with contributions by

Brian D. Andresen
William F. Balistreri
Gerald S. Berenson
Edward W. Bermes
Robert V. Blanke
Robert A. Blouin
Larry D. Bowers
Gregory J. Buffone
Carl A. Burtis
Sati C. Chattoraj
Robert H. Christenson
Walter J. Decker
Franklin R. Elevitch
Emanuel Epstein
Virgil F. Fairbanks
Montserrat deM. Fencl
Donald Fraser
Esther F. Freier
Robert S. Galen
Gregor H. Grant
Michael F. Greene
Mary Alice Heard
A. Ralph Henderson
G. Phillip Hicks
Robert A. Jacob
C. Darrell Jennings
George F. Johnson
Glenville Jones
George G. Klee
Sang Whay Kooh
Robert O. Kringle
Robert F. Labbe

Joel M. Lamon
H. Peter Lehmann
Nancy M. Logan
Ernest Maclin
Donald B. McCormick
Donald W. Moss
Thomas P. Moyer
Dorice M. Czajka-Narins
John O'Brien
Theodore Peters
Charles E. Pippenger
Edward R. Powsner
Ingeborg C. Radde
Alan D. Rinker
Robert C. Rock
Leslie M. Shaw
Ole Siggaard-Andersen
Lawrence M. Silverman
Helge E. Solberg
Sathanur R. Srinivasan
Evan A. Stein
Nancy L. Strang
Thomas O. Tiffany
Dan Tulchinsky
M. David Ullman
Harland L. Verrill
W. Gordon Walker
Nelson B. Watts
James O. Westgard
David M. Wilson
Bethany L. Wise
Donald S. Young

Textbook of CLINICAL CHEMISTRY

Editor:

NORBERT W. TIETZ, Ph.D.

Professor, Department of Pathology,
Director of Clinical Chemistry,
College of Medicine,
University of Kentucky, Lexington, KY

Editorial Committee:

N. V. BHAGAVAN, Ph.D.

WENDELL T. CARAWAY, Ph.D.

REX B. CONN, M.D.

JOHN F. KACHMAR, Ph.D.

ELIZABETH L. PRUDEN, Ph.D., M.T. (ASCP)

RONALD J. WHITLEY, Ph.D.

W. B. SAUNDERS COMPANY

Philadelphia • London • Toronto • Mexico City • Rio de Janeiro • Sydney • Tokyo • Hong Kong

W. B. Saunders Company: West Washington Square
                          Philadelphia, PA 19105

Library of Congress Cataloging in Publication Data

Main entry under title:

Textbook of clinical chemistry.

    1. Chemistry, Clinical. I. Tietz, Norbert W.,
1926–   [DNLM: 1. Chemistry, Clinical. QY 90 T355]
RB40.T44 1986 616.07′56   83-20410
ISBN 0-7216-8886-1

Editor: Dudley Kay
Developmental Editor: Dave Kilmer
Designer: Terri Siegel
Production Manager: Frank Polizzano

Textbook of Clinical Chemistry   ISBN 0-7216-8886-1

Last digit is the print number: 9   8   7   6   5   4   3   2

*Dedicated to our associates
whom we have the privilege of working with
and who made this textbook possible.*

CONTRIBUTORS AND REVIEWERS

BRIAN D. ANDRESEN, Ph.D.

Director, Mass Spectrometry Facilities, Biomedical and Environmental Science Division, University of California, Lawrence Livermore National Laboratory, Livermore, CA. Formerly: Associate Professor, Department of Pharmacology, School of Medicine, Ohio State University, Columbus, OH

Mass Spectrometry

WILLIAM F. BALISTRERI, M.D.

Professor, Departments of Pediatrics and Medicine, College of Medicine, University of Cincinnati; Director, Division of Pediatric Gastroenterology and Nutrition, Children's Hospital Medical Center, Cincinnati, OH

Liver Function

GERALD S. BERENSON, M.D.

Professor, Department of Medicine; Head, Section of Cardiology; Director, Specialized Center of Research—Arteriosclerosis, School of Medicine in New Orleans, Louisiana State University Medical Center, New Orleans, LA

Invited Reviewer

EDWARD W. BERMES, JR., Ph.D.

Professor, Departments of Pathology and Biochemistry, Stritch School of Medicine, Loyola University; Director, Clinical Chemistry, Loyola University Medical Center, Maywood, IL

General Laboratory Techniques and Procedures; Specimen Collection and Processing; Sources of Biological Variation

N. V. BHAGAVAN, Ph.D.

Professor and Chairman, Department of Biochemistry and Biophysics, John A. Burns School of Medicine, University of Hawaii; Consultant Biochemist, Kaiser Medical Center, Honolulu, HI

Editorial Committee

ROBERT V. BLANKE, Ph.D.

Professor, Departments of Pathology, Pharmacology, and Toxicology; Director, Hospital Toxicology Laboratory, Medical College of Virginia, Virginia Commonwealth University, Richmond, VA

Therapeutic Drug Monitoring; Analysis of Toxic Substances

ROBERT A. BLOUIN, Pharm. D.

Associate Professor, College of Pharmacy, University of Kentucky; Director, Clinical Pharmacokinetics Program, University of Kentucky Medical Center, Lexington, KY

Therapeutic Drug Monitoring

LARRY D. BOWERS, Ph.D.

Associate Professor, Department of Laboratory Medicine and Pathology, University of Minnesota; Director, Drug Analysis Section, University of Minnesota Hospitals, Minneapolis, MN

Chromatography

GREGORY J. BUFFONE, Ph.D.

Associate Professor, Department of Pathology, Baylor College of Medicine; Director, Clinical Chemistry, Department of Pathology, Texas Children's Hospital, Houston, TX

Principles of Immunochemical Techniques

CARL A. BURTIS, Ph.D.

Chief, Clinical Chemistry, Health Division, Chemical Technology Division, Oak Ridge National Laboratory, Oak Ridge, TN

Coordinator: Analytical Procedures and Instrumentation; High Performance Liquid Chromatography

WENDELL T. CARAWAY, Ph.D.

Clinical Biochemist, McLaren General Hospital and St. Joseph Hospital, Flint, MI

Editorial Committee; Photometry; Carbohydrates

SATI C. CHATTORAJ, Ph.D.

Associate Professor, Department of Biochemistry; Director, Research Laboratories, Department of Obstetrics and Gynecology, Boston University Medical Center; Director, Gynecological Endocrine Lab, Boston University School of Medicine, Boston, MA

Gas Chromatography; Endocrinology

ROBERT H. CHRISTENSON, Ph.D.

Assistant Professor, Department of Pathology, Duke University Medical Center; Chief, Clinical Chemistry, Durham Veterans Administration Medical Center, Durham, NC

Amino Acids and Proteins

REX B. CONN, M.D.

Professor, Department of Pathology and Laboratory Medicine, Emory University School of Medicine; Clinical Professor of Medical Technology, Georgia State University; Associate Chief, Pathology and Laboratory Medicine, and Director, Clinical Pathology Laboratories, Emory University Hospital, Atlanta, GA

Editorial Committee

WALTER J. DECKER, Ph.D.

Adjunct Associate Professor, Departments of Pharmacology, Toxicology, and Pediatrics, University of Texas Medical Branch, Galveston, TX; Toxicology Consultant Services, Inc., El Paso, TX

Analysis of Toxic Substances

BASIL T. DOUMAS, Ph.D.

Professor, Department of Pathology, Medical College of Wisconsin; Director, Clinical Chemistry, Milwaukee County Medical Complex, Milwaukee, WI

Invited Reviewer

FRANKLIN R. ELEVITCH, M.D.

Associate Clinical Professor, Department of Laboratory Medicine, University of California, San Francisco Medical Center; Director, Clinical Laboratories, El Camino Hospital, Department of Pathology, Mountain View, CA

Microprocessors and Computers in the Clinical Laboratory

EMANUEL EPSTEIN, Ph.D.

Associate Clinical Professor, Department of Pathology, College of Medicine, Wayne State University, Detroit, MI; Medical staff, William Beaumont Hospital, Royal Oak, MI

Electrophoresis

VIRGIL F. FAIRBANKS, M.D.

Professor, Department of Laboratory Medicine, Mayo Medical School; Consultant, Department of Laboratory Medicine, Mayo Clinic, Rochester, MN

Biochemical Aspects of Hematology

MONTSERRAT deM. FENCL, Ph.D.

Associate Director, Laboratory of Reproductive Endocrinology, Brigham and Women's Hospital; Assistant Professor, Department of Obstetrics and Gynecology, Harvard Medical School, Boston, MA

Biochemical Aspects of Pregnancy

DONALD FRASER, M.D., Ph.D.

Professor, Departments of Paediatrics and Physiology, University of Toronto; Senior Physician, Department of Endocrinology, Hospital for Sick Children, Toronto, Ontario, Canada

Calcium and Phosphate Metabolism

ESTHER F. FREIER, M.S., M.T.(ASCP)

Professor, Department of Laboratory Medicine and Pathology, University of Minnesota Medical School; Co-Director, Clinical Chemistry Laboratories, University of Minnesota Hospitals, Minneapolis, MN

Osmometry

ROBERT S. GALEN, M.D., M.P.H.

Chairman, Department of Biochemistry, The Cleveland Clinic Foundation, Cleveland, OH

Analytical Goals and Clinical Relevance of Laboratory Procedures

GREGOR H. GRANT, B.M., F.R.C.Path.

Consultant Chemical Pathologist (Retired), Royal Salop Infirmary, Shrewsbury, Shropshire, England

Amino Acids and Proteins

MICHAEL F. GREENE, M.D.

Instructor, Department of Obstetrics and Gynecology, Harvard Medical School, Cambridge, MA; Obstetrician Gynecologist, Brigham and Women's Hospital, Boston, MA

Biochemical Aspects of Pregnancy

MARY ALICE HEARD, M.A.

Data Coordinator, Department of Pathology, University of Kentucky Medical Center, Lexington, KY

Editorial Assistant

A. RALPH HENDERSON, M.B., Ch.B., Ph.D.

Professor, Department of Biochemistry, University of Western Ontario; Chief, Department of Clinical Biochemistry, University Hospital, London, Ontario, Canada
Enzymes; Gastric, Pancreatic, and Intestinal Function

G. PHILLIP HICKS, Ph.D.

Adjunct Associate Professor, Department of Medicine, University of Wisconsin Medical School; President, Laboratory Consulting, Inc., Madison, WI
Microprocessors and Computers in the Clinical Laboratory

ROBERT A. JACOB, Ph.D.

Project Leader and Research Chemist, Bioanalytical Research Unit, USDA-ARS Western Human Nutrition Research Center, Presidio of San Francisco, CA
Trace Elements

C. DARRELL JENNINGS, M.D.

Assistant Professor, Department of Pathology and Co-Director, Clinical Immunology Laboratory, University of Kentucky Medical Center, College of Medicine; Assistant Chief, Laboratory Service, Veterans Administration Hospital, Lexington, KY
Nitrogen Metabolites and Renal Function

GEORGE F. JOHNSON, Ph.D.

Associate Professor, Department of Pathology, College of Medicine; Director, Special Chemistry Laboratory, University of Iowa, Iowa City, IA
Statistical Procedures

GLENVILLE JONES, Ph.D.

Associate Professor, Departments of Biochemistry and Medicine, Queen's University, Kingston, Ontario. Formerly: Associate Professor, Department of Biochemistry, University of Toronto and Research Institute, Hospital for Sick Children, Toronto, Ontario, Canada
Calcium and Phosphate Metabolism

JOHN F. KACHMAR, Ph.D.

Senior attending scientist, Presbyterian–St. Luke's Hospital; Emeritus Associate Professor of Biochemistry, Rush University, Chicago, IL
Editorial Committee; Enzymes

GEORGE G. KLEE, M.D., Ph.D.

Assistant Professor, Department of Laboratory Medicine, Mayo Medical School; Director, Thyroid Function and Radioimmunoassay Laboratory, Mayo Clinic, Rochester, MN
Quality Assurance; Biochemical Aspects of Hematology

SANG WHAY KOOH, M.D. Ph.D., F.R.C.P.(C)

Associate Professor, Departments of Paediatrics and Physiology, University of Toronto; Senior Physician, Hospital for Sick Children, Toronto, Ontario, Canada
Calcium and Phosphate Metabolism

ROBERT O. KRINGLE, M.S.

Adjunct faculty, Department of Clinical Sciences, College of Science, Rochester Institute of Technology; Biostatistician, Eastman Kodak Company, Rochester, NY
Statistical Procedures

ROBERT F. LABBE, Ph.D.

Professor, Department of Laboratory Medicine, School of Medicine; Head, Clinical Chemistry Division, University of Washington, Seattle, WA
Porphyrins and Disorders of Porphyrin Metabolism

JOEL M. LAMON, M.D.

Clinical Associate Member, Division of Hematology-Oncology; Adjunct Assistant Member, Department of Clinical Research, Scripps Clinic and Research Foundation, La Jolla, CA
Porphyrins and Disorders of Porphyrin Metabolism

H. PETER LEHMANN, Ph.D.

Professor, Department of Pathology, Louisiana State University Medical Center, New Orleans, LA
Invited Reviewer

NANCY M. LOGAN, B.A.

Research Associate, Division of Clinical Chemistry, University of Kentucky Medical Center, Lexington, KY
Appendix

ERNEST MACLIN, P.E.

Vice President, Research and Development, Electro-Nucleonics, Inc., Fairfield, NJ
Automation in the Clinical Laboratory

DONALD B. MCCORMICK, Ph.D.

Fuller E. Callaway Professor and Chairman, Department of Biochemistry, Emory University School of Medicine, Atlanta, GA
Vitamins

DONALD W. MOSS, Ph.D., D.Sc.

Professor of Clinical Enzymology, Department of Chemical Pathology, Royal Postgraduate Medical School, University of London; Honorary Biochemist, Hammersmith Hospital, London, England
Enzymes

THOMAS P. MOYER, Ph.D.

Associate Professor, Department of Laboratory Medicine, Mayo Medical School; Director, Toxicology Laboratory, Mayo Clinic, Rochester, MN
Therapeutic Drug Monitoring

DORICE M. CZAJKA-NARINS, Ph.D.

Guest Lecturer, Chicago Medical School, North Chicago, IL; Formerly: Associate Professor, Section of Clinical Nutrition, Department of Related Health Programs, Rush–Presbyterian–St. Luke's Medical Center, Chicago, IL
Nutrition

JOHN F. O'BRIEN, Ph.D.

Associate Professor, Department of Laboratory Medicine, Mayo Medical School; Director, Enzyme Laboratory, Section of Clinical Chemistry, Mayo Clinic, Rochester, MN
Lysosomal Enzymes

THEODORE PETERS, JR., Ph.D.

Adjunct Associate Professor, Department of Biochemistry, College of Physicians and Surgeons, Columbia University; Research Biochemist, The Mary Imogene Bassett Hospital, Cooperstown, NY
Analytical Goals and Clinical Relevance of Laboratory Procedures; Evaluation of Methods

CHARLES E. PIPPENGER, Ph.D.

Head, Section of Applied Clinical Pharmacology, Department of Biochemistry, Cleveland Clinic Foundation, Cleveland, OH
Therapeutic Drug Monitoring

EDWARD R. POWSNER, M.D.

Director, Division of Nuclear Medicine, Department of Pathology, St. John Hospital, Detroit, MI
Basic Principles of Radioactivity and Its Measurement

ELIZABETH L. PRUDEN, Ph.D., M.T.(ASCP)

Assistant Professor, Department of Pharmacology and Therapeutics, School of Medicine in Shreveport; Co-Director, Clinical Toxicology Laboratory, Louisiana State University, Shreveport, LA
Editorial Committee; Electrolysis, Blood Gases, and Acid-Base Balance

INGEBORG C. RADDE, M.D., Ph.D., F.R.C.P.(C)

Associate Professor, Departments of Paediatrics and Pharmacology, University of Toronto; Associate Professor and Senior Staff Physician, Research Institute, Hospital for Sick Children, Toronto, Ontario, Canada
Calcium and Phosphate Metabolism

ALAN D. RINKER, M.S.

Supervisor, Radioimmunoassay and Immunochemistry Laboratory, Division of Clinical Chemistry, University of Kentucky Medical Center, Lexington, KY
Gastric, Pancreatic, and Intestinal Function; Invited Reviewer

ROBERT C. ROCK, M.D.

Associate Professor, Department of Pathology and Laboratory Medicine, Johns Hopkins University School of Medicine; Director, Department of Laboratory Medicine, Johns Hopkins Hospital, Baltimore, MD
Nitrogen Metabolites and Renal Function

LESLIE M. SHAW, Ph.D.

Associate Professor, Department of Pathology; Director, Toxicology Laboratory, Division of Laboratory Medicine, William Pepper Laboratory, Hospital of the University of Pennsylvania, Philadelphia, PA
Liver Function

OLE SIGGAARD-ANDERSEN, M.D., Ph.D.

Professor, Department of Clinical Chemistry, University of Copenhagen; Director, Clinical Chemistry, Herlev Hospital, Herlev, Denmark
Electrochemistry; Electrolytes, Blood Gases, and Acid-Base Balance

LAWRENCE M. SILVERMAN, Ph.D.

Associate Professor, Department of Pathology and Biochemistry, University of North Carolina; Director, Special Chemistry Laboratory; Associate Director, Clinical Chemistry Laboratories, Department of Hospital Laboratories, North Carolina Memorial Hospital, Chapel Hill, NC
Amino Acids and Proteins

HELGE ERIK SOLBERG, M.D., Ph.D.

Consultant, Department of Clinical Chemistry, Riks-hospitalet, University of Oslo, Oslo, Norway
Establishment and Use of Reference Values

SATHANUR R. SRINIVASAN, Ph.D.

Professor, Departments of Medicine and Biochemistry, Louisiana State University School of Medicine, New Orleans, LA
Invited Reviewer

EVAN A. STEIN, M.D., Ph.D.

Associate Professor, Departments of Pathology and Laboratory Medicine and Internal Medicine, University of Cincinnati Medical Center; Director, Clinical Chemistry and Lipid and Apolipoprotein Laboratories, University Hospital, Cincinnati, OH
Lipids, Lipoproteins, and Apolipoproteins

NANCY L. STRANG, M.S.

Data Coordinator, Department of Pathology, University of Kentucky Medical Center, Lexington, KY
Editorial Assistant

NORBERT W. TIETZ, Ph.D.

Professor, Department of Pathology, College of Medicine; Director, Clinical Chemistry, University of Kentucky, Lexington, KY
Editor; Electrolytes, Blood Gases, and Acid-Base Balance; Gastric, Pancreatic, and Intestinal Function; Appendix

THOMAS O. TIFFANY, Ph.D.

Director, Research and Development, Instrumentation Laboratory, Inc., Spokane, WA
Fluorometry, Nephelometry, and Turbidimetry

DAN TULCHINSKY, M.D.

Associate Professor, Department of Obstetrics and Gynecology, Harvard Medical School; Director, Reproductive Endocrine Laboratory, Brigham and Women's Hospital, Boston, MA
Biochemical Aspects of Pregnancy

M. DAVID ULLMAN, Ph.D.

Associate Research Professor, Department of Psychiatry, Boston University School of Medicine, Boston, MA; Chief, Special Research Laboratory, Veterans Administration Hospital, Bedford, MA
High Performance Liquid Chromatography

HARLAND L. VERRILL, Ph.D.

Assistant Adjunct Professor, Department of Pathology, College of Human Medicine, Michigan State University, East Lansing, MI; Clinical Biochemist, Department of Pathology, Hurley Medical Center, Flint, MI
Invited Reviewer

W. GORDON WALKER, M.D.

Professor, Department of Medicine, Johns Hopkins University School of Medicine; Director, Renal Division, Department of Medicine; Director, Adult Clinical Research Center, and Physician, Johns Hopkins Hospital; Director, Nephrology, and Physician, Good Samaritan Hospital, Baltimore, MD
Nitrogen Metabolites and Renal Function

NELSON B. WATTS, M.D.

Senior Associate, Emory School of Medicine and Emory University Clinic, Section of Internal Medicine, Department of Endocrinology, Atlanta, GA
Carbohydrates; Endocrinology

JAMES O. WESTGARD, Ph.D.

Professor, Department of Pathology and Laboratory Medicine, University of Wisconsin; Associate Director, Clinical Laboratories–Quality Control, University of Wisconsin Hospitals and Clinics, Madison, WI
Evaluation of Methods; Quality Assurance

RONALD J. WHITLEY, Ph.D.

Assistant Professor, Department of Pathology, College of Medicine; Assistant Director, Clinical Chemistry, University of Kentucky, Lexington, KY
Editorial Committee

DAVID M. WILSON, M.D., F.A.C.P.

Associate Professor, Department of Medicine, Mayo Medical School; Chairman, Renal Function Laboratory, Department of Laboratory Medicine, Mayo Clinic, Rochester, MN
Invited Reviewer

BETHANY L. WISE, Ph.D., M.T.(ASCP)

Clinical Chemistry Instructor, Division of Medical Technology, School of Allied Medical Professions, College of Medicine, Ohio State University, Columbus, OH
Mass Spectrometry

DONALD S. YOUNG, M.B., Ph.D.

Professor, Department of Pathology and Laboratory Medicine; Head, Division of Laboratory Medicine, University of Pennsylvania; Director, William Pepper Laboratories, Hospital of the University of Pennsylvania, Philadelphia, PA; Formerly: Professor of Laboratory Medicine, Mayo Medical School; Head, Section of Clinical Chemistry, Department of Laboratory Medicine, Mayo Clinic, Rochester, MN

General Laboratory Techniques and Procedures; Automation in the Clinical Laboratory; Specimen Collection and Processing; Sources of Biological Variation

FOREWORD

Since the publication of *Fundamentals of Clinical Chemistry*, second edition, the field of clinical chemistry has undergone significant changes and has continued its exponential growth. The warm reception that *Fundamentals of Clinical Chemistry* received has encouraged its editors and authors to write a new *Textbook of Clinical Chemistry*, designed to meet the needs created by changes and growth in the field. As with the second edition, this new textbook provides a useful balance between basic biochemistry, the application of laboratory tests in medical diagnosis, and a description of the methods for performing laboratory tests. An important thrust of the *Textbook of Clinical Chemistry* is to bridge the gap between the technology of the laboratory and the bedside application of laboratory information in the patient care process. Several new chapters as well as expanded coverage of sections on clinical significance of laboratory tests are designed to meet this goal. Chapters describing newer instrumental approaches and topics such as nutrition, therapeutic drug monitoring, and monitoring of pregnancy have been added to address these newly developing needs.

It is hoped that the comprehensive and in-depth coverage of the traditional as well as the new areas of clinical chemistry will fill not only contemporary but also future needs of the intermediate and advanced student. The book is also designed to be a resource for clinical chemists and for pathologists and other physicians. Advanced and less important material has been set in smaller-sized type to set it apart from regular text.

The wide use of kit procedures in clinical chemistry laboratories required adjustment in the presentation of methods. To avoid duplication of material that is covered in kit inserts, we have omitted detailed procedures whenever reliable kit procedures are available and in wide use. Instead of a detailed method, in many cases we included a thorough description of the principles of the kit procedures. A list of references and reading material is included at the end of each chapter to guide the reader in further studies.

The work on this book started several years ago. Throughout this time various national and international scientific organizations and many individuals have expended considerable effort to standardize units applicable to clinical chemistry so that these units may be more meaningful and also consistent throughout the world. The SI system has been adopted by the authors, in principle, in conformity with the practices in this country. In many places the older and the SI units are used side by side to aid in the transition until final consensus has been reached in the USA.

The task of selecting material for inclusion in a textbook becomes more and more difficult, because of the rapid expansion of the field and the diversity of interests of the individual members of the clinical chemistry profession. If we have failed in any area to provide adequate coverage, we would appreciate hearing from you so that we can consider your suggestions for inclusion in the next edition.

NORBERT W. TIETZ

ACKNOWLEDGMENTS

Preparation of the manuscripts for this multiauthor book required close cooperation between the contributors and the editor. I am most grateful to all contributors for submitting so willingly to this principle in preparing their manuscripts and for cooperating so well with the review process of manuscripts. My special thanks go to the members of the Editorial Committee for their significant efforts and their many contributions to the editing process. I am especially indebted to Dr. Elizabeth L. Pruden for her dedication, perseverance, and her numerous invaluable contributions.

Nancy L. Strang, M.S., and Mary Alice Heard, M.A., assisted in the editorial process and their efforts and dedication are gratefully acknowledged. Thanks also go to Nancy M. Logan, B.A., Karen D. Reeves, A.A.S., and Marilyn R. Rettaliata, B.S., for proofreading galleys and to Tammy I. Rayburn and Susan C. Blandford for typing manuscripts.

The excellent cooperation of the staff of W. B. Saunders Company, especially of John J. Hanley, formerly President and Medical Editor, Frank Polizzano, Production Manager, David Kilmer, Developmental Editor, Nina Ikeda, Design Department Manager, and Terri Siegel, Book Designer, is also gratefully acknowledged.

NORBERT W. TIETZ
EDITOR

CONTENTS

CHAPTER **1**

LABORATORY PRINCIPLES AND INSTRUMENTATION

General Laboratory Techniques and Procedures

by Edward W. Bermes, Jr., Ph.D., and Donald S. Young, M.B., Ph.D.

The function of the clinical chemistry laboratory is to perform qualitative and quantitative analyses on body fluids such as blood, urine, and spinal fluid as well as feces, tissue, calculi, and other materials. If the results are to be useful to the physician in the diagnosis and treatment of disease, the tests must be performed as accurately as possible. This requires the use of sound analytical methods and good instrumentation. However, underlying all standards of analytical quality are *basic principles and procedures.* Essential factors include purity of reagent solutes and solvents, suitability of containers, reliability and quality of measuring devices and methods, suitability of separative methods and devices, and observance of established safety procedures. There should also be an understanding of the chemical reactions involved in each test and the effect of physical variables on the procedures.

CHEMICALS AND RELATED SUBSTANCES

Laboratory chemicals are available in a variety of grades. The solutes and solvents used in analytical work are *reagent grade chemicals,* among which water is a solvent of primary importance. The classification of *reagent grade chemicals* is quite unofficial, and there is no agreement among manufacturers concerning the designation of the various degrees of purity. The IUPAC (International Union of Pure and Applied Chemistry) has established criteria for "primary standards." The National Bureau of Standards (NBS) has a number of Standard Reference Materials (SRMs) available for the clinical chemistry laboratory, and the College of American Pathologists also supplies some Certified Reference Materials (CRMs). The National Committee for Clinical Laboratory Standards has established a standard for reagent grade water and has published "Guidelines for the Development of Certified Reference Materials for use in Clinical Chemistry" in the "proposed standard" stage.

Reagent Grade Water

The preparation of most reagents and solutions used in the clinical laboratory requires "pure" water. For many years the term "distilled" has been used to mean "pure," since distillation was the principal procedure used to remove the impurities that water contains in its natural state. Now there are alternative methods to produce "pure" water. In fact, single-distilled water fails to meet the specifications for Type I Clinical Laboratory Reagent Water put forward by the National Committee for Clinical Laboratory Standards (NCCLS) in 1980.[8]

NCCLS specifications for reagent grade water (Table 1A-1) have largely replaced those published by the College of American Pathologists in 1978.[2] Although in many laboratories the term "deionized water" may have replaced "distilled water," the use of both terms should be discouraged since they describe methods of preparation and do not in themselves reflect the quality of the final product. The term "Reagent Grade Water," followed by designation of the Type (I through III), better defines the specifications of the water and is independent of the method of preparation.

In general, no single process of purification can produce water that meets the rigid specifications for Reagent Grade Water Type I set forth by NCCLS. The following processes or any combination of these may be utilized in the preparation of reagent grade water as long as the final product meets the specifications. The combination selected is usually dictated in part by the quality of the source water, and in part by the intended use of the water.

Preparation of Reagent Grade Water

Three types of filtration equipment may be used. Most municipal water supplies contain enough solid contaminants to make prefiltration an economically feasible part of a total system. *Prefilters* are usually composed of microfiber glass or cotton; they remove 98% or more of the particulate matter and thus protect the whole water flow and containment system. Prefilters often enhance the economics of the operation, since these filters can usually be cleaned and reused many times.

A second type of filtration is a bed of *activated carbon* through which the water passes. The bed is effective in removing large amounts of organic materials and chlorine. A combination of prefiltration and carbon bed filtration is extensively used to increase the life expectancy of an ion-exchange resin used later in the purification process.

The third type of filtration employs a *submicron filter*. This is typically a Millipore membrane filter which removes all particles or microorganisms larger than the membrane's pore size (usually $0.2 \mu m$) without releasing contaminants that could alter the quality of the water. A submicron filter is usually used as the last stage of a system, i.e., near the point of delivery.

The oldest method of water purification is *distillation*. Because distillation involves phase changes from liquid to vapor to liquid, nonvolatile materials are left behind in the boiling vessel where they may form a crust on the inner surface of the vessel. Volatile impurities may be partially carried over into the distillate. During vigorous boiling a small number of water droplets entrapped in vapor may be carried over and transport sodium, potassium, manganese, carbonates, and sulfates to the distillate. As a result, water treated by distillation alone will not meet the specific

Table 1A-1. NCCLS SPECIFICATIONS FOR REAGENT GRADE WATER[8]

	Type I	Type II	Type III
*Microbiological content,** colony forming units per mL, cfu/mL (maximum)	10	10^3	N.A.
pH	N.A.	N.A.	5.0–8.0
Resistivity,[†] Mohm centimeter (MΩ-cm), 25 °C	10 (in-line)	2.0	0.1
Silicate, mg SiO_2/L (maximum)	0.05	0.1	1.0
Particulate matter[‡]	Water passed through 0.2-μm filter	N.A.	N.A.
Organics[§]	Water passed through activated carbon	N.A.	N.A.

Microbiological content. The microbiological content of viable organisms, as determined by total colony count after incubation at 36 ± 1 °C for 14 h, followed by 48 h at 25 ± 1 °C, and reported as colony forming units per mL (cfu/mL).

[†]*Specific resistance or resistivity.* The electrical resistance in ohms measured between opposite faces of a 1.00 cm cube of an aqueous solution at a specified temperature. For these specifications, the resistivity will be corrected for 25 °C and reported in Mohm/cm. The higher the amount of ionizable materials, the lower the resisitivity and the higher the conductivity.

[‡]*Particulate matter.* When water is passed through an intact membrane filter with a mean pore size of 0.2 μm, it is considered to be free of particulate matter.

[§]*Organic material.* When water is passed through a bed of activated carbon, it is considered to contain minimum organic material.

conductivity requirement of Type I water. Distillation involves a large expenditure of energy and in addition requires a considerable expenditure of time to clean and maintain the still, unless one of the two methods described below precedes the distillation. The amount of maintenance needed for the still is directly related to the quality of the feed water.

Deionized water is water from which the ionized salts have been removed by the process of ion-exchange. It is most conveniently prepared by utilizing commercial deionizing equipment which ranges in size from small disposable cartridges to large tanks containing resin. The resin contained in large tanks is capable of being regenerated. Deionization is accomplished by passing feed water over insoluble resin polymers which exchange H^+ and OH^- ions for the impurities present in ionized form in the water. These insoluble resin polymers are prepared with acid or amine functional groups. The tanks may contain cation-exchangers, anion-exchangers, or a "mixed-bed resin exchanger," which is a mixture of cation- and anion-exchange resins in the same container. A "two-bed deionizer" consists of the separate ion-exchange resins in a series, i.e., cation-exchange followed by anion-exchange or vice versa.

A typical cation-exchange resin reacts as follows:

$$(RSO_3)H + Na^+ \rightarrow (RSO_3)Na + H^+$$

while an anion-exchange resin with quaternary ammonium type of structure acts as follows:

$$(RNR'_3)OH + Cl^- \rightarrow (RNR'_3)Cl + OH^-$$

A single-bed deionizer generally is capable of producing water that has a specific resistance in excess of 1 Mohm/cm. When connected in series, mixed-bed deionizers usually produce water with a specific resistance that exceeds 10 Mohm/cm.

Reverse osmosis is a process by which water is passed through a membrane that acts as a molecular filter. The membrane removes 95–99% of organics, bacteria, and other particulate matter and 90–97% of all ionized and dissolved minerals, but less of the gaseous impurities. Although the process is inadequate for producing reagent grade water for the laboratory, it may be utilized as a preliminary purification method.

Osmosis is the process by which water flows through a membrane from a dilute solution into a more concentrated solution until the osmotic pressure on both sides of the membrane is equal. Reverse osmosis utilizes a pump to apply enough pressure continuously to the side containing the feed water to overcome the osmotic pressure and cause purified water to pass through the membrane. The product (permeate) and the concentrate (brine) are continuously removed. Basic considerations affecting the selection of the type of reverse osmosis equipment are:
1. Quality of raw water
2. Capacity in terms of flow
3. Performance in terms of salt passage compared to water passage
4. Stability in terms of operating life. (Free chlorine will attack many membranes.)

Quality, Use, and Storage of Reagent Water

Type III water may be used for glassware washing. (Final rinsing, however, should be done with the water grade suitable for the intended glassware use.) It may also be used for certain qualitative procedures, such as those used in general urinalysis.

Type II water is used for general laboratory testing not requiring Type I water. Storage should be kept to a minimum, and storage as well as delivery systems should be constructed so as to assure a minimum of chemical or bacterial contamination.

Type I water should be used in test methods requiring minimal interference and maximum precision and accuracy. Such procedures might include trace metal determination, enzyme measurements, electrolyte measurements, and preparation of all standards. This water should be used immediately after production. No specifications for storage systems for Type I water are given because it is not possible to maintain the high resistivity while drawing off water and storing it.

Specific Systems for the Preparation of Reagent Grade Water

Any system may be used as long as the desired specifications are attained. Considerations in the selection of a system to produce Type I reagent water might include quality of feed water, amount of water needed, availability of space near area of use, and electrical energy cost in the area. In general, if the feed water contains more than 1000 ppm of total dissolved solids (TDS), reverse osmosis should be considered as a pretreatment.

Type I reagent water, with a neutral pH and 170 ppm of total dissolved solids, can be produced from municipal feed water by use of a primary dual filtration cartridge to remove particulate matter and organics, followed by two tanks containing a mixed-bed ion-exchange resin. The tanks are followed by a 0.2 μm membrane filter. A conductivity light installed between the two ion-exchange tanks is set to go out when the resistance of the effluent from the first tank goes below 200 000 ohms/cm. The second tank then "polishes" the water to give water with a resistance of up to 15 Mohms/cm. When the deionizers are exchanged, the first tank is removed for regeneration, the second tank is put in its place, and the new tank is put in the second position. The spigot for water delivery should be within a few feet of the final filter.

Testing for Water Purity

The maximum interval in the testing cycle for purity of reagent water should be one week. Specifications measured at the time of production may differ from those at the time and place of use. If the water is piped a long distance, consideration must be given to deterioration en route to the site of use. Thus, a central institution system located on an upper floor may consist of pretreatment (ion exchange or reverse osmosis) followed by distillation, resulting in Type I water, and then delivery to the site several hundred yards away. In this instance it may be necessary to install a mixed-bed ion-exchange resin tank, and a 0.2 μm filter at the use site so that the Type I water produced centrally is still Type I quality when drawn off.

The specific methods used for testing cannot be covered here. At a minimum, water should be tested for microbiological content, pH, resistivity, and soluble silica. The reader is referred to one of several references for specifics.[1,2,8]

Reagent Grade or Analytical Reagent Grade (AR) Reagents

Chemicals which meet specifications of the American Chemical Society are described as reagent or analytical reagent grade. These are available in two forms: (1) lot analyzed reagents, in which each individual lot is analyzed and the actual amount of impurity reported (e.g., arsenic— 0.0005%) and (2) maximum impurities reagents, for which maximum impurities are listed (e.g., arsenic—maximum 0.001%). In the latter instance, the arsenic may actually be only 0.0004%, but the analyst is guaranteed only the maximum limit of 0.001%. Establishment of the ACS specifications marked a turning point in the preparation of high-purity reagents.

The Committee on Analytical Reagents of the American Chemical Society periodically publishes "Reagent Chemicals" listing specifications and now intends to issue supplements between editions. Manufacturers of ACS chemicals check each lot in a control laboratory and place "ACS" on the labels of only those lot chemicals that meet the Society's published specifications. These reagent grade chemicals are of very high purity and are recommended for quantitative or qualitative analyses.

Ultra Pure Reagents

Gas chromatography, high pressure liquid chromatography, fluorometry, and trace metal analyses frequently call for reagents whose purity exceeds the specifications of those described above. Manufacturers now offer selected chemicals that have been especially purified to meet specific needs. There is no uniform designation for these chemicals and organic solvents. Terms such as "Spectrograde," "Nanograde," and "HPLC Pure" have been used. Data of interest to the user, e.g., absorption at a specific ultraviolet wavelength, are supplied with the reagent.

Other Designations of Chemical Purity

There are several other designations of chemical purity. In general, chemicals so designated are not of sufficient purity for use as analytical reagents.

Chemically Pure (CP) is a designation that fails to reveal the tolerance limits of impurities, and the practice followed by different manufacturers in the use of this designation is not uniform. The term *highest purity* is used by manufacturers for organic chemicals that they have purified to as great a degree as they find practical. The purity is usually determined by measurement of melting points or boiling points. However, this group of chemicals may occasionally have to be used in clinical chemical analyses when higher purity biochemicals are not available.

USP and *NF Grade* are chemicals produced to meet specifications set down in the United States Pharmacopeia (USP) or the National Formulary (NF). These designations are of interest primarily to the pharmaceutical chemist, since tolerances specified for impurities are those that will not be injurious to health. In many cases these compounds may be very pure and can be used in chemical analysis as well as in the preparation of various reagents, but high purity cannot be assumed to be the case in all instances; in these categories chemical purity is only incidental.

Purified, Practical, Technical, or *Commercial Grade* chemicals should not be used in clinical chemical analysis without prior purification.

Purity of Organic Reagents

The purity of commercially obtained organic reagents for clinical chemistry purposes is generally inferior to that of inorganic reagents. The majority of impurities in these compounds will have been introduced in their synthesis either along with the starting materials or as by-products; such impurities are presumably more difficult or more expensive to remove than are impurities in inorganic substances. In addition, some organic compounds oxidize or decompose on standing, and the amount of impurities from this cause will depend on how long the bottle of reagent has been opened or stored. Stability is often improved when the compounds are stored in amber bottles and refrigerated; however, phenols and amines oxidize on standing and tend to darken even when refrigerated. Sugars are quite hygroscopic and absorb moisture rapidly unless they are stored in a desiccator.

The presence of impurities in an organic reagent may be a source of difficulty in its use. If the contaminant does not react with the substance being determined, no interference will occur as long as there are no interfering side reactions and there is enough of the original reactant remaining. If a reagent is impure, the net final color developed in a colorimetric reaction may be considerably more than desirable because of a high blank caused by the impurity. The existence of isomers and their presence in a particular lot of an organic reagent may be a cause of analytical problems, since in the rather specific geometrical requirements of a chelate ring only certain isomers may produce the desired colored complex, or in enzyme reactions only one of the isomers may be suitable as substrate.

Organic solvents are available in several grades of purity. The common grades are listed in Table 1A-2. The degree of purity required for the solvent depends on the application. The details of purification procedures are included in most comprehensive analytical chemistry texts. Because of the variety of grades available it is rarely necessary to redistill reagents in the clinical laboratory.

The properties of some of the organic solvents commonly used in the clinical laboratory are listed in Table 1A-3. Organic solvents should generally be considered potentially hazardous to health, and thus inhalation of fumes should be avoided. Unless a solvent is known to be non-flammable, it should be considered a potential fire hazard. Diethyl ether has a particularly high flammability rating. Proper storage conditions are essential to minimize the dangers of organic solvents. Pure solvents are best stored in glass containers under an inert gas either in a sealed container or under positive pressure. With some solvents, sealing the stopper with paraffin wax prolongs their storage life. The maximum quantities of solvents that can be stored in particular types and sizes of containers are described by OSHA regulations. (See section on Safety.)

Table 1A-2. COMMON GRADES OF ORGANIC SOLVENTS

Practical Grade:	Contains some impurities but usually adequate for most organic preparations
USP Grade and NF Grade:	Meets standards established by the US Pharmacopeia (USP) or National Formulary (NF); may contain impurities which have not been tested for
Chemically Pure (CP):	Almost as pure as reagent-grade chemicals
Spectroscopic Grade:	Spectrally pure in the visible, ultraviolet, and near and mid-infrared ranges
Chromatographic Grade:	Minimum purity of greater than 99% as determined by gas chromatography; no single impurity exceeding 0.2%
Reagent Grade:	Certified to contain impurities below levels established by the Committee on Analytical Reagents of the American Chemical Society

(Modified from Shugar, G.J., Shugar, R.A., Bauman, L., et al.: Chemical Technicians' Ready Reference Handbook. 2nd ed. New York, McGraw-Hill, 1981.)

Table 1A-3. PROPERTIES OF ORGANIC SOLVENTS COMMONLY USED FOR EXTRACTION OF AQUEOUS SOLUTIONS

	BP(°C)	Flammability	Lighter/Heavier Than Water	Characteristics
Diethyl ether	34.5	++++*	L†	Absorbs 1.5% water; strong tendency to form peroxides
Methylene chloride	40	−	H	May form emulsions; easily dried
Petroleum ethers	35–60	+++	L	Easily dried; poor solvents for polar compounds
Ethyl acetate	77.1	+++	L	Good for polar compounds; absorbs up to 3.3% water
2-Butanol	99.5	+	L	Good for extraction of highly polar water-soluble compounds from buffer solutions; dries easily
Tetrachloromethane	76.7	−	H	Good for nonpolar compounds; is easily dried
Chloroform	61.7	−	H	Easily dried; tends to form emulsions
Diisopropyl ether	69	+++	L	Tends to form peroxides
Benzene	80.1	+++	L	Prone to form emulsions; good for extraction of alkaloids and phenols from buffered solutions

*− to ++++ indicates degree of flammability.
†L = lighter than; H = heavier than.
(From Shugar, G.J., Shugar, R.A., Bauman, L., et al.: Chemical Technicians' Ready Reference Handbook. 2nd ed. New York, McGraw-Hill, 1981; and Gordon, A.J., and Ford, R.A.: The Chemist's Companion. New York, Wiley-Interscience, 1972.)

Standards

Primary standards are highly purified chemicals that can be directly weighed or measured to produce a solution whose concentration is exactly known. The International Union of Pure and Applied Chemistry has proposed a degree of 99.98% purity for primary standards. This is not achievable (or necessary) for most clinical chemistry laboratory standards.

These highly purified chemicals may be weighed out directly for the preparation of solutions of selected concentration or for the standardization of solutions of unknown strength. They are supplied with a certificate of analysis for each lot. These chemicals must be stable substances of definite composition, which can be dried, preferably at 104–110 °C, without a change in composition. They must not be hygroscopic, so that water is not absorbed during weighing.

Secondary standards are solutions whose concentrations cannot be prepared by weighing the solute and dissolving a known amount into a volume of solution. The concentration of secondary standards is usually determined by analysis of an aliquot of the solution by an acceptable reference method, using a primary standard to calibrate the method.

Reference Standards (Standard Reference Materials, SRMs) for clinical laboratories were developed by the National Bureau of Standards. The first SRM, cholesterol, was issued in 1967. Today the list is extensive (Table 1A-4). Not all standard reference materials have the properties and the degree of purity specified for a primary standard, but each has been well characterized for certain chemical or physical properties and is issued with a certificate that gives the results of the characterization. These may then be used to characterize other materials.

Desiccants and Use of Desiccators

Most published information concerning the comparative efficiency of drying agents is based upon experiments in which the amount of moisture absorbed from air flowing through a bed of desiccant is measured. The limited information that is available concerning the efficacy of drying reagents in a desiccator suggests that the comparative efficiency may not be the same as in flow measurements. Every effort should be made to avoid desiccants that produce dust when used in desiccators; granular calcium chloride, for instance, frequently carries a large amount of dusty "fines" when fresh. Drying agents that incorporate cobalt chloride or some other moisture-sensitive salt to indicate an exhaustion are much preferred to those that do not contain such

Table 1A-4. SOME STANDARD REFERENCE MATERIALS (SRMs) FOR CLINICAL CHEMISTRY

Analyte	Material	SRM Number
Albumin (Bovine)	Powder	926
Angiotensin I (Human)	Powder	998
Bilirubin	Powder	916
Calcium*	Human serum, lyophilized	909
Calcium carbonate	Powder	915
Chloride*	Human serum, lyophilized	909
Cholesterol*	Human serum, lyophilized	909
Cholesterol	Powder	911A
Cortisol	Powder	921
Creatinine	Powder	914
Ethosuximide	Human serum, lyophilized	900
Glucose*	Human serum, lyophilized	909
Glucose	Powder	917
Iron	Metal	937
Lead nitrate	Powder	928
Lithium*	Human serum, lyophilized	909
Lithium carbonate	Powder	924
Magnesium*	Human serum, lyophilized	909
Magnesium gluconate	Powder	929
Mannitol	Powder	920
p-Nitrophenol	Powder	938
Phenobarbital	Human serum, lyophilized	900
Phenytoin	Human serum, lyophilized	900
Potassium*	Human serum, lyophilized	909
Potassium chloride	Powder	918
Potassium dichromate	Crystalline	935
Primidone	Human serum, lyophilized	900
Quinine sulfate	Powder	936
Sodium chloride	Powder	919
Tripalmitin	Crystalline	1595
Urea	Powder	912
Uric acid*	Human serum, lyophilized	909
Uric acid	Powder	913
VMA (4-hydroxy-3-methoxymandelic acid)	Powder	925
Glass filters		930D
Thermometers, clinical laboratory		933, 934

*All analytes are contained in same lyophilized serum.

(From Seward, R.W., and Mavrodineanu, R.: Standard Reference Materials: Summary of the Clinical Laboratory Standards Issued by the National Bureau of Standards. Washington, D.C., National Bureau of Standards, 1981.)

indicators. Silica gel and anhydrous calcium sulfate (Drierite) are sold with indicators included. Although more expensive, such forms are frequently worth the higher cost.

Desiccators should be opened carefully. Ordinary desiccators often contain air at less than atmospheric pressure as a result of cooling the warm air formed from hot samples added prior to closure. If such a desiccator is not opened slowly, the inrush of air may create drafts sufficient to dislodge materials from open vessels or to stir up dust particles from the drying agent that may subsequently settle in the vessels that are being stored. Vacuum desiccators should be provided with a curved inlet tube to deflect incoming air against the lid, and the stopcocks on these should be opened very carefully when restoring the internal pressure to that of the atmosphere.

The properties of various desiccants are described in Table 1A-5; two of these are distinctly alkaline and one is strongly acidic. The choice of the drying agent required for the quantitative absorption of moisture depends on the composition of the gases or materials to be dried, but also on convenience, efficiency, and sometimes cost. For example, magnesium perchlorate quantitatively absorbs ammonia gas, and anhydrous calcium chloride (technical grade) absorbs carbon dioxide and ammonia. These facts should be kept in mind when choosing a desiccant intended for the quantitative removal of water from gases that may also contain ammonia, carbon dioxide, or other reactive substances. Certain drying reagents are deliquescent and, when liquefaction of the drying agent occurs, a decline in drying efficiency results. Calcium chloride and magnesium perchlorate are examples; however, both have a considerable capacity before deliquescence sets

Table 1A-5. CHEMISTRY AND ACTIVITY OF DESICCANTS

Drying Agent	Activity*	Capacity	Deliquescence	Easy Regeneration	Chemical Reaction
Phosphorus pentoxide	0.02	V. low	Yes	No	Acidic
Barium oxide	0.6–0.8	Moderate	No	No	Alkaline
Alumina	0.8–1.2	Low	No	Yes	Neutral
Magnesium perchlorate (anhydrous)	1.6–2.4	High	Yes	No	Neutral
Calcium sulfate (Drierite)	4–6	Moderate	No	Yes	Neutral
Silica gel	2–10	Low	No	Yes	Neutral
Potassium hydroxide (stick)	10–17	Moderate	Yes	No	Alkaline
Calcium chloride (anhydrous)	330–380	High	Yes	No	Neutral

*Micrograms residual water per liter of air at 30 °C.

in. Phosphorus pentoxide is one of the most powerful drying agents in use, but its effective capacity is rapidly reduced by formation of metaphosphoric acid upon uptake of water.

Some drying agents can easily be regenerated, and this is an important consideration when comparing costs. Silica gel can be regenerated by heating in a drying oven at 120 °C, while anhydrous calcium sulfate and activated alumina require temperatures of 275 °C and 175 °C, respectively. Magnesium perchlorate can be regenerated by heating to 240 °C in a partial vacuum.

GENERAL LABORATORY SUPPLIES

Most of the laboratory ware utilized in the clinical chemistry laboratory today is made of either glass or plastic, both of which may be made of one of several different types. Thus, a given item such as a beaker may be made from one of many different materials. The advantages as well as the limitations imposed by the properties of these materials should be appreciated in light of their possible effects upon analyses performed in such containers or devices.

Glassware

Glass is a complex silicate, the properties of which depend on the type of silicate anion in the structure as well as its cation content. The addition of metal cations, such as Fe(III) or Ni(II), to the basic structure of glass can alter its color. The thermal properties of glass can be significantly changed by the addition of boron oxide (B_2O_3). This latter glassware, known as *borosilicate glass,* is extensively used in clinical laboratories.

Glass with high thermal resistance is usually a *borosilicate glass* that has a low alkali content. It is free from zinc-group elements and heavy metals (arsenic and antimony) and resists heat, corrosion, and thermal shock. Since its dimensions change very little with temperature (relatively low coefficient of expansion), this type of glassware should be used whenever heating or sterilization by heat is to be carried out. Some borosilicate glassware, if properly supported and not under internal pressures, can be heated to about 600 °C for a relatively short period of time. If the glass is cooled too quickly, however, it will acquire strains that may affect its future serviceability. The highest safe operating temperature of this glass is its *strain point.* The thermal durability of borosilicate glass is shown in Table 1A-6.

Pyrex (Corning) and *Kimax* (Kimble) brand glass are the most common thermal-resistant borosilicate glassware found in the laboratory. Most clinical laboratory apparatus such as beakers, flasks, and pipets are made of this type of glass. Because the vessel contains so few elements, it contaminates liquids minimally, even when they are hot.

Several years ago a special alumina-silicate glass was developed that is at least six times stronger than borosilicate glass. This *Corex* (Corning) brand laboratory glassware has been strengthened chemically rather than thermally. Corex pipets have a typical impact strength of 30 000 psi compared to a rating of 2000 to 5000 psi for borosilicate pipets. Corex laboratory glassware is harder than conventional borosilicates and better able to resist clouding due to alkali

Table 1A-6. THERMAL DURABILITY OF
BOROSILICATE GLASS

Strain point*	510 °C
Annealing point[†]	555 °C
Softening point	820 °C

*Temperature at which deformation may result due to heat stress.
[†]Temperature to which glass must be heated to relieve strains.

and scratching. This glass is also used for higher temperature thermometers (> 250 °C), graduated cylinders, and centrifuge tubes.

Vycor (Corning) brand laboratory glassware is recommended for use in applications involving high temperatures, drastic heat shock, and extreme chemical treatment with acids and dilute alkalis. This transparent glassware is resistant to attack by all acids except hydrofluoric, and even in the upper temperature range it is more resistant to alkali than is borosilicate glass. Vycor ware is used primarily in ashing and ignition techniques. It can be heated to 900 °C and can withstand the shock of a temperature drop from 900 °C to 0 °C.

The high silica content of *high silica glass* (over 96%) makes it comparable to fused quartz in its thermal endurance, chemical stability, and electrical characteristics. The glass is made by removing almost all elements except silica from borosilicate glass. It is radiation resistant and has good optical qualities and temperature characteristics; it is used for high-precision analytical work and can also be used for optical reflectors and mirrors.

Boron-free glassware has high resistance to alkali and was developed particularly for use with strongly alkaline solutions. Its thermal resistance is much less than that of borosilicate glass, and it must therefore be heated and cooled very carefully. Its primary use should be with solutions or for digestions involving strong alkali. This glass is often referred to as soft glass.

Low actinic glassware contains materials that usually impart an amber or red color to the glass and thereby reduce the exposure of the contents to light. It was developed to provide a highly protective laboratory glassware for handling materials sensitive to light in the 300 to 500 nm range (e.g., bilirubin, carotene, and vitamin A).

Flint glass is a soda-lime glass composed of a mixture of silicon, calcium, and sodium oxides. This type of glass is lowest in cost of all glasses and is readily fabricated in a wide variety of shapes. Such glass has poor resistance to high temperatures and sudden changes of temperature, and its resistance to chemical attack is only fair. Since this glass is relatively easy to melt and shape, it has been used for bottles and some disposable laboratory glassware. Certain manufacturers supply Flint (soft) glass disposable pipets to the clinical laboratory. Users of such pipets should rinse the pipets before use if they have not been water rinsed by the manufacturer. Soda-lime pipets may release alkali into the pipetted liquid and cause considerable errors in certain critical assay procedures (such as trace metal analyses). Organic contaminants found in soft glass and borosilicate glass disposable pipets can also interfere with analytical procedures using UV-absorption or fluorescence techniques.

Colored and opal glasses, used in filters and light bulbs, are made by adding small amounts of metallic oxides to the glass.

Coated glass has a thin, metallic oxide permanently fire-bonded to the surface of glass. It can conduct electricity and has electronic applications as a heat shield to protect against infrared light and as an electrostatic shield to carry off charges.

Optical glass is made of soda-lime, lead, and borosilicate; because of its high optical purity, it is used in making prisms, lenses, and optical mirrors.

Glass-ceramics (Pyroceram) have high thermal resistance, chemical stability, and corrosion resistance, and they are useful for making hot plates, table tops, and heat exchangers.

Radiation-absorbing glass is made of soda-lime and lead and is useful in preventing transmission of high energy radiation (i.e., gamma rays, X-rays).

Plasticware

The introduction of plasticware to clinical chemistry laboratories has greatly enhanced laboratory analysis. Beakers, bottles, flasks, graduated cylinders, funnels, centrifuge tubes, tubing, and pipets now have unique qualities that make them ideal for use when high corrosion resistance and unusual impact and tensile strength are required. Table 1A-7 describes the physical and chemical properties of various resins that are used in the preparation of laboratory ware, while Table 1A-8 lists the resistance of plasticware to various chemicals. The information is intended to be used as a general guide only. Because many factors can affect the chemical resistance of a given product, it is important to test the material under your own conditions. If any doubt exists about specific applications of plastic labware, the manufacturer should be contacted.

The *polyolefins* are a unique group of resins which are chemically relatively inert. Although concentrated sulfuric acid does slowly attack polyethylene at room temperature, the polyolefins as a group are unaffected by acids, alkalis, salt solutions, and most aqueous solutions. Aromatic, aliphatic, and chlorinated hydrocarbons cause moderate swelling at room temperature; organic acids, essential oils, and halogens slowly penetrate these plastics. Strong oxidizing agents attack this group of resins only at elevated temperatures. *Polyethylene* and *polypropylene* are used primarily to fabricate bottles, beakers, jars, carboys, jugs, funnels, pipet jars, pipet baskets, tanks, buret covers, check valves, disconnect valves, twistcock connectors, needle valves, hollow stoppers, dropping pipets, hydrometer jars, stirring rods, tubing, and reagent dispensers. Polyethylene is less expensive than polypropylene and is used in most disposable plasticware. Polypropylene has a distinct advantage in that it withstands higher temperatures and can be sterilized; however, it absorbs pigments and tends to become discolored.

Polycarbonate is twice as strong as polypropylene and may be used at temperatures ranging from -100 to $+160\,^\circ$C; however, its chemical resistance is not as wide as that of the polyolefins. This resin is unsuitable for use with strong acids, bases, or oxidizing agents; it is dissolved by chlorinated aliphatic and aromatic hydrocarbons. Polycarbonate resin is insoluble in aliphatic hydrocarbons, some alcohols, and dilute aqueous acids and salts. Since labware molded from this resin is glass-clear and shatterproof, polycarbonate is used extensively in centrifuge tubes and graduated cylinders.

Fluorocarbon resins (Teflon) have unique qualities that make them almost chemically inert and therefore ideal when high corrosion resistance at extreme temperatures is essential. Because Teflon (fluorinated ethylene propylene, FEP) resists extreme temperatures ranging from $-270\,^\circ$C to $+255\,^\circ$C, bottles and beakers made of this material are suitable for use in cryogenic experiments.

Labware made of fluorocarbon resins is pure, translucent, white, and inert to such corrosive reagents as boiling aqua regia, nitric and sulfuric acids, boiling hydrocarbons, ketones, esters, and alcohols. Because of its unique anti-adhesive properties and its nonwettable surface, Teflon is used for self-lubricating stopcocks, stirring bars, bottle-cap liners, and tubing. It is also quite easy to clean and is fast drying, but it can easily be scratched and warped.

Use and Care of Plastic Laboratory Ware

Whenever possible, plastics should be used in place of glass, because plasticware has the advantage of being unbreakable and does not release ions into a solution as does glass. Two

Table 1A-7. PHYSICAL PROPERTIES OF PLASTICS

	Max. Usable Temp. (°C)	Brittleness Temp. (°C)	Transparency	Autoclavability	Specific Gravity	Flexibility
Low-density polyethylene	80	−100	Translucent	No	0.92	Excellent
High-density polyethylene	120	−100	Translucent	No	0.95	Rigid
Polypropylene	135	0	Translucent	Yes	0.90	Rigid
Polymethylpentene ("TPX")	175	20	Clear	Yes	0.83	Rigid
Teflon FEP (fluorinated ethylene propylene)	205	−270	Translucent	Yes	2.15	Excellent
Tefzel ETFE (ethylene-tetrafluoroethylene)	150	−100	Translucent	Yes	1.70	Moderate
Polycarbonate	135	−135	Clear	Yes*	1.20	Rigid
Polyvinyl chloride (PC)	70†	− 30	Clear	No†	1.34	Rigid
Polyallomer	130	− 40	Translucent	Yes	0.90	Moderate
Polysulfone	165	−100	Clear	Yes	1.24	Rigid

*Sterilizing reduces mechanical strength. Do not use PC vessels for vacuum applications if they have been autoclaved.
†Except for PVC tubing, which will withstand temperatures of 121 °C and which can be autoclaved.
(From Nalgene Labware, 1983–84.)

Table 1A-8. CHEMICAL RESISTANCE OF PLASTICS AT 20 °C

Classes of Substances	Acids, Dilute or Weak	Acids,* Strong and Concentrated	Alcohols, Aliphatic	Bases	Hydrocarbons, Aliphatic	Hydrocarbons, Aromatic	Hydrocarbons, Halogenated	Ketones	Oxidizing Agents, Strong
Low-density polyethylene	E	E	E	E	F	F	N	G	F
High-density polyethylene	E	E	E	E	G	G	F	G	F
Polypropylene/polyallomer	E	E	E	E	G	F	F	G	F
Polymethylpentene ("TPX")	E	E	E	E	F	F	N	G	F
Teflon	E	E	E	E	E	E	E	E	E
Polycarbonate	E	N	G	N	F	N	N	N	N
Polysulfone	E	G	G	E	G	N	N	N	G
Polyvinyl chloride	E	E	E	E	E	N	N	N	G
Polystyrene	E	F	E	E	N	N	N	N	N
Nylon	F	N	G	F	E	E	G	E	N

E = Excellent; G = Good; F = Fair; N = Not recommended.
*Except for oxidizing acids; for oxidizing acids, see "Oxidizing Agents, Strong."
(From Nalgene Labware, 1983–84.)

13

disadvantages of plasticware are a tendency to bind various solutes and to leach surface-bound constituents into subsequent solutions. Polyethylene is permeable to water vapor, and even in tightly stoppered bottles evaporation may occur, resulting in an increased concentration of the reagents and standards. As the volume of remaining liquid diminishes to a small fraction of the total capacity, the ratio of the surface area to volume increases; with the increased surface area of polyethylene available for diffusion, evaporation becomes a significant factor. Therefore, small volumes of reagent should never be stored in oversized plastic bottles for long periods of time.

Polyethylene is not completely inert and may bind or adsorb proteins, dyes, stains, iodine, and picric acid. Colorless reagents may bind to plasticware without being detected, only to manifest themselves in subsequent chemical analyses as a cause of erratic results. Slow reduction of ceric and cuprous ions has been observed in solutions stored in polyethylene bottles. Such solutions may also develop significant fluorescence after being stored in plastic containers for extended periods.

Linear polyethylene, polypropylene, Teflon, polymethylpentene, and polycarbonate plastics can be cleaned in ordinary glassware washing machines. Ultrasonic cleaners may also be used provided the plasticware does not rest directly upon the transducer diaphragm. The use of abrasive cleaners and strong oxidizing agents should be avoided.

Polypropylene, Teflon, and polymethylpentene may be repeatedly autoclaved under normal conditions. Polycarbonate shows some loss of mechanical strength when autoclaved; thus, this procedure should be limited to 20 min at 121 °C. Plasticware should be well cleaned and rinsed with deionized water before sterilizing, since certain chemicals which have no appreciable effect on plastics at room temperature may cause deterioration at autoclaving temperatures. Polystyrene, polyvinyl chloride, styrene acrylonitrile, and conventional polyethylene are not autoclavable, but they may be gas sterilized with ethylene oxide or chemically sterilized by rinsing with benzalkonium chloride. With the exception of Teflon, none of the listed resins should be hot-air sterilized because of the potential for accelerated oxidative degradation.

Some of the transparent plastics (e.g., polycarbonate, polystyrene) may absorb minute quantities of water vapor and appear cloudy after autoclaving. This clouding effect is transient and disappears as the plastic dries. Clearing of the plastic may be accelerated by drying the plasticware in an oven at 110 °C.

Cleaning of Glass and Plasticware

Laboratory glassware should be rinsed immediately after use and placed in a weak solution of a detergent until it is washed. Hazardous materials should be discarded and their containers should be washed immediately, rather than being placed with the glassware that is handled by laboratory washing personnel. In many institutions, the glassware is then washed in an automatic washer, rinsed in a special rinse cycle, and then placed in an automatic dryer. The manufacturer of the equipment should be able to recommend a detergent that is compatible with the equipment and effective with the local water supply. Residual detergent may be detected by measuring the pH of water added to the glassware, or by using a dilute solution of an acid-base indicator that will indicate an alkaline residue.

For general washing without an automatic washer, most laboratories prefer *detergents* which are not highly alkaline, are nonionic, and are metal free. These have largely replaced the harsh alkaline detergents formerly used which, while effective, etched glassware and were hard on the skin of hands. Again, special care must be taken to assure adequate rinsing. All clean laboratory ware should drain with a continuous thin film of water. Imperfect wetting, or the presence of discrete droplets of water, indicates that the vessel is not sufficiently clean.

Ultrasonic cleaners may also be used to supplement the action of detergent. In these baths where very high frequency vibrations cause the soil to be broken free from the walls of the vessel, neutral detergents are usually preferred.

Cleaning solution which consists of sodium dichromate in sulfuric acid is a common cleaning agent. It is prepared by stirring about 20 g of powdered technical grade sodium dichromate with just enough water to make a thick paste. Technical grade concentrated sulfuric acid, 300 mL, is then added slowly and carefully and with constant stirring. This preparation is best carried out in a sink. Clear supernatant solution should be decanted from the top for use. The solution may be used until the reddish color of dichromate has been replaced by the green color of the chromic ion. Because of its strong oxidizing power, this material removes organic material as well as inorganic material; it should not be used on any plasticware except Teflon. Scrupulous rinsing with tap water followed by reagent water must follow the use of this strong chemical cleaner. Because of its corrosive properties, personnel handling cleaning solution must wear safety glasses, rubber gloves, and a rubber or plastic apron.

Nitric acid may also be used as a cleaning agent with the same safety precautions. In addition, strong nitric acid solutions should be handled only in a chemical hood. Although nitric acid effectively removes trace metals, relatively dilute solutions (1 mol/L or less) of hydrochloric acid are equally effective and more easily used since they do not require the use of a hood.

Air drying or oven drying at temperatures below 100 °C, with the laboratory ware bottom up, is the preferred drying method. Occasionally, it is desirable to rinse the glassware with a water miscible organic solvent and then to expose it to a stream of air or nitrogen. To prevent contamination, the solvent must be of high quality and the gas must be pure. Storage should be such that the laboratory ware is protected from dust.

Synthetic and Rubber-Base Tubing

Tygon is a modified polyvinyl plastic substance which is clear and resistant to chemical attack and is used extensively for the manufacture of tubing. It is inert to most chemicals and has been used to transfer liquids, gases, and colloidal suspensions. It comes in a variety of sizes and is specified by inside diameter (I.D.), by outside diameter (O.D.), and by wall thickness. The tubing is rapidly and easily used and can be readily changed to meet specific needs. Since Tygon tubing is flexible, it can curve around corners and avoid obstructions. It occupies a minimum of space and can be used with a variety of positive-action (peristaltic) fluid pumping systems. Because it is available in continuous lengths, Tygon requires few joints or couplings. It is readily cut with a sharp blade and can be coupled to other larger bore Tygon tubings by using a small amount of solvent (cyclohexanone) for softening the tubing ends and as a bond. Careful heating of the tube ends with a hot metal probe can also assist in softening the plastic for the coupling of tubing.

Tygon will sometimes be difficult to slip over metal or glass in making a connection. The use of a soapy solution or a synthetic wetting agent will have the effect of lubricating the tubing, allowing it to slip on more easily. Various Y, H, and T fittings can also be used to join branch lines and make multiple connections with Tygon tubing. Simple collar clamps can serve to ensure a connection on glass or metal.

Tygon is not compatible with polystyrene and acrylics and, when in contact at high temperatures with solutions containing zinc or copper, can become discolored. Since Teflon tubing is chemically inert to synthetic substances, organic solvents, acids, and alkalis and can withstand continuous temperatures to 300 °C, it can be used in place of Tygon in circumstances requiring its special properties.

Amber *latex* rubber tubing is translucent and is recommended for all kinds of glass connections in which a highly elastic tubing with a long life is required.

Neoprene is a synthetic rubber which is soft and pliable with a smooth finish. Neoprene is not affected by oil, alkalis, hot water, and many corrosive substances. It is not recommended for use with chlorinated or aromatic hydrocarbons.

VOLUMETRIC EQUIPMENT AND ITS CALIBRATION

Most clinical chemistry procedures require accurate volumetric measurements to assure accurate results. For precise work only Class A glassware should be used. Class A glassware is certified to conform to the specifications outlined in the National Bureau of Standards (NBS) circular C-602.

Pipets

Two types of pipets are mainly used in clinical chemistry, transfer pipets and measuring pipets (Figure 1A-1). *Transfer pipets* include volumetric and Ostwald-Folin pipets and are designed to deliver a fixed volume of liquid. They consist of a cylindrical bulb joined at both ends to

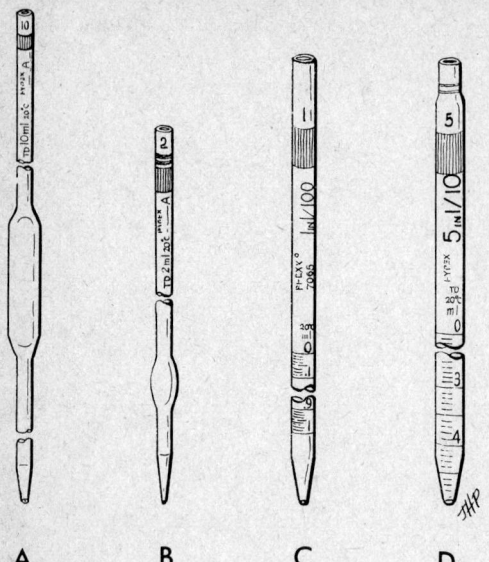

Figure 1A-1. Pipets. *A*, Volumetric (transfer). *B*, Ostwald-Folin (transfer). *C*, Mohr (measuring). *D*, Serologic (graduated to the tip).

narrower glass tubing. A calibration mark is etched around the upper suction tube, and the lower delivery tube is drawn out to a gradual taper. The bore of the delivery orifice should be sufficiently narrow so that rapid outflow of liquid and incomplete drainage cannot cause measurement errors beyond tolerances specified.

A volumetric transfer pipet (Figure 1A-1,*A*) is calibrated to deliver accurately a fixed volume of a dilute aqueous solution such as a standard, a filtrate, or a nonviscous sample. The most commonly used sizes are 1, 2, 3, 4, 5, and 10 mL. Less frequently used sizes delivering 6, 8, 25, 50, and 100 mL can also be obtained. Table 1A-9 shows examples of tolerances of various types of pipets. The reliability of the calibration of the volumetric pipet decreases with a decrease in size and, therefore, special micropipets have been developed for microanalysis.

Ostwald-Folin pipets (Figure 1A-1,*B*) are similar to volumetric pipets but have their bulb closer to the delivery tip, thereby reducing the surface area in contact with the liquid. The commonly used sizes are 0.5, 1.0, 2.0, and 3.0 mL. These pipets are used for the accurate measurement of viscous fluids such as blood or serum. In contrast to a volumetric pipet, an Ostwald-Folin pipet has an etched ring near the mouthpiece, indicating that it is a blow-out pipet. The liquid is blown out of the pipet only after the blood or serum has drained to the last drop in the delivery tip. When filled with opaque fluids such as blood, the top of the meniscus must be read. Controlled slow drainage is required with all viscous solutions so that no residual film is left on the walls of the pipet.

The second type of pipet is the *graduated or measuring pipet* (Figure 1A-1,*C*). This is a piece of glass tubing which is drawn out to a tip and graduated uniformly along its length. Two kinds are available; the Mohr pipet is calibrated between two marks on the stem, whereas the serologic pipet has graduation marks down to the tip. The serologic pipet (Figure 1A-1,*D*) must be blown out to deliver the entire volume of the pipet and has an etched ring (or pair of rings) near the mouth end of the pipet signifying that it is a blow-out pipet. Mohr pipets require a controlled delivery of the solution between the calibration marks. Serologic pipets have a larger orifice than the Mohr pipets, and thus drain faster. Measuring pipets are commonly supplied in 0.1, 0.2, 0.5,

Table 1A-9. TOLERANCES OF VARIOUS TYPES OF PIPETS

Type	Size (mL)	Class	Limit of Error (mL)
Transfer	5	A	0.0100
Ostwald	5	A	0.0100
Serologic	5		0.0400
Mohr	5		0.0400

5.0, and 10.0 mL sizes. In clinical chemistry laboratories, measuring pipets are principally used for the measurement of reagents and are not generally considered accurate enough for measuring samples and standards.

Measuring pipets are made of borosilicate glass; however, constant vigilance is needed to be certain that pipets with broken tips or pipets that have become badly etched are discarded. Pipets are now available that are made from special, chemically tempered glass. Although they cost more, they will outlast their borosilicate counterparts because they break less easily. In addition, they are less susceptible to scratching and chipping because the glass is harder.

There are certain common techniques which apply to all pipets described above. First, mouth pipetting should never be allowed. Instead, pipetting bulbs should be used by each individual in the laboratory. Pipets must be held in a vertical position when adjusting the liquid level to the calibration line and during delivery. The lowest part of the meniscus, when it is sighted at eye level, should be level with the calibration line on the pipet. The flow of the liquid should be unrestricted when using volumetric pipets and the tips should be touched to the inclined surface of the receiving container until 2 seconds after the liquid has ceased to flow.

With graduated pipets, the flow of liquid may have to be slowed during delivery. Serologic pipets are calibrated to the tip, and the etched glass ring on top of the pipet signifies that it is to be blown out. The pipet is first allowed to drain and then the remaining liquid is blown out.

Verification of Pipet Calibration

The following procedure is used to verify calibration of a to-deliver pipet made of borosilicate glass. Note that water, weighing vial, and pipet must all be at room temperature at the time of the procedure.

1. Determine and record the temperature (t) of a large volume of pure water to the nearest whole degree.
2. Weigh a clean, dry, empty, stoppered weighing vial. Handle the vial without touching it with bare fingers. Record weight to nearest 0.1 mg (W_v).
3. Transfer a sample of water to the weighed vial.
4. Restopper and reweigh the vial. Record weight to the nearest 0.1 mg (W_v).
5. Refer to the table titled "*True capacity of glass vessels from the weight of the contained water or mercury when weighed in air*" in Handbook of Chemistry and Physics to obtain the correction factor (F_t) for the water temperature.[9]
6. Calculate

$$(W_f - W_v) \times (F_t) = \text{actual capacity, mL at temperature of manufacturer's calibration}$$

Example for a 10 mL pipet:
$$\begin{aligned} W_f &= 31.9961 \text{ g} \\ W_v &= 22.0391 \text{ g} \\ t &= 24\ ^\circ C \\ F_t &= 1.003\ 771 \text{ (from table)}[9] \end{aligned}$$

$$(31.9961 - 22.0391) \times (1.003\ 771) = 9.9945 \text{ mL, actual capacity}$$

The deviation, or error, from stated capacity is

$$\frac{(10 - 9.9945)}{10} \times 100 = 0.055\%$$

In routine analysis, an error of 0.1% or less may be ignored when using the pipet. With an error in excess of 0.1%, depending upon the application, the stated capacity of a given pipet may require correction to the actual capacity found.

Micropipets

In micro work, the remaining volume that coats the inner wall of a pipet can cause significant error. For this reason most micropipets are calibrated to contain (TC) the stated volume rather than to deliver it. Proper use requires rinsing the pipet with the final solution after delivering the contents into the diluent. Volumes are expressed in microliters (μL); the older term lambda is no longer recommended. (One lambda [λ] = 1 μL = 0.001 mL.) Micropipets are generally available in small sizes, ranging from 1–500 μL.

Semi-Automatic Pipets and Dispensers

Semi-automatic versions of pipets and dispensers are available in sizes from 1 μL to 20 mL. Figure 1A-2,*A* illustrates a positive displacement pipetter. The capillary attachments into which the sample is drawn may be siliconized glass, glass, or plastic. Models that allow for digital adjustment of volume are also available.

Figure 1A-2,*B* illustrates another type of single or multirange micropipetting device. This device will draw up and dispense its predefined volume (from 5–1000 μL) when the plunger is moved through a complete cycle. The fluid containment tip is made of a disposable plastic material which tends to retain less inner surface film than does glass. Such pipets avoid the risk of cross contamination between samples, eliminate the need for washing between samples, and improve the precision of measurements. A macro version of this device can be used to "deliver" or transfer larger volumes such as samples of protein-free filtrates or dilutions. The usable range is 1–5 mL and the device may be adjusted for any volume within those limits.

Figure 1A-3,*A* shows a dilutor that is a semiautomatic sampling, diluting, and dispensing apparatus which measures and dispenses preset volumes of solutions by means of two motor-driven syringes, one for metering the sample and one for metering the diluent. The dilutor can be adjusted to aspirate as little as 10 μL of sample and to deliver it with as much as 10 mL of diluent. The reproducibility of the metered pumps has been claimed to be ±0.5% of delivered volume.

A more automatic and versatile piece of equipment is shown in Figure 1A-3,*B*. This is actually a complete liquid-handling system which can aspirate up to five different fluids and deliver them with diluent. The equipment is microprocessor controlled and is easily programmed. Accuracy is claimed to be better than 0.5% of dispensed volume and reproducibility better than 0.1% when dispensing at least 10% of syringe volume.

Calibration of Micropipets, Dispensers, and Dilutors

Accurate calibration of micropipets, dispensers, and dilutors is as important as the calibration of macro-volumetric pipets. The classic procedure for checking the calibration of micropipets employs gravimetric measurement of mercury as the substance contained by the micropipet.

Measure the weight of mercury contained or delivered by the micropipet and divide by the density of mercury at its observed temperature to find the volume of the pipet. The density of mercury at 25 ±5 °C is 13.5340 ± 0.0123 (± 0.091%); therefore, for temperatures between 20 and 30 °C it is sufficiently accurate to use the density at 25 °C.

Spectrophotometric Calibration of Micropipets

As an alternative to the classic gravimetric calibration using mercury, a colored compound may be used to verify the calibration of micropipets. The following is an example of the use of *p*-nitrophenol in a spectrophotometric verification of the calibration of a 10 μL pipet.

Procedure

1. All glassware used should meet Class A specifications.
2. Reagents required:
   (a) NaOH, 0.01 mol/L
   (b) *p*-nitrophenol, 105 mg/dL

Dissolve 105 mg of high purity *p*-nitrophenol (NBS SRM 938) in deionized water in a 100-mL volumetric flask. Fill to the mark.

3. Preparation of dilutions:

Reference dilutions: Fill three 250-mL volumetric flasks to the mark with 0.01 mol/L NaOH. To each add 1.0 mL of the *p*-nitrophenol solution, using a different volumetric pipet for each addition.

Test dilutions: Add to each of 5 test tubes 2.5 mL NaOH, 0.01 mol/L, using either a calibrated pipet or calibrated dispenser. To each tube, using the test micropipet in the correct

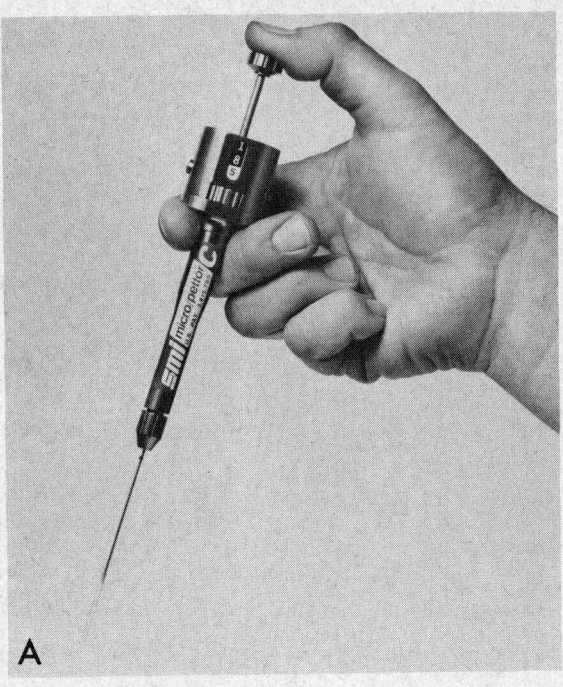

A

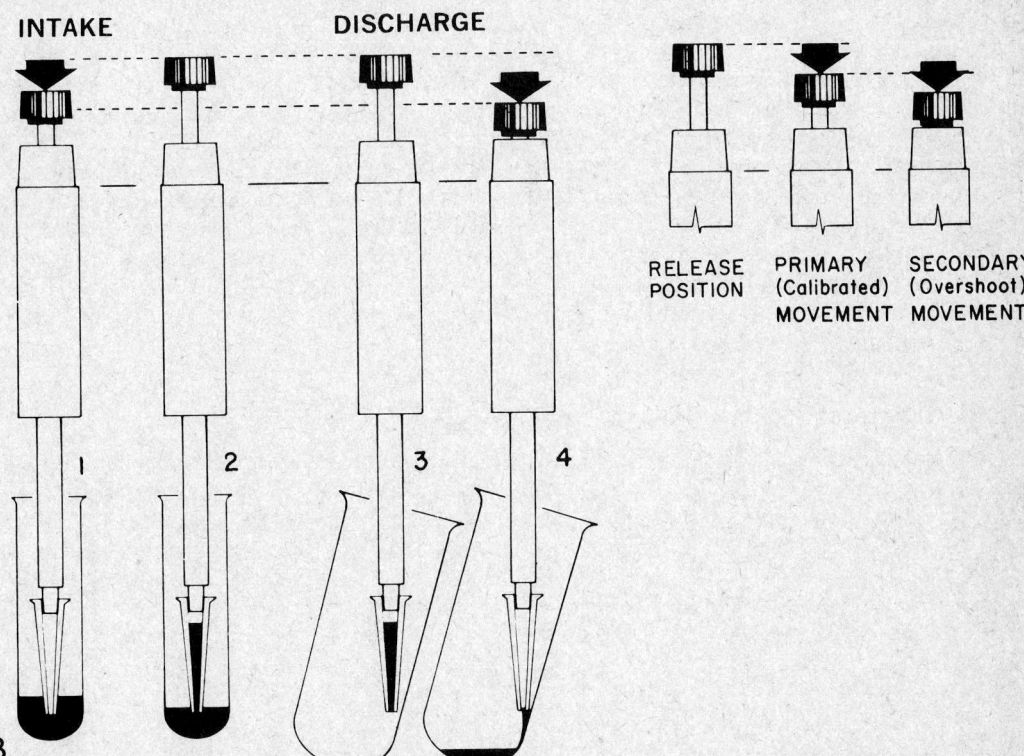

INTAKE   DISCHARGE

RELEASE   PRIMARY   SECONDARY
POSITION   (Calibrated)   (Overshoot)
                MOVEMENT   MOVEMENT

1   2   3   4

B

Figure 1A-2. *A,* Multirange micropipetting device. (Courtesy of Scientific Manufacturing Industries, Inc.) *B,* A semi-automatic micropipetting device which is designated as a TD pipet.

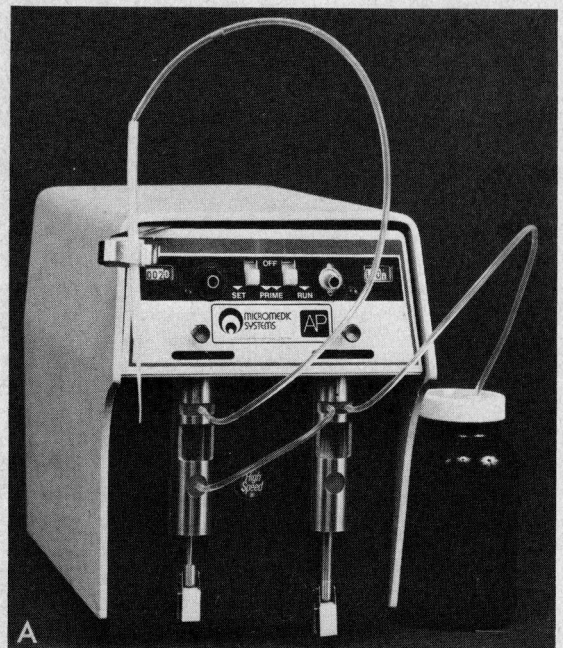
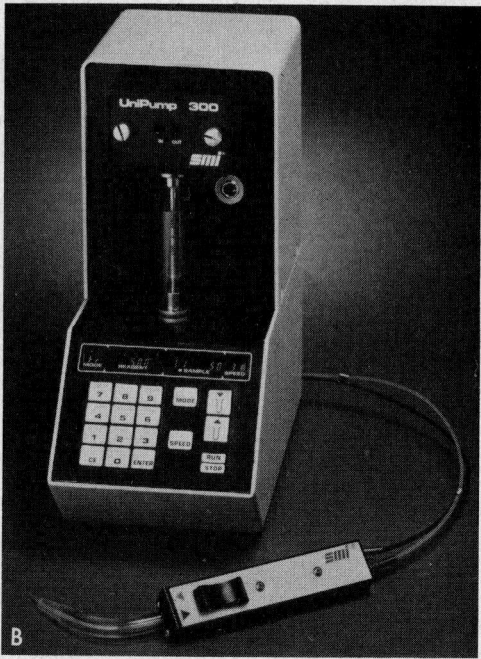

Figure 1A-3. *A*, Semiautomatic sampling, diluting, and dispensing apparatus. (Courtesy of Micromedic Systems, Inc.) *B*, An automatic and versatile sampling, diluting, and dispensing apparatus. (Courtesy of Scientific Manufacturing Industries, Inc.)

delivery mode, add 10 μL of *p*-nitrophenol solution. (If the pipet is the "to deliver" [TD] type, and is being rinsed in receiving solution, set aside a portion of *p*-nitrophenol for rinsing the pipet between each of the five additions.)

4. Read the absorbance of each of the reference and test dilutions, in 10 mm cuvets, at 401 nm on a narrow bandpass spectrophotometer. For NBS *p*-nitrophenol SRM with a "certified specific absorbance" of 131.48 L·g^{-1}·cm^{-1} in NaOH, 0.01 mol/L, A_{401} should be 0.550 for an exactly prepared reference dilution. Average the readings of the three reference dilutions; the mean value (A_1) should be close to 0.550.

Average the five test dilution readings for a mean value (A_2).

5. Calculation:

$$A_2/A_1 \times D \times V = \text{volume, in } \mu\text{L, delivered by pipet}$$

where D is dilution of test solution (1/251 in this example), and V is final volume in μL of the test dilution (2510 μL in this example).

For this example then, if A_1 is 0.550 and A_2 were 0.561, 0.561/0.550 $\times$ 1/251 $\times$ 2510 μL = 10.20 μL delivered by pipet with a nominal capacity of 10 μL. This is a 2% error. The analyst must make a decision as to the acceptability of this error. As supplied by the manufacturer, micropipets of this capacity are capable of delivering between 0.5% and 1% of nominal capacity depending on the type of pipet.

Automatic dilutors can be calibrated utilizing similar techniques.

Calibration of Micropipets and Dilutors Using Radioisotopes

Another technique to verify the calibration of micropipets or dilutors and to estimate the precision of their performance is available to laboratories routinely using gamma emitting radioisotopes. Because of its availability and low radiation hazard the isotope of choice is ^{125}I, and any water-soluble labeled material of high specific activity is acceptable.

To achieve a valid estimate of precision, (1) the volumes of the replicates must be identical in order to avoid variations in counting geometry, (2) the count rate (counts per minute) must be at least 50 times the background count, and (3) the total counts must be sufficiently high to result in a coefficient of variation (CV) of less than 0.5%. The minimal counting time to achieve a given coefficient of variation may be calculated from a formula which assumes a Poisson distribution:

$$CV(\%) = \frac{\sqrt{N}}{N} \times 100 = \frac{\sqrt{RT}}{RT} \times 100$$

where N is the net total count, R is the rate, and T is the time.

The accuracy of the micropipet or dilutor may be ascertained by comparison with a "reference" or "master" pipet whose calibration has been verified utilizing a gravimetric technique.

Burets

Standard burets vary in size from 1–100 mL. A buret of 50 mL capacity is subdivided at 0.1 mL intervals. Burets having a capacity of 10 mL or less are classified as microburets.

The outflow of liquid from the buret is usually controlled by an all-glass or all-Teflon stopcock. The latter type does not require any lubricant and is especially useful when the titrant is an alkali. The all-glass stopcock should be lightly greased with petrolatum-like lubricants. Silicone-containing lubricants are not recommended since they "creep" along the length of the buret with subsequent contamination of the walls. Some burets are equipped with a reservoir and a 2-way stopcock for self-filling.

A buret calibration is verified by first filling the buret to a point just above the zero line with CO_2-free deionized water, and then carefully adjusting the meniscus to the zero line. The drop of water adhering to the buret tip is removed by touching the tip to the inside of a glass vessel. A tared vessel is placed beneath the tip and the buret is fully opened. Delivery should proceed freely with the buret tip not in contact with the vessel wall until the meniscus is about 1 cm above the graduation line of the desired volume. The buret tip should then be touched to the wall of the container and the content allowed to drain into the solution until the meniscus reaches the graduation mark. The tared vessel is then stoppered and reweighed, and the delivered volume corrected for temperature. As in the case with graduated pipets, the accuracy of the overall calibration and mark-to-mark calibration decreases with a decrease in buret size.

Burets used in macroanalysis have major graduation marks completely around the long cylindrical tube and minor graduation marks at least halfway around. Errors of parallax in reading the meniscus are thus minimized at the major graduations but may still occur at the minor graduations.

Microburets

Some microburets are constructed like a syringe. The plunger of the syringe can be machined to an exact diameter. The displacement by the advancing plunger is indicated on a micrometer dial and is proportional to the volume delivered. A typical model is shown in Figure 1A-4.

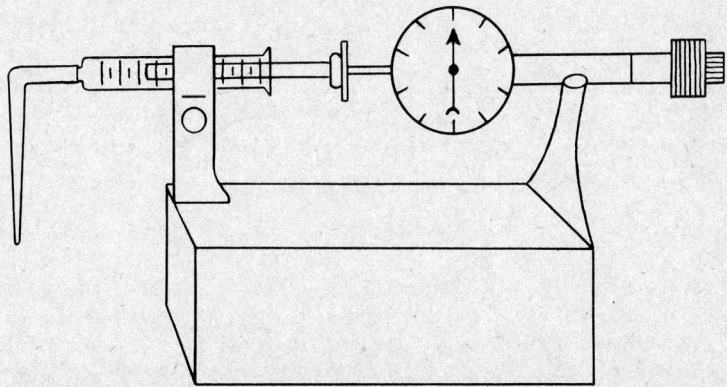

Figure 1A-4. Syringe type microburet.

Various sizes are available that permit accurate delivery of as little as 0.01 μL per scale division. Several points should be noted:

1. The titrant must be prepared at a relatively high concentration to reduce the volume necessary for a titration and to provide sharp end points. For example, if an end point can be determined to ± 1 scale division, the volume of titrant in an average titration should be equivalent to 100 scale divisions to provide a precision of $\pm 1\%$. The concentration of titrant is a compromise between choosing a weaker concentration, which provides a larger volume and therefore a smaller reading error, and a stronger concentration, which minimizes end point error and avoids over-dilution of the sample.

2. The tip of the pipet must be placed beneath the surface of the solution to be titrated to permit even dispensing of reagent and to avoid discrete drop formation. The tip must be drawn to a fine point to prevent diffusion of liquid back into the pipet.

3. Titrations must be performed with good lighting against a white background to permit easier detection of color changes, and the solution must be stirred as the titration proceeds. For titration in test tubes the solution may be mixed with a fine stream of air or bubbles of an inert gas (Figure 1A-5). Vibrators or small cups that are rotated electrically are also means for obtaining mixing during the titration. Magnetic stirrers with very small stirring bars (e.g., 1×5 mm) are available commercially.

4. Blanks and standards must be included routinely and titrated to the same end point as the unknown. Indicator error is more significant in microtitrations; it is therefore advisable to pipet a fixed volume of indicator rather than to add it dropwise. Duplicate determinations are recommended.

Volumetric Flasks

Volumetric flasks (Figure 1A-6) are commonly found in sizes varying from 1–4000 mL. They are primarily used in preparing solutions of known concentration and they are available in various grades. The most accurate are certified to meet standards set forth by the National Bureau of Standards.

An important factor in the use of volumetric apparatus is the need for an accurate adjustment of the meniscus. A small piece of card that is half black and half white is most useful. The card is placed 1 cm behind the apparatus with the white half uppermost and the top of the black area about 1 mm below the meniscus. The meniscus then appears as a clearly defined thin black line. This device is also useful in reading the meniscus of a buret.

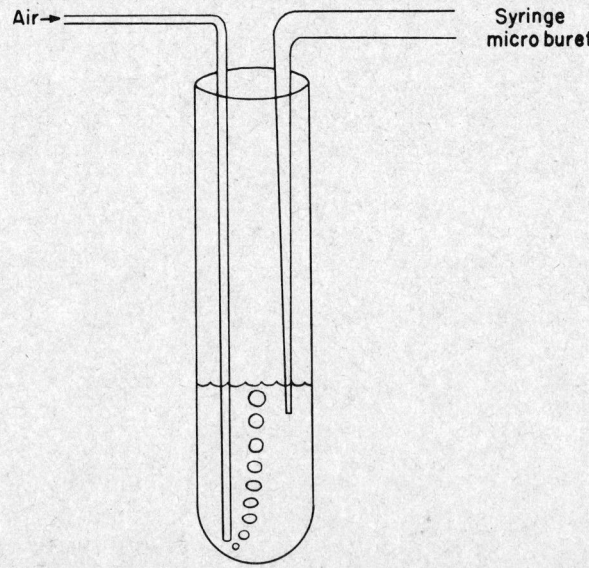

Figure 1A-5. Microtitration in a test tube. Mixing is accomplished with a stream of air bubbles.

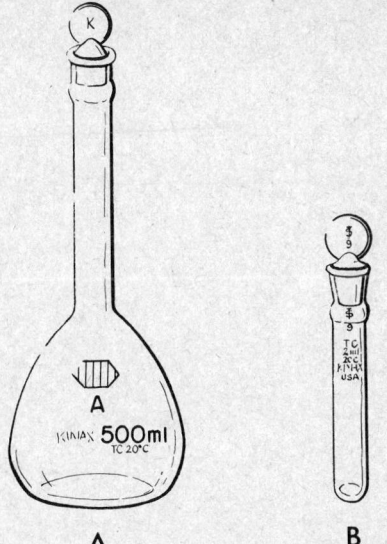

Figure 1A-6. Volumetric flasks. *A*, Macro. *B*, Micro.

Volumetric equipment should be used with solutions equilibrated to room temperature. Solutions diluted in volumetric flasks should be repeatedly mixed during dilution so that the contents are homogeneous before the solution is made up to final volume. Errors due to expansion or contraction of liquids upon mixing are thereby minimized.

Calibration of volumetric flasks. Volumetric flasks should be thoroughly cleaned and dried before calibration. The flask is then weighed and filled with carbon dioxide–free deionized water until just above the graduation mark. The neck of the flask just above the water level should be kept free of water. The meniscus mark is set at the graduation line by removing excess water and the flask is reweighed. The final weight is corrected for the equilibrated water and air temperature to obtain the volume of the flask. Flasks may also be calibrated by the spectrophotometric technique previously described.

Maintenance and Care of Volumetric Glassware

Volumetric glassware and glass apparatus must be absolutely clean to ensure that the volumes measured are accurate and hence chemical reactions are not adversely affected. One gross method that may be used to test for cleanliness is to fill the vessel with distilled water and then empty it and examine the walls to see whether they are covered by a continuous thin film of water. Imperfect wetting or the presence of discrete droplets of water indicates that the vessel is not sufficiently clean. Obviously this criterion cannot be used for nonwettable plastics. A test for residual acid or alkaline detergents is the measurement of the pH of water added to the glassware. A dilute solution of an acid-base indicator applied to the surface is another method to check for residual alkaline detergents. The number of pieces of glassware checked depends on the work load and number of pieces cleaned, but at least one piece of randomly selected glassware should be checked each day.

The cleaning of glassware has been described earlier. Of the various cleaning agents in common use, the National Bureau of Standards prefers fuming sulfuric acid and a chromic sulfuric acid mixture. However, in general the method used to clean glassware should be as mild as possible and appropriate to the type of contamination present.

CENTRIFUGES

A centrifuge is a device to accelerate gravitational separation of substances differing significantly in their masses. In the clinical laboratory centrifugation is used:

1. To separate particles from a solution in which they are suspended. Examples of this application are (a) removing cellular elements from blood to provide cell-free plasma or serum for analysis, (b) concentration of cellular elements and other components of biologic fluids for microscopic examination or chemical analysis, (c) elimination of chemically precipitated protein from an analytic specimen, and (d) separating protein-bound or antibody-bound ligand from free ligand in immunochemical and other assays.

2. To separate two liquid phases of different densities, e.g., (a) extracting solutes in biological fluids from aqueous to organic solvents, and (b) separating lipid components such as chylomicrons from other components of plasma or serum, and lipoproteins from each other.

Types of Centrifuges

Centrifuges may be classified generally into three types: horizontal-head or swinging-bucket, fixed-angle or angle-head, and ultracentrifuge. There are many different models of the fixed-angle and angle-head centrifuges, including benchtop and floor models, as well as refrigerated and nonrefrigerated versions.

Horizontal-head or *swinging-bucket centrifuges* allow the tubes placed in the cups of the rotor to assume a horizontal plane when the rotor is in motion, and a vertical position when it is at rest. During centrifugation, particles travel in a constant manner along the tube while the tube is at right angles to the shaft of the centrifuge; thus the sediment is distributed uniformly against the bottom of the tube. The surface of the sediment is flat (parallel to the shaft of the centrifuge) and remains so, with a column of liquid on top of it when the rotor stops and the tube assumes a vertical position. Supernatant liquid is simply removed by a pipet with negligible disturbance of the packed sediment. If the sediment is well packed, decantation may be used to remove the supernatant.

In an *angle-head rotor,* tubes are held in a fixed position at angles from 25–40 degrees to the vertical axis of rotation. As in the horizontal head centrifuge, particles are driven outward horizontally but strike the side of the tube so that the sediment packs against the side and bottom of the tube with the surface of the sediment parallel to the shaft of the centrifuge. As the rotor slows down and then stops, gravity causes the sediment to slide down the tube; generally a poorly packed pellet is formed.

The aerodynamic shape of a fixed-angle rotor allows more rapid sedimentation of small particles than is usually possible with a horizontal-head rotor. Fixed-angle rotors can be run at a higher speed than swinging-bucket rotors, which offer considerable resistance to rotation and which generate heat as a result of air friction. The resistance of a horizontal-head rotor can be lessened if the swinging buckets are enclosed in a wind shield to improve aerodynamics. The sedimentation of large particles is efficient at low speed, so the horizontal-head rotor is adequate for the separation of erythrocytes from plasma or of a protein precipitate from a supernatant.

Ultracentrifuges are very high-speed centrifuges which usually employ fixed-head rotors. The most common application of an ultracentrifuge in the clinical laboratory is the separation of lipoproteins. Because the separation may require hours or days and may generate considerable heat as a result of friction at high speeds, ultracentrifugation requires a refrigerated chamber. Ultracentrifuges are available in both analytical and preparative models. The tabletop model Airfuge Ultracentrifuge is a miniature air turbine with a small rotor capable of achieving a centrifugal force of 165 000 times gravity ($\times$ g). It is used in the clinical laboratory to clear serum of chylomicrons so that accurate analyses may be performed on the infranatant. The application of ultracentrifuges is discussed in Chapter 7, Lipids.

Components of a Centrifuge

All centrifuges contain a rotor or centrifuge head, a drive-shaft, and a motor. The rotor is enclosed in a chamber that is provided with a cover and latch. Most centrifuges also include a power-switch, timer, speed control, tachometer, and brake. Some include a protective shield to minimize aerosol production if a tube should break, or a refrigerator to reduce the temperature within the chamber. Centrifuges may also include audible or visible alarms to indicate malfunctions such as imbalance of the rotor.

The typical design of a horizontal-head rotor allows the use of four buckets, each able to swing freely between $\times$-shaped metal limbs. The buckets are suspended from support pins placed into grooves in the limbs. The buckets can accept a variety of racks or adapters into which the tubes to be centrifuged are inserted. For any one model of centrifuge, the buckets serve as the tube holders and are interchangeable. Most tube holders contain cushioning pads to lessen the possibility of tube breakage during centrifugation. In some bench-top centrifuges as many as 60 tubes may be inserted into a single rack, allowing a theoretical capability of centrifuging 240 tubes at one time.

The drive-motor of a large centrifuge is usually a high-torque, heavy-duty DC electric motor of up to $3/4$ HP. AC motors are usually used with small centrifuges. Diodes and capacitors are used to rectify the alternating current. Generally the rotor shaft is driven directly or through a gyro system, but occasionally a pulley-system is used. In most centrifuges, sealed bearings are used to reduce the need for lubrication and to minimize vibration. The speed of the centrifuge rotor is governed by a potentiometer, which changes the voltage supplied to the motor. However, differences in the load to be spun cause the rotor to turn at different speeds although the potentiometer setting is the same. The tachometer indicates the speed of the rotor (revolutions per minute, rpm), not the gravitational force applied to the contents of the tubes in the rotor. The brake operates by reversing the polarity of the current to the motor, thus causing rapid deceleration.

The timer on a centrifuge allows a rotor to reach a preprogrammed speed under optimized conditions and then decelerate without braking after the set time. Most centrifuges can be operated independently of a timer.

Centrifuges generate heat. The temperature in the chamber in many centrifuge models may increase by as much as 5 °C following a single run to separate serum or plasma from cells. The change in temperature depends on the initial ambient temperature in the centrifuge chamber, rotor speed, duration of centrifugation, and rotor design. These factors affect the concentration of a specimen due to loss of water caused by evaporation. A large chamber around the rotor dissipates heat. When the material to be centrifuged is temperature labile, a refrigerated centrifuge should be used. In the simplest form a refrigerator unit is mounted beside the centrifuge and cold air is blown into the rotor chamber. This approach is usually inadequate to stabilize the low temperature. In more sophisticated centrifuges, refrigeration coils around the chamber make it possible to maintain a preset temperature within ±1 °C.

Some models of centrifuges are available with a plastic cover over the rotor chamber or over each swinging bucket to reduce aerosol formation when a tube breaks during centrifugation.

Principles of Centrifugation

Speed of centrifugation, measured in revolutions per minute or rpm, does not describe the force required to separate two phases in a centrifuge. The correct term is relative centrifugal force (RCF), also called relative centrifugal field. Units are expressed as number of times greater than gravity, e.g., $5000 \times g$.

Relative centrifugal force is calculated as follows:

$$RCF = 1.118 \times 10^{-5} \times r \times n^2$$

where 1.118×10^{-5} is an empirical factor, r is the horizontal distance, i.e., the radius in centimeters, from the center of rotation to the bottom of the tube in the rotor cavity or bucket during centrifugation, and n is the speed of rotation of the rotor in rpm.

The RCF of a centrifuge may also be determined from the nomogram distributed by manufacturers of centrifuges (see Appendix). RCF is derived from the distance from the rotor center to the bottom of the tube, whether the tube is horizontal to, or at an angle to, the rotor center. The RCF as calculated above is the maximum RCF. Maximum RCF is not the force to which all the contents of a tube are subjected. The minimum RCF, calculated in the same way but from the center of rotation to the *surface* of the liquid, may be several hundred *g* less than the maximum value. Note also that RCF applied to a tube in a fixed-angle head may be much less than that applied to the same tube in a horizontal-head rotor, because the tube in a fixed-angle head is unable to swing outward.

The time required to sediment particles depends on the rotor speed, the radius of the rotor, and the effective path length traveled by the sedimented particles, i.e., the depth of the liquid in the tube. Duplication of conditions of centrifugation is often desirable. The following is a useful formula for calculating speed required of a rotor whose radius differs from the radius with which a prescribed RCF was originally defined.

$$\text{rpm (alternate rotor)} = 1000 \times \sqrt{\frac{\text{RCF (original rotor)}}{11.18 \times r \text{ (cm, alternate rotor)}}}$$

Length of time for centrifugation can also be calculated so that running with an alternate rotor of a different size is equivalent to running with the original rotor.

$$\text{time (alternate rotor)} = \frac{\text{time} \times \text{RCF (with original rotor)}}{\text{RCF (with alternate rotor)}}$$

Note, however, that it may not be possible to reproduce conditions exactly when a different centrifuge is used. Descriptions of times of centrifugation include the time for the rotor to reach operating speed (which may vary from instrument to instrument) and do not include deceleration time, during which sedimentation is still occurring but less efficiently. Even with maximum braking, deceleration may take as long as 3 min in some centrifuges.

Operation of the Centrifuge

Despite years of experience with centrifuges, there are no specific recommendations for RCF or time for centrifugation of blood specimens. NCCLS standard H18-P[5] proposes RCF of 1000–1200 $\times$ g for 10 $\pm$5 min, but clearly, considerable deviations from the recommendation would still allow adequate separation. No standards have been established for centrifugation of other specimens, such as serum to which a protein precipitant has been added.

For proper operation of a centrifuge, only those tubes recommended by its manufacturer should be used. The material used for the tube must withstand the RCF to which the tube is likely to be subjected. Polypropylene tubes are generally capable of withstanding RCFs of up to 5000 $\times$ g. The tubes should have a tapered bottom, particularly if a supernatant is to be removed, and should be of a size to fit securely into the rack to be centrifuged. The top of the tube should not protrude so far above the bucket that the swing into a horizontal position is impeded by the rotor.

For smooth operation of the centrifuge, the rotor must be properly balanced. The weight of racks, tubes, and their contents on opposite sides of a rotor should not differ by more than 1% or by an acceptable limit established by the manufacturer. The smaller the difference, the smoother the centrifugation. Before centrifuging any laboratory specimens, pairs of adapters and specimen tubes should be placed on opposite pans of a balance and the tubes rearranged, so that the weights are equal and the placement of tubes is symmetrical. Tubes filled with water may also be used to equalize the weights. The total weight of each rack should not exceed the limits stated by the centrifuge manufacturer at a rated speed.

Imbalance of the rotor causes vibration, and even when this imbalance is not great enough to trigger an imbalance detector, it will result in increased wear on the centrifuge and more frequent breakage of tubes. Because tubes do not swing fully horizontal, sediments pack poorly during spin and tend to resuspend during deceleration and stopping. Slight vibration of a centrifuge can normally be expected when the rotor is accelerating or decelerating through slow speeds. Vibration is amplified at a critical speed range during the acceleration or deceleration and continuous operation of a centrifuge in this critical range should be avoided. Although the rotor must pass through this range as it decelerates, centrifugal force is still enough to prevent disturbance of a well-packed sediment.

Tubes of collected blood should be centrifuged prior to being unstoppered to reduce the probability of an aerosol being produced when the tube is opened. The practice of using a wooden applicator to release a clot stuck to the top of the tube, or to its stopper, should be avoided; it is a potential cause of hemolysis. Centrifugation at an appropriate RCF will usually ensure that the clot is released from the tube wall and drawn to the bottom of the tube.

Operating Practice

Cleanliness of a centrifuge is important in minimizing the possible spread of infectious agents, such as hepatitis viruses. With proper operation of a centrifuge, few tubes will break. In case of breakage, the racks and chamber of the centrifuge must be carefully cleaned. Any spillage should be considered a possible hepatitis hazard. Gray dust, arising from the sand-blasting of the chamber by fragments of glass, indicates tube breakage and possible contamination which necessitates cleaning of the chamber. Broken glass embedded in cushions of tube holders may be a continuing cause of breakage if cushions are not inspected and replaced in the cleanup procedure.

The speed of a centrifuge should be checked at least once every three months. The procedure should use a stroboscopic light or a vibrating-reed external tachometer of known accuracy. The measured speed should not differ by more than 5% from the rated speed under specified conditions. All the speeds at which the centrifuge is commonly operated should be checked. The centrifuge timer should be checked weekly against a reference timer such as a stop watch and should not be more than 10% in error. The temperature of a refrigerated centrifuge should be measured monthly under reproducible conditions and should be within 2 °C of the expected temperature. Commutators and brushes should be checked at least every three months. Brushes should be replaced when they show considerable wear.

Criteria of satisfactory performance should be established for every centrifuge, just as for other laboratory instruments. Maintenance and repair records should be carefully kept.

PROCEDURES FOR CONCENTRATING SOLUTIONS

Evaporation is a technique most often used to concentrate solutes in a solution, particularly low molecular weight solutes. A technique frequently used for the evaporation of volatile solvents is to direct a stream of air through a fine nozzle onto the surface of liquid in a tube to flush out vapors in equilibrium with the solution. With a suitable manifold this procedure can be applied to many tubes simultaneously, and the procedure can be speeded up by placing the tubes in a heated water bath. The procedure should be carried out in a fume hood to avoid a health hazard resulting from evaporating solvents. If compressed air is used, it should be passed through a filter and over a desiccating agent to prevent particulate matter and moisture from contaminating the solution.

Rapid removal of large volumes of a solvent is best effected through combination of reduced pressure and heat. The solution to be evaporated is placed in a round-bottom flask of a capacity no larger than 100 mL to avoid the danger of implosion. Vacuum is applied to the flask through a safety trap while the flask is swirled in a heated water bath. Swirling gives a large surface area, thus speeding evaporation and reducing bumping. Alternatively, a thick-walled Erlenmeyer flask may be used at ambient temperature. A water aspirator, an institution vacuum line, or a mechanical vacuum pump may be used to reduce the pressure. With any system, a suitable trap is essential to collect the vapors.

To concentrate a larger volume of solution, a rotary evaporator is used. To concentrate many specimens of small volume, vortex evaporators or multiple unit rotary evaporators are available.

Freeze-drying has been extensively used in clinical chemistry for the preparation of control materials, some unstable reagents, and to a lesser extent for the preparation of individual specimens for analysis. More than 99% of the water in a substance to be freeze-dried may be removed in the typical freeze-drying process. The composition of the freeze-dried material is generally unchanged by freeze-drying except for the loss of water or some volatile organic matter. Freeze-drying first entails freezing a material at −40 °C or less and then subjecting it to a high vacuum. Very low temperatures cause the ice to sublime; solid nonsublimable material, initially locked in an ice matrix, remains behind in a dried state.

Commercial freeze-drying instruments allow many specimens to be dried at the same time by means of special manifolds. Such instruments usually combine a freeze unit with a lyophilizer. In the most sophisticated systems, the temperature and vacuum may be adjusted according to the material to be processed. Freeze-drying is most efficient, i.e., drying is most rapid and the material does not thaw, when the material to be dried is spread as a thin film over the inner surface of a flask or other container. This is best accomplished when the container is rotated during freezing. The volume of material to be lyophilized may vary from a few milliliters to several liters.

SEPARATORY FUNNELS AND EXTRACTION PROCEDURES

Separatory funnels are employed in the clinical chemistry laboratory for simple extraction procedures. This involves bringing a given volume of solution into contact with a given volume of solvent (immiscible in the solution) and shaking until equilibrium has been attained, followed by separation of the liquid layers. Most separatory funnels taper off to a narrow bottom, which contains a sealed stopcock. Thus, it is relatively easy to separate two phases for further analysis.

The capacity of the funnel should be such that the liquids occupy no more than approximately one-half the volume of the funnel. If necessary, the extraction procedure may be repeated after the addition of fresh solvent. This type of extraction gives rapid, simple, and clean separations. When extracting from a liquid to a lighter solvent, as in diethyl ether extractions of aqueous solutions, it is necessary to remove the lower phase from the funnel after each extraction before removing the extraction solvent.

Extraction is the process used to separate compounds on the basis of their different solubilities in different solvents. The technique may be used either to purify or to concentrate a material. Application of the technique and the principle on which it is based are discussed in Chapter 1B, Section Six, and in Chapter 9.

The choice of solvent for extraction depends on the solute to be recovered. Generally, it is desirable to use a solvent that does not form emulsions and is easily evaporated. Diethyl ether and ethyl acetate are commonly used, although both tend to absorb water. Diethyl ether, while having the advantage of being easily evaporated, is toxic and is capable of forming explosive peroxides as well as being a fire hazard. In some cases, the emulsion consists of a colloidal suspension of an organic solvent in an aqueous solution or vice versa. Emulsification occurs most often when the solute acts as a detergent or when viscous solutes are present. Emulsification is less likely to occur when the two solvents are swirled together rather than shaken vigorously. An emulsion may be broken by salting out, i.e., by the addition of an inorganic salt soluble in water, to cause the solubility of organic solvents in water to be reduced. Among other techniques that may be effective are the addition of a silicone defoamer, a detergent, an acid, or several drops of ethanol or diethyl ether; the last of these is especially useful when the organic layer is chloroform. An emulsion may also be resolved by allowing it to stand or by centrifugation.

When water and an immiscible solvent have been shaken together, the two phases may be separated by *phase separation paper* (Whatman 1 PS Phase Separator). This paper has been impregnated with a special silicone to make it hydrophobic, so that the paper retains water. When the paper is folded in the conventional way in a conical filter funnel, solvents that are heavier than water will flow through the apex. Solvents that are lighter than water pass through the wall of the folded cone.

LABORATORY MIXERS AND HOMOGENIZERS

Efficient mixing of solutions of very different densities is facilitated by mechanical agitation. Laboratory mixers differ mainly in the design of the motor and the stirrer. Depending on the application, a fixed or variable-speed motor should be selected. For viscous solutions, high-torque, low-speed motors should be used. In the presence of explosive vapors an air-driven motor should replace electrical motors. The speed of rotation of this motor and stirrer is changed by altering the air pressure.

Laboratory mixers are rated according to whether they are for continuous (8 hours uninterrupted) or intermittent duty. The design of the rotating device is dictated by its intended application, such as shearing, mixing, or homogenizing. Most stirrers are made of glass, stainless steel, or Teflon. While most mixers are intended for bench-top use, certain models are driven by 12-V DC motors and may be held by hand above the vessel in which solutions are to be mixed. Solutions in a flask may also be mixed by placing a Teflon-coated stirring bar in the flask and placing it above the stirring module which contains a rotating magnet.

Measurements on cell contents require initial preparation of the tissues. Preparation may involve grinding of the tissue in a ground glass tissue blender using manual force or a rotor driven by a simple electric motor. For large quantities of material, a tissue blender similar to the typical kitchen blender is used to emulsify and pulverize the tissue. Several different designs of blades are available and may easily be exchanged if a different application is required. The speed at which the blades rotate is generally over 15 000 rpm.

FILTRATION

Two types of filtration are used to separate particles from liquids: *surface* filtration and *depth* filtration. *Surface filtration* (screen filtration), common in the clinical laboratory, is performed with filter papers, membranes, or sieves. The process is one of exclusion of the particles having a larger diameter than the pores of the separation medium. Surface filtration retains the solid material on the filter medium and the filtrate passes through the filter, to be collected or discarded as required. The number and size of pores can be accurately controlled in the manufacture of the separation medium so that it is possible to determine the appropriate filter for a given task.

With *depth filtration,* the thick separation medium allows particles to be retained in the body of the filter as well as on the surface. Depth filters are usually made of cotton, fiberglass, asbestos, or other materials such as diatomaceous earth. In depth filters, the matrix of fibers is usually arranged in a randomized manner to eliminate channels through the filters, but still allowing passage of fluid. The lack of uniformity of the matrix does not allow the characteristics of depth filters to be defined precisely. Nevertheless, the filters can be evaluated experimentally after manufacture and a nominal performance rating assigned. Combining depth and surface filters into a single filter is useful for purifying solutions. The depth filter is used as a prefilter to trap large particles, thus lengthening the life of the secondary surface filter. Fiberglass or sintered glass is often used as a prefilter for coarse materials.

Filtration may be conducted under gravity, pressure, or a vacuum. The last may be used to accelerate the filtration rate by combining the influence of gravity, suction, and capillary attraction. The factors affecting gravity filtration include the nature of the filter medium, its surface area, and its pore size. The number and size of particles in solution, and its pH, also may affect the filtration rate.

Many filtrations in the clinical laboratory are carried out with *filter paper.* Different types of filter paper include low-ash or ashless paper as well as various grades related to thickness. Regular filter paper leaves some ash on ignition so it cannot be used in gravimetric analysis. Ashless paper leaves such a low residue on ignition that its mass can be ignored. Vacuum filtration requires paper known as hardened-grade paper, with high, wet tensile strength. Fiberglass papers have the greatest strength and combine the advantages of retention of fine particles with fast filtration. Table 1A-10 describes the grade, flow rate, and retention characteristics of commonly used filter papers. Retention refers to the particle size of precipitate the grade will retain; speed refers to relative mean flow rates.

Much of the filtration in the clinical laboratories is now carried out with *membranes* of controlled pore size. These filters are made from homogeneous polymeric materials, such as polyvinylidene fluoride, cellulose esters, cellulose acetate, polytetrafluoroethylene, and polyvinyl chloride, depending on the intended application of the filter. The most widely used filter is composed of cellulose acetate and cellulose nitrate. The filters contain no loose fibers or particulate matter and are manufactured in a variety of sizes up to almost 30 cm diameter and with pore sizes that vary from 10–0.025 μm, the latter being capable of sterilizing the filtrate by retaining microorganisms. Eighty per cent of the surface area of a typical membrane filter is occupied by pores, so that high flow rates occur through a filter even with pores of the smallest diameter.

Table 1A-10. GRADE, FLOW RATE, AND RETENTION CHARACTERISTICS OF FILTER PAPERS

Grade	Thickness (mm)	Ash %	Wet Strength*	Initial Filtration Speed(s)†	Particle Size Retained (μm)
40	0.20	0.010	L	75	8
41	0.21	0.010	L	12	20–30
42	0.20	0.010	L	240	2.5
43	0.21	0.010	L	40	16
44	0.17	0.010	L	175	3
50	0.12	0.025	H	250	2.7
52	0.17	0.025	H	55	7
54	0.18	0.025	H	10	20–25
540	0.16	0.008	H	55	8
541	0.16	0.008	H	12	20–25
542	0.15	0.008	H	250	2.7

*L = Low; H = High.
†Time taken for 100 mL of clean water to pass through 15 cm quadrant folded circle, according to a modified ASTM method D981-56.
(Information from Whatman Laboratory Products, 9 Bridewell Place, Clifton, N.J.)

The basic structure of membrane filters is hydrophobic, although the surface of the membrane can be modified chemically to make it hydrophilic.

Membrane filters may be used under vacuum, with positive pressure, or with gravity, and some have been included in-line in AutoAnalyzers to filter reagent solutions. Some filter holders have been designed for use with syringes through which pressure is applied to filter a solution. Other membrane filters are designed for ultrafiltration and are available with a variety of pore sizes for selective filtration. *Ultrafiltration* is a technique for removing dissolved particles by an extremely fine filter. It is used to concentrate macromolecules such as proteins, because smaller dissolved molecules pass through the filter.

Many small specimens of biological fluids may be concentrated at the same time using porous *cellulose acetate filter* cones and a centrifuge. The filter cones fit inside centrifuge tubes, and centrifugal force is used to drive water and molecules with a molecular weight of less than 50 000 through the anisotropic membrane filter. This approach is widely used to concentrate proteins in urine or cerebrospinal fluid for electrophoresis.

Fine hollow fiber filters with walls of well-defined porosity are now also used to produce ultrafiltrates or to concentrate proteins. The large filter surface allows rapid concentration.

BALANCES AND WEIGHING

Mass is an invariant property of matter. Weight is a function of mass under the influence of gravity, a relationship expressed by the equation:

$$\text{weight} = \text{mass} \times \text{gravity}$$

Two substances of equal weight and subject to the same gravitational force have equal masses. The determination of mass is made using a balance, by comparing the mass of an unknown against that of a known mass. This comparison is called weighing and the absolute standards against which masses are compared are called weights. In practice, the terms *mass* and *weight* are used synonymously.

The classic form of a balance is a beam poised on an agate knife-edge fulcrum, with a pan hanging from each end of the beam and a rigid pointer hanging from the beam at the poise point. With the object to be weighed on one pan and weights of equal mass on the other pan, the pointer will come to rest at an equilibrium point, i.e., "balance," between the extremes of the path of excursion. The weight required to achieve the equilibrium is therefore equal to the weight of the substance being weighed.

Although the classic form of the balance is of great antiquity, modern balances—both mechanical and electronic—continue to apply the principle of equilibrium in a variety of ingenious ways. More than one type of balance is required for a clinical laboratory, since there is a need to weigh, for example, 3 kg of timed urine output as well as microgram amounts of drug for a standard solution. Coarse balances of large capacity (up to 5 kg) have a sensitivity of 0.1 g. The typical analytical balance has a capacity of 200 g with a sensitivity of about 10 μg. Microbalances may have a maximum capacity of as little as 5 g and sensitivity of 0.1 μg.

All balances require a vibration-free location. The more sensitive a balance, the more protection it needs, not only from vibration but also from air currents that can disturb the equilibrium between the weighed object and weights. The zero or null point of the balance, i.e., the rest point in the absence of either weights or an object to be weighed, must be known or kept adjusted. Scrupulous attention to cleanliness is essential. Chemical substances being weighed should never be placed in direct contact with the pans. Loose crystals of chemicals, or liquids with corrosive vapors, should not be permitted to remain on or around the immediate area of the pans. Good weighing technique at sensitivities under 1 g calls for handling weights with forceps and weighed objects with suitable utensils, to avoid deposition of moisture, oils, or salts from an analyst's skin.

Principles of Weighing

There are two general principles of weighing. In weighing by *substitution,* weights are removed from the side of a balance to which the object to be weighed has been added, to restore equilibrium.

In weighing by *direct comparison,* weights are added to one side of the beam to counterbalance the weight of the object on the other side. This approach is more common.

In practice, two modes of weighing are used. (1) Analytical weights are added to equal the weight of the object being weighed. (2) The material to be weighed is added to a balance pan to achieve equilibrium with a preset weight. This second mode is used more commonly in clinical chemistry, where the major need is to weigh a fixed quantity of chemical so that a standard or reagent solution of known concentration may be prepared. Before weighing a chemical, the weight of the container must be determined or allowed for by taring to reset the equilibrium point. When taring is impractical, the weight of the empty container must be subtracted from the combined weight of container and material, in order to obtain the weight of the material alone.

Types of Balances

A balance may have one or two pans. *Double-pan* balances conform to the classic design with a single beam with arms of equal length. Standard weights are usually added manually by hand to the right-side pan to counterbalance the weight of the object on the other, but in some models a dial or vernier with chain is used to make fine adjustments to the mass associated with the right-side pan. In *single-pan* balances, the arms are of unequal length. The object to be weighed is placed on the pan attached to the shorter arm. A restoring force is applied mechanically or electronically to the other arm to return the beam to its null position. Double- and triple-beam balances are forms of the unequal-arm balance. A single pan on a short arm is balanced about a fulcrum by a larger arm consisting of two or three parallel beams to which weights or poises of different weights are attached. Generally these balances are designed to weigh bulk reagents and have relatively low accuracy.

The *single-pan* balance is now the most common balance in the clinical laboratory. It is most often electronically operated and self-balancing. Such a balance may be coupled directly to a computer or recording device. In the electronic single-pan balance, a load on the pan causes the beam to tilt downward. A null detector senses the position of the beam and indicates when the beam has deviated from the equilibrium point. To ensure that the balance has high sensitivity, the null detector must have little friction; low friction may be achieved through optical, electro-magnetic, or capacitative approaches.

In all electronic balances, an electromagnetic force is applied to return the balance beam to its null position. The electromagnetic force takes the place of weights in a two-pan balance. The restoring force is proportional to the weight on the pan and is applied through a solenoid or torque motor. The current required to produce the force is displayed digitally by liquid crystals or light-emitting diodes, in a form equivalent to the mass on the balance pan. The accuracy of an electronic balance depends on the linearity of both the digital voltmeter and the torque motor.

Almost all electronic balances have a built-in provision for taring so that the mass of the container can be subtracted easily from the total mass measured. Thus, operation of an electronic balance is very simple. It entails switching on the balance, placing a container to hold the material to be weighed on the balance pan, taring off the weight of the container, transferring the material to be weighed to the container, and recording the mass of the material as shown on the digital display.

Analytical Weights

Analytical weights must be used to counterbalance the weight of objects weighed on two-pan balances and to verify the performance of both single and two-pan balances. The National Bureau of Standards recognizes five classes of analytical weights. Class M weights are of primary standard quality and are used only to calibrate other weights. Class S weights are used for calibrating balances. In the clinical laboratory balances should be calibrated at least monthly and prior to very accurate analytical work. Class S-1 weights have greater tolerance than class S weights and are used for routine analytical work. The tolerance of class P weights is even greater. Class J weights are intended for microanalytical work and range from 50–0.05 mg, in contrast to the range of 100 g–1 mg of the other classes.

The integral weights of a set of class S weights are made from brass or stainless steel and are lacquered or plated for protection. The fractional weights of a set of class S standards are usually made of platinum or aluminum. Tolerances of the different weights have been defined by the National Bureau of Standards. For class S weights from 5–1 g the tolerance is ± 0.054 mg, from 500–100 mg ± 0.025 mg, and from 50–1 mg ± 0.014 mg.

CONCEPT OF SOLUTE AND SOLVENT

In clinical chemistry most measurements are concerned with the concentration of substances in solutions, the solutions most often being blood, serum, urine, spinal fluid, or other body fluids. The substance that is dissolved in a solution is a *solute*. Thus the compounds that are typically measured, regardless of whether they are inorganic or organic or of high or low molecular weight, are solutes. In clinical chemistry these are frequently referred to as *analytes*. The substance in which the solute is dissolved is a *solvent*. A *solution* is correctly defined as a physically homogeneous mixture of two or more substances. A solution may be gaseous, liquid, or solid. Again, in the field of clinical chemistry we are concerned primarily with the measurement of solids in liquids, where there is always a relatively large amount of solvent in comparison to amount of solute.

When a solution holds as much of a dissolved solute as it can at a specific temperature, it is said to be *saturated*. Saturation is influenced by the temperature of the solution, the atmospheric pressure, and the nature of the solute and solvent. Solvents generally dissolve more solute at higher temperatures, while the solubility of gases generally decreases with increase in temperature. For those substances that dissolve with absorption of heat, previously dissolved solute may come out of solution when the solution is cooled. An *unsaturated* solution is one that contains less solute than the solvent is capable of holding. A *supersaturated* solution is one that contains more solute than it can hold when the solution is saturated. Addition of undissolved substance, jarring, or stirring the supersaturated solution will cause precipitation of the excess solute and produce a saturated solution.

Compounds of similar chemical composition are usually more soluble in each other than in compounds of very different structure. When two liquids dissolve in each other in any proportion, they are completely *miscible*. If the liquids do not dissolve at all in each other, they are completely *immiscible*. When each of two liquids is partially soluble in the other, they are said to be *partially miscible*.

Solutions

The concentration of a solute in a solution may be expressed in different ways. Traditionally in clinical chemistry, concentration has most often been expressed as mass of solute per volume of solution, i.e., mass concentration. However, the *Système International* (SI, see p. 40) recommends the use of moles of solute per volume of solution for analyte concentrations (i.e., substance concentration), whenever possible, and the use of liter as the reference volume. Concentrations expressed as mass of solute per mass of solvent, as is favored for osmolality, have the advantage that the mass of solute and solvent, in contrast to volume, is independent of the temperature of measurement.

Gases dissolve in liquids to form solutions. Henry's law, which is applicable to gases at low pressures, states that the solubility of a gas in a liquid is directly proportional to the pressure of the gas above the liquid at equilibrium. Thus, as the pressure of a gas is doubled, its solubility is also doubled. The relationship between pressure and solubility varies with the nature of the gas. When several gases are dissolved at the same time in a single solvent, the solubility of each gas is proportional to its partial pressure in the mixture. The solubility of most gases in liquids decreases with an increase in temperature, and indeed a boiling liquid will frequently drive out all dissolved gases.

Traditionally, the unit used to describe the concentration of gases in liquids has been per cent by volume (v/v). Using the SI, gas concentrations are expressed in moles per cubic meter (mol/m^3).

Making Dilutions

With use of proper techniques, errors in the preparation of dilutions should always be less than 0.1%. Most dilutions are made by transferring a precise volume of a concentrated solution into an appropriate flask and then adding water or other diluent to the required volume, with appropriate mixing to ensure homogeneity. Errors may be of two types—*random* and *determinate*. Random errors vary nonreproducibly from one measurement to another, and may thus be minimized by careful technique. A determinate error affects every one of a set of measurements, remaining fixed for all measurements. Such errors may be avoided by using correctly calibrated equipment.

If a solution to be pipetted does not have the same density, viscosity, or surface tension as water, the volume of solution to be delivered may not be the same as it would be with water with which pipets are usually calibrated. For dilute solutions, this error is usually so small that it can be ignored. To avoid errors that arise when two liquids of very different composition are mixed, the technique of diluting to volume is used. Instead of adding 90 mL water to 10 mL concentrated solution, the 10 mL concentrated solution should be pipetted into a 100 mL volumetric flask, and water added to bring the volume to 100 mL.

When a solution is diluted with water, its volume is increased and its concentration decreased, but the total amount of solute remains unchanged. With a constant amount of solute,

$$\text{concentration}_1 \times \text{volume}_1 = \text{concentration}_2 \times \text{volume}_2$$

This formula can be used to determine the volume of a concentrated solution that is required to make a known volume of a solution of a desired lesser concentration.

Gravimetric techniques are inherently more precise than volumetric techniques. Thus, if very accurate dilutions are to be made, an amount of concentrated solution, appropriate to its specific gravity, should be weighed out and transferred to a volumetric flask for dilution. However, in the field of clinical chemistry, satisfactory accuracy can generally be achieved through pipetting. Theoretically, deviations of temperature away from the 20° C at which glassware is calibrated could affect the accuracy of the dilutions made using pipets and calibrated flasks. However, such errors are small, and can be ignored when all solutions and glassware are at room temperature.

Expressing Concentrations of Solutions

The concentration of solutions may be stated in many different ways. These are listed in Table 1A-11. While SI units (see p. 40) are preferred for reporting the concentration of constituents analyzed in the clinical laboratory, use of these units is not universal. Concentrations can be expressed in either chemical or physical terms.

Table 1A-11. CONCENTRATION QUANTITIES AND UNITS

Name of Quantity	Definition	Unit
Substance concentration [of a given solute component]	Amount of substance of a solute divided by volume of solution	mol/m^3; mol/L
Molality [of a given solute component]	Amount of substance of a solute divided by mass of solvent	mol/kg
Mole fraction (or substance fraction) [of a given component]	Amount of substance of a component divided by amount of substance of mixture (i.e., all components of the system)	mol/mol
Mole ratio (or substance ratio) [of a given solute component]	Amount of substance of a solute divided by amount of substance of the solvent	mol/mol
Mass concentration [of a given component]	Mass of a component (e.g., solute) divided by volume of system (e.g., solution)	kg/m^3; kg/L
Mass fraction [of a given component]	Mass of a component divided by mass of system (mixture)	kg/kg
Volume fraction [of a given component]	Volume of a component divided by volume of system (mixture)	m^3/m^3; L/L
Number concentration	Number of specified particles or elementary entities divided by volume of system (mixture)	m^{-3}; L^{-1}
Number fraction	Number of specified particles or elementary entities divided by total number of particles or entities in the system (mixture)	1 (a ratio)
Substance content [of a given component]	Amount of substance of a component divided by mass of system (mixture)	mol/kg

(From The SI for the Health Professions, World Health Organization, 1977.)

PHYSICAL AND CHEMICAL UNITS

Typically laboratory data have been, and generally still are, reported in the United States in terms of mass concentration, i.e., the mass of solute per unit volume of solution. Although the deciliter has been the traditional reference volume, SI specifies the liter. In common parlance, and in some publications, mass concentration has been incorrectly reported in terms of grams per cent or per cent. This terminology indicates an amount of solute per mass of solution, e.g., grams per 100 g, and would be appropriate only if standards against which the unknowns were compared were also measured in the same terms. An exception to the general expression of analyte concentrations in terms of volume of solution is the measurement of osmolality, in which concentrations are expressed in terms of mass of solvent.

When the solution and solvent are both liquids, as in alcohol solutions, the concentration of such a solution is frequently expressed in terms of volume per volume (v/v). By adding 70 mL alcohol to a flask and making it to 100 mL with water, a solution whose concentration is 700 mL/L would be achieved. The expression "700 mL/L" is preferred to the alternatives of 70 volumes per cent or 70% (v/v).

A *molar* solution contains 1 mole of solute in one liter (1 mol/L) of solution; e.g., since the molecular weight of H_2SO_4 is 98.08, a 1 molar solution of H_2SO_4 contains 98.08 g H_2SO_4 per liter of solution. The symbol M, to denote molarity, is no longer acceptable and has been replaced by mol/L. A *molal* solution contains 1 mol of solute in 1 kg of solvent. Molality is properly expressed as mol/kg.

A *normal* solution contains 1 gram equivalent weight of solute in 1 liter solution; i.e., 1 mol HCl, 0.5 mol H_2SO_4, and 0.33 mol H_3PO_4, each in one liter of solution, are 1 normal solutions. The use of normality is limited in that a given solution may have more than one normality, depending on the type of reaction for which the solution is used. The molarity of a solution, however, is a fixed number, since there is only one molecular mass for any substance. Normality is no longer recommended to express concentrations. Nevertheless, the term is included here because it remains in common usage and is related to the equivalent concept, still favored for serum electrolyte concentrations in the United States and some other countries.

The following equations define the expressions of concentrations:

$$Mole = \frac{mass\ (g)}{gram\ molecular\ weight}$$

$$Molarity\ of\ a\ solution = \frac{number\ of\ moles\ of\ solute}{number\ of\ liters\ of\ solution}$$

$$Molality\ of\ a\ solution = \frac{number\ of\ moles\ of\ solute}{number\ of\ kilograms\ of\ solvent}$$

$$Normality\ of\ a\ solution = \frac{number\ of\ gram\ equivalents\ of\ solute}{number\ of\ liters\ of\ solution}$$

$$\begin{matrix} Normality \\ (in\ oxidation\text{-}reduction\ reaction) \end{matrix} = molarity \times \begin{matrix} difference \\ in\ oxidation\ state \end{matrix}$$

$$\begin{matrix} Gram\ equivalent\ weight \\ (as\ oxidant\ or\ reductant) \end{matrix} = \frac{formula\ weight\ (g)}{difference\ in\ oxidation\ state}$$

A milligram equivalent of a substance is its equivalent weight expressed in milligrams. The equivalent mass of H_2SO_4 is 49.04 g. Then one milligram equivalent of H_2SO_4 = 49.04 mg H_2SO_4. Since substances may react on the basis of their valence, one mole calcium (atomic weight = 40), which is bivalent, has twice the combining power of one mole sodium (atomic weight 23). Forty mg Ca is therefore equivalent to two times 23 mg Na.

The unit of measurement commonly used to express the concentration of electrolytes in plasma is the milliequivalent (mEq), which is one-thousandth of an equivalent.

$$Milliequivalents\ (mEq) = \frac{weight\ (g)}{milliequivalent\ mass\ (g)}$$

Milligrams per 100 mL (deciliter) can be converted to mEq per liter using the following formula:

$$mEq/L = \frac{mg/dL \times 10 \times valence}{mg\ atomic\ mass}$$

Example: If the serum sodium concentration is 322 mg/dL, then the serum contains 3220 mg/L. The equivalent mass of sodium is 23, and the valence is 1; therefore

$$mEq/L = \frac{322 \times 10 \times 1}{23} = 140$$

The recommended units for sodium concentration in plasma are millimoles per liter (mmol/L). In the example above, the concentration of sodium is:

$$mmol/L = \frac{mg/L}{mg\ molecular\ mass} = \frac{322 \times 10}{23} = 140$$

In chemical terms the *titer* of a solution is the mass of a substance equivalent to a unit volume of the solution.

$$Titer = \frac{grams\ of\ substance}{liters\ of\ solution} = \frac{mg\ substance}{mL\ solution}$$

However, in clinical laboratory practice a titer is more commonly thought of as the lowest dilution at which a particular reaction takes place. Titer is customarily expressed as ratio, e.g., 1:10 or 1 to 10.

BUFFER SOLUTIONS AND THEIR ACTION

Buffers are defined as substances that resist changes in the pH of a system. All weak acids or bases, in the presence of their salts, form buffer systems. The action of buffers and their role in maintaining the pH of a solution can best be explained with the aid of the Henderson-Hasselbalch equation, which may be derived as follows:

The ionization of a weak acid, HA, and of a salt of that acid, BA, can be represented as:

$$HA \rightleftharpoons H^+ + A^-$$

$$BA \rightleftharpoons B^+ + A^-$$

The dissociation constant for a weak acid may be calculated from the following equation:

$$\frac{[H^+][A^-]}{[HA]} = K_a$$

Thus

$$[H^+] = K_a \times \frac{[HA]}{[A^-]}$$

or

$$\log [H^+] = \log K_a + \log \frac{[HA]}{[A^-]}$$

Now multiplying throughout by -1:

$$- \log [H^+] = - \log K_a - \log \frac{[HA]}{[A^-]}$$

since, by definition, $pH = -\log [H^+]$ and $pK_a = -\log K_a$, we may write

$$pH = pK_a + \log \frac{[A^-]}{[HA]}$$

Since A^- is derived principally from the salt, the equation may, for practical purposes, be written:

$$pH = pK_a + \log \frac{[salt]}{[undissociated\ acid]}$$

or simply:

$$pH = pK_a + \log \frac{[salt]}{[acid]}$$

where $[salt] = [A^-]$ = concentration of dissociated salt and $[acid] = [HA]$ = concentration of undissociated acid.

Consequently, the pH of the system is determined by the pK of the acid and the ratio of $[A^-]$ to $[HA]$. The buffer has its greatest buffer capacity at its pK, i.e., that pH at which the $[A^-] = [HA]$. This entered into the preceding equation gives

$$pH = pK_a + \log 1$$
$$pH = pK_a + 0$$

The capacity of the buffer decreases as the ratio deviates from 1. In general, buffers should not be used at a pH greater than 1 from the pK_a. If the ratio is beyond 50/1 or 1/50, the system is considered to have lost its buffering capacity. This point is approximately 1.7 pH units to either side of the pK of the acid since

$$pH = pK_a + \log 50/1 \qquad\qquad pH = pK_a + \log 1/50$$

or

$$pH = pK_a + 1.7 \qquad\qquad pH = pK_a - 1.7$$

For acetate buffer, the relationship between pH and the ratio of CH_3COONa to CH_3COOH (A^-/HA) is shown in Table 1A-12.

The chemical mechanisms by which buffers exert their effect may be seen by considering the reactions involved upon the addition of base to a buffer solution containing acetate ions, CH_3COO^-, and acetic acid molecules, CH_3COOH.

On addition of NaOH:

$$
\begin{array}{ccc}
CH_3COO^-\ H^+ & & CH_3COO^-Na^+ \\
& + Na^+ + OH^- \longrightarrow & + HOH \\
CH_3COO^-Na^+ & & CH_3COO^-\ Na^+
\end{array}
$$

Table 1A-12. SALT/ACID RATIO AND pH CONCENTRATION

CH_3COONa (mol/L)	CH_3COOH (mol/L)	Ratio Salt/Acid	pH
0.00	0.20	0.00	2.7
0.01	0.20	0.05	3.4
0.05	0.20	0.25	4.1
0.10	0.20	0.50	4.4
0.20	0.20	1.00	4.7
0.40	0.20	2.00	5.0
1.00	0.20	5.00	5.4
2.00	0.20	10.00	5.7

Base OH$^-$ is removed by combining with the hydrogen ion dissociated from acetic acid, thus minimizing pH changes.

The addition of alkali decreases the [CH$_3$COOH] in the buffer and increases the [CH$_3$COONa]. The pH of the solution increases in proportion to the change in ratio of salt to acid in the buffer solution.

On addition of HCl:

$$\begin{matrix} CH_3COO^-\ H^+ \\ \\ CH_3COO^-\ Na^+ \end{matrix} + H^+ + Cl^- \longrightarrow \begin{matrix} CH_3COO^-\ H^+ \\ \\ CH_3COO^-\ H^+ \end{matrix} + Na^+ + Cl^-$$

H$^+$ is removed by combining with acetate to form poorly dissociated acetic acid.

In this case the addition of HCl acts to decrease [CH$_3$COONa] and increase [CH$_3$COOH] in the buffer. The pH of the solution falls in proportion to the change in ratio of salt to acid in the solution; however, since the pH is related to the logarithm of the A$^-$/HA ratio, only a small change in pH occurs.

Other buffers commonly used, and important in clinical chemistry, are phosphate (Sørensen), citrate (Sørensen), carbonate-bicarbonate (Delory-King), acetate, phthalate, boric acid–borate (Palitzsch), veronal–sodium veronal, glycine-glycinate, diethanolamine, and Tris (hydroxymethyl) aminomethane.

Tables for the preparation of commonly used buffers in clinical chemistry laboratories are given in the Appendix.

UNITS OF MEASUREMENT

A meaningful measurement is expressed with both a number and a unit. The unit identifies the dimension—mass, volume, or concentration—of a measured property. The number indicates how many units are contained in the property.

Traditionally measurements in the clinical laboratory have been made in metric units. In the early development of the metric system, units were referenced to length, mass, and time. The first absolute systems were based on the centimeter, gram, and second (CGS) and then the meter, kilogram, and second (MKS). The current system is called the *Système International d'Unités* (SI) and was accepted internationally in 1960. The *units of the system* are called SI units.

Système International d'Unités takes its origin from the Conférence Générales des Poids et Mesures (CGPM), which since 1889 has functioned as the international authority for measurement. Operating under CGPM is the Comité International des Poids et Mesures (CIPM) whose prime responsibilities are to define units and to direct the Bureau International des Poids et Mesures (BIPM). BIPM, established in 1875, serves as the international reference laboratory and has the responsibility for ensuring uniformity of physical measurements.

International System of Units

There are three classes of SI units—base, derived, and supplemental units.

A *base unit* was decreed by the CGPM for each of seven fundamental but dimensionally independent physical quantities. These units, and the provisionally accepted katal, are listed in Table 1A-13. A *derived unit* is derived mathematically from two or more base units. The SI is coherent since derived units relate unequivocally to base units and contain no numerical factor other than one. A *supplemental unit* is a unit that conforms to the SI but which has not been classified as either base or derived. At present only the radian (for plane angles) and the steradian (for solid angles) are classified this way.

The kilogram is the only SI base unit with a prefix. The anomaly exists because historically names of multiples and submultiples of the units of mass were formed by adding prefixes to the word *gram.* Multiples and submultiples of all other base units are denoted by an appropriate prefix to the base unit.

Because derived units consist of base units linked by multiplication or division, some are complex and have been given special names and symbols to simplify their use. Some examples

Table 1A-13. SI BASE UNITS

Quantity	Name	Symbol
Length	meter	m
Mass	kilogram	kg
Time	second	s
Electric current	ampere	A
Thermodynamic temperature	kelvin	K
Amount of substance	mole	mol
Luminous intensity	candela	cd
Catalytic amount	katal	kat

of derived units appear in Table 1A-14. For measurements of properties such as refractive index or relative density that are dimensionless, the appropriate SI unit is 1 (one) because the derived unit is a ratio between two identical base units. However, within the United States the current trend is to present such ratios without any unit.[7]

The SI strictly defines how units and their symbols are to be written. Roman (upright) lower case type is used for symbols of the units, although the first letter of a symbol derived from a proper name is capitalized (e.g., Hz for hertz). Names of units derived from proper names are written in lower case when written in full, except for Celsius, where the first letter is capitalized; in North America the symbol for liter (L) is also capitalized. Thus L is used, where l would be the proper symbol, in order to avoid possible confusion between the letter l and the number 1 which are similar or identical in some type fonts. This deviation from the uniformity of the SI was accepted by the CGPM at its meeting in 1979. Prefix symbols are printed in upright type with no space between the symbols for the prefix and the unit. Symbols, instead of complete unit names, are to be used in conjunction with numerals, e.g., 50 mL rather than 50 milliliters but when numbers are written in full in a text the unit names are to be written out, e.g., the volume of a solution is fifty milliliters (not 50 milliliters). Symbols are not followed by a period except at the end of a sentence, nor do they change in the plural. When the quantity or value of a unit is greater than one, the unit name is made plural, and it is singular when the value is one or less. The product of two or more units is preferably indicated by a dot, although this may be omitted when there is no risk of confusion with other symbols. The dot must be positioned above the line to distinguish it from a decimal point placed on the line. Although in other countries the comma is widely used for the decimal point instead of a dot on the line, the preferred practice in the United States remains the dot. Names and symbols should not be mixed; thus, use m · s or meter second but not m second. Exponents are used to indicate squared or cubic functions, e.g., m^2 and m^3 for square and cubic meter, respectively. Powers of numbers should be written in full in textual prose, but symbols with superscript numbers should be used in formulas and tables. Reciprocals are indicated by s^{-1}, L^{-1}, etc. A solidus (oblique stroke, /), a horizontal line, or negative power may be used to express a unit derived from two others by division, e.g.,

Table 1A-14. EXAMPLES OF SI DERIVED UNITS IMPORTANT IN CLINICAL MEDICINE, EXPRESSED IN TERMS OF BASE UNITS

Quantity	Name	SI Symbol	Expression in Terms of Other SI units	Expression in Terms of SI Base units
Volume	cubic meter	m^3		m^3
Mass density	kilogram per cubic meter	kg/m^3		kg/m^3
Concentration of amount of substance	mole per cubic meter	mol/m^3		mol/m^3
Frequency	hertz	Hz		s^{-1}
Force	newton	N		$m \cdot kg \cdot s^{-2}$
Pressure	pascal	Pa	N/m^2	$m^{-1} \cdot kg \cdot s^{-2}$
Energy, work, quantity of heat	joule	J	$N \cdot m$	$m^2 \cdot kg \cdot s^{-2}$
Power	watt	W	J/s	$m^2 \cdot kg \cdot s^{-3}$
Electric potential, potential difference, electromotive force	volt	V	$W \cdot A^{-1}$	$m^2 \cdot kg \cdot s^{-3} \cdot A^{-1}$

m/s, $\frac{m}{s}$, or m · s^{-1}. A solidus must not be repeated on the same line. Complex units may be simplified by the use of negative powers or parentheses. A full space must be left between a number and the unit or its symbol, e.g., 10 g, not 10g. A zero must be used before the decimal point when numerical values are less than one. Spaces should be used between every three digits in numbers of more than four digits, starting from the decimal point, although the need for this should be made rare by the use of appropriate prefixes in the reporting units. A space is optional with four digit numbers. When numbers are presented in columns in tables, groups of three digits should be aligned vertically.

The CGPM recognizes that some units outside the SI continue to be important and useful in particular applications. An example is the liter as the reference volume in clinical chemistry. Liter is the name of the submultiple (cubic decimeter) of the SI unit of volume, the cubic meter. Considering that one cubic meter represents some 200 times the blood volume of an adult human, the SI unit of volume is neither a convenient nor a reasonable reference volume in a clinical context. Nevertheless, CGPM recommends that such exceptional units as the liter should not be combined with SI units and preferably should be replaced with SI units whenever possible.

The minute, hour, and day have had such longstanding use in everyday life that it is unlikely that new SI units derived from the second could supplant them. Some other non-SI units are still accepted although they are rarely used by most individuals in their daily lives, but have been very important in some specialized fields. Examples of such units are the nautical mile, knot, and hectare. Examples of non-SI units retained for use with the SI are illustrated in Table 1A-15.

Definition of Base Units

Length. The unit of length, the *meter* (in Europe spelled *metre*), was formerly based on the international prototype of platinum-iridium first defined in 1889. It has now been redefined as "the length equal to 1 650 763.73 wavelengths in vacuum of the radiation corresponding to the transition between the levels $2p_{10}$ and $5d_5$ of the krypton-86 atom."

Mass. The *kilogram* is the unit of mass and is equal to the mass of the international prototype of the kilogram. Mass refers to the quantity of matter and is used in place of the ambiguous term, weight, which reflects the force of gravity.

Time. The *second,* the unit of time, was previously defined as the fraction 1/86 400 of the mean solar day. Yet the solar day is not constant so a new definition was created in 1967. The second is now defined as "the duration 9 192 631 770 periods of the radiation corresponding to the transition between two hyperfine levels of the ground state of the cesium-133 atom."

Electric current. The *ampere* is that constant current which, if maintained in two straight parallel conductors of infinite length, of negligible circular cross section, and placed 1 meter apart in vacuum, would produce between these conductors a force equal to 2×10^{-7} newtons per meter of length.

Temperature. The triple point of water is the fundamental fixed point for temperature measurements. The temperature of 273.16 K was assigned to this point. The *kelvin* is defined as the fraction 1/273.16 of the triple point of water. The unit "degree Celsius" is exactly equivalent to the unit kelvin, and an interval or difference of temperature may also be expressed in degrees Celsius. The symbol for degree Celsius is °C, and the degree sign and the C must not be separated. There should be a space between the number and the °C as for other units.

Table 1A-15. NON-SI UNITS RETAINED FOR USE WITH THE SI

Quantity	Unit	Symbol	Value in SI Units
Time	minute	min	1 min = 60 s
	hour	h	1 h = 3 600 s
	day	d	1 d = 86 400 s
Plane angle	degree	°	1° = $\pi/180$ rad
	minute	'	1' = $\pi/10\,800$ rad
	second	"	1" = $\pi/648\,000$ rad
Volume	liter	L	1 L = 1 dm^3 or 10^{-3} m^3
Mass	tonne	t	1 t = 1000 kg

Luminous intensity. The *candela* is the luminous intensity, in the perpendicular direction, of a surface of 1/600 000 square meter of a black-body at the temperature of freezing platinum under a pressure of 101 325 newtons per square meter.

Amount of substance. The *mole* is the amount of substance of a system which contains as many elementary entities as there are atoms in 0.012 kilogram of carbon-12. When the mole is used, the elementary entities must be specified and may be atoms, molecules, ions, electrons, other particles, or specified groups of such particles.

Decimal Multiples and Sub-Multiples of SI Units

In practical application of SI units, certain values are too large or too small to be expressed conveniently in the base or derived units. Numerical values can be brought to convenient size when the unit is appropriately modified by official SI prefixes. Generally, the prefixes are such that the value of the unit changes a thousand-fold. However, certain common previously accepted multiples or submultiples, such as deci- and hecto-, are still accepted within the SI framework. The SI prefixes are listed in Table 1A-16 together with their symbols.

The International Organization for Standards (ISO) recommends that prefix symbols be printed in Roman (upright) type without spacing between the prefix symbol and the unit symbol, thus mL rather than m L. Further, an exponent attached to a symbol containing a prefix indicates that the multiple or submultiple of the unit is raised to the power expressed by the exponent, e.g., cm^3 is $10^{-6}m^3$. Compound prefixes formed by the juxtaposition of two or more SI prefixes are not to be used. Thus 1 nm is appropriate, whereas 1 mμm is not. Likewise, the appropriate unit of mass is the megagram (Mg) rather than the kilokilogram (kkg).

Applications of SI in Laboratory Medicine

Many international clinical laboratory organizations, as well as national professional societies, have accepted the International System of Units in its broad application. The World Health Organization has also recommended its adoption. These organizations have accepted the liter as the preferred unit of volume. For multiples and submultiples of units, including derived units, only one prefix should be used. The prefix should preferably be confined to the numerator, except in the case of kilogram. Units of concentration should use the liter as denominator. Use of the milli-, micro-, or femtoliter for reporting of test values requires little alteration in thinking when liter is used for reporting mass concentrations, provided that the numerator unit is changed by a similar factor. The numerator unit is usually adjusted so that clinically probable values are less than one thousand. When liter is used instead of deciliter for reporting of results, their numerical value is increased by a factor of 10. Such values, when considered in conjunction with the patient's clinical status, should be understandable.

Use of the liter as reference volume causes little confusion in interpretation of test results. However, introduction of the concept of amount of substance expressed as mole (mol) and its

Table 1A-16. PREFIXES AND THEIR SYMBOLS USED TO DESIGNATE DECIMAL MULTIPLES AND SUB-MULTIPLES

Prefix	Symbol	Factor	Numerical Value
tera	T	10^{12}	1 000 000 000 000
giga	G	10^9	1 000 000 000
mega	M	10^6	1 000 000
kilo	k	10^3	1000
hecto*	h	10^2	100
deka*	da	10^1	10
deci*	d	10^{-1}	0.1
centi*	c	10^{-2}	0.01
milli	m	10^{-3}	0.001
micro	μ	10^{-6}	0.000 001
nano	n	10^{-9}	0.000 000 001
pico	p	10^{-12}	0.000 000 000 001
femto	f	10^{-15}	0.000 000 000 000 001
atto	a	10^{-18}	0.000 000 000 000 000 001

Prefixes marked with * do not conform to the systematic pattern with exponents that are simple multiples of three and should be avoided in scientific practice.

decimal fractions of mmol, μmol, nmol, etc., could lead to confusion because of general lack of familiarity with these units in comparison with more traditional units. The SI does not mandate the use of amount of substance (moles) or substance concentration (moles per liter) instead of mass of substance (kilograms) or mass concentration (kilograms per liter). However, in those countries in which the SI has been introduced into medical practice, amounts of constituents of body fluids have generally been reported as substance concentrations rather than mass concentrations, e.g., 2.5 mmol/L instead of 10.0 mg/dL for calcium and 3.9 mmol/L instead of 70 mg/dL for glucose. The rationale for this change is that compounds react on a molar basis and expression of amounts of substances in such terms allows a better understanding of the relative proportion of compounds. Even though physicians initially have little comprehension of the meaning of results when expressed in units of substance concentration, experience in other countries has shown that they soon adjust to the previously unfamiliar units. Nevertheless, during and following the transition from one set of units to another, it is essential that numbers be unequivocally associated with units. There is, therefore, no reason for misinterpretation. A comparison of results of some of the commonly measured serum constituents, at a concentration found in healthy individuals, is shown in Table 1A-17.

There are certain aspects of the SI that have encountered considerable opposition from medical practitioners, although there is theoretically no reason why certain SI units should cause more problems than other units.

Some individuals have advocated reporting of *acidity* of body fluids in terms of hydrogen ion concentration instead of pH. Their rationale is that pH is a measure of the chemical potential of hydrogen ions rather than a measure of the concentration of the ions. Since chemical potential is a function of the activity coefficient, which is not known with certainty, the antilog of pH cannot be assumed to equal the hydrogen ion concentration. pH values should be treated as primary variables and reported as measured.

The proposed base unit katal (symbol kat), mol/s, is the *catalytic amount* of any catalyst, including enzymes, that catalyzes a reaction rate of one mole per second in an assay system. The kind of quantity measured is identified as catalytic amount. There is a constant relationship between the International Unit (1 μmol/min) and the katal (1 mol/s); to convert a value in International Units to nmol/s the value is multiplied by 16.67. Note, however, that dependency on reaction conditions applies to SI units the same as to International Units; therefore, data reported in the same units but obtained under different conditions may not be comparable. Although the katal has been approved by the IUB-IUPAC Joint Commission on Biochemical Nomenclature, it has yet to receive official sanction by CGPM. Replacement of the International Unit for reporting enzyme activity is likely to be slow; even units that antedated the International Unit are still widely used in clinical laboratory practice.

For some *proteins* whose molecular weight is uncertain, debate continues as to the appropriate unit for reporting test results. Nevertheless, as substance concentration gives a better indication of the relative amount of a protein, substance concentration units are still preferred, even if the molecular weight of the protein is not precisely known. The molecular weight actually used for

Table 1A-17. TYPICAL VALUES FOR ANALYTES AND REPORTING INCREMENTS

	Conventional Units		Recommended Units		Rounded Recommended Unit		Smallest Recommended Reporting Increment	
Albumin	3.8	g/dL	550.6	μmol/L	550	μmol/L	10	μmol/L
Bilirubin	0.2	mg/dL	3.42	μmol/L	3	μmol/L	2	μmol/L
Calcium	9.8	mg/dL	2.45	mmol/L	2.45	mmol/L	0.02	mmol/L
Cholesterol	200	mg/dL	5.17	mmol/L	5.2	mmol/L	0.05	mmol/L
Creatinine	0.8	mg/dL	90.48	μmol/L	90	μmol/L	10	μmol/L
Glucose	90	mg/dL	5.00	mmol/L	5.0	mmol/L	0.1	mmol/L
Phosphorus	3.0	mg/dL	0.97	mmol/L	1.0	mmol/L	0.05	mmol/L
Thyroxine	7.0	μg/dL	90.09	nmol/L	90	nmol/L	10	nmol/L
Triglycerides	100	mg/dL	1.14	mmol/L	1.15	mmol/L	0.05	mmol/L
Urea nitrogen*	10	mg/dL	3.57	mmol/L	3.5	mmol/L	0.05	mmol/L
Uric acid	5.0	mg/dL	297	μmol/L	300	μmol/L	10	μmol/L

*Urea nitrogen is reported as urea (mmol/L) when SI units are used.

the substance concentration should be included as part of the test description. The concentration of *hormones* should be reported as substance concentration also. Units of activity, such as international units related to a consensus standard, should be avoided. Again, where uncertainty in the correct molecular weight of the hormone exists, an approximate value may be used without introducing a major error in the reported results. Even though the molecular weight of hemoglobin is known, there is no agreement as to whether the monomer, Hb(Fe), or the tetramer $Hb_4(Fe_4)$ should be used to report values as substance concentration. Until agreement is reached, the International Committee for Standardization in Hematology (ICSH) recommends that hemoglobin results be reported in terms of mass concentration, i.e., g/L.

Although it is probable that *drugs* will be prescribed and administered in mass units for the immediate future, it is still desirable that the concentration of a drug in a body fluid be expressed in terms of substance concentration rather than mass concentration. Substance concentration allows the concentration of a drug to be understood in comparison with the concentrations of proteins, to which drugs are largely bound, and of other compounds with which drugs compete for protein-binding sites. For drugs, it is essential that units be clearly stated to avoid dangerous misinterpretations of test values, which might occur when drugs are administered and measured in body fluids in different units.

The osmole is not an SI unit and therefore theoretically is not suitable for reporting results of measurements of the concentration of osmotically active particles. The appropriate alternatives would be to record the depression of freezing point, whose unit is kelvin or degree Celsius, or the change in vapor pressure in pascals. But, because the units would be method dependent, some of the benefits of using the osmole would be lost. The current recommendation is to report osmolality in osmoles per kilogram. If osmolarity is reported, the preferred units remain osmoles per liter (osmol/L).

Implementation of the SI requires the use of the joule (J) to replace the calorie as the unit of *energy*. The calorie, as now used, is an ill-defined unit. The terms "calorie" and "Calorie," widely used in nutrition practice, actually relate to thermochemical kilocalorie. Thus, replacement of calorie by joule would avoid confusion and introduce a uniform practice. Resistance to conversion from calories to joules is based on the same arguments that have been used against the change from other traditional units to SI units. The arguments, in the case of energy units, also cite the lack of published tables of nutrition information in SI units. It is recommended that dietary allowances be presented in megajoules rather than kilojoules, as the smaller number probably has some psychological benefit to the dieter. A 2000-international-kilocalorie diet seems more acceptable when expressed as 8.37 MJ than as 8370 kJ.

The correct unit for reporting *pressure,* whether it is the partial pressure of blood gases or blood pressure, is the pascal. Because the numerical values expressed in pascals are too small to be used for most clinically important pressure measurements, the kilopascal is a more convenient unit. Thus, kilopascal replaces mm Hg or torr. The present common units for gas concentrations in blood are volume per cent (v/v) or mmol/L. The CO_2 content of blood is expressed in mmol/L. If all the carbon dioxide in the blood were in the form of bicarbonate, mEq/L of bicarbonate and mmol/L of carbon dioxide would be identical. Since carbon dioxide exists in the blood as a mixture of free CO_2, HCO_3^-, and H_2CO_3, concentrations of CO_2 are best reported in mmol/L.

Reporting Increments

Conversion to the unit "mol/L" from g/dL entails first multiplication by 10 (for volume conversion) and then division by the molecular weight of the compound. Depending on the number of decimal places to which the number is taken, different degrees of precision in measurement may be implied. When conversions are made, no greater precision should be implied than was present in the original measurement. While rounding of numerical values is necessary, this should be deferred to the last step to minimize errors.

The data expressed in the second column of Table 1A-17 imply a greater degree of precision than is warranted. The usual increment for reporting albumin in traditional units is 0.1 g/dL. The comparable increment for albumin in SI units is 10 μmol/L. A calcium value of 9.8 mg/dL implies that the measured value is more than 9.75 but less than 9.85 mg/dL. A value of 2.45 mmol/L implies that the measured value is between 2.445 and 2.455 mmol/L or 9.780 and 9.820 mg/dL, a much greater degree of precision than presented in the preceding example. While the

Table 1A-18. ABBREVIATIONS FOR
SYSTEMS AND PREFIXES
RECOMMENDED BY IFCC AND IUPAC

B	blood	a	arterial
Erc	erythrocyte	v	venous
Ercs	erythrocytes	c	capillary
F	feces	f	fasting
Lkc	leukocyte	d	24 hours (diem)
Lkcs	leukocytes		
P	plasma		
Pt	patient		
S	serum		
Sf	spinal fluid		
U	urine		

current reporting increment for calcium is typically 0.1 mg/dL, the increment with the SI should be 0.02 mmol/L, even though it implies slightly greater precision in measurement.

Standardized Reporting of Test Results

To describe test results properly it is important that all necessary information be included in the test description. The International Federation of Clinical Chemistry (IFCC) and the International Union of Pure and Applied Chemistry (IUPAC) recommend that the following items be included:

1. The name of the system or its abbreviation
2. A dash (two hyphens)
3. The name of the component (never abbreviated) with an initial capital letter
4. A comma
5. The quantity name, or its abbreviation
6. An equal sign
7. The numerical value and the unit or its abbreviation.

The following example illustrates a correct description of a test result, in complete and abbreviated forms:

Plasma from fasting patient—Glucose, substance concentration = 5.5 millimoles per liter.
(fPt)P—Glucose, substc. = 5.5 mmol/L.

The *system* represents the material that is examined. It should be described to the greatest extent necessary to make interpretation of results unambiguous. The system may be the patient, a body fluid, or a component of a body fluid. While spelling out the system is preferable, certain abbreviations have been recommended by IFCC and IUPAC. The abbreviations may be combined to build up a complete description, e.g., fPt and aB for fasting patient or arterial blood, respectively. IFCC and IUPAC recommended abbreviations are listed in Table 1A-18. Additional abbreviations are included in Table 1A-19.

Table 1A-19. SUGGESTED
ADDITIONAL
ABBREVIATIONS FOR
SYSTEMS

Amf	amniotic fluid
Df	duodenal fluid
Gf	gastric fluid
Peritf	peritoneal fluid
Plf	pleural fluid
Semf	seminal fluid
Synf	synovial fluid
T	tissue

The *component* that is measured should be described in as complete terms as possible. The component corresponds to what was previously described as the test name. When the molecular mass of a compound is uncertain, as may occur with some proteins or peptide hormones, the test description should include the component followed by its molecular mass in parentheses, e.g., Albumin (69 000). In descriptions of hemoglobin it is necessary to indicate whether data are reported with respect to the monomer or tetramer. The most complete entity that is measured should be reported. Thus, when urea is measured, results should be reported in terms of urea rather than urea nitrogen. While the official recommended name for urea is carbamide, there seems little justification to change test names radically at a time when confusion from change of units may be present. When a component such as iron is measured, the test description (component) should state Iron (II) or (III) to indicate that both ferrous and ferric ions are being measured and reported in terms of Fe. Potassium ions should be listed to indicate unequivocally that the ions were measured.

The *quantity* is the property that is measured. The different kinds of quantities and their accepted abbreviations are listed in Table 1A-20. In any discussion of a test, the quantity must be stated before the numerical value can be interpreted. The "mass concentration" of a given component is its mass divided by the volume of the system. The mass fraction is the mass of a component divided by the mass of the system (mixture). The "substance concentration" is the amount of substance of a solute divided by the volume of the solution. "Number fraction" is the number of specified particles or elementary entities divided by the total number of the system.

Each quantity is associated with a specific unit. The relation between the concentration quantities and their units is illustrated in Table 1A-11.

Rules for Conversion from One Set of Units to Another

To convert from conventional units to SI units the following scheme may be used:
1. Mass concentration (mass units/dL) to mass concentration (mass units/L):

$$\text{Numerical value in mass units/dL} \times 10 = \text{mass units/L}$$
$$\text{e.g., Serum albumin, } 4.0 \text{ g/dL} \times 10 = 40 \text{ g/L}$$

2. Mass concentration to substance concentration:

$$\frac{\text{Numerical value in mass concentration/dL}}{\text{molecular mass}} \times 10 = \text{substance units/L}$$

$$\text{e.g., Serum albumin, } 4.0 \text{ g/dL} = \frac{4.0 \times 10}{69\,000} \text{ mol/L}$$
$$= 580 \ \mu\text{mol/L}$$

Table 1A-20. ABBREVIATIONS FOR NAMES OF QUANTITIES

ams.	amount of substance
diff.	difference
equil.	equilibration
massc.*	mass concentration
massfr.†	mass fraction
molal.	molality
numc.	number concentration
numfr.	number fraction
rel.	relative
substc.	substance concentration
substfr.	substance fraction
vol.	volume
volfr.	volume fraction

*Previously masc.
†Previously masfr.

3. Mass amount to amount of substance:

$$\frac{\text{Numerical value in mass units}}{\text{molecular mass}} = \text{amount of substance}$$

$$\text{e.g., Urine albumin, 150 mg/d} = \frac{150}{69\,000}\text{ mmol/d}$$
$$= 2.2\ \mu\text{mol/d}$$
$$\text{Urine urate, 600 mg/d} = \frac{600}{168}$$
$$= 3.6\text{ mmol/d}$$

In some cases a factor of 1000 may be required so that values in SI units generally fall between 1 and 1000. To convert from SI to conventional units, which is discouraged, the reverse process of that shown above is used.

The factors for converting from traditional to SI units for many of the constituents of body fluids commonly measured in the clinical laboratory are listed in the table of reference ranges included in the Appendix.

SAFETY

Safety in the clinical laboratory in the past often consisted of following practices dictated by common sense rather than following practices which were dictated by an established safety program. Although many large chemical companies had made commendable commitments to employee safety prior to this time, most clinical laboratories all but ignored formal safety programs.

The Federal Occupational Safety and Health Act (OSHA) of 1970 was the beginning of the formal regulatory oversight of employee safety. Consideration for the health and safety of the employees is now accepted as an obligation of all employers and laboratory directors. Clinical laboratories are generally considered safe places to work only because necessary safety rules and procedures have been incorporated into the daily activities of the workers.

Every clinical laboratory must have a formal safety program. Although safety is each person's responsibility, even in a small laboratory it is good practice to assign a specific individual the title of "Safety Officer," with the duties of administering the safety program and keeping it current. One part of a program consists of the education and motivation of the clinical laboratory employees. Each new employee should be given a copy of the general laboratory safety manual as part of his or her orientation. Each employee should know the location of available fire-fighting equipment and how to use it and where the evacuation routes are. The continuing education program of the laboratory should include periodic talks on safety. Several audiovisual aides are available from a variety of sources to support the educational part of the program.

The second part of a safety program relates to ensuring that the laboratory environment meets accepted safety standards. This effort would include, but not be limited to, attention to such items as proper labeling of chemicals, types and location of fire extinguishers, hoods that are in good working order, proper grounding of electrical equipment, and providing means for proper disposal of biological specimens.

Identification of Hazards

Clinical laboratories deal with each of the nine classes of hazardous materials. These are classified by the United Nations (UN) as (1) explosives, (2) compressed gases, (3) flammable liquids, (4) flammable solids, (5) oxidizer materials, (6) toxic materials, (7) radioactive materials, (8) corrosive materials, and (9) miscellaneous materials not elsewhere classified.

Warning labels aid in the identification of chemical hazards during shipment. Under regulations of the U. S. Department of Transportation (DOT), chemicals that are transported in the United States must carry labels based on the UN classification. DOT placards or labels are diamond-shaped with a digit imprinted on the bottom corner that identifies the UN hazard class (1 to 9). The hazard is identified more specifically in printed words placed along the horizontal axis of the diamond. Color-coding and a pictorial art description of the hazard supplement the identification of hazardous material on the label; the art work appears in the top corner of the diamond (see Figure 1A-7,*A*).

The upper half of the label for corrosives shows a metallic bar and a hand, both of which are being eaten away by drops of fluid; the black bottom half of the label is lettered "corrosives" in white. The bottom corner shows the digit "8."

The system is used by the U. S. Department of Transportation for shipping hazardous materials; however, when the hazardous material reaches its destination and is removed from the shipping container, this identification is lost. The laboratory must then label each individual container. Usually, the information needed to classify the contents of the container appropriately is contained on the shipping label and should be noted. Important first aid information is also usually provided on this label.

Even though OSHA prescribes the use of labels or other appropriate warnings at present, no single uniform labeling system for hazardous chemicals exists for clinical laboratories. The National Fire Protection Association (NFPA) developed the 704-M Identification System, which classifies hazardous material from 0–4 (most hazardous) according to flammability and reactivity

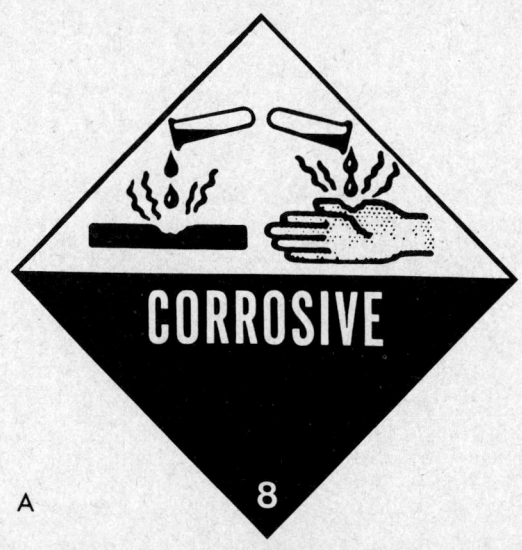

A

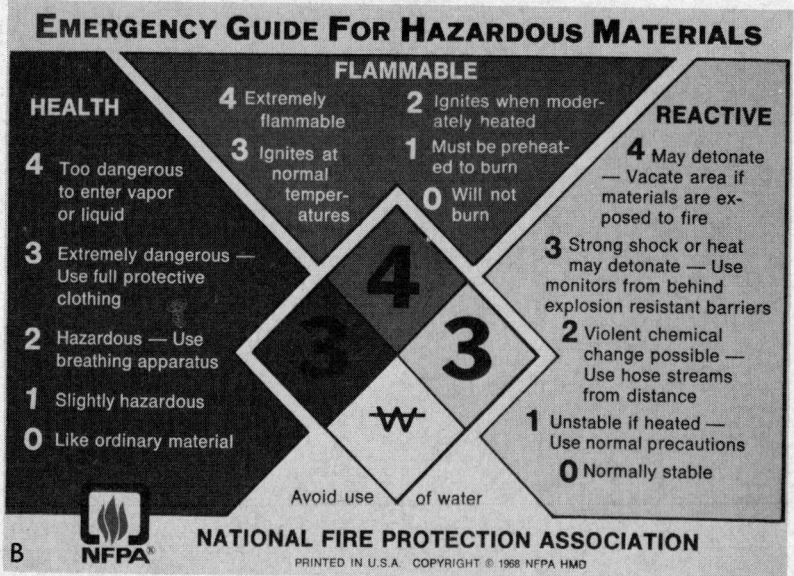

B

Figure 1A-7. *A*, Department of Transportation label for corrosives. *B*, Labeling identification system of National Fire Protection Association.

(instability). This system utilizes diamond-shaped labels which are available from most companies that sell laboratory safety equipment. The labels are color coded and are divided into quadrants. Each of the three quadrants has a characteristic color and represents a type of hazard. A number in the quadrant indicates the degree of the hazard. The fourth (lower) quadrant contains information of special interest to firemen, e.g., "W" indicates a water ignitable material (see Figure 1A-7,*B*). Some chemicals require labels utilizing two or three of the quadrants to convey the necessary information.

Descriptive labels such as "corrosive," "flammable," "poison," and "explosives" may also be used. Obviously, some containers may require two labels or three labels, since several types of hazard are involved.

Kits of assorted safety signs (Figure 1A-8) are available from most laboratory supply dealers. These contain most general labels needed in a clinical laboratory. Some examples of the contents of one kit are "eye wash," "use in hood," "safety shower," "empty cylinder," and "wear eye protection in this area."

Chemical Safety

The proper storage and use of chemicals is necessary to avoid dangers such as burns, explosions, fires, and toxic fumes. Thus, knowledge of the properties of the chemicals in use and of proper handling procedures will greatly reduce dangerous situations. Bottles of chemicals and solutions should also be handled carefully, and a cart should be used to transport a heavy or a multiple number of containers from one area to another. Glass containers with chemicals should be transported in rubber or plastic containers which will protect them from breakage and, in the event of breakage, contain the spill. A manual on the handling of chemicals should be available in each laboratory.

Spattering from acids, caustic materials, and strong oxidizing agents probably represents the greatest hazard to clothing and eyes and is a potential source of chemical burns. A bottle should never be held by its neck, but instead firmly around its body with one or both hands, depending on the size of the bottle. Acids must be diluted by slowly adding them to water, with mixing; *water should never be added to concentrated acid.* When working with acid or alkali solutions, safety glasses should be worn. Acids, caustic materials, and strong oxidizing agents should be mixed in the sink. This will provide water for cooling as well as for confinement of the reagent in the event the flask or bottle were to break.

All bottles containing reagents must be properly labeled. It is good practice to label the container before adding the reagent, thus avoiding the possibility of having an unlabeled reagent.

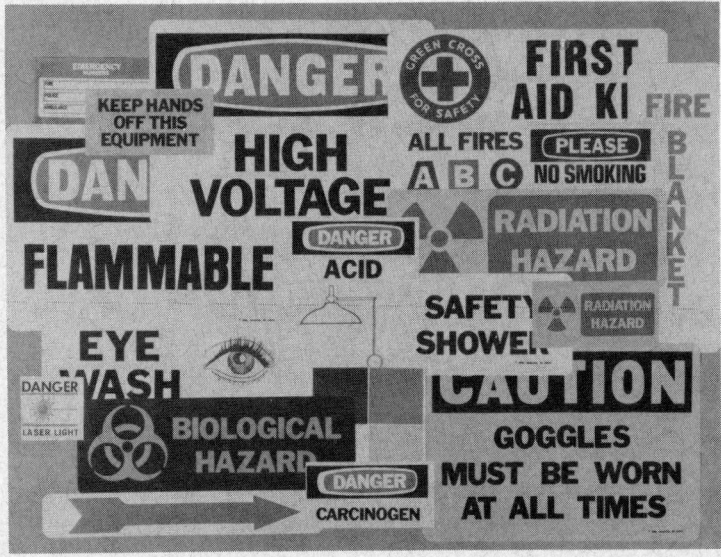

Figure 1A-8. A commercially available kit of OSHA-approved signs. (Courtesy of Fisher Scientific, an Allied company.)

The label should bear the name and concentration of the reagent, the initials of the person who made up the reagent, and the date on which the reagent was prepared. When appropriate, the expiration date should also be included. The labels should be color coded or an additional label added to designate specific storage instructions, such as the need for refrigeration or special storage related to a potential hazard. All reagents found in unlabeled bottles should be discarded using rigid precautions. The reagent should be flushed down the drain with copious amounts of water. In the case of an acid, it must be diluted sufficiently not to harm the plumbing. Other reagents which are not harmful by themselves may be potentially hazardous if poured down a drain in which an acid is subsequently disposed. An example of such a chemical is a salt of cyanide, which releases hydrogen cyanide when in contact with acid. Large amounts of water must be used to flush the drain completely of all reagents, but particularly those that present a danger, either actual or potential.

Strong acids, caustic materials, and strong oxidizing agents should be dispensed by any of a number of commercially available automatic dispensing devices. Under no circumstances is mouth pipetting permitted.

Most of the general precautions just mentioned also hold for automated chemistry equipment such as AutoAnalyzers. In some instances, all waste materials cannot be collected in the same container. With certain pieces of equipment, strong acids or other hazardous materials are pumped directly into the drain. This should always be accompanied by a steady flow of water from the faucet. Safety glasses should be used by instrument operators when acids are pumped under pressure.

Perchloric acid, because it is potentially explosive in contact with organic materials, requires careful handling procedures. Perchloric acid should not be used on wooden bench tops, and bottles of this acid should be stored on a glass tray. Disposal may be accomplished by adding the acid dropwise (utilizing a splatter shield) to at least 100 volumes of cold water and pouring the diluted acid down the drain with large amounts of additional cold water. Special perchloric acid hoods, with special wash-down facilities, should be installed if large amounts of this acid are used.

Special care is needed when dealing with mercury. Many think of mercury as a material used to fill a gasometer and forget that it is also a potential hazard because it is toxic and cumulative. Even small drops of mercury on bench tops and floors may poison the atmosphere in a poorly ventilated room. The element's ability to amalgamate with a number of metals is well known. After an accidental spillage of mercury, the spill area should be cleaned carefully until there are no droplets remaining. All containers of mercury should be kept well stoppered.

The U. S. Environmental Protection Agency (EPA) controls the disposal of nonradioactive hazardous wastes. The Resource Conservation and Recovery Act of 1976 (RCRA) states that disposal of materials classifiable within any of the nine United Nations hazardous materials classes will be enforced in such a way that health and safety professionals involved in the disposal of such materials are personally liable for each individual violation. However, the EPA has not developed standards for nonradioactive hazardous waste disposal based on "best available technology."

Although defined federal standards for hazardous waste disposal are not available, many municipalities have their own regulations. At this time, most clinical laboratories can dispose of aqueous solutions of chemicals down the drain after proper treatment, e.g., neutralization of acids, followed by a generous amount of water. Disposal of volatiles and infectious materials will be covered later.

Hazards from volatiles. The use of organic solvents in a clinical laboratory represents a potential fire hazard as well as hazards to health from inhalation of toxic vapors or skin contact. Use of these solvents should be carried out in a fume hood. Storage of organic solvents is regulated by rules set down by OSHA (Table 1A-21). However, some local fire department rules are more stringent. Solvents should be stored in an OSHA-approved metal storage cabinet that is properly vented. The maximum working volume of flammable solvents allowed outside of storage cabinets is 5 gallons per room. No more than 60 gallons of Types I and II may be stored in a single cabinet. No more than 3 cabinets may be located in each 5000 sq. ft. of laboratory space. Larger amounts than shown in Table 1A-21 must be stored in special refrigerated storage rooms or in outside storage buildings.

Vaporization is the major problem in the ignition and spread of fires. Flammable and combustible liquids and solids are those whose vapor forms a flammable mixture with air. They

Table 1A-21. MAXIMUM ALLOWABLE SIZE OF CONTAINERS FOR STORAGE OF SOLVENTS INSIDE BUILDINGS (OSHA)

	Maximum Allowable Size of Containers			
Container Type	Class IA	Class IB	Class IC	Classes II and III
Glass or approved plastic	1 pt.	1 qt.	1 gal.	1 gal.
Safety cans*	1 gal.	2 gal.	5 gal.	5 gal.
Metal drums (DOT)	60 gal.	60 gal.	60 gal.	60 gal.

	Definition of Classes:
IA	Flash point below 22.8 °C and boiling point below 37.8 °C
IB	Flash point below 22.8 °C and boiling point above 37.8 °C
IC	Flash point between 22.8 °C and 37.8 °C
II	Flash point between 37.8 °C and 60 °C
III	Flash point above 60 °C

*Safety cans are metal or plastic containers having a spring-closing spout cover designed so that it will relieve internal pressure when subjected to the heat of fire and will prevent leakage if tipped over. They are also equipped with a flame-arrester screen. Portable safety cans range in size from 1 pint to 5 gallons and are made in numerous styles with faucets, pouring spouts, or dispensing hoses. Only FM (Factory Mutual Engineering Corporation of the Factory Mutual System) or UL (Underwriters Laboratories, Inc.) recognized safety cans should be used, because these have been tested and approved by these agencies.
OSHA = Occupational Safety and Health Administration.

are characterized by their flash point, where the flash point is defined as the lowest temperature at which a solvent will give off flammable vapors in the close vicinity of its surface. The mixture at its flash point will ignite when exposed to a source of ignition. At temperatures below the flash point, the vapor given off is considered too lean for ignition. The flash points of solvents commonly found in clinical laboratories are listed in Table 1A-22.

Disposal of flammable solvents in storm sewers or sanitary sewers is, in general, not allowed. Exceptions are small amounts of those materials which are miscible with water; but even disposal of these should be followed by large amounts of cold water. Other solvents should be collected in safety cans. Separate cans should be used for ether and for chlorinated solvents; all other solvents may be combined in a third can. The cans should be stored, in keeping with storage quantity rules, in a safety cabinet until pickup by a waste-disposal firm. A more economical approach is to transfer the solvents to larger cans or drums in an outside storage facility so that pickup could be less frequent. Some large institutions have their own in-house disposal facilities.

Hazards from compressed gases. The U. S. Department of Transportation (DOT) regulations cover the labeling of cylinders of compressed gases that are transported by interstate carriers. The diamond-shaped labels described previously are used on all large cylinders and on any boxes containing small cylinders. The OSHA regulations regarding compressed gas in cylinders are based on publications of the Compressed Gas Association, Inc.[3]

Table 1A-22. FLASH POINT OF SOLVENTS COMMONLY FOUND IN CLINICAL LABORATORIES

	°F	°C
Acetone	−4	−20
Butanol	82	28
Cyclohexane	−4	−20
Ethyl alcohol	55	13
Ethyl ether	−49	−45
Ethyl acetate	24	−5
Heptane	25	−4
Hexane	−7	22
Isopropanol	53	12
Petroleum ethers	−70	−56
Propanol	77	26
Toluene	40	14
Xylene	81	27

Some general rules for handling large cylinders of compressed gas are:

1. Always transport cylinders using a hand truck to which the cylinder is secured.

2. Leave valve cap on cylinder until cylinder is ready for use, at which time the cylinder should have been secured by a support around the upper one-third of its body. Disconnect hose or regulator, shut off valve, and replace cap before the cylinder is completely empty to avoid the possibility of the development of a negative pressure. Place "empty" sign or label on cylinder.

3. Chain or secure cylinders at all times even when empty.

4. Always check cylinders for composition of contents before connection.

5. Never force threads; if regulator does not thread readily, something is wrong.

Empty cylinders are usually picked up by gas suppliers for return to the manufacturer for refilling.

Small cylinders. The precautions cited for large refillable gas cylinders also apply to small cylinders which are not refillable. Propane cylinders of the type frequently secured to a flame photometer and cylinders of calibrating gases for blood gas equipment are examples of disposable cylinders. Cylinders in floor-standing base supports require the additional security of a chain or strap attached to a wall or fixed piece of furniture. Local fire department regulations (which vary considerably from place to place) govern disposal of exhausted cylinders.

Electrical Hazards

Wherever there are electrical wires or connections, there is a potential shock or fire hazard. Worn wires on all electrical equipment should be replaced immediately; all equipment should be grounded using three-prong plugs. OSHA regulations stipulate that the requirements for grounding of electrical equipment of the National Electrical Code (published by The National Fire Protection Agency) be met. If grounded receptacles are not available, a licensed electrician should be consulted for proper alternative grounding techniques. Some local codes are more stringent than OSHA requirements and do not allow for 2-pole mating receptacles for a 3-pole plug.

Use of extension cords is prohibited by the 1980 NCCLS standard on Power Requirements for Clinical Laboratories.[6] This standard is more stringent than any other existing regulations. In some instances, an extension cord may have to be used temporarily. In such cases the cord should be less than 12 feet in length, have at least 16 AWG wire, be approved by the Underwriters Laboratory (UL), and have only one outlet at the end. If several outlets are needed in an area, a power strip with its own fuse or circuit breaker may be installed at least three inches above bench-top level. Several manufacturers now sell devices to check for high resistance in neutral or ground wiring or excess voltage in the neutral wiring. One such device is sold by American Scientific Products under the name *Tek Pro Circuit Tester*.

Electrical equipment and connections should not be handled with wet hands, nor should electrical equipment be used after liquid has been spilled on it. The equipment must be turned off immediately and dried thoroughly; a fan or hair dryer will speed up the drying process. In case of a wet or malfunctioning electrical instrument that is used by several people, the plug should be pulled and a note cautioning coworkers against use should be left on the instrument.

Electrical apparatus, especially a motor that is operated in an area where there are flammable vapors, must be explosion-proof; induction-driven motors are well suited for these areas.

Fire Safety

The ideal solution to the problem of fires and indeed to all laboratory accidents is prevention. However, all fires cannot be prevented, so provisions must be made for those that do occur. NFPA (National Fire Protection Association) and OSHA publish standards covering subjects from emergency exits (including means of egress) to safety and firefighting equipment. NFPA also publishes the National Fire Codes. However, these are "standards" without the force of law. Many state and local agencies have adopted these codes (some of which are more stringent than OSHA requirements) and thus make them legally enforceable.

Every laboratory should have the necessary equipment to put out or to confine a fire in the laboratory, as well as to put out a fire on the clothing of an individual. Easy access to safety showers is essential. A safety shower should have a pull chain either attached to the wall at a convenient height or hanging down from the shower head; the chain should have a large ring attached so that the shower may be easily activated, even with eyes closed. Fire blankets for smothering fire on clothing should be available in an easily accessible wall-mounted case. The blanket is unrolled from the case and rolled around the body by taking hold of the rope that is attached to the blanket and turning the body around. The location of this equipment as well as the locations of fire alarms and maps of evacuation routes will be dictated by the local fire marshal.

Various types of *fire extinguishers* are available. The type to use depends upon the type of fire. Because it is impractical to have several types of fire extinguishers present in every area, dry chemical fire extinguishers are among the best all-purpose extinguishers for laboratory areas. An extinguisher should be provided near every laboratory door and, in a large laboratory, also at the end of the room opposite to the door. Everyone in the laboratory should be instructed in the use of these extinguishers and any other firefighting equipment. All fire extinguishers should be tested by qualified personnel at intervals specified by the manufacturer. The three classes of fires and the type of fire extinguisher to be used for each are listed in Table 1A-23. Every fire extinguisher is labeled as to the type of fire it should be used to extinguish.

Two additional types of fires, designated "D" and "E," should be handled only by trained personnel. Type "D" fires include those involving powdered metal materials (e.g., magnesium). A special powder is used to fight this hazard. A type "E" fire is one that cannot be put out or is liable to result in a detonation (such as an arsenal fire). A type "E" fire is usually allowed to burn out while nearby materials are being appropriately protected.

Many clinical laboratories now have a computer which is housed in a temperature- and humidity-controlled room. The most popular automatic fire control system in use for these rooms is Halon 1301 (bromotrifluoromethane). Although this is the least toxic of the halons, NFPA regulations require a warning sign at the entrance to the room and availability of self-contained breathing equipment.

Biological Hazards

Among the most frequent biohazards are (1) accidental puncture with hypodermic needles, (2) spraying of infectious materials by a syringe or spilling and splattering of these materials on bench tops or floors, (3) centrifuge accidents, and (4) cuts or scratches from contaminated glassware.

Viral hepatitis represents a special health hazard not only for clinical laboratory personnel but also for other health care individuals. Although in the past hepatitis B virus presented the

Table 1A-23. CLASSIFICATION OF FIRES AND FIRE EXTINGUISHER REQUIREMENTS

Type of Hazard	Class of Fire	Recommended Extinguisher Agents
Ordinary combustibles: Wood, cloth, paper	A	Water, dry chemical foam, loaded steam
Flammable liquids and gases: Solvents and greases Natural or manufactured gases	B	Dry chemical, carbon dioxide, loaded steam, Halon 1211 or 1301 foam
Electrical equipment: Any energized electrical equipment. If electricity is turned off at source this reverts to a Class A or B	C	Dry chemical, carbon dioxide, Halon 1211 or 1301
Combinations of: Ordinary combustibles and flammable liquids and gases	A & B	Dry chemical, loaded steam, foam
Combinations of: Ordinary combustibles and electrical equipment	A & C	Dry chemical
Combinations of: Flammable liquids and gases and electrical equipment	B & C	Dry chemical, carbon dioxide, Halon 1211 or 1301
Combinations of: Ordinary combustibles, flammable liquids and gases, and electrical equipment	A, B & C	Triplex dry chemical

greatest risk, non-A, non-B hepatitis is now the most common form of post-transfusion hepatitis, and is being seen with increasing frequency among renal dialysis patients.

Techniques and Procedures to Minimize Laboratory Infections

Sound fundamental laboratory techniques, well supervised and conscientiously carried out, can do much to achieve environmental control and reduce the hazards of infection. A list of procedure rules that are widely applicable in clinical laboratories follows:

1. Never perform direct mouth pipetting in the laboratory.
2. Do not blow out pipets that contain infectious material.
3. Do not mix infectious materials by bubbling air through the liquid.
4. Use only needle-locking hypodermic syringes. Avoid using syringes whenever possible. Dispose of needles in special containers. (See Figure 1A-9.)
5. Expel excess fluid and bubbles from a syringe held vertically into a cotton pledget moistened with disinfectant.
6. Sterilize reusable pipets and syringes immediately after use by placing them into a pan containing a phenolic compound such as Staphene.
7. Before centrifuging tubes, inspect them for cracks. Inspect the inside of the trunnion cup for rough walls caused by erosion or adhering matter. Be sure that rubber cushions are free from all bits of glass. A germicidal solution added between the tube and the trunnion cup not only disinfects the surfaces of both but also provides an excellent cushion against shocks that otherwise might break the tube.
8. Avoid decanting centrifuge tubes; if you must do so, wipe off the outer rim afterward with a disinfectant. Avoid filling the tube to the point that the rim becomes wet with a potentially contaminated specimen.
9. Never leave a discarded tube or infected material unattended or unlabeled.
10. Periodically, clean out deepfreeze and dry-ice chests to remove broken ampules or tubes of biological specimens stored there. Use rubber gloves and respiratory protection during this cleaning.
11. Label all samples known to be or strongly suspected of being from a patient with hepatitis upon entry into the laboratory. The label should be placed on all tubes containing aliquots, e.g., all serum aliquots removed from clot and used in various sections of the laboratory.

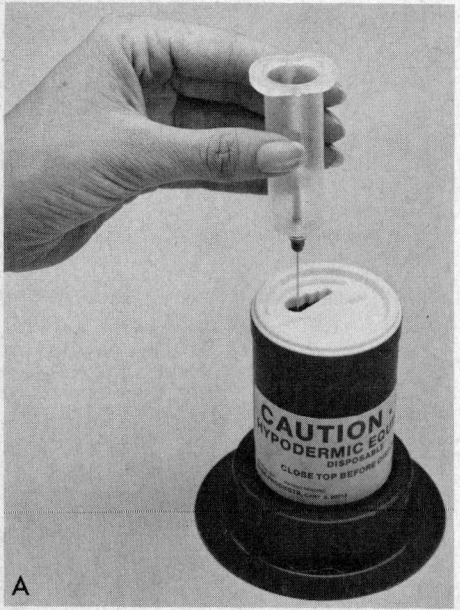

A

B

Figure 1A-9. Convenient needle disposal system is available in various sizes: (A) small for blood-drawing tray, or (B) large. (Courtesy of Sage Products, Inc.)

12. Handle serum specimens which carry a risk of infectious hepatitis with disposable rubber or plastic gloves.

13. Use terminal sterilization of all biological specimens (except urine) by autoclaving or incineration as part of the disposal process.

14. Do not allow smoking or consumption of food or beverages in the laboratory. Food and beverages are never to be stored in any laboratory refrigerator (or freezer) unless it has been specifically and *exclusively* designated for such purpose.

15. Encourage frequent hand washing; it is mandatory any time an employee leaves the laboratory.

16. Develop the habit of keeping your hands away from your mouth, nose, eyes, and any other mucous membranes. This will reduce the possibility of self-inoculation.

Safety Equipment

A large number of safety items are available for the laboratory. Most of these are shown and described in a Manual on Laboratory Safety.[4] A few important safety items will be mentioned briefly here. These safety items should be used in areas where they are appropriate. Eyewashers or facewashers should be available in every chemistry laboratory. Many types are available, and some simply connect to existing plumbing. A hand-held eye/face safety spray is a convenient safety device and can be consistently placed in a position next to each sink while using only a few inches of space. Safety showers, strategically located in the laboratory, should be tested on a regular schedule.

Heat resistant (non-asbestos) gloves should be available for handling hot glassware and for handling dry ice. Safety goggles, glasses, and visors, some that fit conveniently over regular eye glasses, are available in many sizes and shapes. Personnel wearing contact lenses should be aware of the danger of irritants getting under a lens, thus making it difficult to irrigate the eye properly. Shatterproof safety shields should be used in front of systems posing a potential danger because of implosion (vacuum collapse) or pressure explosions. Desiccator guards should be used with vacuum desiccators. Hot beakers should be handled with tongs. Inexpensive polyethylene pumps are available to pump acids from large bottles. Spill kits for acids, caustic materials, or flammable solvents come in various sizes. Such kits and the other appropriate safety materials should be located in convenient sites in the laboratory.

A chemical fume hood is a necessity for every clinical chemistry laboratory. The fume hood is the only safe place to open any container of a material that gives off harmful vapors, or to prepare reagents that produce fumes, or to heat flammable solvents. In the event of an explosion or fire in the hood, closing its window contains the fire.

Safety Inspections

OSHA inspectors have the authority to enter a clinical laboratory unannounced and, upon presentation of credentials, inspect it. The inspection may be regular or as a result of a complaint.

The Commission on Inspection and Accreditation of the College of American Pathologists inspects clinical laboratories and uses various safety check lists (available to the laboratory prior to inspection) when evaluating a laboratory for accreditation. These inspections are carried out on a regular basis.

The Joint Commission on Accreditation of Hospitals (JCAH) does not publish specific safety guidelines for the clinical laboratory. The Commission covers the subject very generally and briefly in its manual. Although JCAH will accept CAP accreditation of a laboratory, it may still conduct a safety inspection of the laboratory when it inspects the hospital.

Inspections may also be made on a regular basis by state or local health departments or by local fire departments to determine conformance to their particular requirements. At the present time, a laboratory that meets Federal or State OSHA requirements is likely to satisfy the standards of any other inspecting agency.

References

1. American Society for Testing and Materials: Water, Annual Book of ASTM Standards, Part 31, Philadelphia, 1981.
2. Commission on Laboratory Inspection and Accreditation: Reagent Water. Skokie, Ill., College of American Pathologists, 1978.

3. Compressed Gas Association, Inc.: Handbook of Compressed Gases. 2nd ed. New York, Reinhold Publishing Corporation, 1981.
4. Fisher Scientific Company: Fisher Safety Manual, Pittsburgh, Fisher Scientific Co., 1982.
5. National Committee for Clinical Laboratory Standards: Centrifugation of Blood, Proposed Standard H18-P. Villanova, Pa., NCCLS, 1981.
6. National Committee for Clinical Laboratory Standards: Power Requirements for Clinical Laboratory Instruments and for Laboratory Power Sources, Approved Standard AS1-5, Villanova, Pa., NCCLS, 1980.
7. National Committee for Clinical Laboratory Standards: Quantities and Units: SI, Committee Report C11-CR. Villanova, Pa., NCCLS, 1979.
8. National Committee for Clinical Laboratory Standards: Specifications for Reagent Water Used in the Clinical Laboratory: Approved Standard ASC-3. Villanova, Pa., NCCLS, 1980.
9. Weast, R. C. (Ed.): The Handbook of Chemistry and Physics. 58th ed. Cleveland, The Chemical Rubber Co., 1980-81, p. D-144.

Additional Readings

College of American Pathologists: Standards, Reference Materials, and Methods. A Practical Guide for the Medical Laboratory. Skokie, Ill., College of American Pathologists, 1983.
Committee on Hazardous Substances in the Laboratory: Prudent Practices for Handling Hazardous Chemicals in Laboratories: National Research Council. Washington, D.C., National Academy Press, 1981.
Gordon, A. J., and Ford, R. A.: The Chemist's Companion, New York, Wiley-Interscience, 1972.
National Bureau of Standards: Standard Reference Materials; Summary of the Clinical Laboratory Standards, NBS Special Publication 260-71. Washington, D.C., U.S. Department of Commerce, 1981.
Shugar, G. J., Shugar, R. A., Bauman, L., et al.: Chemical Technicians' Ready Reference Handbook. 2nd ed. New York, McGraw-Hill, 1981.
World Health Organization: The SI for the Health Professions. Geneva, Switzerland, 1977.

Analytical Procedures and Instrumentation

coordinated by Carl A. Burtis, Ph.D.

PHOTOMETRY

by Wendell T. Caraway, Ph.D.

Many determinations made in the clinical laboratory are based upon measurements of radiant energy emitted, transmitted, absorbed, or reflected under controlled conditions. The principles involved in such measurements will be considered in this section. Details of operation and maintenance for a particular instrument are typically supplied by the manufacturer and should be consulted for further information.

The term "photometric measurement" was defined originally as making a measurement of light intensity independently of wavelength. Most instruments used at present, however, have some means of isolating a narrow wavelength range of the spectrum for measurements. Those that use filters for this purpose are referred to as *filter photometers*, while those that use prisms or gratings are called *spectrophotometers*. Both types will be considered.

Nature of Light

Electromagnetic radiation includes radiant energy from short-wavelength gamma rays to long-wavelength radio waves. The term "light" will be used to describe radiant energy with wavelengths visible to the human eye and with wavelengths bordering on those visible to the human eye. The wavelength of light is defined as the distance between two peaks as the light travels in a wavelike manner. This distance is preferably expressed in nanometers (nm) for wavelengths commonly used in photometry. Other obsolete units that may be encountered are Ångstroms (Å) and millimicrons (mμ):

$$1 \text{ nm} = 1 \text{ m}\mu = 10 \text{ Å} = 10^{-9} \text{ m}$$

The human eye responds to radiant energy with wavelengths between about 380 and 750 nm, but modern instrumentation permits measurements at both shorter wavelength (ultraviolet, UV) and longer wavelength (infrared, IR) portions of the spectrum.

Sunlight, or light emitted from a tungsten filament, is a mixture, or spectrum, of radiant energy of different wavelengths that the eye recognizes as "white." Table 1B-1 shows approximate relationships between wavelengths and color characteristics for the UV, visible, and short IR portions of the spectrum. Thus, a solution will appear green when viewed against white light if it transmits light maximally between 500 and 580 nm but absorbs light at other wavelengths. Similarly, a solid object appears green if it reflects light in this region (500–580 nm) but absorbs

Table 1B-1. UV, VISIBLE, AND SHORT IR SPECTRUM
CHARACTERISTICS

Wavelength (nm)	Region Name	Color Observed*
< 380	Ultraviolet[†]	Not visible
380–440	Visible	Violet
440–500	Visible	Blue
500–580	Visible	Green
580–600	Visible	Yellow
600–620	Visible	Orange
620–750	Visible	Red
750–2000	Short IR	Not visible

*Owing to the subjective nature of color, the wavelength ranges shown
are only approximations.

[†]The ultraviolet portion of the spectrum is sometimes further divided
into "near" UV (220–380 nm) and "far" UV (< 220 nm). This arbitrary
distinction has a practical basis since standard silica used to make cuvets
transmits light effectively at wavelengths ≥ 220 nm.

light at other portions of the spectrum. In general, if we compare the intensity of light transmitted by a colored solution to that of a blank or reference solution over the entire spectrum, we obtain a typical spectral-transmittance curve characteristic for that spectrum. Such curves are shown in Figure 1B-1 for solutions of nickel sulfate (A) and potassium permanganate (B). Inspection of the curves should lead us to predict that the color of solution A is green inasmuch as light is transmitted maximally near the green portion of the spectrum. Curve B, on the other hand, illustrates the spectrum of a solution that transmits light maximally in the blue, violet, and red portions of the spectrum. The eye recognizes this mixture of colors as purple.

The chief advantage of filter photometry or spectrophotometry is that we can isolate and use discrete portions of the spectrum for purposes of measurement. Thus, in the case of the permanganate solution, we may use a relatively pure green light source which will be absorbed selectively, rather than use a white (composite) light source. This results in improved specificity, sensitivity, and linearity.

Beer's Law

Consider an incident light beam with intensity I_O passing through a square cell containing a solution of a compound that absorbs light of a certain wavelength (lambda = λ) (Figure 1B-2). The intensity of the transmitted light beam I_S will be less than I_O and we define the transmittance (T) of light as I_S/I_O. Some of the incident light, however, may be reflected by the surface of the cell or absorbed by the cell wall or by the solvent. In order to focus attention on the compound

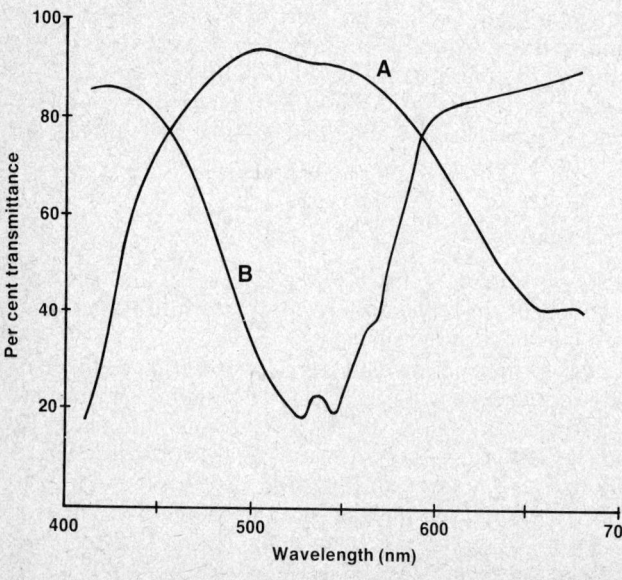

Figure 1B-1. Spectral-transmittance curves of nickel sulfate (A) and potassium permanganate (B). Arbitrary concentrations, read versus water as a blank (Beckman DB-G spectrophotometer).

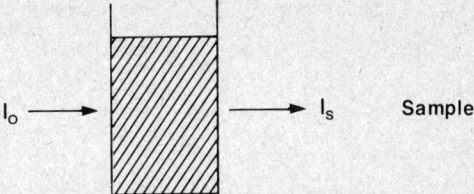

Figure 1B-2. Transmittance of light through sample and reference cells. Transmittance of sample versus reference = I_S/I_R.

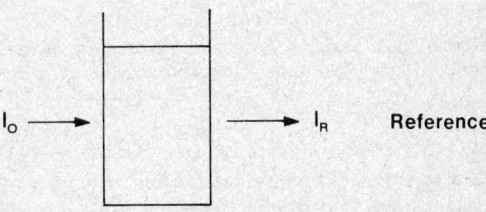

of interest, it is necessary to eliminate these factors. This is done by using a reference cell identical to the sample cell, except that the compound of interest is omitted from the solvent. The transmittance through this reference cell is I_R/I_O; the transmittance for the compound in solution is then defined as I_S/I_R. In practice, the reference cell is inserted and the instrument adjusted to an arbitrary scale reading of 100 (corresponding to 100% transmittance), following which the per cent transmittance reading is made on the sample. As we increase the concentration of the compound in solution, we find that transmittance varies inversely and logarithmically with concentration. Consequently, it is more convenient to define a new term, absorbance (A), that will be directly proportional to concentration.* Hence

$$A = -\log I_S/I_R = -\log T = \log \frac{1}{T} = \log \frac{100\%}{\text{per cent } T}$$

$$= \log 100 - \log \text{ per cent } T = 2 - \log \text{ per cent } T$$

To clarify these relationships more fully, assume that we have a solution of a compound at a concentration of 1 g/L that transmits only half of the incident light ($T = 0.5$) compared to a reference setting of 100, i.e.,

$$T = 50\% \text{ and } A = 2 - \log 50 = 0.301$$

If we increase the concentration to 2 g/L, the light transmitted becomes $0.5 \times 0.5 = (0.5)^2 = 0.25 = 25\%$ T. At 3 g/L, transmittance $= (0.5)^3 = 0.125 = 12.5\%$ T, and so on. It can be seen that as the concentration increases *linearly*, the per cent T decreases *geometrically*. On the other hand, absorbance is *linearly* related (directly proportional) to the concentration (see Figure 1B-3,*A* and *C*):

Concentration (g/L)	Per Cent T	A	A/C
0	100	0.000	
1	50	0.301	0.301
2	25	0.602	0.301
3	12.5	0.903	0.301

Data can also be plotted using semilogarithmic paper as shown in Figure 1B-3,*B*. When per cent *T* is plotted on the log scale, a straight line with negative slope is obtained. Some instruments have scales which show per cent *T* and absorbance. Most instruments now provide direct digital readouts of absorbance.

---

*Another, now obsolete, term sometimes used for absorbance is "optical density" (*OD*). Note that absorbance is *not* 100 − per cent *T*.

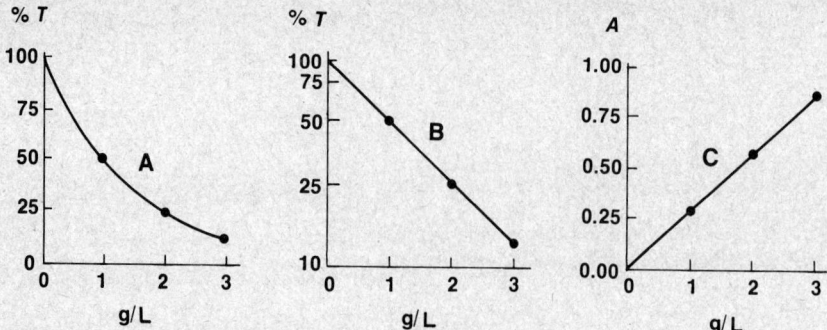

Figure 1B-3. Transmittance and absorbance as a function of concentration. *A*, Per cent *T*, linear scale. *B*, Per cent *T*, logarithmic scale. *C*, Absorbance, linear scale.

Experience has shown that with most photometers the response of the detector to a signal of transmitted light is such that any uncertainty in *T* is constant over the entire *T* scale. The uncertainty derives from electrical and mechanical imperfections in the instrument and from individual variations in the use of the instrument.

A fixed distance on the linear scale (e.g., 1% *T*) represents a greater change in absorbance for low values of per cent *T* than for high values of per cent *T*. For this reason, the *absolute* concentration error or uncertainty is greater when taking readings at high absorbance. However, the *relative* concentration error is greater for readings at either low *or* high absorbances. An uncertainty of 1% in readings on the linear *T* scale results in changes in absorbance (and in apparent concentration) as follows:

Per Cent T	Absorbance	Absolute Error	Per Cent Relative Error
10	1.000		
11	0.959	0.041	4.1
45	0.347		
46	0.337	0.010	2.9
90	0.046		
91	0.041	0.005	10.9
95	0.022		
96	0.018	0.004	18.2

The relative error is actually *minimal* at 36.8% *T*, corresponding to an absorbance of 0.434. Consequently, methods should be designed such that readings fall near the center of the scale, preferably between 20 and 80% *T* (*A* between about 0.7 and 0.1). Judgment must be exercised as to the importance of absolute and relative errors in a particular determination.

When we hold the concentration constant, say at 1 g/L, and double the inside diameter of the cell, the effect on absorbance is the same as doubling the concentration, since we have introduced twice as many absorbing molecules in the light path. From this it follows that absorbance is also directly proportional to the light path through the cell. This relationship is often referred to as Bouguer's law or Lambert's law.

The overall equation relating these variables may be expressed as

$$A = abc \tag{1}$$

where *A* is the absorbance, *a* is a proportionality constant defined as absorptivity, *b* is the light path in cm, and *c* is the concentration of the absorbing compound, usually expressed in g/L. This equation is called *Beer's law* and forms the basis of quantitative analysis by absorption photometry. Absorbance values have no units; hence, the units for *a* are the reciprocal of those for *b* and *c*.

Beer's law may be derived more rigorously as follows: Consider the differential expression

$$\frac{-dI}{dn} = kI$$

where $-dI$ is the increment of light absorbed at intensity level I by an increment dn in the number of absorbing molecules; k is a proportionality constant. Rearrangement followed by integration between limits gives

$$\int_{I_O}^{I} \frac{dI}{I} = -k \int_{O}^{N} dN$$

$$\ln \frac{I}{I_O} = -kN$$

The quantity N is a measure of the number of particles which are effective in absorbing light. A more useful measure of N is the product of the length of light path b and concentration c, or

$$\ln \frac{I}{I_O} = -kbc$$

For convenience, we shall replace k by another constant, a, which includes the factor for conversion of natural to common logarithms and invert the ratio I/I_O to remove the negative sign whence

$$\log_{10} \frac{I_O}{I} = abc$$

The quantity log I_O/I is given a special symbol, A, and is called *absorbance*. Thus, Beer's law is stated as

$$A = abc.$$

When $b = 1$ cm and c is expressed in moles per liter, the symbol ϵ (epsilon) is substituted for the constant a. The value for ϵ is a constant for a given compound at a given wavelength under prescribed conditions of solvent, temperature, pH, and so forth, and is called the *molar absorptivity*. The nomenclature of spectrophotometry is summarized in Table 1B-2. Values for ϵ are useful to characterize compounds, to establish their purity, and to compare sensitivities of measurements obtained on derivatives. Bilirubin, for example, when dissolved in chloroform at 25 °C, should have a molar absorptivity of 60 700 $\pm$ 1600 at 453 nm. The molecular weight of bilirubin is 584. Hence, a solution containing 5 mg/L (0.005 g/L) should have an absorbance of

$$A = (60\ 700)(1)(0.005/584) = 0.520$$

Conversely, a solution of this concentration showing an absorbance of 0.490 could be assumed to have a purity of 0.490/0.520 or 94%.

The molar absorptivity of the complex between ferrous iron and s-tripyridyltriazine is 22 600, while that with 1,10-phenanthroline is 11 000. This illustrates that, for a given concentration of iron, the former reagent produces a complex which has an absorbance about twice that of the complex with the latter reagent. Hence, it imparts greater sensitivity to the measurement of iron.

Table 1B-2. SPECTROPHOTOMETRY NOMENCLATURE

Name	Symbol	Definition
Absorbance	A	$-\log T$ or $\log I_O/I$
Absorptivity	a	A/bc (c in g/L)
Molar absorptivity	ϵ	A/bc (c in mol/L)
Path length	b	Internal cell or sample length, in cm
Transmittance	T	I/I_O*
Wavelength unit	nm	10^{-9} m
Absorption maximum	λ_{max}	Wavelength at which a maximum absorption occurs

*I/I_O is the ratio of the intensity of transmitted light to incident light.

In toxicological work, it is customary to list constants based on concentrations in g/dL rather than in mol/L. This may also be necessary when the molecular weight of a substance is unknown. For $b = 1$ cm and $c = 1$ g/dL (1%), A can be written as $A_{1\,\text{cm}}^{1\%}$. This constant is called the *absorption coefficient*. An older symbol, $E_{1\,\text{cm}}^{1\%}$, now obsolete, was called the *extinction coefficient*.

The direct proportionality between absorbance and concentration must be established experimentally for a given instrument under specified conditions. Frequently there is a linear relationship up to a certain concentration or absorbance. We then say that the solution obeys Beer's law up to this point. Within this limitation, a calibration constant (K) may be derived and used to calculate the concentration of an unknown solution by comparison to a standard. From Equation (1) ($A = abc$)

$$a = \frac{A}{bc} \tag{2}$$

Therefore

$$\frac{A_1}{b_1 c_1} = \frac{A_2}{b_2 c_2} \tag{3}$$

The light path (b) remains constant in a given method of analysis with a fixed cuvet size, and $b_1 = b_2$. Equation (3) then becomes

$$\frac{A_1}{c_1} = \frac{A_2}{c_2} \quad \text{or} \quad \frac{A_s}{c_s} = \frac{A_u}{c_u} \tag{4}$$

where s and u represent standard and unknown, respectively. Solving for the concentration of unknown, we obtain

$$c_u = \frac{A_u}{A_s} \times c_s \tag{5}$$

or the equivalent expression

$$c_u = A_u \times \frac{c_s}{A_s} = A_u \times K \tag{6}$$

where $K = c_s/A_s$. The value of the constant K is obtained by measuring the absorbance (A_s) of a standard of known concentration (c_s).

Certain precautions must be observed in using such calibration constants. Under no circumstances should the constant be used when either the standard or unknown readings exceed the linear portion of the calibration curve, i.e., if the curve no longer obeys Beer's law. Two or more standards should be included in each series of determinations to permit direct comparison of unknown to standard, or to calculate the calibration constant, since variations in reagents, working conditions, cell diameters, deterioration, or change in instruments, and so on, may result in day-to-day changes of the absorbance value for the standard. A nonlinear calibration curve may be used if a sufficient number of standards of varying concentration is included to cover the entire range encountered for readings on unknowns.

In some cases, pure standard may not be readily available and constants may be provided that were obtained on pure standards by others and reported in the literature. In general, the use of published constants should be discouraged unless the method is followed in detail and readings are made on a spectrophotometer capable of providing light of high spectral purity at a verified wavelength. Use of broader-band light sources usually leads to some decrease in absorbance. The absorbance of NADH at 340 nm, for example, is frequently used as a reference for the determination of enzyme activity, based on an assigned molar absorptivity of 6.22×10^3. This value is acceptable only under the carefully controlled conditions just described and should not be used unless these conditions are met. In summary, published values for molar absorptivities and absorption coefficients should be used only as guidelines until they can be verified by readings on pure standards for a given instrument.

COMPONENTS OF SPECTROPHOTOMETERS

The major components of a *single-beam spectrophotometer* are shown schematically in Figure 1B-4. Light is passed through a monochromator to provide selection of the desired region of the spectrum to be used for measurements. Slits are used to isolate a narrow beam of the light and improve its chromatic purity. The light next passes through an absorption cell (cuvet), where a portion of the radiant energy is absorbed, depending on the nature and concentration of the solution. Any light not absorbed is transmitted to a detector (photocell or phototube) which converts light energy to electrical energy that can be registered on a meter or digital readout.

In operation, an opaque block is substituted for the cuvet, so that no light reaches the photocell, and the meter is adjusted to read 0% T. Next, a cuvet containing a reagent blank is inserted and the meter is adjusted to read 100% T (i.e., zero absorbance). The composition of the reagent blank should be identical to that of standard or unknown solutions except for the substance to be measured. Standard solutions containing various known concentrations of the substance are inserted and readings are recorded. Finally, a reading is made on the unknown solution, and its concentration is determined by comparison with the readings obtained on the standards.

Double-beam spectrophotometers are also widely used. One such type of double-beam system is shown schematically in Figure 1B-5 in which all components are duplicated except the light source. This system is referred to as a double-beam-in-*space* spectrophotometer. Another approach is to use a light-beam chopper (a rotating wheel with alternate silvered sections and cut out sections) inserted after the exit slit (Figure 1B-6). A system of mirrors passes the portions of the light reflected off the chopper alternately through the sample and a reference cuvet onto a common detector. This system is referred to as a double-beam-in-*time* spectrophotometer. The chopped-beam approach, using one detector, compensates for light source variation as well as for sensitivity changes of the detector.

Component parts of spectrophotometers will next be considered in more detail.

The Light Source

The light source for measurements in the visible portion of the spectrum is usually a tungsten light bulb. The lifetime of a tungsten filament can be greatly increased by the presence of a low pressure of iodine or bromine vapor within the lamp. An example is the *quartz-halogen* lamp, which has a fused-silica envelope and which provides high intensity light over a wide spectrum. The tungsten lamp is acceptable for making measurements of moderately dilute solutions in which the change in color intensity varies significantly with small changes in concentration. A common disadvantage with some photometers is that a considerable amount of heat is generated by the light source; this heat may cause problems in measurement either by changing the geometry of the optical system or by changing the sensitivity of the photocell.

At the temperature of the tungsten filament, some of the metal vaporizes and condenses on the cooler glass surface of the bulb, thus reducing the intensity of radiant energy. Such coatings may also change the spectrum sufficiently to alter instrument response. Similar changes may be brought about by deposits of chemical fumes or oily particles on the outside of the bulb. These coatings are often uneven and are particularly likely to cause trouble in the double-beam-in-space spectrophotometers. Periodic lamp inspection and replacement are necessary to avoid these difficulties.

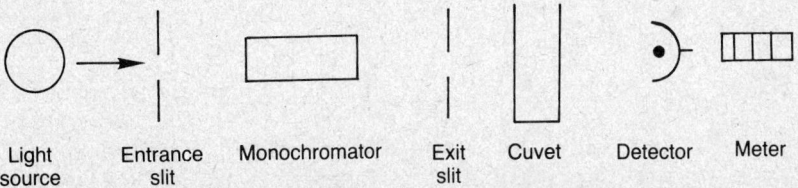

Figure 1B-4. Major components of a single-beam spectrophotometer.

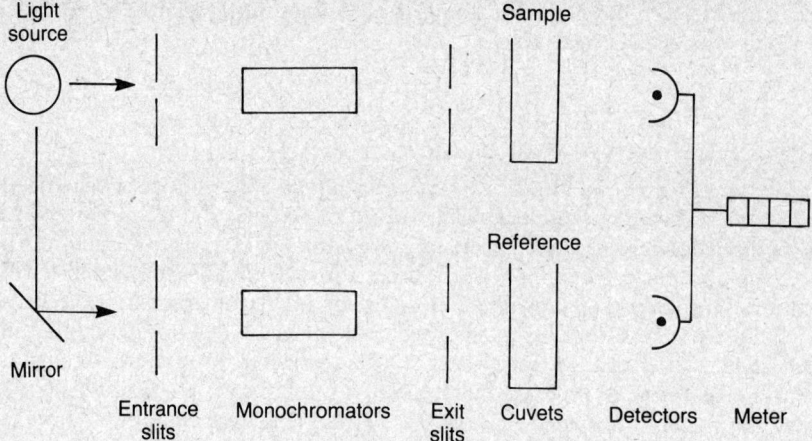

Figure 1B-5. Double-beam-in-space spectrophotometer.

A tungsten light source does not supply sufficient radiant energy for measurements below 320 nm. In the ultraviolet region, a low-pressure mercury-vapor lamp, which emits a discontinuous or line spectrum, is useful for calibration purposes but is not very practical for absorbance measurements, since it can be used only at certain wavelengths. Hydrogen and deuterium lamps provide sources of continuous spectra in the ultraviolet region with some sharp emission lines, as do high pressure mercury and xenon arc lamps. These sources are more commonly used in ultraviolet absorption measurements. A deuterium lamp is more stable and has a longer life than a hydrogen lamp.

Electric current is the energy source for the lamp. Central power stations may vary in their ability to maintain a constant voltage in the lines; in addition, emergency generators may be put into use in case of power failures. Variations in light intensity from the lamp will result in unstable readings on the meter; consequently, suitable voltage-regulation devices are essential for optimum stability. Transformers may also be used to reduce the line voltage for some instruments. A storage battery provides a stable source of current but presents maintenance problems and is susceptible to a slow drift with time.

Spectral Isolation

A system for isolating radiant energy of a desired wavelength and excluding that of other wavelengths is called a *monochromator*. There are various ways of accomplishing this, including the use of filters, prisms, and diffraction gratings. Combinations of lenses and slits may be inserted

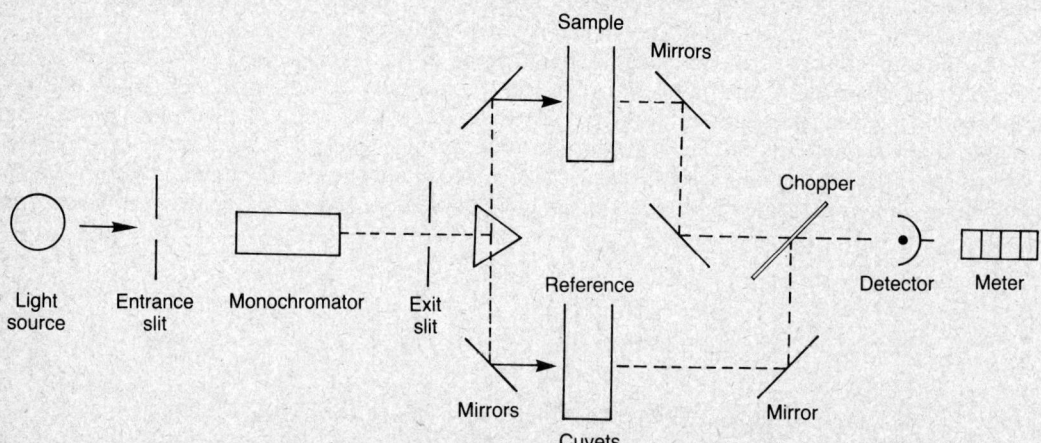

Figure 1B-6. Double-beam-in-time spectrophotometer.

before or after the monochromatic device to render light rays parallel or to isolate narrow portions of the light beam. Variable slits may be used to permit adjustments in total radiant energy reaching the photocell.

The simplest type of filter is a thin layer of colored glass. Certain metal complexes or salts, dissolved or suspended in glass, produce colors corresponding to the predominant wavelengths transmitted. Strictly speaking, a glass filter is not a true monochromator since it transmits light over a relatively wide range of wavelengths. The spectral purity of a filter or other monochromator is usually described in terms of its *spectral bandwidth*. This is defined as the width, in nm, of the spectral transmittance curve at a point equal to one-half the peak transmittance (Figure 1B-7). Commonly used glass filters have spectral bandwidths of approximately 50 nm and are entirely adequate for many purposes. These are referred to as wide-bandpass filters.

Other glass filters include the narrow-bandpass and sharp-cutoff types (Figure 1B-7). As implied, the latter filter typically shows a sharp rise in transmittance over a narrow portion of the spectrum and is used to eliminate light below a given wavelength. Narrow-bandpass filters may be constructed by combining two or more sharp-cutoff filters or regular filters; however, the availability of high-intensity-light sources now favors the use of narrow-bandpass interference filters.

Another approach for construction of narrow-bandpass filters is to use a dielectric material of controlled thickness sandwiched between two thinly silvered pieces of glass. The thickness of the layer determines the wavelength of energy transmitted. Energies of wavelengths that are multiples of this thickness stay in phase as they reflect back and forth through the dielectric materials and finally emerge, whereas other wavelengths will cancel due to phase differences. Light striking the semitransparent film at a given angle θ will be partly reflected and partly passed (Figure 1B-8, point a). The same process occurs at a', b, b', and so on. For reinforcement to occur at point b, the distance traveled by the beam reflected at a' must be some multiple of its wavelength in the medium. For purposes of clarity, the incident beam is shown as arriving at an angle θ from the perpendicular. In ordinary use, however, θ approaches zero and $n\lambda' = 2t$, where λ' is the wavelength of radiation in the dielectric material, t is its thickness, and n is an integer called the order of interference. These filters have narrow spectral bandwidths, usually from 5 to 15 nm, and are referred to as *interference filters*. Since they also transmit harmonics, or multiples, of the desired wavelength, accessory glass filters are required to eliminate these undesired wavelengths. Thus, an interference filter designed for 620 nm will also transmit some radiation at 310 and 1240 nm unless accessory cutoff filters are provided to absorb this undesired stray light.

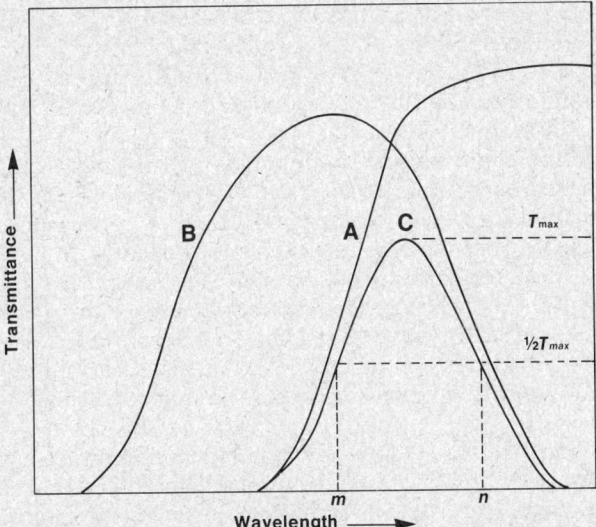

Figure 1B-7. Spectral characteristics of a sharp-cutoff filter (A) and a wide-bandpass filter (B). The narrow-bandpass filter (C) is obtained by combining filters A and B. The spectral bandwidth of filter C (distance $n - m$) is defined as the width in nm of the spectral-transmittance curve at a point equal to one half of maximum transmittance.

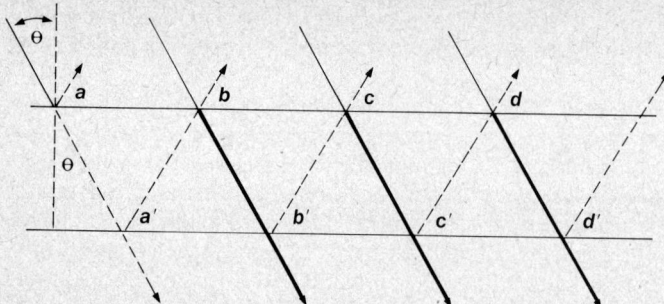

Figure 1B-8. Passage of light through an interference filter. The heavier lines indicate reinforcement of entering light rays.

Prisms and diffraction gratings are also widely used as monochromators. A *prism* separates white light into a continuous spectrum by refraction, i.e., shorter wavelengths are bent, or refracted, more than longer wavelengths as they pass through the prism. This results in a nonlinear spectrum with the longer wavelengths closer together, but with suitable accessories a narrow-bandwidth portion of the spectrum may be isolated. A *diffraction grating* is prepared by depositing a thin layer of aluminum-copper alloy on the surface of a flat glass plate, then ruling many small parallel grooves into the metal coating. Better gratings contain 1000–2000 lines/mm and must be made with great care. These are then used as molds to prepare less expensive replicas for general use in instruments. Rays of radiant energy bend (refract) around a sharp corner, and the extent of refraction varies with the wavelength. Thus, each line ruled on the grating, when illuminated, gives rise to a tiny spectrum. Wave fronts are formed which reinforce those wavelengths in phase and cancel those not in phase. The net result is a uniform linear spectrum. Some instruments contain diffraction gratings that produce spectral bandwidths of 20 nm or more; higher priced instruments may have a resolution of 0.5 nm or less. Good gratings are generally better than prisms for spectral isolation.

The grating discussed above is called a plane *transmission grating*. In most instruments, *reflection gratings* are more common. Lines are engraved on the surface of a mirror, which may be either a polished metal slab or a glass plate on which a thin, metallic film has been deposited. A grating may also be ruled at a specified angle, so that a maximum fraction of the radiant energy is directed into wavelengths diffracted at a selected angle. This type of grating is called an *echelette* and is said to have been given a *blaze* at a particular angle or to have been blazed at a certain wavelength, e.g., 250 nm.

The type of monochromator chosen will depend on the objectives of the analyst. Narrow-bandpass filters are highly desirable in flame photometry in order to isolate emission energy at a given wavelength. Narrow-spectral bandwidths are required in spectrophotometers if one is interested in resolving and identifying sharp absorption peaks that are closely adjacent. Lack of agreement with Beer's law will occur when a part of the spectral energy transmitted by the monochromator is not absorbed at all by the substance being measured. This is more commonly observed with wide-bandpass instruments.

Some increase in absorbance as well as improved linearity with concentration is usually observed with instruments that operate at narrower bandwidths of light. This is especially true for substances that exhibit a sharp peak of absorption. Spectral absorbance curves for a solution of coproporphyrin I (Figure 1B-9) demonstrate the marked decrease in maximum absorbance as the spectral bandwidth is increased from 1 to 20 nm. The *natural bandwidth* of an absorbing substance is defined as "the bandwidth of the spectral absorbance curve at a point equal to one-half of the maximum absorbance." Curve A in Figure 1B-9, scanned at a *spectral* bandwidth of 1 nm, shows a *natural* bandwidth of approximately 10 nm. As a general rule, for peak absorbance readings to be within 99.5% of true values, the spectral bandwidth should not exceed 10% of the natural bandwidth.

Coproporphyrin is an unusual example of a compound showing a very narrow natural bandwidth. Most common clinical chemistry laboratory procedures produce a color for which the natural bandwidth ranges from 40 to over 200 nm.[2] The natural bandwidth of NADH is 58 nm (λ_{max} = 339 nm). Ideally, for accurate measurements of this compound, we should use a spectral bandwidth of 6 nm or less. Actual studies have shown that use of a 10-nm spectral bandwidth at 340 nm produces absorbance values approximately 98% of those obtained with a 1-nm spectral bandwidth.[6]

   The wavelength selected is usually at the peak of maximum absorbance in order to achieve maximum sensitivity; however, it may be desirable to choose another wavelength to minimize interfering substances. For example, turbidity readings on a spectrophotometer are greater in the blue region than in the red region of the spectrum, but the latter region is chosen for turbidity measurements to avoid absorption of light by bilirubin (460 nm) or hemoglobin (417 and 575 nm). The color developed in the alkaline picrate procedure for creatinine produces a relatively flat peak in the visible region at approximately 480 nm, but the reagent blank itself absorbs light strongly below 500 nm. A compromise is made by selecting a wavelength at 520 nm to minimize the contribution of the blank. Blank readings should, of course, be kept to a minimum. A small difference between two large numbers is subject to greater uncertainty; hence, minimizing absorbance of the blank improves precision and accuracy. The linear working range of a method can be expanded also by not measuring at the peak absorbance. Measurements should preferably not be taken on the steep slope of an absorption curve, since a slight error in wavelength adjustment would introduce a significant error in absorbance readings.

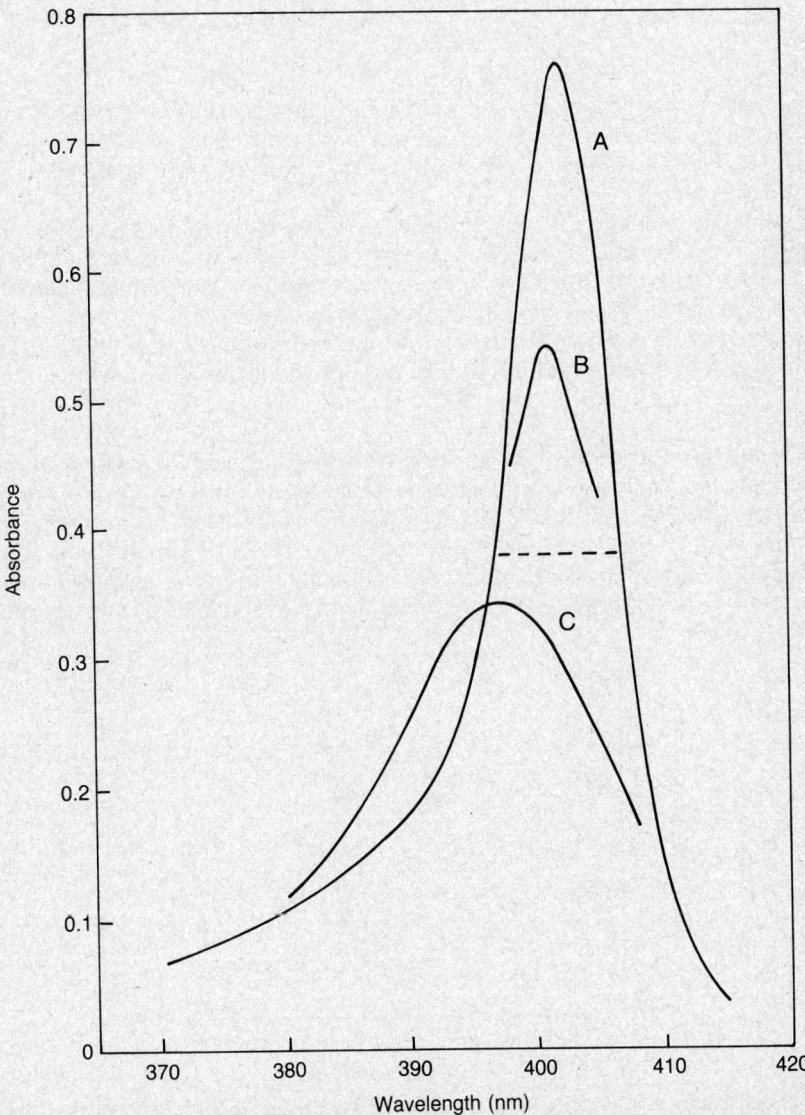

Figure 1B-9. Effect of spectral bandwidth (SBW) on the absorption spectrum of coproporphyrin I. Nominal concentration, 1 μg/mL in HCl, 0.1 mol/L. SBW: Curve A, 1 nm, Beckman DB-G spectrophotometer; Curve B, 10 nm, and curve C, 20 nm, Beckman DB spectrophotometer. The dotted horizontal line shows a natural bandwidth of 10 nm for coproporphyrin I when scanned at a spectral bandwidth of 1 nm. The shift of A_{max} to lower wavelengths as SBW is increased is related to skewness of the absorption spectrum to the left.

Multiple Wavelength Readings

Background interference can often be eliminated or minimized either by inclusion of blanks or by reading absorbance at two or three wavelengths. In one approach, currently used on the DuPont *aca* system, absorbance is measured at two wavelengths, one corresponding to peak absorbance and another at a point near the base of the peak to serve as a baseline. The difference in absorbance at the two wavelengths is related to concentration. In effect, this provides a blank reference point for each individual specimen. Another method to correct for background interference is to measure absorbance at the peak wavelength and at two other wavelengths equidistant from the peak. Values for the latter are averaged to obtain a baseline under the peak, which is then subtracted from the peak reading. The value thus obtained is known as a "corrected" absorbance and can be related to the concentration, provided that the background absorbance is linear with wavelength over the region in which readings are made. This technique of making corrections for interfering substances is called the *Allen correction* and is illustrated in Figure 1B-10. The corrected absorbance at 300 nm is obtained from the Allen equation:

$$A_{corr} = A_{300} - \frac{(A_{280} + A_{320})}{2}$$

Similar corrections are applied in procedures for spectrophotometric determinations of salicylates, porphyrins, steroids, and other compounds. The correction must be applied similarly to standards, since the Allen corrected absorbance is typically less than the total absorbance, even in the absence of interfering background.

Before the Allen correction is used, knowledge of the shape of the absorption curve for the substance of interest and of the interference is required. The linearity of the baseline shift should be verified by measuring the absorption spectrum of commonly encountered interferences. Care should be exercised in the use of the Allen correction because if it is not properly used, it may introduce larger errors than would be observed without correction. For example, such a situation may occur if the background reading is not linear over the region measured.

Cuvets

Cuvets, also known as absorption cells, may be round, square, or rectangular, and may be constructed from glass, silica (quartz), or plastic. (Depending on the material used, the amount of light transmitted by the cell may differ significantly.) For routine colorimetric manual procedures, the round, glass, test tube type is most common. These are inexpensive and satisfactory, provided they are matched; i.e., to be used interchangeably, the cuvets must be of uniform inside diameters so that the absorbance of a solution will be within specified tolerances when measured

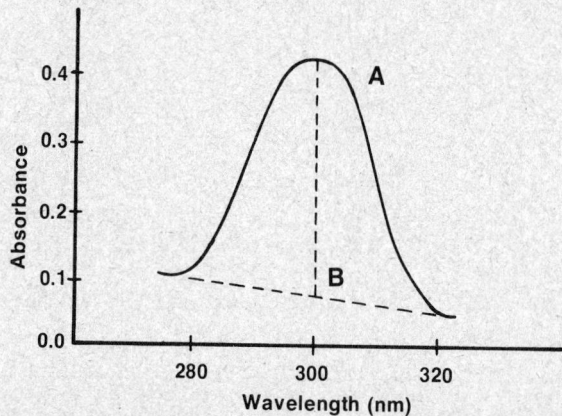

Figure 1B-10. Hypothetical example of an Allen correction. *A* represents the absorption curve of the test mixture. Absorbance readings are taken at the absorption peak (300 nm) and at equidistant points from the peak (280 and 320 nm). *B* represents the background absorbance, which is linear over the measured range. (From Evenson, M. A.: *In*: Fundamentals of Clinical Chemistry. N. W. Tietz, Ed. Philadelphia, W. B. Saunders Company, 1970. Used with permission.)

in different cuvets. This criterion is readily checked by preparing a stable solution, such as cyanmethemoglobin, to read approximately 50% T at 540 nm against a reference blank. This solution is then added to each dry cuvet to be tested and readings are taken. Those cuvets that match within 0.5% T (or other selected tolerance) are reserved for use. Since cuvets may not be perfectly round, they are rotated in the well to observe any changes with position; those showing changes are either discarded or etched to indicate the position of use.

Square or rectangular cuvets have plane-parallel optical surfaces and a constant light path. The most popular have a 1.0-cm light path, held to close tolerances. Compared to round cuvets, they have few optical aberrations such as lens effect and refractive errors. On the other hand, these cuvets are more expensive and are less convenient to handle. Ordinary borosilicate glass cuvets are suitable for measurements in the visible portion of the spectrum. For readings below 340 nm, however, quartz absorption cells are usually required. Some plastic cells have good clarity in both the visible and ultraviolet range but can present problems relating to tolerances, cleaning, etching by solvents, and temperature deformations. Many of the plastic cuvets are designed for disposable, single-use applications.

Cuvets must be clean and optically clear. Etching or deposits on the surface will obviously affect absorbance values. Round cuvets, used in the visible range, are cleaned by copious rinsing with tap water and distilled water. Alkaline solutions should not be left standing in cuvets for prolonged periods since alkali slowly dissolves glass and produces etching. Both round and square cuvets may be cleaned in mild detergent or soaked in a mixture of concentrated HCl:water:ethanol(1:3:4). Cuvets should never be soaked in dichromate cleaning solution since the solution tends to adsorb onto and discolor the glass.

Cuvets used for measurements in the ultraviolet region should be handled with special care. Invisible scratches, fingerprints, or residual traces of previously measured substances may be present and absorb significantly. A good practice is to fill all such cuvets with distilled water and measure the absorbance for each against a reference blank over the wavelengths to be used. This value should be essentially zero. Startling and disturbing inconsistencies will be avoided if this procedure is practiced routinely.

Detectors

The two most commonly used devices for measuring light intensity in the UV and visible regions of the spectrum are barrier layer cells and photomultiplier tubes. The barrier layer cells (also known as photovoltaic cells) are rugged and are used in inexpensive instruments; photomultipliers are almost always used in the higher quality, more expensive spectrophotometers.

Barrier layer cells. These cells operate on the principle that when light falls on certain metals or semiconductors, electrons will flow in proportion to the intensity of the light. The barrier layer cell consists of a thin layer of silver on a layer of the semiconductor selenium. The silver and selenium metals are then mounted on an iron backing or support (Figure 1B-11). When light, passing through the thin metal layer, falls upon the selenium surface, electrons are released and are collected on the silver layer to produce a negative charge. The metal base plate becomes the positive pole. When this cell is connected to a galvanometer, a current will flow which will vary with the intensity of the incident light.

The sensitivity of the barrier layer cell to light of different wavelengths is similar to that of the human eye. The maximum sensitivity of both occurs at 550 nm. Barrier layer cells are usually used at high levels of illumination, and the output from these photocells is generally not amplified. Barrier layer cells are very stable but are slow in responding to changes of light intensity. Because of their slow response time, they are not suitable as detectors in instruments in which interrupted (chopped) light beams fall on the detectors. Another disadvantage of this photocell is that it tends to show fatigue. Fatigue occurs in a barrier layer cell when, at a constant, continuous high level of intensity, the electrical output of the photocell decreases with time. Therefore, barrier layer cells should not be used at extremely high illumination.

A potential problem with barrier layer cells is that their electrical output is very temperature-dependent. If a heat producing, high-intensity lamp or flame is used for the light source, instrument design must be such that thermal stability and rapid temperature equilibrium of the photocell are achieved. This may be accomplished by having the light source far removed from the photocell. Use of heat shields or plastic materials that do not readily conduct heat are other ways of improving thermal stability of instruments.

Photomultiplier tubes. A photomultiplier is an electron tube that is capable of significantly amplifying a current. The photomultiplier tube has as its cathode a light-sensitive metal that absorbs light and emits electrons in proportion to the radiant energy that strikes the surface of

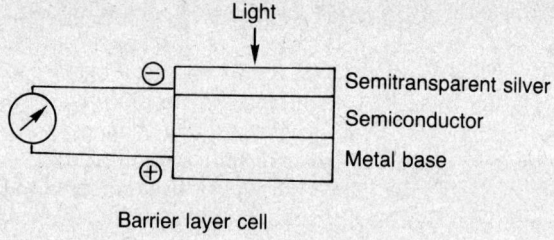

Barrier layer cell

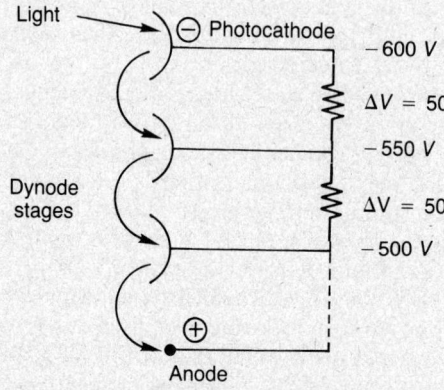

Photomultiplier tube

Figure 1B-11. Schematic diagrams of a barrier layer cell and a photomultiplier tube.

the light-sensitive material. The electrons produced by this first stage go to a secondary stage (surface), where each electron produces between four and six additional electrons. Each of these electrons from the second stage goes on to another stage, again producing four to six electrons. Each electron cascades through the photomultiplier stages; thus, the final current produced by such a tube may be one million times as much as the initial current. As many as 10 to 15 stages or dynodes are present in common photomultipliers.

When operating such a tube, voltage is applied between the photocathode and each successive stage. The normal increment of voltage increase of each photomultiplier stage is from 50–100 V larger than that of the previous stage (Figure 1B-11). A common photomultiplier tube may have approximately 1500 V applied to it.

Photomultiplier tubes have extremely rapid response times, are very sensitive, and do not show as much fatigue as other detectors. Because these tubes have excellent sensitivity and rapid response, they must be carefully shielded from all stray light and daylight. A photomultiplier with the voltage applied should never be exposed to room light because it will burn out. Because of the fast response time of the photomultiplier, this detector is readily used with interrupted light beams such as those produced by choppers, and it therefore has significant advantages when used as a UV-visible detector in spectrophotometers. The rapid response times are needed when a spectrophotometer is being used to determine an absorption spectrum of a compound. The photomultiplier also has adequate sensitivity over a wide wavelength range.

When voltage is applied to photomultipliers and all light has been blocked from them, some current usually will be produced. This current is called *dark current*. It is desirable to have the dark current of photomultipliers at its lowest level, since this current is also amplified and appears as background noise.

Readout Devices

Electrical energy from a detector is displayed on some type of meter or readout system. These may be direct reading or null point systems. In *direct reading systems,* the output of the photocell is used to drive a sensitive meter directly with no further amplification. Other instruments utilize an amplifier to increase the output of the detector. In the *null point system*, the output of the detector is balanced against the output of a reference circuit. The meter may be replaced by a servomotor activated by an imbalance of current, which stops when the two circuits are balanced. Direct digital readouts are thus obtained.

Also popular are digital readout devices that provide a visual numerical display of absorbance or converted values of concentrations. These operate on the principle of selective illumination of portions of a bank of light-emitting tubes or diodes, controlled by the voltage signal generated. Typical examples include visible light-emitting diodes (LEDs). The latter incorporate gallium as the major component. At present, $GaAs_xP_x$ diodes that emit red light are most widely used. Compared to meters, the digital readout devices have faster response, are easier to read, and decrease operator fatigue.

Microprocessors are now widely incorporated in many instruments. Signal output from a standard or calibrator can be stored, signals from blanks can be subtracted from both standards and unknowns, and the concentration of unknowns readily calculated analogous to the application of Beer's law. Data from multiple standards may be used to store a complete calibration curve, display or print out the curve for visible inspection, and calculate results of unknowns based on the curve or some mathematical transformation of it.

Recorders

Spectrophotometers may be equipped with recorders in addition to or instead of the digital display of values. These are synchronized to provide line traces of transmittance or absorbance, as a function of either time or wavelength. Change of absorbance with time is widely used to measure enzyme activity. Thus, in the kinetic determination of lactate dehydrogenase, the rate of disappearance of NADH can be monitored at 340 nm as pyruvate is converted to lactate. Multiple readings at short intervals may also be taken electronically and fed into a microprocessor to identify the linear portion of the curve, calculate the slope as $\Delta A/min$, and multiply by an appropriate constant to provide readout and printout of the final result.

When a continuous tracing of absorbance versus wavelength is recorded, the resultant figure is called an *absorption spectrum*. This type of procedure is especially useful for identification of drugs that absorb in the ultraviolet. Several criteria are used, including determination of those wavelengths showing maximum and minimum absorbance in both dilute acid and alkaline solutions; absorptivity at the wavelength of maximum absorbance; and ratios of absorbance at two wavelengths. Finally, the entire spectrum is compared to that of a known sample of the suspected drug. This sort of information is available in textbooks of toxicology.

STANDARDIZATION OF SPECTROPHOTOMETERS

In most spectrophotometric analytical procedures the absorbance of an unknown is compared directly to that of a standard or series of standards. Under these circumstances, minor errors in wavelength calibration, variation in spectral bandwidths, presence of stray light, and so on, do not usually contribute serious errors. Use of a series of standards covering a wide range of concentrations also provides a measure of linearity, i.e., agreement with Beer's law for a given procedure and instrument. When calculations are based on published or previously determined values for molar absorptivities or absorption coefficients, however, the spectrophotometer must be checked more rigorously. Standardization of spectrophotometers on a periodic basis also improves reliability of routine comparative analyses. The reader is referred to a monograph on standards in absorption spectrometry.[1]

Wavelength Calibration

For many analytical purposes, the wavelength chosen may be satisfactory if it is close to the λ_{max} of the chromogen being measured and the wavelength is reproducible. Most filters fall into this category and are quite satisfactory since unknowns are compared to standards at a fixed wavelength and spectral bandwidth. With prisms and diffraction gratings, however, a continuous choice of wavelengths is available, and it becomes necessary to verify their accuracy and reproducibility. Knowledge of exact wavelength becomes critical when using published molar absorptivities for identification of substances in toxicological studies and in the use of differential absorption techniques. Enzyme assays employing the NAD-NADH reaction, for example, are based on a molar absorptivity constant for NADH of 6.22×10^3 at 340 nm. It is therefore necessary that the wavelength setting be accurate and reproducible and that the instrument show spectrophotometric accuracy if the constant is to be used in calculating results.

For the narrow spectral bandwidth instruments, a *holmium oxide glass* may be scanned over the range of 280–650 nm. This material shows very sharp absorbance peaks at well-defined wavelengths, and the operator may compare the wavelength scale readings which produce maximum absorbance with established values. Should these not coincide, a calibration curve can be constructed to relate scale readings to true wavelengths. A typical spectral-transmittance curve for holmium oxide glass is shown in Figure 1B-12. Selected absorption peaks for this filter, suitable for calibration purposes, occur at the following wavelengths (nm):

279.3	418.5
287.6	536.4
333.8	637.5
360.8	

Solutions of holmium oxide in dilute perchloric acid have also been recommended and may be used with any spectrophotometer.[1]

With broader bandpass instruments a didymium filter may be used to verify wavelength settings. This filter should show a minimum per cent transmittance at 530 nm against an air blank (Figure 1B-13). Since didymium has several absorption peaks, the setting should be verified grossly by visual examination of transmitted light. This light should appear green at 530 nm.

Spectral Bandwidth

The spectral bandwidth (SBW) of a spectrophotometer is customarily stated by the manufacturer. This value is generally accepted without verification because of the lack of a convenient method for directly verifying it. The SBW can be measured by use of a mercury vapor lamp which shows a number of sharp, well-defined emission lines between 250 and 580 nm. The apparent width of an emission band at half-peak height is taken to be the spectral bandwidth of the instrument (see Figure 1B-7). The SBW may also be calculated from the manufacturer's specifications.[4] Interference filters with spectral bandwidths of 1–2 nm are available and may be used to check those instruments with a nominal SBW of 8 nm or more.

Stray Light

Stray light, in general terms, is radiation of wavelengths outside the narrow band nominally transmitted by the monochromator. A perfect monochromator would transmit light only within

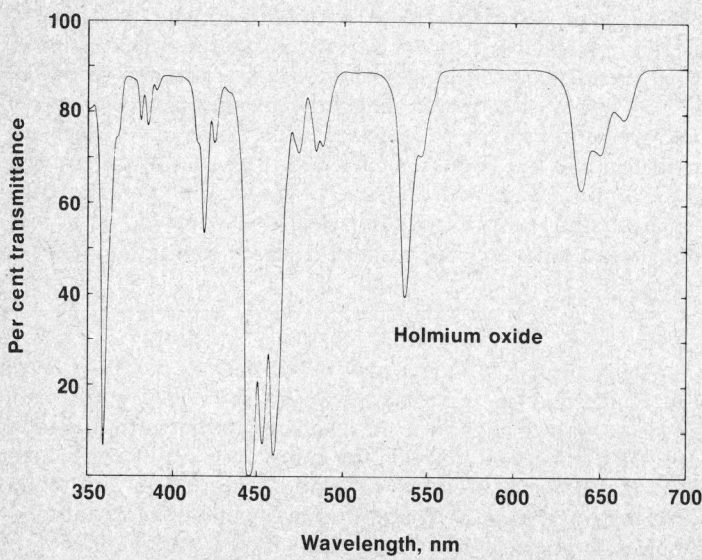

Figure 1B-12. Spectral-transmittance curve of holmium oxide filter. (Courtesy of Beckman Instruments, Inc.)

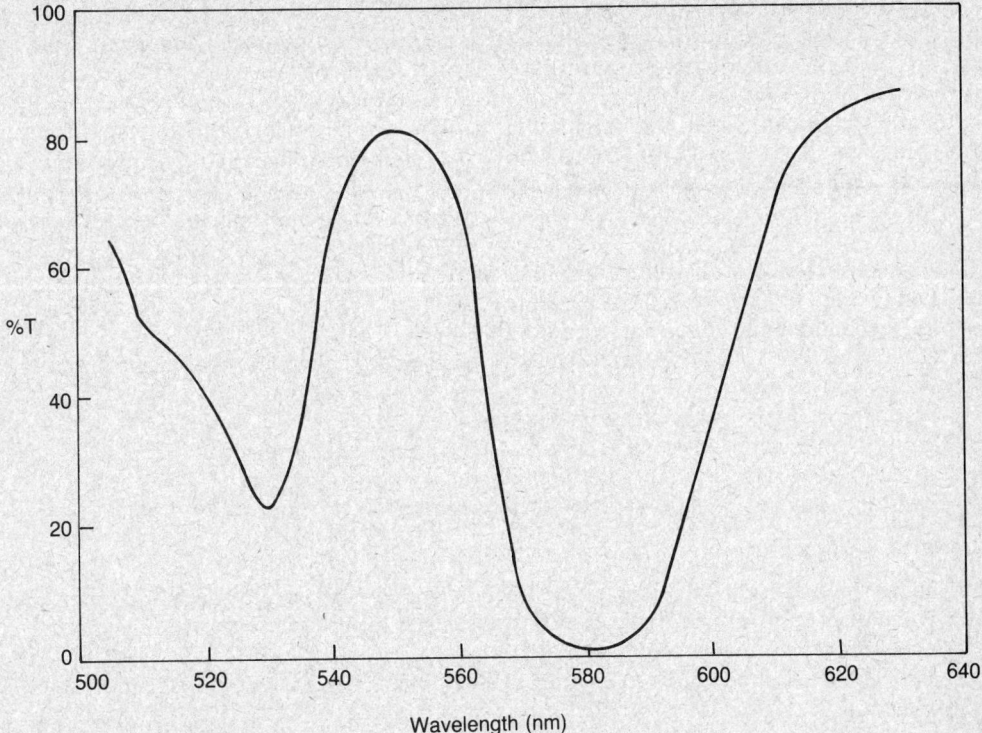

Figure 1B-13. Spectral-transmittance curve of a didymium filter. (Perkin-Elmer Model 35 spectrophotometer, 8-nm nominal spectral bandwidth.)

its passband. In practice, scattering and diffraction inside the monochromator introduce light of other wavelengths into the exit beam. This light is further modified by other components of the spectrophotometer and by the sample itself. Stray light is usually defined as a ratio or per cent of the stray light to the total detected light.

Other sources of unwanted light include light leaks and fluorescence of the sample. Light leaks should be excluded by covering the cell compartments. Light arising from fluorescence can increase the signal to the detector and cause an apparent decrease in absorbance. These sources of light are not included in the usual definition of stray light.

The major effect of stray light on the performance of a spectrophotometer is an absorbance error, especially in the upper end of the absorbance range of the instrument. Most spectrophotometers are equipped with one or more stray-light filters. Thus, a blue filter is used with a tungsten lamp for wavelength settings below about 400 nm. When the spectrophotometer is set to 350 nm, for example, most of the stray light is of wavelengths in the visible range. The blue filter absorbs most of the visible light but transmits well in this UV portion of the spectrum. By analogy, a red filter is used for wavelengths in the range of 650–800 nm.

A cutoff filter is satisfactory for the detection of stray light. These may be of glass, similar to the stray-light filters discussed above, which produce a sharp cut in the spectrum with almost complete absorption on one side and high transmittance on the other. Liquid cutoff filters are satisfactory and convenient in the UV range where stray light is usually more of a problem. A 50 g/L aqueous solution of sodium nitrite should show essentially zero %T when read against water over the range of 300–385 nm. Acetone, read against water, should show zero %T over the range of 250–320 nm.

Linearity

Linearity of calibration curves has been discussed earlier with reference to Beer's law. In order for a spectrophotometer to give accurate absorbance measurements throughout its absorbance range, its response to changes in light intensity must be linear. This means that a linear relationship should exist between the light absorbed and the instrument readout.

Various liquid solutions have been used to check instrument linearity. Several sources of error may be encountered in this type of standard, such as dilution errors and errors from lack of stability, shifts in pH, and temperature effects. An alternative procedure is to use a solid glass filter which, at a given wavelength, has an absorbance that is a small fraction of the total linear range of the instrument.[3] The didymium filter used for Figure 1B-13 has an absorbance of approximately 0.09 at 550 nm and is satisfactory for establishing linearity. The procedure is as follows:

1. Set the wavelength at 550 nm, cover the empty cuvet compartment, and set absorbance to zero.

2. Read absorbance of the didymium filter and record value.

3. Remove the filter and set absorbance to 0.25.

4. Read the didymium filter and record value.

5. Set absorbance to 0.50, 0.75, 1.00, 1.25, etc., with the compartment empty, and read absorbance as above.

6. Calculate ΔA for each increase in settings.

Linearity is satisfactory as long as the ΔA increments remain constant. The same general procedure may be used with a suitable filter on those spectrophotometers in which the blank reference settings can be adjusted in steps up to approximately 1 unit of absorbance.

Photometric Accuracy

Solutions of potassium dichromate are recommended for overall checks on photometric accuracy.[1] These also provide an indication of wavelength calibration, linearity, cuvet light path, and freedom from stray light in the UV region.

Analytical reagent grade $K_2Cr_2O_7$ is dried at 110 °C for 1 h. Solutions in 0.005 mol/L sulfuric acid are prepared:

Solution A: 0.0500 g/L for the absorbance range from 0.2–0.7

Solution B: 0.1000 g/L for the absorbance range from 0.4–1.4

Measurements should be made in 10-mm cells with the temperature controlled in the range of 15–25 °C, using 0.005 mol/L sulfuric acid as the reference. Table 1B-3 gives the expected values for the two absorbance maxima and minima of the solutions based on literature values. Since the natural bandwidth of solution A at 350 nm is approximately 63 nm, the values shown apply strictly to spectrophotometers with a spectral bandwidth of 6 nm or less.

Glass filters made from a moderately neutral glass may also be used to check photometric accuracy. These are available from the National Bureau of Standards (Schott NG-4, SRM 930) and are calibrated on the NBS high-accuracy spectrophotometer at approximately 10, 20, and 30% transmittance levels. A certificate is supplied which describes the filters and gives the calibration values at four wavelengths ranging from 440–635 nm.

FLAME PHOTOMETRY

Flame emission photometry is most commonly used for the quantitative measurement of sodium and potassium in body fluids. Lithium, while not normally present in serum, may also be measured in connection with the therapeutic use of lithium salts in the treatment of some psychiatric disorders.

Table 1B-3. RECOMMENDED ABSORBANCE VALUES FOR ACIDIC POTASSIUM DICHROMATE SOLUTIONS

Wavelength (nm)	Absorbance	
	Solution A	*Solution B*
235 (min)	0.626 ± 0.009	1.251 ± 0.019
257 (max)	0.727 ± 0.007	1.454 ± 0.015
313 (min)	0.244 ± 0.004	0.488 ± 0.007
350 (max)	0.536 ± 0.005	1.071 ± 0.011

Atoms of many metallic elements, when given sufficient energy such as that supplied by a hot flame, will emit this energy at wavelengths characteristic for the element. A specific amount or quantum of thermal energy is absorbed by an orbital electron. The electrons, being unstable in this high energy (excited) state, release their excess energy as photons of a particular wavelength as they change from the excited to their previous or ground state. If the energy is dissipated as light, the light may consist of one or more than one energy level and therefore of different wavelengths. These line spectra are characteristic for each element. Sodium, for example, emits energy primarily at 589 nm, along with other, much less intense emissions (Figure 1B-14). The wavelength to be used for the measurement of an element depends upon the selection of a line of sufficient intensity to provide adequate sensitivity as well as freedom from other interfering lines at or near the selected wavelength.

Alkali metals are comparatively easy to excite in the flame of an ordinary laboratory burner. Lithium produces a red, sodium a yellow, potassium a violet, rubidium a red, and magnesium a blue color in a flame. These colors are characteristic of the metal atoms that are present as cations in solution. Under constant and controlled conditions, the light intensity of the characteristic wavelength produced by each of the atoms is directly proportional to the number of atoms that are emitting energy, which in turn is directly proportional to the concentration of the substance of interest in the sample. Thus, flame photometry lends itself well to direct concentration measurements of some metals.

Other cations, such as calcium, are less easily excited in the ordinary flame. In these cases, the amount of light given off may not always provide adequate sensitivity for analysis by flame emission methods. The sensitivity can be improved slightly by using higher temperature flames.

Of the more easily excited alkali metals like sodium, only 1–5% of those atoms present in solution become excited in a flame. Even with this small percentage of excited atoms, the method has adequate sensitivity for measurement of alkali metals for most bioanalytical applications. Most other metal ions are not as easily excited in a flame, and flame emission methods are not as applicable to their measurement.

Components of a Flame Photometer

Figure 1B-15 shows a schematic diagram of the basic parts of a flame photometer. A tank of gas is required and a two-stage pressure regulator. High-pressure tubing must be used to lead the gases to the flame. An atomizer is needed to spray the sample as fine droplets into the flame. The monochromator, entrance and exit slits, and detectors are similar to those discussed previously for spectrophotometers. In effect, the light source for the spectrophotometer has been replaced with an atomizer-flame combination, and one is measuring emission of light rather than absorption.

Various combinations of gases and oxidants have been proposed and are being used in flame photometry. These include acetylene and oxygen for the hottest flame, and natural gas, acetylene, and propane in combination with either oxygen or compressed air. The choice of flame depends largely on the temperature desired; for sodium and potassium determinations, a propane-com-

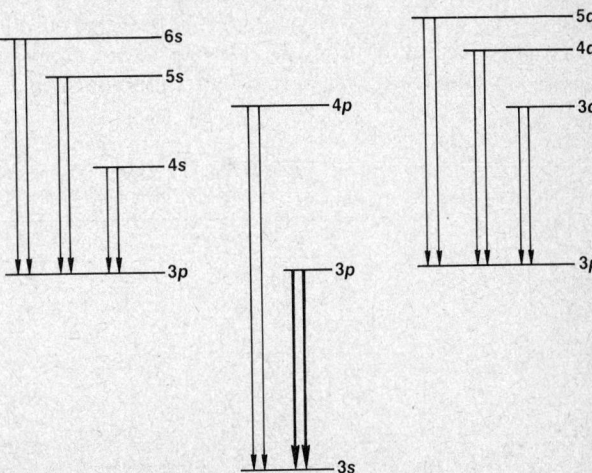

Figure 1B-14. Schematic diagram showing energy levels for certain lines of the sodium spectrum. The major doublet at 589 nm (shown in heavy lines) results when the excited valence electron returns from the 3p orbital to the ground state 3s orbital.

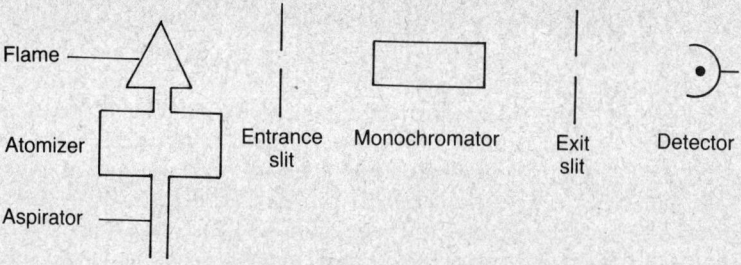

Figure 1B-15. Essentials of a flame photometer.

pressed air flame appears entirely adequate. Typical flame temperatures are shown in Table 1B-4.

The *atomizer* and the *flame* are critical components in a flame photometer. The atomizer provides a means of drawing the sample through the aspirator and converting it into a fine mist, which then enters the flame. This can be done by passing a gas of high velocity over the upper outlet of a capillary tube, the lower end of which is inserted into the sample. Liquid is then drawn up into a chamber and dispersed into small droplets. The larger droplets settle to the bottom and go to waste. The most important variable in the flame itself is the temperature. Frequent standardization of flame photometers is essential because thermal changes do occur and affect the response of the instrument. In addition, temperature changes affect the output of photocell detectors; for this reason, a period of warm-up, with aspiration of water and standards, is required before measurements are taken to establish thermal equilibrium for the flame and the atomizer chamber.

Ideally, monochromators in flame photometers should be of higher quality than those found in absorption spectrophotometers. When nonionic materials are burned, light of varying wavelength is given off. This is known as *continuous emission* and will be added to the *line emission* of the element being measured. For this reason, the narrowest bandpath that is achievable should be used to eliminate as much of the extraneous, continuous emission as possible, but still permit a maximum of the line emission to pass through to the detector. The detectors used in flame photometers operate by the same principle and in the same way as those described for spectrophotometers.

Direct and Internal Standard Flame Photometry

In some of the instruments of earlier designs, standard solutions of sodium or potassium were atomized or aspirated directly into the flame to provide a series of meter readings against which an unknown solution could be compared. This approach, referred to as the *direct reading method*, presents certain problems: (1) Minor fluctuations in air or gas pressure cause unstable response in the instrument and lead to errors. (2) Separate analyses and sometimes separate dilutions must be made for sodium and potassium. (3) The potassium signal is enhanced by the sodium concentration in the specimen. The latter effect, known as *mutual excitation*, results from the transfer of energy from an excited sodium atom to a potassium atom. Consequently, more potassium atoms are excited and light emission is increased. Ideally, then, the concentration of

Table 1B-4. FLAME TEMPERATURES FOR
VARIOUS GAS MIXTURES

Gas Mixture	Flame Temperature, °C
Natural gas–air	1840
Propane-air	1925
Hydrogen-air	2115
Acetylene-air	2250
Hydrogen-oxygen	2700
Natural gas–oxygen	2800
Propane-oxygen	2850
Acetylene-oxygen	3110

sodium and potassium in the standards should closely approximate those in the unknown, a situation that is difficult to achieve when analyzing a specimen such as urine in which these electrolytes show wide variation in concentration.

In the *"internal standard"* method, lithium or cesium is added to all standards, blanks, and unknowns in equal concentrations. Lithium has a high-emission intensity, is normally absent from biological fluids, and emits at a wavelength sufficiently removed from sodium and potassium to permit spectral isolation. The flame photometer makes a comparison of the emission of the desired element (sodium or potassium) with the emission of the reference lithium element. By measuring the ratios of emissions in this way, small variations in atomization rates, flame stability, and solution viscosity are compensated for. Lithium does not function as a "standard" under these conditions but as a reference element. Variable concentrations of sodium and potassium, in the lithium diluent, must be used to establish calibration curves or to verify linearity of response.

Lithium also acts as a *radiation buffer* to minimize the effects of mutual excitation. The final working concentration of lithium is so high, compared to either sodium or potassium, that the same percentage of potassium becomes excited regardless of the sodium concentration in the sample. Serum lithium concentrations in patients receiving lithium salts are maintained at approximately 1 mmol/L. This amount will produce no significant change in final lithium concentrations in samples containing lithium in the diluent.

A wetting agent is freqently recommended for inclusion in standards and sample dilutions. This minimizes changes in atomizer flow rates due to differences in viscosity of the samples. Viscosity effects are further reduced by diluting samples 100- to 200-fold.

ATOMIC ABSORPTION SPECTROPHOTOMETRY

Atomic absorption spectrophotometry in some respects is the inverse of flame emission photometry. In all emission methods, the sample is excited in order to measure the radiant energy given off as the element returns to its lower energy level. Extraneous radiation must be filtered out from the energy of interest if interference by these signals is to be avoided.

In atomic absorption spectrophotometry the element is not appreciably excited in the flame, but is merely dissociated from its chemical bonds and placed in an unexcited or ground state. This means that the atom is at a low energy level in which it is capable of absorbing radiation at a very narrow bandwidth corresponding to its own line spectrum. A hollow cathode lamp, made of the material to be analyzed, is used to produce a wavelength of light specific for the kind of metal in the cathode. Thus, if the cathode were made of sodium, sodium light at predominantly 589 nm would be emitted by the lamp. When the light from the hollow cathode lamp enters the flame, some of it is absorbed by the ground-state sodium atoms in the flame, resulting in a net decrease in the intensity of the beam from the lamp. This process is referred to as atomic absorption.

The process is analogous to absorption spectrophotometry. A specific hollow cathode lamp serves as the light source, and the sample heated in the flame replaces the sample in the cuvet. The pathlength of the flame is analogous to the light path through the cuvet. As noted previously, only a small fraction of the sample in the flame contributes emission energy, and only a fraction of this will be transmitted to the detector. Hence, most of the atoms are in the ground state and are able to absorb light emitted by the cathode lamp. In general, atomic absorption methods are approximately 100 times more sensitive than flame emission methods. In addition, owing to the unique specificity of the wavelength from the hollow cathode lamp, these methods are highly specific for the element being measured.

Components of an Atomic Absorption Spectrophotometer

Figure 1B-16 shows the basic components of an atomic absorption spectrophotometer. The hollow cathode lamp is the light source. A nebulizer sprays the sample into the flame; the monochromator, slits, and detectors have the functions described for flame photometry. The cathode is made of the metal of the substance to be analyzed and is different for each metal analysis. In some cases, an alloy is used to make the cathode, resulting in a multielement lamp.

On most instruments, an electrical-beam chopper and a tuned amplifier are incorporated.

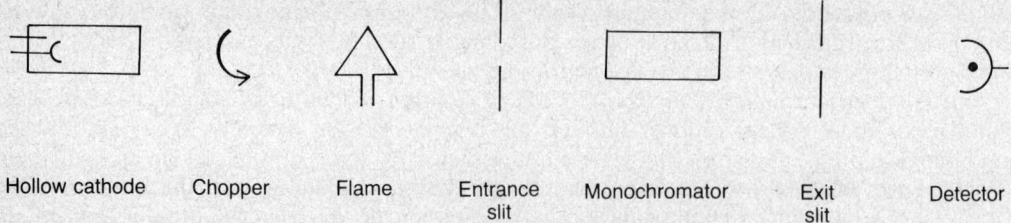

| Hollow cathode | Chopper | Flame | Entrance slit | Monochromator | Exit slit | Detector |

Figure 1B-16. Essentials of an atomic absorption spectrophotometer.

The power to the hollow cathode lamp is pulsed so that the light is emitted by the lamp at a certain number of pulses per second. On the other hand, all of the light coming from the flame is unpulsed. When light leaves the flame, it is composed of pulsed unabsorbed light from the lamp and a small amount of unpulsed flame spectrum and sample emission. The detector senses all light, but the amplifier is electrically tuned to accept only pulsed signals. In this way, the electronics in conjunction with the monochromator remove all of the flame spectrum and sample emission.

The hollow cathode lamp usually contains argon or neon gas at a pressure of a few millimeters of mercury. An argon-filled lamp produces a blue-to-purple glow during operation, and the neon produces a reddish-orange glow inside the hollow cathode lamp. Quartz, or special glass that allows transmission of the proper wavelength, is used as a window. A current is applied between the two electrodes inside the hollow cathode lamp, and metal is sputtered from the cathode into the gases inside the glass envelope. When the metal atoms collide with the neon or argon gases, they lose energy and emit their characteristic radiation. Calcium has a sharp, intense, analytical emission line at 422.7 nm, which is most frequently used for calcium analysis. In an interference-free system, only calcium atoms will absorb the calcium light from the hollow cathode as it passes through the flame.

Two kinds of burners have been used in most clinical applications. One is a *total consumption burner,* as illustrated in Figure 1B-17. With this burner, the gases hydrogen and air mix with the sample within the flame. One disadvantage of the total consumption burner is that relatively large droplets are produced in the flame, which scatter light and thus cause signal noise. Also, the amount of acoustical noise produced is very high and may become uncomfortable after a few hours of operation. An advantage of this type of burner is that the flame is more concentrated and it can be made hotter than other burners, causing molecular dissociation that may be desirable for some chemical systems.

Figure 1B-18 shows a *premix burner* (laminar flow burner) and illustrates how the sample is aspirated, volatilized, and burned. Note that the gases are mixed and the sample is atomized before being burned. An advantage of this system is that the larger droplets go to waste while the fine mist enters the flame, thus producing a less noisy signal. In addition, the pathlength through the flame of the burner is longer than that of the total consumption burner. This produces a greater absorption and increases the sensitivity of the measurement. A disadvantage of the premix burner is that the flame is usually not as hot as that of the total consumption burner, and thus it cannot sufficiently dissociate certain metal complexes in the flame (e.g., calcium-phosphate complexes). *Nitrous oxide premix burners* produce higher temperatures and will dissociate some calcium complexes; however, at these higher temperatures, calcium becomes excited to a significant extent and emits in the flame, thus introducing another problem.

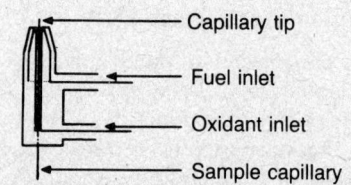

Capillary tip
Fuel inlet
Oxidant inlet
Sample capillary

Figure 1B-17. Total consumption burner.

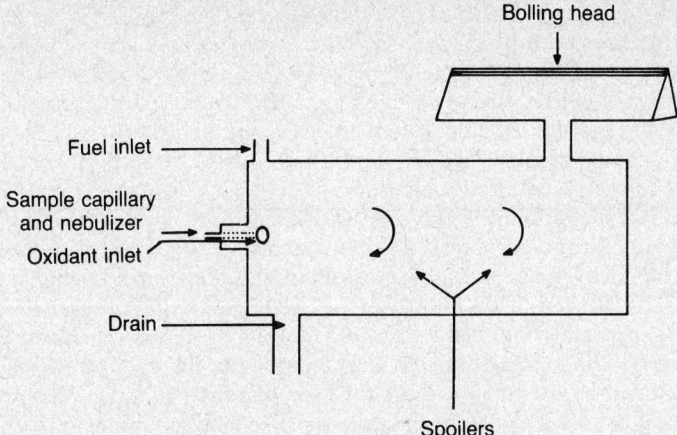

Figure 1B-18. Laminar flow burner.

In *flameless atomic absorption* techniques (carbon rod or "graphite furnace"), the sample is placed in a depression on a carbon rod in an enclosed chamber. Strips of tantalum metal may also be used. In successive steps, the temperature of the rod is raised to dry, char, and finally atomize the sample into the chamber. The atomized element then absorbs energy from the corresponding hollow cathode lamp. This approach is more sensitive than the conventional flame methods and permits determination of trace metals in small samples of blood or tissue. As discussed under Interference, phosphate interferes with calcium determinations by atomic absorption methods. This presents more problems in flameless analysis than in conventional flame atomization; hence, the latter method appears more suitable for the determination of calcium in serum and urine.

With flameless atomic absorption, a novel approach has been used to correct for background absorption. In the presence of an intense magnetic field, the energy levels in the atom that define the wavelength of the emitted radiation are shifted slightly. This is called the *Zeeman effect.*[5] The magnetic field can be pulsed alternately on and off. When the magnetic field is on, the absorbance wavelength of the analyte is shifted slightly away from the resonance line in the source, providing a signal for the background only. The difference between the two signals is background-corrected absorbance. The major advantage is that background correction is accomplished at the same wavelength as that used for the analytical measurement. Accurate background correction can typically be obtained at much higher background absorption levels than is generally possible in other systems.

Interference in Atomic Absorption Spectrophotometry

There are three general types of interferences in atomic absorption spectrophotometry. These are chemical, ionization, and matrix effects.

Chemical interference refers to the situation when the flame cannot dissociate the sample into free atoms so that absorption can occur. An example of this is the phosphate interference in the determination of calcium, caused by the formation of calcium-phosphate complexes. These complexes do not dissociate in the flame unless a special high-temperature burner is used. The phosphate interference is overcome by adding a cation that will compete with calcium for the phosphate. Usually, in the determination of calcium, lanthanum or strontium is added to the dilute sample to replace and release calcium from its phosphate complex. The freeing of calcium occurs because lanthanum and strontium form more stable complexes with phosphate than does calcium. The free calcium is then capable of absorbing the calcium light from the hollow cathode.

Ionization interference results when atoms in the flame become excited, instead of only being dissociated, and then emit energy of the same wavelength that is being measured. This effect can be overcome by adding an excess of a more easily ionized substance that will absorb most of the flame energy so that the substance of interest will not become excited. Ionization interference can also be decreased by reducing the flame temperature.

A third type of interference is the *matrix interference.* One example of a matrix effect is the enhancement of light absorption by organic solvents. An atom may absorb between two and five times more energy when dissolved in an organic solvent than when dissolved in an aqueous solvent. A second kind of matrix effect is the light absorption caused by formation of solids from sample droplets as the solvent is evaporated in the flame. This will usually occur only with solutions of concentrations greater than 0.1 mol/L. Refractory oxides of metals formed in the flame can also lead to matrix interferences.

Atomic absorption spectrophotometry is sensitive, accurate, precise, and highly specific. One of the reasons for these advantages is that the method does not require excitation of the element and thus is less affected by temperature variations in the flame and by the transfer of energy from one atom to another than flame photometry. The high specificity results from the fact that the light used has an extremely narrow bandwidth (0.01 nm) and is selectively absorbed by the atoms being measured. The most significant disadvantage is the problem of interferences. This is being solved, however, by extraction techniques and the introduction of competing cations to release the element to be measured from complexing or chelating anions.

References

1. Burgess, C., and Knowles, A., Eds.: Standards in Absorption Spectrometry. New York, Chapman and Hall, 1981.
2. James, G. P., and DJang, M. H.: Evaluation of clinical laboratory instruments. Part III. Spectral bandwidth and wavelength accuracy. Am. J. Med. Technol., 47:477-483, 1981.
3. Lucas, D. H., and Blank, R. E.: Spectrophotometric standards in the clinical laboratory. Am. Lab., 9:77-83, 1977.
4. Passey, R. B., Gillum, R. L., and Fuller, J. B.: Measurement of spectral bandwidth, as exemplified with the Beckman "Enzyme Analyzer System TR Spectrophotometer." Clin. Chem., 21:1582-1584, 1975.
5. Slavin, W.: Atomic absorption spectroscopy. The present and future. Anal. Chem., 54:685A-694A, 1982.
6. Surles, T., and Erickson, J. O.: Absorbance measurements at various spectral bandwidths. Clin. Chem., 20:1243-1244, 1974.

Additional Reading

Bauer, H. H., Christian, G. D., and O'Reilly, J. E., Eds.: Instrumental Analysis. Boston, Allyn and Bacon, 1978.
Ewing, G.W.: Instrumental Methods of Chemical Analysis. 3rd ed. New York, McGraw-Hill, 1969.
Knowles, A., and Burgess, C., Eds.: Practical Absorption Spectrometry. New York, Chapman and Hall, 1984.

Section Two

FLUOROMETRY, NEPHELOMETRY, AND TURBIDIMETRY

by Thomas O. Tiffany, Ph.D.

FLUORESCENCE AND LIGHT SCATTERING MEASUREMENTS

Definition of Fluorescence and Light Scattering

The interaction of radiant energy with molecules or particles in solution can result in either fluorescence or light scattering. *Fluorescence* occurs when a molecule absorbs light at one wavelength and re-emits light at a longer wavelength. *Light scattering* occurs when radiant energy passing through a solution encounters a molecule in an elastic collision, which results in the light being scattered in all directions. Unlike fluorescence emission, the scattered light is of the same frequency as the incident light. These simplified definitions will be expanded upon in this chapter.

Overview of Fluorescence and Light Scattering Measurements

The use of qualitative and quantitative fluorometric measurements in biological and clinical analysis is relatively new. Although the fluorescence of quinine, chlorophyll, and other natural products was known to Sir G. G. Stokes when he correctly explained the mechanism of absorption and emission of light energy over a century ago, it was not until the collaborative work of Bowman and Udenfriend that qualitative and quantitative fluorometry began to be used for biological applications. Bowman developed a spectrophotofluorometer to broaden the spectral region in which biological compounds could be excited. This more versatile fluorescence-measuring instrument became useful in the study of the dynamics of pharmacology and the metabolism of drugs. Since Udenfriend's classic text in fluorescence was published in 1962 much progress has been made in the development of instrumentation and methods.[9,16-18,20,21] The principal reasons for the interest of analytical and clinical chemists in fluorescence techniques are the potential for increased analytical sensitivity and the possible avenue for new chemistry test methods. Newer automated fluorometric instrumentation and the growing availability of reagents for fluorescence analysis are beginning to justify these interests.

The theory of light scattering from small particles was first developed over one century ago by Lord Rayleigh. The theory was extended to solutions of larger particles and macromolecules in 1947 by Debye. Light scattering measurements have been applied to the determination of molecular weights of macromolecules and to the monitoring of antigen-antibody interactions. More recently, there has been considerable effort to develop instrumentation for the quantitation of large- and small-molecular-weight analytes by light scattering measurements. Light scattering, as measured by turbidimetry or nephelometry, has been used to advantage in the quantitation of immunoglobulins, other proteins (see Chapter 4), and, more recently, therapeutic drugs.

This chapter will discuss the nature of fluorescence and light scattering, their instrumentation, applications, and limitations.

APPROACHES TO FLUORESCENCE MEASUREMENTS

The Nature of Fluorescence

Luminescence is the emission of light or radiant energy when an electron returns from an excited or higher energy level to a lower energy level. There are several types of luminescence phenomena including fluorescence, phosphorescence, and chemiluminescence. Although luminescence phenomena differ in how an electron is activated to the excited state, they result in similar emissions of radiant energy. *Fluorescence luminescence* and *phosphorescence luminescence* occur when radiant energy is absorbed by a molecule. Absorption of light energy by a molecule takes place in discrete units or quanta. The energy of a quantum of light is expressed by the formula $E = h\nu$ or $E = hc/\lambda$, where E is the energy, h is Planck's constant, 6.6×10^{-27} erg s, c the velocity of light, ν the frequency, and λ the wavelength. Each molecule contains a series of closely spaced energy levels. Figure 1B-19 shows diagrammatically the relationship between absorption, fluorescence, and phosphorescence. Absorption of a quantum of light energy by a

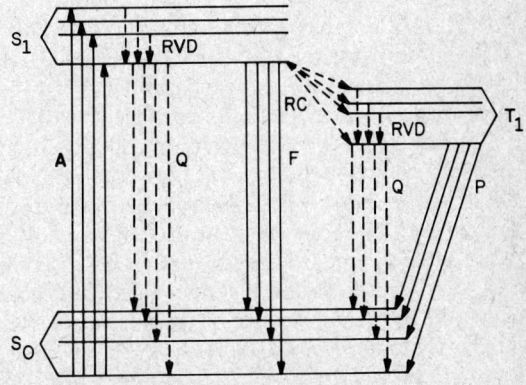

Figure 1B-19. Luminescence energy level diagram of typical organic molecule. S_0 is the ground level singlet state; S_1 is the first excited singlet state; A is the absorption process; T_1 is the first excited triplet state; and RVD is the radiationless vibrational deactivation. Q is quenching of the excited singlet or triplet state. F is the fluorescence process from the first excited singlet state. P is the phosphorescence process from the first excited triplet state. RC is the radiationless crossover from the first excited singlet state to the first excited triplet state.

molecule causes the transition of an electron from the singlet ground state to one of a number of possible vibrational levels of its first excited singlet state. The actual number of molecules in the excited state under typical reaction conditions and excited with a typical 150-W light source is very small and is estimated to be about 10^{-13} moles per mole of fluorophor. Once the molecule is in an excited state there are several ways it can return to its original energy state. These include radiationless vibrational equilibration, the fluorescence process, quenching of the excited singlet state, radiationless crossover to a triplet state, quenching of the first triplet state, and the phosphorescence process. Therefore fluorescence is a similar but competitive process to phosphorescence. The fluorescence process is the most probable process to occur under the typical reaction conditions used by the clinical chemist (i.e., moderately viscous solutions of aromatic organic molecules at room temperature).

There are some important points to learn about *fluorescence.* First, as shown in Figure 1B-19, vibrational equilibration before fluorescence results in some loss of the excitation energy. The emitted fluorescence light is of less energy or has a longer wavelength than the excitation light. Second, the difference between the maximum wavelength of the excitation light and the maximum wavelength of the emitted fluorescence light is a constant referred to as the "Stokes shift." This constant is a measure of the energy lost during the lifetime of the excited state (radiationless vibrational deactivation) before return to the ground singlet level (fluorescence emission). Third, the emission of fluorescence light is characterized by a rapid (10^{-8} s) decay time.

Phosphorescence shows a larger shift in emitted light wavelength than does fluorescence. The decay time of emission of phosphorescence light is longer (10^{-4} to 100 s) than the decay time of fluorescence emission. Decay times are expressed in a time range of several orders of magnitude and they vary with the molecule and its solution environment.

Chemiluminescence and *bioluminescence* differ from the other luminescence phenomena, fluorescence and phosphorescence, in that the excitation event is caused by a chemical reaction and not by photolumination. The physical event of the light emission in chemiluminescence and bioluminescence is similar to fluorescence in that it occurs from an excited singlet state and the light is emitted when the electron returns to the ground state.

Chemiluminescence involves the oxidation of an organic compound, such as luminol, by an oxidant (e.g., hydrogen peroxide, hypochlorite, or oxygen) with the emission of light occurring from the excited product formed in the chemical reaction. These reactions occur in the prescence of a catalyst such as the Cu(II) ion or heme. *Bioluminescence* is the name given to a special form of chemiluminescence found in biological systems in which a catalytic protein, e.g., luciferase, increases the efficiency of the luminescent reaction. Bioluminescence reactions can be classified as (1) pyridine-nucleotide linked systems, (2) adenine-nucleotide linked systems, and (3) enzyme-substrate systems.

Chemiluminescence and particularly bioluminescence are of interest to the clinical chemist today due to the greater sensitivity of the bioluminescence assays compared to most direct fluorescence assays and due to their comparable sensitivity to radioimmunoassays. The discussion of chemiluminescence is purposely limited in scope. More detailed discussions about chemiluminescence, bioluminescence, reaction mechanisms, instrumentation, and applications can be found in the literature.[19,22]

Time Relationships of Fluorescence Emission

The time required for a molecule to absorb radiant energy and to be promoted to an excited state is approximately 10^{-15} s. The length of time for vibrational equilibration to occur to the lowest excited state is of the order of 10^{-14} to 10^{-12} s. The length of time required for fluorescence emission to occur is of the order of 10^{-8} to 10^{-7} s. Relatively speaking, there is a considerable time delay between the absorption of light energy, the return to the lowest excited state, and the emission of fluorescence light. This time relationship is shown in Figure 1B-20. Phase I in Figure 1B-20 represents the time period between absorbance of light energy and radiationless loss of energy during vibrational rearrangement to the lowest excited energy state. This time period is represented by the up and down arrows in the diagram. Phase II shows the emission and decay of a short-lived (b) and a longer-lived fluorophor (a). If the fluorescence emission is measured over time following a pulse of light from an excitation source such as a xenon lamp or laser, the intensity of the emitted light decays as a first-order process similar to radioactive decay (i.e., Phase II of Figure 1B-20). The time required for the emitted light to reach $1/e$ of its initial

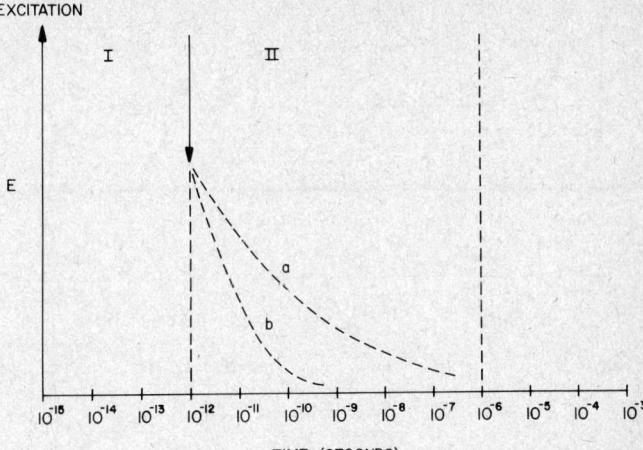

Figure 1B-20. Fluorescence decay process: E is the absorption of energy; I is the vibrational deactivation time phase; II is the fluorescence emission time phase; a is long fluorescence decay time; and b is short fluorescence decay time.

intensity, where e is the Naperian base 2.303, is called the average lifetime of the excited state of the molecule, or the *fluorescence decay time.*

The time delay between absorption of quanta of energy and fluorescence can be used in fluorescence instrumentation designed to function with sensitivity similar to that of a gamma counter.[11,12,23] These instruments are called time-resolved fluorometers. The advantage of a time-resolved fluorometer is the elimination of background light scattering due to Rayleigh and Raman signals and consequent dramatic increase in signal-to-noise and detection sensitivity. These contributors to background scatter and the time-resolved fluorometer will be treated separately in more appropriate sections of this chapter.

Fluorescence Polarization

Light is composed of electric and magnetic waves at right angles to each other. Light waves produced by standard excitation sources have their electric vectors oriented randomly. Light waves, passed through certain crystalline materials (polarizers), have their electric vectors oriented in a single plane and are said to be plane polarized. Fluorophors absorb light most efficiently in the plane of their electronic energy levels. If their rotational relaxation (Brownian movement) is slower than their fluorescence decay time, as is the case for large fluorescent-labeled molecules, the emitted fluorescence light will be polarized. Since small molecules have rotational relaxation times that are much shorter than their fluorescence decay time, their emitted fluorescence light is depolarized. However, if the small fluorescent molecule is attached to a macromolecule or if it is placed in a viscous solution, the small molecule will emit polarized light.

Fluorescence polarization, P, is defined by the following equation:

$$P = (I_v - I_h)/(I_v + I_h) \tag{1}$$

P is the difference of the two observed intensities divided by their sum. Fluorescence polarization is measured by placing a mechanically or electrically driven polarizer between the sample cuvet and the detector. A diagram of a fluorescence polarization measurement system is shown in Figure 1B-21. In the normal instrumentation mode the sample is excited with polarized light to obtain maximum sensitivity. The polarization analyzer is positioned first to measure the intensity of the emitted fluorescence light in the vertical plane (I_v) and then the polarization analyzer is rotated by 90° to measure the emitted fluorescence light intensity in the horizontal plane (I_h). (See also page 90.) P is then calculated manually or automatically by use of equation (1).

Fluorescence polarization is used to quantitate analytes by use of the change in fluorescence depolarization following immunological reactions. Quantitation is accomplished by adding a known quantity of fluorescent-labeled analyte molecules to a reaction solution containing an antibody specific to the analyte. The labeled analyte will bind to the antibody and cause a change in its rotational relaxation time resulting in fluorescence polarization. The addition of nonlabeled

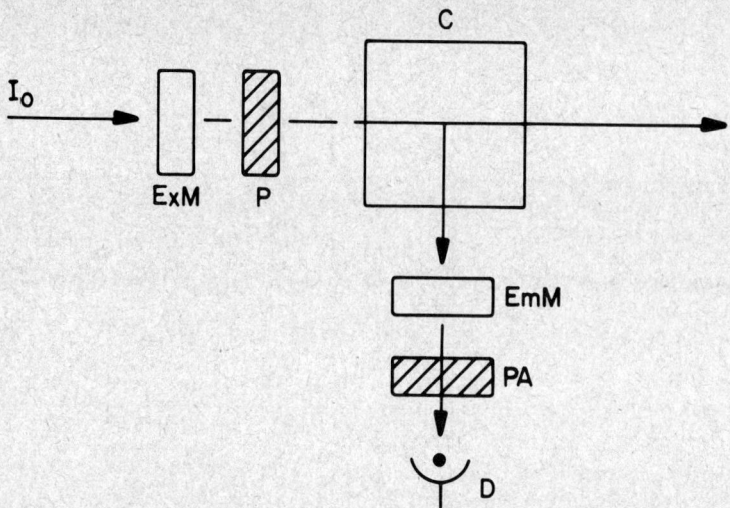

Figure 1B-21. Schematic of a fluorescence polarization analyzer. P is the polarizer to provide polarized excitation light. PA is the polarizer analyzer, which is rotated to provide the measurement of parallel and perpendicular polarized fluorescence-emission intensity. ExM is the excitation monochromator, EmM is the Emission monochromator, D is the detector, and C is reaction cell or cuvet.

analyte such as an unknown quantity of a therapeutic drug in a serum specimen will result in a competition for binding to the antibody with the fluorescent-labeled analyte. This change in binding of the fluorophor-labeled analyte will cause a change in fluorescence polarization that is relative to the amount of analyte contained in a given sample. Because the change in fluorescence polarization is a direct response to the reaction mixture and because bound fluorophor need not be separated from free, fluorescence polarization is applicable to homogeneous assays of small-molecular-weight analytes such as therapeutic drugs.[2,3]

The Relationship of Concentration to Intensity of Fluorescence Emission

The relationship of concentration to intensity of fluorescence emission may be derived from the Beer-Lambert law:

$$I/I_o = 10^{-abc} \qquad (2)$$

where I is the intensity transmitted, I_o intensity of the incident light, a the absorptivity, c the concentration of the absorbing species, and b the path length. The amount of light absorbed is obtained by rearrangement of equation (2):

$$I_o - I = I_o(1 - 10^{-abc}) \qquad (3)$$

The fluorescence intensity F is proportional to the quanta of light absorbed and to the efficiency of the emitted light as expressed by equation (4):

$$F = \phi I_o(1 - 10^{-abc}) \qquad (4)$$

where F is the relative intensity, ϕ is the fluorescence efficiency (i.e., the ratio between quanta of light emitted and quanta of light absorbed), I_o is the initial excitation intensity, a is the molar absorptivity, c is the concentration in mol/L, and b is the volume element defined by the geometry of the excitation and emission slits. Fluorescence is emitted in all directions, but only that passing through the emission slit is measured.

By expansion of equation (4) through a Taylor series, rearrangement, logarithm base conversion, and basic assumptions about dilute solutions, equation (5) can be derived:

$$F = \phi(I_o(2.3abc)) \qquad (5)$$

Equation (5) indicates that fluorescence intensity is directly proportional to the concentration of the fluorophor and the excitation intensity. This relationship holds only for dilute solutions where absorbance is < 2% of the exciting radiation; the fluorescence intensity becomes nonlinear as the absorbance of the solution increases above 2% of the exciting radiation. This phenomenon is called the inner filter effect and it will be discussed in more detail in a later section. Other factors influencing the measurement of fluorescence intensity are the sensitivity of the detector and the degree of background light scatter seen by the detector.

Fluorescence intensity measurements are more sensitive than absorbance measurements. The magnitude of absorbance of a chromophor in solution is determined by its concentration and the path length of the cuvet. The magnitude of fluorescence intensity of a fluorophor is determined by its concentration, the path length, and the intensity of the light source. The sensitivity of fluorescence measurements can be 100–1000 times greater than the sensitivity of absorbance measurements through the use of more intense light sources, by the use of digital signal filtering techniques, and by the use of sensitive emission photometers. All of these are incorporated in conventional spectrofluorometric instrumentation, described further on in this chapter.

Fluorescence measurements are expressed in relative intensity units. The word "relative" is used because the intensity measured is not an absolute quantity. It is a small part of the total fluorescence emission and its magnitude is defined by the instrument slit width, detector sensitivity, monochromator efficiency, and excitation intensity. Since these are instrument-related variables, establishing an absolute intensity unit for a given concentration of a fluorophor that is valid from instrument to instrument is difficult, if not impossible, to achieve.

Fluorometry combines the simplicity of photometry with high specificity and sensitivity. Examples of analytes of clinical chemistry importance that can be determined by the use of fluorescence measurements are given in Table 1B-5.

LIMITATIONS OF FLUORESCENCE MEASUREMENTS

Factors limiting fluorescence measurements include concentration effects on fluorescence (inner filter effect), background effects due to Rayleigh and Raman scattering, solvent effects (e.g., interfering nonspecific fluorescence and quenching), and sample matrix effects (e.g., light scattering, interfering fluorescence, adsorption, and photodecomposition).

The Inner Filter Effect

The linear relationship between concentration and fluorescence emission (equation 5) holds as long as solutions are used that absorb < 2% of the exciting light. As the absorbance of the solution increases above this amount, the relationship becomes nonlinear. In an extreme case the fluorescence emission intensity from very concentrated solutions can become less than the emission intensity from dilute solutions of the same fluorophor. The inner filter effect is caused by a loss of excitation intensity across the cuvet path length as the excitation light is absorbed by the

Table 1B-5. EXAMPLES OF ANALYTES MEASURED
BY FLUORESCENCE

Analytes	Enzymes
Glucose	Oxidoreductases
Bilirubin	Glucosidases
Protoporphyrin	Hydrolases
Magnesium	Proteases
Calcium	*Therapeutic drugs*
Catecholamines	Phenytoin
Bile acids	Phenobarbital
Estrogens (urinary)	Amikacin
*Coagulation**	Tobramycin
Heparin	Gentamicin
Antithrombin III	Theophylline
Plasminogen	Primidone

*Using specific fluorescent substrates for proteolytic enzymes.

fluorophor. Thus as the fluorophor becomes more concentrated, the absorbance of the excitation intensity increases and the loss of the excitation light as it travels through the cuvet increases. This effect is most often encountered with a right angle fluorescence instrument where the emission slits are set to monitor the center of the sample cell where the absorbance of excitation light is greater than at the front surface of the cuvet. Therefore it is less problematic if a front surface fluorescence instrument is used. However, most fluorescent measurements are made on very dilute solutions so that the inner filter effect is not a problem.

Light Scattering

Light is absorbed by molecules at all wavelengths and emitted without loss of energy. This absorption of radiant energy causes changes in vibrational energy but does not cause a change in energy levels. The light is emitted in all directions and its intensity is inversely proportional to the fourth power of the wavelength. The phenomenon was first described by Rayleigh in 1870 and thus is called *Rayleigh scattering*. For fluorophors with small Stokes shifts the excitation and emission spectra overlap and are particularly susceptible to loss of sensitivity due to background light scatter. Rayleigh-type light scatter can be controlled by the use of well-defined emission and excitation interference filters or by appropriate monochromator settings, and by the use of polarizers.

Raman scattering is related to Rayleigh scattering. Rayleigh scattering is the result of an elastic collision of light with a molecule; the molecule's rotational and vibrational energy is changed, but because the collision is elastic no energy is given or taken up. With no energy change, the light emitted occurs with no change in wavelength. The Raman effect, on the other hand, is the result of an inelastic collision where energy is lost or taken up by the molecule. Since Raman scattering by solvents occurs with a loss of excitation energy, the light scattering occurs at longer wavelengths. The Raman light scattering is independent of excitation wavelength and is a property of the solvent. Because Raman light scattering appears at longer wavelengths than the exciting radiation, it can be a difficult interference to eliminate when working at very low fluorophor concentrations. As an example, the wavelength shift in water is ~50 nm at an excitation wavelength of 365 nm and ~75 nm at an excitation wavelength of 436 nm. The shift would represent a problem if the excitation maximum of a fluorophor was 365 nm and the emission maximum was at 415 nm. Raman light scattering can be controlled by setting the excitation and emission wavelengths far enough apart to avoid the Raman scatter. It can also be controlled by narrowing the slit width on the excitation monochromator. However, both options tend to decrease sensitivity.

Solvent and Cuvet Material Effects

Certain quartz glass and plastic materials that contain UV absorbers will fluoresce. Some solvents, such as ethanol, can cause appreciable fluorescence. It is therefore important when fluorescence assays are set up to check the background fluorescence of all components of the reaction mixture. Fluorescence grade solvents and cuvets with minimum fluorescence emission that minimize these types of fluorescence background problems are commercially available.

Quenching by the solvent can also be a problem and should be investigated when setting up a new analytical method. *Quenching* is related to the interaction of the fluorophor with the solvent or with a solute dissolved in the solvent. Such interaction results in a loss of fluorescence due to energy transfer or other mechanisms, but there is no effect on the absorbance spectrum of the fluorophor. An example of quenching is the loss of fluorescence when halides are added to quinine in dilute sulfuric acid. Quenching can also be a useful tool for studying molecular structure, because fluorescence emission is sensitive to and specific for changes in atomic and molecular structure.

Sample Matrix Effects

A serum or urine sample contains many compounds that can fluoresce. Thus, the sample matrix is a potential source of unwanted background fluorescence and must be examined when new methods are developed. The most serious contributors to unwanted fluorescence are proteins and bilirubin. However, since protein excitation maxima are in the range of 260–290 nm, their contribution to overall background fluorescence is minor when excitation occurs above 300 nm.

The light scattering of proteins and other macromolecules in the sample matrix can cause unwanted background. Lipemic samples, for example, are noted for their intense light scattering, and the relative contribution of lipids to the background signal of a fluorescence measurement should be investigated when setting up a new method.

Dilute solutions of fluorophors can exhibit additional problems besides unwanted background interferences. Fluorophors in the concentration range of 10^{-9} mol/L and below can show considerable adsorption to glass containers and other reaction vessels. Also, dilute solutions of fluorophors, when excited over long periods of time, are susceptible to photodecomposition by intense excitation light. These problems can be avoided by proper selection of reaction vessels, by addition of wetting agents, and by minimizing the length of time a sample is exposed to the excitation light.

INSTRUMENTATION FOR MEASUREMENT OF FLUORESCENCE

BASIC INSTRUMENTATION

The instruments used to measure fluorescence are fluorometers or spectrofluorometers. The distinction between the term "fluorometer" and the term "spectrofluorometer" is the way that excitation light and emission light are separated into monochromatic light. The fluorometer employs interference filters or glass filters to produce monochromatic light for sample excitation and for isolation of fluorescence emission, whereas the spectrofluorometer uses a grating or prism monochromator for this purpose.

Many significant advances have been made in the development of fluorometers and spectrofluorometers since Udenfriend's first textbook was published, but the basic components required then for fluorescence spectrometry are similar to those today. These are (1) the excitation source, (2) the excitation monochromator, (3) the sample cell, (4) the emission monochromator, and (5) the detector. These components are shown in Figure 1B-22 as they would be configured in a 90° optical system. It is useful to have some understanding of each of these components and how each functions in a fluorescence instrument.

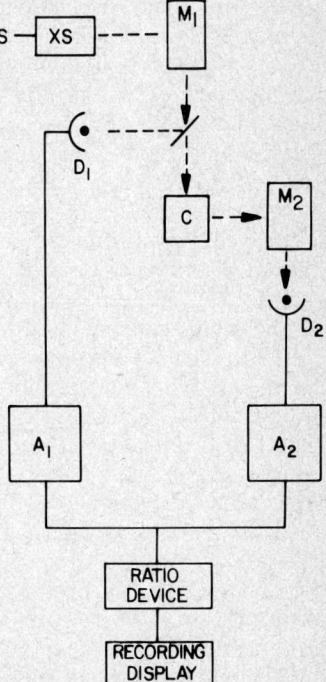

Figure 1B-22. Block diagram of a typical spectrofluorometer: XS is the xenon source; PS is the power supply; M_1 is the excitation monochromator; C is the sample cell; M_2 is the emission monochromator. D_1 and D_2 are detectors; D_1 monitors the variation in excitation intensity and D_2 measures fluorescence emission intensity. A_1 and A_2 are excitation signal and emission signal amplifiers, respectively.

The excitation source. The absorption spectra of most fluorescent compounds of interest are in the range of 300–550 nm. The fluorescence emission intensity is proportional to the initial excitation intensity as well as to concentration and size of the volume element being measured in the sample cell. Therefore, an intense lamp capable of emitting radiant energy over a large spectral region is desirable. The light source that best fits these criteria is the xenon arc lamp, but a quartz halogen cycle lamp of high intensity may also be used for excitation in special applications. A mercury lamp will provide high-intensity line spectra at several wavelengths, but the most useful lamp is one that provides a continuum over the spectral range of interest.

The *xenon lamp* is popular because it provides a continuum of relatively high-intensity radiant energy over the spectral range of 250–800 nm. A limitation of xenon lamps for analytical use is arc wandering or flicker. The arc between the internal electrodes of the xenon lamp tends to move slightly, causing some random variation in the intensity of the light reaching the cuvet. Since the intensity of the light from lamps driven by nonregulated power supplies is modulated by the frequency of the line voltage, the implementation of current-stabilized power supplies has improved the performance of fluorescence instrumentation using xenon lamps.

Excitation and emission monochromator. A monochromator is defined as a device for isolating a narrow portion of the spectrum. Those monochromators used in fluorescence instrumentation are interference filters, colored glass filters, gratings, and prisms. (See also Section One, Photometry.) Most modern analytical instruments using interference filters employ the all-dielectric multicavity filter or a hybrid Fabry-Perot coupled-dielectric-layer filter (i.e., a filter with metal reflective layers). Either type of filter is combined with appropriate sharp cut-off glass filters to form a single filter package, which removes undesired transmission of higher orders and provides narrow bandwidth, higher peak wavelength transmission, and increased band slope. The increased slope of the spectral band makes the transition from peak transmission to nontransmission more abrupt, which is very important for the spectral separation of excitation and emission bands with a small Stokes shift. For additional information on the properties and construction of interference filters, the student should consult chapters on interferometry and thin films contained in most recent optics textbooks.[7]

Colored glass filters selectively absorb certain wavelengths of light. These filters have been used for both excitation and emission wavelength selection, but they are more susceptible to transmitting stray light and they can exhibit unwanted fluorescence.

Grating monochromators are devices that isolate narrow regions of the spectrum. Their function and construction have been described in the Photometry section of this chapter. The spectral resolution of the light at the slit is a function of the slit width and the resolution of the grating. Spectrofluorometers generally use larger slit widths than absorbance spectrophotometers to obtain higher excitation intensities. The advantage the grating monochromator provides is the selectivity of excitation and emission wavelengths that is required when working with new fluorophors with absorbance and emission maxima for which specially fabricated interference filters may not exist. The rotation of the grating can be computer controlled to automate spectral scans of fluorescence excitation and emission. In the conventional operation of a spectrofluorometer either the excitation wavelength or the emission wavelength is held constant and the other is scanned. Newer fluorescence instrumentation, however, allows the analyst to synchronize the excitation and emission monochromators to scan together at programmed rates and to provide a change in emission intensity as a function of change in excitation and emission wavelength. Synchronous scanning gives an additional dimension of specificity to fluorescence measurements.[6]

The sample cell compartment sets the geometry the instrument uses for fluorescence measurements. Fluorescence light is emitted in all directions from a molecule and there are several excitation/emission geometries, illustrated in Figure 1B-23, that can be used to measure fluorescence. Most commercial spectrofluorometers and fluorometers use *the right-angle-detector approach* because it minimizes background signal that limits analytical sensitivity. *The end-on approach* allows the adaptation of a fluorescence detector to existing 180° absorption instruments. Its sensitivity is limited by the quality of the excitation/emission interference filter pair, the excitation/emission spectral band overlap, and the inner filter effect. The *front surface approach* provides the greatest linearity over a broad range of concentration because it minimizes the inner filter effect. The front surface approach shows similar sensitivity to the right angle detectors but is more susceptible to background light scatter. Front surface fluorometry has been most widely applied to heterogeneous solid-phase fluorescence immunoassay systems. As an example, see item 5 in Table 1B-6.

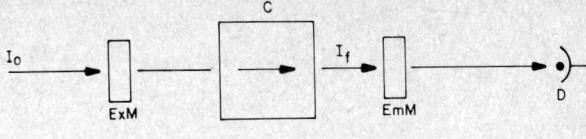

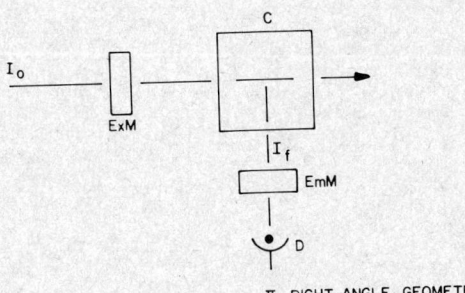

Figure 1B-23. Fluorescence excitation/emission geometries: I_0 is the initial excitation energy, ExM is the excitation monochromator, C is the sample cuvet, I_f is the fluorescence intensity, EmM is the emission monochromator, and D is the detector.

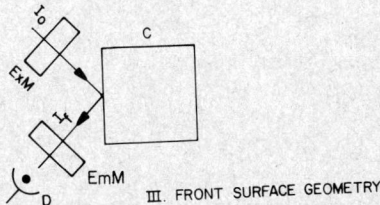

The sample cell can be oriented at different angles in relationship to the excitation source and the detector to accommodate these different geometries. The major concerns related to the geometry of the sample cell are light scattering, the inner filter effect, and the sample volume element seen by the detector. Figure 1B-24 shows the sample cell and slit arrangement for a conventional fluorescence spectrophotometer with the excitation and emission slits oriented at a right angle. S_1 and S_2 designate the excitation and emission slits, respectively, in Figure 1B-24,*A* and *B*. The position of the emission slit and the width of the slit are important. If the emission slit is located near the front edge of the sample cell as shown in Figure 1B-24,*B*, the inner filter effect is minimized. If the emission slit width is increased, sensitivity will increase, but specificity may decrease.

The material of the sample cell is generally fused silica or quartz. Plastic cuvets can be used for certain applications, but certain ultraviolet absorbers in the plastics will fluoresce, causing unwanted background signal and loss of sensitivity.

Table 1B-6. SPECIAL APPLICATIONS FLUOROMETERS

Instrument	Application	Manufacturer
1. Protopath Fluorometer	Coagulation/Specific factor assays	Dade Division, American Hospital Supply Corporation
2. ESA 4000 Hemato-Fluorometer	Bilirubin; protoporphyrin	Environmental Sciences Associates
3. TDx-Fluorescence Polarization Analyzer	Therapeutic drugs	Abbott Laboratories, Diagnostics Division
4. Ortho Spectrum III	Cellular immunofluorescence testing	Ortho Diagnostics
5. Fluorometer-FIAX	Antinuclear antibodies, immunoglobulins (IgG, IgA, IgM), and antibodies to rubella G, cytomegalovirus, herpes simplex virus, rheumatoid factor, *Toxoplasma gondii*	International Diagnostic Technology, Inc.
6. Photon Counting Fluorometer	Thyroxine, triiodothyronine	BioRad Laboratories

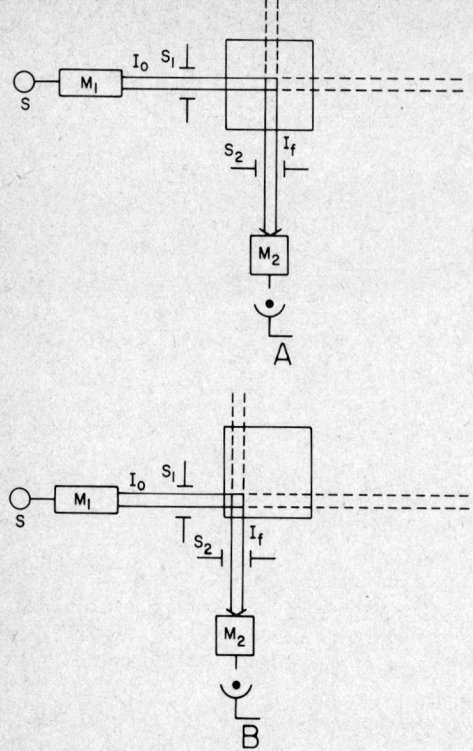

Figure 1B-24. Two right-angle fluorescence sample cuvet positions. *A* is the standard 90° configuration. *B* is the offset positioning of the cuvet to minimize the inner filter effect.

Photodetectors. Since the human eye is a most sensitive detector with a wide range of spectral recognition, qualitative fluorescent thin-layer methods in the clinical laboratory make use not only of a short- and long-wavelength ultraviolet lamp source but also of visual observation. For quantitative assays, the most commonly used detector in fluorometers and spectrofluorometers is the photomultiplier tube (PM), which has been described in some detail in Section One. The important features of the PM tube for fluorescence measurements are a wide choice of spectral responses, rapid photon response time (i.e., nanosecond response time), and sensitivity. Sensitivity is due to the possible gain of 10^6 electrons at the anode of the PM tube for each incident photon hitting the photo cathode.

The measurement of electron flow at the PM tube anode can be accomplished in different ways, depending upon the light level (photon flux) striking the PM cathode and the desired sensitivity. At high light intensities, analog techniques for measurement of PM current are used. The analog signal can then be converted to a digital signal for panel digital display or for computer use. At low light levels, as is the case for photons emitted from a fluorophor at low concentration, photon counting techniques can be used. At low light levels, the photons strike the cathode of the PM tube generating spikes or pulses at the anode of the PM tube. The number of pulses generated is proportional to the number of photons impinging upon the PM tube. The number of pulses that occur per unit time is directly proportional to the intensity of emitted fluorescence light striking the PM tube. This method is called *photon counting.* The use of photon counting increases the signal-to-noise ratio and the sensitivity of the measurement of fluorophors at very low concentrations.

Manual Fluorometers and Spectrofluorometers

There are many commercially available fluorometers and fluorescence spectrophotometers offering a variety of features. These features can include ratio referencing, microprocessor-controlled excitation and emission monochromators, pulsed xenon light sources, photon counting, rhodamine cell for corrected spectra, polarizers, flow cells, front-surface viewing adapters, multiple cell holders, and microprocessor-based data reduction systems. Regardless of the complexity of these instruments at first appearance, they all comprise the basic instrumental components discussed in the previous section.

The simplest type of fluorometer can be discussed by referring to Figure 1B-24,*A.* This fluorometer consists of a light source (S), a sample compartment, excitation and emission filters or monochromators (M_1 and M_2), a detector, and a simple digital or meter display. The operation of this device requires that the fluorometer display first be set to zero by use of a sample blank. A standard solution is then placed into the fluorometer and the range is adjusted. This is usually accomplished by adjusting the high-voltage control of the PM tube, but it may also be controlled by adjustment of the emission slit (S_2). Once the zero and standard settings have been made, unknown solutions can be read. If a number of unknown solutions are to be read, then the zero and the standard must be checked several times to compensate for electronic and light-source drift. This simple single-beam fluorometer, like the ratio-referencing fluorometers described in the next paragraph, can measure concentration over 3 or 4 orders of magnitude.

A typical ratio referencing spectrofluorometer is illustrated in Figure 1B-25. The block diagram of this instrument was shown in Figure 1B-22. This is a basic simple right-angle instrument that uses two monochromators (M_1 and M_2), two photomultiplier detectors (i.e., D_1 and D_2, the reference and sample photomultipliers), and a xenon lamp source. The light from the exciter monochromator (M_1) is split and a small portion (10 %) is directed to the reference photomultiplier (D_1) for ratio referencing purposes. The remaining excitation light is focused into the sample cuvet (C). Emission optics are positioned at a right angle to the excitation optics. An emission monochromator (M_2) is used to select or scan the desired portion of the emission spectra, which is directed to the sample PM tube (D_2) for measurement of the emission intensity. The output signals from the reference and the sample photomultipliers are amplified (A_1 and A_2), and a ratio of the sample to the reference signal is provided by a digital display or a chart recorder. If excitation and emission interference filters are substituted for monochromators (M_1 and M_2), the instrument shown in Figure 1B-22 becomes a simple ratio fluorometer. The operational mode of a ratio fluorometer is similar to that of the spectrofluorometer; however, only discrete excitation and emission wavelengths are available, and the use of this type of instrument is precluded for scanning fluorophors to obtain emission and excitation spectra. The ratio filter fluorometer is most useful for obtaining concentration measurements at defined excitation and emission wavelengths.

The ratio-referencing spectrofluorometer can be operated at fixed excitation and emission wavelength settings for concentration measurements or it can be used to measure the excitation or emission spectrum of a given compound. The measurement of concentration of unknowns is accomplished in a similar manner as with the single-beam fluorometer. A blank and a reference solution are first measured and then the unknown samples. The ratio-referencing spectrofluor-

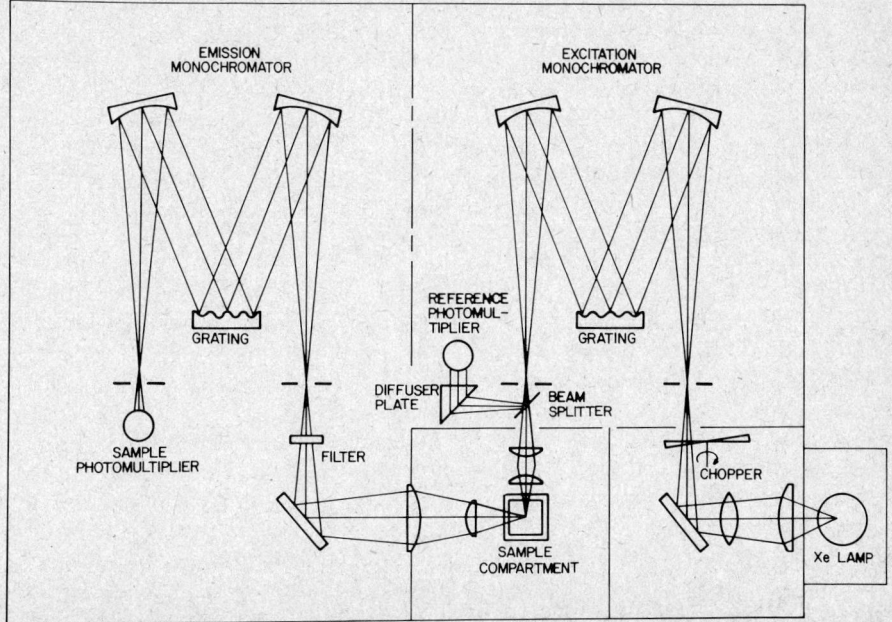

Figure 1B-25. Ray diagram of a typical commercial spectrofluorometer.

ometer in Figure 1B-25 provides two advantages over single-beam spectrofluorometers. First, it eliminates short-term and long-term xenon lamp energy fluctuations, i.e., arc flicker and lamp decay, and thus minimizes the need for frequent standardization of the instrument during analysis. Second, it provides "essentially" corrected excitation spectra by compensating for wavelength dependent energy fluctuations.

Some spectrofluorometers have the capability of synchronous scanning by both the excitation and emission monochromators. The synchronous scanning of the excitation and emission monochromators is performed at a constant wavelength increment of 5–50 nm or more, but intervals in the range of 15–30 nm yield the most distinctive spectra. This technique has been valuable in the area of forensic chemistry. It is another approach to qualitative and quantitative analyses of mixtures of fluorophors, such as drugs and their metabolites in solution. There are other instrumental approaches to the measurement of mixtures of fluorophors. These include the excitation-emission matrix approach (EEM) and excitation scan with modulation of the emission signal.[6,8] These approaches are not commercially available but they warrant the attention of the more interested student of fluorescence.

The time-resolved fluorometer is a newer development in fluorescence instrumentation.[11,12] It is similar to the ratio-referencing fluorometer with the exception that the light source is pulsed and the detector monitors, in a fast photon-counting mode, the decay of the fluorescence signal that occurs after the excitation lamp is off. Therefore, time-resolved fluorometry requires long-lived fluorophors with decay times in the order of 10–100 μs, such as lanthanide chelates. Monitoring only the decaying fluorescence signal from a labeled analyte over a period of several lamp pulses and not the light scattering caused by the light source eliminates a serious limitation to fluorescence measurements at very low fluorophor concentrations. Time-resolved fluorometers are capable of measuring fluorophors at a concentration of 1 pg/mL. The use of a time-resolved fluorometer has been recently reported for quantitating human chorionic gonadotropin by use of a europium-labeled monoclonal antibody to the β-subunit of hCG. The advantage to the use of lanthanide chelates and time-resolved fluorescence detection is a potentially very high detection sensitivity, because the signal/photon emission can be increased by using a stronger excitation light source and the background signal can be discriminated by using time-resolved detection; in addition, lanthanide labels are biochemically inert.[12]

Automated Fluorometers and Spectrofluorometers

There are several automated analyzers with fluorometric and absorbance measurement capabilities for use in the clinical chemistry laboratory. These systems include a centrifugal analyzer that can measure absorbance, fluorescence, or light scattering and a discrete analyzer capable of absorbance or fluorescence measurements.[4,14] A diagram of the optical path of the centrifugal analyzer absorbance fluorescence system is shown in Figure 1B-26. This system uses a xenon light source and a monochromator to isolate the excitation spectrum. The excitation light is directed into the spinning reaction cuvets at a 90° angle by use of a quartz-fiber optic bundle. Interference filters are used to isolate the emission spectrum. A photomultiplier tube is used to measure the emission intensity. A fluorescence compound in the reference cuvet of the rotor is used to provide a stable reference similar to the manual spectrofluorometer. The automated discrete analyzer employs the same lamp and optical path as the absorbance mode. This is an end-on fluorescence analyzer.

FLUOROMETERS FOR SPECIAL APPLICATIONS

Several automated fluorometric analyzers have been developed for special applications in clinical chemistry. These include a fluorescence polarization spectrometer for quantitation of therapeutic drugs, an enzyme-linked fluoro-immunoassay system for measurement of therapeutic drugs; a hematofluorometer for quantitation of protoporphyrin, bilirubin, and reserve bilirubin-binding capacity; a front-surface solid-phase instrument for quantitation of immunoglobulins and antinuclear antibodies; and a flow-cell fluorometer with photon counting for quantitation of thyroxine, free thyroxine, and triiodothyronine. Each of these automated fluorescence instruments makes use of a special principle or process in fluorescence measurements to accomplish a specific type of analysis. The special instruments and their manufacturers are listed in Table 1B-6.

The *fluorescence polarization system* takes advantage of the increased polarization that occurs

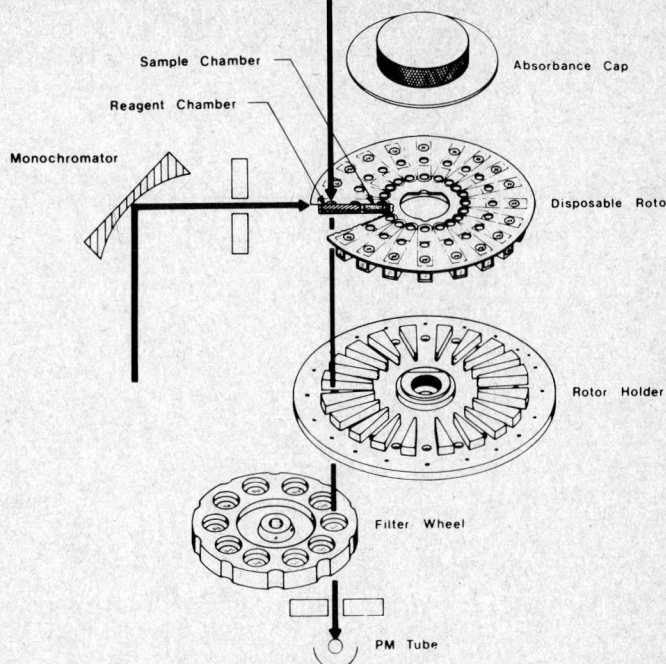

Figure 1B-26. Schematic diagram of a fluorescence/absorbance centrifugal analyzer.

when fluorescence-labeled small molecules are bound by antibodies. (Refer to the fluorescence polarization section, p. 81.) To improve the measurement of fluorescence polarization, an electronically controlled polarizer located in the excitation beam automatically rotates the plane of polarized light. A vertical polarizer is located in the emission optical path. Thus the plane of polarized light can be alternately changed 90° from parallel to perpendicular with less bias than in mechanical systems.

The *hematofluorometer* uses a front-surface cell for quantitation of bilirubin and protoporphyrin in red blood cells by fluorescence measurement. With this instrument the inner-filter effect is minimized by using front-surface fluorescence and small-path-length cuvets. This arrangement allows direct measurement of bilirubin and protoporphyrin in red blood cell hemolysates.

Flow cytometry is becoming more useful in the differentiation of lymphocytes from other peripheral leukocyte types. Flow cytometers have been built with an argon laser that provides excitation light at 488 nm and optics to measure right angle fluorescence, right angle light scatter, and forward light scatter. This powerful new instrument is capable of analyzing a flowing stream of blood cells for size and, by prior admixture of fluorescence-labeled monoclonal antibodies, for subclasses of specific types of cells such as T-lymphocytes.

LIGHT SCATTERING MEASUREMENTS

The Nature of Scattered Light

When electromagnetic radiation, such as light, impinges upon a particle, its electrons become subject to a force in one direction and its nuclei to a force in the opposite direction, causing the electrons about the particle to oscillate in synchronism with the electric field of the incident light. Thus, an oscillating dipole is induced in the particle by the incident light. The magnitude of this dipole moment is proportional to the electric field strength of the incident light. The proportionality constant between the magnitude of the dipole moment and the electric field strength of the light is the polarizability of the electron cloud surrounding the particle. The oscillating dipole becomes a source of electromagnetic radiation, reradiating light at the same wavelength as the incident light and in all directions. This radiation from the oscillating dipole is called *scattered light*. There are several factors to consider and understand about light scattering. These include the effect of particle size, wavelength dependence, distance of observation, effect of polarization of incident light, the concentration of the particles, and the molecular weight of the particles.

Particle Size

The size of the light scattering particle is an important element in the relationship of light scattering theory to observation. The Rayleigh scattering expression (see next section) applies to the scattering of light from small particles whose dimensions are much smaller than the wavelength of incident light (e.g., particle size $< \lambda/10$). When the dimensions of the particle are much smaller than the wavelength of the incident light, then the entire particle is subjected to the same electric field strength at the same time. The reradiated or scattered light waves from the small particle are in phase and reinforce each other. As the particles become larger than the incident light wave, the radiated light waves are no longer all in phase. Reinforcement of radiation occurs in some directions and destructive interference in others. The scattering patterns from these large particles are characteristic of the size and shape of the particle.

Wavelength Dependence of Light Scattering

The experimental measurement of the energy in a light wave is the intensity (i_s) that is defined as the energy that falls on an area of 1 cm^2 per second. Lord Rayleigh, in 1871, derived the first equation for the relationship of the intensity i_s of scattered light to the intensity I_o of the incident light:

$$\frac{i_s}{I_o} = \frac{16 \, \pi^2 a \, \sin^2\theta}{\lambda^4 \, r^2} \tag{6}$$

where i_s is the intensity of scattered light, I_o is the intensity of the excitation light, a is the polarizability of the small particle, λ is the wavelength of the incident light, θ is the angle of observation, and r is the distance from light scattering to the detector. One important observation to be made from equation (6) is the reciprocal fourth-power relationship between light scattering intensity and wavelength. This relationship shows that the intensity of light scatter will increase by the fourth power of the wavelength, as the wavelength of the incident light is decreased. The Rayleigh light scattering equation thus accounts for the fact that sunlight scattered by the earth's atmosphere is greatly enriched in the blue wavelengths. Another useful observation from equation (6) that is important to the design of instruments measuring light scattering is the fact that the light intensity decreases by the square of the distance r from the light scattering particles to the detector. Thus, the detector should be located close to the analytical cell either by the juxtaposition of the cell to the detector or by the use of good collection optics.

Concentration and Molecular Weight Factors in Light Scattering

Equation (6) can be adapted in a nonrigorous way to take into account factors of a defined system of particles suspended in a solvent:

$$\frac{i_s}{I_o} = \frac{4\pi^2 \, (dn/dc)^2 \, Mc \, \sin^2\theta}{N_a \lambda^4 \, r^2} \tag{7}$$

where i_s is the intensity of scattered light from small particles excited by polarized light, I_o is the incident intensity, dn/dc is the change in refractive index of the solvent with respect to change in solute concentration, N_a is Avogadro's number, M is the molecular weight (g/mol), and c is the concentration (g/mL) of the particles. The important observation to be made from equation (7) is the direct relationship of light scattering to the concentration of the particles and to the molecular weight of the particles. For a more detailed discussion of the Rayleigh light scattering formula and its application to macromolecules in solution, consult Kusnetz and Mansberg,[5] Tanford,[14] or Van Holde.[18]

The Effect of Polarized Light on Light Scattering

Equations (6) and (7) are different forms of the Rayleigh expression for light scattering from small particles if excited by polarized light. Figure 1B-27,A shows the effect of polarized and nonpolarized light on light scattering intensity from small particles as a function of scattering angle. Curve 2 shows a spherically symmetrical intensity diagram as predicted by equation (6). Curve 3 is the resultant intensity diagram when curves 1 and 2 are summed and is the scattering angular intensity diagram obtained when light scatters from small particles excited with non-polarized light. Curves 1 and 2 represent intensity diagrams from vertically and horizontally

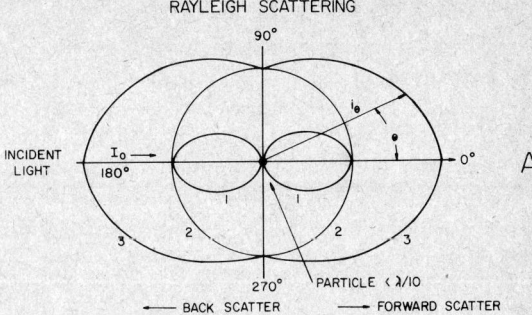

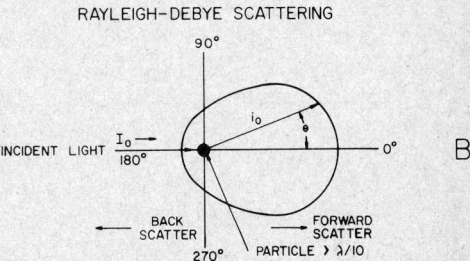

Figure 1B-27. The angular dependence of light scattering intensity with nonpolarized and polarized incident light for small particles (*A*) and the angular dependence of light scattering with nonpolarized light for larger particles (*B*).

polarized light components that can be thought of as comprising nonpolarized light. The Rayleigh light scattering expression for small particles excited by nonpolarized light is given by equation (8):

$$\frac{i_o}{I_o} = \frac{2\pi^2 \, (dn/dc)^2 \, Mc \, (1 + \cos\theta)}{N_a \, \lambda^4 \, r^2} \tag{8}$$

where i_o is the intensity of scattered light by small particles excited with nonpolarized light. There are two important observations to be made from equation (8) and Figure 1B-27,*A*. First, the total light scattered by small particles is less when excited by polarized light than by nonpolarized light, and a reduction of background signal from light scattering in fluorescence measurements can be achieved if an appropriately oriented polarizer is used in front of the emission detector. Second, the light scattering intensity from small particles excited by nonpolarized light shows a symmetrical angular dependence of light scattering about the 90° axis. (Refer to Figure 1B-27,*A*, curve 3.)

The Angular Dependence of Light Scattering

The angular dependence of light scattering from small particles (i.e., size less than $\lambda/10$), as indicated in the previous section, is represented by Figure 1B-27,*A*. Careful examination of Figure 1B-27,*A* (curve 3) shows that the light scatter intensity for forward scatter and back scatter (i.e., i_o at 0° and 180°) from small particles excited by nonpolarized light is equal. However, light scatter intensity at 90° is much less. As the size of particles becomes larger (e.g., $= > \lambda/10$), the angular dependence of light scatter takes on the dissymmetrical relationship shown in Figure 1B-27,*B*. In this case it can be seen that the light scattering intensities at forward and back angles are not equal; the forward scatter intensity is much larger. Also, the light scattering intensity at 90° is much less than the intensity at the forward (0°) angle. As particles become even larger this dissymmetry increases even further. This dissymmetry and the change of angular dependence of light scattering with change in the size of particles is very useful for characterization and differentiation of various classes of macromolecules and cells. As was previously mentioned, this property of light scattering is being used in the design of flow cytometers. These instruments measure near forward light scattering and right angle light scattering from cellular particles flowing through an optical cell and excited by a high-intensity laser. The ratio of the near forward light scattering intensity to the right angle light intensity is used in these instruments to distinguish different cell sizes.

Light Scattering and Plasma Proteins

The expression for light scattering given in equation (8) holds in dilute solution for small particles whose largest dimension is less than one-tenth of the wavelength of the incident light. Thus, the upper limit on size of particles exhibiting Rayleigh scattering is about 40 nm when visible light at 400 nm is used. Many of the plasma proteins such as immunoglobulins, β-lipoproteins, and albumin are below this limit. As the particle becomes larger in size from $\sim$40–400 nm, the angular dependence of the scattered light loses the symmetry around the 90° axis as seen in Figure 1B-27, *A* and *B*, and shows an increase in forward scattering. Some plasma proteins of the IgM class, chylomicrons, and aggregating immunoglobulin/antigen complexes fall into this size category. The scattering from particles in this size range is known as *Rayleigh-Debye* scattering, and the equation for this type of scattering becomes more complex.[5] Particles such as red blood cells and bacteria are larger yet, i.e., 7000–40 000 nm. These particles show a complex angular dependence of light scattering and this type of scattering from very large particles is called *Mie* scattering.[5] These large particles produce a predominance of scattered light in a narrow angular region in the forward direction.

The Rayleigh and Rayleigh-Debye expressions provide useful information about the scattering of light by small and intermediate-size particles and are important in the optimization of analytical instrumentation for measuring light scattering. This information includes the wavelength dependence, the angular dependence, the relationship between intensity of scattering and the detector, and the polarization of the scattered light at a right angle to the incident light. Furthermore the Rayleigh expression, equation (8), shows a direct relationship between concentration of particles and scattered light at fixed angle of observation provided that the particles are uniform in molecular weight and are of the proper size. However, the relationship between light scattering intensity and concentration becomes more complex when applied to aggregating systems such as antigen-antibody reactions, because these reactions involve a change in concentration, a change in molecular weight, and a change in size. Although more difficult to describe by theory, light scattering measurements have proved to be useful for the quantitation of serum proteins as discussed in Chapters 1C and 4.

NEPHELOMETRIC AND TURBIDIMETRIC MEASUREMENTS OF SCATTERED LIGHT

Light scattering is a physical phenomenon resulting from the interaction of light with a particle in solution. The phenomenon should not be confused with turbidimetry and nephelometry, which are methods used to measure scattered light. These two methods will be considered separately in this section.

Turbidimetry

Turbidity causes the attenuation (decrease) of the intensity of the incident beam of light as it passes through a solution of particles. The measurement of this decrease in intensity of the incident light beam that is caused by scattering, reflectance, and absorption of the light is called *turbidimetry*. Analogous to absorption spectroscopy, the turbidity can be defined as:

$$I = I_o e^{-bt} \text{ or } t = (1/b)\ln I_o/I \tag{9}$$

where t is the turbidity, and b is the path length of the incident light through the solution of light scattering particles. Turbidity is measured at 180° from the incident beam, or more simply, in the same manner as absorbance measurements are made in a spectrophotometer. Turbidity can be measured on most spectrophotometers and automated clinical chemistry analyzers. The stability and resolution of modern microprocessor-driven spectrophotometers and photometers have greatly improved the ability to measure turbidity with good accuracy and precision.

Nephelometry

Nephelometry is defined as the detection of light energy scattered or reflected toward a detector that is not in the direct path of the transmitted light.[5] Common nephelometers measure scattered light at right angles to the incident light. The ideal nephelometric instrument would be free of stray light; neither light scatter nor any other signal would be seen by the detector when no particles are present in solution in front of the detector. However, due to stray light generating components in the optics path as well as in the sample cuvet or sample itself, a truly dark field situation is difficult to obtain when making nephelometric measurements. Some nephelometers are designed to measure scattered light at an angle other than 90° in order to take advantage of the increased forward-scatter intensity caused by light scattering from larger particles, e.g., immune complexes. These instruments employ the principle illustrated in Figure 1B-27,B; for a general description, see Light Scattering Instrumentation, below.

Turbidimetry or Nephelometry: Selection of Method

The choice between turbidimetry and nephelometry depends upon the application and the available instumentation. Until recently, the statement was often made that for relatively clear solutions, where the transmission of light in the forward direction is greater than 95%, small changes in absorption due to turbidity were difficult to measure with precision and nephelometry was the method of choice. However, with the advent of stable high-resolution photometric systems, turbidimetric measurements have become competitive in sensitivity with nephelometric methods for immunological quantitation of serum proteins. Nephelometry, however, still offers some advantage in sensitivity when measuring low-level antigen-antibody reactions.[1]

LIMITATIONS OF LIGHT SCATTERING MEASUREMENTS

Antigen excess. Antigen-antibody reactions are complex and appear to result in a mixture of aggregate sizes. As the turbidity increases during addition of antigen to antibody, the signal increases to a maximum value and then decreases. The point at which the decrease begins marks the beginning of the phase of antigen excess; the phenomenon is explained in Chapter 1C, Principles of Immunochemical Techniques. Light scattering methods for quantitation of antigen-antibody reactions, to be fully acceptable, must provide a method for detecting antigen excess. Sternberg[13] and Hills and Tiffany[1] have shown that the kinetics of immune complex formation measured either by nephelometry or turbidimetry are sufficiently different in the three phases—antibody excess, equivalence, and antigen excess—that computer algorithms can be developed to flag antigen excess automatically.[1,13]

Matrix effects. Particles and solvent as well as all serum macromolecules scatter light. Lipoproteins and chylomicrons in lipemic serum provide the highest background turbidity or intensity. With appropriate dilutions, the relative intensity of light scattering from a lipemic sample is less than that of the antiserum blank. However, as the concentration of the antigen in serum decreases and correspondingly less dilute samples are used, the background interference from lipemic samples becomes greater. An effective method for minimizing this background interference is the use of rate measurements, where the initial sample blank is eliminated.

Large particles such as suspended dust can also cause significant background interference. This background interference can be controlled by filtering all buffers and diluted antisera before analysis is attempted.

LIGHT SCATTERING INSTRUMENTATION

Light scattering can be measured with standard clinical fluorometers or photometers. Although intensity of light scattering can be thought of as analogous to fluorescence intensity, the angular dependence of light scattering intensity has resulted in the design of special nephelometers. These devices place the photomultiplier detector at appropriate angles to the excitation light beam. The design principle of a nephelometer is similar to the design principle applied in fluorescence measurements. The major operational difference between the fluorometer and the nephelometer is that the excitation and detection wavelength will be set to the same value. The principal concerns of light scatter instrumentation are excitation intensity, wavelength, distance of the detector from the sample cuvet, sample slits, and minimization of external stray light. A schematic diagram of the basic components of an instrument for measuring light scattering is shown in Figure 1B-28. It consists of a light source, collimating optics, a sample cell, and the collection optics, which include light scattering optics, detector optical filter, and a detector. The schematic also shows the different angles from the incident light beam where the detector, filter,

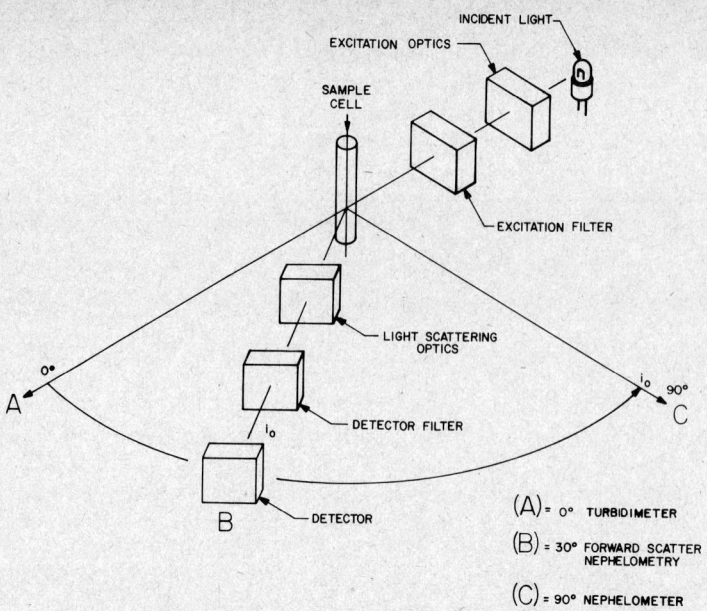

Figure 1B-28. Schematic diagram of light scattering instrumentation showing: *A*, the optics position for a turbidimeter; *B*, the optics position for a forward-scattering nephelometer; and *C*, the optics position for a right-angle nephelometer.

and optics can be placed to measure light scattering. Figure 1B-28,*A* is the straight-through arrangement for turbidimetry, while Figure 1B-28,*B* and 1B-28,*C* are arrangements frequently found in nephelometers. The detector arrangement shown in Figure 1B-28,*B* is for measurement of forward scatter at 30°, the optical arrangement used in the Beckman ICS automated rate nephelometer. Some high-precision research instruments for measuring light scattering provide a detector that is movable about the sample cell to take advantage of the angular dependence of light-scatter intensity. This movable detector is useful in determining the molecular properties of the scattering particle. Most clinical instruments capable of measuring light scattering have a fixed optical arrangement.

The optical components used for the nephelometer or turbidimeter are similar to those in fluorometers or photometers. The light sources commonly used are quartz halogen lamps, xenon lamps, and lasers. The laser beam is used specifically in some nephelometers because of its high intensity; in addition, the coherent nature of laser light makes it ideally suited for nephelometric applications.

Turbidimetric measurements are easily carried out on photometers or spectrophotometers

Table 1B-7. EXAMPLES OF ANALYTES MEASURED BY LIGHT SCATTERING METHOD (NEPHELOMETRY AND TURBIDIMETRY)

Immunoglobulins	*Coagulation factors*
IgG	Antithrombin III
IgA	*Therapeutic drugs*
IgM	Theophylline
IgE	Gentamicin
Specific proteins	Tobramycin
C3	Phenytoin
C4	Phenobarbital
Haptoglobin	
Transferrin	
Alpha$_1$-antitrypsin	
β-Lipoproteins	
Fibronectin	
Human placental lactogen	
Albumin	

and require little optimization. The principal concern of turbidimetric measurements is signal-to-noise ratio. Photometric systems with electro-optical noise in the range of ± 0.0002 absorbance unit or less are useful for turbidity measurements.

Ratio-referencing fluorometers and specifically designed nephelometers are well suited for nephelometric measurements.

Several automated instruments capable of light scattering intensity or turbidimetry measurements are commercially available for the clinical chemistry laboratory. One of these is capable of both. The key to the successful use of any of these systems is the availability of efficient antigen-antibody excess flagging and the provision of automated curve-fitting routines. In less than a decade it has become possible to perform a wide variety of specific protein analyses in as little time as it takes to run a glucose and with similar simplicity. Light scattering measurements are best applied to immunoassays of specific proteins and haptens. Particular applications are described in Chapters 3 and 17. Several examples of the types of analytes measured by light scattering are shown in Table 1B-7.

References

1. Hills, L. P., and Tiffany, T. O.: Comparison of turbidimetric and light-scattering measurements of immunoglobulins by use of a centrifugal analyzer with absorbance and fluorescence/light scattering optics. Clin. Chem., 26:1459-1466, 1980.
2. Jolley, M. E., Stroupe, S. D., Schwenzer, K. S., et al.: Fluorescence polarization immunoassay. III. An automated system for therapeutic drug determination. Clin. Chem., 27:1575-1579, 1981.
3. Jolley, M. E., Stroupe, S. D., Wang, C. J., et al.: Fluorescence polarization immunoassay I. Monitoring aminoglycoside antibiotics in serum and plasma. Clin. Chem., 27:1190-1197, 1981.
4. Khalil, O. S., Routh, W. S., Lingenfelter, K., et al.: Automated in-line ratio-correcting filter fluorometer, Clin. Chem., 27:1586-1591, 1981.
5. Kusnetz, J., and Mansberg, H. P.: Optical considerations: Nephelometry. In: Automated Immunoanalysis, Part 1. R.F. Ritchie, Ed. New York, Marcel Dekker, Inc., 1978.
6. Lloyd, J. B. F., and Evett, I. W.: Prediction of peak wavelengths and intensities in synchronously excited fluorescence emission spectra. Anal. Chem., 49:1710-1714, 1977.
7. Nussbaum, A., and Phillips, R.: Thin Films in Contemporary Optics for Scientists and Engineers. Englewood Cliffs, N.J., Prentice-Hall, 1976.
8. O'Haver, T. C.: Potential clinical applications of derivative and wavelength modulation spectrometry. Clin. Chem., 25:1548-1553, 1979.
9. Schulman, S. G.: Fluorescence and Phosphorescence Spectroscopy: Physiochemical Principles and Practice. Oxford, Pergamon Press, 1977.
10. Shelly, D. C., Warner, I. M., and Quarles, J. M.: Multiparameter approach to the "fingerprinting" of fluorescent pseudomonads. Clin. Chem., 26:1419-1424, 1980.
11. Soini, E., and Hemmila, I.: Fluoroimmunoassay: Present status and key problems. Clin. Chem., 25:353-361, 1979.
12. Soini, E., and Kojola, H.: Time-resolved fluorometer for lanthanide chelates—A new generation of nonisotopic immunoassays. Clin. Chem., 29:65-68, 1983.
13. Sternberg, J.: A rate nephelometer for measuring specific proteins by immunoprecipitin reactions. Clin. Chem., 23:1456-1464, 1977.
14. Tanford, C.: Light scattering. In: Physical Chemistry of Macromolecules. New York, Wiley, 1961.
15. Tiffany, T. O., Huey, E. E., Manning, G. B., et al.: The application of fluorescence measurements to centrifugal analyzers. In: Centrifugal Analyzers in Clinical Chemistry. C. Price and K. Spencer, Eds. Eastbourne, East Sussex, Praeger Publishers, 1980.
16. Udenfriend, S.: Fluorescence Assay in Biology and Medicine. New York, Academic Press, 1962.
17. Udenfriend, S.: Fluorescence Assay in Biology and Medicine, Vol. 2. New York, Academic Press, 1969.
18. Van Holde, K. E.: Scattering in Physical Biochemistry, Vol. 9. Englewood Cliffs, N.J., Prentice-Hall, 1971.
19. Wampler, J. E.: Measurements and physical characteristics of luminescence. In: Bioluminescence in Action. P. J. Herring, Ed. New York, Academic Press, 1978.
20. Wehry, E. L.: Modern Fluorescence Spectroscopy, Vol. 1. New York, Plenum Press, 1976.
21. Wehry, E. L.: Modern Fluorescence Spectroscopy, Vol. 2. New York, Plenum Press, 1976.
22. Whitehead, T. P., Kricka, L. J., Carter, T. J. N., et al.: Analytical luminescence: Its potential in the clinical laboratory. Clin. Chem., 25:1531-1546, 1979.
23. Wieder, I.: Background Rejection in Fluorescent Immunoassay. Proceedings of the VIth International Conference on Immunofluorescence. Amsterdam–New York, Elsevier/North-Holland Biomedical Press, 1978.

Section Three

ELECTROPHORESIS

by Emanuel Epstein, Ph.D.

Electrophoresis refers to the migration of charged solutes or particles in a liquid medium under the influence of an electrical field. While the expression *electrophoresis* applies to the migration of all species of particles, the term *iontophoresis* is specifically limited to the migration of small ions.

The first electrophoresis method used in the study of proteins was the free solution or moving boundary method devised by Tiselius in 1937. The Tiselius moving boundary electrophoresis apparatus is essentially a rectangular sectional U-tube. The U-tube is provided with optical flats so that observation of electrophoretic migration can be made. The serum sample to be analyzed is diluted with buffer and placed in the bottom U of the apparatus. Two vertical limbs containing buffer are placed on top of the U, which is temporarily isolated from the vertical limbs by large ground flats. These vertical limbs are filled with buffer. Other sections connect the vertical limbs to electrode vessels. The ground flats are moved so that contact is made between the liquid in the bottom U and the liquid in the vertical limbs, leaving sharp boundaries between buffer and sample solution. Electrical connections to a direct current power supply are made, and electrophoresis is allowed to proceed for 2 1/2 h at a reduced temperature, e.g., 4 °C. The moving boundary equipment is, by its nature, an analytical tool where measurements are made of the protein constituents of the serum within the arms of the U-tube. Very sensitive Schlieren or interference optics detect the presence of protein zones due to the difference of refraction caused by concentration changes (boundaries) in the buffer solution. The optics are also sensitive to concentration differences of proteins. The term *Schlieren* refers to the plural of the German word *Schliere*, meaning streak. Separated serum protein areas contain mixtures of proteins except for the fastest and the slowest moving components; those in between are contaminated by neighboring proteins. This technique is still used in research for the measurement of electrophoretic mobility and for the study of protein-protein interaction but is not used in clinical laboratories for routine work. A complex apparatus is needed, the technique is difficult, and samples of the order of 0.5 mL of serum are required. Results similar to those obtained with moving boundary electrophoresis can be obtained using agarose gel or cellulose acetate electrophoresis. These techniques, which require only a small sample volume and simple apparatus, will be described later in this section.

The term *zone electrophoresis* is used to refer to the migration of charged macromolecules in a porous supporting medium, such as cellulose paper, cellulose acetate sheets, and agarose gel film. Zone electrophoresis differs from moving boundary electrophoresis in that it generates an *electrophoretogram*, a display of protein zones, each one sharply separated from neighboring zones, on the electrophoretic support material. Solutes of interest in clinical chemistry are mainly macromolecular in size and colloidal in nature and include proteins in serum, urine, CSF, and other physiological fluids, as well as the proteins in erythrocytes and tissues.

THEORY OF ELECTROPHORESIS

Chemical species carrying an electric charge by virtue of protein ionization will move either to the cathode or to the anode in an electrophoresis system, depending on the kind of charge on the molecules. In a solution more acid than the isoelectric point (pI) of the solute, an ampholyte (a molecule which can be either positively or negatively charged, also called a Zwitterion) takes on a positive charge (binds protons) and migrates toward the cathode (the negatively charged electrode). In the reverse situation the ampholyte is in the anionic form and migrates toward the anode (positive electrode). The rate of migration is dependent on such factors as (1) net electric charge of the molecule, (2) size and shape of the molecule, (3) electric field strength, (4) properties of the supporting medium, and (5) the temperature of operation. The equation expressing the driving force which effects migration in such an electrophoresis system is given by

$$F = (X)(Q) = \frac{(EMF)\,(Q)}{d}$$

where

F = the force exerted on an ion
X = the current field strength (V/cm), i.e., voltage drop per unit width of medium
Q = the net charge on the ion
EMF = the electromotive force (voltage [V] applied)
d = the width of, or distance across, the electrophoretic medium (cm)

The steady acceleration of the migrating ion is counteracted by a resisting force, characteristic of the solution in which migration occurs. This force, expressed by Stokes' law, is

$$F' = 6\pi r \eta v$$

where

F' = the counter-force
r = the ionic radius of the solute
η = the viscosity of the buffer solution in which migration is occurring
π = 3.1416, a constant
v = the rate of migration of the solute = velocity, l/t (cm/s)

The force F' counteracts the acceleration which would be produced by F if no counter-force were present, and the resultant of the two forces is a constant velocity. Therefore, when

$$F = F'$$

then

$$6\pi r \eta v = (X)(Q)$$

or

$$\frac{v}{X} = \frac{l \cdot d}{t \cdot E} = \frac{Q}{6\pi r \eta} = \mu$$

where

$\dfrac{v}{X}$ = the rate of migration (cm/s) per unit field strength (E/cm), defined as the electrophoretic mobility. It is expressed by the symbol μ. The units of μ are cm^2/(V)(s).

It can be seen that electrophoretic mobility is directly proportional to the net charge and inversely proportional to the size of the molecule and the viscosity of the electrophoresis medium. As mentioned above, μ = cm^2/(V)(s), and this equation can be used to calculate the electrophoretic mobility in actual practice. To illustrate this, suppose that albumin has traveled 3 cm on a cellulose acetate strip 10 cm in width (d). If it takes 75 min (or 75 × 60 s) to travel this length at a voltage of 250 V, then

$$\mu = \frac{(3)(10)}{(75)(60)(250)} = 2.7 \times 10^{-5} \text{ cm}^2/(V)(s)$$

Since one mobility unit is defined as 10^{-5} cm^2/(V)(s), the above mobility then is 2.7 mobility units. Some literature sources show mobilities as having a negative sign. Assuming that protein migration occurs in a direction opposite to that of the electrophoretic field (i.e., from the anode [positive electrode] to the cathode [negative electrode]), by convention the mobility will have a negative value. In the example just given, electrophoresis was performed at pH 8.6 where proteins have a negative charge; therefore, migration occurred from the negative to the positive electrode, and the mobility value is amended to read −2.7 mobility units.

Mobility values obtained from zone electrophoresis procedures may not exactly compare with values obtained by free (moving boundary) electrophoresis because of possible differences in temperature, ionic strength, rate of endosmotic flow (to be described later), average pore size of support medium, and point of sample application.

Another perturbing factor in electrophoresis is the so-called *wick flow*. When electrophoresis is in progress, heat is evolved, resulting in evaporation of solvent from the electrophoretic support. The drying effect causes buffer to rise into the electrophoresis support from both buffer compartments. This flow of buffer from both directions affects the protein migration and hence the calculated mobility.

Effect of Buffer on Electrophoretic Migration

Buffer ions have a two-fold purpose in electrophoresis; they carry the applied current, and they fix the pH at which electrophoresis is carried out and thus determine the kind of electrical charge on the solute and the extent of ionization of the solute. By these effects the buffer determines the direction of the electrophoretic migration to one of the two electrodes. The buffer ionic strength determines the thickness of the ionic cloud surrounding the charged molecules and, thus, also the rate of migration and the sharpness of the zones. The effect of ionic strength on mobility of ions in the electrophoretic process is due to the concentration of ions (buffer and nonbuffer ions) clustered about the charged molecule. With increasing concentration of ions, the ionic cloud increases, and the molecule becomes more hindered in its movement. High-ionic-strength buffers yield sharper band separations, but the benefits of sharper resolution are diminished by the Joule (heat) effect that leads to denaturation of heat-labile proteins. The ionic strength (I) of an electrolyte (buffer) composed of monovalent ions is equal to its molality (mol/kg). The ionic strength of an electrolyte with one monovalent and one divalent ion is 3 mol/kg, and for a doubly divalent electrolyte it is 4 mol/kg.

Many buffer systems have been used in electrophoretic procedures, but those most widely used are the barbital buffers and the Tris–boric acid–EDTA buffers. Barbital buffers, $I = 0.025$–0.075, and Tris buffers, $I = 0.03$–0.12, are common. Because the base in both of these buffers is monovalent, any association effect between buffer and protein is minimal.

A relatively high ionic strength buffer ($I = 0.075$, pH 8.6), with admixture of calcium lactate, has been used for high-resolution separations of serum proteins in an 8 g/L agarose gel.[8] Under these conditions it is necessary to reduce the temperature of the system to 10–14 °C. Resolution of serum proteins of normal persons into as many as 13 zones is possible, and detection of one or more of an additional 8 zones may be seen in various pathological conditions. These modifications of conventional electrophoresis are collectively referred to as *high-resolution electrophoresis*. This technique also makes use of low electroendosmosis agarose (see below for a discussion of electroendosmosis). Use of calcium ions as the lactate salt aids in resolution of the various globulins in the β-globulin area. A high ionic strength buffer increases resolution of the albumin, α_1, and α_2-globulin fractions.

Electroendosmosis or Endosmosis

An electrophoretic support medium in contact with water takes on a negative charge because of adsorption of hydroxyl ions. Since the ions are fixed to the surface of the electrophoretic support, they are rendered immobile relative to the other ions in solution. Positive ions in solution cluster about the fixed negative charge sites, forming an ionic cloud of mostly positive ions. The number of negative ions increases with increase in distance from the fixed negative charge sites until, eventually, positive and negative ions are present in equal concentration (Figure 1B-29). The potential which exists between the fixed ions and the associated cloud of ions is termed the *electrokinetic potential* or the *zeta potential* (ζ). When a current is applied to such a system, charges attached to the immobile support remain fixed, but the cloud of ions in solution is free to move to the electrode of opposite polarity. Since the ions in solution are highly hydrated, this movement of the ionic cloud results in movement of the solvent as well. This movement of solvent and its solutes relative to the fixed support is referred to as *endosmosis*. Macromolecules in solution which move in the opposite direction must "buck" this flow of hydrated positive ions.

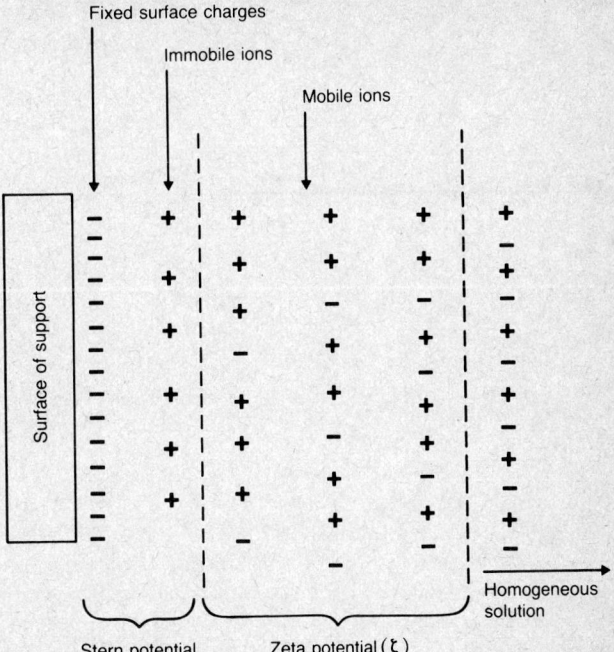

Figure 1B-29. Distribution of + and − ions around the surface of an electrophoretic support. Fixed on the surface of the solid is a layer of − ions. (These may be + ions under suitable conditions.) A second layer of + ions is attracted to the surface. These two layers compose the Stern potential. The large, diffuse layer containing mostly + ions is the electrokinetic or zeta (ζ) potential. Extending further from the surface of the solid is homogeneous solution. The Stern potential plus the zeta potential equals the electrochemical potential or epsilon (ϵ) potential.

If the molecules are insufficiently charged, they may remain immobile or they may even be swept back toward the opposite pole. In media where endosmosis is strong, as is true for cellulose paper, conventional cellulose acetate, and conventional agarose gel, γ-globulins are swept behind the line of application. In electrophoretic media where surface charges are minimal (starch gel or polyacrylamide gel), endosmosis is also minimal.

Most of the electroendosmotic effect observed using agarose is due to the presence of charged groups such as sulfate or carboxylic acid groups. It is possible to reduce endosmotic effects considerably by removing or modifying these charged groups chemically. The further addition of substances such as sucrose or sorbitol increases the osmolality and thus may reduce endosmosis to a point where the agarose may be considered to be electroendosmosis-free.

A General Method for Electrophoresis

Procedure

First, a hydrated support material such as freshly prepared agarose gel or previously wetted cellulose acetate is placed into the electrophoresis chamber. Care should be taken that excess buffer is removed from the support surface and that bubbles are not present. Commercially available agarose gel films may be used directly. Contact of the support is made with buffer previously placed into the electrode chambers. (Some alterations in this general method are common with agarose gel films, and these modifications will be discussed in the appropriate chapters of this text. Methods involving starch gel or polyacrylamide gel electrophoresis cannot be described in these general terms, but mention of methodology will be given in the appropriate sections.) Sample is applied to the support, and electrophoresis is conducted for a determined length of time using either constant voltage or constant current. The support is then removed from the electrophoresis cell and rapidly dried or placed in a fixative to prevent diffusion of sample components. It is then treated with a dye-fixative reagent to locate and visualize the individual protein zones by staining. After washing out excess dye, the support is dried (in the case of paper) or placed in a clearing agent (in the case of cellulose acetate membranes). Agarose gels may be dried with the aid of mild heat. No treatment is necessary to retain transparency. (For details, see Chapter 4.) A schematic of an electrophoresis system is shown in Figure 1B-30. Actual apparatuses may vary in form and detail, but Figure 1B-30 illustrates the essentials common to all designs.

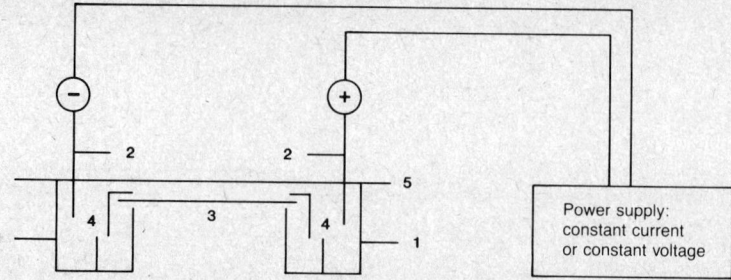

Figure 1B-30. A schematic of a typical electrophoresis apparatus. Two buffer boxes (1) with baffle plates contain the buffer used in the process. In each buffer box is an electrode (2), either platinum or carbon, the polarity of which is fixed by the mode of connection to the power supply. The electrophoresis support (3) on which separation takes place is in contact with buffer by means of wicks (4). In some systems, the support dips directly into the buffer solution. The whole apparatus may be covered (5) to minimize evaporation and protect the system. Direct current power supply may be either constant current (adjustable) or constant voltage (adjustable), or both.

Protein Stains Used in Electrophoresis Procedures

Stains used to visualize and locate the separated protein fractions of the sample differ in accordance with type of application and personal choice. The stain utilized for a specific application is described in connection with the individual procedures detailed in Chapter 4. However, as a general guide, Table 1B-8 lists some dyes commonly used in electrophoresis with suggested wavelengths for quantitation (see below). The amount of dye taken up by the sample is affected by many factors, such as the type of protein and the degree of denaturation of the proteins by the fixing agents. Commonly used protein stains are Amido Black (Naphthol Blue Black), Ponceau S, and Bromophenol Blue. However, some of these stains are not suitable for polyacrylamide gel–isoelectric focusing (PAGE-IEF) because of their reactivity toward carrier ampholytes (vide infra). Members of the Coomassie Brilliant Blue series of dyes, such as Coomassie Brilliant Blue R-250, have been successfully used in PAGE-IEF. More recently, silver nitrate has been used to stain proteins and polypeptides in electrophoretic procedures. Sensitivity of this stain is greater than the sensitivity of dyes used for that purpose.

Table 1B-8. SUGGESTED WAVELENGTHS FOR QUANTITATION OF SERUM PROTEIN ZONES BY EITHER DIRECT DENSITOMETRY OR ELUTION SPECTROPHOTOMETRY

Separation Type		Nominal Wavelength, nm	
		Densitometry	*Elution*
Staining of serum proteins in general	Amido Black (Naphthol Blue Black)	640	600
	Bromophenol Blue	600	595
	Coomassie Brilliant Blue G-250 (Brilliant Blue G)	595	610
	Coomassie Brilliant Blue R-250 (Brilliant Blue R)	560	580
	Nigrosin	540	560
	Ponceau S	520	565 (alkaline) 520 (neutral)
Staining of isoenzymes NAD(P)H → NBTH → formazan	Nitrotetrazolium Blue (as the formazan)	570	—
Staining of lipoprotein zones	Fat Red 7B (Sudan Red 7B)	540	—
	Oil Red O	520	—
	Sudan Black B	600	—

Quantitation of Protein Zones

It is customary to report the results obtained by electrophoretic separation in terms of the percentage of each fraction present or in terms of absolute concentration if the total quantity of protein is known. Quantitation of the dye in the individual zones can conveniently be accomplished either by direct *densitometry* or by *elution* of the dye associated with each protein zone and subsequent spectrophotometric measurement of the eluted dye. In *densitometry*, an electrophoretic strip (or other medium) is moved past the measuring optical system and each fraction in accordance with its location and absorbance value is presented on a recorder chart. In some cases, the area under each peak is automatically integrated. The *elution method* involves cutting the support into zones containing the individual fractions and eluting the adsorbed dye by means of suitable solvents such as basic buffers, weak alkali (e.g., NaOH, 0.1 mol/L), or alcoholic solutions.

The instrumental method of quantitating electrophoretograms, termed *densitometry*, has become the method of choice for the typical clinical laboratory. Although initially expensive (ranging from about $5000–$10 000), a contemporary densitometer may pay for itself in a few years in technologist time and effort. The following features are those to be sought in choosing a densitometer: (1) ability to scan electrophoresis support lengths of 25–150 mm; (2) capability of automatic gain control to prevent the most intense peak or zone of an electrophoretogram from going off-scale; (3) automatic background zeroing capability for background correction, which allows the instrument to choose the lowest background point in the electrophoretogram as baseline so that minor peaks will not be lost or "cut-off"; (4) variable wavelength control, either as a continuously variable monochromator or as variably selectable interference filters, to allow operation in the 400–700 nm range; (5) variable slits from 0.1×2.0 to 0.2×10.0 mm; and (6) an integrating device. The simplest integrator provides a zig-zag integration display where the operator counts peaks and manually computes integration zones.

Microprocessor-containing models automatically compute electrophoretogram zones with great speed. If this feature is desired, one should consider a model that has the capability of integrating up to 30 peaks rather than only the five zones associated with conventional serum protein electrophoresis. Such a choice allows the laboratory to handle advanced techniques that fractionate serum proteins into multiple peaks. Another desirable feature is ultraviolet fluorescence capability. Creatine kinase isoenzymes, for example, are routinely analyzed by making use of the fluorescence of generated NADH or NADPH.

While not essential for routine laboratory use, the following features are desirable: (1) computerized integration and print-out (mentioned above); (2) automatic indexing, a feature that automatically advances the electrophoresis strip which contains multiple sample channels from one channel to the next; (3) built-in diagnostics (microcomputerized) for instrument trouble-shooting; and (4) a choice of one of several (commonly 3) scanning speeds.

A densitometer may be tested for linear response by the use of a standard which is provided by several manufacturers. This standard may be designed in several zones of increasing densities or as a step-wise arrangement of variable densities. The first type of standard causes the pen to come to base line between each dense zone. By the behavior of peak-tracing and base line return, the operator checks the optical, mechanical, and electrical functions of the densitometer. The zones of variable densities are permanently fixed and have expected values of performance.

In paper electrophoresis, accuracy of densitometry is affected by the scattering of incident light by the surface of the paper. In addition, linearity of the photoelectric measurement is affected by the thickness of the dye-protein layer. In such cases, the elution method is likely to be more accurate.

In modern cellulose acetate techniques, an acquired transparency by chemical clearing of the membranes and use of a very small sample size have substantially eliminated the shortcomings that had been associated with densitometry of cellulose paper strips. Similar advantages are common with agarose gel electrophoresis. Nevertheless, problems associated with densitometry persist due to differences in quantity of stain taken up by individual proteins and differences in protein zone sizes, because solutions to such problems are not simple.[10,11]

Power Supply

Commercially available power supplies allow operation at either constant current or constant voltage. The flow of current through a medium which offers electrical resistance is associated with production of heat:

$$Heat = (E)(I)(t)$$

where

E = electromotive force (EMF), in volts (V)

I = current in amperes (A)

t = time in seconds (s)

Heat evolved during electrophoresis increases the conductance of the system (decreases the resistance). With constant-voltage power sources, the resultant rise in current, due to the increase in thermal agitation of all dissolved ions, causes both an increase in the migration rate of the protein and an increase in the rate of evaporation of water from the stationary support medium. The water loss causes an increase in ion concentration and thus a further decrease in resistance (R). To minimize these various effects on the migration rate, it is best to utilize a constant-current power supply. According to Ohm's law:

$$E = (I)(R)$$

Therefore, if R is decreased, the applied EMF also decreases (current remains constant). This in turn decreases the heat effect and thus keeps the migration rate relatively constant. If a constant voltage source is used, the current, and therefore the migration rate, will progressively increase.

For *isoelectric focusing* (IEF), use of a constant-power power supply is advisable. If IEF is carried out with a constant-voltage power supply, frequent adjustments of the voltage may be necessary as current drops during the IEF process. This drop in current is due to lower conductivity of the carrier ampholytes at their isoelectric points, and to the creation of zones of pure water during electrophoresis. With a constant-power power supply frequent resettings of voltage may be avoided or minimized. Constant-current power supplies are not customarily used in IEF.

Paper Electrophoresis (PE)

Paper electrophoresis, once used widely in clinical laboratories for the separation of serum proteins, has largely been replaced by cellulose acetate or agarose gel electrophoresis. Most PE procedures use either Whatman No. 1 or 3 MM paper, or their equivalents from other manufacturers, supplied in the form of precut strips. A paper is chosen on the basis of qualities of separation, handling, dyeing, and destaining. In general, a thick, soft paper (such as Whatman No. 3 MM) is best for resolution of proteins, but electroendosmosis is more prominent with this paper than with a hard paper.

A significant disadvantage in the use of PE is the long separation time needed (14–16 h). Also, in case of serum protein electrophoresis, some albumin is adsorbed to the paper, resulting in excessive background (trailing) and decreased accuracy. The advantages of paper are its high tensile strength, low cost, and ease of handling.

Agarose Gel Electrophoresis (AGE)

Agar gel electrophoresis has been successfully applied to the analysis of serum proteins, hemoglobin variants, lactate dehydrogenase isoenzymes, lipoprotein fractions, and other substances. In fact, this gel medium parallels cellulose acetate in versatility and convenience and competes favorably with other media for applicability to routine clinical laboratory demands.

Impure and even purified agar is composed of at least two fractions: agaropectin and agarose. Agaropectin contains acid sulfate and carboxylic acid groups and accounts for the considerable endosmosis and background staining which are observed with unfractionated agar. Agarose, which is essentially free of ionizable groups, exhibits little endosmosis. It is becoming the agar medium of choice. Some lots of commercially available agarose, however, still contain some residual charged groups. The literature may be consulted for complete procedures for purification of agarose.[4,7]

Use of agarose rather than agar in "agar gel electrophoresis," except for certain special procedures, has become so pervasive in clinical laboratory procedures that the term *agarose gel electrophoresis* will be used in the following text.

The advantage agarose gel has over paper is its lower affinity for proteins and its native clarity, after drying, which permits excellent densitometric examination. Usually 0.5–1.0 g of agarose/dL of buffer provides a gel of desired strength and with good migration properties. In

AGE, the unmodified serum or serum dissolved in warmed agar is applied directly into a precut or precast well. The latter technique is less convenient, but it has the advantage that the agarose sample solution solidifies to become part of the agarose plate. Use of sample wells to which unmodified serum is applied results in an uneven surface of the completed electrophoretogram at the sample application point and causes an artifactual peak in densitometry. More recently, some AGE procedures use a modified sample application technique which avoids the surface artifact. Use is made of a thin plastic template which contains small slots corresponding to sample application points. The template is placed upon the agarose surface and 5 μL samples are placed upon each slot. The serum sample is allowed to diffuse into the agarose for 5 min, excess sample is removed from the template by blotting, and the template removed from the agarose surface. The sample size actually used in the AGE technique is relatively small (0.6–3 μL), and electrophoresis time is relatively short (30–90 min, depending on experimental conditions). Specific applications of this technique are mentioned later in this text.

Cellulose Acetate Electrophoresis (CAE)

If the hydroxyl groups in cellulose are reacted with acetic anhydride, cellulose is acetylated to form the raw material for cellulose acetate membranes. The membranes commercially available contain about 80% air space in the form of pockets within the interlocking cellulose acetate fibers. When purchased, the membranes come as dry, opaque, brittle films which crack easily if not handled gently. However, when the film is placed in buffer, the air spaces fill with liquid and it becomes quite pliable. The characteristics of the membrane will vary with the extent of acetylation, the prewashing procedure employed by the manufacturer, and the additives used, as well as the pore size and the thickness of the membrane. Serum samples (0.3–2.0 μL) are generally applied to cellulose acetate strips (presoaked with buffer) with a twin-wire applicator. The edge of a glass slide or a micropipet has also been used for this purpose. Cellulose acetate membranes may be made transparent (cleared) for densitometry by treatment with a solvent mixture which contains one solvent that tends to dissolve cellulose acetate and another solvent that acts as a vehicle for the first (e.g., 95 parts methanol and 5 parts glacial acetic acid). The cellulose acetate fibers are partially dissolved by the action of the solvent and coalesce so that the original air spaces are eliminated.

An advantage of CAE is the speed of separation (20 min–1 h) and the ability to store the transparent membranes for long periods. Applications of this technique will be discussed in Chapter 4. Cellulose acetate, especially prepared to reduce electroendosmosis, is commercially available. The use of agarose or cellulose acetate in IEF and in high resolution electrophoresis[8] may be of great importance in the clinical laboratory.

Polyacrylamide Gel Electrophoresis (PAGE)

Serum protein zones resolved by ordinary electrophoretic techniques are composed of several proteins possessing the same electrophoretic mobility; they tend to be broad because proteins diffuse during the electrophoresis procedure. *Disk electrophoresis* was introduced in 1964[2,9] to overcome these deficiencies. The term derives from the *discontinuities* in the electrophoretic matrix and from the *discoid* shape of the separated zones of proteins. Protein electrophoresis using paper, cellulose acetate, or agarose gel yields only five zones, namely, albumin and α_1-, α_2-, β-, and γ-globulins. Polyacrylamide or starch gel electrophoresis may yield 20 or more fractions and is therefore widely used in the study of individual proteins in serum, especially in the study of genetic variants and isoenzymes.

The PAGE technique employs layers of gel which differ in composition and pore size. The individual gels are first prepared in situ in glass tubes by polymerizing a gel monomer and a cross-linking agent with the aid of an appropriate catalyst. The first gel to be poured into the tubular-shaped electrophoresis cell is the small-pore *separation gel*. After 30 min, during which gelation takes place, a large-pore gel, the *spacer gel*, is cast on top of the separation gel. Then a large-pore monomer solution containing a small amount of serum, about 3 μL, is polymerized above the spacer gel so that the finished product is composed of three different layers of gel. When electrophoresis begins, all protein ions migrate through the large-pore gels (which do not impair movement of most proteins in serum) and stack up on the separation gel in a very thin zone. This process serves to concentrate protein components at the border (or starting) zone, so that preconcentration of specimens with low protein content (e.g., CSF) may not be necessary.

Separation of the individual protein ions then takes place in the bottom separation gel, not only on the basis of their charge but also on the basis of molecular size. The retardation of some proteins in both polyacrylamide and starch gel electrophoresis (to be described below) is due to a molecular sieve phenomenon. The average pore size in a typical 7.5% PAGE separation gel is about 5 nm (50 Å), large enough to allow most serum proteins to migrate unimpeded. However, proteins whose molecular radius and/or length exceeds critical limits will be more or less impeded in electrophoretic migration. Some of these proteins are fibrinogen, β_1-lipoprotein, α_2-macroglobulin, and γ-globulins. Acrylamide gel is thermostable, transparent, strong, and relatively chemically inert, and can be made in a wide range of pore sizes. Furthermore, these gels are uncharged, thus eliminating electroendosmosis.

In recent years, simplifications of this technique have been introduced. In one such procedure, the use of spacer and sample gel was eliminated and replaced with a continuous buffer system with excellent results. Epstein et al.,[5] in their studies of alkaline phosphatase isoenzymes, used a similar approach. These authors delivered the undiluted serum sample by micropipet through the covering buffer layer directly onto the top surface of the separation gel. A schematic representation of serum protein electrophoresis and haptoglobin typing by PAGE is shown in Figures 1B-31 and 1B-32.

The application of polyacrylamide gel in isoelectric focusing will be presented below.

Starch Gel Electrophoresis

Starch gel electrophoresis also possesses the property of separating macromolecular ions on the basis of both surface charge and molecular size. The starch used is partially hydrolyzed, since native starch does not gel. Like agarose gel, starch gel may be utilized in a horizontal process, but the sample must be applied in such a way that a very fine starting zone results. This can be accomplished by introducing the sample into a slit or trough cut into the gel either directly as a liquid sample or as a paper loaded with specimen. The technique can also be used with migration taking place in the vertical direction. In this case, the liquid sample is applied to a cut or precast slot in the gel and then covered with a warmed wax-petrolatum mixture. The wax hardens and holds the liquid sample in place. Application of electrical current results in compaction of proteins on the surface of the gel so that a thin uniform starting zone is produced. The proper preparation of gels is relatively difficult, requiring considerable skill. Starch gels are used in a concentration of 10–16 g/dL. The pH of the buffer varies according to the specific application and is generally chosen to be between 8.6 and 9.0, but may range between 3 and 11.

Isoelectric Focusing (IEF)

A further advance in separation methods by electrophoresis is *isoelectric focusing (IEF)*. This technique separates amphoteric compounds, such as proteins, by virtue of migration in a medium possessing a stable pH gradient, with pH varying in the direction of the migration. The protein moves to the zone in the medium where the pH is equal to the isoelectric point of the protein. At this pH, the charge becomes zero and migration ceases. Figure 1B-33 illustrates the procedure in a schematic and shows the electrophoretic conditions before and after current is applied. In IEF the protein zones are very sharp since the pI of a protein is confined to a narrow pH range and because diffusion of a protein is counteracted by the acquisition of a charge as it varies from its pI position and by subsequent migration back to this position by electrophoretic forces (Figure

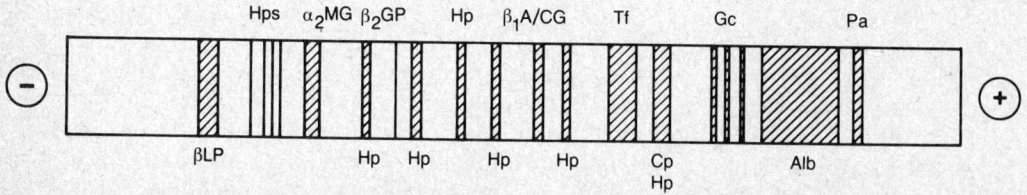

Figure 1B-31. A simplified schematic of a protein pattern from the serum of a subject with haptoglobin type 2–1 (separation by PAGE). Some zones contain more than the one protein shown, as demonstrated by immunological techniques. Abbreviations: Pa, prealbumin; Alb, albumin; Gc, Gc-globulins; Cp, ceruloplasmin; Tf, transferrin; β_1A/CG, β_1A/C globulin; Hp or Hps, haptoglobin(s); β_2GP, β_2-glycoprotein (best shown by PAS staining), α_2MG, α_2-macroglobulin; βLP, β-lipoprotein.

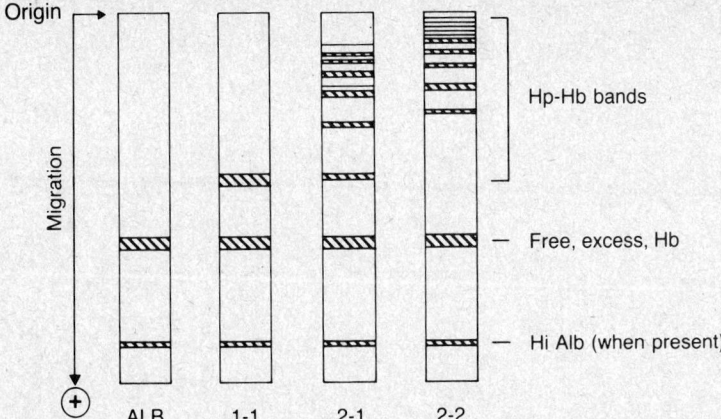

Figure 1B-32. Electrophoretograms of the three main haptoglobin types by a PAGE procedure. Albumin is shown as a blank because albumin is used in the electrophoretic procedure. Excess control hemoglobin (Hb) as a hemolysate is added to the serum and albumin is added to the control hemolysate to increase the viscosity of the solution. Staining is performed by the use of *o*-dianisidine and hydrogen peroxide. Haptoglobin-hemoglobin complexes (Hp-Hb) are readily seen. At times an unstable Hb hemolysate causes the formation of methemalbumin (Hi Alb). [6]

1B-34). Proteins which differ in their pI values by only 0.02 pH unit have been separated by IEF.

The pH gradient is created by the use of amphoteric polyaminocarboxylic acids (*carrier ampholytes*), a group of compounds with molecular weights of 300–1000. The carrier ampholytes are used in mixtures of 50–100 different individual compounds and create under electrophoretic conditions a "natural pH gradient" when the many different carrier ampholytes reach their individual isoelectric points. They create buffered zones with stable but gradually changing pH wherein the slower migrating proteins can separate out at their individual isoelectric points.

As Figure 1B-33 illustrates, the anode is surrounded by a dilute solution of an acid and the cathode by a dilute alkaline solution. After focusing, the most negatively charged carrier ampholytes and proteins will be found at the anodal end and the most positively charged near the cathodal end of the electrophoretic matrix. The other carrier ampholytes and proteins focus at

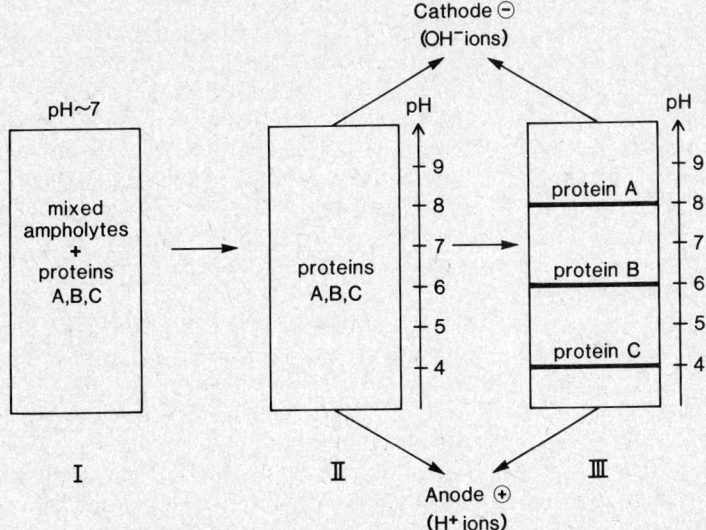

Figure 1B-33. Schematic of an IEF procedure. *I*, A homogeneous mixture of carrier ampholytes, pH range 3–10, to which proteins A, B, and C with pI 8, 6, and 4, respectively, were added. *II*, Current is applied and the carrier ampholytes rapidly migrate to the pH zones where net charge is zero (the pI value). *III*, The proteins A, B, and C migrate more slowly to their respective pI zones where migration ceases. The high buffering capacity of the carrier ampholyte creates stable pH zones where each protein may reach the pH zone where its net charge is zero.

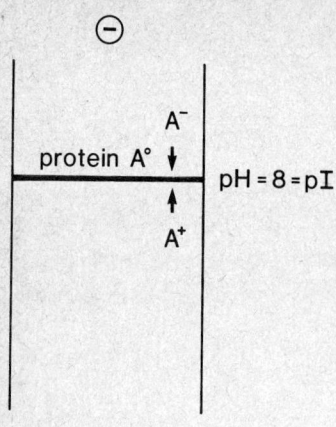

Figure 1B-34. After attaining the pH where protein A has a net charge of zero, A°, diffusion toward the cathode bestows a negative charge on A, A⁻, and migration in the electric field forces A⁻ back to A°. Diffusion toward the anode causes A to take on the opposite charge, A+, and migration is toward the cathode and to the point where A° exists. IEF processes of this kind cause sharp zones to form, i.e., the protein is focused.

intermediate points according to their isoelectric points. Since carrier ampholytes are generally used in IEF in relatively high concentrations, a high-voltage power source (up to 2000 V) is necessary (power is in the vicinity of 2–50 W, depending upon experimental conditions). As a result, the electrophoretic matrix must be cooled.

Polyacrylamide gel–isoelectric focusing (PAGE-IEF) is widely used in analytical work. The polyacrylamide gel must be optically clear and supple, but yet have large enough pore size so that, ideally, protein migration will not be impeded by molecular sieving effects. In real practice impeded migration of some proteins, such as IgM, cannot be avoided. With the availability of electroendosmosis-free materials, IEF methods have been adapted for agarose (AGE) and cellulose acetate electrophoresis (CAE). AGE-IEF and CAE-IEF have the advantages that operating conditions are simple and that large pore sizes make it unlikely that any proteins will be excluded on the basis of molecular weight (size).[1,3]

Limitation and Errors in Routine Electrophoresis

The routine performance of electrophoresis in the routine clinical laboratory demands some knowledge of the constraints and errors involved. The following paragraphs provide some guidelines regarding the materials used in electrophoretic practice.

Buffers. Buffers used in the laboratory are good culture media for growth of microorganisms. For that reason, once a buffer has been made up with water to the desired pH and ionic strength, the buffer should be refrigerated when not in use. Moreover, a cold buffer is preferred in an electrophoretic run since resolution is improved and evaporation from the electrophoretic support is lessened. Buffers used in small-volume apparatuses should be discarded after each run because of pH changes resulting from electrolysis of water that accompanies electrophoresis. It is not necessary to discard large volumes of buffer (from 700 mL–1 L) after each separation if the polarity of the electrophoretic cell is switched after each run. Alternatively, the buffer from both buffer boxes may be poured into a common container which can be stored at 4 °C. Such a practice may be followed for 4 electrophoretic runs.

Stain solution. A typical stain solution may be used several times before being replaced. A good rule of thumb is that a stain solution of 100 mL may be used for a combined total of 387 cm² (60 in²) of cellulose acetate or agarose film. The stain solution may be considered faulty if leaching of stained protein zones occurs in the 5% acetic acid wash solution or in the clearing solution (cellulose acetate only). Stain solution must be stored tightly covered to avoid evaporation.

Sampling. The amount of serum protein which is applied to an electrophoretic support must be optimal to obtain satisfactory results. Albumin is about 10 times more concentrated in serum than the smallest fraction, the α_1-globulins. Therefore, the amount of serum applied should be such that overloading with albumin is avoided, but should be adequate to allow α_1-globulin quantitation. Typical amounts of serum applied to cellulose acetate are 0.3–1.6 μL, depending upon the size of the twin-wire applicator. When procedures call for multiple applications, such as in isoenzyme analysis, the concern over albumin overloading is no longer a factor. In PAGE, a typical sample size is 3 μL (about 210 μg of total protein) for protein electrophoresis, but in applications involving, for example, alkaline phosphatase isoenzymes, as much as 25 μL of a normal serum may be applied. Volumes may be reduced to as little as 10 μL of serum for specimens with high total alkaline phosphatase activity (> 500 U/L).

Volume requirements in agarose electrophoresis generally are from 0.6–2.0 μL, depending upon the test requirements.

Problem Solving in Protein Electrophoresis

Problems may arise from sample application or from electrophoretic support materials. Application problems are especially prominent in both CAE and AGE techniques, and some are listed below.

Discontinuities in sample application may be due to dirty applicators. Use caution in cleaning these applicators because they are easily bent. Twin-wire applicators are best cleaned merely by dipping in water with agitation followed by removal of water and residue by gently pressing the applicators against absorbent paper. It is inadvisable to clean wires by manual wiping.

Distorted protein zones may be due to bent applicators or to excessive drying of portions of the electrophoretic support. Overapplication may also cause zone distortion. Care should be taken not to overwhelm the twin-wire applicator with excessive serum or to overfill the sample well in AGE. Distorted zones may also be caused by excessive drying of the cellulose acetate film before or during electrophoresis. Improper tension on the cellulose acetate film may also cause distortions, often resulting in zones that look "bent over."

Irregularities (other than broken zones) in sample application probably are due to excessively wet cellulose acetate films. Parts of the applied samples may look washed out.

In most cases *unusual bands* are artifacts which may be easily recognized. Hemolyzed samples are frequent causes of an increased β-globulin (where free hemoglobin migrates), or an unusual band between the α_2- and β_2-globulins may be the result of a hemoglobin-haptoglobin complex. A band occurring at the starting point of an electrophoretogram may be fibrinogen. The sample should be verified as being serum before this band is reported as an abnormal protein. It is not unusual to find split α_1-, α_2-, and β-globulins; these are not to be considered errors of electrophoresis. Occasionally a split albumin zone is observed as a result of a rare, benign, genetically related condition. However, a grossly widened albumin zone may well be the result of one of certain medications that are albumin bound and not of faulty electrophoretic practice. Occasionally an irregular but sharp protein zone is seen at the starting point. Unlike fibrinogen or other proteins which may be seen at about the same point, the artifact lacks the regular, somewhat diffuse appearance which proteins normally show; it is actually denatured protein resulting from a deteriorated serum or from damage done to the cellulose acetate by the twin-wire applicator. Finally, it is good laboratory practice to include a control serum with each electrophoretic run, as an aid in the evaluation of the quality of an electrophoretic run.

References

1. Catsimpoolas, N.: Isoelectric focusing and isotachophoresis of proteins. 13th Eastern Analytical Symposium Articles. Separation Sci., 8:71-121, 1973.
2. Davis, B. J.: Disc electrophoresis—II. Method and application to human serum proteins. Ann. N.Y. Acad. Sci., 121:404-427, 1964.
3. Drysdale, J. W.: Isoelectric focusing in polyacrylamide gel. In: Methods of Protein Separation, Vol. 1. N. Catsimpoolas, Ed. New York, Plenum, 1975.
4. Duckworth, M., and Yaphe, W.: Preparation of agarose by fractionation from the spectrum of polysaccharides in agar. Anal. Biochem., 44:636-641, 1971.
5. Epstein, E., Wolf, P. L., Horwitz, J. P., et al.: An indigogenic reaction for alkaline phosphatase in disk electrophoresis. Am. J. Clin. Pathol., 48:530-534, 1967.
6. Epstein, E., and Zak, B.: Separation of haptoglobins by disc electrophoresis. Ann. Clin. Lab. Sci., 2:191-197, 1972.
7. Hegenauer, J. C., and Nace, G. W.: An improved method for preparing agarose. Biochim. Biophys. Acta, 111:334-338, 1965.
8. Jeppsson, J.-O., Laurell, C.-B., and Franzen, B.: Agarose gel electrophoresis. Clin. Chem., 25:629-638, 1979.
9. Ornstein, L.: Disc electrophoresis—I. Background and theory. Ann. N.Y. Acad. Sci., 121:321-349, 1964.
10. Zak, B., Baginski, E. S., and Epstein, E.: Associated problems of protein electrophoresis, staining and densitometry. Ann. Clin. Lab. Sci., 8:385-395, 1978.
11. Zak, B., Epstein, E., and Watkins, R.: Postulated flaw in densitometry. Microchem. J., 23:226-236, 1978.

Additional Reading

Arbuthnott, J. P., and Beeley, J. A., Eds.: Isoelectric Focusing. Boston, Butterworth, 1976.

Deyl, Z., et al.: Electrophoresis: A Survey of Techniques and Applications, Pt. A. E. M. Everaerts, Ed. New York, Elsevier, 1980.

Gaal, O., et al., Eds.: Electrophoresis in the Separation of Biological Macromolecules. New York, Wiley, 1980.

Gordon, A. H., et al., Eds.: Electrophoresis of Proteins in Polyacrylamide and Starch Gels, Pt. 1. 2nd ed. New York, Elsevier, 1975, Vol. 1.

Whitaker, J. R.: Paper chromatography and electrophoresis. *In*: Electrophoresis and Stabilizing Media. G. Zweig, Ed. New York, Academic Press, 1967.

Section Four

ELECTROCHEMISTRY

by Ole Siggaard-Andersen, M.D., Ph.D.

Several types of analytical methods are based on electrochemical phenomena. Those which have been applied to clinical chemical procedures are potentiometry, polarography, amperometry, coulometry, and conductometry.[16]

POTENTIOMETRY

Potentiometry is the measurement of the electric potential difference between two electrodes in an electrochemical cell. An *electrochemical* (galvanic) *cell* always consists of two electrodes (electron or metallic conductors) connected by an electrolyte solution (ion conductor). An *electrode* or *half-cell*, consists of a single metallic conductor in contact with an electrolyte solution. The ion conductors can be composed of one or more phases in direct contact with each other, or they can be separated by membranes which are permeable only to specific cations or anions (Figure 1B-35). One of the electrolyte solutions is the unknown or test solution, which may be replaced by an appropriate standard solution for calibration or quality control purposes. A salt solution or bridge may be interposed in the cell to reduce any liquid-liquid junction potential present (see later). By convention, the cell is shown so that the left electrode (M_L) is the *reference electrode*, while the right electrode (M_R) is the *indicator (measuring) electrode*.[1]

The electromotive force (E) is defined as the maximal difference in potential between the two electrodes (right minus left) obtained when the current drawn from the cell is zero. Measurement of the cell potential is carried out with a *potentiometer*, of which the common pH meter is a special type. Two main classes of potentiometers are available. The *null-point potentiometer* (compensation or balancing potentiometer) applies a potential of increasing magnitude between the cell electrodes until that point at which no current flows through the cell (Figure 1B-35). The *direct-reading potentiometer* is a voltmeter which measures the potential across the cell (between the two electrodes), but in order that such potential measurement be accurate, it is necessary that no current flows through the cell. This is accomplished by incorporating a high resistance within the voltmeter (input impedance $> 10^{12}\ \Omega$). Modern direct-reading potentiometers are equal in accuracy to compensation potentiometers and have largely replaced them. For convenient reading and recording of data, these instruments may be equipped with accessories for direct digital display or printout.

Within any one given conductive phase, the potential is constant as long as the current flow is zero. Between two different phases, however, a potential difference arises. The overall potential of an electrochemical cell is the sum of all the potential jumps existing between different phases of the cell (Figure 1B-35). The potential of a single electrode with respect to the surrounding

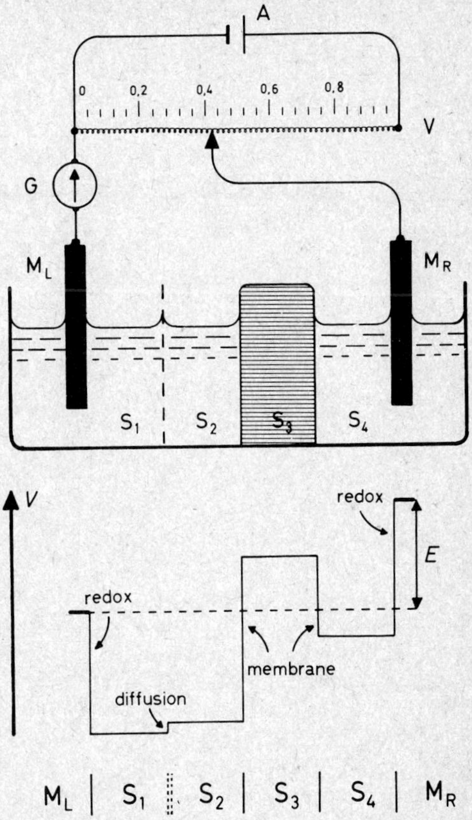

Figure 1B-35. Schematic diagram of an electrochemical cell. M_L and M_R are two metallic conductors. S_1, S_2, S_3, and S_4 symbolize a variable number of *ion conductive* phases. S_1 may be a saturated KCl solution, S_2 a given test solution, S_3 an ion-selective membrane (which need not necessarily be a thin membrane), and S_4 a given reference solution. The liquid-liquid junction between S_1 and S_2 may be an open contact between the two solutions, or a porous membrane, or a fiber junction. The two electrodes (M_L and M_R) are connected externally via a potentiometer, e.g., a null-point potentiometer, consisting of a voltage source (A), a voltage divider (V), and a sensitive galvanometer (G). The voltage divider is adjusted so that the galvanometer shows zero current. The EMF of the cell is then read directly on the calibrated scale.

The lower half of the figure illustrates the potential jumps at the phase boundaries caused by (1) redox potentials, (2) membrane potentials, and (3) diffusion potentials (liquid-liquid junction potentials). At the bottom, the cell is written in symbols, vertical lines indicating phase boundaries, and a double dotted line indicating a liquid-liquid junction.

electrolyte and the absolute magnitude of the individual potential jumps between the phases are actually unknown and cannot be measured. We can measure only the *potential differences* between two electrodes (half-cells). The potential jumps can be classified as (1) redox potentials, (2) membrane potentials, and (3) diffusion potentials. Generally, it is possible to devise a cell in such a manner that all the potential jumps except one are constant. This potential can then be related to the *activity* of some specific ion of interest (H^+, Na^+, and so forth).

MEASUREMENT OF ION ACTIVITY VERSUS ION CONCENTRATION

Most methods, such as flame photometry, atomic absorption spectrometry, and titration, provide the concentration of total ion. Direct potentiometric measurement on the undiluted sample provides a means of determining the concentration of *free unbound ion*, or, more precisely, the *activity of the ion*. This is the relevant quantity for most purposes because chemical equilibria and biological phenomena are dependent on the activity of the ions rather than on the concentration of total ion. But for some purposes, e.g., for balance studies where the input and output of the total amount of substance are of interest, the concentration of *total ion* is more relevant.

The relationships between activity and concentration of free ion and between concentration of free ion and total ion are given in the following.

The *relative activity* (a) equals the numerical value of the *molality* of free ion ($m/(\text{mol}/\text{kg})$)* times the *activity coefficient* (γ), e.g., for Ca^{2+}:

$$a_{Ca^{2+}} = \gamma_{Ca^{2+}} \cdot m_{Ca^{2+}}/(\text{mol}/\text{kg}) \tag{1}$$

*Notice that the symbol m may be employed for *molality* as well as for *mass*. In order to distinguish these, the symbol for mass may be printed as an italic boldface **m**.

Notice that the symbol of a quantity (e.g., mCa^{2+}) represents the numerical value times the unit (e.g., $mCa^{2+} = 1.25 \cdot 10^{-3}$ mol/kg). Both the activity and the activity coefficient are dimensionless quantities.

The activity coefficient depends on the *ionic strength* (I) of the solution $(I = \frac{1}{2}\Sigma m \cdot z^2$, where z is the charge number of the ions) and generally decreases with increasing ionic strength. For dilute aqueous solutions it is possible to calculate the activity coefficients from the ionic strength by means of the Debye-Hückel equation:

$$-\log_{10}\gamma = \frac{z^2 \cdot A \cdot I^{1/2}}{1 + \mathring{a} \cdot B \cdot I^{1/2}} \qquad (2)$$

where $\mathring{a}$ is the mean ion size parameter for the solution [e.g., H^+ (0.9), Na^+ (0.45), K^+ (0.3), Ca^{2+} (0.6), Cl^- (0.3), HCO_3^- (0.45), unit: nm], and A and B are temperature dependent constants [at 37 °C: $A = 0.5213$ (mol $\cdot$ kg^{-1})$^{-1/2}$, $B = 3.305$ nm^{-1} $\cdot$ (mol $\cdot$ kg^{-1})$^{-1/2}$]. At higher ionic strengths $(I > 0.1$ mol/kg) the equation should be extended by two terms that take the hydration of the ions into account (according to the Stokes-Robinson hydration theory).[22]

The equation shows that the decrease in the activity coefficient with increasing ionic strength is more pronounced the higher the charge number (z) of the ion, and hence activity coefficients are generally much lower for divalent than for monovalent ions. For most biological fluids it is difficult to calculate the activity coefficients accurately because the contribution of the protein ions to the ionic strength is highly uncertain. Examples of estimated activity coefficients for normal human blood plasma are Na^+ (0.75), K^+ (0.74), Ca^{2+} (0.31).

The *amount-of-substance concentration* (c), often called the molarity, equals the molality (m) times the mass concentration of water (ρH_2O), e.g., for Ca^{2+}:

$$cCa^{2+} = mCa^{2+} \cdot \rho H_2O \qquad (3)$$

For normal human blood plasma the mass concentration of water is about 0.93 kg/L, but in pathological specimens with lipemia or severe hyperproteinemia the value may be as low as 0.8 kg/L; hence the substance concentration (c) may be 20% lower than the molality (m).

The *concentration of free ion* equals the concentration of total ion times a factor (α) which may be called the degree of dissociation, e.g., for Ca^{2+}:

$$cCa^{2+} = \alpha Ca^{2+} \cdot ctCa \qquad (4)$$

The degree of dissociation is very different for different ions. For Ca^{2+} in normal blood plasma the value is about 0.5, but the value varies with the albumin concentration (decreasing with increasing albumin concentration) and with pH (decreasing with increasing pH). Na^+ and K^+ are almost completely dissociated in normal plasma, i.e., only slightly bound, the value for α being about 0.99 for Na^+ and 0.96 for K^+.

Results of potentiometric measurements on undiluted whole blood, plasma, or serum are usually reported not as activity but as the activity multiplied by an appropriate factor (f), so that the normal reference interval becomes the same as the conventional normal reference interval for substance concentration. For Na^+ the factor may be derived theoretically from the values for ρH_2O, γNa^+, and αNa^+ of normal plasma:

$$f Na^+ = \frac{\rho H_2O}{\gamma Na^+ \cdot \alpha Na^+} \cdot (mol/kg) = \frac{0.93 \cdot kg/L}{0.75 \cdot 0.99} \cdot (mol/kg) = 1.25 \cdot 10^3 \text{ mmol/L}$$

i.e., if the activity is measured as $112 \cdot 10^{-3}$, the result is reported as $112 \cdot 10^{-3} \cdot 1.25 \cdot 10^3$ mmol/L $= 140$ mmol/L. For K^+ the factor is derived similarly, i.e., $fK^+ = (0.93/(0.74 \cdot 0.96))$ $\cdot 10^3$ mmol/L. For Ca^{2+} the factor is $fCa^{2+} = (\rho H_2O/\gamma Ca^{2+}) \cdot (mol/kg) = (0.93/0.31) \cdot 10^3$ mmol/L $= 3.0 \cdot 10^3$ mmol/L; i.e., fCa^{2+} does not include αCa^{2+} because the activity is converted to the concentration of *free* Ca^{2+}, not the concentration of *total* calcium.

With such corrections, major discrepancies between the results obtained by flame photometry and by direct potentiometry for Na^+ and K^+ will occur only when the mass concentration of water deviates significantly from normal (because pathological variations in γ and α are insignificant in this respect). In severe hyperlipemia where the mass concentration of water may be as low as 0.80 kg/L, the result for Na^+ by flame photometry may be 120 mmol/L while direct potentiometry gives 140 mmol/L. The flame photometric value may be misinterpreted as indicating Na^+ deficiency or water intoxication or both. The potentiometric result shows that the Na^+ activity is normal.

In general, *activity determinations* are based on the comparison of the potential of the unknown solution with the potential of several calibration solutions with known activity. The best example is the pH determination, as illustrated in the following equation:

$$pH(X) = pH(S_1) + \frac{pH(S_2) - pH(S_1)}{E(S_2) - E(S_1)} \times [E(X) - E(S_1)] \tag{5}$$

where $E(S_1)$ and $E(S_2)$ are the readings for two different calibration solutions (National Bureau of Standards buffers); pH(X) and $E(X)$ represent the pH and the reading, respectively, for the unknown.

The *precision* of potentiometric measurements is limited by the standard deviation for the measured potentials (about 50 μV). This uncertainty is equivalent to a standard deviation of about 0.001 for log a for a monovalent ion. This corresponds to a coefficient of variation for the activity of a monovalent ion of about 0.2%. For divalent ions, the coefficient of variation will be about twice as high. The *accuracy* of direct potentiometric measurement is limited by the uncertain liquid junction potential, which may vary by about ± 0.5 mV with the composition of the serum. The variation corresponds to a bias in the measured concentration of about $\pm 2\%$ for monovalent ions.

If potentiometric measurements are employed for *the determination of the concentration of total ion* (free + bound), the sample must be diluted by a suitable diluent that liberates the complex-bound ion from its binding agent (e.g., by a pH adjustment). The dilution should at the same time serve to establish a constant ionic strength so that a constant activity coefficient is obtained independently of variations in ionic strength of the original sample. This method is sometimes called *indirect potentiometry* as opposed to *direct potentiometry*, which involves direct measurement in the undiluted sample. A different approach consists of *titration* utilizing a potentiometric end-point detection. In this case, the electrode is used only to sense the sudden change in activity as the end point is reached. This technique is generally considered among the most accurate and precise analytical methods available.

REDOX ELECTRODES

Redox potentials are due to chemical equilibria involving electron transfer reactions:

$$\text{Oxidized form (Ox)} + ze^- \rightleftharpoons \text{Reduced form (Red)} \tag{6}$$

e.g.,

$$Fe^{3+} + e^- \rightleftharpoons Fe^{2+}$$

or

$$2H^+ + 2e^- \rightleftharpoons H_2$$

where z symbolizes the number of electrons involved in the reaction (the numerical stoichiometric number). Any substance which binds (accepts) electrons is an oxidant (Ox) and a substance which gives off electrons is a reductant (Red). The two forms, Ox and Red, represent a redox couple (conjugate redox pair). Usually, redox processes take place only between two redox couples, the electrons being transferred from a reductant (Red_1) to an oxidant (Ox_2). In this process, Red_1 is oxidized to its conjugate Ox_1, while Ox_2 is reduced to Red_2:

$$Red_1 + Ox_2 \rightleftharpoons Ox_1 + Red_2 \tag{7}$$

e.g.,

$$2S_2O_3^{2-} + I_2 \rightleftharpoons S_4O_6^{2-} + 2I^-$$

In an electrochemical cell, electrons may be accepted from, or donated to, an inert metallic conductor (e.g., platinum). A reduction process tends to charge the electrode positively (remove electrons), and an oxidation process tends to charge the electrode negatively. By convention, redox equilibrium (equation 7) is represented by the cell

$$M_L \quad Red_1 - Ox_1 \parallel Ox_2 - Red_2 \mid M_R \tag{8}$$

A positive potential ($E > 0$) for cell (8) signifies that the cell reaction (7) proceeds spontaneously from left to right; $E < 0$ signifies that the reaction proceeds from right to left; and $E = 0$ indicates that the two redox couples are at mutual equilibrium.

The *electrode potential* (reduction potential) for a redox couple is defined as its potential measured with respect to the standard hydrogen electrode which is set equal to zero (see later). This potential, by convention, is the EMF of a cell, where the standard hydrogen electrode is the reference electrode (left electrode) and the given half-cell is the indicator electrode (right electrode). The reduction potential for a given redox couple is given by the Nernst equation (for derivation, see textbooks of physical chemistry):

$$E = E° - \frac{N}{z} \cdot \log \frac{a\text{Red}}{a\text{Ox}} = E° - \frac{0.0592 \text{ V}}{z} \cdot \log \frac{a\text{Red}}{a\text{Ox}} \tag{9}$$

where

E = electrode potential of the half-cell
$E°$ = standard electrode potential when $a\text{Red}/a\,\text{Ox} = 1$
z = number of electrons involved in the reduction reaction
R = gas constant (= $8.3143 \text{ J} \cdot \text{K}^{-1} \cdot \text{mol}^{-1}$)
T = absolute temperature (unit: K, kelvin)
F = Faraday constant (= $96487 \text{ C} \cdot \text{mol}^{-1}$), (C = coulomb)
N = $R \cdot T \cdot \ln 10/F$ (= the Nernst factor if $z = 1$)*
  N = 0.0592 V if $T = 298.15 \text{ K}$ (= 25 °C)
  N = 0.0615 V if $T = 310.15 \text{ K}$ (= 37 °C)
$\ln 10$ = natural logarithm of $10 = 2.303$
a = activity.
$a\text{Red}/a\text{Ox}$ = product of mass action for the reduction reaction.

The redox electrodes presently in use can be divided into two major groups or classes, namely, (1) inert metal electrodes immersed in solutions containing redox couples, (2) metal electrodes where the metal functions as a member of the redox couple.

Inert Metal Electrodes

Platinum and gold are examples of inert metals used to record the redox potential of a redox couple dissolved in an electrolyte solution.†

Not all dissolved redox couples, however, are able to equilibrate with an inert metal like platinum or gold. In many cases the presence of a catalyst is required to establish a reproducible potential. The catalyst can be a small amount of another redox couple, a so-called mediator, which readily equilibrates with both the metal electrode and the more sluggish redox couple of interest. Examples of mediators are methylene blue and quinhydrone. Measurement of the redox

---

*N should not be mistaken for the Avogadro constant (formerly also called the Lochschmidt constant).

†Attempts have been made to measure the redox potential of blood or plasma. However, this potential is an undefined quantity as long as the redox couple is not specified. The different redox couples of blood or plasma are not in thermodynamic equilibrium, in contrast to the equilibrium between the different acid-base pairs. If it were not for this disequilibrium, life would be impossible because all organic substances would rapidly burn to CO_2 and H_2O. The redox potential in blood measured with a gold electrode (using a calomel reference electrode) appears to be the redox potential of the ascorbic acid/dehydroascorbic acid couple, which is of much less clinical interest than that of redox couples such as $NADH/NAD^+$.

potential of dissolved couples may be used for end-point detection in redox titrations, but this procedure rarely finds application in clinical chemistry.

The *hydrogen electrode* is a special redox electrode for pH measurement. It consists of a platinum or gold electrode which is coated with highly porous platinum (platinized) to catalyze the electrode reaction

$$H^+ + e^- \rightleftharpoons 1/2 \ H_2$$

The electrode potential is given by

$$E = E° - N \cdot \log \frac{(aH_2)^{1/2}}{aH^+}$$

or

$$E = E° - N \cdot [\log(aH_2)^{1/2} - \log aH^+]$$

where

$$E° = 0 \text{ at all temperatures (by convention)}$$
$$aH^+ = \text{activity of hydrogen ions}$$
$$-\log aH^+ = \text{negative log of the } H^+ \text{ activity } (paH^+ \text{ or pH})$$

When the partial pressure of H_2 in the solution (and hence aH_2) is maintained constant by bubbling H_2 through the solution, the potential is a linear function of $\log aH^+$ ($= -pH$). In the *standard hydrogen electrode* the electrolyte consists of an aqueous solution of HCl with $aHCl = 1.000$ (or $cHCl \approx 1.2$ mol/L) in equilibrium with a gas phase with $aH_2 = 1.000$ (or $pH_2 = 101.3$ kPa $= 1$ atm).

The *quinhydrone electrode* is another special redox electrode used for pH measurement. It consists of a platinum or gold electrode immersed in a saturated solution of quinhydrone, which is an equimolar mixture of quinone (Q) and hydroquinone (H_2Q). The electrode reaction is

The electrode potential is given by

$$E = E° - N \cdot \log \frac{(aH_2Q)^{1/2}}{(aQ)^{1/2} \cdot aH^+}$$

and because $aH_2Q = aQ$, the electrode potential varies linearly with the negative log aH^+ or pH. The hydrogen electrode and the quinhydrone electrode have both been replaced by the glass electrode for pH measurement, although the use of the quinhydrone electrode has been proposed in connection with a pCO_2 electrode (see later).

Metal Electrodes Participating in Redox Reactions

The *silver electrode* consists of a silver wire immersed in a solution containing silver ions. The electrode process consists of the reduction of silver ions to metallic silver:

$$Ag^+ + e^- \rightleftharpoons Ag$$

The expression for the electrode potential reduces to:

$$E = E° + N \cdot \log aAg^+$$

because the activity of pure silver is unity. According to this equation, the electrode measures the silver ion activity in the solution. An application for this electrode is the determination of chloride by titration with $AgNO_3$. In this method the silver electrode is coupled with a mercurous sulfate reference electrode ($\frac{1}{2}Hg_2^{2+} + e^- \rightleftharpoons Hg$) for the end point determination. When all Cl^- ions are precipitated as $AgCl$, the excess of Ag^+ causes a sudden change in the electrode potential, which indicates the end point of the reaction.

In an analogous fashion, other metal electrodes (e.g., Zn, Cu) measure the activity of their respective ions. When the metal is chemically unstable in pure form, an electrode consisting of an amalgam of the metal (e.g., calcium amalgam) can sometimes be used. Unfortunately, such electrodes are of little value for direct measurements in biological fluids because proteins and lipids tend to coat the amalgam surface, and various redox couples present in biological fluids tend to affect the electrode potential.

The *silver/silver chloride electrode* consists of a silver wire, electrolytically coated with $AgCl$, which dips into a solution containing chloride ions. The electrode process is

$$AgCl \text{ (solid)} + e^- \rightleftharpoons Ag \text{ (metal)} + Cl^-$$

Since $aAgCl$ and aAg are both unity (because both components are present as pure substance), the expression for the electrode potential reduces to

$$E = E° - N \cdot \log aCl^-$$

which shows that the electrode measures chloride activity. This electrode has been utilized for direct measurement of the chloride activity in serum. By placing it directly on the skin surface, it can also be used for the measurement of the chloride activity of sweat in connection with the diagnosis of cystic fibrosis.

When the chloride activity is kept constant, e.g., with $cKCl$ fixed at 0.1 mol/L, the electrode potential is constant, and therefore the electrode is frequently employed as a reference electrode or "inner electrode" in membrane electrodes (see Figure 1B-37).

In complete analogy, a silver/silver bromide electrode, a silver/silver iodide electrode, and a silver/silver sulfide electrode measure bromide activity, iodide activity, and sulfide activity, respectively. Unfortunately, these electrodes are quite sensitive to other redox couples in the test solution and therefore often fail in biological solutions.

The *calomel electrode* consists of mercury covered by a layer of calomel (Hg_2Cl_2), which is in contact with an electrolyte solution containing chloride. The electrode process is

$$\frac{1}{2}Hg_2Cl_2 \text{ (solid)} + e^- \rightleftharpoons Hg \text{ (metal)} + Cl^-$$

Since aHg and aHg_2Cl_2 are both unity (present as pure substance), the electrode potential reduces to

$$E = E° - N \cdot \log aCl^-$$

This shows that the calomel electrode also functions as a chloride electrode, and its potential varies with the chloride activity. The chloride activity is generally maintained constant, either as saturated KCl or at a concentration of 3.5 (or 4.0) mol/L. Calomel electrodes are frequently employed as reference electrodes together with glass electrodes for pH measurement.

ION-SELECTIVE MEMBRANE ELECTRODES[3,5,6,11,13]

Membrane potentials are due to permeability of certain types of membranes to selected anions or cations. Biological membranes (e.g., the glomerular membrane in the kidney) are often impermeable to the high-molecular protein ions; this fact gives rise to the Donnan potential, which in turn leads to an uneven distribution of the diffusible ions on both sides of the membrane. For analytical applications, membranes are required which possess a selective permeability for a single ion species.

The ion-selective membrane, separating the solution on the left side (1) from the solution on the right side (2), can be illustrated as

$$1 \mid \text{Membrane} \mid 2 \tag{10}$$

The membrane potential is conventionally defined as the potential on the right side minus the potential on the left. The potentials of solutions 1 and 2 are recorded with any suitable reference electrode. In the case of a glass membrane, the potential of the reference solution (2) is generally recorded by means of an Ag/AgCl electrode, while the potential of the test solution (1) is recorded by means of a calomel electrode via a liquid-liquid junction between the saturated KCl and the test solution.

If the activity of a diffusible cation is higher in 1 than in 2, a positive membrane potential develops. In most cases the mechanism is as follows: on the left side, where the cation activity is high, cations are bound to the membrane surface by specific binding groups and the membrane is thereby charged positively. This charging of the membrane causes a dissociation of cations from the other side of the membrane into the solution on the right side, which is thereby charged positively with respect to the solution on the left side. Thus, it appears as if the membrane were permeable to the cation only. This process proceeds until an equilibrium is established, which occurs when the electric potential difference across the membrane matches the difference in activity of the diffusible ions on the two sides of the membrane. Attempts to force the ion through the membrane by an electric current often damage the membrane, or at least make it behave differently from when it is at zero current.

The membrane potential is given by the Nernst equation (for derivation see textbooks of physical chemistry):

$$V = -\frac{R \cdot T}{z \cdot F} \cdot \ln \frac{a(2)}{a(1)} \tag{11}$$

where z is the ion charge number (positive for cations, negative for anions), and a is the activity of the diffusible ion (cation or anion). If 2 is a reference solution with constant activity of the diffusible ion [constant $a(2)$], the equation reduces to

$$V = V' + \frac{N}{z} \cdot \log a(1) \tag{12}$$

where V' is constant. In other words, the *membrane potential is directly proportional to the logarithm of the activity of the diffusible ion in the test solution* (Figure 1B-36).

The (theoretical) *sensitivity* of the electrode, $\Delta V / \Delta \log a$, is given by the Nernst slope, N/z. However, the actual sensitivity often deviates slightly, being $s \cdot N/z$, where s is the *relative sensitivity*. For most pH-glass electrodes s is 0.98–1.00. For other ion-selective electrodes s may be as low as 0.90. Even lower values indicate poor electrode performance due to malfunction or poor selectivity. It should be noted that the theoretical sensitivity for divalent ions is only half the value for monovalent ions.

The *selectivity* of the electrode for a single ion species is seldom absolute. If the membrane is "permeable" to several different ions, a diffusion potential may arise, and the theory for the membrane potential becomes more complicated. The following empirical equation (Nicolsky) describes the measurement potential as a function of the activity of the primary ion (1) and an interfering ion (2):

$$V = V' + N \cdot \log (a^{1/z_1} + k_{1,2}^{\text{pot}} \cdot a_2^{1/z_2}) \tag{13}$$

$k_{1,2}^{\text{pot}}$ is the selectivity coefficient which equals the permeability ratio P_2/P_1. The lower the value of $k_{1,2}^{\text{pot}}$, the smaller the interference of 2 when measuring 1. The value for the selectivity coefficient depends to some extent on the method of measurement. One method is based on measurements in solutions containing one of the ions only, and finding the activity of the interfering ion that gives the same electrode potential as a given activity of the primary ion. In this case, $k_{1,2}^{\text{pot}} = a_1/a_2$ (provided $z_1 = z_2$). Better methods are based on measuring $k_{1,2}^{\text{pot}}$ in solutions containing both ions.

The *design of membrane electrodes* is as multifaceted as the human imagination. Electrodes may be bulb-shaped, inverted bulb-shaped, capillary type, needle type, and flow-through. Many

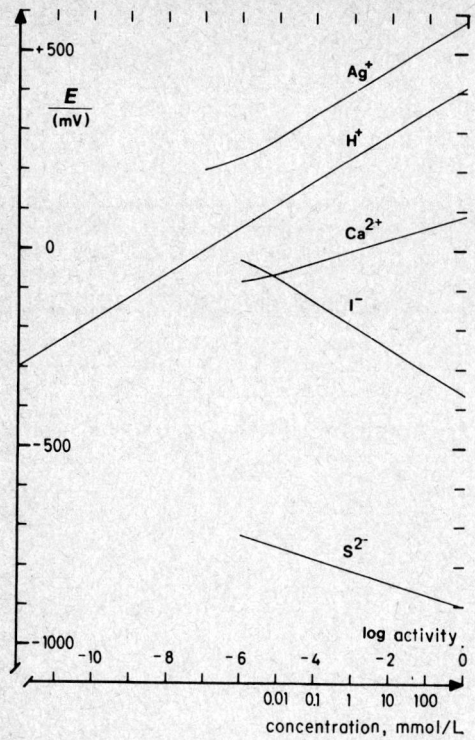

Figure 1B-36. Illustration of the Nernst equation, i.e., the relationship between electromotive force of an ion-selective electrode chain (E) and the logarithm of the activity of the ion (log $\alpha l'$):

$$E = E' + \frac{R \cdot T \cdot \ln 10}{z \cdot F} \cdot \log \alpha l'$$

where E' is constant at constant temperature (dependent upon the type of reference electrodes), and z is the charge number for the ion (positive for cations, negative for anions). The slope for the monovalent ions is ± 59.16 mV; for divalent ions it is ± 29.58 mV ($T = 298.15$ K). The abscissa also indicates the concentration of free ion provided the activity coefficient is taken to be 1.

attempts are being made to miniaturize electrodes for measurement in the blood stream (catheter-tip electrodes) or measurements in single cells (glass capillary microelectrodes).[23]

A special modification of design is the elimination of the "inner" reference solution (symbolized by S_4 in Figure 1B-35) and instead applying the ion-selective membrane as a coating directly on the metallic conductor. These *coated-wire ion-selective electrodes* are especially suitable for miniaturization,[6] but they often suffer from electrochemical instability of the inner reference potential, yielding less reliable results.

Another technical development is the use of field effect transistors (FET) where the gate insulator (often Si_3N_4) is covered by an ion-selective membrane producing an *ion-selective field-effect transistor* (ISFET).[6] In this way an integration of ion-selective membrane and solid state amplifier is accomplished. The main advantage is the possibility of miniaturization, since, in theory, 5–10 different ion-selective membranes may be combined on one integrated circuit within an area of ~ 1 mm². Amplification of the signal in situ obviates the need for heavy shielded leads, so that the probe can be made flexible, an important feature of catheter-tip electrodes.

Available ion-selective membrane electrodes can be classified arbitrarily as (1) glass electrodes, (2) solid-state electrodes, and (3) liquid ion-exchange electrodes.[3,13] By means of additional membranes these electrodes may be modified into (4) gas electrodes or (5) enzyme electrodes.

Glass Electrodes[4]

Glass electrodes are made from specially formulated glass consisting of a melt of SiO_2 with added oxides of various metals. Membranes with varying composition of the glass have been prepared with directed selectivity for H^+, Na^+, K^+, Li^+, Rb^+, Cs^+, Ag^+, Tl^+, and NH_4^+.[5] The membranes generally have a thickness of 10–100 μm. Depending on the type of glass, their electrical resistance is very high, being about 10–800 MΩ at room temperature, increasing considerably with decrease in temperature.

Hydrogen-ion selective glass electrodes (pH electrodes) can be manufactured from the classic Corning 015 glass, consisting of SiO_2, Na_2O, and CaO in the molar ratio of 72.2:21.4:6.4. A newer glass more selective to H^+ consists of SiO_2, Li_2O, and CaO in the ratio 68:25:7, but many other compositions have also been found suitable. With the older glass, significant error in measurement arose in the presence of Na^+ at pH values above 8.0. The sodium error, which is due to a loss in specificity of the electrode, increases greatly with temperature. This error is greatly reduced with newer types of glass which allow pH measurements in NaOH solutions up to about pH 13 with an error of less than 0.02. This corresponds to a selectivity coefficient $k_{H,Na}^{pot}$ of about 10^{-15} at room temperature. The glass membrane of the electrode may be shaped according to the requirements of the application. It is bulb-shaped for most titration purposes, and flat for surface measurements, but it has an inverted bulb-shape for microanalysis. For pH measurements in blood, the thermostatted capillary glass electrode has proved invaluable (Figure 1B-37).

Na+-selective glass electrodes can be prepared from glass consisting of SiO_2, Na_2O, and Al_2O_3 in the ratio of 71:11:18. Lithium aluminum silicates have also been found suitable. The selectivity coefficient $k_{Na,K}^{pot}$ may be as low as 10^{-3}, and the electrode is insensitive to H^+ in the pH range 6–10. Electrodes with a flat surface have been used for the direct measurement of sodium ion activity on the skin surface for the diagnosis of cystic fibrosis. Capillary electrodes have been constructed for measurements of Na^+ in serum or serum dilutions.

K+-selective glass electrodes are as yet less satisfactory than are the Na^+ electrodes, $k_{K,Na}^{pot}$ being about 0.05. Attempts have been made to measure H^+, Na^+, and K^+ in serum simultaneously with three different glass electrodes, utilizing a computer for calculation of the correction for the different electrodes on the basis of the different selectivity constants. Glass electrodes selective to Li^+ and NH_4^+ are also available.

Solid-State Electrodes

Solid-state membranes can be either homogeneous membranes consisting of a "single" crystal, or heterogeneous membranes consisting of an active substance embedded in an inert matrix.[15,18]

The homogeneous-membrane electrodes include those for F^- (lanthanum fluoride crystal), Cl^- (AgCl crystal), Br^- (AgBr crystal), I^- (AgI crystal), S^{2-} (Ag_2S), Cu^{2+} (cupric selenide crystal), and others. The silver salt solid-state membrane electrodes are less susceptible to interference from redox systems than are the equivalent silver/silver salt redox electrodes. The AgCl membrane electrode is used for measurement of the activity of chloride in sweat by direct measurement on the skin surface (Orion electrode; Orion Co., Boston, Mass.). The fluoride electrode has been utilized for measurements of the fluoride concentration in saliva.

Ion-Exchange Electrodes[11]

Liquid ion-exchange membranes consist of an *inert solvent* in which an *ion-selective carrier* substance is dissolved. Both solvent and carrier should be insoluble in water. The membrane solution can be separated from the test solution by means of a collodion membrane, or a porous

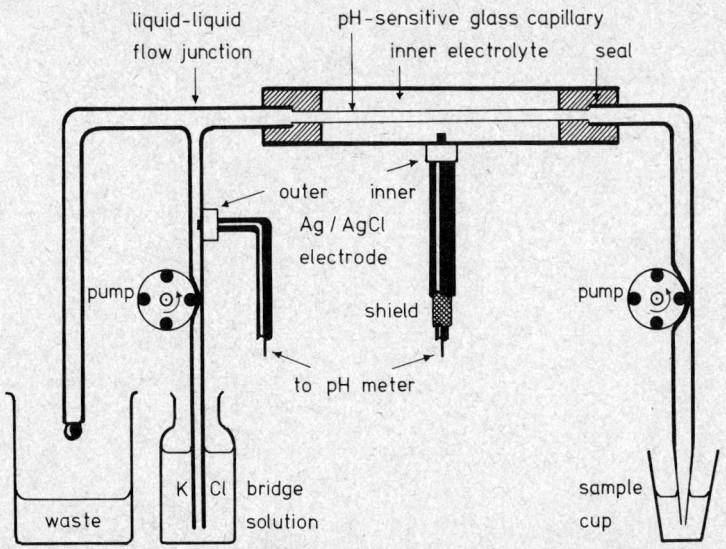

Figure 1B-37. Schematic illustration of a capillary glass electrode for pH measurement in a flow system. The sample and the salt bridge solution are simultaneously pumped through the tubings; where they meet an open liquid-liquid flow junction is established.

matrix can be soaked by the membrane solution. The inert solvent and the ion-selective carrier are often embedded in a matrix of polyvinylchloride (PVC) obtained when solutions of PVC in tetrahydrofurane are evaporated into thin semisolid membranes. Many different membranes have been prepared with selectivity for specific cations or anions, e.g. K^+, NH_4^+, Ca^{2+}, and recently also for H^+.

Potassium-selective membranes can be made by dissolving the antibiotic valinomycin in a suitable solvent.[14] Valinomycin is a neutral carrier which binds K^+ in the center of a ring of oxygen atoms (Figure 1B-38). This membrane is highly selective for potassium, the selectivity ratio $k_{K,Na}^{pot}$ being 2.5×10^{-4}. The electrode is widely used for measurement of potassium ions in serum.

The ammonium-ion-selective membrane is based on a mixture of the antibiotics nonactin and monactin, which are also neutral carriers.

A calcium-selective membrane can be made by dissolving the calcium salt Ca^{2+}-bis(di-*p*-octylphenyl phosphate) in PVC.[17,18] Two di-*p*-octylphenyl phosphate ions bind a Ca^{2+} ion to act as the Ca^{2+} carrier:

Dioctylphenyl phosphonate is added as a plasticizer to the PVC:

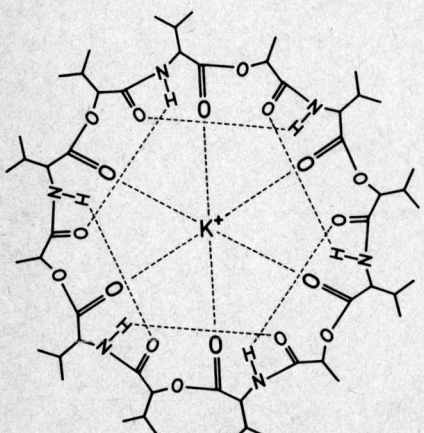

Figure 1B-38. Valinomycin has a heterocyclic structure consisting of alternating peptide and ester linkages. The molecule is cylindrical, stabilized by six hydrogen bridges (from —NH to C=O). The outer surface of the cylinder is strongly hydrophobic, and the interior is strongly hydrophilic. The size of the central cavity nearly equals the diameter of an unhydrated potassium ion.

A neutral carrier has also been described:[20]

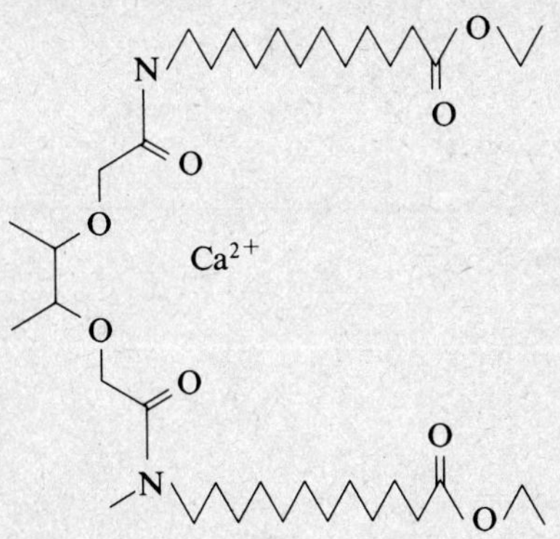

It binds the Ca^{2+} ion in a neutral pocket in a manner similar to the binding of K^+ by valinomycin.

Protein Interference

A protein coating on the ion-selective membrane may cause a lack of reproducibility when measurements are performed alternately in protein-free and protein-containing solutions. This source of bias may be avoided by covering the ion-selective membrane by a dialysis membrane. The ion-selective membrane then measures the activity of the ion in a thin layer of protein-free dialysate under the dialysis membrane. The activity of the ion is not the same in this dialysate as in the original sample because the nondiffusible protein ions cause an uneven distribution of the diffusible ions (Donnan distribution). However, this apparent bias is exactly balanced by the Donnan potential across the dialysis membrane, and this Donnan potential is included in the total cell potential. A disadvantage of using a cellophane membrane is a significant prolongation of the response time of the electrode from less than a second to > 20 s for 98% response.

Gas Electrodes

Gas electrodes are specially designed for the measurement of specific gases in gas mixtures or in solutions. Examples are the CO_2 electrode and the NH_3 electrode, which are both based on potentiometric measurement using a pH-glass electrode. The O_2 electrode is based on amperometric measurement using a polarized platinum cathode (see below).

Usually the gas electrode is separated from the test solution by means of a thin gas-permeable membrane (e.g., polyethylene, polypropylene, Teflon, silicone rubber), but the separation may also simply be a small "air gap."

The *pCO₂ electrode* (or more correctly, the pCO_2 cell) represents a special application of a pH glass electrode (Figure 1B-39). The sample in this case is in contact with a membrane that is permeable to gas but not to solutions. This membrane (e.g., silicone rubber) is separated from the actual glass electrode by a thin film of bicarbonate solution (5 mmol/L). The CO_2 gas diffuses from the sample (or test gas) through the membrane, rapidly enters into equilibrium with the bicarbonate solution, and thus alters its pH. The pH of the bicarbonate solution is a simple function of the pCO_2 obtained by rearrangement of the Henderson-Hasselbalch equation:

$$pH = -s \cdot \log pCO_2 - \log \alpha + pK' + \log cHCO_3^-$$

where s is the relative sensitivity of the electrode, normally 0.95–1.00, α is the solubility coefficient of CO_2 in the bicarbonate solution, K' is the apparent, overall, first dissociation constant of carbonic acid.

The pCO_2 electrode has been used extensively for measurements in arterial blood samples. More recently the application has been extended to *transcutaneous measurements.*[9,19] Good correlation has been demonstrated between the transcutaneous pCO_2 and the arterial pCO_2, especially when the transcutaneous electrode and the skin surface are heated by a small built-in heating unit to a temperature between 42 and 44 °C.

The *NH₃ gas electrode* is similar to the CO_2 electrode except that the bicarbonate solution is replaced by an ammonium chloride solution. Again the measured pH of the ammonium chloride

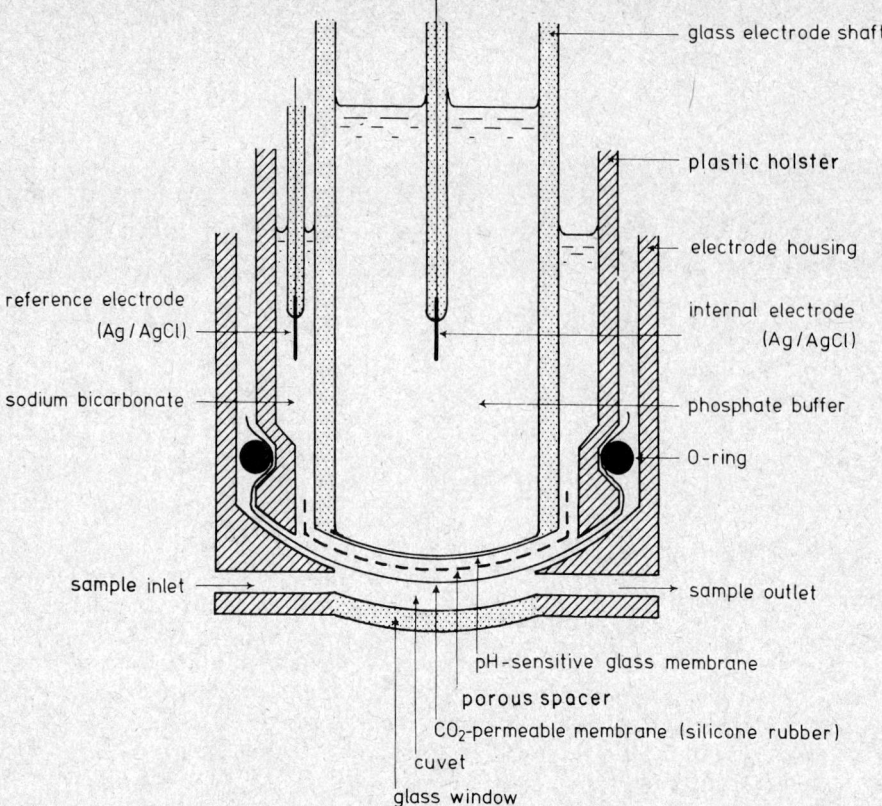

Figure 1B-39. Schematic illustration of a pCO$_2$ electrode.[21]

solution behind the gas-permeable membrane varies linearly with the logarithm of the partial pressure of NH$_3$.

Enzyme Electrodes[7]

The various electrodes may be covered by a layer of immobilized enzymes that catalyze a chemical reaction that can be monitored by the electrode. This principle may be applied to redox electrodes, ion-selective membrane electrodes, and amperometric electrodes (see below).

The response of enzyme electrodes is more complex because it is dependent on the rate of diffusion of substrates into the enzyme layer, the rate of the enzyme-catalyzed reaction, and the rate of back diffusion of the reaction products. The enzyme electrodes therefore do not measure a thermodynamic equilibrium and the sensitivity is not necessarily Nernstian.

The enzymes may be immobilized on the electrode surface by simple adsorption, by entrapment behind a dialysis membrane, or by covalent linkage.

Urea electrodes: urea from the sample diffuses to the membrane consisting of a polyacrylamide matrix containing immobilized urease fixed on a Dacron net. The urea is then hydrolyzed to NH$_4^+$, which is measured by a NH$_4^+$-selective membrane electrode, or the rise in pNH$_3$ is sensed by a NH$_3$ gas electrode (through a gas-permeable membrane or via an air gap), or the rise in pH is sensed with a pH glass electrode.

A *glucose electrode,* based on the use of glucose oxidase in combination with a pH electrode for detection of gluconic acid formed in the enzyme-catalyzed reaction, was the first enzyme electrode proposed. However, amperometric glucose electrodes (see later) have proved to be superior.

A *creatinine electrode* using creatinine-deiminase and a NH$_3$ gas electrode has been described. The response time is 6–10 min and the detection limit about 70 μmol/L.

A *lactate electrode* has been developed based on the following reaction catalyzed by immobilized lactate dehydrogenase:

$$\text{Lactate} + 2\,\text{Fe(CN)}_6^{3-} \rightarrow \text{pyruvate} + 2\,\text{Fe(CN)}_6^{4-} + 2\,\text{H}^+$$

The change in the redox potential of the ferricyanide-ferrocyanide redox couple is sensed by a redox electrode. Unfortunately, the response time is as long as 20–30 min.

In conclusion, it should be noted that enzyme electrodes are much less durable than H^+, K^+, or Ca^{2+} electrodes as immobilized enzymes gradually lose their activity due to surface contamination and aging.

LIQUID-LIQUID JUNCTION POTENTIALS

The variable and generally unknown liquid-liquid junction potential constitutes the most important source of bias in potentiometric measurements of ion activities. Diffusion potentials arise at a liquid-liquid junction $(1 \mid\mid 2)$ where two solutions of different ionic composition are in direct contact. If cations diffuse faster from left to right than anions, a positive liquid junction potential develops at the interface. Diffusion potentials are due to irreversible processes and are therefore more difficult to calculate than the true equilibrium potentials, but they can often be calculated with good approximation by means of the Henderson equation. Generally, the aim is to reduce the liquid junction potential to a value as low and as reproducible as possible. This is achieved by using a concentrated KCl solution as the bridge solution (3.5 mol/L, or saturated = 4.52 mol/L at 37 °C). The potential is then dominated by the large excess of K^+ and Cl^- diffusing at almost the same rate into the test solution, and the potential becomes largely independent of the composition of the test solution.

As an example, the following values may be calculated for the liquid junction potential (E_j) at 37 °C:

KCl (5.32 mol/kg) $\mid\mid$ NBS-phosphate buffer (7.38), $E_j = -3.7$ mV

KCl (5.32 mol/kg) $\mid\mid$ normal plasma, $E_j = -2.5$ mV

The difference between the junction potential with test solution (here plasma) and the calibration solution (here phosphate buffer) is called the *residual liquid junction potential*; in the present example it is $+1.2$ mV. Such a residual liquid junction potential gives a positive bias on the measured hydrogen ion activity of $\approx 5\%$. In order to reduce the residual liquid junction potential, the composition of the calibration solution should match that of the test solution as closely as possible.

Erythrocytes present at the junction with KCl as the bridge solution increase the liquid junction potential ($\approx +0.6$ mV when comparing whole blood and plasma). They do so because precipitated proteins at the liquid junction act as an ion exchanger influencing the diffusion rate of the ions, slowing down Cl^- more than K^+. The effect is reduced to about one third, when a flowing junction is employed. It is also possible to use a different bridge solution. Thus, with sodium formate, 4 mol/kg, as bridge solution, the effect of erythrocytes on the junction potential is negligible, and measurements on whole blood give the same result as measurements on the corresponding plasma.

The potential of the junction: $1 \mid\mid 2$, may be estimated by the Henderson equation:

$$E_j = \frac{R \cdot T}{F} \cdot \frac{f(1) - f(2)}{g(1) - g(2)} \cdot \ln \frac{g(1)}{g(2)}$$

$$f = \Sigma m_i \cdot \lambda_i / z_i \tag{14}$$

$$g = \Sigma m_i \cdot \lambda_i$$

where λ_i is the limiting molar conductance of the ions (unit, mS $\cdot$ m^2/mol), e.g., at 37 °C for K^+ (9.1), Na^+ (6.4), Cl^- (9.6), HCO_3^- (5.9), $HCOO^-$ (6.8), $H_2PO_4^-$ (4.8), HPO_4^{2-} (15.1), Ca^{2+} (14.0). The liquid junction potential is, in principle, inaccessible for direct measurement, and the calculated value is only a rough estimate. Note that *thermal* diffusion potentials may arise at a liquid-liquid junction where two identical solutions of different temperature are in contact. Such thermal diffusion potentials should be avoided by maintaining the whole cell at the same, constant temperature.

POLAROGRAPHY[8]

Polarography is based on the dual measurement of the current flowing through an electrochemical cell and the electrical potential between the two electrodes when the potential is gradually increased at a constant rate by means of an external voltage source.[8] The function between current and potential is called a *polarogram* (Figure 1B-40).

A polarographic cell generally consists of a polarizable indicator electrode and a nonpolarizable reference electrode. The indicator electrode can be a dropping mercury electrode, i.e., a glass capillary tube (50 μm I.D.) filled with mercury that slowly drips in small droplets out of the tip of the tubing into the test solution. In this way the electrode surface is constantly renewed. The reference electrode can be a large pool of mercury in the bottom of the vial.

The cell may be symbolized as follows, where the indicator electrode is the cathode:

$$\text{Hg} \quad | \quad \text{Test solution} \quad | \quad \text{Hg}$$
$$\textbf{(Anode)} \hspace{4.5cm} \textbf{(Cathode)}$$

The cell is symmetrical and therefore shows an electromotive force of zero. When an increasingly negative potential is applied to the cathode at a constant rate, the current at first is almost zero, because the cathode is electrolytically polarized (i.e., the cell becomes asymmetrical and therefore generates a counterelectromotive force that balances the applied potential). If the test solution is oxygen-free water, the electrode processes are

Cathode $\hspace{4cm} H^+ + e^- \rightleftharpoons \tfrac{1}{2}H_2$
Anode $\hspace{4cm} 2Hg \rightleftharpoons Hg_2^{2+} + 2e^-$

Hydrogen is formed at the cathode, which becomes a "hydrogen electrode," and mercury is oxidized to Hg_2^{2+} at the anode. However, due to the large surface of the anode, the concentration changes at the anode are small and insignificant in relation to the concentration changes at the cathode. The anode is therefore considered to be nonpolarized. A calomel electrode or an Ag/AgCl electrode also behaves as a nonpolarizable electrode, because the changes in chloride concentration when a current is flowing are relatively small, and the electrode potential is therefore relatively constant.

The maximal counterelectromotive force that can be obtained is reached when pH_2 at the cathode has reached atmospheric pressure. This is at a potential of about -1.7 V, although the electromotive force of a cell with a hydrogen electrode and a mercury reference electrode is only about -0.8 V. The difference is called the overpotential, which is ascribed to activation energy required for the formation of H_2. If the cathode potential is increased above this so-called

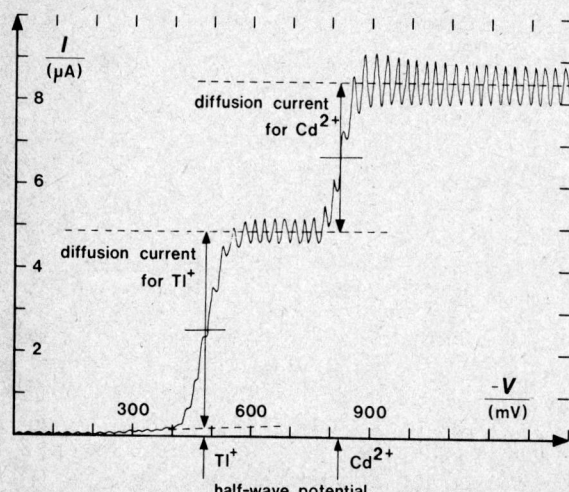

Figure 1B-40. Polarogram of a test solution containing TlCl (1.0 mmol/L) and CdCl$_2$ (0.5 mmol/L) in an ammonia/ammonium chloride buffer (1 mol/L). The abscissa shows the potential of the indicator electrode. The ordinate shows the current through the cell. The current pulsates due to the dropping mercury cathode.

decomposition potential, a current will flow and bubbles of hydrogen will develop at the cathode. When H^+ is continuously reduced, H^+ is said to depolarize the cathode.

If the test solution contains substances that are reduced more easily than H^+, the cathode will be depolarized by these substances and a current will flow through the cell. For example, if the test solution contains oxygen, this will be reduced at the cathode according to the reaction

$$O_2 + 2H_2O + 4e^- \rightarrow 4OH^-$$

When the decomposition potential for this reaction (about -0.3 to -0.4 V) is reached, the current increases until it reaches a plateau, the so-called diffusion current. The height of the plateau is dependent on the rate at which O_2 can diffuse from the surrounding solution to the surface of the cathode where $p\ O_2$ is zero. The diffusion current is therefore directly proportional to the $p\ O_2$ of the test solution. Any substance that is reduced at the cathode is characterized by two parameters: (1) the *half-wave potential* (i.e., the potential where the current is equal to one-half the plateau current) characterizes the kind of substance being reduced, and (2) the *diffusion current* is proportional to the concentration of the substance in the test solution. Polarography may therefore be used for identification as well as for quantitation. In medicine polarographic analysis is especially being used for determination of some metals (copper, lead, zinc, etc.) and to characterize proteins (e.g., the Brdička reaction between cobalt and SH-groups of mucoproteins) in addition to the determination of oxygen (see Amperometry).

A variant of polarography is called *anodic stripping voltammetry.* One of the electrodes is a mercury-coated graphite rod. A negative potential is applied to the electrode. The trace metal ions of the sample are then reduced and plate the electrode. The plating time is usually from 1–30 min, depending on the concentration. A polarogram is then recorded with the plated electrode as the anode and a nonpolarizable cathode. The metals are stripped off the anode by oxidation to the respective cations. The order in which they are stripped off is a function of the metal's unique redox potential. The current flow during the stripping of a given metal is a function of the amount of this metal. The method therefore provides identification as well as quantitative measurement of trace metals. The preconcentration (plating) step permits the analysis of extremely dilute samples.

A modification of anodic stripping is called *potentiometric stripping analysis.*[10] After the initial plating step, the electrode is left undisturbed and the cell potential (the electromotive force) is then measured as a function of time. The electrode reactions proceed in the reverse direction due to the presence of oxidizing species in the solution (Hg^{2+} or O_2). During this process the potential remains almost constant and the length of the plateau on the potential/time curve is proportional to the concentration of the first metal (chronopotentiometry). When the next metal starts stripping, a jump in potential is observed.

AMPEROMETRY

This technique is based on measurement of the current flowing through an electrochemical cell when a constant electric potential is applied to the electrodes. Some examples of this analytical principle are described in the following text.

The *pO₂ electrode* (Clark electrode, Figure 1B-41) is actually a complete electrochemical cell consisting of a small platinum cathode (area about 300 μm^2) and an Ag/AgCl anode in phosphate buffer with added KCl. The platinum cathode, covered by a thin film of electrolyte, is separated from the test solution by a gas-permeable membrane (e.g., polypropylene). The cathode potential is adjusted to -0.65 V. In the absence of oxygen in the test solution, the current is almost zero because the cathode is polarized. In the presence of oxygen, a current is observed that is due to a diffusion of O_2 from the test solution through the membrane to the cathode, where it is reduced. The current is directly proportional to the pO_2 in the test solution. The sensitivity of commercial pO_2 electrodes is on the order of $\Delta I/\Delta pO_2 = 10^{-14}$ A/Pa, depending on the size of the cathode area and the thickness of the gas-permeable membrane.

The purpose of the membrane is two-fold: first, to prevent proteins and other (dissolved) oxidants from gaining access to the cathode surface and, second, to limit the diffusion zone to the membrane and hence prevent variations in the diffusion coefficient of O_2 in the test solution

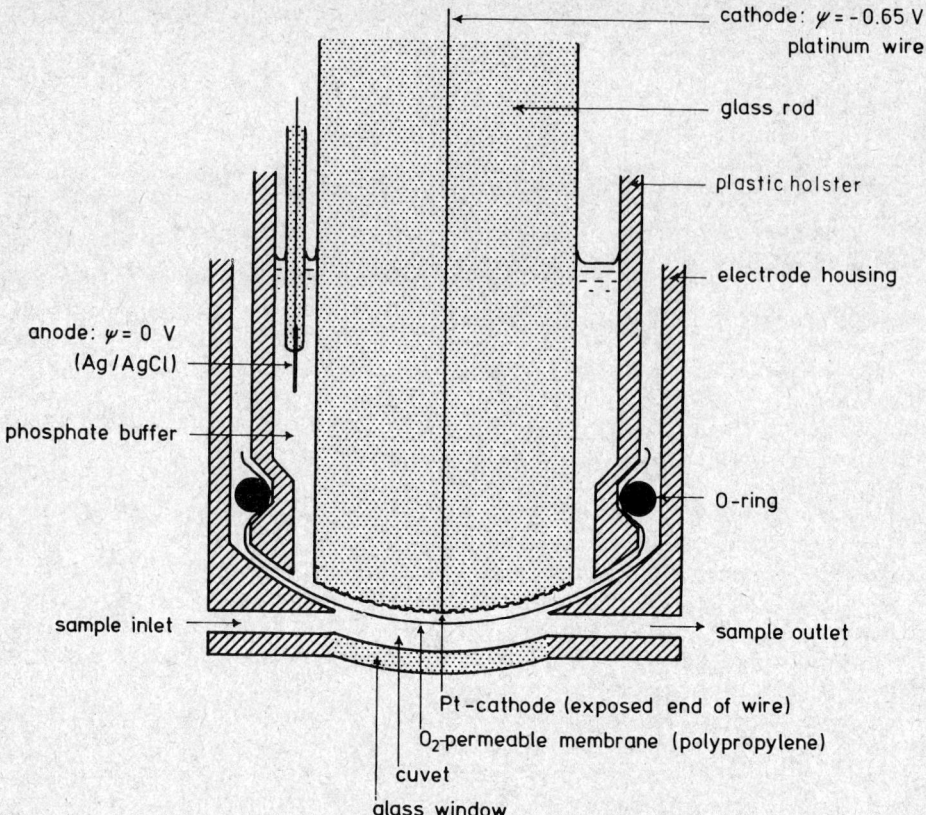

cathode: $\psi = -0.65$ V
platinum wire

glass rod

plastic holster

electrode housing

anode: $\psi = 0$ V
(Ag/AgCl)

phosphate buffer

O-ring

sample inlet

sample outlet

Pt-cathode (exposed end of wire)

O_2-permeable membrane (polypropylene)

cuvet

glass window

Figure 1B-41. Schematic illustration of a pO_2 electrode.[21]

(or gas) from influencing the result. Nevertheless, pO_2 electrodes generally read higher values in gases than in liquids, and it is necessary to take this gas/liquid ratio into account when calibrating the electrodes.

The pO_2 electrode has found widespread application for measurements of pO_2 in arterial or capillary blood. Estimation of the arterial pO_2 by *transcutaneous measurement* is also possible.[9] Special pO_2 electrodes with built-in heating coil have been developed for this purpose. The silver anode forms a ring around the cathode and thereby serves as a circular heating pad. With a temperature of 43–44 °C the skin is sufficiently arterialized to provide a reasonably good correlation between the arterial and the transcutaneous pO_2, especially in newborn infants. Unfortunately, the response is quite sensitive to the local blood flow, and therefore the correlation deteriorates in patients in shock or other conditions with diminished peripheral blood flow.

The pO_2 electrode has also been applied to the measurement of the concentration of *total oxygen* in the blood after liberation of hemoglobin-bound oxygen with ferricyanide or carbon monoxide (by forming methemoglobin and carboxyhemoglobin, respectively).

Glucose determination by means of the pO_2 electrode is based on the glucose oxidase catalyzed reaction:

$$\text{Glucose} + O_2 \rightarrow \text{Gluconic acid} + H_2O_2$$

The rate of fall in pO_2 under standardized conditions is a measure of the glucose concentration. The glucose oxidase may also be trapped in a gel on the outside of the O_2-permeable membrane in front of the cathode so that a diffusion equilibrium involving both glucose and oxygen is established. Under standardized conditions, the electrode response is then directly related to the glucose concentration.

The **peroxidase electrode** consists of a *polarized platinum anode* and a nonpolarized silver/silver chloride cathode. The anode potential is $+0.6$ V. In the presence of peroxide, a current flows owing to oxidation of peroxide at the anode:

$$H_2O_2 \rightarrow 2H^+ + 2e^- + O_2$$

The peroxide electrode covered by an enzyme membrane or enzyme layer has been proposed by Clark[2] for the determination of any substance for which a suitable oxygen oxidoreductase is available. Important examples are glucose, lactate, uric acid, and ethanol, in which case the membrane contains glucose oxidase, lactate oxidase, uricase, or alcohol oxidase, respectively. The substrate diffuses to the enzyme layer, where it is dehydrogenated, and H_2O_2 is formed. The H_2O_2 diffuses to the anode and causes a current proportional to the rate of formation of H_2O_2, i.e., proportional to the diffusion of substrate to the membrane. Further applications of this principle will probably lead to other useful methods in the future. Many examples of such amperometric enzyme electrodes have been published.

Amperometric end-point determination may be applied in the titration of Cl^- with Ag^+. The sample is acidified by means of nitric acid (or HNO_3 plus acetic acid) and the chloride ions are titrated with Ag^+. The silver ions may be added as a solution of $AgNO_3$ (volumetric titration), or Ag^+ may be generated from a silver electrode by means of an electric current (coulometric titration; see later). During titration, the Ag^+ concentration remains low due to the reaction $Ag^+ + Cl^- \rightarrow AgCl$, which causes the precipitation of $AgCl$. At the end point, Ag^+ appears in excess, and the increase in Ag^+ activity may be detected either potentiometrically (with an Ag electrode and a mercury/mercurous sulfate reference electrode), or the Ag^+ may be detected amperometrically. In the latter case, two silver electrodes are employed and a negative potential of 0.15–0.25 V is applied to the cathode. During titration, the cathode is polarized and the current is very low. At the end of the titration, the excess of Ag^+ depolarizes the cathode by the reaction $Ag^+ + e^- \rightarrow Ag$, and a current flows which is proportional to the excess of Ag^+. When the current has reached a preset value, the titration can be automatically stopped. This indicator principle is applied in the Cotlove chloride titrator, which in addition uses coulometric generation of Ag^+.

COULOMETRY

Coulometry is the technique used to measure the *amount* of electricity passing between two electrodes in an electrochemical cell. The amount of electricity is directly proportional to the amount of substance produced or consumed by the redox process at the electrodes. This is called Faraday's first law and may be expressed as

$$z \cdot n \cdot F = Q \tag{15}$$

where z is the numerical stoichiometric number of electrons involved in the reduction (or oxidation) reaction (unit: 1), n is the amount of substance reduced or oxidized (unit: mol), F is the Faraday constant ($= 96487 \text{ C} \cdot \text{mol}^{-1}$), and Q is the amount of electricity (unit: C = coulomb = ampere $\cdot$ second) passing through the cell.

An example of an application is the coulometric titration of chloride (Cotlove titrator), where silver ions are generated by electrolysis from a silver wire used as anode. At the cathode, H^+ is reduced to H_2. The amount of silver ions generated is measured coulometrically. When the current is kept constant, the measurement is reduced to a measurement of time according to

$$Q = I \cdot t \tag{16}$$

where Q is the amount of electricity (unit: coulomb), I is the electric current (unit: ampere), and t is time (unit: s, second). The current may also be gradually decreased as the titration approaches the end point. In this case the amount of electricity is calculated as the integral: $Q = \int_0^t I \cdot dt$, a calculation performed electronically by the coulometer. The end point of the titration may be detected either amperometrically or potentiometrically.

Acid-base titrations can also be performed coulometrically using a platinum generator electrode in the test solution separated from the other electrode by a sintered glass filter. If the generator electrode is the cathode, H^+ is removed ($H^+ + e^- \rightarrow \frac{1}{2}H_2$), which is equivalent to the addition of base. If the generator electrode is the anode, H^+ is added ($H_2O \rightarrow \frac{1}{2}O_2 + 2H^+ + 2e^-$). In either case, two sets of electrodes are necessary: the generator electrodes and the indicator electrodes. The latter may be used for direct potentiometric determination of the end point or in connection with amperometric end-point determination. The position of the indicator electrodes should be specially adjusted in relation to the generator electrodes so that the current through the latter does not disturb the performance of the former. Coulometric titrations are among the most accurate analytical determinations available.

CONDUCTOMETRY

Conductometry is the measurement of the current flow between two nonpolarized electrodes between which a known electrical potential is established. In order to avoid polarization of the electrodes, an alternating potential is applied with a frequency between 100 and 3000 Hz. With increase in conductivity* of the solution, there is less impedance (resistance) and, therefore, increased current flow. (Since the applied potential is alternating, the resulting current is also alternating.) The current is directly proportional to the conductivity of the solution.

The conductivity of aqueous solutions is dependent on the concentration of electrolytes and is closely related to the ionic strength. In the purest available water, the conductivity is $\kappa = 4.9 \cdot 10^{-6}$ S $\cdot$ m^{-1} at 18 °C. In ordinary distilled or deionized water the value is $\kappa < 2 \cdot 10^{-4}$ S $\cdot$ m^{-1}. A higher conductivity indicates the presence of electrolytes, and conductivity measurements therefore serve to monitor the performance of deionizers and to provide a warning that the ion-exchange resin should be regenerated.

Some chemical reactions are associated with a change in conductivity of the reaction medium. For example, urea can be measured rapidly and accurately by the change or the initial rate of change in conductivity that accompanies the urease-catalyzed formation of ammonium ions and bicarbonate ions. However, the high conductivity of biological fluids (due to the presence of the electrolytes) prevents the use of this technique for analytes of low concentrations where the change in conductivity is too small on a high background. Conductivity measurement can also be used for end-point detection in many kinds of titrations (acid-base, precipitation, compleximetric).

The conductivity of whole blood is greatly dependent on the volume fraction of erythrocytes, and attempts have been made to utilize conductivity measurements for determination of the erythrocyte volume fraction (hematocrit).

The Coulter principle for electronic counting of blood cells in suspension utilizes the fact that the conductivity of the blood cells is lower than that of a salt solution employed as suspension medium. The cell suspension is forced to flow through a tiny orifice. Two electrodes are placed on either side of the orifice and a constant current is established between the electrodes. Each time a cell passes the orifice the resistance increases, causing a spike in the electrical potential difference between the electrodes. The pulses are amplified and counted.

References

1. Bates, R. G.: Determination of pH. Theory and Practice. 3rd ed. New York, John Wiley & Sons, Inc., 1973.
2. Clark, L. C., Jr.: A family of polarographic enzyme electrodes and the measurement of alcohol. Biotechnol. Bioeng., 3:377, 1972.
3. Durst, R. A., Ed.: Ion-Selective Electrodes. Washington, D.C., National Bureau of Standards Publication, 314, 1969.
4. Eisenman, G., Ed.: Glass Electrodes for Hydrogen and Other Cations. New York, Marcel Dekker, 1967.
5. Freiser, H., Ed: Ion-Selective Electrodes in Analytical Chemistry, Vol. 1. New York, Plenum Press, 1978.
6. Freiser, H., Ed.: Ion-selective Electrodes in Analytical Chemistry. Vol. 2. New York, Plenum Press, 1980.
7. Guilbault, G. G.: Analytical uses of immobilized enzymes. Biotechnol. Bioeng., 3:361, 1972.
8. Heyrovsky, J., and Kuta, J.: Principles of Polarography. Prague, Publishing House of the Czechoslovak Academy of Science, 1965.
9. Huch, R., and Huch, A. (Eds.): Continuous Transcutaneous Blood Gas Monitoring. New York, Marcel Dekker, 1983.
10. Jagner, D., and Graneli, A.: Potentiometric stripping analysis. Anal. Chim. Acta, 83:19-26, 1976.
11. Kessler, M., Clark, L. C., Lübbers, D. W., et al., Eds.: Ion and Enzyme Electrodes in Biology and Medicine. Munich, Urban & Schwarzenberg, 1976.

---

* *Conductance (G)* is defined as the current (unit: A) divided by the potential difference (unit: V), when no EMF is present, and the unit therefore is ohm^{-1} = siemens, (Ω^{-1} = S).

Conductivity (κ) is defined as the current density (unit: A $\cdot$ m^{-2}) divided by the electric field strength (unit: V $\cdot$ m^{-1}), when no EMF is present, and the unit therefore is: $\Omega^{-1} \cdot$ m^{-1} = S $\cdot$ m^{-1}.

12. Meier, P. C., Amman, D., Morf, W. E., et al.: Liquid-membrane ion-selective electrodes and their biomedical applications. *In*: Medical and Biological Applications of Electrochemical Devices. J. Koryta, Ed. New York, John Wiley & Sons Ltd., 1980, pp. 13-91.
13. Moody, G. J., and Thomas, J. D. R.: Selective Ion Sensitive Electrodes. Watford, England, Merrow Publishing Company, 1971.
14. Pioda, L. A. R., Stankova, V., and Simon, W.: Highly selective potassium responsive liquid membrane electrode. Anal. Letters, 2:665, 1969.
15. Pungor, E., Havas, J., and Toth, K.: Membranes of heterogeneous structure for the determination of the activity of anions. I. Acta Chir. Acad. Sci. Hung., 41:239, 1964.
16. Purdy, W. C.: Electroanalytical Methods in Biochemistry. New York, McGraw-Hill Book Company, 1965.
17. Ross, J. W.: Calcium-selective electrode with liquid ion exchange. Science, 156:1378, 1967.
18. Růžička, J., Hansen, E. H., and Tjell, C.: Selectrode™, the universal ion selective electrode. VI. The calcium(II) selectrode employing a new ion exchanger in a non-porous membrane and a solid state reference system. Anal. Chim. Acta, 67:155, 1972.
19. Severinghaus, J. W., Stafford, M., and Bradley, A. F.: tcP$_{CO_2}$ electrode design, calibration and temperature gradient problems. Acta Anaesth. Scand., Suppl.68:118-122, 1978.
20. Simon, W., Ammann, D., Oehme, M., et al.: Calcium-selective electrodes. Ann. N.Y. Acad. Sci., 307:52, 1978.
21. Siggaard-Andersen, O.: The Acid-Base Status of the Blood. 4th ed. Baltimore, Williams & Wilkins, 1974.
22. Stokes, R. H., and Robinson, R. A.: Ionic hydration and activity in electrolyte solutions. J. Am. Chem. Soc., 78:1870-1878, 1948.
23. Weyer, E. M., Ed.: Bioelectrodes, Ann. N.Y. Acad. Sci., 148:1, 1968.

Section Five

OSMOMETRY

by Esther F. Freier, M.S., M.T. (ASCP)

Osmometry is a technique for measuring the concentration of solute particles. The concentration in turn is related to the osmotic pressure of the solution. The osmotic pressure governs the movement of water (or solvent) across membranes separating two solutions. Different membranes vary in pore size and thus in their ability to select molecules of different size and shape. Those membranes which are permeable only to water are referred to as *strictly semipermeable*. The term *partially semipermeable* is applied to those membranes which permit passage of some selected small molecules or ions in addition to water. Membranes such as those enclosing the glomerular and capillary vessels are permeable to water and essentially to all *small* molecules and ions, but not to macromolecular colloids such as proteins.

Consider an aqueous solution of sucrose placed within a sac made up of a strictly semipermeable membrane, with an open vertical glass tube manometer attached to this sac. If this device is placed into a beaker of distilled water, water will move from the beaker across the membrane into the sucrose solution in an attempt to make the chemical potential of the solvent equal on both sides of the membrane. The sucrose solution will rise some distance up the manometer tube. At equilibrium, the gravitational pressure of the column of solution in the manometer equals the osmotic pressure and thus will prevent further net movement of water from the beaker.

The height of the rise of the sucrose solution in the manometer tube is a measure of the *osmotic pressure* of the sucrose solution. The osmotic pressure of this solution is defined to be equal to the pressure which would have to be exerted on the sucrose side of the membrane to prevent the flow of water across the membrane.

Osmosis

Osmosis[6,10] is that process which constitutes the movement of water across the membrane in response to differences in osmotic pressure across the two sides of the membrane. Water generally migrates from the side of the membrane with the more dilute solute to that containing the more concentrated solute. The water is more concentrated on the pure (100%) water side of the membrane, and therefore water moves down a concentration gradient from its higher concentration to a lower concentration in the sucrose solution side, analogous to the diffusion of molecules and ions.

If the sucrose solution within the membrane sac were replaced with a sodium chloride solution of the same molarity, the solution in the manometer would reach equilibrium at a point almost twice as high as that previously observed. Sodium chloride dissociates into two ions per molecule, and if the ion activity were unrestricted, the sodium chloride solution would have twice as many osmotically active particles for the same molecular concentration as does the sucrose solution. In reality, the number of active particles is less, as explained later. The total number of individual (solute) particles present in the solution per given mass of solvent, regardless of their molecular nature (i.e., nonelectrolyte, ion, or colloid), determines the total osmotic pressure of the solution. In blood plasma, for example, nonelectrolytes such as glucose and urea as well as the electrolytes present either as free ions or in undissociated form, and even the proteins, contribute to the osmotic pressure of this body fluid.

Colligative Properties

If a solute is dissolved in a solvent (water, in the case of biological systems), the following four related phenomena occur:

1. The *osmotic pressure* of the system is increased, as just explained.

2. At any given temperature, the *vapor pressure* of the solution is *lowered* below that of the pure solvent.

3. As a result of the change in vapor pressure, the *boiling point* of the solution, i.e., the temperature at which the vapor pressure equals atmospheric pressure, is *raised* above that of the pure solvent.

4. The *freezing point* of the solution, i.e., the temperature at which the vapor pressure of the solid equals that of the solvent, is *lowered* below that of the pure solvent.

These four properties of solutions, that is, an increase in osmotic pressure and boiling point and a lowering of vapor pressure and freezing point, are called *colligative properties* because they are related and mathematically interconvertible. They all are directly related to the total number of solute particles per mass of solvent. The term *osmolality* expresses concentrations in terms of *mass* of solvent (1 osmolal solution is defined to contain 1 osmol/kg H_2O). In contrast, the term osmolarity expresses concentrations per volume of solution (1 osmolar solution is defined to contain 1 osmol/L solution). Osmolality (osmol/kg H_2O) is a thermodynamically more precise expression because solution concentrations expressed on a weight basis are temperature independent, while those based on volume will vary with temperature in a manner dependent on the thermal expansion of the solution. Since the osmolality of physiological fluids is relatively low, the milliosmol (mOsm) is used as a more convenient unit.

In the ideal case, 1 mole of a nondissociated solute dissolved in 1 kg of water results in a 1 molal solution that contains 6.023×10^{23} particles (Avogadro's number) per kg of water. This solution boils at a temperature 0.52 °C higher and freezes at a temperature 1.858 °C lower than pure water. The vapor pressure of this solution is 0.3 mm Hg lower than the vapor pressure of pure water, which is 23.8 mm Hg at 25 °C. The osmotic pressure of the same solution is increased from zero to 17 000 mm Hg or to 22.4 atmospheres. In this ideal case, all solvent and solute particles have unit activity, i.e., there is no association between any molecules of any species.

A solution of an electrolyte dissociates into two (in the case of NaCl) or three (in the case of $CaCl_2$) particles, and therefore, the colligative effects of such solutions are multiplied by the number (n) of dissociated ions formed per molecule. Because of incomplete electrolyte dissociation, as well as association between solute molecules and between solute molecules and solvent, many solutions do not behave as expected in the ideal case, and a 1 molal solution may give an osmotic pressure higher or (more often) lower than that theoretically expected. The osmotic activity coefficient is a factor used to correct for the deviation from the "ideal" behavior of the system:

$$\text{osmolality} = \text{osmol/kg } H_2O = \phi(n)(C)$$

where

ϕ = osmotic coefficient

n = number of particles into which each molecule in the solution potentially dissociates

C = concentration in mol/kg H_2O

The total osmolality or osmotic pressure of a solution is equal to the sum of the osmotic pressures or osmolalities of all solute species present.

A table of osmotic coefficients of most solutes of biological interest has been compiled by Wolf.[11] The osmotic coefficients were derived experimentally by measuring the osmolality of solutions of known concentration. While glucose and ethanol have osmotic coefficients of 1.00, the ϕ for urea is 0.94 and the ϕ for NaCl is 0.93 at the concentration found in serum and 0.91 at the higher concentrations that can occur in urine. Potassium dihydrogen phosphate (KH_2PO_4) in a 0.5% solution has an osmotic coefficient of only 0.46.[11]

Because of their high molecular weights, serum proteins contribute less than 0.3% of the total osmolality of serum. The electrolytes, Na^+, Cl^- and HCO_3^-, which are present in relatively high concentration, make the greatest contribution to serum osmolality, whereas nonelectrolytes, such as glucose and urea, which are present normally at low concentrations, contribute only to a small extent. For example, the theoretical molality of osmolutes (osmotically active particles) in blood serum is $\sim$307 mmol/kg water divided among the various osmolutes approximately as follows: Na^+, 148; K^+, 4; Ca^{2+}, 2; Mg^{2+}, 1; Cl^-, 108; HCO_3^-, 27; HPO_4^{2-}, 2; SO_4^{2-}, 1; organic anions, 5; glucose, 5; and urea, 4 (all values being molalities; unit: mmol/kg H_2O). The serum proteins contribute only $\sim$1 mmol/kg H_2O. When the osmotic coefficients are applied, this mixture in serum would have an osmolality of only 290 mosmol/kg H_2O.

The Freezing Point Depression Osmometer

Theoretically, any of the four colligative properties could be used as a basis for the measurement of osmolality. The freezing point depression is most commonly used to determine the concentration of osmotically active particles in physiological fluids because its measurement is simple and convenient. Unlike vapor pressure or osmotic pressure measurements, the determination of freezing point depression is independent of changes in ambient temperature. (The vapor pressure of water is 17.5 mm Hg at 20 °C, 23.8 mm Hg at 25 °C, and 47.1 mm Hg at 37 °C.) The instrument used is a freezing point depression osmometer, but it is usually referred to simply as an osmometer or, less commonly, as a cryoscope. The latter term may come into more frequent usage to distinguish this instrument from the vapor pressure osmometer, which is also used to measure total osmolality in serum and urine. The components of a freezing point depression osmometer (see Figure 1B-42) are as follows:

1. A thermostatically controlled cooling bath maintained at -7 °C.
2. A rapid stir mechanism to initiate ("seed") freezing of the sample.

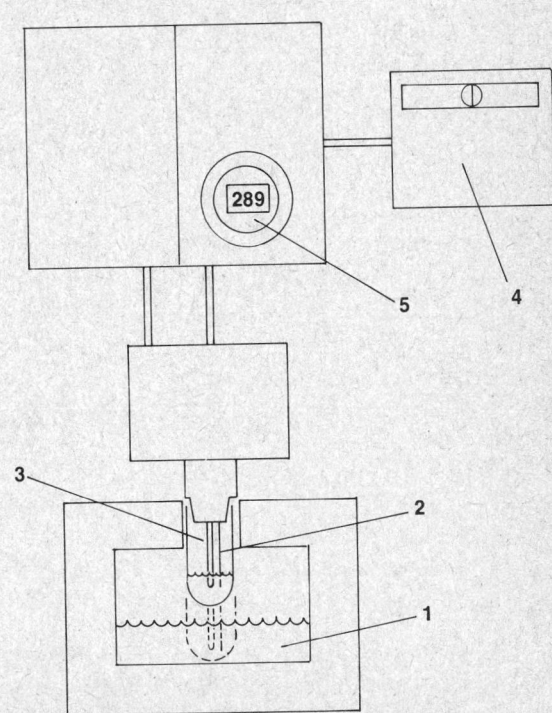

Figure 1B-42. Block diagram of freezing point depression osmometer. *1*, Cooling fluid. *2*, Stirring rod. *3*, Thermistor. *4*, Galvanometer. *5*, Potentiometer with direct readout. Test tube is shown above the liquid in the cooling bath (solid line) and inside the cooling liquid (dashed line).

3. A thermistor probe connected to a Wheatstone bridge circuit to measure the temperature of the sample. (The thermistor is a glass bead attached to a metal stem whose resistance varies rapidly and predictably with temperature.)

4. A galvanometer which displays the freezing curve and which is used as a guide when the measuring potentiometer (5) is used.

5. A measuring potentiometer (variable resistor) is used to null the current in the Wheatstone circuit.

Automatic instruments exist in which components 4 and 5 are replaced by a light-emitting diode (LED) display which both indicates the time course of the freezing curve and locks in the final result in a numerical display.

Operation of an Osmometer[3]

The sample into which the thermistor probe and stirring wire have been centered is lowered into the bath and with gentle stirring is supercooled to a temperature several degrees below its freezing point (-7 °C). When the galvanometer (or LED display) movement indicates that sufficient supercooling has occurred, the sample is raised to a point above the liquid in the cooling bath and the wire stirrer is changed from a gentle rate of stir to a momentary (1 s) vigorous amplitude which initiates freezing of the supercooled solution. This freezing is only to the slush stage, about 2–3% of the solvent solidifying. The galvanometer reverses direction, as the released heat of fusion warms the solution and then remains stationary, indicating the equilibrium temperature at which both freezing and thawing of the solution is occurring. During this equilibrium period of 2–3 min, the balancing potentiometer is adjusted, thus changing the variable resistance of the Wheatstone bridge to bring the galvanometer to the null position. At the end of the equilibrium temperature plateau, the galvanometer again indicates decreasing temperature as the sample freezes further toward a complete solid.

The balancing potentiometer readings can be related to the degree of freezing point depression in °C, but more commonly they are calibrated with suitable standards to read directly in milliosmols. If an instrument is used which measures directly the freezing point depression, the osmolal concentration may be calculated as follows:

$$\text{If the observed freezing point is } -0.53 \text{ °C, the mosmol/kg H}_2\text{O} = \frac{-0.53}{-1.86} \times 1000 = 285$$

where -1.86 °C is the molal freezing point depression of pure water.

If the sample volume available for analysis is limited, a 0.2-mL sample tube may be used instead of the 2-mL sample tube routinely used. After analysis, the thawed sample may be used for other tests. When a measured volume of 0.25 mL is used in the 0.2 mL tubes, equivalent precision to the larger tube is attained. A precision (day-to-day) of ± 2 mosmol/kg H$_2$O can be obtained in a service laboratory. The automatic instruments, by controlling the time at which the sample tube is raised above the bath, the proper time for initiating freezing, and the time at which the automatically balanced final reading is taken, minimize variations between operators of the instrument. The precision of the automatic instrument is essentially the same as noted above for the manual instruments used by trained analysts.

The Vapor Pressure Osmometer

Another type of osmometer intended to measure total osmolality was introduced for use in the clinical laboratory by Wescor Inc., Logan, UT 84321. Though the instrument is called a vapor pressure osmometer, the osmolality measurement is not related directly to a change in vapor pressure (in mm Hg), but to the decrease in the *dew point temperature* of the pure solvent (water) caused by the decrease in vapor pressure of the solvent by the solutes. In this instrument, temperature is measured by means of a thermocouple, which is a device consisting of two different metals joined so that a voltage difference generated between the points of contact (junctions) is a measure of the temperature difference between the points.

The osmolality measurement is made at room temperature by introducing a 7-μL sample into a vaporization-condensation chamber on a 5 mm filter paper disk. The chamber is sealed and time is allowed for thermal and vapor equilibration. An electrical current is fed through the thermocouple junction cooling the thermocouple to a temperature below the dew point, the temperature at which the saturation vapor pressure is equal to the actual vapor pressure of the contained water vapor. Water condenses from the enclosed air in the chamber and forms a thin liquid film on the thermocouple junction. The heat of condensation raises the thermocouple temperature until it reaches the dew point at which no more water condenses. The temperature becomes stabilized at this point because a higher temperature would cause water to evaporate, cooling the thermocouple junction. The response shown on the output meter is proportional to the temperature depression of the thermocouple, which, in turn, is proportional to the depression of the dew point of the water in the chamber. The instrument is calibrated with suitable standards to read directly in milliosmols as is the cryoscopic osmometer. Early models were sensitive to changes in ambient temperature, but a temperature-compensating circuit has reduced such influence to <0.1 mosmol/°C, provided the instrument is not subjected to rapid changes in ambient temperature. The instrument is compact and does not need a refrigerated bath or test tubes. When the chamber and thermocouple hygrometer are kept scrupulously clean, the instrument is capable of a within-run precision of ±3 osmol/kg water at a normal serum level of 290. The technique has been adopted by the Association of Official Analytical Chemists for use by cheese factories to screen milk deliveries for adulteration through addition of water.[9]

In clinical laboratories, the vapor pressure osmolality technique has been reported to be less precise than the freezing point depression method, both in interlaboratory proficiency surveys conducted by the College of American Pathologists (CAP) and in an individual intralaboratory comparison.[4,7] More than 80% of the laboratories in the CAP surveys use freezing point depression osmometers. For serum samples, the coefficients of variation obtained for the vapor pressure osmometer are about twice those obtained for the cryoscopic osmometer. The lesser degree of precision is related to the lower slope of dew point decrease compared with freezing point decrease, i.e., 0.303 °C versus 1.858 °C per osmol/kg H_2O.

Nonlinear response is observed with the vapor pressure osmometer below 200 mosmol/kg H_2O, with deviations from linearity greatest at the zero end of the calibration curve. Because of variation in individual thermocouple response, the measured value for the osmolality of pure water ranges from 20–40 mosmol/kg H_2O. Unless the sample chamber and thermocouple are scrupulously clean, the nonlinearity is even more pronounced. Therefore, measurements below 200 mosmol/kg H_2O should be made with care using a 100 mosmol/kg H_2O standard and water to define the curve.

An important difference between the vapor pressure technique and the cryoscopic osmometer is the failure of the former to include in the measurement of the total osmolality the volatile solutes present in the serum. Substances such as ethanol, methanol, ethylene glycol, and even dissolved CO_2 (dCO_2), which is proportional to pCO_2, are volatile and thus escape from the solution and increase the vapor pressure instead of lowering the vapor pressure of the solvent (water). As a result, the dew point of the remaining solution is lowered to a lesser degree than would have been the case without loss of solute. The error introduced as a result of loss of CO_2 alone may be as much as 7–11 mosmol/kg H_2O.[5] In the freezing point depression method, the loss of CO_2 does not exceed 1.1 mmol/L plasma during the freezing point measurement even if no effort is made to prevent CO_2 loss. Thus the error would be negligible. The osmotic coefficient (ϕ) for CO_2 is 0.9 mosmol/mmol.[5] With either technique, serum specimens should be kept stoppered and analyzed without undue delay, or further loss of CO_2 from bicarbonate may occur before the measurement is made. Lipemic sera have been reported to give higher results on the dew point osmometer compared to the freezing point instrument.[7]

The Colloid Osmotic Pressure Osmometer*

Colloid osmotic pressure (COP) is a direct measure of the contribution of macromolecules to the osmolality. In case of serum, it is a measure of the contribution of the serum proteins to the serum osmolality. Because of their large molecular weight, proteins contribute only about 1 mosmol/kg H_2O to the total serum osmolality measured by the freezing point depression or vapor pressure osmometers. In the COP osmometer (also referred to as a membrane osmometer),

---

*Instruments manufactured by Wescor, Inc., Logan, UT, and Instrumentation Laboratories, Lexington, MA.

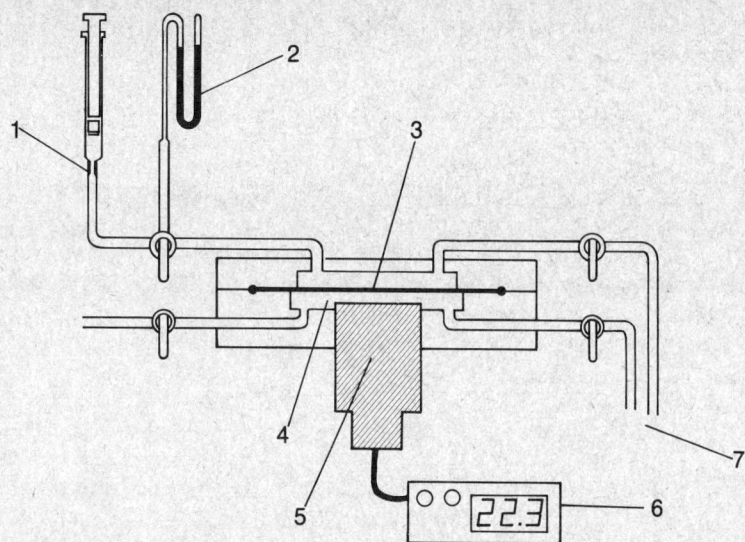

Figure 1B-43. Block diagram of colloid osmotic pressure osmometer. *1*, Sample. *2*, Mercury manometer. *3*, Semipermeable membrane. *4*, Saline side of chamber. *5*, Pressure transducer. *6*, Meter. *7*, Waste.

the measuring cell consists of two chambers of fluid separated by a semipermeable membrane. The components of the instrument are outlined in Figure 1B-43. The sample which contains protein is placed on one side of the membrane and a colloid-free physiological saline solution on the reference side. This saline solution is in contact with a pressure transducer. The instrument is adjusted to zero reference value with saline on both sides of the membrane. When a sample containing protein is introduced on the sample side, fluid moves by osmosis from the saline side across the membrane, causing a negative pressure to develop on the saline (reference) side equal to the COP of the sample. The membrane commonly used for serum is impermeable to particles with molecular weights ≥ 30 000. Human albumin, 50 g/L, or dextran 70, 30 g/L, can be used as reference standards or controls, and a mercury manometer can be used to calibrate the pressure transducer. Less than 500 μL of sample is required, and the measurement can be made in <1 min with a precision of 0.5 mm Hg.

Plasma extenders such as dextrans exhibit two to eight times greater COP than protein solutions of the same concentration. Also, the relationship between albumin concentration and COP is not a linear one. Therefore, the direct measurement of COP is more accurate than a prediction made from serum protein measurements.[2,8] The determination of COP is useful in guiding postoperative intravenous fluid therapy to prevent complications such as pulmonary edema and in the assessment of other abnormalities of water balance or serum protein concentration.

The reference range found for serum is 21 ± 2 mm Hg in supine individuals and 25 ± 2 mm Hg in ambulatory individuals. The higher level in ambulatory individuals correlates with higher blood hemoglobin and higher serum protein levels in ambulatory individuals compared to levels of those who are supine. This is due to higher hydrostatic pressure in the ambulatory state which forces more fluid out of the vascular compartment through the capillary membrane into the interstitial fluid. The COP is a measure of the force drawing this fluid back into the circulation. It has been recommended that specimens on ambulatory patients be drawn after the subject has been in a sitting position for at least 15 min.

References

1. Abel, J. E.: The physical background to freezing-point osmometry and its medical-biological applications. Am. J. Med. Electronics, *2* (Jan-Mar):32-41, 1963.
2. Duncan, A., and Young, D. S.: Measurements of serum colloid pressure are of limited usefulness. Clin. Chem., *28*:141-145, 1982.
3. Johnson, R. B., Jr., and Hoch, H.: Osmolality of serum and urine. *In*: Standard Methods of Clinical Chemistry, Vol. 5. S. Meites, Ed. New York, Academic Press, 1965, pp. 159-168.

4. Juel, R.: Serum osmolality, a CAP survey analysis. Am. J. Clin. Pathol., Suppl. *68*:165-169, 1977.
5. Khosla, S. S., and DuBois, A. B.: Correction for loss of CO_2 from blood during measurement of osmotic pressure, J. Appl. Physiol., *44*:474-478, 1978.
6. Lifson, N., and Visscher, M. B.: Osmosis in living systems. *In*: Medical Physics, Vol. 1. O. Glasser, Ed. Chicago, Year Book Publishers, 1961, pp. 869-892.
7. Mercier, D.E., Feld, R.D., and Witte, D. L.: Comparison of dewpoint and freezing point osmometry. Am. J. Med. Tech., *44*: 1066-1069, 1978.
8. Morissette, M. P.: Colloid osmotic pressure: Its measurement and clinical value. Can. Med. Assoc. J., *116*:897-900, 1977.
9. Richardson, G. H., Mortensen, M. S., and Crockett, R. G.: Quantitation of added water in milk by using vapor pressure osmometry. J. Assoc. Off. Anal. Chem., *61*:1038-1040, 1978.
10. Warhol, R. M., Eichenholz, A., and Mulhausen, R. O.: Osmolality. Arch. Int. Med., *116*:743-749, 1965.
11. Wolf, A. V.: Aqueous Solutions and Body Fluids. New York, Hoeber Medical Division, Harper and Row, 1966.

Section Six

CHROMATOGRAPHY

by Larry D. Bowers, Ph.D.

The association between biochemical analysis and chromatography has been long and productive. The complex nature of plant pigmentation led the Russian botanist Mikhail Tswett to search for a substance which could separate the pigments into their constituent parts. In his landmark paper of 1906, he recognized that the resolution of the plant pigments into a number of colored bands on his column of calcium carbonate was due to adsorption of the chemicals. He coined the terms "chromatogram" and "chromatography" to describe the technique.

The development of partition chromatography in 1941 by Martin and Synge was the outgrowth of an attempt to separate amino acids by countercurrent extraction. They found that the separation of monoamino monocarboxylic acids on a short column of water-coated silica gel far surpassed that obtained with any extraction apparatus. Because silica gel did not always show ideal behavior, Martin and his colleagues also used water-coated paper fibers to separate these compounds. This was the origin of paper chromatography.

In 1952, Martin and James developed gas-liquid partition chromatography to achieve a better separation of fatty acids. The development of thin layer chromatography by Kirchner and by Stahl was inspired by the difficulties encountered in analyzing flavor components and pharmaceutical samples. With this rich heritage, it is not surprising that various modes of chromatography have been used rather extensively in the clinical laboratory.

A chromatographic separation minimally requires that a sample be introduced into a stream of gas or liquid that flows through a bed of support particles. The support particles and any associated solvent are known as the *stationary phase* while the flowing stream is known as the *mobile phase*. The analyte molecules distribute between the stationary and mobile phases according to equilibrium processes. Since a compound can move through the bed only when it is in the mobile phase, a molecule which has greater affinity for the stationary phase would fall behind a molecule which has lesser affinity, and thus it will be separated from it. Once the compounds are separated, chromatography can be used to provide assistance in identifying the analytes, to provide quantitative analysis of the compound, or to provide a quantity of pure chemical for further studies (preparative chromatography).

Chromatographic techniques are often classified by the physical characteristics of their mobile and stationary phases. For example, a column operated with a layer of water coated onto a silica gel support through which an organic solvent is flowing would be referred to as *liquid-liquid chromatography*. If a gas were used as the mobile phase in conjunction with a liquid stationary phase, the technique would be referred to as *gas-liquid chromatography*. A separation system can also be classified by the mechanism of the separation. For example, adsorption of a compound onto a stationary phase particle would be referred to as *adsorption chromatography*. Other commonly used separation modes are *ion exchange, liquid partitioning, steric exclusion,* and *affinity complexation.* This approach to classification can sometimes be ambiguous since more than one type of interaction may be involved in separation. Finally, the chromatographic system may be classified by the form of the bed, as for example in *thin-layer chromatography.*

Chromatographic separations are mainly based on the distribution of the compound of interest between two phases. The role of the analyst is to choose the distribution system which best separates the analytes from each other and from any interferences. For example, glycosylated hemoglobin is formed in erythrocytes from the nonenzymatic reaction between glucose and hemoglobin. To separate glycosylated hemoglobin A_{1C} from hemoglobin A_0, one could take advantage of the fact that the addition of glucose reduces the number of charges on the hemoglobin molecule. However, since hemoglobin has many charges, a chromatographic system that recognizes the added glucose moiety would be easier to implement than a method that attempts to recognize the rather subtle difference in charge.

The selection of the best system for a separation requires an understanding of the interactions involved in a separation. The fact that a flow of mobile phase is essential to effect the separation of the analytes adds a second component to chromatography—a flow dynamics component. This parameter, in conjunction with the equilibrium component, determines the characteristics of a separation technique such as its speed and its selectivity. As analysts, we must be concerned not only with the quality of separation but also with the speed with which the analyses can be performed. This requires an appreciation of the factors which limit or augment the speed of separation.

FORMS OF CHROMATOGRAPHY

In addition to selecting an interaction that will best achieve the separation, the analyst must choose a specific form of chromatography (e.g., liquid column chromatography). This choice is based on the strengths and weaknesses of the techniques and is rarely unequivocal. A simple decision scheme is shown in Figure 1B-44 to illustrate how chromatography might be used to solve an analytical problem. These guidelines are very general and some may not be applicable in the face of particular requirements or problems. Nevertheless, the scheme does provide an approach to applications of chromatography.

The simplest form of chromatographic separation is **extraction**, in which a substance partitions between two immiscible solvents. For example, to extract the class of drugs called barbiturates from serum, we acidify the serum to put the barbiturates in the form most soluble in an organic solvent such as chloroform. A volume of the chloroform is then vigorously shaken with the serum and when the two solvent layers separate, most of the barbiturates are present in the chloroform layer while many troublesome substances such as proteins remain in the water layer. The fraction of the analyte which appears in the organic layer will increase (1) if a large volume of organic solvent relative to serum is used, (2) if the serum is extracted several times with organic solvent and the solvent aliquots are pooled, and (3) if the attractive forces between the selected solvent and the analyte are maximized. Solvent extraction is frequently used as a way to clean up a specimen prior to spectrophotometric, fluorometric, chromatographic, or immunologic assay of nonpolar compounds such as lipids, steroids, and drugs.

In classic **liquid chromatography**, a tube or column is filled with a solid support, such as calcium carbonate, and a liquid is percolated through the column under the influence of gravity. More recently pumps and other means have been used to provide the driving force for the mobile phase. A general schematic of a column chromatograph is shown in Figure 1B-45. The next requirement of a column chromatographic system is a means of introducing the sample into the mobile phase before entry into the column. The common element in column chromatography is that the mobile and stationary phases are in equilibrium and the distribution of the analyte between the two phases causes the separation. After the separation in the column, the sample components are eluted from the downstream end of the column by a continuous flow of mobile phase. The pattern of elution, called the chromatogram, can be detected by an on-line monitor such as a spectrophotometer or by analysis of sequential mobile phase fractions which are collected either manually or automatically with a fraction collector. In a few instances, the stationary phase may be extruded from the column and the analyte band(s) cut out and eluted from the stationary

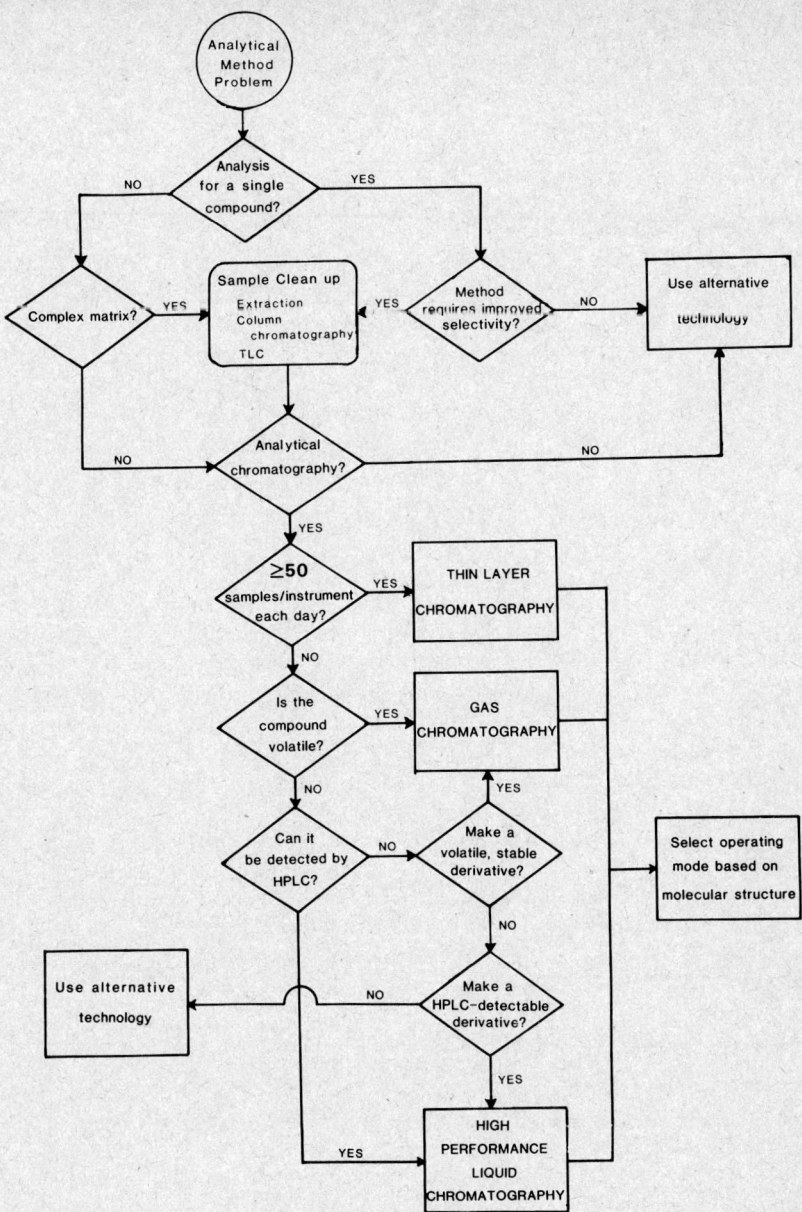

Figure 1B-44. A typical decision tree used to determine the type of chromatography that will solve an analysis or research problem.

phase with solvent. A more detailed discussion of modern gas and liquid column chromatography can be found in Sections Seven and Eight of this chapter.

In **thin-layer chromatography** (TLC), a thin layer of support particles is placed on a flat support such as a glass plate.[3,5] Using a capillary tube, the samples are applied on the plate along a line parallel to the edge but a few centimeters from it. After the sample application spots are dry, the plate is placed in a tank which contains a small amount of mobile phase, as shown in Figure 1B-46. The mobile phase travels up the plate by capillary action, which is a balance between mobile phase surface tension and a retarding force such as viscosity or gravity. Additional separating power can be achieved if the plate is developed in two dimensions as shown in Figure 1B-47. In addition to the ascending technique described above, thin layer plates can be developed in a descending or radial mode. In contrast to column chromatography, the stationary phase is not equilibrated with the mobile phase during the entire development process since the plate is

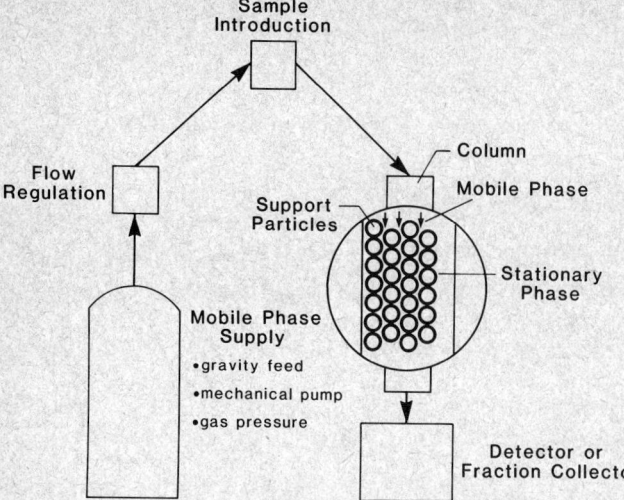

Figure 1B-45. Schematic diagram of a column chromatograph.

dry until the solvent passes over it. Lack of equilibration, along with the inconsistent flow rate obtained from capillary action, the evaporation of solvents from the mobile phase during the run, and varying degrees of "wetting" of the support, makes reproducible movement of an analyte from run to run difficult to achieve. To improve reproducibility, the developing tank is saturated with solvent vapor prior to and during the run. To aid in the identification of the compounds giving rise to the chromatographic spots, standards are usually chromatographed with the unknown sample. If the migrations of the unknown compound and standard are not identical, the compounds are not the same. The converse is not true, however, since more than one compound can have the same migration in a particular chromatographic system. Absolute quantitation is generally difficult, because the size and density of the spots depend on a number of factors, including the application technique. Nevertheless, densitometry has been used to quantitate drugs, lipids from serum and amniotic fluid, and a number of other compounds of clinical interest. The advantages of TLC include the ease with which a compound can be located by chemical or enzymatic reactions which produce color, the potential to chromatograph "dirty" samples without a cleanup step, and the relatively low cost.

Recently the use of smaller stationary phase particles, a thinner layer, and controlled flow rates has led to the development of "high performance thin-layer chromatography." Although more apparatus is required than for classical TLC, the separations are much more rapid, efficient, and reproducible. Details of this technique can be found in reference 7. An intermediate technique using the "high performance" thin-layer plates but retaining the classical development technique has also been introduced with the advantages of increased speed and better resolution. Some problems, however, are still encountered from solvent evaporation and inadequate wetting.

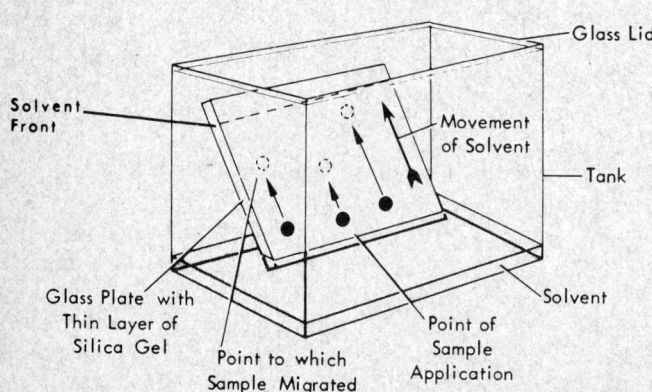

Figure 1B-46. Illustration of thin-layer chromatography. The solvent is drawn up the thin layer of adsorbent by capillary action. Ascending development is commonly used in TLC. (Modified from Bennett, T. P.: Graphic Biochemistry. Vol. 1: Chemistry of Biological Molecules. New York, The Macmillan Company, 1968.)

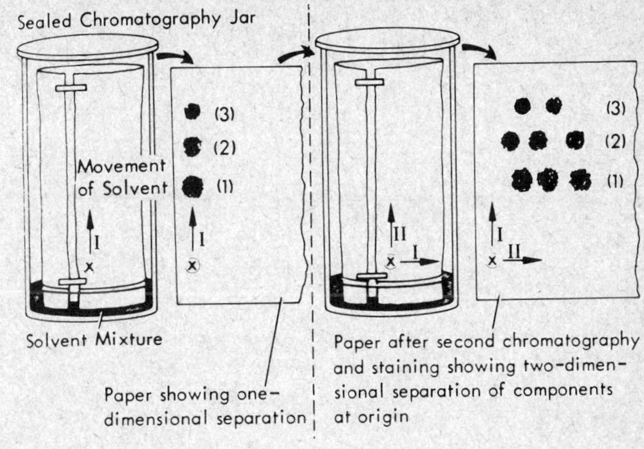

Figure 1B-47. Two-dimensional paper chromatography. Two different solvents are used for development. (From Bennett, T. P.: Graphic Biochemistry. Vol. 1: Chemistry of Biological Molecules. New York, The Macmillan Company, 1968.)

Paper chromatography is operationally quite similar to TLC. The stationary phase, however, is a layer of water coated onto the paper fibers (rather than silica gel, the support particles normally associated with TLC). Cellulose-coated particles can be formed into TLC plates which have retentivity similar to the paper sheets, but have superior separation speed and resolution.

It is important to recognize that any of the aforementioned separation mechanisms (adsorption, ion exchange, partition, steric exclusion, or affinity) can be used with either thin-layer or column techniques.

FUNDAMENTAL PROPERTIES OF CHROMATOGRAPHY

EQUILIBRIUM ASPECTS OF CHROMATOGRAPHY

The basis of any chromatographic separation is the differential migration of two (or more) compounds of interest through the chromatographic bed. For each analyte (solute molecule), the separation is the result of an equilibrium established between the solute in the mobile phase and the solute in the stationary phase. The ratio of the number of moles of solute in the stationary phase to the number of moles of solute in the mobile phase when equilibrium is established is known as the *capacity factor, k'*. The factor k' is also a measure of the amount of time spent by the solute in the stationary phase relative to the time spent in the mobile phase under a particular set of operating conditions. The capacity factor can be related to an equilibrium constant if sufficient information regarding the relative amounts of the mobile and stationary phases is available. Because the equilibrium constant is dependent on the characteristics of the compound involved, the retention behavior of an analyte can assist in the identification of that analyte. It is important to recognize that the retention behavior does not *prove* that a certain compound is present, only that the chromatogram is *consistent* with its presence.

The capacity factor allows us to predict the position in space or time of that particular compound under the separation conditions chosen. In open-bed chromatography, all separation activity must occur within the distance traveled by the solvent. In thin-layer chromatography, the solute migration is described by R_f, which is calculated from the relation:

$$R_f = \frac{\text{distance from application point to spot center}}{\text{distance from application point to solvent front}}$$

The farther the spot moves, the less would be the affinity of the solute for the stationary phase and the smaller would be k'. It is not surprising, then, that R_f and k' are related:

$$R_f = 1/(1 + k')$$

When the compounds of interest are eluted from a column, the *retention volume, V_r,* can be related to k' as follows:

$$V_r = V_m(1 + k')$$

where V_m is the volume of mobile phase required to elute a compound with no affinity for the stationary phase ($k' = 0$) (see Figure 1B-48). *Retention time, t_r,* is a frequently used term which can be related to the retention volume by the relation

$$V_r = t_r \times F$$

where F is the flow rate in mL/min. Thus we can decrease the retention time by increasing the flow rate, but the volume of mobile phase required for the separation remains the same.

It should be apparent that separation of two compounds will require some difference in their equilibrium constants and therefore in the capacity factors. The separation factor or *selectivity,* α, for a pair of compounds A and B can be defined by:

$$\alpha = k'(B)/k'(A)$$

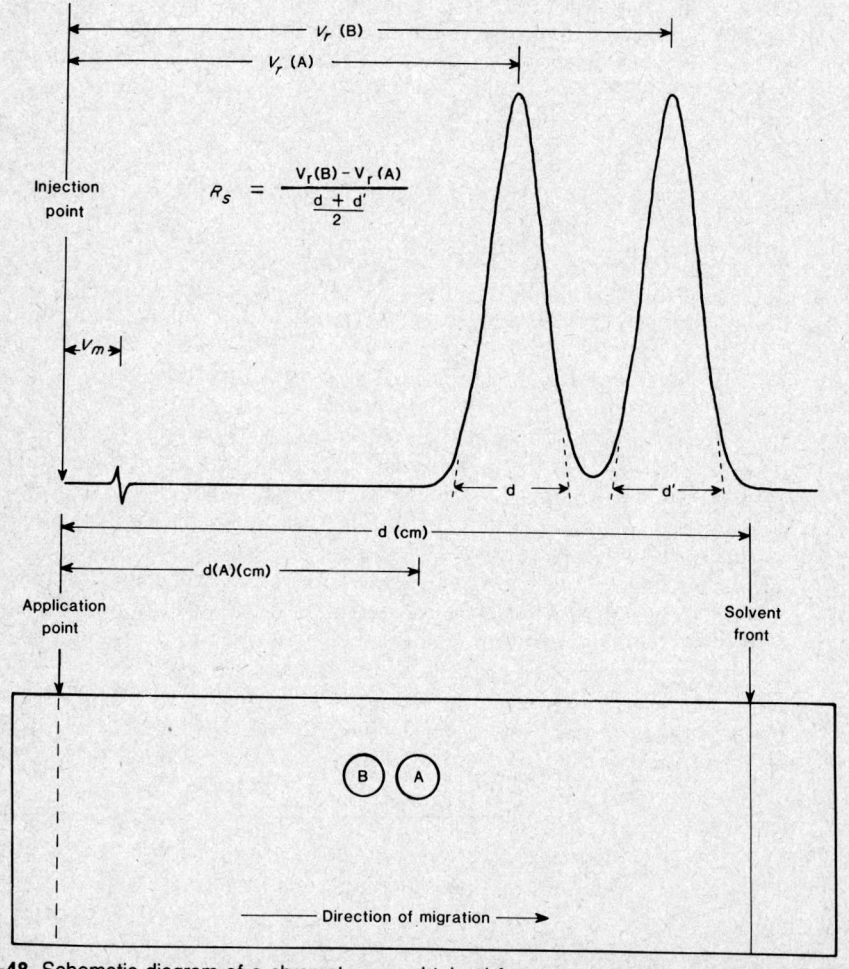

Figure 1B-48. Schematic diagram of a chromatogram obtained from a column and open-bed chromatograph. In open-bed chromatography (bottom of figure), note that strongly retained compounds (B) move slower than less strongly retained compounds, while in column chromatography (top of figure), compound B is eluted later than A, again due to stronger retention.

where α usually has a value greater than 1.00. In chromatography it is necessary to choose a separation system with a large value for α. Using the example of glycosylated hemoglobin again, in the separation mode in which specific retention of the added glucose moiety occurred, only the glycosylated hemoglobin would be retained and the α value would be extremely large. If a separation based on the charge differences were used, the selectivity would be much smaller and the separation conditions would have to be optimized and very rigorously controlled to obtain a good separation. It is technically feasible to separate two compounds having an α of 1.01, reasonably convenient to separate them if α is 1.05, and relatively easy to separate them if α is greater than 1.10.

FLOW DYNAMICS ASPECTS OF CHROMATOGRAPHY

As compounds move through the stationary phase, ideally they migrate as very narrow bands. Unfortunately this is not what is observed. Band broadening or spreading will occur primarily as a result of the flowing mobile phase that is required to effect the separation. Band spreading results in larger spots (in TLC) and broader peaks (in column chromatography). Investigations to explain band spreading resulted in the development of the rapid forms of chromatography that exist today. Martin and Synge recognized that if a solute was retained in a column long enough, the eluting concentration profile could be described by a Gaussian peak. This observation in turn simplified the mathematics of relating peak width to physical events occurring in the separation system.

A characteristic parameter of a chromatographic system is its efficiency or *number of theoretical plates, N*. In theory, a plate is the length of a column needed to allow one equilibration of the solute to occur between the stationary and mobile phases. The plate number can be defined as:

$$N = [V_r(A)/s(A)]^2$$

where $s(A)$ is the standard deviation of the peak (see Figure 1B-48). If the number of plates increases while the retention volume is constant, the peak will become narrower. On the other hand, if N is decreased (by an increase in flow rate, for example), the peak will be broader if the retention volume remains constant. This broadening will affect the resolution of compounds, as we will see below.

Since the standard deviation of a graphical display is difficult to measure, a more easily measured quantity is used to determine the number of theoretical plates. If a tangent is drawn to each of the inflection points of the Gaussian peak, the distance between the points where the tangents intersect the chromatographic baseline will be four standard deviations. Alternatively, the width of a Gaussian peak at one-half its height corresponds to 2.354 standard deviations and therefore $s(A) = w(A)/2.354$. Substituting the latter relationship into the above equation, the number of theoretical plates can be derived by measuring the width of a peak at half height, $w(A)$, and using the relation

$$N = 5.54 \, [V_r(A)/w(A)]^2$$

A second measure of efficiency is the *plate height*, H:

$$H = L/N$$

where L is the length of the column. It is important to note that these characteristic numbers do not correspond to a physical entity. They are, however, widely used to characterize band spreading in chromatographic systems.

In simplest terms, band spreading implies that if two identical molecules start at the same time at the top of a column, they will not necessarily appear at the other end of the column at the same time. Since the equilibria are identical, the phenomenon must be related to flow dynamics. The structure of the column itself has some effect. Some of the mechanisms by which this discrepancy in movement occurs can be explained by referring to Figure 1B-49. Molecule Y has managed to find a direct path through the packing

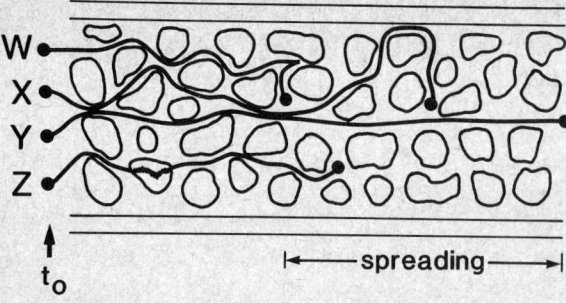

Figure 1B-49. Spreading occurring as a result of molecular flow path tortuosity of eddy diffusion. The dots indicate the position of molecules W, X, Y, and Z at the beginning of flow and after a fixed period of time.

material with no obstructions and therefore would be expected to elute first. Molecule X has had to avoid several obstructions, and thus, even if its velocity were identical to that of molecule Y, the more circuitous route retards its arrival at the end of the column. The third molecule (W) has not only experienced a more tortuous path through the packing material, it has also gotten trapped in some completely obstructed channels and has had to reverse direction and extricate itself by diffusional movement. Obviously, it will lag far behind molecules X and Y. The contribution to band spreading due to the differences in flow velocity just described is frequently called *"eddy diffusion."* This phenomenon occurs independent of the flow rate.

Another factor which affects the spreading of the molecules in the column is the time spent in diffusing into the stagnant fluid in the pores of the packing material and into the stationary phase. As shown by molecule Z, the longer the molecule spends in the support, the further it gets behind a molecule which remains in the flow stream. These processes are strongly dependent on the support particle size and on the thickness of the stationary phase.

A detailed discussion of these parameters is beyond the scope of this section,[1,2,4] but the conclusions can be summarized. For gas chromatography, the relationship between the plate height, H, and flow velocity is given by the van Deemter equation:

$$H = A + B/v + Cv$$

where v is the flow velocity of the mobile phase and A, B, and C are constants related to the mechanistic components of dispersion. The eddy diffusion, which is independent of flow velocity, contributes to the A term. The B term arises from diffusion of the analytes along the long axis of the column and is important only at low flow velocities. The C term includes contributions from radial diffusion in the mobile phase and diffusion in the stationary phase.[1,2] Because of basic differences in the rates of these processes in gases and liquids, the relationship for liquid chromatography is given by the Knox equation:

$$H = Av^{(1/3)} + B/v + Cv$$

For both gas and liquid column chromatography, the efficiency of a chromatographic system is optimal at intermediate flow velocities as shown in Figure 1B-50. Optimum performance is normally not obtained in

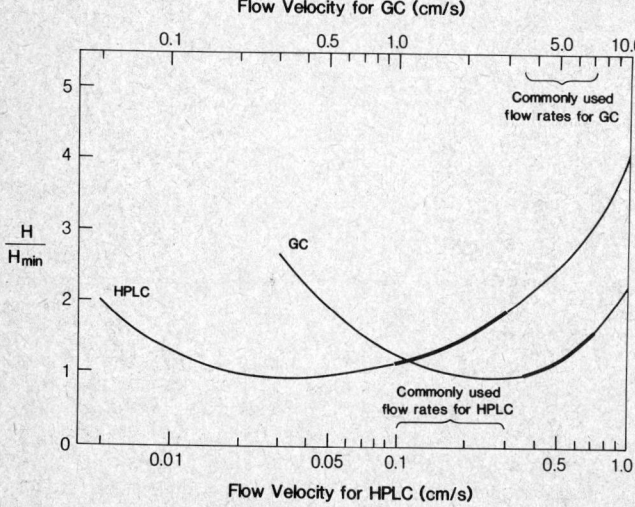

Figure 1B-50. Van Deemter and Knox equations relating a normalized height-equivalent to a theoretical plate to mobile-phase flow velocity. As the flow rate in the normal operating area increases, the height of a plate increases and column efficiency decreases.

practice because of the emphasis on separation speed, which requires the use of flow rates greater than optimal. The factor limiting column efficiency at commonly used flow rates is the homogeneity of the column packing in the column. Since there are strong relationships between the constants and the diameter of the packing material, small particles are more efficient than larger particles and they are more efficient at a higher flow rate. This translates into faster analysis times since the separation time is proportional to the height of a theoretical plate divided by the flow velocity (H/v). Understanding these principles has led to the development of high performance liquid chromatography and capillary gas chromatography, both of which have the kind of speed necessary for clinical analysis. Table 1B-9 compares the relative speeds of the various chromatographic techniques that have been discussed.

PRACTICAL ASPECTS OF CHROMATOGRAPHY

The separation of two compounds depends not only on the distance between the center of the peaks or spots ($V_{r(B)} - V_{r(A)}$) but also on the width of the peaks or spots. *Resolution*, R_s, which is the measure of the relative separation of two peaks, is defined in Figure 1B-48. It is a function of both the distance of band migration and the band spreading. Since the purpose of chromatography is to separate the compounds of interest, resolution is probably the most important aspect of a separation system. In order to separate two compounds such that there is less than 0.1% of compound A present within the peak corresponding to compound B, a resolution (R_s, see below) of 1.5 is required. With this resolution, the signal will almost touch the base line between the peaks as shown in Figure 1B-48. For analytical work where speed is a consideration, a resolution of 1.0 would correspond to an error in the peak areas of two equal-sized peaks of about 2%. The error in the peak heights would be less than 0.1%. Since both peak area and peak height are proportional to the concentration of analyte, this would result in a corresponding error in the analysis. With a resolution of 1.0 for two equal-sized peaks, the valley between the peaks would be a little less than one third the height of the peak maxima. As the relative size of the peaks changes, the relative overlap will change as well, so that when peak A is 16 times larger than peak B the error in the peak areas with a resolution of 1.0 is 0.8% for A and −12% for B. A more detailed discussion of the role of resolution can be found in reference 4.

Since resolution is related to retention and band spreading, it can be controlled by adjusting the equilibrium and flow dynamics of the separation system. For column chromatography, the relationship between these parameters is:

$$R_s = \frac{\sqrt{N}}{4} (\alpha - 1) \left(\frac{k'}{1 + k'} \right)$$

where N is the number of theoretical plates, α is the selectivity, and k' is the capacity factor of the earlier eluting peak. A slightly different relationship exists for thin-layer chromatography. We have discussed the role of system efficiency and its dependence on column characteristics (such as length and particle size) and on flow rate. Often the efficiency is determined by a commercial column purchased for the separation, thus limiting our means to improve the efficiency of the system. Although there is no resolution if there is no retention, retention by itself is of

Table 1B-9. COMPARISON OF THE EFFICIENCY OF SEVERAL CHROMATOGRAPHIC TECHNIQUES

	Performance	Samples/h*
Classical TLC	0.1 plate/s	~10[†]
Circular HPTLC	25–100 plates/s	~80[‡]
Classical LC	0.01 plate/s	1–2
HPLC	25–150 plates/s	2–10
Packed-bed GLC	5–30 plates/s	3–8
Capillary GLC	50–150 plates/s	4–12

*These numbers are representative examples from clinical samples using routine equipment. Higher throughput can be obtained by using column switching and other sophisticated techniques.

†Assumes 6 samples applied to plate concurrently.

‡Assumes 4 samples applied to plate concurrently.

little value. When $k'(B)$ is five, the contribution of retention to resolution is 83% of its maximum. Increasing the capacity factor beyond five, therefore, does little for resolution and creates broader peaks which are more difficult to detect. The important role of selectivity should be readily apparent. If there is sufficient difference in the affinity of the two solutes for the stationary phase, an inefficient column with poor retention will still resolve the two compounds.

The fact that selectivity and retention are both equilibrium properties guides the chromatographer in their adjustment. To a large degree, knowledge of the chemical structure of the compound(s) of interest and the selectivity of the separation modes can be used to choose a separation scheme. (Refer again to the example of separating glycosylated hemoglobin and hemoglobin.) In gas chromatography, the mobile phase is usually inert, and therefore all selectivity and retention modification must be achieved through the choice of a stationary phase and an operating temperature. One might choose, for example, a stationary phase with selectivity for barbiturates for use in a screen for the sedative and hypnotic drug classes.

An important advantage in liquid chromatographic techniques, whether column or thin layer, is the active role that the mobile phase plays in the separation. Secondary equilibria such as the interaction of hydrogen ions in the mobile phase with acidic or basic groups on the analyte, as well as the interaction of silver ions with double bonds in organic molecules or of metal chelates with chiral centers in organic species, or specific interactions between the solvent molecules and the analyte, can be used to assist in the separation of compounds with various types of structural differences.

Although the retention of a compound can be adjusted by various parameters, many samples contain compounds with extreme differences in their structural characteristics. As a result, resolution of two rapidly eluting peaks requires use of conditions which cause excessive retention and resolution of strongly retained peaks. For example, to separate the bile acid conjugates glycodeoxycholate, glycochenodeoxycholate, and taurolithocholate by liquid column chromatography, the mobile phase required to resolve the first two compounds in 5 min causes the last compound to be retained for 40 min. This is known as the *general elution problem*. It often causes unacceptable prolongation of the analysis time, as described above. Generally, the solution is to change the retentivity of the system during the run. In liquid chromatography, the composition of the mobile phase is changed from one composition to another either suddenly, called a step gradient, or continuously, called gradient elution. This change can be accomplished in thin-layer chromatography by two dimensional development as described in Figure 1B-47 and text. In gas chromatography, changes in temperature can vary the elution time.

SEPARATION MECHANISMS

ION-EXCHANGE CHROMATOGRAPHY

In ion-exchange chromatography, differences in the sign and magnitude of ionic charges are the basis of separation. This technique is most useful for separation of inorganic ions, amino acids, nucleotides, and proteins. Although there are inorganic ion-exchange supports (such as Permutit and hydroxyapatite), the greater capacity of polymeric supports such as cellulose, dextran, and polystyrene-divinylbenzene has all but eliminated use of inorganic supports in the clinical laboratory, except for some protein separations.

Materials which can be used to separate cations are known as *cation-exchange resins* and are characterized by the presence of negatively charged groups. These can be either strongly acidic groups, such as sulfonate ions, or weakly acidic groups, such as carboxylate ions or carboxymethyl (CM), phosphate (P), sulfomethyl (SM), sulfoethyl (SE), or sulfopropyl (SP) groups. *Anion-exchange resins* are characterized by the presence of strongly basic quaternary amines such as triethylaminoethyl (TEAE) groups or weakly basic groups such as aminoethyl (AE), diethylaminoethyl (DEAE), guanidoethyl (GE), and epichlorohydrin-triethanolamine (EC-TEOLA) groups which can bear a positive charge. The presence of weakly acidic and basic groups on the resin tends to enhance the acid-base selectivity of the support, since changing the pH will affect the charge not only on the compound of interest but also on the resin. This allows adjustment of retention by changing pH and ionic strength. The role of ionic strength can be viewed simplistically as a competition between the salt ion and the analyte ion for the exchange sites.

For example, a positively charged amino acid can compete better with one sodium atom for a cationic site than it can with ten (see Figure 1B-51). Finally, it should be mentioned that there is some ion-exchange selectivity for inorganic ions even when the net charge on them is the same. For example, the binding of the monovalent cations to a cationic resin increases in the order:

$$Li^+ < H^+ < Na^+ < NH_4^+ < K^+ < Rb^+ < Cs^+ < Ag^+$$

For the anionic resins of the strong type, the selectivity would be:

$$F^- < OH^- < CH_3COO^- < HCOO^- < H_2PO_4^- < HCO_3^- < Cl^- < NO_3^- < HSO_4^- < I^-$$

This is mainly due to the size of the solvated ion and its ability to penetrate the resin. Thus a separation with insufficient resolution using hydrogen ions in the eluent may benefit from the use of a lithium eluent, because the lithium displaces the analytes less vigorously and may allow a difference in their binding to appear.

Ion-exchange resins have been used for a spectrum of applications in the clinical laboratory—from mixed bed resins for removal of ions from water (deionization) to chromatographic separation of hemoglobin variants, isoenzymes of creatine kinase and lactate dehydrogenase, and amino acids. Ion exchangers may also be used to remove an interfering ion. For example, urinary porphobilinogen can be retained on an anion exchange resin while other compounds which might interfere with the color reaction are eluted. Porphobilinogen can then be eluted with acid and reacted with Ehrlich's reagent to obtain a selective quantitative test.

STERIC EXCLUSION CHROMATOGRAPHY

Steric exclusion chromatography (formerly known as gel-filtration, gel-permeation, size exclusion, molecular exclusion, or molecular sieve chromatography) depends primarily on molecular size selectivity to separate molecules, although molecular shape and hydration do exert an effect.

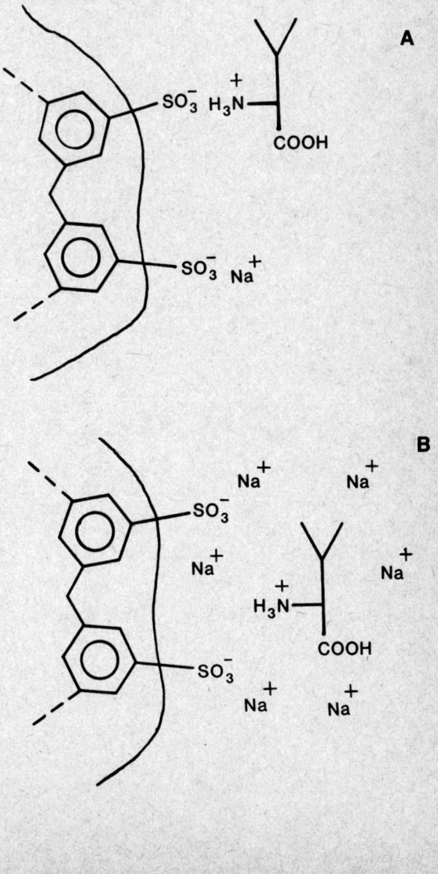

Figure 1B-51. Retention (A) and elution (B) of an amino acid with a cation-exchange resin as a function of sodium ion concentration. A higher ionic strength results in decreased retention. Variation of pH can also be used to adjust retention by changing the charge on the analyte molecule.

The stationary phase consists of a material which has well-defined pore size distribution. In liquid chromatography a variety of materials have been used, including cross-linked dextran (Sephadex), polyacrylamide (Bio-Gel), agarose (Sepharose), polystyrene-divinylbenzene, porous glass, and combinations of the above (Ultrogel, Synchrompak). By choosing a stationary phase with pore sizes that will allow the analyte to penetrate the support, the analyte will be retained, whereas molecules too large to enter the support will be present only in the void volume (V_o) and will be rapidly eluted from the column (see Figure 1B-52). Molecules which are intermediate in size will have access to various fractions of the volume and will elute according to the relation:

$$V_r = V_o + KV_i$$

where V_r is the retention volume, V_o is the volume between the particles, K is the fraction of the pore volume accessible to the molecule, and V_i is the volume within the support particles. Note that the retention is limited to one column volume and therefore resolution is limited. For this reason, steric exclusion is very seldom used in analytical work. Separations made on the basis of size assume there is no interaction between the solutes and the support. Unfortunately, this is often not the case. Hydrophilic gels have been designed to be used in water for separation of polysaccharides as well as enzymes, antibodies, and other proteins, while hydrophobic gels have been developed for separation of nonpolar species such as triglycerides in nonaqueous mobile phases. It should be appreciated that any chromatographic support which has a porous structure may exhibit some steric exclusion behavior.

Several different applications of steric exclusion chromatography have been used in the clinical laboratory. In the chromatographic mode, molecular weight data can be obtained for macromolecules. Small salt or buffer ions can be removed from protein solutions as an alternative to dialysis. One of the more interesting uses of the mode is in the DuPont "*aca*" amylase method. The colorimetric reaction scheme in the amylase method is based on the detection of glucose formed. Endogenous serum glucose, which would interfere, is therefore first separated from the amylase using steric exclusion chromatography. The amount of mobile phase used is just sufficient to elute the enzyme, leaving the glucose trapped in the column.

ADSORPTION CHROMATOGRAPHY

The electrostatic, hydrogen-bonding, or dispersive interactions between a molecule and a solid support or adsorbent are the basis for adsorption chromatography. The practice of *gas chromatography* in the adsorption mode is restricted to the analysis of low-molecular-weight species (e.g., methyl, ethyl, and isopropyl alcohols) and of compounds which are normally gases at room temperature (e.g., CO_2, N_2, and O_2) on stationary phases such as "molecular sieve," alumina, and Porapak polymers. In *liquid chromatography*, three types of adsorbents are generally

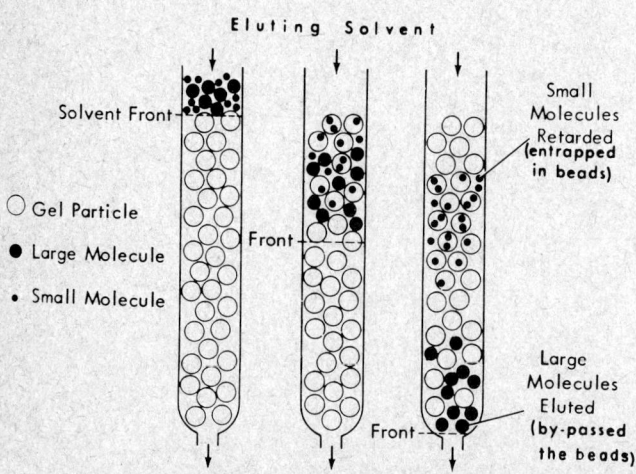

Eluting Solvent

Solvent Front

Gel Particle ○
Large Molecule ●
Small Molecule ·

Front

Front

Small Molecules Retarded (entrapped in beads)

Large Molecules Eluted (by-passed the beads)

Figure 1B-52. Representation of gel-filtration chromatography using a column technique. (Modified from Bennett, T. P.: Graphic Biochemistry. Vol. 1: Chemistry of Biological Molecules. New York, The Macmillan Company, 1968.)

used—nonpolar, acidic polar, and basic polar. The nonpolar adsorbents include charcoal and polystyrene-divinylbenzene. The main acidic polar adsorbent is silica gel. The surface silanol (SiOH) groups can interact well with basic substances, in some cases so well that they cannot be eluted. In these cases, the use of silica gel should be avoided. Alumina is the main basic adsorbent, preferentially retaining acidic substances. Florisil has also been used as a basic adsorbent when catalytic decomposition of the analyte is observed with alumina. Competition between the mobile phase components and the solutes for the adsorption sites on the support is characteristic of *liquid-solid chromatography* (LSC). As a result, a scale of solvent strength, or eluotropic series, can be developed based on the strength of solvent adsorption. For a polar adsorbent like silica gel, one would expect either methanol or water to be strongly adsorbed and therefore to serve as a strong solvent. Pentane would be poorly adsorbed and would thus act as a weak solvent. In contrast, for a nonpolar adsorbent such as charcoal, water would be a weak solvent and hexane would be a strong solvent.

Adjustment of the retention and selectivity are accomplished by judicious choice of a mixture of a weak and a strong solvent, usually with the assistance of solvent strength tables.[4] It should be pointed out that solvent mixtures which have the same eluotropic value in the table frequently show selectivity differences as a result of specific solvent properties.[4] Because of the types of interactions occurring, adsorption chromatography shows strong functional group, planar stereochemical, and isomeric selectivity, but very poor homolog differentiation. For example, mono-, di-, and triglycerides could be separated by LSC, but separation of 1-caproyl-glycerol from 1-stearoyl-glycerol would be impossible. The main drawback to adsorption chromatography is preparation of a support with a homogeneous distribution of adsorption sites. Technical problems associated with the reproducibility of retention have reduced the popularity of this mode of chromatography. Nevertheless, it has been used, for example, to separate steroids or vitamins prior to radioimmunoassay when the antibody cross-reactivity was unacceptable.

PARTITION CHROMATOGRAPHY

The separation force in partition chromatography is the relative solubility of the solute molecule in the mobile and stationary phases. For liquid chromatography, the capacity factor, k', is determined by the distribution coefficient, K_d, and the relative volumes of the stationary and mobile phases:

$$k' = (C_s V_s / C_m V_m) = K_d (V_s / V_m)$$

where C_s and C_m are the concentrations of the analyte in the stationary and mobile phases respectively and V_s and V_m are the volumes of the stationary and mobile phases. For gas chromatography, the same relationship applies, but V_m is the volume of gas. As might be expected, many types of interactions may occur between the solute and the mobile and stationary-phase components, and perhaps even with the support. An eluotropic series can again be developed for liquid chromatography using the Hildebrand solubility parameters or other solubility treatments. For *normal-phase* partition chromatography, a polar solvent such as β,β'-oxydipropionitrile is used as the stationary phase while a nonpolar solvent mixture such as hexane and ethanol is used as the mobile phase. Separation of a nonpolar compound would require a nonpolar stationary phase and a polar mobile phase. An example would be the separation of fatty acids with a squalene stationary phase using a mixture of water and acetonitrile as the eluent. A partition system in which the stationary phase is less polar than the mobile phase is called *reversed-phase* partition chromatography. With the advent of chemically bonded nonpolar supports, as many as 80% of all LC separations are carried out by reversed-phase partition. However, the mechanism of a separation is often not clear cut, with both adsorptive and partitioning forces involved.

The relative polarity of liquid stationary phases used in gas chromatography and their selectivity for specific functional groups can be ascertained from the respective "McReynolds constants," ΔI. These constants are determined by measuring the retention of a series of test compounds relative to a homologous series of normal alkanes. A squalene stationary phase is used as a reference point. The larger the McReynolds constant, the greater the retention of that class of compound. For example, SP-2401 would preferentially retain ketones ($\Delta I = 358$) relative to alcohols ($\Delta I = 238$). This is a unique feature of SP-2401 stationary phases. An extensive tabulation of McReynolds constants has been published by Supina.[6]

The broad selectivity of partition chromatography is due to the physical and chemical functional group selectivity arising from chemical interactions. Although partition chromatography is also good for separating homologs, it is not the method of choice for separating isomers.

AFFINITY CHROMATOGRAPHY

The term *affinity chromatography* has been used to cover a myriad of separation mechanisms involving some aspect of an interaction that occurs between biochemical species. These interactions can range from highly specific enzyme-substrate, hormone-receptor, or antigen-antibody complexes, to less specific charge-transfer complexes. Because of the vagueness of the "biospecificity" criterion, the term "affinity chromatography" has unfortunately been used to describe such varied modes as hydrophobic chromatography, covalent chromatography, metal-chelate chromatography, template chromatography, and charge-transfer chromatography.

The stationary phase in affinity chromatography is prepared by immobilizing a molecule (called a *ligand* because it participates in binding) on a support either directly or through a *spacer* (see Figure 1B-53). Because of the special stereochemical aspects of the interaction, the length and charge of the spacer may be critical to the ability of the stationary phase to bind specifically the analyte and to release it. The support must also be as inert as possible to prevent nonspecific adsorption. These restrictions have led to the general use of agarose, cross-linked dextrans, and polyacrylamide as supports, although cellulose, polystyrene, and controlled-pore glass have also been used. For example, to prepare a system to separate dehydrogenase enzymes which use nicotinamide adenine dinucleotide (NAD) as a cofactor, an NAD analog, adenosine monophosphate (ligand), might be attached to cyanogen bromide–activated agarose support through an amino-carboxy-hexane spacer. Adjustments of pH and ionic strength are required to achieve a strong binding of the analyte to the ligand. If the interaction between analyte and ligand is specific, the analyte may be displaced in a single step by addition of a substrate or an inhibitor, or alternatively by a pH change, by an ionic strength change, or by addition of a hydrogen bond-breaking agent such as guanine hydrochloride, urea, or sulfite. The latter group of compounds

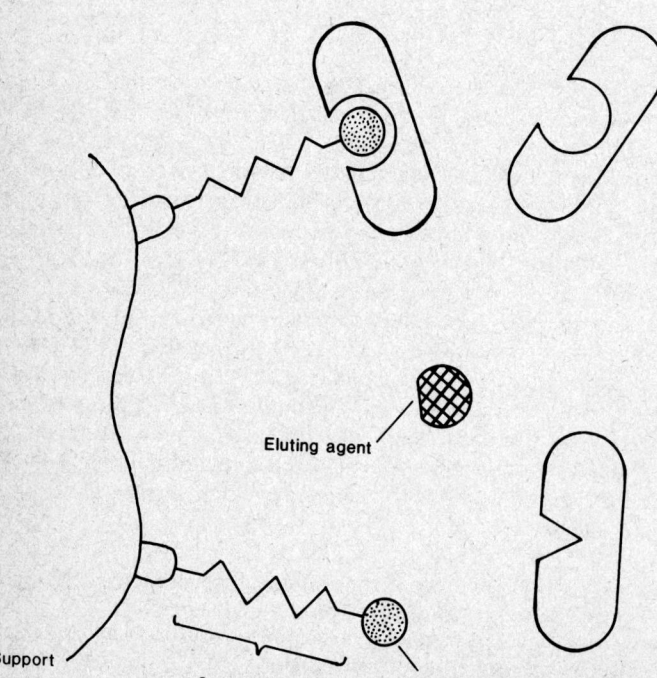

Figure 1B-53. Principle of affinity chromatography separation. The analyte (enzyme, antibody, antigen, tissue receptor, etc.) binds to the support-bound ligand. It can then be eluted with a general eluent, such as a chaotropic agent, a pH change, or with a biospecific eluent, such as an inhibitor or substrate.

Eluting agent

Support

Spacer arm

Ligand

are also known as *chaotropic agents*. If a group separation is required, as for example when using a cofactor as the ligand, gradient elution of the bound solutes, using one of the above-mentioned techniques, may be required to resolve the analyte mixture.

The power of affinity chromatography lies in its selectivity. As mentioned above, absolute selectivity obviates the need for efficient separation systems. For less specific affinity adsorbents, it remains to be seen whether slow equilibration and relatively poor column efficiencies will allow development of high performance affinity chromatography. In the clinical laboratory, use of affinity chromatography has been limited at present to preparative separations of proteins and antibodies. Lectin columns have also been used to separate cells with different surface carbohydrate moieties. More recently, heparin has been used as a ligand to specifically retain low-density and very-low-density lipoproteins. The use of phenyl boronate affinity supports to separate glycosylated hemoglobin and catecholamines may be the first widespread analytical use of affinity chromatography.

References

1. Giddings, J. C.: Dynamics of Chromatography, Volume 1, New York, Marcel Dekker, 1965.
2. Karger, B. L., Snyder, L. R., and Horvath, C.: An Introduction to Separation Science. New York, Wiley-Interscience, 1973.
3. Kirchner, J. G.: Thin Layer Chromatography. 2nd ed. New York, Wiley-Interscience, 1978.
4. Snyder, L. R., and Kirkland, J. J.: Introduction to Modern Liquid Chromatography. 2nd ed. New York, Wiley-Interscience, 1980.
5. Stahl, E.: Thin Layer Chromatography. 2nd ed. New York, Springer-Verlag, 1969.
6. Supina, W. R.: The Packed Column in Gas Chromatography. Bellefonte, Pa., Supelco, 1974.
7. Zlatkis, A., and Kaiser, R. E., Eds.: HPTLC: High Performance Thin Layer Chromatography. New York, Elsevier, 1977.

Additional Reading

Ettre, L. S.: Introduction to Open Tubular Columns. Norwalk, Conn., Perkin-Elmer Corp., 1978.
Ettre, L. S., and Zlatkis, A.: 75 Years of Chromatography—A Historical Dialog. New York, Elsevier, 1979.
Freeman, R. R.: High Resolution Gas Chromatography. 2nd ed. Avondale, Pa., Hewlett-Packard Co., 1981.
Giddings, J. C.: Field flow fractionation. Anal. Chem., *53*:1170A–1178A, 1981.
Jennings, W.: Gas Chromatography with Glass Capillary Columns. 2nd ed. New York, Academic Press, 1980.
Johnson, E. L., and Stevenson, R.: Basic Liquid Chromatography. Palo Alto, Cal., Varian Associates, Inc., 1978.
Turkova, J.: Affinity Chromatography. New York, Elsevier, 1978.

Section Seven

GAS CHROMATOGRAPHY

by Sati C. Chattoraj, Ph.D.

Since James and Martin[3] first reported the separation of volatile fatty acids by gas chromatography (GC), the technique has been widely applied in industrial and biomedical analysis. In clinical chemistry, however, utilization of GC has been less extensive because a great many compounds of analytical interest have a high molecular weight or have ionic character and thus are not readily volatilized. Nevertheless, improved instrument design, availability of more thermostable stationary phases, and introduction of methods to form more volatile derivatives have now made GC a powerful analytical tool for separating and quantitating a large number of compounds of clinical interest (see Chapters 9, 17, and 18).

GC has great advantages over conventional chromatography because of its high resolution, sensitivity, rapid separation, and ability to provide simultaneous quantitation. The following is a brief discussion of the principles, apparatus, and requisites for gas chromatography. For detailed information, the reader is referred to pertinent reviews and monographs.[1,2,4–6,9]

BASIC PRINCIPLES OF TECHNIQUE

Gas chromatography is a process by which a mixture of compounds in volatilized form is separated into its constituent components by moving a mobile (gas) phase over a stationary phase (sorbent or liquid phase). Under controlled conditions, the individual components of the sample, in accordance with their vapor pressure, will be present partially in the stationary phase and partially in the mobile phase. The ratio of the weight of solute per milliliter stationary phase to the weight of solute per milliliter mobile phase is termed the *partition coefficient,* K. A compound with a high vapor pressure will have a low partition coefficient, i.e., more of the compound is present in the gas phase. Therefore, it will be eluted more rapidly than another compound with a lower vapor pressure. If one compound has a selective interaction with the stationary phase and another has no or a lesser degree of interaction, the difference in rate of elution of the two compounds is further enhanced, i.e., the efficiency of separation is increased.

Gas chromatography is divided into two major categories: (1) gas-solid chromatography (GSC), in which the sorbent is a solid of large surface area, and (2) gas-liquid chromatography (GLC), in which a nonvolatile liquid, the stationary phase, is coated on an inert solid support. The mobile phase in both cases is an inert gas such as nitrogen, helium, or argon. The gas carries the solute molecules through the gas chromatographic column, hence the name *carrier* gas. The effluent from the column carries the separated sample constituents to the detector in the order of their elution. The response signal of the detector, after amplification in an electrometer, is fed into a recorder for display. The graphic representation is usually a straight base line upon which is superimposed a number of peaks, one for each of the components of the sample. *Retention time,* the elapsed time between injection of sample and the appearance of a given peak maximum at the detector, is a distinctive property of a particular component in a particular separation system. Retention time is therefore a means of qualitative identification of individual components of the mixture. Peak size (area or height), being proportional to the amount of component detected, is a means for quantitation of the separated components.

In an ideal chromatograph, the rate of transport of the solute molecule is determined by the velocity (flow rate) of the carrier gas and the partition coefficient (equilibrium distribution) of the solute between the mobile and stationary phases. The equilibrium distribution is dependent on the chemical composition of both the samples and the liquid phase, as well as on the column temperature. The rate of movement of the solute band throughout the entire column is uniform under specified experimental conditions. An ideal peak is sharp and needle-like, showing no dispersion. However, in any packed column, solute molecules and carrier gas molecules travel tortuous paths among the particles between the column inlet and the column outlet. Consequently, some solute molecules will leave the column ahead of or behind the center of the solute band, depending on the length of the route they travel. This phenomenon, known as *multiple-path effect,* or *eddy diffusion,* adds to peak broadening. The extent of such a diffusion process influences the *efficiency* (separating power) of the column. Other factors which influence broadening of a peak are the partition coefficient, thickness of the coated film of liquid phase, viscosity of the liquid phase, and carrier gas velocity.

By analogy to distillation and counter-current processes, evaluation of the quality of a column is based on the number of its theoretical plates, *N.* One theoretical plate is defined as a region on a column with perfect equilibrium of the solute between mobile and stationary phases. Stated simply, each plate corresponds to one equilibrium separation in a sequence of multiple extractions.

The calculation of the number of theoretical plates is made with the equation

$$N = 5.54 \ (X/Y)^2$$

where X represents the distance of the peak maximum from the point of injection of the solute and Y is the peak width at half the peak height. This quantity is related to the length of the column, L, by the expression $\overline{H}$, which stands for Height Equivalent to a Theoretical Plate (HETP), which is obtained by the equation

$$\overline{H} = L/N$$

The value of $\overline{H}$ is a useful characteristic for judging the efficiency of a column. The smaller the value of $\overline{H}$, the better is the separatory power of the column. Factors involved in the dispersion of the solute band during chromatography in a packed column are related to $\overline{H}$ by means of the van Deemter equation,[8]

$$\overline{H} = A + B/\mu + C \cdot \mu$$

where A, B, and C are constants related to eddy diffusion, molecular diffusion, and resistance to mass transfer, respectively, and μ is the linear gas velocity (or flow rate) through the chromatographic column. These factors are taken into account in constructing an efficient gas chromatographic column.

THE GAS CHROMATOGRAPH

A gas chromatograph consists of six basic parts: (1) a carrier gas supply with flow control; (2) a sample introduction system; (3) a column, a column oven, and its temperature control; (4) a detector; (5) an electrometer (amplifier); and (6) a recorder (Figure 1B-54). Ancillary attachments include devices for collecting the eluate and instruments such as mass spectrometers and ultraviolet, visible, or infrared spectrophotometers for identification of the eluted components of the sample mixture. In recent years, microcomputer-controlled electronics have made automation possible for the entire chromatography process from injection and separation through peak measurement, calculation, and final report, and have thereby improved the overall accuracy and precision of gas chromatographic analysis. In addition to data handling, the computer system offers the capability of recognizing and quantitating unresolved peaks and correcting for base line drift and nonzero base line. Automated gas chromatographs are now available from many commercial sources.

Carrier Gas Supply with Flow Control

The type of carrier gas is determined by the type of detector utilized in the system. The most frequently used carrier gas is nitrogen, which can be used with flame ionization, electron capture, or thermal conductivity detectors. Helium may be used with flame ionization and thermal conductivity detectors, while nitrogen admixed with argon-methane is used with the electron capture detector.

Precise control of the carrier gas flow rate through the column is important, since the retention time of individual sample components is in part determined by the gas flow rate. Thus, any variations in flow rate will lead to a change in retention time and uncertainty as to the

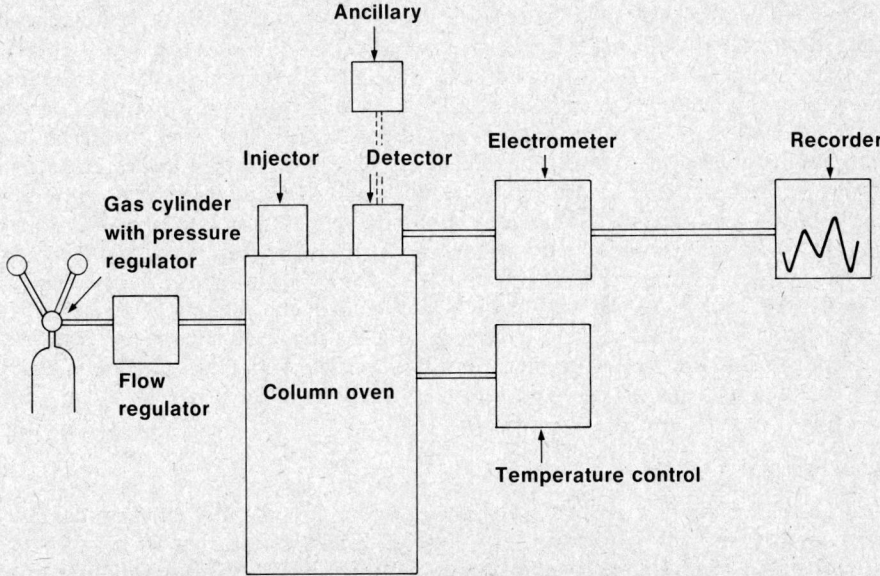

Figure 1B-54. Gas chromatographic system.

identity of the chromatographic peaks. In isothermal operation and in a system free of leaks, constancy of flow rate is adequately maintained by a simple flow-control device. But when temperature-programmed operation is used, maintenance of constant flow rate requires a differential-flow controller to compensate for changes in column pressure caused by varying column temperature and other factors.

Sample Introduction System

The sample to be introduced may be a solid, liquid, or gas. Gases are usually introduced by gas-tight syringes or by gas sampling valves. Liquids are injected into the carrier gas stream with a microliter syringe through a self-sealing septum in the apparatus. Solids may be dissolved in a suitable solvent and injected in liquid form. Injection of liquid sample is the most common introduction mode for clinical use of GC. Recently, devices for the direct injection of solids have also become commercially available.

The sample introduction system is kept at a temperature which will cause quick vaporization of the sample components so that the sample mixture moves onto the column in the form of a narrow band. Excessive dead volume in the injection port results in diffusion of the sample and therefore in increased peak-width and tailing. Efficiency of separation is also affected by the sample volume, which should be kept as small as possible (2–10 μL).

Formation of Derivatives

The high boiling point of some compounds requires that they be injected in the form of a more volatile derivative so that the desired instantaneous vaporization can be accomplished. Additional advantages of derivatization include (1) decrease of polarity, which minimizes "tailing" and adsorption of solute compound on the column; (2) stabilization of thermally reactive structures of the compound, which prevents its thermal breakdown; and (3) alteration of retention behavior, which improves separation of compounds in specific mixtures.

The criteria for the selection of the derivative will depend on the physicochemical characteristics of the compounds under study. Typical examples are as follows: for aliphatic and phenolic hydroxyl groups, derivatives such as trimethylsilyl ether, acetate, heptafluorobutyrate; for aliphatic and aromatic carboxyl groups, trimethylsilyl ester, methyl ester; for aldehyde and keto carbonyl groups, O-methyloxime, trimethylsilyl enol ether, O-(pentafluorobenzyl) oxime; for aliphatic and aromatic primary and secondary amino groups, N-(trimethylsilyl)amine, acetamide, heptafluorobutyramide. Halogen-containing derivatives are suitable for gas chromatographic analysis with electron capture detection (see also p. 155).

Column Oven and Its Temperature Control

The column oven of the gas chromatograph is designed to maintain constancy and uniformity of the column temperature, which are necessities for reproducibility of retention times and for maintaining a constant bleed rate of the column (evaporation of components of the liquid phase itself). Any "hot spots" along the column may cause localized deterioration of the column packing and cause undesirable adsorption of compounds. Any temperature drop either in the column itself or in the connection between column and detector (detector line) may lead to partial condensation and, consequently, to reduced efficiency of separation and to decreased sensitivity.

Isothermal operation limits gas chromatographic analysis to mixtures of a few components all of whose boiling points are within a relatively narrow range. In an isothermal run early peaks, representing low boiling substances, may emerge as sharp overlapping peaks while higher boiling components emerge as flat peaks that are difficult to measure. The use of *temperature programming* (i.e., the controlled change of column temperature) allows the analysis of complex mixtures containing analytes with a wide range of boiling points. At the start, a lower temperature is used and low boiling peaks are well separated. With the rising temperature, each higher boiling component is eluted as a sharp peak, and total analysis time is shortened.

Gas Chromatographic Column

Separation of the sample mixture takes place in the gas chromatographic column. Thus, serious consideration must be given to every detail of column construction in order to ensure efficiency and sufficient sensitivity for quantitative analysis. Packed analytical columns vary from

$1/8$–$1/4$ inch in outer diameter (O.D.) and 4–12 feet in length. Preparative scale separations are run on $3/8$ inch, $1/2$ inch, and larger diameter columns. Longer columns provide more theoretical plates and better resolution while an increase in column diameter increases the column capacity for sample. On the other hand, very long columns require very high carrier-gas pressures at the inlet that pose problems in injection technique and in avoiding gas leaks, and very wide columns lose column efficiency due to diffusion and the multipath effect.

The materials used for the construction of the column are usually glass or stainless steel. The choice of material depends upon the substance being analzyed. Cholesterol and particularly samples containing halogens may react with hot metal surfaces; thus, the use of a glass column in such cases is essential. Stainless steel columns are inexpensive and durable and have a non-adsorptive surface with excellent heat-transfer properties. Glass columns, on the other hand, allow visual inspection of the support during the packing process and detection of discontinuity of packing material or deposition of nonvolatile carbonaceous residues in the top of the column. However, the fragility of the column and the need to deactivate the active sites inside the tubing (silanization) are disadvantages of glass columns. *Capillary columns* are long tubes of small diameter. They range from 0.01–0.03 inch in internal diameter (I.D.) and from 100–500 feet in length. The inside wall of the tubing is coated with a thin film of liquid phase. Capillary columns have high efficiencies but low sample capacity. They are most suitable for the separation of complex mixtures and the resolution of closely related isomers such as closely related saturated hydrocarbons and saturated and unsaturated aromatic compounds.

Solid support material. Ideally, the solid support upon which liquid stationary phase is coated should be inert, with high crushing strength, large surface area, and uniform size. Its sole purpose is to hold the thin layer of the stationary phase. If the support is not inert, interactions of the vapor solute with support will occur, and asymmetrical peaks (tailing) or partial loss of the injected material due to irreversible adsorption or decomposition will result.

Diatomaceous earths (silicates) are widely used for support material. Interaction of polar groups with active sites on silicates is prevented by *silanization,* treatment with dichlorodimethylsilane or hexamethyldisilazane.

Examples of active sites   **After silanization**

The material is then treated with absolute ethanol or methanol in order to place methyl or ethyl radicals at sites where chlorine may have replaced hydroxyl groups during silanization. The treatment prevents reversion of such sites to hydroxyl groups should the support come into contact with water. Among the most popular commercially available silicate supports are Anakrom U, Celite 545, Gas-Chrom P and Q, and Chromosorb P and W.

Other supports include halocarbon types (e.g., Chromosorb T, Fluoropak 80), porous polymer beads (e.g., Tenax-GC, Porapak P), spherical siliceous supports (e.g., Porasil), and siliceous supports with chemically bonded liquid phase (Durapak). Choice of a support depends largely on the nature and type of the substance to be analyzed. The clinical analyst has a wide choice of support materials ready to use from commercial sources.

The efficiency of the column is in part determined by the size and distribution of the supporting particles. By decreasing the particle size of a support of uniform distribution, the multiple-path effect (eddy diffusion) on the transport of the solute is minimized, and thus the separatory power of the column is enhanced. Reduction of particle size, however, adversely affects the permeability of the column for the carrier gas, and therefore the practical limit of particle-diameter is about 150-mesh. Solid supports in the range of 80–100-mesh and 100–120-mesh generally give adequate columns for routine analysis. Supports are usually sized by screening through standard screens. Mesh numbers refer to the number of openings per linear inch. Particles that will pass through 60-mesh but not 80-mesh are referred to as 60/80-mesh.

Stationary phase (liquid phase). The proper choice of a liquid phase is of great importance for the separation, quantitation, and stability of the sample compounds. The nature of the stationary phase to be used is primarily determined by the physical and chemical properties of the compounds to be analyzed. The amount of stationary liquid phase used in relation to the amount of inert support (loading) may vary from less than 1–50% by weight. In general, diffusion phenomena are reduced in columns with relatively low amounts of a liquid phase. However, at extremely low concentrations, the support may possess sufficient residual adsorptivity to cause sample adsorption or tailing of the peaks. From the practical point of view, loadings ranging from 2–10% are suitable for most analytical purposes. However, the volatility of the sample needs to be considered when choosing the amount of liquid phase. As a rule, compounds of low volatility, such as steroids, are best analyzed on relatively low-loaded columns (3% or less), whereas very volatile materials such as light hydrocarbons, because of their poor solubilities, require high concentrations (20–30%) of liquid phase. The retention time is proportional to the amount of liquid phase present; the higher the liquid load, the longer is the retention time for a compound.

Although a wide variety of compounds have been used as liquid phases, a few carefully selected compounds probably suffice to fill the needs of analytical procedures carried out in a clinical laboratory. The ideal liquid phase, under experimental conditions, should be nonvolatile, thermally stable, and chemically inert toward the solutes under study. Methyl silicone polymers (e.g., SE-30, OV-1, DC-200, UCL-45), substituted silicone polymers (e.g., SE-52, OV-17, F-60, XE-60, QF-1, UCW-98), and silicone polyesters (e.g., EGSS-X, EGSP-A, ECNSS-S) fulfill these criteria fairly well. All methyl silicones have the following basic structure:

$$H_3C-\underset{\underset{CH_3}{|}}{\overset{\overset{CH_3}{|}}{Si}}-O-\left[\underset{\underset{CH_3}{|}}{\overset{\overset{CH_3}{|}}{Si}}-O\right]_n-\underset{\underset{CH_3}{|}}{\overset{\overset{CH_3}{|}}{Si}}-CH_3$$

The greater the value of n, the higher the molecular weight and the greater the viscosity. As n increases, the compounds change from oils to gums. When the molecular weight is of the order of 10 000, the vapor pressure of the gum is negligible, even at temperatures of 350 °C. The methyl silicone polymer SE-30 is an excellent example of a thermally stable stationary phase and is therefore used widely. The partition property of the silicone polymers can be modified by replacing the methyl group or silicone part to varying degrees by other groups. Substitution by phenyl as in SE-52, cyanoethyl as in XE-60, and trifluoropropyl groups (QF-1) or the introduction of a polyester ethylene glycol succinate (EGSS-X) has yielded useful stationary phases. Various non-silicone liquid phases such as branched chain C_{87} hydrocarbon (Apolane-87), polyamides (PolyA-135), and polyethylene glycol and its derivatives (Carbowax 1000, Carbowax 20M-TPA) have also found wide application in gas chromatography.

Stationary phases can be separated into nonselective phases (e.g., SE-30) and selective phases (e.g., NGS, XE-60). The former types are suggested for the separation of compounds with differing molecular weights and differing boiling points and for the separation of samples of a relatively nonpolar nature. The selective type phases are more frequently employed for the separation of isomers and compounds of varying polarity. In general, the selective phases are less thermostable than the nonselective phases.

The procedure of coating the solid support with the stationary phase is one of the important facets of column preparation. The stationary phase should consist of a thin uniform film around the solid support. Fragmentation of the very brittle support particles must be avoided to prevent exposure of adsorptive sites.

Column conditioning. Freshly prepared gas chromatographic columns and, especially, their stationary liquid phases may contain impurities such as low-molecular-weight compounds (shorter polymers) of the liquid phase. Such impurities are removed by heating the column, usually to 30–40 °C above the intended operating temperature for the column. This procedure is called column conditioning. In order to avoid contamination of the detector with impurities, detector lines should be disconnected during the process. The conditioning period varies with different liquid phases but is ordinarily 12–24 h. During this time the carrier gas is permitted to sweep

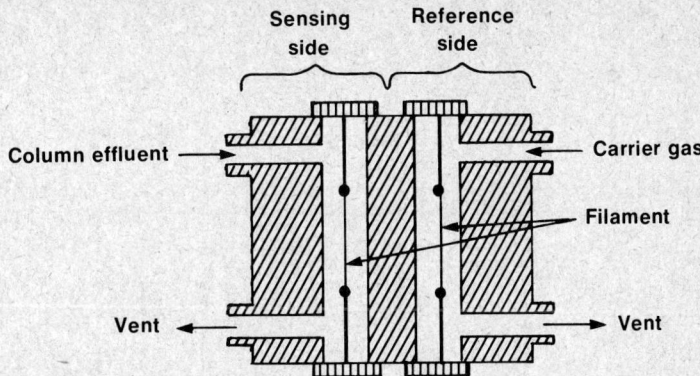

Figure 1B-55. Schematic diagram of a thermal conductivity detector.

through the column to carry off the impurities. Satisfactory conditioning of the column has been accomplished if a gas chromatogram shows a steady base line without any peaks or spikes.

Detectors

The column effluent, containing the individual separated compounds, enters the detector through the detector lines. As the compounds pass through the detector, an electrical signal is generated proportional to the amount of substance present in the carrier gas stream. Many types of detectors have been invented.[7] Among these, the thermal conductivity detector, the electron capture detector, the flame ionization detector, and the recently introduced alkali flame ionization detector are most noteworthy (see Table 1B-10).

Thermal conductivity (TC) detector. As the name implies, the operational principle of this detector is based on the changes in thermal conductivity of the carrier gas caused by the admixture of sample components. Changes in conductivity are proportional to the amount of substance eluted from the column. The detector consists of a metal block with two separate channels through which the column effluent and the pure gas, respectively, flow (Figure 1B-55). Each channel has a wire of tungsten or tungsten-rhenium alloy; these wires are connected to opposite arms of a Wheatstone bridge circuit which is fed from a constant current supply. Both filaments are heated electrically to raise their temperature above that of the detector block.

Since most substances analyzed have low thermal conductivities, sensitivity of the detector is increased by using a gas with high thermal conductivity. Helium is most frequently used. As the column effluent containing the mixture of carrier gas and separated sample components passes through the sample side of the detector, the temperature and therefore the electrical resistance of the sample filament rise, while resistance of the reference filament remains unchanged. Difference in resistance causes a current flow which is amplified by the electrometer and transmitted to the recorder for graphic display.

The TC detectors are nondestructive, and thus sample components may be collected for further studies. Other advantages of TC detectors are their ruggedness and their ability to respond to a wide variety of samples, including respiratory gases such as O_2, N_2, CO_2, CO, and others which cannot be detected with the flame ionization detector. Limitations of TC detectors are their relatively low sensitivity (lowest detection limit approximately 10^{-5} g of methane/mL of detector effluent) and their tendency to give baseline drifts.

Electron capture (EC) detector. These detectors are the most sensitive (lowest detection limit in the vicinity of 1×10^{-13} g/s for lindane), but they are also the most selective, since they detect only compounds with affinity for electrons, that is, those compounds which capture electrons. Thus, application of EC detectors is essentially limited to halogenated compounds. If the compound itself does not contain halogens, appropriate derivatives of the sample compounds can be formed.

The EC detectors contain two electrodes separated by an insulator (Figure 1B-56). A radioactive source such as tritium or nickel-63 is attached to the cathode. The column effluent containing carrier gas and separated sample components enters the detector through a tube leading to the anode, while the gas is exhausted through a hole in the cathode. As the carrier gas (e.g., nitrogen admixed with argon-methane) goes through the detector, it is ionized by the radioactive source. A fixed potential, or a short pulse of potential, just sufficient to collect all

Table 1B-10. SUMMARY OF DETECTOR CHARACTERISTICS

Detector*	Principle of Operation	Selectivity	Minimum Detectable Quantity (g)	Stability	Temp. Limit (°C)	Carrier Gas	Remarks
TCD	Change in thermal conductivity of gases	Universal response to all compounds	10^{-5} of methane per mL of detector effluent	Good	450	He H$_2$ N$_2$	Nondestructive; requires good temperature and flow control; simple, rugged, and inexpensive
FID	H$_2$-O$_2$ flame	Responds to organic compounds, not to inorganic gases or water	1×10^{-12}/s for alkanes	Very good	400	He N$_2$	Destructive; CS$_2$ good solvent because no response
ECD	N$_2$ + $\beta \rightarrow e^- +$ sample $\rightarrow$ decrease of standing current	Responds to electron absorbing compounds especially halogens, nitrates, and conjugated carbonyls				N$_2$ or Ar + CH$_4$	Nondestructive; detector is easily contaminated but easy to clean; sensitive to water; carrier gas must be dry; can be operated in pulsed or DC mode
^{3}H ^{63}Ni			10^{-13}/s for lindane 4×10^{-12}/s for lindane	Fair Fair	225 350		
AFID	Alkali modified H$_2$-O$_2$ flame			Fair	300	N$_2$	
P cmpd.		Enhanced response to phosphorus compounds	2×10^{-12}/s for parathion				Destructive; requires flow controller for hydrogen and air
N cmpd.		Enhanced response to nitrogen compounds	2×10^{-10}/s for azobenzene				Destructive; requires flow controller for hydrogen and air; high sensitivity operating in reduced O$_2$ mode

*TCD, thermal conductivity detector; FID, flame ionization detector; ECD, electron capture detector; AFID, alkali flame ionization detector.

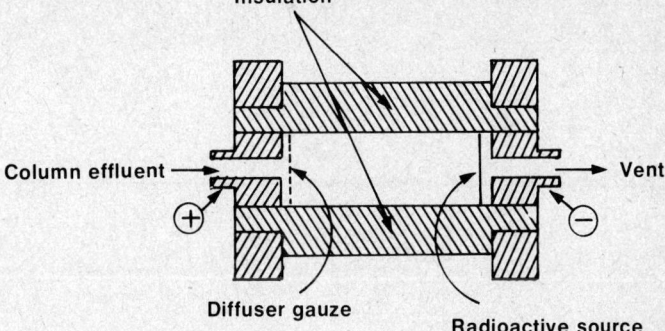

Figure 1B-56. Schematic diagram of an electron capture detector.

ions and electrons is applied to the electrode, yielding a very small but constant current across the detector (standing current). If compounds having an electron-capturing ability enter the detector, some electrons are captured, and as a result the standing current is diminished. The changes in current can be amplified and displayed as negative peaks on the recorder. In normal operation, however, a signal from the detector is inverted in the electrometer so that a positive peak is depicted on the recorder chart.

EC detectors have the great advantage of sensitivity and selectivity, but unfortunately they are also extremely sensitive to impurities in the sample and to changes in the operating conditions. Thus, their use should be considered only when the FI detector proves inadequate.

Flame ionization (FI) detector. This is the most widely used detector because of its simple construction, reliable performance, wide range of linear response, high sensitivity, and ease of operation. FI detectors are more sensitive than TC detectors (detectability up to 10^{-12} g/s for hydrocarbon), and are not subject to corrosion. They can be operated at high temperatures; the stability of the liquid phase is generally the limiting factor. FI detectors can essentially sense all organic substances and are therefore highly suitable for analytical work in a clinical laboratory. Inorganic gases and water, however, cannot be detected. The sample is destroyed during detection in the FI detector, and therefore sample collection is possible only if the column effluent is split prior to entering the detector.

In most FI detectors (Figure 1B-57), the effluent from the column is mixed with hydrogen gas prior to its admission to the detector. The mixture then enters a jet, where it is burned in either air or oxygen atmosphere to produce a small flame. As a sample burns in the flame, ionization occurs and the transmitted electrons are collected either by a collector loop placed around the flame or by collector plates on both sides of the flame. A potential difference (e.g., 300 V) is applied between the jet and the collector loop, each of which serves as an electrode. The degree of thermal ionization, and therefore the current flow, is directly proportional to the amount of sample burned in the flame. The current thus produced is linearly amplified by the electrometer and transmitted to a strip chart recorder for display.

Optimal performance of the FI detector is dependent not only on the applied potential and construction of the detector, but also on the ratio between flow rates of hydrogen and carrier gas. The ratio (rate of H_2 to rate of carrier gas flow) determines flame temperature and, consequently, efficiency of ionization. The usual recommendation for maximum response is for a ratio of 1:1 for hydrogen to carrier gas flow and 10:1 for air to hydrogen flow. Flow rates of carrier gas, hydrogen, and air should be optimized for each experimental condition if maximum detector response is desired.

Alkali flame ionization detector (nitrogen phosphorus detector). The alkali flame ionization detectors (AFID) are very similar in operation and construction to the FID. Indeed, the original construction of the AFID was almost the same as that of the FID, with the difference that a platinum wire coated with alkali sodium hydroxide was placed in the flame. In the presence of alkali in the flame, chlorine-containing compounds have a 15 times greater and phosphorus-containing compounds a 300 times greater response than the corresponding FID response. The principle of the response of the AFID has not yet been fully explained. It is generally considered that the presence of halogens, phosphorus, nitrogen, sulfur, arsenic, tin, silicon, or lead increases

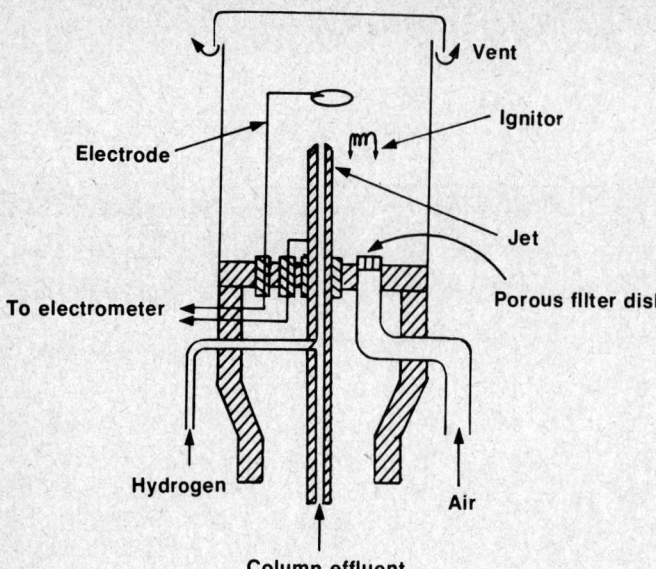

Figure 1B-57. Schematic diagram of a flame ionization detector.

the volatility of the alkali metal salt and thus also the concentration of the alkali metal in the flame. This results in an increase in ionization current with consequent enhanced response for the compounds containing the above elements.

Recent AFID construction differs mainly in the location and the nature of alkali metal in the flame. The most common and suitable construction of AFID is that in which FID is equipped with a burner jet fitted with a cap made from compressed alkali metal salt such as rubidium sulfate. In contrast to FID, AFID operates in a minimal oxygen condition. For example, the air flow rate in a normal FID ranges from 300–400 mL/min, whereas in the alkali detectors the air flow rate is approximately 130 mL/min. Since stable performance of alkali detectors depends on constant air and hydrogen flow rates, the use of flow controllers is important. Recently the use of chlorinated hydrocarbon pesticides has been de-emphasized because of their deleterious effects on health and environment. This has led to the increased use of organophosphate pesticides and to the greater application of selective nitrogen phosphorus detectors. The minimum detectable quantity for phosphorus-containing compounds (e.g., parathion) is approximately 2×10^{-12} g/s, whereas that for nitrogen-containing compounds (e.g., azobenzene) is 2×10^{-10} g/s.

References

1. Eik-Nes, K. B., and Horning, E. C.: Gas Phase Chromatography of Steroids. New York, Springer-Verlag New York Inc., 1968.
2. Gas chromatography in clinical chemistry. Clin. Chim. Acta, 34:129, 1971. 9th West European Symposium on Clinical Chemistry.
3. James, A. T., and Martin, A. J. P.: Gas-liquid partition chromatography: The separation and microestimation of volatile fatty acid from formic acid to dodecanoic acid. Biochem. J., 50:679, 1952.
4. Littlewood, A. B.: Gas Chromatography: Principles, Techniques and Applications. 2nd ed. New York, Academic Press, 1970.
5. Risby, T. H., Field, L. R., Young, F. J., et al.: Gas chromatography. Anal. Chem., 54:410R, 1982.
6. Schomburg, G.: Gas Chromatography. Weinheim, Verlag Chemie, 1977.
7. Sevcik, J.: Detectors in Gas Chromatography. New York, Elsevier Publishing Co., 1976.
8. Van Deemter, J. J., Zniderweg, F. J., and Klikenberg, A.: Longitudinal diffusion and resistance to mass transfer as causes of non-ideality in chromatography. Chem. Eng. Sc., 5:271, 1956.
9. Wotiz, H. H., and Chattoraj, S. C.: The role of gas-liquid chromatography in steroid hormone analysis. J. Chromatogr. Sci., 11:167, 1973.

HIGH PERFORMANCE LIQUID CHROMATOGRAPHY

by M. David Ullman, Ph.D., and Carl A. Burtis, Ph.D.

High performance liquid chromatography (HPLC) is a process that separates components of a mixture more efficiently than normal liquid chromatography. It is a "high performance" process resulting from technological advances in both instrumentation and column design.

Continued developments in HPLC instrumentation have led to flexible, fully automated analytical systems as well as sophisticated components for assembly of specialized systems. Advanced column technology has provided for the chromatographic separation of a broad spectrum of analytes. As a result of this flexibility and separation spectrum, HPLC has found wide use in clinical laboratories.

For the performance and development of HPLC methods, it is necessary to be familiar with the HPLC instrumentation and columns and to understand the basic theory behind the process. This understanding should lead to cost-effective decisions relative to the parameters that must be varied to acquire a desired resolution. Chromatographic theory can also be applied to formulate an efficient protocol for the routine operation of HPLC systems. This section will discuss HPLC instrumentation and columns, chromatographic theory, and some practical suggestions for the routine operation of the HPLC system.

INSTRUMENTATION

The instrument used in HPLC ranges from a single-pump isocratic system to a fully automated quaternary gradient system. Usually there are seven components in the HPLC instrument: a solvent reservoir, a pump, an injector, a chromatographic column, a detector, a data recorder and a microprocessor (Figure 1B-58).

Solvent Reservoir

The solvent reservoir is usually a brown glass bottle or flask which contains the mobile phase.

Pump

The pump drives the mobile phase from the reservoir and through the injector, the column, and the detector(s). In the past, several different types of pumps have been used, including the gas displacement pump, the gas amplified pump, and various types of syringe pumps. The most widely used type today is the *reciprocating pump,* in which an asymmetric cam drives a piston into and out of the pumping chamber (Figure 1B-59). In the fill mode, the piston is withdrawn from the chamber and a volume of liquid proportional to the diameter and stroke of the piston is drawn from the solvent reservoir, through the open inlet check valve, into the pumping chamber. In the pumping mode, the inlet check valve is closed, the outlet valve is opened, the piston is driven into the chamber, and liquid is pumped into the column.

The reciprocating pump has definite advantages over other types of pumps; namely, a small internal volume, modest cost, and simple flow control. However, a major disadvantage is that, due to its reciprocating action, the mobile phase pulsates and causes variations in the flow rate. This phenomenon is known as "pump noise" and is manifested as fluctuations in the output of the detectors. Consequently, pulsations decrease the signal-to-noise ratio of the detector and limit the analytical sensitivity of the entire HPLC system. Thus, most reciprocating pumps utilize either a pulse damper or a multi-head reciprocating pump. In the *multi-head type of pump,* two or more reciprocating pistons (Figure 1B-59) operate out of phase to minimize the flow pulsations.

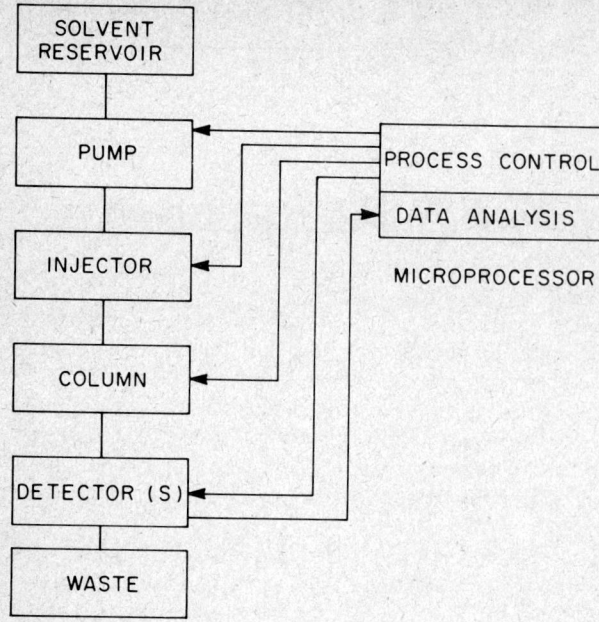

Figure 1B-58. Functional schematic of a high-performance liquid chromatograph.

Reciprocating pumps operate at up to 10 000 psi and generate flow rates between 1 and 10 mL/min.

The pump in a liquid chromatograph operates in two different modes: isocratic or gradient. In the *isocratic* mode the mobile phase composition remains constant throughout the chromatographic run. An isocratic mobile phase can be a single solvent (e.g., methanol) or it can be premixed from several different solvents (e.g., methanol-water) and delivered from a single solvent reservoir. Alternatively, a multisolvent mobile phase can be proportioned from two or more reservoirs.

In the *gradient* mode, the mobile phase composition is changed, either in a stepwise or in a continuous fashion, throughout the run. Many different techniques can be used to generate gradient profiles. In one technique, two pumps are used in parallel. A wide variety of gradient profiles can then be generated by programming the output of each one. Alternatively, the mobile phase can be proportioned on the inlet side of a single pump. For example, three different solvent reservoirs may be connected via proportioning valves to the inlet check valve of a single pump.

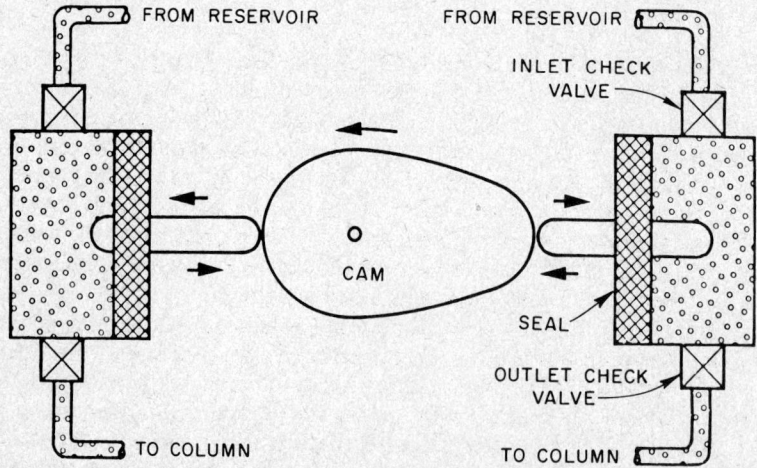

Figure 1B-59. Cross-sectional view of a dual-piston reciprocating pump. (From Walker, J. Q., Jackson, M. T., Jr., and Maynard, J. B.: Chromatographic Systems: Maintenance and Troubleshooting. 2nd ed. New York, Academic Press, 1977. Reproduced with permission.)

The composition of the mobile phase can then be varied by programming the time during which solvent is delivered through each of the proportioning valves.

Injector

An aliquot of sample is introduced into a liquid chromatograph via some type of sample injector. The most widely used type is the loop injector. As shown in Figure 1B-60, in the fill position, an aliquot of sample is introduced with a syringe into an external loop of the injector at atmospheric pressure. In the inject mode, the valve is activated, thereby rotating the sample loop into the flowing stream of mobile phase, which results in the aliquot of sample being flushed into the chromatographic column. Although they are more expensive, loop injectors are precise and can be used at high pressures. In addition, they can be automated, and consequently multi-sample loaders which incorporate a loop injector are commercially available and have been integrated into automated liquid chromatographs.

Chromatographic Column

The chromatographic column consists of a stainless steel tube which contains the packing (stationary phase) and the ancillary fittings used to connect it into the chromatographic system.

Column packings. Numerous materials are used as column packing, including silica, alumina, charcoal, and organic polymers. However, silica is the most widely used material and is produced in three principal types of particles: macroparticulate, microparticulate, and pellicular (Figure 1B-61). Macroparticulate supports have a mean particle diameter larger than 40 μm and are completely porous. Microparticulate supports have a mean particle diameter between 3 and 10 μm, are completely porous, and have either a spherical or irregular shape. Pellicular packing has a thin pellicle or porous layer of stationary phase coated on 20–60 μm solid glass beads. Microparticulate packings have the highest efficiencies as well as the greatest loading capacities and are the most widely used.

Depending on the chromatographic mode in which the column is to be used, the silica-based column packings shown in Figure 1B-61 are used directly or modified by coating or chemically bonding an active phase onto the silica surface. In liquid-solid HPLC, where the separation depends on the differential adsorption and desorption of solutes on the surface of the stationary phase, the particles of silica are used directly. Many lipid separations are performed in this mode.

In partition or liquid-liquid HPLC, the solutes differentially distribute themselves between two immiscible liquids; one of these, attached to particles of silica, serves as the stationary phase. In the past, the attachment process consisted of physically coating the particles with the desired liquid. Since the attachment is a simple physical adhesion, such column packings have a tendency to lose their liquid coating with a concomitant loss in their separating ability. This problem can be partially offset by presaturating the liquid used as the mobile phase with the liquid used as

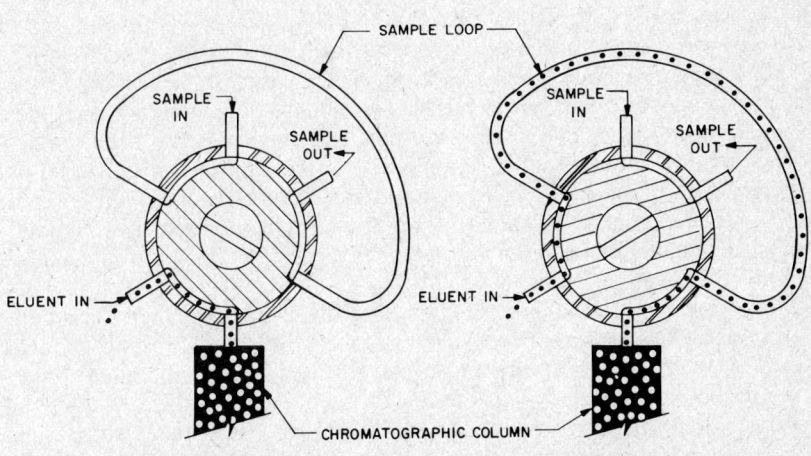

Figure 1B-60. Cross-sectional view of a sample loop injector that is commonly used in a liquid chromatograph.

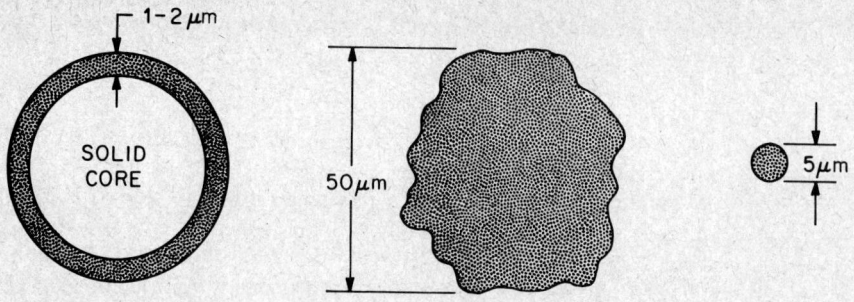

Figure 1B-61. Various types of chromatographic packings used in liquid chromatography.

the stationary phase. However, columns packed with coated column packings have short lifetimes and consequently are not widely used in modern-day HPLC.

A very significant advancement made in column technology has been the development of the "bonded-phase" column packings. With these packings, the molecules of the liquid used as the stationary phase are chemically bonded to the surfaces of the silica particles through a silica ester or, preferably, a silicone polymeric linkage. The resultant packings are mechanically and chemically stable, have long lifetimes, and provide excellent chromatographic performance. The capability of chemically bonding specific functional groups has resulted in the availability of a variety of versatile column packings that can be used for both partition and ion-exchange applications.

Two forms of partition HPLC exist: normal-phase and reversed-phase. In *normal-phase HPLC* the functional groups of the stationary phase are polar, while the mobile phase consists of nonpolar solvents, such as hexane or pentane. Examples of polar functional groups that are used for normal-phase HPLC packings are amino and nitrile groups.

Reversed-phase HPLC is the most widely practiced form of liquid chromatography, and it requires a nonpolar stationary phase. The most popular reversed-phase packing is the octadecyl type, in which a C_{18} hydrocarbon is bonded to silica particles. A column with octadecyl packing is often called an ODS column, which stands for *octadecyl silica*. Reversed-phase column characteristics can be altered by attaching different alkyl chains (e.g., octyl) to the silica. Mobile phases such as acetonitrile, aliphatic alcohols, tetrahydrofuran, water, or mixtures of such solvents are used.

"Ion-pair" HPLC is a version of reversed-phase HPLC and is particularly useful for biological separations. Generally, reversed-phase columns are most effective in separating nonionic compounds; but most biological compounds have ionic functional groups. The ionic nature of weak biological acids and bases (pK_a values between 2 and 8), can be suppressed by buffering the mobile phase. However, for strong acids and bases, pH adjustment cannot be used since silica dissolves at pH values less than 2 and greater than 8. In the ion-pair technique, a second ion (i.e., counter ion) is added to the mobile phase. This counter ion combines with the sample ions and forms a neutral ion-pair. The neutral ion-pair will then undergo the normal partitioning process. In this manner, strongly ionic compounds can be separated by reversed-phase HPLC. When the sample is a strong acid, a basic compound (such as tetrabutyl ammonium phosphate, tributylamine, or tridecylamine) is added to the mobile phase as the counter ion. For strong bases, sodium lauryl sulfonate, sodium dioctylsulfocyanate, or other sulfonic acids are used as the counter ion.

In *ion-exchange HPLC,* the column packings are either cation exchangers or anion exchangers. The cation exchangers have either sulfonic acid (strong cationic) or carboxylic acid (weak cationic) functional groups, while the anion exchangers have a quaternary ammonium functional group bonded to the particles of silica.

The various column packings are contained in a stainless steel tube, which ranges in length from 10–150 cm and has an internal diameter from 1–4.6 mm. These tubes are usually of a straight configuration, although both "U" and coiled shapes have been used. The tubes are packed by either dry- or wet-packing techniques.

The HPLC columns may be operated at ambient temperature or, alternatively, the column temperature may be regulated with a heating oven. Column end-fittings are used to connect the column into the HPLC system. The fittings contain a porous plug which is used to retain the column packing. All fittings and connecting tubing to and from the column must have a minimal dead volume to prevent dilution of the sample and loss of resolution due to remixing of separated components.

In some cases, a precolumn is used to protect the column packing from stripping of the stationary phase or from extraneous materials, such as protein, found in the sample. Most chromatographic packings will irreversibly absorb protein with a concurrent loss in resolution and life of the column.

Detectors

The fifth major component of a liquid chromatograph is the detector. As the name implies, its function is to detect compounds as they elute from the chromatographic column. Many different types of detectors based on a wide range of physical and chemical processes have been developed or proposed for use in liquid chromatographs. However, only a few are commonly used in HPLC, with the ultraviolet and the visible photometers being the most popular, although fluorometers and electrochemical detectors are becoming more widely used for clinical applications.

Ultraviolet and visible photometers. Ultraviolet (UV) and visible photometers are based on absorption of radiant energy by the compounds as they elute from the chromatographic column. The magnitude of the absorption and the wavelengths at which it occurs are a function of the molecular structure of the compounds. Most organic compounds will absorb in the ultraviolet and a few in the visible region of the electromagnetic spectrum. Consequently, UV and visible photometers are used as LC detectors and operate in the radiant energy regions of 190–400 nm and 400–700 nm, respectively.

Photometers operate in either fixed wavelength or variable wavelength modes. The fixed wavelength UV photometer is the most widely used of all LC detectors, and most utilize the intense 254 nm resonance line produced by a mercury arc lamp (Figure 1B-62). The radiation from such a lamp is directed through the reference and sample compartments of the flow cell, and the resultant transmitted radiation is detected and measured by a photo detector. The output from the photo detector is then displayed on a strip-chart recorder or a printer plotter to produce a chromatogram in which information is presented in an absorbance-versus-time format (Figure 1B-63). This type of detector is very sensitive. The absorbance range displayed can be as small

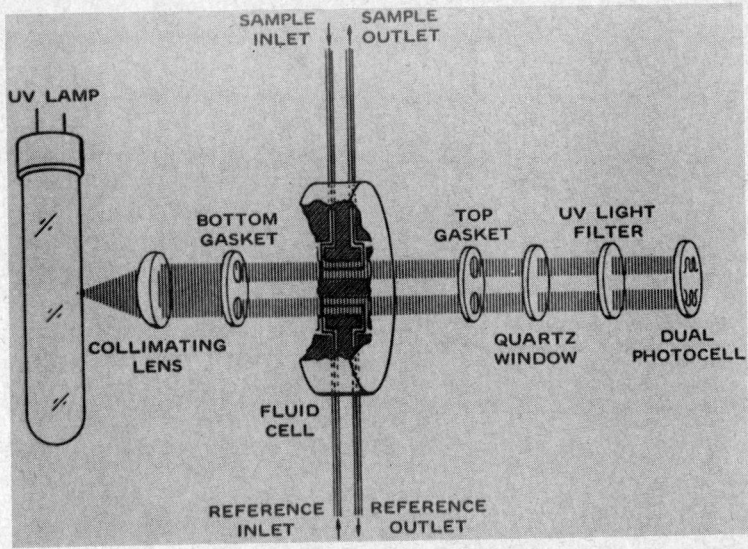

Figure 1B-62. Optical schematic of a simple photometer and flow cell (fluid cell) which constitute the most widely used detector in liquid chromatography. (From Johnson, E. L., and Stevenson, R.: Basic Liquid Chromatography. Palo Alto, Cal., Varian Associates, 1978.)

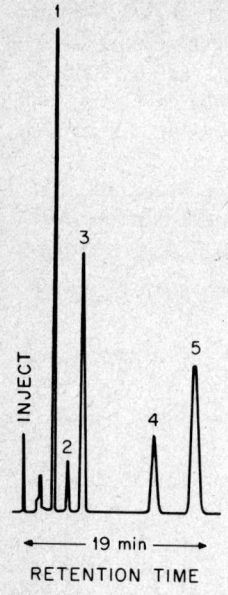

COLUMN: μBONDAPAK C$_{18}$
4 mm ID x 30 cm
SAMPLE: 10 μL
DETECTOR: UV: 254 nm, 0.02 AUFS

(1) ETHOSUXIMIDE
(ZARONTIN)
(2) PRIMIDONE
(MYSOLINE)
(3) PHENOBARBITAL
(LUMINAL)
(4) DIPHENYLHYDANTOIN
(DILANTIN)
(5) CARBAMAZEPINE
(TEGRETOL)

Figure 1B-63. Chromatogram obtained from a reversed-phase separation of anticonvulsants using a UV photometer detector at 254 nm at 0.02 aufs. (Chromatogram courtesy of Waters Associates, Milford, Mass.)

as 0.005 absorbance unit full scale (aufs). Additional advantages of the fixed-wavelength UV detector include simplicity of operation, reasonable cost, low base line noise level, and excellent compatibility with gradient elution. However, its use does require that the compounds of interest absorb energy at the photometer's fixed wavelength.

To provide the fixed wavelength detectors with more flexibility, a few of the other less intense resonance lines of the mercury lamp can be used; alternatively, a phosphor can be placed between the lamp and the flow cell and the emitted fluorescence light resulting from the 254 nm excitation can then be used as the light source. This latter approach is utilized in the dual wavelength photometers that operate at two fixed wavelengths, for example, 254 nm and 280 nm. The intense 214 nm resonance line of a zinc arc lamp can also be used for detection at a lower wavelength, where more compounds absorb.

The second type of photometer is the variable wavelength detector that operates at a wavelength which can be selected from a given wavelength range. Thus, the detector can be "tuned" to operate at the absorbance maximum for a given compound or set of compounds, which greatly enhances the applicability and selectivity of the detector. Another advantage of this detector is its ability to operate at wavelengths down to 190 nm. Since more compounds absorb at the lower wavelengths, this capability enhances the applicability and versatility of these detectors. For example, even a compound such as cholesterol, which has very little UV absorbance at 254 nm, can be detected at 200 nm (Figure 1B-64). It should be noted that at lower wavelengths many solvents absorb, and thus cannot be used as chromatographic eluents. Fortunately, acetonitrile and methanol, two widely used solvents in reversed-phase chromatography, do not absorb above 190 and 205 nm, respectively.

Fluorescence detector. Since many compounds have fluorescent properties, small fluorometers with flow cells have been developed and are increasingly being used as detectors in liquid chromatography. Most are relatively simple in design and construction and are very selective and sensitive if the compounds fluoresce within their operating wavelength range. In addition, either a pre- or postcolumn reactor can be added, which can chemically tag a compound with a fluorescent label. Amino acids or other primary amines are often labeled with either a dansyl or fluorescamine tag, followed by HPLC separation and fluorometric monitoring of the column eluent.

Electrochemical detectors. In this type of detector, the sample is either oxidized or reduced at an electrode surface under a constant potential. For such a detector to be useful, the compounds in question must be electroactive at the available potentials. Electroactive compounds of clinical interest include the urinary catecholamines. Useful methods have been developed for measuring these compounds, using HPLC in conjunction with an electrochemical detector. In addition,

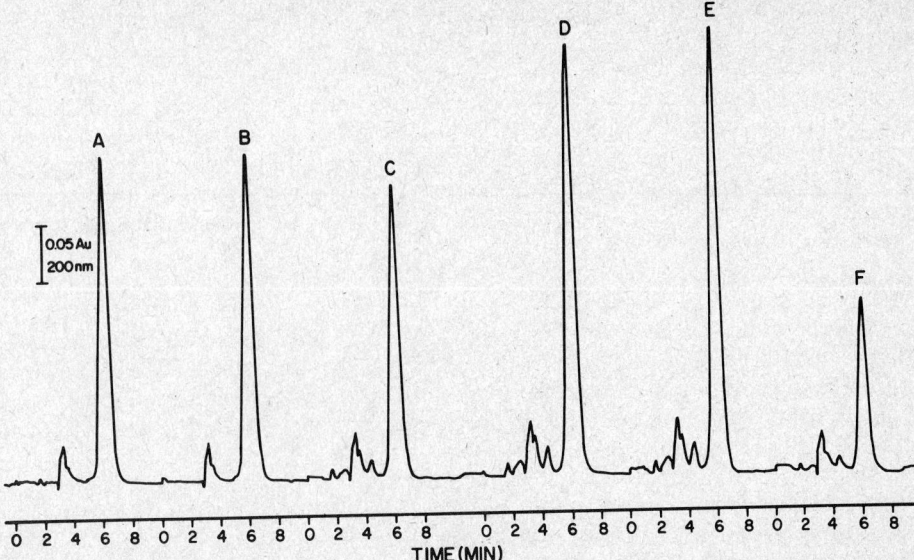

Figure 1B-64. Chromatogram obtained from a reversed-phase separation of free cholesterol using a variable wavelength photometer set at 200 nm at 0.5 aufs. Peak A is a cholesterol standard which contains 3.0 g/L (7.76 mmol/L). Peaks B–F were obtained from analyzing samples of serum. (From Duncan, I. W., Culbreth, P. H., and Burtis, C. A.: J. Chromatog., *162*:281-292, 1979. Reproduced with permission.)

electrochemically active tags (e.g., bromine) can be added to compounds such as unsaturated fatty acids or prostaglandins.

Data Recorder

A liquid chromatograph produces data in a time-versus-response format. The time required for a compound to pass through a liquid chromatograph is called the retention time, and under a given set of chromatographic conditions this time can be used to identify eluting compounds. The detector response is used for quantitative purposes, since the magnitude of the response is proportional to the quantity of the compound passing through it.

A strip chart recorder, one of the earliest data recording devices used in chromatographs, is still widely used for this purpose. A strip chart recorder displays chromatographic data as a sequential series of peaks, called a chromatogram. To utilize quantitatively the data contained in a chromatogram, either the area or height of each individual peak is measured.

A variety of manual, mechanical, and automatic techniques can be used to measure the peak areas. Peak areas can be calculated manually by the triangle formula area $(A) = HW_{1/2}$, where H is the height of the peak and $W_{1/2}$ is its width at half height. The area can also be calculated by triangulation with the formula $A = 1/2\ BH$, where B is the peak width measurement taken at the intersection of the base line with two lines drawn tangent to the sidelines of the peak. Another manual means of measuring peak area is to cut out and weigh each individual peak. Although these manual techniques are simple and inexpensive, they are laborious and less precise, with precision ranging from ±2–4%.

Two mechanical devices which have been used to measure peak areas are the planimeter and the plotting integrator. Both of these devices produce numbers which are proportional to peak area. Although the calculation of peak area with mechanical devices is less laborious than the manual techniques, their precision is still marginal (±1–4%).

Automatic data processing for chromatographic analysis became available in 1961, when digital integrators were used with gas chromatographs. Subsequently, digital integrators were adapted for use with HPLC systems. With these devices, the output of the detector is connected in series with the digital integrator and a strip-chart recorder. As solutes pass through the detector, they produce a signal which is monitored by the integrator and digitally converted to numerical information. As digital integrators became more advanced, they provided the chromatographer with a means to compensate automatically for many chromatographic anomalies, including incompletely separated peaks, tailing peaks, and drifting base lines. The analytical precision of a digital integrator is $\pm 0.5\%$.

Although digital integrators do an excellent job of measuring peak areas, they are still just simple data-acquisition devices. To provide computation and interpretation capabilities with on-line data acquisition, mini- and microcomputers have been interfaced with HPLC systems. These interfaced systems are very useful, since they provide the chromatographer with the practical and analytical advantages that result from a fully automatic data system. However, they are expensive. This economic disadvantage of on-line data systems was minimized when microprocessors became available in the 1970's.

Microprocessor

The incorporation of microprocessor technology into HPLC instrumentation has resulted in cost-effective data processing and fully automatic HPLC systems, which are easy to operate and which provide improved analytical performance. In fact, the advent of microprocessor-based HPLC instruments is one of the more important factors that has led to the acceptance and utilization of HPLC as a routine analytical tool.

In the clinical laboratory, the microprocessor can serve both process control and data processing functions (Figure 1B-65). As a process controller, the microprocessor monitors and controls various parameters such as mobile phase composition and flow rate, column back-pressure, column and detector temperatures, sample injection, detector selection and operation, and the various timing steps necessary for the system to operate. In addition to the control functions, the microprocessor also monitors and displays several system functions such as flow rate, column pressure, temperature, and run time.

For data processing, the microprocessor monitors the system detectors and commands the acquisition and storage of data at specified time intervals. The area or height of each chromatographic peak is determined from these stored data and is then used to compute the concentration of the solute that each peak represents. Available algorithms for this computation include those based on calibration curves or conversion factors using either internal or external standards. A complete report is then prepared for each chromatographic run. Either a cathode ray tube with a keyboard or a printer-plotter with a keyboard is used to interact with the microprocessor. The printer-plotter is of particular importance since it is also used to generate a hard-copy facsimile of the chromatogram (i.e., a response-versus-time plot). This precludes the need for a strip-chart recorder.

Applied Theory

The primary goal of any chromatographic process is to separate a mixture into its individual components, called solutes. For an effective separation, the individual solutes must emerge from the column at different times and their band profiles must be narrow enough to minimize overlap (Figure 1B-66). Resolution (R_s)

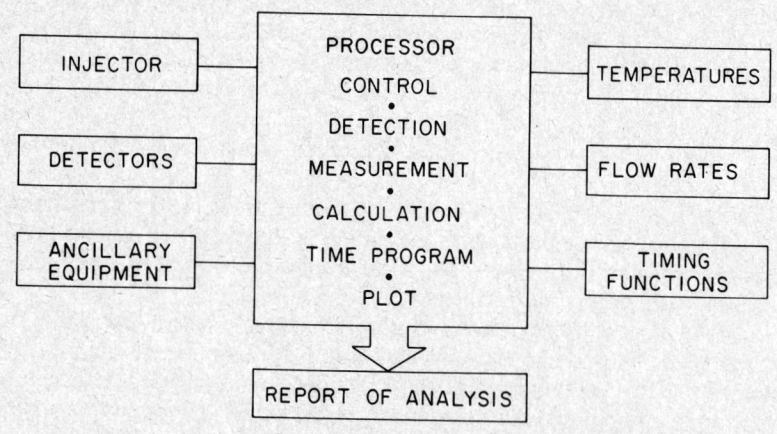

MICROPROCESSOR FUNCTIONS

Figure 1B-65. Functions of a microprocessor in the operation of a liquid chromatograph.

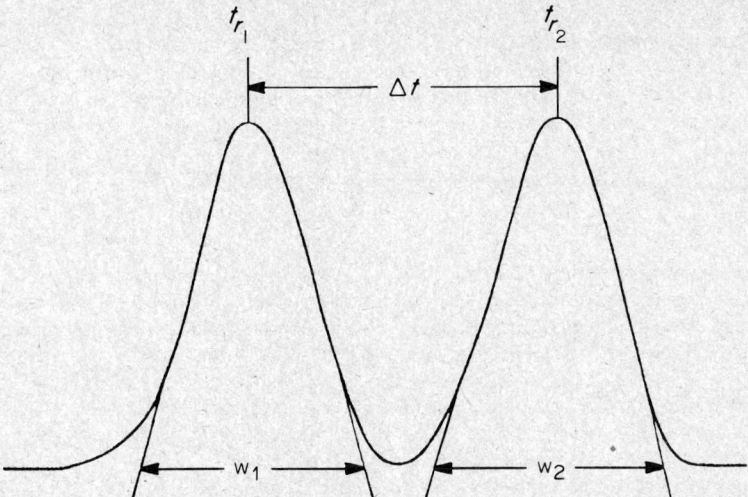

Figure 1B-66. A two-component separation showing the chromatographic parameters that are measured and used to compute resolution.

is a measure of the degree of separation of two solute peaks. By definition, R_s is the distance between the peak centers divided by the average of the peak widths. This is expressed mathematically as:

$$R_s = \frac{t_{r_2} - t_{r_1}}{\dfrac{w_1 + w_2}{2}} = \frac{2\Delta t}{w_1 + w_2} \tag{1}$$

where t_{r_1}, t_{r_2} = retention times for solutes 1 and 2, respectively, and w_1, w_2 = peak widths in units of time measured at the base for solutes 1 and 2, respectively.

Inadequate separations usually occur when the calculated value for R_s is less than 0.8, while base line separation is obtained when R_s is greater than 1.25 (Figure 1B-67). In actual practice, it is seldom necessary to calculate an R_s value, since by simple inspection of the peaks' shapes in a chromatogram, one can determine if the resolution is adequate. If R_s is not adequate, it can be increased by improving the column capacity factor, column selectivity, or column efficiency.

Column Capacity Factor

As a solute is introduced into the HPLC column, one of three things may happen. First, all of the molecules may become attached to the column packing, in which case no migration of the solute occurs, since no molecules enter the mobile phase flow stream. Second, none of the molecules may be retained by the column packing, in which case the solute emerges from the column at t_m, the time from injection until

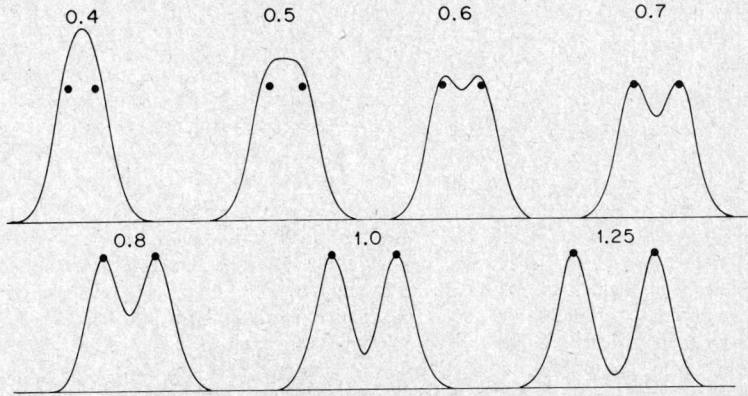

Figure 1B-67. Separation of chromatographic peaks as a function of resolution (R_s). (From Snyder, L. R.: J. Chrom. Sci., *10*:202, 1972. Reproduced with permission.)

unretained materials emerge. Finally, some of the solute molecules may become associated with the column packing, remaining at all times in near equilibrium with those in the mobile phase, in which case migration at a rate different from unretained components occurs. The resolution provided by a chromatographic column is a function of the ratio of solute molecules on the column packing to those in the mobile phase. This ratio is called the capacity factor (k') and can be computed from the following relationship:

$$k' = \frac{t_r - t_m}{t_m} \tag{2}$$

where t_r = retention time of retained solute, and t_m = retention time of unretained solute. The k' is small when few solute molecules are associated with the column packing. Solutes with a k' less than 2 are only slightly retained by the column packing and emerge from the column soon after the unretained materials. This leads to poor resolution because the unretained components may trail through the solute band of interest.

When k' is large, the number of solute molecules which are associated with the column packing is large and the solute band migrates slowly. Large values of k' improve separation, but at the expense of longer elution times and increased broadening of the solute bands. Optimal values of k' lie between 2 and 6, although in actual practice k' values between 1 and 10 can be used.

Column Selectivity

Column selectivity, α, is a measure of the relative separation between solute peak centers. It is a function of the column packing, mobile phase, and solute chemistry. It can be calculated from:

$$\alpha = \frac{t_{r_2} - t_m}{t_{r_1} - t_m} = \frac{k'_2}{k'_1} \tag{3}$$

where t_{r_1} and t_{r_2} = retention times of solutes 1 and 2, respectively; t_m = retention time of unretained solute (often called solvent or injection peak); and k'_1 and k'_2 = capacity factors of solutes 1 and 2, respectively. Since the k' of a solute equals its ratio of molecules on the column packing to those in the mobile phase, column selectivity, α, is improved by changing that ratio for the solutes of interest (see Figure 1B-68).

This change can be induced by altering (1) the composition of the mobile phase; (2) the pH and/or ionic strength of the mobile phase; (3) the stationary phase, e.g., the polarity of a reversed-phase column can be changed by using C_8 instead of C_{18} column packing; (4) the sample chemistry, through derivatization; and (5) the separation temperature.

When calculating α, the larger capacity factor is placed in the numerator; thus, α is always 1.0 or greater. When two solutes have the same k', α will be equal to 1.0 and no separation will occur. In general, solutes are relatively easy to separate if α is 1.1 or greater.

Column Efficiency

A very efficient column is characterized by the appearance of narrow peaks in the chromatogram (Figure 1B-68). The efficiency of a chromatographic column is indicative of its effectiveness in minimizing the spreading of solute bands. Column efficiency is expressed as the number of theoretical plates, N, and can be calculated from:

$$N = \left(\frac{t_r}{\sigma}\right) = 16\left(\frac{t_r}{w}\right)^2 = 5.5\left(\frac{t_r}{w_{\frac{1}{2}}}\right)^2 \tag{4}$$

where t_r = retention time,
$\quad\sigma$ = $\frac{1}{4}$ of the peak width at the base line,
$\quad w$ = peak width at the base line, and
$\quad w_{\frac{1}{2}}$ = the peak width at $\frac{1}{2}$ peak height.
The measurement of $w_{\frac{1}{2}}$ is usually more accurate because it is less arbitrary than base line measurements. Efficient 30-cm HPLC columns frequently have N values of 5000 or greater.

Efficiency is also expressed as the number of *theoretical plates* per unit length (N/L) or the reciprocal (L/N) of this relationship, which is called the height equivalent to one theoretical plate (HETP). Thus smaller HETP's signify more efficient columns. It becomes obvious that N (and therefore column efficiency) can be increased by increasing the length of the column. Other factors affect efficiency: characteristics of the column packing particles, mobile phase viscosity, column temperature, and noncolumn contributions. The most efficient column packings are composed of small (3–10 μ) regularly shaped particles. These particles produce the least number and narrowest alternative flow paths (minimize eddy diffusion). They are also of low porosity, which promotes efficient exchange of solute molecules between the column packing

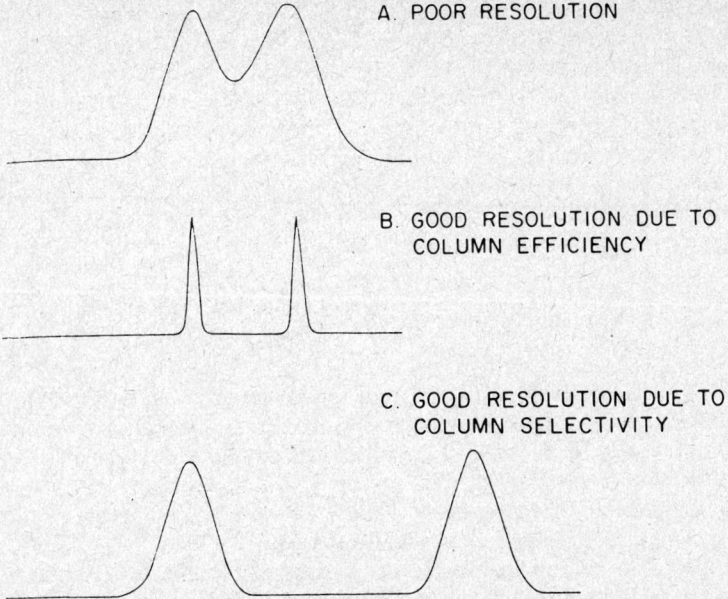

A. POOR RESOLUTION

B. GOOD RESOLUTION DUE TO COLUMN EFFICIENCY

C. GOOD RESOLUTION DUE TO COLUMN SELECTIVITY

Figure 1B-68. Demonstration of the effect of selectivity and efficiency on chromatographic resolution. (From Johnson, E. L., and Stevenson, R.: Basic Liquid Chromatography. Palo Alto, Cal., Varian Associates, 1978.)

and the mobile phase by minimizing mass transfer effects. In fact, any aspect of the chromatographic system which augments the exchange of solute molecules between the column packing and the mobile phase (e.g., low viscosity mobile phase, mildly elevated separation temperature) will provide narrower solute bands and higher efficiencies.

It should be noted that the high efficiency of any HPLC system will be lost if noncolumn contributions to solute band spreading are not properly controlled. The total solute band width (or volume) is the square root of the sum of the squares of the band spreading of each component in the chromatographic system. Pre- and postcolumn tubing, connectors, column end fittings, and the detector cell must be of minimum diameter, length, and volume.

Improvement of resolution. Improvement of resolution of solutes in a chromatographic run may be required for any of three major reasons: inadequate resolution, cost, and time. Cost and time considerations are extremely important to the clinical chemist. Unwarranted expense from a procedure can accrue from excess solvent usage, excess column usage, or wasted work hours due to overly long chromatographic runs. Long chromatographic runs are especially unacceptable if a large number of samples are to be analyzed. To improve resolution within cost and time constraints, it is important to understand the relationship of resolution to the factors that control it, namely capacity, selectivity, and efficiency.

With the appropriate substitution and mathematical manipulation, equations 2, 3, and 4 can be combined to disclose this relationship as:

$$R_s = \left(\frac{k'}{k' + 1}\right)\left(\frac{\alpha - 1}{\alpha}\right)\left(\frac{\sqrt{N}}{4}\right) \tag{5}$$

where k' is the capacity factor for the peak of interest, α is the selectivity factor for the peaks to be resolved, and N is the (column) efficiency factor. Thus, resolution can be improved through an increase in k', α, or N. The most cost- and time-effective approach to improvement of resolution is to adjust first the capacity factor (k'), then the selectivity factor (α), and then the efficiency factor (N).

The k's for the solute zones of interest are first adjusted to between 2 and 6. This adjustment is easy to make, since it entails a simple modification of mobile phase strength. However, R_s will be appreciably improved when the initial value of k' is small, since the term $k'/k' + 1$ is asymptotic with 1.0. Also, an increase in k' requires an increase in the analysis time, a sometimes unacceptable trade-off.

The α value is evaluated after k' is acceptable. If α is less than 1.1, it will have to be increased. The most dramatic improvement in R_s can be obtained with an increase in α, but it is also the most difficult and time consuming factor to vary. The manipulation of α entails essentially a "trial and error" approach to the search for new column-mobile phase-sample chemistry combinations until the α value between the solutes of interest is greater than 1.1.

Finally, if necessary, the chromatographic efficiency should be adjusted. Three parameters have a major influence on chromatographic efficiency: column length, column packing characteristics, and mobile phase flow rate. The column length required to give the desired resolution can be approximated from the following relationship:

$$\text{Column lengths required} = \left(\frac{\text{Desired resolution}}{\text{Observed resolution}}\right)^2$$

For example, if the desired resolution is 1.2 and the observed resolution is 0.6, then the column length must be increased by at least a factor of 4. However, an increase in column length increases the analysis time. It is also expensive, since additional columns are needed.

The efficiency of the chromatographic system can also be improved through a decrease in the mean particle diameter of the column packing. Several manufacturers now make column packings with mean particle diameters as small as 3 μm.

Once the appropriate column length and the packing material have been selected, the efficiency of a column can be improved by decreasing the flow rate of the mobile phase. The relationship between efficiency and flow rate is shown with the van Deemter plot (Figure 1B-69).

There are other approaches to increasing the value of N; and, although they do not have a dramatic effect on its value, they avoid the considerable investment in money (increased column length) and time (decreased flow rate) described above. For example, a slight increase in the column temperature may convert an unacceptable resolution value into an acceptable one if the solute is thermally stable. It is therefore advisable to include a column heating device as part of the high performance liquid chromatography system.

PRACTICAL OPERATION

It is beyond the scope of this text to discuss all elements of the HPLC process. However, there are some rudimentary practices that will augment the routine operation of the HPLC system. These practices pertain to sample preparation, tubing assembly, mobile phase preparation, pumps, injector, column, detector, and laboratory safety.

Sample Preparation

The mixture to be separated by the HPLC process is an integral part of the system. Uninterrupted HPLC operation requires that the dissolved sample be free of any particulate matter,

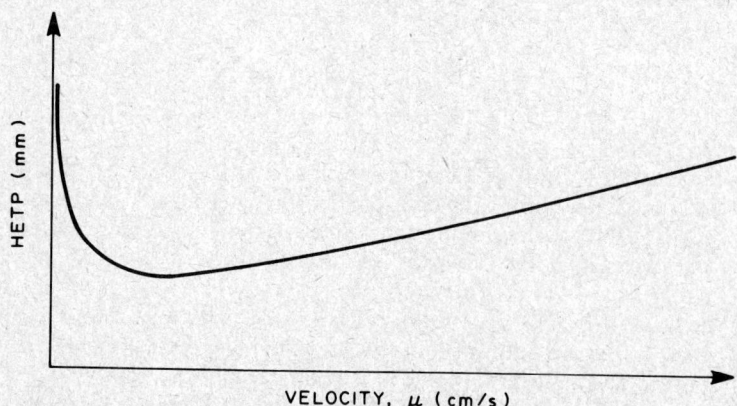

Figure 1B-69. Typical van Deemter plot of mobile phase velocity versus efficiency (HETP).

because the small-diameter tubing (usually 0.009 inch) used to connect the HPLC components plugs easily. Several types of sample filters are available which utilize a 0.5-μm screen to eliminate particles from the sample solutions.

It is advantageous to dissolve the sample in a minimum volume of solvent. This is particularly important in an isocratic elution system, since an excess solvent volume can contribute significantly to solute band spreading. Slightly larger solvent volumes can be used if the solvent is of a lesser strength than the mobile phase, or if the samples are separated in a gradient elution system, since both conditions will allow the solute to concentrate at the column head.

If the sample is to be dissolved in a solvent different from the mobile phase, it must be predetermined that the solute will not precipitate as the solvent and mobile phase mix.

Tubing Assembly

It is important to follow the manufacturer's suggested procedures for fitting and tightening the ferrule assemblies that connect the tubing to the various components of an HPLC system. Overtightening the assemblies may cause irreparable damage, and undertightening or improper alignment of unions may permit leakage. It is possible to have a leak in the chromatographic system that does not discharge mobile phase, but allows air to be absorbed through it. This type of leak will usually be manifested in base line disturbances created by the evolution of dissolved gases, as the mobile phase passes from the high-pressure (column) to the ambient-pressure (detector) side of the HPLC system.

Mobile Phase

Mobile phases should be prepared from HPLC-grade solvents that are free of particulate matter. Fortunately, most commercial-grade HPLC solvents are prefiltered through 0.5-μ screens. Solvents which constitute mobile phase should also be devoid of dissolved gases. One of three procedures is used to avoid the problems that arise from dissolved gases: vacuum degassing, helium or nitrogen purging, or postdetector back pressure.

With vacuum degassing, the mobile phase is stirred in a side-arm vacuum flask under reduced pressure. Nonpolar mobile phases require less than a minute to degas, but polar mobile phases require several minutes.

Helium purging requires that a constant and gentle stream of helium be bubbled through the mobile phase in the reservoir during operation of the HPLC; the relatively insoluble helium extracts the dissolved gases. It is important to place a drying agent in the helium line (between the helium tank and the solvent reservoir) to remove any water that might be in the gas.

Finally, some manufacturers attach a back-pressure valve to the detector outlet (for appropriate detectors) to hold gases in solution until they have passed through the detector.

Special precautions should be taken when buffer or salt solutions are used in the mobile phase. These solutions should never be left static in the pumping system, since the buffers and salts may precipitate. The precipitate can then score the piston(s) in the pump(s) and plug the flow system, or both. The first is costly in repairs and both are costly in time.

Pumps

HPLC pumps are technologically reliable. Little "down-time" is experienced if the manufacturer's suggested maintenance schedule (which is usually printed clearly in the instruction manual) is followed. The pump is composed of parts, such as check valves, pistons, and piston seals, that are subject to wear and dirt accumulation. A few minutes spent on maintenance, as suggested by the manufacturer, more than makes up for the investment in time.

When a different mobile phase is to be used in the HPLC system, it is advisable to pump several times the total pump and mixer volumes with the new mobile phase. When the previous mobile phase is immiscible with the new one, an intermediate or transition solvent, which dissolves in both, must be used. If the transition solvent is not used, the immiscible mobile phases will partition and form a biphasic system. Then, "bleed-off" of the partitioned solvents will cause unstable base lines and lost time.

During gradient mode operation the HPLC solvents which constitute the mobile phase must be adequately mixed. Commonly, one of two types of solvent mixer is used: static or dynamic. Static mixers rely on laminar flow dynamics. Dynamic mixers usually utilize a magnetic stirrer. The viscosity of solvents will affect their mixing characteristics. Inadequate mixing will often cause mobile phase refractive index changes, which can be detected by many UV detectors. These changes will generate unstable base lines.

Injector

Injectors, whether manual or automated, should receive careful and scheduled routine maintenance. Most manual injectors require the use of a sample syringe with a specific needle length and gauge.

Each change in mobile phase requires that the injector be flushed of old and replenished with new mobile phase. Fixed-loop or universal injectors demand that both the sample loop and by-pass loop be flushed with mobile phase. Failure to purge the injector properly can result in aberrant chromatograms.

Column

The HPLC column receives significant abuse by the very nature of its function. The column will accumulate debris, particularly in the clinical laboratory where so many biological materials are analyzed. This accumulation will, in time, alter column capacity, selectivity, and efficiency. The accumulation of debris can be retarded by the use of a guard- or precolumn between the injector and the column. The guard-column will remove many impurities which might otherwise bind irreversibly to the packing material. The avoidance of chemical or mechanical shock will also prolong column life.

The HPLC column will yield better routine performance if it is re-equilibrated with the mobile phase after a shutdown, a flow rate change, or a change in mobile phase composition (including gradient mode operation).

Detector

The UV detector is sensitive to refractive index changes in the mobile phase which will be manifested by base line fluctuations. Therefore, conditions that affect the mobile phase refractive index, namely temperature and pressure, should be carefully stabilized. A piece of rubber tubing which is fitted over the inlet line can be used to insulate the HPLC column and the detector inlet from ambient temperature changes. Pumps with minimum pressure pulsations will guard against pressure changes. The electrochemical detector is susceptible to ambient temperature fluctuations and pump pulsations when it is operated at very high sensitivities.

A log of detector usage should be kept so that lamp failure (UV and fluorescent) or electrode failure (electrochemical) can be anticipated. If it is financially feasible, spare parts such as detector cell windows, seals, and lamps should be kept in stock.

Safety

Toxic and flammable solvents used with the HPLC instrument demand that eating and smoking in the vicinity of the chromatograph be forbidden. When possible, the HPLC system should be located in an exhaust hood. If the laboratory does not have the available hood space, the column effluent, at least, should be directed to a hood and collected in a suitable container prior to proper disposal. The use of noxious solvents also requires that appropriate protective clothing and eye protection be worn.

Finally, the explosive release of pressure in an HPLC system is not a major hazard. Liquids compress only slightly, and therefore they accumulate very little energy. It is imperative, however, that stainless steel unions and columns be utilized. Brass fittings used for GC are *NOT* acceptable for use with the HPLC system, because they can abruptly release at elevated pressure.

Recommended Reading

Dixon, P. F., Gray, C. H., Kim, C. K., et al., Eds.: High Pressure Liquid Chromatography in Clinical Chemistry. New York, Academic Press, 1976.

Johnson, E. L., and Stevenson, R.: Basic Liquid Chromatography. Palo Alto, Cal., Varian Associates, 1978.

Rajcsanyi, P. M., and Rajcsanyi, E.: High-Speed Liquid Chromatography. New York, Marcel Dekker, 1975.

Snyder, L. R., and Kirkland, J. J.: Introduction to Modern Liquid Chromatography. 2nd ed. New York, Wiley Interscience, 1979.

Walker, J. Q., Jackson, M. T., Jr., and Maynard, J. B.: Chromatographic Systems: Maintenance and Troubleshooting. New York, Academic Press, 1977.

Yost, R. W., Ettre, L. S., and Conlon, R. D.: Practical Liquid Chromatography: An Introduction. Norwalk, Conn., Perkin-Elmer Corporation, 1980.

Section Nine

BASIC PRINCIPLES OF RADIOACTIVITY AND ITS MEASUREMENT*

by Edward R. Powsner, M.D.

The value of radionuclides for diagnosis, therapy, and medical research is fully established. Understanding both the risks and benefits of these applications requires knowledge of the basic principles of radioactivity as well as its measurement and of the biologic effects.

ATOMIC STRUCTURE, RADIATION, AND RADIOACTIVITY

The Atom

Atomic nucleus. In the 90 years since J. J. Thomson's suggestion that the atom was like a plum pudding of uniform positive charge with embedded negatively charged electrons, the atomic model has evolved into a small, central nucleus of positive charge surrounded by "orbital" electrons each carrying a single negative charge. In this planetary model, the mass of the atom is concentrated in the dense, central nucleus whose diameter is about 5 pm (picometers) for light atoms and 15 pm for heavy atoms; the overall diameter, including the electron shells, is about 10 000 times greater. Although the composition of the nucleus and the nature of its internal "glue" or binding energy are still not well understood, it will be adequate to assume that the nucleus is composed of *protons* and *neutrons.* The proton carries a single positive charge; the neutron carries no charge. Collectively, protons and neutrons are referred to as *nucleons.*

The masses of atoms and other particles are expressed in *atomic mass units,* amu. By definition, 1 amu is $1/12$ of the mass of one atom of carbon-12. Also, 1 amu is about 1.6604×10^{-27} kg. The mass of a nucleon is slightly more than 1 amu; by contrast, an electron has a mass of about 1/1900 amu. Of the nucleons, the proton is the smaller; it is slightly less massive (1.0072765 amu) than the neutron (1.0086650 amu).

Atomic species. The *atomic number* (Z) is the number of protons in the nucleus; the total number of nucleons, protons plus neutrons, is the *mass number* (A). A *nuclide* is an atomic species with a given atomic number and a given mass number. The generally accepted way of representing a nuclide is to place its mass number as a left superscript to its chemical symbol and its atomic number as a left subscript, e.g., $^{32}_{15}P$, $^{14}_{6}C$. Since both atomic number and chemical symbol identify the chemical species, the subscript is often omitted, e.g., ^{32}P, ^{14}C.

Isotopes are nuclides with the same atomic number but different mass numbers. These represent various nuclear species of the same element. Most elements occurring in nature are mixtures of isotopes. For example, natural carbon is predominantly ^{12}C with about 1% ^{13}C and a trace of ^{14}C. The naturally occurring isotopes of the lighter elements such as carbon and iron are stable, but among the heavy elements, such as lead and bismuth, some or all of the naturally occurring isotopes are unstable. These unstable nuclides undergo spontaneous decay to stable nuclides by a process referred to as radioactive decay. Decay is typically accompanied by the emission of energy in the form of radiation.

In the electrically neutral, un-ionized atom, the number of orbital electrons equals the number of protons. Because chemical properties depend upon the number and arrangement of the orbital electrons and only slightly on the atomic mass, the chemical properties of all isotopes of an element are virtually identical. This identity of chemical properties is the basis for isotopic tracer methodology; the fundamental principle is that the living system does not differentiate between isotopes of the same element.

*Portions of this chapter were retained from "Basic Principles of Radioactivity and its Measurement" by Maurice V. L'Heureux, *Fundamentals of Clinical Chemistry,* 2nd edition. These contributions, as well as additional helpful comments made to this chapter, are thankfully acknowledged.

Atomic binding energy. The mass of the atom is less than the sum of the masses of its components. The difference represents the binding energy of the atom. Thus, the atom of the helium isotope $^4_2$He has a nucleus of two protons and two neutrons plus the requisite two orbital electrons. The mass number of this helium is 4 and its actual mass is 4.0026 amu. Summing the individual components, the mass of the four nucleons is 4.0319 (2 × 1.00728) + (2 × 1.00867), and that of the two electrons is 0.0011 (2 × 0.000549); the total mass is 4.0330 amu, or 0.0304 amu more than the actual mass of the atom. This excess mass represents energy which may be conceptualized as the energy that must be supplied to break the atom into protons, neutrons, and electrons, or conversely, the energy released when the atom was formed. It is customary to assign a negative sign to binding energy to indicate that this energy must be supplied or added to overcome binding. Most of the binding energy of the atom is nuclear; orbital electrons are much less tightly bound to the atom than are nucleons.

Relation of mass to energy. To help explain the meaning of nuclear binding energy and its association with radioactivity, it is necessary to use the relationship between mass and energy. From Einstein's theory,

$$E = Mc^2$$

where E is energy in joules, M is mass in kilograms, and c is the velocity of light in vacuum (3.0 × 10^8 m/s). In these units, the energy equivalent of one kilogram of any material is 9 × 10^{16} J. Perhaps more useful is the corresponding statement in units appropriate to radioactive decay,

$$E_{eV} = 931.4 \times 10^6 \, M_{amu}$$

where E_{eV} is the energy in electron volts (1 eV is the energy gained by one electron falling through a potential difference of 1 V) and M_{amu} is the mass in atomic mass units. To illustrate, one electron, mass 0.911 × 10^{-30} kg or 0.000549 amu, at rest has an energy-equivalent mass,

$$E_{electron} = 0.911 \times 10^{-30} \, kg \times 9 \times 10^{16} \, m^2/s^2 = 8.199 \times 10^{-14} \, J$$

or, expressed in electron volts,

$$E_{electron} = 931.4 \times 10^6 \, eV/amu \times 0.000549 \, amu = 0.511 \times 10^6 \, eV$$

When an electron is annihilated, as it is on colliding with a positron (a particle just like the electron, but with positive charge), the electron and positron disappear and two photons emerge, each with exactly 0.511 MeV. Similarly, the binding energy of a helium nucleus,

$$E_{binding} = (4.0319 - 4.0026) \, amu \times 931.4 \, MeV/amu = 27 \, MeV$$

In nuclear transformations, it is the difference between the masses of parent* and daughter* nuclei which appears as radiation energy.

Nuclear stability. In the neutron-proton model of the nucleus, protons contribute mass and positive charge while neutrons contribute only mass; the combination of protons and neutrons provides stability. Binding energy and nuclear stability depend upon the number of neutrons relative to the number of protons (n/p ratio) in the nucleus. The shaded band of Figure 1B-70 represents stable nuclei. As indicated by the narrowness of this band, only a minority of possible proton-neutron combinations are stable. An excess of either one or the other leads to a redistribution of the particles. As sketched in the small circled diagrams of this figure, the unstable nucleus can adjust its n/p ratio by emission of particles of matter. This readjustment, known as radioactive decay, is discussed below.

A solitary neutron does not form a stable nuclide; in contrast, a solitary proton is the nucleus of ordinary hydrogen ($^1_1$H). The one-proton combinations with neutrons start with stable deuterium ($^2_1$H), one proton with one neutron. The next member of this series is unstable radioactive tritium

---

*Parent: unstable radioactive nucleus; daughter: more stable product nucleus.

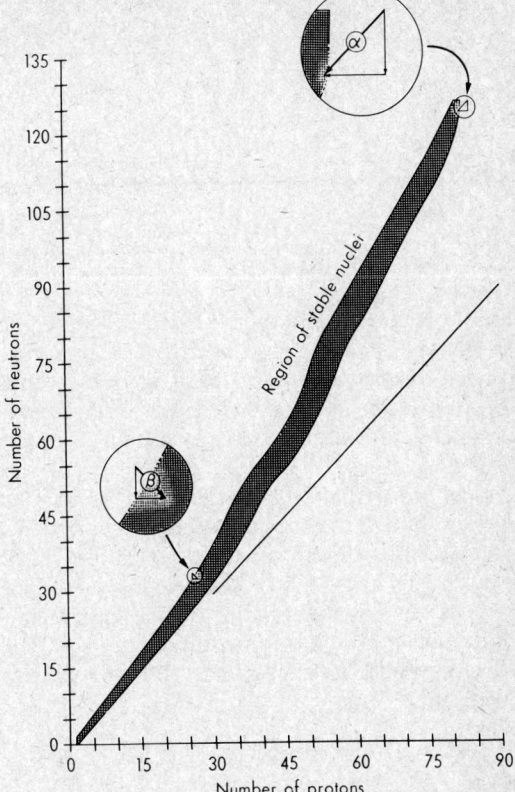

Figure 1B-70. Region of nuclear stability. Combinations of neutrons and protons which fall within the shaded region form stable nuclides; all other combinations are radioactive. The observed radioactive nuclides lie just above or just below the region of stability, and in general, the farther a combination is from this region, the shorter its half-life. For the smaller nuclides within the stable region, the number of neutrons approximates the number of protons; for the larger nuclides, the neutrons always outnumber the protons.

Beta decay (lower inset). In β-decay, the nuclide loses one neutron and gains one proton. The effect of these changes is to bring the nuclide diagonally down and right toward the region of stability.

Alpha decay (upper inset). The nuclide loses two neutrons and two protons. The effect is to bring the nuclide diagonally down and left toward stability.

(From Powsner, E. R., and Raeside, D. E.: Diagnostic Nuclear Medicine. New York, Grune and Stratton, 1971, p. 51.)

($^3_1$H), one proton with two neutrons. One proton with three neutrons, if it exists at all, would be an extremely unstable form of hydrogen. Combinations containing two protons are called helium. There is no stable nuclide of two protons alone; at least one neutron is required. The two-proton combinations with, respectively, one and two neutrons are $^3_2$He and $^4_2$He; both are stable, but helium isotopes with more than two neutrons are unstable.

For light nuclides up to neon (atomic number 10), the numbers of protons and neutrons are equal or approximately equal. Moderately massive nuclides have more neutrons than protons; for example, the stable isotopes of iron (atomic number 26) have between 28 and 32 neutrons. As atomic numbers increase, the difference between the number of neutrons and protons increases (see Figure 1B-70). At the end of the list, bismuth has a stable isotope with 126 neutrons for its 83 protons. No stable nuclides are found with more than 83 protons.

Radiation

Radiations from the atom and its nucleus may be classified as particulate or electromagnetic.

Particulate radiation. As the name suggests, this radiation consists of small bits of high velocity matter, either the so-called elementary particles, such as the electron, positron, proton, neutron, and neutrino (discussed later), or the more complex particles, such as the helium nucleus and larger nuclear fragments. The particles themselves are identified by their measurable physical properties such as rest mass, charge, or "spin." As high velocity particles, these particles also have kinetic energy.

Electromagnetic radiation. Special names have been assigned to electromagnetic radiation on the basis of its energy or, equivalently, of its wavelength or frequency (Table 1B-11). Thus, electromagnetic radiations of very low energy, low frequency, and long wavelength are called radiowaves, while those of higher energy and frequency and shorter wavelength are called light. Those with the highest energy are called X-rays and gamma (γ-) rays. Gamma and X-rays are distinguished as follows: γ-ray implies origin in the de-excitation of an atomic nucleus and X-ray implies origin from the acceleration of orbital or other electrons. Nevertheless, both are

Table 1B-11. A CLASSIFICATION OF
ELECTROMAGNETIC RADIATION

Name	Frequency (Hz)
Radiowaves	$10^4 - 10^{11}$
Infrared light	$10^{12} - 10^{14}$
Visible light	$10^{14} - 10^{15}$
Ultraviolet light	$10^{15} - 10^{17}$
Gamma and X-rays	$10^{17} - 10^{20}$

electromagnetic radiation, and a 35 keV γ-ray is physically indistinguishable from a 35 keV X-ray.

A major tenet of modern physics is that all radiation has properties resembling those of particles and those of waves. For electromagnetic radiation, including γ- and X-rays, *photon* is the name for the particle or quantum of radiation. The photon has no rest mass; indeed, it does not exist except in motion. This property is more easily understood in terms of waves because waves imply motion. On the other hand, electromagnetic waves can travel through a vacuum which appears to contain nothing to support wave motion. This property is more easily understood in terms of a particle moving through the empty space. Electromagnetic radiation has its maximum velocity in a vacuum. This velocity is the same for all electromagnetic radiation; indeed, the velocity in a vacuum is one of the fundamental constants of nature and is usually referred to as the velocity of light (approximately 3×10^8 m/s). The reference to wave-like properties suggests that photons have frequency. For all waves, wavelength times frequency gives velocity; in a vacuum,

$$\nu \lambda = 3 \times 10^8 \, \text{m/s}$$

where ν is the frequency in hertz (1 Hz = 1 cycle per second), λ is the wavelength in meters, and the velocity is that of light in vacuum.

For photons, frequency is also related directly to the energy of the photon. The exact relationship is

$$E = h\nu$$

where E is the energy in joules, ν is the frequency in hertz, and h is a fundamental constant, Planck's constant, 6.6×10^{-34} joule-seconds. Rewritten in terms of wavelength in picometers (pm) and of energy in million electron volts (MeV),

$$E = 1.24 / \lambda$$

For example, the 0.364 MeV gamma radiation emitted after iodine-131 decay has a wavelength of $1.24/0.364 = 3.4$ pm. For comparison, a photon of green light (530 nm or 530 000 pm) has an energy of $1.24/530\,000 = 2.3 \times 10^{-6}$ MeV or 2.3 eV.

Radioactive Decay

Energetics. Radioactive decay is a property of the atomic nucleus and is evidence of nuclear instability. Through radioactive decay the nucleus alters its composition or configuration to increase its stability (Figure 1B-70). Decay is manifested by a spontaneous change within the nucleus that results in loss of mass and the emission of energetic radiations.

Spontaneous decay is possible only if the mass of the daughter nuclide is less than that of the parent. As required for the conservation of mass and energy, the total energy of the radiations equals the energy equivalent to the difference between the masses of the parent and daughter nuclides. For example, $^{74}_{33}\text{As}$ (73.923910 amu) is unstable and decays to either $^{74}_{32}\text{Ge}$ (73.921150 amu) or $^{74}_{34}\text{Se}$ (73.922450 amu) (Figure 1B-71). As can be seen from the masses given, both daughters are lighter than arsenic-74. Differences in the number of orbital electrons, each of mass 0.000549 amu, account for part of the difference, most of which represents nuclear binding energy.

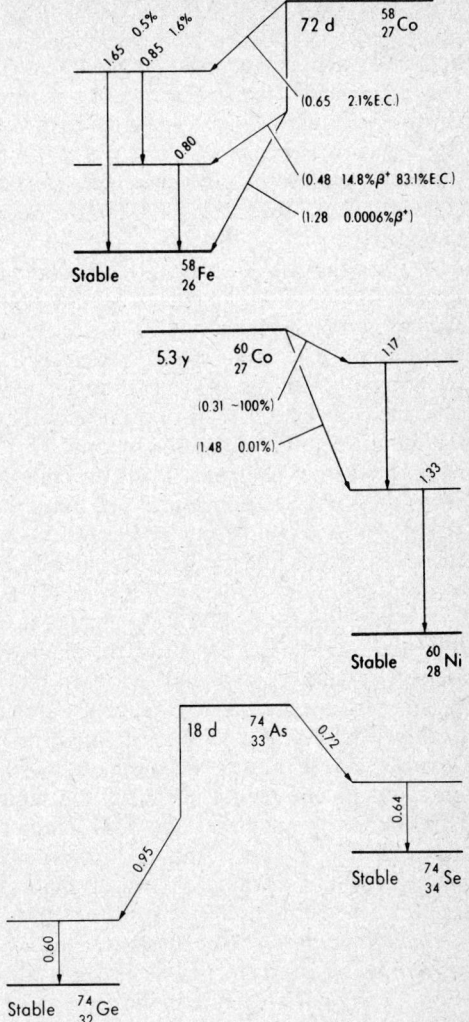

Figure 1B-71. Atomic mass-energy diagrams for positive and negative β-decay. *Top,* Cobalt-58 decays to iron-58 by either electron capture or positron emission. *Center*: Cobalt-60 decays exclusively by negative β-emission to nickel-60. *Bottom,* Arsenic-74 is shown decaying to selenium-74 by negative β-emission or to germanium-74 by positron emission. The frequencies are, respectively, 32% and 29% of all decays. Electron capture, also producing germanium-74, occurs in the remaining 39% of the decays.

In these diagrams, the symbol for the nuclide is given just below the lowest or only horizontal line of each stack. The time to the left of the symbol for a radionuclide is its half-life; nonradioactive nuclides are simply labeled "stable." The height of a horizontal line is proportional to its energy with respect to the ground state of the stable nuclide. The relative position of a line from right to left indicates relative atomic number. Referring to the arsenic diagram, from left to right are the successive positions for atomic numbers 32 (germanium), 33 (arsenic), and 34 (selenium).

Arrows sloping down to the right indicate negative β-emission, vertical arrows indicate gamma emission, and arrows sloping down to the left show electron capture. Positron emission is shown by a dogleg arrow, down vertically and then sloping down left. Positron emission implies electron capture as an alternative; the relative frequency of the two processes may be shown as in the case of the 0.48 MeV positron of cobalt-58. The number assigned to each of the arrows is the energy in million electron volts released during the transition; the second number, where given, is the frequency in percent. For example, the 1.65 MeV γ-ray is emitted following 0.5% of cobalt-58 decays. (Based on data of Way, K., et al.: Nuclear Data Sheets. New York, Academic Press, 1966; from Powsner and Raeside: op. cit.)

In the decay to selenium-74, the products include a negative electron and a characteristic photon of 0.64 MeV. The electron is emitted with a maximum kinetic energy of 0.72 MeV. The total energy released is the difference between the atomic masses of arsenic-74 and selenium-74,

$$(73.923910 \text{ amu} - 73.922450 \text{ amu}) \times 931.4 \text{ MeV/amu} = 1.36 \text{ MeV}$$

which equals the sum of the energies of the electron and photon.

Alpha decay. To achieve stable configurations, heavy elements, particularly those with atomic numbers above 70, may shed some of their nuclear mass by emitting a two-proton two-neutron fragment identifiable after emission as a helium nucleus (Figure 1B-70, upper detail). Because nuclear radiations were observed before their identity was known, this fragment was called an *alpha* (α-) *particle* and its emission was termed α-decay. These names are still used. As an example, radium-226 decays by α-emission to produce radon-222 as follows:

$$^{226}_{88}\text{Ra} \rightarrow {}^{222}_{86}\text{Rn} + {}^{4}_{2}\text{He}$$

For simplicity, the α-particle is shown after it has acquired two orbital electrons to become a helium atom. Most α-emitters are naturally occurring radioisotopes of the heavy elements and have little clinical application.

Beta decay. For some heavy nuclides and for almost all those with atomic numbers below 60, stability is achieved by a rearrangement of the nucleus in which the total number of nucleons and mass number are unchanged. In terms of the neutron-proton model of the nucleus, this rearrangement is the conversion of a neutron to a proton or vice versa. During such conversions, the nucleus emits either a negative electron or its positive equivalent, a *positron.* The emission of the negative electron was discovered first. This was named the *beta (β-) particle,* and unless otherwise indicated, the name still refers to the negative particle. Where necessary, the charge sign can be included to distinguish the negative β-particle or electron from the positive β-particle or positron.

The emission of a negative β-particle leaves the nucleus with one additional positive charge. From our viewpoint, a neutron is converted to a proton, and the nucleus assumes the next higher atomic number (Figure 1B-70, lower detail). Negative β-emission is characteristic of a nucleus that has more neutrons than required by its protons for stability. Examples of nuclides that decay by negative β-emission are carbon-14, hydrogen-3 (tritium), iodine-131, and iron-59. Emission of a positive β-particle or of a positron has the opposite effect: a proton becomes a neutron and the atomic number decreases by one. This process is characteristic of nuclei that have an excess of protons over neutrons. Positron emitters include carbon-11, fluorine-18, and iron-52. Some commonly used radionuclides are listed with their properties in Table 1B-12. A few nuclides, arsenic-74 for example, can achieve stability by either negative or positive β-decay. Atomic mass-energy diagrams illustrating positive β-decay (cobalt-58) and negative β-decay (cobalt-60) as well as the two modes of decay of arsenic-74 are shown in Figure 1B-71.

During β-decay, all of the change in nuclear charge is accounted for by the newly created β-particle. However, not all of the change in energy is accounted for by the β-particle; some is carried by an accompanying *neutrino* (during positron emission) or by an *antineutrino* (during negative β-emission). These neutrinos are uncharged, they have no rest mass, and they react only weakly with matter. As a result, neutrinos have a long range and are very difficult to detect. Because they divide the decay energy with the β-particle in a random fashion, the β-particles from a given emitter are not all of the same energy. Rather, they encompass a spectrum ranging continuously from practically zero up to the maximum energy of the decay process. Both the shape of the spectrum and the maximum energy are characteristic of the emitter. Examples representing a wide range of spectral shapes are shown in Figure 1B-72. Compare these continuous β-spectra with the line spectrum for internal conversion electrons discussed below.

Electron capture. An alternative decay process to the emission of positive β-particles is the capture of an electron. In this process, an orbital electron is absorbed by the nucleus. The end effect on nuclear structure is the same; a proton appears to have changed into a neutron and the

Table 1B-12. RADIATION PROPERTIES OF COMMONLY USED RADIONUCLIDES*

Nuclide	Half-Life		Type of Decay	Maximum Energy of Radiation (MeV)[†]	
				Beta	*Gamma*
^{3}H	12.3	y	β^-	0.018	
^{14}C	5730	y	β^-	0.156	
^{24}Na	15	h	β^-, γ	1.39	1.37, 2.75
^{32}P	14.3	d	β^-	1.71	
^{35}S	87.1	d	β^-	0.167	
^{42}K	12.4	h	β^-, γ	2.00, 3.52	1.52
^{51}Cr	27.8	d	EC[‡]		0.32
^{59}Fe	45	d	β^-, γ	0.273, 0.475	1.095, 1.292
^{57}Co	270	d	EC		0.122, 0.136
^{60}Co	5.26	y	β^-, γ	0.313	1.17, 1.33
^{99m}Tc	6	h	IT[‡]		0.141
^{125}I	60	d	EC		0.036
^{131}I	8.1	d	β^-, γ	0.61	0.364
^{198}Au	2.7	d	β^-, γ	0.962	0.411
^{197}Hg	65	h	EC		0.077
^{203}Hg	46.9	d	β^-, γ	0.213	0.279

*Data taken in part from Dillman, L. T.: Radionuclide decay schemes and nuclear parameters for use in radiation-dose estimation, J. Nucl. Med., Supplement Number 2, March, 1969.
[†]Only the principal energies are given here.
[‡]EC refers to electron capture. IT refers to isomeric transition.

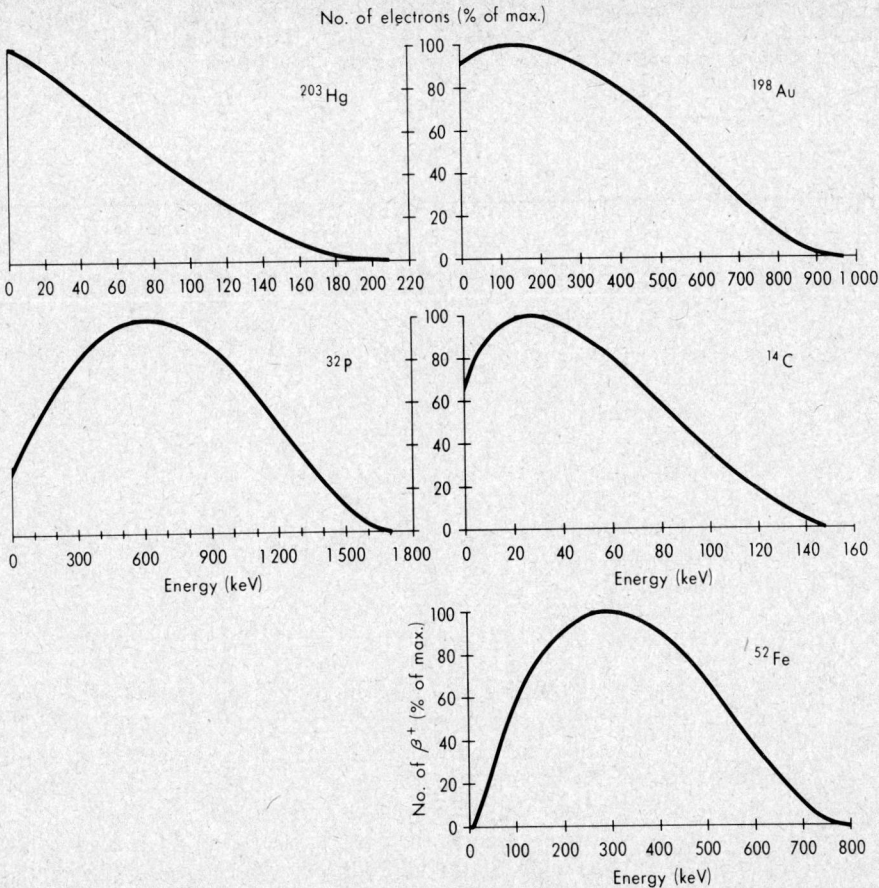

Figure 1B-72. Continuous spectra of β-radiations from representative negative β-emitters (upper four panels) and from a typical positron emitter (lower right panel). Beta particles are emitted with a continuous range of energies from zero to the maximum available; the portion of available energy not carried by the β-particle is carried by the neutrino or antineutrino (see text). (Curves calculated by the method of Widman, J. C., et al.: Intern. J. Appl. Radiation Isotopes, *19*:1-14, 1968; from Powsner and Raeside: op. cit.)

atomic number decreases by one. As an example, Figure 1B-73 shows the decay scheme for iodine-125 which decays exclusively by electron capture. A simple explanation of electron capture is that the orbits of the atomic electrons, particularly those of the inner K-shell, have, as suggested by the wave-like properties of electrons, some probability of overlapping the nucleus and of permitting the nucleus to capture the electron. As with positron emission, a neutrino is emitted.

Following electron capture, the atom is left with a vacancy in one of its inner electron shells. This is an unstable or excited state for the atom. *Atomic de-excitation* is accomplished by rearrangement of the orbital electrons, usually by a jump of an outer-shell electron to the inner-shell vacancy. The energy lost by the electron as it falls to the inner shell may be carried off in the emission of an X-ray or may cause ejection of a more weakly bound orbital electron. An electron ejected by this mechanism is called an *Auger electron* after its discoverer, Pierre Auger (1888–1938). As required by the rules of energy conservation, the kinetic energy of the Auger electron must equal that of the X-ray minus the binding energy of the electron.

Gamma emission and internal conversion. The original observations of naturally occurring radioactive nuclides disclosed a third radiation, the *gamma* (γ-) *ray*. This ray is distinguished by its ability to penetrate materials which block both α- and β-radiation. We have already alluded to the γ-ray as the photon emitted during the decay of arsenic-74, and we stated that γ-radiation is high energy electromagnetic radiation emitted during nuclear de-excitation. After either α- or β-decay or after electron capture, the nucleus may or may not be left in its most stable state, its state of lowest energy, commonly referred to as the ground state. For example, $^3_1$H undergoes β-decay directly to the ground state of $^3_2$He. The decay of $^{32}_{15}$P to $^{32}_{16}$S is another example. These

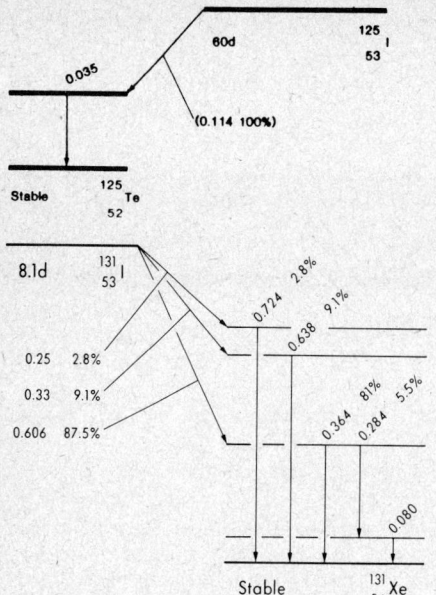

Figure 1B-73. Atomic mass-energy diagram for the decay of two iodine isotopes. *Top,* Electron capture decay of iodine-125 to tellurium-125. *Bottom,* Beta decay of iodine-131 to xenon-131. The symbols used are explained in Figure 1B-71. (In part from Powsner and Raeside: op. cit.; based on data of Way, K., et al.: op. cit.)

nuclides are called pure β-emitters because no other important radiation is produced. On the other hand $^{131}_{53}$I, also a β-emitter, decays by one of several routes to one of the excited levels of $^{131}_{54}$Xe (Figure 1B-73). Other examples are given in Table 1B-12. In these decays, the daughter nuclide goes from the excited to the ground state by shedding the excess energy either through the emission of a γ-photon or through the ejection of an orbital electron.

The energy of the emitted γ-ray is equal to the energy released in de-excitation. Where de-excitation occurs in a single step, a single photon is produced. Where it occurs in multiple steps, each produces a photon of corresponding energy. For every radionuclide, the de-excitation steps and, therefore, the energies of the γ-rays are unique; indeed, the spectrum of γ-ray energies may be used to identify unknown nuclides. When measured with a crystal scintillation detector (discussed later) each photon is represented by a discrete peak, usually referred to as the photopeak. For example, the de-excitation of xenon-131 (produced by the decay of iodine-131) includes many steps; the more frequent of these are shown in Figure 1B-73. The energy spectrum for iodine-131 contains the corresponding photopeaks, each located at the energy level of the photon it represents.

Actual spectra produced in the laboratory also contain features which are related to the size and shape of the crystal and its shielding. Among these are the so-called Compton edge and the backscatter peak. The *Compton edge,* which is found to the left of the corresponding photopeak, appears in the spectrum as a relatively sharp fall in the count rate at an energy equal to the maximum energy that can be imparted to an electron in a Compton interaction with an incoming photon. The difference between the energies of the Compton edge and the photopeak for the incident photon is the energy carried off when the scattered photon escapes from the crystal. Electrons with less than the maximum energy contribute to the count rate at energies to the left of the Compton edge. The *backscatter peak* is caused by Compton scattering at large angles (i.e., backward) into the crystal from interactions within the surrounding lead shield.

The ejection of an orbital electron during nuclear de-excitation is referred to as *internal conversion,* and the ejected electron itself is called a *conversion electron.* Despite the superficial resemblance between conversion electrons and β-particles, internal conversion differs from β-emission in several ways. The most important are that internal conversion, like electron capture, leaves the atom in an excited state and that internal conversion is not accompanied by emission of neutrinos. Because there is no neutrino to share the energy, each electron carries the full energy of de-excitation. A spectrum of the energies of the electrons emitted by internal conversion during the de-excitation of lead-207 is shown in Figure 1B-74. This is referred to as a sharp-line spectrum in contrast to the distributed energy spectrum of electrons emitted in β-decay (Figure 1B-72).

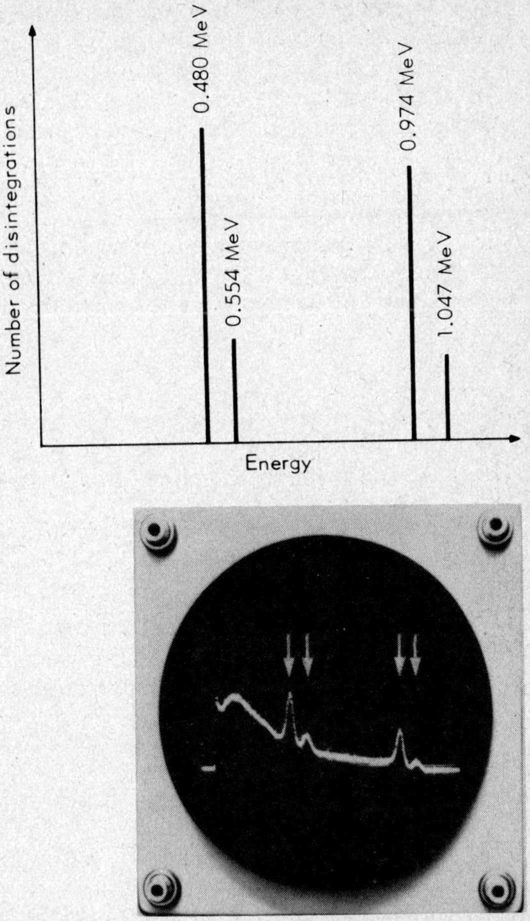

Figure 1B-74. Sharp-line spectrum of internal conversion electrons emitted during the de-excitation of lead-207 following the decay of bismuth-207. *Top,* Schematic representation. *Bottom,* An oscilloscopic trace of the same spectrum. The four energy peaks marked by arrows correspond to the four lines in the top panel. Artifacts of the laboratory technique are responsible for the apparent broadening of the spectral lines and for the slope of the baseline. (Photograph courtesy of N. Horwitz, Ph.D.; data from Way, K., et al.: op. cit.; after Powsner and Raeside: op. cit.)

Gamma emission and internal conversion are competitive processes. Low excitation energy and high atomic number favor internal conversion; high energy and low atomic number favor γ-emission. When the γ-photon is emitted, it carries the full energy of de-excitation; when the internal conversion electron is emitted, it carries the same de-excitation energy minus the energy required to remove the electron from its orbit.

Because the γ-radiations and internal conversion electrons of nuclear de-excitation are typically emitted very promptly, they often appear inseparable from the initial decay and are commonly, but inaccurately, attributed to the parent nuclide. For example, the dominant 0.364 MeV photon observed within a few picoseconds (ps) following the decay of iodine-131 ($t_{1/2} = 8.1$ d) is not from iodine but from its daughter, xenon-131 ($t_{1/2} = 10$ ps) (Figure 1B-73). The β-radiations originate in the initial decay process and come from the iodine-131 nucleus itself. A few nuclides de-excite slowly enough to make the distinction of practical importance. Prior to their de-excitation, such nuclides are said to be in a *metastable state* and are designated by the letter m following the mass number. An important metastable nuclide is technetium-99m ($t_{1/2} = 6$ h), which decays by γ-emission to technetium-99 ($t_{1/2} = 200\ 000$ years). De-excitation of a metastable nuclide to the stable state of the same nuclide is referred to as *isomeric transition.*

Rate of Radioactive Decay

Random nature of decay. The rate of decay is a characteristic of an individual radionuclide; the rate is a constant which is not affected by temperature, pressure, concentration, or any other chemical or physical condition. Ernest Rutherford (1871–1937) observed that the decay of a radionuclide is a random event and that the number of decays in successive intervals of equal time follows the Poisson distribution (see below). The decay of any single atom cannot be predicted, but in a group of millions of atoms the number that will disintegrate during a given time interval can be predicted within statistical limits. It can be expected that the number of disintegrations in a short time interval, the rate of disintegration, is proportional to the number of atoms present at the beginning of the time interval. The proportionality constant, referred to as the decay constant, is a measure of the intrinsic probability of decay. This relationship is expressed by

$$-dN/dt = \lambda N \tag{1}$$

where N is the number of atoms present at time t, dN/dt is the disintegration rate expressed as rate of change in N, λ is the decay constant characteristic of a given nuclide, and the minus sign indicates that the number of atoms present is decreasing with time. Upon integration,

$$N = N_o e^{-\lambda t} \tag{2}$$

where N_o is the number of atoms at $t = 0$, and e is the base of natural logarithms.

Activity and half-life. The strength or rate of decay of a radioactive source is called its *activity*, A, and is simply the rate of loss of atoms, $-dN/dt$. From equations (1) and (2),

$$A = A_o e^{-\lambda t} \tag{3}$$

where A_o equals λN_o, the activity at zero time.

It is more convenient to describe the rate of decay in terms of *half-life*, $t_{1/2}$, the time required for sample activity to decrease to half its initial value. The relationship between $t_{1/2}$ and λ may be derived from the foregoing definition and equation (3),

$$t_{1/2} = (\ln 2)/\lambda = 0.693/\lambda \tag{4}$$

where 0.693 is the natural logarithm of two. Substituting into equation (3)

$$A = A_o e^{-(0.693/t_{1/2})t} \tag{5}$$

This relationship is conveniently pictured by plotting the logarithm of the per cent activity remaining against time (Figure 1B-75). Each half-life reduces the activity by half; the effect is cumulative. In general,

$$A_n = A_o \times 2^{-t/t_{1/2}} \tag{6}$$

or

$$A_n = A_o \times 2^{-n} \tag{7}$$

where n is the number of half-lives. Equation (7) is useful in planning experiments and in the disposal of radioactive waste. For disposal a rule of thumb is that seven half-lives reduces the activity to less than 1% of its original value ($2^{-7} = 1/128 = 0.78\%$) and ten half-lives reduces it to less than 0.1%.

Units. The conventional unit of radioactivity is the *curie* (Ci), defined as 3.7×10^{10} decays per second (dps) or 2.22×10^{12} decays per minute (dpm). This number was originally selected because it approximates the activity of 1 g of radium-226 in equilibrium with its daughters. The new SI unit of activity is the *becquerel* (Bq), equal to 1 dps. One curie equals 3.7×10^{10} Bq; 1 μCi equals 37 kBq.

Relationship of mass to activity. For separate elements, equal activities in curies do not mean equal masses in grams. From the definitions above and Avogadro's number, 6.02×10^{23}, the mass in kilograms of a radionuclide which has an activity of 1 Ci,

$$m_{Ci} = 0.89 \times 10^{-16} M t_{1/2}$$

where M is the gram-atomic weight and $t_{1/2}$ is the half-life in seconds. For example, the mass of

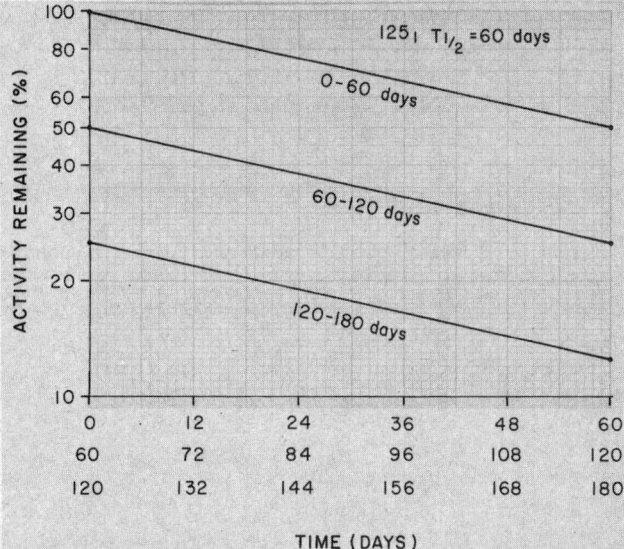

Figure 1B-75. Radioactivity versus time for iodine-125 ($t_{1/2} = 60$ d). The logarithm of activity remaining is plotted against elapsed time over a period of three half-lives.

one curie of cobalt-60 ($t_{1/2} = 5.3$ years) is about 1 mg, whereas that of one curie of uranium-238 ($t_{1/2} = 4.5 \times 10^9$ years) is about 3000 kg.

Specific activity. The term "specific activity" has several meanings; it may refer to any one of the following: radioactivity per unit mass of an element; radioactivity per mass of labeled compound; or radioactivity per unit volume of a solution. The denominator of reference must be specified. In terms of radioactivity per unit mass, the maximum specific activity attainable for each radionuclide is that for the pure radionuclide. For example, pure carbon-14 has a specific activity of 62 Ci/mol or 4400 Ci/kg. As usually available, carbon-14 is a tracer for compounds in which it represents only a small fraction of the total carbon, most of which is the naturally occurring mixture of stable carbon-12 and stable carbon-13. If there is no stable element present, the radionuclide is said to be *carrier free.*

INTERACTION OF RADIATION WITH MATTER

Radioactive emissions possess energy either in the form of kinetic energy of motion, as in the case of α- and β-particles, or in the form of electromagnetic radiation as in the case of γ-rays. In passing through matter these radiations transfer energy to the atoms and molecules encountered, chiefly through excitation and ionization. Alpha and β-particles transfer energy by the interaction of their electric fields. Gamma rays, being uncharged, interact by other processes that will be noted later. The ability of radiation to produce excitation and ionization is one of its most important properties. This property is the basis for the detection of radioactivity and is responsible for the biologic effects of radiation.

Excitation and Ionization

Excitation. In this process, the energy of the incident radiation is transferred to matter by raising the electrons of the irradiated material to higher energy levels.

Ionization. If the energy absorbed from the radiation completely removes an electron from its atom or molecule, the process is called ionization. The resulting positive ion and negative electron are referred to as an *ion pair.* The ejected electron itself is called a *secondary electron.* On the average, the energy required to form one ion pair in water is about 34 eV. Biological effects are dependent on the spatial and temporal distributions of the ionizations as well as on the number of ionizations. The spatial distribution is expressed in terms of *specific ionization,* defined as the number of ion pairs formed per unit length. The amount of energy transferred to the absorber per unit path length is called the *linear energy transfer* (LET). Usual units are keV/μm. Alpha particles have a much higher LET than β-particles.

Following their formation, the excited and ionized atoms and molecules quickly return to the energy of their ground state. The de-excitation of atoms may produce X-rays as the electron vacancies are refilled; the release of energy from excited molecules occurs through a variety of chemical and thermal interactions. X-rays may interact elsewhere in the material to produce further excitation or ionization. Ion pairs contribute to the formation of free ions and free radicals and these, in turn, contribute to the disruption of chemical bonds.

Particulate Radiation

The most important interactions are those of the charged particles, α and β. Both α- and β-particles cause ionization and excitation as a result of their velocity and charge. In mechanical terms, α- and β-particles are rapidly moving, charged bodies capable of forcing electrons from the atoms they pass. At each encounter the particles lose energy, and after many encounters they finally come to rest.

Interaction of alpha radiation. Alpha radiation produces a dense, localized region of ionization over a short, nearly straight path. The straight trail is the result of the relatively large mass of the α-particle (5600 times the mass of an electron) which reduces its deflection at each interaction. The α-particle carries a positive charge which is twice the magnitude of the charge of an electron. The α-particle is relatively slow, e.g., about 3 Mm/s (3×10^6 m/s) for the 4.8 MeV α-particle of radium-226. For comparison, a β-particle of equal energy would travel with nearly the velocity of light. This combination of mass, charge, and velocity of the α-particle is responsible for the dense ionization by which it dissipates its kinetic energy over a short distance. Alpha particles from a source near the surface of the body can penetrate only the superficial layers of the skin. However, if internalized by ingestion or inhalation, α-emitters are a very serious internal hazard because of the dense ionization.

Interaction of β-radiation. On emission, β-particles travel at velocities which approach that of light. For example, the average β-particle of phosphorus-32 (energy 0.70 MeV) is emitted at 270 Mm/s (90% of the velocity of light). In contrast to the α-particle, the lighter β-particle leaves a comparatively sparse, irregular trail of ionization. On the basis of 34 eV per ion pair formed in air, a 0.70 MeV β-particle from phosphorus-32 forms about 20 000 ion pairs before coming to rest. Following their ejection, the secondary electrons themselves cause ionization and excitation. The β-particle can also produce atomic and molecular excitation without ejecting secondary electrons. The overall effect is the transfer of energy from particle to matter. For β-particles in water or soft tissue, this transfer takes place at the rate of about 1 keV/μm; at this rate, the 0.70 MeV β-particle of phosphorus-32 would dissipate all its energy over a range of 0.7 mm. These values are only representative and the results only approximate; however, the general rule is that the greater the energy of the β-particle, the greater its range.

Bremsstrahlung. Charged particles, including β-particles, may also undergo other interactions. Of these, Bremsstrahlung, or braking radiation, is perhaps the most important. The name refers to the original observation that X-rays are emitted when a beam of electrons is stopped. In general, electromagnetic radiation is emitted whenever charged particles are accelerated or decelerated. The medical X-ray tube is a practical application of Bremsstrahlung. Within the tube, a beam of electrons is directed against a dense target; their sudden deceleration upon striking the target produces X-rays. The same phenomenon is encountered in the laboratory when the energetic β-particles (electrons) of a strong β-emitter such as phosphorus-32 strike the walls of a container. The β-particles themselves are not able to penetrate the glass sides of the container, but the X-rays generated penetrate easily.

Electromagnetic Radiation

The nature of the initial interaction of photons, including γ-rays, is energy dependent. At the low end of the energy spectrum, particularly if the energy is insufficient to cause ionization, photons induce electronic and molecular motions which register as *heat*. This is the predominant effect of the radiowaves used in microwave or radar heaters. Photons are higher on the energy spectrum and can eject electrons from matter by several mechanisms, of which the most important are the photoelectric and Compton interactions.

Photoelectric effect. The ejection of an atomic electron (Figure 1B-76) is the most likely interaction of a moderately energetic photon. The photoelectric effect was first observed for photons of visible light and derives its name from this association. The photographic light meter,

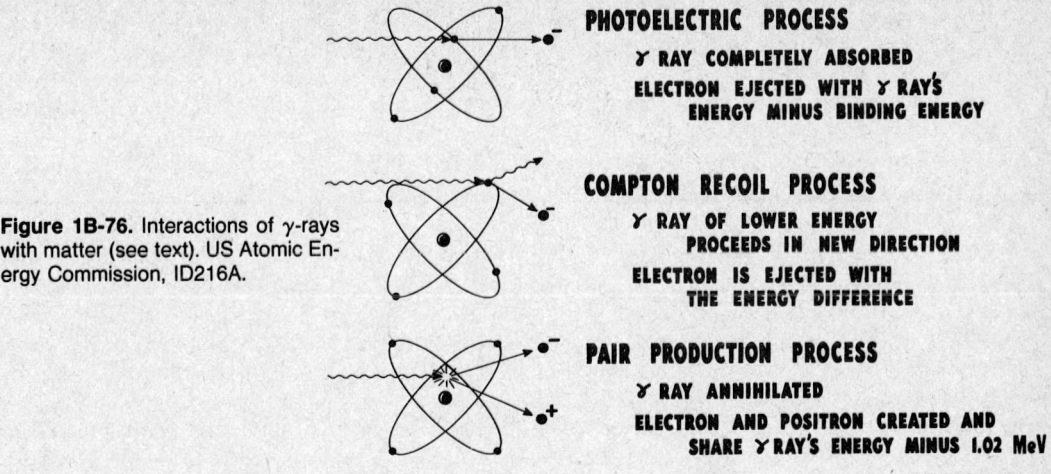

PHOTOELECTRIC PROCESS

ɣ RAY COMPLETELY ABSORBED

ELECTRON EJECTED WITH ɣ RAY'S
ENERGY MINUS BINDING ENERGY

Figure 1B-76. Interactions of ɣ-rays with matter (see text). US Atomic Energy Commission, ID216A.

COMPTON RECOIL PROCESS

ɣ RAY OF LOWER ENERGY
PROCEEDS IN NEW DIRECTION

ELECTRON IS EJECTED WITH
THE ENERGY DIFFERENCE

PAIR PRODUCTION PROCESS

ɣ RAY ANNIHILATED

ELECTRON AND POSITRON CREATED AND
SHARE ɣ RAY'S ENERGY MINUS 1.02 MeV

USAEC-ID-216A

for example, operates by measuring the rate at which electrons are ejected by photoelectric interactions with incident light. The photoelectric effect is possible only if the energy of the incoming photon at least equals the binding energy of the electron. Photon energy in excess of the binding energy imparts kinetic energy to the ejected *photoelectron*; the photon itself is absorbed. Photons of visible light with energies of only a few electron volts can eject only weakly bound electrons such as those found in the alkaline metals. X-ray and ɣ-photons with energies in excess of 150 keV can eject the most tightly bound atomic electrons. The probability that a photoelectric interaction will occur is a function of both the energy of the photon and the binding energy of the electron. In general, interaction is less likely at higher photon energies and more likely with materials of higher atomic number. It is also true that the interaction is most likely for photon energy at, or just above, the binding energy of an atomic electron.

Compton effect. Named after its discoverer, Arthur Compton (1892–1962), the Compton effect (Figure 1B-76) predominates at energies greatly in excess of the binding energy of atomic electrons. In a Compton interaction, the incoming photon interacts with a free or weakly bound, outer-shell electron as though the photon and electron were each billiard balls in collision. The electron is propelled in one direction while the photon is scattered in another direction, simultaneously losing some of its energy to the electron. The electron deposits its energy in the surrounding matter by ionization and excitation of other atoms. The scattered photon has lower energy than the original photon. It may escape from the material without further interaction; it may undergo another Compton interaction; or it may disappear in a photoelectric interaction.

For water or soft tissue, the photoelectric interaction predominates for photons with energies up to 50 or 60 keV; the Compton interaction predominates from ~60 keV to several million electron volts. These are the two most likely and the only important interactions for the counting and detection of ɣ-radiation in the clinical chemistry laboratory.

Pair production. If the energy of the incident photon is high enough, other reactions become possible. Of these, pair production is the most important; it requires a photon of at least 1.02 MeV (Figure 1B-76). In pair production, the photon is absorbed in the vicinity of an atomic nucleus with the formation of an electron-positron pair. The 1.02 MeV is the energy equivalent of the rest mass of the two electrons, 0.51 MeV each. Photon energy in excess of 1.02 MeV is carried off as kinetic energy by the newly created pair. Pair production is the reverse of the annihilation reaction in which a positron plus a negative electron disappears with the production of two 0.51 MeV photons. In fact, the positron created in pair production undergoes annihilation as soon as it comes to rest. As always, the total of mass plus energy is conserved in these reactions.

Range of ɣ-radiation. The distance traveled by a photon traversing matter is best discussed in terms of the attenuation of a beam of photons, that is, the decrease in the intensity of the beam as it passes through matter. There is no definite end point; rather, the intensity of the beam of γ- or X-rays decreases exponentially. This behavior results from the probabilistic nature of photon interactions. Over any stated distance a single photon has a finite chance of interacting with the electrons of the material traversed. This chance remains constant for any equal distance along the path. Equivalently, for a beam of photons, a constant fraction will interact for each unit of distance. As a result, it is useful to speak of a half-value thickness, the thickness of material through which the incoming beam is reduced to half its intensity. For the 28 keV photons of iodine-125 decay, the half-value thickness in water is about 3 cm. For 10 000 photons which enter the water, 5000 will penetrate to 3 cm, 2500 to 6 cm, 1250 to 9 cm, and so on.

Units of Radiation Exposure and Radiation Dose

The units used in quantitation of the interactions of radiation with matter are based either on radiation exposure, that is, on the strength of the radiation field, or on radiation dose, that is, on the energy absorbed by the matter itself.

Exposure. X-ray or γ-ray exposure is usually defined in terms of the ionization produced by the radiation. The conventional unit is the *roentgen* (R); the SI unit is the *coulomb per kilogram* (C/kg). Originally defined together with the instrumentation used for its measurement, the roentgen is the amount of radiation required to produce ions carrying one electrostatic unit of electric charge per cubic centimeter of dry air at standard temperature and pressure. In SI units, 1 R produces ionization equal to 0.258 mC/kg.

Radiation dose. Dose is described by the amount of energy absorbed. The conventional unit is the *rad* (*r*adiation *a*bsorbed *d*ose), which is defined as 100 ergs absorbed per gram of matter. Exposure of dry air to 1 R results in the deposition of 0.87 rad; during the exposure of soft tissue, 1 R deposits 0.93–0.97 rad, depending on photon energy. In part, the rad was chosen because the number of rads is close to the number of roentgens to which soft tissue is exposed. The SI unit of radiation dose is the *gray* (Gy), defined as one joule absorbed per kilogram; 1 Gy equals 100 rad. The relationship between radiation dose and biological effect is presented in a later section.

DETECTION AND MEASUREMENT OF RADIOACTIVITY

The earliest methods for detecting radiation, i.e., darkening of photographic emulsion, ionization of gas, and fluorescent scintillation, are still the basis for most modern techniques. The underlying physical process in all of these methods is the excitation or ionization caused by the radiation as it penetrates the detector. In radiology, exposure and observation of a photographic film is the principal procedure. Measurement of ionization in gas or, more recently, in solids is widely used for environmental monitoring. However, in the clinical chemistry laboratory almost all measurements use scintillation detectors. Photographic and ionization methods will be described only briefly.

Autoradiography

The earliest photographic technique for imaging radiation, autoradiography, uses the radiations from radioactivity within the sample itself in contrast to conventional radiography, which uses an external source of radiation, usually an X-ray tube. In either photographic process, the radiation interacts with an emulsion of silver halide in gelatin to alter the configuration of the electrons in the halide crystals. This forms a latent or undeveloped image made up of microscopic aggregations of silver atoms. Subsequent photographic development, most commonly by an alkaline reducing agent, converts more silver halide to metallic silver. Because reduction is more rapid where previous radiation has formed a latent image, development leaves a greatly intensified, visible image of the original pattern of radiation.

Gas-Filled Detectors

Detectors filled with an inert gas are designed to capture and measure the ions produced by radiation within the detector. The positive and negative ions are moved through the gas in opposite directions by a low intensity electric field; the resulting current is amplified and measured. If the applied field is moderately high, e.g., several hundred volts over a few centimeters, the electrons move with enough velocity to ionize additional gas molecules. This process is referred to as gas amplification. As a result of this mechanism, the initial radiation may trigger a virtual avalanche of ionization and can cause a continuous electrical discharge within the tube. The discharge may be limited by lowering the voltage or by including a small amount of a quenching, energy-absorbing compound in the gas mixture. Properly controlled, gas amplification is advantageous because the current it generates is easier to measure than the small current of the primary ions alone.

Gas-filled detectors are given names that indicate the voltage range in which they operate. The *ionization chamber* is a detector that uses low voltage to measure the primary ionization. The *proportional counter* is a detector that uses a higher voltage with limited gas amplification and special gas mixtures such as neon plus methane. Under these conditions, the charge per ionizing particle remains proportional to the amount of primary ionization produced by the particle. Finally, the *Geiger counter* employs voltages high enough for the radiation to trigger maximum ionization within the tube. The Geiger counter produces a relatively large electric signal, but it cannot distinguish among different types of radiation. The Geiger counter is the only one of the three gas-filled counters commonly seen in the clinical chemistry laboratory; it is used as a portable radiation monitor. The ionization chamber is also used as a radiation monitor; it is particularly useful for measurement of radiation exposure, which is commonly expressed as ionization per unit volume, as will be discussed below. The proportional counter was the principal instrument for counting laboratory samples labeled with weak β-emitters such as carbon-14 and tritium. It has been almost universally replaced for this purpose by the liquid scintillation counter.

Scintillation Detectors

There are two principal types of scintillation detectors found in the clinical chemistry laboratory, the sodium iodide crystal scintillation detector and the organic liquid scintillation detector. The crystal detector is the more widely used because of the ease of its operation and the economy of its sample preparation.

The operating principle for both types of scintillation detectors is similar. The absorption of radiation within the scintillator causes excitation and ionization. This absorbed energy in turn produces a flash of light or scintillation. The distinguishing property of a scintillator is its capacity to convert a significant fraction of the absorbed energy into visible or ultraviolet light.

Crystal scintillation detector. This solid scintillation detector is diagrammed in Figure 1B-77. The usual form of a crystal scintillation detector encountered in the chemistry laboratory is the *well detector.* This has a hole drilled in the end or side of the cylindrical crystal to accept a test tube (Figure 1B-77). Because it is hygroscopic, the crystal is hermetically sealed in an aluminum can with a transparent quartz window at one end through which the blue-violet (420 nm) scintillations can be detected. The can follows the contours of the well. The photons of iodine-125 in the sample easily penetrate the specimen tube and the thin, low-density can and enter the crystal where they are likely to be absorbed in the thick, high-density sodium iodide. Of course, only the energy which is actually absorbed in the crystal can be detected.

The crystal itself is usually a circular cylinder, machined from a single crystal of sodium iodide which was carefully grown to ensure optical clarity for efficient light collection. A small amount of thallium is added to the mother liquor to improve performance. The high atomic number of iodine, the density of the sodium iodide ($3.7 \, g/cm^3$), and the size of the crystal (typically about 5 cm in diameter) make it useful for counting γ-rays of low and moderate energy.

The well detector is the instrument of choice for such γ- and X-ray–emitting nuclides as chromium-51, cobalt-57, iron-59, iodine-125, and iodine-131, and for this reason it is often referred to as a γ-counter. It is not suitable for β-radiation, which usually cannot penetrate the sample container and aluminum lining of the well. For a typical well detector, the counting efficiency for iodine-125, expressed as the percentage of decays that produces counts, approximates 70%.

Liquid scintillation detector. As sketched in Figure 1B-78, this detector measures radioactivity by recording scintillations occurring within a transparent vial that contains the unknown sample and liquid scintillator. Because the radionuclide is intimately mixed with, or actually dissolved in, the liquid scintillator, the technique is ideal for the pure β-emitters of low energy and short range such as tritium (maximum β-energy 19 keV and maximum range $< 10 \, \mu m$ in water) and carbon-14 (156 keV maximum energy and $< 300 \, \mu m$ range). In this respect the liquid scintillation detector differs sharply from the sodium iodide scintillation detector; in the latter the scintillations

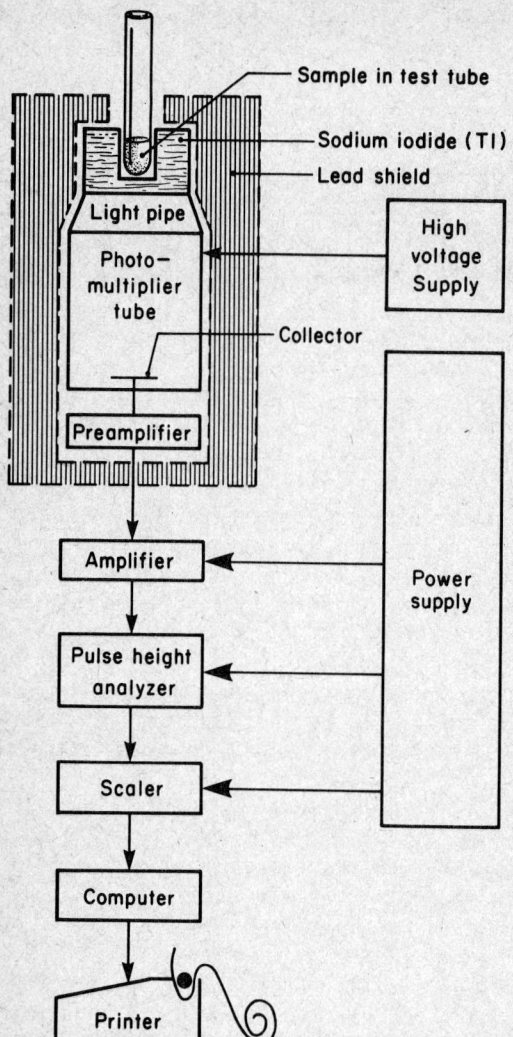

Figure 1B-77. Crystal scintillation well detector.

occur in a crystal that is physically separate from the sample. Typical efficiencies for liquid scintillation counting in the absence of significant quenching are 60% for tritium and 90% for carbon-14.

The liquid scintillator itself, the so-called *scintillation cocktail*, contains at least two components. The first is the *primary solvent*, usually inexpensive, and chosen for its efficiency in absorbing and transferring radiation energy. It is usually one of the aromatic hydrocarbons: toluene, xylene, or pseudocumene (1,2,4-trimethyl benzene). The second necessary component is a *primary scintillator* to absorb energy from the primary solvent and convert it into light. The usual material is PPO (2,5-diphenyl oxazole) used in a concentration of 3–6 g/L. PPO emits ultraviolet light of 380 nm. In addition, the cocktail may contain the following:

1. A *secondary solvent*, usually an alcohol or dioxane, to improve the solubility of aqueous samples, such as serum or urine, or it may contain one of the trademarked surfactants such as Triton X-100 (a polyphenoxy alcohol) to stabilize or emulsify the sample. When a dioxane-naphthalene mixture is used in place of toluene or one of the other common primary solvents, the naphthalene functions as the primary energy converter while either the naphthalene or dioxane may serve as the primary solvent.

2. A *secondary scintillator*, sometimes referred to as a wavelength shifter, to absorb the ultraviolet photons of the primary scintillator and re-emit the energy at a longer wavelength. This shifting was necessary for older photomultiplier tubes because they were insensitive to the

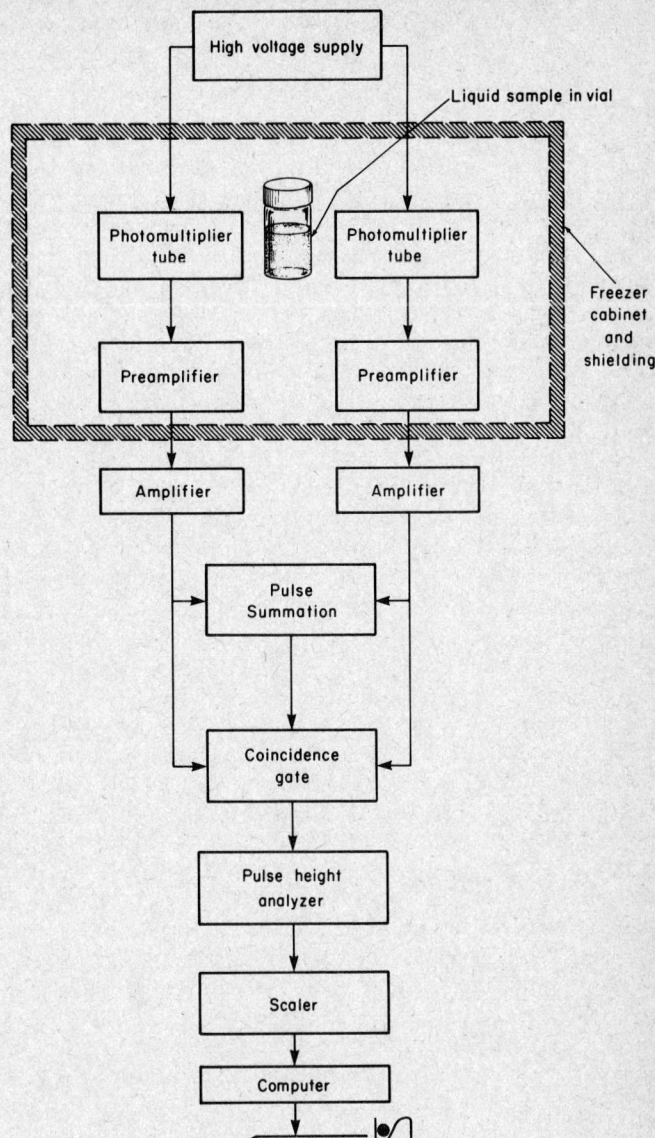

Figure 1B-78. Liquid scintillation detector.

short ultraviolet wavelengths of the primary scintillator. Although it is not necessary for modern photomultipliers, the secondary scintillator is still found in many scintillation cocktails to ensure compatibility with older machines. Examples are POPOP (1,4-bis-2-(5-phenyloxazolyl)-benzene) and dimethyl-POPOP (1,4-bis-2-(4-methyl-5-phenyloxazolyl)-benzene), at a concentration of about 100 mg/L, and bis-MSB (*p*-bis-(*o*-methylstyryl)-benzene) at about 1 g/L. POPOP emits blue light of 432 nm.

3. One or more *adjuvants*. These include suspension agents, typically silicon-based thixotropic gels; solubilizers for biologic tissues, usually strong mineral or organic bases; and antifreezes such as ethylene glycol to prevent freezing and separation of water at low temperatures. Cooling was required for older photomultipliers and is still used to simplify maintenance of temperature or to reduce chemoluminescence.

The components of the liquid scintillation cocktail may be mixed in the laboratory or purchased in a proprietary formulation. Note that the cocktail is flammable and some of its components are toxic. They should be treated with caution during use. They also constitute a hazard on disposal, more because of their toxicity than because of their radioactivity.

Electronics of Scintillation Counting[9]

Following each burst of scintillation, the light must be collected and converted to an electric pulse; the pulses must be amplified, sorted by size, and counted. The components which perform these functions in the crystal scintillation detector and the liquid scintillation detector are sketched in Figures 1B-77 and 1B-78, respectively. The more important components are discussed here.

Photomultiplier. This tube converts scintillations to electrical pulses (Figure 1B-79; see also Chapter 1B, Section One). For well detectors, light is collected through the transparent window in the crystal cover and transmitted to the photomultiplier through a short *light pipe*. For liquid scintillators, the sample vial is placed in a lightproof opening between two photomultipliers. The photons of light enter the photomultiplier through its quartz window. Within the photomultiplier the photons strike the *photosensitive layer* (cathode) and eject outer shell electrons from it. Sensitivity varies from tube to tube; one to several photons are required for each electron ejected. These electrons are accelerated toward the first of several intermediate electrodes, called *dynodes*, each of which is maintained at a progressively higher positive potential relative to the photosensitive cathode. As each electron strikes the dynode, it ejects several additional electrons from the dynode surface, thus effectively multiplying the number of electrons traveling down the tube. These, likewise, are accelerated by the increasing potential differences into successive dynodes where the multiplication is repeated. For dynode surfaces and potentials capable of yielding four electrons per incident electron, the multiplication achievable with 10 dynodes is 4^{10} or 1 048 576. This number of electrons carries 1.7×10^{-13} C (coulombs); for a 1 μs pulse, the average current is 0.17 μA.

Preamplifier. Despite the multiplication of electrons which occurs in the photomultiplier tube, the output of the tube still requires amplification. A preamplifier, typically placed close to the photomultiplier, is used to boost the current before it is analyzed. Note that the size of the current pulses at this point is affected by (1) the energy of the radiation and the efficiency of its absorption and conversion into light photons, (2) the multiplication or gain of the photomultiplier, which in turn depends on the number of dynodes and the voltage between them, and (3) the gain of the preamplifier.

Coincidence circuit. To reduce the effects of electronic noise and small scintillations unrelated to radioactive decay in liquid scintillation counting, the two photomultipliers are connected through an electronic circuit, the coincidence circuit, designed to reject all but significant scintillations. A scintillation in the liquid, caused by the decay of a radionuclide, typically produces several hundred photons, enough to be "seen" by both photomultipliers. On the other hand, electronic noise and the single photon events occurring in the liquid but unrelated to the radiation, the so-called chemoluminescence, produce pulses in the output of only one photomultiplier. Of course, two such events which by chance occur simultaneously, or nearly so, will incorrectly be counted as a decay. Because the probability of this is small compared to the total number of noise pulses, the coincidence circuit greatly reduces the effect of noise on the measured count rate.

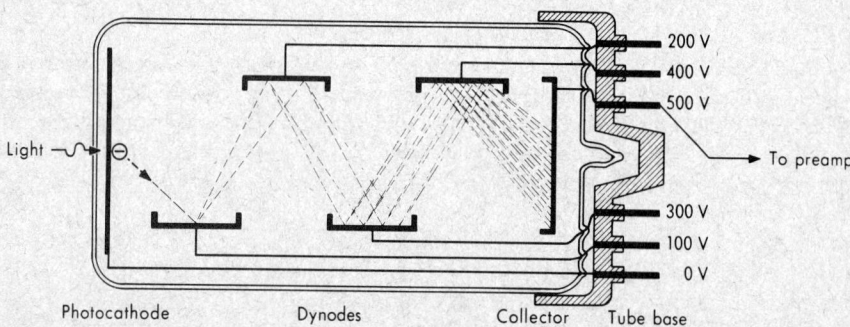

Figure 1B-79. Photomultiplier tube. Light entering the window at the left is shown ejecting a single photoelectron. Two-fold multiplication at each dynode, as sketched, produces 16 electrons at the cathode. In actual tubes, several electrons are ejected by the incident light, and the overall multiplication factor is one million or more. (From Powsner and Raeside: op. cit.)

Pulse-height analyzer. Pulses are sized electronically by an analyzer; its circuit passes only pulses within an acceptable *window* or *channel.* Typically, the window is delimited by a pair of discriminators, one set to the bottom of the window, the other set to the top. Each responds only if the height of the pulse exceeds the value to which it has been set. A pulse is within the window if the *lower discriminator* responds while the *upper discriminator* does not. Conversely, it is not in the window if either both or neither of the discriminators responds. Acceptable pulses are recorded in an electronic register sometimes referred to as a *scaler.* This term originated when mechanical counters were used; these were too slow to count pulses without an intervening circuit to scale down the pulse rate.

The analyzer for a well detector is usually supplied with one or more windows, often factory set, each corresponding to the narrow, γ-energy spectrum of a specific nuclide.* An adjustable window may also be supplied. In liquid scintillation counters, the windows must be wider to accommodate the wider spectrum of β-energies, and the channel width is usually adjusted by the user because the intensity of the scintillation is affected by the sample as well as by the β-energy.

Calibration or "peaking" of an electronic window requires placement of a radionuclide of the appropriate radiation energy in the counter and adjustment of the voltage to the photomultiplier to maximize the count rate in the corresponding window. In older instruments, this adjustment was required daily. In newer instruments, the high voltage adjustment is required infrequently and the adjustment mechanism may be accessible only to a service representative.

Pulses from the amplifier can also be sized by an analyzer with 512 or 1024 closely spaced channels. In this *multichannel analyzer,* each pulse is sized and assigned to the corresponding channel based on its energy. Counts are accumulated separately for each of the channels. The channels are calibrated against the energies of one or more nuclides, such as cesium-137 (photon energy 0.6616 MeV) or cobalt-60 (1.173 and 1.332 MeV), chosen for their well-separated γ-energies and long half-lives. The data from the multichannel analyzer can be displayed on an oscilloscope screen, as shown in Figure 1B-74, or plotted with counts per channel versus channel number, or the total counts can be printed for specified regions of interest.

Efficiency of Scintillation Counting

General factors affecting efficiency. The overall efficiency of a scintillation detector can be defined as follows:

$$\text{Counting efficiency} = (\text{Count rate}/\text{Decay rate})$$

The factors that affect the count rate and, hence, the efficiency are:

1. The fraction of all decays that yield useful radiation, i.e., radiation capable of affecting the scintillator.

2. The fraction of potentially useful radiations that are directed into the scintillator.

3. The fraction of photons entering the scintillator that deposit energy therein.

4. The scintillation efficiency, the ratio of energy emitted as light to the energy deposited as radiation.

5. The detector threshold, the number of light photons (usually 10–20) required to trigger a count in the electronic circuits monitoring the scintillator.

Of the five factors affecting the count rate, scintillation efficiency is the one most likely to require the attention of the laboratory chemist. For solid crystal scintillation detectors, a decrease in efficiency may be an indication that the crystal has cracked or otherwise failed and should be replaced. For liquid scintillation detectors, changes in efficiency are more likely to be related to the sample or its preparation than to the detector. Monitoring the efficiency of a scintillation detector is an important part of the quality control program.

Quenching in liquid scintillation counting. The term "quenching" refers to all the factors which reduce scintillation efficiency. Quenching lowers efficiency by absorbing part of the energy of a photon emitted in the scintillator. This shifts the pulse to the left along the energy spectrum and decreases the number of pulses in the higher energy channels (Figure 1B-80). Quenching

---

*Generally, the window is centered around the photo peak; when multiple photo peaks are used, the window may be set wider to include more than one peak, thus resulting in increased counting efficiency.

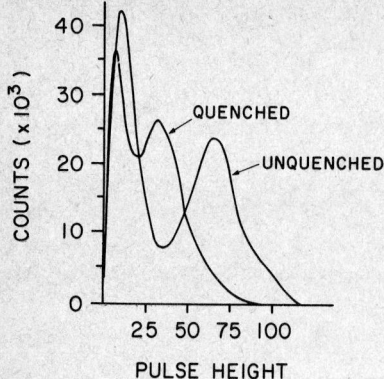

Figure 1B-80. The effect of quenching on the pulse-height spectrum of iodine-125 in a liquid scintillation counter. Chemical quenching caused by the addition of a small amount of nitromethane shifts the spectrum to the left and lowers the total count rate. The nitromethane absorbs energy, reducing that available for production of scintillation photons. In this example, the counting efficiency (counts per decay) for the unquenched sample was 76% and for the quenched sample, 58%. Tripling the amount of nitromethane will shift the spectrum farther to the left and will reduce the efficiency to 34%. (From Horrocks, D. L.: Nucl. Instr. Methods, *133*:293-301, 1976.)

lowers the total count by shifting some pulses so far to the left that they fall below the noise threshold. Quenching can cause batch-to-batch or sample-to-sample variations in efficiency. Nevertheless, laboratories not infrequently ignore quenching in the hope that whatever happened to the unknown samples will also have happened to the standards. Because the factors which cause quenching are so numerous and so difficult to control and because the effect of quenching on counting efficiency is not linear, it is important to monitor and, when necessary, to correct for quenching in every assay.

In homogeneous systems, those for which the sample is soluble in the scintillation cocktail, the two major types of quenching are chemical and optical. *Chemical quenching* is caused by molecules in the sample or the cocktail which compete with the primary scintillator for solvent excitation energy. Most compounds, including water and dissolved oxygen, are chemical quenchers; another is the standard quenching agent, nitromethane (Figure 1B-80). *Optical quenching* refers to the absorption of photons by colored compounds in the scintillator. Heme is an example of a quenching compound with visible color, but colorless compounds may also quench by absorption in the ultraviolet. Although both chemical and optical quench decrease the number of photons and the height of the pulse, the effects of the two can be distinguished to some extent because optical quenching is affected by the length of the photon path through the colored liquid and therefore causes more variation in pulse height than does chemical quenching.

In multiphase systems such as emulsions and suspensions, the minute irregularities in physical contact between sample and solvent may further reduce efficiency by absorbing radiation or by interfering with energy transfer. This is referred to as *phase quenching*.

Quench correction. The original and still the most reliable technique for quench correction is the use of an *internal standard*. This is a liquid of known radioactivity added to blank samples that are presumed to contain the same quenching factors as the unknown samples or added to the unknown samples themselves after measuring the count rate of the sample alone. The efficiency calculated for the internal standard (measured count rate divided by known decay rate) is used to convert the sample count rates to decay rates.

Because of their convenience and adaptability to large automatic counters, *external standards* have largely replaced the internal standard. An external standard is usually a capsule of a long-lived, γ-emitting nuclide such as radium-226, cesium-137, or barium-133. For counting, the capsule is shifted mechanically from its shielded storage to a counting position against the side or end of the sample vial. In these positions, gamma rays from the nuclide produce Compton electrons and some photoelectrons within the liquid which in turn produce scintillations. Quenching also affects the counting efficiency and shifts the spectrum of these electrons. Measurement of the shift in the energy spectrum of the external standard is used to determine and correct for the effects of quenching. Corrected data are expressed as decays per minute, the so-called *absolute count rate*, rather than counts per minute.

Automatic quench correction is commonly provided on the larger counters; machines so equipped can print either the measured count rate or absolute count rate, as desired. Detailed descriptions of quench correction methods are available.[5,6,11] In general, these methods require initial calibration using a set of quenching standards each containing the same, known amount of nuclide but a different amount of quenching compound. Despite their general reliability, methods based on external standards are subject to vagaries and should be monitored with an internal standard.

Counting Statistics

Poisson distribution. Rutherford observed that the number of decays of a radioactive sample in successive, equal time intervals follows a Poisson distribution. For low numbers of decays, particularly if the mean value is below 100, the Poisson distribution differs noticeably from the more familiar, symmetric Gaussian distribution. The tails of the Poisson distribution are more prominent and are asymmetric. Values far removed from the mean are more likely to occur than for the Gaussian distribution. On the other hand, for thousands of counts recorded in a typical assay, the Poisson distribution closely resembles the bell-shaped Gaussian distribution with one provision: the standard deviation of the Poisson distribution is always equal to the square root of the mean.

Answering the common question "How many counts?" serves to illustrate the application of Poisson statistics. How many counts should be accumulated to assure, at least with respect to the randomness of counting, that the measured count rate for a sample can be expected to lie within 1.5% of the mean value for that sample in at least 95% of the times that the sample is counted? To respond, note that for a normal distribution and also for a Poisson distribution with a mean of several hundred or more, the region within two standard deviations to each side of the mean encompasses approximately 95% of the population. Consequently, the requirement that 95% of the counts fall within the mean ± 0.015 times the mean is met if

$$2\,s = 0.015\,\bar{x}$$

where s and $\bar{x}$ are the standard deviation and mean, respectively. For a Poisson distribution,

$$s = \sqrt{\bar{x}}$$

Substituting and squaring,

$$\bar{x} = (2 / 0.015)^2 = 4 / 0.000225 = 17\ 800$$

In other words, each sample must be counted long enough to accumulate at least 17 800 counts. Increasing the number of counts decreases the statistical error of the count. To the extent that background is negligible, quadrupling the number of counts halves the counting error. For the example just used, counting to 71 200 reduces the error from 1.5% to 0.75%. In practical terms, increasing the count requires increasing the activity in the sample or increasing the time for counting the sample.

Effect of background count rate. Samples are always counted in the presence of background. This activity originates from a variety of sources including ever-present, naturally occurring radionuclides in the environment and the radioactive materials in the laboratory. Unless the background count rate is negligible, it must be subtracted from the measured rate to obtain the net activity, and it must be included for calculation of uncertainties in the count, usually referred to as counting error. This calculation follows the general rules for propagation of errors. For the simple case of a sum or difference, the rule states that the square of the error in the result equals the sum of the squares of the individual errors. If sample and background are counted for equal time,

$$C_s = C_{s+b} - C_b$$

where counts, C, are specified by the subscripts: s for sample net, b for background, and s + b for sample with background. From the above rule,

$$s_s^2 = s_{s+b}^2 + s_b^2$$

where the sample standard deviations, s, are specified by the subscripts as before.

Optimization. For most clinical assays, sample activity is high and therefore is not the limiting factor. However, for the laboratory faced with low-activity samples, it is important to achieve a *minimum counting time* while maintaining an acceptably low counting error. This requires selecting conditions and equipment that provide *maximum counting efficiency,* with minimum and stable background count rate. In general, it can be shown that the conditions for maximizing counting efficiency are also the conditions which maximize S^2/B, where S is the net sample count rate, and B is the background count rate.[10] In practice, a sample of moderate activity is counted with each piece of equipment under consideration and at each channel width or other adjusted setting. The equipment and operating conditions used to count the actual samples should be the equipment and conditions for which S^2/B is largest.

RADIATION SAFETY

All individuals working in or frequenting any portion of a restricted area where radioactivity is present are required to have instruction in the health problems associated with radiation exposure.[2] This section is intended to meet this requirement.

Radiation Dose

Absorbed energy. All of the effects of radiation on tissue begin with the deposition of radiation energy and the resultant disruption of chemical bonds. If a large amount of energy is absorbed, cells are killed and the tissue is destroyed almost immediately. If less energy is absorbed, the damage to cells and tissue may not be clinically apparent for hours or even years. At still lower levels there may be no discernible damage either because the damage to cells is too slight to detect or because biologic repair has been complete.

Effective dose. To radiation biologists, the biologic effect of the absorbed radiation energy is more meaningful than the amount of energy itself. The difference arises because some types of radiation produce more tissue damage than others for an equal amount of absorbed energy; for example, α-radiation is more damaging than γ-radiation. *Relative biological effectiveness* (RBE) is the term used by workers in this field to describe the biologic effect for any given type of radiation energy relative to the effect from a standard type of radiation. The standard type of radiation is usually 200–250 keV X-rays. RBE can be calculated following an experiment which measures the amount of each of two types of radiation required to produce the same biologic end point (e.g., cell death, chromosome damage, whole animal death, cancer incidence). Thus,

$$RBE = \frac{\text{Dose of standard radiation for end point}}{\text{Dose of other radiation for same end point}}$$

As might be expected, the value obtained for the RBE depends in part on the end point used for the experiment. After considering the RBE obtained from many radiobiologic experiments, the International Commission of Radiation Protection (ICRP) has determined a *quality factor* (Q) which they recommend for estimating the effects of radiation. Gamma rays, X-rays, and β-radiation are assigned a Q of 1; α-radiation, which as mentioned is more injurious, is assigned a Q of 20. Complete tables of Q values have been prepared by the ICRP.[7]

The *dose equivalent* takes into account both the absorbed energy and the quality factor of the radiation (as well as any other modifying factors). The conventional unit of dose equivalent is the *rem* (roentgen-equivalent-*man*). The newer SI unit is the *sievert* (Sv); 1 Sv equals 100 rem. To summarize, 1 rem of any radiation has the same biologic effectiveness as 1 rad of standard X-rays; likewise, 1 Sv has the effectiveness of 1 Gy of standard X-rays.

Health Risks from Radiation

Types of risk. For the small amounts of radionuclide used and the low exposures received in the clinical biochemistry laboratory, risk is limited to the possible effects of low-level, long-term exposure. *Acute radiation sickness* with immediate organ damage may follow acute exposure greater than 25 rem (0.25 Sv) and usually follows doses greater than 100 rem (1.0 Sv). Doses as large as 25 rem (0.25 Sv) are not encountered in clinical laboratories.

Chronic, low-level exposure is thought to increase the risk of developing *malignant diseases*. It is assumed that the immediate effect of this low-level radiation is damage to nuclear DNA and that this damage results in somatic mutation.

Estimates of risk. Because clinical disease typically fails to appear until years after exposure to low levels of radiation, and because the same malignancies which may follow exposure are known to occur in those who have had no known exposure to radiation, the estimates of risk are extremely crude. Nevertheless, it is illustrative to compare estimates of risk for persons exposed at the maximum rate permitted by regulations to the risks incurred by persons not exposed to radiation. For this comparison, note that the maximum permissible dose equivalent for the whole body of a worker exposed in restricted areas is 1.25 rem for each calendar quarter. Assuming this dose is spread uniformly throughout the work week, it is about 20 mrem daily. Over an adult career of 30 years at the 1.25 rem per quarter rate, the dose will be 150 rem. This exposure has been estimated to entail a risk of about 3.6 cancer deaths per thousand exposed workers (based on a lifetime risk for 1 rem of 120 deaths per million workers).[8,12] For comparison, the risk of developing cancer spontaneously is about 20 times greater. From another perspective, the risk of cancer associated with 30 years of exposure at the maximum permissible rate is lower than the risk of a fatal automobile accident but higher than the risk of a fatal electric shock.[8,12]

Regulatory Requirements

The requirements of state and federal agencies must be followed when using radioactive materials. States differ substantially in their regulations; some have no regulatory agency and depend entirely on federal regulations while others require the user to report to several different agencies at once. The federal regulations are covered in Title 10 of the Code of Federal Regulations (CFR),[1] which discusses radiation protection and licensing requirements. The Nuclear Regulatory Commission (NRC) controls the possession, use, and transfer of licensed radionuclides; the Department of Transportation regulates the shipping of radioactive materials as described in CFR Title 49;[4] and the Food and Drug Administration regulates the manufacture of pharmaceuticals and diagnostic kits, including those containing radionuclides. Users should obtain a current copy of the Code of Federal Regulations and consult a radiation health physicist regarding their own unique requirements for occupational protection from radiation.

In general, the user is required to maintain certain standards of radiation exposure which cannot be exceeded and to maintain a level of occupational exposure as low as reasonably achievable (*ALARA*). Most large institutions employ radiation health officers or radiation health physicists to provide guidance in radiation protection and ensure that the user conforms to current regulations. There is no absolute lower threshold to occupational radiation exposure below which danger of possible injury can be positively excluded. However, existing guidelines for handling radioactive materials will result in risks as low as, and in most cases a risk much lower than, those accepted by workers in other well-regulated industries. This risk is low enough to be acceptable to most persons working with ionizing radiations. *Dosage limits*, as specified in the CFR, represent doses of radioactivity which are not expected to cause appreciable bodily injury to the individual at any time during a lifetime. Some of these limits are listed in Table 1B-13. Doses to workers in clinical laboratories are normally a small fraction of those tabulated in this table.

Table 1B-13. MAXIMUM PERMISSIBLE DOSE
EQUIVALENTS FOR OCCUPATIONALLY EXPOSED
INDIVIDUALS*

	Dose per Calendar Quarter (rem)
Whole body	1.25
Head and trunk	1.25
Active blood-forming organs	1.25
Lens of eye	1.25
Gonads	1.25
Hands and forearms	18.75
Skin of whole body	7.50

*Title 10, Code of Federal Regulations, 20.101, 1983.

Safe Handling of Radionuclides

Radiation monitoring. For individuals working with radioactive materials, it is a good safety precaution to wear some form of *personal radiation monitor*. The most popular devices are the film badge and the thermoluminescent dosimeter, both of which measure absorbed dosage in millirems. In some cases (e.g., when using millicurie quantities of I-125) *bioassay* is appropriate in addition. Useful techniques for bioassay include periodic assay of urine and counting over the thyroid for workers exposed to radioactive iodine and assay of urine and expired air for those exposed to tritium or carbon-14.

For the laboratory, an *area monitor* should be in continuous operation. The usual area monitor is a Geiger counter with an audible signal and a loud alarm to warn of high count rates. In addition, all laboratory work areas should be checked regularly for surface contamination, preferably by *wipe test*. This consists of wiping each surface to be examined with a separate, moistened absorbent material (the swipe) and counting each swipe in a γ-counter for γ-radiation or in a liquid scintillation counter for β-radiation.

Safety precautions. General safety precautions must be practiced in any clinical laboratory where radioisotopes are used. Some of these are as follows:

1. Accurately record the receipt, transfer, and disposal of radioactive materials.

2. Label prominently all substances containing or contaminated with radioactivity.

3. Protect sources and post adequate warning signs in areas in which sources are stored. Do not leave sources of radioactivity unattended; lock unattended storage facilities.

4. Wear a protective laboratory coat at all times in the radiation area.

5. Wear disposable gloves, and thoroughly wash hands after handling radioactive substances and before leaving the work area.

6. Do not eat, store, or prepare food, smoke, or apply cosmetics in a work area.

7. Use and store radioactive materials only in designated, posted, and protected work areas.

8. Plan experiments carefully to minimize the time spent working with high-energy β-emitting or γ-emitting materials.

9. Work behind lead or other shielding and use tongs or remote devices to handle high-energy β-emitting or γ-emitting sources of $> 100 \ \mu$Ci. Radiation intensity is inversely proportional to the square of the distance from a "point" source.

10. Handle millicurie quantities of radioiodine only in a fume hood approved and tested for this purpose.

11. Handle unsealed radioactive materials over removable absorbent covers to simplify decontamination should spillage occur.

12. Pipet radioactive materials only with a pipet bulb or syringe. Do not pipet by mouth.

13. Contain all spills of radioactive materials. When spills are detected, immediately notify the proper safety officers; begin decontamination promptly.

14. Store radioactive waste, liquid or solid, in labeled containers until disposed of by the proper agencies. Do not allow radioactive waste to enter normal routes of trash or sewage disposal.

References

1. Code of Federal Regulations, Title 10 Energy. Office of the Federal Register, General Services Administration, Washington, D.C., January 1, 1983.
2. *Ibid.*, Section 19.12, Instructions to workers, 1983.
3. *Ibid.*, Section 20.101, Radiation dose standards for individuals in restricted areas, 1983.
4. Code of Federal Regulations, Title 49 Transportation, Subchapter C: Hazardous Materials Regulations, Parts 171–179. Washington, D.C., Office of the Federal Register, General Service Administration, October 1, 1982.
5. Gibson, J. A. B.: Modern techniques for measuring the quenching correction in a liquid scintillation counter: A critical review. *In*: Liquid Scintillation Counting, Recent Applications and Development, Vol. 1. C.-T. Peng, D. L. Horrocks, and E. L. Alpen, Eds. New York, Academic Press, 1980, pp. 153-172.
6. Horrocks, D. L.: A new method of quench monitoring in liquid scintillation counting: The H number concept. *In*: Liquid Scintillation Counting, Incorporating Whole-Body Counting and Radioimmunoassay, Vol. 5. M. A. Crook and P. Johnson, Eds. Philadelphia, Heyden, 1978, pp. 145-168.
7. International Commission on Radiological Protection: Recommendations of the ICRP. Publication No. 26 (Annals of the ICRP Vol. 1 No. 3). New York, Pergamon, 1977.
8. International Commission on Radiological Protection: Problems Involved in Developing an Index of Harm. Publication No. 27 (Annals of the ICRP Vol. 1, No. 4). New York, Pergamon, 1977.
9. Krugers, J.: Instrumentation in Applied Nuclear Chemistry. New York, Plenum, 1973.
10. Loevinger, R., and Berman, M.: Efficiency criteria in radioactivity counting. Nucleonics, *9*(1):26-39, 1951.

11. Ring, J. G., Nguyen, D. C., and Everett, L. J.: Liquid scintillation counting from gross counts to spectral analysis. *In*: Liquid Scintillation Counting, Recent Applications and Development, Vol. 1. C.-T. Peng, D. L. Horrocks, and E. L. Alpen, Eds. New York, Academic Press, 1980, pp. 89-104.
12. Whipple, G. H.: Unpublished lecture notes, 1978.

Additional Readings

Bushong, S. C.: Radiologic Science for Technologists. 2nd ed. St. Louis, C.V. Mosby Company, 1980.

Hendee, W. R.: Radioactive Isotopes in Biological Research. New York, John Wiley & Sons, 1973.

Howard, P. L., and Trainer, T. D.: Radionuclides in Clinical Chemistry. Boston, Little, Brown & Company, 1980.

Lederer, C. M., Hollander, J. M., and Perlman, I.: Table of Isotopes. 6th ed. New York, John Wiley & Sons, 1967.

Noz, M. E., and Maguire, G. Q.: Radiation Protection in the Radiologic and Health Sciences. Philadelphia, Lea & Febiger, 1979.

Peng, C.-T., Horrocks, D. L., and Alpen, E. L.: Liquid Scintillation Counting, Recent Applications and Development, Vol. 1, Physical Aspects; Vol. II, Sample Preparation and Applications. New York, Academic Press, 1980.

Pizzarello, D. J., and Witcofski, R. L.: Medical Radiation Biology. Philadelphia, Lea & Febiger, 1982.

Powsner, E. R., and Raeside, D. E.: Diagnostic Nuclear Medicine. New York, Grune & Stratton, 1971.

Section Ten

MASS SPECTROMETRY

by Brian D. Andresen, Ph.D., and Bethany L. Wise, Ph.D., M.T. (ASCP)

Mass spectrometry is a technique that can provide information concerning the elemental composition and structure of organic compounds. The arrangement of functional groups of a molecule may be determined, and the molecular weight may be measured to ten thousandths of an atomic mass unit (amu). Submicrogram quantities of an unknown substance in a complex biological matrix in either a solid, liquid, or gaseous state may be completely characterized and quantitated with extreme specificity and sensitivity. Hence, mass spectrometry has gained increased use in clinical chemistry laboratories for the identification of compounds present in body fluids in health and disease and for the identification of ingested drugs and their metabolites.

In order to perform an analysis utilizing mass spectrometry, compounds (molecules) are ionized by a variety of techniques which will be discussed below. The ionization process allows compounds to be manipulated (filtered) through electrostatic and/or magnetic fields and to be positively identified according to their mass (m) to charge (z) ratio (m/z). Furthermore, sufficient excess energy can be imparted to the molecular ion to generate many fragment ions that can also be separated, measured, and recorded from the smallest fragment to the intact molecular ion, to produce a mass spectrum (Figure 1B-81). A mass spectrum is usually displayed with the most abundant fragment (i.e., the most stable and most-often-recorded ion) normalized to an arbitrary 100% relative intensity. All other fragment ions and molecular ion intensities are presented relative to this major fragment ion (termed the "base peak"). The molecular ion for certain compounds often does not fragment easily, and for stable compounds (e.g., polycyclic aromatics) the molecular ion may be the "base peak" ion. However, most compounds produce a fragment ion which is more stable than the molecular ion.

The mass spectrum of a compound is specific for that particular compound. Reference libraries of mass spectral data are available for use in the identification of an unknown substance. If the compound has not been analyzed previously by mass spectrometry, a search through the library collection of mass spectral data may reveal many compounds that produce similar fragment ions of unique m/z values. Such information identifies specific structural features and aids in supporting a reasonable proposal for the complete identification of the unknown compound.

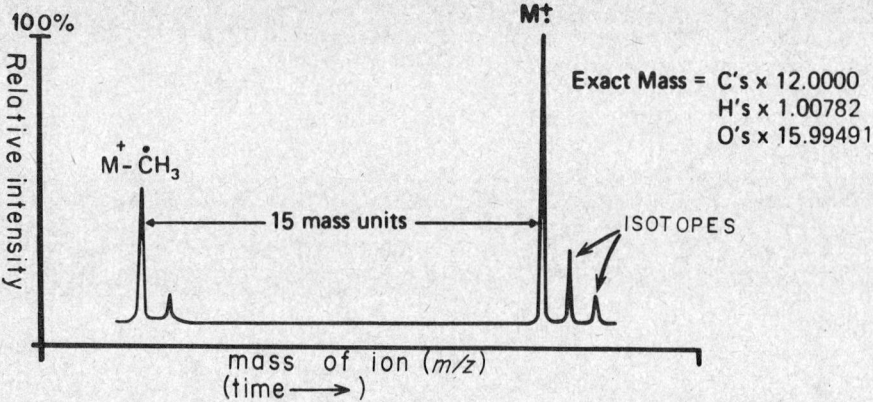

Figure 1B-81. Analog mass spectral data produced by a mass spectrometer. The molecular ion, M⁺, is equal to the molecular weight of a substance and the precursor for all the ions seen in the spectrum. If the mass of the molecular ion can be determined accurately enough, the elemental composition may be determined unequivocally. Stable isotopes of individual elements can also be observed. The x-axis of the plot shows mass while the y-axis reveals abundance of ion formation.

THE INSTRUMENT

The essential components of a mass spectrometer are the vacuum system, ion source, mass filter, and detector, as illustrated in Figure 1B-82.

Vacuum System

Mass spectrometers operate under very high vacuum conditions. Most instruments operate in the range of 10^{-7} to 10^{-8} torr (1 torr = 1.0 mm Hg = 1/760 atmosphere). At this pressure there are approximately 10^8 molecules/cm^3 of background air molecules. If more background air molecules were present in the instrument, ions generated in the ion source would collide with the background molecules and scatter, resulting in decreased sensitivity. Additionally, to avoid contamination and background noise from the condensation of previous samples on the interior of the instrument, the mass spectrometer is continuously heated (in the range of 200–320 °C). In order to maintain an extremely clean interior, the instrument is seldom vented to the atmosphere.

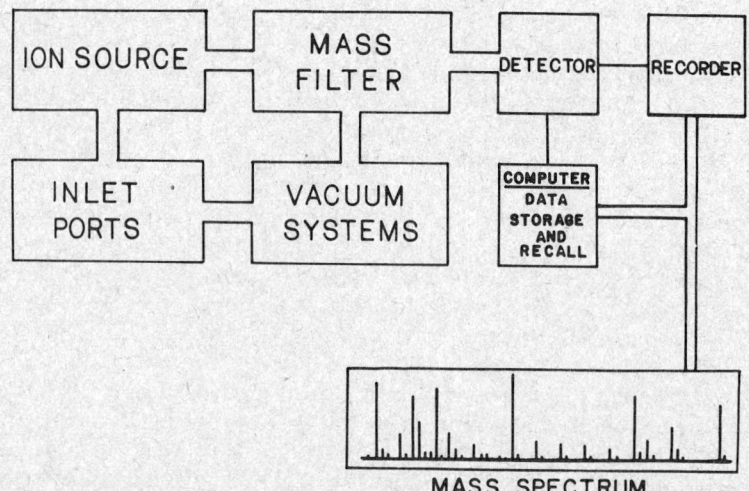

Figure 1B-82. Block diagram of the essential components of a mass spectrometer. The ion source, inlet ports, mass filter, and detector are all under high vacuum (10^{-7} torr).

Ion Source

In mass spectrometry, the sample must be pure and dry, because large amounts of water or other solvents will be detected simultaneously by the instrument. The sample must also be ionized before analysis. Only 10% of a sample which is directed into the ion source is ionized. The remainder of the sample is either removed through the vacuum pumping system or condensed as background residue onto the inside walls of the mass spectrometer. Various ionization techniques have been used. The most common are electron impact, chemical ionization, and field desorption.

Electron impact. The most widely used ionization technique is electron impact, in which an electron gun precisely positioned in the ionization chamber bombards the molecules from the sample. Figure 1B-83 shows the configuration of a typical electron-impact ion source. The sample to be analyzed is either fractionally distilled, sublimed from a probe inserted into the ion source, or eluted from a gas chromatographic column that is directed into the ion source chamber. The ionization beam is produced when electrons are drawn off a heated filament (cathode) toward the walls (anode) of the ionization chamber. The standard potential difference required to generate a target current of 200–300 μamps is 70 electron volts, more than enough energy to ionize organic substances. As the molecules from the compound under investigation are directed into the electron gun, they are exposed to the electron beam, which causes ionization and fragmentation. At the moment of ionization the ion-source chamber contains the molecular ion (parent compound minus an electron) as well as fragments derived from the molecular ion. These fragment ions possess either positive (+), negative (−), or neutral (·) charges. Although quite abundant, neutral fragment ions are not detected by mass spectrometers. For most compounds negative ions are generated approximately 1/100th as often as positive ions. If an opening in the ion source is near a negatively charged draw-out plate, positive molecular ions, and positively charged fragment ions will rush from the ion source in an attempt to neutralize their charge. If the draw-out plate is designed with an escape slit, a great majority of the positively charged fragment ions will be allowed to leave the ion source region in free flight. Focusing plates near the exit of the ion source allow intense total ion beams to enter the analyzer section of the mass spectrometer.

Chemical ionization. In contrast to the rather harsh ionization technique of electron impact, chemical ionization utilizes a reagent gas which ionizes the sample more gently. Reagent gases that may be used include methane, ammonia, water, and isobutane, of which methane is used most frequently. The reagent gas is continuously leaked into the ion source, raising the source pressure to a substantially higher value (0.05–0.1 torr) than the rest of the mass spectrometer (10^{-7} to 10^{-8} torr). Initially, the reagent gas is ionized by the electron gun in a manner similar

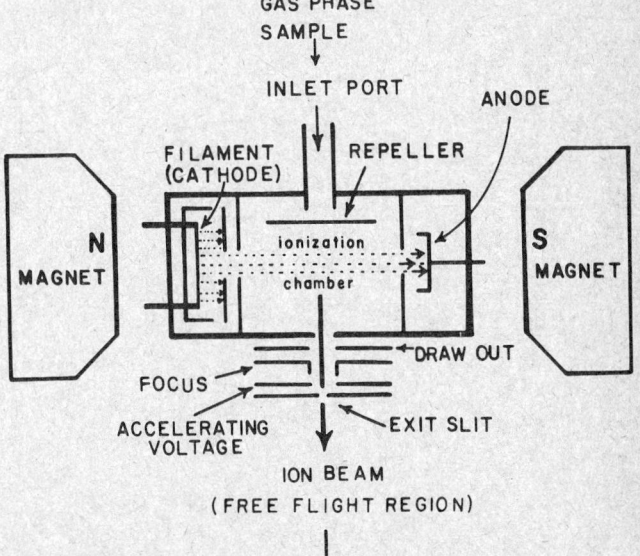

Figure 1B-83. Electron impact ion source. The small magnets are used to collimate a dense electron beam, which is drawn from a heated filament placed at a negative potential. The electron beam is positioned in front of a repeller, which is at a slight positive potential compared to the ion source. The repeller sends any positively charged fragment ions toward the opening at the front of the ion source. The accelerating plates strongly attract the positively charged fragment ions.

to the electron impact technique. Then, due to the higher source pressure, the ionized reagent-gas molecules react chemically in the gas phase with other nonionized reagent molecules to form a highly activated intermediate, which is used to transfer a charge to the sample molecules. The sequence of gas-phase chemical events leading to the transfer of a proton from methane (reagent gas) to the sample molecule is seen below:

$$CH_4 \text{ (bombarded with electron beam)} + e^- \longrightarrow CH_4^+ + 2e^- \text{ (ionization)}$$
$$CH_4^+ + CH_4 \longrightarrow CH_5^+ + CH_3 \cdot \qquad \text{(activation)}$$
$$CH_5^+ + \text{Analyte} \longrightarrow (\text{Analyte-H})^+ + CH_4 \qquad \text{(charge transfer)}$$
$$(M+1)^+$$

Although this process appears more cumbersome than electron impact, chemical ionization provides a unique method for the determination of the molecular weight of compounds which do not form stable molecular ions under electron impact conditions. Substances which produce no molecular ion will, under chemical ionization techniques, produce "quasi" molecular ions (i.e., $M^+ + 1$) and only a few fragment ions. This technique is especially useful for determining the molecular weight of an unknown substance. Figure 1B-84 contrasts the mass spectra of ephedrine under electron impact and chemical ionization (methane) conditions. With electron impact, the molecular ion ($M^+ = 165$) is not evident. However, under chemical ionization conditions the $M^+ + 1$ ion (m/z 166) is quite abundant.

Chemical ionization techniques have been used clinically to screen complex biological samples rapidly in order to identify the molecular weights of unknown drugs. A probe distillation of a crude serum or urine sample extract directly into the chemical ionization source permits the identification of all compounds by their $M^+ + 1$ ions. Relative concentrations of individual drugs can also be determined rapidly by comparisons of peak heights of individual molecular ions.

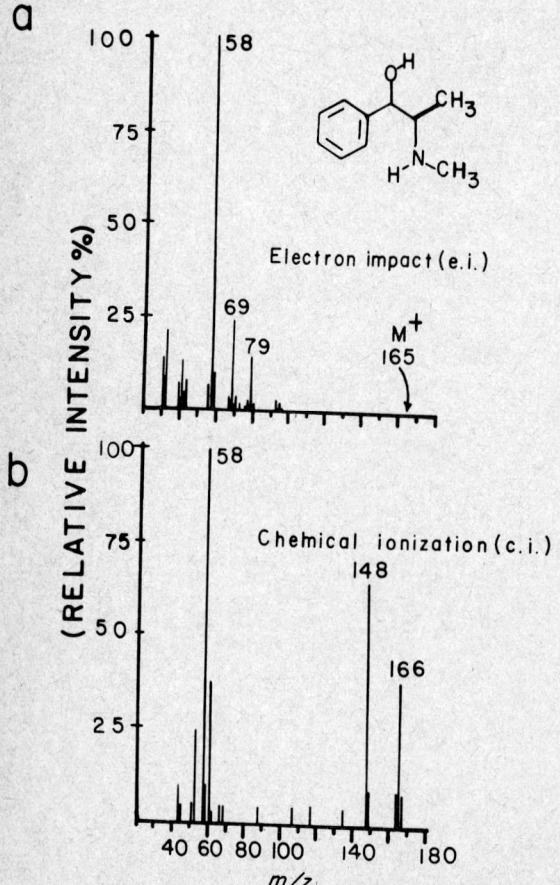

Figure 1B-84. Mass spectral comparisons of ephedrine utilizing (a) electron impact, and (b) chemical ionization (methane) techniques. The ion at m/z 58 corresponds to $(CH_3—CH=NH—CH_3)^+$. The $(M+1)^+$, m/z 166, ion is clearly evident under chemical ionization conditions, yet is less than 0.1% when electron impact is employed.

Field desorption. Mass spectral analysis usually requires that substances should be volatilized directly into the ion source of the mass spectrometer. This requirement would make the analysis of large, polar, and nonvolatile compounds impossible if it were not for the third ionization technique, field desorption. In addition, compounds of high molecular weight often yield relatively weak molecular ions. Field desorption can yield acceptable data for compounds previously not amenable to mass spectrometric analysis.

Field desorption utilizes a probe with a heated filament upon which the sample to be analyzed is placed. The very fine "whisker" type filament, 0.5–1.0 cm long, is placed directly in front of a negatively charged (−8000 to −10 000 V) draw-out plate. When a positively charged voltage is applied to the sample probe, electrons in the molecules of the sample are drawn into the positively charged filament surface (+8000 to +10 000 V), which is simultaneously heated. The ionized molecules are repelled from the surface of the probe and are strongly attracted to the negative field of the draw-out plate. The draw-out plate is constructed with an orifice which allows most of the ionized molecules to enter the mass filter section of the mass spectrometer. The molecular ion is then drawn out of the ion source at a very high velocity. The resultant limited resident time in the ion source will give rise to only a few fragment ions before the molecular ion strikes the detector. Therefore, field desorption primarily yields molecular ions and has been important in the confirmation of molecular weights of very large and polar substances, which are difficult to ionize or which produce no molecular ion with other types of ionization techniques.

Figure 1B-85 contrasts the fragment ions generated for tetrahydrocortisone using the three different ionization techniques. Field desorption clearly produces the simplest mass spectrum and most abundant molecular ion.

Mass Filters

A total ion (TI) beam containing the molecular and fragment ions of the compound under investigation continuously emerges in free flight from the ion source. To obtain a complete mass

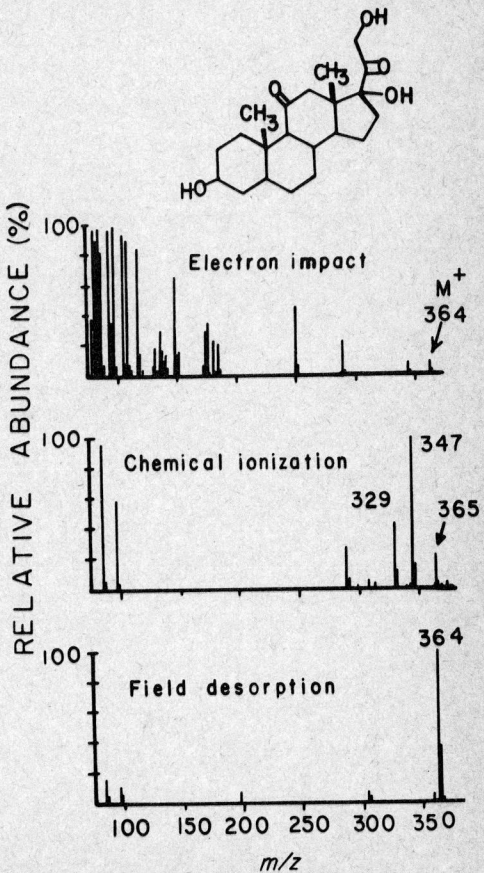

Figure 1B-85. Mass spectral comparisons of tetrahydrocortisone under electron impact, chemical ionization, and field desorption. The molecular ion is clearly evident under field desorption.

spectrum, the ion beam must be separated into its individual ion components according to their masses (m) and charges (z). Two different methods, one using the magnetic sector analyzer and the other the quadrupole analyzer, are found in analytical laboratories to separate the various fragment and molecular ions. Other techniques such as time-of-flight and ion cyclotron resonance are available, but they are usually located only in research or industrial laboratories.

Magnetic sector instruments. As charged particles travel through a magnetic field, they are deflected at right angles to the lines of magnetic flux. If the total ion beam travels in free flight through a strong magnetic field, the fragment ions are deflected differently, depending upon their mass (m) and charge (z). For ions with the same charge, the lightest fragment ions are deflected to the greatest degree and the heavier fragment ions are deflected to a lesser amount. All fragment ions with identical charge and mass are deflected to the same degree and are recorded as a unique signal. The intensity of this signal is dependent upon the number of fragment ions recorded, which in turn is dependent upon the stability of the fragment ion. The more stable an ion is in the gas phase, the more often it is recorded. Utilizing a magnetic filter, the weight and relative abundance for each ion can be recorded. A fragment ion (m) with a charge (z) will be recorded when the following relationship is satisfied:

$$m/z = \frac{H^2 R^2}{2V}$$

where H is the magnetic field strength, R is the radius of curvature of the flight path, and V is the accelerating voltage. Figure 1B-86 shows that the molecular ion (M^+) for acetone is deflected the least in a fixed magnetic field, while the lightest fragment ion shown (CH_3^+) is deflected the most. In this example, only the fragment ion at m/z 43 (CH_3—$C\equiv O^+$) is of the proper mass (possesses the correct momentum based on its mass and velocity obtained from the accelerating voltage) to travel through the flight path (with a radius of curvature R) and strike the detector. To observe all fragment and molecular ions (i.e., a complete mass spectrum of the compound of interest) it is possible to (1) utilize a fixed magnetic field with a photographic plate in the plane of the emerging ions at the exit of the magnetic field, (2) sweep the magnetic field to "fan" individual fragments past the detector, and (3) uniformly increase or decrease the accelerating voltage. Most magnetic sector instruments use the second technique to scan the field and obtain the full spectrum, as each unique mass ion is passed by the entrance to the detector.

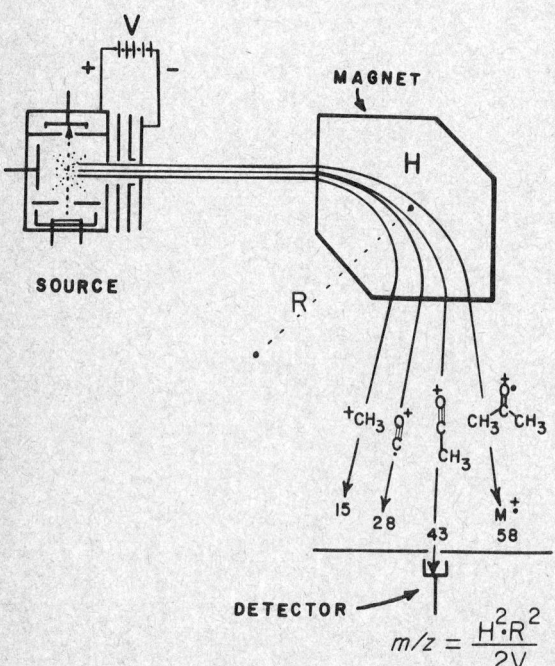

Figure 1B-86. Ion path through a strong external magnetic field. Ion trajectories and radius of curvature, R, are related to the accelerating voltage, V, magnetic field strength, H, and m/z ratio. Only when a specific V, H, and m/z condition is met will an ion follow the unique path of radius R leading to the detector.

A mass spectrum represents all ions which are stable enough to be drawn from the ion source and traverse a distance of 1–2 m before hitting the ion-detection system. The mass spectrum is plotted either electronically or on high-speed chart paper. A completed plot of all masses versus intensity for each compound is unique for each compound. These barplots can be displayed rapidly under computer control and can also be duplicated from one instrument to another.

Scan times for magnetic sector instruments can be very fast (less than one second) but are usually a minimum of 3–5 s because of the time needed between scans for the magnet to settle. Magnetic sector instruments can usually detect high-mass fragment ions in the range of 1000–1500 amu and possess great resolving power. High-resolution instruments are capable of measuring exact masses and determining the molecular formula of each ion recorded, a capability which is often very important for the characterization of a completely unknown compound.

Quadrupole analyzers. In contrast to magnetic sector instruments, quadrupole analyzers are typically smaller in size and separate mass spectral fragment ions in a completely different manner. The quadrupole instrument, as its name implies, utilizes four metallic rods (approximately 10–20 cm long and 1 cm in diameter) which are held in place by ceramic insulators. In operation, an oppositely charged potential is placed on the rods, and the DC voltage is increased while the polarity of each rod is constantly alternated at a high radio frequency (RF). Figure 1B-87 reveals the configuration of the DC voltage, attached frequency generator, and regulation circuits.

Ions leaving the ion source travel into the center of the oscillating electrostatic field produced by the quadrupoles. Mass filtering of the fragment ions is accomplished as ions are attracted to the oscillating charged surface of the rods. Separation of fragment ions is determined by the optimum path which a certain mass ion will take as it travels through the length of the quadrupole.

The paths which the fragment ions follow through the mass filter are very contorted. Each ion with a unique m/z value will follow a specific path as determined by the applied voltage and frequency on the rods. Only when specific conditions are satisfied will an ion emerge at the end of the quadrupole and strike the detector. A complete mass spectrum can be obtained by scanning the quadrupoles through the whole range of voltages and frequencies.

Quadrupole analyzers can produce unit mass resolution of fragment ions to approximately 800–1000 amu, which is adequate for most analytical purposes. The small size of these analyzers, lack of large magnetic power supplies, and low maintenance make them suitable as routine analytical instruments. Although they do not have great resolving power, quadrupole instruments are capable of scanning rapidly (0.5–2 s for complete mass spectral data acquisition) and therefore have been used extensively for combined capillary gas chromatographic–mass spectrometric computer analyses.

Detector

One of the main attributes of mass spectrometry, its great sensitivity, is achieved because fragment ions are recorded with an electron multiplier. The device (Figure 1B-88) resembles a

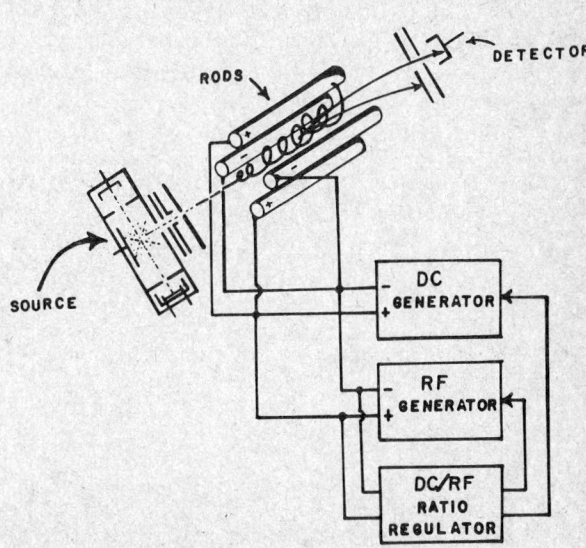

Figure 1B-87. Quadrupole mass spectrometer with associated electronic controls for scanning the DC voltage and radio frequency (RF) of the mass filter rods. Ions in a quadrupole mass filter follow a contorted path and only those which have a specific *m/z* ratio that matches the applied voltage and radio frequency will be detected.

photomultiplier except that ions, rather than photons, set off a large cascade of electron discharges from the surfaces of metal dynodes specifically arranged with high potential differences between them. The electrons boil off the surface of the electron multiplier and amplify the incoming ion signal. The cascade of electrons generated in this manner increases the detected signal 10^6–10^7 times. With an output signal of greater intensity from the electron multiplier, mass spectral data can be recorded directly on an analog recording device, or the signal itself can be passed through an analog-to-digital converter for computer acquisition and storage. This high gain characteristic of electron multipliers permits analysis of compounds present in picogram to femtogram amounts in complex mixtures. Material in nanogram quantities is frequently and readily analyzed by mass spectrometric instruments.

GAS CHROMATOGRAPHY–MASS SPECTROMETRY

The resolving power of gas chromatography, when coupled with the sensitivity and specificity of mass spectrometry, forms a unique tool for the analysis of complex biological samples. Because packed gas chromatographic columns utilize carrier-gas flow rates of 30–40 mL/min, much of the gas must be removed before an eluent from the column can be introduced into the ion source. To analyze the eluents repetitively as they emerge from the end of the gas chromatographic column, various designs (Figure 1B-89) such as membranes, fritted glass, or jet separators have been used to remove the helium carrier gas. Each takes advantage of the fact that helium can be stripped away from larger organic compounds because of its inertness, small molecular size, and high kinetic energy. The jet separator is the most popular and most durable, and has the fewest reactive sites (chemically reactive surfaces) to trap polar compounds.

As an alternative to packed columns, capillary columns with flow rates of only 1–4 mL/min are often attached directly to the ion source. This flow rate is small enough, compared to that from packed columns, to be easily handled by the vacuum system of the mass spectrometer. All interface lines between the gas chromatograph and mass spectrometer are heated to avoid solvent condensation on cool inlet parts.

Sample Preparation

Both gas chromatography (GC) and mass spectrometry (MS) are generally applicable only to volatile compounds. Consequently, many compounds must be derivatized before they are sufficiently volatile for GC, MS, or combined GC-MS analysis. Derivatization becomes especially important when complex biological samples, which contain a broad spectrum of compounds of varying volatility, are to be analyzed. In addition, conversion of polar compounds by chemical derivatization into more volatile substances often enhances the gas chromatographic separation (i.e., causes less tailing and reduces simultaneous elution of compounds), increases the sensitivity of the assay for the individual components, and generates unique mass-spectral data which aid in the identification of unidentified substances in complex biological mixtures. Many derivatization techniques are available. For most biological samples, silylation and methylation are commonly used.

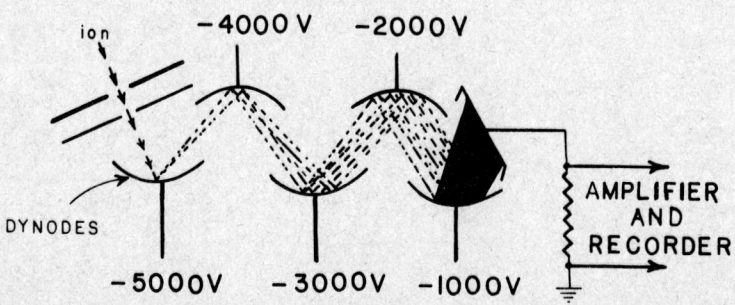

Figure 1B-88. Electron-multiplier detector. As an ion impinges on the first dynode, a cascade of electrons is observed. These are ejected from the surfaces of the remaining dynode stages of the electron multiplier. Gains higher than 10^8 can be achieved with this type of detector.

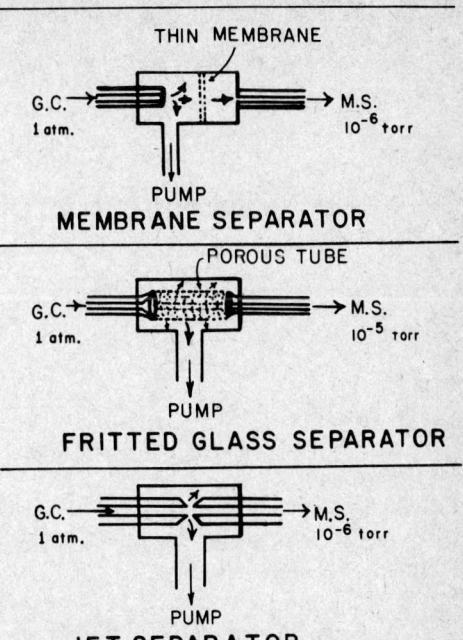

Figure 1B-89. Techniques to remove helium (carrier gas) and enrich the sample as it elutes from the end of the gas chromatographic column. The membrane, fritted glass, and jet separator all remove the helium by the use of an external vacuum pumping system which draws the helium preferentially away from the sample.

Silylation. Commercially available silylation reagents, when added to a completely dried extract of the sample to be analyzed, will quantitatively convert reactive groups (i.e., $=NH$, $-SH$, and $-OH$) to their corresponding trimethylsilyl (TMS, $-Si(CH_3)_3$) derivatives. The reactions often encountered with different samples are seen below:

$$R-CO_2H \quad + \quad \text{TMS-reagent} \quad \longrightarrow \quad R-CO_2-TMS$$
$$R-CH_2OH \quad + \quad \text{TMS-reagent} \quad \longrightarrow \quad R-CH_2O-TMS$$
$$R-NH_2 \quad + \quad \text{TMS-reagent} \quad \longrightarrow \quad R-NH-TMS$$
$$R-NH-R' \quad + \quad \text{TMS-reagent} \quad \longrightarrow \quad R-N(TMS)-R'$$
$$R-SH \quad + \quad \text{TMS-reagent} \quad \longrightarrow \quad R-S-TMS$$

One example of a very common TMS-reagent is $CF_3-C(=N-TMS)(O-TMS)$ [bis-(trimethylsilyl)trifluoroacetamide], often abbreviated BSTFA. This reagent will react rapidly at 50–80 °C with any of the reactive groups shown above. The TMS derivatives, once formed, are generally more volatile for gas chromatographic separations or for probe distillation directly into the ion source of the mass spectrometer.

Methylation. This procedure is useful for the preparation of methyl esters, derivatives which are amenable to gas chromatographic and mass spectrometric analysis. Two methods are available to prepare methylated derivatives. In the first, methanol and dry HCl gas are added to an organic acid. The methanol is used to produce the ester while the HCl gas acts as a dehydrating agent. The reaction is as follows:

$$R-CO_2H + MeOH \cdot HCl \longrightarrow R-CO_2-Me + HCl \cdot H_2O$$

The sample is suitable for analysis after evaporation of the methanol·HCl·H_2O solution.

The second method uses diazomethane (CH_2N_2) for methylation. This reagent is both toxic and explosive but if used in small amounts is relatively safe and is the reagent of choice to generate methyl esters of carboxylic acids. Diazomethane is a gas at room temperature but is easily trapped (dissolved) in ether. When this ether/CH_2N_2 solution is added to a carboxylic acid to be analyzed, methyl esters are quantitatively formed.

Diazomethane is generated from a variety of intermediates, which incorporate N-nitroso-urea and upon treatment with a strong base decompose and liberate CH_2N_2. The sequence of events leading to a methyl ester from diazomethane and a carboxylic acid is outlined below:

$$O=N \quad O$$
$$\quad | \qquad ||$$
$$CH_3-N-C-NH-R + KOH \longrightarrow CH_2N_2 + K^+ \; {}^-O-C-NH-R$$
$$\qquad\qquad\qquad\qquad\qquad\qquad\text{(gas)} \qquad \textit{(remains in the aqueous solution)}$$

$$\downarrow$$

Bubble into ether solution of acids
$$(R-CO_2H)$$

$$\downarrow$$

$$R-CO_2-Me$$
(prepared on the microgram scale)

ISOTOPE DILUTION MASS SPECTROMETRY (IDMS)

Isotopes of various elements (for example, $^{79}Br/^{81}Br$ or $^{35}Cl/^{37}Cl$) can be easily measured with a mass spectrometer. Compounds labeled with stable isotopes (i.e., natural elements replaced with stable ones, such as deuterium for hydrogen, ^{13}C for ^{12}C, ^{15}N for ^{14}N, or ^{18}O for ^{16}O) can also be analyzed by mass spectrometry. These stable, nonradioactively labeled compounds have been used in clinical chemistry for metabolism studies and for quantitative measurements of individual compounds in complex mixtures.

Metabolic Studies

If a compound and its isotopically labeled analog are mixed and analyzed together, data will be generated which show the spectrum of the nonlabeled substance superimposed on the spectrum of the labeled compound. Due to the fact that both compounds are analyzed simultaneously, the molecular ion region appears as an unnatural cluster of ions, separated in mass by the original number of stable isotopes added to the parent compound. Fragment ions also show these unnatural clusters.

If an equal mixture of a nonlabeled and labeled drug or other important chemical containing three or more isotopes is administered to either animals or patients, the starting compound or its metabolites can be readily identified even if present in complex mixtures. The labeled compound can be easily identified because no naturally occurring substance produces a molecular ion region with an equal separation of ions greater than two. Therefore, with biological samples that may contain hundreds of metabolites, only those which produce isotopic doublets are derived from the compound administered. Utilizing this technique, important metabolites of the parent compound can be identified.

In the study of these labeled metabolites, the position which contains the isotope may be eliminated or altered by a metabolic pathway that specifically utilizes the labeled group. The loss of a stable isotope will be reflected in the molecular ion region and in specific fragment ions. This information can then be used to determine unequivocally the location of a metabolic reaction on the molecule of interest.

Quantitation of Individual Compounds

Standard analytical techniques often fail in the determination of a minute amount of a material in a limited amount of complex sample. Because of the discrete separation obtained by gas chromatography and the sensitivity and specificity of the mass spectrometer, low levels of compounds are easily quantitated utilizing analogs labeled with a stable isotope. Typically, the sample to be analyzed is "spiked" with an exact amount of the stable, isotopically labeled analog (internal standard) of the compound to be quantitated. The sample is then extracted, derivatized if necessary, and analyzed by computer-controlled gas chromatography–mass spectrometry. Since the nonlabeled and labeled compounds are extracted equally and subjected to identical isolation and chromatographic procedures, losses during the analysis are similar and thus the ratio of labeled to nonlabeled compounds remains unchanged. As a result, errors in the quantitative measurements are reduced substantially.

During data acquisition the computer can record either selected ions (selected ion monitoring, SIM) or all mass-spectral fragment ions for later recall of plots of selected ions (mass chromatography). The amount of the internal standard can be measured and compared directly to the nonlabeled compound for quantitative analysis. As an example of the technique of isotope dilution, Figure 1B-90 shows the total ion plot (lower trace) of a TMS-derivatized (trimethylsilyl) acid extract of a human urine sample. Mandelic acid–phenyl-D_5 (D=deuterium) was added as an internal standard (1.5 μg/mL). Both mandelic acid and mandelic acid–phenyl-D_5 possess intense fragment ions at m/z 179 and m/z 184, respectively.

Mandelic acid
m/z 179

Mandelic acid-D_5
m/z 184

Mass plots of mandelic acid and mandelic acid–phenyl-D_5 are also shown in Figure 1B-90 (top portion of plots). Although the derivatized urine sample contains many components, plotting only the ions of interest (i.e., m/z 179 and m/z 184) greatly simplifies the analysis. By directly comparing the peak area of only m/z 184 (the internal standard) with the peak area of m/z 179 at the anticipated location for mandelic acid–phenyl-D_5, the amount of mandelic acid in the urine sample can be determined accurately.

Due to the fact that isotopically labeled compounds possess extraction properties, chro-

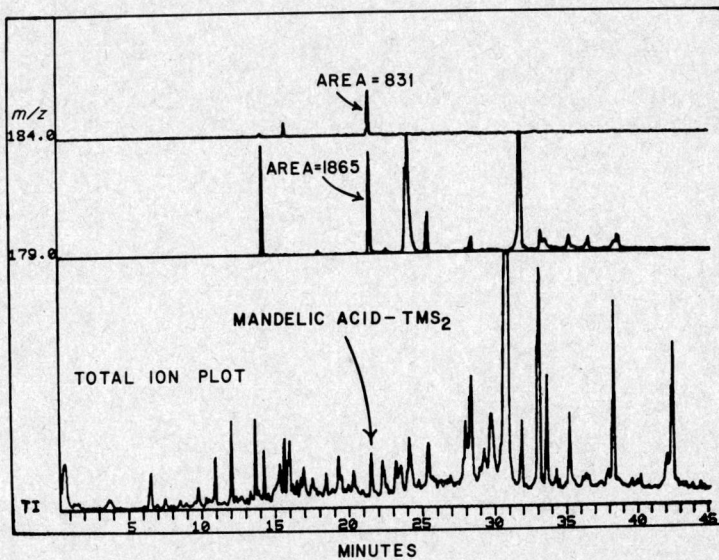

Figure 1B-90. Total ionization (TI) plot of a TMS-derivatized urine extract. Plots of m/z 179 (endogenous mandelic acid) and m/z 184 (internal standard, mandelic acid–phenyl-D_5) are shown above. Although other compounds possess these fragment ions, only mandelic acid is observed at 21.6 min. Comparison of the peak area of m/z 179 (endogenous mandelic acid) with the m/z 184 (the internal standard at 1.5 μg/mL) reveals a value of 3.4 μg/mL in the urine sample.

matographic properties, and ionization characteristics identical to those of the nonlabeled compound, quantitative aspects of IDMS are greatly improved. This labeling technique can be applied to almost any compound which can be prepared with a suitable stable isotope and which will not be exchanged either during the extraction procedures or during derivatization.

A specific application of IDMS analysis that is of particular interest to clinical chemists is the work by scientists at the National Bureau of Standards and other selected institutions in developing definitive methods for analytes of clinical interest. For example, IDMS definitive methods have been developed for several serum electrolytes, glucose, cholesterol, urea nitrogen, uric acid, and creatinine.

Summary

Mass spectrometry is one of the most comprehensive analytical techniques now available. Over the past 20 years outstanding advances have been made in instrument reliability and ease of operation. The mass spectrometer is no longer a specialized research instrument used by only a few specialized laboratories. Today, user-oriented mass spectrometers are available which enable the practicing clinical chemist to utilize this technique for numerous applications.

References

1. Eldjarn, L., Jellum, E., and Stokke, O.: Application of gas chromatography–mass spectrometry in routine and research in clinical chemistry. J. Chromatogr., 91:353–366, 1974.
2. Jellum, E., Stokke, O., and Eldjarn, L.: Application of gas chromatography, mass spectrometry, and computer methods in clinical biochemistry. Anal. Chem., 45:1099–1106, 1973.
3. Lawson, A. M.: The scope of mass spectrometry in clinical chemistry. Clin. Chem., 21:803–824, 1975.
4. Roboz, J.: Mass spectrometry in clinical chemistry. Adv. Clin. Chem., 17:109–191, 1975.

Additional Readings

Beynon, J. H.: Mass Spectrometry and Its Applications to Organic Chemistry. Amsterdam, Elsevier Publishing Co., 1960.
Biemann, K.: Mass Spectrometry, Applications to Organic Chemistry. New York, McGraw-Hill, 1962.
Budzikewicz, H., Djerassi, C., and William, D. H.: Mass Spectrometry of Organic Compounds. San Francisco, Holden-Day, Inc., 1967.
Frigerio, A.: Advances in Mass Spectrometry in Biochemistry and Medicine. New York, Halsted Press, 1977.
Frigerio, A., and Castagnoli, N.: Mass Spectrometry in Biochemistry and Medicine. New York, Raven Press, 1974.
Gudzinowicz, B. J., and Gudzinowicz, M. J.: Analysis of Drugs and Metabolites by Gas Chromatography–Mass Spectrometry, Vols. 1–5. New York, Marcel Dekker, 1977.
McLafferty, F. W.: Interpretation of Mass Spectra. Reading, Mass., W. A. Benjamin, Inc., 1973.
Millard, B. J.: Quantitative Mass Spectrometry. Philadelphia, Heyden Press, 1978.
Roboz, J.: Introduction to Mass Spectrometry, Instrumentation and Techniques. New York, John Wiley and Sons, 1968.
Shrader, R. S.: Introductory Mass Spectrometry. Boston, Allyn and Bacon, Inc., 1974.
Waller, G. W., Ed.: Biochemical Applications of Mass Spectrometry. New York, Wiley Interscience, 1972.
Watson, T.: Introduction to Mass Spectrometry. New York, Raven Press, 1976.

CHAPTER 1C

PRINCIPLES OF IMMUNOCHEMICAL TECHNIQUES

by Gregory J. Buffone, Ph.D.

Any discussion of immunochemistry—the branch of chemistry concerned with detection and quantitation of chemical substances by the measurement of antigen-antibody interactions—requires knowledge of the basic "reactants." In most immunochemical analyses performed in the chemistry laboratory, an antibody is used as a reagent to detect the chemical substance (antigen) of interest. The exquisite specificity and the high affinity of antibodies for specific antigens, coupled with the unique ability of antibodies to cross-link antigens, allows the identification and quantitation of specific substances by a variety of methods. The principles of the methods most commonly used in the laboratory are discussed in this chapter. This introduction is intended to acquaint the reader with the structure and function of antibodies (immunoglobulins) in relation to their use as reagents in immunoanalyses. Immunoglobulins in health and disease will be covered elsewhere in this text.

Antibody. Antibodies are immunoglobulins that are capable of binding specifically to a wide array of natural and synthetic antigens, including proteins, carbohydrates, nucleic acids, lipids, and other chemicals. Immunoglobulins consist of five general classes designated as IgG, IgA, IgM, IgD, and IgE. Immunoglobulin G is the most prevalent immunochemical reagent in use. Immunoglobulins A, M, D, and E, although essential in the immune response for a given host, do not play an important role in immunochemical analysis and will therefore not be discussed here. A schematic diagram of the IgG molecule is shown in Figure 1C-1. IgG is a glycoprotein with a M.W. of about 158 000 and is classified as a 7S globulin by sedimentation velocity analysis. The protein is composed of two duplex chains with each set being made up of a heavy (γ) and light (κ or λ) chain joined by disulfide bonds. Interchain disulfide bonds hold the duplex chains together and create a symmetrical molecule. For more information on immunoglobulins, see Chapter 4.

The variable amino acid sequence at the amino terminal end of each chain determines the antigenic specificity of the particular antibody. Each unique amino acid sequence is a product of a single plasma cell line or clone. The normal host response to an immunogen (any chemical substance capable of inducing an immune response) results in the stimulation of one or two major classes of lymphocytes which are able to divide and produce plasma cells capable of secreting antibodies. Each plasma cell line produces antibodies with a single specificity. A complex antigen is capable of eliciting a multiplicity of antibodies with different specificities which are derived from different cell lines. Antibodies derived in this manner are termed polyclonal and exhibit diverse specificities in their reactivity with the immunogen. Each unique region of the molecular antigen which will bind complementary antibody is termed an *epitope*. Epitope is the more contemporary terminology in use for antigenic determinant. Both terms define a specific physical site on the molecule that reacts with the antibody. The circulating immunoglobulins are collectively referred to as humoral antibodies. Another large group of immunoglobulins, about

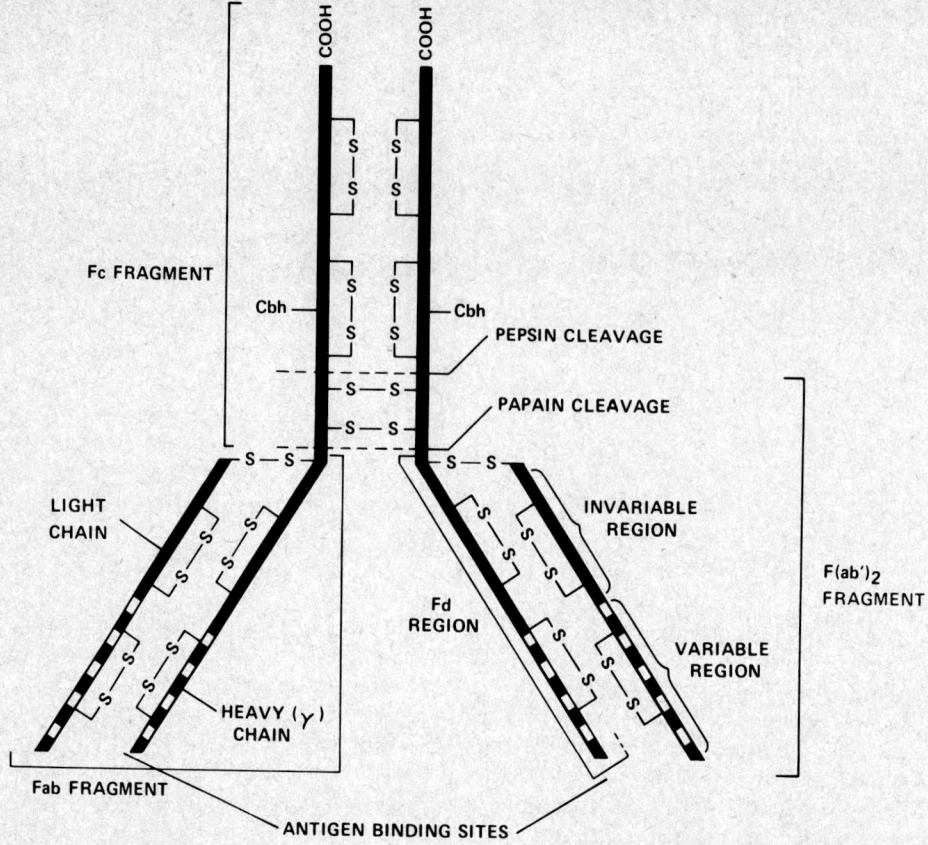

Figure 1C-1. Schematic diagram of IgG antibody molecule showing carbohydrate (Cbh), disulfide bonds (—S—S—) and major fragments produced by proteolytic enzyme treatment (F(ab')₂, Fc, Fab, Fd).

which much less is known, is referred to as cellular antibodies; these are involved in delayed hypersensitivity reactions, rejection of transplants, allergic reactions, tumor immunity, and other phenomena.

Antigens and immunogens. It is extremely difficult to provide a rigid definition of these terms and almost any definition becomes circular. Thus, an *immunogen** (or an immunogenic material) is either a protein or a substance coupled to a carrier, usually a protein, that when introduced into a foreign host is capable of inducing the formation of another protein, an antibody, in the host. This definition is neither precise nor complete. The route of introduction of the immunogen is usually, but not always, intradermal, and the antibody produced may be either circulating (*humoral*) or tissue-bound (*cellular*), as in delayed hypersensitivity reactions or graft reactions.

The types and sources of immunogens vary greatly. If we include those small molecules (haptens) that are capable of inducing specific antibody formation when coupled to a carrier protein, even molecules such as L-tyrosine, penicillin, D-glucose, and arsanilic acid are immunogenic. A *hapten* is a chemically defined determinant that, when conjugated to an immunogenic carrier, stimulates the synthesis of antibody specific for the hapten. It is capable of binding antibody but cannot by itself stimulate an immune response. Natural immunogens are usually more complex than haptens and have several epitopes. Such immunogens are capable of reacting with several antibodies, allowing crosslinking and the formation of large immune complexes. In

---

*The term "immunogen" (instead of the term "antigen") is now used when referring to materials capable of eliciting antibody formation when injected into a host. The term "antigen" is used for any material capable of reacting with an antibody, without necessarily being capable of inducing antibody formation. For example, egg albumin is an immunogen, since it is capable of inducing formation of anti-egg albumin antibody. Morphine is an antigen, since it will react with antimorphine antibodies, but it is not an immunogen since it does not induce antibody formation, unless first conjugated with a protein. In connection with immunochemical laboratory tests, it is current practice to use the broader term "antigen."

addition to proteins and polypeptides, lipids (such as the Wasserman antigen, cardiolipin), nucleic acids, and many other materials can function as antigens. Some general properties requisite for immunogenicity are as follows: (1) areas of structural stability within the molecule, (2) randomness of structure, (3) a minimal M.W. of 4000–5000, (4) ability to be metabolized (a necessary but not sufficient criterion for some classes of antigens), (5) accessibility of a particular immunogenic configuration to the antibody-forming mechanism, and (6) foreignness of structure to the host. The overall shape and charge of the molecule in most cases have very little influence on the ability to induce antibody formation.

Injection of an immunogen into a host animal induces a series of biologic changes leading to markedly increased production of certain serum globulins with specific antibody activity—the immunoglobulins—and to changes in the capability of certain cellular elements to give an immune response. Continuing stimulation by immunogen can result in increasing production of immunoglobulins of different types and of different binding characteristics for antigens. In some cases, excessive stimulation can lead to immunologic paralysis, in which no antibody response is mounted until a major fraction of the immunogen is cleared from the organism. After the first exposure to an immunogen, a latent period (induction) occurs during which no antibody is present in serum; this period may last from 5–10 d. Because antibody formation by sensitized (immunogen-coated) cells can be demonstrated in vitro in a period of minutes, the long delay in antibody build-up in vivo is due to the time required for transport of immunogen to the site of antibody production, for transmission of information, and for actual synthesis of antibody.

A disease known as serum sickness can be induced by a single injection of a large amount of immunogen. In about 5–8 d, sufficient antibody forms so that some appears in the blood. Circulating antibody combines with antigen already present, causing the production of histamine as well as other amines and peptides. The antigen-antibody complexes formed also bind complement and thus deplete the circulating complement levels. The aggregates are removed by the kidney, and this causes an immune-complex nephritis and subsequent albuminuria. This protean disease, encountered frequently in the past as a result of antitoxin therapy, is only one example of the involvement of immunologic reactions in disease.

The strength or energy of interaction between the antibody and antigen is described by two terms. *Affinity* refers to the thermodynamic quantity defining the energy of interaction of a single antibody combining site and its corresponding epitope on the antigen. *Avidity* refers to the overall strength of binding of antibody and antigen and includes the sum of the binding affinities of all the individual combining sites on the antibody. For example, IgG has two affinity-binding sites, whereas IgM has 10 affinity-binding sites per antibody molecule. Thus, affinity is a property of the substance bound (antigen) and avidity is a property of the binder (antibody). For polyclonal antibodies, avidity is difficult to determine, primarily due to the diversity of the antibody population.

Polyclonal antiserum is raised in a normal animal host in response to immunogen administration. In contrast, *monoclonal antiserum* is produced in a very different manner and represents the product of a single clone or plasma cell line rather than a heterogeneous mixture of antibodies produced by many cell clones in response to immunization. A method for their production is briefly outlined in the last section of this chapter.

ANTIGEN-ANTIBODY BINDING—THEORY OF INTERACTION

Several forces act cooperatively to produce antigen-antibody binding. The three major contributing forces are van der Waals–London dipole-dipole interaction, hydrophobic interaction, and ionic coulombic bonding.[39,69] The following review is intended to lay a foundation for the discussion of the antigen-antibody systems, their observed characteristics, and manipulation for analytical and experimental purposes.

Van der Waals–London dipole-dipole interactions are basically electrostatic in nature and are applicable to polarizable, noncharged molecules, that is, those molecules whose structure allows the electron cloud around the molecule to be distorted by outside forces in such a way that a transient dipole is produced. Such polarization results in the formation of an instantaneous dipole moment which in turn induces a dipole moment in adjacent molecules. The induced dipole moment is oriented in an antiparallel fashion with respect to the original dipole moment such that a net attractive force is produced. These forces operate over short distances (4–6 nm) and are more significant for larger molecules. The attractive force is inversely proportional to temperature since polarizability varies inversely with temperature.

Hydrophobic interaction, sometimes referred to as hydrophobic bonding or binding, is not a true bond but an association which results from the natural tendency of a system to seek the lowest possible free energy level. A simple way of stating the second law of thermodynamics is that an isolated system will approach a state of maximum randomness. Hydrophobic molecules or macromolecular species with hydrophobic regions which exist in an aqueous environment necessarily produce an ordering of the normally loosely arranged hydrogen bonding lattice of water, which results in a crystalline or "ice-like" array. Entropy is thereby decreased and an unfavorable situation with regard to overall energy balance is created. If a hydrophobic group exists near the surface of an antigen or antibody-combining site, removal of water molecules from between the antigen-antibody molecule and their return to the bulk water results in a favorable change in free energy. Hydrophobic interaction enhances or stabilizes binding but is not necessarily a driving force in antigen-antibody binding.[66,71]

Coulombic Bonds. Coulombic bonding results from the attraction between charged groups on the antigen and antibody, primarily COO^- and NH_4^+. The attraction between the charged groups is greatest in a medium with a low dielectric constant, due to reduced interaction of the solvent or other solute (salts) with the macromolecular ions. In a medium of high dielectric constant (aqueous solutions containing added salt), a diffuse double layer of charged particles will tend to shield the attraction of the charged species in the reactive sites of the antigen and antibody. This inhibition under certain circumstances can considerably reduce the binding constant for many antigen-antibody systems.

Considering these forces, one would predict that changing pH, temperature, and the ionic strength of the reaction medium should influence the binding of antigen and antibody. However, with the pH between 6 and 8, and the temperature between 25 and 35 °C, only minimal variation in the rate of association and immune complex formation is reported.[55,72] Extremes in pH (≤ 4.0 and ≥ 8.0) can cause inhibition of binding or dissociation of already formed antigen-antibody complexes. Such extremes ultimately result in some degree of denaturation of the proteins involved. Changes in ionic strength have been found to produce a proportionally greater influence on the rate of binding of antigen and antibody than either pH or temperature within the ranges stated. This concept will be examined further in the following sections.

REACTION MECHANISM FOR ANTIGEN AND ANTIBODY BINDING

The binding of antigen to antibody is not static but is an equilibrium reaction which proceeds in three phases. The initial reaction (phase 1) of a multivalent antigen (Ag_n) and a bivalent antibody (Ab) occurs very rapidly in comparison to the subsequent growth of the complexes (phase 2) and is depicted by the following equation:

$$Ag_n + Ab \underset{k_{-1}}{\overset{k_1}{\rightleftharpoons}} Ag_nAb \underset{k_{-2}}{\overset{k_2}{\rightleftharpoons}} Ag_aAb_b \qquad (1)$$

where $k_1 >>> k_2$, n is the number of epitopes per molecule, and a and b are the number of antigen and antibody molecules per complex. The third phase of the reaction involves the precipitation of the complex after a critical size is reached. The speed of these reactions depends on factors such as electrolyte concentration, pH, and temperature, as well as on antigen and antibody types, and the binding affinity of the antibody. The concentration of NaCl is important; in most cases, NaCl, 0.15 mol/L, is used. Greater concentrations of NaCl can lead to smaller amounts of precipitate, which is due not to increased solubility of the antigen-antibody complex but to an equilibrium shift causing a given amount of antigen to combine with smaller amounts of antibody. Decreasing the NaCl concentration can lead to increased precipitation of other proteins. Generally, the presence of divalent cations has no effect on precipitation.

It is best to use dilute solutions for determining the influence of such factors as ionic species, ionic strength, pH, and concentration of soluble linear polymers, or for optical analytical methods. Use of dilute solutions slows the growth of the antigen-antibody complexes and a more stable and a more homogeneous population of complexes results. Most of the discussions presented in subsequent sections are based on dilute systems and may not pertain to solutions where reactants are present in much higher concentrations.

The Precipitin Reaction

If the number of antibody combining sites, [Ab], is significantly greater than the antigen binding sites, [Ag], then antigen binding sites are quickly saturated by antibody before crosslinking can occur and the formation of small antigen-antibody complexes of the composition AgAb results (Figure 1C-2,A). For the case where antibody is in moderate excess, i.e., [Ab] > [Ag], the probability of crosslinking of Ag by Ab is more likely and hence, large complex formation is favored (Figure 1C-2,B). In the case where [Ag] is in great excess, large complexes would be less probable and the theoretical minimum size of complexes would be Ag_2Ab (Figure 1C-2,C). This model describes the results observed when antigens and antibodies are mixed in various concentration ratios. The curve shown in Figure 1C-3 is a schematic diagram of the classical precipitin curve described by Heidelberger and Kendall in the early 1930's. Although the concentration of total antibody is constant, the concentration of free antibody, $[Ab]_f$ (i.e., not bound to antigen), and free antigen, $[Ag]_f$, varies throughout the range for any given Ag/Ab ratio. A low Ag/Ab ratio exists in section A of Figure 1C-3 (zone of antibody excess). Under these conditions, $[Ab]_f$ exists in solution but $[Ag]_f$ does not. As total antigen increases, the size of the immune complexes increases up to equivalence (section B, Figure 1C-3) where little or no $[Ab]_f$ or $[Ag]_f$ exists. This is the zone of maximum immune complex size. This equivalence zone does not represent a ratio of exact molar equivalence of reactants but is the optimal combining ratio for crosslinking in the particular system under examination. As Ag/Ab increases (section C, Figure 1C-3), the immune complex size will decrease and $[Ag]_f$ will increase (zone of antigen excess). No $[Ab]_f$ should exist in this area of the curve. However, for a given Ag/Ab ratio the population of immune complexes formed at equilibrium will be heterogeneous with respect to size and composition.

EFFECT OF IONIC SPECIES, IONIC STRENGTH, AND SOLUBLE LINEAR POLYMERS ON ANTIGEN-ANTIBODY BINDING

Ionic species and ionic strength effects. Experiments performed in the early 1960's by Pressman and others, using haptens of known charge, demonstrated the influence of anionic and cationic species in solution on the binding of antigen and antibody.[25,62] When a cationic hapten was used,

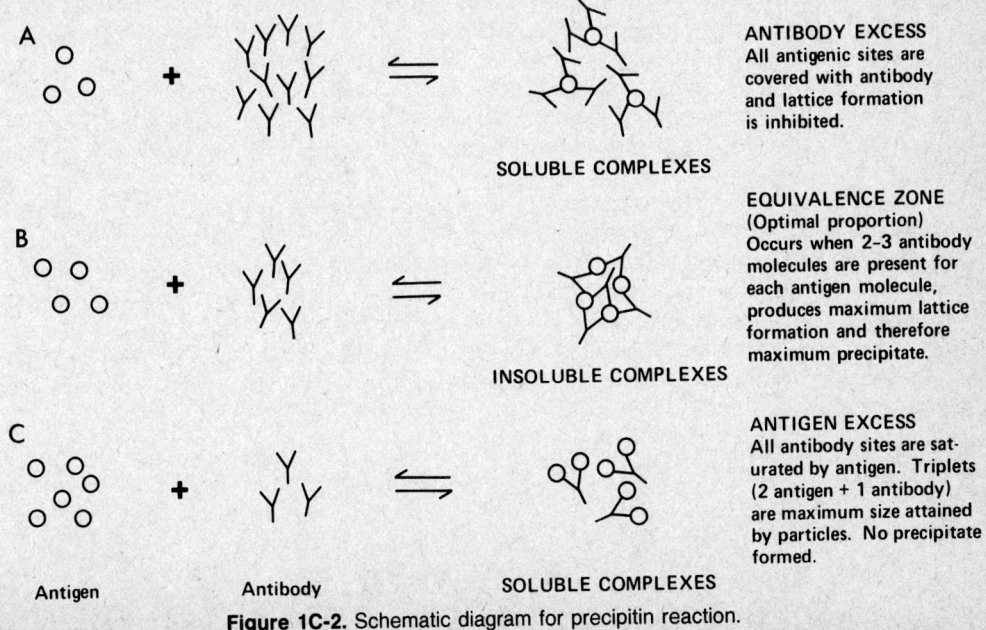

A   ANTIBODY EXCESS
All antigenic sites are covered with antibody and lattice formation is inhibited.

SOLUBLE COMPLEXES

B   EQUIVALENCE ZONE
(Optimal proportion)
Occurs when 2–3 antibody molecules are present for each antigen molecule, produces maximum lattice formation and therefore maximum precipitate.

INSOLUBLE COMPLEXES

C   ANTIGEN EXCESS
All antibody sites are saturated by antigen. Triplets (2 antigen + 1 antibody) are maximum size attained by particles. No precipitate formed.

Antigen   Antibody   SOLUBLE COMPLEXES

Figure 1C-2. Schematic diagram for precipitin reaction.

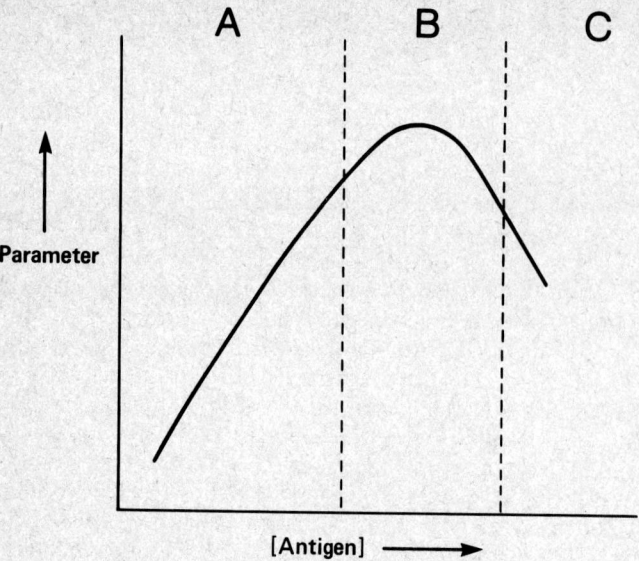

Figure 1C-3. Schematic diagram of precipitin curve illustrating zones of antibody excess (*A*), equivalence (*B*), and antigen excess (*C*). The parameter measured may be quantity of protein precipitated, light scattering, or another measurable parameter. Antibody concentration is held constant in this example.

anionic salts had little or no influence on binding, in contrast to cation salts which produced an inhibition of the antibody-hapten binding. The order of inhibition by various cations is $Cs^+ > Rb^+ > NH_4^+ > K^+ > Na^+ > Li^+$. This order corresponds to the decreasing ionic radius and increasing radius of hydration. Presumably, the lesser degree of hydration permits greater interaction of the salt with an anionic group located in the antibody combining site. Similar results were found for anionic haptens and anionic salts. The order of inhibition of binding is $CNS^- > NO_3^- > I^- > Br^- > Cl^- > F^-$, again in the order of decreasing ionic radius and increasing radius of hydration. If the competition theory as suggested by these experiments is correct, one would expect the degree of inhibition to be a concentration-dependent phenomenon. Marrack and Richards, working with a system of bovine serum albumin (BSA) and anti-BSA, observed the same concentration-dependent inhibition and anion order, that is, $NO_3^- > Br^- > Cl^- > F^-$.[55] The rate of formation of immune complexes (IC) is slower in normal saline (NaCl, 0.15 mol/L) than the same reaction carried out in deionized water. Given the above observation, F^- should be the anion of choice for immunochemical reaction buffers. In fact, F^- does provide a modest improvement over Cl^- but the advantage is so small that laboratories rarely substitute toxic fluoride ion for innocuous chloride ion in buffer solutions. A small but measurable difference in the initial rate of combination of antigen and antibody can be seen for phosphate as compared to Tris buffer, presumably due to the charge difference in the buffer molecules.[67] In general, the most significant differences in reaction rates for the various anionic species examined were seen at $t < 5$ min. When the reactions were examined at longer times (i.e., $t > 5$ min), the difference in rate of antigen-antibody complex formation was relatively small for different anionic species.[55] Some decrease in rates ($\sim$10–15%) was noted on addition of other anions as outlined in the series above.

Polymer effect. More than 10 years ago, several investigators observed that linear molecules such as hyaluronic acid, with a large effective volume, exerted a profound influence on the solubility of proteins in solution.[46,47,60] In the presence of Dextran 500, solubility of different proteins was observed to be in inverse order to their molecular weights, i.e., solubility of α-crystalline $\le$ fibrinogen < γ-globulin < albumin, < < < tyrosine.[46] In general, the solubility of a protein in the presence of different linear polymers was found to be inversely proportional to the molecular weight of the polymer, i.e., the higher the molecular weight of the polymer, the lower the solubility of the protein.

On the basis of these findings, Laurent proposed a steric exclusion mechanism to explain the observations.[45] Assuming a fixed total volume (V_T) of solvent being occupied by both polymer and protein, and defining the volume occupied by polymer as V_E (excluded volume, i.e., volume not accessible to proteins) and the volume occupied by protein as V', then the relation

$$V_T = V' + V_E \qquad (2)$$

implies that any increase in V_E, due to increase in number or size of polymer molecules, forces a decrease in V' and an effective increase in the concentration of protein molecules. Hence, as V_E is increased, the *effective* protein concentration is increased, probability of collision and self-association of protein molecules is increased, and large insoluble aggregates are formed. Hellsing's work on the influence of polymer on the antigen-antibody reaction has provided further support for the steric exclusion model.[28-31] He has shown that the *composition* of the immune complex formed is not affected by the presence of polymer; that there is no complex formed between the polymer and the antigen, antibody, or immune complex; that the polymer effect is dependent on the molecular weight of both antigen and polymer; and that the use of polymer in a reaction mixture can increase the precipitation of immune complex with low avidity antibody.[28-31] Addition of polymer to a mixture of antigen and antibody causes a significant increase in the rate of immune complex growth, especially during the early phase of the reaction.[67] The rate of increase is directly proportional to the polymer concentration and molecular weight. The addition of anionic salts moderates the polymer effect by competitive inhibition as discussed in the previous section. The immune complexes formed in the presence and absence of polymer have also been studied by light scattering methods to determine their molecular weights and approximate radii. These data show that the molecular weights and radii of the immune complexes were independent of the presence of a polymer. Similar results were found for immune complexes formed in solutions containing different concentrations of chloride ions.[12] Given these observations, it appears that neither the polymer nor various ionic species influence the *mechanism* of the antigen-antibody reaction but that they do influence the rate of the reaction by inhibiting or promoting association of antigens and antibodies.

Numerous polymer species have been tested (Table 1C-1) for applications in immunochemical methods. The most desirable characteristics of the polymer are high molecular weight, a high degree of linearity (minimum branching), and high aqueous solubility. Most investigators have found polyethylene glycol 6000 (PEG) in concentrations of 3–5 g/dL to be most useful. Polyethylene glycol can also be used in the fractionation of plasma for protein purification or other applications.

ANTIGEN-ANTIBODY REACTIONS AT A SOLID-LIQUID INTERFACE

If the antigen or antibody of interest is bound to a solid phase such as a cell membrane or to a synthetic particle (polystyrene or cellulose), the protein will exist in a micro-environment that is different from that of a protein which is free in solution. The water surrounding the protein is more highly ordered near the surface of the solid phase and a condition results which is more favorable for van der Waals–London dipole-dipole interaction and coulombic bonding. This situation favors the formation of low avidity as well as high avidity antigen-antibody complexes and hence can provide better sensitivity for analytical applications. Some studies have shown radiometric solid-phase assays to be more sensitive than their counterpart solution assays. In one study, using the same antibody pool of antihuman α_1-antitrypsin, the lower limit of sensitivity for the solution method was 20–30 ng/mL; for the solid-phase method it was 5–10 ng/mL. The difference in sensitivity is most likely a result of additional low avidity antibody binding during the initial incubation in the solid phase system which does not occur as readily in the liquid phase.

QUALITATIVE METHODS FOR DETECTING PROTEINS

Passive Gel Diffusion

Many qualitative and quantitative immunochemical methods are carried out in a semisolid medium such as agar or agarose. The primary advantage in using a gelatinous medium is to stabilize the diffusion process with regard to mixing caused by vibration or convection and to allow visualization of precipitin bands for qualitative and quantitative evaluation of the reaction.

Table 1C-1. LINEAR POLYMERS USED TO ENHANCE ANTIGEN-ANTIBODY REACTION

Polyethylene glycol (Carbowax)	Polypropylene glycol
Dextran	Polyvinyl alcohol
Modified cellulose	Polyvinyl pyrrolidone

Antigen-antibody ratio, salt concentration, and polymer enhancement have the same influence on the antigen-antibody reaction in gels as they have on reactions in solution.

If the matrix does not interact with the molecular species under investigation, passive diffusion of reactants in a semisolid matrix can be described by Fick's equation:

$$\frac{dQ}{dt} = -DA\frac{dC}{dx} \qquad (3)$$

where dQ is the amount of diffusing substance which at time t passes through the area A; dC/dx is the concentration gradient and D is the diffusion coefficient. The diffusion coefficient, D, is a directly proportional function of temperature; it is also a function inversely proportional to hydrated molecular volume of the diffusing species. dQ/dt is clearly a function of dC/dx, the concentration gradient; the amount of diffusing species transferred from the origin to a distant point (over the migration distance) is dependent on the length of time diffusion is allowed to occur.

The initial concentration of antigen and antibody is critical. Each molecule in the system will achieve a unique concentration gradient with time. When the leading fronts of antigen and antibody diffusion overlap, the reaction will begin but formation of a precipitin line will not occur until moderate antibody excess is achieved. A precipitin band may form and be dissolved many times by incoming antigen before equilibrium is established and the position of the precipitin band becomes stable. Because heavier molecules diffuse more slowly, the position of the precipitin band is in part a function of the molecular masses of both antigen and antibody. The precipitin band acts as a specific barrier; neither specific antigen nor antibody can penetrate without being precipitated by the other, but unrelated molecules can cross the band of precipitation freely. However, if both antigen and antibody concentrations are too high, the excessive precipitate formed in the gel can physically block further diffusion of other molecules, including the antigen and antibody of interest, and cause anomalous results.

Two basic approaches to passive diffusion are in common use today. One is simple diffusion in which a concentration gradient is established for only a single reactant. *Single immunodiffusion* usually depends on diffusion of an antigen into agar impregnated with antibody. A quantitative technique based on this principle is radial immunodiffusion (RID) which will be discussed later in this section. The second approach is double diffusion in which a concentration gradient is established for both reactants (antigen and antibody).

Double immunodiffusion in two dimensions (Ouchterlony technique) is a widely used technique in immunology. It allows direct comparison of two or more test materials and provides a simple and direct method for determining if the antigens in the test specimens are identical, cross-reactive, or nonidentical.

The simplest method uses a standard Petri dish filled with agar in saline (2–4 g/dL) to a depth of about 3 mm. Holes are cut in the agar plate with a punch or a sharp cork borer. The plugs of agar are removed by suction, and the bottom of each well is sealed with a drop of agar in saline (1 g/dL) to prevent diffusion under the agar. When the same antigen is in both wells, the lines of precipitation fuse and are continuous—the reaction of identity (Figure 1C-4,*A*). When the precipitin bands cross each other, it is a reaction of nonidentity (Figure 1C-4,*B*); in case the two antigens are related but not identical, a reaction of partial identity is observed (Figure 1C-4,*C*). Here, the cardinal point is that the precipitate serves as a barrier that does not block unrelated diffusing reactants. As shown in Figure 1C-4,*D*, when two related antigens, Ag and Ag$_1$, are in separate wells and the respective antibodies, Ab and Ab$_1$, are in the third well, an AgAb precipitate forms on one side and blocks further diffusion of Ab from the antibody well. However, on the other side the Ag$_1$Ab$_1$ precipitate does not stop Ab from migrating further and forming an AgAb spur.

In an unbalanced, nonequivalence system, formation of a second band is possible while the first is still dissolving in the region of antigen excess. This problem may be resolved by observing the reaction at various time intervals and by setting up a series of plates with different concentrations of both reactants. Still another artifact may be due to precipitation of antibody by organ extracts or, as reported in some cases, by phosphate buffer.

Double diffusion can easily be done on a microscale using small glass slides. Only a small sample (1 μL) is required, and the reaction time is markedly decreased, from as long as a week

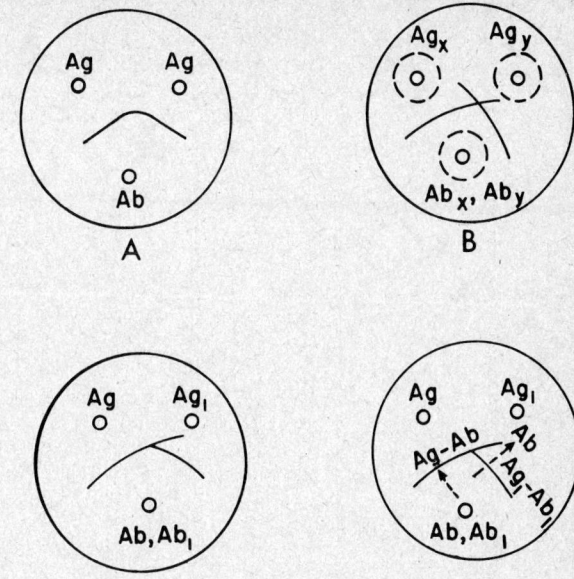

Figure 1C-4. Double immunodiffusion in two dimensions by Ouchterlony technique. Ag = antigen; Ab = antibody. *A*, Reaction of identity. *B*, Reaction of nonidentity. *C*, Reaction of partial identity. *D*, Scheme for spur formation.

to 24 h or less in some cases. Sharp lines are produced because the agar layer is thin and the area from which the reactants diffuse is minute. Finally, the slides can be stained fairly rapidly after the template has been removed, or alternatively, photographed unstained with an oblique light source illuminating the precipitin bands. In some cases, the intensity and number of bands have been measured by using a photoelectric microphotometer and recorder.

Note that a negative reaction does not necessarily imply absence of antibody or antigen. A negative reaction can result from using amounts of material too small for the sensitivity of the method, or alternatively, the antibody may be nonprecipitating.

Immunoelectrophoresis

If several antigens of interest exist in a common solution (e.g., spinal fluid, serum), the various protein species can be separated and identified by immunoelectrophoresis (IEP). This technique, described by Grabar and Williams, has been used extensively in both research and clinical laboratories for the study of antigen mixtures, the evaluation of the specificity of antiserum, and the evaluation of human gammopathies.[24]

The procedure is carried out using an agarose gel medium poured onto a thin plastic sheet. The sample to be analyzed is placed in a reservoir in the gel, and an electrical field is applied across the gel surface. During electrophoresis the proteins in the serum are separated according to their electrophoretic mobilities (Figure 1C-5). Following electrophoresis, an antiserum against the protein of interest is placed in a trough parallel and adjacent to the electrophoresed sample. Simultaneous diffusion of the antigen from the separated sample and antibody from the trough results in formation of precipitin arcs whose shape and position are characteristic of the individual separated proteins in the specimen. The pattern is interpreted after overnight development of the precipitate. By comparison with a known control separated on the same plate, tentative identification of individual proteins can be made. In some cases, the characteristics of the precipitin arc, such as bowing, brushing at the end of the arc, or thickening, may suggest to the experienced observer the presence of special types of proteins. A permanent record can be made either by photographing the electrophoresis film or by staining and retaining the film.

In performing immunoelectrophoresis, a number of factors must be carefully controlled to take full advantage of the sensitivity of the method. The choice of the buffer anion may be more important than selection of the actual pH; for instance, borate buffer promotes electrophoretic migration of polysaccharides with cis-hydroxyl groupings and also alters the supporting medium (agar) to increase electroendosmotic flow. Improved resolution is obtained by decreasing buffer ionic strength at constant current, by varying the duration of electrophoresis, and by adjusting

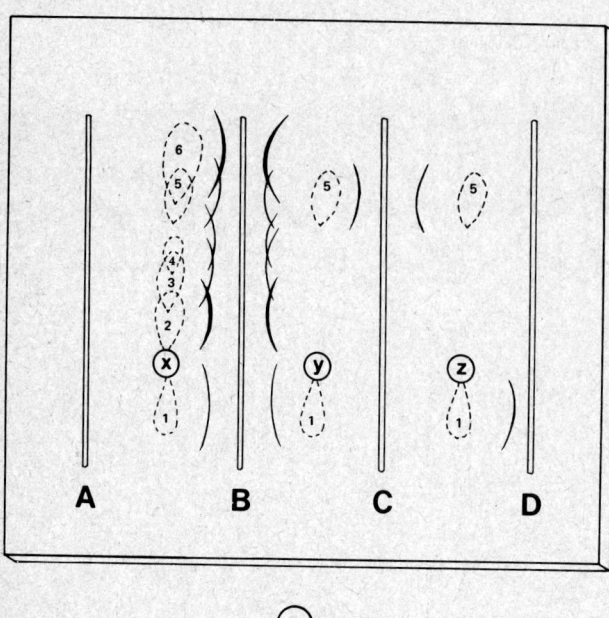

Figure 1C-5. Usual configuration for immunoelectrophoresis. Sample wells are punched in the agar/agarose, sample is applied, and electrophoresis is carried out to separate the proteins in the sample. Antiserum is loaded into the troughs and the gel incubated in a moist chamber at 4 °C for 24–72 h. Track X represents the shape of the protein zones after electrophoresis; tracks Y and Z show the reaction of proteins 5 and 1 with their specific antisera in troughs C and D. Antiserum against protein 1–6 is present in trough B.

the distance between the lateral antibody trough and the antigen locations after electrophoresis. Prolonged electrophoresis results in heating of the agar and limits the resolution unless the procedure is performed at lower temperatures (e.g., in a refrigerated room or with a cooling apparatus).

In clinical laboratory medicine this procedure has primarily been applied to the evaluation of human myeloma proteins. However, the method is slowly being replaced by immunofixation electrophoresis, particularly for the study of protein antigens and their split products, and for evaluation of human myeloma proteins. (See discussion of immunofixation electrophoresis.)

Crossed immunoelectrophoresis (CRIE), also known as two-dimensional immunoelectrophoresis, is a variation of IEP wherein electrophoresis is also used in the second dimension to drive the antigen into a gel containing antibodies specific for the antigens of interest (Figure 1C-6).[16,43] This technique is more sensitive and produces higher resolution than is possible with immunoelectrophoresis. As with immunoelectrophoresis, the specimen is applied to a cylindrical well cut in an agar medium and is then subjected to electrophoresis in the first dimension, followed by electrophoresis at a 90° angle to the first run through the gel which contains antibodies. Precipitin bands form at the zones of equivalence for each antigen. Precipitation occurs along lateral margins as the antigen advances and results in precipitation "peaks" for each antigen. An example of the application of CRIE is shown in Figure 1C-7. A serum specimen treated with two different quantities of trypsin was examined for changes in α_1-antitrypsin. Formation of trypsin-antitrypsin complexes results in a second smaller peak cathodal to the native species. The resolution of the peaks in the second dimension may be improved by the use of a linear application of the sample (i.e., into a rectangular slot instead of a well in the agar). Such application causes the proteins to move from the origin in relatively narrow bands and not in teardrop-shaped zones. Quantities of individual antigens can be estimated by measuring the peak area of the antigen under the precipitin arc and comparing it to the peak area of a standard antigen preparation. This technique has its greatest utility in evaluation of protein mixtures in the study of protein alterations subsequent to activation or interaction with other molecules.

In *counter immunoelectrophoresis (CIE)*[57] two parallel lines of wells are punched in the agar. One row is filled with antigen solution and the opposing row is filled with antibody solution (Figure 1C-8). If the solutions were allowed to diffuse passively over the next 18–24 h, a precipitin line would form between the opposing wells where antigen reacted with specific antibody. In

⊕

1st
dimension

⊖

⊕

Antibody
containing gel

2nd
dimension

1st dimension

B

⊖

Figure 1C-6. Two-dimensional (CRIE) electrophoresis. Diagram *A* shows the configuration for the first dimension of crossed immunoelectrophoresis. The segment of the gel denoted by the dashed lines in diagram *A* is cut out and placed on a second plate. An upper gel containing antibody is added as shown in diagram *B*. Electrophoresis is now carried out at 90° relative to the first dimension run.

CIE, this process is made to occur more rapidly by applying a voltage across the gel so that the antigen and antibody move toward each other at a faster rate. The antigen migrates toward the anode, and the antibody moves in the opposite direction as a result of electroendosmosis, and a precipitin line is formed where they meet. Qualitative information (i.e., identification of antigen) is provided within 1–2 h. This method has found application in the detection of bacterial antigens in blood, urine, and CSF.

Immunofixation (IF), as a method of immunochemical identification of proteins following electrophoretic separation, was first described more than 15 years ago.[1] Only in the last few

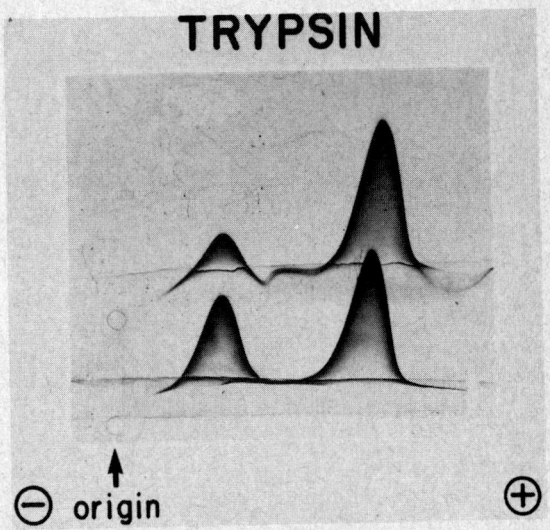

TRYPSIN

⊖ origin ⊕

Figure 1C-7. Crossed immunoelectrophoresis pattern obtained with two different concentrations of trypsin added to normal serum. The first dimension was carried out from left to right and the second dimension from bottom to top. Two separate gels are shown, with the highest trypsin concentration at the bottom. Antibody against α_1-antitrypsin was present in the second dimension gel. The resulting pattern shows two distinct α_1-antitrypsin species, the free protease inhibitor (right) and protease-antiprotease complex (left). This example illustrates the ability of CRIE to evaluate changes in specific protein structure.

⊕

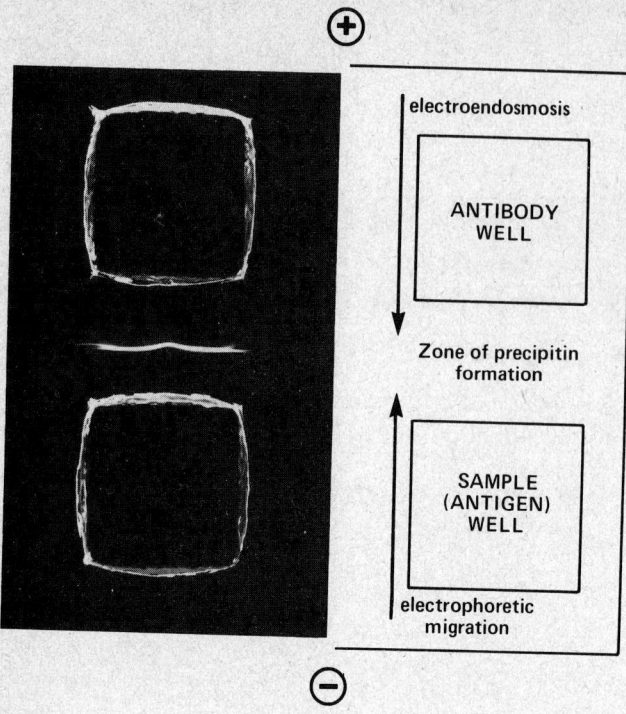

electroendosmosis

ANTIBODY
WELL

Zone of precipitin
formation

SAMPLE
(ANTIGEN)
WELL

electrophoretic
migration

Figure 1C-8. Counter immunoelectrophoresis showing positive reaction between anti–*H. influenzae B* (uper well) and a CSF sample containing *H. influenzae B* (lower well).

⊖

years, however, has this elegant technique gained widespread acceptance. As in IEP and CRIE, a first dimension electrophoresis is performed in agarose gel to separate the proteins in the mixture. Subsequently, antiserum spread directly on the gel causes the protein(s) of interest to precipitate. The immune precipitate is trapped within the gel matrix and all other nonprecipitated proteins can be removed by washing the gel. The gel may then be stained for identification of the proteins. CRIE is more sensitive than IF in terms of detection limit and also shows better resolution. For instance, proteins of closely related or identical electrophoretic mobility can be distinguished better by CRIE since in IF they will appear as a single band. In practice, IF is technically more efficient than either IEP or CRIE and it produces patterns which are more easily interpreted. The utility of immunofixation can readily be demonstrated by examining the conversion of two complement components of the alternative pathway. Both complement component C3 and Factor B are converted into inactive molecules through breakdown of the native protein after activation of the pathway. The activation and rate of breakdown can be easily studied using immunofixation. Examples of C3 and Factor B conversion studied by immunofixation are shown in Figure 1C-9. If the gels are scanned using densitometry, quantitative data can be obtained, allowing a better evaluation of the kinetics of the process (Figure 1C-10). Immunofixation is replacing IEP in many laboratories for the evaluation of myeloma proteins because of the rapidity of the procedure and the ease of interpretation. An excellent summary of the technique has been published by Johnson.[36]

All of the techniques discussed thus far use a direct examination of the separated species in the medium used for electrophoresis, namely, direct immunoprecipitation of the protein in the gel. Certain media, such as polyacrylamide, do not lend themselves to direct immunoprecipitation, nor is there always sufficient antigen concentration to produce an immunoprecipitate which will be retained in the gel during subsequent processing. Under these circumstances transfer of the protein from the medium in which the electrophoresis is done onto a solid phase, such as nitrocellulose paper, cyanogen bromide–activated paper, or cationized nylon-based membrane (Zeta Probe, Bio Rad Laboratories Richmond, CA 94804), will act to fix the proteins after electrophoretic separation.[6,13,21,54] Once the proteins are fixed either through adsorption, in the case of nitrocellulose paper or nylon-based membrane, or through a covalent bonding using *m*-nitrobenzyloxymethyl or cyanogen bromide–activated paper, they can be detected using antibody probes labeled with either radioactive isotopes or enzymes. By using such probes, the

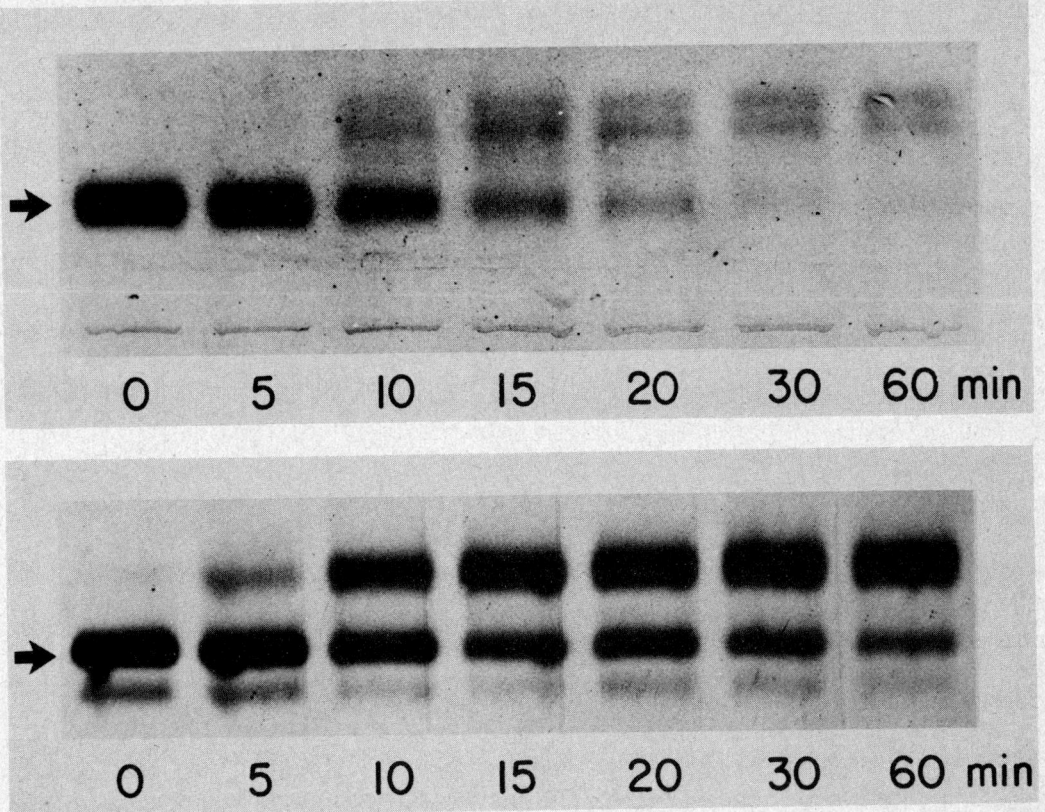

Figure 1C-9. Immunofixation electrophoretic patterns showing the conversion of Factor B (top) and C3 (bottom) after activation of complement by cobra venom factor. Time of sampling after activation is given below each line. Native protein band is indicated by the arrow.

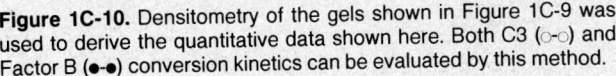

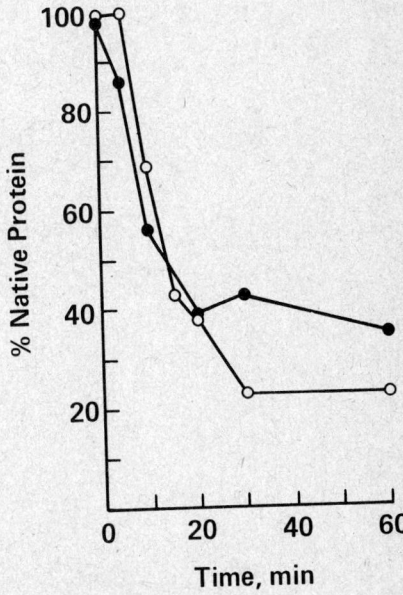

Figure 1C-10. Densitometry of the gels shown in Figure 1C-9 was used to derive the quantitative data shown here. Both C3 (○-○) and Factor B (●-●) conversion kinetics can be evaluated by this method.

sensitivity of detection can be 10–100 times greater than direct immunoprecipitation and staining of proteins.

An example of this technique applied to α_1-antitrypsin Pi typing is shown in Figure 1C-11. The proteins are focused at pH 4–5 in agarose and then transferred to a nylon membrane by blotting. The proteins are transferred by capillary diffusion and bound by nonspecific forces to the nylon membrane. The membrane is then sealed with fetal bovine serum to block all nonspecific binding sites and then reacted with biotin-labeled anti-α_1-antitrypsin. Biotin is a vitamin capable of binding to avidin with a very high binding affinity. Avidin coupled to horseradish peroxidase (HRP) is then added. The avidin-HRP complex binds to any bound biotin–anti-α_1-antitrypsin. The presence of the avidin-HRP complex and hence the protein of interest is detected by addition of 3,3′-diaminobenzidine. HRP will oxidize 3,3′-diaminobenzidine and produce an insoluble product which precipitates at the site where avidin-HRP complex is bound. The precipitation results in the formation of visible brown bands where the protein was originally fixed in the membrane. Concentrations of antigen as low as 500 ng/mL or 2.5 ng per band in the gel can be detected by this method. Immobilization of protein patterns following an analytical separation by electrophoresis, isoelectric focusing, sodium dodecyl sulfate polyacrylamide electrophoresis, or other methods provides a powerful tool for analytical study of proteins present in low concentrations in cell culture or body fluids.

QUANTITATIVE METHODS FOR MEASURING PROTEINS

Radial Immunodiffusion (RID) and Electroimmunoassay (EIA)

Typically, the two most commonly encountered gel-based methods for quantitative immunochemical studies are radial immunodiffusion and electroimmunoassay ("rocket" technique). *RID* is a passive diffusion method in which a concentration gradient is established for a single reactant, usually the antigen. The antibody is uniformly dispersed in the gel matrix. Antigen is allowed to diffuse from a well into the gel until antibody excess exists and immune precipitation occurs; the antigen-antibody interaction is manifested by a well-defined ring of precipitation around the antigen well. The ring diameter will continue to increase until equilibrium is reached. Standards are run at the same time as the sample, and a standard curve of ring area or diameter

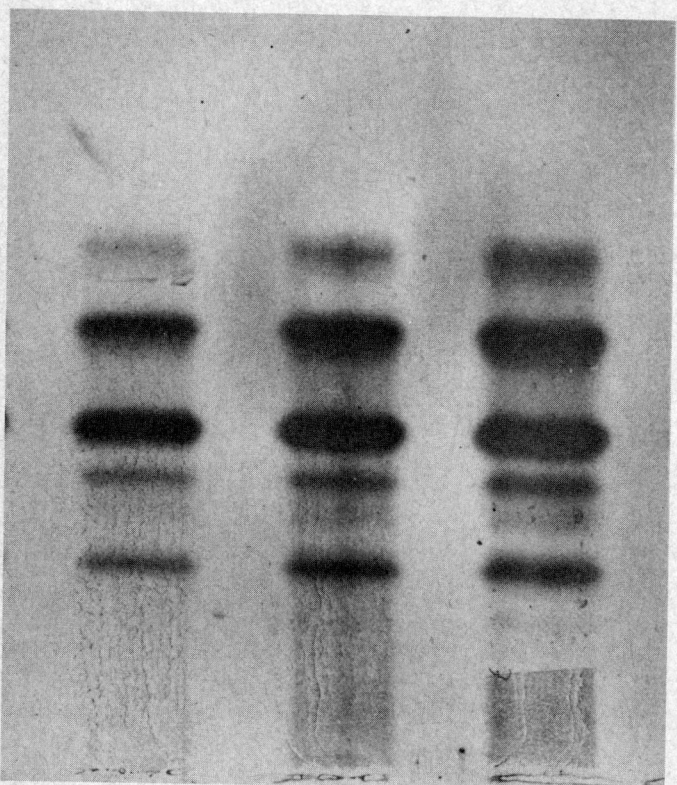

Figure 1C-11. Biotin–avidin–horseradish peroxidase staining of α_1-antitrypsin banding after isoelectric focusing at pH 4–5 and fixation of the protein on a nylon membrane.

versus concentration is plotted. The original procedure described by Mancini[52,53] requires the establishment of equilibrium before measurement of the precipitin ring diameter. Under equilibrium conditions, a linear relation exists between antigen concentration and the square of the precipitin ring diameter. In addition, the precision of the measurement of the ring diameter is better than if the rings are measured before equilibrium is established. According to Fahey,[19] quantitative data can also be derived by reading the ring diameter before equilibrium is established. This approach is often more practical for a clinical laboratory if available time for testing is at a premium.

From Fick's equation, one would predict that the rate of diffusion would be dependent on antigen size, molecular weight, and concentration. While this is indeed the case for the *rate* of diffusion, the ultimate area of the precipitin ring at equilibrium is independent of these factors and dependent instead on the antigen-antibody concentration ratio.[52,53] Antigen concentrations are calculated in both Fahey and Mancini RID methods by plotting the square of the precipitin ring diameter against standard antigen concentrations and interpolating unknowns from this curve. RID can be made more sensitive using polyethylene glycol (PEG) to enhance precipitin line formation or by using ^{125}I-, ^{131}I-, or enzyme-labeled reagents.[2,27,59,65]

In *electroimmunoassay,* as in RID, a single concentration gradient is established for the antigen, but in this case an applied voltage is used to drive the antigen from the application well into a homogeneous suspension of antibody in the gel (Figure 1C-12).[44] This produces, as opposed to RID, a unidirectional migration of antigen and results in increased sensitivity. The height of the resulting rocket-shaped precipitin line is proportional to the antigen concentration. Quantitation is effected by using standards on the same plate along with the unknowns and then estimating the concentrations of unknowns from the heights of the "rockets" obtained. The calibration curve is linear only over a narrow concentration range, so samples may have to be diluted or concentrated as needed. Electroimmunoassay methods produce the best results with antigens having a strong anodic mobility and intermediate to low molecular weight. Proteins such as transferrin, C3, or IgG with low anodal mobility or virtually no net charge at pH 8.6 (the most common pH used for the method) can be modified by carbamylation or run at a lower pH to make their measurements by EIA feasible. One advantage of EIA is that more than one antigen can be measured simultaneously if appropriate conditions are chosen. The simultaneous use of more than one antiserum results in superimposed rockets which can be measured separately. Other modifications, such as the use of an intermediate gel which causes precipitation of C3, allow measurement of C3d in human serum and illustrate the exceptional versatility of this method.[7]

In general, gel-based methods for quantitation of antigens require excellent technical skill, as well as tedious preparation and execution, to produce good results. In many clinical laboratories today, these methods are restricted to qualitative studies or are used as reference methods. Quantitative data are more commonly obtained by turbidimetric and nephelometric methods, radioimmunoassays, enzyme immunoassays, and fluorometric immunoassays.

Turbidimetric and Nephelometric Assays

Turbidimetry and nephelometry are convenient techniques for measuring the rate of formation of immune complexes in vitro. Instrumental principles for these methods are described in Chapter 1B, Section Two. Studies have shown that the reaction between antigen and antibody begins

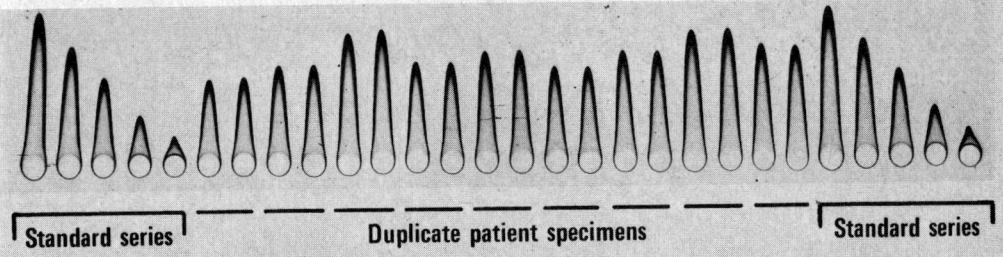

Standard series   **Duplicate patient specimens**   **Standard series**

Figure 1C-12. Rocket immunoelectrophoresis of human serum albumin. Patient samples were applied in duplicate. Standards were placed at opposite ends of the plate.

within milliseconds and continues for hours.[25,67] Both turbidimetric and nephelometric immunochemical methods using rate and pseudo-equilibrium protocols have been described for proteins, antigens, and haptens. In rate assays, measurements are usually made within the first few minutes of the reaction since the largest change (dI_s/dt) in intensity of scattered light (I_s) with respect to time is obtained during this time interval. For so-called equilibrium assays, it is necessary to wait 30–60 min so that the dI_s/dt is small relative to the time required to make the necessary measurements. For the purpose of this discussion, such conditions will be referred to as pseudo-equilibrium because true equilibrium is not reached within the time allowed for these assays. Measurement of the *rate* of immune complex formation can also be used for quantitative immunochemical studies. Either dI_s/dt or the time required to reach peak rate can be related to antigen concentration in a manner analogous to any other rate methodology. Rate nephelometric assays have the advantage that blank correction is not required and that several samples can be assayed in a few minutes instead of the 30–60 min required for pseudo-equilibrium methods.[5,11,32]

The performance of nephelometric or turbidimetric assays can be significantly improved by increasing the reaction rate by addition of water-soluble linear polymers. This allows the use of much lower reactant concentrations and results in a more stable immune complex suspension.[10]

Nephelometric methods in general are more sensitive than turbidimetric assays and have an average sensitivity of ~1–10 mg/L for a serum protein. Better sensitivities are obtained in fluids such as CSF and urine because of their lower lipid and protein concentration which results in a better signal-to-noise ratio.

Nephelometric and turbidimetric assays have also been applied to the measurement of drugs (haptens) using *inhibition techniques*. The reagent is made by attaching the drug of interest to a carrier molecule such as bovine serum albumin. The hapten-bound albumin then competes with free hapten (drug introduced in sample) for antihapten-antibody. In the presence of free hapten, immune complex formation is decreased since more antibody sites are saturated; thus, light scattering is decreased. The decrease of light scattering is related to the concentration of free hapten. Both kinetic and pseudo-equilibrium methods have been described.[15,20] In the absence of free hapten, bound hapten-albumin reacts with available antihapten-antibody sites to form cross-linked immune complexes with high light scattering ability.

Techniques for Detecting Primary Association Between Antigen and Antibody

The methods discussed so far rely on examining immune complex formation as an index of antigen-antibody reaction. Recall, however, that in equation (1) the overall reaction is shown in phases and only the final phase is the formation of immune complex. For measuring the initial phase, i.e., the primary reaction Ag + Ab, different physical and chemical techniques are required. A list of some of these methods is given in Table 1C-2. Many of the methods have provided important information on the mechanism of the antigen-antibody reaction, but are not applicable for use in a clinical laboratory. The following section will discuss selected methods appropriate for clinical laboratory use.

Radioimmunoassays and immunoradiometric assays. Radioimmunoassay (RIA) and immunoradiometric assay (IRMA) methods are capable of measuring the primary reaction between hapten or antigen and a single antibody. In RIA, the hapten or antigen is labeled with a radioactive isotope, whereas in IRMA, the antibody is the labeled species. In RIA or IRMA, we are concerned only with the primary combination of antigen and antibody.

$$Ab + Ag \underset{k_{-1}}{\overset{k_1}{\rightleftharpoons}} AbAg \tag{4a}$$

$$K = \frac{[AbAg]}{[Ab][Ag]} \tag{4b}$$

where k_1 = the rate constant for the forward reaction
k_{-1} = the rate constant for the reverse reaction
K = the equilibrium constant for the overall reaction

As would be predicted from the law of mass action, the concentration of Ab, Ag, and AbAg will be dependent on the magnitude of k_1 and k_{-1}. For polyclonal antiserum, the average avidity

Table 1C-2. METHODS FOR MEASURING THE PRIMARY REACTION
OF ANTIGEN AND ANTIBODY

Equilibrium dialysis	Ultracentrifugation
Fluorescence quenching or enhancement	Stopped-flow analysis
Farr technique	Temperature-jump
Gel permeation chromatography	Isotopic dilution

of the antibody populations will determine K, and the magnitude of k_1 in comparison to k_{-1} will determine the ultimate sensitivity attainable with a given antibody population.

The two standard procedures for RIA are termed competitive and sequential. In a *competitive RIA,* all reactants are mixed together simultaneously. Labeled antigen (Ag*) and unlabeled antigen (Ag) compete for binding to the antibody. In such a system, the avidity of the antibody for both the labeled and unlabeled antigen must be the same. Under these conditions, the probability of the antibody binding the labeled antigen is inversely proportional to the concentration of unlabeled antigen; hence bound counts are inversely proportional to unlabeled antigen concentration. For some antisera, a competitive RIA is not as sensitive as a sequential RIA, especially in cases where the antiserum has different avidities for the labeled and unlabeled antigen.

In the *sequential approach,* unlabeled antigen is first mixed with excess antibody and binding is allowed to achieve equilibrium. Labeled antigen is then added and allowed to equilibrate, and then after separation, the bound and free counts are determined.

$$\text{Step 1} \qquad Ag + Ab \underset{k_{-1}}{\overset{k_1}{\rightleftharpoons}} AgAb + Ab$$

$$\text{Step 2} \qquad AgAb + Ab + Ag^* \rightleftharpoons AgAb + Ag^*Ab + Ag^* \tag{5}$$

With this approach, a higher fraction of the unlabeled antigen can be bound by the antibody than in a competitive assay, especially at low antigen concentrations. Sequential assays can provide a two- to four-fold increase in sensitivity compared to a competitive assay, provided $k_1 >> k_{-1}$. An increase in sensitivity results from an increase in AgAb binding (and thus in a decrease in Ag* binding) which is favored by the sequential addition of Ag and Ag*. If $k_1 \geq k_{-1}$, sensitivity decreases, because dissociation of AgAb becomes more likely and competition between Ag* and Ag results.

Radiolabeling of antigen with an isotope can cause changes in reactivity with the antibody. Therefore, labeled and unlabeled antigens should always be evaluated when a competitive assay is used to assure that the antibody reacts equally with each form.

Separation of free from bound label relies upon some physical method of separation (Table 1C-3). The methods implicitly assume $k_1 >> k_{-1}$; however, use of any separation method requires some sacrifice in sensitivity. Three general types of separation are commonly used. In the first type the removal of the free antigen is achieved by *adsorption.* Many adsorbents are used for this purpose, including activated charcoal, dextran coated charcoal, ion exchange resin, magnesium silicate (Florisil), fuller's earth, and talc. The disadvantage of this method is that the time of contact between the adsorbent and the incubation mixture is critical, especially for the more active adsorbents such as activated charcoal. Sometimes timing has to be controlled to within seconds in order to get reproducible results.

The second type of method involves *precipitation of the bound antigen* from the solution by using a protein precipitant such as $(NH_4)_2SO_4$, ethanol, dioxane, or polypropylene glycol. The bound antigen can also be precipitated immunologically by using a second antibody. For example, if the primary antibody is derived from rabbits, the second antibody can be an antiserum raised

Table 1C-3. PHASE SEPARATION METHODS FOR IMMUNOASSAYS

Electrophoresis	Polymer precipitation
Adsorption	Solvent or salt precipitation
Ion exchange	Protein A
Gel filtration	Biotin-avidin
Double antibody precipitation	Solid phase antibodies

against rabbit γ-globulin in goats or sheep. This method has the advantage that it can be used for practically any assay. It has the disadvantage, however, that it usually requires longer assay times and additional steps.

The third type of method utilizes *solid phase antibodies*. In this method, the binding and competition of antigens occur on a solid surface to which the antibody is attached. The solid surface may be the inside surface of a plastic tube, a polymerized antibody, or antibody conjugated by a covalent bond to an insoluble inert material, such as cellulose or Sephadex. At the end of incubation, the solid phase antibody, along with bound antigen, can be sedimented to the bottom of the tube and the supernatant containing the free antigen can easily be siphoned off or decanted.

A typical RIA calibration curve is shown in Figure 1C-13. The lower limit of sensitivity is usually defined by the inflection point of the curve, i.e., that point at which percentage of bound counts (% B) differs from zero antigen concentration by a specified level of significance. Rodbard has reviewed the mathematical fitting of these data and the reader is referred to his work for more information on this topic.[64]

While RIA can be used for measurement of protein antigen, its primary application has been for quantitating haptens or peptides, particularly hormones. In most cases these molecules have only one antigenic site and therefore are more suitable for RIA. (For additional information on RIA principles, see specific procedures in Chapter 9.)

Labeled antibody assays (IRMA) have the advantage of not requiring a quantity of purified antigen because the antigen need not be labeled. This also obviates potential problems which may be caused by iodination of labile antigens. Antibodies are more stable proteins and are less difficult to label without damaging the protein's function. IRMA's often require additional steps for addition of reactants or washing. Typically, a "sandwich" or two-site IRMA method is used. In these systems, antibody is first attached to a solid phase by passive adsorption or by using reactions which result in a covalent binding of the antibody to the solid phase. Antigen from the sample is then allowed to react with the solid phase antibody, other protein is washed away, and a labeled antibody is added which reacts with the bound antigen through a second and distinct antigenic determinant. After washing again, the bound counts are determined and are directly proportional to the concentration of antigen. IRMA is a common approach for protein antigen measurement, due to its simplicity and ease of application in the routine laboratory.

Indirect IRMA's have also been described. In such a method, the labeled antibody is reacted with the antigen of interest that is bound to a solid phase (e.g., cell membrane). After removal of the bound antibody, the excess labeled antibody, which remains in solution, can then be

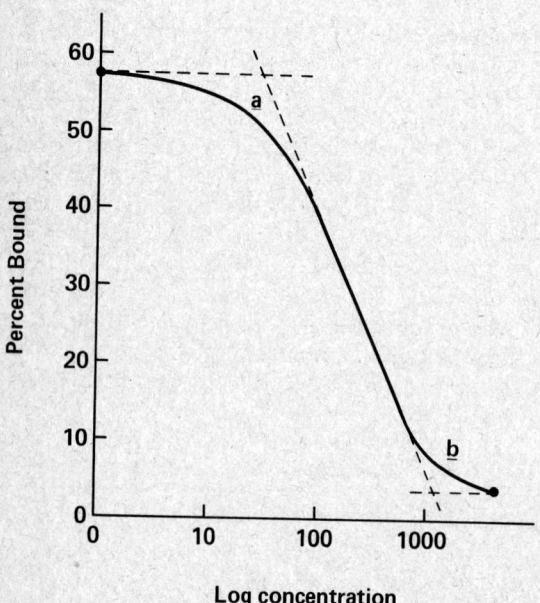

Figure 1C-13. A schematic diagram of the dose response curve for a typical radioimmunoassay. The analytically useful portion of the curve is bracketed by points *a* and *b*.

measured and is inversely related to the antigen concentration. This approach is particularly useful for fragile cells which may lyse during subsequent washing steps or when the cells are required for further studies. Indirect IRMA assays have been applied to measurement of surface-bound C3b on human erythrocytes or microorganisms.[8,9]

Nonisotopic immunoassays. Use of nonisotopic immunoassays has, in the last few years, become increasingly common in both research and clinical laboratories. Although these immunoassays do not employ radioactive labels, they may possess sensitivities similar to those of classic RIA. The immunochemical mechanisms and principles are the same as for RIA; only the label and method of measurement are different. The most common labels are enzymes and fluorescent or fluorogenic conjugates, any of which may be covalently bound to antibodies or haptens. **EMIT** (Enzyme Multiplied Immunoassay Technique) is one example of a nonisotopic immunoassay currently used mainly for drug assays. In this procedure antibody against the drug is added to the patient's sample. The same drug, but labeled with an enzyme, is then added as a reagent; the enzyme-bound drug, reacting with remaining antibody, forms an antigen-antibody complex. The binding of antibody to the drug-enzyme molecule affects enzyme activity by physically blocking access of the substrate to the active site or by changing the conformation of the enzyme molecule and thereby altering its activity. The relative change in activity is proportional to the drug concentration in the patient's sample. Standards are included in the assay, and concentrations in unknown samples can be calculated from a standard curve. The EMIT system is considered a homogeneous system since quantitation can be accomplished without separating the bound and free fractions of the antibody.

A heterogeneous enzyme immunoassay technique is that of *enzyme-linked immunosorbent assay (ELISA)*. One of the reaction components is nonspecifically adsorbed to a solid phase, such as a microtiter tray or a plastic bead. This attachment of one of the reactions facilitates separation of bound and free labeled reactant. The most common approach is as follows. Sample or standard containing the *antigen* to be quantitated is added to, and allowed to bind to, a solid-phase antibody. After washing, enzyme-labeled antibody (different from the bound antibody) is added and forms a "sandwich complex" of solid phase Ab-Ag-Ab-enzyme. Excess (unbound) antibody is then washed away and enzyme substrate is added; the enzyme hydrolyzes (or oxidizes) the substrate, causing color formation which is proportional to the amount of antigen in the sample. *Antibodies* in a sample can also be quantitated using an ELISA system in which antigen instead of antibody is bound to a solid phase and the second reagent is an enzyme-labeled antibody specific for the sample antibody. The procedure is otherwise identical to that described above for the measurement of antigens.

One of the major goals in the design of all nonisotopic immunoassays has been to provide a homogeneous system in which no separation of bound and free antibody or antigen is necessary and therefore, to make these assays technically easier and shorter. An in-depth review of the various approaches for measuring the reaction of antigen and antibody in nonisotopic systems is beyond the scope of this discussion. However, several examples which illustrate the diversity of the methods are given in Table 1C-4, along with references for each. Several considerations regarding the design of nonisotopic immunoassays are worth mentioning here. Whole antibody (7S IgG) is the most commonly used form of immune reagent. Studies[40] have shown that IgG isolated from antiserum by affinity chromatography provides a more sensitive reagent for solid phase immunoassays. The IgG obtained by affinity chromatography displays less nonspecific binding such that signal-to-noise ratio is improved and sensitivity increased. In addition, the use of F(ab')$_2$ or Fab fragments (see Figure 1C-1) has also been shown to provide a more sensitive reagent system than 7S IgG, especially for applications in cytochemistry and immunoelectron microscopy where penetration of the tracer molecule into cells is necessary.

The choice of an *enzyme label* in any nonisotopic immunoassay depends primarily on the application. Alkaline phosphatase, β-glucuronidase, glucose-6-phosphate dehydrogenase, and horseradish peroxidase are commonly used in both commercial and research systems. β-Glucuronidase and alkaline phosphatase, among other enzymes, can be used with either a colorimetric or fluorometric substrate. A significantly more sensitive assay can be achieved with a fluorometric reagent system. Horseradish peroxidase can be used with substrates that produce soluble products (e.g., *o*-phenylenediamine) or insoluble products (3,3'-diaminobenzidine or tetramethylbenzidine). Methods for the preparation of enzyme antibody conjugates are reviewed by Avrameas.[3]

TABLE 1C-4. NONISOTOPIC IMMUNOASSAYS

	Principle	References
Prosthetic-group immunoassay	Competitive assay in which enzyme co-factor is attached to hapten. Antihapten antibody blocks enzyme activity by restricting co-factor availability. Addition of free hapten results in increase in free co-factor and an increase in enzyme activity.	73
Fluorescence polarization immunoassay	Measurement of polarization of fluorescence from ligand attached to hapten as a function of antibody binding. Free drug reduces polarization of fluorescence.	37,38,61
Solid-phase light scattering immunoassay	Indium spheres coated on glass to measure antibody binding to antigen. Binding of antibody to antigen increases dielectric layer thickness which produces a greater degree of scatter than in areas where only antigen is bound. Quantitation is achieved by densitometry.	63
Radial partition immunoassay	Uses solid-phase glass fiber disk to which antibody is attached. Antigen and enzyme labeled antigen are added sequentially with washing between additions. Bound enzyme is developed with 4-methylumbelliferyl phosphate. Requires specialized equipment and reagents; fully automated system.	23
Electrochemical-differential polarographic immunoassay	Electroactive antigen displaced from antibody by addition of unlabeled antigen. Unlabeled antigen concentration is proportional to differential pulse measured at a dropping mercury electrode.	76
Substrate-labeled fluorescent immunoassay	Competitive assay in which free drug blocks binding of antibody to substrate-drug complex, thereby permitting enzymatic action which produces a fluorescent product.	51
Enzyme-enhancement immunoassay	Microenvironment of enzyme is altered by binding of electronegative antibody, thereby modifying enzyme product. Formation of enzyme-antibody-antigen-antibody complex results in a macroaggregate product which can be measured by turbidimetry or light scattering methods.	22
Time-resolved fluorescence	Method based on the use of a flash lamp source and lanthanide-chelate label. This application takes advantage of the longer fluorescence lifetime of the lanthanide-chelate as compared to the native fluorescence of serum. (See also Chapter 1B, Section Two.)	68

Application of indirect detection of bound antibody to nonisotopic immunoassay, qualitative detection of antigen, and cytochemistry has certain advantages. These indirect approaches may be exemplified by what are called *second (or double) antibody systems* often employed in RIA. An antigen is allowed to react with a first antibody to form unlabeled $AgAb_1$. A second antibody (Ab_2), raised against Ab_1 and then labeled, is added and reacts to form a complex of $AgAb_1Ab_2*$. By measuring the label of bound Ab_2, the amount of Ag, Ab_1, or $AgAb_1$ may be determined. If Ab_1 is a rabbit IgG, a labeled goat antirabbit IgG (Ab_2) becomes a reagent capable of reacting with any rabbit Ab_1 specific for a particular Ag. The advantage of such an indirect system is that a single Ab_2-label combination may be used for a wide variety of specific $AgAb_1$ reactions and a single detection system for the label is applicable over the spectrum of Ag's and Ab_1's of the primary reaction. Amplification of the label signal (i.e., an increase in sensitivity) is likely to occur due to the greater number of antigenic sites that exist on the target complex ($AgAb_1$). The binding constants of most antibody systems are 10^5–10^{11} L/mol. By contrast, the binding constant of another type of binding system, the biotin-avidin complex, is extremely high, 10^{15} L/mol; capitalizing on this system allows indirect systems to be devised that are even more sensitive than the double antibody system.

A *biotin-avidin system* uses a biotin-labeled first antibody. Biotin can be attached to the antibody in relatively high proportion without loss of immunoreactivity by the antibody.[4,26] When an avidin-conjugated label is added, a complex of AgAb-biotin:avidin-label is formed. Further amplification can be achieved by a biotin:avidin:biotin linkage since the binding ratio of biotin:avidin is 4:1, e.g., AgAb-biotin:avidin:(4 biotin-label). If the label is an enzyme, then large numbers of enzyme molecules in the complete complex provide a large increase of enzymatic activity coupled to the small amount of Ag being determined, and sensitivity of the Ag assay is correspondingly greater. Recent developments in preparation of avidin-biotin complexes and polymers of enzyme molecules such as horseradish peroxidase and alkaline phosphatase[33,48] hold great promise for increased sensitivity of many immunoassays and wider utilization of nonisotopic immunoassays. One disadvantage of the biotin-avidin system is that it cannot be used in a biological system in which free biotin or avidin might be found in significant concentrations. The reader is referred to listed references for extended discussions of conjugate preparation and applications in enzyme-linked and solid-phase immunoassays.[3,34,35,41,42,49,74]

OTHER IMMUNOCHEMICAL TECHNIQUES

Immunocytochemistry

The use of labeled antibody reagents as specific probes for protein and peptide antigens allows the researcher and pathologist to examine single cells for synthetic capability and for specific markers for identification of various cell lines. Immunochemistry in recent years has been rapidly expanded by immunoenzymatic methods, especially with regard to the use of horseradish peroxidase (HRP)-labeled (immunoperoxidase) assays. Employing enzyme labels provides several advantages over fluorescent labels. They permit the use of fixed tissues embedded in paraffin which provides excellent preservation of cell morphology and eliminates the problem of autofluorescence from tissue. In addition, immunoperoxidase stains are permanent and only a standard light microscope is needed to identify labeled features. The immunoperoxidase methods are also applicable to electron microscopy. Several approaches for immunoenzymatic assay have been used, including direct, indirect, peroxidase-antiperoxidase, and enzyme bridge methods. These techniques are illustrated in Figure 1C-14. The peroxidase-antiperoxidase and indirect immunoperoxidase methods are the most versatile and generally the most sensitive of the four approaches.[23]

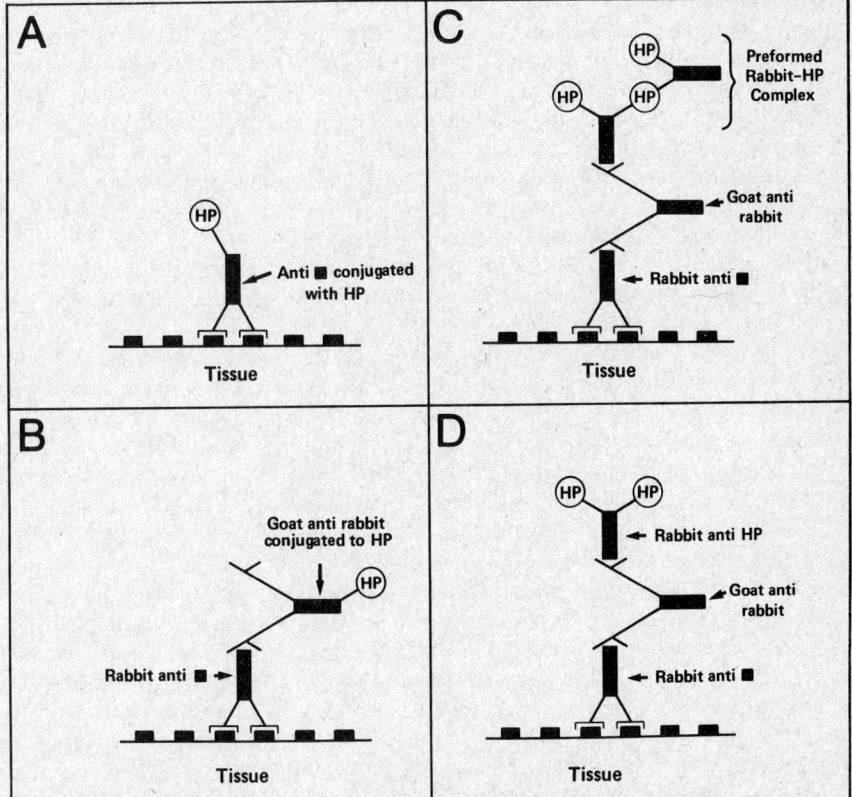

Figure 1C-14. Immunoperoxidase methods. *A,* Direct method requiring only one incubation. *B,* Indirect method requiring two incubations. *C,* The peroxidase-antiperoxidase method (PAP) in which the rabbit antihorseradish peroxidase–horseradish peroxidase complex is formed prior to addition to the tissue, thus requiring three incubation steps. *D,* The enzyme bridge method is similar to *C* but requires an additional incubation step and tends to be less sensitive, since significant amounts of HRP are lost during the final wash.

Agglutination Assays

Agglutination assays have been used for many years for the qualitative and quantitative measurement of antigens and antibodies. In an agglutination method, the visible clumping of particulates such as cells, bacteria, and latex particles is used as an indicator of the primary reaction of antigen and antibody. Agglutination methods require stable and uniform particulates, pure antigen, and specific antibody. IgM antibodies are more likely to produce complete agglutination than are IgG antibodies because of the size and valence of the IgM molecule. Therefore, when only IgG antibodies are involved, it may be necessary to use chemical enhancement or an antiglobulin-agglutination method. As with all immunochemical reactions in which aggregation is the measured endpoint, the ratio of antigen and antibody is critical. Extremes in antigen or antibody concentration will result in inhibition of aggregation.

An incomplete agglutination reaction is one in which the primary reaction occurs, but no or only minimal aggregation of the particles occurs. Many particles such as erythrocytes and bacteria in solution have a net negative charge (zeta potential) causing mutual repulsion.[77] For successful agglutination, the antigen-antibody reaction must overcome this normal resistance. In the case of a weak antigen-antibody reaction or one in which only IgG is involved, this mutual repulsion may be sufficient to inhibit agglutination completely or partially. In systems in which incomplete agglutination results, enhancement may be achieved by lowering the ionic strength or introducing polymeric molecules such as polymerized albumin (5–30%), dextran, polyvrene, polyvinylpyrrolidone, or polyethylene glycol.[77] In some cases, agglutination in the presence of specific antibody can be enhanced by the treatment of cells with enzymes such as bromelin, papain, or ficin.[50]

Antiglobulin methods are also useful for enhancing incomplete agglutination reactions. In this approach, antibody specific to the primary antibody bound to the particulate is added to produce agglutination.

The direct agglutination technique is used for the direct or indirect detection of antigens present on the surface of cells or particles. This technique is generally more sensitive than the precipitin reaction because the bulk of the cell provides added mass for visualization of the reaction. Specific agglutination is the clumping of bacteria, erythrocytes, and so forth in the presence of homologous antibody.

Hemagglutination refers to agglutination reactions in which the antigen is located on an erythrocyte. Erythrocytes are not only good passive carriers of antigen, but are also easily coated with foreign proteins and can be easily obtained and stored.

Direct testing of erythrocytes for blood group, Rh, and other antigenic types is used widely in blood banks; specific antisera, such as anti-A, anti-C, and anti-Kell, are used to detect such antigens on the erythrocyte surface.

In *indirect* or *passive* hemagglutination, the erythrocytes are used as a particulate carrier of foreign antigen (and in some tests, of antibody); this technique has wide applications. Other materials available in the form of fine particles, such as bentonite and latex, also have been used as antigen carriers, but they are more difficult to coat, standardize, and store. In a related variation of this technique, known as *hemagglutination inhibition,* the ability of antigens, haptens, or other substances to inhibit specifically hemagglutination of sensitized (coated) cells by antibody is determined. An application of this technique is a pregnancy test described in Chapter 19.

The classic agglutination test can be conducted either on a slide or in a test tube. Micro procedures that require very small quantities of costly reagents are being used with increasing frequency. These are easier to set up, are more sensitive, and produce settling patterns that can be read easily. The choice of animal for the source of erythrocytes is important; one should use the species that produces the least degree of cross reaction. In practice, sheep erythrocytes are used most frequently. Erythrocytes are relatively unstable after depletion of substrate and accumulation of metabolic products. However, if the cells are collected directly into Alsever's solution (acid-citrate-dextrose solution, pH 6.1, sometimes containing a broad-spectrum antibiotic) and stored at 4 °C, the erythrocytes are usable for up to 30 d. Cells treated with formalin (formalinized cells) can be used for months and may be stored for even longer periods in the frozen or lyophilized state. However, settling patterns may not be as satisfactory as with wet cells stored in the cold.

For use in indirect agglutination tests, erythrocytes can be coated (sensitized) in various ways. The simplest method is by direct adsorption of antigen onto the cell, usually by direct incubation with the desired antigen. The degree of sensitization of the cells thus treated is a function of concentration of the antigen used and time of exposure, but seems to be unrelated to the temperature during adsorption. The erythrocyte membrane possesses many reactive sites and thus it is possible to coat the cell surface sequentially with a number of different antigens, without interference from those initially on the surface. Substances such as polysaccharides and lipopolysaccharides are easily adsorbed onto the cells, whereas more complicated methods are usually necessary for the adsorption of many, but not all, proteins. However, some nonhemagglutinating viruses and antibiotics such as penicillin may be coupled directly to the cells. Some proteins will bind only if the cells are pretreated; the most widely used method for this is the *tanned red cell technique* in which erythrocytes are treated with tannic acid. The exact mechanism by which tannic acid sensitization works is unknown.

The procedure for coating antigens onto tannic acid treated cells is much like that described for simple adsorption. The optimal concentration of antigen to be used may have to be determined in some cases, although only a very small proportion of the antigen ($\sim$1%) is taken up by the cell. Tannic acid also increases cell agglutinability. In some cases, the tanned erythrocytes are so sensitive that they will clump without addition of antibody. If normal serum or serum albumin is added as a stabilizer (if the antigens being used do not cross-react), the suspension not only will give specific agglutination but also will be extremely sensitive to even small amounts of antibody.

The most specific method of sensitization involves the direct covalent attachment of the antigen to the erythrocyte via a chemically reactive grouping. The great hazard in these procedures is the possibility of drastically altering the erythrocyte surface, denaturing or immunologically altering the antigen. There are three generally accepted techniques for the direct attachment of the reactive groupings, based on use of *bis*-diazotized benzidine (BDB), chromic chloride, or 1,3-difluoro-4,6-dinitrobenzene (DFDNB).

The BDB technique couples erythrocyte and antigen through stable, covalent azo bonds. Cells prepared in this fashion are fragile, but if the erythrocytes are first formalinized and then coupled, they can be coated more effectively with larger quantities of antigen and can be stored for longer periods. Coupling with heavy metal cations (for example, chromic chloride) is simple. Chromic chloride, erythrocytes, and antigen are mixed and incubated at room temperature for 1 h; the reactants are then removed by centrifugation. However, chromium-treated cells settle poorly. In another technique, DFDNB reacts with amino groups of the protein and the antigen. When sera are tested with this reagent, they should first be pretreated with DFDNB-treated (nonantigen-conjugated) cells to effect a nonspecific reaction between serum and DFDNB itself. After removal of any precipitate and addition of the coated cells, agglutination is due to action between the antibody and the coated erythrocyte. Erythrocytes sensitized by this or the previous techniques can be used on a micro scale.

In general, the agglutination methods are quite sensitive but are not as quantitative as other immunochemical methods discussed thus far. A biotin-avidin bridging reagent has been described for increasing sensitivity of agglutination reactions.[17] In some cases isotopic and nonisotopic immunoassays, especially enzyme immunoassays, are as convenient as agglutination reactions and are therefore slowly replacing agglutination methods in many laboratories.

MONOCLONAL ANTIBODIES

Monoclonal antibody technology up to this time has had a greater impact on basic research than on clinical or anatomical pathology. The exquisite specificity, referred to by one author as an "immunochemical laser," has made it possible to do molecular mapping of protein antigens.[18] Using a panel of monoclonal antibodies, each directed against a unique determinant, different functions of the complement component C3 have been assigned to specific regions or fragments of the native molecule.[14,70] Similar studies with carcinoembryonic antigen (CEA) and melanoma antigen have been done.[58,75]

The usual method of production of monoclonal antibodies is schematically depicted in Figure 1C-15. Sensitized lymphocytes from the spleens or lymph nodes of mice that have been immunized once or more over a 2–4-week period are fused with a murine myeloma cell line from tissue culture in the presence of polyethylene glycol (PEG) which promotes cell fusion through an unknown mechanism. The murine myeloma cell line is an immortal B-cell line. The murine myeloma cell lines most commonly used are deficient in the enzyme hypoxanthine guanine phosphoribosyl transferase (HGPRT), and therefore cannot synthesize purine bases from thymidine and hypoxanthine in the presence of aminopterin. Following the fusion, the cells are placed into a selection medium containing hypoxanthine, aminopterin, and thymidine (HAT medium) to selectively grow *fused hybrid cell* lines. The fused hybrid cells can survive in a HAT medium since the cells combine the immortality of the myeloma cell with the genetic material of the spleen cell necessary for

HYBRIDOMA PRODUCTION

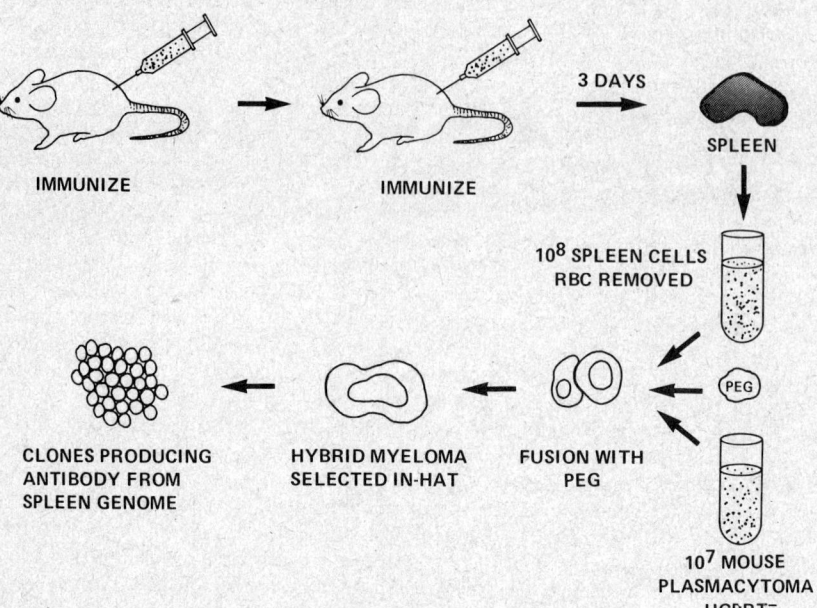

Figure 1C-15. Schematic diagram for production of monoclonal antibody. (Reproduced with permission from Polin, R. A., and Wasserman, R. L.: Monoclonal antibodies: Progress and promise. Laboratory Management, October, 1983, pp. 33–44.)

synthesis of HGPRT. The unfused myeloma cells are killed by the HAT medium since they do not have the HGPRT gene and the unfused spleen cells cannot be maintained in the culture. Colonies arising from the fused cells are then screened for antibody production and those cell lines secreting antibody of the desired specificity are cloned in subcultures. In this way, a single clonal line can be isolated which produces an antibody with a specificity for a single antigen epitope and having a single binding energy or affinity.

The unique ability of a monoclonal antibody to react with a single epitope on a multivalent antigen means that the majority of monoclonal antibodies will not crosslink and precipitate macromolecular antigens. For this reason monoclonal antibodies have not found broad applicability in the clinical laboratory where traditional precipitin methods are in use. Also, until recently monoclonal antibodies have not been available in kit form but are now available for IgE, β-HCG, prostatic acid phosphatase, and other antigens of clinical interest. Both isotopic and nonisotopic solid phase immunoassays using monoclonal antibodies have been described. A practical advantage of using monoclonal antibodies is that two different antibody specificities can be used in a single incubation step. A solid phase antibody specific for a unique epitope and another enzyme or radio-labeled antibody specific for a different epitope can be reacted with antigen in a single step. This eliminates the two-step sequential addition of antigen and labeled antibody to the solid phase, as well as one incubation step and one washing step, which would be necessary when using polyclonal antibodies binding to both sites.

Because monoclonal antibodies can be made against single epitopes on complex multifunctional molecules, they make it possible to measure the appearance or disappearance of fragments of molecules cleaved as a feature of their activation in vivo, e.g., complement components and coagulation factors.

Using biotin-avidin, it is possible to perform solution phase immunoassays using monoclonal antibodies. In a study of CEA, monoclonal anti-CEA labeled with biotin was reacted with CEA and then precipitated after the addition of avidin.[75] With further improvements in methodology and an increase in the range of antibody specificities, monoclonal antibodies will certainly have significant impact on both immunochemistry and immunocytochemistry in the future.

References

1. Alper, C. A., and Johnson, A. M.: Immunofixation electrophoresis: A technique for the study of protein polymorphism. Vox Sang., *17*:445-452, 1969.
2. Arbesman, C. E., Ito, K., Wypych, J. I., et al.: Measurement of serum IgE by a one-step single radial radiodiffusion method. J. Allergy, *49*:72-80, 1972.
3. Avrameas, S., Ternynck, T., and Guesdon, J. L.: Coupling of enzymes to antibodies and antigens. Scand. J. Immunol., *8*(Suppl. 7):7-23, 1978.
4. Bayer, E. A., and Wilcheck, M.: The use of the avidin-biotin complex as a tool in molecular biology. Methods Biochem. Anal., *26*:2-42, 1980.
5. Beck, O. E., and Kaiser, P. E.: Rate nephelometry of human IgE in serum. Clin. Chem., *28*:1349-1351, 1982.
6. Bittner, M., Kupferer, P., and Morris, C. F.: Electrophoretic transfer of proteins and nucleic acids from slab gels to diazobenzyloxymethyl cellulose or nitrocellulose sheets. Anal. Biochem., *102*:459-471, 1980.
7. Brandshund, I., Siersted, S. E., and Teisner, B.: Double-decker rocket immunoelectrophoresis for direct quantitation of complement split products with C3d specificities in plasma. J. Immunol. Methods, *44*:63-71, 1981.
8. Buffone, G. J., Edwards, M. S., Schimbor, C. M., et al.: The role of anticapsular antibody in alternative complement pathway mediated opsonization of *Streptococcus pneumoniae*, type 14. In preparation.
9. Buffone, G. J., Leatherwood, C. M., Person, D. A., et al.: An immunoradiometric assay for erythrocyte complement (C3b) receptor activity applied to a pediatric population with connective tissue disease. Clin. Chem., *29*:1720-1723, 1983.
10. Buffone, G. J., and Lewis, S. A.: Advantages of small angle light scattering measurements in immunonephelometry. *In*: Plasma Protein Pathology. H. Peeters and P. H. Wright, Eds. New York, Pergamon Press, 1979, pp. 55-61.
11. Buffone, G. J., Savory, J., Cross, R. E., et al.: Evaluation of kinetic light scattering as an approach to the measurement of specific proteins with the centrifugal analyzer. I. Methodology. Clin. Chem., *21*:1731-1734, 1975.
12. Buffone, G. J., Savory, J., and Hermans, J.: Evaluation of kinetic light scattering as an approach to the measurement of specific proteins with the centrifugal analyzer. II. Theoretical considerations. Clin. Chem., *21*:1735-1746, 1975.
13. Burnette, W. N.: "Western blotting": Electrophoretic transfer of proteins from sodium dodecyl sulfate–polyacrylamide gels to unmodified nitrocellulose and radiographic detection with antibody and radioiodinated protein A. Anal. Biochem., *112*:195-203, 1981.
14. Burger, R., Deubel, U., Hadding, U., et al.: Identification of functionally relevant determinants on the complement component C3 with monoclonal antibodies. J. Immunol., *129*:2042-2051, 1982.
15. Cheng, A., Bray, K., and Polito, A.: Nephelometric inhibition assay for gentamicin. Clin. Chem., *25*:1078, 1979.
16. Clarke, H. G. M., and Freeman, T.: A quantitative immunoelectrophoresis method (Laurell electrophoresis). *In*: Protides of Biological Fluids, 14th Colloquium. H. Peeters, Ed. Amsterdam, Elsevier, 1967, pp. 503-509.
17. Costello, S. M., Felix, R. T., and Giese, K. W.: Enhancement of immune cellular agglutination by use of an avidin biotin system. Clin. Chem., *25*:1572-1580, 1979.
18. David, G. S., Wang, R., Bartholomew, R., et al.: The hybridoma—an immunochemical laser. Clin. Chem., *27*:1580-1585, 1981.
19. Fahey, J. L., and McKelvey, E. M.: Quantitative determination of serum immunoglobulins in antibody agar plates. J. Immunol., *94*:84-90, 1965.
20. Gauldie, J., and Bienenstock, J.: Automated nephelometric analysis of haptens. *In*: Automated Immunoanalysis, Vol. 7, Part 1. R. Ritchie, Ed. New York, Marcel Dekker, 1978, pp. 321-333.
21. Gershoni, J. M., and Palade, G. E.: Electrophoretic transfer of proteins from sodium dodecyl–sulfate–polyacrylamide gels to a positively charged membrane filter. Anal. Biochem., *124*:396-405, 1982.
22. Gibbons, I., Hanlon, T. M., Skold, C. N., et al.: Enzyme-enhancement immunoassay: A homogeneous assay for polyvalent ligands and antibodies. Clin. Chem., *27*:1602-1607, 1981.
23. Giegel, J. L., Brotherton, M. M., Cronin, P., et al.: Radial partition immunoassay. Clin. Chem., *28*:1894-1898, 1982.
24. Grabar, P., and Williams, C. A.: Méthode permettant l'étude conjuguée propriétés électrophorétiques et immuno-chimiques d'un mélange de protéines. Application au sérum sanguin. Biochim. Biophys. Acta, *10*:193-194, 1953.
25. Grossberg, A. L., Chen, C. C., Rendina, L., et al.: Specific cation effects with antibody to a hapten with a positive charge. J. Immunol., *88*:600-603, 1962.
26. Guesdon, J. L., Ternynck, T., and Avrameas, S.: The use of avidin-biotin interaction in immunoenzymatic techniques. J. Histochem. Cytochem., *27*:1131-1139, 1979.
27. Harrington, J. L., Fenton, J. W., and Pert, H.: Polymer-induced precipitation of antigen-antibody complexes: 'Precipiplex reactions.' Immunochemistry, *8*:413-421, 1971.
28. Hellsing, K.: Immune reactions in polysaccharide media. Experiments with specific antibodies of different affinities for serum albumin. Biochem. J., *114*:151-155, 1969.
29. Hellsing, K.: Immune reactions in polysaccharide media. Polysaccharide-enhanced precipitation reactions with antigens of various sizes. Biochem. J., *114*:145-149, 1969.
30. Hellsing, K.: Immune reactions in polysaccharide media. The composition of antigen-antibody complexes in the precipitin reaction. Biochem. J., *114*:141-144, 1969.
31. Hellsing, K.: Immune reactions in polysaccharide media. I. The effect of dextran on the reaction between ^{125}I-labeled human serum albumin and gamma-G-globulin from rabbit. Acta Chem. Scand., *20*:1251-1262, 1966.
32. Hills, L. P., and Tiffany, T. O.: Comparison of turbidimetric and light-scattering measurements of immunoglobulins by use of a centrifugal analyzer with absorbance and fluorescence-light-scattering optics. Clin. Chem., *26*:1459-1466, 1980.
33. Hsu, S. M., Raine, L., and Fanger, W.: A comparative study of the PAP method and an avidin-biotin complex

method for studying polypeptide hormones with radioimmunoassay antibodies. Am. J. Clin. Pathol., *75*:734-738, 1981.

34. Ishikawa, E., Hamaguchi, Y., and Imagana, M.: An improved preparation of antibody-coated polystyrene beads for sandwich enzyme immunoassay. J. Immunoassay, *1*:385-398, 1980.

35. Ishikawa, E., and Kato, K.: Ultrasensitive enzyme immunoassay. Scand. J. Immunol., *8* (Suppl. 7):43-55, 1978.

36. Johnson, A. M.: Immunofixation electrophoresis and electrofocusing. Clin. Chem., *28*:1797-1800, 1982.

37. Jolley, M. E., Stroupe, S. D., Schwenzer, K. S., et al.: Fluorescence polarization immunoassay. III. An automated system for therapeutic drug determination. Clin. Chem., *27*:1575-1579, 1981.

38. Jolley, M. E., Stroupe, S. D., and Wang, C. J.: Fluorescence polarization immunoassay. I. Monitoring aminoglycoside antibiotics in serum and plasma. Clin. Chem., *27*:1190-1197, 1981.

39. Kabat, E. A.: Structural Concepts in Immunology and Immunochemistry. 2nd ed. New York, Holt, Rinehart and Winston, 1976.

40. Kato, K., Fukui, H., Hamaguchi, Y., et al.: Enzyme-linked immunoassay: Conjugation of the Fab' fragment of rabbit IgG with β-*o*-galactosidase from *E. coli* and its use for immunoassay. J. Immunol., *116*:1554-1560, 1976.

41. King, T. P., and Kochoumian, L.: A comparison of different enzyme conjugates for enzyme linked immunosorbent assay. J. Immunol. Methods, *28*:201-210, 1979.

42. Kricka, L. J., Carter, T. J. N., Burt, S. M., et al.: Variability in the adsorption properties of microtiter plates used as solid supports in enzyme immunoassay. Clin. Chem., *26*:741-744, 1980.

43. Laurell, C. B.: Antigen-antibody crossed electrophoresis. Anal. Biochem., *10*:358-361, 1965.

44. Laurell, C. B.: Electroimmunoassay. Scand. J. Clin. Lab. Invest., *29*(Suppl. 124):21-37, 1972.

45. Laurent, T. C.: *In*: The Chemical Physiology of Mucopolysaccharides. G. Quintarelei, Ed. Boston, Little, Brown & Co., 1968, p. 153.

46. Laurent, T. C.: The interaction between polysaccharides and other macromolecules. 5. The solubility of proteins in the presence of dextran. Biochem. J., *89*:253-257, 1963.

47. Laurent, T. C., and Ogston, A. G.: The interaction between polysaccharides and other macromolecules. 4. The osmotic pressure of mixtures of serum, albumin, and hyaluronic acid. Biochem. J., *89*:249-253, 1963.

48. Leary, J. J., Brigati, D. J., and Ward, D. C.: Rapid sensitive colorimetric method for visualizing biotin-labeled DNA probes hybridized to DNA or RNA immobilized on nitrocellulose: Bio-blots. Proc. Natl. Acad. Sci. USA, *80*:4045-4049, 1983.

49. Lehtonen, O. P., and Viljanen, M. K.: Antigen attachment in ELISA. J. Immunol. Methods, *34*:61-70, 1980.

50. Lewis, A. J.: Papain, ficin, and bromelain in the detection of incomplete rhesus antibodies. Br. J. Haematol., *3*:332-339, 1957.

51. Li, T. M., Benovic, J. L., Buckler, R. T., et al.: Homogeneous substrate labeled fluorescent immunoassay for theophylline. Clin. Chem., *27*:22-26, 1981.

52. Mancini, G., Carbonara, A. O., and Heremans, J. F.: Immunochemical quantitation of antigens by single radial immunodiffusion. Immunochemistry, *2*:235-254, 1965.

53. Mancini, G., Vaerman, J. P., Carbonara, A. O., et al: A single-radial diffusion method for the immunological quantitation of proteins. *In*: Protides of Biological Fluids, 11th Colloquium, Bruges. H. Peeters, Ed. Oxford, Pergamon Press, 1964, pp. 370-373.

54. March, S. C., Parikh, I., and Cuatrecasas, P.: A simplified method for cyanogen bromide activation of agarose for affinity chromatography. Anal. Biochem., *60*:149-152, 1974.

55. Marrack, J. R., and Richards, C. B.: Light scattering studies of the formation of aggregates in mixtures of antigen and antibody. Immunology, *20*:1019-1040, 1970.

56. Mason, D. Y., and Sammons, R.: Alkaline phosphatase and peroxidase for double immunoenzymatic labeling of cellular constituents. J. Clin. Pathol., *31*:454-460, 1978.

57. Nakamura, S.: Cross Electrophoresis: Its Principles and Application. New York, Elservice Publishing Co., 1966.

58. Ng, A.-K., Giacomini, P., Kantor, R., et al.: Molecular heterogeneity and shedding of a high-molecular-mass melanoma-associated antigen identified with monoclonal antibodies. Clin. Chem., *28*:2347-2350, 1980.

59. Nygren, H., and Stenberg, M.: Diffusion-in-gel enzyme-linked immunosorbent assay (DIG-ELISA). Quantification of antigen by diffusion over an antibody-coated surface. Scand. J. Clin. Lab. Invest., *42*:355-359, 1982.

60. Ogston, A. G., and Phelps, C. F.: The partition of solutes between buffer solutions containing hyaluronic acid. Biochem. J., *78*:827-833, 1961.

61. Popelka, B. R., Miller, D. M., Holen, J. T., et al.: Fluorescence polarization immunoassay. II. Analyzer for rapid, precise measurement of fluorescence polarization with use of disposable cuvettes. Clin. Chem., *27*:1198-1201, 1981.

62. Pressman, D., Nisonoff, A., and Radzimski, G.: Specific anion effects with antibenzoate antibody. J. Immunol., *86*:35-41, 1961.

63. Rej, R., Keese, C. R., and Giaever, I.: Direct immunochemical determination of aspartate aminotransferase isoenzymes. Clin. Chem., *27*:1597-1601, 1981.

64. Rodbard, D.: Data processing for radioimmunoassays: An overview. *In*: Clinical Immunochemistry, Vol. 3. S. Natelson, A. J. Pesce, and A. A. Mietz, Eds. Washington, D.C., Am. Assoc. Clin. Chem., 1978, pp. 477-494.

65. Rowe, D. S.: Radioactive single radial diffusion: A method for increasing the sensitivity of immunochemical quantitation of proteins in agar gel. Bull. WHO, *40*:613-616, 1969.

66. Sandler, S. G., Nusbacher, J., and Schanfield, M. S., Eds.: Immunobiology of the Erythrocyte. New York, Alan R. Liss, 1980, pp. 158-164.

67. Savory, J., Buffone, G. J., and Reich, R.: Kinetics of the IgG anti-IgG reaction, as evaluated by conventional and stopped-flow nephelometry. Clin. Chem., *20*:1071-1075, 1974.

68. Siitari, H., Hemmilia, J., Soini, E., et al.: Detection of hepatitis B surface antigen using time-resolved fluoroimmunoassay. Nature, *301*:258-260, 1983.

69. Steward, M. W.: Immunochemistry. New York, John Wiley and Sons, 1977, p. 233.

70. Tamerius, J. D.: Recent advances in monoclonal antibodies to complement components and their application for immune complex assay. Lecture—Clinical Laboratory Tests: New Technology and Future Directions, San Diego, February 2–4, 1983.
71. Tanford, C.: The Hydrophobic Effect: Formation of Micelles and Biological Membranes. New York, John Wiley and Sons, 1980.
72. Tengerdy, R. P.: Reaction kinetic studies of the antigen antibody reaction. J. Immunol., *99*:126-132, 1967.
73. Tyhach, R. J., Rupchock, P. A., Pendergrass, J. H., et al.: Adaption of a prosthetic-group-labeled homogeneous immunoassay to a reagent strip format. Clin. Chem., *27*:1499-1504, 1981.
74. Voller, A., Bartlett, A., and Bidwell, D. E.: Enzyme immunoassays with special reference to ELISA techniques. J. Clin. Pathol., *31*:507-519, 1978.
75. Wagener, C., Clark, B. R., Richard, K. J., et al.: Monoclonal antibodies for carcinoembryonic antigen and related antigens as a model system: Determination of affinities and specificities of monoclonal antibodies by using biotin-labeled antibodies and avidin as a precipitating agent in a solution phase immunoassay. J. Immunol., *130*:2302-2307, 1983.
76. Wehmeyer, K. R., Halsall, H. B., and Heiniman, W. R.: Electrochemical investigation of hapten-antibody interaction by differential polarography. Clin. Chem., *28*:1968-1972, 1982.
77. Williams, C. A., and Chase, M. M., Eds.: Methods in Immunology and Immunochemistry, Vol. 3. New York, Academic Press, 1970, pp. 1-125.

Automation in the Clinical Laboratory

by Ernest Maclin, P.E., and Donald S. Young, M.B., Ph.D.

Increased demand for tests from the clinical laboratory as well as development of automated instruments to perform these tests occurred at about the same time during the 1950's. Automation allowed laboratories to process a much larger workload without a comparable increase in number of staff. The term "automation" has been applied in the field of clinical chemistry to describe the process whereby an analytical instrument performs many tests with only minimal involvement of an analyst. Nevertheless, this use of the term may not be entirely correct, and it might better be described as mechanization. The definition accepted by the International Union of Pure and Applied Chemistry (IUPAC)[12] describes *automation* as the replacement of human manipulative effort and facilities in the performance of a given process by mechanical and instrumental devices which are regulated by feedback of information so that an apparatus is self-monitoring or self-adjusting. By this definition, no currently available clinical chemical instrument could be described as automated, although the term could be applied to certain individual steps in many analytical processes.

In this chapter we will discuss automation in the sense of a mechanized analytical process, as it is traditionally understood by clinical laboratorians. To describe a lesser degree of mechanization than that associated with an instrument's performing a complete chemical analysis, the terms "semi-" or "partial" automation have been applied, even though these terms have not been well defined.

Among the benefits of automation is the elimination of tasks that are repetitive and monotonous for a human, leading to boredom or inattention that may cause variability or errors of analysis. Improvement of reproducibility does not necessarily improve the accuracy of test results, since accuracy is largely influenced by the analytical methods used. Nevertheless, the significant improvements in quality of laboratory tests in recent years owe a great deal to the combination of well-designed automated instrumentation with good analytical methods and effective quality assurance programs.[8] Although initial cost of an automated instrument may be high, it is usually offset by cost savings gained by reduction of staff needed to process heavy workloads.

In this chapter we discuss the principles that are applied in the design of the individual steps of automated analysis. This will be followed by the application of these principles to some representative instruments.* We also include a section to provide guidelines for instrument selection and another section on trends in instrumentation to reflect further potential benefits of automation on laboratory operations.

INSTRUMENTAL CONCEPTS

Automated systems generally incorporate mechanized versions of basic manual laboratory techniques and procedures. Indeed, many manufacturers provide reagents with their systems, as well as instructions to allow the same reagents to be used for manual tests in the event of equipment failure.

---

*To avoid repetition and to make the text easier to read, names and locations and trademark status of manufacturers of instruments noted in the text are listed in Appendix 1D-II. TM and ® notations are similarly defined in Appendix 1D-II.

Since the development of the first successful automated system by Skeggs, introduced as the AutoAnalyzer in 1957, many different concepts and terms have evolved for automated systems. It is difficult to classify automated systems, although manufacturers typically focus on one major process in their automated systems to distinguish one system from another. In fact, many systems incorporate several different concepts into a single unit. Therefore it is useful to define the terms that are commonly used in descriptions of automation in the clinical laboratory, and to provide examples of how these concepts have been integrated into analytical systems.

DEFINITIONS

Batch analysis: a number of specimens are processed in the same analytical session, or "run."

Sequential analysis: each specimen in the batch enters the analytical process one after another, and each result or *set* of results emerges in the same order as specimens are entered.

Continuous flow analysis: each specimen in the batch passes through the same continuous stream and is subjected to the same analytical reactions as every other specimen and at the same rate.

Discrete analysis: each specimen in the batch has its own physical and chemical space, separate from every other specimen.

Single-channel analysis (also known as single test analysis): each specimen is subjected to a single process so that results for a single analyte are produced.

Multiple-channel analysis (also known as multitest analysis): each specimen is subjected to multiple analytical processes so that a set of test results is obtained.

Parallel analysis: all specimens are subjected to a series of analytical processes at the same time in a parallel fashion.

Discretionary multiple-channel analysis: specimens in sequence can be analyzed by any one or more than one of the available processes (methods, channels) as a result of a command to the processing system specific for that specimen.

Random access analysis: any specimen, by a command to the processing system, can be analyzed by any available process, in or out of sequence with other specimens, and without regard to their initial order.

Other definitions of terms used in the automation of clinical chemistry have been published by IUPAC.[12]

Brief descriptions of some systems using different process concepts are presented below. More comprehensive descriptions of several common systems are presented at the end of this chapter.

Several process concepts may be incorporated into a single instrument, as exemplified by centrifugal analyzers. The widely used bench and floor model *centrifugal analyzers* (e.g., GemENI, COBAS-BIO, ENCORE) use discrete pipetting for specimens and reagents. The batch of specimens is sequentially loaded into the discrete vessels in the rotor and is subsequently analyzed in parallel. When a single analyte is measured, results for the entire batch of specimens are available in 2–13 min. When different tests are performed sequentially in a batch mode, from 200–450 results can be produced per hour, depending on the system and the length of time required to measure each analyte. A keyboard entry (for COBAS-BIO, Multistat, or ENCORE) or card entry (for GemENI) is used to instruct the analyzer which test to perform. All specimens within a batch are then tested for the same analyte. Different reagents may then be placed in the analyzer and a different test is programmed into the instrument.

Several benchtop analyzers use a batch, sequential, discrete approach with specimens first being presented in a platter or tray on a *batch* basis; each is then entered into the analyzer sequentially with analyses being performed on a *discretionary* basis. The widely used discrete analyzers, ASTRA 4 and 8 systems, implement *discrete* pipetting of specimens and reagents with discretionary selection of tests from an array of several possible tests (multiple channels) with *sequential* specimen presentation from a batch of specimens. Keyboard commands are used to select the appropriate tests for each specimen.

The *AutoAnalyzer* is a single-channel, continuous flow, batch analyzer providing one result per analyte for each specimen at a rate of 40–60 specimens per hour. The latest version of this system using continuous flow is the SMAC. This is a multiple-channel analyzer capable of performing 23 simultaneous analyses on one specimen of a batch of as many as 168 specimens presented to the analyzer; the SMAC can produce about 2500 test results per hour. This instrument is nondiscretionary, so that all 23 tests are done on every specimen.

When initially introduced, automation mimicked manual test procedures and was applied to those tests requested most often. All procedural steps and their substeps were duplicated. But to reduce the number of manipulative steps, new analytical methods as well as modifications of existing procedures have been developed.

Many of the same analytical principles are used for the quantitation of serum and urine constituents, but automation is more difficult for urine tests than for serum tests because the variable concentration of many urine constituents requires sensitivity to measure constituents at a low concentration as well as expanded linearity to permit measurements of high concentrations without dilution. This requirement, together with the relatively low demand for urine tests compared with serum tests, has restricted the development of analyzers designed specifically for urine constituents. Nevertheless, some urine analyses are performed on serum analyzers in certain institutions.

The *specimen throughput* rate in most multichannel and random access systems is independent of the reaction times of the chemistry procedures; instead, the rate of analysis depends on a fixed sequence of events involving mechanical motions and optical measurement cycles. This is in contrast to batch analyzers, in which throughput rate is affected by the assays performed. In a batch system, the time for a rapid end-point reaction such as that for total protein may be only a fraction of that for a kinetic enzyme assay, such as creatine kinase.

Random access analyzers such as the *aca*, Ektachem, and DACOS perform analyses on a batch of specimens, with each specimen analyzed sequentially but for possibly different tests. The broad array of tests available gives the instruments some of the same capabilities as the multichannel analyzers, but the tests in the random access analyzers are not preordained, since they are selectable through the use of different vials, in the case of liquid reagents (as in the DACOS and DEMAND), or different packs (*aca*), or different thin-film slides (Ektachem). This approach permits measurement of a variable number and variety of analytes in each specimen. As with discretionary multichannel systems, profiles can be defined for a specimen at the time of specimen entry into the analyzer by means of a keyboard, as used in most systems, or by operator selection of appropriate reagent packs as required for the *aca*.

Manufacturers have incorporated low-cost *microprocessors* into many batch analyzers to collate results as well as to perform many statistical and quality control tasks. Individual results may be stored in memory until a previously defined profile for a single specimen is completed and all test results then presented together on a single report form. In some situations, linkage between different analyzers is possible so that a specimen- or patient-oriented, rather than analyte-oriented, report may be produced.

The component steps in automated systems include:
1. Specimen identification
2. Specimen preparation
3. Specimen transport and delivery
4. Removal of protein and other interferents
5. Sample transport and delivery
6. Reagent handling
7. Reagent delivery
8. Chemical reaction phase
9. Measurement approaches
10. Signal processing and data handling.

These steps are usually performed sequentially, but in some instruments they may be combined and occur in parallel.

We describe these processes and methods of implementation in detail in terms of operational and analytical performance. The application of these steps in some representative systems is included at the end of the chapter to illustrate how they are integrated into a single analyzer. Note that the concepts can be applied within instruments that use ion-selective electrodes or fluorometry as well as the more common photometric measurements.

Specimen Identification

Typically, the identifying link between patient and specimen is made at the patient's bedside, and that between the specimen and test results is made in the laboratory. Linkage in the laboratory is done by a variety of techniques ranging from manual transcription of all information to analyzer generation of labels and printing of reports.

In some computerized systems, entry of a test order for a uniquely identified patient generates a specimen label bearing a unique laboratory accession number and establishes a record that remains incomplete until a result (or set of results) is entered into the computer against an accession number. The unique label is affixed to the specimen collection tube when the blood is drawn. Arrival of the specimen in the laboratory is recorded by a manual or computerized log-in procedure. In other systems, the specimen labeled at the patient's bedside with the patient identification and collection information enters the laboratory with a requisition form; there it is assigned an accession number as part of the log-in procedure, which may or may not be computer implemented.

After accessioning, specimens enter the technical handling processes. For those processes requiring physical removal of serum from the original tube, secondary labels bearing the same information as the original label must be affixed to the secondary tube. Some automated analyzers can sample directly from the original collection tube while simultaneously reading the accession number from a secondary label on the tube. The secondary labels may be generated at the time of accessioning by a special, usually computerized device or in some analyzers by a built-in printer that is activated when the analyzer is programmed.

Many methods are used to achieve secondary labeling. A number may be handwritten on the specimen cup or a coded label may be affixed to the original tube or to a specimen cup. The label numbers may have to be correlated with a manually- or computer-generated work list or load list. The load list usually records accession numbers in sequence with the physical positions in the loading zone of the analyzer. This loading zone may be a revolving tray or turntable, a mechanical belt, or a set of racks by which specimens are delivered to the sample aspiration station of the analyzer in a predetermined order.

In those analyzers that do not automatically link specimen identity and sample aspiration, the sequence of results produced must be linked manually with the sequence of entry of specimens. Some analyzers print out or transmit to a host computer each result or set of results from a specimen either by position of the specimen in the loading zone or by the accession number programmed to that position. When results are printed out only by position number, they must subsequently be hand-transcribed and correlated with the work list.

Clearly, many opportunities arise for mismatching specimens and results.[8] The risks begin at the bedside and are compounded with every processing step a specimen undergoes between collection from the patient and the analysis by the instrument. The risks are particularly great when hand transcription is invoked for accessioning, labeling and relabeling, and making load lists. An incorrect accession number or one with a transposition of digits, or a load list with transposed accession numbers, may cause test results to be attributed to the wrong patient. Hand transcription of information may involve errors both in the identifying numbers of patients and in the numerical values of the test results. An additional hazard exists when specimens must be inserted into certain positions in the loading zone defined for them by a load list. Human misreading of either specimen label or loading list may cause misplacement of specimens, calibrators, or controls.

Systematic correlation of results with proper specimen identity can reduce errors in a manual system. In automated analyzers the same objective can be achieved by (1) reading the identification label in the specimen loading zone as in the SMAC (see Figure 1D-1,A); (2) generating coded labels at the time of test request entry as in the Paramax (see Figure 1D-1,B); (3) providing the specimen label with the test requisition as in the Parallel; or (4) reading codes by a separate, off-line light pen in conjunction with separate coded labels, which can be affixed to the specimen container, as in the ACCLAIM.

In all instruments with the capability of scanning coded labels, results are matched with a patient's identity. Typically, the analyzer's computer has enough memory capacity so that when reports are produced the patient demographic data, such as age and sex, may be printed together with the patient's name and identifying number, hospital location, and the physician's name.

Several codes have been used on labels in automated systems. Examples of bar codes that are used in chemistry analyzers are illustrated in Figure 1D-1. All labels must be human readable as well as machine readable to allow specimens to be located rapidly for emergency tests, or, from storage, for retesting.

Guidelines to establish uniformity in blood banking have been published by the American

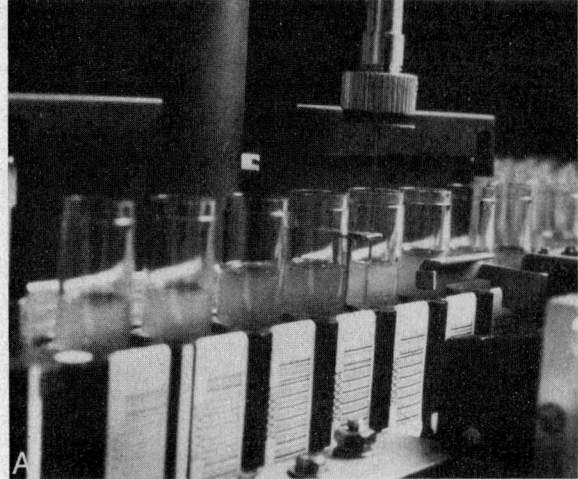
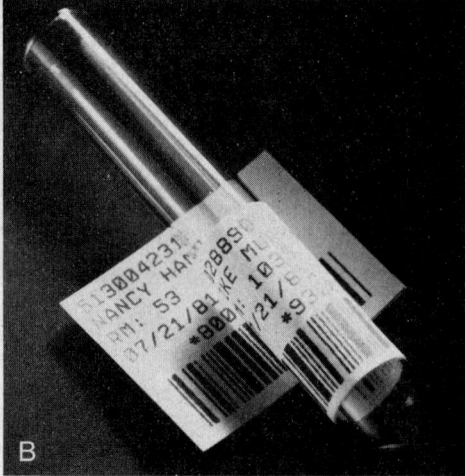

Figure 1D-1. Examples of codes used on labels for specimen identification. Labels, as shown, have human-readable as well as machine-readable coding. *A*, Codabar intermixed on SMAC II. (Courtesy of Technicon Instruments Corp.) *B*, Codell on Paramax. (Courtesy of American Dade, Division of American Hospital Supply Corporation.)

Blood Commission's Committee for Commonality in Blood Bank Automation[1] and by the Food and Drug Administration (FDA) for simplified, uniform, machine- and human-readable labels for blood and blood components. Proposed legislation[6] requires uniform labeling whether or not blood centers use computerized processing. To simplify organization within hospitals, it seems appropriate that identification of all clinical laboratory specimens should be done with the same coded system. Although there is as yet no standard for coded labels within medical practice, efforts are under way to accept codes, known as Universal Product Code (UPC) and Code 39, within the hospital industry; this standard would allow use of either code for specimen identification.

If a bar-code identification can be initiated at a patient's bedside, greater integrity of the identity of a specimen in an analyzer could be assured. Prototype systems to transfer a patient's identity to blood tubes at the patient's bedside have been developed, but none are in routine use. One difficulty in applying these systems is the need to cross over traditional boundaries, which causes conflicts with administrative procedures outside the control of the laboratory staff.

Not all methods that link patient identity and test results require coded labels. The DuPont *aca* uses a unique approach in which specimen and patient identification are handwritten on an ID card that is attached to the specimen container, and a simple photocopier in the analyzer transfers the same information to the test report form. A copy of both documents is illustrated in Figure 1D-2.

Specimen Preparation

The clotting of blood in the specimen collection tubes and the centrifuging of these tubes, with subsequent transfer of serum to secondary tubes, cause delays and expense in the preparation of a specimen for analysis. To date, no system for automatic preparation of a specimen for analysis has been incorporated into any analyzer. However, three ongoing developments are noteworthy.

1. Use of whole blood for analysis. Automated or semiautomated ion-selective electrodes, which measure ion activity rather than concentration in whole blood, can be incorporated into automated systems to provide certain test results for a clinician within minutes of a specimen's being drawn. This approach is already available for measurement of sodium, potassium, and calcium, and extended application of it to other analytes holds considerable promise for more rapid testing. Another approach involves both manual or automated application of whole blood to dry reagent films and visual or instrumental observation of a quantitative change; this approach is exemplified by systems available from Boehringer Mannheim Corporation (BMC) and Ames

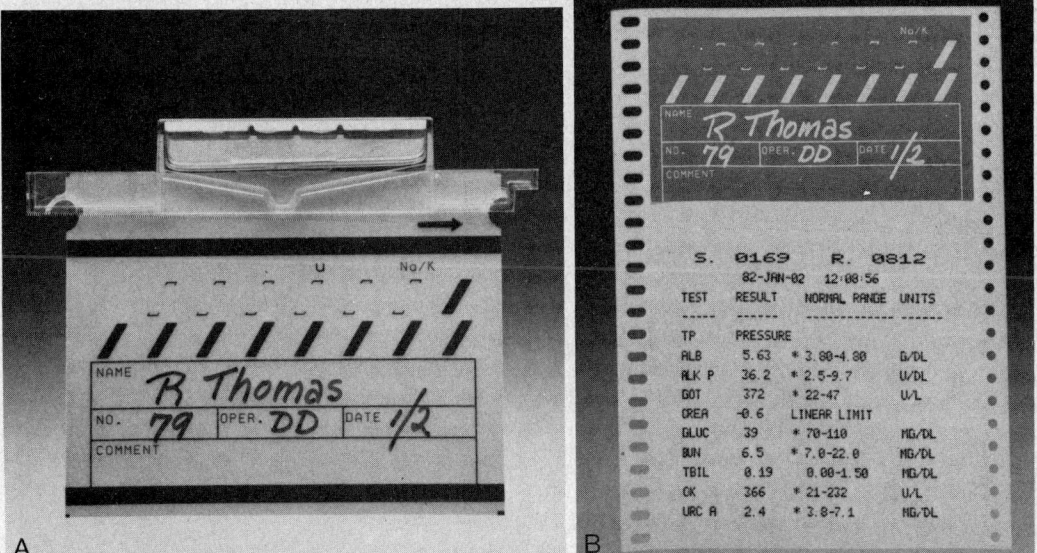

Figure 1D-2. Specimen identification input and output for *aca* II and III discrete clinical analyzer. *A*, Specimen cup and label as annotated by technologist. *B*, Integrated test report for all analytes on the same specimen. (Courtesy of DuPont Company.)

Division, Miles Laboratories, Inc.[27] When whole blood can be added directly to an assay system, specimen preparation time is essentially eliminated.

2. Automation of specimen preparation. Instrumentation Laboratory, Inc., has developed a membrane filtration device in a disposable cartridge (Statsep) to produce unhemolyzed plasma from a specimen of whole blood in 60 s. No clinically significant difference in the concentration of 18 commonly measured analytes between plasma obtained by centrifugation or filtration has been observed. This approach has the potential to reduce the delay between specimen receipt in the clinical laboratory and production of test results, but it has not yet been integrated into any automated system.

3. Use of robotics. Robots can be used to mechanize the specimen preparation steps presently performed by technologists, in a manner analogous to those automated systems that implement manual analytical procedures. A modular robotics system that has "hands" and "fingers," that positions tubes and syringes relative to various specimen racks and centrifuges, and that can pipet, pour, and mix solutions has been offered by Zymark Corporation (Figure 1D-3). Robot actions are controlled by a simple microprocessor. The modules are initially mounted on a laboratory bench by an operator and, when properly programmed, will perform their motions indefinitely. Positioning accuracy of 2.5 mm (0.1 inch) can be attained over distances corresponding to the normal movement of a human arm. Zymark systems have been used hitherto primarily in industrial analytical laboratories and have not been incorporated into any automated clinical laboratory analyzer.

Specimen Handling, Presentation, and Transport

In most situations, the specimen presented to an automated analyzer is serum. A few analyzers, e.g., SMAC and Parallel, have been designed to sample a supernatant serum from the original collection tube. In these cases, collecting tubes frequently contain a separator material that forms a barrier between supernatant and cells. (See Chapter 3.) Most analyzers, however, sample from a cup or tube that is filled with serum transferred from the original specimen tube. The shape and size of the cups or tubes are dictated by the kind, structure, and auxiliary functions (e.g., specimen identification) of the analyzer's presentation device.

Often the design of the sampling cup is unique for a particular analyzer. Each cup should be designed to minimize dead-volume, i.e., the excess serum that must be present in a cup to

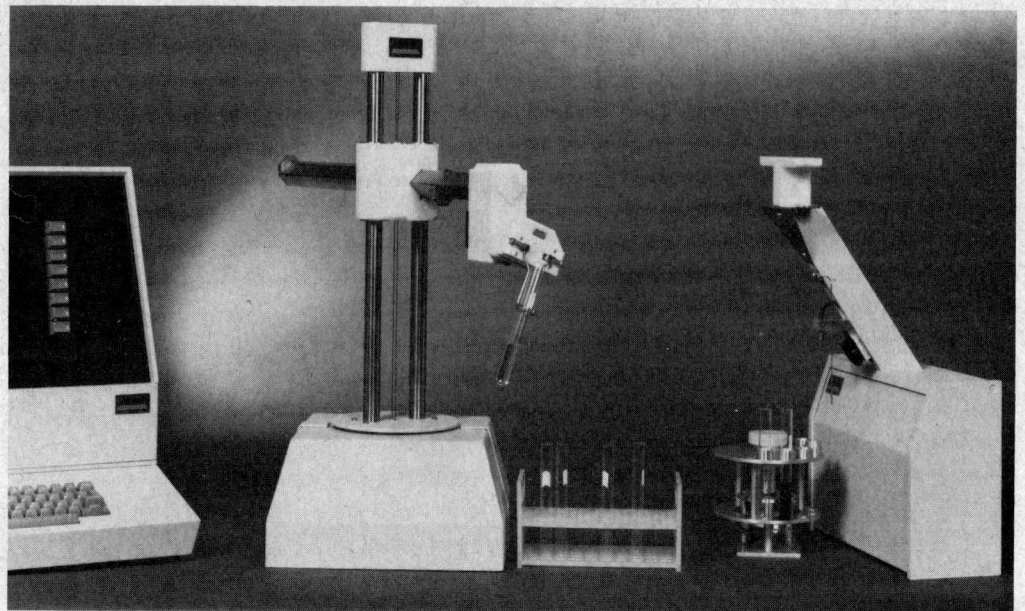

Figure 1D-3. Zymate Laboratory Automation System (robotics module) used for specimen preparation. Modules from left to right are controller, general purpose hand, and master station. (Courtesy of Zymark Corp.)

permit aspiration of the full volume required for testing. Cups must be made of inert material so that no interaction takes place with the analytes to be measured. Hall and Whitehead[10] have shown that calcium may be adsorbed onto the walls of polystyrene cups and thus cause low calcium values. Specimen containers should be made of glass or polyvinyl and, if of polystyrene, should be demonstrated not to remove or add material to the serum that would be measured as the analyte. Specimen cups should also be disposable to minimize cost, and their shape should be such that, even without a cap, little evaporation occurs.

Evaporation of specimen from cups in the loading zone may cause analytical errors as great as 50% over 4 h.[4] The extent of fluid loss depends on environmental factors (temperature, relative humidity, and air flow) and the shape of the cup. A fluid surface area that is small in relation to the depth of the specimen reduces the rate of evaporation. Furthermore, evaporation from serum is less than from water-based standards. The absolute amount of fluid lost from a container tends to be less when the cup is only partially filled, although the relative amount may be more and may have a real effect on the concentration of analytes. Standards and control cups should be freshly filled for every run since the concentration of an analyte evaluated against standard solutions that have become concentrated through evaporation will be falsely low.

All cups or tubes containing solution for analysis should be covered until immediately before the specimen is to be analyzed. Many manufacturers of automated analyzers provide covers for individual cups or for part or the whole of the loading zone to reduce losses due to evaporation. Cups may be covered by Parafilm or by caps, e.g., Pre-vap (Sherman-Boosalis) that are placed over the cups and have crosscuts to permit ready entry of a sample probe.

Specimens may undergo other forms of *degradation* in addition to evaporation. Specimens that contain temperature-labile constituents may undergo degradation of such analytes if held at ambient temperature. Other constituents that are photolabile, such as bilirubin, are destroyed by ultraviolet light. The adverse effects of ambient temperature may be guarded against by holding both specimens and standards in a refrigerated loading zone, but no automated analyzer yet includes such refrigeration as an integral unit. Photodegradation is reduced by the use of semi-opaque cups and smoke- or orange-colored plastic covers placed over the specimen cups. Such covers are often an integral part of the analyzer as in the SMAC and ASTRA 8.

The *loading zone* of an analyzer is the area in which specimens are held in the instrument before they are analyzed. The holding area may be in the form of a circular tray (Hitachi 705 and ASTRA), a rack or series of racks that can be built into a form of cassette (ACCLAIM), or a serpentine chain of containers (DEMAND) into which individual tubes are inserted. When specimens are not identified automatically, they must be presented to the sampling device in the correct sequence, i.e., according to a loading list. The sampling mechanism determines the exact volume of sample removed from the specimen.

It is sometimes important that specimens be advanced into the analyzer at a uniform rate, since in some instruments such as the basic AutoAnalyzers, duration of specimen presentation may influence the volume of sample. The design of the device (tray, rack, or chain) for holding specimens is often specific for a specific instrument. Systems use trays that hold from 20–168 specimens to be analyzed at one time. (See Appendix 1D-I.) Operator intervention is usually required to introduce another tray or to add more tubes.

For most analyzers, specimens for a second run may be prepared on a separate tray while one run is already in progress. This permits machine operation and human actions to proceed in parallel for optimum efficiency. In some analyzers, specimens may be added continuously by the operator as the specimens become available. A desirable feature of any automated analyzer is the ability to insert new specimens ahead of specimens already in place in the loading zone. This feature allows the timely analysis of specimens with a high medical priority when they are received in the clinical laboratory. When specimen identification is machine-read, it is easy for the operator to reposition specimens in the loading zone; however, when specimen identification is tied to a loading list, insertion or repositioning of specimens must be accompanied by revision of the loading list.

Removal of Protein and Other Interferents

Removal of proteins and other interferents is often necessary to assure specificity of an analytical method. Several approaches have been used.

Dialysis is a process by which constituents of low molecular weight are separated across a semipermeable membrane from compounds with a high molecular mass, e.g., proteins. Dialysis is used in some continuous flow analyzers to deproteinize serum and thus prevent precipitation of proteins by denaturing reagents in chemical reactions, to remove optical interference caused by macromolecules, and to extend the life of reusable immobilized enzyme reagents.

In the basic AutoAnalyzer model a Cuprophan sheet is stretched between two plastic plates with matched spiral grooves, thus creating a channel on either side of the membrane. A diluted sample stream ("donor stream") moves on one side of the membrane while a "recipient" stream (generally one of the reagents or a saline solution) circulates through the grooves on the other side of the membrane. To achieve optimum transfer of solute, the two streams must flow concurrently at equal rates.

The quantity of solute that passes through a membrane is determined by the concentration gradient across the membrane, the duration of contact of the two solutions, the area of contact, and the temperature, as well as by the thickness and porosity of the membrane. As solutes cross the membrane, their concentration in the donor stream decreases as does their rate of transfer, so that even if the membrane area is doubled, the quantity of solute transferred is not proportionally increased. Other factors that influence the rate of transfer are the size and shape of the molecules, their electrical charge, and the composition of the fluids on either side of the membrane.

Continuous flow systems are very dependent on maintenance of uniform conditions. When specimens differ in character, as is the case with diluted serum or standard solutions that contain no protein as opposed to serum specimens with relatively high protein content, the rate of dialysis of solutes may also differ.[2]

To increase sensitivity in continuous flow systems, the flow rate of the liquid streams can be decreased, and more concentrated samples and thinner membranes can be used. In contemporary dialyzer plates, the grooves are shallower and shorter (by a factor of 10:1), a design which leads to reduced sample interaction and carryover. Changes in temperature are presumed to be slight during the short transit time of solutions through the new dialyzers, and in the recent generation of continuous flow systems (SMAC II) temperature control of the dialyzer is not even provided.

Membranes age with use and time, and protein deposits reduce the quantity of material that crosses the membrane, so that membranes in the basic AutoAnalyzer formerly had to be changed at intervals of one week. In recent models of the SMAC, its computer alerts the operator to investigate the need for membrane replacement, which is typically required every two weeks.

Column chromatography is used in the DuPont *aca* to remove compounds that would adversely affect test reactions. Gel filters, ion exchangers, DEAE, or resins are used to eliminate organic salts and interfering compounds as well as proteins and to separate isoenzymes. Columns are used for amylase (to remove glucose) and for creatine kinase MB and lactate dehydrogenase LDH-5 (to remove other isoenzymes). In all instances, the specimen and its diluent(s) are injected through the individual columns by the dispensing pump.

Other separation systems are used in the analyzers using dry chemistry. In the Kodak Ektachem, the top spreading layer of the test slide separates low-molecular-weight compounds from large-molecular-weight interferents. In the BMC and Ames dry chemistry devices, a film membrane is used to exclude blood cells that interfere with the analytical procedures.

Sample Transport and Delivery

The method of sample delivery and transport into the analyzer is the major difference between continuous flow and discrete systems. In continuous flow systems, the sample is aspirated through the sample probe into a continuous reagent stream, whereas in discrete analyzers the sample is aspirated into the sample probe and then delivered, often with reagent, through the same orifice into a reaction cup or other container.

Continuous Flow Analyzers

Technicon Instruments Corporation pioneered the use of peristaltic pumps and plastic tubing to advance the sample and reagents. The peristaltic pump is also used in many analyzers with ion-selective electrodes. Peristaltic pumps trap a "slug" of fluid between two rollers that occlude the tubing. As the rollers travel over the tubing, the trapped fluid is pushed forward and, as the leading roller lifts from the tubing, is added to the fluid beyond the pump. To assure proportionality between calibrators, controls, and specimens, the pump must act uniformly on the sample tube and the roller speed must remain constant. Although polyvinyl tubing stretches with use, changes in flow-rate over the time of a typical run are minimal. On a short-term basis, minor changes in proportionality between standards and unknowns are corrected by recalibration approximately every 20 min.

For uniformity of sample aspiration, the length of time the sample probe remains in each specimen cup (dwell time) must be the same, and the wash time, or time between specimen aspirations, must also remain constant. An air bubble is aspirated into the sample line as the sample probe lifts out of the specimen. In AutoAnalyzers, the sample probe then dips into a water reservoir. An air bubble is again aspirated into the sample line as the probe lifts out of the water and before it dips into the next specimen. In all continuous flow systems, the air bubble between specimens and additional bubbles introduced through a separate tubing scour the inside of the tubing, separating specimens and reducing carryover from one specimen to the next specimen. Carryover occurs because the flow of sample in contact with the tube walls is slower than the flow of sample in the center of the tube. The water rinse incorporated in the AutoAnalyzer allows reduction of carryover between serum samples beyond that achieved with an air bubble alone. In the SMAC series of continuous flow analyzers, the sample probe dips three times into a water reservoir between serum specimens, aspirating a bubble of air between each dip. This introduction of air bubbles is very effective at reducing carryover between specimens.

The original AutoAnalyzers used mechanical cams to initiate and control sample aspiration and wash cycle. Current models of the AutoAnalyzer and other continuous flow analyzers use electronic timers. The number of specimens analyzed per hour is governed by the timer. In AutoAnalyzers, the timer determines both the time interval for sample aspiration and the duration of the wash cycle between specimens, i.e., the sample-to-wash ratio. Sample-to-wash ratios of 2:1 are common, although ratios of up to 9:1 are possible. The ability to change the sample-to-wash ratio allows flexibility in setting up parameters for analyses. Generally, a lower ratio allows greater removal of residual sample in the sample line by lengthening the wash cycle and thus reducing specimen interaction and cross contamination.

In the SMA and SMAC series of analyzers, a single sample is aspirated and diluted. The diluted sample is subsequently split into several different sample streams, one for each test or blank. Improvements in the performance of sample aspiration and reduction in carryover in continuous flow analyzers have been obtained by (1) ensuring more uniform delivery of fluids by the use of additional rollers of larger diameter and by the elimination of the rocking motion of the earlier models when the roller lifts from its platen; (2) introducing air segments into the sample line at a rate of one every 0.67 s; and (3) electronic timing of sample, air, and water aspiration. These changes have reduced contamination in SMAC systems to such a degree that 150 specimens per hour may be analyzed.

Sample dilution. The volume of all solutions pumped in a continuous flow analyzer is determined by the inside diameter of the tubing in the pump. In the AutoAnalyzer, most tubing is made of polyvinyl. Although the inside diameters of the available pump tubings range from 0.005–0.110 inch (0.13–2.79 mm) to give flow rates of from 0.015–3.90 mL/min, tubes of different diameters can be used in the same pump; this is possible because the thickness of the tube walls is the same and they are all compressed to the same extent in the pump.

In AutoAnalyzers, a manifold is assembled from tubing of the desired diameters to assure proportional mixing of sample and reagents. The nominal flow rates are used as a guide to select the correct tubes. The manifold is stretched and secured between plastic holders (end-blocks) and fittings. The SMA and AutoAnalyzer II have end-blocks that accommodate up to 28 tubings. For the SMAC series, tubes are mounted on separate pump modules. Stretching of tubings, which occurs as they age, is most marked in large-diameter tubings. Proper tension can be restored by adjusting the end-blocks of the manifold. When this adjustment no longer assures proportionality, the tubings must be replaced, preferably all at one time. An experienced operator can recognize when it is necessary to replace tubings, and this knowledge can be of benefit in the preventive maintenance program for the instrument. Recent models of these systems (SMAC II) provide computerized monitoring to detect declining hydraulic pump performance and to warn the operator to initiate corrective maintenance.

Although standard solutions frequently lack protein and the viscosity of the fluids in the same tube (standards and serum specimens) may be different, flow rates are basically constant since the peristaltic pumps used are positive displacement systems. However, Technicon now provides replacement polyvinyl tubing on the basis of tested flow rates to make calibration easier. Special tubing (Solvaflex) that is chemically resistant must be used to pump organic solvents, and Acidflex tubing must be used to pump strong acid solutions.

Discrete Processing Systems

Positive liquid-displacement pipets are used for sample handling in most discrete automated systems, including centrifugal analyzers. In most batch and random-access systems and in some multitest units, specimens, standards, and controls are delivered by a single pipet to the next stage in the analytical process. An exception to this is the ASTRA, in which two pipets are used, each dedicated to delivering samples to a separate group of test channels. In effect, two multitest systems are incorporated into a single housing.

A positive-displacement pipet may be designed for one of two operational modes—either to dispense only aspirated sample into the reaction receptacle or to flush out sample together with diluent. Both systems use a plastic or glass syringe with a plunger whose tip is usually made of Teflon. Careful control of the speed is required for both aspiration and delivery in these devices and for the transition between the two steps, because abrupt changes in velocity lead to inaccuracy and imprecision. Nevertheless, the cycle time may be as short as 3 s. The cycle time of the sample pipet largely governs the throughput of the analyzer. When a liquid, e.g., diluent or reagent, is used as displacement medium, liquid-displacement pipets are capable of highly reproducible measurement of serum. However, when air is the displacement medium, viscous fluids may be measured with less accuracy. This inaccuracy becomes apparent when sera containing large amounts of lipids or proteins are pipetted and the analytical results are compared against standards that are water based.

Pipets may be categorized as fixed, variable, or selectable volume. Selectable-volume pipets allow the selection of a limited number of predetermined volumes. Variable systems may allow as many as a hundred 1-μL increments to be used in a pipet with a maximum delivery volume of 100 μL. The size of the sample has an important influence on the reagent volume that may be used in a system since the ratio of sample to reagent typically affects the dynamic range (linearity limit) for a particular analyte. Generally, pipets with selectable volumes are used in systems that allow many different applications, whereas fixed-volume pipets are usually used for samples and reagents in instruments dedicated to performing only a small variety of tests.

Inaccuracy and imprecision for delivery of samples from pipets should not exceed 1%. Similar performance is desired for reagent delivery. The accuracy of the pipetting must be verified periodically and such checks are a key factor in assessing the quality of instrument performance. The accuracy and reproducibility of pipets in automated instruments are checked in the same way as they are for pipets in manual systems. (See Chapter 1A.)

Carryover

Carryover between specimens should be kept as small as possible. The extent of carryover can be determined by the procedure described in Chapter 2D, p. 421. The effects of carryover can be reduced by good engineering practices, but it is also possible to correct mathematically for actual interaction. In the SMAC an average carryover factor is provided by the manufacturer at the time of instrument delivery and can be modified by the user under actual operating conditions. Excessive carryover is flagged by the computer as a warning message to prompt corrective action by the operator.

Most manufacturers of discrete systems reduce the carryover through setting an adequate flush-to-specimen ratio and by incorporating wash stations for the sample probe. The ratio of flush to specimen may be as much as 4:1 to limit carryover to < 1%,[5] although recent advances in materials and dispenser velocity control have permitted lower ratios. Appropriate choice of sample probe material, geometry, and surface conditions will minimize imprecision and inaccuracy.

Carryover has been reduced in some systems by flushing the internal and external surfaces of the sample probe with copious amounts of diluent. The outside of the sample probe is wiped in some instruments to prevent transfer of a portion of the previous specimen into the next specimen cup. A recent innovation used in the Technicon RA-1000 uses minute amounts of a fluorocarbon fluid to renew a coating on the probe tip. This coating prevents the sample from adhering to the underlying material. A drop of circulating fluid is dispensed after each sample so that carryover is reduced to 0.0125%.

In discrete systems with disposable reaction vessels and measuring cuvets, any carryover is entirely caused by the pipetting system. In instruments with reusable cuvets or flow cells, carryover may arise at every point through which samples pass sequentially. Disposable sample-probe tips can eliminate both the contamination of one sample by another inside the probe and the carryover of one specimen into the specimen in the next cup. Such an approach is used in the Ektachem analyzers, in which a separate disposable tip for each sample is automatically mounted onto a single, positive-displacement pipet. In these systems, a sample is partially expelled and the resultant pendant drop of serum on the tip is touched to the spreading layer of the test slide, where it spreads and subsequently diffuses into the next layer, consisting of a dry reagent matrix. Although the imprecision and inaccuracy of sample dispensing may be as high as ±2%, the volume of fluid that passes to the area exposed to the photometric light beam remains constant, and the resulting final precision is equivalent to ±0.2%. Since a new pipet tip is used for each specimen, carryover is totally eliminated.

Reagent Handling

Most automated systems use liquid reagents stored in plastic or glass containers. The volumes of reagents stored depend on the number of tests to be performed without operator intervention to refill. Where possible, manufacturers use single reagents for test procedures, although two or more reagents may be required for some tests. Some of the newer analyzers, e.g., *aca* and Paramax, use reagents in dry pill form (Figures 1D-4 and 1D-5). Others, e.g., Ektachem or Seralyzer, use reagent-impregnated slides or sticks.

For many analyzers in which specimens are not processed continuously, reagents are stored in laboratory refrigerators and introduced into the instruments as required. In the larger systems, as in the SMAC II and Parallel, sections of the reagent storage compartments are maintained at 4–8 °C. Many of the reagents delivered in liquid form by the manufacturers of these systems are stable for 2–12 months. Reagent containers in the systems range from 125–1000 mL. Reagents prepared manually or by automated reconstitution from dry pill or lyophilized powder are stable after reconstitution typically for 2–8 d when stored at 4–8 °C. Some reagents using enzymes to

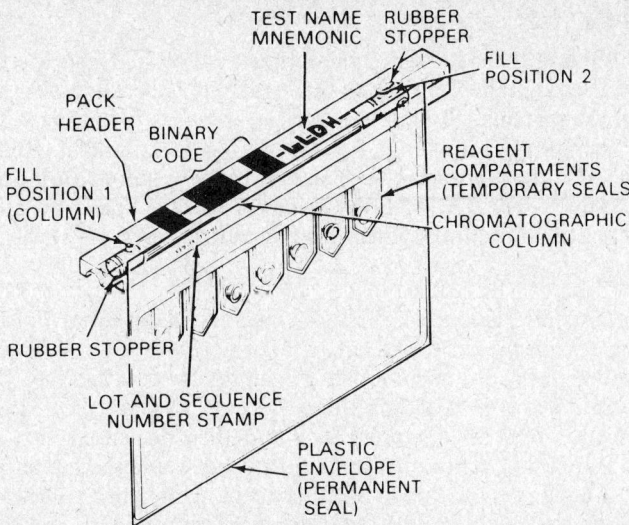

Figure 1D-4. Unit analytical test pack used for *aca* discrete clinical analyzer. Reagents in compartments are in the form of liquids and tablets. (Courtesy of DuPont Company.)

quantitate a substrate, e.g., for glucose and cholesterol, are supplied as single component liquids to eliminate the need for reconstitution. However, typically two separate reagents, each with one year's stability, are provided. When the reagents are mixed, they are usually stable for 10 d. Prepackaging of liquid reagents provides convenience and independence from variations in water quality that may occur with powdered reagents. This development provides economic choice in a laboratory where test volume is low and long-term stability of reagents yields significant cost savings.

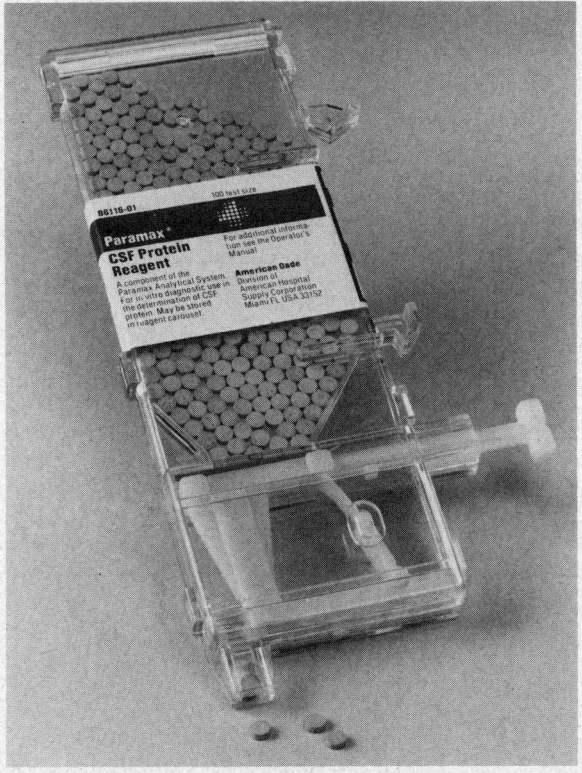

Figure 1D-5. Unit test reagent tablets are stored in a plastic dispenser for use in the Paramax. (Courtesy of American Dade, Division of American Hospital Supply Corporation.)

Liquid Reagent Systems

The ASTRA, Parallel, and SMAC series of continuous flow analyzers are examples of systems using liquid reagents. Large volumes of reagents, adequate for operation of the analyzer for hours or days, are stored in the systems. The containers are reservoirs providing reagent continuously, as in the SMAC, or sample by sample, as for the ASTRA or Parallel. Reagents with limited stability may be prepared in the prerun, set-up phase and may be stored during a run in refrigerated space adjoining the system. The DEMAND conserves storage space by placing concentrated reagents in small containers and diluting them from a central reservoir at the time of dispensing.

Dry Reagent Systems

Reagents for the *aca* and Paramax are initially in a dry tablet form. For the *aca*, tablets are in compartments of a test pack, as illustrated in Figure 1D-4. After sample and diluent have been injected, mechanical breaker-mixers crush the compartments holding the reagent tablets, thus allowing the reagents to mix with the diluted serum to start the reaction. The Paramax tablet is held in a plastic dispenser (Figure 1D-5) and then dispensed into a one-test reaction vessel; fluid is added and mixed by an ultrasonic horn to dissolve the tablet, and sample is then added. Both analyzers illustrate the "unit dose" concept. Unit dose reagents use little storage space and operator time and avoid test-to-test reagent carryover. Dry reagents, although more costly than liquid reagents, have the advantage of long stability and are less likely to be wasted. That is because expiration dates are long and every unit in a batch is used.

Nonliquid Reagent Systems

Unit dose application is carried to a further level of sophistication in systems where no, or very little, liquid is used in the analytic process, as in the Ektachem, Seralyzer, and Stratus. For "color reactions," the Ektachem uses a multilayered, 16-mm square slide (see Figure 1D-6) in which reagents dispersed in emulsions are activated by diffusion of the sample fluid into the layers. From 3–7 layers containing reagents are used for each of the different tests now available. The Ektachem also uses slides for electrometric assays that incorporate miniature ion-selective

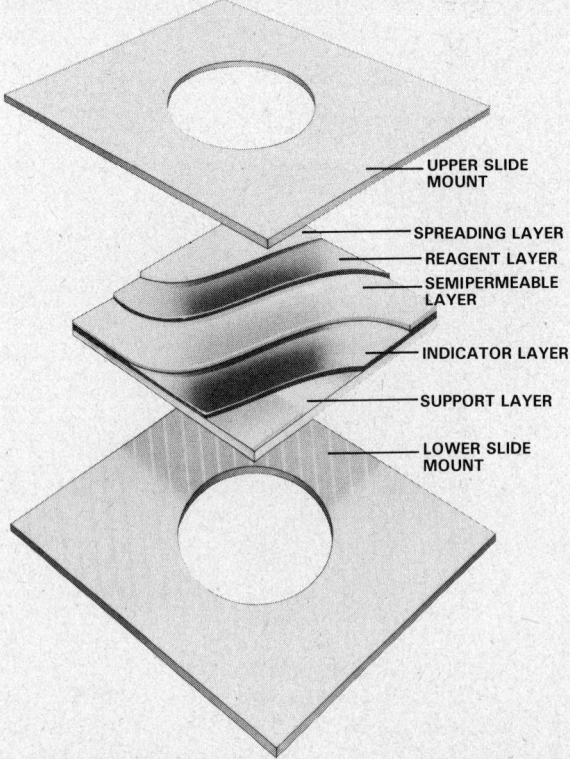

UPPER SLIDE MOUNT

SPREADING LAYER

REAGENT LAYER

SEMIPERMEABLE LAYER

INDICATOR LAYER

SUPPORT LAYER

LOWER SLIDE MOUNT

Figure 1D-6. Schematic, exploded view of the multilayer dry-film reagent used in Kodak Ektachem analyzers. (Courtesy of Eastman Kodak Company.)

electrodes. On these, a reference solution and patient specimen provide fluid that turns the electrodes into electrochemical half-cells. (See Chapter 10.) Slides of this type that measure sodium, potassium, carbon dioxide, and chloride are in current use.

The Seralyzer system has dry reagents impregnated in a spongelike matrix at the end of a plastic strip. Serum, which is diluted for some tests, activates the reaction when it is deposited on the fiber matrix. The fiber sticks contain little residual water, so they can be packaged in glass or plastic jars without special precautions until opened.

The immunoassays of the Stratus use dry reagents in fiberglass "tabs." The analyzer permits both sequential and competitive immunoassays. In the sequential methodology, antigen in the sample, when deposited on the tab, binds with antibody immobilized in a glass fiber layer. An enzyme-labeled conjugate (second antigen) subsequently added to the fiber layer binds with the antibody sites not bound by antigen in the patient sample. Substrate for the enzyme, added at a third station, washes excess labeled antigen and proteins to the outer portion of the tab and reacts with the conjugate, producing a fluorescence signal.

Systems Using Recyclable Reagents

The protein immobilization used in the Stratus has an analogy in immobilization of reagent enzymes in continuous flow (SMAC) and stopped flow (ASTRA) systems and of reagent antibodies in continuous flow immunoassay (ARIA II). In these systems, the fixed enzyme or antibody reagent is presented to the analyte for each successive sample. In the SMAC glucose assay, hexokinase and glucose-6-phosphate dehydrogenase are bound to the inner wall of a coiled tube. Several thousand tests can be performed with one coil during a 4–5-week period. Coils for enzymatic measurement of uric acid are also available. In the YSI Glucose Analyzer, enzymes for measurement of glucose are immobilized on membranes of sensing electrodes. The reaction products are then measured by the sensing device.

In the ARIA II, antibody is covalently bound to a chamber through which antigen (analyte) and competitive antigen pass. Separation of free from bound antigen is accomplished by a buffer in the first phase; then the immobilized antibody is cleaned of bound antigen by another buffer before the next sample enters the chamber.

Reagent Delivery

Liquid reagents are taken up and delivered to mixing and reaction chambers either by pumps through tubes or by positive displacement syringe devices. In continuous flow systems such as SMAC, a peristaltic pump and manifold both deliver fluids and determine the proportion of reagents to sample. Delivery is accomplished by intermittent pressure of forward-driving rollers of the pump against a manifold of tubes lying against a platen, as explained previously (p. 244). In the ASTRA, a stepper-motor is used to drive a pump that functions in the same way as the peristaltic pump of the AutoAnalyzer by alternately compressing and releasing a single reagent tube. In most high-throughput automated analyzers, reagents and diluent are drawn up from bulk containers through tubes and the sample from the specimen cup through the aspirating probe.

Syringe devices are common to many automated systems for both reagent and sample delivery. They are usually positive-displacement devices and the volumes of reagents delivered may be changed. Reproducibility of delivered volume is frequently 1% with an accuracy within 1%. In those analyzers where more than one reagent is acquired and dispensed by the same syringe, washing or flushing of the probe is essential to prevent reagent carryover that may be deleterious to successive analytic steps in the same assay or to successive and different assays.

Chemical Reaction Phase

Concerns related to the chemical reaction phase and the measurement of the reaction are addressed in the design of every automated analyzer, but the manner in which each aspect is handled is ordinarily and intimately dependent on one or more of the other factors involved. Five concerns are to be considered: (1) container in which the reaction occurs, (2) mixing and transport of reactants, (3) thermal conditioning of fluids, (4) timing of reaction(s), and (5) measurement of the reaction.

In *continuous flow systems*, the reactants are contained in the tubing; the *mixing* is achieved by tumbling of air-segmented slugs of fluids passing through helical coils; *thermal conditioning* occurs as the stream passes through tubing enclosed in baths set at particular temperatures; *timing of the reaction period* is determined by the distance the stream travels in the tubing and coils between reagent-sample mixing and the measurement station; the *measurement* of the reaction occurs in a stationary flow cell. Serum blanking, if required, is done simultaneously in a parallel channel.

Discrete analyzers may use as *reaction containers* either (1) individual (disposable or reusable) reaction vessels that are transported through the system after sample and reagent have been dispensed (*aca*, DEMAND, Parallel) (p. 269), or (2) a stationary reaction chamber as in the ASTRA (p. 261). *Mixing of reactants* is accomplished by a variety of means—forceful dispensing (ABA 100, Abbott VP), magnetic stirring (ASTRA or ACCLAIM), vigorous lateral displacement (RA-1000), vibrating rod (DEMAND, Hitachi 705), or probe (DACOS) to name a few. *Thermal conditioning* is achieved by temperature-controlled air or liquid baths or by heating blocks placed around vessels or chambers. In the case of *aca* and DEMAND, *reaction time* is dependent on the rate of transport through the system to the measurement station or on timed events of reagent addition (or activation) relative to measurement, or on both. For the ASTRA, reaction time is either a function of length of stay in the chamber for the sample-reagent mixture prior to measurement or of the time interval over which continuous measurement is made. In transported-vessel systems, measurement may be made on fluid in the original vessel that has been transported into the measurement station or in a stationary flow cell to which the fluid has been transferred at the measurement station. If the first approach is used, the optical quality of the reaction vessel is an important consideration; if the reaction vessel is reusable, it must be cleaned well before it is recycled. Obviously, cleansing of a stationary flow cell or reaction chamber is equally important to minimize carryover.

Mixing, so easy as a manual process, is a difficult process to automate. The tumbling action of an air-segmented stream in the helical coil (SMAC) is analogous to mixing by repeated manual inversion. Vibrating rods (DEMAND) mimic glass stirring rods in manual use; however, they must be cleaned and dried before transfer to the next reaction vessel. Magnetic-bar stirring is conveniently adaptable only to stationary reaction chambers (ASTRA) or to reusable reaction vessels (ACCLAIM). Vigorous lateral displacement (RA-1000) or forced fluid dispensing (ABA 100 and VP) requires special design of recipient reaction vessels to maximize the mixing. The *aca*, with its unique, flexible test pack, utilizes breaker-mixers that can selectively break open, by compression, internal reagent compartments and mix by patting the fluid contents. The nature of systems such as Ektachem, Seralyzer, and Stratus obviates the need for mixing since the serum completely interacts with the dry chemicals as it flows through the matrix of the reaction unit.

Thermal conditioning, overall, is essentially a matter of establishing a controlled temperature environment in close contact with the reaction container and efficient heat transfer from the environment to the reaction mixture. In continuous flow systems (SMAC), the environment is a solid (aluminum powder) in intimate contact with glass coils through which the stream flows. Temperatures of 37, 45, or 90 °C are used, depending on the analyte measured in the channel. In the *aca* the analysis compartment is a large air bath heated at 37 °C. In the ACCLAIM and DEMAND it is a water bath at 37 °C, while in the Seralyzer and Ektachem the reaction system is kept in close contact with temperature-controlled metal surfaces.

In current models of centrifugal analyzers, the reaction containers become, in effect, a multicuvet insert to the rotor. A disposable type (for COBAS-BIO) is the model for discussion. The ring consists of 30 optical-quality cuvets projecting radially from the circular collar. The whole assembly fits closely against rotor surfaces that serve as the thermal environment that is established by heaters built into the rotor body and by a temperature-controlled air bath around the rotor. *Reagents* as well as *samples* are dispensed by a syringe device, one cuvet at a time, to separate compartments in the innermost portion of the rotor. *Mixing* of individual reagents and of the reagent mixture with the sample is the result of centrifugal force exerted when the rotor spins. The centrifugal force will cause the contents of the innermost compartment to move and mix with the contents of the next chamber and ultimately to move into the individual cuvets. *Reaction time* is dictated by the length or kind of measurement interval. The *measurement* is generally photometric but may be turbidimetric or fluorometric; the signal from each cuvet is generated with each revolution of the rotor and is subsequently analyzed, integrated, and processed by an internal microprocessor.

Mixing and Incubation in Centrifugal Analyzers

In centrifugal analyzers, the functions of mixing, thermal equilibration, transport, and measurement are integrated into a rotor assembly. The basic principle of operation is illustrated in Figure 1D-7. Within a rotor, a set of cavities radially aligned to a corresponding cuvet creates a set of discrete reaction units. Each specimen of a batch is processed in a separate reaction unit. A light source and a monochromator (or interference filter) are arranged on one side of the rotor so that light passes through the cuvets to a photo-detector on the other side, while the cuvets spin through the light beam (Figures 1D-8 and 1D-9).

Serum and reagents are placed in the innermost cavities of the rotor disk. In most systems, fluids are pipetted into compartments of the rotors outside the centrifugal analyzer and are subsequently mounted into the instrument. As the disk is rotated rapidly about its axis in the analyzer, the reagents and the sample flow upward and over the walls of their cavities and come together in the outer chamber of the rotor. Partial mixing occurs with this transfer of fluids. The rotor is abruptly braked and then accelerated so that the fluids are mixed completely by agitation.

Three recent versions of centrifugal analyzers, COBAS-BIO, Multistat, and GemENI, use disposable rotors with from 20–30 positions. The diameter of the rotors varies from 7.5–15 cm. As in the early centrifugal analyzers, reagents and sera are held in separate sections of a rotor until the rotor is spun at 500–1000 rpm to combine the fluids in the outermost position (cuvet). The rotor unit is mounted and removed from the centrifuge manually. Cuvet pathlengths are 1.0 and 0.5 cm in the GemENI and Multistat, respectively. The pathlength of the COBAS-BIO is not fixed, since it is determined by the volume of dispensed fluids rather than by the separation of cuvet walls. With precise control of the volume of the cuvets by injection molding, the pathlength depends on the fluid volume along the longitudinal axis of the cuvet. (See Figure 1D-9.)

Methods of thermal equilibration differ in each system. To ensure rapid transfer of heat into the fluids, rotors have been manufactured of aluminum or graphite with the basic materials covered with Teflon. The usual heat sources are air-baths or electric heaters mounted directly on the rotor and controlled through slip rings. In GemENI, fluids are heated by intimate contact of the plastic disk with the thermal mass to which it is pressed prior to spinning. In the Multistat, the disk is rotated slowly and heated by an infrared lamp controlled by a thermistor set in the

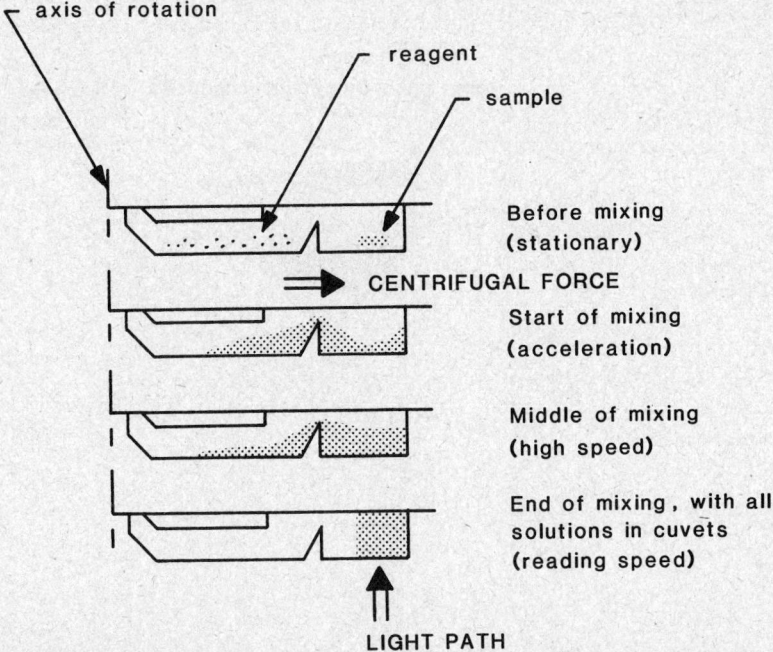

Figure 1D-7. Sequence of transfer of solutions and mixing in a disposable rotor of a centrifugal fast analyzer. (Courtesy of Electro-Nucleonics, Inc.)

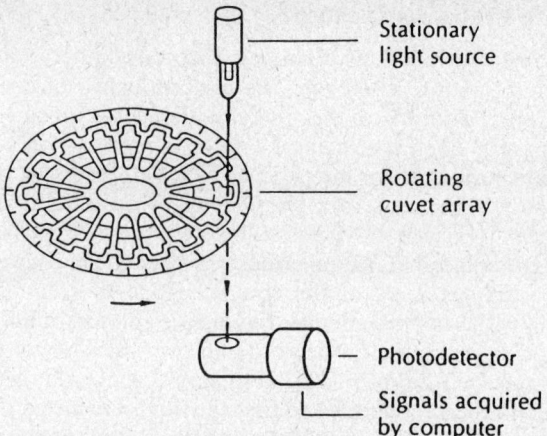

Figure 1D-8. Schematic view of operation of centrifugal analyzer. The cuvet array rotates through a stationary light beam that monitors absorbance of each cuvet at every revolution. One cuvet with a water blank permits setting to 100% transmission while the solid portion of the rotor between every cuvet allows setting of 0% transmission. Absorbance measurements in each cuvet are acquired every 100 milliseconds and averaged to enhance precision. (Courtesy of Electro-Nucleonics, Inc.)

base of the centrifuge rotor. Angular speed is kept low to provide even heating and to prevent the fluids from being inadvertently mixed. In the COBAS-BIO, reagents are warmed by an aluminum rotor/baffle whose temperature is controlled by the flow of thermally modulated air. In the COBAS-BIO, Multistat, and GemENI, the reaction mixture is maintained at operating temperature, while the rotor is spinning, by heated air and by conduction from the mass on which the cuvets rest.

Reaction Vessels and Cuvets in Discrete Systems

In some discrete systems, reaction vessels are reused; in others they are discarded after use. In some systems, the reaction vessel serves as a cuvet, whereas in others solutions are aspirated from the reaction vessel into a separate cuvet for measurement.

The use of inexpensive disposable cuvets followed the development of improved plastics (notably acrylic and polyvinylchloride) and manufacturing technology. Large-scale production

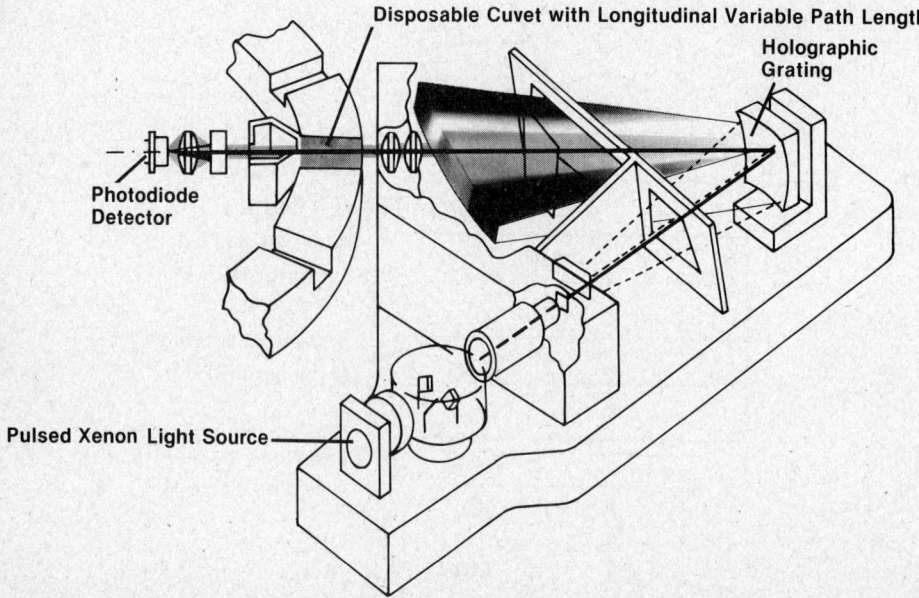

Figure 1D-9. Optical arrangement for COBAS-BIO centrifugal analyzer including a view of the variable-pathlength cuvet, where pathlength depends on volume of dispersed fluids as well as geometry of cuvet. (Reprinted with permission of Roche Diagnostic Systems.)

of cuvets with excellent dimensional tolerances was an essential requirement. The cuvets must be transparent in the spectral range of interest, and some permit measurement at wavelengths as low as 290 nm. The use of disposable cuvets has simplified automation, with elimination of carryover in the cuvets and maintenance of flow cells. There is increasing use of disposable components in discrete automated systems including more complex, higher-throughput, random access systems such as the Paramax, RA-1000, and DEMAND. Some examples of systems using combinations of reusable or disposable reaction vessels and cuvets are discussed below.

In the ACCLAIM, reagents are dispensed into one of 75 polyvinyl reaction cups in a circular table sitting in the same heated water bath as the reagent storage cups. (See Figure 1D-10.) Undiluted serum specimens are added to the reaction cups from 45 s to 11 min prior to aspiration into a series of 4 photometer flow cells. In each cup, a magnet, attracted sequentially by electromagnetic coils at each of the 4 reagent pipetting stations, is used to mix the reagents. Several reagents can be added to each reaction cup so that substrate-initiated reactions can be performed. After the reaction solution has been aspirated into a cuvet, a wash cup descends into the reaction vessel to flush it with a high-pressure, conical water jet oriented to scour the sides of the cup. The wash fluid is aspirated and the cup is air-dried, allowing its reuse. At the end of each day's operation, the reagent storage containers are also flushed with deionized water to prevent crystallization of reagents. Reaction vessels are also reused in the DACOS, but here the reaction vessel is also used as the measuring cuvet (Figure 1D-11). One hundred twenty reaction cups/cuvets are arranged in a large, circular rotor that sits in an air bath at 37 °C. The probe used to dispense reagents into the cuvet is oscillated to mix the reactants.

In the SBA 300 batch analyzer the reaction fluids are aspirated into the flow cell of a spectrophotometer. Since the reaction mixtures pass through the same flow cell, potential for variability due to differences in the pathlength of individual cuvets is eliminated. Carryover is minimized by careful design of the flow cell and by passage of a portion of the reaction mixture, equal to 5–10 times the internal volume of the flow cell, through the cell prior to the actual photometric measurement.

Figure 1D-12 illustrates the Technicon RA-1000 in which an array of 100 cuvets is used. The unit is slued rapidly to mix reagents in the cuvets. Each cuvet is designed so that the fluorocarbon fluid dispensed through the sample probe to prevent carryover (see p. 246) does not interfere with the optical path. At the end of each group of up to 100 reactions, the rotor is manually removed and replaced. Other instruments, such as the Abbott ABA 100 and VP, also use a disposable circular array of reaction chambers/cuvets; in these, high-velocity injection of a reagent into the cuvets is used to mix solutions. In these instruments, optical measurements are made through the water bath in which the cuvet holder sits. The water/cuvet interface reduces some reflections that might otherwise increase imprecision.

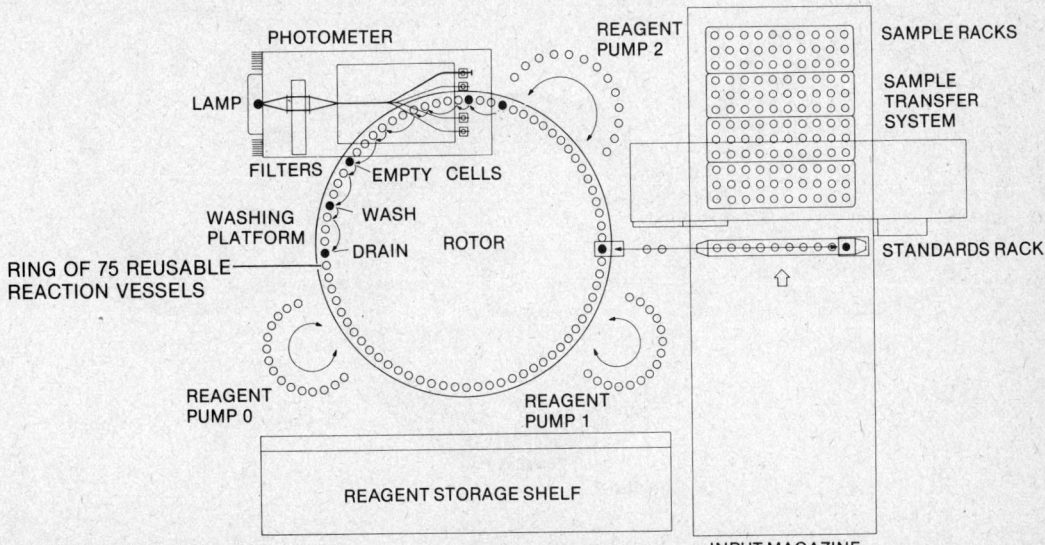

Figure 1D-10. Schematic plan for ACCLAIM system, which employs reusable reaction vessels with reaction mixtures transported to a photometer. (Courtesy of Electro-Nucleonics, Inc.)

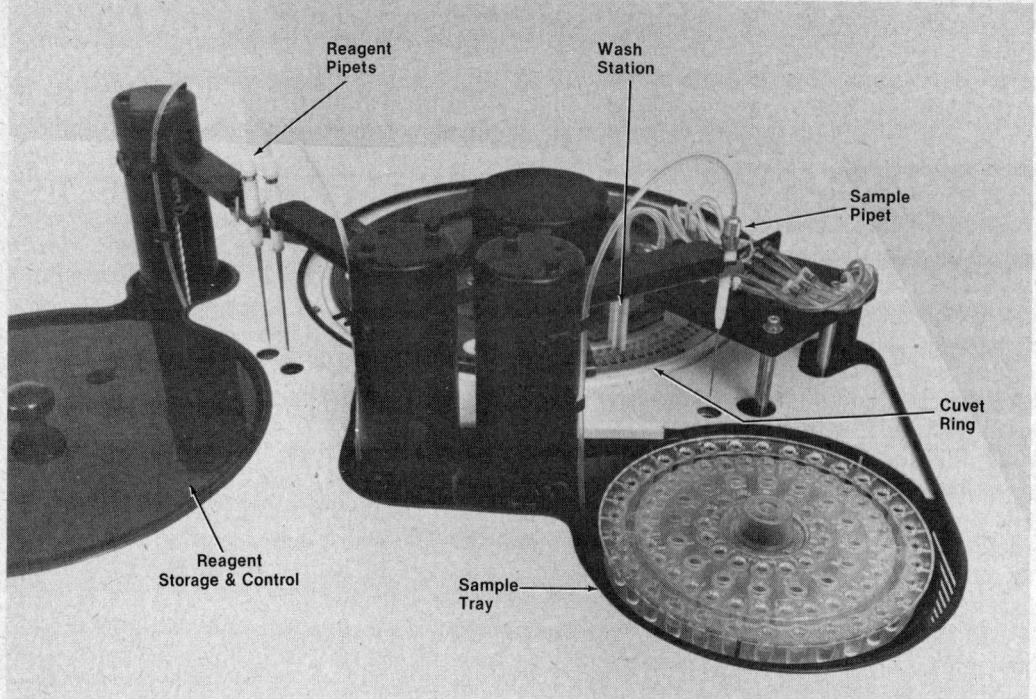

Figure 1D-11. Major components of the DACOS, a random access analyzer. (Courtesy of Coulter Electronics, Inc.)

The unit test pack of the *aca*, made of a flexible ionomer (Surlyn), acts as a reagent container, mixing vessel, and measuring cuvet. The sample is flushed into the pack by one of six diluents dispensed at the initial filling station. The pack is then attached to a continuous chain to transport it through the instrument. A breaker-mixer breaks the sealed reagent compartments and kneads the pack for 25 s to mix reagents with sample and diluent. The pack is transported to a measurement station where a cuvet with a precise 1-cm pathlength is formed by hydraulic pistons

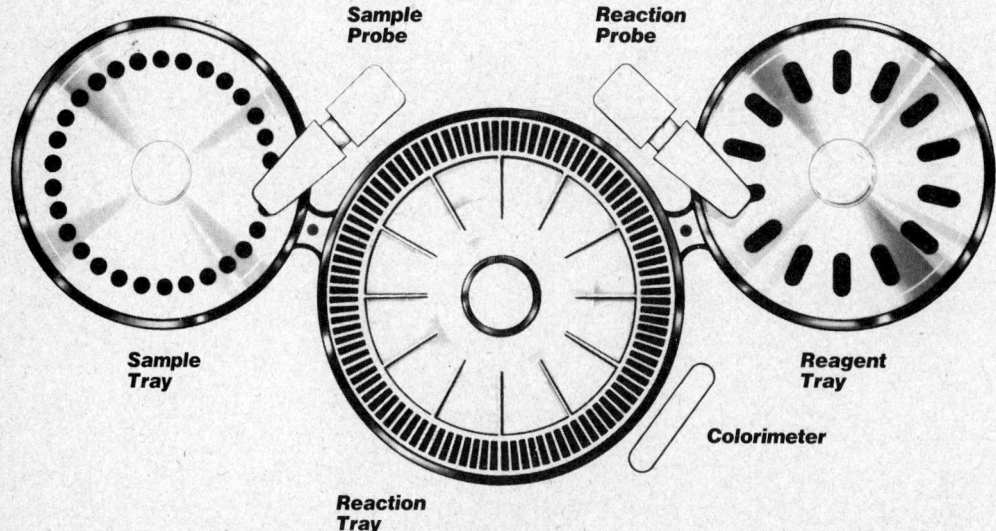

Figure 1D-12. Schematic, plan view of the RA-1000, a random access instrument that employs a 100-cuvet disposable reaction tray. (Courtesy of Technicon Instruments Corp.)

that squeeze the flexible pack between quartz windows. A wetting solution applied to the surface of the windows eliminates distortion of the optical interface at the pack walls. After the photometric measurement is made, the entire pack is discarded automatically.

The dry reagent systems such as the Ektachem, which uses multilayer film slides, and the Seralyzer, which uses impregnated fiber strips, eliminate dispensing and mixing of liquid reagents. Nevertheless, these instruments still require a mechanism to maintain a stable temperature and to provide accurate positioning of the reaction unit for optical measurements. Thermal control in the Kodak instruments is provided through intimate contact of the slides with a heated carousel that supports the slide and rotates it to the measurement station.

Measurement Approaches

Automated analyzers have traditionally relied on principles of absorbance/transmittance photometry for measurement of analytes. However, in recent years several alternative photometric approaches have been adapted to automated analysis: reflectance photometry (Ektachem), fluorometry (Optimate, Advance), nephelometry (ICS), and fluorescence polarization (TDx). Nonphotometric analysis by means of ion-selective electrodes and other electrochemical techniques is becoming increasingly popular. Principles of these measurement techniques are discussed elsewhere (Chapter 1B, Sections Two and Four). Here, the special features and applications of the various approaches to automated analysis are reviewed.

Systems Using Absorbance/Transmittance Photometry

The measurement of absorbance requires three basic components—a radiant energy source, a means of spectral isolation, and a detector. (See also Chapter 1B, Sections One and Two.)

The *radiant energy sources* used in automated systems include tungsten, quartz halogen, deuterium, mercury, and xenon lamps, as well as lasers. In the quartz halogen lamp, low-pressure halogen vapor (such as iodine or bromine) is enclosed in a fused silica envelope in which a tungsten filament serves as an incandescent light source. The spectrum produced includes wavelengths from ~300–700 nm. In the COBAS-BIO, a pulsed xenon lamp is used to provide a high-intensity, long-lived source. Some nephelometric instrument systems use lasers to provide monochromatic light at discrete wavelengths. (See Appendix 1D-I, B.) The ICS uses a tungsten lamp, however.

Spectral isolation is most commonly achieved with interference filters in automated systems. Such filters are now quite inexpensive and only a few are needed in any one instrument, since only a limited number of wavelengths are required for analysis of a large number of absorbing species. Typical interference filters have peak transmittances of 30–80% and bandwidths of 5–15 nm. (See Chapter 1B, Section One.) Although the bandpass of an interference filter can be made as narrow as 1 nm, such narrow bandpasses are not usually necessary, even for kinetic enzyme assays that require high spectral accuracy.[24] In several of the multitest analyzers in which a single cuvet is used for all photometric measurements, the required filters are mounted in a filter wheel and the appropriate filter is moved into place under command of the microprocessor. Monochromators for spectral isolation are more often used in European than in American instruments. Since a monochromator provides a continuous choice of wavelengths and therefore greater flexibility, instruments with monochromators are especially well suited for the development and addition of new assays.

The most popular *detector* used in automated systems is the photomultiplier tube (PMT), although photodiodes have adequate sensitivity for measuring most reactions. PMT's, however, will probably continue to be required to give sufficient sensitivity for fluorescent and chemiluminescent reaction measurements and for measurements that require a fast detector response time.

The combination of components to be incorporated into an automated analyzer must be appropriate to the intended throughput, precision, range of wavelengths, and cost of the instrument. Several approaches have been used for the *electro-optical integration and packaging* in different analyzers. In centrifugal analyzers, a double-beam spectrophotometric system is, in effect, created by constant monitoring of a reference solution (usually water) set at 100% transmittance against the zero transmittance obtained as solid portions of the rotor between cuvets block the light path (Figure 1D-8). Absorbance readings are made at very short intervals, such as every 0.2 ms. A logarithmic amplifier or microprocessor software routine converts transmittance to absorbance. Low-cost analog-to-digital converters with conversion times of fractions of milliseconds, together with low-cost oscillators with timing accuracy of microseconds, are coupled with high-performance computers to accumulate, store, and process optical signals on a real-time basis. The rapid, continuous sampling of optical signals in centrifugal and other analyzers allows optimal timing of measurements, especially when following kinetic reactions.[25]

The DACOS (Discrete Analyzer with Continuous Optical Scanning) uses a concept similar to that used in a centrifugal analyzer to obtain 100% and 0% transmittance. However, instead of cuvets being rapidly rotated with respect to a stationary light beam, in the DACOS the light beam is rapidly rotated relative to a series of 120 slowly rotating cuvets. Eight filters, each with its own detector, rotate around the quartz halogen light source, creating, in effect, eight double-beam photometers. Slip rings are used to provide electrical connections to and from the optical system on the rotor; to reduce electrical interference, the analog signal is immediately digitized.

The Paramax uses liquid-filled light guides to convey energy to and from remotely located reaction cuvets. Light from a single tungsten quartz halogen source is focused by eight radially arranged condensing lenses onto a spinning filter wheel containing several filters (Figure 1D-13). The resulting pulses of light of seven different wavelengths are conveyed by light guides to each of eight photometric read stations. During the 5-s period that each disposable cuvet is at the read station, it receives 150 pulses of light of each wavelength. The transmitted energy is directed through a second spinning filter wheel to eight radially arranged PMT's. For each measured analyte the microprocessor chooses the appropriate pair of wavelengths that have previously been defined for the bichromatic reading and then processes the signals.

Systems that use time-sharing of components of the optical system tend to be less costly and show good performance and reliability. In general, an electro-optical package should offer an absorbance range of up to 2.5 A in order to permit the extended linearity desirable for some reactions. Detection sensitivity should be 0.001 absorbance unit at 1 A, and the noise level should not exceed 0.0005 A at 1 A, in order to assure precision adequate for measuring kinetic reactions. The spectral range typically required is 340–650 nm, but spectral isolation need be no more exact than is necessary for enzyme determinations. For analytes with broad absorbance peaks, interference filters with a bandpass up to 20 nm are satisfactory.

Proper alignment of cuvets with the light path(s) is as important in automated as in manual photometry; stray energy and internal reflections must be kept to acceptable levels. If the light path is not perpendicular to the cuvet, inaccuracy and imprecision may occur, particularly in kinetic analyses since these are not referenced to a standard. Proper alignment is achieved by fixing lenses and light guides rigidly to the analyzer structure while cuvets are transported past the light beam in a uniform manner. Stray energy is reduced to 0.2% or less by careful design of the wavelength isolation filters or the monochromator, or by use of dual filters (as in the

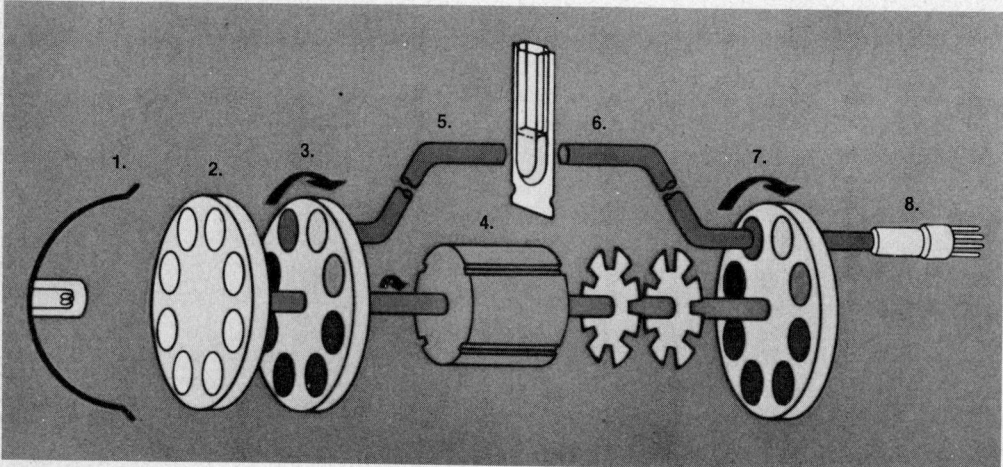

Figure 1D-13. Optical arrangement for Paramax using multiple wavelength. (1) Source with reflector, (2) fixed focusing lens, (3) filter wheel, (4) double-shifted motor, (5) fiber optic bundle, (6) cuvet, (7) filter wheel, (8) photomultiplier tube. (Courtesy of American Dade, Division of American Hospital Supply Corporation.)

Paramax, Figure 1D-13) to increase rejection of stray light. To reduce external reflections and enhance precision, the ABA 100 and Paramax both transport cuvets through liquid while photometric measurements are made.

Other Forms of Photometry

Several forms of photometry other than absorbance photometry are now widely used in automated systems either to increase the sensitivity of measurement or to implement newly developed measuring techniques that generate signals that do not conform to Beer's law. Among these techniques are reflectance photometry, fluorometry, fluorescence polarization, and nephelometry.

In *reflectance photometry*, diffuse reflected light is measured. The reflected light results from illumination with the diffused light of a reaction mixture in a carrier or from the diffusion of light by a reaction mixture in an illuminated carrier. The intensity of the reflected light from the reagent carrier is compared to the intensity of light reflected from a reference surface. Intensity of reflected light is not linear with concentration of the analyte. Two algorithms are commonly utilized to linearize the relation of reflectance to concentration. The Kubelka-Munk equation is applied in the Seralyzer, while the Clapper-Williams transformation is applied in the Ektachem.[27] The electro-optical components used in reflectance photometry are essentially the same as those required for absorbance photometry. In the Seralyzer, a xenon flash tube located in an integrating sphere is used to provide high-intensity pulses of white light. Light reflected from the sphere uniformly illuminates the top of the reagent pad in which the chemical reaction occurs. The reflected light is conducted by a collimating tube through an optical interference filter to a photodiode. At the same time, reference light from the integrating sphere is conducted through the same interference filter to a reference photodiode. The microprocessor then applies the Kubelka-Munk equation to relate reflective measurement to concentration. A description of the Ektachem application of reflectance photometry will be found in the section Examples of Automated Clinical Chemistry Systems, below. At present, reflectance photometry is primarily applied in analyzers using dry, unit dose reagent systems.

Fluorescence is the emission of electromagnetic radiation by a species that has absorbed exciting radiation from an outside source. Intensity of emitted (fluorescent) light is directly proportional to concentration of the excited species, and mathematical conversion of the signal prior to calculation of the concentration is not required. (See Chapter 1B, Section Two.)

Fluorometry is widely used for automated immunoassay. Sensitivity is about 1000 times greater than for comparable absorbance spectrophotometry, but background interference due to fluorescence of native serum may be a major problem. This interference can be minimized by careful design of the filters used for spectral isolation and by selection of a fluorophore whose emission spectrum is distinct from those of interfering compounds. Examples of different optical arrangements are found in the Optimate, Stratus, and the VP series of analyzers.

Right-angle fluorescence measurement is a feature of the Optimate. The system can be used for both absorbance spectrophotometry and fluorometry. The primary path, used for photometric measurements, consists of a quartz halogen lamp, interference filters mounted in a stepper motor-driven filter wheel, the flow cell, and a photomultiplier. A secondary light path, used in fluorometric measurements, starts at right angles to the flow cell, passes through another interference filter mounted in a four-position wheel, and to a second photomultiplier. Current applications of the analyzer utilize a 400-nm primary filter and a 450-nm secondary filter; the additional filter spaces in the secondary path remain open for future applications.

Front-surface reflectance fluorometry is used in the Stratus, where light passed through a 360-nm filter is reflected from a dichroic mirror onto the front of the reagent tab on which the reaction of the antigen-conjugated enzyme occurs. The light emitted from the front surface of the tab travels back through the dichroic mirror and to a photomultiplier through an interference filter that selects only 450-nm light. Fluorescence intensity is monitored over a 30-s interval and rate of fluorescence development is calculated by the microprocessor. The rate of fluorescence change is then compared to the standard curve stored in the microprocessor, and the test value is computed.

Forward-emitted fluorescence is measured in the VP series II-F analyzer with a 55-W tungsten-halogen lamp, a photomultiplier tube, and four filters (excitation, reference, emission, reference) mounted on a rapidly spinning filter wheel.[13] The wheel is coupled with a fixed prefilter with the same bandpass as the excitation filter. By careful design of filter blocking and by comparison of fluorescence to reference signals, a ratio-correcting, in-line fluorometer with a detection limit of 192 pmol of fluorescein is obtained. This sensitivity is adequate for the quantitation of therapeutic drugs by fluorescence immunoassay. As with the Optimate, the approach combines elements for both absorbance and fluorescence detection in a single instrument.

In *fluorescence polarization*, the radiant energy is in the form of polarized light. Measurement is then made of the change in the angle of rotation of polarized light emitted by a fluorescent molecule as described in Chapter 1B, Section Two. Fluorescence polarization is applied in the TDx for homogeneous assays of small-molecular-weight compounds. Radiant energy in the form of polarized light is alternately pulsed vertically and horizontally to impinge on a reaction mixture containing antibody, specimen, and tracer. Free tracer not bound to antibody depolarizes the emitted light, while light emitted from bound tracer is highly polarized. As a result, in this system polarization is inversely proportional to concentration. The optical system consists of a 50-W projector bulb, an interference filter, a fixed horizontal polarizer, and a photomultiplier tube. An electrically switched liquid crystal is located between the polarizer and the reaction mixture. Absence or presence of voltage on the crystal changes the orientation of the polarized light. Because variability of pathlength is immaterial to the analytical principle used, inexpensive borosilicate tubes are satisfactorily used for reaction cells. Use of a liquid crystal instead of mechanical devices to modulate the polarized light significantly improves the reliability of the optical system for this sensitive technique.

Turbidimetry and nephelometry are particularly applicable to methods measuring the precipitate formation in antigen-antibody reactions. *Turbidimetry* measurements have been adapted to centrifugal analyzers because the parallel nature of the analytical process permits simultaneous evaluation of standards and unknowns under identical conditions. Details of turbidimetry and *nephelometry* have been discussed in Chapter 1B, Section Two.

The ICS nephelometric system employs a tungsten-iodide source lamp, with a series of focusing lenses and a photomultiplier tube located at a 70° forward angle to detect scattering of light in the 400–500 nm range. Rate of change of scattered light intensity is monitored; the nonlinear signal is converted mathematically in the microprocessor to achieve a linear relation of rate to concentration.

In the PDQ Laser Nephelometer, a highly collimated beam from a 0.5-mW He-Ne laser is directed to the antigen-antibody complexes. A photomultiplier tube measures the scattered-light intensity passing through apertures placed at an angle of 31° above the incident-light beam.

Systems Using Electrochemical Measurements

A variety of electrochemical methods has been incorporated into automated systems. Perhaps the most widely used electrochemical approach involves *ion-selective electrodes.* These electrodes have replaced flame photometry for the determination of sodium and potassium in many analyzers and have lately found direct application in the measurement of other electrolytes and indirect application in the analysis of several other serum constituents. The operating principle of ion-selective electrodes is given in some detail in Chapter 1B, Section Four. The relationship between ion activity and the concentration of ions in the specimens must be established with standard solutions, and frequent recalibration must be done to compensate for alterations of electrode response.

Electrodes in automated systems may be either of *direct dip* or *flowthrough design.* In the direct dip design, as used in the DYNALYTE and the IONETICS Dual Analyzer, miniature electrodes may be clustered around a reference electrode, and when dipped into the specimen the whole assembly may be agitated to assure contact of each electrode with sample. In the more widely used flowthrough systems (ORION 1020, NOVA), peristaltic pumps are used to move the sample into chambers containing fixed sample and reference electrodes. Whatever the design, the electrodes must remain in contact with the specimen for a sufficiently long time (typically, 20–30 s) to reach steady-state conditions. Response time, therefore, determines throughput of an automated system, and care must be taken to monitor response time and to restore it, when it degrades, by appropriate maintenance procedures.

There remains some controversy in respect to validity of direct (using undiluted sample) and indirect (using diluted sample) measurements with ion-selective electrodes. (See Chapter 1B,

Section Four, and Chapter 10.) Due to differences in electrode design (membrane material as well as type and composition of the liquid junction) and in calibrator employed, results of two different potentiometric systems may not correlate. For particular applications of ion-selective electrodes, see descriptions in Examples of Automated Clinical Chemistry Systems, below.

Ion-selective electrodes coupled with immobilized enzymes represent an extended application. In such systems, the enzymatic reaction with the analyte involves a product that can be measured electrochemically. An example is the YSI Glucose Analyzer. In this system glucose oxidase is immobilized between two membrane layers and catalyzes oxidation of glucose in the fluid, in contact with the membrane, to produce hydrogen peroxide. The hydrogen peroxide in turn is oxidized at a platinum anode. The oxidation results in an electrical current directly proportional to hydrogen peroxide concentration and therefore proportional to glucose concentration.

Coulometry is used in the ASTRA's to measure chloride concentration. The specimen is introduced into an acid solution in the reaction cup, which contains a silver ion–generating electrode and detector electrodes. Solid state electronic circuits monitor the amount of silver ion required to titrate the chloride in solution, and the amount of silver consumed is directly proportional to the concentration of chloride.

Conductimetry is also an electrochemical approach used in the ASTRA systems, specifically for the measurement of urea nitrogen. The specimen is introduced into the reaction cup along with a buffered urease solution. The rate of change of conductivity across gold electrodes embedded in the cup, as urea is converted to ammonium and carbonate ions, is proportional to the concentration of urea nitrogen.

Signal Processing, Data Handling, and Microprocessors

Inclusion of digital computers in automated analyzers has had a major impact on the acquisition and processing of analytical data. Analog signals from detectors are routinely and rapidly (10^{-3} to 10^{-5} s) converted to digital form by analog-to-digital converters. The computer then processes the digital data, by means of algorithms, into immediately useful and meaningful output. Data processing can be as sophisticated as computer hardware and software allow. Real-time acquisition and processing of data by microprocessors have allowed automation of such procedures as nonisotopic immunoassays and reflectance spectrometry because algorithms can readily transform complex, nonlinear standard responses into linear calibration curves. The nature and character of data output can also be as varied as output software and hardware allow and as analytical purposes require. Some functions performed by microprocessors incorporated into automated analyzers are listed in Table 1D-1.

Table 1D-1. SOME SIGNAL AND DATA PROCESSING FUNCTIONS PERFORMED BY COMPUTERS OF AUTOMATED ANALYZERS

DATA ACQUISITION AND CALCULATION
  Acquisition of response signal and signal averaging
  Subtraction of blank response
  Correction of response of unknown for interferences (e.g., Allen-type corrections)
  First-order linear regression for determining slope
    ($\Delta A / \Delta t$) of rate reactions; ($\Delta A / \Delta C$) of absorbance/concentration relation;
    ($\Delta R / \Delta C$) of any response parameter to concentration
  Statistics (mean, SD, CV) on patient or control values
  Mathematical transformation of nonlinear relations to linear counterparts

MONITORING
  Test for fit of data to linearity criteria for calibration curves or rate reactions
  Test of patient result against reference range criteria
  Test of control result against criteria of a quality control standard of performance

DISPLAY
  Accumulation of sets of patient results
  Collation of results for patient-oriented printout
  Provide warning messages to alert operator to instrument malfunction, need for maintenance, or unusual clinical
    situation

1. Microprocessors *command* and *phase* electromechanical operation of the analyzer, thus ensuring that all functions are performed uniformly, repeatably, and in the correct sequence (e.g., transfer of solutions, placement of proper filters, regulation and change in speed of rotation). So important is a microprocessor to the operation of a centrifugal analyzer that benchtop versions of these analyzers could not have been implemented without integration of a microprocessor component. Microprocessor control of operational features of automated equipment, calculation of results, and monitoring of operation are a source of increased reproducibility and consistency of results. These combined features, theoretically at least, allow less-skilled operators to produce results of equal quality to those obtained by more highly skilled and highly paid workers.

2. Microprocessors *acquire, assess, process,* and *store* operational data from the analyzers. Many microprocessors include BITE (Built In Test Equipment) that monitors instrument functions for correct execution and that reacts to improper function by recording the site and nature of the malfunction.

3. Microprocessors enable important *communication interactions between the analyzer and the operator.* Alphanumeric displays on a CRT or on light-emitting diode arrays provide operator-readable information and signals. The signals may instruct the operator to intervene to replenish reagent supply or to empty waste containers, or the signals may warn of operating problems that can compromise test results. The message describing the site and type of problem, together with the manufacturer's service manuals, enables quick identification of problems and their prompt correction. Output data tested against preset criteria can be flagged and displayed for the operator's evaluation and judgment. Such information may specify that linearity of a reaction has been exceeded; that a reaction is nonlinear; that substrate exhaustion has occurred; that absorbance of a reagent is too high or too low; or that baseline drift is excessive. The microprocessor can also take commands from the operator through the keyboard or a touch-sensitive screen. These commands may reprogram certain of the analyzer's functions (e.g., timing interval for a kinetic reaction, or a temperature set point); enter certain values such as standard concentrations; display stored information in raw or processed form; or define the format of printed output.

4. Microprocessors can *communicate to main-frame computers.* A typical interface implemented in microprocessors is a UART (Universal Asynchronous Receiver and Transmitter), which is in turn connected to an RS-232 or RS-422 interface to permit interactive communication to other computer systems. The same interface capability permits transmission by modem (modulator-demodulator) over telephone lines. The telephone line may lead to the manufacturer's central service department, as for Ektachem and DACOS, and thus enhance ability of the on-site operator to service and repair the analyzer. With a bidirectional interface, it is possible to "download" from the mainframe computer such data as specimen identification that govern specimen location in the analyzer's loading zone and to send test results from the analyzer back to the main-frame, either in real-time or by dump from disk storage.

EXAMPLES OF AUTOMATED CLINICAL CHEMISTRY SYSTEMS

The field of automation is changing rapidly. New instruments are introduced continuously, replacing older systems. Any attempt, therefore, to describe all currently used automated systems is unrealistic. However, to provide a perspective on different approaches to the implementation of automation, six representative systems are described. These systems exemplify the various processes required for automation that have been previously discussed. These examples include instruments that employ continuous flow, parallel, batch, discrete, random access, and unit dose concepts in their design. Note, however, that these concepts are not exclusive, and some manufacturers utilize several of these characteristics in one system.

From 23 to 61 assays can be performed on the six systems described. Most measure the same analytes and the methodology is quite similar. The key differences are the way in which the test assays are automated and the throughput, response time, and operating economics of the systems, factors which are briefly discussed in a later section on Instrument selection.

ASTRA

The ASTRA systems are discrete, multitest, selective, sequential analyzers. The ASTRA 4 is a benchtop unit with slots for four analytical modules, while the ASTRA 8 has eight available slots. Most analytical modules can perform one test procedure, but enzyme modules and the cholesterol–triglyceride–uric acid module are designed to perform three different tests and the Na/K module can perform two tests. A possible configuration of 13 channels can be assembled that might consist of two modules for enzymes (three tests per module), one module for concurrent sodium and potassium analyses, and five modules for single assays selected by the purchaser. Reconfiguration must be made by the manufacturer's service staff. Throughput is dependent on the test-mix selection and may be between 585 and 720 tests/h. The recently introduced ASTRA IDEAL consists of two ASTRA 8 units with a communication module link, which allows keyboard selection of an entire 23-test panel, of individual tests, or of any desired test combination. Storage of data and facilities for automatic printout of test results or panels are among the features of this unit.

The ASTRA 8 is described here. It consists of a specimen changer, sample assay compartment, video display, printer, programming keyboard, and microprocessor with a disk operating system. (See Figure 1D-14.) Thirty-eight specimens and 2 calibrators can be placed in disposable cups on the 40-position specimen changer for presentation to a dual sample probe. One of three possible operating modes is selected at the keyboard: (1) a basic mode in which all specimens are assayed on all channels; (2) a program mode in which the operator determines any combination of assays on an individual specimen or group of specimens; and (3) a stat mode that allows interruption of other modes for analysis of a specimen with high priority. The system performs analysis of

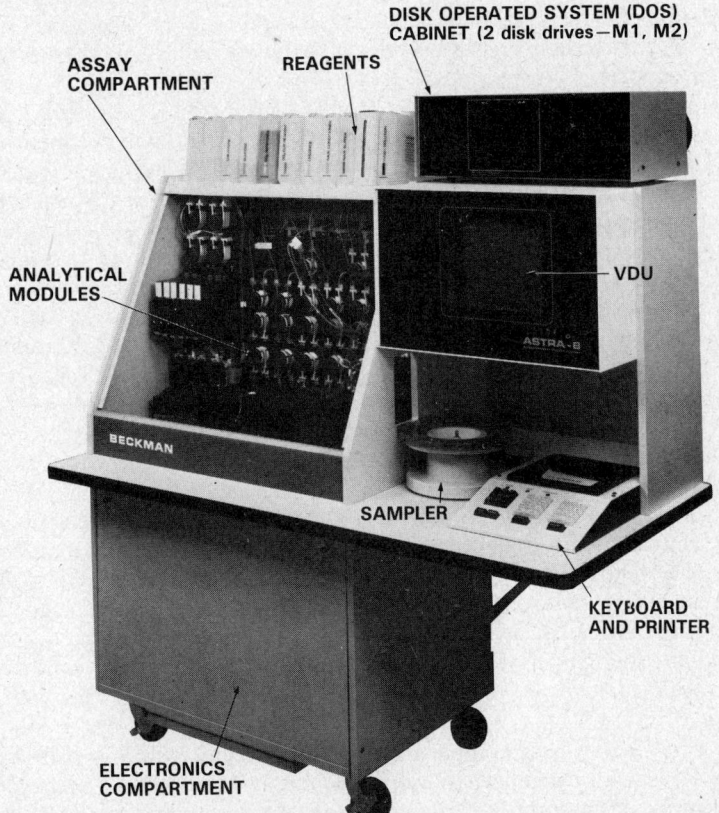

Figure 1D-14. The ASTRA 8 is a discrete, discretionary multitest analyzer in which replaceable analytical modules with similar components are used. Detection methods of modules in the assay compartment differ for different analytes. (Courtesy of Beckman Instruments, Inc.)

the stat specimen, placed in a tray position defined by the operator, as soon as a current measuring cycle is completed and then returns to the operating mode that was interrupted.

A plastic cover placed over the specimen tray reduces evaporation and protects specimens and calibrators from foreign particles. The tray accommodates specimen cups with capacities of 0.25–4.0 mL. Absence of cups and contents, or insufficient specimen, is detected by a test of conductivity between the two elements of the dual sample probe. Absence or insufficient specimen is reported on the printout. The system aborts operation when the calibrator is absent or low in volume. The specimen identification system is specimen oriented rather than patient oriented; tray and cup numbers, date and time, and specimen identification numbers can be entered on the keyboard, but demographic data cannot. The system is, however, fitted with an interface to a main laboratory or hospital computer, so that transmission to computer devices that have capture and merge functions relative to specimen number identification can be usefully implemented. Keyboard entry also allows identification of control specimens, reference ranges, instructions for flagging patient data, and some other functions.

Two probes, moving as a single unit, drop through a hole in the specimen tray cover to acquire specimens. Motor-driven syringes sip the required volume of specimen for the tests programmed plus a small excess, as well as two air slugs that are used to minimize cross contamination. The dual probes move into the analytical section of the instrument and separate to dispense the required specimen volume into the individual analytical modules. The probes traverse the assay compartment and stop to dispense the specimen at two adjacent analytical modules, provided both tests have been requested. The time required for sample acquisition and dispensing is reduced by the use of the dual probe. All of the sample required for all tests is aspirated at the same time; the sample is then dispensed into the reaction chambers without flush. The probes return to a wash cup, and pumps purge the inside and the outside of the sample probes with deionized water containing cleaning solution to prepare the probes for the next sample. The reaction chambers are fixed, cuplike elements that are repetitively filled and drained by peristaltic pumps for successive assays. Specimen handling and chemistry measurement modules are similar in appearance.

Modular design of analytical (and other) units of the ASTRA systems has the advantage that the user can minimize downtime by removal of a malfunctioning element and replacement with a spare. Among Beckman service options are replacement packages, called "Rapid Kits," that a user can stock as his needs seem to justify. The chemistry measurement modules, with the exception of the integrated sensor devices and reaction chambers, are essentially similar. Each has 2–4 peristaltic pumps; one or two to fill the reaction chamber with reagent(s); one to sip excess reagent; and one to drain the chamber after measurement is completed. Accuracy of reagent delivery is not dependent on the delivery pump since reagent volume is fixed by the chamber size and by an overflow device that removes excess reagent with a sipping pump.

Mixing in the reaction chamber is accomplished by a Teflon-coated magnetic bar in the cup that is driven by a stirrer motor (drive magnet) under the cup. For some assays a heater, also under the cup, provides thermal control of 37 ± 0.1 °C. Only cuvets used for kinetic assays are flushed after they are drained; they are flushed twice to prevent carryover. Total carryover in the sample probe and reaction chamber is $<0.5\%$.

Six of the 23 assays available are performed in modules utilizing well-tried electrochemical principles of measurement and engineering that have been offered by Beckman in smaller systems for many years. The remaining available assays are spectrophotometric. Operators of the equipment can maintain the modules very easily.

Liquid reagents and wash solutions are located either on top of or in the sample assay compartment but are not temperature conditioned. Reagents for electrochemical assays are stable at room temperature for over 21 d, for most colorimetric assays for over 5 d, and for kinetic assays for 3 d.

Electronic units in the base compartment of the original ASTRA systems used PROM's (Programmable Read Only Memory) to store machine instruction and user-defined parameters in the 128K memory of an Intel 8080 microprocessor. In addition, recent models have floppy disks to improve user convenience and data backup capacity.

The video display provides communication between the operator and the microprocessor and is used in conjunction with the system keyboard to display instructions as they are programmed. It is at this keyboard that chemistry measurement selection and numerical patient data are entered (i.e., tray number to be programmed, patient ID number, and other data), and a variety of auxiliary functions may be called. An RS-232 port is provided to enable communication with an external computer. The LINK computer interface module for the ASTRA IDEAL also provides a bidirectional interface to allow downloading of test requests and patient data from a host computer.

COBAS-BIO

The COBAS-BIO is a discrete, batch, centrifugal, parallel, fast analyzer. The instrument is a tabletop system that provides a throughput of 200–400 results per hour. It utilizes 30 discrete, disposable, reaction cuvets assembled into a rotor, a grating monochromator, and a single Intel 8080A microprocessor to control the total system (Figure 1D-15).

Specimens are loaded synchronously, with a printout to identify their positions in the circular specimen tray. Translucent polyethylene cups have a snap top that prevents evaporation of specimen; the sampling probe penetrates the top. Distilled or deionized water, the volume of which is regulated by internal valving, flushes the sample from the sample probe into the rotor. Carryover, which can occur only in the sample probe, is negligible.

Specimen cups can be placed into one of two positions. If cups are in the down position, specimens will be aspirated; no aspiration will take place if specimen cups are in the up position. This permits a degree of selectivity for running different batches of tests on specimens on the same tray. Partial batches can be run, sequentially, on the same cuvet rotor, as long as an adequate number of unused positions are available for blanks, standards, and specimens.

Tests are selected by number at the keyboard of the control panel from a list of as many as 40 tests previously programmed by the user. A specific test is programmed by selection from a menu, displayed on the front panel, of suitable parameters (up to 19) for that test. Parameters include linearity limits, operating temperature (which can be selected from 25–50 °C in 0.1 °C increments), incubation time, pipettor settings, and one of seven report formats. The formats

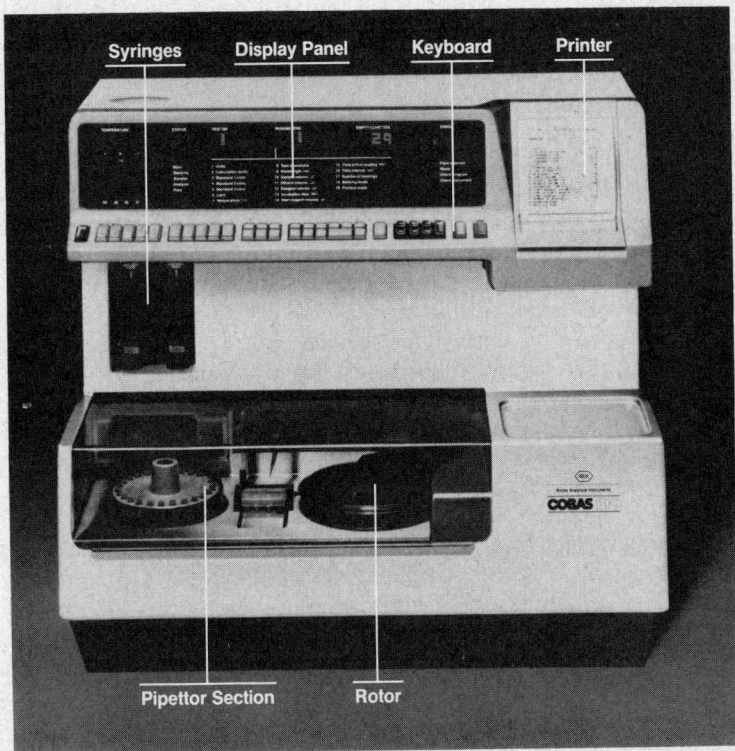

Figure 1D-15. COBAS-BIO—a centrifugal, batch, parallel fast analyzer. (Reprinted with permission of Roche Diagnostic Systems.)

range from those that print final test results to those that print raw analytical data (e.g., absorbance vs time).

Reagents and standards are drawn up from a storage tray placed on the instrument at the time a test is initiated. The tray has wells for two different reagent solutions and three standard containers.

The 30-position disposable cuvet ring is manually mounted on the rotor in the analysis station. Twenty-eight of these positions are available for unknowns as well as standards or controls, while the other two cuvets are reserved for water and reagent blanks. Sequential positions not used in a preceding run are available for use in the next run so long as a sufficient number of them remain for the blanks, standards, and specimens of the next run. Inside the analysis station, an air bath heats and cools the metal surface on which the cuvet ring is mounted. Temperature of the reaction mixture is maintained to ± 0.1 °C. The photodetector is on a hinged arm at the center of rotation in the analysis station.

Sample transfer and reagent transfer are made by two positive-displacement Hamilton syringes. Ranges of adjustment for sample and diluent volumes are 2–80 μL in 1-μL increments. The range for the primary reagent volume is 5–370 μL in 5-μL increments; the range for the secondary reagent volume is 2–80 μL in 1-μL increments. To minimize carryover during fluid aspiration, an air bubble is maintained between the sample and a column of deionized water. Sample washout occurs with the delivery of deionized water, and external cleaning is accomplished by rinsing the sample probe at a wash station. The reagent dispenser requires that a disposable tip be affixed each time a different reagent is used. Prior to the transfer of fluid from the specimen ring and reagent trays, the cuvet ring is spun briefly and optical quality of the cuvets is verified; fluid transfer is then completed within 3.5 min for the first reagent transfer and within an additional 1.5 min for the second reagent transfer.

Different analytical modes, selected in the initial programming of the test configuration, include fixed-point and kinetic modes. A substrate-initiated mode is implemented by automatic stop of the rotor and the subsequent transfer, by the sample pipettor, of the trigger reagent into the side compartment of the cuvet. When spin is resumed, the reaction is initiated through transfer by centrifugal force of the trigger reagent to the reaction cuvet.

Wavelength selection is provided by a holographic grating monochromator with a spectral range of 285–750 nm; wavelength is selectable in 1-nm increments. A pulsed, xenon, flash tube provides a high-intensity, long-lived light source; pulsing is phased to those times when readings are required. A semiconductor photodiode serves as detector.

In contrast to most spectrophotometers, measurement in the COBAS does not depend on fixed, optical path length; the optical path is through the longitudinal axis of the fluid contained in the cuvet (Figure 1D-9) so that the length is defined by the contained volume. The plane of the meniscus is made flat by the centrifugal field and is parallel to the center of rotation. In the application of Beer's law that is invoked, the relationship of absorbance to analyte concentration becomes a function *only* of the volume of sample in which the analyte was added; i.e., calculation factors do not change with change in reagent volume.[3] Furthermore, only minimal error is introduced from reagent- and diluent-volume delivery.[23] An optimum range of absorbance can be achieved by varying either sample, diluent, or reagent volumes.

The loader unit is an integral part of the system. While this arrangement has some advantages in regard to convenience and space saving, it has the disadvantage that only one tray can be loaded in advance of the analytical run.

The microprocessor monitors many system features, such as fluid volume in the diluent reservoir, and alerts the operator to potential errors or to such problems as incorrect reaction direction (e.g., increasing vs decreasing absorbance change), substrate exhaustion, excessive electro-optical noise, or nonlinearity. Coded computer messages can be used for trouble-shooting in conjunction with the service manual and a test board. The operator can do this either independently or together with the manufacturer's service personnel.

Results are printed on a 48-column, thermal strip-printer. The Intel computer has a 64K memory for function control and programming of the system. An RS-232C port allows communication to a central computer.

EMIT technologies are practical on the instrument either with an optional board for 4-parameter, logistic data reduction or through the interface with a microcomputer having appropriate programs for data reduction.

DEMAND

The DEMAND system is a random-access analyzer that utilizes disposable reaction cuvets and concentrated liquid reagents. The system, housed in a floor-model console (Figure 1D-16), performs up to 23 different on-line assays at a throughput of 400 results per hour (up to 600 per hour with the ion-selective electrode module).

More than 100 specimens may be presented for sampling by placing them in cups in numbered positions on a serpentine chain. The chain is enclosed in a compartment kept at high relative humidity to reduce evaporation. Color-coded carriers can be used to identify locations into which standards or controls are placed. The color codes are machine readable and serve as a signal to the analyzer to implement automatically a previously defined calibration procedure or to analyze a control. Specimen identification occurs at the time of entry of test requests on the CRT keyboard by assignment of the specimen to a specific position on the specimen chain.

The system is "specimen oriented" in that every test requested is performed on each specimen before the next specimen is sampled and tested. Profiles with 2–23 components can be preprogrammed; up to 10 profiles can be defined and stored on an 8-inch floppy disk. Specific profiles can then be selected, specimen by specimen, from a menu presented on the CRT display at the time of specimen entry.

Emergency tests can be performed for any one analyte or any group of analytes available on the system by placing the stat specimen in a special position above the specimen loading compartment and then instructing the system, by means of the keyboard, to aspirate that sample as soon as the specimen in process has been completed. Results for up to 23 analytes can be available within 10 min of the stat specimen entry.

Disposable, 10-mm rectangular reaction cuvets, 100 to a pack, are loaded into the lower section of the console. Since up to 10 packs may be loaded, ~2.5 h of operation will elapse

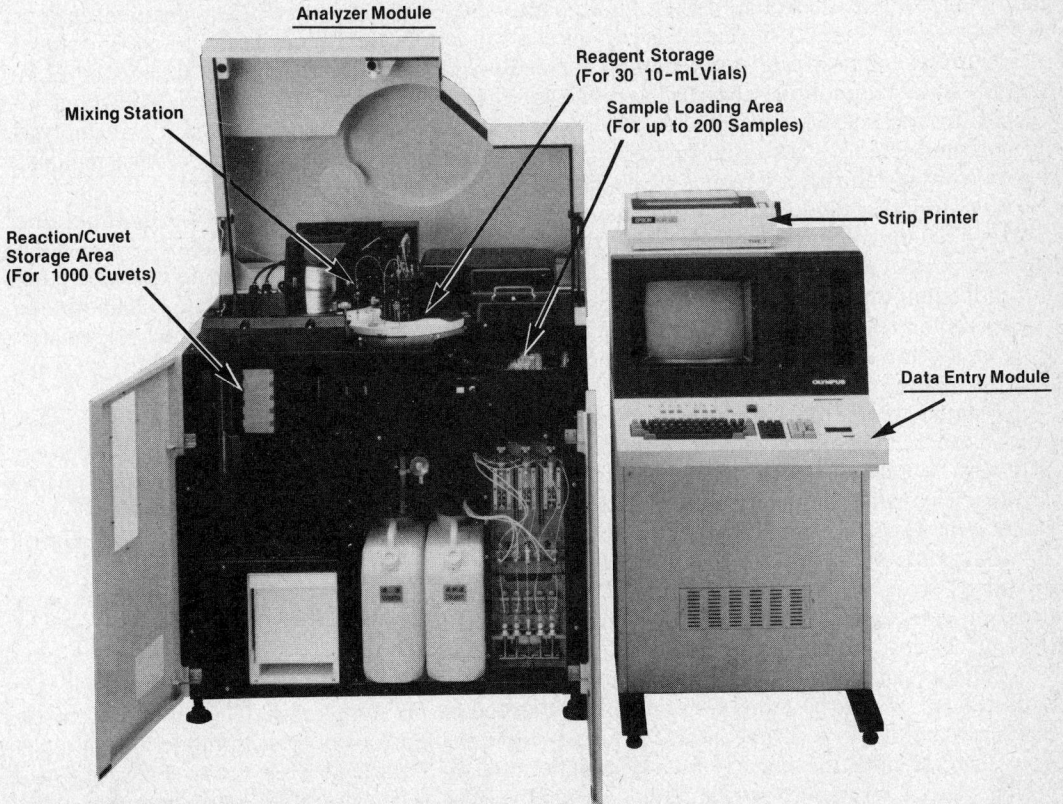

Figure 1D-16. DEMAND—a random access analyzer with major system components shown. (Courtesy of Cooper Biomedical, Inc.)

before the computer signals the operator to replenish the supply. Cuvets are fed by an electrically driven push-bar onto a rotating reaction ring that holds 72 cuvets and indexes every 9 s. At the end of one rotation, used cuvets drop through a chute into a disposal bag. A water bath monitored by thermistors at 37 $\pm$ 0.1 °C maintains the temperature of an aluminum ring in which the reaction cuvets sit.

Sample volumes, which can be programmed from the keyboard, range from 5–25 μL in 1-μL steps and are metered by a digitally controlled positive-displacement syringe. Deionized water is used to flush the probe. A separate wash position, located on the path from sample cup to cuvet ring, supplies deionized water for internal and external washing of the probe. The arrangement reduces carryover to <1.0%. Delivery volumes for reagent concentrate, from 25–100 μL in 5-μL increments, are also set at the keyboard for each of two dispensers. Concentrated reagents packaged in uniform 18-mL vials and stored in a refrigerated compartment on the console may provide for up to 300 assays. Stability of most of the reagent concentrates in the refrigerated compartment is at least 5 d.

In the dispensing step, the concentrate is automatically diluted 3–5 times with prewarmed, deionized water. Reagent containers are rotated on a removable circular tray to the pipetting position, where a cap that has been mounted on each vial to reduce evaporation and contamination is automatically lifted for the aspiration. The physical arrangement eliminates the need for extra tubes, pumps, or valves. Dilution of the reagent with preheated diluent provides reaction aliquots at operating temperatures. To eliminate carryover of reagent, the reagent dispenser probe, like the sample probe, stops at wash stations. Oscillating metal rods provide mixing after each reagent and sample delivery. These rods are also washed and are subsequently dried with forced air at separate stations.

Fifteen fixed photometer stations located around the rotating cuvet ring acquire absorbance data on each cuvet every 36 s. The output of a single lamp, located above the plane of the turntable, is directed down through one of 10 selectable filters in a wheel. The monochromatic beam is sequentially reflected to each of the 15 silicon photodiode detectors by a digitally controlled, rotating, reflecting prism. All 15 photometer stations receive light signals through all 10 filters every 9 s. One of seven microprocessors monitors the filter wheel and cuvet position to identify the appropriate wavelength(s) for analysis and to select only the data required for analysis from the multiple absorbance readings obtained. Bichromatic absorbance readings are selected from all of those acquired to correct for several types of interference, such as hemolysis, lipemia, and icterus. Correction for serum blanks is possible, although serum blanks require a separate cuvet and thereby reduce throughput.

As with other random-access systems, the throughput rate is determined by the fixed indexing time and fixed photometer stations rather than by the length of time for completion of a reaction. Chemistry reactions are monitored every 36 s, an interval that appears adequate to encode data for both equilibrium and kinetic analyses. Up to 14 bichromatic measurements can be made after triggering the reaction. The start-and-stop reading times are user-adjustable in 36-s increments. Rate assays use a multipoint, fixed-time approach,[22] and a line of best fit is calculated by least-squares regression.

Manufacturer's test parameters for each methodology or parameters selected by the laboratorian can be used and stored for each chemistry on an 8-inch floppy disk. The disk also permits long-term storage of patient and QC data. An on-line, multirule Shewhart algorithm provides real-time evaluation of quality control data. Shewhart (Levey-Jennings) plots and quality control summary reports in various formats can be printed out for each test performed by the analyzer.

Data output is available on the CRT or from one of two printers that provide hard copy of various types of information. A simple, 24-column impact strip-printer at the keyboard entry panel provides raw data such as absorbance and reagent blank readings and an "audit trail" of the run. A separate line printer produces patient reports.

Patient demographic information may be entered through the keyboard at the time tests are requisitioned or may be edited into patient files stored on the floppy disk. Chartable reports can be prepared either in real-time, i.e., as results become available, or in a batch mode, i.e., dumped from the disk after the run has been completed and the data have been reviewed on the strip printer output. A report-writing utility program allows the laboratorian to design his own chartable report form.

A bidirectional RS-232C port is available that will accept test requisitions from, and transmit data to, another computer.

EKTACHEM 700

The Ektachem 700 is a random-access system that integrates a multilayer, film-slide technology for endpoint, electrolyte, and enzyme rate measurements within one housing. One slide is used per assay and then discarded. The system is a floor model (Figure 1D-17) with a maximum throughput of 540 results per hour; it uses a CRT with "touch screen" display/input, a keyboard for operator entry, a printer, and the analyzer module. (See Figure 1D-18.)

Ten specimens are placed in a tray quadrant and four of these are assembled to make a complete unit. Specimens are identified at the CRT screen by their position in the quadrant and by labeling the quadrant. Up to 400 test requests can be entered, on a continuous basis, for the specimens mounted in a quadrant prior to the time that the quadrant is transferred to the loading zone. Each quadrant contains disposable pipet tips mounted concentrically with the specimens. An arm in the metering tower that contains the fluid dispensers for specimens picks up each pipet tip as required. (See Figure 1D-18.)

A positive-displacement pump in the metering tower aspirates as many aliquots of specimen as required for the number of tests requested. Endpoint assays require 10 μL and enzyme assays 11 μL of serum. The aspirator dispenser has a positive-displacement piston with an air space between the piston and the actual fluid so that the piston never touches the specimen. Since the metering tips are disposable, carryover is eliminated. Reactions are initiated following diffusion of the specimen into the film layers so that no mechanical agitation is required to accomplish mixing.

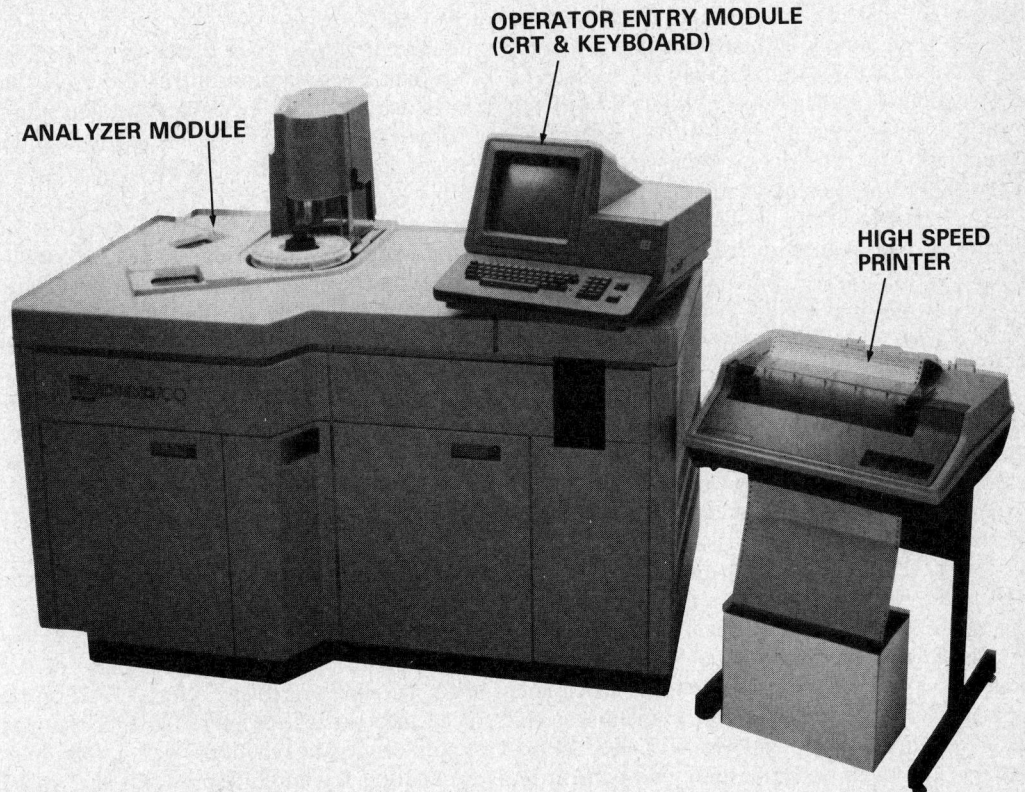

OPERATOR ENTRY MODULE (CRT & KEYBOARD)

ANALYZER MODULE

HIGH SPEED PRINTER

Figure 1D-17. Main modules of the Ektachem 700, a unit dose, random access analyzer. (Courtesy of Eastman Kodak Company.)

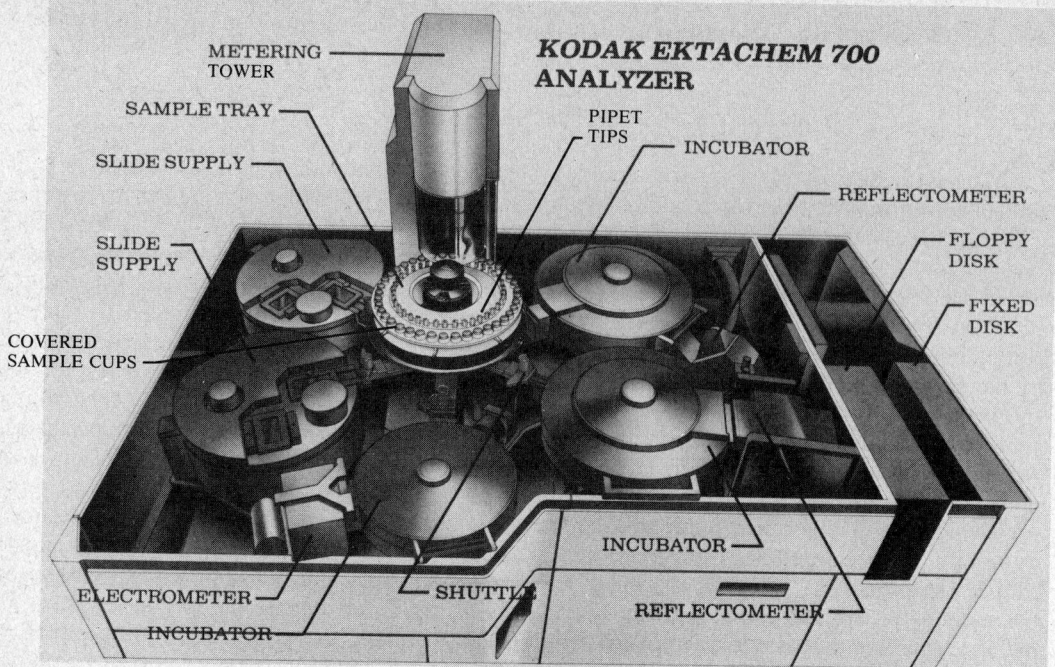

Figure 1D-18. Analyzer module for the Ektachem 700. (Courtesy of Eastman Kodak Company.)

Each specimen cup is covered with a flexible plastic cap to prevent evaporation. The cap is cross-cut so that the sample pipet tip penetrates easily.

The slide supply compartments provide space for 60 cartridges, each containing 50 slides, so that a total of 3000 slides may be in place in the system. These compartments are accessible for reloading from the outside. Some 25 different assays are presently available, and depending upon the workload mix, ~2.5–6 h of operation are to be expected before replacement of slides is required. The cartridges are bar-coded and are automatically identified by an optical reader in the slide supply compartments.

As individual tests are called up for each specimen, a slide drawn from the cartridge in the housing is placed in a distributor arm that moves the slide to a metering station below the specimen tray, where a metered drop of sample is injected on the slide. The potentiometric slides receive, in addition to the patient specimen, a 10-μL drop of reference fluid for a reference electrode; reference fluid is deposited by a separate dispenser in the metering tower. (See Chapter 10.)

After metering, the slide is moved by the distributor arm to the proper incubator station, where it is transferred into the incubator. Each incubator has a different capacity—27 slides for endpoint reactions, 17 for ion-selective electrode measurements, and 24 for kinetic determinations. Time in each incubator is set to be consistent with the analytical requirements; endpoint reactions require 5 min and electrolyte determinations 3 min. Enzyme reactions are monitored approximately every 6 s for 5 min.

Slides that have the built-in ion-selective electrodes are maintained at 25 °C and monitored by an electrometer. Endpoint measurements and kinetic enzyme measurements are monitored by separate reflectometers, each with a permanent reflectance standard for reference. Thermal control for all slides is obtained through contact of the slide with the rotating thermal mass of the incubator; the enzyme and endpoint measurement disks are maintained at 37 ± 0.1 °C. Slides are removed from the colorimetric incubator and placed into the reflectometer so that the first reading occurs at 5 min and every 12 s thereafter. For some endpoint reactions, such as albumin, readings are taken at 2.5 min. A two-point rate determination for amylase uses readings at 2.5 and 5 min.

At the enzyme incubator, slides are read, in situ, by a reflectometer with a 1-σ precision of 0.0004 arbitrary reflectance transmission units to provide high precision. Up to 54 readings can be taken for each slide during a 5-min incubation. The adaptive algorithms that define the number of points selected for calculation of enzyme activity are, in essence, a variable-time analysis as defined by Pardue.[22] This algorithm permits selection of the portion of the reaction-rate curve with maximum rate and linear reaction kinetics as well as monitoring of reactions over a wide analytical range. After the test readouts are completed, all slides drop down chutes into disposal boxes.

Standardization and calibration are required only once per week since temperature and relative humidity controls integrated in the slide supply compartments maintain stable environmental conditions.

Primary instrument controls are activated when the operator touches an appropriate area of the touch-sensitive CRT of the operator entry module. Tests are ordered by means of the same mechanism. The software of the Ektachem 700 permits the operator to program up to 14 customer-selected test panels, configure special report formats, edit test results, and perform diagnostic trouble shooting when the system is inoperative.

A system of 18 microprocessors permits independent control of various functions such as incubator motion, data acquisition, thermal control, and power requirements. Two hierarchical systems form a master computer. They share more than 1 000 000 bytes of random access memory as well as a 20-megabyte Winchester fixed disk, and a 1-megabyte flexible disk. The computer monitors instrument functions and environmental conditions and provides self-diagnostic capability in the event that repair is necessary. Diagnostic messages are displayed on the CRT screen to guide the operator through the problem. If further help is required, a direct telecommunications link to the manufacturer is available. The link permits additional electronic analysis through an analyzer-to-computer connection and facilitates prompt identification and correction of problems. A bidirectional interface for intercommunication with a house computer is an optional feature.

PARALLEL

The Parallel analytical system is a floor-model, 30-channel selective and discrete analyzer (Figure 1D-19). It has multiple analytical stations and generates results at a maximum rate of 7200 per hour. As with many discretionary systems, actual throughput depends on the test repertoire.

The system consists of five major components: the chemistry console, the electronics console, the database console, the video terminals, and a printer. Each of the individual channels on the chemistry console contains its own sample and reagent dispensers, mixer(s), measurement station(s), and wash position. The system has substantial computer capacity to accumulate, process, and display data. User-friendly menu screens contribute to ease of operation.

A key feature of the analyzer is its test selectivity. Any one or a combination of the 30 available tests can be requested as a test schedule. A desired test schedule can be entered into the system by one of four options: (1) readout of test schedule and patient's accession number from a bar-code label affixed to the specimen; (2) download of accession numbers and corresponding test schedules from an external laboratory computer; (3) direct entry of the test schedule from the chemistry console's STAT pad; or (4) keyboard entry of the test schedule at the video display unit (VDU).

Option 1 provides the highest throughput. The required bar-code labels must be generated external to the Parallel System, e.g., by a laboratory computer or special printer. A labeled specimen is centrifuged and the tube is placed in the specimen transport chain or serum is transferred to an analyzer cup, which is then set into the top of the original 10-mL specimen tube in the instrument's specimen transport chain. When the tube's label is read at the bar-code reader in the chemistry console, the test schedule is passed to the analyzer. (See Figure 1D-1,C.) Under Option 2, patient accession numbers and test schedules are sent from the laboratory computer directly to the computer on the Parallel analyzer. Compared to continuous operation (Option 1), the system in this mode causes specimens to be analyzed as groups. The group size can range from 1 to 450 specimens depending on the schedules downloaded from an external laboratory computer. Option 3 allows the operator either to order a STAT schedule or to override an encoded test schedule from his keypad. Option 4 is used when the analyzer is not linked to a laboratory computer. Additional computer software is provided to permit input of patient demographic information. A collated final report is printed directly from the Parallel analytical system.

Specimens identified by bar-code can be placed randomly on the analyzer's specimen transport

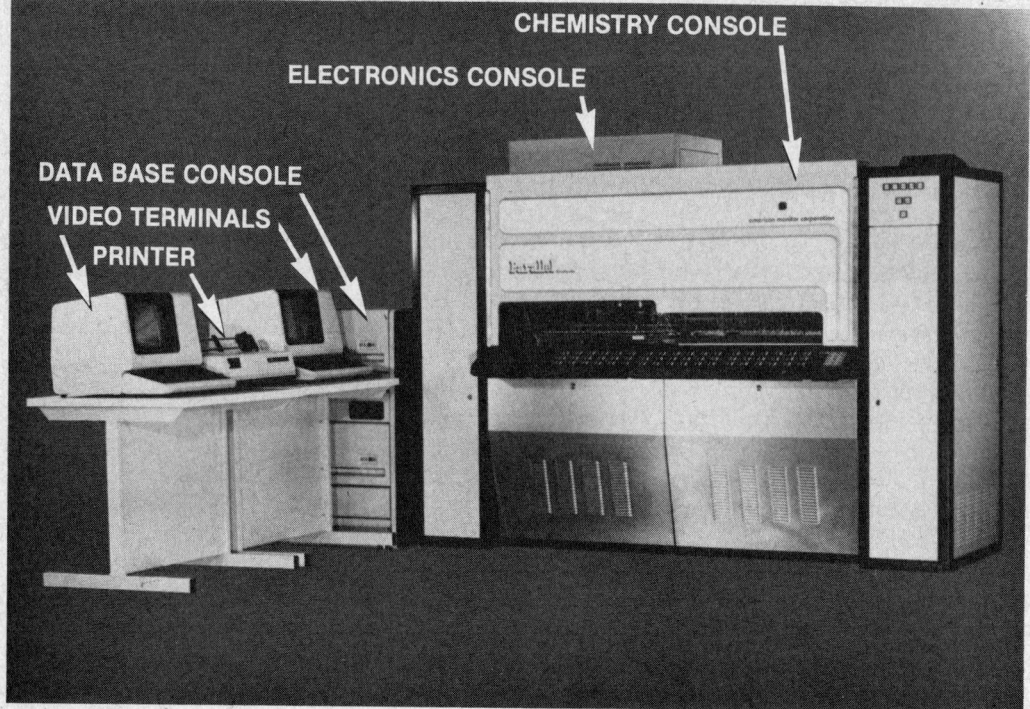

Figure 1D-19. Main modules of the Parallel, a multichannel, discretionary, highly computerized system. (Courtesy of American Monitor Corp.)

chain. If bar-code reading is not used, the keyboard is used to inform the system of the location of a specimen. The specimen chain accommodates 150 specimens in a continuous loop, and once in place the specimen is available for processing. For each assay, an aliquot of the specimen is transferred to one of 30 washable, polypropylene reaction vessels, each with a 2.0-mL capacity. Each group of 30 reaction vessels is carried by a reaction vial bar; there are 68 carrier bars on the system so that a total of 2040 reaction vessels are available. A schematic of the main elements of the analyzer's chemistry console is shown in Figure 1D-20.

Sampling for all or any of the 30 channels occurs according to a fixed cycle time of 15 s. Sample and reagents are aspirated and dispensed for a specific channel only if the test has been

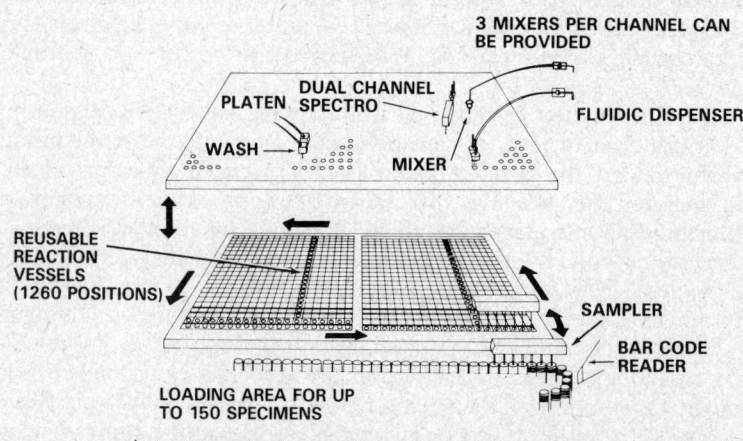

Figure 1D-20. Schematic of the main elements of the Parallel chemistry console. (Courtesy of American Monitor Corp.)

requested. Sampling is by positive displacement; a deionized water column is present in the probe between the syringe piston and the aspirated sample. Dispensing is accomplished by passing deionized water through a bypass valve and flushing the specimen from the sample aspiration tip. This sample-dispense design reduces syringe motion, thus minimizing wear, and avoids serum contact with the syringe piston. Sample volumes up to 100 μL can be dispensed in 1-μL increments; volumes used by the instrument range from 30–70 μL. Sample carryover is eliminated by an additional flush of the inside of the sample probe and the washing of the outside with deionized water.

A fiberglass platen contains fluid-dispensing heads, motorized stirrers, transfer probes, photometers, and wash probes, all of which move up and down above the plane of the reaction vessels once every index cycle. Up to 6 reagents can be dispensed per channel; reagent volumes dispensed range from 50–1300 μL. The configuration of devices on the platen is fixed by the manufacturer; however, configuration can be altered to permit changeover of channels from one analyte to another.

Bulk liquid reagents are stored in 32 containers at ambient temperature and in 28 containers at 10–14 °C. Stability of the manufacturer's working reagents is typically one month. Capacity of bulk storage containers ranges from 125–1000 mL.

A motorized stirrer for each channel is mounted on the platen between the reagent dispenser and the photometer. Up to three mixers can be placed in one channel, although for most chemistries only one is required. Twenty-nine transfer probes aspirate reaction mixture to the 10-mm flow cells of 29 photometers. The location of a transfer probe is determined by the reaction requirements of a channel. One channel is devoted to monitoring specimen absorbance at 500, 546, and 650 nm for subsequent correction of icterus, hemolysis, or turbidity. The thirtieth probe leads to a flame photometer for determination of sodium and potassium. Each flow cell has a measurement volume of 76 μL and is flushed continuously as fresh reagents are aspirated. If no test has been requested on a particular channel, its reaction vial is filled with deionized water which, when aspirated into the flow cell, assures that it is never empty.

Each optical measuring station is a dual-channel filter photometer with a water-cooled tungsten-halogen lamp and an interference filter. The interference filters cover the range of 340–750 nm; they have bandpasses of 7–19 nm. Within each photometer a beam-splitter creates two separate light paths. One beam illuminates a photodetector that regulates output of the light source through a negative feedback network. The other beam passes through an interference filter and then through the flow cell to a second detector for measurement of absorbance of the reaction mixture.

The spectrophotometric read process begins after transfer of the reaction mixture is complete. Readings occur over a period of 10.5 s, during which interval the photometers are scanned eight times with four readings taken during each scan. Readings are processed through a log amplifier and A/D converter and sent to the database console, where various mathematical routines correct for carryover and spectral interference and reduce absorbance data to concentration values. Analytical modes for standardizing include linear regression, third-degree polynomial fits, and multiple kinetic standardization relative to a specific reference material. Nonlinear, log-logit transformations for nonisotopic immunoassays are also supported.

The last step in the assay sequence is the wash station, where reaction vessels are cleansed by dispensing and aspiration of 1.8 mL of deionized water. A final vacuum aspiration is applied to the reaction vials, leaving them clean and dry for the next assay cycle.

The chemistry console is directed by the electronics console, a 64K minicomputer that controls not only the mechanical functions but also the transfer of spectrophotometric data to the database console. The database console consists of a 256K minicomputer with two 10-megabyte hard-disk drives; it executes all mainline programs and provides order entry, result reporting, and data storage and retrieval.

The database includes the last 10 standardization curves for each analyte as well as quality control data. Using the VDU the operator may make a historical review of any particular analyte, including plots of absorbance against concentration and tabulations of blank values, slope, intercept, and certain other derived coefficients. The same data may be obtained in hard copy.

Fifteen different quality control materials and their expected means and ranges can be identified to the system. Four of them can be uniquely specified as on-line monitors for a multirule data-acceptance routine. Quality control data can be stored and Levey-Jennings plots, together with statistical summaries, may be produced on a daily or monthly basis.

The database console also offers certain computer utility programs for defining up to 99 different test profiles, age-adjusted reference ranges, and "alert limits." Additionally, as many as 10 different report formats may be designed by the operator.

The Parallel system contains an RS-232C data communications package supporting a variety of communication messages. Data can be transmitted from the analyzer to the laboratory computer in real time or in batch mode. Complete or partial reports may be transmitted to the host computer; partial report transmission is advantageous when one or more assays of a set require repetition before results can be reported.

SMAC II

The Technicon SMAC II system (Sequential Multiple Analysis plus Computer) is the most recent model of the continuous flow systems applying principles devised by Skeggs in the 1950's. Samples, sequentially presented, are split automatically into multiple, parallel, analytical pathways for the simultaneous assay of 23 serum components. Throughput of the system is 150 specimens per hour. Two host computers and ten microprocessors control the mechanical, analytical, and data handling functions of the system.

The system consists of two major subsystems. The Analytical Processor (AP) instructs and monitors analysis; the Results Processor (RP) provides data handling of results and release of an edited report, complete with appropriate demographic information. The five operating units of the system are shown in Figure 1D-21.

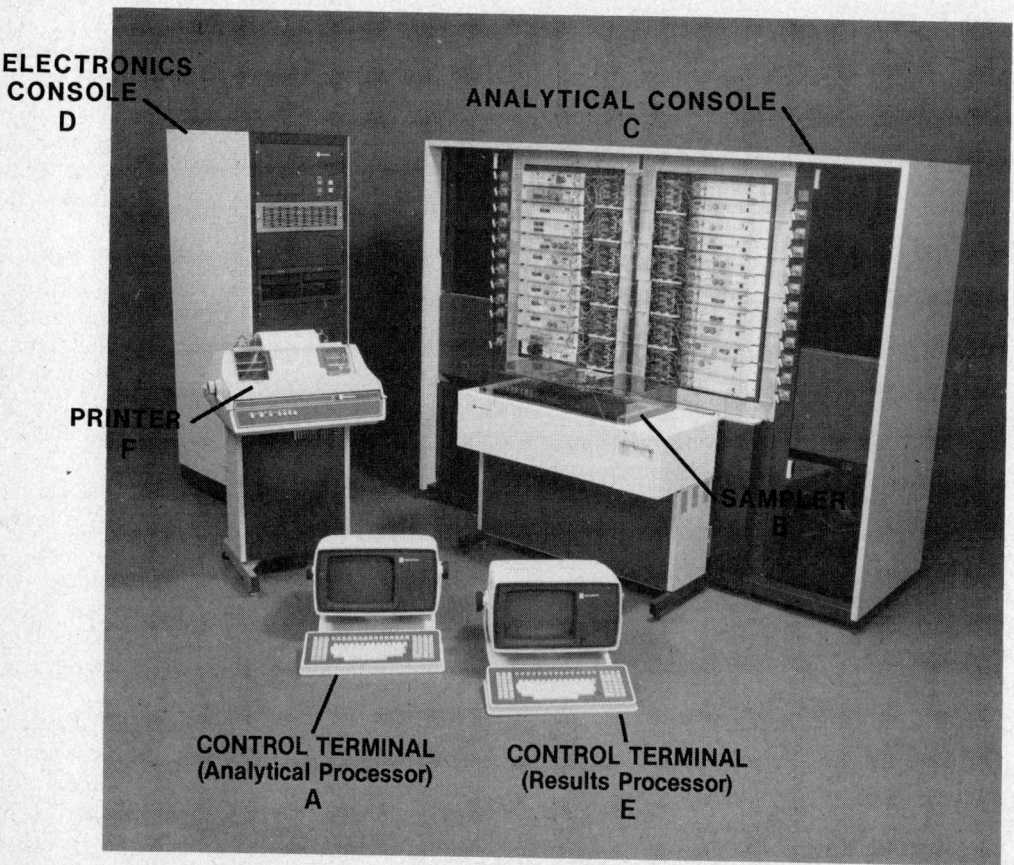

Figure 1D-21. Main module of the SMAC II, a sequential, multichannel, nondiscretionary, continuous flow analyzer. (Courtesy of Technicon Instruments Corp.)

The operator initiates analysis through the keyboard of the AP control terminal (Figure 1D-21,*A*) by use of command codes and dedicated keys. Startup procedures such as automatic delivery of water and reagents are directed from this terminal. The operator also defines through keyboard entries cycles for calibration and frequency of control specimens and intervals for readjustment of reagent baselines. During the run, the operator can display analytical curves of 16 channels at one time on the CRT or can call up management reports of system parameters.

The sampler (Figure 1D-21,*B*) has the capacity for 18 carrier blocks, each with space for eight specimen collection tubes or polystyrene specimen cups. Each position in a block may be assigned to a specimen by the operator's entering specimen number versus position at the keyboard, or the position may be identified by an IDee bar-coded label affixed to a specimen collection tube or to a specimen position in a block. The IDee bar code, read at the sampling station, allows specimens to be presented randomly; when specimen numbers are entered at the keyboard, however, a load list is required to assure correct sequencing. A fixed module in the center of the sampler holds wash solution, control solutions, calibrators (these in a refrigerated portion), and a position for placing a priority (STAT) specimen. Sampling of the STAT specimen is directed from the keyboard and occurs without impact on sequence of specimens already in the loading zone. Calibrators and controls are sampled in the pattern previously defined by the operator.

Samples are taken into the analytical console (Figure 1D-21,*C*) by a probe dipping into and aspirating from specimens in appropriate locations in carrier blocks and the central module. Approximately 600 µL of specimen is required for the sampling process. The probe is connected by tubing to the dilution cartridge on the analytical console. The major portion of aspirated sample is diluted with distilled water and segmented with air bubbles to minimize sample-to-sample interaction. Diluted sample then travels up a riser in the center of the console past a series of analytical cartridges. Each cartridge is a miniature, single-channel analyzer configured in a uniform module but made of different components. A typical cartridge includes a peristaltic pump, pump tubes, mixing coils, injection blocks, and flow cells. When required, an on-line dialyzer for protein separation or an on-line heating bath for color development or both are present.

Each cartridge draws from the central riser an aliquot of diluted sample; the amount is dependent on the internal diameter of the connected sampling tube. Reagents are drawn through tubes from an array of reagent containers mounted on both sides of the analytical console. Refrigerated compartments provide stability for labile reagents for up to 5 d. Other reagents not requiring refrigeration are stable for 1 month or up to 1 year. Capacities of reagent containers range from 20–1000 mL, and reagent consumption is ~30–60 mL per channel per hour. In a routine maintenance program, pump tubes are normally changed after every 200 h of operation.

Quantitation of analytes is by spectrophotometry except for sodium and potassium, which are measured by in-line ion-selective electrodes. Blanking, when necessary, is achieved by a parallel channel on an analytical cartridge. Methods for glucose and uric acid use specific enzymes immobilized in nylon coils and indicator reagents for spectrophotometry. Kinetic assays are implemented by placing multiple flow cells on an analytical cartridge and making multipoint measurements.

Optical fibers transmit light from one of two light sources (UV and visible) to each flow cell and from each flow cell to a single photomultiplier tube detector. A computer synchronizes selected wavelengths provided by filters on a rotating wheel with the appropriate flow cell. The transmittance signal of the photomultiplier is converted to absorbance by a logarithmic amplifier and the analog signal is digitized through an analog-to-digital converter. Digital data are then checked by a peak monitor program and tested for abnormal wave forms against method-specific curve characteristics stored in memory. Atypical curve shapes due to insufficient sample, poor mixing, or irregularities of flow are detected. The AP control system alerts the operator to problems and presents a message defining probable cause and suggested remedial action. Validation of curves by the monitor allows corresponding patient data to be released by the operator from the AP. Pending release, up to 2048 data files can be stored under AP control.

Both the AP and the RP interact with the electronics console (Figure 1D-21,*D*), which contains two disk drives for AP control and a 10-megabyte hard disk for RP control. Data released from the AP system flow, together with IDee number or specimen number, to the RP where the IDee number is the key to linking analytical data to demographic information. Demographic information can be entered into the RP either through the keyboard on the RP control terminal (Figure 1D-21,*E*) or by download from an external laboratory computer. Up to four RP terminals may be tied to one SMAC II system. The RP terminal and its subsystem manage the review and editing of patient and control data received from the AP through an RS-232 port, as well as the release of reports as hard copy on the printer (Figure 1D-21,*F*), or as data stream to an external computer, or as both. Daily patient reports may be printed in up to 76 different, user-defined formats, or a 7 day cumulative report may be printed for one patient in a selected format. The system can store up to 7500 test files. Quality control programs apply Westgard multirule algorithms[28] for testing control data (see Chapter 2E); they also store and summarize control data and print performance reports.

SELECTION OF INSTRUMENTS

The demand for increased efficiency in health care has prompted laboratory directors to provide services more rapidly and at lower cost. These general requirements must be met in the face of the demand for a greater variety of tests with increased specificity, a decreased supply of skilled technologists, higher personnel turnover, and services provided at multiple as well as remote locations. In the context of the many operational and technical demands, clearly one system in and by itself will not fill all of the various needs. Equally clearly, automation cannot be an efficient and effective response to all needs and demands in laboratories. Automated systems will always need to be judiciously selected to fit into that combination of automated, semiautomated, and manual modes of analysis that is, in any one laboratory, optimal for the kind and volume of work performed there.

No one instrument can serve all the needs of different-sized facilities and all concerns of the physician, the laboratorian, and the administrator. Presently, there are no standardized approaches to the selection of instrumentation, but useful checklists, workshop manuals, and charts have been published[7,14,21] to aid the laboratorian in the selection and evaluation process. This process is necessarily iterative with many of its steps performed in parallel.

A two-phase effort is suggested: the first, selection; the second, evaluation. In both phases the steps are the same, but priority of each step differs.

In the *selection* phase, the priority is: first, systematic analysis of the role of an automated system in the workflow and identification of candidate systems; second, detailed analysis of cost-effectiveness of rival candidate systems; third, assessment of acceptability to operators ("user friendliness," serviceability, and similar factors); fourth, characterization of analytical performance. At this stage, manufacturers' claims may be assumed to be valid because of today's environment of competitive pressures and government regulations. In the *evaluation* phase, the priority is exactly reversed.

A preliminary screening method for candidate systems has been proposed.[9] In this approach a matrix is set up with one axis listing parameters to be considered, and the other listing candidate systems. The list of parameters may include any operational, technical, or financial elements of interest. Table 1D-2 is an example of such a matrix.[9] In the table are shown values assigned on an arbitrary scale of 0–2 that indicate relative importance of a parameter according to individual judgment or to a consensus. This approach

Table 1D-2. PRELIMINARY 'WANT' LIST FOR THE EVALUATION OF AUTOMATED ANALYZERS[9]

Instrument Designation	Evaluation Categories*													Total Points
	1	*2*	*3*	*4*	*5*	*6*	*7*	*8*	*9*	*10*	*11*	*12*	*13*	
A	1.0	0.0	0.0	1.0	1.0	1.0	0.5	0.0	0.0	1.0	0.5	0.0	2.0	8.0
B	1.5	0.0	0.0	1.5	1.5	1.0	0.5	0.5	2.0	2.0	2.0	0.0	1.5	14.0
C	0.5	0.0	0.0	0.0	2.0	0.5	1.0	2.0	0.5	2.0	2.0	1.5	2.0	14.0
D	1.0	2.0	2.0	1.5	2.0	0.5	2.0	1.5	0.0	0.0	0.5	0.0	2.0	15.0
E	1.0	2.0	0.0	1.5	2.0	1.5	2.0	0.0	2.0	2.0	0.5	1.0	1.5	17.0
F	2.0	1.5	0.0	1.5	2.0	1.5	1.5	1.0	1.5	2.0	0.5	0.0	2.0	17.0
G	2.0	1.5	0.0	1.5	2.0	2.0	1.5	1.5	0.0	0.0	0.5	0.0	2.0	14.5
H	2.0	2.0	0.0	1.5	1.0	2.0	2.0	0.5	1.5	0.0	0.5	0.0	2.0	15.0
I	1.0	2.0	2.0	0.5	1.0	1.0	2.0	1.5	0.0	0.0	0.5	0.0	2.0	13.5
J	2.0	1.0	0.0	1.5	1.0	0.5	1.0	0.0	0.5	2.0	0.5	2.0	2.0	14.0
K	2.0	0.0	0.0	1.0	1.5	1.0	1.5	0.0	0.0	2.0	0.5	0.0	2.0	11.5

*Categories: 1. sample consumed per test; 2. throughput (testing rate); 3. stat capability; 4. methodology; 5. accuracy and precision; 6. reagent cost/test; 7. manpower requirements; 8. personnel training requirements; 9. computer compatibility; 10. capability to flag instrumental anomalies; 11. automatic deproteinization; 12. sample identification system; 13. selectable tests for profiling.

(Reproduced with permission from Gudzinowicz, B. J., et al.: Chem. Biomed. Environ. Instrument., *11*:107-126, 1981, by courtesy of Marcel Dekker, Inc.)

enables a semiquantitative estimate of a candidate system's agreement with targeted goals. This screening mechanism forces systematic definition of areas of concern in the selection process and provides a first estimate of the strongest contenders among competing systems.

Every resource for obtaining information, including manufacturers' literature, should be utilized. Published articles that describe system configuration and operation in depth and from a user's perspective should be consulted. Two useful journals are the *Journal of Clinical Laboratory Automation* and *Clinical Instrument Systems*. Performance evaluations using recognized protocols (such as the NCCLS EP5-T[19]) and published in peer review journals provide additional assessments of analytical performance. Reviews of this type appear regularly in *Clinical Chemistry* and *The Journal of Automatic Chemistry*. On-site demonstrations, visits to laboratories that use the systems, conversations with colleagues, and viewing of instruments at professional meetings provide additional opportunities to gain information.

When range of choice among systems has been narrowed by the screening process, detailed analyses of costs of acquiring and running candidate systems are required. Several references[7,9,21] are useful in systematizing the process of cost analysis of acquiring capital equipment. Table 1D-3 lists the economic considerations involved in a selection process.[9] Oszustowicz[21] defines 33 steps in making financial comparisons necessary to the process.

The next step in the selection process is a detailed analysis of costs and cost-effectiveness of candidate systems.

Acquisition costs must be considered in terms of possible options of capital investment (purchase), lease, straight rental, or reagent rental. In the case of purchase, amortization, depreciation, and, in particular instances, the cost of borrowing purchase money or the rate and volume of cash flow, are matters of major concern. In the case of the other options, length of time and other dimensions of a contractual commitment require evaluation in connection with both costs and expected revenue. In every cost analysis, both fixed and variable costs must be estimated. Since variable costs are intimately dependent on work volume and cost/benefit ratios on expected revenue, good workload and financial data are needed to make rational projections.

Operating costs are another aspect of the analysis. The chief determinants are labor demand of a system both in operation and maintenance; supply cost, which is a function of both work volumes and the nature of assays on the system; and service requirements, which may be supplied in-house or by a service contract. Systems with attractively low cost per test may have unattractive total operating costs if using them is labor-intensive in setup and maintenance or if service arrangements are expensive. Low cost per test may not be a financial advantage if a large proportion of tests must be dedicated to nonrevenue-producing assays (controls, calibrators, and the like).

Manufacturers offer a variety of arrangements, which may or may not be consonant with institutional fiscal and financial policy. The overall view of acquisition and operating costs is highly dependent on administrative stance with respect to payback period, present value, and return on investment. For this reason, institutional management and laboratory directors must work together before embarking on (and during) this cost analysis of candidate systems.

The success of many automated systems depends heavily on the interface between the machine and the human operator. Those systems deemed "user friendly" obtain greatest acceptance by laboratory workers;

Table 1D-3. ECONOMIC ASPECTS OF INSTRUMENT EVALUATION[9]

1. Workload (yearly) projected for a given test over a specific instrument amortization period.
2. Manpower needs (yearly) for a given test over the given instrument amortization period.
3. Manpower costs (yearly) for a given test over the given amortization period.
4. Preventive maintenance costs (yearly) for a given test over the given instrument amortization period.
5. Down-time costs (yearly) for the instrument over the given instrument amortization period.
6. Reagent and disposable item costs (yearly) for a given test over the given instrument amortization period.
7. Service contract costs (yearly, if any) for the instrument over the given instrument amortization period.
8. Instrument cost (yearly) adjusted for interest charges over its amortization period.
9. Cost adjustments (yearly) for wage-scale adjustments, cost-of-living adjustments, effect of inflation on reagent and disposable item costs, etc., over the given instrument amortization period.
10. Total direct costs for a given test over the given instrument amortization period.
11. Total indirect costs for a given test over the given instrument amortization period.
12. Total economic cost (sum of total direct and indirect costs) for a given test over the given instrument amortization period.
13. If the instrument selected for evaluation is a multicomponent analyzer, steps 1, 2, 3, 6, 9, 10, 11, and 12 must be determined for each test performed by the analyzer.
14. For a single-element analyzer, the total economic cost (item 12) must be compared over the given instrument amortization period to a similar total derived from an economic evaluation of the cost of the laboratory's present operation for the given element or component. The same is true for multielement analyzers.

(Reproduced with permission from Gudzinowicz, B. J., et al.: Chem. Biomed. Environ. Instrument., 9:315-333, 1979, by courtesy of Marcel Dekker, Inc.)

acceptance, in turn, leads to more rapid and less disruptive introduction into the laboratory workflow. It is highly desirable for technologists who will eventually use the systems under consideration to operate them for some period of time in other laboratories, in a demonstration session, or in a preliminary on-site trial. During a trial period, detailed evaluation of performance is not suggested, although basic checks outlined in NCCLS EP5-T[19] can be made as personnel are becoming familiar with the system.

A useful approach to aid the final choice[7] is to send uniform bid questionnaires to various manufacturers. The information returned is helpful in the economic analysis and also in reaching an understanding of the manufacturer's claims for performance and ease of operation. Adoption of a standard method for describing specifications, as suggested in IFCC provisional guidelines,[20] will aid assessment of many variables of concern in the selection of potential equipment. These specifications outline general information, detailed operational specifications, and description of technical details. As manufacturers begin to present information according to the guidelines outlined by IFCC and in NCCLS I1-A2,[17] screening and selection will become easier.

Once systems have been delivered, *performance evaluation* must assure that the sensitivity, specificity, accuracy, and precision claimed by the manufacturer are achieved. The NCCLS user-evaluator protocol EP5-P provides a standard approach to assess precision and carryover for a wide range of clinical values and within a reasonable time period. Since many manufacturers utilize NCCLS protocols EP2,[16] EP4,[15] and EP3[18] for verification of claims and for documentation, agreement between the manufacturer's claims and the user's findings is a reasonable expectation. Use of NCCLS evaluation protocols will not, however, identify specificity problems or effects of different lots of calibrating materials, nor will they detect variations of operator skills, preanalytical error,[26] or environmental conditions. In this context, a major concern is the generation of a falsely negative result that fails to identify patients with an existing disease state. Gudzinowicz et al.[9] deal with this concern by testing the ability of the overall system (instrument and reagents) to identify the maximum separation between false negative and false positive test results. They use a statistical method employing a Gram-Charlier dissection algorithm. While use of these protocols will assess technical performance, it will not determine how reliably and rapidly the analytical equipment will provide results to the physician.

Instrument systems in use should be demonstrably rugged enough to withstand misuse. Sufficient training and adequate manuals should be provided for operators so that they can detect ordinary malfunctions and make ordinary on-site repairs. Microprocessor technology has reduced dependency on manufacturers' service engineers either by incorporating diagnostic routines on software and making available 24-h telephone consultation or by speedy dispatch of replacement or substitution boards. Overall, the best recommendation for equipment reliability is a well-founded reputation for satisfying past and current users.

TRENDS IN INSTRUMENTATION

Current trends in laboratory operation provide an indication of the probable directions automation will take. The drive for greater economy in hospital practice will reduce the duration of hospitalization and thereby dictate greater emphasis on outpatient practice. These trends will stimulate the need for a more rapid response within a centralized laboratory as well as fostering satellite laboratories. Infrequently requested tests will probably be referred more often to reference laboratories, where the large volume of work may prompt automation of tests that were previously performed manually. Esoteric tests previously performed in research or special laboratories will be transferred to the chemistry laboratory. Immunoassays requiring radioactivity counting systems will be performed with increasing frequency by nonisotopic methods using automated analyzers.

Technological advances of the past decade are now being put to work. Nonisotopic antibody labels with spectrophotometric, fluorometric, and chemiluminescent detection systems are achieving sufficient sensitivity so that they provide viable alternatives to radioisotopes for many measurements. Improved purification and production of monoclonal antisera have enhanced analytical specificity. Many reagents can now be delivered in solid as well as liquid forms. Reusable, immunoreactive reagents coupled to ion-selective electrodes as detectors have been developed. As analyzers with solid reagent systems are implemented, control of the analytical process will be more in the manufacturer's and less in the laboratorian's hands. This is already true for tests as evidenced by analytical systems such as the DuPont *aca*, Paramax, Seralyzer, and Ektachem for routine assays and Stratus, TDx, and Advance for immunoassays.

The integration of such systems into laboratory operations reinforces the role of the laboratory as a producer of good-quality, clinically useful information on a timely basis. It also redirects emphasis away from the major operational problems that surfaced in laboratories in the mid-60's—collection and processing of specimens, recording and transcription of data, and production of reports of test results[11]—problems that, if not solved, have at least been ameliorated by more extensive use of computer technology.

Availability of microprocessors in the 1970's permitted their incorporation into analytical systems, allowing the laboratory to become independent of central multiuser computers. Hicks[11] suggests that a dual role for computers—one for the clinician and one for the laboratorian—will evolve in the future. The focus for the physician will be primarily on cumulative patient reports while that of the laboratory will be on workload analysis and quality control summaries with displays. The transfer of information from one domain to the other will be accomplished by increased use of cost-effective, dedicated computer hardware and software. This conceptual model is equally applicable for operations in a large or small, central or satellite laboratory, or at the patient's bedside. Indeed, many of the automated analytical systems described earlier, whether batch, multitest, or random access, will use their computers to function in both domains. They will interact with a host computer, delivering patient-collated reports to the physician, or they will communicate to a personal computer for the reporting function as well as for the quality control data processing to produce Levey-Jennings charts and other tools of analytical control. The computer may also direct further testing based on results of the tests initially requested.

Concurrent improvements in computer technology and analytical techniques will continue the transfer of information in a more cost-effective and timely manner. Computer capability will be enhanced by improvements in chip technology out of proportion to the concomitant modest increase in cost. This trend can be expected to persist although the cost of software development may increase.

In the context of this scenario, experts perceive several trends. The number of manipulative steps in analytical systems is being reduced. This reduction has been assisted by use of nonliquid reagents in tablets or in thin-films or fiber matrices, bound enzymes, ion-selective electrodes coupled to reagent-active membranes, or reusable solid substrates. Single-vial liquid reagents simplify the operation of random access analyzers, especially when the reagent is contained in a disposable tube that also serves as cuvet. Use of solid-state reusable reagents reduces the flexibility of an instrument because it forces a fixed set of tests, but there is the potential benefit of relatively low operating costs. Increased use of robotic systems to separate serum from cells can be expected.

Microprocessors can be assumed to allow true automation of analytical systems with feedback control of processing steps. Furthermore, they are likely to provide interpretation of some test results and real-time quality assurance. A more intimate connection to central computer systems via direct modem, RS-232 or RS-422 ports, or telemetry will greatly reduce production and transfer of paper documents and the time and effort required for handling them. This intimate integration of the microprocessor into central computer systems will further reduce the distinction between instrument information and laboratory information.

Innovations in expanded-function microprocessors will bring instrumental analysis closer, both in time and space, to the patient's bedside. Miniaturization of memory, with bubble chips capable of storing one million bytes of information, as well as development of input/output devices such as voice control and response or flat-screen and touch-screen display, will affect the practice of laboratory medicine. Less-skilled operators could be taught by voice output to operate and troubleshoot the instrument or could themselves direct testing through voice input. With development of whole-blood assay technologies, miniaturized, rapid plasma-separation techniques, and compact, dry-reagent systems, reduction of manipulative steps on random-access instruments should lead to faster turnaround of results, reduction of the number of errors, and the reduction of costs. It is the combination of a compact, almost fully self-contained analysis system with a powerful data-handling capability that will bring the assay process to the patient's bedside.

Due to the low cost of microprocessors, repair of systems may become feasible by simple disposal of damaged or nonfunctioning subsystems. Manufacturers will provide service and updates by means of PROM's (Programmable Read Only Memory), cartridges, updated disks, remote modem transmission, or data dump to the manufacturer's computer.

Although decision-making by instruments, including ability to base a decision on results obtained at another analytical station, is already possible, these functions have yet to be implemented on a large scale. Thus instruments, in the future, can be expected to provide better quality assurance as well as greater control of tests to be performed through automatic interpretation of results and decision-making capabilities. Furthermore, instruments are likely to be simpler to service and to interact with the operator in a user-friendly manner.

Although computerization and miniaturization of components will have considerable impact on laboratory testing, these same features are likely to allow more extensive use of continuous monitoring of certain analytes in vivo. Currently, systems are used to monitor pCO_2 and pO_2 transcutaneously. It is probable that intravascular miniature electrodes will allow blood pH, electrolytes, and substances such as glucose and urea to be monitored continuously in acutely ill patients with little risk. These electrode systems, in turn, could be coupled to infusion pumps to regulate the circulating concentration of the measured analytes. The ability to monitor some analytes in vivo and others at the patient's bedside may well widen the role of the phlebotomist to one of an analyst.

Whether these changes in practice, while feasible, will actually be implemented will largely depend on their impact on the quality of health care as well as their cost-effectiveness. Nevertheless, it is certain that many developments in the traditional practice of clinical chemistry will result from changes in technology. Clinical chemists must be ready to adapt to these changes.

Appendix 1D-I. SELECTED PARAMETERS OF REPRESENTATIVE INSTRUMENTS

A. PHOTOMETRIC INSTRUMENTS

Name and Categories	Manufacturer	Specimen Loading Zone Capacity	Number of Channels	Approximate Throughput (Results/h)	Cabinet Type	Type of Reagents Used	Measurement Method(s)	Comments
ABA VP Batch, sequential, discrete	Abbott Laboratories	32	N/A	300–465	Table top	Liquid, bulk	Photometric	Disposable cuvet array
aca Discrete, selective random access	DuPont Company	34	N/A	100	Floor console	Tablet and liquid	Photometric and ion-selective electrodes	Disposable, unit dose, cuvet pack; integral columns for separation for some assays
ACCLAIM Discrete, selective random access	Electro-Nucleonics, Inc.	150	N/A	360	Floor console	Liquid	Photometric	Reusable reaction vessels, flow-through cuvets
ASTRA 8 Multichannel, discretionary, discrete	Beckman Instruments, Inc.	40	8	580–720	Floor console	Liquid, bulk	Photometric, ion-selective electrode, coulometric, and conductometric	Reusable reaction vessels
AUTOANALYZER I Single channel, sequential batch, continuous flow	Technicon Instruments Corp.	40	N/A	60	Table top	Liquid, bulk	Photometric	The basic, initial, continuous flow, single channel
COBAS-BIO Discrete, batch, centrifugal, parallel, fast analyzer	Roche Diagnostic Systems	28	N/A	200–400	Table top	Liquid	Photometric, fluorescence polarization	Disposable cuvets
DACOS Discrete, selective random access	Coulter Electronics, Inc.	40+	N/A	600	Floor console	Liquid	Photometric	Semireusable cuvets
DEMAND Discrete, selective random access	Cooper Biomedical, Inc.	100	N/A	400	Floor console	Liquid	Photometric	Disposable cuvet packets
EKTACHEM 400 Discrete, selective random access	Eastman Kodak Company	40	N/A	300	Floor console	Film, unit dose	Reflectometric	Disposable slides
ENCORE Batch, centrifugal, fast analyzer	Baker Instruments Corp.	96	N/A	400	Floor console	Liquid, bulk	Photometric, fluorometric	Reusable rotor; 6 separate analytes may be run simultaneously on one rotor

Instrument / Type	Manufacturer							
GemENI Batch, centrifugal, fast analyzer	Electro-Nucleonics, Inc.	20	N/A	200–400	Table top	Liquid, bulk	Photometric	Disposable rotor
HITACHI 705 Discrete, selective random access	Boehringer Mannheim	40	N/A	180 (300 with ISE)	Floor console	Liquid; ion-selective electrodes	Photometric	Reusable cuvets
HITACHI 737 Discrete, selective random access	Boehringer Mannheim	60	N/A	1200	Floor console	Liquid	Photometric	Reusable cuvets
IL Multistat III Plus Batch, centrifugal, fast analyzer	Instrumentation Laboratory, Inc.	20	N/A	400	Floor console	Liquid, bulk	Photometric, fluorometric, and nephelometric	Disposable rotor
IMPACT 400 Batch, sequential, discrete	Gilford	60	N/A	218–480	Table top	Liquid, bulk	Photometric	
PARALLEL Multichannel, discretionary, discrete	American Monitor Corp.	150	30	7200	Floor console	Liquid, bulk	Photometric and ion-selective electrode	
PARAMAX Discrete, selective random access	American Dade	—	N/A	720	Floor console	Tablet, unit dose	Filter photometer	Disposable cuvets in reels of 2100
RA-1000 Discrete, selective random access	Technicon Instruments Corp.	30	N/A	240	Table top	Liquid	Photometric	Disposable cuvet ring
SMA 12/60 Multichannel, sequential, nondiscretionary, continuous flow	Technicon Instruments Corp.	40	12	720	Floor console	Liquid, bulk	Photometric	
SMAC Multichannel, sequential, nondiscretionary, continuous flow	Technicon Instruments Corp.	144	23	3400	Floor console	Liquid, bulk	Photometric	
SPIRIT Discrete, selective random access	Baker Instruments Corp.	40	N/A	320	Table top	Liquid	Photometric and ion-selective electrodes	Disposable reaction vessel; reusable flow-through cuvet; maximum of 12 reagents can be used at one time
System 508 Multichannel, discretionary discrete	Instrumentation Laboratory, Inc.	100	8	800	Floor console	Liquid, bulk	Photometric, ion-selective electrode, and manometric	Reusable reaction vessels

Table continued on following page

Appendix 1D-I. SELECTED PARAMETERS OF REPRESENTATIVE INSTRUMENTS (*Continued*)

B. INSTRUMENTS WITH NONPHOTOMETRIC MEASUREMENT PRINCIPLES

Name and Category	Manufacturer	Measurement, Techniques, Energy Source, and Details	Approximate Throughput Results/Hour	Comments
Auto ICS Sequential, batch	Beckman Instruments, Inc.	Rate nephelometry; quartz iodide lamp, 450–620 nm; PMT at 70° to incident light	25–55	Reusable cuvet; automated 40-position sampler
IL System 504 Discrete 4-channel (selectable for 3 channels)	Instrumentation Laboratory, Inc.	Ion-selective electrodes for Na, K; mercuric thiocyanate for Cl (photometric); direct pressure for tCO_2 (manometric)	400	Automated 100-position, linear sampler
Model 420 Sequential, batch, discrete	Baker Instruments Corp.	Rate and endpoint nephelometry; helium-neon laser at 632.8 nm; dual diode mounted at 37.8° to incident light	25–30	Flow-through cuvet; automated 25-specimen sampler
Model 1020 Semiautomated discrete, dual channel	Orion Research, Inc.	Ion-selective electrodes for Na, K	60	Calibration is automatic; specimen: whole blood, plasma, serum
NOVA 4 + 4 Discrete, 4-channel, nondiscretionary	NOVA Biomedical	Ion-selective electrodes for Na, K, Cl, and CO_2	85	Automated 40-position sampler
PDQ Laser Nephelometer Sequential, batch, discrete	Hyland Diagnostics	Endpoint nephelometry; using a helium-neon laser at 632.8 nm, with PMT mounted at optimum forward detection angle of 31°	20–30	Automated pipettor-dilutor available for up to 150 specimens
YSI Model 23A Blood Glucose Analyzer Discrete, single channel	Yellow Springs Instrument Co., Inc.	Immobilized enzyme (glucose oxidase) on membrane converts glucose to H_2O_2, monitored amperometrically	60	Specimen: whole blood, plasma

Appendix 1D-II. NAMES AND ADDRESSES OF MANUFACTURERS OR DISTRIBUTORS OF PRODUCTS CITED IN TEXT

Abbott Laboratories 1921 Hurd Drive Irving, TX 75061	ABA 100™, TDx™, VP™
American Can—Parafilm Div. American Lane Greenwich, CT 06830	Parafilm®
American Dade Division American Hospital Supply Corp. P.O. Box 520672 Miami, FL 33152	Paramax®, Stratus®
American Monitor Corp. P.O. Box 68505 Indianapolis, IN 46268	Parallel®
Ames Company Division of Miles Laboratories Elkhart, IN 46514	Optimate®, Seralyzer®
Baker Instruments Corp. 100 Cascade Drive Allentown, PA 18103	Encore™, Spirit™, Model 420™
Beckman Instruments, Inc. Healthcare Products 200 South Kramer Blvd. Brea, CA 92621	ASTRA™, ICS™
Becton Dickinson Rutherford, NJ 07070	ARIA II™
Boehringer Mannheim Diagnostics 9115 Hague Road Indianapolis, IN 46250	Hitachi 705, Hitachi 737
Cooper Biomedical, Inc. One Technology Court Malvern, PA 19355	DEMAND™
Coulter Electronics, Inc. 590 West 20th Street Hialeah, FL 33010	DACOS™
DuPont Instruments Quillen Bldg./Concord Plaza Wilmington, DE 19898	*aca*™ (Teflon®, Surlyn®)
Eastman Kodak Co. 343 State Street Rochester, NY 14650	Ektachem
Electro-Nucleonics, Inc. Separation and Analytical Systems Div. 368 Passaic Avenue Fairfield, NJ 07006	ACCLAIM™, GEMENI®
Gilford Instruments 132 Artino Street Oberlin, OH 44074	SBA 300, IMPACT® 400E
Hyland Diagnostics Division of Cooper Biomedical, Inc. One Technology Court Malvern, PA 19355	PDQ™

Table continued on following page

Appendix 1D-II. NAMES AND ADDRESSES OF MANUFACTURERS OR DISTRIBUTORS OF PRODUCTS CITED IN TEXT (*Continued*)

Instrumentation Laboratory, Inc. 113 Hartwell Avenue Lexington, MA 02173	MULTISTAT®, STATSEP™, SYSTEM 504™, SYSTEM 508™
Ionetics, Inc. 3020 Enterprise Street Costa Mesa, CA 92626	IONETICS Dual Analyzer
Nova Biomedical 20 Ossipee Road Newton, MA 02164	Nova 4 + 4
Orion Research, Inc. 840 Memorial Drive Cambridge, MA 02139	ORION Model 1020©
Roche Diagnostic Systems 340 Kingsland Nutley, NJ 07110	COBAS-BIO®
Sequoia-Turner Corp. 755 Ravendale Avenue Mountain View, CA 94043	DYNA-LYTE 100, 200
Sherman-Boosalis Corp. 2716 Humboldt Avenue S. Minneapolis, MN 55408	Pre-Vap™
Syva Corporation 3181 Porter Drive Palo Alto, CA 94304	Advance™
Technicon Instruments Corp. 511 Benedict Avenue Tarrytown, NY 10591	AutoAnalyzer®, RA-1000®, SMA®, SMAC®
Yellow Springs Instrument Co. P.O. Box 279 Yellow Springs, OH 45387	Model 23A
Zymark Corporation Zymark Center Hopkinton, MA 01748	Zymate

References

1. American Blood Commission Committee for Commonality in Blood Bank Automation: Guidelines for the uniform labeling of blood and blood components. Arlington, Va, American Blood Commission, 1980.
2. Babson, A. L., and Kleinman, N. M.: A source of error in an AutoAnalyzer determination of serum iron. Clin. Chem., 13:163-166, 1967.
3. Billmeyer, F. W.: Principles of Color Technology. New York, Interscience Publishers, 1966, pp. 61-63.
4. Burtis, C. A.: Factors influencing evaporation from sample cups, and assessment of their effect on analytical error. Clin. Chem., 21:1907-1917, 1975.
5. Burtis, C. A., Johnson, W. F., Mailen, J. C., et al.: Automated sample-reagent loader for use with the GeMSAEC fast analyzer. Clin. Chem., 18:433-439, 1972.
6. Food and Drug Administration: Changes in proper names of certain biological products; current good manufacturing practice for blood and blood components; uniform blood labeling; additional standards for human blood and blood products; reorganization and revision of regulations. Federal Register, 45:72404, 81065, 1980.
7. Gochman, N., and Bowie, L. J.: Selection factors for automated analytical instrumentation. Workshop No. 201, American Association for Clinical Chemistry National Meeting. Clin. Chem., 28:1675, 1982.
8. Grannis, G. F., Grümer, H.-D., Lott, J. A., et al.: Proficiency evaluation of clinical chemistry laboratories. Clin. Chem. 18:222-236, 1972.

9. Gudzinowicz, B. J., Driscoll, J. L., Martin, H. C., et al.: Economic and technical aspects of instrument evaluation. Chem. Biomed. Environ. Instrument., 4:1-25, 107-166, 315-333, 1980.
10. Hall, R. A., and Whitehead, T. P.: Absorption of serum calcium by plastic cups. J. Clin. Pathol., 23:323-336, 1970.
11. Hicks, G. P.: Microcomputer applications in the clinical laboratory of the 1980's. J. Clin. Lab. Auto., 1:185-186, 1981.
12. International Union of Pure and Applied Chemistry, Commission on Automation: Characteristics and attributes of instruments intended for automated analysis in clinical chemistry. IUPAC Inf. Bull., No. 3, pp. 233-240, 1978.
13. Khalil, O. S., Routh, W. S., Lingenfelter, K., et al.: Automated in-line ratio-correcting filter fluorometer. Clin. Chem., 27:1586-1591, 1981.
14. Krieg, A. F., and Baer, D.: Sharpen your instrument-shopping savvy. Med. Lab. Obs., April 1974, pp. 70-82.
15. National Committee for Clinical Laboratory Standards: Establishing performance claims for clinical chemical methods; comparison of methods experiment. NCCLS Standard EP4-T. Villanova, Pa., National Committee for Clinical Laboratory Standards, 1979.
16. National Committee for Clinical Laboratory Standards: Establishing performance claims for clinical chemical methods; replication experiment. NCCLS Standard EP2-T. Villanova, Pa., National Committee for Clinical Laboratory Standards, 1983.
17. National Committee for Clinical Laboratory Standards: Preparation of manuals for installation, operation and repair of laboratory instruments. NCCLS Standard I1-A2. Villanova, Pa., National Committee for Clinical Laboratory Standards, 1979.
18. National Committee for Clinical Laboratory Standards: Establishing performance claims for clinical chemical methods; replication experiment. NCCLS Standard EP3-T. Villanova, Pa., National Committee for Clinical Laboratory Standards, 1982.
19. National Committee for Clinical Laboratory Standards: User evaluation of precision performance of clinical chemistry devices. NCCLS Standard EP5-T. Villanova, Pa., National Committee for Clinical Laboratory Standards, 1982.
20. Okuda, K., Bonini, P., Burtis, C. A., et al.: Provisional guidelines (1982) for listing specifications of clinical chemical analyzers. International Federation of Clinical Chemistry, IFCC Document Stage 3, Draft 1: 1982-01-10.
21. Oszustowicz, R. J.: Following the equipment acquisition process. J. Hosp. Fin. Manage. Assoc., April 1982, pp. 12-33.
22. Pardue, H. L.: A comprehensive classification of kinetic methods of analysis used in clinical chemistry. Clin. Chem., 23:2189-2201, 1977.
23. Price, C., and Spencer, K.: Centrifugal analyzers in clinical chemistry. London, Praeger, 1980.
24. Rains, S.: Effect of spectral bandwidth and wavelength accuracy on the measurement of reduced NAD. Clin. Chem., 19:647, 1977.
25. Renoe, B. W., Savory, J., and Wills, M.: Measurement principles for centrifugal analyzer methods. J. Clin. Lab. Auto., 1:47-54, 1981.
26. Statland, B. E., and Winkel, P.: Physiologic variation of the concentration values of selected analytes as determined in healthy young adults. Proceedings of the 1976 Aspen Conference on Analytical Goals in Clinical Chemistry. Chicago, College of American Pathologists, 1976, pp. 94-101.
27. Walter, B.: Dry reagent chemistries in clinical analysis. Anal. Chem., 55:498A-514A, 1983.
28. Westgard, J. O., Barry, P. L., Hunt, M. R., et al.: A multi-rule Shewhart chart for quality control in clinical chemistry. Clin. Chem., 27:493-501, 1981.

CHAPTER 2

ACQUISITION, MANAGEMENT, AND APPLICATION OF LABORATORY DATA

Statistical Procedures

by Robert O. Kringle, M.S., and George F. Johnson, Ph.D.

Analytical work in the clinical chemistry laboratory consists of the determination of constituents of body fluids. The information is used in making decisions regarding the health and treatment of patients. In order to do so effectively, analytical work is also conducted with patient and synthetic fluids to make decisions regarding the day-to-day analytical stability of laboratory methods, the acceptability of the analytical performance of new methods and instruments, and the improvement of existing laboratory procedures and methods. All the data, and all the quantities calculated from the data, upon which these decisions and conclusions are based, are subject to many sources of variation. While some of these sources are known to the analyst, others are unknown. Some are controllable, while others cannot be controlled. The task confronting the analyst is to make the required decisions in the face of this uncertainty in the data. The science of statistics provides methods for dealing with these many sources of variation, both in the conduct of the experimental work and in the analysis and interpretation of the data. The proper use of statistical methods in analytical work is essential, since reliable inferences and conclusions can be made only when the experimental work and the statistical analyses are properly performed.

This introductory section provides a discussion of typical sources of variation and the role of statistics in dealing with variation in analytical work. Other sections of the chapter introduce and discuss specific statistical concepts and techniques important in analytical work in clinical chemistry: fundamental statistical concepts, sampling distributions, inferences about population means, inferences about population variances, the study of variation, regression analysis, total analytical error, quality control, outliers, and simulations. These sections also provide a statistical foundation for topics covered in succeeding chapters: 2B, *Establishment and Use of Reference Values*; 2C, *Analytical Goals and Clinical Relevance of Laboratory Procedures*; 2D, *Evaluation of Methods*; 2E, *Quality Assurance*; and 2F, *Microprocessors and Computers in the Clinical Laboratory*. A list of symbols appears at the end of each section for easy reference.

SOURCES OF VARIATION IN ANALYTICAL DATA

The many sources of variation or uncertainty in analytical data can be classified into four broad categories: inter-individual, intra-individual, preanalytical, and analytical.

Inter-individual variation refers to differences in the true level of an analyte between individuals. This variation may occur as a result of differences in age, sex, race, genetics, and long-term health status of individuals.

Intra-individual variation refers to differences in the true level of an analyte within an individual over time. Temporal intra-individual variation occurs because of differences in season, activity, emotional state, and short-term health. Intra-individual variation also occurs due to differences in patient preparation with respect to diet, exercise, drugs, sleep, posture, time of venipuncture, and the length of time the tourniquet is applied.

Preanalytical variation refers to differences in the analyte level of a specimen due to handling of the specimen before it enters an analyzer system. Differences in transport, standing time before separation, centrifugation time, and storage conditions are typical sources of preanalytical variation.

Analytical variation refers to differences in the analytical measurements of a specimen after it has been prepared for analysis. While discussing analytical variation, it will be convenient to introduce and illustrate the concepts of *systematic* and *random* variation. A source of variation, however, may not be uniquely systematic or random; many sources of variation can be systematic in nature in one particular experimental situation and random in another.

Different laboratories, methods, instruments, and technicians are common sources of *systematic* analytical variation, also called analytical *bias*. For example, consider an infinite number of measurements by two laboratories on the same lyophilized control specimen as depicted graphically in Figure 2A-1. The systematic analytical variation or analytical bias of laboratory A relative to the true concentration is the difference $C_A - C_T$ between the average C_A of the laboratory A measurements and the true concentration C_T. Similarly, the analytical bias of laboratory A relative to laboratory B is the difference $C_A - C_B$ between the averages of the laboratory A and laboratory B measurements. Thus, a source of variation is *systematic* if it influences all measurements in the same direction and to the same magnitude. In this example, the systematic variation or bias of laboratory A influences all measurements by $C_A - C_T$ units relative to the true concentration. If there were no other sources of variation, then all laboratory A measurements on this specimen would in fact be low by exactly $C_A - C_T$ units. In practice, however, many sources of variation within laboratory A will cause these measurements to be different from one another.

Aging phenomena can also be sources of systematic variation. Chemicals, reagents, standards, and instrument components may deteriorate with time causing an increasing or decreasing trend in laboratory results.

Sources of *random* analytical variation influence each measurement differently, in either a positive or negative direction, and to a different extent in magnitude. For example, multiple determinations on the same specimen on the same system in the same run vary in an unpredictable manner due to random fluctuations in the electro-optical mechanism, the fluid dispensing of the sample and reagent, the temperature of the instrument, and the evaporation of sample and reagent. Other sources, such as random variation associated with the act of calibration, cause determinations to vary from run to run and day to day in a random manner. Separation and estimation of components of random variation are discussed in a later section.

THE ROLE OF STATISTICS IN ANALYTICAL WORK

The purpose of analytical work in the clinical laboratory is to learn about the nature and magnitude of particular sources of variation. For instance, investigations are commonly conducted to evaluate and/or compare the systematic variation (i.e., bias) of one or more particular methods, laboratories, instruments, or preanalytical procedures. Sources of variation other than those being investigated which affect the experimental results are called *experimental error*. The term *error* in statistics refers to variation in measurements due to random and/or systematic sources, and does not refer to a mistake. Whether the sources of experimental error are known or unknown to the analyst, their existence causes problems in interpretation of the experimental results. They may obscure real effects or differences that the analyst is attempting to detect or measure, or conversely, they may mislead the analyst into concluding that some nonexistent effect or difference is real.

Example 1. Consider the following experiment to compare the determinations on specimens stored for 1 day at 0 °C with those on fresh specimens. Twenty specimens are split into two parts to be analyzed fresh on day 1 and again after 24-h storage at 0 °C on day 2. Suppose that the average of the values on the stored specimens is 5% lower than that on the freshly drawn specimens.

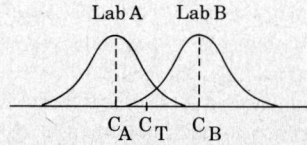

Figure 2A-1. The systematic analytical variation of laboratory A, relative to the true concentration C_T, is $C_A - C_T$. The systematic analytical variation of laboratory A, relative to laboratory B, is $C_A - C_B$.

Can the analyst infer that storage for one day at 0 °C results in a 5% reduction in concentration? Obviously not. The effect or variation being investigated in this example is the storage effect on concentration, but this effect is inseparable from the run-to-run and day-to-day experimental errors from day 1 to day 2. Thus, the 5% reduction may be partially or wholly due to the experimental errors associated with these two particular runs on these two particular days. Similarly, if there had been no reduction in the average result from day 1 to day 2, the analyst could not conclude that there was no storage effect. Typical sources of these unavoidable experimental errors include random error associated with the act of calibrating each run, random error associated with new vials of calibrators and reconstitution of these vials each day, and random/systematic error associated with environmental conditions each day.

The object of statistical methods is to foresee such problems and to take steps to eliminate or minimize them. First and foremost, precautions should be taken in the design of the experimental work to minimize the influence of experimental error. Secondly, statistical analyses provide the most informative ways to summarize the results, including statements regarding the analyst's confidence in the existence or magnitude of an effect or difference.

Although proper statistical design and analysis are both important in analytical work, the design is clearly more important. If the experimental error is not adequately controlled or minimized through proper statistical design, no analysis, no matter how sophisticated, will be able to extract useful information from the data. Alternatively, simple examination and analysis of the data are often all that is needed for interpreting the results of a well-designed experiment.

FUNDAMENTAL STATISTICAL CONCEPTS

In this section fundamental statistical concepts and techniques will be introduced in the context of several typical analytical investigations. The basic concepts of populations, samples, parameters, statistics, probability distributions, and random sampling will be defined and illustrated. Two important probability distributions, the Gaussian and Student's t, will be introduced and discussed.

BASIC CONCEPTS

It will be instructive to introduce the basic statistical concepts of frequency distribution, population, sample, probability distribution, parameter, statistic, and random sample in the context of realistic analytical investigations. Suppose an analyst is interested in a new kit for the determination of serum albumin with a centrifugal analyzer system. Of particular interest is the sensitivity as defined by the true slope β of the calibration curve, $A = \alpha + \beta$ (C), relating absorbance A to concentration C. Ten consecutive runs are made with two calibration standards, at 0 and 4.6 g/dL, yielding the following slopes: 0.1505, 0.1486, 0.1500, 0.1472, 0.1528, 0.1514, 0.1479, 0.1521, 0.1493, and 0.1475.

The Dot Diagram and the Frequency Distribution

A *dot diagram* is a simple but informative device for displaying the results of small experiments like the albumin sensitivity study. The dot diagram in Figure 2A-2, displaying the ten albumin slopes, is simple to construct yet it provides useful information about the general location ($\sim$0.150) and general spread ($\sim$0.147–0.153) of the results.

A corresponding graphical device for displaying a larger set of data is the *frequency distribution*, also called a *histogram*. Figure 2A-3 shows a frequency distribution displaying the results of serum GGT measurements of 100 apparently healthy 20 to 29 year old males. The frequency distribution is constructed by dividing the measurement scale into cells of equal width, counting the number n_i of values that fall within each cell, and drawing a rectangle above each cell whose area (and height since the cell widths are all equal) is proportional to n_i. In this example, the

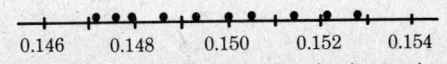

0.146	0.148	0.150	0.152	0.154

Figure 2A-2. Dot diagram of albumin slope values.

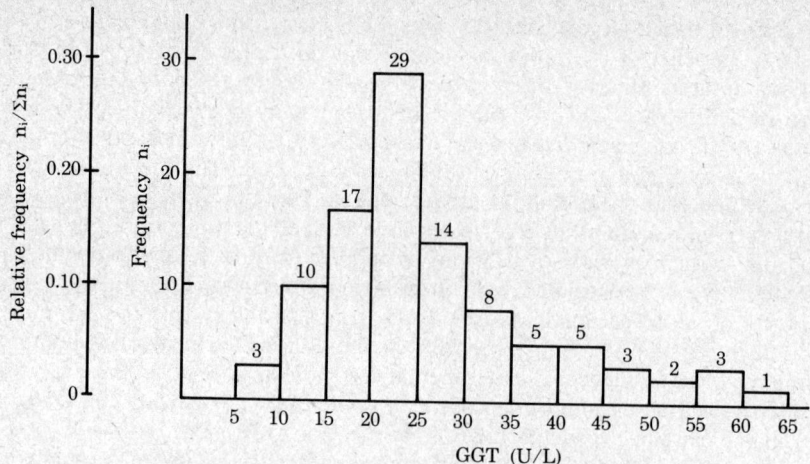

Figure 2A-3. Frequency distribution of 100 GGT values.

selected cells were 5 to 9, 10 to 14, 15 to 19, 20 to 24, 25 to 29, and so on, with 60 to 64 being the last cell. The ordinate axis of the frequency distribution gives the number of values falling within each cell. When this number is divided by the total number of values in the data set, the *relative frequency* in each cell is obtained. Inspection of the GGT frequency distribution reveals an approximate average of 20 or 25 U/L with a spread of values from about 5–65 U/L.

These and other graphical summaries should always be the first step in any analysis of data. In addition to providing information about the central location and variation of the data, they often reveal values that do not appear to belong with the majority of the data. Such values should be investigated before further data analysis proceeds. Frequently, clerical mistakes will be discovered to be the cause of atypical results. Sometimes, errors or unintentional changes in procedures will be discovered, in which case the atypical result may provide more interesting information than the majority of the data. When no assignable cause can be found for a suspicious-looking measurement, statistical tests can be performed to determine the probability that such a data point belongs with the majority of the data. Such statistical tests are discussed in the section on outliers.

Population and Sample

The purpose of analytical work is to obtain information and draw conclusions about characteristics of one or more populations of values. For example, in the albumin sensitivity investigation, the analyst seeks information about the central location and spread of the population of calibration slopes on a particular instrument with a particular reagent kit. In the GGT example, the interest is in the location and spread of the population of GGT values on 20 to 29 year old healthy males visiting a certain clinical laboratory. Thus, a working definition of a *population* is the complete set of all observations that might occur as a result of performing a particular procedure according to specified conditions.

As in the two examples above, most populations of interest in clinical chemistry are infinite in size, and so are impossible to study in their entirety. Usually a subgroup of observations is taken from the population as a basis to form conclusions about the population characteristics. The group of observations that has actually been selected from the population is called a *sample*. For example, the ten albumin slopes and the 100 GGT values are each a sample from their respective populations as defined above. However, a sample can be used to study the characteristics of a population only if it has been properly selected. For instance, if the analyst is interested in the population of albumin slopes with a particular calibration protocol and various lots of calibrator material, the sample must be selected in such a way as to be representative of these various lots of material and this particular protocol. Similarly, if the analyst is interested in the population of GGT values over various lots of materials and some time period, the sample must be selected to be representative of these factors as well as in the age, sex, and health factors. Consequently, exact specification of the population(s) of interest is necessary before designing a plan for obtaining the sample(s).

Probability and Probability Distributions

Consider again the frequency distribution in Figure 2A-3. In addition to the general location and spread of the GGT determinations, other useful information is easily extracted from this frequency distribution. For instance, 96% (96 of 100) of the determinations are less than 55 U/L, and 91% (91 of 100) are greater than or equal to 10 but less than 50 U/L. Because the cell interval is 5 U/L in this example, statements like these can be made only to the nearest 5 U/L. A larger sample would allow a smaller cell interval and more refined statements. For a sufficiently large sample, the cell interval can be made so small that the frequency distribution can be approximated by a continuous smooth curve like that shown in Figure 2A-4. In fact, if the sample is large enough, we can consider this a close representation of the true *population frequency distribution*. In general, the functional form of the population frequency distribution curve of a variable x is denoted by f(x).

The *population frequency distribution* allows us to make probability statements about the GGT values x of a randomly selected member of the population of healthy 20 to 29 year old males. For example, the probability $Pr(x > x_a)$ that the GGT value x of a randomly selected 20 to 29 year old healthy male is greater than some particular value x_a is equal to the area under the population frequency distribution to the right of x_a. If $x_a = 58$, then from Figure 2A-4, $Pr(x > 58) = 0.05$. Similarly, the probability $Pr(x_a < x < x_b)$ that x is greater than x_a but less than x_b is equal to the area under the population frequency distribution between x_a and x_b. For example, if $x_a = 9$ and $x_b = 58$, then from Figure 2A-4, $Pr(9 < x < 58) = 0.90$. Because the population frequency distribution provides all the information about probabilities of a randomly selected member of the population, it is called the *probability distribution* of the population. Although the true probability distribution is never exactly known in practice, it can be approximated with a large sample of observations.

Parameters: Descriptive Measures of a Population

Any population of values can be described by measures of its characteristics. A *parameter* is a constant that describes some particular characteristic of a population. Although most populations of interest in analytical work are infinite in size, for the following definitions we shall consider the population to be of finite size N, where N is very large.

One important characteristic of a population is its **central location.** The parameter most commonly used to describe the central location of a population of N values is the *population mean μ* (mu):

$$\mu = \frac{\sum_{i=1}^{N} x_i}{N}$$

The population mean is also called the *expected value* of x and is then denoted by E(x). Thus, $\mu = E(x)$.

Another important characteristic of a population is the **dispersion** of the values about the population mean. A parameter very useful in describing this dispersion of a population of N

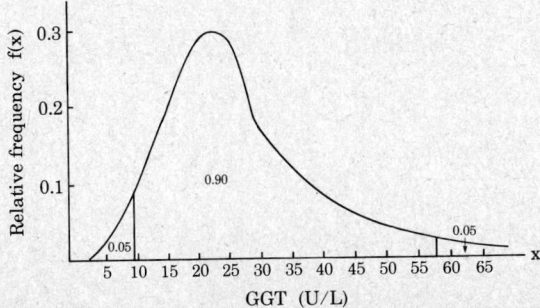

Figure 2A-4. Population frequency distribution of GGT values.

values is the *population variance* σ^2 (sigma squared):

$$\sigma^2 = \frac{\sum_{i=1}^{N} (x_i - \mu)^2}{N}$$

The population variance is often denoted by $V(x)$. Thus $\sigma^2 = V(x)$.

The *population standard deviation* σ, the positive square root of the population variance, is a parameter frequently used to describe the population dispersion in the same units (e.g., mg/dL) as the population values.

In analytical work, it is frequently the case that the dispersion of a population of analytical values is larger if the mean of the population of values is larger. For instance, the dispersion of the population of analytical values on a specimen with a high concentration is usually greater than that on a specimen of lower concentration. Consequently, the dispersion of a population is often described relative to its mean by the *population coefficient of variation*, defined as $(\sigma/\mu)(100)$.

Statistics: Descriptive Measures of the Sample

As noted earlier, the analyst usually has at hand only a sample of observations from the population of interest. A *statistic* is a value calculated from the observations in a sample to describe a particular characteristic of that sample. Corresponding to the population parameters defined above, the following statistics are commonly used to describe the central location and dispersion of a sample of n observations, $x_1, x_2, \ldots, x_n$.

$$\text{Sample mean: } \bar{x} = \frac{\sum_{i=1}^{n} x_i}{n}$$

$$\text{Sample variance: } s^2 = \frac{\sum_{i=1}^{n} (x_i - \bar{x})^2}{n-1} = \frac{n \sum_{i=1}^{n} x_i^2 - \left(\sum_{i=1}^{n} x_i\right)^2}{n(n-1)}$$

$$\text{Sample standard deviation: } s = \sqrt{s^2}$$

$$\text{Sample coefficient of variation: } CV = (s/\bar{x})(100)$$

The denominator $n - 1$ in the definitions of s^2 and s is known as the *degrees of freedom* associated with these statistics, and will be denoted by the Greek letter ν (nu) in this chapter. The loss of one degree of freedom from the total of n observations in the sample is due to the necessity of replacing the unknown population mean μ by the sample mean $\bar{x}$. The second expression for s^2 is easier for hand calculation, while the first will minimize round-off problems in computer algorithms.

Example 2. For the sample of ten slopes from the conceptual population of all possible slopes for the new albumin kit, these statistics are:

$$\bar{x} = \frac{\sum_{i=1}^{10} x_i}{10} = \frac{0.1505 + 0.1486 + \ldots + 0.1475}{10} = 0.1497$$

$$s^2 = \frac{\sum_{i=1}^{10} (x_i - 0.1497)^2}{10 - 1}$$

$$= \frac{(0.1505 - 0.1497)^2 + (0.1486 - 0.1497)^2 + \ldots + (0.1475 - 0.1497)^2}{10 - 1}$$

$$= 0.3853 \, (10^{-5})$$

$$s = \sqrt{0.3853 \, (10^{-5})} = 0.0020$$

Example 3. For the sample of 100 GGT observations in Table 2A-1 from the conceptual population of 20 to 29 year old healthy males, these statistics are:

$$\bar{x} = \frac{\sum_{i=1}^{100} x_i}{100} = 25.9 \text{ U/L}$$

$$s^2 = \frac{\sum_{i=1}^{100} (x_i - 25.9)^2}{100 - 1} = 133.5 \text{ (U/L)}^2$$

$$s = \sqrt{133.5} = 11.6 \text{ U/L}$$

Random Sampling

A *random selection* from a population is one in which each member of the population has an equal chance of being selected. A *random sample* is one in which each member of the sample can be considered to be a random selection from the population of interest. Although much of statistical analysis and interpretation depends on the assumption of a random sample from some fixed population, actual data collection often does not satisfy this assumption. In particular, for sequentially generated data it is often true that observations adjacent to each other tend to be more alike than observations separated in time. A sample of such observations cannot be considered a sample of random selections from a fixed population. A later example will illustrate the possible problems in conducting a common statistical procedure with a sample that is incorrectly assumed to be a random sample from a fixed population. Fortunately, precautions can usually be taken in the design of an investigation to approximately validate the random sampling assumption.

THE GAUSSIAN PROBABILITY DISTRIBUTION

The Gaussian probability distribution, illustrated in Figure 2A-5, is of fundamental importance in statistics for several reasons. As mentioned earlier, a particular analytical value x will not usually be equal to the true value μ of the specimen being measured. Rather, associated with this particular value x there will be a particular measurement error $\epsilon = x - \mu$, which is the result of many contributing sources of error. These measurement errors tend to follow a probability distribution like that shown in Figure 2A-5, where the errors are symmetrically distributed with

Table 2A-1. SAMPLE OF 100 GGT OBSERVATIONS FROM THE CONCEPTUAL POPULATION OF 20 TO 29 YEAR OLD HEALTHY MALES

26	18	51	14	25
14	27	21	33	25
13	19	38	25	39
24	45	26	35	24
27	35	22	21	16
22	17	21	34	21
19	25	14	23	22
20	24	18	22	27
15	23	21	17	19
9	42	45	51	23
21	34	19	24	19
24	14	21	63	44
44	19	41	24	22
14	22	57	10	18
30	56	20	8	22
15	27	14	33	12
30	48	23	25	37
29	23	44	8	15
18	26	24	31	18
31	55	12	21	28

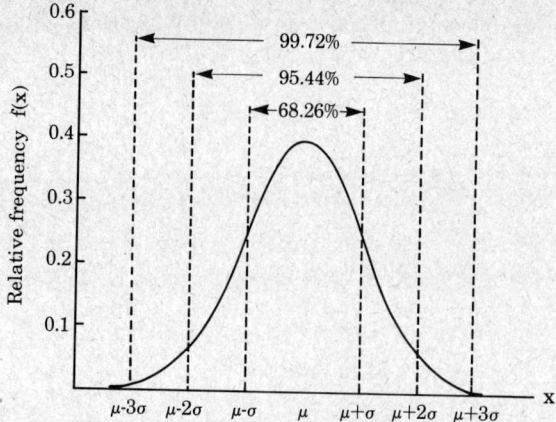

Figure 2A-5. The Gaussian probability distribution.

smaller errors occurring more frequently than larger ones, and with an expected value of 0. This important fact is known as *the central limit effect* for distributions of errors: if a measurement error ϵ is the sum of many independent sources of error, $\epsilon_1, \epsilon_2, \ldots, \epsilon_k$, several of which are major contributors, the probability distribution of the measurement error ϵ will tend to be Gaussian as the number of sources of error becomes large.

Another reason for the importance of the Gaussian probability distribution is that many statistical procedures are based on the assumption of a Gaussian distribution of values. Furthermore, these procedures are usually not seriously invalidated by departures from this assumption.

The magnitude of the uncertainty associated with sample statistics can be ascertained based on the fact that many sample statistics computed from large samples have a Gaussian probability distribution.

Characteristics of the Gaussian Probability Distribution

The Gaussian probability distribution is completely characterized by its mean μ and variance σ^2. The notation $N(\mu, \sigma^2)$ is often used for the distribution of a variable that is Gaussian with mean μ and variance σ^2. The following facts provide further insight into the Gaussian probability distribution.

1. 15.87% of all observations are positive deviations from the mean exceeding one standard deviation. Likewise, due to symmetry, 15.87% of all observations are negative deviations from the mean exceeding one standard deviation. Thus, 31.74% of the observations are deviations in either direction exceeding one standard deviation. Consequently, as illustrated in Figure 2A-5, 68.26% or roughly two-thirds of the observations deviate in either direction from the mean by less than one standard deviation.

2. Similarly, 4.56% or roughly 1/20 of the observations are deviations in either direction from the mean exceeding two standard deviations, and 95.44% or roughly 19/20 of the observations are deviations in either direction within two standard deviations from the mean.

3. Similarly, 0.28% of the observations deviate in either direction from the mean by more, and 99.72% by less, than three standard deviations.

These facts about the variable x that follows a $N(\mu, \sigma^2)$ distribution can be expressed by the following probability statements.

$$
\begin{aligned}
1.\ \Pr(x > \mu + \sigma) &= 0.1587 \\
\Pr(x < \mu - \sigma) &= 0.1587 \\
\Pr(|x - \mu| > \sigma) &= 0.3174 \\
\Pr(|x - \mu| < \sigma) &= 0.6826
\end{aligned}
$$

$$
\begin{aligned}
2.\ \Pr(x > \mu + 2\sigma) &= 0.0228 \\
\Pr(x < \mu - 2\sigma) &= 0.0228 \\
\Pr(|x - \mu| > 2\sigma) &= 0.0456 \\
\Pr(|x - \mu| < 2\sigma) &= 0.9544
\end{aligned}
$$

3. $\Pr(x > \mu + 3\sigma) = 0.0014$
   $\Pr(x < \mu - 3\sigma) = 0.0014$
   $\Pr(|x - \mu| > 3\sigma) = 0.0028$
   $\Pr(|x - \mu| < 3\sigma) = 0.9972$

The Standard Gaussian Distribution

Probability statements about a variable x that follows a $N(\mu, \sigma^2)$ distribution are usually made by considering the variable

$$z = \frac{x - \mu}{\sigma} \qquad (1)$$

which is called the *standard Gaussian variable*. The variable z has a Gaussian probability distribution with $\mu = 0$ and $\sigma^2 = 1$, i.e., z is $N(0, 1)$. Thus, the previous probability statements can be rewritten as follows:

1. $\Pr(x > \mu + \sigma) = \Pr((x - \mu) > \sigma) = \Pr((x - \mu)/\sigma > 1) = \Pr(z > 1) = 0.1587$
   $\Pr(z < -1) \quad = 0.1587$
   $\Pr(|z| > 1) \quad = 0.3174$
   $\Pr(|z| < 1) \quad = 0.6826$

2. $\Pr(z > 2) \quad = 0.0228$
   $\Pr(z < -2) = 0.0228$
   $\Pr(|z| > 2) = 0.0456$
   $\Pr(|z| < 2) = 0.9544$

3. $\Pr(z > 3) \quad = 0.0014$
   $\Pr(z < -3) = 0.0014$
   $\Pr(|z| > 3) = 0.0028$
   $\Pr(|z| < 3) = 0.9972$

Probability Statements for Any Gaussian Distribution

The procedure for determining the probability $\Pr(x > x_a)$ that a randomly selected value of the variable x with a $N(\mu, \sigma^2)$ distribution will exceed some value x_a is as follows:

1. Compute the standard Gaussian value $z_a = (x_a - \mu)/\sigma$.

2. Obtain $\Pr(z > z_a)$ from the table of probability points of the standard Gaussian distribution in Statistical Table 20-21, in the Appendix.

3. Then $\Pr(x > x_a) = \Pr(z > z_a)$.

Example 4. Suppose that the distribution of fasting serum glucose values in 20 to 29 year old healthy males is known to be $N(90, 10^2)$. To find the probability that the fasting serum glucose value of a randomly selected member of this population will exceed 105 mg/dL, i.e., $\Pr(x > 105)$, we proceed as follows (refer to Figure 2A-6):

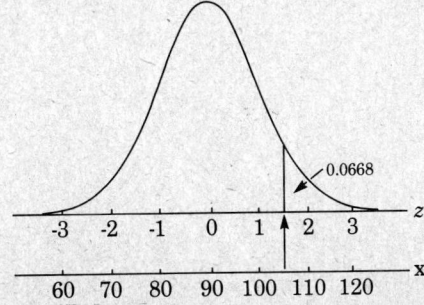

Figure 2A-6. Illustration for Example 4. $\Pr(x > 105) = \Pr(z > 1.5) = 0.0668$.

1. $z_a = (x_a - \mu)/\sigma = (105 - 90)/10 = 1.5$

2. $\Pr(z > z_a) = \Pr(z > 1.5) = 0.0668$ from the table of probability points of the standard Gaussian distribution in Statistical Table 20-21, in the Appendix.

3. $\Pr(x > 105) = 0.0668$

STUDENT'S t PROBABILITY DISTRIBUTION

In order to determine probabilities associated with a Gaussian distribution by the procedure just illustrated, it is necessary to know the population standard deviation σ. In actual practice σ is often unknown so we cannot calculate $z_a = (x_a - \mu)/\sigma$. However, if a random sample can be taken from the Gaussian population, we can calculate the sample standard deviation s, substitute s for σ, and compute the value $t_a = (x_a - \mu)/s$ of the variable

$$t = \frac{x - \mu}{s} \tag{2}$$

Under these conditions the variable $t = (x - \mu)/s$ has a probability distribution called the *Student's t distribution*. The t distribution is really a family of distributions depending on the degrees of freedom ν for the sample standard deviation. Several t distributions from this family are shown in Figure 2A-7. When the size of the sample and the degrees of freedom for s are infinite, there is no uncertainty in s, and so the t distribution is identical to the standard Gaussian distribution. However, when the sample size is small, the uncertainty in s causes the t distribution to have greater dispersion and heavier tails than the standard Gaussian distribution as illustrated in Figure 2A-7. A table of probability points for the t distribution is given in Statistical Table 20-22, in the Appendix.

Example 5. Returning to the fasting serum glucose example, suppose that σ is unknown and that a random sample of size 20 from the $N(90, \sigma^2)$ distribution of fasting serum glucose values in 20 to 29 year old healthy males yielded a sample standard deviation s = 10.0 mg/dL. Then to find the probability $\Pr(x > 105)$, we proceed as follows (refer to Figure 2A-8):

1. $t_a = (x_a - \mu)/s = (105 - 90)/10 = 1.5$
2. $\Pr(t > t_a) = \Pr(t > 1.5) = 0.08$, approximately, from the table of probability points of the t distribution with $\nu = n - 1 = 19$ degrees of freedom in Statistical Table 20-22, in the Appendix.
3. $\Pr(x > 105) = 0.08$

List of Symbols for Fundamental Statistical Concepts Section (in order of appearance)

β slope of calibration curve
α intercept of calibration curve
A absorbance
C concentration

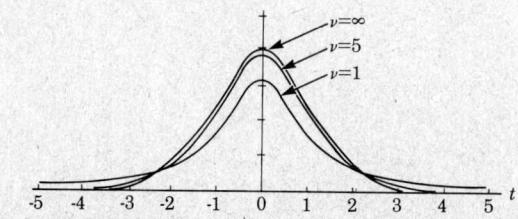

Figure 2A-7. The t distribution for $\nu = 1$, 5 and ∞.

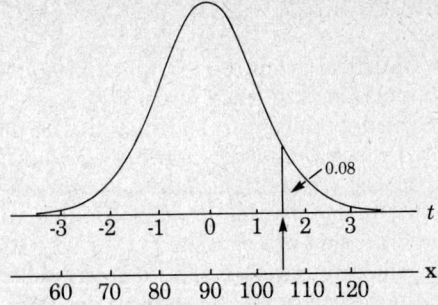

Figure 2A-8. Illustration for Example 5. $Pr(x > 105) = Pr(t > 1.5) = 0.08$.

n_i	number of observations in the ith cell of a frequency distribution
x	variable
f(x)	functional form of the population frequency distribution curve for variable x
x_a	a particular value of the variable x
x_b	a particular value of the variable x
Pr( )	probability that the variable specified takes on a value from those specified in ( )
N	number of members of a finite population
x_i	ith member of a population or a sample
μ	population mean for variable x
Σ	summation of specified observations of a specified variable; $\sum_{i=1}^{N} x_i$ is the sum of all the values of variable x from the first to the last for a finite population
E(x)	expected value of variable x; $E(x) = \mu$
σ^2	population variance for variable x
V(x)	population variance for variable x; $V(x) = \sigma^2$
σ	population standard deviation for variable x
n	number of members in a sample
$\bar{x}$	sample mean for variable x
s^2	sample variance for variable x
s	sample standard deviation for variable x
CV	sample coefficient of variation for variable x
ν	degrees of freedom associated with s^2
ϵ	measurement error variable defined as the difference between the values of variable x and the expected value of x; $\epsilon = x - \mu$
$N(\mu, \sigma^2)$	probability distribution that is Gaussian with mean μ and variance σ^2
|x|	absolute value of number x; the positive part of x; $\|-2.3\| = \|2.3\| = 2.3$
z	variable with the standard Gaussian N(0, 1) probability distribution
z_a	a particular value of the variable z
N(,)	probability distribution that is Gaussian with mean and variance specified in (,)
t	variable with Student's t probability distribution
t_a	a particular value of the variable t

SAMPLING DISTRIBUTIONS

In this section we introduce the important concept of sampling distributions, which forms the basis for the statistical inference procedures presented in the remaining sections. First we discuss the critical assumption of random sampling from a fixed population, and at the end of this section we give an example of the potential consequences of applying a common statistical procedure in violation of this assumption. The two general areas of statistical inference, namely estimation and significance testing, are developed through use of the sampling distribution.

INDEPENDENCE AND RANDOM SAMPLING

It has been stated that statistical procedures for making inferences about a quantity x that varies according to a fixed population frequency function f(x) generally require a random sample $x_1, x_2, \ldots, x_n$ from that fixed population. The quantity x that varies due to chance factors according to f(x) is called a *random variable*, and the values $x_1, x_2, \ldots, x_n$ of the random sample are observed values of the random variable x. *Statistical independence* of these values $x_1, x_2, \ldots, x_n$ means that the ith observation x_i of the sample varies according to $f(x_i) = f(x)$ regardless of the value of the preceding observation x_{i-1}, or any other observation in the sample. For example, statistical independence means that x_i is equally likely to be on either side of the mean μ regardless of where the preceding observation x_{i-1} is with respect to μ. When a sample consists of statistically independent observations from a fixed population, then that sample can be considered a random sample. As stated earlier, often in analytical work the sources of experimental error tend to persist over time causing the ith observation x_i to vary high or low compared to μ if x_{i-1} was high or low. In this situation the observations are *statistically dependent*, i.e., $f(x_i)$ *depends* on x_{i-1} such that $E(x_i) > \mu$ if $x_{i-1} > \mu$, and $E(x_i) < \mu$ if $x_{i-1} < \mu$. A sample of such observations cannot be considered a random sample from a fixed population.

SAMPLING DISTRIBUTIONS

Statistical inference is basically concerned with extracting information about unknown population parameters from a sample of observations, and making decisions about these parameters based on this information and its inherent uncertainty. For example, in the initial stages of a method evaluation study, we might be interested in the within-run standard deviation σ to determine whether or not it meets our requirements. For illustrative purposes, consider a simple experiment where we make 15 randomly spaced determinations on a particular fluid within a single run and calculate the statistic s as the basis for our decision. Our decision about σ will obviously be based on this particular value of s. However, a different random sample of 15 within-run determinations would likely produce a different value of s. Clearly, inference about the population parameter σ from the sample statistic s can be made with confidence only if we understand the fluctuating behavior of the statistic s when computed from different random samples of size 15. This uncertain "behavior" of the statistic s, and of any other statistic, can be characterized by its probability distribution, called the *sampling distribution* of the statistic. Formally, the *sampling distribution* of a statistic is the probability distribution of all possible values which can be assumed by that statistic, computed from samples of the same size from the same population. A clear understanding of sampling distributions is essential, since this is the concept most basic to an understanding of statistical inference.

There are two ways of approaching the study of sampling distributions. One, based on actual or simulated repeat samples, leads to what is called the *experimental sampling distribution*. The other, based on mathematical theory, leads to what is called the *theoretical sampling distribution*. The first sampling distribution we will study is that of the sample mean $\bar{x}$.

Experimental Sampling Distribution of the Mean

We begin our discussion of the distribution of the sample mean by constructing an experimental sampling distribution. Such an exercise provides insight into the "behavior" of the sample mean $\bar{x}$.

Suppose that an analyst wants to learn about the mean fasting serum glucose, μ, of the population of diabetics who visit a certain clinic. A random sample of twenty-five diabetic patients visiting the clinic over several weeks produces a mean fasting serum glucose $\bar{x} = 152.8$ mg/dL. Realizing that a different random sample of twenty-five diabetics from this same clinic would likely produce a sample mean different from 152.8 mg/dL, the analyst would like to know how much the sample means will vary when computed from different random samples of size 25 from this particular population.

Simulation methods can be used to determine how much such means will vary as a result of pure chance or random sampling variation. Suppose that for this population of diabetics,

Table 2A-2. MEANS OF 50 RANDOM SAMPLES, EACH OF SIZE 25, FROM THE CONCEPTUAL POPULATION OF DIABETICS

Sample Number	$\bar{x}$	Sample Number	$\bar{x}$	Sample Number	$\bar{x}$	Sample Number	$\bar{x}$	Sample Number	$\bar{x}$
1	160.2	11	157.9	21	154.4	31	157.4	41	159.5
2	164.2	12	157.7	22	163.8	32	170.6	42	163.8
3	159.6	13	159.5	23	156.7	33	163.2	43	165.9
4	160.5	14	158.0	24	154.1	34	162.1	44	158.6
5	165.6	15	151.2	25	159.1	35	160.0	45	161.2
6	158.4	16	169.0	26	157.4	36	163.9	46	162.3
7	159.1	17	161.8	27	166.2	37	157.2	47	152.5
8	161.7	18	153.9	28	165.1	38	163.6	48	158.0
9	160.9	19	152.7	29	157.0	39	167.3	49	158.7
10	160.8	20	159.1	30	162.4	40	165.4	50	161.7

fasting serum glucose is a random variable following a Gaussian distribution with mean $\mu = 160$ mg/dL and standard deviation $\sigma = 20$ mg/dL. We can then simulate the drawing of a large number of random samples of size 25 from this distribution to study the characteristics of the distribution of sample means. We simulated the drawing of fifty such random samples and discuss how this can be done in the last section of the chapter. The resultant fifty sample means listed in Table 2A-2 were used to construct the experimental sampling distribution plotted in Figure 2A-9.

Inspection of the experimental sampling distribution of $\bar{x}$ provides a great deal of information about how the sample means in different random samples vary due to pure chance. For instance, the smallest mean is 151.2 mg/dL and the largest is 170.6 mg/dL. Furthermore, 70% (35 of 50) of the sample means are between 156.0 mg/dL and 164.0 mg/dL. Since we know that the true population mean is 160 mg/dL, we can say that 70% of the sample means differ, or are "off" by less than 4 mg/dL. Also, 94% (47 of 50) of the sample means are between 152.0 mg/dL and 168.0 mg/dL, and so are "off" by less than 8 mg/dL.

As this example has shown, we can gain valuable insight into the distribution of the sample mean by taking many random samples of the same size and constructing an experimental sampling distribution. However, in actual practice, we ordinarily take only a single random sample. Fortunately, the sampling distribution of the mean has been derived by mathematical theory, thus allowing us to determine how close a sample mean may be to the mean of the population from which it came.

Theoretical Sampling Distribution of the Mean

Provided the random sampling model is appropriate, it can be shown through mathematical theory that the sampling distribution of the mean $\bar{x}$ has the following characteristics:

1. The mean of the distribution of $\bar{x}$ will be equal to the mean of the population being sampled, i.e., $\mu_{\bar{x}} = E(\bar{x}) = \mu$.

2. The variance of the distribution of $\bar{x}$ will be equal to the variance of the population being sampled divided by the sample size, i.e., $\sigma_{\bar{x}}^2 = V(\bar{x}) = \sigma^2/n$, and $\sigma_{\bar{x}} = \sigma/\sqrt{n}$.

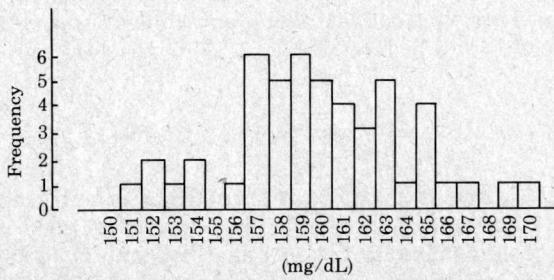

Figure 2A-9. Frequency distribution of means of 50 random samples, each of size 25, from a hypothetical population of fasting serum glucose concentrations in diabetics.

3. The form of the distribution of $\bar{x}$ will be Gaussian
   a. if the population being sampled is Gaussian or
   b. if the sample size is large, regardless of the functional form of the population being sampled.

Result (3b) is due to the central limit effect discussed earlier, and is known as the *central limit theorem*. This theorem is of fundamental importance in statistics, since it justifies the use of a Gaussian sampling distribution in many types of problems as long as the sample size is large. The sample size necessary for the central limit theorem to apply depends upon the degree to which the population being sampled deviates from Gaussian form. Generally, a sample size of 30 is adequate except in cases of very unusual distribution shapes. Additionally, the theorem states that the larger the sample size, the more closely the distribution of the sample mean will approximate the Gaussian distribution.

Example 6. In the example of fasting serum glucose values from diabetics, we initially took a random sample of size 25 from a Gaussian population having a mean $\mu = 160$ mg/dL and standard deviation $\sigma = 20$ mg/dL. The sampling distribution of $\bar{x}$ therefore is a Gaussian distribution with mean and standard deviation given by

$$\mu_{\bar{x}} = \mu = 160 \text{ mg/dL and}$$

$$\sigma_{\bar{x}} = \frac{\sigma}{\sqrt{n}} = \frac{20}{\sqrt{25}} = 4 \text{ mg/dL}$$

It is interesting to see how close the mean and standard deviation of the 50 means in the experimental sampling distribution are to these theoretical values. The 50 means have an average of 160.4 mg/dL and a standard deviation of 4.2 mg/dL. These values are close enough to the theoretical values above to provide experimental verification of the theorem for the mean and standard deviation of the sampling distribution of $\bar{x}$.

We can also check the agreement between the proportion of sample means falling between 156.0 mg/dL and 164.0 mg/dL in the experimental and theoretical sampling distributions. In the experimental sampling distribution, 70% of the sample means were in this range. The theoretical sampling distribution of $\bar{x}$ is Gaussian with a mean $\mu_{\bar{x}} = 160$ mg/dL and a standard deviation $\sigma_{\bar{x}} = 4$ mg/dL. Since 156 corresponds to $\mu_{\bar{x}} - 1\sigma_{\bar{x}}$ and 164 corresponds to $\mu_{\bar{x}} + 1\sigma_{\bar{x}}$, the proportion of values between 156 and 164 is 68% from Gaussian distribution probabilities and agrees closely with 70% obtained from the experimental distribution. Similarly, the 94% of sample means between 152 and 168 in the experimental sampling distribution is in close agreement with the 95% of sample means expected to fall between $\mu_{\bar{x}} - 2\sigma_{\bar{x}}$ and $\mu_{\bar{x}} + 2\sigma_{\bar{x}}$ based on the theoretical distribution of $\bar{x}$.

Estimation of Population Parameters

Under the random sampling model, the sample mean $\bar{x}$ has an expected value μ and varies about μ with standard deviation $\sigma/\sqrt{n}$, which can be decreased by increasing the size of the random sample. In other words, $\bar{x}$ will tend to be closer to μ as the sample size increases. Thus, it is reasonable to think of $\bar{x}$ as an *estimate* of the unknown population mean μ. Similarly, it can be shown that the sampling distribution of s^2 has σ^2 as its mean with a standard deviation proportional to $1/\sqrt{n}$. Thus, we can think of s^2 as an *estimate* of the unknown population variance σ^2. Estimation procedures will be discussed in detail in the following sections.

Significance Testing

The sampling distribution of a statistic provides the basis for a common statistical procedure, the significance or hypothesis test. For illustration, consider again the population of GGT values in 20 to 29 year old healthy males, and assume that σ is known to be 11 U/L. Now suppose it is of interest to test the hypothesis or claim that μ is equal to 25 U/L. Recall that a sample of size n = 100 values produced $\bar{x} = 25.9$ U/L. Under the assumption of random sampling, the sampling distribution of $\bar{x}$ is $N(\mu, \sigma^2/n)$ or $N(\mu, 1.21)$. Now if the mean μ of the original population is really equal to the hypothesized value of 25 U/L, then the observed $\bar{x} = 25.9$ U/L should look like a typical random selection from the sampling distribution $N(25, 1.21)$ shown in Figure 2A-10. Consequently, if 25.9 U/L appears to be a typical random observation from the $N(25, 1.21)$ distribution, there is no evidence to doubt the hypothesis that $\mu = 25$ U/L. Alternatively, if 25.9 appears to be an atypical random observation from the $N(25, 1.21)$ distribution, then this is evidence to doubt that $\mu = 25$ U/L. In fact, the observed $\bar{x} = 25.9$ appears to be a typical random selection from the $N(25, 1.21)$ distribution in Figure 2A-10, so there is no evidence to reject the hypothesis that $\mu = 25$ U/L. A more objective method of comparing the result $\bar{x} = 25.9$ to its sampling distribution under the hypothesis that $\mu = 25$ is demonstrated in the following example.

Example 7. To determine just how usual or unusual a result $\bar{x} = 25.9$ U/L is from the $N(25, 1.21)$ distribution, we calculate the probability $Pr(\bar{x} \geq 25.9)$ of a random selection as large as or larger than that observed, $\bar{x} = 25.9$.

1. $z_o = \dfrac{\bar{x} - \mu_{\bar{x}}}{\sigma_{\bar{x}}} = \dfrac{25.9 - 25}{\sqrt{1.21}} = 0.82$

2. $Pr(z \geq z_o) = Pr(z \geq 0.82) = 0.21$ from the table of probability points of the standard Gaussian distribution in Statistical Table 20-21, in the Appendix.

3. $Pr(\bar{x} \geq 25.9) = 0.21$. In other words, the probability is 0.21 that a sample mean $\bar{x}$, computed from a random sample of 100 observations from a hypothetical distribution with $\mu = 25$ and $\sigma = 11$, would be as large as or larger than 25.9.

Thus, $\bar{x} = 25.9$ U/L is not an especially unusual random observation from the $N(25, 1.21)$ distribution since a mean this large or larger would occur in 21% of random samples of size 100. The conclusion is that there is not sufficient evidence to reject the hypothesis that the mean of the population of GGT values in 20 to 29 year old healthy males is 25 U/L.

The utilization of the sampling distribution of $\bar{x}$ for the above significance test of the population mean requires three assumptions: (1) that the sample is a random sample, (2) that the population sampled is Gaussian or that the sample is large enough so that the distribution of $\bar{x}$ is approximately Gaussian, and (3) that σ^2 is known. In the GGT example, the random sampling assumption should be approximately validated by the collection of GGT values over several weeks or months, and the sample size of $n = 100$ ensures the Gaussian form of the distribution of $\bar{x}$ even though the distribution of GGT values itself is likely to be non-Gaussian. Although σ was assumed to be known, in reality it is often unknown. If so, the above significance testing procedure can still be used by replacing the unknown population variance $\sigma_{\bar{x}}^2 = \sigma^2/n$ of the sampling distribution by $s_{\bar{x}}^2 = s^2/n$ where s^2 is the sample variance computed from the random sample. To determine the likelihood associated with the observed sample mean $\bar{x}$, we use $t = (\bar{x} - \mu_{\bar{x}})/s_{\bar{x}}$, distributed according to the t distribution with $\nu = n - 1$ degrees of freedom, as in the following example.

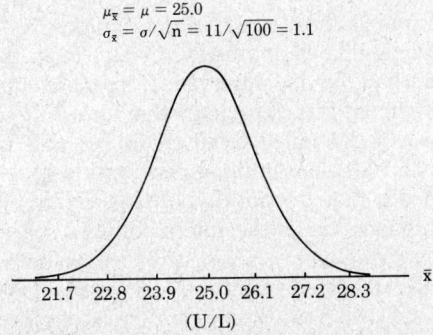

$$\mu_{\bar{x}} = \mu = 25.0$$
$$\sigma_{\bar{x}} = \sigma/\sqrt{n} = 11/\sqrt{100} = 1.1$$

| 21.7 | 22.8 | 23.9 | 25.0 | 26.1 | 27.2 | 28.3 |
(U/L)

Figure 2A-10. Sampling distribution of $\bar{x}$ for random samples of 100 observations from a population with $\mu = 25$ U/L and $\sigma = 11$ U/L.

Example 8. If σ is unknown in the GGT example, the significance test based on the random sample of n = 100 values with $\bar{x}$ = 25.9 U/L and s = 11.6 U/L would proceed as follows:

1. $t_o = \dfrac{\bar{x} - \mu_{\bar{x}}}{s_{\bar{x}}} = \dfrac{25.9 - 25}{11.6/\sqrt{100}} = 0.78$

2. $\Pr(t \geq t_o) = \Pr(t \geq 0.78) = 0.22$, approximately, from the table of probability points of the t distribution with $\nu = 99$ degrees of freedom. (See Statistical Table 20-22, in the Appendix.)

3. $\Pr(\bar{x} \geq 25.9) = 0.22$. In other words, the probability is 0.22 that a sample mean $\bar{x}$, computed from a random sample of 100 observations from a hypothetical distribution with $\mu = 25$ and σ estimated by s = 11.6, would be as large as or larger than 25.9.

The conclusion again is that there is not sufficient evidence to reject the hypothesis that $\mu = 25$ U/L.

Significance Testing Strategy

The GGT example of a significance test about a single population mean has illustrated the general strategy of a statistical significance test.

1. State the possible decisions to be made in the form of a pair of hypotheses called the *null* and *alternative hypotheses.* In the GGT example, the null hypothesis is H_o: $\mu = 25$ U/L and the alternative is H_a: $\mu \neq 25$ U/L. The alternative hypothesis may be two-sided as in this example ($\mu < 25$ or $\mu > 25$), or it may be one-sided if there is only one alternative direction that is plausible or of interest. The alternative hypothesis should be stated in advance of the data collection.

2. State what the relevant statistic is for testing the null hypothesis against the alternative. In the GGT example of a significance test about a single population mean, the relevant statistic is $\bar{x}$.

3. State and roughly sketch the sampling distribution showing how the statistic would behave if the null hypothesis were true. In the GGT example with σ^2 known this is shown in Figure 2A-10.

4. Compute the observed value for this statistic from the data. In the GGT example, the observed value for the statistic $\bar{x}$ is 25.9 U/L.

5. Assuming that the null hypothesis is true, calculate the probability of observing by chance (i.e., random sampling variation) a result for the statistic that differs, in the direction indicated by the alternative hypothesis, as much as or more than the one actually obtained. In the GGT example with σ^2 known, the probability of observing a sample mean by chance as large or larger than 25.9 U/L if $\mu = 25$ U/L is 21%. This probability is a measure of the degree of evidence against H_o: $\mu = 25$ and in favor of $\mu > 25$. Since the alternative hypothesis H_a: $\mu \neq 25$ is two-sided, it is customary to report twice this probability, or 2(21%) = 42%, as the probability of observing a sample mean differing by at least 0.9 U/L in either direction from the expected value $\mu = 25$ U/L.

6. The probability determined above is called the *significance level* or *p-value* of the observed result. It is defined as the probability of an observed result from the sampling distribution under H_o as much or more extreme than the one observed. It measures the degree of agreement between the observed value of a statistic and its expected value under the null hypothesis. If the *p*-value is moderate or large, the data have failed to discredit the null hypothesis. In addition, if the sample size is large enough so that the sampling distribution is relatively narrow in width, then the data have provided evidence that the null hypothesis is true or almost true. If the *p*-value is small, the observed result is unusual under the null hypothesis, so the null hypothesis is discredited or rejected and a *statistically significant result* is said to have been obtained.

In practice, it is generally accepted that the observed result is statistically significant if the *p*-value is less than 0.05. This means that there is a 5% risk of rejecting the null hypothesis when in fact it is true. Depending upon the situation, a larger or smaller risk may be appropriate. In any case, it should be noted that it is much more informative to report the actual *p*-value than merely to state that the observed result is or is not statistically significant at the 0.05 level. For instance, while there would be no practical difference between a result with $p = 0.06$ and a result with $p = 0.04$, dichotomizing the results at the 0.05 level would report one of these results as statistically significant and the other as not.

The *power* of a statistical significance test provides the investigator with the answer to an important question: if it is necessary to know that the null hypothesis is false when some specific alternative situation is true, what is the likelihood that the significance test procedure will reject H_o in favor of H_a? For example, suppose it is important for the analyst to reject H_o: $\mu = 25$ in favor of H_a: $\mu > 25$ if in fact μ is really 30 in the GGT example. Now if μ is really 30 U/L, then the sample mean $\bar{x}$ from a random sample of size 100 is likely (with 95% probability) to fall between $30 \pm 2(1.1)$, or 27.8 and 32.2 U/L. Since no such sample mean would appear to be a typical random selection from the sampling distribution in Figure 2A-10 under H_o: $\mu = 25$, the significance test procedure is likely to lead to rejection of H_o: $\mu = 25$ in favor of H_a: $\mu > 25$, as required. This significance test with n = 100 would be said to have good power for the alternative $\mu = 30$ U/L. Clearly, the power of a statistical significance test depends on the alternative situation and the width of the sampling distribution. Since the sampling distribution becomes narrower as the sample size increases, the power will increase as n increases for a fixed alternative. Further discussion of the power of statistical significance tests can be found in many textbooks.[23,25]

In many investigations the decision to be made is a practical one rather than a statistical one. Therefore, it is important to recognize that statistical significance does not necessarily imply *practical significance*. However, the magnitude of the *p*-value and the magnitude of the difference between the observed result and its expected value under H_o provide objective information as part of the basis for the decision. Additionally, it should be noted that statistical nonsignificance does not imply practical nonsignificance if the sample size is small and the power is poor for an alternative of practical importance.

Misapplication of the Significance Test for μ

As an example of a situation where the random sampling model is often assumed but likely invalid, consider the albumin sensitivity experiment with ten slopes from ten consecutive runs. Suppose we conduct a significance test of the null hypothesis H_o: $\mu = 0.1480$ against the alternative H_a: $\mu > 0.1480$, where 0.1480 is the average slope of 120 runs over the previous three months using the standard kit. In other words, we would like to test the hypothesis that the sensitivity of the new kit is the same as that of the standard kit against the alternative that its sensitivity is better. The sample of ten slopes has a mean $\bar{x} = 0.1497$ and a standard deviation s = 0.0020.

1. $t_o = \dfrac{\bar{x} - \mu_{\bar{x}}}{s_{\bar{x}}} = \dfrac{0.1497 - 0.1480}{0.0020/\sqrt{10}} = 2.69$

2. $Pr(t \geq t_o) = Pr(t \geq 2.69) = 0.01$, approximately, from the table of probability points of the t distribution with $\nu = 9$ degrees of freedom. (See Statistical Table 20-22, in the Appendix.)

3. $Pr(\bar{x} \geq 0.1497) = 0.01$

Since the *p*-value is 0.01, the result $\bar{x} = 0.1497$ is statistically significant, and there is sufficient evidence to conclude that the sensitivity of the new kit is better than that of the standard kit *if* the assumptions of the significance test are not in question. In fact, the assumption that the ten slopes are independent observations and constitute a random sample from the conceptual population of all possible slopes with the new kit is highly suspect. Inspection of the past 120 consecutive slopes with the standard kit in Figure 2A-11 reveals a high degree of dependency. Clearly any sample of ten consecutive slopes with a particular kit from this system cannot be considered a random sample of ten independent slopes from the conceptual population of all possible slopes. Thus, the significance test and the conclusion based on the sample of ten consecutive slopes is inappropriate. In fact, inspection of the ten slopes with the new kit compared to the previous 120 slopes with the standard kit in Figure 2A-11 reveals no evidence that the new kit has better sensitivity.

In comparing the sensitivity of the new albumin kit to the assumed average sensitivity 0.1480

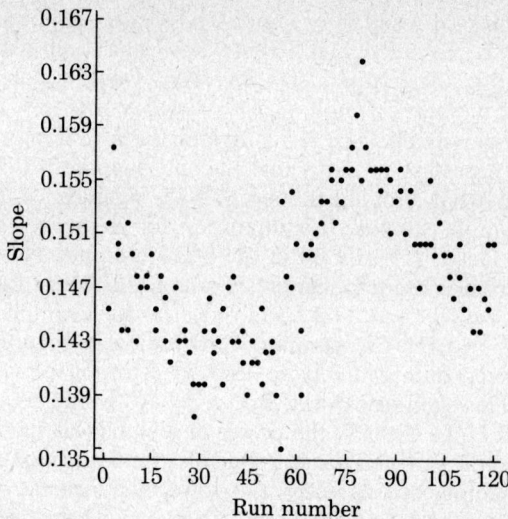

Figure 2A-11. Slopes of 120 consecutive runs with the standard albumin kit.

of the standard kit, ten slopes from ten runs over an extended period of time may approximately validate the random sampling model, but would clearly be very inefficient in respect to time. A better experimental design for this investigation is a randomized comparison design, which will be discussed in the next section.

List of Symbols for Sampling Distributions Section (in order of appearance)

x	random variable
f(x)	functional form of the population frequency distribution curve for the variable x
n	number of members in a sample
x_i	*i*th member of a sample
f( )	functional form of the population frequency distribution curve for the variable specified in ( )
μ	population mean for the variable x
E( )	expected value (i.e., population mean) of the variable specified in ( )
σ	population standard deviation for the variable x
s	sample standard deviation for the variable x
$\bar{x}$	sample mean for the variable x
$\mu_{\bar{x}}$	population mean for the variable $\bar{x}$
$E(\bar{x})$	expected value of the variable $\bar{x}$; $E(\bar{x}) = \mu_{\bar{x}} = \mu$
$\sigma_{\bar{x}}^2$	population variance for the variable $\bar{x}$
$V(\bar{x})$	population variance for the variable $\bar{x}$; $V(\bar{x}) = \sigma_{\bar{x}}^2 = \sigma^2/n$
$\sigma_{\bar{x}}$	population standard deviation for the variable $\bar{x}$
N(,)	Gaussian probability distribution with mean and variance specified in (,)
z	variable with the standard Gaussian N(0, 1) probability distribution
z_o	observed value of the variable z under the null hypothesis H_o
Pr( )	probability that a specified variable takes on a value from those specified in ( )
$s_{\bar{x}}^2$	sample variance for the variable $\bar{x}$
$s_{\bar{x}}$	sample standard deviation for the variable $\bar{x}$; also frequently called the standard error of the mean
t	variable with the Student's t probability distribution
t_o	observed value of the variable t under the null hypothesis H_o
ν	degrees of freedom associated with s^2
H_o	null hypothesis
H_a	alternative hypothesis
p	p-value of the observed result in a statistical significance test; also called the significance level of the observed result

INFERENCE ABOUT POPULATION MEANS

We are now ready to begin discussion of inferences about population means based on random samples drawn from the populations of interest. In addition to extending the significance testing procedures discussed in the previous section, this section covers interval estimation of population means, including sample size determination. It will also be shown how experimental design techniques such as randomization and blocking can be used to ensure the validity and enhance the sensitivity of analytical investigations. A brief introduction to the Poisson distribution and an application in radioimmunoassay is included. Table 2A-5 at the end of this section contains a summary of significance tests and confidence intervals for population means.

THE RANDOMIZED COMPARISON

Consider the comparison of the new and standard albumin kits discussed at the end of the previous section. A better experiment to evaluate the sensitivity of the new kit with respect to the standard kit would be to run tests with both kits several times during the same short time period. We would gain the advantage of comparing the two kits under like conditions, and thus eliminate the possibility of instrument or system changes over time which would cause confusion in our comparison of the average sensitivities of the two kits. Other applications of randomized comparisons in analytical work include comparisons of the population means of measurements on specimens from males and females and of measurements on a split-serum pool where one part is neat and the other is spiked with an interfering substance of interest. The design and analysis of such experiments will be considered after developing the sampling distribution theory a little further.

Sampling Distribution of $\bar{x}_A - \bar{x}_B$

The following mathematical results are necessary for determining the sampling distribution of the difference between two sample means. If x_A and x_B are two independent random variables with means μ_A and μ_B and variances σ_A^2 and σ_B^2, then the random variable defined as $y = x_A + x_B$ has mean and variance given by

$$E(y) = E(x_A) + E(x_B) = \mu_A + \mu_B \tag{3}$$

$$V(y) = V(x_A) + V(x_B) = \sigma_A^2 + \sigma_B^2 \tag{4}$$

Also, the random variable $y = x_A - x_B$ has mean and variance given by

$$E(y) = E(x_A) - E(x_B) = \mu_A - \mu_B \tag{5}$$

$$V(y) = V(x_A) + V(x_B) = \sigma_A^2 + \sigma_B^2 \tag{6}$$

Now assume we have two independent random samples of size n_A and n_B from two populations with means μ_A and μ_B and variances σ_A^2 and σ_B^2. Then the sampling distributions of $\bar{x}_A$ and $\bar{x}_B$ are $N(\mu_A, \sigma_A^2/n_A)$ and $N(\mu_B, \sigma_B^2/n_B)$. Using equations (5) and (6) we have

$$E(\bar{x}_A - \bar{x}_B) = E(\bar{x}_A) - E(\bar{x}_B) = \mu_A - \mu_B \text{ and} \tag{7}$$

$$V(\bar{x}_A - \bar{x}_B) = V(\bar{x}_A) + V(\bar{x}_B) = \sigma_A^2/n_A + \sigma_B^2/n_B \tag{8}$$

so that the sampling distribution of $\bar{x}_A - \bar{x}_B$ is

$$N(\mu_A - \mu_B, \sigma_A^2/n_A + \sigma_B^2/n_B) \tag{9}$$

Thus, to conduct a significance test about two population means,

$$z = \frac{(\bar{x}_A - \bar{x}_B) - (\mu_A - \mu_B)}{\sqrt{\sigma_A^2/n_A + \sigma_B^2/n_B}} \tag{10}$$

is distributed as a standard Gaussian variable and can be used to determine the significance of an observed result $\bar{x}_A - \bar{x}_B$ for any null hypothesis about $\mu_A - \mu_B$.

In practice it is often the case that the variances σ_A^2 and σ_B^2 are unknown. In such cases, it is usually reasonable to assume that $\sigma_A^2 = \sigma_B^2 = \sigma^2$, and then a pooled or weighted average

$$s_p^2 = (\nu_A s_A^2 + \nu_B s_B^2)/(\nu_A + \nu_B) \tag{11}$$

of the two sample variances provides an estimate of the unknown common variance σ^2. The degrees of freedom associated with the estimate s_p^2 are $\nu_A + \nu_B = (n_A - 1) + (n_B - 1)$. Then

$$t = \frac{(\bar{x}_A - \bar{x}_B) - (\mu_A - \mu_B)}{\sqrt{s_p^2/n_A + s_p^2/n_B}} \tag{12}$$

is distributed according to the t distribution with $\nu_A + \nu_B$ degrees of freedom and can be used to determine the significance level of an observed result $\bar{x}_A - \bar{x}_B$ for any null hypothesis about $\mu_A - \mu_B$. For the case $\sigma_A \neq \sigma_B$, the corresponding formula for equation (12) is given in Table 2A-5 at the end of this section.

Design of the Experiment

Consider the comparison of the new and standard albumin kits. How should the experiment be designed so that the assumption of two independent random samples is approximately validated, allowing use of expression (9) for the sampling distribution of $\bar{x}_N - \bar{x}_S$ (N = new, S = standard)? Suppose ten runs are to be carried out with each kit. The twenty runs must be assigned to the two kits in such a manner that the experimental errors are likely to contribute equally to the sample slopes of the new and standard kits. Clearly, ten consecutive runs with one kit followed by ten runs with the other would not satisfy this requirement. However, a *random allocation* of one of the two kits to each of the twenty runs would ensure, at least in probability, that the experimental errors contribute equally to the two kits, thereby validating the assumption of two independent random samples. This could be accomplished, for example, by randomly drawing a card for each run from an intital set of twenty cards, ten marked with "N" and ten with "S." In general, this random assignment of the experimental treatments (kits in this example) to the experimental material (runs in this example) is called *randomization.* Thus, randomization is a sampling procedure used in the actual performance of an experiment to ensure that the assumption of independent random sampling is valid.

Note that in the above example, it is not necessary to have random samples from the conceptual populations of all possible slopes with each kit, since the objective is comparison of the two kits under like conditions. It is only necessary to ensure that the experimental errors throughout the duration of the experiment are randomly allocated to the two kits. Of course all other experimental factors, such as the calibrator material and the operator, should be identical for the two kits and held constant for the duration of the experiment.

Procedure for the Significance Test

Suppose the data in Table 2A-3 were obtained by performing the experiment as discussed. The significance test comparing the sensitivities of the two kits as defined by the calibration slope proceeds as follows:

1. The null hypothesis is H_o: $\mu_N = \mu_S$, or $\mu_N - \mu_S = 0$. The alternative hypothesis is H_a: $\mu_N > \mu_S$, or $\mu_N - \mu_S > 0$.
2. The test statistic is $\bar{x}_N - \bar{x}_S$.
3. The sampling distribution of $\bar{x}_N - \bar{x}_S$ is $N(\mu_N - \mu_S, \sigma_N^2/n_N + \sigma_S^2/n_S)$, where $\mu_N - \mu_S = 0$ under the null hypothesis.
4. The observed result is $(\bar{x}_N - \bar{x}_S) = (0.1496 - 0.1486) = 0.0010$.
5. The significance level or p-value of this result is $\Pr(\bar{x}_N - \bar{x}_S > 0.0010)$, which is obtained as follows:

    a. $t_o = \dfrac{(\bar{x}_N - \bar{x}_S) - (\mu_N - \mu_S)_o}{\sqrt{s_p^2/n_N + s_p^2/n_S}} = \dfrac{(0.1496 - 0.1486) - 0}{\sqrt{0.0025^2/10 + 0.0025^2/10}} = 0.89$

    b. $\Pr(t \geq t_o) = \Pr(t \geq 0.89) = 0.20$, approximately, from a table of probability points of the t distribution with $\nu_N + \nu_S = 18$ degrees of freedom.

    c. $\Pr(\bar{x}_N - \bar{x}_S \geq 0.0010) = 0.20$

Table 2A-3. RESULTS OF THE RANDOMIZED
COMPARISON OF THE STANDARD
AND NEW ALBUMIN KITS

Run	Kit	Slope	
1	S	0.0147	$n_S = 10$
2	N	0.0151	$\bar{x}_S = 0.1486$
3	N	0.0153	$s_S^2 = 0.0022^2$
4	S	0.0151	
5	N	0.0149	
6	S	0.0150	
7	S	0.0152	$n_N = 10$
8	S	0.0147	$\bar{x}_N = 0.1496$
9	N	0.0144	$s_N^2 = 0.0028^2$
10	S	0.0145	
11	N	0.0150	
12	N	0.0149	
13	S	0.0148	$s_p^2 = \dfrac{9s_N^2 + 9s_S^2}{18} = 0.0025^2$
14	N	0.0149	
15	N	0.0147	
16	S	0.0147	
17	S	0.0150	
18	N	0.0154	
19	S	0.0149	
20	N	0.0150	

6. Since the significance level is $p = 0.20$, there is not sufficient evidence to reject the null hypothesis, i.e., there is not sufficient evidence that the sensitivity of the new kit is better than that of the standard kit.

Another application of the randomized comparison experiment is the traditional interferent experiment in which half of an initial serum pool is spiked with the potential interfering substance and the other half is not. Randomization in this case consists of the random order in which the repeat determinations on the two pools are made during a single run, thereby forcing the experimental errors within this run to contribute equally in probability to the determinations on the interferent and control pools. For example, if evaporation was occurring throughout this run and all control pool determinations were made first, followed by all interferent pool determinations, then the experimental errors would tend to be negative for the control pool determinations and positive for the interferent pool determinations. The observed difference in means would reflect the evaporation effect as well as the interferent effect.

The randomized comparison experiment can easily be extended to comparing more than two treatment or population means. The statistical data analysis is usually conducted by analysis of variance (ANOVA) techniques discussed in many statistical textbooks.[3,23,25]

THE RANDOMIZED PAIRED COMPARISON

When the experimental material for the comparison of two treatments is known or expected to be nonhomogeneous, the comparison can often be conducted with increased sensitivity by comparing the treatments within *matched pairs* of experimental material. For example, suppose that to compare glucose plasma and serum values on a particular analytical system, both a plasma and a serum specimen are collected from each of ten subjects. The results shown in Figure 2A-12 and Table 2A-4 appear to reveal no difference between the two population means. In fact, conducting a significance test for a randomized comparison yields

$$t_o = \frac{(\bar{x}_P - \bar{x}_S) - (\mu_P - \mu_S)_o}{\sqrt{s_p^2/10 + s_p^2/10}} = \frac{(91.3 - 89.6) - 0}{\sqrt{7.32^2/10 + 7.32^2/10}} = 0.52$$

which is clearly not statistically significant. However, we have not taken advantage of the fact that the comparisons were done in pairs. By considering the ten paired differences $x_d = x_{Plasma} - x_{Serum}$ shown in Figure 2A-13, the biological variability of the glucose concentration

Table 2A-4. RESULTS OF THE RANDOMIZED PAIRED COMPARISON OF PLASMA AND SERUM GLUCOSE CONCENTRATIONS

Subject	Plasma	Serum	Difference
1	92	90	2
2	98	95	3
3	86	87	−1
4	90	89	1
5	82	81	1
6	101	97	4
7	90	90	0
8	104	102	2
9	90	87	3
10	80	78	2
Average	$\bar{x} = 91.3$	$\bar{x} = 89.6$	$\bar{x}_d = 1.7$

between the ten subjects is eliminated. Such elimination of biological variability is possible because the plasma and serum specimens were collected in a paired manner. Only random analytical variation remains when these differences are considered, permitting a much more sensitive comparison. From the dot diagram of the paired differences in Figure 2A-13 it is clear that the plasma values are higher than the serum values by approximately 1–2 mg/dL on average. This will be confirmed by the appropriate significance test for the randomized paired comparison that follows. Randomization of (1) the order of drawing the serum and plasma specimen on each subject and (2) the order of measuring the serum and plasma specimen for each subject will ensure that the experimental errors are randomly allocated to the two treatments, thereby validating the significance test that follows.

Procedure for the Significance Test

The significance test for this randomized paired comparison is based on the ten differences $x_d = x_P - x_S$, shown in Table 2A-4 and Figure 2A-13, which are assumed to be a random sample from the Gaussian distribution $N(\mu_d, \sigma_d^2)$ of all possible differences. The null hypothesis is $H_o: \mu_d = 0$ and the alternative is $H_a: \mu_d > 0$. The sampling distribution of $\bar{x}_d$ is $N(\mu_d, \sigma_d^2/n)$. Since σ_d^2 is likely unknown, the variance σ_d^2/n can be replaced by $s_d^2/n = 1.49^2/10$, where $s_d^2 = 1.49^2$ is computed from the sample of $n = 10$ differences. Thus, to determine the significance level of the observed result $\bar{x}_d = 1.7$, we proceed as follows:

1. $t_o = \dfrac{\bar{x}_d - \mu_{do}}{\sqrt{s_d^2/n}} = \dfrac{1.7 - 0}{\sqrt{1.49^2/10}} = \dfrac{1.7}{0.47} = 3.61$

2. $\Pr(t \geq t_o) = P(t \geq 3.61) < 0.005$ from a table of probability points of the t distribution with $\nu = 10 - 1 = 9$ degrees of freedom.

3. $\Pr(\bar{x}_d \geq 1.7) < 0.005$

Since the *p*-value is so small, there is substantial evidence to reject the null hypothesis and to conclude that plasma values are higher on the average than serum values. The estimated difference is $\bar{x}_d = \bar{x}_P - \bar{x}_S = 1.7$ mg/dL. Later in this section we will discuss how large the true difference $\mu_d = \mu_P - \mu_S$ may actually be.

Further Considerations

In comparing two treatments, the true difference between the treatment means is frequently dependent on the concentration of the constituent of interest. Thus, when subjects are used for pairing in the comparison of two treatments, and the subjects actually used for the experiment cover a broad concentration range, the paired differences should be plotted against the concen-

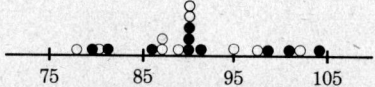

Figure 2A-12. Dot diagram of plasma (●) and serum (○) glucose concentrations in mg/dL.

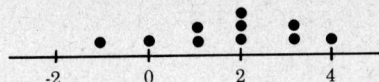

Figure 2A-13. Dot diagram of paired differences between plasma and serum glucose concentrations in mg/dL for 10 subjects.

tration of the constituent and inspected for a possible dependency. If the mean difference between treatments appears to change as a function of concentration, then the statistical summarization and significance test should be conducted on subsets of subjects such that the change in the mean difference appears to be negligible for the subjects in a particular subset. In the plasma versus serum example, the range of fasting glucose concentrations of the subjects actually used was narrow enough to believe the mean difference μ_d would be constant over the range studied, or at least that the change in μ_d would be negligible over this range. The difference between plasma and serum values may in fact be different at concentrations lower and higher than the concentration range covered in this experiment.

A potential misapplication of the significance test for the randomized paired comparison is in the analysis of method comparison data where determinations are made on split samples by two analyzer systems. The first problem relates to the fact stated above, namely that the true mean difference between the values on the two systems is likely to change as a function of the concentration of the constituent being measured. Additionally, the variation of the differences usually increases as the concentration increases. Separating the differences into two or three groups as suggested above will usually validate within these groups the assumption of a population of differences with a single mean and variance. The second problem is a result of the fact that this experiment is usually conducted over several days, since the results from a single day would confuse or confound the random day error of each system on that particular day with the difference between the two systems. This makes the experimental design more complicated than a simple paired comparison, since there is now replication both between days and within days. In other words, all of the differences are not independent of each other. Those differences on a particular day will tend to be more like each other than those on another day because of the particular random day error associated with each day for each system. The design and analysis of method comparison experiments will be discussed in detail later in this chapter.

The pairing concept can easily be extended to the comparison of more than two treatments, in which case the term *blocking* is usually applied. A *block* is a portion of the experiment that is expected to be more homogeneous than all portions together. For example, two or more treatments may be compared on the same specimens, during the same runs, or on the same days to obtain more sensitive comparisons. The statistical analysis of a blocked design is conducted according to two-way and higher-way ANOVA for blocked designs discussed in many statistical textbooks.[3,23,25]

DESIGNING EXPERIMENTS IN ANALYTICAL WORK

Blocking and randomization provide for more valid and sensitive comparisons in analytical work. Blocking takes advantage of known or expected variability among the experimental material to obtain greater sensitivity in comparing treatments. Randomization ensures the validity of the statistical significance test by forcing the unknown sources of variability to contribute equally to the treatments.

In addition to employing blocking and randomization for dealing with unavoidable sources of variation, the analyst should think about sources of variability that are avoidable beforehand and eliminate these sources from the comparison of treatments within blocks. However, it is useful and desirable to have variation between blocks that is representative of the range of blocks of interest. For example, in the comparison of treatments using subjects as blocks, a wide range of concentrations is desirable to determine how the treatments compare at different concentrations.

In summary, the analyst should deal with known sources of variability by either avoiding or blocking them, and should force unknown sources of variability to influence the treatments equally in probability by randomization.

CONFIDENCE INTERVALS FOR POPULATION MEANS

So far we have emphasized one of two general areas of statistical inference about population means, namely significance testing, and included only a brief introduction to the second area, estimation, which we will now consider. Recall that uncertainty associated with a particular sample mean $\bar{x}$ causes this "point" estimate to be different from the exact value of μ. A more useful form of estimation provides an "interval" within which the true unknown population mean μ almost certainly lies. For example, it would be informative to know that the interval from 24.2 U/L to 28.4 U/L includes the true mean μ of a certain population of GGT values with 95% probability or confidence. Similarly, in comparing the means of the populations of plasma and serum glucose values, it would be informative to know that the interval from 1.0 mg/dL to 3.0 mg/dL includes the true difference $\mu_P - \mu_S$ with 95% confidence. Such interval estimates of population parameters are called *confidence intervals*.

Confidence Intervals for a Single Population Mean

Consider again estimation of the mean μ of the population of GGT values in 20 to 29 year old healthy males, and suppose that $\sigma = 11$ is known. The random sample of n = 100 values produced the point estimate $\bar{x} = 25.9$ U/L. As in the case of the significance test about μ, the basis for interval estimates or confidence intervals for μ is the sampling distribution of $\bar{x}$ shown in Figure 2A-14. Since the sampling distribution of $\bar{x}$ is $N(\mu, 1.1^2)$ for random samples of size n = 100, 95% of sample means computed from random samples of size 100 will differ from the unknown mean μ by at most $2\sigma_{\bar{x}}$ or 2.2 U/L. Thus, if $2\sigma_{\bar{x}} = 2.2$ is added to and subtracted from each sample mean $\bar{x}$ from this distribution (i.e., $\bar{x} \pm 2.2$), then 95% of these intervals will in fact include the unknown mean μ, and 5% will not. Also, for a particular random sample with mean $\bar{x}$, the probability is then 95% that this $\bar{x}$ differs from μ by at most $2\sigma_{\bar{x}}$ or 2.2, so that $\bar{x} \pm 2.2$ U/L includes μ with 95% probability or confidence. For our example with $\bar{x} = 25.9$ U/L, a 95% confidence interval for μ is computed as follows:

$$25.9 \pm 2\sigma_{\bar{x}} = 25.9 \pm 2.2 = 23.7 \text{ to } 28.1 \text{ U/L}$$

The 95% confidence interval for a population parameter can be interpreted as follows: if the same experiment is repeated many times, and a 95% confidence interval is constructed for the parameter each time, then 95% of these intervals will actually include the true population parameter, and 5% will not. It follows that for the one confidence interval at hand from a single particular experiment, there is a 95% chance that it really includes the true population parameter, and a 5% chance that it does not.

The procedure illustrated above can be used to construct intervals with confidence or prob-

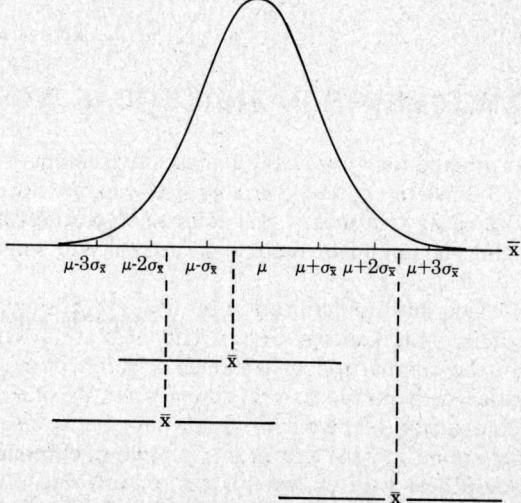

Figure 2A-14. Ninety-five per cent confidence intervals $\bar{x} \pm 2\sigma_{\bar{x}}$ for the population mean μ. The bottom 95% confidence interval does not include μ because $\bar{x}$ is more than $2\sigma_{\bar{x}}$ away from μ.

ability of including μ other than 95%, in which case the factor 2 in the formula $\bar{x} \pm 2\sigma_{\bar{x}}$ would depend upon the degree of confidence. For example, for a 95% confidence interval the factor is 1.96 or roughly 2; for a 90% confidence interval the factor is 1.645; and for a 99% confidence interval the factor is 2.576. The general formula for a $(1 - \alpha)100$ confidence interval for μ when σ is known is

$$\bar{x} \pm z_{\alpha/2} (\sigma_{\bar{x}}) \quad \text{or} \quad \bar{x} \pm z_{\alpha/2} (\sigma/\sqrt{n}) \tag{13}$$

where $z_{\alpha/2}$ is the value of a standard Gaussian variable such that a proportion $\alpha/2$ of the values exceed $z_{\alpha/2}$. For instance, for a 95% confidence interval $\alpha = 0.05$, $\alpha/2 = 0.025$, and $z_{0.025} = 1.96$ or approximately 2.0 as often used in practice.

If σ is unknown, a $(1 - \alpha)100$ confidence interval for μ is

$$\bar{x} \pm t_{\alpha/2} (s_{\bar{x}}) \quad \text{or} \quad \bar{x} \pm t_{\alpha/2} (s/\sqrt{n}) \tag{14}$$

where s is an estimate of σ with ν degrees of freedom and $t_{\alpha/2}$ is the value of the t distribution with ν degrees of freedom such that a proportion $\alpha/2$ of the values exceed $t_{\alpha/2}$.

Example 9. If σ is unknown in the GGT example, a 95% confidence interval for μ based on the random sample of $n = 100$ values with $\bar{x} = 25.9$ U/L, s $= 11.6$ U/L, and $t_{0.025} = 1.98$ for $\nu = 99$ is computed as follows:

$$25.9 \pm 1.98 \, (11.6/\sqrt{100}) = 25.9 \pm 2.3 = 23.6 \text{ to } 28.2 \text{ U/L}$$

Since both confidence intervals and significance tests for μ are based on the sampling distribution of $\bar{x}$, there is a direct relationship between confidence intervals and significance tests. In the above example, the lower confidence limit for μ of 23.6 U/L is the value of μ below the observed result $\bar{x} = 25.9$ that would produce a p-value of 0.025. Similarly, the upper confidence limit for μ of 28.2 U/L is the value of μ above the observed result $\bar{x} = 25.9$ that would produce a p-value of 0.025. Any hypothesized value of μ between 23.6 and 28.2 will result in a significance test p-value greater than 0.025, while any hypothesized value below 23.6 or above 28.2 will result in a p-value less than 0.025. Thus, a confidence interval for μ can be interpreted as including all the values of μ that are compatible with the observed data in significance testing about μ, and it is therefore much more informative than a single significance test for a particular hypothesized value of μ.

If the uncertainty in estimating μ is too great in that the confidence interval for μ is too wide, one can determine approximately how large a sample is needed to reduce the width of the confidence interval to any desired value.

Example 10. Suppose that it is desirable to have a 95% confidence interval of approximately 2 U/L in width rather than the 4.6 U/L width in Example 9. Then from equation (14) the quantity $t_{0.025} (s/\sqrt{n})$ must be approximately equal to 1 U/L. Substituting $t_{0.025} \simeq 2$ and s $= 11.6$ into $t_{0.025} (s/\sqrt{n}) = 1$, the required sample size is

$$n = (t_{0.025}(s)/1)^2 = ((2)(11.6))^2 = 538.2, \text{ or } \simeq 539$$

Determination of the sample size is best conducted before the data are collected but requires a reasonable idea of the magnitude of σ, the standard deviation of the experimental error. The sample size required for obtaining a $(1 - \alpha)100$ confidence interval of approximate width w is determined by setting w/2 equal to $z_{\alpha/2} (\sigma/\sqrt{n})$ in equation (13) and solving for n:

$$n = \left[\frac{z_{\alpha/2} (\sigma)}{w/2} \right]^2 \tag{15}$$

The quantity w/2 can also be thought of as the maximum error, with $1 - \alpha$ probability, in $\bar{x}$ as an estimate of μ.

Confidence Intervals for $\mu_A - \mu_B$: Paired Design

In a randomized paired comparison of two treatments, recall that the sampling distribution of $\bar{x}_d$, the mean of the paired differences, provides the basis for inference about the difference $\mu_A - \mu_B$. The n paired differences $x_d = x_A - x_B$ are assumed to be a random sample from a population of differences whose distribution is $N(\mu_d, \sigma_d^2)$, and the sampling distribution of $\bar{x}_d$ is $N(\mu_d, \sigma_d^2/n)$. Thus, this is simply the case of constructing a confidence interval for the mean of a single Gaussian population, so a $(1 - \alpha)100$ confidence interval for μ_d (and $\mu_A - \mu_B$) is

$$\bar{x}_d \pm z_{\alpha/2}\,(\sigma_{\bar{x}_d}) \qquad \text{or} \qquad \bar{x}_d \pm z_{\alpha/2}\,(\sigma_d/\sqrt{n}) \tag{16}$$

If σ_d is unknown, as is often the case, then a $(1 - \alpha)100$ confidence interval for μ_d is

$$\bar{x}_d \pm t_{\alpha/2}\,(s_{\bar{x}_d}) \qquad \text{or} \qquad \bar{x}_d \pm t_{\alpha/2}\,(s_d/\sqrt{n}) \tag{17}$$

Example 11. In the randomized paired comparison of glucose plasma and serum measurements with $\bar{x}_d = 1.7$, $s_d = 1.49$, and $t_{0.025} = 2.262$ for $\nu = 9$, a 95% confidence interval for $\mu_P - \mu_S$ is computed as follows:

$$1.7 \pm 2.262\,(1.49/\sqrt{10}) = 1.7 \pm 1.1 = 0.6 \text{ to } 2.8 \text{ mg/dL}$$

Thus, the true difference $\mu_P - \mu_S$ lies between 0.6 and 2.8 mg/dL with 95% certainty. Expressed as a percentage of the approximate 90 mg/dL glucose concentration, it is 95% certain that the true difference $\mu_P - \mu_S$ is no less than 0.7% and no more than 3.1%. In particular, since zero is not included in the interval it is not a plausible value for $\mu_P - \mu_S$.

The number n of paired differences required to obtain a confidence interval of width w can be determined by setting $w/2 = z_{\alpha/2}\,(\sigma_d/\sqrt{n})$ in equation (16) and solving for n:

$$n = \left[\frac{z_{\alpha/2}\,(\sigma_d)}{w/2}\right]^2 \tag{18}$$

To obtain an initial guess for σ_d, recall that since $x_d = x_A - x_B$, we have $\sigma_d^2 = \sigma_A^2 + \sigma_B^2$ where σ_A^2 and σ_B^2 are the variances of the experimental errors influencing measurements for treatments A and B.

Confidence Intervals for $\mu_A - \mu_B$: Unpaired Design

Confidence intervals for the difference of two population means in an unpaired randomized comparison are based on the sampling distribution of $\bar{x}_A - \bar{x}_B$, which is $N(\mu_A - \mu_B, \sigma_A^2/n_A + \sigma_B^2/n_B)$. Employing the same logic used to derive confidence intervals for a single population mean, a $(1 - \alpha)100$ confidence interval for $\mu_A - \mu_B$ is

$$(\bar{x}_A - \bar{x}_B) \pm z_{\alpha/2}\sqrt{\sigma_A^2/n_A + \sigma_B^2/n_B} \tag{19}$$

If σ_A^2 and σ_B^2 are unknown, as is often the case, a $(1 - \alpha)100$ confidence interval for $\mu_A - \mu_B$ is

$$(\bar{x}_A - \bar{x}_B) \pm t_{\alpha/2}\sqrt{s_p^2/n_A + s_p^2/n_B} \tag{20}$$

where $t_{\alpha/2}$ is the value of the t distribution with $\nu = n_A + n_B - 2$ degrees of freedom such that a proportion $\alpha/2$ of the values exceed $t_{\alpha/2}$, and s_p^2 is the pooled estimate in equation (11) of the common unknown population variance. For the case $\sigma_A \neq \sigma_B$, the corresponding formula for equation (20) is given in Table 2A-5 at the end of this section.

Example 12. In the randomized comparison of the albumin kits with $\bar{x}_N = 0.1496$, $\bar{x}_S = 0.1486$, $s_p = 0.0025$, and $t_{0.025} = 2.101$ for $\nu = 18$, a 95% confidence interval for the true difference $\mu_N - \mu_S$ in the slope means is

$$(0.1496 - 0.1486) \pm 2.101\sqrt{0.0025^2/10 + 0.0025^2/10}$$
$$\text{or } -0.0013 \text{ to } 0.0033$$

Table 2A-5. SIGNIFICANCE TESTS AND CONFIDENCE INTERVALS FOR POPULATION MEANS

H_o	H_a	Design	Test Statistic	p-Value	$(1 - \alpha)100$ Confidence Interval
$\mu = \mu_o$	$\mu < \mu_o$ $\mu > \mu_o$ $\mu \neq \mu_o$	Single random sample; σ known	$z_o = \dfrac{\bar{x} - \mu_o}{\sigma/\sqrt{n}}$	$\Pr(z < z_o)$ $\Pr(z > z_o)$ $2\Pr(z > |z_o|)$	$\bar{x} \pm z_{\alpha/2}\,(\sigma/\sqrt{n})$
$\mu = \mu_o$	$\mu < \mu_o$ $\mu > \mu_o$ $\mu \neq \mu_o$	Single random sample; σ unknown	$t_o = \dfrac{\bar{x} - \mu_o}{s/\sqrt{n}}$ $\nu = n - 1$	$\Pr(t < t_o)$ $\Pr(t > t_o)$ $2\Pr(t > |t_o|)$	$\bar{x} \pm t_{\alpha/2}\,(s/\sqrt{n})$
$\mu_A - \mu_B = (\mu_A - \mu_B)_o$	$\mu_A - \mu_B < (\mu_A - \mu_B)_o$ $\mu_A - \mu_B > (\mu_A - \mu_B)_o$ $\mu_A - \mu_B \neq (\mu_A - \mu_B)_o$	Randomized comparison; σ_A and σ_B known	$z_o = \dfrac{(\bar{x}_A - \bar{x}_B) - (\mu_A - \mu_B)_o}{\sqrt{\sigma_A^2/n_A + \sigma_B^2/n_B}}$	$\Pr(z < z_o)$ $\Pr(z > z_o)$ $2\Pr(z > |z_o|)$	$(\bar{x}_A - \bar{x}_B) \pm z_{\alpha/2}\sqrt{\dfrac{\sigma_A^2}{n_A} + \dfrac{\sigma_B^2}{n_B}}$
$\mu_A - \mu_B = (\mu_A - \mu_B)_o$	$\mu_A - \mu_B < (\mu_A - \mu_B)_o$ $\mu_A - \mu_B > (\mu_A - \mu_B)_o$ $\mu_A - \mu_B \neq (\mu_A - \mu_B)_o$	Randomized comparison; $\sigma_A = \sigma_B$ but unknown	$t_o = \dfrac{(\bar{x}_A - \bar{x}_B) - (\mu_A - \mu_B)_o}{\sqrt{s_p^2/n_A + s_p^2/n_B}}$ $\nu = n_A + n_B - 2$	$\Pr(t < t_o)$ $\Pr(t > t_o)$ $2\Pr(t > |t_o|)$	$(\bar{x}_A - \bar{x}_B) \pm t_{\alpha/2}\sqrt{\dfrac{s_p^2}{n_A} + \dfrac{s_p^2}{n_B}}$
$\mu_A - \mu_B = (\mu_A - \mu_B)_o$	$\mu_A - \mu_B < (\mu_A - \mu_B)_o$ $\mu_A - \mu_B > (\mu_A - \mu_B)_o$ $\mu_A - \mu_B \neq (\mu_A - \mu_B)_o$	Randomized comparison; $\sigma_A \neq \sigma_B$ and unknown	$t_o = \dfrac{(\bar{x}_A - \bar{x}_B) - (\mu_A - \mu_B)_o}{\sqrt{s_A^2/n_A + s_B^2/n_B}}$ $\nu = \dfrac{(s_A^2/n_A + s_B^2/n_B)^2}{\dfrac{(s_A^2/n_A)^2}{(n_A - 1)} + \dfrac{(s_B^2/n_B)^2}{(n_B - 1)}}$	$\Pr(t < t_o)$ $\Pr(t > t_o)$ $2\Pr(t > |t_o|)$	$(\bar{x}_A - \bar{x}_B) \pm t_{\alpha/2}\sqrt{\dfrac{s_A^2}{n_A} + \dfrac{s_B^2}{n_B}}$
$\mu_d = \mu_{do}$	$\mu_d < \mu_{do}$ $\mu_d > \mu_{do}$ $\mu_d \neq \mu_{do}$	Randomized paired comparison; σ_d known	$z_o = \dfrac{\bar{x}_d - \mu_{do}}{\sigma_d/\sqrt{n}}$	$\Pr(z < z_o)$ $\Pr(z > z_o)$ $2\Pr(z > |z_o|)$	$\bar{x}_d \pm z_{\alpha/2}(\sigma_d/\sqrt{n})$
$\mu_d = \mu_{do}$	$\mu_d < \mu_{do}$ $\mu_d > \mu_{do}$ $\mu_d \neq \mu_{do}$	Randomized paired comparison; σ_d unknown	$t_o = \dfrac{\bar{x}_d - \mu_{do}}{s_d/\sqrt{n}}$ $\nu = n - 1$	$\Pr(t < t_o)$ $\Pr(t > t_o)$ $2\Pr(t > |t_o|)$	$\bar{x}_d \pm t_{\alpha/2}(s_d/\sqrt{n})$

Thus, the true difference $\mu_N - \mu_S$ between the sensitivities of the albumin kits lies between -0.0013 and 0.0033 with 95% certainty. In particular, since zero is in the interval, $\mu_N = \mu_S$ is compatible with the data. Furthermore, it is 95% certain that the new kit is not more than 0.9% ($0.0013/0.149 = 0.9\%$) less sensitive than the standard kit, and no more than 2.2% ($0.0033/0.149 = 2.2\%$) more sensitive.

The sample size $n = n_A = n_B$ required for each treatment in a randomized comparison to obtain a <u>confidence</u> interval of width w can be determined by setting $w/2$ equal to $z_{\alpha/2}\sqrt{\sigma^2/n + \sigma^2/n}$ in equation (19), where σ^2 is the common population variance, and then solving for n:

$$n = \left[\frac{\sqrt{2} \cdot z_{\alpha/2}(\sigma)}{w/2} \right]^2 \tag{21}$$

THE POISSON DISTRIBUTION

The *Poisson distribution* serves as a model for many random time-related phenomena. The disintegration of radioactive atoms from which radioimmunoassay derives its analytical signal is an example of this random process. From some simple assumptions and basic probability considerations, the discrete probability distribution that describes the number of radioactive disintegrations in a given interval can be derived. In general, if x is the number of occurrences of some random event in an interval of time or space T, the probability that x will occur is

$$f(x) = \frac{(\lambda T)^x \cdot e^{-\lambda T}}{x!}, x = 0,1,2,... \tag{22}$$

The Greek letter λ (lambda) is the occurrence rate, and both the theoretical mean and variance of x are given by λT. In the radioactive disintegration example $f(x)$ is the probability that exactly x disintegrations will be observed in the time interval T. The theoretical mean number of disintegrations will be λT and the theoretical variance will also be λT.

The Poisson distribution is thus a discrete distribution that has the interesting property that the mean and variance of the distribution are identical. In radioimmunoassay, we use the observed radioactive counts in a given time interval to estimate the total number of radioactive atoms present in the assay system. Because of the nature of radioactive disintegration there is statistical variation in this measurement process. The Poisson nature of the process tells us that the standard deviation of the observed counts can be estimated from the square root of the observed counts.

Example 13. If the number of observed counts in a time interval is N, then the estimated standard deviation of the underlying Poisson distribution will be $\sqrt{N}$, and the coefficient of variation due to counting error will be $(\sqrt{N}/N)100\%$ or $(1/\sqrt{N})100\%$. For 10 000 counts we would observe a CV of 1% due to counting error, while for 900 counts a CV of 3.3% due to counting error would be obtained.

For the radioactive counting example, λT corresponds to the mean number of counts. As λT becomes large, the Poisson distribution is approximated by the continuous Gaussian distribution with $\mu = \lambda T$ and $\sigma^2 = \lambda T$, and statistical methods based on the Gaussian distribution can be used to analyze data from radioactive counting experiments, and to construct confidence intervals on the counts observed. For isotope samples used for the quality control of a counting instrument, the observed variance can be compared to the theoretical variance. Instrumental drift would cause a greater observed variance than predicted from the Poisson distribution.

List of Symbols for Inference About Population Means (in order of appearance)

x_A	random variable
x_B	random variable
μ_A	population mean for random variable x_A
μ_B	population mean for random variable x_B
σ_A^2	population variance for random variable x_A
σ_B^2	population variance for random variable x_B
y	random variable
E( )	expected value of the random variable in ( )

$V(\ )$	variance of the random variable in $(\ )$
n_A	size of random sample from population A
n_B	size of random sample from population B
$\bar{x}_A$	mean of sample from population A
$\bar{x}_B$	mean of sample from population B
$N(\ ,\ )$	probability distribution that is Gaussian with mean and variance specified in $(\ ,\ )$
z	the standard Gaussian variable with $N(0, 1)$ probability distribution
σ^2	population variance
s_A^2	sample variance for sample from population A
s_B^2	sample variance for sample from population B
ν_A	degrees of freedom for s_A^2
ν_B	degrees of freedom for s_B^2
s_p^2	pooled sample variance $s_p^2 = (\nu_A s_A^2 + \nu_B s_B^2)/(\nu_A + \nu_B)$
t	a variable that has the Student's t distribution
H_o	null hypothesis
H_a	alternative hypothesis
$Pr(\ )$	probability that the specified variable takes on a value from those specified in $(\ )$
$(\mu_A - \mu_B)_o$	hypothetical value of $(\mu_A - \mu_B)$ under the null hypothesis H_o
t_o	observed value of the random variable t under the null hypothesis H_o
p	p-value or significance level of the observed result
x_d	random variable that is the difference between two other random variables
μ_d	population mean for random variable x_d
σ_d^2	population variance for random variable x_d
$\bar{x}_d$	sample mean for random variable x_d
s_d^2	sample variance for random variable x_d
μ_{do}	hypothetical value of μ_d under the null hypothesis H_o
ν	degrees of freedom for s_d^2
μ	population mean for random variable x
σ	population standard deviation for random variable x
X	sample mean for random variable x
$\sigma_{\bar{x}}$	population standard deviation for the random variable $\bar{x}$
$1 - \alpha$	degree of confidence for a confidence interval; e.g., for a 95% confidence interval $\alpha = 0.05$ and $1 - \alpha = 0.95$
$z_{\alpha/2}$	value of the standard Gaussian variable z such that a proportion $\alpha/2$ of the values exceed it; e.g., for a 95% confidence interval $\alpha/2 = 0.025$ and $z_{\alpha/2} = 1.96$
s	sample standard deviation for random variable x
$s_{\bar{x}}$	sample standard deviation for the random variable $\bar{x}$
$t_{\alpha/2}$	value of the Student's t variable t such that a proportion $\alpha/2$ of the values exceed it
ν	degrees of freedom for s
w	width of a confidence interval
$\sigma_{\bar{x}_d}$	population standard deviation for the random variable $\bar{x}_d$
$s_{\bar{x}_d}$	sample standard deviation for the random variable $\bar{x}_d$
λ	rate parameter for a Poisson random variable
T	time interval for a Poisson random variable
e	the constant 2.7183
x!	the product $x(x - 1)(x - 2) \ldots (2)(1)$
N	observed value of a Poisson random variable

INFERENCES ABOUT POPULATION VARIANCES

In addition to estimating and comparing population means, analytical work is often conducted to estimate and compare population variances. For example, an experiment is usually conducted with a new analyzer system to determine how repeatable the determinations on a particular specimen are within a run, from run to run within a day, and from day to day. In this section the treatment of random analytical variation will be limited to inferences about the within-run variance σ_{WR}^2 based on quite simple experiments. Table 2A-6 contains a summary of these significance test and confidence interval procedures. Extension of these statistical procedures to inferences about within-run variance and total variance based on more elaborate experimentation will be discussed in the next section.

INFERENCES ABOUT A SINGLE POPULATION VARIANCE

Consider an experiment where twenty determinations are made on a particular fluid within a single run to estimate the within-run variance σ_{WR}^2 of a new glucose analyzer system and to check the manufacturer's claim that the within-run variation is no larger than 1% CV. Although a much more elaborate experiment is recommended to evaluate the random analytical variation of an analyzer system, this simple example will serve to illustrate the general procedures for significance tests and confidence intervals about variances. Whatever the size of the experiment, the determinations within a run on a particular fluid should be spaced randomly throughout a run of length typically encountered in daily operation. The determinations can then be considered a random sample from a Gaussian population of all possible determinations with variance σ_{WR}^2, an assumption necessary for inferences about variances.

Suppose that the twenty values yield the statistics $\bar{x} = 120.4$ mg/dL and $s = 1.4$ mg/dL. The point estimate of the within-run population variance σ_{WR}^2 is $s^2 = 1.96$ (mg/dL)^2. Since the manufacturer's claim of a CV of 1% is equivalent to $\sigma_{WR}^2 = 1.2^2 = 1.44$ at a concentration of 120 mg/dL, the point estimate s^2 of σ_{WR}^2 is in fact higher than the claim. However, whether this result is statistically significant or not requires a significance test of the hypothesis $H_0: \sigma_{WR}^2 = 1.2^2$ against the alternative $H_a: \sigma_{WR}^2 > 1.2^2$. The sampling distribution of s^2 will be introduced next, since it forms the basis for significance tests and confidence intervals for the population variance σ^2.

The Sampling Distribution of s^2

In situations like the above, where $x_1, x_2, \ldots, x_n$ represent a random sample of n values from a Gaussian population with variance σ^2, it can be shown that the variable

$$\chi^2 = (n - 1)s^2/\sigma^2 \tag{23}$$

has a distribution called the *chi-square* (χ^2) *distribution* with $\nu = n - 1$ degrees of freedom. The chi-square distribution has a mean ν, variance 2ν, and is skewed with a long right tail. The skewness decreases as ν increases until at $\nu = 50$ the distribution is approximately Gaussian. The percentage points of the χ^2 distribution are given in Statistical Table 20-23, in the Appendix.

Significance Test for a Population Variance σ^2

Since the sampling distribution of $(n - 1)s^2/\sigma^2$ is a χ^2 distribution with $\nu = n - 1$ degrees of freedom, a significance test can be conducted by calculating

$$\chi_0^2 = (n - 1)s^2/\sigma_0^2 \tag{24}$$

for any hypothetical value σ_0^2 of σ^2 and the observed value of s^2, and referring this to the percentage points of the χ^2 distribution with $\nu = n - 1$ degrees of freedom. If this observed value χ_0^2 can be considered a typical random observation from the χ^2 distribution, then there is no evidence to reject the hypothesized value σ_0^2 of σ^2. Conversely, if it cannot be considered a typical random observation, then there is evidence to reject the hypothesized value of σ^2.

Example 14. To test the null hypothesis $H_o: \sigma^2 = 1.2^2$ against the alternative hypothesis $H_a: \sigma^2 > 1.2^2$ in the glucose example, the observed value of $(n - 1)s^2/\sigma^2$ is $\chi_0^2 = (20 - 1)1.4^2/1.2^2 = 25.9$. From the table of percentage points of the χ^2 distribution with $\nu = 19$ degrees of freedom, in Statistical Table 20-23, in the Appendix, the value 25.9 is exceeded by chance about 10% of the time; i.e., the *p*-value is $p = 0.10$. Thus, there is not sufficient evidence to reject the manufacturer's claim.

Table 2A-6. SIGNIFICANCE TESTS AND CONFIDENCE INTERVALS FOR POPULATION VARIANCES

H_o	H_a	Design	p-Value Calculation		$(1-\alpha)100$ Confidence Interval
$\sigma^2 = \sigma_o^2$	$\sigma^2 < \sigma_o^2$ $\sigma^2 > \sigma_o^2$	Single random sample from one population	$\chi_o^2 = \dfrac{(n-1)s^2}{\sigma_o^2}$ $\nu = n - 1$	$\Pr(\chi^2 < \chi_o^2)$ $\Pr(\chi^2 > \chi_o^2)$	$\dfrac{(n-1)s^2}{U}$ to $\dfrac{(n-1)s^2}{L}$ U = upper tail $\alpha/2$ value of χ^2 L = lower tail $\alpha/2$ value of χ^2
	$\sigma^2 \neq \sigma_o^2$			2(minimum of above)	
$\sigma^2 = \sigma_o^2$	$\sigma^2 < \sigma_o^2$ $\sigma^2 > \sigma_o^2$	k random samples from one population	$\chi_o^2 = \dfrac{\nu_p s_p^2}{\sigma_o^2}$ $\nu_p = \nu_1 + \cdots + \nu_k$	$\Pr(\chi^2 < \chi_o^2)$ $\Pr(\chi^2 > \chi_o^2)$	$\dfrac{\nu_p s_p^2}{U}$ to $\dfrac{\nu_p s_p^2}{L}$ U = upper tail $\alpha/2$ value of χ^2 L = lower tail $\alpha/2$ value of χ^2
	$\sigma^2 \neq \sigma_o^2$			2(minimum of above)	
$\dfrac{\sigma_A^2}{\sigma_B^2} = 1$	$\dfrac{\sigma_A^2}{\sigma_B^2} > 1$	Independent random samples from two populations	$F_o = \dfrac{s_A^2}{s_B^2}$ $s_A^2 > s_B^2$ $\nu_A = n_A - 1$ $\nu_B = n_B - 1$	$\Pr(F > F_o)$	$\dfrac{s_A^2/s_B^2}{U}$ to $\dfrac{s_A^2/s_B^2}{L}$ U = upper tail $\alpha/2$ value of F L = lower tail $\alpha/2$ value of F

Confidence Interval for σ^2

In order to construct a $(1 - \alpha)100$ confidence interval for σ^2 based on an estimate s^2 with $\nu = n - 1$ degrees of freedom, let L and U be the lower and upper tail $\alpha/2$ percentage points of the χ^2 distribution with $\nu = n - 1$ degrees of freedom. Then

$$\Pr\left(L < \frac{(n - 1)s^2}{\sigma^2} < U\right) = 1 - \alpha \tag{25}$$

and upon algebraic rearrangement

$$\Pr\left(\frac{(n - 1)s^2}{U} < \sigma^2 < \frac{(n - 1)s^2}{L}\right) = 1 - \alpha \tag{26}$$

Thus, the lower and upper $(1 - \alpha)100$ confidence limits for σ^2 are

$$(n - 1)s^2/U \text{ and } (n - 1)s^2/L \tag{27}$$

Example 15. Continuing with the glucose system example, in order to construct a 95% confidence interval for σ_{WR}^2 based on the random sample of twenty observations with $s^2 = 1.96$, we first obtain $L = 8.91$ and $U = 32.9$ as the lower and upper $0.05/2 = 0.025$ probability points of the χ^2 distribution with $\nu = 19$ degrees of freedom. Then the 95% confidence limits for σ_{WR}^2 are $19(1.96)/32.9$ and $19(1.96)/8.91$ or 1.13 and 4.18 $(mg/dL)^2$. Furthermore, by taking the square root of each limit for σ_{WR}^2, a 95% confidence interval for σ_{WR} is 1.06–2.04 mg/dL; and by expressing these limits for σ_{WR} as a percentage of 120 mg/dL, a 95% confidence interval for the population CV_{WR} is 0.9–1.7%.

To determine the sample size n necessary to obtain a confidence interval for a population variance σ^2 of sufficiently narrow width, we first select a reasonable value for s^2, and then calculate confidence limits for various values of n, with L and U obtained from the table of percentage points of the χ^2 distribution with $\nu = n - 1$ degrees of freedom. The rate at which we reduce the uncertainty about s^2 decreases as n increases. For example, the reduction in uncertainty is greater as n increases from 2 to 20 than from 20 to 40.

The purpose of this simple example has been to illustrate the statistical procedures for inferences about a population variance, and it is not typical of the recommended experimental protocols for estimation of system random analytical variation.[11] Next we discuss a frequently used statistical design and analysis for obtaining a more representative estimate of σ_{WR}^2, and shortly we will discuss more elaborate protocols for the study of components of variation in addition to σ_{WR}^2.

Pooling Several Estimates of σ^2

It is often useful to combine several estimates of σ^2, each with a few degrees of freedom, into a single estimate of σ^2 with degrees of freedom equal to the total of the degrees of freedom of the individual estimates. For example, suppose that for each of k patient specimens two determinations have been obtained, both in the same run. If the two determinations on each specimen are randomly spaced, an estimate s^2 of σ_{WR}^2 with a single degree of freedom can be obtained for each of the k specimens. For the ith specimen, denote the estimate of σ_{WR}^2 as s_i^2 with $\nu_i = 1$ degree of freedom. By averaging or *pooling* all k estimates of σ_{WR}^2, we obtain

$$s_p^2 = \frac{s_1^2 + s_2^2 + s_3^2 + \ldots + s_k^2}{k} \tag{28}$$

as an estimate of σ_{WR}^2 with degrees of freedom $\nu_p = \sum_{i=1}^{k} \nu_i$. The subscript p is used to indicate that the estimate and its associated degrees of freedom have been obtained by pooling together several independent estimates. The above formula for s_p^2 applies as long as the number of random observations n_i, and thus the degrees of freedom $\nu_i = n_i - 1$, are equal for all k estimates. For example, if three determinations were made on each of k specimens, producing k individual estimates s_i^2 each with $\nu_i = 2$ degrees of freedom, then the pooled estimate s_p^2 of σ_{WR}^2 is obtained by equation (28) with degrees of freedom $\nu_p = \Sigma\nu_i = 2k$. When the degrees of freedom ν_i, and thus the uncertainty associated with the individual estimates s_i^2 are not all equal, the pooled estimate s_p^2 of σ_{WR}^2 is given by

$$s_p^2 = \frac{\nu_1 s_1^2 + \nu_2 s_2^2 + \ldots + \nu_k s_k^2}{\Sigma\nu_i} \tag{29}$$

where the individual estimates s_i^2 are weighted by their respective degrees of freedom ν_i. The degrees of freedom are given by $\nu_p = \Sigma\nu_i$.

In the same manner as above, a pooled estimate of σ_{WR}^2 can be obtained from multiple determinations of k specimens of the same control where the determinations on the ith specimen are all in the same run. Again, the multiple determinations on a particular specimen should be randomly spaced throughout the run containing that specimen.

When obtaining a pooled estimate of σ_{WR}^2 based on patient specimens one must consider the fact that random analytical variation usually increases as a function of concentration for most analyzer systems. Consequently, if the patient specimens cover a broad range of concentration, then they should be separated into several (usually two or three will suffice) groups such that the change in σ_{WR}^2 over the concentration range for each group is negligible. A scatterplot of the individual estimates s^2 or s against concentration is often useful to help select appropriate groups.

When obtaining an estimate of σ^2 by pooling, the confidence interval and significance test procedures illustrated earlier apply, with the degrees of freedom ν_p replacing the degrees of freedom $\nu = n - 1$ in equations (23)–(27).

Example 16. Suppose that twenty-five patient specimens with glucose concentrations between 80 and 120 mg/dL have been analyzed in duplicate over several days, where the two determinations for a particular specimen were randomly spaced throughout that particular run. Then twenty-five single degree of freedom estimates of σ_{WR}^2 can be obtained and combined into a pooled estimate with $\nu_p = \Sigma\nu_i = 25$ degrees of freedom. If the twenty-five individual estimates s_i^2 have a sum $\Sigma s_i^2 = 38.4$, then

$$s_p^2 = \frac{s_1^2 + s_2^2 + \ldots + s_{25}^2}{25} = 1.54$$

is an estimate of σ_{WR}^2 with $\nu_p = 25$ degrees of freedom. The corresponding estimates of the population standard deviation and CV are 1.24 and 1.2%, respectively. A 95% confidence interval for σ_{WR}^2 based on $s_p^2 = 1.54$, $\nu_p = 25$, and $L = 13.1$ and $U = 40.6$ from the table of χ^2 probability points with $\nu_p = 25$, is

$$\frac{\nu_p \cdot s_p^2}{U} \text{ to } \frac{\nu_p \cdot s_p^2}{L} = \frac{25(1.54)}{40.6} \text{ to } \frac{25(1.54)}{13.1} = 0.95 \text{ to } 2.94 \ (mg/dL)^2$$

Corresponding 95% confidence intervals for the population parameters σ_{WR} and CV_{WR} are 0.98–1.71 mg/dL and 1.0–1.7%, respectively.

INFERENCE ABOUT THE RATIO σ_A^2/σ_B^2 OF TWO POPULATION VARIANCES

In analytical work it is sometimes of interest to compare the variances of two Gaussian populations of values. For example, one may wish to compare the variances σ_A^2 and σ_B^2 of values produced by a new and an experienced laboratorian, or by two different analyzer systems. Again, we will introduce the statistical procedures for comparing two variances through a simple example dealing with within-run variances only. Extension of these procedures for comparison of within-run variances and/or total variances based on more elaborate experimentation will be discussed in the next section.

The Sampling Distribution of s_A^2/s_B^2

Suppose an inexperienced laboratorian makes 25 determinations randomly spaced in a single run on a particular control fluid yielding $s_A^2 = 2.3^2$ with $\nu_A = 24$ degrees of freedom. Also, suppose an experienced laboratorian makes 21 determinations randomly spaced in a single run on the same fluid yielding $s_B^2 = 1.5^2$ with $\nu_B = 20$ degrees of freedom. Then s_A^2 and s_B^2 are estimates of the within-run variances σ_A^2 and σ_B^2 for these two laboratorians with ν_A and ν_B degrees of freedom, respectively. Inferences about the ratio σ_A^2/σ_B^2 can be made by considering the variable

$$F = \frac{s_A^2/\sigma_A^2}{s_B^2/\sigma_B^2} \tag{30}$$

which has a sampling distribution called the *F distribution* with numerator degrees of freedom ν_A and denominator degrees of freedom ν_B. Tables of the percentage points of the *F* distribution are given in Statistical Table 20-24, in the Appendix. These tables are arranged assuming that the numerator in equation (30) is larger than the denominator.

Significance Test for a Hypothetical Ratio σ_A^2/σ_B^2

In the example above, it may be of interest to test the null hypothesis that $\sigma_A^2 = \sigma_B^2$, or $\sigma_A^2/\sigma_B^2 = 1$, against the alternative that $\sigma_A^2 > \sigma_B^2$ or $\sigma_A^2/\sigma_B^2 > 1$. Under the null hypothesis, the variable $F = (s_A^2/\sigma_A^2)/(s_B^2/\sigma_B^2)$ is equal to s_A^2/s_B^2 and has an *F* distribution with 24 and 20 degrees of freedom. Thus, if the observed result $F_o = s_A^2/s_B^2 = 2.3^2/1.5^2 = 2.35$ appears to be a typical random observation from the *F* distribution with 24 and 20 degrees of freedom, then there is no evidence to reject the null hypothesis. However, in comparing 2.35 to the table of probability points of the *F* distribution with 24 and 20 degrees of freedom, 2.35 is exceeded by chance approximately 3% of the time. Thus, since the significance level of the observed result is $p = 0.03$, there is sufficient evidence to suspect that the inexperienced laboratorian has a larger within-run variance.

Confidence Intervals for σ_A^2/σ_B^2

To construct a $(1 - \alpha)100$ confidence interval for the ratio σ_A^2/σ_B^2 based on estimates s_A^2 and s_B^2 with ν_A and ν_B degrees of freedom, let L and U be the lower and upper tail $\alpha/2$ percentage points of the *F* distribution with ν_A and ν_B degrees of freedom. Then

$$\Pr\left(L < \frac{s_A^2/\sigma_A^2}{s_B^2/\sigma_B^2} < U \right) = 1 - \alpha \tag{31}$$

and upon algebraic rearrangement

$$\Pr\left(\frac{s_A^2/s_B^2}{U} < \frac{\sigma_A^2}{\sigma_B^2} < \frac{s_A^2/s_B^2}{L} \right) = 1 - \alpha \tag{32}$$

giving the lower and upper $(1 - \alpha)100$ confidence limits for the ratio σ_A^2/σ_B^2.

Example 17. To obtain a 95% confidence interval for the ratio σ_A^2/σ_B^2 of the within-run variances of the two laboratorians, we first obtain $L = 0.43$ and $U = 2.41$ as the lower and upper tail 0.025 probability points of the *F* distribution with $\nu_A = 24$ and $\nu_B = 20$ degrees of freedom. Note that since lower tail probability points of the *F* distribution are not usually tabled, it is necessary to use the fact that the lower tail $\alpha/2$ probability point of the *F* distribution with ν_A and ν_B degrees of freedom can be obtained by taking the reciprocal of the upper tail $\alpha/2$ probability point with the degrees of freedom reversed. Thus, $L = 1/2.33 = 0.43$ where 2.33 is the upper tail 0.025 probability point of the *F* distribution with $\nu_A = 20$ and $\nu_B = 24$ degrees of freedom. Then the 95% confidence limits for the ratio σ_A^2/σ_B^2 are

$$\frac{2.3^2/1.5^2}{2.41} \quad \text{and} \quad \frac{2.3^2/1.5^2}{0.43}$$

or 0.98 and 5.47. Recall that the $(1 - \alpha)100$ confidence limits are the hypothetical values that would give a significance test p-value of $\alpha/2$. The fact that the lower 95% confidence limit just includes 1.0 agrees with the significance test p-value just over 0.025 obtained earlier.

List of Symbols for Inference about Population Variances (in order of appearance)

σ^2_{WR}	population within-run variance
CV	population or sample coefficient of variation
$\bar{x}$	sample mean
s	sample standard deviation
s^2	sample variance
H_o	null hypothesis
H_a	alternative hypothesis
x_i	ith member of a sample
n	number of members in a sample
χ^2	variable with the chi-square probability distribution
ν	degrees of freedom for s^2
σ^2_o	hypothetical value of σ^2 under the null hypothesis H_o
χ^2_o	observed value of the variable χ^2 under the null hypothesis H_o
p	p-value or significance level of the observed result in a statistical significance test
$1 - \alpha$	confidence level for a $(1 - \alpha)100$ confidence interval
L	lower tail $\alpha/2$ percentage point of the probability distribution of interest
U	upper tail $\alpha/2$ percentage point of the probability distribution of interest
$\Pr(\ )$	probability that the variable specified takes on a value from those specified in $(\ )$
s^2_i	sample variance for ith random sample
ν_i	degrees of freedom for s^2_i
s^2_p	pooled sample variance $s^2_p = (\nu_1 s^2_1 + \ldots + \nu_k s^2_k)/(\nu_1 + \ldots + \nu_k)$
ν_p	degrees of freedom for s^2_p; $\nu_p = \nu_1 + \ldots + \nu_k$
n_i	size of ith random sample
σ^2_A	population variance for population A
σ^2_B	population variance for population B
s^2_A	sample variance for sample from population A
s^2_B	sample variance for sample from population B
ν_A	degrees of freedom for s^2_A
ν_B	degrees of freedom for s^2_B
F	variable with the F probability distribution
F_o	observed value of the variable F under the null hypothesis H_o

THE STUDY OF VARIATION IN ANALYTICAL WORK

The study of variation in analytical work often involves more than one source of variation. For example, we may be interested in determining the variation of some final result that is a combination or function of several measurements with known variances, or conversely, we may be interested in separating the final variation into components associated with individual causes. In this section we present techniques for handling these situations by discussing the variance of functions of random variables and nested ANOVA for estimating components of variance.

THE EXPECTED VALUE AND VARIANCE OF A LINEAR COMBINATION OF RANDOM VARIABLES

Often in clinical chemistry we are interested in a random variable that is a function of other variables. An important case we will consider is when a random variable can be expressed as a linear combination of independent random variables. For example, consider the variable $y =$

$\Sigma a_i x_i$ where the x_i are random variables and the a_i are constants. Denote the expected value of each x_i by μ_i and the variance of each x_i by σ_i^2. Then the expected value and variance of y can be expressed in terms of the means and variances of the x_i as follows:

$$E(y) = \sum_{i=1}^{n} a_i \mu_i = a_1\mu_1 + a_2\mu_2 + \ldots + a_n\mu_n \tag{33}$$

$$V(y) = \sum_{i=1}^{n} a_i^2\sigma_i^2 = a_1^2\sigma_1^2 + a_2^2\sigma_2^2 + \ldots + a_n^2\sigma_n^2 \tag{34}$$

For equation (34) to be valid, the x_i must be *independent random variables*, which means that the expected value of any product $(x_i - \mu_i)(x_j - \mu_j)$ equals zero when $i \neq j$. Furthermore, if the distribution of each variable x_i is Gaussian, then the distribution of y is also Gaussian.

Example 18. The anion gap, defined as $AG = (Na) - [(Cl) + (HCO_3)]$, is the sum of three analytically independent random variables. From equation (34), the analytical variance of the anion gap (σ_{AG}^2) is equal to the sum of the variances of the individual values, $\sigma_{Na}^2 + \sigma_{Cl}^2 + \sigma_{HCO_3}^2$. For example, if the values for the analytical standard deviations for serum sodium, chloride, and bicarbonate are 1.3, 1.2, and 0.7 mmol/L, the standard deviation of the AG determination is:

$$\sigma_{AG} = \sqrt{(1.3)^2 + (1.2)^2 + (0.7)^2} = 1.9 \text{ mmol/L}$$

If the three analytical measurement errors are assumed to be drawn from three Gaussian distributions and the calculated AG is 12 for a particular specimen, then a 95% confidence interval for the true mean μ_{AG} of the values on that specimen is $AG \pm (2.0)\,\sigma_{AG}$, or $12 \pm (2.0)(1.9)$, or 8.2–15.8 mmol/L.

Example 19. Let x_1 and x_2 be two separate analytical values for a patient on specimens drawn on different days. Assume that both values are drawn from the same population with unknown mean μ and known variance σ^2. Consider the random variable

$$y = (x_2 - x_1)/\sqrt{2}\sigma = a_1 x_1 + a_2 x_2 \tag{35}$$

where

$$a_1 = -1/(\sqrt{2}\sigma) \text{ and } a_2 = 1/(\sqrt{2}\sigma)$$

Since y is a linear combination of x_1 and x_2, from equation (33) we have

$$E(y) = a_1\mu_1 + a_2\mu_2 = a_1\mu + a_2\mu = 0(\mu) = 0$$

and from equation (34) we have

$$V(y) = a_1^2\sigma_1^2 + a_2^2\sigma_2^2 = (a_1^2 + a_2^2)\sigma^2 = \sigma^2/\sigma^2 = 1$$

The random variable y is an example of a standardized random variable with a mean of zero and a variance of one. If x_1 and x_2 have Gaussian distributions, then the distribution of y is also Gaussian, permitting simple procedures for statistical inference about μ_y. In practice σ is usually unknown but can be approximated from repeated measurements of control fluids over many days.

Example 20. Suppose we have obtained successive serum creatinine values of 9 and 12 mg/L for a particular patient on different days. Should we conclude that serum creatinine is increasing, or that these values represent the same underlying value μ in the presence of analytical variability? Assume that the total analytical standard deviation of serum creatinine measurements at this concentration is known from previous studies of analytical variation to be about 0.7 mg/L. Ideally, the standard deviation σ in the denominator of equation (35) should include the intra-individual biological variance component in addition to the analytical variance component, if it is desired to allow for typical intra-individual variation. The calculation of y is as follows:

$$y = (x_2 - x_1)/(\sqrt{2}\sigma) = +3/(0.7\sqrt{2}) = 3.0$$

Since the distribution of y is $N(0, 1)$ under the hypothesis that $\mu_2 = \mu_1$, we can test this hypothesis by finding the probability that a random selection from the $N(0, 1)$ distribution would be as large as or larger than 3.0. From Statistical Table 20-21, in the Appendix, this probability is 0.0013. Thus, we conclude that it is unlikely that x_1 and x_2 have the same population mean μ, and that $\mu_2 > \mu_1$. A 95% confidence interval for the true change $\mu_2 - \mu_1$ in serum creatinine is

$$x_2 - x_1 \pm 2\sqrt{\sigma_1^2 + \sigma_2^2} =$$

$$12 - 9 \pm 2\sqrt{0.7^2 + 0.7^2} =$$

$$3.0 \pm 2\sqrt{0.98} =$$

$$1.0\text{--}5.0 \text{ mg/L}$$

TRANSMISSION OF ERROR

Frequently a calculated quantity in clinical chemistry is related to multiple random analytical variables in a complex fashion. Consider a calculated variable g which is a function of the random variables x, y, and z, i.e., $g = f(x, y, z)$. If the variances of x, y, and z are known, what is the variance of g? The answer to this question when the variables are independent and combined linearly has just been discussed. For more complex functions of random variables, an approximate solution can be obtained by using the *Taylor series expansion* of the function and disregarding all but the first-order terms. If $g = f(x, y, z)$ is approximately linear with respect to x, y, and z in the region of interest, and σ_x^2, σ_y^2, and σ_z^2 are known, then the approximate variance of g is

$$\sigma_g^2 \simeq \left(\frac{\partial g}{\partial x}\right)^2 \sigma_x^2 + \left(\frac{\partial g}{\partial y}\right)^2 \sigma_y^2 + \left(\frac{\partial g}{\partial z}\right)^2 \sigma_z^2 \tag{36}$$

The mean values of the variables are substituted into the expressions for the partial derivatives of g with respect to the random variables x, y, and z. The variables x, y, and z are assumed to be independent, which eliminates considerations of covariance terms.

Example 21. The formula for the measurement of creatinine clearance C_{cr} is

$$C_{cr} = (U)(V)/P$$

where the variable U is the urine creatinine value, the variable V is the measured volume (in mL) of urine produced per minute, and P is the serum creatinine value. When applied to this example, the approximation formula yields the following equation for the variance of the calculated clearance:

$$\sigma_{C_{cr}}^2 = \left(\frac{\partial C_{cr}}{\partial U}\right)^2 \sigma_U^2 + \left(\frac{\partial C_{cr}}{\partial V}\right)^2 \sigma_V^2 + \left(\frac{\partial C_{cr}}{\partial P}\right)^2 \sigma_P^2$$

$$= \left(\frac{V}{P}\right)^2 \sigma_U^2 + \left(\frac{U}{P}\right)^2 \sigma_V^2 + \left(\frac{-UV}{P^2}\right)^2 \sigma_P^2$$

If we divide both sides of this equation by $C_{cr}^2 = U^2V^2/P^2$ and cancel common terms, we have the following expression:

$$\sigma_{C_{cr}}^2/C_{cr}^2 = \sigma_U^2/U^2 + \sigma_V^2/V^2 + \sigma_P^2/P^2 \quad \text{or}$$

$$CV_{C_{cr}}^2 = CV_U^2 + CV_V^2 + CV_P^2$$

Thus, we have shown that the coefficient of variation of the measured clearance is approximately the sum of the coefficients of variation of the three component variables. A CV of 7% for U and P and 10% for V, for example, would yield a CV for the clearance of

$$CV_{C_{cr}} = \sqrt{7^2 + 10^2 + 7^2} = 14\%$$

ESTIMATING VARIANCE COMPONENTS BY NESTED ANOVA

In studying variation, the objective often is to find the magnitudes of individual or component variances that cannot be found directly. *Variance component analysis* provides estimation of these component variances. For illustration, consider a typical analytical process. Usually a volume of sample is mixed sequentially with volumes of reagents, and a single analytical response is measured at a given temperature. Random variation in these volumetric, spectrophotometric, and other processes results in what is known as within-run variation. A standard sample used to calibrate an analytical procedure is subject to the same random analytical variation as any other sample. For this reason small random shifts in an analytical system occur when the analytical process is recalibrated in order to run additional samples. This source of variation contributes to run-to-run variation within a day. In the same manner, we can identify sources of variation that cause random shifts on a day-to-day basis, such as the volumetric reconstitution of a freeze-dried secondary standard material for calibration. A sample from a specimen that arrives in the laboratory for single aliquot analysis on a particular day in a particular run is subject to the total analytical variation from all three of these components, as illustrated in Figure 2A-15. A common variance component study is the generation and analysis of data to separate the total analytical variation into parts assignable to the temporal components: within-run, run-to-run within-a-day, and day-to-day. Because each specific source of system variation can be linked to one of these temporal components, an unacceptably large temporal variance component implicates a set of system sources of variation as potential problems to be corrected.

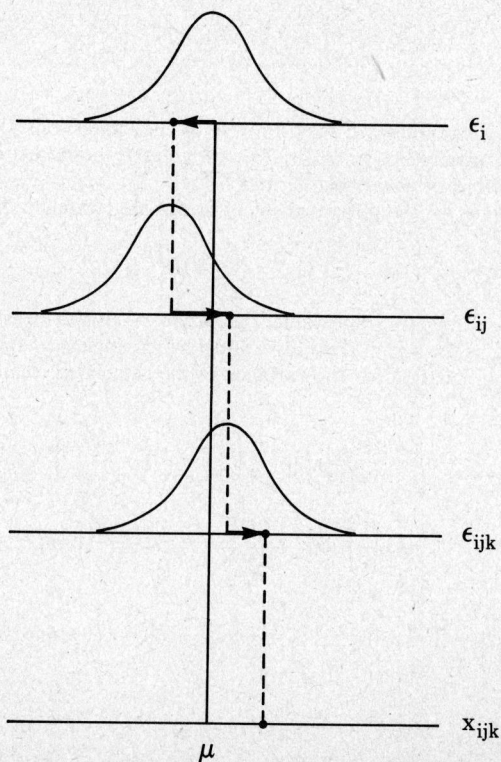

Figure 2A-15. The two-stage nested model: the random error associated with an observation x_{ijk} is the sum of three components of random error.

Other applications of variance component analysis include the separation and estimation of inter-individual, intra-individual, and analytical components of variance, and the estimation of the vial-to-vial variation of control and calibrator fluids.[15,27] In all these applications of variance component analysis, exactly which components are estimable depends upon the way the experiment is designed.

The One-Stage Nested Design, Model, and ANOVA

Estimation of variance components requires an experimental design for generating the observations, a statistical model that can explain the variation in the observations, and a subsequent analysis of variance (ANOVA). We will first describe the *one-stage nested* design, model, and ANOVA. As an example, consider an experiment where an analytical process is run for I days with J replicates of a particular fluid in a single run each day. Analytical variation can then be partitioned into two components, within-run and day-to-day. Since there is only one run per day, the day-to-day component will actually be the sum of the run-to-run within day and the day-to-day components. The run-to-run component can be separated from the day-to-day component only if multiple runs are performed on some days, as in the two-stage nested design. The statistical model for the one-stage nested design is

$$x_{ij} = \mu + \epsilon_i + \epsilon_{ij} \tag{37}$$

where $i = 1, \ldots, I$ days and $j = 1, \ldots, J$ replicates/day.

The random variable x_{ij} refers to the analytical result for replicate j on day i, and has an expected value of μ. The random variable ϵ_{ij} is the analytical measurement error for replicate j on day i due to the within-run sources of variability. It has an expected value of zero and a variance σ_{WR}^2, the within-run component of variance. The random variable ϵ_i is the analytical measurement error influencing all observations for the ith day due to the day-to-day sources of variability. It has an expected value of zero and a variance σ_D^2, the day-to-day component of variance. Since ϵ_{ij} and ϵ_i are independent, according to equation (34) the variance of $x_{ij} = \mu + \epsilon_i + \epsilon_{ij}$ is $\sigma_{WR}^2 + \sigma_D^2$, which we refer to as the total analytical variation σ_{TOT}^2.

The calculations for estimation of the components of variance for the one-stage nested design are presented in Table 2A-7. It should be noted that this ANOVA table is a slight modification of that usually presented in statistical textbooks. The modification is that we have *not* multiplied the day-to-day sum of squares by J, the number of replicates, to obtain SSD, and we have therefore adjusted the quantity estimated by the day-to-day mean square in the last column. We feel that these equations are more intuitive for estimating the variance components.

The estimation procedure in Table 2A-7 can be used only when the number J of replicates per day is the same for each of the I days. The remainder of this section also assumes that J is constant. If a computer is not available and only a small number of days do not have the intended number of replicates J, then it may be wise to discard and rerun the tests on those days to produce a *balanced* (i.e., equal J's) design. The *unbalanced* (i.e., unequal J's) case is discussed in various statistical textbooks.[23,25] Commercially available statistical software packages have procedures to estimate variance components for both balanced and unbalanced designs.[20,21]

As shown in Table 2A-7, the quantity s_{WR}^2 directly estimates σ_{WR}^2. The estimate s_{WR}^2 is calculated as the average of the I individual estimates of σ_{WR}^2, which is equivalent to equation (28) given earlier. A convenient formula exists for the special case $J = 2$ in which there are only two replicates in each daily run. If the two replicates in the ith run are designated x_{i1} and x_{i2}, then using the fact that $\bar{x}_{i\cdot} = (x_{i1} + x_{i2})/2$, we have

$$s_{WR}^2 = \frac{\sum_{i=1}^{I} (x_{i1} - x_{i2})^2}{2I}$$

The quantity s_D^2 is calculated from the I daily means using the standard formula for the sample variance of I data values. It should be noted that s_D^2 does not directly estimate σ_D^2. From the postulated model (37) and equation (34) for the variance of a sum of random variables, it can be shown that the variance of the daily means is $\sigma_D^2 + \sigma_{WR}^2/J$. Thus s_D^2 estimates $\sigma_D^2 + \sigma_{WR}^2/J$, which will be σ_D^2 only when J is infinite. To estimate σ_D^2 we use

Table 2A-7. MODIFIED ANOVA TABLE FOR ESTIMATING THE COMPONENTS OF VARIANCE IN THE ONE-STAGE NESTED DESIGN

Source of Variation	Sum of Squares (ss)	Degrees of Freedom (df)	Mean Square (ss/df)	Quantity Estimated by Mean Square
Day-to-day	$SSD = \sum_{i=1}^{I} (\bar{x}_{i.} - \bar{x}_{..})^2$	$I - 1$	$s_D^2 = \dfrac{SSD}{(I-1)}$	$\sigma_D^2 + \dfrac{\sigma_{WR}^2}{J}$
Within-run	$SSW = \sum_{i=1}^{I} \sum_{j=1}^{J} (x_{ij} - \bar{x}_{i.})^2$	$I(J-1)$	$s_{WR}^2 = \dfrac{SSW}{I(J-1)}$	σ_{WR}^2
Total	$SST = \sum_{i=1}^{I} \sum_{j=1}^{J} (x_{ij} - \bar{x}_{..})^2$	$IJ - 1$	$s_T^2 = \dfrac{SST}{IJ-1}$	$\sigma_{WR}^2 + \sigma_D^2 \dfrac{(IJ-J)}{(IJ-1)}$

Estimates of Variance Components

$$\hat{\sigma}_{WR}^2 = s_{WR}^2$$

$$\hat{\sigma}_D^2 = s_D^2 - \frac{s_{WR}^2}{J}$$

$$\hat{\sigma}_{TOT}^2 = \hat{\sigma}_{WR}^2 + \hat{\sigma}_D^2 = s_D^2 + \frac{(J-1)}{J} s_{WR}^2$$

x_{ij} = observation on day i, replicate j
$i = 1, \ldots, I$ days
$j = 1, \ldots, J$ replicates per run

$\bar{x}_{i.} = \dfrac{\sum_{j=1}^{J} x_{ij}}{J}$ = average of all replicates on day i

$\bar{x}_{..} = \dfrac{\sum_{i=1}^{I} \sum_{j=1}^{J} x_{ij}}{IJ} = \dfrac{\sum_{i=1}^{I} \bar{x}_{i.}}{I}$ = average of all results over all days

$$\hat{\sigma}_D^2 = s_D^2 - s_{WR}^2/J \qquad (38)$$

where the symbol " ^ " stands for "estimator of." Sometimes equation (38) yields a negative value for $\hat{\sigma}_D^2$. The causes may be (1) outliers in the data that inflate s_{WR}^2, (2) a small true σ_D^2 in comparison to σ_{WR}^2, or (3) small sample sizes I and/or J. If $\hat{\sigma}_D^2$ is negative when the data are free of obvious outliers, we suggest replacing it with zero to calculate $\hat{\sigma}_{TOT}^2$ in

$$\hat{\sigma}_{TOT}^2 = \hat{\sigma}_D^2 + \hat{\sigma}_{WR}^2 \qquad (39)$$

By ignoring the nested nature of the experimental design, the quantity s_T^2 can be calculated as the sample variance of all the IJ data values. It should be noted from Table 2A-7 that s_T^2 has an expected value $\sigma_{TOT}^2 = \sigma_{WR}^2 + \sigma_D^2$ only when J = 1. When J > 1 the expected value of s_T^2 is less than σ_{TOT}^2. This underestimation decreases as I increases relative to J and as σ_D^2 decreases relative to σ_{WR}^2.

Example 22. A control sample is analyzed in duplicate in each analytical run with one run performed daily for ten days. Although the NCCLS has recommended that variance component studies be conducted over 20 days to ensure adequate representation of day-to-day variation, we have chosen this smaller example for ease of illustration.[11] The data, ANOVA table, and estimated variance components are presented in Table 2A-8.

The Two-Stage Nested Design, Model, and ANOVA

The two-stage nested model simply contains another component for the hierarchy of variance. In the components of analytical variance example, we can partition the total variance into another component if multiple runs are performed daily. The expanded variance model, illustrated in

Table 2A-8. ESTIMATION OF VARIANCE COMPONENTS BY ANOVA FOR ONE-STAGE NESTED DESIGN IN EXAMPLE 22

		Measurements (x_{ij}) Days (i = 1, . . . , 10)									
		1	2	3	4	5	6	7	8	9	10
Replicates (j = 1,2)	1	7.6	7.4	7.6	7.6	7.7	7.7	7.9	7.6	7.5	7.8
	2	7.5	7.6	7.5	7.7	7.6	7.6	7.8	7.6	7.3	7.6
$\bar{x}_{i.}$		7.55	7.5	7.55	7.65	7.65	7.65	7.85	7.6	7.4	7.7

$$\bar{x}_{..} = \frac{\sum_{i=1}^{10} \bar{x}_{i.}}{10} = 7.61$$

Modified ANOVA Table

Source of Variation	Sum of Squares (ss)	Degrees of Freedom (df)	Mean Square (ss/df)	Quantity Estimated by Mean Square
Day-to-day	$SSD = \sum_{i=1}^{10}(\bar{x}_{i.} - \bar{x}_{..})^2 = 0.1340$	9	$s_D^2 = 0.0149$	$\sigma_D^2 + \sigma_{WR}^2/2$
Within-run	$SSW = \sum_{i=1}^{10}\sum_{j=1}^{2}(x_{ij} - \bar{x}_{i.})^2 = 0.0900$	10	$s_{WR}^2 = 0.0090$	σ_{WR}^2
Total	$SST = \sum_{i=1}^{10}\sum_{j=1}^{2}(x_{ij} - \bar{x}_{..})^2 = 0.3580$	19		

Estimated Variance Components

$\hat{\sigma}_{WR}^2 = s_{WR}^2 = 0.0090$

$\hat{\sigma}_D^2 = s_D^2 - s_{WR}^2/2 = 0.0104$

$\hat{\sigma}_{TOT}^2 = \hat{\sigma}_{WR}^2 + \hat{\sigma}_D^2 = 0.0194$

$\hat{CV}_{WR} = (100)\hat{\sigma}_{WR}/\bar{x}_{..} = 1.25\%$

$\hat{CV}_D = (100)\hat{\sigma}_D/\bar{x}_{..} = 1.34\%$

$\hat{CV}_{TOT} = (100)\hat{\sigma}_{TOT}/\bar{x}_{..} = 1.83\%$

Figure 2A-15, can be written as follows:

$$x_{ijk} = \mu + \epsilon_i + \epsilon_{ij} + \epsilon_{ijk} \tag{40}$$

where i $= 1, \ldots ,$ I days
$\quad$ j $= 1, \ldots ,$ J runs per day
$\quad$ k $= 1, \ldots ,$ K replicates per run.

As before μ is the mean concentration. The random variable ϵ_{ijk} is the analytical measurement error for the measurement on the ith day, jth run, and kth replicate. It has a zero expected value and variance σ^2_{WR} that reflects the within-run sources of variation. The random variable ϵ_{ij} is the analytical error attributable to, and impacting all observations in, the jth run on the ith day. It has a zero expected value and variance σ^2_R that reflects within-day run-to-run sources of variation. The random variable ϵ_i is the analytical error attributable to, and impacting all observation in, the ith day. It has zero expected value and variance σ^2_D that reflects day-to-day sources of variation. The variance of the random variable x_{ijk} is the sum of these variances, $\sigma^2_{TOT} = \sigma^2_D + \sigma^2_R + \sigma^2_{WR}$.

The modified ANOVA table for estimating the components of variance in the completely balanced two-stage nested design is given in Table 2A-9. Again the calculations involve only means and variances that can be obtained with use of a scientific pocket calculator. Negative values for $\hat{\sigma}^2_D$ and $\hat{\sigma}^2_R$ are possible. After checking to eliminate the possibility of obvious outliers as the cause, the negative values should be replaced by zero for calculating $\hat{\sigma}^2_{TOT}$. As was the case with the one-stage model, s^2_T will in general underestimate σ^2_{TOT} to a degree that depends upon the relative sample sizes and the relative magnitudes of the true variance components.

Another application of the two-stage nested analysis of variance model in clinical chemistry is the determination of the intra-individual and inter-individual variances of a biochemical constituent present in human specimens. An experimental design to determine these variances might involve selecting I individuals from a population, and obtaining their sera for J days at the same time each day under the same experimental conditions. All the sera could then be analyzed in one analytical run with the replication of each serum sample K times. From the K repeated analyses of each of the IJ samples, a pooled analytical within-run variance could be obtained analogous to the within-run variance s^2_{WR} in Table 2A-9. An intra-individual variance could be calculated for each individual using the mean assayed concentrations from the different daily specimens, and these variances averaged over all individuals to give s^2_{INTRA}. The variances of grand individual means over all individuals would be calculated to obtain s^2_{INTER}. In analogy with the equations in Table 2A-9, we could calculate the following estimates:

$$\hat{\sigma}^2_{WR} = s^2_{WR}$$

$$\hat{\sigma}^2_{INTRA} = s^2_{INTRA} - s^2_{WR}/K$$

$$\hat{\sigma}^2_{INTER} = s^2_{INTER} - s^2_{INTRA}/J$$

INFERENCES CONCERNING VARIANCES ESTIMATED IN ONE-STAGE AND TWO-STAGE NESTED ANOVA

In the previous section, we discussed statistical procedures for constructing confidence intervals for, and conducting significance tests about, population variances. In this subsection we will extend these statistical procedures to variance components estimated by ANOVA from nested designs. For example, a new analyzer system can be tested to determine whether or not it meets a manufacturer's claims for within-run and/or total analytical variation. Provided the distribution of the measurement errors in the one-stage and two-stage nested models are Gaussian, the distributions of the estimates $\hat{\sigma}^2_{WR}$ and $\hat{\sigma}^2_{TOT}$ are approximately chi-square distributions. Therefore, to make inferences about the population variances σ^2_{WR} and σ^2_{TOT} we can use the procedures outlined in equations (23)–(27). However, we must replace the degrees of freedom $(n - 1)$ in equations (23)–(27) with the degrees of freedom associated with the estimates $\hat{\sigma}^2_{WR}$ and $\hat{\sigma}^2_{TOT}$. The sampling distributions of the other variance component estimators are *not* approximately chi-square. Statistical procedures for inference about these other components are discussed by Snee.[24]

Table 2A-9. MODIFIED ANOVA TABLE FOR ESTIMATING THE COMPONENTS OF VARIANCE IN THE TWO-STAGE NESTED DESIGN

Source of Variation	Sums of Squares (ss)	Degrees of Freedom (df)	Mean Square (ss/df)	Quantity Estimated by Mean Square
Day-to-day	$SSD = \sum\limits_{i=1}^{I} (\bar{x}_{i..} - \bar{x}_{...})^2$	$I - 1$	$s_D^2 = \dfrac{SSD}{(I-1)}$	$\sigma_D^2 + \dfrac{\sigma_R^2}{J} + \dfrac{\sigma_{WR}^2}{JK}$
Run-to-run	$SSR = \sum\limits_{i=1}^{I}\sum\limits_{j=1}^{J} (\bar{x}_{ij.} - \bar{x}_{i..})^2$	$I(J-1)$	$s_R^2 = \dfrac{SSR}{I(J-1)}$	$\sigma_R^2 + \dfrac{\sigma_{WR}^2}{K}$
Within-run	$SSW = \sum\limits_{i=1}^{I}\sum\limits_{j=1}^{J}\sum\limits_{k=1}^{K} (x_{ijk} - \bar{x}_{ij.})^2$	$IJ(K-1)$	$s_{WR}^2 = \dfrac{SSW}{IJ(K-1)}$	σ_{WR}^2
Total	$SST = \sum\limits_{i=1}^{I}\sum\limits_{j=1}^{J}\sum\limits_{k=1}^{K} (x_{ijk} - \bar{x}_{...})^2$	$IJK - 1$	$s_T^2 = \dfrac{SST}{IJK-1}$	$\sigma_{WR}^2 + \dfrac{K(IJ-1)}{IJK-1}\,\sigma_R^2 + \dfrac{JK(I-1)}{IJK-1}\,\sigma_D^2$

Estimates of variance components

$$\hat{\sigma}_{WR}^2 = s_{WR}^2$$

$$\hat{\sigma}_R^2 = s_R^2 - \frac{s_{WR}^2}{K}$$

$$\hat{\sigma}_D^2 = s_D^2 - \frac{s_R^2}{J}$$

$$\hat{\sigma}_{TOT}^2 = \hat{\sigma}_{WR}^2 + \hat{\sigma}_R^2 + \hat{\sigma}_D^2$$

x_{ijk}
$i = 1, \ldots, I$ days
$j = 1, \ldots, J$ runs per day
$k = 1, \ldots, K$ replicates per run

$$\bar{x}_{ij.} = \frac{\sum\limits_{k=1}^{K} x_{ijk}}{K} \qquad \bar{x}_{i..} = \frac{\sum\limits_{j=1}^{J}\sum\limits_{k=1}^{K} x_{ijk}}{JK} \qquad \bar{x}_{...} = \frac{\sum\limits_{i=1}^{I}\sum\limits_{j=1}^{J}\sum\limits_{k=1}^{K} x_{ijk}}{IJK}$$

Inference about σ_{WR}^2

The number of degrees of freedom for the estimate $\hat{\sigma}_{WR}^2$ of the within-run variance is simply obtained from the ANOVA tables in Tables 2A-7 and 2A-9, and this is equivalent to the pooling procedure discussed in the previous section. Thus the number of degrees of freedom ν_{WR} for $\hat{\sigma}_{WR}^2$ in the one-stage and two-stage nested designs are $I(J - 1)$ and $IJ(K - 1)$, respectively. Equation (27), with ν_{WR} replacing $(n - 1)$, can be used to construct confidence intervals for σ_{WR}^2 based on an estimate $\hat{\sigma}_{WR}^2$. Equation (24), with ν_{WR} replacing $(n - 1)$, can be used to conduct significance tests about σ_{WR}^2 based on the estimate $\hat{\sigma}_{WR}^2$.

Example 23. In the one-stage nested design in Example 22, duplicate determinations were made in a single run each day for 10 days. The number of degrees of freedom for $\hat{\sigma}_{WR}^2$ is $\nu_{WR} = 10(2 - 1) = 10$. A 95% confidence interval for σ_{WR}^2 based on $\hat{\sigma}_{WR}^2 = 0.009$, and $L = 3.25$ and $U = 20.5$ from the table of chi-square probability points with $\nu_{WR} = 10$, is

$$\frac{(10)\,(0.009)}{20.5} \quad \text{to} \quad \frac{(10)\,(0.009)}{3.25}, \quad \text{or } 0.004 \text{ to } 0.28$$

Taking the square root of these limits and dividing by the mean of 7.61, the corresponding interval for the population CV_{WR} would be 0.9–2.2%.

Example 24. Suppose σ_{WR}^2 is estimated from two runs with two replicates each day for I days. Then there are $\nu_{WR} = (I)(2)(2 - 1)$ or $2I$ degrees of freedom associated with this estimate, where $2I$ is simply the total number of runs.

Inference about σ_{TOT}^2

The number of degrees of freedom for the estimate $\hat{\sigma}_{TOT}^2$ in the one-stage and two-stage nested designs is *not* given by $IJ - 1$ and $IJK - 1$, because all the observations are not independent of each other. An approximation for ν_{TOT} given by Satterthwaite takes into account this lack of independence in nested and other designs.[9] The approximate number of degrees of freedom ν_{TOT} appropriate for $\hat{\sigma}_{TOT}^2$ in the one-stage nested design is

$$\nu_{TOT} = \frac{(\sigma_{WR}^2 + \sigma_D^2)^2}{(J - 1)s_{WR}^4/(J^2 \times I) + s_D^4/(I - 1)} \tag{41}$$

where the various quantities are defined in Table 2A-7. The population variances in equation (41) can be replaced by their estimates.

Example 25. In the one-stage nested design in Example 22, $J=2$, $I=10$, $\hat{\sigma}_{WR}^2 + \hat{\sigma}_D^2 = 0.019$, $s_{WR}^2 = 0.009$, and $s_D^2 = 0.015$, so

$$\nu_{TOT} = \frac{(0.019)^2}{(0.009)^2/40 + (0.015)^2/9} = 14.1$$

A 95% confidence interval for σ_{TOT}^2 based on $\hat{\sigma}_{TOT}^2 = 0.019$ and $L = 5.63$ and $U = 26.1$ from the table of chi-square probability points with $\nu_{TOT} = 14$ is

$$\frac{(14)\,(0.019)}{26.1} \quad \text{to} \quad \frac{(14)\,(0.019)}{5.63}, \quad \text{or } 0.010 \text{ to } 0.047$$

Taking the square root of these limits and dividing by the mean of 7.61, the corresponding 95% interval for the population CV_{TOT} is 1.3–2.9%.

The appropriate formula for the approximate number of degrees of freedom ν_{TOT} for $\hat{\sigma}_{TOT}^2$ in the two-stage nested design shown in Table 2A-9 is

$$\nu_{TOT} = \frac{(\sigma_{WR}^2 + \sigma_R^2 + \sigma_D^2)^2}{\dfrac{(K - 1)s_{WR}^4}{K^2(I)(J)} + \dfrac{(J - 1)s_R^4}{(J^2)(I)} + \dfrac{s_D^4}{I - 1}} \tag{42}$$

with all quantities as defined in Table 2A-9. The population variances in equation (42) can be replaced by their estimates.

Example 26. The total analytical variation of an instrument is often estimated by running a sample in duplicate (K = 2) in two runs per day (J = 2) over I days. The formula for ν_{TOT} is then

$$\nu_{TOT} = \frac{(\sigma_{WR}^2 + \sigma_R^2 + \sigma_D^2)^2}{\dfrac{s_{WR}^4}{8I} + \dfrac{s_R^4}{4I} + \dfrac{s_D^4}{I-1}}$$

Specifically, consider the case where I = 20 and the other quantities are as follows:

$$s_{WR}^2 = 7.90, \; s_R^2 = 7.02, \; s_D^2 = 5.48$$

$$\hat{\sigma}_{WR}^2 = 7.90, \; \hat{\sigma}_R^2 = 3.07, \; \hat{\sigma}_D^2 = 1.96, \; \hat{\sigma}_{TOT}^2 = 12.93$$

Then

$$\nu_{TOT} = \frac{(7.9 + 3.07 + 1.96)^2}{\dfrac{(7.9)^2}{160} + \dfrac{(7.02)^2}{80} + \dfrac{(5.48)^2}{19}} = 64.6$$

Using $\hat{\sigma}_{TOT}^2 = 12.93$ with $\nu_{TOT} = 64$, we can compute confidence intervals for, and conduct significance tests about, σ_{TOT}^2.

List of Symbols for Study of Variation Section (in order of appearance)

x_i	random variable
μ_i	population mean for x_i
σ_i^2	population variance for x_i
y	random variable
E(y)	expected value (i.e., population mean) for y
V(y)	population variance for y
N(0, 1)	Gaussian probability distribution with mean of zero and variance of one
x,y,z,	random variables
g = f(x,y,z)	random variable g is a function of x, y, and z
$\sigma_x^2, \sigma_y^2, \sigma_z^2$	population variances for x, y, and z
σ_g^2	population variance for g
$\simeq$	approximately equal to
$\partial g/\partial x$	partial derivative of the function g with respect to x; similarly for y and z

The one-stage nested model

x_{ij}	ijth observation in the one-stage nested model
I	number of levels of the first factor (e.g., day)
J	number of levels of the second factor (e.g., replication within-run)
ϵ_i	random error attributable to the ith level of factor one (e.g., ith day)
ϵ_{ij}	random error attributable to jth observation for ith level of factor one
σ_D^2	component of variance for factor one (e.g., day-to-day)
σ_{WR}^2	component of variance for factor two (e.g., replicate within-run)
σ_{TOT}^2	total variance
$\bar{x}_{i.}$	sample mean of the observations in ith level of factor one (e.g., ith day)
$\bar{x}_{..}$	sample grand mean
SSD	sum of squares for factor one in Table 2A-7
SSW	sum of squares for factor two in Table 2A-7
SST	sum of squares total
s_D^2	mean square for factor one in Table 2A-7
s_{WR}^2	mean square for factor two in Table 2A-7
s_T^2	mean square total

$\hat{\sigma}_D^2$ estimate of variance component for factor one (e.g., day-to-day)
$\hat{\sigma}_{WR}^2$ estimate of variance component for factor two (e.g., within-run)
$\hat{\sigma}_{TOT}^2$ estimate of total variance

The two-stage nested model

x_{ijk} ijkth observation in the two-stage nested model
I number of levels of the first factor (e.g., day)
J number of levels of the second factor (e.g., run within-day)
K number of levels of the third factor (e.g., replicate within-run)
ϵ_i random error attributable to ith level of factor one
ϵ_{ij} random error attributable to jth level of factor two for ith level of factor one
ϵ_{ijk} random error attributable to kth observation for jth level of factor two and ith level of factor one
σ_D^2 component of variance for factor one (e.g., day-to-day)
σ_R^2 component of variance for factor two (e.g., run-to-run within-day)
σ_{WR}^2 component of variance for factor three (e.g., replicate within-run)
σ_{TOT}^2 total variance
$\bar{x}_{ij.}$ sample mean of observations in ith level of factor one and jth level of factor two
$\bar{x}_{i..}$ sample mean of observations in ith level of factor one
$\bar{x}_{...}$ sample grand mean
SSD, SSR,   sums of squares in modified ANOVA table, Table 2A-9
SSW, SST
$s_D^2, s_R^2,$ mean squares in modified ANOVA table, Table 2A-9
s_{WR}^2, s_T^2
$\hat{\sigma}_D^2$ estimate of variance component for factor one (e.g., day-to-day)
$\hat{\sigma}_R^2$ estimate of variance component for factor two (e.g., run-to-run)
$\hat{\sigma}_{WR}^2$ estimate of variance component for factor three (e.g., replicate within-run)
$\hat{\sigma}_{TOT}^2$ estimate of total variance
ν_{WR} degrees of freedom for estimate $\hat{\sigma}_{WR}^2$
ν_{TOT} approximate degrees of freedom for the estimate $\hat{\sigma}_{TOT}^2$

REGRESSION ANALYSIS

In many analytical investigations we are interested in the relationship between two variables y and x. Mathematically y is often called a *function* of x, but statistically the term *regression* is used to describe the relationship between y and x. To distinguish between the two variables, x has traditionally been called the *independent* or *predictor* variable, and y has been called the *dependent* or *response* variable. In some applications these terms are appropriate and in others they are less suitable.

Regression methods are useful in many different types of investigations including the following:

1. Determining if any relationship between y and x exists
2. Determining the shape of the relationship
3. Determining a regression model that will describe this relationship, and the number of parameters necessary for this model
4. Estimating the unknown parameters in a specified model
5. Estimating or predicting the mean value of y for a particular value of x
6. Estimating the error or variation in y after adjustment for x

In this section we first discuss the classical simple linear regression model and the associated regression analysis. This is followed by the application of regression analysis for calibration of an analyzer system, a short introduction to nonlinear regression models, and a brief discussion of the commonly used regression approach for comparing two methods.

THE CLASSICAL SIMPLE LINEAR REGRESSION MODEL

Consider an investigation to determine the interference caused by ascorbate on glucose measurements. The purpose of the investigation is to determine if a relationship exists between the glucose measurement y and the amount x of ascorbate in a specimen. Additionally, if a relationship exists, we would like to (1) determine an appropriate model for the relationship, (2) estimate the unknown parameters of the model, and (3) estimate or predict y at different concentrations of ascorbate to determine how ascorbate affects glucose measurements.

To conduct the experiment, a high-normal glucose serum pool with no ascorbate was obtained. This pool was split into four parts and spiked with 0, 2, 4, and 6 mg/dL ascorbic acid. Eight samples were prepared from each of the four new pools. The thirty-two samples were assayed for glucose in a random order in a single run. The thirty-two pairs $(x_1, y_1), \ldots, (x_{32}, y_{32})$ are given in Table 2A-10 and plotted in Figure 2A-16. As well as can be determined from this plot, a negative linear relationship appears to exist between the glucose measurement y and the amount of ascorbate x. We present next a discussion of the classical simple linear regression model and its assumptions, followed by regression analysis of these data.

Assumptions

The assumptions underlying the *classical simple linear regression model* are described below and illustrated in Figure 2A-17.

1. The values of the variable x are "fixed" or preselected by the investigator.

2. The values of the variable x are known or measured without error, or measured with error that is negligible.

3. For each value of x there is a population of y values. Additionally, these populations are assumed to be Gaussian for validity of the statistical inferential procedures that we will illustrate.

Table 2A-10. ASCORBATE INTERFERENCE FOR GLUCOSE CONCENTRATION

Observation i	Ascorbate Concentration x_i	Glucose Concentration y_i	Predicted Glucose $\hat{y}_i$	Residual $\hat{\epsilon}_i = y_i - \hat{y}_i$
1	0	113.9	115.53	−1.63
2	0	116.8	115.53	1.27
3	0	115.6	115.53	0.07
4	0	115.7	115.53	0.17
5	0	115.1	115.53	−0.43
6	0	115.2	115.53	−0.33
7	0	116.3	115.53	0.77
8	0	115.5	115.53	−0.03
9	2	114.0	114.04	−0.04
10	2	112.9	114.04	−1.14
11	2	115.2	114.04	1.16
12	2	113.1	114.04	−0.94
13	2	114.8	114.04	0.76
14	2	114.6	114.04	0.56
15	2	114.1	114.04	0.06
16	2	113.8	114.04	−0.24
17	4	112.8	112.56	0.24
18	4	112.4	112.56	−0.16
19	4	112.1	112.56	−0.46
20	4	112.5	112.56	−0.06
21	4	113.5	112.56	0.94
22	4	113.7	112.56	1.14
23	4	111.1	112.56	−1.46
24	4	112.5	112.56	−0.06
25	6	111.1	111.08	0.02
26	6	110.0	111.08	−1.08
27	6	112.3	111.08	1.22
28	6	110.6	111.08	−0.48
29	6	110.3	111.08	−0.78
30	6	110.9	111.08	−0.18
31	6	111.4	111.08	0.32
32	6	111.9	111.08	0.82

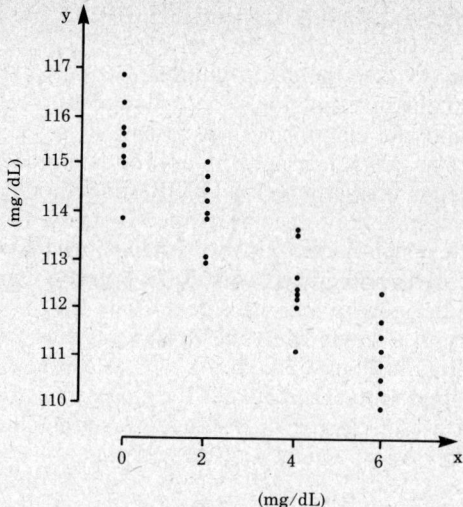

Figure 2A-16. Scatterplot of glucose concentrations y in mg/dL, against ascorbate concentrations x in mg/dL.

4. The variances of these populations are all equal, with the common variance denoted by $\sigma^2_{y|x}$.

5. The means of these populations of y values lie on a straight line, which is expressed as

$$\mu_{y|x} = \alpha + \beta x \tag{43}$$

where $\mu_{y|x}$ is the mean of the population of y values for a particular value of x, and α and β are the population regression parameters. Geometrically, α is the y-intercept and β is the slope of the line in the (x, y) plane on which the population means lie.

6. The y values sampled from these populations are assumed to be statistically independent. This means that the value of any y measurement does not depend upon the value of any other y measurement.

These assumptions can be summarized by what we call the regression model

$$y = \alpha + \beta x + \epsilon \tag{44}$$

where y is a random observation from one of the populations of y values, $\alpha + \beta x$ is the mean of this population, and ϵ is the random measurement error associated with this random y observation. As defined previously in this chapter, the measurement error

$$\epsilon = y - (\alpha + \beta x) \tag{45}$$

associated with a random measurement y is a random variable representing the difference between the random measurement y and the mean of the population from which y is drawn. The mea-

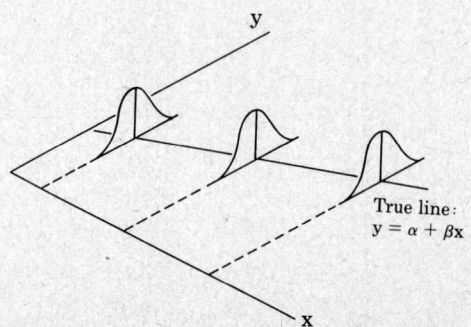

Figure 2A-17. The simple linear regression model. The Gaussian distributions of y values, $N(\alpha + \beta x, \sigma^2)$, are shown for three selected values of x.

surement error ϵ for a measurement y at a particular value of x is assumed to have a Gaussian distribution with a mean of zero and a variance equal to the common variance $\sigma^2_{y|x}$ of the populations of y values.

Estimating the Simple Linear Regression Line

In order to estimate the population parameters α and β that determine the line of population means defined by equation (43), we draw random samples from some of the populations of y values. In our glucose example we have drawn random samples of size eight from each of four populations of y values. The pattern of these plotted data in Figure 2A-16 suggests a linear relationship between y and x. Although it would be possible to draw by hand a straight line y = a + bx through the sample data that somewhat describes the relationship between y and x, this approach is clearly too subjective, and an objective approach is necessary for determining the "best fitting" line. The intercept a and slope b of this best fitting line will provide the estimates of the parameters α and β.

The method usually employed for obtaining the best fitting regression line is called the *method of least squares,* and the resultant line y = a + bx is called the *least squares line.* The sample coefficients a and b are called the *least squares estimates* of α and β, and are given by

$$b = \frac{\Sigma x_i y_i - (\Sigma x_i)(\Sigma y_i)/n}{\Sigma x_i^2 - (\Sigma x_i)^2/n} = \frac{\Sigma(x_i - \bar{x})(y_i - \bar{y})}{\Sigma(x_i - \bar{x})^2} \qquad (46)$$

$$a = \bar{y} - (b)(\bar{x}) \qquad (47)$$

where all summations are from i = 1 to n.

For the glucose example in Table 2A-10 we have

$$b = \frac{10758.41 - (96)(3625.7)/32}{448 - (96)^2/32} = -0.74$$

$$a = 113.30 - (-0.74)(3) = 115.53$$

giving the estimated regression line

$$\hat{y} = 115.53 - (0.74)(x)$$

where $\hat{y}$ is used to denote the estimated population means calculated from this line. This least squares line, plotted in Figure 2A-18, is best in the sense that the sum of the squared vertical distances $\Sigma(y_i - \hat{y}_i)^2$ between the observed data points y_i and the least squares line is smaller than that for any other line.

The interpretation of the coefficient a = 115.53 is that the estimated mean of the population of y values, or glucose concentrations, is 115.53 mg/dL when x, or the concentration of ascorbate, is 0 mg/dL. The coefficient b = −0.74 is the estimated increase (decrease in this case since the coefficient is negative) in the mean glucose concentration in mg/dL for each one mg/dL increase in ascorbate.

Checking the Adequacy of the Assumed Model

So far we have only conducted some algebraic computations to get the estimated or sample regression equation for the assumed simple linear model. Next it is important to determine if the simple linear model adequately describes the relationship between the variables. If the adequacy is verified, we will then determine if the sample regression line can be used effectively for estimation and prediction.

Consider for each observation y_i the true measurement error

$$\epsilon_i = y_i - (\alpha + \beta x_i) \qquad (48)$$

which is unknown. These errors are assumed to have a $N(0, \sigma^2_{y|x})$ distribution. The estimated measurement error is

$$\hat{\epsilon}_i = y_i - (a + bx_i) = y_i - \hat{y}_i \qquad (49)$$

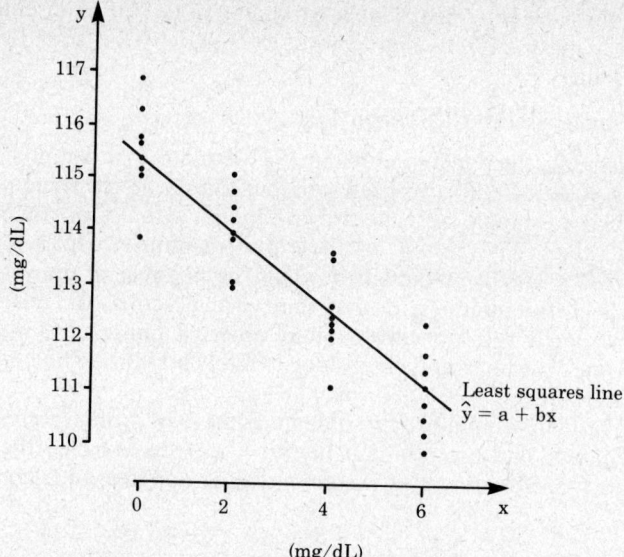

Figure 2A-18. The least squares line is the line which minimizes the sum of the squares of the vertical distances $y_i - \hat{y}_i$.

the vertical distance of y_i from its estimated mean given by the regression line. These $\hat{\epsilon}_i$, given in Table 2A-10 for the glucose example, are called the *residuals* of the fitted regression, and are assumed to have a $N(0, \sigma^2_{y|x})$ distribution when the assumed regression model correctly describes the true relationship. Examination of these residuals provides a useful check on the adequacy of the simple linear model and the validity of the underlying assumptions. In order to check model adequacy by examination of the residuals, the number n of pairs of data values (x, y) must be greater than the number p of model parameters. When n is equal to p there will be an exact fit of the model to the data and each of the n residuals $\hat{\epsilon}_i$ will be zero. For the simple linear model p is 2, and n = 2 data points will determine a unique straight line with $\hat{\epsilon}_1 = \hat{\epsilon}_2 = 0$.

Below we give a few ways to examine the residuals. More extensive discussion, including statistical tests of model "lack-of-fit," can be found in various textbooks.[3,6] We leave it to the reader to verify that examination of the residuals in Table 2A-10 in fact provides no evidence to doubt the adequacy of the simple linear model and its assumptions in the glucose example.

1. A dot diagram or histogram of the residuals $\hat{\epsilon}_i$ is useful to check roughly the assumption that the ϵ_i have a Gaussian distribution. Gross outliers will be easily revealed. If the residuals show evidence of skewness, transforming the y values, e.g., ln(y), should be considered. Such transformation will also affect the form of the model being considered.

2. Scatterplots of the $\hat{\epsilon}_i$ against x_i and $\hat{y}_i$ are very informative. First, if the $\hat{\epsilon}_i$ appear to have a mean of zero at all x and $\hat{y}$, then there is no evidence from these data of inadequacy of the assumed linear relationship. A curved pattern for the means of the $\hat{\epsilon}_i$ when plotted against x_i or $\hat{y}_i$ indicates that the assumed linear relationship between y and x is not correct, and that a model with curvature would provide a better fit to the data. One such curvilinear model is the *quadratic model*

$$y = \alpha + \beta x + \gamma x^2 + \epsilon \tag{50}$$

Such models and the corresponding regression procedures are discussed in the regression references.[3,6] Before entertaining such curvilinear models, one should examine the magnitude of the mean residual at different values of x. If this "model-lack-of-fit" is not practically important for the intended use of the regression line, then it may be decided that the simple linear model provides an adequate approximation to the true curvilinear relationship. The fitting of simple linear models where x and/or y have been transformed, such as

$$y = \alpha + \beta \ln(x) + \epsilon \tag{51}$$

can also be attempted before considering curvilinear models.

3. It is also informative to inspect the variability of the $\hat{\epsilon}_i$ as a function of $\hat{x}$ or y in the scatterplot above. For example, increasing variation of the $\hat{\epsilon}_i$ as a function of x or $\hat{y}$ is evidence of violation of the assumption of a constant variance $\sigma_{y|x}^2$ for the populations of measurement errors and y values. In such cases, estimation of the regression equation by *weighted least squares* should be considered.[3,6,7,19] Ordinary least squares will still yield a sample regression equation that has the population regression as its expected value, but its uncertainty will be greater than that of the equation estimated by weighted least squares.

Estimate of $\sigma_{y|x}^2$ and Confidence Interval for β

Given that the residuals have shown no reason to doubt the adequacy of the simple linear model and its underlying assumptions, we can construct confidence intervals for α, β, and the population means $\mu_{y|x} = \alpha + \beta x$, as well as prediction intervals for new observations.

When the assumed regression model is correct, the estimator of the common variance $\sigma_{y|x}^2$ of the populations of y values is

$$s_{y|x}^2 = \frac{\Sigma \hat{\epsilon}_i^2}{n - 2} = \frac{\Sigma (y_i - \hat{y}_i)^2}{n - 2} \tag{52}$$

The denominator is n − 2 because two quantities, a and b, have to be calculated from the n sample values to estimate the population means $\hat{y}_i = a + bx_i$. The quantity $s_{y|x}$ is often called the *standard error of estimate*, although *standard deviation of the residuals* seems a more appropriate term.

A $(1 - \alpha)100$ confidence interval for β is given by

$$b \pm t_{\alpha/2} (s_b) \tag{53}$$

where the estimated standard deviation of b is

$$s_b = \frac{s_{y|x}}{\sqrt{\Sigma (x_i - \bar{x})^2}} \tag{54}$$

and $t_{\alpha/2}$ is the value of the variable with the t distribution, with n − 2 degrees of freedom, such that a proportion $\alpha/2$ of the values exceed it.

Example 27. For the glucose example in Table 2A-10, the estimate of the variance $\sigma_{y|x}^2$ of the population of glucose measurements at any fixed value of x is

$$s_{y|x}^2 = \frac{\Sigma (y_i - \hat{y}_i)^2}{n - 2} = \frac{18.59}{32 - 2} = 0.62$$

The estimated standard deviation is $s_{y|x} = 0.79$. A 95% confidence interval for β is then given by

$$-0.74 \pm 2.04 \frac{0.79}{\sqrt{159.99}}, \text{ or } -0.87 \text{ to } -0.61$$

Thus, we are 95% certain that the true average effect on glucose measurements is a loss of between 0.61 and 0.87 mg/dL for every one mg/dL ascorbate.

Confidence Interval for α

A $(1 - \alpha)100$ confidence interval for α is given by

$$a \pm t_{\alpha/2} (s_a) \tag{55}$$

where the estimated standard deviation of a is

$$s_a = s_{y|x} \sqrt{\frac{\Sigma\, x_i^2}{n\, \Sigma\, (x_i - \bar{x})^2}} \tag{56}$$

Confidence Interval for the True Mean Value $\mu_{y|x}$

A $(1 - \alpha)100$ confidence interval for the true mean value of y for a particular value x_o of x is given by

$$\hat{y}_o \pm t_{\alpha/2}\,(s\hat{y}_o) \tag{57}$$

where $\hat{y}_o = a + bx_o$ is the estimated mean of the distribution of y values and

$$s\hat{y}_o = (s_{y|x}) \sqrt{\frac{1}{n} + \frac{(x_o - \bar{x})^2}{\Sigma\, (x_i - \bar{x})^2}} \tag{58}$$

is the estimated standard deviation of $\hat{y}_o$.

It should be noted that the width of this confidence interval depends on the distance of x_o from $\bar{x}$. In particular, the confidence interval will be narrowest when $x_o = \bar{x}$, and will get progressively wider as x_o departs from $\bar{x}$ in both directions.

Prediction Interval for a Single New Observation y

A $(1 - \alpha)100$ prediction interval for a single new observation y from the population of y values at a particular value x_o of x is given by

$$\hat{y}_o \pm t_{\alpha/2}\,(s_{y|x}) \sqrt{1 + \frac{1}{n} + \frac{(x_o - \bar{x})^2}{\Sigma\, (x_i - \bar{x})^2}} \tag{59}$$

where $\hat{y}_o = a + bx_o$. Note that this interval is similar to that in equation (57) except that an extra $s_{y|x}^2$ is added to the variance because the actual observed value of y varies about its true mean value with variance $\sigma_{y|x}^2$.

Squared Multiple Correlation Coefficient

Another informative regression statistic is the *squared multiple correlation coefficient*

$$R^2 = \frac{\Sigma\, (\hat{y}_i - \bar{y})^2}{\Sigma\, (y_i - \bar{y})^2} \tag{60}$$

which measures the proportion of the total variation of the y values about the grand mean $\bar{y}$ that is explained by the linear regression. This definition applies to models other than the simple linear model as well. R^2 takes on values between 0 and 1. It is equal to 1 only when all the (x_i, y_i) observations fall exactly on the regression line. It is usually expressed as a percentage by multiplying by 100.

Example 28. In the glucose example in Table 2A-10, we have

$$R^2 = \frac{88.04}{106.63} = 0.826, \text{ or } 82.6\%$$

Thus, the estimated regression equation $\hat{y} = 115.53 - 0.74(x)$ explains 82.6% of the total variation of the y values about the grand mean $\bar{y}$.

The Correlation Coefficient

We now introduce a measure of the *linear* association between two variables x and y where both x and y are random variables, together having some bivariate probability distribution. This measure of linear association is defined as

$$\rho_{xy} = \frac{E[(x - \mu_x)(y - \mu_y)]}{(\sigma_x)(\sigma_y)} \tag{61}$$

and is called the *population correlation coefficient*. It can be shown that $-1 \leq \rho_{xy} \leq 1$. If $\rho_{xy} = 1$, then x and y are positively correlated and all values of (x, y) lie on a straight line with positive slope. If $\rho_{xy} = -1$ then x and y are negatively correlated and all values of (x, y) lie on a straight line with negative slope. If $\rho_{xy} = 0$, then x and y are uncorrelated in that there is no linear association between them, although x and y may still be nonlinearly associated.

For a random sample of n pairs of values $(x_1, y_1) \ldots (x_n, y_n)$ from the bivariate probability distribution of x and y, the *sample correlation coefficient* r_{xy} that estimates ρ_{xy} is given by

$$r_{xy} = \frac{\Sigma (x_i - \bar{x})(y_i - \bar{y})}{\sqrt{\Sigma (x_i - \bar{x})^2} \sqrt{\Sigma (y_i - \bar{y})^2}} \tag{62}$$

The statistic r_{xy} takes on values between -1 and 1 also, and is typically used to measure the linear association between two variables x and y even when one or both are not random variables. The squared multiple correlation coefficient R^2, in the case of the simple linear regression model only, is equal to the square of the sample correlation coefficient r_{xy}^2.

CALIBRATION

Most analytical procedures require *calibration* to transform the instrument response measurement to predicted concentration, activity, etc. Calibration involves measurement of the instrument response y to special samples called *calibrators,* whose concentrations x are known, and subsequent estimation of the calibration regression model. The calibration procedure is performed periodically to adjust for system drift. Whatever calibration procedure is chosen, it will have an impact on the analytical performance of the system. Some of the important aspects of the calibration procedure are discussed briefly below and more extensively in the literature.[1,4]

1. When first evaluating an analyzer system, an investigation should be conducted to determine the shape of the underlying relationship between the instrument response measurements y, e.g., absorbance, and the true analyte values x, e.g., concentration. A set of calibrators at many concentrations spanning the range of the analyzer system should be run several times, and preferably over several days, to study this relationship. These data will help determine the appropriateness of the manufacturer's recommended calibration model. An incorrect calibration model will result in a constant bias in predicted concentration that will be different at each concentration. Using the fitted calibration model, the concentrations of the calibrators can be predicted from their instrument response measurements, and these can be compared to their known concentrations to describe the bias pattern in concentration units.

2. The number and concentration values of the calibrators to be used in routine calibration should be carefully selected because the random calibration variation, and therefore the total analytical variance of the system, will depend upon these factors. In addition, if a curved relationship between absorbance and concentration is being approximated by a straight line because of economic or system constraints, this model lack-of-fit will cause the measurements to be biased in comparison to the measurements with a better fitting model. The bias at a particular concentration will depend upon the number and concentration values of the calibrators.

3. If the variance of the measured instrument response increases as the analyte concentration increases, then weighted least squares should be considered to fit the calibration model. This will reduce the calibration-to-calibration contribution to the total analytical variation of the system. However, if the calibration procedure has been limited to a single instrument response measurement on each of p calibrators to fit a calibration model with p parameters, then there is only one possible fitted equation in the (x, y) plane. For example, a single response measurement for each of two calibrators exactly determines the fitted line for the simple linear model.

4. There is a choice of conducting regression of y on x, called *classical regression,* or x on y, called *inverse regression.*

For the simple linear model, the classical regression equation, estimated by minimizing $\Sigma(y_i - \hat{y}_i)^2$, is

$$y_i = a + b(x_i) \tag{63}$$

Subsequent prediction of the unknown concentration x of a sample based on an observed instrument response measurement y_o is given by

$$\hat{x}_o = \frac{y_o - a}{b} \tag{64}$$

The inverse regression equation, estimated by minimizing $\Sigma(x_i - \hat{x}_i)^2$, is

$$x_i = a^* + b^*(y_i) \tag{65}$$

Subsequent prediction of the unknown concentration x of a sample with observed instrument response y_o is given by

$$\hat{x}_o = a^* + b^*(y_o) \tag{66}$$

When calibrating a simple linear model with one measurement from each of two calibrators, the two approaches are identical, since there is only one possible fitted line to relate y and x in the (x, y) plane. However, when there are more calibration points than there are model parameters, the two approaches lead to different fitted calibration lines in the (x, y) plane, and thus different concentration predictions. In such cases, the inverse approach in equations (65) and (66) has been shown to give predicted concentration values $\hat{x}$ that are closer to the true unknown concentrations when interpolating between the extremes of the calibrators.[22] Additionally, when the calibration model being fitted is mathematically more complex than the simple linear model in equation (63), the classical approach requires solving the classical regression equation for x in terms of y to obtain a concentration prediction equation similar to equation (64). This may be difficult or impossible, while the inverse approach lends itself directly to prediction of unknown concentrations.

Parameter Estimation in Nonlinear Models

So far we have discussed least squares estimation of the parameters for the simple linear model

$$y = \alpha + \beta x + \epsilon \tag{67}$$

The other models we have mentioned were the quadratic polynomial model

$$y = \alpha + \beta x + \gamma x^2 + \epsilon \tag{68}$$

and the simple linear model in y and ln(x)

$$y = \alpha + \beta \ln(x) + \epsilon \tag{69}$$

All of these models are *linear in the parameters* since they are of the form

$$y = \beta_0 + \beta_1 z_1 + \beta_2 z_2 + \ldots + \beta_p z_p + \epsilon \tag{70}$$

where the z_i are predictor variables (or functions of predictor variables) and the β_i are the parameters to be estimated by least squares. Although models that are linear in the parameters can be used to represent many types of relationships, there are also many instances where they are not the best models to use. For example, when definite chemical and physical information is available about the functional form of the relationship between the response and predictor variables, it is often preferable to fit this "mechanistic" model, since (1) its parameters usually have chemical or physical interpretations and (2) it generally has fewer parameters to be estimated. However, the mechanistic model often cannot be written in the form of

Table 2A-11. THE FOUR-PARAMETER LOGISTIC MODEL EXPRESSED IN THREE DIFFERENT FORMS

Algebraic Form	Variables*	Parameters[†]
$y = \dfrac{a - d}{1 + (x/c)^b} + d$	(x, y)	a, b, c, d
$R = R_o + \dfrac{K_c}{1 + \exp\left[-(a + b \log(C))\right]}$	(C, R)	R_o, K_c, a, b
$y = y_o + \dfrac{(y_\infty - y_o)(x^d)}{b + x^d}$	(x, y)	y_o, y_∞, b, d

*Concentration and instrument response variables shown in parentheses.
[†]Equivalent letters do not necessarily denote equivalent parameters.

equation (70). Any such model that cannot be written in the form of equation (70) is *nonlinear in the parameters* and will be called a *nonlinear model*.

Example 29. An apparent first order reaction is monitored spectrophotometrically and absorbance y is measured as a function of time x. The model to be fitted to the data is

$$y = (\theta_1 - \theta_2) \exp(-\theta_3 x) + \theta_2$$

where the parameters θ_1, θ_2, and θ_3 have the following interpretations:

θ_1 is the absorbance at time 0

θ_2 is the absorbance at time infinity

θ_3 is the first order rate constant.

This is a nonlinear model because it cannot be written in the form of equation (70). It is linear in the parameters θ_1 and θ_2, but it is nonlinear in the parameter θ_3.

Parameter Estimation

The approach to parameter estimation for nonlinear models is similar to that for linear models in that the least squares method is also used. If we have a nonlinear model with p parameters to be estimated from n observed data points, we minimize the sum of the residuals $\Sigma(y_i - \hat{y}_i)^2$ as before. However, in the case of nonlinear parameters, we will not be able to find the solution for the least squares estimates of the p parameters in closed form. Iterative numerical techniques are necessary which take several steps before an adequate approximation to the correct parameter estimates is obtained. An introduction to the theory of nonlinear estimation can be found in Draper and Smith.[6] The availability of microprocessor-based laboratory instrumentation has made it possible to implement iterative solutions for estimation in nonlinear models.

The Logistic Model

A useful application of nonlinear models in clinical chemistry is the fitting of immunoassay calibration curves. The four-parameter logistic model has been used for both radioimmunoassay and enzyme immunoassay techniques, and can be written in several forms shown in Table 2A-11.[16] All the forms are identical from a model viewpoint, and relationships between the parameters in the different equations can be easily derived. Because there are four parameters to be estimated in the logistic model, there must be a minimum of four distinct analyte values in the calibrator set.

As shown in Table 2A-12, the logistic model can be transformed with the logit function to obtain a

Table 2A-12. LOGIT TRANSFORMATION OF A FOUR-PARAMETER LOGISTIC MODEL USING THE FIRST FORM FROM TABLE 2A-11

$$y = \frac{a - d}{1 + (x/c)^b} + d$$

$$y' = \frac{y - d}{a - d} = \frac{1}{1 + (x/c)^b}$$

$$Y = \text{logit}(y') \equiv \log \frac{y'}{(1 - y')} = b \log(x) - \log(c^b) = \alpha + \beta \log(x)$$

$$V(Y) \simeq \frac{\sigma_{y'}^2}{[(y')(1 - y')]^2}$$

simple linear model. In order to use this approach two parameters must be assumed known from prior study: the analytical responses "*a*" and "*d*" at zero and infinite analyte concentration, respectively. If these assigned values are grossly incorrect, a linear transformation will not be obtained. The transformation also causes greater nonhomogeneity of variance in the transformed response variable than that in the original response variable; thus use of weighted least squares is recommended.[13] The appropriate weighting factor is the reciprocal of the standard deviation of the response variable, $w_i = 1/\sqrt{V(y_i)}$. The Taylor expansion (see equation 36) was used to obtain the approximate variance $V(y_i)$ given in Table 2A-12.

METHOD COMPARISON

In method comparison studies using patient samples, one of the objectives is to compare the "test" method measurements y to those of the "comparative" method x. When the comparative method can be assumed to measure "truth" with negligible measurement error and negligible nonspecificity, we can completely characterize the distributions of measurement errors for the test method through the concept of total analytical error, the topic of the next section. When the comparative method cannot be assumed to yield truth, all we can practically do is compare the means of the two methods. Although the paired difference approach discussed in the next section is recommended for both situations, we give a short discussion here of the commonly used simple linear regression approach for the situation where the comparative method has measurement error. Discussion of both the regression and paired difference approaches in the former situation will be deferred until the next section.

Suppose that we have a method comparison where both the test and comparative methods are field methods, and that we have conducted a simple linear regression analysis of the data. The estimated simple linear regression equation $\hat{y} = a + bx$ is often used to estimate values of the test method that correspond to certain values, for example, the 95% reference values of the comparative method. Before using the estimated regression equation for this or any other decisions about the test method, the adequacy of the assumed simple linear model and its assumptions should be checked. Only when these assumptions have been shown to be satisfied should the simple linear regression approach be used instead of the straightforward paired difference approach discussed in the next section.

1. The assumption of a linear relationship between y and x can be checked only roughly by inspection of a scatterplot of y against x. This assumption can be checked with better resolution by inspection of a scatterplot of the residuals $y_i - \hat{y}_i$ against $\hat{y}_i$. Evidence of curvilinearity indicates that a model that allows for curvature would provide a better fit to the data. Alternatively, the paired difference approach discussed in the next section can be used.

2. The assumption of known fixed values of x, without measurement error, is violated by definition of the comparative method as another field method. As a result, it can be shown that the estimated slope b of the simple linear model has an expected value that is not β, but is biased toward zero. The magnitude of this bias depends on the magnitude of the variance of the errors in measuring the x values relative to the variation of the concentrations of the x values themselves. In a method comparison for an analyte where the patient specimens span a broad concentration range, this problem is usually not of great practical significance. However, for analytes where this is not possible, e.g., sodium, calcium, or carbonate, the estimated slope b is usually biased by a practically significant amount. Various approaches for unbiased estimation of the slope have been discussed in the literature.[5] Alternatively, the paired difference approach discussed in the next section can be used.

3. The variances of the populations of y measurements are assumed to be equal, which is often not the case when the specimens cover a broad concentration range. As discussed earlier, ordinary least squares yields an estimated slope with more sensitivity to the highly variable measurements than does weighted least squares. Alternatively, the paired difference approach discussed in the next section can be used.

List of Symbols for Regression Analysis (in order of appearance)

x	dependent or predictor variable; not a random variable
y	independent or response variable; a random variable

(x_i, y_i)	randomly sampled value from the population of y values at a particular value of x
$\sigma^2_{y|x}$	common variance of the populations of y values
$\mu_{y|x}$	mean of the population of y values at a particular value of x; $\mu_{y|x} = \alpha + \beta x$
α	y-intercept of the population regression line in the simple linear model
β	slope of the population regression line in the simple linear model
ϵ	measurement error variable associated with measurement variable y: $\epsilon = y - (\alpha + \beta x)$
n	number of values (x_i, y_i) sampled from the populations of y values
a	least squares estimate of α
b	least squares estimate of β
$\bar{x}$	sample mean for the variable x
$\bar{y}$	sample mean for the variable y
$\hat{y}$	estimated means of the populations of y values: $\hat{y} = a + bx$
$\hat{y}_i$	estimated mean of the population of y values at a particular value of x: $\hat{y}_i = a + bx_i$
ϵ_i	true unknown measurement error associated with a particular measurement y_i
$\hat{\epsilon}_i$	estimated measurement error associated with a particular measurement y_i: $\hat{\epsilon}_i = y_i - (a + bx_i) = y_i - \hat{y}_i$
p	number of parameters in the regression model
$s^2_{y|x}$	estimated common variance of the populations of y values
$1 - \alpha$	degree of confidence for a confidence interval
$t_{\alpha/2}$	value of the Student's t variable such that a proportion $\alpha/2$ of the values exceed it
s_b	estimated standard deviation of b
s_a	estimated standard deviation of a
x_o	a particular value of the variable x
$\hat{y}_o$	estimated mean of the population of y values at a particular value x_o of x; $\hat{y}_o = a + bx_o$
$s\hat{y}_o$	estimated standard deviation of $\hat{y}_o$
R^2	squared multiple correlation coefficient
ρ_{xy}	population linear correlation coefficient between random variables x and y
μ_x, μ_y	population means for random variables x and y
σ_x, σ_y	population standard deviations for random variables x and y
r_{xy}	sample correlation coefficient between random variables x and y
$\hat{x}_o$	predicted value of the variable x based on an observed measurement y_o
a*	estimated intercept of the inverse simple linear regression model
b*	estimated slope of the inverse simple linear regression model
β_i	ith population parameter for the general linear regression model: $\beta_o, \beta_1, \ldots, \beta_p$
z_i	ith independent or predictor variable in the general linear regression model: $z_1, z_2, \ldots, z_p$
θ_i	ith population parameter in the general nonlinear (in the parameters) regression model: $\theta_1, \theta_2, \ldots, \theta_p$
w_i	weight for the ith observation (x_i, y_i) for weighted regression analysis
$V(y_i)$	variance of the population of y values at a particular value x_i of x

THE TOTAL ANALYTICAL ERROR OF AN ANALYZER SYSTEM

Various approaches have been recommended for defining and estimating the analytical performance of analyzer systems in clinical chemistry. An appropriate approach should describe the magnitude and likelihood of errors in measurements on patient specimens. In this section, we discuss such an approach through the concept of total analytical error. After defining total analytical error, we discuss the design and analysis of an experiment to estimate the total analytical error of an analyzer system, comparing a simple and straightforward estimation procedure with the popular regression procedure. Also included is a short discussion about judging the acceptability of total analytical performance. Total analytical error, as well as other aspects of analytical performance, is also discussed in Chapter 2D, Evaluation of Methods.

DEFINITION OF TOTAL ANALYTICAL ERROR

In the evaluation of a particular analyzer system, consider the probability distribution of measurements x on patient specimens all having the same true concentration C_{true}, e.g., 100 mg/dL. Suppose this distribution of values is Gaussian with mean $\mu_x = 103$ mg/dL and standard deviation $\sigma_x = 2$ mg/dL as shown in Figure 2A-19. If we subtract the true value 100 mg/dL from each of these values, we obtain the probability distribution of the analytical measurement errors $\epsilon = x - 100$, which is Gaussian with mean $\mu_\epsilon = 3$ mg/dL and standard deviation $\sigma_\epsilon = 2$ mg/dL. This probability distribution contains all the information about the magnitude and likelihood of the analytical errors of measurements on patient specimens with a true concentration of 100 mg/dL.

1. The *systematic analytical error* of measurements on patient specimens with a true concentration of 100 mg/dL is $\mu_\epsilon = 3$ mg/dL. Thus, on average, the measurement on a patient specimen with a true concentration of 100 mg/dL is 3 mg/dL high.

2. The *random analytical error* of measurements on patient specimens with a true concentration of 100 mg/dL is $\sigma_\epsilon = 2$ mg/dL. It should be noted that two distinct sources contribute to this random analytical error. The first is the random error associated with repeated measurements over time on the same specimen, which was discussed in a previous section, and will hereafter be called *imprecision*. The second is the *random interference* of measurement errors on patient specimens with the same true concentration C_{true}. These random interference errors are specimen-specific errors that vary from one patient specimen to another depending upon the method's sensitivity to the various physical and chemical characteristics of the specimen other than the analyte of interest. These errors can be either positive or negative and are unknown, just like errors due to imprecision. However, the standard deviation σ_{RI} of these errors cannot be estimated by repeated measurements on a serum pool or on the same specimen. Rather, nested designs can be used with patients' specimens followed by nested ANOVA to separate and estimate this variance component as well as the temporal variance components associated with system imprecision. This will be illustrated shortly.

3. The central 95% of the distribution of measurement errors is $\mu_\epsilon \pm 2(\sigma_\epsilon)$, or $3 \pm 2(2)$, or -1 mg/dL to $+7$ mg/dL. Thus, for specimens at $C_{true} = 100$ mg/dL the maximum absolute error for 95% of the patient specimen measurements with this system is 7 mg/dL. The *total analytical error* of a system at a particular concentration is defined as this maximum absolute error for 95% of measurements on patient specimens. Thus, the total analytical error at $C_{true} = 100$ mg/dL is 7 mg/dL for this system.

In general, the distribution of measurement errors at a concentration C_{true} other than 100 mg/dL will be different from that at $C_{true} = 100$ mg/dL. The distributions of measurement errors over the whole concentration range can be represented in general as illustrated in Figure 2A-20, where distributions of measurement errors at three particular concentrations are shown. The systematic, random, and total analytical errors of an analyzer system are generally concentration dependent.

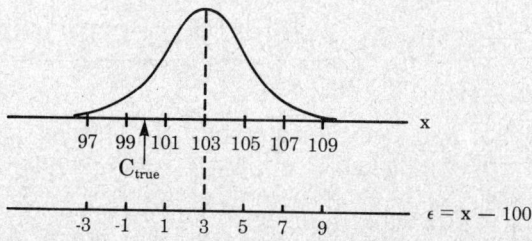

Figure 2A-19. Probability distributions of measurements (upper scale) and measurement errors (lower scale) on patient specimens with $C_{true} = 100$ mg/dL.

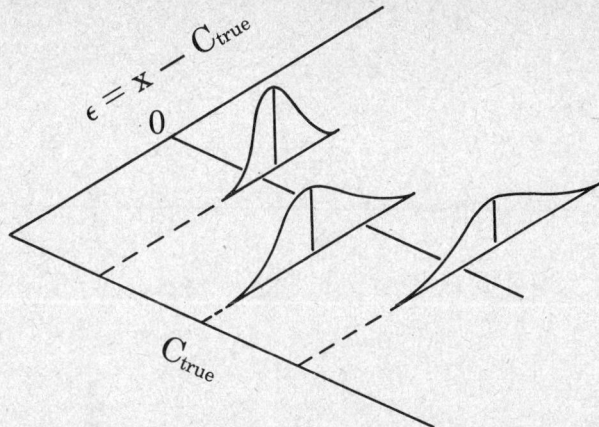

Figure 2A-20. Probability distributions of measurement errors at three particular values of C_{true}.

We have introduced the concepts of systematic, random, and total analytical error of measurements on patient specimens in this discussion, while purposely avoiding use of the terms accuracy or inaccuracy. These terms are used inconsistently throughout the clinical literature, so one must always read carefully to determine which concept these terms refer to. For example, the term accuracy or inaccuracy is commonly used in referring to the systematic error, which we call bias, as well as to the total error.

ESTIMATION OF TOTAL ANALYTICAL ERROR

Experimental Design

Estimation of the parameters of the distributions of analytical measurement errors requires random samples from these distributions. Such samples can be obtained by measuring split samples from patient specimens by (1) the analyzer system being tested and (2) a comparative analyzer system with little or no systematic or random analytical error (e.g., a reference method), and then calculating the paired differences $x_d = x_{test} - x_{comp}$. Under the assumption that x_{comp} is "truth," the computed values of $x_d = x_{test} - x_{comp}$ are the observed measurement errors for the test method. A large number of patient specimens spanning a wide range of concentration and analyzed over several days will provide adequate random samples for estimating the parameters of the distributions of measurement errors. The NCCLS has recommended a minimum of 40 specimens analyzed over a minimum of five days, with suggested distributions of specimens over the clinically meaningful ranges for many analytes.[12] A nested design (e.g., two measurements per specimen, specimens nested within-run) will permit estimation of the variance components associated with imprecision and random interference.

The example we will use for illustration is an evaluation of an automated Jaffe creatinine system in comparison to an HPLC creatinine system. Split samples from 206 patient specimens were analyzed in duplicate over 14 days with a single run-per-day per system.

Graphical Analysis

Proper data analysis begins with graphical display and visual inspection of the data. A simple but high resolution graphical display of the observed measurement errors is a scatterplot of the individual measurement errors $x_d = x_{test} - x_{comp}$ against x_{comp}. Such a plot is shown in Figure 2A-21 for the creatinine example. Simple inspection of this plot of individual measurement errors reveals the general magnitude of the systematic, random, and total analytical error over the whole range of concentration. For example, the systematic error is approximately +0.1 mg/dL at concentrations of 2 mg/dL or less, and then becomes zero and slightly negative as the concentration increases. The random analytical error is smallest between 0 and 1 mg/dL and then increases slightly between 1 and 2 mg/dL, and dramatically beyond 2 mg/dL. Analytical performance is of most importance near the medical decision concentration, which is approximately 1.5 mg/dL for creatinine. At this concentration, the majority of the measurement errors appear

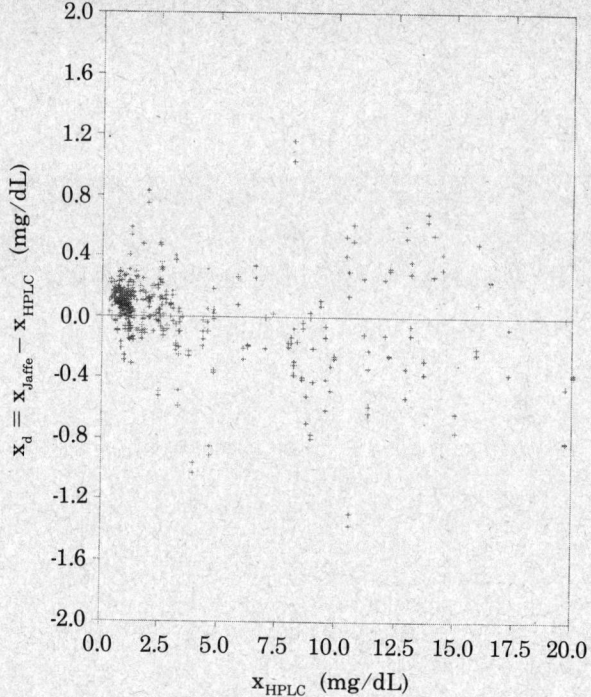

Figure 2A-21. Scatterplot of the individual measurement errors $x_d = (x_{Jaffe} - x_{HPLC})$ against x_{HPLC}.

to be between -0.2 mg/dL and $+0.4$ mg/dL so that the total analytical error appears to be about 0.4 mg/dL. Several high measurement errors in this region indicate a very undesirable sensitivity to interfering substances in these specimens.

It is interesting to compare this simple informative graphical display with the popular regression-type scatterplot shown in Figure 2A-22. Because the concentration range is so large in comparison to the measurement errors, the resolution in this type of plot is usually inadequate to reveal any useful information about the magnitude of the analytical measurement errors.

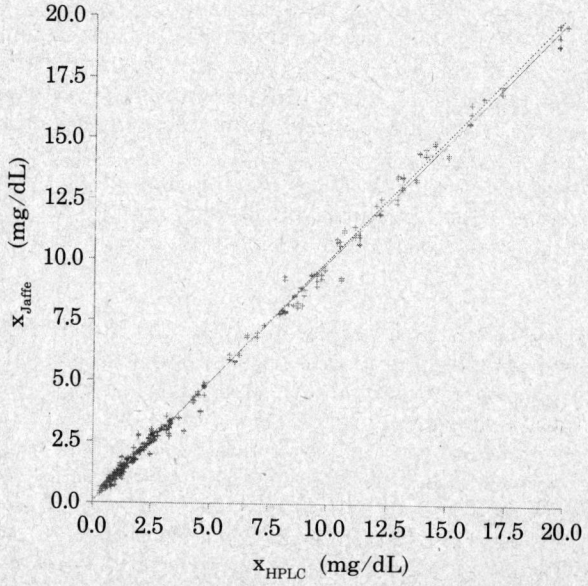

Figure 2A-22. Scatterplot of the individual measurements x_{Jaffe} against x_{HPLC}, with least squares line (——) and line identity (···).

Statistical Estimation: Paired Differences Approach

Several approaches have been used to estimate the parameters of the distributions of measurement errors. One easy and direct approach begins with partitioning the concentrations x_{comp} into several narrow concentration ranges in such a way that within each range the changes in both the systematic and random error are negligible in comparison to their own magnitudes. For example, for the individual measurement errors in Figure 2A-21, the following concentration ranges appear to be reasonable: 0–1 mg/dL, 1–2 mg/dL, 2–4 mg/dL, and 4–21 mg/dL. Note that although the systematic error does not appear to change from the first group of measurement errors to the second, the random error appears to be much smaller in the first group than in the second. The numbers of specimens and measurements within each range are shown in columns (2) and (3) of Table 2A-13.

This method of inspecting the scatterplot of the paired differences to define the ranges can cause bias in estimation of the systematic error discussed below. A preferable method may be to define ranges based on natural separation, e.g., below, within, and above the health-associated concentration range, provided the systematic and random error are relatively constant in each range.

After appropriate concentration ranges have been defined, the systematic error is estimated by calculating the sample mean $\bar{x}_d$ of the observed measurement errors or differences within each group. These estimates are given in column (4) of Table 2A-13. For example, in the 1–2 mg/dL range, the estimated systematic error is $\bar{x}_d = +0.09$ mg/dL.

Because the observed measurement errors $x_d = x_{test} - x_{comp}$ in a method comparison are not independent, and because it will be informative to look at the contributions of the variance components to the total variation of the measurement errors, the standard deviation σ_{TOT} of the measurement errors should be estimated by considering the experimental design, which in this case is a two-stage nested design. Therefore nested ANOVA can be conducted on the observed differences x_d, according to Table 2A-9, to separate and estimate the variance components for the test method:

$$\sigma_{TOT}^2 = \sigma_D^2 + \sigma_{RI}^2 + \sigma_{WR}^2 \tag{71}$$

where σ_D^2 is the day-to-day variance component
$\quad\sigma_{RI}^2$ is the random interference variance component
$\quad\sigma_{WR}^2$ is the within-run variance component
$\quad\sigma_{TOT}^2$ is the total variance.

The nested design in these situations will in general be unbalanced, requiring a statistical computer program package, e.g., SAS, to obtain the estimates of the variance components in equation (71).[20,21] Alternatively, the sample standard deviation of all the observed differences will not usually be a serious underestimate of σ_{TOT} when the numbers of patients and days are large as in this example. These estimates are given in column (5) of Table 2A-13 and agree quite closely with the nested ANOVA estimates of σ_{TOT} in Table 2A-14. However, nested ANOVA provides valuable information about the relative magnitudes of the random error contributors. For example, from Table 2A-14 it is clear that the major contributor to the random analytical error of this Jaffe creatinine system is the random interference, accounting for approximately 75% ($\hat{\sigma}_{RI}^2 / \hat{\sigma}_{TOT}^2 \simeq 0.75$) of the total random variation. This separation of the random error components helps identify those sources in need of improvement when total random error is judged to be unacceptable.

Table 2A-13. ESTIMATES OF SYSTEMATIC, RANDOM, AND TOTAL ANALYTICAL ERROR FOR CREATININE EXAMPLE

1 Concentration Range (mg/dL)	2 Number of Specimens	3 Number of Points	4 Systematic Error (mg/dL)	5 Random Error (mg/dL)	6 Central 95% Measurement Errors (mg/dL)	7 Total Error (mg/dL)
0–1	54	108	+0.10	0.10	−0.10 to +0.30	0.30
1–2	56	112	+0.09	0.18	−0.27 to +0.45	0.45
2–4	39	78	−0.01	0.26	−0.53 to +0.51	0.53
4–20	57	114	−0.11	0.43	−0.97 to +0.75	0.97

Table 2A-14. ESTIMATED COMPONENTS OF RANDOM ANALYTICAL
ERROR (mg/dL) FOR THE CREATININE EXAMPLE

Concentration	$\hat{\sigma}_{WR}$	$\hat{\sigma}_{RI}$	$\hat{\sigma}_D$	$\hat{\sigma}_{TOT}$
0–1	0.04	0.09	0	0.10
1–2	0.06	0.15	0.07	0.18
2–4	0.08	0.23	0.12	0.27
4–20	0.17	0.37	0.16	0.44

The central 95% of the distribution of measurement errors is estimated as $\bar{x}_d \pm 2(\hat{\sigma}_{TOT})$ and is given in column (6) of Table 2A-13. The total analytical error is estimated as the maximum absolute error for 95% of this distribution of measurement errors and is given in column (7).

If the comparative method is not a reference method, interpretation of the estimates of systematic, random, and total error must acknowledge the fact that these estimates reflect the error in the comparative method as well as error in the test method. For example, the systematic error must then be interpreted as the difference between the systematic errors of the two systems, and so the systematic error describes how well the systems agree, on average, relative to each other. The imprecision components will be the sum of the imprecisions of the two methods, as will the random interference if the two systems are very different in the substances and conditions causing interference. When the two systems are very similar in the substances and conditions causing interference, then the estimated random interference component will be very small, reflecting little interference of the test method relative to the comparative method. The estimated total analytical error describes the maximum difference between the two methods for 95% of patient specimen measurements.

Alternative Estimation Procedures

Simple linear regression analysis is a popular statistical procedure for estimating analytical performance from method comparison data. However, the assumptions of the simple linear regression model are often not satisfied in the method comparison situation. In review, these assumptions are:

1. The values of x are known without error.
2. There is a Gaussian population of y values at each particular value of x.
3. The means of these populations of y values are given by a linear function of x, i.e., $\mu_{y|x} = \alpha + \beta x$.
4. There is a common variance $\sigma^2_{y|x}$ for these populations of y values.

For illustrating some of the potential problems with simple linear regression analysis, we have applied this procedure to the creatinine method comparison data.

Example 30. The simple linear regression model for the creatinine method comparison is $y = \alpha + \beta x + \epsilon$ where y represents the Jaffe measurements and x represents the HPLC measurements. The estimated parameters are a = 0.086, b = 0.983, and $s_{y|x}$ = 0.27 mg/dL. Thus $\hat{y} = 0.086 + 0.983(x)$. At the medical decision concentration, i.e., at x_c = 1.50 mg/dL, the estimated mean of the population of y values is $\hat{y}$ = 0.086 + 0.983(1.50) = 1.56 mg/dL, so that an estimate of the systematic error is $\hat{y} - x_c$ = +0.06 mg/dL. The random error is estimated by $s_{y|x}$ = 0.27 mg/dL. The central 95% of the distribution of measurement errors at x_c = 1.50 mg/dL is estimated by

$$(\hat{y} - x_c) \pm 2(s_{y|x}) \text{ or } 0.06 \pm 2(0.27) \text{ or } -0.48 \text{ mg/dL to } +0.60 \text{ mg/dL}$$

Thus, the total analytical error is estimated to be 0.60 mg/dL.

The estimates of analytical performance based on the regression approach are not in agreement with those based on the paired differences approach. This often happens in the method comparison situation because the assumptions of simple linear regression are frequently not satisfied.

1. Because the x values are measurements by the comparative method they cannot be considered fixed known quantities as assumed in regression analysis. As mentioned in the regression section, the consequence is that the statistic b for estimating the true slope β is biased toward zero. Thus, the regression estimate of systematic error will be incorrect in the general method comparison situation. The amount of bias is a function of the magnitude of the total

random analytical error $\sigma^2_{TOT(comp)}$ (imprecision plus random interference) in the comparative method relative to the variance of the concentrations of the specimens used in the study. Therefore, the random analytical error in the comparative method can be compensated to some extent by using patient specimens that span a wide range of concentration. However, for some tests (e.g., sodium and calcium) this is not possible, and the computed slope can be as much as 25% too low when the comparative method is another field method.

2. Both regression analysis and the paired difference analysis assume that the distribution of y measurements at any fixed x is Gaussian. Although this is likely to be valid for the variation in y measurements caused by imprecision contributors, it may not be true for the variation caused by random interference, especially when there are one or two major interfering substances or conditions that do not vary biologically according to a Gaussian distribution. Inspection of the paired difference plot should reveal any gross violation of this assumption.

3. The linear regression estimate of systematic error is often affected by curvilinearity in the relationship between the two systems. Although a quadratic regression model can be fit to determine if departure from linearity exists, this is generally a poor check because the curvilinearity may not be quadratic in nature. For example, if curvilinearity exists mainly at the lower concentrations, the quadratic model may not be a significant improvement over the linear model since the fit of both models is dominated by the data at the higher concentrations where the variation is usually larger.

4. The variance of the distributions of errors about the regression line is usually not constant, but rather increases as the concentration increases. The first and obvious problem here is that the statistic $s_{y|x}$ estimates some type of "average" random error over the whole concentration range, which may or may not be appropriate at the medical decision concentration(s). For example, the estimate $s_{y|x} = 0.27$ mg/dL is 50% higher than the estimate 0.18 mg/dL from the direct paired differences analysis at x = 1.50 mg/dL in the Jaffe creatinine evaluation. One solution to this problem is to calculate the standard deviation of the regression residuals in a narrow range centered at the medical decision concentration. The second problem is that the nonhomogeneous variance causes the observations at high concentrations to have excessive influence on the least squares estimates a and b. The estimates are still unbiased estimates of α and β, but weighted regression can be conducted to obtain estimates that have less sensitivity to the highly variable measurements.

Another approach to estimating total analytical error incorporates estimates of systematic error from regression analysis and random error from repeated measurements over time on a control fluid. In addition to the problems related to the systematic error estimate discussed above, the estimate of the random analytical error does not include the random interference component, which can be a major contributor as demonstrated in the Jaffe creatinine example.

Judging the Acceptability of Analytical Performance

Various individuals and organizations have recommended limits for the magnitude of analytical error that can be tolerated from a medical usefulness perspective. However, there is little or no consistency regarding the type of analytical error for which these limits apply. Furthermore, for a particular type of error the recommended limits may vary considerably.

Of the various recommendations for analytical performance, that of Tonks[26] can be interpreted as an allowable limit for total analytical error. Also, since Gilbert[8] provides separate limits for random error and systematic error, a limit for total error (although it excludes allowance for random interference) can be obtained as the systematic error limit plus two times the random error limit. (See Chapter 2D, Figure 2D-2.) These recommendations for total analytical performance should be used to judge the performance of the test method only when the comparative method is a reference method.

Example 31. At a true creatinine concentration of 1.5 mg/dL, Tonks' limit for total analytical error is 10% or 0.15 mg/dL while Gilbert's limit is 0.10 + 2(0.10) or 0.30 mg/dL. The estimated total analytical error for the automated Jaffe creatinine system at 1.5 mg/dL is 0.45 mg/dL. Based on this large data set, the value of 0.45 has little uncertainty, although demonstrating it is beyond the scope of this text. Since this estimate is statistically and clinically significantly larger than the allowable limit of both Tonks and Gilbert, the analytical performance of this system is judged to be unacceptable. Comparing the estimates of the systematic error (0.09 mg/dL), random interference (0.15 mg/dL), and imprecision (0.07 mg/dL day-to-day and 0.06 mg/dL within-run), we conclude that random interference is the major source of the unacceptable analytical performance, and that the systematic error is also a large contributor.

List of Symbols for Total Analytical Error (in order of appearance)

x	measurements of concentrations on patient specimens
C_{true}	true concentration
μ_x	population mean for x
σ_x	population standard deviation for x
ϵ	measurement errors for values x compared to the true values
μ_ϵ	population mean for ϵ
σ_ϵ	population standard deviation for ϵ
σ_{RI}	population standard deviation of measurement errors due to interfering characteristics of samples with the same C_{true}
x_{test}	test method measurements of concentrations on patient specimens
x_{comp}	comparative method measurements of concentrations on patient specimens
x_d	paired differences between x_{test} and x_{comp}
$\bar{x}_d$	sample mean of paired differences
σ_D^2	day-to-day variance component of measurement errors
σ_{RI}^2	random interference variance component of measurement errors
σ_{WR}^2	within-run variance component of measurement errors
σ_{TOT}^2	total variance (imprecision plus random interference) of measurement errors
x	comparative method measurements (regression approach)
y	test method measurements (regression approach)
α	y-intercept of population regression line
β	slope of population regression line
$\mu_{y|x}$	mean of the population of y values at a particular value of x
$\sigma_{y|x}^2$	common variance of the populations of y values
a	least squares estimate of α
b	least squares estimate of β
$\hat{y}$	estimated mean of the population of y values at a particular value of x
$s_{y|x}^2$	estimated common variance of the populations of y values
x_c	medical decision concentration

STATISTICAL QUALITY CONTROL

The procedure used to monitor analytical performance in most clinical laboratories consists of regularly making determinations on a stable, homogeneous control material and tracking these measurements over time on some type of control chart. In this section we describe the statistical concepts underlying the construction of several types of control charts and the application of several statistical control rules. Among other aspects of quality assurance, Chapter 2E describes how to evaluate and compare the performance of statistical control procedures.

CONTROL PROCEDURES FOR INDIVIDUAL MEASUREMENTS

The Control Chart

To construct a control chart for individual measurements on a particular control fluid, we must estimate the mean μ and standard deviation σ_{TOT} of the probability distribution of analytical measurements on this fluid. This is typically accomplished by making many determinations on this fluid over a period of time sufficient to obtain adequate representation of the major variance components of the analyzer system. This period of time is usually called the *baseline period*. From these measurements we obtain the desired estimates $\bar{x}$ of μ and $\hat{\sigma}_{TOT}$ of σ_{TOT}, and construct a *control chart* based on the assumption of a Gaussian distribution of values, as illustrated in Figure

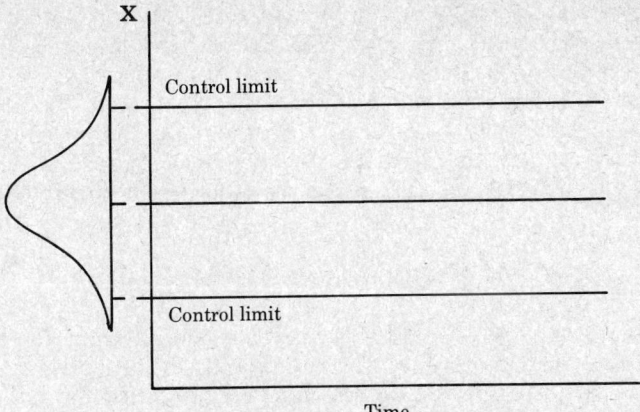

Figure 2A-23. Quality control chart.

2A-23. The analytical performance of the analyzer system is then monitored by plotting subsequent values on this chart, and subjecting these measurements to one or more statistical control rules.

Statistical Control Rules

Ideally we would like to have a control rule(s) that would always provide a "signal" when analytical performance has significantly changed or deteriorated and that would never provide a false signal. Unfortunately this is not possible. For any particular control rule, whenever we increase the probability of a signal for a true analytical performance change, we also increase the probability of a false signal. For example, the $\bar{x} \pm 2\hat{\sigma}_{TOT}$ control limits will result in a greater probability of true signals compared to the $\bar{x} \pm 3\hat{\sigma}_{TOT}$ limits for a change in analytical performance of a given magnitude, but the probability of a false signal with the $\bar{x} \pm 2\hat{\sigma}_{TOT}$ limits ($\simeq 4.6\%$) will also be higher than that with the $\bar{x} \pm 3\hat{\sigma}_{TOT}$ limits ($\simeq 0.3\%$). Chapter 2E discusses the evaluation and comparison of statistical control rules based on these probabilities of error detection (i.e., power) and false rejection. An acceptable control rule is one that has an adequately large probability of error detection for clinically important changes in analytical performance and a tolerably small probability of false rejection, as well as being reasonable from practical and economic viewpoints.

Alternate QC Procedures

Consider individual values plotted on the control chart as discussed above. Such values typically do not behave like independent observations from a single Gaussian population. The reason for this lies in the fact that individual measurements over time can be explained by nested models like those in equations (37) and (40) where there are several variance components, e.g.,

$$x_{ijk} = \mu + \epsilon_i + \epsilon_{ij} + \epsilon_{ijk} \tag{72}$$

Suppose that several measurements are made on a control fluid in each run, with several runs per day. The measurements in any given run will tend to cluster because of the random shift introduced at the beginning of the run, which will persist throughout that run. For example, the act of calibrating each run will contribute to this random run-to-run error. Similarly, measurements in any given day will have a small random shift because of the unique random error for that day, due to such sources as vial-to-vial variation in the manufacture and reconstitution of the control fluid. Most statistical control procedures, like those discussed above, do not consider this realistic model of the nature of the analytical system. Rather, they are based upon the inappropriate simple model

$$x_i = \mu + \epsilon_i \tag{73}$$

where each individual measurement x_i is assumed to be the result of the mean level μ and a unique random error ϵ_i that is independent of all the other random errors $\epsilon_j, j \neq i$.

The statistical control procedures usually applied in the clinical laboratory are based upon an inappropriate simplified model of the analyzer system characteristics, and therefore are not optimal. Use of more realistic models like that in equation (72) could improve the performance of statistical control procedures. One such control procedure that has been used by engineers since the early 1960's for modeling process control systems is the *Kalman Filter*.[10,14] System drift in the mean level μ, due to such factors as control material deterioration over time, can also be incorporated into these models. With the increased utilization of microprocessor-based laboratory instrumentation, more frequent application of such improved control techniques will be possible.

CONTROL PROCEDURES FOR MEANS

Consider monitoring system performance by use of a control chart for the mean of n measurements within a run. The general control chart procedures for individual measurements apply to this situation as well. Consider the construction and interpretation of the control chart for this case. Recall that the control chart reflects the probability distribution of the data to be subsequently plotted on it. The data in this case are run means, where each run mean is the average of the same number n of measurements on the same control fluid. Thus, the mean $\mu_{\bar{x}}$ and standard deviation $\sigma_{\bar{x}}$ for the population of run means must be estimated. A straightforward approach is to take n measurements of the control fluid in each run with an experimental design that will provide adequate representation of the run-to-run variation. For example, if the control chart is intended to be used for monitoring run means over many calibrations, then n measurements should be made in each of I (at least 10, preferably 20) individually calibrated runs in the baseline period. Then the estimates of $\mu_{\bar{x}}$ and $\sigma_{\bar{x}}$ from the individual run means $\bar{x}_i$, i = 1, . . . , I, are given by

$$\hat{\mu}_{\bar{x}} = \frac{\sum_{i=1}^{I} \bar{x}_{i.}}{I} = \bar{x}.. \tag{74}$$

$$\hat{\sigma}_{\bar{x}} = \sqrt{\frac{\sum_{i=1}^{I} (\bar{x}_{i.} - \bar{x}..)^2}{I - 1}} \tag{75}$$

From Table 2A-7 for the one-stage nested design, it can be seen that the standard deviation $\sigma_{\bar{x}}$ being estimated in equation (75) is

$$\sigma_{\bar{x}} = \sqrt{\sigma_R^2 + \sigma_{WR}^2/n} \tag{76}$$

Thus, the standard deviation of the distribution of run means reflects the fact that the uncertainty in each run mean is the result of a single random run-to-run error and the average of n random within-run errors. However, as mentioned in the discussion of alternate QC procedures, if there are multiple run means within each day or within each calibration, then the run means will not behave independently.

OUTLIERS

An *outlier* in a sample is an observation that belongs to a population other than the one to which the majority of the observations belong. Outliers in a sample can distort the computed values of statistics and can cause incorrect inferences to be made about the population parameters of interest. As mentioned throughout this chapter, the first step in any data analysis should consist of plotting and inspecting the observations in the sample(s). Any suspicious observations are potential outliers and should be investigated before statistical procedures are performed. Those potential outliers for which no assignable cause can be found should be subjected to a statistical outlier test.

Most statistical tests for outliers assume that the population under study is of Gaussian form. When the observations vary due to analytical measurement variation only, and not due to biological variation as well, this assumption is usually satisfied. When the observations vary due to biological variation, such as in reference value studies, this assumption is usually not satisfied, and transformation of the sample data is usually necessary before outlier tests can be validly performed. (See also Chapter 2B.)

Most tests for outliers also require specification of the particular observation(s) that are to be tested. One procedure requires specification of only the maximum number of potential outliers in the sample.[17,18]

A recent book reviews tests for outliers and discusses their advantages and disadvantages.[2] Below we describe one significance test for outliers, and then discuss how it can be applied in the various types of experiments discussed in this text.

Whatever method is used to identify observations as outliers, a complete data analysis should always include documentation of the observations omitted from the statistical estimates and significance tests. It is usually wise to compare and report the statistical and practical conclusions obtained with the unexplainable outliers (a) included and (b) excluded.

SIGNIFICANCE TEST FOR OUTLIERS

Suppose we have a sample of n observations, intended to be from a single population, and that we want to test this sample for outliers. A test of significance can be conducted as follows.[23]

1. Identify the observation that is most extreme, i.e., furthest from the sample mean. For illustration, consider the sample of 10 observations from the Gaussian population of differences in Figure 2A-13 and Table 2A-4 with the sample mean 1.7. The observation furthest from the mean is $x = -1$. We will test the hypothesis that this observation is an outlier.

2. Compute $\bar{x}$ and s with this observation omitted. For the nine remaining observations we have $\bar{x} = 2.0$ and $s = 1.2$.

3. Determine the probability of observing this value as a new randomly sampled observation by calculating

$$t_o = \frac{x - \bar{x}}{s} \tag{77}$$

and referring t_o to the table of probability points of the Student's t distribution in Table 20-22 in the Appendix. For our example

$$t_o = \frac{-1 - 2.0}{1.2} = -2.46$$

On the low end of the t distribution with $\nu = 9 - 1 = 8$, t_o falls between $t = -2.896$ ($p = 0.010$) and $t = -2.306$ ($p = 0.025$). Linear interpolation yields $p = 0.021$. However, this test assumes that the observation $x = -1$ is a new randomly selected observation, when in fact it was selected as the most extreme observation in a sample of size 10. It can be shown that the correct p-value for the observation selected as the most extreme observation in a sample of size n is approximately equal to $(n)(p)$ for small values of p, where p is the p-value for t_o as a random observation from the t distribution. Thus for our example, the significance test p-value is approximately $(10)(0.021) = 0.21$, and there is not sufficient evidence to say that the observation $x = -1$ is an outlier.

When the most extreme observation is actually found to be an outlier by this test, the test can be repeated to check the next most extreme observation. It should be noted that when there are several outliers in the sample, this and most other outlier tests may in fact fail to find that the most extreme observation is an outlier, especially when the sample size is small.

OUTLIERS IN OTHER EXPERIMENTS

Outliers in Two Sample Tests for Means or Variances

In two sample tests for means or variances, we have two samples that can be checked for outliers. The outlier test can be applied to each sample separately. To get a better estimate of the variance in tests for means with a common population variance, the two sample variances can be pooled together after omitting the most extreme observation to be tested.

Outliers in Nested Designs

Consider the one-stage nested design in Table 2A-7. In this situation, we have a random sample of size J from each of I populations, and another random sample of size I from a single population. The I random samples of size J should be checked for outliers first. When J = 2, plotting and outlier testing should be applied to the set of I differences $x_{i1} - x_{i2}$, which should behave like a random sample of size I from a Gaussian population if there are no outliers. The outlier test will indicate an outlying difference (i.e., a pair of values) in this situation. When $J \geq 3$, plotting and outlier testing should be applied to the total set of IJ differences $x_{ij} - \bar{x}_{i.}$. Finally, the group of I differences $\bar{x}_{i.} - \bar{x}_{..}$ should be plotted and tested for outliers. The two-stage nested design can be handled similarly.

Outliers in Regression

In the classical regression situation, plotting and outlier testing should be applied to the n residuals $\hat{\epsilon}_i = y_i - \hat{y}_i$. In the method comparison situation, the design is actually a two-stage nested design (days, patients nested within days, and measurements nested within patients) for each of the two analyzer systems. For each system, the replicate measurements on the specimens should be treated as discussed above. If desired, the patient means $\bar{x}_{ij.}$ and day means $\bar{x}_{i..}$ can also be treated as discussed above.

SIMULATION OF RANDOM EVENTS

The *computer simulation* of random events allows study of the properties of statistical procedures which are too complex to analyze theoretically. Simulation allows "what if" experiments to be designed and conducted on the computer to test outcomes under a given set of assumptions. In clinical laboratory applications of statistics, important aspects of a statistical procedure such as its power, robustness, and bias can be investigated by generating data with known characteristics. For example, if one wished to study the bias of the estimated slope of a least squares straight line in a comparison-of-methods experiment, data from variance models typical of both methods could be generated. Complex statistical quality control procedures can be evaluated and compared by generating random data having defined changes in systematic and/or random analytical error and then estimating the power functions of these procedures, as discussed in Chapter 2E.

A particularly useful technique for educational and research simulations in the clinical laboratory is the generation of random Gaussian data. Many statistical computer software packages have a *random number generator* for the Gaussian distribution.[19-21] If such packages are not available, Gaussian random numbers can still be generated with most microcomputers using BASIC computer languages. The random number generated with BASIC is between zero and one has a *continuous uniform distribution*, which means that for the variable x, the probability of observing a value between 0 and x is the number x. The probability frequency function of x is simply f(x) = 1 for x between 0 and 1 and f(x) = 0 for all other values of x.

Once a source of uniformly distributed random numbers between 0 and 1 is available, random Gaussian numbers can be generated. Two uniform numbers u_1 and u_2 are generated and then substituted into the following formulas:

$$z_1 = -2 \cdot \ln(u_1) \cdot \cos(2\pi \cdot u_2) \tag{78}$$

$$z_2 = -2 \cdot \ln(u_1) \cdot \sin(2\pi \cdot u_2) \tag{79}$$

The values z_1 and z_2 are independent random values of the standard Gaussian random variable z with zero mean and unit variance. These values can be transformed into random values of a Gaussian random variable x with mean μ and standard deviation σ by using

$$x_i = \mu + (\sigma)(z_i) \tag{80}$$

In this way random Gaussian data with any mean and standard deviation can be generated with a microcomputer system or programmable calculator.

The output of most random number generators is usually pseudo-random because a series of numbers is generated that repeats after a very large number of generated numbers. Nonetheless, the characteristics of the serial numbers are those of random numbers to an acceptable approximation. An application of simulation was given in the section on the experimental sampling distribution of the mean. Other applications of simulation in clinical chemistry include investigations of calibration procedures,[1] method comparison procedures,[5] and quality control procedures in Chapter 2E.

References

1. Aronsson, T., deVerdier, C., and Groth, T.: Factors influencing the quality of analytical methods—a systems analysis, with computer simulation. Clin. Chem., 20:738-748, 1974.
2. Barnett, V., and Lewis, T.: Outliers in Statistical Data. New York, John Wiley and Sons, 1978.
3. Box, G. E. P., Hunter, W. G., and Hunter, J. S.: Statistics for Experimenters. New York, John Wiley and Sons, 1978.
4. Brauer, G. A., and Sylvestre, E. A.: Techniques for specifying calibration functions and calibrator analyte values. J Qual. Tech. (in press).
5. Cornbleet, P. J., and Gochman, N.: Incorrect least-squares regression coefficients in method-comparison analysis. Clin. Chem., 25:432-438, 1979.
6. Draper, N., and Smith, H.: Applied Regression Analysis. 2nd ed. New York, John Wiley and Sons, 1981.
7. Garden, J. S., Mitchell, D. G., and Mills, W. N.: Nonconstant variance regression techniques for calibration-curve based analysis. Anal. Chem., 52:2310-2315, 1980.
8. Gilbert, R. K.: Progress and analytical goals in clinical chemistry. Am. J. Clin. Pathol., 63:960-973, 1975.
9. Graybill, F. A.: Theory and Application of the Linear Model. Section 15.6: Additional Results on Components-of-Variance Models. North Scituate, Mass., Duxbury Press, 1976.
10. Meinhold, R. J., and Singpurwalla, N. D.: Understanding the Kalman Filter. The American Statistician, 37:123-127, 1983.
11. National Committee for Clinical Laboratory Standards: Proposed Guidelines for User Evaluation of Precision Performance of Clinical Chemistry Devices. NCCLS Document EP5. Villanova, Pa., 1982.
12. National Committee for Clinical Laboratory Standards: Proposed Guidelines for User Comparison of Quantitative Clinical Laboratory Methods Using Patient Samples. Draft #7. NCCLS Document EP9. Villanova, Pa., 1983.
13. Odell, W. D., and Doughaday, W. H.: Principles of Competitive Protein-Binding Assays. Philadelphia, J. B. Lippincott, 1971.
14. Phadke, M. S.: Quality audit using adaptive Kalman filtering. San Francisco, ASQC Quality Congress Transactions, 1981.
15. Pickup, J. F., Harris, E. K., Kearns, M., et al.: Intra-individual variation of some serum constituents and its relevance to population-based reference ranges. Clin. Chem., 23:842-850, 1977.
16. Rodbard, D., and McClean, S. W.: Automated computer analysis for enzyme-multiplied immunological techniques. Clin. Chem., 23:112-115, 1977.
17. Rosner, B.: On the detection of many outliers. Technometrics, 17:221-227, 1975.
18. Rosner, B.: Percentage points for the RST many outlier procedure. Technometrics, 19:307-312, 1977.
19. Ryan, T. A., Joiner, B. L., and Ryan, B. F.: Minitab Student Handbook. North Scituate, Mass., Duxbury Press, 1976.
20. SAS User's Guide: Basics, 1982 ed. Cary, N.C., SAS Institute Inc., 1982.
21. SAS User's Guide: Statistics, 1982 ed. Cary, N.C., SAS Institute Inc., 1982.
22. Shukla, G. K.: On the problem of calibration. Technometrics, 14:547-553, 1972.
23. Snedecor, G. W., and Cochran, W. G.: Statistical Methods. Ames, Iowa, The Iowa State University Press, 1967.
24. Snee, R. D.: Computation and use of expected mean squares in analysis of variance. J. Qual. Tech., 6:128-137, 1974.
25. Steel, R. G. D., and Torrie, J. H.: Principles and Procedures of Statistics. New York, McGraw-Hill, 1980.
26. Tonks, B. D.: A study of the accuracy and precision of clinical chemistry determinations in 170 Canadian laboratories. Clin. Chem., 9:217-233, 1963.
27. Williams, G. Z., et al.: Biological and analytical components of variation in long-term studies of serum constituents in normal subjects. I. Williams, G. Z., Young, D. S., Stein, M. R., and Cotlove, E.: Objectives, selection of subjects, laboratory procedures, and estimation of analytical variance. Clin. Chem., 16:1016-1021, 1970; II. Harris, E. K., Kanofsky, P., Shakarji, G., and Cotlove, E.: Estimating biological components of variation. Clin. Chem., 16:1022-1027, 1970; III. Cotlove, E., Harris, E. K., and Williams, G. Z.: Physiological and medical applications. Clin. Chem., 16:1028-1032, 1970.

Establishment and Use of Reference Values

by Helge Erik Solberg, M.D., Ph.D.

THE CONCEPT OF REFERENCE VALUES

Medicine is an art and a science in the service of fellow human beings. To increase the health of their patients, physicians collect empirical data, interpret these data using scientific knowledge and professional experience, make decisions concerning diagnoses, recommend preventive measures, and execute therapeutic actions. There is a fundamental limitation in this activity; absolute health does not exist. Health is necessarily a relative concept.

Interpretation by Comparison

To say that health is *relative* implies that the condition of individuals must be related to something. Data collected during the medical interview, clinical examination, and supplementary investigations must be interpreted by *comparison* with reference data. The physician does this when making a diagnosis. If the condition of the patient resembles what is considered typical of a particular disease, the physician may base his diagnosis on this observation (positive diagnosis). This diagnosis is made more likely if observed symptoms and signs do not fit the patterns characterizing a set of alternative diseases (diagnosis by exclusion). Such disease patterns are examples of reference data needed for the medical interpretation. Also, the different degrees of health have their set of characteristics, which provide reference sources for the judgment of how healthy an individual is.

The process of medical interpretation by comparison may be more or less formalized. Some diagnoses are recognized by an intuitive assessment based on "clinical experience." Others are based on reasoning using advanced knowledge of normal and pathologic anatomy, physiology, and biochemistry and of other relevant areas of medical science. Sometimes the evaluation is of a qualitative nature, and in other cases it may be quantitative. The decision making may even be computer-assisted, using rules based on the laws of probability and statistical techniques.

The approaches are different, but the basis is in all cases the same: to relate, in one way or another, observed data to reference data.

The interpretation of medical laboratory data is just a special case of this decision-making process by comparison. For this decision process, we need *reference values* for all tests performed in the clinical chemistry laboratory, not only from healthy individuals but also from patients with relevant diseases. Ideally, observed values should be related to several collections of reference values: values from healthy persons, from the undifferentiated hospital population, from persons with typical diseases, and from ambulatory individuals, as well as previous values from the same subjects.[21] A patient's laboratory result simply is not medically useful if appropriate data for comparison are lacking. The establishment and use of such reference values are the topics of this chapter.

Certain conditions are mandatory to make comparison of a patient's laboratory results with reference values possible and valid:[15]

1. All groups of reference individuals used should be clearly defined.
2. The patient examined should sufficiently resemble the reference individuals (in all groups selected for comparison) in all respects other than those under investigation.

3. The conditions under which the specimens were obtained and processed for analysis should be known.

4. All quantities compared should be of the same type.

5. All laboratory results should be produced by adequately standardized methods under sufficient analytical quality control.

To these general requirements one may add others that become necessary when the more advanced techniques for decision making are applied:[62]

1. The stages in the pathogenesis of the diseases that are the objectives for diagnosis should be demarcated.

2. The diagnostic sensitivity, the diagnostic specificity, the prevalence, and the clinical costs of misclassification should be known for all laboratory tests used.

"Normal Values"—an Obsolete Term

The term "normal values" has frequently been used in the past. This term is roughly equivalent to "reference values obtained from healthy individuals."

The concept of reference values was introduced to avoid the ambiguities inherent in the term "normal values."[19] The substitution of "reference values" for "normal values" at first glance appears to be an unimportant one, but on closer scrutiny this change in nomenclature can be seen as an important step toward establishing a scientific basis for clinical interpretation of laboratory data.[62]

The term "normal values" seems to be more precise than is actually the case. Confusion arises because the word "normal" has several very different connotations.[45] Three important and very different meanings of the word "normal" are given below:

1. Values are often qualified as "normal" if their observed distribution seems to follow closely the theoretical "normal distribution" of statistics, i.e., the Gaussian probability distribution. Somehow this use of "normal" has misled people to believe that the distribution of biological data is symmetrical and bell-shaped like the Gaussian distribution. But on closer examination this is usually found not to be the case. In order to exorcise the "ghost of Gauss," it has been recommended that the term "normal limits" not be used.[16] For a similar reason the term "normal distribution" should also be avoided and replaced by the term "Gaussian distribution."[45]

2. Another meaning of "normal" is met in the following statement: it is "normal" to find that the concentration of triglyceride in serum is between 0.6 and 2.4 mmol/L, while it is considered "abnormal" to have a serum triglyceride concentration outside these limits. Here a more precise statement would read as follows: Approximately 95% of the values obtained, when measuring the concentration of triglyceride in sera collected from individuals considered to be healthy, is included in the interval 0.6–2.4 mmol/L. The obsolete concept of "normal values" partly carried this meaning. Alternative terms for "normal" in this sense are "common," "frequent," "habitual," "usual," and "typical."[45]

3. The term "normal" is also often used to indicate that values show the absence of certain diseases or the lack of risks for the development of diseases. In this sense a "normal value" is considered as a sign of health. Better terms are "healthy," "nonpathological," or "harmless" values.[45]

In summary, the word "normal" is used in relation to clinical chemistry values with at least three different meanings:

Statistical sense: "Gaussian"

Epidemiological sense: "common"

Clinical sense: "nonpathological."

Concept of Reference Values

The best way to avoid the problems related to the term "normal values" is to use the concept of reference values.[19] This concept was purposely introduced as a very vague term, to force us to define in each case what is actually meant. Reference values are results of a certain type of quantity obtained from a single individual or a group of individuals corresponding to a stated description, which must be spelled out and made available if others are to use these values.

The International Federation of Clinical Chemistry (IFCC) recommends the term "reference values" and related terms like "reference individual," "reference limit," "reference interval," and "observed values."[38] The definitions given below and the presentation in the following sections of this chapter are in accordance with the IFCC recommendations.*

The first term defined in the IFCC recommendation is the **reference individual:**[38]

> *an individual selected for comparison using defined criteria.*

For the interpretation of values obtained from an individual under clinical investigation, we need appropriate comparison values. To provide such values, we must select suitable individuals. The characteristics of the individuals in each group chosen for comparison should be clearly defined. We must specify whether they should be healthy or have a certain disease, their age and sex, and the conditions for the specimen collection. The definition of a reference individual also covers cases in which the individual under clinical investigation is her/his own reference, as discussed in the section on subject-based reference values.

The definition of a **reference value** is based on that of the reference individual:[38]

> *a value obtained by observation or measurement of a particular type of quantity on a reference individual.*

If, for example, we measure the concentration of triglyceride in sera collected from a group of reference individuals selected for comparison according to a sufficiently precise set of criteria, the triglyceride results are our reference values.

The definition of an **observed value** is as follows:[38]

> *a value of a particular type of quantity, obtained by observation or measurement and produced to make a medical decision. Observed values can be compared with reference values, reference distributions, reference limits, or reference intervals.*

Or rephrased: an observed value is the laboratory result obtained by analysis of specimens collected from an individual under clinical investigation. Some call such values "test values," but the word "test" in this term is ambiguous (laboratory test? statistical test?) and it should be avoided.

The IFCC also defines other terms related to the concept of reference values: reference population, reference sample group, reference distribution, reference limit, and reference interval.[38] We shall, however, postpone the discussion of these terms until the section on the statistical treatment of reference values.

Types of Reference Values

It is convenient, and frequently sufficient, to give a short description associated with the term reference values, such as "health-associated reference values" (close to what was understood by the obsolete term "normal values"). Other examples of such qualifying words are "diabetic," "hospitalized diabetic," and "ambulatory diabetic." These short descriptions prevent the common misunderstanding that reference values are associated only with health.

We may further distinguish between subject-based and population-based reference values. *Subject-based reference values* are previous values from the same individual, obtained when she/he was in a defined state of health. *Population-based reference values* are those obtained from a group of well-defined reference individuals and are usually the type of values referred to when the term "reference values" is used without any qualifying words. The present chapter deals predominantly with population-based values.

The body of the chapter discusses population-based *univariate reference values* and quantities derived from them. If we, for example, produce, treat, and use separately reference values of cholesterol and triglyceride in serum, we have two sets of univariate reference values. The term *multivariate reference values* denotes that results of two or more analytes obtained from the same set of reference individuals are treated in combination. The serum cholesterol and triglyceride

---

*A note on the literature: The Expert Panel on Theory of Reference Values of the IFCC has produced a series of six recommendations on the establishment and use of reference values.[38] The recent textbook on reference values edited by Gräsbeck and Alström is the best available general source of information on the topic.[22] The French committee on reference values has also prepared a set of valuable recommendations (with translation into English).[53]

values may, for example, be used to define a bivariate reference region. We shall return to this subject in a later section.

SELECTION OF REFERENCE INDIVIDUALS

A set of *selection criteria* determines which individual should be included in the group of reference individuals.[38] Such selection criteria include statements describing the source population, specifications of criteria for health, or the disease of interest.

Often separate reference values for each sex, different age groups, and other criteria are needed. Our group of reference individuals may therefore have to be divided into more homogeneous subgroups. For this purpose we need to specify rules for the division, called stratification or *partition criteria*.

It is important to distinguish between selection and partition criteria. First, we apply the selection criteria to obtain a group of reference individuals. Thereafter, we may divide this group into subgroups using partition criteria. Whether a specific criterion, e.g., sex, is a selection or a partition criterion depends on the purpose of the actual project. Sex is a selection criterion if reference values from females only are needed.

Concept of Health in Relation to Reference Values

There is an obvious need for health-associated reference values for quantities measured in the clinical chemistry laboratory. But the concept of health is problematic; much confusion may arise if the selection criteria for health are not clearly stated for a specific project.[20]

The World Health Organization has defined health as "a state of complete physical, mental and social wellbeing and not merely the absence of disease or infirmity."[20] This is an attempt to define *absolute health*, but as such, absolute health is never attained.

Thus, in the context of reference values we need a more modest concept of health. Past experience has taught us that health is a *relative concept*.[20,38] It is possible to be ill in one respect and well in another; what is considered healthy in a developing country may be judged rather unhealthy in Western Europe and North America, and so on.

Furthermore, the diagnosis of health cannot be based solely on excluding pathology. This fact, which has been named the *privative concept of health*,[20] may cause difficulties. If no signs of disease can be demonstrated, uncertainty remains, since such signs might be detected on closer examination. The "feeling" of health is not a reliable criterion because of its subjectivity. In addition, an individual may try to conceal an illness for various reasons (e.g., to qualify for life insurance).

When producing reference values, we should ask ourselves, why do we need these values? How are we going to use them? To what extent does the intended purpose of the project determine how we should identify health? In short, we need a *goal-oriented concept of health*.[20,38]

The need for goal-oriented and relative concepts of health may be obvious if we consider some contexts where reference values may be required:

Clinical diagnosis and decisions concerning therapy
Clinical follow-up
Health screening
Identification of people at risk
Epidemiological evaluation of the state of health of a population.

Gräsbeck suggested the following **general definition of health**, which summarizes the relative, privative, and goal-oriented aspects discussed above:[20]

Health is characterized by a minimum of subjective feelings and objective signs of disease, assessed in relation to the social situation of the subject and the purpose of the medical activity, and it is in the absolute sense an unattainable ideal state.

Strategies for Selection of Reference Individuals

Several methods have been suggested for the selection of reference individuals. Table 2B-1 shows three pairs of concepts that may be used to describe a sampling scheme. The concepts of each pair are mutually exclusive. The sampling is, for example, either direct or indirect. One may,

Table 2B-1. STRATEGIES FOR SAMPLING OF REFERENCE INDIVIDUALS

Direct	*Individuals* are selected from a parent population using defined criteria.
Indirect	The individuals are not considered, but defined rules are applied to a data base of analysis results to obtain a subset of *values* with the required characteristics.
A priori	A direct method (see above) where individuals are selected for specimen collection and analysis if they fulfill defined inclusion criteria.
A posteriori	A direct method using a data base containing both analysis results and information on a large number of individuals. Values of individuals fulfilling defined inclusion criteria are selected.
Random	A process of selection giving each item (individual or test result) an equal chance of being chosen.
Nonrandom	A process of selection giving each item an unequal chance of being chosen.

however, combine one concept from several pairs to obtain a more precise description. An example: the selection may be direct, a posteriori, and nonrandom.

The merits and disadvantages of these strategies will be described below. It is not possible to recommend one sampling scheme that is superior in all respects and applicable to all situations. One must choose the optimal approach for a given project and state clearly what has been done.

Direct or Indirect Sampling?

Direct selection of reference individuals (Table 2B-1) is the only method that agrees with the reference values concept as recommended by the IFCC, and it is the basis for the presentation here.[38] Its only disadvantages are the problems and costs of obtaining a representative group of reference individuals.

These practical problems have led to the search for simpler and less expensive approaches. Hoffmann suggested an *indirect* method that has become rather popular.[37] It is based on the observation that the majority of analysis results produced in the clinical laboratory seems to be "normal." Figure 2B-1 shows one example from the author's laboratory. As can be seen, the values of the serum sodium concentration have a distribution with a preponderant central peak and a shape not too far from that of the Gaussian distribution. According to Hoffmann's method, the dominant peak is composed mainly of "normal values." The advocates of the method, therefore, claim that it is possible to estimate the "normal interval" if we extract the distribution of "normal values" from this part of the distribution. "Normal limits" determined by the indirect method on the basis of the distribution shown in Figure 2B-1 would, however, be obviously biased as compared with the shown health-associated reference limits. Note that the term "normal" is here

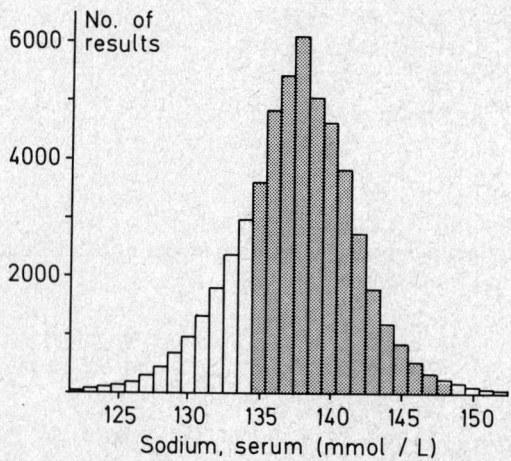

Figure 2B-1. Distribution of sodium concentrations in serum obtained in a routine laboratory. The histogram shows the distribution of 53 128 serum sodium concentrations measured in consecutive clinical specimens during a six months' period in 1982 at Rikshospitalet, Oslo. The shaded area is within reference limits as determined by a direct method (193 healthy adults of both sexes).

used intentionally to distinguish between the concepts of "normal values" and reference values.

Several mathematical methods have been used to extract the distribution of "normal values" from routine laboratory data.[22,37,42] The methods have become popular, since computer programs make the computation simple and fast.

The indirect method, however, has several important deficiencies. The two major ones are as follows:

1. The estimates of the lower and upper "normal limits" depend heavily on the particular mathematical method used and its underlying assumptions.

2. According to Harris,[22] the indirect method destroys the scientific basis for obtaining and comparing reference values. The results for each hospital would depend on the characteristics of the hospital's patient group at that particular time. These results would vary not only across hospitals but also for the same hospital at different times. The outcome would be a mass of unstable values for each analyte.

A Priori or A Posteriori Sampling?

When carefully performed, both mentioned methods may result in reliable reference values. The choice is often a question of practicality. They require the same set of successive steps, but the order of some of these operations differs depending on the mode of selection, a priori or a posteriori, as pointed out by Siest.[22]

The first step in the process of producing reference values for a laboratory test should always be the collection of quantitative information about the sources of biological variation for the analyte studied. A search through relevant literature may give the required data (see Chapter 3).[60] If the relevant information cannot be found in the literature, pilot studies may be necessary before the selection of reference individuals is planned in detail.

Serum sodium is an example of a biological quantity that is little affected by biological variation. The list of factors may be rather long for other analytes, such as serum enzymes, proteins, and hormones.

It is important to distinguish between controllable and noncontrollable sources of biological variation. Some factors may be controlled by the standardization of the procedure for preparation of reference individuals and specimen collection. Other factors, such as age and sex, may be relevant partitioning criteria. The remaining sources of variation should be considered when defining the criteria for the selection of reference individuals.

The a priori strategy is best suited for smaller studies. Possible reference individuals from the parent population are interviewed and examined clinically and by selected laboratory methods to decide whether or not they fulfill the defined inclusion criteria. If the decision is positive, specimens for analysis are collected by a standardized procedure (including the necessary preparation of the individuals prior to the collection).

The a posteriori method is based on the availability of a large collection of data on medically examined individuals and measured quantities. Well-planned studies by centers for health screening or preventive medicine may provide such data. It is important that the data be collected by a strictly standardized and comprehensive protocol concerning the sampling from the parent population, the registration of demographic and clinical data on participating individuals, the preparation for and execution of specimen collection, and the handling and analysis of the specimens. If these requirements are met, values may be selected after application of the defined inclusion criteria to the individuals included in the data base.

A study performed in Kristianstad, Sweden,[4] highlights a practical problem often met when selecting reference individuals: the number of subjects fulfilling the inclusion criteria may be too small. In that study only 17% of the participants were accepted by the criteria used, leaving an insufficient reference sample group. The frequency of exclusion was higher among females and in the higher age groups.

There are several solutions to this problem:

1. The exclusion criteria may be relaxed. As already discussed, the set of relevant sources of biological variation differs among different analytes. One may define a minimum set of exclusion criteria for a given laboratory test. In the Kristianstad study, the complete group of individuals could probably be used for establishment of reference values for serum sodium, for example, and a majority of the individuals would be acceptable for the determination of reference values for several other analytes.[4]

2. Another design of the sampling procedure could reduce the practical problems and costs of obtaining a sufficiently large group of reference individuals. The Kristianstad study showed that 75% of the excluded subjects could be identified by using only a simple questionnaire.[4] In the higher age group this percentage was even higher. Therefore, a preliminary screening of a large number of individuals from the parent population, using a carefully designed autoanamnestic questionnaire, would result in a much smaller sample of individuals to be examined clinically and by laboratory methods. Using the results of the Kristianstad study, it may be estimated that if 3000 individuals are prescreened, and if the individuals remaining in the reduced sample are subjected to a closer examination, a group of 240 reference individuals may be obtained.

The two modifications of the protocol may also be combined.

Random or Nonrandom Sampling?

Ideally, the group of reference individuals should be a random sample of all the individuals fulfilling the defined inclusion criteria in the parent population. Statistical estimation of distribution parameters and their confidence intervals and statistical hypothesis testing require this assumption.

For several reasons, most collections of reference values are, in fact, obtained by a nonrandom process.[29] That means that all possible reference individuals in the entire population under study do not have an equal chance of being chosen for inclusion in the usually much smaller sample of individuals studied. A strictly random sampling scheme is in most cases impossible for practical reasons. It would imply the examination of and application of inclusion criteria to the entire population (thousands or millions of persons) and then the random selection, e.g., by raffling, of a subset of individuals from among those accepted.

It is important to realize that we do not obtain a random sample, in the strict sense, if we start by selecting individuals randomly from the entire population and then apply inclusion criteria to sort out the subset of individuals fulfilling these criteria, even though this may be the best approximation we can hope to obtain. Usually the situation is less satisfactory. A sample of reference individuals obtained by selecting among blood donors, persons working in a factory, or the hospital's staff, is definitely not random sampling of the possible reference individuals in the general population. Data obtained by the indirect method are even more nonrandom.

The conclusions are obvious: (1) We have to use the best reference sample we can possibly get, all practical considerations taken into account. (2) Data should be used and interpreted with due caution, remembering the possible bias introduced by the nonrandomness of the sample selection process.

Selection Criteria and Evaluation of Subjects

The selection of reference individuals consists essentially of applying defined criteria to a group of examined candidate persons.[38] The required characteristics of the reference values determine which criteria should be used in the selection process. Table 2B-2 shows a list of important criteria to use when the production of health-associated reference values is the aim.

The Scandinavian Committee on Reference Values has published a list of diseases to consider for exclusion of individuals.[2] The complete list is too comprehensive for application in all situations,[4] but it may prove valuable as a checklist when defining the criteria for an actual project.

Among the *risk factors* (Table 2B-2) obesity and hypertension deserve a few comments.

The definition of *obesity* is problematic. It might be based on a known or assumed contribution to the risk of developing a specified disease. Scientific data of this type are, however, seldom available for the studied population. Another possibility for establishing obesity is to use upper limits based on weight measurements in different age, sex, and height groups of the general population. The Scandinavian Committee recommends the use of the national age, sex, and height-specific mean weight +20% as the upper limit.[2] National differences are, however, great. For example, a male of 180 cm height and 55 years old is considered to be overweight at 89 kg in Norway, at 97 kg in Finland, at 104 kg in Denmark, and at 111 kg in Sweden.[2] Tables of optimum or ideal weights have been published by life insurance companies. They may be more appropriate for demarcation of obesity.

Similar problems affect the definition of *hypertension* in relation to the establishment of health-associated reference values. The Scandinavian Committee[2] recommends the criteria shown in Table 2B-3. This Committee has also recommended exclusion criteria based on *laboratory examinations* (see Table 2B-4).

Table 2B-2. EXAMPLES OF EXCLUSION CRITERIA FOR HEALTH-ASSOCIATED REFERENCE VALUES

The table lists only some major classes of criteria. It should be supplemented with other relevant criteria based on known sources of biological variation. (See Chapter 3.)

Disease [2]
Risk factors
  Obesity
  Hypertension
  Risks from occupation or environment
  Genetically determined risks
Intake of pharmacologically active agents [63,68]
  Drug treatment for disease or suffering
  Oral contraceptives
  Drug abuse
  Alcohol
  Tobacco
Specific physiological states
  Pregnancy
  Stress
  Excessive exercise

It has been argued that we might be lost in a circular process when we use laboratory tests to assess the health of subjects who are subsequently to be used as healthy controls for laboratory tests. But actually there is no difference, in this context, between measuring the height, weight, and blood pressure and performing selected laboratory tests, provided that these laboratory tests are not those for which we shall produce reference values. [20]

Usually the individuals should be evaluated clinically by an anamnestic interview or questionnaire and by a physical examination. Published *anamnestic and examination forms* may be tailored to the requirements of the actual project. [2]

Partitioning of the Reference Group

It may also be necessary to define *partition criteria* for the subclassification of the set of selected reference individuals into more homogeneous groups (Table 2B-5). [38] We shall return in later sections to the question of determining when stratification of the reference sample group is necessary and justified. The number of partition criteria should usually be kept as small as possible in order to obtain sufficient sample sizes to derive valid data.

Age and *sex* are the most frequently used criteria for subgrouping, since several analytes vary significantly among different age and sex groups. (See Chapter 3.) Age may be categorized by equal intervals, for example by decades, or the intervals may be narrower in the periods of life where greater variation is observed. It is also often convenient to use qualitative age groups such as postnatal, infancy, childhood, prepubertal, pubertal, adult, premenopausal, menopausal, or geriatric. Height and weight could be used as criteria for categorizing children.

Additional factors are discussed in Chapter 3.

Table 2B-3. CRITERIA FOR BLOOD PRESSURE IN THE SELECTION OF REFERENCE INDIVIDUALS (SCANDINAVIAN RECOMMENDATIONS[2])

Criteria		
Age (Years)	*Blood pressure (mm Hg)* Diastolic	Systolic
16–39	95	150
> 39	100	160

Interpretation
1. The blood pressure is considered acceptable if neither the diastolic nor the systolic pressure is above the stated limits.
2. If the specifications are not met, three additional measurements of blood pressure should be taken after 5 min of resting (sitting) with intervals of at least 1 h.
3. The individual should be excluded if the mean values of the replicate measurements exceed the stated limits.

Table 2B-4. EXCLUSION CRITERIA BASED ON LABORATORY EXAMINATIONS[2]

Candidate subjects should be rejected as reference individuals if one or more values are as stated below.

Test	Result		
Albumin, urine	"Positive"		
Glucose, urine	"Positive"		
Erythrocyte sedimentation rate (1 h), venous blood	*Age (y)*	*Females*	*Males*
	≤ 50	> 21 mm	> 14 mm
	> 50	> 28 mm	> 20 mm
Hematocrit, venous blood	*Females*	*Males*	
	< 31%	< 36%	
Serum cholesterol, fasting	> 9.1 mmol/L (> 350 mg/dL)		
Serum triglyceride, fasting	> 2.4 mmol/L (> 195 mg/dL)		

SPECIMEN COLLECTION

Several preanalytical factors influence the values of biological quantities such as the concentration of components in blood and in other specimens and the amount excreted in feces, urine, or sweat. This topic is thoroughly covered elsewhere (Chapter 3).[17,60] In this discussion we shall only highlight aspects of special relevance to the generation of reliable reference values.[38]

Preanalytical standardization, such as the preparation of individuals before sample collection, the procedure of sample collection itself, and the handling of the specimen prior to analysis may eliminate or minimize bias or variation from these factors. This will reduce biological "noise" that might otherwise conceal important biological "signals" of disease, risk, or treatment effect.

Two Philosophies of Standardization

Preanalytical procedures used prior to determining patient and reference values should be as similar as possible. In general, it is much easier to standardize the routines for studies of reference values than in the daily clinical setting, especially when collecting specimens in emergency or other unplanned situations. Thus, two approaches have been suggested:

1. ". . . Only such factors which may be relatively easily controlled in the clinical setting should be controlled when the reference values are produced. If they cannot be controlled, they should be allowed to vary in the same way as they vary in the clinical setting."[65]

2. The Scandinavian Committee on Reference Values[2] recommends a set of rules for preanalytical standardization when producing reference values (Table 2B-6). These rules should also serve as ideal standards for the clinical situation. Some hospitals in Scandinavia have, in fact, shown that it is possible to apply these rules rather closely in the clinical setting for both hospitalized and ambulatory patients. The same philosophy is the basis for recent recommendations concerning routine blood specimen collection by venipuncture[46] and skin puncture.[22,43,47]

Either philosophy is, however, concordant with the concept of reference values, provided that the conditions under which reference values are produced are clearly stated.

Table 2B-5. EXAMPLES OF PARTITION CRITERIA TO BE USED FOR POSSIBLE SUBGROUPING OF THE REFERENCE GROUP

Age (not necessarily categorized by equal intervals)
Sex
Genetic factors
    Race (ethnic origin)
    Blood groups (ABO)
    Histocompatibility antigens (HLA)
Physiological factors
    Stage in menstrual cycle
    Stage in pregnancy
    Physical condition
Other factors
    Socioeconomic
    Environmental
    Chronobiological

Table 2B-6. STANDARDIZATION OF PREANALYTICAL FACTORS IN THE ESTABLISHMENT OF REFERENCE VALUES FOR ADULT INDIVIDUALS (A SUMMARY OF THE RECOMMENDATIONS PUBLISHED BY THE SCANDINAVIAN COMMITTEE ON REFERENCE VALUES[2])

The day before specimen collection

Food	Ordinary intake; last meal before 2200 h.
Alcohol	Maximum of one small bottle of beer (or equivalent of other beverage) taken with a meal.
Abstinence	No solid food or tobacco and maximum of one glass of water after 2200 h.

Ambulatory subjects; collection in the morning

Rise	1–3 h before collection (record time).
Transport	Public or car transport for maximum of 45 min; walking a maximum of 500 m (≈550 yards) at moderate speed.
Rest	Sitting for at least 15 min; arm muscle work not allowed.
Collection	Between 0800 and 1000 h (record time); sitting position with the arm approximately 45° below the horizontal position.

Subjects lying in bed; collection in the morning

Rest	Bed rest from 2200 h until collection; a short visit to the toilet allowed, but minimum of 1 h before collection.
Collection	Between 0700 and 0900 h (record time); supine position with the arm approximately in the horizontal plane.

Ambulatory subjects; collection in the afternoon

Breakfast	A light meal in the morning (approximately 310 kcal, 1300 kJ) composed of milk, coffee, or tea (maximum two cups), two open sandwiches with butter, slices of lunch meat/cheese or marmalade.
Activity	No exercise or heavy work.
Rest	Sitting at least 15 min; arm muscle work not allowed.
Collection	Between 1300 and 1500 h (record time); minimum of 4 h after breakfast; otherwise as above.

Collection and handling of specimen

Venipuncture	In the cubital fold; no tourniquet; finger pressure proximal to the site allowed.
Difficulties	A new attempt on opposite arm after 15 min rest.

(Consult Chapter 3 for a discussion of other requirements for the collection and handling of specimens.)

Standardization Schemes

The magnitudes of preanalytical sources of variation are clearly not equal for different analytes (see Chapter 3).[17,60] It could therefore be argued that we should consider only those factors that cause unwanted variation for the biological quantity for which we want to produce reference values.[67] Body posture during specimen collection is, for instance, highly relevant for the establishment of reference values for nondiffusible analytes, such as albumin in serum, but irrelevant for serum sodium.[17]

On the other hand, several constituents are usually analyzed in the same clinical specimen. It would therefore be impractical to devise special systems for every single type of quantity.[19] For that reason three standardized procedures for blood specimen collection by venipuncture have been recommended:[2,22] (1) collection in the morning from hospitalized patients, (2) collection in the morning from ambulatory patients, and (3) collection in the afternoon from ambulatory patients. Table 2B-6 summarizes these procedures. In addition, the Scandinavian Committee (Moe[22]) has produced preliminary recommendations for collection of blood by venipuncture from infants and children that are similar to American proposals.[22,43]

Such schemes of standardization have to be modified depending on local conditions and needs and the intended use of the produced reference values.

A special problem is caused by drugs taken by individuals prior to specimen collection. We may distinguish between indispensable and dispensable medication. The latter-type drugs should always be avoided for at least two days. The use of indispensable drugs such as contraceptive pills or essential medication may be a criterion for exclusion or partition.

In emergency or other unplanned clinical situations, even a partial application of the standardized procedure for specimen collection has been shown by Gräsbeck and coworkers to be of great value.[22]

The Need for Additional Information

The clinical situation often is different from a controlled research situation; specimens have to be taken during operations, in emergency situations, or when patients are unwilling to follow the rules. Therefore the clinician needs additional information for the interpretation of a patient's values in relation to reference values obtained under fairly standardized conditions.

An *empirical approach*[19] is to produce other sets of reference values, such as postprandial values, post-exercise values, or post-party values.[22] Such a method, however, is very expensive and cannot cover all situations that could possibly arise.

There is another and more general solution to the problem, called the *predictive approach.*[19] Starting from a set of ordinary reference values and using quantitative information on the effect of various factors, such as intake of food, alcohol and drugs, exercise, stress, or posture (Chapter 3) we could estimate "expected reference values" which fit the actual clinical setting.

More studies of such effects are needed, especially for the combined effect of two or more sources of variation. Is, for example, the combined effect of alcohol and contraceptive drugs on γ-glutamyltransferase activity in serum less than, equal to, or greater than the sum of their individual effects?

IMPORTANCE OF ANALYTICAL PROCEDURES AND QUALITY CONTROL

Essential components of the required definition of a set of reference values are specifications concerning:[22,38]

Analysis method (including information on equipment, reagents, calibration standards, type of raw data, and calculation method)

Quality control (see Chapter 2E)

Reliability criteria (see Chapter 2C).

Specifications should be so carefully described that another investigator can reproduce the study, and that the user of reference values can evaluate their comparability with values obtained with the methods used for producing the patient's values in a routine laboratory. To ensure comparability between reference values and observed values, the same analytical method should preferably be used.

It is often claimed that the analytical quality should be better when determining reference values than when producing routine values. The claim may be true for accuracy; all measures should be taken to eliminate bias. The question of imprecision is more difficult because it is partly dependent on the intended use of the reference values. Increases in analytical random variation result in widening of the reference interval.[10,22] For some special uses of reference values, the narrower reference interval obtained by a more precise analytical method may be appropriate. However, this is usually not true for routine clinical use of reference values. The interpretation is simplest if a patient's values and reference values are comparable with regard to both biological variation and analytical imprecision. For the same reason it is advisable to analyze specimens from reference individuals in several runs to include the between-run components of variation. A safe way to obtain comparability is to include these specimens in routine runs together with the patient's specimens.

STATISTICAL PROCEDURES USED IN ESTABLISHING REFERENCE VALUES

This section deals with two main topics: the partitioning of reference values into more homogeneous classes and the determination of reference limits and intervals.[29,38,42,49,55,67] The subject matter is presented in the order in which the data are best treated. Figure 2B-2 gives an outline of an approach and refers to the corresponding sections in the text. Prior to the presentation of the methods we shall briefly state some statistical concepts used. Further details of statistical concepts and methods are given in Chapter 2A.

Statistical Concepts

The first step in the establishment of reference values is the selection of a group of reference individuals. It is usually not feasible to obtain observations on all possible reference individuals of a certain category of the general population. We therefore hope that the smaller group examined, the *subset* (sometimes called the reference *sample* group),[38] can give us the desired information about the characteristics of the *complete set* of individuals (the reference *population*).[38]

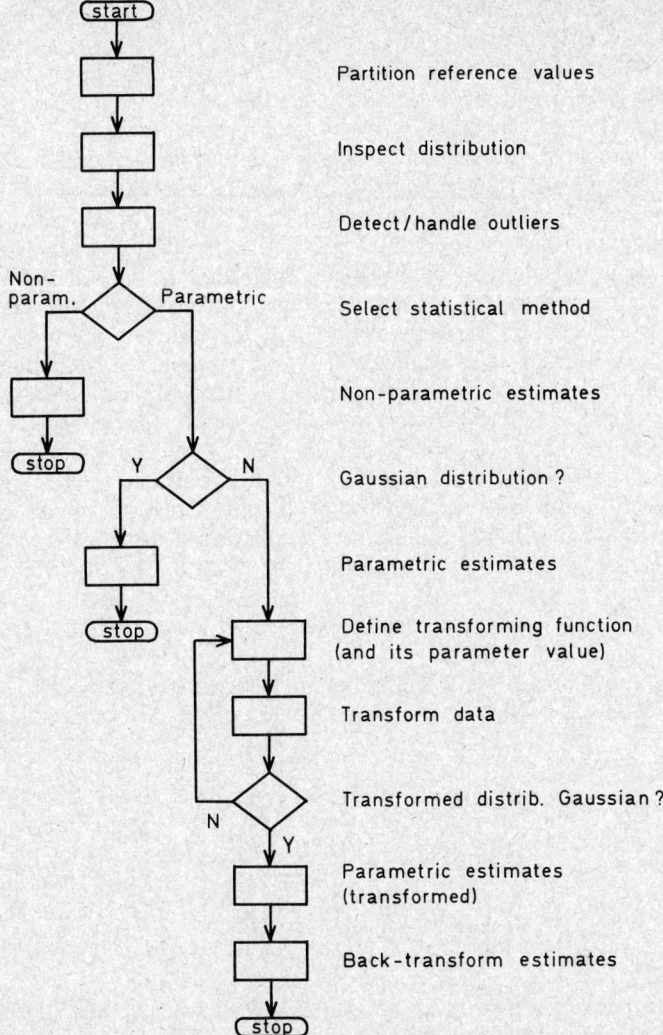

Figure 2B-2. Procedure for the statistical treatment of reference values. The "boxes" in the flow-chart refer to sections in the text. Y = "yes"; N = "no."

The larger set is often considered *hypothetical* because its characteristics are not observed directly; we know neither the number (the set size) nor the properties of all its individuals. We therefore want to infer from observations made on the subset to the hypothetical set. An obvious requirement is that the individuals in the subset are typical of those in the complete set. Statistical theory usually assumes that the items in the subset are selected at *random* from among those in the set; otherwise the subset may be biased. If the items are not randomly selected, we can still use the statistical techniques, but only with due caution and remembering the possible bias introduced.

There are two main types of *inferences* made from values obtained from the subset (sample group) to the set (total reference population).

1. We can *estimate properties* of the set. A reference limit (a percentile) of a biological quantity, such as the concentration of serum triglyceride, based on subset reference values, is an example of a *point estimate* (a single value). We consider it representative of the property found if we had observed all possible values in the set. If we examine many randomly selected subsets from the same set, we obtain several estimates with some variation around the "true" value of the set. It is possible to produce an *interval estimate* bounded by limits within which the "true" value is located with a specified confidence: the confidence interval. The *confidence* is expressed as a number in the interval 0–1, indicating the degree on the scale between "never" and "always."

Our reference limit for serum triglyceride could thus be associated by a confidence interval showing its region of uncertainty.

The "true" (unknown, hypothetical) value of a property of the set is often called a *parameter* and given a lower-case Greek letter symbol (e.g., the standard deviation of the population is symbolized by σ_x). The corresponding property determined in the subset is named a *statistic* and symbolized by a lower case Latin letter (s_x for the standard deviation of the subset).

2. Another inference is to *test hypotheses* regarding properties of the set. We might, for example, state the hypothesis that the distribution of values for serum triglyceride concentrations is of the Gaussian type (the "null" hypothesis). If the deviations of the subset values from the Gaussian distribution are small, we could ascribe them to variation due to chance alone. In that case, we should be permitted to use statistical methods based on the Gaussian distribution. However, we must reject the hypothesis if it is unlikely that the observed deviations from the Gaussian distribution are caused by chance alone. *Statistical tests* provide quantitative approaches to these types of decisions: the hypothesis is rejected if the test shows that the probability of the hypothesis being true is less than a *stated significance level*. The *probability* is a number in the interval 0–1, indicating the degree on the scale between "unlikely" and "certain." If we state a significance level of 0.05 when we test our Gaussian hypothesis for the distribution of serum triglyceride values, we should reject it if the probability obtained by the test is, for example, $p = 0.01$. Then we have to accept the alternative hypothesis that the distribution is non-Gaussian. The *power* of a statistical test is the probability of rejection when the "null" hypothesis is, in fact, false.

In the following sections we shall use the term *reference distribution*[38] for the distribution of reference values.

The statistics, arithmetic *mean* $\bar{x}$ and *standard deviation* s_x, are used as measures of the location of the reference distribution and the dispersion of the reference values, respectively. They are defined as

$$\bar{x} = \frac{\Sigma x}{n}$$

$$s_x = \sqrt{\frac{\Sigma(x - \bar{x})^2}{n - 1}} = \sqrt{\frac{\Sigma x^2 - \frac{(\Sigma x)^2}{n}}{n - 1}}$$

where x is any of the n reference values in the subset (or a subclass of it). Standard deviations are commonly computed as shown in the second part of the equation above. The result may, however, be grossly inaccurate due to rounding errors introduced by calculators and computers. A more accurate algorithm has been published.[13,59]

The variable x in the formulas above denotes a quantity that varies. In our context it signifies a reference value. If the variable by chance may take any one of a specified set of values, we use the term *variate*, i.e., a random variable. In this section we consider distributions of single variates, that is, *univariate* distributions. In a later section we shall also discuss the joint distribution of two or more variates, *bivariate* or *multivariate* distributions.

An observed distribution may be presented as a table or graph (histogram) showing the number of observations in small intervals (Figure 2B-3, *a*). The number of observations in an interval divided by the total number of observations in the distribution (its size) is an estimator of the probability of finding a value in the corresponding interval of the hypothetical *probability distribution* of the population (assuming random sampling). By consecutive summing of all these ratios, starting with the leftmost interval of the observed distribution, we get estimates of cumulative probabilities of the hypothetical *cumulative probability distribution* (Figure 2B-3,*b*).

Partitioning of Reference Values

The subset of reference individuals and the corresponding reference values may be partitioned according to sex, age, and other characteristics. The process of partitioning is also called stratification, categorization, or subgrouping, and its results are called partitions, strata, categories, classes, or subgroups. We shall use the pair of terms *partitioning* (for the process) and *(sub-)classes* (for its result).

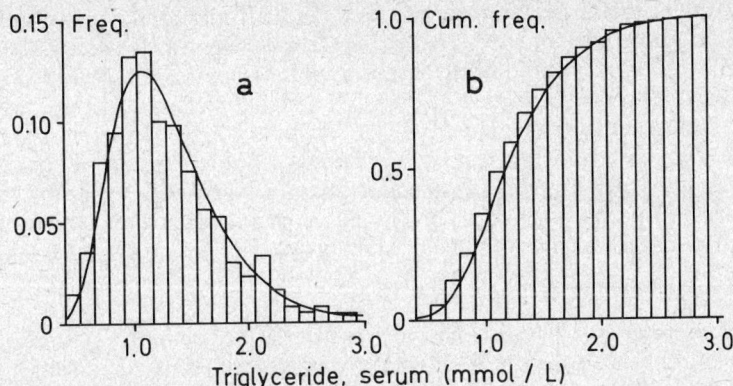

Figure 2B-3. Observed and hypothetical distributions of 500 triglyceride values in serum (in mmol/L). *Part a*, The vertical bars of the histogram show the number of observations in the interval divided by the total number of observations. The curve is the estimated probability distribution of the population, assuming random sampling and a log-Gaussian distribution. *Part b* shows the cumulated ratios (bars) and the estimated cumulative probability distribution (curve). The data were computer generated.

The aim of the partitioning is to reduce, if possible and necessary, variation among subjects so that there will be less biological "noise." Less intraclass variation gives narrower and more sensitive reference intervals. We call such intervals *class-specific*, for example, age- and sex-specific reference intervals.

The distribution of reference values in the classes may show different locations (mean values may vary) or different intraclass variations (standard deviations may vary). Both contribute to the larger total variation of reference values before partitioning.

In general, reference values may be partitioned when the differences between the classes are statistically significant (rejection of the "null" hypothesis of equal distributions). *Differences in location* may be tested by the Student's *t*-test (two classes) or by analysis of variance (several classes), or by nonparametric alternatives to these tests (Wilcoxon's rank-sum test and Kruskal-Wallis' multi-sample test). There are also parametric tests of *differences in intraclass variation*: Fisher's *F*-test (ratio test of two variances), Bartlett's test (several variances), and Levene's test (which is less sensitive to the assumption of Gaussianity).[13] These tests are not described here. The reader is referred to Chapter 2A and to standard textbooks of parametric[14,52] and nonparametric statistics.[8] Most computer centers have packages of statistical programs that make the testing easy, for example, BMDP[13] and SAS.[36]

It has been shown that when the intended purpose of the reference interval is to detect individual changes in biochemical status, differences among means of class-specific reference values may be statistically significant and still too small to justify replacing a single total reference interval with several age- and sex-specific intervals.[27,28] In such cases subject-based reference values are more appropriate for interpretation. We shall return to this problem in a later section, where a criterion for the necessary between-class difference will also be given.

In the following sections we assume that we have a homogeneous reference distribution, either the complete subset distribution (if partitioning has been shown to be unnecessary) or the distributions of subclasses following partitioning.

Inspection of Distribution

It is always advisable to display the reference distribution graphically and to inspect it. A *histogram*, as shown in Figure 2B-3,*a*, is easily prepared manually or by using computer programs and is the data display best suited for visual inspection. The examination of the histogram is a safeguard against the misapplication or misinterpretation of statistical methods, and it may give valuable information about the data. We should look for the following characteristics of the distribution:

1. Highly deviating values (*outliers*) may represent aberrant values (see next section).

2. *Bimodal* or *polymodal* distributions have more than one peak and may indicate that the distribution is nonhomogeneous because of the mixing of two or more distributions. If so, the criteria used to select reference individuals should be re-evaluated, or partitioning of the values according to age, sex, or other relevant factors should be tried.

3. The shape of the distribution should be noticed. It may be asymmetrical or more or less peaked than the symmetrical and bell-shaped Gaussian distribution (Figure 2B-4). The asymmetry most frequently observed with clinical chemistry data is positive *skewness* (Figure 2B-4,*a*). A symmetrical distribution with positive *kurtosis* has a high and slim peak and more values in both tails than the Gaussian type (Figure 2B-4,*c*). Conversely, negative kurtosis indicates that the distribution has a broad and flat top with relatively few observations in the tails (Figure 2B-4,*d*). Asymmetry and non-Gaussian peakedness may be combined.

4. The visual inspection may also provide initial estimates of the location of reference limits that are useful as checks on the validity of computations.

Identification and Handling of Aberrant Values

An *aberrant value* can be traced to a gross deviation from the prescribed procedure for establishment of reference values.[29] Such values may either deviate significantly from the proper reference values (*outliers*) or be hidden in the reference distribution. Only a strict experimental protocol, with adequate controls at each step, can eliminate the latter type of aberrant values.

Visual inspection of a histogram is a reliable method for identification of possible outliers. It is important to keep in mind, however, that values far out in the long tail of a skewed distribution may easily be misinterpreted as outliers. If the distribution is positively skewed, inspection of a histogram displaying the logarithms of the values may aid in the identification of outliers.

Some outliers may also be identified by *statistical tests,* but no single method is capable of detecting outliers in every situation that may occur. The number of techniques suggested or recommended is, for that reason, very large. Two recent books review statistical tests for identification of outliers and discuss their merits and disadvantages.[3,34] The two main problems encountered are discussed below.

1. Many of the tests assume that the type of the true distribution is known prior to using the test. Some of these specifically require that the distribution be Gaussian. However, biological distributions are very often non-Gaussian and their types are seldom known in advance. Furthermore, statistical tests of distribution type are unreliable in the presence of outliers. We then have an impossible dilemma; some tests for outliers assume that the type of distribution is known, but tests for determining the type of distribution require that outliers be absent! As a consequence, we cannot transform the distribution to Gaussian form before the identification of outliers by statistical tests. We therefore should resort to tests that are relatively insensitive to departures from Gaussianity.[3] The following *range test* is relatively robust.[14,49] Identify the extreme value as an outlier if the difference between the two highest (or lowest) values in the distribution exceeds one-third of the range of all values.

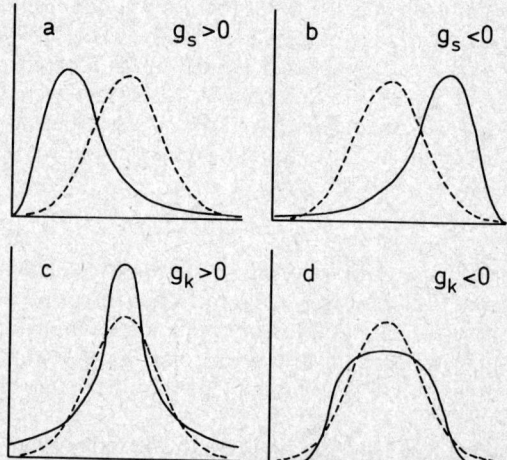

Figure 2B-4. Skewness and kurtosis. The two upper figures show asymmetric distributions; *a*, positive skewness, and *b*, negative skewness. The two lower figures show distributions with non-Gaussian peakedness; *c*, positive kurtosis, and *d*, negative kurtosis. The Gaussian distribution (dashed curve) is shown in all graphs for comparison. The values of the coefficients of skewness (g_s) and kurtosis (g_k) are also shown.

2. Several tests for outliers assume that only one outlier is present in the data. The limitation of these tests is obvious. Some tests may detect a specified number of outliers, or they may be run several times discarding one outlier in each pass of data. The range test, however, usually fails in the presence of several outliers. It is also possible to estimate the standard deviation using data remaining after "*trimming*" of both tails of the distribution by a specified percentage of observations.[35] A formula and tables for the estimation of the standard deviation in "trimmed" data has been published.[35] Outliers could be identified by this method as the values lying three or four standard deviations from the arithmetic mean. The method assumes, however, that the true distribution is Gaussian.

The author's personal experience with several statistical tests for identification of outliers indicates that none of the tests is universally more sensitive or more reliable than simple visual inspection of a histogram.

Deviating values identified as possible outliers should not be discarded automatically. Values should be included or excluded on a rational basis. Check the records of the dubious values and correct errors. In some cases, deviating values should be rejected because noncorrectable causes have been found, such as previously unrecognized conditions qualifying individuals for exclusion from the group of reference individuals.

Determination of Reference Limits: General Considerations

In clinical practice one usually compares an observed patient's value with the corresponding **reference interval** (sometimes called reference range).[38] This interval may be defined in different ways (see below). The reference interval is a useful condensation of the information carried by the total set of reference values. The interval is bounded by a pair of **reference limits,**[38] which should be clearly distinguished from clinical decision limits that provide optimum separation between clinical categories.[22,62]

Three kinds of reference intervals have been suggested: tolerance interval, prediction interval, and interpercentile interval.[15,38,67]

1. The simple *tolerance interval,* without specified confidence, is defined as an interval bounded by upper and lower limits between which on the average a stated fraction of the reference values is contained.[12] This interval is used less than the wider tolerance interval with specified confidence, which will contain at least the specified fraction of values with a stated degree of confidence.[12,25,26,67] One may, for example, estimate an interval containing at least 95% of the population's serum triglyceride concentrations with a confidence of 0.90. The use of tolerance intervals is limited to situations where the assumption of random sampling is fulfilled.

2. The *prediction interval* is defined as an interval bounded by upper and lower limits between which a future observation is expected to fall with a specified degree of probability.[25,26,67] A future value of serum triglyceride concentration will, for example, be within the 95% prediction interval with a probability of 0.95. This type of interval is based on two assumptions. First, as for the tolerance interval, random sampling is presupposed. Second, the future observation should be a true member of the set from which we obtained a subset of reference values. The latter assumption may be false for observations obtained at a later time.

The tolerance and prediction intervals are usually determined by parametric techniques using estimates of the mean and the standard deviation, the only difference being the size of the factor k used to compute the two limits of the interval: $\bar{x} \pm k \cdot s_x$. The values of k for tolerance and prediction intervals depend on the sample size n and the percentage, $100(1-2\alpha)$, of reference values to be contained in the interval, and, when required, the confidence β. Details on the estimation of these types of reference intervals are given in the referenced literature.[12,25,26]

3. The *interpercentile interval* is defined as an interval bounded by two percentiles of the reference distribution. It is the most frequently used type of reference interval. A *percentile* denotes a value that divides the reference distribution such that a specified percentage of its values have magnitudes less than or equal to the limiting value. For example, if 2.32 mmol/L is the 97.5 percentile of serum triglyceride, 97.5% of the concentration values are equal to or below this value. This kind of reference interval has been chosen for detailed discussion here since it is recommended by the IFCC.[38]

It is an arbitrary but common convention to define the reference interval as the *central 95% interval* bounded by the 2.5 and 97.5 percentiles (Figure 2B-5), that is, 2.5% of the values is cut

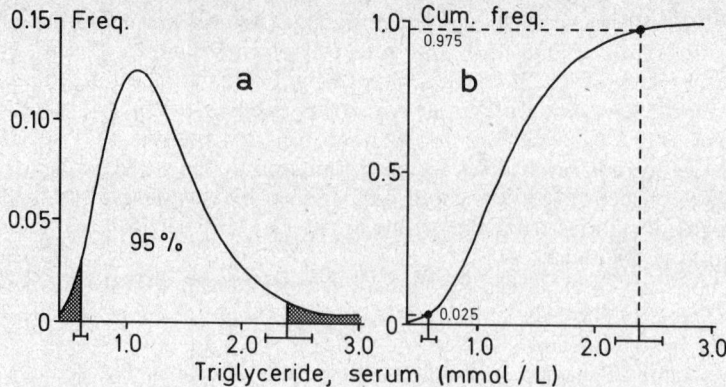

Figure 2B-5. Central 95% reference interval. The 2.5 and 97.5 percentiles of the 500 serum triglyceride concentrations (see Figure 2B-3). The percentiles and their 0.90 confidence intervals were determined by the parametric method (see the text). The curves are the estimated probability distributions (see Figure 2B-3).

off in both tails of the reference distribution.[38] Another size or an asymmetric location of the reference interval may be more appropriate in particular cases. To avoid ambiguity, the definition of the interval should always be stated. The estimation of percentiles presented in the following sections is based on the conventional central 95% interval, but the techniques are easily adapted to other locations of the limits.

The percentiles are point estimates of population parameters. Accordingly, they are unbiased estimates only if the subset of values was selected randomly from the population. But, as was discussed, random sampling is often difficult to achieve. The interpercentile interval may always be used, however, as a summary or description of the subset reference distribution.

The precision of a percentile as an estimate of a population value is dependent on the size of the subset, being less precise with few observations. If the assumption of random sampling is fulfilled, we may determine the *confidence interval* of the percentile, that is, the limits within which the true percentile is located with a specified degree of confidence (Figure 2B-5). The 0.90 confidence interval of the 97.5 percentile (upper reference limit) for serum triglyceride may, for example, be 2.22–2.62 mmol/L. We would expect to find the true percentile in this interval with a confidence of 0.90 if we measured all serum triglyceride concentrations in the total reference population.

A further advantage of the interpercentile interval is that it is easily determined by both parametric and nonparametric statistical techniques, while only parametric methods are in common use for estimating the tolerance and prediction intervals.

The **parametric method** for the determination of percentiles and their confidence intervals assumes a certain type of distribution, and it is based on estimates of population parameters such as the mean and the standard deviation. We are, for example, using a parametric method if we believe that the true distribution is Gaussian and then we determine the reference limits (percentiles) as the values located two standard deviations on each side of the mean. The majority of the parametric methods are, in fact, based on the Gaussian distribution. If the reference distribution shows another shape, we may use mathematical functions that transform data to approximately Gaussian shape. Some positively skewed distributions (Figure 2B-4,*a*) may, for example, be made symmetric by using logarithms of the data values.

The **nonparametric method** makes no assumptions concerning the type of distribution and does not use estimates of distribution parameters. The percentiles are simply determined by cutting off the required percentage of values in each tail of the subset reference distribution.

When we compare the results obtained by these two methods, we usually find that the estimates of the percentiles are very similar, the only difference being that the parametric estimates of percentiles are theoretically more precise (narrower confidence intervals) than those obtained by the nonparametric method, especially with smaller sample sizes. The real difference in precision, however, is probably minimal, since the type of distribution is not known with certainty.

Sample size. In general, the theoretical lower limit of the sample size required for the estimation of the 100α and $100(1-\alpha)$ percentiles is equal to $1/\alpha$. Thus, the estimation of the 2.5 percentile needs at least $1/0.025 = 40$ observations. The precision of percentiles increases with an increasing number of observations, as shown by the narrowing of their confidence intervals. A sample size of at least 120 reference values has been recommended when nonparametric confidence intervals of the percentiles are needed.[38,49]

Conclusions. The practical differences between the three types of reference intervals (tolerance, prediction, and interpercentile intervals) are small when based on at least 100 reference values. The tolerance and prediction intervals assume random sampling, while the percentiles may be useful descriptors of the subset reference distribution also in situations where this assumption is violated. The simple nonparametric method available for the determination of percentiles is an advantage. We do not, in fact, need the more complex parametric method. We shall still present it here in detail because of its popularity and frequent misapplication. The theoretic minimum of the sample size is 40 observations for the estimation of the central 95% interval but the sample size should preferably be much larger. It is a good practice to compute the confidence intervals of the percentiles to get an estimate of their imprecision.

Nonparametric Method

This method consists essentially of cutting off a specified percentage of the values from each tail of the reference distribution. Three techniques may be used:

1. The percentiles may be determined *graphically* by plotting the cumulative distribution on ordinary graph paper (Figure 2B-5,*b*) or on Gaussian probability paper.

2. A *mathematical function* may be fitted to the reference distribution using, for example, statistical nonlinear regression methods. The percentiles are then determined using the fitted function. A computerized method by which a percentile is determined after fitting a second-degree polynomial to the values in a tail of the distribution has been published.[50]

3. Experience has shown that the method described below,[38] which is the simplest of all nonparametric methods, is reliable. It is based on *rank numbers* and also allows nonparametric estimation of the confidence intervals of the percentiles.[49] The method can easily be applied manually or implemented on a computer.[36,56] Proceed as follows:

a. Sort the n reference values in ascending order of magnitude. When dealing with large amounts of data, considerable labor can be saved by identifying and sorting only values in the tails of the distribution, each containing, for example, 7–10% of the values.

b. Rank the values. The minimum value has rank number 1, the next value has rank number 2, and so on until the maximum value, which has rank number n. Consecutive rank numbers should be given to two or more values that are equal ("ties").

c. Compute the rank number of the 100α and $100(1-\alpha)$ percentiles as $\alpha \cdot (n+1)$ and $(1-\alpha) \cdot (n+1)$, respectively. Thus, the limits of the conventional 95% reference interval have rank numbers equal to $0.025 \cdot (n+1)$ and $0.975 \cdot (n+1)$.

d. Determine the percentiles by finding the original reference values that correspond to the computed rank numbers, provided that the rank numbers are integers. Otherwise interpolate between the two limiting values.

e. Finally, determine the confidence interval of each percentile using the binomial distribution.[49] Table 2B-7 facilitates this for the 0.90 confidence interval of 2.5 and 97.5 percentiles. Just enter the table and look up the bounding rank numbers for each percentile.

Table 2B-8 shows an example of the nonparametric determination of percentiles using the serum triglyceride values shown in Figure 2B-3.

Parametric Method

The parametric method is much more complicated and often needs a computer when large samples are to be processed;[38,42,55] computer programs are, however, available.[7,31,56,57] The method is presented under separate headings for testing of distribution type, transformation of data, and the estimation of percentiles and their confidence intervals. The section on data transformation has two parts: the first describes a simple procedure, while the second presents a more comprehensive method that requires a computer.

Testing Fit to Gaussian Distribution

The parametric method for estimating percentiles assumes that the true distribution is Gaussian. This fact was frequently ignored in the past and caused Elveback to warn against "the ghost of Gauss."[16] Negligence often results in seriously biased estimates of reference limits. If,

Table 2B-7. NONPARAMETRIC CONFIDENCE INTERVALS OF REFERENCE LIMITS

The table shows the rank numbers of the 0.90 confidence interval of the 2.5 percentile for samples with 119–1000 values. To obtain the corresponding rank numbers of the 97.5 percentile, subtract the rank numbers in the table from $(n + 1)$ where n is the sample size.

Sample Size	Rank Numbers		Sample Size	Rank Numbers	
	Lower	*Upper*		*Lower*	*Upper*
119–132	1	7	556–574	8	22
133–160	1	8	575–598	9	22
161–187	1	9	599–624	9	23
188–189	2	9	625–631	10	23
190–218	2	10	632–665	10	24
219–248	2	11	666–674	10	25
249–249	2	12	675–698	11	25
250–279	3	12	699–724	11	26
280–307	3	13	725–732	12	26
308–309	4	13	733–765	12	27
310–340	4	14	766–773	12	28
341–363	4	15	774–799	13	28
364–372	5	15	800–822	13	29
373–403	5	16	823–833	14	29
404–417	5	17	834–867	14	30
418–435	6	17	868–871	14	31
436–468	6	18	872–901	15	31
469–470	6	19	902–919	15	32
471–500	7	19	920–935	16	32
501–522	7	20	936–967	16	33
523–533	8	20	968–970	17	33
534–565	8	21	971–1000	17	34

The table is taken from Part 5 of the series of recommendations on reference values prepared by the IFCC[38] (with permission from the IFCC).

Table 2B-8. NONPARAMETRIC DETERMINATION OF REFERENCE INTERVAL

The table shows a worked-out example using the 500 serum triglyceride concentrations displayed in Figure 2B-3. See the text for a description of the nonparametric method. The unit of all concentrations in the table is mmol/L.

Sorted and ranked serum triglyceride values in the left tail of the distribution:

Values:	0.41	0.43	0.45	0.46	0.47	0.49	0.51	0.55	0.55	0.55
Ranks:	1	2	3	4	5	6	7	8	9	10

Values:	0.56	0.58	0.58	0.61	0.62	0.62	0.64	0.64	0.65	0.65
Ranks:	11	12	13	14	15	16	17	18	19	20

Sorted and ranked serum triglyceride values in the right tail of the distribution:

Values:	2.21	2.22	2.26	2.27	2.27	2.28	2.30	2.31	2.34	2.35
Ranks:	481	482	483	484	485	486	487	488	489	490

Values:	2.48	2.50	2.55	2.62	2.63	2.65	2.72	2.78	2.90	2.91
Ranks:	491	492	493	494	495	496	497	498	499	500

Calculation of rank numbers of the percentiles:
Lower: $0.025 \cdot (500 + 1) = 12.5$
Upper: $0.975 \cdot (500 + 1) = 488.5$

Finding the original values corresponding to these rank numbers:
Lower reference limit (2.5 percentile): 0.58
Upper reference limit (97.5 percentile): 2.32 (by interpolation)

Rank numbers (see Table 2B–7) and values of the 0.90 confidence limit of the lower reference limit:
Rank numbers: 7 and 19
Confidence limits: 0.51 and 0.65

Rank numbers (see Table 2B–7) and values of the 0.90 confidence limit of the upper reference limit:
Rank numbers: $500 + 1 - 19 = 482$
$500 + 1 - 7 = 494$
Confidence limits: 2.22 and 2.62

Summary:
Lower reference limit: 0.58 (0.51–0.65) mmol/L
Upper reference limit: 2.32 (2.22–2.62) mmol/L

for example, we compute the mean and standard deviation of the 500 serum triglyceride values displayed in Figure 2B-3, and then calculate the reference interval as $\bar{x} \pm 1.960 \cdot s_x$, we would find the following limits: 0.4 and 2.2 mmol/L (correct values: 0.6 and 2.3 mmol/L). A highly positively skewed distribution may even result in a negative value for the lower reference limit!

A critical phase in the parametric method is, therefore, to test the goodness-of-fit of the reference distribution to a hypothetical Gaussian distribution. If we must reject the Gaussian hypothesis at a specified significance level, we have two alternatives (see Figure 2B-2): either we can use the nonparametric method or we can try to apply mathematical transformation of data to approximate the Gaussian distribution. Only when the Gaussian hypothesis is not rejected by the test can we pass directly to parametric estimation of percentiles and their confidence intervals (see Figure 2B-2).

Goodness-of-fit tests have recently been reviewed by Mardia.[41] We shall discuss only a few simple tests that have proved valuable for our purpose.

1. A **graphic procedure** consists of plotting the cumulative distribution (Figure 2B-3,b) on Gaussian probability paper, which has a nonlinear vertical axis based on the Gaussian distribution.[38] The plot should be close to a straight line if the distribution is Gaussian. It is, however, very difficult to evaluate visually the deviations from the straight line because of the nonlinearity of the vertical distances in the graph. (See, however, the Kolmogorov-Smirnov test below.)

2. The **coefficient-based tests** use statistical measures of skewness and kurtosis.[38,52] These measures are computed from the subset second, third, and fourth moments about the mean (m_2, m_3 and m_4, respectively):[11,52]

$$m_k = \frac{\Sigma(x - \bar{x})^k}{n} \qquad k = 2, 3, \text{ or } 4$$

These formulas give slightly biased estimates of population moments, but they suffice for practical use. Cramér[11] may be consulted for correction factors to reduce the bias.

The subset *coefficient of skewness*, g_s, and its asymptotic standard deviation, s_s (it approaches the true standard deviation as the number of observations increases), are computed by these formulas:[52]

$$g_s = m_3/(m_2\sqrt{m_2})$$

$$s_s = \sqrt{6/n}$$

A more accurate estimate of the true standard deviation of g_s is described in Mardia[41] and shown in Table A6 in Snedecor and Cochran.[52] The coefficient g_s is zero for the Gaussian and other symmetrical distributions. The sign of a nonzero coefficient indicates the type of skewness present in the data (Figure 2B-4,a and b).

The subset *coefficient of kurtosis*, g_k, and its standard deviation, s_k, are computed as follows:[52]

$$g_k = (m_4/m_2^2) - 3$$

$$s_k = \sqrt{24/n} = 2\sqrt{6/n}$$

Mardia[41] may be consulted for a more accurate estimate of the standard deviation of g_k. The coefficient g_k is approximately zero for the Gaussian distribution. The sign of a nonzero coefficient indicates the type of kurtosis present in the data (Figure 2B-4,c and d).

The statistical significance of these two coefficients may be found by referring to Table A6 in Snedecor and Cochran,[52] which presents critical values for two-sided tests at the significance levels 0.02 and 0.10. A rough test is obtained by dividing the coefficient by its standard deviation and then referring to a table of the Gaussian distribution. The latter procedure is only reasonably accurate for sample sizes above 1000 (for g_s) and 2000 (for g_k). Computer programs for calculation and testing of these coefficients are available.[13,31,36,56]

Example: The coefficients computed for the serum triglyceride data shown in Figure 2B-3 are $g_s = 0.873$ and $g_k = 0.677$. In the table given by Snedecor and Cochran,[52] we find their critical values for n = 500 at the 0.02 level of significance: 0.255 and 0.60, respectively. Both coefficients are thus statistically significant and the Gaussian hypothesis should be rejected (the data was, in fact, approximately log-Gaussian). We arrive at the same conclusion by the approximate tests using the ratio between a coefficient and its approximate standard deviation:

$$g_s/s_s = 0.873/\sqrt{6/500} = 7.97$$

$$g_k/s_k = 0.677/\sqrt{24/500} = 3.09$$

Both coefficients exceed three standard deviations. By a computer program[56] these probabilities were found: $p < 0.001$ for g_s and $p = 0.011$ for g_k.

3. Several tests of goodness-of-fit based on the empirical distribution function (*EDF-tests*) have been suggested.[41,61] Three of these tests are good alternatives for our purpose: the Kolmogorov-Smirnov test, the Cramér-von Mises test, and the Anderson-Darling test.[38] Detailed descriptions of these tests and tables of critical values are found in a paper by Stephens.[61] Computer programs for all three tests are available.[56]

It is recommended that *allowances for rounding of data* be made when using these tests, for example by adding random "noise" to the values.[7,38,56]

$$x_i' = x_i' + L \cdot (r_i - 0.5) \quad (i = 1, \ldots, n)$$

Here x_i' is the value with random "noise" added, x_i is the original data value, L is the step size of the value scale (least significant digit, e.g., $L = 0.1$ if laboratory results are reported to one decimal place), and r_i is a random number between 0.0 and 1.0.

The **Kolmogorov-Smirnov test** may be regarded as a test of the vertical distances between the plot of the cumulative distribution and the straight line on Gaussian probability paper (graphic test, see above).[38] The test statistic, D_{max}, is equal to the greatest absolute difference found between the cumulative distribution of reference values and the cumulative Gaussian distribution. The description of the test and tables of critical values of D_{max} given by Stephens[61] are recommended. The data should be corrected for the effect of rounding, for example by adding and subtracting half the step size of the value scale (i.e., $0.5 \cdot L$, see above) to each value and then performing the test with the new set of values. (The number of values becomes two times the sample size.) A simpler procedure is to use adding of random "noise" (see above).

The **Anderson-Darling test** is slightly more powerful than the Kolmogorov-Smirnov test.[61] The test is recommended by the IFCC[38] for use with reference values. The procedure for the Anderson-Darling test is as follows:[38,56,61]

a. Make allowances for rounding of data; for example, add random "noise" (see above).
b. Sort the data values in an ascending order of magnitude.
c. Compute the deviations $v_i = (x_i - \bar{x})/s_x$, where $\bar{x}$ is the mean and s_x is the standard deviation of the data values.
d. Now find the values w_i of the cumulative Gaussian distribution corresponding to the deviations v_i by using statistical tables or a computerized algorithm.
e. Calculate the test statistic A^2 by this formula:

$$A^2 = -n - [\Sigma(2i - 1) \cdot \{\ln(w_i) + \ln(1 - w_{n+1-i})\}]/n$$

f. Reject the hypothesis that the data distribution fits the Gaussian distribution at the 0.01 level of significance if the size-adjusted test statistic A^{2*} is greater than 1.092:[61]

$$A^{2*} = A^2 \cdot \left(1 + \frac{4}{n} - \frac{25}{n^2}\right)$$

Transformation of Data, Simple Method

We saw in the previous section that $\bar{x} \pm 1.960 \cdot s_x$ of the serum triglyceride data in Figure 2B-3 resulted in biased reference limits (too low values), as was to be expected with this positively skewed distribution. The data were, in fact, approximately log-Gaussian; that is, the logarithms of the reference values had a near-Gaussian distribution. Neither the coefficient-based test nor any of the three EDF-tests mentioned above showed significant deviations of logarithmically transformed values from the Gaussian distribution.

It is frequently observed that *logarithmically transformed* values, $y = \ln(x)$, of a positively skewed distribution fit the Gaussian distribution rather closely. In other cases *square roots* of the values, $y = \sqrt{x}$, result in a better approximation to the Gaussian distribution. This is the basis for the common use of the logarithmic and square root transformations when estimating reference limits. The method is applicable only to positively skewed distributions. Computer programs implementing the method are available,[31] but the method is easily performed manually when the sample size is not too large. The procedure is as follows:

1. Test the fit of the distribution of original data to the Gaussian distribution. If the distribution has approximately Gaussian shape, the 2.5 and 97.5 percentiles can be calculated directly as $\bar{x} \pm 1.960 \cdot s_x$. Otherwise continue with the steps described below.

2. Transform data by the logarithmic function $y = \ln(x)$ or the square root function $y = \sqrt{x}$ and then test the fit to the Gaussian distribution. If the transformed distribution is significantly different from Gaussian shape, try another transformation, or estimate the percentiles by the nonparametric method (see above). Continue with the next step if the transformation to Gaussianity was successful.

3. Compute the mean $\bar{y}$ and the standard deviation s_y of transformed data. Then estimate the 2.5 and 97.5 percentiles in the transformed data scale as $\bar{y} \pm 1.960 \cdot s_y$.

4. The final step is reconversion of these percentiles to the original data scale. The inverse functions of the two transformations described here are:

$$\text{Inverse logarithmic function: } x = e^y = \exp(y)$$

$$\text{Inverse square root function: } x = y^2$$

It is also possible to estimate the confidence limits of percentiles determined by the parametric method. The method is presented in a later section.

Example: The mean and standard deviation of the serum triglyceride values of Figure 2B-3 following logarithmic transformation are $\bar{y} = 0.172$ and $s_y = 0.357$, respectively. The transformed 2.5 percentile is $0.172 - 1.960 \cdot 0.357 = -0.528$. Upon reconversion to the original data scale we get $\exp(-0.528) = 0.59$. The lower reference limit of serum triglyceride is thus 0.59 mmol/L. Similarly, we find that the upper reference limit is 2.39 mmol/L. These values are in close agreement with those found by the nonparametric method: 0.58 and 2.32 mmol/L (Table 2B-8).

Transformation of Data; Comprehensive Method

The logarithmic and square root functions often fail to transform data to fit a Gaussian distribution. Then more general transformations can be used. The complete adjustment frequently requires a *two-stage procedure*: transformation to symmetry followed by transformation to compensate for non-Gaussian kurtosis.[7,31,38,55,56]

The following functions are examples of functions that may transform *skewed distributions* to symmetry: the modified logarithmic function $y = \ln(x + C)$,[7,31,55] the power function of Box and Cox,[5,55,57] and the exponential function described below. Adjustment of *non-Gaussian kurtosis* can be obtained by the hyperbolic sine function and its inverse function,[7,31,55] the power function of Boyd and Lacher,[7] and the modulus function described below.

Here we shall present the procedure recommended by the IFCC.[38,56] It is based on the exponential function suggested by Manly[40] for correction of skewness (positive or negative), and on the modulus function of John and Draper[39] for correction of non-Gaussian kurtosis (positive or negative). A computer is necessary, since the estimation of the parameters, γ and λ, of the two transforming functions (see below) requires a lot of computing. Computer programs are available.[56] The procedure is as follows:

1. Compute the mean and standard deviation of the original data. Then standardize data to zero mean and unit variance by subtracting the mean from each value and dividing each difference by the standard deviation.

2. Determine the value of γ that gives a distribution without skewness ($g_s = 0$, see below) when data are transformed by the *exponential function*:

$$y = \begin{cases} \dfrac{\exp(\gamma \cdot x) - 1}{\gamma} & \gamma \neq 0 \\ x & \gamma = 0 \end{cases}$$

3. Compute the mean and standard deviation of the exponentially transformed data. Then standardize transformed data to zero mean and unit variance by subtracting the mean from each value and dividing each difference by the standard deviation.

4. Determine the value of λ that gives a distribution with zero kurtosis ($g_k = 0$, see below) when the exponentially and standardized data are transformed by the *modulus function*:

$$z = \begin{cases} \text{sign} \left[\dfrac{(|y| + 1)^{\lambda} - 1}{\lambda} \right] & (\lambda \neq 0) \\[2em] \text{sign} \{ \ln(|y| + 1) \} & (\lambda = 0) \end{cases}$$

The *sign* in the formulas above is the algebraic sign associated with each input value y.

5. If data now fit a Gaussian distribution, determine the percentiles and their confidence intervals as described in the next section. Then reconvert the percentiles and the confidence limits to the original data scale by performing the above-mentioned steps in reverse order and using the inverse functions at each step.

There are several methods for estimating the parameters γ and λ of the two transforming functions. A particularly simple procedure is to determine each parameter by an iterative routine.[56] Starting with an initial guess of the parameter's value, one performs repetitiously the following steps in sequence until the coefficient of skewness (when determining γ) or the coefficient of kurtosis (for λ) is sufficiently close to zero:

1. Transform data using the current value of the parameter.

2. Determine the relevant shape coefficient (skewness or kurtosis) for the distribution of transformed values.

3. Stop the process if the guiding coefficient is sufficiently close to zero. Otherwise, adjust the value of the function parameter so that the absolute value of the guiding coefficient is expected to be further diminished during the next iteration.

Parametric Estimates of Percentiles and Their Confidence Intervals

General estimates for the 100α and $100(1-\alpha)$ percentiles and their β confidence intervals can be determined by the following method, provided that data (original or transformed) fit the Gaussian distribution:[38]

The 100α and $100(1-\alpha)$ *percentiles* are determined as

$$(\text{mean}) \pm c \cdot (\text{standard deviation})$$

where c is the $(1-\alpha)$ standard Gaussian deviate, as can be found in statistical tables.[12,52] For the 2.5 and 97.5 percentiles we must find the $(1 - 0.025) = 0.975$ standard Gaussian deviate: $c = 1.960$.

The β *confidence interval* of a percentile is estimated as

$$(\text{percentile}) \pm u_{0.5 \cdot (1+\beta)} \cdot \sqrt{\frac{(2 + c_{1-\alpha}^2) \cdot s_y^2}{2n}}$$

where u is the $0.5 \cdot (1+\beta)$ standard Gaussian deviate ($u = 1.645$ for $\beta = 0.90$), c is the standard Gaussian deviate used above, and s_y is the standard deviation of data (original or transformed). For the common situation where $\alpha = 0.025$ and $\beta = 0.90$ we get the simplified formula:

$$(\text{percentile}) \pm 2.81 \cdot s_y / \sqrt{n}.$$

Example: The parametric estimate of the 2.5 percentile of serum triglyceride was determined above by the logarithmic transformation. The $\beta = 0.90$ confidence limits of the lower percentile are then

$$-0.528 - 2.81 \cdot 0.357 / \sqrt{500} = -0.573 \qquad \exp(-0.573) = 0.56$$

$$-0.528 + 2.81 \cdot 0.357 / \sqrt{500} = -0.483 \qquad \exp(-0.483) = 0.62$$

Thus, the complete estimate of the 2.5 percentile (and its $\beta = 0.90$ confidence interval) is 0.59 (0.56–0.62) mmol/L. The 97.5 percentile is, by the same method, found to be 2.39 (2.29–2.50) mmol/L.

PRESENTATION OF AN OBSERVED VALUE IN RELATION TO REFERENCE VALUES

An observed value (patient's value) is interpreted by comparison with reference values. This comparison often looks like hypothesis testing, but it is seldom a statistical test in the strict sense. Ideally, the patient and the reference individuals should match; that is, we might state the hypothesis that they were all picked from the same set (population). Often, however, this is not the case. Thus, it is advisable to consider the reference values as the yardstick for a less formal assessment than hypothesis testing.

The clinician should always be supplied with as much information about the reference values as he needs for the interpretation. Reference intervals for all laboratory tests may be presented to the physicians in a booklet together with information about the analysis methods, their imprecision, and descriptions of the reference values. The aim must be to present to the physicians enough information for rational clinical judgments.

In addition, a convenient presentation of an observed value in relation to reference values may be a great help for the busy clinician.[15,21,38,58]

The presentation of the observed value together with a *complete set of reference values* for the corresponding test is a feasible procedure only when few reference values are available. When we have many reference values, is it more convenient to present the *reference distribution* in a table or graphically in the form of a histogram (Figure 2B-3,*a*) or by a plot of the cumulative distribution (Figures 2B-3,*b* and 2B-5,*b*)? A very informative presentation of the observed value is to show its location on the graph. This is particularly suited for display on a computer's data screen.[21]

A more condensed presentation of the relationship between the observed value and the *reference interval* is their presentation on the same report sheet. The reference intervals may be preprinted on report forms, or the computer system may select the appropriate age- and sex-specific reference interval from a file and print it next to the test result. This type of presentation may also be graphical.[21] Several multichannel analyzers, e.g., the older models of Technicon's SMA, produce a trace of the results across scales on which the reference intervals are indicated as shaded areas.

It is also possible to use various *mathematical indices* or *flagging* with convenient symbols to show the relation of the observed value to the reference distribution. When using such presentation methods, the original observed value should also be reported, to allow comparison with results of other laboratory tests and metabolic calculations.

An observed value may be classified as "low," "usual," or "high" (three classes) depending on its *location in relation to the reference interval*. On reports printed by a computer it is convenient to flag unusual results with symbols, such as L and H.[21]

A more detailed division of the value scale has also been advocated.[15] The regions outside the reference interval may be subdivided in order to indicate how "unusual" the observed value is. The reference interval may also be subclassified. The advantages are doubtful, however, since the shape of the reference distribution is not taken into account.

Another popular method is to express the observed value by a *mathematical distance measure*. Such distances are ratios of the following type:

$$\frac{\text{(observed value)} - \text{(measure of location)}}{\text{(measure of dispersion)}}$$

These measures are also linear transformations that do not take into account the shape of the reference distribution. The well-known "*SD-unit*" or "normal equivalent deviate" is the difference between the observed value and the mean of the reference values divided by their standard deviation.[24] Several similar ratios have been suggested,[15] but none has significant advantages over the other. All produce very confusing values if the reference distribution is very skewed. An observed value with a "SD-unit" of 2.2, for example, would be above the 97.5 percentile if the reference distribution has a Gaussian shape, but might be well below the upper reference limit of a positively skewed distribution. Mathematical transformation of the reference distribution to the Gaussian shape may solve this problem.[58]

To report the observed value as a *percentile* of the reference distribution provides a very accurate measure of the relation.[15,50] The observed serum triglyceride value 2.4 mmol/L may, for example, be reported as 2.4 mmol/L (98 percentile). Alternatively, we can estimate the probability of finding a value closer to the mean than the observed value, the "*index of atypicality.*"[1,58]

When observed values of several analytes are reported simultaneously it is possible to use multivariate analogs of the "SD-unit" and the "index of atypicality."[1] (See the next section.)

MULTIVARIATE, POPULATION-BASED REFERENCE REGIONS

The topic of the previous sections has been univariate population-based reference values and quantities derived from them. Such values do not, however, fit the common clinical situation in which observed values of several different laboratory tests are available for interpretation and decision making. For example, the average number of individual laboratory tests requested on one sample received in the author's laboratory is 5.4 (range: 1–35). There are two models for interpretation by comparison in this situation. We can compare each observed value with the corresponding reference values or interval, i.e., we perform multiple, univariate comparisons; or we can consider the set of observed values as a single multivariate observation and interpret it as such by a multivariate comparison. In this section we shall discuss the relative merits of these two approaches and present methods for the latter type of comparison.

The multivariate concept.[54,64] A univariate observation, such as a single laboratory result, may be represented graphically as a point on a line, the axis or scale of values. The results obtained by two different laboratory tests performed on the same specimen, a bivariate observation, may be displayed as a point in a plane defined by two perpendicular axes. With three results we have a trivariate observation and a point in a space defined by three perpendicular axes, and so on. We lose the possibility for direct graphical representation of multivariate observation with more than three dimensions. Still, we can consider the multivariate observation as a point in a multidimensional hyperspace with as many mutually perpendicular axes as there are results of different tests. The prefix "hyper-" signifies, in this context, "more than three dimensions." Such multivariate observations are also called patterns or profiles. A multivariate distribution is thus represented by a cluster of points on a plane or in a space or in a hyperspace depending on the dimensionality of the observation. Several statistical methods are based on multivariate methods, and some of them are straightforward extensions of well-known univariate methods. A number of good and not too difficult textbooks on multivariate statistical methods exist.[33,44] These books also have simple introductions to matrix algebra, the mathematics used in multivariate analysis.

The multiple, univariate reference region. The univariate reference interval is bounded by two reference limits on the result axis (see Figure 2B-5). Figure 2B-6 shows that the univariate reference intervals for two laboratory tests describe a square in the plane of the two axes. Similarly, three or more univariate reference intervals define boxes or hyperboxes in the (hyper-)space. The multiple, univariate comparison referred to above consists of deciding whether a multivariate observation point lies inside or outside this square, box, or hyperbox. This method has, however, two very serious deficiencies;[18,51,66] an observation may lie outside the limits of the region without being unusual (Figure 2B-6, point A), or it may be found on the inside and still be a very atypical observation (Figure 2B-6, point B). If we use the central 95% interval, 5% of the values are, by definition, expected to be located in the two tails of the univariate reference distribution. But more than 5% of the values would be located outside the square or (hyper-)box created by several 95% intervals. To be accurate, $100 \cdot (1 - 0.95^m)$ per cent of multivariate reference values would be excluded by the method of multiple, univariate comparison (m being the number of different tests or number of dimensions). We would, for example, expect to find a fraction of $100 \cdot (1 - 0.95^{10}) = 40\%$ of "false positives" when using 10 laboratory tests! This discouraging result has been verified in several multiphasic screening programs.[51] We therefore need a better method.

The multivariate reference region.[1,6,18,64,66] It is possible to define a common multivariate reference region based on the joint distribution of the reference values for two or more laboratory tests. This multivariate region is not a right-angled area or (hyper-)box, but more like an ellipse in the plane (Figure 2B-6) or an ellipsoid (hyper-)body in (hyper-)space. This region may be a straightforward extension of the univariate 95% interval to the multivariate situation; it may be

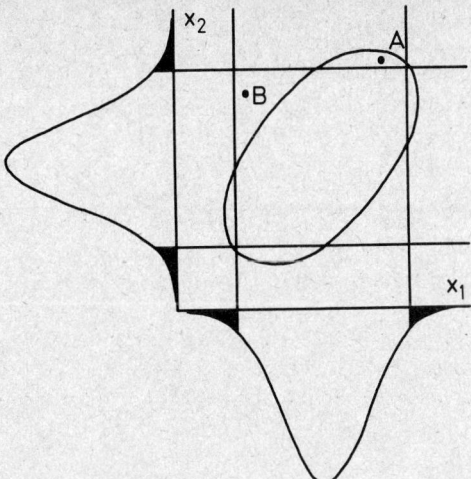

Figure 2B-6. Bivariate reference region (ellipse) compared with the region defined by the two univariate reference intervals (box).

set to enclose 95% of the central multivariate reference data points. In that case we would expect to find only 5% "false positives." Multivariate regions analogous to the univariate tolerance and prediction intervals can also be defined and estimated.[9]

The use of multivariate reference regions usually requires the assistance of a computer. The computer program takes a set of results obtained by several laboratory tests on the same clinical specimen and calculates an index. The interpretation of a multivariate observation in relation to reference values is then the task of comparing the index with a critical value estimated from the reference values. This, obviously, is much simpler than comparing each result with its proper reference interval.

The index is essentially a distance measure (Mahalanobis' squared distance, D^2) expressing the multivariate distance between the observation point and the common mean of the reference values, taking into account the dispersion and correlation of the variables (see below).[1,44] More interpretational guidance may be obtained from this distance by expressing it as a percentile analogous to the percentile presentation of univariate observed values.[6] Also, the index of atypicality has a multivariate counterpart.[1]

The *Mahalanobis' squared distance*[1,6,44,54] of an observation is the multivariate analog of the square of the "SD-unit," d^2:

$$d^2 = \{(x - \bar{x})/s_x\}^2 = (x - \bar{x}) \cdot (s_x^2)^{-1} \cdot (x - \bar{x})$$

The second form of the formula above is given to show the analogy to the Mahalanobis' distance, D^2, which is defined as

$$D^2 = (\mathbf{x} - \bar{\mathbf{x}})' \cdot \mathbf{S}^{-1} \cdot (\mathbf{x} - \bar{\mathbf{x}})$$

where $(\mathbf{x} - \bar{\mathbf{x}})'$ and $(\mathbf{x} - \bar{\mathbf{x}})$ are row and column vectors of the differences between each variable and its mean, and $\mathbf{S}^{-1}$ is the inverse matrix of variances and covariances. (According to the notation of matrix algebra, a lower case boldface letter denotes a vector [an array of numbers]. The symbol for a matrix [a table of numbers] is a boldface capital letter.) The diagonal elements of the matrix $\mathbf{S}$ before inversion are the variances, i.e., measures of the dispersion of individual variables. The covariances are measures of their interrelationship (off-diagonal elements of $\mathbf{S}$). The D^2 for a set of m-dimensional observations follows approximately a χ^2 distribution with m degrees of freedom. The critical value of D^2 for a 95% reference region is thus approximately the $(1 - 0.95)$ fractile of this distribution.[1,6] An observed pattern is located outside the multivariate region if D^2 exceeds the critical value. If we want to report the D^2 as a percentile, we also refer to the same χ^2 distribution.[6]

Although the theory of multivariate reference regions has been known for a long time, surprisingly few applications have been reported in the literature. A recent report reviews the topic and presents the results of a very careful study on the multivariate 95% region for a 20-test chemistry profile.[6] Some of the findings are important to summarize here.

1. Sixty-eight per cent of the subjects had at least one test result outside univariate reference intervals, which was close to what was theoretically expected: $100 \cdot (1 - 0.95^{20}) = 64\%$.

2. By contrast, only 5% of the patterns were outside the multivariate reference region (as expected).

3. Transformation to the approximately Gaussian shape of the univariate distributions was necessary.

4. A test profile may be distinctly unusual in the multivariate sense even though each individual result is within its proper reference interval (as, for example, point B in Figure 2B-6).

5. The multivariate reference region could detect minor deviations of multiple analytes.

6. Conversely, it could also be quite insensitive to highly deviating results for a single analyte.

7. The sensitivity could be increased by defining multivariate reference regions for subsets of physiologically related tests.

SUBJECT-BASED REFERENCE VALUES

Figure 2B-7 depicts the inherent problem associated with population-based reference values. It shows two hypothetical reference distributions. The one represents the common reference distribution based on single specimens obtained from a group of several different reference individuals. It has a true (hypothetical) mean μ and standard deviation σ. The other distribution is based on several specimens collected over time in a single individual, the ith individual. Its hypothetical mean is μ_i and the standard deviation σ_i.

If an observed value is located outside the subject's 2.5 and 97.5 percentiles, the personal or subject-based reference interval, the cause may be a change in the biochemical status, suggesting a disease. Figure 2B-7 shows that such an observed value may still be within the population-based reference interval. The sensitivity of the latter interval to changes in a subject's biochemical status is accordingly dependent on the location of the individual's mean μ_i relative to the common mean μ and to the relative magnitudes of the corresponding standard deviations σ_i and σ. A mean μ_i close to μ and a small σ_i relative to σ may conceal the individual's changes entirely within the population-based reference interval.

Harris[27,28] analyzed this topic and found that the ratio R of intraindividual (personal) variation over interindividual (among subjects) variation provides a criterion of the usefulness of the population-based reference interval. The average value of this ratio is defined as $R = \sqrt{E(\sigma_i^2)/Var(\mu_i)}$ where $E(\sigma_i^2)$ is the average value of individual variances σ_i^2, and $Var(\mu_i)$ is the variance of the individual mean values μ_i (Figure 2B-7). Estimates of $E(\sigma_i^2)$ and $Var(\mu_i)$ are obtained by collecting several specimens over time from each individual in the subset examined, and then treating the results by statistical analysis of variance.[32]

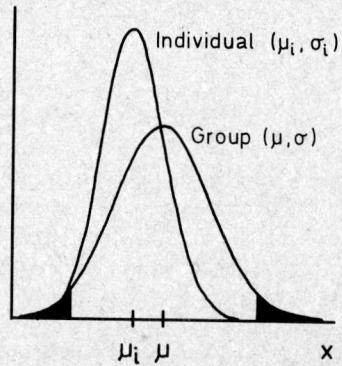

Figure 2B-7. The relationship between population-based and subject-based reference distributions and reference intervals. The example is hypothetical, and the two distributions are, for simplicity, Gaussian. (After Harris.[27])

It has been shown that the population-based reference interval has less than the desired sensitivity to changes in biochemical status if the ratio value is $R \le 0.6$.[27,28] This reference interval is a more trustworthy reference if $R > 1.4$, at least for the individual whose standard deviation σ_i is close to the average value $\sqrt{E(\sigma_i^2)}$. Published data usually show that homeostatically tightly-controlled quantities such as serum electrolytes have high ratio values.[27,48,65] The population-based reference intervals of such analytes suffice for clinical use. Serum proteins and enzymes, for example, have very low ratios, because they are not under the same degree of metabolic control. Here subject-based reference intervals seem more appropriate.

There are two possible solutions to the problem of the clinical insensitivity of population-based reference intervals.

1. We can try to reduce the variation in the reference values by *partitioning* into more homogeneous subclasses. But to increase the ratio R, for example, from 0.6 to 1.4 by partitioning, requires that we can obtain the rather dramatic reduction of 37% in the standard deviation.[27] This often is difficult to attain in practice.

2. The other possibility is to use the subject's previous values, obtained when the subject was in a well-defined state of health, as the reference for any future value. The application of *subject-based reference values* becomes more feasible as "health-screening" by laboratory tests and as computer storage of results become available to large segments of the general population.

There are two not completely separated classes of models for construction of subject-based reference intervals: statistical and physiological models.

1. Harris has developed several models based on statistical *time series analysis*.[27–30] At the one extreme we have a stationary or *homeostatic model* suitable for analytes showing relatively fast, random fluctuations around a constant mean (set point). The set point is estimated from past values that are given equal weights. Another model, the nonstationary *random-walk model,* allows a changing set point over time in healthy subjects. Then the more recent values are given heavier weights during the estimation of the current set point. Intermediate as well as more or less complex models exist. Some of these data-following methods are suited for adaptive forecasting in situations where the time intervals are short, as, for example, during hospitalization.[29] They might thus be implemented on a computer as part of a laboratory cumulative reporting system. The reader is referred to papers by Harris for details on statistical time series models.[28–30]

2. It is also possible to construct *physiological models* that utilize known physiological and biochemical time-dependent relationships. Winkel has developed a time-series model for monitoring plasma progesterone in pregnancy using the assumption of a simple exponential growth curve for the size of the placenta.[65]

TRANSFERABILITY OF REFERENCE VALUES

The determination of reliable reference values for each test on the laboratory's repertoire is a major task, often far beyond the capabilities of the individual laboratory. It would therefore be convenient if reference values generated in another laboratory could be used. Petitclerc and Kelly[22] studied the transfer of reference values within the same population (in Sherbrooke, Canada) and between two populations (in Canada and France). They found that "even with the same population, the same instrument, with methodologies under good quality control and some standardization of blood collection, it is difficult to reproduce identical reference intervals." Still, they concluded, it is possible for a laboratory that is unable to produce its own reference values to use those established elsewhere, provided that certain conditions were fulfilled.

1. The populations should be adequately described and matching.

2. Subsets of data should be compared to check for bias arising mainly from analytical factors.

3. Analytical performance in both laboratories should agree.

4. Preparation of individuals and specimen collection should follow the same standardization scheme.

Another possibility, not yet fully explored, is to transfer the "true" biological distribution, making adjustments for the effects of analytical variation. The procedure might be as follows:

1. Estimate the systematic components of analytical variation (bias) of the analytical methods used by the two laboratories by comparing results obtained on the same (human) specimens or by referring to a highly accurate method.

2. Estimate the imprecision of both methods by adequate experimental designs, preferably by using human specimens.

3. Adjust the reference distribution obtained in the donor laboratory for the effects of bias (subtraction)

and imprecision (reduction of the standard deviation) to provide an estimate of the "true" biological distribution. This is fairly simple if the shape of data is approximately Gaussian. Otherwise a mathematical transformation may be necessary.

4. Perform the converse adjustments in the receiving laboratory to obtain an estimate of the reference distribution believed to be found if the reference values were produced locally.

DYNAMIC VS. STATIC INTERPRETATION OF CLINICAL CHEMISTRY DATA

The interpretation of observed values by comparison with population-based reference values or intervals is not the only way clinical data may be used. Often dynamic approaches to data interpretation are more appropriate. Time-dependent variation may provide important information. The time series analysis of consecutive values from the same individual is one example. Other examples are the dynamic analysis of kinetic processes in the organism, such as intermediary metabolism, and the exchange of substances between metabolic pools. It is, for example, possible to design a model for urea turnover in the body.[23] The model defines the rates of urea input from various sources to the extracellular fluid, the exchange of urea across cell membranes, the urea degradation in the gut, the handling of urea by the kidneys, and so forth. Such a model may facilitate the interpretation of serum urea values with the purpose of detecting hemorrhage or necrosis following major surgery and the evaluation of the magnitude of these complications. Biochemical model building, estimation of model parameters from observed values, and computer simulation of the models may add greatly to our understanding. This rapidly expanding field of data interpretation has been reviewed by Groth and de Verdier.[23]

References

1. Albert, A., and Heusghem, C.: Relating observed values to reference values: The multivariate approach. *In*: Reference Values in Laboratory Medicine. R. Gräsbeck and T. Alström, Eds. Chichester, England, John Wiley and Sons, Ltd., 1981, pp. 289–296.
2. Alström, T., Gräsbeck, R., Hjelm, M., et al.: Recommendations concerning the collection of reference values in clinical chemistry. Scand. J. Clin. Lab. Invest., *35*, Suppl. 144:1–45, 1975.
3. Barnett, V., and Lewis, T.: Outliers in Statistical Data. Chichester, England, John Wiley and Sons, Ltd., 1978.
4. Berg, B., Nilsson, J.-E., Solberg, H. E., et al.: Practical experience in the selection and preparation of reference individuals: Empirical testing of the provisional Scandinavian recommendations. *In*: Reference Values in Laboratory Medicine. R. Gräsbeck and T. Alström, Eds. Chichester, England, John Wiley and Sons, Ltd., 1981, pp. 55–64.
5. Box, G. E. P., and Cox, D. R.: An analysis of transformations. J. R. Statist. Soc., Ser. B, *26*:211–252, 1964.
6. Boyd, J. C., and Lacher, D. A.: The multivariate reference range: An alternative interpretation of multiple profiles. Clin. Chem., *28*:259–265, 1982.
7. Boyd, J. C., and Lacher, D. A.: A multi-stage algorithm for clinical laboratory data. Clin. Chem., *28*:1735–1741, 1982.
8. Bradley, J. V.: Distribution-Free Statistical Tests. Englewood Cliffs, N.J., Prentice-Hall, Inc., 1968.
9. Chew, V.: Confidence, prediction and tolerance regions for the multivariate normal distribution. J. Am. Statist. Assoc., *61*:605–617, 1966.
10. Cotlove, E., Harris, E. K., and Williams, G. Z.: Biological and analytical components of variation in long-term studies of serum constituents in normal subjects. III. Physiological and medical implications. Clin. Chem., *16*: 1028–1032, 1970.
11. Cramér, H.: Mathematical Methods of Statistics. Princeton, N.J., Princeton University Press, 1946.
12. Diem, K., and Seldrup, J.: Introduction to Statistics, Statistical Tables, Mathematical Formulae. *Vol. 2 of*: Geigy Scientific Tables. C. Lentner, Ed. 8th ed. Basle, Ciba-Geigy, 1982.
13. Dixon, W. J., and Brown, M. B., Eds.: BMDP Statistical Software 1981. Berkeley, University of California Press, 1981.
14. Dixon, W. J., and Massey, F. J.: Introduction to Statistical Analysis. 3rd ed. New York, McGraw-Hill Book Co., 1969.
15. Dybkaer, R.: Observed values related to reference values. *In*: Reference Values in Laboratory Medicine. R. Gräsbeck and T. Alström, Eds. Chichester, England, John Wiley and Sons, Ltd., 1981, pp. 263–278.
16. Elveback, L. R., Guillier, C. L., and Keating, F. R.: Health, normality and the ghost of Gauss. JAMA, *211*:69–75, 1970.
17. Felding, P., Tryding, N., Hyltoft Pedersen, P., et al.: Effects of posture on concentration of blood constituents in healthy adults: Practical application of blood specimen collection procedures recommended by the Scandinavian Committee on Reference Values. Scand. J. Clin. Lab. Invest., *40*:615–621, 1980.
18. Grams, R. G., Johnson, E. A., and Benson, E. S.: Laboratory data analysis system: Section III—Multivariate normality. Am. J. Clin. Pathol., *58*:188–200, 1972.
19. Gräsbeck, R.: Terminology and biological aspects of reference values. *In*: Logic and Economics of Clinical Laboratory Use. E. S. Benson and M. Rubin, Eds. New York, Elsevier, 1978, pp. 77–90.
20. Gräsbeck, R.: Health as seen from the laboratory. *In*: Reference Values in Laboratory Medicine. R. Gräsbeck and T. Alström, Eds. Chichester, England, John Wiley and Sons, Ltd., 1981, pp. 17–24.

21. Gräsbeck, R.: Display of results with reference values. *In*: Reference Values in Laboratory Medicine. R. Gräsbeck and T. Alström, Eds. Chichester, England, John Wiley and Sons, Ltd., 1981, pp. 279–288.

22. Gräsbeck, R., and Alström, T., Eds.: Reference Values in Laboratory Medicine. Chichester, England, John Wiley and Sons, Ltd., 1981.

23. Groth, T., and de Verdier, C.-H.: The potential use of biochemical-physiological simulation models in clinical chemistry. Scand. J. Clin. Lab. Invest., *39*:103–110, 1979.

24. Gullick, H. D., and Schauble, M. K.: SD unit system for standardized reporting and interpretation of laboratory data. Am. J. Clin. Pathol., *57*:517–525, 1972.

25. Hahn, G. J.: Statistical intervals for a normal population, part I. Tables, examples and applications. J. Qual. Techn., *2*:115–125, 1970.

26. Hahn, G. J.: Statistical intervals for a normal population, part II. Formulas, assumptions, some derivations. J. Qual. Techn., *2*:195–206, 1970.

27. Harris, E. K.: Effects of intra- and interindividual variation on the appropriate use of normal ranges. Clin. Chem., *20*:1535–1542, 1974.

28. Harris, E. K.: Some theory of reference values. I. Stratified (categorized) normal ranges and a method for following an individual's clinical laboratory values. Clin. Chem., *21*:1457–1464, 1975.

29. Harris, E. K.: Statistical aspects of reference values in clinical pathology. *In*: Progress in Clinical Pathology, Vol. VII. M. Stefanini and E. S. Benson, Eds. New York, Grune and Stratton, Inc., 1981, pp. 45–66.

30. Harris, E. K., Cooil, B. K., Shakarji, G., et al.: On the use of statistical models of within-person variation in long-term studies of healthy individuals. Clin. Chem., *26*:383–391, 1980.

31. Harris, E. K., and DeMets, D. L.: Estimation of normal ranges and cumulative proportions by transforming observed distributions to Gaussian form. Clin. Chem., *18*:605–612, 1972.

32. Harris, E. K., Kanofsky, P., Shakarji, G., et al.: Biological and analytical components of variation in long-term studies of serum constituents in normal subjects. II. Estimating biological components of variation. Clin. Chem., *16*:1022–1027, 1970.

33. Harris, R. J.: A Primer of Multivariate Statistics. New York, Academic Press, 1975.

34. Hawkins, D. M.: Identification of Outliers. London, Chapman and Hall, 1980.

35. Healy, M. J. R.: Outliers in clinical chemistry quality-control schemes. Clin. Chem., *25*:675–677, 1979.

36. Helwig, J. T., and Council, K. A., Eds.: SAS User's Guide. Cary, N.C., SAS Institute Inc., 1979.

37. Hoffmann, R. G.: Statistics in the practice of medicine. JAMA, *185*:864–873, 1963.

38. IFCC Expert Panel on Theory of Reference Values: The theory of reference values. *Part 1.* The concept of reference values. J. Clin. Chem. Clin. Biochem., *17*:337–339, 1979. *Part 2.* Selection of individuals for the production of reference values. J. Clin. Chem. Clin. Biochem., *22*:203–208, 1984. *Part 3.* Preparation of individuals and execution of blood specimen collection for the production of reference (and observed) values. In preparation. *Part 4.* Control of analytical variation in the production, transfer and application of reference values. In preparation. *Part 5.* Statistical treatment of collected reference values. Determination of reference limits. J. Clin. Chem. Clin. Biochem. *21*:749–760, 1983. *Part 6.* Presentation of observed values related to reference values. J. Clin. Chem. Clin. Biochem. *20*:841–845, 1982.

39. John, J. A., and Draper, N. R.: An alternative family of transformations. Appl. Statist., *29*:190–197, 1980.

40. Manly, B. F. J.: Exponential data transformations. The Statistician, *25*:37–42, 1976.

41. Mardia, K. V.: Tests of univariate and multivariate normality. *In*: Handbook of Statistics, Vol. I, Analysis of Variance. P. R. Krishnaiah, Ed. Amsterdam, North-Holland Publishing Co., 1980, pp. 279–320.

42. Martin, H. F., Gudzinowicz, B. J., and Fanger, H.: Normal Values in Clinical Chemistry. A Guide to Statistical Analysis of Laboratory Data. New York, Marcel Dekker, 1975.

43. Meites, S., and Levis, M. J.: Skin-puncture and blood-collecting techniques for infants. Clin. Chem., *25*:183–189, 1979.

44. Morrison, D. F.: Multivariate Statistical Methods. New York, McGraw-Hill Book Co., 1976.

45. Murphy, E. A.: The normal, and the perils of the sylleptic argument. Persp. Biol. Med., *15*:566–582, 1972.

46. NCCLS: Standard Procedures for the Collection of Diagnostic Blood Specimens by Venipuncture (TSH-3). Villanova, Pa., National Committee for Clinical Laboratory Standards, 1979.

47. NCCLS: Standard Procedures for the Collection of Diagnostic Blood Specimens by Skin Puncture (TSH-4). Villanova, Pa., National Committee for Clinical Laboratory Standards, 1979.

48. Pickup, J. F., Harris, E. K., Kearns, M., et al.: Intra-individual variation of some serum constituents and its relevance to population-based reference ranges. Clin. Chem., *23*:842–850, 1977.

49. Reed, A. H., Henry, R. J., and Mason, W. B.: Influence of statistical method used on the resulting estimate of normal range. Clin. Chem., *17*:275–284, 1971.

50. Rossing, R. G., and Hatcher, W. E.: A computer program for estimation of reference percentile values in laboratory data. Computer Progr. Biomed., *9*:69–74, 1979.

51. Sackett, D. L.: The usefulness of laboratory tests in health-screening programs. Clin. Chem., *19*:366–372, 1973.

52. Snedecor, G. W., and Cochran, W. G.: Statistical Methods. 6th ed. Ames, Iowa State University Press, 1967.

53. Société Francaise de Biologie Clinique, Comité Scientifique, Commission "Valeurs de référence": Documents. Ann. Biol. Clin., *39*:381–384, 1981; *37*:119–124, 1979; *37*:125–126, 1979; *37*:229–239, 1979; *38*:251–265, 1980; *39*:235–244, 1981.

54. Solberg, H. E.: Discriminant analysis. CRC Crit. Rev. Clin. Lab. Sci., *9*:209–242, 1978.

55. Solberg, H. E.: Statistical treatment of collected reference values and determination of reference limits. *In*: Reference Values in Laboratory Medicine. R. Gräsbeck and T. Alström, Eds. Chichester, England, John Wiley and Sons, Ltd., 1981, pp. 193–205.

56. Solberg, H. E.: REFVAL. Technical Report, February 1983. Department of Clinical Chemistry, Rikshospitalet, N-Oslo 1, Norway, 1983. (The technical report, including computer programs, is available on request to the author.)

57. Solberg, H. E.: Statistical treatment of reference values. Bull. Molec. Biol. Med., *8*:13–19, 1983.
58. Solberg, H. E.: Presentation of observed values in relation to reference values. Bull. Molec. Biol. Med., *8*:21–26, 1983.
59. Solberg, H. E.: Inaccuracies in computer calculation of standard deviation. Anal. Chem., *55*:1611–1614, 1983.
60. Statland, B. E., and Winkel, P.: Sources of variation in laboratory measurements. *In*: Clinical Diagnosis and Management by Laboratory Methods. 16th ed. J. B. Henry, Ed. Philadelphia, W. B. Saunders Co., 1979, pp. 3–28.
61. Stephens, M. A.: EDF statistics for goodness of fit and some comparisons. J. Am. Statist. Assoc., *69*:730–737, 1974.
62. Sunderman, F. W.: Current concepts of "normal values," "reference values," and "discrimination values" in clinical chemistry. Clin. Chem., *21*:1873–1877, 1975.
63. Tryding, N., and Lindblad, C.-G.: Drug Effects in Clinical Chemistry. Stockholm, Apoteksbolaget, 1981.
64. Winkel, P.: Patterns and clusters—multivariate approach for interpreting clinical chemistry results. Clin Chem., *19*:1329–1338, 1973.
65. Winkel, P.: The use of the subject as his own referent. *In*: Reference Values in Laboratory Medicine. R. Gräsbeck and T. Alström, Eds. Chichester, England, John Wiley and Sons, Ltd., 1981, pp. 65–78.
66. Winkel, P., Lyngbye, J., and Jörgensen, K.: The normal region—a multivariate problem. Scand. J. Clin. Lab. Invest., *30*:339–344, 1972.
67. Winkel, P., and Statland, B. E.: Reference values. *In*: Clinical Diagnosis and Management by Laboratory Methods, 16th ed. J. B. Henry, Ed. Philadelphia, W. B. Saunders Co., 1979, pp. 29–52.
68. Young, D. S., Pestaner, L. C., and Gibberman, V.: Effects of drugs on clinical laboratory tests. Clin. Chem., *21*: 1D-432D, 1975.

Analytical Goals and Clinical Relevance of Laboratory Procedures

by Robert S. Galen, M.D., M.P.H., and Theodore Peters, Jr., Ph.D.

Clinical laboratory information is used by the physician to answer specific and usually quite limited questions about a patient in order to make medical decisions. Valid medical decisions require valid information. Accuracy, precision, analytical sensitivity, analytical specificity, and factors such as patient preparation and sample handling determine the validity of information provided by the laboratory. To be useful, the information must also be relevant to the clinical problem. This chapter will examine these considerations.

Laboratory information should be distinguished from laboratory data. Numerical figures or qualitative descriptions represent data. To be of optimum use, other data must also be available, such as values to be expected in a healthy individual (reference ranges), precision of quantitative measurements, and values characteristically found in various pathological states. When these data are properly integrated, they constitute laboratory information.

Laboratory information is used in the medical decision making process in the same way other clinical information is used. For example, splenomegaly (enlargement of the spleen) is a significant finding. The normal spleen cannot be palpated (felt) through the abdominal wall, slight enlargement frequently cannot be distinguished from normal on physical examination, and gross enlargement is easily detected; splenomegaly occurs in a large number of disorders ranging from infectious mononucleosis to leukemia. Similarly, a slight abnormality of a quantitative test may be difficult to identify with certainty, while gross abnormalities are easily detected even by an imprecise method, and there are usually multiple potential causes for the abnormality.

What are the questions to be answered with laboratory and other clinical information? Most fall into one of the following categories:

1. Is inapparent disease present (screening)?
2. What is the disease process (pathophysiology)?
3. How can we increase confidence in the tentative diagnosis (confirmation)?
4. How severe is the disease process (prognosis)?
5. Has a change occurred since the last observation (monitoring)?

Note that the same laboratory test may be used for any or all of these purposes under different circumstances. Analytical attributes of a test are critically important in determining whether the test will be useful in answering one or all of the questions; analytical precision will directly affect our ability to distinguish a normal from an abnormal result or to determine whether a day-to-day change in test results is significant. The diagnostic attributes of a test, diagnostic sensitivity and diagnostic specificity (which will be discussed later in this chapter), are important in assessing whether a test is useful, e.g., relevant in screening, differential diagnosis, or confirmation of a tentative diagnosis.

Medical decision making is an iterative process, and it does not consist of merely attaching the proper diagnostic label to a patient. The first steps usually consist of taking a medical history and performing a physical examination. Even these processes are iterative; if a patient complains of chest pain, the physician can immediately ask about the nature of the pain and whether it is precipitated by exercise. Usually a tentative diagnosis, or "clinical impression," can be made after the medical history and physical examination. Laboratory tests and other diagnostic procedures are selected on the basis of that information. An important decision point is reached when the available information is adequate to decide whether or not to treat the patient. Under some circumstances, treatment may be part of the diagnostic process, a "therapeutic trial."

Some diagnostic procedures carry a significant risk to the patient, and the information to be obtained must be sufficiently important in subsequent decisions to warrant the risk involved. An example of such a procedure is needle biopsy of the liver. Although it would provide definitive diagnostic information in acute viral hepatitis, this diagnosis can be made on the basis of clinical and laboratory information with sufficient confidence that the risk of biopsy outweighs the importance of the additional information that might be obtained. While the direct risks involved in performing a chemical analysis are negligible, there is always a possibility that a misleading result will indicate a need for further diagnostic procedures that do involve real risks to the patient. However, even without further risk to the patient, misleading or inaccurate chemical tests usually result in a financial cost to the patient, to the insurance carrier, or to the laboratory.

Disease should be considered a process, not a condition. Diagnostic considerations include the cause (etiology), the destructive and reparative processes involved (pathogenesis), the abnormalities that can be observed to establish presence of the disorder (diagnostic criteria), and the probable outcome (prognosis). With these considerations, it is easily recognized that a clinical situation in which a single test provides a definitive understanding of a disease process is a rarity. These considerations also emphasize the necessity for understanding the chronological component in application of laboratory tests to the study of disease and in using laboratory tests in the diagnostic process. A good example is the use of serum creatine kinase (CK) measurements in the diagnosis of myocardial infarction (heart attack) discussed elsewhere in this book. Unless there is another pathological process present, the serum CK is not increased immediately following a myocardial infarction, and the test is virtually useless in the diagnostic process. However, serum activities become elevated within a few hours, peak at ∼18 h, and return to usual levels within about 3 days. Consideration of this chronological sequence is an essential element in interpretation of CK measurements. Further consideration of the chronological component will disclose that myocardial infarction, in most cases, is the result of coronary atherosclerosis. This condition in many cases is the result of a disorder of lipid metabolism, which in some cases is a genetically determined abnormality. In this context, a heart attack is simply one event in a rather prolonged metabolic disorder that can be defined only by chemical measurements and identification of genetic determinants.

The impressive resources of the clinical laboratory can be applied most effectively in the patient care process if they are directed toward supplying information that cannot be obtained by talking to or examining the patient. The most effective instrument for detecting disease of the heart valves is the stethoscope; the most effective approach to detecting and defining biochemical disorders is appropriate use of quantitative chemical measurements. Graves' disease (a type of hyperthyroidism) and myxedema (a type of hypothyroidism) were accurately identified and described before clinical laboratories existed. But laboratory tests can now detect both of these conditions long before patients develop the distressing and destructive physical signs characteristic of these disorders. It is the role of the clinical chemistry laboratory to detect biochemical disorders in their early, clinically inapparent, and more easily treatable stages. Perhaps a more instructive example is the measurement of serum potassium concentration. There are no physical signs that are definitively characteristic of hyperkalemia or hypokalemia (electrocardiography can be helpful), but either can result in death due to cardiac arrest. Thus, measurement of serum potassium can provide information unobtainable by other means that is essential to selection of proper treatment of the patient.

The following discussion will address analytical requirements for laboratory tests and some approaches to evaluating their clinical relevance.

ANALYTICAL GOALS

Valid data are essential in making medical decisions. The two most important concepts used in judging analytical performance, and thus to judge validity, are:

analytical accuracy: the agreement between the best estimate of a quantity and its "true" value, and

analytical precision: the agreement between replicates.

Two other considerations are:

analytical sensitivity: the ability of an analytical method to detect small quantities of the measured component, and

analytical specificity: the ability of an analytical method to determine solely the component(s) it purports to measure.

These definitions are enlarged upon in Chapter 2D, Evaluation of Methods, which also discusses other means of evaluating performance of methods. The distinction between analytical and diagnostic sensitivity and analytical and diagnostic specificity should be noted carefully, and will be considered further in a later section.

SETTING ANALYTICAL GOALS

Analytical goals should be appropriate for their intended medical use, and hence will vary in different clinical situations as well as among different analytes. Although one intuitively assumes that a laboratory test is usually obtained for the purpose of diagnosing disease, most laboratory tests are in fact obtained to monitor treatment. There is a basic distinction in performance requirements for the two purposes. For monitoring therapy or following the course of disease, the concentration of one or more substances is compared with values obtained previously on the same patient. In this case the day-to-day observations must be reliable, and precision, or reproducibility, is the prime concern. For diagnosis, on the other hand, the concentration of one or more substances is compared with the established reference range or discrimination values to determine if the value for the subject is suggestive of a suspected illness. In a diagnostic application not only precision but also accuracy is important, since comparison must be made with results from other patients, often from many different institutions.

When laboratory analysis is used to screen for disease, an abnormal value may not lead to immediate therapy; it may be instead the signal for further, more extensive testing. Results must be both accurate and precise in order to minimize the expense of the additional tests and the effect on the subject of a false positive finding. At the same time, the chance that the presence of a disease would be missed (false negative result) must be minimized. In a later part of this chapter the implications of false negative and false positive results are more fully described.

Testing to determine *prognosis* is an extension of the diagnosis into the future. Accuracy requirements are not as rigid as for the initial diagnosis, since less is known about the relationship of concentration to prognosis than about reference ranges. For example, whether the serum prostatic acid phosphatase is 5 or 6 U/L would make little difference in predicting the course of a prostatic carcinoma, but an equal error in activity of 1 U/L could be critical in the initial detection of this tumor.

The *assessment of risk factors* is a form of prognosis. An example is the assay of serum HDL-cholesterol to predict the likelihood of atherosclerotic changes and associated coronary heart disease. The information may be sought in order to recommend a change in dietary or living habits, or merely for reassurance of a concerned patient.

In *emergencies* the timeliness of a report may be of such importance that a somewhat lower standard of performance in terms of accuracy and precision is acceptable. Merely knowing whether the blood glucose is high or low would justify initiation of appropriate treatment of a comatose, insulin-dependent, diabetic patient who might be in hypoglycemic coma or in ketoacidosis. Various "quick" tests for blood glucose, amylase, and urea have proved to be useful for certain applications, but have less precision than the conventional methods. The accuracy and precision of these tests must still be known and the analytical runs must be under control. Fortunately, emergency tests can now be performed rapidly and at any time of day on many automated instruments without significant interruption of routine testing.[1] About two-thirds of laboratories in the U.S.A. now use the same instrumentation for stat and routine testing. These test results then carry the same accuracy and precision as a routine test result, a welcome development.

Whether it is feasible or even appropriate to set different analytical goals for all of the above clinical situations has been questioned in a review by Fraser in 1981.[13] This author points out the advantages of delineating a single set of goals, which would be the most stringent of those for different situations, and describes the difficulties in deciding how to define goals. There are few objective studies presently available.

Goals for Analytical Accuracy

Criteria for medical needs in terms of accuracy have seldom been defined. The reasons are several. Laboratories have been in the habit of defining the reference ranges believed applicable to their own locale; there was little movement of patients around the country and consequently little need for common standards for the "true value." Furthermore, on a practical scale, there was, for many analytes, no agreement on the "true value" among laboratories, nor was there agreement on reference methods to obtain such values. It cannot now be maintained that accuracy is less important than precision. Patients are shifting geographical locations frequently, and there is a need for standardization of both analytical results and reference ranges among laboratories. Studies of a patient or of groups of patients often extend over long periods of time, requiring that the laboratory results be comparable from lab to lab.

Since there has been less experience in comparing bias of interlaboratory results than in comparing precision, there are few published analytical goals for accuracy of clinical laboratory tests. Gilbert in 1975 published accuracy goals as percentages, based on his experience with the evaluation of College of American Pathologists (CAP) survey results.[17] These goals are shown as the second column of figures in Table 2C-1. They are stated to be approximations, and to be nearly the same as one coefficient of variation for precision.

Gilbert's approach reflects trust in the mean of values from many participating laboratories as an approach to the establishment of the "true" value for an analyte. This confidence has been strengthened by agreement between such interlaboratory means with values established by definitive methods (Table 2C-1). Here the bias for each of the clinical laboratory values, except that for creatinine, is less than Gilbert's accuracy goal.

Expanding on Gilbert's approach, the College of American Pathologists has combined the assessment of accuracy and precision into a "total error," which is the sum of bias and imprecision as described in Figure 2D-2 (p. 414). This approach has the advantage of presenting clinicians with only a single term for total analytical error and has been suggested as more medically useful.

Since 1978 the College of American Pathologists has employed the total error concept to evaluate survey results for analytes which yield close agreement among participating laboratories. Values for potassium which deviate by more than 0.5 mmol/L from the mean of all participating laboratories, for calcium which deviate by more than 1.0 mg/dL, for sodium by 4 mmol/L, or for chloride by 5 mmol/L are flagged as unsatisfactory.

Table 2C-1. ACCURACY GOALS FOR CLINICAL LABORATORY TESTS AND PERFORMANCE OF CLINICAL LABORATORIES

Substance (in Serum)	Definitive Value*	Accuracy Goal, %[c]	Clinical Laboratory Values	
			Bias, %	CV, %
Calcium[a]	10.0 mg/dL	2.0	−1.1	4.2
Chloride[a]	105 mmol/L	1.0	−0.8	2.4
Iron[a]	80 μg/dL	5.0	−3.5	12.7
Lithium[a]	1.00 mmol/L		+0.2	13.6
Magnesium[a]	2.00 mg/dL	5.0	−0.7	12.9
Potassium[a]	4.0 mmol/L	2.0	+1.0	2.4
Sodium[a]	145 mmol/L	0.7	−0.5	1.5
Cholesterol[b]	173 mg/dL	4.0	+1.1	5.8
Glucose[b]	104 mg/dL	5.0	−1.2	7.0
Urea-nitrogen[b]	17 mg/dL	5.0	+2.8	7.8
Urate[b]	4.8 mg/dL	3.8	−1.8	4.7
Creatinine[b]	0.89 mg/dL	6.7	+17.8	11.2
Cortisol[b]	15.8 μg/dL		+29.1	14.3

*Definitive values on two serum pools were obtained by isotope dilution-mass spectrometry at the U.S. National Bureau of Standards or at the Swedish Karolinska Institute[b] and compared to the mean values found by clinical laboratories on the same pools.
[a]Mean of 1100–4200 U.S. laboratories in CAP survey.[18]
[b]Mean of 18–64 Swedish clinical laboratories.[5]
[c]Gilbert.[17]

Goals for Analytical Precision

Two basic approaches have been used to set analytical goals for precision in terms of medical needs: questionnaires and consensus. Neither of these attempts defines a requirement for accuracy. The first and more widely used approach is to poll a large number of practicing physicians concerning how large a change in the concentration of an analyte would be considered medically significant, i.e., real, and requiring action. For example, would a further workup be indicated if the fasting serum glucose were 145 rather than 140 mg/dL? Precision criteria derived in this fashion are listed in column 3 of Table 2C-2.

Precision, expressed as coefficient of variation, generally varies with the concentration of the analyte. A 5 mg/dL error in glucose measurement would be 10% at a concentration of 50 mg/dL, but <4% at a concentration of 140 mg/dL. Hence goals should be specified at several levels, particularly at the medically important "decision levels" mentioned earlier.

The process of defining precision limits admittedly involves a process of circular reasoning. The true clinical utility of a certain precision level is unknown, and will become known only when clinicians have had experience with that level of precision. To find an approach which avoids such subjectivity, a group of clinical chemists, pathologists, statisticians, and other laboratory scientists met in Aspen, Colorado, in 1976 to discuss ways to set analytical goals. They reached a tentative consensus which was set forth in their report, "Analytical Goals in Clinical Chemistry,"[8] and which has been approved by the Subcommittee on Analytical Goals in Clinical Chemistry of the World Association of Societies of Pathology. The chief recommendations of this report were:

1. Analytical goals can only be defined in terms of the needs for patient care. Any other basis is irrelevant.

2. Goals for precision and accuracy of quantitative methods, where the analyte is well defined, should be at least as stringent as the current performance by well-managed laboratories, or the "state of the art."

3. For group screening, in which an individual is to be selected from a population, a goal for an analytic coefficient of variation (CV_a) is defined as:

$$CV_a \leq \frac{1}{2}\sqrt{(CV_{\text{Intra-individual}})^2 + (CV_{\text{Inter-individual}})^2}$$

4. For testing to evaluate diagnosis or monitor treatment of an individual, a goal for an analytic coefficient of variation is defined as:

$$CV_a \leq \frac{1}{2}CV_{\text{Intra-individual}}$$

5. There is a major need for further study and understanding of the statistical relationships that should be used to derive a clinical diagnosis from an analytical value.

6. When the physiological and pathophysiological mechanisms which produce variations of particular analytes in health are understood, it may be possible to reduce the effects of the biologic variance on total observed variability.

Rather than relying on subjective replies to a questionnaire to set precision goals, the Aspen group took into account the variability of analyte concentrations in healthy individuals (inter-individual or between-person variability). There is no utility in measuring triglycerides in serum to ±1%, for instance, if their concentration in an individual varies 15% or more during a single day or from day to day; thus, the precision criteria for the determination of triglycerides can be less rigorous than those for the determination of sodium or calcium, which are in the body under close homeostatic control and vary only about ±2% in the course of several months.[7] Precision criteria based on intra-individual variations are listed in column 4 of Table 2C-2. These values were calculated as one-half of the measured day-to-day or intra-individual variability.

The most stringent of the precision goals set by clinical surveys (Table 2C-2, column 3) are those for calcium, sodium, chloride, and osmolality (CV's of ≤1.4%). These substances are under tight homeostatic control; note that they show day-to-day variability within the same individual of ≤1.1% (column 4). Day-to-day individual variability is also small for albumin, alkaline phosphatase, cholesterol, creatinine, glucose (fasting), phosphate, potassium, total protein, thyroxine, and uric acid (column 4). A weakness of this approach is that these values are set from observations on healthy individuals; day-to-day variability may be appreciably higher in pathological states.

Table 2C-2. ANALYTICAL PERFORMANCE AND MEDICAL NEEDS FOR FREQUENTLY ANALYZED SERUM CONSTITUENTS

Analyte	Mean Value Found for Analyte	Estimate of Precision Required for Medical Needs (CV, %)		"State-of-the-Art" Analytical Precision (CV, %)	
		Based on Clinical Surveys	Based on Biological Variability	Average Laboratory	Laboratory in 0–10th Percentile
1	2	3	4	5	6
Acid phosphatase	1.2 U/L			9.3	
Alanine aminotransferase	32 U/L	10.3		7.9	
Albumin	3.5 g/dL	3.7	1.4	3.4	5.7
Alkaline phosphatase	280 U/L	4.3	1.8	3.4	
Amylase	117 U/L	6.0		7.6	
Aspartate aminotransferase	35 U/L			7.5	
Bilirubin, direct	0.3 mg/dL			22.7	
	1.0	10.5		9.8	15.4
Bilirubin, total	1.0 mg/dL			8.3	17.9
Calcium, total	11.0 mg/dL	0.8	0.9	2.0	3.5
Carbon dioxide, total	20 mmol/L	9.0		4.9	
Chloride	90 mmol/L	1.3	1.1	1.5	2.8
Cholesterol	250 mg/dL	2.8	2.4	3.2	6.5
Cortisol	20 µg/dL		13.0	10.1*	16.1*
Creatine kinase	100 U/L	5.1	13.0	9.0	
Creatinine	1.5 mg/dL	15.4	2.2	5.6	10.0
Digoxin	2.5 ng/mL			7.0	10.4
Fibrinogen	200 mg/dL			6.0	10.0
Glucose	50 mg/dL	3.2		3.9	6.0
	120	2.7	2.2	2.8	4.9
γ-Glutamyl transferase	30 U/L			5.0	
Iron	80 µg/dL		13.0	4.0	8.6
Iron binding capacity, total	250 µg/dL			6.2*	10.4*
Lactate	1.5 mmol/L			11.6	
Lactate dehydrogenase	155 U/L	7.9		3.5	
Lipase	10.1 U/L			27.5	
Lithium	1.5 mmol/L			3.8	6.9
Magnesium	1.5 mg/dL			9.0	14.8
Osmolality, total	280 mOsmol/kg	1.3		1.8*	3.6*
pCO_2	32 mm Hg	4.4		5.0	
pO_2	75 mm Hg	4.8		7.2	
pH	7.30	2.8		2.7	
Phosphate-P	4.5 mg/dL	4.0	2.9	3.0	5.3
Potassium	3.0 mmol/L	3.7	2.2	2.4	4.0
Protein, total	6.0 g/dL	2.9	1.5	2.2	3.7
Salicylate	15.0 mg/dL			3.5	5.4
Sodium	130 mmol/L	1.4	0.4	1.1	1.8
Thyroxine	12 µg/dL	3.7	3.8	7.2	10.1
Triglycerides	130 mg/dL	10.5	13.0	5.2	9.2
Urea nitrogen	27 mg/dL	15.0	6.2	3.5	6.1
Uric acid	6.0 mg/dL	3.1	3.7	3.0	5.3

*Manual methods.

Estimates of medical needs were obtained either from questionnaires received from clinicians (column 3)[2,9] or from the biological variability of the concentration of an analyte as evaluated at a conference on analytical goals held in Aspen, Colorado, in 1976 (column 4).[8,17]

State-of-the-art figures (columns 5 and 6) were obtained from cumulative results of large-scale survey and quality control programs.[20,22,23,30]

When the Aspen group considered precision goals for screening populations, where the between-day variability among subjects is much larger than is the variability among individuals, its approach resulted in unusually high allowable imprecisions: a CV of 37% for triglycerides and a CV of 26% for creatine kinase. For such analytes the group recommended that the analytical goals should be at least as stringent as the current performance of laboratories, or the state of the art, which is considered later in this section.

ANALYTICAL PERFORMANCE COMPARED WITH ANALYTICAL GOALS

Accuracy. The lack of objective data for goals for analytical accuracy and the paucity of defined standard materials have prompted most organizations that assess performance to include bias together with imprecision as total error. Thus bias is not identified, but is considered to be less than the allowable total error.

With the development of *definitive methods* for certain analytes in serum specimens, however, it has been possible to assess accuracy of performance of clinical laboratories for these analytes by applying the definitive methods to serum pools used in nationwide survey programs. The definitive methods usually employ isotope dilution–mass spectrometry, and furnish a value that is as close to a "true value" for the analyte in the serum matrix as is obtainable with current technology. Two such studies are shown in Table 2C-1. Gilbert compared the mean of results obtained by 4200 CAP Survey participants with definitive values for 7 analytes assigned by the U.S. National Bureau of Standards,[18] and Bjorkhem et al.[5] compared the findings of 64 Swedish laboratories with values assigned for 6 other analytes by the Karolinska Institute.

Accuracy performance as judged by the clinical laboratory values for "Bias" in Table 2C-1 was within 2% for 9 of the 13 substances tested at the levels shown. For urea and iron it was about 3%. At the extremes of the ranges of concentration tested (data not shown), the bias increased to about 4% for most analytes.

The larger error for iron is probably caused by a fundamental difference in methodology, since the National Bureau of Standards' method measures total iron after digestion and therefore includes heme iron. The error for creatinine of +17.8% is probably due to fundamental problems with interfering substances in the Jaffe reaction used in many laboratories, as well as to the difficulties of measuring creatinine precisely at concentrations near the decision point of 1 mg/dL. The high bias and CV for cortisol reflect problems in measuring such materials in low concentrations.

The bias in the averaged results of surveys for many frequently assayed substances is seen to be small—less than the CV of the assay (Table 2C-1). This also indicates that the mean value generated in a large number of participating laboratories for a common pool is a reasonable estimate of the true value for these substances. This conclusion cannot be applied as yet to other analytes, particularly enzymes, for which reaction conditions vary among laboratories, nor can it be extended to specimens containing interfering substances.

There are obvious weaknesses of the survey approach to the determination of analytical accuracy. A more serious drawback is that the encouragingly low figures for bias shown by survey results do not necessarily apply to the performance of individual laboratories; the dispersion of individual results can be quite large. Laboratories should determine their own bias from their survey results.

Precision. More quantitative data are on hand to assess precision than accuracy. The large quality assurance programs sponsored by commercial firms or by the College of American Pathologists provide an assessment for participating laboratories in the U.S.A. Columns 5 and 6 of Table 2C-2 show recent findings from three such sources, mainly from the CAP Quality Assurance Service report in 1982.

The CV's given in Table 2C-2, columns 5 and 6, are a measure of day-to-day precision for individual laboratories, a summation of over 250 daily determinations of two control preparations by about 1800 laboratories. Most of the data are for automated procedures, which in the U.S.A. now outnumber manual tests by about three to one. The CV's for manual tests, indicated by an asterisk, are about one-third higher than those for automated tests, but for some there is no difference. The reported CV's are for several operators for the same type of instrument or method.

The precision for the average participating laboratory is given in column 5 of Table 2C-2. In order to show the spread of performance about this average, results for laboratories with the poorest performance in terms of precision (i.e., the 0–10th percentile) are listed in column 6 of the table. These laboratories show CV's which are 1.5–2 times those of average laboratories (column 5). For laboratories in the top 10% in performance (data not shown) the CV's are one-half to two-thirds of the CV's of the average laboratory.

For direct bilirubin, two concentrations are listed in Table 2C-2. At the lower value, which is near the medical decision range for direct bilirubin, the distinction between relative and absolute variability is important. The relative precision of 22.7% at the direct bilirubin value of

0.3 mg/dL is an absolute variability (SD) of only $0.227 \times 0.3 = 0.07$ mg/dL; the SD at the value of 1.0 mg/dL is 9.8%, which is $0.098 \times 1.0 = 0.10$ mg/dL. Hence the analyte is actually measured more accurately at the lower value, but the relative error is greater.

When the performance of the average laboratory in terms of precision (column 5, Table 2C-2) is measured against the goals set by surveys of physicians (column 3), performance is seen to meet or exceed the goal for 19 of the 25 analytes listed. Only for amylase, calcium, cholesterol, creatine kinase, pO_2, and thyroxine does the measured precision fall short of the estimated needs. By their very nature, however, the analytical goals set by surveys of physicians (column 3) may be overly permissive.[29] The allowable CV for creatinine at 1.5 mg/dL of 15.4% is a good example. At this concentration a change of 15.4%, or only 0.23 mg/dL, means a change of 15.4% in creatinine *clearance*, the usual index of renal function. For a normal female patient, with serum creatinine of 0.92 mg/dL, a similar error of 0.23 mg/dL would mean a 25% change in renal function. Surely a change of this magnitude is of concern to nephrologists, and this consideration shows that there is fallacious or weak logic in setting analytical goals by clinical surveys before the benefits of using more precise results have been realized.

Comparison of the data of column 5 (Table 2C-2) with goals established by the Aspen Conference on the basis of intra-individual day-to-day variability (column 4) shows that laboratory performance falls short of the goal not only for calcium, cholesterol, and thyroxine but also for albumin, alkaline phosphatase, chloride, creatinine, glucose, protein, and sodium. This means that improving precision performance for these analytes is a worthwhile endeavor, since the laboratory is not yet able to measure them, on the average, with the same precision as physiological control mechanisms can maintain their concentrations in the blood.

Although the precision of calcium measurements has improved, calcium remains the analyte which, according to both criteria of Table 2C-2, is in greatest need of improvement. This is a reflection of its biological control and its importance in detecting hyperparathyroidism. Creatine kinase, on the other hand, is measured with a precision of 9.0% (column 5), better than its day-to-day variation of 13.0% (column 4), while its medical need based on surveys of 5.1% (column 3) appears to have been set unrealistically low.

The average laboratory in 1980–1982 (column 5) is seen to have met, or nearly met, the precision recommended by physicians for most analytes listed in column 3. The lowest 10% of laboratories (column 6), however, met the goals only for creatinine and triglycerides. When judged by the ability to detect day-to-day variations in an individual, the average laboratory performed satisfactorily only for cortisol and iron (which have large diurnal changes), creatine kinase, triglycerides, urea, and uric acid, whereas the laboratories below the 10th percentile met the requirement only for iron, triglycerides, and urea. There is obviously need for improvement.

Note also that the comparisons in Table 2C-1 and Table 2C-2 have been made using pools of control serum. Major errors could still affect precision if tests were carried out with actual specimens and if these contained interfering substances such as hemoglobin, bilirubin, or drugs; if there were specimen carryover in the course of an automated batch procedure; if a clerical mistake were made; or if any other variables, such as technologist error, were introduced.

The trend toward improved precision over the last decade is encouraging.[30] CV's for determination of glucose, calcium, cholesterol, electrolytes, urea, uric acid, and protein have fallen from about 8 to 3%. Even CV's for assays of enzymes have improved markedly to the 4–10% range. The improvement springs both from the increasing use of automated equipment and from the increased effectiveness and growth of quality control programs.

CLINICAL RELEVANCE OF LABORATORY PROCEDURES

To be relevant in the patient care process, a laboratory test must provide useful information that cannot be obtained by other means or must provide similar information at lower cost or at less risk to the patient. Many laboratory tests now performed probably do not meet any of these criteria and should be considered redundant. Generally speaking, a relevant laboratory test is one that detects or confirms a disorder in circumstances when other approaches are ineffective, defines a pathological process, or quantitates an abnormality already identified.

Although it has been studied extensively, the exact process used by a physician in arriving at a diagnosis has eluded explanation. If the process could be defined precisely, there is little doubt that modern computers, with their capability for accessing enormous databases, could perform this task more effectively. This is partly due to our unsystematic, perhaps illogical, classification of disease, a classification which changes as new concepts and facts are discovered.

Some diagnostic terms are purely descriptive; for example, hyperthyroidism, hypokalemia, and chronic alcoholism. Other terms refer to a cluster of findings, not all of which must be present to identify the disorder. Clearly, many laboratory tests identify conditions characterized by descriptive terms such as hypercalcemia, acidosis, hypoxia, and hypercapnia. However, many disorders are defined in terms of morphological (anatomical) changes, and laboratory tests are used to identify these conditions in order to avoid other diagnostic procedures that may be difficult or impossible to perform, or that pose significant risk to the patient. An extreme, but illustrative, example is the use of serum enzyme measurements in the diagnosis of myocardial infarction. An infarct is a morphological abnormality: necrotic myocardial tissue. However, it is not feasible to examine microscopically myocardial tissue from every patient with chest pain to identify those who have infarcts. Instead, it is necessary to utilize information from the medical history and physical examination, electrocardiography, and serum enzyme and isoenzyme measurements for this purpose.

To be useful, i.e., relevant, in the diagnostic process, a laboratory test abnormality must have a high association with the disorder in question and should be infrequently encountered in other conditions. A variety of statistical methods can assess relevancy in these terms. The most widely used is the predictive value model. It should be emphasized that while this type of analysis is valuable, even essential, in evaluating the clinical usefulness of a laboratory test, the techniques are of limited applicability in the diagnostic process because they only give the probability that a patient has the disease under consideration.

THE PREDICTIVE VALUE OF LABORATORY TESTS[16,31,32]

The diagnostic performance of a laboratory test can be defined by its diagnostic sensitivity, diagnostic specificity, predictive value, and efficiency.

Diagnostic sensitivity is a measure of the frequency of a positive test when a particular disease is present. *Diagnostic specificity* is a measure of frequency of a negative test in the absence of a particular disease; both are expressed in percent. A perfect laboratory test would have 100% sensitivity and 100% specificity, and an abnormal result would unfailingly identify those with the disease in question. Such tests, except those used in a purely descriptive situation, do not exist.

When a test is applied to a population, not all of whom have the disorder being considered, some patients having the disorder will have negative tests (false negatives) and some not afflicted with the disorder will have positive tests (false positives). The probabilities of these outcomes can be calculated for any defined population and are expressed as predictive values. The *predictive value* of a positive result is the percentage of all positive results that are true positives, or it is the frequency of the disease in question in all patients with positive test results. The predictive value of a negative result is the percentage of all negative results that are true negatives, or it is the frequency of patients not having the disease in question in all patients with negative test results. The *efficiency* of a test indicates the percentage of patients who are correctly classified by the test result as having the disease or not having the disease in question. The relationships of these terms are shown in detail in Table 2C-3.

In most clinical applications, we are interested primarily in the predictive value of a positive result. The predictive value of a positive test identifies the percentage of patients who will be presumptive positives for a particular disease and who, after detailed diagnostic work-ups, turn out to have the disease. Those patients who turn out not to have the disease will be defined as false positives. In the discussion that follows, unless otherwise indicated, predictive value will be used to refer to the predictive value of the positive test result.

The Effect of Disease Frequency

The *incidence* of a disease is the number of cases of the disease which arise during a specified period of time (e.g., number of patients per 100 000 population who develop the disease in a given year). The *prevalence* is the frequency of the disease at a designated point in time. In contrast to incidence rates, which measure events, prevalence measures what prevails or exists.

Table 2C-3. PREDICTIVE VALUE OF TEST APPLIED TO HEALTHY AND DISEASED POPULATIONS[14,16]

	Number with Positive Test Result	Number with Negative Test Result	Totals
Number with disease	TP	FN	TP + FN
Number without disease	FP	TN	FP + TN
Totals	TP + FP	FN + TN	TP + FP + TN + FN

Definitions:

TP = true positives: number of diseased patients correctly classified by the test.

FP = false positives: number of nondiseased patients misclassified by the test.

FN = false negatives: number of diseased patients misclassified by the test.

TN = true negatives: number of nondiseased patients correctly classified by the test.

Sensitivity = positivity in disease, expressed as per cent $= \dfrac{TP}{TP + FN} \times 100$.

Specificity = negativity in health, or absence of a particular disease, expressed as per cent $= \dfrac{TN}{FP + TN} \times 100$.

Predictive value of positive test = per cent of patients with positive test results that are diseased $= \dfrac{TP}{TP + FP} \times 100$.

Predictive value of negative test

= per cent of patients with negative test results that are nondiseased $= \dfrac{TN}{TN + FN} \times 100$.

Efficiency of test

= per cent of patients correctly classified as diseased and nondiseased $= \dfrac{TP + TN}{TP + FP + FN + TN} \times 100$.

The particular application determines whether prevalence or incidence rates are more appropriate for calculating the predictive value of a test. In the discussion that follows, prevalence will be used to refer to the frequency of disease in a population. It should be noted that the relationships described for disease prevalence would be the same for disease incidence (the frequency of disease during a specified time period).

A marked change in predictive value occurs when there is a change in the prevalence of the disease in the population under study. For example, for a hypothetical test, we might know from the value for sensitivity and specificity that the test was positive in 95% of diseased patients and negative in 95% of nondiseased patients, i.e., it has a sensitivity of 95% and a specificity of 95%. Because of the relationship between these two parameters and the prevalence of disease, predictive values will vary according to how many persons in the population being studied have the disease. In other words, the number of false positives and false negatives will vary between populations having differing prevalence of the disease. Table 2C-4 demonstrates the change in predictive value that occurs for this test with changing prevalence of disease. A particular test will have a higher predictive value when it is applied to a population with a higher prevalence of the disease being studied. This explains why a good diagnostic test frequently fails as a screening test when applied to a population in which the prevalence of a disease is very low. Usually laboratory tests are obtained on the basis of previous clinical information about the patient, which has the effect of selecting a population for testing that has a higher prevalence of the disorder in question. For example, the prevalence of liver disease in a clinic population might be 1%, and even without a laboratory test (random guessing) the diagnosis would be correct 1% of the time. If a test with 95% sensitivity and 95% specificity, as described above, were done on all patients, the probability of a patient with a positive test having liver disease (the predictive value of a positive test) would be 16.1%. However, if within this clinic population one fourth of patients who admit to excess ethanol consumption have liver disease, the predictive value of a positive test within this selected group of heavy drinkers increases to 86.4%. (See Table 2C-4.)

Another example: if the hypothetical test were a serum enzyme determination used in the diagnosis of myocardial infarction, we would know from its sensitivity and specificity that it was positive in 95% of patients with a myocardial infarction (MI) and negative in 95% of non-MI patients. We can then compare the predictive value of this enzyme test in patients in the Outpatient Clinic and the Coronary Care Unit (CCU). Let us assume that the prevalence of MI in the clinic in all patients with "chest pain" was 5%, whereas in the CCU it was 50% in patients admitted to rule out MI. When used in the clinic, the enzyme test would have a predictive value of 50%,

Table 2C-4. PREDICTIVE VALUE AS A FUNCTION OF
DISEASE PREVALENCE[14] FOR A LABORATORY TEST
WITH 95% SENSITIVITY AND 95% SPECIFICITY

Prevalence of Disease (%)	Predictive Value (%)
1	16.1
2	27.9
5	50.0
10	67.9
15	77.0
20	82.6
25	86.4
50	95.0

i.e., 50% of the patients with positive results would have MI's, the remaining 50% would not and would therefore be false positives. This same enzyme test when used in the CCU would have a predictive value of 95%, i.e., 95% of the patients with positive results would have MI, with only 5% of positive results being false positives. (See Table 2C-5.)

The Effect of Referent Value

While an abnormal laboratory test result can be defined as one falling outside the limits of a predefined reference range, this distinction may not be the most effective for making medical decisions since it may produce an unacceptably high frequency of false positive or false negative results. By using an empirically selected referent value or "cut-off point" rather than a reference range for a normal population, it is possible to eliminate or decrease false positive tests (at the cost of more false negative tests) or to eliminate or decrease false negative results (at the cost of more false positive results).

Figure 2C-1 illustrates how sensitivity and specificity are altered by the selection of a referent value for a particular test. If we studied patients admitted to the coronary care unit with the possible diagnosis of "myocardial infarction," divided the population into those with MI and those without MI based on history and electrocardiographic (ECG) findings, and then expressed the enzyme results for each group as a frequency distribution, we would find the classical overlapping distributions seen in Figure 2C-1. It is impossible to select a referent value for enzyme activity that would afford complete discrimination between the two groups. At any cut-off point, one must sacrifice sensitivity for specificity or vice versa. Herein lies the major flaw of diagnosis from laboratory tests, in that tests are not sensitive *and* specific at the same time. Use of multiple laboratory tests improves the predictive value somewhat, but the tradeoff between sensitivity and specificity always remains. If we wanted to select a cut-off point for the test illustrated in Figure 2C-1, we would choose any point in the overlapping region. Several referent values are listed in Table 2C-6 with the sensitivity, specificity, and predictive value associated with each. For any laboratory test studied in any disease, the predictive value of the test in diagnosing the disease can be calculated at any referent value.

Table 2C-5. PREDICTIVE VALUE OF AN ENZYME ASSAY IN HIGH- AND LOW-PREVALENCE
SITUATIONS (MYOCARDIAL INFARCTION, MI)[14]

	Low Prevalence				High Prevalence		
Test	+	−	*Total*	*Test*	+	−	*Total*
MI	48	2	50	MI	475	25	500
Non-MI	47	903	950	Non-MI	25	475	500
Total	95	905	1000	Total	500	500	1000

Clinic: prevalence of MI = 50/1000 = 5%

Sensitivity = 48/50 = 96%
Specificity = 903/950 = 95%
Predictive value = 50%

Coronary Care Unit: prevalence of MI = 500/1000
= 50%

Sensitivity = 475/500 = 95%
Specificity = 475/500 = 95%
Predictive value = 475/500 = 95%

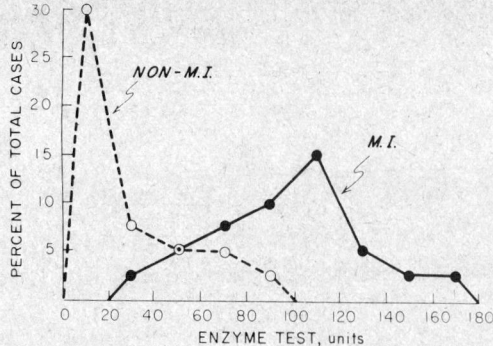

Figure 2C-1. Selection of a referent value based on the sensitivity and specificity of a laboratory test.[14] Prevalence of myocardial infarction = 50%. Mean enzyme activity, MI = 97 U/L. Mean enzyme activity, non-MI = 27 U/L. Upper limit of normal = 20 U/L, sensitivity = 100%, specificity = 60%. Referent value = 100 U/L, sensitivity = 50%, specificity = 100%. (From Galen, R. S.: Orthop. Clin. North Am., *10:*293, 1979.)

The relationships described above can be expressed algebraically by letting the disease prevalence equal p, the test sensitivity equal a, and the test specificity equal b. The predictive value is shown calculated in Table 2C-7.

$$\text{Predictive value} = \frac{pa}{pa + (1 - p)(1 - b)}$$

Converting back to our starting terminology,

$$\text{Predictive value} = \frac{(\text{prevalence})(\text{sensitivity})}{(\text{prevalence})(\text{sensitivity}) + (1 - \text{prevalence})(1 - \text{specificity})}$$

This formula is frequently referred to as Bayes' formula and was derived from Bayes' theorem.[3] Before a test result for a random patient is evaluated, it is known from the figure for prevalence that the patient has a certain probability of having the disease in question, the a priori probability. After the test result is evaluated, the probability of the patient's having the disease is known as the a posteriori probability, which is the predictive value of the positive test result. If the laboratory test is useful, the predictive value will be greater than the disease prevalence.

Application of the Model—Single Test

In 1977, Foti and co-workers[11] reported on a radioimmunoassay (RIA) procedure for prostatic acid phosphatase (PAP). Foti's findings are summarized in Table 2C-8. (Stages I through IV describe the extent of spread of the tumor.) Foti's study, and an editorial by Gittes[19] that followed, suggested acid phosphatase be used not only as a tumor marker but as a cancer screening test.

Table 2C-6. REFERENT VALUES AND ASSOCIATED PREDICTIVE VALUES IN ENZYME TESTING[14]

Referent Value(U/L)	Sensitivity (%)	Specificity (%)	Predictive Value of a Positive Test (%)	Predictive Value of a Negative Test (%)	Efficiency (%)
20	100.0	60.0	71.4	100.0	80.0
40	95.0	75.0	79.2	93.8	85.0
50	90.0	80.0	81.8	88.9	85.0
60	85.0	85.0	85.0	85.0	85.0
80	70.0	95.0	93.3	76.0	82.5
100	50.0	100.0	100.0	66.7	75.0
Chance test*	50.0	50.0	50.0	50.0	50.0
Perfect test*	100.0	100.0	100.0	100.0	100.0

Disease prevalence = 50%.
*Hypothetical situation.
Referent values and associated predictive values and efficiencies for data illustrated in Figure 2C-1.

Table 2C-7. ALGEBRAIC EXPRESSION OF PREDICTIVE VALUE TABLE[14]

	Number with Positive Test Result	Number with Negative Test Result	Totals
Number with disease	pa	p(1 − a)	p
Number without disease	(1 − p)(1 − b)	(1 − p)b	(1 − p)
Totals	pa + (1 − p)(1 − b)	p(1 − a) + (1 − p)b	1

Sensitivity = a
Specificity = b
Prevalence = p

Predictive value of positive = $\dfrac{pa}{pa + (1 - p)(1 - b)}$

Foti and co-workers determined the sensitivity of the radioimmunoassay of PAP (based on current clinical staging criteria) to be 33, 79, 71, and 92% in Stages I to IV, respectively. The specificity of radioimmunoassay of PAP has been measured in specific subgroups, but is not available for the at-risk population as a whole. However, on the basis of specificity of 100% in normal subjects, 94% in patients with benign prostatic hypertrophy, and 89% in patients with other carcinomas, a reasonable estimate of specificity would be 95%. The third factor, prevalence, is a function of age and the method of diagnosis. An overall estimate of prevalence of 25% for men over age 60 seems more than reasonable.

Finally, the cases must be staged. Clinical data, probably biased toward advanced stages, combined with autopsy data, were used to estimate that 15, 25, 35, and 25% of the patients with carcinomas would fall into stages I to IV, respectively.[10] Data showing how this test can be evaluated using the predictive value model are presented in Table 2C-9. On the basis of these estimates, the predictive value of the radioimmunoassay for PAP can be calculated to be 83% with an overall sensitivity of 73%.

These calculations are representative of an unscreened population. If the population had been previously screened for prostatic carcinoma, the test would be detecting the incident, rather than the prevalent, cases. An incidence rate of 5% per year (5 to 10 times the maximal reported clinical incidence rate) is more than adequate to explain the increase in prevalence with age. This rate yields a predictive value of a positive test of 43%. Therefore, follow-up studies (perhaps needle biopsy of the prostate) on 57% of the patients would be done unnecessarily. With this frequency of false positives, it would be neither medically prudent nor cost-effective to use this test for screening purposes. With a lower incidence, the test would perform even worse. With a 1% incidence, the predictive value would be only 13%.

These considerations demonstrate the importance of evaluating laboratory tests in terms of sensitivity, specificity, and predictive values in the population in which they will be used.

Table 2C-8. SENSITIVITY AND SPECIFICITY OF RIA TEST FOR PROSTATIC ACID PHOSPHATASE*

Group	Number of Patients	Sensitivity (%)
Patients with prostate cancer	113	70
Stage I	24	33
Stage II	33	79
Stage II	31	71
Stage IV	25	92

Group	Number of Patients	Specificity (%)
Patients without prostate cancer	217	94
Normal controls	50	100
Benign prostatic hyperplasia	36	94
Total prostatectomy	28	96
Other cancers	83	89
Gastrointestinal disorders	20	95

*Data summarized from Table 2, Foti et al.[11]

Table 2C-9. TESTING FOR CARCINOMA OF THE PROSTATE: DISTRIBUTION OF RESULTS IN A HYPOTHETICAL UNSCREENED POPULATION, n = 10 000; PREVALENCE = 25%; SPECIFICITY = 95%[10]

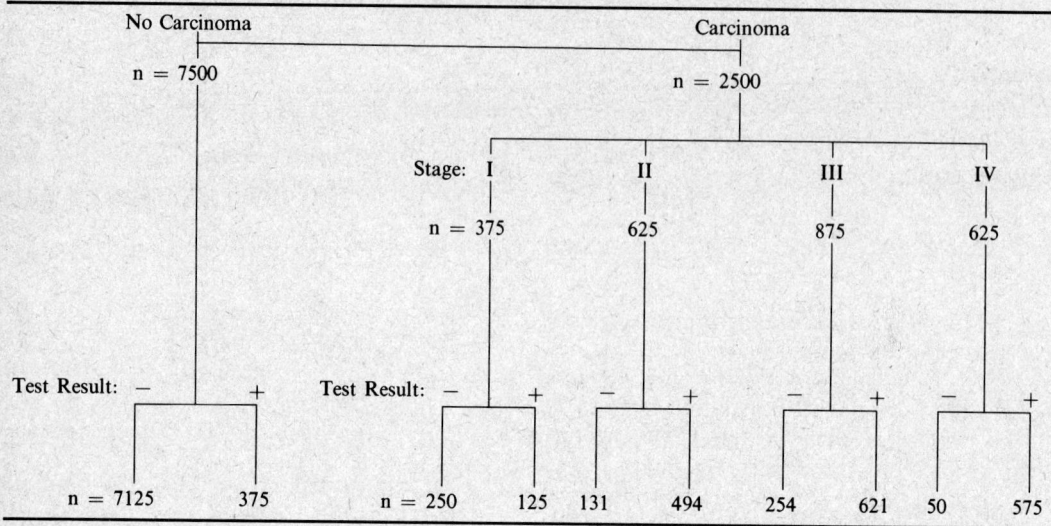

Predictive Value of Laboratory Tests—Combination Testing

So far we have discussed the predictive value of a single test result. But in fact it is rare to use the result of a single test as the final arbiter of a medical decision. Therefore, let us explore some actual examples of multiple test situations to illustrate the benefit of using more than one test to resolve a clinical problem.

Two independent tests in a screening or diagnostic situation can be used in three different ways:

1. Test A is applied first and all those with a positive result are retested with test B. (Series approach.)

2. Test B is applied first and all those with a positive result are retested with test A. (Series approach.)

3. Tests A and B can be used together and all those with positive results for either or both tests are considered to be positives. (Parallel approach.)

Which approach or sequence is best depends on the testing situation and the sensitivity and specificity of the individual tests and their combinations. With parallel testing, the combined sensitivity is greater than the individual sensitivities of the contributing tests.

While parallel testing results in the highest sensitivity but the lowest specificity, series testing results in the lowest sensitivity but highest specificity.

The *series approach* can be taken as either A then B or B then A. If two tests are going to be used in series, the optimal sequence can be determined using the predictive value model. If *parallel testing* is used, tests A and B are applied simultaneously. If either test A or B or both are positive, the result of the test is considered to be positive; a negative test result requires that A *and* B be negative.

The parallel approach has the highest sensitivity. It detects the greatest number of diseased patients in the population. However, it also produces the greatest number of false positives. Of all the approaches, the parallel approach requires the most laboratory work since both tests are performed on all patients in the population.

Application of the Model—Multiple Tests

The above relationship can be demonstrated with actual data from a study of haptoglobin, total LDH, and LDH isoenzymes in the diagnosis of hemolytic disorders.[24] One hundred hospitalized patients were evaluated based on the physician's request for a serum haptoglobin. The laboratory then performed LDH and LDH isoenzyme determinations. Without knowledge of the haptoglobin, LDH, or LDH isoenzyme data, a hematologist classified these 100 cases into two

Table 2C-10. HEMOLYTIC DISEASE STUDY[24]

Test	Sensitivity (%)	Specificity (%)	Predictive Value (%)	Efficiency (%)
Haptoglobin	83	96	87	93
Lactate dehydrogenase	83	61	40	66
Lactate dehydrogenase isoenzymes	58	93	74	85
Haptoglobin *and* LDH isoenzymes*	50	100	100	88
Haptoglobin *or* LDH isoenzymes†	92	89	73	90

*Series interpretation.
†Parallel interpretation.

major groups: hemolytic disease and other (nonhemolytic disease). Twenty-four patients were classified as having hemolytic disease, including autoimmune hemolytic anemia (Coombs' positive), autoimmune hemolytic anemia (Coombs' negative), pernicious anemia, mechanical hemolytic anemia, and hypersplenism. The remainder, or 76 patients, were classified as having other diseases including "nonhemolytic" hematologic disorders such as iron deficiency anemia and bleeding. This remainder represented a broad spectrum of disease states. Table 2C-10 summarizes the sensitivity, specificity, predictive value, and efficiency of these procedures used as single tests and in combination with each other. Haptoglobin was determined nephelometrically and considered positive for hemolytic disease at a concentration of ≤ 25 mg/dL (3 μmol/L). LDH activity was considered abnormal at ≥ 250 U/L. LDH isoenzymes were separated on cellulose acetate and scanned fluorometrically. LDH isoenzymes were considered positive when LDH_1 activity exceeded LDH_2 activity (a "flipped" LDH pattern).

The highest efficiency (overall correct classification) is achieved by simply using haptoglobin alone. The predictive value for haptoglobin is presented in Table 2C-11. It is interesting to explore the four cases of hemolytic disease that were "false negatives," that is, patients without depressed haptoglobin levels. These turned out to be two cases of hypersplenism and two cases of pernicious anemia. On the other hand, there were three "false positives," that is, nonhemolytic patients with depressed haptoglobin levels. These turned out to be cases of ovarian carcinoma, cirrhosis, and sarcoidosis.

The contrast between series and parallel testing is demonstrated quite nicely by the combined use of haptoglobin and LDH isoenzymes. While the efficiency is essentially the same, a series approach requiring both tests to be positive has a predictive value of 100%. The sensitivity, however, is only 50% and half of the cases would go undetected. A parallel approach requiring either test to be positive actually has the highest sensitivity of the tests listed (Table 2C-10).

The effect of increasing specificity with the series approach can be demonstrated with actual data from a study of enzymes and isoenzymes in the diagnosis of acute myocardial infarction.[14] Using the final clinical diagnosis as the ultimate criterion, the sensitivity and specificity of single and multiple enzyme determinations were calculated. Positive results were defined for creatine kinase and aspartate aminotransferase on day one, and for lactate dehydrogenase on day two, following chest pain; the laboratory's usual upper limit was used for each test. The results are summarized in Table 2C-12.

It can be seen that interpreting results in a series fashion always decreases sensitivity and increases specificity relative to single tests or smaller test combinations. The overall effect of the

Table 2C-11. HEMOLYTIC DISEASE STUDY[24]

	Haptoglobin Results		
	≤ 25 *mg/dL*	> 25 *mg/dL*	Total
Hemolytic disease	20	4	24
Other	3	73	76
Totals	23	77	100

*Sensitivity = 83%; specificity = 96%; predictive value = 87%; efficiency = 93%.

series approach is that more cases will be missed, but there will be a higher predictive value of the positive result, i.e., fewer false positive results.

Biochemical profiles or batteries of laboratory tests interpreted in a parallel fashion have extremely high sensitivity, but low specificity. Because of the high sensitivity, these profiles, when negative, have a very high negative predictive value. For that reason, they are used by clinicians to exclude or rule out a variety of diagnoses with a high degree of certainty.

Later on in the diagnostic work-up, series testing is performed. This is highly predictive of the presence of a particular disease because of its high specificity. As the clinician follows his own algorithm for working up a patient, the probability of a diagnosis continues to increase as he receives more information.

It is essential that laboratorians and clinicians be able to use the predictive value model for evaluating the diagnostic accuracy of laboratory tests as well as other procedures. This will facilitate rational test selection and anticipation of outcomes. Cut-off points (referent values) for laboratory tests can be selected to regulate the frequency of false positive tests to keep within medically or economically derived guidelines. Furthermore, multiple laboratory tests for a single disease can be evaluated and one can decide whether adding a test to a profile or series increases the predictive value enough to justify the extra cost.

There are several caveats in using the predictive value model, and in reading journal articles in which it is used, that should be kept in mind. To calculate the sensitivity and specificity of a diagnostic test it is necessary to perform the test on two (preferably large) populations. All members of one population should unequivocally have the disorder being tested for, and all members of the other should unequivocally be free of the disease, but can have other diseases. Sometimes, however, it is difficult to be certain that a patient does or does not have the disorder, and classification may be uncertain. If these indeterminate patients make up a sizable fraction of the patients being tested, figures for sensitivity and specificity will be affected by how they are ultimately classified. Needless to say, the results of the test being evaluated cannot be used to decide into which group a subject should be placed. In order to avoid these difficulties, some investigators have selected patients with far advanced disease for one group and completely healthy young subjects for the other. Not surprisingly, figures for sensitivity and specificity are much higher than in most clinical situations where it is necessary to distinguish a disease in its early stages from an unknown number of other diseases, some of which also affect the test being used. Ransohoff and Feinstein[28] have pointed out that this type of evaluation may be a reason for enthusiastic initial articles describing a diagnostic test which subsequently is shown to produce many false positive and false negative results.

Another problem with the predictive value model when applied to medical diagnosis is that it requires knowledge of the prevalence of the disorder in the population that includes the patient being tested. This information is rarely available. While prevalence can be estimated, this approach adds an element of uncertainty to the figure indicating probability that the patient with a positive test does have the disease.

APPROPRIATE USE OF THE LABORATORY

The function of a well-run laboratory extends beyond producing reliable results on a given specimen. Results should be reported promptly, so that treatment will not be delayed; assistance should be provided to the clinician on the judicious use of tests; advice should be available on

Table 2C-12. SENSITIVITY AND SPECIFICITY OF SINGLE AND COMBINATION* ENZYME TESTS PERFORMED ON PATIENTS IN THE CCU[14]

Test	Sensitivity (%)	Specificity (%)
CK	95.7	57.1
AST	91.4	73.5
LDH	87.2	90.5
AST, then LDH	82.2	92.9
CK, then LDH	82.2	92.9
CK, then AST	87.2	83.7
CK, AST, then LDH	77.7	95.2

*All combination results interpreted in series fashion as listed.

selecting the most appropriate tests for a particular patient, and on interpreting the test results. The need for judicious use of the laboratory has been heightened with the continuing growth in volume, complexity, and resultant cost of laboratory testing. Several symposia have been held on this topic;[4,15] all concluded that clinicians and laboratory scientists share responsibility to work together to optimize the use of the laboratory and to minimize the cost of this component of health care to the patient or his insurance carrier.

GOALS AND OBJECTIVES

Selection of laboratory tests should be appropriate; the tests should be obtainable at reasonable cost, and results should be reliable, prompt, and useful. All of these results can best be achieved by close communication between the practicing physician and the laboratory scientist.

Eight goals and objectives have been proposed for effective use of the laboratory:[6]

1. Clinicians should select tests which are optimal for the particular patient's situation. Test selection is discussed in the following section.

2. Orders for tests should be clear and unambiguous. Errors in ordering all too often result in the wrong test being performed, e.g., total rather than free thyroxine, or the results being reported on the wrong patient, e.g., Richard Smith instead of Robert Smith.

3. Specimens should be collected appropriately for the test and handled promptly and with care. The importance of this aspect cannot be overemphasized, since the accuracy of any test result depends on the appropriateness of the specimen submitted to the laboratory. Recommendations on this topic are found in Chapter 3.

4. The analytical methods in use should be accurate and precise. Selection and evaluation of methods are discussed in Chapter 2D.

5. Test results should be reported promptly to the clinician who ordered the test or to the appropriate nursing unit. Prompt reporting can be a significant factor to avoid repeated testing or to shorten hospital stay. The last section of this chapter offers some criteria for speed of reporting and advice on report forms and procedures.

6. The clinician should take action appropriate to the medical significance of the test result. Ways to judge medical significance are discussed in the preceding section. Aids from the laboratory, such as reference ranges, are considered in Chapter 2B and the Appendix.

7. Charges for laboratory tests should be based only on actual laboratory costs; total costs to the patient should not be inflated to underwrite other functions of the hospital.

8. Guidelines regarding appropriate laboratory use should be furnished to clinicians, and procedures for the proper collection and handling of specimens should be provided to nursing unit personnel. Many laboratories issue a booklet or pamphlet listing tests available, requirements for preparation of the patient and specimen collection, reference ranges, and sometimes brief discussions of the significance of abnormal values. Information on reference ranges may be found in the Appendix of this book.

TEST SELECTION

Tests for Monitoring Disease

Optimal selection of laboratory tests is a challenge to the physician. Needless tests increase costs and may cause discomfort. Test selection for following the course of a disease and monitoring therapy is probably the most straightforward. Often the severity of a disease can be characterized by the concentration of a single constituent of serum, such as glucose for diabetes, or creatinine for renal failure. During therapy only a single therapeutic drug may need to be monitored. The chief performance requirement for the laboratory in such monitoring is that the precision of the selected test be better than the day-to-day variability of the analyte concentration, so that trends are readily apparent.

Tests for Diagnosis

The use of the laboratory to establish a diagnosis is more challenging. If the disease is severe, its diagnosis and its remedy are often obvious to the clinician. But the more vague the symptoms and the less severe the disease the more difficult is the test selection. For the difficult diagnosis the clinical chemist should be able to assist the physician both in selecting tests which help to define the biochemical nature of the disease and in furnishing information regarding analytical performance of these tests in the laboratory.

The predictive value model, discussed earlier in this chapter, is one basis for test evaluation. Use of this model requires knowledge of diagnostic sensitivities and specificities of tests, of the prevalence of diseases, and of optimal referent values or cut-off points for test results. Note also that the severity of a disease and its potential for treatment can affect the relative importance of sensitivity and specificity.

The analytical performance of the test can be a crucial factor in diagnosis. Calcium is one of the most closely regulated serum constituents, and the ability to detect a significant increase in its concentration in suspected hyperparathyroidism can be limited by inadequate analytical performance. The key to resolving such deficiencies is close communication between the clinician and the laboratory staff; and the laboratory should respond with replicate analyses of the specimen to improve the precision of the results and with the use of additional control procedures to confirm accuracy.

Combination testing. Results from additional tests will often increase the likelihood of reaching a diagnosis. When more than one test is applied to a clinical problem, the principles of combination testing (section on Clinical Relevance) should be followed. Performing several tests at the same time (parallel testing) will increase the overall sensitivity. Performing several tests sequentially (series testing) will lower sensitivity but will improve specificity. (See also discussion on p. 400.)

Tests for Screening: Profiles

For use in parallel testing, laboratories usually offer profiles appropriate to frequently encountered clinical problems.[6] Advantages accrue in several ways from the use of profiles; for example, reduced cost per test owing to decreased personnel and specimen requirements, reduced clerical errors due to decreased paperwork, and even saving of clinicians' time in reading reports through more efficient reporting. Some typical profiles are listed in Table 2C-13.

Testing asymptomatic persons for evidence of disease or for risk factors for disease places critical demands on laboratory testing. The prevalence of the disease and the possible benefits of therapy must justify the cost of the screening; i.e., both the analytical and clinical performance of the test must be evaluated. Several criteria should be considered before undertaking a screening program:[12] (1) Does the disease have a significant effect on the quality or length of life? (2) Are acceptable methods of treatment available? (3) Does the disease have an asymptomatic period during which it can be detected by testing, and in which treatment is of greater benefit than the treatment which would be supplied when symptoms appear?

Few diseases meet these criteria, and follow-up treatment is often not pursued even when indicated.[33] Mass screening for diabetes, for instance, has not proved effective, since it is a matter for debate whether treatment of an asymptomatic diabetic has reduced either morbidity or mortality.[21]

One beneficial effect of screening may be the psychological reassurance given a subject by learning that he or she is *not* in a publicized "risk zone" for a debilitating disease, such as atherosclerosis. It is impossible to evaluate this benefit objectively, but the reassurance of a

Table 2C-13. SOME EXAMPLES OF CHEMICAL PROFILES FOR TESTING SERUM

Acute care	*Hepatic*	*Evaluation of nutritional status*
Sodium	Alkaline phosphatase	Albumin
Potassium	Aspartate aminotransferase	Transferrin
Chloride	Alanine aminotransferase	Prealbumin
CO_2, total	γ-Glutamyl transferase	Vitamin A
Glucose	Bilirubin, total	Urea
Creatinine	Albumin	Triglycerides
Cardiac evaluation		Magnesium
Cholesterol		Glucose
HDL cholesterol		*Cardiac injury*
Triglycerides		Creatine kinase
Glucose		Lactate dehydrogenase
Uric acid		CK-MB isoenzyme
Free thyroxine		LDH isoenzymes

favorable test result has been claimed to improve attendance at work and to decrease use of medications.[33]

Screening profiles often provide more results than are needed for a clinical situation.[4] Redundant or useless tests may complicate the diagnostic process and adversely affect medical decision making. The clinician may feel that any value outside the reference range, whether related to the situation under investigation or not, requires follow-up treatment. This can lead to "treating a value" rather than treating a disease. It rarely leads to the discovery of an unsuspected disease, and far more frequently means conducting a further series of more complex tests to explain false positive results. In one study of serum uric acid in over 4000 men, only 1% had values over 9 mg/dL which were felt to warrant treatment for hyperuricemia although they showed no symptoms of gout.[26] Another 1.5% had a history of gout, but over 10% more had uric acid levels about 7 mg/dL which on investigation were chiefly attributable to azotemia, acidosis, or the use of diuretics.

The challenge in structuring profiles is to avoid unnecessary tests, so that a cost/benefit analysis would show that the increased information is beneficial and not detrimental to the patient.

OPTIMAL USE OF THE LABORATORY

Overuse of the laboratory has several causes. Standing orders for profiles are one abuse. Consultation may convince the ordering physician that only selected tests are needed daily, and others may be ordered at longer intervals, perhaps weekly. Serial testing without appreciation of the physiological day-to-day variation of concentrations of serum constituents is wasteful. Day-to-day variability can be calculated from column 4 of Table 2C-2; the figures given are one-half the day-to-day variability of individuals. It can be seen, for instance, that the normal daily variation of the urea concentration in serum is 2×6.2 or 12.4%. Hence it would be nonsensical to base a medical decision on a day-to-day change of urea-nitrogen from 20 to 22 mg/dL.

Newly introduced tests are particular problems. These tend to be ordered without adequate knowledge of their efficiency. The laboratory staff should provide information not only on the properties of the new test but on its appropriate application in clinical situations.

Ideally, physicians should select laboratory services with an understanding of the inherent statistical limitations of *all* procedures, including history and physical diagnosis. Few medical schools provide formal education in this area, and there is little applied research on the clinical value of annual expenditures on laboratory tests. The ordering of laboratory tests by physicians merits serious consideration, worthy of the same thought as a history and physical examination.

Education in the appropriate use of the laboratory should logically begin in medical school, where suitable strategies have proved effective in reducing improper use of the laboratory. For house staff and junior physicians, concurrent review of charts by senior staff has proved the most practical approach.[25] Intrinsic to this teaching is the awareness that every test has a price which includes both monetary cost and morbidity, or at least discomfort, to a patient and which must be weighed against possible benefits of resultant therapy. Instruction in test selection is usually welcomed by physicians, and has been part of medicine for over half a century, as reflected in the following quotation from Dr. Francis W. Peabody:[27]

"Good medicine does not consist in the indiscriminate application of laboratory examinations to a patient, but rather in having so clear a comprehension of the probabilities of a case as to know what tests may be of value . . . it should be the duty of every hospital to see that no house officer receives his diploma unless he has demonstrated . . . a knowledge of how to use the results in the study of his patients."

ORDERING TESTS

Test Priorities: Stat, ASAP, Routine

There are generally only two categories of priority for performing laboratory tests. *Routine tests* are those which are performed during regular working hours and on fixed schedules. The schedule of operation may include accumulating batches of samples which are analyzed once a day, weekly, or on certain days of the week. Tests for which the results are needed urgently and at any hour of the day are designated *"stat"* (abbreviation for the Latin *statim,* immediately). To the laboratory the priority "stat" means that results will be obtained and reported as quickly as possible—usually within 15–20 min, but certainly no longer than 1 h—even if routine work must be set aside temporarily. A truly stat situation is one in which results will immediately determine or influence a therapeutic decision, such as in treatment of suspected diabetic coma or hypoglycemia. Unfortunately, a stat request has other shades of meaning, and is often assigned merely to avoid delays, for instance, to write an order before a change of nursing shifts or to allow a patient to leave who has been waiting in a clinic.

Since stat tests invariably cause interruption of laboratory operations, they mean higher costs; thus their use should be minimized. Restricting the stat designation to true emergencies can be encouraged by careful selection of the tests which are offered with this priority, through consultation with the clinical staff. Many of the pseudo-stat requests can be avoided by publicizing the schedule for reporting of routine test results and adhering to this schedule.

Another expedient which may be employed to combat overuse of the stat priority is to offer an intermediate priority, sometimes termed *"as soon as possible"* or "ASAP." ASAP tests are performed with the next batch of specimens on a particular instrument, or as soon as they can be performed without interrupting assays in progress. A typical result is reported in less than three hours, but the time will vary greatly with the workload and the hour of the day. ASAP tests often satisfy the need of the clinician for prompt reporting.

Which tests should be offered on a stat basis? The list varies from laboratory to laboratory and hospital to hospital, depending on clinical needs. Thus, firm recommendations cannot be offered. Many hospital laboratories offer most of the chemical tests shown in Table 2C-14. This list has grown in recent years due to increased stat capability of newer instruments. Often such an instrument can handle stat tests even in the middle of a batch of routine specimens with little interruption of service. These instruments are a welcome development, since the treatment of emergency situations is a crucial purpose of hospitalization, and any good laboratory takes pride in contributing to this critical care. Indeed, stat tests are considered one of the chief reasons for maintaining a laboratory on the premises of a hospital; otherwise tests might be sent to a referral laboratory.

Table 2C-14. CHEMICAL TESTS FREQUENTLY
OFFERED ON A STAT BASIS

Serum	*Cerebrospinal fluid*
Sodium	Total protein
Potassium	Glucose
Chloride	*Urine*
CO_2, total	Protein
Glucose	Glucose
Amylase	Specific gravity
Creatinine	*Blood*
Urea	pH
Calcium (total)	pCO_2
Calcium (ionized)	pO_2
Bilirubin (neonatal)	
Osmolality	
Aspartate aminotransferase	
Creatine kinase	
Acetone	
Ammonia	
Alcohol	
Acetaminophen	
Barbiturate	
Salicylate	
Therapeutic drugs (various)	

Table 2C-15. EXAMPLES OF CRITICAL VALUES AND ALLOWED DIFFERENCES FROM PRIOR VALUES*

	Critical Values		Allowed Differences from Prior Values		
Analyte	Lower	Upper	Low Range	Division Point	High Range
Glucose (serum)	40 mg/dL	400 mg/dL	100 mg/dL	300 mg/dL	50% of prior value
Glucose (CSF)	40 mg/dL	—	—	—	20% of prior value[a]
Urea-N	3 mg/dL	50 mg/dl	10 mg/dL	25 mg/dL	25% of prior value
Creatinine	0.3 mg/dL	7.5 mg/dL	0.5 mg/dL	2 mg/dL	1.5 mg/dL
Calcium, total	6.0 mg/dL	13.0 mg/dL	1.5 mg/dL	—	—
Calcium, ionized	3.0 mg/dL	6.5 mg/dL	—	—	—
Magnesium	1.0 mg/dL	—	1.0 mg/dL	—	1.0 mg/dL
Phosphate-P	1.0 mg/dL	8 mg/dL	2.0 mg/dL	12 mg/dL	25% of prior value
Uric acid	1.0 mg/dL	12 mg/dL	2.0 mg/dL	5 mg/dL	50% of prior value
Bilirubin (adult)	—	—	2.5 mg/dL	10 mg/dL	25% of prior value
Bilirubin (neonatal)	—	12 mg/dL	—	—	—
Albumin	1.5 g/dL	6 g/dL	1.0 g/dL	—	1.5 g/dL
Protein	4 g/dL	9 g/dL	1.5 g/dL	—	1.5 g/dL
Protein (CSF)	—	100 mg/dL	—	—	—
Sodium	120 mmol/L	160 mmol/L	10 mmol/L	—	10 mmol/L
Potassium	2.5 mmol/L	6.5 mmol/L	1.2 mmol/L	—	1.2 mmol/L
Lithium	—	2.0 mmol/L	—	—	—
Chloride	85 mmol/L	115 mmol/L	10 mmol/L	—	10 mmol/L
CO$_2$ content	10 mmol/L	40 mmol/L	6 mmol/L	—	6 mmol/L
Alkaline phosphatase	—	—	50[b] U/L	125 U/L	50% of prior value
Aspartate aminotransferase	—	—	100[b] U/L	500 U/L	40% of prior value
Alanine aminotransferase	—	—	100[b] U/L	500 U/L	40% of prior value
Creatine kinase	—	—	200[b] U/L	600 U/L	40% of prior value
Lactate dehydrogenase	—	—	200[b] U/L	600 U/L	40% of prior value
Amylase	—	—	100[b] U/L	500 U/L	25% of prior value
pH	7.20	7.65	—	—	—
pO$_2$	40 mm Hg	—	—	—	—
pCO$_2$	—	50 mm Hg	—	—	—

*From references 34, 35, and the authors' experience.
[a]Not applicable if patient is undergoing renal dialysis.
[b]Method-dependent values.

Critical values are those which should be reported immediately to the clinician responsible for the patient as well as verified in the laboratory. Values showing changes greater than the *Allowed differences from prior values* should be verified and then flagged to indicate their possible importance to the clinician.

REPORTING RESULTS

EVALUATION OF TEST RESULTS BY THE LABORATORY

Critical values. The laboratory staff has the responsibility to evaluate a test result before reporting it. First, it must be confirmed that the test is within control as judged by quality control results from that assay run (Chapter 2E). Next, results should be screened for values which are so far from the norm as to indicate a potentially dangerous condition requiring prompt action by the clinician. Such critical results should be reported immediately, even if they were not requested on a priority basis. Thus, a potassium level above 6.5 mmol/L requires action to avoid myocardial conduction problems, and a serum glucose concentration below 40 mg/dL is a danger signal for cerebral damage. Table 2C-15 lists some typical "critical values," sometimes unfortunately called "panic values"; certainly prompt and thoughtful action is indicated—not panic. Note that critical values may be either below or above the reference range. On seeing a result outside these ranges, the laboratory supervisor should notify the responsible clinical staff by the most rapid method and immediately verify the result with the technologist performing the test, repeating the assay if there is any question regarding its validity.

Cumulative reports. If serial analyses are performed on specimens taken at different times from the same patient, it is helpful to the clinician if these results are presented as a table on a single sheet. Such *cumulative reports,* often generated by a computer, enable the physician to see at a glance the progress of his patient (see also Chapter 2F) and allow the laboratory staff to make additional checks on the validity of values, as discussed below.

Checks with prior values. The first check should confirm that a pattern of results for different analytes is reasonable, considering the suspected diagnosis. Random errors such as an albumin concentration higher than the total protein or transposition of urea and creatinine can readily be seen by an experienced laboratorian.

The second additional check compares the new value for an analyte with the value obtained last for the same analyte. Such comparisons are termed *prior value tests* or *delta checks*[35] (see also Chapter 2E). Examples of the allowable differences between two subsequent test results are given in Table 2C-15. A difference greater than the figures given implies either that the patient has experienced an acute change in condition or that there is a laboratory error in one of the results. Typically 5–25% of results failing a delta check are attributable to laboratory errors.[34]

It is also helpful to the clinician to receive calculated data such as urea nitrogen/creatinine ratio, anion gap, and base excess, which are further discussed in Chapters 2E, 10, and 11.

References

1. Barnett, R. N., McIver, D. D., and Gorton, W. L.: The medical usefulness of stat tests. Am. J. Clin. Pathol., *69*:520-524, 1978.
2. Barrett, A. E., Cameron, S. J., Fraser, C. G., et al.: A clinical view of analytical goals in clinical biochemistry. J. Clin. Pathol., *32*:893-896, 1979.
3. Bayes, T.: An essay toward solving a problem in the doctrine of change. Philos. Trans. R. Soc. Lond., *53*:370-418, 1763.
4. Beck, J. R., Cornwell, G. G., III, French, E. E., et al.: The "iron screen": Modification of standard laboratory practice with data analysis. Hum. Pathol., *12*:118-126, 1981.
5. Björkhem, I., Bergman, A., Falk, O., et al.: Accuracy of some routine methods used in clinical chemistry as judged by isotope dilution–mass spectrometry. Clin. Chem., *27*:733-735, 1981.
6. Cole, H. W.: Biochemical test profiles and laboratory system design. Hum. Pathol., *11*:424-434, 1980.
7. Cotlove, E., Harris, E. K., and Williams, G. Z.: Biological and analytic components of variation in long-term studies of serum constituents in normal subjects. III. Physiological and medical implications. Clin. Chem., *16*:1028-1032, 1970.
8. Elevitch, F. R., Ed.: Proceedings of the 1976 Conference on Analytical Goals in Clinical Chemistry, College of American Pathologists. Skokie, Ill., College of American Pathologists, 1977.
9. Elion-Gerritzen, W. E.: Analytic precision in clinical chemistry and medical decisions. Am. J. Clin. Pathol., *73*:183-197, 1980.
10. Fink, D. J., and Galen, R. S.: Immunologic detection of prostatic acid phosphatase: Critique II. Hum. Pathol., *9*:621-623, 1978.
11. Foti, A. G., Cooper, J. F., Herschman, H., et al.: Detection of prostatic cancer by solid-phase radioimmunoassay of serum prostatic acid phosphatase. N. Engl. J. Med., *297*:1357-1361, 1977.
12. Frame, P. S.: Screening in a rural practice. Family Pract., *9*:57-64, 1979.
13. Fraser, C. G.: Analytical goals in clinical biochemistry. Prog. Clin. Pathol., *8*:101-122, 1981.

14. Galen, R. S.: Selection of appropriate laboratory tests. *In*: Clinician and Chemist. D. S. Young, J. Hicks, H. Nipper, et al., Eds. Washington, D.C., American Association for Clinical Chemistry, 1979, pp. 69-105.
15. Galen, R. S., and Burke, M. D., Eds.: Symposium on the role of the laboratory in clinical decision making. Hum. Pathol., *11*:406-448, 1980; *12*:106-144, 1981.
16. Galen, R. S., and Gambino, S. R.: The Predictive Value and Efficiency of Medical Diagnoses. New York, John Wiley and Sons, 1975.
17. Gilbert, R. K.: Progress and analytic goals in clinical chemistry. Am. J. Clin. Pathol., *63*:960-973, 1975.
18. Gilbert, R. K.: Accuracy of clinical laboratories studied by comparison with definitive methods. Am. J. Clin. Pathol., *70*:450-470, 1978.
19. Gittes, R.: Acid phosphatase reappraised. N. Engl. J. Med., *297*:1398-1399, 1977.
20. Hyland SCAN Service, Hyland Diagnostics, Deerfield, Ill. 60015, January-December, 1982.
21. Ingelfinger, F. J.: Debates on diabetes. N. Engl. J. Med., *296*:1228-1230, 1977.
22. Itano, M.: CAP blood gas survey—first year's experience. Am. J. Clin. Pathol., *74*:535-541, 1980.
23. Lohff, M. R., DiSilvio, T. V., Ross, J. W., et al.: Analytic clinical laboratory precision. State of the art for selected enzymes. Am. J. Clin. Pathol., *78*(Suppl.):634-643, 1982.
24. Marchand, A., Galen, R. S., and Van Lente, F.: The value of serum haptoglobin in hemolytic disease. JAMA, *243*:1909-1911, 1980.
25. Martin, A. R., Wolf, M. A., Thibodeau, L. A., et al.: A trial of two strategies to modify the test-ordering behavior of medical residents. N. Engl. J. Med., *303*:1330-1336, 1980.
26. Paulus, H. E., Coutts, A., Calabro, J. J., et al.: Clinical significance of hyperuricemia in routinely screened hospitalized men. JAMA, *211*:277-281, 1970.
27. Peabody, F. W.: The physician and the laboratory. Boston Med. Surg., *187*:324-328, 1922.
28. Ransohoff, D. F., and Feinstein, A. R.: Problems of spectrum and bias in evaluating the efficiency of diagnostic tests. N. Engl. J. Med., *299*:926-930, 1978.
29. Rosano, T. G., and Brown, H. H.: Analytical and biological variability of serum creatinine and creatinine clearance: Implications for clinical interpretation. Clin. Chem., *28*:2330-2331, 1982.
30. Ross, J. W., and Fraser, M. D.: Clinical laboratory precision. The state of the art and medical usefulness based internal quality control. Am. J. Clin. Pathol., *78*(Suppl.):578-586, 1982.
31. Vecchio, T. J.: Predictive value of a single diagnostic test in unselected populations. N. Engl. J. Med., *274*:1171-1173, 1966.
32. Weinstein, M. C., and Fineberg, H. V.: Clinical Decision Making. Philadelphia, W. B. Saunders Company, 1980.
33. Werner, M., and Altshuler, C. H.: Cost effectiveness of multiphasic screening: Old controversies and a new rationale. Hum. Pathol., *12*:111-117, 1981.
34. Wheeler, L. A., and Sheiner, L. B.: A clinical evaluation of various delta check methods. Clin. Chem., *27*:5-9, 1981.
35. Whitehurst, P., DiSilvio, T. V., and Boyadjian, G.: Evaluation of discrepancies in patients' results—an aspect of computer-assisted quality control. Clin. Chem., *21*:87-92, 1975.

Evaluation of Methods

by Theodore Peters, Jr., Ph.D., and James O. Westgard, Ph.D.

Before a new or improved method is introduced into a laboratory, it must be selected with care and its performance evaluated under laboratory conditions. Method selection and evaluation are important steps in establishing high quality laboratory services. The purpose of this chapter is to outline some of the techniques that are utilized for selecting and evaluating analytical methods, whether these are manual or automated, and whether the reagents are prepared in the laboratory or obtained from a manufacturer.

Objective schemes for selecting and evaluating laboratory methods were first introduced by Barnett in the 1960's.[5] These schemes have been further refined, particularly the statistical analysis of the method evaluation studies. Logan's[14] recent review of the literature in this area, while emphasizing kit methods, provides a thorough discussion of current practices in selecting and evaluating methods. The review by Westgard[24] focuses more on the concepts and principles of assessing the performance of analytical methods. Practical applications of these ideas along with examples and problem sets are found in a monograph.[27] Documents developed by the International Federation of Clinical Chemistry (IFCC) provide a definition of terms[6] and a general discussion of the philosophy of method evaluation[7] as it relates to an overall program in quality control. More detailed evaluation guidelines are being developed by the National Committee for Clinical Laboratory Standards (NCCLS). One series of documents[16-18] provides guidelines to be used by manufacturers when making performance claims for analytical methods. Another series[19] provides guidelines for the clinical laboratory users who want to verify a manufacturer's claims.

The selection and evaluation of a new method require considerable judgment. The analyst must be able to understand the claims of manufacturers and the studies reported in the literature. He must also be able to perform evaluation studies to make sure that the desired performance is achievable under the conditions prevailing in his laboratory.

DEFINITIONS AND GOALS

The first step in selecting a method is to define the need for the assay and the goals which are to be accomplished. Pertinent properties relate to medical usefulness (see Chapter 2C) and include the choice of analyte to be measured, the diagnostic sensitivity and specificity of the procedure, and the precision and accuracy required for effective performance. Other related characteristics are the type of specimen to be analyzed, the hours the test should be available, and required turnaround time for results. The importance of a careful definition of requirements should be apparent. It is this list of requirements that guides the whole process of selecting and evaluating analytical methods.

CHARACTERISTICS OF ANALYTICAL METHODS

Practicality characteristics. These are factors (other than analytical performance) which determine whether the method can be implemented in the laboratory. They include the required equipment, workload, specimen handling, run size, personnel skill, cost per test, methods of standardization and quality control, space needs (including reagent storage), and precautions and procedures required for safety. Most practicality requirements can be defined by the laboratory personnel through discussion among the analysts, supervisors, and directors. Practicality characteristics can often be assessed from the technical literature which describes the analytical method.

Reliability characteristics. These properties relate to the performance of the method, including the precision, accuracy, analytical sensitivity, analytical specificity, recovery, interference, blank readings, linear range, sample interaction, and reagent stability. A "rugged" method should provide consistently reliable performance with different operators and different batches of reagents over a long period of time. The assessment of reliability characteristics requires experimental studies to estimate the method's performance, followed by judgments on whether the observed performance is satisfactory for the medical use of the test results. The definition of "medical usefulness requirements" may require conferences between laboratory and clinical personnel. Requirements will vary with the intended use of the test results or the medical application, thus yielding different goals or specifications for the medically useful accuracy and precision. The definition of goals for analytical performance will be discussed more fully after defining the characteristics to be tested during method evaluation.

Analytical range. This is the "range of concentration or other quantity in the specimen over which the method is applicable without modification."[6] It is tested by a "linearity experiment" in which a series of solutions, usually standards representing a wide concentration range, are analyzed by the analytical method. Ideally the standard curve (plot of response versus analyte concentration) should be linear and pass through the origin. The analytical range should be wide enough to include most (95%) of the expected clinical specimens without pre-dilution. When linearity is assumed in the calibration process, the linear range of the standard curve is the critical characteristic. If linear response is not obtained, calibration procedures should employ a higher number of standard solutions to define the response curve adequately, and unknowns should be bracketed by the standard solutions. These requirements may be somewhat reduced when nonlinear standard curves are very stable. The difficulties in using nonlinear standard curves are minimized when microprocessors are employed for data reduction.

Analytical sensitivity. This is defined as a measure of "the ability of an analytical method to detect small quantities of the measured component."[6] Sensitivity is related to precision. When the concern is performance at concentrations above the detection limit, it is best to determine precision at the particular concentrations of interest. When the concern is performance at very low concentrations, for example in measuring trace materials, then it is useful to determine the detection limit.

Detection limit. This is defined as "the smallest single result which, with stated probability (commonly 95%), can be distinguished from a suitable blank. The limit may be a concentration or an amount and defines the point at which the analysis becomes just feasible."[6] The detection limit is related to precision and also depends on the amplitude of blank readings. A more quantitative definition is that the detection limit is equal to 3 times the standard deviation of the blank, or is located 3 standard deviations above the measured average blank.[2]

Blank readings. These are the responses observed by the measurement procedure due to reagent and sample constituents, not including the desired analyte. It may be useful to quantify the blank readings directly by making measurements of the reagent solutions without sample present (reagent blanks) and sample dilutions without at least one reagent which initiates the reaction (sample blanks). Low blank readings are of course desirable and provide the best possibilities for precision and accuracy. For high blank readings, the performance of a method may still be satisfactory if the blank measurements are stable and precise, therefore permitting small differences between measurements to be meaningful. The analyst should be aware of the magnitude of blank readings so that he can appreciate the requirements for precision of measurement.

Analytical specificity. This is "the ability of an analytical method to determine solely the component(s) it purports to measure."[6] Specificity is related to accuracy. Lack of cross-reactivity with closely related substances is one aspect of specificity; an example is the ability to measure glucose in the presence of a similar hexose such as mannose, or the ability to measure chloride in the presence of bromide. Other common examples are the potential errors due to frequently encountered substances such as bilirubin, hemoglobin, and lipids, which may cause errors by virtue of their color, turbidity, or other characteristics.

Interference. This is another aspect of specificity, namely, "the effect of a component, which does not by itself produce a reading, on the accuracy of measurement of another component."[6] An example is the measurement of glucose by a glucose oxidase reaction, where the intermediate product hydrogen peroxide may react with uric acid, rather than with the intended chromogen. Interferences may be very subtle, such as the unsuspected presence of a drug which exhausts a diazo reagent. It is difficult to perform exhaustive experiments which test for all possible interferences when the causes may not even be suspected.

Recovery. This is "the ability of an analytical method to correctly measure pure analyte when added to the samples routinely analyzed." Recovery is related to accuracy and it tests whether the analytical method can measure the desired substance in the presence of all of the other materials appearing in the matrix of the real specimen. It also tests for competitive interferences, for example, the competition of protein for calcium in certain compleximetric reactions used in the determination of calcium. Addition of the analyte in vitro is obviously an artificial approach; it cannot be assured that the physical state of the analyte, its solubility, and its binding to serum proteins are the same as would be attained in vivo. Nor can the possibility of a change in chemical form in vivo (such as conjugation or modification in the liver) be taken into account. But in the absence of analytical reference methods and reference materials, recovery experiments may be one of the few ways of assessing accuracy. On the other hand, recovery studies are often poorly performed and the data improperly calculated and presented; thus the full value of such studies frequently is not realized.

Precision. This refers to the "agreement between replicate measurements"[6] and is estimated by a "replication experiment" in which the same sample material is analyzed several times (usually a minimum of 10–20 measurements). Different components of precision may be estimated depending on how the experiment is performed. *Within-run precision* is the variability found when the same material is analyzed repeatedly in the same analytical run (or alternatively when duplicate analyses are made within a run on a series of clinical specimens and the SD of the duplicates is calculated). It usually provides an optimistic estimate of the expected performance since there has been minimal opportunity for conditions to change during the time between replicate analyses. *Within-day precision* is estimated when the same material is analyzed repeatedly in several different runs on the same working day. The variability is usually somewhat higher than observed for within-run replicates. *Day-to-day precision* (sometimes called *between-day precision*) is the variability found when the same material is analyzed repeatedly on different days. This last estimate is the most realistic assessment of the performance that will be observed by the clinicians in their use of the test because it includes the variable factors of different operators, pipets, and other conditions in the laboratory.

The term *imprecision* has been recommended as more appropriate for emphasizing the lack of agreement.[6] Imprecision has been defined as the "standard deviation or coefficient of variation of the results of a set of replicate measurement."[6] This can also be described as random analytical error, "an error which can be either positive or negative, whose direction and exact magnitude cannot be predicted."[27]

Accuracy—the systematic error concept. This is a measure of the "agreement between the best estimate of a quantity and its true value."[6] It is usually estimated by a "comparison of methods experiment" in which clinical specimens are analyzed by the method under evaluation and by another method whose accuracy has already been established. The term *inaccuracy* has been recommended to emphasize lack of agreement and is defined as the "numerical difference between the mean of a set of replicate measurements and the true value."[6] This disagreement can also be described as *systematic error*—"an error that is always in one direction."[27] Systematic errors can further be divided into two types—constant and proportional. *Constant systematic error* is "an error that is always in the same direction and of the same magnitude even as the concentration of analyte changes."[27] *Proportional systematic error* is "an error which is always in one direction and whose magnitude is a percentage of the concentration of analyte being measured."[27]

In addition to these types of errors, there is random bias (discussed in Chapter 2A) which is a measure of systematic error which varies from sample to sample. Random bias can occur when there is an interfering material which varies from specimen to specimen. This error is estimated from a comparison of methods experiment, and its magnitude will depend on the patient population tested.

The nature of these different types of errors can be seen by plotting the results of a test method versus the true values for a group of specimens. As shown in Figure 2D-1, random error (including random bias) shows up as scatter in the data about the line of best fit. Constant error causes a shift of the line in one direction and is most easily seen by the point at which the line

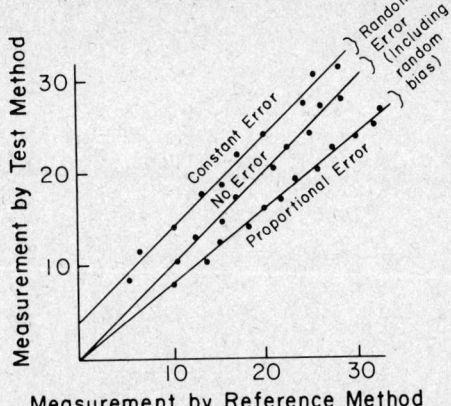

Figure 2D-1. Appearance of different types of analytical errors when observed values (test method measurements) are plotted on the y-axis versus some estimate of the true values (measurements by a reference method) plotted on the x-axis. (From Westgard, J. O., de Vos, D. J., Hunt, M. R., et al.: Method Evaluation. Houston, American Society for Medical Technology, 1978.)

intersects the ordinate. Proportional error causes the angle of the line to change. Thus, careful inspection of this plot can provide information about the random, constant, and proportional components of analytical error.

Another useful graph is the plot of the difference between the test and reference methods on the y-axis versus the value by the reference method on the x-axis. This "scatterplot" expands the scale for displaying the random error in the data and is particularly useful for close inspection of the between-method differences. However, it is not as easy to interpret the constant and proportional errors using this graph.

Information on the components of error is particularly useful to the analyst when attempting to identify the source of errors in order to reduce their magnitude. On the other hand, information on the overall effect of the components of error, or the total error, is of interest when determining the analytical quality that is achievable and its acceptability for the intended medical purposes and applications of the analytical method. Many of the components of error are additive; thus the final test results may be in error by an amount greater than any one of the components. Because of this, many analysts prefer to employ a total error concept of accuracy.

Accuracy—the total error concept. Another view of accuracy considers all types of error, both random and systematic, as causing an observed measurement to deviate from its true value. This total error concept and its relationship to the random and systematic components are illustrated in Figure 2D-2. The distribution of values around a central value represents random error. The shift of the central value of the distribution from the true value represents systematic error. The total error shows how large the errors can be when the random and systematic components occur in the same direction. This worst case estimate of the errors can be seen to be larger than either the random or systematic components alone, and therefore has been argued to be a safer estimate for judging the quality of analytical methods.[26] Total error is actually a more comprehensive concept of accuracy than the traditional systematic error concept. It seems particularly appropriate for use in applications when a single measurement is performed and the test results are affected by both random and systematic errors.

SETTING QUALITY GOALS FOR ANALYTICAL PERFORMANCE

Performance goals should be expressed as specifications for the amount of analytical error that can be permitted without invalidating the medical usefulness of test results. Separate specifications can be defined as goals for precision and accuracy, or a single specification can be defined as a goal for total error.

Precision goal. This goal can be stated as an allowable standard deviation (s_A). For a urea method, for example, the allowable standard deviation could be set as ± 1 mg/dL at a concentration of 15 mg/dL and as ± 5 mg/dL at a concentration of 100 mg/dL.

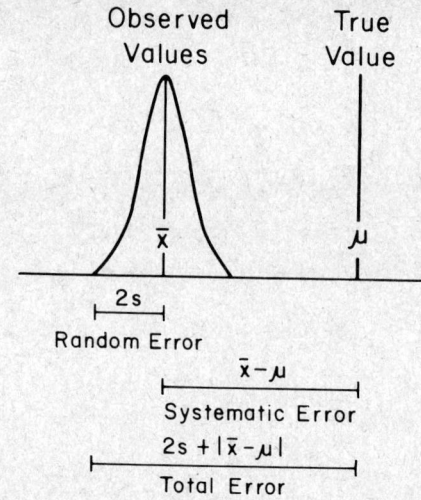

Figure 2D-2. Illustration of concepts of accuracy: random error, systematic error, and total error. (From Westgard, J. O., de Vos, D. J., Hunt, M. R., et al.: Method Evaluation. Houston, American Society for Medical Technology, 1978.)

Accuracy goal. This can be stated as an allowable systematic error or allowable bias (B_A). For a urea method, the allowable bias could be defined as 0.8 mg/dL at a concentration of 15 mg/dL, and 4 mg/dL at 100 mg/dL.

Total error goal (TE_A). This goal can be stated as a 95% limit of the allowable analytical error, i.e., a 2 standard deviation limit which includes the effects of both random error (RE) and systematic error (SE), as seen in Figure 2D-2.

$$TE = SE + RE$$

where $RE = 2 \times$ the day-to-day SD. The 95% in the definition means that only one patient test result out of twenty would be permitted to have an error greater than the amount specified. For the urea example, the total allowable error would be $(2 \times 1) + 0.8 = 2.8$ mg/dL at 15 mg/dL, and $(2 \times 5) + 4 = 14$ mg/dL at 100 mg/dL.

Medical decision level. All of the error specifications should include some statement of the concentration or medical decision level (X_C) at which the desired performance is to be achieved. This medical decision level can be any concentration where test results are critically interpreted for purposes of diagnosis, monitoring, or therapeutic decisions. Several levels can be defined when there are different medical applications. Method evaluation studies should be designed to estimate the errors at these critical concentrations.

RECOMMENDATIONS FOR QUALITY GOALS

Specifications for allowable analytical errors should be judgments based on experience with the medical use of the laboratory tests. There are no specifications that are universally applicable. What is appropriate depends on the medical mission of the health care facility, the population being served, the particular application of the test, and the way in which the physician interprets the test result. This means that the performance goals may differ from laboratory to laboratory.

Some general guidelines are available in the literature. Barnett's early discussion of this subject is widely referenced.[4] There are other recommendations by Tonks,[23] Cotlove et al.,[8] Gilbert,[13] Elion-Gerritzen,[10] and the CAP Aspen Conference.[9] Even though most of the literature references discuss only precision goals or allowable standard deviations (random error), these can be interpreted as total error goals by multiplying them by 2 to provide a 95% limit for analytical error. There is little discussion of accuracy goals or allowable biases except in the review by Frazer.[11] Allowable bias can be added to 2 times the allowable standard deviation to give the total error specification. In the absence of such information, it would be safest to consider the total error specification to be equal to 2 times the allowable standard deviation. Chapter 2C presents a more thorough discussion of the medical usefulness of laboratory tests and the implications for quality goals. Table 2C-2 lists precision requirements for medical needs which have been derived from some of the above references.

At the outset, making judgments on allowable analytical errors may appear difficult; however, this judgment is at least made in the subject area that is understood by laboratory personnel. The alternative is to try to make judgments on what statistical results are acceptable. This has led to the use of statistics such as t-value, F-value, and r (correlation coefficient) for judging the acceptability of analytical performance. Judgments based on such values are much more difficult and usually not very objective.[28] Thus, it is preferable for laboratory personnel to adopt the approach of defining allowable analytical errors and to judge analytical performance relative to medical needs.

JUDGING PERFORMANCE BASED ON QUALITY GOALS

When a specification for the allowable analytical error is given, the acceptability of a method's performance can be judged as follows:

1. The observed errors are compared to the specified allowable errors.
2. When an observed error is less than the allowable error, performance is acceptable.
3. When an observed error is greater than the allowable error, performance is not acceptable; the errors of the method must be reduced by modification of the procedure, or else the method must be rejected.

Quantitative criteria can be developed for each experiment to relate the observed errors to the specified allowable errors. The observed errors are estimates based on the use of certain statistics. For example, the standard deviation is used to estimate precision; therefore, a criterion for judging precision can compare the observed standard deviation with the specified allowable standard deviation. Or, when a random error goal has been specified, the observed standard deviation can be multiplied by 2 to provide a 95% limit which can then be compared with the allowable total error. Similar criteria can be developed for interference, recovery, and comparison of methods experiments,[26] and are shown in Table 2D-1.

These criteria are based on point estimates or single-value estimates of the observed errors. It is assumed that sufficient data are collected to provide a reliable estimate of the observed errors by a single number of single value. A more objective set of criteria can be developed by estimating the confidence interval for the observed errors, thereby taking into account the uncertainty due to the amount of data collected.[26] An observed error is then estimated by a range of values, or by the upper and lower limits for the error. When the upper limit or larger estimate of the observed error is less than the allowable error, the method's performance is judged acceptable. When the lower limit or smaller estimate of the errror is greater than the allowable error, the

Table 2D-1. CRITERIA FOR JUDGING THE ACCEPTABILITY OF ERRORS AS ESTIMATED FROM DIFFERENT EVALUATION EXPERIMENTS

Type of Error	Experiment	Criteria
Random error	Replication	$s_{obs} < s_A,$ or $2s_{obs} < TE_A$
Proportional error	Recovery	$\left\lvert \dfrac{(\bar{R} - 100)}{100} \right\rvert X_C < B_A$
Constant error	Interference	$\lvert Bias \rvert < B_A$
Systematic error	Comparison	$\lvert (a + bX_C) - X_C \rvert < B_A$
Total error	Replication and comparison	$2s_{obs} + \lvert (a + bX_C) - X_C \rvert < TE_A$

s_{obs} is the standard deviation determined in the replication experiment.
$\bar{R}$ is the average recovery (in percent) determined in the recovery experiment.
s_A, B_A, and TE_A are the allowable standard deviation, allowable bias, and allowable total error, respectively.
X_C is the decision level concentration where medical interpretation is critical.
Bias is the average difference determined in the interference experiment.
a and b are the y-intercept and slope, respectively, determined by regression analysis of the comparison of methods data.

method's performance is judged unacceptable. When neither of the above criteria is fulfilled, the uncertainty in the data does not permit an objective decision to be made. Either more data must be collected or the method is judged to have "borderline" performance.

Other criteria may also be developed based on the way the different errors are estimated. For example, in the total error criterion in Table 2D-1, the estimate of random error is obtained from a replication experiment. This has some advantages in that the test materials can be chosen to have concentrations corresponding to particular medical decision levels of interest, thus providing good estimates of precision as a function of medical decision points. However, this total error criterion does not include the random bias component of error (see discussion in Chapter 2A). To take random bias into account, the data from the comparison of methods experiment can be used to estimate both the random error of the analytical method and the random interference from specimen to specimen. This requires different statistical calculations, and gives a somewhat different and more complicated total error criterion. For sake of simplicity, we illustrate the use of the simple single-value criteria in Table 2D-1 in the subsequent discussions in this chapter.

SELECTING AN ANALYTICAL METHOD

After setting goals for a method, the next step is to survey the technical literature and decide which procedures best fulfill the requirements. This literature should include both professional journals and manufacturers' reports and brochures. Then, before proceeding to obtain the reagents, kit, or instrument system, the pertinent information should be assembled in a tabular form. A qualitative judgment can often be made from review of this information, but it is also possible to assign weights or points to the different requirements based on their relative importance, thus providing a more quantitative selection process.

Some guidelines for reviewing prospective methods have been established by both the American Association for Clinical Chemistry[1] and the International Federation of Clinical Chemistry.[21] These are oriented to kit methods, but provide generally useful guidelines. Attention is directed to the following features:

1. Principle of the assay, with original references.

2. Composition of reagents and standards, quantity provided (equivalent number of tests), and storage requirements (space, temperature, light, humidity) applicable both before and after opening the original container.

3. Stability of reagents and standards (shelf life).

4. Possible hazards and appropriate safety precautions.

5. Specimen requirements, i.e., conditions for collection, specimen volume, anticoagulants, preservatives, storage conditions.

6. Anticipated analytical performance, i.e., accuracy, precision, analytical range, linear range, specificity using biological samples.

7. Reference range with information on how it was derived; typical values obtained in health and disease; whether there will be a need to determine a reference range for one's own institution.

8. Complete and detailed protocol for performing a test.

9. Instrumental requirements and limitations.

10. Availability of technical support, supplies, and service.

These points should be considered in relation to the situation in the laboratory. Is the optical equipment suitable in terms of spectral range and bandwidth? Is there sufficient space? How much technologist time will be required, and at what level of skill? If training of the entire staff in a new technique such as micropipetting is required, is such training worth the possible benefit? What is the estimated cost of an assay, including calibrators and quality control specimens? How would the data be handled, from the aspects of report format to compatibility with data processing equipment already in use? From consideration of such factors as these, a decision should be made on the analytical method which best fits the laboratory requirements and has the potential for achieving the necessary analytical quality.

EVALUATING AN ANALYTICAL METHOD

Following the definition of needs and quality goals as well as the careful selection of an analytical method, evaluation studies are performed to test the characteristics defined above. In general, a preliminary period should be allowed for familiarization with the analytical method, equipment, and procedural steps. This may involve trial runs to determine within-run precision and analytical range. Once the analyst is satisfied that the method can be performed without any procedural difficulties, studies on recovery and interference can be started. Precision should then be studied over at least a one-month period. Accuracy should be tested by a comparison of methods experiment, using a reference method with established and satisfactory accuracy and precision. The acceptability of the new method is determined from the experimental results by estimating the analytical errors, then comparing these estimates with the defined error specifications. When the observed errors are less than the specified allowable errors, the method can be implemented for routine service.

STEP-BY-STEP OUTLINE OF A METHOD EVALUATION STUDY

1. Plan in advance to assign several technologists, not just the best-trained one, to carry out the testing. If the test is not frequently performed, it will save time later to stockpile frozen clinical specimens which have been analyzed by the existing procedure; specimens with possible interference from bilirubin, hemolysis, or turbidity are of special interest. If the sample volume required per test is high, pool clinical specimens with high and low concentrations, respectively, for later use in testing precision. Directions for preparing such pools are given in Chapter 2E. Any frozen specimens must be thawed and mixed completely before use.

2. Obtain the reagents, kit, instrument, and all necessary materials and supplies. Study the instructions, then set up the method and perform some initial runs on control sera and random clinical specimens, carefully following the directions of the literature reference or of the manufacturer. Do this for several days to become familiar with the method.

3. Determine the within-run precision by analyzing 10–20 aliquots of a control specimen or a clinical specimen. Alternatively, perform duplicate analyses on a variety of clinical specimens. Do this at the different decision level concentrations of interest, or select groups of clinical specimens near the different decision levels of interest. Calculate the mean and standard deviation for each data set.

Exclusion of outliers. In assembling data, results which are questionable should be excluded only under the following rules: (1) those related to documented errors, such as results obtained when an automated instrument is indicating an "Error" status; (2) those which can arbitrarily be excluded as undocumented errors, such as decimal point or transcription errors; and (3) statistical outliers, defined as values lying more than 3 times the SD from the mean. After excluding the outliers, the mean and SD are recalculated for each data set.

To judge acceptability, compare the observed standard deviation (s_{obs}) to the allowable standard deviation (s_A). If the precision meets this requirement, proceed to further testing as described below. If not, attempt to locate the sources of imprecision, or abandon this procedure in favor of another. Steps at which imprecision frequently arises are dispensing of specimens or reagents, mixing operations, and reading of outputs such as optical absorbance.

4. Determine the analytical range by analyzing a set of standards or a series of dilutions of either control materials or an elevated specimen pool. (In some assays, the choice of diluent may be important; for example, a protein diluent may be necessary to maintain albumin or total protein concentrations relatively constant. In some instances, the only suitable diluent may be serum known to be free of the analyte of interest.) Plot the observed value versus the concentrations of the samples, or versus the relative dilutions for the series of samples. Estimate the analytical range by visual inspection of the graph. Compare the observed value with the necessary analytical range. When applicable, also include measurements of the reagent blanks and sample blanks (if any) versus water, and note the magnitude of the blank values compared with the read-out value for a low concentration of analyte.

5. Determine recovery by adding known amounts of the analyte in concentrated form to

actual clinical specimens to prepare a test sample. Add a similar volume of solvent without analyte to provide a baseline sample. Analyze the test and baseline samples, calculate the difference to determine the amount recovered, then calculate the ratio of the amount recovered divided by the original amount added, and multiply by 100 to give the recovery in per cent. Do this for several specimens and average the per cent recoveries to provide the final estimate. The difference between this average recovery and the ideal 100% recovery provides an estimate of proportional error, which can then be multiplied by a decision level of interest to determine the magnitude of the error at that critical concentration. Compare this estimate of systematic error with the allowable bias (B_A) or with the allowable total error (TE_A) to judge acceptability (see Table 2D-1).

6. Test the interference of common substances, such as bilirubin, hemoglobin, lipids, anti-coagulants, and common drugs, and of specific substances known to cause problems in the determination of the particular analyte being measured. The concentrations of added drugs should be in the usual therapeutic or toxic range. The effects of anticoagulants are best treated on paired specimens from a blood sample. Interference can be tested by an experiment analogous to the recovery experiment, except that the material added is the suspected interfering substance, rather than the sought-for analyte. The results can be expressed in terms of the difference in concentration between the sample containing the interferent and the baseline sample. Interference can also be tested by comparing results with an analytical method known to be free of interference, again taking the difference between the observed values as the estimate of interference. This difference, or bias, can then be compared with the allowable bias (B_A) or the allowable total error (TE_A) to judge its acceptability (Table 2D-1).

7. Determine other specific characteristics which may be critical for the particular test method. For example, sample interaction may occur with certain automated analytical instruments owing to slow washout of pipetting devices or cuvets. The experimental design will depend on the type of instrument system, but generally will involve introduction of a high concentration specimen followed by two or more aliquots of a low concentration specimen—for example, the series H_1, L_1, L_2. The difference between the L_1 and L_2 results is taken as the error due to interaction. This is a systematic error, at least in the sense that it will be the same when the same concentrations of materials are involved; thus acceptability of performance can be judged by comparing the observed error to the allowable bias (B_A).

8. Determine day-to-day precision by analyzing serum pools or control materials each working day over a one-month period. Select materials at two or three decision levels, or one near the upper limit of the analytical range. Calculate the mean and standard deviation for each data set. To judge acceptability, compare the observed standard deviation with the allowable standard deviation (s_A).

9. Compare the results by the test method with those obtained by a reference analytical method. The differences between methods can be interpreted as analytical errors, but there are several factors which are critical for the performance of the experiment and the interpretation of the results:

Reference analytical method. To assign the cause of a systematic error to the test method assumes that the comparison method is of high quality and free from analytical errors. There are some high quality analytical methods that have been established as reliable methods for use in validating new analytical methods,[22] but these are often not readily accessible in service laboratories. There are other recommendations concerning the preferred methodological principles for the measurement of different analytes.[15] Most often the method chosen is one that is in routine use in the laboratory. In such a case, there exists the possibility that any differences observed may actually be caused by the limitations of the comparison method. When these differences are large, additional data from interference and recovery experiments may be necessary to determine which method is at fault.

Clinical specimens. Specimens from 40–200 patients should be analyzed. The number of specimens is less important than the concentration range covered by the specimens and the variety of diseases and medical problems they represent. The specimens should be distributed over the entire analytical range of interest, with about a fifth of the total near each medical decision level and a fifth near the upper limit of the analytical range. Specimens should be fresh, if possible, but if not, then preserved under conditions which are known not to alter the results. It is desirable to obtain duplicate results by both analytical methods, preferably from two different analytical runs. The specimens should be analyzed by the two methods within a 4-h period if possible, to prevent any changes due to instability of the specimens themselves. They should be analyzed in small groups over a period of several days to provide data representative of the performance of both analytical methods.

Data analysis. The results should first be plotted, the test results on the y-axis and the comparative results on the x-axis. The data may be subjected to a variety of statistical analyses. (See Chapter 2A for details.) Although the use of t-values and correlation coefficients has been popular, they are of little value for assessing performance relative to the medically based performance goals.[27,28] It is best to subject the data to regression analysis to determine the slope, y-intercept, and standard deviation of the residuals or points about the regression line. The slope (b) provides an estimate of proportional analytical error, the intercept (a) an estimate of constant analytical error, and the standard deviation ($s_{y/x}$) an estimate of the random error between the methods. The regression equation ($Y = a + bX$) is used to estimate the systematic error at the decision level concentrations. The Y-value (Y_C) corresponding to the X-value of interest (X_C) is calculated ($Y_C = a + bX_C$); then systematic error is estimated as the difference between X_C and Y_C (SE $= |Y_C - X_C|$) (see Table 2D-1). SE can be compared to the specification for allowable bias (B_A) or allowable total error (TE_A) to judge acceptability. A total error criterion can be employed by adding the observed systematic error to 2 times the observed standard deviation (TE $=$ SE $+ 2 s_{obs}$), then comparing this estimate to the allowable total error (TE_A).

10. Determine the stability and uniformity of reagents and standards by comparing results using different batches (lots) of materials, and by using batches (lots) that have been stored for varying periods of time. The variation of results should be much less than the defined error specifications.

11. Establish or verify the reference ranges (intervals). If one accurate method is replacing another, there should be no change from the previously established reference intervals. In other instances where no reference ranges have been established, it will be necessary to analyze a group of specimens from healthy persons, following the guidelines in Chapter 2B.

12. Prepare a report of the evaluation studies, summarizing the experimental results and data analysis. Carefully review the data and determine whether the method is acceptable on an overall basis. For a newly introduced analyte, a conference with clinicians may be useful to help assess the performance of the method and may also be helpful to introduce the new test to the clinicians.

13. Document the procedure to be followed in the routine operation of the analytical method. Prepare in-service training materials and instruct personnel how to perform the test. Implement quality control procedures and establish control limits and actions.

14. Prepare documentation for the clinical staff, with instructions on specimen collection, hours of test availability, expected turnaround time, reference ranges, charge for the test, and a short description of its clinical utility.

15. Inform the clinical staff of the test's availability. Encourage staff feedback during the introductory phase of the new test. Monitor performance carefully during the first several months by use of statistical control procedures.

AN EXAMPLE EVALUATION STUDY: GLUCOSE IN SERUM

Analytical needs. A rapid procedure is needed to determine glucose at times outside of regular working hours. A sample volume of 0.2 mL or less is desired, an analytical range of 0–500 mg/dL, and a turnaround time of 30 min for individual specimens ordered on an emergency basis. The short turnaround time means that loss of glucose by glycolysis would not be significant.

Quality goals. The medical decision levels of interest for the glucose analysis are taken as 50 mg/dL and 200 mg/dL to indicate hypo- and hyperglycemia, respectively. Since the procedure is intended for use outside of first shift working hours, the decision level usually applied for screening (140 mg/dL) is not included.

Precision goals for glucose are defined to be 3 mg/dL at 50 mg/dL and 6 mg/dL at 200 mg/dL. Allowable bias is specified to be 4 mg/dL at 50 mg/dL and 8 mg/dL at 200 mg/dL. Calculating total error as twice the SD plus the bias (Figure 2D-2) gives total error goals (TE_A) of 10 mg/dL at 50 mg/dL and 20 mg/dL at 200 mg/dL.

Method selection. After surveying available methods, a test method is selected which is available in kit form and can be set up on an existing laboratory spectrophotometer. One technologist is assigned for the evaluation study, with a second being assigned time to participate in the experiments.

Familiarization period. After the kit materials and primary standard materials are obtained, the method is set up and tried. Primary standard solutions are prepared for calibration. Several runs are performed using control materials and randomly selected clinical specimens.

Within-run precision. A within-run replication experiment is performed by analyzing 20 aliquots each of a low-abnormal control and a moderately high abnormal control. The respective means and standard deviations are 56.5 mg/dL and 0.7 mg/dL for the low material and 182.6 mg/dL and 2.1 mg/dL for the high material. These SD's are judged acceptable when compared with the allowable SD's of 3.0 and 6.0 mg/dL. Although the observed performance is better than the analytical goals, these estimates only include within-run variation. As a rule of thumb, it may be useful to require within-run precision to be a factor of two better than the specified allowable standard deviation, especially for nonautomated methods which are subject to variations in technique by different analysts.

Analytical range. The range of linearity is determined by duplicate analyses on a series of glucose standards prepared from a stock glucose standard of 1000 mg/dL. Results are presented graphically in Figure 2D-3. The plot shows excellent linearity to 600 mg/dL, which meets the specification for linearity to 500 mg/dL. The absorbance at zero glucose concentration is the reagent blank value, which is reproducible and satisfactorily low so that omission of this blank absorbance should not be a significant source of error.

Recovery. Two pools of sera are prepared as baseline specimens. A concentrated glucose standard is prepared so that the standard addition will not dilute the serum appreciably. Two different amounts of glucose are added. To 9.6 mL of pooled serum is added 0, 100, or 400 μL of glucose standard having a concentration of 10 000 mg/dL, plus 400, 300, or 0 μL, respectively, of sodium chloride, 0.15 mol/L, to give a final volume of 10.0 mL in each case. Quadruplicate assays are performed and the four individual values for each specimen are averaged (Table 2D-2). The amount recovered is determined by subtracting the amount originally present in the respective pools (61 and 171 mg/dL) from the amount measured or "found." Per cent recovery is obtained by dividing this figure by the amount added and multiplying by 100. These individual recoveries are averaged to give an estimate of 98% recovery, which corresponds to a 2% proportional error, or actual errors of 1.0 mg/dL at 50 mg/dL and 4.0 mg/dL at 200 mg/dL. These errors are less than the allowable biases (4.0 and 8.0 mg/dL), and therefore do not invalidate the analytical goals.

Interference. The potential interference of visibly abnormal specimens is tested by analyzing a series of icteric, turbid, and hemolyzed sera by the new method and an established method known to be free of such interferences (Table 2D-3). (Note that this is not the classical experiment for testing interferences by adding the interfering material directly and measuring its effects; the comparison approach is employed here because of the limitations of adding these materials to serum pools.) The differences between the glucose values by the two procedures are from 1–4 mg/dL at 100–140 mg/dL, and are judged to be acceptably small compared to the allowable biases of 4–8 mg/dL.

Other possible interferences are tested by adding small amounts of highly concentrated solutions to portions of a large pool of serum. The average differences from the baseline value are 0–3 mg/dL at 120 mg/dL (Table 2D-4). These differences are smaller than the allowable biases; thus they do not limit the usefulness of the method. Whether the procedure would be applicable to heparinized or other types of plasma may also be tested by splitting some blood samples appropriately.

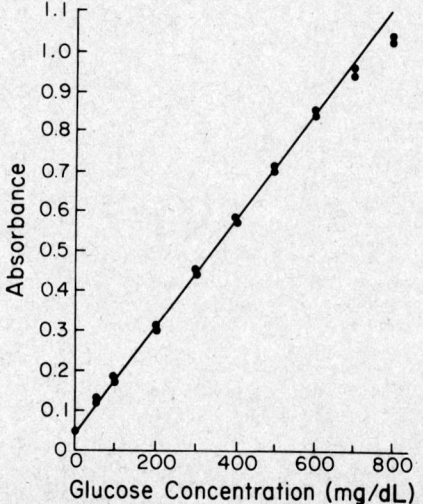

Figure 2D-3. Plot of observed response versus standard concentration to determine linear analytical range.

Table 2D-2. RECOVERY RESULTS—EXAMPLE EVALUATION DATA

Serum Pool	Glucose Added		Glucose Found* mg/dL	Glucose Recovered mg/dL	Recovery %
	μL/10mL	mg/dL			
A	0	0	61	—	—
A	100	100	159	98	98
A	400	400	457	396	99
B	0	0	171	—	—
B	100	100	268	97	97
B	400	400	562	391	98
				(Average)	98

*Average of quadruplicate assays.

Table 2D-3. INTERFERENCE RESULTS (BILIRUBIN, HEMOGLOBIN, LIPEMIA)—EXAMPLE EVALUATION DATA

Serum	Interferent			Method for Glucose	
	Type	Conc.	n	Comparative (mg/dL)	Test (mg/dL)
Normal	None		10	112 ± 4.4	113 ± 5.6
Icteric	Bilirubin	6.5 mg/dL	8	98 ± 5.3	97 ± 6.4
Hemolyzed	Hemoglobin	96 mg/dL	11	122 ± 3.2	120 ± 7.6
Turbid	$A_{660} = 0.45$		6	144 ± 5.5	140 ± 7.9

Table 2D-4. INTERFERENCE RESULTS (OTHER SUBSTANCES)—EXAMPLE EVALUATION DATA

Substance	Conc. (mg/dL)	n	Glucose (mg/dL)	
			Mean	±SD
None (control)	—	10	122	3.2
Ascorbic acid	14	6	119	4.1
Sodium salicylate	18	7	121	4.4
Trisodium citrate	—	8	122	3.5
Heparin	—	6	121	2.8
Disodium EDTA	—	9	120	3.3

Interaction. This parameter should be tested when the instrumentation used is subject to contamination of one sample by another, as for example in continuous flow systems, with use of flow through cuvets. An analytical run is made with the following sequence of low (L, 53 mg/dL) and high (H, 500 mg/dL) pools: $L_1L_2L_3$ $H_1H_2L_4$ $H_3H_4L_5$ $L_6L_7L_8$ $H_5H_6L_9$ $H_7H_8L_{10}$ $H_9H_{10}L_{11}$. Results of the five L pools which follow other L pools (L_2, L_3, L_6, L_7, L_8) give a mean of 52.3 mg/dL (SD 1.2 mg/dL). This average is compared with the average of 54.3 mg/dL (SD 1.5 mg/dL) obtained from the five L pools which follow the H pools (L_4, L_5, L_9, L_{10}, L_{11}). The difference of 2.0 mg/dL is the interaction or carryover error, which is less than the allowable biases and does not by itself limit the usefulness of the test results.

Day-to-day precision. Two quality control pools were analyzed daily for 20 days, giving the following means and standard deviations: low pool mean 52.5 mg/dL, SD 1.5 mg/dL; high pool mean 184.5 mg/dL, SD 5.1 mg/dL. The observed precision is acceptable when compared with the allowable standard deviations of 3.0 and 6.0 mg/dL, respectively.

Comparison of methods. A method of proven reliability is available for comparison purposes. Serum specimens from 114 patients are analyzed in duplicate by both the test and comparison methods. The duplicates are averaged for each method and the averages are plotted with the test method as the ordinate and the comparison method as the abscissa (Figure 2D-4). The plot shows that the glucose concentrations have a reasonable distribution over the analytical range of interest.

Regression analysis gives the following statistical data: slope 0.982, y-intercept 1.2 mg/dL, standard error ($s_{y/x}$) 4.2 mg/dL, correlation coefficient 0.98. These statistics are interpreted as follows: The slope is close to 1.00, showing that the proportional analytical error is 1.8%, which corresponds to -0.9 mg/dL at 50 mg/dL and -3.6 mg/dL at 200 mg/dL. The y-intercept is close to zero, indicating a small constant

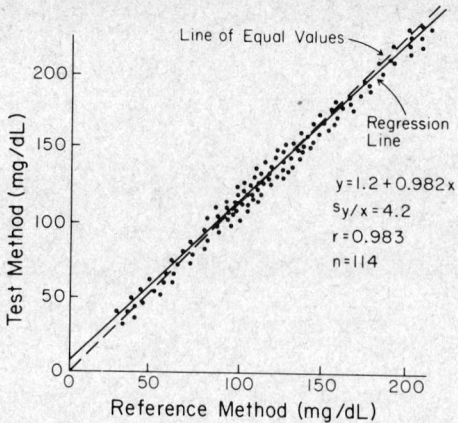

Figure 2D-4. Comparison of serum glucose values obtained with a test method versus those obtained on the same specimens with a reference method.

systematic error of 1.2 mg/dL. The scatter in the data, or the size of random differences (between the averages of duplicates by the two methods) is estimated by the standard deviation of 4.2 mg/dL for the points about the regression line. This shows that the differences between the averages of duplicates cover a range of 8–9 mg/dL, but differences between single measurements would be expected to be somewhat higher. The high value for the correlation coefficient confirms that a wide range of concentrations has been studied and indicates that the data should provide reliable estimates of the regression statistics (i.e., that simple linear regression should be satisfactory for analyzing this set of data).

To judge the acceptability of the method based on these data, the systematic errors must be estimated at the decision levels of interest. For example, the systematic error at the decision level of 50 mg/dL is only 0.3 mg/dL, $Y_C = 1.2 + 0.982 (50) = 50.3$. The systematic error at the decision level of 200 mg/dL is 2.4 mg/dL, $Y_C = 1.2 + 0.982 (200) = 197.6$. Note that the constant and proportional components are in opposite directions and their effects are somewhat balanced out at the concentrations of interest. Thus, systematic error by itself is quite small, less than the specified allowable biases. The performance of the method is judged acceptable when the systematic error concept of accuracy is employed.

To apply a total error criterion, the standard deviations from the day-to-day precision study are also required (TE = SE + RE, where RE is estimated as 2 times the day-to-day SD). The total error at 50 mg/dL is estimated as 3.3 mg/dL [0.3 + 2 (1.5)] and at 200 mg/dL as 12.6 mg/dL [2.4 + 2 (5.1)]. These observed total errors are also acceptably small compared with the allowable total errors (TE_A) of 10.0 and 20.0 mg/dL.

After determining that analytical performance is acceptable, then additional studies can be carried out to determine the stability and reproducibility of reagents. Reference intervals can be verified for selected populations, and would be expected to be very close to those of the reference method since there is little systematic error.

Evaluation reports which have been published in the scientific literature provide more detailed descriptions of the evaluation procedure and experiments, the data analysis and interpretation, and the final judgments on the acceptability of analytical methods. Some examples are the reports by Bandi et al.,[3] Passey et al.,[20] Westgard et al.,[25] and Garber et al.[12] Review of these reports will provide the analyst with real evaluation studies which can serve as models for developing evaluation protocols for one's own laboratory.

For assays other than glucose and specimens other than serum the above evaluation procedure may be too rigid. Recovery studies will seldom be feasible when assaying enzymes, for example. With expensive tests such as immunoassays or complex procedures such as electrophoresis, the analyst may have to settle for smaller numbers of test specimens. With specimens such as cerebrospinal fluid, which are seldom available in excess of needs, some testing for precision, interference, and sample interaction can be performed with dummy solutions made to simulate the biological materials. But regardless of the modifications of these guidelines which are applied, the essential caveat remains that the reliability of a method must be established by actual testing before it is introduced into the laboratory.

References

1. American Association for Clinical Chemistry, Committee on Standards: AACC policy regarding reagent sets and kits. Clin. Chem., *12*:43-44, 1966.
2. American Chemical Society, Committee on Environmental Improvement (D. MacDougall, Chairman), Subcommittee on Environmental Analytical Chemistry (W. B. Crummett, Chairman): Guidelines for data acquisition and data quality evaluation in environmental chemistry. Anal. Chem., *52*:2242-2249, 1980.

3. Bandi, Z. L., Fuller, J. B., Bee, D. E., et al.: Extended clinical trial and evaluation of urea nitrogen determination with the Ektachem GLU/BUN analyzer. Clin. Chem., *27*:480-485, 1981.

4. Barnett, R. N.: Medical significance of laboratory results. Am. J. Clin. Pathol., *50*:671-676, 1968.

5. Barnett, R. N., and Youden, W. J.: A revised scheme for the comparison of quantitative methods. Am. J. Clin. Pathol., Suppl., *54*:454-462, 1970.

6. Büttner, J., Borth, R., Boutwell, J. H., et al. (IFCC Committee on Standards): Provisional recommendation on quality control in clinical chemistry. I. General principles and terminology. Clin. Chem., *22*:532-539, 1976.

7. Büttner, J., Borth, R., Boutwell, J. H., et al. (IFCC Committee on Standards): Provisional recommendation on quality control in clinical chemistry. II. Assessment of analytical methods for routine use. Clin. Chem., *22*:1922-1932, 1976.

8. Cotlove, E., Harris, E. K., and Williams, G. Z.: Biological and analytic components of variation in long-term studies of serum constituents in normal subjects. III. Physiological and medical implications. Clin. Chem., *16*:1028-1032, 1970.

9. Elevitch, F. R., Ed.: CAP Aspen Conference 1976: Analytical goals in clinical chemistry. Skokie, Ill., College of American Pathologists, 1977.

10. Elion-Gerritzen, W. E.: Analytical precision in clinical chemistry and medical decisions. Am. J. Clin. Pathol., *73*:183-195, 1980.

11. Fraser, C. G.: Analytical goals in clinical biochemistry. *In*: Progress in Clinical Pathology. M. Stefanini and E. Benson, Eds. New York, Grune & Stratton, 1981, Chapter 5.

12. Garber, C. C., Westgard, J. O., Milz, L., et al.: DuPont *aca* III performance as tested according to NCCLS guidelines. Clin. Chem., *25*:1730-1738, 1979.

13. Gilbert, R. K.: Progress and analytic goals in clinical chemistry. Am. J. Clin. Pathol., *63*:960-973, 1975.

14. Logan, J. E.: Criteria for kit selection in clinical chemistry. *In*: Clinical Biochemistry: Contemporary Theories and Techniques, Vol. 1. H. Spiegel, Ed. New York, Academic Press, 1981, p. 43.

15. NCCLS Document C5-T: Methodological principles for establishing principal assigned values to calibrators. Villanova, Pa., National Committee for Clinical Laboratory Standards, 1977.

16. NCCLS Document EP2-T: Protocol for establishing performance claims for clinical chemistry methods—Introduction and performance check experiment. Villanova, Pa., National Committee for Clinical Laboratory Standards, 1979.

17. NCCLS Document EP3-T: Protocol for establishing performance claims for clinical chemical methods—Replication experiment. Villanova, Pa., National Committee for Clinical Laboratory Standards, 1979.

18. NCCLS Document EP4-T: Protocol for establishing performance claims for clinical chemical methods—Comparison of methods experiment. Villanova, Pa., National Committee for Clinical Laboratory Standards, 1979.

19. NCCLS Document EP5-T: User evaluation of precision performance of clinical chemistry devices—Tentative Guidelines. Villanova, Pa., National Committee for Clinical Laboratory Standards, 1982.

20. Passey, R. B., Gillum, R. L., Fuller, J. B., et al.: Evaluation and comparison of 10 glucose methods and the reference method recommended in the proposed product class standard (1974). *In*: Selected Methods of Clinical Chemistry, Vol. 8. American Association for Clinical Chemistry, 1977, p. 9.

21. Rubin, M., Barnett, R. N., Bayse, D., et al. (IFCC Expert Group on Diagnostic Kits and Reagents): Provisional recommendation (1978) on evaluation of diagnostic kits. Part 1. Recommendation for specifications of labelling of clinical laboratory materials. Clin. Chem., *25*:1499-1502, 1979; Revised recommendation (1983); Clin. Chim. Acta, *137*:371F-379F, 1984.

22. Tietz, N. W.: A model for a comprehensive measurement system in clinical chemistry. Clin. Chem., *25*:833-839, 1979.

23. Tonks, D. B.: A study of the accuracy and precision of clinical chemistry determinations in 170 Canadian laboratories. Clin. Chem., *9*:217-233, 1963.

24. Westgard, J. O.: Precision and accuracy: Concepts and assessments by method evaluation testing. CRC Crit. Rev. Clin. Lab. Sci., *13*:283-330, 1981.

25. Westgard, J. O., Carey, R. N., Feldbruegge, D. H., et al.: Performance studies on the Technicon SMAC analyzer: Precision and comparison of values with methods in routine laboratory service. Clin. Chem., *22*:489-496, 1976.

26. Westgard, J. O., Carey, R. N., and Wold, S.: Criteria for judging precision and accuracy in method development and evaluation. Clin. Chem., *20*:825-833, 1974.

27. Westgard, J. O., de Vos, D. J., Hunt, M. R., et al.: Method evaluation. Houston, American Society for Medical Technology, 1978.

28. Westgard, J. O., and Hunt, M. R.: Use and interpretation of common statistical tests in method comparison studies. Clin. Chem., *19*:49-57, 1973.

Quality Assurance

by James O. Westgard, Ph.D., and George G. Klee, M.D., Ph.D.

ESTABLISHING QUALITY GOALS

All laboratory personnel must be conscious of how the quality of their work affects the medical diagnosis and treatment of patients. Along with a general concern for accuracy, there must also be a thorough understanding of the essential quality characteristics of the analytical services they provide. Laboratory analysts, supervisors, and directors must have concern for the scope of testing that is offered, for the delivery of analytical services in a useful and convenient manner, and for ensuring that the analytical quality is appropriate for the medical application and interpretation of the test results.

It is important for laboratories to define their service objectives and to establish goals for quality. Without quality goals, there is no objective way to determine whether acceptable quality is being achieved, to plan effective strategies for improving quality, or to design procedures which assure that a specified level of quality will be attained.

Quality goals cannot be set on an absolute basis. They will vary from laboratory to laboratory, depending on the medical missions of the health care facilities and professional interests of the physicians utilizing the laboratory tests. Quality goals must also be considered in relation to cost. A goal of achieving the highest possible quality is not appropriate or practical when costs are being curtailed. In establishing quality goals, it is therefore more realistic to specify the quality that is necessary or adequate for the medical applications of the laboratory test results to be produced.

This chapter will focus primarily on analytical quality and the procedures by which it can be monitored. Goals for analytical quality can be established in the same way they can be established for purposes of method evaluation. The philosophy is to define an "allowable analytical error" based on "medical usefulness" requirements. A "total error" specification is useful because it will permit the calculation of the sizes of random and systematic errors that need to be detected to maintain performance within the allowable error limit. "Medical decision concentrations," i.e., the concentrations where medical interpretation of laboratory test results is particularly critical, are important in establishing the analytical concentrations at which analytical performance needs to be most carefully monitored. Thus, analytical goals can be established by specifying the allowable analytical error and the critical medical decision concentration. Method evaluation is only the first step in validating that analytical performance satisfies those goals. Quality control procedures should provide for the continuing verification that those goals are being achieved during routine service.

ELEMENTS OF A QUALITY ASSURANCE PROGRAM

The attainment of quality goals in a service laboratory requires a comprehensive quality assurance program. "Quality assurance" is used here to represent those practices which are generally recommended for assuring that desired quality goals are achieved. Quality assurance is a broad spectrum of plans, policies, and procedures which together provide an administrative structure for a laboratory's efforts to achieve quality goals. The term "quality control" (QC) is often used to represent those techniques and procedures that monitor performance parameters. Generally these are quantitative techniques that monitor particular sources of errors, estimate the magnitude of the errors, and alert laboratory personnel when there are indications that quality has deteriorated.

A quality assurance program involves virtually everything and everybody in the clinical laboratory. An error in any one step during the acquisition, processing, and analysis of a specimen and the reporting of a laboratory test result can invalidate the quality of the analysis and cause the laboratory to fall short of its quality goals. There are several essential elements of a quality assurance program.

Commitment. Dedication to quality service must be central; otherwise quality goals are not likely to be achieved. Quality must be a major consideration in all management decisions because any single decision may compromise other plans and practices for attaining quality goals. A true commitment is required by laboratory directors, managers, and supervisors if the efforts of other laboratory personnel are to be successful.

Facilities and resources. Laboratories must have the administrative support necessary to provide the quality of services that is desired. This means having adequate space, equipment, materials, supplies, staffing, supervision, and budgetary resources. These resources provide the basis upon which quality services can be developed and maintained.

Technical competence. High quality personnel are essential for high quality services. The educational background and experience of all personnel are important, as is the capability of providing the in-service training which can develop and maintain skills. In-service training can be a mechanism for assuring the competency of laboratory personnel, instilling quality goals, implementing quality control procedures, and providing for the continuing development of laboratory personnel, both technically and intellectually.

Quality assurance procedures. Good technical procedures are necessary to provide quality laboratory services. Three groups of procedures are discussed in this chapter:

1. The control of pre-analytical conditions or variables such as test requests, patient preparation, patient identification, specimen acquisition, specimen transport, specimen processing, specimen distribution, preparation of worklists and logs, and maintenance of records.

2. The control of analytical variables, which includes analytical methodology, standardization and calibration procedures, documentation of analytical protocols and procedures, and the monitoring of critical equipment and materials.

3. The monitoring of analytical quality by the use of statistical methods and control charts.

Problem solving mechanism. Although it is a particularly critical element in a quality assurance program, the need for a mechanism for problem solving is often underemphasized. Such a mechanism provides the link between the identification of a problem and the implementation of a solution to that problem. It is a "feedback loop" which responds to an error signal by making adjustments to reduce the size of the error or to prevent its recurrence. For problems limited to individual methods or instrument systems, the delegation of responsibility for the systems may provide the corrective mechanism. Specialized trouble-shooting skills can be developed and improved; preventive maintenance programs can be instituted. For problems occurring more generally, the in-service training program can be an important part of the mechanism but will often require additional input from a quality control technologist or supervisor to initiate the use of this mechanism and to help define its objectives. A different mechanism, not widely used in this country at present but receiving increased attention, is the use of "quality control circles."[54] These are small discussion groups which meet regularly to discuss problems and suggest solutions. By involvement of personnel, quality control circles heighten the interest and commitment to quality and also provide a creative feedback mechanism.

The comprehensive nature of quality assurance programs and their missions, goals, and activities have been discussed in greater detail by Eilers.[24] There are also detailed outlines of the elements of cost management for quality assurance,[25] literature reviews of the recent trends and developments,[31,35,40] detailed recommendations by professional organizations such as the International Federation for Clinical Chemistry[10-15] and the College of American Pathologists,[16] and entire books devoted to quality assurance practices in clinical laboratories.[22,38,52,75]

CONTROL OF PRE-ANALYTICAL VARIABLES

SYSTEMS ANALYSIS OF THE PROCESS

The responsibility for accurate and timely test reports generally lies with the laboratory, but many problems can arise prior to the analysis of the submitted specimens. In order to monitor and control these errors, it is essential to understand the process and identify the potential sources of error. Table 2E-1 shows the steps involved between the physician's initial request for a test and the arrival of a specimen for analytical testing. This is a "systems analysis" that identifies most of the critical steps for a typical laboratory, but each laboratory situation will be somewhat different and there may be additional sources of error. It is important for each laboratory to do this kind of systems analysis of its own specimen acquisition process in order to identify those areas where errors may occur.

Once the process has been documented, those steps which are most susceptible to error should be identified and should receive the greatest attention. Many times the steps of the process which cause the greatest number of complaints, such as lost specimens or delayed results, are judged to be most important, even though other steps, such as the appropriateness of test selection and the acceptability of a specimen, may be of greater importance for optimal medical care. Guidelines describing procedures for specimen handling are available from organizations such as the National Committee for Clinical Laboratory Standards (NCCLS). Documents by accrediting agencies such as the College of American Pathologists,[16] Centers for Disease Control,[22,52] and state regulatory agencies are also helpful in this regard.

Table 2E-1. SYSTEMS ANALYSIS OF STEPS IN ORDERING A TEST AND COLLECTING AND PROCESSING A SPECIMEN

Step	Potential Errors
1. Physician orders test	Inappropriate test
	Handwriting not legible
	Request written on wrong form
	Wrong patient identification
	Special requirements not listed
2. Nurse reviews order	Form lost or delayed
	Improper patient preparation
3. Clerk processes form	Form lost or delayed
	Wrong Addressograph plate used
4. Lab prepares tube	Wrong collection tube
	Wrong labels
	Wrong time scheduled
5. Phlebotomist draws blood	Wrong patient
	Sample hemolyzes
	Inadequate volume of specimen
	Tourniquet on too long
	Blood diluted with intravenous fluids
6. Transport to lab	Specimen lost or delayed
	Delivered to wrong area
	Unsuitable conditions
7. Separation of serum	Wrong centrifuge speed
	Tube breaks in centrifuge
	Aliquot tubes contaminated
	Wrong name on aliquot tube
	Blood hemolyzes
8. Storage before analysis	Wrong temperature
	Contamination from stoppers
	Precipitate forms
	Improper light exposure
9. Log in specimen	Transcription error
	Placed on wrong list
10. Prepare worksheets	Transcription error
	Specimen missed
	Specimens out of sequence

PROCEDURES FOR MONITORING PRE-ANALYTICAL VARIABLES

It is very difficult to establish effective methods for monitoring and controlling pre-analytical variables because many of the variables are outside the traditional laboratory areas. Monitoring pre-analytical variables requires the coordinated effort of many individuals and hospital departments, each of which must recognize the importance of these efforts in maintaining a high quality of service. Accomplishing such monitoring may require support from outside the laboratory, particularly from the institution's clinical practice committee or some similar authority.

Test utilization. Traditionally, laboratory test utilization has always been monitored or controlled to some degree, but current emphasis on the cost of medical care and government regulation of medical care may increase the importance of this factor. Clinical practice committees may decide that only certain tests are necessary for emergency care, and may therefore limit the availability of kind and volume of "stat" requests. Peer review audits may lead to other guidelines concerning the appropriate use of diagnostic tests in different clinical situations. Careful monitoring of test requests and their appropriateness is likely to increase in importance, and the laboratory will likely have a role in identifying situations where test utilization can be optimized and in providing in-service education to cause changes in ordering patterns.

Patient identification. Correct identifications of patients and specimens are a major concern for laboratories. The highest frequency of errors occurs with the use of handwritten labels and request forms. One method for checking identification is to compare identifiers, such as the patient's name and his unique hospital number. The identification on the specimen label should also check with the identification on the requisition form. The use of plastic embossed patient identification cards to imprint the patient's name on test request forms and on blood collection labels can eliminate transcription and identification errors but does not guarantee that the patient name on the labels correctly identifies the donor of the specimen. The greatest promise for correct identification of patient specimens is the introduction of the laser beam light wand which can read bar code or optical characters from labels, combined with printers that can generate matching labels. The use of reliable, machine-readable labels should markedly reduce identification problems.

Turnaround time. Delayed and lost test requisitions, specimens, and reports can be major problems for laboratories. An essential feature for monitoring the cause of delays is the recording of the actual times of specimen collection, receipt in the laboratory, and reporting of test results. This can be done manually by placing time stamps in key locations, such as blood-drawing centers, specimen-processing stations, result-reporting areas, and wards or chart-posting areas. It can be done more effectively by having computer systems which automatically document the times of test requests, specimen acquisition, processing, analysis, and reporting. Turnaround time can be monitored like any other quality control variable, and limits can be established to flag "out-of bounds" specimens. Lists of delayed specimens can also provide a powerful mechanism for detecting lost specimens or reports. Resolution of problems in this area can be aided by a systems analysis of the laboratory operations, which would help identify those steps and areas which cause delays and disruptions in service. A good system for monitoring patient, specimen, and information flow may be obtained by integration of the light wand identification system with a computer that could automatically track each specimen at each of the steps from test request to result posting.

Patient preparation. Laboratory tests are affected by many factors, such as recent intake of food, alcohol, or drugs, as well as by smoking, exercise, stress, sleep, posture during specimen collection, and other variables. (See Chapter 3 on specimen collection.) Proper patient preparation is essential for the test results to be meaningful. Although responsibility for this usually resides with personnel outside the laboratory, the laboratory must define the instructions and procedures for patient preparation and specimen acquisition. These procedures should be included in hospital procedure manuals and should be transmitted to patients by both oral and written instructions. Compliance with these instructions can be monitored directly when the laboratory employs its own phlebotomists. Specific inquiry should be made regarding patient preparation before specimens are collected, and efforts should be made to correct noncompliance. For tests in which standardization is very important (such as plasma catecholamines), it is best to collect the specimens in a controlled environment such as a clinical testing unit.

Specimen collection. The techniques used to acquire a specimen can affect many laboratory tests. Prolonged tourniquet application causes local anoxia to cells and excessive venous back pressure. The anoxia causes small solutes such as potassium to leak from cells, and the venous pressure concentrates cells, proteins, and substances bound to proteins (such as calcium). Blood collected from an arm into which an intravenous infusion is running can be diluted or contaminated. Hemolysis during and after collection alters the concentration of any analyte which has a red cell/plasma concentration differential. Improper containers and incorrect preservatives can greatly affect test results and make them inappropriate. The best mechanism to monitor and control this aspect of laboratory processing is to have a specially trained laboratory team assigned to specimen collection. All members of the team should be given explicit instruction in the proper methods of specimen collection. The identification of the person collecting a specimen should be maintained. Individuals who process the specimens should be trained to look for and document collection problems. Physicians should be encouraged to report clinically inconsistent results. Similarly, errors detected by limit checks, delta checks (differences between consecutive results on individual patients), or other algorithms should be recorded. Any collection problems should be reviewed with the individuals collecting the specimens. Pride of workmanship should be encouraged, and quality performance should be rewarded.

Specimen transport. The stability of specimens during transport from the patient to the laboratory is seldom monitored, though this aspect may be critical for some tests when performed locally, and for most tests when sent to regional centers and commercial laboratories.[49] Most laboratories have recommendations for specimen storage and transport, but many of these are empirical and lack adequate scientific documentation. Even the definition of "stability" is not well agreed upon, with some investigators accepting changes less than 10%,[44,77] while others relate stability to medically significant changes[57] or a percentage of the established analytical variability.[61]

In controlling specimen transport, the essential feature is the authority to reject specimens which arrive in the laboratory in an obviously unsatisfactory condition (such as a thawed specimen which should have arrived frozen). As with other QC procedures, a small number of problems are expected, but if the error rate gets too large, there is a need for systems analysis of transport procedures, with resulting modifications. In tests where stability is a major problem, the design of specific control procedures appropriate for those tests may be necessary.

Specimen separation and aliquoting. The separation and aliquoting of blood specimens is more directly under the control of the laboratory. The main variables are the centrifuges, the containers used, and the personnel.

Centrifuge performance. The care and maintenance of centrifuges has been discussed in Chapter 1A. For quality control purposes, centrifuges should be monitored by checking the speed, timer, and temperature.[34]

Container monitoring. Collection tubes, pipets, and aliquot tubes are sources of contamination by calcium and trace metals. Also, glass beads or other materials added to blood specimens to aid in the separation of serum from cells may cause contamination. Each lot number of materials used should be tested for contamination by calcium and possibly other elements. Cork stoppers should not be used on specimens intended for calcium determinations since false elevations of 10–50% may occur.[2] Some of the plasticizers used in making plastic containers interfere with drug analyses.[7,62] Also, some plastic materials adsorb trace amounts of some analytes and should not be used for substances in low concentration such as parathyroid hormone.[39] Because of the intricate relationships between specimen processing and analytical testing, supplies and processing procedures should not be changed without consulting the personnel responsible for the analytical testing. (See also Chapter 3.)

Personnel monitoring. As in other areas of the laboratory, the personnel who process the laboratory specimens should be carefully trained and supervised. A written procedure manual should be available in the area. "Fast-track" mechanisms should be established for the processing of emergency specimens. Quality control mechanisms should monitor both personnel performance and personnel safety. An important part of performance is throughput time, which can be calculated if one records the specimen arrival time and the time when processing is completed. Weekly or monthly summary statistics are an effective method for monitoring processing time. Also, the identification of the technicians should be recorded to facilitate the detection and correction of problems.

In optimizing the efficiency of a specimen processing laboratory, there is a trade-off between the time it takes to record and check parameters and the error rates or inconsistencies of the function. However, these records are necessary for documenting the daily operation and for helping to identify systematic problems.

Laboratory logs. When the serum aliquot tubes arrive in the laboratory, various logging and monitoring systems are necessary. In laboratories without computerized reporting, a request/report form generally accompanies the specimens. One should check that the patient name, identification number, and tests requested on the form match the information on the label of the specimen tube. The specimen should be inspected to confirm adequacy of volume and freedom from problems which would interfere with the assay, such as lipemia or hemolysis. The specimens are then stored appropriately, and the identification information and arrival time are recorded in a master log. If the analyses are performed in batches, specimen identification generally is recorded in specific locations on the worksheets using the information on the tube labels. After analysis, the results are recorded on the worksheet, and if both the assay and the individual test results pass the quality control criteria, the test results are transferred to the result forms for reporting. However, prior to reporting the results, a second technologist should verify the adequacy of the quality control and should check for transcription errors by comparing the results on the report forms with those on the master log. Specimens which require further analysis due to dilution or assay problems should be indicated either on the master log or on a delayed report log.

Transcription errors. A substantial risk of transcription error exists in the laboratory even with the double checking of results. Reduction of this type of transcription error is one of the theoretical advantages of computerization. However, unless patient identification and test results are entered directly from instrumentation that is interfaced to the computer, manual entry of the patient identification and test results can still result in errors. Some computer systems have error detection routines programmed into the terminal entry functions, such as check digits, limit checks, test correlation checks, and verification checks with master hospital files, but these routines are not well developed on most laboratory systems.

An elegant system for monitoring manual clerical functions has been developed to detect errors in blood banking records,[60] and it should be directly applicable to clinical chemistry. In this system, known errors are discretely introduced into the system using factitious patients. The types of errors introduced are chosen to represent errors likely to occur or errors which cause major problems. The factitious reports are routed to dummy locations and eventually are returned to the quality control technologist. One can calculate the efficiency of the laboratory error detection program by comparing the number of factitious errors discovered with the number introduced. This efficiency factor then can be used to estimate the actual laboratory error rate based on the number of true errors discovered in the laboratory. The implementation of this scheme requires the cooperation of laboratory personnel and the undercover work of a quality control person. By a combination of paranoia and pride of workmanship, most personnel are more conscientious when they know they are being monitored. There was a substantial increase in the rate of error detection shortly after this plan was implemented in the blood bank. This effect gradually faded over the next several months, but the rate remained better than prior to implementation. Public display of individual error rates encourages pride of workmanship.

CONTROL OF ANALYTICAL VARIABLES

There are many analytical variables which must be carefully controlled in order to assure accurate measurements by analytical methods. Reliable analytical methods are obtained by a careful process of selection, evaluation, implementation, maintenance, and control. Smooth and uninterrupted laboratory service requires many procedures aimed at preventing the occurrence of problems. Laboratories may experience different problems with the same analytical methods, owing to different amounts of effort being allocated to the care and support of those methods.

There are certain variables—water quality, calibration of analytical balances, calibration of volumetric glassware and pipets, stability of electrical power, and the temperature of heating baths, refrigerators, freezers, and centrifuges—that should be monitored on a laboratory-wide basis, since they will affect many of the methods in the laboratory. (These are discussed in other chapters.) In addition, there will be certain variables that relate more directly to individual analytical methods, and these require that procedures be developed to deal specifically with the characteristics of those methods.

Choice of analytical methodology. The selection and evaluation of analytical methods has been discussed in an earlier chapter, but it is important to recognize that the initial evaluation of a method often takes place in a setting somewhat more idealized than the production setting. It is therefore desirable to have a start-up period in the service laboratory before test results are to be reported. This period allows time to discover any additional problems, to develop maintenance programs which alleviate those problems, and to train a sufficient number of analysts to support the routine service operation.

Standards and calibration. The reliability of the analytical values obtained with a procedure often depends on the quality of the standards and the calibration procedure employed. Some knowledge of the overall structure of an accuracy-based measurement system is necessary to appreciate fully the problems in standardization and calibration. Uriano and Cali[64] have defined a hierarchy of analytical methods and reference materials which shows the relationship between reference materials and analytical methods of different quality. Figure 2E-1 illustrates how "field methods" are related to "reference methods" by "secondary reference materials," and how reference methods are related to "definitive methods" by "primary reference materials."

The practical meaning of this measurement structure has been discussed by Tietz.[63] The highest quality methods, the *definitive methods,* are to be used to validate reference methods and primary reference materials. The highest quality reference materials, namely *primary reference materials,* should be used in the development and validation of reference methods, the calibration of definitive and reference methods, and the production of secondary reference materials. *Reference methods* should be used to validate field methods. *Secondary reference materials* should be used to provide working standards for field methods and to assign values to control materials. *Control materials* are used only to monitor field methods.

To help implement this conceptual structure in clinical chemistry, the following definitions of reference materials have been proposed:[48]

Reference materials (RM): "A material or substance one or more properties of which are sufficiently well established to be used for the calibration of an apparatus or for the verification of a measurement method."

Certified reference material (CRM): This is a primary material described as a "reference material accompanied by, or traceable to, a certificate stating the property value(s) concerned, issued by an organization, public or private, which is generally accepted as technically competent." (For example, standard reference materials, SRM's, from the U. S. National Bureau of Standards.)

Calibration and test material (CTM): This is a secondary material described as "a reference material or solution with which the test sample is compared in order to determine the concentration of analytes or other quantities."

Control material: A "specimen, or solution, which is analyzed solely for quality control purposes and is not used for calibration purposes."

To illustrate the use of these materials, a field (routine) method would be traceable to a reference method when it has been evaluated against that method or when it has been calibrated with a secondary reference material or CTM whose value was established by analysis with a reference method. The reference method itself should have been validated against a definitive method and calibrated with a primary reference material or CRM whose value was established by the definitive method. Proper choice of calibrating materials establishes the "traceability" of the analytical results to the "true value" through the structure of the measurement system, as illustrated in Figure 2E-1. Note also that the use of control materials solely for internal quality control does not establish traceability, but use of such materials in external quality assurance may validate field methods against reference methods and thereby may establish traceability.

The selection of reference materials is therefore critical in validating the performance of an analytical method. CRM's should be used at least in the initial evaluation studies and in establishing the acceptability of CTM's used in routine service. The specific characteristics of CTM's should be documented, along with the number of different concentrations of standard solutions and the frequency of their use. These latter choices depend on the characteristics of the analytical method, particularly the stability, reproducibility, and linearity.

Documentation of analytical protocols. Step-by-step procedures for performing analytical determinations are critical if the methods are to provide the same results when used by different analysts over a long period of time. To maintain such consistency requires written protocols, or method and procedure manuals. The essential information to be contained in such documents is outlined in Table 2E-2. More detailed guidelines are provided in an NCCLS recommendation.[50] Method manuals should be revised and updated whenever changes occur. While this has been a

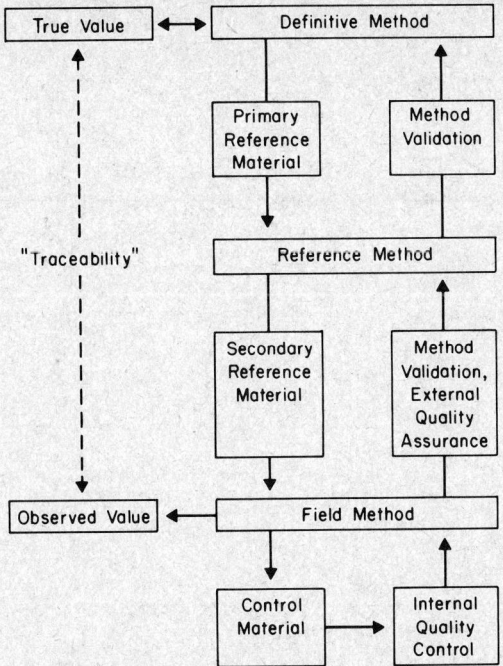

Figure 2E-1. Structure of an accuracy-based measurement system showing relationships between reference methods and materials.

problem because of the time required to retype the directions, the advent of word processors should alleviate this difficulty and permit rapid revisions whenever changes occur.

Verification of reference intervals. The details for the establishment of reference values have been presented in an earlier chapter. Such studies are generally performed for most methods during their development, and the results are available in the technical literature from manufacturers and in published scientific reports. However, these intervals may be dependent on variables such as subject population, which may not adequately represent the performance of a method in another laboratory. It is therefore recommended that reference intervals be verified for normal subjects representing the population encountered in a respective laboratory. It is also helpful to compare the distribution of patient test results with the expected intervals. When possible, the medical records of patients having test results near the limits of the reference intervals should

Table 2E-2. OUTLINE FOR A PROCEDURE MANUAL

A. *Procedure name*: List principal name of procedure first, alternative names next. List commonly used abbreviations.

B. *Clinical significance*: Give a brief explanation of how the test is used in clinical medicine. Include reference intervals for specific diseases and recommended diagnostic and therapeutic action limits.

C. *Principle of method*: Give a brief statement of the principles on which the method is based.

D. *Specimen*: List type of specimens that can be used, recommended volume, as well as minimum volume. Indicate conditions that render the specimen unacceptable, such as hemolysis or lipemia. List patient preparation procedures. Provide instruction for specimen handling prior to testing.

E. *Reagents and equipment*: Provide a list of reagents in order of their use, including standards. Indicate the names and addresses of suppliers and detailed instructions for preparation, including checks to be performed prior to use. List equipment used and special precautions required.

F. *Procedure*: This should be a step-by-step description, including calibration and quality control procedures. There should be enough detail so that the assay could be performed by a person unfamiliar with the test. The necessary calculations should be included.

G. *Reference values*: List the reference ranges for healthy subjects. Indicate parameters such as age, sex, or race that affect reference values. Include the nature of the population studied, the number of subjects, and the date or reference for the original work.

H. *Comments*: Include any special analytical variables affecting the test, such as pH or temperature. Include effects of commonly used drugs, any dangers or personal hazards in the procedure, and any special safety precautions and procedures.

I. *References*: Give the primary literature references which describe the method, or the references upon which the method is based.

be reviewed to confirm these ranges. It is also desirable to check for sex and age differences, using distributions of patient results normally encountered at the institution.

Monitoring technical competence. The personal characteristics and techniques of individual analysts may greatly affect certain analytical methods, particularly manual methods. Proper training of laboratory personnel to achieve uniformity in technique is important, as is scheduling sufficient routine service to maintain proper techniques. In training personnel on new analytical methods, it is helpful to have written lists of objectives that outline the critical tasks and knowledge that must be mastered. Having such objectives assures systematic instruction which covers the critical points. Prior to performing analyses for clinical use, the technical competence of the personnel should be checked and practice runs should be performed. Periodic monitoring of competency may be difficult, but incident reports and results from internal and external quality control checks will identify specific problems, which should be discussed directly with the personnel involved. In-service and continuing education programs will help maintain and improve competence. Employee conferences will also help uncover nontechnical problems which may affect work quality.

Inventory control of materials. Stable operation of laboratory methods is vitally dependent on the materials utilized with the methods. Procedures are needed to inventory materials and initiate orders when supplies are low. These procedures must be tailored to the particular reagents and supplies that are in use. In general, when materials are stable and changes in lot numbers cause problems, large stocks should be maintained, though this may be limited by economic considerations and storage space. Adequate inventory should be maintained to allow time for additional shipment and testing of additional supplies if the current shipment should not be satisfactory. Formal inventory control procedures will help minimize required storage by initiating orders when stock reaches a certain predetermined level. Many companies will ship reagents on a standing order once usage rates are established.

Input control of materials. Along with inventory management, the quality of materials should be monitored when they are received. Interruptions in service can be avoided by testing the adequacy of materials prior to their introduction for routine use. It is particularly important that patient results from new lots of reagents and standards be compared with those obtained with previous lots. In radioimmunoassays, for example, many problems are related to changes in radio-labels, standards, or second stage antibodies. New lots of blood-drawing tubes, processing tubes, and the like should receive some parallel or comparison testing prior to their routine use. In establishing pretesting procedures for monitoring incoming materials, it should be recognized that this kind of testing can become expensive.

Another part of input control is the proper labeling of reagents and materials. Proper identification should include the name, lot number (particular manufacturing batch), concentration, date received, date prepared, date opened, date of expiration, and recommended storage conditions.

Monitoring method changes. An essential tool for solving analytical problems is a record or log of all changes and problems occurring with a method. Detailed examples of instrument logs and check lists have been developed by Winstead.[78] These logs should include the date, time, analyst, and any changes in lots of reagents, materials, or standards. All instrument maintenance should be recorded, including all work performed by service personnel from outside the laboratory. The occurrence of control problems should be indicated, along with actions employed to resolve the problems. Periodic review of these logs should lead to implementation of preventive maintenance programs, based on the frequency of occurrence of particular problems.

CONTROL OF ANALYTICAL QUALITY USING STABLE CONTROL MATERIALS

The performance of analytical methods can be monitored by analyzing specimens whose concentrations are known, then comparing the observed values with the known values. The known values are usually represented by a range of acceptable values, or upper and lower limits for control (control limits). When the observed values fall within the control limits, this should assure the analyst that the analytical method is working properly. When the observed values fall outside the control limits, the analyst should be alerted to the possibility of problems in the analytical determination.

CONTROL MATERIALS

The known specimens that are analyzed for quality control purposes are called "control materials." They need to be stable materials, available in aliquots or vials, that can be analyzed periodically over a long time. There should be little vial-to-vial variation so that differences between repeated measurements can be attributed to the analytical method alone. The control material should preferably have the same matrix as the test specimens of interest; for example, a protein matrix may be best when serum is the test material to be analyzed by the analytical method. Materials from human sources have generally been preferred, but because there is some risk of hepatitis infection, bovine materials offer a certain advantage in safety and are more readily available. The concentration of analyte should be in the normal and abnormal ranges, corresponding to concentrations which are critical in the medical interpretation of the test results.

Control materials can be prepared in the laboratory from left-over sera. Bowers et al.[6] provide directions for preparing 25-L pools of human control material. Excess sera are collected daily in polyethylene containers and stored at -20 °C in a freezer. After 25 L has been collected, the frozen material is thawed, refrozen, and stored for one month at -20 °C. After this stabilization period, the pool is thawed again and the concentrations of electrolytes are adjusted by additives. Glacial acetic acid is added to adjust the pH; enzyme activities are adjusted by addition of high activity concentrates, and the pool is mixed after each addition. The entire pool is then filtered through a series of seven different Millipore filters and finally is aliquoted into test tubes which are stored in a freezer at -20 °C for one month before use.

Many laboratories choose to purchase materials from one of several companies that manufacture control sera or "control products." These are generally supplied as lyophilized or freeze-dried materials which are reconstituted by adding water or a specific diluent solution. Also available are materials having matrices representing urine, spinal fluid, and whole blood. Some liquid control materials are available and have the potential advantage of eliminating errors due to reconstitution. However, the matrices of these liquid materials contain other materials which are a potential source of error with some analytical methods and instruments. Studies on the effects of the matrix of control materials, such as the comparison of intra-method variability of patient sera and control sera, can be found in the literature.[65]

When selecting commercial control materials, there are several other considerations in addition to the matrix of the product. Stability is critical because it is often desirable to purchase a year's supply of one manufacturing lot or batch. Different batches (or lot numbers) of the same material will have different concentrations, which require new estimates of the mean and standard deviation. The size of the aliquots or vials should be convenient for the analytical methods to be monitored. Larger-sized vials are generally less expensive (on a per mL basis), but unused materials may eliminate any savings. Two or three different materials should be selected to provide concentrations which monitor performance at different medical decision levels. There may be some advantage in selecting materials from different manufacturers to minimize possible problems with a given manufacturing process. On the other hand, a series of materials designed by one manufacturer to have a certain relationship between concentrations can be used to gain additional information regarding linearity and accuracy.[32]

Control products can be purchased as assayed or unassayed materials. Assayed materials come with a list of values for the concentrations that are expected for that material. This list often includes both the mean and standard deviation. Values may even be specified for several of the common analytical methods and preferably for a reference method used to measure a particular analyte. Because of the work required to determine these values, the assayed materials are more expensive. While the stated assay values are useful in selecting the desired materials, it is advisable to determine the mean and standard deviation in the user's laboratory, because this process improves the performance characteristics of statistical control procedures.

GENERAL PRINCIPLES OF CONTROL CHARTS

The most common way of comparing the values observed for control materials with their known values is through the use of control charts. These are simple graphical displays in which the observed values are plotted versus the time when the observations were made. The known values are represented by an acceptable range of values, as indicated on the chart by lines for upper and lower control limits. When the plotted points fall within the control limits, this is generally interpreted to mean that the method is performing properly. When points fall outside the control limits, problems may be developing.

The control limits are usually calculated from the mean ($\bar{x}$) and standard deviation (s) obtained from repeated measurements on the known specimens by the particular analytical method that is to be controlled. The mean and standard deviation are calculated from the following equations:

$$\bar{x} = \frac{\Sigma x_i}{n}$$

$$s = \sqrt{\frac{n\Sigma x_i^2 - (\Sigma x_i)^2}{n(n-1)}}$$

where x_i is an individual control observation and n is the number of observations in the time period being monitored. The initial estimate should be based on measurements obtained over a period of at least one month when the method is working properly. These conditions are necessary because the measurements are used to characterize the distribution of values that is expected during stable routine operation of the analytical method. The initial estimate may not be entirely reliable, due to the low number of data points and possible outliers in the data. The estimates can be revised when more data have been accumulated by recording n and the summations of x_i and x_i^2, then using the cumulative totals in the equations above to give cumulative means and standard deviations. The effects of outliers can be minimized by eliminating values exceeding the mean by more than 3.1 to 3.8 s's (where the exact factor depends on the total number of data points: 3.14 for n=30; 3.22, n=40; 3.33, n=60; 3.41, n=80; 3.47, n=100; 3.66, n=200; 3.83, n=400).[9]

It is assumed that error distribution of the analytical method is Gaussian. The control limits are set to include most of the control values, usually 95 to 99.7%, which correspond to the mean ± 2 or 3 standard deviations (s). Since it should be a relatively rare occurrence to observe a value in the tails of the distribution (only 1 out of 20 times for 2 s limits, 3 out of 1000 for 3 s limits), such an observation is suspect and suggests that something may have happened to the analytical method. Such an occurrence could have caused a shift in the mean (an accuracy problem), which would result in a higher probability for exceeding the limits, or it could have caused an increase in the standard deviation (a precision problem), which would widen the distribution and also result in a higher probability for exceeding the limits.

Figure 2E-2,*A* illustrates how the distributions of control values will appear for three different situations: (a) stable performance where only an occasional observation exceeds the control limits; (b) occurrence of a systematic error which shifts the mean of the distribution and causes a much higher expectation or probability of observing control values outside one of the control limits; and (c) occurrence of an increase in random error or imprecision, which widens the distribution and causes a much higher probability of observing a control value outside either of the control limits.

Control charts are utilized to compare the observed control values to the control limits and to provide a visual display that can be quickly inspected and reviewed. These charts have the concentration or observed value plotted on the y-axis versus time of observation on the x-axis. It has been common practice to plot one month's data on a chart, usually one or two points a day, but the time axis should be chosen to be appropriate for the method being monitored. An example of a control chart is shown in Figure 2E-2,*B*, where the control values represent the three situations in Figure 2E-2,*A*, with 10 values per situation for a total of 30 values. When the analytical method is operating properly, the control values fall predominantly within the control limits. When there is an accuracy problem, the control values are shifted to one side, and there may be several values in a row falling outside one of the limits. When there is a precision problem, the control values fluctuate much more widely, and there may be values exceeding both the upper and lower control limits.

Interpretation of the control data can be guided by certain decision criteria or control rules, which define when an analytical run is judged "in-control" (acceptable) or "out-of-control" (unacceptable). These control rules are given symbols as A_L or n_L, where A is the abbreviation for a statistic, n is the number of control observations, and L refers to the control limits. For example, 1_{3s} refers to a control rule where 1 observation exceeding the mean ± 3 s control limits is the criterion for rejecting the analytical run.

"Analytical run" is used here to refer to that segment of data for which a decision on acceptability is to be made. This is the group of patient results that are to be reported, based on the control results available for inspection at that time.

"n" is the number of control observations available for inspection when a decision is to be made on the acceptability of an analytical run. For example, when there is one control observation

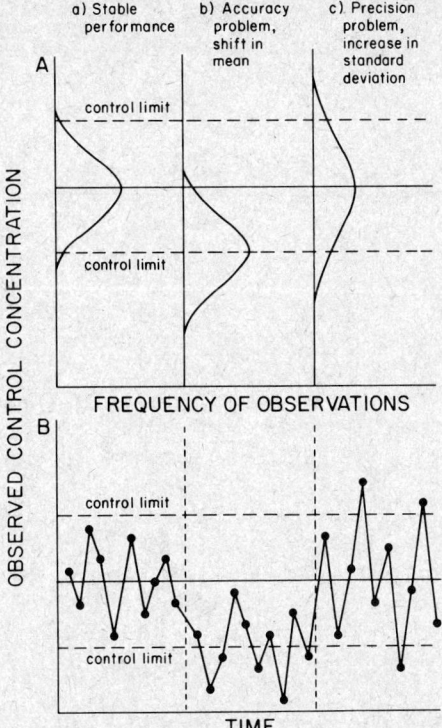

Figure 2E-2. Conceptual basis of control charts: *A*, Frequency distributions of control observations for different error conditions. *B*, Display of control values representing those distributions when concentration is plotted versus time on a control chart.

preceding and one following a group of ten patient samples whose results are to be reported, there are two control observations in that analytical run.

The different control procedures discussed here have different performance capabilities, depending on the control rules and the number of control observations that are chosen. These choices should be related to the quality goals set by the laboratory. Many of the procedures in use today have not been chosen for best performance, but rather for ease of use when implemented manually. Knowledge of the performance characteristics of control procedures is necessary to select procedures that will detect relevant laboratory problems without causing too many "false alarms" (rejections when there are no analytical problems). Experienced analysts often employ a series of informal rules or judgments to reduce the number of false alarms, without knowing their effects on the detection of real problems, or true alarms. There should be some quantitative assessment of these two characteristics, false alarms and true alarms, whenever capabilities of new control procedures are assessed or established control procedures are reviewed.

PERFORMANCE CHARACTERISTICS OF A CONTROL PROCEDURE

The performance of a control procedure can be described by its probability for rejection.[72] Probability refers to the likelihood that an event will occur; in this case, the event is the rejection of an analytical run. Ideally, there should be no rejections of analytical runs when the analytical method is performing properly. On the other hand, when analytical errors occur, the control procedure should provide a rejection signal. This corresponds to having a probability of 0.00 in the first situation and a probability of 1.00 in the second.

The term "probability for false rejection" (p_{fr}) is used to describe the first situation where there are no analytical errors present, except for the inherent imprecision or inherent random error of the analytical method. (There is always some random error associated with an analytical method, even when it is working properly. This is the random error that is estimated by the replication experiment during method evaluation studies.) When only this inherent random error is present, without any additional errors, the probability for false rejection should be zero. The frequency of false rejections is very critical, because false rejections are like false alarms. Too many false alarms cause the analyst to disregard the alarm system, even when the alarm is occurring as a result of real errors that should be corrected.

The term "probability for error detection" (p_{ed}) is used to describe the second case, where there is an analytical error in addition to the inherent or background random error. The probability for error detection should be high (near 1.00) when these errors are large enough to invalidate the quality goals for the analytical method. Unfortunately, this is difficult to achieve because the control procedure is attempting to detect a signal (additional error) in the presence of noise (inherent random error).

These critical performance characteristics—probability for false rejection and probability for error detection—can be summarized by graphs of probability for rejection (y-axis) versus the size of analytical error (x-axis). Graphs such as that shown in Figure 2E-3 are called power functions because they describe the statistical power of the control procedure.[70] The different lines on the graph represent different numbers of control observations, for example, n = 1, 2, and 4 for the three lines on this graph. The probability for false rejection is given by the y-intercept, the point where each line intersects the y-axis. For this control procedure with n from 1–4, p_{fr} is 0.01–0.02, or 1–2%. The probability for error detection can be determined for any size error by first locating the size of the error on the x-axis, drawing a line up to intersect the power function curve, drawing a horizontal line from the point of intersection to the y-axis, and reading the probability for rejection from the scale on the y-axis. For this control procedure, p_{ed} is 0.13 for n = 1, 0.22, for n = 2, and 0.48 for n = 4. Thus, there is a 13–48% chance of detecting an error of this size with this control procedure, depending on the number of control observations used.

Two power function graphs are necessary, one to describe the performance for random error and the other the performance for systematic error. For random error (RE), as shown in Figure 2E-4,A, the x-axis is labeled ΔRE. A value of 1.0 corresponds with the original standard deviation of the analytical method, a value of 2.0 to a doubling of that standard deviation, 3.0 to a tripling, and so on. For systematic error (SE), the x-axis is labeled ΔSE (Figure 2E-4,B). A value of 1.0 s corresponds to a systematic shift equivalent to the size of the standard deviation, a value of 2.0 s to a shift equivalent to two times s, and so on.

Power functions can be determined by mathematical calculations or by computer simulation studies, the latter having been used in clinical chemistry to characterize many of the commonly used control procedures.[70] Power functions are useful for evaluating the performance capabilities of individual control procedures, for comparing the performance of different control procedures, and for designing new procedures with improved performance characteristics. The best control procedure is the one with the lowest probability for false rejection and the highest probability for detecting those errors that are large enough to invalidate the analytical quality goals for the method.

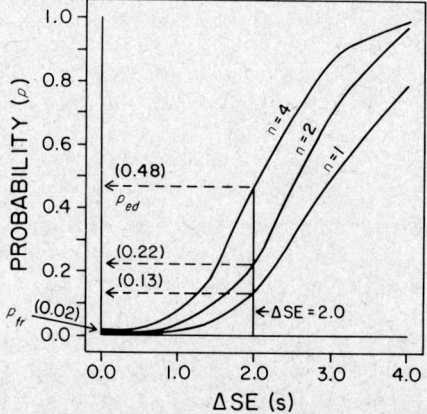

Figure 2E-3. Power function graph presenting probability for rejection on the y-axis versus size of error on the x-axis. The different lines represent different numbers of control observations, in this case n = 1, 2, and 4.

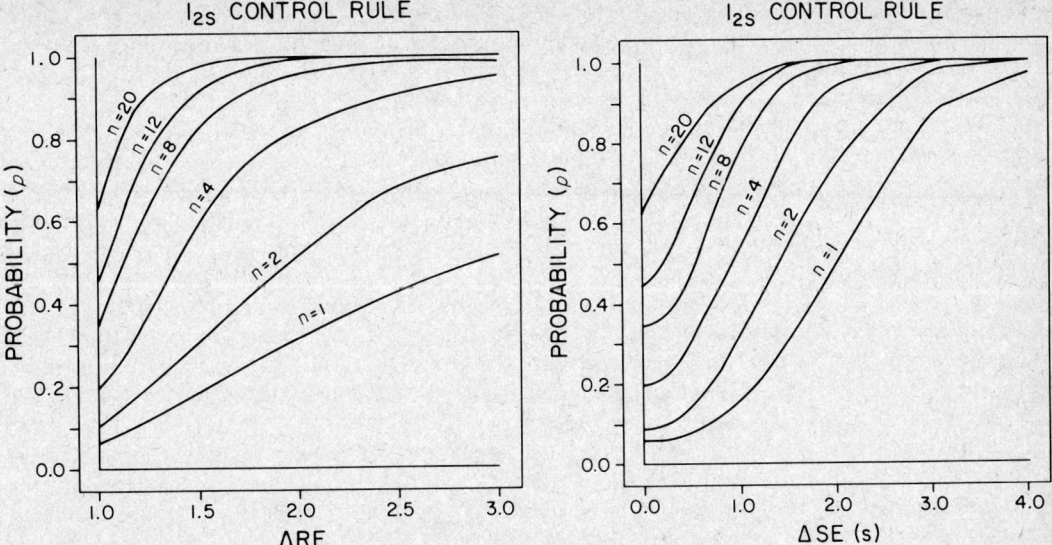

Figure 2E-4. Power functions for 1_{2s} control rule. A, Random error. B, Systematic error. (From Westgard, J. O., and Groth, T.: Clin. Chem., *25*:863-869, 1979.)

The size of the critical errors that need to be detected if the control procedure is to maintain the specified quality can be calculated from the following equations:

$$\Delta SE_{critical} = \frac{AAE}{s} - 1.65$$

$$\Delta RE_{critical} = \frac{AAE}{1.96\ s}$$

The error specification is represented by the term AAE, which is the allowable analytical error, defined as a 95% limit for total error. $\Delta SE_{critical}$ is the smallest systematic error that would invalidate the quality of the analytical results. $\Delta RE_{critical}$ is the smallest increase in random error that would invalidate the quality of the analytical results.

The rationale for these equations is shown in Figure 2E-5. For systematic error, it is necessary

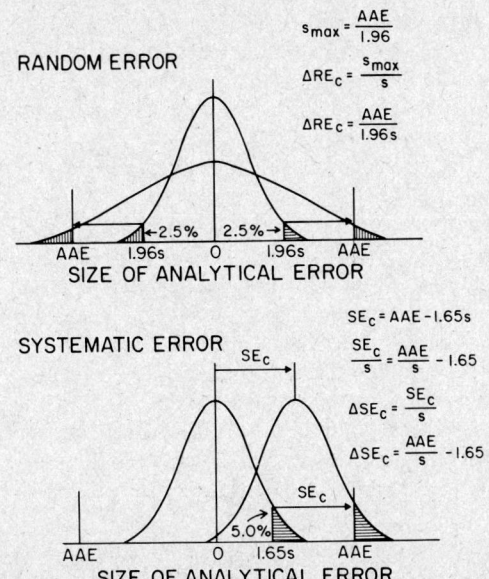

Figure 2E-5. Rationale for calculation of errors that must be detected in order to achieve a quality goal.

to keep the 5% tail of the distribution inside the AAE limit to maintain the specified quality. The 5% point in the tail of the error distribution occurs at 1.65 s from the mean; thus the difference between AAE and 1.65 s is the critical systematic error. Because the units of the two terms are different, it is convenient to divide by s and then define SE/s as $\Delta SE_{critical}$. This term expresses the critical systematic error as a multiplier of the standard deviation (SE equals $\Delta SE_{critical}$ $\times$ s). For random error, it is necessary to keep both of the 2.5% tails inside the AAE limits to maintain the specified quality. The 2.5% tails occur at ± 1.96 s from the mean; thus AAE divided by 1.96 would be the maximum standard deviation that could be allowed. $\Delta RE_{critical}$ is the ratio of this maximum s to the stable s, and expresses the increase in the method's standard deviation that must be detected.

For example, if AAE were specified to be 10 mg/dL at a concentration of 120 mg/dL, and s were estimated as 2.0 mg/dL and bias as 0.0 mg/dL, ΔSE would be 3.4 s and ΔRE is 2.5. This means that the control procedure needs to be able to detect a systematic shift equivalent to 6.8 mg/dL (3.4 s) and a random error equivalent to a standard deviation of 5.0 mg/dL (2.5 s).

With knowledge of these values for critical errors and the power functions for a control procedure, it is possible to make a critical assessment of the performance achievable by different statistical control procedures. The quality of the control procedures themselves can be evaluated and related to the quality goals defined for the laboratory.

Levey-Jennings Control Chart

Control charts were first introduced into the clinical chemistry laboratory by Levey and Jennings in 1950.[46] They demonstrated how the industrial control procedures developed by Shewhart[59] could be used with the mean and range of duplicate measurements from clinical chemical methods. In 1952 Henry and Segalove[36] further popularized the concepts of quality control in clinical chemistry and described an alternate chart where single control values were plotted directly. This "single value" chart was adopted by most laboratories because data calculations were not required prior to plotting the control results. Today this single value chart is commonly known as a "Levey-Jennings" chart, even though Levey and Jennings recommended plotting the mean and range of replicate measurements.

To use a Levey-Jennings control chart, follow these steps:

1. Analyze samples of the control material by the analytical method to be controlled on at least 20 different days. Calculate the mean and standard deviation for those results.

2. Construct a control chart on graph paper. Label the y-axis "control value" and set the range of concentrations to include the mean ± 4 s. Draw horizontal lines for the mean and the upper and lower control limits. Set the control limits as the mean ± 3 s when the number of control observations, n, is 2 or greater. When n is 1, control limits may be set as the mean ± 2 s. (See following discussion of performance characteristics.) Label the x-axis in terms of time, using day, run number, control observation number, or whatever is most appropriate for recording the relative time of the control observations.

3. Introduce control specimens into each analytical run, record the values, and plot each value on the control chart.

4. When the control values fall within the control limits, interpret the run as being "in-control" and report the patient results. When a single control value exceeds the control limits, stop the method. Do not report patient results. Inspect the method to determine the cause for the errors. Solve the problem, then repeat the entire run—specimens and control samples. Determine control status for the new run in the same manner.

An example of a Levey-Jennings chart is shown in Figure 2E-6, where control limits have been set as the mean ± 3 s. Power functions for a Levey-Jennings chart having 3 s control limits, or a 1_{3s} control rule, are shown in Figure 2E-7. The probability for false rejection is seen to be less than 0.05 or 5% even when n is very large. The probability for error detection increases as n increases, but for n's of 2–4 the procedure is not very sensitive for either random or systematic errors.

Figure 2E-4 shows the power functions for a Levey-Jennings chart having 2 s control limits, or the 1_{2s} control rule. Observe that the y-intercept increases rapidly as n increases. The probabilities for false rejection are as follows: n=1, 0.05; n=2, 0.09; n=4, 0.18; n=8, 0.33; n=12,

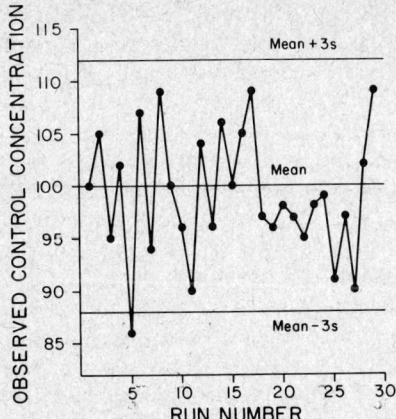

Figure 2E-6. Levey-Jennings control chart having control limits set as the mean $\pm 3s$. Concentration is plotted on the y-axis versus time (run number) on the x-axis.

0.46; and n = 20, 0.64. Since probabilities for false rejection greater than 0.05 or 5% are not desirable, the use of the 1_{2s} rule is limited to the case where n = 1.

The 1_{2s} rule has actually been widely used with higher n's. Such use causes a false rejection problem which may compromise the usefulness of the control procedure. There is no easy way to tell whether the rejection signal is due to the background random error (false rejection), or whether an additional error has occurred (true rejection). This difficulty has led to a more qualitative interpretation of the control results and often to routine repetition of the control measurements (and sometimes patient samples as well) whenever a control value exceeds a 2 s limit. When the second or repeated control value is observed to be outside the 2 s control limits, this provides evidence for a true rejection, and problem-solving procedures should be started.

A similar false rejection problem arises when multi-channel instrument systems are controlled using the 1_{2s} rule on each of several channels. For 1 control material being analyzed by 4, 8, 12, and 20 channels, the chances that the control value on at least one channel exceeds its 2 s limits are 18%, 33%, 46%, and 64%, respectively. Such a high rate of values exceeding the reference range may cause the same percentage of work to be routinely repeated, obviously compromising the efficiency of the laboratory.

It is important to recognize the seriousness of the false rejection problem and its relationship to the control limits that are chosen for the Levey-Jennings chart. These false rejections are in

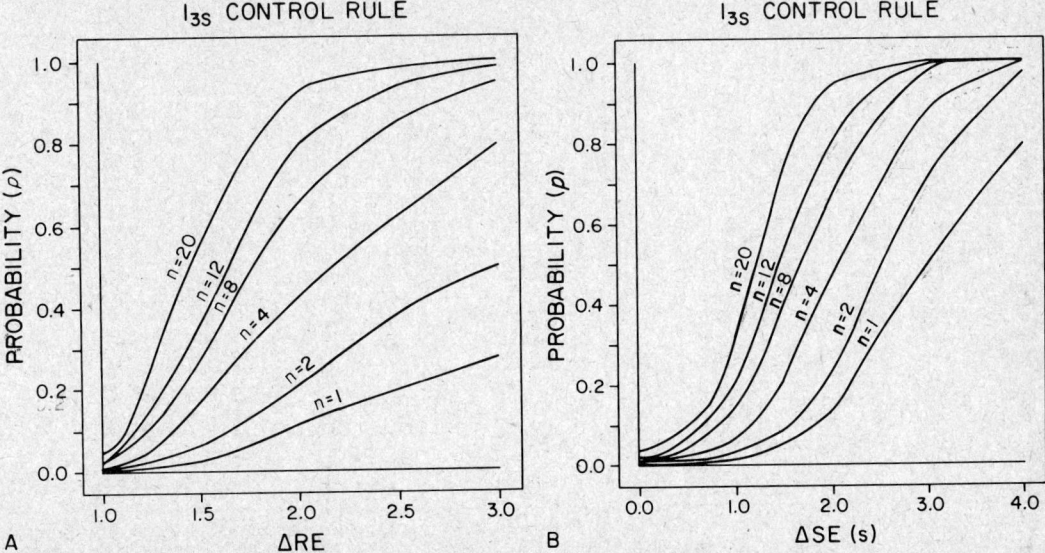

Figure 2E-7. Power functions for the 1_{3s} control rule. A, Random error. B, Systematic error. (From Westgard, J. O., and Groth, T.: Clin. Chem., *25*:863–869, 1979.)

effect an inherent property of the control procedure. They occur because of the control limits that have been selected, not because of any problems with the analytical method. The use of 2 s control limits cannot therefore be generally recommended. With the use of 3 s control limits, the false rejection problem is eliminated, but unfortunately error detection is also reduced.

An alternative approach to having to choose between 2 s and 3 s limits is to use control limits that increase as n increases, thus maintaining a specified probability for false rejection. For example, a $1_{0.05}$ control rule can be employed, setting the control limits to maintain a probability for false rejection of 0.05. Factors for calculating the control limits, as a function of n, are given in Table 2E-3 for false rejection probabilities of 0.05, 0.01, and 0.002. As an example of using these limits, a method having 4 control observations per analytical run could have control limits set as the mean ±2.50 s in order to maintain a 0.05 probability for false rejection. To monitor an instrument system having 6 simultaneous channels with 1 control measurement per channel could employ control limits set as the mean ±3.13 s to maintain a 0.01 probability for false rejection. A 4-channel instrument having 2 control measurements per channel and per run could have control limits set as the mean ±3.59 s to maintain a 0.002 probability for false rejection.

The use of these control rules where the actual control limits vary with n has not been very widespread. This is probably due to the practical difficulties of preparing control charts having different limits for different methods, and also due to the difficulties in teaching laboratory analysts the use of these more complicated control procedures. The advantage of these rules over the 1_{2s} rule is that false rejections can be maintained at a specified acceptable level. Their advantage over the 1_{3s} rule is that error detection can be somewhat improved for low n's when the false rejection level is set at 0.05, and to a lesser extent when set at 0.01.

Westgard Multi-Rule Chart

The "multi-rule" procedure developed by Westgard et al.[67] utilizes a series of control rules for interpreting control data. The probability for false rejections is kept low by selecting only those rules whose individual probabilities for false rejection are very low (0.01 or less). The

Table 2E-3. FACTORS FOR CALCULATING CONTROL LIMITS

Control Rules	Number of Control Observations								
	2	3	4	6	8	10	12	16	20
A. Factors for calculating control limits from the standard deviation									
$1_{0.05}$*	2.24	2.39	2.50	2.64	2.74	2.81	2.86	2.94	3.02
$1_{0.01}$	2.81	2.93	3.01	3.13	3.21	3.27	3.31	3.38	3.43
$1_{0.002}$	3.27	3.36	3.44	3.52	3.59	3.64	3.66	3.72	3.75
$2_{0.05}$*	1.01	1.22	1.33	1.47	1.56	1.62	1.67	1.74	1.80
$2_{0.01}$	1.47	1.64	1.74	1.86	1.93	2.00	2.03	2.09	2.14
$2_{0.002}$	1.86	2.01	2.09	2.19	2.26	2.30	2.34	2.40	2.44
$\bar{x}_{0.05}$*	1.39	1.13	0.98	0.80	0.69	0.62	0.57	0.49	0.44
$\bar{x}_{0.01}$	1.82	1.49	1.29	1.05	0.91	0.82	0.74	0.65	0.58
$\bar{x}_{0.002}$	2.19	1.78	1.54	1.26	1.09	0.98	0.89	0.77	0.69
$R_{0.05}$*	2.77	3.31	3.63	4.03	4.29	4.47	4.62	4.84	5.01
$R_{0.01}$	3.64	4.12	4.40	4.76	4.99	5.16	5.29	5.50	5.65
$R_{0.002}$	4.37	4.80	5.05	5.37	5.58	5.75	5.80	6.06	6.20
$s_{0.003}$†	1.84	1.86	1.81	1.71	1.64	1.58	1.54	1.48	1.43
B. Factors for calculating control limits from the average range									
$R_{0.003}$‡	3.27	2.57	2.28	2.00	1.86	1.78	1.72	1.64	1.59
$\bar{x}_{0.003}$‡	1.88	1.02	0.73	0.48	0.37	0.31	0.27	0.21	0.18
C. Critical values for χ^2 test									
$\chi^2_{0.05}$	5.99	7.81	9.49	12.59	15.51	18.31	21.03	26.30	31.41
$\chi^2_{0.01}$	9.21	11.34	13.28	16.81	20.09	23.21	26.22	32.00	37.57
$\chi^2_{0.005}$	10.59	12.83	14.86	18.54	21.95	25.18	28.30	34.26	39.99

*From Table 1 in Westgard, J. O., Groth, T., Aronsson, T., et al.: Performance characteristics of rules for internal quality control: Probabilities for false rejection and error detection. Clin. Chem., 23:1857–1867, 1977.

†From Table E, page 647, in Grant, E. I., and Leavenworth, R. S.: Statistical Quality Control. 4th ed. New York, McGraw-Hill, 1972.

‡From Table C, page 645, in Grant and Leavenworth.

probability for error detection is improved by selecting those rules that are particularly sensitive to random and systematic errors. The procedure requires a chart having lines for control limits drawn at the mean ±1 s, 2 s, and 3 s, and can be adapted to existing Levey-Jennings charts by addition of one or two sets of control limits.

The following control rules are used:

1_{2s}—one control observation exceeding the mean ±2 s—is used only as a "warning" rule which initiates testing of the control data by the other control rules.

1_{3s}—one control observation exceeding the mean $\pm\,3$ s—is a rejection rule that is primarily sensitive to random error.

2_{2s}—two consecutive control observations exceeding the same mean plus 2 s or mean minus 2 s limit—is a rejection rule that is sensitive to systematic error.

R_{4s}—one observation exceeding the mean plus 2 s and another exceeding the mean minus 2 s—is a rejection rule that is sensitive to random error.

4_{1s}—four consecutive observations exceeding the mean plus 1 s or the mean minus 1 s—is a rejection rule that is sensitive to systematic error.

$10_{\bar{x}}$—ten consecutive control observations falling on one side of the mean (above or below, with no other requirement on size of the deviations)—is a rejection rule that is sensitive to systematic error.

The use of the multi-rule procedure is similar to the use of a Levey-Jennings chart, but the data interpretation is more structured. To employ the multi-rule procedure, follow these steps:

1. Analyze samples of the control material by the analytical method to be controlled on at least 20 different days. Two different materials having appropriate concentrations are recommended, but a single material can be used. Calculate the mean and standard deviation for the results for each control material being used.

2. Construct a control chart for each of the control materials being used. The observed concentration or control value should be plotted on the y-axis, setting the range of concentrations to include the mean ±4 s. Draw horizontal lines for the mean, the mean ±1 s, the mean ±2 s, and the mean ±3 s. It may be desirable to use different colors for these lines, perhaps blue, yellow, and red for the 1 s, 2 s, and 3 s limits, respectively. The x-axis should be scaled for time, day, or run number, and labeled accordingly.

3. Introduce two control specimens into each analytical run, one for each of the two concentrations when two different materials have been selected. Record the control values and plot each on its respective control chart.

4. When both control observations fall within the 2 s limits, accept the analytical run and report the patient results. When one of the control observations exceeds the 2 s limits, hold the patient results. Inspect the control data using the 1_{3s}, 2_{2s}, R_{4s}, 4_{1s}, and $10_{\bar{x}}$ rules. When any one of these rules indicates the run is out-of-control, reject the analytical run and do not report the patient results. When all of these rules indicate that the run is in-control, accept the analytical run and report the patient results.

5. When a run is out-of-control, determine the type of error occurring based on the control rule that has been violated. Look for sources of that type of error. Correct the problem, then reanalyze the entire run—controls and patients.

An example application of the multi-rule procedure is shown in Figure 2E-8, where the top chart is for a high concentration control material and the bottom chart is for a low concentration material. Table 2E-4 summarizes the interpretation of the charted data, providing the run number, the accept/reject decision, control rules violated, and the type of error suspected based on the rule violations. It is important to note that the R_{4s} rule is applied only within a run, so that between-run systematic errors are not wrongly interpreted as random errors. However, the rule may be applied "across" materials, meaning that one of the observations can be on the low material and the other on the high material, as long as they are within the same run. On the other hand, note that the 2_{2s}, 4_{1s}, and $10_{\bar{x}}$ rules can be applied across runs and materials. This effectively increases n and improves the error detection capabilities of the procedure.

Power functions for the multi-rule procedure are shown in Figure 2E-9 for n from 2–6. The probability for false rejection will be much lower at these n's than for the Levey-Jennings chart having 2 s control limits. However, the false rejections do increase as n increases, limiting n to a maximum of 4–6. For larger n's, it would be necessary to modify the procedure by eliminating the R_{4s} rule, or by replacing this approximate range rule with an exact range rule. (See discussion of Shewhart range procedure.)

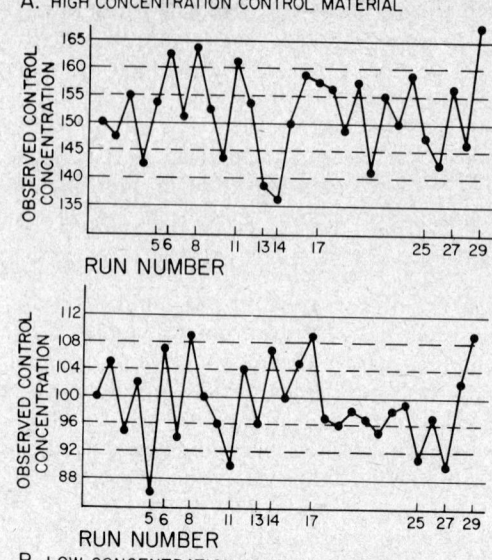

A. HIGH CONCENTRATION CONTROL MATERIAL

B. LOW CONCENTRATION CONTROL MATERIAL

Figure 2E-8. Westgard multi-rule control chart having control limits drawn at the mean ± 1 s, 2 s, and 3 s. Concentration is plotted on the y-axis versus time (run number) on the x-axis. *A*, Chart for high concentration control material. *B*, Chart for low concentration control material. (From Westgard, J. O., et al.: Clin. Chem., *27*:493-501, 1981.)

Comparison of the probability for error detection between the multi-rule procedure and the Levey-Jennings chart having 3 s limits shows improved error detection for the multi-rule procedure. The R_{4s} rule improves the detection of random error and the 2_{2s}, 4_{1s}, and $10_{\bar{x}}$ rules improve the detection of systematic error. Elimination of the $10_{\bar{x}}$ rule does not cause much loss in error detection, but does considerably reduce the amount of control data that must be inspected; thus this simplification may make the multi-rule procedure easier to use. The 4_{1s} rule could possibly be eliminated to further simplify the procedure, but this would be at an appreciable loss in power for detecting systematic errors.

Cumulative Sum (Cusum) Control Chart

The single value control charts discussed in the previous sections provide a display of the differences between the observed values and the expected mean. Control rules such as 2_{2s}, 4_{1s}, and $10_{\bar{x}}$ provide one way of determining when these successive differences no longer appear to be random (too many in a row on one side of a limit). A more exact and quantitative method is the cumulative sum control procedure or cusum chart. This chart was introduced in industry in the late 1950's[27,53] and has found limited use in clinical laboratories since the early 1960's.

Table 2E-4. INTERPRETATION OF EXAMPLE CONTROL DATA USING THE WESTGARD MULTI-RULE PROCEDURE

Run Number	Decision on Run		Control Rule Violated					Type of Error	
	Accept	Reject	1_{3s}	2_{2s}	R_{4s}	4_{1s}	$10_{\bar{x}}$	RE	SE
5		x	x					x	
6	x								
8		x		x					x
11		x			x			x	
13	x								
14		x		x					x
17		x							x
25	x					x			x
27		x					x		
29		x	x	x					x

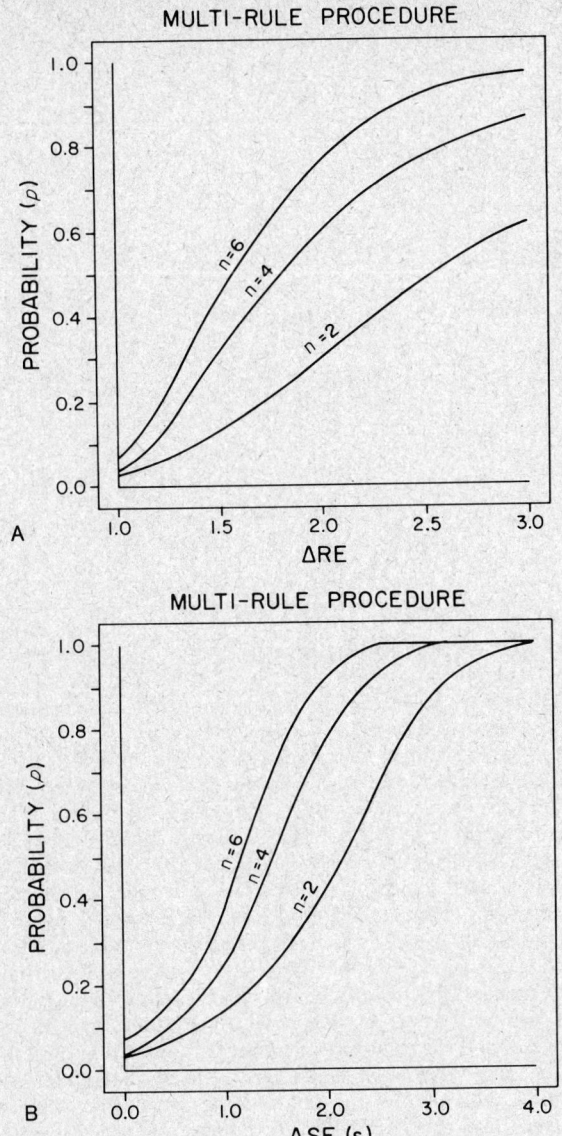

Figure 2E-9. Power functions for Westgard multi-rule control procedure. *A*, Random error. *B*, Systematic error. (From Westgard, J. O., et al.: Clin. Chem., *27*:493-501, 1981.)

To set up a cusum chart, follow these steps:

1. Analyze the control material by the analytical method to be controlled on at least 20 different days, and calculate the mean and standard deviation of those results. (This is the same as the initial step required for a Levey-Jennings chart or for a multi-rule chart.)

2. Construct a control chart. Label the y-axis "cusum." Draw a horizontal line at the midpoint of the y-axis to represent a cusum of zero. Set the range of values above and below to be about 10 times the standard deviation. Label the x-axis in terms of time, using day, run number, control observation number, or whatever is most appropriate for recording the relative time of the control observation.

3. Introduce control specimens into each analytical run and record the value obtained.

4. Calculate the difference between the value and the expected mean. Obtain the cusum by adding this difference to the cumulative sum of the previous differences. Plot the cusum on the control chart and inspect the plot.

5. Interpret the charted data by examining the slope of the cusum line. A steep slope suggests that a systematic error is present and that the run is out-of-control.

Example calculations are shown in Table 2E-5. These cusum values are plotted versus observation number in Figure 2E-10. When control values scatter on both sides of the mean, giving both positive and

Table 2E-5. EXAMPLE CUSUM CALCULATIONS
AND TABULAR RECORD FOR V-MASK CUSUM
CHART (FOR CONTROL MATERIAL WITH
$\bar{x} = 100$, $s = 5.0$)

Control Observation Number	Control Value	d_i	CS_i
1	110	+10	+10
2	100	0	+10
3	108	+ 8	+18
4	105	+ 5	+23
5	105	+ 5	+28
6	101	+ 1	+29
7	96	− 4	+25
8	105	+ 5	+30
9	101	+ 1	+31
10	101	+ 1	+32
11	111	+11	+43
12	102	+ 2	+45
13	110	+10	+55
14	107	+ 7	+62
15	107	+ 7	+69
16	107	+ 7	+76

negative differences, the cusum will alternate in sign and the plotted values will wander back and forth across the zero line on the control chart. When the control values fall predominantly on one side of the mean so that most of the differences have the same sign, the cusum value increases in magnitude and the plotted values will move away from the zero line of the control chart.

It is more difficult to judge the control status from cusum charts than from Levey-Jennings or multi-rule charts. The approach that has been used most extensively is to make the judgment based on the slope of the cusum line. In industry, this has often been done by constructing templates having a V-shaped section removed from a rectangular sheet of clear plastic. This V-shaped cutout establishes the angle which is the control limit and gives the technique its name of *"V-mask" cusum*. The apex of the V-mask is positioned on the control chart at a specified distance in front of the most recent cusum observation. If all of the plotted values are contained within the angle of the V-mask, the method is judged to be in-control. If any of the plotted values fall outside the angle of the V-mask, the method is judged to be out-of-control. Although this technique for interpreting cusum data is very objective, V-mask overlays have not been very widely used or recommended for clinical laboratories; one notable exception is the V-mask scheme of Kemp et al.[41] Instead, interpretation has been based on visual inspection and judgment of the angle of the cusum line, sometimes aided by use of special graph paper having an underlying pattern of 45° angles (∠∠∠∠∠∠ lines across the chart). When using this special graph paper, the convention has been to scale the graph so that a change of 2 s on the y-axis is the same distance as the distance between two points on the x-axis. The 45° angle then represents the slope expected when the observed mean is approximately 2 s from the expected mean.

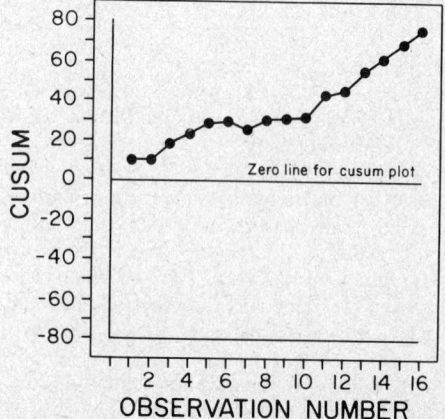

Figure 2E-10. V-mask cumulative sum control chart for example data in Table 2E-5. The cumulative sum of the differences from the mean is plotted on the y-axis versus time (control observation number) on the x-axis.

Table 2E-6. EXAMPLE CUSUM CALCULATIONS AND TABULAR RECORD FOR DECISION LIMIT CUSUM (FOR CONTROL MATERIAL WITH $\bar{x}$ = 100, s = 5.0; FOR CONTROL CHART WITH k_u = 105, k_l = 95, h_u = 13.5, h_l = 13.5)

Control Observation Number	Control Value	d_i	CS_i	Comment
1	110	+5	+ 5	Start cusum calculation
2	100	−5	0	
3	108	+3	+ 3	
4	105	0	+ 3	
5	105	0	+ 3	
6	101	−4	−1	End cusum calculation
7	96			
8	105			
9	101			
10	101			
11	111	+6	+ 6	Start cusum calculation
12	102	−3	+ 3	
13	110	+5	+ 8	
14	107	+2	+10	
15	107	+2	+12	
16	107	+2	+14	Out-of-control

An alternative way of interpreting cusum data is to employ a numerical limit for the cusum value itself, a technique known as *"decision limit"* cusum. Particular decision limit schemes are characterized by two parameters: k, a factor for calculating the threshold, and h, a factor for calculating the decision limit or control limit. The cusum calculations do not start until a control value exceeds a certain threshold above (k_u) or below (k_l) the expected mean ($\bar{x}$). Once exceeded, the differences from the threshold are calculated and summed for successive observations to provide the cusum. This process continues until the cusum either exceeds the control limits (h_u and h_l) and the method is judged out-of-control, or until the cusum changes sign, in which case the cusum calculation is stopped and the method judged to be in-control. An example of the cusum calculations for the decision-limit technique is given in Table 2E-6.

This example data is plotted in Figure 2E-11. One advantage of the decision limit cusum chart is that it has horizontal lines for control limits, permitting the cusum chart to be interpreted in a manner similar to a Levey-Jennings chart. This form of interpretation requires less experience and judgment, therefore providing more consistent interpretation when many analysts use the control procedure. Another advantage of the decision-limit technique is that it can be implemented using only tabular operations (without charting), thus making it very easy to program for computerized handling of control data.

Although interpretation is easier and more objective, selecting the threshold and control limits is somewhat difficult. This selection is done with the aid of nomograms, which are quite complicated to understand and use, and therefore it is initially perhaps best to use one of the decision-limit schemes recommended in the clinical chemistry literature.[71] For manual implementation, set the threshold at the $\bar{x}$ ±1.0 s and control limits at 2.7 s. For computer implementation, set the threshold at the $\bar{x}$ ±0.5 s and

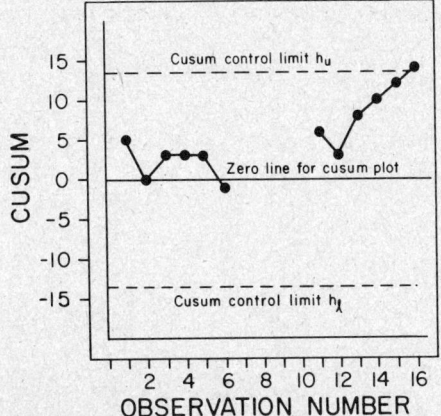

Figure 2E-11. Decision limit cumulative sum control chart for example data in Table 2E-6. The cumulative sum of the differences from the threshold is plotted on the y-axis versus time (control observation number) on the x-axis.

Table 2E-7. POWER (PROBABILITY FOR REJECTION) FOR VARIOUS CONTROL PROCEDURES[70]

Control Rules	n	ΔRE (Increases in Random Error)							ΔSE (Systematic Error)						
		1.00	1.25	1.50	1.75	2.00	2.50	3.00	0.0 s	0.5 s	1.0 s	1.5 s	2.0 s	3.0 s	4.0 s
1_{3s} and cusum combined	1	0.00	0.02	0.02	0.07	0.12	0.10	0.27	0.00	0.00	0.03	0.04	0.13	0.50	0.79
	2	0.00	0.03	0.06	0.15	0.23	0.41	0.52	0.00	0.01	0.04	0.14	0.34	0.85	0.99
	4	0.01	0.06	0.21	0.35	0.46	0.63	0.81	0.01	0.04	0.14	0.48	0.81	1.00	1.00
	8	0.02	0.14	0.37	0.58	0.68	0.88	0.96	0.02	0.09	0.38	0.83	0.95	1.00	1.00
$\bar{x}_{0.05}$	2	0.05	0.13	0.18	0.27	0.33	0.45	0.49	0.05	0.10	0.32	0.54	0.78	0.98	1.00
	4	0.05	0.12	0.22	0.31	0.31	0.38	0.51	0.05	0.20	0.52	0.82	0.98	1.00	1.00
	8	0.05	0.12	0.20	0.27	0.32	0.43	0.55	0.05	0.32	0.82	0.99	1.00	1.00	1.00
	12	0.05	0.13	0.20	0.27	0.31	0.45	0.55	0.05	0.44	0.93	1.00	1.00	1.00	1.00
	20	0.05	0.10	0.18	0.25	0.32	0.42	0.54	0.05	0.61	1.00	1.00	1.00	1.00	1.00
$R_{0.05}$	2	0.05	0.12	0.18	0.27	0.28	0.45	0.55	0.05	0.05	0.06	0.05	0.04	0.05	0.04
	4	0.05	0.17	0.35	0.47	0.59	0.70	0.85	0.05	0.05	0.04	0.07	0.06	0.03	0.06
	8	0.05	0.23	0.47	0.65	0.80	0.92	0.96	0.05	0.05	0.06	0.05	0.04	0.06	0.05
	12	0.05	0.31	0.53	0.81	0.90	0.98	0.99	0.05	0.07	0.06	0.07	0.05	0.03	0.05
	20	0.05	0.35	0.71	0.88	0.97	0.99	1.00	0.05	0.07	0.06	0.06	0.06	0.06	0.06
$\chi^2_{0.05}$	2	0.05	0.11	0.17	0.27	0.28	0.44	0.55	0.05	0.05	0.06	0.05	0.04	0.05	0.04
	4	0.05	0.18	0.35	0.47	0.58	0.72	0.84	0.05	0.05	0.04	0.07	0.06	0.03	0.06
	8	0.05	0.28	0.50	0.70	0.83	0.94	0.97	0.05	0.04	0.06	0.07	0.04	0.06	0.04
	12	0.05	0.36	0.64	0.87	0.95	0.99	0.99	0.05	0.07	0.05	0.05	0.05	0.03	0.05
	20	0.05	0.44	0.83	0.95	0.99	1.00	1.00	0.05	0.06	0.04	0.05	0.05	0.04	0.05
$\bar{x}_{0.01}$ and $R_{0.01}$	2	0.02	0.08	0.12	0.26	0.34	0.51	0.63	0.02	0.04	0.16	0.33	0.61	0.95	1.00
	4	0.02	0.08	0.25	0.44	0.53	0.68	0.83	0.02	0.08	0.30	0.65	0.93	1.00	1.00
	8	0.02	0.12	0.35	0.57	0.57	0.88	0.96	0.02	0.14	0.58	0.95	1.00	1.00	1.00
	12	0.02	0.16	0.40	0.65	0.82	0.96	0.99	0.02	0.22	0.80	1.00	1.00	1.00	1.00
	20	0.02	0.17	0.53	0.75	0.94	0.99	1.00	0.02	0.34	0.97	1.00	1.00	1.00	1.00
$\bar{x}_{0.01}$ and $\chi^2_{0.01}$	2	0.02	0.08	0.12	0.26	0.34	0.51	0.63	0.02	0.04	0.16	0.33	0.61	0.95	1.00
	4	0.02	0.08	0.24	0.45	0.56	0.68	0.84	0.02	0.08	0.31	0.65	0.93	1.00	1.00
	8	0.02	0.15	0.41	0.62	0.77	0.91	0.97	0.02	0.14	0.59	0.95	1.00	1.00	1.00
	12	0.02	0.22	0.49	0.78	0.91	0.98	1.00	0.02	0.22	0.80	1.00	1.00	1.00	1.00
	20	0.02	0.26	0.71	0.89	0.98	0.99	1.00	0.02	0.34	0.97	1.00	1.00	1.00	1.00

the control limits at 5.1 s. This latter procedure will be sensitive to smaller errors because of the choice of the lower threshold. However, the charting is not as convenient when performed manually, particularly when the cusum is used in combination with the 1_{3s} control rule to provide better detection of random errors.[71]

When compared with the Levey-Jennings chart with 3 s limits, the cusum procedure will provide better detection of systematic errors but less sensitivity for random errors. Because of the low sensitivity to random errors, cusum should not be used alone, but rather should be included with a Levey-Jennings procedure on a combined chart[71] or employed as a separate chart along with a Levey-Jennings chart. Power functions for a combined 1_{3s} and cusum procedure are given in Table 2E-7, where probability for rejection is tabulated as a function of the size of analytical error. Performance is similar to the multi-rule procedure for low n's, at least as far as detecting systematic errors. However, the multi-rule procedure will be more sensitive to random errors because of the added R_{4s} rule.

Shewhart Mean and Range (or Standard Deviation) Control Charts

Statistical quality control charts were first introduced in industry during the 1930's by Shewhart,[59] who recommended obtaining a group of control measurements, determining the average and standard deviation of the group, then plotting the average on an "x-bar" or mean chart and the standard deviation on an S-chart. For n's less than 10, the S-chart is usually replaced by a "range" or R-chart on which is plotted the range, or difference between the highest and lowest observations of the group.

To set up mean and range charts, initial estimates must be obtained for the mean and standard deviation, or the mean and average range, as determined for the control material to be used and the analytical method to be controlled. Control limits for the charts are calculated using the factors given in Table 2E-3. A group of control measurements is then collected in each analytical run, and the mean and range (or standard deviation) are calculated and plotted. An example set of data is given in Table 2E-8 and plotted on the mean and range charts in Figure 2E-12.

Table 2E-8. SHEWHART MEAN AND RANGE CALCULATIONS AND TABULAR RECORD

| Run | Individual Control Measurements | | | | | | Group | |
	1	2	3	4	5	6	x̄	R
1	96	98	101	98	95	102	98.33	7
2	99	100	105	97	100	103	100.67	8
3	101	100	113	95	106	101	102.67	18
4	98	93	96	104	101	102	99.00	11
5	101	103	106	100	99	98	101.17	8
6	95	109	99	103	105	101	102.00	10
7	103	94	92	94	101	98	97.00	11
8	101	105	104	108	103	104	104.17	7
9	96	105	92	105	101	104	100.50	13
10	101	99	93	107	102	95	99.50	14
11	102	88	92	101	99	98	96.67	14
12	92	97	107	95	99	106	99.33	15
13	94	105	96	103	104	100	100.33	11
14	101	96	101	101	95	95	98.17	6
15	107	99	101	96	102	97	100.33	11
16	102	97	98	98	99	97	98.50	5
17	101	112	104	102	107	108	105.67	11
18	97	99	104	96	95	109	100.00	14
19	98	95	101	91	97	96	96.33	10
20	97	96	96	94	96	98	96.17	4
21	98	95	92	99	95	93	95.33	7
22	96	89	95	94	93	94	93.50	7
23	95	100	93	89	98	101	96.00	12
24	96	104	103	102	101	107	102.17	8
25	92	89	94	94	90	90	91.50	5
26	92	101	93	102	102	100	98.33	10
27	100	101	105	101	104	95	101.00	10
28	112	91	97	97	97	102	99.33	21
29	103	102	95	92	80	88	93.33	23

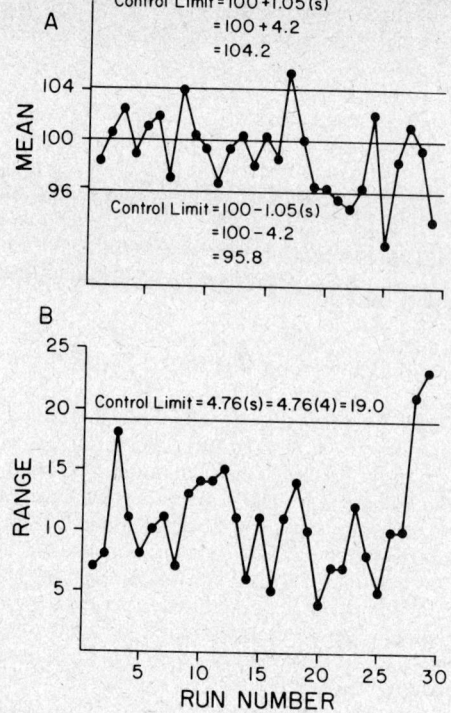

Figure 2E-12. Shewhart mean and range control charts. *A,* Mean chart. *B,* Range chart.

Implementation by use of the Z and chi-square (X^2) tests. Charts for mean, range, and standard deviation are seldom found in clinical laboratories due to the time required for calculating the control statistics. These control procedures are probably practical only when computerized data handling is available. An easy way of implementing these procedures on computers is to employ statistical tests of significance. A Z-test can be used to determine whether the mean has changed from its original value, and a χ^2 test to determine whether the standard deviation has changed from its original value.

The values for Z and χ^2 are determined from the following equations:

$$Z_{obs} = \frac{\overline{x}_{obs} - \overline{x}_{exp}}{\dfrac{s_{exp}}{\sqrt{n}}}$$

$$\chi^2_{obs} = \frac{(s_{obs})^2 \, (n-1)}{(s_{exp})^2}$$

where $\overline{x}_{obs}$ is the mean observed in the analytical run being tested, $\overline{x}_{exp}$ is the mean expected when the method is working properly, n is the number of control measurements in the group, s_{obs} is the standard deviation observed in the analytical run being tested, and s_{exp} is the standard deviation expected when the analytical method is in stable operation.

To determine the control status of an analytical run when using the Z-test, the observed Z-value is compared to the critical Z-value for a selected probability for false rejection. The critical values are 1.96, 2.58, and 3.09 for probabilities of false rejection of 0.05, 0.01, and 0.002, respectively. When the observed Z-value is less than the critical Z-value, the run is judged to be in-control. When it exceeds the critical Z-value, the run is judged to be out-of-control.

To determine control status using the χ^2 test, the observed χ^2 value is compared to the critical χ^2 value (given in Table 2E-3). When the observed value is less than the critical value, the run is judged to be in-control; when it exceeds the critical value, the run is judged to be out-of-control.

Alternative calculations. Moving averages and moving standard deviations are in use in industry and should find application in clinical laboratories. With these procedures, n is chosen, measurements are collected until n is attained, and then the mean and standard deviation are calculated. With each additional measurement, the oldest measurement is dropped, the newest is added, and new estimates of the mean and standard deviation are obtained after each new control measurement, rather than waiting for new groups of n measurements. Similar updated means and standard deviations can be obtained from exponential smoothing, a calculation procedure which provides an average that weighs recent observations more heavily than older observations.[19,69] The "smoothed mean" is sensitive to trends which are just beginning to develop, and therefore it is often referred to as a trend analysis procedure. The calculations are too complicated to be implemented manually, but the availability of computers makes these calculations feasible in clinical laboratories.

Implications for future applications. Power functions for mean (or Z-test), range, and χ^2 (or standard deviation) control procedures are given in Table 2E-7. When compared with those for previous control procedures, they show higher probabilities for error detection, particularly at larger n's. The probability for false rejection can be set at a suitably low level by proper choice of control limits. Thus these control procedures appear to offer better performance characteristics than single-value control charts, having higher error detection and lower false rejection as n increases.

The use of mean and standard deviation control procedures (or related procedures) to monitor accuracy and precision is a direct extension of the practices of method evaluation, the mean being used to monitor systematic error and the standard deviation to monitor random error. This relationship helps explain why present control procedures having n's from 1–4 have relatively low power or error-detection capability. Not surprisingly, it is more difficult to estimate precision and accuracy using quality control procedures with n's of 1–4 than with method evaluation procedures using n's of 20–40. For low n's, all control procedures will have relatively low power. As n increases, power will increase, particularly for procedures utilizing mean, range, S-charts, Z- and χ^2 tests, moving averages and moving standard deviations, and trend analysis. Future practices in statistical quality control in clinical laboratories will undoubtedly make more use of these procedures.

The widespread availability of microprocessors and microcomputers makes it practical to use complicated control procedures on a routine basis in clinical laboratories. Calculations on control data, once properly programmed, can be performed quickly and without mistakes. Graphical displays can be generated, with color added to aid the interpretation of control data. Specific control rules can be programmed to provide "accept" and "reject" signals or printouts. Records can be maintained to document instrument changes, preventive maintenance, and control problems. Monthly summary statistics and quality control reports can be generated. Many of these features, along with many of the statistical control procedures that have been described, are found in microcomputer quality control programs that are commercially available today.

CONTROL OF ANALYTICAL QUALITY USING PATIENT DATA

PROCEDURES USING INDIVIDUAL PATIENT RESULTS

The patient test results are the final product of most laboratory procedures, and the monitoring of these results is the most direct form of quality control. Unfortunately, procedures for monitoring results are not very sensitive and have low probabilities for error detection. The most effective procedure is the clinical correlation of test results with other information related to the patient, especially surgical findings, response to therapy, and autopsy data. Less sensitive but easier to implement are comparisons with previous test values and correlation with related test results. The easiest procedure is the comparison of test results with physiologic or probabilistic limits.

Clinical correlation. In modern high-volume laboratories, it is impractical for laboratories to correlate all test results with the clinical status of the patients. In general, the clinicians ordering the tests are in a better position to examine the appropriateness of test results. However, clinicians often order a test because they are uncertain of the exact clinical status of their patient and base their diagnoses heavily on the laboratory test results. In such cases, it is circular reasoning to correlate clinical diagnoses with laboratory test results. Clinical correlation studies are best done retrospectively when laboratory test results can be directly related to other evidence, such as surgical findings. Unfortunately, most laboratory tests do not correlate perfectly with disease states, and one must examine multiple cases before determining the diagnostic efficiency of a given test (as discussed further in the next section). In an individual patient, clinical correlation can identify impossible or highly unlikely test results that cannot be flagged in the laboratory, such as a normal serum bilirubin concentration in a highly jaundiced patient. Clinicians should be encouraged to report these discrepancies to the laboratory, and a mechanism should be available to follow up these problems. Also, after using a test for a period of time, many clinicians can detect alterations in test results, and although this is not reliable for documenting problems, informal discussions with clinicians frequently using the test can aid in identifying aspects of laboratory tests which should be further investigated.

Correlation with other laboratory tests. As in clinical correlation, there are a limited number of situations where a single test result is plausible, but the combination of several test results is impossible or very unlikely. If the tests involved in these comparisons are performed at the same time, one often can identify errors and correct problems prior to reporting the test results to the clinicians. Even though very few tests have exact clinical relationships, and one often must consider statistical percentages rather than the results of a single patient to identify malfunctions, there are several relationships that provide some possibilities for monitoring an individual patient's results, at least for purposes of initiating further review. These are:

1. *Blood typing.* There is a close association between the ABO blood antigens on erythrocytes and isoantibodies found in plasma. Exceptions can be found in patients recently transfused with whole blood.

2. *Anion gap.* To maintain electrical neutrality, the sum of the charges of anions in a blood sample must equal the sum of the charges of cations when expressed in molar concentrations. An anion gap (AG) can be calculated as:

$$AG = (Na^+ + K^+) - (Cl^- + HCO_3)$$

Values less than about 10 mmol/L or greater than about 20 mmol/L may indicate error.[73,76] Elevated values may be found in patients with renal failure, diabetic acidosis, cardiac failure, anoxia, and other conditions (see Chapter 10). Low values occur in hypoproteinemia and with intravenous hydration. The capabilities of anion gap control procedures have been studied by Cembrowski,[20] who recommends that the average anion gap of groups of 8 or more patients be used to provide a more sensitive statistical control.

3. *Osmolar gap.* Serum osmolality can be estimated from the concentrations of a few major blood constituents and can also be measured using colligative properties. The osmolar gap can be calculated as:

$$OG = osmolality - 1.86 (Na^+ + K^+) + \frac{glucose}{18} + \frac{urea\ nitrogen}{2.8}$$

where osmolality is in mOsm/kg, sodium and potassium are in mmol/L, and glucose and urea nitrogen are in mg/dL. Normally this gap is near zero. Alcohol consumption or the ingestion of toxins such as salicylates or ethylene glycol can cause an increase. Alterations of unmeasured electrolytes and dehydration may also increase the gap.

4. *Acid-base balance.* The Henderson-Hasselbalch equation can be used to calculate theoretical bicarbonate and total CO_2 concentrations when pH and pCO_2 are measured. Theoretical and measured results generally agree within 2.0 mmol/L.[76]

5. *Thyroxine—TSH.* Normally, thyroxine produces a negative feedback to the pituitary and reduces the secretion of thyroid stimulating hormone (TSH). Patients with serum thyroxine concentrations above 8.0 μg/dL seldom have elevated concentrations of TSH.[21] Exceptions occur in secondary and tertiary hyperthyroidism.

Intra-laboratory duplicates. Samples can be divided into two aliquots and analyzed, and the duplicates used for control purposes.[5] This is a simple quality control procedure that does not require stable control materials and therefore can be used when stable materials are not available, or as a supplemental procedure when stable control materials are available. The differences between duplicates can be plotted on a range type of control chart having limits calculated from the standard deviation of the differences (using factors for the Shewhart range chart in Table 2E-3). When the duplicates are obtained from the same method, this range chart monitors only random error and thus is not adequate for assuring the accuracy of the analytical method. When the duplicates are obtained from two different laboratory methods, then the range chart actually monitors both random and systematic errors but cannot separate the two types of errors. The interpretation becomes more difficult, particularly when there are stable systematic differences or biases between the two analytical methods. Multiplicative factors may be necessary to deal with proportional differences, and additive factors may be necessary to allow for constant differences. Interpretation of observed differences becomes more qualitative; nevertheless, this procedure still provides a useful way of monitoring the consistency of the data being generated by the laboratory.

Table 2E-9. RECOMMENDED LIMITS FOR
DELTA CHECKS[45]

Test	Delta Check Limit
Albumin	20%
Bilirubin, total	50%
Calcium, total	15%
Creatine kinase	99%
Creatinine	50%
Phosphorus	20%
Potassium	20%
Protein, total	20%
Sodium	5%
Thyroxine	25%
Urea nitrogen	50%
Uric acid	40%

Delta checks with previous test. Certain errors, particularly errors in specimen identification, can be detected by comparing laboratory test results with values obtained on previous specimens from the same patient.[51,58] The expected variability of test results depends on both the analyte and the time interval between determinations. Ladenson[45] has defined delta check limits based on a three-day interval in terms of a percentage change from the initial value. His check limits for some common tests are shown in Table 2E-9.

In identifying specimen mix-ups, it is helpful to examine multiple test parameters, including hematologic tests, since differences in a profile of test results often are more obvious and more statistically significant than changes in individual tests.[51,74] Sheiner et al.[58] have evaluated the performance of several delta check methods, including two discriminant functions, for detecting mislabeled specimens; they observed a false positive rate of 5% and an error detection of about 50% for mislabeled specimens.

Limit checks. A patient's test results should be reviewed to check that they are within the physiologic ranges compatible with life. These limit checks are helpful for detecting clerical errors such as transposed digits or misplaced decimal points. This checking can be combined with warning limit checks for detecting and verifying possible, but infrequently occurring, test results. These warning limits are dependent on test methodology and the characteristics of the patient population being tested. Table 2E-10 shows limits for some common tests as recommended by Whitehurst et al.[76]

The efficiency of each of these methods for detecting errors is relatively low. The review limits must be set wide to select only a manageable number of records for review. Using the Whitehurst system in a series of 30 870 test results, 1254 or 4.1% were flagged for lack of laboratory test correlation, but only 179 or 14.2% were found to be errors, although about 10% were not rechecked. A comparison of test results done by multiple methods flagged another 389 test results, and again only about 14% were found to be in error. Delta checks with previous

Table 2E-10. RECOMMENDED RANGES FOR LIMIT CHECKS[76]

Test	Low Warning	High Warning
Acid phosphatase* (U/L)	0.1	10
Albumin (g/dL)	1.5	6
Alkaline phosphatase* (U/L)	5	300
Amylase* (U/L)	20	1000
Bilirubin (mg/dL)	0.2	10.0
Calcium (mg/dL)	6.5	13.0
Creatine kinase (U/L)	5	1500
Creatinine (mg/dL)	0.3	7.5
Phosphorus (mg/dL)	1.0	8.0
Potassium (mmol/L)	3.0	6.0
Sodium (mmol/L)	120	150
Urea nitrogen (mg/dL)	3	50
Uric acid (mg/dL)	1.0	12.0

*Values are method dependent.

determinations and limit checks each flagged 450 more test results. Only about 2% of the data identified by delta checks and 3% by the limit checks were determined to be real errors. In total, about 8.4% of the test results were reviewed, and 0.8% were found to be in error. Obviously, there is a trade-off between the percentage of results retested and the detection of errors. Ironically, the lower the laboratory error rate, the lower the efficiency of the mechanisms for error detection. Thus, these methods cannot replace conventional quality control procedures, but they should be used for supplemental purposes because they can identify errors which are not detected by other methods of quality control.

PROCEDURES BASED ON RESULTS FROM MULTIPLE PATIENTS

Test distribution statistics. Statistics based on distributions of test results from large numbers of patients are useful for detecting systematic errors (shifts and drifts), but are of no value for detecting random errors (increased variability or scatter). They are useful adjuncts to the fundamental control procedures which employ stable control materials but should not be substituted for them. Patient values include numerous sources of variation—demographic, biologic, pathologic, and pre-analytic—in addition to the analytic variation caused by the analytical method. Because of this, individual test values have too much variability to have any utility for quality control; however, the mean of multiple test values or groups of patients is more stable and therefore may be useful for control purposes.

The variability of the mean test result from a group of n patients decreases proportionally to the square root of the number of determinations. The variability of the estimated mean is given by the standard error of the mean (SEM), which is equal to the standard deviation of the patient population divided by the square root of n, the number of patient measurements included in the group mean. Consider the following example, where the s of the patient values is 100 units. As n increases to 16, 25, 100, and 400, the standard error for the patient mean becomes 25, 20, 10, and 5 units, respectively. If the s of the analytical method were 10 units, the distribution of patients would be 10 times as large; however, if one considers the mean values of groups of 100 patients, the variability (SEM) of the group mean is approximately equal to the variability observed on individual observations on stable control materials. For groups of 400 patients, the variability of the mean is half that for individual observations on the stable control material.

Changes in the means of patient populations may be caused by multiple variables. Changes in the demographic and clinical characteristics of the patients, such as the ratio of males to females, the ratio of hospitalized patients to outpatients, or the presence of many specimens from a specialty clinic can alter the mean value. Similarly, changes in pre-analytic conditions, such as tourniquet time and specimen storage, can alter patient population means, and can therefore be monitored by use of patient means. These variables are not monitored by those control procedures employing stable materials; thus the "patient mean procedures" provide additional capabilities and should be used in conjunction with other control procedures.

Statistical methods for monitoring patient means. The "mean of normals" approach calls for establishing limits, usually the limits for reference values, for "trimming" the patient data. Values outside these limits are eliminated from the calculations, thus reducing the response to outliers and subpopulations, and also to real errors.[37,56] An alternative approach has been to employ the median as the control statistic, and this may be a more sensitive indicator of change than the mean.[29] More complicated estimates have been recommended to account for hospitalized patients having slightly different test values from those of ambulatory patients. A weighted mean based on the percentage of patients in each category has greater sensitivity for error detection, particularly for measurements of the serum concentration of total protein, albumin, and calcium.[23]

A particular algorithm, called *Bull's algorithm,* has been widely employed for on-line monitoring of automated hematologic cell counters. Bull et al.[8,42] evaluated six statistics for monitoring erythrocyte indices: the sample mean, two moving average means, the mean of a truncated sample, the John's mean, and the median. They assessed the usefulness of these different statistics by mathematically introducing both abrupt bias changes and cyclical time changes. The moving average statistics were found to have superior error detection. The batch size used in calculating the moving average regulates the smoothing of the data. Large batch sizes not only smooth out undesired individual patient effects but also increase the number of specimens (and time) needed to detect instrument malfunction. A weighted moving average based on a batch size of 20 specimens was recommended.

Although control statistics based on patient data can be readily available in computerized laboratories, their utility is still controversial. Several authors have shown that these control procedures are not as sensitive as procedures using stable reference material.[1,4,23,56] All of these statistics are designed to detect systematic errors and have virtually no power for detecting random error. The relative sensitivity of these statistics for detection of systematic errors, as compared with stable reference controls, is dependent on three factors: variation of patient test values (and the test statistics derived from them), analytic variation, and the ratio of the number of specimens to the number of controls. Low patient variability, high analytic variability, and a high ratio of specimens to controls favor patient-based quality control procedures. Patient test result variability depends not only on the analyte measured but also on the population being tested. Populations consisting mainly of healthy individuals generally have lower test variability than populations of medical specialty clinics. A quantitative study of these variables and their effects on the statistical power of procedures utilizing patient means has been performed by Cembrowski et al.[18]

Clinical correlation studies. Retrospective correlation studies relating laboratory test results to surgical findings or therapeutic changes in patients are effective mechanisms for long-term quality control. Two separate patient populations are needed to efficiently monitor both false positive and false negative laboratory test results. To monitor false positives, the records of patients with positive results should be randomly sampled and reviewed for confirmatory evidence of disease. For many laboratory tests, substantial medical follow-up is necessary to confirm or refute test results, and, unfortunately, medical records often do not contain this information. Again, stated diagnoses cannot be used to confirm laboratory test results, since the laboratory test results may have been the primary evidence for the diagnosis. To monitor false negatives, the records of patients with objective clinical confirmation of the disease should be randomly sampled and reviewed. Examples of objective confirmation are surgical discovery of a tumor, improvement in physical signs and symptoms following pharmacologic or surgical therapy, and autopsy findings. It is difficult to obtain unbiased record selection, since test results often alter clinical decisions regarding surgery or therapy. There is less selection bias in autopsy data, but the disease process generally is more advanced in these patients.

In *summary,* quality control mechanisms based on patient data can provide additional information useful in monitoring the quality of laboratory analyses. These procedures are often time-consuming and generally are not sensitive enough to serve as the only means of quality control. On the other hand, many of the control problems detected with these techniques may not be evident with conventional quality control systems. A quality assurance program should make appropriate use of "patient data procedures," but it must be recognized that these procedures are not as well developed, not as widely employed, and not as well understood in terms of their strengths and weaknesses.

ROLE OF EXTERNAL QUALITY ASSURANCE (QA) PROCEDURES

All of the control procedures described so far have focused on monitoring a single laboratory. These procedures comprise what is called *internal quality assurance* to distinguish them from procedures used to compare the performance between different laboratories, the latter being known as *external quality assurance.* The two are complementary activities, internal QA being necessary for the daily monitoring of the precision and accuracy of the analytical method, and external QA being important for maintaining the long-term accuracy of the analytical methods.[13,15]

Limitations of internal QA. It is important to recognize that internal QA procedures only detect changes in performance between the present operation and the "stable" operation that was characterized during the baseline period when the analytical method was thought to be working properly. Although the procedures can detect systematic errors as well as random errors, the only systematic errors detected are those changes from the original baseline. If the method actually had some undetected systematic errors during the baseline period, those systematic errors would be included in the mean that was used to calculate the control limits for the procedure.

Thus, only systematic changes from this original mean will be detected by internal QA procedures.

Initial method evaluation studies are essential to assure that systematic errors are not present prior to the baseline period and the determination of the mean and the control limits. The accuracy of the method should be initially established by comparison with other analytical methods (as well as recovery and interference studies) and should continue to be monitored by comparison with other analytical methods. Ongoing comparison-of-methods studies are desirable to assure that systematic errors do not slowly increase and go undetected by internal QA procedures. These ongoing comparison studies are provided by the external QA programs.

Features of external QA programs. There are several external quality assurance programs that are available to laboratories, some sponsored by professional societies and others by manufacturers of control materials. Detailed descriptions of the objectives and design are available in the literature.[30,32,35] The basic operation of these programs involves having all the participating laboratories analyze the same lot of control material, usually daily as part of the internal QA activities. The results are tabulated monthly and sent to the sponsoring group for the data analysis. Summary reports are prepared by the program sponsor and are distributed to all participating laboratories. This reporting takes time for processing all the data from a large number of laboratories; thus the data analysis is not available in real time and therefore is useful only for monthly reviews and periodic problem-solving activities.

The reports often include extensive data analysis, statistical summaries, and plots. The overall mean of all laboratories in the program, or alternatively the mean of values of all reference laboratories, is taken as the "true" or correct value, and is used for comparison with the individual laboratory's mean. Different programs do this in different ways. For example, the statistical significance of any difference between an individual laboratory's observed mean and the group mean can be tested by use of the t-test. When the difference is significant, the laboratory can be alerted that its results are biased as compared with the results of most of the other laboratories. Another approach is to divide the difference by the overall standard deviation of the group, and then to express the difference in terms of the number of standard deviations:

$$\text{SDI} = \frac{\text{Lab mean} - \text{Group mean}}{\text{Group s}}$$

where SDI is the abbreviation for standard deviation interval or index, and Group s is the standard deviation for the group or a selected subset of the group. Differences greater than 2 indicate that a laboratory is not in very good agreement with the rest of the laboratories in the program. These calculations reduce all the test results to the same values, which makes it possible to interpret the data without reference to the exact mean and s for each analytical method. For example, a value of $+2.0$ has the same meaning for any test, indicating that the value is 2 s above its established mean.

Some additional information about the nature of the systematic error can be obtained when there are two different control materials analyzed by each laboratory. The laboratory's observed mean for material A can be plotted on the y-axis versus its observed mean for material B on the x-axis. These graphs are called Youden plots, and their use is described in detail by Elion-Gerritzen.[26] Ideally, the point for a laboratory should fall at the center of the plot. Points falling away from the center, but on the 45 degree line, suggest a proportional analytical error. Points falling away from the center, but not on the 45 degree line, suggest either an error that is constant for both materials or an error which occurs with just one material.

The report may also include Levey-Jennings plots of the data, but because this information is not available in real time, it cannot effectively serve the purposes of internal quality control. Blank control charts which are set up for each analyte and each control material can save the laboratory the time required when these charts are prepared manually.

The operation of external quality assurance programs is greatly improved by the incorporation of a microcomputer. The external comparisons can be more quickly performed by using the microcomputer as a terminal for transmitting data via telephone to the central computer. The processing of the data to compare the results from the many different laboratories can then be done, and the reports returned via telephone. Since the individual control observations are entered into the microcomputer, the control data can be immediately tested by internal control procedures to determine control status. Thus, the microcomputer when used in this way will integrate the internal and external procedures into a more efficient program for quality assurance.

IDENTIFYING THE SOURCES OF ANALYTICAL ERRORS

The control procedures discussed in the previous sections provide a way of alerting the analyst to analytical problems which may cause the quality of analytical performance to fall short of the quality goals set for the laboratory. However, these control procedures do not identify the sources of the analytical errors and solve the control problems. Additional procedures are often needed to identify the sources.

PHYSICAL INSPECTION

When alerted to a control problem, the first step should be to inspect carefully the analytical method, equipment, reagents, and specimens. Does everything look, feel, smell, and sound correct? An inspection may seem to be a very qualitative and sensory technique, but it can be very powerful when performed with checklists developed for specific analytical methods. This inspection should include a review of records documenting changes occurring with the instrument and reagents. Brief instrument function checks can often be performed to verify proper system performance and separate chemical and instrumental sources of errors. An experienced analyst can often spot the problem by making this kind of inspection, while inexperienced analysts will be aided by formal checklists.

RELATIONSHIP OF TYPE AND SOURCE OF ERROR

The type of error itself provides a clue about the source of the error. For example, systematic errors are often related to calibration problems, e.g., impure calibration materials, improper preparation of standard solutions, erroneous set point and assigned values, unstable standard solutions, contaminated solutions, inadequate calibration techniques, nonlinear or unstable calibration functions, unstable reagent blanks, and inadequate sample blanks. Random errors are more likely due to lack of reproducibility in the pipetting of samples and reagents, to dissolving of reagent tablets and the mixing of sample and reagents, and to the lack of stability of temperature baths, timing regulation, and photometric and other sensors. A specific example of the instrument variables affecting kinetic assays has been provided by Maclin et al.[47] Individual analytical methods may not be subject to all of these possible sources of error; rather, there may be only a few plausible sources for a particular type of error. Experienced analysts often know what these common sources are for their particular analytical methods and can quickly identify the sources once the type of error is known.

Some information about the type of error occurring can be obtained by knowing which errors the control procedures respond to. Different control rules have different sensitivities to detect random and systematic errors, as illustrated by the multi-rule control procedure. It is often possible to make a good judgment as to the type of error occurring based on the control procedure that provides the alert. Control procedures which employ patient samples rather than stable control materials can help identify pre-analytic sources of errors such as sample handling and processing. External control procedures may provide more extensive information about systematic errors than what is available from internal procedures. The information from all of these procedures is complementary and when used together provides a more complete assessment of the types of errors occurring, and their possible sources.

METHOD EVALUATION EXPERIMENTS

In cases when the type of error cannot be easily identified and quantitated from available control data, it may be necessary to utilize the experiments employed in method evaluation studies. The "comparison-of-methods experiment" may be employed to estimate systematic errors when there is another routine method available for measuring the analyte of interest. Interpretation of the results must consider that the observed differences between two methods could be caused by either method, and cannot therefore be assigned to the "test" method unless the quality of the "comparison" method is well documented. Recovery, interference, and linearity experiments may be more specific in estimating systematic errors, including the constant or proportional nature of the systematic errors.

To identify sources of random error, the "replication experiment" can be designed to estimate the contributions from several different components.[3] For example, when duplicates are analyzed within a run for several different runs, the data can be used to determine the components of within-run and between-run variation. This process isolates the errors occurring in different time periods, which in turn may help identify the sources of the random error. Replication experiments of this kind, using analysis of variance statistical techniques, can be tailored to particular analytical systems to identify the major factors contributing to the random error of those analytical methods. The performance of control procedures also depends on the different components of variance, and it is particularly important that the between-run component be kept as small as possible.[68]

OPTIMIZATION TECHNIQUES

More complicated experimental techniques such as factorial experiments, "simplex" optimization, and "response surface" optimization could be used to solve complex control problems (see Chapter 5). However, these are usually beyond the scope of most clinical laboratories, and are employed primarily in industrial and research settings. Youden[79] described the use of a series of experiments in which several factors could be varied simultaneously, thus reducing the total number of experiments necessary to determine how these factors affect the performance of the analytical method. These "factorial" experiments can be used for testing the "ruggedness" of an analytical method, i.e., its sensitivity to any disturbance in the chosen experimental conditions. A properly optimized method should be rugged, meaning that small changes in the analytical conditions should not greatly affect analytical performance. Formal optimization strategies can be employed to decide on those analytical conditions. Simplex optimization has been applied to clinical chemistry methods by Krause and Lott.[43] More sophisticated "response surface" optimization techniques have been employed by Rautela et al.[55] These techniques provide systematic approaches for selecting the analytical conditions which minimize errors and reduce control problems.

References

1. Amador, E., Bartholomew, P. H., and Massod, M. F.: An evaluation of the "average of normals" and related methods of quality control. Am. J. Clin. Pathol., 50:369–378, 1968.
2. Baer, D. M., and Krause, R. B.: Spurious laboratory values resulting from simulation mailing conditions. Am. J. Clin. Pathol., 50:111–119, 1968.
3. Bauer, S., and Kennedy, J. W.: Applied statistics in the clinical laboratory: II. Variability over time. J. Clin. Lab. Autom., 2:35–40, 1982.
4. Begtrup, H., Leroy, S., Thyregod, P., et al.: Average of normals used as control of accuracy, and a comparison with other controls. Scand. J. Clin. Lab. Invest., 27:247–253, 1971.
5. Bokelund, H., Winkel, P., and Statland, B. E.: Factors contributing to intraindividual variation of serum constituents: 3. Use of randomized duplicates to evaluate sources of analytic error. Clin. Chem., 20:1507–1512, 1974.
6. Bowers, G. N., Burnett, R. W., and McComb, R. B.: Preparation and use of human serum control materials for monitoring precision in clinical chemistry. Selected Methods for Clinical Chemistry, 8:21–27, 1977.
7. Brunswick, D. J., and Mendels, J.: Reduced levels of tricyclic antidepressants in plasma from vacutainers. Commun. Psychopharmacol., 1:131–134, 1977.
8. Bull, B. S., Elashoff, R. M., Heilbron, D. C., et al.: A study of various estimators for the derivation of quality control procedures from patient erythrocyte indices. Am. J. Clin. Pathol., 61:473–481, 1974.
9. Burnett, R. W.: Accurate estimation of standard deviations for quantitative methods used in clinical chemistry. Clin. Chem., 21:1935–1938, 1975.
10. Büttner, J., Borth, R., Boutwell, J. H., et al.: International Federation of Clinical Chemistry provisional recommendation on quality control in clinical chemistry. I. General principles and terminology. Clin. Chem., 22:532–539, 1976.
11. Büttner, J., Borth, R., Boutwell, J. H., et al.: International Federation of Clinical Chemistry provisional recommendation on quality control in clinical chemistry. II. Assessment of analytical methods for routine use. Clin. Chem., 22:1922–1932, 1976.
12. Büttner, J., Borth, R., Boutwell, J. H., et al.: International Federation of Clinical Chemistry provisional recommendation on quality control in clinical chemistry. III. Calibration and control materials. Clin. Chem., 23:1784–1789, 1977.
13. Büttner, J., Borth, R., Boutwell, J. H., et al.: International Federation of Clinical Chemistry approved recommendation (1983) on quality control in clinical chemistry. V. External quality control. J. Clin. Chem. Clin. Biochem., 21:885–892, 1983.

14. Büttner, J., Borth, R., Boutwell, J. H., et al.: International Federation of Clinical Chemistry provisional recommendation on quality control in clinical chemistry. VI. Quality requirements from the point of view of health care. Clin. Chim. Acta, 74:F1–F9, 1977.

15. Büttner, J., Borth, R., Broughton, P. M. G., et al.: International Federation of Clinical Chemistry approved recommendation (1983) on quality control in clinical chemistry. IV. Internal quality control. J. Clin. Chem. Clin. Biochem., 21:877–884, 1983.

16. CAP Standards for Accreditation of Medical Laboratories. Skokie, Ill., College of American Pathologists, 1974.

17. Caragher, T. E., and Grannis, G. F.: Performance evaluation of multi-channel analyzers by use of linearly-related survey specimens. Clin. Chem., 24:403–413, 1978.

18. Cembrowski, G. S., Chandler, E. P., and Westgard, J. O.: Assessment of "average of normals" quality control procedures and guidelines for implementation. Am. J. Clin. Pathol., 81:492–499, 1984.

19. Cembrowski, G. S., Westgard, J. O., Eggert, A. A., et al.: Trend detection in control data: Optimization and interpretation of Trigg's technique for trend analysis. Clin. Chem., 21:1396–1405, 1975.

20. Cembrowski, G. S., Westgard, J. O., and Iyama-Kurtycz, D. F.: Use of anion gap for the quality control of electrolyte analyzers. Am. J. Clin. Pathol., 79:688–696, 1983.

21. Cotton, G. E., Gorman, C. A., and Mayberry, W. E.: Suppression of thyrotrophin (h-TSH) in serums of patients with myxedema of varying etiology treated with thyroid hormones. N. Engl. J. Med., 285:529–533, 1971.

22. Dharan, M.: Total Quality Control in the Clinical Laboratory. St. Louis, The C. V. Mosby Company, 1977.

23. Dixon, K., and Northam, B. E.: Quality control using the daily mean. Clin. Chim. Acta, 30:453–461, 1970.

24. Eilers, R. J.: Quality assurance in health care: Missions, goals, activities. Clin. Chem., 21:1357–1367, 1975.

25. Elin, R. J.: Elements of cost management for quality assurance. Pathologist, 34:182–183, 1980.

26. Elion-Gerritzen, W. E.: Quality control in clinical chemistry—the two-sample plot and improvement of laboratory performance. Am. J. Clin. Pathol., 67:91–96, 1977.

27. Ewan, W. D.: When and how to use cu-sum charts. Technometrics, 5:1–22, 1963.

28. Garg, A. K., and Nanji, A. A.: The anion gap and other critical calculations. Diagn. Med., 5:32–43, 1982.

29. Glick, J. H.: Statistics of patient test values: Application to indirect normal range and to quality control. Clin. Chem., 18:1504–1513, 1972.

30. Grannis, G. F.: Interlaboratory survey of enzyme analyses: I. Preliminary studies. Am. J. Clin. Pathol., 66:206–222, 1976.

31. Grannis, G. F., and Caragher, T. E.: Quality-control programs in clinical chemistry. CRC Crit. Rev. Clin. Lab. Sci., 8:327–364, 1977.

32. Grannis, G. F., and Miller, W. G.: On the design of clinical chemistry quality-control sera. Clin. Chem., 22:500–512, 1976.

33. Groth, T., Falk, H., and Westgard, J. O.: An interactive computer simulation program for the design of statistical control procedures in clinical chemistry. Comput. Programs Biomed., 13:73–86, 1981.

34. Hamlin, W. B., Duckworth, J. K., Gilmer, P. R., et al.: Laboratory Instrument Maintenance Manual. Skokie, Ill., College of American Pathologists, 1977.

35. Haven, G. T., Lawson, N. S., and Ross, J. W.: Quality control in the 1980s. In: Clinical Laboratory Annual, 1982. H. A. Homburge and J. G. Batsakis, Eds. New York, Appleton-Century-Crofts, 1982, pp. 209–225.

36. Henry, R. J., and Segalove, M.: The running of standards in clinical chemistry and the use of the control chart. J. Clin. Pathol., 5:305–311, 1952.

37. Hoffman, R. G., and Waid, M. E.: The "average of normals" method of quality control. Am. J. Clin. Pathol., 43:134–141, 1965.

38. Inhorn, S. L., Ed.: Quality Assurance Practices for Health Laboratories. Washington, D.C., American Public Health Association, 1978.

39. Juppner, H., Mohr, H., and Hesch, R. D.: Adsorption of parathyrin: Pitfall for solid phase assays using radiolabelled antibodies? J. Clin. Chem. Clin. Biochem., 18:585–590, 1980.

40. Kelly, D. T., and Kelly, M. E.: Proposals and current practices in quality control. Am. J. Med. Tech., 47:957–964, 1981.

41. Kemp, K. W., Nix, A. B. J., Wilson, D. W., et al.: Internal quality control of radioimmunoassays. J. Endocrinol., 76:203–210, 1978.

42. Korpman, R. A., and Bull, B. A.: The implementation of a robust estimator of the mean for quality control on a programmable calculator or a laboratory computer. Am. J. Clin. Pathol., 65:252–253, 1976.

43. Krause, R. D., and Lott, J. A.: Use of the Simplex method to optimize analytical conditions in clinical chemistry. Clin. Chem., 20:775, 1974.

44. Kubasik, N. P., Ricotta, M., Hunter, T., et al.: Effect of duration and temperature of storage on serum analyte stability: Examination of 14 selected radioimmunoassay procedures. Clin. Chem., 28:164–165, 1982.

45. Ladenson, J. H.: Patients as their own controls: Use of the computer to identify "laboratory error." Clin. Chem., 21:1648–1653, 1975.

46. Levey, S., and Jennings, E. R.: The use of control charts in the clinical laboratories. Am. J. Clin. Pathol., 20:1059–1066, 1950.

47. MacLin, E., Rohlfing, D., and Ansour, M.: Relationship between variables in instrument performance and results of kinetic enzyme assays—a system view. Clin. Chem., 19:832–837, 1973.

48. National Committee for Clinical Laboratory Standards: Nomenclature and definitions for use in the national reference system in clinical chemistry. NCCLS Document PSC-13. Villanova, Pa., 1979.

49. National Committee for Clinical Laboratory Standards: Standard procedures for the handling and transport of domestic diagnostic specimens and etiologic agents. NCCLS Document ASH-5. Villanova, Pa., 1980.

50. National Committee for Clinical Laboratory Standards: Tentative guidelines for clinical laboratory procedure manuals. NCCLS Document L2-T. Villanova, Pa., 1981.

51. Nosanchuk, J. S., and Gottmann, A. W.: CUMS and delta checks. Am. J. Clin. Pathol., 62:707–712, 1974.

52. Octaviano, P. J., and DiSalvo, A. F.: Quality Control in the Clinical Laboratory—A Procedural Text. Baltimore, University Park Press, 1977.
53. Page, E. S.: Continuous inspection schemes. Biometrika, 41:100–114, 1954.
54. QC Circles: Applications, Tools, and Theory. Milwaukee, American Society for Quality Control, 1976.
55. Rautela, G. S., Snee, R. D., and Miller, W. K.: Response-surface cooptimization of reaction conditions in clinical chemical methods. Clin. Chem., 25:1954, 1979.
56. Reed, A. H.: Use of patient data for quality control of clinical laboratory tests. Clin. Chem., 16:129–134, 1970.
57. Rossing, R. G., and Foster, D. M.: The stability of clinical chemistry specimens during refrigerated storage for 24 hours. Am. J. Clin. Pathol., 73:91–95, 1980.
58. Sheiner, L. B., Wheeler, L. A., and Moore, J. K.: The performance of delta check methods. Clin. Chem., 25:2034–2037, 1979.
59. Shewhart, W. A.: Economic Control of Quality of the Manufactured Product. New York, Van Nostrand, 1931.
59a. Slockbower, J. M., and Blumenfeld, T. A.: Collection and Handling of Laboratory Specimens. Philadelphia, J. B. Lippincott Company, 1983.
60. Taswell, H. F., Smith, A. M., Sweatt, M. A., et al.: Quality control in the blood bank—a new approach. Am. J. Clin. Pathol., 62:491–495, 1974.
61. Thiers, R. E., Wu, G. T., Reed, A. H., et al.: Sample stability: A suggested definition and method of determination. Clin. Chem., 22:176–183, 1976.
62. Thomas, J. A., Darby, T. D., Wallin, R. F., et al.: A review of the biological effects of di-(2-ethylhexyl)phthalate. Toxicol. Appl. Pharmacol., 45:1–17, 1978.
63. Tietz, N. W.: A model for a comprehensive measurement system in clinical chemistry. Clin. Chem., 25:594, 1979.
64. Uriano, G. A., and Cali, J. P.: Role of reference materials and reference methods in the measurement process. In: Validation of the Measurement Process, J. R. DeVoe, Ed. ACS Symposium Series. Washington, D.C., American Chemical Society, 1977, p. 140.
65. Van Helden, W. C. H., Visser, R. W. J., Van den Bergh, F. A. J-T. M., et al.: Comparison of intermethod analytical variability of patient sera and commercial quality control sera. Clin. Chim. Acta, 93:335–347, 1979.
66. Westgard, J. O.: Better quality control through microcomputers. Diagn. Med., 5:60–74, 1982.
67. Westgard, J. O., Barry, P. L., Hunt, M. R., et al.: A multi-rule Shewhart chart for quality control in clinical chemistry. Clin. Chem., 27:493–501, 1981.
68. Westgard, J. O., Falk, H., and Groth, T.: Influence of a between-run component of variation, choice of control limits, and shape of error distribution on the performance characteristics of rules for internal quality control. Clin. Chem., 25:394–400, 1979.
69. Westgard, J. O., and Groth, T.: Design and evaluation of statistical control procedures: Applications of a computer "QC Simulator" program. Clin. Chem., 27:1536–1545, 1981.
70. Westgard, J. O., and Groth, T.: Power functions for statistical control rules. Clin. Chem., 25:863–869, 1979.
71. Westgard, J. O., Groth, T., Aronsson, T., et al.: Combined Shewhart-cusum control chart for improved quality control in clinical chemistry. Clin. Chem., 23:1881–1887, 1977.
72. Westgard, J. O., Groth, T., Aronsson, T., et al.: Performance characteristics of rules for internal quality control: Probabilities for false rejection and error detection. Clin. Chem., 23:1857–1867, 1977.
73. Wheeler, L. A., and Sheiner, L. B.: Delta check tables for the Technicon SMA 6 continuous-flow analyzer. Clin. Chem., 23:216–219, 1977.
74. Wheeler, L. A., and Sheiner, L. B.: A clinical evaluation of various delta check methods. Clin. Chem., 27:5–9, 1981.
75. Whitehead, T.: Quality Control in Clinical Chemistry. New York, John Wiley & Sons, 1977.
76. Whitehurst, P., DiSilvio, T. V., and Boyadjian, G.: Evaluation of discrepancies in patients' results—an aspect of computer-assisted quality control. Clin. Chem., 21:87–92, 1975.
77. Wilding, P., Zilva, J. F., and Wilde, C. E.: Transport of specimens for clinical chemistry analysis. Ann. Clin. Biochem., 14:301–306, 1977.
78. Winstead, M.: Instrument Check Systems. Philadelphia, Lea and Febiger, 1971.
79. Youden, W. J.: Statistical Techniques for Collaborative Tests. Washington, D.C., Association of Official Analytical Chemists, 1969.

Microprocessors and Computers in the Clinical Laboratory

by Franklin R. Elevitch, M.D., and G. Phillip Hicks, Ph.D.

Although it was recognized in the 1950's that computers could have a significant impact on health care, early investigators were hindered by the limited and expensive hardware available. Computers were unable to communicate directly (on-line) with laboratory instruments or personnel and required use of off-line methods of data input with cards punched by nonlaboratory personnel. In the early 1960's the development of integrated solid state electronic circuits made possible the production of more sophisticated computer equipment at lower cost. By the late 1960's and early 1970's a new generation of hardware allowed the development of highly efficient "interactive systems" which could communicate directly with laboratory personnel and instrumentation. By the late 1970's the large scale integration (LSI) of electronic circuits permitted miniaturized components to be integrated as an entire computer in a single wafer-thin electronic chip. The chip computer, called the microprocessor or microcomputer, has become the most common computer component of the 1980's. Even large computers may be built from many LSI chips combined to form a single computer.

In the mid-1970's clinical laboratories found themselves at the forefront of analytical mechanization but constrained by manual data acquisition, data processing, and communications systems. For example, the emerging areas of immunodiagnostics and therapeutic drug monitoring were dependent upon nonlinear calibration models which in turn required rapid data acquisition and data reduction technologies. At the same time, a public clamor for improved productivity and stricter accreditation standards created the need to make conventional instrumentation more efficient while maintaining and documenting analytic quality.

Laboratory instrument manufacturers turned to microprocessor technology as a solution to both of these problems. Because of their miniature size, as well as their speed and low cost, microprocessors can be incorporated into laboratory instruments. Adding a keyboard and display device for communication with the analyst was a practical next step. The result was a new generation of instruments with built-in capabilities to identify specimens, request tests, dispense samples, diluents and reagents, monitor reaction kinetics, optimize calibration curves (and retain them for later retrieval), correct baseline drift and sample carryover, flag unacceptable control values, and diagnose instrument malfunctions. Specimen throughput was increased by replacing the slower electromechanical switches and timing devices of the past with faster microprocessor-controlled circuits. Until recently, computer hardware was so expensive that a single computer had to be used for as many applications as possible to justify its cost. However, the much less expensive microcomputer permits the utilization of numerous individual units for use in quality control, on-line monitoring, data interpretation, communications, and inventory control.[7,8] Integrated through a common database and communications networks, the modern microcomputer will rival the power of some computers sold as complete systems over the past two decades. In the future, the scope of computer applications will increase as the laboratory utilizes data to produce additional information for improved patient care, such as lists of possible diagnoses, suggestions for further tests, and the identification of drug and test interactions.[18]

To appreciate fully the impact and benefits of the computer, one must thoroughly understand the laboratory's problem of managing specimens and data. The increase in laboratory tests per se, as well as the trend to centralize laboratory services, has led to significant management problems of handling and identifying specimens submitted for analysis. Furthermore, the critical nature of the data involved requires an accurate flow of laboratory information between other areas of the hospital, including nursing services, the business office, the medical record room, and physicians' offices. Thus, it is necessary to consider the entire patient care process and the impact the computer has on all associated areas as well as the place it occupies in the entire hospital communications network (Figure 2F-1).

In view of the space limitations, this chapter provides only an introduction to basic computer information and concepts, as well as an overview of the application of the computer in the laboratory. This material is supplemented with references to more detailed sources of information.[1-3,10,11]

THE COMPUTER SYSTEM

There are two components of a computer system, *hardware* and *software*. These two components are analogous to the division of the analytical process into instrumentation and methodology. Just as the best instrumentation in the laboratory cannot produce good results without a method designed especially for accurate analysis, neither can the best computer (hardware) be used effectively without programs (software) specifically designed for clinical laboratory applications.

When an LSI chip is used as a component of equipment, such as a home appliance or a laboratory instrument, it is referred to as a *microprocessor*. A system designed as a general purpose computer which uses the same chip is called a *microcomputer*. The *minicomputer* is manufactured from the same technology as the microcomputer but is generally designed to utilize several terminals and programs simultaneously (multi-user environment), while the microcomputer is generally intended for one user at a time (single-user environment). Some general purpose microcomputers now rival the capacity of the commonly known minicomputers and the distinction between them is becoming less clear. For more detailed information dealing with the basic terminology of computer systems, the following references can be consulted.[9,12,13]

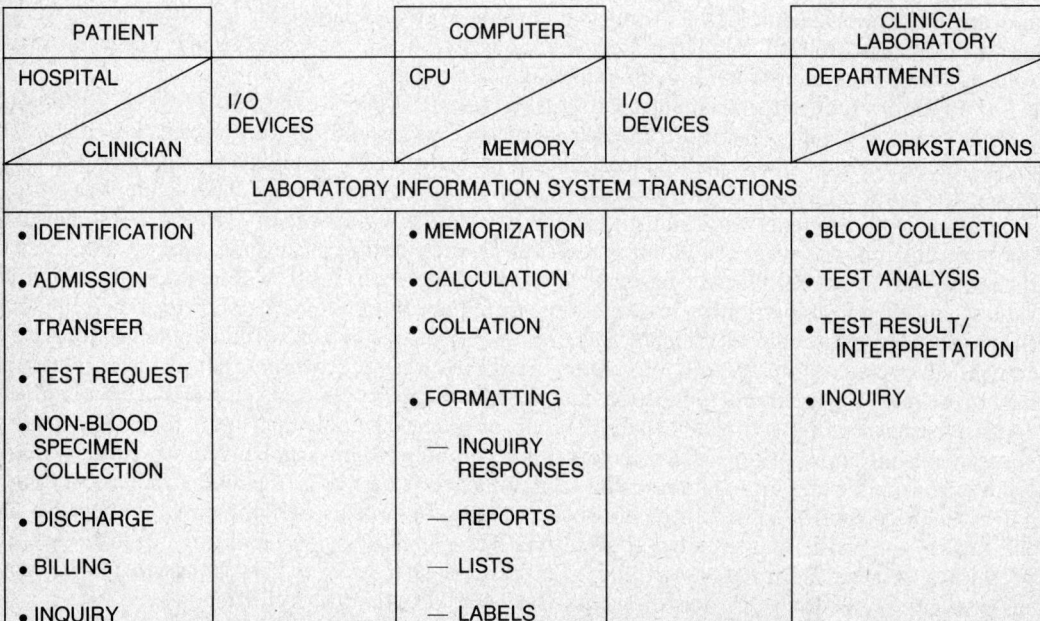

PATIENT		COMPUTER		CLINICAL LABORATORY
HOSPITAL / CLINICIAN	I/O DEVICES	CPU / MEMORY	I/O DEVICES	DEPARTMENTS / WORKSTATIONS
LABORATORY INFORMATION SYSTEM TRANSACTIONS				
• IDENTIFICATION		• MEMORIZATION		• BLOOD COLLECTION
• ADMISSION		• CALCULATION		• TEST ANALYSIS
• TRANSFER		• COLLATION		• TEST RESULT/ INTERPRETATION
• TEST REQUEST				
• NON-BLOOD SPECIMEN COLLECTION		• FORMATTING		• INQUIRY
		— INQUIRY RESPONSES		
• DISCHARGE		— REPORTS		
• BILLING		— LISTS		
• INQUIRY		— LABELS		

Figure 2F-1. Overview of a computerized laboratory information system.

COMPUTER HARDWARE

Peripheral Devices

The computer hardware can be divided into two categories—the *central processing unit* (CPU), which is the computer proper, and *peripheral devices* (peripherals), which are external to the CPU. The various peripheral devices provide the means for transferring information into and out of the CPU and for storing data and programs for use in the CPU. Peripherals combined with the CPU make up what is known as a *configuration.*

There are two types of peripheral devices: (1) those which allow the transfer of information between the computer and the laboratorian (user) or laboratory instrumentation, and (2) those which store and retrieve information.

Input and Output Devices

Cathode Ray Tube (CRT)

Devices that communicate with the user or instrumentation are referred to as I/O (Input/ Output) devices. One of the most common peripheral I/O devices is the cathode ray tube (CRT), or monitor, which is very much like a television set with a keyboard. In fact, personal micro-computer systems frequently use a television set as the monitor.

The CRT terminal is used to transfer information between the user and the computer. It can be used in a conversational mode at a high speed of output from the computer to the screen (30 to 9600 characters/second). Since the CRT has no moving parts, it is silent. The CRT allows for observation on the screen of the keyboard entries as well as for program-controlled edits of the data entered. However, as the screen fills up, the information is either erased or disappears off the top of the screen. Consequently, there is no permanent record (audit trail) of the transaction between the user and the computer. It is for this reason that laboratories may prefer to use the CRT for interactive functions, such as inquiry, and to use printer terminals for the output of data for laboratory reports.

CRTs may be "intelligent" or "dumb." An *intelligent terminal* looks very much like any other CRT terminal in outward appearance. However, it has one or more internal microprocessors with programs to perform specific tasks at the work station. The program may be permanently contained in the terminal or it may be loaded from the main computer. The latter technique is called *down loading.* In either case, the intelligent terminal performs functions that would otherwise be required of the main computer. For example, the intelligent terminal may preformat the CRT screen with instructions and blank fields for the input of data, allowing entries only into the appropriate field and editing the inputs for proper format such as decimal point position or use of number versus letters where required. Input to the intelligent terminal can continue even when the main computer is not operating or available to receive the data. In this case, the "intelligent" terminal stores the information until the main computer is operational and ready to receive the information. A "dumb" terminal, in contrast, requires that the main computer be operating to accept each character or each line and to control the input process.

Terminals

The second major type of I/O terminal is a printer which has a keyboard, similar to a typewriter, allowing keyboard entry of information or queries. The computer in turn responds by typing information back on either the same or another terminal. Output from any printing terminal is referred to as a *hard copy,* and has the advantage of providing a permanent record (called *audit trail*) of every transaction, thus allowing retrospective examination of entries. There is a variety of hardcopy terminals which range in speeds from 30 characters/second up to 240 characters/second. Hardcopy terminals are used to communicate with central processing units in minicomputer systems.

Card Reader

Another type of device, which is suitable for input of information only, is the card reader. It can "read" both pencil marks and holes on a card (thus the name mark-sense reader) by measuring the light reflected from or transmitted through the card. The light pattern corresponds to logical characters (numbers, letters, decimal points) which are transmitted to the computer. During the 1970's the mark-sense card became a standard way of entering data in areas such as hematology and urinalysis. The mark-sense document continues to be a common and inexpensive form of input for most computer systems. However, the advent of optical scanners may replace the "card" approach.

Optical Scanners

Optical scanners may read and recognize either a bar code or a printed character directly. (OCR refers to optical character recognition.) The optical scanner shown in Figure 2F-2 is a hand-held microcomputer-based device that is widely used in supermarkets and industry inventory systems. Such a unit may interact with the user, read bar codes, and communicate directly with the main central computer. Optical bar code scanners which use laser beams are used, for example, in blood banks to read identification labels. Laser technology permits a code to be read on curved as well as smudged surfaces, thereby improving reliability.

Digital Matrix Devices

The incorporation of the microprocessor into peripheral devices has provided many options for data input that are beyond the scope of this chapter, but a few are worth mentioning here. *Digital pads* allow the direct input of a "drawing" or "graphics" into the computer through a pen. *Light pen* and *touch sensitive screens* can be used to select options on a CRT by pointing to the location on the screen or to input graphic data by "drawing" on the CRT screen.

Voice Input and Output

Recognition of voice as input for the computer is a new developmental area. Considerable computing power is required for such techniques, but the internal microprocessor in an input device can provide such computing power. One of the limitations of current technology is the range of vocabulary for reliable input. However, the laboratory could utilize a device that recognizes only numbers from 0–9, and words such as point, start, stop.

Voice is also an output mode for computers and is used in devices that talk as they teach

Figure 2F-2. Handheld microcomputer with optical scanner. (Courtesy of MSI.)

spelling and mathematics. Voice input devices for computers that are controlled by computer programs are available at relatively low cost. Additional research is needed to determine the best application of such audio-output systems in the clinical laboratory.

Printers

Low speed printers. There are a variety of flexible and inexpensive *dot matrix* and *character* printers available as output devices with speeds of 30–240 characters/second. These types of printers are also commonly used with microcomputer systems and may have capabilities for several character sizes and line widths, and full graphics. Such printers also use an internal microprocessor with programs to provide printing flexibility. Despite their low speed, these printers can be an effective ouput backup to higher speed printers.

High speed printers. The high speed printer is an output device only, usually capable of printing at speeds from 300–1100 lines/minute. This device is often referred to as a *lineprinter* because it prints one whole line at a time. There is a complete character set for each space on the lineprinter, and when a complete line has been received by the printer from the CPU, the entire line is printed in one process. Because of their high printing speed, lineprinters are suitable for the output of bulky reports such as cumulative reports for all patients in the hospital, and long administrative reports such as patient directories and master work logs for the entire laboratory. Since the high speed lineprinter is expensive compared to other printing terminal devices, usually only one or two such printers are used in a laboratory configuration. Lineprinters are expensive since technology for high speed printed output has not advanced at the same rate as electronic circuit technology. New developments, however, such as laser beam printing, are becoming available. The direct interfacing of the laboratory system to the central hospital information system may reduce the need for high speed printing in the laboratory, but does not eliminate it. Such an interface also allows the laboratory system to send all laboratory results to a central computer on a timely basis for distribution by the central system.

Color Graphics

It is common for microcomputers to use a color monitor or television set for output. This makes the use of color graphics practical and has already been applied in such areas as quality control.[16] Color printers are also available, and although they are more expensive than noncolor printers, their costs are becoming competitive. Presently, color printers are much slower than conventional printers because a line must be printed once for each color. In terms of output, however, a single screen or page of color graphics can often convey information more clearly than many pages of standard printing.

Instrument Interfaces

The instrument interface is a peripheral device which allows the direct electronic connection of a laboratory instrument to the CPU. There are two major types of interfaces, *analog* and *digital*. One must understand the different characteristics of an analog signal and a digital signal in order to understand these interfaces and how they are used with laboratory devices.

Figure 2F-3 illustrates the two types of signals. An *analog signal* varies continuously with time. A recorder tracing of a chromatographic separation provides a good example of a graphic demonstration of an analog signal. Since a digital computer can process only discrete numbers, it is necessary to convert the continuously changing voltage generated by the detector into a series of discrete numbers. This conversion is made by an *analog to digital converter* (ADC) which "reads" the analog signal and converts the value of the signal height into a discrete number proportional to the voltage of the original signal. A series of numbers taken at sufficiently frequent intervals (e.g., each second) permits the computer to construct a corresponding curve and to find the highest point on this curve. Other examples of analog instruments are continuous-recording densitometers for electrophoresis and recording spectrophotometers.

A *digital signal* typically consists of a series of equal height voltage "pulses." The magnitude or "value" of the signal is indicated by the presence or absence of pulses in the pulse pattern, or sometimes by the interval between pulses, but not by the height of the voltage pulses. Figure 2F-3 illustrates four sets of digital pulses which correspond to the four analog peaks. Since the digital pulses are already discrete, a digital interface need only "decode" the pulse pattern to obtain the discrete number representing the signal value. Most laboratory instruments utilize a microprocessor to preprocess data into a digital format before output.

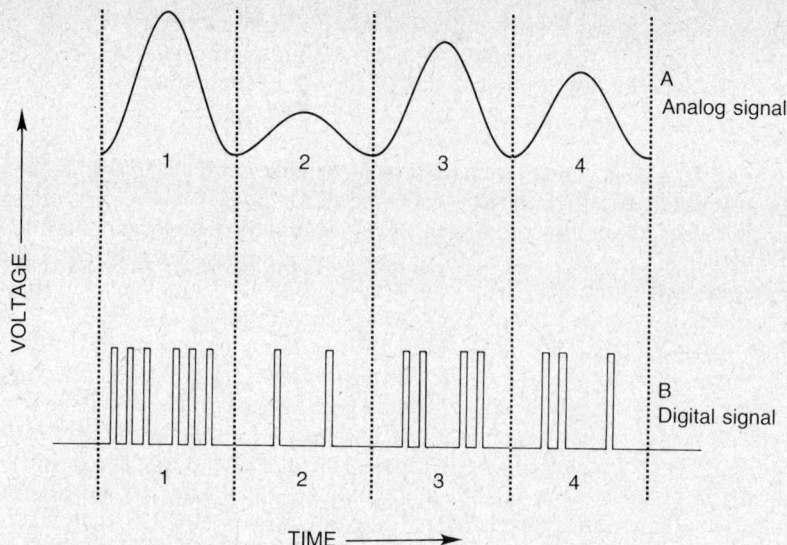

Figure 2F-3. Representation of analog and digital signals.

Because digital pulses contain more high frequency components compared with the most common analog signals, digital instruments must be closer to the computer (25–200 feet) than most analog instruments (25–1000 feet) unless the pulses are amplified.

Storage Devices

Disks

Perhaps the most common storage device external to the CPU is the disk. This is literally a platter on which magnetized spots or *bits** can be deposited (or erased). A disk can be thought of as a phonograph record with grooves which correspond to magnetic tracks. The heads within the disk drive read and write information on the disks, and correspond to a phonograph pickup and stylus.

Figure 2F-4 illustrates magnetic bits of data on the tracks of each disk. Each bit produces an electronic pulse when it passes the read head. A bit of data is determined to be "0" or "1" by the presence or absence of an electronic pulse. A number of bits together produce a pattern of pulses called a *word*. For example, a 16-bit word consists of a pattern of 16 pulses which give the numeric value of the word in a manner similar to that described for digital signals in Figure 2F-3. Each word has a specific location on the disk assigned to it; this location is called the *address* of the word. One half of a word, generally 8 bits, is called a *byte*. A byte or one half word is generally used to store the equivalent of a single character such as "A", "1", etc., and has become one of the most common ways of expressing memory and file sizes. Thus, 64 000 bytes of memory would be 32 000 words or 64 000 characters.

The disk platter is divided into tracks and the tracks are divided into *sectors* and *blocks* which may be designated to store specific files of information or programs. Some disk drives have a read head for each track, called a "fixed head," but most disk drives have a single read head for each disk platter which must move to a track and then read the block of interest. Since the head has to move to the appropriate location and then read the sector to find the information, the average access time is typically 50–80 milliseconds. The moving head disk, however, usually has a high capacity, namely in the range of 2.5–50 million words per platter.

The moving head disks are further categorized as *hard* and *floppy*. The *hard disk* is made of a rigid aluminum platter which has a magnetized-coated surface. The *floppy disk* is similar to magnetic tape except that it has the shape of a flexible disk and is contained in a rigid paper

---

*All information handled by the computer is contained in a numerical code whose fundamental unit is called a *bit*. Since the values of a bit are limited to zero and one, all calculations and data are contained as coded numbers in base 2 or "binary." Word sizes for typical laboratory computers range from 8 to 32 bits.

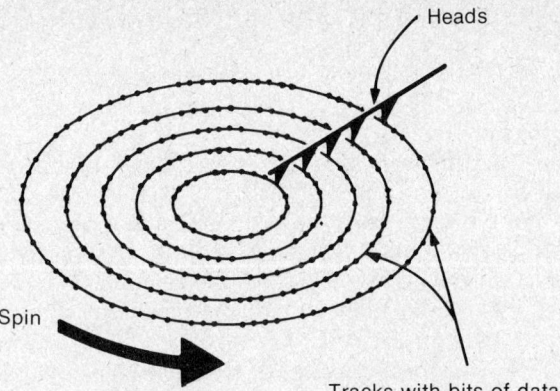

Figure 2F-4. Representation of disk.

jacket for protection. Although both floppy and hard disks may be used on all types of computers from the microcomputer to the largest mainframes, the floppy disk is more commony used with the microcomputer.

Floppy disks are available in three sizes: $3^1/_2$, $5^1/_4$, and 8-inch diameters. Higher capacity floppy disks, i.e., double or quad density, and dual side formats are also available. Although relatively inexpensive (approximately $500 for the disk drive and $5 for each floppy diskette with a typical capacity of 160 000 bytes), the capacity for storage on floppy diskettes is fairly small, and they need to be changed frequently in order to load programs or data into the computer. Hard disks, on the other hand, are significantly more expensive, but offer greater storage capacity (upwards of $3000 for a capacity of 5 million bytes or more).

Floppy disks and some hard disks may be removed from the disk drive for storage at a site remote from the computer. The Winchester, a nonremovable hard disk, provides high-capacity storage at low cost, and information is easily retrievable.

The storage device with the fastest access time is the *mass memory*. These forms of memory are accessible at the same speed as the main memory of the computer and may be constructed by LSI chip technology. Decreasing manufacturing costs have made these "bulk" or "mass" memories competitive with fixed head disks and by comparison they are inexpensive to maintain. Mass memories are provided in modules of 250 000 to 500 000 bytes and are usually "plug-to-plug compatible" with the disk controllers provided by major manufacturers. Large systems use up to 4 million bytes of this type of storage. Typical access times are 3 microseconds, a thousand times faster than fixed-head disks with access times of 4 milliseconds. Therefore, manufacture of fixed-head disks generally is being discontinued.

In some systems, high-speed mass memories equivalent to disk memory are used to store and access information most frequently needed, such as programs. The moving-head disk, which is much slower, might be used for storage of patient data that are accessed less frequently.

Magnetic Tape

Magnetic tape is the slowest storage medium. It also retains information in the form of magnetic bits. The retrieval time depends on the length of the tape. Small tapes, used on laboratory computers, typically contain only 100 000 to 200 000 words and the computer can scan such a tape within 20 to 30 seconds. While this is extremely slow compared to a disk, it may be quite acceptable for storing information which is to be retrieved infrequently, e.g., billing information that is generated once a day only.

Large amounts of information may be stored on "industry-compatible tape," sometimes referred to as IBM-compatible tape. These tapes may be 2400 feet in length and may have 20 times or more the capacity of small tapes, but the access time may be as long as 20 minutes. Use of industry-compatible tape has the great advantage that information generated by the laboratory computer can be transferred to any other computer equipped with an industry-compatible tape drive. This may be helpful for transmitting billing information or information related to admissions, transfers, and discharges from one computer to another. Such transfer of information by tape is less expensive than direct computer interfaces by wire or telephone.

Magnetic tape is also a more economical medium for the long-term storage of patient data than disk or mass memory.

The video tape recorder used in the home is an example of a *high-density digital tape storage* device. Video tape cassette systems are used as a backup storage for some of the medium-capacity disks in the 10 million byte range.

The *video disk* is a new development that uses a laser beam in a manner analogous to the stylus of a phonograph; it has the potential for storage of billions of bytes of data. Video disks presently allow for information retrieval only; however, research is under way to develop a video disk that will have capabilities for both reading and writing. If successful, video disks will be one of the highest capacity storage media available for the computer.[17] For example, a single side of a credit card size video disk presently can store one megabyte of information.

Central Processing Unit (CPU)

Computers used for the laboratory typically are classified as minicomputers and microcomputers. In the 1970's this classification was meaningful to distinguish the limited capability of the microcomputer from the more powerful minicomputer. Today, this distinction is less clear as the microcomputer approaches the capacity of the minicomputer and because the minicomputer is manufactured with the same chip technologies used in the microcomputer.

The CPU is the "electronic logic" of the computer configuration. In the case of the microcomputer, the CPU might be a single chip on the main electronic board. Each CPU has its own set of instructions which it executes in the appropriate order as specified by the software. The number of different instructions in an *instruction set* ranges from six to several hundred. The CPU has its own storage device, the main memory, which may contain from 4000 (4 K) to one million (1 mega) or more bytes or instructions, depending upon the specific requirements. The CPU electronic logic will typically fetch instructions one-by-one from the main memory, then recognize and execute them appropriately. The time required for these functions is generally 1 microsecond or less.

The fast access time of the main memory and the high-speed circuitry of the CPU in general are very seldom the limiting factors in the speed of the overall system. Peripheral devices, such as disks, terminals, and tape units are many orders of magnitude slower than the CPU, and therefore determine the overall speed of the system. The central processor is capable of executing many thousands of instructions in the same time in which it takes a terminal to print or display one character.

COMPUTER SOFTWARE

Software Programming Languages

Computer software programs provide detailed instructions to the electronic hardware, and thus enable the overall system to function. Programs can be constructed in several "levels" of languages. Those languages relating to the detailed machine structure (binary words, bytes, bits, etc.) are called *machine* or *assembly languages* and are considered to be "low level." Languages which relate to commands which provide operations directly for the user (C=A+B, PRINT, etc.) are referred to as "high level."

Machine instruction is the most fundamental means of telling the computer what function to perform. An instruction is expressed as one or more discrete numbers which are decoded by the electronic logic of the computer. For example, the number coded 2000 may instruct the computer to ADD a number (xx) to another number (xy). The address (location) of the number to be added (xy) might also be included within the basic instruction. Thus, the number coded 2100 may instruct the computer to ADD the number (xy) in location 100 to the other number (xx). A series of such discrete instructions which are executed in sequence constitutes a *program*.

Constructing a program with such a series of discrete numbers is very tedious work, and therefore techniques have been developed to allow programmers to substitute English-type *mnemonic codes* for instructions, such as ADD, LOAD, MULT, or STORE, in place of their numeric equivalents. The computer program that converts the mnemonic codes to the equivalent numeric instructions which are decoded by the computer is called an *assembler.* Programming in this manner is called *machine language programming* or *assembly programming.*

A useful program, for example, to print a patient's name from a computer file may require hundreds or even thousands of discrete machine instructions and can be tedious to construct even with assembly language. Once a function is written in machine instructions, it may be "called" or used by many programs without the necessity of writing the same function many times in one system. An even higher level of mnemonics can be implemented to call a series of functions. For example, the mnemonic "PRINT NAME 178756" may mean to print the name of the patient with the hospital number 178756. This simple mnemonic may in turn call a corresponding machine program consisting of hundreds of instructions. Such a technique, referred to as high level programming, allows the programmer to construct programs more easily and at a "higher level." Similarly, a series of high-level mnemonics is referred to as a *high-level language.* Examples of some well-known high-level languages are FORTRAN (FORmula TRANslation), which is especially useful for mathematics, COBOL (COmmon Business Oriented Language), useful for business applications, BASIC (developed by Dartmouth University as a simple beginning language), PASCAL (structured language developed by the University of California), and MUMPS (designed at Massachusetts General Hospital for medical applications).

High-level and machine-level programming both have advantages and disadvantages. A high-level language is easier to learn and to use, requires less knowledge of the computer, and is therefore preferred by many programmers. While more difficult to use, machine language is more efficient in terms of speed and size of programs. A single high-level instruction may use hundreds of machine instructions generalized to carry out a function with a wide range of conditions and applications, while a machine program can be tailored for a specific application. A well-designed system usually will mix the higher level functions with specific machine instructions to optimize the advantages of both types of programming. Since memory is becoming less expensive, more programming is being done in high-level languages. However, certain applications such as interfacing to instrumentation or communications require the use of machine language to attain sufficient speed and control.

A program or system cannot simply be rated as "good" or "bad" based on the language in which it is written. Rather, one must be concerned with the overall functions performed, the speed of execution of each function, the response time, and the ease with which laboratory functions are changed or developed within the system.

For functional purposes, a program is classified either as an *operating system* or as an *application program.* The group of programs called the operating system is analogous to the central nervous system. Written for specific central processing units, these programs "manage" all of the jobs and the devices on the computer. Application programs may be written in assembler or higher level languages, and are designed to interrelate the computer to its user.

In the 1970's the operating system and the application programs resided in a single computer. Today microcomputers are usually interfaced with the operating system and database in a main minicomputer. Data "downloaded" from the minicomputer are for use in the microcomputers. The advantage of an efficient operating system and comprehensive database is thereby combined with the relative simplicity of microcomputer programming. This "networking" of microcomputers with database systems is a growing trend and will be an important factor in the evolution of new laboratory systems.

Database

The organization of data in the computer ultimately determines the function and flexibility of the system software. New approaches for structuring data have recently been developed and promise to increase greatly the utility of computer information systems. A detailed discussion of this new approach, called database, is beyond the scope of this chapter; however, key features to database systems are mentioned to encourage further study by the reader.[15]

Traditionally, computer files were designed to resemble the processes which they were intended to serve. In the 1970's, for example, laboratory systems typically were designed with "request files," "worksheet files," etc. Structuring data in the computer files, analogous to the processes in the laboratory, provided a convenient way to access the data for each function. This approach, however, assumes that the computer system will be utilized only in the manner intended by the designer. When the system is used in ways not anticipated by the designer, the specific structure of the data files imposes limitations on the number of ways that the data can be used by programs; i.e., it is very inflexible.

Programs using structured file systems typically are "data dependent." This means that each program is written to utilize the files exactly as they are designed by referring to the specific location and structure of data within each file. With this approach, it is necessary to rewrite every program each time the data files are altered by adding new types of data or by changing its structure. Thus, both process-oriented file designs and data-dependent programs have contributed to the inflexibility of laboratory systems.

On the other hand, database designs promise to provide dramatic improvements in flexibility. Briefly stated, a database system is designed to file or store data without a programming structure related to the specific process it initially intended to serve; i.e., all data are accessible without limitations set by previous assumptions regarding the use of data. The data are organized in logical groupings which provide the computer programs essentially equal access to all data. This allows new programs to use data in ways not anticipated by the designer of the system, and makes the programs *data independent*.

COMPUTERIZED LABORATORY SYSTEM

BACKGROUND

In a computerized laboratory information system (LIS) the flow of information passes through the central processing unit and then into memory, where it is available for inquiry and reports (Figure 2F-1). All information is entered into the computer through terminals or machine readable forms. In this interactive mode the computer is used to manage the data transactions through appropriate software-controlled prompts and edits. Requisitions, logs, and workbooks are used as a backup on those occasions when the computer is inoperative for prolonged periods of time.

Entry of information related to patient identification, admission/transfer/discharge, test requests, and some specimen collections is a function performed by nonlaboratory personnel. In institutions that have a hospital information system (HIS), this information is automatically transmitted to the LIS through an electronic interface and communications protocol. Usually test results are returned through LIS to HIS for inquiry and reporting on the nursing stations. Billing information is transferred to the hospital financial management system (FMS) either directly or more commonly by magnetic tape.

Messages between LIS and HIS are transmitted through an interface according to a transmission protocol with specific characteristics:

1. *Electrical interface standard*, e.g., Electronic Industries Association (EIA) RS-232. This standard specifies voltages, connectors, and pin assignments for interconnecting data communications devices over limited distances.

2. *Mode of transmission*, either *parallel* or *serial*. In parallel transmission, all bits in a character, i.e., a byte, are transmitted simultaneously. Each bit in an eight bit byte requires a separate wire for data transmission. In serial transmission, binary bits are transmitted sequentially and require one pair of wires for transmission. In serial *asynchronous* data transmission, timing is not critical. A start and stop bit is added to each character so that the receiving computer can identify when the character starts and stops. In serial *synchronous* transmission, timing is derived through synchronizing characters at the beginning of each message.

3. *Transmission character set*, e.g., ASCII (American Standard Code for Information Interchange). Each ASCII character is represented by seven binary bits giving a total of 128 (2^7) possible characters.

4. *Parity check*, a means of checking the correctness of data transmission. An additional bit is added to each character or word which represents the sum of 1's or 0's. Parity may be odd or even. The rule for odd parity: if the sum of 1's is odd, the parity bit is 0; if the sum is even, the parity bit is 1.

5. *Transmission rate* in bits per second is called *baud*. The baud rate in a clinical setting is usually 100, 300, 1200, 2400, or 9600.

6. *Record length* is number of characters, e.g., 512, of which 499 may be data with the balance of characters used for message timing and parity checking.

PLANNING A LABORATORY INFORMATION SYSTEM

A LIS integrates all aspects of the laboratory, medical staff, and hospital that are concerned with laboratory testing.[6] Because a LIS commits an organization to a literal, formalized system of policies and procedures embodied within the computer system, a pre-existing harmonious relationship among all of the parties is essential. In this atmosphere, a long-term institutional commitment can be made to planning, implementing, operating, and maintaining a LIS.

Since the installation of a LIS is a major commitment, a feasibility study should first be conducted by the laboratory, hospital administration, and medical staff to determine the technical and economic practicality of replacing the existing manual system with a computer system.[5] The feasibility study may show that improvements in the manual system are sufficient to correct existing problems and bring about enhanced service, in which case the plans for a LIS would be abandoned. On the other hand, if a LIS appears preferable, the next step is writing specifications for the LIS. These specifications must be detailed and specific, addressing the needs of the laboratory as well as the rest of the hospital and medical staff. If a HIS already exists (or if one is anticipated imminently), a computer interface specification is included in the planning process. The hospital may then choose to develop its own LIS or purchase one from the LIS vendor which meets the required specifications. In either case, a lifelong relationship (at least for the life of the LIS) is established with those who provide the computer, its peripheral devices, and its programs. The mutual obligations of this relationship are described as clearly as possible in a legal contract signed by the hospital and the LIS vendor.

DATA TRANSACTIONS IN THE LABORATORY

For purposes of this discussion, the flow of clinical laboratory information has been divided into seven steps. Each step is characterized by a *data transaction* in which there is a unidirectional transfer of data from a *data provider* to a *data receiver*. Although laboratory personnel are involved at each step, the patient, medical staff, and five other hospital services participate in the overall process (Table 2F-1).

Depending on the admission circumstances, i.e., inpatient, emergency, or outpatient, a patient

Table 2F-1. CLINICAL LABORATORY INFORMATION DATA TRANSACTION

Transaction	Data Provider	Data Receiver
Patient identification admission transfer	Patient Nurse	Admission clerk ER clerk Nurse Lab clerk
Test request	Clinician	Lab phlebotomist Nurse
Specimen collection	Lab phlebotomist Nurse	Lab technologist Nurse Clinician
Test analysis	Lab instrument	Lab technologist
Test result/interpretation	Lab technologist Lab director	Clinician Nurse
Patient discharge	Clinician	Admission clerk Business clerk Lab clerk Medical records Nurse
Patient/guarantor billing	Lab clerk	Business clerk

provides identification information to a clerk in the admission office, emergency room, or clinical laboratory. When a patient is transferred or discharged, nursing service or admission office personnel provide the pertinent information to the laboratory. The patient's physician requests specific tests in longhand on order sheets in the patient's chart. These orders are then transcribed by a ward clerk onto a requisition form. Specimens such as urine or sputum are collected and labeled by either nursing or laboratory staff. Other specimens, such as wound or body fluid, which are collected and labeled by physicians, are also sent to the laboratory accompanied by a requisition. Receipt of a labeled specimen and requisition in the laboratory is acknowledged by a time stamp on the requisition, followed by an entry in a manual accession log book. Accessioned specimens are distributed to departments where they are processed through centrifugation and aliquotting if necessary. The specimens are then distributed to work stations for analysis. Analytical results are transcribed by laboratory technologists onto report forms which are distributed for interpretation and action by the medical or nursing service.

At some point during this process, a charge slip is sent to the business office for patient and/or guarantor (insurance company, Medicare, Medicaid) billing.

Communication throughout this entire data transaction process is both verbal and written, typically involving several different types of forms, embossed plates, labels, requisitions, reports, and statements. Mistaken data transcription, delivery, or filing at any of the various steps can result in erroneous laboratory service. The problem is compounded by the fact that a single patient usually has multiple specimens within a hospital admission and numerous patients are receiving service each day.

THE LABORATORY DATABASE

A major advantage of a LIS in a medical facility is that it establishes standard laboratory nomenclature and standardized information processing procedures. The basic unit of information in a LIS file is called a *data element*. All the aspects of clinical laboratory testing are described by the appropriate combinations of fewer than 50 data elements grouped to define a patient, a specimen, or a test (see Table 2F-2). Actually, all the transactions shown in Table 2F-1 translate into data element processing in LIS: adding, deleting, editing. LIS facilitates not only data entry but also their storage, collation, formatting for retrieval, inquiry, and reporting (see Table 2F-3).

Data elements are organized in LIS into two functional types of files: definitional and descriptive. *Definitional files* or dictionaries are intended for long-term repetitious use. They include data elements which constitute the basic services of the laboratory and whose definitions

Table 2F-2. DATA ELEMENTS OF A LABORATORY INFORMATION SYSTEM

Patient	Specimen	Test
1. Number	1. Number	1. Number
2. Name	2. Status	2. Name(s)
3. Birthdate	a. Ordered	3. Mnemonic
4. Sex	b. Collected	4. Source
5. Admission date	c. Received	5. Turnaround time
6. Admission status	d. Reported	6. Container/volume
7. Location	3. Date	7. Special requirements
8. Room/bed	a. Year	8. Department
9. Physician(s)	b. Month	9. Worklist
10. Diagnosis(es)	c. Day	10. Method
11. Discharge date	4. Time	11. Units of measurement
12. Financial status	a. 12-hour clock	12. Numeric?
13. Guarantor(s)	b. 24-hour clock	13. Required?
14. Billing number	5. Worker identification	14. Delta check
		15. Controls/standards
		16. Reference intervals
		17. Panic value(s)
		18. Charge code
		19. Charge
		20. Workunits

Table 2F-3. LABORATORY INFORMATION SYSTEM INTERACTIONS

Transaction	Inquiry/Reports
Patient ID, admission/discharge/transfer	Daily listings
Test request	Clinicians' orders
	Uncollected specimens
Specimen collection	Collection lists
	Labels
	Master log
	Outstanding specimens
	Tests pending
	Worklists
	QNS (quantity not sufficient) report
	ND (not done) report
Specimen analysis	QC listings/graphs
Test result/interpretation	Stat
	Interim
	Cumulative
	Exception reports
	Abnormal values
	Panic values
	Delta check
Patient/guarantor billing	Billing listing
	Productivity/workload

can be changed only by management. These files include (1) test list; (2) worklist formats; (3) standards and controls; (4) coded comments; (5) laboratory departments; (6) hospital, emergency room, and clinic locations; (7) physicians' names, addresses, and phone numbers; (8) diagnostic classifications; and (9) automated instrument interface specifications. *Descriptive files* retain the historical events in LIS, e.g., patient identification/admission/discharge/transfer, test requests, specimens. Relevant definitional data elements (Table 2F-2) are included in descriptive files which make up each patient's record. Each LIS will have its own file structure, which may be structured, hierarchical, network, relational, or some combination of these.

In the hierarchical file structure, directories are arranged in a branching tree structure. Pointers link data elements together in a network structure, while two-dimensional tables connect data elements in a relational structure.

For every transaction in LIS, the appropriate data elements are incorporated into a descriptive file. Table 2F-4, the LIS database, illustrates the logical accumulation of patient (P), specimen (S), and test (T) data elements from patient identification through billing. With each new transaction, data elements are added. Because of the cumulative process, it is unnecessary to repeatedly store data elements which are already filed with each new test request. All transactions relate specific test numbers to specific specimen numbers for a specific patient number.

Patient Identification

All hospital services for each patient are related to a unique patient number. On admission the patient's name, birthdate, and sex are recorded by the admission clerk. If there is a HIS, a unique patient number is assigned by the HIS computer, and the four data elements, the patient's number, name, birthdate, and sex, are transmitted to the LIS. Otherwise, the admission clerk assigns a unique patient number from a nonrepeating list of serial numbers, and the four data elements are delivered to a laboratory clerk for transcription into LIS.

The patient number, or unit number, should be unique and permanent during a lifetime of medical care. This number typically has six or more digits (e.g., Social Security number). Sometimes an alpha prefix "I" or "O" may be used for quick recognition of inpatient or outpatient status or a check digit may be added. Other times a numerical suffix identifies each of multiple admissions, or an additional separate admission number may be assigned from a nonrepeating serial sequence of numbers. In an emergency room admission of a patient in a life-threatening circumstance, time

Table 2F-4. LABORATORY INFORMATION SYSTEM
DATABASE

Transaction		Data Element
Patient identification admission transfer	P	1. Patient number
		2. Patient name
		3. Birthdate
		4. Sex
		5. Admission date
		6. Admission status
		7. Location
		8. Room/bed
		9. Clinician(s)
		10. Diagnosis(es)
Test request	S	1. Specimen number
		2. Specimen status
		3. Date
		4. Time
		5. Worker ID
	T	1. Test number
		2. Test name(s)
		3. Mnemonic
		4. Source
		5. Turnaround time
Specimen collection		6. Container volume
		7. Special requirements
Specimen analysis		8. Department
		9. Worklist
		10. Method
		11. Units of measure
		12. Numeric?
		13. Required?
		14. Delta check
		15. Controls/standards
Test result/interpretation		16. Reference intervals
		17. Panic value(s)
Patient discharge	P	11. Discharge date
Patient/guarantor billing	T	18. Charge code
		19. Charge
		20. Work units
	P	12. Financial status
		13. Guarantor(s)
		14. Billing number
		15. Addresses

may be insufficient to obtain a formal patient number from HIS or the Admission Office. In such cases, a separate manual system consisting of unique, nonrepeating numbered labels and matching wristbands which can be worn by the patient is essential. The temporary number is transcribed into LIS so that testing can proceed without delay. When a permanent patient number is assigned by LIS, the information under the temporary number is merged into the file under the permanent number. All test results will then be associated with the single correct number. Merging of files is also helpful when a patient is admitted as an inpatient from the emergency room, and allows all temporarily related data to be kept combined in a single patient report. When merging takes place, the patient number eliminated by merging should become one of the patient identifiers which are displayed on patient reports as a reference to previous transactions. Lifetime physical attributes, such as blood type, may be retained in LIS for quick retrieval in emergency situations.

Age is usually derived from the patient's birthdate and is used in LIS for assigning age-related reference intervals to the respective test results. The birthdate may also help to discriminate between patients with identical names. Sex-related reference intervals may be edited by gender. If a patient fraudulently borrows a friend's medical eligibility identification card, blood type may be a helpful distinguishing feature.

In a multiprocessor network, the patient number is recognized by each processor so that there is always a way to relate a specimen to the patient from whom it was obtained.

Once a patient is properly identified, this information is associated with all subsequent laboratory transactions. Although LIS-printed labels correct the illegibility problem sometimes associated with embossed plates, it is still possible to introduce an error by putting the wrong label on a specimen. Use of bar code and optical character readers reduces transcription and identification errors in specimen processing.

Data are retrieved in LIS by either patient name or patient number. When a patient name is used for retrieval, it is convenient if a Soundex (sounds like) code or some sort of partial lookup routine using the patient's initials is available, in case the patient's name is misspelled or illegible.

Admission/Transfer/Discharge

A patient's demographic data elements may differ from one hospital stay to another. These data elements include (1) admission date, (2) admission status, (3) location, (4) room/bed, (5) clinician(s), (6) diagnosis(es), (7) discharge date, (8) financial status, (9) guarantor(s), and (10) billing number.

Admission statuses are outpatient, inpatient, preadmission, ambulatory surgery, etc. In LIS, these classifications initiate report routing, define the length of time data are retained in LIS, and organize pricing schedules. For example, inpatients may be retained in LIS for 15 days after discharge until all testing is completed, whereas the retention time may be 60 days for outpatients. Charges for laboratory testing may also be related to admission status if there are cost differences between inpatient and outpatient services.

Location, such as room and bed, is used for inpatient specimen collection rounds and report delivery. Outpatient reports are routed to the emergency room, clinics, or clinicians' offices, respectively.

Responsibility for documenting the patient's diagnosis(es) is shared between the hospital staff and the medical staff. While physicians make diagnoses, hospitals maintain the medical records. Physicians' diagnoses follow a variety of classification systems which hospitals have the responsibility to translate into a single uniform system of coding. Hospitals currently use a coding system called ICD9-CM (International Classification of Diseases, Clinical Manifestations). These codes are collated on a regular basis into reports of the hospitals' activities.

In LIS, patient identification and admission/transfer/discharge data elements must be entered before tests can be ordered and processed. If there is a HIS, these data elements are transmitted automatically to the LIS. If HIS assigns patient numbers but is temporarily inoperable, a system of temporary numbers must be available for manual entry into LIS. Temporary numbers are merged into permanent numbers when the HIS comes back up. If there is no HIS, all data elements are manually entered into LIS.

The billing process is initiated by acquisition of the patient's financial status and guarantor as specified on admission. (See Patient/Guarantor Billing, page 475.) Hospital accounting systems are becoming increasingly complex because of the growing number of reimbursement categories, such as self pay, private insurance, part pay, Medicare, and others. The guarantor is the specific insurance company or government agency to which the statement of charges will be sent.

Test Request

Nursing service is the communications hub for patient care. All clinician orders, including those for laboratory testing, pass through nursing service. Even if a clinician places an order directly into a HIS, which is interfaced with LIS, these orders are documented in the patient's chart. Nursing service has the responsibility to assure that all specimens are collected and that all results return to the patient's chart. Nonblood specimens, i.e., urine, sputum, are collected by the nursing service. The specimen together with a requisition (transmittal slip) is then sent to the laboratory. In a manual system, this usually means that a multipart form is imprinted with the patient's identifying information, and the appropriate tests are either checked off or written on the requisition. With an interfaced HIS, transmittal slips are printed at the nursing station for nurse-collected specimens. For routine and timed specimens collected by the laboratory, either collection lists and labels are printed in the laboratory through HIS, or manual requisitions are sent to the laboratory for transcription into LIS. For STAT orders in a manual system, the laboratory is called by phone and the requisition is held at the nursing station until the laboratory arrives. In a HIS ordering system, STAT orders for blood collection should be printed closest to the location of the phlebotomist, either in the laboratory or on the nursing unit. When tests are ordered in HIS, a specimen number is assigned by the computer to all of the tests ordered at that time. This number is the unique identifier which both HIS and LIS use for inquiry and reporting. If this specimen number is also used as an accession number in the laboratory, the numerical series in each department will not be continuous. This makes it somewhat difficult at a later time to look for retained specimens, but with specimen lists and machine-readable labels, the specimens are identifiable.

A record of cancellations is maintained by entering into LIS the specimen number, test identification, date cancelled, reason for cancellation, and user identification. Duplicate test orders within a specimen should be rejected automatically.

Specimen Collection

The receipt of a collected specimen in the laboratory is recorded in LIS by changing the specimen status from "ordered" or "collected" to "received." This specimen status change automatically completes the order in the clinicians' HIS record, and can be used to assign an accession number in LIS. The latter action has the advantage of maintaining consecutive accession numbers exclusive of the cancelled orders that occurred before the specimen was collected. In this case, accession numbers relate to the year, month, day, laboratory department, and sequence in which the specimen was received. For all departments other than Microbiology, it is convenient to assign numbers serially within a calendar day, with the clock starting at midnight. Microbiology conveniently runs on an annual calendar beginning 0001 hours January 1st.

If test requests are added on to a specimen which has been received, it is useful to be able to assign the previous accession number to that test. Quality control specimens, both controls and standards, should have an accession numbering sequence analogous to but distinguishable from patient specimens. Phlebotomist and nursing service identification should also be stored in LIS.

Specimen Analysis

For specimen processing, LIS groups tests with the appropriate standards and controls by workstation or reference laboratory. These groupings are called worklists and may be reviewed on either a video terminal or printed report. The format for worklists is specifically designed for each work area. A worklist may be a loading list for an automated instrument or it may provide a format for recording data if tests are done manually. When a worklist is partially completed, incomplete tests are automatically advanced to the next worklist. In this way, worklists serve as "conveyor belts" for work in progress in the laboratory. In order to keep track of uncompleted work, LIS produces the following operational reports on request throughout the day: a cumulative list of specimens received in the laboratory but not completed; a list of received specimens which were not processed because of insufficient quantity or any other reason; a list of tests ordered but cancelled; daily master logs of all specimens collected; and a cumulative summary of all tests ordered for each patient during an admission.

Test results are produced by a variety of laboratory instruments. With built-in microprocessor capabilities, many instruments display preformatted results together with limit checks and results of control samples. Once matched by a laboratory technologist with the proper patient identifiers, these values are transmitted directly to LIS and HIS. Many instruments contain sufficient memory to retain a large number of patient and control values in the event LIS is inoperable. Alternatively, supplemental memory storage devices (buffers) are commercially available as backup for those instruments that do not have this capability.

LIS stores all control results and uses these results for several quality control routines, including the multi-rule Shewhart, a daily listing of observed raw values, expected values, and a calculated running mean and standard deviation. A cumulative summary of observed control values is displayed monthly in a standard graphical format (see Chapter 2E, Quality Assurance), together with a display of the expected mean and standard deviation and a list of observed means and their standard deviations for at least the five previous months.

Test Results

Laboratory technologists review and verify all results entered manually or through on-line instrument interfaces before the results are available for inquiry or reporting. LIS retains the identification of the laboratory technologists who performed and verified the results. Free text and coded comment entries are also included with results. The LIS test list definitions (Table 2F-2) provide on-line editing capabilities for numeric/non-numeric results, required results, and results outside of reference, panic, or delta check limits. (See also Chapter 2E, Quality Assurance.) Coded comments are also edited by user-definable tables of allowable responses. LIS may record the time at which results were verified and use this time as a test completion time. LIS also may record the start, stop, and elapsed time of automated instrument runs.

In addition to patient stat, interim, or cumulative reports, LIS provides lists throughout the day of all patients whose results are abnormal, panic, and out-of-delta check limits for review by the laboratory director.

Test Interpretation

Ideally, computer-generated *interpretive reports* should assist clinicians in evaluating patient problems and arriving at medical decisions with accuracy and efficiency. The first step in report design is to format data in such a way that abnormal results and pathophysiological relationships are highlighted. This can be accomplished by grouping physiologically related analytes in the print collation sequence of the cumulative report, and by flagging abnormal and panic values with distinctive symbols. For analytes whose trend may be meaningful on a daily basis, e.g., calcium, a vertical format with analytes listed in rows and with dates in columns is appropriate. For analytes where date and time are more meaningful, e.g., blood gases or isoenzymes, a vertical format is preferable. Of course, graphic displays are ideal in either circumstance but the technology is just coming of age.

The next step in interpretive reporting is the *explanatory comment*. This can be added by the laboratory director to emphasize an abnormal finding or caution against false positive results. An elaboration of this is the categorical listing of disease entities that might be associated with abnormalities.

Computer-assisted interpretive reports have taken several different approaches.[4] The report may list possible causes of a single abnormal result, differential possibilities offered by multiple abnormal results, and suggestions for further laboratory testing. This approach has been useful for analytes such as calcium, thyroid hormones, and serum proteins, where the pathophysiology is defined.

Application of *statistical protocols* to interpretive reports is in its infancy. Such protocols require thoughtful design. Not only must study and test populations be appropriately selected, but the ratio of patients to test variables must be adequate. This means that the laboratory must have the capability of monitoring the prevalence of patient diagnoses while maintaining consistent test standardization (accuracy) over extended periods of time. In practice, probabilistic approaches will require information and biostatistical expertise which is not now commonplace in clinical laboratories.

The process of developing interpretive laboratory reports requires a professional team consisting of clinicians, administrators, computer scientists, biostatisticians, and laboratorians. Whether the approach is to be categorical, probabilistic, a combination of both, or something else (e.g., symbolic reasoning), there must be "infrastructure" supporting the database that perpetuates the relevance of interpretive laboratory reports to individual patients.[14]

Patient/Guarantor Billing

The hospital business office is responsible for processing laboratory charges. Together with the laboratory, the business office and hospital administration develop a price list which includes test name, charge code, and charge. The charge code has a single dollar value and identifies one or more tests in LIS and the hospital financial management system (FMS). For example, blood and urine sodium will have the same charge code because the amount of work involved in the analysis is comparable. On the other hand, the charge code for serum and cerebrospinal fluid protein electrophoresis may be different because of the extra concentrating step required for spinal fluid. The laboratory is assigned a unique series of charge code numbers from the hospital's chart of accounts. The laboratory must then maintain both the LIS and FMS price lists when charges are revised or tests added.

Outpatient laboratory services demand more accounting time than inpatient services. Patient identification and admission information must be gathered efficiently so that patients are registered promptly. Missed addresses, guarantor names, or zip codes will result in wasted time and inefficiency. If there is a sizable repeat outpatient population, it is helpful to register patients at the outpatient laboratory reception area where the patient information need only be edited rather than completely entered at each visit. In this instance, it is also helpful to have patients' accounts on line with current postings, because patients are as interested in their bills as they are in laboratory results. If HIS, FMS, and LIS are not interfaced, a terminal from each system may be required. An interim solution is to have microfiche available for HIS and FMS information, although the accounting information in such case may be several days old.

Accurate billing information is provided by a list which identifies the patient by an appropriate hospital number and lists the tests that were performed, together with billing codes and charges. The LIS can provide billing information on magnetic tape to a central data-processing computer. LIS billing can minimize two clerical problems in the laboratory: namely, double charges and missed charges.

If a laboratory test, once requested, cannot be performed, proper credit can be given to the patient or the charge can be deleted automatically if the billing report has not yet been issued. Certain codes, such as "QNS" (quantity not sufficient) or "ACC" (laboratory accident) are "crediting statements" which cause the computer to generate automatically a credit to the patient. Such control over the billing (and crediting) process, without human intervention, greatly improves the timeliness of billing while assuring that the patient is charged only for services rendered.

FUTURE USE OF COMPUTERS

In the past decade the computer has come of age. While retaining its clerical functions of accepting requests, collecting data, and generating reports, the computer is actively participating in analytical procedures and in the activities of the nursing unit. Microprocessors are now being used in laboratory instruments to control the analytical process. By placing sample aspiration, reagent pipetting, temperature control, calibration, phasing, drift, and carryover correction under control of microprocessors, analytical quality and efficiency have been improved.

Computer technology is now turning to the challenge of interpretive reporting and data utilization. However, as has been pointed out, clinical laboratory data are only a portion of the total database required for making accurate patient-specific diagnostic, therapeutic, and prognostic decisions. This will require computer access to clinical information, as well as development of rational models of medical decision-making processes.

Many hospitals and companies are in the process of developing total information systems where computers will control the information flow throughout the entire hospital. Such information systems can be interfaced with the laboratory computer, so that laboratory data can be transmitted automatically to the central communications system. Such development is certainly technically feasible, and routine implementation can be expected in the next decade. It is likely that large central computers will be connected to various configurations of smaller laboratory computers. The performance of functions unique to the laboratory and the integration of laboratory instrumentation with small computers make the presence of the latter in the laboratory highly desirable, if not essential.

As a result of new and significant developments in electronics, the price and size of the computer hardware have decreased steadily. Thus, it becomes apparent that the most limiting and most costly aspect of a computer system is the software. In order to reduce the high cost of individualized programs, and to avoid staffing each laboratory with its own programming experts, new approaches in the development of more flexible software are also necessary and can be anticipated.

References

1. Aller, R. D., and Elevitch, F. R., Eds.: Computers in the clinical laboratory. Clin. Lab. Med., *3*:1–254, 1983.
2. American Society for Testing and Materials Subcommittee: Standard Guide for Computer Automation in the Clinical Laboratory E792-81. Philadelphia, American Society for Testing and Materials, 1981.
3. Davis, W. S.: Information Processing Systems. Reading, Mass., Addison-Wesley, 1978.

4. Elevitch, F. R.: Computer-assisted clinical laboratory interpretive reports. Lab. Med., *13*:45–47, 1982.
5. Elevitch, F. R., and Aller, R. D.: Planning for a Computer in Your Laboratory. Chicago, Am. Soc. of Clin. Path., 1983, p. 208.
6. Grams, R., Ed.: Medsy Report: Review of Laboratory Systems. Gainesville, Fla., University of Florida, 1981.
7. Hicks, G. P.: Chip technology: Its influence on the distribution of laboratory data and procedures in the 80's. Arch. Pathol. Lab. Med., *105*:341–342, 1981.
8. Hicks, G. P.: Microcomputer applications in the clinical laboratory of the 1980's. J. Clin. Lab. Automation, *1*:185–196, 1981.
9. Hosty, A. T., Lundberg, G. D., Kreig, A. F., et al.: So a computer system sounds like a good idea? Pathologist, *33*:293–296, 1979.
10. Johnson, J. L.: Achieving the Optimum Information System for the Laboratory. Northbrook, Ill., JLJ Associates, 1975 Update, 1976.
11. Krieg, A. F., Johnson, T. J., McDonald, C., et al.: Clinical Laboratory Computerization. Baltimore, University Park Press, 1971.
12. Krieg, A. F., Lundberg, G. D., Hosty, T. A., et al.: A definition and classification of clinical laboratory data processing. Pathologist, *33*:78–83, 1979.
13. Lewis, J. W., and Jarett, L.: Computer application in laboratory medicine. *In*: Gradwohl's Clinical Laboratory Methods and Diagnosis. 8th ed. A. Sonnenwirth and L. Jarett, Eds. St. Louis, The C. V. Mosby Company, 1980.
14. Miller, R. A., Pople, H. E., Jr., and Myers, J. D.: *Internist–I,* an experimental computer-based diagnostic consultant for general internal medicine. N. Engl. J. Med., *307*:468–476, 1982.
15. Weiderhold, G.: Database technology in health care. J. Med. Sys., *5*:175–196, 1981.
16. Westgard, J.: Improving quality control using microcomputers. Diag. Med., January-February, 1982, p. 60.
17. White, R. M.: Disk-storage technology. Sci. Am., *243*:138–148, 1980.
18. Winrich, L. B., and Hobbie, R. K.: Computer interpretation of laboratory test results—an annotated bibliography. J. Med. Sys., *5*:219–241, 1981.

CHAPTER 3

SPECIMEN COLLECTION AND PROCESSING; SOURCES OF BIOLOGICAL VARIATION

by Donald S. Young, M.B., Ph.D., and Edward W. Bermes, Jr., Ph.D.

Physicians use laboratory tests to diagnose disease, to monitor its progress or its response to treatment, and to screen for disease in seemingly healthy individuals. Many physicians consider disease as the only possible cause of abnormal test results. Yet, many factors besides disease affect the composition of body fluids; these factors may be either preanalytical or analytical. Whenever possible, preanalytical variability should be controlled so that correct interpretation of test results is facilitated. The variability of test results due to biological factors is often greater than the variability due to analytical factors. Control of biological variability begins with proper preparation of an individual prior to specimen collection. When factors are not controllable, as is the case with genetic and other long-term influences, their possible effect on test values should be recognized and considered in the evaluation of laboratory data.

In this chapter, we discuss proper specimen collection techniques, short-term biological variables that should be controlled, and long-term biological variables that affect test values and their interpretation. For more detailed reviews of the influence of biological variability the reader should consult articles by Ladenson[28] and Young.[61]

SPECIMEN COLLECTION

COLLECTION OF BLOOD

Blood for analysis may be obtained from veins, arteries, or capillaries. Venous blood is usually the specimen of choice and *venipuncture* is the method for obtaining this specimen. In young children *skin puncture* is frequently used to obtain what is predominantly capillary blood; *arterial puncture* is used mainly for blood gas analyses. A syringe or an evacuated blood tube is used to obtain blood specimens. The blood is usually drawn through a needle, although an indwelling intravenous line with tap or an indwelling butterfly-winged needle for administration of medications such as heparin may occasionally be used.

Before collecting a specimen of any type, a phlebotomist verifies the patient's identity. Specimen containers must be labeled at least with the patient's name, hospital or identification number, location within the hospital, and date and time of collection.

Venipuncture

Before performing a venipuncture, the phlebotomist should verify that the patient is fasting, should fasting be necessary to ensure medically useful results. The patient should be comfortably seated or supine, if sitting is not feasible, and should have been in this position for 20 minutes before the specimen is drawn. This standardization minimizes differences in concentrations of blood constituents due to variations in blood volume (hemoconcentration). The patient's arm should be extended in a straight line from the shoulder to the wrist. Which arm is used is not important, although many patients prefer to spare the one with which they write. An arm with an inserted intravenous line should be avoided, as should an arm with extensive scarring or a hematoma at the intended collection site. If a woman has had a mastectomy, arm veins on that side of the body should not be used since the surgery may have caused lymphostasis affecting the blood composition.

The median cubital vein in the antecubital fossa, or crook of the elbow, is the preferred site for collecting venous blood in adults, since the vein is both large and close to the surface of the skin. Veins on the back of the hand or at the ankle may be used, although these are less desirable and should be avoided in diabetics and other individuals with poor circulation. In severely ill individuals or those requiring many intravenous injections, as for example patients undergoing chemotherapy, an alternative blood drawing site should be chosen to preserve the good veins for the patient's treatment. An arm containing a cannula or arteriovenous fistula should not be used without consent of the patient's physician. If fluid is being infused intravenously into a limb, the fluid should be shut off for 3 min before a specimen is obtained and a suitable note made in the patient's chart and on the result report form. (See also Collection of Blood from Limbs Containing Intravenous Lines, p. 487.) Specimens obtained from the opposite arm or below the infusion site in the same arm may be satisfactory for most tests except for those analytes that are contained in the infused solutions (e.g., glucose or electrolytes).[42]

Selection of a vein for puncture is facilitated by palpation. The area around the intended puncture site should be cleaned with a prepackaged alcohol swab or with a gauze pad saturated with 70% isopropanol. Cleaning of the puncture site should be done with a circular motion and from the site outward. The skin should be allowed to dry in the air. No alcohol should remain on the skin, because traces may cause hemolysis and invalidate test results. When specimens are to be collected for ethanol determinations, the skin should be cleaned with a benzalkonium chloride solution (Zephiran chloride solution, 1:750) that is free of alcohol. Povidone-iodine should be avoided as a cleaning agent because it may interfere with several chemistry procedures. Once the skin has been cleaned, it should not be touched until after the venipuncture has been completed.

When the skin is cleaned, a tourniquet is applied 4–6 inches (10–15 cm) above the intended puncture site to obstruct the return of venous blood to the heart and to distend the veins. Tourniquets are best made from precut soft-rubber strips about 1 inch wide and 15 inches long that are commercially available, or from Velcro-type bands (Propper Manufacturing Co., Long Island City, NY) designed for this purpose. Alternatively, a blood pressure cuff may be used, but it should be inflated only enough to impede venous return to the heart. A tourniquet should not be left in position for more than one minute and the patient should not be allowed to pump his fist while the tourniquet is in place.

Before performing a venipuncture the phlebotomist should estimate the volume of blood to be drawn and select the appropriate tubes for the plasma or serum tests requested. An appropriate needle must also be selected. The most commonly used sizes are gauges 19–22 (1.06–0.71 mm in outside diameter). The larger the gauge size, the smaller the bore. The usual choice for an adult with normal veins is gauge 20; if veins tend to be thready and collapse easily, a size 21 is preferred. For volumes of blood from 30–50 mL, an 18-gauge needle (O.D. 1.24 mm) may be required to ensure adequate blood flow. A needle is typically 1.5 inches (3.7 cm) long, but 1-inch (2.5-cm) needles are also available and may be easier for an inexperienced phlebotomist to use. All needles must be sterile, sharp, and without barbs. If blood is drawn for trace element measurements, the needle should be stainless steel and known to be free from contamination. For trace element determinations, all apparatus coming in contact with the specimen should be acid washed or known to be free of trace metal contamination. (See Chapter 8C.)

The time at which a specimen is obtained is important for those blood constituents that undergo marked diurnal variation, e.g., corticosteroids and iron, and for those used to monitor drug therapy, e.g., digoxin or prothrombin time, since the interval after drug administration affects the drug concentration. Furthermore, special precautions in the collection and handling of specimens are required for tests for alcohol and for tests of medicolegal importance for which a chain of custody for the specimen must be established.

When blood collection is complete, the tourniquet should be released, and then the needle withdrawn. When the needle has been removed, the patient should hold a dry gauze pad over the puncture site, with the arm raised to lessen the likelihood of leakage of blood. The pad can subsequently be held in place by a bandage, which can be removed after 15 min.

The techniques for venipuncture in children and adults are similar. However, *children* are likely to make unexpected movements, and assistance in holding them still is often desirable. Either a syringe or evacuated blood tube system may be used to collect specimens. A syringe should be either the tuberculin type or a 3-mL-capacity syringe, except when a large volume of blood is required for analysis. A 21–23 gauge needle, or 20–23 gauge butterfly needle with attached tubing, is appropriate to collect specimens.

A comprehensive description of the proper technique for venipuncture is contained in NCCLS approved standard ASH-3.[39] The standard contains additional information regarding the performance of difficult venipunctures, collection procedures for blood culture and other sterile techniques, and for specimen collections from patients in isolation.

Venous Occlusion

A tourniquet is invariably used to obstruct the return of venous blood to the heart, as explained earlier. When a blood pressure cuff is used as a tourniquet, it is usually inflated to ~60 mm Hg (8.0 kPa). When a Velcro cuff is used, venous obstruction is less. It is rarely necessary to leave a tourniquet in place for longer than one minute, but even within this short time the composition of blood changes. When the flow of blood in the large veins is obstructed by a tourniquet, the filtration pressure across the capillary walls is increased. The increased filtration pressure causes fluid and low-molecular-weight compounds to pass through the capillary wall. The situation is analogous to that caused by the change of posture from lying to standing, i.e., a relative hemoconcentration. Although the changes that occur in 1 min are slight, marked changes have been observed after 3 min. (See Table 3-1.) Note that these changes often exceed the analytical variability of the methods used to measure the constituents of the blood.

The composition of blood drawn first, i.e., that blood closest to the tourniquet, will be most representative of the composition of circulating blood. The first-drawn specimen should therefore be used for those tests pertinent to critical medical decisions, e.g., calcium. Blood drawn later, which was contained in the small veins and capillaries when the tourniquet was applied, shows a greater effect of venous stasis. Thus, the first tube may show a 5% increase of protein, whereas the third tube may show a 10% change.[35] The concentration of protein-bound constituents is also influenced by stasis, and, indeed, one study has demonstrated that 3-min stasis may be enough to increase the concentration of protein or protein-bound constituents by 15%. A uniform procedure for order of draw for tests should therefore be established.

Table 3-1. CHANGES* IN COMPOSITION OF SERUM WHEN
VENOUS OCCLUSION IS PROLONGED FROM 1 MIN TO 3 MIN[49]

Increase	%	Decrease	%
Total protein	4.9	Potassium	6.2
Iron	6.7		
Total lipids	4.7		
Cholesterol	5.1		
Aspartate aminotransferase	9.3		
Bilirubin	8.4		

Mean values obtained from 11 healthy individuals

*In order to estimate the probable effect of a factor on results, relate per cent increase or decrease shown (or intimated) in table to analytical variation ($\pm$ % CV) routinely found for analytes.

The increase in activity of creatine kinase and aspartate aminotransferase in serum may be greater than would be anticipated for hemoconcentration alone. The added increase may be due to slight trauma to tissue as the needle pierces the skin and to stasis of blood in the tissues.

Pumping of the fist before venipuncture should be avoided because it causes an increase in the plasma potassium, phosphate, and lactate concentrations. The lowering of the blood pH by accumulation of lactate causes the plasma ionized calcium concentration to increase. The ionized calcium concentration reverts to normal 10 min after the tourniquet is released.

Stress associated with blood collection can have effects in patients at any age. Plasma concentrations of cortisol and growth hormone may increase. Particularly in young children who are frightened, struggling, and held in physical restraint, collection may cause adrenal stimulation leading to an increased plasma glucose concentration or create increases in the serum activities of enzymes that originate in skeletal muscle.

Blood Collection with Syringe

Evacuated blood tubes are generally considered to be less expensive and more convenient to use than syringes, but syringes are invariably used for patients with difficult veins. If a syringe is used, the needle is placed firmly over the nozzle of the syringe and the cover of the needle is removed. If the syringe has an eccentric nozzle, the needle should be arranged with the nozzle downward but the bevel of the needle upward. The syringe and needle should be aligned with the vein to be entered and the needle pushed into the vein at an angle to the skin of ~15 °. When the initial resistance of the vein wall is overcome as it is pierced, forward pressure on the syringe is eased, and the blood is withdrawn by gently pulling back the plunger of the syringe. When an appropriate size of syringe has been selected, rarely will a second syringe be required to complete the draw. Should a second syringe be needed, a gauze pad may be placed under the hub of the needle to absorb spill; the first syringe is then quickly disconnected and the second put in place to continue the draw. After removal of the needle from the syringe, drawn blood should be quickly transferred by gentle ejection into tubes prepared for its receipt. The tubes should then be capped; if they contain an additive or anticoagulant, they should be gently mixed by 5–10 inversions.

Vigorous suction on a syringe during collection or forceful transfer from the syringe to the receiving vessel may cause hemolysis of blood. Hemolysis is generally less when blood is drawn through a small-bore needle because turbulence of the blood is less than when a larger-bore needle is used.

Blood Collection with Evacuated Blood Tube

Typical systems for collecting blood in evacuated tubes are shown in Figure 3-1. The needle is screwed into the collection tube holder and the tube is then gently inserted into this holder. Before use, the tube should be gently tapped to dislodge any additive from the stopper before the needle is inserted into a vein. This will prevent aspiration of the additive into the patient's vein.

After the skin is cleaned, the needle should be guided gently into the patient's vein, and once the needle is in place, the tube should be pressed forward into the holder to puncture the stopper and release the vacuum. When blood begins to flow into the tube, the tourniquet should be released without moving the needle. The tube is filled until the vacuum is exhausted and the tube can then be withdrawn from the holder and replaced by another tube, if this is necessary. Other tubes may be filled, if required, using the same technique with the holder in place.

When several tubes are required for a single blood collection, a shut-off valve, consisting of rubber tubing that slides over the needle opening inside the tube, is used to prevent spillage of blood during exchange of tubes. (See Figure 3-1.)

Evacuated blood tubes. Most phlebotomists prefer evacuated blood tubes to syringes because evacuated tubes are easy to use and there is less likelihood of contamination of their outside with blood. There are two main classes of evacuated blood tubes, those containing a serum-separating material and those without. Various additives, including anticoagulants and preservatives, may be used with these tubes. Stopper color denotes the type of additive as shown in Table 3-2. Evacuated blood tubes are siliconized to reduce adhesion of clots to walls or stoppers and to decrease risk of hemolysis. Stoppers of the tubes may be siliconized or coated with glycerin to facilitate their removal. Siliconized stoppers are generally preferred because there is less interference

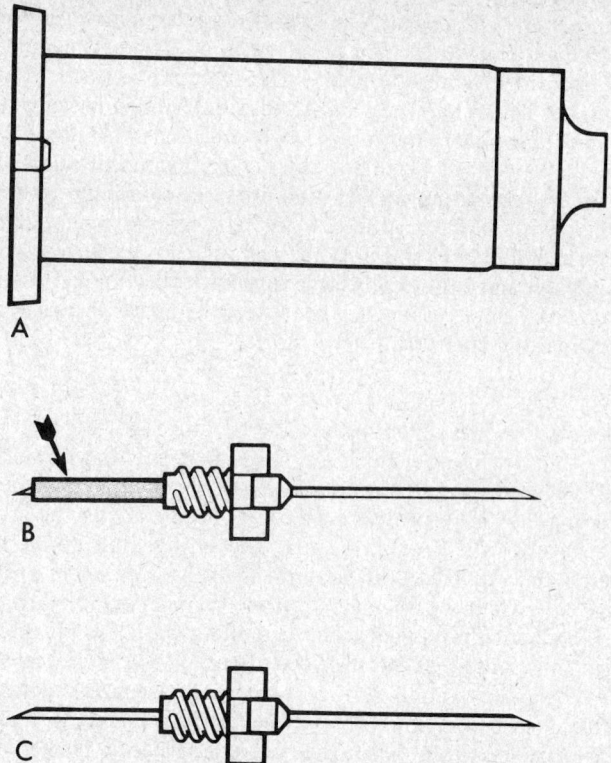

Figure 3-1. Blood collection tube holder (Becton Dickinson) (*A*), with needles for use with multiple tubes (*B*) and single tubes (*C*). Arrow indicates rubber sleeve.

by silicone with test procedures. Tubes containing glycerin should not be used when lipid measurements are to be made on the specimen. The additives listed in Table 3-2 must fall within certain tolerances as listed in NCCLS-approved standard, ASH-1.[38] Thrombin has been added to some tubes to accelerate clotting. However, the silicone-coated wall may also activate and accelerate the clotting mechanism. The NCCLS standard applies to a limited number of tube sizes with maximum draws of from 3–15 mL. The tube sizes listed are adequate for all volumes likely to be required in a clinical chemistry laboratory.

Serum separator tubes contain an inert, thixotropic, polymer gel material with a specific gravity of ~1.04. Silica or glass particles that accelerate clotting may be associated with it. Aspiration of blood into the tube and subsequent centrifugation displace the gel, which settles like a disk between cells and supernatant when the tube is centrifuged. RCF must be at least $1100 \times g$ for gel release and barrier formation. Release of intracellular components into the supernatant is prevented by the barrier for several hours or, in some cases, for a few days. These separator tubes may be used as primary containers from which serum can be directly aspirated by a number of analytical instruments. Several studies indicate that the separation material does not contaminate the serum,[30] but others demonstrate that free gel particles may persist and can affect dialysis membranes or ion-selective electrodes.

Table 3-2. CODING OF STOPPER COLOR TO INDICATE ADDITIVE IN EVACUATED BLOOD TUBE[38]

Color	Use	Additive
Gray	Plasma or whole blood with glycolysis inhibition	Oxalate (Na or K), fluoride (Na), iodoacetate (Na)
Yellow	Sterile interior of tube	
Green	Plasma or whole blood	Heparin (Na, Li, or NH_4)
Red	Serum	
Blue	Plasma or whole blood	Citrate (Na)
Lavender	Plasma or whole blood	EDTA (Na_2 or K_2)

Evacuated blood tubes may be made of soda-lime or borosilicate glass. Tubes made from the former may release trace elements, particularly calcium and magnesium, into solutions.

Stoppers may contain zinc, invalidating the use of evacuated blood tubes for zinc measurement, and TBEP (tris-butoxyethyl phosphate), a constituent of rubber, which may interfere with the measurement of certain drugs. As evacuated tubes age, the vacuum fails and their effective draw diminishes. The silicone coating also tends to break down with age. Therefore, the stock of these tubes should be rotated and careful attention paid to the expiration date. Problems with evacuated blood tubes also arise when too little blood is collected into the tubes (short draw). Although an incorrect draw is of little consequence when clotted blood is required, a disproportionate ratio of blood to anticoagulant may affect some hematology and other special tests. Therefore, check some tubes from each batch for volume of draw before the lot is put in use.

In a few patients, backflow from blood tubes into veins occurs due to a decrease in venous pressure. The dangerous consequences of this occurrence may be avoided if only sterile tubes are used for collection of blood. Backflow is minimized if the arm is held downward and blood is kept from contact with the stopper during the collection procedure.

Skin Puncture

If only a small volume of blood is required for a blood test such as blood glucose, venipuncture is unnecessary. However, a skin puncture is time consuming, and there is a greater risk of infection than from a venipuncture since the fingertip is more difficult to sterilize than the antecubital fossa. In an *adult or grown child*, blood may be obtained by puncturing the tip of a finger or by piercing an earlobe. The phlebotomist thoroughly cleans the skin of the center of the palmar side of the tip (distal phalanx) of the third or fourth finger of the nonwriting hand with a gauze pad saturated with 70% isopropanol. All alcohol must be allowed to evaporate from the skin so that hemolysis does not occur. When the skin is dry, it is quickly punctured by a sharp stab with a lancet. The depth of the incision should be < 2.5 mm to avoid contact with bone.[5] To minimize the possibility of infection, a different site should be selected for each finger puncture. The finger should be held in such a way that gravity assists the collection of blood on the finger tip. Massage of the finger to stimulate blood flow should be avoided, because it causes the outflow of debris and of tissue fluid, which does not have the same composition as plasma. To improve circulation of the blood, the finger may be warmed by application of a warm, wet washcloth for 3 min prior to pricking. The first drop of blood is wiped off and subsequent drops are transferred to the appropriate collection tube by gentle contact. Filling should be done rapidly to prevent clotting, and introduction of air bubbles should be avoided. This is best done by holding the collection tube at only a slight angle to the skin surface. However, even with the most careful technique it is not always possible to avoid air bubbles.

Blood may be collected into capillary blood tubes by capillary attraction. A variety of collection tubes is commercially available. (See Figure 3-2.) The Natelson tubes (280-μL capacity, 1.5–1.6 mm I.D.) and Caraway tubes (350 μL, 2.5–2.7 mm I.D.) are widely used. Such containers can be purchased with different anticoagulants, such as sodium and ammonium heparin, and some can be obtained in brown glass for collection of light-sensitive analytes such as bilirubin. There are also many small plastic or glass tubes that can be used to collect blood. These have the disadvantage that blood tends to pool in the mouth of the tube and must be flicked down the tube with a risk of hemolysis. Drop-by-drop collection should be avoided since it increases hemolysis.

Several devices are available for pricking skin to obtain capillary specimens. These include the Autolet (Ulster Scientific, Inc., Highland, NY 12528), a spring-loaded device that can be used by an adult without assistance. The lancet blade is retracted by cocking the device; when the blade is released, it penetrates the skin to a depth of 2 mm.

In an *infant* less than one year old, the lateral or medial plantar surface of the foot should be used for skin puncture; suitable areas are illustrated in Figure 3-3. In older children the plantar surface of the big toe may also be used, although blood collection on ambulatory patients from anywhere on the foot should be avoided. Blumenfeld et al.[5] have described in detail the suitable locations for drawing blood from the feet of infants. The proper procedure for collecting blood from infants is also contained in an NCCLS standard.[40] The technique for filling and processing tubes is similar in children and adults.

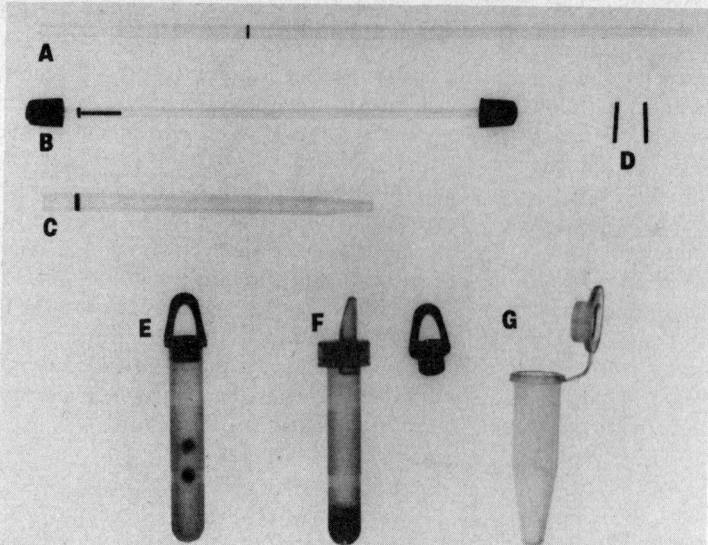

Figure 3-2. Representative devices for collection of capillary specimens. *A,* Natelson micro collection tube. *B,* Heparinzed capillary tube containing metal mixer, used for specimens for blood gas analysis. *C,* Caraway tube. *D,* Metal mixers for mixing blood in capillary tubes. *E,* Plastic micro tube containing anticoagulant and two inert mixing beads. *F,* Plastic micro tube with serum separator gel and collecting device to improve blood flow into tube. *G,* Plastic micro centrifuge tube.

For the collection of blood specimens on filter paper for neonatal screening, the skin is cleaned and punctured as described previously. Then the filter paper is gently touched against a large drop of blood, which is allowed to soak into the paper to fill the marked circle. Only a single application per circle should be made. The paper is examined to verify that there has been complete penetration of the paper. The procedure is repeated to fill all the circles. Avoid milking or squeezing of the foot since this procedure contributes tissue fluids. The filter papers should be air-dried. Blood should not be transferred onto filter paper after it has been collected in capillary tubes since partial clotting may have occurred.

Arterial Puncture

Arterial punctures require considerable skill and are usually performed only by physicians or specially trained technicians or nurses. The preferred sites of arterial puncture are, in order, the radial artery at the wrist, the brachial artery in the elbow, and the femoral artery in the groin. Since bleeding from the femoral artery tends to be greater, especially in the elderly, sites in the arm are used most often.

The artery to be punctured is identified from its pulsation and thick wall. The skin surrounding the puncture site should be cleaned as for a venipuncture. No tourniquet is required for arterial punctures. An 18- or 20-gauge needle should be used for one of the larger arteries, but a smaller needle, e.g., 23–25 gauge (O.D. 0.64 mm to 0.51 mm), should be used to collect the blood from

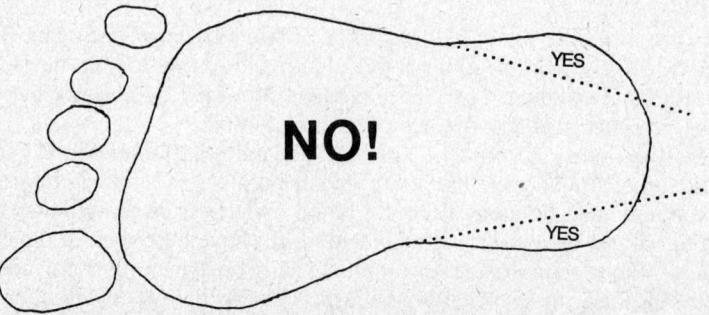

Figure 3-3. Acceptable sites for skin puncture to collect blood from an infant's foot. (Modified from Blumenfeld, T. A., Turi, G. K., and Blanc, W. A.: Lancet, *1*:230-233, 1979.)

smaller arteries. The needle and syringe should be flushed out with a heparin solution both to ensure adequate anticoagulation and to eliminate trapping of air in the needle and in the dead space of the nozzle. Plastic syringes should be avoided because, unlike glass syringes, the plunger will not rise due to the arterial blood pressure alone. Some plastics may also be permeable to gases. Evacuated blood tubes should not be used for the collection of specimens for blood gas analysis because the residual air in the tube may cause erroneous results when it equilibrates with the blood. Once an arterial puncture has been performed, firm pressure should be applied over the puncture site for at least 5 min to minimize bleeding.

If the collected specimen is intended for blood gas analysis, the nozzle of the syringe containing the blood should be sealed and the syringe placed in ice-water for immediate transport to the clinical laboratory. Analyses should be performed within 15 min because chilling will not completely inhibit the metabolic activity of white cells. (See Chapter 10.) Generally, heparin has no effect on the blood pH, but an excess may cause a reduction of up to 15% in the pCO_2.

In the neonate, an indwelling catheter in the umbilical artery is best to obtain specimens for blood gas analysis. In the older child or adult in whom it is impossible to perform an arterial puncture, a capillary puncture may be performed to obtain *arterialized capillary blood*. Such a specimen yields acceptable values for pH and pCO_2, but not always for pO_2. (See Chapter 10.) In the older child or adult the preferred puncture site is the earlobe; in the young child or infant it is the heel. Capillary blood specimens are particularly inappropriate when the blood circulation is poor and thus should be avoided when a patient has reduced cardiac output, hypotension, or vasoconstriction. For each capillary puncture, the skin should first be warmed with a hot, moist towel to improve the circulation. The puncture itself should be performed as described previously; a free flow of blood is essential. Heparinized capillary tubes containing a small metal bar are used to collect the blood. Tubes should be sealed quickly and the contents mixed well by using a magnet to move the metal bar up and down in the tube so that a uniform specimen is available for analysis. (See Figure 3-2.)

Anticoagulants and Preservatives for Blood

If whole blood or plasma is desired for testing, an anticoagulant must be added to the specimen during the collection procedure. Whole blood is rarely required for clinical chemistry tests; indeed, only for blood gas and ammonia determinations is it the preferred specimen, although it may also be used for glucose, urea nitrogen, and lactate determinations. Serum from coagulated blood is the specimen of choice for many assay systems, but plasma obtained with an appropriate anticoagulant may be an equally valid specimen, and, in certain circumstances, preferable to serum. Because harvest of serum must wait 15–30 min for completion of coagulation before centrifugation, use of plasma expedites analysis in medical emergencies. Furthermore, plasma yield from a given volume of whole blood is always greater than the yield of serum. A disadvantage is the formation of fibrin clots or fragments when plasma is stored and the subsequent risk of clogging sample probes of automated analytical instruments. Plasma is also not a suitable specimen for electrophoretic analyses since the presence of fibrinogen can confuse interpretation of electrophoretic patterns.

Heparin is the most widely used anticoagulant for clinical chemical analyses. It causes the least interference with tests. It is a mucoitin polysulfuric acid and is available as sodium, potassium, lithium, and ammonium salts. This anticoagulant probably acts as an antithrombin to prevent the transformation of prothrombin into thrombin and thus the formation of fibrin from fibrinogen.

About 20 units of heparin are usually required to anticoagulate 1 mL of blood; thus, most blood tubes are prepared with ~0.2 mg heparin for each mL of blood to be collected. The heparin is usually present as a dry powder. It is hygroscopic and dissolves rapidly. Although ammonium heparin has some effect on erythrocyte volume, other heparin salts appear to have none. Nonetheless, the sodium and potassium salts should be avoided when plasma is obtained for analysis of sodium and potassium; the positive error in sodium analysis is ~2 mmol/L. Although lithium is often used as an internal standard for flame photometry, the amount of lithium added as lithium heparinate has an insignificant effect on sodium and potassium measurements. Heparin was once considered to be contaminated with phosphate; this is no longer the case and heparin salts may be used safely in conjunction with tests for phosphorus.

Heparin has the disadvantage of high cost and temporary action, and it produces a blue background in blood smears that are stained with Wright's stain. In addition, heparin is said to inhibit acid phosphatase activity and to interfere with the binding of calcium to EDTA in calcium methods involving complexing with EDTA. It has also been reported to affect the binding of triiodothyronine and thyroxine to their carrier proteins, thus producing higher free concentrations of these hormones. Heparin has been demonstrated to cause significant inactivation of hydroxybutyrate dehydrogenase, and possibly of lactate dehydrogenase.

Ethylenediaminetetraacetic acid (EDTA). The chelating agent EDTA is particularly useful for hematologic examinations because it preserves the cellular components of blood. It is used as the disodium or dipotassium salt, the latter being more soluble. It is effective at a final concentration of 1–2 mg/mL of blood. Higher concentrations do not appear to affect the red cells. EDTA prevents coagulation by binding calcium, which is essential for the clotting mechanism. Blood collection tubes are prepared by adding a 0.1% solution of an EDTA salt followed by evaporation of water at room temperature.

EDTA, probably by chelation of metallic cofactors, inhibits alkaline phosphatase, creatine kinase, and leucine aminopeptidase activities. Its chelation of calcium makes EDTA unsuitable for specimens for calcium and iron analyses using colorimetric or titrimetric techniques. As an anticoagulant, it has little effect on other clinical chemistry tests.

Sodium fluoride is usually considered as a preservative for blood glucose; however, it also acts as a weak anticoagulant. As a preservative, together with another anticoagulant such as potassium oxalate, it is effective at a concentration of ~2 mg/mL blood. It exerts its preservative action by inhibiting the enzyme systems involved in glycolysis. When sodium fluoride is used alone for anticoagulation, 3–5 times greater concentrations than the usual 2 mg/mL are required. The high concentration and the inhibition of the glycolytic cycle are likely to cause fluid shifts. Fluoride is a potent inhibitor of many serum enzymes and in high concentrations also affects urease, used to measure urea nitrogen in many analytical systems. There is little justification for its use as an anticoagulant for clinical chemistry tests.

Fluoride at a concentration of ~2 mg/mL is the best available preservative for glucose. Most specimens will be preserved at 25 °C for 24 h or at 4 °C for 48 h. Without an antiglycolytic agent, the blood glucose concentration decreases ~10 mg/dL (0.56 mmol/L) per hour at 25 °C. The rate of decrease is faster in newborns because of the increased metabolic activity of their erythrocytes and in leukemic patients because of the high metabolic activity of the white cells.

Sodium fluoride is poorly soluble and blood must be well mixed before effective antiglycolysis occurs.

Citrate. Sodium citrate solution, at a concentration of 3.4 or 3.8 g/dL in a ratio of 1 part to 9 parts of blood, is widely used for coagulation studies since the effect is easily reversible by addition of Ca(II). It appears to preserve labile procoagulants, but it has little application in clinical chemistry. Since citrate chelates calcium, it is unsuitable as an anticoagulant for specimens for measurement of this element. It inhibits aminotransferase and alkaline phosphatase but stimulates acid phosphatase when phenylphosphate is used as substrate. Since citrate complexes molybdate, it decreases the color yield in phosphate measurements and thus produces low results.

Oxalates. Sodium, potassium, ammonium, and lithium oxalates inhibit blood coagulation by forming rather insoluble complexes with calcium ions. Potassium oxalate ($K_2C_2O_4 \cdot H_2O$) at a concentration of ~1–2 mg/mL blood is the most widely used oxalate. If tubes are prepared in the laboratory, the potassium oxalate may be added as a 30 g/dL solution and dried in an oven. The oxalate should be dried as a thin film over the sides of the tube; drying temperature should not exceed 100 °C in order to avoid decomposition of the oxalate to carbonate, which has no anticoagulant activity. At concentrations of > 3 mg oxalate/mL, hemolysis is likely to occur.

Although the combined ammonium/potassium oxalate used in hematology before the introduction of EDTA did not cause shrinkage of erythrocytes, other oxalates cause shrinkage by drawing water into the plasma. Reduction in hematocrit may be as much as 10%, causing a reduction in the concentration of plasma constituents of 5%. As fluid is lost from the cells, an exchange of electrolytes and other constituents across the cell membrane occurs. Oxalate inhibits several enzymes, including acid and alkaline phosphatases, amylase, and lactate dehydrogenase, and may cause precipitation of calcium as the oxalate salt.

Iodoacetate. Sodium iodoacetate at a concentration of 2 g/L is an effective antiglycolytic agent and a substitute for sodium fluoride. Since it has no effect on urease, it can be used when glucose and urea are performed on a single specimen. It inhibits creatine kinase but appears to have no significant effects on other clinical chemistry tests.

The differences between the plasma and serum concentrations of commonly ordered analytes are shown in Table 3-3. Triglyceride concentration is also higher in serum than plasma.[34]

Table 3-3. DIFFERENCE IN COMPOSITION OF PLASMA AND SERUM[29]

Plasma Value Greater Than Serum Value (%)		No Difference in Serum and Plasma Value	Plasma Value Less Than Serum Value (%)	
Calcium	0.9	Bilirubin	Albumin	1.3
Chloride	0.2	Cholesterol	Alkaline phosphatase	1.6
Lactate dehydrogenase	2.7	Creatinine	Aspartate aminotransferase	0.9
Total protein	4.0		Bicarbonate	1.8
			Creatine kinase	2.1
			Glucose	5.1
			Phosphorus	7.0
			Potassium	8.4
			Sodium	0.1
			Urea	0.6
			Uric acid	0.2

See footnote, Table 3-1.

Influence of Site of Collection on Blood Composition

Blood obtained from different sites differs in composition. Skin puncture blood is more like arterial blood than venous blood. Thus, there are no clinically significant differences between freely flowing capillary blood and arterial blood in pH, pCO_2, pO_2, and oxygen saturation. The pCO_2 of venous blood is up to 6–7 mm Hg (0.8–0.9 kPa) higher. Venous blood glucose is as much as 7 mg/dL (0.39 mmol/L) less than the capillary blood glucose as a result of tissue utilization.

Skin puncture blood obtained from the earlobe or nonwarmed skin sites may show changes characteristic of stasis. These include increased serum protein and protein-bound constituents. The serum potassium may also be increased by 0.2 mmol/L, but most of the changes, although statistically significant, are not clinically significant. Warming of the skin surface to improve the circulation lessens the differences due to stasis.

Blood obtained by skin puncture is contaminated to some extent with interstitial and intracellular fluids. The potassium concentration of interstitial fluid is as much as 0.3 mmol/L less than that of venous serum. Despite the lower potassium concentration in interstitial fluid, potassium in capillary blood may be higher than in venous blood; this may be due to contamination with tissue debris. Plasma total protein concentration is more than three times the concentration of interstitial fluid. The low capillary blood protein may therefore reflect mixing of interstitial fluid with capillary blood. Such dilution also reduces the concentration of protein-bound constituents such as calcium and bilirubin.

The major differences between venous serum and capillary serum are illustrated in Table 3-4.

Collection of Blood From Limbs Containing Intravenous Lines

Proper collection of blood from a central venous catheter is necessary to ensure that the composition of specimen is not affected by the fluid that is infused into the patient. The fluid is shut off using the stopcock on the catheter and 10 mL blood is aspirated through the stopcock and discarded before the specimen for analysis is withdrawn. Blood properly collected from a central venous catheter, and compared with blood drawn from a peripheral vein at the same

Table 3-4. DIFFERENCE IN COMPOSITION OF CAPILLARY AND VENOUS SERUM[27]

Capillary Value Greater Than Venous Value (%)		No Difference in Capillary and Venous Value	Capillary Value Less Than Venous Value (%)	
Glucose	1.4	Phophorus	Bilirubin	5.0
Potassium	0.9	Urea	Calcium	4.6
			Chloride	1.8
			Sodium	2.3
			Total protein	3.3

See footnote, Table 3-1.

time, shows significant differences in composition. A comparison of arterial blood with central and peripheral venous blood is illustrated in Table 3-5.

Blood may be collected from the veins of an arm below an intravenous line without interference from the fluid being infused, because retrograde blood flow does not occur in the veins and the fluid that is infused must first circulate through the heart and be returned to the tissues before it reaches the sampling site.

Hemolysis

Serum shows visual evidence of hemolysis when the hemoglobin concentration exceeds 20 mg/dL. Slight hemolysis has little effect on most test values. Severe hemolysis causes a slight dilutional effect on those constituents present at a lower concentration in the erythrocytes than in plasma. However, a marked effect may be observed on those constituents that are present at a higher concentration in erythrocytes than in plasma. Thus, plasma concentrations or activities of aldolase, total acid phosphatase, lactate dehydrogenase, isocitrate dehydrogenase, potassium, magnesium, and phosphate are particularly increased by hemolysis. The inorganic phosphate in serum increases rapidly as the organic esters in the cells are hydrolyzed. Aspartate aminotransferase activity is increased by 2% for each 10 mg hemoglobin/dL. When colorimetric procedures without extraction are used, 10 mg hemoglobin/dL will raise the apparent cholesterol concentration by 5 mg/dL. Ten mg hemoglobin/dL will increase serum lactate dehydrogenase by ~10%, and serum potassium by ~0.6%. An additional band due to hemoglobin may be observed on serum protein electrophoresis.

Although the amount of free hemoglobin could be measured and a calculation made to correct test values affected by hemoglobin,[9] this practice is undesirable because other factors than hemoglobin could contribute to the altered test values, and it would be impossible to assess their impact.

URINE COLLECTION

The type of urine specimen to be collected is dictated by the tests to be performed. Untimed or random specimens are suitable for only a few chemical tests; generally urine specimens must be collected over a predetermined interval of time, such as 1, 4, or 24 h. A *clean, early-morning, fasting specimen* is generally the most concentrated specimen and thus is preferred for microscopic examinations and for the detection of abnormal amounts of constituents such as proteins, or of unusual compounds such as β-chorionic gonadotropin. The *clean timed specimen* is one obtained at specific times of the day or during certain phases of the act of micturition. Bacterial examination of the first 10 mL of urine voided is most appropriate to detect urethritis, while the midstream specimen is best for investigating bladder disorders. The *double voided specimen* is the urine excreted during a timed period following a complete emptying of the bladder; it is used, for

Table 3-5. INFLUENCE OF COLLECTION SITE ON COMPOSITION OF PLASMA[46]

	Arterial	Central Venous	Peripheral Venous
Alanine aminotransferase (U/L)	62	61	81
Albumin (g/dL)	3.6	3.7	3.9
Alkaline phosphatase (U/L)	114	113	107
Amylase (U/L)	149	148	177
Aspartate aminotransferase (U/L)	20	20	21
Calcium (mg/dL)	8.1	8.2	8.3
Chloride (mmol/L)	99	97	101
Creatine kinase (U/L)	82	73	91
Creatinine (mg/dL)	1.4	1.3	1.2
Gamma-glutamyltransferase (U/L)	13	14	14
Potassium (mmol/L)	4.0	3.9	3.8
Sodium (mmol/L)	144	145	144
Total protein (g/dL)	6.6	6.8	7.7
Urea N (mg/dL)	32	31	25
Uric acid (mg/dL)	8.1	8.1	7.9

See footnote, Table 3-1.

example, to assess glucose excretion during a glucose tolerance test. Its collection must be timed in relation to the ingestion of glucose.

Catheter specimens are used for microbiological examination in critically ill patients or in those with urinary tract obstruction but should not normally be obtained just for examination of chemical constituents. The *suprapubic tap specimen* is a useful alternative, because the tap is unlikely to cause infection. Following appropriate cleaning of the skin over the full bladder, a 22-gauge spinal needle is passed through a small wheal made by a local anesthetic. The bladder is penetrated and the urine withdrawn into the syringe.

Even though tests in the clinical chemistry laboratory are not generally affected by lack of sterile collection procedures, the patient's genitalia should be cleaned before each voiding to minimize the transfer of surface bacteria to the urine. Cleansing is essential if the true concentration of white cells is to be obtained. Details of collection of urine specimens are contained in an NCCLS standard currently in preparation.

Timed Urine Specimens

The collection period for timed specimens should be long enough to minimize the influence of short-term biological variations. When specimens are to be collected over a specified period of time, the patient's close adherence to instructions is important. The bladder should be emptied at the time the collection is to begin and this urine is discarded. Thereafter, all urine should be collected until the end of the scheduled time. If a patient has a bowel movement during the collection period, precautions should be taken to prevent fecal contamination. If a collection has to be made over several hours, urine should be passed into a separate container at each voiding and then emptied into a larger container for the complete specimen. This two-step procedure prevents the danger of a patient's splashing himself with a preservative such as acid. The large container should be stored at ~4 °C in a refrigerator during the entire collection period.

Before beginning a timed collection a patient should be given written instructions with regard to diet or drug ingestion, if appropriate, to avoid interference of ingested compounds with analytical procedures. Thus, instructions for collection of specimens for 5-hydroxyindoleacetic acid measurements should specify avoidance of avocados, bananas, plums, walnuts, pineapples, and eggplant, as well as acetaminophen and cough syrups containing glyceryl guaiacolate (guaifenesin). The dietary components are sources of 5-hydroxytryptamine and should be avoided for this reason; the other compounds interfere with certain analytical procedures but may not interfere with highly specific analytical methods. The staff of each laboratory should determine its own requirements. See also specimen information for specific analytes in the respective chapters.

For 2-h specimens, a prelabeled 1-L bottle is adequate. For a 12-h collection a 2-L bottle usually suffices; for a 24-h collection a 3- or 4-L bottle is appropriate for most patients. A single bottle allows adequate mixing of the specimen and avoids possible loss of some of the specimen if a second container does not reach the laboratory. Urine should not be collected at the same time for two or more tests for which different preservatives are required. Aliquots for such an analysis as a microscopic examination should not be removed while a 24-h collection is in process. Removal of aliquots is not permissible even when the volume removed is measured and corrected, since the excretion of most compounds varies throughout the day, and results will be affected. Appropriate information regarding the collection, including warnings with respect to handling of the specimen, should appear on the bottle label. An example appears in Figure 3-4.

Collection of Urine From Children

Collection of a timed specimen from a baby is difficult, but fortunately such specimens are rarely required. The scrotal or perineal area is first cleaned and dried and any natural or applied skin oils are removed. For an untimed specimen, a plastic bag (U-bag, Hollister Inc., Chicago, IL, or Tink-Col, C.R. Bard Inc., Murray Hill, NJ) is placed around the infant's genitalia and left in place until urine has been voided.

A metabolic bed is used to collect timed specimens from infants. The infant lies on a fine screen above a funnel-shaped base containing a drain under which a container is placed to receive urine. The fine screen retains fecal material. Nevertheless, the urine is likely to be contaminated, to some extent, by such material.

To obtain a sterile urine specimen for culture from an infant, a suprapubic tap is performed. The collection of specimens from older children is done as in adults, using assistance from a parent when this is necessary.

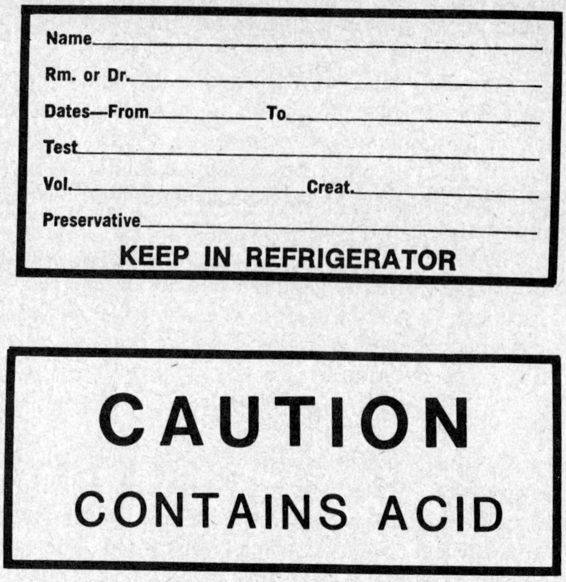

```
Name_____

Rm. or Dr._____

Dates—From_____To_____

Test_____

Vol._____Creat._____

Preservative_____
```

KEEP IN REFRIGERATOR

CAUTION
CONTAINS ACID

Figure 3-4. Examples of labels for urine containers.

Urine Preservatives

The most common preservatives and the tests for which the preservatives are required are listed in Table 3-6. Preservatives have different roles but are usually added to reduce bacterial action or chemical decomposition or to solubilize constituents that might otherwise precipitate out of solution. Another application is to decrease atmospheric oxidation of unstable compounds. Some specimens should not have *any* preservatives added because of the possibility of interference with analytical methods.

One of the most satisfactory forms of preservation of urine specimens is *refrigeration* during collection; it is even more successful when combined with chemical preservation. Urinary preservative tablets that contain a mixture of chemicals, such as potassium acid phosphate, sodium benzoate, benzoic acid, methenamine, sodium bicarbonate, and red mercuric oxide (Metropolitan Life Insurance Co.), have been used for chemical and microscopic examination. Since these tablets contain sodium and potassium salts, among others, they cannot be used for analysis of these analytes. The preservative tablets act mainly by lowering the pH of the urine and by releasing formaldehyde. *Formalin* has also been used for preserving specimens, but in large amounts it will precipitate urea and inhibit certain reactions, e.g., the dipstick esterase test for leukocytes. Acidification to below pH 3 is widely used to preserve 24-h specimens and is particularly useful for specimens for calcium, steroids, and vanillylmandelic acid determinations (10 mL HCl, 6 mol/L, per 24-h excretion). However, precipitation of urates will occur, thereby rendering a specimen unsuitable for measurement of uric acid. *Sulfamic acid* (10 g/L urine) has also been used to reduce pH. *Boric acid* (5 mg/30 mL) has been used, but it too causes precipitation of urates. Although *thymol* and *chloroform* were widely used in the past to preserve specimens for chemical and microscopic urinalysis, it is now recognized that specimens for these tests should be analyzed immediately and that the addition of preservatives is both largely ineffective and a source of interference with several methods. *Toluene* is the only organic solvent that is still used as a preservative. When present in a large enough amount, it acts as a barrier between the air and the surface of the specimen. Toluene, however, will not prevent the growth of anaerobic microorganisms and, because of its flammable nature, it is a safety hazard. *Sodium carbonate* is used to preserve porphyrins and urobilinogen. Five grams is used for a 24-h collection.

When a timed collection is complete, the specimen should be delivered without delay to the clinical laboratory, where the volume should be measured. This may be done by using graduated cylinders or by weighing the container and urine when preweighed or uniform containers are used. The mass in 1 g may be reported as if it were the volume in mL. There is rarely a need to measure the specific gravity of a weighed specimen since errors in analysis usually exceed the error arising from failure to correct the volume of urine for its mass. Nevertheless, it is relatively

Table 3-6. COMMONLY USED URINE PRESERVATIVES

Analyte	None	Freeze	Glacial Acetic Acid	Boric Acid	Toluene	Concentrated Hydrochloric Acid	Sodium Bicarbonate	Petroleum Ether	Nitric Acid
Aldosterone		X	X	X					
Amino acids					X	X			
δ-Aminolevulinic acid			X			X			
Calcium						X			
Catecholamines			X			X			
β-hCG	X			X					
Citrate					X	X			
Copper	X					X			
Cortisol (free)	X		X	X					
Cystine		X			X	X			
Estrogens			X	X					
Homogentisic acid	X								
Homovanillic acid			X	X		X			
5-Hydroxyindoleacetic acid				X		X			
Hydroxyproline					X	X			
17-Ketosteroids			X	X					
Magnesium	X					X			
Mercury	X					X			X
Metanephrines			X			X			
Nitrogen	X					X			
Osmolality	X	X							
Oxalate	X					X			
Porphyrins							X		
Pregnanetriol			X	X					
Uric acid	X				X				
Urobilinogen	X	X						X	
Vanillylmandelic acid				X		X			
Zinc	X					X			

easy to apply a simple mathematical correction based on an average specific gravity, or to measure the specific gravity and to derive a correct volume for each specimen.

Before a specimen is transferred into small containers for each of the ordered tests, it must be thoroughly mixed to assure homogeneity since the specific gravity, volume, and composition of the urine may all vary throughout the collection period. The small container into which an aliquot is transferred should not be a plastic bottle if toluene or another organic compound has been used as a preservative; metal-free containers must be used for trace metal analyses.

COLLECTION OF FECES

Small aliquots of feces are frequently analyzed to detect the presence of "hidden" blood, so-called *occult blood*, which is recognized as one of the most effective clues to the presence of a bleeding ulcer or malignant disease in the gastrointestinal tract. The utility of screening for occult blood is such that it is included as part of many periodic health examinations. Tests for occult blood should be done on aliquots of excreted stools rather than on material obtained on the glove of a physician doing a rectal examination, since this procedure may cause enough bleeding to produce a positive result. In other instances, the small amount of stool present on the glove may not be representative of the whole, so that bleeding may not be recognized. Patients have even been encouraged to recover, by means of a spatula, small parts of feces that have been excreted into a toilet bowl and apply them directly to the reagent-impregnated slides used for the detection of blood. These specimens are then mailed or delivered to a laboratory for analysis. (See also Chapter 15.)

Feces from children may be screened for tryptic activity to detect cystic fibrosis. In the infant, fecal material for these tests is usually recovered from the child's diaper. See Chapter 5 for a discussion of the measurement of trypsin in feces.

In adults, measurement of fecal nitrogen and fat in 72-h specimens is used to assess the severity of malabsorption; measurement of fecal porphyrins is occasionally required to characterize the type of porphyria. (See Chapters 14 and 16.) Generally no preservative is added to the feces, but the container should be kept refrigerated throughout the collection period and care should be taken to avoid contamination from urine. When the collection is complete, the container and feces are weighed and the mass of excreted feces is calculated. The specimen is homogenized and aliquotted so that the amount of fat or nitrogen excreted per day and the proportion of dietary intake excreted can be calculated.

For metabolic balance studies, collections of stool are usually made over a 72-h period. (See also Chapter 8A.) Many balance studies are carried out in conjunction with research on the metabolism of such elements as calcium. It is important for such studies that a patient be on a controlled diet for a sufficiently long time prior to the commencement of the study, so that a steady state has been attained.

COLLECTION OF SPINAL FLUID

Spinal fluid is normally obtained from the lumbar region, although a physician may occasionally request analysis of fluid obtained during surgery from the cervical region or from a cistern or ventricle of the brain. Spinal fluid is examined when there is a question as to the presence of a cerebrovascular accident, meningitis, demyelinating disease, or meningeal involvement in malignant disease. Lumbar punctures should always be performed by a physician. The physician thoroughly cleans the skin of the lumbar region below the termination of the spinal cord where the cauda equina goes through the spinal canal. He makes a small bleb in the skin over the space between the third and fourth or fourth and fifth lumbar vertebrae with 2% procaine and then introduces a spinal needle (22-gauge, 3.5 inches long) through the bleb into the spinal canal. He measures the pressure with a manometer and then allows 3–4 mL of fluid to drip into plain tubes. The tubes should be sterile, especially if microbiological tests are required. Since the initial specimen may be contaminated by tissue debris or skin bacteria, the first tube should be used for chemical or serological tests, the second for microbiological tests, and the third for microscopic and cytologic examination. The same procedure is used for infants and children but the volume of fluid withdrawn should be the minimum for the requested tests.

Up to 20 mL of spinal fluid can be safely removed from an adult, although this amount is not usually required. Antiglycolytic agents are generally not added to the tube for glucose measurement; rapid processing of specimens, a clinical requirement for tests on spinal fluid, ensures that little metabolism of glucose occurs even in the presence of many bacteria. To allow proper interpretation of spinal fluid glucose values, a simultaneous blood specimen should be obtained.

COLLECTION OF OTHER FLUIDS AND TISSUES FOR ANALYSIS

Synovial Fluid Aspiration

The technique of obtaining synovial fluid for examination is called *arthrocentesis*. Synovial fluid is withdrawn from joints to aid characterization of the type of arthritis and to differentiate noninflammatory effusions from inflammatory fluids. Normally, only a very small amount of fluid is present in any joint but this volume is usually very much increased in the presence of inflammatory conditions. Arthrocentesis should be performed by a physician using sterile procedures and the technique must be modified from joint to joint depending on the anatomic location and size of the joint. The skin over the joint is cleaned with an antiseptic such as iodine or hexachlorophene (pHisoHex) and then anesthetized with ethyl chloride. A needle of appropriate size is introduced into the joint and the required amount of fluid aspirated into the syringe. The physician should establish priorities for the tests to be performed in case the available volume is insufficient for all tests. Sterile plain tubes should be used for culture and for glucose and protein measurements; an EDTA tube is needed for a total leukocyte, differential, and erythrocyte count. Microscopic slides are prepared for staining with Gram's or other stains indicated, and for visual inspection.

Amniocentesis

Amniocentesis is performed by a physician for prenatal diagnosis of congenital disorders, to assess fetal maturity, or to look for Rh isoimmunization or intrauterine infection. Although ultrasound is not essential, amniocentesis is best performed with its assistance to aid localization of the placenta and to determine the presentation of the fetus. The best sites for obtaining amniotic fluid are behind the neck of the fetus, below its head, or from other unoccupied areas of the amniotic cavity.

The skin is cleaned and anesthetized as for other similar procedures and 10 mL of fluid is aspirated into a syringe connected to the spinal needle that is typically used. Glass containers are used to transport the fluid to the laboratory. If a specimen is for the determination of the lecithin/sphingomyelin (L/S) ratio, the container is immediately placed in ice; if it is for spectrophotometric analysis, the specimen should be transferred to a brown tube or bottle to prevent photodegradation of bilirubin. Alternatively, the specimen container may be wrapped in aluminum foil.

There are few complications from amniocentesis. Occasionally a bloody tap is made. The blood may come from the uterine wall, the placenta, or even the fetus. Determination of fetal hemoglobin can be used to help ascertain the source, if it is important to do so.

Pleural, Pericardial, and Ascitic Fluid Collection

The pleural, pericardial, and peritoneal cavities normally contain a small amount of serous fluid that lubricates the opposing parietal and visceral membrane surfaces. Inflammation or infections affecting the cavities cause fluid to accumulate. The fluid may be removed to determine if it is an effusion or an exudate, a distinction possible by protein or enzyme analysis. The collection procedure is called *paracentesis*. When specifically applied to the pleural cavity, the procedure is a *thoracentesis*; if applied to the pericardial cavity, a *pericardiocentesis*. Paracenteses should be performed only by skilled and experienced physicians. Pericardiocentesis has now been largely supplanted by echocardiography.

The skin over the intended puncture site should be cleaned with 70% isopropanol and then allowed to dry in the air. A spinal needle is then inserted into the body cavity through a small bleb in the skin raised by injection of a local anesthetic. Fluid is then withdrawn by a syringe and transferred to appropriate tubes for analysis. Paracentesis is rarely associated with complications. Occasionally, blood-stained fluid is obtained because of puncture of a small blood vessel. If adhesions are present between the intestine and abdominal wall, a part of the intestine could be perforated by a peritoneal tap. With a thoracentesis, pneumothorax and bronchopleural fistulas are potential complications.

Collection of Saliva

While measurements of concentrations of certain analytes in saliva have been advocated, there is presently no clinical justification for such assays, with the possible exception of blood group substances to determine secretor status and Le genotype. Measurement of a drug in saliva has been suggested to estimate the free, pharmacologically active concentration of the drug in serum. There is, however, a considerable difference in pH between saliva and serum, and ratios of bound/free drug would not be the same. Fortunately, ultrafiltration techniques are now available that facilitate the processing of serum for free drug analysis.

Several slightly different techniques have been devised for the collection of saliva. Generally, an individual is asked to rinse out his mouth with water and then chew an inert material such as a piece of rubber or paraffin wax from 30 seconds to several minutes. The first mouthful of saliva is discarded; thereafter the saliva is collected into a small glass bottle.

Preparation of Solid Tissue

The solid tissue most often analyzed in the clinical chemistry laboratory is malignant tissue from the breast for estrogen and progesterone receptors. During surgery, at least 0.5–1 g tissue should be removed and trimmed of fat and nontumor material. The tissue should be quickly frozen, preferably in liquid nitrogen or in a mixture of dry ice and alcohol. The time between collection and freezing should be less than 20 min. A histologic section should always be examined at the time of analysis of the specimen to confirm that the specimen is indeed malignant tissue. The same procedure may be used to obtain and prepare solid tissues for toxicological analysis; but when trace element determinations are to be made, all materials used in the collection or handling of the tissue should be made of plastic or materials known to be free of contaminating trace elements. (See also Chapter 8C.)

HANDLING OF SPECIMENS FOR TESTING

MAINTENANCE OF SPECIMEN IDENTIFICATION

Valid test results require a representative, properly collected, and properly preserved specimen. Proper identification of the specimen must be maintained. Every specimen container must be adequately labeled even if the specimen must be placed in ice or if the container is so small that a label cannot be placed along the tube, as might happen with a capillary blood tube. Direct labeling of a capillary blood tube by folding the label like a flag around the tube is preferred; a less satisfactory alternative is to label a larger container into which the capillary tube can be placed. The minimum information on a label should include the patient's name, location, and identifying number, as well as the date and time of collection. A system of color-coded tags should also be used to identify tubes that contain specimens known to come from patients with an infectious disease. Either one warning label for all diseases or labels specific for different diseases may be used. The same information should be attached to all tubes to which the specimen is transferred during its processing. Examples of labels that are used in daily practice are included in Figure 3-5.

Preservation of Specimens in Transit

Although delays of a specimen in transit from a patient to the laboratory are usually short, the time elapsing from the separation of serum and cells until analysis may be considerable. The specimen must be properly treated both during its transport to the laboratory and from the time the serum has been separated until it is analyzed. For some tests, specimens must be kept at ~4 °C from the time the blood is drawn until the specimens are analyzed or until the serum or plasma is separated from the cells. Examples are specimens for ammonia and blood gas determinations such as pCO_2, pO_2, and blood pH. (See Chapter 10.) Transfer of these specimens to the laboratory must be done by placing the specimen container in ice water. Specimens for acid phosphatase, lactate and pyruvate, and certain hormone tests, e.g., gastrin, renin activity, and parathyroid hormone, should be treated the same way. A significant decrease in pyruvate and increase in lactate concentration occurs within a few minutes at ambient temperature. (See Chapter 6.)

For all test constituents that are thermally labile, serum and plasma should be separated from cells in a refrigerated centrifuge. Specimens for bilirubin or carotene must be protected from

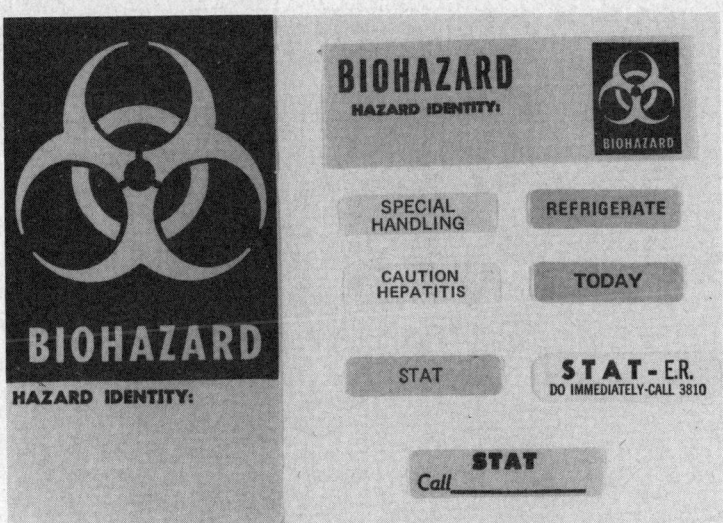

Figure 3-5. Examples of labels for blood tubes or containers to indicate special handling or possible hazardous specimens.

both daylight and fluorescent light to avoid photodegradation. Some special handling requirements are listed in Table 3-7.

Although transport of specimens from the patient to the clinical laboratory is usually done by messenger, pneumatic tube systems may be used to move the specimens more rapidly over long distances. Hemolysis may occur in these systems unless the tubes are completely filled and movement of the blood tubes inside the specimen carrier is prevented.[52] The pneumatic tube system should be designed to eliminate sharp curves and sudden stops of the specimen carriers, since these factors are responsible for much of the hemolysis that may occur. With many systems, however, the plasma hemoglobin concentration may be increased and the serum activity of red cell enzymes, e.g., lactate dehydrogenase, may also be increased. Nonetheless, the amount of hemolysis is usually so small that it can be ignored.

Separation and Storage of Specimens

Plasma or serum should be separated from cells as soon as possible, and certainly within 2 h.[31] Premature separation of serum, however, may permit continued formation of fibrin and lead to obstruction of sample probes in testing equipment. Coagulation in plain or silicone-coated glass tubes is usually complete in 20–30 min but is prolonged in plastic containers. If it is impossible to centrifuge a blood specimen within 2 h, the specimen should be held at room temperature rather than at 4 °C to decrease hemolysis. If the specimens cannot be analyzed at once, the separated serum should generally be stored in capped tubes at 4 °C until analysis, both to maintain stability of the specimen and to reduce evaporation. If a specimen for a particular test is sufficiently unstable at 4 °C, the serum specimen should be held at −20 °C in a freezer capable of maintaining this temperature. Frost-free freezers should be avoided because they have a wide temperature swing during the freeze/thaw cycle. Note, however, that 4 °C or −20 °C is not the optimal storage temperature for all tests; some lactate dehydrogenase isoenzymes, for instance, are more stable at room temperature than at 4 °C. Although changes in concentration of test constituents have been observed when serum is stored in a serum separator tube in a refrigerator for 24 h, the changes do not appear to be large enough to be of clinical significance.

Specimen tubes should be centrifuged with stoppers in place. Closure reduces evaporation, which occurs rapidly in a warm centrifuge with the air currents set up by centrifugation. Stoppers also prevent aerosolization of infectious particles. Specimen tubes containing volatiles such as ethanol *must* be stoppered while they are spun. Centrifuging specimens with the stopper in place maintains anaerobic conditions that are important in the measurement of carbon dioxide and ionized calcium. Removal of the stopper before centrifugation allows loss of carbon dioxide and

Table 3-7. SELECTED BLOOD CONSTITUENTS REQUIRING SPECIAL COLLECTION AND STORAGE CONDITIONS[59]*†

Constituent	Anticoagulant	Handling Requirements
S-Acetone, acetoacetate†‡		Freeze; stopper
S-Acid phosphatase		Add citrate (10 mg/mL); freeze
P-ACTH (corticotropin)	Heparin, 0.2 mg/mL	Freeze within 15 min of collection
S-Alcohol		Add NaF (10 mg/mL) to serum
S-Aldolase		Freeze
P,S-Aldosterone		Add boric acid (25 mg/mL) or freeze
P-Amino acids	Heparin, 0.2 mg/mL	Freeze
S-Androstenedione		Draw sample in A.M. (0000–1200 h)
S-Ascorbic acid		Freeze
S-Barbiturates		Do not use heparin
S-Bile acids		Freeze
S-C-peptide		Draw from fasting patient; freeze
S-Calcitonin		Freeze
P-Carcinoembryonic antigen (CEA)	EDTA, 2 mg/mL	
B-Cholinesterase	Heparin, 0.2 mg/mL	
S-Citric acid		Freeze
S-Complement		Freeze
P-Cortisol	Heparin, 0.2 mg/mL	Separate immediately
S-Creatine		Freeze
S-Creatine kinase isoenzymes		Freeze
S-Creatinine		Freeze
S-Cryoglobulins		Keep above 20 °C
P-11-Deoxycortisol	Heparin, 0.2 mg/mL	Separate immediately
P,S-Digitoxin		Draw 6–12 h post-administration
S-Digoxin		Draw 8 h post-administration
P-Estradiol	Heparin, 0.2 mg/mL	Freeze
P-Fatty acids (free and esterified)	Heparin, 0.2 mg/mL	Freeze
P-Fibrinogen	Citrate, ~4 mg/mL	Do not use heparin
S,B-Fluoride		Do not collect in glass container
P-Folate (tetrahydrofolate)		Freeze
B-Galactose	Heparin, 0.2 mg/mL	NaF (10 mg/mL) added with heparin
S-Gastrin		Collect from fasting patient; freeze
B-Glucose-6-phosphate dehydrogenase†	EDTA, 2 mg/mL	
P-HDL-cholesterol	EDTA, 2 mg/mL	Collected after 12–14 h fast; freeze
B-Hemoglobins (for quantitation and electrophoresis)	EDTA, 2 mg/mL	
S-Histidine		Freeze
S-17-Hydroxyprogesterone		Draw between 0900 and 1100 hours
S-Insulin		Collect from fasting patient; freeze
S-Isocitrate dehydrogenase		Freeze
B-Lactate		Immediately dilute with an equal volume of 5% (50 g/L) perchloric acid; shake and mix
B-Lead	Heparin, 0.2 mg/mL	Collect in lead-free tube containing heparin
S-Lipoprotein phenotyping		Freeze
S-Lysozyme		Freeze
S-Magnesium		Separate immediately
S-Parathyroid hormone (PTH)		Freeze
S-Pepsinogen		Freeze
S-Placental lactogen		Freeze
S-Prolactin		Freeze
S-Prostaglandin $F_{2\alpha}$		Freeze
P-Pyridoxal phosphate (vitamin B_6)	EDTA, 2 mg/mL	Protect from light
B-Pyruvate		Immediately dilute with an equal volume of 5% (50 g/L) perchloric acid; shake and mix
P-Renin	EDTA, 2 mg/mL	Chill during collection, centrifugation
S-Vitamin A		Protect from light
S-Zinc		Use only acid-washed glass; avoid hemolysis

P = plasma, S = serum, B = whole blood, EDTA = ethylenediaminetetraacetate.

*All specimens listed, except those to be assayed for cryoglobulins, should be transported at temperatures below 15 °C, except as noted.

†A sealable Styrofoam container with "freezer packs" is usually sufficient to keep specimen frozen for 12 h. Solid CO_2 is necessary for longer periods.

‡There is some question as to whether this constituent remains stable even with this procedure.

an increase in blood pH. Control of pH is especially important for the enzymatic measurement of acid phosphatase, which is labile under alkaline conditions engendered by CO_2 loss. Indeed, once the serum is separated for acid phosphatase tests, a tablet of disodium citrate should be added to stabilize the pH at about 6.2.

Transport of Specimens to Referral Laboratories

Before a referral laboratory is used for any tests, the quality of its work should be verified by the referring laboratory. The specimen handling requirements of the referral laboratory must be observed, and test results reported by a referral laboratory must be identified as such when they are filed in a patient's chart.

It should be assumed that transport from a referring laboratory to a referral laboratory may take as long as 72 h. A referring laboratory should retain enough specimen under optimum conditions for retesting should an unanticipated problem arise during shipment. The tube used for holding a specimen (*primary container*) should be so constructed that the contents do not escape if the container is exposed to extremes of heat, cold, or sunlight.[60] Reduced pressure of 0.50 atmosphere (50 kPa) may be encountered during air transportation, together with vibration, and specimens should be protected by a suitable container from these adverse conditions. Variability of temperature is the biggest factor in causing instability of test constituents.[16]

Polypropylene and polyethylene containers are usually suitable for specimen transport. Glass should be avoided because of its tendency to break. Polystyrene is also unsuitable as it may crack when frozen. The containers must be leak-proof and should have a Teflon-lined screw cap that does not loosen under the variety of temperatures to which the container may be exposed. The materials of both stopper and container must be inert and must not have any effect on the concentration of the analyte.

The shipping or *secondary container* used to hold one or more specimen tubes or bottles must be constructed to prevent the tubes from knocking against each other. Corrugated, fiberboard, or Styrofoam boxes designed to fit around a single specimen tube may be used. A padded shipping envelope provides adequate protection for shipping single specimens. When specimens are shipped as drops of blood on filter paper, e.g., for neonatal screening, the paper should be enclosed in a plastic bag which then can be mailed in a regular envelope.

For transportation of *frozen* or *refrigerated* specimens, a Styrofoam container should be used. The container walls should be 1 inch (2.5 cm) thick to provide effective insulation. The container should be vented to prevent buildup of carbon dioxide under pressure and a possible explosion. Solid carbon dioxide (dry ice) is the most convenient refrigerant material for keeping specimens frozen, and temperatures as low as -70 °C can be achieved. The amount of dry ice required in a container depends on the size of the container and the efficiency of its insulation, as well as the time for which the specimens must be kept frozen. One piece of solid dry ice (about 3 inches $\times$ 4 inches $\times$ 1 inch) in a container with 1-inch Styrofoam walls and a volume of 125 cubic inches (2000 cm³) will maintain a single specimen frozen for 48 h.

Various laws and regulations apply to the shipment of biological specimens. While these theoretically apply only to etiologic agents (known infectious agents), all specimens should be transported as if the same regulations applied. Airlines have rigid regulations covering the transport of specimens. The airlines deem dry ice a hazardous material; therefore, the transport of most clinical laboratory specimens is affected by the regulations.

The various modes of transport of specimens influence the shipping time and cost. Priority mail service is usually adequate within the United States if a delay of up to 3 d is acceptable. Express mail service assures delivery within 24 h, but it is available only between certain cities. Small-package service is available from many airlines, as is service by bus lines and Amtrak.

The cost of specimen transport is often included in the fee of a referral laboratory for each test. Many referral laboratories either provide their own courier service or contract with large national companies to provide such services.

The director of a referring laboratory has the responsibility to ensure that specimens will be adequately transported to the referral laboratory. He should determine the benefits of different services, and should keep in mind that the fastest service is generally the most expensive. He should also know that specimens should not be sent to a referral laboratory at the end of the week; since more delays in transit occur during weekends than during the working week, deterioration of specimens is more likely.

PHYSIOLOGICAL FACTORS AFFECTING THE COMPOSITION OF BODY FLUIDS

Physiological factors that affect laboratory test values fall into two categories—those that can be controlled and those that cannot. Those that can be controlled have short-lived effects. Duration of the other factors is much longer. Standardization of specimen collection practices minimizes the variables that cause changes in test values within one day or from one day to another and thereby reduces the difficulty in interpretation of values. However, in hospital practice standardization is not always possible; thus, one must understand the effects of both controllable and uncontrollable variables on the composition of body fluids. Most of the available information has been gathered under conditions where only one experimental variable was studied at a time. Few studies have addressed the interaction of several variables and their possible effects.

Many factors that affect specimen collection procedures have already been considered in this chapter. However, the controllable effects discussed in the following section may also have a profound influence on the composition of body fluids.

Controllable Biological Variables

Posture

The blood volume of an adult in an upright compared to a recumbent position is typically 600–700 mL less. Change from a lying to upright position equates to a reduction of about 10% in the blood volume, but because only protein-free fluid passes through the capillaries to the tissues, the reduction of the plasma volume is greater than that of the blood volume. Fluid reduction in plasma is associated with a comparable increase in the plasma protein concentration. The concentrations of all proteins, including enzymes and protein hormones, and of such compounds as drugs, calcium, and bilirubin that circulate partly bound to protein are also affected.

The typical pressure at the arterial end of a capillary is 24 mm Hg (3.2 kPa) and at the venous end 10 mm Hg (1.3 kPa), although it varies with the distance of the capillary from the heart. Transfer of fluid and solute across a capillary wall depends on a complex interaction of hydrostatic and osmotic pressures of the capillary and interstitial fluids. Fluid moves into the interstitial space at the arteriolar end of the capillary and returns to the capillary at the venular end. A greater volume of fluid leaves the capillary at the arteriolar end than is returned to it at the venous end. The excess drains into the lymphatic system. When an individual lies down, more fluid returns to the capillaries because the capillary pressure is reduced. The volume of fluid returning to capillaries progressively declines when an individual is recumbent for a long time. Diuresis serves as a mechanism to prevent overloading of the vascular system. The pressure relationships involved in fluid transfer across capillaries are illustrated in Figure 3-6.

Heart rate and systolic and diastolic blood pressures are greater in the upright than in the recumbent individual. The change in posture from lying to standing increases the secretion of catecholamines, aldosterone, angiotensin II, renin, and antidiuretic hormone. Epinephrine and norepinephrine concentrations in serum may double within 10 min, but there is no change in their urinary excretion. The increase of plasma aldosterone and plasma renin activity is slower, but their concentrations may still double within 1 h. Significant changes also take place with a change from the lying to the sitting position.[47]

The reduction of the extracellular fluid volume with standing reduces the renal blood flow and causes a reduction of the glomerular filtration rate and urine production. The changes are apparent in 1 h. Lithium, used therapeutically to treat some forms of schizophrenia, responds in the same way as sodium to postural changes. Sodium, potassium, and lithium excretions are reduced in response to increased hormone secretion, but the normal diurnal variation persists. When an individual stands, his urinary pH decreases and excretion of bicarbonate is reduced as hydrogen ions are exchanged for sodium. The excretion of protein is reduced in most individuals with the reduction of the glomerular filtration rate that occurs with standing. Orthostatic proteinuria is a condition in which protein is present when people are standing but essentially absent when they are recumbent. This phenomenon is possibly due to increased glomerular permeability from increased venous pressure. The incidence of orthostatic proteinuria is probably < 5%.

Normally, the alteration of blood volume that takes place with a change from standing to

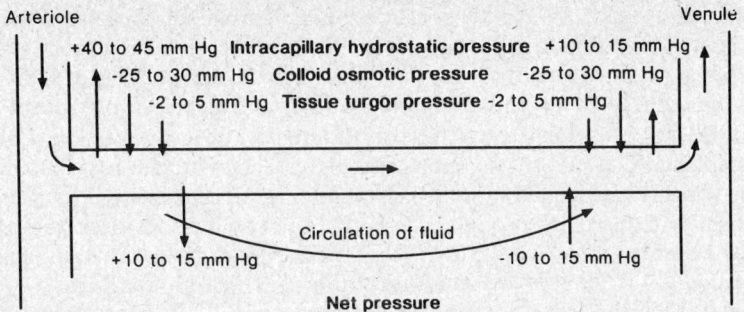

Figure 3-6. Pressure relationships across blood capillary affecting the transfer of fluids and solutes. (Reproduced with permission from Pitts, R. F.: Physiology of the Kidney and Body Fluids. 3rd ed. Chicago, Year Book Medical Publishers, Inc., 1974.)

lying is complete in 30 min. The decrease with the change from lying to standing is complete in 10 min. Changes in concentration of proteins and protein-bound constituents in serum are greater in hypertensives than normotensives, in individuals with a low plasma protein concentration than in those with a normal concentration, and in the elderly when compared with the young. Most of the plasma oncotic pressure is attributable to albumin because of its high concentration, so that protein malnutrition, with its associated reduction of plasma albumin concentration, will reduce the retention of the fluid within the capillaries. Conversely, the impact of postural changes is less in individuals with abnormally high concentrations of protein, such as those with a monoclonal gammopathy (multiple myeloma).

In general, the concentrations of freely diffusible constituents with molecular weights of < 5000 are unaffected by postural changes. However, a significant increase in potassium (about 0.2–0.3 mmol/L) occurs with 30 min standing. This increase in K^+ has been attributed to the release of intracellular potassium from muscle. Although postural changes affect urinary sodium excretion, sodium concentration in plasma is only slightly affected.

Changes in the concentration of some major serum constituents with change in posture are listed in Table 3-8.

Hospitalization and Immobilization

The plasma and extracellular fluid volumes decrease within a few days of bed rest. Consequently the blood hematocrit may increase by as much as 10% within 4 d. There is usually a slight reduction of total body water.

With prolonged bed rest, fluid retention occurs and serum protein and albumin concentrations may be decreased by an average of 0.5 g/dL and 0.3 g/dL, respectively. The concentrations of protein-bound constituents are also reduced, although mobilization of calcium from bones with an increased ionized fraction compensates for the reduced protein-bound calcium, so serum total calcium is less affected. Serum aspartate aminotransferase activity is usually slightly less in

Table 3-8. CHANGE IN CONCENTRATION OF SERUM CONSTITUENTS WITH CHANGE FROM LYING TO STANDING[17]

Constituent	Average Increase (%)
Alanine aminotransferase	7
Albumin	9
Alkaline phosphatase	7
Amylase	6
Aspartate aminotransferase	5
Calcium	3
Cholesterol	7
IgA	7
IgG	7
IgM	5
Thyroxine	11
Triglycerides	6

See footnote, Table 3-1.

individuals confined to bed than in those undertaking normal physical activity. Initially and paradoxically, creatine kinase activity is increased as a result of its release from skeletal muscles, but ultimately creatine kinase activity may be less than in active healthy individuals. Serum potassium may be reduced by up to 0.5 mmol/L because of reduction of skeletal muscle mass.

Prolonged bed rest is associated with increased urinary nitrogen excretion. Calcium, sodium, potassium, phosphate, and sulfate excretions are increased; hydrogen ion excretion is reduced, presumably due to decreased metabolism of skeletal muscle.[14] The amplitude of circadian variation of plasma cortisol is reduced by prolonged immobilization, and the urinary excretion of catecholamines may be reduced to one third of the concentration in an active individual. Vanillyl-mandelic acid excretion is reduced by one fourth after 2–3 weeks of bed rest.

When an individual becomes active after a period of bed rest, more than three weeks are required before calcium excretion reverts to normal, and another three weeks before positive calcium balance is achieved. Several weeks are required before positive nitrogen balance is restored.

Exercise

The influence of exercise on the composition of body fluids is related to the duration and intensity of the activity.

With *moderate exercise*, the provoked stress-response causes an increase in the blood glucose, which stimulates insulin secretion. The arteriovenous difference in glucose concentration is increased by the greater tissue demand for glucose. Plasma pyruvate and lactate are increased by the increased metabolic activity of skeletal muscle. Even mild exercise may increase the plasma lactate two-fold. Arterial pH and pCO_2 are reduced by exercise. Reduced renal blood flow causes a slight increase in the serum creatinine concentration. Competition between uric acid and lactate and products of increased tissue catabolism for renal excretion cause the serum urate concentration to increase. Exercise causes a reduction of cellular ATP, which increases cellular permeability. The increased permeability causes slight increases in the serum activities of enzymes originating from skeletal muscle, e.g., aspartate aminotransferase, lactate dehydrogenase, creatine kinase, and aldolase.[54] As little as 5 min of walking will increase the activity of these enzymes in plasma. Mild exercise produces a slight decrease in the serum cholesterol and triglyceride concentrations that may persist for several days.

Generally, the effects of *strenuous exercise* are exaggerations of those occurring with mild exercise. Thus, hypoglycemia and increased glucose tolerance may occur. The plasma lactate may be increased ten-fold. Severe exercise increases the concentration of plasma proteins due to an influx of protein from interstitial spaces, which occurs after an initial loss of both fluid and protein through the capillaries. Glycoproteins, transferrin, and α_2-macroglobulin concentrations are typically increased.[22] Fibrinolytic activity is also increased. Strenuous exercise may more than double creatine kinase activity, but the activity of enzymes with primarily liver or kidney origins is little changed, although both hepatic and renal blood flow are reduced.

Heavy exercise for 10 min increases plasma renin activity by 400%. Cortisol secretion is stimulated and the normal diurnal variation may be abolished.[13] Urinary free cortisol excretion and the plasma concentrations of aldosterone, growth hormone, and prolactin are also increased by exercise. Strenuous exercise increases both the plasma and urinary concentrations of catecholamines.

The blood pH, oxygen saturation, and venous bicarbonate concentrations are decreased by strenuous exercise. The concentration of triglycerides is reduced briefly by exercise, but the free fatty acid concentration is greatly increased; serum creatinine and urea nitrogen concentrations are also increased. Although the creatinine concentration returns rapidly to normal on the cessation of exercise, the increased urea nitrogen concentration persists for some time. Hematuria and proteinuria occur with exercise and worsen in proportion to the extent of the exercise.

Some representative changes in concentration or activity of serum constituents induced by exercise are listed in Table 3-9.

Physical Training

Athletes generally have a higher serum activity of enzymes of skeletal muscular origin than do nonathletes. However, the response of these enzymes to exercise is less in athletes than in other individuals. Serum concentrations of urea, urate, creatinine, and thyroxine are higher in

Table 3-9. EFFECT OF STRENUOUS EXERCISE ON SELECTED SERUM CONSTITUENTS[51]

Constituent Value	% Increase	Constituent Value	% Decrease
Acid phosphatase	11	Albumin	4
Alanine aminotransferase	41	Bilirubin	4
Alkaline phosphatase	3	Iron	11
Aspartate aminotransferase	31	Lactate dehydrogenase	1
Calcium	1	Potassium	8
Chloride	1	Sodium	1
Cholesterol	3	Total lipids	12
Creatinine	17		
Phosphorus	12		
Total protein	3		
Urea N	3		
Uric acid	4		

See footnote, Table 3-1.
Changes were determined 15 min after conclusion of 20 min exercise.

athletes than in comparable untrained individuals.[61] This is probably related to the increased muscle mass and a good turnover of muscle mass in athletes.

The total serum lipid concentration is reduced by physical conditioning; serum cholesterol may be lowered by as much as 25%. HDL-cholesterol, however, is increased. Thus, the decrease is mostly due to a reduction in LDL-cholesterol. The serum triglyceride concentration may be reduced by up to 20 mg/dL (0.23 mmol/L), but the free fatty acid concentration is higher in fit individuals than others. Generally, the same exercise produces a less-marked biochemical response in the fit person than in the unfit person.

Circadian Variation

Many constituents of body fluids exhibit cyclical variations throughout the day. Factors contributing to such variations include posture, activity, food ingestion, stress, and daylight or darkness, as well as sleep or wakefulness. These cyclical variations may be quite large, and therefore the drawing of the specimen must be strictly controlled. The concentration of serum iron, for example, may change by as much as 50% from 0800–1400 h, and that of cortisol by a similar amount between 0800 and 1600 h. Serum potassium has been reported to decline from 5.4 mmol/L at 0800 h to 4.3 mmol/L at 1400 h.[50] The typical total variation of several commonly measured serum constituents over 6 h is illustrated in Table 3-10. The total variation is listed together with analytic error.

Table 3-10. TOTAL AND ANALYTICAL VARIATION FOR SERUM TESTS ON SPECIMENS OBTAINED AT 0800 AND 1400 H[58]

Constituent	Mean	Total Variation (%)	Analytical Variation (%)
Sodium (mmol/L)	141	1.9	1.8
Potassium (mmol/L)	4.4	7.1	2.8
Calcium (mg/dL)	10.8	3.2	2.7
Chloride (mmol/L)	102	3.8	3.4
Phosphate (mg/dL)	3.8	10.7	2.4
Urea N (mg/dL)	14	22.5	2.5
Creatinine (mg/dL)	1.0	14.5	6.3
Uric acid (mg/dL)	5.6	11.5	2.6
Iron (μg/dL)	116	36.6	3.4
Cholesterol (mg/dL)	193	14.8	5.7
Albumin (g/dL)	4.5	5.5	3.9
Total protein (g/dL)	7.3	4.8	1.7
Total lipids (g/L)	5.3	25.0	3.6
Aspartate aminotransferase (U/L)	9	25	6
Alanine aminotransferase (U/L)	6	56	17
Acid phosphatase (U/L)	3	15	8
Alkaline phosphatase (U/L)	63	20	3
Lactate dehydrogenase (U/L)	195	16	12

11 male subjects, age 21–27 years, studied at 0800, 1100, 1400 h.

See footnote, Table 3-1.

Hormones are secreted in bursts, and this, together with the cyclical variation to which most hormones are subject, may make it very difficult to interpret their serum concentration properly. Corticotropin secretion is influenced by cortisol-like steroids, but it is also affected by posture as well as by light and darkness and stress. Its secretion is increased three- to five-fold from its minimum between afternoon and midnight to its maximum around waking. Cortisol concentrations are greatest around 0600–0800 h.[55]

Maximum renin activity normally occurs early in the morning during sleep; its minimum occurs late in the afternoon. Glomerular filtration rate (GFR) varies inversely with the secretion of renin; GFR is least at the time of maximum renin secretion and ~20% greater in the afternoon when renin activity is at a minimum. The excretion of 17-ketosteroids and 17-hydroxycorticosteroids is low at night and reaches a maximum about midafternoon.

There is no circadian variation in the plasma concentrations of FSH and LH in men, but a 20–40% increase of plasma testosterone occurs during the night. Prolactin is secreted, like other hormones, in multiple bursts; prolactin concentration is greatest during sleep.

The serum TSH is at a maximum between 0200 and 0400 h and at a minimum between 1800 and 2200 h. The variation is of the order of 50%. There are also variations in the serum thyroxine concentration, but these appear to be related to the changes in concentration of binding protein brought about by changes in posture. These variations are maximal between 1000 h and 1400 h.

Growth hormone secretion is greatest shortly after sleep commences. Conversely, basal plasma insulin is higher in the morning than later in the day, and its response to glucose is also greatest in the morning and least about midnight. When a glucose tolerance test is given in the afternoon, higher glucose values occur than when the test is given early in the day. The higher plasma glucose occurs in spite of a greater insulin response which is nevertheless delayed and less effective.

Urinary excretion of catecholamines and their metabolites is less at night than during the day. The effect is related to activity, since in night workers excretion is less during the day.

The peak urinary excretion of sodium and potassium is about noon, whereas the excretion of calcium and magnesium is greatest during the night. Urinary phosphate excretion is low at night, with the result that serum phosphate is as much as 30% higher at night than during the morning. Urinary volume and creatinine excretion are low during the night. The creatinine clearance may be reduced by up to 10% during the night. Night urine contains much ammonium and its titratable acidity is high.[53]

Blindness

Although not controllable, the influence of blindness is best considered here because of its effect on circadian variation. With blindness the normal stimulation of the hypothalamic-pituitary axis is reduced. Consequently, certain features of hypopituitarism and hypoadrenalism may be observed.[6] In some blind individuals, the normal diurnal variation of cortisol may persist; in others it does not. Urinary excretion of 17-ketosteroids and 17-hydroxycorticosteroids is reduced. Plasma sodium and chloride are often low in blind individuals, probably as a result of reduced aldosterone secretion. Plasma glucose may be reduced in blind people, and insulin tolerance is often less. The excretion of urate is reduced. Renal function may be slightly impaired as evidenced by slight increases in serum creatinine and urea nitrogen.

Negative nitrogen balance may occur in blind people and the serum protein concentration may be reduced. The serum cholesterol is frequently increased and bilirubin concentration may also exceed the upper limit of normal. The diurnal variation of serum iron is often lost.

Travel

Travel across several time zones affects the normal circadian rhythm. Five days are required to establish a new stable diurnal rhythm after travel across 10 time zones. The changes in laboratory test results are attributable to altered pituitary and adrenal function. Urinary excretion of catecholamines is usually increased for two days; serum cortisol is reduced. During a flight, serum glucose and triglyceride concentrations increase, while glucocorticoid secretion is stimulated. During a prolonged flight fluid and sodium retention occur but urinary excretion returns to normal after two days.[10]

INFLUENCE OF FOOD AND STIMULANTS

Recent Food Ingestion

The concentration of certain plasma constituents is affected by the ingestion of a meal. The biggest increases in serum concentrations occur for glucose, iron, total lipids, and alkaline phosphatase. The increase in alkaline phosphatase (mainly intestinal isoenzyme) is greater when a fatty meal is ingested and is influenced by the blood group of the individual and the substrate used for the enzyme assay. Lipemia may affect some analytical methods used to measure serum constituents. Ultracentrifugation or the use of serum blanks can reduce the adverse analytical effects of lipemia.

The effects of a meal may be long lasting. Thus, ingestion of a protein-rich meal in the evening may cause increases in the serum urea nitrogen, phosphorus, and urate concentrations that are still apparent 12 h later. Nevertheless, these changes may be less than the typical intraindividual variability. Large protein meals at lunch or in the evening also increase the serum cholesterol and growth hormone concentrations for at least 1 h after a meal. The effect of carbohydrate meals on blood composition is less than that of protein meals. No change in the cortisol concentration is noted when breakfast is taken, probably because cortisol occupies completely all cortisol-binding sites on its binding protein in the early morning. Glucagon and insulin secretions are stimulated by a protein meal, and insulin is also stimulated by carbohydrate meals.

In response to a meal, the stomach secretes hydrochloric acid, causing a reduction in the plasma chloride concentration. The venous blood from the stomach contains an increased amount of bicarbonate. This condition reflects a mild metabolic alkalosis ("alkaline tide") and an increased $p\text{CO}_2$. The metabolic alkalosis is sufficiently marked to reduce the serum ionized calcium by 0.2 mg/dL (0.05 mmol/L). Following ingestion of a meal the liver becomes the prime site for metabolism of the ingested substances. This metabolic activity may impair its secretory capability and thus account for the observed increase in postprandial bilirubin concentration, Bromsulfophthalein (BSP) retention, and serum enzyme activity.

The effects of ingestion of a 700 kcal (2.93 MJ) meal on some commonly measured blood constituents are illustrated in Table 3-11. The effects differ with different meals. Thus, glucose increase is often greater and phosphate usually decreases after a carbohydrate meal.

The ingestion of one glass of water has been observed to be followed by statistically significant alterations in the concentration of several of the commonly measured test constituents. When 75 g glucose is ingested with water, as in a glucose tolerance test, the concentration of glucose is increased. This stimulates the secretion of insulin. Insulin causes the release of sodium from cells and stimulates the transport of potassium into the cells.

Table 3-11. INFLUENCE OF A STANDARD 700 KCAL MEAL ON SERUM CONSTITUENTS[53]

Constituent	Before Meal	2 h After Meal
Alanine aminotransferase (U/L)	31	33
Albumin (g/dL)	4.5	4.6
Alkaline phosphatase (U/L)	46	46
Aspartate aminotransferase (U/L)	22	28
Bilirubin (mg/dL)	0.7	0.8
Calcium (mg/dL)	9.9	10.0
Cholesterol (mg/dL)	220	220
Glucose (mg/dL)	71	82*
Lactate dehydrogenase (U/L)	198	198
Phosphorus (mg/dL)	3.1	3.6*
Potassium (mmol/L)	3.8	4.0*
Sodium (mmol/L)	140	141
Total protein (g/dL)	7.8	7.9
Urea N (mg/dL)	16	16
Uric acid (mg/dL)	6.0	6.2

Results are mean values in 200 healthy individuals.

*Note also that other studies have reported greater increases in the glucose concentration and a reduction of the phosphorus and potassium concentrations depending on the type of meal.

See footnote, Table 3-1.

Ingestion of Specific Foods and Beverages

Caffeine, which is contained in many beverages including coffee, tea, and colas, has considerable effect on the concentration of blood constituents. Caffeine stimulates the adrenal medulla, causing an increased excretion of the catecholamines and their metabolites and a slight increase in the plasma glucose concentration with impairment of glucose tolerance.[3] The adrenal cortex is also affected; plasma cortisol is increased, accompanied by increased excretion of free cortisol, 11-hydroxycorticoids, and 5-hydroxyindoleacetic acid. The effect of caffeine may be so marked that the normal diurnal variation of plasma cortisol may be eliminated.

Caffeine has a marked effect on lipid metabolism. Ingestion of 2 cups of coffee may increase the plasma free fatty acid concentration by as much as 30% and glycerol, total lipids, and lipoproteins to a lesser extent. Prolonged ingestion of caffeine, e.g., over several weeks, causes a slight reduction of the serum cholesterol concentration but an increase in the serum triglyceride concentration.

Caffeine is also a potent stimulant of gastric juice, hydrochloric acid, and pepsin secretion. The serum gastrin concentration may be increased by as much as 5 times after the ingestion of 3 cups of coffee. Coffee has a diuretic effect and also increases the excretion of erythrocytes and renal tubular cells in the urine.

Habitual ingestion of *bran* impedes the absorption of certain compounds, including calcium, cholesterol, and triglycerides from the gastrointestinal tract. The concentration of calcium may be reduced by as much as 0.3 mg/dL (0.08 mmol/L) and that of triglycerides by 20 mg/dL (0.23 mmol/L), especially if triglycerides were high initially. Pectin and dietary fibers reduce the serum apolipoprotein-B and cholesterol concentrations. Many fruits and vegetables that contain 5-hydroxytryptamine (serotonin), e.g., bananas, cause an increased excretion of 5-hydroxyindole-acetic acid. Avocados impair glucose tolerance by affecting insulin secretion. Onions reduce both the plasma glucose and insulin response to glucose.

Smoking

Smoking, through the action of nicotine, may affect several laboratory tests. The extent of the effect is related to the number of cigarettes smoked and to the amount of smoke inhaled.

Through stimulation of the adrenal medulla, nicotine increases the concentration of epinephrine in the plasma and the urinary excretion of catecholamines and their metabolites.[11] Glucose concentration may be increased by 10 mg/dL (0.56 mmol/L) within 10 min of smoking a cigarette. The increase may persist for 1 h. Plasma lactate is increased, and since the pyruvate concentration is reduced, the lactate/pyruvate ratio is increased. Plasma insulin concentration shows a delayed response to the increased blood glucose, rising about 1 h after a cigarette is smoked. Typically, the plasma glucose concentration is higher in smokers than in nonsmokers, and glucose tolerance is mildly impaired in smokers. The plasma growth hormone concentration is particularly sensitive to smoking. It may increase ten-fold within 30 min after an individual has smoked a cigarette.

The plasma β-lipoprotein, cholesterol, and triglyceride concentrations are higher in smokers than in nonsmokers. Free fatty acid concentration tends to be variable, but inhalation during smoking produces an immediate increase of free fatty acids of about 30%. Some of the effects of smoking on serum constituents are listed in Table 3-12.

Smoking affects the adrenal cortex as well as the medulla; plasma 11-hydroxycorticosteroids may be increased by 75% with heavy smoking. In addition, the plasma cortisol concentration may increase by as much as 40% within 5 min of the start of smoking, although the normal

Table 3-12. REPORTED CHANGES IN SERUM COMPOSITION IN SMOKERS[48]

Constituent	% Change
Albumin	3
Cholesterol	4
Glucose	10
Phospholipids	5
Triglycerides	20
Urea N	10

See footnote, Table 3-1.

diurnal rhythmicity of cortisol is unaffected. Smokers excrete more 5-hydroxyindoleacetic acid than do nonsmokers.

The blood erythrocyte count is increased in smokers. The amount of carboxyhemoglobin may exceed 10% of the total hemoglobin in heavy smokers, and the increased number of cells compensates for impaired ability of the red cells to transport oxygen. The blood pO_2 of the habitual smoker is usually about 5 mm Hg (0.7 kPa) less than in the nonsmoker, while the pCO_2 is unaffected. The blood leukocyte concentration is increased by as much as 30% in smokers, and the leukocyte concentration of ascorbic acid is greatly reduced.

Fluid retention caused by nicotine causes a mild decrease in the plasma protein concentration but without demonstrable effect on the calcium concentration or on the activity of serum enzymes. The plasma urate concentration is less in smokers than in nonsmokers, probably as a result of lessened intake of food by smokers. Both the serum urea and creatinine concentrations tend to be less in smokers than in nonsmokers.

Nicotine is a potent stimulant of the secretion of gastric juice. Both volume and acid secretion are increased within 1 h of smoking several cigarettes. In contrast, the bicarbonate concentration and volume of pancreatic juice are reduced.

Smoking affects the body's immune response. Serum IgA, IgG, and IgM are generally lower in smokers than in nonsmokers, whereas the IgE concentration is higher. Smokers, more often than nonsmokers, may show the presence of antinuclear antibodies and weakly positive tests for carcinoembryonic antigen.

The serum vitamin B_{12} concentration is often markedly reduced in smokers, and the decrease is in inverse proportion to the serum concentration of thiocyanate.

Alcohol Ingestion

A single moderate dose of alcohol has few effects on laboratory tests. Ingestion of enough alcohol to produce mild inebriation may increase the blood glucose concentration by 20–50%. The increase may be even more marked in diabetics. More commonly, inhibition of gluconeogenesis occurs and becomes apparent as hypoglycemia and ketonemia. Hypoglycemia is most common in children, alcoholics, and the malnourished. Lactate accumulates and competes with urate for excretion in the kidneys so that the serum urate is also increased. Marked hypertriglyceridemia following alcohol ingestion is due to a combination of increased triglyceride formation in the liver and impaired removal of chylomicrons and VLDL (very-low-density lipoproteins) from the circulation. The effect is most noticeable when alcohol is ingested with a fatty meal. The effect may persist for > 12 h. When moderate amounts of alcohol are ingested for 1 week, the serum triglyceride concentration is increased by > 20 mg/dL (0.23 mmol/L).

Intoxicating amounts of alcohol stimulate the release of cortisol, although the effect is more related to the intoxication than to the alcohol per se. Sympathico-medullary activity is increased by acute alcohol ingestion but without detectable effect on the plasma epinephrine concentration and only a mild effect on norepinephrine. With intoxication, plasma concentrations of catechol-amines are markedly increased. Acute ingestion of alcohol leads to a sharp reduction in the plasma testosterone in men, with an increase in the plasma luteinizing hormone (LH) concentration.

Chronic alcohol ingestion affects the activity of many serum enzymes. Gamma-glutamyl-transferase activity has been most studied, and increased activity of the enzyme is used as a marker of persistent drinking. A single acute ingestion of alcohol has been reported to increase the serum activity of the enzyme, a manifestation of hepatic microsomal enzyme induction. Chronic alcoholism is associated with many characteristic biochemical abnormalities, including abnormal pituitary, adrenal cortical, and medullary function. Acute alcohol ingestion has been reported to increase the activity of several serum enzymes, including γ-glutamyltransferase, isocitrate dehydrogenase, and ornithine carbamoyl transferase.[18,20]

Drug Administration

It is rare for a patient to be hospitalized without receiving some drugs. For certain medical conditions more than 10 drugs may be administered at one time. Even many healthy individuals take several drugs regularly, such as vitamins, oral contraceptives, or sleeping tablets. Individuals with chronic diseases often ingest drugs on a continuing basis. Drugs may have both in vivo and in vitro effects on laboratory tests. The in vivo effects arise from the therapeutic intent of drugs, their side effects, and patient idiosyncrasies. Effects on the composition of body fluids are likely to be more apparent when large doses of a drug are administered for a long time than when administration of a single dose occurs on an isolated occasion.

Comprehensive listings of the effects of drugs on laboratory tests have been published by Young et al.[63] and Young.[62] Only a few representative effects are discussed here.

Many drugs, when administered intramuscularly, cause sufficient muscle irritation to increase amounts of enzyme released into the serum. The activities of creatine kinase, aldolase, and the skeletal muscle component of lactate dehydrogenase are increased in the serum. The increased activities may persist for several days after a single injection, and consistently high values may be observed during a course of treatment. *Penicillin* derivatives given intramuscularly are particularly likely to increase the activity of these enzymes, although any drug given intramuscularly appears capable of increasing enzyme activity. Drugs that have been demonstrated to increase serum creatine kinase activity following intramuscular injection are listed in Table 3-13. *Opiates* such as morphine or meperidine can cause spasm of the sphincter of Oddi. The spasm transmits pressure back to the liver, causing release of liver and pancreatic enzymes into the serum. Increases in aspartate aminotransferase activity may be so large that they may be suggestive of a myocardial infarction.

Oral contraceptives affect many different constituents measured in the clinical laboratory. Tests are affected by both the progestin and estrogen components. The overall effect depends on the proportion of the two components.

Diuretic drugs often cause a mild reduction of the plasma potassium concentration; hyponatremia may be observed. Hypercalcemia may occur with hemoconcentration, but occasionally the ionized as well as the protein-bound fraction is increased. Thiazides cause hyperglycemia and reduce glucose tolerance, especially in diabetics. Thiazides may cause prerenal azotemia with hyperuricemia due to decreased renal blood flow and glomerular filtration rate as a result of reduced blood volume.

The broad range of possible effects of a drug on clinical laboratory tests is exemplified by *phenytoin.* With long-term treatment, many patients have reduced serum calcium and phosphate concentrations and increased activity of alkaline phosphatase. Phenytoin induces the synthesis of bilirubin-conjugating enzymes in the liver. Consequently, the serum bilirubin concentration is reduced, serum γ-glutamyltransferase activity is increased, and urinary glucaric acid excretion is augmented. A few cases of increased serum aminotransferase activity have been reported, together with prolongation of the prothrombin time. Occasionally, cholestatic, cytotoxic, or mixed hepatic injury may occur. The overall incidence of slight alteration of liver function is about 25%.

Phenytoin decreases the urinary excretion of 17-ketosteroids and 17-hydroxycorticosteroids by stimulating the conversion of cortisol to 6-β-hydroxycortisol; it also diminishes serum FSH and the sperm count in semen, and thereby reduces fertility. Phenytoin also lowers the serum thyroxine concentration, probably by competitive displacement of thyroxine from its protein binding sites; free T_4 also tends to be low. Serum T_3 is low, probably as a result of stimulated metabolism in the liver, but the concentration of TSH is unaffected by the altered thyroxine metabolism.

Many epileptics receiving anticonvulsants excrete increased amounts of copper and zinc in their urine. Increased serum ceruloplasmin also increases the total serum copper concentration. In 20–30% of epileptic children receiving anticonvulsant therapy, erythrocyte aspartate aminotransferase activity is low, indicating a lowered pyridoxal (vitamin B_6) status. In as many as 50% of the adults receiving phenytoin for some time, there will be folate deficiency, manifested by reduced erythrocyte and serum folate concentrations. The mechanism for the deficiency has not yet been established conclusively. In about 10% of adults taking phenytoin the serum B_{12} is low.

T-cell function is reduced in many epileptics receiving anticonvulsants, and systemic lupus

Table 3-13. DRUGS ASSOCIATED WITH INCREASED SERUM CREATINE KINASE ACTIVITY FOLLOWING INTRAMUSCULAR INJECTION

Ampicillin	Chlorpromazine	Meperidine
Analgesics	Clindamycin	Morphine
Antibiotics	Digoxin	Penicillin
Barbiturates	Diuretics	Phenothiazines
Carbeicillin	Lidocaine	Tubocurarine

erythematosus occurs in about 10% of treated individuals. Antinuclear antibodies have been reported in 55% and low serum IgA and IgG concentrations in 25% of the patients. Other reported biochemical abnormalities include reduced serum concentrations of HDL-cholesterol, uric acid, and urea.

Underlying Medical Conditions

Some general clinical conditions have an effect per se on the composition of body fluids. These conditions may exist in addition to the primary complaint that prompted a patient's admission to the hospital.

Fever

Fever provokes many hormone responses.[45] Hyperglycemia occurs early and stimulates the secretion of insulin, which improves glucose tolerance; but insulin secretion does not necessarily reduce the blood glucose concentration, since increased secretion of growth hormone and glucagon also occurs. Fever appears to reduce the secretion of thyroxine, as do acute illnesses even without fever. In response to increased corticotropin secretion, the plasma cortisol concentration is increased and its normal diurnal variation may be abolished. The urinary excretion of free cortisol, 17-hydroxycorticosteroids, and 17-ketosteroids is increased. As acute fever subsides, or if it lessens but persists for a prolonged period, the hormone responses diminish.

Glycogenolysis and a negative nitrogen balance occur with the onset of fever. These are prompted by the typically decreased food intake and wastage of skeletal muscle that accompany fever. Although there is usually an increase in the blood volume with fever, the serum concentrations of creatinine and uric acid are usually increased. Aldosterone secretion is increased with retention of sodium and chloride. Secretion of antidiuretic hormone also contributes to the retention of water by the kidneys. Increased synthesis of protein occurs in the liver, and the plasma concentrations of acute-phase reactants and glycoproteins are increased.

Fever accelerates lipid metabolism. The serum concentrations of cholesterol, nonesterified fatty acids, and the other lipids may decrease initially, but within a few days the free fatty acid concentration may increase markedly. Fever is often associated with a respiratory alkalosis caused by hyperventilation. This pH increase causes a reduction of the plasma phosphate concentration, with an increased excretion of phosphate and other electrolytes.[2] Serum iron and zinc concentrations decline with accumulation of both elements in the liver. The copper concentration increases because of increased production of ceruloplasmin by the liver. Some representative changes in serum composition induced by fever are listed in Table 3-14.

Shock and Trauma

Regardless of the cause of shock or trauma, certain characteristic biochemical changes ensue.[12] Corticotropin secretion is stimulated to produce a three- to five-fold increase in the serum cortisol concentration. The 17-hydroxycorticosteroid excretion is greatly increased, although the excretion of 17-ketosteroids and of metabolites of adrenal androgens may be unaffected. Aldosterone secretion is stimulated. Plasma renin activity is increased, as are the secretions of growth hormone,

Table 3-14. EFFECT OF FEVER ON COMPOSITION OF SERUM[2]

Constituent	Baseline Value	Concentration After Induction of Fever			
		18 h	*48 h*	*72 h*	*96 h*
Sodium (mmol/L)	141	130	130	132	135
Chloride (mmol/L)	99	91	89	92	94
Potassium (mmol/L)	3.6	3.5	3.0	3.4	3.6
Calcium (mg/dL)	9.7	8.4	8.5	9.0	9.1
Phosphate (mg/dL)	3.3	2.3	3.2	3.2	3.7
Magnesium (mg/dL)	1.85	1.62	1.73	1.78	1.70
Creatinine (mg/dL)	1.10	1.03	1.04	1.00	1.09
Urea N (mg/dL)	13.4	14.0	15.2	18.5	17.4
Uric acid (mg/dL)	5.0	5.5	5.7	6.2	6.2

See footnote, Table 3-1.

glucagon, and insulin. Anxiety and stress increase the excretion of catecholamines. The stress of surgery has been shown to reduce the serum triiodothyronine by 50% in patients without thyroid disease.

The general metabolic response to shock includes the normal response to stress by mobilization of lipids, although the serum triglyceride concentration is not usually affected. Plasma glucose concentration is increased and glucose tolerance is reduced.

Immediately following an injury there is loss of fluid to extravascular tissues with the result of a decreased plasma volume. If the decrease is enough to impair circulation, glomerular filtration is diminished. Diminished renal function will lead to the accumulation of urea and other end products of protein metabolism in the circulation. In burned patients, serum total protein concentration falls by as much as 0.8 g/dL because of both loss to extravascular spaces and catabolism of protein. Serum α_1-, α_2-, and γ-globulin concentrations increase but not enough to compensate for the reduced albumin concentration. The plasma fibrinogen concentration responds dramatically to trauma and may double in 2–8 d following surgery.

With tissue destruction, there is increased urinary excretion of the major components of skeletal muscle. The muscle damage associated with the trauma of surgery can markedly increase the serum activity of enzymes originating in skeletal muscle, and this increased activity may persist for several days. Typical alterations in activity of serum enzymes following surgery are illustrated in Table 3-15. Increased tissue catabolism requires increased oxygen consumption and also leads to the production of acid metabolites. Thus, blood lactate may increase two- to three-fold. With tissue anoxia and impairment of renal and respiratory function, a metabolic acidosis develops.

Transfusion

The protein-rich fluid lost from the intravascular space following trauma is replaced with protein-poor fluid from the interstitial spaces. Subsequently, this is replaced by a fluid similar in composition to plasma. Transfusion of whole blood or plasma raises the plasma protein concentration; the amount of increase depends on the amount of blood administered. Serum lactate dehydrogenase activity, primarily LDH-1 and LDH-2, is increased by the breakdown of transfused erythrocytes. Transfusions to replace blood lost due to injury reduce sodium, chloride, and water retention precipitated by the injury. Serum iron and transferrin concentrations are reduced immediately following an injury, but extensive blood transfusions can lead to siderosis and an increased serum iron concentration. Serum potassium may increase with transfusion of stored blood.

Infusions of glucose solutions usually result in a reduction of both the plasma phosphate and potassium concentrations as these compounds are taken up by the erythrocytes. Infusions of solutions of albumin may increase plasma alkaline phosphatase activity if the albumin has been prepared from placentas.

LONG-TERM BIOLOGICAL INFLUENCES ON BODY FLUIDS

Published reference ranges usually are broad; however, the intra-individual variability is significantly less than the group or population variability. Indeed, a situation can arise in which the values obtained in one individual are very different from those found within another individual,

Table 3-15. INCIDENCE OF INCREASED ACTIVITY OF SERUM ENZYMES AND ISOENZYMES FOLLOWING SURGERY[26]

Enzyme	%
Creatine kinase	76
CK-MB isoenzyme	6
Aspartate aminotransferase	50
α-Hydroxybutyrate dehydrogenase	28
LDH-1 isoenzyme	18
LDH-1 > LDH-2	10
LDH-5 isoenzyme	20

See footnote, Table 3-1.

yet both sets of values could fall within a predetermined reference range. Not only does the set-point for different individuals appear to be different, but the extent by which values change from one occasion to another, even when specimens are obtained under standardized conditions, appears to differ among individuals. Both of these factors are probably determined genetically, and the difference in set-points underscores the fallibility of using a population reference range to associate small changes in laboratory data with a deviation from health for a given individual.

There is better concordance between the serum concentrations or activities of several constituents in monozygotic twins than in dizygotic twins.[23] This evidence indicates the importance of genetic constitution in determining the concentration of blood constituents. An influence of heredity on the plasma concentrations of cholesterol, glucose, urea nitrogen, urate, and bilirubin has been substantiated.

An association of blood type with concentration of certain constituents, e.g., uric acid, α_1-antitrypsin, cholesterol, and alkaline phosphatase, has been established.

The influences of gender and age as they affect the results of individual laboratory tests are discussed in various chapters of this book and are given in the Appendix, Table 20-24. Age, second only to gender, influences reference values, but the most important influences that determine the overall effect of age are the degree of sexual maturity and the amount of skeletal muscle mass of the individual. Typical changes in serum composition occurring with age are listed in Table 3-16, although the degree of changes differs in various reports. Generally, individuals can be considered in four groups—the newborn, the older child to puberty, the sexually mature adult, and the elderly adult.

INFLUENCES OF AGE, GENDER, AND RACE

Newborn. The body fluids of the newborn infant reflect both the trauma of birth and the changes related to the infant's adaptation to an independent existence. The composition of the blood is affected by the maturity of the infant at birth. In the mature infant most of the hemoglobin is the adult form, hemoglobin A, whereas in the immature infant much of the hemoglobin may be the fetal form, hemoglobin F. In both the mature and immature infant, the arterial blood oxygen saturation is very low initially. Newborns develop a metabolic acidosis that results from the accumulation of organic acids, especially lactic acid. The acid-base status, however, reverts to normal within 24 h.

Within a few minutes of an infant's birth, fluid passes from the blood vessels into the extravascular spaces. This fluid is similar to plasma except that plasma protein increases because the fluid lost from the intravascular space contains no protein.

The serum activities of several enzymes, including creatine kinase, γ-glutamyltransferase, and aspartate aminotransferase, are high at birth, but the degree of increase of alanine aminotransferase activity is less.

Table 3-16. INFLUENCE OF AGE ON MEAN CONCENTRATION OF SERUM CONSTITUENTS IN MALES[32]

	Measured Value <29 y	Change Compared with <29 y Value			
		30–39 y	40–49 y	50–59 y	60–69 y
Albumin (g/dL)	4.6	0.2↓	0.3↓	0.4↓	0.6↓
Alkaline phosphatase (U/L)	51	3↓	1↓	1↑	4↑
Aspartate aminotransferase (U/L)	41	3↑	3↑	1↑	1↑
Bilirubin (mg/dL)	0.4	0.1↑	0	0	0
Calcium (mg/dL)	9.8	0.1↓	0.2↓	0.2↓	0.3↓
Cholesterol (mg/dL)	211	29↑	43↑	48↑	36↑
Creatinine (mg/dL)	1.1	0	0.1↑	0.1↑	0
Glucose (mg/dL)	108	1↑	6↑	2↑	9↑
Phosphorus (mg/dL)	4.0	0.1↓	0.3↓	0.2↓	0.2↓
Total protein (g/dL)	7.6	0.1↓	0.2↓	0.2↓	0.2↓
Urea N (mg/dL)	15	1↑	1↑	2↑	3↑
Uric acid (mg/dL)	5.9	0	0.2↑	0.1↓	0.2↓

See footnote, Table 3-1.

In infants, even in the absence of disease, the concentration of bilirubin rises following birth and peaks about the third to fifth day of life. The physiological jaundice of the newborn rarely produces serum bilirubin values > 5 mg/dL (85 μmol/L). Distinguishing this naturally occurring phenomenon from other conditions that produce neonatal hyperbilirubinemia may be difficult, and the chronological course of the hyperbilirubinemia is important.

The blood glucose concentration is low in newborns because of their small glycogen reserves, although some attribute the low glucose to adrenal immaturity. Blood lipid concentrations are low but reach ∼80% of the adult values after two weeks. The plasma sodium concentration in an infant at birth is slightly higher than in the adult; at 12 h it decreases to below the adult value before rising to a value slightly greater than in the adult. The chloride concentration changes similarly, and the changes are largely related to fluid transfer in and out of the blood capillaries. The plasma potassium concentration may be as high as 7 mmol/L at birth, but it falls rapidly thereafter. Plasma calcium is also high initially but falls by as much as 1.4 mg/dL (0.35 mmol/L) during the first day of life.

The plasma urea nitrogen concentration decreases following birth as the infant synthesizes new protein, and the concentration does not begin to rise until tissue catabolism becomes prominent. The plasma amino acid concentration is low as a result of synthesis of tissue protein, although urinary excretion of amino acids may be quite high because of immaturity of the tubular reabsorptive mechanisms. The plasma urate concentration is high at birth, but the high clearance of urate soon reduces the plasma concentration below the adult value.

The serum thyroxine concentration of the healthy newborn, like that in the pregnant woman, is considerably higher than in the nonpregnant adult. Following its birth, an infant secretes thyroid-stimulating hormone (TSH), which causes a further increase in the serum thyroxine concentration. The physiological hyperthyroidism gradually declines over the first year of life.

Childhood to puberty. Many changes take place in the composition of body fluids between infancy and puberty. Most of the changes are gradual and there are rarely abrupt changes to adult concentrations.

The plasma protein concentrations increase after infancy and adult concentration values are attained by the age of 10 years. Serum IgG increases slightly out of proportion to the increase in concentration of α_2-globulin. The serum activity of most enzymes decreases during childhood to adult values by puberty or earlier, although the activity of alanine aminotransferase may continue to rise, at least in men, until middle age. Serum alkaline phosphatase activity is high in infancy but decreases during childhood, and rises again with growth before puberty. The activity of the enzyme is better correlated with skeletal growth and sexual maturity than with chronological age; it is greatest at the time of maximal osteoblastic activity occurring with bone growth. The activity decreases rapidly after puberty, especially in girls.

The serum creatinine concentration increases steadily from infancy to puberty parallel to development of skeletal muscle; until puberty there is little difference in the concentration between sexes. The serum uric acid concentration decreases from its high at birth until age 7–10 years, at which time it begins to increase, especially in boys, until about age 16 years.

The adult. Adult values are usually taken as the reference to which those of young and elderly persons are compared. The concentrations of most test constituents remain quite constant between puberty and menopause in women and between puberty and middle age in men.

During the midlife years, serum total protein and albumin concentrations decrease slightly. There may be a slight decrease in the serum calcium concentration in both sexes. In men, the serum phosphorus decreases markedly after age 20 years; in women, the phosphorus also decreases until menopause, when a marked increase takes place. The serum alkaline phosphatase begins to rise in women at the menopause so that in elderly women activity of this enzyme may actually be higher than in men.

Serum uric acid concentrations peak in men in their twenties and in women during middle age. Urea concentration increases in both sexes in middle age. Age does not affect the serum creatinine concentration in men but does increase the concentration in women. The serum total cholesterol and triglyceride concentrations increase in both men and women at a rate of ∼2 mg/dL (0.02 mmol/L) per year to a maximum between ages 50 and 60 years. The activity of most enzymes in serum is less during adult life than during adolescence. This increased enzyme activity presumably reflects the greater physical activity of the adolescents. The concentration of glucose in plasma 1 h after a loading dose rises ∼8 mg/dL (0.44 mmol/L) per decade.

Table 3-17. CHANGES IN COMPOSITION OF
SERUM WITH MENOPAUSE[48]

Constituent	% Increase
Alanine aminotransferase	12
Albumin	2
Alkaline phosphatase	25
Apolipoprotein AI	4
Aspartate aminotransferase	11
Cholesterol	10
Glucose	2
Phospholipids	8
Phosphorus	10
Sodium	1.5
Total protein	0.7
Uric acid	10

See footnote, Table 3-1.

The elderly adult. Significant increases in the plasma concentrations of many constituents occur in women after the menopause.[57] Some of these changes are listed in Table 3-17.

Renal concentrating ability is reduced in the elderly adult, so that creatinine clearance may decline by as much as 50% between the third and ninth decades. This decreased clearance is caused more by a decrease in urinary creatinine excretion as a result of decreased lean body mass than by renal problems. The tubular maximum capacity for glucose is reduced. The blood urea concentration rises with age, as does the urinary excretion of protein.

Hormone concentrations are also affected by aging. However, changes in concentration are much less pronounced than an endocrine organ's response to stimuli. Triiodothyronine concentration decreases by up to 40% in people over 40 years of age. Although thyroxine secretion is reduced, the thyroxine concentration is not changed, because its degradation is also reduced. Plasma parathyroid hormone concentration decreases with age. Cortisol secretion is reduced, although the serum concentration may not be affected. The reduced secretion leads to a reduction in the urinary excretion of 17-hydroxycorticosteroids. 17-Ketosteroid excretion in the elderly adult is about half of that of the younger adult. The secretion and metabolic clearance of aldosterone are decreased with a reduction of ∼50% in the plasma concentration. The aldosterone response to sodium restriction is diminished. Basal insulin concentration is unaffected by aging but its response to glucose is reduced. In the male, the secretion rate and concentration of testosterone are reduced after age 50 years. In women, the concentration of pituitary gonadotropins, especially FSH, is increased in the blood and urine.

Estrogen secretion in women begins to decrease before the menopause and continues at a greater rate after the menopause while gonadotropins show a feedback-mediated reciprocal rise. Serum concentrations of estrogens decrease by 70% or more, and urinary excretion of estrogens is decreased comparably. The decreased estrogen secretion may be responsible for the increase of serum cholesterol that occurs up to age 60 years in women. Estrogen secretion in men, although always less than in women, declines with age.

Gender. Until puberty there are few differences in laboratory data between boys and girls. After puberty, the serum activities of alkaline phosphatase, the aminotransferases, creatine kinase, and aldolase are greater in men than in women. The higher activity of enzymes originating from skeletal muscle in men is related to their greater muscle mass. After the menopause, the activity of alkaline phosphatase increases until it is higher in women than in men. Although total lactate dehydrogenase activity is similar in men and women, the LDH-1 and LDH-3 activities are higher and LDH-2 is less in young women than in men. These differences disappear after the menopause.

The concentrations of albumin, calcium, and magnesium are higher in men than women, but the concentration of γ-globulin is less. Blood hemoglobin concentrations are less in women; thus the serum bilirubin concentrations are also slightly lower. Serum iron is low during a woman's fertile years, and her plasma ferritin may be only one third the concentration in males. The reduced iron concentration in women is attributable to menstrual blood loss. Cholesterol concentration is typically higher in men than women while the α-lipoprotein concentration is less. The plasma amino acid concentrations as well as the concentrations of creatinine, urea, and uric acid are higher in males than in females. The effect of age on the difference in concentrations of serum constituents between men and women is illustrated in Table 3-18.

Race. Differentiation of the effects of race from those of socioeconomic conditions is often difficult. Nevertheless, the total serum protein concentration is known to be higher in Blacks than in Whites. This is largely attributable to a much higher γ-globulin, although usually the concentrations of α_1- and β-globulins are also increased. The serum albumin is typically less in Blacks than Whites. In black men, serum IgG is often 40% higher and serum IgA may be as much as 20% higher than in white men.[8]

The activity of creatine kinase and lactate dehydrogenase is usually much higher in both black men and women than in Whites. This is presumed to be related to the amount of skeletal muscle, which tends to be greater in Blacks than Whites. Due to their greater skeletal development, black children generally have a higher serum alkaline phosphatase at puberty than do white children.

Carbohydrate and lipid metabolism differ in Blacks and Whites.[4] Glucose tolerance is less in Blacks, Polynesians, American Indians, and Eskimos than in comparable age- and sex-matched Whites. This is substantiated by a plasma glucose 1 h after glucose challenge that may be as much as 15–25 mg/dL (0.8–1.4 mmol/L) higher in the Blacks than Whites. After age 40 years, the serum cholesterol and triglyceride concentrations are consistently higher in both white men and women than in Blacks. These may be dietary rather than racial factors since the concentration of plasma lipids has been shown to be different for the same racial group in different parts of the world. The blood hemoglobin concentration is as much as 1.0 g/dL higher in Whites than Blacks. Some of the indigenous groups of the Pacific (e.g., Maoris of New Zealand) have significantly higher mean serum urate concentrations than Caucasian populations.

EFFECT OF ENVIRONMENTAL FACTORS

In individuals living at a high *altitude* the blood hemoglobin is markedly increased due to reduced atmospheric pO_2.[15] Erythrocyte 2,3-diphosphoglycerate is also increased, and the oxygen dissociation curve is shifted to the right. The increased erythrocyte concentration leads to an increased turnover of nucleoproteins and excretion of urate. The fasting, basal concentration of growth hormone concentration is high in individuals living at a high altitude.

Environmental temperature affects the composition of body fluids. Acute exposure to *heat* causes the plasma volume to expand by an influx of interstitial fluid into the intravascular space, and by reduction of glomerular filtration. The plasma protein concentration may decrease by up to 10%. Salt and water may be lost by sweating, but generally there are no changes in the plasma sodium and chloride concentrations. Plasma potassium concentration may decrease by as much as 10% as potassium is taken up by the cells. If sweating is extensive, hemoconcentration rather than hemodilution may occur.

Table 3-18. INFLUENCE OF GENDER ON COMPOSITION OF SERUM AT DIFFERENT AGES[32]

Constituent	Male-Female Difference				
	< 29 y	30–39 y	40–49 y	50–59 y	60–69 y
Albumin (g/dL)	0.1	0.1	0	0	−0.1
Alkaline phosphatase (U/L)	14	12	−8	2	−1
Aspartate aminotransferase (U/L)	5	8	8	1	−1
Bilirubin (mg/dL)	0.1	0.1	0.1	0.1	0.1
Calcium (mg/dL)	0.1	0.1	0.1	−0.1	−0.2
Cholesterol (mg/dL)	−14	2	6	−16	−34
Creatinine (mg/dL)	0.2	0.2	0.2	0.2	0.1
Glucose (mg/dL)	5	3	6	0	6
Phosphorus (mg/dL)	0.1	0.1	0	−0.1	−0.2
Total protein (g/dL)	−0.1	−0.1	−0.1	−0.1	−0.2
Urea N (mg/dL)	3	3	3	2	0
Uric acid (mg/dL)	1.5	1.7	1.7	1.0	0.5

See footnote, Table 3-1.
Male values are higher than female except where indicated by a minus sign.

The *geographical location* in which an individual lives may affect the composition of body fluids. Thus, a statistically significant increase in the serum concentrations of cholesterol, triglycerides, and magnesium has been observed in people living in areas with hard water. Trace element concentrations are also affected by locale; e.g., in areas where there is much ore smelting, serum concentrations of the trace elements involved may be increased. Blood lead and carboxyhemoglobin concentrations are higher in areas where there is much heavier automobile traffic than in rural areas.

LONG-TERM CYCLICAL CHANGES

Seasonal Influences

Seasonal influences on the composition of body fluids are small in comparison to those related to changes in posture or effected by misuse of a tourniquet.[33] Probable factors are dietary changes as different foods come into season and altered physical activity as more or different forms of exercise become feasible. Evaluations of seasonal variation are difficult since they depend on the definition of a season and on the magnitude of temperature change from one season to another. Day-to-day variability in the composition of body fluids is greater in summer than winter. Nevertheless, biological variability is generally only little greater than analytical variability.

In summer in the northern hemisphere, the γ-globulin may increase by as much as 50%. Serum urate concentration appears to be ~5–7% higher in summer than winter. Serum triglyceride concentration is up to 10% higher in summer, whereas the serum cholesterol has been reported to be up to 50 mg/dL (1.3 mmol/L) higher in men and 30 mg/dL (0.7 mmol/L) higher in women in winter than in summer. Activities of serum enzymes arising from skeletal muscle are higher in summer than winter, presumably as a result of increased physical activity. The increase of serum lactate dehydrogenase may be as much as 20%.

Calcium metabolism is affected by an individual's exposure to sunlight. Dehydrocholecalciferol in the skin is converted by UV irradiation to cholecalciferol, which is further metabolized in the liver and kidney to 1,25-dihydroxycholecalciferol. The calcium concentration in serum is increased, as is its elimination in urine.[21] Seasonal changes appear to affect other endocrine systems as well. A lower plasma glucose concentration may be observed and glucose tolerance is improved during summer. Serum concentrations of thyroid hormones are unaffected but the urinary excretion of triiodothyronine is increased in summer. The excretion of metabolites of adrenal hormones is generally greater in summer than winter. This could be attributable to greater physical activity.

Exposure to sunshine for a weekend during summer may cause enough photodegradation of bilirubin to reduce the serum concentration by 20%. Some seasonal effects on the composition of body fluids are listed in Table 3-19.

Influence of Menstrual Cycle

The plasma concentrations of many female sex hormones, as well as other hormones, are affected by the menstrual cycle.[61] (See also Chapter 9.) Thus, the plasma corticosterone concentration is as much as 50% higher in the luteal phase than in the follicular phase. The urinary

Table 3-19. SEASONAL EFFECTS ON COMPOSITION OF SERUM[33]

Constituent	Concentration Highest	Concentration Lowest	% Difference Between High and Low
Alanine aminotransferase	Winter	Spring, summer	5.0
Albumin	Fall	Summer	1.2
Aspartate aminotransferase	Spring	Fall	11.7
Calcium	Fall	Winter	1.0
Creatinine	Summer	Winter	4.7
Glucose	Fall	Spring	1.5
Lactate dehydrogenase	Summer	Winter	1.8
Triglycerides	Spring	Fall	5.4
Urea N	Fall	Spring, summer	3.2
Uric acid	Summer	Winter	4.3

See footnote, Table 3-1.

excretion of 17-hydroxycorticosteroids reaches a peak at midcycle. Plasma androstenedione concentration and plasma aldosterone concentration increase from the follicular phase to the luteal phase of the menstrual cycle. On the preovulatory day the aldosterone concentration may actually be twice that of the early part of the follicular phase. The change in renin activity is almost as great. These changes are usually more marked in women who retain fluid prior to menstruation. Urinary catecholamine excretion increases at midcycle and remains high throughout the luteal phase.

The plasma cholesterol concentration is least at ovulation, corresponding to the time of maximum estrogen secretion, and rises immediately prior to menstruation, when it remains high for one week. The cyclical variation in cholesterol is not observed with anovulatory cycles. The total protein and albumin concentrations decrease at the time of ovulation but then increase again. The plasma fibrinogen concentration decreases markedly at menstruation. The serum calcium correlates with changes in albumin. Serum phosphate is reduced at the time of menstruation although the creatinine and urate concentrations are highest at this time and are lowest toward the end of the intermenstrual period.

The plasma iron concentration may be very low with the onset of menstruation; the magnesium concentration is least at this point of the cycle. Plasma sodium and chloride concentrations increase up to the onset of menstruation but may fall by 2 mmol/L with the postmenstrual diuresis.

Plasma ascorbic acid is low at the time of ovulation, whereas the concentration of folate is unaffected by the menstrual cycle. Serum creatine kinase activity may be slightly reduced at the time of ovulation, but the activities of other enzymes appear to be unaffected by the menstrual cycle.

BODY HABITUS

The serum concentrations of cholesterol, triglycerides, and β-lipoproteins are positively correlated with obesity.[48] The serum urate concentration is also correlated with body weight, especially in individuals weighing over 80 kg.[37] Serum lactate dehydrogenase activity and glucose concentration increase in both sexes with increasing body weight.[48] In men serum aspartate aminotransferase, creatinine, and total protein increase with increasing body weight, as does the blood hemoglobin concentration. In women serum calcium increases with increasing body weight. In both sexes serum phosphate decreases with increased body mass.

Cortisol production is increased in obese individuals. However, increased metabolism maintains the serum concentration unchanged so that urinary excretion of 17-hydroxycorticosteroids and 17-ketosteroids is increased. Since growth hormone concentration is reduced in obese individuals, it responds poorly to the normal challenges. Plasma insulin concentration is increased but glucose tolerance is impaired in the obese. (See Chapter 6.) Although the serum thyroxine concentration is unaffected by obesity, the serum triiodothyronine correlates significantly with body weight and increases further with overeating. In obese men, the serum testosterone concentration is reduced.

The fasting concentrations of pyruvate, lactate, citrate, and unesterified fatty acids are higher in obese individuals than in those of normal body weight. Serum iron and transferrin concentrations are low.

Gastric juice volume and acid output are increased in obese individuals.

INFLUENCE OF DIET

An individual's typical diet has considerable influence on the composition of his plasma. Studies with synthetic diets have shown that day-to-day changes in the amount of protein are reflected within a few days in the composition of the plasma and in the excretion of end products of protein metabolism.

Four days after the change from a normal diet to a high-protein diet, a doubling of the plasma urea concentration occurs with an increase in its urinary excretion.[7] Serum cholesterol and phosphate concentrations are also increased. A high-protein intake increases both serum and urinary urea and urate. A high-fat diet, in contrast, depletes the nitrogen pool because of the requirement for excretion of ammonium ions to maintain acid-base homeostasis. A high-fat diet also reduces serum urate. Reduction of fat intake reduces serum lactate dehydrogenase activity. The ingestion of very different amounts of cholesterol has little effect on the serum cholesterol concentration; an increase in intake of 50% may only affect the serum concentration by 5–10 mg/dL (0.13–0.26 mmol/L).[24] Ingestion of unsaturated fat does have a hypocholesterolemic effect.

If dietary carbohydrate consists mainly of starch or sucrose rather than other sugars, the serum activities of alkaline phosphatase and lactate dehydrogenase are increased. Asparate aminotransferase activity is influenced by the type of sugar ingested. The plasma triglyceride concentration is reduced when sucrose intake is reduced. Flatter glucose tolerance curves are observed with a bread diet than when a high-sucrose diet is ingested. A high-carbohydrate diet decreases the serum concentrations of very-low-density lipoprotein, triglyceride, cholesterol, and protein.

Vegetarianism

In longstanding vegetarians, the concentrations of low-density and very-low-density lipoproteins are low. The total lipid and phospholipid concentrations are reduced, and the concentrations of cholesterol and triglyceride may be only two-thirds of those in people on a mixed diet. The effects are less marked in individuals who have been on a vegetarian diet for only a short time. The lipid concentrations are also less in individuals who eat only a vegetable diet than in those who consume eggs and milk as well. When individuals previously on a mixed diet begin a vegetarian diet, their serum albumin concentration may fall by 10% and their urea concentration by 50%. However, there is little difference in the concentration of protein or of activities of enzymes in the serum of longstanding vegetarians and individuals on a mixed diet.

Urinary pH is usually higher in vegetarians than in meat-eaters because of reduced intake of precursors of acid metabolites. The plasma vitamin B_{12} may be reduced in vegetarians to a concentration approaching deficiency. Vegetarians tend to have a higher serum bilirubin concentration than do meat-eaters. An explanation for the low vitamin B_{12} and high bilirubin has still to be established. Differences in the composition of serum of vegetarians and nonvegetarians are listed in Table 3-20.

Malnutrition

In malnutrition, total serum protein, albumin, and β-globulin concentrations are reduced. The increased concentration of γ-globulin does not fully compensate for the decrease in other proteins. The concentrations of complement C3, retinol-binding globulin, transferrin, and prealbumin decrease rapidly with the onset of malnutrition[41] and are measured to define the severity of the condition. (See Chapter 8A.) The plasma concentrations of lipoproteins are reduced and serum cholesterol and triglycerides may be only 50% of the concentrations in healthy individuals. In spite of severe malnutrition, glucose concentration is maintained close to that in healthy individuals. However, the concentrations of serum urea and creatinine are greatly reduced as a result of decreased skeletal mass, and creatinine clearance is also decreased.

Plasma cortisol concentration is increased due to decreased metabolic clearance. The plasma concentrations of total triiodothyronine, thyroxine, and TSH are considerably reduced with the

Table 3-20. COMPARISON OF BLOOD CONSTITUENTS BETWEEN VEGETARIANS AND NONVEGETARIANS[19]

Constituent	Vegetarians	Nonvegetarians
S-Albumin (g/dL)	4.2	4.2
P-Calcium (mg/dL)	9.4	9.7
P-Cholesterol (mg/dL)	213	252
P-HDL-cholesterol (mg/dL)	66	66
B-Glucose (mg/dL)	90	101
B-Hemoglobin (g/dL)	13.9	14.3
P-Triglycerides (mg/dL)	106	124
B-Urea N (mg/dL)	14	16
P-Uric acid (mg/dL)	5.3	5.8

See footnote, Table 3-1.

thyroxine concentration being most affected. This is partly due to reduced concentrations of thyroxine-binding globulin and prealbumin.

Erythrocyte and plasma folate concentrations are reduced in protein-calorie malnutrition, but the serum vitamin B_{12} concentration is unaffected or may even be slightly increased.[25] The plasma concentrations of vitamins A and E are much reduced. Although the blood hemoglobin concentration is reduced, the serum iron concentration is initially little affected by malnutrition.

The activity of most of the commonly measured enzymes is reduced but increases with restoration of good nutrition.

Fasting and Starvation

Withdrawal of most or all caloric intake has been used to treat certain cases of obesity. Such withdrawal provokes many metabolic responses. The body attempts to conserve protein at the expense of other sources of energy such as fat. The blood glucose concentration decreases by as much as 18 mg/dL (1 mmol/L) within the first three days of the start of a fast in spite of the body's attempts to maintain glucose production.[36] Insulin secretion is markedly reduced, while glucagon secretion may double in an attempt to maintain normal glucose concentration. Lipolysis and hepatic ketogenesis are stimulated. Ketoacids and fatty acids become the principal sources of energy for muscle. In addition, the concentrations of ketone bodies, fatty acids, and glycerol in serum rise considerably. Serum triglycerides increase by 20% after 48 h of fasting but decline thereafter; the cholesterol concentration also decreases. Amino acids are released from skeletal muscle and the plasma concentration of the branched-chain amino acids may increase by as much as 100% with 1 d of fasting.

The breakdown of fat leads to a transient increase in body water. Normally, however, an osmotic diuresis soon reduces the blood volume even though glomerular filtration may be impaired, as demonstrated by increased serum creatinine concentration and reduced creatinine clearance. Hepatic blood supply may also be reduced. Bromsulfophthalein (BSP) retention is increased and the serum bilirubin rises; the unconjugated bilirubin more than doubles within 48 h.[1] Increased serum activities of aspartate and alanine aminotransferase and of lactate dehydrogenase are observed. The enzyme changes may be linked more to focal necrosis of the liver than to general circulatory impairment.

In spite of the catabolism of tissue induced by starvation, the serum protein concentration is little affected initially; ultimately, a reduction occurs. However, from the beginning, the catabolism of nucleoproteins causes an increased serum urate. Rise in serum urate is exacerbated by the reduced glomerular filtration rate and the competition for excretion from lactate and ketoacids. A metabolic acidosis is common with associated reduction of the blood pH and pCO_2; often the blood pO_2 is also reduced.

With the onset of starvation, aldosterone secretion increases with the results of increased urinary excretion and decreased plasma concentration of potassium. Magnesium, calcium, and phosphate are affected similarly, although the urinary excretion of phosphate gradually declines.

Plasma growth hormone concentration may rise by as much as 15 times at the start of a fast but may return to normal after 3 d. Free and total triiodothyronine decrease by up to 50% within 3 d of the start of a fast. Free thyroxine concentration is also affected, but to a lesser extent; total thyroxine is little changed. Urinary free cortisol is decreased by fasting and the plasma cortisol concentration (free and total) shows a slight increase along with loss of the normal diurnal variation.

Early in refeeding, sodium retention occurs as a result of decreased sodium and chloride excretion in the urine.[44] The reduction in potassium excretion takes longer. These events are associated with an even greater secretion of aldosterone than occurs during the period of fasting. The abnormal concentrations of most constituents rapidly revert to normal. Nitrogen balance soon becomes positive, especially if the nonprotein calories are mainly from carbohydrate.

References

1. Barrett, J. V. D.: Hyperbilirubinemia of fasting. JAMA, *217*:1349-1353, 1971.
2. Beisel, W. R., Goldman, R. F., and Joy, R. J. T.: Metabolic balance studies during induced hyperthermia in man. J. Appl. Physiol., *24*:1-10, 1968.
3. Bellet, S., Roman, L., DeCastro, O., et al.: Effect of coffee ingestion on catecholamine release. Metabolism, *18*:288-291, 1969.

4. Benedek, T. G., and Sunder, J. H.: Comparison of serum lipid and uric acid content in white and Negro men. Am. J. Med. Sci., *260*:331-340, 1970.

5. Blumenfeld, T. A., Turi, G. K., and Blanc, W. A.: Recommended site and depth of newborn heel skin punctures based on anatomic measurements and histopathology. Lancet, *1*:230-233, 1979.

6. Bodenheimer, S., Winter, J. S. D., and Faiman, C.: Diurnal rhythms of serum gonadotropins, testosterone, estradiol and cortisol in blind men. J. Clin. Endocr. Metab., *37*:472-475, 1973.

7. Brohult, J.: Effects of high protein and low protein diets on ornithine carbamoyl transferase activity in human serum (S-OCT). Acta Med. Scand., *185*:357-362, 1969.

8. Buckley, C. E., and Dorsey, F. C.: Serum immunoglobulin levels throughout the life-span of healthy men. Ann. Intern. Med., *75*:673-682, 1971.

9. Caraway, W. T.: Chemical and diagnostic specificity of laboratory tests. Am. J. Clin. Pathol., *37*:445-464, 1962.

10. Carruthers, M., Arguelles, A. E., and Mosovich, A.: Man in transit: Biochemical and physiological changes during intercontinental flights. Lancet, *1*:977-981, 1976.

11. Cryer, P. E., Haymond, M. W., Santiago, J. V., et al.: Norepinephrine and epinephrine release and adrenergic mediation of smoking-associated hemodynamic and metabolic events. N. Engl. J. Med., *295*:573-577, 1976.

12. Cuthbertson, D. P., and Tilstone, W. J.: Metabolism during the post-injury period. Adv. Clin. Chem., *12*:1-55, 1969.

13. Davies, C. T. M., and Few, J. D.: Effects of exercise on adrenocortical function. J. Appl. Physiol., *35*:887-891, 1973.

14. Deitrick, J. E., Whedon, G. D., and Shorr, E.: Effects of immobilization upon various metabolic and physiologic functions of normal men. Am. J. Med., *4*:3-36, 1948.

15. Eaton, J. W., Brewer, G. J., and Grover, R. F.: Role of Red cell 2,3-diphosphoglycerate in the adaptation of man to altitude. J. Lab. Clin. Med., *73*:603-609, 1969.

16. Felding, P., Hyltoft Petersen, P., and Horder, M.: The stability of plasma and serum constituents during simulated transport. Scand. J. Clin. Lab. Invest., *41*:35-40, 1981.

17. Felding, P., Tryding, N., Hyltoft Petersen, P., et al.: Effects of posture on concentrations of blood constituents in healthy adults: Practical application of blood specimen collection procedures recommended by the Scandinavian Committee on Reference Values. Scand. J. Clin. Lab. Invest., *40*:615-621, 1980.

18. Freer, D. E., and Statland, B. E.: The effects of ethanol (0.75 g/kg body weight) on the activities of selected enzymes in sera of healthy adults: 1. Intermediate-term effects. Clin. Chem., *23*:830-834, 1977.

19. Gear, J. S., Mann, J. I., Thorogood, M., et al.: Biochemical and haematological variables in vegetarians. Br. Med. J., *280*:1415, 1980.

20. Goldberg, D. M., and Watts, C.: Serum enzyme changes as evidence of liver reaction to oral alcohol. Gastroenterology, *49*:256-261, 1965.

21. Green, A. G.: Circannual excretory patterns in man. J. Clin. Pathol., *27*:932, 1974.

22. Haralambie, G.: Serum glycoproteins and physical exercise. Clin. Chim. Acta, *26*:287-291, 1969.

23. Jensen, J., Blankenhorn, D. H., Chin, H. P., et al.: Serum lipids and serum uric acid in human twins. J. Lipid Res., *6*:193-204, 1965.

24. Keys, A., Anderson, J. T., Mickelsen, O., et al.: Diet and serum cholesterol in men. Lack of effect of dietary cholesterol. J. Nutr., *59*:39-56, 1956.

25. Khalil, M., Tanios, A., Moghazy, M., et al.: Serum and red cell folates, and serum vitamin B_{12} in protein calorie malnutrition. Arch. Dis. Child., *48*:366-369, 1973.

26. Krafft, J., Fink, R., and Rosalki, S. B.: Serum enzymes and isoenzymes after surgery. Ann. Clin. Biochem., *14*:294-296, 1977.

27. Kupke, I. R., Kather, B., and Zeugner, S.: On the composition of capillary and venous blood serum. Clin. Chim. Acta, *112*:177-185, 1981.

28. Ladenson, J. H.: Nonanalytical sources of variation in clinical chemistry results. *In*: Gradwohl's Clinical Laboratory Methods and Diagnosis. A. C. Sonnenwirth and L. Jarett, Eds. 8th ed. St. Louis, Mosby, 1980, pp. 149-192.

29. Ladenson, J. H., Tsai, L-M. B., Michael, J. M., et al.: Serum versus herparinized plasma for eighteen common chemistry tests. Am. J. Clin. Pathol., *62*:545-552, 1974.

30. Laessig, R. H., Hassemer, D. J., Westgard J. O., et al.: Assessment of the serum separator tube as an intermediate storage device within the laboratory. Am. J. Clin. Pathol., *66*:653-657, 1976.

31. Laessig, R. H., Indriksons, A. A., Hassemer, D. J., et al.: Changes in serum chemical values as a result of prolonged contact with the clot. Am. J. Clin. Pathol., *66*:598-604, 1976.

32. Leonard, P. J.: The effect of age and sex on biochemical parameters in blood of healthy human subjects. *In*: Reference Values in Human Chemistry. G. Siest, Ed. Basel, Karger, 1973, pp. 134-140.

33. Letellier, G., and Desjarlais, F.: Study of seasonal variations for eighteen biochemical parameters over a four-year period. Clin. Biochem., *15*:206-211, 1982.

34. Lum, G., and Gambino, S. R.: A comparison of serum versus heparinized plasma for routine chemistry tests. Am. J. Clin. Pathol., *61*:108-113, 1974.

35. McNair, P., Nielsen, S. L., Christiansen, C., et al.: Gross errors made by routine blood sampling from two sites using a tourniquet applied at different positions. Clin. Chim. Acta, *98*:113-118, 1979.

36. Merimee, T. J., and Fineberg, E. S.: Homeostasis during fasting. II. Hormone substrate differences between men and women. J. Clin. Endocr. Metab., *37*:698-702, 1973.

37. Munan, L., Kelly, A., PetitClerc, C., et al.: Association with body weight of selected chemical constituents in blood. Clin. Chem., *24*:772-777, 1978.

38. National Committee for Clinical Laboratory Standards: Standard for Evacuated Tubes for Blood Specimen Collection. 2nd ed. NCCLS Standard ASH-1. Villanova, Pa., National Committee for Clinical Laboratory Standards, 1980.

39. National Committee for Clinical Laboratory Standards: Standard Procedure for the Collection of Diagnostic Blood Specimens by Venipuncture. NCCLS Standard ASH-3. Villanova, Pa., National Committee for Clinical Laboratory Standards, 1977.

40. National Committee for Clinical Laboratory Standards: Approved Standard Procedures for the Collection of Diagnostic Blood Specimens by Skin Puncture. NCCLS Standard H4-A. Villanova, Pa., National Committee for Clinical Laboratory Standards, 1982.

41. Olusi, S. O., McFarlane, H., Osunkoya, B. O., et al.: Specific protein assays in protein calorie malnutrition. Clin. Chim. Acta, 62:107-116, 1975.

42. Ong, Y. Y., Boykin, S. F., and Barnett, R. N.: You can draw blood from the "IV arm" below the intravenous needle if you put a tourniquet in between. Am. J. Clin. Pathol., 72:101-102, 1979.

43. Pitts, R. F.: Physiology of the Kidney and Body Fluids. 3rd ed. Chicago, Year Book Medical Publishers, 1974.

44. Rapoport, A., From, G. L. A., and Husdan, H.: Metabolic studies in prolonged fasting. I. Inorganic metabolism and kidney function. Metabolism, 14:31-46, 1965.

45. Rayfield, E. J., Curnow, R. T., George, D. T., et al.: Impaired carbohydrate metabolism during a mild viral illness. N. Engl. J. Med., 289:618-621, 1973.

46. Rommel, K., Koch, C-D., and Spilker, D.: Einfluss der Materialgewinnung auf klinisch-chemische Parameter in Blut, Plasma und Serum bei Patienten mit stabilem und zentralisiertem Kreislauf. J. Clin. Chem. Clin. Biochem., 16:373-380, 1978.

47. Sassard, J., Vincent, M., Annat, G., et al.: A kinetic study of plasma renin and aldosterone during changes of posture in man. J. Clin. Endocr. Metab., 42:20-27, 1976.

48. Siest, G., Henny, J., and Schiele, F., Eds.: Interpretation des examens de laboratoire. Basel, Karger, 1981.

49. Statland, B. E., Bokelund, H., and Winkel, P.: Factors contributing to intraindividual variation of serum constituents. 4. Effects of posture and tourniquet application on variation of serum constituents in healthy subjects. Clin. Chem., 20:1513-1519, 1974.

50. Statland, B. E., Winkel, P., and Bokelund, H.: Factors contributing to intraindividual variation of serum constituents: 1. Within-day variation of serum constituents in healthy subjects. Clin. Chem., 19:1374-1379, 1973.

51. Statland, B. E., Winkel, P., and Bokelund, H.: Factors contributing to variation of serum constituents in healthy subjects. In: Organisation des Laboratoires. Biologie Perspective. G. Siest, Ed. Paris, L'Expansion Scientifique Francaise, 1975, pp. 717-750.

52. Steige, H., and Jones, J. D.: Evaluation of pneumatic tube system for delivery of blood specimens. Clin. Chem., 17:1160-1164, 1971.

53. Steinmetz, J., Panek, E., Sourieau, F., et al.: Influence of food intake on biological parameters. In: Reference Values in Human Chemistry. G. Siest, Ed. Basel, Karger, 1973, pp. 193-200.

54. Thomson, W. H. S., Sweetin, J. C., and Hamilton, I. J. D.: ATP and muscle enzyme efflux after physical exertion. Clin. Chim. Acta, 59:241-245, 1975.

55. Weitzman, E. D.: Circadian rhythms and episodic hormone secretion. Ann. Rev. Med., 27:225-243, 1976.

56. Wesson, L. G., Jr.: Electrolyte excretion in relation to diurnal cycles of renal function. Medicine, 43:547-592, 1964.

57. Wilding, P., Rollason, J. G., and Robinson, D.: Pattern of change for various biochemical constituents detected in well-population screening. Clin. Chim. Acta, 41:375-387, 1972.

58. Winkel, P., Statland, B. E., and Bokelund, H.: The effects of time of venipuncture on variation of serum constituents. Am. J. Clin. Pathol., 64:433-447, 1975.

59. Winsten, S., and Gordesky, S. E.: Transportation of specimens. In: Selected Methods of Clinical Chemistry, Vol. 9. W. R. Faulkner, Ed. Washington, D.C., AACC, 1982, pp. 11-15.

60. Wollner, G. C.: Handling, storing and transporting diagnostic specimens. Lab. Med., 11:87-91, 1980.

61. Young, D. S.: Biological variability. In: Chemical Diagnosis of Disease. S. S. Brown, F. L. Mitchell, and D. S. Young, Eds. Amsterdam, Elsevier, 1979, pp. 1-13.

62. Young, D. S.: The effects of frequently prescribed drugs on common laboratory procedures. In: Practice of Medicine. J. A. Spittell, Jr., Ed. Philadelphia, J. B. Lippincott Co., 1984. Vol. 1, chapter 26, pp. 1-21.

63. Young, D. S., Pestaner, L. C., and Gibberman, V.: Effects of drugs on clinical laboratory tests. Clin. Chem., 21:1D-432D, 1975.

Additional Reading

AMA Division of Drugs: AMA Drug Evaluations. 5th ed. Chicago, AMA, 1983.

Brown, S. S., Mitchell, F. L., and Young, D. S., Eds.: Chemical Diagnosis of Disease. Amsterdam, Elsevier, 1979.

Calam, R. R.: Reviewing the importance of specimen collection. J. Am. Med. Technol., 39:297-302, 1977.

Gräsbeck, R., and Alström, T., Eds.: Reference Values in Laboratory Medicine. New York, Wiley, 1981.

Henry, J. B., Ed.: Clinical Diagnosis and Management by Laboratory Methods. 17th ed. Philadelphia, W. B. Saunders Co., 1984.

Johnson, T. R., Moore, W. M., and Jeffries, J. E., Eds.: Children are Different. 2nd ed. Columbus, Ohio, Ross Laboratories, 1978.

Meites, S., Ed.: Pediatric Clinical Chemistry. 2nd ed. Washington, D.C., AACC, 1981.

Slockbower, J. M., and Blumenfeld, T. A., Eds.: Collection and Handling of Laboratory Specimens. Philadelphia, J. B. Lippincott Co., 1983.

Statland, B. E., and Winkel, P.: Effects of preanalytical factors on the intraindividual variation of analytes in the blood of healthy subjects: Consideration of preparation of the subject and time of venipuncture. CRC Crit. Rev. Clin. Lab. Sci., 8:105-144, 1977.

Tietz, N. W., Ed.: Clinical Guide to Laboratory Tests. Philadelphia, W. B. Saunders Co., 1983.

Wilding, P., Zilva, J. F., and Wilde, C. E.: Transport of specimens for clinical chemistry analysis. Ann. Clin. Biochem., 14:301-306, 1977.

CHAPTER 4

AMINO ACIDS AND PROTEINS

by Lawrence M. Silverman, Ph.D., Robert H. Christenson, Ph.D., and Gregor H. Grant, B.M., F.R.C. Path.

BASIC CHEMISTRY OF AMINO ACIDS AND PROTEINS

Amino acids are organic compounds containing both an amino group and a carboxyl group. Those occurring in proteins are called α-amino acids and have the empirical formula $RCH(NH_2)COOH$. Beta-amino acids and γ-amino acids also occur in nature but not as components of proteins. As shown below, the core of an α-amino acid is the carbon atom next to the carboxylic acid group, the α-carbon. With the exception of glycine, all α-amino acids are asymmetric, i.e., four different groups are bonded to the α-carbon—a hydrogen atom, a carboxyl group, an amino group, and a distinctive R group. These four groups can be bonded in two different ways, which are mirror images of one another, and rotate polarized light in opposite senses. They are called D and L optical isomers; in natural proteins of higher organisms, only the L isomer is found. In the illustration below, **A** is the conventional way to depict the L configuration, but **B** is the more convenient form for writing and printing.

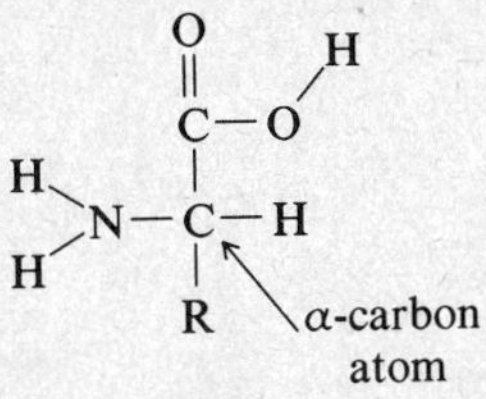

Table 4-1 lists the L-α-amino acids of biological importance. Interestingly, only 20 amino acids are used to build the enormous number of biologically active peptides and proteins that exist in nature. What makes peptides and proteins so versatile in terms of structure and function are the characteristic acid-base properties of individual amino acids, as well as the variety of possible R group interactions.

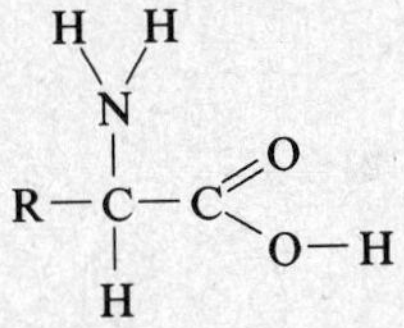

Acid-base properties of amino acids depend on the amino and carboxyl groups attached to the α-carbon and on the basic or acidic functional groups which may be present in an R group. In the illustration above, the amino acid is shown in an uncharged form. In aqueous media, however, very little amino acid exists in an un-ionized form. The acidic carboxyl group is a proton donor and the basic amino group is a proton acceptor. In the physiological pH range of 7.35–7.45, the carboxyl group is dissociated and the amino group is protonated to give the following structure:

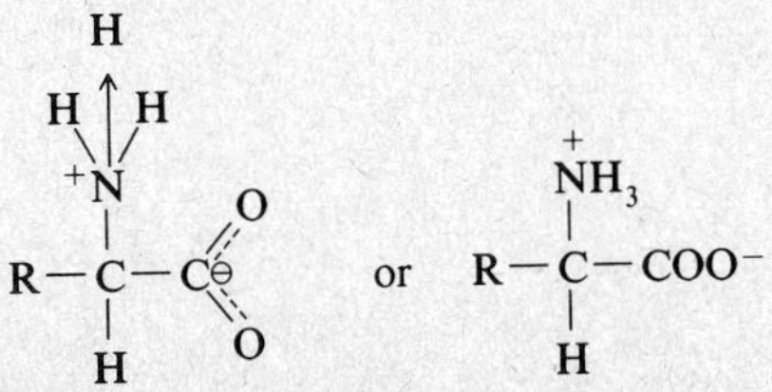

Text continued on page 524

Table 4-1. AMINO ACIDS

Name and Abbreviation	M.W.	Structure at pH 6-7	Comments
		I. AMINO ACIDS FOUND IN MOST PROTEINS	
		Hydrophobic Amino Acids; Nonpolar R Groups	

Alanine
Ala — 89 — Substrate for ALT (alanine transaminase); least hydrophobic of the group

Leucine
Leu — 131 — Branched-chain R group; essential; ketogenic; metabolism is faulty in maple syrup urine disease

Isoleucine
Ile — 131 — Essential; partly ketogenic; see Leu

Valine
Val — 117 — Essential; partly ketogenic; see Leu

Proline
Pro — 115 — Important constituent of connective tissue proteins (e.g., collagen and elastin); some hydroxylated to Hyp during collagen synthesis; destabilizes α-helical and β-structures; contains an α-imino group

Methionine
Met — 149 — Essential; important in transfer of methyl groups; provides sulfur for other sulfur-containing compounds

Phenylalanine
Phe — 165 — Essential; elevated levels in phenylketonuria

Tryptophan
Trp — 204 — Essential; metabolites found in carcinoid disease; contains indole ring system; precursor of serotonin and melatonin

Table 4-1. AMINO ACIDS (*Continued*)

Name and Abbreviation	M.W.	Structure at pH 6-7	Comments
		Hydrophilic Amino Acids; Uncharged Polar Groups	
Glycine Gly	75		Simplest amino acid; optically inactive; placed in this group because its R group (single H) is unable to affect polarity of the rest of the molecule; used in biosynthesis of purines and porphyrins; used in vitro as a buffer
Serine Ser	105		Constituent in active center of many enzymes; hydroxyl group can be phosphorylated
Threonine Thr	119		Essential
Cysteine Cys	121		Sulfhydryl group functional in the activity of many enzymes; is responsible for disulfide bridges in peptides and proteins; cystine is dicysteine, Cys—S—S—Cys; homocysteine has one carbon more than cysteine and forms homocystine (dihomocysteine)
Tyrosine Tyr	181		Usually nonessential; intermediate in synthesis of catecholamines, thyroxine, and melanin; functional phenolic group; reacts with Folin's reagent in quantitative protein assay
Glutamine Gln	146		Storage form of ammonia in tissue; supplies the amido nitrogen used in purine and pyrimidine biosynthesis
Asparagine Asn	132		Storage form of ammonia in tissues
Hydroxyproline Hyp	131		Constituent of collagen—the only human protein to contain appreciable amounts; urinary output is used as an indicator of bone matrix metabolism; contains an α-imino group

Table continued on following page

Table 4-1. AMINO ACIDS (*Continued*)

Name and Abbreviation	M.W.	Structure at pH 6-7	Comments
Dicarboxylic Amino Acids; Acidic R Groups			
Aspartic acid Asp	133		Co-substrate with Glu for AST; used in pyrimidine biosynthesis
Glutamic acid Glu	147		Co-substrate with Ala for ALT and with Asp for AST
Basic Amino Acids; Basic R Groups			
Lysine Lys	147		Essential; terminal NH_2 called ϵ-amino
Arginine Arg	174		Involved in urea synthesis; the basic group is a guanidinium group
Histidine His	155		The imidazole group of histidine is the most important buffer group in the physiological pH range
II. MISCELLANEOUS AMINO ACIDS			
Thyroxine T_4	777		Thyroid hormone
Triiodo-thyronine T_3	650		Thyroid hormone; more active than T_4
β-Alanine β-Ala	89		Constituent of the vitamin pantothenic acid

Table 4-1. AMINO ACIDS (*Continued*)

Name and Abbreviation	M.W.	Structure at pH 6-7	Comments
Dihydroxy-phenyl-alanine Dopa*	197		Intermediate in catecholamine synthesis
γ-Amino-butyric acid GABA*	103		Metabolite of Glu; a neurotransmitter
Ornithine Orn*	132		Intermediate in urea synthesis
Citrulline Citr*	175		Intermediate in urea synthesis
Phosphoserine	185		In casein and other phosphoproteins
Pyrrolidine carboxylic acid	129		Cyclized form of Glu, rare; used to terminate peptide chains, as at *N*-terminal end of L-chains in γ-globulins
Taurine	125		Forms conjugates with bile acids; inhibits nerve impulse transmission
β-Aminoiso-butyric acid β-AIB*	103		Present in urine; a metabolite of pyrimidines

*Abbreviation useful but not official.

This kind of ionized molecule with coexistent negative and positive charges is called a dipolar ion or ampholyte (in older terminology, a zwitterion). At low pH, an amino acid is in its cationic form, i.e., both the amino and carboxyl groups are protonated as $-N^+H_3$ and $-COOH$. As the pH rises, the carboxyl group loses its proton, and the ampholyte form appears at about pH 6. With further increases in pH, the amino $-N^+H_3$ is also deprotonated and the anionic form of the molecule is formed. The process can be shown for a monoamino and monocarboxylic amino acid as follows:

| Cation at pH < pI | Ampholyte at pH = pI | Anion at pH > pI |

The dissociation constants, K_1 and K_2, are usually expressed logarithmically as pK_1 and pK_2, where $pK = -\log K$, in a manner analogous to the notation for pH. A pK is the pH at which equal quantities of the protonated (associated) and unprotonated (dissociated) forms are present. The isoelectric point, pI, is that pH at which all molecules exist in the ampholyte form, i.e., with net charge = 0. At pH > pI, the amino acid is negatively charged, and at pH < pI, it is positively charged. The isoelectric point of a neutral amino acid can be calculated from the pK's of its amino and carboxyl groups, i.e., $pI = 1/2(pK_1 + pK_2)$. The concept of an ampholyte and its dissociation characteristics as described here for amino acids can be applied to proteins as well; in physiologic ranges of pH, a majority of proteins are negatively charged, but manipulation of the pH as well as the ionic strength of the medium may change their net charge to positive or zero.

Glycine, with a pK_1 of 2.34 and pK_2 of 9.60, exemplifies the acid-base behavior of that group of amino acids whose R groups have no ionizable substituents. Glycine is sometimes used in vitro to prepare buffers at pH 1.0–3.8 or 8.1–11.0. Ionization constants in amino acids are given in Table 4-2.

The R groups of individual amino acids are responsible for their special properties. Note that Table 4-1 is organized in terms of the differences in R groups. Some R groups are nonpolar and therefore hydrophobic; others, although uncharged, are polar and hydrophilic. Still others

Table 4-2. IONIZATION CONSTANTS OF IONIZABLE GROUPS IN FREE AMINO ACIDS AND IN PROTEINS*

Ionizing Group	Range of pK Values	
	Free Amino Acids	Proteins
Principal carboxyl ($-COOH$) = pK_1	1.7 – 2.6	3.0 – 3.2
α-Amino ($-N^+H_3$) = pK_2	9.0 – 10.8	7.6 – 8.4
Second carboxyls of Glu and Asp ($-COOH$)	3.8 – 4.3	3.0 – 4.5
Imidazole nitrogen of His $\left(\underset{\diagdown}{\overset{\diagup}{>}}\overset{+}{N}-H \right)$	6.0	5.6 – 7.0
Sulfhydryl of cysteine ($-SH$)	8.3	9.1 – 10.8
Phenolic hydroxyl of Tyr ($-OH$)	10.1	9.2 – 9.8
ε-Amino of Lys ($-N^+H_3$)	10.5	9.4 – 10.6
Guanidinium group of Arg $\left(-N-C{\overset{NH_2}{\underset{N^+H_2}{\diagup}}} \right)$	12.5	11.5 – 12.6

* Given are the pK values for proton-donating charged groups when present in free amino acids and when present in peptide-linked amino acid residues in proteins. The pK value for the primary carboxyl varies from 1.71 for Cys to 2.63 for Thr. Similarly, the pK for the α-ammonium group varies from 8.95 for Lys to 10.78 for Cys. In protein chains the proximity of other amino acid residues and charged groups may modify the pK for any given ionizable group. The amino acid symbols are those listed in Table 4-1.

can become charged, either negatively (the acidic amino acids) or positively (the basic amino acids). R groups may be linear as in valine or cyclic as in proline, small as in glycine or bulky as in tryptophan. Electron density may be low as in aliphatic chains or high as in aromatic rings. This variation in R group structure makes possible several kinds of interaction between them. As will be apparent later, this is of great importance in determining protein structure.

Some of the amino acids have R groups that contain charged or ionizable substituents. These substituents have their own pK's, which are also shown in Table 4-2. At pH 7, the acidic amino acids, Glu and Asp, have their second carboxyl group fully ionized and are therefore negatively charged in the physiologic pH range. Except for histidine, most basic amino acids are positively charged at physiologic pH's; < 10% of histidine is positively charged in this range. Charges that reside or develop on R groups are responsible for electrostatic bonding that may occur in proteins or between a protein and a ligand. Amino acids, therefore, are distinguishable by their general and individual characteristics. Their differing solubilities and acid-base properties provide the basis for their separation by partition chromatography, ion-exchange chromatography, or electrophoresis. Differences in the chemical nature of their R groups permit, in some cases, their identification or quantitation by color reactions. The aromatic amino acids—tyrosine and tryptophan—absorb ultraviolet light at 280 nm; all amino acids absorb UV radiation at 220 nm. The absorption at 280 nm can be used to estimate the concentration of an individual protein in solution, such as fibrinogen. Aside from the distinctive properties used for analyses, the general and special properties of amino acids are critical to the role they play in determining the structure and function of the proteins into which they are incorporated.

Polymerization of Amino Acids

Amino acids can be linked to one another through a *peptide bond* formed between the α-amino group of one and the α-carboxyl group of another.

| First amino acid | Second amino acid | Dipeptide |

The peptide bond is described by the structure in the shaded area. It may be formed between the —NH_2 of the first and the —COOH of the second or between the —COOH of the first and the —NH_2 of the second. For example, glycine and alanine can react to form two different dipeptides, either glycyl-alanine or alanyl-glycine:

| Glycine | Alanine | Glycyl-alanine |

| Alanine | Glycine | Alanyl-glycine |

In glycyl-alanine, the *carboxyl* group of alanine is not engaged in the peptide bond and is therefore terminal to the molecule. Alanine is called the *C*-terminal residue of the peptide; glycine is the *N*-terminal residue because its *amino* group is free. In alanyl-glycine, the designations are reversed.

The C- and N-terminal designations shown here for dipeptides apply also to polypeptides and proteins which have amino acids linked by peptide bonds to form chains,

$$\ldots -\overset{\overset{H}{|}}{\underset{\underset{R_1}{|}}{C}}-\overset{\overset{O}{\|}}{C}-\overset{}{\underset{\underset{H}{|}}{N}}-\overset{\overset{H}{|}}{\underset{\underset{R_2}{|}}{C}}-\overset{\overset{O}{\|}}{C}-\overset{}{\underset{\underset{H}{|}}{N}}-\overset{\overset{H}{|}}{\underset{\underset{R_3}{|}}{C}}-\overset{\overset{O}{\|}}{C}-\overset{}{\underset{\underset{H}{|}}{N}}-\overset{\overset{H}{|}}{\underset{\underset{R_4}{|}}{C}}-\ldots$$

where R_1, R_2, R_3, and so on may be the R groups of any of the 20 amino acids. Very short chains are often designated as tri-, tetra-, or pentapeptides and so on. Glutathione, for instance, is a tripeptide of Glu, Cys, and Gly. The hormones oxytocin and vasopressin are nonapeptides. Chains up to five residues are called oligopeptides. Longer chains (6–30 residues) are referred to as *polypeptides*. When the number of amino acids linked together exceeds 40 (M.W. $\sim$5000), the chain takes on the physical properties associated with proteins. The terms proteose and peptone refer to protein breakdown products, large polypeptides, that differ from true proteins in that they are not coagulated by heat. A wide variety of peptides are formed when proteins are hydrolyzed in the digestive tract or in the laboratory.

Strictly speaking, the peptide bond should be depicted as:

$$-\overset{}{\underset{\underset{|}{|}}{C_\alpha}}-\overset{\overset{O}{\|}}{C}=\overset{}{\underset{\underset{H}{|}}{N}}-\overset{}{\underset{\underset{|}{|}}{C_\alpha}}-$$

where the dashed line indicates partial double-bond character imposed over three components of the bond. This feature of the peptide bond freezes the amide group in a single plane, so that R groups attached to the α-carbons on either side of the peptide bond are restricted in the positions they can assume relative to the peptide bond and to one another. The polypeptide chain, therefore, takes on a three-dimensional structure that is governed by the specific arrangement of R groups around the backbone of peptide bonds.

From the structures of the amino acids in Table 4-1, it is clear that in an appropriate three-dimensional orientation, interactions of one R group with another are possible, including disulfide bond formation, hydrogen bond formation, hydrophobic interaction, interaction between charged R groups, and steric effects. The disulfide bond is the only covalent bond, other than the peptide bond, that contributes to polypeptide structure, RSH + HSR → R-S-S-R. Other types of interactions are weaker, but are often so numerous that their collective strength is large. *Hydrogen bonds* result from the sharing of a hydrogen atom between two electronegative atoms (e.g., N or O) which have unbonded electrons. In proteins, groups having a hydrogen atom which can be shared include $>$N—H (peptide nitrogen, imidazole, and indole); —OH (serine, threonine, tyrosine, and hydroxyproline); —NH_2 and —N^+H_3 (arginine, lysine, and α-amino); and —CONH (carbamino). Groups which can accept the sharing of a hydrogen atom include —COO^- (aspartate, glutamate, and α-carboxylate); —S—S— (disulfide); and $>$C=O (in peptides and ester linkages). Examples of hydrogen bonds linking the $>$CO of one peptide bond to the $>$NH of another, or the —CO— of one R group to the —OH of another R group, are shown below.

Hydrophobic interactions result from the tendency of nonpolar groups (e.g., methyl or phenyl) to associate with one another in aqueous or other polar solutions. In proteins, this association

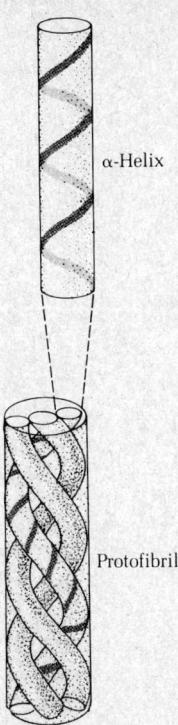

α-Helix

Protofibril

Figure 4-1. The structure of an α-keratin protofibril. This protofibril consists of three right-handed α-helices arranged in a left-handed superhelix coil. (From Rawn, J. D.: Biochemistry. New York, Harper and Row Publishers, 1983, p. 85.)

serves to bend and fold a molecule in such a way as to bring nonpolar R groups inside to the less polar interior and polar R groups outside to the more polar aqueous environment.

In recent years much effort has been devoted to investigating protein structures by physical methods, especially X-ray diffraction of crystalline proteins. Although the soluble globular proteins are especially important in clinical chemistry, thorough investigation of the insoluble protease-resistant fibrous proteins has revealed many of the basic principles of the three-dimensional structures of proteins. The most familiar fibrous proteins are the α-keratins, which form the principal structural proteins of epidermis, hair, and nails. It was in hair keratin that the presence of the α-helix was first proved. As illustrated in Figure 4-1, α-keratin is composed of three right-handed α-helix coils supercoiled around each other. Each coil is held together by hydrogen bonding between peptide $>$CO and $>$NH groups of one coil and those of the next or previous coil. The amino acids of these coils include many cysteine residues which bind the separate α-helices to one another by disulfide bonds aided by hydrogen and electrovalent bonds. This structure gives α-keratins their high tensile strength, flexibility, and elasticity.

Another type of fibrous protein structure, found in silk, is β-sheet conformation, a term coined because of the rippled or pleated structure of these molecules. This is due to the polypeptide chains of the extended coils that are bound together laterally by disulfide covalent bonds and hydrogen bonds. The latter connect the peptide bonds of different chains, instead of connecting different coils of the same chain as in an α-helix. The β-sheet is strong and inflexible when force is applied parallel to the covalently bonded or long axis, whereas an α-helical structure, such as hair, elongates when tension is applied from an α-helix to a β-sheet structure.

When describing the structure of a protein, particularly a globular protein, it is convenient to consider its complexity in four stages.

1. *Primary structure* refers to the identity and specific order of amino acid residues in the polypeptide chain. This sequence, which depends exclusively on covalent (peptide) bonds, is predetermined by the DNA coding; the remarkable fact is that the three-dimensional structure and any special biological properties of the protein follow automatically from this amino acid sequence, which folds itself into the most stable structure possible under physiological conditions (e.g., of temperature, pH).

2. *Secondary structure* may be defined as a regular recurring arrangement in space of the primary structure extending along one dimension. The secondary structure of a globular protein usually has stretches of α-helix, β-pleated sheet, and random coils, all dependent on numerous hydrogen bonds and occasional disulfide covalent bonds.

3. *Tertiary structure* involves the intramolecular folding of the polypeptide chain into a compact three-dimensional structure with a specific shape. This structure is maintained by electrovalent linkages, hydrogen bonds, disulfide bridges, van der Waal's forces, and hydrophobic interactions. Hydrophobic interactions are considered to be a major force in maintaining the unique tertiary structure of proteins.

4. *Quaternary structure* refers to the association of several polypeptide chains or subunits into a larger "oligomeric" aggregate unit. This structure depends on the close fit of the polypeptide subunits through interactions at their contact surfaces and with any prosthetic groups.

Some proteins contain nonamino acid components known as *prosthetic groups*. Such proteins are often referred to as *conjugated proteins* and are classified according to the nature of their prosthetic groups as metalloproteins, lipoproteins, glycoproteins or mucoproteins, phosphoproteins, and so on. Glycoproteins and mucoproteins both have covalently linked carbohydrate prosthetic groups, but the amount of carbohydrate varies—in glycoproteins it may be ~5–15%, in mucoproteins ~15–75%. Conjugated proteins freed of their prosthetic groups are called *apoproteins*.

Nearly all proteins of interest in clinical chemistry are soluble globular proteins such as hemoglobin (see Figures 15-4 and 15-5), enzymes, and plasma proteins (with the exception of fibrinogen). The complex bending and folding of their polypeptide chains are a result of the numerous interactions of their R groups as indicated above. Globular proteins are compact with little or no space for water in the interior of the molecule where most of the hydrophobic R groups are located. On the other hand, most polar R groups are located on the surface of the protein, where they exert a substantial influence on protein solubility, acid-base behavior, and electrophoretic mobility.

Most globular proteins retain their biological activity only within very limited ranges of temperature and pH, at which their native structural arrangements are the most stable. Even short periods of exposure to temperatures of 65–70 °C or to extremes of pH will cause "denaturation" with loss of solubility and biological activity; enzymes, for example, lose their catalytic powers after denaturation occurs. (See Chapter 5.) The amino acid sequence remains unchanged, but the polypeptide chain becomes unfolded, or the quaternary structure is lost.

After mild denaturation by pH change or by the action of agents such as urea, many protein molecules are able to revert spontaneously to their most stable or native globular forms and regain their biological functions when physiologic conditions are restored, good evidence that the amino acid sequence alone determines their normal complex shapes. If, however, there is a high activation energy barrier between the two states, the denatured state will not revert to the native state.

The relationship between the structure of a protein and its biological function is slowly beginning to be understood. For example, in enzyme molecules there are clefts in the surface where their active centers are situated (see Chapter 5), and in the hemoglobin molecule each globin chain is looped about itself to form a cleft lined by hydrophobic R groups in which the heme group lies. (See Chapter 15.)

Biological functions of proteins are numerous, varied, and important. Enzymes catalyze biochemical transformations essential to metabolism; protein, poly-, and oligopeptide hormones regulate metabolism; antibodies and components of the complement system protect against infection. Plasma proteins maintain the oncotic pressure of plasma; they transport hormones, vitamins, metals, and drugs, often serving as reservoirs for their release and utilization; apolipoproteins solubilize lipids; hemoglobin carries oxygen; protein coagulation factors affect hemostasis. In addition, there are structural proteins such as the keratins, collagen, and elastin; contractile proteins such as myosin; storage proteins such as ferritin; chromosomal proteins such as the basic histones; and many others.

Many of the properties of proteins can be utilized for their separation, identification, and assay.[19,23,66] Among these properties are:

1. *Molecular size.* A majority of proteins are macromolecules, compounds of large size and consequently of high molecular weight. Because of their size, proteins can be separated from smaller molecules by dialysis or ultrafiltration across membranes of specified pore size, or by

molecular exclusion chromatography (gel filtration). Because of their differing molecular weights, they can be separated by density-gradient ultracentrifugation.

2. Differential solubility. Loss of solubility is essentially a loss of affinity of the solute for the solvent. The solubility of a protein is affected by the pH, ionic strength, temperature, and dielectric constant of the solvent in which it is placed. Changing the pH of the solvent affects the surface charges of the polar R groups exposed to the solvent; at its isoelectric point (net charge zero), a protein in a polar solvent is in its least soluble state because it has the least affinity for the solvent. Manipulation of ionic strength of the solvent affects the degree of hydration of the protein; in high concentrations, ionic salts compete with protein for water and thus decrease the amount of water available for hydration of the protein molecules. This behavior is the basis of "salting-in" and "salting-out" procedures which historically were the earliest methods for separating and characterizing different proteins and groups of proteins. Raising the temperature affects solubility by decreasing the intramolecular bonding of a protein; loss of secondary and tertiary structure exposes the less polar interior of the protein molecule to the polar medium. Water-miscible, neutral organic solvents have dielectric constants that are less than that of water; they suppress the ionization of the surface R groups on the protein and thereby decrease its solubility. Manipulation of solvent character and composition in such ways is often useful in analyses of biological, protein-containing fluids.

3. Electrical charge. The effect of pH to confer, enhance, or change the surface charges on a protein can create, in a mixture of different proteins, a variety of species of different charge/mass ratios, which migrate at different rates in an electrical field; electrophoresis and isoelectric focusing are based on this behavior. Ion-exchange chromatography is based on electrostatic interactions of charged proteins with oppositely charged solid media.

4. Adsorption on finely divided inert materials. These materials offer large surface areas for interaction of protein with the material. Adsorption of protein on charcoal, a nonpolar substance, occurs presumably by hydrophobic interaction and can be used to separate proteins from a fluid matrix. Adsorption on polar substances such as silica, alumina, or hydroxyapatite depends on ionic interactions or hydrogen bonding; these substances can be used with buffer *elution* in chromatographic separations.

5. Specific binding to antibodies, co-enzymes, or hormone receptors. The unique property of a protein to recognize and bind to a complementary compound with high specificity has been valuable in immuno- and competitive protein binding assays of all types. Proteins can also be separated by affinity chromatography, where a ligand attached to a solid medium provides high selectivity for a protein that binds it. Because immunochemical methods do not require separation of the protein to be determined from all the others in a biological mixture such as serum, specific antisera are now the reagents most commonly used for the determination of individual proteins.

DISORDERS OF AMINO ACID METABOLISM

In this section we first consider the normal metabolism of amino acids and their concentrations in body fluids; this is followed by a discussion of the principal disorders of amino acid metabolism together with the laboratory methods used for their diagnosis and screening.

In health, the primary supply of amino acids for endogenous protein synthesis is provided by dietary proteins. However, eight to ten of the twenty common amino acids cannot be synthesized by most mammals, and these are therefore considered "essential" constituents of the diet for maintenance of health or growth or both. (See Table 4-1, where essential amino acids are identified in the Comment column.) Proteolytic enzymes in the GI tract, acting on proteins, release amino acids, which are then absorbed from the jejunum into the blood and subsequently become part of the body pool of amino acids. The liver and other tissues draw on this pool for synthesis of plasma and intracellular proteins. The liver and kidneys are also actively involved in interconverting amino acids by transamination and degrading them by deamination. Deamination produces ammonium ions, which are consumed in the synthesis of urea. Urea, in turn, is excreted by the kidneys. (The major pathways of amino acid metabolism are outlined in Figure 4-2.)

Amino acids in blood are filtered through the glomerular membranes but are normally reabsorbed in the renal tubules by saturable transport systems. Hence, high blood levels of amino acids result in significant renal excretion (*aminoaciduria*). In normal subjects, aminoaciduria is transient and is associated with protein intake in excess of amino acid requirements for repletion

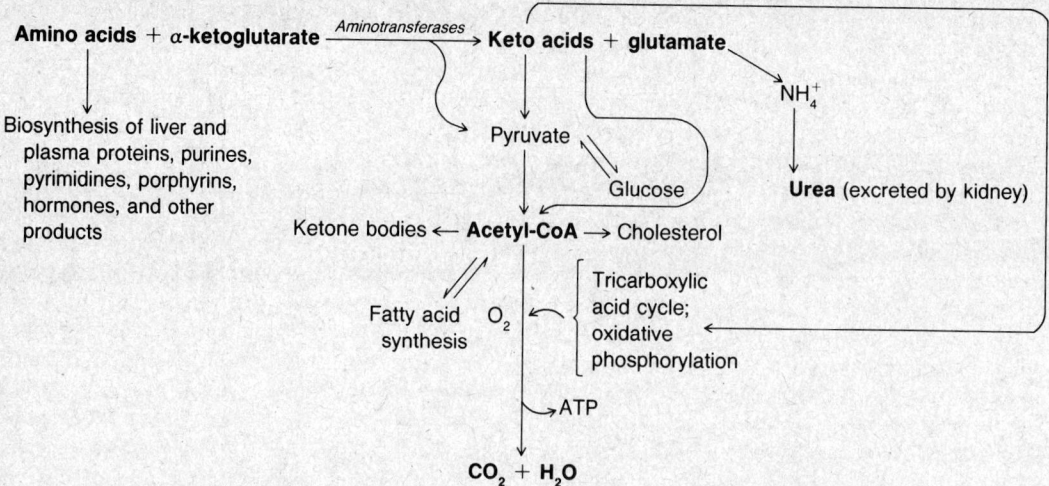

Figure 4-2. A generalized scheme of amino acid metabolism in the liver.

of the body pool. The mechanism of reabsorption is still obscure, although it is known to be an active transport system dependent upon membrane-bound carriers as well as intraluminal cNa$^+$. Four different transport systems have been identified: one for neutral amino acids; one for basic amino acids; one for proline, hydroxyproline, and glycine; and one for dicarboxylic amino acids. For some amino acids, transport appears to occur by two systems: one of low capacity but of high specificity, and the other of high capacity but of low specificity. A few amino acids such as cystathionine and homocystine are not efficiently reabsorbed by renal tubular mechanisms, but are cleared by the kidney and apparently excreted into the urine at the glomerular filtration rate.

Much that is known about renal reabsorption mechanisms has been learned from the study of various forms of aminoaciduria. Three types of aminoaciduria have been identified: (1) *overflow aminoaciduria* occurs when the plasma level of one or more amino acids exceeds the renal threshold (tubular capacity for reabsorption); (2) *renal aminoaciduria* occurs when plasma levels are normal but there is a congenital or acquired defect of the renal transport system; (3) *no-threshold aminoaciduria* occurs when excessive amounts of an amino acid, arising from an inherited metabolic block, are present in urine, while plasma levels are essentially normal because all the amino acid is excreted. Note that no-threshold aminoacidurias, such as homocystinuria, are not due to congenital or acquired kidney defects, but are solely due to inefficient reabsorption by normal renal tubular mechanisms.

Amino Acid Concentrations in Body Fluids[91]

Plasma amino acid concentrations vary during the day by about 30%; therefore, blood specimens should be collected at the same time each day. Values are highest in mid-afternoon and lowest in early morning. This diurnal variation is particularly important when specimens are analyzed for detection of heterozygotic states of defective metabolism.

Plasma amino acid concentrations are high during the first days of life, especially in premature neonates, but they tend to be low in babies with birth weights low for their gestational age; malnutrition due to placental insufficiency is the cause. Maternal values are low in the first half of pregnancy.

In a chromatogram of normal *adult urine,* glycine is usually the dominant fraction, with alanine, serine, glutamine, and, in heavy meat eaters, histidine and 1-methyl histidine present in smaller quantities; in some normal urines, taurine is prominent; in others, β-aminoisobutyric acid is seen. The renal threshold for many substances is lowered during pregnancy, and amino acids such as histidine, phenylalanine, lysine, and tyrosine are commonly seen in urine. Premature babies, especially during the first week, have a physiological generalized renal aminoaciduria; even at full term, aminoaciduria is more marked than in normal adults.

CSF concentrations of most amino acids are less than in plasma. The plasma/CSF ratio varies for individual amino acids, suggesting that CSF concentrations are dependent on other factors besides diffusion from the choroid plexuses.

Cells have amino acid concentrations about ten times those of plasma.

THE AMINOACIDURIAS

Aminoacidurias[2,8,37,99] may be primary or secondary. Primary disease is due to an inherited enzyme defect, also called an inborn error of metabolism. The defect is located either in the pathway by which a specific amino acid is metabolized or in the specific renal tubular transport system by which the amino acid is reabsorbed. Secondary aminoaciduria is due to disease of an organ such as the liver, which is an active site of amino acid metabolism, to generalized renal tubular dysfunction, or to protein-energy malnutrition.

The *primary aminoacidurias* are a group of "inborn errors of metabolism," a term first used by A. E. Garrod 75 years ago following his study of alkaptonuria. Alkaptonuria is a rare, inherited condition in which the urine darkens on standing because of the presence of homogentisic acid, a substance not normally found in urine. Garrod suggested that alkaptonuria, as well as a number of other congenital conditions, might be due to a deficiency of one of the enzymes involved in amino acid metabolism, and that the various biochemical, pathological, and clinical manifestations of these inherited disorders were secondary to the absence of the enzyme. Fifty years later, the enzyme deficiency in alkaptonuria was directly demonstrated. By then the concept of "one gene–one enzyme or other protein" had become firmly established, and direct evidence of a human mutation actually producing an alteration in an amino acid sequence of a protein had been obtained in sickle cell anemia, where valine replaces glutamic acid at the sixth position of the β-chain of hemoglobin.

Increasing knowledge in molecular biology[29] has made clear that an inborn error of metabolism is the result of a mutation in the DNA base sequence that codes for the specific amino acid sequence of a particular enzyme protein. The defect in the enzyme that derives from the altered gene is reflected as diminished or absent biological activity, the equivalent of a block in the metabolic path of that enzyme's substrate. The substrate accumulates or is diverted into alternative paths, so that products of the normal path are not made at all or are made in smaller amounts, or products of the alternative path are present in much larger than normal amounts. The metabolic defect is then expressed as abnormal concentrations of normal metabolites or as the appearance of abnormal metabolites associated with characteristic symptoms. There are, therefore, three levels at which it should be theoretically possible to diagnose such an inherited disease: (1) the DNA abnormality, (2) the enzyme defect, and (3) the metabolic abnormalities that are due to the defect. Presently, diagnosis is heavily dependent on recognizing clinical signs and symptoms as well as characterizing the chemical nature of these disorders. Demonstration of an enzyme defect is confined to prenatal diagnosis or to confirmation of a postnatal diagnosis by testing cultured cells. Approaches are still under development for demonstrating DNA abnormalities.

Although an enormous number of congenital errors exist, many of them are harmless and often go unrecognized; nevertheless, more than 1350 human diseases, including the primary aminoacidurias, are now known to be caused by alteration of one single gene out of the 50 000 or more in each human cell. The locations of more than 350 of those genes in the chromosome are also now known.[99] Most of the serious inherited diseases are rare, but collectively they pose an increasingly important diagnostic problem because some are now treatable, if diagnosed sufficiently early.

The symptoms and prognosis of the primary aminoacidurias which cause disease may vary from being almost benign, as in alkaptonuria, to being often lethal, as in maple syrup urine disease. Their causes are usually a block in a major catabolic pathway, and their expression often indicates the diversion of substrate to an alternative, metabolically ineffective minor path. As a result, substrates and precursor substrates behind the block accumulate in the blood. When renal transport mechanisms for these substances are saturated, these products spill into the urine. Accumulation of keto acids in maple syrup urine disease and of phenylalanine in phenylketonuria (PKU) are examples of precursors and substrates with toxic effects. Past the block, products and intermediates are decreased, for example, tyrosine in PKU. If an alternative path is used as it is for phenylalanine in PKU, products and intermediates of the alternative path are increased in blood and urine; in PKU these substances include phenylpyruvate and phenyllactate. (See page

546.) Occasionally, an enzyme block occurs in an anabolic pathway, as in the synthesis of melanin where the consequence is albinism. In some cases, substrates do not accumulate in the blood because there is no reabsorption in the kidneys (no-threshold aminoaciduria); blood levels are low and urine levels are increased.

The *renal* types of primary aminoacidurias (e.g., cystinuria) are not characterized by high levels of amino acids in blood because the defective protein coded by the altered gene is an element of the tubular reabsorption mechanism for that amino acid. The nature of the defect is uncertain; either a membrane-fixed enzyme or some carrier protein essential to the transport mechanism is missing or defective. The transport defect may affect a single amino acid or a group of amino acids; in cystinuria, for instance, it is the dibasic amino acid transport system which is impaired.

The *primary aminoacidurias* listed in Table 4-3 are all inherited as autosomal recessive disorders. The overflow types (Table 4-3, *A*) probably involve a change in only one amino acid residue of the enzyme concerned. For an autosomally recessive disease to occur in an offspring, the gene defect must be present in the chromosomes of both parents, i.e., the offspring's genome must be homozygous for the defect. The homozygote therefore fails to make the critical protein with a correct amino acid sequence; in this case normal enzyme activity is either absent or significantly reduced. In the heterozygote, enzyme activity is decreased (e.g., half-normal) and the block in the metabolic path is only partial. Defective proteins are usually detectable immunologically even when they are inactive as enzymes. Occasionally they can be completely absent as in half the cases of homocystinuria.

The *secondary aminoacidurias* affect many amino acids simultaneously; as in primary aminoacidurias, these defects may be of an "overflow" or "renal" type. An example of a disorder that results in secondary overflow aminoaciduria is fulminant hepatic failure (acute massive necrosis or acute yellow atrophy of the liver).[97] This condition may suddenly overwhelm the young and healthy person, usually as a result of an acute viral hepatitis or self-induced acetaminophen poisoning that progresses to acute hepatic encephalopathy and coma. The mortality rate is 80–90%; however, if the patient recovers, due to the regenerative powers of the liver, he usually recovers completely without liver scarring (cirrhosis). If patients could be supported temporarily by an "artificial liver," many more might recover. Present attempts have met with only partial success because the cause of the fatal encephalopathy is not fully understood. It appears to be due to a number of small-molecular water-soluble "toxins," both free and bound to proteins, which result in the failure of metabolic and detoxification processes normally carried out by the liver. These toxins are thought to include the characteristically high concentrations of methionine, phenylalanine, tyrosine, and tryptophan found with these disorders.

Generalized secondary *renal aminoaciduria* is due to progressive damage to the renal tubules and has a variety of causes, acquired or inherited, all of which cause proximal renal tubular dysfunction. If rickets is also present, the condition is called Fanconi's syndrome;[87] in childhood it is often secondary to cystinosis, an autosomal recessive disorder causing widespread intracellular deposits of cystine but without cystinuria. Generalized renal aminoacidurias may be caused by poisons, especially heavy metals; by wasting from starvation or disease; by acute tubular necrosis; or by congenital diseases such as galactosemia and Wilson's disease. In some patients the tubular defect involves amino acids only. In others, all the functions carried out by the proximal tubule are affected, as in Fanconi's syndrome where glucosuria and hyperphosphaturia are also common.

The Diagnosis of the Inherited Aminoacidurias

In the study and treatment of inherited defects in amino acid metabolism, one can list three separate problems: the diagnosis of the sick infant or child, routine neonatal screening, and the possibility of prenatal diagnosis.

The Sick Infant or Child.[15] An inherited aminoaciduria must be diagnosed rapidly before irreparable damage is done; however, the diseases are uncommon and their symptoms are often nonspecific. Consequently, inherited aminoacidurias are sometimes not recognized sufficiently early. An inborn error should be suspected when there has been a previous unexplained neonatal death, when the parents are biologically related, if symptoms appear when feedings are changed, or if the condition improves when food is withheld and the patient maintained on a glucose/saline drip. An inherited aminoaciduria should also be suspected when an infant or young child shows clinical symptoms of persistent vomiting, failure to thrive, or neurological or liver abnormalities or both; in some cases massive ketosis, unexplained acute acidosis, urinary stones, or a

Table 4-3, A. PRIMARY OVERFLOW AMINOACIDURIAS (AUTOSOMAL RECESSIVE DISORDERS*)

Disorder	Prevalence	Abnormal Enzyme(s) (or other defects)	Excesses in Blood	Excesses in Urine	Clinical Features	Treatment
Hyperphenylalaninemia *Classic PKU* (Type I)	1:10 000	Phenylalanine hydroxylase (absent)	Phenylalanine	Phenylalanine and its metabolites (phenylpyruvate, phenyllactate, o-hydroxyphenylacetate)	Mental retardation; seizures; eczema	Dietary restriction of Phe
Variant PKU (Type II)	1:14 000	Phenylalanine hydroxylase (deficient)	Phenylalanine	Variable	Mild retardation	Dietary restriction of Phe
Transient neonatal (Type III)	1:30 000	Phenylalanine hydroxylase (deficient)	Phenylalanine	Phenylalanine metabolites usually not found	Normal	None required
Type IV	Rare	Dihydropteridine reductase (absent)	Phenylalanine	Variable	Neurological disorders	DOPA, 5-OH-tryptophan
Type V	1:30 000	Defect in biopterin synthesis	Phenylalanine	Variable	Neurological disorders	DOPA, 5-OH-tryptophan
Tyrosinemia *Type I* (Tyrosinosis)	1:100 000	Fumarylacetoacetate hydrolase? (absent)	Tyrosine; methionine	Tyrosine and its metabolites (PHPPA, PHPLA, PHPAA); DOPA; generalized aminoaciduria (e.g., Fanconi's syndrome)	Hepatic cirrhosis; renal damage	Dietary restriction of Phe, Tyr, Met (treatment does not cure liver disease)
Type II	Rare	Tyrosine aminotransferase (absent)	Tyrosine	Tyrosine and its metabolites; tyramine	Eye and skin lesions; mental retardation	Dietary restriction of Tyr, Phe
Transient neonatal	Full term, 1:10; premature, 1:3	Liver immaturity	Tyrosine; phenylalanine	Tyrosine and its metabolites; tyramine	None, long term	Vitamin C; reduced protein intake
Alkaptonuria	1:250 000	Homogentisic acid oxidase (absent)	Homogentisic acid (slight)	Homogentisic acid	Degenerative arthritis, cartilage pigmentation	None available
Homocystinuria	1:200 000	Cystathionine β-synthase (absent or deficient)	Homocystine; methionine	Homocystine; methionine and its sulfoxide	Ocular, skeletal, vascular effects	Pyridoxine; low methionine diet supplemented with cystine
	Rare	Methylenetetrahydrofolate reductase (absent or deficient)	Homocystine with normal methionine	As in blood	Mental retardation	Folate
	Rare	Methyltransferase	Homocystine with normal methionine	Homocystine; methylmalonic acid	Failure to thrive; mental retardation	Vitamin B_{12}

Table continued on following page

533

Table 4-3, A. PRIMARY OVERFLOW AMINOACIDURIAS (AUTOSOMAL RECESSIVE DISORDERS*) *Continued*

Disorder	Prevalence	Abnormal Enzyme(s) (or other defects)	Excesses in Blood	Excesses in Urine	Clinical Features	Treatment
Histidinemia	1:20 000	Histidase (absent)	Histidine; alanine	Imidazole; pyruvic acid and other histidine metabolites	Sometimes normal; sometimes neurological symptoms, such as mental retardation or speech defects	Dietary restriction of His
Branched-chain ketoaciduria (Maple syrup urine disease)	1:250 000	Branched-chain ketoacid decarboxylase (deficient)	During acute attacks: leucine, isoleucine, alloisoleucine, valine, and corresponding ketoacids	During acute attacks: as in blood; also odor	Overwhelming acidosis; vomiting and CNS symptoms; sometimes mental retardation; respiratory failure; may be fatal	Dietary restriction of Leu, Ile, Val
Nonketotic hyperglycinemia	1:150 000	Block in glycine cleavage enzyme system	Glycine	Glycine (also in CSF)	Seizures, hypotonia, no ketosis; if severe, mental retardation; fatal within 2 years	None available
Propionic acidemia	Rare	Propionyl CoA carboxylase (deficient)	Glycine	Glycine, propionate, hydroxypropionate, methylcitrate	Metabolic ketoacidosis, developmental retardation	Low-protein diet
Methylmalonic acidemia	1:20 000	Methylmalonyl CoA mutase (absent or deficient)	Glycine; methylmalonic acid	Glycine; methylmalonic acid (also in CSF); ketonuria	Metabolic ketoacidosis, developmental retardation	Vitamin B_{12}
Cystathioninuria	1:70 000	γ-Cystathionase (absent or deficient)	Cystathionine	Cystathionine (also in CSF) and cystathionine metabolites	Benign	None required
Carnosinemia	1:500 000	Carnosinase (deficient)	Carnosine	Carnosine (carnosine and homocarnosine also in CSF)	Severe neurologic disease	None available
Hyperprolinemia I & II	1:300 000	Type I: proline oxidase (deficient) Type II: Δ⁵-pyrroline-5-carboxylic acid dehydrogenase (deficient)	Proline	Proline; hydroxyproline; glycine	Probably benign	None required
UREA CYCLE DISORDERS						
Citrullinemia	Rare	Argininosuccinate synthetase (deficient)	Citrulline; ammonia; alanine	Citrulline; glutamine	Vomiting; seizures; coma; mental retardation; hepatomegaly	Low protein diet, arginine supplements

Disorder	Prevalence	Enzyme Defect	Excess in Blood	Excess in Urine/CSF	Clinical Symptoms	Treatment
Argininosuccinic aciduria	1:75 000	Argininosuccinate lyase (deficient)	Argininosuccinic acid; citrulline; ammonia after meals	Argininosuccinic acid and its anhydride; citrulline; increased argininosuccinic acid in CSF	Vomiting; growth failure; neurologic dysfunction; trichorrhexis nodosa	Low protein diet; arginine supplements
Argininemia	Rare	Arginase (deficient)	Arginine; ammonia after meals	Normal or excess arginine, cystine, and ornithine	Vomiting; coma; spastic diplegia	Low protein diet
Hyperornithinemia	Rare	Ornithine decarboxylase (deficient)	Ornithine, glutamine, alanine, ammonia after meals	Ornithine, homocitrulline	Vomiting, lethargy, coma, retardation	Low protein diet
Ornithine transcarbamylase deficiency	Rare	Ornithine transcarbamylase (deficient, X-linked dominant)	Ammonia, glutamine	Orotic acid, uridine, uracil	Lethargy, irritability, vomiting, convulsions, coma, apnea, death	Low protein diet
Carbamoylphosphate synthetase deficiency	Rare	Carbamoylphosphate synthetase (absent or deficient)	Ammonia, glycine, glutamine	Glycine, glutamine	Vomiting, lethargy, hypotonia, coma, pulmonary and GI hemorrhages, developmental retardation	Low protein diet

Table 4-3, B. PRIMARY RENAL AMINOACIDURIAS (AUTOSOMAL RECESSIVE DISORDERS*)

Disorder	Prevalence	Excess Amino Acids in Urine	Clinical Symptoms†	Treatment
Cystinuria, classic	1:13 000	Lysine; ornithine; arginine; cystine	Cystine renal calculi	High fluid intake; keep urine alkaline; D-penicillamine
Hypercystinuria	Rare	Cystine	Cystine renal calculi	High fluid intake; keep urine alkaline; D-penicillamine
Dibasic aminoaciduria and lysinuric protein intolerance	Rare	Ornithine; lysine; arginine	Vomiting, hepatomegaly; growth failure; protein intolerance; retardation	Low protein diet
Hartnup disease	1:18 000	All neutral amino acids	May be symptomless; usually pellagra-like dermatitis; neurological and psychiatric symptoms	Adequate protein diet; nicotinamide
Iminoglycinuria	1:12 000	Glycine; proline; hydroxyproline	Benign	
Dicarboxylicaminoaciduria	Rare	Glutamic acid; aspartic acid	Probably benign	
Methionine malabsorption	Rare	Methionine; also tyrosine, phenylalanine, and branched-chain amino acids; α-hydroxybutyric acid; urinary odor	White hair; seizures; mental retardation	Low methionine diet

*In some probable, rather than proved.
†Affected subjects usually have a transport defect in the jejunal mucosa, as well as in the proximal renal tubules.

characteristic odor of the urine or body will suggest an inborn error. A metabolic cause should be suspected in babies who appear well initially, but who deteriorate after a few days as their protein intake increases. If coma is present from birth, however, it is more likely to be due to such common causes as intracranial hemorrhage and hypoxia.

The determination of *total* amino acids in plasma or urine is seldom useful clinically. (For methods for the determination of total amino acids, see references 57 and 111.) What *is* required is the separation, detection, and estimation of *individual* amino acids. Thin-layer chromatography is ideal for this purpose since it is relatively simple and inexpensive, and a large number of specimens can be processed simultaneously. This technique, however, is only semiquantitative and is difficult to automate. Simple chemical or microbiological tests for individual amino acids and metabolites are often used to supplement TLC.

An attempt should be made to reach, as soon as possible, a provisional diagnosis that can be confirmed later if necessary. Failure to do so carries the risk that the child may die or develop irreversible problems before any information is obtained from the laboratory. Ideally the laboratory should carry out a battery of tests to exclude inherited metabolic diseases in general, not only the aminoacidurias.[36]

Collection of specimens. The child should be on a normal diet, and blood and urine specimens should be collected simultaneously. Plasma should be harvested promptly; plasma and urine specimens should be stored frozen if analysis is delayed. Hemolysis should be avoided in the blood collection procedure because the concentration of amino acids is much greater in erythrocytes than in plasma. Specimens, together with some skin for fibroblast culture, should be collected even if the child is likely to die. Diagnosis of an aminoaciduria cannot be made on post mortem specimens, and it is important for genetic counseling of parents who plan a future pregnancy. If neurological symptoms necessitate a lumbar puncture, some CSF should be preserved frozen; the CSF/plasma glycine level is diagnostic in nonketotic hyperglycinemia. It should be remembered that drugs given either to the mother prior to the baby's birth or to the infant may produce unknown metabolites which may complicate interpretation of the chromatogram. This is the case, for example, with ampicillin.

Neonatal Screening. Genetic screening of newborns is commonly practiced in order to make the earliest possible diagnosis of a potentially treatable inherited aminoaciduria. Most screening programs have been directed to the early detection of phenylketonuria (PKU). One of the most widely used tests for PKU screening has been the *Guthrie test.* Since its introduction in 1961, the test has also been applied to the diagnosis of some of the other, rarer aminoacidurias.[34] Typical Guthrie test systems are shown in Table 4-4.

The Guthrie test is a semiquantitative microbiological assay. Bacterial spores, usually *B. subtilis,* are incorporated into an agar medium to which has been added a competitive growth inhibitor specific for the amino acid to be determined. The inhibitor often has a molecular structure similar to the amino acid of interest; comparison of the L-amino acid and inhibitor columns of Table 4-4 illustrates this point. Blood or urine from the patient is spotted onto a piece of soft filter paper; a standardized portion of the paper with the specimen is punched out and laid on the agar surface. The agar plate is then incubated and later observed for bacterial growth. In the presence of elevated concentrations of the amino acid of interest, the effect of the growth inhibitor is diminished or overcome, and zones of bacterial growth will be observed. The test system is designed to show growth only when the concentration of the amino acid of interest exceeds its upper reference limit.

In order to avoid false negative results in the Guthrie test, the specimen must be free of drugs such as antibiotics or endogenous substances that inhibit bacterial growth. False positive results may also occur; therefore, every positive result should be confirmed with a quantitative

Table 4-4. GUTHRIE TEST SYSTEMS

L-Amino Acid	Inhibitor	Disease	Reference Limits mg/dL
L-Phenylalanine	β-2-Thienylalanine	Phenylketonuria	4
L-Leucine	4-Azaleucine	Maple syrup urine disease	4
L-Methionine	Methionine sulfoximine	Homocystinuria	2
L-Tyrosine	D-Tyrosine	Tyrosinemia	8
L-Lysine	S(β-Aminoethyl) cysteine	Hyperlysinemia	4

chemical test or by chromatography. Positive results obtained on premature infants must be interpreted in conjunction with additional tests, since immaturity of the liver, as well as inherited disease, may be a cause of increased amino acid levels. Specimens should not be obtained until 48 h after birth, since 2 d of protein intake are required to accumulate an abnormal concentration of an amino acid, e.g., phenylalanine. The patient should be on a normal diet for at least 24 h prior to having a specimen drawn.

Prenatal Diagnosis. Prenatal diagnosis of an inherited aminoaciduria is particularly desirable in the case of a mother who has had a previous child with a severe inherited defect. Preliminary study of parents, affected relatives, or the previous affected child, preferably by skin fibroblast cultures, is essential for defining the defect to be excluded.

The usual chemical tests for metabolites are not applicable to prenatal diagnosis. Although amniotic fluid is easily obtained, it may not reflect a metabolic defect in the fetus. Fetal blood, on the other hand, can reflect the metabolic defect reliably but few physicians are willing to perform the high-risk procedures to obtain the specimen. Prenatal diagnosis, therefore, must be based on identification either of primary enzyme deficiency or of abnormal DNA in fetal cells. The most common approach is to collect amniotic fluid by amniocentesis at 14–16 weeks of gestation, culture the fetal cells, and then examine the cells for enzyme activity. Unfortunately, 3–6 weeks may be required to grow enough cells for the enzyme assays. The ideal prenatal test would lead to a diagnosis during the first three months of pregnancy. Such a test, however, would require direct gene analysis of tissue obtained by biopsy of trophoblastic villi, or of fetal cells obtained by amniocentesis.[67] If amniotic fluid is used, fetal cells have to be cultured to supply enough DNA; this causes delay so that it has no advantages over the present enzyme determinations, unless enzyme assays are impractical as in PKU.

Direct gene analysis has already been applied to the diagnosis of thalassemia and may be used for the severe aminoacidurias. This technique involves cutting up DNA with restriction endonucleases, separating the fragments by gel electrophoresis, and identifying the genes with radioactive gene probes which are exact complementary copies of the DNA sequence being sought.[29] Unless the restriction or cutting site is altered by the mutation, single base changes cannot be identified. Detection of such changes requires the development of shorter gene probes. In the meantime, gene-mapping techniques can be used to explore differences in nucleotide sequences that flank the relevant gene. These "nucleotide sequence polymorphisms" may alter the restriction cleavage site or change the size of the restriction fragment. The relevant fragment can then be detected. Such an analysis has recently been used to detect phenylketonuria.[116]

AMINO ACID ANALYSIS

Three groups of tests for amino acid analysis may be considered.

1. The initial diagnostic screening tests referred to above include thin layer chromatography, urine color tests, and Guthrie microbiological tests.

2. Quantitative tests for monitoring treatment or confirming the initial diagnosis. For these estimations, ion exchange or high performance liquid chromatography is preferred; these procedures are complex and expensive, but can be automated. Some individual amino acids may also be determined by quantitative chemical methods.

3. Exact identification of an unknown amino acid or metabolite. Since chromatographic separation gives only one identifying characteristic, the retention time, other properties such as the infrared spectrum may need to be examined; the most discriminating automated system is a combination of gas liquid chromatography and mass spectrometry, available in selected medical centers.

Screening Tests

Thin Layer Chromatography (TLC)

There are three stages in any chromatographic analysis of amino acids: preparation of the sample, chromatographic separation, and identification of the separated amino acids. Chromatography is a versatile technique, and details can be varied to suit specific problems. (See Chapter 1B, Section Six.) For the analysis of amino acids in body fluids and tissues, pretreatment of the sample is often necessary in order to remove proteins, lipids, inorganic salts, or other substances that will interfere with chromatographic resolution. Plasma, serum, and indeed most other biological fluids are first deproteinized with sulfosalicylic acid or some other appropriate protein-precipitating agent. Following centrifugation to remove precipitated proteins, the supernatant is then desalted by means of solvent extractions, ion-exchange resins, or electrodialysis. Relatively dilute specimens such as CSF or urine may not require extensive desalting pretreatment.

The amount of amino acids visible in a chromatogram is influenced not only by the disease process but also by the volume of fluid applied to the chromatogram. It is desirable, therefore, that the sample volume be standardized by reference to its total nitrogen content or the amount of creatinine in a specified volume of the specimen. A convenient and sufficiently accurate way to estimate total amino acid nitrogen is by the method of Pope and Stevens.[77] This method is based on the reaction between copper and amino nitrogen to form Cu(II) complexes. The extent of Cu(II) complex formation is assessed by iodometric titration with thiosulfate. Other methods include direct spectrophotometry[98] and atomic absorption spectroscopy.[7]

Of the possible choices of *stationary phases*, paper and thin layers of cellulose continue to be favored by many investigators. Procedures using cellulose have the advantage of giving superior chromatographic resolution and reducing the time required for solvent development. Procedures using paper, however, are very useful when blood or urine samples are collected on filter paper disks.

A large number of *solvent systems*[97a] have been proposed for the separation of amino acid mixtures. One of these is described in detail in a later section of this chapter. One-dimensional chromatography is favored by some laboratories because of its simplicity—multiple standards and samples can be easily run on a single plate. Two-dimensional TLC is similar to one-dimensional methods except that, after the first migration, the chromatogram is rotated 90° and then transferred to another solvent system for a second migration. As a result, amino acids are separated in 2 dimensions. When selective staining reagents are used in conjunction with two-dimensional solvent systems, identification of more than 75 compounds of biochemical interest is possible.

Many *staining reagents* can be used to visualize amino acids separated by TLC. The most widely used reagent for both qualitative and quantitative assessment of amino acids is ninhydrin. A number of colored products are formed, but the major one is presumed to result from deamination and condensation as follows:

Ninhydrin   Amino acid   Hydrindantin   Aldehyde

Ninhydrin   Hydrindantin   Colored Product

Note that only the α-nitrogen from the original amino acid molecule is ultimately part of the measured product. The ninhydrin reaction has historically been and still remains the most useful amino acid staining procedure for paper and thin layer chromatography as well as for the detection of amino acids in column effluents.

Most amino acids react with ninhydrin at ambient temperatures to form a blue color that becomes purple upon heating. However, proline and hydroxyproline yield yellow compounds

Table 4-5. REAGENT SYSTEMS FOR STAINING CHROMATOGRAPHIC PLATES

Staining Method	Use	Reagent Composition	Procedures and Comments
Ninhydrin–isatin reagent	Large number of amino acids commonly encountered in biological fluids	a. Ninhydrin, 0.25 g/dL in acetone b. Isatin, 10 mg/dL in acetone c. Lutidine	1. Mix 50 mL of reagent (a) and 50 mL of reagent (b); then add 1 mL of lutidine. 2. Dip or spray chromatograms; dry at 40 °C for 20–30 min.
Ninhydrin–collidine reagent	Polychromatic reagent; facilitates identification of individual amino acids	a. Ninhydrin, 0.2 g/dL in isopropanol b. Collidine (2,4,6-trimethylpyridine)	1. Mix 2.5 mL of reagent (b) with 100 mL of reagent (a). 2. Dip or spray chromatograms and heat for 10 min in a 70 °C humidified oven.
Chlorine–o-tolidine	General reagent for amino acids; requires higher concentrations for detection	a. Saturated solution of o-tolidine in 2% acetic acid b. KI, 0.85 g/dL in H_2O c. Sodium hypochlorite, 2 g/dL in H_2O	1. Mix equal volumes of reagents (a) and (b) just prior to use (solution 1). 2. Spray chromatograms lightly with solution (c) and dry for 1–1.5 h; overspray the chromatograms with solution 1.
Pauly reagent	Histidine and other imidazoles, tyrosine and other phenolic compounds form red, brown, and yellow azo dyes	a. Sodium nitrite, 5 g/L in H_2O b. Sulfanilic acid, 9 g in 90 mL concentrated HCl and 900 mL H_2O c. Sodium carbonate, anhydrous, 10 g/dL in H_2O	1. Mix 1 volume of solution (a) and 1 volume of solution (b). After 5 min, add 2 volumes Na_2CO_3. *Caution:* CO_2 is liberated vigorously. 2. Chromatograms may be dipped or sprayed. Dry chromatograms at 100 °C for 4–5 min.
Ehrlich's reagent	Indoles: purple; hydroxy-indoles: blue; aromatic amines: yellow	a. p-Dimethylaminobenzaldehyde, 10 g in 100 mL concentrated HCl b. Acetone	1. Mix 1 volume of reagent (a) with 4 volumes of reagent (b) prior to use.
Fluorescamine	Very sensitive general reagent for amines, amino acids, peptides	a. Triethylamine, 10 g/dL in methylene chloride b. Fluorescamine, 20 mg/dL in anhydrous acetone	1. Dry chromatograms at 100 °C, then spray with reagent (a) and allow to dry for a few seconds. 2. Spray reagent (b) and allow to dry. Apply reagent (a) again; amino acid spots are visible at 360 nm. Sensitivity is very good at 250–300 pmol/L.
Iodoplatinate	Unoxidized sulfur amino acids form white spots on purple background	a. Chloroplatinic acid ($H_2PtCl_6 \cdot 6\ H_2O$), 1.0 g/L, in hydrochloric acid, 0.2 mol/L b. KI, 16.7 g/L in H_2O	1. Just before use, combine 0.6 volume of reagent (a) with 10 volumes of reagent (b) and mix; subsequently add 40 volumes of acetone. 2. Color development may take 24 h; plastic or glass plates *must* be used; chromatograms must be stored in a plastic bag in the dark.
Sakaguchi reaction	Arginine and other substituted guanidines form deep red or orange spots (fades)	a. 8-Hydroxyquinoline, 0.1 g/dL in acetone b. Bromine liquid, 0.3 mL in 100 mL sodium hydroxide, 0.5 mol/L	1. Prepare fresh. 2. Dip chromatograms into solution (a) and evaporate acetone. 3. Dip into reagent (b).

that are less satisfactory for visual observation; consequently, additional stains such as isatin (indole-2,3-dione) are often used. Isatin converts proline and hydroxyproline to blue compounds that are detectable against a yellow background. The addition of various divalent metal ions to ninhydrin or isatin staining solutions is frequently used to help stabilize the color complexes that are formed. With ninhydrin-Cd(II), all spots are given a uniform, bright, red-orange color, which facilitates comparisons of relative concentrations of the different amino acids. With isatin, Cd(II) produces a polychromatic reagent that reacts with amino acids to give a wide variety of colors. The addition of organic bases, such as collidine, to ninhydrin solutions also produces polychromatic staining, which facilitates identification of individual amino acids; the reader is referred to the article by Saifer for further information.[89] A brief summary of some useful stains is shown in Table 4-5. Stained chromatograms can be sprayed with acrylic solution and covered with protective plastic, both to preserve the quality and to facilitate storage.

One-Dimensional Chromatography

The principle of one-dimensional chromatography is described in Chapter 1B, Section Six. One-dimensional thin-layer chromatography of plasma, serum, or urine specimens can detect most aminoacidopathies. Relatively inexpensive commercial kits (e.g., Helena Laboratories, Beaumont, TX 77704, catalog no. 544) are available and well-suited for use in smaller laboratories with limited resources. In these methods, specimens, standards, and controls are applied to precoated-cellulose thin layer plates. Development is carried out in an appropriate solvent-vapor system,[8] followed by drying and staining of the chromatogram. The standard is a mixture of amino acids of known concentration. Examples of one-dimensional chromatograms are shown in Figure 4-3. By running several samples of the same specimen on a single plate and then staining each with different specific stains, it is often possible to identify an individual amino acid without resorting to the two-dimensional technique.

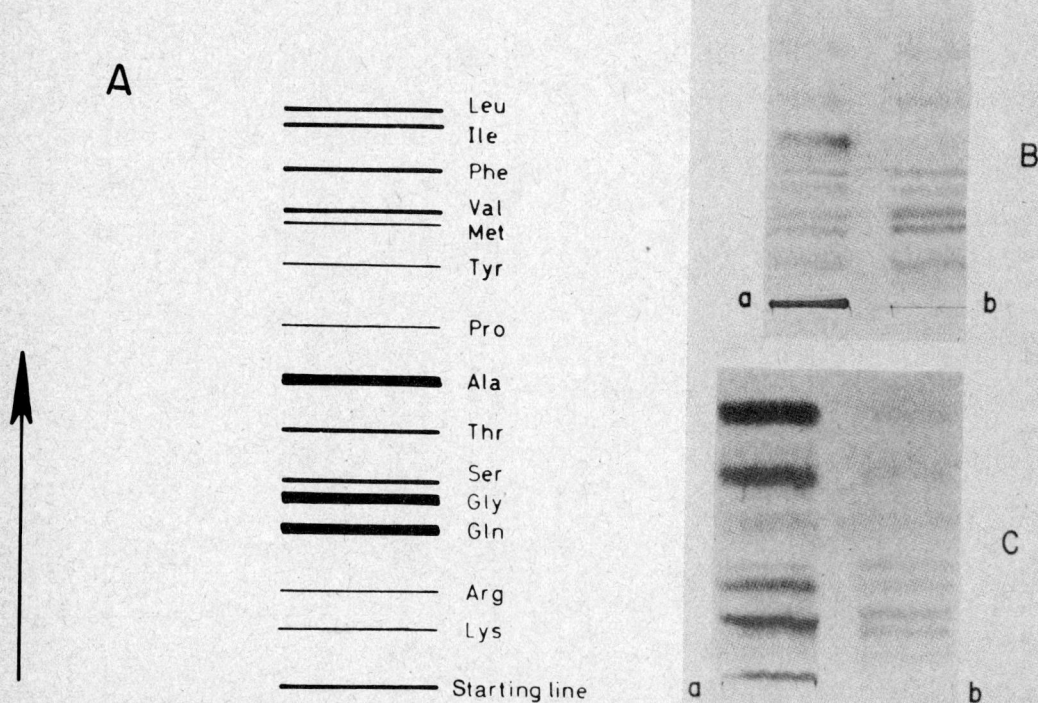

Figure 4-3. One-dimensional chromatography of amino acids, in a solvent system consisting of n-butanol:acetone: acetic acid:water (35:35:10:20, V|V). *A*, Positions of several amino acids after their separation from a standard mixture. *B*, Separation of plasma amino acids in transient neonatal tyrosinemia. Note tyrosine band a; **b** = amino acid standard. *C*, Separation of plasma amino acids in untreated maple syrup urine disease; **b** = amino acid standard. (From Bremer, H. J., Duran, M., Kamerling, J. P., et al.: Disturbances of Amino Acid Metabolism: Clinical Chemistry and Diagnosis. Baltimore, Urban and Schwarzenberg, Inc., 1981.)

Two-Dimensional Thin Layer Chromatography

Specimen. Two-dimensional thin layer chromatography can be used to identify free amino acids in blood, urine, cerebrospinal fluid, and other biological fluids and tissues. For the qualitative identification of amino acids in urine, a freshly voided early morning specimen is collected without preservatives. If analysis is not begun within 3 h, the specimen should be frozen until the test can be performed.

Principle.[89] Following protein precipitation with absolute ethanol, the urine sample is extracted with chloroform to remove urea and other organic and inorganic substances. Amino acids in the aqueous layer are then separated by thin layer chromatography on precoated cellulose plates. Two developing solvents are used: the first contains ammonia to increase the mobility of amino acids with basic side chains; the second contains formic acid to increase the mobility of amino acids with acidic side chains.

After development, individual amino acids are visualized with ninhydrin-collidine, a polychromatic staining reagent. Presumptive identification is made by comparing R_f values and characteristic colors of unknowns with those of standard mixtures run at the same time. Urine specimens containing high concentrations of inorganic salts, drug metabolites, or other interfering anions and neutral substances will generally require desalting by an ion-exchange technique to improve chromatographic resolution.

Reagents

1. Absolute ethanol, reagent grade.
2. Chloroform, reagent grade.
3. Dowex 50W X-8 ion-exchange resin, 200–400 mesh, hydrogen form. Suspend the commercial product overnight in hydrochloric acid, 2 mol/L. Filter and wash with distilled water until the eluate is neutral to pH paper. Dry the resin in a 38 °C oven for 24 h, and store the dried material in a dark bottle at room temperature.
4. Solvent System #1. Mix 40 mL pyridine, 30 mL acetone, 5 mL concentrated ammonium hydroxide, and 20 mL water in a flask. Volumes are sufficient for one TLC tank. Prepare fresh daily.
5. Solvent System #2. Mix 66 mL 2-propanol, 15 mL concentrated formic acid, and 15 mL water in a flask. Volumes are sufficient for one TLC tank. Prepare fresh daily.
6. Staining Reagent. Dissolve 0.2 g ninhydrin monohydrate in 100 mL isopropanol. Add 2.5 mL 2,4,6-trimethylpyridine (collidine) and mix. Prepare fresh daily.
7. Standards.
Solution A. Weigh and dissolve amino acids in hydrochloric acid, 0.1 mol/L, to give 5 mmol/L of each of the following: alanine, arginine hydrochloride, aspartic acid, cystine, glutamic acid, glycine, histidine hydrochloride, hydroxyproline, isoleucine, leucine, lysine hydrochloride, methionine, ornithine hydrochloride, phenylalanine, proline, serine, threonine, tryptophan, and tyrosine.
Solution B. Weigh and dissolve amino acids in hydrochloric acid, 0.1 mol/L, to give 5 mmol/L of each of the following: alanine, β-aminoisobutyric acid monohydrate, ethanolamine hydrochloride, glutamic acid, glutamine, glycine, histidine hydrochloride, lysine hydrochloride, serine, taurine, and threonine.
Note: Store reference standards at 4 °C where they are stable for at least six months. Neutralize before use by adding an equal volume of sodium hydroxide, 0.1 mol/L.

Supplies

Merck precoated 0.1 mm cellulose TLC plastic sheets (without indicator), 20 × 20 cm (Bodman Chemicals, Media, PA 19063, catalog no. 5577-7).

Procedure

Sample preparation

1. Transfer 1.0 mL of urine to a glass test tube. Add 3.0 mL of absolute ethanol and mix for 15 s. Allow the mixture to stand for 10 min and then centrifuge at 2000 × g for 10 min.
2. Decant the supernatant into another glass test tube. Add 3.0 mL of chloroform and mix vigorously. Allow the mixture to stand for 5 min, and then centrifuge at 2000 × g for 5 min.
3. Using a disposable glass pipet, transfer the upper aqueous layer to a labeled glass test tube. Cap the tube and use the extract for TLC analysis. Extracts may be stored at 4 °C for one week.
Note: In preparing the extract, about 20% of the total amino acids are lost, including approximately half the phenylalanine, leucine, and isoleucine.

Resin Desalting

4. If necessary, remove interfering anions and neutral substances as follows:
a. Transfer 1.0 mL of urine extract to a glass test tube. Add 0.5 g of Dowex 50W X-8. Mix vigorously for 1 min.
b. Centrifuge the mixture at 2000 × g for 5 min to precipitate the resin. Discard the supernatant.

c. Add 10 mL of deionized water to the resin and mix again for 1 min. Centrifuge as above and discard the supernatant.

d. Add 1 mL of ammonium hydroxide, 2 mol/L, to release amino acids from the ion-exchange resin. Centrifuge as above and save the supernatant. Repeat twice more for a total of 3 mL of NH$_4$OH.

e. Evaporate the combined eluates to dryness under a gentle stream of nitrogen. The temperature of the water bath should not exceed 40 °C.

f. Reconstitute the dried residue with 1 mL of water. Cap the tube and use the solution for TLC analysis. Reconstituted solutions may be stored at 4 °C for one week.

5. Divide a 20 × 20 cm cellulose TLC sheet into four quadrants by marking with a pencil. Position a straight-edged ruler along the penciled line and remove about 2 mm of the cellulose layer with the flat end of a small screwdriver.

Note: Top edges are scraped to prevent the cellulose surface from coming in contact with the solvent-saturated paper wicks during chromatographic development.

6. Using a 10-μL syringe, apply 4 μL of each urine extract and reference standard to a discrete point about 1 cm from each lower left corner of each quadrant of the 20 × 20 cm sheet. Keep the spot as small as possible by applying the sample in small amounts and drying with a heat gun between applications.

Note: The amount of urine extract applied to the chromatogram may be related to either a definite amount of amino acid nitrogen or to a definite amount of creatinine. Do not apply >8 μL.

Table 4-6. COLOR REACTION AND R$_f$ VALUES OF VARIOUS
AMINO ACIDS ON CELLULOSE TLC PLATES STAINED
WITH NINHYDRIN-COLLIDINE

Amino Acid	Color	R$_f$ × 100	
		1st Dimension	*2nd Dimension*
1. Cystine	Gray	15	10
2. Cysteine	Gray	15	10
3. Cystathionine	Violet	8	13
4. Cysteic acid	Violet	15	14
5. Phosphoethanolamine	Violet	4	21
6. Argininosuccinic acid	Violet	7	27–40
7. Ornithine	Violet	23–28	23
8. Asparagine	Beige	21	24
9. Homocystine	Violet	18	33
10. Arginine	Violet	11	23–37
11. Lysine	Violet	30	23–35
12. 1-Methylhistidine	Gray	29	26–39
13. Histidine	Gray	36	19
14. Carnosine	Tan	34	20–30
15. Taurine	Violet	46–60	23
16. Glutamine	Violet	21	34
17. Aspartic acid	Aquamarine	8	37
18. Citrulline	Violet	22	41
19. Methionine sulfoxide	Violet	26	39
20. Methionine sulfone	Violet	40	40
21. Serine	Violet	40	40
22. Glycine	Rose	24	43
23. Hydroxyproline	Orange	28	43
24. Glutamic acid	Violet	9	47
25. Sarcosine	Gray-pink	31	51
26. Threonine	Violet	65	44
27. β-Alanine	Aquamarine	33	54
28. Alanine	Violet	31	57
29. Proline	Yellow	36	57
30. Tyrosine	Gray-brown	53	55
31. Tryptophan	Yellow-gray	53	55
32. Ethanolamine	Violet	80	47
33. γ-Aminobutyric acid	Violet	35	65
34. α-Aminobutyric acid	Violet	38	58
35. β-Aminoisobutyric acid	Violet	37	66
36. Methionine	Violet	49	65
37. Phenylalanine	Gray	59	69
38. Valine	Violet	44	72
39. Leucine	Violet	54	78
40. Isoleucine	Violet	54	78

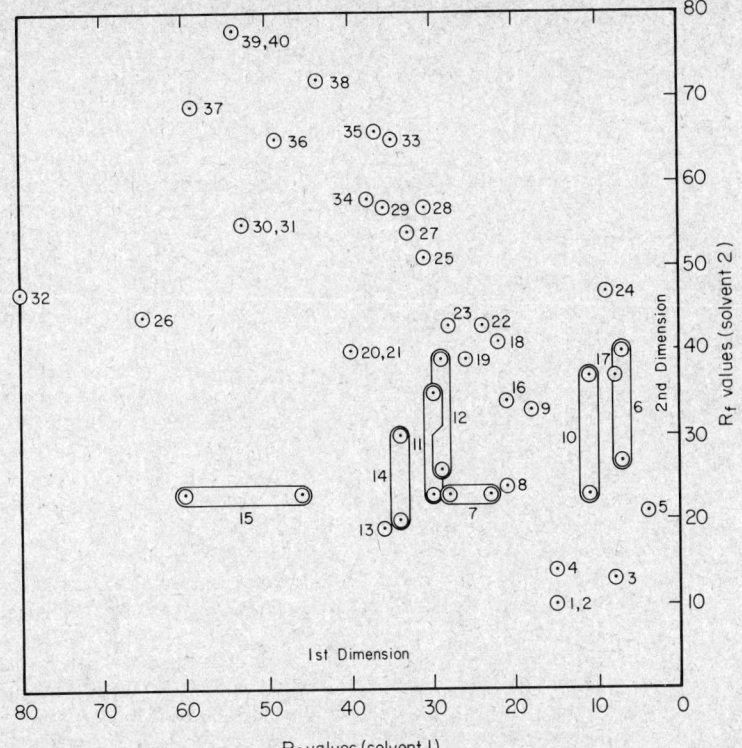

Figure 4-4. Composite map of R_f values ($\times 100$) of amino acids obtained with two-dimensional chromatography on cellulose. First dimension solvent—pyridine:acetone:concentrated ammonium hydroxide:water (40:30:5:20 by volume). Second-dimension solvent—2-propanol:concentrated formic acid:water (66:15:15 by volume). O = origin. The numbers correspond to the amino acids listed in Table 4-6.

Chromatographic Development

7. Cut the plastic TLC sheet into four 10×10 cm quadrants. Place each section into a tank previously equilibrated with solvent system #1. Solvents should be used only once. Tanks may accommodate more than one 10×10 cm section at a time.

8. Develop vertically for approximately 45 min, allowing the solvent front to reach the top of each sheet.

9. After development, remove sheets from the tanks and air dry at room temperature in a fume hood for 4 h.

10. Rotate plate origins 90° counterclockwise so that the application point is now at the right bottom corner. Place the TLC sheets in tanks previously equilibrated with solvent system #2.

11. Develop vertically for ~90 min, allowing the solvent front to reach the top of each sheet.

12. After development, remove sheets from the tanks and air dry at room temperature in a fume hood overnight.

13. Place sheets in a 4 °C refrigerator for 30 min. Then proceed immediately to the next step.

Visualization

14. In a well-vented fume hood, spray the chromatograms with the ninhydrin-collidine reagent until evenly moistened.

15. Support the plastic TLC sheets on cool Pyrex glass plates and heat in a 70 °C humidified oven for 10 min to visualize the colored spots. A damp towel may be placed in the oven 30 min prior to color development in order to saturate the air space with water vapor.

Results and Interpretations

1. Compare R_f values and characteristic colors of patients' chromatograms with those of standards. A fluorescent lightbox may be conveniently used to view each translucent chromatogram.

2. Trace patients' chromatograms onto clear vinyl sheets with a wax crayon. Then align each tracing over chromatograms of standards run at the same time. Identify each amino acid and record results. Expected color reactions and R_f values for a number of amino acids are listed in Table 4-6. R_f values will vary slightly with each individual run. A composite map is presented in Figure 4-4.

3. Examine each chromatogram for deviations from normal. Increases or decreases in any particular amino acid are noted and recorded. A "normal" chromatogram is characterized by the presence of five prominent amino acids: alanine, glutamine, glycine, histidine, and serine. Moderate amounts of β-amino-isobutyric acid, ethanolamine, glutamic acid, lysine, taurine, and threonine also occur in most normal pediatric urine samples. Traces of asparagine, cystine, or methylhistidine are only infrequently found.

4. Report results as positive or negative according to the following criteria:

Negative: Normal amino acid pattern—no detectable increase in any particular amino acid (or group of amino acids).

Positive: Abnormal amino acid pattern—detectable increase in a particular amino acid (or group of amino acids). Positive findings are reported with a suggestion that a second sterile morning urine be obtained to confirm the abnormal pattern. (The patient should be off any drug therapy for at least 3 d prior to specimen collection). A positive finding in this instance should be verified by quantitative amino acid analysis.

5. Consult Table 4-3 for interpretation of positive results. Transient aminocidurias in infants, due probably to developmental delays in enzymes or transport mechanisms, often present dilemmas in the interpretation. Many of the diagnostically useful amino acids, such as tyrosine, proline, and hydroxyproline, may or may not be elevated in newborns. Only when a hyperaminoaciduria has been demonstrated to be *persistent* can a reliable diagnosis be made. In addition, certain drugs can result in metabolic interferences or in the appearance of anomalous spots on the chromatogram. Review of the patient's history and close cooperation with the physician are, therefore, especially important.

6. For reliable interpretation, chromatograms must be of high quality. Visual inspection by an experienced analyst is presently more satisfactory than automated scanning/quantitation techniques. Considerable experience with normal and abnormal patterns and with artifacts must be gained before clinical interpretation is attempted.

7. Sensitivity: This TLC method will detect at least 0.1–0.5 μg of each amino acid applied to the plate.

8. In general, urine is more useful than plasma or serum for evaluation of a suspected metabolic disorder. However, urine chromatography may not be effective in screening for disorders such as PKU or MSUD during the neonatal period. If desired, two dimensional TLC can be used to identify amino acids in blood. If < 1.0 mL of plasma or serum is available, then the volumes of extraction solvents must be reduced proportionately. Approximately 2.5 μL of the extract is applied to the chromatogram.

Colorimetric Screening Tests for Urine

A variety of qualitative tests (described later in detail) can be employed for screening, spot checking, or supplemental information. These tests are summarized in Table 4-7.

Table 4-7. COLOR TESTS ON URINE

Test and Color	Metabolites	Disorder
*Ferric chloride** (p. 550)		
Dark blue-green (persistent)	Phenylpyruvate	Phenylketonuria
Green (transient)	p-Hydroxyphenylpyruvate	Tyrosinuria
Blue (transient)	Homogentisic acid	Alkaptonuria
Gray-green	Imidazolepyruvate	Histidinemia
Gray-blue	Branched-chain keto acids	Maple syrup urine disease
Blue-green	5-Hydroxyindoleacetic acid	Carcinoid
Purple	Salicylates	(Interferent)
Purple-brown	Phenothiazines	(Interferent)
Cyanide/nitroprusside (p. 556)		
Cherry-red	Cystine	Cystinuria; generalized aminocidurias
	Homocystine	Homocystinuria
	Cysteine-homocysteine disulfide	Homocystinuria
	Penicillamine-cysteine disulfide	(Treatment)
2,4-Dinitrophenylhydrazine (p. 557)		
Yellow-white (ppt)	Branched-chain keto acids; phenyl-pyruvate; p-hydroxyphenylpyruvate	Maple syrup urine disease; phenylketonuria; tyrosinosis
Nitrosonaphthol (p. 554)		
Orange-red	Tyrosine and its metabolites (e.g., p-hydroxyphenyl-pyruvate, -lactate, -acetate)	Tyrosinosis; tyrosinemia

*Phenistix (Ames Co., Division of Miles Laboratories, Elkhart, IN 46514) is a commercial version of the ferric chloride test (p. 550).

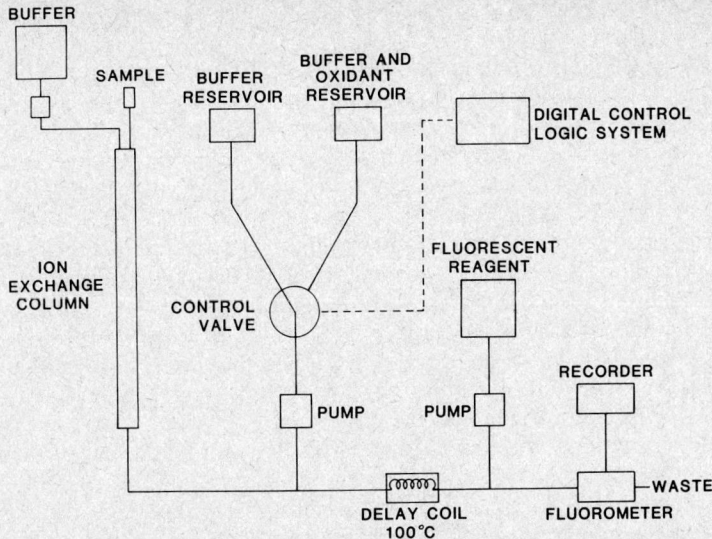

Figure 4-5. Automated amino acid analyzer equipped for fluorescent detection of amino acids. As proline and hydroxyproline emerge from the ion-exchange column, the digital control logic system introduces a reagent that oxidizes these imino acids to compounds capable of reacting with fluorescent detection reagents. The control valve is returned to the buffer reservoir when the other amino acids are eluted. (From Hammond, J. E., and Savory, J.: Advances in the detection of amino acids in biological fluids. Ann. Clin. Lab. Sci., 6:158–166, 1976.)

Quantitative Tests for the Separation of Amino Acids

Ion-exchange chromatography. The principle of the ion-exchange chromatographic separation of amino acids has been extensively applied in research laboratories. (See Chapter 1B, Section Six.) Following column separation of the compounds, ninhydrin or some other indicator is combined with the column effluent so that amino acids can be detected using an online spectrophotometer, fluorometer, or other detection device (Figure 4-5). The amino acids are identified by relating the *retention times* of the components in the specimen to those of standards. Quantitation is made by comparison of specimen peak areas with those from sets of standards or, alternatively, by using an internal standardization technique. Fluorometric detection with fluorescamine and *o*-phthalaldehyde have improved sensitivity, while shorter, narrower columns packed with smaller particles operating at higher flow rates have improved the rate of analysis.

Ion-exchange chromatography, however, is not suitable for the average clinical laboratory. The analyzer is a large, expensive instrument dedicated solely to assays that are infrequently needed. Furthermore, completion of an analysis of one specimen requires at least 3 h.

High performance liquid chromatography (HPLC). HPLC has emerged as an attractive means of amino acid analysis because of its great sensitivity, high resolution, and relatively short analysis time. (See Chapter 1B, Section Eight.) The major advantage of HPLC over GLC is that the relatively high temperatures needed for sample volatilization by GLC are unnecessary with this technique. Thus, the problem of amino acid decomposition at high temperatures is avoided. HPLC has the additional advantage of avoiding the time-consuming derivatization reactions that are necessary for GLC analysis. Precise solvent delivery systems, microbore separation columns, and automated instrumentation continue to create widespread interest in HPLC. Reference 49 serves as a good source for more detailed information.

Detection of amino acids separated by HPLC is ordinarily achieved by ultraviolet or visible spectrophotometry, fluorometry, or electrochemical detection; however, absorption by amino acids of ultraviolet radiation between 250 and 280 nm is not sufficiently great for many applications. A number of interfering, nonamino acid substances also absorb in this wavelength range so that an initial cleanup of the specimen is generally necessary. Post-column derivatizations with ninhydrin[64] or fluorogenic reagents,[6,88] such as *o*-phthaldialdehyde or fluorescamine, have been successfully used for detection purposes. More recently, precolumn derivatization techniques using *o*-phthaldialdehyde (OPA)[50] or dansyl derivatives[22,110] have been used with reversed-phase

HPLC. Electrochemical detection has also been coupled with derivatization methods to enhance analytical sensitivity.

Gas-liquid chromatography. Separations using gas-liquid chromatography (GLC) are based on the interaction of volatilized sample compounds with a relatively nonvolatile liquid phase which has been coated onto a solid column support. Separation on the GLC column allows identification and quantitation of sample compounds by appropriate detection techniques (see Chapter 1B, Section Seven).

Advantages of GLC include small sample size, sensitivity, and speed, but a major limitation is the relatively low volatility of amino acids at temperatures conventionally used in this technique. In addition, many amino acids tend to decompose rather than to volatilize upon heating. Fortunately, these compounds can be reacted with derivatizing agents to increase volatility and to enhance their chromatographic and detection characteristics. Single derivatization reactions, however, have not proved very successful and multiple derivatization reactions have to be used. These reactions often involve formation of amino acid alkyl esters with subsequent acetylation reactions involving the amino and other functional groups.[96]

The necessity for an initial "cleanup" step, usually on an ion-exchange column, followed by derivatization makes GLC unattractive to the routine clinical laboratory and thus it is rarely used. There are, however, two applications for which it is the method of choice, namely, the investigation of organic acidurias and, in combination with mass spectrometry, the exact identification of an unknown amino acid.

SELECTED DISORDERS OF AMINO ACID METABOLISM

Hyperphenylalaninemias

The hyperphenylalaninemias are a group of disorders resulting from impaired conversion of phenylalanine (Phe) to tyrosine. Normally, the major metabolic pathway for phenylalanine involves an enzyme complex known as phenylalanine hydroxylase (PH).

This catalytic system is found in appreciable amounts only in the liver and kidney. Tetrahydro-biopterin is an obligatory cofactor; another enzyme, dihydropteridine reductase, keeps this cofactor in an active reduced state.

The hyperphenylalaninemias are nearly always due to deficiencies of phenylalanine hydroxy-lase, although some 1–3% of cases are due to a defect in either dihydropteridine reductase or one of the enzymatic steps involved in biopterin synthesis. As a group, the hyperphenylalaninemias have an incidence of about 1:10 000 live births; rates are as high as 1:5400 in Ireland and as low as 1:16 000 in Switzerland.

Approximately half of all cases are *classic phenylketonuria* (PKU), in which PH activity is almost totally absent. As a direct consequence of impaired parahydroxylation, Phe accumulates in blood, urine, and CSF. In contrast to blood Phe levels found in normal subjects (adults <2 mg/dL), levels in classic PKU are >20 mg/dL. As a result of these high concentrations of blood Phe, normally minor pathways of Phe metabolism are activated, leading to increased production of phenylketones (e.g., phenylpyruvate) and other metabolites. These metabolites are rapidly cleared by the kidney and excreted into the urine.

Less severe forms of PKU, often referred to as *PKU variants*, are also observed and result from partial deficiencies of PH. In the atypical variant, PH activities may be up to 6% of normal, and ketonuria may or may not be present; blood Phe levels are generally in the 8–20 mg/dL range. In the benign persistent variant, enzyme activities are about 9% of normal, and blood Phe levels generally do not exceed 8 mg/dL. Another form of hyperphenylalaninemia is referred to as *transient neonatal hyperphenylalaninemia*. This disorder is caused by delayed hepatic maturation of the PH enzyme system. This condition is not an inherited defect, and blood Phe levels may be >12 mg/dL initially but will progressively decline toward normal as the neonate matures.

Untreated PKU causes severe mental retardation. Affected children, however, appear normal at birth, and the earliest symptoms are usually nonspecific—delayed development, feeding dif-ficulties, and vomiting sometimes sufficiently severe to suggest pyloric stenosis. In some children, an unusual but characteristic musty odor may be observed in urine or sweat, due to increased production of phenylpyruvate. Without treatment, retardation progresses. Older patients fre-quently show hyperactivity, seizures, eczema, and hypopigmentation.

Hypopigmentation in PKU relates to the hyperphenylalaninemia since Phe is a competitive inhibitor of tyrosinase, an enzyme which initiates melanogenesis. Accumulation of Phe is also known to reduce myelin, norepinephrine, and serotonin synthesis, and may, directly or indirectly, cause severe neurologic symptoms, although the mechanism is not known. Actual injury to brain tissue begins within the second or third week of life and becomes maximal at about eight or nine months. Without therapy, the child may develop with an IQ of less than 20; <4% of untreated cases reach an IQ of more than 60.

Treatment consists of restricting dietary Phe before the onset of brain damage so that the serum Phe concentration is kept at 3–8 mg/dL. Even when diagnosis is made as late as 4–6 months of age, institution of diet therapy will decrease the rate of further mental deterioration. It used to be thought that dietary restriction could be lifted after age five, but further experience has shown that monitoring of Phe blood levels and dietary control should continue indefinitely. Monitoring and control are especially important for females (see maternal hyperphenylalanemia, below). The patient on dietary therapy is best monitored by blood analysis using quantitative methods, but the Phenistix or ferric chloride test for phenylpyruvate in urine can be used more frequently and inexpensively as an early warning test, analogous to dipstick testing in control of diabetes mellitus.

Because *early diagnosis* is essential in order to avoid the severe effects of PKU, neonatal screening has become widespread in the U.S., Australia, Great Britain, and other European countries. Ideally, prenatal diagnosis would be best, but PH is a hepatic enzyme and, therefore, is not present in blood or fibroblasts. DNA analysis is promising. The human gene for PH has now been cloned and can be used to analyze the PH locus. Results suggest that prenatal diagnosis from cultured amniotic cells or trophoblast biopsy should be possible in 75% of PKU families by taking advantage of harmless polymorphisms that alter positions at which endonucleases cut the DNA chain.[116] Considerable advantage can be gained by examining parents and siblings for heterozygosity. About 2% of the Caucasian population in the United States are *heterozygote carriers* of the defective gene. Carriers usually have blood Phe levels within the normal range, but hepatic PH activity is subnormal so that limited capacity to metabolize Phe can be dem-onstrated by a load test (see below).

Maternal hyperphenylalaninemia is a condition that is likely to occur in the adult phenyl-ketonuric pregnant female who has been successfully treated by dietary control since infancy. More and more instances of this situation are arising as treated PKU children become adults. High maternal levels of Phe cross the placenta and have disastrous effects on the fetus—cardiac defects, intrauterine growth retardation, anencephaly, and mental retardation. To reduce or prevent some or all of these adverse effects, a restricted diet must be followed before conception and throughout pregnancy. This also applies to the consumption of the synthetic sweetener Aspartame, which is a dipeptide of phenylalanine and aspartic acid.

Hyperphenylalaninemia due to tetrahydrobiopterin deficiency is caused by a defect in *dihydropteridine reductase* or one of the enzymatic steps involved in *biopterin synthesis*. The disorder results in severe progressive neurological disease, probably the result of failure to synthesize the neurotransmitters serotonin, dopamine, and norepinephrine. Dietary control of Phe blood levels usually has no beneficial effect, but replacement therapy with precursors of the neurotransmitters that appear after the block in the pathway, namely 5-hydroxytryptophan and dihydroxyphenyl-alanine (DOPA), has some benefits. In one of the variants, treatment required dietary restriction of Phe as well as supplementation of tetrahydrobiopterin.

Test for Heterozygote or Carrier State

The suspected carrier is given an oral dose of 100 mg of phenylalanine per kg body weight. Blood specimens are then obtained at hourly intervals for 4 h and assayed for phenylalanine. In a noncarrier, phenylalanine concentrations will rise from a fasting value of about 1.4 mg/dL to a value of 9 mg/dL at 1 and 2 h, and then drop to 5 mg/dL at 4 h. In a heterozygote, the phenylalanine concentration will rise from normal to a value of about 19 mg/dL at the first hour and fall much more slowly than in the normal person. A phenylalanine tolerance index is calculated by summing the 1-, 3-, and 4-h values. The values of the index for normal individuals and carriers are sufficiently disparate to distinguish clearly between the two. In phenylketonurics the rise in serum level is higher and more prolonged, reaching 30 mg/dL at 1 h and 40–59 mg/dL at 2–5 h.

Alternatively, the elimination rate of phenylalanine may be observed following an intravenous loading dose of 300 μmol/kg. Distinction of normal from homozygotes and heterozygotes has been achieved with only a 2% overlap when an elimination rate of 4.34 mmol/h per m^2 was used as a discriminatory level.[42]

Methods for the Determination of Serum Phenylalanine

Fluorometric Determination of Serum Phenylalanine[28]

Principle. Phenylalanine is reacted with ninhydrin in the presence of the dipeptide L-leucyl-L-alanine to form a fluorescent product which is proportional to the phenylalanine concentrations. The reaction is carried out at pH 5.88. An alkaline copper tartrate reagent is added to stabilize the fluorescent product.

Specimen. Blood specimens are drawn from infants into microcapillary tubes. If the assay must be delayed, the plasma or serum is separated and frozen until the test is begun (50 μL required for the procedure may be stored in a plastic microcentrifuge tube). Phenylalanine values increase during storage of whole blood; when phenylalanine levels are within the reference range, the increase may be as much as 30–50% in the course of 5–7 d. This increase does not occur in frozen serum or plasma.

Reagents

1. Succinate buffer, 0.6 mol/L, pH 5.88 $\pm$ 0.03. Dissolve 70.9 g of succinic acid in about 600 mL of water; using a pH meter, carefully add about 200 mL of NaOH, 5 mol/L, with thorough stirring, until a pH of 5.88 is obtained (see Comments). Store at 0–5 °C. Stable for 2–4 months. Recheck pH.
2. Ninhydrin solution, 0.03 mol/L. Dissolve 2.67 g of reagent grade ninhydrin in water to make 500 mL of solution. Store at 0–5 °C. Stable for 7–10 d.
3. L-Leucyl-L-alanine, 5 mmol/L. Dissolve 101 mg in 100 mL of water. Aliquot and store frozen. Stable for six months at −20 °C.
4. Buffered ninhydrin-peptide mixture. Prepare fresh daily by mixing 5 volumes of succinate buffer, 2 volumes of ninhydrin solution, and 1 volume of peptide reagent 3.
5. Copper sulfate solution, 0.8 mmol/L. Dissolve 50 mg of $CuSO_4 \cdot 5 H_2O$ in 250 mL of water. Stable indefinitely.
6. Alkaline tartrate solution. Dissolve 1.33 g of anhydrous Na_2CO_3 and 57 mg Rochelle salt, potassium sodium tartrate tetrahydrate ($KNaC_4H_4O_6 \cdot 4 H_2O$), in 500 mL of water. Stable if stored in a polyethylene bottle at room temperature.

7. Alkaline copper reagent. Mix 3 volumes of alkaline tartrate with 2 volumes of copper solution. Prepare this reagent fresh daily.

8. Trichloroacetic acid (TCA), 0.6 mol/L. Dissolve 98.0 g TCA, fluorometric grade, in 1 L of water. Store refrigerated. Stable for six months.

9. Phenylalanine standards.

a. Stock standard, phenylalanine 200 mg/dL, in water. Use only fluorometric grade reagents. Store frozen. Stable for 1–2 months.

b. Working standards, in water. Dilute 25 μL and 50 μL of stock standard to 4.0 mL with TCA, 0.3 mol/L. Store the working standards frozen. Treated in the same manner as sample filtrates, these standards represent concentrations equivalent to plasma levels of 2.5 and 5.0 mg/dL. Working standards in water are used only to assay normal pooled serum that is used to prepare standards in serum.

c. Working standards, in serum. Pool normal sera free of hemolysis, icterus, and turbidity. Use as a diluent to prepare assay standards as follows:

Standard Number	Stock Standard, μL	Serum Pool, mL	Phenylalanine Concentration, mg/dL
1	0	5.0	A*
2	100	4.9	0.98A + 4.0
3	200	4.8	0.96A + 8.0
4	300	4.7	0.94A + 12.0
5	400	4.6	0.92A + 16.0

Microprocedure

In this method a set of microliter pipets, a microcentrifuge, and plastic microcentrifuge tubes (0.3–0.5 mL capacity) should be used.

1. Into microcentrifuge tubes, pipet 50 μL of specimen or standards, and add 50 μL of TCA, 0.6 mol/L. Mix the tubes by vigorous hand-tapping or with a mechanical mixer. Allow them to stand for 10 min and then centrifuge in a microcentrifuge.

2. Pipet 50 μL of each filtrate to a set of 13 × 100 mm test tubes. Also set up a blank containing 50 μL of TCA, 0.30 mol/L.

3. Add 0.80 mL of the buffered ninhydrin-peptide mixture to each tube and thoroughly mix the tubes. Cover the tubes with small marbles, or otherwise cap, transfer to a water bath at 60 ± 1 °C, and heat for 2 h.

4. While the tubes are incubating, turn on the fluorometer. Set the excitation (primary) wavelength at 365 nm, the emission (secondary) wavelength at 515 nm.

5. After the 2-h incubation period, remove the tubes from the bath and place into a cooling bath at about 20 °C.

6. Add 5.0 mL of fresh alkaline copper reagent to each tube. Mix the contents of the tubes and transfer to fluorometer cuvets. Zero the fluorometer and choose the aperture or slit settings so that the highest standard will give a scale reading of 60–80. Record the fluorescence readings of the blank, standards, and specimens. Correct all readings for the reading of the TCA blank.

7. Draw a calibration curve relating the corrected fluorometer scale readings to the concentrations of the serum phenylalanine standards, and read the concentrations of the unknowns from the standard curve. The curve should be linear to 14–16 mg/dL.

8. Run all standards in duplicate. Repeat any test with a value >4 mg/dL to verify the result. The physician should be informed immediately and asked to provide a fresh, new specimen to confirm the abnormal result.

Comments

1. The above microprocedure can be scaled down further: filtrates are made with 20 μL each of serum and of TCA; then 20 μL of filtrate is used for the reaction; 300 μL of buffered ninhydrin-peptide and 2.0 mL of alkaline copper reagent are used in place of the quantities previously mentioned.

2. The pH of the succinate buffer must be 5.88 ± 0.03. If the pH is lower, a loss of sensitivity occurs, and at higher pH values other serum amino acids may give significant fluorescence. At pH 5.88, only tyrosine, leucine, and perhaps arginine give measurable readings, but these are <5% of an equimolar concentration of phenylalanine. If these amino acids are present in substantially increased amounts, they

---

*A stands for the concentration of phenylalanine in the serum pool as determined against working standards in water or against a previous, reliable set of serum standards. If A is found, for instance, to be 2.1 mg/dL, the serum standard set would have values of 2.1, 6.1, 10.0, 14.0 and 17.9 mg/dL. Serum standards may be kept frozen for 4–6 weeks. They are treated in the assay procedure exactly like the unknowns.

will give an apparent increase in the phenylalanine value. The use of standards in serum corrects for the presence of normal levels of other amino acids and assures the same pH in unknowns and standards.

Reference Ranges

	mg/dL
Adults:	0.8–1.8
Newborns, full term and normal weight:	1.2–3.4
Newborns, low birth weight or premature:	2.0–7.5
Newborns, phenylketonuric: 2–3 days after birth	>4.5
untreated, at 10 days	15–30

Determination of Phenylalanine by Chromatography

Automated ion-exchange column chromatography[61] may be used as a confirmatory test for specimens yielding positive phenylalanine screening results and for monitoring the dietary treatment of patients with PKU. GLC methods[44] using various derivatives and one-dimensional paper chromatography are also available.

Spectrophotometric Methods for the Determination of Phenylalanine

Estimation of phenylalanine using the enzyme phenylalanine ammonia lyase to convert phenylalanine to measurable transcinnamic acid has been developed.[95] Older spectrophotometric methods are based on conversion of phenylalanine to phenylpyruvate by L-amino acid oxidase.[55] Tyrosine, tryptophan, and histidine interfere with the L-amino acid oxidase methods; in addition, these methods require a high resolution spectrophotometer and considerable technical skill.

Ferric Chloride Test for Phenylpyruvic Acid (PPA) in Urine

This test is both qualitative and nonspecific; however, the ferric chloride test is useful in monitoring the effectiveness of dietary therapy of PKU patients. This test should not be used for screening purposes (see below).

Reagents

1. $FeCl_3$ solution. Dissolve 10 g $FeCl_3 \cdot 6\ H_2O$ in water to give 100 mL solution.
2. Phosphate precipitating agent. Dissolve 2.2 g $MgCl_2 \cdot 6\ H_2O$, 1.4 g NH_4Cl, and 2.0 mL concentrated NH_4OH in water to give 100 mL of solution.

Procedure

Add, with mixing, 1.0 mL of phosphate precipitant to 4.0 mL of fresh urine. Filter the mixture and acidify the filtrate with 2–3 drops of concentrated HCl. Add 2–3 drops of $FeCl_3$ solution; observe the filtrate after each drop for any color formation. A dark green to blue-green color that persists for 2–4 min indicates a positive test. Very rapidly fading greens suggest homogentisic acid (HGA) or p-hydroxyphenylpyruvic acid (PHPPA). Imidazolepyruvic acid gives a positive test in histidinemia, identical with that for PPA, but it is encountered only rarely. Bilirubin may give a false positive reaction, but would normally be absent. Other colors than those given in Table 4-7 are read as negative. The sensitivity for PPA is about 10 mg/dL. By using serial dilutions of the urine filtrate, a semiquantitative measure of the PPA concentration may be made.

PPA usually cannot be detected in infant urines until the serum phenylalanine concentration has reached levels between 12 and 15 mg/dL; this does not usually occur until 10–14 d after birth. Infants affected with phenylketonuria will excrete up to 2 g PPA/d, and urinary concentrations will vary from 50–100 mg/dL. PPA is unstable.

A commercial product called Phenistix (Ames Co., Division of Miles Laboratories, Elkhart, IN 46515) may also be used to test for PPA. In this procedure, also based on the ferric chloride reaction, the presence of PPA will change the color of the reagent-impregnated paper from a light gray to a gray-green or green-blue. Light, medium, and high readings correspond to PPA levels of 15, 40, and 100 mg/dL respectively.

Tyrosinemia and Related Disorders

There are several forms of tyrosinemia, each of which is accompanied by tyrosinuria and phenolic aciduria. Tyrosine is essential for protein synthesis and serves as a precursor for thyroxine, melanin, and catecholamines. The body obtains tyrosine from dietary proteins and partly, as we have seen, from hydroxylation of phenylalanine to tyrosine in vivo. The metabolism of tyrosine may follow a number of different pathways. The major pathway leads to p-hydroxyphenylpyruvic acid (PHPPA) and then to homogentisic acid (HGA). The aromatic ring in the latter is opened by homogentisic acid oxidase to form maleylacetoacetic acid (MAA), which is then isomerized and hydrolyzed to fumarate and acetoacetate. These four-carbon containing compounds are then metabolized in the appropriate pathways.

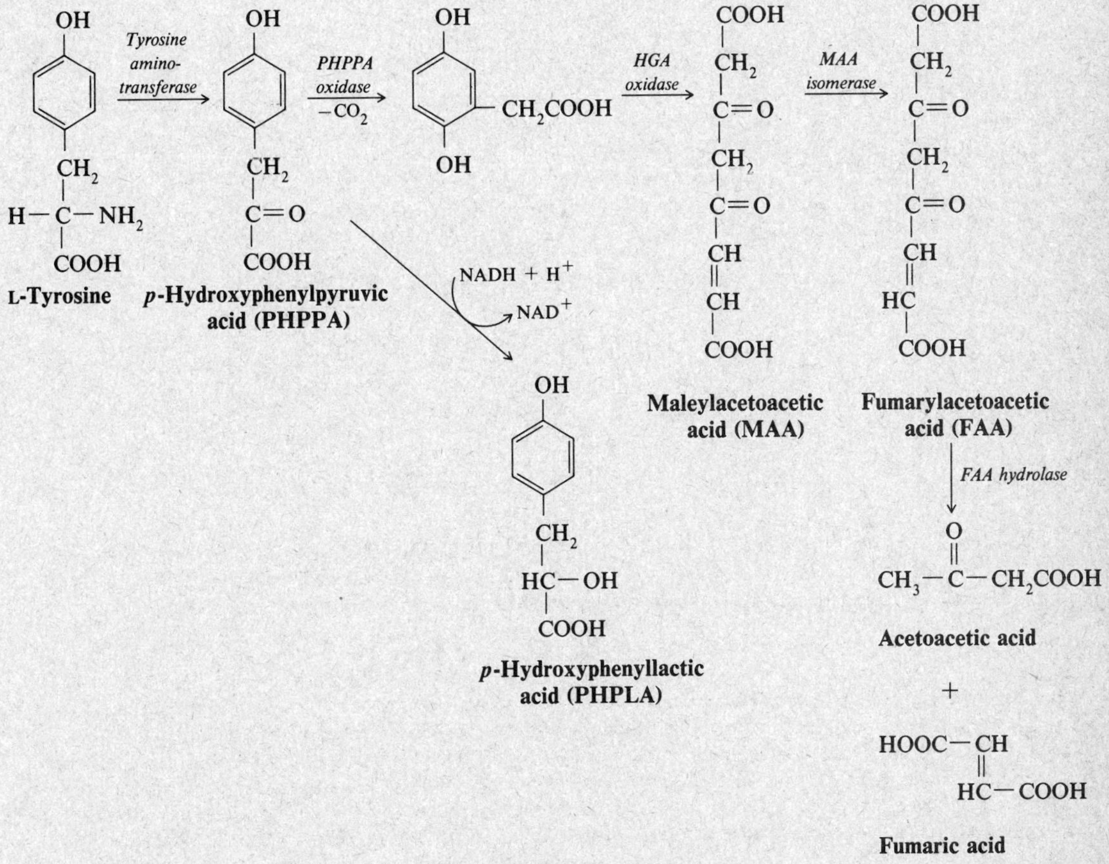

In a minor pathway, tyrosine is converted to dihydroxyphenylalanine (DOPA) and then to the pressor hormones norepinephrine and epinephrine. The reaction is as follows:

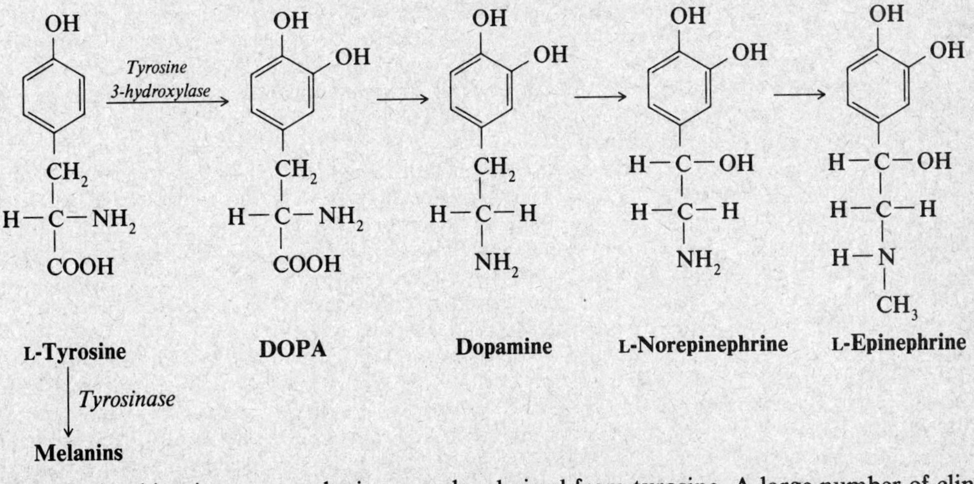

As noted, the skin pigments, melanins, are also derived from tyrosine. A large number of clinical syndromes exhibiting hypomelanosis have been described. Those that result from inherited defects in melanin synthesis are collectively known as *albinism*. Tyrosinase activity is frequently absent or deficient in many of these disorders.

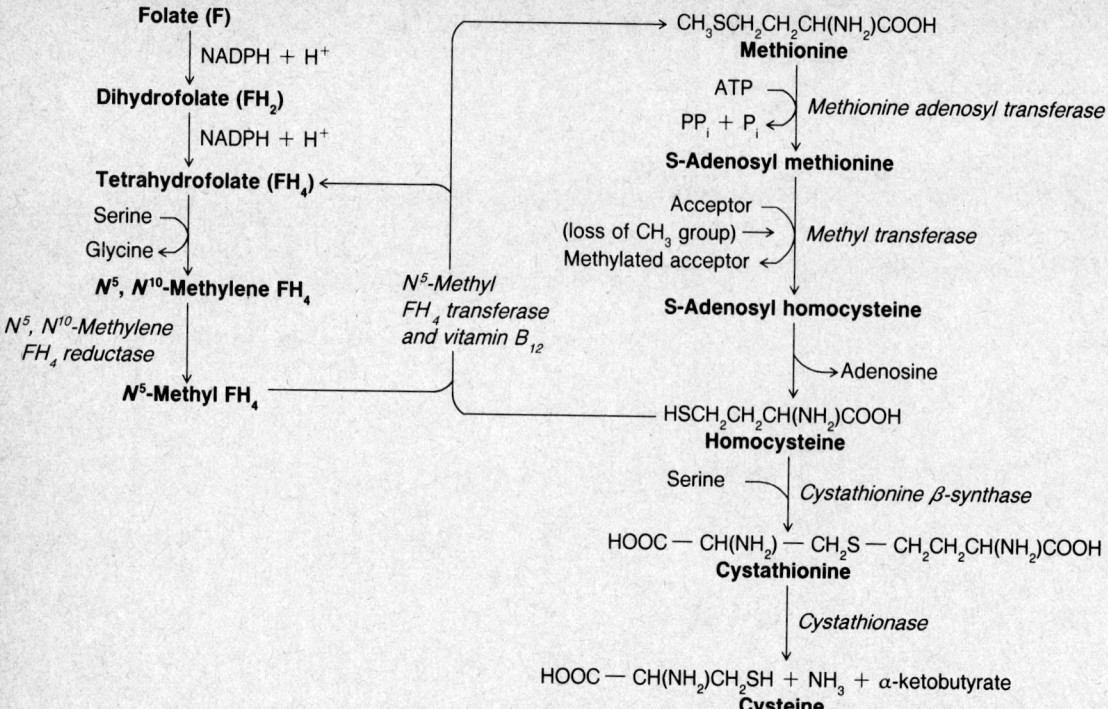

Figure 4-6. Biochemical pathways of the conversion of methionine to homocysteine and cysteine.

Tyrosinemia I (tyrosinosis, hepatorenal tyrosinemia). Reduced activity of fumarylacetoacetate (FAA) hydrolase is thought to be the primary defect in tyrosinosis, but the activity of PHPPA oxidase has also been found to be reduced. The enzyme block causes increased tyrosine levels in blood and urine, elevated methionine levels in blood, and excretion of large amounts of DOPA and other tyrosine metabolites in urine. The high blood concentration of methionine appears to be due to an ill-understood inhibitor of methionine adenosyltransferase. (See Figure 4-6.) Increased DOPA is the result of increased activity of the minor pathway caused by the enzyme block. Other distinctive biochemical findings include increased serum α-fetoprotein, sometimes to very high levels, and increased urinary excretion of δ-aminolevulinic acid.

The primary effects of the disease are on the liver and kidneys. The liver damage varies clinically from acute hepatic failure and death in infancy to a chronic condition with cirrhosis later in life. A generalized transport failure (Fanconi's syndrome) develops in the proximal tubules of the kidney with resulting vitamin D resistant rickets (sometimes the presenting symptom), hyperphosphaturia, glucosuria, and aminoaciduria. Unfortunately, this generalized aminoaciduria tends to complicate the specific urine amino acid findings associated with hepatorenal tyrosinemia, and liver failure complicates the plasma findings. Most cases of Type I tyrosinemia have been found in an isolated French Canadian population in Quebec. Elsewhere, the disease has an incidence of about 1:100 000. Dietary restriction will correct the biochemical change, but unfortunately not the progressive liver damage.

Tyrosinemia II. The deficiency of the hepatic enzyme, tyrosine aminotransferase, which catalyzes the first stage of tyrosine catabolism, is the cause of tyrosinemia II. This form is rare but better understood than tyrosinemia I. Distinctive clinical features of Type II tyrosinemia include eye lesions, such as corneal erosions, and skin lesions of the palms and soles. These eye and skin lesions are probably secondary to intracellular formation of tyrosine crystals, which cause inflammation. There is also occasional mental retardation. Elevated levels of tyrosine are found in blood and urine, as well as increased concentrations of phenolic acids and tyramine in urine; plasma methionine, however, is not elevated.

Transient neonatal tyrosinemia. Serum tyrosine levels are high in premature infants and in full-term infants of low birth weight, due to the immaturity of the liver and its decreased ability to synthesize the appropriate enzymes. As the liver matures, the accumulated tyrosine is metabolized and serum levels decrease to adult levels within 4–8 weeks. Immaturity of an infant's liver may also be associated with phenylalanine hydroxylase deficiency and, therefore, a temporary increase in serum phenylalanine levels. Furthermore, phenylalanine hydroxylase may be inhibited as a result of the accumulation of tyrosine. Care must be taken in such cases to avoid a possible erroneous diagnosis of phenylketonuria. *Neonatal screening* for tyrosinemia can be carried out by a Guthrie test (Table 4-4) but is impractical except in areas of known high frequency.

Methods for the Determination of Tyrosine in Serum

Ion-exchange column chromatography will detect a substantial increase in the serum level of tyrosine and is regarded as a reference method. The most convenient method for the quantitative assay of tyrosine in serum is the fluorometric procedure of Udenfriend and Cooper.[107] A TCA filtrate of the serum is treated with a mixture of α-nitroso-β-naphthol (ANBN) and nitrite in the presence of nitric acid. Initially the ANBN and nitrite react with tyrosine to form a pink-colored complex. Treatment with nitric acid converts this pink complex to a yellow material that fluoresces. This yellow pigment is extracted into ethylene dichloride in order to remove excess ANBN, and its fluorescence is measured at 570 nm after excitation at 460 nm. Simple micro- and semi-micro modifications of the Udenfriend and Cooper procedure have been reported by Hsia and Inouye.[41] These fluorometric methods are not specific; tyramine and other p-hydroxyphenyl compounds can also react but are generally not present in sufficient quantity to interfere in serum assays. The method is unsuitable for urine, where interfering compounds are present in greater concentration.

Enzymatic methods employ tyrosinase (from mushrooms), L-amino acid oxidase (from snake venom[55]), and ammonia lyase (from yeast).[95] Tyrosinase catalyzes the oxidation of tyrosine to dopaquinone, and consumption of oxygen has been measured amperometrically. The method is sensitive, requires only 10–50 μL of sample, and does not demand deproteinization, extraction, or incubation. Products formed in the other enzymatic assays can be measured spectrophotometrically.

Reference Ranges

	mg/dL[105]	mg/dL
Premature, 1 d:	2.17 ± 1.81	7–24
Newborn, 1 d:	0.76 – 1.79	1.6–3.7
1–3 months:	1.48 ± 0.47	
2–6 months:	1.30 – 3.91	
9 months–2 years:	0.20 – 2.21	
3–10 years:	1.01 – 3.22	
6–18 years:	0.78 – 1.59	
Adults:	0.40 – 1.57	0.8–1.3
		(mean: 1.1)

Tyrosinemia I (Tyrosinosis)	mg/dL
Children:	?
Adults:	4 – 10

Tyrosinemia II	mg/dL (SD)
Younger patients (up to 7 years):	43.4 ± 10.97
Older patients (8–55 years):	23.0 ± 5.3

Phenylketonuria	mg/dL
Children:	Low adult range
Adults:	0.53 – 0.57

Detection of p-Hydroxyphenylpyruvic Acid (PHPPA) in Urine

Tyrosinemia can be diagnosed most reliably by chromatographic isolation and identification of PHPPA from urine. Urinary excretion of PHPPA in affected patients may reach 1.6 g/d, about 25 times that excreted by normal subjects. More conveniently, the Millon reaction or the nitrosonaphthol test can be used to detect a wide variety of substituted phenolic compounds in urine. These tests, however, are not specific for PHPPA; positive reactions are also given by tyrosine and tyrosine metabolites, such as p-hydroxyphenyllactate (PHPLA) and p-hydroxyphenylacetate (PHPA), that accompany the increased urinary output of p-hydroxyphenylpyruvic acid (PHPPA).

Modified millon reaction. Mix 2.5 mL of urine and 2.0 ml of a solution of 15 g $HgSO_4$ in H_2SO_4, 2.5 mol/L; allow to stand for 1 h and then centrifuge. Transfer the supernatant to a 50-mL beaker, add 10 mL of H_2SO_4, 1 mol/L, and gently boil the mixture for 10–15 min. Add an additional 2 mL of sulfuric acid, 1 mol/L, and cool for 30 min. Then add 1 mL of fresh $NaNO_2$, 2 g/dL. The formation of an orange color indicates a positive test.

Nitrosonaphthol test. To 1 mL nitric acid (one volume of concentrated acid to five volumes H_2O), add 1 drop of a freshly prepared aqueous solution of sodium nitrite, 2.5 g/dL. Mix well. Then add 10 drops nitrosonaphthol solution, 0.1 g/dL, in 95% ethanol and 3 drops of urine. Mix well. Formation of an orange-red color within 3–5 min is reported as a positive result.

Alkaptonuria

Alkaptonuria is a rare hereditary metabolic disease resulting from deficient activity of homogentisic acid oxidase in the major catabolic pathway that converts tyrosine to fumarate and acetoacetate (see above). Reduced activity of this oxidase causes the accumulation of homogentisic acid (HGA) in cells and body fluids. Polymers of HGA, as well as the compound itself, bind to collagen in cartilage, intervertebral disks, and other connective tissue, eventually causing degenerative arthritis and pigmentation of cartilage. Pigment is often first apparent in the ears (ochronosis). Urine darkens on standing, on exposure to air, or upon addition of alkali. Alkaptonuria is not usually diagnosed until middle age when ochronosis and arthritis lead to suspicion, but it can be diagnosed in neonates if the dark stain in an unwashed diaper is noticed and investigated.

There is no satisfactory treatment for alkaptonuria. If the diagnosis is made early, dietary restriction of tyrosine or its precursor phenylalanine may be beneficial. Another suggested therapy is administration of vitamin C, which is known to be required for maximum activity of HGA oxidase. Additionally, vitamin C may prevent in vivo oxidation and polymerization of HGA.

Darkening of urine on exposure to air and sunlight, though characteristic of alkaptonuria, takes many hours and must be distinguished from darkening due to other substances. Melanuria, phenols, gentisic acid (a salicylate metabolite), and indoxyl sulfate (indican, a metabolite of tryptophan, formed in stagnant intestinal contents) may also cause darkening of urine.

Reduction Tests for Homogentisic Acid

HGA reduces ammoniacal silver nitrate very rapidly. Add 5.0 mL of an aqueous $AgNO_3$ solution, 3 g/dL, to 0.50 mL of urine, followed by a few drops of dilute ammonia solution, NH_4OH (1 volume of concentrated base to 9 volumes H_2O). If HGA is present, a brown-black to black precipitate of reduced elemental silver will be formed immediately, often even before the addition of the NH_4OH. Melanogens also react in this test but very slowly and only if there is an excess of ammonia. Urine containing HGA, when heated with Benedict's qualitative glucose reagent, will develop a dark supernatant and a yellow precipitate of cuprous oxide. Melanogens do not reduce the reagent except when present in very large quantity; even then, the supernatant is not dark.

Melanuria

The skin pigment melanin is produced in specialized cells, melanocytes, and is the end product of one metabolic pathway from tyrosine and DOPA. Melanomas (tumors consisting of melanocytes) of the skin or retina may sometimes produce sufficient amounts of pigments or colorless precursors to be excreted and detected in urine, particularly in cases with metastases in the liver. Like HGA, melanogens in urine darken upon exposure to air and sunlight.

Tests for Melanogens in Urine

Ferric chloride test. Add about 1 mL of a solution of $FeCl_3 \cdot 6 H_2O$ (10 g/L in HCl, 1.2 mol/L) to about 5 mL of urine. If melanogens are present, the color changes rapidly to a dark brown.

Thormählen test. Add 5–6 drops of fresh sodium nitroferricyanide solution (5.0 g/dL in water) to 5 mL of urine in a test tube, then add 0.5 mL of NaOH, 10 mol/L, and mix vigorously. Rapidly cool the tube under cold tap water and acidify with acetic acid solution (glacial acetic acid, 33 mL/dL). Normal urine gives an olive or brownish-green color; if melanogens are present, the color varies from greenish-blue to bluish-black, depending on the amount present. Although high levels of creatinine give a brown color, the Thormählen test is probably the most specific and sensitive test available for melanin or melanogens.

Ammoniacal silver nitrate test. Add 0.5 mL of urine to 5.0 mL of an aqueous $AgNO_3$ solution, 3 g/dL, followed by dilute ammonia solution (concentrated NH_4OH, 2 mL/dL) until the AgCl precipitate is almost dissolved. The solution will darken as a result of the formation of both melanin and colloidal silver. The reaction develops slowly. Homogentisic acid, on the other hand, rapidly darkens in the silver solution even before the addition of the ammonia. Used in conjunction with the Thormählen test, the ammoniacal silver nitrate test assists in discriminating between melanuria and alkaptonuria.

Disorders of Sulfur-Containing Amino Acids

The homocystinurias are a group of disorders each characterized by increased concentrations of homocysteine in body tissues.

Homocystinuria (Incidence 1:200 000)

Although homocysteine is an amino acid, it is not a building block for proteins. It lies, however, at an important metabolic branch point in the pathway from methionine to cysteine, where there is an alternative sulfur-conserving pathway back to methionine (Figure 4-6). Homocysteine does not normally accumulate in plasma because it is very unstable in aqueous solution, and when present in excess, undergoes oxidation to homocystine.

$$2 \; HS-CH_2-CH_2-CH(NH_2)COOH \quad \xrightarrow{-2\,H} \quad \begin{array}{l} S-CH_2-CH_2-CH(NH_2)COOH \\ | \\ S-CH_2-CH_2-CH(NH_2)COOH \end{array}$$

Homocysteine   **Homocystine**

Concentrations of homocystine in normal urine are too low for detection, but decreased rates of homocysteine conversion to cystine (by cystathionine β-synthase) or back to methionine (by N^5-methyltetrahydrofolate transferase) may cause homocystinuria.

Cystathionine β-synthase deficiency is the most common cause of homocystinuria. Deficient activity of this enzyme places a block between homocysteine and cystathionine with the result that homocysteine and its precursor accumulate and cysteine and cystine decrease in body fluids. Biochemical abnormalities include detectable plasma levels of homocystine and increased plasma methionine levels; the latter are normally $\sim$0.45 mg/dL, and may increase up to 30 mg/dL. The urine contains homocystine and other sulfur-containing amino acids such as methionine, methionine sulfoxide, and the mixed disulfide of cysteine and homocysteine.

There are no symptoms in the newborn, but ocular, skeletal, and vascular symptoms gradually develop with age. Dislocation of the ocular lenses appears in the first few years, often followed by myopia, glaucoma, and retinal detachment. Skeletal abnormalities include osteoporosis, thinning and lengthening of the long bones, genu valgum (knock knee), and frequent chest, vertebral, and foot deformities. The most serious symptoms are due to arterial or venous thromboses, which may be lethal even at an early age. Mental retardation occurs in less than half the cases, probably as a consequence of thrombotic complications.

Homocystinuria due to cystathionine β-synthase deficiency is genetically heterogeneous; enzyme activities measured in fibroblasts or lymphocytes vary from 0–10% of normal. Patients in whom the enzyme defect is incomplete will respond biochemically and clinically to large doses of pyridoxine, the cofactor needed to activate the enzyme. Those with complete enzyme deficiency should be treated with a diet low in methionine and supplemented with cystine.

Defects in the N^5-methyltetrahydrofolate-dependent methylation of homocysteine. The methylation of homocysteine to form methionine is mediated by a folate-dependent system (upper left, Figure 4-6). Folate is a required substrate of the cycle, and vitamin B_{12} is a required cofactor for N^5-methyltetrahydrofolate transferase activity. Accumulation of homocysteine leads to homocystinuria as a result of either a dietary deficiency of folate or vitamin B_{12}, an inherited inability to metabolize these vitamins to their biochemically effective forms, or defects in the reductase or transferase enzymes of the cycle. *5,10-Methylenetetrahydrofolate reductase deficiency* is rare. When it occurs, death in infancy or childhood is probable; surviving adults are mentally retarded with cerebral atrophy. Increased levels of homocysteine in urine and normal plasma methionine concentrations are diagnostic of these cases. Treatment consists of administration of folate. Patients with *N^5-methyltetrahydrofolate transferase deficiency* present with similar signs and symptoms, namely failure to thrive and mental retardation. In this form of homocystinuria, there is a primary genetic defect in the synthesis of methylcobalamin, a vitamin B_{12} coenzyme required for enzyme activity. Characteristically, methylmalonic acid accumulates in body tissues and is excreted in urine. Unlike other types of homocystinuria, plasma folate levels are also increased. Treatment is administration of vitamin B_{12}.

Neonatal screening is practical only for cystathionine β-synthase deficiency. A Guthrie test for increased plasma *methionine* is performed. A positive result, however, should be interpreted with care since it may be transient or due to liver damage, tyrosinosis, or a deficiency of hepatic S-adenosyl methionine synthetase. The incidence of transient hypermethioninemia has decreased recently, probably because infant milk preparations now have lower protein content. In the absence of neonatal screening, homocystinuria is unlikely to be diagnosed until symptoms appear and the urine is tested.

Cyanide-nitroprusside test for cystine and homocystine. This test is based on the reaction of sodium nitroprusside with sulfhydryl compounds such as cysteine and homocysteine to produce a red-purple color. Oxidized disulfides (e.g., cystine and homocystine) will not react but must first be reduced to their free thiol forms with alkaline sodium cyanide prior to color development.

To 5 mL urine add 5 drops of concentrated NH_4OH followed by 2 mL of an aqueous solution of NaCN, 5 g/dL. *Caution*: NaCN is poisonous. Allow the mixture to stand for 10 min. Then add, dropwise, a freshly prepared aqueous solution of sodium nitroprusside, 5 g/dL. Normal urines show a persistent yellow or pale brown color. Formation of a red-purple color is a positive result. The use of NH_4OH instead of NaOH prevents interference from creatinine. Acetoacetic acid and acetone give negative results with this particular version of the nitroprusside test.

Silver nitroprusside test for homocystine. A modification of the cyanide nitroprusside test is used to differentiate between cystine and homocystine. By substituting silver nitrate for sodium cyanide as the reducing agent, homocystine can be reduced to its thiol form (homocysteine) while cystine remains in the nonreactive oxidized form. Homocysteine then reacts with sodium nitroprusside to yield a pink-purple color.

Procedure. Saturate 2 mL of urine with solid sodium chloride. Then add 4 drops of an ammoniacal solution of silver nitrate solution, 1 g/dL (10 mL concentrated NH_4OH per 100 mL). Allow the mixture to stand for 1 min, and then add 4 drops of a freshly prepared aqueous solution of sodium nitroprusside, 1 g/dL, and 4 drops of an aqueous solution of NaCN, 0.7 g/dL. Immediate development of a pink or purple color represents a positive test for homocystine. Cystine will undergo a slow reaction with excess cyanide, eventually resulting in a positive test.

Cystinuria

Classic cystinuria is the most common inborn error of amino acid transport. This disease is characterized by massive urinary excretion of the dibasic amino acids: cystine, lysine, arginine, and ornithine. Normally these amino acids are filtered by the glomerulus and reabsorbed in the proximal renal tubule, the latter by specific carrier molecules which are probably membrane-fixed enzymes. There are probably three specific classes of carriers: one for cystine; another for lysine, arginine, and ornithine; and a third that transports all four amino acids. Cystinuria is a transport defect of the third carrier system.

Because cystine is the least soluble of the naturally occurring amino acids, its overexcretion often leads to the formation of cystine calculi in the renal pelves, ureters, and bladder; obstruction, infection, and renal insufficiency occasionally result. Some 1–2% of all urinary tract stones are composed of cystine. Treatment involves reducing the concentration of cystine in urine by drinking large amounts of water, increasing cystine solubility by maintaining an alkaline urine, and, if necessary, reducing cystine excretion by using D-penicillamine.

Histidinemia

Histidine is normally largely catabolized to urocanic acid, which is converted to a product that is utilized in the synthesis of folic acid derivatives.

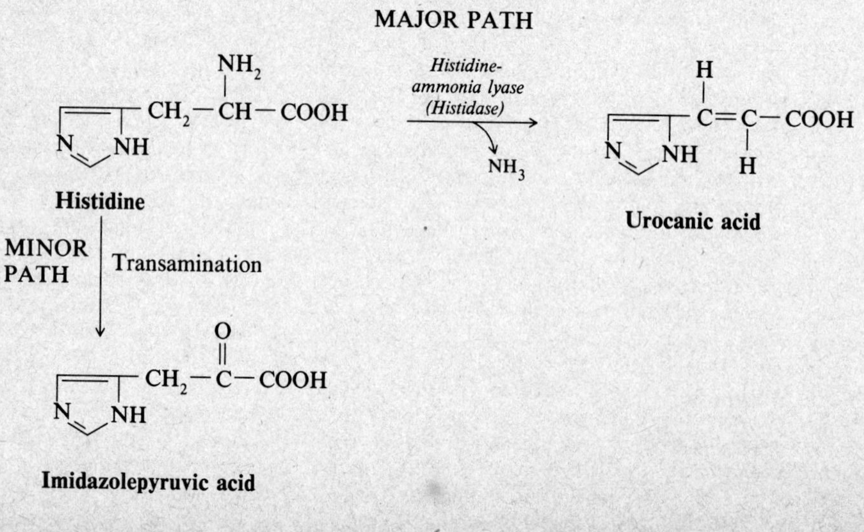

MAJOR PATH

Histidine

Urocanic acid

Imidazolepyruvic acid

An inherited defect in the enzyme histidase causes plasma levels of histidine to increase; thus, imidazole pyruvate and other histidine metabolites appear in the urine. The overall incidence of histidinemia is 1:20 000, but in Japan it is 1:8000. A majority of patients are clinically normal, although some display abnormalities such as mental retardation and speech defects. The enzyme deficiency has been demonstrated in skin and liver biopsies. Although neonatal screening with a Guthrie test is feasible, there remains some doubt as to the seriousness of the disease; thus, currently, screening is not recommended.

Organic Acidurias

The group of organic acidurias is heterogeneous in both expression and causation.[27,102] They are characterized by excessive urinary excretion of aliphatic and aromatic organic acids and may be congenital or acquired. Clinically, organic acidurias typically present as unexplained severe ketoacidosis; symptoms of the various types are similar—vomiting, lethargy, seizures, coma, and, in survivors, mental retardation. Since most of the acids cannot be visualized by TLC, diagnosis within this group of aminoacidurias is chiefly dependent (with the exception of maple syrup urine disease) on analysis by gas chromatography–mass spectrometry.

Maple Syrup Urine Disease (MSUD)

The branched-chain amino acids, leucine, isoleucine, and valine, are normally converted by transamination to their corresponding α-keto acids, which, by oxidative decarboxylation, are then converted to acyl-CoA derivatives. An inherited defect of the decarboxylase step results in the accumulation of the branched-chain amino acids and their corresponding α-keto acids in blood, urine, and CSF. Overall, the incidence of the disease is 1:250 000. Several types of MSUD have been identified clinically and biochemically. In the classic type, the infant appears normal at birth but soon begins to vomit and fails to thrive. Acute ketoacidotic episodes, often triggered by recurring infections, are observed as a result of increased production of organic keto acids. Severe neurologic dysfunction leads to seizures, coma, respiratory failure, and death in many patients. Survivors are usually mentally retarded. Milder variants of the disease, identified by enzyme analysis of leukocytes or fibroblasts, are known.

MSUD takes its name from the characteristic odor of the urine of affected individuals; the maple syrup or burnt sugar odor is due to the high concentration of aliphatic keto acids. Amino acid analyses of blood and urine show high levels of leucine, isoleucine, and valine. Allo-isoleucine, which is not present in normal patients, is also typically found in MSUD. Abnormal amounts of keto acids in urine are detected by the dinitrophenylhydrazine test. A gray-blue color is given with the urine ferric chloride test.

Neonatal screening with a Guthrie test is possible, though uncommon in practice. Antenatal diagnosis may be made by assay of decarboxylase activity in cultured cells from amniotic fluid. Treatment is by dietary restriction of branched-chain amino acids. Daily analysis of urine with dinitrophenylhydrazine and monthly quantitation of plasma amino acid levels are used to monitor effectiveness of the diet.

Dinitrophenylhydrazine Test for Keto Acids

2,4-Dinitrophenylhydrazine (DNPH) reacts with a variety of aliphatic, cyclic, and aromatic carbonyl compounds to form relatively insoluble crystalline hydrazones. Mix equal quantities of filtered urine and 2,4-dinitrophenylhydrazine solution, 0.4 g/dL in HCl, 1 mol/L. A yellow to yellow-white precipitate forming within 5 min indicates a large quantity of keto acids. A slight precipitate or turbidity forming later than 1 min may be normal. Interference contributed by acetone can be eliminated by heating the urine briefly at 100 °C before testing.

Disorders of Propionate and Methylmalonate Metabolism

Among the organic acidurias, disorders of propionic acid and methylmalonic acid metabolism are worthy of brief mention. Infants with these disorders usually present with overwhelming illness early in life. Intolerance to dietary protein, particularly amino acids such as leucine, isoleucine, valine, threonine, and methionine, is characteristic. It is now recognized that these disorders involve deficiencies of propionyl CoA carboxylase and methylmalonyl CoA mutase. Chemically, massive ketonuria and elevated levels of glycine in body fluids are noted. High concentrations of methylmalonic acid can also be demonstrated in methylmalonic aciduria but not in propionic aciduria. Simple screening tests for methylmalonic acid in urine are available. Definitive diagnosis of propionic acidemia can be made by demonstrating the presence of methyl-citrate in urine, or by analyzing for organic acids in urine by GC-MS. Since propionyl CoA carboxylase (EC 6.4.1.3) is a biotin-containing enzyme, some patients respond to biotin treatment. On the other hand, methylmalonyl CoA mutase requires vitamin B_{12} as its coenzyme, and some patients with this enzyme deficiency respond to vitamin B_{12} therapy. Dietary treatment involves restriction of branched-chain amino acids. Prenatal diagnosis for both disorders can be made by enzyme analysis on cultured cells from amniotic fluid.

Nonketotic Hyperglycinemia

Incidence of this inborn error of metabolism is 1:150 000. The metabolic block is due to a defect in glycine cleavage which normally forms carbon dioxide, ammonia, and $N5,N10$-methy-lenetetrahydrofolate in the liver and brain.

High levels of glycine accumulate in liver, brain, blood, and urine; the most characteristic finding, however, is an increase in CSF glycine and a low plasma/CSF glycine ratio. High concentrations of glycine in CSF in the absence of ketosis distinguish the disease from those organic acidurias with severe ketosis that may present clinically with the same symptoms. Glycine is a potent inhibitory neurotransmitter, and symptoms of hyperglycinemia appear related to high concentrations of glycine in brain tissues. The prominent symptoms are epileptic attacks and hypotonia, which may lead soon to fatal respiratory failure. There is no satisfactory treatment, and patients with mild cases who survive are usually severely retarded.

Defects of Urea Cycle Enzymes

Metabolic blocks in the urea cycle lead to accumulation of ammonia and its intermediates. Symptoms vary in severity and in the age of presentation. Three types of disease—fulminant neonatal, subacute, and late onset—are distinguished. In the fulminant-neonatal type, failure to thrive is evident within a week, followed by vomiting and seizures from encephalopathy, and finally by respiratory or liver failure, or pulmonary or intracranial hemorrhage.

Any of six enzymes of the urea cycle may be deficient (see Figures 11-5 and 13-22): argi-ninosuccinate synthetase, argininosuccinate lyase, arginase, ornithine decarboxylase, ornithine transcarbamylase, or carbamoyl phosphate synthetase. *Citrullinemia* is caused by deficiency of argininosuccinate synthetase. Only about 50 cases have been described. Plasma citrulline is increased but not argininosuccinate. *Argininosuccinic aciduria* is due to deficiency of arginino-succinate lyase; the incidence of the disease is 1:75 000. Large quantities of argininosuccinate (ASA) are excreted in urine; during analysis, ASA tends to form stable anhydrides. Because little ASA is reabsorbed by the renal tubules, plasma levels of this intermediate may not be substantially increased. Citrulline is slightly increased in both blood and urine, and arginine and ornithine are decreased. Plasma ammonia increases substantially after a meal. In the less severe cases, treatment by dietary protein restriction has given encouraging results. *Argininemia* occurs because of arginase deficiency. Only 13 cases, mainly in Spanish Americans, have been reported. Plasma arginine is increased and plasma ammonia rises after protein intake. No acute neonatal variant of this disorder has been described. In the hyperammonemic form of *hyperornithinemia*, there is a deficiency of ornithine decarboxylase.

Diagnosis of urea cycle disorders can be made from their amino acid patterns on thin layer chromatography of plasma or urine. Prenatal diagnosis of deficiency of argininosuccinate lyase and synthetase can be made by enzyme analysis on cultured cells from amniotic fluid. Blood urea in these disorders remains normal or only slightly decreased. Treatment is by dietary restriction, but prognosis is generally poor.

Patients with hyperammonemia due to ornithine transcarbamylase deficiencies or carbamoyl-phosphate synthetase deficiencies can be distinguished by the pronounced oroticaciduria present in the former but not the latter disorder. (Oroticaciduria also occurs as a hereditary disorder, transmitted as an autosomal recessive trait, in which a defect in the metabolism of pyrimidines is associated with excessive excretion of orotic acid in the urine. It is characterized by megaloblastic anemia, crystalluria, and frequently physical and mental retardation.) Impaired urea formation in these groups of patients leads to overwhelming illness and coma. Organic acid analysis by gas chromatography–mass spectrometry allows testing for these disorders.

More common than these urea cycle enzyme defects is *transient hyperammonemia of the newborn.* In these cases, initially high ammonia levels in the newborn go down in a few days and do not go up again despite protein challenges.

Other aminoacidurias. Numerous other aminoacidurias have been described;[2,99] for some only one case was reported. Several aminoacidurias are summarized in Table 4-3 together with those described above.

PROTEINS OF BODY FLUIDS

The human body contains countless different proteins. Some investigators have estimated that there are 30 000–50 000 structural genes that code for the synthesis of human proteins; others estimate the figures at twice these numbers. The number of distinct proteins within one cell is estimated at 3000–5000. With state-of-the-art methodology, more than 300 different proteins can presently be identified in plasma alone. (See Figure 4-7.) Some proteins are elaborated or circulate only at certain stages of growth or under particular physiologic or pathologic circumstances. Many proteins are structural elements of cells or organized tissues and can be examined only after their separation from the tissue where they reside; others are soluble, i.e., they are free molecules moving in intracellular or extracellular fluids. Some soluble proteins which are normally intracellular escape into extracellular and intravascular fluids when cells are damaged; some, which as a rule reside on cell surfaces, may be shed into blood or urine in detectable quantities. The variety due to the number of genes that code for the synthesis of specific proteins is increased by post-synthetic modifications of some proteins that make them selectively functional or degradable. Not only is the variety of proteins seemingly infinite, so are their variations of concentration in health and disease, their distribution within the body, their functions, their compositions, and their structures. The proteins most amenable to routine laboratory evaluation are those in blood, urine, spinal fluid, amniotic fluid, saliva, feces, and peritoneal or pleural fluids. With a few exceptions, which will be noted, proteins found in these biological materials were originally contained in blood plasma. In the following discussion, we must necessarily limit our consideration to the most abundant *plasma* proteins, to changes of their concentrations in the most accessible body fluids, and to but a few of the possible methods for their evaluation.

Most plasma proteins, with the exception of immunoglobulins and protein hormones, are synthesized in the liver. They are secreted by the hepatocytes into the space of Disse and then move into the bloodstream through the hepatic sinusoids which open into the central veins of the liver. Plasma proteins circulate in the blood and between the blood and the extracellular tissue spaces; their movement occurs not only by passive diffusion through junctions between capillary endothelial cells but also by active transport mechanisms. Because of this movement, most extravascular fluids normally contain small amounts of plasma proteins. Differences in molecular size among individual proteins and the specificity of some of their transport mechanisms account for differences in kind and proportions of individual proteins in different fluids. Disease often alters the amount and proportions of plasma proteins in body fluids in characteristic ways. Most plasma proteins are catabolized in the liver; for some, the signal that marks them for degradation appears to be the loss of part or all of their sialic acid content.

Table 4-8 lists the principal plasma proteins, together with the adult reference ranges, the half-lives ($t_{1/2}$), isoelectric points (pI), molecular weights (M.W.), carbohydrate contents, and acute phase behaviors; the individual proteins are listed in the order of their electrophoretic mobilities on agarose. These proteins are described below, both in groups and individually.[112] The immunoglobulins and the complement factors, as distinct groups of closely related proteins, are considered first. Other proteins are considered in the order in which they appear in Table 4-8. Chapters elsewhere in this book describe many more proteins: enzymes (Chapter 5); lipoproteins (Chapter 7); hormones (Chapter 9); hemoglobin, fibrinogen, and other coagulation proteins (Chapter 15). Additional references should be consulted for a more complete listing of plasma proteins,[86] for reference ranges,[105] and for physiologic variations.[52]

As methods have evolved[85] from the Kjeldahl determinations of total protein and albumin to immunochemical analyses of individual proteins, our knowledge of body fluid proteins and

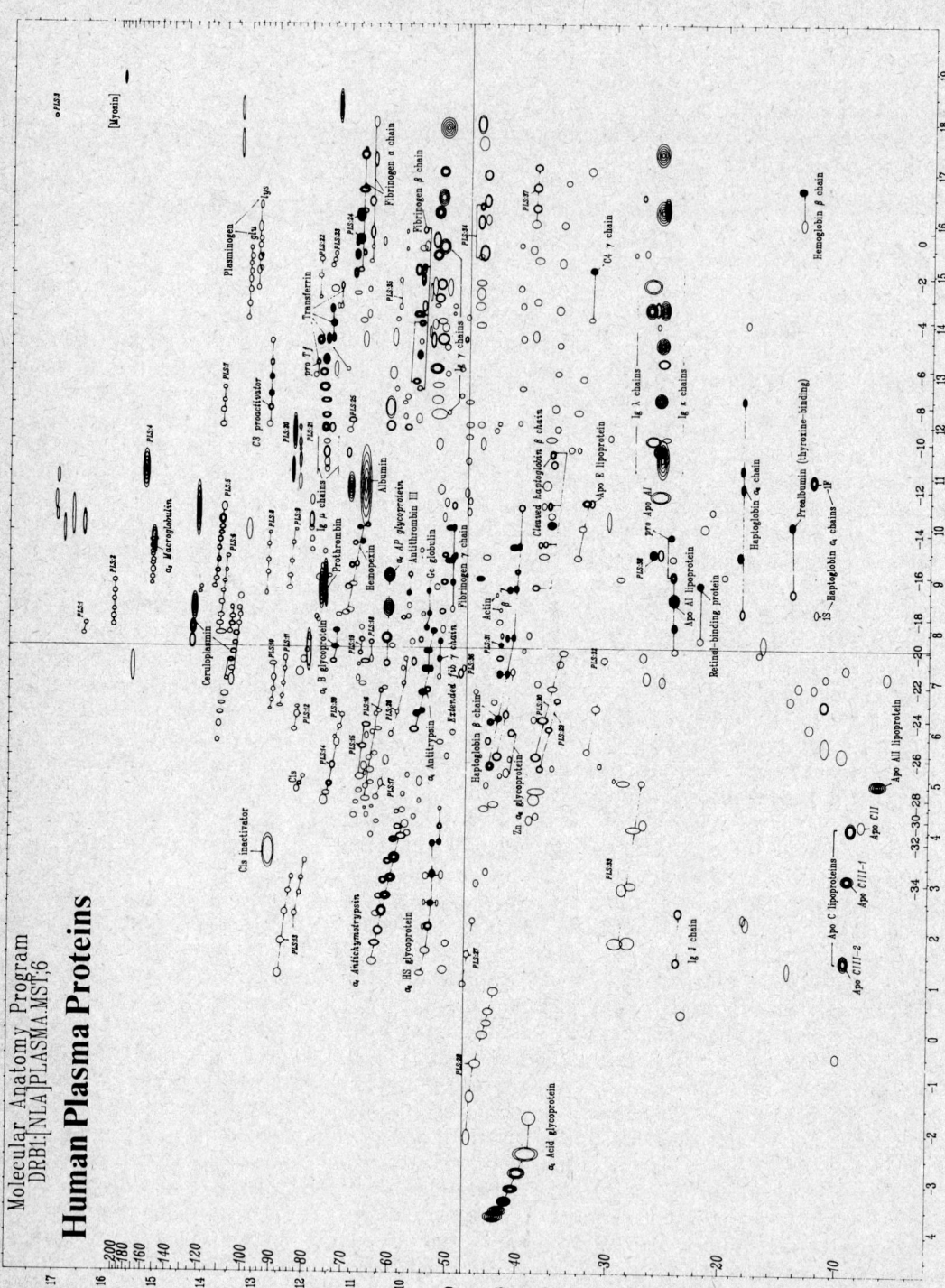

Figure 4-7. Two-dimensional electrophoretic map of human plasma proteins, obtained by combining isoelectric focusing in the presence of urea in one horizontal dimension, and electrophoresis in the presence of sodium dodecyl sulfate in the second dimension. (From Anderson, N. L., Tracy, R. P., and Anderson, N. G.: High resolution electrophoretic mapping of human plasma proteins. *In:* The Plasma Proteins, Vol. 4. 2nd ed. F. Putnam, Ed. New York, Academic Press, 1984.)

Table 4-8. PROPERTIES OF SELECTED PLASMA PROTEINS

Protein	Adult Reference Ranges, mg/dL	$t_{1/2}$	pI	M.W.	Carbohydrate Content, %	Acute Phase Behavior	Comments
Prealbumin (PA)	20–40	12 h	4.7	54 400	0	Neg APR*	Indicator of nutrition
Albumin	3500–5000	15–19 d	4–5.8	66 000	0	Neg APR	Reduced in many diseases; a general transport protein
Electrophoresis, α_1-region							
α_1-Antitrypsin (AAT)	78–200	4 d	4.8	55 000	12	APR	Congenital deficiency may cause emphysema or cirrhosis
α_1-Acid glycoprotein (AAG, orosomucoid)	50–150	5 d	2.7–4	40 000	45	APR	Function obscure
α_1-Lipoprotein (Apoprotein A)	170–325			200 000			Transports lipids
α_1-Fetoprotein (AFP)	0.003			69 000			Principal fetal protein
Electrophoresis, α_2-region							
Haptoglobin (HAP, Hp)	30–215	2 d	4.1†	85 000–1 000 000+	12	APR	Binds Hb; reduced by hemolysis
α_2-Macroglobulin (AMG)	125–410	5 d	5.4	800 000	8		Increased in nephrotic syndrome
Ceruloplasmin (CER)	20–50	4.5 d	4.4	160 000	7	APR	Decreased in Wilson's disease; contains Cu
Electrophoresis, β_1-region							
Transferrin (TRF, siderophilin)	200–350	7 d	5.7	77 000	6	Neg APR	Transports Fe, increased in hypochromic anemia
Hemopexin (Hx, Hpx)	50–115			57 000			Binds heme
β-Lipoprotein (Apoprotein B)	60–155			~3 000 000			Transports lipids
C4	10–40			206 000	7	APR	Complement factor
Electrophoresis, β_2-region							
Fibrinogen	200–400	2.5 d	5.5	340 000	3	APR	Precursor of fibrin clot
C3	70–150			180 000	2	APR	Complement factor
β_2-Microglobulin (BMG)	0.1–0.2			11 800			Used to test renal tubular function
Electrophoresis, γ-region							
IgG	525–1650	24 d	6–7.3	160 000	3		Antibodies increase in immune reactions; monoclonal increases in B cell tumors
IgA	40–390	6 d		170 000	8		
IgM	25–310	5 d		900 000	12		
C-reactive protein (CRP)	<0.8		6.2	~120 000	0	APR ++	Probably a nonadaptive defense protein

*APR: Acute phase reactant.

†For HAP 1-1 phenotype.

Note: The adult reference ranges reported are established by nephelometry and may vary with differences in technique. Other reference ranges in this chapter may differ according to age and methodology used.

their alterations in health and disease has grown rapidly. This knowledge, however, is far from complete. Some of the alterations have a genetic origin; many more reflect physiological or pathological processes. Presumably, every protein exists for a reason, which is often not understood. In some cases, apparent association of a specific disease with a specific pattern of altered proteins remains empirical, simply because the cause and effect relationship remains unknown. Variations in the amount or kinds of protein found in plasma or extravascular fluids depend on many factors—genetic, physiologic, or pathologic. In many instances, laboratory analyses detect and evaluate these variations reliably, but they must always be considered in the context of clinical findings and in terms of all the factors, technical as well as biological, which may affect the results.

Table 4-8 lists several principal plasma proteins that can be grouped as "*acute phase reactants*" (APR). Concentrations of these proteins rise significantly during acute inflammation due to causes such as surgery, myocardial infarction, infections, and tumors. Presumably, they all play a part in the very complex defensive process of inflammation. As will be seen later, some aspects of their role are now partly understood. This acute phase reaction is a general reaction to inflammation, comparable to the increase in temperature or leukocyte count, and is not specific for any given disease; all these changes are probably triggered by small proteins, known as leukocytic endogenous mediators (LEM), released from the site of injury.[65] Plasma levels of the individual acute phase proteins rise at different rates—first, C-reactive protein and α_1-antichymotrypsin; then, within the first 12 h α_1-acid glycoprotein; then, α_1-antitrypsin, haptoglobin, C4, and fibrinogen; and finally C3 and ceruloplasmin; all reach their maxima within 2–5 d. The sequence for C-reactive protein, α_1-antitrypsin, and C3 increases is illustrated in Figure 4-8. These changes, caused by increased synthesis in the liver, do not aid in the diagnosis of the cause of inflammation, but measurement of those proteins with the largest and earliest rises (e.g., C-reactive protein) can be useful in monitoring the progress of the inflammation or its response to treatment. Increase in synthesis of APR is accompanied by a decrease in the synthesis of prealbumin, albumin, and transferrin (the so-called *negative acute phase reactants*), so that only a slight rise in total plasma protein occurs. The inflammatory process therefore causes nonspecific changes in levels of individual proteins, which may mask changes attributable to a specific disease. Laboratory findings must be interpreted accordingly.

Rates of hepatic synthesis of many plasma proteins are affected by the patient's endocrine status. The reader is referred to Chapter 9 for specific details. The effects of steroid hormones on individual plasma protein levels are given in Table 4-9. The plasma protein levels characteristic of a specific disease can therefore be complicated by the steroid status of the patient as well as by an inflammatory acute phase reaction. The abnormal steroid status may be the result of an intrinsic hormonal disorder or of treatment with steroid hormones, as in inflammation.

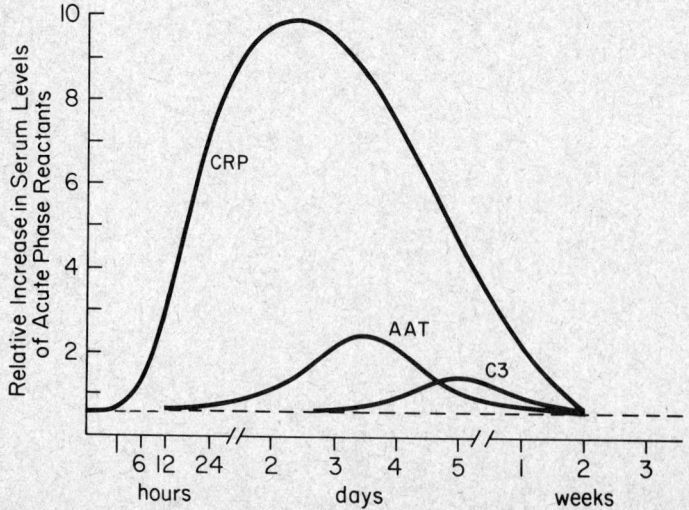

Figure 4-8. Relative increases of acute phase reactants: C-reactive protein (CRP), α_1-antitrypsin (AAT), complement factor 3 (C3). Concentrations are expressed as multiples of the upper limit of the reference range. The dotted line represents the upper reference limit.

Table 4-9. EFFECTS OF STEROIDS ON PLASMA PROTEIN
CONCENTRATIONS

	Corticoid	Androgen	Estrogen
Prealbumin	−	+	−
Albumin	−	N	−
α_1-Lipoprotein	N	−	+ +
α_1-Acid glycoprotein	+	+	−
α_1-Antitrypsin	N	+	+ +
Haptoglobin	+	+ +	−
Ceruloplasmin	N	N	+ + +
Transferrin	N	+	+
β-Lipoprotein	+ +	+	N
IgG	−	N	−

+: Increase; N: No change; −: Decrease.

Adapted from: Whicher, J. T.: Abnormalities of plasma proteins. *In* Biochemistry in
Clinical Practice. D. L. Williams and V. Marks, Eds. London, William Heinemann, 1983,
pp. 221-250.

ANALYSIS OF PROTEINS

Methods for the analysis of proteins in body fluids can be roughly grouped as follows:

1. Quantitative measurements of total protein and albumin; these are described below under the sections of the different body fluids.

2. Separation by electrophoresis, which provides semiquantitative estimations of the main classes of proteins present in fairly high concentrations. (See Chapter 1B, Section Three.)

3. Specific quantitative assays of particular proteins by immunochemical methods (see Chapter 1B, Section Two, and Chapter 1C) using specific antisera and measurement of the antigen-antibody (Ag-Ab) complexes by nephelometry, turbidimetry, radial immunodiffusion, or electroimmunoassay; or, if present in very low concentrations, by radioimmunoassay (RIA) or enzyme immunoassay (EIA).

4. Detection and identification of abnormal proteins by radial immunodiffusion (RID), immunoelectrophoresis (IEP), and immunofixation electrophoresis (IFE). (See Chapter 1C.)

Immunochemical Methods for the Determination of Individual Proteins

Since these methods are generally applicable to any of the proteins described below, they will be discussed more fully here.

Nephelometry and turbidimetry, because of their speed and ease, are most widely used.[30] These techniques can be performed either by measuring the amount of Ag-Ab complex formation (endpoint methods) or by measuring the rate of complex formation (kinetic methods). The kinetic methods are more rapid since measurements are accomplished within 20 s, and are more precise since sample blanks are not necessary. Kinetic assays are, however, somewhat less sensitive since low-affinity antibodies do not have time to react.

Several factors must be considered in the selection of an appropriate method.

Sensitivity. Sensitivities of ~10 μg/mL are attained with nephelometry; turbidimetric methods are slightly less sensitive, the detection limit being 20–30 μg/mL. RID methods are sensitive to a minimum of 10–20 μg/mL; the detection limit can be lowered by using special "low level" plates (Calbiochem-Behring Corp., La Jolla, CA 92037). Nephelometry, turbidimetry, and RID are all sufficiently sensitive for quantitation of proteins present in concentrations of 20 μg/mL or greater. The most sensitive method, RIA, is required for quantitation at the ng/mL level, e.g., for serum IgE.

Precision. With nephelometry, within-run CV's of < 5% are usual; the endpoint methods, however, tend to be less precise than the kinetic methods because sample blanks are required. RID and EIA systems have higher run-to-run CV's, usually in the range of 5–15%. RIA, measuring much smaller amounts of protein, generally has within-run CV's of 5–10%.

Turnaround time. Nephelometry is fast; kinetic methods give results within minutes,[10,100] although endpoint methods take up to 1 h. RID usually requires 24–48 h of incubation, an unacceptable period for urgent tests. Depending upon the procedure, turnaround time for RIA is usually several hours.

Instrumentation and equipment. Although relatively expensive nephelometers[83] and spectro-photometers are required for the light-scattering and light-absorbing methods, these instruments are often equipped with microprocessors that generate standard curves and calculate patient results. RID requires no instrumentation other than pipets, although some type of illuminated plate reader is advantageous. RIA requires radiation scintillation counters.

Standardization of Immunochemical Methods

Reference materials containing accurately determined quantities of purified individual proteins seldom exist. For many proteins of interest, there are no well-defined purified protein preparations available. Depending on their sources, antisera may vary greatly in specificity and sensitivity. Preparation of a pure protein for raising a highly specific antiserum is not easily accomplished. The problems of standardization of methods and preparation of antisera are reflected in a recent survey of the College of American Pathologists (CAP). Identical samples were assayed for α_1-antitrypsin by 510 laboratories using five different methods; the means for the five methods were as low as 160 and as high as 234 mg/dL. On the same survey, the C3 complement component, assayed by eight methods, gave values ranging from 14.9–28.2 mg/dL.

The World Health Organization (WHO) has available reference preparations for IgG, IgA, IgM, IgD, IgE, α-fetoprotein, carcinoembryonic antigen (CEA), albumin, C3, ceruloplasmin, and transferrin. The values assigned are, however, in International Units (IU); factors have not been defined for conversion of IU to mass units. Another preparation, the United States National Reference Preparation (USNRP) for Specific Human Proteins, is available from the Centers for Disease Control, Atlanta, GA; it contains 13 individual proteins and the values in IU are assigned against WHO standards. A reference preparation whose assigned values are in mass units has been developed by the CAP. The 12 individual proteins in this preparation were assayed by a number of experts in specific protein assay and by manufacturers of diagnostic products for such assays. The methods used included nephelometry, RID, electroimmunodiffusion, and turbidimetry. The arithmetic mean value for each protein was calculated from the values of all the methods and the mean was defined as the assigned value for the specified analyte.

The inconsistent character of available antisera and lack of well-defined reference materials makes it imperative that each laboratory establish its own reference ranges for its own reagent systems and methods.

Determination of Proteins by Rate Nephelometry

A rate nephelometric procedure for quantitation of IgG, IgA, and IgM in serum or CSF is described below. The procedure utilizes a rate nephelometer and reagents supplied by the manufacturer (Beckman Instruments, Inc., Brea, CA 92621). This specific assay is included as an example of rate nephelometry as it is applied to the assay of numerous specific proteins. Endpoint nephelometric and turbidimetric techniques are also commonly used to quantitate immunoglobulins; these methods are thoroughly described in the literature made available by manufacturers of analytical systems, as well as in analytical journals such as Clinical Chemistry. Their principles are discussed in Chapter 1B, Section Two, and Chapter 1C.

Principle

Specific antiserum against the protein to be determined is added to a reaction cuvet containing a serum sample in a buffered solution, to allow the formation of antigen-antibody complexes. The measurement is made by passing an incident light beam through the cuvet containing the reaction mixture. The light beam is scattered by the suspended antigen-antibody complexes, and the intensity of scattered light, measured at an angle of 70° to the incident beam, is proportional to the number of suspended particles in solution. As antigen and antibody react, the rate of formation of particles progressively increases until the rate reaches a maximum level. The *rate-of-change* of the light scatter is monitored and the maximum rate during the reaction is selected, and is directly proportional to the concentration of antigen in the reaction mixture. When standards of known concentrations are used, maximum rate versus concentration can be plotted to form a standard curve. The concentration of antigen in an unknown sample is then calculated from the standard curve. This method can be performed on the Beckman Immunochemistry System (ICS).

Specimen Collection and Storage

Test specimens must be nonhemolyzed, cell-free serum, urine, or cerebrospinal fluid (CSF). CSF specimens should be collected by a physician and need no additional processing except for centrifugation if cells are present. Serum and CSF samples may be stored at 2–8 °C for up to 3 d, or at −20 °C for longer periods. Repeated freezing and thawing of specimens may cause deterioration of immunoglobulins and should be avoided. Minimum amount of specimen required is 100 μL.

Reagents and Preparations

Reagents are obtained in packaged form from the manufacturer and include the following:

1. ICS Calibrator Serum. Each bottle contains 1 mL of processed human serum with known amounts of IgG, IgA, and IgM and a preservative (sodium azide, 0.1 g/dL). This calibrator should be stored at 2–8 °C and is stable until the expiration date stated on the vial.

2. ICS Buffer Solution. Each bottle contains phosphate-buffered saline containing a polymer and sodium azide, 0.1 g/dL, as a preservative. Store at room temperature; stable up to the expiration date stated on the container.

3. ICS Diluent Solution. Same as ICS Buffer Solution, but with no polymer.

4. Antiserum (mammalian) against specific protein to be determined. Store at 2–8 °C; stable up to the expiration date stated on the vials.

Controls. Beckman ICS Control Serum (nephelometric grade, order no. 662950); Kallestad Quantitrol Serum (Kallestad Laboratories, Inc., Austin, TX 78701).

Procedure

1. Prepare the rate nephelometer for operation according to the manufacturer's instructions.

2. Allow calibrators, antiserum solutions, controls, and patient samples to warm to room temperature prior to use.

3. Using the diluter provided with the instrument, serially dilute 100 μL of ICS calibrator, controls, and samples as required for the immunoglobulin being measured. For most proteins, a 1:6 serial dilution sequence provides enough different dilutions for the diverse concentration ranges encountered for different serum proteins.

Dilution sequence: A—Undiluted
B—1:6
C—1:36
D—1:216
E—1:1296

This dilution scheme allows the analyst to be prepared for the occasional out-of-range specimens. After the calibrator serum and all samples are serially diluted, begin assay with the appropriate dilution (A–E) called for by the alphanumeric message displayed on the instrument for the particular immunoglobulin being assayed. Mix all samples thoroughly before dilutions are made. Mix each specimen after dilution, and again immediately before assaying.

4. Insert the appropriate calibrator card attached to the calibrator vial into the instrument card reader. The coded information on this card is used for calibration of the instrument. (For more explanation, see Calculations.)

5. Remove calibrator card and insert card that is attached to the appropriate antibody vial and allow it to remain in the card reader. *Note*: Calibration of the system requires that the lot number of the calibrator card be the same as the lot number of the calibrator serum (printed on the bottle label). It is also essential that the lot number on the antibody card be the same as the lot number printed on the label of the antiserum bottle. Use of reagents with cards bearing lot numbers different from those appearing on the attached reagent bottle will result in erroneous test results because of improper instrument calibration.

6. Dispense 600 μL of ICS buffer solution into an ICS reaction cell containing a stirring bar. Avoid placing fingers on the cell window.

7. Insert cell into the instrument cell compartment and close cell cover.

Calibration

8. After the instrument indicates which dilution to use (depending on the immunoglobulin being measured), pipet 42 μL of this dilution of calibrator serum into the reaction cell through the cell cover opening.

9. Within 20 s after calibrator addition, pipet 42 μL of the appropriate antiserum into the reaction cell.

10. After the peak rate has been determined and verified, the analyzer will store the rate value in the microcomputer and the display will alternately present the rate value and a series of messages directing the removal of the cell, insertion of a new cell, and closing the cell cover.

11. Repeat steps 8 and 9. The analyzer will recall the previously stored peak rate value from the microcomputer, compare it with the new value, and determine if both values are within a programmed per cent of one another. If the first and second peak rate values fall within the required range, the two values will be averaged and the peak rate signal will be internally adjusted to correspond with the calibrator's target value. If the first and second peak rate values are out of range, another series of messages on the alphanumeric display will direct the analyst through a third assay of the calibrator dilution.

12. After calibration is completed, remove and discard the cell. At this time, the appropriate calibrator dilution should be recycled as an unknown; this value *must be* within ±5% of the target value before the calibration can be accepted.

Patient Testing

13. Once calibration is acceptable, inject the appropriate dilution of control (in the same manner that the calibrators were injected).

14. If the controls are not within acceptable quality control ranges, new dilutions should be made and assayed. If they still remain out of range, the system may need to be recalibrated. If the recycled calibrator is acceptable and the controls are within range, proceed with the analysis of patient samples.

15. After all measurements on patient samples have been completed, the controls are assayed again to confirm that the assay is still within control.

Calculations

A calibration curve consisting of 8–15 points is prepared by the reagent manufacturer for each lot of calibrator serum and specific antiserum for each analyte. These values are coded onto the antibody program card which provides the microprocessor with all the parameters needed to perform a specific protein assay and calculate results for patients' specimens. The derived constants define the shape of the calibration curve and are used by the microprocessor to convert peak rate values for unknown samples to actual concentration values. Since the microprocessor recognizes the calibration curve by reading the antibody card, it is necessary to measure and verify only *one* point on the curve. Once the instrument verifies the known target value (calibration point), all unknown samples may be assayed and their results interpolated from the calibration curve for that particular antiserum.

Limitations of the Procedure

1. Lipemic sera should not be assayed by nephelometric methods due to extreme light scattering by the lipid particles in the samples. These sera must be centrifuged in a high speed centrifuge (e.g., Airfuge, Beckman Instruments, Inc., Brea, CA 92621) to separate lipids and obtain clear sera prior to assay.

2. Test tubes containing samples, controls, and calibrator must remain covered during assay to prevent dust and dirt particle contamination. The reaction cells must also remain covered during the assay and when stored. Dust particles and other particulate matter in the reaction mixture can result in extraneous light scattering signals and lead to erroneous results.

3. Adverse storage conditions or use of outdated reagents may cause erroneous results.

IMMUNOGLOBULINS (HUMORAL ANTIBODIES)

Immunoglobulins (Igs) are unique in their heterogeneity, in their sites of synthesis, and in the fact that their synthesis is an adaptive response to antigenic stimulation. They have two functions: recognition of antigens and initiation of effector mechanisms to destroy antigens. Their basic characteristics and the definitions of some terms commonly used in immunology are described in Chapter 1C.

Whereas most plasma proteins are synthesized in the liver, immunoglobulins are synthesized in and secreted by plasma cells which are the end products of a line of cell types that start with the B lymphocyte stem cell in bone marrow (Table 4-10). More mature (early and late) B lymphocytes, found mainly in lymph nodes and in blood, develop numerous receptor immunoglobulins on their surface membranes. Upon encountering antigen, these B lymphocytes proliferate and develop into plasma cells. Each of these secretes into the blood a highly specific antibody capable of binding additional antigen. The stimulating antigens are normally foreign, but may be on host cell surfaces and cause autoimmune disease.

All immunoglobulin molecules consist of two or more basic units built of two identical heavy

Table 4-10. THE B LYMPHOCYTE SERIES AND ASSOCIATED MALIGNANT NEOPLASMS

| Stages in Maturation and Proliferation | Principal Site | Immunoglobulins | | Associated Malignant Neoplasms |
		Surface Receptor	Secreted into Blood	
Stem cell	Bone marrow	None	None	Acute lymphocytic leukemia
↓				
Early B lymphocyte	Lymph nodes	IgM, IgD	None	Lymphoma, chronic lymphocytic leukemia (85%)
Antigen → *Primary immune response*				
↓				
Late B lymphocyte	Lymph nodes	IgM	IgM	Lymphoma, chronic lymphocytic leukemia (15%), and Waldenström's macroglobulinemia
Antigen → *Secondary immune response*		*H-chain*	*changes*	
↓		↓		
Plasma cell	Lymphatic tissue, bone marrow	IgG, IgA	IgG, IgA	Multiple myeloma

(H) chains and two identical light (L) chains. This structure is illustrated in Figure 1C-1. The amino acid sequences of the variable regions at the N-terminal ends of the four chains determine the antigenic specificity of the particular antibody molecules produced by a single plasma cell or by a "clone" of identical plasma cells. The remainder of the molecule, the "constant" part, is the same for every immunoglobulin molecule of a given class and carries the effector sites. The two antigen-binding sites are at the ends of each identical light and heavy chain pair (see Fab, in Figure 1C-1). A single bacterium will have numerous surface antigenic proteins; each protein will have many determinants, and each determinant will stimulate production of plasma cells that produce an antibody for that determinant. This is the source of the heterogeneity of immunoglobulins, a heterogeneity illustrated by the diffuse bands, particularly for IgG, seen on electrophoresis.

The effector sites of an Ig molecule that interact with cells (e.g., mast cells) and with complement are on the constant (Fc) parts of the heavy chain. Variations in these parts of the molecule are responsible for the classes and subclasses into which Igs are grouped: IgM, IgG with four subclasses, IgA with two subclasses, IgD, and IgE. (Their respective heavy chains are called μ, γ, α, δ, and ϵ.) The hinge region between the Fc and Fab portions, the site of proteolytic cleavage, controls the interaction between the Fab and Fc parts. The hinge region contains one or more half-cystines which provide the interchain disulfide bridges. It is thought to be extended in IgG_3 and in IgD to allow their easy binding to C1q and thus activation of complement by the classical pathway. These structural variations in Ig classes are the reasons for the differences in their functions.[79]

Light chains, which are produced independently and in slight excess of their incorporation into Igs, are of two types, κ and λ; their constant regions have different structures. Both occur in all Igs in the proportion $\kappa:\gamma = 2:1$; the two halves of a given molecule always have the same type. There are four subclasses of λ-chains.

Properties of Immunoglobulin Classes

IgM. IgM is the most primitive and least specialized Ig and the only Ig that a newborn synthesizes. In adult serum it is the third most abundant Ig and accounts for 5–10% of the total circulating Ig. IgM as a membrane receptor molecule is monomeric, but most of the serum IgM is a pentamer of five IgM monomers; each monomer is very similar to the IgG molecule shown in Figure 1C-1, except that IgM has attached to it a small glycopeptide, the J chain. IgM contains about 10% carbohydrate; its high M.W. (900 000) prevents its passage into extravascular spaces. B lymphocytes at first have IgM surface receptors (Table 4-10), and secrete IgM in the first or "primary" response to an antigen. The H chains of the IgM surface receptor molecules then are modified in situ to IgG or IgA H chains, while the variable regions remain unchanged; as the cells change into plasma cells, a second dose of the same antigen causes a larger, "secondary" response, now of IgG secretion. IgM continues, however, to be synthesized against antigens confined to the blood, such as erythrocyte surface antigens and tropical parasites. IgM is not transported across the placenta and is therefore not involved in hemolytic disease of the newborn. It is an efficient complement activator, the Fc chains being spaced at the correct distance to match the C1q binding sites.

IgG. The major Ig produced by plasma cells is IgG, which makes up 70–75% of the total IgG. Of this amount, 65% is extravascular; the remainder is mainly present in plasma. Its major function appears to be neutralization of toxins in tissue spaces. Antibodies of the IgG class are produced in response to most bacteria and viruses; they aggregate and coat small soluble foreign proteins such as bacterial toxins. IgG consists of two heavy and two light chains (see Figure 1C-1.); its M.W. is 160 000 and it contains <3% carbohydrate. During cellulose acetate or agarose electrophoresis, IgG usually migrates in the γ- and slow-β-regions; the heterogeneity of the IgG antibody molecules synthesized by different plasma cells causes the region to stain diffusely. The broad response to various antigens is called a polyclonal response, i.e., numerous subsets or clones of plasma cells are producing and secreting IgGs that are slightly different in the structure of their variable regions and consequently in electrophoretic mobility.

IgG has four subclasses: IgG_1, IgG_2, IgG_3, and IgG_4. Structurally these subclasses differ in the hinge region. In IgG_3, the hinge is extended by up to 15 half-cystines, which allows binding on C1q. IgG_1 and IgG_3 both bind firmly to the Fc receptors of phagocytic cells, activate killer monocytes (K cells), and cross the placenta by an active transport process dependent on Fc binding. IgG_1 is the principal IgG to cross the placenta and to protect the newborn for the first three months (Figure 4-9). The $t_{1/2}$ of IgG_1, like those of IgG_2 and IgG_4, is 22 d, much longer than that of IgG_3 (7 d).

IgA. Approximately 10–15% of serum Ig is IgA. IgA contains 10% carbohydrate, has a M.W. of 160 000, and a $t_{1/2}$ of 6 d. In its monomeric form, its structure is similar to that of IgG, but 10–15% of IgA in serum is polymeric, particularly IgA_2, which is more resistant to destruction by some pathogenic bacteria than IgA_1. On electrophoresis, IgA migrates in the beta-gamma region, ahead of most of the IgG.

Another and probably more important form of IgA is called *secretory IgA*. It is found in tears, sweat, saliva, milk, and colostrum, and in gastrointestinal and bronchial secretions. Secretory IgA has a M.W. of 380 000 and consists of two molecules of IgA, a secretory component whose M.W. is 70 000, and a J chain of M.W. 15 600. It is synthesized mainly by plasma cells in the mucous membranes of the gut and bronchi and in the ductules of the lactating breast. The secretory component makes secretory IgA more resistant to enzymes and protects the mucosa

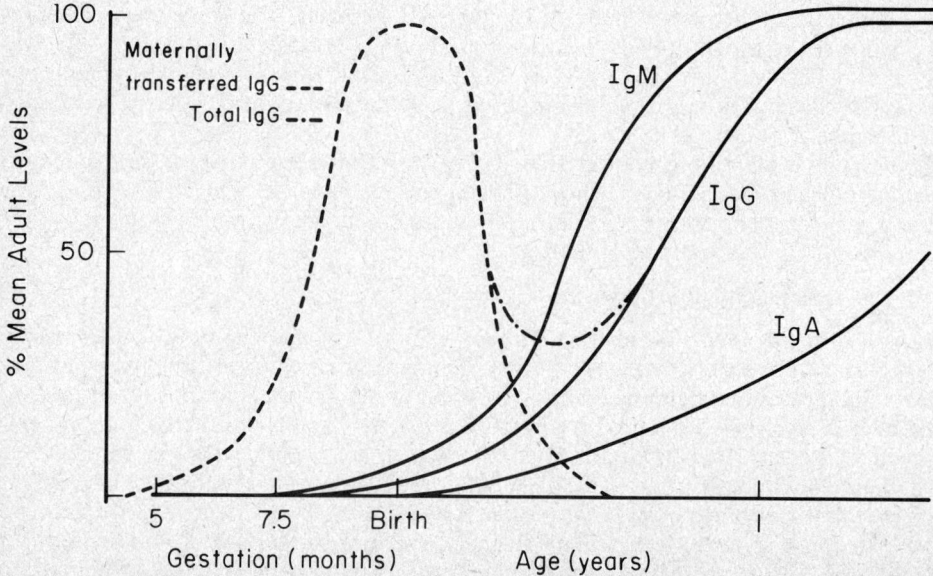

Figure 4-9. Serum immunoglobulin levels as per cent of adult levels before birth and for the first year of life.

from bacteria and viruses. Its presence in colostrum and milk probably protects the newborn from intestinal infections. IgA can activate complement by the alternative pathway (see page 576) but the exact role of IgA in serum is not clear.

IgD. IgD accounts for <1% of serum Igs. It is monomeric, contains about 12% carbohydrate, and has a M.W. of 184 000. Its structure is similar to that of IgG. Like IgM, IgD is a surface receptor for antigen in B lymphocytes, but its primary function is unknown.

IgE. IgE is so rapidly and firmly bound to mast cells that only trace amounts of it are normally present in serum. IgE contains 15% carbohydrate and has a M.W. of 188 000; its structure is similar to that of IgG. Many IgE molecules are attached to mast cell surfaces. Each molecule may be a different antibody produced by a different variable region; the IgE is attached through binding sites on its Fc chains. When antigen (allergen) combines with one of the attached IgE molecules, the mast cell is stimulated to release histamine and other vasoactive amines. These vasoactive amines are responsible for the vascular permeability and smooth muscle contraction occurring in such allergic reactions as hay fever, asthma, urticaria, and eczema.

"Normal" Immunoglobulin Levels

Reference ranges for IgG and IgM in various human adult populations differ around the world because levels depend on the extent of antigenic stimulation coming from the environment. IgA levels, however, are unaffected by environmental factors. Newborn infants have B lymphocytes with antigen receptors of the IgM type, but do not have significant rates of Ig synthesis. The IgG level in a neonate represents IgG transferred across the placenta from the mother. Immunoglobulin synthesis is stimulated by environmental antigens so that serum IgM reaches adult levels at about 9 months (see Figure 4-9), IgG in 3 years, IgA, IgE, and IgD in about 14 years. IgA in saliva, however, reaches adult levels in about 6 weeks.

Clinical Significance of Immunoglobulins

When we estimate the concentration of a single class of Ig, we are evaluating a polyclonal mixture of antibodies that are "idiotypes," i.e., the products of many different clones of plasma cells, each clone producing Ig molecules with a different variable region. Benign or malignant proliferation of one such clone will produce a high concentration of a single idiotype (a monoclonal antibody) that may appear as a sharp, narrow band on protein electrophoresis. If a few clones proliferate, there are several sharp bands; for example, the oligoclonal bands seen in electrophoresis of CSF in demyelinating diseases such as multiple sclerosis. The changes found in disease may therefore be a decrease or an increase in the normal polyclonal mixture of serum Igs, or an increase in one or more monoclonal idiotypes (in multiple myeloma usually one, rarely two).

Immunoglobulin deficiency. Immune defense depends on four complex, interactive systems: humoral antibodies (Igs) of the B lymphocyte series; cell-mediated immunity of the T lymphocytes; the phagocytic system; and the complement system. The last two systems are nonspecific in that they have no immunologic memory for antigen. Only the first and fourth systems are composed of plasma proteins. Immunodeficient states characterized by recurrent infections may be the result of a defect in any of these systems.

Marked reduction or absence of the γ-band on electrophoresis indicates deficiency of IgG antibodies. IgG deficiency is commonly secondary to protein loss or to failure of synthesis, but may be due to a primary congenital disorder. (Table 4-11 summarizes some causes and effects of Ig deficiencies.) The diagnosis of a deficiency state is important because replacement therapy with γ-globulin can be provided. Presence of a normal-appearing γ-band on protein electrophoresis does not rule out Ig deficiency. Some primary deficiencies involve only one or two Ig classes; if the total Ig level is not greatly affected, the deficiency (e.g., IgM) may not be suspected from the electrophoretic pattern. Furthermore, some clinically immunodeficient patients have normal IgG levels, yet the antibodies appear incompetent when the patient is challenged with antigen.

Immunodeficiency is a risk for infants. Figure 4-9 shows how levels of maternal IgG, transferred across the placenta, rise in the fetus during the last three months of pregnancy. Contact of the neonate with environmental antigens causes B lymphocytes to begin to multiply, IgM levels to start to rise, and plasma cells producing IgG and IgA to increase in number. These developments, however, are paralleled by a decrease of maternal IgG, so that in the infant's blood, IgG falls to a minimum at about three months of age. Two groups of newborns are at risk: premature babies, because they start with less than the full-term amount of maternal IgG,

Table 4-11. CAUSES OF IMMUNOGLOBULIN DEFICIENCY

Secondary causes
  Defective synthesis (IgM falls first, then IgA, finally IgG)
    Lymphoid malignancy, multiple myeloma, lymphoma, chronic lymphocytic leukemia
    Toxic reaction, e.g., renal failure, diabetes mellitus
    Drugs, e.g., phenytoin, penicillamine
    In newborn only: prematurity, transient delay in initiation of synthesis
  Abnormal loss of proteins
    Nephrotic syndrome, burns, protein-losing enteropathy
Primary or inherited causes
  Failure of antibody production
    Generalized (severe pyogenic infections occur)
      Infantile X-linked Bruton type
      Acquired, variable, unclassifiable, occurring at any age
    Selective immunoglobulin deficiency of
      IgA: Most common (1:700), symptomless but those affected tend to suffer allergic or autoimmune disease
      IgG and IgA (IgM increased): recurrent pyogenic infections
      IgA and IgM: Giardiasis common
      IgG: Recurrent pyogenic infections
      IgM: Susceptibility to autoimmune disease and to septicemia following splenectomy
  Combined failure of antibody and cell-mediated immunity
    Severe combined immunodeficiency. Swiss and sex-linked types, death in infancy from fungal or viral infections
    Associated with thymoma, achondroplasia, or thrombocytopenia and eczema (Wiskott-Aldrich syndrome)

and babies in whom initiation of IgG synthesis is transiently delayed. IgG determinations are invaluable in these cases since levels may fall dangerously low if the baby is not treated. Rising IgM and normal salivary IgA concentrations at six weeks of age suggest a good prognosis.

Polyclonal hyperimmunoglobulinemia. Polyclonal increases in serum Igs are the normal response to infections. IgG tends to predominate in autoimmune responses; IgA in skin, gut, respiratory, and renal infections; and IgM in primary viral infections and bloodstream infections such as malaria. *Chronic bacterial infections* cause an increase in serum levels of all Igs. In such cases, estimations of the individual Igs seldom provide more information than protein electrophoresis. They are of value, however, in the differential diagnosis of liver disease and of intrauterine infections. In *primary biliary cirrhosis,* the IgM level is markedly increased; in *chronic active hepatitis,* IgG and sometimes IgM are increased, and in *portal cirrhosis,* IgA and sometimes IgG are increased. In *intrauterine infections,* production of IgM by the fetus increases and at birth the IgM level in cord blood is increased. Estimations of IgE are used in the management of asthma and other allergic conditions, especially in children.

Monoclonal Immunoglobulins (Paraproteins)

A single clone of plasma cells produces immunoglobulin molecules with identical structures. If the clone is permitted to multiply, the concentration of its particular protein in the patient's serum becomes so great that on electrophoresis it often produces a narrow, sharply discrete spike. These monoclonal immunoglobulins, which are also called paraproteins, may be polymers, monomers, or fragments of Ig molecules; if fragments, they are usually light chains (Bence Jones proteins) or, rarely, heavy chains or half molecules; both monomers and fragments may be polymerized. About 60% of paraproteins are due to multiple myeloma or to a solitary plasmacytoma; about 15% are due to overproduction of B lymphocytes, mainly in lymph nodes, lymphomas, chronic lymphocytic leukemia, Waldenström's macroglobulinemia, or (rarely) in heavy chain disease; probably up to 25% of paraproteins are benign, and many are never discovered.[54]

Multiple myeloma is a malignant neoplasm of a single clone of plasma cells of the bone marrow. The plasma cells most often proliferate diffusely throughout the marrow but occasionally they form a solitary tumor called a *plasmacytoma.* Osteolytic bone lesions are produced and the other bone marrow cells are reduced so that thrombocytopenia, anemia, and leukopenia develop. At the same time, development of normal clones of plasma cells is inhibited; consequently, synthesis of the other Igs is reduced and a syndrome of recurrent infections occurs. The incidence of multiple myeloma is low in individuals of less than age 60, but rises rapidly with age. The patient may present with local symptoms of a bone lesion, but more often presents with nonspecific symptoms such as weight loss, obscure anemia, hemorrhages, repeated infections, or renal failure.

A highly suggestive laboratory finding is a normal serum alkaline phophatase level in a patient with destructive bone lesions. Cardinal diagnostic features of the disease are the findings of neoplastic plasma cells in bone marrow aspirate, radiologic demonstration of osteolytic lesions, and identification of a paraprotein in serum or in concentrated urine. All patients who could conceivably have the disease should be screened for paraproteins since <1% with the disease fail to have detectable paraproteins. Table 4-12 lists the paraproteins which may be associated with multiple myeloma and shows some characteristic findings for them.

Should a paraprotein be identified in blood, or urine, or both, its heavy and light chains should be typed and the concentrations of polyclonal IgG, IgA, and IgM should be determined. These studies confirm whether the spike on the electrophoretic pattern is indeed a paraprotein, they help to decide the probable prognosis, and they show whether the polyclonal Igs are so low as to make the patient vulnerable to infections. If the paraprotein proves to be IgM, the diagnosis is probably Waldenström's macroglobulinemia and not multiple myeloma. Prognosis is based on the class of the paraprotein found, its concentration at the time of diagnosis, and the rate at which its concentration increases. The concentration at the time of diagnosis must correlate with the current extent of the disease process. The rate of increase in concentration, when compared with the average doubling time for the concentration of the particular class of paraprotein, should correlate with the rate of growth of the neoplasm.

When a paraprotein is identified, but the condition is clinically benign with normal bone marrow and bone X-rays, the following findings suggest that the condition will eventually become malignant:

1. IgG > 2 g/dL; or either IgA or IgM > 1 g/dL; or the finding of an IgD or IgE paraprotein at any concentration.

2. Immunoglobulin fragments in urine or serum—in urine, usually Bence Jones protein; in serum, occasionally other fragments such as monomeric IgM.

3. Progressive increase in the concentration of the paraprotein.

4. Low levels of polyclonal Igs.

The converse of these criteria suggests that the condition is benign. Even then the patient should be followed for at least 5 years.

Table 4-12. MONOCLONAL IMMUNOGLOBULINS (PARAPROTEINS) IN MULTIPLE MYELOMA

Plasma Paraprotein	Incidence,* %	Age of Occurrence,* Mean	Average Doubling Time for Concentration, Months	Incidence of Bence Jones Proteinuria, %	Comments
IgG	50	65	10	60	Patients more susceptible to immunodeficiency; paraproteins reach highest levels
IgA	25	65	6	70	Tend to have hypercalcemia and amyloidosis
Bence Jones protein only	20	56	3.5	100	Often renal failure; bone lesions; amyloidosis. Poor prognosis.
IgD	2	57	—	100	90% λ type. Often have extraosseous lesions, amyloidosis, renal failure; 50% have enlarged lymph nodes, liver, spleen. Poor prognosis.
IgM	1	—	—	100	May or may not have viscosity syndrome
IgE	0.1	—	—	Most	—
Biclonal	1	—	—	—	—
None detected	<1	—	—	0	Usually reduction of normal immunoglobulins

*Approximate.

Lymphoid tumors, presenting as lymphomas or chronic lymphocytic leukemias, arise from less mature stages in B lymphocyte development; about 1 in 5 produce paraproteins, usually of the IgM class. *Waldenström's macroglobulinemia,* however, arises from the most mature B lymphocytes and invariably produces IgM; in fact, it is the presence of this very high molecular weight protein that produces the salient symptom of the disease—an increase in viscosity of the blood. Bence Jones proteinuria occurs in 80% of these cases, but the condition is much less malignant than multiple myeloma. Lymph nodes and spleen are enlarged, but the lymphoid infiltration is slow-growing and the symptoms are treatable by exchange transfusion. Rarely, unusual forms of IgG, IgA, or light chains polymerize and cause a similar syndrome of high blood viscosity. *Heavy chain diseases* are rare conditions also associated with lymphoid infiltration in which the paraprotein consists only of a heavy chain, usually incomplete. The most common is *α-chain disease,* in which the intestine is infiltrated and a severe malabsorption syndrome is produced; but *γ-chain disease* and *μ-chain disease* also occur.

Two other disorders are sometimes characterized by paraproteins: *cryoglobulinemia* and *amyloid disease.* A cryoglobulin is a serum protein which precipitates at temperatures lower than body temperature. Most cryoglobulins are polyclonal immunoglobulin complexes, but nearly half are monoclonal, usually IgM. For cryoglobulin examinations, a temperature of 37 °C must be maintained for blood collection and for serum separation and storage in order to keep the cryoglobulin from precipitating out of the serum. Amyloid disease is characterized by deposits of insoluble, fibrillar protein complexes in various tissues; with special staining, the deposits are easily seen in biopsy sections. Some of the deposits contain fragments of L chains, especially from the variable region. Amyloid deposits may also occur in multiple myeloma. (See section on Amyloid.)

Methods for Analysis of Immunoglobulins[114]

Only immunochemical methods are sensitive enough to detect or quantitate individual Igs at normal levels. Although RID or electroimmunoassay gel techniques may be used, nephelometry or turbidimetry is preferred now that very specific antisera with high titer and affinity are available. The latter methods require few manipulations and are more rapid and precise. They are applicable to serum and CSF IgG, IgA, and IgM assays, but for IgD and IgE the greater sensitivity of radioimmunoassay or enzyme immunoassay is required. Remarkably good precision can be obtained for Igs, but because reference materials containing precisely determined quantities of purified proteins are not presently available, it is impossible to say how accurate the methods are.

Assays of "individual" polyclonal Igs involve determining the concentration of a mixture of protein molecules of different sizes having similar constant regions but different variable regions (idiotypes). Reagent antisera and reference Ig standards used in most immunochemical assays have been generated against normal human sera containing a mixture of Ig subclasses. Despite the uncertainties inherent in these assay systems, determination of individual polyclonal Igs, by either an individual or a series of techniques, is quite reliable. By contrast, the determination of monoclonal proteins is not. A monoclonal immunoglobulin will have only some of the determinants with which the antibodies in the antiserum usually react and thus the standard dilution curve will not apply. The few antibodies in the antiserum which *do* react are soon precipitated and the paraprotein overestimated.[114] Furthermore, if a new batch of antiserum is introduced, the relationship between paraprotein concentration and the standard curve is likely to change. For these reasons, the absolute concentration of a paraprotein must be estimated by electrophoresis and densitometry, but subsequent *changes* of concentration of the *same* paraprotein can be determined immunochemically provided that the *same* batch of antiserum is used for all the determinations.

Reference preparations of immunoglobulins are available from the World Health Organization (WHO); the International Federation of Clinical Chemistry (IFCC); the Centers for Disease Control, which provide the United States National Reference Preparations (USNRP); and the American College of Pathologists (CAP). The preparations are usually standardized in International Units (IU) and there are no agreed factors for the conversion of IU to mass units. (See p. 564.)

Reference Ranges

By nephelometry:

Age	IgG (mg/dL)	IgA (mg/dL)	IgM (mg/dL)	IgD (mg/dL)	IgE (IU/mL)
Serum					
Newborn (4 d):	700 – 1480	0 – 2.2	5 – 30		
16–60 years:	650 – 1500	76 – 390	40 – 345	0 – 8	0 – 380
> 60 years:	600 – 1560	90 – 410	30 – 360	(RID)	(RIA)
Spinal fluid:	0 – 5.5	0 – 0.6	0 – 1.3		
Saliva:		~11			

Figure 4-9 shows in graphic form changes in IgA, IgG, and IgM levels in childhood. The reader is cautioned, however, that the data given in the figure and the above listing are only illustrative. Reference ranges are related to age, geographic, and methodological differences, and vary widely in different circumstances. (See also Chapter 20.) Consult reference 105 for more extensive listings for different ages, methods, and body fluids.

Investigation of Paraproteins

The first step consists of agarose electrophoresis of serum and of concentrated urine. Serum should be collected and separated at 37 °C to avoid loss of cryoglobulins. An estimate of the paraprotein should then be made from the spike identified on the electrophoretic plate by densitometric scan and by determination of total protein in the sample. Bence Jones protein, when it is present only in urine, is the paraprotein most often missed.

The second step is to identify heavy and light chains by immunoelectrophoresis or, more simply, by immunofixation.[84] Specific antisera against γ-, α-, and μ-heavy chains and against κ- and λ-light chains are used; antisera against IgD and IgE are seldom required. Reaction of the paraprotein with only one heavy chain and one light chain antiserum confirms that an anomalous spike on protein electrophoresis is indeed a paraprotein, and not due to denaturation, fibrinogen, hemoglobin, or some other artifact. Nonparaprotein immunoglobulins should then be determined by nephelometry or other immunochemical methods. If a paraprotein does not react with any of these heavy chain antisera, it should be tested with anti-delta and anti-epsilon before assuming it is a Bence Jones protein.

The patient should subsequently be monitored for changes in the paraprotein's concentration using a quantitative immunochemical assay, conducted at appropriate intervals. Monitoring of other nonparaprotein Igs may also be desirable.

Immunofixation Electrophoresis

Immunofixation electrophoresis (IFE) is gradually replacing immunoelectrophoresis (IEP) because of its rapidity and ease of interpretation. Several procedures, including commercial kits, are available for both IFE and IEP. While these procedures may differ in detail, their principles are the same as those described here.

Specimen Collection and Storage. Fresh serum is the specimen of choice. Blood is collected in the fasting state without anticoagulants and then centrifuged immediately for 15 min at ~900 × g. Serum can be stored at 2–8 °C up to 5 d, or at −20 °C for longer periods. The minimum amount of specimen required is 100 μL. Cerebrospinal fluid, urine, and other biological fluids may be used, but concentrating these specimens prior to analysis may be necessary.

Principles and Interpretation. Samples of the patients' specimens are placed in six separate wells on an agarose gel and their major protein groups separated by electrophoresis. One of these tracks is then treated with a chemical fixative solution to fix all proteins in the agarose and create an electrophoresis reference pattern for the specimen. The other five tracks are treated with specific heavy chain and light chain antisera which react with individual immunoglobulins in the specimens, causing them to become immunofixed in the agarose. All unreacted proteins in these five tracks are then washed out of the gel, and all six tracks are stained to visualize the fixed protein bands.

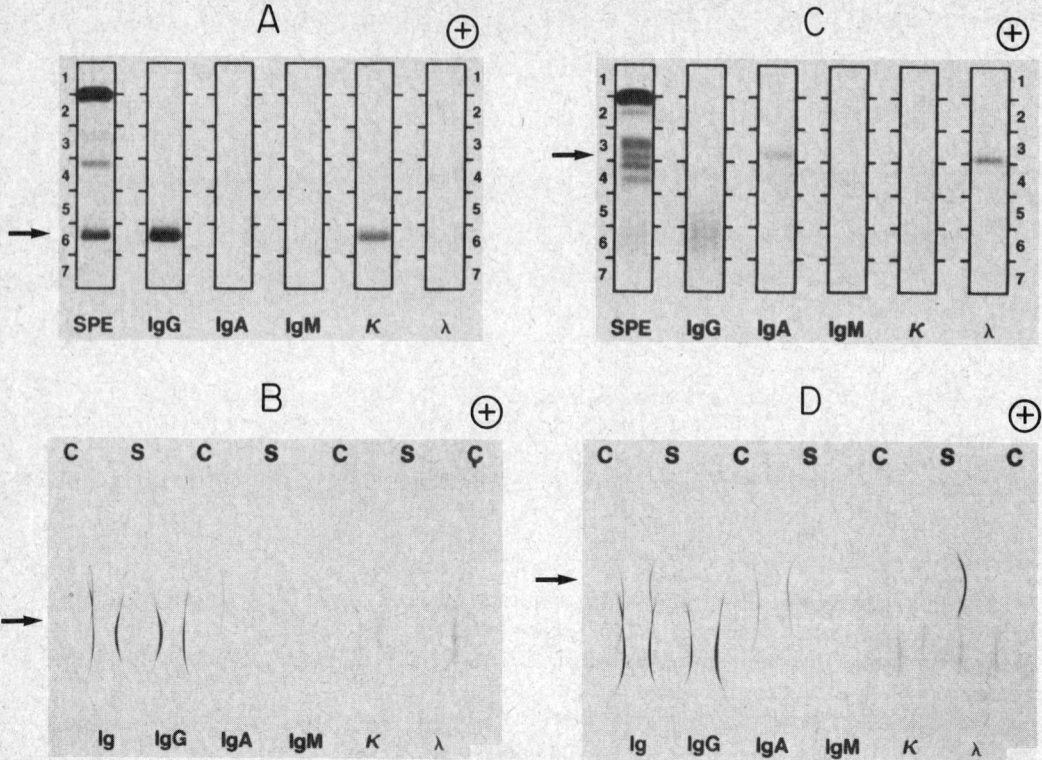

Figure 4-10. Comparison of IFE and IEP for two patients with monoclonal gammopathies.

A, Patient specimen with an IgG (kappa, κ) monoclonal protein as identified by IFE. Arrow indicates position of monoclonal protein. After electrophoresis, each track except SPE is reacted with its respective antiserum, then all tracks are stained to visualize the respective protein bands. SPE = chemically fixed serum protein electrophoresis; IgG, IgA, IgM, κ and λ indicate antiserum used on each track.

B, Same specimen as in A, with proteins identified by IEP. Arrow indicates position of monoclonal protein. Normal control (C) and patient serum (S) are alternated. After electrophoresis, antiserum is added to each trough as indicated by the labels Ig (polyvalent Ig antiserum), IgG, IgA, IgM, κ and λ. The antisera react with separated proteins in the specimens to form precipitates in the shape of arcs. The IgG and kappa arcs are shorter and thicker than those in the normal control, showing the presence of the IgG (kappa) monoclonal protein. The concentrations of IgA, IgM, and λ-light chains are also reduced.

C, Patient specimen with an IgA (lambda, λ) monoclonal protein identified by IFE procedure as described in A.

D, Same specimen as in C with proteins identified by IEP as described in B. The abnormal IgA and λ-arcs for the patient specimen indicate an elevated concentration of a monoclonal IgA (lambda) protein. All separations were performed using the Beckman Paragon system.

By comparing the locations of the stained immunofixed bands with a band of the same location in the reference pattern, the identification of a specific protein may be made.

In monoclonal gammopathies, the IFE patterns yield a distinct, sharply defined, precipitin band with one heavy chain and one light chain antiserum. These bands match the location of the particular Ig in the reference pattern. (See Figure 4-10.) In polyclonal gammopathies, a diffuse precipitin band will occur with the specific antiserum, in contrast to the sharp band observed in monoclonal gammopathies.

IFE patterns should always be confirmed by quantitating the immunoglobulins (IgG, IgA, IgM) in the specimen. This can be done by nephelometry as described earlier. Elevations of specific immunoglobulins should correspond to more intensely stained bands on the IFE pattern, but if a monoclonal protein is present, the result will be an approximation only. In order to determine proper dilutions of the specimen to be used, it is helpful to perform these quantitations before IFE is performed.

Reagents. Agarose gels for IFE may be prepared by the laboratory. A more common and convenient practice, however, is to purchase ready-made gels from commercial sources (e.g., Beckman Instruments, Inc., Brea, CA 92621; or Worthington Diagnostics, Freehold, NJ 07728). These gels are often supplied as part of a kit which contains most other necessary materials for performance of IFE (e.g., antisera, protein stains, electrophoresis buffer). Antisera to human μ-, α-, and γ- heavy chains and to κ- and λ-light chains may be obtained from commercial sources as part of a kit (see above) or as individual reagents (e.g., from Atlantic Antibodies, Scarborough, MA 04074; or Kallestad Laboratories, Inc., Austin, TX 78701). Several different electrophoresis buffers may be used; the most common one is barbital buffer, 0.05 mol/L, pH 8.6. Likewise, various protein stains are available; Coumassie Brilliant Blue stain is commonly used. A protein fixative solution is required for fixing proteins onto the agarose gel after electrophoresis. This solution often consists of glacial acetic acid, 5 mL/dL in water or acetic acid, 5 mL/dL in aqueous ethanol (60 mL/dL in water).

Procedure

1. Apply 3–5 μL of test specimen into each of the six specimen wells on the agarose gel. *Note*: Some procedures require a specimen *dilution* to be made before application of specimen to the gel. The amount of dilution depends on the sensitivity of the staining reagent and resolution of the major protein groups on the agarose gel after electrophoresis. In general, for normal levels of proteins in the patient specimen, a two-fold dilution should be used for the reference pattern and a ten-fold dilution should be used for the immunofixation patterns. These dilutions may have to be adjusted, depending on the actual concentrations of immunoglobulins in the specimen. For urine and CSF samples, it may be necessary to concentrate the specimens.
2. Place gel in an electrophoresis chamber and electrophorese the specimens for 30 min at 100 V (time and voltage may differ with various procedures).
3. Remove gel from chamber and gently blot gel surface.
4. Apply each antiserum to a corresponding track on the gel, beginning with anti-lambda, then anti-kappa, anti-IgM, anti-IgA, and anti-IgG. Add to the remaining track the protein fixative solution. For this one track only, all separated proteins will be fixed in the gel and will not be washed out in the subsequent washing steps. After staining, this track is then used as a reference to which immunofixed specific proteins on the other tracks can be compared.
5. Incubate the gel in a moist chamber at 45 °C for 30 min to allow the antisera to bind the specific proteins in the specimen.
6. Wash the gel in saline, blot, and repeat the wash procedure.
7. Dry gel in an oven at 90 °C until completely dry.
8. Place gel in protein stain for 3–5 min, then into acetic acid, 5 mL/dL, until background is clear.
9. Dry gel in oven until completely dry.

A comparison of IFE and immunoelectrophoresis (IEP) for two patients with monoclonal gammopathies is shown in Figure 4-10. The principles of IEP are described in Chapter 1C.

COMPLEMENT PROTEINS

The complement proteins are a group of at least 20 immunologically distinct components in blood and tissue fluids.[1,18,63] They are able to interact sequentially with antigen-antibody complexes, with each other, and with cell membranes in a complex but adaptable way to destroy viruses and bacteria and, pathologically, even the host's own cells. Complement proteins are synthesized by the liver and are normally present in the blood as functionally inactive molecules. A sequence of interactions can be activated by antigen-antibody complexes through the "classical" pathway from C1 to C5, or even by nonantibody initiating factors through the "alternative" or nonclassical pathway from C3 to C5. (See Figure 4-11.) Either path leads to the final membrane attack sequence (C5 to C9 and lysis). The whole process is controlled by the very short life of many of the intermediates and by the presence of specific inhibitor proteins which are an integral part of the system.

The proteins of the classical and final pathways are numbered from C1 to C9; with the exception of C4, numbering is in the order of their reaction. Proteins of the alternative path are designated by single capitals (B, D, P), except for C3 which is common to both paths. Lower case suffixes (a,b) indicate fragments into which a component has been split. An enzymatically active component is designated by a bar over the symbol. For example, $\overline{D}$ and $\overline{C4b,2a,3b}$ are enzymes; the latter is a complex of fragments derived from C4, C2, and C3.

Activation of the classical pathway is usually initiated by IgG or IgM bound to antigen or by C-reactive protein. In the activation, IgG_3 is more active than IgG_1 or IgG_2. Antibody is not essential to initiation of the alternative pathway, even though IgA and IgG can initiate it. The

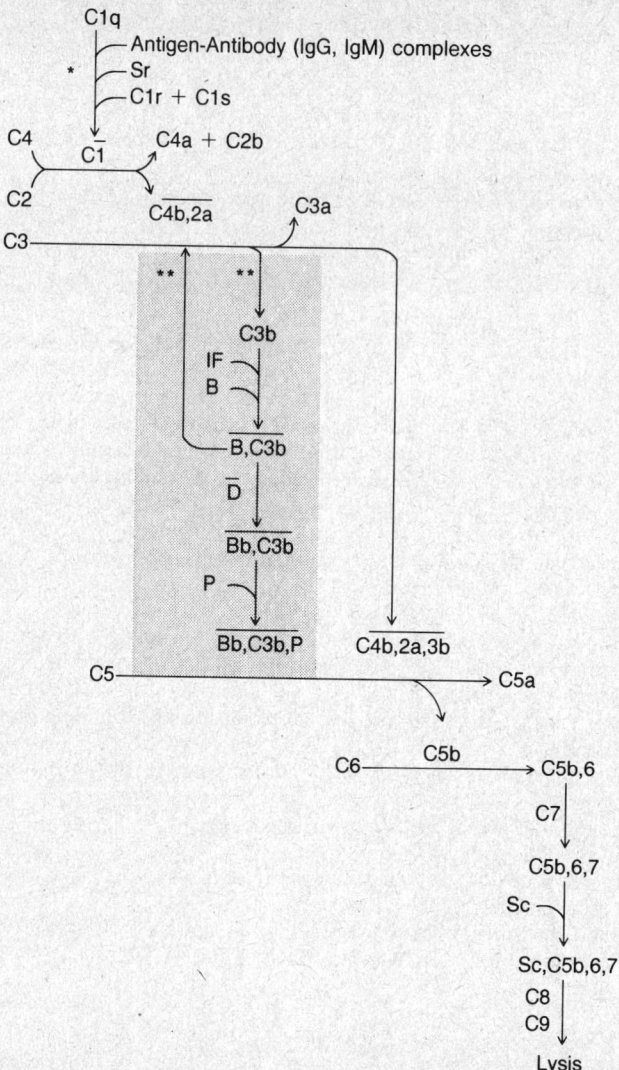

Figure 4-11. The classic and alternative pathways of complement activity. The shaded area is the alternative pathway. The bar over certain components indicates that they have enzymatic activity. Note that each pathway *produces* an enzyme that acts on C5, the classic pathway C4b,2a,3b, and the alternative pathway Bb,C3b,P. Sr = Recognition site on cell membrane; Sc = cytolytic site; C1q, C1r, C1s = subunits of C1; IF = initiating factor, alternative pathway. Initiating factors include zymosan in yeast cell wall, endotoxic lipopolysaccharide from gram-negative bacteria, certain kinds of immune complexes, and a factor from cobra venom. P = Factor P, originally called "properdin." * = Site of inhibition of C1 esterase inhibitor (C1INA). ** = Site of inhibition by C3bINA. (Modified from Mayer, M. M.: Complement and lysis. *In:* Principles of Immunology. N. R. Rose, F. Milgrom, and C. J. van Oss, Eds. New York, Macmillan Publishing Co., 1979.)

alternative pathway is therefore responsible for an immediate, nonspecific, nonantibody defense against invading organisms. Proteolytic cleavage of complement proteins C3, C4, and C5 (Figure 4-11) produces a small fragment, a peptide, designated by the suffix a, and a major fragment with suffix b. The major fragment possesses an enzymic site for attacking the next component of the sequence. It has also a binding site for attachment to a cell membrane or to an immune complex, so that subsequent effects are localized. The peptides C3a and C5a cause release of histamine from mast cells (anaphylatoxic effect); histamine increases vascular permeability, contraction of smooth muscle, and, hence, edema and stasis. Immune complexes bound with complement will cause platelet aggregation and thrombosis so that vessels surrounding the infected area are sealed off. C3a, C5a, and C5b67 are chemotaxins; they attract phagocytes to the infected area. C3b acts as an opsonin by attaching to immune complexes. This newly formed complex binds to receptor sites on the phagocyte surface (immune adherence) and thus promotes phagocytosis; phagocytosis leads to proteolytic digestion of the foreign material. Destruction of the targeted cell's membranes is mediated by the C5-9 complex and results in lysis.

Since pure preparations and specific antisera are now available, complement components can be determined either immunologically or by a functional assay (hemolytic complement assay). Here we shall consider C1, C1 esterase inhibitor (C1 inhibitor, C1INA), C3, C3 proactivator (Factor B), and C4, since estimations of them and their breakdown products have proved most useful clinically.

C1 consists of three distinct protein molecules—C1q, C1r, and C1s—combined in a molar ratio of 1:2:2. These components are held together by a Ca^{2+}-dependent interaction and dissociate only under pathological conditions. C1q has a hexameric structure; each monomer contains a globular head so that the overall structure resembles a bunch of six tulips. The monomers are connected by a radial strand to a common stem region which has a collagen-like structure. This common stem region provides the Ca^{2+}-attachment site for C1r and C1s. Binding sites for the Fc region of IgG_1, IgG_2, IgG_3, and IgM reside on the globular heads of the C1q moiety. When bound with antigen-antibody, the C1 complex interacts with a recognition site on a cell to activate the complement cascade.

C1 esterase inhibitor (C1INH) is an α_2-globulin with a M.W. of 105 000. It controls the first stage of the classical pathway and also inhibits plasmin, thrombin, and kallikrein. Its absence causes inherited angioedema. C1 esterase inhibitor is also an acute phase reactant.

C3 and *C4* are so similar in their properties that they may have a common evolutionary origin. C3 has a M.W. of 180 000 and C4 has a M.W. of 206 000. Both are glycoproteins which migrate on electrophoresis in the β-region on cellulose acetate or agarose. In fresh serum, the concentration of C3 is sufficient to produce a β_2-band on electrophoresis; C4 is usually not visible. Both molecules are polymers; C3 contains two subunits and C4 three. Genetic polymorphism has been observed but without clinical abnormalities. C3 and C4 polymers dissociate upon storage. Their electrophoretic mobilities are altered and the β_2-band will disappear; immunochemical recognition, however, is retained. Both C3 and C4 are acute phase reactants, and both take part in the classical pathway, but only C3 participates in the alternative pathway.

C3 proactivator (Factor B) is a β_2-globulin with a M.W. of 95 000. Its serum concentration normally is about 20 mg/dL.

Clinical Significance

Abnormal serum levels of complement proteins may be due to either inherited or acquired diseases.[31] Inherited deficiencies of most complement proteins have been described.[32]

Inherited deficiencies. Inherited angioedema is an autosomal dominant disease that results from a deficiency of C1 esterase inhibitor. It is characterized by attacks of subcutaneous, bronchial, and gastrointestinal edema. Spontaneous activation of uninhibited C1 leads to consumption of C4 and C2 but not of C3 or later components. A peptide breakdown product of C2 is responsible for the vascular permeability that leads to edema. Laboratory findings include low levels of C4 and C1 esterase inhibitor, but normal levels of C3. In about 15% of the cases, levels of C1 inhibitor found are normal, but the protein is nonfunctional. In such cases, a functional assay for the inhibitor is essential to diagnosis. Diagnosis is important since mortality is high if the disorder is untreated. Treatment with the anabolic steroid danazol is effective. Therapy may be monitored by assays of C1 esterase inhibitor.

Inherited deficiencies of C1, C4, and, most commonly, *C2* also occur. These deficiencies are associated with increased incidence of immune complex diseases such as systemic lupus erythematosus, Henoch-Schönlein purpura, polymyositis, and glomerulonephritis. *Deficiency of C3* results in recurrent infections and impaired leukocyte mobilization. *Deficiency of C3b inhibitor* has a similar outcome since activation of the alternative pathway is not controlled and C3 and Factor B deficiencies ensue. *Deficiencies in* components *C5 to C9* are associated with recurrent and persistent Neisserial infections because the system fails to lyse cell membranes of the invading organisms.

Acquired disease in serum complement levels may occur in either direction. Levels increase in acute phase reactions; they decrease in immune complex diseases because the classical pathway is activated so that consumption of complement proteins exceeds their synthesis. C3 and C4 are sometimes called "subacute phase" proteins because the levels of these proteins rise at a later time than other acute phase proteins (Figure 4-8). Circulating levels of C3 and C4 reflect a

Table 4-13. CONDITIONS IN WHICH ESTIMATES OF COMPLEMENT FACTORS
ARE USEFUL FOR DIAGNOSIS

Systemic lupus erythematosus	C4 low, C3 sometimes low*
Rheumatoid vasculitis	C3 low*
Subacute bacterial endocarditis	Both C3* and C4 low
Shunt nephritis	Both C3* and C4 low
Post-streptococcal glomerulonephritis	C3 low,* returns to normal in three months
Mesangio-capillary glomerulonephritis	C3 low* (persistent), C4 normal, Factor B also found
Polymyalgia rheumatica	Only C3 conversion products found
Mixed cryoglobulinemia	Only C3 conversion products found
Gram-negative bacteremic shock (early diagnosis)	C3 low,* C4 normal, Factor B low
Gram-positive bacteremia	Both C3* and C4 low
Disseminated cytomegalovirus infection	C4 very low, C3* normal or increased

*C3 conversion products are also present.

balance between complement consumption due to immune complex formation and increased synthesis due to the acute phase response.

Immune complex diseases are disorders in which antigen-antibody complexes either are formed and deposited in tissues or are formed in the circulation and then become permanently attached to tissue cells. Measurement of complement components can be useful in the diagnosis of these diseases, because the Ag-Ab complexes activate the complement system. Hence, changes in the concentration of the complement components reflect the activity of the deposition process. Evidence for an active process consists of decreasing levels of C3 and C4 as well as the presence of C3 fragments in serum. C1q, often a part of antigen-antibody complexes, can also be measured. Laboratory results, however, must be interpreted with caution. For example, inadequate preservation of the specimen will allow dissociation of C3 and C4 polymers. As a result, serum levels of C3 or C4 will be overestimated since RID assays also measure the low molecular weight breakdown fragments. A concurrent acute phase reaction in the patient will also affect C3 and C4 levels. In this case, extensive tissue damage releases proteolytic enzymes that cleave complement components that become detectable in the assays. In some cases, immune complexes are produced intermittently, and clinical symptoms may not have appeared or testing may not have been ordered until after the complement levels in serum have returned to normal.

Table 4-13 lists conditions in which estimates of complement factors are useful for diagnosis.

Methods for the Determination of Selected Complement Proteins

Specimen. Serum is preferred; however, plasma from EDTA-anticoagulated blood may be suitable for assay of most complement factors. The most important requirements are prompt separation of plasma or serum from cells and prompt analysis or storage at −70 °C, in order to preserve the very labile components.

Immunochemical Assays. Nephelometry or RID can be used for assays of C3, C4, and Factor B, although there may be overestimation of C3 and C4 by RID when breakdown products are present. Breakdown products can be readily detected and estimated by immunofixation,[113] as illustrated in Figures 1C-9 and 1C-10. C1q and C1 esterase inhibitor are also estimated immunochemically.

Reference ranges[105] for complement factors are age dependent and method dependent. Selected data are presented below.

Factor	Method	Newborn mg/dL	Adult mg/dL
C1q	RID	4.7± 5.1*	6.7± 0.7*
C1 esterase inhibitor	RID	17.4 – 24.0	17.4 – 24.0
C3	RID	88.4± 11.7*	141.2± 14.9*
	Nephelometry	58 – 120	80 – 155
C4	RID	16 – 39	15 – 45
	Nephelometry	10 – 26	13 – 37
Factor B	RID	11.8± 20*	24.1± 4.7*
	Nephelometry	14 – 33	20 – 45

* ±SD.

Functional assays depend on observing the final stage of the final pathway of complement activity as indicated by the hemolysis of erythrocytes coated with antibody. Individual components are tested by adding all the other components in excess. Sheep erythrocytes coated with rabbit antisheep antibody are used; the added components must be pure. If test serum added to the complement/indicator system does not activate complement-mediated lysis, then the test serum is functionally defective for the factor excluded from the assay system. Various modifications allow examination of both classical and alternative pathway factors. Quantitative estimates can be made by titering the test serum and comparing endpoint lysis to the lytic behavior when a standard amount of the pure factor is present in the system. Test results are often expressed relative to the amount of factor required to cause 50% lysis (CH50).

Functional assays are used for evaluating patients whose symptoms imply deficiency of a complement factor but whose serum levels of complement factors are apparently normal by immunochemical assays. In such cases, these assays are essential. They are time-consuming and expensive to perform and are optimally performed in a specialized laboratory.

METHODS FOR THE DETERMINATION OF PROTEINS IN SERUM AND PLASMA

The overall protein composition of a patient's plasma or serum should be studied first by determining its total protein content and by examining its composition by cellulose acetate or agarose gel electrophoresis.

Serum Total Protein

Allowing freshly drawn blood to clot removes fibrinogen, which is present in plasma at levels of 200–400 mg/dL, as well as some other trace proteins. The total protein concentration of serum obtained from a healthy, ambulatory adult is 6.3–8.3 g/dL, and from an adult at rest, 6.0–7.8 g/dL.

There are two general causes for alterations of serum total protein: change in the volume of plasma water, and change in the concentration of one or more of the specific proteins in the plasma. Decrease in the volume of plasma water (*hemoconcentration*) is reflected as relative hyperproteinemia; concentrations of all the individual plasma proteins are increased to the same degree. *Hyperproteinemia* is seen in dehydration due to inadequate water intake or to excessive water loss as in severe vomiting, diarrhea, Addison's disease, or diabetic acidosis. *Hemodilution* (increase in plasma water volume) is reflected as relative hypoproteinemia; concentrations of all the individual plasma proteins are decreased to the same degree. Hemodilution occurs with water intoxication or salt retention syndromes, during massive intravenous infusions, and physiologically when a recumbent position is assumed. A recumbent position decreases total protein concentration by 0.3–0.5 g/dL.

Of the individual serum proteins, albumin is present in such high concentrations that low levels of this protein alone may cause *hypoproteinemia*. Such hypoproteinemia is common and has many causes. See page 588. A mild *hyperproteinemia* may be caused by an increase in the concentration of specific proteins normally present in relatively low concentration, as for example, increases in acute phase reactants and polyclonal immunoglobulins, as a result of infection. Marked hyperproteinemia may be caused by high levels of the *monoclonal immunoglobulins* produced in multiple myeloma and other malignant paraproteinemias.

Serum Protein Electrophoresis (SPE)

Separation of proteins on cellulose acetate or agarose gel is almost always performed on serum in order to avoid the complication of a fibrinogen band in a region where proteins of greater interest will migrate. The principles of electrophoresis are described in Chapter 1B, Section Three. In practice, most SPE is performed using systems that integrate apparatus, materials, and reagents from a single supplier (Beckman Instruments, Inc., Palo Alto, CA 94304; Helena Laboratories, Beaumont, TX 77704). The standard buffer is barbital with an ionic strength of 0.05 and pH 8.6. The usual sample is 3–5 μL, applied with a mechanical device to obtain an even stripe of sample across the width of the support medium or over a track of the medium. Typical parameters for the run are 1.5 mA per 2 cm width of cellulose acetate medium, 10 mA per 1 cm width of agarose medium, and a run time of 40–60 min producing a 5–6 cm migration

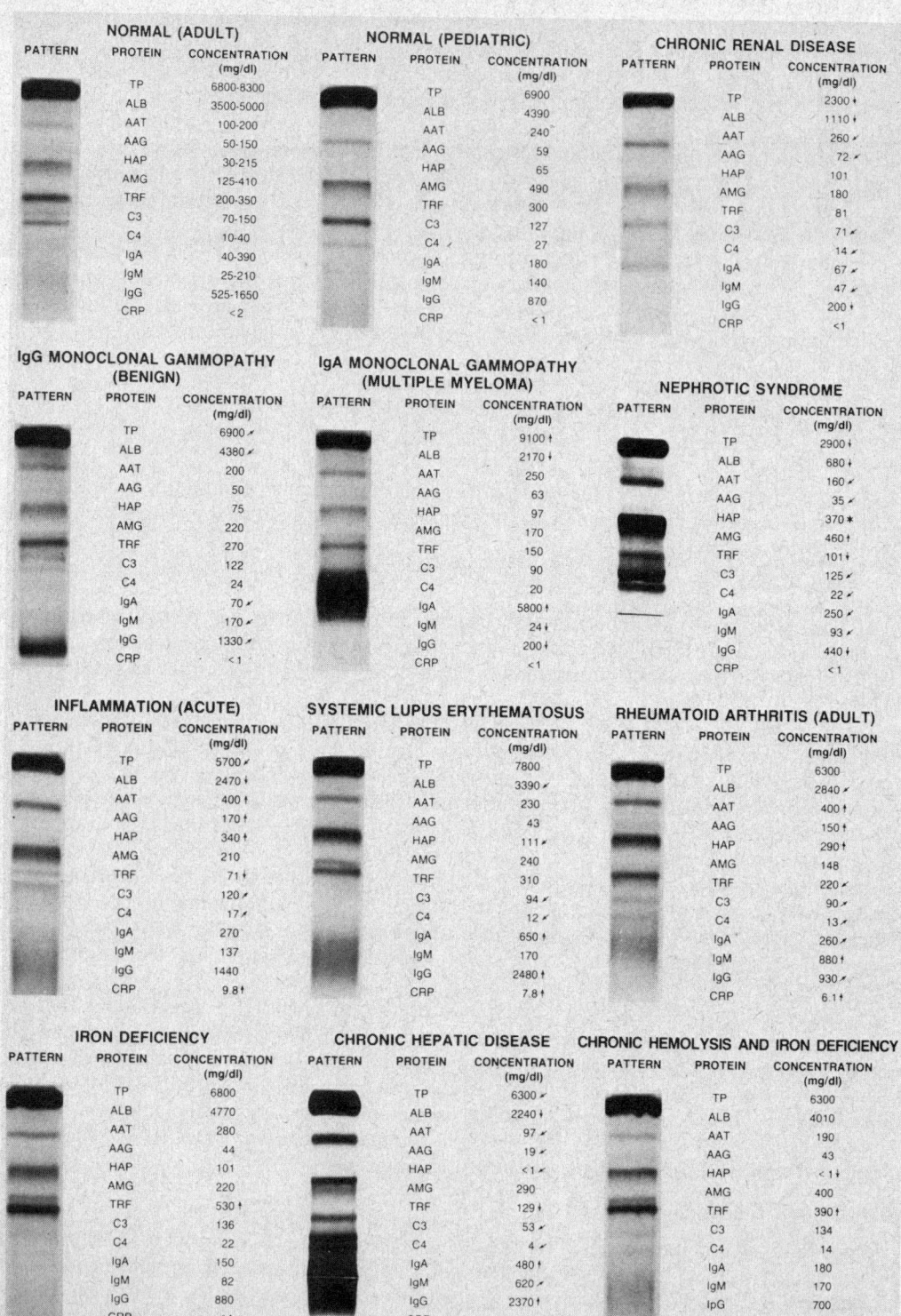

Figure 4-12. Electrophoretic patterns typical of normal and of some pathological conditions (agarose gel). Upright and down arrows (↑↓) indicate increase and decrease from the reference range, respectively. Slanting right (↗) or left (↙) arrows indicate variation from normal to an increase or from normal to a decrease from the reference range, respectively. *Hp 2-2 phenotype.

distance for albumin. Figure 4-12 illustrates some serum separations typical of normal and pathological conditions. Although regions and typical quantities of individual proteins are indicated on the figure, blurring and overlap of migrating species do not allow a clear distinction and quantitation of any individual protein except albumin. Consequently, many consider that the inspection of an electrophoretogram by a trained observer gives more reliable information than densitometric measurements. Changes in certain bands, however, are clearly associated with particular disorders, making serum protein electrophoresis a valuable screening device. The normal serum presents usually only five bands (albumin, α_1, α_2, β, and γ), although a sixth band (β_2, C3 complement component) may be seen if the serum is fresh and a buffer containing Ca(II) ions is used. Densitometry may be used for rough quantitation of individual bands and for graphic displays of stained electrophoresis patterns. Many of the individual proteins have too low concentrations to manifest as distinct stained bands, or they are overshadowed by proteins of higher concentrations that migrate near them (e.g., ceruloplasmin is masked by haptoglobin and α_2-macroglobulin). Some proteins stain poorly because they contain high proportions of lipid (lipoproteins) or carbohydrate (α_1-acid glycoprotein). A variety of stains may be used to visualize bands (Table 1B-8). Amido Black and Ponceau S have perhaps been most popular in the past; Coomassie Brilliant Blue is more sensitive. After visual inspection of the separated bands, the electrophoresis strips can be cleared to allow semiquantitative evaluation of the bands by densitometry, and then dried to preserve them in a permanent record. Most laboratories provide a chart report bearing an image of the densitometric scan, or of the stained strip, or of both. Special fat stains are needed to visualize lipoproteins that migrate in bands of somewhat variable mobility, in the fast α_1-region (α_1-lipoprotein), the α_2 or pre-β-region (very low density lipoprotein), the β_1-region (β-lipoprotein), or remain at the origin (chylomicrons). (See also Chapter 7.)

The patient's serum should always be run in parallel with a normal control serum; the two patterns are then compared, and the following changes noted. (See also reference 46.)

Intensely stained bands occurring from the α- to γ-regions. These suggest monoclonal immunoglobulins (paraproteins).

Multiple bands, absent bands, or different mobility of normal bands. These could be due to genetic variants such as haptoglobin in the α_2-region, α_1-antitrypsin in the α_1-region, and transferrin in the β-region.

Other causes of altered mobility. Increased mobility of albumin occurs when it is bound to penicillin or salicylates, or to greater than normal amounts of bilirubin or fatty acids. Decreased mobility of α_1-antitrypsin occurs when it binds thiol groups or Bence Jones protein.

Appearance of a band not normally seen. The concentration of a normal protein may increase to such a level as to become visible as a line. For instance, a faint sharp band may appear between albumin and the α_1-region as a result of a hundred-fold increase of α-fetoprotein from certain tumors. Similarly, a large increase of CRP in a severe acute phase reaction may generate a faint band in the γ-region, or an increase of lysozyme in monocytic leukemia may produce a band in the post-γ-region.

The stained strip should also be observed for changes in relative concentrations of the major plasma proteins. Such changes can suggest the presence of certain pathological conditions:

Decreased albumin and γ-bands in conjunction with an increased α_2-band suggest selective proteinuria such as is displayed in the *nephrotic syndrome*. Note, however, that the albumin concentration must fall by at least one third of its normal level before its decrease will be evident on the electrophoretic strip.

An increase in the α_1-band (α_1-antitrypsin and α_1-acid glycoprotein) and α_2-band (haptoglobin) suggests an *acute phase reaction*. An increase in α_1-components only may be seen in chronic hepatitis and in acute phase reactions accompanied by hemolysis, as well as in estrogen therapy or pregnancy. In diseases characterized by vasculitis (such as rheumatoid arthritis) or in immune complex diseases, there may be a predominant increase in the α_2-band.

An increase in the β_1-band suggests *iron deficiency anemia* (transferrin increase) or high levels of estrogen.

Fusion or bridging of β- and γ-bands suggests an increase in IgA such as occurs in cirrhosis, respiratory or skin infections, and rheumatoid arthritis.

An increase in the γ-band suggests a polyclonal γ-globulin increase associated with an immune reaction, chronic inflammatory disease, liver disease, or disseminated neoplasms. Oligoclonal bands are occasionally seen in chronic aggressive hepatitis and chronic viral infections.

An absence or decrease of the γ-band suggests immune deficiency, either congenital or acquired. (See Table 4-11.)

Determination of Total Protein in Serum

When the total protein concentration of serum is measured, two assumptions are arbitrarily made: (1) all protein molecules are pure polypeptide chains, containing on the average 16% by weight of nitrogen; and (2) each of the several hundred individual proteins present in serum reacts chemically like every other protein. Clearly, the first assumption is not true at all, and the second cannot always be true. Nevertheless, these simplifying assumptions make measurement of total protein a practical, though empirical, procedure.

Historically, total protein was first determined by the *Kjeldahl method*, in which acid digestion was employed to convert nitrogen in the protein to ammonium ion. The concentration of ammonia nitrogen was then evaluated by titration or nesslerization, a correction was made for nitrogen contributed by nonprotein compounds also present in serum, and the ammonia nitrogen value was multiplied by 100%/16% (i.e., the factor 6.25) to express protein nitrogen as total protein. The method was well defined and reproducible but so time consuming and inconvenient as to be impractical for widespread routine use. The Kjeldahl determination, however, remains a means of defining reference standards for the biuret method in use today.

The *biuret method* depends on the presence of peptide bonds in all proteins. When a solution of protein is treated with Cu(II) ions in a moderately alkaline medium, a colored chelate is formed between the Cu(II) ion and the carbonyl oxygen ($>$C=O) and amide nitrogen ($>$NH) atoms of the peptide bond.[53] An analogous reaction occurs between cupric ion and the organic compound biuret,

$$\underset{\displaystyle O}{\overset{\displaystyle H}{HN-C}}-\underset{}{\overset{\displaystyle H}{N}}-\underset{\displaystyle O}{\overset{\displaystyle H}{C-NH}}$$

hence the name. The reaction occurs with any compound containing at least two H_2N—C—, H_2N—CH_2—, H_2N—CS—, or similar groups joined together directly or through a carbon or nitrogen atom. One copper ion is probably linked to six nearby peptide linkages by co-ordinate bonds. Amino acids and dipeptides do not react, but tri-, oligo-, and polypeptides do react to give pink to reddish-violet products. The intensity of the color produced is proportional to the number of peptide bonds which are reacting, and therefore to the number of protein molecules present in the reaction system. Thus, the biuret reaction with protein is suitable for the quantitative determination of total protein by spectrophotometry.

Although small peptides present in serum react, their concentration in serum is so low that they contribute little to the biuret color. Ammonium ions interfere but not at concentrations that can occur in serum. Most biuret methods can detect between 1 and 15 mg of protein in the aliquot being measured, an amount present in 15–200 μL of a serum containing protein at 7 g/dL. Numerous versions of the biuret method have been reported; the method given in detail below is an example. All the methods are simple and, when used with contemporary technology and automation, are sufficiently precise for clinical use.

Refractometry is a quick and reasonably accurate alternative to chemical analysis for serum total protein, when a rapid estimate is required. Some laboratories find it a convenient way to determine total protein prior to serum protein electrophoresis. The refractive index of water at 20 °C is 1.330. If solute is added to the water, the refractive index of a dilute solution is increased linearly and proportionately to the solute concentration; at higher concentrations of dissolved solids (5–20 g/dL), the increase is nearly linear. The relationship of refractive index to concentration holds for a mixture of solutes so long as their individual refractive indices are similar in magnitude. The refractive index of a solution is appreciably affected by its temperature, but refractometers intended for clinical laboratory use have been designed to compensate for effects of different temperatures in a range between 15 and 37 °C. Clinical instruments are precalibrated for total solids expressed either as concentration of protein in g/dL of serum or as specific gravity for urine.

Serum contains dissolved solids in concentrations of 8–10 g/dL, most of which is protein. In the refractometry of serum, it is assumed that the concentrations of inorganic electrolytes and nonprotein organic compounds do not vary appreciably from serum to serum, and that differences in the refractive index reflect primarily differences in protein concentrations. The assumption has been shown to be reliable for clear, nonpigmented samples with normal levels of glucose and urea nitrogen. However, large positive errors occur with azotemia and hyperglycemia; at a level of ~700 mg/dL for blood glucose, or ~300 mg/dL for urea nitrogen, the error in total protein value is 0.6 g/dL. Alterations in albumin/globulin ratio, hyperbilirubinemia, and especially lipemia cause positive errors of up to 5%; errors are even greater with highly lipemic sera.

The TS Meter (Model 10401, American Optical Instrument Co., Buffalo, NY 14215) is the refractometer most often used in clinical laboratories. It is frequently mounted on a stand with a built-in light source. A large drop of serum (or urine) is allowed to spread in a thin film between a cover plate and a glass plate located over the measuring prism. Light enters the instrument in a beam parallel to the prism, is refracted by the solution, and is then projected against the eyepiece, which is marked with scales calibrated for g/dL of serum protein and for specific gravity of urine. The refracted rays light a segment of the field viewed in the eyepiece, the field being separated by a sharp demarcation into a light and a dark area. The scale is read where it is cut by the boundary line between the two areas. Distilled water (specific gravity 1.000) is used to verify the zero position on the protein and specific gravity scale.

At protein concentrations <3.5 g/dL, refractometric results are likely to be inaccurate. At a level >11.0 g/dL, a valid result can be obtained by diluting the serum with equal parts of water and reading the dilution. A day-to-day CV of <2.0% is acceptable precision.

Determination of Total Protein (Biuret Method)

Specimen. Either serum or plasma may be used but serum is preferred. A fasting specimen is not required but may be desirable to decrease lipemia. Hemolysis should be avoided. Tightly stoppered samples of serum are stable for one week or more at room temperature and for one month at 2–4 °C. Specimens that have been frozen and thawed should be thoroughly mixed prior to assay.

Principle. Peptide bonds of proteins react with Cu(II) ions in alkaline solutions to form a colored product whose absorbance is measured spectrophotometrically at 540 nm.

The biuret reagent contains sodium potassium tartrate to complex cupric ions and maintain their solubility in alkaline solution. Iodide is included as an antioxidant.

Reagents

1. NaOH, 6.0 mol/L. Dissolve 240 g NaOH in ~800 mL of water. Cool and dilute to 1 L. Use a newly opened bottle of NaOH in order to minimize carbonate contamination. Store solution in a tightly closed polyethylene bottle at room temperature.

2. Biuret reagent. Dissolve 3.00 g $CuSO_4 \cdot 5 H_2O$ in ~500 mL of water. Add 9.0 g sodium potassium tartrate [$KOOC-(CHOH)_2-COONa \cdot 4 H_2O$] and 5.0 g KI. When the solution is clear, add 100 mL NaOH, 6 mol/L, and dilute to 1 L with water.

3. Biuret blank reagent. Prepare exactly as the biuret reagent but add no $CuSO_4$.

4. Protein standard, bovine albumin, 6–7 g/dL. The most convenient standard is one obtained with a concentration value assigned by some unequivocal means such as Kjeldahl nitrogen content (Armour Pharmaceutical Co., Chicago, IL 60690) or weighed in and verified by reference laboratories (College of American Pathologists, Skokie, IL 60076). A standardized bovine serum albumin solution, Standard Reference Material (SRM) #927, is available from the National Bureau of Standards, Washington, D.C. It may also be used for assigning protein concentrations to secondary standards, i.e., pooled sera or bovine albumin solutions. Standardization with bovine albumin is less costly than with human albumin; its use by a majority of laboratories is an advantage to laboratories that participate in regional and national quality control programs and proficiency surveys.

Procedure

1. Set up a pair of tubes, one labeled B (blank) and the other U (unknown) for each standard, control, and unknown sample. Also prepare a tube labeled RB (reagent blank).

2. To B tubes, add 5.0 mL biuret blank reagent.

3. To the RB and U tubes, add 5.0 mL biuret reagent.

4. To the RB tube, add 100 μL of water. To each tube of each pair, add 100 μL of appropriate sample. Mix.

5. Let stand at room temperature for 30 min.

6. Set zero absorbance with the biuret blank reagent (without copper). Then read and record absorbances of RB, B, and of U tubes.

Calculation

$$\text{Total protein, g/dL} = (A_U - A_B - A_{RB})/(A_S - A_B - A_{RB}) \times C_S$$

where C_S is the concentration of protein in the standard in g/dL.

Comments

1. The length of time in step 5 may be shortened to 10 min by incubating reagent-sample mixtures at 37 °C.

2. If the calculated result exceeds 12.0 g/dL, repeat the test after dilution of the sample with an equal volume of NaCl, 0.15 mol/L, and multiply the result by 2.

3. The method has been shown to be linear over the concentration range of 1.0–12.0 g/dL.

4. Hyperbilirubinemia is corrected by use of a blank. Hemoglobin introduces no error up to 300 mg/dL. Blank absorbance for clear specimens is usually < 0.010. With extraordinarily lipemic sera, the following pretreatment is recommended: Into each of two tubes place 100 µL serum, 0.5 mL water, and 10 mL acetone; mix by inversion, then centrifuge. Pour off the supernatant, invert the tube, and touch the lip to absorbent paper. Immediately, add 5.0 mL biuret blank reagent to one tube and 5.0 mL biuret reagent to the other, and mix. The precipitate should dissolve promptly. Then proceed from step 5 as usual. Alternatively, the specimen may be centrifuged at high speed (e.g., Airfuge, Beckman Instruments, Inc.).

5. A day-to-day CV of < 2% is acceptable.

Reference Ranges.[105] The range of total protein in serum of the ambulatory adult is 6.3–8.3 g/dL, for the recumbent adult 6.0–7.8 g/dL. Past age 60, levels are ~0.2 g/dL lower.

Reference

Peters, T., Jr., Biamonte, G. T., and Doumas, B. T.: Protein (total protein) in serum, urine, and cerebrospinal fluid; albumin in serum. *In*: Selected Methods of Clinical Chemistry, Vol. 9, W. R. Faulkner, and S. Meites, Eds. Washington, D.C., American Association for Clinical Chemistry, 1982.

Other Methods for the Determination of Total Protein

The methods discussed here are dependent on four general properties of proteins, namely:

1. Reaction of tyrosine and tryptophan residues with the Folin-Ciocalteu (phenol) reagent.
2. Characteristic absorption of ultraviolet (UV) light at 200–225 nm and 270–290 nm.
3. Precipitation (for measurement by turbidimetry or nephelometry).
4. Ability to bind colored dyes.

Although any of these methods could be and have been used for total serum protein quantitation, none is more convenient or practical than the biuret method. These methods are reviewed here predominantly as background information for the assay of total protein in urine or cerebrospinal fluid (CSF).

Most proteins contain tyrosine or tryptophan or both, but each protein contains a unique proportion of them. Albumin, for instance, has only 0.2% tryptophan by weight, while the tryptophan content of individual globulins varies between 2 and 3%. These amino acids, either free or in an unfolded polypeptide chain, reduce phosphotungstic-phosphomolybdic acid (Folin-Ciocalteu) reagent to give a blue color. This property is more advantageous for assaying a pure protein whose composition and relative reactivity are known (e.g., fibrinogen), than it is for a mixture of individual proteins with different concentrations and reactivities. However, this reaction has been applied to the assay of total protein in the *Lowry method*[39] as an auxiliary to the biuret reaction. In this method, proteins are allowed first to react with Cu(II) in alkaline solution to form copper–peptide bond–protein complexes. When the Folin-Ciocalteu reagent is added, the copper-protein complexes join with tyrosine and tryptophan residues in the reduction process. About 75% of the color produced depends on the Cu(II) complex; the absorbance of the colored complex is measured at a wavelength between 650 and 750 nm; less chromogenic complexes are also formed with histidine and cysteine. The Lowry method is sensitive to protein at a concentration of 10–60 µg/mL, i.,e., it is ~100 times more sensitive than the biuret reaction alone. The sensitivity is an advantage for measurement of very low concentrations of protein, and the Lowry method is widely used in research to quantitate tissue proteins and enzyme protein in purified preparations. Despite the desirability of such sensitivity in measuring total protein in urine or

CSF, application of the method to these fluids is not wholly satisfactory, because the reagents react nonspecifically with nonprotein compounds, particularly in spinal fluid, to give a positive error of 3–9 mg/dL; gel filtration can be used to remove interferents from urine or CSF but this process is time consuming and not always feasible when the specimen volume of CSF is limited. Drugs such as salicylates, chlorpromazine, tetracyclines, and some sulfa drugs also give a positive interference. For all these reasons, the Lowry method is not widely used in clinical laboratories.

Absorption of UV light at 280 nm depends chiefly on the aromatic rings of tyrosine and tryptophan at pH 8. Accuracy and specificity suffer from an uneven distribution of these amino acids among individual proteins in a mixture, as well as from the presence in body fluids of free tyrosine and tryptophan, uric acid, and bilirubin which also absorb light near 280 nm. At 200–225 nm, peptide bonds are chiefly responsible for UV absorption (70% at A_{205}); specific absorption by proteins at these shorter wavelengths is 10–30 times greater than at 280 nm. Interference from free tyrosine and tryptophan is significant at these short wavelengths, although it can be made negligible for serum by dilution of 1:1000 or 1:2000 in NaCl, 0.15 mol/L. The method has been used for CSF after removal of small interfering molecules by gel filtration. This approach is sensitive and simple, but requires the use of an appropriate spectrophotometer and expensive cuvets with high transmission of light at 220 nm.

Precipitation of protein for *turbidimetric* or *nephelometric assays* can be achieved with sulfosalicylic acid alone, or sulfosalicylic acid in combination with sodium sulfate or trichloroacetic acid, or with trichloroacetic acid alone. Precipitation methods for total protein assay depend on formation of a fine precipitate of uniform, insoluble protein particles which, in suspension, scatter incident light (Chapter 1B, Section Two). Critical features of these methods are:

1. The choice of conditions of reagent and temperature (or both) at which globulins precipitate as efficiently as albumin.

2. The formation of small, well-dispersed particles of consistent size so that light-scattering is a reproducible phenomenon.

3. A homogeneous distribution of suspended particles throughout the medium during the measurement step.

4. The choice of a standard material that has as nearly as possible the precipitation behavior of normal and abnormal mixtures of plasma proteins.

Specific turbidimetric precipitation methods for total protein in urine and CSF are described later in this chapter.

Another approach to analyze proteins of biologic fluids with low concentrations is trichloroacetic acid precipitation of a fairly large volume of specimen containing mixed proteins followed by decantation of the precipitant solution and addition of biuret reagent to the precipitate.[106] While appropriate for urine specimens, this approach is not helpful for assay of proteins in CSF because CSF is usually obtained only in small volumes that often must be divided among multiple analyses.

Dye-binding methods are based on the ability of proteins to bind dyes, such as Amido Black 10B and Coomassie Brilliant Blue. This property is invoked to stain protein bands after electrophoresis and for assaying serum albumin; the dyes can also be utilized in spectrophotometric methods for total protein. The unequal affinities and binding capacities of individual proteins for dyes is a limitation in all these applications and is complicated further by the inability to define a consistent material for use as a standard. The dye-binding method of greatest contemporary interest, particularly for assay of total protein in urine, employs Coomassie Brilliant Blue G-250. The Coomassie Brilliant Blue (CBB) method was originally described for CSF[48] and subsequently adapted to urine.[58] CBB binds to protonated amine groups of amino acid residues in the polypeptide chain, and the absorbance maximum for the bound species of the dye decreases at 465 nm and increases at 595 nm (Figure 4-13).

The reagent is prepared by dissolving 100 mg of CBB G-250 (CI number, 42655) in 50 mL of 95% ethanol. After addition of 100 mL of phosphoric acid (85%), the reagent is diluted to 1 L with water; it should have an A_{465} of 0.800 ± 0.100 when used. To 5.0 mL of reagent is added 50 μL of standard, sample, or control, and 5 min later the absorbance is measured at 495 and 595 nm (a bichromatic mode). The method is simple, fast, and linear up to 150 mg/dL. Run-to-run coefficients of variation with automated analysis were 5.7% at 24 mg/dL and 2.9% at 84 mg/dL. Positive interferences were noted in urine specimens containing tolbutamide or very high concentrations of urea. Very high concentrations of NaCl and HCl, the latter used to preserve specimens, caused very large negative interferences. If color obtained with albumin is

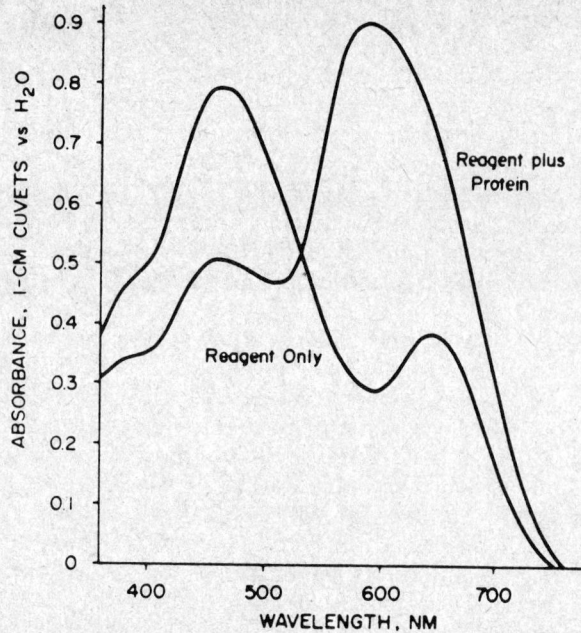

Figure 4-13. Absorption spectrum of Coomassie Brilliant Blue with and without added protein.[58] (From Lott, J. A., Stephan, V. A., and Pritchard, K. A.: Evaluation of the Coomassie Brilliant Blue G-250 method for urinary protein. Clin. Chem., *29*:1946-1950, 1983.)

taken as 100%, hemoglobin and transferrin give a color intensity equivalent to that for albumin but globulins and κ and λ chains give only 60%.[82] The method has yet to be widely implemented. A manual version is available in kit form from Quantimetrix, Hawthorne, CA 90250.

Standardization of Total Protein Methods

Use of bovine or human albumin to standardize *biuret methods* is well established. Albumin can be prepared consistently in high purity, it contains only amino acids, its nitrogen content is a constant fraction of its molecular mass, and the number of peptide bonds per molecule is known. Since the peptide bond is the biuret-reacting unit in all proteins, and the number of peptide bonds determines the absorbance of the colored product, albumin is a reasonable peptide-bond standard for all the proteins in the mixture.

For the standardization of *precipitation and dye-binding methods,* the recommendation is usually to use a suitable dilution of a serum (or serum pool) with a normal albumin/globulin ratio, obtained from a healthy subject, and quantitated for total protein with a correctly standardized and well-controlled biuret or Kjeldahl method. The choice is dictated by the intent to provide in the standard a mixture of proteins whose precipitation and dye-binding behavior are similar to that of proteins in the sample to be analyzed. Use of BSA or HSA as standards prohibits the use of sulfosalicylic acid because these pure proteins give about 2.5 times the turbidity given by serum globulins. With these pure proteins, TCA is the reagent of choice.

INDIVIDUAL PLASMA PROTEINS

PREALBUMIN AND RETINOL-BINDING PROTEIN

Prealbumin (PA), M.W. 54 000, and retinol-binding protein (RBP), M.W. 21 000, are both synthesized by the liver. During electrophoresis, both migrate ahead of albumin. Their $t_{1/2}$ are extremely short, probably < 12 h; consequently, measurement of their serum levels may provide a more timely and sensitive assessment of protein malnutrition or liver dysfunction than transferrin or albumin. (See also Chapter 8A.)

Both PA and RBP are transport proteins. PA binds thyroxine and triiodothyronine, the latter with greater affinity. RBP, in the form of a 1:1 complex with PA, transports vitamin A (retinol); at the target cell, uptake of retinol is followed by dissociation of the PA-RBP complex to its separate proteins. PA is a negative acute phase reactant; the serum level falls in inflammation and malignancy, and also in cirrhosis of the liver and protein-wasting diseases of the gut or kidney, due to decreased synthesis. In Hodgkin's disease, however, the level rises. Serum RBP rises in chronic renal disease, especially if there is tubular proteinuria. Decreases are associated primarily with liver disease and protein malnutrition. Zinc is required for RBP synthesis and zinc deficiency states are characterized by low serum levels of both RBP and total vitamin A.

Determination of PA and RBP can be made by nephelometry[100] or RID. The adult reference range for PA if measured by RID is 10–40 mg/dL and for RBP 3.5–9.0 ($\pm$SD) mg/dL.[105] The level in children is about half the adult level; a sharp increase occurs at puberty.[105]

ALBUMIN

Albumin, from week 20 of gestation and continuing throughout life, is the most abundant protein in human plasma, representing 40–60% of the total protein.[74] It is synthesized in the liver at a rate that is dependent on protein intake but subject to feedback regulation by the plasma albumin level. The molecule contains no carbohydrate and is not stored to any extent in parenchymal cells. The $t_{1/2}$ of albumin has been estimated at 15–19 d. Normally, traces of albumin can be found in almost all extravascular body fluids, and little is lost from the body by excretion. Presently, there is some controversy over the amount of albumin filtered through the glomerulus. The albumin concentration in the glomerular filtrate is very small, probably no more than 0.04% of the plasma albumin concentration, but the glomerular filtration rate is so great that even this represents a urinary albumin excretion rate of 3.6 g/d, 36 times the observed rate. Excess albumin is probably reabsorbed by the proximal tubular cells and degraded by their lysosomal enzymes into fragments, which are returned to the circulation in a manner similar to those of small molecular weight proteins. Albumin is catabolized in various tissues, where it is taken up by cells by pinocytosis. Its constituent amino acids are released by intracellular proteolysis and returned to the body pool.

Albumin has a M.W. of ~66 000. Because its pI lies between 4 and 5.8 (depending on the method of determination), albumin is an anion at pH 7.4 with > 200 negative charges per molecule. The chief biological functions of albumin are to transport and store a wide variety of ligands, to maintain the plasma oncotic pressure, and to serve as a source of endogenous amino acids.

The vast capacity of albumin for ligand binding is due to the large number of charges on each molecule as well as the very large number of molecules available. For example, albumin binds (and solubilizes) nonpolar compounds such as plasma bilirubin and long-chain fatty acids. Albumin also binds a proportion of many hormones, among them thyroxine, triiodothyronine, cortisol, and aldosterone, thus acting as a reservoir in which these potent compounds are stored in an inactive form, but from which they are readily mobilized. Some 40% of serum calcium is bound to albumin (see Chapter 12) and many drugs are also strongly bound—phenylbutazone, warfarin, clofibrate, and salicylate, to mention only a few. (See Chapter 17.)

Albumin concentration is the major determinant of plasma oncotic pressure, one of the factors that regulate partition of water between intra- and extravascular compartments. Because oncotic pressure depends on the number of large molecular weight particles, albumin, which provides a high concentration of relatively small protein molecules, contributes significantly to oncotic equilibria. In certain diseases, abnormally low plasma albumin concentrations allow water to move out of the vascular bed and into the tissues (edema). In protein-losing syndromes, such as the nephrotic syndrome, increases in other plasma proteins may partially compensate for the loss. During the first 20 weeks of fetal gestation α-fetoprotein may serve as the osmotic equivalent of albumin.

Because albumin is a relatively small molecule, its level in extravascular body fluids is a useful indicator of the integrity of various membrane barriers, as described in the sections on body fluids.

Clinical Significance

Plasma levels of albumin, because they depend on protein intake, are frequently used to assess nutritional status. Moderate to large changes in plasma concentration of albumin have significant effects on the relative amounts of the bound and free concentrations of the ligands it carries; since free ligands are those which interact with tissue receptor sites and which can be excreted, plasma albumin levels have important influences on the metabolism of endogenous substances such as calcium, bilirubin, and fatty acids, and on the effects of drugs and hormones.

Albumin levels, although important for management and follow-up, have very little value in diagnosis. *Hyperalbuminemia* is of little diagnostic significance except in dehydration. *Hypoalbuminemia,* however, is very common in many illnesses and results in most instances from one or more of the following factors:

1. Impaired synthesis, either primary as in liver disease or secondary to diminished protein intake.

2. Increased catabolism as a result of tissue damage and inflammation.

3. Reduced absorption of amino acids caused by malabsorption syndromes or malnutrition.

4. Protein loss: in urine, due to nephrotic syndrome, chronic glomerulonephritis, diabetes, or systemic lupus erythematosus; in feces, due to protein-losing enteropathy arising from inflammatory or neoplastic disease; or from the skin through burns.

5. Altered distribution that may sequester large amounts of albumin in an extravascular compartment, as for instance in ascites where high pressure in the portal circulation drives albumin into the peritoneal fluid.

The most severe hypoalbuminemia is caused by protein loss by way of urine or feces; when plasma albumin levels are below 2.0 g/L, edema is usually present.

There are more than 20 *genetic variants* of albumin, which are not associated with disease, but which cause two bands or a single, wide band in the albumin region on electrophoresis. The condition is called *bisalbuminemia*. There is also a transient form, sometimes caused by drugs such as penicillin or cephalothin.[103] Surprisingly, congenital absence of albumin, or *analbuminemia,* is asymptomatic except for occasional slight edema.

Methods for the Determination of Albumin

Determination of albumin[75] in serum or plasma is usually based on the binding behavior of the protein with the anionic dyes bromcresol green (BCG) or bromcresol purple (BCP) in a manual or automated procedure. Earlier dye-binding methods using 2-(4'-hydroxyazobenzene)-benzoic acid (HABA) or methyl orange have fallen into disuse because of undesirable interferences. A BCG method is described in detail later.

Electrophoretic and immunochemical methods are also available but are more appropriate for body fluids in which albumin concentration is normally low, i.e., urine and cerebrospinal fluid. The albumin band, separated electrophoretically on cellulose acetate or agarose gel, is well resolved and usually homogeneous. After staining with Amido Black 10B or Coomassie Brilliant Blue, the band can be estimated by scanning in a densitometer, together with a total protein estimation, or can be cut out and eluted for spectrophotometric quantitation. Elution is preferred by some since it provides a direct estimate of the albumin-stain complex and avoids the inaccuracy caused by nonlinearity. However, for the albumin band to be visible, the fluid applied must contain albumin at concentrations of 30–50 mg/dL and a preliminary procedure to concentrate urine and CSF is usually necessary. The extra handling takes time and may increase the imprecision and inaccuracy of the method. Immunochemical methods such as RID and nephelometry are highly specific and much more sensitive. Albumin assay by an immunochemical method does not pose the same problems as immunochemical methods for other specific proteins. Good, consistent antisera can be generated because highly purified protein is available and the pure protein can also be used for unequivocal standardization of methods. These methods are, however, relatively expensive and unlikely to be applied except for special diagnostic purposes.

Dipstick tests for *urinary* albumin are probably performed more often than any other test and are discussed later under the section on "Proteins in Urine."

Determination of Albumin with Bromcresol Green

The requirements of a dye-binding method for albumin include specific binding of the dye to albumin in the presence of other plasma or serum proteins; high binding affinity between the dye and albumin so that small changes in ionic strength and pH or the presence of competing ligands will not break the dye-protein complex; a substantial shift in the absorption wavelength of the dye in the bound form so that it will be spectrally distinct from the free form present in excess; and an absorption maximum for the bound form at a wavelength distinct from those where bilirubin and hemoglobin can interfere. The manual method given here for albumin by BCG binding is adaptable to automated analysis.

Principle. Albumin and bromcresol green are allowed to bind at pH 4.2 and absorption of the BCG-albumin complex is determined spectrophotometrically at 628 nm. At pH 4.2, albumin acts as a cation to bind the anionic dye.

Specimen. Serum is preferred. A fasting specimen is not required, although it may be desirable since marked lipemia interferes in the assay. Venostasis should be avoided in specimen collection because hemoconcentration increases the apparent concentrations of albumin and other plasma proteins.

Reagents

1. BCG reagent.* Dissolve 105.0 mg bromcresol green (3,3′,5,5′-tetrabromo-m-cresolsulfonphthalein) or 108.0 mg of its sodium salt; 8.85 g of succinic acid; 100 mg sodium azide; and 4 mL Brij-35, 300 g/L (Technicon Instrument Corp., Tarrytown, NY 10591) in ∼950 mL of water. Adjust the pH of the solution to 4.15–4.25 with NaOH, 6 mol/L, and dilute to 1 L with water. Store at room temperature in a tightly closed polyethylene bottle; reagent is stable for at least six months.

2. BCG blank solution. Prepare exactly as for BCG reagent, but do not add any BCG.

3. Albumin (human) standard,* ∼6.0 g/dL. Dissolve 6 g human serum albumin (Cohn Fraction V) and 50 mg sodium azide in 100 mL of water in a small beaker with gentle stirring. Measure total protein concentration by a correctly standardized and well controlled biuret method for total protein. The standard, stored in a tightly closed bottle at 2–4 °C, is stable for up to six months.

Procedure

1. Set spectrophotometer to zero absorbance at 628 nm with water. Read absorbance of BCG reagent (usually ∼0.150 A) and reset absorbance to zero against the BCG reagent.

2. Into a set of tubes, labeled for standard, control, and unknowns, add 5.0 mL BCG reagent.

3. Add 20 μL of appropriate specimen to each tube, one at a time, and mix without delay. Read and record A_{628nm} at 30 ($\pm$3) s after the addition.

Calculation

$$\text{Albumin, g/dL} = A_u/A_s \times C_s, \text{g/dL}$$

Comments

1. Hyperbilirubinemia or hemolysis does not interfere. Marked lipemia causes positive interference but the interference can be corrected. Add 20 μL of sample to 5.0 mL of BCG *blank solution*; set to zero absorbance with blank solution; read absorbance of the serum-blank mixture (A_b). Then subtract A_b from A of serum-reagent mixture and calculate with corrected A_u.

2. Absorbance is a linear function of albumin concentration from 1.0–6.0 g/dL but should be confirmed by suitable dilutions of standard. Acceptable within-run CV is < 3%, between-run CV < 4%. Careful pipetting techniques are required.

3. The binding of BCG to albumin is not wholly specific; some dye binds also to α_2- and β_2-globulins, although more slowly than with albumin. Nonspecific binding is minimized in the assay by reading the absorbance at 30 s after mixing dye and serum. Reading at 10 s is also acceptable, but one or the other time should be selected and consistently observed in routine use. Nevertheless, at very low serum concentrations of albumin, nonspecific binding causes significant positive interference and an immunochemical method of quantitation will be more accurate.

Reference Ranges

In the recumbent adult, the reference range for serum albumin by the BCG method is 3.5–5.0 g/dL. In the upright position, the level is ∼0.3 g/dL higher because of hemoconcentration. In ambulatory adult males, the range is 4.2–5.5 g/dL. For ambulatory adult females, the range is 3.7–5.3 g/dL; it is lower in the last two trimesters of pregnancy. Serum albumin concentration

---

*BCG reagent similar to the one described here is available from commercial sources. Another buffer ion (citrate or lactate) is used in some preparations and linearity with alternative buffers may not be the same as for the method given here. An alternative standard, commercially prepared, is available from a number of reliable suppliers (e.g., Sigma Chemical Co., St. Louis, MO 63178).

varies with age, particularly in the neonatal period when physiologic shifts of body water are common.[105]

Reference ranges determined by nephelometry are:[105]

	g/dL
Newborn, 2–4 d:	2.8–4.4
Adult:	3.4–5.0
> 60 years:	3.4–4.8

Reference

Peters, T. Jr., Biamonte, G. T., and Doumas, B. T.: Protein (total protein) in serum, urine, and cerebrospinal fluid; albumin in serum. *In*: Selected Methods of Clinical Chemistry, Vol. 9. W. R. Faulkner, and S. Meites, Eds. Washington, D.C., American Association for Clinical Chemistry, 1982.

Determination of Albumin with Bromcresol Purple

Bromcresol purple (BCP) can be used for assay of serum albumin in a manner similar to BCG.[59] Yellow BCP dye, buffered at pH 5.2 with acetate, turns green when complexed with albumin. Absorbance of the green complex is measured at 603 nm. A manual method, in which absorbance is read at 1 min, is available in kit form from Quantimetrix, Hawthorne, CA 90250. The method has been adapted to continuous flow[76] and centrifugal analyzers,[38] and to the DuPont *aca*. Excellent specificity is claimed by proponents[26] for assay of albumin in human serum and human serum-based controls, when human albumin standards are used. Method characteristics and performance are not as well documented as they are for BCG since so far the method has not come into wide use.

α_1-ANTITRYPSIN

Alpha$_1$-antitrypsin (AAT) is an acute phase reactant with antiprotease activity,[56] whose deficiency is associated with lung and liver disease. Since its isolation and description in 1955, interest in this glycoprotein has centered on its genetic polymorphism and the diseases associated with its biological function as a protease inhibitor.

Properties. Alpha$_1$-antitrypsin has a M.W. of $\sim$55 000 and a pI of 4.8, and contains 10–12% carbohydrate. On agarose or cellulose acetate electrophoresis it migrates in the α_1-region where it appears to account for $\sim$90% of the protein in the band. The other two proteins of the band stain only faintly—α_1-acid glycoprotein because of its high carbohydrate content, and α_1-lipoprotein because of its high lipid content. As a *protease inhibitor*, AAT acts against chymotrypsin, kallikrein, renin, urokinase, plasmin, and possibly thrombin, but the inhibition of greatest clinical significance is directed against neutrophil elastase and collagenase. The name α_1-*antitrypsin* is almost a misnomer since AAT is relatively inactive toward trypsin, although responsible for 90% of serum antitrypsin activity. The inhibitory activity of AAT is maximal at the neutral to slightly alkaline pH of blood; at pH 4.5 its activity is negligible. This pH-dependence is responsible for the lesser role AAT plays in the inhibition of intestinal enzymes such as trypsin and the larger role it plays in the respiratory tract.

Function. The function of AAT is to neutralize lysosomal elastase released upon phagocytosis of particles by polymorphonuclear leukocytes. AAT, being a relatively small molecule, can pass from capillaries into tissue fluid, bind protease, and pass back into the intravascular fluid. It may even transfer bound protease to α_2-macroglobulin, which, because of its larger size, cannot leave the intravascular compartment; by this mechanism, protease is rapidly transported to the reticuloendothelial system for rapid degradation.

Genetic Variants. AAT has significant microheterogeneity; depending on buffer pH and type of support for electrophoresis, a number of bands may be seen. Heterogeneity remains after sialidase treatment, an indication that AAT phenotypes are probably not due to differences in the sialic acid moieties. Typing of multiple forms is empirically based on differences in electrophoretic mobilities.[45,47] Some 33 allotypes have been described. The protein produced by 95% of the U.S. population, whose genotype is Pi^{MM} (homozygotic for M protease inhibitor), is called M protein. Two other proteins, Z and S, are found in genotypes Pi^{ZZ}, Pi^{SS}, Pi^{SZ}, Pi^{MZ}, and Pi^{MS}; taken all together, these genotypes have a frequency of <1% in the U.S. population. Z and S proteins differ from one another and from M protein by only one amino acid residue. Ability to inhibit proteases is directly related to the circulating level of M protein. If normal mean activity in the MM phenotype is taken as 100%, the level of activity in ZZ is 15%, in SS 60%, in MZ 57.5%, and in MS 80%. A rare null genotype (Pi⁻) produces no AAT at all.[101]

Clinical Significance. Low levels of AAT are found in neonatal respiratory distress syndrome, in severe protein-losing disorders, and in congenital deficiency. Increased levels are more common since AAT is an acute phase reactant. AAT deficiency of the ZZ, SS, and possibly MS phenotypes is associated with pulmonary emphysema of early onset (20–30 years). Inhaled particles and bacteria are continually being removed in the lungs by polymorphonuclear leukocytes. This phagocytosis gives rise to the release of the lysosomal enzyme neutrophil elastase. When the M protein of AAT is deficient, the uninhibited enzyme attacks the elastin of the alveolar wall, whose springy recoil after distention by inhaled air is normally a mechanism to drive out exhaled air. The loss of elasticity of the lung tissue results in emphysema with impaired ventilation and vulnerability to serious respiratory infections. The liver is also affected, most seriously in the ZZ phenotype, where the Z protein accumulates in hepatic cells. The mechanism of hepatic damage is not as well understood as the mechanism in the lung, but the result is cirrhosis. In the first few weeks of life, some 10–20% of neonates with ZZ phenotype develop hepatitis which may end fatally in juvenile cirrhosis. Some adults also develop liver damage but the considerable proportion who are not affected suggests that an additional, perhaps environmental, factor is required for damage to occur. Whether SZ, MS, and SS phenotypes are also at risk is not yet clear.

Methods of Analysis. The functional assay, called trypsin inhibitory capacity (TIC),[24] has been virtually replaced by immunochemical (chiefly nephelometric) assays for the M protein. The *reference ranges* for the nephelometric method are:[105]

	mg/dL
Newborn:	145–270
Adult:	78–200
> 60 years:	115–200

In the absence of an acute phase reaction to inflammation, a level of < 50 mg/dL is presumptive evidence for a variant phenotype; testing should be pursued with typing by isoelectric focusing[45] or starch gel electrophoresis. In the presence of clinical infection and demonstrable increases in other acute phase reactants, a level > 50 mg/dL does not exclude deficiency. A sensitive method for typing α_1-antitrypsin is illustrated in Figure 1C-11.

α_1-ACID GLYCOPROTEIN

Schmid's excellent review[90] covers historical aspects of work that dates back to the 1940's when the seromucoid proteins were first described. The major portion of plasma seromucoids has since been recognized as α_1-acid glycoprotein.

Properties. Alpha₁-acid glycoprotein (AAG, orosomucoid) is unique among the plasma proteins because of its low pI, 2.7–3.5. Its M.W. is ~40 000; 45% of it is carbohydrate with hexose, hexosamine, and sialic acid in equal proportions. The antigenic determinant (epitope) utilized in immunochemical assays, however, is on the polypeptide moiety. AAG has excellent solubility in water and polar solvents and has a high negative charge density at pH 7.4. Treatment with neuraminidase followed by isoelectric focusing reveals polymorphism in the α_1-region. Polymorphism has no known clinical significance, but because of it AAG migration on agarose or cellulose acetate electrophoresis may differ slightly from one specimen to another. Different but related molecular forms have also been isolated from extravascular fluids such as sputum.

Although the exact role of AAG is unknown, it must be associated with inflammation; increased levels are observed following an acute inflammatory episode, i.e., AAG is an acute phase reactant. Some evidence points to a role for AAG in inactivation of progesterone. A certain degree of homology exists among the amino acid sequences of AAG, the immunoglobulins, and the haptoglobin α-chain, suggesting a common ancestry or role in the immune system. Functions in coagulation, collagen formation, and hemagglutination of influenza virus have been proposed.

Early work indicated that the liver was the only site of synthesis of AAG. More recent studies have shown that under certain well-defined conditions, some tumors are able to synthesize AAG. Catabolism of AAG proceeds first by desialation and then rapid (within minutes) degradation in the liver. AAG binds lidocaine and propranolol, drugs frequently used in managing a patient after myocardial infarction. The post-infarction increase in AAG as an acute phase reactant is a complicating factor in adjusting doses of the drugs; regimens appropriate to the immediate post-infarct period, when AAG concentrations are lower, may not be adequate as AAG concentration rises and more drug is sequestered in protein-bound form.

Clinical Significance. The clinical value of AAG determination is currently limited to monitoring the acute phase reaction. However, increases occur also in rheumatoid arthritis, systemic lupus erythematosus, Crohn's ileitis, malignant neoplasms (especially those with metastases and large tumor mass), and in myocardial infarction. Decreases occur in malnutrition, severe hepatic damage, and severe protein-losing gastroenteropathies.

Methods of Analysis. Since antibodies to AAG appear to recognize all the polymorphic forms, immunochemical methods such as RID and nephelometry are most commonly employed for assay. The *reference range* by RID is 55–140 mg/dL[105] and 50–150 mg/dL by nephelometry.[100]

α_1-FETOPROTEIN

Synthesized in the fetal liver, α_1-fetoprotein (AFP) has a M.W. of 69 000. In the thirteenth week of fetal life, AFP forms one third of the total fetal plasma protein. The maximum level is reached at week 30 of gestation; the level then begins to decline. At birth, the level is ~1% of the earlier peak, and at 1 year and into adult life it is < 30 ng/mL. AFP has no known function.

Clinical Significance. In obstetrics, determination of AFP levels in amniotic fluid or maternal serum is utilized for antenatal screening for *neural tube defects* (NTD) in the fetus. (See also Chapter 19, Biochemical Aspects of Pregnancy.) In NTD, AFP leaks from the plasma through the open tube into the amniotic fluid so that the fluid has grossly elevated AFP levels in open spina bifida and anencephaly. Intrauterine death, congenital nephrosis, teratoma, exomphalos, and some other fetal defects also cause increases. From amniotic fluid, some AFP passes into the maternal circulation where the level will rise above the very low normal adult concentration. In about 85% of cases of open spina bifida and anencephaly, maternal AFP concentrations become diagnostic at 16–18 weeks of pregnancy. Use of AFP determinations in screening pregnancies for NTD is controversial because of the serious consequences of a false positive finding. At present, it is recommended that determinations of AFP for antenatal screening be performed only in laboratories where personnel have considerable interpretative experience.

Later in life gross elevations of AFP serum levels are found in ~80% of patients with hepatocellular carcinoma, in 50% of those with *germ cell tumors,* and in all children with *hepatoblastoma.* Sequential assays are particularly useful for prognosis and monitoring treatment. A lesser elevation of AFP occurs in some cases of carcinoma of the pancreas, gut, or lung, and in some nonmalignant diseases such as cirrhosis of the liver, tyrosinemia, and ataxia-telangiectasia.

Methods of Analysis.[92] The choice of method is determined by the specimen. For amniotic fluid, RID or electroimmunoassay (rocket technique) is adequate, but radioimmunoassay on a suitable dilution of fluid is also used. It is essential to avoid contamination of the sample with fetal blood which has an AFP concentration up to 200 times that of amniotic fluid. For serum, radioimmunoassay or enzyme-labeled immunoassay is required. Counterimmunoelectrophoresis may be used to screen patients with hepatic disease for increased AFP due to hepatocellular carcinoma. This application is useful in those parts of the world where hepatitis B, an etiological factor in the occurrence of hepatocellular carcinoma, is prevalent.

Reference ranges for amniotic fluid and maternal serum vary with weeks of gestation.[105] Here we present some selected data.

Amniotic fluid, 20 weeks:	5–25 mg/L
Serum, Maternal, 20 weeks:	21–96 µg/L
Newborn:	< 5 mg/dL
> 1 year:	< 30 µg/L

HAPTOGLOBIN

Haptoglobin (HAP, Hp) binds free hemoglobin in plasma; it is an acute phase reactant.

Properties.[20,80] Haptoglobins are glycoproteins which migrate on agarose and cellulose acetate electrophoresis in the α_2-region. The molecule consists of two pairs of nonidentical chains, α and

β, linked by disulfide bonds. The subunit structure can therefore be designated as $\alpha_2\beta_2$. The α-chains are of two types, α^1- and α^2-; variants of α^1- are α^{1F}- and α^{1S}-. Three phenotypes can be identified by electrophoresis: HAP 1-1, containing only α^1- and β-chains; HAP 2-1, containing α^1-, α^2-, and β-chains; and HAP 2-2, containing only α^2- and β-chains. The variants of HAP 1-1 are $\alpha_1^{1F}\beta_2$ and $\alpha_2^{1S}\beta_2$; F and S indicate their relatively fast or slow electrophoretic migration. Oligomers of HAP 2-1 and HAP 2-2 with full or half molecules of themselves or of one another can also form; HAP 1-1 does not polymerize in this manner. The molecular weight of monomeric HAP 2-1, per se, is 120 000 but in oligomeric form may be 200 000; M.W. of HAP 2-2 is 160 000 but $\sim$400 000 as an oligomer; M.W. of HAP 1-1 is 80 000–85 000. HAP appears to be synthesized in significant quantity only in the liver.

The relative frequency of the α^1-chain in populations varies in different parts of the world, from 0.77 in the Congo to 0.10 in parts of India; it is 0.31–0.45 in European populations. Incidence of phenotypes in Europe is greatest for HAP 1-1, followed by HAP 2-1, and then by HAP 2-2. A small percentage of Black and Oriental adults have no detectable HAP (phenotype HAP 0-0). The phenotypes have no clinical importance, except that HAP 0-0 (anhaptoglobinemia) may be an unrecognized factor in the diagnosis of diseases which deplete HAP plasma levels.

The primary function of HAP appears to be the binding of free oxyhemoglobin in plasma; one molecule of HAP binds two molecules of hemoglobin. The binding is irreversible and occurs between HAP and the α-chain of the globin portions of hemoglobins A, F, S, or C; HAP does not bind methemoglobin, heme, or unusual forms of Hb in which the α-chain is missing. The oxygen-binding characteristics of hemoglobin in the HAP-Hb complex are altered, but the nature of the reaction and bonding between HAP and Hb is still unknown. The complex is removed from the plasma within minutes by the reticuloendothelial system, where its components are metabolized to free amino acids and iron within a few hours. HAP thus acts to prevent loss of hemoglobin to urine and to conserve iron. In an acute episode of intravascular hemolysis, the total circulating HAP pool is capable of binding about 3 g of Hb. A single severe episode of hemolysis may deplete HAP entirely. After such depletion plasma levels return to normal in about one week.

Hemoglobin which has been broken into heme and globin is further handled by *hemopexin*, a β_1-globulin (M.W. 70 000) that binds free heme. The heme-hemopexin complex is taken up by the liver, where iron from heme is bound to ferritin and the remainder of the heme moiety is converted into bilirubin. Unlike HAP, hemopexin is not an acute phase protein. In conditions where hemolysis and an acute phase reaction are concurrent, it may be preferable to follow the changes in hemopexin concentration rather than those of HAP. After hemolysis, a third complex may also appear—*methemalbumin*, a combination of oxidized heme with albumin. The formation of methemalbumin and the hemopexin-heme complex may explain why anhaptoglobinemia (HAP 0-0) is symptomless. Mechanisms for the capture and handling of free hemoglobin are so efficient that only after they are exhausted are hemoglobin or its degradation products excreted by the kidney. Even when hemolysis is extensive enough to cause hemoglobinuria, some of the filtered Hb is reabsorbed in the renal tubules.

Clinical Significance. Isolated estimations of plasma HAP have limited value because of the wide normal range. Serial determinations, however, are used to detect and monitor acute phase reactions and hemolytic states.

Increases in plasma HAP concentrations occur in acute phase reactions. Levels may appear to be increased in conditions such as burns and nephrotic syndrome, when large amounts of the lower molecular weight plasma proteins have been lost. An acute phase response may be confirmed and monitored by assay of other acute phase reactants such as CRP and AAT. *Low serum HAP* levels are most frequently associated with conditions of increased intravascular hemolysis or Hb turnover such as occur in hemolytic anemias, transfusion reactions, and malaria. Chronic intravascular hemolysis causes a small but continuing drop in HAP levels. A typical "hemolysis panel" ordered for evaluation of such cases might include HAP, lactate dehydrogenase, and plasma Hb assays. Hemolysis in the extravascular compartment does not usually affect plasma HAP concentration. Other causes of low levels are anhaptoglobinemia (HAP 0-0) and severe liver diseases where protein synthesis is diminished. HAP is detectable in the plasma of only 10–20% of neonates, but at six months, upon maturation of the liver, adult levels are achieved. Participation in sports such as running, aerobic dancing, and weight lifting cause repetitive trauma to erythrocytes, and may be a cause of a sustained lower HAP level.

Methods of Analysis. Immunochemical methods are preferred for quantitation of HAP. Antisera raised against HAP react with each of the proteins of the three phenotypes. Nephelometry[108] is often more satisfactory than RID because the great difference in the molecular weight of the various proteins and the effect of molecular weight on diffusion rate cause an undesirable delay in obtaining results. RID plates for the highly polymerized HAP 2-2 protein cannot be read until several days have been allowed for diffusion and reaction with antiserum. Other methods are based on the peroxidase activity of HAP-Hb complex separated by electrophoresis or gel filtration. Typing of haptoglobins is performed by starch gel electrophoresis or by polyacrylamide gel electrophoresis (PAGE). See Figure 1B-32.

Reference ranges for HAP using nephelometry are given here; consult reference 105 for more extensive data.

	mg/dL
Newborn:	5–48
Adult:	30–215

α_2-MACROGLOBULIN

The biological functions of α_2-macroglobulin (AMG) are poorly understood,[4,56] but because of its structure and high molecular weight, AMG has some limited clinical significance. AMG inhibits proteases in a manner different from that of other protease inhibitors (see below). It is not an acute phase reactant.

Properties. AMG is one of the largest plasma proteins; its M.W. has been estimated as 625 000 to over 800 000. Because of its very large size, AMG tends to remain in the intravascular compartment. The protein contains ~8% carbohydrate and has a pI of 5.4. Four subunits have been described and evidence exists for a dimeric structure. Three polymorphic forms are known.

The outstanding characteristic of AMG is its ability to bind a wide variety of molecules and ions, particularly endopeptidases, including the carboxyl-, thiol-, serine, and metal proteases. Proteases such as plasmin, pepsin, trypsin, chymotrypsin, and cathepsin D are bound irreversibly, but the active sites of these enzymes are not masked. Changes in the molecular structure of the AMG-protease complex, seen by electron microscopy, suggest that the AMG molecule only partly surrounds the protease molecule, allowing interaction with its smaller substrates and inhibitors to continue. Such a mechanism would explain how AMG binding inhibits protease activity against large substrate molecules, while the AMG-protease complex retains proteolytic activity against certain low molecular weight substrates. There is considerable additional evidence that AMG has a vital but unclear role in immunological and inflammatory processes.[43]

AMG is synthesized in the liver and reticuloendothelial system. In the free form, AMG has a $t_{1/2}$ of $5^1/2$ d; when complexed with proteases, the $t_{1/2}$ is ~10 min since these complexes are rapidly cleared by the reticuloendothelial system. Clearance of the AMG-protease complex is much faster than clearance of the AAT-protease complex ($t_{1/2}$ ~3 h), from which the protease molecule may have been transferred.

Clinical Significance. Estimation of AMG is seldom of clinical value. In states such as the nephrotic syndrome, in which selective loss (sieving) of protein occurs, AMG is retained because of its large size. There is some evidence that hepatic synthesis of AMG increases in order to compensate partially for the decrease in albumin normally active in maintaining oncotic pressure. Increased levels of AMG are also associated with estrogen stimulation, as in pregnancy or the use of oral contraceptives, but the mechanism is unclear. Children display higher levels than adults until the age of 15, an observation that has not been explained. Increased levels have been reported in some liver diseases, diabetes mellitus, neural tube defects, ataxia-telangiectasia, atopic dermatitis, and Down's syndrome.

Methods of Analysis. Immunochemical methods are the methods of choice; nephelometry[100] is commonly employed, although RID is acceptable.

Reference Ranges. In the newborn, the level is 1.5 times that of adult males; the level reaches a maximum at 1–3 years, with subsequent slow decline until stabilization at about age 25. Levels rise again after age 70. For adults, reference intervals by RID are 150–350 mg/dL for males and 175–240 mg/dL for females.[105]

CERULOPLASMIN

Ceruloplasmin (CER) [35,78] is a late acute phase reactant; it is the principal copper-containing protein of plasma. Its main clinical importance is in the diagnosis of Wilson's disease (hepato-lenticular degeneration).

Properties. Ceruloplasmin is an α_2-glycoprotein with a M.W. between 120 000 and 160 000, probably ∼134 000. Because of its lability, CER has been difficult to isolate, but there is now general agreement that it is a single polypeptide chain containing 6–7 copper atoms per molecule. Because of its copper content, the pure protein has a blue color, hence the name. CER contains ∼10% carbohydrate as terminal sialic acid chains attached to the polypeptide. With a pI of 4.4, CER migrates on agarose or cellulose electrophoresis in the fast α_2-region, but it makes no visible contribution to the α_2 band except when the concentration in plasma is greatly increased. Like AAG and AMG, desialized CER is rapidly taken up and catabolized by the liver. Genetic polymorphism occurs and can affect levels found in serum; however, abnormalities associated with a specific phenotype have not been identified.

The function of CER is still debated. Its chief functions would appear to be as an enzyme and probably as a copper donor. Soon after its discovery 35 years ago, CER was shown to possess oxidase activity toward many polyamine and polyphenol substrates. CER was, in fact, originally called "copper oxidase," and its oxidase behavior (reduction of O_2 to water) was used for its identification and assay. Recent evidence has shown that CER rapidly catalyzes oxidation of Fe(II) to Fe(III), and that its "ferroxidase" activity is essential to transformation of Fe(II) at the cell surface as a preliminary step for the binding of iron to transferrin. Ferroxidase activity of CER may answer longstanding questions about iron metabolism—how, for instance, Fe(II) can be converted to Fe(III) without the formation of insoluble ferric hydroxide and toxic oxygen atoms that would be produced upon auto-oxidation of Fe(II).

Ceruloplasmin was originally thought to be the major transport protein for copper in plasma, but studies with copper radioisotopes have shown negligible turnover of copper in CER. Once CER is secreted by the liver, it neither gains nor loses copper; it can donate copper only if some process exists to internalize the CER molecule into a cell where the copper can be removed from the protein. These findings indicate that CER does not act in respect to copper as transferrin does in respect to iron. Plasma copper consists of a nondialyzable fraction (95%) attached to CER and of a dialyzable fraction (5%) loosely bound to albumin and histidine. Copper is believed to be transported in the dialyzable form from the gut to the liver, where copper is incorporated into the CER apoprotein, which is then sialized and released into the bloodstream. Animal experiments have shown that copper cannot combine with apoprotein in the circulation. Increased absorption of copper leads to increased synthesis of CER and increased excretion of nonresorbable copper-protein complexes in the bile. Synthesis of CER thereby provides a first line reaction to potential copper toxicity. In the plasma, circulating CER serves as a stable pool of copper in nontoxic form. Cells probably take up CER and degrade it to obtain the copper essential for synthesis of such enzymes as monoamine oxidase, diamino oxidase, and ascorbate oxidase, each of which contains one copper atom per molecule.

A possible role for CER is as an antioxidant. Spontaneous oxidation of many organic substances in contact with oxygen is a constant danger to life but is prevented by the presence of antioxidants in plasma and tissues. Ceruloplasmin and transferrin have been shown in vitro to account for most of the antioxidant activity of plasma. CER may prevent lipid peroxidation and free radical production in inflammatory states; this is perhaps its role in an acute phase reaction.

Clinical Significance. CER plasma level is increased in infections, malignancy, and trauma. Increases are particularly notable in diseases of the reticuloendothelial system such as Hodgkin's disease. The level is also increased in infection or obstruction of the biliary tract.

The most important clinical application of CER assay is in the diagnosis of Wilson's disease.[21] Typically the plasma CER is reduced while dialyzable copper concentration is increased. Unless treated with copper chelators such as D-penicillamine, the disease is always progressive and fatal. Prompt diagnosis is important since the treatment takes 3–6 months to have the desired effect. When treated, most patients lead a normal life. Unfortunately, probably only a quarter of the cases are correctly diagnosed. A majority of cases present with liver dysfunction; any patient between the age of 5 and 50 years with unexplained liver disease, enlarged liver and spleen, a history of attacks of jaundice, or signs of unexplained brain damage should be screened for Wilson's disease.[69] All liver biopsies obtained for diagnosis of a chronic liver disease should be tested for excess copper content. Neurological symptoms are more common in adults, while symptoms of liver dysfunction are more common in children.

The disease is rare (incidence 1:50 000–1:100 000); it is due to an autosomal recessive defect in the regulation of copper metabolism. The cause cannot be a molecular defect in CER because a small number of patients have normal CER levels. The two fundamental disturbances are a gross decrease in the rate of incorporation of copper into apoprotein and a marked reduction in the biliary excretion of copper. Although the exact defect is unknown, it is probably lack of an enzyme or a carrier protein required for the incorporation of copper into proteins. Recent findings have shown increased levels of copper in fibroblasts cultured from affected individuals, an indication that the defect is not confined to the liver. Copper deposits in the kidneys, in the liver where it causes cirrhosis, and in the brain where it damages the basal ganglia. The combination of liver and neuron damage gives the disease its alternative name, hepatolenticular degeneration, since the neural structure involved is the nucleus lentiformis. Copper also deposits in the cornea, giving rise to the characteristic Kayser-Fleischer rings. The finding of Kayser-Fleischer rings, together with plasma CER < 10 mg/dL, high plasma levels of dialyzable copper, and increased urinary copper, is diagnostic in all cases with neurological involvement, but is diagnostic in only 80% of the cases with liver disease. The other 20% of adult cases with liver disease have normal CER levels, but not > 30 mg/dL. In such cases, increased copper content in a liver biopsy sample is a more reliable diagnostic finding. The most reliable test for Wilson's disease is injection of radioactive copper, followed by timed observations of plasma radioactivity. In normal patients, the radioactivity first increases, then falls, then rises again as CER is synthesized. The secondary rise will not occur in Wilson's disease because of the defect in incorporating copper into CER.

Low plasma levels of CER are also found in malnutrition, malabsorption, nephrosis, and severe liver disease, particularly primary biliary cirrhosis. In the last instance, the radiocopper test will probably be required to rule out Wilson's disease.

Methods of Analysis.[100] Currently, CER is most readily measured immunochemically by nephelometry. RID and spectrophotometric measurement of oxidase activity are alternatives.

Reference Ranges. CER levels vary with age. Exercise, pregnancy, and estrogen administration increase CER concentration. In the absence of severe liver disease, serum levels < 10 mg/dL are suggestive of Wilson's disease. Reference intervals given below are for the nephelometric assay.[105]

		mg/dL
Infant,	1–3 months:	5–18
	6–12 months:	33–43
	13–36 months:	26–55
Child,	4–5 years:	27–56
	6–7 years:	24–48
	7 years:	20–54
Adult:		18–45

TRANSFERRIN

Transferrin (TRF, siderophilin)[9,81] is the principal plasma protein for transport of iron. Its concentration correlates with the total iron binding capacity (TIBC) of serum. (See Chapter 15.) TRF is a negative acute phase reactant.

Properties. TRF has a M.W. of 77 000; it is a glycoprotein migrating in the β-region on cellulose acetate and agarose electrophoresis. It has a single polypeptide chain with ~6% carbohydrate; pI is between 5.5 and 5.9. At least 20 variants of TRF have been reported, but polymorphism has no clinical significance except as the rare congenital defect atransferrinemia.

TRF reversibly binds numerous polycations—iron, copper, zinc, cobalt, and calcium—although only iron and copper binding appear to have physiological significance. One molecule of TRF binds two ferric ions and an associated anion which is, in vivo, usually bicarbonate. There are two binding sites for iron, probably with different affinities, which are strong at physiological pH but decrease with decreasing pH. The iron-TRF complex has an absorbance maximum at 470 nm. Ceruloplasmin is believed to mediate the oxidation of Fe(II) to Fe(III) as a preliminary step for binding of Fe(III) to TRF.

TRF is synthesized in the liver and to a small extent in the reticuloendothelial system and in endocrine glands such as testes and ovaries. It has a $t_{1/2}$ of ~7 d. Plasma levels are apparently regulated by availability of iron, since in iron deficiency states plasma levels rise and, upon successful treatment with iron, return to normal levels. The site of catabolism is unknown, although some TRF is lost from the body in exfoliated intestinal mucosal and other cells. About half the extracellular transferrin exists outside the vascular compartment in body fluids such as lymph and cerebrospinal fluid. There appears to be some involvement of transferrin in intracellular iron metabolism, particularly in intestinal mucosal cells.

Iron metabolism is discussed in Chapter 15; thus, only a brief discussion of the role of TRF is presented here. The apoprotein binds iron released from catabolism of hemoglobin or absorbed from the intestine. The TRF-Fe(III) complex in plasma then carries iron to storage sites (the liver and reticuloendothelial system) where it is incorporated into ferritin and hemosiderin, and to cells that synthesize the iron-containing compounds such as hemoglobin, myoglobin, and cytochromes. Recent work suggests that rapidly proliferating cells such as those of the erythroid bone marrow or of the placenta have surface receptors for the TRF-iron complex, and that receptor binding may be a significant step in cellular uptake or iron. How the iron is released from the complex is not known; however, the complex is disrupted by cellular uptake of the iron and the apoprotein is freed to bind more iron.

Clinical Significance. Evaluation of plasma transferrin levels are useful for the differential diagnosis of anemia and for monitoring its treatment. In the very common disease iron deficiency or hypochromic anemia, the transferrin level is increased due to increased synthesis, but the protein is less saturated with iron because plasma iron levels are low. On the other hand, if the anemia is due to a failure to incorporate iron into erythrocytes, the transferrin level is normal or low but the protein is highly saturated with iron. In iron overload, transferrin concentration is normal but saturation (normally 30–38%) is > 55%, and may be as high as 90%.

Transferrin is a negative acute phase reactant, and *low levels* will occur, along with low levels of albumin, prealbumin, and β-lipoprotein, in inflammation or malignancy. The cause of this decreased synthesis is still obscure. Other, more obvious causes of decreased synthesis and low plasma levels are chronic liver disease and malnutrition.[16] Transferrin levels may, in fact, be used for assessing nutritional status (Chapter 8A). Protein loss, as in the nephrotic syndrome or protein-losing enteropathies, also causes low levels. In the congenital defect atransferrinemia, a very low level of transferrin is accompanied by iron overload and a severe hypochromic anemia which is resistant to iron therapy. *High levels* of transferrin occur in pregnancy and during estrogen administration.

Methods of Analysis. Immunochemical methods, nephelometry or RID, are preferred. Abnormally low levels may first have been suspected from a routine electrophoretic separation. Because of the convenience of simultaneous measurement of serum iron and TIBC, and the desirability of knowing per cent saturation of TRF, TRF is sometimes estimated indirectly from the TIBC value by the equation:

$$\text{TRF, mg/dL} = 0.70 \times \text{TIBC, } \mu g/dL \text{ or TIBC} = 1.43 \times \text{TRF}$$

This procedure, however, overestimates the TRF concentration by 16–20% because it assumes that all plasma iron is bound to TRF. In fact, when TRF is more than half saturated, some of the iron is bound to other plasma proteins such as albumin.

Reference ranges for nephelometry are:[105]

	mg/dL
Newborn:	130–275
Adults:	220–400
> 60 years:	180–380

β_2-MICROGLOBULIN

Beta$_2$-microglobulin (BMG) is a low molecular weight (11 800) protein found on the cell surfaces of all nucleated cells and shed into the blood, particularly by lymphocytes and tumor cells. The protein is the light or β-chain of the human leukocyte antigen (HLA) on cell surfaces, and consists of a single polypeptide chain with one intrachain disulfide bridge. BMG does not contain carbohydrate. It has a $t_{1/2}$ of 107 min. The small size of the molecule allows BMG to pass through the glomerular membrane, but normally < 1% of the BMG filtered is excreted in the urine; the remainder is reabsorbed and catabolized in the proximal tubules of the kidney.

High plasma levels occur in renal failure, inflammation, and neoplasms, especially those associated with B lymphocytes. The principal clinical value of BMG assay is to test renal tubular function,[115] particularly in kidney transplant patients in whom rejection of the allograft will manifest as diminished tubular function. Serial assays of β_2-microglobulin are also useful to monitor B cell tumors (see Table 4-8). In acute leukemia and lymphoma with central nervous system involvement, the level of BMG is increased in cerebrospinal fluid; and in Sjögren's syndrome with lymphoid infiltration of salivary glands, its level rises in saliva.

Radioimmunoassay is the usual method of analysis for BMG.

Reference intervals are:[105]

Serum:	0.10–0.26 mg/dL
Urine:	0.03–0.37 mg/d
CSF:	1.5 ± 0.2 mg/L (1 SE)

C-REACTIVE PROTEIN[68,72]

A substance present in the sera of acutely ill patients and able to bind the C-polysaccharide on the cell wall of *Streptococcus pneumoniae* was first described in 1930. In 1941, it was shown to be a protein which was given the name C-reactive protein (CRP). The earliest assays for CRP were semiquantitative precipitin tests, but subsequent characterization of the protein and production of reliable antisera have led to highly specific, sensitive, and reproducible quantitative methods. CRP, the first acute phase reactant to be discovered, is also the one exhibiting the most dramatic increases in concentration.

Properties. CRP consists of five identical, nonglycosylated, polypeptide subunits noncovalently linked to form a disk-shaped cyclic polymer with a M.W. of 115 000–140 000. The protein contains little or no carbohydrate and migrates on either cellulose acetate or agarose electrophoresis anywhere from the slow γ- to the mid-β-region; the extent of migration depends on various factors such as the concentration of calcium ions in the electrophoretic buffer. Because CRP is normally present in plasma at a mean concentration < 800 μg/dL, sensitive immunochemical methods are required for its detection. The protein is synthesized in the liver.

CRP binds not only the polysaccharides present in many bacteria, fungi, and protozoal parasites, but also—in the presence of calcium ions—phosphorylcholine, phosphatidylcholines such as lecithin, and polyanions such as nucleic acids. In the absence of calcium ions, CRP binds polycations such as histones. Once complexed, CRP becomes an activator of the classical complement pathway starting at C1q. (See Figure 4-11.) Like antibodies, CRP can initiate opsonization, phagocytosis, and lysis of invading cells, as a response to the inflammatory reaction. Perhaps the main role of CRP is to recognize potentially toxic autogenous substances released from damaged tissue, to bind them, and then to detoxify them or clear them from the blood. CRP is metabolized in the opsonization process, but whether it is metabolized by any other route is not clear. Serum proteins capable of Ca(II)-dependent binding to C-polysaccharide occur in fish and birds, as well as in mammals. Their appearance is similar to CRP by electron microscopy and they have similar amino acid sequences. The group of C-reacting proteins is closely related to serum amyloid P (see below). No genetic abnormalities have been associated with CRP, although a genetic deficiency has been suggested.

Clinical Significance. CRP has long been recognized as one of the most, if not the most, sensitive of the acute phase reactants. CRP levels in plasma can rise dramatically following myocardial infarction, stress, trauma, infection, inflammation, surgery, or neoplastic proliferation. The increase occurs within 24–48 h and the level may be 2000 times normal. But, because the increase is nonspecific, it cannot be interpreted without a complete clinical history, and even then only by comparison with previous values. Cord blood normally has low CRP concentrations (1–35 μg/dL) but in intrauterine infection levels may be as high as 26 000 μg/dL. Determination of CRP is clinically useful for screening for organic disease; for assessing the activity of an inflammatory disease such as rheumatoid arthritis; for detecting intercurrent infections in systemic lupus erythematosus, in leukemia, or after surgery (secondary rise in plasma level); for detecting rejection in renal allograft recipients; as well as for managing neonatal septicemia and meningitis when the collection of specimens for bacteriological investigations may be difficult.

For reasons unknown, the degree of CRP response varies in some diseases that are otherwise apparently similar. For example, the CRP response in systemic lupus and ulcerative colitis, even where there are obvious signs and symptoms of inflammation, is slight in contrast to its very large response in rheumatoid arthritis and Crohn's disease.

Methods of analysis.[14,100] A variety of immunochemical methods can be used, among them RID, radioimmunoassay, rate nephelometry, and homogeneous enzyme immunoassay (Syva Co., Palo Alto, CA 94303).

The *reference range* for both RID and rate nephelometry is 80–800 μg/dL.

AMYLOID

Amyloid (Greek: resembling starch, because of its staining properties) is a pathological extracellular deposit that appears translucent and structureless. It stains with iodine and a variety of dyes, but its affinity for Congo red is the basis of its definition and identification in biopsy specimens. All types of amyloid bind Congo red, which emits an apple-green fluorescence under polarized light. The affinity of amyloid for Congo red is also the basis of a now obsolete test for amyloidosis. (Injected intravenously, the dye is rapidly removed from plasma by amyloid-containing tissues.) Under the electron microscope, all types of amyloid are seen to share two fibrillar components with characteristic periodicity and a minor rodlike "P-component," a glycoprotein that, viewed on end, has the appearance of a pentamer with a hollow core. This component circulates normally in plasma. The fibril deposit, as determined by X-ray, consists of a characteristic β-pleated sheet that can be extracted from tissue with water but not physiologic salt solutions.

Since amyloid stains with Congo red, the formation of this insoluble variety of protein, resistant to normal proteolysis owing to its β-conformation, appears to be the one characteristic common to all types of amyloid, whatever their composition.[32a] Amyloid fibrils have diverse soluble plasma protein precursors that are either increased in quantity or modified by proteolysis to make them insoluble. The result is a disease group with different etiologies, called *amyloidosis.* The deposits can be local or systemic. They exert pressure on vital structures and eventually cause death. No details are known as to the local mechanism of formation of these deposits or the determinant for the site of deposition.

On clinical grounds amyloidosis can be classified (omitting some rare forms) into five main groups: (1) primary amyloidosis, (2) amyloidosis associated with multiple myeloma, (3) secondary amyloidosis associated with inflammatory or infectious diseases, (4) a form associated with aging, and (5) familial amyloidosis. Increased knowledge of the chemical structure of the deposits has led to a revised classification[11] that takes into consideration the type of amyloid protein present. This classification is used here. (See Table 4-14.)

The first two clinical groups are associated with **AL amyloidosis.** Immunoglobulin light chain amyloid (AL) protein contains fibrillar proteins made up of homogeneous Ig light polypeptide chains mainly of the λ (predominantly λ_{VI}) type or their amino terminal variable fragments, or both. AL proteins have a M.W. of 14 000–22 000 and have an affinity to clotting factor X, which may be responsible for the decrease in plasma level of this factor and for associated bleedings. Since they always include the variable part of the Ig light chain, they differ from one case to another. Not all light chains seem capable of precipitation in the β-conformation which constitutes "amyloid."

AL amyloidosis is most common. Bone marrow plasmacytosis and excess plasma cell production of antigenically identical monoclonal light chains are common to primary amyloidosis and multiple myeloma. Thus, a clear distinction between these two conditions is not possible chemically. They appear to differ only in the presence or absence of osteolytic lesions. AL deposits may occur in the tongue, heart, lymph nodes, spleen, joints, peripheral nerves, and skin.

The third clinical group (secondary amyloidosis) is associated with AA amyloid deposits and is called **AA amyloidosis.** In this form amyloid A (AA) proteins with a MW of 5000–9000 form the main component and λ-light chains a minor one. AA proteins are amino terminal fragments of serum amyloid A protein (SAA protein) that is composed of monomers of M.W. 12 500. It circulates as a molecule of M.W. 220 000–235 000 that is complexed mostly to HDL-protein but at times to albumin. It increases rapidly in infections or noninfectious inflammation and thus behaves like an acute phase protein. AA protein can be distinguished from AL protein by the ability of permanganate to bleach the Congo red staining. AL protein staining is resistant to permanganate.

Table 4-14. CHEMICAL CLASSIFICATION OF THE AMYLOID DISEASES

Clinical Syndrome	Fibril Precursor	Fibril	Common Term	Chemical Description
Primary myeloma with amyloid	Ig L-chain	L-chain or V_L fragment	AL	A_λ $(_{1-n})$ or A_κ $(_{1-n})$
Secondary (inflammation-associated)	SAA	AA	AA	$AA_{prototype}$ $AA(_{trp})$ var†
Localized				
Endocrine			AE	
Thyroid medullary carcinoma*	Procalcitonin	?Procalcitonin	AE_t	$A_{procalcitonin}$
Pancreatic islet	—	—	AE_i	—
Skin—Papular, macular, nodular	—	—	AD	—
Senile				
Cardiac	Prealbumin	Prealbumin	AS_{c1} AS_{c2}	$A_{prealbumin}$ —
Brain	—	—	AS_b	—
Pancreas	—	—	AS_p	—
Familial				
Neuropathic				
Portuguese	Prealbumin	Prealbumin	AF_p	$A_{prealbumin}$ (val30→met)
Swedish	Prealbumin	Prealbumin	AF_{sw}	$A_{prealbumin}$ (val30→met)
Israeli	Prealbumin	Prealbumin	AF_{Is}	$A_{prealbumin}$ (thr49→gly)
Japanese	—	—	AF_j	$A_{prealbumin}$ (val30→met)
Indiana-Swiss	—	—	AF_{i-s}	$A_{prealbumin}$ (var)
Iowa	—	—	AF_i	—
Cardiomyopathic				
Danish	Prealbumin	Prealbumin	AF_{da}	$A_{prealbumin}$ (var)
Nephropathic				
FMF (North African)	SAA	AA	AF_{FMF}	AA
Irish-American	SAA	AA	AF_{IR}	AA
Urticaria, deafness (England)	—	—	AF_{de}	—
Polish (hypertensive)	—	—	AF_{po}	—
German (renal failure)	—	—	AF_g	—
Vascular				
HCHWA (Iceland)§	Gamma trace‡	Gamma trace fragment polymer	AF_{HCHWA}	$A_{\gamma\ trace}$

*Also found in other APUD tumors.
†This designation is used to denote amino acids that depart from the prototype sequences.
§Hereditary cerebral hemorrhage with amyloidosis.
‡Gastroenteropancreatic neuroendocrine protein is a hormonelike peptide with no known function found to date in peptidergic CNS neurons, the anterior pituitary, and pancreas. Its concentration is high in cerebrospinal fluid.
(From Buxbaum, J.N.: The amyloid diseases. *In*: Cecil Textbook of Medicine. 17th ed. J.B. Wyngaarden, and L.H. Smith, Jr., Eds. Philadelphia, W.B. Saunders Company, 1985, p. 1169. Updated.)

AA protein is often deposited in chronic, noninfectious inflammatory diseases, such as rheumatoid arthritis (incidence up to 20%) and other inflammatory joint diseases, as well as in chronic suppurative and granulomatous infections, such as tuberculosis and osteomyelitis. Deposits of AA protein are also observed in nonlymphoid tumors such as renal and gastric carcinomas, and in Hodgkin's disease. Deposits of AA protein are most often found in the kidney, liver, and spleen, usually causing a nephrotic syndrome and hepatosplenomegaly.

Senile amyloidosis. Senile amyloid (AS) protein has been found most often in the heart (senile cardiac amyloid, SCA) but also in the pancreas and brain. There seems to be an independent pathogensis for these three forms. Nodular or infiltrative amyloid deposits may also be seen in skin, lung, and trachea and may involve endocrine organs such as the pancreas (in longstanding diabetes) or the thyroid (in medullary carcinoma). These forms are generally asymptomatic, except for the cardiac form.

Familial amylodosis. Several genetically transmitted forms of amyloid disease have been reported. These have predominantly neuropathic but also nephropathic and vascular symptoms. (See Table 4-14.) The fibrils of the Portuguese, Swedish, and polyneuropathic amyloid syndrome of Ashkenazic Jews have monomers of M.W. 14 000 that share antigenic determinants and amino acid homology with prealbumin.

Other forms of amyloidosis. Other less common forms of amyloidosis are known but are not well characterized.

OTHER PLASMA PROTEINS

The protease inhibitors α_1-antichymotrypsin (α_1AC) and *inter-α-trypsin inhibitor* (IαTI), together with the vitamin D–binding protein (DBP; Gc globulin), form two very faint bands between the α_1- and α_2-regions on cellulose acetate or agarose electrophoresis. α_1AC has a M.W. of 68 000, is a rapidly responsive acute phase protein, and occurs normally in relatively high concentrations in bronchial secretions. IαTI has a M.W. of 160 000, and, because it breaks easily into fragments, probably serves as a parent of protease inhibitors of smaller size. DBP has a M.W. of 59 000 and functions in vitamin D transport; it is measured by RIA. Both protease inhibitors can be assayed by RID; the *reference range* for α_1AC is 30–60 mg/dL, for IαTI 20–70 mg/dL, and for DBP 525 $\pm$ 24 μg/mL.

A number of *placental proteins* have been found, aside from human chorionic gonadotropin (M.W. 40 000) and placental lactogen (M.W. 27 000). These include *Schwangerschaft's protein* (SP) with M.W. 90 000; the *pregnancy-associated plasma proteins A* and *B* (PAPP-A, M.W. 750 000; PAPP-B, M.W. 1 000 000); and placental protein 5 (PP5, M.W. 36 000). Assay of members of this group of proteins may have applications in pregnancy testing, for evaluation of feto-placental function, or as tumor markers.

Ferritin, the chief storage form of iron in the body, is discussed in Chapter 15. The minute amount normally present in serum (12–200 ng/mL) is determined by either immunoradiometric assay or radioimmunoassay. The level of ferritin falls in iron deficiency anemia; it rises in iron storage diseases, in an acute phase reaction, and in liver disease. The principal clinical value of the serum ferritin assay lies in its use to estimate bone marrow iron stores without subjecting the patient to bone marrow aspiration.

Lysozyme (muramidase) has a M.W. of 15 000; it occurs in lysosomes intracellularly and in most extracellular fluids, especially in exocrine secretions. It is bactericidal by virtue of its enzymatic activity against the constituents in bacterial cell walls. Lysozyme is produced by granulocytes and monocytes but not by lymphocytes. The plasma level rises in tuberculosis, sarcoidosis, and other granulomatous diseases. The enzyme is filtered by the glomerulus, but almost all of it is resorbed and catabolized by the tubules; hence, its measurement in urine may be employed to test renal tubular function. Lysozyme is produced in such large quantities in monocytic leukemia that it may be visible as a post-γ band in an electrophoretic separation of serum or urine concentrate. Its concentration, determined by nephelometry,[105] is normally 0.36–0.78 mg/dL of serum; urinary excretion is 1.3–3.6 mg/d.

Carcinoembryonic antigen (CEA)[33] is a glycoprotein with a M.W. of ~200 000. Its level in serum is increased when there are malignant tumors of the colon, lung, pancreas, stomach, or breast. The finding has little diagnostic value since the CEA levels may also be increased by smoking, pulmonary infections, hepatitis, ulcerative colitis, Crohn's disease, and some other nonmalignant conditions. Assay of CEA has, however, great clinical value in prognosis and monitoring the treatment of carcinoma of the colon or rectum.[5] Determination of CEA prior to surgical removal of the tumor has prognostic value since its level reflects tumor volume. After surgery, levels may be used to judge the completeness of resection; effective removal is indicated by finding a normal CEA level within one month. Serial determinations at monthly intervals thereafter are helpful in detecting a recurrence of the tumor because the level of CEA increases 4–6 months before a recurrence becomes apparent clinically. In the case of an inoperable tumor, CEA levels may also be used to judge the effectiveness of treatment by chemotherapy. CEA is determined by radioimmunoassay or enzyme immunoassay; values in excess of 10 ng/mL are rarely found in healthy subjects. Values in healthy nonsmokers are ordinarily <5 ng/mL.

PROTEINS IN URINE[17,51]

The glomeruli behave as ultrafilters for the plasma proteins. The degree to which individual proteins are normally filtered through the membrane is a function both of their molecular size and of their plasma concentration. In general, transport of protein molecules through the glomerular membrane progressively diminishes as protein size increases. Normally, high molecular weight proteins such as IgM (M.W. 900 000) do not appear in glomerular filtrate except in trace amounts. Relatively small yet significant amounts of albumin (M.W. 66 000) are passed into the filtrate as a result of its high plasma concentration and relatively low molecular weight. Proteins with M.W. 15 000–40 000 filter more readily but in lesser quantities because of their low plasma concentrations. The proportions of individual proteins excreted in the urine depend on the extent of their absorption by the renal tubules; albumin represents ∼60% of total protein excreted because it is not completely removed from the filtrate by the tubular cells. The low molecular weight proteins are actively reabsorbed from the filtrate and catabolized in the proximal tubule. Tubular reabsorption involves binding of the protein to specific receptors on the tubule epithelial cells, uptake into the cells by pinocytosis, fusion of uptake vacuoles with lysosomes, and hydrolysis of the protein by lysosomal enzymes, followed by return of the constituent amino acids to the body pool. Very little, if any, of the total urinary protein normally excreted consists of these small proteins. Only a small amount of protein is excreted (20–150 mg/d), and the majority of it is albumin; the remainder is almost entirely the Tamm-Horsfall protein uromucoid, a constituent of urinary casts, probably secreted by the distal tubules. Increased permeability of the glomeruli is therefore indicated first by increased amounts of albumin in the urine, and increasing permeability is indicated by the appearance in urine of proteins with increasingly greater molecular weights. Diminished or diminishing tubular reabsorption is indicated by the appearance of increasing concentrations of small molecular weight proteins in urine. (See also Chapter 11.)

Clinical Significance

There are four ways in which *proteinuria* (increased amounts of protein in urine) can occur; increased glomerular permeability (*glomerular proteinuria*) in which the urinary protein is mainly albumin; defective tubular reabsorption (*tubular proteinuria*) in which the urinary proteins are mainly normal, low molecular weight plasma proteins; increased concentration in the plasma of an abnormal, low molecular weight protein such as immunoglobulin light chains (*overload proteinuria*); and abnormal secretion of protein into the urinary tract (*post-renal proteinuria*). The last two are the least common and will be considered first. For further information, see Chapter 11.

Overload proteinuria includes hemoglobinuria, myoglobinuria, and Bence Jones proteinuria, which is due to high plasma concentrations of immunoglobulin light chain paraproteins, e.g., in multiple myeloma. Detection of light chains depends on electrophoretic and immunochemical testing. (See page 573.) Hemoglobin or myoglobin can be detected by tests discussed in Chapter 15.

Post-renal proteinuria refers to protein arising from the urinary tract below the kidneys and is usually due to inflammation or malignancy. It can be diagnosed by microscopic examination of the urinary sediment for inflammatory (pus) cells and malignant cells. The presence of erythrocytes or pus cells in casts in such a centrifuged urinary sediment is valuable proof that their origin is from the kidney and *not* extrarenal.

Glomerular proteinuria is by far the most common and serious type of proteinuria. Patients are routinely screened for this disorder by a simple dipstick test for albumin. If the dipstick test is negative, clinically significant glomerular proteinuria is excluded; if the test is positive, further investigation such as a quantitative evaluation of protein excretion is indicated. Since most of the excreted protein is albumin, glomerular proteinuria is often labeled "albuminuria." Increase in glomerular permeability occurs in numerous conditions characterized by diffuse injury to the kidney. (See also Chapter 11.) The types of proteinuria listed in Table 11-4 are all of the glomerular type except Bence Jones proteinuria, an overload proteinuria. In diabetes, vascular permeability increases and albuminuria appears when metabolic regulation is poor.[70] Immune complex disease often affects the glomerulus; in nephritis associated with systemic lupus erythematosus and in most cases of glomerulonephritis, granular deposits, containing immunoglobulins and complement components, have been demonstrated on the glomerular membrane. In such progressive diffuse renal diseases, the ability to restrict filtration of the smallest of the larger proteins is lost first;

thus, albumin appears first in the urine. So long as the molecular weight determines which proteins are filtered, the proteinuria is called *selective*. Progressively severe glomerular lesions produce less selective proteinuria until proteins of all sizes pass the glomerulus (*nonselective proteinuria*). Changes in selectivity parallel histologic changes in renal biopsy material; earliest changes can be diagnosed by electron microscopy. In the final stages of disease, as glomeruli are destroyed and large numbers become functionless, proteinuria decreases and renal failure is established. Evaluation of selectivity of the glomerular membrane is an aid to differential diagnosis of these diseases and their prognosis.

Protein excretion rate is ordinarily determined from a 24-h urine collection because random specimens vary considerably in protein concentration. However, a more convenient alternative is to determine, in random samples, the ratio of protein to creatinine (protein-creatinine index).[3,94] Creatinine concentration is relatively constant in any one subject and the index correlates well with the 24-h total excretion of protein. Total protein determination, however, does not discriminate between individual proteins, and when 24-h excretion or protein-creatinine index values are near normal, estimation of one or two individual, representative proteins is preferred. When protein excretion is >1 g/d, the proteinuria is likely to be glomerular. When excretion is >2 g/d, heavy loss of plasma albumin may cause generalized edema (e.g., in the nephrotic syndrome). In these cases, selectivity should be assessed by electrophoresis of the urine and comparison of clearance of proteins of differing molecular weights (e.g., IgG/albumin ratio). When selectivity is high (IgG/albumin ratio <0.16), as it often is in children, treatment with corticosteroids is likely to be successful. Effect of treatment can be monitored by measuring excretion rates of total or individual proteins.

Functional or benign proteinuria is a form of glomerular proteinuria that is probably due to changes of blood flow through the glomeruli. It is associated with exercise, pyrexia, exposure to cold, congestive heart failure, hypertension, or arteriosclerosis. Protein excretion rates are <1 g/d. *Postural or orthostatic proteinuria*, associated with the upright position, is also a functional proteinuria but excretion may exceed 1 g/d. Orthostatic proteinuria complicates assessment of proteinuria in otherwise symptomless patients. If transient, it is probably benign; but if chronic or not entirely related to posture, protein excretion rates should be checked at six-month intervals, preferably by a quantitative test. Persistence of proteinuria suggests underlying renal disease. In normal pregnancy, protein excretion may increase harmlessly to 200–300 mg/d. This slight increase is in contrast to the proteinuria of pre-eclamptic toxemia, a glomerular proteinuria of up to 3 g/d, and to the proteinurias of latent renal disease or of urinary infection.

Tubular proteinuria is characterized by the appearance of low molecular weight proteins in the urine because of defective reabsorption of these compounds in the proximal renal tubules. It may occur alone but is more commonly associated with glomerular proteinuria. When tubular proteinuria occurs alone, albumin excretion is only slightly increased but not enough to give a positive dipstick reaction. More specific tests are required to detect simple tubular proteinuria or to identify it in the presence of glomerular proteinuria. Agarose electrophoresis of urine gives a characteristic pattern—prominent α- and β-bands, a relatively faint albumin band, and sometimes a post-γ-band. SDS-PAGE (sodium dodecyl sulfate–polyacrylamide gel electrophoresis) is even more useful since it can demonstrate tubular proteinuria in the presence of glomerular proteinuria. Urinary proteins are first heated with SDS to which they bind and thus acquire a uniform negative charge. Electrophoresis on 10% polyacrylamide gel (Chapter 1B, Section Three) will then separate the proteins by their molecular weights (size) rather than by their charges. (See Figure 4-14.) The proteins typically excreted in tubular proteinuria are low molecular weight proteins, such as β_2-microglobulin (M.W. 11 800), lysozyme (M.W. 14 500), retinol-binding protein (M.W. 21 000), α_1-microglobulin (M.W. 27 000), and α_1-acid glycoprotein (M.W. 40 000), as well as various polypeptide hormones and enzymes. A simple way to screen for tubular proteinuria is to measure one of them, such as β_2-microglobulin or lysozyme, quantitatively. The protein-creatinine clearance ratio for the marker protein is a useful index of excretion (for example, the ratio for lysozyme is increased 100 times in Fanconi's syndrome) and is sufficiently reliable for most diagnostic and prognostic purposes.

Acute tubular proteinuria may occur in severe disturbances of metabolism that are associated with burns, acute pancreatitis, or administration of renotoxic drugs; it may later clear up completely. *Chronic tubular proteinuria* is usually irreversible. Its etiology may be hereditary as in Fanconi's syndrome, or a kidney disease such as chronic pyelonephritis, or a systemic disease such as cirrhosis or sarcoidosis, or acquired due to drugs such as phenacetin or to poisons such

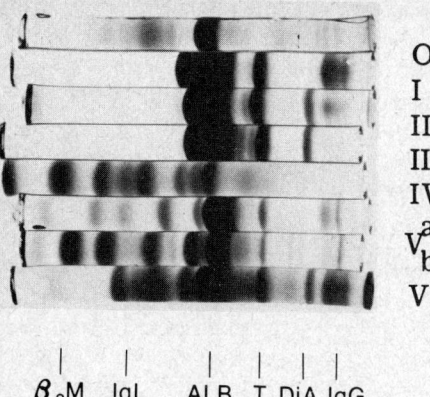

Figure 4-14. Classification of renal proteinuria by SDS poly-acrylamide rod gel electrophoresis. 0, physiological; I, unse-lective glomerular; II, intermediate selective glomerular; III, se-lective glomerular; IV, micromolecular tubular; Va and Vb, mixed types of tubular; VI, mixed glomerular and tubular (DIA = albumin dimer; IgL = immunoglobulin light chains; T = transferrin). In sample IV all the low molecular weight proteins of the glomerular filtrate are being excreted. In Va and b, more albumin is being excreted but the tubules are now absorbing the smallest microglobulins, especially Va, a pattern found dur-ing recovery from acute tubular proteinuria. (Courtesy of Prof. Dr. med. W. H. Boesken.)

as cadmium. In such instances, slight tubular proteinuria may be the only sign of progressive renal damage. Tests for tubular proteinuria are therefore now being used to monitor renal allograft rejection, aminoglycoside and cadmium toxicity, and chronic pyelonephritis; a favorite marker protein is β_2-microglobulin.[115]

LABORATORY STUDIES OF PROTEINURIA

Qualitative detection of excess protein in urine is largely based on use of dipsticks. The reactive portion of the stick is coated with a buffered indicator which develops color in the presence of protein. The principle is often called the "protein error of indicators." A typical example is Albustix (Ames, Elkhart, IN 46514), in which bromphenol blue, buffered to pH 3 with citrate, is present predominantly in the protonated, yellow form. When protein is added, the affinity of the anionic form of the indicator dye for protein causes a shift of the equilibrium between anionic and protonated forms of the indicator toward formation of the blue anionic species. The intensity of blue color produced is then proportional to the concentration of protein in the specimen. The Combur 8 strips (Boehringer Mannheim Diagnostics, Inc. Indianapolis, IN 46250) are said to be less subject to drug interferences. Detection limit is 7 mg/dL.

Like all dye-binding techniques, the dipstick methods are more sensitive to albumin than to other plasma proteins. They are therefore excellent screening tests for glomerular proteinuria, but unsatisfactory for detection of tubular proteinuria or overload proteinuria of the Bence Jones type. Most tests are sensitive to protein at > 10 mg/dL; they are only roughly quantitative and their use should be limited to screening and to rough estimates required before concentrating the specimen for electrophoresis or diluting it for quantitative assay. A first morning urine specimen is best, since it tends to be concentrated and unaffected by postural factors.

Quantitative assay for total protein or for individual proteins is usually performed on timed collections. Periods of 4, 8, and 12 h may be appropriate for following a renal transplant patient or a patient whose acute renal losses of albumin are being compensated with closely regulated replacement therapy. In most cases, however, a 24-h collection is chosen, both for quantitative total or specific protein assay and for electrophoretic separation. An alternative approach is to measure the protein/creatinine ratios of random specimens.

Quantitative assay of specific proteins is necessary for estimates of glomerular selectivity and for evaluation of tubular proteinuria. Nephelometry, radial immunodiffusion, and rocket im-munoelectrophoresis are sufficiently sensitive for most of the analytes. But for quantitating β_2-microglobulin, whose excretion normally is ∼0.1 mg/d, radioimmunoassay is required. Because this protein is susceptible to degradation at low pH, specimens should be stored at a pH > 6.0 during the collection period. In assessment of the nephrotic syndrome, glomerular selectivity is estimated usually by measuring two individual proteins, one which in health is just retained by the glomerulus (e.g., albumin) and another which is well retained (e.g., IgG). A random urine sample and a corresponding serum are collected and analyzed for the two proteins. A calculation of the ratio of the individual clearances is then made:

$$\left(\frac{UV}{P} \text{ for IgG}\right) \bigg/ \left(\frac{UV}{P} \text{ for albumin}\right)$$

Thus "V" cancels out and

$$\text{Ratio} = \frac{[\text{Serum albumin}] \times [\text{Urine IgG}]}{[\text{Urine albumin}] \times [\text{Serum IgG}]}$$

In this form of the ratio, a value above the upper limit of the reference interval would imply loss of selectivity. *Clearance* of individual proteins may also be evaluated by the classic equation:

$$\text{Clearance, mL/min} = \frac{U \times V}{P}$$

where U and P represent the concentration of the compound in urine and plasma, and V is the volume of excreted urine per minute. For evaluation of tubular proteinuria, one may use either a 24-h specimen to calculate the excretion rate of the protein per day, or a random specimen. In the latter case, the concentrations of both creatinine and the specific protein (e.g., lysozyme) are determined and the protein-creatinine index is calculated, thus:

$$\text{Lysozyme-creatinine Index} = \frac{[\text{Lysozyme}]}{[\text{Creatinine}]}$$

To evaluate selectivity using a random specimen, one would calculate the protein-creatinine indices for the two proteins in the urine specimen and substitute them for urine concentration terms in the equation given above.

Methods for the Determination of Total Protein in Urine

The choice of a manual method for quantitation of urinary total protein is not easy. Recent comparisons of methods have been reported in the literature.[25,60] The biuret method applied to acid-precipitated protein[106] or to a concentrate obtained by membrane filtration has the advantage of equal sensitivity to each of the individual proteins in the mixture. Many laboratories, however, find this approach too time-consuming for routine use and prefer turbidimetric and dye-binding methods because they are fast and simple. Of the dye-binding methods, Ponceau S and Coomassie Brilliant Blue (CBB) are the most popular; the CBB method is probably the better of the two and is still being improved. It is sensitive and has a precision of 6–7% (CV). The CBB method has, like turbidimetric methods, limited linearity so that urines must be diluted to varying degrees for assay, and like turbidimetric and other dye-binding methods, it has unequal sensitivities for individual proteins. CBB underestimates low molecular weight proteins in tubular proteinuria and immunoglobulin light chains in overload proteinuria. Conflicting opinions and claims as to which method is best lead us to believe none is wholly satisfactory. A method using a TCA-Ponceau dye reagent has been published by Pesce and Strande.[73]

The method chosen for detailed description here is simply one which has been evaluated and reported as a Selected Method in Clinical Chemistry.[75] Recent automation of a similar method for the *aca* (EI duPont de Nemours and Co., Wilmington, DE), in which TCA is replaced by benzethonium chloride has relieved many laboratories of constant worry with methodology, not because the method is beyond reproach, but because it has been qualified under Food and Drug Administration standards as being a state-of-the-art technique.

Determination of Total Protein by TCA Precipitation

Principle. Proteins in urine are precipitated as fine particles with trichloroacetic acid and quantitated by turbidimetry at 620 nm. Precipitation under these conditions is a general reaction of proteins, but standardization is empirical since there is some difference in the behavior of individual proteins in a mixture. A recent modification of the method uses the supernatant of TCA-treated urine for blanking and absorbance is read at 420 nm after 35 min.[93]

Specimen. Random specimens may be used but timed 24-h specimens are preferred so that the excretion rate can be judged against established reference ranges. No preservation is required during collection, but thereafter urine aliquots should be stored at 2–4 °C or frozen. Any urine collected with strong acid preservatives for other analytes is not suitable for assay.

Reagents

1. Trichloroacetic acid (TCA), 0.76 mol/L. Dissolve 125 g TCA in ~500 mL water and dilute to 1 L volumetrically. Stable at room temperature for several months and indefinitely at 2–4 °C.

2. NaCl-azide reagent. Dissolve 8.5 g NaCl and 0.5 g NaN_3 in 1 L of water. Stable under the same conditions as the TCA reagent.

3. Standards

   a. *Stock.* Obtain a single fresh serum specimen or a pool of serum. It should be free of icterus, hemolysis, and turbidity, and should have a normal albumin/globulin ratio (1.3–1.8). Add NaN_3 to serum in a proportion of 1 mg to 2 mL. Determine the total protein concentration of the serum by a correctly standardized and carefully controlled biuret method. Store tightly closed at 2–4 °C and reanalyze every two months. The serum is stable ~6 months but should be replaced if sediment or mold appears.

   b. *Dilute standard.* Dilute 3.00 mL of stock serum volumetrically to 200 mL with NaCl-azide reagent. Take precautions in mixing to minimize foaming. Calculate the total protein concentration of this preparation.

$$\text{Dilute total protein standard, mg/dL} = 0.015 \times \text{Serum total protein, g/dL} \times 1000$$

Store dilute standard at 2–4 °C and prepare fresh each month.

   c. *Working standards.* Make fresh daily or for each assay. Dilute 1.0 and 5.0 mL of each dilute standard to 10.0 mL with NaCl-azide. The concentration of these standards are, respectively, $1/10$ and $1/2$ the concentration of the dilute standard. Keep solution at 2–4 °C when not in use.

Procedure

1. Allow reagents to come to room temperature before use in the assay.

2. Allow refrigerated urine specimen to come to room temperature before analysis. Mix the specimen gently but thoroughly, measure and record its total volume, and filter an aliquot of ~100 mL through Whatman No. 1 paper.

3. Estimate the protein concentration of a filtered aliquot by a dipstick method. If the concentration is > 100 mg/dL, dilute a portion with NaCl, 0.15 mol/L, to bring it into the range of 50–100 mg/dL. Given the uncertainty in a dipstick estimate (especially at very high concentrations of protein), time is often saved by bracketing the projected dilution with two others and retesting all of them with the dipstick before selecting one for analysis. Record dilution factor(s).

4. Set up a pair of tubes for each standard, control, and unknown. Label one tube B and the other T. To all tubes add 4.00 mL of appropriate sample.

5. To B tubes, add 1.00 mL of NaCl-azide reagent. Mix gently by inversion.

6. To T tubes, at timed intervals, add 1.00 mL TCA, 0.76 mol/L. Upon addition, mix each by inversion.

7. Allow to stand for 10 min at room temperature.

8. Observing the time interval used in addition of TCA, gently mix a pair of B and T tubes. Immediately, set the spectrophotometer to zero absorbance with the B tube at 620 nm, and read and record the absorbance of the corresponding T tube.

Calculations

$$\text{Total protein, mg/dL} = A_u/A_s \times C_s, \text{mg/dL}$$

or

$$\text{Total protein, mg/d} = A_u/A_s \times C_s, \text{mg/dL} \times \frac{\text{24-h volume, mL}}{100 \text{ mL/dL}}$$

where A_u and A_s are absorbance of T tubes for unknown and standard and C_s is concentration of a standard as evaluated in item 3c of the reagent descriptions. Should the calculated value exceed 1000 mg, report it in grams.

Comments

1. Linearity of the system should be checked every few months with a set of 6 standards that span a concentration range of 10–100 mg/dL. The relation of absorbance to concentration has been found linear up to 110 mg/dL.

2. Large errors can be introduced by using cold specimens or reagents (or both) in the assay. Positive interference occurs in the presence of drugs that precipitate in acid, among them cephaloridine, cephalothin, chlorpromazine, promazine, penicillin in high dosage, sulfamethoxazole, and many radiographic dyes.

3. Day-to-day CV's are related to the concentration of the control. Evaluators of the method reported CV of 27.7% at 19 mg/dL and 9.0% at 41 mg/dL. In a 1979 CAP survey, with 112 laboratories using a human albumin standard supplied by CAP, the interlaboratory CV was ~30% at concentrations of 105–120 mg/dL. By comparison, the CV for the acid precipitation/biuret method was ~15%.

4. The method is also applicable to spinal fluid. With proper concern for spectrophotometer design and cuvet volume, the method can be scaled down to require only 0.5 mL of sample.

Reference Ranges

The range for concentration is 1–14 mg/dL.[105] The excretion rate at rest is 50–80 mg/d;[59] many laboratories indicate the reference value as < 100 mg/d (< 150 mg/d in pregnancy). The concentration may reach 300 mg/d in urine of healthy subjects after exercise.

Reference

Peters, T. Jr., Biamonte, G. T., and Doumas, G. T.: Protein (total protein) in serum, urine, and cerebrospinal fluid; albumin in serum. *In*: Selected Methods of Clinical Chemistry, Vol. 9. W. R. Faulkner, and S. Meites, Eds. Washington D.C., American Association for Clinical Chemistry, 1982.

Electrophoretic Separation of Urinary Proteins

The procedure for electrophoresis of urine on agarose gel or cellulose acetate is identical to that for serum, with the exception that urine must be concentrated prior to application. Concentration is most conveniently achieved with the Minicon B15 concentrators (Amicon Corporation, Danvers, MA 01923). Up to 5 mL of a sample is placed in a chamber of the concentrator. The inner surface of the chamber is a membrane of selective permeability that is backed by absorbent pads. Water from the specimen crosses the membrane and is absorbed by the pads; proteins left behind on the specimen side of the membrane are concentrated to a degree dependent on the time allowed for the process. Up to 100-fold concentration is possible. This particular concentrator retains in the specimen proteins with M.W.'s > 15 000, which includes Bence Jones proteins and albumin.

Simple electrophoresis of urine concentrated to a protein concentration of ~3 g/dL can identify the presence of immunoglobulin light chains (Bence Jones protein) or other low molecular weight proteins typical of tubular proteinuria. Comparison of a urine separation with a corresponding serum separation may also indicate the degree of selectivity in glomerular proteinuria. Separations of a urine concentrate on SDS-PAGE (Figure 4-14) are particularly informative.

PROTEINS IN CEREBROSPINAL FLUID[51,109]

Cerebrospinal fluid (CSF) is secreted by the choroid plexuses, around the cerebral vessels, and along the walls of the ventricles of the brain. It fills the ventricles and cisternae, bathes the spinal cord, and is reabsorbed into the blood through the arachnoid villi. More than 80% of its protein content originates from plasma by ultrafiltration through the walls of capillaries in the meninges and choroid plexuses; the remainder originates from intrathecal synthesis. The lowest concentration of total protein and the smallest proportion of the larger protein molecules are in the ventricular fluid; as the CSF passes down to the lumbar spine (from which site specimens are usually collected), the protein concentration increases. This difference in concentration at different levels of the CNS is illustrated by the following reference ranges:[105]

	Total Protein, mg/dL
Ventricular fluid	5–15
Cisternal fluid	15–25
Lumbar fluid	15–45

In CSF from premature and full-term neonates, levels are considerably higher (up to 130 mg/dL); in healthy elderly adults, concentrations are up to 60 mg/dL.

Because CSF is mainly an ultrafiltrate of plasma, low molecular weight plasma proteins such as prealbumin, albumin, and transferrin normally predominate. No protein with a molecular weight greater than that of IgG is present in sufficient concentration to be visible on electrophoresis. The electrophoretic pattern of normal CSF after concentrating the fluid, has two striking features—a prominent prealbumin band and two transferrin bands. The second of the transferrin bands is called the τ band; it is produced or transformed intrathecally and, by comparison with plasma transferrin, is deficient in sialic acid content.

Clinical Significance of CSF Protein Analysis

Examination of CSF total protein and specific proteins is used chiefly to detect increased permeability of the blood/brain barrier* to plasma proteins or to detect increased intrathecal secretion of immunoglobulins (Igs). (See Table 4-15.)

The permeability of the blood/brain barrier to plasma proteins is increased by high intracranial

---

*The blood/brain barrier is a concept rather than an anatomical structure. The barrier is defined by the many complex factors that govern the distribution of compounds other than water, CO_2, and O_2 between the blood and the extracellular fluid of the brain and its accessory structures. In the present discussion, the term "blood/brain barrier" is used, admittedly inaccurately, as a synonym for the capillary endothelium of vessels of the central nervous system.

Table 4-15. CEREBROSPINAL FLUID PROTEIN IN VARIOUS DISEASES

Clinical Condition	Appearance and Cells × 10⁶/L	Total Protein (mg/dL)	Pandy Test* for Immunoglobulins
Normal	Clear, colorless; 0–5 lymphocytes	15–45†	Negative
Increased admixture of proteins from blood			
Increased capillary permeability:			
Bacterial meningitis	Turbid, opalescent, purulent, usually >500 polymorphs	80–500	1+ to 3+
Cryptococcal meningitis	Clear or turbid; 50–150 polymorphs or lymphocytes	25–200	1+ to 3+
Leptospiral meningitis	Clear to slight haze; polymorphs early, then 5–100 lymphocytes	50–100	
Viral meningitis	Clear or slight haze, colorless; usually up to 500 lymphocytes	30–100	Usually negative
Encephalitis	Clear or slight haze, colorless; usually up to 500 lymphocytes	15–100	Usually negative
Poliomyelitis	Clear, colorless; up to 500 lymphocytes	10–300	Usually negative
Brain tumor	Usually clear; 0–80 lymphocytes	15–200 (usually normal)	Usually negative
Mechanical obstruction:			
Spinal cord tumor‡	Clear, colorless or yellow	100–2000	1+ to 3+
Hemorrhage:			
Cerebral hemorrhage	Colorless, yellow or bloody; blood cells	30–150	Negative or 1+
Local immunoglobulin production:			
Neurosyphilis	Clear, colorless; 10–100 lymphocytes	50–150	3+ to 4+
Multiple sclerosis§	Clear, colorless; 0–10 lymphocytes	25–50	Negative or 1+
Both increased capillary permeability and local immunoglobulin production:			
Tuberculous meningitis	Colorless, fibrin clot, or slightly turbid; 50–500 lymphocytes	50–300 (occ. up to 1000)	1+ to 3+
Brain abscess	Clear or slightly turbid	20–120	Usually negative
After myelography (inflammatory reaction)		Slight increase	

*One drop of CSF is added to 1.0 mL of saturated aqueous phenol solution.
†Premature infant: up to 400 mg/dL. Children: 30–100 mg/dL. Old age: up to 60 mg/dL.
‡Froin's syndrome: lumbar fluid values are much higher than cisternal fluid values.
§Similar values may occur in certain other chronic inflammatory conditions of the nervous system (see text).

pressure due to brain tumor or intracerebral hemorrhage, or by inflammation due to bacterial or viral meningitis, encephalitis, or poliomyelitis. Striking elevations of CSF total protein are seen in bacterial meningitis; smaller elevations occur in the other inflammatory diseases and with tumor or hemorrhage. Lumbar CSF protein is increased when the CSF circulation is mechanically obstructed above the puncture site (as by a spinal cord tumor) and plasma proteins equilibrate across the walls of meningeal capillaries into the stagnant CSF. The effect of any of these conditions is that the proportions of specific proteins in CSF come increasingly to resemble those characteristic of serum.[13] The degree of permeability of the blood/brain barrier can be evaluated by immunochemical measurements of albumin in CSF and in serum specimens obtained at the same time. (Note that, for this purpose, determination of CSF albumin must be made by a quantitative immunochemical method. Electrophoretic separation, although generally informative of increased amounts of specific plasma proteins in the CSF, is not useful for an objective definition of degree of permeability.) Albumin is a particularly suitable indicator protein because it is neither synthesized nor metabolized intrathecally. In CSF that is free of contaminating blood, albumin must necessarily have come from plasma through the blood/brain barrier.

Using the results of the albumin assays, a ratio can be calculated:

$$\frac{\text{Albumin}_{CSF}, \text{ g/dL}}{\text{Albumin}_{SER}, \text{ g/dL}}$$

But since the quotient obtained by dividing 0.015–0.045 g/dL for albumin$_{CSF}$ by 4.0–5.5 g/dL for albumin$_{SER}$ would be a very small decimal value, this ratio is renamed "CSF/Serum Albumin Index" and is arbitrarily calculated as follows:

$$\text{CSF/Serum Albumin Index} = \frac{\text{Albumin}_{CSF}, \text{ mg/dL}}{\text{Albumin}_{SER}, \text{ g/dL}}$$

An index value of < 9 is considered consistent with an intact barrier. Values of 9–14 are interpreted as slight impairment, of 14–30 as moderate impairment, and of 30–100 as severe impairment. Values > 100 indicate complete breakdown of the barrier.

Demonstration of increased *intrathecal synthesis of immunoglobulins,* particularly IgG, has great importance in the diagnosis of demyelinating diseases of the central nervous system (CNS), especially multiple sclerosis.[109] In multiple sclerosis, patchy deterioration of myelin sheaths of axons in the CNS profoundly affects conduction of nerve impulses. The cause of demyelination is unknown. Sites of the lesions are unpredictable and the types of symptoms produced vary widely. B lymphocytes that infiltrate the lesions synthesize IgG and occasionally other Igs. Since axons of the CNS are in intimate contact with CSF, the Igs produced in the lesion appear in the CSF.

Determination of a protein component of the fatty myelin sheath, *myelin basic protein* (MBP), by radioimmunoassay would be a more direct diagnostic approach, but assays have been beset with difficulties, because of the fragmentation of MBP in vivo by proteinases, and because of the difficulties of making a specific antiserum against MBP. The problem may not be solved until the fragments can be synthesized and monoclonal antisera made against them. The evidence suggests that the assay is helpful in assessing the activity of demyelination because MBP levels rise and fall with exacerbation and remission of multiple sclerosis.[12]

Increased intrathecal synthesis of IgG causes an increase in the CSF/serum immunoglobulin concentration ratio and the appearance of oligoclonal Ig bands on electrophoretic separations of CSF.[104] Why synthesis is restricted to only a few clones is not known. The investigation of these changes has now become routine when multiple sclerosis is suspected, but their presence is not diagnostic. At least 90% of cases of multiple sclerosis give positive findings, but increased Ig and oligoclonal Ig may also be found in some other chronic inflammatory diseases of the CNS such as chronic meningoencephalitis due to bacteria, viruses, fungi, or parasites; subacute sclerosing panencephalitis; and Guillain-Barré syndrome. These can be ruled out clinically or by finding a CSF total protein > 100 mg/dL, a CSF leukocyte count > 50/μL, or a positive test for neurosyphilis. The CSF albumin concentration in 70% of cases of multiple sclerosis is within the reference range.

A complicating factor in testing for intrathecal IgG is that an increase in CSF IgG concentration or in the CSF/serum IgG concentration ratio may occur either because of increased permeability of the blood/brain barrier, or increased local production of IgG, or both. In order to identify intrathecal production specifically, correction for increased permeability is necessary. Corrections use *CSF and serum* albumin and IgG concentrations in one of several ways.

1. Concentrations *in CSF* of IgG and albumin are measured and a ratio is calculated:

$$\text{Ratio} = \frac{\text{IgG}_{\text{CSF}}, \text{mg/dL}}{\text{Albumin}_{\text{CSF}}, \text{mg/dL}}$$

A ratio > 0.27 is considered indicative of increased synthesis; in about 70% of cases of multiple sclerosis, the ratio is > 0.27.

2. Concentrations *in CSF and serum* of IgG and albumin are measured and the "CSF immunoglobulin index" is calculated from CSF/serum IgG concentration divided by CSF/serum albumin concentration so that

$$\text{Index} = \frac{\text{IgG}_{\text{CSF}}, \text{mg/dL} \times \text{Albumin}_{\text{SER}}, \text{g/dL}}{\text{Albumin}_{\text{CSF}}, \text{mg/dL} \times \text{IgG}_{\text{SER}}, \text{g/dL}}$$

The reference range for the index is 0.3–0.77. Values > 0.77 are considered evidence of increased IgG synthesis; in over 90% of cases of multiple sclerosis, the index exceeds 0.77. This estimator is presently the one most often used.

3. Concentrations *in CSF and serum* of IgG and albumin are measured and the intrathecal rate of IgG synthesis is calculated by an empirical (Tourtellotte's) formula, based on the known rate of CSF production, by assessing the increase above normal of IgG in the patient's CSF and deducting from it the amount of leakage through the blood/brain barrier as indicated by the excess CSF albumin.

$$\text{Synthesis rate,} \\ \text{mg IgG/d} = 5 \text{ dL/d} \left[\left\{ \text{IgG}_{\text{CSF}} - \frac{\text{IgG}_{\text{SER}}}{369} \right\} - \left\{ \left(\text{Albumin}_{\text{CSF}} - \frac{\text{Albumin}_{\text{SER}}}{230} \right) \times \frac{0.43 \, (\text{IgG}_{\text{SER}})}{\text{Albumin}_{\text{SER}}} \right\} \right]$$

All protein concentrations are expressed in mg/dL. The first bracketed term represents the difference between IgG found in CSF and the IgG expected if the blood/brain barrier is intact. The second bracketed term represents the same for albumin, but is corrected by a ratio, M.W. albumin/M.W. IgG = 0.43, in order to express the difference in albumin as if it were a difference in IgG, assuming that one mole of IgG accompanies every mole of albumin that passes the blood/brain barrier. The term 5 dL/d converts mg/dL units in the other terms and relates IgG to the average volume of CSF, 5 dL/d, normally formed in 24 h. The number 369 is the ratio of Serum/CSF IgG concentrations in normal subjects; the number 230 is the same ratio but for albumin concentrations. The reference range for the synthesis rate is −9.9 to +3.3 mg/d. Values > 8 mg/d are interpreted as increased rate. Although more experience with this estimator is needed, it presently appears that values > 8 mg/d are currently found in ~95% of cases of multiple sclerosis.

Method for the Determination of Total Protein in CSF

Methods for total protein are reviewed on page 584. The small volumes of CSF available limit the choices. Turbidimetric methods and versions of the Coomassie Brilliant Blue (CBB) dye-binding method[40,48] remain the most popular. The most serious disadvantage of turbidimetric methods is the requirement for 0.2–0.5 mL of sample. CBB methods are sensitive enough for use with samples of as little as 25 μL, but they underestimate globulins. Since albumin is the predominant protein of CSF, this underestimation may often not be serious enough to preclude the use of a CBB method. It seems probable, however, that immunochemical methods that use small sample volumes and determine individual proteins rapidly, specifically, precisely, and quantitatively may eventually replace determinations of total protein in CSF.

A *turbidimetric method* is presented in which the precipitating reagent has been shown to be equally efficient in precipitating albumin and globulins.[71]

Principle. Proteins in CSF are precipitated by a combination of sulfosalicylic acid and sodium sulfate and are quantitated turbidimetrically at 620 nm.

Specimen. A CSF specimen is very precious since only a small volume is generally obtained (5–10 mL or less), the protein concentration is low, the specimen often must be shared with the microbiology and cytology departments, and spinal puncture may be painful for the patient. Repeated puncture must also be avoided because total protein determined on specimens obtained by a second puncture may reflect trauma created by the first, rather than a change in the patient's condition. Assay techniques must therefore be chosen to give maximum information on the least volume of sample and any sample remaining should be refrigerated or frozen in case additional testing is required. Blood in a CSF specimen may be present because of bleeding within the cord, ventricles, or cisternae, or as the result of a "bloody tap." Blood in the specimen invalidates the protein values, since the protein concentration of whole blood is ~1000 times that of normal CSF. Total protein measurement, even on clear supernatant of a bloody specimen, inevitably reflects contamination with plasma proteins, and the proportions of specific proteins characteristic of spinal fluid are deranged by the presence of proteins characteristic of plasma.

Reagents

1. Precipitating reagent. Dissolve 30.0 g of sulfosalicylic acid (dihydrate) and 70.0 g of anhydrous sodium sulfate in water and dilute to 1 L volumetrically. Mix and filter until the filtrate is clear. Store the clear reagent in a brown bottle at room temperature. The reagent is stable but should be discarded if discoloration or turbidity develop.

2. NaCl-azide reagent. See page 606, Method for Total Protein in Urine.

3. Standards. See page 606.

Procedure

1. First confirm clarity of precipitating reagent. To 0.5 mL of NaCl-azide reagent, add 2.0 mL precipitating reagent. Mix and read absorbance at 620 nm against NaCl-azide reagent set to zero absorbance. If A is greater than 0.010, prepare new precipitating reagent.

2. Set up a pair of tubes, one labeled B and the other T, for each standard, unknown, or control.

3. If CSF contains cells or particles, remove these by centrifugation. Add 0.5 mL sample to corresponding B and T tubes.

4. Add 2.00 mL NaCl-azide reagent to B tube and mix.

5. At timed intervals, add 2.00 mL precipitating reagent to T tubes, mixing each upon addition.

6. Let stand 10 min at room temperature.

7. While observing time interval, mix separately the B and T tubes of a pair and immediately read A for the T tube against B set to zero absorbance at 620 nm. Record A for T.

Calculation

$$\text{Total protein, mg/dL} = A_u/A_s \times C_s, \text{mg/dL}$$

where A_u and A_s are readings for unknown and standard and C_s is concentration of standard.

Comments

1. Linearity should be checked regularly in a manner similar to that described for the Method of Total Protein in Urine.

2. Run-to-run CV of <5% is acceptable but a CV of ~10% can be expected at control levels <15 mg/dL.

3. Interferences include hemoglobin and drugs that precipitate in acid, as well as radiographic contrast medium.

4. The blank and choice of wavelength compensates for the yellow pigments of xanthochromia. If the specimen is colorless, one may conserve a scanty specimen by substituting NaCl, 0.15 mol/L, for sample in the blank tube without serious risk of error.

Reference

Peters, T., Jr., Biamonte, G. T., and Doumas, B. T.: Protein (total protein) in serum, urine, and cerebrospinal fluid; albumin in serum. *In*: Selected Methods of Clinical Chemistry, Vol. 9, W. R. Faulkner, and S. Meites, Eds. Washington, D.C., American Association for Clinical Chemistry, 1982.

Methods for Specific Proteins in CSF

Rate nephelometry, electroimmunodiffusion, and radial immunodiffusion are most often used for measurements of albumin and IgG. When diffusion methods are employed, CSF and serum concentrations of a protein can be directly compared on the same plate; electroimmunodiffusion even allows all four estimations needed for calculating the CSF/serum immunoglobulin index to be performed on the same plate, since albumin and IgG migrate in opposite directions. Apparent absence of IgG may be due to its degradation by proteinases in the specimen. Radioimmunoassay (RIA) will be required for determination of specific proteins present in very low concentrations, e.g., IgM.

Reference ranges.[105] The range for albumin levels in lumbar spinal fluid by RID is 17.7–25.1 mg/dL. In normal CSF, IgA, IgD, and IgM, measured by RIA, are each less than ~0.2 mg/dL. Reference ranges for IgG, also measured by RIA, are age related and their means rise from 3.5 mg/dL in the 15–20-year age group to 5.8 in adults 60 years or older.

Electrophoretic Separation of CSF Proteins

With sufficient concentration of CSF, the same electrophoretic procedure used for serum can be employed. Concentration requires a considerable volume of specimen and always involves some loss of proteins. Electrophoresis techniques known as high-resolution electrophoresis have also been applied to the separation of CSF protein. This technique and instrumentation are commercially available (e.g., Beckman Instruments, Brea, CA 92621; Helena Laboratories, Beaumont, TX 77704) and involve electrophoresis on thin agarose gels using high voltage separation with concurrent cooling. A 100-fold concentration is necessary but only 3 µL of concentrate is used for application; oligoclonal bands in only 5 µg of IgG are clearly demonstrated with Coomassie Brilliant Blue. Recently, two methods sensitive enough to use unconcentrated CSF have been developed. If silver staining is used, an unconcentrated sample of 10 µL that contains only 0.4–0.8 µg of IgG is sufficient.[62] Comparisons of silver and CBB staining for CSF separations are shown in Figure 4-15. If immunofixation is utilized before silver staining, the technique is highly sensitive and IgG bands can be identified with certainty.

Polyacrylamide gel electrophoresis is another technique applicable to unconcentrated CSF and suitable for routine work. A sample of 100–200 µL is applied directly to the gel and concentration occurs in the medium (see Chapter 1B, Section Three). After electrophoresis and staining, oligoclonal bands can be clearly seen. If the sample is pretreated with sodium dodecyl sulfate to invoke separation by molecular size, then the SDS-PAGE procedure can be used for evaluating the permeability of the blood/brain barrier in a manner analogous to its use for investigating proteinuria. (See p. 603.) Electrofocusing is an even more sensitive approach but is not as suitable for routine use; with this technique, the pattern of bands appears to be unique for each patient.

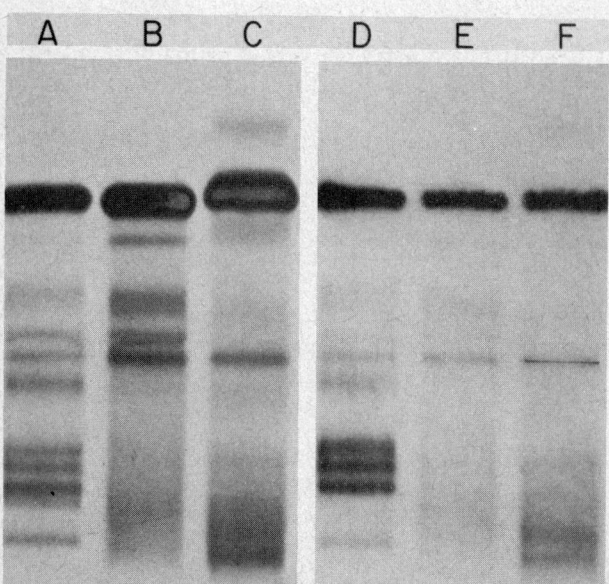

Figure 4-15. Comparison of staining methods for cerebrospinal fluid proteins. A–C, Coomassie Brilliant Blue stain. D–F, Silver stain. Samples were run for 40 min at 100 V, on Beckman SPE-II agarose gel in a Paragon electrophoresis system. Lanes A, D: Control Mixture—serum from multiple myeloma patients, diluted 100-fold, 2 µL applied. Lanes B, E: Serum from a multiple sclerosis patient, diluted 1:100, 2 µL applied. Lane C: CSF from a multiple sclerosis patient, concentrated 80-fold in an Amicon B-15 concentrator; 1 µL applied. Lane F: Same CSF, unconcentrated, 1 µL applied. (Photograph courtesy of Dennis B. Lubahn, Ph.D., Department of Hospital Laboratories, North Carolina Memorial Hospital, Chapel Hill, N.C.)

With any of the electrophoretic approaches, CSF and serum should be collected from the patient at the same time and the two specimens should be run in parallel, after dilution of the serum to a protein concentration similar to that of CSF. Although the main purpose of electrophoretic separations of CSF is to detect oligoclonal IgG bands, a number of other abnormalities may be seen and matched to the serum pattern. Some are secondary to changes in the proportions of the plasma proteins due, for example, to an acute phase reaction or to the presence of Bence Jones proteins in plasma. Others are the result of intrathecal desialization of glycoproteins. Whether or not the presence of desialized protein is indicative of degenerative disease is not known. In such cases, the τ-band, believed to arise by desialization of transferrin, becomes more prominent, whereas increased permeability of the blood/brain barrier decreases the prominence of the τ-band.

Silver Staining of Protein Fractions after Electrophoresis

After electrophoretic separation, protein fractions are visualized by staining with a suitable dye. Ideally, a dye should have equal affinity for all proteins and sufficient sensitivity to detect each protein. However, the most commonly used dye for cellulose acetate separations, Ponceau S, has 1.5 times greater affinity for albumin than for γ-globulins. Other widely used dyes behave in a similar manner, for example, Amido Black 10B, also known as Amidoschwarz 10B; Naphthol Blue Black, or Buffalo Black, a stain commonly employed for agarose separations; and Coomassie Brilliant Blue, a dye widely used for agarose and polyacrylamide gel separations. In terms of sensitivity, Ponceau S on cellulose acetate or Amido Black on agarose can detect proteins whose concentrations in the sample exceed 30–50 mg/dL. Coomassie Brilliant Blue is slightly more sensitive and is suitable for detection of oligoclonal bands in concentrated CSF.

Silver staining binds to specific amino acid groups in proteins in a manner independent of the net protein charge and can routinely detect proteins at concentrations as low as 100 μg/dL, a sensitivity some 20–50 times greater than that of Coomassie Brilliant Blue. Silver staining is therefore an attractive choice for cerebrospinal and any other body fluids when both a low protein concentration and a small volume are available. Its disadvantage is its greater complexity. In addition, the technique is less suited for serum and other body fluids.

To quantitate each electrophoretic protein band, the plate may be scanned with a densitometer. Each fraction is then multiplied by the total protein concentration to yield the concentration of each fraction.

Reagents

Use distilled water throughout.

1. Sodium chloride, 9.0 g/L in water.
2. Glutaraldehyde, 2.5 g/dL. Dilute 10 mL of glutaraldehyde, 25 g/dL, to 100 mL with water. Prepare fresh at time of use.
3. Triton X-100, 0.5 mL/dL. Add 10 mL Triton X-100 to 2 L of water. Stir until solution is complete and store in refrigerator.
4. Dithiothreitol, 0.01 g/dL. Add 200 mg dithiothreitol to 2 L of water and stir until dissolved. Store in refrigerator; stable one month.
5. Sodium carbonate, 5 g/dL. Add 100 g Na_2CO_3 to 2 L of water and stir until dissolved. Store at room temperature.
6. Fixative. Mix 1 L methanol, 1 L water, and 200 mL glacial acetic acid. Store at room temperature.
7. Formaldehyde, 37 g/dL.
8. Silver nitrate, 10 g/dL.
9. Ammonium nitrate, 12 g/dL.
10. Silicotungstic acid, 50 g/dL.
11. Silver stain solution. This preparation serves for one gel; quantities may be doubled if two gels are handled in the same session. *Caution*: Silver nitrate stains hands. Wear protective gloves and exercise care.
   a. Add 94 mL of water to a Styrofoam cup which has been rinsed with water. Add a magnetic stirring bar which has been washed with nitric acid and then rinsed with water.
   b. Add 2 mL $AgNO_3$, 10 g/dL, while stirring.
   c. Add 2 mL NH_4NO_3, 12 g/dL.

    d. Add 2 mL silicotungstic acid, 50 g/dL. Shake the silicotungstic acid solution well before pipetting it.

    e. Mix for 30 s, then add 1.4 mL of formaldehyde, 37 g/dL.

Specimen

CSF should be applied directly to the agarose gel without prior concentration. If serum is used, it must be free of fibrinogen which will obscure the beta-gamma zone. Serum and other biological fluids should be diluted (with saline) or concentrated such that the *total protein* is at least 30–50 mg/dL. Specimens may be stored in covered containers for 48 h at 2–6 °C.

Procedure

    1. Fill each side of an electrophoresis cell with an appropriate buffer, e.g., barbital, 0.05 mol/L.

    2. Gently blot the surface of the agarose gel and then discard the gel blotter. *Note:* Do not touch the gel with bare hands. Always use forceps for movement of the gel.

    3. If using a template, rinse it with water and dry with a Kimwipe. Then align the template on the gel.

    4. Preparation and application of sample.

    a. Cerebrospinal fluid. If the specimen has not been concentrated, apply 3 μL (less if total protein is greatly elevated).

    b. Serum. Dilute 5 μL serum with 0.5 mL saline, 0.9 g/dL. Apply 3 μL.

    5. Wait 3–5 min to allow the sample to diffuse into the gel.

    6. Remove and discard template. Place gel onto the gel bridge and insert the bridge into the electrophoresis cell. Cover the cell and plug it into the power supply.

    7. Electrophorese at 100 V for 45 min.

    8.* Pour ~100 mL of fixative solution into each of 2 plastic weighing boats. Upon completion of electrophoresis, place the gel into one of the boats for 2.5 min. Then place the gel into the second boat for an additional 2.5 min.

    9. Allow the gel to drain on filter paper to remove excess fixative, then place the gel in a forced air draft oven at 65 °C for 20 min. The positive (anodal) end of the gel should be tilted upward while it dries. Once dried, the gel may be kept in dust-free storage and stained at a later time.

    10.* To stain, place the gel in a weighing boat containing ~100 mL glutaraldehyde, 2.5 g/dL, for 10 min. If the gel was stored after drying, soak it for 15–20 min in this solution to remove any crystals.

    11.* Rinse the gel in Triton X-100, 0.5 g/dL, for 5 min. Repeat.

    12.* Place the gel in a weighing boat containing ~100 mL of dithiothreitol, 0.01 g/dL, for 10–30 min, as needed.

    13. While the gel soaks in dithiothreitol, prepare a silver developing solution.

    a. Into a Styrofoam cup previously rinsed with water and containing an acid-washed magnetic stirring bar, transfer ~50 mL of Na_2CO_3, 5 g/dL.

    b. While stirring rapidly, gradually add 50 mL of the silver solution prepared in step 7, above. Immediately pour this new-made solution into a plastic weighing boat. Be sure that the solution is clear initially. If the solution is cloudy, discard it since it may cause black precipitation in the staining procedure. This solution may be stored up to 12 h in the dark at 15–30 °C.

    14.* Place the gel into the developing solution and watch for appearance of protein bands. A fine brown precipitate, which will eventually turn black, will also start to form in the solution. After the albumin bands begin to form (~3 min), transfer the gel to a boat containing 100 mL fresh fixative, in order to arrest the development.

    15.* Allow the gel to soak in fixative for 2 min, then transfer the gel to another boat containing fresh fixative for an additional 2 min.

    16. Remove the gel and allow excess fixative to drain off. Dry gel in oven for 5 min.

    17.* Soak gel in water to wash off excess fixative. While it soaks, repeat step 13 to make a fresh batch of developing solution.

    18.* Place the wet gel in the freshly prepared developing solution and watch it carefully for rapid development of the pattern.

    **Note:* In steps marked with an asterisk, agitation of the liquid is performed with the aid of a horizontally moving rotator.

19.* When the pattern is sufficiently dark, place the gel in fresh fixative for 2 min. Drain and dry the gel for a permanent record. *Note*: For more or less sensitivity, increase or decrease the staining time, respectively. Excessive staining will cause the plate to turn black.

20. For visual interpretation, inspect the gel for the presence of the protein bands. If scanning is desired, a white light (no filter) should be used on a densitometer. The stained and dried gel is stable indefinitely if stored in the dark.

Figure 4-15 shows a comparison of proteins visualized with Coomassie Brilliant Blue and with the silver stain.

Comments

1. The silver stain involves an accelerated reaction such that the higher the protein concentration, the more rapidly the staining procedure proceeds. Therefore, optimal staining time is variable. When running various types of specimens on the same agarose gel, care must be taken that the specimens with lower protein concentrations stain sufficiently.

2. It is very important not to touch the plates. Chloride and amino acids from the skin will contaminate the agarose and cause background staining which can interfere with the interpretation of the stained protein bands.

3. The silver stain should be stored in the dark. Excessive exposure to light will cause the reagent to darken and become unusable.

PROTEINS IN OTHER BODY FLUIDS

Investigation of proteins in other body fluids is seldom of clinical value, although there are exceptions to this statement. *Amniotic fluid* is analyzed for α-fetoprotein for detection of neural tube defects. *Saliva* is tested for secretory IgA in infantile hypogammaglobulinemia and for β_2-microglobulin in Sjögren's syndrome. Assay of *feces* for α_1-antitrypsin is sometimes used in the diagnosis of exudative enteropathy, since, unlike other plasma proteins, AAT appears not to be hydrolyzed in the gut. However, a better test is intravenous injection of ^{52}Cr-labeled albumin, followed by measurement of the radioactive label in feces.

Pathological accumulations of fluid in the peritoneal and pleural cavities or elsewhere vary greatly in protein content; they may be ultrafiltrates with low protein concentrations and scant amounts of high molecular weight proteins, or they may be serous fluids with high protein concentrations and significant amounts of large proteins. These fluids are divided arbitrarily according to their protein concentration into *transudates* with total protein below $\sim$3 g/dL and *exudates* with total protein concentrations above $\sim$3 g/dL. Transudates ordinarily reflect changes in permeability of filtering membranes, while exudates usually result from infection or malignancy; the latter may contain large numbers of leukocytes or malignant cells. Both transudates and exudates usually contain enough protein to allow the biuret method to be used for the determination of total protein.

References

1. Alper, C. A.: Complement. *In*: Structure and Function of Plasma Proteins. A. C. Allison, Ed., VI. New York, Plenum Publishing Corp., 1974.
2. Ampola, M. G.: Metabolic Diseases in Pediatric Practice. Boston, Little, Brown and Company, 1982.
3. Barratt, M.: Proteinuria. Br. Med. J., *287*:1489-1490, 1983.
4. Barrett, A. J., Starkey, D. M., and Munn, E. A.: The unique nature of the interaction of α_2-macroglobulin with proteases. *In*: Bayer Symposium V Protease Inhibitors. H. Fritz, H. Tschesche, and L. J. Greene, Eds. Berlin, Springer-Verlag, 1974, p. 72.
5. Begent, R. H. J.: The value of carcinoembryonic antigen measurement in clinical practice. Ann. Clin. Biochem., *21*:231-238, 1984.
6. Benson, J. R., and Hare, P. E.: *o*-Phthalaldehyde: Fluorogenic detection of primary amines in the picomole range. Comparison with fluorescamine and ninhydrin. Proc. Natl. Acad. Sci. USA, *72*:619-622, 1975.
7. Bonsquet, B., Bonvier, J. L., and Dreux, C.: Amino acidurie globale determinée par spectrométrie d'absorption atomique. Clin. Chim. Acta, *42*:327-334, 1972.
8. Bremer, H. J., Duran, M., and Kamerling, J. P. et al.: Disturbances of Amino Acid Metabolism: Clinical Chemistry and Diagnosis. Baltimore, Urban and Schwarzenberg, Inc., 1981.
9. Buffone, G. J., Lewis, S. A., Iosefsohn, M., et al.: Chemical and immunochemical measurement of total iron-binding capacity. Clin. Chem., *24*:1788-1791, 1978.
10. Buffone, G. J., Savory, J., Cross, R. E., et al.: Evaluation of kinetic light scattering as an approach to the measurement of specific proteins with the centrifugal analyzer. I. Methodology. Clin. Chem., *21*:1731-1734, 1975.
11. Buxbaum, J. N.: The amyloid diseases. *In*: Cecil Textbook of Medicine. 17th ed. J. B. Wyngaarden and L. H. Smith, Jr., Eds., Philadelphia, W. B. Saunders Company, 1985.

12. Calbreath, D. F.: Analysis of multiple sclerosis and related disorders. Part 1: Analysis of CSF proteins. Washington, D.C., American Clinical Products Review. November/December, 1983, pp. 14-19.

13. Christenson, R. H., Behlmer, P., Howard, J. F., et al.: Interpretation of cerebrospinal fluid protein assays in various neurological diseases. Clin. Chem., 29:1028-1030, 1983.

14. Claus, D. R., Osmand, A. P., and Gewurz, H.: Radioimmunoassay of human C-reactive protein and levels in normal sera. J. Lab. Clin. Med., 87:120, 1976.

15. Clayton, B. E., and Round, J. M.: Chemical Pathology and the Sick Child. Oxford, Blackwell, 1984.

16. Collins, J. A.: Clinical judgment versus the laboratory. N. Engl. J. Med., 306:987, 1982.

17. Cooper, E. H., and Morgan, D. B.: Proteinuria. Am. Assoc. Clin. Chem., 1(10):1-11, 1984.

18. Cooper, N. R.: The complement system. In: Basic and Clinical Immunology. 4th ed. D. P. Stites, J. D. Stobo, H. H. Fudenberg, et al., Eds. Lange Medical Publishers, Los Altos, CA, 1982, pp. 124-135.

19. Cooper, T. G.: The Tools of Biochemistry. New York, Wiley, 1977.

20. Daniels, J. C.: Carrier protein abnormalities. In: Serum Protein Abnormalities: Diagnostic and Clinical Aspects. S. E. Ritzmann, and J. C. Daniels, Eds. Boston, Little, Brown and Company, 1975, pp. 213-242.

21. Danks, D. M.: Hereditary disorders of copper metabolism in Wilson's disease and Menkes' disease. In: The Metabolic Basis of Inherited Disease. 5th ed. J. B. Stanbury, J. B. Wyngaarden, D. S. Fredrickson, et al., Eds. New York, McGraw-Hill, 1983, pp. 1251-1268.

22. De Jong, C., and Hughes, G. J.: Amino acid analyses by high-performance liquid chromatography. An evaluation of the usefulness of pre-column DNS derivatization. J. Chromatogr., 241:345-359, 1982.

23. Dickerson, R. E., and Geis, I.: Proteins, Structure, Function and Evolution. 2nd ed. Menlo Park, CA, Benjamin/ Cummings Publishing Company, 1983.

24. Dietz, A. A., Rubinstein, H. M., and Hodges, L.: Measurement of α_1-antitrypsin in serum by immunodiffusion and by enzymic assay. Clin. Chem., 20:396-399, 1974.

25. Dilena, B. A., Penberthy, L. A., and Fraser, C. G.: Six methods for determining urinary protein compared. Clin. Chem., 29:1533-1557, 1983.

26. Duggan, J., and Duggan, P. F.: Albumin by bromcresol green—a case of laboratory conservatism. Clin. Chem., 28:1407-1408, 1982. Letter.

27. Duran, M., and Wadman, S. K.: Organic acidurias. In: Recent Advances in Clinical Biochemistry, Number 2. K. G. M. Alberti and C. P. Price, Eds. Edinburgh, Churchill Livingstone, 1981, pp. 103-127.

28. Faulkner, W. R.: Phenylalanine. In: Standard Methods of Clinical Chemistry, Vol. 5. S. Meites, Ed. New York, Academic Press, 1965, pp. 199-209.

29. Fincham, J. R. S.: Genetics. Bristol, John Wright, 1983.

30. Finley, P. R.: Nephelometry: Principles and clinical laboratory applications. Lab. Management, 20:34-45, 1982.

31. Gaither, T. A., and Frank, M. M.: Complement. In: Clinical Diagnosis and Management by Laboratory Methods. 17th ed. J. B. Henry, Ed. Philadelphia, W. B. Saunders Co., 1984, pp. 879-892.

32. Glass, D. N., Fearon, D. T., and Austen, K. F.: Inherited abnormalities of the complement system. In: The Metabolic Basis of Inherited Diseases. 5th ed. J. B. Stanbury, J. B. Wyngaarden, D. S. Fredrickson, et al., Eds. New York, McGraw-Hill, 1983, pp. 1934-1955.

32a. Glenner, G. G.: Amyloid deposits and amyloidosis. N. Engl. J. Med., 302:1283-1292, 1333-1343, 1980.

33. Gold, P., and Freedman, S. O.: Carcinoembryonic antigen (CEA) in clinical medicine. Cancer, 42:1399-1402, 1965.

34. Guthrie, R., and Susi, A.: A simple phenylalanine method for detecting phenylketonuria in large populations of newborn infants. Pediatrics, 32:338-343, 1963.

35. Gutteridge, J. M. C., and Stocks, J.: Ceruloplasmin: Physiological and pathological perspectives. CRC Crit. Rev. Clin. Lab. Sci., 14:257-329, 1981.

36. Haan, E. A., and Danks, D. M.: Clinical investigation of suspected metabolic disease. In: Laboratory Investigation of Fetal Disease. A. J. Basson, Ed. Bristol, John Wright, 1981, pp. 410-428.

37. Harrison, T. R.: Principles of Internal Medicine. 8th ed. New York, McGraw-Hill, 1977.

38. Haythorn, P., and Sheehan, M.: Improved centrifugal analyzer assay of albumin. Clin. Chem., 25:194, 1979. Letter.

39. Henry, R. J., Cannon, D. C., and Winkelman, J. W., Eds.: Clinical Chemistry: Principles and Techniques. 2nd ed. New York, Harper and Row, 1974, pp. 424-428.

40. Hische, E. A. H., van der Helm, H. J., van Meegen, M. T., et al.: Protein estimations in cerebrospinal fluid with Coomassie Brilliant Blue. Clin. Chem., 28:1236-1237, 1982.

41. Hsia, D. Y. Y., and Inouye, T.: Inborn Errors of Metabolism. 2nd ed. Chicago, Year Book Medical Publishers, 1966, pp. 69, 71, 221.

42. Jagenburg, R., and Rodjer, S.: Detection of heterozygotes for phenylketonuria by constant intravenous infusion of L-phenylalanine. Clin. Chem., 23:1661-1665, 1977.

43. James, K.: Alpha$_2$-macroglobulin and its possible importance in the immune system. In: Trends in Biochemical Sciences, 1980, pp. 43-47.

44. Jellum, E., Stokke, O., and Eldjarn, L.: Combined use of gas chromatography, mass spectrometry, and computer in diagnosis and studies of metabolic disorders. Clin. Chem., 18:800-809, 1972.

45. Jeppsson, J. O., and Franzen, B.: Typing of genetic variants of α_1-antitrypsin by electrofocusing. Clin. Chem., 28:219-225, 1982.

46. Jeppsson, J. O., Laurell, C. B., and Franzen, B.: Agarose gel electrophoresis. Clin. Chem., 25:629-638, 1979.

47. Johnson, A. M.: Genetically determined variations in plasma proteins. In: Protein Abnormalities: Proteins in Body Fluids, Amino Acids, and Tumor Markers: Diagnostic and Clinical Aspects, Vol. 1. S. E. Ritzmann and L. M. Killingsworth, Eds. New York, Alan R. Liss, Inc., 1983, pp. 53-100.

48. Johnson, A., and Lott, J. A.: Standardization of the Coomassie Brilliant Blue method for cerebrospinal fluid proteins. Clin. Chem., 24:1931-1933, 1978.

49. Jones, B. N., and Gilligan, J. P.: Amino acid analysis by o-phthaldialdehyde precolumn derivatization and reversed-phase HPLC. Am. Lab., 15(12):46-51, 1983.

50. Jones, B. N., and Gilligan, J. P.: o-Phthaldialdehyde precolumn derivatization and reverse-phased high-performance liquid chromatography of polypeptide hydrolysates and physiological fluids. J. Chromatogr., 266:471-482, 1983.

51. Killingsworth, L. M.: Clinical applications of protein determinations in biological fluids other than blood. Clin. Chem., 28:1093-1103, 1982.

52. Killingsworth, L. M.: Plasma protein patterns in health and disease. CRC Crit. Rev. Clin. Lab. Sci., 11:1-30, 1979.

53. Kingsley, G. R.: The direct biuret method for the determination of serum proteins as applied to photoelectric and visual colorimetry. J. Lab. Clin. Med., 27:840-845, 1942.

54. Kohn, J.: Monoclonal proteins. In: Immunochemistry in Clinical Laboratory Medicine. A. M. Ward and J. T. Whicker, Eds. Lancaster, England, MTP Press, Ltd., 1979.

55. La Du, B. N., and Michael, P. J.: An enzymatic spectrophotometric method for the determination of phenylalanine in blood. J. Lab. Clin. Med., 55:491-496, 1960.

56. Laurell, C. B., and Jeppsson, J. O.: Protease inhibitors in plasma. In: The Plasma Proteins, Vol. 1. 2nd ed. F. Putnam, Ed. New York, Academic Press, 1975, p. 229.

57. Lorentz, K., and Flatter, B.: Simplified colorimetry of L-amino nitrogen in plasma, serum, or urine. Clin. Chem., 20:1553-1554, 1974.

58. Lott, J. A., Stephan, V. A., and Pritchard, K. A.: Evaluation of the Coomassie Brilliant Blue G-250 method for urinary protein. Clin. Chem., 29:1946-1950, 1983.

59. Louderback, A., Mealy, E. H., and Taylor, N. A.: A new dye-binding technique using bromcresol purple for determination of albumin in serum. Clin. Chem., 14:793-794, 1968. Abstract.

60. McEldeny, L. A., Tarbit, I. F., and Cassells-Smith, A. J.: Six methods for urinary protein compared. Clin. Chem., 28:256-260, 1982.

61. Mechanic, G., Efron, M. L., and Shih, V. E.: A rapid quantitative estimation of tyrosine and phenylalanine by ion exchange chromatography. Anal. Biochem., 16:420-426, 1966.

62. Mehta, P. D., Mehta, S. P., and Patrick, B. A.: Silver staining of unconcentrated cerebrospinal fluid in agarose gel (Panagel) electrophoresis. Clin. Chem., 30:735-736, 1984.

63. Müller-Eberhard, H. J.: Complement. Annu. Rev. Biochem., 45:697-724, 1975.

64. Murayama, K., and Sugawara, T.: Resolution of 52 ninhydrin positive compounds with a high-speed amino acid analyzer. J. Chromatogr., 224:315-321, 1981.

65. Nagashima, M., and Schreiber, G.: Changes in specific proteins during acute inflammation. Am. Assoc. Clin. Chem., 1(9):1-8, 1984.

66. Neurath, H., and Hill, R. L.: The Proteins. 3rd ed. New York, Academic Press, 1976.

67. Old, J. M., Ward, R. H. T., Petrou, M., et al.: First-trimester fetal diagnosis. Lancet, 2:1413-1416, 1982.

68. Osmand, A. P., Friedenson, B., Gewurz, H., et al.: Characterization of C-reactive protein and the complement subcomponent C1t as homologous proteins displaying cyclic pentameric symmetry (pentraxins). Proc. Natl. Acad. Sci. USA, 74:739-743, 1977.

69. Parkes, D.: Wilson's disease. Br. Med. J., 288:1180-1181, 1984.

70. Parving, H. H., Noer, I., Decrert, T., et al.: The effect of metabolic regulation on microvascular permeability to small and large molecules in short-term diabetes. Diabetologia, 12:161-166, 1976.

71. Pennock, C. A., Passant, L. P., and Bolton, F. G.: Estimation of cerebrospinal fluid protein. J. Clin. Pathol., 21:518-520, 1968.

72. Pepys, M. B.: C-reactive protein. Fifty years on. Lancet, 1:653-657, 1981.

73. Pesce, M. A., and Strande, C. S.: A new micromethod for determination of protein in cerebrospinal fluid and urine. Clin. Chem., 19:1265-1267, 1973.

74. Peters, T., Jr.: Serum albumin. In: The Plasma Proteins, Vol. 1. 2nd ed. F. Putnam, Ed. New York, Academic Press, 1975, p. 133.

75. Peters, T., Jr., Biamonte, G. T., and Doumas, B. T.: Protein (total protein) in serum, urine, and cerebrospinal fluid; albumin in serum. In: Selected Methods of Clinical Chemistry, Vol. 9. W. R. Faulkner, and S. Meites, Eds. Washington, D.C., American Association of Clinical Chemistry, 1982.

76. Pinnel, A. E., and Northam, B. E.: New automated dye-binding method for serum albumin determination with bromcresol purple. Clin. Chem., 24:80-86, 1978.

77. Pope, C. G., and Stevens, M. F.: The determination of amino nitrogen using a copper method. Biochem. J., 33:1070-1077, 1933.

78. Poulik, M. D., and Weiss, M. L.: Ceruloplasmin. In: The Plasma Proteins, Vol. II. 2nd ed. F. Putnam, Ed. New York, Academic Press, 1975, pp. 52-109.

79. Pumphrey, R. S. H.: Structure and function of immunoglobulin. In: Immunochemistry in Clinical Laboratory Medicine. A. M. Ward and J. T. Whicker, Eds. Lancaster, England, M.T.P. Press, 1979, pp. 85-98.

80. Putnam, F. W.: Haptoglobin. In: The Plasma Proteins, Vol. 2. 2nd ed. F. W. Putnam, Ed. New York, Academic Press, 1975, pp. 2-51.

81. Putnam, F. W.: Transferrin. In: The Plasma Proteins, Vol. 1. 2nd ed. F. W. Putnam, Ed. New York, Academic Press, 1975, pp. 265-316.

82. Ramakers, J. M.: Coomassie Blue: An alternative procedure for proteins. Clin. Chem., 30:1433-1434, 1984.

83. Ritchie, R. F.: Automated immunoprecipitation analysis of serum proteins. In: The Plasma Proteins. 2nd ed. F. W. Putnam, Ed. New York, Academic Press, 1975, pp. 376-425.

84. Ritchie, R. F., and Smith, R.: Immunofixation III. Application to the study of monoclonal proteins. Clin. Chem., 22:1982-1985, 1976.

85. Ritzmann, S. E., and Daniels, J. C.: Introduction In: Serum Protein Abnormalities: Diagnostic and Clinical Aspects. S. Ritzmann and J. Daniels, Eds. Boston, Little, Brown and Company, 1975.

86. Ritzmann, S. E., and Finney, M. A.: Proteins—synopsis of characteristics and properties. In: Proteins in Body Fluids, Amino Acids and Tumor Markers. Diagnostic and Clinical Aspects. New York, Alan R. Liss, Inc., 1983, pp. 415-455.

87. Roth, K. S., Foreman, J. W., and Segal, S.: The Fanconi syndrome and mechanisms of tubular transport dysfunction. Kidney Int., *20*:705-716, 1981.
88. Roth, M.: Fluorescence reaction for amino acids. Anal. Chem., *43*:880-882, 1971.
89. Saifer, A.: Rapid screening methods for the detection of inherited and acquired aminoacidopathies. Adv. Clin. Chem., *14*:145-218, 1971.
90. Schmid, K.: Alpha₁-acid glycoprotein. *In*: The Plasma Proteins, Vol. 1. 2nd ed. F. W. Putnam, Ed. New York, Academic Press, 1975, pp. 184-228.
91. Scriver, C. R., and Rosenberg, L. E.: Amino Acid Metabolism and Its Disorders. Philadelphia, W. B. Saunders Company, 1973.
92. Sell, S., and Becker, F. F.: Alpha-fetoprotein, I. Natl. Cancer Inst. Monogr., *60*:19-26, 1978.
93. Shahangian, S., Brown, P. I., and Ash, K. O.: Turbidimetric measurement of total urinary proteins: A revised method. Am. J. Clin. Pathol., *81*:651-654, 1984.
94. Shaw, A. B., Risdon, P., and Lewis-Jackson, J. D.: Protein creatinine index and Albustix in assessment of proteinuria. Br. Med. J., *287*:929-932, 1983.
95. Shen, R. S., and Abell, C. W.: Phenylketonuria: A new method for the simultaneous determination of plasma phenylalanine and tyrosine. Science, *197*:665-667, 1977.
96. Siezen, R. J., and Mague, T. H.: Gas-liquid chromatography of the *N*-heptafluorobutyryl isobutyl esters of fifty biologically interesting amino acids. J. Chromatogr., *130*:151-160, 1977.
97. Silk, D. B. A., and Williams, R. Clinical biochemistry of the "artificial liver." *In*: Biochemistry in Clinical Practice. D. L. Williams and V. Marks, Eds. London, Heinemann, 1983, pp. 164-173.
97a. Smith, I.: Chromatographic and Electrophoretic Techniques, Vol. 1. 3rd ed. London, Heinemann Medical Publishers, 1969.
98. Spies, J. R., and Chambers, D. C.: Spectrophotometric analysis of amino acids and peptides with their copper salts. J. Biol. Chem., *191*:787-797, 1951.
99. Stanbury, J. B., Wyngaarden, J. B., Fredrickson, D. S., et al.: The Metabolic Basis of Inherited Disease. 5th ed. New York, McGraw-Hill, 1983.
100. Sternberg, J. C.: A rate nephelometer for measuring specific proteins by immunoprecipitation reactions. Clin. Chem., *23*:1456-1464, 1977.
101. Talamo, R. C., Langley, C. E., and Reed, C. E.: Alpha₁-antitrypsin deficiency: A variant with no detectable α_1-antitrypsin. Science, *181*:70, 1973.
102. Tanaka, K.: Disorders of organic acid metabolism. *In*: Biology of Brain Dysfunction, Vol. 3. G. E. Gaull, Ed. New York, Plenum Press, 1975, pp. 145-214.
103. Tarnaky, A. L.: Genetic and drug-induced variation in serum albumin. *In*: Advances in Clinical Chemistry, *21*:101-146, 1980.
104. Thompson, E. J., Kaufmann, P., Shortman, R. C., et al.: Oligoclonal immunoglobulins and plasma cells in spinal fluid of patients with multiple sclerosis. Br. Med. J., *1*:16-17, 1979.
105. Tietz, N. W., Ed.: Clinical Guide to Laboratory Tests. Philadelphia, W. B. Saunders Co., 1983.
106. Tietz, N. W., Ed.: Fundamentals of Clinical Chemistry. 2nd ed. Philadelphia, W. B. Saunders Co., 1976.
107. Udenfriend, S., and Cooper, J. R.: Assay of L-phenylalanine as phenylethylamine after enzymatic decarboxylation; application to isotopic studies. J. Biol. Chem., *203*:953-960, 1953.
108. Van Lente, F., Marchand, A., and Galen, R. S.: Evaluation of a nephelometric assay for haptoglobin and its clinical usefulness. Clin. Chem., *25*:2007-2010, 1979.
109. Walsh, M. J., and Tourtellotte, W. W.: The cerebrospinal fluid in multiple sclerosis. *In*: Multiple Sclerosis. J. F. Hallpike, C. W. M. Adams, and W. W. Tourtellotte, Eds. London, Chapman and Hall, 1983, pp. 275-358.
110. Weidmeier, V. T., Porterfield, S. P., and Hendrick, C. E.: Quantitation of DNS-amino acids from body tissues and fluids using high-performance liquid chromatography. J. Chromatogr., *231*:410-417, 1982.
111. Wells, M. G.: Simple rapid method for determination of α-amino acids in urine. Clin. Chim. Acta, *25*:27-29, 1969.
112. Whicher, J. T.: Abnormalities of plasma proteins. *In*: Biochemistry in Clinical Practice. D. L. Williams and V. Marks, Eds. London, William Heinemann, 1983, pp. 221-250.
113. Whicher, J. T., Higginson, J., Riches, P. G., et al.: Clinical applications of immunofixation: detection and quantitation of complement activation. J. Clin. Pathol., *33*:781-785, 1980.
114. Whicher, J. T., Warren, C., and Chambers, R. E.: Immunochemical assays for immunoglobulins. Ann. Clin. Biochem., *21*:78-91, 1984.
115. Woo, J., Floyd, M., and Cannon, D. C.: Albumin and β₂-microglobulin radioimmunoassays applied to monitoring of renal-allograft function and in differentiating glomerular and tubular diseases. Clin. Chem., *27*:709-713, 1981.
116. Woo, S. L. C., Lidsky, A. S., Güttler, F., et al.: Cloned human phenylalanine hydroxylase gene allows prenatal diagnosis and carrier detection of classical phenylketonuria. Nature, *306*:151-155, 1983.

Additional Reading

Dickerson, R. E., and Geis, I.: Proteins, Structure, Function and Evolution. 2nd ed. Menlo Park, CA Benjamin/Cummings Publishing Company, 1983.
Ritzmann, S. E.: Protein Abnormalities. Volume 1: Physiology of Immunoglobulins: Diagnostic and Clinical Aspects. New York, Alan R. Liss, Inc., 1982.
Ritzmann, S. E.: Protein Abnormalities. Volume 2: Pathology of Immunoglobulins: Diagnostic and Clinical Aspects. New York, Alan R. Liss, Inc., 1982.
Ritzmann, S. E., and Killingsworth, L. M.: Protein Abnormalities. Volume 3: Fluids, Amino Acids, and Tumor Markers: Diagnostic and Clinical Aspects. New York, Alan R. Liss, Inc., 1983.
Stanbury, J. B., Wyngaarden, J. B., Fredrickson, D. S., et al.: The Metabolic Basis of Inherited Disease. 5th ed. New York, McGraw-Hill Book Company, 1983.

CHAPTER 5

ENZYMES

by Donald W. Moss, Ph.D., D.Sc., A. Ralph Henderson, M.B., Ch.B., Ph.D., and John F. Kachmar, Ph.D.

Clinical enzymology is the application of the science of enzymes to the diagnosis and treatment of disease. It is one of the most important fields of contemporary clinical chemistry. Today, in the larger hospital laboratories, enzyme assays may account for as much as 20–25% of the total workload, with as many as 12–15 different enzymes being estimated routinely. Measurements of the activity of digestive enzymes in body fluids as an aid to diagnosis date back to the early 1900's, and some of these earliest observations (e.g., those on amylase in urine, first studied by Wohlgemuth in 1908) are still useful. Measurements of enzyme activity in serum began in the 1920's and 1930's with the studies of Kay, King, Bodansky, and Roberts on alkaline phosphatase in bone and liver disease. Shortly afterward, Kutscher and Wolbergs, as well as Gutman and Gutman, recognized the value of changes of acid phosphatase activity in serum in the diagnosis of prostatic cancer.

An important phase of serum enzymology began in 1943, with the observation of Warburg and Christian that increased activities of glycolytic enzymes were present in the sera of tumor-bearing rats. The measurement of cellular enzymes released into plasma as a consequence of tissue damage received an even greater stimulus in 1955 when La Due, Wroblewski, and Karmen reported the transitory rise of glutamic-oxaloacetic transaminase activity (now called aspartate aminotransferase) in serum following an acute myocardial infarction. This observation marks the beginning of the modern phase of diagnostic enzymology, which has been characterized by a search for methods of greater analytical and clinical sensitivity and for enzymes whose changes are more specific for particular disease states and organs.

Enzymes are proteins with catalytic properties due to their powers of specific activation of their substrates. This definition indicates the characteristic properties of enzymes, which in turn govern the principles of methods of enzyme analysis. This chapter surveys these principles, outlines the main diagnostic interpretations of enzyme assays, and describes in detail the most common and clinically useful methods for the determination of enzyme activities.

Enzyme Nomenclature

By the turn of the century it had become customary to identify individual enzymes by using the name of the substrate or group on which the enzyme acts and then adding the suffix -*ase*. Thus the enzyme hydrolyzing urea was ure*ase*, that acting on starch, amyl*ase* (from amylum, starch), and that acting on phosphate esters, phosphat*ase*. In a few instances, for purposes of clarity, the type of reaction involved was also identified, as in carbonic anhydrase, D-amino acid oxidase, and succinic dehydrogenase. Enzymes known prior to this attempt at systemization had already been given empirical names. Some trivial names of this type are trypsin, diastase, ptyalin, pepsin, and emulsin.

This combination of a few trivial, common names and the larger number of semisystematic names was originally found serviceable. However, many enzymes were found to act on the same substrate while catalyzing different or related reactions. Other enzymes showed a specific requirement for a single substrate or pair of substrates. Thus, the need for a definitive and standardized system of identifying enzymes was recognized. In 1955 the International Union of Biochemistry appointed a commission to study the problem of enzyme nomenclature. Its proposals have been accepted by all workers in the field. With the periodic updating undertaken by the Nomenclature Committee of IUB, the proposals provide a rational and practical basis for identifying all enzymes now known and for those enzymes which will be discovered in the future.[110]

Two names are provided for each enzyme: (1) a systematic name, which clearly describes the nature of the reaction catalyzed, and with which is associated a unique numerical code designation; and (2) a trivial or practical name, which may be identical with the systematic name but is often a simplification of it, suitable for everyday use. The unique numerical designation for each enzyme consists of four numbers, separated by periods, as for example 2.2.8.11. The number is prefixed by the letters "EC," denoting "Enzyme Commission." The first number defines the class to which the enzyme belongs. All enzymes are assigned to one of six classes, characterized by the type of reaction they catalyze: (1) oxidoreductases, (2) transferases, (3) hydrolases, (4) lyases, (5) isomerases, and (6) ligases. The next two numbers indicate the subclass and sub-subclass to which the enzyme is assigned. For example, these may differentiate the amino-transferring subclass from the phosphate-transferring category, or the ethanol acceptor sub-subclass from that accepting acyl groups. The last number is the specific serial number given to each enzyme within its sub-subclass.

The systematic name of each enzyme consists of two parts; the first gives the name of the substrate or substrates acted upon, and the second, a word ending in -ase, indicates the type of reaction catalyzed by all enzymes in the group. If two substrates are involved, both names are used and are separated by a colon; e.g., L-lactate:NAD oxidoreductase. Occasionally an expression in parentheses, such as (decarboxylating), may be inserted to identify the reaction further. Because of the precise rules governing terminology, any enzyme can be identified by both its code number and its systematic name. Table 5-1 lists some selected enzymes of clinical interest, identified by trivial, practical, and systematic names and also by code numbers.

Table 5-1. ENZYME COMMISSION (EC) NUMBERS, SYSTEMATIC AND TRIVIAL NAMES, TOGETHER WITH FREQUENTLY ADOPTED ABBREVIATIONS OF ENZYMES OF MAJOR DIAGNOSTIC IMPORTANCE

EC Number	Systematic Name	Trivial Name	Abbreviation
1.1.1.27	L-Lactate:NAD$^+$ oxidoreductase	Lactate dehydrogenase	LDH, LD
1.1.1.42	Threo-D$_s$-isocitrate:NAD(P)$^+$ oxidoreductase (decarboxylating)	Isocitrate dehydrogenase	ICD
1.4.1.3	L-Glutamate:NAD(P)$^+$ oxidoreductase (deaminating)	Glutamate dehydrogenase	GLDH
2.3.2.2	(5-Glutamyl)-peptide:amino-acid 5-glutamyl-transferase	γ-Glutamyltransferase	GGT
2.6.1.1	L-Aspartate:2-oxoglutarate aminotransferase	Aspartate aminotransferase (transaminase)	AST
2.6.1.2	L-Alanine:2-oxoglutarate aminotransferase	Alanine aminotransferase (transaminase)	ALT
2.7.3.2	ATP:creatine N-phosphotransferase	Creatine kinase	CK
3.1.1.3	Triacylglycerol acylhydrolase	Lipase	Lip
3.1.1.7	Acetylcholine acetylhydrolase	Acetylcholinesterase, true cholinesterase, choline esterase I	
3.1.1.8	Acylcholine acylhydrolase	Pseudocholinesterase, benzoyl cholinesterase, choline esterase II (serum cholinesterase)	ChE (SChE)
3.1.3.1	Orthophosphoric-monoester phosphohydrolase (alkaline optimum)	Alkaline phosphatase	ALP
3.1.3.2	Orthophosphoric-monoester phosphohydrolase (acid optimum)	Acid phosphatase	ACP
3.1.3.5	5'-Ribonucleotide phosphohydrolase	5'-Nucleotidase	5NT
3.2.1.1	1,4-α-D-Glucan glucanohydrolase	Amylase	Amy
3.4.21.4		Trypsin	—
4.1.2.13	D-Fructose-1,6-bisphosphate D-glyceraldehyde-3-phosphate-lyase	Aldolase	ALD

It is a common and convenient practice to use capital letter abbreviations for the names of certain enzymes, such as ALT (formerly GPT) for alanine aminotransferase (EC 2.6.1.2); other examples are AST (GOT), LD(H), and CK, as illustrated in Table 5-1. This practice is not recommended by Commission rules. However, it is so well established, and the convenience in oral and printed communication is so real, that the practice persists despite the formal disapproval. Abbreviations will be used in this chapter when appropriate and after being clearly defined.

Enzymes as Proteins

All enzyme molecules possess the three levels of structure characteristic of proteins in general. The *primary structure*, the linear sequence of amino acids linked through their α-carboxyl and α-amino groups by peptide bonds, is specific for each type of enzyme molecule. Each polypeptide chain is coiled up into three-dimensional *secondary* and *tertiary* levels of structure. Secondary structure refers to the conformation of limited segments of the polypeptide chain, namely α-helices, β-pleated sheet, random coils, and β-turns. The complex, three-dimensional structure assumed by the whole chain is referred to as its tertiary structure. Since the amino acid sequence, the primary structure, determines the conformation of the higher levels of molecular structure that are assumed spontaneously by the growing chain, the secondary and tertiary structures are as characteristic of a particular type of protein molecule as its primary structure is. A further level of structure exists in protein molecules which contain two or more polypeptide chains (subunits). This *quaternary* level of structure consists of the association of small groups of subunits (also called protomers), each consisting of one polypeptide chain in its characteristic conformation, to make up an oligomeric protein molecule; e.g., creatine kinase is a dimer, lactate dehydrogenase a tetramer. Biological activity, such as the catalytic activity of enzymes, is often found to be a property of the oligomeric molecule, so that activity is lost under conditions in which the subunits separate from each other. The protomers that compose a particular type of enzyme molecule are frequently identical (e.g., as in the MM isoenzyme of CK, or the H_4 lactate dehydrogenase isoenzyme). However, association of unlike subunits can occur and gives rise, for example, to the heteropolymeric isoenzymes described later.

Although the techniques needed to define the structure of a protein molecule in all its levels are laborious, considerable progress has been made in determining the structures of enzymes, particularly at the primary structural level. Some important generalizations have emerged from these studies.

There is no feature of primary structure (such as repetition of particular amino acid sequences) that is common to all enzyme molecules. However, considerable homologies of sequence are found between enzymes that appear to share a common evolutionary origin, such as the proteases trypsin and chymotrypsin, and similarities of sequence are even more marked between the members of a family of isoenzymes. The amino acid sequence in the immediate neighborhood of the active center of the enzyme (see below) is often closely similar in enzymes of related function; for example, the "serine proteases" are so called because they all have this amino acid in the active center.

Enzyme molecules differ in the proportion of secondary-level structures such as α-helices that they contain, although no enzyme molecule so far studied approaches the large proportion of α-helices found in myoglobin and hemoglobin. The tertiary structures of different types of enzyme molecules are as individually characteristic as their primary structures; nevertheless, some common features can be discerned at this structural level. Enzyme molecules are roughly globular in overall shape, with a preponderance of polar amino acid side chains on the outside of the molecule and nonpolar side chains in the interior. The ionizable residues in contact with the surrounding medium are responsible for many of the properties of the enzyme molecules in solution, such as their migration in an electric field and their solubility. A few covalent disulfide bridges link different parts of the polypeptide chains in some enzyme molecules, but the three-dimensional structure is mainly stabilized by the large number of hydrophobic interactions which are formed between the nonpolar side chains in the interior of the molecule.

The biological activity of a protein molecule—its catalytic activity in the case of an enzyme molecule—depends generally on the integrity of its structure. Therefore, any disruption of the structure is accompanied by a loss of activity, a process that is known as *denaturation*. If the process of denaturation has not gone too far it may be reversed, with recovery of activity, when the denaturing agent is removed. This is because of the tendency of the enzyme molecule to resume its usual conformation. However, prolonged or severe denaturing conditions result in an irreversible loss of activity. Denaturing conditions include elevated temperatures, which weaken the stabilizing bonds by increasing the vibration of the constituent atoms. Heat inactivation of most enzymes takes place at an appreciable rate at room temperatures and, since the temperature coefficients* of inactivation have values in the order of tens or hundreds, the rate of inactivation increases rapidly as temperature rises, becoming almost instantaneous in most cases above about 60 °C. Low temperatures are therefore needed to preserve enzyme activity, especially in aqueous solutions such as serum. The implications of denaturation for the storage of serum samples while awaiting analysis for enzyme activities are discussed more fully in a later section. Extremes of pH also cause unfolding of enzyme molecular structures and, except in a few cases, should be avoided when preserving enzyme samples. Urea and related compounds disrupt hydrogen bonds and hydrophobic interactions, so that exposure of enzymes to strong solutions of these reagents results in inactivation. Because the sensitivity of an enzyme molecule to denaturation is a reflection of its three-dimensional structure, differences in rates of inactivation provide a sensitive indication of structural differences between enzyme molecules. Therefore, heat inactivation has been used extensively in the analysis of mixtures of isoenzymes. However, since the temperature coefficients of inactivation are so high, slight changes in experimental conditions can have a profound effect on the validity of the results, and the conditions of inactivation must be strictly controlled.

Specificity and the Active Center

With the exception of enzymes such as proteases, nucleases, and amylases, which act on macromolecular substrates, enzyme molecules are considerably larger than the molecules of their substrates. Nevertheless, interaction between the enzyme and its substrate involves the combination of one molecule of enzyme with one substrate molecule (or two, in the case of bisubstrate reactions). The reaction involves the attachment of the substrate molecule to a specialized region of the enzyme molecule, its *active center*. Studies of the three-dimensional structures of enzyme molecules show that the active center is located in a pocket or cleft in the surface of the enzyme molecule (Figure 5-1). The various groups that are important in substrate-binding are brought together at the active center, and it is there that the processes of activation and transformation of the substrate take place. The composition and spatial arrangement of the active center are also the basis of what is perhaps the most outstanding characteristic of enymes, their specificity.

Early in the study of enzymes it was appreciated that each enzyme catalyzed only one reaction or at most a limited range of chemical reactions. The degree of specificity varies from one enzyme to another. Some enzymes show absolute specificity—they catalyze a unique reaction, and no others. Pyruvate kinase, for example, mediates the transfer of a phosphate group between phosphoenol pyruvate and ADP and can function in no other reaction.

A somewhat lesser degree of substrate specificity is found in hexokinase. This enzyme transfers a phosphate group from ATP to D-glucose, but it will also phosphorylate D-fructose, D-mannose, and 2-deoxy-D-glucose at almost equivalent rates. It will not act, however, on D-galactose or on a variety of other hexoses or pentoses, although some of these are bound to the enzyme and can competitively inhibit enzyme activity. Disaccharides, methylated sugars, and sugar alcohols neither bind to the enzyme nor inhibit its activity.

The phosphatases are examples of enzymes with group specificity. These enzymes split phosphate from any of a large variety of organic phosphate esters, although at somewhat different rates. Substances as varied as glucose-6-phosphate, phenyl phosphate, and β-glycerophosphate can serve as substrates. The esterases are enzymes with even lower specificity. They hydrolyze esters to alcohols and carboxylic acids. A considerable range of chain lengths in both the alkyl (alcohol) and acyl (acid) portions of the esters is permitted.

Stereoisomeric specificity is characteristic of many enzymes. The enzymes involved in glycolysis act only on the D-stereoisomers of glucose and its derivatives and never on the L-forms. The transaminases convert oxo-acids only to the L-isomers of the amino acids, and fumarase hydrates

---

*The temperature coefficient of a reaction is usually expressed as the ratio of the reaction rate at a temperature $(t + 10)$ °C to that at temperature t, and is given the symbol Q_{10}.

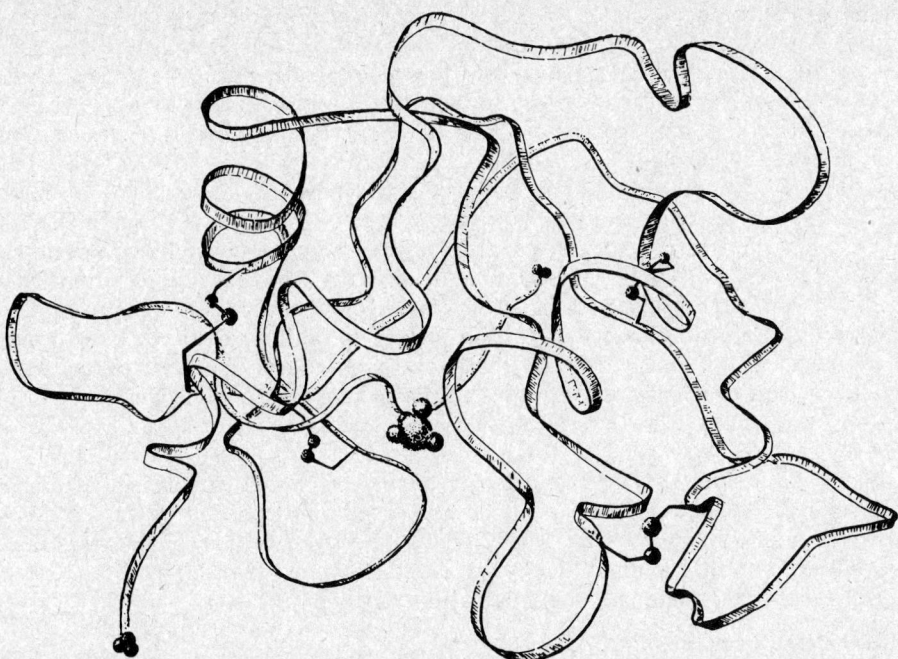

Figure 5-1. Plane projection of the three-dimensional structure of ribonuclease A, as proposed by Harker from X-ray studies. The phosphate group lies in the "active center" of the enzyme, situated in a cleft in the three-dimensional structure. (From Kartha: Accounts Chem. Res., *1*:374, 1968.)

fumarate to the L-form of malate, rather than to the D-glucose–related mirror image (D-) form. Human α-amylase hydrolyzes only the linear segments of starches in which the D-glucose residues are linked by α-1,4-linkages. It is inactive toward cellulose, in which the sugar residues are connected by β-1,4-linkages, and toward the branch points (α-1,6-linkages) in glycogen and amylopectin.

Enzymes as Antigens

Like other proteins, enzymes usually elicit the production of antibodies when they are injected into animals of a species other than that in which they originate. The antigenicity of individual proteins is mainly due to the three-dimensional configuration of relatively small regions of their molecules. Even small structural differences between closely similar molecules, such as the members of a family of isoenzymes, are often sufficient to render them antigenically distinct. Thus, in many cases, antibodies specific for a single type of molecule can be produced. The availability of enzyme-specific antisera opens up a wide range of methods in enzyme analysis, some of which, such as radioimmunoassay, do not depend on the catalytic activity of the enzyme molecules that are being determined and thus can measure inactive enzyme protein. The availability of immunochemical methods is particularly important in the analysis of isoenzyme mixtures. (See later sections.)

ENZYMES AS CATALYSTS

A catalyst may be defined as a substance which increases the rate of a particular chemical reaction without itself being consumed or permanently altered. In other words, at the end of a catalyzed reaction the catalyst appears unchanged in form and quantity, whereas the main reaction materials have undergone transformation into new products. Enzymes are protein catalysts of biological origin. Virtually all the chemical reactions that take place in living matter are catalyzed by specific enzymes, so that life itself can be regarded as an integrated series of enzymatic reactions and some diseases as a derangement of the normal pattern of metabolism. Apart from these fundamental considerations, it is the catalytic property of enzymes that makes them such sensitive indicators of pathological change.

Because of their remarkable catalytic activity, a given number of enzyme molecules will convert an enormously greater number of substrate molecules to products within a short space of time. Therefore, the appearance of increased amounts of enzymes in the blood stream can be detected with great sensitivity, although the amount of enzyme protein released from damaged cells is negligible compared with the background level of nonenzymic proteins in blood. For example, the amount of acid phosphatase *protein* present in normal serum is of the order of a few $\mu g/L$, compared with a total protein concentration of 60–80 g/L. Since enzymes are specific in their actions, a particular enzyme, such as acid phosphatase, can be recognized by its characteristic effect on a given chemical reaction, in spite of the presence of a vast excess of other proteins.

Like other catalysts, an enzyme changes only the rate at which equilibrium is established between reactants and products; it does not alter the equilibrium constant of the reaction. In a reaction in which only one set of products is chemically possible, the catalyst cannot effect any change in the nature of the products, but when several different possible pathways exist the enzyme will direct the reaction along only one pathway.

The Enzyme-Substrate Complex

A reaction involving the transformation of a substrate, S, into a product, P, can proceed spontaneously only if there is a decrease in free energy or chemical potential in the course of the reaction. Stated simply, chemical reactions proceed downhill, from the energy viewpoint. The symbol for Gibbs' free energy is G, and ΔG represents the difference between the chemical potential of the products, P, and that of the initial reactants, S. Thus, a chemical reaction will tend to proceed spontaneously to completion or to equilibrium only if ΔG has a negative value. This concept is illustrated in Figure 5-2. If ΔG is zero, no reaction will occur; the system is in equilibrium, and P and S have the same free energy content. If ΔG is positive, P will tend to react to form S; the uphill reaction will take place only if energy is provided from the outside to push the reaction uphill.

Even though a chemical reaction is thermodynamically possible (i.e., ΔG is negative), the reaction may not proceed spontaneously, since only those molecules that are "excited" or "activated" will undergo reaction. The "active" molecules are those that have absorbed extra energy, with the result that the bonds linking some or all of the atoms in the molecule are weakened. Most substances are stable at ordinary conditions because only a very small fraction of the

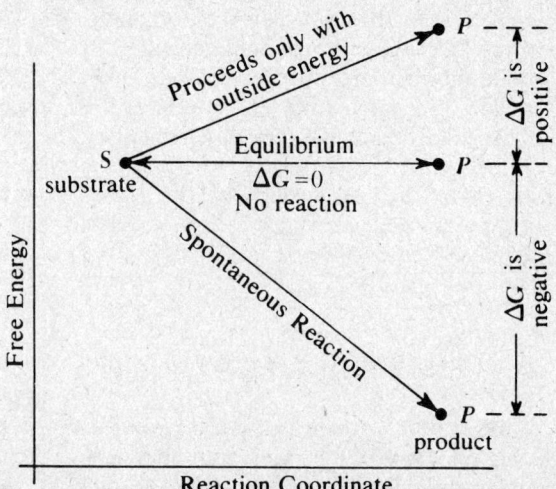

Figure 5-2. Free energy change and the course of chemical reactions.

molecules are in an "activated" state, that is, have enough extra energy to cross the "activation barrier." In the spontaneous reaction

$$S \rightleftharpoons S^* \rightleftharpoons P^* \rightleftharpoons P \qquad\qquad (1)$$

the molecules of S must first absorb ΔA units of activation energy to form the "activated" molecules, S^*, as shown schematically in Figure 5-3 and in equation (1). The activated S^* molecules then undergo reaction to P^* and P. Many reactions require only a relatively low activation energy, and the energy of the thermal motion of the solute and solvent molecules is sufficient to activate enough molecules to initiate the reaction. The energy released in the reaction then makes possible the activation of all the reactants and allows the reaction to proceed to completion within a comparatively short time period. Many reactions with high activation barriers can be initiated by heating the reactants, as is often done in the laboratory.

If the activation energy needed for a given reaction pathway is too high to permit initiation of the reaction, the reactant may still be able to undergo chemical change if some other reaction pathway having a lower activation energy requirement is available. Catalysts provide such alternative paths.

Enzymes act through the formation of an enzyme-substrate complex, ES, in which a molecule of substrate is bound to the specialized substrate-binding region (the *active center*) of the enzyme molecule. The binding process transforms the substrate molecule to its activated state. The energy required for this transformation is provided by the free energy of binding of S to E. Therefore, activation takes place without the addition of external energy, so that the energy barrier to the reaction is lowered and breakdown to products is speeded up (Figure 5-3). The ES complex breaks down to give the reaction products (P) and free enzyme (E):

$$E + S \rightleftharpoons ES \rightarrow P + E \qquad\qquad (2)$$

Part of the free-energy change on binding produces a "strain" in the substrate molecule. The "strained" state is stabilized by the specific binding site of the enzyme; otherwise, it would

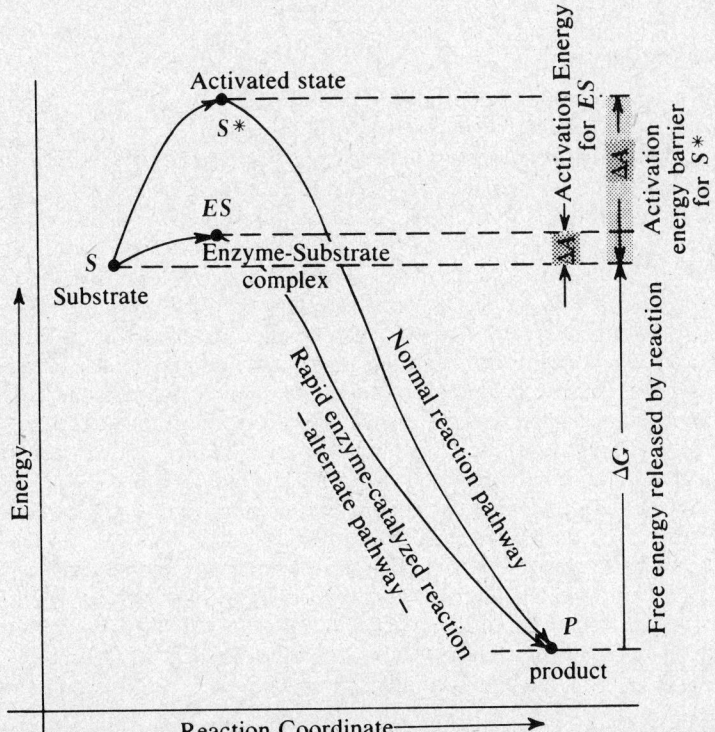

Figure 5-3. Activation energy barrier and reaction course, with and without enzyme catalysis.

return to its ground state without undergoing reaction. The binding of the "strained" or transition state of the substrate can be studied in some cases (e.g., by synthesizing analogs of the molecule in its transition state). This approach shows that the binding of the transition-state molecule is very strong—much stronger than that of the substrate in its ground state. This helps to explain the reversibility of enzymatic reactions; if the substrate molecule were the most strongly bound form, it would be difficult to see how product molecules could be bound, as they would have to be to allow the reaction to proceed in the reverse direction.

Factors Governing the Rate of Enzyme-Catalyzed Reactions

Enzyme Concentration

The overall reaction of conversion of substrate to products in an enzymatic reaction, with the intermediate formation of an enzyme-substrate complex, is the resultant of several individual reactions, each with its characteristic rate constant:

$$E_f + S \underset{k_{-1}}{\overset{k_1}{\rightleftharpoons}} ES \overset{k_2}{\longrightarrow} E_f + P \tag{3}$$

In these equations E_f refers to free enzyme, and k_1 and k_{-1} to the rate constants for the association and dissociation of the complex. The complex breaks down into P and free enzyme (E_f) at a rate governed by k_2. Michaelis and Menten assumed that equilibrium is attained rapidly between E, S, and ES; i.e., the effect of product formation $(ES \rightarrow P)$ on the concentration of ES is negligible. Therefore, the overall rate of the reaction under otherwise constant conditions is proportional to the concentration of the ES complex.

Provided that an excess of free substrate molecules is present, addition of more enzyme molecules to the reaction system will increase the concentration of the ES complex and thus the overall rate of reaction. This accounts for the observation that the rate of reaction is generally proportional to the concentration of enzyme present in the system and is the basis for the quantitative determination of enzymes by measurement of reaction rates. Reaction conditions are selected to ensure that the observed reaction rate is proportional to enzyme concentration over as wide a range as possible.

Substrate Concentration

In addition to explaining the dependence of reaction rate on enzyme concentration under conditions in which excess substrate is present, the formation of an enzyme-substrate complex also accounts for the typical relationship between reaction velocity and substrate concentration. If the enzyme concentration is fixed and the substrate concentration is varied, the rate of reaction is almost directly proportional to substrate concentration at low values of the latter; i.e., the reaction is essentially first order with respect to substrate concentration. At high substrate concentrations, on the other hand, variation in substrate concentration has no effect on rate, and the reaction is zero order with respect to substrate concentration (Figure 5-4). At low concentrations of substrate, only a fraction of the enzyme is associated with substrate and the rate observed reflects the low concentration of the enzyme-substrate complex. At very high substrate concentrations, all the enzyme is bound to substrate, and a much higher rate of reaction is obtained. Moreover, since all the enzyme is now present in the form of the complex, no further increase in complex concentration and no further increment in reaction rate are possible. The maximum possible velocity for the reaction has been reached.

The significance of substrate-rate curves was first emphasized by Michaelis and Menten, and such curves are referred to as Michaelis-Menten plots.

Referring again to equation (3), if we denote the total enzyme concentration by $[E_t]$, then at any time $[E_f] = [E_t] - [ES]$. The concentration of free substrate can be equated with total substrate concentration, since the amount combined as ES will be a negligible fraction of the total. (The concentration of E is very small compared to that of S.) Then the dissociation (Michaelis) constant (K_m) of the equilibrium reaction $(ES \rightleftharpoons E + S)$ is given by:

$$K_m = \frac{[E_f] \times [S]}{[ES]} = \frac{([E_t] - [ES]) \times [S]}{[ES]} = \frac{k_{-1}}{k_1} \tag{4}$$

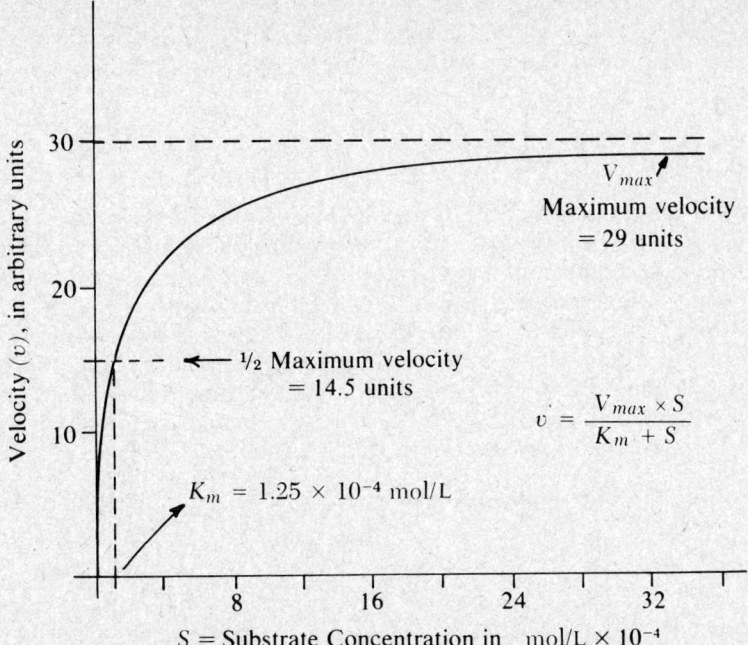

Figure 5-4. Michaelis-Menten curve relating velocity (rate) of an enzyme-catalyzed reaction to substrate concentration. The value of K_m is given by the substrate concentration at which one-half of the maximum velocity is obtained.

The measured velocity of the reaction (v) is determined by the rate of decomposition of ES into P and E_f, i.e.,

$$v = k_2 \times [ES] \tag{5}$$

Rearranging equation (4) and eliminating $[ES]$ from equation (5),

$$v = k_2 \cdot \frac{[E_t] \times [S]}{K_m + [S]} \tag{6}$$

For a given amount of enzyme, the maximum reaction velocity (V_{max}) is reached when all of the enzyme is saturated with substrate, i.e., $[ES] = [E_t]$, and therefore, $V_{max} = k_2 \times [E_t]$. Substituting this in equation (6) gives:

$$v = \frac{V_{max}[S]}{K_m + [S]} \tag{7}$$

Further rearrangement gives:

$$\frac{V_{max} - v}{v} = \frac{K_m}{[S]} \tag{7a}$$

A plot of v against $[S]$ should give a section of a rectangular hyperbola (Figure 5-4) and this is the shape of the curve that is found experimentally for most enzymes. When $[S] = K_m$, manipulation of equation (7) yields $v = \frac{1}{2}(V_{max})$; i.e., the Michaelis constant K_m is that particular substrate concentration at which the reaction proceeds at one half of its maximum velocity.

In Michaelis and Menten's original scheme, K_m was a true equilibrium constant; a smaller value indicated a greater affinity of the enzyme for its substrate and vice versa. In introductory derivations such as that presented here, the symbol K_m for the Michaelis constant is usually used in this sense. However, now that it is possible to measure rates of formation and breakdown of ES in several cases (e.g., by fast-reaction absorption spectrophotometry), it has been found that the rate of breakdown of *ES* into products may not be negligible compared to that of the dissociation of *ES* into *E* and *S*. The Michaelis constant thus ceases to be a true equilibrium constant and now contains a velocity term, becoming a steady-state constant,

$$K_m = \frac{k_{-1} + k_2}{k_1}$$

Nevertheless, the shape of the curve relating v and $[S]$ remains the same. In precise demonstrations, it is now customary to restrict the term Michaelis-Menten constant (K_m) to the experimentally determined substrate concentration at which $v = \frac{1}{2}V_{max}$, and to use the symbol K_s to represent the true enzyme-substrate association constant, where this is known.

While it is quite simple to set up an experiment to determine the variation of v with $[S]$, the exact value of V_{max} cannot be evaluated easily from the hyperbolic curve. Furthermore, many enzymes deviate from ideal behavior at high substrate concentrations and indeed may be inhibited by excess substrate, so that the calculated value of V_{max} cannot be achieved in practice. The Michaelis-Menten equation (7) is therefore usually transformed into one of several reciprocal forms [equations (8) and (9)], and either $\frac{1}{v}$ is plotted against $\frac{1}{[S]}$, or $\frac{[S]}{v}$ is plotted against $[S]$.

$$\frac{1}{v} = \left(\frac{K_m}{V_{max}} \times \frac{1}{[S]}\right) + \frac{1}{V_{max}} \tag{8}$$

$$\frac{[S]}{v} = \left(\frac{1}{V_{max}} \times [S]\right) + \frac{K_m}{V_{max}} \tag{9}$$

Equation (8), for example, when plotted, gives a straight line with intercepts at $\frac{1}{V_{max}}$ on the ordinate and $-\frac{1}{K_m}$ on the abscissa. (The graph is usually referred to as the Lineweaver-Burk plot.) For illustrative purposes, the data for Figure 5-4 are recast in Lineweaver-Burk form in Figure 5-5. The Michaelis-Menten curve and its linearized plots are altered in the presence of inhibitors in a way that is described in a later section.

The value of the Michaelis constant has been used to compare the binding of homologous or related substrates to the same enzyme. Also, if measured against the same substrate under defined conditions, the K_m value can be used to compare the properties of similar enzymes from different sources. Isoenzymes determined by distinct genetic loci typically differ in their Michaelis constants, as has been shown, for example, for the isoenzymes of lactate dehydrogenase.

It is necessary to explore the relationship between reaction velocity and substrate concentration over a wide range when setting up methods of enzyme assay, to determine K_m, and to detect any inhibition at high substrate concentrations. Zero-order kinetics will be maintained if the substrate is present in large excess, i.e., at concentrations at least 10 and preferably 100 times that of the value of K_m. When $[S] = 10 \times K_m$, v is approximately 91% of the theoretical V_{max}. The K_m values for the majority of enzymes are of the order of 10^{-5} to 10^{-3} mol/L; therefore, substrate concentrations are usually chosen to be in the range of 0.001–0.10 mol/L. There are occasions when optimal concentrations of substrate cannot be used, for example, when the substrate has limited solubility, or when the concentration of a given substrate will inhibit the activity of another enzyme needed in a coupled reaction system.

Allosteric enzymes. Instead of the usual hyperbolic dependence of reaction velocity on substrate concentration, some polymeric enzymes show a sigmoid relationship between these variables (Figure 5-6). Thus the increase in v for a given increase in $[S]$ is lower than that expected from the Michaelis-Menten equation at low values of $[S]$ but greater than predicted at higher values of $[S]$. The nearly flat portion of the curve at low substrate concentrations essentially constitutes a threshold of substrate concentration, below which v is low and is little affected by changes in $[S]$. Around a critical concentration of substrate the enzyme responds markedly to changes in $[S]$, switching from very low to pronounced activity. Not surprisingly, therefore, enzymes that occupy rate-limiting positions in metabolic reaction sequences are often found to exhibit sigmoid kinetics.

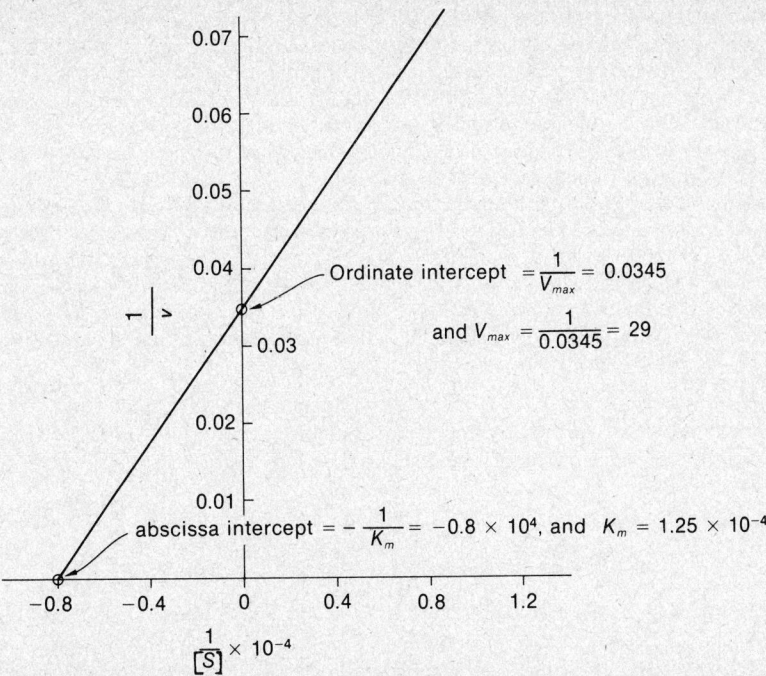

Figure 5-5. Lineweaver-Burk transformation of the curve in Figure 5-4, with $1/v$ plotted on the ordinate (Y-axis), and $1/[S]$ on the abscissa (X-axis). The indicated intercepts permit calculation of V_{max} and K_m. The units of v and $[S]$ are those given in Figure 5-4.

In general, the existence of a sigmoid curve for v against $[S]$ implies that more than one molecule of substrate binds to each molecule of enzyme and that binding of the second molecule is affected by the presence of the substrate molecule already bound, and so on until all the substrate-combining sites are filled. When the attachment of succeeding molecules is facilitated by those bound earlier, the interaction between them is described by the term "cooperativity."

The degree of sigmoidicity of the curve relating v and $[S]$ may be altered by the presence of an appropriate modifier. An inhibitor tends to make the curve more sigmoidal, increasing the concentration of substrate needed to produce a significant increase in v. On the other hand, an activator may bring the curve to a more hyperbolic shape (Figure 5-6). These modifiers combine with the enzyme at sites distinct

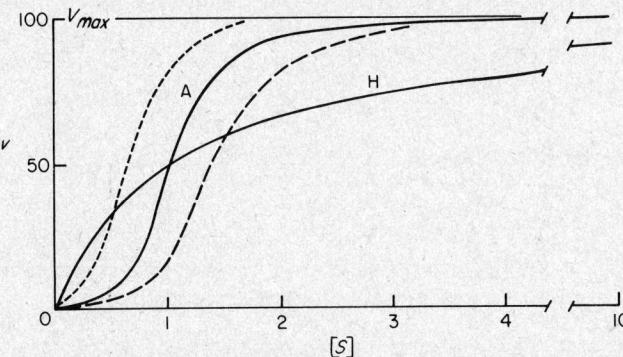

Figure 5-6. Dependence of v upon $[S]$ for an allosteric enzyme (curve A) compared with that for an enzyme showing typical Michaelis-Menten kinetics (H). Both have similar V_{max} values, but for H, velocity is greater at low substrate concentrations. Above a threshold value of $[S]$ of about 0.6 mmol $\times$ L^{-1}, however, velocity approaches V_{max} much more rapidly in the case of A as substrate concentration increases. Activators of A reduce the sigmoidicity of curve A (short-dashed curve); i.e., they lower the substrate concentration threshold, whereas inhibitors (long-dashed curve) increase the sigmoidicity and raise the threshold value. (From Moss, D. W.: Measurement of enzymes. *In*: Enzymes in Cardiology: Diagnosis and Research. D. J. Hearse and J. de Leiris, Eds. New York, John Wiley & Sons, 1979.)

from the substrate-binding site and have been termed allosteric modifiers to emphasize their supposed effects on the conformation of the enzyme molecules. The existence of different binding sites, with the implication of different specificity requirements for substrate and modifiers, has obvious advantages for the operation of a regulatory enzyme in a metabolic sequence. The modifier need not resemble the substrate of the enzyme in structure, as a competitive inhibitor would have to do; it could be the product of a later enzyme in the same sequence or even a molecule produced in a different sequence of reactions. Allosteric modifiers thus offer increased possibilities for control of metabolism.

The separate but identical multiple binding sites for substrate and the different binding sites for activators or inhibitors could arise by repeated sequences of amino acids in the primary structure of a single polypeptide chain. It is easier to envisage a set of protein subunits that associate with each other to form the active enzyme oligomer, each of which carries a substrate-binding site as well as activator and/or inhibitor-binding sites. All the enzymes exhibiting cooperativity that have so far been investigated have been found to be oligomeric, and most of the models proposed for the action of allosteric enzymes have assumed the existence of this quaternary level of protein structure.

Two-substrate reactions. Although the effect of changes in the concentration of only a single substrate on the rate of reaction has been discussed so far, most enzymatic reactions are of the type:

$$\overset{E}{\text{Substrate 1 + Substrate 2} \rightleftharpoons \text{Product 1 + Product 2}}$$
$$S_1 \qquad\qquad S_2 \qquad\qquad P_1 \qquad\qquad P_2$$

When one of the substrates is water (i.e., when the process is one of hydrolysis) with the reaction taking place in aqueous solution, only a fraction of the total number of water molecules present participates in the reaction. The small change in the concentration of water has no effect on the rate of reaction. More generally, however, the concentrations of both substrates may be variable and both may affect the rate of reaction. Among the bisubstrate reactions important in clinical enzymology are the reactions catalyzed by dehydrogenases, in which the second substrate is a specific coenzyme, NAD(H) or NADP(H), and the amino-group transfers catalyzed by the aminotransferases.

If a bisubstrate reaction proceeds by way of intermediate enzyme-substrate complexes, so that

$$E + S_1 \rightleftharpoons ES_1$$

followed by

$$ES_1 + S_2 \rightleftharpoons ES_1S_2 \rightarrow P_1 + P_2 + E$$

and if S_1 and S_2 combine with separate sites on the enzyme molecule, the rate of the reaction from left to right is given by

$$v = \frac{V_{max} \times [S_1][S_2]}{[S_1][S_2] + [S_2]K_m^1 + [S_1]K_m^2 + K_s^1K_m^2} \tag{10}$$

K_m^1 and K_m^2 are the Michaelis constants for the two substrates, and $[S_1]$ and $[S_2]$ are their concentrations. K_s^1 is the equilibrium constant for the reversible reaction between the enzyme and S_1. If the equation is rearranged into the double reciprocal form,

$$\frac{1}{v} = \frac{1}{[S_1]}\left(\frac{K_m^1}{V_{max}} + \frac{K_m^2 K_s^1}{[S_2]V_{max}}\right) + \frac{1}{V_{max}}\left(1 + \frac{K_m^2}{[S_2]}\right) \tag{11}$$

it can be seen that a plot of $1/v$ against $1/[S_1]$ will give a straight line, but both the slope of the line and its intercept on the ordinate are affected by $[S_2]$, the concentration of the second substrate (Figure 5-7,*a*). Similarly, a plot of $1/v$ against $1/[S_2]$ is rectilinear, but with the slope and intercept

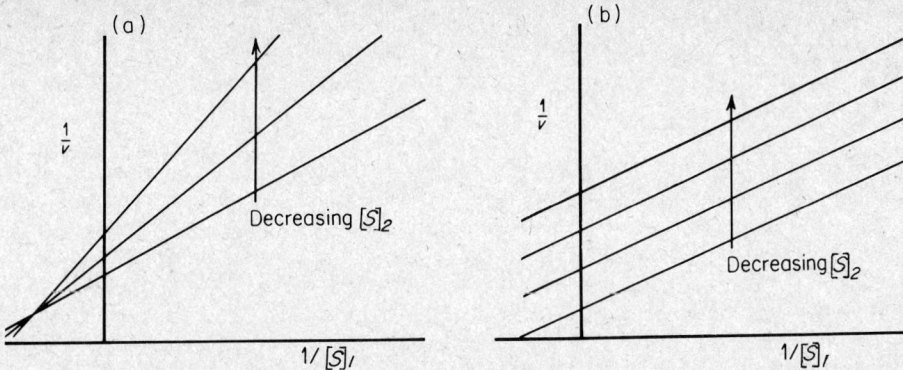

Figure 5-7. Double-reciprocal plots of $1/v$ against $1/[S_1]$ for two-substrate reactions, showing the effect of falling concentration of the second substrate, $[S_2]$. (a) In a dehydrogenase reaction in which a ternary complex is formed, and (b) in a ping-pong bi-bi reaction mechanism (e.g., aminotransferase) in which no ternary complex is formed. (From Moss, D. W.: Measurement of enzymes. *In*: Enzymes in Cardiology: Diagnosis and Research. D. J. Hearse and J. de Leiris, Eds. New York, John Wiley & Sons, 1979.)

dependent on $[S_1]$. If, however, the concentration of one substrate (e.g., S_2) is infinitely great, the rate equation reduces to the usual form for one-substrate reactions:

$$v = \frac{V_{max} \times [S_1]}{[S_1] + K_m^1} \tag{12}$$

Values of K_m and V_{max} for each substrate can be derived from experiments in which the concentration of the first substrate is held constant at saturating levels while the concentration of the second substrate is varied, and vice versa. There is of course no reason why the K_m values for the two substrates should be the same, or even similar; e.g., pyruvate and NADH, the two-substrate pair in the reaction catalyzed by lactate dehydrogenase of beef heart, have K_m values of 2×10^{-5} mol/L^{-1} and 3×10^{-6} mol/L^{-1}, respectively.

In some bisubstrate reactions no ternary complex, ES_1S_2 is formed, because the binding of the first substrate is followed by release of the first product before the second substrate is bound and the second product is released. This sequence is described as a "ping-pong bi-bi" type of reaction. It occurs in reactions catalyzed by aminotransferases.

The relationship between reaction velocity and the concentrations of the two substrates in ping-pong bi-bi reactions reduces to the form

$$v = \frac{V_{max} \times [S_1][S_2]}{[S_1][S_2] + [S_2]K_m^1 + [S_1]K_m^2} \tag{13}$$

The reciprocals of v and $[S_1]$ are related by the equation

$$\frac{1}{v} = \frac{1}{[S_1]} \cdot \frac{K_m^1}{V_{max}} + \frac{1}{V_{max}}\left(1 + \frac{K_m^2}{[S_2]}\right) \tag{14}$$

so that a plot of $1/v$ against $1/[S_1]$ is unchanged in slope by variation in $[S_2]$, but the intercept on the ordinate, and therefore the value of V_{max}, changes as $[S_2]$ is varied (Figure 5-7,b). Similar equations describe the variation of V_{max} with $[S_1]$ when $1/v$ is plotted as a function of $1/[S_2]$.

The selection of reaction conditions for the measurement of enzymatic activity involving two substrates can be approached empirically by varying the concentration of the first substrate while keeping the concentration of the second substrate constant, until maximum activity is reached. The process is then repeated with the concentration of the first substrate held at the value thus determined, while the concentration of the second substrate is varied. Many attempts have been made to define optimal substrate concentrations in this way for the measurement of clinically important enzymes, such as the aminotransferases. However, the recommended concentrations frequently differ, and indeed there is no reason why this experimental approach should give a single solution to the problem of optimization of substrate concentrations in two-substrate reactions.

Inspection of the rate equation for a ping-pong bi-bi reaction shows that a particular observed reaction velocity, expressed as a fixed fraction of the theoretical V_{max} value, will be given by an infinite number of pairs of values of the substrate concentrations $[S_1]$ and $[S_2]$. If the two Michaelis constants of the reaction, K_m^1 and K_m^2, are known, a plot of the concentration of the first substrate, S_1, against that of the second substrate, S_2, for all values of v which are the same fraction of V_{max} will be a hyperbola (Figure 5-8).

In practice the choice of substrate concentrations is limited by such considerations as the solubility of the substrates, the viscosity and high initial absorbance of concentrated solutions, and the relative costs of the reagents. Furthermore, the selection of appropriate substrate concentrations is only one of the factors to be considered in formulating an optimal assay system for the measurement of a specific enzyme activity: critical choices must also be made with respect to other, frequently interdependent factors that affect reaction rate, such as the concentrations of activators and the nature and pH of the buffer system. In recent years the traditional empirical approach to optimization has been replaced by newer techniques of simplex co-optimization or response-surface methodology.

Consecutive enzymatic reactions. All reactions catalyzed by enzymes are in theory reversible, the enzyme catalyzing the forward and backward reactions equally. However, in practice, the reaction is usually found to be more rapid in one direction than the other, so that an equilibrium is reached in which the product of either the forward or the backward reaction predominates, sometimes so markedly that the reaction is virtually irreversible.

If the product of the reaction in one direction is removed as it is formed, e.g., because it is the substrate of a second enzyme present in the reaction mixture, the equilibrium of the first

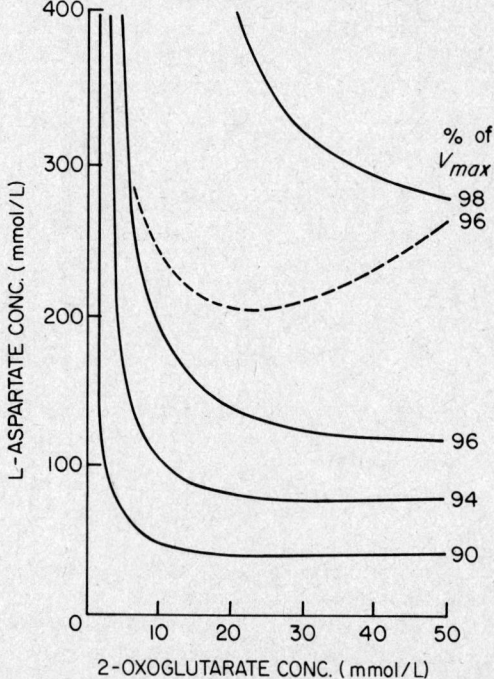

Figure 5-8. Relationship between calculated reaction velocity (expressed as a percentage of the theoretical V_{max}) and concentration of the two substrates, L-aspartate and 2-oxoglutarate, for human heart aspartate aminotransferase. The solid curves make no allowance for inhibition by 2-oxoglutarate; when this is taken into account, the broken line (showing the effect on the 96% V_{max} curve as an example) is obtained. (From Moss, D. W.: Measurement of enzymes. *In*: Enzymes in Cardiology: Diagnosis and Research. D. J. Hearse and J. de Leiris, Eds. New York, John Wiley & Sons, 1979.)

enzymatic process will be displaced so that the reaction may go to completion in that direction. Reaction sequences in which the product of one enzyme-catalyzed reaction becomes the substrate of the next enzyme and so on, often through many stages, are characteristic of biological processes. In the laboratory also, several enzymatic reactions may be linked together in order to provide a means of measuring the activity of the first enzyme or the concentration of the initial substrate in the chain. For example, the activity of creatine kinase is usually measured by a series of linked reactions, and glucose can be determined by consecutive reactions catalyzed by hexokinase and glucose-6-phosphate dehydrogenase.

For such a series of consecutive reactions, a Michaelis-Menten curve can be constructed for each enzyme taking part. The maximum velocity (V_{max}) of each stage in the overall process will depend on the amount of the particular enzyme catalyzing that stage that is present in the system. Therefore, however much substrate is introduced into the system at the beginning of the sequence, the rate of appearance of end-products will not increase beyond the limit set by the lowest of the several values of V_{max}. The overall rate of conversion by the system will increase only if the value of V_{max} for the rate-limiting enzyme is raised by adding more enzyme. However, if the concentration of the first substrate and the rates of the succeeding reactions are such that none of the enzymes is saturated with its substrate, an increase in the concentration of the primary substrate will result in a greater rate of appearance of products.

When a secondary enzyme-catalyzed reaction, known as an indicator reaction, is used to determine the activity of a different enzyme, it is essential that the primary reaction should be the rate-limiting step. For example, in the determination of aspartate aminotransferase activity, the indicator reaction is the reduction of the oxaloacetate formed in the aminotransferase reaction to malate by malate dehydrogenase and NADH. The activity of the indicator enzyme must be sufficient to ensure the virtually instantaneous removal of the product of the first reaction, to prevent significant reversal of the first reaction. The measured enzyme is typically acting under conditions of saturation with respect to its substrate; however, the concentration of the substrate of the indicator enzyme (i.e., the product of the first reaction) remains in the region of the Michaelis-Menten curve in which v is directly proportional to $[S]$. Therefore, the rate of reaction catalyzed by the indicator enzyme is directly proportional to the rate of product formation in the first reaction.

A lag period occurs after the start of the first reaction during which the concentration of its product reaches a steady state. Since the rate of the second reaction depends on the activity of the indicator enzyme as well as on the concentration of its substrate (i.e., the product of the primary reaction), the duration of the lag period is reduced by increasing the concentration of the indicator enzyme, thus lowering the steady-state concentration of the product of the primary reaction.

The rate of the indicator reaction, v_i, is related to substrate concentration and therefore to the product concentration, $[P]$, by the Michaelis-Menten equation

$$v_i = \frac{V^i_{max} \times [P]}{[P] + K^i_m} \tag{15}$$

in which V^i_{max} and K^i_m are the maximum velocity and Michaelis constant of the indicator enzyme. For the rate of the indicator reaction to match that of the primary reaction, v_i must equal the limiting velocity of the primary reaction, v_t, which the assay system is expected to measure. Therefore, the activity of indicator enzyme needed is given by

$$v_t = \frac{V^i_{max} \times [P]}{[P] + K^i_m} \tag{16}$$

or, rearranged,

$$V^i_{max} = v_t \left(1 + \frac{K^i_m}{[P]}\right) \tag{17}$$

For example, consider a coupled assay designed to measure up to 400 μmol min^{-1} $\times$ L^{-1} of aspartate aminotransferase activity in serum, corresponding to 48 μmol min^{-1} $\times$ L^{-1} when diluted in the assay mixture. If a steady-state concentration of oxaloacetate of 1 mol $\times$ L^{-1} is chosen for [P], the appropriate activity of the indicator enzyme malate dehydrogenase (K_m^i ~15 μmol $\times$ L^{-1} oxaloacetate) would be

$$V_{max}^i = 48\left(1 + \frac{15}{1}\right) = 768 \ \mu\text{mol min}^{-1} \times \text{L}^{-1} \text{ of assay mixture.}$$

The ratio of activities of the indicator and primary enzymes will vary from one assay method to another depending on the range of activity which it is desired to measure, the Michaelis constant of the indicator enzyme, and the lag period which is considered acceptable. Nevertheless, the catalytic concentration of the indicator enzyme in the reaction mixture must always be much greater than that of the enzyme being determined.

Effect of pH on Reaction Rate

If an enzyme-catalyzed reaction is carried out at a series of different pH values with the concentration of enzyme and substrate kept constant, a marked dependence of reaction rate on pH is observed. When enzyme activity (rate) is plotted against pH, curves of the type presented in Figure 5-9 are obtained. Similar curves relating reaction rate and pH can be obtained for all enzymes, although the pH at which maximum activity occurs and the shape of the curve vary from one enzyme to another and with such conditions of measurement as choice of buffer species (e.g., Figure 5-9). Many of the enzymes in blood plasma show maximum activity in vitro somewhere in the pH range from 7–8. However, activity can be observed at pH values as low as 1.5 (pepsin) and as high as 10.5 (alkaline phosphatase). The optimal pH for a given reaction may be different from the optimal pH found for the reverse reaction.

The form of the pH-dependence curve is a result of a number of separate effects. The enzyme may catalyze the reaction only if the substrate is in either the undissociated or dissociated form, and the extent of dissociation is dependent on the pH of the reaction system and the pK of the dissociating acid or base group of the substrate. The activity of the enzyme itself is also affected by the extent of dissociation of certain key amino acid side-chains in the protein molecule, both

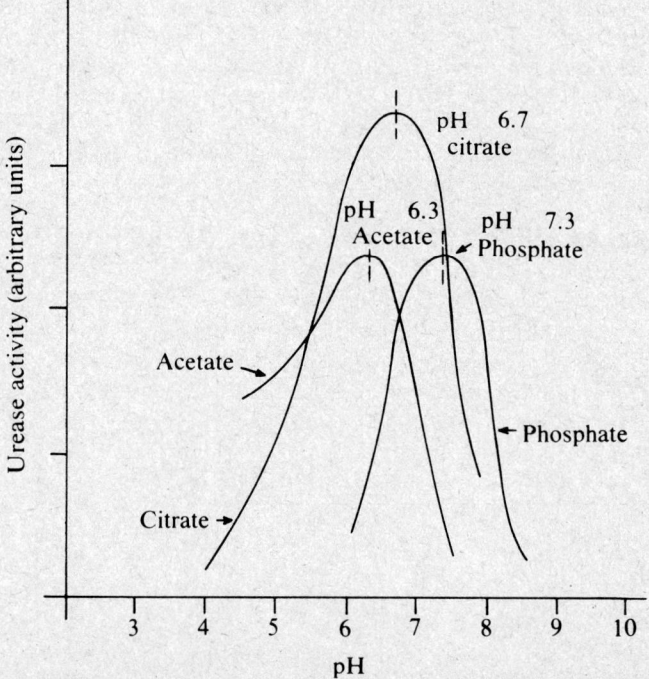

Figure 5-9. pH activity curves for urease, showing effect of buffer species on pH optimum. (Adapted from Howell and Sumner: J. Biol. Chem., *104*:619, 1934.)

at the "active center" and elsewhere in the molecule. Both pH and ionic environment will have an effect on the three-dimensional conformation of the protein and therefore on enzyme activity. At extreme values of pH, enzymes may be irreversibly denatured. The extent of denaturation depends on the temperature and the period of exposure to the extreme pH. The active forms of many enzymes consist of associations of two or more individual peptide chains; at extreme values of pH these associations are disrupted, with ensuing loss of catalytic activity.

The pronounced effects of pH on enzyme reactions emphasize the need always to control this variable by means of adequate buffer solutions or by use of a pH-stat. Enzyme assays should be carried out at the pH of optimal activity because the sensitivity of the measurement is maximal at this pH, and because the pH-activity curve usually has a minimum slope or plateau near this pH so that a small variation in pH will cause a minimal change in enzyme activity. The buffer system must be capable of counteracting the potential effect on pH caused by addition of the enzyme sample (e.g., serum itself is a powerful buffer) to the assay system, as well as the effects of acids or bases formed during the reaction (e.g., formation of fatty acids by the action of lipase, or formation of ammonia from urea by urease). Since buffers have their maximum buffering capacity close to their pK_a values, whenever possible a buffer system should be chosen with a pK_a value within 1 pH unit of the desired pH of the assay. Interaction between certain buffer ions and other components of the assay system (e.g., activating metal ions) may eliminate these buffers from consideration. Although substrates such as amino acids in aminotransferase reactions have buffering properties, their concentrations are usually too low to control the pH in the complete assay system. Exceptions are peptides and amino alcohols that act as amino acid or phosphate acceptors in the reactions catalyzed by γ-glutamyltransferase and alkaline phosphatase, respectively, at concentrations high enough to provide most or all of the necessary buffering capacity.

When plots of the variation of velocity with substrate concentration are made at different pH values for a given amount of an enzyme, it is usually found that the Michaelis constant, as well as the maximum velocity, is markedly dependent on pH. Thus both binding of the substrate by the enzyme and the rate of breakdown of the enzyme-substrate complex are affected by changes in pH. The probable identity of groups present in the active center has been inferred from a study of pH-activity relationships for a number of enzymes. When K_m has a marked dependence on pH, i.e., when the curves of v against $[S]$ vary in shape with pH, the observed pH optimum depends on the substrate concentration. Therefore, the optimum pH for an enzyme assay must be determined or redetermined after the appropriate substrate concentration has been selected.

Effect of Temperature on Reaction Rate

The rate of any chemical reaction increases as the temperature at which the reaction is taking place increases. Enzymatic reactions are no exception. For most enzymatic reactions values of Q_{10} vary from 1.7–2.5. The relation between the rate of a chemical reaction and the absolute temperature is given by the empirical equation of Arrhenius:

$$2.303 \frac{d(\log k)}{dt} = \frac{E}{RT^2} \qquad (18)$$

in which k is the rate constant for the reaction, T the absolute temperature, and R the gas constant ($1.987 \text{ cal} \times \text{deg}^{-1} \times \text{mol}^{-1}$). E is also a constant for the particular reaction, called the energy of activation. Integrating the Arrhenius equation,

$$\log k = \frac{-E}{2.303RT} + \text{constant} \qquad (19)$$

Since the maximum velocity is given by $k \times [E]$, a plot of $\log V_{max}$ against $1/T$ is linear, of slope $-E/2.303R$. The energy of activation of the reaction can thus be calculated. Comparison of the values of E obtained for an enzymatic reaction in the presence and absence of the catalyst (if the reaction proceeds at a measurable rate without the enzyme) shows that the energy of activation is much lower when the enzyme is present. Therefore, the enzyme increases the rate of reaction by lowering the energy barrier that otherwise separates the reactants and the products.

However, an increase in the rate of the catalyzed reaction is not the only effect of increasing temperature on an enzymatic reaction. The initial rate of reaction measured instantaneously goes on increasing with rising temperature, at least in theory. In practice, however, a finite time is needed in all methods to allow the components of the reaction mixture, including the enzyme solution, to reach temperature equilibrium and to permit the formation of a measurable amount of product. During this period the enzyme is undergoing thermal inactivation and denaturation. This process has a very large temperature coefficient for most enzymes and thus becomes virtually instantaneous at temperatures in the region of 60–70 °C.

The counteracting effects of the increased rate of the catalyzed reaction and more rapid enzyme inactivation as the temperature is raised account for the existence of an apparent "optimum temperature" for enzyme activity (Figure 5-10). The apparent optimum temperature depends on the time taken to make the activity measurement. With older fixed-time assay methods that require lengthy periods of incubation of enzyme with substrate, enzyme inactivation takes effect at lower temperatures and the phenomenon is more easily seen.

Thermal inactivation of enzymes is influenced by other factors. These include the presence of substrate and its concentration, the pH, and the nature, as well as the ionic strength, of the buffer. The presence of other proteins, as in serum samples, may help to stabilize the enzyme. Storage of serum samples at low temperatures is necessary to minimize loss of enzyme activity while awaiting analysis, although repeated freezing and thawing should be avoided. However, individual enzymes vary in their stability characteristics and appropriate storage conditions vary correspondingly. Amylase, for example, is stable at room temperature (22–25 °C) for 24 h, whereas acid phosphatase is exceedingly unstable, even when refrigerated, unless kept at a pH below 6.0. Alkaline phosphatase exhibits an unusual property: the tendency for the activity of frozen, partially purified preparations of the enzyme to increase after thawing over a period of 24 h or longer. This effect is shared by reconstituted, lyophilized preparations of the enzyme and affects their use for quality-control purposes. A few enzymes are inactivated at refrigerator temperatures; clinically important examples are the liver-type isoenzymes of lactate dehydrogenase, LDH-4 and LDH-5, which appear to be less stable at lower temperatures. As a result, sera for lactate dehydrogenase determinations should be kept at room temperature and not refrigerated.

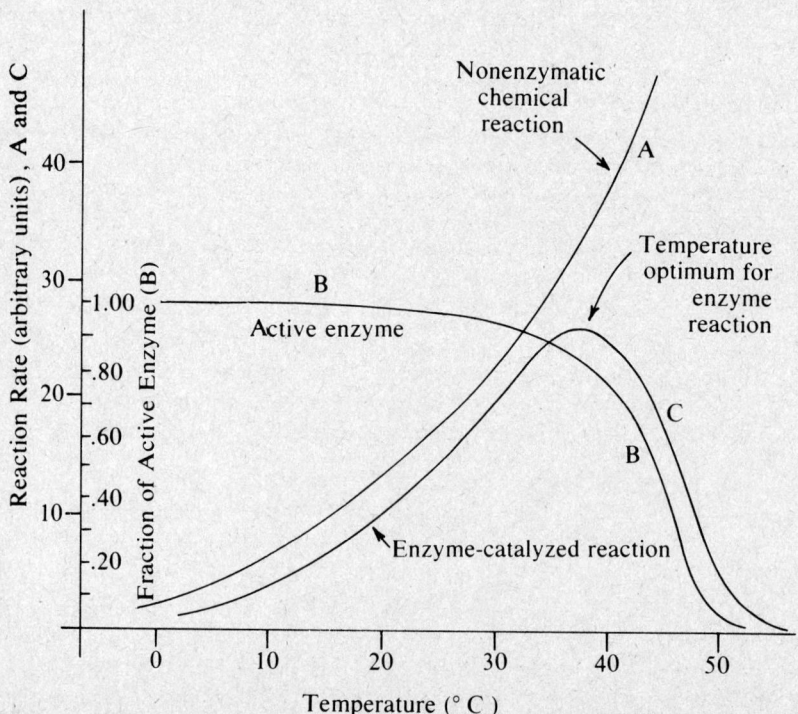

Figure 5-10. Schematic diagram showing effect of temperature on rate of nonenzyme-catalyzed and enzyme-catalyzed reactions.

The choice of temperature for the assay of enzymes of clinical importance has been the subject of extensive debate. Temperatures close to the ambient temperature, e.g., 25 or 30 °C, are more suitable in manual or semiautomated continuous-monitoring enzyme assays than a higher temperature such as 37 °C. In such methods, temperature fluctuations are magnified and equilibration periods are prolonged during the necessary repeated openings of the cuvet compartment of the photometer when a large difference exists between the ambient and measurement temperatures. This objection to the use of higher temperatures does not usually apply to automated methods of assay in which samples are processed within a controlled environment, nor in the analysis of batches of specimens by fixed-time methods. Therefore, 37 °C is often chosen in these circumstances. The greater rate of reaction at 37 °C compared with 25 or 30 °C increases the sensitivity of the assay, shortening the time required for measurement, and this also favors the use of the higher temperature. However, enzyme inactivation and deterioration of substrates and coenzymes are also more rapid.

Whatever temperature is chosen, precise temperature control to within ± 0.1 °C during the enzymatic reaction is essential. Heat is lost more rapidly to the surrounding air at 37 °C than at lower temperatures, and this heat loss is frequently regarded as an advantage. However, cycles of passive heat loss alternating with periods of heating are unlikely to meet tightening specifications for temperature control. Active cooling by means of a refrigeration unit or by the Peltier effect will therefore become increasingly necessary.

It is necessary that the chosen assay temperature should be accurately set, as well as closely maintained, during enzyme assay. An accurate temperature standard that can readily be transferred from one laboratory to another is provided by the gallium melting-point cell.[91] Pure gallium melts at 29.77 °C.

The Expert Panel on Enzymes of the IFCC has recommended 30 °C as the standard temperature for *reference* methods in clinical enzymology, principally because this temperature is close to the standard temperatures at which other physicochemical calibrations are carried out.[36] Although referring primarily to reference methodology, this recommendation has influenced various national bodies to suggest that 30 °C should become a standard temperature for routine analysis also. These suggestions have not found universal support, however, and differences in assay temperature continue to be a major cause of lack of agreement between results for a particular enzyme reported from different laboratories.

It seems probable that universal use of a single assay temperature in routine clinical enzymology will never be achieved, because of the different practical considerations that apply in different laboratories. However, the choice of assay temperature will probably be restricted in any laboratory to one of two alternatives, 30 or 37 °C. This development directs attention to the validity of temperature conversion factors as a means of comparing results obtained at one temperature with those at the other temperature.

The temperature coefficients of enzyme activity can be determined accurately for a single sample, such as a single serum. However, the question then arises: can this same factor be applied to *all* serum samples containing that enzyme? Attempts to answer this question have led to conflicting results, some of which are due to the use of inappropriate experimental design or of methods that were not optimal at one or the other of the chosen temperatures. However, recent work shows that, provided optimized methods are used and the temperature difference is not too great, temperature coefficients valid for human sera can be determined for enzymes of clinical interest.[118] Coefficients for 30–37 °C conversions are therefore given in this chapter.

It remains preferable whenever possible to report results of enzyme assays as obtained without the application of conversion factors, unless comparison with other results is required. Such comparisons are further facilitated by including appropriate reference ranges.

Inhibition and Activation of Enzyme Activity

The rates of enzymatic reactions are often found to be affected by changes in the concentrations of substances other than the enzyme or substrate. These modifiers may be activators, i.e., they increase the rate of reaction; or they may be inhibitors, i.e., their presence reduces the reaction rate. Activators and inhibitors are usually small molecules (compared with the enzyme itself) or even ions. They vary in specificity from modifiers that exert similar effects on a wide range of different enzymic reactions at one extreme, to substances that affect only a single reaction at the other extreme. Reagents such as strong acids or multivalent anions and cations, which denature or precipitate proteins, destroy enzyme activity and thus may be regarded as extreme examples of nonspecific enzyme inhibitors. However, these effects, which depend on the properties of enzymes as proteins rather than as catalysts, are not usually included in discussions of enzyme inhibition, although they have obvious practical implications in the treatment and storage of specimens in which enzyme activity is to be measured. The activity of some enzymes depends on the presence of particular chemical groups, such as reduced sulfhydryl (-SH) groups, in the active center. Reagents that alter these groups (e.g., oxidants of SH groups) therefore act as general inhibitors of such enzymes.

Some phenomena of enzyme activation or inhibition are due to interaction between the modifier and a nonenzymic component of the reaction system, such as the substrate. In most cases, however, the modifier combines with the enzyme itself, in a manner analogous to the combination of enzyme and substrate.

Inhibition of Enzyme Activity

Inhibitors are classified into reversible and irreversible types. *Reversible inhibition* implies that the activity of the enzyme is fully restored when the inhibitor is removed from the system in which the enzyme acts by some physical separative process such as dialysis, gel filtration, or chromatography. An *irreversible inhibitor,* on the other hand, combines covalently with the enzyme so that physical methods are ineffective in separating the two. For example, organophosphorus compounds are extremely potent irreversible inhibitors of esterases, including acetylcholinesterase. The enzyme breaks one of the bonds in the inhibitor, but part of the molecule is left bound to the active center of the enzyme, preventing further activity. In some cases enzymes that have combined with irreversible inhibitors can be reactivated by a chemical reaction which removes the blocking group; e.g., the phosphoryl enzymes formed with organophosphorus compounds can sometimes be reactivated by treatment with oximes or hydroxamic acids.

A physiologically important category of irreversible enzyme inhibition is produced by the *anti-enzymes,* exemplified by a variety of trypsin inhibitors. These are proteins that bind to trypsin irreversibly, nullifying its proteolytic activity. One such inhibitor is present in the α_1-globulin fraction of serum proteins; others are found in soy beans and lima beans. Similar proteolysis inhibitors present in plasma prevent the accumulation of excess thrombin and other coagulation enzymes, thus keeping the coagulation process under control.

Characteristics of irreversible inhibition. An irreversible inhibitor is not in equilibrium with the enzyme. Its effect is progressive with time, becoming complete if the amount of inhibitor present exceeds the total amount of enzyme. The rate of the reaction between enzyme and inhibitor is expressed as the fraction of the enzyme activity that is inhibited in a fixed time by a given concentration of inhibitor. The velocity constant of the reaction of the inhibitor with the enzyme is a measure of the effectiveness of the inhibitor.

When the inhibitor is added to the enzyme in the presence of its substrate, the reaction between the enzyme and inhibitor may be delayed because some of the enzyme molecules will be combined with the substrate and therefore protected from reacting with the inhibitor. However, as the substrate molecules are broken down, the active centers become available for combination with the inhibitor. Thus inhibition will eventually become complete even though an excess of substrate may initially have been present. Furthermore, addition of more substrate is ineffective in reversing the inhibition, in contrast to its effect on reversible competitive inhibition discussed later.

Kinetics of reversible inhibition. Reversible inhibition is characterized by the existence of an equilibrium between enzyme, E, and inhibitor, I:

$$E + I \rightleftharpoons EI$$

The equilibrium constant of the reaction, K_i (the *inhibitor constant*), is a measure of the affinity of the inhibitor for the enzyme, just as the value of the Michaelis constant, K_m, generally reflects the affinity of the enzyme for its substrate.

A *competitive* inhibitor is usually a structural analog of the substrate and binds to the enzyme at the substrate-binding site, but because it is not identical with the substrate, breakdown into products does not take place. When the process of inhibition is fully competitive, the enzyme can combine with either the substrate or the inhibitor but not with both simultaneously. Two equilibria are therefore possible:

$$E + S \rightleftharpoons ES \rightarrow E + \text{products}$$

and

$$E + I \rightleftharpoons EI$$

An equation can be derived that relates the observed reaction velocity to the concentrations of substrate, $[S]$, and inhibitor, $[I]$:

$$v = \frac{V_{max}[S]}{[S] + K_m\left(1 + \dfrac{[I]}{K_i}\right)} \tag{20}$$

This is the Michaelis-Menten equation, but with K_m modified by a term including the inhibitor concentration and inhibitor constant. V_{max} is unaltered. Therefore, curves of v against $[S]$ in the presence and absence of inhibitor reach the same limiting value at high substrate concentrations, but when the inhibitor is present, K_m is apparently greater. Plots of $1/v$ against $1/[S]$ with and without inhibitor cut the ordinate at the same point but have different slopes and intercepts on the abscissa (Figure 5-11).

Competitive inhibition accounts for the toxic effect of various compounds; e.g., sulfonamides, which are structural analogs of p-aminobenzoic acid, prevent the growth of bacteria that need the latter compound by a process of competitive inhibition of enzymes required for the metabolism of p-aminobenzoic acid. Competitive inhibition is also responsible for the inhibition of some enzymes by excess substrate, because of competition between substrate molecules for a single binding site. In two-substrate reactions, high concentrations of the second substrate may compete with the binding of the first substrate. For example, aspartate aminotransferase is inhibited by

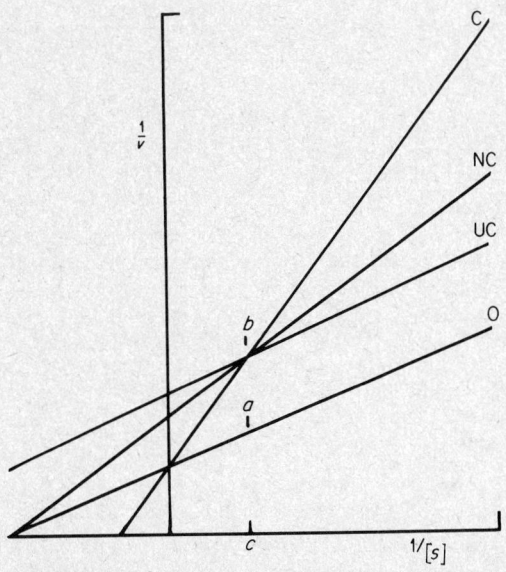

Figure 5-11. Effects of different types of inhibitors on the double-reciprocal plot of $1/v$ against $1/[S]$. Each of the inhibitors has been assumed to reduce the activity of the enzyme by the same amount, represented by the change in $1/v$ from a to b at a substrate concentration of c. Line O is the plot for enzyme without inhibitor, C with a competitive inhibitor, NC with a noncompetitive inhibitor, and UC with an uncompetitive inhibitor. (From Moss: D. W.: Measurement of enzymes. *In*: Enzymes in Cardiology: Diagnosis and Research. D. J. Hearse and J. de Leiris, Eds. New York, John Wiley & Sons, 1979.)

excess concentrations of the substrate 2-oxoglutarate and this inhibition is competitive with respect to L-aspartate. Therefore, to maintain a given velocity at high 2-oxoglutarate concentrations, the concentration of L-aspartate has to be increased above the value needed at lower concentrations of 2-oxoglutarate. This requirement has the effect of altering the shape of the line linking pairs of substrate concentrations that correspond to a given reaction velocity from a hyperbola to a parabola (Figure 5-8).

Competitive inhibition can also contribute to the reduction of the rate of an enzymatic reaction with time. For example, a rate reduction can occur because increasing concentrations of reaction products tend to drive the reaction backward, if it is freely reversible. A product may itself be an inhibitor of the forward reaction so that, even if the reaction is not readily reversible, it proceeds against a rising concentration of inhibitor. A familiar example of "product inhibition" is the release of the competitive inhibitor, inorganic phosphate, by the action of alkaline phosphatase on its substrates. In this case both organic phosphates and inorganic phosphate bind to the active center of the enzyme with similar affinities; i.e., K_m and K_i are of the same order of magnitude.

Product inhibition is a cause of nonlinearity of reaction progress curves during fixed-time methods of enzyme assay. For example, oxaloacetate produced by the action of aspartate aminotransferase inhibits the enzyme, particularly the mitochondrial isoenzyme. The inhibitory product may be removed as it is formed by a coupled enzymatic reaction: malate dehydrogenase converts the oxaloacetate to malate, and at the same time oxidizes NADH to NAD^+.

Competitive inhibition by metal ions can arise when two metal ions compete for the same binding site on the enzyme. Thus, Ca^{2+} is an inhibitor for some enzymes that depend on Mg^{2+} activation. An example of this occurs with creatine kinase: inhibition by the Ca^{2+} ions of serum can be reduced by the addition of EDTA to assay mixtures. Because of the lower affinity of EDTA for Mg^{2+}, activation by this ion still takes place. Sodium and lithium are potent inhibitors of pyruvate kinase, for which potassium is an obligatory activator.

A *noncompetitive* inhibitor usually is unlike the substrate in structure. It is assumed to bind at a site on the enzyme molecule different from the substrate-binding site; thus there is no competition between inhibitor and substrate, and a ternary enzyme-inhibitor-substrate complex can form. Attachment of the inhibitor to the enzyme does not alter the affinity of the enzyme for its substrate, but the enzyme-inhibitor-substrate complex does not break down to give products. The following equilibria are therefore possible:

$$E + S \rightleftharpoons ES \rightarrow E + \text{products}$$

$$E + I \rightleftharpoons EI$$

$$EI + S \rightleftharpoons EIS$$

$$\text{and } ES + I \rightleftharpoons EIS$$

Since S and I do not interfere with the binding of each other, the rate constants for the combination of S with EI are the same as those for the binding of S with E. Similarly, the same rate constants apply to the reaction of I with ES as with E.

Derivation of an expression for v gives the following:

$$v = \frac{V_{max} \times [S]}{(K_m + [S])\left(1 + \dfrac{[I]}{K_i}\right)} \tag{21}$$

The effect of a noncompetitive inhibitor is therefore to divide V_{max} by a factor of $(1 + [I]/K_i)$. Plots of v against $[S]$ have the same shape in the presence or absence of the inhibitor, but do not reach the uninhibited value of V_{max}, however great the substrate concentration. Double-reciprocal plots of $1/v$ against $1/[S]$ are altered in both slope and intercept on the ordinate by the presence of the inhibitor (Figure 5-11).

An example of noncompetitive enzyme inhibition is the inactivation of enolase by fluoride (F^-) ion. In one step in the metabolism of glucose, enolase catalyzes the conversion of 2-phosphoglycerate to phospho-enolpyruvate. By inhibiting this reaction, F^- stops the glycolytic breakdown of glucose by preventing formation of pyruvate. This is the biochemical rationale for the use of F^- as a preservative in specimens to be used for glucose determinations. Inhibition of enzymes by heavy metal ions is usually noncompetitive in character.

A rather unusual type of reversible inhibition has been named *uncompetitive inhibition*. In this type of inhibition, parallel lines are obtained when plots of $1/v$ against $1/[S]$ with and without the inhibitor are compared (Figure 5-11), i.e., the slope remains constant but the intercept on the ordinate is altered by the presence of the inhibitor. The effect on the rate equation is to divide both V_{max} and K_m by the factor $(1 + [I]/K_i)$.

Uncompetitive inhibition is rare in single-substrate reactions. An example of importance in clinical enzymology is the inhibition of intestinal and placental alkaline phosphatases by L-phenylalanine. It can occur in such reactions when the inhibitor binds to the *ES* complex but not to the free enzyme. A ternary complex *ESI* can form, but only after the enzyme has combined with the substrate. Uncompetitive inhibition is more common in two-substrate reactions.

Plots of $1/v$ against $1/[S]$ for the inhibited and uninhibited reactions are often found not to fit any of the three patterns described above. Instead, the lines intersect to the left of the ordinate but above the abscissa. These cases are usually interpreted as implying that the inhibitor interferes with both the binding of the substrate to the enzyme and its subsequent breakdown; the effect is therefore referred to as *mixed inhibition*.

Inhibition by antibodies. Combination of enzyme molecules with specific antibodies often has no effect on catalytic activity, which is retained by the enzyme-antibody complex. However, in some cases reaction of the enzyme and antibody reduces or even abolishes enzymatic activity. The most probable explanation for this type of inhibition is that the antibody molecule restricts access of the substrate molecules to the active center by steric hindrance, or in extreme cases completely masks the substrate-binding site. However, it appears that some examples of enzyme inhibition by combination with antibodies are due to a conformational change induced in the enzyme molecule.

Inhibition of the activity of an enzyme molecule labeled with a hapten (e.g., morphine) as a result of combination with a specific antibody is the basis of homogeneous enzyme immunoassay (EMIT, Syva Corp., Palo Alto, CA).

Enzyme Activation

Activators are considered to increase the rates of enzyme-catalyzed reactions by promoting formation of the most active state either of the enzyme itself or of other reactants such as the substrate. This generalization covers a wide variety of mechanisms of activation.

Many enzymes contain metal ions as an integral part of their structures, e.g., zinc in alkaline phosphatase and carboxypeptidase A. The function of the metal may be to stabilize tertiary and quaternary protein structure. Removal of the divalent metal ions by treatment with an appropriate concentration of EDTA solution is accompanied by conformational changes with inactivation of the enzyme. The enzyme can often be reactivated by dialysis against a solution of the appropriate metal ion, or simply by adding the ion to the reaction mixture. Reactivation may take some time, because rearrangement of the polypeptide chains into the active conformation is not instantaneous.

The metal ion components of many enzymes appear to play a direct part in catalysis, in addition to any possible structural role they may fulfill. A metal ion may function in catalysis, e.g., by providing an electropositive center in the enzyme with which negatively charged groups in the substrate can form coordinate links.

When the activator ion is an essential part of the functional enzyme molecule, whether as a purely structural element or with an additional catalytic role, it is usually incorporated quite firmly into the enzyme molecule. Therefore, it is not usually necessary to add the activator to reaction mixtures and excess of the ion may even have an inhibitory effect. However, in many cases the activating ion is attached only weakly or transiently to the enzyme (or its substrate) during catalysis. Enzyme samples may therefore be deficient in the ion, so that addition of the ion increases the reaction rate or indeed may be essential for the reaction to take place. For example, all phosphate transfer enzymes (kinases) such as creatine kinase require the presence of Mg^{2+} ions. Without magnesium, kinases are inactive; i.e., the metal ion is an essential activator. Other common activating cations are Mn^{2+}, Fe^{2+}, Ca^{2+}, Zn^{2+}, and K^+. More rarely, anions may act as activators. Amylase will function at its maximal rate only if Cl^- or other monovalent anions such as Br^- or NO_3^- are present. This process probably involves a reversible combination of chloride with the enzyme. Addition of 5 mmol $\times$ L^{-1} of chloride increases amylase activity almost three-fold, at the same time shifting the pH optimum from 6.5 to 7.0. The chloride ion may combine with a positively charged group in the enzyme, changing the ionization constant of a group important in catalysis. However, other anions such as bromide are less effective activators of amylase, so that some degree of specificity is involved in the process of activation. Some enzymes require the obligate presence of two activating ions. K^+ and Mg^{2+} are essential for the activity of pyruvate kinase, and both Mg^{2+} and Zn^{2+} are required for alkaline phosphatase activity.

The velocity of the reaction depends on the concentration of a reversible activator in a fashion similar to its dependence on substrate concentration, and an activator constant K_a, analogous to the Michaelis constant K_m, can be determined from data relating enzyme activity to increasing activator concentration in the presence of excess substrate. The simplest interpretation of K_a is that it is the dissociation constant of the equilibrium between E and the activator, A. However, this is only true when the combination of enzyme and activator is independent of the reaction between E and S, and the same value for K_a is obtained at all concentrations of the substrate. If the free enzyme and the enzyme-substrate complex have different affinities for the activator, the value for K_a varies with $[S]$.

It is important that activator-dependent enzyme reactions be performed in the presence of both excess activator and excess substrate. In many cases, however, the addition of activator beyond a certain optimal concentration may result in a decrease in reaction rate (inhibition by excess activator). In such cases the optimal concentration of activator must be used.

The mechanisms by which cations and anions activate enzymes vary from case to case. For example, the ion may alter the spatial configuration of the protein to allow for proper binding of the substrate to the enzyme. This is the most likely explanation of the activity of monovalent ions such as K^+ and Cl^-. Other metal ions are involved in electron transfer reactions and undergo oxidation and reduction, as in the case of iron ($Fe^{2+} \rightleftarrows Fe^{3+}$) in cytochrome C oxidase. For some enzymes combination of the apparent substrate with a metal ion may be necessary before full, or even any, catalytic activity is observed. In these cases the true substrate of the enzyme is the metal-substrate complex, and it is the concentration of this complex which influences the rate of reaction. In many reactions catalyzed by ATPases and other phosphate-transferring enzymes, e.g., creatine kinase, there seems to be obligatory formation of complexes between the substrate and magnesium ions. In these cases the metal ion may act as a bridge between the substrate and the enzyme, or it may alter the configuration of the substrate, or it may neutralize ionic charges in the substrate that would otherwise hinder the approach of the substrate to the active center.

In a broader sense, the term "activation" can be applied to any process whereby an inactive enzyme is made catalytically active. The proteolytic enzymes are synthesized in the body in the form of inactive precursors, termed "proenzymes" or "zymogens," which are then transformed by chemical agents into the active enzymes. Trypsin is formed from its precursor, trypsinogen, when a hexapeptide fragment is split from the amino terminal end of the protein by the action of enterokinase or by active trypsin. Similarly, H^+ ions in the stomach convert pepsinogen to the active gastric enzyme, pepsin.

Apparent activation of an enzyme may be observed whenever a substance is added that can counteract the presence of some inhibiting agent.

Coenzymes and Prosthetic Groups

Coenzymes are usually more complex molecules than activators, although smaller molecules than the enzyme proteins themselves. Some compounds that are classified as coenzymes (e.g., the dinucleotides NAD and NADP) are specific substrates in two-substrate reactions, and their effect on the rate of reaction follows the Michaelis-Menten pattern of dependence on substrate concentration. The structures of these two coenzymes are identical except for the presence of an additional phosphate group in NADP (Figure 5-12); nevertheless, individual dehydrogenases, for which these coenzymes are substrates, are predominantly or even absolutely specific for one or the other form. Oxidation or reduction of the coenzymes involves the removal or addition of two electrons at the C-4 atom of the pyridine ring, with an accompanying shift of electrons away from or toward the N-1 atom of the ring.

Coenzymes such as NAD and NADP are bound only momentarily to the enzyme during

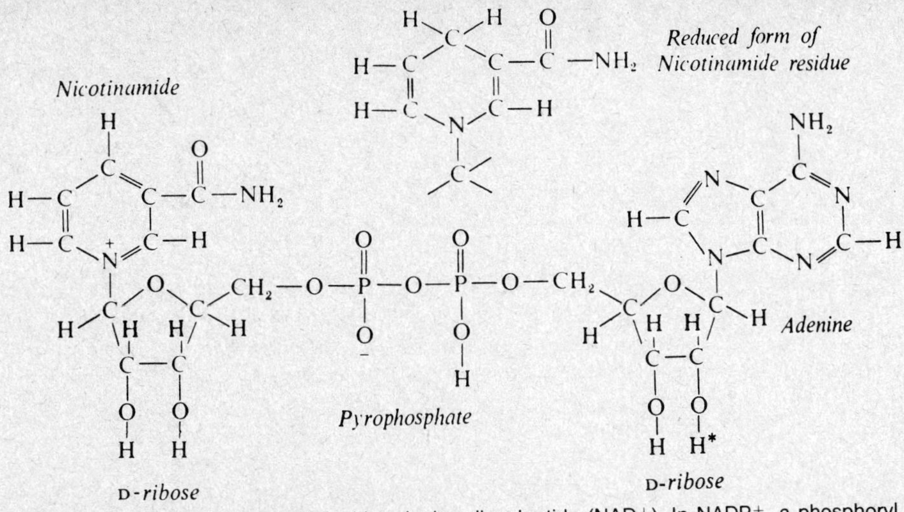

Figure 5-12. Structural formula for nicotinamide adenine dinucleotide (NAD$^+$). In NADP$^+$, a phosphoryl residue, $-PO_3^{2-}$ H, replaces the (H) at the position indicated by the star (*). The structure of the nicotinamide portion in the reduced form of the coenzyme is also shown.

the course of reaction, as is the case for substrates in general. Therefore, no reaction takes place unless the appropriate coenzyme is present in solution, e.g., by adding it to the reaction mixture in the assay of dehydrogenase activity. Among other examples of coenzymes in this category is adenosine triphosphate (ATP), which takes part in phosphate-transfer reactions catalyzed by kinases. In contrast to these entirely soluble coenzymes, a number of coenzymes are more or less permanently bound to the enzyme molecules, where they form part of the active center and undergo cycles of chemical change during the reaction.

The active holo-enzyme results from the combination of the inactive apo-enzyme with the *prosthetic group*, as such a bound coenzyme is called. An example of a prosthetic group is pyridoxal phosphate, a component of aspartate and alanine aminotransferases. The pyridoxal phosphate prosthetic group undergoes a cycle of conversion of the pyridoxal moiety to pyridoxamine and back again during the transfer of an amino group from an amino acid to an oxo-acid. Prosthetic groups, like activators with a structural role, do not usually have to be added to elicit full catalytic activity of the enzyme unless previous treatment has caused the prosthetic group to be lost from some enzyme molecules. However, both normal and pathological serum samples contain appreciable amounts of apo-aminotransferases, which can be converted to the active holo-enzymes by a suitable period of incubation with pyridoxal phosphate. This treatment increases the aspartate aminotransferase activity of serum specimens by an average of about 50%, but the effect on alanine aminotransferase levels is considerably less. There is considerable variation from sample to sample in the degree of activation and also, in the case of aspartate aminotransferase, between sera from patients with different diseases. In order to ensure that assay methods measure only differences in the total enzyme content of specimens, the most recent descriptions of methods for aminotransferase assay specify conditions for activation of any apo-enzyme which may be present. This principle applies also to known cofactors in other enzyme assays.

A study of the formulas of coenzymes and prosthetic groups shows that many contain structures derived from the vitamins. Thus, the nicotinamide portion of NAD and NADP derives from the vitamin niacin (a mixture of nicotinamide and nicotinic acid), while the pyridoxal phosphate prosthetic group of the aminotransferases is a derivative of pyridoxine, vitamin B$_6$. Other derivatives of the B-group vitamins that participate in enzymatic reactions are thiamin pyrophosphate (derived from vitamin B$_1$), the coenzyme involved in decarboxylation of pyruvate as well as in other reactions of carbohydrate metabolism, and the flavin mono- and dinucleotides (formed from riboflavin, vitamin B$_2$) that act as the prosthetic groups of electron-transferring enzymes in the respiratory chain. Biotin (vitamin H) is the prosthetic group of carboxylation reactions. Folic acid, in the form of various derivatives, takes part in the transfer of one-carbon units, and cobamide derivatives of vitamin B$_{12}$ participate in isomerization reactions.

Progress of Enzymatic Reactions and the Measurement of Reaction Rates

Since the rate of an enzyme-catalyzed reaction is directly proportional to the amount of active enzyme present in the system, determination of the rate of reaction under defined and controlled conditions provides a very sensitive and specific method for the measurement of enzymes in samples such as serum. The sensitivity of rate determinations in quantitative analysis of enzymes is approached only by a few immunological techniques; therefore, almost all determinations of enzymes in clinical chemistry are made by measuring the reaction rate.

All determinations of reaction rate involve the measurement of the amount of change produced in a defined time interval; i.e., all measurements of reaction rate are kinetic measurements. This is just as true when the amount of change produced by the enzyme is measured after stopping the reaction at the end of a fixed-time interval as it is when the progress of the reaction is monitored continuously (although the term "kinetic method" has often been limited erroneously to the latter procedure). These two analytical approaches are referred to respectively as *fixed-time* and *continuous-monitoring* methods. They have different advantages and limitations. To appreciate these it is necessary to consider the way in which the rate of an enzymatic reaction varies with time.

The progress of conversion of the substrate into products in the presence of an enzyme can be followed by measuring the decreasing concentration of the substrate or the increasing concentration of the products. Measurement of product formation is preferable where possible, since determination of the increase in concentration of a substance above an initially zero or low level is analytically more reliable than measurement of a fall from an initially high level. However, certain enzymes of importance in clinical chemistry are usually measured in terms of a fall in substrate concentration.

At the moment when the enzyme and substrate are mixed, the rate of the reaction is zero or virtually so. Typically, the rate then rises rapidly to a maximum value, which remains constant for a period of time (Figure 5-13). During the period of constant reaction rate, which may be

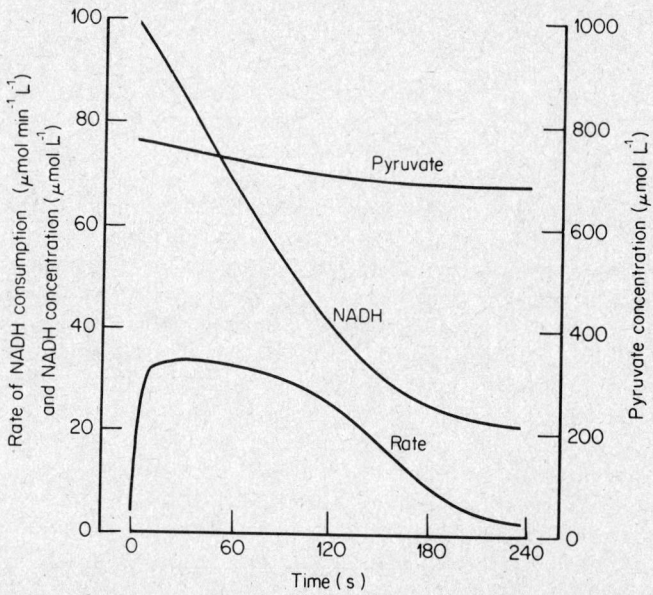

Figure 5-13. Changes in substrate concentrations and rate of reaction during an assay of lactate dehydrogenase activity at 37 °C in phosphate buffer, with pyruvate and NADH as substrates. The reaction is followed by observing the fall in absorbance at 340 nm as NADH is oxidized to NAD$^+$. The rate of reaction rises rapidly to a maximum value, from which it declines only slightly until about half the NADH has been used up. During this phase of the reaction the rate is essentially zero order with respect to substrate concentration. At the point at which the rate falls below about 90% of its maximum value, NADH concentration is approximately $10 \times K_m$. K_m for NADH is of the order of 5×10^{-6} mol $\times$ L^{-1}, whereas for pyruvate it is 9×10^{-5} mol $\times$ L^{-1}. Thus an initial pyruvate concentration approximately 10 times that of NADH is used. (Concentrations are per liter of reaction mixture.) (From Moss, D. W.: Measurement of enzymes. *In*: Enzymes in Cardiology: Diagnosis and Research. D. J. Hearse and J. de Leiris, Eds. New York, John Wiley & Sons, 1979.)

short or long in different circumstances, the rate depends only on enzyme concentration and is completely independent of substrate concentration. The reaction is said to follow zero-order kinetics; i.e., its rate is a function of the zero power ($S^0 = 1$) of the substrate concentration. Ultimately, however, as more and more substrate is consumed, the reaction rate declines and enters a phase of first-order dependence on substrate concentration. Other factors that contribute to the decline in reaction rate include accumulation of products that may be inhibitory, the growing importance of the reverse reaction, and even enzyme denaturation.

Although the rates of reaction produced by different amounts of an enzyme can be compared under first-order conditions, it is obviously easier to standardize such comparisons when the enzyme concentration is the only variable that influences the reaction rate. Therefore, enzyme assays are almost always made under conditions which are initially saturating with respect to substrate concentration. The rate of reaction during the zero-order phase can be determined by measuring the product formed during a fixed period of incubation, but this assumes that the rate has, in fact, remained constant during this period. If it has not, the apparent rate of reaction will no longer be proportional to enzyme concentration. This is illustrated in Figure 5-14. Measurement of reaction rates at any portion of curve A will give results that will be identical to the true "initial rate." However, curve B deviates from linearity over its entire course, and rates fall off with time. At no time will it give a measure of the "initial rate." From curve C, correct results can be obtained only if the rate is measured along Segment II. Incorrect results are obtained if the rate is measured during the lag phase (I), or during phase III.

Careful selection of reaction conditions, such as the concentrations of substrates and cofactors, can often improve the reaction progress curves, eliminating lag phases and prolonging the period of linearity, so that fixed-time methods of analysis become feasible. Improvements in photometry leading to more sensitive measurement of product formation have also allowed the duration of incubation to be shortened compared with older assays. This has resulted in a corresponding increase in the range over which enzyme activity can be measured (Figure 5-15). Nevertheless, an upper limit of activity exists in all fixed-time methods, above which progress curves will no longer be linear, in which case the amount of change measured over the fixed-time interval no longer represents the true zero-order rate.

The existence of an upper limit of enzyme activity implies that samples with activities above this limit must receive special attention. The upper limit of activity acceptable in the unmodified method must be chosen so that samples with activities below it can be presumed with a high degree of certainty to give linear progress curves; on the other hand, if the limit is set too low,

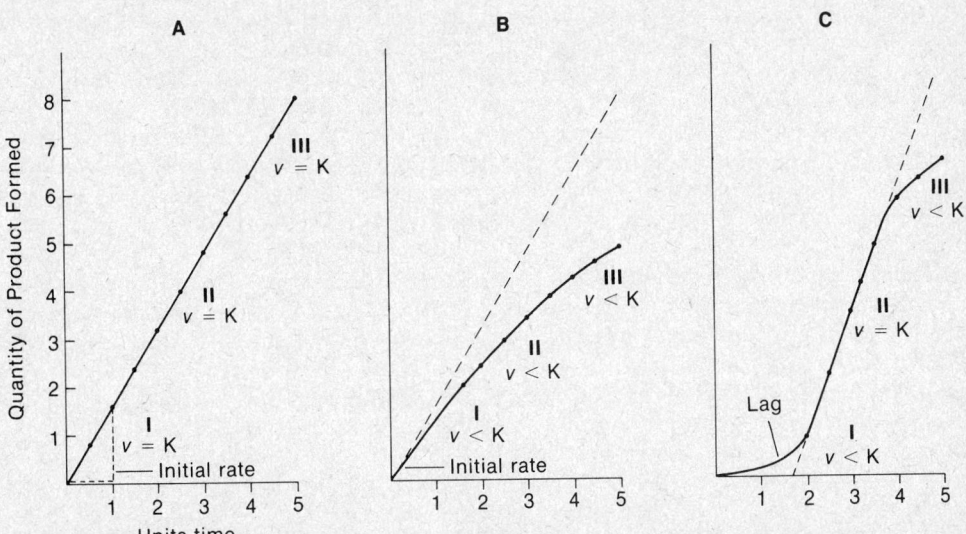

Figure 5-14. Forms of graphs showing change in enzyme reaction rate as a function of time. In *A*, the rate is constant during the entire run, and rates calculated as I, II, and III will be identical to the initial rate. In *B*, the rate falls off continuously; rates calculated at I, II, and III will be different, and less than the true initial rate. In *C*, a measurement at II will be representative of the maximal rate, but as I (lag period) and III (substrate depletion) it will be less than at II.

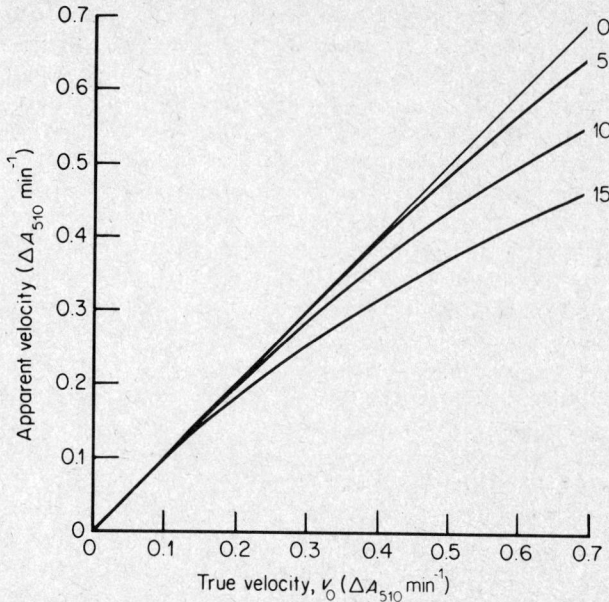

Figure 5-15. Apparent velocity of hydrolysis of phenyl phosphate by alkaline phosphatase measured over a fixed-time interval, as a function of the true initial velocity, v_o. The curves 5, 10, and 15 correspond to incubation periods of 5, 10, and 15 min and show the increasing deviation of the apparent velocity from the initial rate as the incubation period is lengthened. Initial substrate concentration 4.75 mmol $\times$ L^{-1} of reaction mixture in carbonate-bicarbonate buffer, pH 9.9 at 37 °C. (From Moss, D. W.: Measurement of enzymes. *In*: Enzymes in Cardiology: Diagnosis and Research. D. J. Hearse and J. de Leiris, Eds. New York, John Wiley & Sons, 1979.)

many samples will be reanalyzed unnecessarily. Samples that are above the limit should ideally be assayed by shortening the incubation period until a constant reaction rate is obtained. However, this is difficult or impossible in some automated methods, in which the duration of incubation is fixed by the configuration of the apparatus. It then becomes necessary to dilute the specimen; however, dilution may not always result in a proportionate change in activity.

The initial rate of reaction theoretically increases without limit as enzyme concentration increases, as long as no other factor such as substrate concentration becomes limiting. In practice, the reaction becomes so rapid at high enzyme activities that it is impossible to measure the initial rate of reaction, even with continuous-monitoring methods. Therefore, an upper limit of activity that can be accepted without modification of the assay procedure exists even in continuous-monitoring methods, but this limit is usually much higher than that applicable in corresponding fixed-time methods. Fewer samples therefore require special treatment. Furthermore, continuous monitoring allows identification of the appropriate zero-order portion of the progress curve for each sample, as well as identification of samples that require special treatment. Continuous-monitoring methods (or methods in which the progress of reaction is monitored at frequent intervals) therefore possess a decisive advantage in enzyme assay and should be used whenever possible.

The relative merits and disadvantages of fixed-time and continuous-monitoring methods are also relevant in the automation of enzyme assays. The balance between these two approaches may be altered by the constraints introduced by different types of automation, as is discussed in a later section.

Enzyme activity can also be measured by noting the time required to consume all of a fixed amount of substrate, since the amount of substrate transformed, Q, is a function of $[E]$:

$$Q = k_1 \times [E] \times t$$

and

$$[E] = \frac{Q}{k_1} \times \frac{1}{t} \qquad (22)$$

$[E]$ is the concentration of active enzyme, k_l is the rate constant, and t is the elapsed time. Therefore, $[E] \propto 1/t$. This is the principle of the Wohlgemuth method of amylase estimation, for example. However, the end-point of the reaction is approached asymptotically, making its precise timing difficult, and each sample requires a different time for analysis. Methods of this type have therefore largely been discontinued.

Units for Measuring Enzyme Activity

Since enzymes are almost always measured by their catalytic activities, the results of such determinations are expressed in terms of the number of activity units present in a convenient volume or mass of specimen. The unit of activity is the measure of the rate at which the reaction proceeds, e.g., the quantity of substrate consumed or product formed in a chosen unit of time. The quantity of substrate may be given in any convenient unit—milligrams, micromoles, change in absorbance, change in viscosity, or microliters of gas formed; time may be expressed in seconds, minutes, or hours. Since the rate of the reaction will depend on experimental parameters such as pH, type of buffer, temperature, nature of substrate, ionic strength, concentration of activators, and other variables, these parameters must be specified in the definition of the unit. Any methodological modification, however slight, that involves a change in the values chosen for these variables may alter the significance of the units associated with that method.

In the course of many decades a multiplicity of units for expressing enzyme activity have been introduced. Even for the same or similar enzymes, each investigator defined his unit in terms of quantities analytically or otherwise convenient for him at the time. Thus, a series of eponymous enzyme units evolved, e.g., the Bodansky or King-Armstrong units for phosphatase, the Somogyi unit for amylase, and others.

The concentration of an enzyme is often expressed in terms of activity units per volume or mass of specimen, e.g., per mg of a specimen of liver tissue, per mg of protein, or per mg of fat-free protein-nitrogen in the specimen. In clinical work, the concentration is generally reported in terms of some convenient unit of volume, such as activity per 100 mL or per liter of serum, or per 1.0 mL of packed erythrocytes.

Just as a multiplicity of units for reporting activity developed, a similar multiplicity of units of volume evolved. The Commission on Enzymes proposed that the unit of enzyme activity be defined as that quantity of enzyme that will catalyze the reaction of one micromole (μmol) of substrate per minute, and that this unit be termed the *International Unit* (U). Catalytic concentration is to be expressed in terms of U/L (or mU/L), whichever gives the more convenient numerical value. In this chapter the symbol U is used to denote the International Unit. In those instances in which there is some uncertainty as to the precise nature of the substrate or where there is difficulty in calculating the number of micromoles reacting (as with macromolecules such as starch, protein, and complex lipids), the unit is to be expressed in terms of the chemical group or residue measured in following the reaction (e.g., glucose units, or amino acid units formed).

Although the proposals of the Enzyme Commission have been accepted by the majority of scientists working with enzymes, many laboratories using well-established enzyme procedures continue to report enzyme values in terms of the older eponymous units to which they have become accustomed. But as new methods are devised, it is anticipated that the activity units will be established in accordance with the recommendations of the Commission. However, it cannot be emphasized too strongly that the International Unit itself is a function of the method used and is not an absolute quantity. Therefore, methods with different reaction conditions will give different numerical values for activity even when all values are expressed in International Units.

The International Unit itself may soon be replaced by a new unit. Both IUPAC and IUB recommend[70] that enzyme activity be expressed in mol/s, given the name katal, and that the enzyme concentration be expressed in terms of katals per liter (kat/L). This is consistent with the SI (Système Internationale) scheme of units in that the mole is the measure of substrate transformed and the second is the unit of time. Thus, $1\ U = 10^{-6}\ mol/60\ s = 16.7 \times 10^{-9}$ mol/s, or 1.0 nkat/L = 0.06 U/L.

Measurement of Chemical Changes in Enzymatic Reactions

The amount of substrate transformed into products during an enzyme-catalyzed reaction can be measured with any appropriate analytical method. However, the method most commonly used is photometric analysis. The reaction may be accompanied by a change in the absorbance characteristics of some component of the assay system, in either the visible or ultraviolet spectrum, which can be observed while it is proceeding. "Self-indicating" reactions of this type are particularly valuable since they make possible continuous monitoring in a recording spectrophotometer. Important examples of self-indicating reactions are the determination of dehydrogenase activity by following the change in absorbance at 339 (340) nm of the coenzymes NAD(H) or NADP(H) during oxidation or reduction (Figure 5-16) and the measurement of alkaline phosphatase activity by the liberation of the yellow p-nitrophenate ion from the substrate p-nitrophenyl phosphate in alkaline solution. (Acid phosphatase cannot be determined by continuous monitoring with this substrate, since the spectral differences between the phosphate ester and its parent phenol are present only in alkaline solution.) The advantages of continuous monitoring are so marked that coupled reactions are frequently used to provide an observable change in absorbance accompanying a primary reaction in which such a change is not present.

The photometric requirements of enzymatic analysis are not different from those of photometric quantitative analysis in general. However, the photometer used must be provided with means to maintain the contents of the cuvet at a constant temperature during the reaction, and provision of a chart recorder simplifies monitoring of the progress of reaction.

The introduction of prism or diffraction-grating spectrophotometers capable of isolating a narrow beam of monochromatic light in the ultraviolet or visible spectrum and with stable and sensitive photomultipliers as detectors has greatly improved the reproducibility of photometric measurements. Consequently, it has become customary to make use of the known molar absorptivity of well-defined reaction products (such as NADH) when calculating changes in their concentrations based on measurements made with spectrophotometers. However, this procedure can lead to serious errors if the assumptions on which it is based are not justified. Therefore, the absorbance and wavelength accuracy of the spectrophotometer should be checked regularly.

Automation of Enzyme Activity Measurements

As with other types of analysis, the increasing demand for enzyme estimations in diagnosis and treatment has been met by the introduction of automated methods of analysis. Two separate lines of development of automatic enzyme analyzers can be distinguished.[101]

The first of these is the mechanization of fixed-time methods of analysis, e.g., with the Technicon Corporation's continuous-flow AutoAnalyzer. Provided that good fixed-time methods are chosen, such automated methods perform at least as well as and often better than their manual counterparts, because of the improved reproducibility of measurement of sample and reagent volumes achieved by automated analysis. However, automated fixed-time procedures share the same limitations as manual methods of this type; i.e., they are valid only over a limited

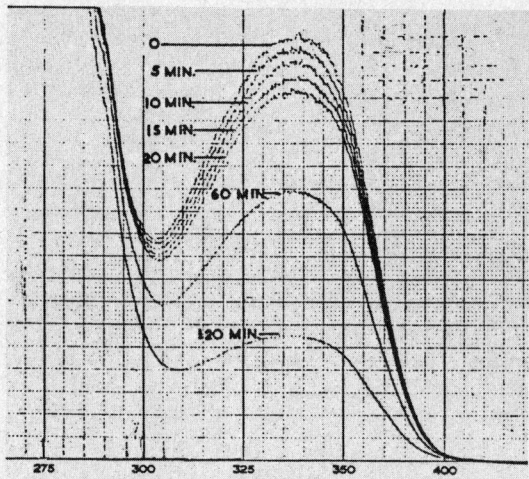

Figure 5-16. Progressive reduction in absorbance at 339 nm during the oxidation of NADH and NAD$^+$ with transfer of hydrogen to pyruvate to form lactate, catalyzed by lactate dehydrogenase. (From Moss, D. W.: Enzymes as tools in the laboratory. *In*: 2nd Symposium on Advanced Medicine. London, Pitman Medical, 1966.)

range of enzyme activity. Because chemical reactions do not go to completion in the AutoAnalyzer, the use of *calibrators* (i.e., solutions with assigned values of enzyme activity) is always necessary. The precise times at which reagent streams mingle to initiate and terminate enzymatic reactions may be difficult to determine exactly and the precise duration of the enzymatic reaction may not be known. Therefore, calibration with serum or other samples of known enzymatic activity, determined by a different assay method, is preferable to the use of standard solutions of the reaction product.

The second approach to automated enzyme analysis is represented by the many instruments now available that are designed to mechanize the successive stages of sample and reagent volume measurement, preincubation and temperature control, initiation of the enzymatic reaction, monitoring of absorbance changes, and calculation of enzyme activity, which together constitute an enzyme measurement by the continuous-monitoring procedure. These automatic enzyme analyzers are capable of providing satisfactory analytical data with greatly increased throughput of samples. However, in their design, as in the design of all such apparatus, choices and compromises may have been made in the interests of simplicity, reliability, or reduced capital cost. These compromises may have introduced limitations into the analytical process which were not present in the original manual method. For example, the infinite number of points represented by the analog signal of changing absorbance with time drawn on a recorder chart may be replaced by three, or even two, digitized readings during the progress of the reaction. This simplification of the reaction-monitoring process may result in failure to detect nonlinear progress curves or to take account of lag phases. Furthermore, the mode of operation of some automatic analyzers requires the enzymatic reaction to be initiated by the addition of serum to the reaction mixture. Therefore, preincubation to allow sample-dependent side reactions to be completed before starting the reaction with substrate is impossible. However, the difficulties caused by the absence of preincubation or excessively long lag phases can be overcome in some cases by a modification of the composition of the reagents.

Optimization, Standardization, and Quality Control of Enzyme Assays

In all analytical methods, a difference in the quantity of the analyte being measured should be the only variable that causes the difference in results obtained. In the case of enzyme activity measurements, it is obvious that all the factors that affect the reaction rate, other than the concentration of active enzyme, must be rigidly controlled. However, some slight variation in such factors as pH or temperature between samples or between analytical runs is inevitable, and this contributes to the imprecision of the analysis. The effect of these slight variations is least under optimal conditions, e.g., of pH or substrate concentration, since at the top of the pH optimum curve or at high substrate concentration the dependence of reaction velocity on these variables is least marked. Furthermore, since the reaction velocity is at or near its maximum under optimal conditions, a larger analytical signal is obtained which can be measured more easily and precisely than the smaller signal obtained under suboptimal conditions. Much effort has therefore been devoted to determining optimal conditions for measuring the activities of enzymes of clinical importance.

Optimal conditions are not necessarily those that provide the maximum attainable rate of reaction. Sometimes the highest rate can be obtained only by the use of solutions that are so concentrated that they have inconveniently high absorbances, or are too viscous for accurate dispensing, or are prohibitively expensive to prepare. Optimal conditions are those that combine a high rate of reaction with the least marked dependence on variables other than enzyme concentration and that are practicable in the routine laboratory.

Optimization of reaction conditions for enzyme assays has traditionally been carried out by varying a single factor and studying its effect on the reaction rate, then repeating the experiment with a second factor, and so on until effects of all the variables have been tested. An optimal combination of variables is selected on the basis of these experiments and the validity of the chosen conditions is verified. Not only is this approach laborious, but it also is not well adapted to situations in which the effects of different variables are interdependent, as is frequently the case in enzyme analysis. Therefore, newer techniques in which the effect of changes in more than one variable are studied simultaneously are used more and more frequently. Simplex co-optimization[77] and response-surface methods[116] are typical of these newer strategies.

Standardization. The large number of methods used for the measurement of the activities of enzymes in clinical laboratories, and the multitude of different units of activity and reference ranges to which this lack of standardization gives rise, form a serious barrier to the interpretation of data obtained in different laboratories, both in the records of individual patients and in reports in the scientific literature.

Two approaches to standardization of enzyme assays are possible. The first is to provide enzyme preparations of stated catalytic activity that can be used as calibration standards (calibrators), in the way in which solutions of accurately known concentration are used to standardize determinations of other substances. However, this approach requires the availability of enzyme preparations of defined and reliable properties, particularly with regard to stability. The susceptibility of enzymes to denaturation has not so far been reliably overcome, although there are hopeful signs that stable enzyme calibrators may soon become available.

The second approach is to standardize the conditions of assay, thus reducing the number of methods in use. This has been the objective of various working-parties and committees whose specific approaches have not always been identical. Methods recommended by these groups range from methods intended for use in small groups of neighboring laboratories to rigorously specified reference methods intended to provide a criterion of analytical performance against which routine methods can be judged. Between these extremes are the methods proposed by national or international enzyme committees and designed to combine analytical reliability with applicability to current standards of instrumentation and skill in well-equipped laboratories.

Many laboratories now recognize the advantages of being able to refer their enzyme results to a particular set of recommended methods, to provide at least a partial solution to the problems of enzyme standardization. Recommended methods are being used increasingly with a marked improvement in interlaboratory agreement as shown in external quality assessment schemes.

The role proposed by the Expert Panel on Enzymes of the International Federation of Clinical Chemistry (IFCC) for reference methods in enzyme standardization is different from that of the recommended methods proposed by national enzyme committees. IFCC reference methods are intended to take account of all the factors that are known to affect the measurement of a particular enzyme's activity, and thus to give the most reliable and unbiased estimate that can be obtained in the present state of knowledge. With their rigorous specifications for reagents and equipment and their careful exclusion of possible interfering reactions by the use of control and blank estimations, such reference methods are not immediately suitable for daily analysis of large numbers of specimens. Their influence on routine practice is therefore indirect and is exerted in several ways.

The reference methods set standards of precision and accuracy (in the sense of reaction rates unbiased by side-reactions) against which the relative performances of methods intended for routine use can be judged. Sources of error identified in developing reference methods can be taken into account in routine methods. It is also probable that, as suitable enzyme calibrators become available, their activities will be assigned in terms of the reference method. These preparations will then be used to calibrate routine methods, so that, even if several routine methods are in use, the results obtained by all of them can be expressed in a common set of units.

Quality control. As with calibration, the application of quality control procedures to ensure that satisfactory analytical performance of enzyme assays is maintained on a day-to-day basis is also complicated by the tendency of enzyme preparations to undergo denaturation with loss of activity. These various factors make it difficult to distinguish between poor analytical performance and denaturation as possible causes of a low result obtained for a control sample introduced into a batch of analyses. Assured stability within a defined usable time-span is therefore the prime requirement for enzyme control materials as it is for enzyme calibrators. However, specifications for the two types of materials can differ in other respects. Since the function of a calibration standard is to provide a stated activity under defined assay conditions, it is not necessary for the calibrator to show a sensitivity to changes in the assay system identical to that of the samples under test; within certain limits, therefore, enzymes from various sources can be considered in the search for stability. On the other hand, it is the function of a *control* to reveal small variations in reaction conditions, so it must be at least as sensitive to these variations as the samples being analyzed. The preparation of enzymes from human sources is not by itself a guarantee of an effective control. For example, human placental alkaline phosphatase is very stable, but it differs significantly in kinetic properties from the liver and bone enzymes that contribute most of the alkaline phosphatase activity of human serum samples; it is therefore not an ideal enzyme for use in control material for the determination of alkaline phosphatase.

Lyophilized preparations containing various enzymes are available from commercial sources and these have a useful function in quality control. Serum pools prepared in the laboratory (with care to exclude any specimens contaminated with hepatitis virus), then assayed for enzyme activity and stored in the deep-freeze in small portions for daily use, are also valuable. However, the use of both types of control materials is attended with uncertainties for the reasons mentioned above. Furthermore, it is often difficult to reproduce the values for activity stated by the manufacturers for freeze-dried sera, even when the same methods are apparently in use, so that recalibration in the laboratory is necessary.

Another useful procedure in the control of enzyme analysis is to carry forward one or two samples from one batch to the next, so that the successive results can be compared. Some loss of activity usually takes place between the first and second analysis, but for many enzymes this is less than 10% with suitable interim storage. When large batches of specimens (50–100 or more) drawn from a constant type of population are routinely analyzed, comparison of the daily batch-means provides a useful additional means of quality control in enzymology and one which makes no assumptions about the stability of enzymes in control materials. However, it may be difficult to decide how much day-to-day variation in batch mean arises from analytical variability and how much from changes in the composition of the population from which the samples come.

In spite of these various difficulties, systematic application of quality control programs is as essential in enzyme analysis as in other forms of clinical analysis if the results obtained are to be useful in diagnosis and treatment. When well-designed assay methods are in use, the reproducibility of results of enzyme assays on a day-to-day basis is usually less than $\pm 10\%$ coefficient of variation for activities within the normal range and less than $\pm 5\%$ CV for increased activities.

Measurement of Enzymes in Terms of Mass

The problems of standardization of enzyme assays would be greatly reduced or even abolished if the results could be expressed in terms of the mass of enzyme present rather than in method-dependent activity units. Results of activity measurements can be converted to mass units if the molecular activity of the enzyme is known, since

$$\text{Amount of enzyme (moles)} = \frac{\text{Catalyzed rate of reaction (mol substrate transformed/s)}}{\text{Molecular activity (mol substrate transformed/s per mol enzyme)}} \quad (23)$$

The molecular activity depends on the reaction conditions, as all measurements of catalytic activity do. However, the molecular activity is constant under defined conditions; thus, if its value is known for an enzyme acting in a particular assay method, the number of activity units obtained by that method for a given volume of sample can be converted to the corresponding molar concentration of the enzyme. If the molecular activities of the enzyme in several different assay methods are known, the results of each can be converted to molar terms by the use of the appropriate factor.

Few human enzymes of diagnostic importance have so far been obtained in the state of complete purity that is necessary for the preparation of solutions of known molarity with which molecular activity can be determined. However, the molarities of some impure enzyme solutions can be determined by titrating them with a reagent that combines specifically with the enzyme in a detectable way. For example, alkaline phosphatase incorporates inorganic phosphate irreversibly at its active center at acid pH. If radioactive phosphate is used, the molarity of an impure solution of the enzyme can be deduced from the amount of labeled phosphate that is bound. The molarity of a solution of an aminotransferase can be determined by observing the spectral change that occurs as the apoenzyme is titrated with its prosthetic group, pyridoxal phosphate. These methods actually determine the concentration of catalytic centers in the solution; to obtain the concentration of enzyme molecules it is necessary to divide the result by the number of active centers on each molecule of enzyme when this number is more than one.

Conversion of enzyme activity to mass by use of molecular activity factors has so far received little attention from clinical chemists, and attempts to avoid the difficulties of method-dependent activity units have concentrated mainly on the direct measurement of enzymes as proteins. Only a few immunological methods of protein analysis can approach the sensitivity required for clinical purposes, while at the same time having sufficient specificity to be applicable to impure solutions of enzymes such as serum.

A considerable number of radioimmunoassays for human enzymes have now been described. Some purified enzyme has to be prepared to act as a calibration standard, to be radioactively labeled, and to be used to raise the enzyme-specific antibody. These methods determine all molecules with the antigenic determinants necessary for recognition by the antibody, so that inactive enzyme molecules that are immunologically unaltered are measured as well as active molecules. This potentially increases the sensitivity of RIA over that of assays of catalytic activity, since it is possible that inactive as well as active enzyme molecules are present in the circulation. This advantage of RIA has not been found to be significant except in the determination of some digestive enzymes such as trypsin when inactive precursors as well as inhibitors of catalytic activity are present in plasma. In some cases, however, the amount of *active* enzyme is clinically the more significant quantity, e.g., when serum cholinesterase is inhibited by drugs or poisons.

Other immunologically based enzyme assays include such techniques as radial immunodiffusion and "rocket" electroimmunoassay. In these methods the activity of the enzyme being determined is used to locate the precipitin arcs or peaks by reactions analogous to those used in histochemistry. Again, pure enzyme is needed as a standard to enable results to be expressed in mass units.

RIA and other immunoassays have been little used for the determination of total enzyme activities for diagnostic purposes. In general, these assays cannot compete in speed and precision with measurements of catalytic activity. Furthermore, several enzyme activities in serum are due to mixtures of immunologically distinct isoenzymes, so that an assay using a single type of antibody will usually determine only one of the enzyme forms. However, this disadvantage in the determination of total enzyme activity becomes a marked advantage in the measurement of a specific isoenzyme, and immunological methods have assumed great importance in isoenzyme analysis for diagnostic purposes.

Enzymes as Analytical Reagents

The use of enzymes as analytical reagents offers the advantage of great specificity for the substance being determined. This high specificity typically removes the need for preliminary separation or purification stages, so that the analysis can be carried out directly on complex mixtures such as serum. Enzymes with absolute specificity for the substance being estimated are clearly preferable for analytical use. Uricase (urate oxidase), urease, and glucose oxidase are examples of highly specific enzymes used in clinically important assays. However, this ideal cannot always be achieved in practice, and a knowledge of the substrate specificities of reagent enzymes is therefore essential to allow possible interferences with the assay to be anticipated and corrected. Coupled reactions are often used to construct an enzymatic analytical system for determining a particular compound, as they are in measurements of enzyme activity, and the specificity of the coupled reactions may modify the specificity of the overall process. An example of this is the determination of glucose by the hexokinase reaction. Hexokinase will convert sugars other than glucose to their 6-phosphate esters. However, the indicator reaction used to monitor this change is catalyzed by glucose-6-phosphate dehydrogenase, an enzyme that is highly specific for its substrate, so that the overall process is highly specific for glucose.

The principle that is most widely used to determine the amount of a substance enzymatically is to allow the reaction to go to completion, so that all the substrate has been converted into a measurable product. These methods are called "end-point" or, more correctly, "equilibrium" methods, since the reaction ceases when the equilibrium is reached. Reactions in which the equilibrium point corresponds to virtually complete conversion of the substrate are obviously preferable for this type of analysis. However, unfavorable equilibria can often be displaced in the desired direction by additional enzymatic or nonenzymatic reactions that convert or "trap" a product of the first reaction; e.g., in measuring lactate with lactate dehydrogenase, the pyruvate formed can be trapped by the addition of hydrazine with which it forms a hydrazone.

From equation (22) we can see that the time required to transform a fixed quantity, Q, of substrate into products is inversely proportional to the amount of enzyme, E, present. Equilibrium methods may therefore require the use of appreciable amounts of enzyme for each sample to avoid inconveniently long incubation periods. As the substrate concentration falls to low levels toward the end of the reaction, the K_m of the enzyme becomes important in determining the reaction rate. Enzymes with high affinities for their substrates (low K_m values) are therefore most suitable for equilibrium analysis. Equilibrium methods are largely insensitive to minor changes in reaction conditions. It is not necessary to have exactly the same amount of enzyme in each reaction mixture, or to maintain the pH or temperature absolutely constant, provided that the variations are not so great that the reaction is not completed within the fixed time allowed.

The amount of reagent enzyme required for each analysis can be reduced and the time shortened by the use of kinetic methods, i.e., methods in which the amount of change produced in a fixed time interval ($=$ rate) is measured. As already described, the rate of an enzyme-catalyzed reaction with initially high substrate concentration first follows zero-order kinetics, then, as the substrate concentration declines, passes into a first-order kinetic phase.

For any first-order reaction, the substrate concentration $[S]$ at a given time t after the start of the reaction is given by

$$[S] = [S_0] \times e^{-kt} \tag{24}$$

where $[S_0]$ is the initial substrate concentration, e is the base of the natural log, and k is the rate constant.

The change in substrate concentration $\Delta[S]$ over a fixed time interval t_1 to t_2 is related to $[S_0]$ by the equation

$$[S_0] = \frac{-\Delta[S]}{e^{-kt_1} - e^{-kt_2}} \tag{25}$$

That is, the change in substrate concentration over a fixed time interval is directly proportional to its initial concentration. This is a general property of first-order reactions.

For an enzymatic reaction, first-order kinetics are followed when $[S]$ is small compared to K_m. Equation (7) then reduces to

$$v = \frac{V_{max}}{K_m} \times [S] \text{ or } v = k[S]$$

Thus, the first-order rate constant, k, is equal to V_{max}/K_m.

Methods in which some property related to substrate concentration (such as light absorbance) is measured at two fixed times during the course of the reaction are known as "two-point" kinetic methods. They are theoretically the most accurate for the enzymatic determination of substrates. However, these methods are technically more demanding than equilibrium methods. Since reaction rate is being measured in the two-point methods, all the factors that affect reaction rate such as pH, temperature, and amount of enzyme must be kept constant from one assay to the next, as must the timing of the two measurements. These conditions can now readily be achieved in automatic analyzers. A standard solution of the analyte (substrate) must be used for calibration. To ensure first-order reaction conditions the substrate concentration must be low, of the order of less than $0.2 \times K_m$. Enzymes with high K_m values are therefore preferred for kinetic analysis to give a wider usable range of substrate concentration. Introduction of a competitive inhibitor has been suggested as a way of increasing the apparent K_m of a reagent enzyme. The low substrate concentrations needed for kinetic analysis also require the measurement of small changes, e.g., in absorbance. In spite of their technically demanding nature, however, two-point kinetic methods are increasingly used for substrate determinations.

Immobilized Enzymes

The consumption of relatively expensive enzymes, one of the disadvantages of enzymatic analysis, can be reduced by the use of immobilized enzymes that can be recovered and reused for several analyses. Thus, the enzyme fulfills its true role as a catalyst, instead of being a consumable reagent. Immobilized enzymes have been chemically bonded to adsorbents such as microcrystalline cellulose, diethylaminoethyl cellulose, carboxymethyl cellulose, and agarose. Diazo, triazine, and azide groups are used to join the enzyme protein to the insoluble matrix, forming either particles in contact with the substrate solution or a surface in contact with substrate solution, such as a membrane or a coating on the inner surface of a vessel holding the substrate solution. Among enzymes available in such immobilized form are urease, hexokinase, α-amylase, glucose oxidase, trypsin, and leucine aminopeptidase. Stability to heat and other forms of inactivation is considerably increased compared to enzymes in solution. Immobilized proteolytic enzymes are not subject to autodigestion. However, some properties of the enzyme, such as its Michaelis constant or its pH optimum, may be altered.

Immobilized enzymes are mainly useful in analytical systems in which the reaction products can be detected directly, or at least without the use of coupled enzymatic reactions, since such reactions would require additional enzymes and coenzymes that would negate the original cost savings. Techniques such as potentiometry, polarography, and microcalorimetry are therefore frequently chosen in exploiting the benefits of immobilized enzymes. Enzymes immobilized on the inner walls of plastic tubing are suitable for use in continuous-flow analyzers. Enzymes incorporated into membranes form part of enzyme electrodes. The surface of an ion-sensitive electrode is coated with a layer of porous gel in which an enzyme has been polymerized. When the electrode is immersed in a solution of the appropriate substrate, the action of the enzyme produces ions to which the electrode is sensitive. For example, an oxygen electrode coated with a layer containing glucose oxidase can be used to determine glucose by the amount of oxygen consumed in the reaction, and urea can be estimated by the combination of a selective ammonium ion-sensitive electrode and a urease membrane.

Enzyme Immunoassays

The ability to conjugate enzymes with other proteins without significant loss of catalytic activity has opened up the possibility of using enzyme labels in place of radioactive isotopes in various types of immunoassays. The principles of enzyme immunoassays are described in Chapter 1C.

ISOENZYMES AND OTHER MULTIPLE FORMS OF ENZYMES

Improved techniques for analyzing mixtures of proteins, developed over the past twenty years, have shown that a particular type of catalytic activity within a single species is frequently due to the existence of several distinct forms of an enzyme rather than to only one type of molecule.[103] These enzyme variants may occur within a single organ or even within a single type of cell. The forms can be distinguished on the basis of differences in various physical properties, such as electrophoretic mobility or resistance to chemical or thermal inactivation. Often there are significant quantitative differences in catalytic properties between them. However, all the forms of a particular enzyme retain the ability to catalyze its characteristic reaction. The multiple molecular forms of an enzyme are often described as "isoenzymes" (or isozymes), although it is now recommended by IUB that this term should be restricted to those forms which originate at the level of the genes that encode the structures of the enzyme proteins in question.[23]

The existence of multiple forms of enzymes in human tissues has important implications in the study of human disease. The presence in different organs of isoenzymes with distinctive properties helps in understanding organ-specific patterns of metabolism, while genetically determined variations in enzyme structure between individuals account for such characteristics as differences in sensitivity to drugs, as well as those differences in metabolism which manifest themselves as hereditary metabolic diseases. For the diagnostic enzymologist, the existence of multiple forms of enzymes, whether due to genetic or nongenetic causes, provides opportunities to increase the diagnostic specificity and sensitivity of enzyme assays carried out on samples of body fluids.

Genetic Origins of Enzyme Variants

True isoenzymes may be due to the existence of more than one gene locus coding for the structure of the enzyme protein. A substantial proportion of human enzymes (perhaps more than one third) seem to be determined by more than one structural gene locus. The structural genes at the different loci have undergone differential modifications during the course of evolution, so that the enzyme proteins coded by them no longer have identical structures, although they are recognizably similar; in other words, they are isoenzymes. *Multiple gene loci* have become disseminated throughout a whole species during the course of evolution, so that these genes and their resultant isoenzymes typically are present in all individuals of that species.

The several genes that determine a particular group of isoenzymes are not necessarily closely linked on one chromosome; indeed, they are often located on different chromosomes. For example, the structural genes that code for human salivary and pancreatic amylases are both located on chromosome 1, whereas the genes that code for mitochondrial and cytoplasmic malate dehydrogenase are carried on chromosomes 7 and 2, respectively. Among the enzymes of clinical importance that exist as isoenzymes because of the presence of multiple gene loci are lactate dehydrogenase, creatine kinase, and some forms of alkaline phosphatase.

A large number of human enzymes exist in multiple molecular forms, which differ in type from one individual to another. Family studies show that the various forms of a given enzyme are inherited according to Mendelian laws. In these cases the different enzyme forms originate from modified genes, or alleles. Allelic genes are alternative forms of the gene that occur at a particular locus and that give rise to gene products with the same function. The isoenzymes that result from the existence of allelic genes are termed *"allelozymes."* The proportion of human gene loci subject to allelic variation is considerable, and the probability that individual human beings will differ to some degree in their isoenzyme patterns is correspondingly high. The likelihood of different isoenzyme patterns is further increased by the possibility that individuals may be heterozygous for particular alleles; i.e., they have inherited different alleles at a particular locus.

The number of allelic variants and the frequency with which particular variants occur within the population vary considerably from one enzyme to another. For example, mutations at either of the two principal loci that determine human lactate dehydrogenase are extremely rare, but there is a high incidence of mutant alleles at the single locus that determines the structure of placental alkaline phosphatase. More than 150 distinct forms of the glucose-6-phosphate dehydrogenase of human erythrocytes have now been identified, together with a further 50 or so awaiting verification, each form being determined by a different allele at the locus on the X chromosome that codes for this enzyme. Some of these alleles are extremely rare, whereas others occur with appreciable frequencies in particular populations or geographical locations. When isoenzymes due to variation at a single locus occur with appreciable frequencies in a human population, the population is said to be polymorphic with respect to the isoenzymes in question.

Another category of multiple molecular forms can arise in the case of enzymes that are oligomeric, i.e., that consist of molecules made up of subunits. The association of different types of subunits in various combinations gives rise to a range of active enzyme molecules. When the different subunits derive from distinct structural genes, either multiple loci or multiple alleles, the hybrid molecules so formed are included within the definition of isoenzymes and are called *hybrid isoenzymes.* The ability to form hybrid isoenzymes is evidence of considerable structural similarities between the different subunits. Hybrid isoenzymes can be formed in vitro, but they are also formed in vivo in cells in which the different types of constituent subunits are present in the same subcellular compartment.

The number of different hybrid isoenzymes that can be formed from two nonidentical protomers depends on the number of subunits in the complete enzyme molecule. For a dimeric enzyme, one mixed dimer (hybrid isoenzyme) can be formed. If the enzyme is a tetramer, three heteropolymeric isoenzymes may be formed (Figure 5-17). Examples of hybrid isoenzymes are the mixed MB dimer of creatine kinase (p. 679) and the three hybrid isoenzymes, LDH-2, LDH-3 and LDH-4, of lactate dehydrogenase (p. 691).

To summarize, "true isoenzymes" are multiple forms of an enzyme that all possess the ability to catalyze the enzyme's characteristic reaction, but that differ in structure because they are encoded by distinct structural genes. The structural variations of the isoenzymes are reflected to various degrees in their properties.

The different structural genes from which the isoenzymes arise may be due to the existence of multiple gene loci (typically possessed by all individuals), or of allelic genes at a particular locus (giving rise to interindividual variation in isoenzymes).

In the case of oligomeric isoenzymes, hybrid isoenzymes may be formed by the combination of unlike subunits. The unlike subunits may be derived from the existence of multiple gene loci,

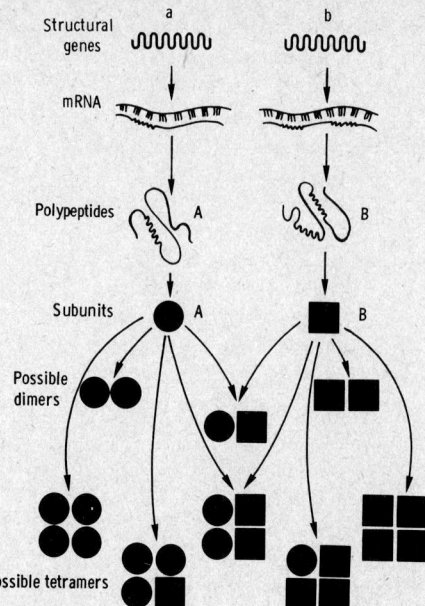

Figure 5-17. Diagram showing the origin of isoenzymes, assuming the existence of two distinct gene loci. When the active enzymes are polymers containing more than one subunit, hybrid isoenzymes consisting of mixtures of different subunits may be formed. One such isoenzyme can be formed in the case of a dimeric enzyme such as creatine kinase, and three if the enzyme is a tetramer (e.g., lactate dehydrogenase). In both cases two homopoly-meric isoenzymes can also exist. (From Moss, D. W.: Isoenzyme Analysis. London, The Chemical Society, 1979.)

multiple alleles, or both. The hybrid isoenzymes thus formed are also included in the definition of true isoenzymes.

It should be added that while by strict definition the term isoenzymes should be restricted to the three categories listed above, "isoenzymes" is still often used in an operational sense to describe any multiple enzyme forms, whatever their origins.

Distribution of Isoenzymes and Other Multiple Forms of Enzymes

The existence of multiple gene loci and the isoenzymes derived from them has presumably conferred an evolutionary advantage on the species and has thus become part of its normal metabolic pattern. Some of these adaptations are related to the division of function between and within different types of specialized cells and tissues. It is to be expected, therefore, that the distribution of isoenzymes will not be uniform throughout the body, and wide variations in the activity of different isoenzymes do indeed occur between organs, between the cells which compose a particular organ, and even between the structures that constitute a single cell. Tissue-specific differences are also found in the distributions of some multiple forms of enzymes that are not due to the existence of multiple gene loci. The tissue-specific distribution of isoenzymes and other multiple forms of enzymes provides the basis for organ-specific diagnosis through isoenzyme measurements.

Certain gene loci may be expressed almost exclusively in a single tissue, perhaps at a particular stage in development. In addition to the two gene loci that determine the two most common subunits of lactate dehydrogenase, a third lactate dehydrogenase locus is active only in mature testis. It determines the structure of a third type of subunit, X or C, which makes up a specific isoenzyme, LDH-X or LDH-C, found only in testis. The isoenzyme of alkaline phosphatase that occurs in the human placenta is the product of a single structural gene locus, distinct from loci that specify the structures of other forms of alkaline phosphatase, and the product of the placental phosphatase locus is normally detectable only in the placenta.

A particularly striking example of the localized expression of multiple gene loci is provided by distinct isoenzymes that occur exclusively in specific subcellular organelles. Differences between mitochondrial isoenzymes and their functionally analogous counterparts in the cytoplasm have been demonstrated in several cases. That the mitochondrial and extramitochondrial enzyme forms are indeed determined by separate gene loci (i.e., that they are true isoenzymes) has been shown for both human aspartate aminotransferases and malate dehydrogenases by the discovery of rare variants of the respective isoenzymes from mitochondria or cytoplasm. The variants are inherited in a Mendelian manner without corresponding changes in the isoenzymes located elsewhere in the cell.

The mitochondrial isoenzyme of aspartate aminotransferase accounts for more than three-quarters of this activity in the parenchymal cells of the liver. Although isoenzymes of nonmitochondrial organelles have been less well characterized than the enzyme variants found in mitochondria, differences in properties do exist between catalytically similar enzymes prepared from various other subcellular fractions, e.g., cytosol and cell membranes. It is probable that some of these variations also arise from the presence of multiple structural gene loci.

Changes in Isoenzyme Distribution During Development

Multiple gene loci and their dependent isoenzymes provide means for the adaptation of metabolic patterns to the changing needs of different organs and tissues in the course of normal development, or in response to environmental change. Pathological conditions may also be associated with alterations in the activities of specific isoenzymes.

Changes during normal development. The patterns of several sets of isoenzymes change during normal development in tissues from many species.

Changes in the relative proportions of several isoenzymes are seen during the embryonic development of skeletal muscle. The proportions of the electrophoretically more-cathodal isoenzymes of both lactate dehydrogenase and creatine kinase increase in this tissue, so that the qualitative patterns associated with the differentiated muscle are present by about the sixth month of intrauterine life. Smaller, quantitative changes in isoenzyme distribution may continue to birth and into early postnatal life. An increased proportion of anodal lactate dehydrogenase isoenzymes has been noted in muscle tissue of normal subjects over 60 years of age; thus, the pattern tends toward that seen in young children.

As might be expected from the highly specialized nature of the metabolism of the liver, this tissue shows characteristic changes in the patterns of several isoenzymes during embryogenesis. In early fetal development, three aldolase isoenzymes, A, B, and C, together with the various hybrid tetramers, can be detected in extracts of liver. However, at birth aldolase B is the predominant isoenzyme, as in adult liver. Striking changes in the distribution of isoenzymes of alcohol dehydrogenase also occur in human liver during prenatal development.

The changes in isoenzyme patterns during development result from changes in the relative activities of gene loci within developing cells of a particular type, e.g., muscle cells. Other alterations in the balance of isoenzymes within the whole organism may derive from changes in the number or activity of cells that contain large amounts of a characteristic isoenzyme. An example of this is the increased number and activity of the osteoblasts that are responsible for mineralization of the skeleton between the early postnatal period and the beginning of the third decade of life. The excess of alkaline phosphatase from the active osteoblasts enters the circulation, where its presence can be recognized by its characteristic properties and where it elevates the total serum alkaline phosphatase activity of young persons above that of skeletally mature adults (Figure 5-18). An alkaline phosphatase from the liver also contributes to the total activity of this enzyme in normal plasma, and the amount of this isoenzyme in plasma shows a small, progressive increase with age (Figure 5-18). The reason for the latter age-dependent change is not known, but it may result from increased synthesis of the isoenzyme by hepatocytes in response to continuing exposure to inducing factors.

Changes in isoenzyme distribution in disease. Certain diseases, such as the progressive muscular dystrophies, appear to involve a failure of the affected tissues to mature normally or to maintain a normal state. Cancer cells show a progressive loss of the structure and metabolism of the healthy cells from which they arise. Therefore, the pattern of isoenzymes of mature, differentiated tissue may be lost or modified if normal differentiation is arrested or reversed, and many examples of isoenzyme changes accompanying such processes have been reported.

The distributions of isoenzymes of aldolase, lactate dehydrogenase, and creatine kinase in the muscles of patients with progressive muscular dystrophy have been found to be similar to those seen in the earlier stages of development of fetal muscle. The isoenzyme abnormalities seen in dystrophic muscle have been interpreted as a failure to reach or maintain a normal degree of

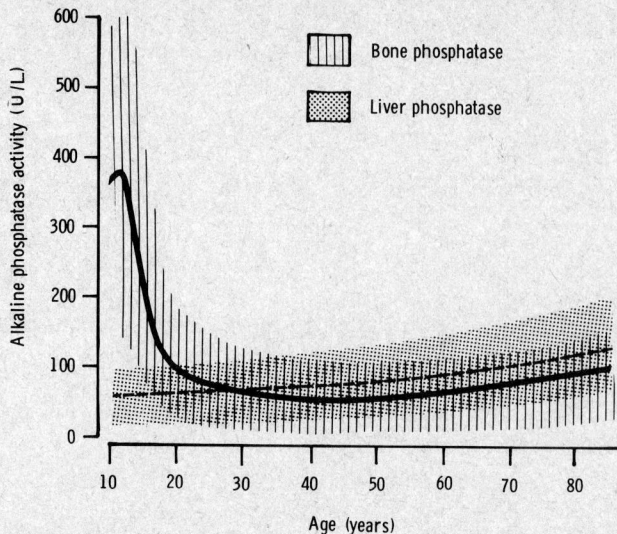

Figure 5-18. Age-dependent changes in the activities of bone and liver alkaline phosphatases in serum, determined by a selective heat-inactivation method.[107] The activities were determined at 37 °C by the Scandinavian recommended method. (Reproduced with permission from Moss, D. W.: Isoenzymes. London, Chapman & Hall, 1982.)

differentiation. Isoenzyme patterns in regenerating tissues may also show some tendency to approach fetal distributions. This tendency may result from relaxation or modification of control systems in rapidly dividing cells and may account for some of the isoenzyme changes seen, for example, in muscle in acute polymyositis.

A re-emergence of fetal patterns of isoenzyme distribution is also a feature of malignant transformation in many tissues. This phenomenon was first studied extensively in the case of lactate dehydrogenase isoenzymes. Malignant tumors in general show a significant shift in the balance of isoenzymes toward the electrophoretically more-cathodal forms, LDH-4 and LDH-5. The decline in activity of the LDH-1 and LDH-2 isoenzymes results in patterns which are reminiscent of those occurring in embryonic tissues. Tumors of prostate, cervix, breast, brain, stomach, colon, rectum, bronchus, and lymph nodes are among those that show this transformation. In contrast, comparatively benign gliomas show a relative increase in anionic isoenzymes. A relative increase in the proportion of cathodal isoenzymes of lactate dehydrogenase has also been observed in tissues adjacent to malignant tumors, e.g., of the colon, although the cells in these regions are morphologically normal.

The isoenzyme patterns of aldolase, pyruvate kinase, and hexosaminidase have also been shown to undergo a change toward fetal-like patterns in hepatoma. Other isoenzyme systems that differ in neoplastic cells from their counterparts in normal tissues include malate dehydrogenase and aspartate aminotransferase, in which the balance between cytoplasmic and mitochondrial isoenzymes is altered.

In 1968, Fishman and his collaborators reported the identification of an alkaline phosphatase in the serum of a patient with metastatic squamous cell carcinoma of the lung that was biochemically and immunologically identical with the alkaline phosphatase of normal placenta.[40] The newly discovered isoenzyme was termed the Regan isoenzyme after the patient in whom it was discovered. The Regan isoenzyme has been detected in tumor tissues and in sera of patients with many types of malignant disease, and also in some patients with nonmalignant diseases. An incidence of the isoenzyme of 3–15% in sera of cancer patients has been estimated, but this varies with the sensitivity of the methods used for its detection. Other variant forms of alkaline phosphatase have since been discovered in tumor tissues. These variants also show many similarities to both normal placental alkaline phosphatase and the Regan isoenzyme but differ from them in such properties as response to some inhibitors.

Nongenetic Causes of Multiple Forms of Enzymes

Many different types of post-translational modification of enzyme molecules can give rise to multiple forms (Figure 5-19). Several of these processes have been shown to cause the heterogeneity of various enzymes, either in living matter or as a result of changes taking place during extraction or storage.

Modification of the residues in the polypeptide chains of enzyme molecules can take place in living cells to give multiple forms; e.g., removal of amide groups accounts for some of the heterogeneity of amylase and carbonic anhydrase. (These enzymes also each exist as true isoenzymes.) Modification can also take place as a result of extraction procedures; many erythrocyte enzymes, including adenosine deaminase, acid phosphatase, and some forms of phosphoglucomutase, contain sulfhydryl groups that are susceptible to oxidation. In hemolysates, oxidation may be brought about by the action of oxidized glutathione, although in the intact cells this compound is present in its reduced form. Thus, variant enzyme molecules with altered molecular charge may be generated.

Modifications affecting nonprotein components of enzyme molecules may also lead to molecular heterogeneity. Many enzymes are glycoproteins, and variations in carbohydrate side chains are a common cause of nonhomogeneity of preparations of these enzymes. Some carbohydrate moieties, notably N-acetyl neuraminic acid (sialic acid), are strongly ionized and consequently have a profound effect on some properties of enzyme molecules in which they occur. For example, removal of terminal sialic acid groups from human kidney alkaline phosphatase with neuraminidase greatly reduces the electrophoretic heterogeneity of the enzyme.

Aggregation of enzyme molecules with each other or with nonenzymic proteins may give rise to multiple forms that can be separated by techniques that depend on differences in molecular size. A number of examples of enzyme heterogeneity caused in this way have been described. For example, four catalytically active cholinesterase components with molecular weights ranging from about 80 000–340 000 are found in most sera, with the heaviest component, C_4, contributing most of the enzyme activity. Other enzyme forms are also present occasionally, but it appears that the principal serum cholinesterase fractions can be attributed to different states of aggregation of a single monomer.

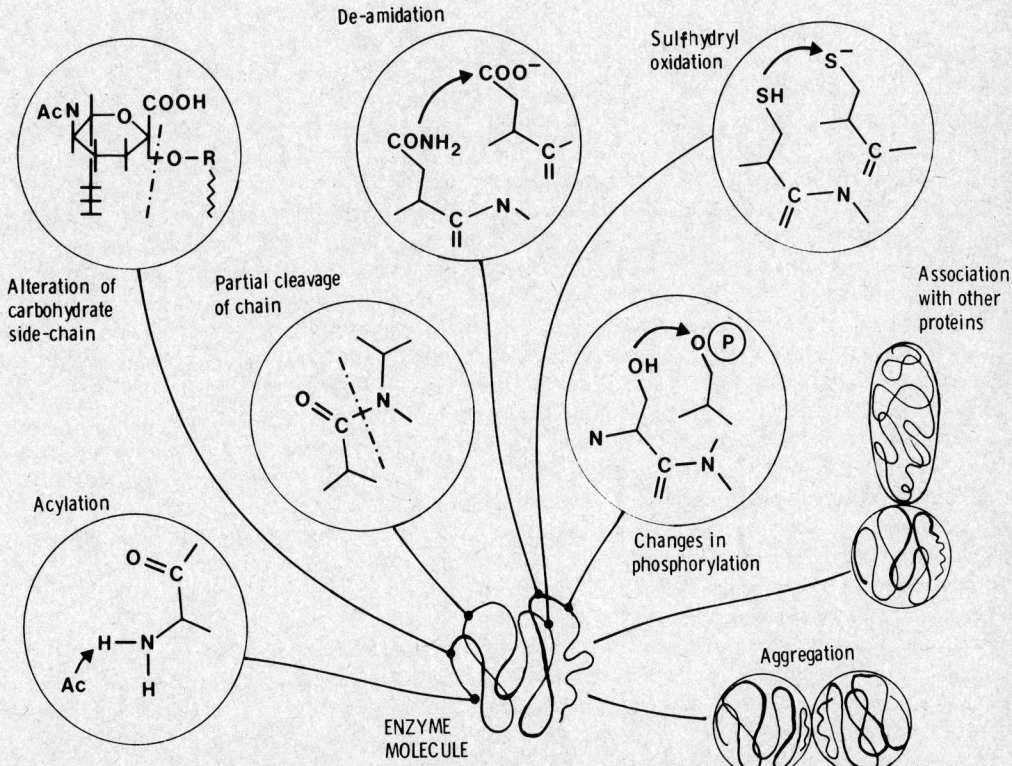

Figure 5-19. Nongenetic modifications which may give rise to multiple forms of enzymes. (From Moss, D. W.: Isoenzymes. London, Chapman & Hall, 1982.)

A specific form of interaction between enzymic and nonenzymic proteins is the cause of unusual enzyme components seen when some samples of human plasma are fractionated by electrophoresis or chromatography. These components are due to combination of apparently normal enzyme or isoenzyme molecules with plasma immunoglobulins. The enzyme-protein complexes thus formed may themselves be heterogeneous. Since the identification of "macroamylase," the first such enzyme-immunoglobulin complex to be identified, similar complexes involving lactate dehydrogenase, alkaline phosphatase, and creatine kinase have been observed.

A single polypeptide chain can in theory exist in an infinite number of different conformations. However, one specific *conformation* generally appears to be the most stable for any given sequence of amino acids, and this conformation is assumed by the chain as it is synthesized within the cell. Thus, the primary structure of the polypeptide chain also determines its three-dimensional secondary and tertiary structures. It is conceivable that, in some cases, there may be several alternative conformations ("conformers") of a single chain that are of nearly equal stabilities and, therefore, that these alternative forms may coexist. This possibility was first suggested to account for the heterogeneity that is seen in preparations of the cytoplasmic and mitochondrial isoenzymes of malate dehydrogenase and has also been proposed as an explanation of the multiple electrophoretic zones of erythrocyte acid phosphatase. However, no multiple enzyme forms have been shown unequivocally to be due to conformational isomerism.

Differences in Properties Between Multiple Forms of Enzymes

The structural differences between the multiple forms of an enzyme give rise to greater or lesser differences in physicochemical properties such as electrophoretic mobility, resistance to inactivation, or solubility, or in catalytic characteristics such as ratio of reaction with substrate analogs or response to inhibitors. Methods of isoenzyme analysis are therefore designed to investigate a wide range of catalytic and structural properties of enzyme molecules.[102] However, it is usually possible to make only limited deductions as to the nature of the underlying structural differences between isoenzymes that are responsible for the dissimilar properties. Equally, the changes in catalytic and other properties that may result from specific structural alterations in enzyme molecules can rarely be predicted from current theoretical knowledge of the relationship between structure and function of proteins.

Isoenzymes due to the existence of multiple gene loci usually differ quantitatively in catalytic properties. These differences may be seen in such characteristics as molecular activity, Michaelis constants for the substrate or substrates, sensitivity to various inhibitors, and relative rates of activity with substrate analogs (when the specificity of the isoenzymes allows the substrate to be varied), and it is presumably these differences that are the basis of the biological importance of this type of isoenzymic variation. In contrast, multiple enzyme forms that arise by such posttranslational modifications as aggregation usually have similar catalytic properties.

Multilocus isoenzymes also usually differ in antigenic specificity, although there are exceptions to this generalization among isoenzymes that have presumably emerged relatively recently in evolutionary history and are closely related in structure. Immunological cross-reaction is also not uncommon among multilocus isoenzymes. Multiple enzyme forms due to postsynthetic modification frequently have common antigenic determinants. Isoenzymes derived from allelic genes ("allelozymes") are often antigenically similar, even to the extent that substances that cross-react with antisera to the common isoenzyme may be detectable when mutation has abolished enzyme activity altogether. However, the greater specificity of monoclonal antibodies seems capable of revealing differences between otherwise antigenically similar isoenzyme molecules.

Differences in resistance to denaturation, e.g., by heat or concentrated urea solutions, are commonly found between true isoenzymes, whether these are the products of multiple loci or multiple alleles. Other multiple forms of enzymes often do not differ, or differ only slightly, in this respect. The most commonly exploited difference between isoenzymes is the difference in net molecular charge that results from the altered amino acid compositions of the molecules; this forms the basis of separation by zone electrophoresis, ion-exchange chromatography, or isoelectric focusing. Separation methods that depend on differences in molecular size, such as gel filtration, do not distinguish between the small differences in size which often exist between true isoenzyme molecules but are important in the detection of multiple forms that involve aggregation, or association of enzyme molecules with other proteins.

Zone electrophoresis is the most generally useful technique of isoenzyme analysis. When suitably sensitive methods are used to detect and to measure the separated enzyme zones, electrophoresis requires only small volumes of the specimen. Certain supporting media, notably starch and polyacrylamide gels, have pores of sizes that approach the molecular dimensions of protein molecules. This introduces an element of segregation according to size into the resolution of protein mixtures on these media. When the members of the isoenzyme system do not differ markedly from each other in molecular size, as is the case for lactate dehydrogenase isoenzymes, the use of gel media does not significantly affect the electrophoretic separation of the isoenzymes, and similar patterns are obtained with both nonsieving supports such as cellulose acetate film and sieving media such as starch gel. However, the restricted diffusion of protein molecules and reaction products in the gels tends to preserve discrete enzyme zones during staining after electrophoresis. Against this advantage must be set the greater technical difficulty of preparing and handling gels compared with films, the relative impermeability of some gels to reagents used for enzyme location, and difficulties that may be encountered in densitometric scanning of stained gels.

The methods used to locate isoenzyme zones after electrophoresis are usually adaptations of those used in histochemistry to demonstrate enzyme activities in tissue sections by forming a colored precipitate. For example, in locating zones of lactate dehydrogenase activity, oxidation of the substrate, lactate, is coupled through several stages to the reduction of a tetrazolium salt:

After a suitable interval the reaction is stopped and the colored bands are fixed by treating the strip with dilute acetic acid. With suitable modifications, this sequence of reactions can be applied to the location of other enzymes that directly, or indirectly through coupled reactions, effect the reduction of NAD^+ or $NADP^+$.

When the purpose of isoenzyme separation is to distinguish between possible alternative sources of a raised enzyme activity, as for instance between heart and liver as sources of lactate dehydrogenase in serum, inspection of the relative intensity of the colored isoenzyme zones may provide sufficient information. However, in order to fully exploit the potential diagnostic sensitivity of isoenzyme separation, quantitative determination of the relative proportions of the different fractions is necessary.

Lactate dehydrogenase isoenzymes are well separated by zone electrophoresis. Densitometric scanning of stained electrophoretic strips therefore gives a good estimate of the activities of the individual isoenzymes, provided the total activity applied to the strip and the staining conditions generate an intensity of the color proportional to the enzymatic activity. Creatine kinase isoenzymes also separate well on electrophoresis, but the need to detect and measure low activities of the MB isoenzyme calls for the use of sensitive techniques, such as measurement of the fluorescence of NADPH in a suitable fluorescence scanner. The fluorescence of NAD(P)H can similarly be used in the measurement of other isoenzymes after electrophoretic separation.

Elution of electrophoretically separated isoenzyme zones, followed by measurement of the enzymatic activities of the eluates, provides an alternative to densitometric methods of isoenzyme quantitation. This approach has been applied to creatine kinase. However, the volume of sample that can be applied to some media, such as cellulose acetate, is limited. Therefore, sensitive methods are needed to assay the activity eluted from the strip. Recoveries from other media, such as starch gel, are low or variable.

The qualitative separation of isoenzymes by zone electrophoresis is not in general affected by changes in composition of the serum itself. Dissociation and regrouping of subunits during storage of serum samples, or the selective inactivation or alteration of certain isoenzymes, have been reported to change the electrophoretic patterns of both lactate dehydrogenase and creatine kinase isoenzymes. While these possibilities must be kept in mind, isoenzyme patterns determined by electrophoresis are, in the main, characteristic and reproducible. Consequently, a considerable advantage of electrophoresis is the high degree of certainty that it brings to the identification of specific isoenzymes.

Ion-exchange chromatography also makes use of differences in net molecular charge at a given pH to separate isoenzymes. A typical ion-exchange material is DEAE-cellulose, in which ionizable diethyl-aminoethyl groups are attached to an inert cellulose matrix. Ion-exchange chromatography is not in general as highly resolving of closely similar proteins as is zone electrophoresis, but relatively large amounts of proteins can be separated with good recoveries of enzymatic activity, so that the method is of great value in enzyme purification.

Measurement of the enzymatic activity of each of the multiple fractions eluted by the gradient method from an ion-exchange column is laborious and elution may take several hours. When the isoenzymes to be separated are quite distinct in their elution characteristics, stepwise elution or batch processing may be used. In the first technique, a small ion-exchange column is washed with a given volume of eluent of constant composition. The total eluate is collected for measurement of enzymatic activity. The column is then washed with a further fixed volume of eluent of a different composition, so as to elute a second isoenzyme, and the whole eluate is collected. Further washings may be interposed between the two elution steps. In batch processes, the sample and successive eluents are shaken with solid ion-exchange material in which the inert support may be beads of glass or macroporous resin.

The simplicity of stepwise ion-exchange methods makes them suitable for routine analysis of serum samples, and they have been widely applied to the quantitation of creatine kinase isoenzymes. However, these methods are susceptible to error if all the activity of a particular isoenzyme is not eluted by the appropriate eluent, or by succeeding washes, so that a fraction of its activity is eluted with a second isoenzyme whose activity is consequently overestimated. Errors of this type are particularly likely to affect estimates of the second isoenzyme when the isoenzyme eluted in the first step is present in the original mixture in large excess compared with the second isoenzyme. This is the case in stepwise elution methods for the determination of the MB isoenzyme of creatine kinase, where incomplete removal of the MM isoenzyme causes overestimation of the MB form.

Other forms of chromatography that have been applied to fractionation of isoenzyme mixtures include high-performance liquid chromatography and affinity chromatography. The latter makes use of differences between isoenzymes in their affinities for a specific ligand that is attached to an inert insoluble support used as the stationary phase in a chromatography column or in a batch technique.

Probably the most highly resolving of all protein separation methods that depend on differences in ionization characteristics is isoelectric focusing, in which protein molecules migrate through a stabilized pH gradient under the influence of a potential difference applied across it and concentrate as discrete zones at the pH regions corresponding to their isoelectric points. However, the time required to reach equilibrium is lengthy, enzymes are susceptible to inactivation at their isoelectric pH values, and interaction may occur with the carrier ampholytes used to produce the pH gradient.

Selective inactivation under controlled conditions has become an important technique in isoenzyme characterization. The method is based on differences in stability that result from small changes in the structure of protein molecules. Elevated temperatures or concentrated solutions of urea or other reagents are frequently chosen to denature the enzyme. Rates of enzyme inactivation by these agents are critically dependent on the conditions of the experiment, which must therefore be strictly controlled if reliable comparisons between samples are to be made. For example, temperature coefficients of inactivation by heat are of the order of 10–100 or more, so that variations in temperature of fractions of a degree markedly alter rates of inactivation. For heat inactivation studies, therefore, large, well-stirred water baths with high-performance thermostats are needed to minimize temperature fluctuations, and the duration of inactivation must be accurately timed. When urea is used as the inactivating agent, freshly prepared solutions must be used since cyanate is formed on storage of urea solutions and may itself act as an enzyme inhibitor.

Selective inactivation is a useful semiquantitative or quantitative method for determining the composition of isoenzyme mixtures in serum. However, the results are less easy to interpret when more than two isoenzymes are present, especially when hybrid isoenzymes form part of the mixtures, since these forms possess stability characteristics intermediate between those of their parent homopolymeric isoenzymes.

Differences in catalytic properties, such as differences in Michaelis constants, relative rates of reaction with substrate analogs (when the specificity of the enzyme allows for variation in the structure of the substrate), pH optima, and response to inhibitors, typically exist between isoenzymes that are the products of multiple gene loci. These differences can be made the basis of methods of identification and measurement of particular isoenzymes. Under the most favorable circumstances, this approach to isoenzyme measurement has the advantage that only slight changes in the usual method of measuring enzyme activity may be needed to provide information on the isoenzyme composition of a sample. Differences in catalytic properties have particularly been used in isoenzyme studies of lactate dehydrogenase and acid phosphatase.

Immunochemical methods of isoenzyme analysis are particularly applicable to isoenzymes derived from multiple gene loci, since these are usually antigenically distinct. Some of these methods make use of catalytic activity of the isoenzymes. For example, residual activity may be measured after reaction with antiserum. Alternatively, enzymatic activity may be used to locate the enzyme-antigen precipitate in immunodiffusion or "rocket" electroimmunoassays if the reaction with antibody does not inhibit activity. Radioimmunoassays, in which isoenzyme labeled with a radioactive tracer competes with unlabeled isoenzyme for antibody-binding sites, have also been applied to isoenzyme measurement, and these methods do not depend on the catalytic activity of the isoenzyme being determined.

Immunoinhibition has become a popular method for the determination of MB creatine kinase in serum, and various methods have been developed for the measurement of prostatic acid phosphatase. Radioimmunoassays have been developed for both these isoenzymes and for placental alkaline phosphatase.

The choice and application of various methods of isoenzyme analysis in clinical enzymology are discussed later in relation to specific isoenzyme systems.

PRINCIPLES OF DIAGNOSTIC ENZYMOLOGY

All of the hundreds of different enzymes present in the human body are synthesized intracellularly, and most of them carry out their functions within the cells in which they are formed. However, certain enzymes are secreted, usually in an inactive form, and, after activation, function within the extracellular fluids. The most obvious examples of these enzymes are the proteases and other hydrolases secreted into the gastrointestinal tract, some of which (e.g., pancreatic amylase) may find their way into the blood. Other examples of secreted enzymes include the enzymes of the blood-clotting mechanism and those concerned with fibrinolysis, which become functionally active in the plasma and are referred to as plasma-specific enzymes (Table 5-2). Cholinesterase is secreted in an active form into the plasma by the liver, but its function remains unclear.

However, with a few exceptions, such as serum cholinesterase, the clinical chemist is principally concerned with changes in the activity in serum or plasma of enzymes that are predominantly intracellular and that are normally present in the serum at low activities only. By measuring changes in the activities of these enzymes in disease, he attempts to infer the location and nature of pathological changes in the tissues of the body. Therefore, it is necessary to understand the factors that affect the rate of release of enzymes from their cells of origin and the rate at which

Table 5-2. CLASSIFICATION OF ENZYMES IN BLOOD

Classification	Examples
Plasma-specific enzymes	Serine protease procoagulants: thrombin, factor XII (Hageman factor), factor X (Stuart-Prower factor), and others
	Fibrinolytic enzymes or precursors: plasminogen, plasminogen proactivator
Secreted enzymes	Lipase, α-amylase (from salivary glands and pancreas), trypsinogen, cholinesterase, prostatic acid phosphatase
Cellular enzymes	Lactate dehydrogenase, aminotransferases, alkaline phosphatases, and others

they are cleared from the circulation, so that changes in activity in disease can be interpreted correctly.

Factors Affecting Enzyme Levels in Plasma or Serum

The measured level of activity of an enzyme in blood is the result of the balance between the rate at which it is entering the circulation from its cells of origin and the rate at which it is being inactivated or removed.

Clearance of enzymes.[43,78] Little is known about the way in which enzymes are cleared from the circulation. Few enzyme molecules are small enough to pass through the healthy glomerulus of the kidney, and therefore urinary excretion is not a major route for elimination of enzymes from the circulation. An exception to this is amylase; increased levels of this enzyme in the blood, e.g., following acute pancreatitis, are accompanied by increased excretion in the urine. Older views that enzymes are excreted through the liver and into the bile have largely been abandoned, and present evidence suggests that inactivation begins in the plasma and that inactivated enzymes are rapidly removed, probably by the reticuloendothelial system. The half-lives of enzymes in plasma vary from a few hours to several days, but in most cases an average half-life of 24–48 h can be expected (Table 5-3).

The existence of circulating inhibitors or activators of enzymes similarly has little effect on activities measured in the laboratory. Where these inhibitors are of a reversible nature, the relatively high dilutions of serum used in most modern enzyme assays are sufficient to eliminate any possible effects from this cause. Among the few exceptions to this rule that have been noted is the accumulation of inhibitors of lactate dehydrogenase that may occur in the sera of patients with chronic renal disease.

Entry of enzymes into the blood. Since the foregoing effects can be regarded as minor as far as the measured levels of enzymes in serum are concerned, the remaining and by far the most important factors that affect enzyme activities in serum or plasma are those that influence the rate at which enzymes enter the circulation from the cells. These factors can be divided into two main categories: those that affect the rates at which enzymes leak from cells; and those that reflect altered rates of enzyme production, due either to increased synthesis of a particular enzyme by individual cell types or to proliferation of a particular type of enzyme-producing cell.

Leakage of enzymes from cells.[43,78] Enzymes are retained within their cells of origin by the plasma membrane surrounding the cell. The plasma membrane is a metabolically active part of the cell, and its integrity depends on the cell's energy production. Any process that impairs

Table 5-3. RATES OF DECAY OF COMMON SERUM ENZYMES IN MAN (SOME REPORTED VALUES FROM THE LITERATURE)

Enzyme	$t_{1/2}$ (Hours)	Molecular Weight
Alanine aminotransferase	50	110 000
	151	
Aspartate aminotransferase		
Mixed isoenzymes, following myocardial infarction	12	—
Cytoplasmic isoenzyme	20	120 000
Mitochondrial isoenzyme	35	100 000
Lactate dehydrogenase		
LDH-1	113	135 000
	163	
LDH-5	10	135 000
	34	
Creatine kinase		
Mixed isoenzymes, following myocardial infarction	12	—
CK-3 (CK-MM)	13	85 000
CK-2 (CK-MB)	6	87 000
CK-1 (CK-BB)	2	88 400
Alkaline phosphatase		
Intestinal	<1	—
Bone	40	—
Placenta	170	120 000

energy production, either by depriving the cell of oxidizable substrates or by reducing the efficiency of energy production by restricting the access of oxygen, will promote deterioration of the cell membrane. The membrane will become leaky and, if cellular injury becomes irreversible, the cell will die. Small molecules are the first to leak from damaged or dying cells, followed by larger molecules such as enzymes, and ultimately the whole of the contents of necrotic cells are discharged.

Because of the very high concentrations of enzymes within the cells—thousands or even tens of thousands times greater than the concentrations in extracellular fluid—and because of the sensitivity with which small amounts of enzyme can be detected by their catalytic activity, an increase of enzyme activity in the extracellular fluid or plasma is an extremely sensitive indicator of even minor cellular damage, some causes of which are listed in Table 5-4.

Direct attack on the cell membranes by such agents as viruses or organic chemicals is an obvious cause of enzyme release, and one which is particularly important in the case of the liver. However, a reduction in the supply of oxygenated blood perfusing *any* tissue will promote enzyme release. The most obvious clinical condition in which such a reduction occurs is myocardial infarction, when occlusion of a branch of a coronary artery, usually by a blood clot, deprives that segment of the heart served by the artery of its supply of oxygen and oxidizable metabolites. The cells of the affected region rapidly begin to deteriorate and die, releasing their enzyme contents. Transfer of the enzymes to the extracellular fluid and thence to the systemic circulation accounts for the rapid rise in serum enzyme activity that is characteristic of this condition.

Hepatic infarcts are rare because of the dual blood supply of the liver from the hepatic artery and the portal system. Blood from these two sources transverses the sinusoids and drains into the central veins (in the centrilobular area), which in turn drain into the hepatic veins. The hepatic veins join the inferior vena cava, which goes to the right atrium of the heart.

The liver is very sensitive to hypoxia, which can result from a reduction in arterial blood flow due to diminished cardiac output (heart failure). Another factor in heart failure is hepatic venous hypertension due to a backup of blood from the failing right side of the heart. The hypoxia due to both of these factors is most severe in the centrilobular areas of the liver, and cellular necrosis with release of cellular enzymes then occurs. A third factor that might contribute to centrilobular hypoxia is arterial hypoxemia. These effects account for the common observation that increased activities of hepatocellular enzymes in the blood accompany a wide variety of conditions such as *congestive heart failure, shock,* and *hypoxemia*. Severe congestive heart failure

Table 5-4. CAUSES OF CELL DAMAGE OR DEATH

Category	Examples
Hypoxia (an extremely common accompaniment of clinical disease)	Loss of blood supply due to narrowing (atheromatous plaques) or blocking (thrombosis) of artery or vein; inadequate oxygenation due to cardiorespiratory failure; loss of oxygen-carrying capacity (anemia)
Chemicals and drugs (an important cause of cellular damage)	Environmental pollutants—lead, mercury; drugs—use and abuse; alcohol; tobacco
Physical agents	Trauma; extremes of heat and cold; radiation; electrical energy
Microbiological agents	Bacteria, viruses, fungi, protozoa, and helminths
Immune mechanisms	Immune disorders can cause tissue damage by a number of mechanisms: 1. Anaphylaxis (causing release of vasoactive amines) 2. Cytotoxicity (causing the target cell to be lysed) 3. Immune complex disease (leading to release of lysosomal enzymes) 4. Cell-mediated hypersensitivity (leading to cytotoxicity)
Genetic defects	Disorders with polygenic inheritance—diabetes mellitus, gout Mendelian disorders—X-linked disorders, autosomal dominant and recessive disorders, disorders with variable modes of transmission
Nutritional disorders	Protein-calorie malnutrition, vitamin deficiencies, mineral deficiencies

After the classification of Robbins, Angell, and Kumar. *In*: Basic Pathology, 3rd ed. Philadelphia, W.B. Saunders, 1981.

may result in central hemorrhagic necrosis, which gives the cut surface of the liver a "nutmeg" appearance. If heart failure is prolonged, this process eventually leads to scarring around the central veins of the liver lobules, a condition known as *cardiac sclerosis* or *cardiac cirrhosis.*

Other tissues that may contribute enzymes to the blood, as a result of enzyme leakage, include skeletal muscle. Again, the cause can often be traced to poor perfusion of the muscles by oxygenated blood, or to hypothermia, with reduced energy production, as well as to direct trauma of the muscles such as results from crush injuries. Disorders of skeletal muscle due to infection, inflammation (e.g., *polymyositis*), degenerative changes (e.g., *dystrophies*), drugs, and alcohol (e.g., *alcoholic myopathy*) will also cause enzyme leakage from myocytes. Enzyme release from muscles and other tissues also occurs as a result of anesthesia, probably due to a combination of impaired perfusion of the tissues and a direct toxic effect of anesthetic agents on cell membranes.

Efflux of enzymes from damaged cells. Once the conditions for leakage of enzymes from cells have become established, the speed and extent with which the process is reflected in enzyme changes in the blood depend on several factors.

The driving force of enzyme release is the steep concentration gradient of enzyme activity that exists between the interior and the exterior of the cells. The rate of escape of enzyme molecules is presumably controlled to some extent by diffusion; therefore, smaller enzyme molecules might be expected to appear in the extracellular fluid earlier than larger ones. Although this is probably true to some extent, the range of molecular weights covered by enzymes of diagnostic interest is rather narrow, so that this factor may be of lesser importance.

The way in which released enzyme molecules are transferred from the interstitial fluid to the blood probably varies from one tissue to another. Direct transfer probably occurs to a large extent in liver, which is a highly vascular tissue with many permeable capillaries. On the other hand, the capillaries of skeletal muscle are relatively impermeable, and in this tissue it is probable that a large proportion of the released enzyme reaches the circulation by way of the lymph. This can be demonstrated in experiments in which the rate of entry of enzymes released from an ischemic limb is delayed if the limb is immobilized, but is accelerated if the limb is exercised, thus stimulating lymph drainage. Lymph drainage is also important in transporting enzymes released from damaged myocardial cells to the circulation. In this case, the pumping action of the heart maintains a fairly constant lymph flow.

The *intracellular location* of the leaking enzymes also affects the rates at which they will appear in the circulation. As would be expected, the most sensitive indicators of cell damage are the enzymes that are present in the soluble fraction of the cell. Enzymes associated with subcellular structures such as mitochondria are less readily released into the circulation; this fact has been used in attempts to distinguish reversible leakage, presumed to reflect damage only to the cell membrane, from necrotic lesions in which the intracellular structures are destroyed. Distinction is made by comparing the relative amounts of cytoplasmic and mitochondrial enzymes or isoenzymes released into the circulation. However, these attempts have not been translated into general practical use.

The relation between *tissue injury* and the appearance of enzymes in the circulation is most clearly seen in the condition of myocardial infarction, in which a relatively short episode of damage is followed by a rapid transfer of enzymes to the circulation. About 24 h after a myocardial infarction, the pattern of relative activity of various enzymes in the circulation comes to resemble closely that in the myocardial tissue. These relationships are less clearly seen in other conditions, such as chronic liver disease, in which enzyme release is a process that continues over a period of time. The pattern of relative enzyme activities in serum in chronic disease may also become distorted by differential rates of removal of enzymes from the circulation, and possibly also by differential changes of rates of enzyme synthesis in the affected tissue.

Enzyme changes reflecting altered enzyme production. Some of the small amounts of intracellular enzymes normally present in the plasma can be assumed to result from wear and tear of cells or overflow of enzyme from healthy cells. This contribution of enzymes to the circulating blood may fall, either as the result of a genetic deficiency of enzyme production (e.g., as is the case for alkaline phosphatase in hypophosphatasia or in individuals homozygous for the "silent" gene for serum cholinesterase), or when enzyme production is depressed as a result of disease. (For example, the production of serum cholinesterase is often depressed in liver disease.) However, cases in which enzyme production is *increased* are of more general interest in diagnostic enzymology.

An increase in the number and activity of the alkaline phosphatase-producing cells of bone (the osteoblasts) is responsible for the increased level of alkaline phosphatase in the serum of normally growing children. Increased osteoblastic activity also accounts for the increased levels of this enzyme in serum in various types of bone disease. Toward the end of normal pregnancy, the placenta constitutes a new source of alkaline phosphatase, which contributes its characteristic isoenzyme to the maternal circulation.

It is now recognized that biliary obstruction stimulates the production of alkaline phosphatase by the liver, and this example of **enzyme induction** is the source of much if not all of the raised activity of this enzyme in hepatobiliary disease. However, the nature of the inducer is not known in this case. Enzyme induction by other agents, such as drugs or alcohol, may also increase enzyme production; a now familiar example of such induction is the increased activity of γ-glutamyltransferase in serum that may result from the administration of drugs such as the barbiturates or phenytoin and from an excessive intake of ethanol.

Proliferation of acid phosphatase–producing cells accounts for the increased levels of prostatic-type acid phosphatase that are characteristic of about three-quarters of all cases of metastatic cancer of the prostate. As the malignant cells spread beyond the capsule of the prostate gland, they are more favorably placed to contribute their characteristic prostatic acid phosphatase isoenzyme to the circulation. However, in this case, the amount of enzyme produced by each cell is often less than that produced by a normal prostatic cell and indeed may decline to zero as the cancer cell becomes less and less like the prostatic cell from which it has originated.

Other mechanisms. Release of enzymes from damaged or dying cells or changes in the rate of enzyme production constitute the most important mechanisms by which changes in enzyme activity in the serum or plasma are produced. However, other possibilities exist and appear to account for some changes of diagnostic importance. For example, much of the γ-glutamyltransferase activity of liver cells is located on their exterior surfaces. It is possible that ectoenzymes such as this may be eluted from the surfaces, especially where there is increased detergent action of the blood through the accumulation of bile salts. This process does not involve cell damage, in the sense of increased membrane permeability, as evidenced by the lack of correlation between the activities of γ-glutamyltransferase and aminotransferases in serum in liver diseases of different types.

Selection of Enzyme Tests

The selection of which enzyme to measure in serum for diagnostic or prognostic purposes depends on a number of factors. An important factor is the distribution of enzymes among the various tissues, shown, for example, for aspartate and alanine aminotransferase and creatine kinase in Figure 5-20. The main enzymes of established clinical value together with their tissues or origin and their clinical applications are listed in Table 5-5.

Not all intracellular enzymes are equally valuable as indicators of cellular damage. For example, isocitrate dehydrogenase activity is high in heart muscle but, following a myocardial infarction, it is rapidly inactivated on entering the vascular compartment. Ornithine carbamyltransferase is an enzyme of the urea cycle and it is therefore almost totally liver-specific with a liver:blood ratio of about 10^5:1; thus, even minor degrees of hepatocyte damage should create a readily detectable elevation in blood levels. Yet this enzyme has not found particular favor in clinical enzymology, possibly because of the inconvenience of the assay and a consequent lack of wide clinical experience with the results of the assay.

The mass of the damaged or malfunctioning organ, together with the enzyme cell:blood gradient, obviously has a profound influence on the resulting elevation of enzyme activity in blood. Thus, the gradient of activity of prostatic acid phosphatase between prostate and blood is about 10^3:1 and the mass of that organ is 20 g. By contrast, the cell:blood gradient of alanine aminotransferase in the liver cell is 10^4:1 and the mass of the liver can exceed 1000 g. Obviously, fewer cells have to be damaged in the liver than in the prostate for the abnormality to be detected by an enzyme elevation in blood. If, on the other hand, there is total organ involvement, then clearly the vast number of affected liver cells will elevate blood levels of any liver enzyme quite markedly. It has been estimated that if only one liver cell in every 750 is damaged there would be a detectable elevation in the blood level of alanine aminotransferase.

Pathological damage to a tissue can embrace a wide spectrum of effects. Thus a mild, reversible, viral inflammation of the liver—such as a mild attack of viral hepatitis—is likely to

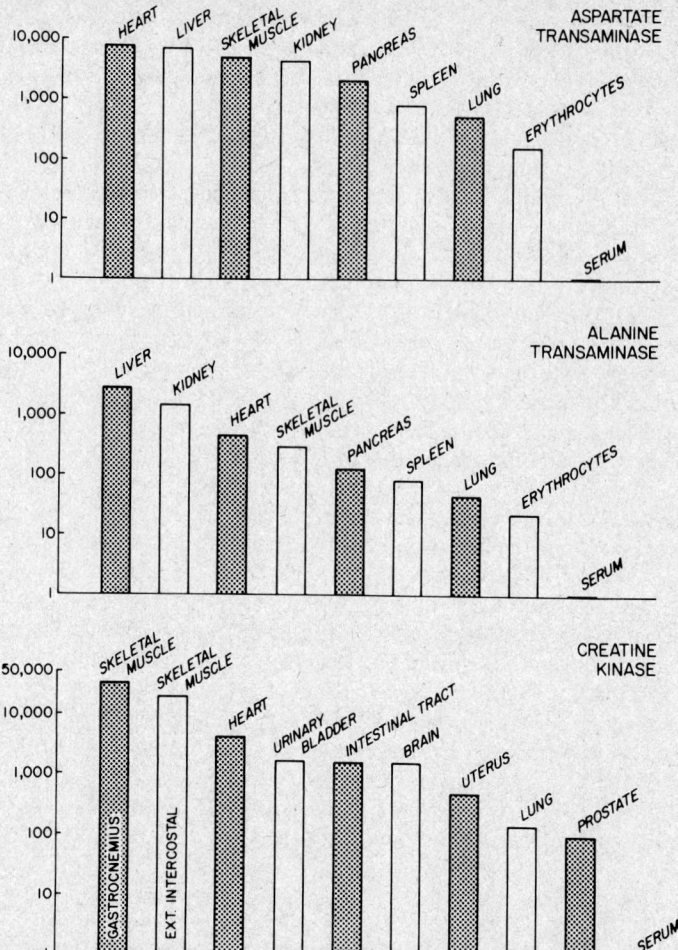

Figure 5-20. The concentration gradients between some human tissues and serum for aspartate transaminase, alanine transaminase, and creatine kinase. The concentration gradient axis is logarithmic.

Table 5-5. DISTRIBUTION OF DIAGNOSTICALLY IMPORTANT ENZYMES

Enzyme	Principal Sources	Principal Clinical Applications
Acid phosphatase	Prostate, erythrocytes	Carcinoma of prostate
Alanine aminotransferase	Liver, skeletal muscle, heart	Hepatic parenchymal disease
Aldolase	Skeletal muscle, heart	Muscle diseases
Alkaline phosphatase	Liver, bone, intestinal mucosa, placenta, kidney	Bone diseases, hepatobiliary diseases
Amylase	Salivary glands, pancreas, ovaries	Pancreatic diseases
Aspartate aminotransferase	Liver, skeletal muscle, heart, kidney, erythrocytes	Myocardial infarction, hepatic parenchymal disease, muscle disease
Cholinesterase	Liver	Organophosphorus insecticide poisoning, suxamethonium sensitivity, hepatic parenchymal diseases
Creatine kinase	Skeletal muscle, brain, heart, smooth muscle	Myocardial infarction, muscle diseases
Glutamate dehydrogenase	Liver	Hepatic parenchymal diseases
γ-Glutamyltransferase	Liver, kidney	Hepatobiliary disease, alcoholism
Lactate dehydrogenase	Heart, liver, skeletal muscle, erythrocytes, platelets, lymph nodes	Myocardial infarction, hemolysis, hepatic parenchymal diseases
5′-Nucleotidase	Hepatobiliary tract	Hepatobiliary disease
Sorbitol dehydrogenase	Liver	Parenchymal hepatic diseases
Trypsin(ogen)	Pancreas	Pancreatic diseases

increase only the permeability of the cell membrane and allow cytoplasmic enzymes to leak out into the blood, whereas a severe attack causing cell necrosis will also disrupt the mitochondrial membrane and both cytoplasmic and mitochondrial enzymes will be detected in blood. Thus, knowledge of the intracellular location of enzymes can assist in determining the nature and severity of a pathological process if suitable enzymes are assayed in the blood. However, as already mentioned, this possibility remains in an exploratory stage.

Finally, in selecting a suitable enzyme to assay in blood for diagnostic purposes, the rate at which its activity will disappear from blood is of some significance. An enzyme with a very short half-life in blood has little or no diagnostic value. Table 5-3 indicates that the most commonly assayed enzymes are those with half-lives in the range of 6 h or greater.

It must be emphasized that diagnostic decisions are made by integrating the history of the patient's illness, the results of clinical examination, and the analytical findings. Therefore, the results of enzyme analyses should always be assessed alongside the results of other analyses (such as bilirubin in acute hepatitis, for example) and the clinical findings.

Individual enzymes of clinical importance are discussed in detail in the following sections in the order shown below:

> Aspartate aminotransferase
> Alanine aminotransferase
> Creatine kinase
> Aldolase
> Lactate dehydrogenase
> Isocitrate dehydrogenase
> Glutamate dehydrogenase
> Alkaline phosphatase
> Biliary tract enzymes
>   5'-Nucleotidase
>   Gamma glutamyltransferase
> Digestive enzymes of pancreatic origin
>   Amylase
>   Lipase
>   Trypsin
>   Chymotrypsin
> Cholinesterase
> Acid phosphatase
> Terminal deoxynucleotidyl transferase

The selected order has a certain functional significance. Thus, enzymes used in the investigation of diseases of cardiac and skeletal muscle, liver parenchyma, biliary tract, and pancreas are associated together in the text. Clearly, there is considerable overlap in this classification; many enzymes are used for the investigation of disease in several organs (e.g., aspartate aminotransferase, alkaline phosphatase), but it must be evident that no single classification is entirely satisfactory.

THE AMINOTRANSFERASES (TRANSAMINASES)

ASPARTATE AMINOTRANSFERASE; ASPARTATE TRANSAMINASE

(EC 2.6.1.1; L-Aspartate:2-Oxoglutarate Aminotransferase, AST or AspAT; Formerly Glutamate Oxaloacetate Transaminase, GOT)

ALANINE AMINOTRANSFERASE; ALANINE TRANSAMINASE

(EC 2.6.1.2; L-Alanine:2-Oxoglutarate Aminotransferase, ALT or AlaAT; Formerly Glutamate Pyruvate Transaminase, GPT)

The aminotransferases constitute a group of enzymes that catalyze the interconversion of amino acids and α-oxoacids by transfer of amino groups. Three-letter abbreviations, AST and ALT, have been suggested for the two enzymes of greatest clinical significance. The term "transaminase" is accepted as an alternative to aminotransferase by the Enzyme Commission, and this term, together with the old abbreviations GOT and GPT, still remains in general use. Distinct isoenzymes of AST are present respectively in the cytoplasm and the mitochondria of cells. In conditions associated with a mild degree of tissue injury, the predominant form in serum is that from the cytoplasm, although some mitochondrial enzyme is also present. Severe tissue damage results in the release of much mitochondrial enzyme as well.

The α-oxoglutarate/L-glutamate couple serves as one amino group acceptor and donor pair in all amino-transfer reactions; the specificity of the individual enzymes derives from the particular amino acid that serves as the other donor of an amino group. Thus, aspartate aminotransferase catalyzes the reaction shown in equation (26). Alanine aminotransferase catalyzes the analogous reaction presented in equation (27).

$$
\begin{array}{ccccc}
\text{COO}^- & \text{COO}^- & & \text{COO}^- & \text{COO}^- \\
| & | & & | & | \\
\text{H}-\text{C}-\text{NH}_2 & \text{C}{=}\text{O} & \xrightarrow{\text{AST, P-5'-P}} & \text{C}{=}\text{O} & \text{H}-\text{C}-\text{NH}_2 \\
| & | & & | & | \\
\text{CH}_2 & \text{CH}_2 & + & \text{CH}_2 & + \quad \text{CH}_2 \\
| & | & & | & | \\
\text{COO}^- & \text{CH}_2 & & \text{COO}^- & \text{CH}_2 \\
& | & & & | \\
& \text{COO}^- & & & \text{COO}^-
\end{array}
\tag{26}
$$

L-Aspartate   α-Oxoglutarate   Oxaloacetate   L-Glutamate

$$
\begin{array}{ccccc}
\text{COO}^- & \text{COO}^- & & \text{COO}^- & \text{COO}^- \\
| & | & & | & | \\
\text{H}-\text{C}-\text{NH}_2 & \text{C}{=}\text{O} & \xrightarrow{\text{ALT, P-5'-P}} & \text{C}{=}\text{O} & \text{H}-\text{C}-\text{NH}_2 \\
| & | & & | & | \\
\text{CH}_3 & \text{CH}_2 & + & \text{CH}_3 & + \quad \text{CH}_2 \\
& | & & & | \\
& \text{CH}_2 & & & \text{CH}_2 \\
& | & & & | \\
& \text{COO}^- & & & \text{COO}^-
\end{array}
\tag{27}
$$

L-Alanine   α-Oxoglutarate   Pyruvate   L-Glutamate

The reactions are reversible, but the equilibria of the AST and ALT reactions favor formation of aspartate and alanine, respectively.

Pyridoxal-5'-phosphate (phosphopyridoxal, P-5'-P) and its amino analog, pyridoxamine-5'-phosphate, function as coenzymes in the amino transfer reactions. The P-5'-P is bound to the apoenzyme, and serves as a true prosthetic group. The pyridoxal-5'-phosphate bound to the apoenzyme accepts the amino group from the first substrate, aspartate or alanine, to form enzyme-bound pyridoxamine-5'-phosphate and the first reaction product, oxaloacetate or pyruvate, respectively. The coenzyme in amino form then transfers its amino group to the second substrate, α-oxoglutarate, to form the second product, glutamate. Pyridoxal-5'-phosphate is thus regenerated.

Serum or crude tissue extracts contain mainly holoaminotransferases, i.e., enzymes with bound coenzyme, so that it is not necessary to add pyridoxal phosphate in order to demonstrate some aminotransferase activity. However, the prosthetic group is lost progressively throughout the purification of tissue extracts, more easily in the case of AST than ALT, so that greater activity results when pyridoxal phosphate is added. This increased activity can sometimes be seen in the case of control materials containing partially purified aminotransferases; the presence of pyridoxal phosphate may produce a marked increase in enzyme activity. Addition of the coenzyme to such materials is now becoming a more common practice.

Both holoaminotransferases and the coenzyme-deficient apoenzymes may be present in serum. Therefore, addition of pyridoxal phosphate under conditions that allow recombination with the enzymes usually produces a marked increase in aminotransferase activity. This increase ranges from zero to three- or four-fold, and averages approximately 50% for AST in normal serum and 20% for ALT. It has not been customary to add pyridoxal phosphate when assaying aminotransferase activity for routine diagnostic purposes, but the reference method drawn up by the Expert Panel on Enzymes of IFCC specifies this addition, in accordance with the principle that all factors affecting the rate of reaction must be optimized and controlled. Addition of pyridoxal phosphate is also recommended in some "national" methods intended for routine use. It is not known whether the presence of apoenzyme in serum results from the release of enzyme molecules in the apo form, or whether the prosthetic group is lost during passage from the cell to the bloodstream.

Transaminases are widely distributed in animal tissues. Both AST and ALT are normally present in human plasma, bile, cerebrospinal fluid, and saliva, but none is found in urine unless a kidney lesion is present. Transaminase activities in various tissues, relative to that in serum, are shown in Table 5-6.

Clinical Significance

In *viral hepatitis* and other forms of liver disease associated with hepatic necrosis, serum AST and ALT are elevated even before the clinical signs and symptoms of disease (such as jaundice) appear. Levels for both enzymes may reach values as high as 100 times the upper reference limit, although 20- to 50-fold elevations are most frequently encountered. Peak values of transaminase activity are seen between the seventh and twelfth days; activities then gradually decrease, reaching normal levels by the third to fifth week if recovery is uneventful. *Alcoholic hepatitis* has more modest elevations.

In *infectious hepatitis* and other inflammatory conditions affecting the liver, ALT is characteristically as high as or higher than AST, and the ALT/AST (De Ritis) ratio, which normally and in other conditions is < 1, becomes greater than unity. The use of different assay methods for the two enzymes may alter the activities of the two enzymes relative to each other, and hence the numerical values of the ratio observed may differ between laboratories. Nevertheless, the principle that hepatitis is associated with comparable elevations of the two activities remains valid. The relatively similar elevations of AST and ALT in hepatitis have been attributed to the release of only the cytoplasmic isoenzyme of AST into the circulation from reversibly damaged parenchymal cells. When necrosis of cells occurs, considerable amounts of mitochondrial AST are also released, depressing the ALT/AST ratio.

The picture in *toxic hepatitis* is similar to that seen in infectious hepatitis, with very high ALT and AST activities being observed in severe cases. Elevations up to 20 times the upper reference limit may be encountered in *infectious mononucleosis* with liver involvement and somewhat lower values in *intrahepatic cholestasis*. Increased levels may also be observed in *extrahepatic cholestasis*, with levels tending to be higher the more longstanding the obstruction. The aminotransferase levels observed in *cirrhosis* vary with the status of the cirrhotic process; they range from upper normal to some four to five times normal, with the level of AST activity higher than that of ALT activity (De Ritis ratio < 1). Elevations probably indicate continuing cellular necrosis.

Five- to ten-fold elevations of the two enzymes occur in patients with primary or metastatic *carcinoma of the liver*, with AST usually being higher than ALT, but levels are often normal in the early stages of malignant infiltration of the liver. Slight or moderate elevations of both AST and ALT activities may be observed after intake of alcohol, in delirium tremens, and after administration of a variety of drugs, such as opiates, salicylates, or ampicillin.

Although serum levels of both AST and ALT become elevated whenever disease processes affect liver cell integrity, ALT is the more liver-specific enzyme.[32] Serum elevations of alanine

Table 5-6. TRANSAMINASE ACTIVITIES IN HUMAN TISSUES, RELATIVE TO SERUM AS UNITY

	AST	ALT
Heart	7800	450
Liver	7100	2850
Skeletal muscle	5000	300
Kidney	4500	1200
Pancreas	1400	130
Spleen	700	80
Lung	500	45
Erythrocytes	15	7
Serum	1	1

Reproduced with permission from King, J.: Practical Clinical Enzymology. London, D. Van Nostrand Co., Ltd., 1965.

transaminase activity are rarely observed in conditions other than parenchymal liver disease. Moreover, elevations of ALT activity persist longer than do those of AST activity. Measurement of both AST and ALT has some value in distinguishing hepatitis from other parenchymal lesions.

Following *myocardial infarction* an increased level of AST activity appears in serum, as might be expected from the relatively high AST concentration in heart muscle (Table 5-6). On average, serum levels do not become abnormal, however, until 6–8 h have elapsed after the onset of chest pain. Abnormal AST levels are observed in more than 97% of cases of myocardial infarction when correctly timed blood specimens are analyzed. Peak values of AST activity are reached after 18–24 h, and the activity values fall to within the normal range by the fourth or fifth day, provided no new infarct has occurred. The peak values of AST activity are roughly proportional to the extent of cardiac damage. Average increases are of the order of four to five times the upper limit of normal; levels of 10–15 times normal are frequently associated with fatal infarcts. However, small elevations in serum levels do not necessarily indicate a favorable prognosis.

Alanine transaminase (ALT) levels are within normal limits, or are only marginally increased, in uncomplicated myocardial infarction, since the concentration of ALT activity in heart muscle is only a fraction of that of AST activity.

AST (and occasionally ALT) activity levels are increased in *progressive muscular dystrophy* and *dermatomyositis,* reaching levels up to eight times normal; they are usually normal in other types of muscle diseases, especially in those of neurogenic origin. *Pulmonary emboli* can raise AST levels to two to three times normal, and slight to moderate elevations (two to five times normal) are seen in *acute pancreatitis,* crushed *muscle injuries, gangrene,* and *hemolytic disease.*

Changes in the level of aspartate aminotransferase activity in *cerebrospinal fluid* have been found to be relatively nonspecific and are of little diagnostic value. Substantial elevations of AST activity have been found in the spinal fluids of patients who suffered *cerebrovascular accidents.* If the enzyme level is elevated both in serum and in spinal fluid, massive *parenchymal brain damage* is suggested and the prognosis is poor; if only the spinal fluid level is raised, recovery is probable. Both primary and secondary *malignancies* involving the brain or spinal cord are accompanied by varied degrees of elevation of AST activity in the spinal fluid. Normal values are usually found in patients with bacterial meningitis; raised levels suggest complications.

METHODS FOR THE MEASUREMENT OF TRANSAMINASE ACTIVITY

As is evident from equations (26) and (27), the assay system for measuring transaminase activity contains two amino acids and two oxoacids. This presents certain problems in measuring enzyme activity. There is no convenient method available for assaying either of the amino acids in the reaction system. Therefore, formation or consumption of the oxoacids is measured.

Colorimetric methods. The oxoacids can be assayed colorimetrically by coupling with 2,4-dinitrophenylhydrazine. There are several disadvantages to this approach, however. In each aminotransferase reaction, two oxoacids are present, one on each side of the equation. In the AST reaction, these are α-oxoglutarate and oxaloacetate. Both can react to give dinitrophenylhydrazones. As the reaction proceeds, one oxoacid is increasing in concentration, while the other is decreasing. The reaction begins with an oxoacid substrate, and hence reagent blanks have a high absorbance. If absorbance is kept low by using low substrate concentrations, the latter may become a factor limiting the reaction rate. In the AST reaction, one of the products, oxaloacetate, is relatively unstable. Furthermore, oxoacids such as pyruvate, which are normal components of serum, also produce phenylhydrazones and contribute to the high blanks.

Despite these limitations, the colorimetric approach is still feasible, because the phenylhydrazones of the products of both the AST and ALT reactions (oxaloacetate and pyruvate, respectively) are considerably more chromogenic than is the phenylhydrazone of α-oxoglutarate. Colorimetric methods based on dinitrophenylhydrazone formation are relatively simple and have limited but acceptable accuracy. Assay methods based on this approach, particularly that described by Reitman and Frankel, have been widely used but must now be considered obsolete.

In the case of AST, an alternative type of colorimetric procedure was introduced by Babson and associates. These workers measured the oxaloacetate formed, by combining it in a specific reaction with Azoene Fast Violet B (6-benzamide-4-methoxy-*m*-toluidine diazonium chloride).

Similar diazonium salts (e.g., Azoene Fast Red and True Ponceau L) have been introduced by Sax and Moore and others. These procedures are sensitive enough to permit use of a 20-min reaction time; reagent blank absorbances are minimal compared to those with the phenylhydrazone methods, and a wider range of activities can be assayed without specimen dilution. The methods have the disadvantage that they must be standardized against preassayed serum specimens; alternatively, the oxaloacetate standard used must be assayed enzymatically.

Both of these types of colorimetric methods are two-point methods. As well as sharing in the general disadvantages of two-point methods, those for AST suffer additionally from the fact that the enzyme is inhibited by its product, oxaloacetate, which accumulates during the reaction, causing deviation from linearity. Although oxaloacetate inhibits both isoenzymes of AST, the mitochondrial enzyme is inhibited to a greater degree.

Continuous-monitoring methods. It is not possible to monitor transaminase reactions directly, but the advantages of continuous-monitoring assays can be obtained by coupling the transaminase reactions to specific dehydrogenase reactions. The oxoacids formed in the transaminase reaction are measured indirectly by enzymatic reduction to the corresponding hydroxyacids, the accompanying change in NADH concentration being monitored spectrophotometrically. Thus, oxaloacetate, formed in the AST reaction, is reduced to malate in the presence of malate dehydrogenase (MDH), as shown in equation (28); pyruvate formed in the ALT reaction is reduced to lactate by lactate dehydrogenase (LDH). The substrate, NADH, and the auxiliary enzymes, MDH and LDH, must be present in sufficient quantity so that the reaction rate is limited only by the amounts of AST and ALT, respectively. As the reactions proceed, NADH is oxidized to NAD$^+$. The disappearance of NADH per unit time is followed by measuring the decrease in absorbance at 339 (340) nm* for several minutes, either continuously or at frequent intervals. The change in absorbance per minute (ΔA/min) may be related directly to micromoles of NADH oxidized and, in turn, to micromoles of substrate transformed per minute (International Units).

$$
\begin{array}{cccc}
\alpha\text{-Oxoglutarate} & \text{L-Aspartate} & \text{L-Malate} & \text{NAD}^+ \\
\searrow \nearrow & & \searrow \nearrow & \\
\text{AST} & & \text{MD} & \quad (28) \\
\nearrow \searrow & & \nearrow \searrow & \\
\text{L-Glutamate} & \text{Oxaloacetate} & & \text{NADH}
\end{array}
$$

Aminotransferase reaction	Dehydrogenase reaction
(Formation of oxaloacetate)	(Quantitation of oxaloacetate)
Assay reaction	**Indicator reaction**

Another type of coupled enzyme system has been proposed for use in assaying the transaminases. In this instance, the indicator enzyme is glutamate dehydrogenase, EC 1.4.1.3, which uses NAD as its coenzyme. The glutamate formed in the transaminase reaction is oxidatively deaminated; i.e., NH$_4^+$ is removed and α-oxoglutarate is regenerated. At the same time NAD is reduced to NADH. The rate of formation of NADH is a measure of the transaminase reaction.

Standardization. Because of the large numbers of AST and ALT activity measurements performed daily in many laboratories throughout the world, the development of standard or reference methods for these two enzymes has been given priority by national and international

---

*The wavelength of maximum absorbance of NADH and NADPH has been quoted as 340 nm and their molar absorptivities at this wavelength as 6220 L $\times$ mol^{-1} $\times$ cm^{-1} ever since photometric measurements involving these compounds were first introduced into clinical chemistry. However, it is now apparent that a number of variables affect the molar absorptivities of these coenzymes. The wavelength of maximum absorbance decreases slightly with increasing temperature, and the absorption peaks also broaden, so that the absorbance measured at a fixed wavelength falls as the temperature is increased. Molar absorptivities also increase as pH rises, whereas these values are lower in solutions of higher ionic strengths. The molar absorptivities and wavelengths of maximum absorbance of the coenzymes have been redetermined under conditions more usual in clinical enzyme analysis. As a result, it is suggested that measurements should be made at 339 nm and that a value of 6300 L $\times$ mol^{-1} $\times$ cm^{-1} for molar absorptivity should be used. These values have been incorporated into the newest recommended methods for enzyme assay. However, the older values continue to be used, and the difference in results is apparent only in measurements of the highest degree of accuracy. (See McComb, R. B., Bond, L. W., Burnett, R. W., et al.: Determination of the molar absorptivity of NADH. Clin. Chem., *22*:141–150, 1976; and Ziegenhorn, J., Senn, M., and Bücher, T.: Molar absorptivities of β-NADH and β-NADPH. Clin. Chem., *22*:151–160, 1976.)

groups. All of these have chosen the coupled-reaction approach with malate or lactate dehydrogenases as the indicator enzymes. However, the methods proposed differ in several details, such as substrate concentrations, nature of buffer (e.g., Tris or phosphate), and assay temperature. The technique of response-surface methodology has been used in developing the most recent of these methods to select values for the many variables that affect the rate of reaction. As already mentioned, supplementation with pyridoxal phosphate ensures that all the transaminase activity of the sample is measured.

Continuous Monitoring Method for the Measurement of Aminotransferase Activity

Measurements should be made in a spectrophotometer with established spectrophotometric and wavelength accuracy and with good resolution at 339 (340) nm. The temperature of the reaction mixture in the cuvet must be controlled at a constant known level by means of a thermostatically controlled cuvet compartment. A preliminary incubation period is necessary to ensure that NADH-dependent reduction of endogenous oxoacids in the sample is completed before adding 2-oxoglutarate to start the transaminase reaction. After a brief lag phase, seen with most coupled reactions, the ΔA is monitored, preferably with a recorder; alternatively, several readings (five to eight at 1-min intervals) are taken to establish the linear portion of the curve. Because the reaction mixture has a rather high initial absorbance, it is customary to use a solution such as potassium dichromate as a blank reference solution so as to produce an initial absorbance for the test of about 0.7–0.9; readings within this absorbance range can also be accomplished with a zero offset (zero suppression) facility if the spectrophotometer is fitted with such a device.

Determination of Aspartate Transaminase (AST) Activity

In the procedure described, Tris is used as buffer in place of the earlier choice of phosphate because phosphate appears to increase the rate of NADH decomposition and to inhibit association of P-5′-P with the transaminase apoenzyme. The coenzyme is added to ensure that all transaminase is fully active. Addition of lactate dehydrogenase to the coupled enzyme system accelerates the endogenous side reactions and thus shortens the preincubation period. A pH of 7.8 is used because the pH optimum for the coupled enzyme system appears to be between 7.7 and 7.9; also, the stability of NADH is greater at this pH than at pH 7.4. Stock preparations of both MDH and LDH are diluted with glycerol rather than with $(NH_4)_2SO_4$ to avoid introducing ammonium ions, thereby to eliminate a possible side reaction catalyzed by glutamate dehydrogenase in which NADH would be oxidized. The reaction temperature is 30 °C.

Reagents

1. Tris base stock, 1.0 mol/L. Dissolve 121.1 g Tris(hydroxymethyl)aminomethane in water to make 1 L. Store in a polyethylene bottle; the reagent is stable.

2. Tris buffer, pH 7.8, 0.10 mol/L. Dilute 50 mL of the Tris base, 1.0 mol/L, with ~300 mL of water. Add ~5.5 mL of HCl, 6 mol/L, adjust temperature to 30 °C, and with the aid of a pH meter adjust pH to exactly 7.8 by the addition of NaOH, 1 mol/L. Dilute with water to 500 mL. The reagent is stable and may be stored in a polyethylene bottle at 4 °C or at room temperature.

Note: Electrodes used for pH measurement must have a ceramic junction and a negligible sodium error; certain pH electrodes do not give correct readings in Tris solutions. Refer to the instructions describing the properties of the glass and reference electrodes being used.

3. L-Aspartate, 228 mmol/L, in 0.10 mol/L Tris buffer, pH 7.8. Dissolve 7.59 g of L-aspartic acid in a solution containing 25.0 mL of Tris base and 175 mL of water, with the addition of about 25 mL of NaOH, 2.5 mol/L, and the application of heat, if necessary. Cool to 30 °C, and with the aid of a pH meter adjust the pH to 7.8 by the addition of 1 mol/L NaOH or HCl. Add water to make 250 mL of solution. Store in refrigerator; stable for 6 months.

4. 2-Oxoglutarate, 225 mmol/L, in Tris buffer, 0.10 mol/L, pH 7.8. Dissolve 1.65 g of 2-oxoglutaric acid in ~25 mL of water, add 5.0 mL of Tris base, warm to 30 °C, and, as before, with the aid of a pH meter, adjust the pH to exactly 7.8 by the addition of NaOH, 2.5 mol/L. Add water to make 50 mL of solution. Keep refrigerated; stable for 2 weeks.

5. NADH solution, 4.2 mmol/L, in Tris buffer, 0.10 mol/L, pH 7.8. Dissolve 3.2 mg of β-NADH Na$_2$ · 4 H$_2$O, $\geq$ 98% grade, in 1.0 mL of the Tris buffer, 0.10 mol/L, pH 7.8. If stored at refrigerator temperature, the reagent is stable for one working week; however, it is recommended to prepare just enough reagent for one day's work. It is convenient to use commercially available vials containing preweighed quantities of the reduced coenzyme. Keep cold during use, e.g., in ice water.

6. Malate dehydrogenase suspension, 36 000 U/L in glycerol-water (1/1). Dilute the commercially available stock suspension of MDH with glycerol-water (1/1) to a concentration of 36 000 U/L. The stock suspension should be free of aminotransferase, apoaminotransferase, and L-glutamate dehydrogenase.

7. Lactate dehydrogenase suspension, 72 000 U/L in glycerol-water (1/1). Prepare this reagent as described for MDH. This enzyme suspension should also be essentially free of aminotransferase, apoaminotransferase, and L-glutamate dehydrogenase activity. Both enzyme reagents are stable for about six months if stored in a refrigerator and kept on ice during use.

8. Pyridoxal-5'-phosphate, 3.4 mmol/L, in Tris buffer, 0.10 mol/L, pH 7.8. Dissolve 11.1 mg of pyridoxal-5'-phosphoric acid in 10 mL of Tris buffer, 0.10 mol/L, pH 7.8. Keep refrigerated; stable for one week.

9. Dichromate solutions (blanks). Prepare a stock solution of 30 mg of K$_2$Cr$_2$O$_7$/100 mL of water containing 3 drops of concentrated H$_2$SO$_4$. Dilute this solution further with water as needed in the procedure. The reagent is stable.

10. The actual concentrations of the various components in the complete reaction mixture are as follows: Tris buffer, pH 7.8, 90 mmol/L; L-aspartate, 175 mmol/L; 2-oxoglutarate, 15 mmol/L; NADH, 0.15 mmol/L; MDH, 600 U/L (25 °C); LDH, 1200 U/L (25 °C); and P-5'-P, 0.12 mmol/L. The ratio of serum volume to total reaction volume (serum dilution) is 1/15. The substrate concentrations are in agreement with those recommended by the Enzyme Study Group of AACC. The substrate concentrations chosen by the Expert Panel on Enzymes are slightly different: aspartate, 240 mmol/L; 2-oxoglutarate, 12 mmol/L. However, in both methods the substrate concentrations support a reaction rate equivalent to about 96% of the theoretical V_{max} value.[117]

Procedure

1. Add the following reaction components to a cuvet that has a 1.0-cm light path:

Reagent	mL
L-Aspartate, 228 mmol/L	2.30
NADH, 4.2 mmol/L	0.10
P-5'-P, 3.4 mmol/L	0.10
MDH, 36 000 U/L	0.05
LDH, 72 000 U/L	0.05
Serum	0.20
Total volume	2.80

If many determinations are to be done, it is convenient to mix a batch of these five reagents sufficient for one day's work and to pipet 2.6 mL of the composite reagent into individual cuvets. The reagent mixture is stable for several days if kept refrigerated. It should be kept in ice water on the bench or in the refrigerator when not being used to prepare reaction tubes. The serum specimens are then pipeted into the cuvets and the contents mixed.

2. Preincubate the cuvets and contents in a water bath or in the thermostatted cuvet compartment of the spectrophotometer at 30 °C to ensure temperature equilibration and to permit side reactions to proceed to completion. Reduction of endogenous oxoacids takes place rapidly in the presence of added LDH.

3. Initiate the AST enzyme reaction by adding 0.20 mL of the 225 mmol/L oxoglutarate solution that has been prewarmed to 30 °C. Rapidly mix the cuvet contents; place the cuvet in the spectrophotometer, and measure the absorbance at 339 (340) nm for a period of 5–8 min against a dichromate blank. Choose the blank or the zero suppression so that the initial absorbance readings are between 0.70 and 0.90.

4. Calculate the average absorbance change per minute (ΔA/min), using either the recorder tracing or the measured absorbance values. Use only the linear portion of the rate curve or those consecutive individual values of ΔA/min that are essentially constant.

Calculation

$$\text{AST activity in U/L } (\mu\text{mol/min/L}) = \frac{\Delta A}{\text{min}} \times \frac{1}{6.3 \times 10^{-3}} \times \frac{3.0}{0.20}$$

$$= \frac{\Delta A}{\text{min}} \times 2381$$

where
- 3.0 = total volume in cuvet, in mL
- 0.2 = volume of serum specimen, in mL
- 6.3×10^{-3} = micromolar absorptivity of NADH at 339 nm

The average rate of the AST reaction in human serum at 37 °C is 1.59 times the rate at 30 °C.[118] If the activity is measured at 37 °C, the value obtained can be converted to that which would be obtained at 30 °C by dividing the measured activity by this factor. However, if local circumstances dictate the selection of a temperature other than 30 °C for measurement, results should preferably be reported at that temperature, with appropriate reference values.

Comments

1. Elevated levels of L-glutamate dehydrogenase (GLDH, EC 1.4.1.3) may be encountered in some sera from patients with parenchymal liver disease. If ammonia is also present in the serum specimen or in one of the reagents, the GLDH reaction will proceed and interfere with the AST determination by consuming oxoglutarate and NADH. Thus, use of ammonia-free reagents is recommended.

2. Occasionally, the side reactions in the preincubation stage may be quite extensive, thus consuming a large fraction of the available NADH. If the ΔA/min in the test decreases with time (i.e., the rate is nonlinear), suboptimal NADH concentration is indicated. In this event, the test must be repeated with a higher initial concentration of NADH or a smaller volume of specimen (diluted specimen).

3. If the activity in a given serum specimen is high (ΔA/min more than 0.10), the serum will need to be diluted. Saline may be used, but an albumin solution, 5 g/dL, is preferred.

4. Hemolyzed serum specimens should not be used. As shown in Table 5-6, AST and ALT activities in erythrocytes are some 15 and 7 times higher, respectively, than those in sera.

5. Authorities disagree as to the stability of AST activity in serum. Specimens are best stored frozen if they are to be kept more than 3–4 d. Minimal loss of activity occurs at 0–4 °C over 1–3 d.

Reference Ranges

Since methods for the determination of AST activity have recently been under intense investigation directed toward establishing conditions that will provide the most accurate results, any suggested range of normal values is still subject to review. The following values, in U/L at 30 °C, reflect current and past experience.[143]

Specimen	Adults	Infants	Newborn
Serum (no P-5'-P added)	8–20	15–60	25–75
M, > 60 years:	10–25		
Serum (P-5'-P added)	10–30		
Cerebrospinal fluid (no P-5'-P added)	3–10	2–10	

The distribution of values is skewed to the right; the skewness may be real or may reflect the inclusion in the normal population of sera with nonvisible hemolysis and sera from persons with possible occult tissue damage. Individuals in the fasting state may show slightly lower values. Values for men are slightly higher than those for women.

Measurements of AST Activity by Double-Beam Spectrophotometry

It is possible to correct effectively for the GLDH, LDH, and ALT side reactions by using a double-beam spectrophotometer in which the ΔA/min is measured in a specimen and in a reference cuvet at the same time. The specimen cuvet contains the complete reaction mixture, whereas the reference cuvet lacks the aspartate. The reaction is initiated with either serum or 2-oxoglutarate. The absorbance change in the specimen cuvet represents both the transaminase and the concurrent side reactions, whereas that in the reference cuvet represents the side reactions only. The difference in ΔA/min values between the two cuvets is directly measured in the double-beam spectrophotometer and gives a more accurate measure of the AST activity. Alternatively, *dynamic blanking* may be used; the ΔA/min is measured in two separate cuvets in the usual spectrophotometer, with and without aspartate present. Oxoglutarate is preferably included in both the specimen and reference mixtures, in place of aspartate, because it is required for the GLDH and ALT side reactions. If this technique is used, care must be taken that identical time intervals elapse between initiation of the reaction and the start of absorbance measurements for each cuvet. The rates of the side reactions, in general, are not linear in time.

In principle, this technique can also be applied to measuring the activity of other enzymes.

Determination of Alanine Transaminase (ALT) Activity

The procedure used is identical to that for measuring AST activity, except that L-alanine replaces aspartate as the amino group donor, and LDH replaces MDH as the indicator enzyme [equation (27)]. The concentration of alanine needed to give an optimal rate of ALT activity is much greater than that of aspartate in measuring AST activity. The added LDH both speeds up the side reaction and serves as the indicator enzyme.

Reagents

1. Tris base, 1.0 mol/L, as in the AST procedure.
2. NADH solution. Refer to the AST procedure.
3. L-Alanine, 610 mmol/L, in Tris buffer, 0.10 mol/L, pH 7.3. Dissolve 13.7 g of L-alanine (99+% purity) in about 200 mL of water and 25 mL of the Tris base, 1.0 mol/L. Warm the solution to 30 °C and, with the aid of a pH meter, adjust the pH to 7.3 with HCl, 1 mol/L, or NaOH, 1 mol/L. Dilute with water to a volume of 250 mL. Stable for 4–6 weeks, if stored at 4–8 °C.
4. 2-Oxoglutarate solution, 225 mmol/L, in Tris buffer, 0.10 mol/L, pH 7.3. Refer to the AST procedure.
5. Pyridoxal-5'-phosphate, 3.4 mmol/L in Tris buffer, 0.10 mol/L, pH 7.3. Refer to the AST procedure.
6. Lactate dehydrogenase solution. 72 000 U/L, in glycerol-water (1/1). This is the same preparation as used in the AST procedure.

Procedure

Pipet the following components into a 10-mm cuvet:

Reagent	mL
L-Alanine, 610 mmol/L	2.30
NADH, 4.2 mmol/L	0.10
P-5'-P, 3.4 mmol/L	0.10
LDH, 72 000 U/L	0.10
Serum	0.20
Total volume	2.80

It is convenient to prepare a mixture of the four reagents, as suggested in the AST procedure, and to pipet 2.60 mL of this composite into a cuvet (or a 12 × 100-mm test tube), followed by 0.20 mL of the serum specimen. From this point on, the procedure and calculations are identical to those outlined for the AST procedure.

With the large quantity of LDH added, the endogenous (side) reactions are completed within 2 min.

The ALT reaction is initiated by adding 0.20 mL of the 225 mmol/L oxoglutarate solution.

The composition of the complete reaction mixture is as follows: Tris buffer, 90 mmol/L, pH 7.3; L-alanine, 500 mmol/L; oxoglutarate, 15 mmol/L; NADH, 0.15 mmol/L; LDH, 2400 U/L; and P-5'-P, 0.12 mmol/L. The serum dilution is 1/15. The substrate concentrations are in agreement with recommendations of the Enzyme Study Group of the AACC and the Expert Panel on Enzymes of IFCC and support a reaction rate of approximately 92% of the theoretical V_{max} value.[65]

Comments

1. As indicated in the procedure, the reaction components and serum may be added to a test tube or directly into a cuvet, and the side reactions permitted to reach completion while temperature equilibrium is being reached. If a test tube is used, the oxoglutarate is rapidly added to the tube contents, which, after mixing, are poured into cuvets prewarmed to 30 °C. The absorbance readings of the solutions are measured and recorded.

2. Very turbid or icteric sera may be diluted in order to avoid working with very high absorbances, although this problem is usually resolved by selecting a $K_2Cr_2O_7$ blank of appropriate concentration.

3. The comments discussed under the AST procedure are also pertinent to the determination of ALT.

Reference Ranges

Reference ranges for ALT activity in U/L in serum and spinal fluid measured at 30 °C are as follows:[143]

Specimen	Adults	Newborn/ Infants
Serum (no P-5'-P added)	8–20	5–28
> 60 years,		
M:	6–24	
F:	7–16	
Serum (P-5'-P added)	Up to 30	
Cerebrospinal fluid (no P-5'-P added)	0–4	

Values in men are slightly higher than in women.

CREATINE KINASE

(EC 2.7.3.2; Adenosine Triphosphate:Creatine *N*-Phosphotransferase; CK)

Creatine kinase, also incorrectly referred to as creatine phosphokinase (CPK), catalyzes the reversible phosphorylation of creatine by adenosine triphosphate (ATP), as shown in equation (29). The optimal pH values for the forward (Cr + ATP → ADP + CrP) and reverse (CrP + ADP → ATP + Cr) reactions are 9.0 and 6.7, respectively. The equilibrium position for the reaction is dependent on pH. At neutral pH, CrP has a much higher phosphorylating potential than does ATP; this higher potential favors the reverse reaction, with ATP being formed from CrP. The reverse reaction proceeds two to six times faster than the forward reaction, depending on the reaction conditions.[80] See equation (29).

As is true for all kinases, Mg(II) is an obligate activating ion, functioning with ADP and ATP. The optimal concentration range for Mg(II) is quite narrow, and excess Mg(II) is inhibitory. Many metal ions, such as Mn(II), Ca(II), Zn(II), and Cu(II), inhibit enzyme activity as do iodoacetate and other sulfhydryl-binding reagents. Activity is inhibited by excess ADP and by citrate, fluoride, nitrate, acetate, iodide, bromide, malonate, and L-thyroxine. Urate and cystine are potent inhibitors of the enzyme in serum. Even Cl⁻ and SO_4^{2-} inhibit activity to some degree, and thus the concentrations of these ions should be kept low in any enzyme assay system based on the CrP + ADP (reverse) reaction. The enzyme in serum is relatively unstable, activity being lost as a result of sulfhydryl group oxidation at the active site of the enzyme. Activity can be partially restored by incubating the enzyme preparation with sulfhydryl compounds such as N-acetylcysteine, thioglycerol, dithioerythritol, dithiothreitol (Cleland's reagent), thioglycolic acid, mercaptoethanol, or cysteine. The extent of activity recovered by this procedure is inversely proportional to the time delay in adding these reagents after collection of the blood specimen and the type of SH reagent. The current agents of choice are N-acetylcysteine and thioglycerol, both used at a final concentration of 20 mmol/L. The former is soluble, stable, odorless, and inexpensive; N-acetylcysteine can be lyophilized, thioglycerol cannot.

$$
\text{Creatine (Cr)} \;+\; \text{Adenosine triphosphate (ATP)} \;\underset{\substack{\text{pH, 6.7}}}{\overset{\substack{\text{pH, 9.0}\\ \text{CK, Mg}^{2+}}}{\rightleftharpoons}}\;
$$

Creatine (Cr)   Adenosine triphosphate (ATP)

(29)

$$
\text{Phosphocreatine (Creatine phosphate, CrP)} \;+\; \text{Adenosine diphosphate (ADP)} \;+\; \text{H}^+
$$

Phosphocreatine
(Creatine phosphate, CrP)   Adenosine diphosphate
(ADP)

CK activity is greatest in striated muscle, brain, and heart tissue, which contain some 2500, 550, and 470 U/g protein, respectively. Other tissues, such as the kidney and the diaphragm, contain significantly less activity (< 30 U/g protein), and the liver and erythrocytes are essentially devoid of activity.[80]

Creatine kinase is a dimer composed of two subunits, each with a molecular weight of about 40 000. These subunits (B or brain and M or muscle) are the products of two distinct structural genes, and since the active form of the enzyme is a dimer, only three different pairs of subunits can exist: BB (or CK-1), MB (or CK-2), and MM (or CK-3). The Commission on Biochemical Nomenclature has recommended that isoenzymes should be numbered on the basis of their electrophoretic mobility with the most anodal form receiving the lowest number. Accordingly, the CK isoenzymes should, like the LD isoenzymes, be numbered CK-1, CK-2, and CK-3, and this convention will be followed here. The distribution of these isoenzymes in the various tissues of the human are shown in Table 5-7; the sometimes wide divergence of data shown in this table for the same tissue illustrates the confusing nature of much of the available information. CK-1 predominates in brain, prostate, gut, lung, bladder, uterus, placenta, and thyroid while CK-3 predominates in skeletal and cardiac muscle. CK-2 is present, to varying degrees, in heart muscle (25–46% of CK activity) and also to a minor degree in skeletal muscle (< 5%).

All three of these isoenzyme species are found in the cell in the cytosol or associated with myofibrillar structures. However, there exists a fourth form that differs from the others both immunologically and by electrophoretic mobility. This isoenzyme, CK-Mt, is located between the inner and outer membranes of mitochondria and it constitutes, in the heart, for example, up to 15% of the total CK activity. CK activity may also be found in macromolecular form—the so-called "macro CK." This exists in two forms, types 1 and 2.[81] Type 1 is CK-1 associated with immunoglobulin G or sometimes CK-3 with IgA. Type 2 appears to be oligomeric CK-Mt. CK-3 is now known to undergo post-translational modification in serum after release from tissue. Thus, in addition to the two subunits in CK-3, there exist at least two other M subunits, each of which is capable of hybridizing with other M or B subunits to form active enzyme species with electrophoretic mobilities slightly different from the original, unmodified subunit.[80,81]

The physiological role of creatine kinase can best be understood by inspecting equation (29). Creatine phosphate, the major phosphorylated compound in muscle, is present in about an eight-fold excess over ATP. When muscle contracts, ATP is consumed (to form ADP) and creatine kinase catalyzes the rephosphorylation of ADP (to form ATP), using creatine phosphate as the phosphorylation reservoir.

Table 5-7. CREATINE KINASE ISOENZYME PATTERNS OF HUMAN TISSUES[80]

Tissue	CK Activity (U/g Wet Weight)	CK-3 (%)	CK-2 (%)	CK-1 (%)
Skeletal muscle	2500	98.9	1.1	0.06
Rectus abdominis		81	19	
Rectus abdominis		94	5	
Pectoralis major		100	0	
Gastrocnemius		70–82	18–30	
Brain	555	0	2.7	97.3
Heart	473	78.7	20.0	1.3
Left ventricle		54±6	41±7	
Papillary muscle		52±4	46±5	
Stomach	190	4.3		95.7
		10	0	90
Small intestine	112	1.2	0	98.8
		11–13	7–9	78–80
Colon	138	2.1	0	97.8
		3–4	0–1	96
Rectum	267	1.2	0	98.8
Kidney	32	2.8	0	97.2
		8–12	0	88–92
Bladder	145	6.6	0	93.4
		2	6	92
Prostate	114	6	0	94
		34–39	2–6	59–60
Lung		27–72	0–4	18–69
Liver	0.6	0	0	100
Uterus	115	2.3	0	97.4
		5–16	2–20	64–93
Placenta		0	0	100
Thyroid		4–26	0–1	73–96

Serum CK activity is subject to a number of physiological variations. For example, the activity in serum appears to be a function of the muscle mass of the individual. This is presumably the basis for the finding that females have lower serum activities than males, and that slightly built individuals often have lower serum CK activities than more heavily built members of the same sex. It has also been found that ethnic origin is important; thus, a black North American female may have higher serum CK activities than a white male.

Clinical Significance[46,80,81]

Diseases of skeletal muscle. Serum CK activity is greatly elevated, at some time during the course of the disease, in all types of muscular dystrophy, and especially so in the Duchenne type, in which levels up to 50 times the upper limit of normal may be encountered. In *progressive muscular dystrophy* (particularly the Duchenne sex-linked muscular dystrophy), enzyme activity in serum is highest in infancy and childhood (7–10 years of age) and may be elevated long before the disease is clinically apparent. Serum CK activity characteristically falls as the patient gets older and as the mass of functioning muscle diminishes with the progression of the disease. About 50–80% of the asymptomatic female carriers of Duchenne dystrophy show three- to six-fold elevations of CK activity, but values may be normal if specimens are obtained after patients have experienced a period of physical inactivity. Quite high values of CK are seen in *viral myositis, polymyositis,* and similar muscle diseases (see also necrotizing polymyopathy, below). However, in *neurogenic muscle diseases,* such as *myasthenia gravis, multiple sclerosis, poliomyelitis,* and *parkinsonism,* serum enzyme activity is normal. Very high activity is also seen in *malignant hyperthermia,* a familial disease characterized by high fever and brought on by administration of inhalation anesthesia (usually halothane) to the affected individual. Apparently, CK-2 replaces part of the CK-3 form in the muscles, and this results in impaired storage of creatine phosphate in muscle. In dystrophies and myopathies, usually only CK-3 is present in serum; but if total CK activity is high, some CK-2 can also be detected (<6%) since diseased skeletal muscle may contain a significant proportion of this isoenzyme due to the phenomenon of fetal reversion in which fetal patterns of protein synthesis reappear. In *necrotizing polymyopathy* (acute rhabdomyolysis) due, for example, to crush injuries, muscle infarction, viral infections, toxins, and malignant hyperthermia, there is severe muscle destruction; serum CK activities > 200 times the upper reference limit may be found. An interesting, but not completely explained, association between elevated serum CK activity (as CK-3) and acute psychotic states has been repeatedly reported. It has been suggested that such increase is a manifestation of skeletal muscle abnormalities in these states; alternatively, it may be a reflection of increased unphysiological motor activity.

Diseases of the heart. Following a *myocardial infarction,* total CK activity in serum (which is mostly due to CK-3) begins to rise within 4–6 h, reaches a peak value between 18 and 30 h, and then rapidly returns to normal by the third day. The average elevation of total CK activity is seven to 12 times the upper reference limit. Invariably, the rise in total serum CK activity is accompanied by a simultaneous rise in serum CK-2 activity that reaches a maximum about 10–20 h after the onset of chest pain, i.e., usually peaking up to 12 h earlier than the total CK activity. Since CK-2 has a shorter serum half-life than serum CK-3 (see Table 5-3), it decreases faster than the total CK; thus CK-2 may not be increased 48 h post-infarct. There is, however, considerable variation between patients. The average elevation of serum CK-2 after a myocardial infarction is 10–25 times the upper reference range. Serum CK-2 values may also be reported as a percentage of total CK activity; usually these values are < 3–6%, but following an infarction values can rise anywhere from 10–30%, depending on the methodology in use, the extent of myocardial damage, and the location of the infarct. In myocardial infarction the diagnostic sensitivity for both total serum CK and serum CK-2 has been variously reported as being from 80–100% with the latter test usually being superior. Other cardiac conditions—*angina pectoris, cardiogenic shock, electrical countershock, tachycardia, myocarditis,* and *congestive cardiac failure*—have variously been reported as having a low occurrence of elevated total serum CK and/or CK-2. In the few positive cases, particularly those with CK-2 elevations, it seems reasonable to suspect cardiac damage severe enough to cause changes in the cell membranes of the myocytes and to allow intracellular CK to diffuse into the extracellular space. Obviously, *cardiac trauma* following heart surgery will also cause an elevation of total serum CK and CK-2 sufficient to mask elevations subsequent to intraoperative myocardial infarction.

The mere presence of CK-2 activity in serum does not necessarily indicate myocardial damage. Many methodologies are now sensitive enough to detect serum CK-2 in health, and CK-2 can arise from tissues other than the myocardium. In fact, CK-2 (< 6%) activity can be detected in serum in inflammatory and degenerative muscle diseases, traumatic lesions including shock, intoxications, delirium tremens, hypothyroidism, and acute psychosis, and in women immediately after obstetric delivery. Thus the diagnosis of myocardial damage must be based on both clinical findings and a consideration of the extent of CK-2 elevation. Myocardial damage is usually excluded if the CK-2 is < 5 or 6% of the total CK activity.

Diseases of the liver. Because liver contains a negligible quantity of CK, patients with *primary liver disease* (but see Reye's syndrome, below) and *cirrhosis* have normal CK activity in their sera. For the same reason *hepatic congestion* and *hypoxia,* which may accompany cardiac disease, do not usually effect an elevation in serum CK values, although they often contribute to the elevations seen in serum ALT and LDH activity.

Diseases of the central nervous system. Serum CK activity may increase in patients with acute *cerebrovascular disease* and with *cerebral ischemia.* Isoenzyme studies show that the increase is entirely in the CK-3 isoenzyme; no CK-1 isoenzyme increase is demonstrable. By contrast, following *head injury,* serum CK-1 activity can readily be detected in many of these patients, and the extent of elevation may correlate with the severity of the injury and, of course, the prognosis. Paradoxically, serum CK-2 can often be detected in these head injury patients and also in cases of *subarachnoid hemorrhage*; this appearance of CK-2 suggests myocardial damage following the cerebral accident and indicates the need to monitor these patients by following the serum CK isoenzyme levels. In *Reye's syndrome* (a childhood disorder characterized by acute brain swelling with fatty infiltration and nonicteric dysfunction of the liver) there is an elevation of total serum CK activity, up to 70-fold, with just detectable CK-1 present in serum; the extent of the total CK elevation appears to be an indicator of the severity of the encephalopathy.

Diseases of the thyroid. Serum CK activity demonstrates an inverse relationship with *thyroid activity.* About 60% of hypothyroid subjects show an average elevation of CK activity five-fold over the upper reference limit; elevations of as high as 50-fold may also be found. The major isoenzyme present is CK-3, although up to 13% of CK activity may be present as CK-2, suggesting possible myocardial involvement. In any case, hypothyroidism predisposes to ischemic heart disease. By contrast, in *hyperthyroidism,* the serum CK activity tends to be at the low end of the reference range.

Clinical Conditions Associated with the Presence of CK-1 in Serum[80,81]

During *normal childbirth,* when there can be a six-fold elevation in maternal total serum CK activity, the uterus and possibly the placenta are the sources of the CK-1 (which can be as high as 10% of the total CK) found in serum; surgical intervention during labor will further increase the serum activity of CK-1. The presence of CK-1, usually at low levels, has also been reported in a wide variety of disorders such as *multisystem insult,* in many critical care patients, following *aortocoronary bypass* operations, and in *hypothermia.* Certain *gastrointestinal disorders,* such as infarction or adenocarcinoma, as well as a variety of *lung tumors,* may cause elevated serum CK-1 activities. *Tumors* of the prostate, bladder, kidney, breast, and ovary as well as *leukemias* and *lymphomas* have been found to be associated to varying degrees with elevations of serum CK-1; these findings have led to the suggestion that serum CK-1 might be of use as a tumor marker.

Creatine Kinase Activity in Cerebrospinal Fluid

Assay of CK activity in cerebrospinal fluid has only limited value in clinical diagnosis because any changes observed are irregular and often nonspecific in nature. Elevations can be seen in some specimens from *epileptic patients* and from patients with *brain tumors* or *cerebral infarcts,* but the elevations are not consistent with the degree of pathologic change. Similarly, spinal fluid specimens from a majority (but not all) of patients with either bacterial or nonbacterial *meningitis* and of those with *autism* have increased levels of CK.

METHODS FOR THE DETERMINATION OF CREATINE KINASE ACTIVITY[80]

Numerous colorimetric, fluorimetric, and coupled enzyme methods have been developed for the assay of CK activity, using either the forward (Cr → CrP) or the reverse (CrP → Cr) reaction. The reverse reaction is preferred because it proceeds faster than the forward reaction, although the cost of the starting chemicals, creatine phosphate and ADP, is greater than the cost of creatine and ATP.

Most commonly, the ATP produced in the reverse reaction is measured by coupling the hexokinase (HK) and the glucose-6-phosphate dehydrogenase (GPD) reactions to the CK reaction [equation (30)]. The HK catalyzes phosphorylation of glucose by ATP to form glucose-6-phosphate (G-6-P) and regenerates ADP for the CK reaction. The G-6-P is then oxidized with $NADP^+$ to form 6-phosphogluconic acid and NADPH. The rate of NADPH formation is a measure of the CK activity, provided that the concentrations of all other components in the three-enzyme system are present in suitable excess, so that the CK activity is the only limiting factor.

$$\text{Creatine phosphate} + \text{ADP} \xrightarrow[\text{pH 6.7}]{\text{CK}} \text{creatine} + \text{ATP}$$

$$\text{ATP} + \text{glucose} \xrightarrow{\text{HK}} \text{glucose-6-phosphate} + \text{ADP} \tag{30}$$

$$\text{Glucose-6-phosphate} + NADP^+ \xrightarrow{\text{GPD}} \text{6-phosphogluconate} + \text{NADPH} + H^+$$

The coupled enzyme system is completely "downhill"; i.e., all reactions proceed in a favorable direction. The pH optimum for the entire system is 6.8–6.9.

There are a number of sources of interference in this assay system. The *glutathione reductase effect* is significant only when reduced glutathione (GSH), contaminated with oxidized glutathione (GSSG), is used as the thiol agent. Under these circumstances any glutathione reductase (GR) activity present in serum will catalyze the reaction (31):

$$\text{GSSG} + \text{NADPH} + H^+ \xrightarrow{\text{GR}} \text{GSH} + NADP^+ \tag{31}$$

The consumption of NADPH results in an apparent reduction in CK activity [see reaction (30)]. This effect can be avoided by using an alternative thiol agent such as N-acetylcysteine. The *adenylate kinase effect,* caused by an enzyme found in fairly high concentration in nearly all tissues including erythrocytes, is due to the presence of adenylate kinase (AK) in the serum to be tested. AK catalyzes the reaction (32):

$$2ADP \xrightarrow{AK} ATP + AMP \tag{32}$$

which, in the presence of ADP (derived from the CK assay mixture), will produce ATP; ATP is also the product of the CK reaction. [See reaction (30).] Therefore, serum AK activity results in apparent increase in CK activity. AK activity can be inhibited by adding fluoride ion, AMP, or diadenosine pentaphosphate (Ap_5A) to the CK assay. Although fluoride ion is an effective inhibitor of AK, it may form insoluble MgF_2 with the magnesium ion in the CK assay. The presence of AMP, a competitive inhibitor of AK, in the CK assay mixture causes product inhibition of reaction (32). Ap_5A competitively inhibits the AK of muscle and erythrocytes but has less effect on the liver and kidney enzymes. The combination in an assay of AMP, 5 mmol/L, and Ap_5A, 10 mmol/L, is an extremely effective inhibitor of erythrocyte and liver AK activity. Alternatively, AK activity can be measured directly by omitting phosphocreatine from the CK assay to obtain a "blank" activity, which is then subtracted from the value for total CK activity in a complete assay system.

CK assays demonstrate a *lag phase,* i.e., a slow increase in reaction rate after the addition of substrate or serum to the assay mixture. Typical lag phases in an optimized assay system are 110 s at 25 °C, 90 s at 30 °C, and 60 s at 37 °C. This phenomenon has important practical implications since inappropriate timing of the measurement period may cause incorrect and low estimations of the reaction rate.

The presence of *endogenous inhibitors* in serum has been referred to earlier. Ca(II) is an important competitive inhibitor for Mg(II), which is required by CK for full catalytic activity. The Ca(II) effect can be nullified by the addition of EDTA, 2 mmol/L, provided the Mg(II) concentration is increased to 10 mmol/L to compensate for some binding of Mg(II) by the EDTA. The presence of endogenous inhibitors in serum means that dilution of serum in the CK assay mixtures will have a significant effect on the apparent CK activity. Decreasing the serum volume fraction from 0.260 (dilution of 1:3.84) to 0.022 (dilution of 1:45) results in a 162% increase in the observed CK activity. It is therefore essential to define the actual serum volume fraction in use in an assay. In a well-optimized CK assay a two-fold dilution (using NaCl, 150 mmol/L) of serum with high activity produces about a 10% increase in CK activity. *CK stability* in stored serum is discussed in more detail under Procedure for the Determination of Creatine Kinase Activity.

Currently, the most satisfactory formulation of the *reverse reaction* assay is the method recommended by the Scandinavian Society for Clinical Chemistry and Clinical Physiology.[24,46] This method will be described under Procedure for the Determination of Creatine Kinase Activity. The Scandinavian method has been adopted by many European professional societies and is increasingly accepted in North America. Despite the growing acceptance of the assay, however, the use of an imidazole buffer has been questioned by Morin, who proposed a Bis-Tris-based assay. Recently, Morin's original formulation has been shown to be capable of further optimization, and response-surface and simplex maximization techniques have demonstrated that CK activities can be increased up to 11% using improved assays.[38] In addition to these formulations, several others exist;[28] for example, the DuPont *aca* assay was used by a quarter of all participants in the College of American Pathologists' Special Enzyme Survey in 1980.

The forward reaction [see equations (29) and (33)], in which CrP and ADP are produced, has been applied in a number of methods. The most popular version involves the following reactions:

$$Creatine + ATP \xrightarrow[pH\ 9.0]{CK} CrP + ADP$$

$$ADP + phosphoenolpyruvate \xrightarrow{PK} ATP + pyruvate \tag{33}$$

$$Pyruvate + NADH + H^+ \xrightleftharpoons{LDH} lactate + NAD^+$$

Unfortunately, this approach produces only one sixth the CK activity of the reverse reaction and is therefore little used nowadays.

Bioluminescent methods have greater sensitivity than spectrophotometric methods.[46] They are likely to become increasingly popular now that luminometers are commercially available. In these methods the reverse reaction is linked to the luciferin/luciferase reaction [equation (34)]:

$$CrP + ADP \xrightarrow{\text{CK}} \text{creatine} + ATP \tag{34}$$

$$ATP + \text{luciferin} + O_2 \xrightarrow{\text{luciferase}} AMP + \text{oxiluciferin} + PP_i + CO_2 + \text{light}$$

The light produced in this reaction is a measure of enzyme activity.

Procedure for the Determination of Creatine Kinase Activity

Specimens[24,46]

Serum is the preferred specimen, although heparinized plasma may be used. Citrate or fluoride inhibits CK activity. CK activity in serum is unstable and is rapidly lost during storage. Full activity may persist at ambient temperatures for 4 h, at 4 °C for ~8–12 h, and for 2–3 d, when frozen; but the degree of stability varies with the individual specimen and also with the activating efficiency of the assay mixture. Creatine kinase is susceptible to thermal denaturation; the degree of inactivation corresponds to the degree of temperature increase and is not reversed by addition of sulfhydryl (thiol) reagents. Creatine kinase is also inactivated by oxidation of sulfhydryl groups at the active site; this inactivation is reversible by the addition of thiol agents. Therefore the serum specimen should be chilled to 4 °C as rapidly as possible after collection. It is probably not necessary to add any thiol agent for storage because an optimized assay formulation, containing EDTA, 2 mmol/L, and N-acetylcysteine, 20 mmol/L, will reactivate creatine kinase in serum to the extent of 99% after it has been stored for one week at 4 °C.

A slight degree of hemolysis can be tolerated because erythrocytes contain no CK activity. However, moderately or severely hemolyzed specimens are unsatisfactory because enzymes and intermediates (adenylate kinase, ATP, glucose-6-phosphate) liberated from the erythrocytes may affect the lag phase and the side reactions occurring in the assay system.

Principle

The reverse reaction [see equation (30)], based on the original formulation of Oliver, is used; the hexokinase and glucose-6-phosphate dehydrogenase reactions are coupled to the production of ATP. Rosalki subsequently modified and improved the method by including AMP to inhibit adenylate kinase and cysteine to activate creatine kinase. Subsequently, Szasz and his colleagues optimized the assay by the incorporation of N-acetylcysteine to activate CK, of EDTA to bind Ca(II) and to increase the stability of the reaction mixture, and of AP₅A and AMP to inhibit adenylate kinase.

Reagents

1. Stock imidazole acetate buffer, 128 mmol/L, pH 7.0 at 25 °C, with magnesium acetate, 12.8 mmol/L and EDTA, 2.55 mmol/L. Prepare by dissolving 8.27 g of imidazole in about 950 mL of purified water, adding 0.95 g of EDTA and 2.75 g of magnesium acetate. When solution is complete, add sufficient acetic acid, 1 mol/L, to bring the pH to 8.0 at 25 °C. Make up to 1 L. The solution is stable for two months at 4 °C.

2. First working reagent. Imidazole acetate, 115 mmol/L; EDTA, 2.3 mmol/L; magnesium acetate, 11.5 mmol/L; N-acetylcysteine, 23.0 mmol/L; ADP, 2.3 mmol/L; AMP, 5.8 mmol/L; AP₅A, 11.5 μmol/L; D-glucose, 23.0 mmol/L; NADP, 2.3 mmol/L; hexokinase, 4000 U/L; D-glucose-6-phosphate dehydrogenase, 2300 U/L; pH 6.5 at 37 °C or pH ~6.7 at 25 °C.

Prepare by adding 90 mL of stock solution (reagent 1) to 98 mg ADP, 211 mg AMP, 1.1 mg AP₅A, 414 mg D-glucose, 181 mg NADP (disodium salt), and 375 mg N-acetylcysteine. Allow solution to equilibrate at the selected temperature and then adjust pH by the addition of acetic acid, 1 mol/L. Finally, add 260–290 U of hexokinase and ~175 U of D-glucose-6-phosphate dehydrogenase to give final catalytic enzyme concentrations of 4000 U/L (hexokinase) and

2300 U/L (G-6-P dehydrogenase) when these enzymes are assayed by optimized assays at 25 °C. Make up the solution to a final volume of 100 mL with water. After preparation the absorbance of the solution should be <0.350 at 339 (340) nm. The reagent is stable for 5 d at 6 °C and for 30 h at room temperature.

3. Second working reagent. Creatine phosphate, 345 mmol/L. Dissolve 1.25 g of creatine phosphate (disodium salt) in water and make up to 10 mL. After preparation the absorbance of the solution at 339 (340) nm should be <0.150. Stable for three months at 4 °C.

Instrument

A narrow-bandpass spectrophotometer equipped with a device to maintain the temperature of the cuvet compartment at 30 ± 0.2 °C, is required; set wavelength at 339 (340) nm.

Procedure

1. Pipet 2.0 mL reagent 2 into a test tube and add 100 μL serum. Mix and place tube into 30 °C bath for at least 5 min.

2. Warm a portion of reagent 3 in the 30 °C bath for at least 5 min.

3. Add 200 μL warm reagent 3 to contents of tube. Mix and transfer contents to a 3-mL cuvet (10-mm light path). Place cuvet into well of spectrophotometer.

4. Allow 2 min for lag phase, then continuously record changing absorbance and use linear section of reaction rate curve for calculation.

Calculation

$$\text{U/L } (\mu\text{mol/min/L}) = \frac{\Delta A/\text{min}}{6.3 \times 10^{-3}} \times \frac{2.30}{0.10} = \Delta A/\text{min} \times 3651$$

where:

$$2.30 = \text{total volume in cuvet, in mL}$$
$$0.10 = \text{volume of serum specimen, in mL (volume fraction} = 0.04348)$$
$$6.3 \times 10^{-3} = \text{micromolar absorption coefficient of NADPH at 339 nm}$$
$$\Delta A/\text{min} = \text{average absorbance change per min}$$

Comments

1. This assay is linear up to at least 3000 U/L. If dilution of sera with very high CK activities is necessary, use NaCl, 150 mmol/L. A two-fold dilution will produce an apparent increase of CK activity of about 10%.

2. This assay has mainly been used at a reaction temperature of 37 °C but performs well at 30 °C.

3. The reagent blank rate (ΔA/min) should be <0.001 (i.e., <3.7 U/L).

4. The stability of the first working reagent is extended because of the presence of EDTA. This reagent is stable for five days. EDTA appears to protect N-acetylcysteine from oxidation, a process known to be catalyzed by divalent ions.

Reference

Committee on Enzymes of The Scandinavian Society for Clinical Chemistry and Clinical Physiology: Recommended method for the determination of creatine kinase in blood modified by the inclusion of EDTA. Scand. J. Clin. Lab. Invest., *39*:1-5, 1979.

Reference Ranges[46,143]

The CK levels in the sera of normal newborns are elevated during the first 24 h postpartum to about three-fold adult values, and a slight elevation remains throughout the first year of life.

In the male, serum CK activity remains constant after the first year until 12 years of age; it then increases at about 15 years of age, a reflection of the increase in muscle mass occurring during puberty. CK activity thereafter decreases slightly until 20 years of age. In the female, activity is stable from the first year until 12 years of age and then rises to the time of menstruation. Thereafter, activity falls, particularly during pregnancy, so that the mean level during pregnancy is less than half that of premenarchal teenagers. After the menopause, the serum CK level rises. For both sexes, there is a distinct seasonal variation with CK activities being higher in the summer months.

Exercise and *muscle trauma* (contact sports, traffic accidents, intramuscular injections, surgery, convulsions, wasp or bee stings) can elevate serum CK values. It has recently been found that serum CK-2 activity is also significantly elevated by regular jogging or after participation in a marathon (i.e., 26-mile) race. Many *drugs*, including the commonly used hypnotic agents, sedatives, tranquilizers, antidepressants, quinidine, and alcohol, can cause elevations of serum CK activity because of their effect on skeletal muscle to allow CK to diffuse from the myocytes. Exposure to *carbon monoxide* can have similar effects.

Examples of adult reference ranges (U/L) for serum CK, using the CK-NAC-EDTA method described above, are:

	Male	Female
25 °C	10–65	7–55
30 °C	15–105	10–80
37 °C	38–174	26–140

It is sometimes convenient to convert enzyme activities at one assay temperature to another, although there is no substitute for the actual determination at each assay temperature. If CK activity at 30 °C is taken as 1.00, activities at other temperatures, are ~0.68-fold at 25 °C and ~1.72-fold at 37 °C.

Caution is necessary when interpreting published data regarding reference ranges or physiological variation, because they were obtained with a wide variety of CK assays containing different activators, varying extents of CK inhibition, and varying serum volume fractions. In addition, the physiological state of the subjects in many studies is ill-defined. It has been established, for example, that bed rest, even overnight bed rest, can lower serum creatine kinase levels by 20% or more. This is presumably part of the explanation for the reported difference between CK values of hospital in- and outpatients.

Methods for the Separation and Quantitation of Creatine Kinase Isoenzymes[46,80,81]

The three techniques most commonly used are electrophoresis, ion-exchange chromatography, and a variety of immunological methods.

Electrophoresis on agar, agarose, or cellulose acetate can separate the CK isoenzymes satisfactorily. The isoenzyme bands are then visualized by incubating the support with a concentrated CK assay mixture using the reverse reaction [see equation (29)]. The NADPH formed in this reaction may be detected by observing the bluish-white fluorescence after excitation by long-wave (360 nm) ultraviolet light. NADPH may be quantitated by fluorescent densitometry, which is capable of detecting 2–10 U/L. Typical examples of results obtained by this technique on a serum sample of a healthy adult and for a patient who has suffered a myocardial infarction 24 h previously are shown in Figure 5-21,*A*. Alternatively, the NADPH can reduce a tetrazolium salt (NBT, nitroblue tetrazolium; MTT, 3-(4,5-dimethylthioazolyl-2)-2,5-diphenyl-2H-tetrazolium bromide; or INT, 3-(4-iodophenyl)-2-(4-nitrophenyl)-5-phenyl-2H-tetrazolium chloride) to form a colored formazan. The discriminating power of electrophoresis allows the detection of many abnormal bands, some of which are shown in Figure 5-21,*B*. In patients on maintenance hemodialysis, for end-stage renal disease, albumin becomes endogenously fluorescent and may simulate the CK-1 band. This possibility can be readily excluded by examining the electrophoretic strip, under fluorescent light, before and after staining. The macro-CK's have been described earlier (p. 679). The postsynthetic bands result from post-translational modifications occurring to the M subunit in blood; these sub-bands will appear between CK-3 and CK-2 and between CK-2 and CK-1 (Figure 5-21,*B*) but are not readily detectable by routine methodologies.

Ion-exchange techniques have been used to separate the CK isoenzymes either by batch adsorption or by column (or minicolumn) chromatography. A variety of media have been used (DEAE-Sephadex A50, DEAE-Cellulose, DEAE-Glycophase, DEAE-Biogel A), but DEAE-

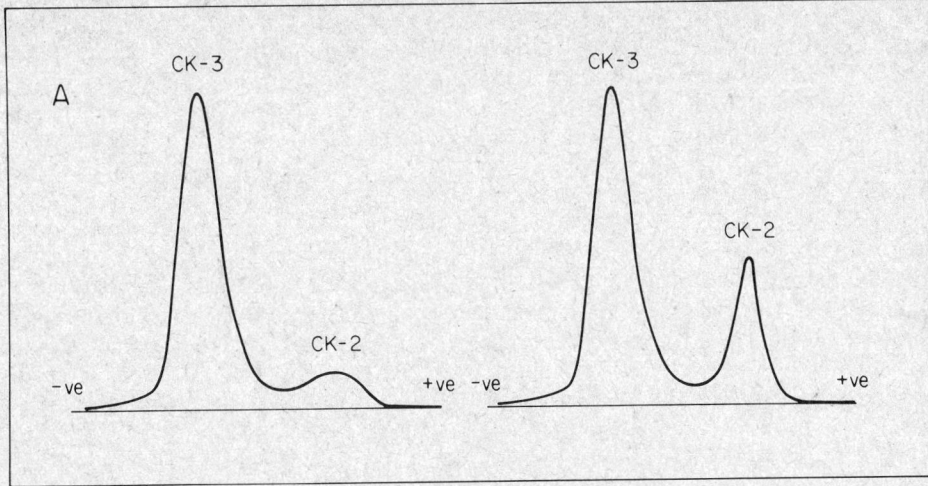

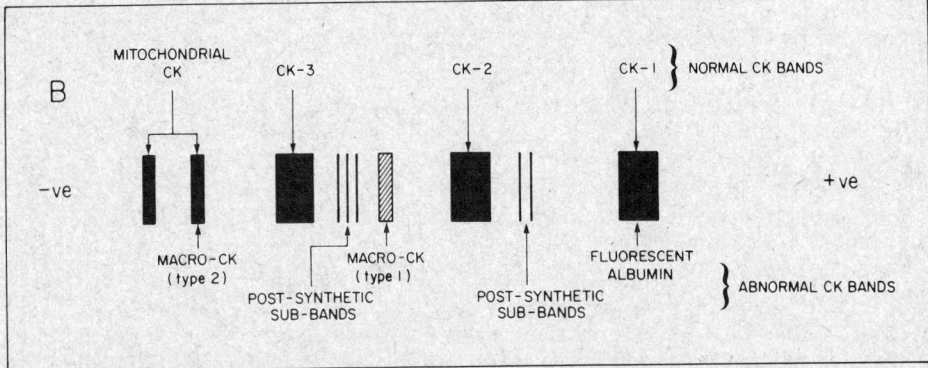

Figure 5-21. *A,* The separation of serum CK isoenzymes from a healthy adult (left) and from a patient (right) who had a myocardial infarction 24 h previously. *B,* A diagrammatic representation of the naturally occurring CK isoenzymes (some of which are seen only in disease) and some of the reported anomalous isoenzymes seen on agarose gels.

Sephadex A50, usually in the form of a minicolumn (0.5 × 6.0 cm) holding about 60 mg of gel, has been most favored. The CK isoenzymes, as well as other proteins, are adsorbed onto the gel, the gel is then washed, and the CK-3 is eluted with Tris buffer, 50 mmol/L, pH 8.0, containing NaCl, 100 mmol/L. CK-2 is next eluted with the same Tris buffer containing NaCl, 200 mmol/L. Finally, CK-1 is eluted with Tris buffer, 50 mmol/L, pH 7.0, containing NaCl, 500 mmol/L. The lower limit of detection is about 1–5 U/L. The elution process considerably dilutes the CK isoenzymes. It is therefore important to reduce this dilutional effect by using small volumes of eluting buffer, or by concentrating each fraction after collection, or by using the eluate as part solvent for the CK reagents. There are several published procedures based on this technique and many commercially available kits. The method is relatively simple and rapid to perform, but precautions must be taken to ensure that there is no carryover between the different isoenzymes eluted from the gel. Carryover is the major limitation of this technique; this risk is greatest for samples with high CK-3 activities.

Immunological methods for measuring the CK isoenzymes require specific antisera against the M and B subunits. These are obtained by injecting pure CK-1 (BB) or CK-3 (MM) into either rabbits or goats. The resulting rabbit antisera can be used to precipitate specific isoenzymes out of solution (immunoprecipitation) or they can competitively bind specific isoenzymes (immunoassay). Goat antisera appear to be more effective inhibitors of subunit activity (immunoinhibition).

In the procedure based on *immunoprecipitation,* anti-CK-B sera precipitate (after centrifugation) all CK-1 (BB) and CK-2 (MB) from a serum sample, leaving only the CK-3 (MM) activity to be detected. Similarly, anti-CK-M sera will precipitate all CK-3 and CK-2, leaving only CK-1 (if present) to be detected. The serum content of CK-2 can thus be measured by calculating the difference between the total serum CK activity and the sum of CK-1 and CK-3 activities. The detection limit of this assay is about 4 U/L; the coefficient of variation of the CK-1 and CK-3 determinations is 4% and that of CK-2 between 10 and 20%. The greater CV for CK-2 is due to the summation of errors in the separate determinations of total CK, CK-1, and CK-3 activities.

Immunoprecipitation is a time-consuming procedure, whereas the technique of *immunoinhibition* is simpler and faster. The inhibiting anti-CK-M sera will inhibit both M subunits of CK-3 and the single M subunit of CK-2 and thus will allow the determination of the enzyme activity of the B subunit of CK-2 and the B subunits of CK-1, if present. To be an effective method for determining CK-2, this technique therefore assumes the absence of CK-1 from the tested serum, a circumstance that does not always occur. However, when CK-1 is present in a specimen, it usually has an activity of <5 U/L, which is below the detection limit of the immunoinhibition technique anyway. Therefore CK-1 interference in this assay is generally acceptable. The detection limit for CK-2, using the *N*-acetylcysteine-activated assay described earlier, is 6–7 U/L, the coefficient of variation is 10%, and the discrimination (threshold) value is 15 U/L at 30 °C.

By contrast with immunoprecipitation and immunoinhibition, which measure the isoenzymes by determinations of enzyme *activity,* *immunoassays* such as RIA measure the enzyme mass whether or not the enzyme molecule is catalytically active. While the RIA of CK-1 and CK-3 is readily achieved, with <2% cross-reactivity, one of the CK-2 subunits will be included in any determination of either the M or the B subunits. The cross-reactions have been reported by different groups as from 35–59%, 7–12%, 3%, or none. CK-2 may be measured specifically by application of the "sandwich" technique in which the B subunits of CK-2 are bound by anti-CK-B sera rendered immobile on a matrix, and then the M subunits associated with the B subunits (MB or CK-2) are estimated by use of a ^{125}I-labeled anti-CK-M serum. The analytical sensitivity of RIA exceeds that of methods that measure enzyme activity, but the RIA technique cannot replace the activity measurement because results of these two methods sometimes differ. There is usually a concordance between these assays for the measurement of serum CK-3, but when serum CK-2 (or CK-1) is measured, enzyme activity disappears faster from serum than does the immunologically reactive enzyme molecule—an observation that suggests intravascular inactivation as a first stage in the metabolism of the enzyme molecule.

A wide spread of *reference ranges* has been reported: for serum CK-3, adult males 102–1688 μg/L and adult females 36–487 μg/L; for CK-2, 2–49 μg/L; and for CK-1, <6.2 μg/L. The detection limit for these assays varies between 0.5 and 10 μg/L, and the precision between 5 and 11%.

The Centers for Disease Control in Atlanta, Georgia, reported in 1982 on the methods used by 280 US laboratories to determine CK isoenzymes. Twenty-nine different methods or kits were used; the most frequent techniques were electrophoresis on cellulose acetate (45%), electrophoresis on agarose (39%), ion-exchange column chromatography (13%), and selective activation (3%). Over one fifth of the laboratories using electrophoresis determined the presence or absence of CK-2 and CK-1 by visual examination under ultraviolet illumination after isoenzyme separation. Of the 158 participants who quantitated the isoenzymes, 77% used electrophoresis, 20% ion-exchange chromatography, and 3% selective activation; with these techniques the survey materials were reported to contain CK-2 ranging from within the laboratory's own reference range to 19-fold greater than the range, a very wide divergence in the reporting of an analyte of some clinical importance.

Are there preferred methods? Morin, in 1977,[99] reported on an extensive study of 18 methods (including three electrophoretic and twelve chromatographic techniques) using four criteria: resolution, reliability, sensitivity, and efficiency.* Electrophoretic methods were found to excel in resolution and reliability but were relatively inefficient and insensitive. By contrast, ion-exchange methods were shown to have good resolution and reliability but, because of variation in factors affecting column performance, both resolution and reliability could degrade without detection.

---

*Resolution—ability to accurately and precisely measure each fraction distinctly, without cross-contamination. Reliability—ability to maintain resolution repeatedly over a broad range of specimens. Sensitivity—limit of detectability. Efficiency—ratio of capacity (number of samples) to simplicity (number of manipulations) and rapidity (time taken).

Morin concluded that ion-exchange methods are potentially more sensitive and efficient than electrophoretic methods.

The US Food and Drug Administration, in a contract study reported in 1981, after an evaluation of 17 CK isoenzyme methods for precision, clinical correlation, and clinical consensus, ranked ion-exchange chromatography more useful than immunoinhibition and immunoinhibition more useful than electrophoresis.[46]

Representative CK isoenzyme results are given in Table 5-8 for a variety of clinical conditions and techniques.

ALDOLASE

(EC 4.1.2.13; D-Fructose-1,6-bisdiphosphate D-Glyceraldehyde-3-phosphate-lyase)

The enzyme aldolase (ALD) belongs to the class of enzymes called lyases, which reversibly cleave substrates without hydrolysis into two units, one or both of which contain a double-bonded carbon atom. Carbonic anhydrase and the many decarboxylases are included in the class. Aldolase catalyzes the splitting of D-fructose-1,6-diphosphate (FDP) to D-glyceraldehyde-3-phosphate (GLAP) and dihydroxyacetone-phosphate (DAP), an important reaction in the glycolytic breakdown of glucose to lactate. The reaction equilibrium favors the formation of fructose diphosphate [equation (35)].

D-Fructose-1,6-diphosphate (FDP)   **Dihydroxyacetone phosphate (DAP)**   **D-Glyceraldehyde-3-phosphate (GLAP)**

$$\text{(35)}$$

Aldolase is a tetrameric enzyme with subunits determined by three separate gene loci.[66] Only two of these loci, those producing A and B subunits, appear to be active simultaneously in most tissues, so that the most common isoenzyme pattern consists of varying proportions of the components of a five-membered set of isoenzymes, of which two members correspond to the A and B homopolymers. The locus which determines the structure of the C subunit is active in brain tissue, as is the A locus, so that this tissue contains aldolases A and C together with the three corresponding heteropolymers.

Table 5-8. CREATINE KINASE ISOENZYME RESULTS IN SERA OF PATIENTS WITH A VARIETY OF CLINICAL CONDITIONS, AND OBTAINED BY SEVERAL DIFFERENT TECHNIQUES

Clinical Condition	CK Activity, U/L			Per cent MB Activity	Technique[†]
	Total	*MM*	*MB*		
Hospitalized noncardiac patients	35±13	33±12	2±1	3–9	E
Myocardial infarctions	860±120	797±112	63±43	2–12	E
Patients receiving morphine IM	347±89	345±89	3±2	0.3–0.9	E
Dermatomyositis	1500	1410	90	6	E
Polymyositis	1725	1052	673	39	E
Normal	36–277	—	0–2.6	0–1	C
Myocardial infarctions	84–236	—	4.6–28	6–11	C
Patient A, myocardial infarct, 24 h	82	53	29	35	I
Patient B, myocardial infarct, 48 h	416	449	67	16	I
Myocardial infarctions (35 patients)					
Total CK, 35–100 U/L	60*	51	9	15	I
Total CK, over 200 U/L	416*	339	67	16	I

*Average values.
[†]Symbols: E = enzymoelectrophoresis; C = column chromatography; I = immunoprecipitation.

The properties of aldolases A, B, and C from several species have been studied extensively. The catalytic differences between the A and B isoenzymes are consistent with the patterns of metabolism in the tissues in which they mainly occur. Both isoenzymes will cleave fructose-1,6-diphosphate or fructose-1-phosphate, but aldolase A shows a 50-fold greater activity toward fructose-1,6-diphosphate than toward fructose-1-phosphate. This is in keeping with its part in the glycolytic metabolism of skeletal muscle, in which it is the predominant isoenzyme, since cleavage of fructose-1,6-diphosphate to triosephosphate is a key reaction in glycolysis. Aldolase B, the main isoenzyme of liver, shows no marked preference for the diphosphate substrate and this and other properties indicate that it is better adapted to utilization of fructose and to gluconeogenesis. The properties of aldolase C are intermediate between those of the A and B isoenzymes, but the function of this isoenzyme is uncertain.

Clinical Significance

Serum aldolase determinations have been of greatest clinical interest in primary diseases of *skeletal muscle*.[140] Elevations of 10–50 times the upper normal level may be seen, the highest levels occurring in progressive *Duchenne-type muscular dystrophy*. The greatest serum increases occur early in the disease, but as the capacity of the cells to synthesize enzyme decreases, serum levels also decrease. Slight to moderate increases in ALD activity can often be demonstrated in the sera of female carriers of the gene causing the disease. Lesser degrees of ALD elevation are encountered in *dermatomyositis, polymyositis,* and *limb-girdle dystrophy,* but normal values are observed in *poliomyelitis, myasthenia gravis, multiple sclerosis,* and muscle diseases of neurogenic origin.

Increases in serum aldolase activity to 5–8 times normal are also observed with *myocardial infarctions;* the pattern of rise and fall parallels that of AST. Increases of 7–20 times normal are associated with *viral hepatitis,* the pattern of rise and fall being similar to that for ALT; normal values are reached 15–20 days after the first rise occurred. Levels of ALD in *chronic hepatitis, portal cirrhosis,* and *obstructive jaundice* are normal or only marginally raised. In liver diseases it is the B isoenzyme, the so-called fructose-1-phosphate aldolase, which is increased.

Other disease states in which increased levels of aldolase may be encountered are *trichinosis* (moderate to high values), *gangrene, prostatic tumors,* some *carcinomas metastatic to the liver, granulocytic leukemia* (about 6 times normal), *megaloblastic anemia* (10–13 times normal), *delirium tremens,* and 60–80% of patients with *acute psychoses* and *schizophrenia*. Activity is normal in *hypothyroid* patients and normal or slightly depressed in persons affected with fructose intolerance.

Injections of cortisone and ACTH will raise serum ALD to levels between 10 and 18 U/L. This physiological response to hormone therapy must be kept in mind when interpreting elevated aldolase values.

In general, however, measurement of aldolase activity in serum does not provide information that is not available more readily by measurement of other, more easily assayed enzymes, such as AST, LDH, and especially CK. Because of its greater diagnostic sensitivity and ease of measurement, CK is generally regarded as the enzyme of choice in the investigation of disorders of skeletal muscle.

Assay Methods

All assay methods are based on the forward reaction, as written in equation (35). Both colorimetric fixed-time and continuous-monitoring procedures have been developed. In the colorimetric methods, hydrazine is added to the reaction mixture as a trapping agent to force the reaction to completion, by forming the hydrazones of DAP and GLAP. These are then hydrolyzed with NaOH to form free dihydroxyacetone and glyceraldehyde, which are then treated with 2,4-dinitrophenylhydrazine. The colored dinitrophenylhydrazones are then measured at 540 nm.[115]

Two continuous-monitoring procedures have been proposed. In the Bruns procedure, on which all the commonly used procedures and kits are based, the aldolase reaction is coupled with two other enzyme reactions. Triosephosphate isomerase (EC 5.3.1.1) is added to ensure rapid conversion of all GLAP to DAP. Glycerol-3-phosphate dehydrogenase (EC 1.1.1.8) is added to reduce the DAP to glycerol-3-phosphate, with NADH acting as hydrogen donor. The decrease in NADH concentration is then measured.[114]

In the other coupled enzyme approach, the GLAP is oxidized by NAD^+ to 3-phosphoglyceric acid, and the rate of formation of NADH is measured. This reaction is catalyzed by the added enzyme, glyceraldehyde-3-phosphate dehydrogenase. Triosephosphate isomerase and arsenate are also added to force the reaction to completion. This coupled system is subject to many interferences, and has fallen into disfavor.

The ALD activity in serum is quite stable. Activity is unchanged at ambient temperatures for up to 48 h and at 4 °C for several weeks. Hemolyzed specimens should not be used, and plasma is preferred over serum because of the possible release of platelet enzyme during the clotting process.

Reference Ranges

The accepted range of values for the activity of ALD in serum in adults is 1.0–7.5 U/L, measured at 30 °C. However, there is a definite sex difference. Pinto and coworkers give 8.0 ± 4.0 U/L for men and 4.7 ± 3.2 U/L for women, for activity measured at 37 °C (5.2 ± 2.6, and 3.0 ± 2.1 U/L, respectively, at 30 °C). Experience has shown that ALD values for inactive persons and patients at bed rest are only some 50–70% of the values obtained for active individuals.

The level at birth is twice as high as that of adults; values double in early childhood, and then slowly fall to the adult range by 18–20 years of age.

LACTATE DEHYDROGENASE

(EC 1.1.1.27; L-Lactate:NAD$^+$ Oxidoreductase; LDH)

Lactate dehydrogenase is a hydrogen transfer enzyme that catalyzes the oxidation of L-lactate to pyruvate with the mediation of NAD$^+$ as hydrogen acceptor. The reaction is reversible [equation (36)] and the reaction equilibrium strongly favors the reverse reaction, namely the reduction of pyruvate to lactate (P→L).[92]

$$
\begin{array}{c}
CH_3 \\
| \\
H-C-O-H \\
| \\
C=O \\
| \\
O^-
\end{array}
\quad + \quad NAD^+
\quad
\underset{\text{pH 7.4-7.8}}{\overset{\substack{LDH \\ \text{pH 8.8-9.8}}}{\rightleftharpoons}}
\quad
\begin{array}{c}
CH_3 \\
| \\
C=O \\
| \\
C=O \\
| \\
O^-
\end{array}
\quad + \quad NADH \quad + \quad H^+ \qquad (36)
$$

L-Lactate   Pyruvate

The pH optimum for the lactate to pyruvate (L→P) reaction is 8.8–9.8 and an optimal assay mixture at 30 °C contains NAD$^+$, 5 mmol/L, and L-lactate, 50 mmol/L; for the P→L assay, at 30 °C, the pH optimum is 7.4–7.8, NADH 150 μmol/L, and pyruvate 1.5 mmol/L. The optimal pH varies with the source of enzyme (i.e., with the predominant isoenzymes in the sample) and depends on the temperature, as well as on substrate and buffer concentrations.[92] The specificity of the enzyme extends from L-lactate to a variety of related α-hydroxy acids and γ-oxoacids. The catalytic oxidation of α-hydroxybutyrate, the next higher homolog of lactate, to α-oxobutyrate is referred to as α-hydroxybutyrate dehydrogenase (HBDH) activity. LDH does not act on D-lactate, and only NAD$^+$ will serve as coenzyme. The enzyme has a molecular weight of 134 000 and is composed of four peptide chains of two types: M (or A) and H (or B), each under separate genetic control. The structures of LDH-M and LDH-H are determined by loci on human chromosomes 11 and 12, respectively. The subunit compositions of the five isoenzymes, in order of decreasing anodal mobility in an alkaline medium, are LDH-1 (HHHH; H$_4$); LDH-2 (HHHM; H$_3$M); LDH-3 (HHMM; H$_2$M$_2$); LDH-4 (HMMM; HM$_3$); LDH-5 (MMMM; M$_4$). A different, sixth lactate dehydrogenase isoenzyme, LDH-X (also called LDH$_c$) composed of four X (or C) subunits, is present in postpubertal human testis.

Lactate dehydrogenases are inhibited by reagents with reactivity against thiol groups such as mercuric ions and p-chloromercuribenzoate, the inhibition being reversed by the addition of cysteine or glutathione. Borate and oxalate inhibit by competing with lactate for its binding site on the enzyme; similarly, oxamate competes with pyruvate for its binding site. Both pyruvate and lactate in excess inhibit enzyme activity, although the effect of pyruvate is greater. Inhibition by either substrate is greater for the H form than for the M form and substrate inhibition decreases with increase in pH. EDTA inhibits the enzyme perhaps by binding Zn(II); however, the postulated activator role for zinc ions is not fully established.

Distribution and Clinical Significance[151]

LDH content in various tissues and serum. LDH activity is present in almost all cells of the body and is invariably found only in the cytoplasm of the cell. Enzyme levels in various tissues (in U/g) are very high compared to those in serum: liver, 145; heart, 124; kidney, 106; skeletal muscle, 147; and erythrocytes (U/g hemoglobin), 36. Thus, tissue levels are about 500-fold higher than those normally found in serum, and leakage of the enzyme from even a small mass of damaged tissue can increase the observed serum level of LDH to a significant extent.

In addition to their higher enzyme concentration, many of these tissues show different isoenzyme composition. In cardiac muscle, kidney, and erythrocytes the electrophoretically faster-moving isoenzymes LDH-1 and LDH-2 predominate, whereas in liver and skeletal muscle the more cathodal LDH-4 and LDH-5 isoenzymes predominate. Isoenzymes of intermediate mobility account for the lactate dehydrogenase activity of many tissues, e.g., endocrine glands, spleen, lung, lymph nodes, platelets, and nongravid uterine muscle (Figure 5-22).

Serum Levels of Lactate Dehydrogenase in Disease States

Myocardial infarction may be associated with elevations of total LDH by as much as ten times the upper reference limit, but usually only three to four times. The rise in serum level begins 8–12 h after onset of pain, reaches a maximum 24–48 h after the episode, and remains elevated for some 7–12 d. The pattern should be contrasted with creatine kinase (p. 681), which begins to rise about 6 h after the onset of chest pain, peaks in about 24 h, and generally returns to the reference range by the third day, and with aspartate aminotransferase (p. 672), which begins to rise at 6–8 h, peaks at 18–24 h (at four to six times the upper reference range), and returns to the reference range by the third or fourth day. Diagnostic sensitivity is enhanced by determining the LDH isoenzyme pattern by electrophoresis or by estimating the level of LDH-1 by immunoprecipitation. A pattern of LDH-1 > LDH-2, the so-called "flipped" pattern, is obtained in more than 80% of all patients at some time after the infarction (Figure 5-23). Since a pattern similar to that observed in myocardial infarction can also be observed following *hemolysis* (see below), it is important to be able to exclude causes of in vivo hemolysis before assuming that such a pattern is due to myocardial necrosis.

Following a myocardial infarction, total serum activities greater than seven to eight times the upper reference limit suggest a poor prognosis. Values may be moderately elevated in *myocarditis* and in *cardiac failure* with *hepatic congestion* (see Figure 5-23), but are normal in *angina* and in *pericarditis.* Enzyme levels may be moderately elevated in *severe shock* and in *anoxia.*

As observed above, almost any cause of *hemolysis,* if sufficiently severe, can produce an LDH isoenzyme pattern similar to that seen in myocardial infarction. An increasingly important cause of hemolysis is the presence of a *prosthetic heart valve. Megaloblastic anemias,* usually resulting from the deficiency of folate or vitamin B_{12}, can cause the erythrocyte precursor cell to break down in the bone marrow (ineffective erythropoiesis), resulting in the release of large quantities of LDH-1 and LDH-2 isoenzymes. Marked elevations of the total serum LDH activity— up to 50 times the upper reference limit—may be observed in the megaloblastic anemias. These elevations rapidly return to normal following appropriate treatment.

Elevations of LDH activity are observed in *liver disease,* but these elevations are not as great as the increases seen in aminotransferase activity. Elevations are especially high (ten times normal)

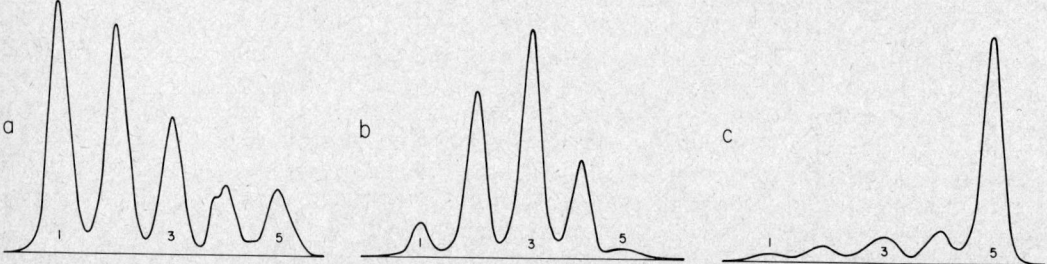

Figure 5-22. Tissue LDH isoenzyme patterns obtained with a thin-layer agarose gel electrophoresis system. The isoenzymes were detected by incubating the gel with a liquid overlay of NAD^+ and L-lactate. The generated NADH was detected by fluorescent scanning. Pattern *a*, myocardial tissue; pattern *b*, platelets; pattern *c*, liver tissue.

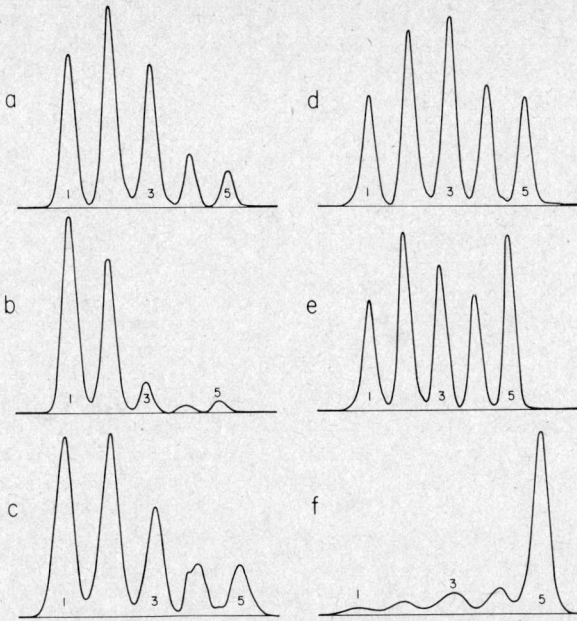

Figure 5-23. Serum LDH isoenzyme patterns obtained with a thin-layer agarose gel electrophoresis system. The isoenzymes were detected by incubating the gel with a liquid overlay of NAD$^+$ and L-lactate. The generated NADH was detected by fluorescent scanning. Pattern *a*, normal serum. Pattern *b*, acute myocardial infarction (common pattern showing "flipped" LDH-1). Pattern *c*, acute myocardial infarction (showing elevated LDH-1 that has not "flipped"). Pattern *d*, involvement of platelets or lymphatic tissue. This pattern was obtained from a patient with infectious mononucleosis. Pattern *e*, congestive cardiac failure showing elevated LDH-5 as a result of hepatic anoxia. Pattern *f*, acute circulatory shock showing very severe hepatic anoxia.

in *toxic hepatitis with jaundice;* slightly lower values are observed in *viral hepatitis* and in *infectious mononucleosis,* often associated with elevations of LDH-3 (see Figure 5-23). LDH activity is normal, or at most twice the upper limit of normal, in *cirrhosis* and in *obstructive jaundice.* Serum LDH-5 is often markedly elevated (Figure 5-23) in patients with either *primary liver disease* or *liver anoxia* secondary to decreased oxygen perfusion.

Increased levels of the enzyme in serum are also found in about one third of patients with *renal disease,* especially those with *tubular necrosis* or *pyelonephritis.* However, these elevations do not correlate well with proteinuria and other parameters of renal disease. The LDH isoenzyme pattern seen in renal disease is very similar to a normal serum pattern except for the higher absolute values. In *renal infarction* the serum LDH pattern can mimic that seen in myocardial infarction.

Patients with *malignant disease* show increased LDH activity in serum; up to 70% of cancer patients with liver metastases and 20–60% of cancer patients who do not have hepatic metastases have elevated total LDH activity. Especially high values are associated with *Hodgkin's disease* and with *abdominal* and *lung cancers.* The isoenzyme pattern of the elevated LDH may occasionally reflect the organ affected by the malignancy, but most often it shows only a nonspecific increase in the slow-moving forms (LDH-4 and LDH-5), suggesting that the tissue has regressed into synthesizing the more embryonic, anaerobic LDH types. One unusual exception to this observation is the markedly elevated LDH-1 seen in *germ cell tumors* such as *teratomas, seminoma of the testis,* and *dysgerminoma of the ovary.* As a rule, the elevations of LDH seen in cancer patients are too erratic to be of use in clinical diagnosis, although serum levels have been assayed to follow changes in tumor burden after chemotherapy. More helpful are measurements of LDH activity in *exudative effusions* obtained from areas near or adjacent to malignancies. In these fluids LDH levels are often higher than in serum, whereas the opposite is true for fluids bathing healthy tissues. In exudates caused by malignancies, LDH-2 is found to be unusually high in 33% of cases, and in 50% of cases LDH-2 was > 35% or LDH-5 was < 12%. *Leukemias* are associated with only moderate elevations of serum LDH.

Moderately increased LDH activity is found in the sera of all patients affected by *progressive muscular dystrophy,* especially in the early and middle stages of the disease. The observed increase is confined to LDH-5, the isoenzyme form characteristic of striated muscle. In the later stages of the disease, after a large mass of the tissue containing the LDH-5 has been lost, the observed LDH levels in serum may even drop to normal levels, with the predominant isoenzyme forms now being LDH-1 and LDH-2. Only occasionally is an elevated serum LDH level found in persons with other forms of neuromuscular disorders.

Elevated values of the enzyme in serum are also seen in cases of *pulmonary embolism;* on occasion, a raised LDH level may be the only evidence to suggest the presence of a hidden embolus. The serum LDH-3 is elevated probably due to the massive destruction of platelets following the formation of an embolus (Figure 5-23).

α-Hydroxybutyrate Dehydrogenase

As an alternative to LDH isoenzyme determinations, Rosalki and Wilkinson advocated measurement of *α-hydroxybutyrate dehydrogenase* (HBDH) activity, and calculation of the LDH/HBDH ratio. LDH-1 and LDH-2 (the two isoenzyme forms with greater proportions of monomer H) are relatively more active with α-oxobutyrate as substrate than with pyruvate. With serum from healthy individuals, the LDH/HBDH ratio varies from 1.2–1.6; in *parenchymal liver disease,* the ratio is increased and ranges from 1.6–2.5. In *myocardial infarction,* with elevated LDH-1 and LDH-2 activity (and greater HBDH activity), the ratio is decreased to between 0.8 and 1.2. The exact ratios will depend on the conditions chosen for the assay of the two activities.

Lactate Dehydrogenase in Urine

Elevations of LDH activity in urine to three to six times normal are associated with *chronic glomerulonephritis, systemic lupus erythematosus, diabetic nephrosclerosis,* and *bladder and kidney malignancies.* Determination of LDH activity in urine is affected by uncertainties arising from the presence of inhibitors such as urea and small peptides, and from the possible inactivation of LDH under adverse pH conditions in the urine.

Lactate Dehydrogenase in Spinal Fluid

Elevations of LDH activity in spinal fluid (CSF) are associated with *subarachnoid hemorrhage* and with *cerebrovascular thrombosis and hemorrhage.* CSF-LDH is usually normal in patients with brain or meningeal tumors (although LDH-5 has been reported as raised in this latter condition), but may be elevated in cases of invasive cancers originating from primary sources elsewhere.

Methods for Lactate Dehydrogenase Activity Measurements[92,129,148]

A multiplicity of procedures have been introduced over the last 25 years; the procedures use the forward (L→P) and reverse (P→L) reactions in almost equal numbers. The L→P assay has the following advantages: substrate inhibition by lactate is less than that produced by pyruvate, NAD preparations (used in the L→P reaction) appear to contain less endogenous LDH inhibitors than NADH preparations (used in the P→L reaction), and the reaction linearity of the L→P assay is more prolonged than that of the P→L assay. However, the converse disadvantages of the P→L reaction seem to be less severe than was formerly thought; for example, commercial preparations of NADH are now substantially free of LDH inhibitors. Other advantages of the P→L assay include a less expensive assay formulation (because of the much lower concentration of reactants), the greater change in absorbance with time, thus allowing more precise measurements, and greater stability of the working reagents once they are prepared as assay solutions. Both assays enjoy significant popularity. The colorimetric methods are largely discontinued in favor of the "kinetic" methods because of the lower reliability of the colorimetric methods and the almost universal availability of spectrophotometers capable of precise measurement at 339 or 340 nm.

The College of American Pathologists reported that, among those using one of their 1981 Special Enzyme Surveys, 80% of 653 participants used an assay temperature of 37 °C; the remainder used 30 °C. Of the 22 different methods used, two methods were used by 50% of the participants—

DuPont *aca* (235 participants) or the Technicon SMAC (93 participants). Both of these instruments used the L→P assay formulation at 37 °C, with detection of NADH production [see equation (36)] at 339 nm. The operating manuals of these analyzers should be consulted for further details of these assays.

In this chapter we will describe a P→L continuous-monitoring procedure that is widely used throughout Europe and that was developed by the Committee on Enzymes of the Scandinavian Society for Clinical Chemistry and Clinical Physiology.

Determination of LDH Activity by Measurement of NADH Consumption[129]

Specimens

Serum or heparinized plasma samples are satisfactory; however, serum or plasma should be separated from the clot as soon as possible after the specimen has been obtained. Plasma containing other anticoagulants, especially oxalate, should not be used. Hemolyzed serum or plasma must not be used since erythrocytes contain 150-fold more LDH activity (particularly LDH-1 and 2) than serum.

Purified LDH is unstable when dissolved in water, but it is quite stable in concentrated ammonium sulfate, in glycerol-water (1:1), and in the presence of other proteins. The different isoenzymes vary in their sensitivity to cold, LDH-4 and LDH-5 being especially labile. In tissue extracts, all activity of LDH-4 and -5 is lost if the extracts are stored at −20 °C overnight. Loss of activity may be prevented by addition of NAD^+ or glutathione. Both types of monomers bind a molecule of NAD^+, but the binding of NAD to the M form is weaker and some dissociation occurs, with concomitant exposure of sulfhydryl groups to oxidation. In serum, the sulfhydryl in albumin and other proteins retards inactivation of the M-rich isoenzymes (LDH-4 and LDH-5). Serum specimens should be stored at room temperature, at which no loss of activity will occur for 2–3 d. If specimens of sera must be stored for longer periods, they should be kept at 4 °C with NAD^+ (10 mg/mL) or glutathione (3.1 mg/mL) added to decrease the rate of inactivation of LDH-4 and -5. Note that there is considerable disagreement over the essential storage conditions for the LDH isoenzymes; some workers claim that 4 °C storage is perfectly adequate. Therefore each laboratory should determine suitable storage conditions.

Principle

Pyruvate is reduced to lactate at pH 7.4 and 30 or 37 °C in the presence of LDH. The progress of the accompanying oxidation of NADH to NAD^+ is monitored continuously by measuring the rate of absorbance decrease at 339 nm in a spectrophotometer [see equation (36)].

Reagents

1. Tris-EDTA buffer, pH 7.4 (37 °C). Weigh out 6.8 g of Tris (hydroxymethyl)aminomethane (56 mmol/L), and 2.1 g of ethylenediaminotetraacetate, disodium salt (5.6 mmol/L), dissolve in 900 mL of water, warm to 37 °C, and carefully add HCl, 1 mol/L, to adjust the pH to 7.4 at 37 °C. Make the solution up to 1 L. This reagent is stable for up to six weeks when stored at 4 °C in a tightly capped dark bottle.

2. NADH solution. Weigh out 11.7 mg (170 µmol/L) of β-NADH (disodium salt, trihydrate) and dissolve in 90 mL of the Tris-EDTA buffer, pH 7.4 (37 °C). Stable for at least 72 h when stored in a tightly capped dark bottle at 4 °C.

3. Pyruvate solution. Weigh out 149 mg (14 mmol/L) of sodium pyruvate and dissolve in 100 mL of water. Stable for at least 20 days when stored at 4 °C in a tightly capped bottle.

Procedure

1. Use a narrow-bandpass (≤ 8 nm) spectrophotometer equipped with a circulating constant temperature bath or equivalent temperature controlling device. Set the thermostat to maintain the temperature of the cuvet compartment at 37 ± 0.2 °C.

2. Pipet 2 mL of Tris-EDTA-NADH buffer (reagent 2) and 50 µL of serum into a test tube. After mixing the contents, place the tube in the 37 °C water bath for 5–15 min. This incubation permits a reduction by the NADH of any pyruvate and other oxoacids present in the serum. At the end of the incubation period add 200 µL of the pyruvate solution—which has been prewarmed to 37 °C—mix rapidly, and transfer to a 3-mL cuvet, with 10-mm path length. Insert cuvet in

the spectrophotometer and maintain at 37 °C. Determine the initial reaction velocity by continuous recording of the absorption at 339 (340) nm.

The actual concentration of reaction components in the cuvet is as follows: Tris buffer, 50 mmol/L (pH 7.4 at 37 °C); EDTA, 5 mmol/L; pyruvate, 1.2 mmol/L; NADH, 150 μmol/L; volume fraction of serum = 0.023 (1:44).

Calculation

The LDH activity in international units at 37 °C is obtained as follows:

$$\text{U/L (μmol/min/L)} = \frac{\Delta A/\text{min}}{6.3 \times 10^{-3}} \times \frac{2.250}{0.05} = \frac{\Delta A}{\text{min}} \times 7143$$

where

2.250 = total volume in cuvet, in mL

0.05 = volume of serum specimen, in mL

6.3×10^{-3} = micromolar absorption coefficient of NADH at 339 nm

$\dfrac{\Delta A}{\text{min}}$ = average absorbance change (decrease) per min

Calculate the average absorbance change per minute ($\Delta A/\text{min}$), using either the recorder tracing or the measured absorbance values. Use only the linear portion of the rate curve or those consecutive individual values of $\Delta A/\text{min}$ that are essentially constant.

Comments

1. Reaction temperature. Most experience with this assay has been obtained at 37 °C. However, satisfactory results can be obtained at 30 °C provided appropriate reference ranges are established.

2. Linearity. This assay will give linear reaction rates up to ~3000 U/L at 37 °C. If activity is in excess of this level, the serum should be diluted 10-fold with assay buffer (or preferably with an albumin solution, 50 g/L) and the assay repeated.

3. Reaction blank. The reaction is carried out without any blank. If the intrinsic absorption of the serum is high, it is convenient to compensate with a blank cuvet using serum and buffer only. Alternatively, a dichromate blank can be used. Prepare a stock solution containing 300 mg $K_2Cr_2O_7$ in 1 L of water containing three or four drops of concentrated H_2SO_4. The stock solution is diluted four-fold, then seven-fold and so on, so that the test cuvet read against the dichromate blank gives an initial absorbance reading of between 0.55 and 0.7 before pyruvate is added.

4. Some analysts omit the preincubation step, claiming that the endogenous reactions do not significantly alter the value of $\Delta A/\text{min}$; others use only short preincubation periods of 3–5 min. Each worker should determine his or her approach by experimentation.

5. Tris buffer was selected because it has adequate buffering capacity at 37 °C and pH 7.4 and because NADH is more stable in Tris buffer than in phosphate buffer. When NADH (10 mmol/L) was stored in the Tris-EDTA-HCl buffer at −20 °C for two weeks, at 4 °C for one week, or at 25 °C for 24 h, initial assay reaction rates remained constant.

Reference Ranges[143]

Values for LDH activity in serum vary considerably, depending on the direction of the enzyme reaction, the type of method used, and the experimental parameters. For the P→L reaction at 30 °C and at pH 7.4, a range of 95–200 U/L represents the experience of most workers. Using the Scandinavian assay described above, reference ranges from 200–380 U/L (at 37 °C) are obtained. For the L→P reaction at pH 8.8–9.0 at 30 °C, the range of 35–88 U/L represents the consensus of values proposed by various investigators.

The reference values for spinal fluid LDH are 7–30 U/L for the P→L reaction at 30 °C. In urine, a reference range of 42–98 U/L has been reported for the L→P reaction at 25 °C.

Reference

Scandinavian Society for Clinical Chemistry and Clinical Physiology: Recommended methods for the determination of four enzymes in blood. Scand. J. Clin. Lab. Invest., *33*:291, 1974.

Method for Demonstrating or Measuring Changes in LDH Isoenzyme Patterns[58,94]

Electrophoretic separation on agarose gels or cellulose acetate membranes is still the procedure most commonly used to demonstrate LDH isoenzymes. Many procedures and specialized pieces of equipment are available commercially to assist those wishing to use this technique. In outline, the serum sample is inserted into a well in the gel surface or is touched onto the surface of the membrane. After the isoenzymes have been separated by electrophoresis, a reaction mixture is layered over the separation medium. The mixture (typically D,L-lactate, 500 mmol/L, and NAD$^+$, 13 mmol/L, often dissolved in a suitable pH 8.0 buffer) may be applied as a liquid or in a gel. The overlay and medium are incubated at 37 °C. The NADH generated over the LDH zones is detected either by its fluorescence, when excited by long-wave ultraviolet light (365 nm), or by its reduction of a tetrazolium salt [NBT: nitroblue tetrazolium; MTT: 3-(4,5-dimethylthiazolyl-2)-2,5-diphenyl-2H-tetrazolium bromide; INT: 3-(4-iodophenyl)-2-(4-nitrophenyl)-5-phenyl-2H-tetrazolium chloride] to form a colored formazan.

Reference Ranges

Using an agarose gel technique with fluorimetric quantitation of the generated NADH, the following reference range for isoenzymes was obtained for a healthy population (n = 250), expressed as per cent of total LDH: LDH-1, 14–26; LDH-2, 29–39; LDH-3, 20–26; LDH-4, 8–16; LDH-5, 6–16. The LDH-1/LDH-2 ratio was 0.45–0.74. The sensitivity and specificity of this ratio test as determined in the Coronary Care Unit were 100% and 90%, respectively. A cellulose acetate method using NBT staining produced the following reference range, expressed as per cent of total LDH: LDH-1, 27–35; LDH-2, 34–44; LDH-3, 16–22; LDH-4, 4–8; LDH-5, 3–7.

Heat Stability

A less precise but often clinically useful evaluation of a serum LDH pattern may be obtained by determining the LDH activity remaining after heating serum at 57 °C for 30 min and the activity remaining after 30 min incubation at 65 °C. If the total activity (at normal assay temperature) is denoted by T, the activity remaining after heating at 57 °C by L, and that remaining after heating at 65 °C by H, then $(T-L)$ will represent the heat-labile (LDH-4 and LDH-5) fractions, with a value of 10–25% in normal sera, increasing to 33–80% in patients with liver disease. The stable fraction is given by H, with a normal range from 20–40%, rising to 45–65% in patients with myocardial infarction.

HBDH Activity[58]

When α-oxobutyrate is used as enzyme substrate in place of pyruvate, the reduction of substrate proceeds at an appreciable rate only when LDH-1 and LDH-2 are present, the other LDH isoenzymes being much less active. α-Hydroxybutyrate dehydrogenase (HBDH), present in serum, represents the LDH activity of, mostly, the LDH-1 and LDH-2 isoenzymes. HBDH measurement can then be a measure of the cardiac LDH isoenzymes. British workers advocated the measurement of HBDH activity and the calculation of the LDH/HBDH ratio (see p. 694). In general, the reaction systems are identical, except that five to ten times as much α-oxobutyrate is used in place of pyruvate. This approach, however, is not much used in the USA.

Ion-Exchange Chromatography

Efforts have also been made to separate the LDH isoenzymes by use of chromatography. Mercer[95] has described an ion-exchange minicolumn chromatographic separation of the LDH isoenzymes, which allows the discrete estimation of serum LDH-1 and LDH-2 activities and thus the determination of LDH-1/LDH-2 ratio. The separation is carried out in three steps. The first step elutes LDH-3, -4, and -5, the second LDH-2, and the third LDH-1. The LDH-1/LDH-2 ratio was calculated and it was shown to have a diagnostic sensitivity for myocardial infarction of 96% and a specificity of 97% when a decision threshold of 0.76 was used.

Immunoprecipitation

The most important recent advance in methodology is undoubtedly the immunoprecipitation technique for measuring serum LDH-1. Usategui-Gomez and her colleagues[146] prepared a goat antiserum to purified rhesus monkey LDH-5. This antiserum binds all human LDH isoenzymes containing one or more M subunits (i.e., LDH-2 through LDH-5). The technique of assay requires that the LDH-5 antiserum be added to a patient's serum specimen. This antiserum binds all LDH isoenzymes except LDH-1. These bound isoenzymes are precipitated by the addition of a second antibody. Since this second antibody is conjugated to polyvinylidene fluoride particles, the particles with the immune complex are readily centrifuged out of solution, leaving only LDH-1 in the supernatant. The LDH-1 activity is assayed by reaction rate methods. LDH-1 results correlate well with those obtained with the electrophoretic assay (r = 0.983). Reported diagnostic sensitivity (for myocardial infarction) of this technique is 94%. It seems likely that this convenient assay will displace the more demanding electrophoretic and chromatographic techniques described above but only for the diagnosis of myocardial infarction.

Reference

McKenzie, D., and Henderson, A. R.: Electrophoresis of lactate dehydrogenase isoenzymes. *In* Selected Methods of Clinical Chemistry, *10*:59-67. G.R. Cooper, Ed. Washington, D.C., American Association for Clinical Chemistry, 1983.

ENZYME TESTS IN THE DETERMINATION OF MYOCARDIAL INFARCTION[1]

The key roles played by CK, AST, and LDH in the investigation of suspected myocardial infarction have already been discussed separately. It is customary to measure two or even all three of these enzyme activities, as well as the isoenzymes of CK and LDH, in the investigation of this disorder. Therefore, this section considers these enzyme changes collectively and discusses the relative advantages and disadvantages of each type of measurement.

The first symptom of a myocardial infarction is usually chest pain (due to ischemia of the cardiac muscle), often severe and frequently described as "crushing" or "tightness." Characteristically, the patient is sweating markedly, and nausea as well as vomiting is frequent. There may also be abnormalities of blood pressure or heart rhythm and the patient may be in shock, i.e., may have inadequate tissue perfusion and oxygenation leading to impaired cellular metabolism. The patient frequently does die if the abnormality of heart rhythm or the shock-like state remains untreated; this is the reason for admission to special coronary care units (CCU) where intensive and continuing electrocardiographic monitoring and therapy for arrhythmias are readily available.[1]

About a quarter of all myocardial infarctions may be clinically "silent," associated either with atypical symptoms or with no symptoms at all. Many of these patients with silent infarcts are diabetic, and the lack of pain is attributed to the autonomic neuropathy that accompanies longstanding diabetes. Often it is difficult to diagnose a myocardial infarction after general surgery because of the effects of the surgery, the anesthetic, and the patient's illness. In the elderly, a myocardial infarction often presents with the onset of sudden breathlessness, acute confusion, fainting, or even a stroke. Chest pain, per se, may be due to causes other than myocardial infarction.[1] Among these are *nonischemic cardiac pathology,* such as *aortic dissection, mitral valve prolapse,* or *pericarditis*; and pulmonary pathology, such as *pulmonary embolism, pneumothorax,* or *pneumonia. Esophageal spasm* or *rupture* or *gastric ulcer* may cause chest pain, as may *degenerative arthritis* of cervical or thoracic vertebrae and *herpes zoster.* Chest pain may have an emotional origin such as *depression, anxiety, or malingering.*

With many patients the examining physician is fairly certain after obtaining the patient's history and completing a physical examination that a myocardial infarction has occurred. In these circumstances the existence of either typical electrocardiographic (ECG) changes or typical serum enzyme changes may be regarded as establishing the diagnosis. The ECG can show changes that are quite characteristic of myocardial infarction (pathologic Q waves, ST segment, and T-wave changes), provided the ECG had previously been normal.[1] These specific patterns can sometimes be obscured by previous *heart block, cardiac muscle hypertrophy,* or *infarction,* or they may not appear at all. (Q waves do not develop in subendocardial infarctions.) This fact accounts for the diagnostic sensitivity of the ECG ranging from only 63–82%, although the diagnostic specificity is about 100%. If the ECG pattern is equivocal, then the physician must depend on the serum enzyme patterns.

Total Creatine Kinase and CK-2[90,136]

After a myocardial infarction there is an initial lag phase, during which all serum enzyme activities remain within their reference ranges. The duration of the lag phase is usually about 6 h but is often less, particularly when the infarct is large. Thereafter, activities rapidly rise to peaks whose magnitude is proportional to the extent of myocardial necrosis (Figure 5-24). The first enzyme to rise is CK-3, roughly equivalent to the total CK activity, which reaches a peak between 18 and 30 h after the onset of chest pain and returns to the reference range by the third or fourth day. CK-2 rises about the same time but it reaches a peak up to 12 h earlier than CK-3. CK-2 then returns, because of its shorter half-life in blood, to preinfarction levels by or before the third day. While there are many noncardiac causes for elevations of CK-3 (described in the section on Creatine Kinase), the rise and subsequent fall of *both* CK-3 and CK-2 is nearly always indicative of a myocardial infarction. Diagnostic sensitivity of total CK elevation is 93–98%, of CK-2 nearly 100%. Diagnostic specificity for total CK elevation is 75–85%, for CK-2 nearly 100%.

Experimental work has shown that the magnitude of enzyme elevation in blood following a myocardial infarction is directly proportional to the size of infarct in the affected tissue. Because of their early rate of rise, both total CK and CK-2 activities in blood have been used for the purpose of infarct sizing.[127] The rationale for early infarct sizing is that prompt therapeutic intervention (drugs or surgery) might reduce the eventual size of the developing infarct. Unfortunately, the varying proportions of CK-3 and CK-2 in different regions of the myocardium introduce significant uncertainties into the calculation of infarct size; therefore the early promise of this aproach has not been achieved.

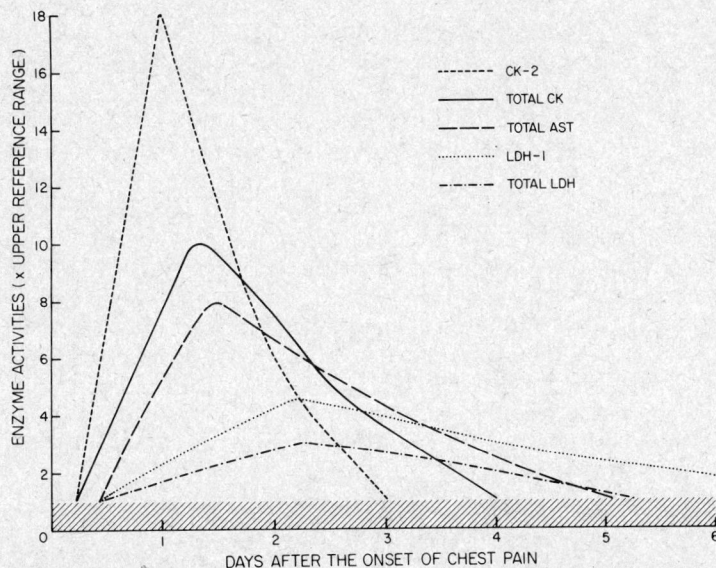

Figure 5-24. Typical pattern of changes in serum enzyme activities following an uncomplicated myocardial infarction. These patterns and the magnitude of the peaks can vary quite markedly among patients, depending on the location and size of the infarct and the enzyme assay methodology in use. The hatched area indicates the reference range of all enzymes.

Aspartate Aminotransferase

Tracing the trajectory of the CK's, aspartate aminotransferase (AST) rises 6–8 h after the onset of chest pain, peaks at ~18–24 h, and returns to preinfarction levels by the fourth or fifth day. AST has poorer diagnostic performance because it is less cardiac specific. Recall that AST rise may reflect diseases of liver, lung, and skeletal muscle as well. Although AST was the first serum enzyme to be used for the diagnosis of myocardial infarction by serum enzyme changes, it tends to be less commonly requested nowadays than are the isoenzymes of CK and LDH.

Total LDH and LDH-1[87,90]

LDH-1, often paralleled by total LDH activity, rises 8–12 h after the onset of chest pain (although it may rise concomitantly with CK-3). The isoenzyme peaks at 24–48 h and returns to preinfarction levels after 7–12 days. LDH-1 is therefore an extremely useful marker of myocardial infarction because of its prolonged presence in serum. The diagnostic sensitivity of LDH-1 estimations in myocardial infarction approaches 90% and diagnostic specificity is 90–99%. The test sensitivity may be further increased by using the LDH-1/LDH-2 ratio. About 80% of all myocardial infarctions show, at some time, a "flipped" ratio, i.e., LDH-1 > LDH-2. However, it may be more useful to use as a diagnostic index elevations of the LDH-1/LDH-2 ratio above the laboratory's reference range, since the frequency of a flipped ratio in myocardial infarction tends to be method-dependent.[58] A common cause of false positive results with LDH-1 or LDH-1/LDH-2 values is the presence of hemolysis. It must be emphasized that myocardial infarction and *hemolysis* produce exactly the same effect on LDH-1 and LDH-1/LDH-2 values. Therefore the finding of an elevated LDH-1 or LDH-1/LDH-2 value in suspected myocardial infarction is meaningful only if the possibility of hemolysis can be ruled out. Usually it can, but the presence of a *prosthetic heart valve,* which can cause hemolysis, should always be considered.

Test Sequence

Appropriate timing of blood sampling for the cardiac enzymes and isoenzymes is essential (Figure 5-24), because the demonstration of both a rise and a subsequent fall in the activities of these enzymes increases the predictive value of the test for myocardial infarction. Clearly then, the most effective test sequence for confirming or excluding a diagnosis of myocardial infarction is the serial and parallel measurements of CK and LDH isoenzymes. A variety of sampling sequences have been suggested: every 12 h for the first 48 h of hospital admission; or three samples within the first 36 h of admission (at admission, 6–12 h later, and 24–36 h later); or sampling every 6–12 h as an aid in the detection of even small subendocardial infarcts; or every 8 h for the first 48 h. In practice, use depends on the local availability of service. Isoenzyme tests are rarely available at night—except under special circumstances—and specimens obtained for these tests will be held until the next working day; by contrast, total CK and LDH assays, or alternatively aspartate aminotransferase estimations, are usually available around the clock. The last two tests are usually entirely adequate for making any decisions necessary for immediate care of the patient.

A patient with a suspected myocardial infarction is admitted to the CCU and treated as such until infarction can be ruled out. The expense of a CCU bed compared to the cost of a bed in a general medical floor mandates that a diagnosis of myocardial infarction be confirmed or excluded as soon as possible. To rule out the diagnosis requires tests with high diagnostic sensitivity. Undoubtedly, CK-2 and, later in time after admission, LDH-1 and the LDH-1/LDH-2 ratio are most useful in this respect. Alternatively, to rule in (confirm) the diagnosis requires tests with high diagnostic specificity; again CK-2, LDH-1, and the LDH-1/LDH-2 ratio are all valuable. However, the use of these tests for rule-in or rule-out purposes presupposes that the physician is knowledgeable about the need to establish prior decision thresholds for these purposes; this is a situation in which the laboratory can provide valuable advice.

Isocitrate Dehydrogenase

(EC 1.1.1.42; Threo-D_s-isocitrate:NADP$^+$ Oxidoreductase, Decarboxylating)

Isocitrate dehydrogenase (ICD) catalyzes the oxidative decarboxylation of isocitrate to α-oxoglutarate [equations (37) and (38)]. The enzyme is substrate specific; only D_s-isocitrate is acted on by the enzyme (both *cis*-aconitate and L-isocitrate are inert), and only NADP can serve as the hydrogen transfer coenzyme in the reaction.

$$
\underset{\text{D}_s\text{-Isocitrate}}{
\begin{array}{c}
\text{COO}^- \\
| \\
\text{H}-\text{C}-\text{OH} \\
| \\
^-\text{OOC}-\text{C}-\text{H} \\
| \\
\text{CH}_2 \\
| \\
\text{COO}^-
\end{array}}
+ \text{NADP}^+
\xrightleftharpoons{\text{Isocitrate dehydrogenase}}
\text{NADPH} + \text{H}^+ +
\underset{\text{Oxalosuccinate}}{
\begin{array}{c}
\text{COO}^- \\
| \\
\text{C}=\text{O} \\
| \\
^-\text{OOC}-\text{C}-\text{H} \\
| \\
\text{CH}_2 \\
| \\
\text{COO}^-
\end{array}}
\qquad (37)
$$

Hydrogen Transfer Reaction

$$
\underset{\text{Oxalosuccinate}}{
\begin{array}{c}
\text{COO}^- \\
| \\
\text{C}=\text{O} \\
| \\
^-\text{OOC}-\text{C}-\text{H} \\
| \\
\text{CH}_2 \\
| \\
\text{COO}^-
\end{array}}
+ \text{H}^+
\xrightleftharpoons[\text{Mn}^{2+}]{\text{Isocitrate dehydrogenase}}
\underset{\alpha\text{-Oxoglutarate}}{
\begin{array}{c}
\text{COO}^- \\
| \\
\text{C}=\text{O} \\
| \\
\text{H}-\text{C}-\text{H} \\
| \\
\text{CH}_2 \\
| \\
\text{COO}^-
\end{array}}
+ \underset{\substack{\text{Carbon} \\ \text{dioxide}}}{\text{CO}_2}
\qquad (38)
$$

Decarboxylation Reaction

It is not clear whether the oxidation and decarboxylation reactions occur simultaneously or in separate steps. The enzyme that is found in serum is primarily of liver origin, although it is found to some extent in all cells. Genetically distinct cytoplasmic (soluble) and mitochondrial isoenzymes exist.[61] As with other isoenzymes determined by separate gene loci, the isoenzymes of NADP-dependent ICD differ in quantitative catalytic properties and in physical properties.

The ICD discussed in this section must be distinguished from the NAD-dependent isocitrate dehydrogenase (EC 1.1.1.41) that participates in the Krebs tricarboxylic acid cycle, a distinct enzyme with quite different properties. NAD-dependent ICD is also located in the mitochondria.

The NADP-dependent enzyme is found in high concentrations not only in the liver but also in heart, skeletal muscle, kidney, and adrenal tissue, as well as in platelets and red cells. It has a M.W. of 64 000 and requires Mn(II) as an activator, the ion being essential for the decarboxylation reaction. The concentration giving optimal activation is in the range of 0.5–1.5 mmol/L. The Mn(II) can be replaced by Mg(II) or Co(II); however, only 60–80% of the activity possible with Mn(II) is obtained, even at the optimal concentration range of 1.7–3.0 mmol/L. The pH range for optimal activity is broad (pH 7.0–7.8).

Sulfhydryl binding reagents such as Cu(II), Hg(II), iodoacetate, and *p*-chloromercuribenzoate (COOH-C$_6$H$_4$-HgCl) are potent inhibitors, and the inhibition can be only partly reversed by addition of sulfhydryl compounds such as glutathione. Cyanide, azide, and *N*-ethylmaleimide are also potent inhibitors of enzyme activity. Inhibition occurs in the presence of NaCl at concentrations as low as 50 mmol/L; at 300 mmol/L the degree of inhibition is 50%. The effect appears to be due to the Na ion, inasmuch as similar inhibition is obtained if other sodium salts are used in place of NaCl. Decrease in activity is also observed if the concentration of a variety of buffers is increased above 0.10 mol/L. This decrease suggests that enzyme activity is depressed in the presence of increasing ionic strength. Anticoagulants such as oxalate, EDTA, and F ions also inhibit activity, perhaps by removing the activating Mn(II).

The existence and properties of mitochondrial and cytoplasmic isoenzymes of NADP-dependent ICD explain the different responses of the serum levels of ICD in liver disease and myocardial infarction, discussed below. The cytoplasmic isoenzyme moves faster toward the anode on electrophoresis and is more stable to heat than the slower-moving mitochondrial isoenzyme. The cytoplasmic form is the predominant form in liver extracts, whereas in heart extracts most of the activity is contributed by the mitochondrial isoenzyme. Minor electrophoretic zones of ICD occur that are presumed to be due to post-translational modifications of the principal isoenzymes.

Clinical Significance

The level of ICD activity found in the sera of healthy adults is quite low (1.2–7.0 U/L, measured at 30 °C). Increases in ICD activity are very sensitive indicators of *parenchymal liver disease,* and permit early detection, inasmuch as they can be seen even in the early, incubation phase of the disease. As prognostic tools, they are of less value. The highest values are found in *viral hepatitis* (10–40 times the upper limit of normal). In *chronic hepatitis,* increased levels of enzyme activity can persist for several months, although serum ICD values usually return to within the normal range within 14–20 d. A sudden large drop in activity following an appreciable initial rise suggests that massive cell necrosis may have occurred, with a poor prognosis. Elevations as high as those seen in viral hepatitis may also be seen in cases of *infectious mononucleosis,* but in patients with *serum* or *toxic hepatitis* the elevations observed are only of the order of three to eight times normal.

Normal or slightly increased levels of ICD activity are encountered in uncomplicated *obstructive jaundice,* and the degree of increase is independent of the severity, duration, and etiology of the obstructive process.[150] However, very high levels of activity are observed in cases of *acute inflammation* of the biliary tract. High levels of the enzyme are also common in neonatal *biliary duct atresia.* In *cirrhosis,* serum enzyme levels will vary from normal to four times the upper normal value. If persistent and more elevated enzyme values are seen, the prognosis is poor. *Hepatic neoplasms* and carcinoma metastatic to the liver are also associated with slight to moderate elevations of serum ICD activity.

Normal enzyme values are encountered in a variety of diseases involving the heart, lungs, kidney, skeletal muscle, and other tissues. Despite the high concentrations of ICD activity in heart tissue, no elevations are observed in *myocardial infarctions,* unless the infarct is accompanied by *congestive failure* resulting in *hepatic ischemia.*[6] Since the heart contains mainly the heat-labile mitochondrial isoenzyme, it is presumed that ICD released from heart muscle does not persist in the circulation, as does the more stable cytoplasmic isoenzyme from hepatocytes.

Ingestion of *alcohol* causes a rise in ICD activity in serum, and it has been suggested that ICD values can be used as a measure of the reaction of the liver to ethanol consumption. However, the much more sensitive and useful response of γ-glutamyltransferase to alcohol ingestion has superseded this application of ICD measurements. Similarly, a number of *drugs,* such as *p*-aminosalicylic acid, may also cause a rise in the ICD level. This rise probably reflects the toxic effect of these drugs on the liver cells.

A sudden rise in the ICD level during *pregnancy* is suggestive of possible placental damage or degeneration. However, in uncomplicated pregnancies, ICD levels remain within normal limits.

Red cells contain some 100 times more enzyme than does serum. However, normal values for serum ICD activity are seen in all forms of anemias except those due to vitamin B_{12} or folate deficiency (the megaloblastic anemias), in which activity may range from normal to eight times the upper limit of normal.

Very low levels of ICD activity are encountered in *cerebrospinal fluid* (up to 0.3 U/L). These levels are too low to measure with the usual procedures used for serum. Moderate elevations of ICD in cerebrospinal fluid have been reported in cases involving acute *bacterial meningitis, vascular cerebral lesions,* and *tumors* primary or metastatic to the cerebrospinal system.

The main diagnostic use of ICD measurements has thus been in the investigation of *hepatobiliary disease,* in which ICD is placed in the same category as other enzymes such as the transaminases, which indicate the leakage of enzymes from damaged or dying cells. In this respect, ICD has been perceived to have an advantage of sensitivity, in that levels in hepatocellular disease are relatively much increased from the very low levels normally present in serum. However, this large relative increase is offset by the low absolute levels of the enzyme, with the resultant difficulties of rapid and accurate measurement. The high degree of organ specificity of ICD has also been seen as an advantage, but equivalent specificity can be obtained by combinations of other enzyme tests when an equivocal clinical presentation makes this necessary.

The distribution of enzymes throughout the zones that make up the liver parenchyma is not uniform. Intracellular concentrations of ICD are higher in cells in the region of the central veins;

ICD is therefore potentially a more sensitive indicator of centrilobular necrosis, in particular, than some other enzymes. However, the localization of intrahepatic lesions by the interpretation of multiple enzyme tests has not entered general clinical chemical practice. Therefore, ICD must be placed in the second rank of tests of hepatocellular damage, since it is essentially equaled in diagnostic sensitivity and surpassed in ease and reliability of measurement by alanine aminotransferase.

Assay Methods

Wolfson and Williams-Ashman first quantitated ICD activity in serum by using a manual spectrophotometric procedure and measuring the increase in NADPH concentration (ΔA_{340}) during the course of the reaction.[153] This procedure was modified to provide the continuous-monitoring methods in current use. The various authors differ in the choice of buffer, temperature, and NADP and Mn(II) concentration; no "optimized" procedure has as yet been developed.

Some two-point, colorimetric methods have also been worked out.[151] In one approach, the oxoglutarate formed in the reaction is measured in the form of its colored 2,4-dinitrophenylhydrazone (ΔA_{390}).

A detailed continuous-monitoring procedure[31,50] is given in *Fundamentals of Clinical Chemistry*, 2nd edition.

Reference Ranges

With continuous-monitoring methods at 30 °C, the ICD activity of sera of normal adults is 1–7 U/L (2–13 U/L at 37 °C). There are no significant age- or sex-related variations, except that levels in newborns may be up to four times adult levels, falling to normal within 2 weeks.

GLUTAMATE DEHYDROGENASE

(EC 1.4.1.3; L-Glutamate:NAD(P)⁺ Oxidoreductase, Deaminating)

Glutamate dehydrogenase (GLDH or GDH) is a mitochondrial enzyme found mainly in the liver, heart muscle, and kidney, but small amounts occur in other tissues, including brain, skeletal muscle, and leukocytes. It is a zinc-containing enzyme. The smallest active molecule has a M.W. of ~350 000, consisting of 6 polypeptide chains, but larger polymers are also found. The enzyme catalyzes the removal of hydrogen from L-glutamate to form the corresponding ketimino-acid that undergoes spontaneous hydrolysis to 2-oxoglutarate:

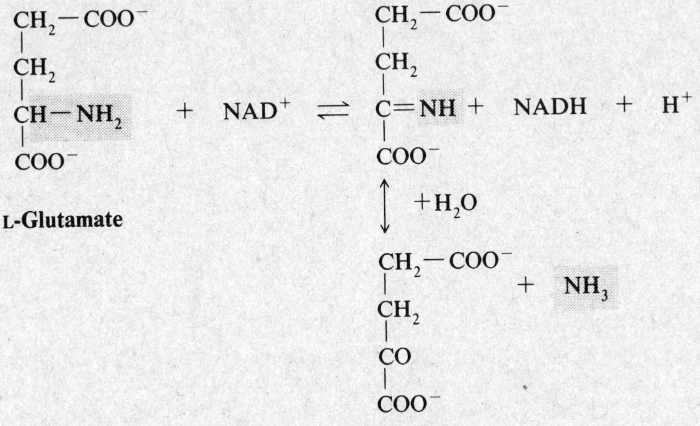

$$
\begin{array}{l}
CH_2-COO^- \\
| \\
CH_2 \\
| \\
CH-NH_2 \\
| \\
COO^-
\end{array}
\; + \; NAD^+ \; \rightleftharpoons \;
\begin{array}{l}
CH_2-COO^- \\
| \\
CH_2 \\
| \\
C=NH \\
| \\
COO^-
\end{array}
\; + \; NADH \; + \; H^+
$$

L-Glutamate

$$\updownarrow \; +H_2O$$

$$
\begin{array}{l}
CH_2-COO^- \\
| \\
CH_2 \\
| \\
CO \\
| \\
COO^-
\end{array}
\; + \; NH_3
$$

2-Oxoglutarate

Although NAD is the preferred coenzyme, NADP can also act as the hydrogen acceptor. Similarly, specificity for L-glutamate is also not absolute; a number of other amino acids, including L-norvaline and L-2-aminobutyrate, can act as substrates. Glutamate dehydrogenase is inhibited by metal ions such as Ag(I) and Hg(I), by several chelating agents, and by L-thyroxine. ADP activates glutamate dehydrogenase above pH 7 and also reverses the inhibition caused by thyroxine and diethylstilbestrol.

The glutamate dehydrogenases of human tissues may be separated by electrophoresis into several forms probably resulting from different polymeric states.

Clinical Significance

Glutamate dehydrogenase is present in normal serum only in trace amounts, but increased activities are observed in liver disease.[130] Activity in serum is increased in all conditions in which hepatocellular damage is present. However, the degree of elevation in uncomplicated *viral hepatitis* is relatively much less than that of the transaminases. Four- or five-fold elevations are seen in *chronic hepatitis* (similar to those for the transaminases); in *cirrhosis* increases are only up to two-fold. Very large rises in serum glutamate dehydrogenase occur in *halothane toxicity,* and marked increases are also seen in response to some other *hepatotoxic agents.* Compared with other hepatocellular enzymes such as the transaminases, increases in the activity of glutamate dehydrogenase in serum are disproportionately large, and may reach 10–20 times normal.

Glutamate dehydrogenase can thus offer some differential diagnostic potential in the investigation of liver disease, particularly when interpreted in conjunction with other enzyme tests. This is often done by calculating the transaminase/glutamate dehydrogenase ratio. The key to this differential diagnostic potential is to be found in the intraorgan and intracellular distribution of the enzyme.

As an exclusively mitochondrial isoenzyme, glutamate dehydrogenase is released from cells that are necrotic; therefore, release is less in diffuse inflammatory processes such as acute hepatitis, and in these conditions the release of cytoplasmic enzymes such as alanine transaminase is quantitatively more pronounced. Glutamate dehydrogenase is more easily measured as an index of cellular destruction than the mitochondrial isoenzyme of aspartate transaminase.

Glutamate dehydrogenase is more concentrated in the central areas of the liver lobules than in the periportal zones. This pattern of distribution is the reverse of that of alanine transaminase. Pronounced release of glutamate dehydrogenase is therefore to be expected in conditions in which centrilobular necrosis occurs, e.g., as a result of ischemia or in halothane toxicity. The relatively greater release of glutamate dehydrogenase in these conditions is emphasized by calculating transaminase/glutamate dehydrogenase ratios.

The difficulties of making use of the additional diagnostic information potentially offered by glutamate dehydrogenase are similar to those discussed under isocitrate dehydrogenase. Absolute levels of activity in serum are low; precision of assay, particularly the precision of ratios, is correspondingly difficult to achieve and to maintain. The use and interpretation of this enzyme test thus remain mainly in specialist laboratories of diagnostic enzymology.

Assay Methods[131]

Continuous-monitoring methods for the determination of glutamate dehydrogenase can be based upon both forward and reverse reactions. The equilibrium favors the formation of glutamate, and higher reaction rates are observed when 2-oxoglutarate is employed as substrate. Serum is added to a solution of NADH, an ammonium salt, and ADP in buffer at pH 7.4–8.0, and the reaction is initiated by the addition of the substrate, 2-oxoglutarate. Optimum conditions for 25 °C have been published by the Deutsche Gesellschaft für Klinische Chemie[29] and for 37 °C by Jung et al.[72] L-Leucine is incorporated into the reaction mixture since this amino acid, like ADP, activates the enzyme. The rate of decrease in absorbance at 340 nm is measured.

Reference Ranges

Many of the data have been collected at 25 °C. Values are up to 0.9 U/L with earlier methods, and up to 3 U/L (women) and up to 4 U/L (men) by optimized methods at this temperature. The upper level is increased to 7.5 U/L when a method optimized at 37 °C is used. No data are yet available at 30 °C.

ALKALINE PHOSPHATASE

(EC 3.1.3.1; Orthophosphoric-Monoester Phosphohydrolase (alkaline optimum) ALP)

The isoenzymes of alkaline phosphatase exhibit optimum activity at a pH of about 10 in vitro, but the optimum observed varies with the nature and concentration of the substrate acted upon, the type of buffer or phosphate acceptor present, and, to some extent, the nature of the isoenzymes. Although alkaline phosphatase displays considerable inter- and intratissue heterogeneity, there are rarely more than two or three forms in any one serum sample. The forms present in sera from patients with various diseases have the characteristics of the specific forms present in liver, bone, intestinal, placental, and, very rarely, renal tissue. The predominant forms present in normal serum can be characterized as the liver and bone varieties.

Alkaline phosphatases act on a large variety of naturally occurring and synthetic substrates, but the natural substrates on which they act in the body are not known.[93] The fact that individuals with an apparent inborn absence of the enzyme excrete large quantities of ethanolamine phosphate suggests that this (or perhaps phosphatidyl ethanolamine) may be one of the true physiologic substrates of the enzyme.

Some divalent ions such as Mg(II), Co(II), and Mn(II) are activators of the enzyme, and Zn(II) is a constituent metal ion.[93] The correct ratio of Mg(II)/Zn(II) ions is necessary to avoid displacement of Mg(II) and to obtain optimal activity. Phosphate, borate, oxalate, and cyanide ions are inhibitors of all forms of the enzyme. The individual multiple forms are inhibited to different extents by L-phenylalanine, urea, excess Zn(II), or AsO_4^{3-}. Variations in Mg(II) and substrate concentrations change the pH optimum. The type of buffer present (except at low concentrations) affects the rate of enzyme activity. Buffers can be classified as inert (carbonate, barbital), inhibiting (glycine, propylamine), or activating (2-amino-2-methyl-1-propanol, AMP, tris(hydroxymethyl)aminomethane, Tris, and diethanolamine, DEA). Glycine inhibits apparently by complexing the activating Mg(II) ion.

The enzyme is present in practically all tissues of the body, especially at or in the cell membranes, and it occurs at particularly high levels in intestinal epithelium, kidney tubules, bone (osteoblasts), liver, and placenta.[93] Although the precise metabolic function of the enzyme is not yet understood, it appears that the enzyme is associated with lipid transport in the intestine and with the calcification process in bone.

The form present in the sera of normal adults probably originates mainly in the liver or the biliary tract with up to half the total activity coming from the skeleton. The respective contributions of these two forms to the total activity is markedly age-dependent. There is also a significant difference between the sexes in serum alkaline phosphatase activity at some ages, although this is less important for interpretation of values than the dependence on age. A small amount of intestinal alkaline phosphatase may also be present, particularly in the sera of individuals of blood groups B or O who are secretors of blood-group substances. The enzyme found in urine is probably derived from renal tissue and does not represent serum enzyme cleared by the kidney. Alkaline phosphatase in serum is rapidly denatured at 56 °C but is relatively stable at lower temperatures. (The placental isoenzyme is most stable.) Sera kept at room temperatures usually show a slight but real increase in activity, which varies from 1% over a 6-h period to 3–6% over a 1- to 4-d period. Even in sera stored at refrigerator temperature, activity increases slowly (2%/day). In frozen sera, there is a decrease in activity, which is slowly recovered after thawing the serum.

A similar enhancement of activity, but of greater magnitude, occurs with reconstituted lyophilized preparations, such as those available as commercial "control sera" or "reference materials." In reconstituted material stored at 37 °C, the increase in activity of some materials can be as high as 50–100% over a 24-h period; and the increases with storage at 4 and 20 °C are about 10 and 30%, respectively. Enhancement of activity continues for several days, but at a decreasing rate. The cause of this phenomenon is not known, but may be due to renaturation of partially denatured enzyme or to dissociation, upon warming, of a phosphate-lipoprotein complex or a multimer of the enzyme that was formed in the freeze-drying process.

Clinical Significance[93]

Serum alkaline phosphatase measurements are of particular interest in the investigation of two groups of conditions: *hepatobiliary disease* (Chapter 13) and *bone disease* associated with increased osteoblastic activity (Chapter 12). For many years, it was believed that alkaline phosphatase reaching the liver from other tissues (especially bone) was excreted into the bile, and that the elevated serum enzyme activity found in hepatobiliary disease was a result of a failure to excrete the enzyme through the bile. However, more recent studies suggest that the response of the liver to any form of *biliary-tree obstruction* is to synthesize more alkaline phosphatase; i.e., the effect is one of enzyme induction. The main site of new enzyme synthesis is the hepatocytes adjacent to the biliary canaliculi. Some of the newly formed enzyme enters the circulation to raise the enzyme level in serum. The elevation tends to be more marked (more than three-fold) in extrahepatic obstruction (e.g., by stone or by cancer of the head of the pancreas) than in intrahepatic obstruction and is greater the more complete the obstruction. Serum enzyme activities may reach 10–12 times the upper limit of normal, returning to normal on surgical removal of the obstruction.

Intrahepatic obstruction of the bile flow (e.g., by invading cancer tissue or by drugs such as chlorpromazine that affect the biliary tree) also raises serum alkaline phosphatase, but usually to a lesser extent (up to 2.5 times the upper normal limit). Liver diseases that principally affect parenchymal cells, such as *infectious hepatitis,* typically also show only moderately elevated or even normal serum alkaline phosphatase levels. The degree of elevation is usually less than three-fold and depends on the degree of biliary stasis. These distinctions between extrahepatic and intrahepatic obstructive jaundice and parenchymal jaundice are clinically very useful, but it must be remembered that they are based on averages of many cases and that exceptions to the rules may be encountered in individual instances.

Among the *bone diseases* the highest levels of serum ALP activity are encountered in *Paget's disease* (osteitis deformans) as a result of the action of the osteoblastic cells as they try to rebuild bone that is being resorbed by the uncontrolled activity of osteoclasts. Values from 10–25 times the upper limit of normal are not unusual. Only moderate rises are observed in *osteomalacia,* the levels dropping slowly in response to vitamin D therapy. Levels are generally normal in *osteoporosis.* In *rickets,* levels two to four times the normal may be observed, and these fall slowly to normal on treatment with vitamin D. Slight to moderate elevations are seen in the *Fanconi syndrome. Primary hyperparathyroidism* and *secondary hyperparathyroidism* are associated with slight to moderate elevations of alkaline phosphatase activity in serum, the existence and degree of elevation reflecting the presence and extent of skeletal involvement. Very high enzyme levels are present in patients with *osteogenic bone cancer.* Transient elevations may be found during healing of *bone fractures.* Physiological *bone growth* elevates alkaline phosphatase in serum, and this accounts for the fact that in the sera of growing children one finds enzyme activity some 1.5–2.5 times that present in normal adult serum.

An increase of up to two to three times normal may be observed in women in the third trimester of *pregnancy* although the range is very wide and levels may not exceed the upper limit of the reference range in some cases. The additional enzyme is of placental origin. Upward or downward trends in placental alkaline phosphatase may presage complications of pregnancy, such as *hypertension* or *pre-eclampsia.* However, because of the wide range of placental alkaline phosphatase activity in serum in normal pregnancy, single estimations are of little diagnostic value.

An unexpected result of the application of the techniques of isoenzyme analysis to the characterization of alkaline phosphatase in serum was the discovery that forms of the enzyme that are essentially identical with the normal placental isoenzyme may appear in the sera of some patients with malignant diseases. These carcinoplacental or "Regan" isoenzymes appear to result from the derepression of the placental phosphatase gene. In some cases, modification of the gene or its product may cause the appearance of isoenzymes that differ in some respects from the normal placental isoenzyme while still retaining its general characteristics. Tumors may also produce alkaline phosphatases that appear to be modified forms of nonplacental isoenzymes.

METHODS FOR THE DETERMINATION OF ALKALINE PHOSPHATASE ACTIVITY

Methods of determining alkaline phosphatase activity have a long history, and numerous methods have had a more or less wide clinical use.[93] This variety of methods contributes to the difficulties of comparing alkaline phosphatase results from different laboratories or reported in the literature. Methodological developments have been directed toward increasing the speed and sensitivity of the assay by selecting readily hydrolyzed substrates and phosphate-accepting buffers and toward the use of continuous-monitoring methods based on "self-indicating" substrates.

The earliest methods of alkaline phosphatase assay involved the measurement of inorganic phosphate released by the enzyme. In 1930, Kay introduced the use of β-glycerophosphate as a substrate for alkaline phosphatase, and Bodansky used this as the basis for his classic procedure in 1932. In 1934 King and Armstrong proposed the use of phenyl phosphate as substrate. The rate of reaction was determined by measuring the phenol formed in a fixed time. Phenyl phosphate has the advantage of being hydrolyzed more rapidly than β-glycerophosphate; moreover, phenol is a reactive compound that can be determined by several sensitive colorimetric methods, thus permitting shorter incubation times. The King-Armstrong method is readily adaptable to some forms of automation, such as the Technicon AutoAnalyzer, and is still in use at the present time.

The most popular of the chromogenic or self-indicating substrates for alkaline phosphatase is 4-nitrophenyl phosphate (usually abbreviated 4-NPP or PNPP from the older name, para-nitrophenyl phosphate), a chromogenic, substituted phenyl phosphate introduced in 1946 by Bessey, Lowry, and Brock. This ester is colorless, but the reaction product is colored at the pH of the reaction; thus the enzyme reaction can be followed continuously by observing the rate of formation of the yellow color of the 4-nitrophenoxide ion. With improvements in the reaction conditions, this reaction forms the basis of current recommended and standard methods of alkaline phosphatase assay. Other self-indicating substrates include phenolphthalein monophosphate, which avoids the nonlinear kinetics of the diphosphate introduced earlier, thymolphthalein phosphate, and α-naphthyl phosphate. However, methods based on these substrates are relatively little used, and the older fixed-time methods such as those of Bodansky and King must now be regarded as obsolete. Fixed-time methods for alkaline phosphatase assay are still carried out in considerable numbers because of their suitability for automatic analyzers such as the Technicon AutoAnalyzer. However, 4-nitrophenyl phosphate is used increasingly as the substrate in these assays.

In all alkaline phosphatase methods discussed so far, the liberated phosphate group is transferred to water; i.e., the reaction is primarily hydrolytic. The rate of phosphatase action is much enhanced, however, if certain amino alcohols are used as buffers. Among these apparent activators are compounds such as 2-amino-2-methyl-1-propanol (AMP), diethanolamine (DEA), tris(hydroxymethyl)aminomethane (Tris), and ethylaminoethanol (EAE). These materials are derivatives of aliphatic amines and function as buffers by binding protons at the nitrogen atom. Being hydroxyl compounds, however, they can act as phosphate group acceptors, and the enzymatic rate-enhancement observed in their presence derives from their participation in the phosphate-transfer reaction. Not all compounds containing NH_2 and OH groups can serve as activating buffers; the greatest degree of phosphorylation is observed when the OH group is separated from the N atom by two carbon atoms (e.g., in DEA and EAE), and a one- or two-carbon alkyl group is attached to one of the other nitrogen bonds:

$$-(C)-C-N-C-C-OH$$
$$|$$
$$H$$

Enzyme activity in the presence of optimal concentrations of these buffers can be two- to six-fold greater than in the presence of a nonactivating buffer such as carbonate. The ALP methods described here employ such phosphate-accepting buffers.

The improved sensitivity and the considerably shortened reaction period associated with the use of a phosphate-accepting buffer were demonstrated by Bowers and McComb in their procedure using 4-NPP as substrate.[15] Many other amino-alcohols, such as DEA, EAE, and MAE (2-methylaminoethanol), are effective in the pH range from 8.5–10.5. The use of such buffers complicates the kinetics of the enzyme reaction, since the reaction rate will depend not only on the substrate concentration but also on the concentration of the buffer-acceptor that behaves as cosubstrate. Rather high concentrations of buffers (0.9–2.0 mol/L) are needed for maximum enzyme activity.

A large number of manual, automated, and commercial kit procedures that use these buffers and chromogenic substrates have been introduced. Both the "standard" procedure for the assay of alkaline phosphatase proposed by clinical chemists in Germany and the recommended method put forward by their Scandinavian colleagues use DEA buffer, pH 9.8, at 1.0 mol/L and PNPP as substrate at 10 mmol/L. Standardizing committees in the USA and France as well as the Expert Panel on Enzymes of the IFCC have preferred to recommend AMP. Both buffers may be contaminated with inhibiting or inactivating impurities, so that careful selection and checking of reagent batches is essential. DEA gives higher rates of activity than AMP, but the viscosity of concentrated solutions of DEA can cause pipetting problems, while the additional sensitivity of methods using DEA necessitates the use of inconveniently small volumes of serum, thus increasing the imprecision.

Determination of Alkaline Phosphatase Activity Using a Continuous-Monitoring Procedure (Bowers and McComb)[15]

Principle

4-Nitrophenyl phosphate (4-NPP; PNPP) is colorless. The enzyme splits off the phosphate group to form free 4-nitrophenol (4-NP, PNP), which in dilute acid solutions is also colorless. Under alkaline conditions, however, 4-NP is converted to the 4-nitrophenoxide ion which, in assuming a quinonoid form, develops a very intense yellow color.

4-Nitrophenyl phosphate
(colorless)

4-Nitrophenoxide
(colorless benzenoid form)

4-Nitrophenoxide
(yellow, quinonoid form)

At the pH of the enzymatic reaction, most of the 4-NP is present in the yellow-colored quinonoid form. Thus, the rate of formation of 4-nitrophenol by the action of the enzyme on 4-nitrophenyl phosphate at 30 °C can be monitored with a recording spectrophotometer.

The concentrations of substrate, buffer/phosphate-acceptor, and Mg(II) in the reaction mixture are 14×10^{-3}, 0.75, and 1.0×10^{-4} mol/L, respectively.

The method can readily be adapted to a two-point procedure in which the reaction is allowed to proceed for a predetermined time (e.g., 15 min) and is then stopped by addition of 10 mL NaOH, 0.05 mol/L, to a 1-mL reaction mixture. The added alkali raises the pH to 11.5–12.0, inactivates the enzyme, converts the phenoxide to the colored quinonoid form, and dilutes the yellow color to a measurable intensity.

Specimens

Only serum or heparinized plasma, preferably free of hemolysis, should be used.

Reagents

1. Buffer, 2-amino-2-methyl-1-propanol (AMP), 0.84 mol/L, pH 10.30 at 30 °C. In a 2000-mL volumetric flask, mix 300 mL of HCl solution, 1.0 mol/L, and 150 g of AMP dissolved in 1000 mL of H_2O, and dilute to volume with additional water. Adjust pH to 10.30 ± 0.02, measured at 30 °C, with the aid of a pH-meter. Use CO_2-free water; in preparing, handling, and storing the buffer, take care that contact with air (CO_2 absorption) is minimal. The buffer is stable at room temperature for about 60 d.

Note: Electrodes used for pH measurement must have a ceramic junction and a negligible sodium error; certain pH electrodes do not give correct readings in amino-alcohol solutions. Refer to the instructions describing the properties of the glass and reference electrodes being used.

2. Stock $MgCl_2$ solution, 150 mmol/L. Dissolve 3.0 g of $MgCl_2 \cdot 6\ H_2O$ in water to make 100 mL of solution. This reagent is stable at room temperature for 30 days.

3. Substrate solution, 215 mmol/L of 4-nitrophenyl phosphate (4-NPP) in $MgCl_2$ solution, 1.5 mmol/L. Prepare a fresh working solution of $MgCl_2$ by diluting 1.00 mL of the stock $MgCl_2$ solution to 100 mL. Weigh out 800 mg of high-purity disodium hexahydrate salt of 4-NPP (correct for hydration if other than 6 H_2O) and dissolve it in 10.0 mL of this working $MgCl_2$ solution.

This solution should preferably be used fresh, but solutions will keep with negligible decomposition for about 7 d if kept at 4 °C and in the dark.

4. 4-NP stock standard, 1.00 mmol 4-nitrophenol/L. Dissolve 139.1 mg of high-purity or recrystallized 4-NP in water to make 1000 mL of solution. This solution is stable if kept in the dark. High-purity 4-NP is available commercially, but a batch of 4-NP may be purified by recrystallization from hot water and drying overnight in a vacuum desiccator over silica gel.

5. 4-NP working standard solution, 4.00×10^{-2} mmol/L in AMP buffer solution, 0.84 mol/L. Pipet 10.0 mL of the stock standard into a 250-mL volumetric flask and dilute to volume with the AMP buffer. Mix thoroughly. Check the absorbance of this solution by measuring its value in a narrow-beam spectrophotometer at 404 nm in a 10-mm cell at 30 °C. The reading should be 0.751 ± 0.002 ($\epsilon = 18\ 700$–$18\ 800$ L $\times$ mol^{-1} $\times$ cm^{-1}). A similar dilution of the stock 4-NP in NaOH, 0.05 mol/L, measured at 401 nm at 25 °C, should give a value of 0.735 ($\epsilon = 18\ 300$–$18\ 400$ L $\times$ mol^{-1} $\times$ cm^{-1}).

Procedure

1. A spectrophotometer having a thermostatted cell compartment and connected to a recorder is required. Alternatively, an automatic analyzer capable of repeated, timed absorbance readings and with a suitable rate-evaluation program can be used. Set the wavelength to 404 nm. Adjust the temperature of the cell compartment to 30 ± 0.1 °C. Place cuvets in the compartment or an incubator at 30 °C to prewarm to temperature.

2. Pipet 2.70 mL of buffer into a 12×100-mm test tube.

3. Add 100 µL of serum to the tube containing the buffer. Mix gently but well. Deliver the serum with an Eppendorf or similar micropipet with disposable tips. If a glass micropipet (TC) is used, rinse the pipet twice with buffer before proceeding to the next specimen.

4. Place the tubes containing buffer and serum into a 30 °C water bath for at least 5 min to equilibrate to temperature. At the same time, place a tube containing an aliquot of substrate into the bath to bring it to temperature.

5. Initiate the reaction by adding 200 µL of warmed substrate to the tube containing the buffered serum. Immediately transfer the tube contents to a prewarmed cuvet, and place the cuvet into the cell compartment of the spectrophotometer. In many instruments it is possible to read several (4–7) cuvets serially in a single assay run. The changes in the absorbances of the solutions in the cuvet(s) with time, measured at 404 nm, are recorded for a period of from 2–5 min. The trace of absorbance versus time should be a straight line or should include a linear segment over a 2–3-min period.

6. The rate of change of absorbance with time $= \dfrac{\Delta A_{404}}{\text{time}} = \Delta A/\text{min}$ is calculated, using data obtained from a linear portion of the reaction progress-curve.

Calculations

1. If $\epsilon = 18\ 750$, then a solution containing 1.0 mmol 4-NP/L will have an absorbance of 18.75, and the measured $\Delta A/\text{min}$ will correspond to $\dfrac{\Delta A/\text{min}}{18.75} \times$ mol 4-NP/mL. This quantity of 4-NP is present in a 3.0-mL volume when the sample size used is 0.1 mL. Thus,

$$\text{U/L} = \frac{\Delta A_{404}/\text{min}}{18.75} \times 1.0 \times 3.0 \times \frac{1000}{0.10} = (\Delta A/\text{min}) \times 1600$$

where
1.0 = mmol 4-NP/mL
3.0 = total volume in cuvet
0.1 = volume of serum in cuvet
1000 = converts mmol to µmol

Comments

1. If a recorder is not available, the absorbances of the cuvets can be read manually every 15 or 30 s and ΔA_{404} can be plotted against time on graph paper.

2. If the spectrophotometer does not give an absorbance of 0.751 for the 4.0×10^{-2} mmol/L

working standard, calculate the "apparent molar absorptivity" (AMA) from the actually observed value:

$$AMA = \frac{(\text{measured absorbance of working standard}) \times 10^3}{4 \times 10^{-2}}$$

The formula for calculating activity in U/L then becomes:

$$U/L = \frac{\Delta A}{\min} \times 1600 \times \frac{18\ 750}{AMA}$$

3. Temperature coefficient. When enzyme activity at 30 °C is taken as 1.0, relative enzyme activities have been reported as 0.76 at 25 °C and 1.47 at 37 °C. The Q_{10} value for the system is 1.7.

Reference Ranges for ALP Activity in Serum

		U/L
Females,	1–12 years:	< 350
	> 15 years:	25–100
Males,	1–12 years:	< 350
	12–15 years:	< 500
	> 20 years:	25–100

Determination of Alkaline Phosphatase Activity Using a Continuous-Monitoring Procedure (AACC and Provisional IFCC Method)[37,145]*

Principle

The method developed by the Expert Panel on Enzymes of the IFCC and the Alkaline Phosphatase Study Group of the AACC also uses 4-NPP as first substrate and 2A2M1P as phosphate-acceptor buffer. The concentrations of Mg(II) and Zn(II), excesses of which are inhibitory, are controlled at optimal levels by addition of Mg(II) and Zn(II) as well as the chelating agent, N-hydroxyethylethylenediaminetriacetic acid (HEDTA). Although Zn(II) ions are present in a total concentration of 1 mmol/L, most are bound to HEDTA leaving only a small, experimentally determined, optimal concentration of free ions. A similar situation exists for Mg(II) ions. Thus, HEDTA acts as a metal ion buffer, maintaining optimal concentrations of the two ions. The temperature range of measurement is 30 ± 0.1 °C and the acceptable wavelength range is 405 ± 2 nm.

Specimens

Serum or heparinized plasma, free of hemolysis, should be used. Complexing anticoagulants such as citrate, oxalate, and EDTA must be avoided. Freshly collected serum samples should be kept at room temperature and assayed as soon as possible, but preferably not later than 4 h after collection. Frozen specimens should be thawed and kept at room temperature (20–26 °C) for 18–24 h before measurement to achieve full enzyme reactivation.

Optimal Conditions for Measurement

The concentrations in the complete reaction mixture are:

2-Amino-2-methyl-1-propanol (2A2M1P)	0.35	mol/L
4-Nitrophenyl phosphate	16.0	mmol/L
Magnesium acetate	2.0	mmol/L
Zinc sulfate	1.0	mmol/L
N-hydroxyethylethylenediaminetriacetic acid (HEDTA)	2.0	mmol/L
Volume fraction of sample	0.0196	(1:51)

---

* Reproduced with permission of the American College of Pathologists, Skokie, IL.

Reagents

1. Metal ion buffer, stock solution, N-hydroxyethylethylenediaminetriacetic acid, trisodium salt, dihydrate (HEDTA), 112.2 mmol/L; zinc sulfate ($ZnSO_4 \cdot 7 H_2O$), 56.1 mmol/L; magnesium acetate [Mg $(C_2H_3O_2)_2 \cdot 4 H_2O$], 112.2 mmol/L. The order of addition of the individual reagents is extremely important to avoid precipitation. Dissolve 4.266 g of HEDTA in ~70 mL of reagent grade water. Then add 1.613 g of $ZnSO_4 \cdot 7 H_2O$ and allow it to dissolve completely. Next, add 2.407 g of Mg $(C_2H_3O_2)_2 \cdot 4 H_2O$, dissolve completely by mixing, and dilute the solution to exactly 100 mL with reagent grade water. This solution is stable at least three months when stored at 2–6 °C.

2. Buffer/metal ion buffer. 2-Amino-2-methyl-1-propanol (2A2M1P), 0.393 mol/L; HEDTA, 2.24 mmol/L; $ZnSO_4$, 1.12 mmol/L; Mg $(C_2H_3O_2)_2$, 2.24 mmol/L; pH (30 °C) 10.40 ± 0.05. Warm the 2A2M1P to ~37 °C until it is completely liquefied. Mix 17.52 g of 2A2M1P with ~400 mL of reagent grade water. Adjust pH to 10.40 at 30 °C by addition of HCl, 1 mol/L. While stirring, slowly add 10.0 mL of metal ion buffer stock solution (solution 1) with a volumetric pipet and readjust to pH (30 °C) 10.40 ± 0.05 by addition of HCl, 1 mol/L. Dilute to exactly 500 mL with reagent grade water and store in a closed container to minimize CO_2 absorption. This buffer solution is stable at least three months if stored at 2–6 °C.

3. 4-Nitrophenyl phosphate (4-NPP), 179.5 mmol/L. The 4-nitrophenyl phosphate reagent must be of the highest possible purity and must meet the following criteria: (a) enzymatic conversion of 4-NPP to 4-NP should result in a hydrolysis of > 98.0%; (b) the molar absorptivity of 4-NPP at 311 nm in NaOH, 10 mmol/L, at 25 °C should be 9867 ± 76 $L \cdot mol^{-1} \cdot cm^{-1}$; (c) the free 4-NP must be < 0.3 mmol per mol of 4-NPP; and (d) inorganic phosphate must be < 10 mmol per mol of 4-NPP. Dissolve 0.666 g of 4-NPP (disodium salt, hexahydrate) in ~8 mL of reagent grade water and make up to 10.0 mL. The 4-NPP solution should be prepared fresh each day and is stable up to 8 h at 20–26 °C.

4. 4-Nitrophenol (4-NP), 1 mmol/L, stock standard. The crystalline 4-nitrophenol preparation selected for use must meet the following specifications: (a) color: colorless to slightly yellow; (b) melting point, 113–114 °C; (c) water content, < 0.10 g per 100 g 4-NP; and (d) molar absorptivity at 401 nm in NaOH, 10 mmol/L, at 24 °C: 18 380 ± 90 $L \cdot mol^{-1} \cdot cm^{-1}$. Acceptable 4-NP can be prepared by recrystallization or sublimation techniques, although several commercial preparations will meet these criteria. 4-NP is also available as a clinical standard (SRM 938) from the National Bureau of Standards in Washington, D.C.

Place ~400 mg of high-purity 4-NP into a 10-mL beaker and put into an oven at 50 °C for 24 h. To prevent water absorption, quickly weigh and dissolve 139.1 mg of 4-NP in reagent grade water and dilute to 1000 mL in a volumetric flask. Mix thoroughly. This solution is stable for at least three months if protected from light and evaporation.

5. 4-Nitrophenol (4-NP), 0.04 mmol/L, in 2A2M1P buffer/metal ion buffer (solution 2). Using a Class A volumetric pipet, transfer exactly 10.00 mL of 4-NP, 1 mmol/L (solution 4), to a 250-mL volumetric flask. Dilute to the mark with 2A2M1P, 0.393 mol/L (solution 2), and mix thoroughly. After allowing sufficient time for temperature equilibration, measure the absorbance of this buffered solution in a 10-mm cuvet at 405 nm at 30 ± 1 °C against a buffer (solution 2) blank using the same instrument and bandpass as for the enzyme assay procedure. This net absorbance reading is used as an instrumentation calibration factor for the calculation of alkaline phosphatase activities and the solution should be prepared at least twice with absorbance readings determined in triplicate. All readings should agree within ±1.0% of the mean.

6. Alkaline buffered substrate solution. Combine 10 volumes of solution 2 with one volume of solution 3 making a total volume that is adequate for the intended use. The absorbance of this solution at 405 nm should not exceed 0.6 A. This solution may be used for up to five working days if stored at 2–6 °C. Alternatively, this reagent may be stored frozen (−20 °C) in convenient aliquots; it is stable at least three months if thawed only once.

7. Alkaline buffered solution for individual sample blank. Combine 10 volumes of solution 2 with one volume of reagent grade water to make a total volume that is adequate for the intended use.

Note: Reagents from the following manufacturers have been found to be suitable for alkaline phosphatase activity measurements: Sigma Chemical Co., St. Louis, MO 63178 (2A2M1P, 4-NPP, magnesium acetate, zinc sulfate, 4-NP); Aldrich Chemical Co., Milwaukee, WI 53233 (2A2M1P, HEDTA, 99+%, Gold Label).

Procedure

1. A spectrophotometer with a thermostatted cell compartment (30 ± 0.1 °C) and connected to a recorder is required. Set the wavelength to 405 ± 2 nm.

2. Pipet 2.50 mL of buffered substrate solution (solution 6) into the reaction cuvet and bring to 30 °C.

3. Initiate the reaction by adding 50 μL of sample to the cuvet and mix thoroughly.

4. Immediately record the change in absorbance of the reaction mixture at 405 nm at 30 °C for up to 5 min after initiation of the reaction. Calculate the change in absorbance per time ($\Delta A / \Delta t$) from the linear portion of the curve. Extension of the reaction time beyond the 5-min period may result in deviations from linearity for some samples.

Note: Various preparations of 2A2M1P have been found to contain inhibitors of alkaline phosphatase activity that cause nonlinear reaction rates after extended incubation periods. Therefore, activities must be calculated during the first 5 min of the reaction to prevent significant decreases in the reaction rate relative to the initial rate. Also, new lots of buffer should be checked against acceptable lots to ensure maximum activity.

5. A reagent blank activity must be determined once for each batch of reagent mixture (solution 6) to account for any nonenzymatic hydrolysis of the substrate. Repeat steps 2 through 4 of the above procedure, substituting 50 μL of distilled water for sample. This blank activity, if present, must be subtracted from all alkaline phosphatase activity measurements determined with the same batch of reagents. Observed reagent blank activities are, however, generally <1 U/L.

6. Sample blank activities must be determined for lipemic and icteric samples by substituting solution 7 for solution 6 and following the procedure as above. Sample blank reactions may result in absorbances that are either increasing ($+\Delta A / \Delta t$) or decreasing ($-\Delta A / \Delta t$). With the stated measurement conditions, sample blank reactions have been observed that range from +0.0017 A/min (4.7 U/L) for a lipemic sample to −0.0203 A/min (−56.1 U/L) for a severely icteric sample. Most serum samples have blank reactions less than ±0.0007 A/min (±1.9 U/L).

7. The values of $\Delta A / \Delta t$ for the alkaline phosphatase reaction are constant for the above stated reaction periods for sera with activities up to 800 U/L (0.291 ΔA/min). If the value of ΔA/min is >0.291 or decreases during monitoring, the sample should be diluted five- to tenfold with sodium chloride, 154 mmol/L, and the measurement repeated.

Calculations

$$\text{Catalytic concentration (U/L)} = \frac{V}{\epsilon \times 1 \times v} \times (\Delta A / \Delta t)$$

where:

V = the reaction volume (L),

ϵ = the micromolar absorption coefficient, L × μmol^{-1} × cm^{-1},

   as determined by measuring A of reagent (5) × 25,

1 = the pathlength of the cuvet (cm),

v = the sample volume (L),

ΔA = the change in absorbance ($\Delta A = \Delta A_{\text{test}} - \Delta A_{\text{reagent blank}} - \Delta A_{\text{sample blank}}$), and

t = the reaction time (min).

For the described method:

$$t = 1 \text{ min}$$

$$V = 2.55 \times 10^{-3} \text{ L}$$

$$v = 0.05 \times 10^{-3} \text{ L}$$

$$1 = 1 \text{ cm}$$

$$U/L = \frac{2.55 \times 10^{-3}}{\epsilon \times 1 \times 0.05 \times 10^{-3}} \times (\Delta A/min)$$

$$= \frac{51}{\epsilon} \times (\Delta A/min)$$

If ϵ, measured in the laboratory, is 18.45×10^{-3}, then

$$U/L = \frac{51}{18.45 \times 10^{-3}} \times (\Delta A/min) = 2764 \times (\Delta A/min)$$

Notes

Comments on the development of the method are outlined elsewhere.[145]

Reference Ranges

Using this method, the following reference ranges have been established for adults in the fasting state:

Males, < 60 years: 30–90 U/L
Females, < 60 years: 20–80 U/L
Adults, > 60 years: 30–90 U/L

Multiple Forms of Alkaline Phosphatase

Alkaline phosphatases prepared from different human tissues are not identical in all properties, although they are similar in many respects.[100] The nature and origin of these tissue-specific differences are not completely understood, but in the case of the placental isoenzyme a distinct structural gene determines the synthesis of this particular form. Available evidence strongly suggests that intestinal alkaline phosphatase is also the product of a unique gene. There is considerable electrophoretic heterogeneity of alkaline phosphatase within each tissue. However, this probably results from the formation of various complexes containing the enzyme as well as from variations in carbohydrate components. The alkaline phosphatase activity of tissues such as liver, bone, or kidney seems to be due almost entirely to the presence in each tissue of a form of the enzyme that is characteristic of that tissue (Figure 5-25).

Criteria that have been used to differentiate the isoenzymes of alkaline phosphatase include differences in relative rates of reaction with various substrates, differences in response to the presence of selected inhibitors, variations in stability to denaturation by heat or urea, differences in electrophoretic mobility, and differences in immunochemical characteristics.

The tissue-specific characteristics of the alkaline phosphatase isoenzymes are retained when the enzymes are released into the circulation and can be used to identify the tissue responsible for the elevation of the serum level of the enzyme. However, the differences in properties between alkaline phosphatase isoenzymes are in some cases quite small (this is particularly true for bone and liver phosphatases), so that a combination of more than one technique may be needed to ensure reliable discrimination.

Electrophoretic separation. When serum samples containing different alkaline phosphatase isoenzymes are separated by electrophoresis at alkaline pH, the liver phosphatase typically moves most rapidly toward the anode (Figure 5-26). Bone phosphatase, which typically gives a more diffuse zone than the liver isoenzyme, has a slightly lower anodal mobility, although the two zones usually overlap to some extent. Intestinal phosphatase also migrates diffusely but more slowly than the bone enzyme, and kidney phosphatase, which occurs very rarely in serum, migrates even more slowly. The placental isoenzymes have mobilities of the same order as those of liver and bone, depending on the phenotype.

Additional minor phosphatase zones are also present in extracts of tissues and occasionally in serum. One such zone contains a high-molecular-weight form of alkaline phosphatase but is also strongly negatively charged. Therefore, it moves slowly in starch gel or may even fail to enter polyacrylamide gel, but it migrates more anodally than the main liver zone on nonsieving media such as cellulose acetate. This zone has been named the "fast liver" fraction because of the latter property and because it has been observed more frequently in serum in hepatobiliary

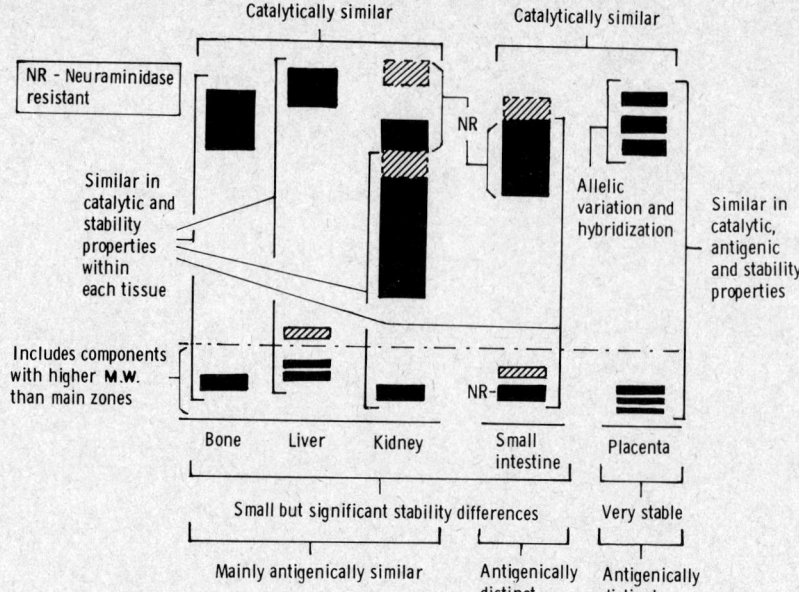

Figure 5-25. Diagrammatic summary of differences in properties between multiple forms of human alkaline phosphatase. Some of these (e.g., placental vs nonplacental phosphatases) originate at the level of the structural gene; others are probably the result of post-translational modifications. The forms shown are based on those separated by starch-gel electrophoresis. Cross-hatching indicates less prominent zones. (Reproduced with permission from Moss, D. W.: Scientific foundations of the estimation of isoenzymes in diagnosis. *In*: Multiple Forms of Enzymes. 1st Leonor-Michaelis-Symposium, Weimar, 1980. Adv. Clin. Enzymol. *2*, Karger, Basel, 1982.)

disease of various types. Complexes between alkaline phosphatase and immunoglobulins occur occasionally in serum, giving rise to abnormally migrating zones; however, these do not provide specific diagnostic information in the present state of knowledge.

Although the difference in electrophoretic mobility between bone and liver phosphatases is small, it can be exploited to distinguish between these two tissues as the possible source of increased phosphatase levels (hyperphosphatasemia). However, it is advisable to run reference sera (markers) of known isoenzyme composition (e.g., samples from cases of proven bone or liver disease) along with the unknown specimens. Comparison is easier when the electrophoresis of sample and markers is done on the same support medium at the same time (e.g., on gel slabs

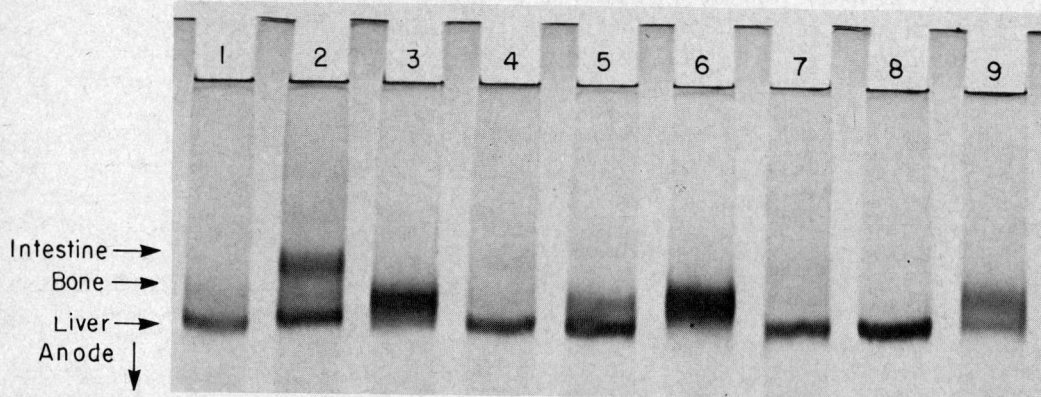

Figure 5-26. Alkaline phosphatase zones in human serum separated by vertical polyacrylamide gel electrophoresis at pH 9.5. The compact zone of liver phosphatase is prominent in (from the left) samples 1, 2, 4, 7, and 8. The more diffuse, less anodal zone of bone phosphatase can be seen in samples, 3, 6, and 9. Sample 2 displays a prominent zone of intestinal phosphatase, migrating more slowly than either the liver or bone forms. The anodal direction is downward.

rather than in individual tubes). It is also useful to compare the electrophoretic patterns obtained on specimens before and after they have been subjected to heat denaturation. (See below.)

Recently, two methods have been reported to greatly improve the electrophoretic separation between bone and liver alkaline phosphatases. The improved separation not only makes visual assessment of the liver and bone zones easier but also allows quantitative estimations to be made by densitometric scanning. Both methods exploit differences in the carbohydrate portions of the two forms of alkaline phosphatase.

In one method, electrophoresis is carried out in the presence of wheat-germ lectin, which retards bone alkaline phosphatase to a greater extent than the liver enzyme.[124] An even simpler method is to treat the serum briefly with neuraminidase.[18,104] Prolonged treatment with neuraminidase considerably reduces the anodal electrophoretic mobilities of both bone and liver alkaline phosphatases as terminal sialic acid residues are removed, until both enzymes reach identical isoelectric points and have identical mobilities. However, in the early stages of treatment, the sialic acid residues of bone phosphatase are more readily attacked than those of liver phosphatase, so that the mobility of bone phosphatase is reduced more than that of liver phosphatase.

Treatment of serum samples with neuraminidase from *Vibrio cholerae* (5 U per 50 μL serum) for 15 min at 37 °C, followed by electrophoresis, gives patterns in which the bone and liver alkaline phosphatase zones are sufficiently well separated to allow the amount of, e.g., liver phosphatase to be determined from a densitometric scan of the pattern (Figure 5-27). Measurements of liver phosphatase made in this way agreed well with results of measurements by a quantitative selective heat-inactivation technique,[107] and showed good reproducibility.[104]

Starch, agar, and polyacrylamide gels, as well as cellulose acetate, have been used for the separation of alkaline phosphatase isoenzymes. The choice of a particular system depends on the interest and resources of individual laboratories, but it is important that the chosen method be used regularly and consistently to ensure repeatable and reliable results.

The same electrophoretic techniques are used for the separation of alkaline phosphatase isoenzymes in serum as for separations of serum proteins. However, more careful attention to

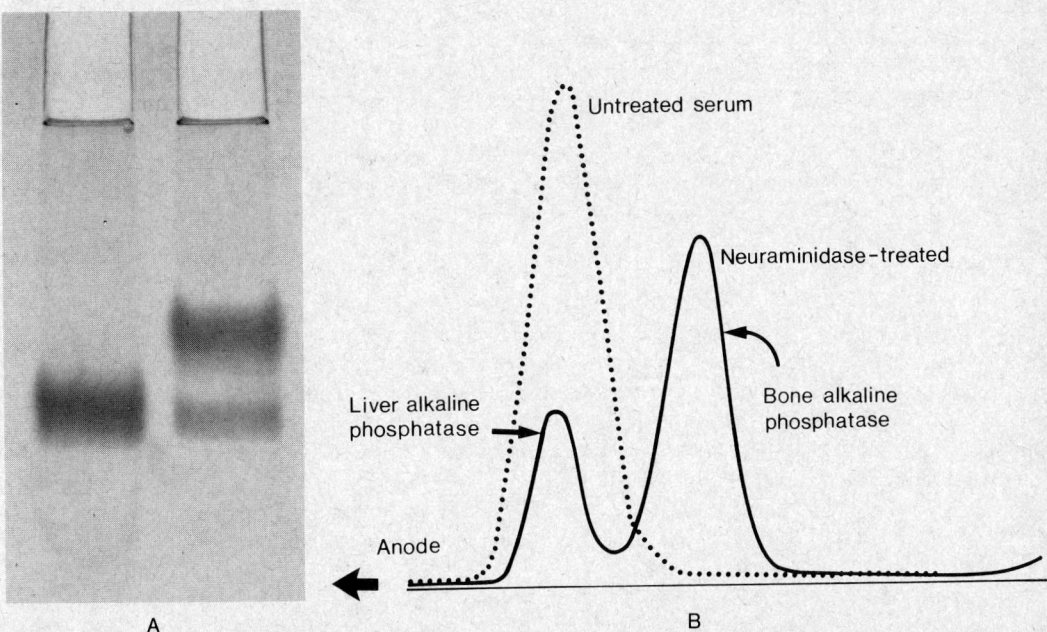

Figure 5-27. *A*, Polyacrylamide-gel electrophoresis of bone and liver alkaline phosphatases in human serum. Left: mixture of two sera containing, respectively, entirely bone phosphatase and entirely liver phosphatase. Right: mixture of the same two sera after each had been treated with neuraminidase for 10 min at 37 °C. The anodal direction is downward. The more anodal zone is liver phosphatase. *B*, Densitometric scans of electrophoretic patterns shown in *A*. Broken line: scan of mixture of untreated sera. Solid line: scan of mixture of sera treated briefly with neuraminidase. The anode is to the left. (From Moss, D. W., and Edwards, R. K.: Clin. Chim. Acta, *143*:177-182, 1984.)

cooling (e.g., by carrying out the procedure in a cold room) is needed in isoenzyme separation, to minimize enzyme inactivation. After electrophoresis, alkaline phosphatase zones are made visible by incubating the gel in a solution of buffered substrate to which a chromogenic system is added. (In the case of electrophoresis on cellulose acetate, the strips can be covered with an agar-gel layer containing the staining system.) A suitable staining mixture consists of 1-naphthyl phosphate (50 mg) and Fast Blue BB salt (70 mg) in Tris-borate-$MgCl_2$ buffer (70 mL) at pH 9.5, or in carbonate-bicarbonate buffer at pH 10.0.[102]

When electrophoresis suggests the presence of intestinal phosphatase, confirmatory evidence can be obtained by repeating the separation after prolonged incubation of the sample with neuraminidase. This treatment, by removing negatively charged sialic acid (N-acetyl-neuraminic acid) residues, reduces the anodal mobility of all phosphatase isoenzymes except that of intestinal origin, which is neuraminidase-resistant since terminal sialic acid residues are not present in the molecule.[121]

Heat-inactivation analysis. The most remarkable property of placental alkaline phosphatase is its pronounced stability to heat.[109] Incubation of the enzyme at a temperature as high as 65 °C for 30 min therefore provides a convenient and specific test for the presence of this isoenzyme, e.g., in late pregnancy. Pronounced heat stability is also shown by the "Regan isoenzyme," a placental-like fetal form of alkaline phosphatase that occurs in 5–15% of specimens from patients with cancers of various types.[39] Thus, heating the sample at 65 °C can also be used to detect this abnormal isoenzyme. The similarity between these two enzyme forms is evidenced by the fact that the "Regan isoenzyme" will react with antibodies raised against the placental isoenzyme.

Other isoenzymes of alkaline phosphatase can also be differentiated on the basis of their stability at temperatures lower than 65 °C.[105] At 56 °C, for example, liver ALP is more stable than the bone isoenzyme. Thus, if after 10 min at 56 °C <20% of the serum alkaline phosphatase activity remains, the enzyme present is probably largely of bone type; residual activities of between 25 and 55% support electrophoretic evidence that liver phosphatase is the predominant isoenzyme. At present, however, there is no agreement on a standard temperature or time of heating. Many workers prefer 56 °C, but time intervals vary from 10–30 min. For this test to be meaningful, very careful control of the experimental conditions is essential. The temperature coefficients for inactivation of enzymes by heat are high, so that a slight variation in the temperature produces a large change in the rate of inactivation.[106] The use of a water bath that is capable of precise and accurate control of temperature within narrow limits is therefore required, and the duration of incubation must be timed exactly. At the end of the heating period, specimens must be immediately immersed in ice water.

The following procedure for heat inactivation of alkaline phosphatase in serum has been found to give reproducible results provided the precautions mentioned above are followed carefully:

1. Put ∼0.5 mL of serum in a small, thin-walled glass tube (Dreyer tube) and seal with plastic film (Parafilm).

2. Place the tube into a thermostatically controlled water bath that is already stabilized at 56 °C so that the surface of the serum in the tube is below the level of the water, and simultaneously start the stopwatch. The water bath should have good temperature stability and be of large volume and well stirred to minimize temperature fluctuations on introducing the specimens.

3. After *exactly* 10 min, remove the serum tube rapidly and place it in ice water.

4. Determine the alkaline phosphatase activity of the heated specimen; express it as a percentage of the activity of an unheated specimen of the same serum that has been kept in ice water but not frozen.

Residual activities of 20% or less suggest that the predominant isoenzyme is of the bone type. Values between 25 and 55% are associated with sera in which the predominant isoenzyme(s) is of liver or intestinal origin, or both.

If the presence of placental or Regan isoenzymes is suspected, the incubation should be done at 65 °C for 30 min. Any remaining activity is due to one of these two isoenzymes.

Urea inhibition. Urea at high concentrations inhibits the enzyme, the inhibition being irreversible and varying with the tissue origin of the ALP.[9] This urea inhibition provides an alternative to heat inactivation in isoenzyme analysis. Again, no agreement exists as to the concentration of urea to be used or the duration of treatment. Bone isoenzyme is most susceptible to urea denaturation (16% residual activity after treatment at 37 °C for 18 min in urea, 3 mol/L). The liver enzyme has intermediate resistance (44% residual activity), whereas the intestinal and placental enzymes are most resistant under these conditions (69% activity left).

Chemical inhibition. Specific chemical inhibitors have also been applied to the characterization of alkaline phosphatase isoenzymes in serum. L-Phenylalanine markedly inhibits intestinal, placental, and Regan isoenzymes when present at a concentration of 5 mmol/L but has less effect on the isoenzymes of bone or liver.[39] Levamisole preferentially inhibits bone and liver phosphatases and is effective at much lower concentrations.[147]

Immunological techniques. Another approach to aid discrimination between the alkaline phosphatase isoenzymes in serum includes the use of antisera to specific tissue phosphatases. Monospecific antisera to placental and intestinal alkaline phosphatase have been prepared, but antisera to either liver or bone phosphatases cross-react completely.[86] Immunological methods provide the best measurements of placental or intestinal phosphatases. In other cases, however, the highest degree of certainty in the identification of ALP isoenzymes can be obtained if a combination of two or more of the following techniques is employed: electrophoresis, heat-stability measurements, and inhibition by L-phenylalanine or urea.

BILIARY TRACT ENZYMES

The activities of several enzymes in serum change in a manner similar to that of alkaline phosphatase in hepatobiliary disease. Since these changes are most pronounced in such diseases as intra- or extrahepatic cholestasis that focus principally on the biliary tract rather than in diseases chiefly affecting parenchymal cells, such as infectious hepatitis, these diagnostically useful enzymes are often collectively referred to as "biliary tract enzymes." This term should not be taken to imply that the enzymes come exclusively from biliary tract cells, although it is likely that some of the increased enzyme activity is from this source; rather, the description reflects their major diagnostic value. Nor should it be assumed that the levels of all the enzymes change in parallel in every case of biliary tract disease, as the comparison of three biliary tract enzymes in Table 5-9 shows. Evidence from experimentally induced biliary obstruction indicates that rather different processes apparently result in elevations of the three enzyme activities in serum; the rapid increase in activity of alkaline phosphatase in obstructed livers is not paralleled by a similar increase in 5'-nucleotidase or γ-glutamyltransferase activities.

Nevertheless, this group of enzymes is conveniently considered together, since their measurement is requested as a diagnostic aid in essentially similar conditions and their activities are interpreted in a broadly similar manner.

Apart from alkaline phosphatase, the biliary tract enzymes include arylamidase (often erroneously called leucine aminopeptidase, LAP), 5'-nucleotidase, and γ-glutamyltransferase. 5'-Nucleotidase and γ-glutamyltransferase are discussed in more detail in this chapter. Currently, use of 5'-nucleotidase is declining while that of γ-glutamyltransferase is increasing because of its greater diagnostic sensitivity and ease of assay.

Table 5-9. COMPARISON OF RESULTS OF ALKALINE PHOSPHATASE (ALP), 5'-NUCLEOTIDASE (NTP) AND γ-GLUTAMYLTRANSFERASE (GGT) IN 174 SERA FROM PATIENTS WITH HEPATOBILIARY DISEASE

	Average Activities (Multiples of Upper Reference Limits)			Incidence of Abnormal Results (%) in All Specimens		
	ALP	*NTP*	*GGT*	*ALP*	*NTP*	*GGT*
Biliary tract disease (primary biliary cirrhosis, carcinoma of pancreas, biliary stricture, cholangitis)	4.0	6.2	11.9	57	43	74
Parenchymal cell disease (various forms of acute and chronic hepatitis)	1.5	1.1	2.3			

5'-NUCLEOTIDASE

(EC 3.1.3.5; 5'-Ribonucleotide Phosphohydrolase)

5'-Nucleotidase (NTP; 5'NT*) is a phosphatase that acts only on nucleoside-5'-phosphates such as adenosine-5'-phosphate (adenosine-5'-monophosphate, AMP; adenylic acid), releasing inorganic phosphate. The enzyme is widely distributed throughout the tissues of the body and appears to be localized in the cytoplasmic membrane of the cells in which it occurs. Its pH optimum is between 6.6 and 7.0.

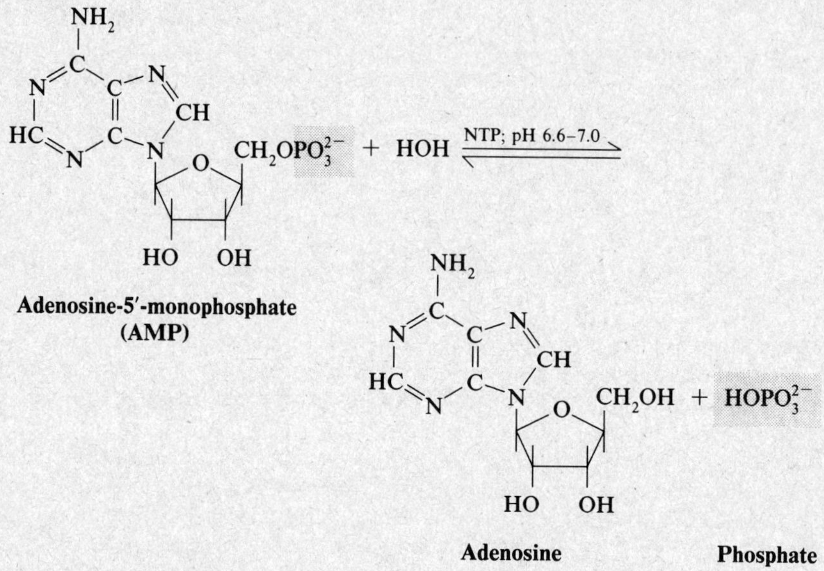

Adenosine-5'-monophosphate (AMP)

Adenosine   **Phosphate**

(39)

Clinical Significance[11]

The activity of 5'-nucleotidase in serum is increased two- to six-fold in those *hepatobiliary diseases* in which there is interference with the secretion of the bile. This may be due to extrahepatic causes (a *stone* or *tumor* occluding the bile duct), or it may arise from intrahepatic conditions such as cholestasis caused by *chlorpromazine, malignant infiltration* of the liver, or *biliary cirrhosis.* When parenchymal cell damage is predominant, as in early *infectious hepatitis,* serum 5'-nucleotidase is normal or only moderately elevated. Thus, as already noted, both 5'-nucleotidase and alkaline phosphatase behave similarly in hepatobiliary disease, except that elevations of 5'-nucleotidase are somewhat more pronounced and persist longer than those of alkaline phosphatase, particularly in cases of chronic liver disease.

In cases of *skeletal disease,* increases in the concentration of 5'-nucleotidase are rarely observed, and when they do occur, they are marginal. Thus, assays of 5'-nucleotidase are valuable in differentiating between elevations in alkaline phosphatase caused by hepatobiliary diseases and those caused by disease involving the skeletal system. In fact, any rise in NTP activity that is more than marginal is virtually specific for hepatobiliary disease.

Normal levels of 5'-nucleotidase are seen in all forms of *pancreatic disease* provided the common duct is not involved and hepatic metastases are not present. No increase or, at best, only minimal increases are seen in *Hodgkin's disease, lupus erythematosus,* a large variety of *infectious diseases,* and diseases involving organs other than the liver. 5'-Nucleotidase is also normal in *late pregnancy* and during *childhood,* conditions in which alkaline phosphatase is generally elevated.

---

* 5'NT is the more literal abbreviation form often seen in the literature; however, NTP has the advantage of indicating the enzyme's membership of the class of phosphatases.

Methods for the Determination of NTP Activity

The substrate most generally used in measuring the activity of 5′-nucleotidase is adenylate (adenosine 5′-monophosphate, AMP). However, this substrate is an organic phosphate ester and thus can also be hydrolyzed to an appreciable degree by other nonspecific (alkaline) phosphatases, even at a pH as low as 7.5, the pH previously assumed to be optimal for NTP activity. Methods for the estimation of 5′-nucleotidase in serum must therefore incorporate some means for correcting for the hydrolysis of the substrate by the nonspecific phosphatases.

5′-Nucleotidase differs from alkaline phosphatase in being inhibited by nickelous ions, and this property is used to distinguish the two enzymes in the method of Campbell described here.[19] The mechanism of inhibition is not entirely clear. NTP is present in serum in a high molecular weight, highly charged form, which Ni(II) ions may precipitate. Some isoenzymes of alkaline phosphatase may also be inhibited by Ni(II). In another approach used by Belfield and Goldberg[5] and by Persijn et al.,[113] a large excess of some phosphate ester, such as β-glycerophosphate or phenylphosphate, is added to the reaction mixture. These materials are substrates for alkaline phosphatase but not for 5′-nucleotidase, and by forming substrate complexes with the former enzyme, they reduce the proportion of the total alkaline phosphatase activity that is directed to the hydrolysis of the NTP substrate, adenosine 5′-monophosphate. This approach cannot be used when the measurement of the reaction rate is based on the determination of inorganic phosphate, since phosphate is a product of the reaction catalyzed by both enzymes. However, methods have been described in which adenosine (the second product of the 5′-nucleotidase reaction) is determined colorimetrically or spectro-photometrically by measuring the amount of ammonia formed by the action of added adenosine deaminase. Radiometric determination of isotopically labeled reaction products has also been described. A recent method of avoiding interference from serum alkaline phosphatase uses concanavalin A, which selectively inhibits NTP, allowing an estimate of the interference from ALP to be made.[156]

Automated methods for the estimation of 5′-nucleotidase are based on nickel inhibition or on the addition of an excess of a nonspecific substrate, as outlined above.

Determination of 5′-Nucleotidase Activity in Serum (Campbell Method)[19]

Principle

Serum is incubated with adenosine 5′-monophosphate at pH 7.5 and 37 °C, with and without added nickel ions. After 30 min the amount of inorganic phosphate liberated is determined. Phosphate produced in the absence of nickel represents the combined activities of ALP and NTP, whereas that produced in the presence of nickel is due to the activity of ALP alone. Thus, the difference between these two values for liberated phosphate corresponds to the activity of 5′-nucleotidase in the serum sample. Manganese ions serve as an activator of 5′-nucleotidase. The presence of copper in the acetate buffer accelerates color development.

Specimens

Serum should be used for analysis. Plasma may cause turbidity and, although EDTA is said to be without effect, this and other metal-binding anticoagulants could interfere with activation by manganese.

Reagents

1. Barbiturate buffer, 0.04 mol/L, pH 7.5.* Dissolve 8.25 g of sodium diethylbarbiturate (sodium barbital) in 140 mL of HCl, 0.2 mol/L, and dilute to 1 L with water. Adjust to pH 7.5 at 37 °C if necessary.

2. Substrate, 10 mmol/L. Dissolve 0.365 g of adenosine 5′-phosphoric acid monohydrate in 18 mL of NaOH, 0.1 mol/L, and dilute to 100 mL with water. Alternatively, use 0.391 g of disodium adenosine 5′-phosphate and dilute to 100 mL with water.

3. Manganese ion solution, 20 mmol/L. Dissolve 0.338 g of $MnSO_4 \cdot H_2O$ or 0.396 of $MnCl_2 \cdot 4 H_2O$ in 100 mL water.

4. Nickel chloride solution, 0.1 mol/L. Dissolve 2.4 g of $NiCl_2 \cdot 6 H_2O$ in 100 mL water.

5. Trichloroacetic acid, 100 g/L.

6. Acetate buffer, 2.4 mol/L, pH 4.0. Dissolve 2.5 g of $CuSO_4 \cdot 5 H_2O$ and 46 g of sodium acetate trihydrate in 1 L of acetic acid, 2 mol/L (115 mL glacial acetic acid diluted to 1 L). Adjust to pH 4.0 if necessary.

7. Ammonium molybdate solution, 5 g/dL. Dissolve 5 g of $(NH_4)_6Mo_7O_{24} \cdot 4 H_2O$ in 100 mL water.

---

*Newer work establishes the optimal pH to be in the range from 6.6–7.0. As alternatives to barbital ($pK'_a = 7.43$), PIPES piperazine-N,N′-bis(2-ethanesulfonic acid), with $pK'_a = 6.80$, and HEPES (N-2-hydroxyethylpiperazine-N-2-ethanesulfonic acid), with $pK'_a = 7.55$, have been proposed as useful buffer materials. Changes in buffer type, as well as pH, affect the measured enzyme activity.

8. **Reducing agent.** Dissolve 2 g of *p*-methylaminophenol monohydrogen sulfate ("Rhodol," "Elon," or "Metol") in 80 mL of water. Add 5 g of Na_2SO_3, anhydrous, dilute to 100 mL with water, and filter. Store in a dark bottle at 4 °C.

9. **Stock phosphate standard, 1.00 mg P/mL.** Dissolve 2.193 g of dry KH_2PO_4, anhydrous, reagent grade, in 500 mL water. Add 0.20 mL of concentrated sulfuric acid as preservative before diluting to the mark. Store at 4 °C.

10. **Working phosphate standard, 10 µg P/mL.** Measure 1.00 mL of stock standard into a 100-mL graduated flask, and dilute to 100 mL with trichloroacetic acid, 100 g/L. Stable for 4 weeks if stored at 4 °C.

Procedure

1. The reaction is carried out in 12-mL conical centrifuge tubes (borosilicate glass). Two tubes are needed per specimen, labeled T (test) and C (control). To the T tube add 1.5 mL of buffer and 100 µL of $MnSO_4$ solution; to the C tube add 1.3 mL buffer, 100 µL $MnSO_4$, and 200 µL of $NiCl_2$ solution. Mix tube contents well.

2. Add 200 µL of serum to both reaction tubes, mix, and place them into a water bath at 37 ± 0.2 °C. Allow 5 min for the tubes and contents to attain 37 °C, and then add 200 µL of substrate solution to each tube and mix without removing the tubes from the water bath.

3. After exactly 30 min, stop the enzymatic reaction by the addition of 2.0 mL of trichloroacetic acid (100 g/L) to each tube. Mix the tube contents thoroughly, remove tubes from the water bath, and centrifuge them at ~3000 × g for 15 min.

4. Pipet 2.0 mL of the clear supernatant from the test (T) and control (C) mixtures into clean borosilicate tubes. All tubes should be washed with dilute HCl and thoroughly rinsed with distilled water to ensure freedom from detergent contamination; similarly, pipets should be rinsed with reagent before delivery.

5. Prepare a standard (S) by pipetting 1.0 mL of working standard and 1.0 mL of water into an acid-washed test tube and prepare a reagent blank (B) by mixing 1.0 mL water and 1.0 mL of trichloroacetic acid.

6. To all tubes (T, C, S, B) add 3.0 mL of acetate buffer, followed by 0.50 mL of ammonium molybdate solution and 0.50 mL of Elon solution. Mix well after each addition.

7. After 5-min color development, transfer the tube contents to cuvets and measure the absorbance of all solutions against the blank in a spectrophotometer at some wavelength between 680 and 880 nm. (The higher the wavelength, if available on the instrument, the higher the absorbance for any given quantity of phosphorus produced.) The color is stable for at least 30 min.

Calculation

The standard tube contains 10 µg of phosphate (as P). Therefore, the amount of phosphate-phosphorus produced by the action of the 5′-nucleotidase is given by

$$\mu g \ P = \frac{A_T - A_C}{A_S - A_B} \times 10$$

or

$$\mu mol \ P = \frac{A_T - A_C}{A_S - A_B} \times \frac{10}{31}$$

This amount of phosphate-phosphorus is released in 30 min by the enzyme present in 0.10 mL of serum since half of the total reaction mixture is taken for estimation of phosphate. Therefore, serum 5′-nucleotidase activity in µmoles of substrate hydrolyzed per min per liter of serum (U/L) is given by

$$U/L = \frac{A_T - A_C}{A_S - A_B} \times \frac{10}{31} \times \frac{1}{30} \times \frac{1000}{0.1}$$

or

$$U/L = \frac{A_T - A_C}{A_S - A_B} \times 108$$

If the activity is greater than 150 U/L, repeat the assay using a shorter incubation time and make corresponding adjustments in the calculations.

Reference Ranges

The normal range for 5'-nucleotidase activity in sera from normal men and normal nonpregnant and pregnant women is from 2–17 U/L. Lower values have been reported in children.

GAMMA GLUTAMYLTRANSFERASE[49,123]

(EC 2.3.2.2; γ-Glutamyl-Peptide:Amino Acid γ-Glutamyltransferase)

Peptidases are enzymes that catalyze the hydrolytic cleavage of peptides to form amino acids or smaller-sized peptides or both. They constitute a broad group of enzymes of varied specificity, and some individual enzymes catalyze the transfer of amino acids from one peptide to another amino acid or peptide; i.e., they act as amino acid transferases. Included among the latter is γ-glutamyltransferase (GGT), which transfers the γ-glutamyl group from peptides and compounds that contain it to some acceptor. This enzyme was originally termed a "transpeptidase," but a more appropriate term is "transferase." The γ-glutamyl acceptor can be the substrate itself, some amino acid or peptide, or even water, in which case a simple hydrolysis takes place. The enzyme acts only on peptides or peptide-like compounds containing a terminal glutamate residue joined to the remainder of the compound through the terminal (-γ-) carboxyl [see equation (40)]. Beyond the fact that glycylglycine is five times more effective as an acceptor than is either glycine or the tripeptide (gly-gly-gly), little is known about the optimal properties of the acceptor cosubstrate. The rate of the peptidase transfer reaction is considerably faster than that of the simple hydrolysis reaction. An example of a reaction catalyzed by the enzyme is presented in equation (40). This is the reaction used in measuring enzyme activity.

| γ-Glutamyl-p-nitroanilide | Glycylglycine | γ-Glutamylglycylglycine | p-Nitroaniline |
| Substrate (donor) | Acceptor | Transfer product | Donor residue |

(40)

The enzyme was first identified in kidney tissue, where its concentration is quite high, but was later shown to be present in serum and in all cells except those in muscle. Some enzyme is present in the cytosol, but the larger fraction is located in the cell membrane and may act to transport amino acids and peptides into the cell across the cell membrane in the form of γ-glutamyl peptides. It may also be involved in some aspects of glutathione metabolism.

Clinical Significance

Even though renal tissue has the highest level of GGT, the enzyme present in serum appears to originate primarily from the hepatobiliary system, and GGT activity is elevated in any and all forms of liver disease. It is highest in cases of *intra-* or *posthepatic biliary obstruction,* reaching levels some 5–30 times normal. It is more sensitive than alkaline phosphatase, 5'-nucleotidase (Table 5-9), leucine aminopeptidase, and the transaminases in detecting *obstructive jaundice, cholangitis,* and *cholecystitis*; its rise occurs earlier than with these other enzymes and persists longer. Only moderate elevations (2–5 times normal) are seen in *infectious hepatitis,* and in this condition GGT determinations are less useful diagnostically than are measurements of the transaminases. High elevations of GGT are also seen in patients with either primary or secondary (metastatic) *neoplasms*; again changes occur earlier and are more pronounced than those with the other liver enzymes. Small increases (2–5 times normal) of GGT activity are observed in patients with *fatty livers,* and similar, but transient, increases are seen in cases of *drug intoxication.* In acute and chronic *pancreatitis,* and in some *pancreatic malignancies* (especially if associated with hepatobiliary obstruction), enzyme activity may be 5–15 times the upper limit of normal. In summary, GGT is the most sensitive enzymatic indicator of hepatobiliary disease available at present. Normal values are rarely found in the presence of liver disease, although, as mentioned below, elevations occur in response to certain stimuli when disease is absent. However, GGT is of little value in attempting to discriminate between different kinds of liver disease.

Normal levels of the enzyme are seen in cases of *skeletal disease* (Paget's disease, bone neoplasms), in *children* above 1 year of age, and in healthy *pregnant women*—conditions in which alkaline phosphatase is elevated. Thus, measurement of GGT levels in serum can be used to ascertain whether observed elevations of ALP are due to skeletal disease or reflect the presence of hepatobiliary disease. 5′-Nucleotidase determinations can provide the same information, but GGT has the advantage of becoming elevated earlier in liver disease and of rising to greater levels than does 5′-NT. However, elevations of 5′-nucleotidase have sometimes been more useful clinically, because they are seen only in hepatobiliary disease.

Normal levels of GGT are encountered in various *muscle diseases* and in *renal failure,* but mild elevations may be seen in untreated *lipoid nephrosis.* In *myocardial infarctions,* GGT is usually normal; if there is a rise, it occurs at about the fourth day and reaches a maximum value in another four days and probably implies *liver damage* secondary to cardiac insufficiency.

Elevated levels of GGT not only are seen in the sera of patients with *alcoholic cirrhosis,* but are also seen in the majority of sera from persons who are *heavy drinkers.* Rosalki stresses the value of serum GGT levels in detecting *alcohol-induced liver disease.* In patients receiving *drugs* such as Dilantin (phenytoin) and phenobarbital, raised levels of the enzyme are found in serum specimens but not in specimens of spinal fluid. Such an increase of activity may reflect induction of new enzyme activity by the action of the anticonvulsant drugs. The release of GGT into serum reflects the toxic effects of alcohol and other drugs on microsomal structures in liver cells. The enzyme level found correlates well with the duration of the drug action. Hepatic complications occurring in *cystic fibrosis* (mucoviscidosis) also lead to elevations of GGT.

High levels of GGT are present in the *prostate,* and this may account for the fact that the activity of GGT in sera of males is ~50% higher than that seen in sera from females. *Prostatic malignancy* may at times be the source of an elevated GGT activity in serum. The *irradiation of tumors* in cancer patients may be accompanied by a rise in GGT activity, although lactate dehydrogenase activity in the course of such treatment remains unchanged. However, in malignant disease in general, an increased serum GGT activity must first arouse suspicion that the disease is metastatic to the liver.

The enzyme that is found in *urine* probably originates in the kidney and genitourinary tract. Elevated enzyme activity is found in the urine of patients with acute urorenal infections and with diseases involving renal tissue destruction. However, in chronic renal disease and in older individuals, urine enzyme levels may be depressed.

METHODS FOR THE DETERMINATION OF γ-GLUTAMYLTRANSFERASE

Although GGT was a subject of research interest for many years, its widespread and systematic use in diagnostic enzymology dates from the introduction in 1969 of a convenient method of assay by Szasz. This method and modifications of it use L-γ-glutamyl-*p*-nitroanilide [equation (40)] as the substrate, with glycylglycine serving as the γ-glutamyl residue acceptor. Both two-point and continuous monitoring methods have been described. Other substrates that have been investigated include the γ-glutamyl derivatives of aminopropionitrile, α-naphthylamine, and aniline, but γ-glutamyl-*p*-nitroanilide has been found to be most convenient because it is most sensitive and the *p*-nitroaniline formed can be directly measured. Unfortunately, this substrate also possesses several negative features. Its solubility in water at pH 8.0 is limited; to approach concentrations in the reaction mixture of some 10 or 20 times the value of the K_m requires that the concentration in the substrate reagent preparation be in the supersaturation range. To obtain a substrate reagent concentration high enough so that the addition of 50–100 μL of the preparation could be used to initiate the transferase reaction, Rosalki and Tarlow[125] dissolved the anilide in HCl and used a stronger buffer to overcome the added acid. The substrate is somewhat unstable and does undergo nonenzymatic hydrolysis, even at pH 8. Moreover, it is even more subject to rapid decomposition at 30 and 37 °C and in the presence of HCl. An appropriate blank may be used to correct for this hydrolysis, but this is not necessary for most purposes. The substrate is dissolved in acid in the method recommended by the Scandinavian Committee on Enzymes,[25] described below.

An alternative procedure has become available in recent years with the production of derivatives of γ-glutamyl-p-nitroanilide in which various groups have been introduced into the benzene ring to increase solubility in water. The most useful of these substrates is L-γ-glutamyl-3-carboxy-4-nitroanilide.[112] This compound is readily soluble in water and is split by GGT at a rate comparable to that observed with L-γ-glutamyl-p-nitroanilide. Activities in serum are higher with the carboxyl derivative than with the noncarboxylated substrate; this is partly due to the higher substrate concentrations attainable with the former.

Reagent kits for GGT determination that incorporate the carboxylated substrate are available commercially. This substrate is recommended in the reference method currently being developed by the Expert Panel on Enzymes of the IFCC.[69] However, most estimations are at present made with the unsubstituted substrate, with which most clinical experience has been gained.

Determination of γ-Glutamyltransferase (GGT)[25]

The substrate employed in this method is γ-glutamyl-p-nitroanilide (GGPNA), with glycylglycine serving as acceptor. Buffering is provided by tris(hydroxymethyl)aminomethane (Tris) and by glycylglycine ($pK_2 = 8.25$). Serum may be added to the buffer-acceptor solution, and the reaction may be initiated by addition of substrate in HCl solution. Alternatively, a serum-start procedure may be used. The increase in absorbance at 405 nm due to the p-nitroaniline formed in the reaction is measured spectrophotometrically. The temperature used is 30 °C, although 37 °C is the temperature preferred by the Scandinavian Committee.

Specimens

Serum free from hemolysis is the preferred specimen, but EDTA-plasma (up to 1 mg/mL blood) can be used. Heparin produces turbidity in the reaction mixture; citrate, oxalate, and fluoride depress activity by 10–15%.

Reagents

(a) Substrate-start procedure
1. Buffer-glycylglycine reagent, pH 8.2 at 25 °C, 120 mmol Tris/L and 90 mmol glycylglycine/L, containing 12 mol/L MgCl$_2$. Dissolve 14.54 g of Tris (base), 11.89 g of glycylglycine, and 2.44 g MgCl$_2$ · 6 H$_2$O in ~800 mL of water. Adjust the pH to 8.2 at 25 °C by the addition of HCl, 1 mol/L, and then dilute to 1 L. The reagent is stable for at least three months if refrigerated.
2. GGPNA substrate, 48 mmol/L in HCl, 0.15 mol/L. Prepare a volume sufficient for a day's work, by dissolving 1.37 g of L-γ-glutamyl-p-nitroanilide, monohydrate (or 1.28 g non-hydrated GGPNA) per 100 mL of HCl, 0.15 mol/L. Use within 2 h.

(b) Serum-start procedure
The reagents are slightly modified, as follows:

Buffer-substrate reagent, Tris, 110 mmol/L; glycylglycine, 8.25 mmol/L; GGPNA, 4.4 mmol/L. Dissolve 13.33 g Tris, 2.34 g MgCl$_2$ · 6 H$_2$O, and 10.90 g glycylglycine in ~800 mL of water. Adjust pH to 8.0 (25 °C) with HCl, 1 mol/L. Dissolve 1.26 g of GGPNA monohydrate (or 1.18 g of the anhydrous compound) in this solution by heating carefully to about 60 °C with constant stirring. Cool and make up to 1 L with water. The solution is stable for a few hours at room temperature and for several weeks at −20 °C.

Procedure

(a) Substrate-start procedure
1. Pipet 2.0 mL of buffer-glycylglycine and 200 μL of specimen into a 1.0-cm cuvet. Prewarm the cuvet to 30 °C.
2. After temperature equilibration has been achieved, initiate the enzyme reaction by adding 200 μL of GGPNA solution. Mix cuvet contents rapidly and then monitor the change in absorbance at 405 nm continuously, or at 1.0-min or 0.5-min intervals, for a period of several minutes. From these absorbance values calculate the ΔA/min, and the activity in the serum specimen in U/L.

The composition of the final reaction mixture is as follows: Tris buffer, 100 mmol/L; glycylglycine, 75 mmol/L; GGPNA, 4 mmol/L, pH 8.0; volume fraction of serum = 0.083.

(b) Serum-start procedure

Pipet 2.0 mL buffered substrate solution into a 1.0-cm cuvet. After temperature equilibration, add 200 μL serum, mix, and monitor the reaction as before.

Calculations

The millimolar absorptivity for p-nitroaniline is reported to be 9.87 L $\times$ mmol^{-1} $\times$ cm^{-1} at 405 nm. (Since 405 nm is on the side of the absorption peak, the absorbance values are markedly affected by such factors as the bandwidth of the instrument. The absorption coefficient should therefore be checked in the laboratory.) Therefore, enzyme activity is given by

$$U/L = \frac{\Delta A}{min} \times \frac{10^3}{9.87} \times \frac{V}{v}$$

where $\dfrac{\Delta A}{min}$ = observed rate of reaction, V is the total reaction volume, and v the serum volume, each in mL; 10^3 converts mmol to μmol.

Substituting the appropriate values of V and v, GGT activity (U/L) is $\dfrac{\Delta A}{min} \times 1215$ for the substrate-start procedure, and $\dfrac{\Delta A}{min} \times 1114$ for the serum-start procedure.

Reference Ranges

	U/L, 30 °C	U/L, 37 °C
Males:	≤ 40	≤ 50
Females:	≤ 25	≤ 30

Proposed IFCC Method

An IFCC reference method for GGT is at present available as a consultative document.[69] It proposes the use of L-γ-glutamyl-2-carboxy-4-nitroanilide as the donor substrate, and retains glycylglycine as the acceptor. A higher concentration of the latter provides sufficient buffering capacity without the need for added Tris. $MgCl_2$, which aids the solubility of the noncarboxylated substrate, is also considered unnecessary. The conditions in the reaction mixture, based on a response-surface optimization, are as follows: temperature, 30 °C; pH (30 °C) 7.90; L-γ-glutamyl-3-carboxy-4-nitroanilide, 6 mmol/L; glycylglycine, 150 mmol/L; volume fraction of serum, 0.091. The wavelength of measurement is 410 nm, at which the product, 5-amino-2-nitrobenzoate, has an absorption coefficient of 7.908 L $\times$ mmol^{-1} $\times$ cm^{-1}.

Multiple Forms of GGT

GGT in serum is heterogeneous with respect to both net molecular charge (e.g., shown by electrophoresis) and size. These forms appear to be due to post-translational modifications of a single type of enzyme molecule rather than to the existence of true isoenzymes. For example, high-molecular-weight forms may represent the release of cell-membrane fragments into the circulation. In spite of numerous investigations, clear correlations between patterns of multiple forms and particular diseases cannot be discerned.

DIGESTIVE ENZYMES OF PANCREATIC ORIGIN

Assays of serum *amylase, lipase, and trypsin* are almost exclusively applied to investigation of pancreatic disease.[48] Although pancreatic function is treated in Chapter 14, a brief outline is included here to orient the reader to clinical applications for individual digestive enzymes.

Pancreatitis is classified as *acute*[135] or *chronic*[52] by clinical or pathological criteria. Recovery from an attack of *acute pancreatitis* leaves the patient asymptomatic. *Chronic pancreatitis* leaves the patient with persistent pain or evidence of endocrine or exocrine insufficiency of the pancreas.

Relapsing is a term characterizing recurrent attacks of either acute or chronic pancreatitis. The milder form of acute pancreatitis is *edematous;* mortality in the edematous form, though significant, is much less than in the *hemorrhagic* form, where it has been reported to be from 20–50%. *Ascites* and *pleural* (usually left-sided) *effusion,* either or both, are frequent complications of acute pancreatitis. Persistent ascites may cause elevation of digestive enzymes in serum as a result of absorption from the ascitic fluid into the circulation. Some 50% of all cases of acute pancreatitis go on to form a pseudocyst. *Pseudocyst* (i.e., not contained by a true capsule) is a collection of enzyme-rich pancreatic fluid encapsulated by inflammatory tissue and occurring as a consequence of autodigestion and liquefaction of pancreatic tissue after hemorrhage and necrosis. Pseudocyst formation is suspected if, one week after onset of an attack, clinical improvement has not occurred. In 50–85% of patients with pseudocyst, serum levels of digestive enzymes are increased.

In chronic pancreatitis, marginal elevations of serum digestive enzymes may be observed during attacks when the disease is in its early stages; eventually severe destruction of acinar tissue reduces the amount of enzymes released into the circulation, and normal or even low levels of serum enzymes can be expected as the disease progresses.

Trauma to the pancreas—as from automobile accidents, physical assaults, surgical procedures, and diagnostic radiologic procedures (e.g., endoscopic retrograde cholangiopancreatography, ERCP)—may cause transient or persistent elevation of serum enzymes, depending on the severity of the injury.

Carcinoma[20] of the pancreas with obstruction of the pancreatic duct makes elevation of serum enzymes likely; in the absence of obstruction, the carcinoma is "silent" in terms of altered serum enzyme levels or may even be associated with decreased serum enzyme levels.

AMYLASE

(EC 3.2.1.1; 1,4-α-Glucan,D-Glucanohydrolase)

Amylases are a group of hydrolases that split complex carbohydrates constituted of α-D-glucose units linked through carbon atoms 1 and 4 located on adjacent glucose residues.[96] Both straight-chain (linear) polyglucans, such as amylose, and branched polyglucans, such as amylopectin and glycogen, are hydrolyzed, but at different rates. In the case of amylose, the enzyme splits the chains at alternate α-1,4-hemiacetal (—C—O—C—) links, forming maltose and some residual glucose; maltose, glucose, and a residue of limit dextrins are formed if branched-chain polyglucans are used as substrate. The α-1,6-linkages at the branch points are not attacked by the enzyme. (See Figure 6-7.)

Two types of amylases are recognized. *Beta-amylase* (e.g., plant and bacterial exoamylase) acts only at the terminal reducing end of a polyglucan chain; it splits off two glucose units (maltose) at a time. Animal amylases, including those present in human tissues, are α-*amylases.* They are also called endoamylases because they can attack α-1,4-linkages in a random manner anywhere along the polyglucan chain. Large polysaccharide molecules are thus rapidly broken down into small units, for example, dextrins, maltose, and some glucose units. Since both maltose and glucose are reducing sugars, the course of the hydrolytic reaction is paralleled by an increase in soluble reducing materials.

Linear starch chains in helical form react with molecular iodine to form the well-known, deep blue starch-iodine complex. The endoamylolytic hydrolysis (dextrinization) of starch to dextrins and oligosaccharides is thus also paralleled by a gradual loss of ability to bind iodine; the hue of the glucan-iodine complex slowly changes to light blue, then to violet, and finally to red. No iodine color is formed when the chain size is six glucose units or less. As dextrinization proceeds, the turbidity and viscosity characteristics of starch sols fall off very rapidly, much faster than the number of freed reducing groups increases. The turbidimetric and nephelometric procedures for measuring amylase are based on this decrease in turbidity as a result of the dextrinization (amyloclastic) reaction.

Amylase in human serum has a moderately sharp pH optimum at 6.9–7.0. The enzyme is customarily assayed at 37 °C, although it is active at 50 °C, and some automated procedures employ this higher temperature. The temperature coefficient is $+6\%/°C$ increase in temperature ($Q_{10} = 1.6$). α-Amylases are calcium metalloenzymes, with the calcium absolutely required for functional integrity; saturation with Ca(II) is obtained at ~1.0 mmol/L. However, full activity is displayed only in the presence of a variety of anions, such as chloride, bromide, nitrate, cholate, or HPO_4^{2-}. The first two anions are the most effective activators; for example, optimal activity is obtained at Cl ion concentrations of 10 mmol/L. The amylases normally occurring in human plasma are small molecules with molecular weights varying from 40 000–50 000. The enzyme is thus small enough to pass through the glomeruli of the kidney, and amylase is the only plasma enzyme normally found in urine. Serum and urine amylases migrate electrophoretically with the β- and γ-globulins.

Sources of Amylase

In the body, amylase is present in a number of organs and tissues.[4,96] The greatest concentration is present in the pancreas, where the enzyme is synthesized by the acinar cells and then secreted into the intestinal tract by way of the pancreatic duct system. The salivary glands also secrete a potent amylase to initiate hydrolysis of starches while the food is still in the mouth and esophagus. The action of the salivary enzyme, once referred to as *ptyalin,* is terminated by acid in the stomach. In the intestinal tract, effective action of pancreatic and intestinal amylase is favored by the mildly alkaline conditions in the duodenum. Intestinal maltase then further hydrolyzes maltose to glucose.

Most of the pancreatic amylase is destroyed by trypsin activity in the lower portions of the intestinal tract, although some amylase activity is present in feces. Amylase activity is also found in extracts from semen, testes, ovary, Fallopian tubes, striated muscle, lung, and adipose tissue. The enzyme is also found in colostrum, tears, and milk. Tumors of lung and ovary may also contain considerable amylase activity. The enzyme present in normal serum and urine is predominantly of pancreatic (P-type) and salivary gland (S-type) origin. Amylase activity is not markedly diminished after pancreatectomy. Thus, the remaining activity is derived from the salivary glands and perhaps from some as yet unidentified source. There is little or no amylase activity in the liver. The enzyme found in urine is derived from the plasma. Ascitic and pleural fluids may contain amylase as a result of the presence of a tumor or pancreatitis.

The existence of at least seven serum amylase isoenzymes can be demonstrated by gel filtration, ion-exchange chromatography, isoelectric focusing, or electrophoretic techniques. Multiple forms are found in salivary and pancreatic extracts and three forms in urine, but usually only the S- and P-types are found in serum.

Macroamylases[41,57] are sometimes present in sera; these rare forms are probably complexes between ordinary amylase (usually S-type) and IgA, IgG, or other normal or abnormal high-molecular-weight plasma proteins. These macroamylases cannot be filtered through the glomeruli of the kidney because of their large size (M.W. > 200 000) and are thus retained in the plasma, where their presence may increase amylase activity to some six- to eight-fold over that observed in health. In macroamylasemia, amylase activity in the urine is lower than normal, since less amylase is cleared by the kidneys. No clinical symptoms are associated with this disorder; however, many of these cases have been detected following investigation of abdominal pain.

Clinical Significance[128]

Assays of amylase activity in serum and urine are largely of use in the diagnosis of diseases of the pancreas and in the investigation of pancreatic function. (See also Chapter 14.)[7] In *acute pancreatitis,* there is a transient rise in serum amylase activity within 2–12 h of the onset; return to normal levels occurs by the third or fourth day. There is usually a four-to six-fold elevation in amylase activity above the reference limit (although a significant number of subjects show a lesser elevation and sometimes none), with maximal levels attained in 12–72 h. The magnitude of the elevation of serum enzyme activity is not related to the severity of pancreatic involvement; however, the greater the rise, the greater the probability of acute pancreatitis. In acute pancreatitis associated with hyperlipemia, serum amylase activity may be normal. A significant amount of the serum amylase is excreted in the urine, and therefore elevation of serum activity is reflected in the rise of urinary amylase activity. Urine amylase, as compared with serum amylase, appears to be *more* frequently elevated, reaches *higher* levels, and persists for *longer* periods. The urinary clearance of amylase is markedly increased in acute pancreatitis; this finding will be discussed in some detail later in this section. In quiescent *chronic pancreatitis,* both serum and urine activities are usually within the reference ranges.

The diagnosis of acute pancreatitis is sometimes difficult to make, since it must be differentiated from other acute intra-abdominal disorders (see Lipase, p. 736), and since an increase in serum amylase activity may not necessarily be due to pancreatitis. Table 5-10 lists some of the many causes of hyperamylasemia and hyperamylasuria. The organ source can sometimes be identified by determining whether the major isoenzyme present is P-type or S-type.

Amylase assays are also of value in detecting the development of complications, such as *pseudocyst, ascites,* and *pleural effusion* following *acute pancreatitis* (see p. 725). It is always wise to examine pleural fluid for amylase activity; activities can be 100-fold the serum reference range following an attack. In *pancreatic abscess* serum amylase is occasionally elevated. In addition, *traumatic lesions* to the pancreas including surgical trauma and radiological investigations may all cause a transient hyperamylasemia; serious injury will often lead to persisting elevations of the enzyme. If there is obstruction of the pancreatic duct by *carcinoma of the pancreas,* then there is likely to be elevation of serum amylase activity. In an *acute relapse of chronic pancreatitis,* the serum amylase may be minimally elevated; more often the activity remains within the reference range.

In some important groups of *nonpancreatic disorders,* reasons for hyperamylasemia are known. Although normally about 25% of the serum amylase is eliminated in the urine, in *renal insufficiency* the serum amylase activity is increased up to two-fold of the upper reference limit and in proportion to the extent of renal impairment. *Neoplastic hyperamylasemia* is an increasingly recognized entity for which the clinical chemist must be constantly alert. Additionally, *tumors of the lung* and *serous tumors of the ovary* can produce hyperamylasemia (with an S-type isoenzyme mobility although it may not be identical) with elevations as high as 50-fold the upper reference limit. Both kinds of tumor can produce *pleural effusion;* in the case of ovarian tumors, the result has

Table 5-10. CAUSES OF HYPERAMYLASEMIA AND HYPERAMYLASURIA*

Pancreatic disease (P-type↑)[†]
   Pancreatitis
      Acute
      Chronic
      Complications
         Pseudocyst
         Ascites and pleural effusion
         Abscess
   Pancreatic trauma, including investigative maneuvers
   Pancreatic carcinoma
Disorders of nonpancreatic origin (mechanism unknown)
   Renal insufficiency (mixed↑)
   Neoplastic hyperamylasemia—usually bronchogenic or ovarian (usually S-type↑)
   Salivary gland lesions, e.g., mumps, calculus disease (S-type↑)
   Macroamylasemia (predominantly S-type)
Disorders of complex origin (mechanism unknown or uncertain)
   Biliary tract disease
   Intra-abdominal disease (other than pancreatic diseases; see above)
      Perforated peptic ulcer (P-type↑)
      Intestinal obstruction (P-type↑)
      Mesenteric infarction (P-type↑)
      Peritonitis (mixed↑; depends on cause)
      Acute appendicitis
      Ruptured ectopic pregnancy (S-type↑)
      Aortic aneurysm with dissection
   Cerebral trauma (type depends on other organ damage)
   Burns and traumatic shock
   Postoperative hyperamylasemia (usually S-type↑)
   Diabetic ketoacidosis (mixed↑)
   Renal transplantation (S-type↑)
   Acute alcoholism (mixed↑)
   Drugs
      Medicinal opiates (P-type↑)
      Heroin addiction →? heroin lung (S-type↑)

*After Salt and Schenkers' classification.[128]
[†]Predominant isoenzyme type is shown in parentheses: P-type (= pancreatic); S-type (= salivary); mixed (= either or both isoenzymes may be present).

been called pseudo-Meigs' syndrome. (Meigs' syndrome: association of ascites and pleural effusion with fibroma of the ovary.) The amylase activities in these effusions can be >200-fold the upper reference limit for serum. Salivary gland lesions caused by infection, irradiation, obstruction, surgery, and tumor have all been reported as producing a significant S-type hyperamylasemia. *Mumps* (infective parotitis) and maxillofacial surgery can cause a two-fold elevation, while *salivary gland irradiation* can produce a transient 9- to 18-fold elevation of serum amylase activity.

Biliary tract diseases such as *cholecystitis* can cause up to four-fold elevations of the serum amylase activity, due to either primary or secondary pancreatic involvement. A variety of *intra-abdominal events* (see Table 5-10) can lead to a significant increase in serum amylase activities up to, and sometimes beyond, a four-fold elevation. Such increases may be due to leakage of the P-type amylase from the intestine into the peritoneal cavity and thence into the circulation. *Peritonitis* and *acute appendicitis* have been previously reported to produce a slight elevation (up to two- and three-fold) of serum amylase activity. Clearly, both the extent of the elevation and the isoenzyme type will depend on the underlying pathology. In *ruptured ectopic pregnancy*, serum amylase activities can be elevated as high as eight-fold the upper reference limit; the amylase isoenzyme is the S-type. *Dissecting aortic aneurysms* can produce elevated serum amylase activities, but the mechanism is uncertain.

Cerebral trauma has also been associated with hyperamylasemia, but the cause is uncertain. Associated trauma to salivary glands and to the abdomen may be responsible. Likewise, in *traumatic shock* a combination of trauma and hypoxia (creating transient pancreatic ischemia) may cause these elevations. *Postoperative* hyperamylasemia occurs in about 20% of all patients subjected to a wide variety of surgical interventions, including extra-abdominal procedures. The elevation can be about four-fold, and the principal isoenzyme is usually S-type and less often P-type. This finding suggests that the source is usually extrapancreatic. In *diabetic ketoacidosis*, elevations of around four-fold the upper reference limit are found in as many as 80% of these patients. These abnormalities are more frequent when the blood glucose exceeds 500 mg/dL and the onset of ketoacidosis is relatively acute. The source of the amylase is uncertain. After *renal transplantation* about one fifth of the patients develop hyperamylasemia; only a small proportion of these instances are due to the occurrence of pancreatitis and the remainder are asymptomatic. In *pneumonia* and other *non-neoplastic diseases* of the lung, hyperamylasemia is increasingly being recognized; although its exact incidence is uncertain, the isoenzyme appears likely to be S-type. In *acute alcoholic intoxication* about 10% of subjects have a three-fold elevation; the isoenzyme type is equally divided between the P- and S-types. Finally, a wide variety of *drugs* must always be considered as possible causes of hyperamylasemia.

Renal clearance of amylase, as related to the reasonably constant clearance of creatinine, has been found useful as a diagnostic concept. A ratio, amylase/creatinine clearance ratio (ACCR), has been defined and expressed as a percentage:

$$\text{ACCR (\%)} = \frac{\text{urinary clearance of amylase}}{\text{urinary clearance of creatinine}} \times 100$$

The ACCR is readily calculated from amylase activity and creatinine concentration determined on the same urine and on a single serum sample obtained at the time of urine collection. The calculation simplifies to:

$$\text{ACCR (\%)} = \frac{\text{urine amylase (U/L)} \times \text{serum creatinine (mg/L)}}{\text{serum amylase (U/L)} \times \text{urine creatinine (mg/L)}} \times 100$$

because volume/time terms in the respective clearance equations are identical and thus cancel out. Note that exactly timed urine collection is unnecessary, and the only constraint is an adequate volume of specimen. Therefore random or short (2–4 h) collections are adequate. The reference range for ACCR is ~2–5%. Note, however, that the ACCR is affected by the type of amylase assay used. Thus, it is imperative to establish a reference range for the assay method currently in use.

In *acute pancreatitis*, tubular reabsorption of amylase and other proteins is reduced (probably due to competition from other low-molecular-weight proteins) and ACCR is increased; values

> 8% are not uncommon. Caution must be exercised in interpreting increased ACCR values,[57] because elevations have been observed also in *burns, ketoacidosis, renal insufficiency, myeloma, light-chain proteinuria, and march hemoglobinuria*, and following *extracorporeal circulation, large intravenous doses of corticosteroids, duodenal perforations*, and *extraperitoneal surgical procedures*. It is possible that the increased clearance of amylase in these conditions is due to the release of S- or P-type amylase from other tissues. In *macroamylasemia*, ACCR is usually < 2%.

METHODS FOR THE DETERMINATION OF AMYLASE ACTIVITY[4,73,89]

Specimen

The serum enzyme is quite stable; activity loss is negligible at room temperature in the course of a week or at refrigerator temperatures over a two-month period. In urine, an acid pH may make the enzyme less stable; therefore, pH should be adjusted to ∼7.0 before storage. With the exception of heparin, all common anticoagulants inhibit amylase activity because they chelate Ca(II); citrate, EDTA, and oxalate inhibit it by as much as 15%. As a consequence, amylase assays should be performed only on serum or heparinized plasma.

Starch-Based Methods[73]

It has been estimated that some 200 methods for the assay of amylase activity have been described, based on nine different principles and various substrates. However, until a decade ago, three main methods (each with several variations) predominated throughout the world, namely, the *saccharogenic*, the *amyloclastic*, and the *chromolytic*. These techniques will be outlined, but first it is necessary to mention the difficulties associated with the amylase substrate. Different starches vary considerably in their proportion of amylose and amylopectin, and the average chain length of the starch molecules will depend on the manner of preparation of the starch and the method used to prepare a solution. Starch does not disperse in water to form a true molecular solution but forms instead a colloidal sol containing hydrated starch micelles of various sizes. The degree of dispersion varies with temperature; at lower temperatures amylase chains aggregate into large micelles. Potato, corn, and Lintner's "soluble" starch have been most commonly used, although pure amylose, amylopectin, and glycogen were preferred by some. Starch sols deteriorate rather rapidly as a result of mold contamination; benzoic acid, sodium azide, or *p*-hydroxy-propylbenzoate may be added as a preservative. Sterile (autoclaved) sols keep well for several months. Obviously, starch substrates must be prepared, treated, and stored in a consistent manner.

Saccharogenic Assays

In saccharogenic assays the course of the enzyme reaction is followed by measuring the quantity of reducing materials (sugars, dextrins) formed. Any of the common procedures for measuring reducing substances, such as the Folin-Wu or Somogyi-Nelson methods, may be used. Methods based on the reduction of picrate, ferricyanide, and 3,5-dinitrosalicylic acid and on the anthrone-sugar reaction have also been advocated. The results obtained with these methods are reported in terms of milligrams of total reducing substances formed, expressed as "apparent glucose." The chief reducing sugar present is maltose, which, mass for mass, has about 40% of the reducing capacity of glucose. Quantities of reducing sugars produced are determined in protein-free filtrates of the reaction mixtures. With some starch preparations, it is difficult to obtain clear, nonopalescent filtrates. Opalescence does not affect the assay, provided that the turbidity is due to starch and not due to incompletely precipitated proteins; turbidity due to starch clears up at a later stage of the procedure.

Several automated forms of saccharogenic procedures have been described. In one such method glucose oxidase and catalase are used to destroy endogenous glucose; starch is then added, and after incubation the reducing sugars are determined using the cupric-neocuproine reaction. This type of assay is the most reliable of the older assays but it is very time consuming, the preparation of reproducible starch solutions is difficult (see above), and sample blanks are frequently high. These factors probably account for the 1978 finding that < 3% of all amylase assays done in the United States are performed using the saccharogenic method. The reader is referred to a detailed description of the Somogyi saccharogenic method as modified by Henry and Chiamori.[62] In the last decade, reaction rate saccharogenic assays have been devised; these will be described later in this section along with more recent assay formulations.

Amyloclastic Assays

Amylase activity can also be evaluated by following the decrease in starch substrate concentration rather than by measuring the product formed. Methods based on this approach are referred to as *amyloclastic* methods. In *chronometric* procedures, the time required for amylase to hydrolyze completely all the starch present in a reaction mixture is measured. The end point is reached when there is absence of any substrate capable of forming the blue starch-iodine color. In the Wohlgemuth method, serial dilutions of the enzyme preparation are added to a fixed quantity of starch, and that dilution is found that is just able to hydrolyze all the starch present in a fixed time period. The *amylometric* procedures of Van Loon and others measure the amount of starch hydrolyzed in a fixed period of time, using the intensity of the blue starch-iodine color as the means for quantitating unhydrolyzed starch.

Some sera and urines contain material that apparently inhibits formation of the starch-iodine color and thus erroneously indicates high amylase activity. The nature of this material is not known; some reports state that it is dialyzable and others that it is some type of paraprotein. Therefore, the use of amylase procedures based on measuring starch by the starch-iodine color should be discouraged. However, these procedures are simple and rapid and are still in use.

In most assay methods the pH is maintained at ~6.9–7.0 by the use of phosphate buffer; in some methods, a Cl ion is added, but in others, the Cl^- added with the serum or urine is assumed to be sufficient (10 mmol/L) to provide maximum activation. The calcium requirement has been ignored, and it is possible that, at times, the Ca(II) in the reaction mixture may not be at optimal levels. In 1978, the amyloclastic group of methods was estimated to constitute ~50% of all amylase assays done in the United States, but the use of the test is now rapidly declining.

Chromolytic Assays

The last group of well-established assays to be described is the *chromolytic* assays, which use dye-labeled amylase substrate materials. Most of these substrates are synthesized by linking amylose or amylopectin, through ether or ester bonds, to a variety of reactive dyes. The dyes are derivatives of triazine, in which R_1 and R_2 (Figure 5-28) are aryl chromophores linked to the triazine ring through their amino groups. The starches are coupled to the dye at C-2 of the glucose residues.

Roche Diagnostics (Nutley, NJ) makes available Amylochrome, amylose bonded to Cibachron Blue F3GA. Amylopectin Azure, available from Calbiochem (Los Angeles, CA), was described by Rinderknecht, Wilding, and Haverback and consists of amylopectin coupled to Remazol Brilliant Blue R. Phadebas is marketed by Pharmacia Laboratories, Inc. (Piscataway, NJ); this substrate is made up of an unidentified dye bonded to starch by cross-linking with diepoxide. All three dye substrates are water insoluble. Their suspensions, in buffer solutions, are attacked by the enzyme at the α-1,4 bonds to produce small-sized dye-containing fragments that are water soluble. These can be measured colorimetrically after being separated by centrifugation or filtration from the insoluble, unreacted substrate. However, D-Amyl (General Diagnostics, Morris Plains, NJ) contains a red dye, Reactone Red 2B, coupled to amylopectin, and is water soluble. The soluble, red hydrolytic products are measured after precipitating the unreacted substrate with tannic acid and alcohol. With the single exception of Phadebas, the absorbance of the color of the soluble split products formed from these dye-labeled starches is proportional to enzyme activity over a practical range of enzyme concentration. In the case of Phadebas, it appears that the logarithm of absorbance is proportional to the logarithm of enzyme activity.

Even though insoluble substrates are used, results obtained in most of these procedures generally appear to be comparable with those obtained with the more classical procedures using natural glucans. Procedures using these dye-coupled starches have been calibrated either empirically or by using enzyme preparations evaluated by the classical methods. These methods are convenient and, in 1978, represented ~40% of routine amylase assays performed in the United States. However, it is observed with increasing frequency that results, especially in the abnormal range, do not always correlate well with those of more established procedures.

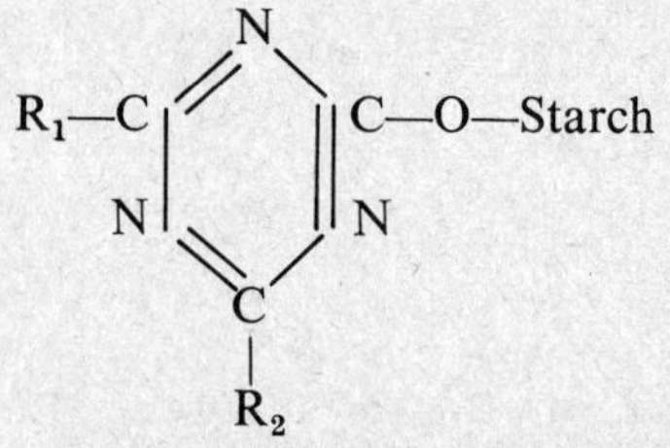

Figure 5-28. Thiazine-labeled starch.

Units for Expressing Results of Conventional Amylase Assays

Somogyi defined a unit of amylase activity in the saccharogenic assay as that quantity of enzyme that is able to liberate reducing substances with a reducing value equivalent to 1 mg of glucose in the course of a 30-min reaction at 40 °C and at a pH of 6.9–7.0. The concentration is expressed as the number of units per 100 mL of specimen. Most methods based on measuring the formation of reducing sugars report results in terms of Somogyi units, although not all measurements are performed at 40 °C.

There is not such uniformity, however, with the *amyloclastic* methods. The Wohlgemuth *diastatic index* is defined as the number of milligrams of starch digested by the enzyme present in 1.0 mL of specimen at 37 °C in 30 min. The Huggins and Russell unit is the quantity of enzyme that will hydrolyze 1.0 mg of starch to the dextrin stage in 60 min at 37 °C, whereas one Street and Close unit is that amount that converts 20 mg of starch to dextrin in 15 min. There is no direct relationship between any one of these units and the Somogyi unit. The latter has been so generally accepted in North America that values obtained by other types of procedures are usually multiplied by an appropriate factor to convert them into equivalent Somogyi units. In the chronometric methods, amylase units are calculated from the formula, units $= C/t$, where t is the time required to decolorize the starch preparation, and C is a factor so chosen that the values obtained are again comparable to those obtained with saccharogenic methods. Inasmuch as neither the substrate nor the products of the amylolytic reaction are single, well-defined chemical entities, and a variety of temperatures are used, there is no real advantage to be gained in converting common amylase units into International Units. To convert from Somogyi units measured at 40 °C to International Units at 37 °C, the appropriate combined factor is 1.6 (1.0 Somogyi unit/100 mL = 1.85 U/L, temperature factor is 0.85, therefore $1.85 \times 0.85 = 1.6$).

DEFINED-SUBSTRATE METHODS FOR AMYLASE ACTIVITY MEASUREMENTS[4,76,89]

The recent use of both well-defined amylase substrates and auxiliary and indicator enzymes in the amylase assay has improved the reaction stoichiometry and has led to more controlled, and therefore more consistent, hydrolysis conditions.

Experimental work has shown that α-amylase hydrolysis of small oligosaccharides gives better-defined products than starches. For example, both maltopentaose and maltotetraose are excellent α-amylase substrates with good stability, consistent hydrolysis products, and unambiguous reaction stoichiometry. Other useful substrates are small oligosaccharides (of 4–7 glucose units) with the 4-nitrophenyl group covalently bound to the reducing end of the oligosaccharide. Examples of these different substrate formulations will be described. The use of coupled enzyme assay (i.e., an assay using auxiliary and indicator enzymes) was first proposed in practical form by Tietz and his colleagues,[76] who used α-glucosidase (EC 3.2.1.20; maltase) to hydrolyze to glucose the di- and trisaccharides that are produced by the action of amylase [Equation (41)]. The rate of glucose production was determined with glucose oxidase (EC 1.1.3.4) and an oxygen electrode to measure the decrease in pO_2:

$$\text{Starch} \xrightarrow{\alpha\text{-amylase}} \text{maltose} + \text{maltotriose} + \text{dextrins}$$

$$\text{Maltose} + \text{maltotriose} \xrightarrow{\alpha\text{-glucosidase}} \text{glucose} \tag{41}$$

$$\text{Glucose} + \tfrac{1}{2}\,O_2 \xrightarrow{\text{glucose oxidase}} H_2O_2 + \delta\text{-gluconolactone}$$

For each mole of glucose oxidized, 0.5 mol of O_2 is consumed. Thus the rate of oxygen removal is directly related to the amylase activity in the sample.

An alternative method, an extension of the above, involves the same principle but requires the addition of peroxidase (EC 1.11.1.7) in an indicator reaction using the diammonium salt of 2,2'-azine-di-(3-ethylbenzthiazoline)-6-sulfonic acid (ABTS); ABTS has an absorbance maximum of 410 nm when oxidized. Hydrogen peroxide formed from endogenous glucose during preincubation [see equation (42)] is decomposed by catalase (EC 1.11.1.6); at the end of the preincubation period, catalase activity is blocked and the indicator reaction shown in equation (42) is used:

$$H_2O_2 + \text{ABTS (reduced: colorless)} \xrightarrow{\text{peroxidase}} H_2O + \text{ABTS (oxidized: colored)} \tag{42}$$

The increase in absorbance at 410 nm is proportional to the amylase activity of the specimen.

Both of these methods require a 10-min preincubation period to consume the endogenous glucose and to overcome the lag phase of the total reaction sequence.

Several variations of this reaction rate formulation have been devised. For example, the detection of glucose can be achieved by the hexokinase (EC 2.7.1.1) reaction as used in the α-Amyl-Harleco and *aca* amylase assays. Maltose generation can be detected by the maltose phosphorylase (EC 2.4.1.8) reaction (Beckman Amylase DS assay) that avoids interference by the endogenous glucose of the sample.

The DuPont *aca* Amylase Assay

In a 1980 College of American Pathologists Quality Assurance Program this assay constituted 45% of all amylase assays used. The basic reaction is shown in equation (43):

$$\text{Maltopentaose} \xrightarrow{\alpha\text{-amylase}} \text{maltotriose} + \text{maltose}$$

$$\text{Maltotriose} + \text{maltose} \xrightarrow{\alpha\text{-glucosidase}} 5 \text{ glucose}$$

$$\text{Glucose} + \text{ATP} \xrightarrow{\text{hexokinase}} \text{G-6-P} + \text{ADP}$$

$$\text{G-6-P} + \text{NAD}^+ \xrightarrow{\text{G-6-P dehydrogenase}} \text{6-P-gluconolactone} + \text{NADH} + \text{H}^+$$

(43)

Endogenous glucose in the sample does not interfere in this assay because it is removed before the reaction by a gel-filtration column in the test pack. Hydrolysis of maltopentaose produces 5 molecules of glucose (and hence NADH) for every single saccharide bond hydrolyzed by amylase. This assay has a rather substantial blank reaction due to contamination by trace amounts of shorter-chain oligosaccharides; the blank reaction displays nonlinear kinetics as do patient samples. However, measurements are reproducible because of the accurate and consistent timing of the preincubation and measurement intervals on the DuPont analyzer.

The Beckman DS Amylase Assay

This method uses a short-chain oligosaccharide as substrate and a more complex auxiliary and indicator enzyme system for measuring the reducing groups produced [equation (44)]:

$$\text{Maltotetraose} \xrightarrow{\alpha\text{-amylase}} 2 \text{ maltose}$$

$$\text{Maltose} + \text{P}_i \xrightarrow{\text{maltose phosphorylase}} \text{glucose} + \text{glucose-1-P}$$

$$\text{G-1-P} \xrightarrow{\beta\text{-phosphoglucomutase}} \text{G-6-P}$$

$$\text{G-6-P} + \text{NAD}^+ \xrightarrow{\text{G-6-P dehydrogenase}} \text{6-P-gluconolactone} + \text{NADH} + \text{H}^+$$

(44)

Therefore, for each bond hydrolyzed by amylase, two molecules of NADH are produced. This stoichiometric assumption holds true because neither P- nor S-type amylase hydrolyzes malto-tetraose to glucose and maltotriose. Note that endogenous glucose does not interfere with the estimation of amylase activity. However, lactate dehydrogenase contamination, from both the specimen and the assay reagents, and the presence of pyruvate cause negative interference because LDH activity, in the presence of pyruvate (from the specimen) and NADH (generated by G-6-P dehydrogenase activity), [equation (44)], will produce lactate and NAD$^+$. The degree of this interference is therefore dependent on the lactate dehydrogenase activity, the pyruvate concentration, and the amylase activity of the individual specimen.

Amylase Assays Using 4-Nitrophenyl-Glycoside Substrates[55]

Substrates for this assay are prepared by bonding 4-nitrophenol (NP) to the reducing end of a defined oligosaccharide. If the oligosaccharide is maltoheptaose (G7), the substrate is then

4NP-G7. Amylase splits this substrate to produce free oligosaccharides (G5, G4, G3) and 4NP-G2 (9%), 4NP-G3 (31%), and 4NP-G4 (60%). Note that G6, G2, G1, 4NP-G6, and 4NP-G5 are not produced in appreciable quantities. If α-glucosidase is added, some of the 4NP-G4 will be hydrolyzed to free NP and free oligosaccharides; 4NP-G3 and 4NP-G2 will be hydrolyzed to free NP and glucose. The result of the combined hydrolysis by amylase in the specimen and by the reagent α-glucosidase is that $> 30\%$ of the product is free NP. The free NP is detected by its absorbance at 405 nm. Alpha-glucosidase does not react with the original substrate or with any oligosaccharide containing more than four glucose molecules in the chain; G-4 is hydrolyzed only very slowly.

The Boehringer Mannheim assay utilizes the 4NP-G7 example substrate (above) and α-glucosidase. The mixture is preincubated for a few minutes at the assay temperature, 50 or 100 μL of specimen is added, and after a further incubation of 2 or 3 min, ΔA/min is determined at 405 nm. The ΔA/min is proportional to the amylase activity of the specimen [equation (45)]:

$$
\begin{aligned}
\text{4-NP-(glucose)}_7 &\xrightarrow{\text{α-amylase}} \text{4-NP-(glucose)}_{4,3,2} + \text{(glucose)}_{5,4,3} \\
\text{4-NP-(glucose)}_{4,3,2} &\xrightarrow{\text{α-glucosidase}} \text{4-NP-(glucose)}_4 + \text{x-glucose} + \text{NP}
\end{aligned}
\tag{45}
$$

The kit insert of this and other similar assays may be consulted for further details.

Other procedures utilize 4NP-G5 and 4NP-G6 as substrates. In the presence of α-glucosidase, these substrates are also hydrolyzed to free 4NP- and 4NP-glycosides with short chain length. An example is the method by Calbiochem-Behring (San Diego, CA). In other respects, the method is similar to that described above.

Reference Ranges[143]

Starch-Based Methods

Saccharogenic method. The reference range for *serum* of healthy persons for the Somogyi method, as modified by Henry and Chiamori, is 60–160 Somogyi units/dL or 111–296 U/L. The actual lower and upper limits vary with the methods used and reflect the poor precision and accuracy of these assay techniques. The reference range in terms of Street and Close units is 6–33 units.

In the assay for *urinary amylase* on 1-, 2-, or 24-h timed specimens, the lower range of normal is between 1500 and 1800 Somogyi units/d and the upper range is between 6000 and 7500 units/d. The average daytime hourly output may be taken as between 70 and 275 units. Amylase levels in *duodenal fluid* specimens may vary depending on the different assay conditions; concentrations of the order of 50 000–80 000 Somogyi units/dL are not uncommon for persons with normal pancreatic function.

Dye-coupled starch methods. Using the Amylochrome (Roche) procedure, the following values (at 37 °C) have been reported:

Serum: 450–2000 dye units/L
Urine: 40–330 dye units/h

Using the Phadebas (Pharmacia) insoluble blue-starch polymer the following values (at 37 °C) have been reported:

Serum: 70–300 U/L
Urine: 170–2000 U/L

Defined-Substrate Methods

DuPont aca amylase assay (at 37 °C):
Serum: 23–85 U/L
Urine: 4–37 U/2 h

Beckman DS amylase assay (at 37 °C):

Serum: newborn, 2–4 d: 5–65 U/L
adult: 25–125 U/L
> 70 years 21–160 U/L
Urine: 1–17 U/h

α-Amylase 4-NP (Boehringer Mannheim) method:

Serum (at 30 °C): ≤ 130 U/L
Urine (at 30 °C): ≤ 650 U/L

Note: As only one out of the three substrate molecules actually cleaved produces free 4-nitrophenyl, a stoichiometric factor of three is used to express amylase activity in International Units.

AMYLASE ISOENZYMES[57,134]

Human amylase isoenzyme (isoamylase) methods include use of electrophoresis, ion-exchange chromatography, isoelectric focusing, and selective inhibition of the S-type isoamylase by a wheat germ inhibitor. These will be discussed in turn.

Electrophoresis. This technique is probably most commonly used, although amylase isoenzyme separation is more difficult than the separation of LDH or CK isoenzymes.

Skude[134] used 0.8% agarose gels containing 6% linear polyacrylamide polymer to reduce endosmotic flow during electrophoresis. Separation of the isoenzymes in serum and urine was achieved at 4 °C. The isoenzymes were then visualized by incubating with the gel a layered suspension of Phadebas blue starch tablets at 56 °C for 2.5 h. This technique detects amylase fractions with activities 2–3% of the total serum activity.

Skude reported the existence of one main pancreatic isoenzyme (P-isoamylase) together with an anodally migrating fraction (P_o-isoamylase) that constitutes < 20% of the main fraction. One other cathodally migrating zone is found in P-type heterozygotes and constitutes ~70% of the main, P-type, fraction. There is also one main salivary isoenzyme (S-isoamylase) together with some minor, anodally migrating variants. Some phenotypes show only presence of these anodally migrating S-type forms without the major S-type fractions. These phenotypes are the most common cause of low total S-type activity.

Legaz and Kenney[85] separated the amylase isoenzymes in serum and urine by electrophoresis on cellulose acetate membranes at 4 °C for 1.25 h in a discontinuous buffer system. The isoamylases were detected by incubating the membrane for 20 min at 37 °C with an agar gel overlay containing a suspension of Phadebas tablets. Six isoenzymes were resolved; the faster migrating isoenzymes were of salivary origin (S1, S2, and S3) and the slower ones were of pancreatic origin (P1, P2, and P3). Legaz and Kenney were unable to provide a reference range for each isoenzyme because of wide variation in isoenzyme distribution between healthy individuals. P2, S1, and S2 isoamylases were most commonly observed in serum patterns. Less than 2% of all patients studied possessed the P1 isoenzyme (a possible genetic variant). S3 was not consistently present in sera; P3 only appeared in the sera of patients with *acute* or *chronic pancreatitis,* or of those with *renal transplants.*

Ion-exchange chromatography. Berk's group[42] has described a column chromatographic technique for examining the amylase isoenzymes in hyperamylasemia. The technique is not applicable to serum of healthy persons because amylase levels are too low; it may, however, be used on serum with elevated levels and on urine with either normal or elevated levels. Serum or urine is first passed through a superfine grade of a G-75 Sephadex gel. Two peaks of amylase activity are obtained. Each peak is then subsequently chromatographed on a DEAE-Sephadex A-50 ion-exchange gel. Two P-type and three S-type isoamylases are obtained by this procedure. The entire procedure is laborious, however; Berk's group used an amylase assay on an AutoAnalyzer to process the large number of column fractions.

Štěpán and Škrha[138] described a 1 × 0.9 cm minicolumn DEAE-cellulose technique much more suited to the needs of the routine hospital laboratory. The serum or urine sample is 100 μL and a single peak of P-type isoamylase is eluted first. The proportion of S-type isoamylase is calculated by difference from the total amylase activity of the specimen and from the activity of the P-type fraction.

Isoelectric focusing. Levitt's group[88] used isoelectric focusing for the separation of the amylase isoenzymes. A 3-mL serum specimen was focused in an isoelectric focusing column for 72 h at 4 °C, although other workers have used shorter time intervals with a corresponding increase in electrical power applied to the column. Sera from healthy persons had three or four major isoamylase peaks, one P-type fraction, and two or three S-type fractions. This technique suffers from the disadvantage that the collection of many fractions from the preparative column may require several amylase assays to detect the separated zones of isoamylase.

Wheat inhibitor. S-type amylase is preferentially inhibited by a protein isolated from wheat.[67,111] The amylase activity of the serum specimen is determined in the presence and absence of the inhibitor. The extent of inhibition is proportional to the S-type isoenzyme content of the specimen. The P-type/S-type isoenzyme ratio can be obtained by comparing the degree of inhibition to a standard calibration curve obtained by assaying mixtures of pure human salivary and pancreatic isoamylases in the presence and absence of the inhibitor. Huang and Tietz[67] adapted this approach to the Beckman DS assay and the Boehringer Mannheim Diagnostics (BMD)-UV method.

Other assays. Radioimmunoassays have been described for both the S- and P-type isoamylases, but some cross-reactivity has been observed between the isoenzymes. However, this technique has not received much attention.

Reference Ranges[143]

Skude, using the electrophoretic technique described earlier, showed that after birth serum S-type activity increases steadily with age and reaches normal adult levels of 82 U/L (range 35–155 U/L) or 48% of total activity at the age of 5 years. The serum P-type activity develops later. The majority of children below 1 year have no demonstrable pancreatic serum isoamylase activity; activity rises slowly thereafter to reach adult levels (88 U/L, with a range of 45–180 U/L, or 52% of total activity) at the age of 10–15 years. Štěpán and Škrha, using a minicolumn ion-exchange technique, found in sera from adults similar ratios (48% and 52%) for the serum S- and P-type activities. Levitt's group obtained values of 68% and 32% for the S- and P-type isoamylases in serum with their isoelectric focusing method. O'Donnell and colleagues,[111] using the wheat inhibitor technique, obtained the following values: males, 38% S-type and 62% P-type; females, 34% S-type and 66% P-type.

Huang and Tietz,[67] for a modified technique, found values for P-type amylase of 27–70% of total amylase activity.

LIPASE

(EC 3.1.1.3; Triacylglycerol Acylhydrolase)

Human lipase is a glycoprotein with a molecular weight of 48 000, and an isoelectric point of about 5.8. For full catalytic activity and greatest specificity, presence of bile salts and a cofactor, called colipase,[14] is required. Human lipase can also be fully activated by colipases from other species, e.g., porcine colipase.

Lipases are defined as enzymes that hydrolyze glycerol esters of long-chain fatty acids. Only the ester bonds at carbons 1 and 3 (α-positions) are attacked, and the products of the reaction are two moles of fatty acids and one mole of 2-acylglycerol (β-monoglyceride) per mole of substrate. The latter is resistant to hydrolysis, due probably to steric hindrance, but it can spontaneously isomerize to the α-form (3-acylglycerol). This isomerization permits the third fatty acid to be split off, but at a much slower rate. A scheme for the steps in the complete hydrolysis of a fat to glycerol and three fatty acids is given in equation (46).

$$
\begin{array}{c}
\mathrm{CH_2OFA'} \\
| \\
\mathrm{CHOFA''} \\
| \\
\mathrm{CH_2OFA'''}
\end{array}
\xrightarrow[\substack{\mathrm{HOH} \\ \mathrm{OH^-}}]{\mathit{Lipase}}
\begin{array}{c}
\mathrm{CH_2OH} \\
| \\
\mathrm{CHOFA''} \\
| \\
\mathrm{CH_2OFA'''}
\end{array}
\xrightarrow[\substack{\mathrm{HOH} \\ \mathrm{OH^-}}]{\mathit{Lipase}}
\begin{array}{c}
\mathrm{CH_2OH} \\
| \\
\mathrm{CHOFA''} \\
| \\
\mathrm{CH_2OH}
\end{array}
\xrightarrow{\mathit{Isomerization}}
\begin{array}{c}
\mathrm{CH_2OH} \\
| \\
\mathrm{CHOH} \\
| \\
\mathrm{CH_2OFA'''}
\end{array}
\xrightarrow[\substack{\mathrm{HOH} \\ \mathrm{OH^-}}]{\mathit{Lipase}}
\begin{array}{c}
\mathrm{CH_2OH} \\
| \\
\mathrm{CHOH} \\
| \\
\mathrm{CH_2OH}
\end{array}
\quad (46)
$$

Triglyceride α,β-**Diglyceride** + β-**Monoglyceride** + α-**Monoglyceride**   **Glycerol** +
                                FA'OH   **FA'''OH**   **FA''OH**
                           (Fatty acid I)   **(Fatty acid III)**   **(Fatty acid II)**

Lipase acts only at the interface between water and the substrate, i.e., when the substrate is present in an emulsified form. The rate of lipase action depends on the surface area of the dispersed substrate. Bile acids appear to be necessary to ensure that the surface of the dispersed substrate remains free of other proteins, including lipolytic enzymes (e.g., lipoprotein lipase), yet available to pancreatic lipase. Lipase seems to gain access to the substrate surface in the following manner: colipase attaches to a micelle of bile salts, thus forming a colipase–bile salt complex that attaches to the substrate surface. This complex binds lipase with high affinity, thus allowing lipase action to proceed.

Lipase may be activated by low concentrations (80–140 mmol/L) of NaCl and other ions, but there is controversy about this effect as well as the benefit of Ca(II) addition.

Lipase and colipase are secreted by the pancreas in roughly equimolar quantities—about 10 μmol/L. This equimolar relationship also applies at the time lipase and colipase enter the circulation. However, the smaller molecular weight of colipase (about 11 000) permits its glomerular excretion and urinary elimination. During an attack of acute pancreatitis, it therefore appears likely that the serum colipase/lipase ratio will drop during the course of the episode. Consequently, true estimates of serum lipase activity may not be obtained unless lipase assay mixtures are supplemented with colipase.

There are at least three related enzymes (aliesterase, arylesterase, and lipoprotein lipase) in tissues and serum whose activities may be included in the measurement of "lipase" activity unless there is careful selection of suitable assay conditions for true lipase activity.[73] Aliesterases (carboxylic-ester hydrolase or carboxylesterase; EC 3.1.1.1) hydrolyze glycerol esters of short-chain fatty acids, such as tributyrin, as well as esters of monohydric alcohols, such as ethyl acetate, with dibasic acids. Aryl esterases (aryl-ester hydrolase; EC 3.1.1.2) hydrolyze such esters as phenyl acetate and β-naphthyl butyrate. Lipoprotein lipase (triacylglyceroprotein acylhydrolase; EC 3.1.1.34), known also as clearing factor, is activated in vivo by heparin; it hydrolyzes protein-bound triglycerides to form free fatty acids and monoglycerides.

Most of the lipase found in serum is produced in the pancreas but some is also secreted by gastric, pulmonary, and intestinal mucosa. Lipase activity can also be demonstrated in leukocytes, in adipose tissue cells, and in milk. After pancreatectomy, lipase activity in blood is maintained at a lower level by the lipase from these other tissues. In the older literature, lipase activity was reported to be present in the urine. These findings were probably due to errors in methodology; it is now fairly well established that urine contains little or no true lipase activity.

Clinical Significance

Lipase measurements on serum, plasma, and ascitic and pleural fluid are used exclusively to investigate pancreatic disorders, usually pancreatitis.

Following an attack of *acute pancreatitis* the serum lipase activity rises within 2–12 h to two- to four-fold (sometimes as much as 10-fold) of the upper reference limit with the 2-point assay and 2- to 50-fold with the kinetic assay. Values may return to normal within 48–72 h but may stay elevated for considerably longer periods (10–14 d) than amylase values. Lipase elevations usually parallel those of amylase, but increases in lipase activity may occur sooner or later than increases in amylase activity and lipase may rise to a greater extent. For this reason it is suggested that the two assays complement and not exclude each other and that both enzymes be assayed. The increase in serum lipase activity is not necessarily proportional to the severity of the attack.

Acute pancreatitis may produce *ascitic fluid* or *pleural fluid* (usually on the left side), or both. These fluids may contain lipase activity. About 50% of patients with acute pancreatitis develop a pseudocyst; its presence is usually suspected when there is no clinical improvement within a week of the acute attack. About 50% of patients with *pseudocyst* formation show persistent serum amylase elevations, and it is suggested that lipase assays may also be useful under these circumstances.

The diagnosis of acute pancreatitis is sometimes difficult to make since this disorder must be differentiated from other acute intra-abdominal disorders with similar clinical findings, such as perforated gastric or duodenal ulcer, intestinal obstruction, mesenteric vascular obstruction, and many others. Elevation of serum lipase activity is probably a more specific diagnostic test in these cases than serum amylase activity, because many of these conditions are less likely to cause increases in lipase activity than in amylase activity.

Serum lipase assays may also be of value in the diagnosis of *chronic pancreatitis*, but in the later stages of the disease there is severe destruction of the acinar tissue resulting in a reduction of the amount of enzyme that can enter the circulation. Marginal or no increase of serum lipase activity is therefore not unusual in this disease. *Obstruction* of the pancreatic duct by a calculus or by carcinoma of the pancreas may cause an increase in serum lipase activity depending on the location of the obstruction and the amount of remaining functioning tissue.

In *acute* and *chronic renal disease,* serum lipase activity is increased, although this elevation is neither as frequent nor as pronounced as that seen with serum amylase activity. Thus, care should be exercised in the interpretation of elevated serum lipase values in the presence of renal disease.

In contrast to amylase, which is present in both the pancreas and the parotid glands, lipase is not present in the parotid gland. Therefore, in *mumps* (acute parotitis) without pancreatic involvement, serum lipase activity is usually not elevated while serum amylase activity is. Finally, investigation of the biliary tract by *retrograde pancreatography* (ERCP) or treatment with opiates (which causes the sphincter of Oddi to contract) may cause elevation of serum lipase activity.

METHODS FOR MEASURING LIPASE ACTIVITY

In 1932 Cherry and Crandall first recognized the clinical value of blood lipase measurement in following pancreatic injury.[73] The method bearing their names employed a buffered, stabilized 50% emulsion of olive oil as substrate; lipase activity was defined by the amount of free fatty acids released in a 24-h incubation at 37 °C and estimated by titration to neutrality with dilute alkali. Since 1932 many lipase methods have been described; they have utilized both triglyceride and nontriglyceride substrates in titrimetric, turbidimetric, spectrophotometric, fluorometric, and immunologic techniques.[152] In this brief survey we will devote the most attention to methods using long-chain triglyceride substrates. Over the years, these methods have displayed a correlation of results with the clinical state that is superior to that seen with other methods.

Long-Chain Triglyceride Substrates

Titrimetric Methods

Olive oil, the substrate of the Cherry and Crandall procedure,[73] is a mixture of long-chain triglycerides. One of many modifications of the method has been in the length of the incubation period; since 50% of the hydrolysis was shown to occur in the first several hours, investigators have found that reduction of the incubation time to six, four, three, or even one hour is feasible, depending on the degree of optimization of reaction conditions. Phenolphthalein, the original acid-base indicator of the titration step, has been replaced by thymolphthalein, whose endpoint at pH 10.5 confers superior stoichiometry. Dependence on a visual endpoint was obviated by Tietz and Fiereck,[142] who demonstrated the effectiveness of potentiometric titration. The character of the olive oil emulsion has been considerably improved in respect to stability, uniformity, and droplet size. Triolein, a pure long-chain triglyceride, has been shown to be an alternative substrate. Although triolein has the advantage of being a single molecular species, the low cost and availability of olive oil, specially purified to remove endogenous fatty acids and other impurities, will no doubt contribute to its continued use for routine analyses.

The relatively long incubation times of the titrimetric methods have always posed a risk of enzyme denaturation due to long exposure of enzyme to the relatively high reaction temperature. The much greater sensitivity of kinetic methods using a pH-Stat makes shorter incubation periods possible, and thus inactivation of the enzyme during the incubation period is greatly reduced. The method of Tietz and Repique[144] is an example of such a continuous monitoring-type technique. During the reaction period, the pH is kept constant at 8.8 at 30 °C (or 8.6 at 25 °C) by addition of dilute alkali. The amount of alkali used is recorded as a function of time and serves as a measure of fatty acid produced during the reaction. Good precision is obtained with a serum sample of 200–500 μL and a reaction time of 3–8 min. When olive oil substrate is used at 30 °C, the reported reference range is < 160 U/L; with triolein substrate, slightly higher activities are obtained and the upper limit of the reference range increases to 200 U/L.[144]

Colipase, until recently, has not been added to reaction mixtures. Older methods therefore determine both lipase and lipoprotein lipase activities, particularly in post-heparin specimens. Recently, however, Hockeborn and Rick[63] have described a continuous monitoring (pH-Stat) method with addition of colipase and bile salts to triolein substrate, but a later communication advised against the use of colipase at this time. Tietz and Astles,[141] using 20 mmol/L of sodium glycocholate and colipase at 3 mg/L, were able to stabilize the enzyme and eliminate post-heparin lipoprotein lipase activity.

Kinetic titrimetric assays employing a pH-Stat are potentially the most reliable means of defining serum lipase activity. They are, unfortunately, demanding of special skills and equipment and thus probably not well suited to routine laboratory use. They mainly have been, and will probably continue to be, used as reference methods for evaluating other lipase assays and for research purposes.

Turbidimetric Methods

Emulsions of fats in water have a milky appearance because the micelles absorb and scatter incident light. Clearing occurs as lipase hydrolyzes the triglycerides in the micelles and micellar disruption takes place. The rate of micellar disintegration as measured by decreases in turbidity or light scattering, however, is not necessarily equal to the rate of triglyceride hydrolysis as measured by the appearance of free fatty acids. A turbidimetric method must therefore be calibrated against a titrimetric method or with a serum calibrator with a lipase activity assigned to it by a titrimetric technique. Stable, reproducible substrates with appropriate initial turbidity are difficult to prepare and store; high readings for initial absorbance result in a high degree of analytical imprecision. Turbidities have occasionally been observed to increase rather than to decrease during the reaction period. Increase may be due to different rates of micelle degradation in blank and test mixtures or to precipitation of abnormal serum proteins in the reaction mixture; such increases have frequently been observed in specimens containing rheumatoid factor. Despite these theoretical and practical objections, turbidimetric assays for lipase activities gain popularity, especially because of recent improvements in techniques as described below.

The method of Vogel and Zieve[152] uses an olive oil emulsion and measures change in turbidity at 400 nm over a 20-min reaction period. An optimal deoxycholate concentration and pH are defined, but the method is relatively insensitive. Shihabi and Bishop[152] have modified the procedure by increasing the reaction temperature and reading absorbance at a lower wavelength. The DuPont *aca* lipase assay is adapted from both methods but uses triolein as substrate.

Recent improvements of the turbidimetric assay incorporate a high concentration (19 mmol/L) of deoxycholate and porcine colipase at 3 mg/L.[64] The deoxycholate makes the small amount (300 mmol/L) of emulsified triolein almost inaccessible to lipoprotein lipase and thus eliminates the overestimation of pancreatic lipase activity. The colipase mediates specific interaction of pancreatic lipase and substrate; $CaCl_2$ is present at 100 μmol/L. The upper limit of the reference range for this assay is 150–200 U/L at 25 or 30 °C.

An interesting variant of the turbidimetric approach utilizes radial diffusion of enzyme into buffered agarose gel containing deoxycholate and olive oil emulsion.[51] Lipase activity is proportional to the zone of clearing observed after incubation of 10 μL of serum for 2 h at 37 °C.

Spectrophotometric and Fluorometric Methods[73,152]

Spectrophotometric methods are based on extraction of free fatty acids released by lipolysis from the reaction mixture, followed by conversion of the acids to their cupric salts and re-extraction into an organic phase containing diethyldithiocarbamate (DTC). The copper-DTC complex is brown in color and suitable for spectrophotometric quantitation. Versions of this approach concentrate on minimizing the number of extractions and facilitating phase separations. Myrtle and Zell[108] found with their modification a reference range of 20–180 U/L at 37 °C, which compares well with that found by Tietz and Repique for the 30 °C kinetic assay. These methods, however, are not susceptible to optimization by deoxycholate since bile salts produce unacceptably high blanks in copper-soap procedures.

A fluorometric method by Rietz and Guilbault[119] uses triolein substrate and estimates fatty acids released from their quenching of fluorescence of 4-methylumbelliferone.

Short-Chain Triglyceride Substrates[73,152]

Short-chain triglycerides have the analytical advantage of greater solubility in aqueous media; they have the theoretical disadvantage of being unphysiologic substrates for lipase. Tributyrin emulsions have been suggested as substrates for assays of lipase activity. Erlanson and Bergström,[34] with a system fortified with taurodeoxycholate and colipase, found that tributyrin was hydrolyzed 12 times faster than triolein by human pancreatic juice or duodenal fluid. Recently a tributyrin derivative, 2,3-dimercaptopropan-1-ol tributyrate (BALB), has been used in a spectrophotometric method.[79] Hydrolysis of BALB by lipase in the presence of the surfactant sodium dodecyl sulfate produces 2,3-dimercaptopropan-1-ol. This product reacts with 5,5'-dithiobis-(2-nitrobenzoic acid), DTNB, to form 5-thio-2-nitrobenzoate ions, which absorb strongly at 412 nm. Both serum arylesterase and carboxylic esterase hydrolyze BALB, but their effect can be eliminated by pretreatment of serum with the potent esterase inhibitor phenylmethylsulfonylfluoride.

Nontriglyceride Substrates[73,152]

The original Cherry-Crandall assay, because of its long incubation time and difficulties of substrate preparation and endpoint detection, stimulated a search for alternative methods. Inevitably, a variety of nontriglyceride substrates came under consideration, among them β-naphthyl laurate, β-naphthyl myristate, phenyl laurate, and sorbitan esters. Most of these are hydrolyzed not only by lipase but by other serum enzymes as well. Attempts to incorporate selective inhibitors into reaction mixtures have not been successful, and methods based on the substrates noted above are not recommended.

Another category of nontriglyceride substrates includes compounds which, upon lipolysis, release fluorophores. Examples are methylumbelliferone and N-methylindoxyl myristate, the monodecanyl ester of fluorescein. However, contemporary experimental work largely favors triglyceride substrates, and methods employing fluorogenic substrates are not likely to become popular in clinical laboratories.

Determination of Lipase Activity in Serum

At present it is difficult to suggest one particular assay formulation for the determination of lipase activity in the service laboratory. Accordingly, one example each of titrimetric and turbidimetric assays is briefly outlined.

Specimen

Lipase activity in serum is stable at room temperature for one week; sera may be stored for three weeks in the refrigerator, and for several months if frozen. Bacterial contamination of the specimen may result in an increase in lipase activity.

Principle of Titrimetric Assay (Tietz and Fiereck)[142]

An aliquot of serum is incubated in a stabilized olive oil emulsion at a reaction pH of 7.8 for 3 h at 37 °C. The liberated fatty acids are then titrated to pH 10.5 with NaOH, 50 mmol/L, either potentiometrically or to a light blue color with thymolphthalein as indicator.

Reagents and Procedure

The details of the reagents and procedure are described in Chapter 12 of *Fundamentals of Clinical Chemistry*[73] and in the original communication.[142] This assay is also available commercially from the Sigma Chemical Company, St. Louis, MO.

Comments

1. The reference range is 28–280 U/L or up to 1.0 (conventional) unit; the CV of the assay is <12% at the upper reference limit but approaches 4% at higher lipase activities.

2. Considerable enzyme inactivation may be expected over a 3-h incubation period at 37 °C; therefore, 30 °C may be a preferable reaction temperature although there is a consequent loss of assay sensitivity.

3. The enzyme reaction is not linear with time over the 3-h period, although the deviation is not pronounced; lack of linearity is significantly less than in methods using 6- or 12-h incubation periods.

4. This assay does not include deoxycholate or colipase. Therefore it is likely to be less specific for pancreatic lipase than assay formulations that incorporate these reagents into the reaction mixture.

Reference

Tietz, N. W., and Fiereck, E. A.: Measurement of lipase activity in serum. Standard Methods of Clinical Chemistry, Vol. 7. New York, Academic Press, 1972, pp. 19-31.

Principle of Turbidimetric Assay (Ziegenhorn et al.,[155] Hoffman and Weiss[64])

An aliquot of serum is added to a preincubated stabilized triolein emulsion containing Na deoxycholate (17 mmol/L), $CaCl_2$ (0.1 mmol/L) and porcine colipase (3 mg/L) at pH 9.2 with Tris buffer at 25 mmol/L and triolein at 0.29 mmol/L. Assay temperature can be 25 °C or 30 °C. Absorbance at 340 nm is read after 4 min and again after an additional 5 min. The ΔA/min is taken as a measure of lipase activity. This method requires the use of a calibrator (a standard lipase preparation). Addition of deoxycholate and colipase increases the specificity and sensitivity of this assay. This method is available as a kit from Boehringer Mannheim Diagnostics; a modified version will be released soon and will allow assay at 37 °C.

Reagents and Procedure

The details of the reagents and procedure are described in a number of publications to which the reader is referred. The assay is also available commercially from Boehringer Mannheim Diagnostics and is described in detail in the package insert.

Comments

1. High concentrations of deoxycholate (19 mmol/L) and colipase (3 mg/L) ensure that only the activity of pancreatic lipase is measured and that neither lipoprotein lipase nor nonspecific esterases are detected.

This assay, as well as the commercial version (BMD), still has the disadvantage of giving negative results (increase in absorbance) with some sera, predominantly from patients with rheumatoid factor and other disorders of active immune processes. These difficulties can be overcome by monitoring the reaction with a chart recorder and selecting the linear segment of the reaction rate curve (between 11 and 19 min, depending on sample) or by preincubating the serum for 10 min with polyethyleneglycol at a final concentration of 55 g/L. This treatment results in the precipitation or dissociation of the interfering factors (e.g., IgM).

2. The presence of a high concentration of deoxycholate also ensures that the reaction lag-phase is minimized.

3. Heating the serum sample for 3 h at 56 °C totally destroys all pancreatic lipase activity as measured by this turbidimetric assay. This assay therefore appears to be more specific for pancreatic lipase than are current titrimetric assays that detect residual "activity" in heated serum.

Immunoassay of Lipase

Recently an analytically sensitive and specific immunoassay for serum lipase has been reported.[53] The assay is based on the sandwich technique. A specific antilipase antibody, conjugated to horseradish peroxidase, binds pancreatic lipase in serum; peroxidase activity measured is proportional to the amount of lipase bound. Detection limit of the assay is 300 ng/L; reference range in a healthy adult population is 7.7–56 μg/L (median 23). A particular advantage of the assay is its ability to determine low lipase concentrations characteristic of cystic fibrosis, chronic pancreatitis, and some cases of carcinoma of the pancreas.

Lipase in Duodenal Fluid

Although amylase and trypsin in duodenal fluid are routinely measured, measurement of lipase in duodenal fluid has been discouraged by the unsatisfactory nature of lipase methods. Duodenal specimens may need to be diluted before being measured. Physiological saline may be used as diluent, although a solution containing 50 mg albumin/dL is preferable.

TRYPSIN

(EC 3.4.21.4; No Systematic Name)

Trypsin is a serine proteinase characterized by the presence at the active site of serine and histidine, both of which participate in the catalytic process. Trypsin hydrolyzes the peptide bonds formed by the carboxyl groups of lysine or arginine with other amino acids, although esters and amides involving these amino acids are actually split more rapidly than peptide bonds. This specificity is illustrated in Figure 5-29, which presents the structures of three synthetic substrates, benzoyl-L-lysineamide (BLA), *p*-toluenesulfonyl-L-arginine methyl ester (TAME), and benzoyl-L-arginine-*p*-nitroanilide (BAPNA); all are hydrolyzed rapidly by trypsin.

The acinar cells of the human pancreas synthesize two different trypsins (I and II) in the form of the inactive proenzymes (or zymogens), trypsinogens I and II. These zymogens are stored in zymogen granules and are secreted into the duodenum under the stimulus of either the vagus nerve or the intestinal hormone cholecystokinin-pancreozymin. The two trypsinogens represent ~19% of the total protein in pancreatic juice; trypsinogen I is present at about twice the concentration of trypsinogen II. In the intestinal tract, the trypsinogens are converted to the active enzyme trypsin by the intestinal enzyme enterokinase, or by preformed trypsin (autocatalysis).

Trypsin I is also described as cationic, trypsin II as anionic, because of their differing electrophoretic mobilities; the cationic form predominates and is the better documented enzyme.[57] Trypsin I and trypsin II have molecular weights of 25 800 and 22 900 and pI values of 4.6–6.5 and >6.5, respectively. The pH optimum for trypsin I for natural substrates is in the range of 8.0–9.0, but for synthetic substrates such as TAME or BAPNA it is 7.8; the pH optimum for trypsin II is 8.0–10.0. Trypsin II differs from trypsin I in that it undergoes autolysis rapidly at neutral or alkaline pH values and it is not stabilized against autolysis by Ca(II). Because the two trypsins show little immunologic cross-reactivity, specific immunoassay of each of them is possible.

Trypsin activity is stimulated by calcium and magnesium ions and to a lesser extent by cobalt and manganese ions and by aliphatic alcohols. Cyanide, sulfide, citrate, fluoride, and heavy metals inhibit activity, as do those organic phosphorus compounds that combine with serine at the active site.

Materials such as soy beans, lima beans, and egg white contain natural trypsin inhibitors—small polypeptides—which combine irreversibly with trypsin and inactivate it by blocking the active center. Similar nondialyzable trypsin inhibitors are present in pancreatic juice, in serum, and in urine. The plasma inhibitors, α_1-antitrypsin and α_2-macroglobulin, and their immuno-

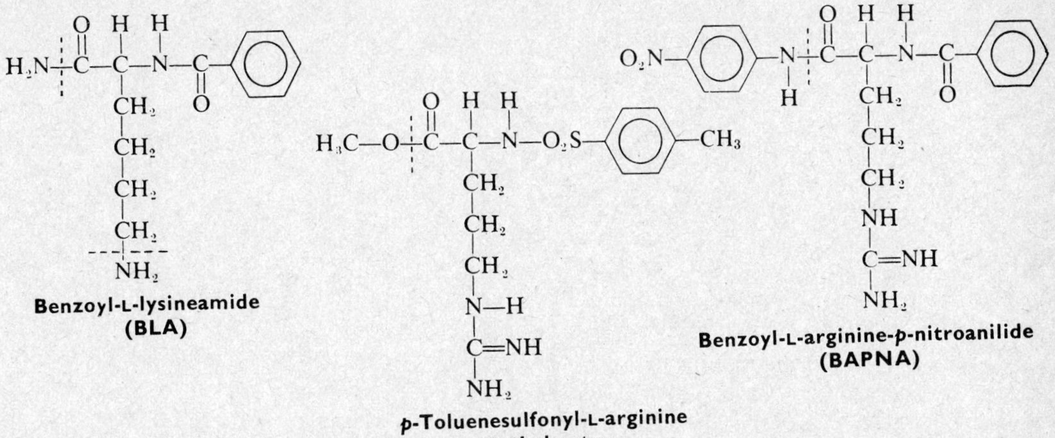

Benzoyl-L-lysineamide (BLA)

***p*-Toluenesulfonyl-L-arginine methyl ester (TAME)**

Benzoyl-L-arginine-*p*-nitroanilide (BAPNA)

Figure 5-29. Formulas for three synthetic substrates for trypsin, illustrating the bond specificity of the enzyme, which attacks bonds involving the —COOH of lysine and arginine. The dashed line indicates the bond which is hydrolyzed. Hydrolysis of TAME is an example of esterase activity; hydrolyses of BLA and of BAPNA are examples of amidase activity.

chemical assay are discussed in Chapter 4. These inhibitors protect plasma and other proteins against hydrolysis by trypsin and other proteases if for some reason any appreciable quantity of the enzyme enters the vascular system. The absence of α_1-antitrypsin is associated with an increased tendency toward panlobular emphysema in early life; this example nicely illustrates the effects of uninhibited proteases on organ function.

Clinical Significance[48]

Trypsin in Duodenal Content and Feces[97]

Determinations of trypsin in gastrointestinal secretions are used as an aid in the evaluation of pancreatic function and in the diagnosis of *chronic pancreatitis* and *fibrocystic disease*, as discussed in Chapter 14. Trypsin assays can be performed either on aspirated specimens of duodenal content or on fresh feces. Most investigators agree that determinations of trypsin in feces are now of limited value, since much of the pancreatic trypsin may be destroyed by proteases as the intestinal contents pass through the tract, and the remainder cannot easily be differentiated from proteases associated with intestinal bacteria.

Fibrocystic disease (cystic fibrosis, mucoviscidosis) in children is accompanied by deficient secretion of trypsin by the pancreas. The disease is diagnosed on clinical grounds and by the measurement of sweat electrolytes. Either at the time of presentation or at the stage when pancreatic insufficiency develops, little or no tryptic activity in fecal specimens (or, better, in duodenal contents) is usually evident. Decrease in activity is conveniently demonstrated with the X-ray film test or with one of the previously mentioned quantitative methods employing synthetic substrates.

Immunoreactive Trypsin in Serum[12,44,84,137]

The recent availability of immunoassays for the quantitation of trypsin in blood is an important technical advance. In the case of trypsin I, radioimmunoassay will detect trypsinogen I, trypsin I, and the trypsin I:α_1-antitrypsin complex; it does not detect the trypsin I:α_2-macroglobulin complex. Free trypsin I is not usually found in serum; it is always complexed. Trypsin II exists, in health, as trypsinogen II and it is also found, in disease, complexed to α_1-antitrypsin and α_2-macroglobulin. Unfortunately, while assays of trypsin II have been reported there is little clinical experience with this enzyme.

Trypsin I (Cationic Trypsin)

In *healthy* individuals, free trypsinogen I is the major form found in serum. Following an attack of *acute pancreatitis*, serum immunoreactive trypsin rises in parallel with serum amylase activity to peak values ranging from 2–400 times the upper reference range. The distribution of the different forms of trypsin appears to be related to the type and severity of the acute pancreatitis. Thus, in *edematous pancreatitis*, the mildest form of acute pancreatitis, 80–99% of the immunoreactive trypsin exists as free trypsinogen I with smaller proportions existing as bound trypsin I. In the more serious *hemorrhagic pancreatitis*, in which the mortality ranges from ~20% to >50%, the proportion of free trypsinogen I may be as low as 30% of the total with appreciable proportions existing as the α_1-antitrypsin and α_2-macroglobulin bound trypsin I.[16]

Immunoreactive trypsin I in serum is elevated in *chronic renal failure*, as is serum amylase. Thus, renal failure must be excluded when interpreting elevated levels.[137] In *chronic pancreatitis* without steatorrhea, plasma levels of immunoreactive trypsin I do not differ from those found in health; when steatorrhea is present, however, fasting levels are extremely low. In the *relapsing phase* of chronic pancreatitis, plasma immunoreactive trypsin may be considerably elevated. In *carcinoma of the pancreas* immunoreactive trypsin levels may be high, normal, or even low, which limits the usefulness of its measurement in this condition.

In *fibrocystic disease*, plasma immunoreactive trypsin levels have been reported to be high in neonates; as the disease progresses, the level falls. Dried blood specimens have been suggested for use in screening tests.

In *diabetes*, both juvenile (Type I insulin-dependent diabetes mellitus) and maturity-onset (Type II non-insulin-dependent diabetes mellitus), significantly lower than normal plasma immunoreactive trypsin levels have been reported with a graded reduction in levels related to the degree of insulin failure.

Increased plasma levels have been reported in virologically confirmed cases of *mumps* and

Coxsackie B infections and also in children with clinical diagnoses of mumps, *pyrexia of unknown origin*, and *meningitis*, as well as in adults with *Bornholm disease* or *cardiac* and *respiratory infections*. It was suggested that in these infections a silent invasion of the exocrine pancreas may have occurred.

Methods for the Determination of Trypsin in Duodenal Fluid and Stool

Semiquantitative Determination of Trypsin by the X-Ray Film Test[27]

The test specimen may be either duodenal fluid or a fresh stool specimen. Serial, two-fold dilutions of the specimen in barbital buffer, 100 mmol/L (pH 8.0), are prepared. For a duodenal fluid specimen, the dilution may range from 1/2 to 1/256, and for a stool specimen from 1/5 to 1/320. A drop of each dilution is spotted onto a small piece of unexposed X-ray film, along with a drop of barbital buffer to act as a control. The X-ray film with the series of specimen and control drops is then incubated for 30 min at 37 °C, cooled in a refrigerator to harden the gelatin, and washed with a slight stream of water to remove any loose or hydrolyzed gelatin. The piece of film is then examined for the presence of clear (digested) areas, indicating hydrolysis of the gelatin layer by the action of the enzyme; the control spot should show no evidence of digestion. The greatest dilution of the specimen giving a cleared area is taken as the measure of the trypsin activity present. The test is not recommended for stool specimens from adults because of the high rate of false negative results.

Reference Ranges

The stool of a normal infant under one year of age will show tryptic activity through a dilution of 1/80 or higher. With older children, activity may be evident only through a dilution of 1/20 or 1/40. Infants with fibrocystic disease will rarely give positive tests beyond the 1/10 dilution. In the case of duodenal contents, normal infants give positive tests through a dilution of 1/32 or 1/64, fibrocystic children usually in dilutions of 1/4 or less.

Determination of Trypsin in Stool or Duodenal Fluid Using TAME as Substrate[56]

Trypsin attacks denatured proteins at a more rapid rate than native proteins, and the former have therefore been used as substrates for the enzyme. However, these substrates are also hydrolyzed by chymotrypsin and other proteases and the trend is now toward the use of synthetic peptide substrates such as BAEE (benzoyl-L-arginine ethyl ester), BLA, BAPNA, and TAME, which are specific for trypsin and are not attacked by other proteases found in *duodenal contents* and *stool specimens*. In addition, these synthetic substrates have only one bond that can be split so that expression of activity in International Units is possible.

Procedures using these well-defined substrates are simple and convenient, although on occasion specimen color may be a source of interference. Where possible, the change in absorbance at some fixed wavelength is followed during the course of the reaction. For example, since benzoyl arginine, a reaction product, has a greater absorbance at 253 nm than does the substrate BAEE, the reaction may be followed by measuring the rate of increase in absorbance at 253 nm. If amides are used, the ammonia formed can be measured by the Berthelot method. With the use of any synthetic substrate, the reaction can be monitored with a pH-Stat by measuring the carboxyl H^+ produced. With BAPA as a substrate, the yellow color of the *p*-nitroaniline split off in the reaction permits the measurement of trypsin activity by a continuous-monitoring procedure.

A procedure using TAME in Tris buffer and based on measuring tryptic activity by titrating the carboxyl hydrogen ion released on hydrolysis of the methyl ester has been described by Haverback and associates.[56] Tryptic activity was recorded as mmol of NaOH required to maintain a constant pH and was converted to μg of crystalline trypsin by comparison with a standard curve.

Reference Ranges for TAME Method

Normal children and adults give values of 40–760 μg trypsin/g stool; adults with (alcoholic) pancreatic insufficiency give values of 0–33 μg/g stool. Fibrocystic children have concentrations of <20 μg/g stool.

Determination of Immunoreactive Trypsin in Blood Serum or Plasma by RIA[12,44,137]

Several groups have purified human pancreatic trypsin I and trypsin II and used them to produce antitrypsin sera. All these assays detect, by mass, trypsin, trypsinogen, and trypsin bound to α_1-antitrypsin. The trypsin bound to α_2-macroglobulin cannot be detected by RIA unless the complex is isolated and dissociated. The trypsin I:α_2-macroglobulin complex is isolated from serum by gel filtration and trypsin I is dissociated by treatment of the complex with acid, after which the freed trypsin I may be assayed. Trypsin I bound to α_2-macroglobulin can consist of > 30% of the total serum trypsin in disease, but many assays do not measure this particular trypsin fraction.

There are several commercial assays available for measurement of trypsin I. In the Behring Diagnostics method (RIA-gnost Trypsin; Calbiochem-Behring, P.O. Box 12087, San Diego, CA 92112) the sample is incubated with ^{125}I trypsin and the antitrypsin serum for ~16 h at room temperature; addition of a precipitating second antibody is followed by centrifugation, and measurement of bound ^{125}I trypsin in the pellet by means of a gamma scintillation counter. A standard curve is established using dilutions of a standard with known amounts of trypsin I instead of a patient sample. The CIS method (Trypsin Radioimmunoassay Kit; marketed by Damon Diagnostics, Needham Heights, MA 02194) uses the same formulation as the Behring Diagnostics kit for the initial assay mixture, but the assay mixture is incubated at room temperature for only 1 h. The immunocomplex is then precipitated with a solution of second antibody and polyethylene glycol. After centrifugation, the bound ^{125}I in the pellet is assayed as in the previous method.

Reference Ranges

Users of the RIA-gnost assay report values of 140–400 μg/L of trypsin I in serum, although others using this method report mean values as low as 87 μg/L. Users of the CIS assay report values of 10–60 μg/L for adults and 8.4–26 μg/L for children aged 2–12 years.

Groups using their own assays—derived from different preparations of human trypsin—report serum trypsin, in fasting males, of 25 $\pm$ 5 g/L ($\overline{X} \pm$ SD) and 26 $\pm$ 9 g/L. Each laboratory should establish its own reference range for the method in use.

It has recently become apparent that serum trypsin I levels show a positive correlation with age, that there are higher values in females (~50% higher than males), and that serum concentrations increase as renal function decreases.

CHYMOTRYPSIN

(EC 3.4.21.1; No Systematic Name)

Chymotrypsin, like trypsin, is a serine proteinase. It hydrolyzes peptide bonds involving carboxyl groups of Trp, Leu, Tyr, or Phe with preference for the aromatic residues. Specificity of chymotrypsin is therefore in contrast to that of pepsin, which splits bonds involving amino groups of the aromatic amino acids. Chymotrypsin also demonstrates hydrolytic activity for other types of bonds in the following order: esters (especially N-substituted tyrosine esters) > amides > peptides. N-acetyl-L-tyrosine ethyl ester is an example of a synthetic substrate, used for chymotrypsin assays in duodenal aspirates and feces, that contains an ester bond and that is not significantly hydrolyzed by trypsin (Figure 5-30).

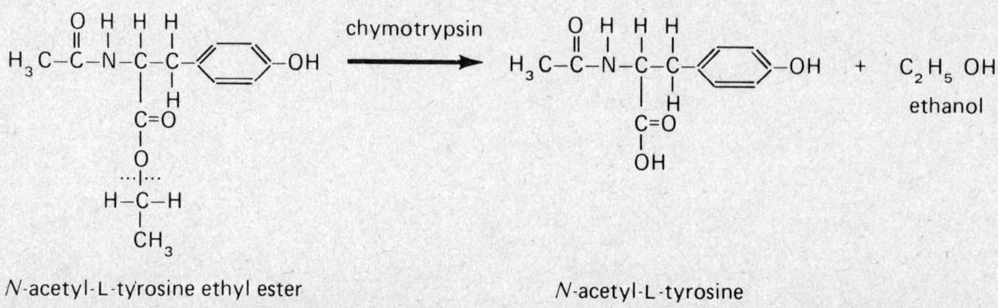

Figure 5-30. N-acetyl-L-tyrosine ethyl ester, a synthetic substrate for chymotrypsin.

The acinar cells of the human pancreas synthesize two different chymotrypsins (I and II, the latter being the major species) in the form of the inactive proenzymes (or zymogens), chymotrypsinogens I and II. These zymogens are stored in granules and are secreted like trypsinogen into the pancreatic duct. In the intestinal tract, the chymotrypsinogens are converted to chymotrypsin by the action of trypsin. Chymotrypsin is more resistant than trypsin to degradation in the intestine; it is therefore the enzyme of choice for assay in feces.

Chymotrypsin I is also described as anionic and chymotrypsin II as cationic because of their differing electrophoretic mobilities; the cationic form predominates. The molecular weight of both forms is ~25 000. There is close immunological identity between the chymotrypsins and chymotrypsinogens; immunoassay for chymotrypsin II reacts totally with chymotrypsinogen II and strongly with both chymotrypsin I and chymotrypsinogen I. Chymotrypsin, like trypsin, is bound in blood plasma by α_1-antitrypsin and α_2-macroglobulin. The complex of chymotrypsin I (or II): α_1-antitrypsin can be detected by the chymotrypsin II immunoassay; however, the chymotrypsin I (or II): α_2-macroglobulin complex cannot be detected with this assay.

Clinical Significance[2,82]

The predominant form of chymotrypsin found in blood is chymotrypsin II. Although RIA's for immunoreactive chymotrypsin II were described in the 1970's, little clinical experience with these assays is available.

The major application of these assays is in the investigation of pancreatic disease. In the newborn, chymotrypsin II in cord blood is 2.2–7.5 times higher in infants with *fibrocystic disease* than in healthy infants. In the adult, *acute pancreatitis* can elevate immunoreactive chymotrypsin II to 35 times the levels found in health. In *pancreatic carcinoma*, levels may be normal or slightly (2.5-fold) elevated. Following *total pancreatectomy*, no chymotrypsin II is detectable in serum.

In *gastric carcinoma*, levels may be slightly elevated (up to 2.5-fold) but are more usually within the reference range. In *hepatobiliary disease*, elevations up to two-fold the upper reference range are found. In *renal failure*, as with trypsin and amylase, there is elevation of serum chymotrypsin II up to eight-fold the reference range.

In duodenal aspirates (or pancreatic juice) and in feces, chymotrypsin levels are variably reduced in pancreatic dysfunction. For further details recent reviews should be consulted. (See Chapter 14.)

Methods for the Determination of Chymotrypsin in Blood

Synthetic Peptides

Methods for the determination of chymotrypsin that employ synthetic substrates or peptides are no longer preferred for use in the clinical laboratory and have been replaced by immunoassays.

Immunoassay[13,45,71]

Several investigators have purified chymotrypsin II and prepared antichymotrypsin sera for immunoassay of chymotrypsin. These assays detect, by mass, chymotrypsinogen II, chymotrypsin II, and, to a lesser extent, chymotrypsin II bound to α_1-antitrypsin. There is a significant cross reaction with chymotrypsinogen I (55–85%), chymotrypsin I (25–50%), and chymotrypsin I: α_1-antitrypsin (20–32%) in these assays; however, the complex, chymotrypsin I (or II): α_2-macroglobulin, is not detected.

This type of assay is not yet commercially available. Therefore chymotrypsin II must be purified from human pancreas and antisera developed against the purified enzyme. Refer to the referenced papers for further details of the assay.

Methods for the Determination of Chymotrypsin in Duodenal or Pancreatic Aspirates and Feces[48]

Synthetic substrates are commonly used for these applications. Recently a sensitive kinetic assay using Succ-Ala-Ala-Pro-Phe-*p*-nitroanilide has been described. Duodenal or pancreatic aspirates are obtained following stimulation with test meals or by administration of secretin or secretin and cholecystokinin-pancreozymin. (See Chapter 14.) Stool specimens are either random or 24-h collections. When chymotrypsin is measured in stool specimens, prior treatment of the specimen with detergent to release particle-bound chymotrypsin[75] is now thought essential.

An *indirect* estimate of chymotryptic activity in the gut can be made by giving the compound *N*-benzoyl-L-tyrosyl-*p*-aminobenzoic acid (Bz-Tyr-PABA) orally, usually with a Lundh meal. Bz-Tyr-PABA is specifically hydrolyzed by chymotrypsin; in the gut, the hydrolysis releases free *p*-aminobenzoic acid (PABA), which is absorbed, conjugated in the liver, and excreted in the urine where it can be readily measured. Details of these assays can be obtained from current reviews.[2,82] See also Chapter 14.

The obvious applications of chymotrypsin *immunoassay* to duodenal or pancreatic aspirate and to feces have not yet been reported. Whether immunoassay will prove superior to current approaches remains to be seen.

Reference Ranges

Serum. In neonates (cord blood) total chymotrypsin is <24 μg/L. In adults total chymotrypsin ranges from 15–78 μg/L (mean of 37.5 μg/L) have been reported, although an earlier report suggested ~10 μg/L.

Duodenal aspirates, pancreatic juice, and feces. Because of the many and varied techniques used for these assays, reference ranges should be obtained from current reviews or should be established in the laboratory performing the test.

CHOLINESTERASES

(EC 3.1.1.7; Acetylcholine Acetylhydrolase)

(EC 3.1.1.8; Acylcholine Acylhydrolase, SChE)

There are two related enzymes that have the ability to hydrolyze acetylcholine. One is *acetylcholinesterase,* which is called "true" cholinesterase, or choline esterase I. True cholinesterase is found in erythrocytes, in lung and spleen, in nerve endings, and in the gray matter of the brain. It is responsible for the prompt hydrolysis of acetylcholine released at the nerve endings to mediate transmission of the neural impulse across the synapse. The degradation of acetylcholine is necessary to the depolarization of the nerve so that it can be repolarized in the next conduction event.

The other cholinesterase is *acylcholine acylhydrolase;* it is usually called pseudocholinesterase, benzoyl cholinesterase, or choline esterase II. Although it is found in liver, pancreas, heart, the white matter of the brain, and serum, its biological role is unknown. The serum enzyme is the one whose assay is clinically useful.

The type reaction catalyzed by both cholinesterases is:

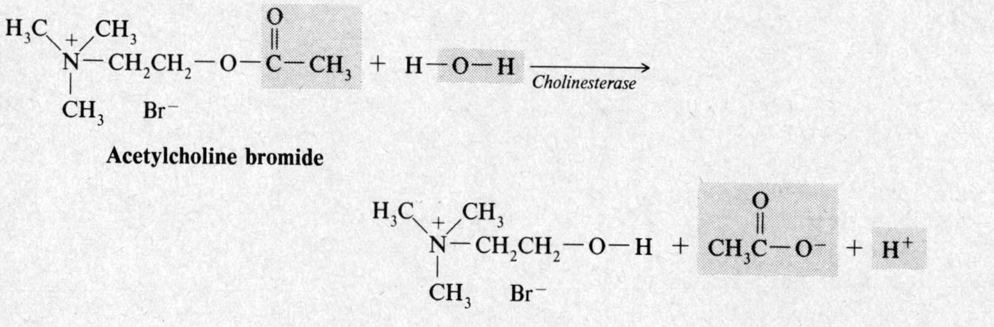

Acetylcholine bromide

Choline bromide   **Acetate ion**   **Hydrogen ion**

The two enzymes differ in specificity toward some substrates while behaving similarly toward others. The serum enzyme acts on benzoylcholine, but cannot hydrolyze acetyl-β-methylcholine; the red cell enzyme acts on the latter but not on the former. Only choline esters are split by the red cell enzyme; aryl or alkyl esters are not attacked. The red cell enzyme is inhibited by its substrate, acetylcholine, if present at concentrations about 10^{-2} mol/L; the serum enzyme is not inhibited by this substrate.

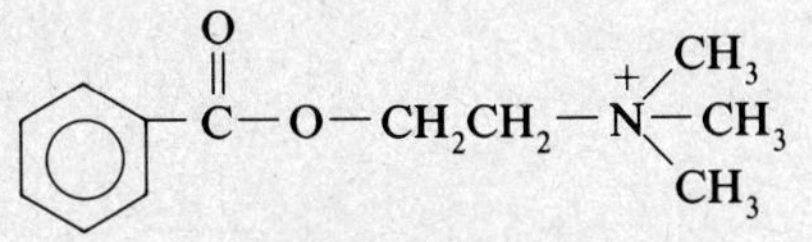

Benzoylcholine

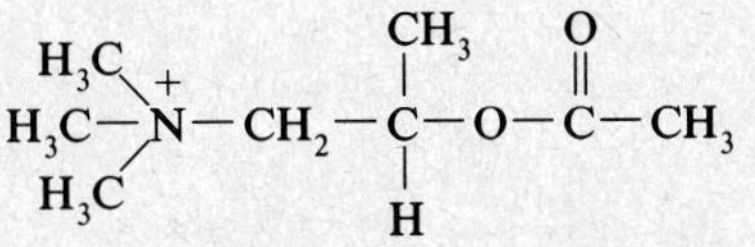

Acetyl-β-methylcholine

Both enzymes are inhibited by the alkaloids prostigmine and physostigmine, both of which contain quaternary nitrogen (also present in choline) in their structures. These two compounds are typical competitive inhibitors, competing with the choline residue of acetylcholine for its binding site on the enzyme surface. Both enzymes are irreversibly inhibited by some organic phosphorus compounds, such as diisopropylfluorophosphate. The phosphoryl group binds very tightly to the enzyme site at which binding of the acyl group normally occurs, thus preventing attachment of the acetylcholine. Both enzymes are also inhibited by a large variety of other compounds, among which are morphine, quinine, tertiary amines, phenothiazines, pyrophosphate, bile salts, citrate, fluoride, and borate.

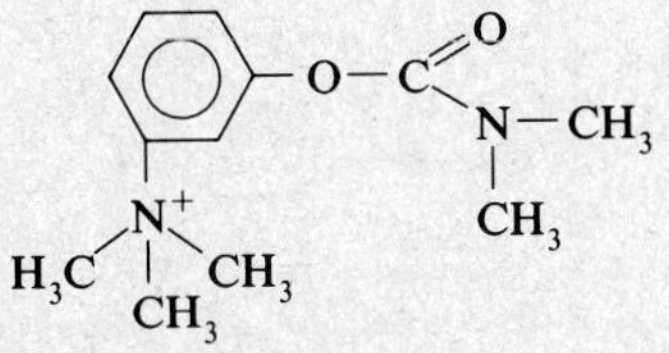

Prostigmine

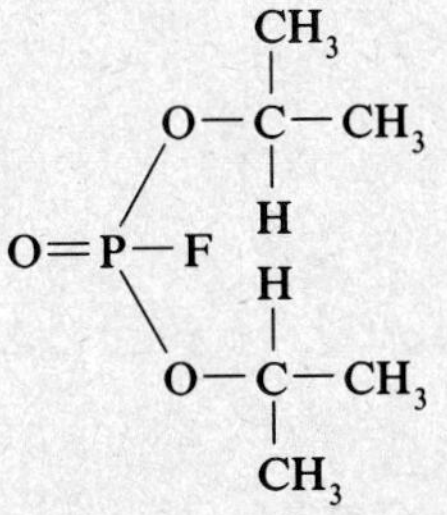

Diisopropylfluorophosphate

The cholinesterase present in normal sera can be separated by electrophoresis into seven to twelve bands, the number obtained depending on the experimental technique used. The isoenzymes of serum cholinesterase (SChE) differ in molecular size and appear to be aggregates of different numbers of the same basic unit. Of more interest are the atypical (genetic) variants of the enzyme, characterized by diminished activity against acetylcholine and other substrates, which are found in the sera of a small fraction of apparently healthy persons.

The gene controlling the synthesis of SChE can exist in at least four allelic forms, designated as E_1^u, E_1^a, E_1^f, and E_1^s. Several other allelic forms may also exist, and another gene locus is recognized (E_2). The normal, most common phenotype is designated as $E_1^u E_1^u$, or UU. The gene E_1^a is referred to as the atypical gene; the sera of persons homozygous for this gene ($E_1^a E_1^a$ = AA) are only weakly active toward most substrates for cholinesterase and possess increased resistance to inhibition of enzyme activity by dibucaine. The E_1^f gene also gives rise to a weakly active enzyme but with increased resistance to fluoride inhibition. The E_1^s gene (s for silent) is associated with absence of enzyme or the presence of a protein with minimal or no catalytic activity. The mutations that give rise to the atypical and fluoride-resistant SChE variants involve a change in the structure of the active center. The variant isoenzymes (allelozymes) are less effective catalysts than the usual form; the affinity of the enzymes for substrates is reduced (i.e., K_m is increased), and affinity for competitive inhibitors such as dibucaine or fluoride is similarly decreased. This gives rise to the characteristic dibucaine- or fluoride-resistant properties of the genetic variants that are exploited in their characterization.

The homozygous forms, AA or FF, are found in only 0.3–0.5% of the white population; their incidence among blacks is even lower. Inheritance of increased serum cholinesterase activity has also been reported in a few families. This is apparently due to increased production of the usual allelozyme.

Clinical Significance

Cholinesterase levels in serum are useful as a test of *liver function,* as an indicator of possible *insecticide poisoning,* or for the detection of patients with *atypical forms* of the enzyme.[17] The spread of values encountered in apparently healthy people is rather wide, ranging between 4000 and 12 000 U/L at 37 °C. The level in any given person is fairly constant; a fall in enzyme level of greater than 670 U/L is significant at the 95% confidence level. Levels at birth are only one-fourth of those of adults but reach adult levels by the second month of life. No enzyme is found in urine.

Measurements of serum cholinesterase activity can serve as sensitive measures of the synthetic capacity of the liver if the patient's normal (baseline) level is known, which unfortunately is rarely the case. In the absence of known inhibitors, any decrease in activity in serum reflects impaired synthesis of the enzyme by the liver. A 30–50% decrease in level is observed in *acute hepatitis,* and in *chronic hepatitis* of long duration. Decreases of 50–70% occur in advanced *cirrhosis* and *carcinoma* with metastases to the liver. Essentially *normal* levels are seen in *chronic hepatitis, mild cirrhosis,* and *obstructive jaundice.*

Decreased levels of serum enzyme are also found in patients with *acute infections, pulmonary embolism,* and *muscular dystrophy,* and after *surgical procedures.* After a *myocardial infarction,* the enzyme level decreases until the fifth day and then begins a slow rise to normal. Decreased levels are also seen in *chronic renal disease* and in *pregnancy.*

A marginal increase in enzyme levels may be observed in patients with *nephrotic syndrome.* Synthesis of albumin to replace that lost in the urine is accompanied by some synthesis of additional cholinesterase, although syntheses of the enzyme and albumin by the liver are independent of each other. Marginal increases in enzyme are also seen in *thyrotoxicosis* and *hemochromatosis,* in *obese diabetics,* and in patients with *anxiety* and other *psychiatric states.*

Among the organic phosphorus compounds that inhibit cholinesterase activity are many organic insecticides, such as Parathion, Sarin, and tetraethyl pyrophosphate. Workers engaged in agriculture and those working in organic chemical industries may be subject to *poisoning* by inhalation of these materials or by contact with them. Obviously, if enough material is absorbed to inactivate all the acetylcholinesterase of nervous tissue, death will result. Both cholinesterases are inhibited, but the activity of the serum enzyme falls more rapidly than does that of the red cell enzyme. A 40% drop in serum enzyme activity occurs before the first symptoms are felt, and a drop of 80% is required before serious neuromuscular effects become apparent. Near-zero or zero levels of enzyme activity require emergency treatment of the patient with such enzyme reactivators as pyridine-2-aldoxime.

Succinyldicholine (suxamethonium) is a drug used in surgery as a muscle relaxant. Because it is very similar to acetylcholine, it is also hydrolyzed by cholinesterase, and its physiological effect persists only long enough (30–50 min) to meet the needs of the surgical procedure. In patients with low levels of enzyme activity, or in those with the atypical, weakly active enzyme variant, this destruction of the drug will not occur rapidly enough, and the patient may enter a period of *prolonged apnea* requiring mechanical ventilation until the drug is eliminated by other routes. Preoperative screening has been advocated to identify patients in whom suxamethonium administration may lead to complications. However, more often the laboratory is asked to investigate patients and their relatives in whom suxamethonium sensitivity has been found.

The degree of sensitivity varies with the phenotype of the patient. There is a decline in total activity from individuals who are homozygous for the usual allele, through those who are heterozygous for the usual and a variant allele, those who are homozygous or heterozygous for variant alleles, to zero in subjects in whom two "silent" alleles are paired. Subjects who possess one normal allele (i.e., who are heterozygous for the normal and a variant allele) usually produce enough enzyme to protect them against suxamethonium sensitivity, whereas patients with paired variant alleles (either as homozygotes or heterozygotes) show various degrees of sensitivity.

Measurements of total serum cholinesterase activity as well as determination of the "dibucaine number" and "fluoride number" are needed to characterize cholinesterase variants fully. The latter parameters indicate the percentage inhibition of enzyme activity toward specified substrates in the presence of a standard concentration of these reagents. The average values of the dibucaine numbers (DN) for normals, heterozygotes, and homozygotes (E_1^a gene) are 78, 60, and 16%, respectively, when benzoylcholine is used as substrate. Dietz and associates reported values of 81–86, 67–80, and 8–35%, respectively, with propionyl thiocholine (PTC) as substrate.[30] Analogous fluoride numbers (FN) with PTC as substrate are 77–81, 67–69, and 54–65%, respectively. The phenotypes most susceptible to apnea after succinylcholine administration are AA, AS, FF, FS, SS, AF, and, to some extent, UA.

Cholinesterase levels in *cerebrospinal fluid* are very low. The total activity present in fluids from healthy individuals is about 17 ± 4 U/L. Both the serum and red cell enzyme forms can be detected, with the serum form present in the greater concentration. Appreciable elevations of SChE activity are seen in diseases involving *destruction of brain parenchyma*, in *brain tumors* and *brain abscesses*, in *hydrocephalus* (to 40 U/L), in *Guillain-Barré disease* (to 60 U/L), and often in *meningitis* and *multiple sclerosis*. In other neurological and non-neurological diseases, enzyme levels are normal or any changes seen are too nonspecific and inconsistent to be of much value in diagnostic work. SChE activity is independent of protein concentration and of leukocyte count, the leukocytes containing no enzyme activity.

METHODS FOR THE DETERMINATION OF CHOLINESTERASE ACTIVITY[3,133]

The hydrolysis of acetylcholine, the substrate used in older methods, results in the formation of one mole of hydrogen ion for each mole of substrate reacted. In manometric methods this hydrogen ion reacts with HCO_3^- buffer to release CO_2, which is then measured. In the potentiometric procedure of Michel, the decrease in pH during the course of a fixed reaction period is measured. The acid formed can be titrated with a pH-Stat apparatus. If pH indicators that change color in the pH range of 6.0–8.5 (bromothymol blue, *m*-nitrophenol) are incorporated into the reaction mixture, the change in color, measured spectrophotometrically, can serve as a measure of enzyme activity. A number of "kits" and test papers devised for testing for enzyme activity in the field are based on this principle.

In the procedure of de la Huerga et al., the quantity of unreacted acetylcholine remaining after a 60-min reaction period at 37 °C is measured. The ester is reacted with hydroxylamine to form the hydroxamate derivative; the complex forms an orange-brown complex with ferric ion in acid solution that can be measured colorimetrically at 540 nm. Kalow and Genest proposed benzoylcholine as substrate and measured the decrease in absorbance of the substrate at 240 nm.

Contemporary methods employ acylthiocholine esters as substrates. These substrates are hydrolyzed at approximately the same rate as choline esters, and the thiocholine formed can be measured by reaction with chromogenic disulfide agents such as DTNB [5,5′-dithiobis(2-nitrobenzoic acid), Ellman's reagent] or 4,4′-dithiodipyridine. The iodide salts of acetyl-, propionyl-, and butyrylthiocholine have all been used as substrates. The reactions for butyrylthiocholine, with DTNB as the chromogen, are given in equations (47) and (48).

$$\text{Butyrylthiocholine ion} + \text{HOH} \xrightarrow{\text{SChE}} \text{Thiocholine ion} + \text{Butyrate} + \text{Hydrogen ion} \qquad (47)$$

Butyrylthiocholine ion

| Thiocholine ion | DTNB (colorless) | Mixed disulfide | 5-Mercapto-2-nitro-benzoic acid; 5-MNBA (colorless) |

$$(48)$$

The DTNB reaction produces a color that can be measured at 410 nm. If 4,4′-dithiodipyridine is used as a chromogen, mercaptopyridine is formed; this immediately tautomerizes to thiopyridine, which absorbs at 344 nm.

Mercaptopyridine   Thiopyridine

Determination of Serum Cholinesterase[30]

Principle

Activity of SChE is determined from the rate of hydrolysis of propionylthiocholine in the presence of DTNB. The reaction of the thiocholine product with colorless DTNB [equation (48)] forms colored 5-MNBA, which is measured spectrophotometrically at 410 nm. Dibucaine or fluoride inhibition can be estimated by performing concurrent assays in which dibucaine or fluoride is present in the substrate mixture. Per cent inhibition is evaluated by comparison of activity in the inhibited system with that in the uninhibited system.

The method described here is considerably condensed. Readers should consult the original paper[30] for important details of procedure and interpretations.

Specimen

Serum is the sample of choice. Enzyme activity in serum is stable for several weeks whether specimen is stored at room temperature or under refrigeration. Moderate hemolysis does not interfere if separated serum has been centrifuged to remove red cell ghosts. Specimens from patients who have displayed apnea after succinylcholine treatment should not be obtained until after paralysis has passed; metabolites of the drug appear to interfere with the assay.

Reagents

1. Phosphate buffer, pH 7.6, ionic strength 0.1 in water.
2. Propionylthiocholine iodide (PTCI), 20 mmol/L in water.
3. Dibucaine, 0.3 mmol/L in water.
4. Sodium fluoride, 40 mmol/L in water.
5. DTNB-buffer (color reagent). DTNB, 0.423 mmol/L in phosphate buffer.

6. Substrates:
   a. Uninhibited reaction. Mix equal parts of PTCI (reagent 2) and water.
   b. Dibucaine-inhibited reaction. Mix equal parts of PTCI and dibucaine solution (reagent 3).
   c. Fluoride-inhibited reaction. Mix equal parts of PTCI (reagent 2) and fluoride solution (reagent 4).
7. Quinidine sulfate, 0.5 g/dL in water.

Final concentrations in assay mixtures are, in mmol/L: PTCI, 2; DTNB, 0.254; phosphate, 25; and (when present) dibucaine, 0.03, or fluoride, 4.

Procedure

1. Dilute serum 100-fold with water.
2. For each inhibited or uninhibited assay, prepare two 16 × 125 mm tubes, labeling one B (blank) and the other U (unknown).
3. Add, to U and B tubes, 3.0 mL DTNB-buffer and 1.0 mL of the appropriate substrate. Mix.
4. Warm tubes at 37 °C for 5 min.
5. Without removing tube from bath and starting timer upon addition, add to the U tube 1.0 mL of diluted serum. Mix and allow to remain at 37 °C for 3 min.
6. Add, to U and B tubes, 1.0 mL quinidine sulfate. Mix.
7. Add, to tube B, 1.0 mL diluted serum and mix.
8. Transfer contents of corresponding U and B tubes to 1-cm cuvets and proceed promptly to read absorbance (color fades).
9. Read absorbance of U against B set to zero absorbance at 410 nm. Under these circumstances, absorbance of U represents $\Delta A/3$ min.

Calculations

Uninhibited reaction:

$$\text{U/L} = \frac{\Delta A/\text{min}}{13.6} \times \frac{6.0}{0.01} = \Delta A/\text{min} \times 14\ 710$$

where:

$$\Delta A/\text{min} = \frac{A_U}{3\ \text{min}}$$

$13.6 = \text{L} \times \mu\text{mol}^{-1} \times \text{cm}^{-1}$, the micromolar absorption coefficient of 5-mercapto-2-nitrobenzoic acid (5-MNBA)

$6.0 = $ total volume in reaction system

$0.01 = $ volume of serum in reaction system

Percent of inhibition:

$$\% \text{ inhibition} = \frac{A_U \text{ of inhibited system}}{A_U \text{ of uninhibited system}} \times 100$$

Comments

The original paper describes alternatives of timing, temperature, and standardization. Reference ranges for genetic variants are also presented.[30]

Reference Range

For the genotype $E_1^u E_1^u$, the mean $\pm$ SD for the above method is 8440 $\pm$ 1780 U/L.

Reference

Dietz, A. A., Rubinstein, H. M., and Lubrano, T.: Colorimetric determination of serum cholinesterase and its genetic variants by the propionylthiocholine-dithiobis(nitrobenzoic acid) procedure. Clin. Chem., *19*:1309-1313, 1973.

ACID PHOSPHATASE

(EC 3.1.3.2: Orthophosphoric-Monoester Phosphohydrolase (acid optimum); ACP)

Under the name of acid phosphatase are included all phosphatases with optimal activity below a pH of 7.0. Thus, the name refers to a group of similar or related enzymes rather than to one particular enzyme species.[10] However, the acid phosphatase of greatest clinical importance, that derived from the prostate, has a pH optimum in the range of pH 5–6.

Acid phosphatase is present in lysosomes, which are organelles present in all cells with the possible exception of erythrocytes. Extralysosomal acid phosphatases are also present in many cells. The greatest concentrations of acid phosphatase activity occur in liver, spleen, milk, erythrocytes, platelets, bone marrow, and the prostate gland. The last is the richest source, and it contributes about one-third to one-half of the enzyme present in sera from healthy males. The source of the remainder of the acid phosphatase in sera from healthy males and females is not known, but there is some evidence that it derives from disintegrated platelets, erythrocytes, the osteoclasts of bone, and leukocytes. The number of genes responsible for the various human acid phosphatases is unknown; however, the red cell enzyme is certainly the product of a unique gene.

The optimal pH for the individual acid phosphatases varies, depending on the tissues from which they are obtained. The observed pH optimum also varies with the substrate on which the enzyme acts; the more acidic the substrate, the lower the pH at which maximum activity is obtained. The enzymes can hydrolyze a variety of phosphate esters, and indeed every substrate utilized in measuring alkaline phosphatase activity in serum has also been used to determine acid phosphatase activity.

The acid phosphatases are unstable, especially at temperatures above 37 °C and at pH levels above 7.0. Some of the enzyme forms in serum (especially the prostatic enzyme) are particularly labile, and over 50% of the ACP activity may be lost in 1 h at room temperature. Acidification of the serum specimen to a pH below 6.5 aids in stabilizing the enzymes.

Because of the clinical importance of increased serum acid phosphatase levels in the diagnosis and monitoring of prostatic cancer, it is desirable to be able to differentiate specifically between increases in the concentrations of the prostatic and nonprostatic forms. Certain inhibitors can enhance the discrimination between prostatic and nonprostatic acid phosphatases. The prostatic enzyme is strongly inhibited by dextrorotatory tartrate ions, whereas the red cell isoenzyme is not. Red cell acid phosphatase is inhibited by formaldehyde and by cupric ions, to which prostatic acid phosphatase is resistant. Thus, these inhibitors, particularly tartrate, allow a distinction to be made between prostatic and erythrocyte acid phosphatases. However, it should be recognized that interference by the erythrocyte isoenzyme in the measurement of prostatic acid phosphatase in serum is not a significant problem in blood specimens taken with precautions against hemolysis and from which serum has been separated without delay.

An alternative approach to increase the specificity of the assay is the use of substrates that are hydrolyzed rapidly by the prostatic enzyme but at a significantly slower rate by the other forms of the enzyme. For example, α-naphthyl phosphate and thymolphthalein monophosphate are relatively less sensitive to the action of nonprostatic acid phosphatases than are such substrates as phenylphosphate and *p*-nitrophenylphosphate.

Clinical Significance[54]

Determination of acid phosphatase activity in serum is almost always directed toward the prostatic enzyme with the intent of detecting or monitoring *carcinoma of the prostate*. This important disease is currently the second most frequent cause of cancer deaths (17 000/year) in American males. Carcinoma of the prostate is graded by clinical examination—a process that is often supplemented by a variety of radiological examinations and histological examination of biopsy tissue.[60] A typical grading scheme is shown in Table 5-11.

It will be noted from Table 5-11 that stage A carcinoma cannot be detected by rectal palpation; in practice it is detected by histological examination of tissue removed to relieve urinary symptoms such as obstruction of the urinary tract. Clinical staging is an inexact process; recent work has shown that the use of radioisotopic bone scan, lymphangiography, and lymph node dissection of the pelvis led to nearly 50% of patients in clinical stages A, B, and C being regraded to a higher stage. Thus, reports that attempt to relate staging of tumor to serum prostatic acid phosphatase levels are valid only when a range of staging procedures has been used.

Table 5-11. CLINICAL STAGING OF PROSTATIC CANCER*

| Clinical Stage | Information Used to Assign Clinical Stage | | | Information Not Currently Required for Staging | |
	Digital Rectal Examination	Serum Acid Phosphatase Activity	Bone Scan Survey	Usual Histological Pattern	Frequency of Lymph Node Metastases (%)
A1	Tumor not detected	N	N	Well differentiated	5
A2	Tumor not detected	N	N	Poorly differentiated	25
B1	Tumor <1.5 cm in diameter or confined to one lobe	N	N	Well differentiated	8–21
B2	Tumor >1.5 cm in diameter or more than one lobe involved	N	N	Moderately to poorly differentiated	14–45
C	Disease locally extensive	N/↑	N	Moderately to poorly differentiated	40–80
D1	Any size of tumor which has spread beyond gland to regional lymph nodes only	N/↑	N	Any	100
D2	Any size of tumor which has spread to distant nodes or bone or viscera	↑	N/+	Any	100

*Reproduced by permission from N. Engl. J. Med., 300:824-833, 1979.

Elevations of the enzymatic activity of prostatic acid phosphatase (and thus, generally, of total acid phosphatase activity also) are found in the sera of ~75% of males with prostatic cancer with metastases. Total activities may reach 40–50 times the upper limit of normal, nearly all of the activity being inhibited by tartrate. However, when the carcinoma remains localized in the prostate gland, normal or only slightly raised levels of enzyme activity are found. After surgery or estrogen therapy, the levels slowly approach normal, with a subsequent rise if the treatment is unsuccessful. In patients with benign (nodular) hypertrophy of the prostate, the enzyme level in serum is within normal limits, even after prostatic manipulation, except in rare instances.

It is still widely held that rectal palpation of the prostate elevates serum prostatic acid phosphatase. There is now much evidence that this is not so. However, transient elevations can occur in certain clinical circumstances. These include *prostatic needle biopsy, cystoscopy, prostatic infarction* (caused either by the act of catheterization or by the presence of an indwelling catheter) and, finally, the very rare case of rupture of a *prostatic cyst.*

In recent years several immunological methods based on the use of specific antisera to prostatic acid phosphatase have been introduced. The potentially greater sensitivity and specificity of these methods, especially RIA, have prompted a re-examination of the value of acid phosphatase measurements in the determinations of the various stages of prostatic cancer.[54,122,132]

A slight advantage in diagnostic sensitivity is gained when analytically sensitive and specific immunologic methods are compared with enzyme activity measurements. The advantage is, however, marginal, particularly in the early stages of prostatic carcinoma. Many of the reports on the results of these immunological assays, up to the end of 1980, have been reviewed.[59,60] Present experience shows that although the number of early-stage cases that can be detected is increased by the use of immunological methods, the cases thus found remain a minority of all cases. This fact, together with the low prevalence of the disease in the population as a whole, does not justify programs to screen healthy subjects by acid phosphatase measurements.

Slight or moderate elevations in total ACP activity often occur in *Paget's disease,* in *hyperparathyroidism* with skeletal involvement, and in the presence of *malignant invasion* of the bones by cancers such as female breast cancer. The enzyme in these cases typically is not inhibited by tartrate and is thought to come from osteoclasts. In some cases of breast cancer, however, elevations of the tartrate-inhibited ACP have been observed. The osteoclasts are also the probable source of the increased tartrate-resistant acid phosphatase activity of growing children. Increased concentrations of nonprostatic acid phosphatase have also been observed in patients with *Gaucher's* and *Niemann-Pick diseases,* in *myelocytic leukemia,* and in some other *hematological disorders.*[120,154]

Acid phosphatase is present in very high concentrations in *semen,* a fact utilized in forensic medicine in investigations of rape and similar offenses.[47,83] It has been suggested[83] that swabs taken from alleged rape victims should be preserved by immersion in 2.5 mL of a protective broth (containing, per liter, 50 g of bovine albumin, 0.2 g NaN_3, 10 mmol phosphate buffer at pH 7.4, 9 g NaCl) and stored at either 4 °C or room temperature. Specimens stored in this fashion retain acid phosphatase activity for at least one month. Using this technique, and a thymolphthalein monophosphate substrate, it was found that vaginal acid phosphatase in noncoital women is < 10 U/L of broth, whereas in recently postcoital women it is > 50 U/L (mean ± SD; 242 ± 104 U/L).[83] In vivo degradation of vaginal postcoital acid phosphatase activity follows a logarithmic course so that by 4 d after intercourse it reaches noncoital levels.[83] An alternative preservative procedure of placing swabs in 1 mL of isotonic saline has been proposed.[47] Samples are then frozen and are allowed to thaw at 2–4 °C for 24 h before assay. The original publication[47] should be consulted for details of the range of values obtained. It should be realized that as male sexual dysfunction is common during rape, acid phosphatase activity may not be detected in samples from alleged rape victims.

METHODS FOR THE DETERMINATION OF ACID PHOSPHATASE ACTIVITY IN SERUM

Methods for acid phosphatase activity measurements are essentially adaptations of those developed for alkaline phosphatase. Phenyl phosphate and 4-nitrophenyl phosphate, for example, have frequently been used in an acid pH medium for acid phosphatase estimation. However, when these methods are used to measure acid phosphatase they must generally be performed as fixed-time procedures. This is because the spectral differences between the phosphate ester and its parent phenol which are exploited in continuous-monitoring procedures for the assay of alkaline phosphatase are distinctive only at alkaline pH. Therefore, the acid phosphatase reaction must first be stopped by making the reaction mixture alkaline before the amount of product formed can be measured.

Continuous-monitoring methods for assay of acid phosphatase activity can be based on the principle introduced by Hillmann, in which α-naphthol released from its phosphate ester forms a colored product with the stabilized diazonium salt of 2-amino-5-chlorotoluene-1,5-naphthalene disulfonate (Fast Red TR). Fast Red TR does not react significantly with bilirubin and coupling with α-naphthol takes place at the pH of enzyme action. The method is not without practical difficulties; the rate of coupling is dependent on pH, and pronounced lag-phases may occur before zero-order kinetics are established. In addition, the colored product is not very stable. Nevertheless, the method has been successfully used in manual and automated procedures. A recent development is the introduction of alcohols, such as 1,5-pentanediol, which accelerate the reaction and increase sensitivity by acting as phosphate acceptors in transfer reactions.

As already mentioned, nonprostatic acid phosphatases hydrolyze certain substrates less readily than does the prostatic enzyme. Since the object of diagnostic assays is almost always to determine the prostatic phosphatase, the substrates preferred by the enzyme are now usually chosen for the assay systems. The two most popular of these are thymolphthalein monophosphate and α-naphthyl phosphate, the latter being favored for continuous-monitoring procedures. Although neither substrate is completely specific, tartrate inhibition to enhance specificity is seldom considered advantageous for methods using the mentioned substrates.

Obviously, these substrates that are relatively insensitive to acid phosphatases of nonprostatic origin must be replaced with another substrate, such as 4-nitrophenyl phosphate, if changes in activities of nonprostatic enzyme are to be examined. With 4-nitrophenyl phosphate substrate, then, the assay system is frequently set up to determine total and tartrate-*un*inhibitable phosphatase activity, and consequently to determine tartrate-inhibitable prostatic enzyme activity by difference.

Immunological methods such as RIA, counterimmunoelectrophoresis, and immunoprecipitation derive their specificity from the specificity of the antiserum, which is high. The potential increase in sensitivity offered by RIA methods relates to the ability of antiserum to recognize *all* molecules of the enzyme, whether or not they are catalytically active. The specific antiserum may be combined with a colorimetric assay of activity; if the antiserum is used to "capture" or precipitate prostatic acid phosphatase activity, determination of total and residual activity again provides measurement, by difference, of the prostatic enzyme's activity.

Determination of Acid Phosphatase with Thymolphthalein Monophosphate as Substrate (Roy, Brower, and Hayden,[126] Modified by Ewen and Spitzer[35])

Specimens

Separate serum immediately from erythrocytes and stabilize by the addition of disodium citrate monohydrate at a level of 10 mg/mL of serum. Pellets (10 mg) are available from several commercial sources. Alternatively, add 50 μL of acetic acid (5 mol/L) per mL of serum to lower the pH to 5.4, at which the enzyme is stable. Under these conditions activity is maintained at room temperature for several hours and for up to a week if the serum is refrigerated. Although the substrate is relatively insensitive to erythrocyte acid phosphatase, hemolyzed serum specimens are contaminated with considerable amounts of this isoenzyme and should be rejected. Chylous sera should be avoided because of possible interference with measurement due to turbidity.

RIA determines enzymatically inactive as well as active molecules, provided that they are immunologically recognizable. The antigenic identity of acid phosphatase molecules is lost at a slower rate than catalytic activity. Nevertheless, since loss of antigenic identity takes place at appreciable rates at room temperature in nonacidified sera, the precautions of acidification and refrigeration should be applied to specimens for the determination of acid phosphatase by RIA as well as to those for enzymatic methods.

Principle

Thymolphthalein monophosphate is hydrolyzed by prostatic acid phosphatase at pH 5.4 and 37 °C. The reaction is stopped after 30 min by addition of NaOH-Na_2CO_3 solution. This develops the alkaline color of the liberated thymolphthalein, which is measured at 595 nm.

The conditions for the complete incubation mixture are:

Reaction temperature	37 °C
pH	5.4
Acetate buffer	0.15 mol/L
Thymolphthalein monophosphate (TMP)	1.0 mmol/L
Brij-35	1.5 g/L
Volume fraction (sample/total)	1:12 (0.083)

Reagents

1. Acetic acid, 5 mol/L. Measure 28.87 mL of glacial acetic acid into a 100-mL volumetric flask and dilute to volume with water. Mix.

2. Sodium acetate, 5 mol/L. Dissolve 68.0 g of sodium acetate trihydrate ($CH_3COONa \cdot 3 H_2O$) in water in a 100-mL volumetric flask and add water to the mark. Mix.

3. Acetate buffer, 5 mol/L, pH 5.4 at 25 °C. Add sufficient reagent 2 to acetic acid, 5 mol/L (reagent 1), to adjust to pH 5.4 at 25 °C.

4. Acetate buffer, 0.25 mol/L, pH 5.4 at 25 °C. Pipet 5 mL of reagent 3 into a 100-mL volumetric flask. Dilute to volume with water. Mix.

5. Buffered substrate. Dilute a solution of Brij-35, 300 g/L, to a concentration of 3.24 g/L. Dissolve 82.8 mg disodium thymolphthalein monophosphate $\cdot$ 11 H_2O, M.W. 752.65, in 50 mL of this solution and adjust volume to 100 mL. If a substrate preparation with a different water of crystallization is used, adjust the weight of salt appropriately. Add 1.92 g of sodium acetate $\cdot$ 3 H_2O to the solution, mix to dissolve, adjust the pH of the solution to 5.4 at 25 °C with HCl, 0.1 mol/L, and adjust volume to 100 mL with water. Store in the refrigerator. The solution is stable for up to 1 month.

6. Alkaline reagent, Na_2CO_3, 0.1 mol/L, in NaOH, 0.1 mol/L. Dissolve 10.6 g of anhydrous Na_2CO_3 and 4.0 g of NaOH in distilled water. Dilute to 1000 mL and mix. The reagent is stable at room temperature.

7. Thymolphthalein stock standard, 3 mmol/L. Dissolve 129.2 mg of thymolphthalein in n-propanol:water (70:30, v/v) and make up to 100 mL. Mix well. Store at 4 °C.

Procedure

1. For each specimen, pipet 0.55 mL buffered substrate (reagent 5) into a labeled test tube and place in a water bath at 37 °C to reach temperature equilibrium.

2. At zero time, add 50 µL serum. Rinse pipet, mix well, and return tube rapidly to water bath.

3. While incubation is proceeding, prepare a "blank" tube for each specimen, containing 1.0 mL of alkaline reagent (6). Add 0.55 mL of buffered substrate (5) followed by 50 µL of the serum specimen to the appropriate "blank" tube. Mix. (The acid phosphatase cannot act at the high pH of the alkaline reagent; the "blank" tube thus compensates for endogenous color of both specimen and reagents.)

4. After exactly 30 min, add 1.0 mL of alkaline reagent (6) to each "test" tube. Mix well to stop the reaction and develop the color.

5. Read absorbances of "test" and "blank" tubes for each specimen at 595 nm, with water as a reference. Subtract the "blank" reading from the "test" to give the absorbance change, ΔA in 30 min due to acid phosphatase activity.

6. If a well-calibrated, narrow-bandpass spectrophotometer is used for measurement, the molar absorptivity of thymolphthalein (39.20×10^{-3} L $\times$ µmol^{-1} $\times$ cm^{-1}) can be used to calculate the catalytic activity concentration of the specimen from the following formula:

$$\text{Acid phosphatase (U/L)} = \frac{\Delta A}{30} \times \frac{1}{39.2 \times 10^{-3}} \times \frac{1.6}{0.05}$$

$$= \Delta A \times 27.2$$

since the final 1.6 mL reaction mixture contains 0.05 mL of the serum sample.

7. Alternatively, a calibration curve can be prepared from the stock standard solution (7) by diluting 1.0, 2.0, 3.0, 4.0, 5.0, and 6.0-mL volumes of the standard to 10 mL with *n*-propanol:water. Working standards consist of 50 μL of each of the respective standard dilutions plus 0.55 mL buffered substrate (5) plus 1 mL alkaline reagent (6). The absorbances of these solutions at 595 nm correspond to those produced by 10, 20, 30, 40, and 50 U/L of acid phosphatase.

Comment

The incubation period can be varied in the enzymatic method to allow for differing levels of activity; e.g., if activity is very high, a 5-min incubation may be sufficient, in which case the result calculated from the formula given above is multiplied by 6.

Reference Ranges

The method is intended for the measurement of prostatic acid phosphatase. In the sera of healthy adult men, the range of activity is from 0.5–1.9 U/L (mean $\pm$ 2 SD). Results of RIA or other immunologic methods such as counterimmunoelectrophoresis for the determination of prostatic acid phosphatase are usually expressed in mass concentrations. Reference ranges are method dependent;[60] representative values are <3 μg/L for RIA[59] or <20 μg/L for counterimmunoelectrophoresis.[149]

Terminal Deoxynucleotidyl Transferase[*]

(EC 2.7.7.31; DNA Nucleotidylexotransferase, TdT)

Terminal deoxynucleotidyl transferase (TdT) is a unique DNA polymerase in that DNA is not required as a template for nucleoside triphosphate addition to the 3′ terminal end of a primer DNA chain. The biological distribution of TdT in normal individuals is generally restricted to thymocytes and to precursor cells of lymphocytes in bone marrow. An ontological association between the appearance of TdT and immunocompetent T-cells has been described by Sujimoto and Bollum.[139] Thus, TdT seems to be present early in lymphocyte differentiation and may persist somewhat longer in those precursors destined to mature as T-cells. The physiological function of this enzyme is not known.

Clinical Significance

The presence of TdT in T-cell precursors suggests that it should be found in T-cell–derived lymphoblastic leukemias and lymphomas. In fact, the most significant elevations of TdT are found in *acute lymphoblastic leukemia*. Virtually all patients with "T" and "null" cell acute lymphoblastic leukemia (ALL) and *lymphomas,* as well as many patients with pre-B cell lymphoblastic leukemia, will have TdT in their malignant cells. Some adult non-Hodgkin's lymphoma lymphocytes show lesser, yet significant, elevations. Approximately 50% of patients with *acute undifferentiated leukemia* and 30% with *chronic myelogenous leukemia* in blast crisis will have elevated TdT activity in their leukemic blasts. In such cases, the acute phase of chronic granulocytic leukemia is a lymphoblastic leukemia. A small number of other unusual acute leukemias may also demonstrate TdT activity. A compilation of data derived from a large patient population has been published.[22]

Because of the association of TdT with immature T-cell malignancies, assay of TdT activity is a useful adjunct in determining the type of leukemia, for selecting appropriate treatment, and for monitoring the effectiveness of the treatment.

---

*Contributed by John F. O'Brien, Ph.D.

TdT may not be a dependable predictor of relapse in children with common ALL. This is because TdT-positive cells may constitute up to 10% of the nucleated cells from normal pediatric marrows and because *marrow regeneration, idiopathic thrombocytopenic purpura,* and *neuroblastoma* may be causes for TdT-positive marrows.[68]

Specimens

Bone marrow aspirates and peripheral blood are the usual specimens used for TdT assays, although pleural fluid, spinal fluid, lymph nodes, and testicular tissue may also be used. For semiquantitative methods, a slide smear prepared from the specimen is necessary, while quantitative methods require separation and concentration as well as lysing of lymphocytes or tissue cells to release TdT. Preparation of specimens may vary; therefore, appropriate references should be consulted.[8,21,26,74,98]

Principles of TdT Methods

TdT catalyzes the addition of deoxynucleoside triphosphates to the 3'-OH end of a polydeoxynucleotide primer. TdT activity is determined by incubating the enzyme with ^{3}H-deoxynucleoside triphosphate, a divalent cation such as Mg(II), and an appropriate primer DNA polymer in a suitable buffer system. The extent of polymerization is determined by isolating the newly formed polymer and measuring its radioactivity. The amount of radioactivity incorporated in the polymer during a defined reaction time is used to calculate TdT enzyme activity.

The general reaction catalyzed by TdT is:

$$\text{n (Deoxynucleoside triphosphate)} + \text{(oligodeoxynucleotide)}_m \xrightarrow{\text{TdT}} \text{(oligodeoxynucleotide)}_{m+n} + \text{n (pyrophosphate)}$$

where m = the number of nucleotides in the primer DNA chain and n = the number of nucleotides added to the DNA chain.

Methods for the Determination of TdT Activity

Published methods for TdT activity measurements demonstrate a large range of variability in both assay components and in the TdT activity found in lymphocytes. The most important variables and their reported uses are pH of reaction (6.8, 7.5, 7.8); the size of the primer DNA used (dA_8, dA_{10}, dA_{12-18}, dA_{50}); method of isolation of the product (glass fiber filters, ion exchange filter paper, TCA precipitation); and the β-scintillation counting procedure. Adequate sensitivity required for detecting low levels of enzyme activity is often a problem, as is the apparent nonlinearity of the reaction kinetics observed with the human enzyme as compared to purified calf thymus preparations. For detailed descriptions of reported methods, the appropriate references should be consulted.[21,26,98]

An immunological method for the *semiquantitative* determination of terminal deoxynucleotidyl transferase has also been described.[21] This immunofluorescent (IF) method, in theory, has advantages over the quantitative radiometric assay in that fewer cells are necessary for detection of TdT and no preliminary preparation of lymphocytes is necessary. It may also permit identification of those cell types which are TdT positive. However, this method is not as sensitive as the radiometric method, and problems with preparation of slide smears do occur. The only extensive comparative study to date used anti-calf TdT as primary antibody followed by application of the fluorescence-labeled second antibody (goat anti-rabbit IgG) to the slide smear.[68] For cases in which a positive qualitative result is sufficient, this shorter, more convenient method may be the assay of choice.

A second immunological method that yields *quantitative* results is the solid-phase enzyme immunoassay described by Kaneda et al.[74] More extensive clinical studies using this technique must be done to confirm the utility of this assay. Also, an enzyme immunoassay for TdT activity may soon be marketed by Abbott Laboratories (North Chicago, IL 60064). This method is based on the sandwich principle in which TdT in the specimen is incubated with anti-TdT immobilized on polystyrene beads. The beads are then washed to remove any excess TdT, and anti-TdT with an enzyme label is added. After another incubation period, the beads are washed again, and enzyme substrate is added. The enzyme-labeled anti-TdT that is bound to the TdT in the sandwich-complex on the bead converts the substrate to a colored product. The intensity of the color is measured spectrophotometrically and is proportional to the amount of TdT in the sample.

References

1. Alpert, J. S., and Braunwald, E.: Acute myocardial infarction: Pathological, pathophysiological, and clinical manifestations. *In*: Heart Disease: A Textbook of Cardiovascular Medicine. 2nd ed. E. Braunwald, Ed. Philadelphia, W. B. Saunders Co., 1984, pp. 1262-1300.

2. Arvanitakis, C., and Cooke, A. R.: Diagnostic tests of exocrine pancreatic function and disease. Gastroenterology, *74*:932-948, 1978.

3. Augustinsson, K. B.: Determination of activity of cholinesterases. *In*: Methods of Biochemical Analysis, Vol. 19. D. Glick, Ed. New York, John Wiley & Sons, 1971, pp. 217-273.

4. Bayse, D. D., Duncan, P. H., and Mather, A.: Amylase in serum and urine. *In*: Clinical and Analytical Concepts in Enzymology. H. A. Homburger, Ed. Skokie, Ill., College of American Pathologists, 1983, pp. 173-194.

5. Belfield, A., and Goldberg, D. M.: Inhibition of the nucleotidase effect of alkaline phosphatase by β-glycerophosphate. Nature (London), *219*:73-75, 1968.

6. Bell, J. L., Shaldon, S., and Baron, D. N.: Serum isocitrate dehydrogenase in liver disease and some other conditions. Clin. Sci., *23*:57-77, 1962.

7. Berk, J. E., and Fridhandler, L.: Hyperamylasemia: Interpretation and newer approaches to evaluation. Adv. Intern. Med., *26*:235-264, 1980.

8. Beutler, E., and Kuhl, W.: An assay for terminal deoxynucleotidyl transferase in leukocytes and bone marrow. Am. J. Clin. Pathol., *70*:733-737, 1978.

9. Birkett, D. J., Conyers, R. A. J., Neale, F. C., et al.: Action of urea on human alkaline phosphatases with a description of some automated techniques for the study of enzyme kinetics. Arch. Biochem. Biophys., *124*:470-479, 1967.

10. Bodansky, O.: Acid phosphatase. Adv. Clin. Chem., *15*:44-136, 1972.

11. Bodansky, O., and Schwartz, M. K.: 5′-Nucleotidase. Adv. Clin. Chem., *11*:277-328, 1968.

12. Borgström, A., and Ohlsson, K.: Radioimmunological determination and characterization of cathodal trypsin-like immunoreactivity in normal human plasma. Scand. J. Clin. Lab. Invest., *36*:809-814, 1976.

13. Borgström, A., Sveger, T., Lindberg, T., et al.: Immunoreactive trypsin, chymotrypsin, and pancreatic secretory trypsin inhibitor in cord blood from infants with cystic fibrosis. Acta Pediatr. Scand., *70*:619-621, 1981.

14. Borgström, B., Erlanson-Albertsson, C., and Wieloch, T.: Pancreatic colipase: Chemistry and physiology. J. Lipid Res., *20*:805-816, 1979.

15. Bowers, G. N., Jr., and McComb, R. B.: A continuous spectrophotometric method for measuring the activity of serum alkaline phosphatase. Clin. Chem., *12*:70-89, 1966.

16. Broderick, J. W., Geokas, M. C., Largman, C., et al.: Molecular forms of immunoreactive pancreatic cationic trypsin in pancreatitis patient sera. Am. J. Physiol., *237*:E474-E480, 1979.

17. Brown, S. S., Kalow, W., Pilz, W., et al.: The plasma cholinesterases: A new perspective. Adv. Clin. Chem., *22*:1-123, 1981.

18. Bücher, T., Dunnweld, M., Fuhrer, S., et al.: Quantitative evaluation of serum alkaline phosphatase isozyme patterns. Abstracts of communications, joint meeting of Assoc. of Clin. Biochemists, U.K., Netherlands Soc. for Clin. Chem., and German Soc. for Clin. Chem., Newcastle upon Tyne, U.K., April, 1983.

19. Campbell, D. M.: Determination of 5′-nucleotidase in blood serum. Biochem. J., *84*:34P, 1962.

20. Cello, J. P.: Carcinoma of the pancreas. *In*: Gastrointestinal Disease: Pathophysiology, Diagnosis, Management. 3rd ed. M. H. Sleisenger and J. S. Fordtran, Eds. Philadelphia, W.B. Saunders Co., 1983, pp. 1514-1527.

21. Cibull, M. L., Coleman, M. S., Nelson, O., et al.: Evaluation of methods of detecting terminal deoxynucleotidyl transferase in human hematologic malignancies. Am. J. Clin. Pathol., *77*:420-423, 1982.

22. Coleman, M. S., and Hutton, J. J.: Terminal transferase. *In*: Methods in Hematology, Vol. 2, The Leukemic Cell. D. Catovsky, Ed. Edinburgh, Churchill Livingstone, 1981.

23. Commission on Biochemical Nomenclature, IUPAC-IUB: Nomenclature of multiple forms of enzymes. J. Biol. Chem., *252*:5939-5941, 1977.

24. Committee on Enzymes of the Scandinavian Society for Clinical Chemistry and Clinical Physiology: Recommended method for the determination of creatine kinase in blood modified by the inclusion of EDTA. Scand. J. Clin. Lab. Invest., *39*:1-5, 1979.

25. Committee on Enzymes of the Scandinavian Society for Clinical Chemistry and Clinical Physiology: Recommended method for the determination of γ-glutamyl transferase in blood. Scand. J. Clin. Lab. Invest., *36*:119-125, 1976.

26. Deibel, M. R., Jr., and Coleman, M. S.: Biochemical properties of purified human terminal deoxynucleotidyl transferase. J. Biol. Chem., *255*:4206-4212, 1980.

27. Demetriou, J. A., Drewes, P. A., and Gin, J. B.: Enzymes. *In*: Clinical Chemistry, Principles and Technics. 2nd ed. R. J. Henry, D. C. Cannon, and J. W. Winkelman, Eds. Hagerstown, Md., Harper and Row, 1974, pp. 815-1101.

28. Desjarlais, F., Morin, L. G., and Daigneault, R.: In search of optimum conditions for the measurement of creatine kinase activity: A critical review of nineteen formulations. Clin. Biochem., *13*:116-121, 1980.

29. Deutsches Gesellschaft für Klinische Chemie: Standardisierung von Methoden zur Bestimmung von Enzymaktivitäten in biologischen Flüssigkeiten. Z. Klin. Chem. Klin. Biochem., *10*:182-192, 1972.

30. Dietz, A. A., Rubinstein, H. H., and Lubrano, T.: Colorimetric determination of serum cholinesterase and its genetic variants by the propionylthiocholine-dithiobis (nitrobenzoic acid) procedure. Clin. Chem., *19*:1309-1313, 1973.

31. Ellis, G., and Goldberg, D. M.: An improved manual and semiautomatic assay for NADP-dependent isocitrate dehydrogenase activity, with a description of some kinetic properties of human liver and serum enzyme. Clin. Biochem., 2:175-185, 1971.

32. Ellis, G., Goldberg, D. M., Spooner, R. J., et al.: Serum enzyme tests in diseases of the liver and biliary tree. Am. J. Clin. Pathol., 70:248-258, 1978.

33. Elser, R.: The measurement of creatine kinase. In: Clinical and Analytical Concepts in Enzymology. H. A. Homburger, Ed. Skokie, Ill., College of American Pathologists, 1983, pp. 141-156.

34. Erlanson, C., and Borgström, B: Tributyrine as a substrate for determination of lipase activity of pancreatic juice and small intestinal content. Scand. J. Gastroenterol., 5:293-295, 1970.

35. Ewen, L. M., and Spitzer, R. W.: Improved determination of prostatic acid phosphatase (sodium thymolphthalein monophosphate substrate). Clin. Chem., 22:627-632, 1976.

36. Expert Panel on Enzymes, Committee on Standards (IFCC): Approved recommendations of IFCC methods for the measurement of catalytic concentrations of enzymes, Part 1. General Considerations. Clin. Chim. Acta, 98:163F-174F, 1979.

37. Expert Panel on Enzymes, IFCC: IFCC methods for the measurement of catalytic concentration of enzymes. Part 5. IFCC method for alkaline phosphatase. Clin. Chim. Acta, 135:339F-367F, 1983; and J. Clin. Chem. Clin. Biochem., 21:731-748, 1983.

38. Fast, D. M., Sampson, E. J., Whitner, V. S., et al.: Creatine kinase response surfaces explored by use of factorial experiments and simplex maximization. Clin. Chem., 29:793-799, 1983.

39. Fishman, W. H., Inglis, N. I., and Krant, M. J.: Serum alkaline phosphatase of intestinal origin in patients with cancer and with cirrhosis of the liver. Clin. Chim. Acta, 12:298-303, 1965.

40. Fishman, W. H., Inglis, N. R., Green, S., et al.: Immunology and biochemistry of Regan isoenzyme of alkaline phosphatase in human cancer, Nature (London), 219:697-699, 1968.

41. Fridhandler, L., and Berk, J. E.: Macroamylasemia. Adv. Clin. Chem., 20:267-286, 1978.

42. Fridhandler, L., Berk, J. E., Montgomery, K. A., et al.: Column-chromatographic studies of isoamylases in human serum, urine and milk. Clin. Chem., 20:547-552, 1974.

43. Friedel, R., Diederichs, F., and Lindena, J.: Release and extracellular turnover of intracellular enzymes. In: Advances in Clinical Enzymology. E. Schmidt, F. W. Schmidt, I. Trautschold, et al., Eds. Basel, Karger, 1979, pp. 70-105.

44. Geokas, M. C., Largman, C., Brodrick, J. W., et al.: Determination of human pancreatic cationic trypsinogen in serum by radioimmunoassay. Am. J. Physiol., 236:E77-E83, 1979.

45. Geokas, M. C., Largman, C., Brodrick, J. W., et al.: Immunoreactive forms of human pancreatic chymotrypsin in normal plasma. J. Biol. Chem., 254:2775-2781, 1979.

46. Gerhardt, W., and Wulff, K.: Creatine kinase. In: Methods of Enzymatic Analysis. 3rd ed. H. U. Bergmeyer, J. Bergmeyer, and M. Grassl, Eds. Weinheim, Verlag-Chemie, 1983, Vol. 3, pp. 508-539.

47. Gohara, W. F.: Rate of decrease of glutamyltransferase and acid phosphatase activities in the human vagina after coitus. Clin. Chem., 26:254-257, 1980.

48. Goldberg, D. M.: Enzymes and isoenzymes in the evaluation of diseases of the pancreas. In: Clinical and Analytical Concepts in Enzymology. H. A. Homburger, Ed. Skokie, Ill., College of American Pathologists, 1983, pp. 31-55.

49. Goldberg, D. M.: Structural, functional and clinical aspects of gamma-glutamyltransferase. CRC Crit. Rev. Clin. Lab. Sci., 12:1-58, 1980.

50. Goldberg, D. M., and Ellis, G.: Isocitrate dehydrogenase. In: Methods of Enzymatic Analysis. 3rd ed. H. U. Bergmeyer, J. Bergmeyer, and M. Grassl, Eds. Weinheim, Verlag Chemie, 1983, Vol. 3, pp. 183-190.

51. Goldberg, J. M., and Pagast, P.: Evaluation of lipase activity in serum by radial enzyme diffusion. Clin. Chem., 22:633-637, 1976.

52. Grendell, J. H., and Cello, J. P.: Chronic pancreatitis. In: Gastrointestinal Disease: Pathophysiology, Diagnosis, Management. 3rd ed. M. H. Sleisenger and J. S. Fordtran, Eds. Philadelphia, W. B. Saunders Co., 1983, pp. 1485-1514.

53. Grenner, G., Deutsch, G., Schmidtberger, R., et al.: A highly sensitive enzyme immunoassay for the determination of pancreatic lipase. J. Clin. Chem. Clin. Biochem., 20:515-519, 1982.

54. Griffiths, J. C.: The laboratory diagnosis of prostatic adenocarcinoma. CRC Crit. Rev. Clin. Lab. Sci., 19:187-204, 1983.

55. Hägele, E-O., Schaich, E., Rauscher, E., et al.: Mechanism of action of human pancreatic and salivary α-amylase on α-4-nitrophenyl maltoheptaoside substrate. Clin. Chem., 28:2201-2205, 1982.

56. Haverback, B. J., Dyce, B. J., Gutentag, P. J., et al.: Measurement of trypsin and chymotrypsin in stool. Gastroenterology, 44:588-597, 1963.

57. Henderson, A. R.: Clinical enzymology. In: Clinical Biochemistry Reviews. D. M. Goldberg, Ed. New York, John Wiley & Sons, 1981, Vol. 2, pp. 185-219.

58. Henderson, A. R.: Lactate dehydrogenase isoenzymes. In: Methods of Enzymatic Analysis. 3rd ed. H. U. Bergmeyer, J. Bergmeyer, and M. Grassl, Eds. Weinheim, Verlag-Chemie, 1983, Vol. 3, pp. 138-155.

59. Henderson, A. R., and Nealon, D. A.: Clinical enzymology. In: Clinical Biochemistry Reviews. D. M. Goldberg, Ed. New York, John Wiley & Sons, 1982, Vol. 3, pp. 187-234.

60. Henderson, A. R., and Nealon, D. A.: Enzyme measurements by mass: An interim review of the clinical efficacy of some mass measurements of prostatic acid phosphatase and the isoenzymes of creatine kinase. Clin. Chim. Acta, 115:9-32, 1981.

61. Henderson, N. S.: Intracellular location and genetic control of isozymes of NADP-dependent isocitrate dehydrogenase and malate dehydrogenase. Ann. N.Y. Acad. Sci., 151:429-440, 1968.

62. Henry R. J., and Chiamori, N.: Study of the saccharogenic method for the determination of serum and urine amylase. Clin. Chem., 6:434-452, 1960.

63. Hockeborn, M., and Rick, W.: Determination of lipase by continuous titration. J. Clin. Chem. Clin. Biochem., 20:773-785, 1982.

64. Hoffmann, G. E., and Weiss, L.: Specific serum pancreatic lipase determination, with use of purified colipase. Clin. Chem., 26:1732-1733, 1980.

65. Hørder, M., and Rej, R.: Alanine transaminase. In: Methods of Enzymatic Analysis. 3rd ed. H. U. Bergmeyer, J. Bergmeyer, and M. Grassl, Eds. Weinheim, Verlag-Chemie, 1983, Vol. 3, pp. 444-456.

66. Horecker, B. L.: Biochemistry of isozymes. In: Isozymes. C. L. Markert, Ed. New York, Academic Press, 1975, Vol. 1, pp. 11-38.

67. Huang, W. Y., and Tietz, N. W.: Determinations of amylase isoenzymes in serum by use of a selective inhibitor. Clin. Chem., 28:1525-1527, 1982.

68. Hutton, J. J., Coleman, M. S., Moffit, S., et al.: Prognostic significance of terminal transferase activity in childhood acute lymphoblastic leukemia: A prospective analysis of 164 patients. Blood, 60:1267-1276, 1982.

69. IFCC Expert Panel on Enzymes: IFCC methods for the measurement of the catalytic concentration of enzymes, Part 4. IFCC method for γ-glutamyl transferase. J. Clin. Chem. Clin. Biochem., 21:633-646, 1983.

70. IUPAC Commission on Quantities and Units and IFCC Expert Panel on Quantities and Units: Approved Recommendations (1978). Quantities and units in clinical chemistry. Clin. Chim. Acta, 96:157F-183F, 1979.

71. Iwaki K., Ogawa, M., Tanaka, S., et al.: Radioimmunoassay for human pancreatic chymotrypsin and measurement of serum immunoreactive chymotrypsin contents in various diseases. Res. Commun. Chem. Pathol. Pharmacol., 40:489-496, 1983.

72. Jung, K., Sokolowski, A., and Egger, E.: An optimized assay of human serum glutamate dehydrogenase activity. Enzyme, 14:44-54, 1973.

73. Kachmar, J. F., and Moss, D. W.: Enzymes. In: Fundamentals of Clinical Chemistry. 2nd ed. N. W. Tietz, Ed. Philadelphia, W. B. Saunders Co., 1976, pp. 565-698.

74. Kaneda, T., Kuroda, S., Hirota, Y., et al.: Highly sensitive solid-phase enzyme immunoassay for terminal deoxynucleotidyl transferase. Anal. Biochem., 126:327-334, 1982.

75. Kaspar, P., Moeller, G., Wahlefeld, A. W., et al.: A new photometric method for the determination of chymotrypsin in stool. Fresenius Z. Anal. Chem., 311:391-392, 1982.

76. Kaufman, R. A., and Tietz, N. W.: Recent advances in measurement of amylase activity—a comparative study. Clin. Chem., 26:846-853, 1980.

77. Krause, R. D., and Lott, J. A.: Use of the simplex method to optimize analytical conditions in clinical chemistry. Clin. Chem., 20:775-782, 1974.

78. Kristensen, S. R., and Hørder, M.: Release and turnover of intracellular enzymes. In: Clinical and Analytical Concepts in Enzymology. H. A. Homburger, Ed. Skokie, Ill., College of American Pathologists, 1983, pp. 1-13.

79. Kurooka, S., and Kitamura, T.: Properties of serum lipase in patients with various pancreatic diseases. J. Biochem., 84:1459-1466, 1978.

80. Lang, H., Ed.: Creatine Kinase Isoenzymes. Berlin, Springer-Verlag, 1981.

81. Lang, H., and Wurzburg, U.: Creatine kinase, an enzyme of many forms. Clin. Chem., 28:1439-1447, 1982.

82. Lankisch, P. G.: Exocrine pancreatic function tests. Gut, 23:777-798, 1982.

83. Lantz, R. K., and Eisenberg, R. B.: Preservation of acid phosphatase activity in medico-legal specimens. Clin. Chem., 24:486-488, 1978.

84. Largman, C., Brodrick, J. W., Geokas, M. C., et al.: Demonstration of human pancreatic anionic trypsinogen in normal serum by radioimmunoassay. Biochim. Biophys. Acta, 543:450-451, 1978.

85. Legaz, M. E., and Kenney, M. A.: Electrophoretic amylase fractionation as an aid in diagnosis of pancreatic disease. Clin. Chem., 22:57-62, 1976.

86. Lehmann, F. G.: Immunological relationship between human placental and intestinal alkaline phosphatase. Clin. Chim. Acta, 65:257-269, 1975.

87. Leung, F. Y., and Henderson, A. R.: Thin-layer agarose electrophoresis of lactate dehydrogenase isoenzymes in serum: A note on the method of reporting and on the lactate dehydrogenase isoenzyme-1/isoenzyme-2 ratio in acute myocardial infarction. Clin. Chem., 25:209-211, 1979.

88. Levitt, M. D., Ellis, C., and Engel, R. R.: Isoelectric focusing studies of human serum and tissue isoamylases. J. Lab. Clin. Med., 90:141-152, 1977.

89. Lorentz, K.: α-Amylase assay: Current state and future development. J. Clin. Chem. Clin. Biochem., 17:499-504, 1979.

90. Lott, J. A., and Stang, J. M.: Serum enzymes and isoenzymes in the diagnosis and differential diagnosis of myocardial ischemia and necrosis. Clin. Chem., 26:1241-1250, 1980.

91. Mangum, B. W., and Thornton, D. D., Eds.: Standard Reference Materials: The Gallium Melting Point Standard. Washington, D.C., NBS Special Publication 481, 1977.

92. McComb, R. B.: The measurement of lactate dehydrogenase. In: Clinical and Analytical Concepts in Enzymology. H. A. Homburger, Ed. Skokie, Ill., College of American Pathologists, 1983, pp. 157-171.

93. McComb, R. B., Bowers, G. N., Jr., and Posen, S.: Alkaline Phosphatase. New York, Plenum Press, 1979.

94. McKenzie, D., and Henderson, A. R.: Electrophoresis of lactate dehydrogenase isoenzymes. In: Selected Methods of Clinical Chemistry. G. R. Cooper, Ed. Washington, D.C., American Association for Clinical Chemistry, 1983, Vol. 10, pp. 58-67.

95. Mercer, D. W.: Improved column method for separating lactate dehydrogenase isoenzymes 1 and 2. Clin. Chem., 24:480-482, 1978.

96. Merritt, A. D., and Karn, R. C.: The human α-amylases. Adv. Hum. Genet., 8:135-234, 1977.

97. Meyer, J. H.: Pancreatic physiology. In: Gastrointestinal Disease: Pathophysiology, Diagnosis, Management. 3rd ed. M. H. Sleisenger and J. S. Fordtran, Eds. Philadelphia, W. B. Saunders Co., 1983, pp. 1426-1436.

98. Modak, M. J., et al.: A micro method for determination of terminal deoxynucleotidyl transferase (TdT) in the diagnostic evaluation of acute leukemias. J. Cancer Res. Clin. Oncol. *98*:91-104, 1980.
99. Morin, L. G.: Evaluation of current methods for creatine kinase isoenzyme fractionation. Clin. Chem., *23*:205-210, 1977.
100. Moss, D. W.: Alkaline phosphatase isoenzymes. Clin. Chem., *28*:2007-2016, 1982.
101. Moss, D. W.: Automatic enzyme analyzers. Adv. Clin. Chem., *19*:1-56, 1977.
102. Moss, D. W.: Isoenzyme Analysis. London, The Chemical Society, 1979.
103. Moss, D. W.: Isoenzymes. London and New York, Chapman and Hall, 1982.
104. Moss, D. W., and Edwards, R. K.: Improved electrophoretic resolution of bone and liver alkaline phosphatases resulting from partial digestion with neuraminidase. Clin. Chim. Acta, *143*:177-182, 1984.
105. Moss, D. W., and King, E. J.: Properties of alkaline phosphatase fractions separated by starch-gel electrophoresis. Biochem. J., *84*:192-195, 1962.
106. Moss, D. W., Shakespeare, M. J., and Thomas, D. M.: Observations on the heat-stability of alkaline phosphatase isoenzymes in serum. Clin. Chim. Acta, *40*:35-41, 1972.
107. Moss, D. W., and Whitby, L. G.: A simplified heat-inactivation method for investigating alkaline phosphatase isoenzymes in serum. Clin. Chim. Acta, *61*:63-71, 1975.
108. Myrtle, J. F., and Zell, W. J.: Simplified photometric copper soap method for rapid assay of serum lipase activity. Clin. Chem., *21*:1469-1473, 1975.
109. Neale, F. C., Clubb, J. S., Hotchkis, D., et al.: Heat stability of human placental alkaline phosphatase. J. Clin. Pathol., *18*:359-363, 1965.
110. Nomenclature Committee, IUB: Enzyme Nomenclature, 1978. Recommendations of the Nomenclature Committee of IUB on the Nomenclature and Classification of Enzymes. New York, Academic Press, 1979.
111. O'Donnell, M. D., Fitzgerald, O., and McGeeney, K. F.: Differential serum amylase determination by use of an inhibitor, and design of a routine procedure. Clin. Chem., *23*:560-566, 1977.
112. Persijn, J. P., and van der Slik, W.: A new method for the determination of γ-glutamyltransferase in serum. J. Clin. Chem. Clin. Biochem., *18*:421-427, 1976.
113. Persijn, J. P., van der Slik, W., Kramer, K., et al.: A new method for the determination of serum nucleotidase. Z. Klin. Chem. Klin. Biochem., *6*:441-446, 1968.
114. Pinto, P. V. C., Kaplan, A., and Van Dreal, P. A.: Aldolase. II. Spectrophotometric determination using an ultraviolet procedure. Clin. Chem., *15*:349-360, 1969.
115. Pinto, P. V. C., Van Dreal, P. A., and Kaplan, A.: Aldolase. I. Colorimetric determination. Clin. Chem., *15*:339-348, 1969.
116. Rautela, G. S., Snee, R. D., and Miller, W. K.: Response-surface co-optimization of reaction conditions in clinical chemical methods. Clin. Chem., *25*:1954-1964, 1979.
117. Rej, R., and Hørder, M.: Aspartate aminotransferase. *In*: Methods of Enzymatic Analysis. 3rd ed. H. U. Bergmeyer, J. Bergmeyer, and M. Grassl, Eds. Weinheim, Verlag-Chemie, 1983, Vol. 3, pp. 416-433.
118. Rej, R., and Vanderlinde, R. E.: Effects of temperature on the steady-state kinetics and measurement of aspartate aminotransferases. Clin. Chem., *27*:213-219, 1981.
119. Rietz, B., and Guilbault, G. G.: Fluorimetric method for measuring serum lipase activity. Clin. Chem., *21*:1788-1790, 1975.
120. Robinson, D. B., and Glew, R. H.: Acid phosphatase in Gaucher's disease. Clin. Chem., *26*:371-382, 1980.
121. Robinson, J. C., and Pierce, J. E.: Differential action of neuraminidase on human serum alkaline phosphatase. Nature (London), *204*:472-473, 1964.
122. Romas, N. A., Rose, N. R., and Tannenbaum, M.: Acid phosphatase: New developments. Hum. Pathol., *10*:501-512, 1979.
123. Rosalki, S. B.: Gamma-glutamyl transpeptidase. Adv. Clin. Chem., *17*:53-107, 1975.
124. Rosalki, S. B., and Foo, Y. A.: Two new methods for separating and quantifying bone and liver alkaline phosphatase isoenzymes in plasma. Clin. Chem., *30*:1182-1186, 1984.
125. Rosalki, S. B., and Tarlow, D.: Optimized determination of γ-glutamyl transferase by reaction-rate analysis. Clin. Chem., *20*:1121-1124, 1974.
126. Roy, A. V., Brower, M. E., and Hayden, J. E.: Sodium thymolphthalein monophosphate: A new acid phosphatase substrate with greater specificity for the prostatic enzyme in serum. Clin. Chem., *17*:1093-1102, 1971.
127. Rude, R. E., Muller, J. E., and Braunwald, E.: Efforts to limit the size of myocardial infarcts. Ann. Intern. Med., *95*:736-761, 1981.
128. Salt, W. B., and Schenker, S.: Amylase—its clinical significance: A review of the literature. Medicine, *55*:269-289, 1976.
129. Scandinavian Society for Clinical Chemistry and Clinical Physiology: Recommended Methods for the Determination of Four Enzymes in Blood. Scand. J. Clin. Lab. Invest., *33*:291-306, 1974.
130. Schmidt, E., and Schmidt, F. W.: Enzyme diagnosis in diseases of the liver and biliary system. *In*: Advances in Clinical Enzymology. F. W. Schmidt, E. Schmidt, I. Trautschold, et al., Eds. Basel, Karger, 1979, pp. 239-293.
131. Schmidt, E., and Schmidt, F. W.: Glutamate dehydrogenase. *In*: Methods of Enzymatic Analysis. 3rd ed. H. U. Bergmeyer, J. Bergmeyer, and M. Grassl, Eds. Weinheim, Verlag Chemie, 1983, Vol. 3, pp. 216-227.
132. Shaw, L. M.: The specificity of the immunochemical determination of prostatic acid phosphatase. *In*: Clinical and Analytical Concepts in Enzymology. H. A. Homburger, Ed. Skokie, Ill., College of American Pathologists, 1983, pp. 57-68.
133. Silk, E., King, J., and Whittaker, M.: Assay of cholinesterase in clinical chemistry, Ann. Clin. Biochem., *16*:57-75, 1979.
134. Skude, G.: Human amylase isoenzymes. Scand. J. Gastroent., *12*: Suppl. 44, 1-36, 1977.
135. Soergel, K. H.: Acute pancreatitis. *In*: Gastrointestinal Disease: Pathophysiology, Diagnosis, Management. 3rd ed. M. H. Sleisenger and J. S. Fordtran, Eds. Philadelphia, W. B. Saunders Co., 1983, pp. 1462-1485.

136. Speicher, C. E., and Smith, J. W.: Cardiovascular Subproblems. *In*: Choosing Effective Laboratory Tests. Philadelphia, W. B. Saunders Co., 1983, pp. 155-182.
137. Stagg, B. H., and Wood, T. P.: Radioimmunoassay of trypsin: A new aid in the assessment of pancreatic function. Ann. Clin. Biochem., *16*:147-151, 1979.
138. Štěpán, J., and Škrha, J., Measurement of amylase isoenzymes in human sera and urine using a DEAE-cellulose mini-column method. Clin. Chim. Acta, *91*:263-271, 1979.
139. Sujimoto, M., and Bollum, F. J.: Terminal deoxynucleotidyl transferase (TdT) in chick-embryo lymphoid tissues. J. Immunol., *122*:392-397, 1979.
140. Thompson, W. H. S.: The clinical chemistry of muscular dystrophies. Adv. Clin. Chem., *7*:138-197, 1964.
141. Tietz, N. W., and Astle, J. R.: Personal communication, 1984.
142. Tietz, N. W., and Fiereck, E. A.: Measurement of lipase activity in serum. *In*: Standard Methods of Clinical Chemistry. G. R. Cooper, Ed. New York, Academic Press, 1972, Vol. 7, pp. 19-31.
143. Tietz, N. W., and Finley, P. R., Eds.: Clinical Guide to Laboratory Tests. Philadelphia, W. B. Saunders Co., 1983.
144. Tietz, N. W., and Repique, E. V.: Proposed standard method for measuring lipase activity in serum by a continuous sampling technique. Clin. Chem., *19*:1268-1275, 1973.
145. Tietz, N. W., and Rinker, A. D.: The measurement of alkaline phosphatase (ALP). *In*: Analytical Concepts in Enzymology. H. A. Homburger, Ed. Skokie, Ill., College of American Pathologists, 1983, pp. 195-203.
146. Usategui-Gomez, M., Wicks, R. W., and Warshaw, M.: Immunochemical determination of the heart isoenzymes of lactate dehydrogenase (LDH_1) in human serum. Clin. Chem., *25*:729-734, 1979.
147. Van Belle, H.: Alkaline phosphatase, 1. Kinetics and inhibition by levamisole of purified isoenzymes from humans. Clin. Chem., *22*:972-976, 1976.
148. Vassault, A., Wahlefeld, A. W., and Deneke, U.: Lactate dehydrogenase. *In*: Methods of Enzymatic Analysis, Vol. 3. 3rd ed. H. U. Bergmeyer, J. Bergmeyer, and M. Grassl, Eds. Weinheim, Verlag-Chemie, 1983, pp. 118-138.
149. Wajsman, Z., Chu, T. M., Saroff, J., et al.: Two new, direct, and specific methods of acid phosphatase determination. Urology, *13*:8-11, 1979.
150. Watts, C.: Serum isocitrate dehydrogenase in obstructive jaundice. Clin. Chim. Acta, *14*:177-184, 1966.
151. Wilkinson, J. H.: The Principles and Practice of Diagnostic Enzymology. London, Edward Arnold, 1976.
152. Williamson, T.: The estimation of pancreatic lipase—a brief review. Med. Lab. Sci., *33*:265-279, 1976.
153. Wolfson, S. K., and Williams-Ashman, H. G.: Isocitric and 6-phosphogluconic dehydrogenases in human blood serum. Proc. Soc. Exp. Biol. Med., *96*:231-234, 1957.
154. Yam, L. T.: Clinical significance of the human acid phosphatases: A review. Am. J. Med., *56*:604-616, 1974.
155. Ziegenhorn, J., Neumann, U., Knitsch, K. W., et al.: Determination of serum lipase. Clin. Chem., *25*:1067, 1979.
156. Zygowicz, E. L., Sunderman, F. W., Jr., Horak, E., et al.: Inhibition by concanavalin A as the basis for a specific assay of serum 5'-nucleotidase activity. Clin. Chem., *23*:2311-2323, 1977.

THE LYSOSOMAL ENZYMES

by John F. O'Brien, Ph.D.

"Inborn errors of metabolism" is a general term that is applied to numerous genetic disorders whose pathology is usually attributable to excessive tissue stores or circulating concentrations of a specific undegraded metabolite. The lysosomal storage diseases, which are summarized in Table 5-12, are a distinct subset of inborn errors. These diseases result from accumulation in lysosomes of metabolites that would normally be degraded by one of the many hydrolytic enzymes that reside in these subcellular organelles. The lysosomal enzymes are unique inasmuch as they have acid pH optima. The specific enzyme deficiencies of the lysosomal storage diseases have been elucidated by identifying products stored in tissues and by their metabolites found in urine. The nomenclature of enzymes, substrates, storage products, and metabolites is not always systematic. The general names for the storage products are sphingolipids or glycosaminoglycans, the latter having been previously known as mucopolysaccharides.

In most cases, demonstration of enzyme deficiency by an assay using an appropriate non-physiological substrate is sufficient for diagnostic purposes. Some lysosomal enzyme deficiencies can be detected by assay of a serum specimen; others require a specimen consisting of cells that contain lysosomes, such as leukocytes, fibroblasts, or amniotic fluid cells. Erythrocytes lack lysosomes and thus are unsuitable for assay purposes. In affected cells, the deficiency of a lysosomal

Table 5-12. LYSOSOMAL STORAGE DISORDERS

Storage Disease	Clinical Features
I. SPHINGOLIPIDOSES	
Niemann-Pick disease (*Sphingomyelin lipidosis*) Deficient enzyme: Sphingomyelinase Method reference: 6 *Comments*: Types C and D are grouped with Niemann-Pick largely on clinical and histologic grounds; an absolute deficiency of sphingomyelinase has not been demonstrated; type C may result from a deficiency or structural alteration in a sphingomyelinase activator protein	*Type A*: Foamy histiocytes in bone marrow; acute neuropathic progressive loss of motor and intellectual capacity early in life; hepatosplenomegaly; often a cherry-red macula; commonly fatal in infancy *Type B*: Foamy histiocytes in bone marrow; chronic form with no nervous system involvement; hepatosplenomegaly *Type C*: Similar to type A except later onset *Type D*: Nova Scotia variant similar to type C Autosomal recessive inheritance
Gaucher's disease (*Glucosyl ceramide lipidosis*) Deficient enzyme: β-Glucocerebrosidase Method reference: 21	*Type 1*: Adult (chronic nonneuropathic): hepatosplenomegaly; Gaucher's cells; anemia; elevated acid phosphatase; bone pain and lytic lesions *Type 2*: Infantile (acute neuropathic): rapidly degenerative central nervous system manifestations; peripheral symptoms similar to type 1 but greatly exaggerated; death usually occurs by 1 year of age Autosomal recessive inheritance
Krabbe's disease (*Galactosylceramide lipidosis, globoid cell leukodystrophy*) Deficient enzyme: Galactosylceramide-β-galactosidase Method reference: 23	A number of forms are recognized, differing largely in age at onset and severity of symptoms; progressive psychomotor retardation; globoid cells in the central nervous system; spastic quadriparesis; hypertonicity; hyperthermia; elevated cerebrospinal fluid protein Autosomal recessive inheritance
Metachromatic leukodystrophy (*Sulfatide lipidosis*) Deficient enzyme: Arylsulfatase A Method references: 1,2	Several closely related disorders with differing ages at onset from 1 year of age well into adulthood; peripheral neuropathy; intermittent pain in arms and legs with eventual difficulty sitting; gait disturbances; absence of deep tendon reflexes; plantar flexion of feet; adult form—slowly progressive dementia; often confused with nonorganic psychoses Autosomal recessive inheritance
Fabry's disease (*Angiokeratoma corporis diffusum universale*) Deficient enzyme: α-Galactosidase A Method reference: 4	Severe pain in extremities; angiokeratoma on buttocks and around navel; tortuous, dilated conjunctival and retinal venules; neuropathy; hypertension; myocardial ischemia; female carriers may show manifestations in later life X-linked inheritance
G_{M2} gangliosidoses: **Tay-Sachs disease** (*Infantile onset*) Deficient enzyme: β-N-Acetylglucosaminidase A	Early motor weakness; psychomotor deterioration after 1 year of age; progressive deafness; blindness; startle response; red macula; primarily found in families with Ashkenazi Jewish ancestry
Bernheimer-Seitelberger disease (*Juvenile onset*) Deficient enzyme: β-N-Acetylglucosaminidase A	Similar to Tay-Sachs in expression except for later age of onset and infrequency of red macula
Adult—Chronic gangliosidosis (*Juvenile—adult onset*) Deficient enzyme: β-N-Acetylglucosaminidase A	Slowly progressive deterioration of gait; muscle atrophy; ataxia of limbs; no dementia; unlike other forms of Tay-Sachs, this form has lateness of onset and absence of dementia
Sandhoff's disease (*A B variant*) Deficient enzyme: β-N-Acetylglucosaminidase A *and* B Method reference: 16	Similar to Tay-Sachs but with mild peripheral neuropathy, mild organomegaly, lipid accumulation in viscera, particularly the asialia form of G_{M2} (globoside), no predilection toward Jewish ancestry Autosomal recessive inheritance

Table 5-12. LYSOSOMAL STORAGE DISORDERS *Continued*

Storage Disease	Clinical Features
I. SPHINGOLIPIDOSES (*continued*)	
Generalized gangliosidosis (*G$_{MI}$ gangliosidosis*) Deficient enzyme: β-Galactosidase Method references: 9,12	*Type 1*: Severe mental retardation; seizures and muscle hypotonicity apparent in first few months of life; dysostosis multiplex; foam cells in marrow *Type 2*: Regression of normal motor development apparent after age 1; seizures and rapidly progressive spasticity follow after first symptoms become apparent; mild dysostosis multiplex Autosomal recessive inheritance
Ganglioside sialidase deficiency Deficient enzyme: Gangliosidesialidase (tentative)	Progressive psychomotor retardation, corneal opacities Autosomal recessive inheritance
II. MUCOPOLYSACCHARIDOSES (MPS)*^{5}	
Hurler's syndrome (*MPS I*) Deficient enzyme: α-L-Iduronidase Method reference: 10	Progressive mental and physical debilitation beginning at age 1; corneal opacities; coarse facies; gingival hyperplasia; dysostosis multiplex; stiff joints (clawhands); dwarfing; organomegaly Autosomal recessive inheritance
Scheie's syndrome (*MPS Ia*) Deficient enzyme: α-L-Iduronidase Method reference: 10	A mild form of MPS I with corneal opacity; mild or absent mental retardation; clawhand deformity; aortic stenosis Autosomal recessive inheritance
Hunter's syndrome (*MPS II*) Deficient enzyme: Iduronate sulfatase Method reference: 10	Dysostosis multiplex essentially the same as in MPS I; mental retardation in the severe forms (syndrome runs gamut from severe to mild); along with the genetics, the distinguishing characteristic from MPS I is lack of corneal clouding, and there is a longer life span of even the severest form X-linked inheritance
Sanfilippo's syndrome (*MPS IIIa*) Deficient enzyme: Heparan *N*-sulfatase Method reference: 8	Behavioral problems progressing to severe mental retardation; comparatively mild or nonexistent connective tissue abnormalities; pronounced hirsutism Autosomal recessive inheritance
Sanfilippo's syndrome (*MPS IIIb*) Deficient enzyme: α-*N*-Acetylglucosaminidase (α-hexosaminidase) Method reference: 24	Indistinguishable clinically from MPS IIIa Autosomal recessive inheritance
Sanfilippo's syndrome (*MPS IIIc*) Deficient enzyme: α-Glucosaminide-*N*-acetyltransferase	Indistinguishable clinically from MPS IIIa Autosomal recessive inheritance
Sanfilippo's syndrome (*MPS IIId*) Deficient enzyme: *N*-Acetylglucosamine-6-sulfate sulfatase	Indistinguishable clinically from MSP IIIa Autosomal recessive inheritance
Morquio's disease (*MPS IVa*) Deficient enzyme: Galactosamine-6-sulfate sulfatase	Pronounced skeletal anomalies with small stature (short trunk dwarfism); short neck; prominent lower ribs; odontoid anomalies; normal intellect Autosomal recessive inheritance

Table continued on following page

---

*All the mucopolysaccharidoses (except for Morquio's disease) and mucolipidoses II and III show delayed breakdown of mucopolysaccharides in cultured skin fibroblasts. This is a helpful screening adjunct, which can be followed by specific enzyme assays if available.

Mucopolysaccharides are now called glycosaminoglycans. However, the term mucopolysaccharidosis (MPS) is retained because of its frequent use in the clinical description of these diseases.

Table 5-12. LYSOSOMAL STORAGE DISORDERS *Continued*

Storage Disease	Clinical Features
II. MUCOPOLYSACCHARIDOSES (MPS) (*continued*)	
Morquio's disease (*MPS IVb*) Deficient enzyme: β-Galactosidase Method reference: 9	A mild form of IVa; spondyloepiphyseal dysplasia; short stature; cloudy corneas; normal intellect Autosomal recessive inheritance
Maroteaux-Lamy disease (*MPS VI*) Deficient enzyme: Arylsulfatase-B Method reference: 1	Severe to mild dysostosis multiplex; gross corneal opacity; retardation of growth; normal intellect Autosomal recessive inheritance
Glucuronidase deficiency (*MPS VII*) Deficient enzyme: β-Glucuronidase Method reference: 7	Mild mental retardation; somewhat coarse facies; gingivitis; organomegaly; sometimes with corneal clouding Autosomal recessive inheritance
III. DISORDERS OF UNDEFINED GLYCOPROTEIN AND/OR LIPID STORAGE PRODUCTS	
Fucosidosis Deficient enzyme: α-Fucosidase Method reference: 3	*Type 1*: Frequent respiratory infections; progressive psychomotor retardation; dysostosis multiplex ("Hurler-like"); thick skin with hypersecretion of sweat with elevated salinity; cardiomegaly *Type 2*: Distinguished from type 1 by its milder clinical presentation and absence of unusual sweating; skin lesions (angiokeratoma) are present Autosomal recessive inheritance
Mannosidosis Deficient enzyme: α-Mannosidase Method reference: 7	Mild hepatosplenomegaly; mild radiologic bone changes; psychomotor retardation; nerve deafness Autosomal recessive inheritance
IV. DISORDERS PRESUMED TO INVOLVE MULTIPLE STORAGE PRODUCTS	
Mucolipidosis I (*ML I, sialidosis*) Deficient enzyme: Neuraminidase Method reference: 17	*Type 1 (dysmorphic)*: Dysostosis multiplex; mental retardation; myoclonus; cherry-red spots of the macula; both infantile and juvenile onsets *Type 2 (normomorphic)*: Cherry-red spots of the macula; myoclonus; normal intelligence Autosomal recessive inheritance
Mucolipidosis II (*ML II*) (*I-cell disease*) Deficient enzyme: Biochemical defect is post-transcriptional modification which leads to absence of mannose-6-phosphate on a number of acid hydrolases which are required for proper sequestration into lysosomes	Early onset of Hurler-like symptoms (often noted at birth); gingival hyperplasia; thoracic deformities; hepatosplenomegaly Autosomal recessive inheritance
Mucolipidosis III (*ML III*) (*pseudo-Hurler polydystrophy*) Deficient enzyme: See ML II	Mild variant of ML II; slowly progressing Hurler-like features, particularly clawhand deformity; mild growth and mental retardation; joint stiffness with reduced mobility Autosomal recessive inheritance
Multiple sulfatase deficiency (*MSD*) Deficient enzyme: Deficiency of all lysosomal sulfatases to greater or lesser degrees (arylsulfatase A < arylsulfatase B) Method reference: 1	Similar to infantile metachromatic leukodystrophy but with the skeletal anomalies of mucopolysaccharidoses; hepatosplenomegaly; thickening of the skin; ichthyosis Autosomal recessive inheritance

enzyme is constant; it is not inducible or episodic. Multiple allelic forms exist for many of the lysosomal enzymes, and these may have differing kinetic or other functional properties that may be associated with variable expression of the clinical disease. One must expect that for each of these proteins there may be as many as or more variants than have been described for hemoglobin or glucose-6-phosphate dehydrogenase, and that the severity of disease may vary accordingly. The rare nature of the lysosomal storage diseases has made it impractical for most clinical laboratories to establish the technical expertise for performance of all the lysosomal enzymes.

Among the lysosomal storage diseases, *Tay-Sachs disease* (TSD) has been the most thoroughly studied. The biochemical defect responsible for TSD is a near total deficiency of the enzyme N-acetyl-β-hexosaminidase A (β-NAGS-A; Hex A).[18] This enzyme is responsible for the hydrolysis of the $\beta(1 \rightarrow 3)$-glycosidic bond between N-acetyl-galactosamine and galactose in the G_{M2} ganglioside (Figure 5-31). The absence of enzyme activity results in the neuronal storage of G_{M2} ganglioside and the attendant clinical description given in Table 5-12. Because of the storage of G_{M2} ganglioside, this disease, along with at least three other variants, is classified as a G_{M2} gangliosidosis.

The isoenzymes of β-NAGS that are present in highest concentration in human tissues and blood are commonly designated A and B. They differ in subunit structure; the A isoenzyme is a mixed multimer of α- and β-subunits ($\alpha_2\beta_2$ the most likely structure);[15] isoenzyme B is a homomultimer composed of four β-subunits. Because of this subunit structure, mutations in the α-chain give rise only to isoenzyme A deficiency while mutations in the β-subunit result in both A and B deficiency. Deficiency of both A and B isoenzyme results in the more rare *Sandhoff variant* of G_{M2} gangliosidosis.

The difference in structure of β-NAGS isoenzymes results in distinct physical properties of the two isoenzymes. β-NAGS A has an isoelectric point (pI) of 5.4, while β-NAGS B has a pI of 7.9. This property allows the separation of the two isoenzymes by electrophoresis and ion-exchange chromatography. The isoenzymes also differ in terms of heat stability. It is the latter difference that is most often used to estimate the activity of each of the isoenzymes in a mixture. (See method below.)

Recent developments in the area of lysosomal storage diseases have focused more on allelic mutations within a specific enzyme deficiency than with diseases caused by the deficiency of a unique system; deficiency of β-NAGS is a good example of this. Although Table 5-12 describes only four subtypes of G_{M2} gangliosidosis, at least six α-chain and three β-chain mutations have been suggested. In some instances, the variants can be distinguished by quantitative comparison of β-NAG A levels or by the fact that the mutation also involves the B isoenzyme. There are, however, instances when patients present with strikingly different symptoms yet the usual enzymatic investigation finds nothing quantitatively unique about the β-NAG deficiency. An example of this phenomenon is found by comparing typical infantile-onset G_{M2} gangliosidosis (Tay-Sachs disease) with adult, chronic G_{M2} gangliosidosis. In contrast, patients with neuronal G_{M2} ganglioside accumulation but with normal β-NAGS activity toward artificial substrates have been described. In these rare instances, the natural substrate must be used to prove the deficiency, which is in an activator subunit of β-NAGS A. The existence of numerous mutations in β-NAGS also probably results in compound heterozygotes who are deficient in functional β-NAGS. The complexity caused by allelism requires that laboratories involved in such testing be aware that in some instances assays using artificial substrates will be inadequate to assign an individual to

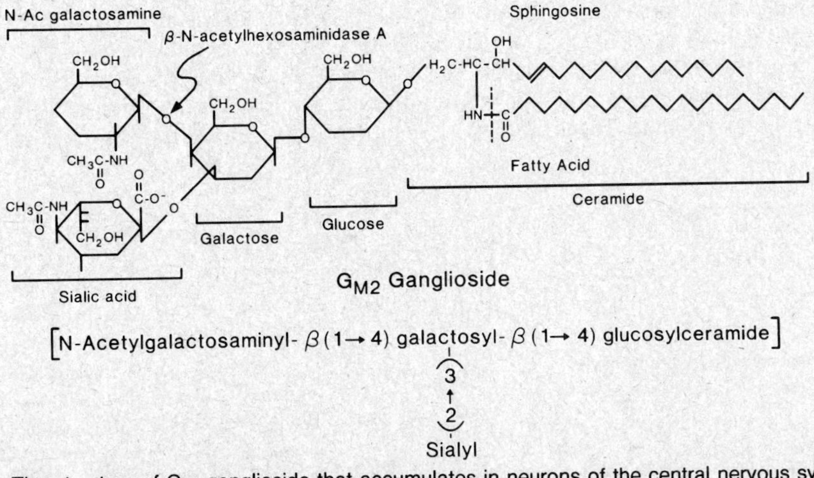

Figure 5-31. The structure of G_{M2} ganglioside that accumulates in neurons of the central nervous system in Tay-Sachs disease. The terminal monosaccharides are N-acetylgalactosamine and sialic acid. Both these monosaccharides are linked to galactose, which is, in turn, linked to the ceramide moiety through a glucose unit. Ceramide is a combination of sphingosine and an amide-linked fatty acid.

carrier or even homozygotic-affected status. This is particularly important in prenatal diagnosis for variants other than those resulting in total enzyme deficiency.

In the recessively inherited disorders (except for X-linked Fabry's disease), clinical manifestation is apparent only in homozygotes. However, unaffected heterozygotes can often be detected by enzyme assay. In laboratories where such assays are performed, detection of heterozygotes of some of these disorders has become an important function. The activities of lysosomal enzymes in heterozygotes are usually below the reference range but higher than those obtained in specimens from affected homozygotes. A frequently cited example is the identification of heterozygotes for the Tay-Sachs disease gene. Tay-Sachs disease is rare in the general population, but its increased incidence in Ashkenazi Jews—about 1 in 28 are carriers—makes screening for carriers in the subpopulation both practical and desirable. Genetic counseling of mates, both of whom have been identified as carriers, clarifies the risk of an affected infant. If the couple chooses to have natural children, fetal Tay-Sachs can be diagnosed using cells either cultured from amniotic fluid or obtained by biopsy of chorionic villi. Identification of carriers of Tay-Sachs disease has therefore become the most frequently requested function of laboratories engaged in enzymatic diagnosis of lysosomal storage disease.

It is more difficult to identify heterozygotes of X-chromosome–linked disorders than to identify those due to autosomal mutations. Because of the random inactivation of one or the other of the two X chromosomes in each cell nucleus early in embryonic life, assay of lysosomal enzyme activity may not distinguish from normals some women who are carriers for X-linked disorders. The only method by which such heterozygotes might be detected is by identifying coexistence of abnormal and normal cells by cloning cultured cells or by enzymatic analysis of hair follicles. These follicles are small clumps of cells that are purported to be in vivo clones. At present, such methods are not widely practiced, but it is conceivable that they may become practical in the future. Carrier identification of any of the lysosomal storage diseases may be advanced in the future by use of specific gene probes, as has been done for β-globin genes. This may be especially useful in carrier identification of X-linked disease.

Clinical Significance

Each disorder has its own characteristic clinical features; these cannot be briefly stated. A synopsis of clinical indications is presented in Table 5-12. Most of these disorders are expressed early in life. Generally, delayed or regressing psychomotor performance is among the first features noticed by the parents of the affected child. Bone development may be abnormal. Significant ophthalmological abnormalities occur in some of the lysosomal storage diseases. For a more complete description of the clinical presentation, biochemistry, and genetics of the lysosomal storage diseases, the reader is referred to reference 22.

METHODS

Lysosomal enzyme quantitation most often uses artificial glycosides or sulfate derivatives of 4-methylumbelliferone or p-nitrophenol as substrates. The former substrate yields a fluorescent product (Figure 5-32) and the latter yields the colored p-nitrophenol product that has long been common in enzymatic determinations. In addition to the specific enzyme tests that are now available for many of the lysosomal storage disorders, the demonstration of elevated urinary

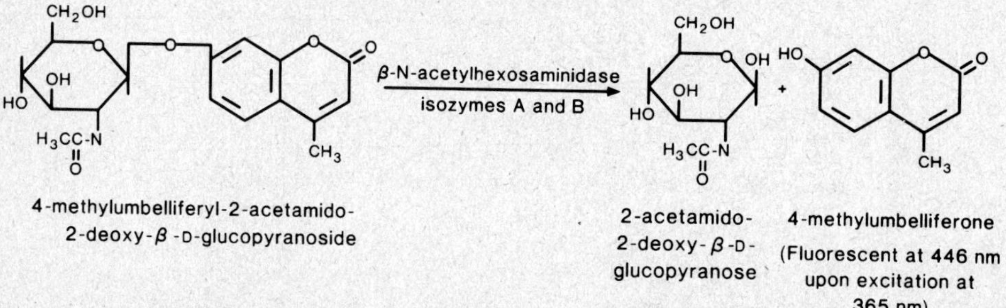

4-methylumbelliferyl-2-acetamido-
2-deoxy-β-D-glucopyranoside

β-N-acetylhexosaminidase
isozymes A and B

2-acetamido-
2-deoxy-β-D-
glucopyranose

4-methylumbelliferone
(Fluorescent at 446 nm
upon excitation at
365 nm)

Figure 5-32. The most frequently used reaction for the determination of β-N-acetylhexosaminidase involves hydrolysis of the 4-methylumbelliferyl derivative of N-acetylglucosamine (shown) or N-acetylgalactosamine. The free 4-methylumbelliferone product is quantitated by spectrophotofluorometry.

glycosaminoglycans and thereafter the evaluation of the rate of turnover of ^{35}S-labeled glycosa-minoglycans in cultured skin fibroblasts can be used. These adjunct tests provide a general approach to identify any of the glycosaminoglycan storage disorders with the exception of the Morquio syndrome.

Because carrier testing for Tay-Sachs disease is relatively frequently requested and the assay for β-D-N-acetylgalactosaminidase has been quite rigorously standardized, only the assay for serum β-NAGS will be presented here. Table 5-12 may be consulted for methods used in the detection of other lysosomal enzyme deficiencies.

Method for the Determination of β-N-acetylglucosaminidase Isoenzymes (Hexosaminidase A and B) in Serum

Principle

Hexosaminidase activity is determined from the rate of hydrolysis of nonfluorescent 4-methylumbelliferone derivative of N-acetylglucosamine to produce fluorescent 4-methylumbel-liferone (4-MU). The isoenzymes are distinguished by their relative heat stabilities. Heating the A isoenzyme at 50–52 °C for 1–3 h reduces its activity by 95%; the B isoenzyme activity is reduced by only 5%. Isoenzyme A activity is therefore evaluated as the difference between total activity (A plus B in unheated sample) and B activity (in heated sample).

Specimen

A minimum of 0.5 mL of serum is required; collect blood without anticoagulant. Heparinized plasma has slightly lower activity than does serum, probably due to heparin-hexosaminidase interaction.

Reagents

1. Citrate-phosphate buffer, 0.05 mol/L, pH 4.4. Partially fill a 500-mL volumetric flask with water and dissolve in it 4.2 g citric acid (M.W. 210.0) and 3.55 g dried Na_2HPO_4 (M.W. 142). Check pH, which is usually ~4.05. Adjust to pH 4.4 with NaOH or HCl, 2 mol/L. Fill to the mark with water. The final solution is 0.05 mmol/L in phosphate and 0.04 mmol/L in citrate. Keep refrigerated up to 2 months.

2. Substrate, 4-methylumbelliferyl-2-acetamido-2-deoxy-β-D-glucopyranoside, 2.5 mmol/L in citrate-phosphate buffer. Dissolve 9.48 mg of compound (M.W. 379.88; Sigma Chemical Co., St. Louis, MO) in 10 mL buffer (reagent 1). Stable for 6 months if frozen.

3. Glycine-hydroxide buffer, 0.2 mol/L, pH 10.7. Dissolve 30.02 g of glycine in 1500 mL water in a 2-L flask. Add ~170 mL NaOH, 2 mol/L, and adjust pH to 10.7 with additional NaOH. Fill to the mark with water. Keep refrigerated except for 1 week's supply, which should be kept at room temperature.

4. Standard, 4-methylumbelliferone (4-MU) (Sigma Chemical Co., St. Louis, MO). Store at room temperature.

    a. Stock standard, 5 mmol/L in ethanol. Dissolve 44 mg 4-MU in a 50-mL volumetric flask containing ethanol. Fill to the mark with ethanol. Stable for 3 months if kept in dark at 8 °C.

    b. Diluted stock, 0.05 mmol/L. Place 0.5 mL of reagent 4a in a 50-mL volumetric flask and fill to the mark with glycine buffer (reagent 3). Stable for 4 d at 8 °C.

    c. Working standard, 0.5 μmol/L. Place 0.5 mL of reagent 4b in a 50-mL volumetric flask and fill to the mark with glycine buffer (reagent 3). Keep at room temperature before use.

    Note: This concentration for a working standard is suggested, since its fluorescence is normally between that obtained for total β-NAGS and isoenzyme B. A 1.0 μmol/L standard is prepared if β-NAGS is increased.

Procedure

1. Make a five-fold serum dilution by mixing 0.5 mL serum with 2.0 mL buffer (reagent 1). Assay positive controls with each analytical run.

2. Label two tubes, one B_2, the other B_3 (Isoenzyme B activities). Into each place 0.5 mL of diluted serum. Cover tubes with rubber stoppers and heat tube B_2 for exactly 2 h and tube B_3 for exactly 3 h at 51 °C. Keep another test tube, labeled T for total hexosaminidase, containing 0.5 mL serum, at room temperature.

3. During this heat treatment, add 3.0 mL of glycine buffer, 0.2 mol/L, pH 10.7 to three 10×75-mm tubes labeled B_2, B_3, and T.

4. After the 2-h heat treatment of tube B_2, submerge test tube in ice water and then place in test tube rack with the T tube. After exactly 3 h treat B_3 tube like the B_2 tube.

5. Remove 50-μL aliquots from tubes B_2, B_3, and T and place into different test tubes; prepare duplicates. Keep the remainder of the three buffered serum solutions in the refrigerator in the event that the test needs to be repeated.

6. Initiate the enzyme reaction by adding 200 μL of substrate (reagent 2) into one B_2, B_3, and T tube. Mix and incubate the three reaction tubes at 37 °C. The three duplicate tubes are reaction blanks for B_2, B_3, and T and are kept at room temperature during the reaction. These tubes are used in step 8.

7. After exactly 30 min incubation, transfer 200 μL from each of the incubated tubes into correspondingly labeled tubes prepared in step 3 and quickly add 3.0 mL of glycine-hydroxide buffer to stop the reaction.

8. To the second set of B_2, B_3, and T tubes (step 6) add 200 μL chilled substrate-buffer and immediately add the 200 μL reaction aliquot to glycine-hydroxide buffer. These tubes serve as blanks.

9. Adjust the spectrophotofluorometer by zeroing the instrument with glycine buffer and adjusting the sensitivity to $\sim$75% full scale with the total reaction tube (T). The 0.5 μmol/L working standard (reagent 4c) should read about 30–40% full scale. Read and record the values for the 0.5 μmol/L standard; B_2 and B_2 blank; B_3 and B_3 blank; and T and T blank.

Calculations

$$\text{Sample dilution:} \quad 5 \quad \times \quad \frac{250}{50} \quad \times \quad \frac{3200}{200} \quad = 400$$

| original dilution of serum | dilution in reaction | dilution with glycine buffer |

$$\text{Total hexosaminidase, U/L} = T - T_b \times \frac{0.5\ \mu mol/L}{Fl_s} \times 400 \times \frac{1}{30\ min}$$

where 0.5 μmol/L = concentration of 4-methylumbelliferone standard and Fl_s = fluorescence of standard.

The same calculation is repeated for B_2 and B_3.

The % isoenzyme A activity $= \dfrac{T - B}{T} \times 100$ and is calculated for both 2- and 3-h inactivation periods.

Comments on the Procedure

1. Considerable disagreement in respect to time and temperature of inactivation appears in the literature. The procedure described here is a compromise in large measure consistent with the method of J. S. O'Brien et al.[16] Examination of B_2 and B_3 activities and of %A activity in control serum is a check to assure that isoenzyme A inactivation is complete and isoenzyme B inactivation is insignificant. When B_2 and B_3 activities are about the same and %A is consistently less than 55 with 2-h inactivation, then samples are insufficiently inactivated; the procedure should be repeated after raising the temperature by 1 °C. Conversely, a significant difference between B_2 and B_3, with %A consistently 70–85, indicates that a significant amount of B isoenzyme is being inactivated and the increase in %A is a mathematical consequence. In this case, lowering the temperature is appropriate corrective action.

2. Good quality control over this assay has important implications. Enzyme tests often provide the only laboratory evidence for identifying the asymptomatic carriers as well as affected fetuses with enzyme deficiencies. The laboratory's potential liability is thus greater, since the consequences of errors may be more drastic than in conventional clinical enzymology, which reflects aspects of organ system function. It is therefore incumbent on the laboratory to ensure accuracy in the β-NAGS assay. Maintenance of such accuracy can be monitored by laboratory quality control procedures, such as inclusion of carrier or indeterminate range samples in every

assay series. Participation in an outside quality control program, e.g., that conducted by the National Tay-Sachs and Allied Disease Association, Harbor-U.C.L.A. Medical Center, Torrance, CA, is also extremely worthwhile if not obligatory.

3. Effective use of this assay requires good communication and interaction between laboratory and physicians. The testing laboratory, ordering physician, and tested individual must be aware that testing in itself does not prevent Tay-Sachs disease. A method should be in place for immediate notification of the ordering physician of ambiguous results or results that indicate a carrier. The meaning of such results in terms of testing specimens or family genetic counseling should be provided by the laboratory or a medical geneticist affiliated with the laboratory. The most frequent cause for repeating the test is that its result is not valid when obtained during pregnancy. This can be circumvented in many instances by obtaining serum from the pregnant woman's mate. If the mate does not have clearly normal β-NAGS-A levels, leukocytes from the pregnant woman can be assayed because the interfering heat-stable form of the enzyme that is found in serum from pregnant individuals is not expressed in leukocytes. For this reason, leukocytes have often been suggested to be the definitive source of β-NAGS for carrier detection. Unfortunately, the variability of results in leukocytes, due among other things to incomplete and unequal solubilization of both isoenzymes from the leukocytes, has proved to be a problem unless the laboratory is thoroughly familiar with this procedure. Reference laboratories can be called on to do the assay and may be sent either prepared leukocytes or anticoagulated whole blood.

Other nuances in Tay-Sachs screening have been discussed elsewhere[11,13,20] and should be thoroughly studied before a laboratory engages in Tay-Sachs carrier screening.

Reference Range

Total β-NAGS: 16.9 $\pm$ 8.2 U/L (2 SD); n = 47
β-NAGS A: 56–76% of total activity

Preparation of Leukocytes (Percy and Brady,[19] Modified)

Leukocytes contain a relatively high amount of lysosomal enzymes and thus constitute a good specimen for the detection of lysosomal enzyme deficiencies. Leukocytes can easily be isolated, and such preparations are often the specimen of choice. Use of the buffy coat is not recommended since lysosomal enzyme activity is usually expressed per gram of protein, and crude specimens would show greater variation in protein content compared to enzyme activity. A description of the method for the preparation of leukocytes follows.

Principle

Whole blood is allowed to stand at room temperature for 30–60 min. Erythrocytes settle first; leukocytes, whose buoyant density is less than that of red cells, sediment more slowly and form a "buffy coat" between the erythrocytes and the supernatant plasma. If the density of the medium through which sedimentation occurs is increased, e.g., by addition of dextran, leukocytes remain suspended. Recovery of the supernatant and centrifugation at low centrifugal force result in a pellet made up largely of leukocytes. Contaminating erythrocytes can be eliminated by selective shock treatment, since they are less resistant than white cells to lysis in hypotonic solutions.

Specimen

Ten milliliters of blood collected in acid-citrate-dextrose (ACD-solution B, Becton-Dickinson, Parsippany, NJ) is the usual source of white blood cells. Leukocytes in such specimens remain viable for up to 24 h of storage or shipment. A considerably lesser yield of leukocytes is obtained from heparinized blood unless the isolation is carried out within a few hours after the sample is drawn. Furthermore, the inclusion of glucose (dextrose) in ACD solution B helps maintain leukocyte viability during prolonged storage or shipment times.

Reagents

1. Dextran solution, 5 g/dL. Into 16 × 125-mm tubes, weigh 0.5-g aliquots of dextran, grade HH, average M.W. 254 000 (Pharmachem Corp., Bethlehem, PA). Immediately before use, dissolve one aliquot in 10.0 mL of NaCl, 0.7 g/dL.

2. NaCl, 0.7 g/dL. Place 0.7 g NaCl into a 100-mL volumetric flask and dilute to mark with water. Stable for two months at 8 °C.

3. NaCl, 0.9 g/dL. Place 0.9 g NaCl into a 100-mL volumetric flask and dilute to mark with water. Stable for two months at 8 °C.

4. NaCl, 1.8 g/dL. Place 1.8 g NaCl into a 100-mL volumetric flask and dilute to mark with water. Stable for two months at 8 °C.

5. ISOTON II (Coulter Electronics, Inc., Hialeah, FL).

6. Na_2CO_3, stock 2 g/dL, in NaOH, 0.1 mol/L. Place 20 g Na_2CO_3 into a 1-L flask and dilute to mark with NaOH, 0.1 mol/L.

7. Sodium potassium tartrate stock, 2 g/dL. Place 2 g of $NaKC_4H_4O_6$ into a 100-mL flask and dilute to mark with distilled, deionized water. Stable for at least one month at room temperature.

8. $CuSO_4$ stock, 1 g/dL. Place 1 g $CuSO_4$, anhydrous, into a 100-mL flask and dilute to mark with distilled, deionized water. Stable for at least one month at room temperature.

9. Phenol reagent stock solution, 2 mol/L. (Fisher Scientific Co., Pittsburgh, PA; Cat. no. SO-P-24.)

10. Working reagents

Reagent A. Mix in the following proportions: 98 parts Na_2CO_3 stock; 1 part NaK tartrate stock; and 1 part $CuSO_4$ stock. Prepare sufficient quantity for daily need (2.5 mL/sample).

Reagent B. Dilute 1 part of phenol stock reagent with 1 part distilled, deionized water. Prepare sufficient quantity for daily need (0.25 mL/sample).

Procedure

1. Place 10 mL of anticoagulated blood into a 15-mL plastic centrifuge tube.

2. Add 2 mL of a freshly prepared solution of dextran, 5 g/dL, in NaCl, 0.7 g/dL. Mix gently by inversion.

3. Allow to stand for 45 min for sedimentation of cells.

4. Draw off supernatant with a plastic disposable pipet and discharge it into another plastic centrifuge tube.

5. Centrifuge at $500 \times g$ for 10 min in the cold.

6. Draw off supernatant and discard.

7. Resuspend the button of white blood cells in 1.0 mL of *cold* NaCl, 0.9 g/dL.

8. Shock treat the cells as follows: Add 3.0 mL of ice-cold distilled H_2O and mix gently for 45 s. Immediately add 3.0 mL of *cold* NaCl, 1.8 g/dL, and mix.

9. Centrifuge at $500 \times g$ for 10 min.

10. Draw off supernatant and discard.

11. Follow steps 7 through 10 for second shock treatment.

At this point, it is possible either to count the white cells or to determine the protein concentration in the preparation. These determinations are eventually used to determine the specific activity of the lysosomal enzyme. Circumstances in each laboratory may dictate which of these alternatives to use. If a Coulter Counter is readily available and the white cell preparation is freshly prepared, counting the cells provides narrower reference limits, probably because small cells and membrane contaminants are excluded from the count. When frozen pellets of white cells are used for the assay, counting is precluded, and the protein concentration must be determined. To determine protein, see page 773.

Determination of Cell Count on Sample

1. Resuspend cells in 1.0 mL of cold NaCl, 0.9 g/dL. Then add an additional 4 mL of NaCl, 0.9 g/dL, to give a total volume of 5 mL (= V).

Note: If there are only a few WBC's present, add less saline, but record the total volume, since it will enter into the final calculations.

2. Allow to stand 30–45 min at 4 °C and resuspend the cells.

3. Make a 500-fold dilution, i.e., 40 μL WBC suspension + 20 mL ISOTON, and count cells in this suspension on the Coulter Counter at the WBC settings. Record the count (= W).

4. Centrifuge the WBC suspension at $600 \times g$ for 10 min.

5. Draw off and discard supernatant. Resuspend cells in 1.0 mL ice-cold, double-distilled H_2O in a graduated, conical test tube, and record the volume. This final WBC suspension = v.

6. If needed, quick-freeze in dry ice-methanol mixture. (See below.)

7. Calculate and record the number of leukocytes per L of preparation. This is called the "liter count" or LC.

$$LC, 10^{10} \text{ cells/L} = W/0.5 \times 500 \times V/v \times 10^3$$

where: W = cell count

V = total volume in mL of WBC suspension from step 1

v = final volume in mL from step 5

500 = Coulter dilution factor

0.5 = volume in mL counted on Coulter

10^3 = conversion factor from mL to L

Example: V = 4.96 mL; v = 1.1 mL; W = 3330 cells

$$LC = 3330/0.5 \times 500 \times 4.96/1.1 \times 10^3 \text{ or } 1.64 \times 10^{10} \text{ cells/L}$$

Note: For consistency, always use the 10^{10} multiplier even when the count is low. For instance, 0.93×10^{10} cells/L rather than 9.3×10^9 cells/L.

Preparation of Sample for Protein Determination

1. Quick-freeze the sample obtained at conclusion of steps 1 through 11 in a dry ice–methanol mixture. Store at $-70\ °C$ until the enzyme assay is to be performed.

Note: Certain enzyme assays cannot be performed on samples that have been frozen, notably that for galactocerebroside β-galactosidase. Check the restrictions on freezing for each assay method.

2. Allow to thaw at room temperature and then completely suspend the sample by sonication, three bursts of 10 s each, in a Heat Systems sonicator cup (Heat Systems, Ultrasonics, Inc., Plainview, NY) filled with ice water.

Determination of Protein in Sample[14]

1. Add 90 μL of water to each of two 10-μL aliquots of sonicate in 12 $\times$ 75-mm tubes.

2. Prepare a blank of pure water, 100 μL, in duplicate.

3. Standardize the protein assay monthly, using bovine albumin, 2.0 g/dL, obtained from New England Reagent Laboratory, East Providence, R.I. Dilute the solution 40-fold with water to obtain a 50 mg/dL solution. Prepare standards as follows:

Standard, μg	Dilute Albumin Solution, μL	Water, μL
2.5	5	95
5.0	10	90
10.0	20	80
25.0	50	50
50.0	100	0

4. Add 100 μL of each standard to 12 $\times$ 75-mm tubes, in duplicate.

5. To water blank, standards, and diluted sonicate, add 2.5 mL reagent A. Mix and let stand 10 min.

6. Add 2.5 mL reagent B to each tube. Mix each tube immediately after reagent addition. Let stand 30 min.

7. Read absorbances of standards and unknowns against water blank, set to zero absorbance, at 660 nm.

Calculation

Determine the relation of absorbance to protein concentration of the standards either by calculation or graphically. The slope should be about 0.007 $A/\mu g$. Calculate protein concentration of unknown:

$$\text{Protein, g/L} = \frac{A_{\text{unknown}} \times 10^6\ \mu L}{A/\mu g \times 10^6\ \mu g/g \times 10\ \mu L}$$

where 10 μL is volume of sonicate taken for assay. The method is nonlinear at high protein concentrations; thus, a dilution of homogenate is required if $A_{\text{test sample}}$ exceeds 0.40. Protein concentrations of homogenates prepared in the manner described above vary from 1–6 g/L.

References

1. Baum, H., Dodgson, K. S., and Spencer, B.: The assay of arylsulphatases A and B in human urine. Clin. Chim. Acta, *4*:453-455, 1959.
2. Beratis, N. G., Aron, A. M., and Hirschhorn, K.: Metachromatic leukodystrophy: Detection in serum. J. Pediatr., *83*:824-827, 1973.
3. Beratis, N. G., Turner, B. M., Labadie, G., et al.: α-L-Fucosidase in cultured skin fibroblasts from normal subjects and fucosidosis patients. Pediatr. Res., *11*:862-866, 1977.
4. Desnick, R. J., Allen, K. Y., Desnick, S. J., et al.: Fabry's disease: Enzymatic diagnosis of homozygotes and heterozygotes. J. Lab. Clin. Med., *81*:157-171, 1973.
5. Fratantoni, J. C., Hall, C. W., and Neufeld, E. F.: The defect in Hurler's and Hunter's syndromes: Faulty degradation of mucopolysaccharide. Proc. Natl. Acad. Sci. U.S.A., *60*:699-706, 1968.
6. Gal, A. E., Brady, R. O., Hibbert, S. R., et al.: A practical chromogenic procedure for the detection of homozygotes and heterozygous carriers of Niemann-Pick disease. N. Engl. J. Med., *293*:632-636, 1975.
7. Gehler, J., Cantz, M., Tolksdorf, E., et al.: Mucopolysaccharidosis VII: β-glucuronidase deficiency. Humangenetik, *23*:149-158, 1974.
8. Hall, C. W., Liebaers, I., Di Natale, P., et al.: Enzymic diagnosis of the genetic mucopolysaccharide storage disorders. Meth. Enzymol., *50*:439-456, 1978.
9. Ho, M. W., and O'Brien, J. S.: Differential effect of chloride ions on β-galactosidase isoenzymes: A method for separate assay. Clin. Chim. Acta, *32*:443-450, 1971.
10. Hopwood, J. J., Muller, V., Smithson, A., et al.: A fluorometric assay using 4-methylumbelliferyl α-L-iduronide for the estimation of α-L-iduronidase activity and the detection of Hurler and Scheie syndromes. Clin. Chim. Acta, *92*:257-265, 1979.
11. Kaback, M. M., Zeiger, R. S., Reynolds, L. W., et al.: Approaches to the control and prevention of Tay-Sachs disease. Prog. Med. Genet., *10*:103-134, 1974.
12. Kint, J. A.: Fabry's disease: Alpha-galactosidase deficiency. Science, *167*:1268-1269, 1970.
13. Lowden, J. A.: Role of the physician in screening for carriers of Tay-Sachs disease. Can. Med. Assoc. J., *119*:575-578, 1978.
14. Lowry, O. H., Rosebrough, N. J., Farr, A. L., et al.: Protein measurement with the Folin phenol reagent. J. Biol. Chem., *193*:265-275, 1951.
15. Mahuran, D., and Lowden, J. A.: The subunit and polypeptide structure of hexosaminidase from human placenta. Can. J. Biochem., *58*:287-294, 1980.
16. O'Brien, J. S., Okada, S., Chen, A., et al.: Tay-Sachs disease: Detection of heterozygotes and homozygotes by serum hexosaminidase assay. N. Engl. J. Med., *283*:15-20, 1970.
17. O'Brien, J. S., and Warner, T. G.: Sialidosis: Delineation of subtypes by neuraminidase assay. Clin. Genet., *17*:35-38, 1980.
18. Okada, S., and O'Brien, J. S.: Tay-Sachs disease: Generalized absence of a beta-D-N-acetylhexosaminidase component. Science, *165*:698-700, 1969.
19. Percy, A. K., and Brady, R. O.: Metachromatic leukodystrophy: Diagnosis with samples of venous blood. Science, *161*:594-595, 1968.
20. Potter, J. L., and Robinson, H. B., Jr.: Screening for the Tay-Sachs carrier: A compromise program. Clin. Chem., *27*:523-525, 1981.
21. Raghaven, S. S., Topol, J., and Kolodny, E. H.: Leukocyte β-glucosidase in homozygotes and heterozygotes for Gaucher's disease. Am. J. Hum. Genet., *32*:158-173, 1980.
22. Stanbury, J. B., Wyngaarden, J. B., Fredrickson, D. S., et al., Eds.: Part 5: Disorders of lysosomal enzymes. *In*: The Metabolic Basis of Inherited Disease. 5th ed. New York, McGraw-Hill, 1983.
23. Suzuki, K.: Enzymatic diagnosis of sphingolipidoses. *In*: Methods in Enzymology. V. Ginsburg, Ed. New York, Academic Press, 1978, Vol. 50.
24. von Figura, K., Logering, M., Mersmann, G., et al.: Sanfilippo B disease: Serum assays for detection of homozygous and heterozygous individuals in three families. J. Pediatr., *83*:607-611, 1973.

CHAPTER 6

CARBOHYDRATES

by Wendell T. Caraway, Ph.D., and Nelson B. Watts, M.D.

Glucose is the major energy source for the human body and is derived primarily from dietary carbohydrates (grains, starchy vegetables, and legumes), from body stores of carbohydrates, and from the synthesis of glucose from protein and the glycerol moiety of triglycerides. When the total daily energy intake exceeds the daily expenditure, the excess is converted to fat and stored in adipose tissue. When daily energy needs are not met by sufficient calorie intake, glucose is produced from the breakdown of carbohydrate stores and from noncarbohydrate sources (e.g., gluconeogenic amino acids, lactate, and glycerol).

The glucose level in blood is kept within a fairly narrow range through a variety of influences. While there is some variation in blood glucose as circumstances change (feeding, prolonged fasting), levels above or below the normal range usually indicate disease.

High blood glucose due to diabetes mellitus is the most commonly encountered disorder of carbohydrate metabolism. Low blood glucose is an uncommon cause of serious disease. There are a number of rare conditions that cause hypoglycemia in the neonatal period and early childhood. In adults, low blood glucose in the fasting state is almost always due to a serious underlying condition; a transient fall of blood glucose several hours after eating is part of normal physiology. Pathological hypoglycemia in the postprandial state occurs infrequently, except after gastrointestinal surgery.

Measurement of glucose is the most frequently performed procedure in most hospital chemistry laboratories. Glucose determinations may be ordered for the diagnosis and follow-up of abnormalities of carbohydrate metabolism (diabetes mellitus, hypoglycemia) and other conditions (such as cerebrospinal fluid glucose in suspected meningitis). Many factors influence the glucose level, and the knowledge of these factors is just as important as the laboratory determination. The meaning of any glucose value, whether "normal" or "abnormal," cannot be known without understanding of the circumstances in the subject at the time the specimen was collected.

CHEMISTRY OF CARBOHYDRATES

The term "carbohydrate" refers to hydrates of carbon and is derived from the observation that the empirical formulas for these compounds contain approximately one molecule of water per carbon atom. Thus glucose, $C_6H_{12}O_6$, and lactose, $C_{12}H_{22}O_{11}$, can be written as $C_6(H_2O)_6$ and $C_{12}(H_2O)_{11}$, respectively. These compounds are not hydrates in the usual chemical sense, however, and noncarbohydrate compounds such as lactic acid, $CH_3CH(OH)COOH$ or $C_3(H_2O)_3$, can have similar empirical formulas. In more descriptive terminology, the carbohydrates are defined as the aldehyde and ketone derivatives of polyhydric alcohols. The simplest carbohydrate is glycol aldehyde, the aldehyde derivative of ethylene glycol. The aldehyde and ketone derivatives of glycerol are, respectively, glyceraldehyde (glycerose) and dihydroxyacetone (Figure 6-1).

Monosaccharides

Sugars containing three, four, five, and six or more carbon atoms are known, respectively, as trioses, tetroses, pentoses, hexoses, and so on, and are classified as monosaccharides. Aldehyde derivatives are called aldoses and ketone derivatives are called ketoses, as shown in Figure 6-2.

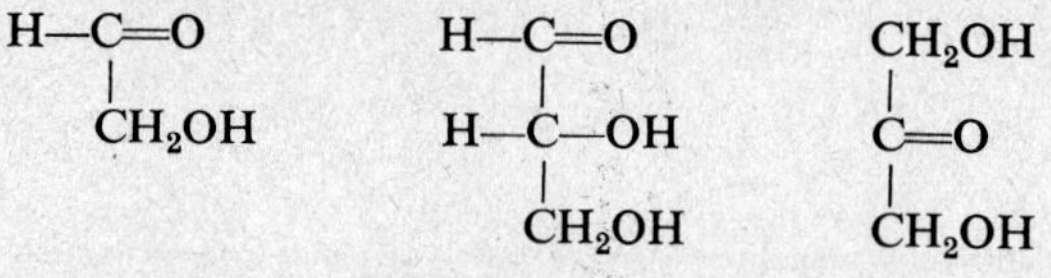

$$H-C=O \qquad H-C=O \qquad CH_2OH$$
$$\ \ \ \ CH_2OH \qquad H-C-OH \qquad C=O$$
$$\qquad\qquad\qquad\qquad CH_2OH \qquad CH_2OH$$

Glycol aldehyde **Glyceraldehyde** **Dihydroxyacetone**
           (glycerose)

Figure 6-1. 2- and 3-Carbon carbohydrates.

The carbon atoms in the chain are numbered 1 to 6 as shown by the numbers at the left of the formula for D-glucose. The designation D- or L- refers to the position of the hydroxyl group on the carbon atom next to the last (bottom) -CH$_2$OH group. In general, the designation of D- and L- for a sugar molecule refers to the stereoisomeric forms of the highest-numbered asymmetric carbon atom. By convention, the D-sugars are written with the hydroxyl group on the right and the L-sugars are written with the hydroxyl group on the left. Compounds that are identical in composition and differ only in spatial configuration are called stereoisomers. The majority of the sugars occurring in the body are of the D-configuration. A number of different structures exist, depending on the relative positions of the hydroxyl groups on the carbon atoms. The D-hexose series is shown in Figure 6-3.

The formula for glucose can be written in either aldehyde or enol form. Shift to the enol anion is favored in alkaline solution:

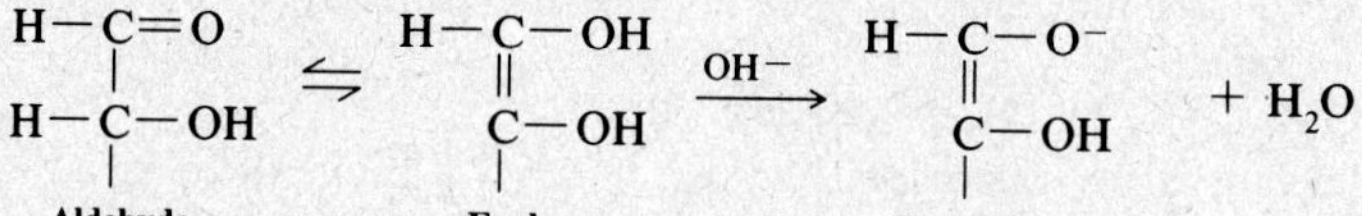

$$H-C=O \qquad H-C-OH \qquad\qquad H-C-O^- $$
$$H-C-OH \ \rightleftharpoons \ C-OH \ \xrightarrow{OH^-} \ C-OH \quad + H_2O$$
  Aldehyde   Enol   Enol anion

The presence of a double bond and a negative charge in the enol anion form makes glucose an active reducing substance and provides a basis for its analytical determination. Thus, glucose in hot alkaline solution readily reduces metallic ions such as cupric to cuprous ions, and the color change can be used as a presumptive indication for the presence of glucose. Sugars capable of reducing cupric ions in alkaline solution are commonly known as reducing sugars.

Aldehyde and alcohol groups can react to form hemiacetals. In the case of glucose, the aldehyde group reacts with the hydroxyl group on carbon 5 as shown in Figure 6-4. With this ring structure the hydroxyl group on the first carbon can be written to the right or to the left. By convention, the form with the hydroxyl group on the right is called α-D-glucose and the form with the hydroxyl group on the left is called β-D-glucose. The common anhydrous crystalline glucose is in the α-D-form. The β-D-form is obtained by crystallization from acetic acid. The two forms differ with respect to optical rotation of polarized light. The specific rotation, $[\alpha]_D^{25}$, for the α-D-form is +113° and for the β-D-form it is +19.7°. As a result of mutarotation, either form in aqueous solution gives rise to an equilibrium mixture that has a specific rotation of +52.5°. The equilibrium established at room temperature is such that about 36% of the glucose exists in the α-form and 64% in the β-form; only a trace remains in the free aldehyde form. The enzyme glucose oxidase reacts only with β-D-glucose. For this reason, standard solutions to be used in glucose oxidase methods for glucose determinations should be permitted to stand at least 2 h in order to obtain equilibrium comparable with that in the test samples to be analyzed.

From the ring structures shown in Figure 6-4, it is not apparent why the aldehyde group should react

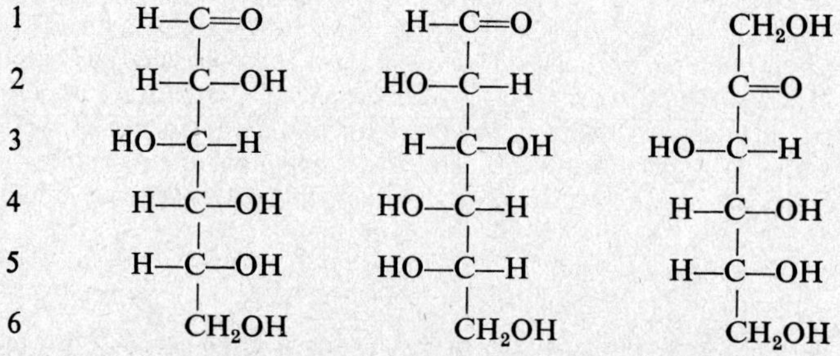

1	H—C=O	H—C=O	CH$_2$OH
2	H—C—OH	HO—C—H	C=O
3	HO—C—H	H—C—OH	HO—C—H
4	H—C—OH	HO—C—H	H—C—OH
5	H—C—OH	HO—C—H	H—C—OH
6	CH$_2$OH	CH$_2$OH	CH$_2$OH
	D-**Glucose**	L-**Glucose**	D-**Fructose**

Figure 6-2. Typical 6-carbon sugars.

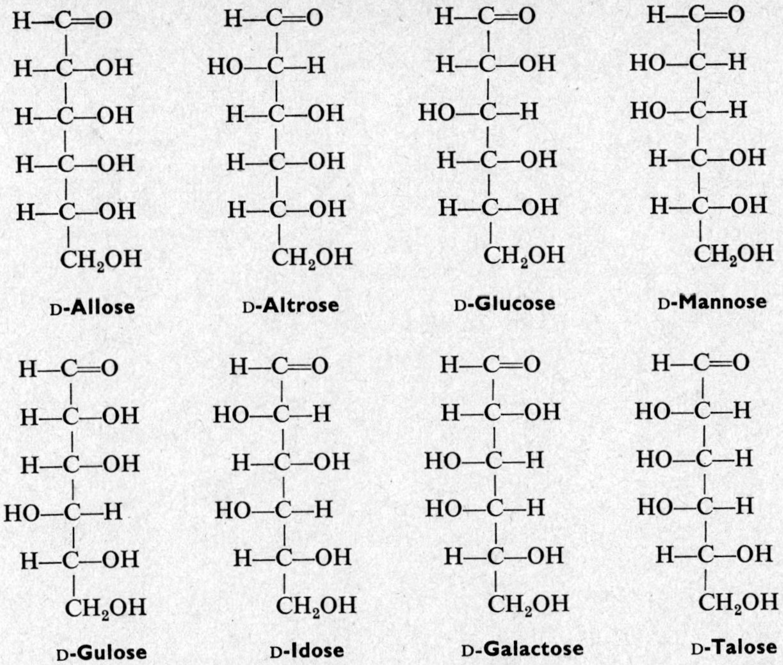

Figure 6-3. The D-hexose sugar series.

with the distant hydroxyl group on carbon 5. The spatial arrangement of the atoms is better represented by a symmetrical ring structure, depicted by the Haworth formula, in which glucose is considered as having the same basic structure as pyran (Figure 6-5). In this formula the plane of the ring is considered as perpendicular to the plane of the paper, with the heavy lines pointing toward the reader. Hydroxyl groups in position 1 are then below the plane (α configuration) or above the plane (β configuration). A six-membered ring sugar, containing five carbons and one oxygen, is a derivative of pyran and is called a pyranose. When linkage occurs with formation of a five-membered ring, containing four carbons and one oxygen, the sugar has the same basic structure as furan and is called a furanose. Representative formulas are shown in Figure 6-5. Fructose is shown in two cyclic forms. Fructopyranose is the configuration of the free sugar and fructofuranose occurs whenever fructose exists combined in disaccharides and polysaccharides, as in sucrose and inulin.

Disaccharides

Chemical interaction of groups between two monosaccharides, with loss of a molecule of water, leads to the formation of disaccharides. The chemical bond between the saccharides always involves the aldehyde or ketone group of one monosaccharide joined to an alcohol group (e.g., maltose) or to an aldehyde or ketone group (e.g., sucrose) of the other monosaccharide. The linkage of the oxygen on C-1 of sugars

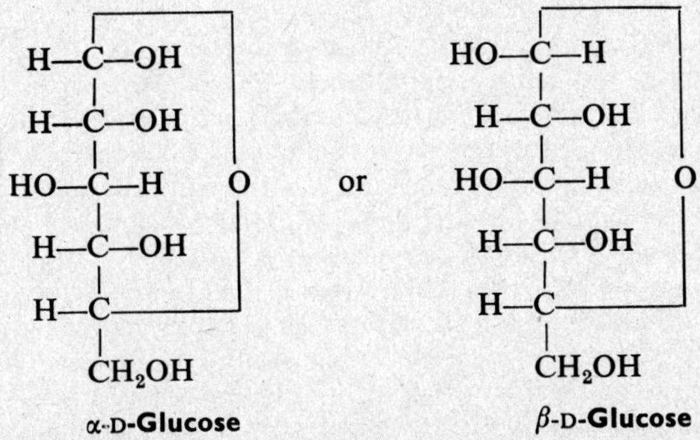

Figure 6-4. Structure of D-glucose (hemiacetal form).

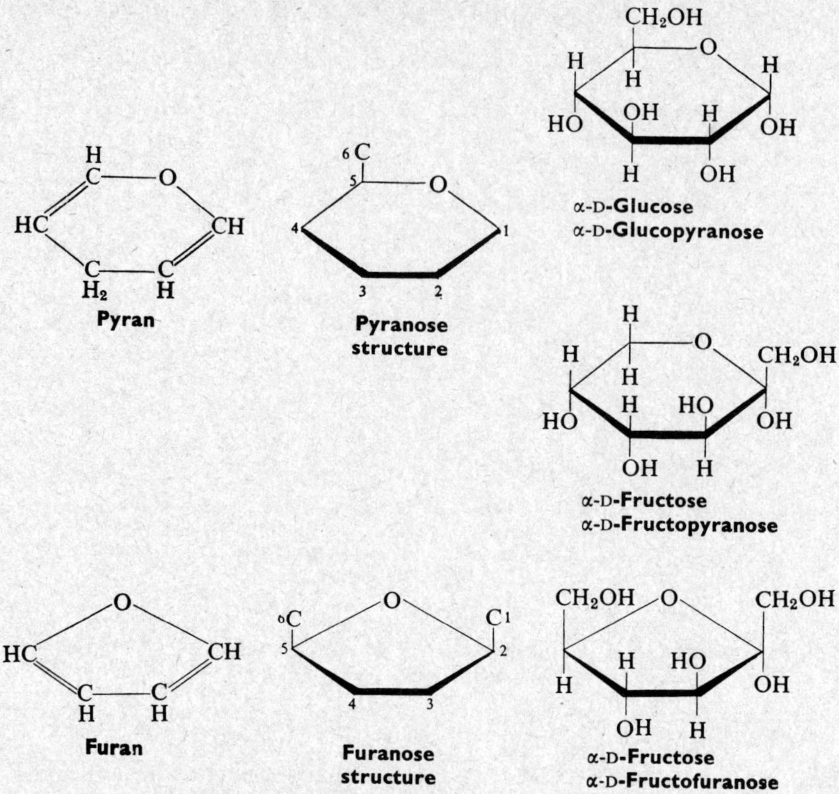

Figure 6-5. The Haworth formula for sugars.

(C-2 in the case of fructose) with an alcohol is called a glycosidic linkage. The most common disaccharides are

Maltose (glucose + glucose)
Lactose (glucose + galactose)
Sucrose (glucose + fructose)

Structural formulas and chemical names are shown in Figure 6-6.

If the linkage between two monosaccharides is between the aldehyde or ketone group of one molecule and a hydroxyl group of another molecule (as in maltose and lactose), there remains one potentially free ketone or aldehyde group on the second monosaccharide. Consequently, this sugar will be a reducing sugar and also can undergo mutarotation. The reducing power, however, is only approximately 40% of the reducing power of the two single monosaccharides added together, since one of the reducing groups is not available. On the other hand, if the linkage between two monosaccharides involves the aldehyde or ketone groups of both molecules (as in sucrose), a nonreducing sugar results since there is no remaining free aldehyde or ketone group.

Polysaccharides

The linkage of many monosaccharide units together results in the formation of polysaccharides. Each molecule of starch or glycogen, the chief reserve carbohydrates of plants or animals, respectively, typically contains 25–2500 glucose units. The suffix *-an* attached to a name of a monosaccharide indicates the main type of sugar present in the polysaccharide. Starch and glycogen, for example, are glucosans, since they are composed of a series of individual glucose molecules. Inulin, a polysaccharide found in the tubers of the dahlia and the Jerusalem artichoke, consists largely of fructose units and is known as a fructosan.

Nearly all starches are composed of a mixture of two kinds of glucosans called amyloses and amylopectins. The relative proportions of these two glucosans in a starch vary from approximately 20% amylose and 80% amylopectin in wheat, potato, and ordinary corn starch, to nearly 100% amylopectin in the starch of waxy corn. On the other hand, a few corn starches are known that contain up to 75% amylose. Although both amylose and amylopectin are made up of glucose molecules, there is one significant difference in their structure. Amylose, which has a molecular weight from 4000 to about 50 000, consists of one long, unbranched chain of 25–300 units of glucose. These units are linked together by 1,4-α-linkages with only the terminal aldehyde group free. In amylopectin the majority of the units are also connected with 1,4-α links, but, in

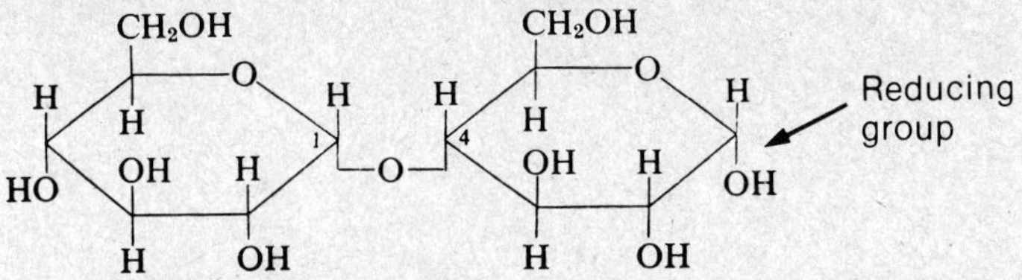

α-D-Maltose
α-D-**glucopyranosyl-4-**α-D-**glucopyranose**

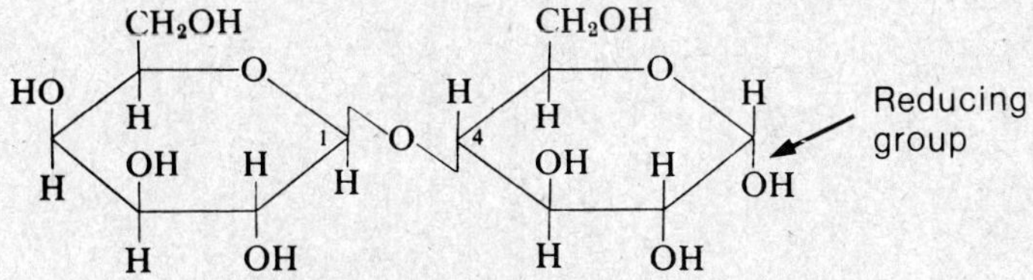

α-D-Lactose
β-D-**galactopyranosyl-4-**α-D-**glucopyranose**

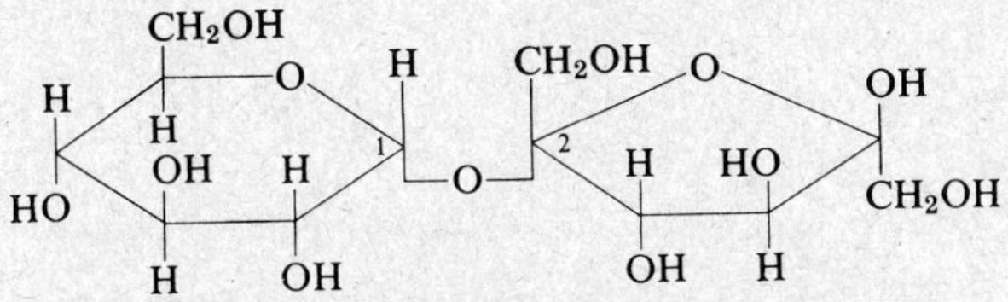

Sucrose
α-D-**glucopyranosyl-**β-D-**fructofuranose**

Figure 6-6. Structural formulas of disaccharides.

addition, there are 1,6-α-glycosidic bonds (amounting to about 4% of the total) that form side chains. The structure is a branch-on-branch arrangement of 1000 or more D-glucopyranose units with a molecular weight for amylopectin ranging from 50 000 to about 1 000 000. Glycogens have structures similar to amylopectins except that branching is more extensive. The average length of a branch in a glycogen molecule is usually 12 or 18 D-glucopyranose units, compared to about 25 units in amylopectin. Examples of polysaccharide linkages are shown in Figure 6-7.

The difference in structure between amylose and amylopectin becomes important in the proper selection of the starch substrate for amylase determinations. Any differences in the structure of starch will affect the rate of hydrolysis. The so-called α-amylase of pancreatic origin hydrolyzes the 1,4-glucoside linkage with special preference for the more central internal linkages. This results initially in the production of some maltose and a mixture of dextrins, which are subsequently also hydrolyzed to maltose. The 1,6-glucoside linkages are not attacked by α-amylase, and relatively large molecules of so-called residual (limit) dextrins are left after the action of the enzyme on amylopectin. Iodine gives a deep blue color with amylose and a red-to-violet color with amylopectin. These characteristic colors disappear when the respective glucosans are hydrolyzed to smaller units such as dextrins and maltose. This disappearance of the starch-iodine color is utilized in some quantitative methods for amylase determination.

Dextrins are the products of partial hydrolysis of starch. They are a complex mixture of molecules of different sizes. Those formed from amylose are unbranched chains, and those formed from amylopectin are branched chains of glucose molecules. Erythrodextrins are larger, branched dextrins that produce a reddish color with iodine.

METABOLISM OF CARBOHYDRATES

Starch and glycogen ingested as food are partially digested by the action of salivary amylase to form intermediate dextrins and maltose. Amylase activity is inhibited by the acid pH of the stomach. In the small intestine the pH is increased by alkaline pancreatic juice, and the amylase of the pancreas completes digestion of starch and glycogen to maltose. The latter, along with any ingested lactose and sucrose, is split by the disaccharidases (maltase, lactase, and sucrase) from the intestinal mucosa to form the monosaccharides glucose, galactose, and fructose.

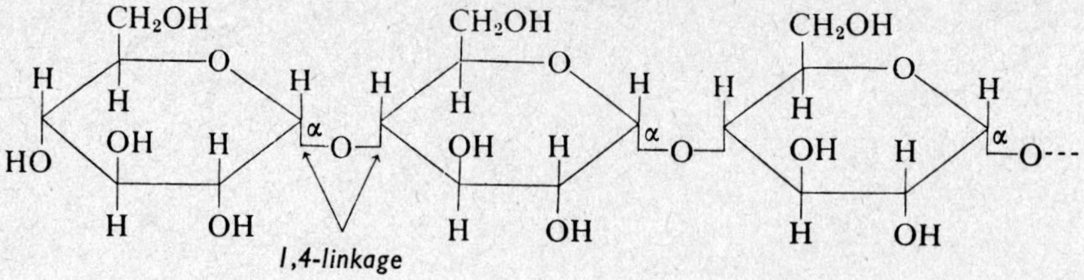

Chain of glucose molecules linked by 1,4-linkages as found in amylose

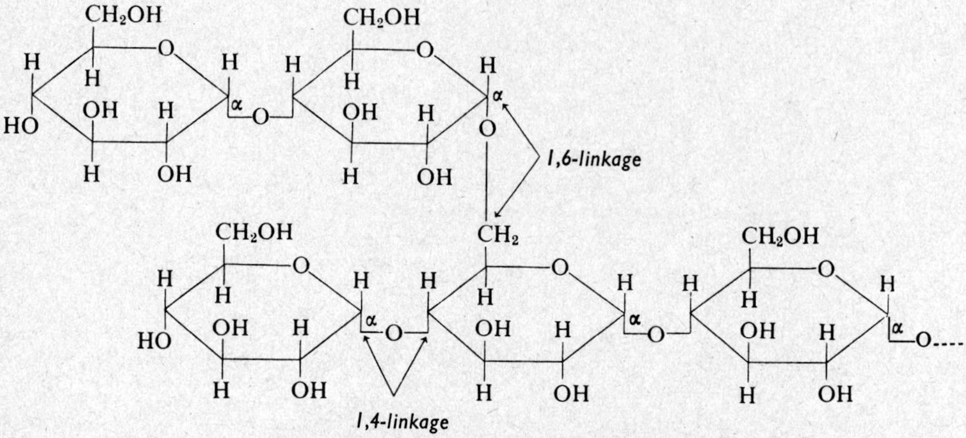

Chain of glucose molecules linked by 1,4-linkages and side chains
linked by 1,6-linkage as found in amylopectin

Figure 6-7. Structures for amylose and amylopectin.

Absorption of these monosaccharides is fairly complete and appears to occur by an active carrier-mediated transfer process. The rate of absorption for glucose and galactose is several times greater than for similar molecules absorbed by passive diffusion (e.g., xylose). Some conversion of fructose to glucose may occur during the process of absorption, and the interconversion can be visualized in terms of the enediol form common to both, as shown in Figure 6-8. Fructose is absorbed more slowly than glucose and galactose by a carrier-mediated process that is independent of both glucose and galactose transport mechanisms.

Following absorption into the portal vein, the hexoses are transported to the liver. Depending on the needs of the body, the carbohydrates are converted to and stored as liver glycogen; or metabolized completely to carbon dioxide and water to provide immediate energy; or converted to keto acids, amino acids, and protein; or converted to fat (triglyceride) and stored in adipose tissue. Fructose and galactose are phosphorylated and eventually enter the same metabolic pathway as glucose.

Some steps in the intermediary metabolism of glycogen and hexoses are shown in the partial outline in Figure 6-9. Each step is catalyzed by a specific enzyme and, in some cases, different enzymes may be responsible for a given step, depending on the direction in which the reaction proceeds. For example, the initial phosphorylation of glucose is mediated by glucokinase, but the reverse reaction depends upon glucose-6-phosphatase.

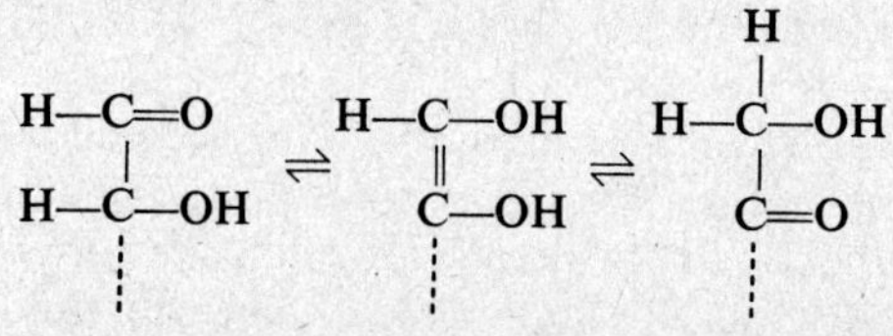

Glucose   **Enediol**   **Fructose**

Figure 6-8. Interconversion of glucose and fructose.

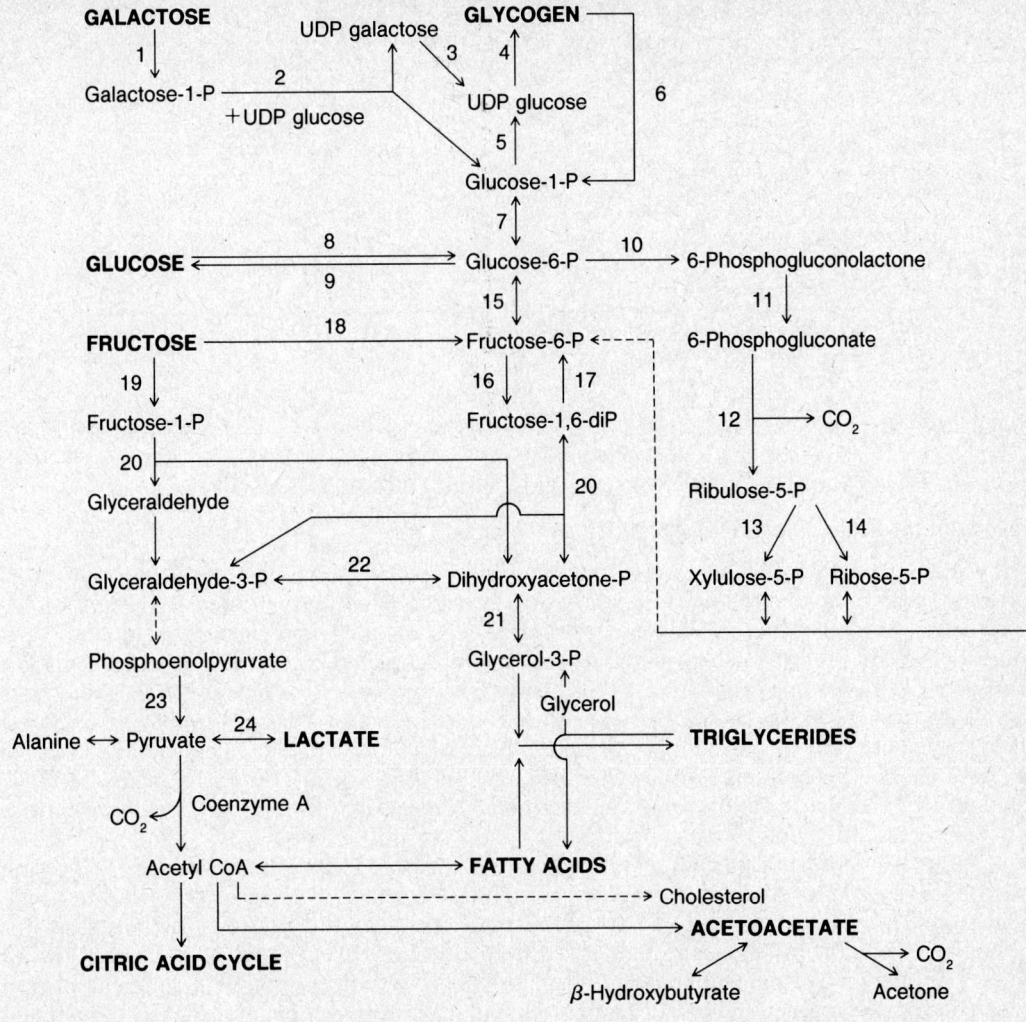

Figure 6-9. Major steps in the intermediary metabolism of carbohydrates. Numbers shown refer to specific enzymes. (------) Multistep pathway. (———) Single step pathway.

1. Galactokinase	13. Ribulose 5-P-epimerase
2. Galactose-1-P-uridyl transferase	14. Ribose 5-P-isomerase
3. UDP-galactose-4-epimerase	15. Phosphohexose isomerase
4. Glycogen synthetase (plus branching enzyme)	16. Phosphofructokinase
5. UDP-glucose pyrophosphorylase	17. Fructose-1,6-diphosphatase
6. Glycogen phosphorylase	18. Hexokinase (extrahepatic)
7. Phosphoglucomutase	19. Fructokinase
8. Glucokinase (and hexokinase)	20. Aldolase
9. Glucose-6-phosphatase	21. Glycerol phosphate dehydrogenase
10. Glucose-6-phosphate dehydrogenase	22. Triose-P-isomerase
11. 6-Phosphogluconolactonase	23. Pyruvate kinase
12. 6-Phosphogluconate dehydrogenase	24. Lactate dehydrogenase

Of clinical interest are metabolic disease states based on a deficiency of one or more enzymes. A number of these, keyed to Figure 6-9, are listed in Table 6-1. Figure 6-9 also shows relationships of carbohydrate metabolism to the production of lactate, ketone bodies, and triglycerides. The pentose phosphate pathway (PPP), also known as the hexose monophosphate shunt (HMPS), is an alternative pathway for glucose metabolism that is important in generating NADPH, which is used in maintaining the integrity of red blood cell membranes, in lipid and steroid biosynthesis,

Table 6-1. ENZYME DEFICIENCIES AND METABOLIC DISEASE STATES

Enzyme Deficiency	Disease State
Glucose-6-phosphatase (9)*	Type I GSD[†] (von Gierke's disease)
Muscle phosphorylase	Type V GSD[†] (McArdle's disease)
Hepatic phosphorylase	Type VI GSD[†] (Hers' disease)
Galactose-1-P-uridyl transferase (2)	Galactosemia
Galactokinase (1)	Galactosemia
Hepatic fructokinase (19)	Essential fructosuria
Hepatic fructose-1-P aldolase (20)	Hereditary fructose intolerance
Pyruvate kinase (23)	Hemolytic anemia
Glucose-6-P dehydrogenase (10)	Hemolytic disease

*Numbers shown refer to enzymes in Figure 6-9.
[†]GSD = Glycogen storage disease.

in hydroxylation reactions, and in other anabolic reactions. The complete picture of intermediary metabolism of carbohydrates is rather complex and interwoven with the metabolism of lipids and amino acids. For details, the reader should consult textbooks of biochemistry.

Regulation of Blood Glucose Concentration

Various terms are applied to describe general processes in carbohydrate metabolism. *Glycogenesis* refers to the conversion of glucose to glycogen, and *glycogenolysis* refers to the breakdown of glycogen to form glucose and other intermediate products. The formation of glucose from noncarbohydrate sources, such as amino acids, glycerol, or lactate, is called *gluconeogenesis.* The conversion of glucose or other hexoses into lactate or pyruvate is called *glycolysis.* Glucose is completely oxidized to CO_2 and H_2O via the mitochondrial electron transport coupled to oxidative phosphorylation, providing ATP. The oxidation of glucose to CO_2 and H_2O also occurs via the hexose monophosphate shunt pathway, providing NADPH, which is used for many anabolic reactions (e.g., fatty acid synthesis, steroid synthesis). The net result of all factors affecting these various processes determines the concentration of glucose in the blood.

During a brief fast, a precipitous drop in the level of blood glucose is avoided by drawing upon the glycogen stores of the liver. A slight amount of glucose may also be derived from the kidney. Both of these organs contain the specific enzyme, glucose-6-phosphatase, necessary for the conversion of glucose-6-phosphate, derived from either gluconeogenesis or glycogenolysis, to glucose. During long-term fasting, gluconeogenesis plays a significant role in maintaining plasma glucose levels. Glycogen stored in skeletal muscle cannot directly contribute glucose to the blood since muscle lacks this enzyme. On the other hand, as blood glucose levels increase, usually by absorption of carbohydrates from the intestine, glycogenolysis is replaced by glycogenesis, whereby excess blood glucose is converted into liver and muscle glycogen. A number of hormones are important in the regulation of blood glucose concentration.

Hormones Which Affect Blood Glucose Concentration

The concentration of glucose in blood is normally maintained in a narrow range despite changes in feeding and fasting. This is accomplished by the coordinated effects of **insulin** on one hand, which tends to lower blood glucose, and several **"counterregulatory" hormones** on the other hand (glucagon, cortisol, epinephrine, and growth hormone), which tend to elevate glucose levels (Figure 6-10).[58] Insulin enhances the entry of glucose into liver, muscle, and adipose tissues, and promotes the storage of energy substrate in the form of glycogen, fat, and protein.[103]

The absolute concentration of insulin required for a specific decrease of blood glucose varies considerably, depending on circumstances. It has been demonstrated that overweight individuals with normal carbohydrate metabolism require much higher levels of insulin than do normal-weight individuals with the same carbohydrate response, both in the basal state and after eating.[31] The problem does not seem to be with the insulin molecule itself; there appears to be a resistance to the effect of insulin in the overweight individual.[75]

Insulin action at the cell level depends on several events. First, the insulin molecule combines with a hormone-specific receptor which is part of the cell membrane. Then the hormone-receptor combination triggers a "second messenger"; this chemical change within the cell leads to changes in cellular activity that can be identified as the result of insulin action (Figure 6-11).[87] For many polypeptide hormones, this "second messenger" is 3′5′-adenosine monophosphate (cyclic AMP),

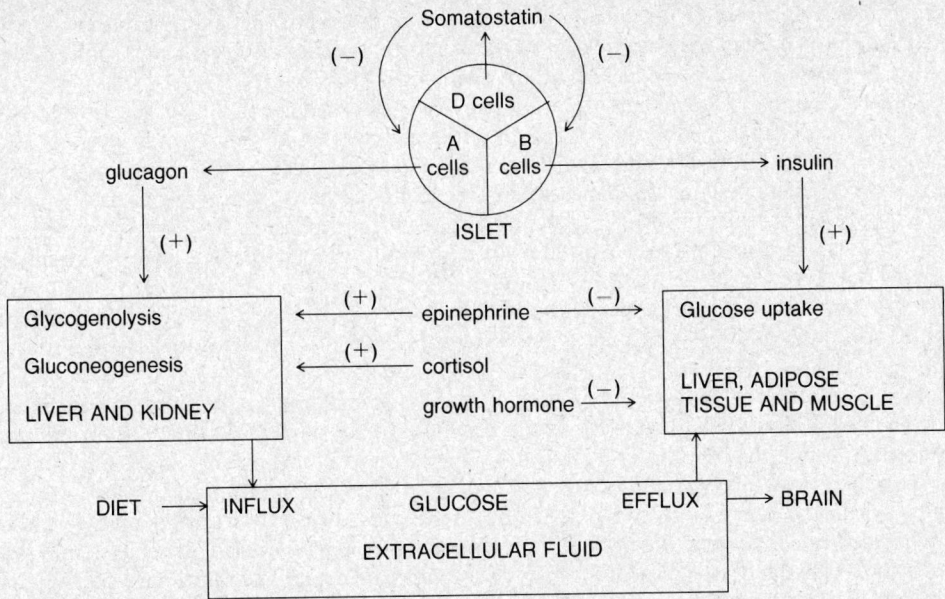

Figure 6-10. Hormonal influences on glucose homeostasis (see text). (+) Stimulation. (−) Inhibition.

but the exact nature of the "second messenger" for insulin activity is not known. There is indirect evidence that the "second messenger" is a small peptide formed from cell-membrane proteins.[53a] Decreases in insulin receptor number have been demonstrated which seem to account for most of the insulin resistance seen in obesity. Fasting serum insulin concentrations under basal conditions are increased in obesity, and the insulin response to a glucose load given orally is much greater than in normal-weight individuals.[84] The elevation of insulin causes a decrease in the number of insulin receptors per cell area (perhaps as a protective response). This "down-regulation" of insulin receptor number causes the target cells to be less sensitive to the effects of insulin and will lead to a sustained elevation of blood glucose if the compensatory increase in insulin is insufficient.[118] The down-regulation of insulin receptor number is a reversible event. Reduction in body weight will cause a fall in serum insulin concentration, an increase in insulin

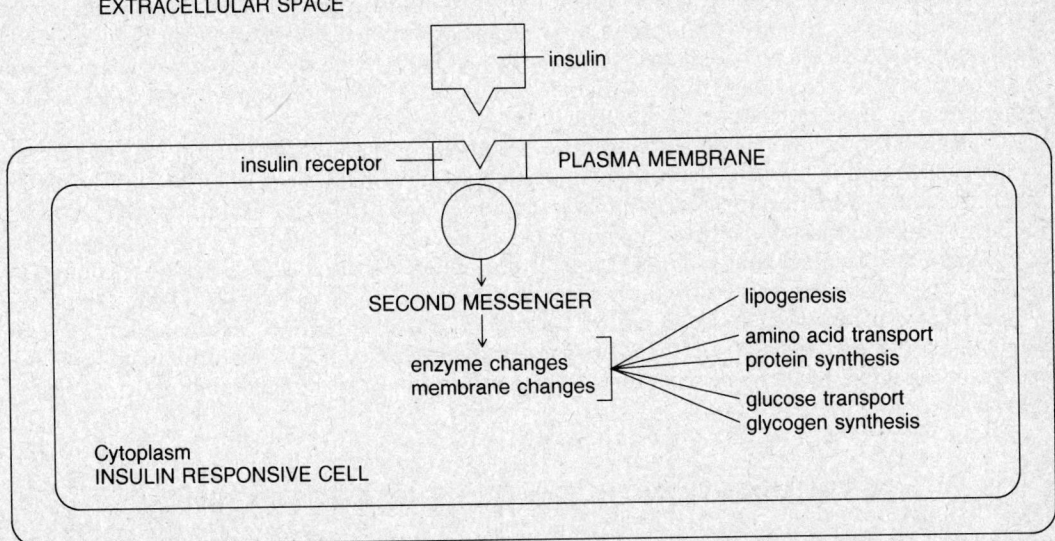

Figure 6-11. Mechanism of insulin action at the cell level. The insulin molecule combines with the specific insulin receptor on the plasma membrane; this combination initiates changes within the cell that become evident as "insulin action."

receptor number, and a lowering of blood glucose. Similar results can be shown with increases in physical activity, even in the absence of weight reduction. Further discussion of insulin, including C-peptide, proinsulin, and assay techniques, will be found in Chapter 9.

"Counterregulatory" hormones serve as substrate providers, enhancing the breakdown of glycogen to glucose (glycogenolysis) and stimulating the synthesis of glucose from glycerol and propionic acid derived from fat, and from protein (gluconeogenesis).[16, 114] Somatostatin (produced in the specialized delta cells of the pancreatic islets of Langerhans, where the beta cells produce insulin and the alpha cells produce glucagon) modulates the relationship between glucose and insulin.[32] Changes in blood glucose, as well as other stimuli, will trigger the appropriate hormone response.

Growth hormone and **adrenocorticotropic hormone** (ACTH) are polypeptides secreted by the anterior pituitary. Both have an antagonistic action to insulin and tend to raise blood glucose concentration.

Cortisol and other 11-oxysteroids secreted by the adrenal cortex stimulate gluconeogenesis. The effect of ACTH is mediated by its action on the adrenal cortex to stimulate production of 11-oxysteroids. Since the 11-oxysteroids tend to increase the blood glucose level and are antagonistic to insulin, they are sometimes referred to as diabetogenic hormones. In individuals with Cushing's syndrome, there is an overproduction of steroids owing to a tumor or hyperplasia of the adrenal cortex, and these individuals tend to show hyperglycemia. Conversely, individuals with Addison's disease (destruction or atrophy of the adrenal cortex) have a primary adrenocortical insufficiency and show moderate hypoglycemia.

Epinephrine, a catecholamine secreted by the adrenal medulla, stimulates glycogenolysis with a resultant increase in blood glucose levels. Physical or emotional stress causes increased production of epinephrine and an immediate increase in production of blood glucose for energy requirements. Tumors of adrenal medullary tissue, known as pheochromocytomas, secrete excess epinephrine or norepinephrine and produce moderate hyperglycemia as long as glycogen stores are available in the liver.

Glucagon, a polypeptide secreted by the alpha cells of the pancreas, increases blood glucose concentration by stimulating hepatic glycogenolysis and gluconeogenesis. Glucagon has no effect on muscle glycogen, as shown by the lack of elevation of blood lactate and pyruvate levels following glucagon administration.

Thyroxine, secreted by the thyroid gland, is a tetraiodinated amino acid derived from tyrosine. Thyroxine stimulates glycogenolysis, often leading to depletion of glycogen stores in the liver. Rate of gastric emptying and rate of glucose absorption from the intestine may be increased. These factors often contribute to a mildly abnormal, diabetic type of glucose intolerance in thyrotoxic individuals, although the fasting plasma glucose is normal in most cases.

Human placental lactogen (HPL), also known as human chorionic somatomammotropin (HCS), is a polypeptide with anti-insulin activity and is secreted by the placenta. It is believed that the action of HPL may be responsible for the development of diabetic ketoacidosis in pregnant women who were not known to have diabetes before they became pregnant and who do not require insulin after termination of pregnancy.

Somatostatin is a polypeptide found in several areas of the body but is concentrated in the hypothalamus and in the D cells of the pancreatic islets. It inhibits release of growth hormone by the pituitary. Somatostatin also inhibits secretion of glucagon and insulin, thus modulating the reciprocal relationship of these two hormones.

Somatomedins are a family of small peptide hormones. Somatomedin A has been implicated as one of the substances responsible for nonsuppressible insulin-like activity (NSILA). The nature of NSILA is discussed in more detail in Chapter 9. Parallel changes in the serum level of somatomedin A and NSILA are found in patients with acromegaly, hypopituitarism, and hypoglycemia caused by non-islet-cell tumors. Somatomedin C is a potent inhibitor of insulin degradation.

DETERMINATION OF GLUCOSE IN BODY FLUIDS

Many analytical procedures have been developed to measure blood glucose. In the past, analyses were often performed with relatively nonspecific methods which resulted in falsely elevated values. Furthermore, whole blood was often the sample of choice for analysis. However, values for glucose in whole blood are less than in plasma since red cells contain only about 80% as much water as an equivalent volume of plasma, this despite the identical concentration of glucose in the *water phase* of both cells and plasma. At present, plasma or serum, rather than whole blood, is used for nearly all glucose determinations. In the fasting state, whole blood glucose concentration is approximately 90% of that found in plasma. During fasting, arterial blood glucose is only about 5 mg/dL higher than that of venous blood. Following glucose challenge, however, the concentration of glucose in arterial blood ranges from 20–70 mg/dL above that of venous blood drawn at the same time.[24,54]

Procedures in common use can be classified as enzymatic (hexokinase, glucose oxidase, and glucose dehydrogenase) or colorimetric (*o*-toluidine). An oxygen rate method, which measures consumption of oxygen, depends on the use of glucose oxidase and a pO_2 electrode. Alternatively, the enzyme can be immobilized on a membrane and the rate of production of hydrogen peroxide measured with a hydrogen peroxide electrode as described below. Older methods, based on oxidation-reduction techniques, are now considered obsolete.

The procedures most widely used in the USA for measuring serum or plasma glucose may be judged by inspection of quality control surveys conducted by the College of American Pathologists. Results from 6567 laboratories reported in one survey conducted in late 1981 are tabulated in Table 6-2. From these data we may conclude, based on participating laboratories only, that automated hexokinase methods are most widely used. This is a reflection of the procedures adopted by the manufacturers of automated equipment. Glucose oxidase and *o*-toluidine methods show about equal frequency of overall use. Automation is favored in the glucose oxidase method by 3 to 1. The most commonly used manual procedure is the *o*-toluidine method.

We should emphasize that these data apply to this CAP survey only and are weighted to laboratories participating either on a voluntary basis or to comply with state regulatory agencies. Many reagent sets (kits) are commercially available for measuring glucose and are widely used, especially in smaller laboratories. Reference to CAP surveys shows that reasonable agreement among the more commonly used methods is obtained for glucose values on lyophilized serum.

Methods for the determination of glucose in blood have been reviewed by Cooper.[15] Ten glucose methods have been evaluated and compared with a generally accepted hexokinase reference method.[79]

Collection and Handling of Specimens

When normal *blood* is drawn and permitted to clot and to stand uncentrifuged at room temperature, the average rate of decrease in serum glucose is approximately 7% in 1 h (5–10 mg/dL).[117] This decrease is the result of glycolysis. Other workers have observed a decrease of

Table 6-2. PROCEDURES USED FOR GLUCOSE ANALYSES IN 6567 LABORATORIES

Method	Number	Per cent of Total	Mean, mg/dL*
Hexokinase, no blank	2867[†]	44	91.1
Multianalyzer (2667)[†]			91.0
Hexokinase, with blank	267	4	89.2
Multianalyzer (253)			89.1
Glucose oxidase	1252	19	
MBTH/DMA (116)			91.6
PAP (529)			90.1
Multianalyzer (425)			90.6
Other (607)			88.7
Multianalyzer (409)			89.2
Oxygen rate method	930	14	87.5
Multianalyzer (875)			87.4
o-Toluidine	1018	16	87.7
Multianalyzer (251)			90.5
Manual (765)			86.7
Other	233	3	
Total	6567	100	

*Results are based on CAP 1981 Set C-D, Specimen C-16 (Copyright, 1981, College of American Pathologists; data used with permission). This specimen contained 1.0 mg/dL of bilirubin. See text for discussion of methods.
†Number, and numbers in parentheses, refer to the number of laboratories using the respective method.

only 1.9 mg/dL per hour when blood was drawn into sterile Vacutainer-brand tubes.[90] This lower rate of glucose loss can be attributed in part to prevention of bacterial contamination and to lower ambient temperatures in the latter study. In separated, unhemolyzed *serum*, the glucose concentration is generally stable up to 8 h at 25 °C if kept free of bacterial contamination, and up to 72 h at 4 °C. Variable stability (related to bacterial contamination) is observed after longer storage periods. Plasma, removed from the cells after moderate centrifugation, contains leukocytes that also metabolize glucose, although cell-free, sterile plasma shows no glycolytic activity. It follows that plasma or serum from blood without preservative must be separated from the cells or clot within 1 h after the blood is drawn if glucose values within 10 mg/dL of the original value are to be obtained consistently. Some vacuum collecting tubes contain an inert material which forms an effective barrier between the cells and serum after centrifuging.

Glycolysis can be prevented and glucose stabilized up to 24 h at room temperature by adding sodium fluoride to the specimen. Fluoride ions prevent glycolysis by inhibiting enolase, an enzyme which requires Mg(II). The inhibition is due to the formation of an ionic complex consisting of Mg(II), inorganic phosphate, and fluoride ions; this complex interferes with the combination of the enzyme with its substrate. Fluoride ions also inhibit coagulation by binding calcium; however, clotting may occur after several hours, and it is therefore advisable to use a combined fluoride-oxalate mixture, such as 2 mg of potassium oxalate and 2.5 mg of sodium fluoride per mL of blood, to prevent such late clotting. Fluoride ions in high concentration inhibit the activity of urease and certain other enzymes; consequently, the specimens are unsuitable not only for determination of urea in some procedures that require urease but also for direct assay of some serum enzymes.

Cerebrospinal fluids are frequently contaminated with bacteria or other cellular constituents and should be analyzed for glucose without delay. Glucose may be preserved in 24-h collections of urine by adding 5 mL of glacial acetic acid to the container before starting the collection. The final pH of the urine is usually between 4 and 5 and bacterial activity is inhibited at this level of acidity. The use of 5 g of sodium benzoate per 24-h specimen is also effective in preserving the urine.

Hexokinase Methods

Principle

Hexokinase is an enzyme that catalyzes the phosphorylation of glucose by adenosine triphosphate (ATP) to form glucose-6-phosphate and adenosine diphosphate (ADP). To follow the reaction, a second enzyme, glucose-6-phosphate dehydrogenase (G-6-PD), is used to catalyze the oxidation of glucose-6-phosphate by nicotinamide adenine dinucleotide phosphate (NADP$^+$) to form NADPH in direct proportion to the amount of glucose originally present. The increase in absorbance at 340 nm due to NADPH provides a measure of glucose content. G-6-PD derived from yeast is used in the assay with NADP as the cofactor. NAD$^+$ is used if the source of the dehydrogenase is bacterial (*Leuconostoc mesenteroides*).

$$\text{Glucose} + \text{ATP} \xrightarrow{\text{Hexokinase}} \text{Glucose-6-phosphate} + \text{ADP}$$

$$\underset{\text{(or NAD}^+\text{)}}{\text{Glucose-6-phosphate} + \text{NADP}^+} \xrightarrow{\text{G-6-PD}} \underset{\text{(or NADH)}}{\text{6-Phosphogluconate} + \text{NADPH} + \text{H}^+}$$

A generally accepted reference method based on this principle has been developed and validated.[70] Serum or plasma is deproteinized by adding solutions of barium hydroxide and zinc sulfate. The clear supernatant is mixed with a reagent containing ATP, NAD$^+$, hexokinase, and glucose-6-phosphate dehydrogenase. The mixture is incubated at 25 °C until the reaction is complete and the absorbance related to the production of NADH is measured at 340 nm. Standards and blanks are carried through the entire procedure, including the deproteinization step. Detailed specifications are given for the equipment, materials, and reagents, including tests of enzyme reagent adequacy.

Although capable of high accuracy and precision, the reference method is too exacting and time-consuming to serve as a routine method in the clinical laboratory. An alternative approach is to apply the reaction directly to serum or plasma and use a specimen blank to correct for interfering substances that absorb at 340 nm.[69] Since most methods in use rely on commercially prepared reagents supplied in lyophilized form, only a general discussion of the direct procedure will be presented.

Either serum or plasma may be used. Sodium fluoride, 2 mg/mL of blood, is the recommended preservative for collecting blood and must be combined with an anticoagulant such as EDTA·Na$_2$, 1 mg/mL of blood. Other satisfactory anticoagulants are heparin, oxalate, and citrate. Hemolyzed specimens containing more than 0.5 g hemoglobin/dL are unsatisfactory owing to the release from red cells of phosphate esters and enzymes that interfere with the measurements.

Nineteen substances were tested for interference with the reaction. Only fructose, per se, affected the blanked direct hexokinase method in serum. Only small amounts of fructose are usually present in normal fasting serum. After ingestion of 2 g sucrose/kg of body weight, serum fructose increases up to 8–10 mg/dL within 1 h and persists for 2 h. Solutions administered during glucose tolerance testing should not contain fructose or fructose-containing saccharides.

Absorbances of samples or standards are measured against corresponding blanks after the reaction has gone to completion. Although glucose concentrations may be calculated directly, based on the molar absorptivity of NADPH or NADH, inclusion of a set of standards is recommended to detect possible deterioration of enzymes, ATP, NADP$^+$, or NAD$^+$, all of which are unstable. Reagents may also contain substances that react with the coenzymes. Presence of these substances can be evaluated by measuring the increase in absorbance observed in a reagent blank. The highest standard provides a check on linearity of response and the adequacy of the enzyme reagent.

Since a buffer solution for preparing specimen blanks is not generally supplied in commercially available reagent sets, the user will need to prepare this separately. It should be of the same composition, concentration, and pH as the buffer used in the commercial reagent. As may be seen in Table 6-2, the serum blank is generally omitted. In one study, the average correction for the blank was found to be equivalent to a glucose concentration of 7 mg/dL. Errors as great as 20–30 mg/dL could result if the serum blank is omitted. With grossly hemolyzed, turbid, or icteric specimens (which were excluded from this study), blank corrections would be expected to be higher.

Hexokinase procedures are also available in which indicator reactions produce colored products so that absorbance may be measured in the visible range.[119] An oxidation-reduction system containing phenazine methosulfate (PMS) and a substituted tetrazolium compound, 2-(p-iodophenyl)-3-p-nitrophenyl-5-phenyltetrazolium chloride (INT), is reacted with NADPH formed in the reaction. The reduced INT is colored with maximum absorbance at 520 nm. The PMS-INT color developer must be refrigerated when not in use and must be protected from exposure to light to retard autoreduction. Co-immobilized hexokinase and G-6-PD on the inner surface of small bore plastic tubing are used in certain continuous-flow automated systems with considerable savings on reagent cost.[56]

Glucose Oxidase Methods

Principle

The enzyme glucose oxidase catalyzes the oxidation of glucose to gluconic acid and hydrogen peroxide:

$$\text{Glucose} + O_2 \xrightarrow{\text{Glucose oxidase}} \text{Gluconic acid} + H_2O_2$$

Addition of the enzyme peroxidase and a chromogenic oxygen acceptor, such as o-dianisidine, results in formation of color which can be measured:

$$\underset{\text{(colorless)}}{o\text{-Dianisidine}} + H_2O_2 \xrightarrow{\text{Peroxidase}} \underset{\text{(colored)}}{\text{Oxidized } o\text{-dianisidine}} + H_2O$$

Glucose oxidase is highly specific for β-D-glucose. As noted earlier, glucose in solution exists as 36 % α- and 64% β-form. Complete reaction of glucose, therefore, requires mutarotation of the α- to β-form. Some commercial preparations of glucose oxidase contain an enzyme, mutarotase, that accelerates this reaction. Otherwise, extended incubation time allows for spontaneous conversion.

The second step, involving peroxidase, is much less specific than the glucose oxidase reaction. Various substances, such as uric acid, ascorbic acid, bilirubin, and glutathione, inhibit the reaction, presumably by competing with the chromogen for hydrogen peroxide, which results in lower values. Most interfering substances can be eliminated by use of a Somogyi filtrate. (See discussion under oxidation-reduction methods.) Acid filtrates cannot be used since peroxides which cause positive errors may be released. Most modern methods omit the preparation of protein-free filtrates in order to make the procedure faster and simpler. Some glucose oxidase preparations contain catalase as a contaminant; catalase activity decomposes peroxide and decreases the final color obtained. Standards and unknowns should be analyzed simultaneously under conditions such that the rate of oxidation is proportional to glucose concentration.

In some methods the final mixture is acidified slightly to stop the reaction, and the yellow color is measured at 400 nm. In stronger acid solution the color becomes pink, with maximum absorbance at 540 nm, and both sensitivity and stability are improved. Other approaches to measurement of the hydrogen peroxide produced include the peroxide-mediated oxidative coupling of 3-methyl-2-benzothiazolinone hydrazone (MBTH) with N,N-dimethyl-aniline (DMA) catalyzed by peroxidase[33] or the oxidative coupling of p-aminophenazone (PAP) to phenol.[109] Both procedures have been adapted to automation. The MBTH/DMA and PAP procedures are not affected by high concentrations of creatinine, uric acid, or hemoglobin and are performed directly on serum.

Glucose oxidase methods are suitable for measuring glucose in *cerebrospinal fluid*. Such methods, however, are not directly applicable to urine specimens, owing to the high concentration of substances which interfere with the peroxidase reaction and produce falsely low results. Uric acid is the major interfering substance in urine. A method has been described in which the urine is first pretreated with an ion-exchange resin to remove interfering substances.

Some instruments employ a *polarographic oxygen electrode* which measures the rate of oxygen consumption after addition of the sample to a solution containing glucose oxidase.[49] Since this measurement involves only the first reaction shown above, interferences encountered in the peroxidase step are eliminated. To prevent formation of O_2 from H_2O_2 by catalase present in some preparations of glucose oxidase, H_2O_2 is removed by inclusion of two additional reactions:

$$H_2O_2 + C_2H_5OH \xrightarrow{Catalase} CH_3CHO + 2 H_2O$$
ethanol \qquad acetaldehyde

$$H_2O_2 + 2 H^+ + 2 I^- \xrightarrow{Molybdate} I_2 + 2 H_2O$$

The latter reaction is effective even when catalase activity has diminished upon storage of reagents. The procedure can be applied directly to urine, serum, plasma, or cerebrospinal fluid. Since blood cells consume oxygen, this approach cannot be used for the determination of glucose in whole blood.

In the YSI Model 23A (Yellow Springs Instrument Co., Yellow Springs, OH 45387), glucose oxidase is immobilized in a thin layer of resinous material sandwiched between two membranes. When a buffered sample is introduced, glucose diffuses through the first polycarbonate membrane and reacts with the enzyme to produce H_2O_2. The latter then diffuses through the second smaller-pore cellulose acetate membrane and contacts a platinum anode where oxidation takes place. The current thus created is directly proportional to the glucose concentration in the diluted sample.

$$H_2O_2 \longrightarrow 2 H^+ + O_2 + 2 e^-$$

The circuit is completed by a silver cathode at which oxygen is reduced to water.

$$4 e^- + 4 H^+ + O_2 \longrightarrow 2 H_2O$$

Any H_2O_2 diffusing back into the sample chamber is destroyed by catalase to prevent interference with the analysis. Determinations may be performed on 25 μL of plasma, serum, or whole blood. Good precision and correlation with an oxygen-consumption rate analyzer have been reported.[12,100]

The *Kodak Ektachem system* (Eastman Kodak Co., Rochester, NY 14650) makes use of dry multilayer films for chemical analyses.[17] Glucose is measured by a glucose oxidase procedure. A 10-μL sample of serum is placed on a porous film on top of the layer containing the reagents. Glucose diffuses through the film and reacts with the reagents to produce a colored end product or dye. The intensity of this dye is measured through a lower transparent film by reflectance spectrophotometry. Advantages of this system include small sample size, no liquid reagents, and improved storage requirements.

The colorimetric method presented below may be applied to plasma, serum, whole blood, and cerebrospinal fluid. The method is not applicable for the determination of glucose in urine because of the high concentration of enzyme inhibitors present. Since only a small amount of sample is used, the reaction may be applied directly to normal-appearing serum, plasma, and cerebrospinal fluid. For whole blood and for serum or plasma that is appreciably hemolyzed, icteric, or lipemic, a deproteinized filtrate should be prepared. Heparin, oxalate, or EDTA is a satisfactory anticoagulant. Sodium fluoride (2.5 mg/mL of blood) may be used as a preservative, but not thymol, which inhibits the enzyme reaction.

The procedure is based on that of Raabo and Terkildsen modified for optional use with deproteinized filtrates. Reagents may also be obtained separately or in kit form from Sigma Chemical Co., St. Louis, MO 63178.

Reagents

1. Zinc sulfate solution, 50 g/L. Dissolve 50 g of $ZnSO_4 \cdot 7H_2O$ in water and dilute to 1 L.
2. Barium hydroxide solution, approximately 0.15 mol/L. Dissolve 50 g of $Ba(OH)_2 \cdot 8H_2O$ in water and dilute to 1 L. Let stand for 2 d in a covered container, then decant or filter. Store in a polyethylene bottle and protect from carbon dioxide in the air. This solution must be balanced against the zinc sulfate solution as follows: Pipet 10.0 mL of the zinc sulfate solution into a flask and dilute with 25 mL of water. Add phenolphthalein indicator and titrate slowly with the barium hydroxide solution, using vigorous mixing, to a definite permanent pink color. Dilute the stronger of the two reagents, if necessary, such that 10.0 mL of zinc sulfate solution requires 10.0 $\pm$ 0.1 mL of barium hydroxide solution for neutralization.
3. Phosphate buffer, 67 mmol/L, pH 7.0. Dissolve 5.81 g of anhydrous Na_2HPO_4 and 3.55 g of KH_2PO_4 in water and dilute to 1 L. Store in a polyethylene bottle at room temperature. The pH of the buffer is not critical since the color development in the procedure has a rather wide optimum from pH 6.5–7.5.
4. Enzyme solution. This reagent is prepared to contain 500 International units of glucose oxidase (*Aspergillus niger*) and 100 Purpurogallin units of peroxidase (horseradish) in 100 mL of phosphate buffer. The enzyme units are those defined by Sigma. Preweighed capsules containing the enzymes and buffer salts are available.
5. *o*-Dianisidine reagent. Dissolve 50 mg of *o*-dianisidine dihydrochloride in 20 mL of water. Stable for 3 months in a brown bottle at refrigerator temperature.
6. Working reagent. Add 1.6 mL of *o*-dianisidine reagent to 100 mL of enzyme solution. The working reagent is stable up to 1 month at refrigerator temperature.
7. Benzoic acid solution, 2 g/L. Dissolve 2 g of benzoic acid in water, with warming, and dilute to 1 L.
8. Stock standard glucose, 1 g/dL. Transfer 1.000 g of dry, reagent grade glucose to a 100-mL volumetric flask. Add benzoic acid solution, mix to dissolve, then dilute to the mark with benzoic acid solution. Mix thoroughly and store in a tightly stoppered bottle in the refrigerator. This solution is stable indefinitely, if kept from evaporation, due to the bacteriostatic effect of benzoic acid. Standard solutions prepared from dry glucose should stand at least 2 h to ensure that mutarotation has reached a state of equilibrium.
9. Working glucose standard, 100 mg/dL. Warm a portion of the stock standard to room temperature. Pipet 10.0 mL of stock standard into a 100-mL volumetric flask and dilute to the mark with benzoic acid solution. Additional working standards may be prepared by suitable dilution of the stock standard.

Procedure Using the Direct Reaction

1. Pipet 25 μL of working standard, serum, plasma, or cerebrospinal fluid into respective test tubes.
2. Add 0.5 mL of water to each tube and to a separate tube for a reagent blank. *Note*: Alternatively, 0.5 mL of a 1:20 aqueous dilution of standards and samples may be used.
3. Add 5.0 mL of working reagent to each tube and mix.
4. Incubate all tubes at 37 °C for 30 min or at room temperature for 45 min. Color is now fully developed. Avoid exposure of the tubes to direct sunlight during incubation.

5. Measure absorbance at 500 nm within the next 30 min against the reagent blank.

$$\frac{A_{sample}}{A_{standard}} \times 100 \text{ mg/dL standard} = \text{mg glucose/dL sample}$$

6. If test result is greater than 300 mg/dL, dilute the original specimen 2-fold or 4-fold with water and repeat the assay, including the standard and reagent blank.

Procedure Using Deproteinized Samples

1. Pipet 0.20 mL of working standard and samples into respective test tubes.
2. Add 1.8 mL of water to each.
3. Pipet 2.0 mL of water into a separate tube for a reagent blank.
4. Add 1.0 mL of barium hydroxide solution to each tube and mix.
5. Add 1.0 mL of zinc sulfate solution to each tube and mix well.
6. Centrifuge to obtain a clear supernatant.
7. Transfer 0.5 mL of supernatant from each tube to corresponding test tubes.
8. Proceed as described under the direct procedure, starting with step number 3.

Comments on the Procedure

The absorbance is usually linear with glucose concentrations up to 300 mg/dL; however, linearity should be checked with each new lot of reagent, using standards of 100, 200, and 300 mg/dL.

A new reagent blank must be prepared for each series of tests because the color of the blank gradually increases with time.

If glassware becomes stained by the reagent, the stain may be removed by rinsing with acetone followed by water.

Rubber tubing, used for dispensing solutions, has been found to interfere with adequate color development; therefore, rubber tubing connections on dispensers and automatic sampling or diluting devices should be kept to a minimum and should be suspect if standards yield unexpectedly low absorbance values.

References

Raabo, E., and Terkildsen, T. C.: On the enzymatic determination of blood glucose. Scand. J. Clin. Lab. Invest., *12*:402-407, 1960.

The enzymatic colorimetric determination of glucose in whole blood, plasma or serum. Sigma Technical Bulletin No. 510, revised January 1982. Sigma Chemical Co., St. Louis, MO 63178.

Glucose Dehydrogenase Methods

Principle

The enzyme glucose dehydrogenase (GDH; β-D-glucose:NAD oxidoreductase, EC 1.1.1.47) catalyzes the oxidation of glucose to gluconolactone:

$$\text{Glucose} + \text{NAD}^+ \xrightarrow{\text{Glucose dehydrogenase}} \text{D-glucono-}\delta\text{-lactone} + \text{NADH} + \text{H}^+$$

Mutarotase is added to shorten the time necessary to reach the end point. The amount of NADH generated is proportional to the glucose concentration.

Glucose dehydrogenase for this assay is isolated from *Bacillus cereus*. The reaction appears to be highly specific for glucose, shows no interference from common anticoagulants and substances normally found in serum, and provides results in good agreement with hexokinase procedures. Methods have been adapted to continuous flow analyzers,[8] including the use of immobilized enzyme,[105] and to a centrifugal analyzer.[61] At this writing, the glucose dehydrogenase procedure is not widely used in the USA, although some commercial reagent sets are now available. One of the authors (WTC) has had experience with a kit marketed by Harleco as UltraChem Gluc-DH. The procedure which follows is based largely on this product. A rapid "kinetic" modification is also presented.

Reagents

1. Buffer. Contains phosphate, 0.12 mol/L, sodium chloride, 0.15 mol/L, sodium azide, 1.0 g/L, adjusted to pH 7.6 at 25 °C. Supplied in bottles containing 100 mL of buffer.

2. Enzyme-coenzyme mixture. This lyophilized mixture, when reconstituted, provides glucose dehydrogenase ≥ 4500 U/L, mutarotase ≥ 90 U/L, and NAD, 2.2 mmol/L. One unit of GDH catalyzes the oxidation of 1 micromole of substrate per minute at pH 7.6 and 25 °C. One unit of mutarotase catalyzes the mutarotation of 1 micromole of α-D-glucose per minute at pH 7.2 and 25 °C.

3. Working reagent. Transfer a portion of buffer from one bottle to the vial of enzyme mixture and let stand 15 min with occasional swirling. Quantitatively transfer the contents of the vial to the bottle of buffer to supply 100 mL of working reagent. Reconstituted reagent is stable for 12 weeks when stored at 2–8 °C or up to 4 weeks when stored at room temperature. An absorbance of the reagent greater than 0.4 A (10-mm light path against water) at 340 nm may be indicative of decomposition.

4. Stock standard glucose, 1 g/dL. Refer to reagents under the glucose oxidase method.

5. Working standard, 200 mg/dL. Warm a portion of the stock standard to room temperature. Pipet 20.0 mL of stock standard into a 100-mL volumetric flask and dilute to the mark with benzoic acid solution. Additional working standards may be prepared by suitable dilution of the stock standard.

Procedure

A spectrophotometer capable of measuring absorbance at 340 nm is required. Cuvets should have a 10-mm light path and permit measurements on 2 mL of solution. The procedure may be applied to serum, plasma, or urine.

1. Label cuvets or test tubes for reagent blank, standard, and sample.

2. Pipet 2.0 mL of working reagent into each cuvet.

3. Pipet 10 μL of water into the blank cuvet and 10 μL of each sample or standard to the appropriate cuvets. Mix thoroughly by inversion or on a vortex mixer.

4. Incubate all cuvets in a constant temperature bath or at room temperature for an appropriate time. Recommended times are:

Temperature, °C	20	25	30	37
Reaction time, min	10	8	7	7

5. Remove cuvets after incubation, tap gently to remove bubbles, and wipe dry.

6. Set spectrophotometer with the reagent blank to zero absorbance at 340 nm. Measure absorbance of standard and sample.

$$\frac{A_{sample}}{A_{standard}} \times 200 \text{ mg/dL standard} = \text{mg glucose/dL sample}$$

Comments on the Procedure

Absorbance is usually linear with glucose concentrations up to 600 mg/dL; however, linearity should be checked using standards of both 200 and 600 mg/dL with each new lot of reagent. When the sample glucose concentration is too high for accurate measurement, dilute the sample 2-fold with saline (9 g NaCl/L) and repeat the analysis.

Lipemia interferes by contributing to the absorbance reading at 340 nm. Set up a sample blank by adding 10 μL of serum or plasma to 2.0 mL of saline. Measure absorbance against saline and subtract this value from the observed absorbance for the unknown in the test cuvet before making calculations.

Anticoagulants or preservatives reported not to interfere at usual concentrations include heparin, EDTA, citrate, oxalate, fluoride, and iodoacetate. The following substances were reported to have no effect on the performance of the method at the concentrations stated: bilirubin, 5 mg/dL; hemoglobin, 95 mg/dL; ascorbic acid, 200 mg/dL; glutathione, 20 mg/dL; uric acid, 10 mg/dL; urea, 2000 mg/dL; creatinine, 25 mg/dL. In our experience, bilirubin at 20 mg/dL and hemoglobin at 100 mg/dL increased the apparent glucose concentration by 13 and 4 mg/dL, respectively.

Relative rates of reaction of other sugars with GDH compared to glucose are:

2-Deoxyglucose	125%
Glucose	100%
D-Glucosamine	31%
D-Xylose	15%
D-Mannose	8%
Fructose	None
Galactose	None

2-Deoxyglucose does not occur in the mammalian organism. Concentrations of other reacting sugars should be negligible; however, the procedure should not be used for blood glucose determinations during oral xylose absorption tests.

Reference

UltraChem Gluc-DH product No. 65048. R7050 issued 4/81 by Harleco, Division of EM Industries, Inc., 480 Democrat Road, Gibbstown, NJ 08027.

Manual Kinetic Procedure Using Glucose Dehydrogenase

The following modification, with use of the above reagent, has been found to be reliable and eliminates the need for either a reagent blank or sample blank. A spectrophotometer capable of rapid automatic sampling and temperature equilibration is required. Time between readings must be exact. The change in absorbance (ΔA) between the two readings is printed out or multiplied by a calibration constant stored in a processor to provide results in concentration units. A Gilford Stasar III spectrophotometer attached to a SYVA CP 5000 processor and a Beckman Model 42 spectrophotometer have been used in this modification (WTC). Other instrumentation may be equally satisfactory. This procedure with the Gilford spectrophotometer will be described.

Settings on Gilford:

Absorbance mode; wavelength: 340 nm; temperature: 30 °C. Set absorbance to zero with water.

Settings on Processor:

Assay: ΔA; delay time: 5 s; read interval: 30 s; number of readings: 1.

Enter $K = 1$ (for absorbance readings) or the value of the calibration constant (for printout in mg/dL.)

Procedure

1. Pipet 20 μL of standard (200 mg/dL) or sample to be analyzed into appropriate small test tubes (e.g., 12 × 75 mm).

2. Sample 1000 μL of reagent with an automatic syringe-type pipet and deliver with one stroke into the tube so as to obtain good mixing.

3. Immediately aspirate the reaction mixture into the spectrophotometer. *Note*: The time between adding the reagent and sampling should be kept to a minimum. The reaction rate is nonlinear and has no lag phase.

4. Results will be printed 35 s after sampling.

Comments on the Procedure

To derive a calibration constant, set $K = 1$ so that ΔA will print in milliabsorbance units. Analyze glucose standards with concentrations of 100, 200, 300, 400, and 500 mg/dL. Calculate K for each standard:

$$K = \frac{\text{mg/dL of standard}}{\Delta A}$$

A linear relationship between ΔA and concentration is usually observed to 400 mg/dL. Alternatively, a calibration curve may be prepared by plotting ΔA versus concentration on linear graph paper.

An automatic sampler-dilutor is recommended for multiple analyses. The reaction mixture may also be dispensed into small plastic cups to provide adequate mixing.

Reference

Caraway, W. T.: Unpublished observations.

o-Toluidine Methods

Principle

Various aromatic amines react with glucose in hot acetic acid solution to produce colored derivatives. Among those used are aniline, benzidine, 2-aminobiphenyl, and o-toluidine. Since the others are (or are suspected) carcinogens, only o-toluidine finds wide use. The latter condenses initially with the aldehyde group of glucose to form an equilibrium mixture of a glycosylamine and the corresponding Schiff base, as illustrated in Figure 6-12. Further rearrangements and reactions take place after the original condensation to produce a mixture of green chromogens with an absorption maximum at 630 nm.[121]

Dubowski applied this reaction to trichloroacetic acid filtrates of serum and demonstrated good specificity for glucose.[23] Negligible values were obtained on serum, cerebrospinal fluid, and urine following yeast fermentation, which destroys glucose. Other workers have noted that the reaction can be applied directly to serum without removal of protein.

Sugars other than glucose produce variable amounts of color in the reaction. The relative absorbance observed, compared to the same quantity of glucose, appears to depend somewhat on reaction conditions and should be determined by the individual laboratory interested in applying such data. The following absorbance ratios relative to glucose were found in our laboratory, using the method to be presented, for standards containing 200 mg/dL of the respective sugars. Of the sugars tested, only glucose, fructose, and galactose occur naturally in serum.

Sugar	Relative Absorbance
Glucose	1.00
Fructose	0.06
Mannose	0.96
Galactose	1.42
Sucrose	0.16
Maltose	0.09
Lactose	0.39
Xylose	0.12

Pentoses react with o-toluidine to produce an orange color with maximum absorbance near 480 nm. By reading at two wavelengths, xylose may be measured in the presence of glucose. A wavelength is chosen (near 680 nm) where the absorbance of the color produced by glucose is the same as that observed at 480 nm but where the absorbance of the color produced by xylose is negligible. The absorbance near 680 nm is then subtracted from the total absorbance at 480 to provide the net absorbance contributed by xylose.

Reagents

1. o-Toluidine, 40 mL/L. Transfer 3.0 g of thiourea to a 3-L Erlenmeyer flask. Add 1920 mL of glacial acetic acid and 80 mL of o-toluidine. Mix until the thiourea is dissolved. Store in a brown bottle at room temperature. Contact with skin should be avoided. The reagent should be dispensed from an all-glass automatic pipet such as a Repipet (Labindustries, Berkeley, CA 94710).

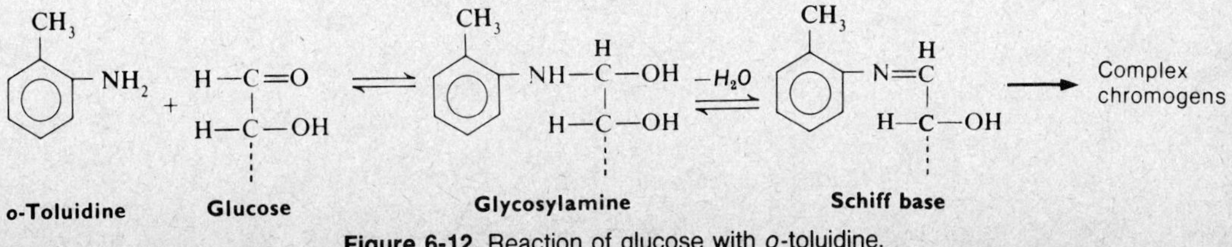

Figure 6-12. Reaction of glucose with o-toluidine.

This reagent may be used for months, although some variability occurs between batches, and older reagents tend to produce increased absorbance values with glucose. Thus, a standard should be included with each series of determinations. Thiourea decreases the color of the reagent blank to an absorbance of about 0.01 or less. Blanks can usually be omitted and the standards and unknowns measured against a distilled water blank; however, each new lot of o-toluidine should be checked by substituting water for serum in the method. If the absorbance of the blank is appreciable, it will be necessary to include a blank with each series of determinations.

2. Standard glucose solution, 200 mg/dL in benzoic acid solution. Prepare from stock 1.000 g/dL standard described for the glucose oxidase procedure.

Procedure

1. Pipet 100 μL of serum, plasma, cerebrospinal fluid, urine, or standard solution into respective large test tubes (e.g., 19 × 150 mm for ease of mixing).
2. Add 7.0 mL of o-toluidine reagent, mix, and place in a 100 °C water bath for 10 min.
3. Cool in a cold water bath for 2 or 3 min; remix.
4. Measure the absorbance within the next 30 min against a water blank at 630 nm.

$$\frac{A_{sample}}{A_{standard}} \times 200 \text{ mg/dL standard} = \text{mg glucose/dL sample}$$

Comments on the Procedure

For whole blood, prepare a protein-free supernatant by mixing 0.20 mL of specimen with 1.80 mL of trichloroacetic acid solution (3 g/dL). Allow to stand 5–10 min, centrifuge, and pipet 1.0 mL for analysis. This procedure should also be followed for grossly hemolyzed or lipemic sera. If protein-free supernates are used, the 200 mg/dL standard must also be diluted 10-fold with the trichloroacetic acid solution, since the acid reduces the absorbance obtained in the reaction. Add 7.0 mL of o-toluidine reagent and proceed as above.

The color follows Beer's law with most spectrophotometers, but this should always be checked for a given instrument by analyzing standards ranging from 100–500 mg/dL. Sufficient reagent is present to permit simple dilution of the final reaction mixture with acetic acid for values up to 2000 mg/dL.

Occasionally the final solution may show some turbidity. Dextran, used as a plasma expander, produces turbidity in the reaction and leads to falsely elevated values. Turbidity can occur if the cooling bath is too cold but usually is encountered with serum or plasma specimens having a high lipid content. In this event, add 2.0 mL of 2-propanol to 4.0 mL of final reaction mixture, mix, measure the absorbance, and multiply the result by 1.5. 2-Propanol is used to solubilize lipids. If the mixture is still cloudy, repeat the test on a protein-free supernatant.

Moderate hemolysis does not interfere significantly. Each 100 mg hemoglobin/dL increases the apparent glucose concentration by 2 mg/dL. Bilirubin contributes some absorbance because of partial conversion to the green pigment biliverdin. A commercial control containing 20 mg/dL of bilirubin in 5.5 g human albumin/dL produced an absorbance in the above method equivalent to 25 mg glucose/dL. A protein-free supernatant produced zero absorbance. This limitation of the direct method must be considered, especially when measuring blood glucose in infants with elevated serum bilirubin levels. EDTA in concentrations greater than 1 mg/mL and sodium fluoride at levels greater than 5 mg/mL in the specimen will cause some increase in color. Thymol preservative should be avoided, since this inhibits color formation. Rarely, certain lots of glacial acetic acid have been found to produce a brown color reaction with o-toluidine.

Reference ranges for the o-toluidine method are essentially the same as those for the glucose oxidase and hexokinase procedures, since all three methods give similar results. In patients with uremia, somewhat higher values are obtained with the o-toluidine method.

Reference

Feteris, W. A.: A serum glucose method without protein precipitation. Am. J. Med. Technol., 31:17-21, 1965.

Oxidation-Reduction Methods

Many older procedures for measuring glucose were based on its property of acting as a reducing agent in hot alkaline solution. Some of these procedures will be discussed briefly because of their historical interest.

In hot alkaline solution the aldehyde group of glucose readily reduces cupric ion to cuprous ion. Under these conditions cuprous ion forms mainly cuprous oxide (Cu_2O). The reaction is not stoichiometric, being dependent on the alkalinity, the time and temperature of heating, and the concentration of reagents. Under carefully controlled conditions the reaction is reproducible and provides quantitative results when standards are analyzed in the same manner as protein-free filtrates. Reoxidation of cuprous ion by oxygen from the air is prevented by using a constricted tube to minimize the surface area or by incorporating sodium sulfate (18 g/dL) in the reagent to decrease the solubility of oxygen. Phosphomolybdic (or arsenomolybdic) acid, Mo(VI), added in a second step is reduced by the cuprous ion to form compounds with lower oxidation states of molybdenum. These products have a blue color and are suitable for photometric measurements.

In the *Folin-Wu* procedure, proteins are precipitated with tungstic acid and the water-clear, protein-free filtrate is used in the reaction with cupric ion. The method lacks specificity, owing to the presence of nonglucose reducing substances in the filtrate, and should now be considered obsolete.

In the *Somogyi-Nelson* procedure, proteins are precipitated by the addition of barium hydroxide and zinc sulfate. Protein is removed as zinc proteinate, sulfhydryl compounds as zinc salts, and the remaining zinc and barium ions as zinc hydroxide and barium sulfate:

$$ZnSO_4 + Ba(OH)_2 \longrightarrow Zn(OH)_2 \downarrow + BaSO_4 \downarrow$$

Uric acid and some creatinine are also precipitated and adsorbed on barium sulfate so that the resultant filtrate is virtually free of nonsugar reducing substances. When the reagents are properly balanced, the filtrate has a pH of approximately 7.4. In spite of its relative specificity, this method is also considered obsolete. Preparation of a Somogyi filtrate is described above under glucose oxidase methods.

In the *neocuproine* method, cupric ion is reduced to cuprous ion by glucose in hot alkaline solution. Cuprous ion then forms an orange-colored complex with 2,9-dimethyl-1,10-phenanthroline (neocuproine), which is suitable for photometric measurement. The procedure is highly sensitive but is relatively nonspecific. Major problems encountered are high blanks, increasing baseline values in automated systems, and interference by uric acid and ascorbic acid. Careful control of temperature and heating time is vital. Application of the method was limited essentially to continuous-flow automated systems and has been discontinued.

In hot alkaline solution the aldehyde group of glucose reduces *ferricyanide* ion (yellow) to ferrocyanide ion (colorless).

$$\text{Glucose} + Fe^{3+}(CN)_6^{3-} \longrightarrow Fe^{2+}(CN)_6^{4-} + \text{oxidized products of glucose}$$
$$\qquad\qquad \text{Ferricyanide} \qquad\qquad \text{Ferrocyanide}$$
$$\qquad\qquad \text{(colored)} \qquad\qquad\quad \text{(colorless)}$$

The amount of decrease in color of ferricyanide, measured at 420 nm, is proportional to the increase in glucose concentration (inverse photometry). The reaction blank, without glucose, has the greatest absorbance, and measurements in the low or normal range are inherently less accurate since the results are based on a small difference between large absorbance values. In addition, the maximum amount of glucose that follows Beer's law is severely restricted by the maximum color that can be measured accurately in the reagent blank. Precision is improved by automating the method. The alkaline ferricyanide method was in wide use when glucose determinations were performed on the Technicon AutoAnalyzer. Other reducing substances interfere, e.g., 1 mg of creatinine reacts the same as 1 mg of glucose, and 1 mg of uric acid reacts the same as 0.5 mg of glucose. Thus, serum from a patient with uremia in which both creatinine and uric acid are markedly elevated gives a falsely elevated value for glucose. The high concentration of creatinine and uric acid found in normal urine renders the alkaline ferricyanide method unsuitable for the determination of glucose in urine. The large amounts of reducing substances liberated by hemolysis make hemolyzed specimens unacceptable.

HOME MONITORING OF BLOOD GLUCOSE

Persons with diabetes, especially those with Type I insulin-dependent diabetes, require some type of monitoring to maintain better control of blood glucose concentrations. Testing urine for glucose is a simple but unreliable guide since the renal threshold must be exceeded before glucose appears in the urine. Renal thresholds vary considerably and tend to be higher in diabetic individuals, especially in elderly persons who have decreased glomerular filtration rates associated with arteriosclerosis and glomerulosclerosis. Insulin pumps that now replace daily injections for many patients should not be used without frequent monitoring of blood glucose. It is impractical for patients themselves to perform blood or plasma glucose determinations by the methods given above, but simple test strips are now available that permit rapid and reasonably accurate measurements on a drop of whole blood. Colors developed on the strips may be evaluated visually or measured quantitatively on meters designed especially for such strips.

In general, home monitoring of blood glucose is recommended for the following: diabetic women during pregnancy; insulin infusion pump users; patients with renal disease; patients with unstable, insulin-dependent diabetes; and diabetics who are color blind and who must use meters to obtain readings.[104] Two methods will be described briefly, both of which depend on the glucose oxidase–peroxidase chromogenic reaction. These reagents are combined in dry form on a small surface area of test strips.

With Dextrostix (Ames Co., Elkhart, IN 46514) a drop of blood is placed on the reagent pad, let stand exactly 60 s, then washed off for 2 s with water from a wash bottle. The strip is read on a reflectance colorimeter (several models are available) which displays results on a digital display screen with a range of 0–399 mg/dL. The meter must be calibrated and controlled daily prior to use. To obtain valid results, the test must be performed exactly as directed. Both the 60-s incubation and 2-s wash times are crucial. The strips may not be held under tap water, as this tends to result in overwashing, removal of chromogen, and thus in falsely low results. Readings must be made immediately, since the color is unstable; hence, the strip may not be kept for later validation.[85] Visual reading of Dextrostix with a color chart is not accurate enough for most clinical circumstances.[106]

The bG Chemstrips (Bio-Dynamics, Indianapolis, IN 46250) have two adjacent reagent zones. Each zone is covered simultaneously with one drop of blood. After 60 s, the blood is wiped off with a cotton ball. After a further 60 s, colors are compared with a color chart. The two zones appear as different colors and provide a total range of 20–800 mg/dL of blood glucose. The final color is usually stable for several days, depending on temperature and humidity; thus, results can be checked by a trained observer. The Accucheck bG meter (Bio-Dynamics) can be used for a digital reading with bG Chemstrips. The Stat-Tek meter may be used with different test strips to provide a readout of results on a scale. One strip covers a range of 50–350 mg/dL of glucose; the other provides an expanded scale over a range of 10–150 mg/dL and is designed to provide better monitoring in newborns or in adults with hypoglycemia.

With careful attention to instructions, results obtained by self-monitoring correlate quite well with those obtained by conventional laboratory procedures. Results with the strip tests obtained on whole blood are approximately 10% less than results obtained on plasma or serum. The reader is referred to a series of nine papers presented as part of a symposium on blood glucose self-monitoring.[106]

Reference Ranges

The normal adult fasting range for *plasma* or *serum* glucose, determined by highly specific methods, is 70–105 mg/dL. For whole blood the range is 60–95 mg/dL. There is no sex difference and the ranges given are applicable to children after the first few weeks of life. In the premature newborn, plasma levels may range from 25–80 mg/dL; in the full-term infant, expected values are 30–90 mg/dL. Fasting plasma glucose values increase about 2 mg/dL per decade in the adult years. Postprandial values increase at the rate of 4 mg/dL per decade and those following a glucose challenge as much as 8–13 mg/dL per decade.[76]

Cerebrospinal fluid glucose values are usually about 60% of the plasma values (fasting range 40–75 mg/dL) and should be compared with plasma values for adequate clinical interpretation. The amount of glucose excreted in the urine, determined by highly specific enzymatic methods, is less than 500 mg/d. Random specimens show an upper limit of normal of approximately 30 mg/dL. Less specific methods for measuring glucose in urine are discussed in the section on urinary sugars.

DIABETES MELLITUS

Diabetes mellitus, as presently understood, is almost certainly more than a single disorder. Of particular importance, yet poorly understood, is the propensity for persons with diabetes mellitus to develop specific complications (retinopathy with blindness, kidney failure with uremia, nerve damage, and circulatory problems which predispose to tissue damage and may lead to amputations, heart disease, and stroke). Diabetes mellitus is a common disorder, but the actual prevalence of this disease depends on the criteria used for its diagnosis. Using criteria that will be discussed in detail later in this section, there are perhaps as many as 10 million persons with diabetes mellitus in the United States; half of them are not aware of the diagnosis.

Hyperglycemia is the hallmark of diabetes mellitus. The common underlying defect is a *deficiency of insulin action which leads to the development of hyperglycemia*. This may be an absolute deficiency of insulin production, typical of type I diabetes, or a relative deficiency of insulin action, commonly seen in type II diabetes. In both types, insulin production by the β-cells of the pancreas is impaired. In most persons with type II diabetes, there is also resistance to the peripheral action of insulin. This insulin resistance may be due to changes in insulin receptor number or affinity, or in post-receptor events.

CLASSIFICATION OF HYPERGLYCEMIC DISORDERS

In an effort to standardize terminology applied to diabetes mellitus, the National Diabetes Data Group has proposed a classification for hyperglycemic disorders which has been widely adopted.[68] This classification has the advantage of being fairly simple, and it fits well with current concepts of the different disorders which may cause hyperglycemia. Familiarity with this classification is essential to the proper use of diagnostic tests in diabetes mellitus and is helpful in formulating treatment programs for persons with this disease. An outline of this classification is shown in Table 6-3.

Insulin-dependent diabetes mellitus (IDDM, type I). Persons with this type of diabetes mellitus are usually acutely ill at the time of diagnosis, with severe hyperglycemia and other metabolic derangements. These individuals produce very little insulin and thus have an *absolute deficiency of insulin*. Without insulin treatment, these metabolic changes will cause death in a short time. With insulin treatment, satisfactory control of blood glucose can be obtained, yet many of these individuals will develop significant microvascular changes with time. Diagnosis of this type of diabetes is fairly straightforward when the patient presents in the acute stages, but earlier diagnosis, with our present understanding, does not seem possible in most cases. Type I, insulin-dependent diabetes mellitus makes up about 10% of all patients with diabetes mellitus.

Non-insulin-dependent diabetes mellitus (NIDDM, type II). While plasma insulin responses for individuals in this group range from low to considerably above normal, most individuals with NIDDM have plasma insulin concentrations that are higher than normal but not sufficient to maintain a normal blood glucose concentration in the face of insulin resistance. Thus, most persons with this form of diabetes mellitus have a *relative deficiency of insulin activity*. Obesity is commonly associated with this form of diabetes, and improvement in the obesity usually brings about an improvement in the hyperglycemia and often normal glucose concentrations. This is not always the case, however, and some persons with NIDDM may require insulin injections, or an oral hypoglycemic agent, to control symptomatic hyperglycemia. NIDDM is subdivided into (A) those patients who are not obese and (B) those who are obese; these subgroups are further divided into (1) those who require insulin for control of hyperglycemia and (2) those who do not require insulin. NIDDM is the most common form of diabetes mellitus (80–90% of patients).

Table 6-3. CLASSIFICATION OF DIABETES MELLITUS
AND OTHER TYPES OF GLUCOSE INTOLERANCE
(FROM THE NATIONAL DIABETES DATA GROUP)[68]

Idiopathic diabetes mellitus
   I. Insulin-dependent type (IDDM)
  II. Non-insulin-dependent type (NIDDM)
     A. Nonobese NIDDM
       1. Requiring insulin for hyperglycemia
       2. Not requiring insulin
     B. Obese NIDDM
       1. Requiring insulin for hyperglycemia
       2. Not requiring insulin
Gestational diabetes (GDM)
Impaired glucose tolerance (IGT)
Previous abnormality of glucose tolerance (Prev AGT)
Potential abnormality of glucose tolerance (Pot AGT)
Glucose intolerance associated with certain conditions
   and syndromes

Most of the patients in this group are obese and do not require insulin treatment. Early diagnosis in this type of patient is difficult, since the disease is usually slow to develop and progress, and precise separation of all persons with NIDDM from normal may not be possible.

Impaired glucose tolerance (IGT). This category identifies individuals who are not clearly normal at the time of testing, but are not sufficiently abnormal to be certain that they have diabetes mellitus. In this group there is a greater than usual risk for the later development of diabetes mellitus; there is a slow rate of progression to overt diabetes (1–5% a year) but a high likelihood that results will revert to normal or remain in the borderline range on repeat testing. In this group, there is a greater than normal risk for the development of cardiovascular disease, but the predictive risk value for a given individual is limited. Treatment with oral hypoglycemic agents has little, if any, protective effect on the later progression to more serious hyperglycemia in this group. Microvascular disease is quite rare in this group, even in patients in whom more definite hyperglycemia appears with time. This category is useful for identifying persons whose glucose tolerance is not clearly normal, but it avoids the inappropriate label of diabetes mellitus for persons in this group.

Gestational diabetes mellitus (GDM). This diagnosis identifies a pregnant woman at high risk for fetal morbidity, where therapeutic intervention has been demonstrated to improve the outcome. Early diagnosis of gestational diabetes is important, so the patient can be monitored and treatment can be instituted when necessary. Gestational diabetes is diagnosed when the abnormality of glucose handling is discovered for the first time during pregnancy; this is not the same as a known diabetic who becomes pregnant. There is a high probability that a woman with gestational diabetes will develop diabetes mellitus later in life (30–60%), but carbohydrate tolerance may revert to normal after delivery and may remain normal.

Previous abnormality of glucose tolerance (Prev AGT). This category includes persons known to have had an abnormality of glucose tolerance at one time, but who have been shown to have normal results on repeat testing. For example, this would include someone with type II diabetes mellitus, obese, not insulin treated, who after weight reduction had a normal glucose tolerance test on follow-up, or a woman with gestational diabetes mellitus who had a normal glucose tolerance test after delivery. This category has no definite prognostic significance and is useful primarily for epidemiological studies.

Potential abnormality of glucose tolerance (Pot AGT). This would be the person at greater than average theoretical risk for the later development of diabetes mellitus, for example, the identical twin of a patient with type II diabetes, or the child of parents who each had type II diabetes. Even more than the category of Prev AGT, this category has no specific prognostic significance and is mainly useful for epidemiological studies.

Glucose intolerance associated with certain conditions and syndromes. This category includes the uncommon patient in whom hyperglycemia is due to a specific underlying disorder, such as chronic pancreatitis with endocrine and exocrine insufficiency, Cushing's disease, acromegaly, glucagon-producing islet-cell tumor of the pancreas, other diseases, and certain drugs which are known to cause hyperglycemia.

The terms "prediabetes," "latent diabetes," "chemical diabetes," "subclinical diabetes," "borderline diabetes," and "asymptomatic diabetes" do not appear in this classification, but these conditions would probably be covered in the categories of Pot AGT, Prev AGT, and IGT. Juvenile-onset or ketosis-prone diabetes would now be classified as type I diabetes, and maturity-onset or ketosis-resistant diabetes would be classified as type II diabetes. While insulin-dependent diabetes is most commonly seen in young persons and non-insulin-dependent diabetes usually appears in middle age, these disorders are not age-specific. The terms "mild" and "severe" are not a part of this classification.

DIAGNOSTIC STRATEGIES

The demonstration of significant hyperglycemia is the key to the diagnosis of diabetes mellitus. While there are abnormalities of insulin secretion and insulin action that precede the development of definite hyperglycemia, and other early abnormalities may also exist, these cannot be quantitated with enough certainty to arrive at the early diagnosis of diabetes mellitus in an individual case. For type I diabetes, the diagnosis is usually simple, since hyperglycemia appears abruptly, is severe, and is accompanied by serious metabolic derangements. In gestational diabetes, criteria set forth by O'Sullivan in 1964 have been shown to identify clearly women with high-risk pregnancies, and effective screening and testing protocols are widely accepted. (See later section on GDM.) It is in type II diabetes mellitus (NIDDM) that early diagnosis becomes troublesome.

Defining normal. The diagnosis of diabetes mellitus must be made using criteria which are arbitrary. Thus, it seems reasonable to evaluate the advantages and limitations of various options for population screening and individual testing, and consider alternatives which may be selected for use in varying circumstances.

Some difficulty exists in setting limits of hyperglycemia that separate normal individuals from those with diabetes mellitus, either in the fasting state or after a glucose challenge administered orally. Most populations show a unimodal distribution for plasma glucose results, that is, a distribution curve skewed to the high end which tends to become bell-shaped on a logarithmic axis (Figure 6-13). Early population studies which were used to set standards for normal and abnormal responses to an oral glucose load were done on young, healthy individuals—a group not particularly likely to develop type II diabetes mellitus. For these reasons, some well-accepted standards for interpretation of an oral glucose tolerance test are unreasonably low when dealing with a population that is at high risk for the development of type II diabetes mellitus, i.e., middle-aged and overweight individuals.

There are some populations in which the distribution of plasma glucose fits a bimodal curve, i.e., two curves with a point of intersection that separates the normal population from those with diabetes. In two of the best-studied groups (the Pima Indians of the United States and the Nauruans, a Micronesian tribe) the prevalence of diabetes mellitus type II approaches half of the adult population. In these groups, separation of normal and abnormal subgroups is marked by a fasting glucose around 140 mg/dL and a value of more than 200 mg/dL, 2 h after a glucose load (Figure 6-13).[29]

The risk for the later development of microvascular disease makes it important to identify patients with type II diabetes. While it seems that bringing glucose levels to normal or near normal will help prevent or minimize microvascular disease, the evidence is inconclusive. In the Pima Indians, the prevalence of retinopathy increases dramatically in persons who have a 2-h plasma glucose over 200 mg/dL.[22] Studies done in England[46] showed that microvascular disease is rare in persons with 2-h plasma glucose values under 200 mg/dL. Even though some English

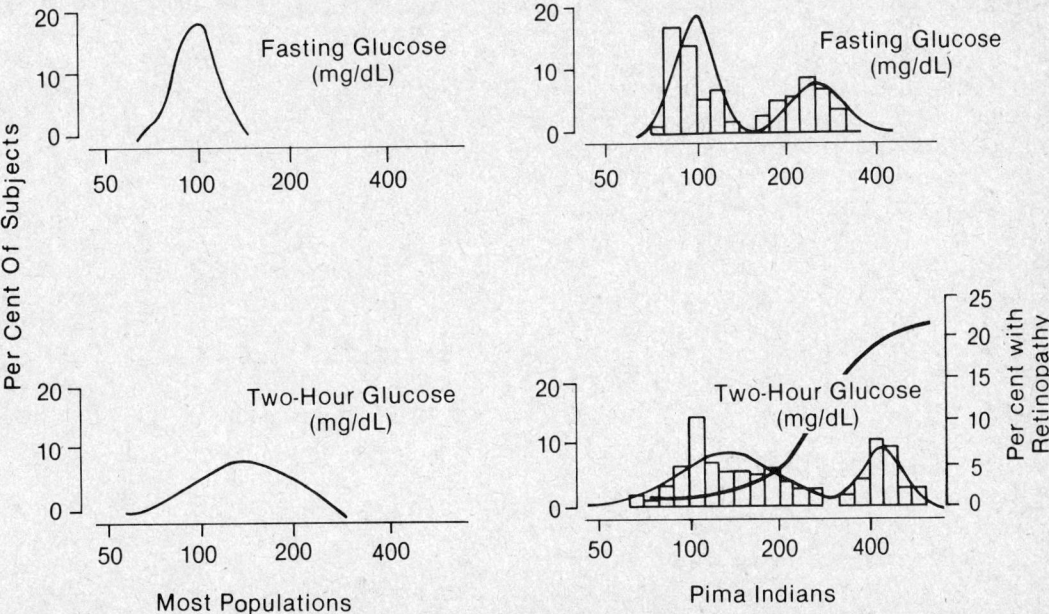

Figure 6-13. Distribution of plasma glucose values in subjects fasting and 2 h after a 75-g oral carbohydrate load. A unimodal distribution is seen in most populations (idealized data, left panel), while a bimodal distribution is found in the Pima Indians (right panel).[29] The prevalence of retinopathy is strikingly increased in subjects with 2-h glucose values over 200 mg/dL (shown as a continuous line superimposed on the 2-h plasma glucose distribution).[22]

subjects with lower values later developed significant hyperglycemia, the development of micro-vascular disease in those persons with initial 2-h plasma glucose values under 200 mg/dL was unusual.

Postprandial plasma glucose. Demonstration of unequivocal hyperglycemia (plasma glucose over 200 mg/dL) 2 h or more after a mixed meal would be considered diagnostic for diabetes mellitus. In a person with a clear elevation of plasma glucose, no further diagnostic testing would be necessary. In a person with a completely normal value (under 140 mg/dL, though most people are usually less than 120 mg/dL) the diagnosis of diabetes mellitus cannot be completely excluded, but for practical purposes, no specific treatment could be recommended for diabetes mellitus at such an early stage. Weight reduction should be recommended for an overweight person regardless of glucose tolerance results, and moderation in dietary sugar intake should be a general rule. For diagnosis in an individual case, the postprandial plasma glucose either will be normal and exclude the diagnosis of diabetes mellitus or will quickly establish the diagnosis where it is clear cut, leaving only a small number of individuals with borderline values who need to be studied further.

Post-challenge plasma glucose. Giving a standard oral glucose load* (75 g is recommended, though 50- and 100-g loads have been used)[68] and determining the plasma glucose concentration 2 h later allows for more objective conditions of testing and a more precise comparison of one individual's response with a large group. This test has the advantage of being more precise than a postprandial glucose, where the meal content, time used to consume a meal, and absorption of the meal cannot be completely controlled. However, the procedure requires more preparation in that the subject must have had unrestricted activity and a diet containing at least 150 g of carbohydrate daily for 3 days before the test; must fast for a specified time (10–16 h); must begin the test in the morning; and must be available for sampling 2 h after ingestion of the glucose load. For the 2-h post-challenge glucose, values over 200 mg/dL would be considered diagnostic of diabetes mellitus.

Effect of age. Postprandial and post-challenge glucose values tend to increase with age.[76] For 2-h values, this increase may be as high as 10 mg/dL per decade after age 40. Thus, cut-off values higher than 200 mg/dL may be found in elderly individuals who do not have diabetes mellitus.

Fasting plasma glucose. An elevation of fasting plasma glucose values is highly indicative of diabetes mellitus. A value over 140 mg/dL would be diagnostic. However, fasting hyperglycemia occurs relatively late in type II diabetes mellitus, so that in a population where the frequency of diabetes mellitus is low, waiting for fasting hyperglycemia to appear would delay the time of diagnosis in a specific case. This would underestimate the prevalence of diabetes mellitus in that population.[107]

Oral glucose tolerance test. The serial measurement of plasma glucose before and after a specific amount of glucose given orally should provide a standard method to evaluate individuals and establish values for normal and disease states. Unfortunately, there are a number of factors which may affect glucose tolerance that should be controlled or eliminated before such a test, and other conditions which will alter the glucose response need to be controlled during the test (Table 6-4). As a result, abnormal values may occur in the absence of diabetes mellitus, and repeat testing in the same individual does not always give reproducible results, even under the same circumstances. When an oral glucose tolerance test is ordered, the following conditions should be met: omit medications known to affect glucose tolerance; perform in A.M. after 3 d of unrestricted diet and activity; and perform after a 10–16 h fast. Plasma glucose should be measured fasting, then every 30 min for 2 h after an oral glucose load. For adults, the recommended load is 75 g; for children, 1.75 g/kg, up to 75 g maximum. There is still some question as to what constitutes the ideal glucose load. In adults, 50 g is submaximal, and 100 g frequently causes nausea and vomiting; 75 g is the consensus but may not be a maximal stimulus.[20,38,68]

Many older but well-accepted standards for the interpretation of an oral glucose tolerance test were set too low because the normative data had been derived from an atypical population, leading to overdiagnosis of diabetes mellitus.[99,116] For an individual who is improperly labeled as diabetic, serious consequences are possible, ranging from anxiety over having a chronic disease to limitations on jobs or insurability and to inappropriate treatment with insulin or oral hypo-glycemic medication.

---

*A commercially prepared and flavored carbohydrate load equivalent to this glucose dose is also acceptable.[68]

Table 6-4. FACTORS THAT AFFECT GLUCOSE TOLERANCE

Factors During the Test That Affect Glucose Tolerance	
Posture	Cigarettes
Nausea	Time of day
Anxiety	Activity
Coffee	Amount of glucose ingested

Factors Before the Test That Affect Glucose Tolerance	
Carbohydrate intake	Propranolol
Time of previous food intake	Corticosteroids
GI surgery and malabsorption	Age
Thiazides	Inactivity
Estrogens	Weight
Diphenylhydantoin	Stress (surgery, infection)

From Watts, N. B. (used with permission).[116]

The National Diabetes Data Group has proposed that a standard set of criteria be adopted for use in the diagnosis of diabetes mellitus (Table 6-5).[68] These criteria are higher than most earlier criteria and are more specific for the diagnosis of diabetes mellitus, though somewhat less sensitive. Using these criteria, some individuals with diabetes mellitus in its early stages will not be identified at that point, but persons with abnormal results can be diagnosed with relative certainty as having diabetes mellitus. These criteria have been widely accepted and should be used to replace previous criteria for interpretation of glucose tolerance test results.

An oral glucose tolerance test is not always necessary for the diagnosis of diabetes mellitus. A clearly elevated plasma glucose after a meal, a high value 2 h after a glucose load, or fasting hyperglycemia will establish the diagnosis without any need for further diagnostic testing, particularly in a person with classic symptoms of diabetes mellitus (thirst, frequent urination, unexplained weight loss). The oral glucose tolerance test is indicated in the following situations:

1. Diagnosis of gestational diabetes mellitus
2. Further evaluation of an individual with a borderline elevation of fasting or postprandial plasma glucose
3. Risk counseling in an individual with a previously abnormal glucose tolerance test under suboptimal conditions (repeat results will often be normal)
4. Risk counseling in persons at high risk for the development of diabetes mellitus, but with normal fasting and postprandial plasma glucose values (a normal result, however, does not mean that diabetes mellitus will not develop in the future)
5. Evaluation of a patient with unexplained nephropathy, neuropathy, or retinopathy. Abnormal results in this setting do not necessarily indicate a cause-and-effect relationship, and further studies will be needed to exclude conclusively other diseases.
6. Population studies for epidemiologic data

Evaluation for gestational diabetes mellitus. Clues from the history of a patient that would raise the suspicion of gestational diabetes include a strong family history of diabetes mellitus, a history of unexplained stillbirth or neonatal death, a history of an infant with a congenital anomaly, the delivery of an infant weighing 4000 g or more, or a poor reproductive history. Suspicious clinical findings include obesity, recurrent monilial infections, hydramnios, or glycosuria. Pregnant women with symptoms of hyperglycemia (thirst, frequent urination, unexplained weight loss) should have a serum glucose measured without delay to look for clear-cut and serious

Table 6-5. CRITERIA FOR DIAGNOSIS OF DIABETES MELLITUS WITH ORAL GLUCOSE TOLERANCE TEST (FROM THE NATIONAL DIABETES DATA GROUP)[68]

Glucose tolerance test is not necessary if fasting plasma glucose exceeds 140 mg/dL or if 2-h postprandial value exceeds 200 mg/dL.

Diabetes mellitus:
  Two-h plasma glucose over 200 mg/dL *and* at least one other value over 200 mg/dL
Impaired glucose tolerance:
  Two-h plasma glucose between 140 mg/dL and 200 mg/dL *and* at least one other value over 200 mg/dL

hyperglycemia; pregnant women with some other reason to suspect gestational diabetes can be evaluated in a more systematic way.

The O'Sullivan screen for gestational diabetes is a simple way to evaluate the significance of a suspicious point from the history or clinical presentation.[77] The woman is given a glucose load after fasting 10–16 h, and plasma glucose is measured afterward. A plasma glucose over 150 mg/dL 1 h after a 50-g glucose load, or a plasma glucose over 140 mg/dL 2 h after a 100-g glucose load would indicate a high probability of gestational diabetes, and these women should be tested further with an oral glucose tolerance test. If the initial screen is normal, the test should be repeated in the third trimester, since carbohydrate tolerance is known to deteriorate in the second half of pregnancy.

When an oral glucose tolerance test is done for the diagnosis of gestational diabetes, the conditions and criteria for diagnosis are different from those when the test is done for the diagnosis of idiopathic diabetes mellitus (Table 6-6).[77]

A 2-h glucose tolerance test is appropriate for the diagnosis of non-insulin-dependent diabetes mellitus. A 3-h glucose tolerance test is used for the diagnosis of gestational diabetes. Longer testing periods or variations on sample times are *not* indicated for the diagnosis of diabetes mellitus or gestational diabetes. The role of the glucose tolerance test and the diagnosis of "hypoglycemia" are discussed later in this chapter.

GLYCOHEMOGLOBINS

Human adult hemoglobin (Hb) typically consists of Hb A (97% of the total), Hb A_2 (2.5%) and Hb F (0.5%). Chromatographic analysis of Hb A has shown that it contains a number of minor hemoglobins identified as Hb A_{1a}, Hb A_{1b}, and Hb A_{1c}. Collectively, they are referred to as Hb A_1, "fast hemoglobins," glycosylated hemoglobins, or simply glycohemoglobins. The Joint Commission on Biochemical Nomenclature of the International Union of Pure and Applied Chemistry recommends the term *neoglycoprotein* for such derivatives, and the new term *glycation* to describe this process. Hemoglobin A consists of four polypeptide chains, two α-chains and two β-chains. Hemoglobin A_{1c} is formed by the condensation of the N-terminal valine amino acid of each β-chain with glucose to form an unstable Schiff base (aldimine, pre-A_{1c}) which then undergoes an Amadori rearrangement to form a stable ketoamine, Hb A_{1c} (Figure 6-14). Of the other minor glycohemoglobins, Hb A_{1a} has been separated further into Hb A_{1a1}, in which the moiety attached to the amino terminal is fructose-1,6-diphosphate, and Hb A_{1a2}, in which the moiety is glucose-6-phosphate. The exact structure of Hb A_{1b} remains uncertain. Hb A_{1c} is the major fraction (about 80%) of Hb A_1.

The formation of the glycohemoglobins is nonenzymatic, occurs over the life span of the red cell (average 120 days), and is proportional to the concentration of glucose in the blood. The amount of total Hb A_1 (or Hb A_{1c}) is related to the time-averaged glucose concentration over the two or three months prior to the measurement. This value provides a single objective parameter reflecting control of diabetes, unrelated to short-term fluctuations in plasma glucose levels. Thus, measurement of glycohemoglobins supplements other more traditional methods of assessing control. It becomes a means for judging control when urine glucose records are inadequate; judging control when blood glucose levels vary markedly through the day or from day to day;

Table 6-6. DIAGNOSIS OF GESTATIONAL DIABETES (GDM)

Oral Glucose Tolerance Test (OGTT) for GDM
Omit medications known to affect glucose tolerance.
Perform in A.M. after 3 days of unrestricted diet and activity, and after 10–16-h fast.
Measure fasting plasma glucose.
Give 100 g of glucose or equivalent orally.
Measure plasma glucose hourly for 3 h.
At least 2 values must exceed the following values:
Fasting: 105 mg/dL
1 h: 190 mg/dL
2 h: 165 mg/dL
3 h: 145 mg/dL
If results are normal in a clinically suspicious situation, repeat OGTT during the third trimester.

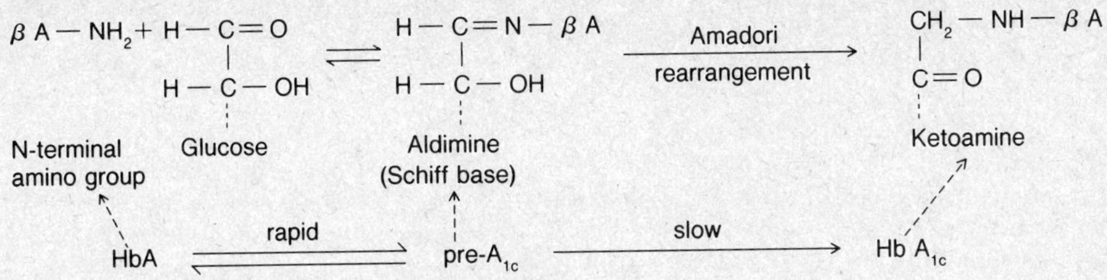

Figure 6-14. Formation of Hb A_{1c}.

judging control in a new patient with known diabetes; providing additional confirmation of the clinical impression of control to reassure the well-controlled patient; urging improvement in the poorly controlled patient; and monitoring patients during pregnancy when close control is especially important.

The interpretation of Hb A_{1c} values is based on the assumption of a normal red cell life span. Patients with hemolytic disease, or other conditions with shortened red cell survival, have a significant reduction in Hb A_{1c}. The effect of other hemoglobin variants (such as F, S, and C) will be considered later with respect to specific methods of analysis.

Most methods for measurement of glycohemoglobins yield values for total Hb A_1, rather than Hb A_{1c} alone. This is of little clinical significance, since the two values have been shown to have a high degree of correlation.[45] Of more concern is the inclusion of the labile intermediate (pre A_{1c}) in most assays, including those by chromatography, HPLC, and electrophoresis.[66] The labile fraction changes rapidly with acute changes in blood glucose concentration; hence, values for Hb A_1 or Hb A_{1c} may show changes not truly reflecting longer time-averaged glucose concentrations. The labile fraction amounts to 5–8 % of total Hb A_1 in normal individuals and ranges from 8–30% in patients with diabetes, depending on the degree of control of blood glucose levels.[43] Reference to Figure 6-14 suggests that pre-A_{1c} should revert to glucose and Hb A in the absence of excess glucose. This is the basis for some procedures in which washed red cells are incubated in saline solution to eliminate the labile fraction. Thus, if the analytical method includes both fractions, the labile pre-A_{1c} should first be removed to avoid factitiously high results.

Nonenzymatic attachment of sugars to amino groups of other proteins also occurs. A stable *glycosylated form of serum albumin* has been described and found to be elevated in humans with diabetes.[21] Since the circulating half-life of human serum albumin is about 19 days, this suggests that monitoring of glycosylated albumin would provide a more recent time-averaging of blood glucose than would Hb A_1.

On a more speculative level, the same chemical reactions that are involved in the glycosylation of hemoglobin may cause similar modifications of tissue proteins related to the complications of diabetes. In addition to serum albumin, examples of nonenzymatic glycosylation include other serum proteins, the red cell membrane, collagen, and crystalline lens. Increased levels of ketoamine-linked glucose have been found in the collagen of glomerular basement membranes from diabetic rats. The fact that such linkages are nearly irreversible may lead to a cumulative effect. The reader is referred to a review on the present status of research on glycoproteins.[96]

METHODS FOR THE DETERMINATION OF GLYCOHEMOGLOBINS

Currently available methods for the determination of glycohemoglobins include ion exchange chromatography, high-performance liquid chromatography, affinity chromatography, colorimetry, spectrophotometry, radioimmunoassay, electrophoresis, and isoelectric focusing. A comparison of six methods has been reported.[73]

Ion-exchange chromatography. In its original form, ion-exchange chromatography for the separation and measurement of Hb A_{1c} was complex and time consuming.[3] Subsequent modifications of hydrolysate preparations, buffers, pH, and resins provided enhanced separation and reduced the time for analysis, but routine laboratory application still did not seem practical.[110] Further study was directed to the use of short columns containing weakly acidic cation-exchange resin or negatively charged carboxymethyl cellulose resin. Glycosylated hemoglobins at a selected ionic strength and pH of the eluent buffer are less positively charged than Hb A and are therefore eluted first. A second buffer may be used to elute Hb A and the absorbance of the two eluates used to calculate per cent of total glycosylated hemoglobin, i.e., $A_{1a} + A_{1b} + A_{1c}$, expressed collectively as Hb A_1. Alternatively, only the Hb A_1 is eluted and a separate dilution of the original hemolysate is made against which the Hb A_1 is compared. A number of commercial modifications have appeared. In some microcolumns, flow rates are accelerated by centrifugation. Simple agitation of resin preparations with hemolysates (batch technique) to adsorb Hb A have also been described. In the latter approach the supernatant solution containing the Hb A_1 fraction is removed by filtration or centrifugation.

In all ion-exchange column methods it is important to control temperature of reagents and columns to obtain accurate and reproducible results. This is best done by thermostatting the columns. Otherwise, a "temperature correction" is applied if the room temperature is different from that of a specified optimum. In addition, rigid control of pH and ionic strength must be maintained.[98] Results with three commonly used commercially available column-chromatographic methods were found to correlate satisfactorily with a high-performance liquid chromatographic method.[37]

The labile Hb pre-A_1 fractions elute with the stable forms and will produce elevated results as discussed above unless some pretreatment of the red cells is included to destroy the labile forms. Hemoglobin F (fetal hemoglobin) also typically elutes with Hb A_1 and produces falsely elevated results. On the other hand, Hb S, Hb C, and their glycosylated derivatives, if present, do not elute; hence, misleading low values for Hb A_1 would be obtained in the presence of Hb S and Hb C since the relative concentration of Hb A is decreased.[1]

High-performance liquid chromatography (HPLC). HPLC has been introduced as a method for separating and quantitating Hb A_{1c} as well as other hemoglobin fractions. Cole et al., by varying the elution rates with phosphate-cyanide buffer at differing pH values, obtained good separation of Hb $A_{1a + b}$ from Hb A_{1c} and Hb A.[14] A more rapid system was also developed in which all the Hb A_1 fractions eluted as one peak. The authors employed Bio-Rex 70 as the resin bed and used a pressure of 400 psi. Eluates were evaluated with a dual wavelength detector at 405 nm for Hb A_1 and 546 nm for Hb A. Replicate analyses from a single normal volunteer over a one-month period showed a coefficient of variation of 2.5%. Further validation of the HPLC procedure has been reported[91] and the system has been recommended as a reference method.[73]

Colorimetry. A colorimetric method has been devised based on the observation that Hb A_{1c}, when subjected to mild acid hydrolysis, releases 5-hydroxymethylfurfural (5-HMF). Oxalic acid is added to an aliquot of hemolysate, the mixture is heated at 100 °C for 5 h and cooled, and proteins are precipitated with trichloroacetic acid. The supernatant is reacted with thiobarbituric acid to produce a final colored solution. Absorbance is measured at 443 nm. Because the reaction is specific for ketoamine-linked glucose, it is unaffected by the presence of Hb F, hemoglobin variants, and labile Hb pre-A_{1c}. However, the test is difficult to standardize because the yield of 5-HMF from Hb A_{1c} is only about 30%. An evaluation of the colorimetric assay and comparison with an ion-exchange chromatographic method have been reported.[80] Under carefully controlled conditions, the assay appears to provide reliable results.

Spectrophotometry. When inositol hexaphosphate (phytic acid) is added to a solution of hemoglobin, a shift in the absorption spectrum occurs, as phytic acid binds to the N-terminal amino groups of the β-chains. Absorbance at 560 nm decreases while that at 633 nm increases. The spectrum of glycosylated hemoglobin is not changed because of the blocking effect of the glucose moiety. The change in absorbance induced by phytic acid is thus inversely proportional to the percentage of glycosylated hemoglobin. This observation forms the basis of a spectrophotometric assay.

One study with a then commercially available kit showed reasonably good correlation for Hb A_1 values between the spectrophotometric method and a microcolumn technique, but absolute values were significantly higher.[115] 2,3-Diphosphoglycerate (DPG), normally present in red cells, binds to the same region of the hemoglobin molecule as phytic acid. This endogenous binding reduces the available sites for phytic acid and leads to higher results for Hb A_1. When 1.8 mmol/L of 2,3-DPG was added to the standards, results showed good agreement. However, addition of higher concentrations of 2,3-DPG caused a progressive increase in apparent Hb A_1. Since endogenous 2,3-DPG concentrations are variable and unknown for a given sample, the method in its present form should be used with caution. Heparin also increases the apparent Hb A₁ values. Another study showed no significant correlation of values obtained by the spectrophotometric assay with either a microcolumn procedure or the colorimetric thiobarbituric acid assay.[73]

Radioimmunoassay. Sheep antiserum against Hb A$_{1c}$ has been prepared and used as the basis for a radioimmunoassay.[47] Partial cross-reactivity to Hb A$_{1a}$ was noted. At present, the antiserum is not commercially available and the method has not been fully evaluated.

Electrophoresis. Agar gel electrophoresis on whole blood hemolysates at pH 6.3 has been shown to provide good resolution of Hb A and Hb A$_1$.[2,63] In addition, Hb S and Hb C and their glycosylated components are resolved and do not interfere. Results agreed well with those obtained by HPLC and by column chromatography. Minor variations in pH, ionic strength, or temperature had no significant effect on results. Hb F migrates to the same region as Hb A$_1$ and would cause a falsely elevated Hb A$_1$ value. This method includes the labile pre A$_{1c}$ aldimine as part of the total Hb A$_1$ fraction.

Isoelectric focusing. Spicer et al. have used isoelectric focusing as a method of quantitating Hb A$_{1c}$.[101] Hemolysates were prepared from washed red cells and diluted to a hemoglobin concentration of 10 g/L with a phosphate-glycerol reagent. Ampholines in the pH range of 6–8 were used to establish the gradient in 1-mm-thick acrylamide gel slabs. Up to 35 samples per gel slab were individually applied to filter paper tabs. On completion of isoelectric focusing, the gels were fixed in 12.5% trichloroacetic acid for 20 min and then scanned on a high-resolution, integrating microdensitometer. Hemoglobin A$_{1c}$ was adequately resolved from hemoglobins A$_{1a}$, A$_{1b}$, S, and F. Results showed good agreement with a column chromatography method. The procedure was also adapted for use with capillary blood samples.

REMOVAL OF LABILE GLYCOHEMOGLOBIN FROM RED BLOOD CELLS

The concentration of the labile form of Hb A$_{1c}$ fluctuates rapidly in response to changes in plasma glucose levels. To reflect accurately long-term blood glucose regulation, especially in poorly controlled diabetics, the labile form should be removed prior to assay for Hb A$_{1c}$ or total Hb A$_1$. This may be accomplished by incubating red cells in saline or in buffer solutions at pH 5–6. Either procedure results in a suspension of red blood cells suitable for use in all procedures discussed above.

Incubation with saline.[34] Collect blood with EDTA anticoagulant, centrifuge, and remove plasma and buffy coat. To 0.5 mL of packed cells add 5 mL of saline (9 g NaCl/L), mix, and centrifuge. Aspirate and discard supernatant. Add 5 mL of saline, mix, and incubate in a 37 °C water bath for 5 h. If more convenient, the mixture may be incubated overnight at room temperature (approximately 14 h at 21–25 °C).[67] Centrifuge and remove supernatant. Add 0.5 mL of saline to the cells, mix, and process for glycohemoglobin assay as for whole blood.

Incubation at pH 5.[5] Collect blood with EDTA anticoagulant, centrifuge, and remove plasma and buffy coat. To 0.5 mL of packed cells add 5 mL of citrate buffer (0.1 mol/L, pH 5.0), mix, and incubate in a 37 °C water bath for 15 min. Centrifuge and remove supernatant. Add 0.5 mL of saline to the cells, mix, and process for glycohemoglobin assay as for whole blood.

To prepare the citrate buffer, dissolve 8.62 g of citric acid monohydrate and 17.35 g of trisodium citrate dihydrate in water and dilute to 1 L.

Determination of Glycosylated Hemoglobin by Affinity Chromatography

Principle

Affinity gel columns are used to separate bound, glycosylated hemoglobin from the nonglycosylated fraction. The gel contains immobilized *m*-aminophenylboronic acid on cross-linked, beaded agarose. The boronic acid reacts with the *cis*-diol groups of glucose bound to hemoglobin to form a reversible 5-membered ring complex, thus selectively holding the glycosylated hemoglobin on the column (Figure 6-15). The nonglycosylated hemoglobin is eluted. The complex is next dissociated by sorbitol, which permits elution of the glycosylated hemoglobin. Absorbances of the bound and nonbound fractions, measured at 415 nm, are used to calculate the per cent of glycosylated hemoglobin.

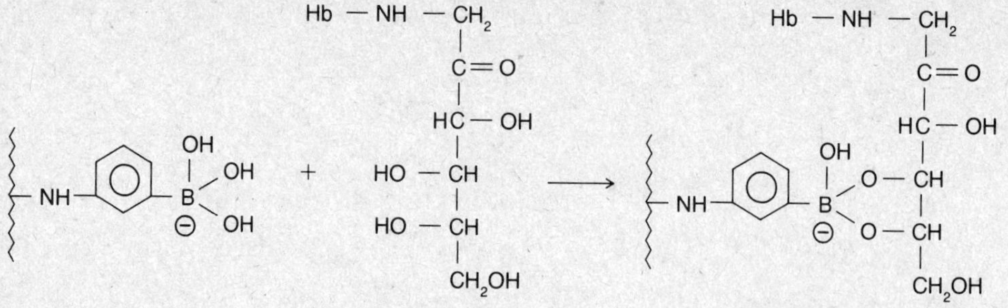

Figure 6-15. Reaction of glycosylated hemoglobin with immobilized boronic acid.

The major advantages of affinity chromatography are (1) no interference from nonglycosylated hemoglobins, (2) negligible interference from the labile intermediate form of Hb A_{1c}, and (3) minimal dependence on variations in ambient temperature in contrast to ion-exchange chromatography. The method has been evaluated in some detail. Columns and reagents are available from Pierce Chemical Co., P. O. Box 117, Rockford, IL 61105.

Specimen

Draw venous blood in blood collection tubes containing EDTA, heparin, or fluoride, and mix well. Whole blood may be stored at 4 °C for 1 week.

Reagents

1. Equilibration wash buffer (WB). One liter of solution contains 250 mmol of ammonium acetate, 50 mmol magnesium chloride, and 0.2 g of sodium azide, adjusted to pH 8.0. Store at room temperature.
2. Elution buffer (EB). One liter of solution contains 200 mmol of sorbitol, 100 mmol of Tris, and 0.2 g of sodium azide, adjusted to pH 8.5. Store at room temperature.
3. HCl solutions, 0.1 mol/L and 0.001 mol/L.

Preparation of Hemolysate

1. Centrifuge the blood specimen and remove the plasma and buffy coat by aspiration.
2. Pipet 100 μL of packed cells to a small test tube. This measurement is not critical.
3. Add 2.0 mL of water and mix well.
4. Let stand about 5 min, remix, and centrifuge. The supernatant should be clear.

Preparation of Columns

The gel columns contain 0.5 mL of immobilized gel (GLYCO-GEL B) and should be stored at 4 °C and protected from direct sunlight. If the gels become darkly colored (red-purple), they should be discarded.

1. Bring columns to room temperature.
2. Remove top stopper only, then pour off and discard the liquid in the column.
3. Remove bottom cap and place column in a suitable rack or test tube.
4. Add 2.0 mL of equilibration wash buffer (WB), let drain, and discard eluate. Column flow stops when the liquid level reaches a disk on the surface of the gel.

Procedure

1. Place the washed column in a clean 16 × 125 mm tube marked *NB* (for *nonbound* fraction).
2. Add 50 μL of clear hemolysate to top of disk. Let drain.
3. Add 0.5 mL of wash buffer (WB) and let drain. This serves to ensure complete transfer of the sample through the disk and onto the gel.
4. Add 5.0 mL of WB and let drain. Total volume of eluate is 5.55 mL.
5. Transfer column to a clean tube marked *B* (for *bound*, or glycosylated fraction).

6. Add 3.0 mL of elution buffer (EB) and let drain.

7. Mix the contents of the NB and B tubes respectively and transfer to corresponding cuvets.

8. Measure the absorbance of NB and B against water at 415 nm.

$$\frac{3.0\,A_{\text{B}}}{5.55\,A_{\text{NB}} + 3.0\,A_{\text{B}}} \times 100 = \text{per cent glycosylated hemoglobin}$$

Regeneration of Columns

This procedure should be done without delay.

1. To the used column add 5 mL of 0.1 mol/L HCl, let drain, and discard eluate.

2. Add 3 mL of HCl, 0.001 mol/L, let drain, and discard eluate.

3. Add 3 mL of HCl, 0.001 mol/L, pick up the column and insert the top stopper. Then place bottom cap over the column tip.

4. Label the column "1" to show it has been used once, "2" if it has been used twice, etc. Columns may be used 5 times and are then discarded.

5. Store all columns in the dark and refrigerated when not in use.

Comments

Values in the reference (normal) range tend to agree well with other procedures. At higher values results are generally greater, presumably owing to the inclusion of glucose attached to free amino groups in hemoglobin as well as to the N-terminal amino groups.

Reference Ranges

Values for glycohemoglobins are usually expressed as a percentage of total blood hemoglobin. Reference (normal) ranges vary, depending on the method, the subfractions measured (Hb A_1 or Hb A_{1c}), and whether the labile fraction is included in the assay. The latter is included in either subfraction for most reports in the literature. A consensus, based on several studies of normal subjects, suggests the following reference ranges.

	Mean, %	Range, %
Hb A_1 (A_{1a+b+c})	6.5	5.0–8.0
Hb A_{1c} only	4.5	3.0–6.0

Reference ranges show some increase with age, in agreement with similar observations of fasting blood glucose levels. In poorly controlled patients with diabetes mellitus, values may extend to twice the upper limit of normal or more but rarely exceed 20%. Values over 20% should prompt further studies to determine the possible presence of Hb F if this is known to interfere in the method.

Reference

Klenk, D. C., Hermanson, G. T., Krohn, R. I., et al.: Determination of glycosylated hemoglobin by affinity chromatography: Comparison with colorimetric and ion-exchange methods, and effects of common interferences. Clin. Chem., 28:2088–2094, 1982.

KETONE BODIES

The metabolism of fatty acids results in the formation of a small amount of acetoacetate, which is subsequently metabolized in the peripheral tissues. In conditions in which there is carbohydrate deprivation (e.g., starvation) or decreased utilization of carbohydrates (e.g., diabetes mellitus), there is an increased production of acetoacetate, and the quantity present may exceed the capacity of the peripheral tissues to metabolize this compound. Thus, the acetoacetate accumulates in the blood; a small part is converted to acetone by spontaneous decarboxylation, whereas the greater part is converted to β-hydroxybutyrate in accordance with the following reactions:

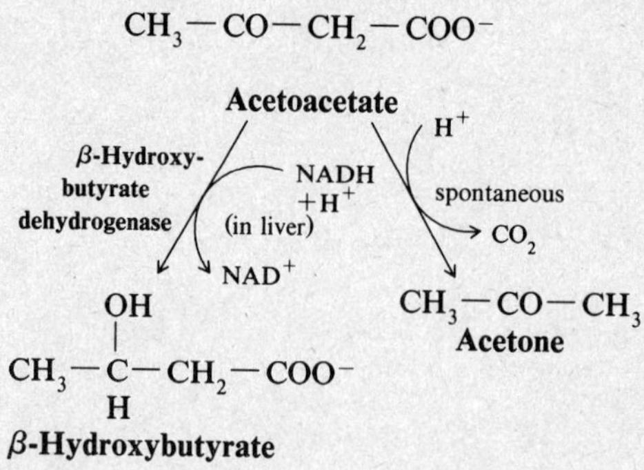

$$CH_3-CO-CH_2-COO^-$$

Acetoacetate

β-Hydroxy-butyrate dehydrogenase   NADH $+H^+$ (in liver) H^+ spontaneous $\rightarrow CO_2$

NAD$^+$

$$CH_3-CO-CH_3$$
Acetone

$$CH_3-\underset{\underset{H}{|}}{\overset{\overset{OH}{|}}{C}}-CH_2-COO^-$$

β-Hydroxybutyrate

The relative proportions in which the three ketone bodies are present in blood may vary; average figures are 78% β-hydroxybutyrate, 20% acetoacetate, and 2% acetone. Of the most commonly used methods for the detection and determination of ketone bodies in serum or urine, none reacts with all three ketone bodies. Gerhardt's ferric chloride test reacts with acetoacetate only; the various tests employing nitroprusside are 15–20 times more sensitive for acetoacetate than for acetone and give no reaction at all with β-hydroxybutyrate. Thus, the tests to be described essentially detect or measure acetoacetate only. Tests for β-hydroxybutyrate are indirect; they require brief boiling of the urine to remove acetone and acetoacetate by evaporation (acetoacetate first breaks down spontaneously to acetone), followed by gentle oxidation of β-hydroxybutyrate to acetoacetate and acetone with peroxide, ferric ions, or dichromate. The acetoacetate thus formed may be detected with Gerhardt's test or one of the procedures employing nitroprusside (see procedure). Determination of β-hydroxybutyrate in urine is not considered to be a routine procedure.

Clinical Significance

Excessive formation of ketone bodies results in increased blood levels (ketonemia) and increased excretion in the urine (ketonuria). This process is observed in conditions associated with a decreased intake of carbohydrates, such as starvation, digestive disturbances, dietary imbalance, and frequent vomiting. A more frequent cause of increased production of ketone bodies is decreased utilization of carbohydrates, such as is found in diabetes mellitus. Both glycogen storage disease (von Gierke's disease) and alkalosis, as a result of some obscure mechanism involving decreased carbohydrate utilization in the liver, may also result in excessive production of ketone bodies.

Semiquantitative determination of ketone bodies in blood is an extremely helpful guide, more so than the determination of these compounds in urine, in the treatment of ketonemia associated with diabetes. In fact, some authors claim that knowledge of the degree of ketonemia is the most valuable guide to insulin therapy and offers more information than knowledge of the degree of ketonuria or of the blood glucose levels.[55] Although this viewpoint is opposed by others, it still points out the great importance of detecting and measuring ketone bodies in serum. Excess production of ketone bodies in the diabetic leads to ketoacidosis with a decrease in plasma pH and bicarbonate. Treatment of such patients must be directed both to the metabolic abnormalities and to the associated acid-base electrolyte problems.

Determination of Ketone Bodies in Serum

Although a number of quantitative and semiquantitative methods for the estimation of ketone bodies have been devised, it is generally agreed that the use of the semiquantitative Acetest and Ketostix reagents (Ames Co., Div. Miles Laboratories, Elkhart, IN 46514) offers information sufficient for clinical purposes.

Specimens

The serum should be free of visible hemolysis, since discoloration of the tablet or reagent strip may occur if an excessive amount of hemoglobin is present. If there is any significant delay in performing the determination, the specimen should be kept well stoppered at refrigerator temperature.

Detection of Ketone Bodies by Acetest

Principle

The Acetest tablets contain a mixture of glycine, sodium nitroprusside, disodium phosphate, and lactose. Acetoacetate or acetone in the presence of glycine will form a complex of lavender-purple color with nitroprusside. The disodium phosphate provides an optimum pH for the reaction, and lactose enhances the color.

Procedure

A detailed procedure for the detection of ketone bodies by Acetest is supplied by the manufacturer with each package of tablets, and the reader is referred to these instructions. Acetest was designed mainly for the detection of ketone bodies in urine. If serum is used, the tablets should be crushed and a drop of serum should be added to the powder. Failure to do so will result in false low results. It has been reported that this procedure is more reliable than is the use of Ketostix described later.[44]

Urine specimens containing very high quantities of phenylketones may give false positive results, as will specimens preserved with 8-hydroxyquinoline. L-Dopa metabolites may produce an atypical reaction which could be interpreted as a positive result.

A positive reaction (appearance of a purple-lavender color) indicates the presence of 5–10 mg or more of ketone bodies per dL. A color chart provided with the package may be used to estimate actual concentrations of the ketone bodies. Approximate values assigned to the color blocks representing increasingly more positive reactions are 20 mg/dL for "small," 30–40 mg/dL for "moderate," and 80–100 mg/dL for "large." If desirable, dilutions of serum with saline can be prepared to measure levels of ketone bodies above 80 mg/dL.* Since a "large" reaction in an undiluted sample corresponds to approximately 80 mg/dL, a "large" reaction in a 2-fold dilution corresponds to approximately 160 mg/dL. Similar calculations can be performed if other dilutions are used.

Detection of Ketone Bodies by Ketostix

Ketostix is a modification of the nitroprusside test in which a reagent strip is used instead of a tablet. The Ketostix test gives a positive reaction within 15 s with a specimen containing 5–10 mg acetoacetate/dL. Approximate serum acetoacetate values assigned to the color blocks representing increasingly more positive reactions are 15 mg/dL for "small," 40 mg/dL for "moderate," and 80 mg/dL for "large." Acetone reacts also, but to a considerably lesser extent.

Determination of Ketone Bodies in Urine

Acetest and Ketostix are also suitable for the detection of ketone bodies in urine. The sensitivity and specificity of the tests are the same as outlined for serum. The original test by Rothera has been essentially replaced by these two modifications.

Gerhardt's test is based on the reaction of ferric chloride with acetoacetate, resulting in the production of a wine red color. Other compounds such as salicylates, phenol, and antipyrine give a similar color;[108] thus, a positive reaction merely indicates the possible presence of acetoacetate. To confirm its presence, urine is heated to decompose acetoacetate to acetone and to drive off the acetone. The test is then repeated. If it is now negative, it can be assumed that the original color was due to acetoacetate. This test also has been replaced by the Acetest and Ketostix procedures.

---

*Since the reaction is affected by proteins, any dilution with saline introduces a certain error.

References

Fraser, J., Fetter, M. C., Mast, R. L., et al.: Studies with a simplified nitroprusside test for ketone bodies in urine, serum, plasma, and milk. Clin. Chim. Acta, *11*:372–378, 1965.

Free, A. H., and Free, H. M.: Nature of nitroprusside reactive material in urine in ketosis. Am. J. Clin. Pathol., *30*:7–10, 1958.

HYPOGLYCEMIA

Hypoglycemia as a laboratory diagnosis has no specific defined limits. It is not uncommon for a plasma glucose concentration as low as 50 mg/dL to be seen several hours after the ingestion of an oral glucose load; even in the fasting state, extremely low blood glucose values may occasionally be seen without symptoms or evidence of underlying disease. The interpretation of a single low blood glucose value is impossible without knowledge of the clinical setting in which it occurs.

There are no symptoms that are specific for hypoglycemia. A rapid fall in plasma glucose will usually trigger the release of epinephrine, and it is epinephrine that accounts for the signs and symptoms which are most commonly attributed to "hypoglycemia":[18] weakness, shakiness, sweating, nausea, rapid pulse, lightheadedness, hunger, and epigastric discomfort. Identical signs and symptoms often occur with other conditions (hyperthyroidism, pheochromocytoma, anxiety). The symptoms may occasionally be triggered by a rapid drop in blood glucose even though the blood glucose itself may not drop below the "normal" range.[19] These symptoms can be termed "adrenergic" symptoms.

A gradual fall in plasma glucose often is not accompanied by epinephrine release or associated adrenergic symptoms. A very low level of plasma glucose (less than 20 or 30 mg/dL) causes impairment of central nervous system function: confusion, lethargy, seizures, loss of consciousness; these symptoms are known as "neuroglycopenia." It is extremely important to differentiate adrenergic signs and symptoms from neuroglycopenia, and to realize that all of these symptoms may be caused by problems other than low blood glucose concentrations.

When an abnormally low blood glucose is found after a period of fasting, there is usually serious underlying organic disease. Such an individual should be studied intensively, and the work-up is best done in the hospital because of the chance that serious permanent neurological damage may occur from severely low blood glucose levels.

Adrenergic symptoms that occur 2–3 h after eating are common. These symptoms are usually mild, and often occur when the blood glucose is either normal or just slightly below normal. In these patients, it is unusual to find any serious or progressive underlying condition to account for such symptoms.

Hypoglycemia in Infancy and Childhood

It is generally accepted that values for plasma glucose below 30 mg/dL in a term infant and below 20 mg/dL in a premature infant are abnormal. There are several comprehensive review articles on the topic of hypoglycemia in the pediatric age group.[13,36,78,94] A list of the common and unusual conditions that need to be considered in the differential diagnosis of hypoglycemia in the neonate or infant is shown in Table 6-7.

Oral glucose tolerance testing is not generally indicated in childhood. If spontaneous hypoglycemia can be demonstrated, appropriate studies for the suspected diagnosis should be undertaken.

Hypoglycemia in Adults

It is extremely helpful when planning an evaluation to consider whether hypoglycemia occurs spontaneously in the fasting state or seems to be triggered by the ingestion of food. Appropriate diagnostic strategies would be considerably different.

Fasting Hypoglycemia

Fasting hypoglycemia in an adult is rare. As previously mentioned, a precise chemical point for separation between the low range of normal and abnormally low is not possible, and values as low as 30 mg/dL may be seen in healthy premenopausal women after a 72-h fast.[64] A fasting

Table 6-7. CAUSES OF HYPOGLYCEMIA IN
NEONATES AND CHILDREN

Transient
  Small for age/prematurity
  Maternal diabetes/toxemia
  Perinatal asphyxia
  Cold stress
  Polycythemia
  Respiratory distress syndrome
Persistent
  Deficient glucose production
    "Ketotic" hypoglycemia
    Glycogen storage disorders
    Deficiency in gluconeogenic enzymes
    Abnormalities of counterregulatory hormones
  Galactosemia
  Hereditary fructose intolerance
  Leucine hypersensitivity
  Endogenous hyperinsulinism
  Idiopathic hypoglycemia

plasma glucose concentration below 50–60 mg/dL is uncommon; a value at or below this level should be viewed with suspicion, and underlying disease should be considered.

Patients with hypoglycemia (or symptoms which suggest hypoglycemia) which occurs in the fasting state will generally have serious organic disease as the cause. Since profound or prolonged hypoglycemia can cause severe and permanent neurologic damage, the evaluation of such patients should be done in the hospital. The initial part of the work-up should be directed at demonstrating the fact that hypoglycemia does occur in the fasting state. Samples for plasma glucose should be obtained frequently during such a fast to anticipate a dangerously low value before it occurs. The patient should be allowed a liberal intake of non-glucose-containing fluids, and samples should be drawn for plasma glucose every 4 h. Most patients with spontaneous hypoglycemia will show an abnormally low value within 12 h of beginning a fast.[95] If the fast has progressed to 48 h with no hypoglycemia demonstrated, the patient should be exercised and final samples drawn before the fast is terminated. It is uncommon to find significant spontaneous hypoglycemia in a patient who tolerates a 48-h fast without a fall in blood glucose.

Conditions to be considered in patients with fasting hypoglycemia are shown in Table 6-8. The demonstration of a low plasma glucose in the presence of an abnormally high plasma insulin value would be highly suggestive of an insulin-producing pancreatic islet cell tumor. Most insulin assays are not sensitive or specific enough to get a precise diagnosis by relying on the insulin value alone, but the finding of a plasma glucose below 30 mg/dL with a plasma insulin above 15 μU/mL would be highly suspicious.[27] Other ways of expressing this relationship (insulin/glucose ratio,[28] amended insulin/glucose ratio[113]) may help in some unusual or borderline situations. Measurement of C-peptide or proinsulin may establish the diagnosis in the rare insulinoma patient in whom the tumor produces more proinsulin than insulin.[4,42,112] Provocative tests (glucagon,[53] tolbutamide,[28,95] or calcium[52]) or suppression tests[111,112] (infusion of pork or beef insulin and measuring C-peptide, or an infusion of fish insulin and measuring insulin) are not generally necessary.

Table 6-8. CAUSES OF SPONTANEOUS (FASTING)
HYPOGLYCEMIA IN ADULTS

Medications (insulin, oral hypoglycemic agents, alcohol)
Severe hepatic dysfunction
Deficiency of adrenal glucocorticoids or other counterregulatory hormones[6]
Extrahepatic neoplasms
Insulin-producing pancreatic tumors
Sepsis

Appropriate historical information should be taken to exclude the possibility of drug-induced hypoglycemia.[93] The C-peptide assay can help in establishing surreptitious injection of insulin (C-peptide should be suppressed).[42] Sepsis as a cause of hypoglycemia should be readily apparent.[65,74] The mechanism for hypoglycemia with sepsis is not well defined. Depleted glycogen stores, impaired gluconeogenesis, and increased peripheral utilization of glucose may all be contributing factors. Laboratory testing can confirm the suspicion of hepatic dysfunction or glucocorticoid deficiency.

Nonpancreatic neoplasms which cause hypoglycemia represent a heterogeneous group.[10,50] Often, these neoplasms are extremely large mesenchymal neoplasms which appear to overutilize glucose and may also have an inhibitory effect on glucose mobilization.[97] Occasionally, small tumors have been reported which cause hypoglycemia, apparently through the production of substances known as insulin-like growth factors or nonsuppressible insulin-like activity (NSILA).[35] Searching for the tumor is the most direct way to arrive at this diagnosis. There are no specific tests to allow the definition of the mechanism of hypoglycemia in most of these individuals under routine clinical circumstances.

The oral glucose tolerance test is not an appropriate study in evaluating a patient suspected of having fasting hypoglycemia.

Postprandial Hypoglycemia[40,57,82]

It is unusual to encounter neuroglycopenia in the postprandial state. In most cases, the question of hypoglycemia will be raised in the patient who complains of adrenergic symptoms occurring approximately 2 h after eating and who seems to obtain relief, lasting 30–45 min, by food intake. There are specific situations where these symptoms are likely to be related to low blood glucose (early diabetes mellitus,[120] where insulin release is exaggerated and delayed, or after gastrointestinal surgery with rapid gastric emptying[92]), but many individuals with adrenergic complaints in the post-meal setting will have these symptoms at a time when their blood glucose is clearly normal. A 5- or 6-h glucose tolerance test has been the standard procedure to establish the presence of postprandial hypoglycemia, but it is not an ideal test.[48] The test is not highly reproducible in any particular individual,[83] and low values for plasma glucose may be seen in the absence of symptoms, while symptoms may accompany normal or near-normal levels.[11] In addition, patients who show a low blood glucose with adrenergic symptoms 3 or 4 h after an oral glucose load may have identical symptoms with a normal blood glucose after a mixed meal.[11,51] Other tests of carbohydrate handling are not particularly helpful in looking at this particular group of patients.

The best diagnostic strategy when the question of postprandial hypoglycemia is considered is to obtain a blood sample under random conditions at the time the patient has symptoms. This can be done if symptoms occur with enough predictability that the patient can be close to a laboratory when symptoms begin, or if the patient has been instructed in a technique of self blood glucose determination and can do a capillary blood glucose when symptoms begin. Finding a normal blood glucose at a time when symptoms are present is strong evidence that the symptoms are not causally related to hypoglycemia.

If it is not possible to obtain a blood glucose at the time symptoms occur with the patient at normal diet and activity, a 5-h meal tolerance test[11,51] has a specific advantage over a 5-h glucose tolerance test in that the meal simulates the composition of a normal diet. Normal and abnormal glucose values are hard to define, but a blood glucose value less than 50 mg/dL with associated adrenergic symptoms would suggest a cause-and-effect relationship.

"Nonhypoglycemia"[9]

In the past, the diagnosis of hypoglycemia has been used to explain a wide variety of disorders which appear to have no direct connection with blood glucose abnormalities. Complaints that may raise the question of hypoglycemia include fatigue, muscle spasms, palpitations, numbness, tingling, pain, sweating, mental dullness, sleepiness, weakness, fainting, and a host of additional nonspecific complaints. Behavior abnormalities, poor school performance, and delinquency have in the past been misattributed to low blood glucose. The widespread use of the insensitive and nonspecific 5-h glucose tolerance test caused a fad of overdiagnosis of hypoglycemia and led the American Diabetes Association to publish a statement[102] to downplay the inappropriate use of the oral glucose tolerance test for the diagnosis of hypoglycemia when applied to these individuals.

Lay publications[26] have added support to this, but it is still important for the medical community to reassure such patients that low blood glucose is not the cause for their symptoms, and to deal with specific abnormalities that might underlie the patient's complaints or problems.

OTHER TOLERANCE TESTS

Intravenous Glucose Tolerance Test

Poor absorption of orally administered glucose may result in a "flat" tolerance curve. Some patients are unable to tolerate a large carbohydrate load orally (vomiting) or may have altered gastric physiology (e.g., gastric resection). In these patients, an intravenous glucose tolerance test may be performed to eliminate factors related to rate of absorption. Plasma insulin assays may also be requested in conjunction with the test.

The dose of glucose is 0.5 g/kg of body weight, given as a 25 g/dL solution. The dose is administered intravenously within 2–4 min and blood is collected every 10 min for 1 h. If insulin assays are performed, a specimen is also obtained 5 min after the start of the injection. Blood glucose levels decrease in an exponential manner, and the rate of glucose disappearance can be calculated from the formula $K = 70/t_{1/2}$, where $t_{1/2}$ is the number of minutes required for the blood glucose to fall to one-half of the 10-min level, and K is the rate of disappearance of blood glucose, expressed as per cent per minute of the 10-min level. To determine $t_{1/2}$, the glucose values are plotted on the log scale of semilog paper versus time on the abscissa. A line drawn through the points is extrapolated to a glucose concentration equal to half the 10-min level. The time interval between 10 min and the extrapolated point is equal to $t_{1/2}$. In normal individuals, K usually exceeds 1.5%; values below 1.0% are considered diagnostic of diabetes. In at least 50% of subjects, results of intravenous glucose tolerance tests do not correlate with the results of oral glucose tolerance tests.[25,30]

In the formula $K = 70/t_{1/2}$, the value of 70 is derived from the logarithmic nature of the decrease in glucose concentration with time. The concentration of glucose at 10 min will be twice that of the value obtained from the plot at $t_{1/2}$. Using natural logarithms, the rate of decrease in glucose concentration, expressed as per cent per min (K), is given by $K = 100 (\ln 2 - \ln 1)/t_{1/2} = 69.3/t_{1/2} \simeq 70/t_{1/2}$.

Insulin Tolerance Test

This test is sometimes used to evaluate patients with resistance to administered insulin or with certain endocrine disorders. The patient is placed on a diet containing at least 300 g of carbohydrate daily for 2 or 3 d before the test. With the patient in the fasting state, blood is taken for a baseline glucose level, after which regular insulin is injected intravenously by a physician in an amount corresponding to 0.1 unit/kg of body weight. Blood specimens are then taken for glucose determinations at 20, 30, 45, 60, 90, and 120 min after the insulin is given. A syringe containing 50 mL of 50 g glucose/dL should be available for intravenous injection. The patient should be observed closely and a physician should be available to make the injection and to terminate the test should a hypoglycemic reaction occur.

Normally the blood glucose decreases to about 50% of the fasting level within 30 min and then returns to normal fasting limits by 90–120 min. There are two types of abnormal response. The insulin-resistant type shows only slight or delayed decrease in blood glucose, and this occurs with adrenal cortical hyperfunction (Cushing's syndrome), in acromegaly, and in some cases of diabetes. In the second type of response the blood glucose falls normally, but the subsequent rise is delayed or does not occur at all. This situation occurs with hypofunction of the anterior pituitary or of the adrenal cortex (Addison's disease), and in hyperinsulinism. In cases of suspected pituitary or adrenal insufficiency it is recommended that half the usual dose of insulin be given and that the patient be watched carefully for signs of hypoglycemia. With severe insulin intolerance an excessive hypoglycemic response with detrimental results (convulsions) can occur. Glucose solutions or fruit juice should normally be given to patients at the end of insulin tolerance tests.

Tolbutamide Tolerance Test

Tolbutamide, 1-butyl-3-(p-tolylsulfonyl)urea (Orinase), is a compound that stimulates the normal pancreas to produce insulin. Following intravenous injection, the normal response is similar to that observed with the insulin tolerance test; the blood glucose decreases to about 50% of the fasting level by 30 min, then returns to normal. If the blood glucose level at 20 min is between 80 and 84% of the fasting value, the patient is said to have a 50% probability of having diabetes. In more severe cases the response will be even less, inasmuch as the pancreas is unable to secrete adequate insulin. The test is also claimed to be valuable in evaluating hypoglycemic states caused by insulinomas. In this condition, injection of tolbutamide results in marked decrease in blood glucose to values in the range of 20–30 mg/dL and persistent hypoglycemia up to 3 h. If the insulin response to tolbutamide is measured, diagnostic data can usually be obtained by 60 min, thus reducing the risk of severe hypoglycemia.[7] As with the insulin tolerance test, patients must be watched carefully for hypoglycemic reactions and the test terminated, if necessary, by intravenous administration of glucose.

Epinephrine Tolerance Test

This test is used to evaluate one form of glycogen storage disease (type I, von Gierke's), a condition in which there is a deficiency or absence of the enzyme glucose-6-phosphatase in the liver. This enzyme is the catalyst for the final step in the formation of blood glucose from hepatic glycogen. Individuals with von Gierke's disease have low blood glucose, increased liver glycogen, but decreased *availability* of liver glycogen as shown by less than normal or no increase in blood glucose following administration of epinephrine. In a normal person, after intramuscular injection of 1 mL of a 1/1000 (1 g/L) solution of epinephrine hydrochloride, the blood glucose increases 35–45 mg/dL in 40–60 min and returns to the fasting level by 2 h. Blood specimens are taken at 30, 45, 60, 90, and 120 min after injection.

Lactose Tolerance Test

A lactose tolerance test can be used to evaluate a deficiency of small bowel mucosal lactase.[72] This has been found to be a rather common condition in healthy adults. Such deficiency may be associated with intolerance to lactose manifested by diarrhea and other symptoms following ingestion of milk. The diarrhea will usually disappear if lactose is eliminated from the patient's diet. The diagnosis is often apparent from the clinical history.

To perform the test, prepare a solution of 50 g of lactose in about 200 mL of lemon-flavored water. For children, a suitable dose is 2 g of lactose/kg of body weight but not to exceed 50 g. Draw a fasting blood specimen, then have the patient consume the lactose over a 5-min period. Draw blood specimens at 15, 30, 45, 60, and 90 min after the lactose is consumed. Analyze plasma for glucose by any method specific for glucose, e.g., hexokinase or glucose oxidase; o-toluidine methods are not recommended, since this reagent also reacts with galactose.

An increase of 30 mg/dL over the fasting glucose level is considered normal. A rise of 20–30 mg/dL is inconclusive. An increase of less than 20 mg/dL is evidence for a deficiency of intestinal lactase. Any abnormal test must be followed by a control test in which 25 g each of glucose and galactose are given to ensure that the individual has normal tolerance to the constituent monosaccharides.

LACTATE AND PYRUVATE

Clinical Significance

Lactic acid, present in *blood* entirely as lactate ion (pK = 3.86), is an intermediary product of carbohydrate metabolism and is derived mainly from muscle cells and erythrocytes. It is normally metabolized by the liver. The blood lactate concentration is, therefore, affected by the rate of production as well as the rate of metabolism. During exercise, lactate levels may increase significantly, from an average normal concentration of about 0.9 mmol/L to about 12 mmol/L. However, pyruvate levels increase under these conditions as well, and the normal ratio of lactate:pyruvate remains approximately 6 or 7:1. Some investigators have reported higher normal ratios, e.g., 9:1 or even higher, owing to variations in technique and sample collection, especially when working under routine conditions.

Severe oxygen deprivation of tissues results in a blockage of aerobic oxidation of pyruvic acid in the tricarboxylic acid cycle and subsequent glycolytic reduction of pyruvate to lactate. This leads to a severe acidosis, called "lactic acidosis," which is associated with a significant increase in the lactate:pyruvate ratio in blood and lactate levels as high as 25 mmol/L (225 mg/dL) or more. Such extreme findings signal deterioration of the cellular oxidative process and are associated with marked hyperpnea, weakness, fatigue, stupor, and finally coma. Conditions of this type are frequently irreversible even when treatment for acidosis and hypoxia is instituted; examples are irreversible stage of shock, diabetic coma without ketosis, and a variety of illnesses in the terminal stage.

Hypoxia with hypoxemia is frequently seen in shock, cardiac decompensation, hematologic disorders, and pulmonary insufficiency. Such conditions are also associated with increases in blood lactate, but the latter are frequently reversible after treatment for hypoxia and the primary condition.

The liver can normally metabolize significantly more lactate than is produced. In the case of decreased perfusion of the liver, however, removal of lactate by the liver may be significantly reduced. Thus, this organ may play an important role in the production of lactic acidosis.

Lactate in cerebrospinal fluid normally parallels blood levels. In case of biochemical alterations in the central nervous system, however, CSF lactate values change independently of blood values. Increased CSF levels are seen in cerebrovascular accidents (CVA), intracranial hemorrhage, bacterial meningitis, epilepsy, and other CNS disorders. In aseptic meningitis, lactate levels in CSF are not usually elevated; hence, CSF lactate has been used to help discriminate between viral and bacterial meningitis. Recent reports suggest, however, that this test has little diagnostic value.[88]

Determination of Lactate in Whole Blood[59,60,62]

Specimens

Collection of a satisfactory specimen for lactate analysis requires special procedures to prevent changes in lactate while and after the specimen is drawn. The patient should be fasting and at complete rest.

If a venous specimen is desired, the specimen is best drawn without the use of a tourniquet or immediately after the tourniquet has been applied. If there is any delay in obtaining the specimen, the tourniquet should be removed after the puncture has been performed, and the blood should be allowed to circulate for at least 2 min before the sample is withdrawn.

Both venous and arterial blood may be collected conveniently in heparinized syringes and immediately delivered into a premeasured amount of chilled protein precipitant, such as metaphosphoric acid or perchloric acid. Alternatively, the blood may be allowed to flow directly into the protein precipitant, preferably after discarding the first few milliliters of blood. The clear supernatant, after centrifugation, is stable at refrigerator temperature for up to 8 d. If blood is not preserved as indicated, lactate will increase rapidly in blood as a result of glycolysis. Increases may be as great as 20% within 3 min and 70% within 30 min at 25 °C. Specimens collected as described are also suitable for the determination of pyruvate.

If plasma is required as specimen, blood is best collected in a container with 10 mg sodium fluoride (NaF) and 2 mg potassium oxalate ($K_2C_2O_4$) per mL of blood, followed by *immediate* chilling of the specimen and separation of the cells *within 15 min.*[89]

Principle

Lactate in the presence of NAD^+ and LDH is oxidized to pyruvate. The NADH formed in this reaction is measured spectrophotometrically at 340 nm, and serves as a measure of the lactate concentration.

$$\text{L-Lactate} + NAD^+ \xrightleftharpoons[\text{pH } 9.0-9.6]{\text{LDH}} \text{Pyruvate} + NADH + H^+$$

The equilibrium of the reaction normally lies far to the left. However, by using a pH of 9.0–9.6 and an excess of NAD^+ and by trapping the reaction product pyruvate with hydrazine, the equilibrium can be shifted to the right. Pyruvate can also be removed by reacting it with L-glutamate in the presence of alanine aminotransferase (ALT). Use of Tris buffer results in faster completion of a side reaction between NAD^+ and hydrazine and prevents the "creeping" of blank values observed when glycine buffer is used.[59]

Metaphosphoric acid is preferred by some as the protein precipitating agent, since perchloric acid does not precipitate mucoproteins and it interferes with the enzymatic method for pyruvate (if this analyte is determined on the same filtrate as lactate). Also, the enzymatic LDH reaction proceeds much more slowly in the presence of perchloric acid.

Because of its high specificity and simplicity, the enzymatic method is the method of choice for measuring lactate, although other methods may also be used, e.g., gas chromatography[89] and colorimetry.[39]

Collection of Specimens

To an appropriate number of tared test tubes (15 × 100 mm) add 6 mL of MPA (5 g/dL). Weigh and place the tubes into ice water. Run assays in duplicate. Collect about 5 mL of arterial or venous blood in a heparinized syringe without production of air bubbles and add 2 mL of the blood to each of two tubes within 15 s of withdrawal.

Mix the blood with the MPA solution by three inversions of the tubes. Once the proteins have been precipitated, there is no further need to chill the tubes. Allow the tubes to warm to room temperature and determine their weights. For complete protein precipitation, allow the tubes to stand for at least 15 min. Centrifuge for 15 min at 3000 × g or for at least 30 min at 1500 × g.[71] The supernatant *must* be clear.

Determine the dilution factor (D) as follows:

$$D = \frac{W_b - W_t}{W_b - W_m}$$

where W_t is the weight of the empty tube, W_m is the weight of the tube plus MPA solution, and W_b is the weight of the tube plus the MPA and blood. If the method is followed as described, the dilution factor (D) is about 4.

Reagents

1. Metaphosphoric acid (MPA), 5 g/dL. Dissolve 5 g of MPA in water and dilute to 100 mL. Prepare fresh daily. (See Comments.)

2. Metaphosphoric acid, 3 g/dL. Dissolve 3 g of MPA in water and dilute to 100 mL. Prepare fresh daily.

3. Tris-hydrazine buffer mixture, pH 9.6 (Tris, 79 mmol/L; hydrazine, 400 mmol/L). To about 700 mL of NaOH, 1 mol/L, add 9.57 g of tris(hydroxymethyl)-aminomethane, 52 g of hydrazine sulfate, and 1.85 g of EDTA · Na_2 (to chelate metal ions and prevent inhibition of lactate dehydrogenase). Adjust the pH to 9.6 with NaOH, 1 mol/L, and dilute the solution to 1000 mL with distilled water. This buffer is stable for 8 d at 4 °C.

4. NAD^+ solution, 27 mmol/L. Dissolve NAD^+ in distilled water to obtain a 20 mg/mL solution. This solution is stable for about 48 h at 4 °C.

5. Lactic dehydrogenase solution. Dilute the stock solution (Sigma, Type III, crystalline bovine heart LDH, 10 mg protein/mL) with saline to a protein concentration of 3 mg/mL (0.3 mL of stock diluted to 1.0 mL). Keep refrigerated and use within 48 h.

6. Lactate standard, 1 mmol/L (9 mg/dL). Dissolve 9.60 mg lithium L-lactate (Calbiochem, A Grade) in water in a 100-mL volumetric flask. Add 25 μL concentrated H_2SO_4 and bring to volume with water. Stable indefinitely at 4 °C. Lactate stock standards are also commercially available.

Procedure

1. To each of three cuvets labeled test, standard, and blank, respectively, add 2.0 mL of Tris-hydrazine buffer.

2. Add 0.1 mL of supernatant solution to the test cuvet, 0.1 mL of lactate standard to the standard cuvet, and 0.1 mL of MPA (3 g/dL) to the blank cuvet.

3. Mix the solutions and add 30 μL of LDH solution and 0.2 mL of NAD^+ solution to each cuvet.

4. Mix the contents again; after approximately 15 min at room temperature, measure the absorbance at 340 nm against the blank. A separate reagent blank and working lactate standard should be determined with each run.

Results obtained with the lactate method are linear up to concentrations of approximately 5.6 mmol/L (50 mg/dL).

Calculation

$$\frac{A_t}{A_s} \times 1 \times D = \text{lactate (mmol/L)}$$

where t and s stand for the test and standard, respectively, and D is the dilution factor.
Alternately, utilizing the absorption coefficient of NADH,

$$A_t \times \frac{2.33}{6.22} \times \frac{D}{0.1} = \text{lactate (mmol/L)}$$

where
A_t = absorbance of the test
2.33 = total volume of the test
6.22 = millimolar absorption coefficient for NADH
0.1 = volume of supernatant
To convert mmol/L to mg/dL, multiply by 9.

Comments

Metaphosphoric acid is generally composed of a variable mixture of HPO_3 and $NaPO_3$. MPA in aqueous solution forms several polymers, $(HPO_3)_x$, and, catalyzed by hydrogen ions, it hydrolyzes to the ortho acid as follows: $HPO_3 + H_2O \rightarrow H_3PO_4$. Since orthophosphoric acid does not precipitate proteins, MPA solutions retain their protein precipitating activity for only about one week at refrigerator temperature.

Reference Ranges

Fasting *venous* blood has a lactate concentration of 0.5–1.3 mmol/L (5–12 mg/dL). *Arterial* blood contains 0.36–0.75 mmol/L (3–7 mg/dL) if the patient is in a state of complete rest. Patients under usual hospital conditions show a wider range, up to 0.9–1.7 mmol/L (8–15 mg/dL) for venous blood and up to 1.25 mmol/L for arterial blood. Sudden, severe exercise increases lactate levels dramatically. Even movement of leg muscles of patients at bed rest may result in significant increases above the normal. Plasma values are about 7% higher than those seen in whole blood, although differences are dependent on the procedure employed. CSF values are normally close to those found in blood but may change independently in CNS disorders. Normal 24-h urine output of lactate is 5.5–22 mmol/d.

Determination of Pyruvate in Whole Blood[60,62]

Principle

The reaction involved in the determination of pyruvate is essentially the reverse of the reaction used in the lactate procedure.

$$\text{Pyruvate} + \text{NADH} + H^+ \xrightleftharpoons{\text{LDH, pH 7.5}} \text{Lactate} + NAD^+$$

At about pH 7.5, the equilibrium constant strongly favors the reaction to the right. The method is very specific, and α-oxoglutarate, oxaloacetate, acetoacetate, and β-hydroxybutyrate do not interfere as is the case with colorimetric methods.

Specimen

Pyruvate in blood is extremely unstable, and the same precautions detailed for lactate should be observed. A protein-free filtrate prepared with metaphosphoric acid (MPA) is suitable for both lactate and pyruvate determinations. The collection of blood and the preparation of a protein-free filtrate are discussed under lactate.

Reagents

1. Tris buffer, 0.75 mol/L. Dissolve 90.8 g of tris(hydroxymethyl)-aminomethane in distilled water and bring to a total volume of 1 L.

2. NADH solution, 0.013 mol/L. Dissolve 10 mg of β-nicotinamide adenine dinucleotide, reduced form, disodium salt, in 1 mL of NaHCO$_3$ solution (1 g/dL). The solution should be refrigerated and used within 48 h.

3. Lactate dehydrogenase solution. Dilute the stock solution (Sigma Type III, crystalline bovine heart LDH, 10 mg protein/mL) with NaCl, 0.15 mol/L, to a protein concentration of 3 mg/mL (e.g., 0.3 mL stock diluted to 1.0 mL). Keep refrigerated and use within 48 h.

4. Pyruvate standards.

    a. Stock pyruvate standard, 0.1 mol/L. Dissolve 1.101 g of sodium pyruvate (Sigma, Type II) in 0.1 mol/L HCl and dilute to a volume of 100 mL. Store in refrigerator.

    b. Working pyruvate standard, 0.05 mmol/L. Dilute the stock solution 2000-fold with MPA, 3 g/dL (e.g., 50 µL diluted to 100 mL). Prepare fresh daily.

Procedure

1. To each of three appropriately labeled cuvets, add 1.0 mL of supernatant solution (test), pyruvate working standard, or MPA, 3 g/dL (blank), respectively, followed by 0.5 mL of Tris buffer, 0.75 mol/L, and 30 µL of NADH solution, 0.013 mol/L. Mix after each addition.

2. Measure absorbances at 340 nm against water.

3. Add to each cuvet 30 µL of LDH solution (3 mg protein/mL), mix, and incubate at room temperature for 2 min.

4. Read absorbances again after 2 min and at additional 1-min intervals thereafter until stable readings are obtained.

Calculation

$$\frac{\Delta A_t - \Delta A_b}{\Delta A_s - \Delta A_b} \times 0.05 \times D = \text{pyruvate (mmol/L)}$$

where t and s indicate test and working standard, respectively, and b indicates blank. D is the dilution factor.

Alternatively, utilizing the absorption coefficient of NADH,

$$(\Delta A_t - \Delta A_b) \times \frac{1.56}{6.22} \times \frac{D}{1.0} = \text{pyruvate (mmol/L)}$$

where

    ΔA_t = change in absorbance of test
    1.56 = total volume in mL
    6.22 = millimolar absorption coefficient for NADH
    1.0 = volume of supernatant in mL
    D = dilution factor

To convert mmol/L to mg/dL, multiply by 8.8.

Reference Ranges

Fasting *venous* blood, drawn with the patient at rest, has a pyruvate concentration of 0.03–0.10 mmol/L (0.3–0.9 mg/dL). *Arterial* blood contains 0.02–0.08 mmol/L (0.2–0.7 mg/dL). Values for cerebrospinal fluid are 0.06–0.19 mmol/L (0.5–1.7 mg/dL).[86] Normal 24-h urine output of pyruvate is approximately 1 mmol/d or less.

Comments

1. Pyruvate in blood is extremely unstable, a significant decrease being observable as early as 1 min after withdrawing the blood. Pyruvate in metaphosphoric acid filtrates of blood is stable for 6 d at room temperature and for 8 d at 4 °C.

2. Pyruvate standard solutions should be prepared fresh as needed. Pyruvic acid polymerizes in solution and may then behave differently in enzymatic reactions.

URINARY SUGARS

Occurrence of Sugars in Urine

Urine is examined routinely to detect or determine the presence or amount of *glucose*; this is done either as a screening procedure or as a guide to insulin therapy. Other sugars may also appear in the urine in certain conditions. The sugars of clinical interest are all reducing sugars; that is, they readily reduce cupric ion in hot alkaline solution. Except for galactose, glucose is the only sugar found in urine that is of pathological significance.

Galactose appears in the urine of infants with galactosemia, a condition characterized by inability to metabolize galactose. Such infants fail to thrive on milk since half of the milk sugar, lactose, is galactose. Failure to discover and control this condition can result in liver disease, mental retardation, and cataracts. *Lactose* is sometimes found in urine of women during lactation and occasionally toward the end of pregnancy. The laboratory may be required to differentiate this sugar from glucose. *Fructose* may appear in the urine after eating fruits, honey, and syrups, but has no significance. Fructosuria also occurs in a rare and harmless congenital defect that should not be confused with diabetes. *Pentoses* may occur in urine after eating such fruits as cherries, plums, or prunes, or as a harmless congenital anomaly and, as with fructose, must be distinguished from glucose. *Maltose* has been reported to occur along with glucose in the urine of some patients with diabetes.

Many *reducing substances* other than sugars may also occur in urine. A partial list of the more important reducing substances is shown in Table 6-9. Ascorbic acid, especially, may be ingested in large amounts or be present in antibiotic preparations administered intravenously. In either case, excess concentrations usually appear in the urine and contribute significantly to the total reducing substances present.

Qualitative Methods for Total Reducing Substances

Principle

Benedict's qualitative reagent contains cupric ion complexed with citrate in alkaline solution. Glucose or other reducing substances reduce cupric ion to cuprous ion with resultant formation of yellow cuprous hydroxide or red cuprous oxide.

Reagent

Dissolve 17.3 g of $CuSO_4 \cdot 5H_2O$ in 100 mL of hot water. Dissolve separately, with heating, 173 g of sodium citrate ($Na_3C_6H_5O_7 \cdot 2H_2O$) and 100 g of Na_2CO_3 in 800 mL of water. Allow to cool, then add the citrate-carbonate solution, with mixing, to the copper sulfate solution. Dilute to 1 L with water. This reagent is stable.

Procedure

Add 8 drops (0.4 mL) of urine to 5 mL of reagent in a test tube. Mix and place in a boiling water bath for 3 min. Remove and examine immediately. Report as 0 to 4+ according to the following criteria:

Appearance	Report	Approximate Glucose Concentration g/dL
Blue to green, no precipitate	0	0–0.1
Green with yellow precipitate	1+	0.3
Olive green	2+	1.0
Brownish-orange	3+	1.5
Brick red	4+	2.0 or more

A convenient adaptation of the above procedure is marketed in tablet form (Clinitest; Ames Co., Elkhart, IN 46514). The tablets contain anhydrous cupric sulfate, sodium hydroxide, citric acid, and sodium bicarbonate. Five drops (0.25 mL) of urine are mixed with 10 drops of water in a test tube. One tablet is added and the mixture is allowed to stand undisturbed for 15 s, remixed, and observed for color. A chart provided by the manufacturer is used to interpret the

Table 6-9. REDUCING SUBSTANCES IN URINE

Fructose	Ketone bodies
Lactose	Sulfanilamide
Galactose	Oxalic acid
Maltose	Hippuric acid
Arabinose	Homogentisic acid
Xylose	Glucuronic acid
Ribose	Formaldehyde
Uric acid	Isoniazid
Ascorbic acid	Salicylates
Creatinine	Cinchophen
Cysteine	Salicyluric acid

result. Heat is generated by contact of sodium hydroxide and water. The initial reaction between citric acid and sodium bicarbonate causes the release of carbon dioxide, which blankets the mixture and reduces contact with oxygen from the air to prevent reoxidation of cuprous ions.

Quantitative Methods for the Determination of Glucose in Urine

Applications of various procedures for the quantitative determination of glucose in urine have been discussed under methods for the determination of glucose in body fluids. The hexokinase or glucose dehydrogenase procedures are recommended for greatest accuracy and specificity. Glucose oxidase procedures that depend only on the consumption of oxygen or the production of hydrogen peroxide are also reliable. Glucose oxidase procedures that include the hydrogen peroxide–peroxidase reaction are not acceptable. The o-toluidine procedure is a satisfactory method for most routine applications.

Separation and Identification of Sugars

Techniques for separating and identifying sugars have included fermentation, optical rotation, osazone formation with phenylhydrazine, specific chemical tests, and paper or thin-layer chromatography. The availability of glucose oxidase test strips has greatly simplified the differentiation of glucose from many other reducing substances. For practical purposes, the urinary sugars of clinical interest are glucose and galactose. Galactose, pentoses, and other reducing sugars do not react with the glucose oxidase strips. Urine from infants and children should be tested routinely by both the glucose oxidase and copper reduction tests to identify individuals with congenital anomalies who might otherwise be diagnosed as diabetics. Nonglucose reducing substances should be further identified by chromatographic procedures.

Qualitative Tests for Individual Sugars

Glucose

A convenient paper test strip is commercially available (Clinistix; Ames Co., Elkhart, IN 46514). The filter paper is impregnated with glucose oxidase, peroxidase, and the dye o-tolidine and provides a simple color test according to principles discussed earlier. The test end is moistened with urine and examined after 10 s. A blue color develops if glucose is present. The sensitivity of the strip has been adjusted to take into account the presence of enzyme inhibitors normally occurring in urine. Thus, a positive test will be obtained with lower concentrations of glucose in water as compared with urine. For the same reason, misleading high results may be obtained with very dilute specimens.

In one study of 2000 urine specimens, 11 false negative enzyme paper tests were encountered. Among the inhibitors identified were ascorbic acid, dipyrone, and meralluride sodium (Mercuhydrin). Several antibiotics contain ascorbic acid as a preservative; the ascorbic acid is largely excreted unchanged and can cause false negative results. Contamination of urine with hydrogen peroxide or a strong oxidizing agent such as hypochlorite (bleach) produces false positive results. For routine examinations, a negative result by the strip test is usually interpreted to mean that the urine specimen is negative for glucose.

Another strip test (Diastix; Ames Co., Elkhart, IN 46514) is designed for the semiquantitative estimation of both glucose and ketone bodies. The glucose portion of the strip utilizes the glucose oxidase–peroxidase method. The peroxide produced oxidizes iodide to iodine, yielding varying intensities of brown color which correspond to the concentration of glucose in the urine. Compared with Clinistix, the Diastix glucose test is much less sensitive to inhibition by ascorbic acid. Other glucose test strips are also available (e.g., Chemstrip; Bio-Dynamics/bmc, Indianapolis, IN 46250).

Seliwanoff's Test for Fructose

Hot hydrochloric acid converts fructose to hydroxymethyl furfural, which links with resorcinol to produce a red-colored compound. To make the reagent, dissolve 50 mg of resorcinol in 33 mL of concentrated hydrochloric acid and dilute to 100 mL with water. Add 0.5 mL of urine to 5 mL of reagent in a test tube and bring to a boil. Fructose produces a red color within 1/2 minute. The test is sensitive to 100 mg fructose/dL provided excess glucose is absent. A 2 g/dL solution of glucose will produce about the same color as 100 mg/dL of fructose after 1/2 min of boiling. A solution of fructose (0.5 g/dL) should be used as a control. With high concentrations of fructose, a red precipitate forms.

Bial's Test for Pentoses

By heating with hydrochloric acid, pentoses are converted to furfural, which reacts with orcinol to form green-colored compounds.

Dissolve 300 mg of orcinol in 100 mL of concentrated hydrochloric acid and add 0.25 mL of ferric chloride solution (10 g/dL). Glucose, if present in the urine, should be removed by fermentation with yeast. Add 0.5 mL of urine to 5 mL of reagent in a test tube and bring to a boil. Pentoses produce a green color. The test is sensitive to 100 mg pentose/dL. A solution of xylose (0.5 g/dL) should be used as a control. Glucuronates will produce a similar color if the boiling is prolonged. Fructose, as with Seliwanoff's reagent, produces a red color.

Identification of Urinary Sugars by Paper Chromatography

Principle

Sugars can be separated by ascending or descending chromatography on paper and located after color development with dinitrosalicylic acid. The variable rates of migration depend upon the solubility of the sugars in the particular solvent system. Presumptive identification is made by comparison of the R_f value of the unknown with those of standards. The following procedure may be performed conveniently in a 6×18 inch Pyrex jar with a tightly fitting cover.

Reagents

1. Solvent. Perform the following procedure under a hood. Mix 60 mL of n-butanol, 40 mL of pyridine, and 30 mL of water. The mixture is completely miscible. Pour into the bottom of the jar, cover, and allow to equilibrate at least 30 min before use.

2. Spray reagent. Dissolve 0.5 g of 3,5-dinitrosalicylic acid in100 mL of sodium hydroxide solution (4 g/dL).

3. Reference sugar solutions. Prepare solutions of glucose, fructose, galactose, maltose, lactose, and xylose to contain 1.75 g of each in 100 mL of benzoic acid solution (2 g/L). These solutions are stable for months.

Procedure

1. Estimate the concentration of sugar in the urine by means of one of the qualitative copper reduction tests. Dilute the specimen, if necessary, to a sugar concentration of approximately 1 g/dL. If the concentration is only 0.5 g/dL, use twice as much sample in the test.

2. Draw a pencil line 1 inch from and parallel to the 10-inch side of a 10×14 inch section of Whatman No. 1 filter paper. Place pencil marks 1 inch apart on the line to indicate starting positions for each reference sample and for the urine specimen.

3. Apply approximately 10 μL of each solution to its respective point. Half of this amount should be applied and permitted to dry before adding the remainder in order to keep the diameter of the spots as small as possible. Allow all samples to dry completely.

4. Arrange the sheet into a 14-inch-high cylinder so that the line of application is at the

bottom. Staple the edges together. Insert the paper into the chromatography jar, tape on the cover, and allow to stand undisturbed for about 16 h (overnight) at room temperature.

5. Remove the sheet and mark with a pencil along the solvent front. Allow to air dry under a hood.

6. Spray the sheet with dinitrosalicylic acid reagent from an atomizer and allow to air dry.

7. Heat the paper at 100 °C for 10 min in a drying oven. The reducing sugars appear as brown spots against a yellow background.

8. Measure the distance from the starting line to the edge of the solvent front. Calculate the ratio of fronts, R_f:

$$R_f = \frac{\text{distance traveled by solute spot}}{\text{distance traveled by solvent front}}$$

The R_f values vary slightly from run to run, and for this reason known reference samples should be included each time. Average values are as follows:

Sugar	R_f
Lactose	0.22
Maltose	0.28
Galactose	0.36
Glucose	0.41
Fructose	0.46
Xylose	0.52

The unknown sugar in the urine is presumed to be the same as a known reference standard when both migrate to the same distance under the test conditions. For confirmation, the urine specimen may be mixed with an equal volume of the known standard and rechromatographed. Only one sugar spot will appear on the paper if the two sugars are identical. Dinitrosalicylic acid is a highly specific reagent for reducing sugars.

Other reagents to detect reducing substances in urine have also been described.[41] Phloroglucinol reagent, for example, produces a faint brown color with galactose, orange-brown with fructose, and greenish-blue with pentoses. No color is produced with lactose, maltose, or glucose. The reagent is prepared by dissolving 0.2 g of phloroglucinol in 80 mL of ethanol (900 mL/L) and diluting to 100 mL with trichloroacetic acid solution (250 g/L).

Identification of urine sugars can also be made by using thin-layer chromatographic techniques as described by Young and Jackson.[122] When frequent chromatographic separations are necessary, this method is preferred over paper chromatography, because of the shorter time period required. If such studies are performed infrequently, paper chromatography is simple, is adequate for most separations, and requires little actual working time.

Reference

Sophian, L. H., and Connolly, V. J.: Chromatographic identification of reducing sugars in urine. Am. J. Clin. Pathol., 22:41–45, 1952.

MUCOPOLYSACCHARIDES

The mucopolysaccharide storage diseases are clinically progressive, hereditary disorders characterized by the accumulation of mucopolysaccharides (MPS, glycosaminoglycans) in various tissues. This accumulation is due to deficiencies of one of a group of enzymes which degrade three classes of mucopolysaccharides: dermatan sulfate, heparan sulfate, and keratan sulfate. The general MPS phenotype includes coarse facies, corneal clouding, hepatosplenomegaly, joint stiffness, hernias, dystosis multiplex, MPS excretion in the urine, and metachromatic staining in peripheral leukocytes and bone marrow.

The mucopolysaccharides are complex molecules which contain repeating disaccharide units (Figure 6-16). *Dermatan sulfate* is composed of alternating sulfated N-acetylgalactosamine and

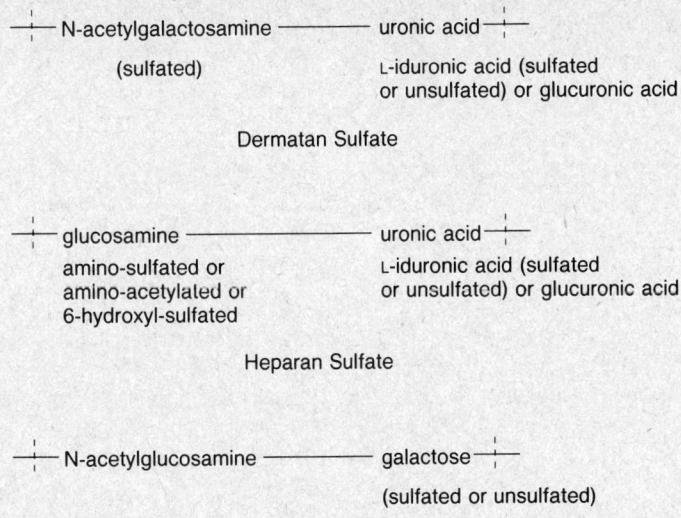

Figure 6-16. Schematic representation of major repeating units of dermatan sulfate, heparan sulfate, and keratan sulfate.

uronic acid residues. The uronic acid may be glucuronic acid, or sulfated or unsulfated L-iduronic acid. *Heparan sulfate* is composed of alternating glucosamine and uronic acid residues. The glucosamine may be sulfated or acetylated on the amino nitrogen as well as sulfated on the 6-hydroxyl group. The uronic acid may be glucuronic acid, or sulfated or unsulfated L-iduronic acid. *Keratan sulfate* is composed of alternating N-acetylglucosamine and galactose residues. The galactose may be sulfated or unsulfated. *Chondroitin sulfates* are composed of alternating sulfated N-acetylgalactosamine and galactose residues.

The sulfated mucopolysaccharides are covalently linked to proteins to form macromolecules, which may be parts of larger aggregates in tissues. The mucopolysaccharidoses are inherited by a pattern of simple Mendelian genetics and, with the exception of the Hunter syndrome, which is an X-linked recessive disease, are autosomal recessive diseases. Table 6-10 indicates the classification of MPS storage diseases, products appearing in the urine in increased amounts, and enzymatic deficiencies. The major MPS excreted by normal individuals is chondroitin-6-sulfate.

Laboratory diagnostic aids for the detection and evaluation of MPS storage diseases have been reviewed by Pennock.[81] A simple screening test for urinary MPS should be performed first. If this is positive, further tests could include the quantitative estimation of MPS by measurement of hexuronic acid content and identification of the MPS excretion pattern by electrophoresis on cellulose acetate or by thin-layer chromatography.

Table 6-10. MUCOPOLYSACCHARIDE STORAGE DISEASES

	Designation	Excess Urinary MPS	Enzyme Deficiency
MPS IH	Hurler syndrome	Dermatan sulfate (DS) Heparan sulfate (HS)	α-L-Iduronidase
MPS IS	Scheie syndrome*	DS and HS	α-L-Iduronidase
MPS II	Hunter syndrome	DS and HS	Iduronate sulfatase
MPS IIIA	Sanfilippo syndrome A	HS	Heparan N-sulfatase
MPS IIIB	Sanfilippo syndrome B	HS	N-Acetyl-α-D-glucosaminidase
MPS IV	Morquio syndrome	Keratan sulfate	N-Acetylgalactosamine-6-sulfatase
MPS VI	Maroteaux-Lamy syndrome	Dermatan sulfate	Arylsulfatase B
MPS VII	β-Glucuronidase deficiency	DS and HS	β-Glucuronidase

*Formerly MPS V.

Screening Test for Mucopolysaccharides in Urine

Specimen

Early morning specimens should not be used for screening tests, and bacterially infected specimens collected at any time are unsuitable for analysis. Either single specimens passed between 0900 and 1800 h or complete 24-h specimens should be collected without preservative and stored frozen until analyzed.

Principle

Under controlled conditions of pH and electrolyte concentration, cetylpyridinium chloride (CPC) reacts with MPS to form an insoluble precipitate. In the presence of citrate buffer at pH 4.8 this precipitate is sufficiently stabilized and dispersed so that the absorbance of the turbid solution may be compared to a similarly treated standard of chondroitin sulfate.

Reagents

1. Citrate buffer, 0.1 mol/L, pH 4.8. Dissolve 9.68 g of citric acid monohydrate and 15.88 g of trisodium citrate dihydrate in water and dilute to 1 L.
2. CPC reagent, 1.0 g/L. Dissolve 1.0 g of cetylpyridinium chloride in 1 L of citrate buffer. This reagent is stable at least one year at room temperature.
3. Standards. Prepare aqueous solutions of chondroitin sulfate to contain 50 and 100 mg/L. Store at 4 °C.

Procedure

1. Measure the specific gravity of the urine and adjust to less than 1.020 with water if necessary. Centrifuge the specimen and use the supernatant for the test. Let standards and urine specimens reach room temperature (20–25 °C) before testing. This is important since CPC precipitates at low temperatures and will do so if added to a cold sample.
2. Measure the creatinine concentration on the supernatant and express as g/L.
3. Mix 1 mL of urine with 1 mL of citrate buffer for a blank.
4. Mix 1 mL of urine with 1 mL of CPC reagent for the test.
5. Mix 1 mL of standard with 1 mL of CPC reagent.
6. Let stand at room temperature for at least 30 min, remix, and measure absorbance of blank, tests, and standards against water at 680 nm. Lower wavelengths may be used but preferably not less than 500 nm. Absorbance (A) is linear up to a concentration of 100 mg/L.

If the absorbance of the unknown exceeds that of the 100 mg/L standard, the urine specimen should be further diluted with water and the entire procedure repeated, including the determination of creatinine on the diluted specimen.

Calculations

$$\frac{A_{test} - A_{blank}}{A_{standard}} \times \frac{\text{concentration of standard (mg/L)}}{10} = \text{CPC units/L}$$

One CPC unit is equivalent to the absorbance given by a solution containing 10 mg of chondroitin sulfate/L. The results are not expressed in mg since different MPS give slightly different turbidity readings at the same concentrations. Calculate CPC units/g of creatinine:

$$\frac{\text{CPC units/L}}{\text{creatinine, g/L}} = \text{CPC units/g creatinine}$$

The upper limits of the normal range for the CPC screening test are given in Table 6-11.

Table 6-11. UPPER LIMITS OF NORMAL FOR CPC
SCREENING TEST FOR MUCOPOLYSACCHARIDES
IN URINE

Age	CPC units/g Creatinine
0–6 months	300
6–12 months	280
1–2 years	262
2 3	244
3–4	228
4–5	213
5–6	198
6–7	185
7–8	173
8–9	161
9–10	150
10–11	140
11–12	131
12–13	122
13–14	114
14–18	100
Adults	50

The upper limit in the first year of life is an arbitrary one chosen to ensure detection of abnormal specimens but which gives a slightly higher percentage of false positive results at this age than is usually found later in life.

Reference

Pennock, C. A.: A review and selection of simple laboratory methods used for the study of glycosaminoglycan excretion and the diagnosis of the mucopolysaccharidoses. J. Clin. Pathol., *29*:111-123, 1976.

References

1. Aleyassine H.: Low proportions of glycosylated hemoglobin associated with hemoglobin S and hemoglobin C. Clin. Chem., *25*:1484-1486, 1979.
2. Aleyassine, H., Gardiner, R. J., Blankstein, L. A., et al.: Agar gel electrophoretic determination of glycosylated hemoglobin: Effect of variant hemoglobins, hyperlipidemia, and temperature. Clin. Chem., *27*:472-475, 1981.
3. Allen, D. W., Schroeder, W. A., and Balog, J.: Observations on the chromatographic heterogeneity of normal adult and fetal human hemoglobin. J. Am. Chem. Soc., *80*:1628-1634, 1958.
4. Alsever, R. N., Roberts, J. P., Gerber, J. G., et al.: Insulinoma with low circulating insulin levels: The diagnostic value of proinsulin measurements. Ann. Intern. Med., *82*:347-350, 1975.
5. Bannon, P.: Effect of pH on the elimination of the labile fraction of glycosylated hemoglobin. Clin. Chem., *28*: 2183, 1982.
6. Boden, G., Reichard, G. A., Hoeldtke, R. D., et al.: Severe insulin-induced hypoglycemia with deficiencies in the release of counterregulatory hormones. N. Engl. J. Med., *305*:1200-1205, 1981.
7. Boehm, T. M., and Lebovitz, H. E.: Statistical analysis of glucose and insulin responses to intravenous tolbutamide: Evaluation of hypoglycemic and hyperinsulinemic states. Diabetes Care, *2*:479-490, 1979.
8. Bush, J. L., Campbell, J., and Sanderson, J. A.: Performance of a glucose procedure based on the glucose dehydrogenase reaction on Technicon continuous flow equipment. Clin. Chem., *27*:1050, 1981.
9. Cahill, G. R., Jr., and Soeldner, J. S.: "A non-editorial on non-hypoglycemia." N. Engl. J. Med., *291*:905-906, 1974.
10. Chandalia, H. B., and Boshell, B. R.: Hypoglycemia associated with extrapancreatic tumors. Report of two cases with studies on its pathogenesis. Arch. Intern. Med., *129*:447-456, 1972.
11. Charles, M. A., Hofeldt, F., Shakelford, A., et al.: Comparison of oral glucose tolerance tests and mixed meals in patients with apparent idiopathic postabsorptive hypoglycemia. Absence of hypoglycemia after meals. Diabetes, *30*:465-470, 1981.
12. Chua, K. S., and Tan, I. K.: Plasma glucose measurement with the Yellow Springs Glucose Analyzer. Clin. Chem., *24*:150-152, 1978.
13. Coates, J. R.: Neonatal hypoglycemia. Med. J. Australia, *1*:11-14, 1974.
14. Cole, R. A., Soeldner, J. S., Dunn, P. J., et al.: A rapid method for the determination of glycosylated hemoglobin using high pressure liquid chromatography. Metabolism, *27*:289-301, 1978.
15. Cooper, G. R.: Methods for determining the amount of glucose in blood. Crit. Rev. Clin. Lab. Sci., *4*:101-145, 1973.
16. Cryer, P. E.: Glucose counterregulation in man. Diabetes, *30*:261-264, 1981.

17. Curme, H. G., Columbus, R. L., Dappen, G. M., et al.: Multilayer film elements for clinical analysis: General concepts. Clin. Chem., 24:1335-1342, 1978.

18. DeFronzo, R. A. Andres, R., Bledsoe, T. A., et al.: A test of the hypothesis that the rate of fall in glucose concentration triggers counterregulatory hormonal responses in man. Diabetes, 26:445-452, 1977.

19. DeFronzo, R. A., Hendler, R., and Christensen, N.: Stimulation of counterregulatory hormonal response in diabetic man by a fall in glucose concentration. Diabetes, 29:125-131, 1980.

20. de Nobel, E., and van't Laar, A.: The size of the loading dose as an important determinant of the results of the oral glucose tolerance test: A study in subjects with slightly impaired glucose tolerance. Diabetes, 27:42-48, 1978.

21. Dolhofer, R., and Wieland, O. H.: Increased glycosylation of serum albumin in diabetes mellitus. Diabetes, 29:417-442, 1980.

22. Dorf, A., Ballintine, E. J., Bennett, P. H., et al.: Retinopathy in Pima Indians. Diabetes, 25:554-560, 1976.

23. Dubowski, K. M.: An o-toluidine method for body-fluid glucose determination. Clin. Chem., 8:215-235, 1962.

24. Duffy, T., Phillips, N., and Pellegrin, F.: Review of glucose tolerance—a problem in methodology. Am. J. Med. Sci., 265:117-133, 1973.

25. Elahi, D., Andersen, D. K., Tobin, J. D., et al.: Discrepant performance on oral and intravenous glucose tolerance tests: The role of gastric inhibitory polypeptide. J. Clin. Endocrinol. Metab., 52:1199-1203, 1981.

26. The fad disease. Hypoglycemia is being diagnosed too often. Time, April 7, 1980, p. 71.

27. Fajans, S. S., and Floyd, J. C., Jr..: Fasting hypoglycemia in adults. N. Engl. J. Med., 294:766-772, 1976.

28. Fajans, S. S., Floyd, J. C., Jr., and Vij, S. K.: Differential diagnosis of spontaneous hypoglycemia. In Endocrinology and Diabetes: The thirtieth Hahnemann symposium. Kryston, L. J., and Shaw, R. A., Eds. New York, Grune & Stratton, 1975, pp. 453-472.

29. Flock, E. V., Bennett, P. H., Savage, P. J., et al.: Biomodality of glycosylated hemoglobin distribution in Pima Indians. Diabetes, 28:984-989, 1979.

30. Ganda, O. P., Day, J. L., Soeldner, J. S., et al.: Reproducibility and comparative analysis of repeated intravenous and oral glucose tolerance tests. Diabetes, 27:715-725, 1978.

31. Genuth, S. M.: Plasma insulin and glucose profiles in normal, obese, and diabetic persons. Ann. Intern. Med., 79:812-822, 1973.

32. Gerich, J. E.: Somatostatin and diabetes. Am. J. Med., 70:619-626, 1981.

33. Gochman, N., and Schmitz, J. M.: Application of a new peroxide indicator reaction to the specific automated determination of glucose with glucose oxidase. Clin. Chem., 18:943-950, 1972.

34. Goldstein, D. E., Peth, S. B., England, J. D., et al.: Effects of acute changes in blood glucose on Hb A_{1c}. Diabetes, 29:623-628, 1980.

35. Gordon, P., Hendricks, C. M., Kahn, C. R., et al.: Hypoglycemia associated with non-islet-cell tumor and insulin-like growth factors. N. Engl. J. Med., 305:1452-1455, 1981.

36. Gutberlet, R. L., and Cornblath, M.: Neonatal hypoglycemia revisited, 1975. Pediatrics, 58:10-17, 1976.

37. Hammons, G. T., Junger, K., McDonald, J. M., et al.: Evaluation of three minicolumn procedures for measuring hemoglobin A_1. Clin. Chem., 28:1775-1778, 1982.

38. Halsbeck, M.: Diagnostiche Probleme bei Diabetes mellitus. Der Internist, 22:187-196, 1981.

39. Henry, R. J., Cannon, D. C., and Winkelman, J. W., Eds.: Clinical Chemistry. Principles and Technics. 2nd ed. Hagerstown, Md., Harper and Row, Publishers, 1974.

40. Hofeldt, F. D.: Reactive hypoglycemia. Metabolism, 24:1193-1208, 1975.

41. Horrocks, R. H., and Manning, G. B.: Partition chromatography on paper. Identification of reducing substances in urine. Lancet, 1:1042-1045, 1949.

42. Horwitz, D. L., Kuzuya, H., and Rubenstein, A. H.: Circulating serum C-peptide. A brief review of diagnostic implications. N. Engl. J. Med., 295:207-209, 1976.

43. Huisman, W., Kuijken, J. P. A. A., Tan-Tjiong, H. L., et al.: Unstable glycosylated hemoglobin in patients with diabetes mellitus. Clin. Chim. Acta, 118:303-309, 1982.

44. James, R. C., and Chase, G. R.: Evaluation of some commonly used semiquantitative methods for urinary glucose and ketone determinations. Diabetes, 23:474-479, 1974.

45. James, T. M., Davis, J. E., McDonald, J. M., et al.: Comparison of hemoglobin A_{1c} and hemoglobin A_1 in diabetic patients. Clin. Biochem., 14:25-27, 1981.

46. Jarrett, R. J., and Keen, H.: Hyperglycemia and diabetes mellitus. Lancet, 2:1009-1012, 1976.

47. Javid, J., Pettis, P. K., Koenig, R. J., et al.: Immunologic characterization and quantification of hemoglobin A_{1c}. Br. J. Haematol., 38:329-337, 1978.

48. Johnson, D. D., Dorr, K. E., Swenson, S. M., et al.: Reactive hypoglycemia. JAMA, 243:1151-1155, 1980.

49. Kadish, A. H., Little, R. A., and Sternberg, J. C.: A new and rapid method for the determination of glucose by measurement of rate of oxygen consumption. Clin. Chem., 14:116-131, 1968.

50. Kahn, C. R.: The riddle of tumor hypoglycemia revisited. Clin. Endocrinol. Metab., 9:335-360, 1980.

51. Kansal, P. C., Buss, R. W., Pino, J. A., et al.: Glucose tolerance test (GTT) vs. meal test (MT) in reactive hypoglycemia. Diabetes, 26(Suppl. 2):403, 1977.

52. Kaplan, E. L., Rubenstein, A. H., Evans, R., et al.: Calcium infusion. A new provocative test for insulinomas. Ann. Surg., 190:501-507, 1979.

53. Kumar, D., Mehtalia, S. D., and Miller, L. V.: Diagnostic use of glucagon-induced insulin response. Ann. Intern. Med., 80:697-701, 1974.

53a. Larner, J.: Mediators of postreceptor action of insulin. Am. J. Med., 74,1A:38-51, 1983.

54. Larsson-Conn, U.: Differences between capillary and venous blood glucose during oral glucose tolerance tests. Scand. J. Clin. Lab. Invest., 36:805-808, 1976.

55. Lee, C. T., and Duncan, G. G.: Diabetic coma; the value of a simple test for acetone in the plasma—an aid to diagnosis and treatment. Metabolism, 5:144-149, 1956.

56. León, L. P., Chu, D. K., Snyder, L. R., et al.: Continuous flow analysis for glucose in serum, with use of hexokinase and glucose-6-phosphate dehydrogenase co-immobilized in tubular form. Clin. Chem., 26:123-129, 1980.

57. Lev-Ran, A., and Anderson, R. W.: The diagnosis of postprandial hypoglycemia. Diabetes, *30*:996-999, 1981.

58. Levine, R., and Haft, D. E.: Carbohydrate homeostasis. N. Engl. J. Med., *283*:175-183, 237-246, 1970.

59. Livesley, B., and Atkinson, L.: Accurate quantitative estimation of lactate in whole blood. Clin. Chem., *20*:1478, 1974.

60. Lubran, M.: Measurement of lactic and pyruvic acid in biological fluids. *In*: Laboratory Diagnosis of Endocrine Diseases. F. W. Sunderman and F. W. Sunderman, Jr., Eds. St. Louis, Warren H. Green, Inc.,1971, pp. 401- 408.

61. Lutz, R. A., and Flückiger, J.: Kinetic determination of glucose with the GEMSAEC (ENI) centrifugal analyzer by the glucose dehydrogenase reaction, and comparison with two commonly used procedures. Clin. Chem., *21*: 1372-1377, 1975.

62. Marbach, E. P., and Wells, M. H.: Rapid enzymatic measurement of blood lactate and pyruvate. Clin. Chem., *13*:314-325, 1967.

63. Menard, L., Dempsey, M. E., Blankstein, L. A., et al.: Quantitative determination of glycosylated hemoglobin A_1 by agar gel electrophoresis. Clin. Chem., *26*:1598-1602, 1980.

64. Merimee, T. J., and Tyson, J. E.: Stabilization of plasma glucose during fasting. N. Engl. J. Med., *291*:1275-1278, 1974.

65. Miller, S. I., Wallace, R. J., Musher, D. M., et al.: Hypoglycemia as a manifestation of sepsis. Am. J. Med., *68*: 649-654, 1980.

66. Nathan, D. M.: Labile glycosylated hemoglobin contributes to hemoglobin A_1 as measured by liquid chromatography or electrophoresis. Clin. Chem., *27*:1261-1263, 1981.

67. Nathan, D. M., Avezzano, E. S., and Palmer, J. L.: A rapid chemical means for removing labile glycohemoglobin. Diabetes, *30*:700-701, 1981.

68. National Diabetes Data Group: Classification and diagnosis of diabetes mellitus and other categories of glucose intolerance. Diabetes, *28*:1039-1057, 1979.

69. Neese, J. W.: Glucose, direct hexokinase method. Selected Methods. Clin. Chem., *9*:241-248, 1982.

70. Neese, J. W., Duncan, P., Bayse, D., et al.: Development and evaluation of a hexokinase/glucose-6-phosphate dehydrogenase procedure for use as a national glucose reference method. HEW Publication No. (CDC) 77-8330. Atlanta, Center for Disease Control, 1976.

71. Neville, J. F., Jr., and Gelder, R. L.: Modified enzymatic methods for the determination of L-(+)-lactic and pyruvic acids in blood. Am. J. Clin. Pathol., *55*:152-158, 1971.

72. Newcomer, A. D., McGill, D. B., Thomas, P. J., et al.: Prospective comparison of indirect methods for detecting lactase deficiency. N. Engl. J. Med., *293*:1232-1236, 1975.

73. Niederau, C. M., and Reinauer, H.: Analyseverfahren für glykosidierte Hämoglobine. Ein Methodenvergleich. J. Clin. Chem. Clin. Biochem., *19*:1097-1101, 1981.

74. Nouel, O., Bernuau, J., Rueff, B., et al.: Hypoglycemia. A common complication of septicemia in cirrhosis. Arch. Intern. Med., *141*:1477-1478, 1981.

75. Olefsky, J. M.: Insulin resistance and insulin action. Diabetes, *30*:148-162, 1981.

76. O'Sullivan, J. B.: Age gradient in blood glucose levels. Diabetes, *23*:713-715, 1974.

77. O'Sullivan, J. B., Mahan, C. M., Charles, D., et al.: Screening criteria for high-risk gestational diabetic patients. Am. J. Obstet. Gynecol., *116*:895-900, 1973.

78. Pagliara, A. S., Karl, I. E., Haymond, M., et al.: Hypoglycemia in infancy and childhood. J. Pediatr., *82*:365-379, 558-577, 1973.

79. Passey, R. B., Gillum, R. L., Fuller, J. B., et al.: Evaluation and comparison of 10 glucose methods and the reference method recommended in the proposed product class standard (1974). Clin. Chem., *23*:131-139, 1977.

80. Pecoraro, R. E., Graf, R. J., Halter, J. B., et al.: Comparison of a colorimetric assay for glycosylated hemoglobin with ion-exchange chromatography. Diabetes, *28*:1120-1125, 1979.

81. Pennock, C. A.: A review and selection of simple laboratory methods used for the study of glycosaminoglycan excretion and the diagnosis of the mucopolysaccharidoses. J. Clin. Pathol., *29*:111-123, 1976.

82. Permutt, M. A.: Postprandial hypoglycemia. Diabetes, *25*:719-733, 1976.

83. Permutt, M. A., Delmez, J., and Stenson, W.: Effects of carbohydrate restriction on the hypoglycemic phase of the glucose tolerance test. J. Clin. Endocrinol. Metab., *43*:1088-1093, 1976.

84. Pfeifer, M. A., Halter, J. B., and Porte, D., Jr.: Insulin secretion in diabetes mellitus. Am. J. Med., *70*:579-588, 1981.

85. Plasse, N. J.: Monitoring blood glucose at home. A comparison of three products. Am. J. Nursing, *81*:2028-2029, 1981.

86. Pryce, J. D., Gant, P. W., and Saul, K. J.: Normal concentrations of lactate, glucose, and protein in cerebrospinal fluid, and the diagnostic implications of abnormal concentrations. Clin. Chem., *16*:562-565, 1970.

87. Roth, J.: Insulin receptors in diabetes. Hospital Pract., *15*(5):98-103, 1980.

88. Rutledge, J., Benjamin, D., Hood, L., et al.: Is the CSF lactate measurement useful in the management of children with suspected bacterial meningitis? J. Pediatr., *98*:20-24, 1981.

89. Savory, J., and Kaplan, A.: A gas chromatographic method for the determination of lactic acid in blood. Clin. Chem., *12*:559-569, 1966.

90. Sazama, K., Robertson, E. A., and Chester, R. A.: Is antiglycolysis required for routine glucose analysis? Clin. Chem., *25*:2038-2039, 1979.

91. Schifreen, R. S., Hickingbotham, J. M., and Bowers, G. N., Jr.: Accuracy, precision, and stability in measurement of hemoglobin A_{1c} by "high-performance" cation-exchange chromatography. Clin. Chem., *26*:466-472, 1980.

92. Schultz, K. T., Neelon, F. A., Nilsen, L. B., et al.: Mechanisms of postgastrectomy hypoglycemia. Arch. Intern. Med., *128*:240-246, 1971.

93. Seltzer, H. S.: Drug-induced hypoglycemia: A review based on 473 cases. Diabetes, *21*:955-966, 1972.

94. Senior, B.: Neonatal hypoglycemia. N. Engl. J. Med., *289*:790-793, 1973.

95. Service, F. J., Dale, A. J. D., Elveback, L. R., et al.: Insulinoma: Clinical and diagnostic features of 60 consecutive cases. Mayo Clin. Proc., *51*:417-429, 1976.

96. Sharon, N., and Lis, H.: Glycoproteins: Research booming on long-ignored ubiquitous compounds. Chem. Eng. News, 59(13):21-44, 1981.

97. Silbert, C. K., Rossini, A. A., Ghazvinian, S., et al.: Tumor hypoglycemia: Deficient splanchnic glucose output and deficient glucagon secretion. Diabetes, 25:202-206, 1976.

98. Simon, M., and Eissler, J.: Critical factors in the chromatographic measurement of glycohemoglobin (Hb A$_1$). Diabetes, 29:467-474, 1980.

99. Siperstein, M. D.: The glucose tolerance test: A pitfall in the diagnosis of diabetes mellitus. Adv. Intern. Med., 20:297-323, 1975.

100. Spencer, W. W., Sylvester, D., and Nelson, G. H.: Evaluation of a glucose method in which a hydrogen peroxide electrode is used. Clin. Chem., 24:386-387, 1978.

101. Spicer, K. M., Allen, R. C., and Buse, M. G.: A simplified assay of hemoglobin A$_{1c}$ in diabetic patients by use of isoelectric focusing and quantitative microdensitometry. Diabetes, 27:384-388, 1978.

102. Statement on hypoglycemia. Diabetes, 22:137, 1973.

103. Steiner, D. F.: Insulin today. Diabetes, 26:322-340, 1977.

104. Stevens, A. D.: Monitoring blood glucose at home. Who should do it. Am. J. Nursing, 81:2026-2027, 1981.

105. Sundaram, P. V., Blumenberg, B., and Hinsch, W.: Routine glucose determination in serum by use of an immobilized glucose dehydrogenase nylon-tube reactor. Clin. Chem., 25:1436-1439, 1979.

106. Symposium on blood glucose self-monitoring. Diabetes Care, 4:392-426, 1981.

107. Taylor, R., and Zimmet, P.: Limitation of fasting plasma glucose for the diagnosis of diabetes mellitus. Diabetes Care, 4:556-568, 1981.

108. Thomas, G. H., and Howell, R. R.: Selected Screening Tests for Genetic Metabolic Diseases. Chicago, Year Book Medical Publishers, Inc., 1973.

109. Trinder, P.: Determination of glucose in blood using glucose oxidase with an alternative oxygen acceptor. Ann. Clin. Biochem., 6:24-27, 1969.

110. Trivelli, L. A., Ranney, H. M., and Lai, H.-T.: Hemoglobin components in patients with diabetes mellitus. N. Engl. J. Med., 284:353-357, 1971.

111. Turner, R. C., and Harris, E.: Diagnosis of insulinomas by suppression tests. Lancet, 2:188-190, 1974.

112. Turner, R. C., and Heding, L. G.: Plasma proinsulin, C-peptide and insulin in diagnostic suppression tests for insulinoma. Diabetologia, 13:571-577, 1977.

113. Turner, R. C., Oakley, N. W., and Nabarro, J. D. N.: Control of basal insulin secretion, with special reference to the diagnosis of insulinomas. Br. Med. J., 2:132-135, 1971.

114. Unger, R. H., and Orci, L.: Glucagon and the A cell. N. Engl. J. Med., 304:1518-1524, 1575-1580, 1981.

115. Wålinder, O., Ronquist, G., and Fager, P.-J.: New spectrophotometric method for the determination of hemoglobin A$_1$ compared with a microcolumn technique. Clin. Chem., 28:96-99, 1982.

116. Watts, N. B.: Oral glucose tolerance test: Nondiagnostic. Lab '79, 2:28-33, 1979.

117. Weissman, M., and Klein, B.: Evaluation of glucose determinations in untreated serum samples. Clin. Chem., 4:420-422, 1958.

118. Wigand, J. P., and Blackard, W. G.: Downregulation of insulin receptors in obese man. Diabetes, 28:287-291, 1979.

119. Wright, W. R., Rainwater, J. C., and Tolle, L. D.: Glucose assay systems: Evaluation of a colorimetric hexokinase procedure. Clin. Chem., 17:1010-1015, 1971.

120. Yalow, R. S., and Berson, S. A.: Dynamics of insulin secretion in hypoglycemia. Diabetes, 14:341-349, 1965.

121. Yee, H. Y., and Goodwin, J. F.: Evaluation of some factors influencing the o-toluidine reaction with glucose. Anal. Chem., 45:2162-2165, 1973.

122. Young, D. S., and Jackson, A. J.: Thin-layer chromatography of urinary carbohydrates. A comparative evaluation of procedures. Clin. Chem., 16:954-959, 1970.

CHAPTER 7

LIPIDS, LIPOPROTEINS, AND APOLIPOPROTEINS*

by Evan A. Stein, M.D., Ph.D.

Of the numerous different lipids known to exist in humans, only a limited number are of clinical and analytical importance. Lipids are ubiquitous in the body tissues and play an important role in virtually all aspects of biological life—serving as hormones or hormone precursors, aiding in digestion, providing energy storage and metabolic fuels, acting as functional and structural components in biomembranes, and forming insulation to allow nerve conduction or to prevent heat loss.

In the last decade, however, the use of the term lipids in clinical chemistry has become virtually synonymous with lipoprotein metabolism and a form of arterial disease known as atherosclerosis (see p. 853). Much of this association has been due to the large national and international collaborative analytical, epidemiological, and clinical studies conducted by the Lipid Metabolism Section of the National Heart, Lung and Blood Institute (NHLBI) under the Lipid Research Clinics (LRC) Program.

CHEMISTRY AND PHYSIOLOGY OF LIPIDS

The term lipid applies to a class of compounds which are soluble in organic solvents and nearly insoluble in water. Chemically, lipids are either compounds that yield fatty acids on hydrolysis or complex alcohols that can combine with fatty acids to form esters. Some lipids are more complex, containing nonlipid groups such as sialic, phosphoryl, amino, or sulfate groups. These groups tend to increase the solubility of lipids in polar solvents. Lipids can be broadly subdivided into five groups based on their chemical structure (Table 7-1).

CHOLESTEROL

Although every living organism examined has been found to contain sterols, cholesterol is found almost exclusively in animals and man, where it is also the main sterol. Virtually all cells and body fluids contain some cholesterol. Like other sterols, cholesterol is a solid alcohol of high molecular weight and possesses the tetracyclic perhydrocyclopentanophenanthrene skeleton. The molecule contains 27 carbon atoms, numbered as shown in Figure 7-1. Knowledge of this sterane skeleton and numbering system is important not only to the clinical chemist but also to the practicing clinician because cholesterol is the initial starting point in many metabolic pathways. These include vitamin D synthesis (Chapter 12), steroid hormone synthesis (Chapter 9), and bile acid metabolism (Chapter 13). Since the enzymes modifying the sterane ring or its radicals in

*Portions of this chapter have been adapted from "Lipids and Lipoproteins" by Ralph D. Ellefson, Ph.D., and Wendell T. Caraway, Ph.D., *Fundamentals of Clinical Chemistry*, 2nd edition. Helpful contributions by Ronald J. Whitley, Ph.D., University of Kentucky Medical Center, are also gratefully acknowledged.

Table 7-1. CLASSIFICATION OF CLINICALLY IMPORTANT LIPIDS

Sterol derivatives	*Glycerol esters*
Cholesterol and cholesteryl esters	Triglycerides (triacylglycerols)
Steroid hormones	Phosphoglycerides
Bile acids	*Sphingosine derivatives*
Vitamin D	Sphingomyelin
Fatty acids	Glycosphingolipids
Short chain (2 to 4 carbon atoms)	*Terpenes (isoprene polymers)*
Medium chain (6 to 10 carbon atoms)	Vitamin A
Long chain (12 to 26 carbon atoms)	Vitamin E
Prostaglandins	Vitamin K

each metabolic pathway are known by both their site and type of reaction (e.g., 21-hydroxylase in cortisol synthesis), the diagnosis of many disease states consequently depends on isolating the site of enzyme dysfunction (e.g., 21-hydroxylase deficiency in adrenogenital syndrome).

Cholesterol Absorption

Cholesterol is presented to the intestinal wall from three sources: the diet, bile and intestinal secretions, and cells. Animal products, especially meat, egg yolk, seafood, and whole-fat dairy products, provide the bulk of dietary cholesterol. Although cholesterol intake varies considerably according to the dietary intake of animal products, the average American diet is estimated to contain ~400–700 mg of cholesterol per day. A similar amount of cholesterol is present in the gut from biliary secretion and the turnover of mucosal cells. Practically all cholesterol in the intestine is present in the unesterified (free) form. Esterified cholesterol in the diet is rapidly hydrolyzed in the intestine to free cholesterol and free fatty acids by cholesterol esterases in pancreatic and small intestinal secretions.

In order to be absorbed, unesterified cholesterol must first be solubilized. This solubility is achieved by formation of mixed micelles containing the unesterified cholesterol, fatty acids, monoglycerides (monoacylglycerols), phospholipids (lysolecithin), and conjugated bile acids. Due to their amphipathic properties, the bile acids are the most important factor affecting micelle formation and therefore cholesterol absorption. In the absence of bile acids, digestion and absorption of both cholesterol and triglyceride are severely impaired. The quantity of dietary cholesterol that can be absorbed appears to be dependent on the amount that can be solubilized by micelles. On the average, 30–60% of dietary and intestinal cholesterol is absorbed daily. With increments in dietary cholesterol, additional cholesterol is absorbed to a maximum of ~1 g/d when the oral intake reaches 3 g/d. Cholesterol absorption is also affected by the form in which cholesterol enters the digestive system. The crystalline form has a lower absorption than cholesterol occurring in a natural state (egg yolks) or dissolved in oil. The ability of cholesterol to enter micelles is also influenced by the quantity, but not the degree of saturation, of dietary fat. Increased amounts of fat in the diet (98% triglycerides) result in expansion of mixed micelles which, in turn, allows for more cholesterol to be solubilized and absorbed.

Maximum absorption of cholesterol occurs in the small intestine (middle and terminal ileum), where micelles contain significant amounts of fatty acids and monoglycerides. As absorption of fat and cholesterol occurs in the small intestine, the micelles break up, thus reducing further cholesterol absorption. Formation of mixed micelles aids cholesterol absorption by both solu-

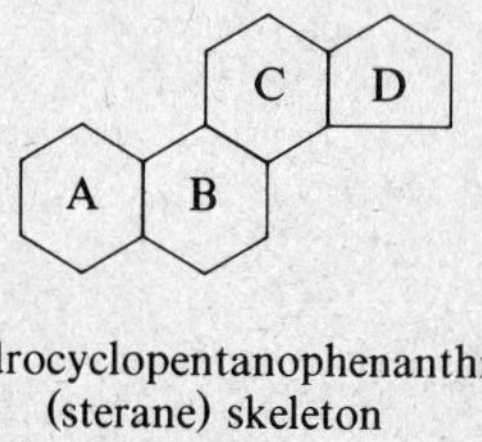

Perhydrocyclopentanophenanthrene
(sterane) skeleton

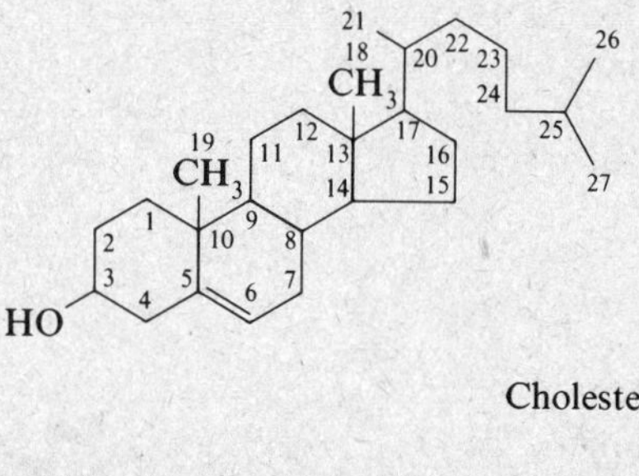

Cholesterol

Figure 7-1. Structure of cholesterol.

bilizing the cholesterol and facilitating its transport across the surface of the luminal cell. Although the specific role of micelles in cholesterol transfer across the cell membrane is not entirely clear, it is probably mediated by bile acids.

In addition to animal cholesterol, ~200–300 mg of plant sterols are ingested daily. The most common plant sterol is β-sitosterol. Plant sterols differ from cholesterol only by small variations on the sterol side chain. Despite their close similarity to cholesterol, plant sterols are poorly absorbed. When plant sterols are administered in amounts of 5–15 g/d, they significantly inhibit the absorption of cholesterol. Although the exact mechanism for reducing cholesterol absorption has not been determined, plant sterols have been used therapeutically in patients with elevated plasma cholesterol levels.

Following its absorption into the mucosal cell, cholesterol is reassembled along with triglycerides, phospholipids, and specific apoproteins into a large micelle called a chylomicron. The apoprotein component is vital to the formation of chylomicrons, and in subjects with a rare deficiency of apolipoprotein B, chylomicron formation and consequently fat and cholesterol absorption are severely impaired. Chylomicrons enter the lymphatics, which empty into the thoracic duct, and eventually enter the systemic venous circulation at the junction of the left subclavian and left internal jugular vein.

Cholesterol Synthesis

Although a portion of the body's cholesterol is derived from dietary intake, the majority of tissue and plasma cholesterol is synthesized endogenously by the liver and other tissues from simpler molecules, particularly acetate. Knowledge of the endogenous cholesterol synthetic pathway has assumed great significance in the last decade as agents have been sought to suppress or decrease endogenous cholesterol synthesis. The need for clinical medicine to understand the fundamental biochemistry of this pathway was underscored by the triparanol disaster of 1960. Triparanol is a drug which inhibits the last step in the endogenous cholesterol synthetic path (desmosterol to cholesterol), but it does not inhibit the rate-limiting step which is now known to be 3-hydroxy-3-methylglutaryl-CoA (HMG-CoA) reductase. When triparanol was used to treat hypercholesterolemia in 1960, the drug caused tissue accumulation of desmosterol and appearance of cataracts, alopecia, and atherosclerosis in treated patients. More recently two drugs, Compactin and Mevinolin, have been shown to suppress HMG-CoA reductase selectively and thereby to lower serum cholesterol levels significantly. The pathway for cholesterol synthesis is shown in Figure 7-2; note that the precursor, acetyl-CoA, is a pivotal compound in intermediary metabolism.

Although essentially all cells have the capacity to synthesize cholesterol from acetyl-CoA, almost 90% of synthesis occurs in the liver and gut; peripheral cells and other organs depend on cholesterol delivery from the circulation. Hepatic synthesis of cholesterol appears to be inhibited by newly absorbed cholesterol which reaches the liver in the chylomicron remnant. This feedback mechanism assists in control of the cholesterol body pool by adjusting the rate of endogenous synthesis against the rate of dietary absorption. It is uncertain whether cholesterol produced by endogenous synthesis and transported by lipoproteins has a similar feedback effect on the liver.

Stage 1

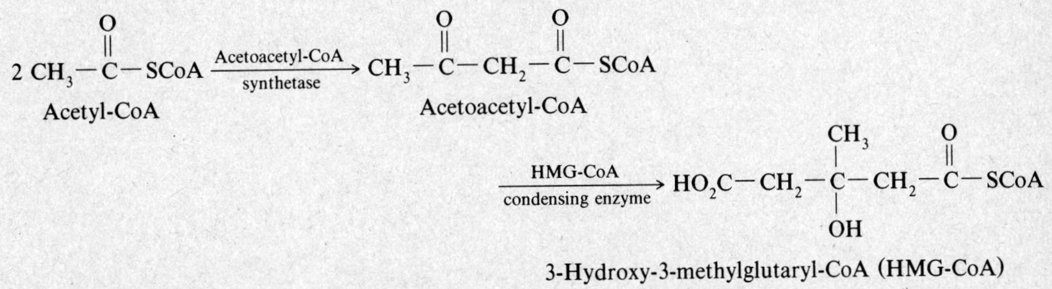

Figure 7-2. Cholesterol biosynthesis.

Illustration continued on following page

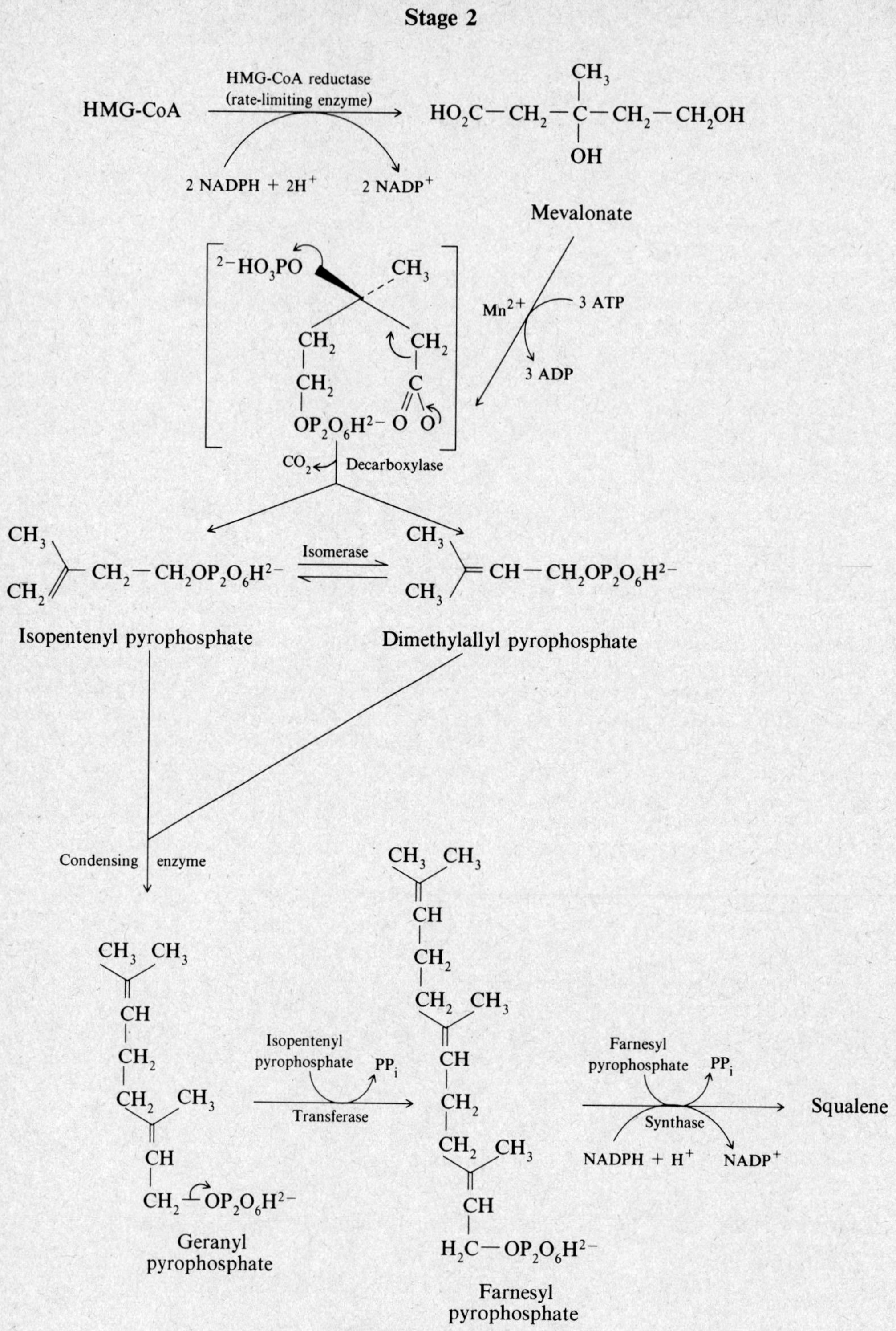

Stage 2

Figure 7-2. *Continued*

Cholesterol biosynthesis is best conceptualized as occurring in three stages (Figure 7-2). In the first stage, acetyl-CoA moieties derived from carbohydrates, amino acids, and fatty acids are condensed to form the 6-carbon thioester, HMG-CoA. In the second stage, HMG-CoA is reduced and decarboxylated to 5-carbon isoprene units. These isoprene units are condensed to form first a 10-carbon and then a 15-carbon intermediate, the latter being farnesyl pyrophosphate. Two of these C_{15} molecules combine to produce the

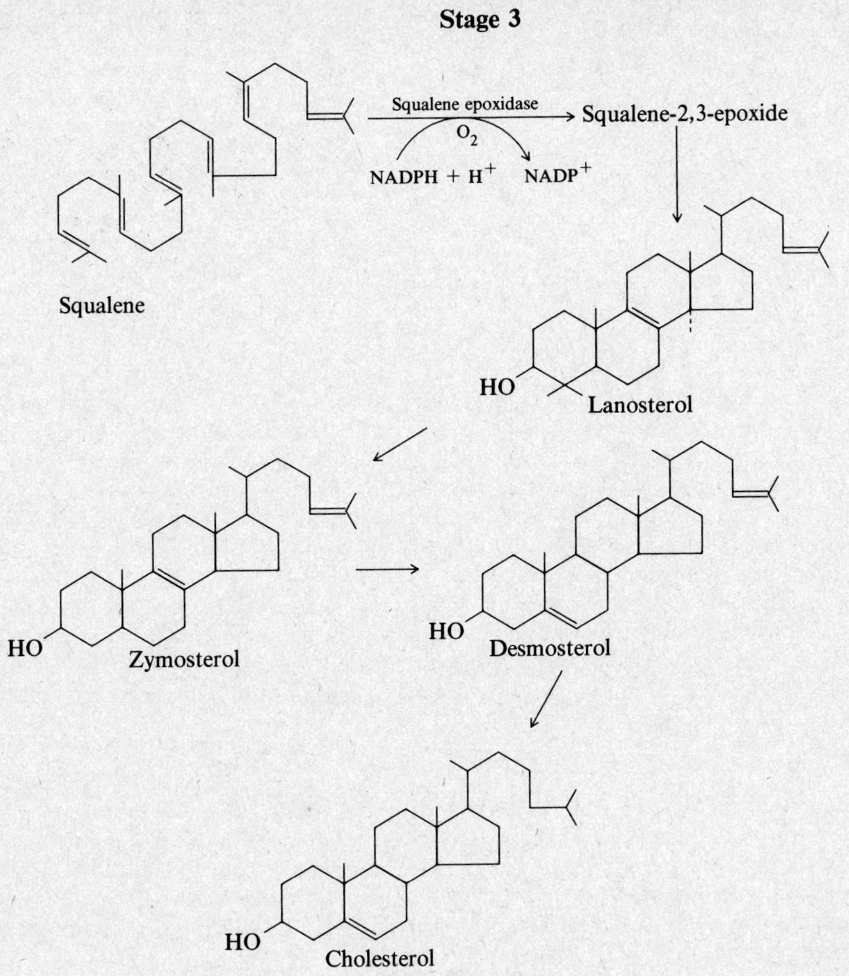

Stage 3

Figure 7-2. *Continued*

final product of the second stage—squalene, a 30-carbon acyclic hydrocarbon. The second stage is important since it contains the step involving the microsomal enzyme HMG-CoA reductase—the step which is *rate-limiting* in cholesterol biosynthesis. The third and final stage of synthesis occurs in the endoplasmic reticulum with many of the intermediate products being bound to a specific carrier protein. Squalene, after an initial oxidation, undergoes cyclization to form the 4-ringed, 30-carbon intermediate, lanosterol. In a series of oxidation-decarboxylation reactions, a number of side chains are removed from the pentanophenanthrene structure to form the 27-carbon molecule of cholesterol.

Until recently, the cholesterol synthetic reactions after HMG-CoA conversion to mevalonate were considered irreversible. However, dimethylallyl pyrophosphate, one of the isoprene molecules derived from mevalonate, can be dephosphorylated and the resulting dimethylallyl alcohol converted back to HMG-CoA. HMG-CoA can be catabolized by the action of HMG-CoA lyase to acetoacetate and acetyl-CoA, which can enter other metabolic pathways. The discovery of this pathway, termed the mevalonate shunt, has potential clinical applications since it may be possible to develop drugs which could enhance this alternate pathway and thus reduce cholesterol synthesis.

Cholesterol Esterification

Once synthesized, cholesterol is released into the circulation for transport in combination with specific apoproteins, the apolipoproteins, in complexes known as lipoproteins. Minimal cholesterol esterification occurs within the liver prior to its release, and cholesterol is mainly esterified within the vascular compartment. Esterification is important since it serves to enhance the lipid-carrying capacity of the lipoproteins. The reaction is catalyzed by the enzymes lecithin-cholesterol-acyltransferase (LCAT) in the plasma, and acyl-cholesterol-acyltransferase (ACAT) intracellularly. The intracellular ACAT pathway is the major pathway in liver, intestine, adrenal cortex, and probably in the arterial wall. The initial reaction (see Figure 7-3) involves activation

Intracellular:

$$\text{Fatty acid} + \text{CoASH} \xrightarrow[\text{ATP} \quad \text{PP}_i + \text{AMP}]{\text{Acyl-CoA} \atop \text{synthetase}} \text{Acyl-CoA}$$

$$\text{Acyl-CoA} + \text{cholesterol} \xrightarrow{\text{ACAT}} \text{Cholesteryl ester} + \text{CoASH}$$

Intravascular:

$$\text{Lecithin} + \text{cholesterol} \xrightarrow{\text{LCAT}} \text{Cholesteryl ester} + \text{lysolecithin}$$

Figure 7-3. Intracellular and intravascular esterification of cholesterol mediated by acylcholesterol acyltransferase (ACAT) and lecithin cholesterol acyltransferase (LCAT), respectively.

of a fatty acid with CoASH to form an acyl-CoA. This, in turn, reacts with cholesterol to form an ester. The LCAT reaction does not involve CoASH and results in fatty acid transfer from the second carbon position of lecithin to cholesterol. LCAT activity is responsible for nearly all the esterified cholesterol in plasma. These cholesterol esters account for about 70% of the total cholesterol in plasma; the remaining 30% exists in free form. LCAT is synthesized in the liver, released into the circulation, and appears to be activated by apolipoprotein AI and perhaps apolipoprotein CI. (See section on Apolipoproteins.)

The plasma LCAT activity may have some significance analytically since the enzyme will continue to esterify plasma cholesterol even after the blood sample is drawn and stored at room temperature. Consequently, analysis requiring free and/or esterified cholesterol should be carried out as soon as possible or the samples should be quickly cooled to 4 °C or frozen, preferably at −70 °C.

Cholesterol Catabolism

Once cholesterol enters the cell, the esters are hydrolyzed by the action of specific esterases. Although there are probably a number of cholesterol esterases, the important isoenzyme appears to reside in the lysosomes. This enzyme, currently indistinguishable from lysosomal acid lipase, functions optimally at an acid pH. The lack or malfunction of this lysosomal enzyme results in intracellular accumulation of cholesterol esters and produces a clinical disorder known as *cholesterol ester storage disease.*

Cholesterol reaching the liver is either secreted unchanged into bile or metabolized to bile acids. Approximately one third of the daily production of cholesterol is catabolized into bile acids. (See Figure 7-4.) The bile acid synthesis rate averages 200–400 mg/d. The first step in bile acid synthesis involves the rate limiting step, 7α-hydroxylation. Two bile acids, cholic and chenodeoxycholic, constitute the primary bile acids. They are conjugated with either glycine or taurine and enter the bile canaliculi. After reaching the small intestine, the conjugated bile acids play an active part in cholesterol and fat absorption, as has been discussed previously. Some of the bile acids are deconjugated and converted by bacteria in the intestine to secondary bile acids. Cholic acid is converted to deoxycholic acid and chenodeoxycholic acid is metabolized to lithocholic acid. Virtually all bile acids except lithocholic are reabsorbed in the lower third of the ileum and returned to the liver via the portal vein, thus completing the enterohepatic circulation. (See also Chapter 13.)

A significant amount of cholesterol is excreted directly into the biliary system where it is normally solubilized by mixed micelles of bile acids and phospholipids. If, however, the amount of cholesterol exceeds the capacity of the solubilizing agents, a supersaturated state can occur (lithogenic bile). This excess cholesterol can precipitate and result in the development of cholesterol gallstones. In westernized societies, ~80% of all gallstones are cholesterol stones.

FATTY ACIDS

The fatty acids (FA) are one of the simpler molecular forms of lipids. They are depicted by the chemical formula RCOOH, where R stands for an alkyl chain. The chain length varies, and the most common classification is thus based on the number of carbon atoms present. The three arbitrary fatty acid groups consist of 2–4 carbon atoms (short chain), 6–10 carbon atoms (medium chain), and 12–26 carbon atoms (long chain). Those of importance in human nutrition and metabolism are of the long chain class and contain even numbers of carbon atoms.

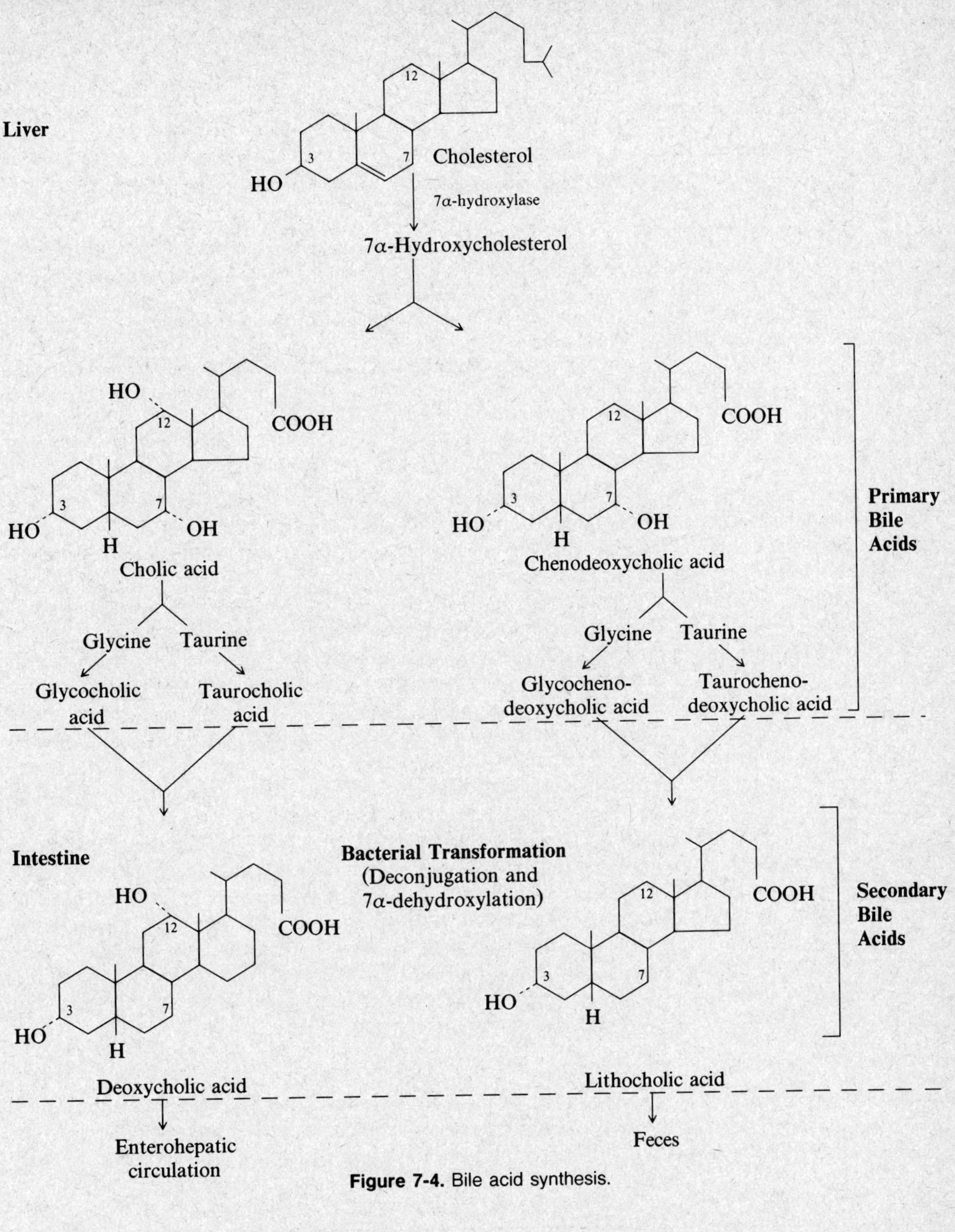

Figure 7-4. Bile acid synthesis.

Fatty acids can be further classified according to their degree of saturation. Saturated fatty acids contain an alkyl chain without any double bonds between carbon atoms, monounsaturated fatty acids contain one double bond, and polyunsaturated fatty acids contain more than one double bond (Figure 7-5). The double bonds in polyunsaturated fatty acids of plant origin are generally three carbon atoms apart, a feature which tends to prevent oxidation (auto-oxidation or peroxidation). Some oils from marine fish, such as salmon, living in deep, cold waters, possess numerous (up to 6) unsaturated bonds and are usually > 20 carbon atoms long. These oils can undergo oxidation due to their many double bonds.

The labeling of the carbon atoms in fatty acids can be either from the carboxyl terminal (Δ-numbering system) or from the methyl-terminal (*n*- or *ω*-numbering system). In addition, the carbon atoms may be labeled with Greek symbols, with α being adjacent to the carboxyl group and *ω* being farthest away. In

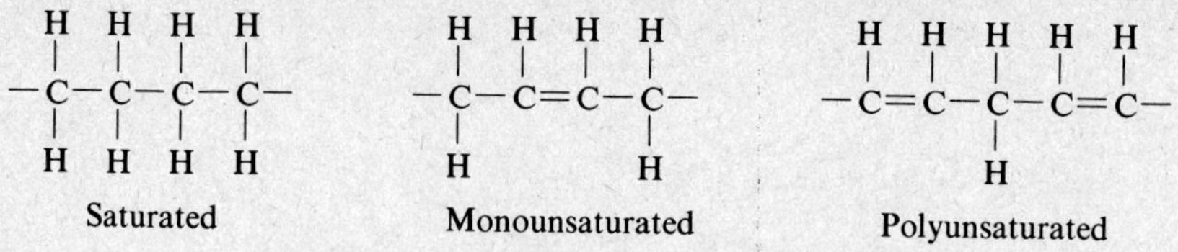

Figure 7-5. Saturated and unsaturated fatty acids.

the Δ-system, fatty acids are abbreviated according to the number of carbon atoms, number of double bonds, and the position(s) of double bond(s). For example, linoleic acid, which contains 18 carbons and two unsaturated bonds between carbons 9 and 10 and between carbons 12 and 13, could be written as $C_{18}{:}2^{9,12}$. Using the n- or ω-system, linoleic acid would be abbreviated to $C_{18}{:}2n\text{-}6$, where only the first carbon forming the unsaturated pair is written.

Knowledge of both classifications is necessary because the Δ-nomenclature is used in prostaglandin metabolism, lipid nutrition, and β-oxidation, while the n-system is used for discussing lengthening and desaturation reactions. A further system of nomenclature is known as the Geneva or systematic classification. In this system, linoleic acid would become 9,12-octadecadienoic acid. Table 7-2 lists a number of the fatty acids commonly found in human tissues.

In saturated fatty acids, the chain is extended and flexible, i.e., the carbon atoms can rotate freely around the longitudinal axis. Unsaturated fatty acids, however, have fixed 30° bends in their chains at each double bond. Depending on the plane in which this bend occurs, either the *cis*- or *trans*-isomer is produced. In mammals, all naturally occurring unsaturated fatty acids are of the *cis*-variety. *Trans*-fatty acids are a result of catalytic hydrogenation, a process used to "harden" fats in the manufacture of certain foods such as margarine. Most fats in the human body are derived from our diet, which on the average contains 40% fat, 90% of which is triglyceride. In addition, man can synthesize most fatty acids including saturated, mono-, and some polyunsaturated fats. But one fatty acid in particular, linoleic acid ($C_{18}{:}2^{9,12}$), cannot be synthesized by mammals; it is found only in plants. This fatty acid is vital for maintenance of health, as well as growth and development, and is termed an essential fatty acid. It is probable that linoleic acid plays an important role in prostaglandin synthesis and perhaps in myelinization of the central nervous system.

The fatty acid carboxyl group has a pK_a of ~4.8; thus, free fatty acid molecules in both plasma and intracellular fluid (pH of 7.4 and 7.0, respectively) exist in an ionized form. The ability of fatty acids to ionize is reduced by the fact that they exist as either esters with cholesterol or glycerol, or bound to albumin and pre-albumin. One molecule of albumin can carry as many as 20 molecules of fatty acid. The normal level of free FA in human blood is only 0.30–1.10 mmol/L, amounting to about 8–31 mg/dL of plasma, but the flux is very large and quite sensitive to exercise and physical work, to the level of blood glucose, and to excitement or other psychologic stress that causes liberation of epinephrine.

Fatty Acid Catabolism

Long chain fatty acids are oxidized for the production of energy in the mitochondria of cells by a series of reactions that operate in a repetitive manner to shorten the chains by two carbon atoms at a time. This process is known as β-oxidation. For example, each C_{16} molecule is converted

Table 7-2. FATTY ACIDS COMMONLY FOUND IN HUMAN TISSUE

Common Name	Systematic Name	Δ-Numbering	$n\text{-}(\omega)$ Numbering
Lauric	Dodecanoic	12:0	12:0
Myristic*	Tetradecanoic	14:0	14:0
Palmitic	Hexadecanoic	16:0	16:0
Palmitoleic	9-Hexadecenoic	$16{:}1^9$	16:1n–7
Stearic	Octadecanoic	18:0	18:0
Oleic*	9-Octadecenoic	$18{:}1^9$	18:1n–9
Linoleic*	9,12-Octadecadienoic	$18{:}2^{9,12}$	18:2n–6
Linolenic	9,12,15-Octadecatrienoic	$18{:}3^{9,12,15}$	18:3n–3
Arachidic	Eicosanoic	20:0	20:0
Arachidonic	5,8,11,14-Eicosatetraenoic	$20{:}4^{5,8,11,14}$	20:4n–6

*Most common fatty acids.

to eight molecules of acetyl-CoA. Acetyl-CoA does not normally accumulate in the cell, but is enzymatically condensed with oxaloacetate, a substance derived largely from carbohydrate metabolism (Figure 7-6). The product of the condensation reaction is citrate, which is a major component of the Krebs cycle (the citric acid cycle, or the tricarboxylic acid cycle). The Krebs cycle serves as a common pathway for the final oxidation of nearly all food material, whether derived from carbohydrate, fat, or protein. It is important to bear in mind that the smooth operation of the Krebs cycle depends on the availability of sufficient oxaloacetate to serve as acceptor for acetyl-CoA.

From each fatty acid molecule a large quantity of energy can be produced; for instance, the complete oxidation of a mole of palmitic acid to carbon dioxide and water produces:

$$CH_3(CH_2)_{14}\,COOH + 23\,O_2 \rightarrow 16\,CO_2 + 16\,H_2O + 2340\,Cal*$$

By means of suitable enzyme reactions, the chemical energy may be released for metabolic processes or stored in the form of high energy compounds, such as adenosine triphosphate (ATP). Triglyceride esters containing three fatty acid molecules, therefore, make an excellent storage form for reserve energy. The amount of energy produced by metabolizing one mole of palmitic acid (16 carbon atoms) is approximately twice that produced by metabolizing an equivalent

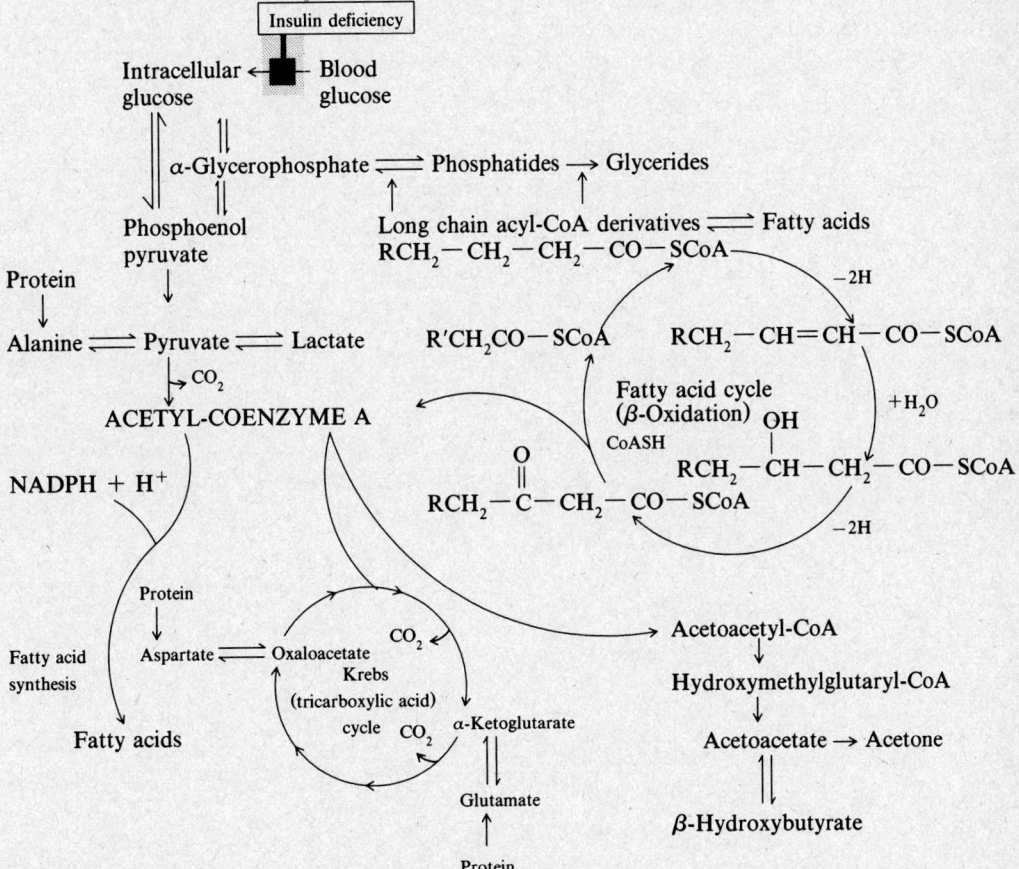

Figure 7-6. Metabolic relations among intermediates of carbohydrate, fat, and protein metabolism. Note that acetyl-CoA is produced from both carbohydrate and fat. The glucogenic amino acids, derived from protein metabolism, enter glycolytic paths as α-keto acids. Ketogenic amino acids enter as acetyl-CoA.

---

*The unit employed in discussing the energy value of food is the Calorie (Cal), equal to 1000 calories or 1 kilocalorie. In the SI system, the unit of energy is the joule (J), and 1 calorie = 4.1868 J.

amount (2.5 moles) of glucose (6 carbon atoms per molecule). Carbohydrate storage requires water of hydration; triglyceride storage does not. In addition to their high intrinsic energy content, triglycerides have a low density (< 1 g/mL) and, due to their hydrophobic property and peripheral distribution, provide excellent insulation.

Ketone Formation

During prolonged starvation or when carbohydrate metabolism is severely impaired, as in uncontrolled diabetes mellitus, the formation of acetyl-CoA exceeds the supply of oxaloacetate. The abundance of acetyl-CoA results from excessive mobilization of fatty acids from adipose tissue and an excessive degradation of the fatty acids by β-oxidation in the liver. The resulting acetyl-CoA excess is diverted to an alternative pathway in the mitochondria which forms acetoacetic acids, β-hydroxybutyric acid and acetone, three compounds known collectively as ketone bodies. This path is shown in Figure 7-7. Presence of ketone bodies is a frequent finding in severe uncontrolled diabetes mellitus.

The first product, acetoacetyl-CoA, condenses with a third molecule of acetyl-CoA to yield β-hydroxy-β-methylglutaryl CoA. This product is cleaved enzymatically to yield acetoacetate and acetyl-CoA. Some of the acetoacetate formed in liver cells usually is reduced to β-hydroxybutyrate. Acetoacetate is unstable and some of it decomposes to form carbon dioxide and acetone. This decarboxylation reaction accounts for the formation of the third ketone body, acetone, frequently observed in severe untreated diabetes mellitus. Ketosis, therefore, develops from excessive production of acetyl-CoA as the body attempts to obtain necessary energy from stored fat in the absence of an adequate supply of carbohydrate metabolites.

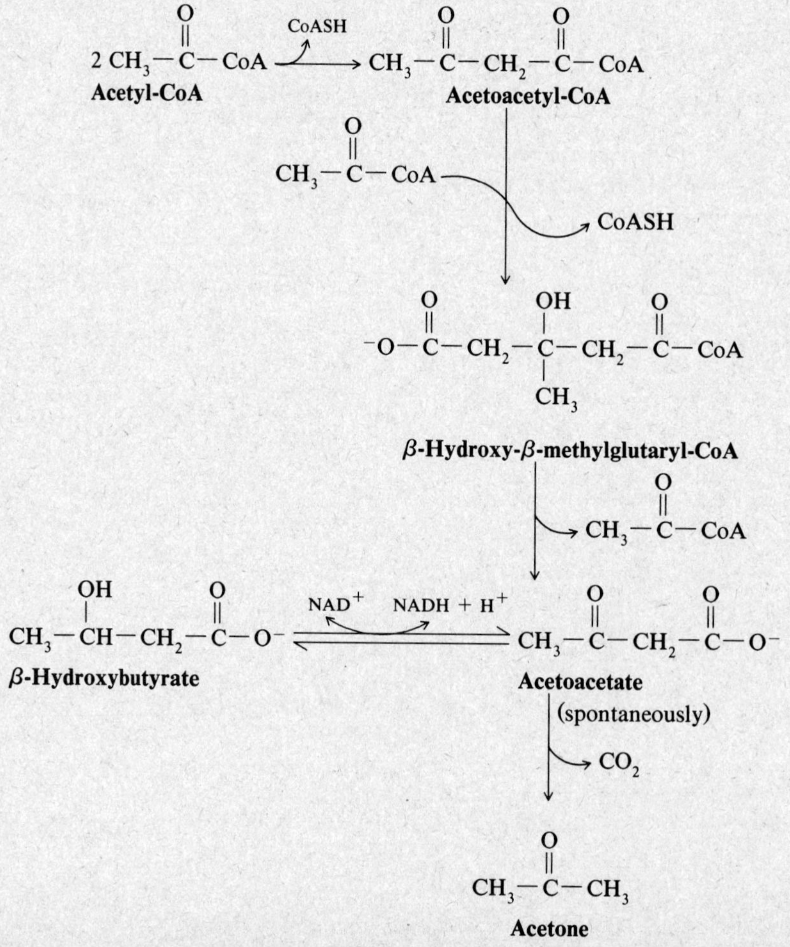

Figure 7-7. Formation of ketone bodies.

Inadequate incorporation of acetyl-CoA into the Krebs cycle may be further aggravated by inhibition of the oxaloacetate-generating enzyme system by excess accumulation of palmitic-CoA and other long chain fatty acid-CoA derivatives in the liver. Skeletal muscle and heart (and brain in prolonged fasting) can use ketone bodies by resynthesizing their CoA derivatives of the acids and subsequently oxidizing them for the production of energy. While liver cells are largely responsible for converting fatty acids, they cannot metabolize acetoacetate, since liver lacks the enzyme required for transferring CoA from succinyl-CoA.

The entire process of ketosis can be reversed by restoring an adequate level of carbohydrate metabolism. In starvation, restoration consists of adequate carbohydrate ingestion; in diabetes mellitus, ketosis can be reversed by insulin administration, which permits circulating blood glucose to be taken up by the cells. With production of oxaloacetate, the acceptor of acetyl-CoA, normal metabolism is restored, and the release of fatty acids from adipose tissues slows and is finally reversed. A graphic view of these metabolic reactions is outlined in Figure 7-6, which shows the interrelationship between carbohydrate, fatty acid, and protein metabolism.

PROSTAGLANDINS

Prostaglandins and related compounds are derivatives of fatty acids. The group consists of prostaglandins, thromboxanes, some hydroperoxy and hydroxy fatty acid derivatives, and leukotrienes. Although their full physiological role is not completely known, they exert diverse biological actions. They are extremely potent, producing physiological actions at concentrations of 1 μg/L.

The prostaglandins are a series of C_{20} unsaturated fatty acids containing a cyclopentane ring; the parent fatty acid has been given the trivial name prostanoic acid.

The carbon chain linked to C-8 of prostanoic acid projects below the plane of the ring, while the chain attached to C-12 projects above the ring. Since the 1960's when prostaglandins were first isolated and their structure identified, their number has grown considerably.

By convention, prostaglandins are abbreviated PG, with the class designated by a capital letter (A, B, E, F, G, H, and I), followed by a number, and then in some cases a Greek letter. (See Figure 7-8.) With the exception of PGG and PGH, which have the same ring structure (cyclopentane endoperoxide), the letters refer to different ring structures. PGA and PGB have keto groups at C-9, with the A series having a double bond between C-10 and C-11 and the B series having a double bond between C-8 and C-12. PGE also has a C-9 keto bond but has a hydroxyl group at C-11. The F series has hydroxyl groups at both C-9 and C-11. The difference between PGG and PGH, which have identical ring structures, occurs in the side chain at C-15; the G series has a peroxide group, while the H series has a hydroxyl group. The I series has a double ring formation, C-9 of the cyclopentane ring being linked to C-6 of the side chain by an oxygen molecule to form a second five-sided ring. (See Figure 7-8.) The endoperoxide PGs (G and H series) are intermediates in the formation of other PGs such as the A, B, E, F, and I series.

The number after the capital letter is usually written as a subscript and is used to designate the number of unsaturated bonds in the PG side chains (i.e., not within the ring structure itself). In PGE_1, for example, a double bond exists between C-13 and C-14; in the 2 series (PGE_2) a double bond exists between C-13 and C-14 and between C-5 and C-6; and in the 3 series (PGE_3) an additional double bond occurs between C-17 and C-18. The bond between C-13 and C-14 is always *trans*, while those between C-5 and C-6 and between C-17 and C-18 are always *cis*. All naturally occurring prostaglandins have a hydroxyl group at C-15 that projects below the plane of the ring. The use of the Greek letter (α or β) applies only to the F series and refers to the hydroxyl group found at C-9. In the α-series, the hydroxyl group projects below the ring plane in the same direction as the C-11 hydroxyl group. The β-series denotes that the hydroxyl at C-9 is above the plane of the ring.

While prostaglandins appear hormone-like in action, they are different from hormones in at least one respect—they are synthesized at the site of action and in almost all tissues. Linoleic acid ($C_{18}:2^{9,12}$) is the precursor of two of the three 20-carbon fatty acids which form prostaglandins. The other precursor is linolenic acid ($C_{18}:3^{9,12,15}$). Both of these fatty acids are considered *essential*, as they cannot be synthesized in the body and therefore must be present in our diet. The three

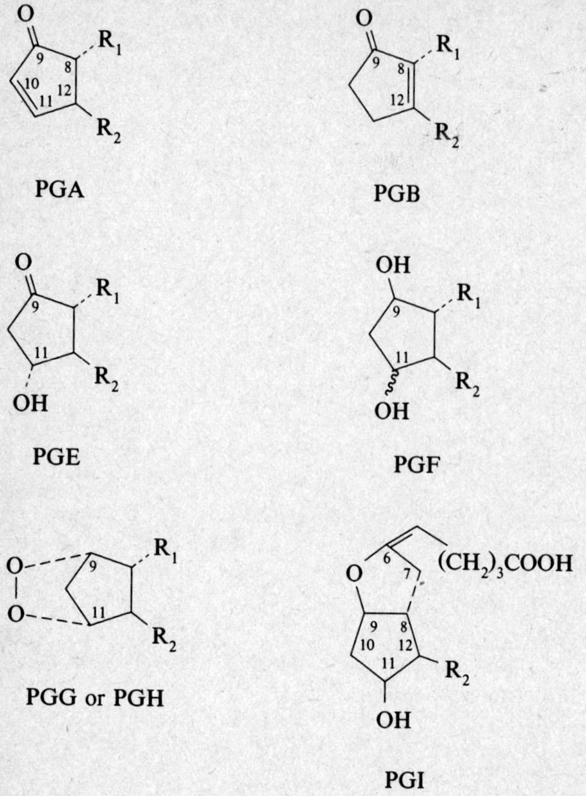

PGA

PGB

PGE

PGF

Figure 7-8. Major prostaglandin classes (series). R_1, R_2 = Prostaglandin side chains. (See text for details.)

PGG or PGH

PGI

C_{20} fatty acids subsequently formed are $C_{20}:3^{5,8,11}$ (eicosatrienoic acid), $C_{20}:4^{5,8,11,14}$ (eicosatetraenoic or arachidonic acid), and $C_{20}:5^{5,8,11,14,17}$ (eicosapentaenoic acid). These three fatty acids form the PG_1, PG_2, and PG_3 series, respectively.

Once formed, prostaglandins exert a very short-lived effect and are rapidly catabolized. They are not stored preformed; instead the precursor C_{20} fatty acids are present in tissues attached to C-2 (see glycerol esters) of the phosphoglycerides. On demand, the C_{20} precursor is hydrolyzed by phospholipase A_2, which is specific for the C-2 atom of the phosphoglyceride. The release of the C_{20} fatty acid, which may be mediated by the effect of bradykinin, thrombins, or angiotensin II, appears to be the rate-limiting step in PG synthesis.

While it is probable that all PG's follow a similar synthetic pathway, $C_{20}:4$ (arachidonic acid) has been the most intensively studied and is used to illustrate the pathway. (See Figure 7-9.) Once released, arachidonic acid can follow one of two pathways. The lipo-oxygenase route produces 12-L-hydroperoxy-5,8,10,14 eicosatetraenoic acid (HPETE); HPETE becomes 12-L-hydroxy-5,8,10,14 eicosotestraenoic (HETE) acid, which is thought to be a chemotactic agent. The alternative pathway is that mediated by cyclo-oxygenase to produce the endoperoxides PGG_2 and PGH_2. The latter can be degraded to 12-L-hydroxy-5,8,10 heptadecatrienoic acid (HHT). What controls the entry into a specific pathway remains speculative; however, it is known that anti-inflammatory drugs (aspirin, acetaminophen, and indomethacin) inhibit the cyclo-oxygenases, thereby decreasing PG synthesis. Inactivation of PG is better known and appears to be mediated by two enzymes, 15α-hydroxy-prostaglandin dehydrogenase and Δ^{13}-prostaglandin reductase.

Although 17 naturally occurring prostaglandins have been described (Table 7-3), only seven are commonly found throughout the body. These are termed the primary prostaglandins.

PGI_2, or *prostacyclin*, is derived from arachidonic acid (Table 7-2) in the vascular endothelium. It has a powerful vasodilatory action, especially on the coronary arteries, and is also responsible for inhibiting platelet aggregation. Thromboxane A_2 (TXA_2) is synthesized from arachidonic acid but is also produced by platelets, where its release has the exact opposite effect to prostacyclin; it contracts arterial smooth muscle and enhances platelet aggregation. It has a very short $t_{1/2}$ of about 30 s, and is rapidly converted to its inactive metabolite, thromboxane B_2 (TXB_2). The thromboxanes are slightly different from the other prostaglandins in that they contain six-sided

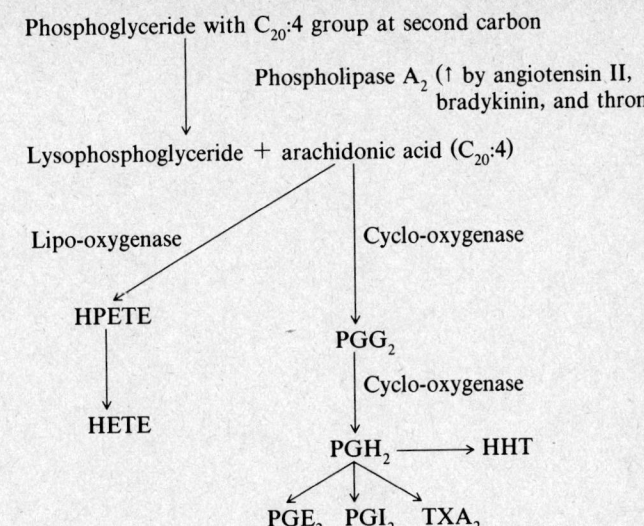

Phosphoglyceride with C_{20}:4 group at second carbon

Phospholipase A_2 (↑ by angiotensin II,
bradykinin, and thrombin)

Lysophosphoglyceride + arachidonic acid (C_{20}:4)

Lipo-oxygenase   Cyclo-oxygenase

Figure 7-9. Synthesis of prostaglandins from arachidonic precursor. For abbreviations, see text.

HPETE

PGG$_2$

Cyclo-oxygenase

HETE

PGH$_2$ ⟶ HHT

PGE$_2$ PGI$_2$ TXA$_2$

PGF$_{2\alpha}$ TXB$_2$

rings of five carbon atoms and one oxygen atom (Figure 7-10). Table 7-4 lists some of the reported functions of the various prostaglandins. With increasing knowledge of the physiological role of the prostaglandins, as a group and individually, it is likely that discrete disorders of prostaglandin metabolism will be discovered and that prostaglandins, prostaglandin analogs, or prostaglandin agonists will eventually be used in clinical practice.

GLYCEROL ESTERS (ACYLGLYCEROLS)

The complex lipids are virtually all fatty acid derivatives; in most cases they are covalently linked to an alcohol. One of the most common alcohols found in human metabolism is glycerol, a three carbon molecule containing three hydroxyl groups.

$$H_2 - C_\alpha - OH$$
$$H - C_\beta - OH$$
$$H_2 - C_{\alpha'} - OH$$

The two terminal carbon atoms in the molecule are chemically equivalent and are designated α and α' (α prime). The center carbon is labeled β. An alternative labeling system commonly employed uses the numeral 1 for the α-carbon, 2 for the β-carbon, and 3 for the α'-carbon.

The class of acylglycerol (glyceride) is determined by the number of alcohol groups that are

Table 7-3. NATURALLY OCCURRING
PROSTAGLANDINS (PG)

Primary PG	Other PG
PGE$_1$	PGA$_1$
PGF$_{1\alpha}$	PGA$_2$
PGE$_2$	19α-OHPGA$_1$
PGF$_{2\alpha}$	19α-OHPGA$_2$
PGI$_2$	PGB$_1$
Thromboxane A$_2$	PGB$_2$
Thromboxane B$_2$	19α-OHPGB$_2$
	PGE$_3$
	PGF$_{3\alpha}$

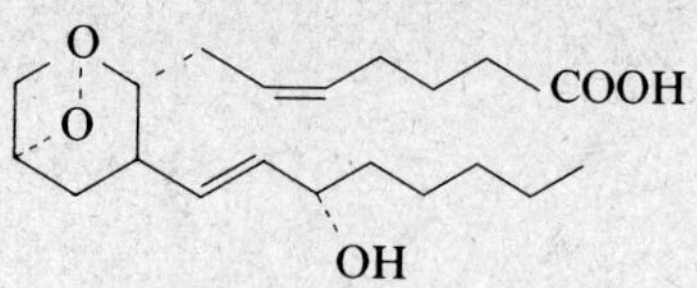

Thromboxane A$_2$ (TXA$_2$)

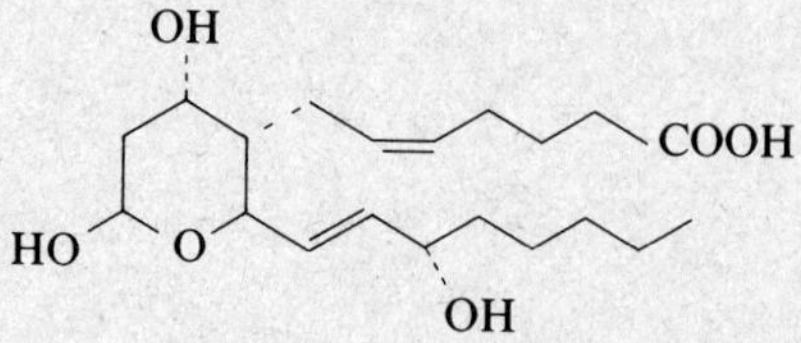

Thromboxane B$_2$ (TXB$_2$)
Figure 7-10. Structures of thromboxanes.

esterified to produce monoacylglycerols (monoglycerides), diacylglycerols (diglycerides), or triacylglycerols (triglycerides). In a monoacylglycerol, the fatty acid may be linked to either of the two terminal carbons (α or α'), or the central (β) carbon atom. By convention, the number system is used to indicate the carbon position, e.g., 1-monoglyceride indicates a fatty acid attachment to the first α-carbon. This numbering system applies to all acylglycerols including the phosphoglycerides, as shown later. Diglycerides can be either 1,2- or 1,3-diglycerides (Figure 7-11).

In human nutrition, *triglycerides* are the most prevalent glycerol esters encountered. They constitute 95% of tissue storage fat and are the predominant form of glycerol ester found in plasma. The fatty acid residues found in mono-, di-, or triglycerides vary considerably, and usually include combinations of the long chain fatty acids as shown in Table 7-2. Triglycerides from plants (e.g., corn, sunflower seed, and safflower oils) tend to have large amounts of C_{18}:2 or linoleic residues. They are termed polyunsaturated fats and are liquid even at 4 °C. Triglycerides from animals, especially ruminants, tend to have C_{12}:0 through C_{18}:0 fatty acid residues (saturated fats), and are solid even at room temperature. Some plant triglycerides, such as coconut oil, are highly saturated and may be solid at room temperature.

Triglycerides undergo digestion in the duodenum and proximal ileum. Through the action of lipases and bile acids, triglycerides are hydrolyzed into glycerol and fatty acids. Following absorption, triglycerides are resynthesized in the epithelial cells and combined with cholesterol and a number of apolipoproteins to form chylomicrons. The chylomicrons travel through the lymphatic system to the thoracic duct and eventually to the jugular vein.

Another major class of glycerol esters consists of those containing phosphoric acid at the third (α') carbon atom; these esters are called *phosphoglycerides* (Figure 7-12). In the simplest form, the A group is an H atom and the molecule is, therefore, a diacylphosphoglyceride. Usually, however, the A is substituted by an alcohol derived group such as choline, serine, inositol, or ethanolamine (Figure 7-12). If choline is the attached group, it is referred to as phosphatidylcholine, and if it is ethanolamine, as phosphatidylethanolamine, and so on. A more common term for

Table 7-4. PROSTAGLANDIN-MEDIATED EFFECTS

Site of Action	Physiological Response
Arterial smooth muscle	Alterations of blood pressure
Uterine muscle	Induces labor, therapeutic abortion
Lower gastrointestinal tract	Increases motility
Bronchial smooth muscle	Bronchospasm
Platelets	Increases coagulability
Stomach	Enhances gastric acid secretion
Capillaries	Increased permeability with flushing
Adipose tissue	Inhibits triglyceride lipolysis

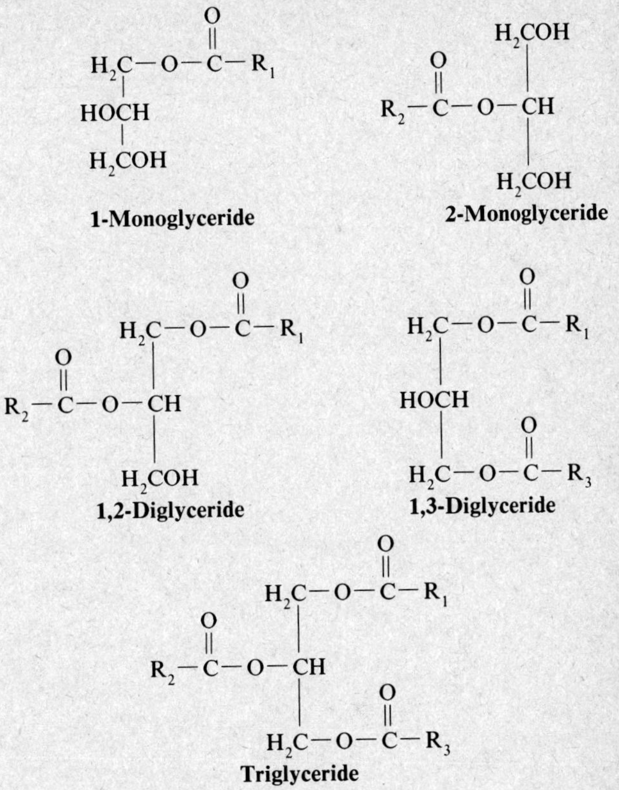

$$\begin{array}{c} O \\ \parallel \\ H_2C-O-C-R_1 \\ | \\ HOCH \\ | \\ H_2COH \end{array}$$

1-Monoglyceride

$$\begin{array}{c} H_2COH \\ O \\ \parallel \\ R_2-C-O-CH \\ | \\ H_2COH \end{array}$$

2-Monoglyceride

Figure 7-11. Structure and classification of glycerol esters (acylglycerols). R_1, R_2, R_3 = Fatty acid(s) of varying carbon atom lengths.

$$\begin{array}{c} O \\ \parallel \\ H_2C-O-C-R_1 \\ O \\ \parallel \\ R_2-C-O-CH \\ | \\ H_2COH \end{array}$$

1,2-Diglyceride

$$\begin{array}{c} O \\ \parallel \\ H_2C-O-C-R_1 \\ | \\ HOCH \\ O \\ \parallel \\ H_2C-O-C-R_3 \end{array}$$

1,3-Diglyceride

$$\begin{array}{c} O \\ \parallel \\ H_2C-O-C-R_1 \\ O \\ \parallel \\ R_2-C-O-CH \\ O \\ \parallel \\ H_2C-O-C-R_3 \end{array}$$

Triglyceride

$$\begin{array}{c} O \\ \parallel \\ H_2C-O-C-R_1 \\ O \\ \parallel \\ R_2-C-O-CH \quad O \\ | \quad\quad \parallel \\ H_2C-O-P-O-A \\ | \\ O^- \end{array}$$

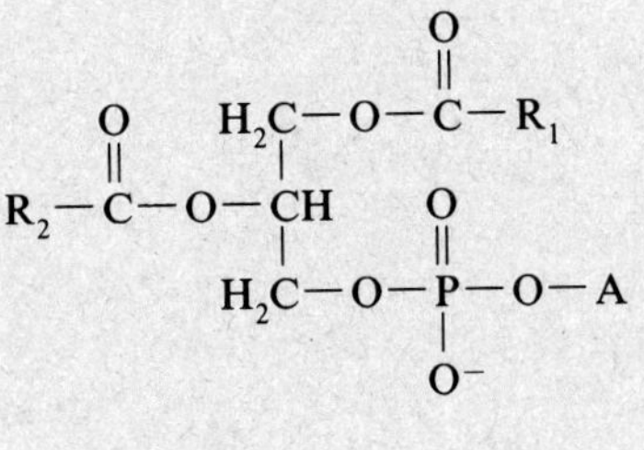

Figure 7-12. Structures of phosphoglycerides and common alcohol groups associated with them. R_1, R_2 = Fatty acid(s) of varying carbon atom lengths.

Where A = $-H$ — Phosphatidic acid

A = $-CH_2CH_2-\overset{+}{N}H_3$ — Phosphatidyl ethanolamine*

A = $-CH_2CH_2-\overset{+}{N}(CH_3)_3$ — Phosphatidyl choline (lecithin)

A = $-CH_2CH-COO^-$ Phosphatidyl serine*
$\quad\quad\quad\quad | \overset{+}{}$
$\quad\quad\quad\quad NH_3$

A = [inositol ring structure] — Phosphatidyl inositol*

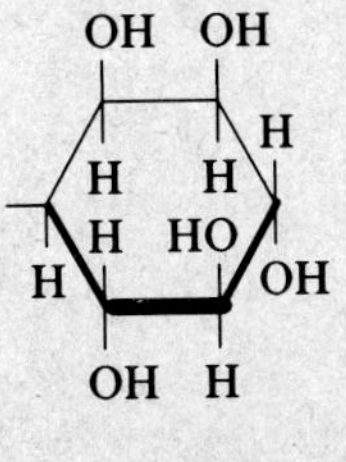

*Commonly known as cephalins.

phosphatidylcholine is *lecithin*, while phosphatidylethanolamine, -serine, and -inositol are referred to as *cephalins*. As the fatty acid residues (R_1 and R_2 in Figure 7-12) vary, several different lecithins and cephalins are formed. These phosphoglycerides are named according to the fatty acid acyl ester attached at C-1 and C-2. For the most part, the most saturated fatty acid attaches to the first carbon atom.

In inner mitochondrial membranes, more complex phosphoglycerides known as *cardiolipins* can be found. They are derived from two phosphoglyceride molecules joined by a glycerol bridge.

SPHINGOLIPIDS

The fourth class of lipids found in humans is derived from the amino alcohol sphingosine. (See Figure 7-13.) This dihydric 18-carbon alcohol contains an amino group at C-17. Sphingosine binds to one fatty acid containing 18 or more carbon atoms via the amino group. This sphingosine amide-acyl molecule is known as a *ceramide* and forms the intermediary step in the formation of three important sphingolipids: *sphingomyelin, galactosylceramide,* and *glucosylceramide* (Figure 7-13). The sugar-containing ceramides can have a sulfate group attached (usually to the 2 position of the galactose residue) to form *sulfatides.* The glycosyl ceramides can also have additional monosaccharide moieties (such as galactose, *N*-acetyl galactosamine, and *N*-acetyl-neuraminic acid) to form complex *globosides* and *gangliosides.* These complex sphingolipids form the major

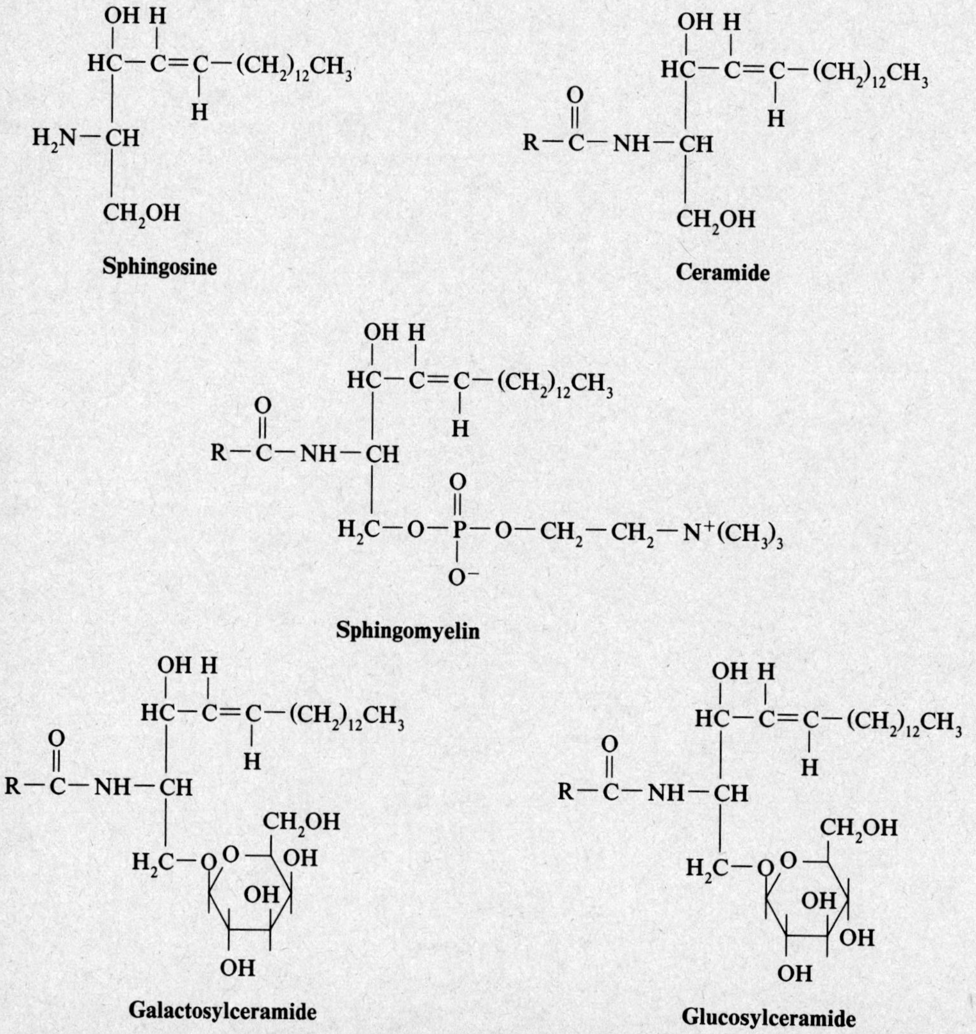

Figure 7-13. Structures of sphingolipids.

lipids of cell membranes and the central nervous system. Gangliosides are particularly prevalent in the gray matter of the brain, while membrane glycosphingolipids play a major role in cell recognition and blood typing.

TERPENES

These are polymers of the five-carbon isoprene unit, and include vitamins A, E, and K. (See Chapter 8B.)

APOLIPOPROTEINS

It is apparent from the previous discussion on lipids that, while they are synthesized in the intestine or liver, they need to be transported to many distant tissues and organs in order to complete their varied metabolic functions. Given the hydrophobic nature of the neutral fats, triglycerides, and cholesteryl esters, lipid transport and delivery via plasma would not be possible without some form of hydrophilic adaptation. The lipids are consequently transported by means of a series of rather complex micellar structures that consist of an outer monolayer of protein (an apolipoprotein) and polar lipids (phospholipid and unesterified cholesterol), and an inner core of neutral lipids (triglycerides and cholesteryl esters). A schematic diagram of a micelle, known as a lipoprotein, is shown in Figure 7-14. The core neutral lipids are by and large inactive components or "passengers," while the apolipoproteins are mainly responsible for the further metabolism and catabolism of the particle. While the core size is dependent on the amount of neutral sterol, the surface monolayer is unchanged in all lipoprotein classes and consists of a layer of protein, phospholipid, and unesterified cholesterol. The previous section dealt with the lipid moieties and this section discusses the protein components. The two components are very difficult to separate in a physiological sense and are therefore considered together in detail in the next section on lipoproteins.

While plasma and tissue lipids have been extensively studied for over 50 years, it was not until the mid-1950's that scientists started viewing the plasma lipids as part of an overall transport system called lipoproteins. By using a variety of separation techniques such as ultracentrifugation and electrophoresis, the lipoproteins were first classified, then quantitated, and eventually correlated with various disease states. The major advances in our knowledge of lipid transport and lipoprotein metabolism, in both health and disease, have come with the enormous amount of work carried out mainly in the 1960's and 1970's on the structure and function of the apolipoproteins, the protein moieties of the lipoproteins. The initial classification divided the lipoproteins into five main density classes: chylomicrons, very low density lipoprotein (VLDL), intermediate density lipoprotein (IDL), low density lipoprotein (LDL), and high density lipoprotein (HDL).

Associated with these lipoproteins, at least five major apolipoproteins have been described and have been labeled A through E. However, a number of subgroups within each major group have been differentiated (Table 7-5). The major groupings and their subgroups differ in their primary, secondary, and tertiary structures, in their physicochemical behavior, and in their function and distribution in the various lipoprotein species.

Figure 7-14. Schematic diagram of a lipoprotein molecule. (Modified and published with permission from M. L. Kashyap, M.D.)

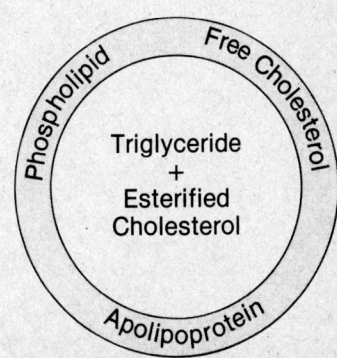

Table 7-5. MAJOR AND MINOR APOLIPOPROTEINS

Apolipoprotein	Function	M.W.	Site of Synthesis
AI	LCAT activation	28 300	Liver, intestine
AII	LCAT inhibition; lipid transport	17 000	Liver, intestine
AIV	Chylomicron triglyceride transport		Intestine
B100	Lipid transport and clearance	8000–275 000	Liver
B48	? Chylomicron transport		Intestine
CI	? LCAT activation	6331	Liver
CII	Lipoprotein lipase activation	8837	Liver
CIII	? Lipoprotein lipase inhibition	8764	Liver
D	? LCAT activation; lipid transfer	22 100	?
EII		38 000	Liver
EIII		38 000	Liver
EIV	IDL clearance	39 500	Liver

APOLIPOPROTEIN A

The A apolipoproteins[42,65] form the major proteins found in HDL, but they are also found in chylomicrons. Approximately 50% of HDL mass is protein, with Apo AI and AII constituting nearly 90%. The ratio of AI to AII is roughly 3:1, with average plasma concentrations of Apo AI being 121 mg/dL and AII being 37 mg/dL (Table 7-6). Both proteins have been sequenced; Apo AI and AII contain 243 and 154 amino acids, respectively. In the absence of lipid, both proteins self associate and form higher molecular weight aggregates. Apo AII appears as two identical chains linked by a disulfide bond at residue 6. While the exact site of *synthesis* is still unclear, experimental evidence suggests that Apo AI and AII originate in the intestine or liver or both. Once in the circulation, after release from the intestine as part of chylomicrons, Apo AI and AII have been shown to transfer and accumulate in HDL.

Apolipoprotein AI plays a *role* in the activation of LCAT and removal of free cholesterol from extrahepatic tissues. For maximal activation of LCAT, the Apo AI:phospholipid ratio must be 6:1. While the function of Apo AII is not entirely clear, it appears to play a structural role in HDL and may inhibit LCAT, possibly by dissociating Apo AI from the Apo AI–phospholipid complex. Apo AII may also play a role in HDL catabolism, perhaps by stimulating hepatic lipase.

Since it appears that apolipoproteins are fairly freely exchangeable between lipoprotein classes, it has been difficult to study apolipoprotein A *catabolism*. The tissue site of degradation is probably the liver or kidney or both. Apo AI and AII have almost identical rates of catabolism.

With the advent of immunological assays which permit quantitation of specific apolipoproteins in general, and Apo AI in particular, a number of disease states have been associated with deficiencies of individual or combinations of apolipoproteins. *Clinical classification* of disease states will be discussed later.

A variant of Apo A, apolipoprotein AIV, is not normally found in plasma but is a constituent of lymph chylomicrons. In certain patients with dys-β-lipoproteinemia (Type III HLP), Apo AIV has been found in significant concentrations in circulating IDL and LDL.[57]

Table 7-6. APOLIPOPROTEIN REFERENCE RANGES
FOR PLASMA[60]

Apolipoprotein	Mean (mg/dL)	Mean ± 2 SD
AI	121	73–169
AII	37	19–55
B	98	58–138
CI	7	3–11
CII	3.7	0.5–6.9
CIII	13	3–23
E	4	2–6

APOLIPOPROTEIN B

Due to its insolubility in the delipidated state, Apo B is the least well characterized apolipoprotein.[42,65] It is, however, the major protein moiety of all lipoproteins other than HDL. The molecular weight of Apo B remains speculative; estimates range from 8000–275 000.

Recently it has been shown that at least two forms of Apo B can be found in humans. The most abundant form is known as large B or B-100 and constitutes the Apo B found in lipoproteins synthesized from the liver. The other form of Apo B, small B or B-48, is believed to be synthesized in the intestinal wall and is found in chylomicron-rich lymph draining into the thoracic duct. In normal subjects, very little if any B-48 is found in fasting plasma; however, in subjects with defects in chylomicron remnant clearance (for example, chronic renal failure), levels of B-48 are increased.[57] Apo B, like the other apolipoproteins, has distinct immunological properties, and these have led to the production of a variety of polyvalent antisera and monoclonal antibodies which have subsequently been used to identify and quantitate circulating levels of Apo B (Table 7-6). More recently, monoclonal antibodies acting at differing sites on the Apo B molecules have been used to elucidate the structure and functional regions of Apo B.[68]

Very little information is available on the synthesis, secretion, and metabolic control involved in Apo B and subsequent lipoprotein production. *Synthesis* of Apo B is thought to occur on membrane-bound polyribosomes, followed by release into the cisternae of the endoplasmic reticulum. The synthesis of Apo B and its incorporation into chylomicrons and VLDL is essential to the formation and release of these lipoproteins into the plasma. Hepatic synthesis of Apo B, which eventually leaves the liver as VLDL, is about 10 mg/kg/d. While Apo B is a component of the VLDL molecule, crucial parts of the Apo B are masked and the apolipoprotein plays a passive role. With delipidation of VLDL and gradual conversion to the relatively triglyceride-poor LDL molecule, Apo B assumes a major role in ensuring lipid delivery to the extrahepatic cells. With unmasking of specific sites, Apo B assumes the role of a recognition protein on the surface of the lipoprotein, allowing the recognition and binding of LDL to specific high affinity receptors on the cell membrane. With the exception of the nervous system and red blood cells, virtually all tissues have receptors for Apo B in the LDL form.

The *catabolism* of Apo B occurs mainly via the receptor-mediated pathway. Once internalized in the cell, Apo B is degraded to its constituent amino acids by lysosomal enzymes. Some Apo B is also catabolized by nonreceptor-mediated processes that include a scavenger pathway. These latter processes may be important in the development of atherosclerosis, as will be discussed later. Some Apo B is also catabolized as part of VLDL and IDL.

Both excess and absence of Apo B in the circulation have been associated with various *disease states*. These situations are discussed in greater detail under the section on hyper- and hypolipoproteinemias.

APOLIPOPROTEIN C

This group of low-molecular-weight apolipoproteins[42,65] has been described as three different proteins: Apo CI, CII, and CIII, consisting of 57, 78, and 79 amino acids, respectively. Apo CIII exists in three isoforms, each with one sialic acid difference, i.e., $CIII_0$, $CIII_1$, $CIII_2$. The C apolipoproteins are soluble in aqueous solution and bind avidly to phospholipids to form protein-lipid complexes. The $t_{1/2}$ of Apo CII is 10–18 h, and daily synthesis has been calculated at 100–400 mg.

While the site of *synthesis* is not known with certainty, current evidence suggests that the liver is the primary organ. In fasting subjects, most of the C apoproteins are found in the VLDL and HDL fractions. The plasma concentrations of Apo CII and CIII have been reported to be ~3.7 and ~14 mg/dL, respectively (Table 7-6). The *function* of Apo CII appears well established. Apo CII has been conclusively demonstrated to be a cofactor of extrahepatic lipoprotein lipase, the catalytic enzyme responsible for hydrolysis of the triglyceride-rich lipoproteins, chylomicrons, and VLDL. In the absence of Apo CII, lipoprotein lipase activity is impaired. In patients with Apo CII deficiency, infusion of fresh plasma containing VLDL or HDL with normal amounts of Apo CII immediately activates their lipoprotein lipase and rapidly results in VLDL breakdown. Apo CII also inhibits binding of VLDL and chylomicrons to hepatic receptors, thereby targeting them for extrahepatic clearance. The presence of a lipoprotein lipase activated by Apo CI has also been reported, and CI has also been implicated in the activation of LCAT. The third

Apo C, CIII, is also thought to play a role in triglyceride hydrolysis, possibly as an inhibitor of lipoprotein lipase. Because of the prominent role of the C apolipoproteins in triglyceride-rich lipoprotein metabolism, they have been referred to as the functional apolipoproteins.

Little is known about the *catabolism* of the C apolipoproteins. There is virtually no Apo C found in IDL or LDL. There is, however, well documented transfer and redistribution of the C apolipoproteins between VLDL, HDL, and chylomicrons. Apo C from HDL is transferred to newly secreted chylomicrons and VLDL, while chylomicron and VLDL hydrolysis results in movement of Apo Cs, especially CII, from these lipoproteins to HDL. Due to the relative short $t_{1/2}$ of Apo C (10–18 h), compared to 2–5 d for Apo A and B, it is speculated that Apo C is cleared independently, perhaps at the stage of interlipoprotein transfer.

APOLIPOPROTEIN D

This apolipoprotein,[42,65] also known as thin-line peptide, was first isolated as part of a subclass of HDL_3. It has since been shown to be a glycoprotein with a M.W. of 22 100. Virtually nothing is known of Apo D synthesis or *catabolism*. It is suggested that Apo D may *function* as a transfer protein, assisting in the movement of cholesterol esters and triglycerides between lipoprotein species, especially from VLDL to HDL and vice versa.

APOLIPOPROTEIN E

Apo E[42,65] was first isolated from plasma in 1973 and originally was known as arginine-rich apolipoprotein. It is a glycoprotein with at least four polymorphic forms known as EI, EII, EIII, and EIV. The isoforms are separated at their isoelectric points, 5.3, 5.5, 5.6, and 5.75, respectively. The major isoforms differ from one another by one charge only. Recent studies have determined that EII has two cysteine residues, EIII one cysteine, and EIV no cysteine. The substitute amino acid for cysteine appears in most cases to be arginine, and the points of substitution have been found to be positions 4 and 33 in the 17 and 89 residue fragments. Apo EII therefore contains cysteine at both these sites, while EIV has arginine at both locations and EIII has cysteine at position 4 and arginine at position 33. Each of these major isoforms of Apo E can exist with minor modifications, probably as a result of post-translational glycosylation.

The three major isoforms (EII, EIII, and EIV) are the result of three independent alleles acting at a single gene locus. Thus, there exist three homozygous and three heterozygous states. Apo E is most likely synthesized in the liver and initially enters the plasma as part of nascent HDL. Under the influence of LCAT, HDL accumulates cholesterol and Apo E is rapidly transferred to VLDL and chylomicrons, where Apo E remains as the lipoprotein is catabolized to IDL by the lipoprotein lipase system. Apo E levels in plasma have been reported to range from 2–6 mg/dL (Table 7-6).

Apo E plays a significant *role* in chylomicron remnant and IDL recognition and catabolism via specific receptors in hepatic cells. It is mainly the EIII and EIV isoforms which interact with the hepatic receptors while EI and EII are poorly recognized and slowly catabolized. In subjects who produce only EII, there is a strong association with impaired clearance of chylomicron remnants and IDL, producing a disorder known as dys-β-lipoproteinemia or Type III hyperlipoproteinemia. In addition to the specific Apo E hepatic receptors, there is a strong affinity between Apo E and the LDL receptor. This is especially true for a form of Apo E–enriched HDL known as HDL_c, which is produced by excessive dietary cholesterol feeding. This HDL_c not only binds 10–20 times more readily then LDL-Apo B but also with higher affinity.

LIPOPROTEIN LITTLE A [Lp(a)] ANTIGEN

In addition to the above apoproteins, another apoprotein known as Lp(a) antigen occurs in a variant form of LDL termed "sinking pre-β-lipoprotein." This apoprotein is relatively rich in carbohydrate residues, especially sialic acid. Its function is not known but it has been independently associated with increased atherosclerosis risk.

LIPOPROTEINS

The rapidly changing knowledge of the apolipoproteins and their functional interrelationships among the various lipoproteins does not permit a definitive classification of the lipoproteins at present. However, the varied physical and chemical properties of these lipoproteins do allow for an operational definition based on their sizes, floating densities in an ultracentrifugal field, or electrophoretic mobilities. For purposes of clarity and consistency, the classification based on ultracentrifugal flotation density will be used, and the reader is referred to Table 7-7 for comparisons with classifications by other isolation techniques.

As knowledge of the *physiological function* and *degradation* of the lipoproteins continues to expand, it is probable that their role will not be restricted solely to the transport of water-insoluble lipids. Lipoproteins also play an important role in cholesterol and triglyceride metabolism. Unlike many other serum components that are mostly transferred from one point to another as metabolites, lipoproteins undergo a series of complex metabolic processes in which changes and exchanges occur continuously in and between the various lipoproteins. This is referred to as the *lipoprotein cascade*.

Chylomicrons

Chylomicrons, responsible for transporting dietary fat, are synthesized by and released from the intestinal epithelial cells. The lipid content is derived predominantly from the alimentary tract and is termed exogenous. The major lipid fraction is triglyceride, which constitutes over 80% of the total particle by weight. Chylomicrons contain a variety of apolipoproteins, but only 1–2% of the total particle is made up of apolipoproteins (Table 7-7). The most notable apolipoproteins found are B-48, AI, AII, AIV, and the C apolipoproteins. Chylomicrons *do not* enter the portal venous system but instead traverse the lymphatics into the thoracic duct, eventually entering directly into the jugular vein and the main systemic circulation. While traveling through the lymphatic system, chylomicrons acquire significant quantities of the C apolipoproteins, presumably as a result of transfer from HDL which has filtered into the lymph from plasma. After entry into the systemic circulation, chylomicrons meet with other triglyceride-rich lipoproteins (TRL) and are eventually acted on by a number of lipoprotein lipases. These enzymes hydrolyze the

Table 7-7. CLASSIFICATION PROPERTIES AND COMPOSITION OF HUMAN SERUM LIPOPROTEINS

Parameter	Chylomicron	VLDL	IDL	LDL	HDL
Hydrated density (g/mL)	0.93	0.97	1.003	1.034	1.121
Solvent density for isolation (g/mL)	<1.006	<1.006	1.006–1.019	1.019–1.063	1.063–1.21
Molecular weight	$(0.4–30) \times 10^9$	$(5–10) \times 10^6$	$(3.9–4.8) \times 10^6$	2.75×10^6	$(3.6–1.75) \times 10^5$
Diameter (nm)	>70.0	25.0–70.0	22.0–24.0	19.6–22.7	4–10
Electrophoretic mobility (paper, agarose)	Origin	Pre-β	Broad β (between β and pre-β)	β	α
Composition (% by weight)					
Cholesterol, unesterified	2	5–8	8	13	6*
Cholesterol, esterified	5	11–14	22	49	13*
Phospholipid	7	20–23	25	27	28*
Triglyceride	84	44–60	30	11	3*
Protein	2	4–11	15	23	50*
Apoproteins (% total apolipoprotein)					
AI	7.4	Trace	—	—	67
AII	4.2	Trace	—	—	22
B-100	Trace	36.9	50–70	98	Trace
B-48	22.5	Trace	Trace	—	—
CI, CII, CIII	66	49.9	5–10	Trace	5–11
EII, EIII, EIV	—	13.0	10–20	Trace	1–2
D	—	—	—	—	Trace
Synthesis	Intestine	Liver, intestine	Intravascular	Intravascular	Intestine, liver

*From Jackson, R. R., Morrisett, J. D., and Gotto, A. M.: Lipoproteins and lipid transport: Structural and functional concepts. *In*: Hyperlipidemia: Diagnosis and Therapy. Rifkind, B. M., and Levy, R. I., Eds. New York, Grune & Stratton, 1977, pp. 1–16.

triglyceride component to monoglycerol, glycerol, and free fatty acids which can then be taken up at the cellular level for energy metabolism or for resynthesis of triglyceride for storage. The Apo AI, AII, and C are transferred to HDL together with a small amount of lipid. The remainder of the chylomicron, known as the chylomicron *remnant*, now falls into the density region between VLDL and IDL, with its main apolipoproteins being B-48 and E. The remnant particle is 30–80 nm in diameter and retains most of the cholesteryl esters found in the chylomicron. Chylomicron remnants are rapidly removed after entering the circulation by the liver (usually in <10 min), with the catabolic process being mediated, in all likelihood, by the Apo E receptor in the liver. The remnants are rapidly internalized by receptor-mediated endocytosis and degraded in the hepatic lysosomes, thus delivering dietary cholesterol to the liver. The exact mechanism in humans has not been well studied. Some estrogen preparations stimulate these hepatic receptors and facilitate remnant clearance. Estrogen therapy has, therefore, been used for patients with remnant accumulation. In rare patients with impaired remnant removal but normal chylomicron catabolism, there is an increase in plasma Apo B-48 and Apo AIV, two lymph chylomicron apolipoprotein constituents not usually found circulating in normal plasma. In such subjects, the impaired catabolism occurs due to an abnormality in the Apo E moiety, with a lack of either or both EIII and EIV, the isomorphic forms responsible for interaction with the hepatic receptor.

Very Low Density Lipoproteins

The other triglyceride-rich lipoprotein, very low density lipoprotein (VLDL), is synthesized in and released from the liver (Figure 7-15). VLDL transports hepatic-synthesized triglyceride and cholesterol that are probably derived from dietary precursors, such as free fatty acids, glycerol, and carbohydrates. VLDL synthesis is closely related to energy availability, being stimulated by energy excess without regard to the calorie source. Besides triglyceride, VLDL contains about 10% cholesterol and a number of functionally important apolipoproteins. The Apo B present (B-100) is of high molecular weight, and although identical to the Apo B found on LDL, its behavior is significantly different. The Apo B on VLDL does not bind significantly to the Apo B (LDL or B-E) receptor, probably due to some form of steric hindrance. In recent studies using a monoclonal antibody specific for the receptor region of Apo B, the antibody failed to bind to VLDL–Apo B, confirming that the Apo B site in VLDL is antigenically nonfunctional. With delipidation of VLDL, the Apo B binds to cell receptors and the monoclonal antibody.[68]

Apo CII is as vital to the catabolism of VLDL as it is to chylomicrons. Figure 7-15 schematically depicts apolipoprotein CII in VLDL as a cofactor of extrahepatic lipoprotein lipase (LPL) which removes most of the triglyceride from VLDL. A *lack of Apo CII*, as has been reported in a number of subjects, leads to lack of LPL activation and accumulation of chylomicrons and VLDL.[9] The addition of even small amounts of Apo CII (by infusions of fresh plasma from a non-CII deficient subject) results in rapid LPL activation and subsequent reductions in circulating triglyceride-rich lipoproteins. Although Apo CII deficiency is rare, with only about ten such reported families in the world, Apo CII deficiency should be considered and differentiated from other causes of TRL elevation, such as deficiency or absence of LPL itself. As VLDL is

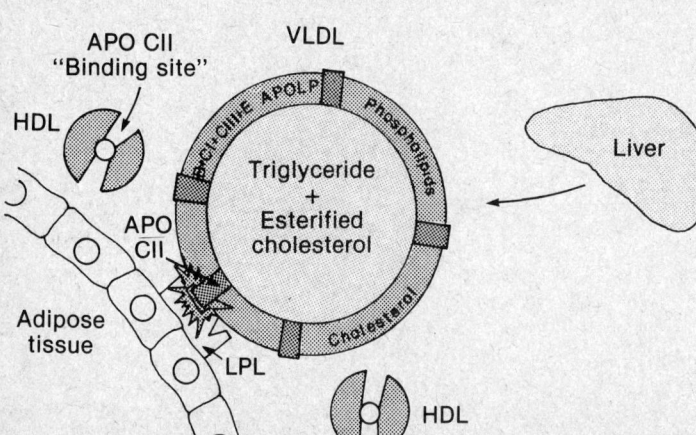

Figure 7-15. Schematic diagram of VLDL catabolism. Activation of lipoprotein lipase (LPL) by Apo CII leads to removal of triglycerides from VLDL. (Modified and published with permission from M. L. Kashyap, M.D., et al.: Lipids, *13:* 933–942, 1978.)

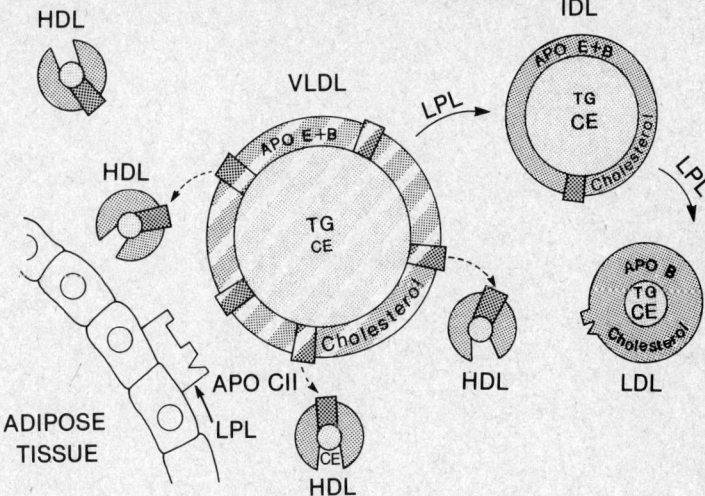

Figure 7-16. Catabolism of VLDL to LDL. According to the postulated "lipoprotein cascade" pathway, hydrolysis of VLDL triglycerides by lipoprotein lipase (LPL) leads to a short-lived lipoprotein termed intermediate-density lipoprotein (IDL). IDL is further catabolized by LPL to form the cholesterol-rich LDL particle. During VLDL catabolism, Apo CII is transferred to HDL; all the remaining apolipoproteins except Apo B are lost as LDL is formed. (Modified and published with permission from M. L. Kashyap, M.D.)

catabolized (see Figure 7-16), the C apolipoproteins together with some lipid material are transferred to HDL. The Apo C on HDL is later transferred back to chylomicrons and VLDL to complete the cycle. The $t_{1/2}$ for VLDL in serum is 1–3 h.

Intermediate Density Lipoproteins

Following VLDL hydrolysis by lipoprotein lipase, a short-lived intermediate density lipoprotein (IDL), partly depleted of TG, is formed. IDL contains about equal amounts of cholesterol and triglyceride (see Table 7-7), and its major apolipoproteins are Apo B and E. It is Apo E that determines the continued catabolic process (cascade) of IDL to LDL for hepatic uptake and degradation. The *role of Apo E* here may be viewed as similar to its role in respect to chylomicron remnants. The disease processes characterized by defective metabolism of chylomicron remnants and of IDL metabolism stem from a defect in Apo E.

Low Density Lipoproteins

In normal circumstances, IDL is further delipidated by hepatic lipoprotein lipase to form low density lipoprotein (LDL). The precursor-product relationship of Apo B in VLDL and LDL is now well established.

LDL catabolism takes place in the liver and the peripheral tissues. The proposed sequence of events is as follows (see also Figure 7-17):

1. LDL interacts with high-affinity receptor sites located in regions of the cell membrane called "coated pits."

2. The bound LDL is then internalized by invagination of these pits into the cell where the pits pinch off to form endocytic vesicles.

3. These vesicles fuse with intracellular lysosomes and the LDL moiety is subjected to a series of hydrolytic enzymatic degradations that ultimately hydrolyze the Apo B to amino acids.

4. The esterified cholesterol in LDL is hydrolyzed by lysosomal cholesterol esterase, and the free cholesterol enters the cytoplasm.

The release of free cholesterol is responsible for three regulatory responses which assist in cholesterol homeostasis: (a) suppression of the rate-limiting enzyme, HMG-CoA reductase, and of new cholesterol synthesis; (b) activation of ACAT activity to esterify excess cholesterol for intracellular storage as cholesteryl ester droplets; and (c) modulation of the number of LDL receptors on the plasma membrane to prevent the over-accumulation of intracellular cholesterol through the receptor pathway. In culture, cells appear to adjust the number of receptors to provide sufficient cholesterol to support cell growth and membrane turnover. But as the rate of growth decreases, the number of receptors formed declines. High-affinity LDL receptors have been found on most cells; however, they are more numerous in certain cell types such as the adrenocortical cells where LDL-derived cholesterol provides the major source of substrate for synthesis of steroid hormones.

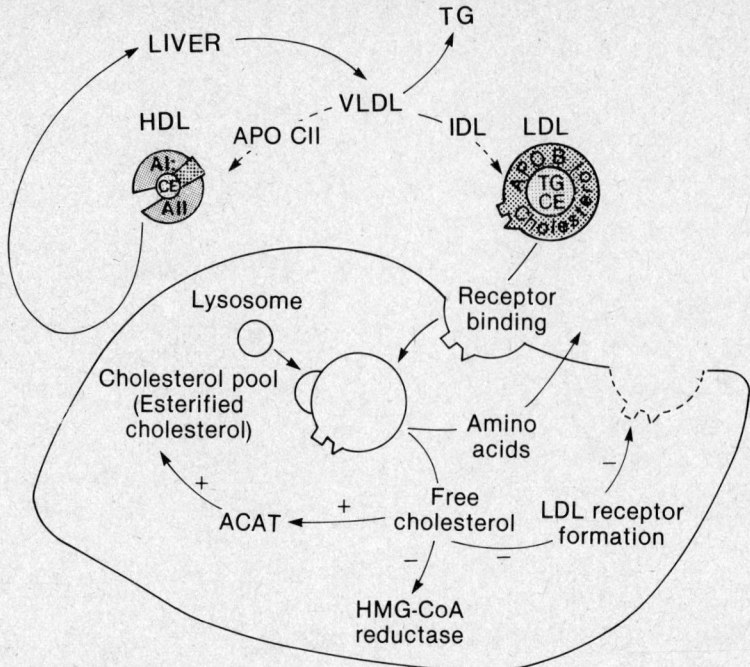

Figure 7-17. LDL catabolism via high-affinity receptor pathway. (See text for details.)

In addition to catabolism via the high affinity Apo B LDL-receptor pathway, plasma LDL can be degraded by nonspecific pathways not subject to feedback regulation. These modes are less efficient and require elevated plasma levels to achieve significant rates of removal. One of these pathways, known as bulk endocytosis, occurs in the scavenger cells or macrophages of the reticuloendothelial system. With increasing plasma levels of LDL, the scavenger cells take up larger amounts of the circulating lipoprotein for degradation. At some point, they become overloaded with cholesterol esters and take on the appearance of "foam cells." In the arterial wall, both macrophages and smooth muscle accumulate cholesterol esters by this mechanism; foam cells therefore become the hallmark of the atherosclerotic plaque. It has been estimated that normally 35–64% of LDL is degraded by the high affinity receptor pathway, with the remainder removed by the scavenger cell system.

The significance of the LDL receptor becomes evident in patients with the genetic disorder *familial hypercholesterolemia* (FH). These subjects have a defect in the gene coding for the LDL receptor. There are three defects so far described: the receptor may be absent; it may be deficient; or it may be present and bind normally, but be unable to internalize the LDL particle. In subjects who are homozygous for the gene defect, functional receptors are few or absent. Heterozygote subjects have approximately half the normal number of LDL receptors. The inability of the LDL moiety to deliver cholesterol to the cells of FH patients results in increased HMG-CoA reductase activity and increased cholesterol synthesis. Reduction in the removal of LDL from plasma causes elevated plasma levels of LDL-cholesterol, increased uptake of LDL by macrophages and smooth muscle cells, cholesterol deposits in arterial walls, and, consequently, premature atherosclerosis.

High Density Lipoproteins

While the metabolism of chylomicrons, remnants, VLDL, IDL, and LDL is reasonably well understood, our knowledge about "high density lipoprotein" (HDL) is relatively new and still growing. HDL consists of a number of polydisperse and heterogeneous particles which vary with respect to size and content of lipid and apolipoprotein. HDL can be separated not only by ultracentrifugation and electrophoresis but also by polyanionic precipitation. At least three well defined HDL subgroups have been studied. These include HDL$_c$, HDL$_2$ and HDL$_3$. About 50% of HDL mass is protein, 30% is phospholipid, and 20% is cholesterol (Table 7-7). The ratio of phosphatidylcholine to sphingomyelin is 5:1, and the ratio of esterified to free cholesterol is about

3:1. The major apolipoproteins found in HDL are AI and AII and constitute about 90% of total HDL protein. The ratio of Apo AI to Apo AII is ∼3:1 by weight.

HDL is usually divided into two density classes, HDL_2 ($d = 1.063–1.125$ g/L) and HDL_3 ($d = 1.125–1.21$ g/L). HDL_2 is the larger of the two particles with an estimated M.W. of 360 000 compared with 175 000 for HDL_3. HDL_2 is also associated with a larger lipid mass: 60% compared to 45% in HDL_3. Reports of the relative amounts of Apo AI and AII in HDL_2 compared with HDL_3 are contradictory. However, on average, HDL_2 contains higher Apo AI levels. The other significant apolipoproteins found in HDL are the C apoproteins, which play an integral role in TRL metabolism. The only other apolipoprotein found to any significant degree in HDL is Apo E, which occurs especially in a variant form of HDL, known as Apo E-HDL_c. This lipoprotein is thought to develop following ingestion of large amounts of cholesterol; it may be responsible for transporting dietary cholesterol. Because of its Apo E content, HDL_c has an extremely high affinity for the LDL receptor, 10- to 25-fold greater than the receptor's affinity for LDL. However, at saturation the number of HDL_c particles bound is only one-fourth the number of LDL particles bound, suggesting that each Apo E-HDL_c particle binds to four LDL receptors while each LDL particle binds to a single LDL receptor site.

Both the liver and the intestine are involved in the *production* of HDL, although the exact roles and relative importance of each are not fully understood. Current evidence indicates that the mature spherical form of HDL is not secreted directly into the plasma but is derived from a discoidal shaped precursor form consisting of Apo AI, AII, lecithin, and free cholesterol. Transformation of this "nascent" or discoidal form of HDL into the spherical lipid transporting moiety depends to a great extent on the action of the plasma enzyme lecithin-cholesterol acyl transferase (LCAT). LCAT is the only factor catalyzing sterol ester synthesis in plasma, and its activity is increased by Apo AI and inhibited by Apo AII. The initial complex of Apo AI, lecithin, and free cholesterol appears to stimulate LCAT activity to form cholesterol ester; with association of APO AII the neutral sterol then enters the core of the particle (Figure 7-18). It is also speculated that HDL components may arise from VLDL and chylomicrons during their catabolism. Transfer of both apolipoproteins and lipids from VLDL and chylomicrons to HDL has been demonstrated. The $t_{1/2}$ of HDL in plasma in normal subjects is ∼4 d. Little is known about the sites of HDL *catabolism*; however, the liver and kidney are probably involved.

Although our knowledge is still limited, HDL appears to play an important *role* in cholesterol efflux from tissues, thereby reducing the amount of cholesterol stored there. HDL also plays a role in returning cholesterol from the periphery to the liver for removal as bile acids, a process known as reverse cholesterol transport. Recent evidence tends to support the suggestion that HDL plays a major role as a scavenger of lipid and apolipoprotein during the normal catabolism of chylomicrons and VLDL. HDL acquires free cholesterol released from these molecules, and plasma LCAT converts this free cholesterol to its esters (with FA derived from lecithin). These esters are later transferred back to VLDL and IDL by way of Apo D or a protein known as lipid transfer protein. HDL is also known to play an important role as a plasma reservoir for Apo CII. The relationship of HDL and VLDL with chylomicrons is exemplified by the fact that defects in TRL catabolism are commonly associated with marked reduction in HDL levels. Recent epidemiological studies have suggested that HDL protects against cardiovascular disease, and a significant amount of work has been done on HDL in order to demonstrate its role in reverse cholesterol transport.

Figure 7-18 attempts to summarize the complex interrelationships between the various lipoproteins and provides a starting point from which a coherent and systematic *classification of lipoprotein disorders* can be derived. The preceding discussion shows that excess circulating levels of any lipoprotein can be caused by one of two factors, either excess production or decreased catabolism. To date, most abnormalities have been related to impaired catabolism. Because each catabolic process involves a number of steps, increased knowledge of these steps will probably reveal more than one catabolic abnormality.

CLINICAL INTERRELATIONSHIPS OF LIPIDS, LIPOPROTEINS, AND APOLIPOPROTEINS

Epidemiology of Lipoproteins and Disease

Lipoprotein metabolism is of particular interest to clinicians concerned with the diagnosis and treatment of atherosclerosis. Atherosclerosis results from thickening of the inner layer of the arterial wall. This thickening is caused by cellular material and deposits of several substances,

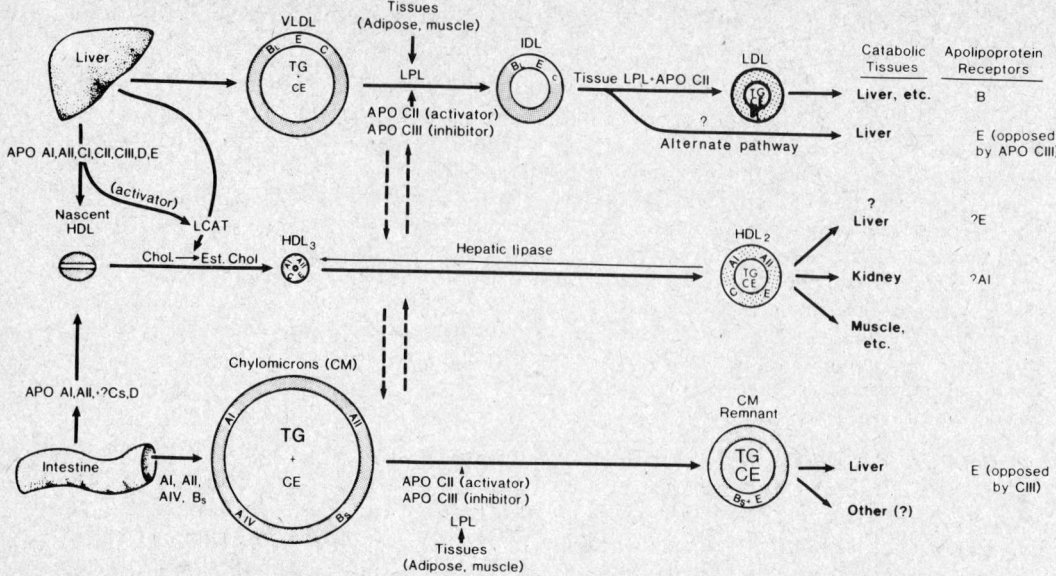

Figure 7-18. Overview of lipoprotein interrelationships. (From Vascular Medicine, *2:*16–21, 1984. Reproduced with permission of copyright owner: Westminster Publications, Inc., Roslyn, New York 11576. All rights reserved.)

Upper path: Endogenous lipoprotein metabolism of VLDL. Hepatic synthesized triglyceride and cholesterol are transported by VLDL. Apo CII catalyzes and Apo CIII inhibits lipoprotein triglyceride hydrolysis by LPL. As VLDL is catabolized to IDL and then to LDL, the C-apoproteins are transferred to HDL (indicated by the broken arrow). IDL may also be catabolized via the Apo E receptor pathway in the liver. LDL is largely catabolized via the Apo B receptor. Degradation of LDL results in the release of esterified cholesterol which is later hydrolyzed to free cholesterol.

Middle path: HDL formation and lipid transfer. Apo AI and AII are synthesized in the liver and intestine, enter the plasma (probably as part of VLDL or chylomicrons), and then come together to form nascent HDL in plasma where Apo AI activates LCAT. Free cholesterol is then esterified and enters the core of the HDL particle. Thus, removal of cholesterol from cell surfaces and other lipoproteins is facilitated. These esters are later transferred from HDL to VLDL for transport to the liver for catabolism. Little is known about the sites of HDL catabolism. As indicated by the broken arrows, the Apo C on HDL is transferred to VLDL and chylomicrons.

Lower path: Exogenous lipoprotein metabolism of chylomicron. Dietary triglyceride constitutes over 80% of these particles. As chylomicrons are catabolized, Apo AI, II, and C are transferred to HDL (indicated by the broken arrow). The remaining chylomicron remnant is removed by the Apo E receptor. Degradation of these remnants results in delivery of dietary cholesteral to the liver.

TG = triglycerides; CE = cholesteryl ester; B$_L$ = large form of Apo B; B$_s$ = small form of Apo B.

particularly lipids. Despite a recent and continuing decline in the incidence of death from atherosclerotic coronary artery disease (CAD), the disease still accounts for more than 50% of all deaths in the USA.[49] Of the many risk factors so far implicated in the causation of CAD, three are accepted as being of major importance: dyslipoproteinemia, hypertension, and cigarette smoking. There are numerous minor risk factors including diabetes, inactivity, obesity, and family history.

The importance of serum lipoprotein disturbances as an etiological factor in the development and potentiation of atherosclerosis is now supported by a considerable body of evidence, direct as well as circumstantial, amassed from epidemiological and population studies.[3,19,52,82] Although work by Gofman and associates[34] in the early 1950's drew attention to the ratios of various lipoproteins as indicators of risk for atherosclerosis, their use lagged since the easily analyzed plasma total cholesterol was thought to be a more sensitive indicator. Early studies by Barr and colleagues[7] and by Nikkila and co-workers[58] demonstrated relative or absolute decrease in the cholesterol carried in the HDL fraction of plasma of atherosclerotic patients. However, it was not until the mid-1970's that population studies again showed that levels of HDL-cholesterol (HDL-C) were powerfully, inversely, and independently related to the risk of CAD.[31] Studies of kindreds with familial elevations of HDL-C revealed lower incidence, morbidity, and mortality for CAD and longer life spans for more individuals in the kindred. Thus, many studies through different approaches have suggested that elevated HDL levels are "anti-atherogenic," while reduced levels are associated with increased risk for coronary artery disease.

Our previous discussion of lipoprotein metabolism and the summary of Figure 7-18 point out the positive association of impaired catabolism of chylomicron remnants, IDL, and LDL with atherosclerosis. Increased HDL appears to retard or prevent the development of atherosclerosis, while altered metabolism of chylomicrons and VLDL does not appear directly related to the disease. Neither the manner in which chylomicron remnants, IDL, and LDL cause atherosclerosis nor how HDL prevents it is known. It is known, however, that the disease involves more than just circulating lipoproteins and that cellular enzyme systems (such as LPL), platelets, arterial smooth muscle and endothelium, and macrophages play significant roles. To these cellular and humoral factors can be added a host of genetic, environmental, and local factors.

The three atherogenic lipoproteins were previously shown to be dependent on either hepatic or extrahepatic receptors for their ultimate catabolism. The chylomicron remnant, taken up via hepatic receptors, permits delivery of large amounts of dietary cholesterol to the liver, which in turn modulates endogenous cholesterol synthesis. In some circumstances, however, this clearance system is disrupted. In dys-β-lipoproteinemia, where major isoforms of Apo E (EIII and EIV) seem not to be produced, remnants and IDL fail to be removed by the liver and serum cholesterol rises due mainly to an increase in cholesterol of dietary origin. In these patients, dietary restriction in cholesterol and saturated fat is often effective in lowering serum cholesterol levels.

Large amounts of endogenously synthesized cholesterol can be transported from the liver to the periphery by means of the LDL receptor mechanism without the need for high levels of LDL in plasma. However, this endogenous transport system can also be disrupted, as seen in familial hypercholesterolemia (FH). With a defect in the LDL receptor mechanism, circulating LDL increases as does the plasma cholesterol content. This LDL-cholesterol has been derived predominantly from endogenous synthesis in the liver, and consequently patients with severe FH maintain high serum cholesterol and LDL-cholesterol levels even when they consume a virtually cholesterol-free diet.

Although hyperlipoproteinemia in the majority of individuals in western countries has not been associated with a major genetic receptor abnormality, LDL levels are apparently above those which appear to be appropriate for the receptor system.[11] These inappropriately elevated LDL levels are believed to be in a range that predisposes to, or accelerates the development of, atherosclerosis. The increase in LDL and LDL-cholesterol suggests elevation of endogenous production, whereas epidemiologic evidence points toward excessive dietary intake of fat, cholesterol, and calories. Dietary intake would impact more on the exogenous cholesterol-carrying lipoproteins, chylomicrons, and their remnants. The epidemiological evidence is tied-in with increased LDL levels through the ability of excess dietary fat and calories to stimulate overproduction of VLDL in the liver. With subsequent VLDL catabolism to LDL, the levels of circulating LDL-cholesterol increase and potentially saturate the LDL receptor system.

Even in subjects without major genetic defects in the LDL receptor system, the ability to cope with excess LDL production as a consequence of diet may be modulated by minor genetic factors and possibly hormonal factors which affect the number of lipoprotein receptors. It is speculated that the more receptors an individual possesses, the lower the LDL-cholesterol level will be, and if all other factors known to affect atherosclerosis are held constant, subjects with the most lipoprotein receptors will be least susceptible when faced with dietary excess of cholesterol and total fat.

Thus, the interaction of genetic and environmental factors in controlling the number of lipoprotein receptors could explain, in part, the most common forms of hypercholesterolemia that beset the western population. These "polygenic" interactions may also explain the significant differences within populations to similar dietary constituents.

The mechanism by which the anti-atherogenic lipoprotein HDL decreases the atherosclerotic process is also speculative. It may function by removing free cholesterol from the peripheral cell, esterifying it, transporting it in the neutral lipid core, and transferring it to other lipoproteins; it could increase the rate of catabolism of the triglyceride-rich lipoproteins by enhancing the interaction of Apo CII with lipoprotein lipase; alternatively, HDL may inhibit the uptake and degradation of LDL in smooth muscles by competitively inhibiting the LDL receptor-mediated pathway.

From the above discussion, it should be clear that understanding of the lipoprotein transport system, its receptors, and the enzymes involved in its catabolism should facilitate the development of methods to predict which individuals are at risk from atherosclerosis and, ideally, to delineate rational lines of therapy.

Apolipoproteins

In recent years an increasing body of evidence has accumulated linking apolipoproteins to atherosclerosis and suggesting that their measurement may be as good as, if not better than, measurements of lipoproteins for assessment of risk for CAD. Numerous studies[4,5,72,83,90] indicate that patients with atherosclerosis are more exactly discriminated from patients without atherosclerosis by the finding of increased plasma Apo B levels than by the findings of decreased HDL-C and increased LDL-C. Atherosclerosis in conjunction with very low HDL-C and deficiency of Apo AI has also been reported,[60,66] and another study[55] shows plasma Apo AI to be the best of all lipid and lipoprotein fractions at predicting CAD in subjects requiring angiography. Apo A levels have also been shown to be better correlated with peripheral vascular disease (PVD) than lipid parameters presently measured.[25]

Although the traditional marker for familial hypercholesterolemia (FH) has been, and will remain, the increase of LDL-cholesterol, some affected subjects undoubtedly present with normal LDL-C but would be found to have increased Apo B if that testing were done; and some unaffected subjects with increased LDL-C would be found to have normal levels of Apo B. Type III disease (dyslipoproteinemia), due to improper catabolism of IDL and remnant particles, now requires difficult, tedious, and expensive ultracentrifugal isolation and measurement of IDL. While Type III hyperlipidemia is relatively rare, it must still be ruled out in all patients who display combined hyperlipidemia (increased cholesterol and triglyceride), a lipid phenotype found in 2–3% of the population. Measurement of Apo E levels and Apo E isomorphic forms appears potentially more specific and perhaps less difficult.[94]

Besides the evident value of measurement of the four major lipoproteins—AI, AII, B, and E—there appears to be a developing need to measure a number of potentially important minor apolipoproteins. Apo CII has been very well documented as the prime cofactor of extrahepatic lipoprotein lipase. Apo CII deficiency, as first described by Breckenridge and colleagues[9] in 1978, if identifiable, is a potentially treatable disease. Recent studies of patients with chronic renal failure, and consequently at increased risk for coronary artery disease, seem to indicate that Apo AIV and B-48, as well as abnormal distribution of Apo C and E, play a role in the atherogenic process.[57]

The plasma apolipoproteins AI and AII have been used as markers in the control of diabetes and liver disease.[45,92] With the wider availability of methods to measure apolipoproteins, more clinical correlations and a better understanding of atherosclerotic mechanisms are likely to be found.

CLASSIFICATION OF CLINICAL LIPOPROTEIN DISORDERS

A contemporary classification of dyslipoproteinemias, organized according to hydrated density of characteristic lipoproteins, is shown in Tables 7-8 and 7-9. It is an extension of the first, comprehensive, and still valuable classification that was formulated in 1965 and 1966. This first classification, known as the *Fredrickson* or *NHLBI classification*[27] and adopted by the World Health Organization (WHO), described hyperlipoproteinemias as plasma phenotypes, i.e., particular patterns of lipid and lipoprotein levels with strong associations to clinical disease. Fredrickson's types continue to be important tools because they (1) allow sharp focus on a diverse group of metabolic abnormalities related to the hyperlipoproteinemias; (2) identify hyperlipoproteinemias, not as specific disease states, but as disorders which affect concentrations of particular lipoproteins in a similar way; (3) associate a majority of lipoprotein types with certain distinctive clinical features; and (4) provide a basis for successful approaches to diet and drug therapy of a majority of lipoprotein disorders, regardless of their etiology.

The major limitation of the Fredrickson system is its tendency to "lump" etiologically heterogeneous disease states together. For example, hyperchylomicronemia (Fredrickson Type I) can, as is now known, be due either to LPL deficiency or to Apo CII deficiency. The system also does not assign roles or a place to HDL or individual apolipoproteins, nor does it take into account significant changes of plasma phenotypes that occur with alterations in diet, use of drugs or alcohol, or general health. Despite these shortcomings, however, the underlying rationale of the Fredrickson system, with its emphasis on biochemical findings of plasma lipid and lipoprotein concentrations as a point of departure for diagnosis of dyslipoproteinemias, remains an important element of any classification, old or new.

Advances in our understanding of normal and pathological lipoprotein metabolism, and elucidation of the lipoprotein "cascade" in terms of interactions among lipoproteins and apolipoproteins, are beginning to lead to a more logical pathophysiologic classification. For instance, hyper-β-lipoproteinemia (Fredrickson Type II) is clearly not a single entity. We know that

Table 7-8. CLASSIFICATION OF HYPERLIPOPROTEINEMIAS

Lipoprotein Abnormality (Increase)	Plasma Appearance*	Total Cholesterol	Triglyceride	LDL-Cholesterol	HDL Cholesterol	Apolipoprotein	Lipoprotein Electrophoresis	Lipoprotein Phenotype	Clinical Association
Chylomicrons	Cream layer, infranate clear or slightly turbid	Normal to moderately elevated	Markedly elevated	Normal	Normal to decreased	↑B-48 ↑A-IV ↑↓CII	Intense band at origin	Type I†	Acute abdomen, pancreatitis
Low density lipoprotein (LDL)	Clear, possible increase in yellow-orange tint	Usually elevated, occasionally within normal range	Normal	Elevated	Normal to decreased	↑B-100	Increased band in β-region	Type IIA†	Markedly increased risk of CAD
Low density lipoprotein (LDL); very low density lipoprotein (VLDL)	Clear to slightly turbid	Elevated, occasionally marginally so	Elevated	Elevated	Normal to decreased	↑B-100	Increased β- and pre-β-band	Type IIB†	Increased risk of CAD
Intermediate density lipoprotein (IDL)	Turbid to opaque with thin creamy layer occasionally present	Elevated	Elevated	Normal to decreased	Normal to decreased	↑EII ↓EIII ↓EIV	Broad β-band	Type III†	Increased risk of CAD
Very low density lipoprotein (VLDL)	Turbid to opaque	Normal to slightly elevated	Moderately to markedly elevated	Normal	Normal to decreased	↓CII ↑B-100	Increased pre-β-band	Type IV†	Increased risk of CAD
Very low density lipoprotein (VLDL); chylomicrons	Creamy layer, infranate turbid to opaque	Slightly to moderately elevated	Markedly elevated	Normal	Normal to decreased	↓↑CII ↑B-48 ↑B-100	Intense band at origin plus increased pre-β-band	Type V†	Pancreatitis, increased risk of CAD
High density lipoprotein (HDL)	Clear	Normal to moderately elevated	Normal	Normal	Elevated	↑AI ↑AII	Increased α-band	Hyper-α-lipoproteinemia	Decreased risk of CAD

*After 16 h at 4 °C.
†Fredrickson type.

Table 7-9. CLASSIFICATION OF HYPOLIPOPROTEINEMIAS

Lipoprotein Abnormality	Plasma Appearance*	Total Cholesterol	Triglyceride	LDL-Cholesterol	HDL-Cholesterol	Apolipoprotein	Electrophoresis	Lipoprotein Phenotype	Clinical Association
Low density lipoprotein (LDL)	Clear	Markedly decreased	Decreased	Absent	Normal	$\downarrow\downarrow$B-100	Absent β-band and decreased pre-β-band	A-β-lipoproteinemia	Malabsorption; mental retardation; growth failure
Low density lipoprotein (LDL)	Clear	Mildly to markedly decreased	Normal	Decreased	Normal	$\downarrow$B-100	Decreased β-band	Hypo-β-lipoproteinemia	Decreased risk of CAD
High density lipoprotein (HDL)	Clear	Normal to decreased	Normal	Normal	Absent	$\downarrow\downarrow\downarrow$AI $\downarrow\downarrow$AII ?$\downarrow$CIII	Absent α-band	A-α-lipoproteinemia	Increased risk of hypersplenism, CAD
High density lipoprotein (HDL)	Clear	Normal, decreased, or increased	Normal to increased	Normal to increased	Decreased	$\downarrow$AI $\downarrow$AII	Decreased α-band	Hypo-α-lipoproteinemia	Increased risk of CAD

*After 16 h at 4 °C.

Adapted from Stein, E. A., and Glueck, C. J.: Hyperlipoproteinemia: Implications, diagnosis and therapy. *In*: Cardiac Diagnosis and Treatment. 3rd ed. N. O. Fowler, Ed. New York, Harper and Row, 1980, pp. 677-695.

elevations of circulating LDL may be due to a number of defects in the high-affinity receptor-mediated LDL pathway and that these defects include both a binding defect and an internalization defect. Further identification of specific defects will eventually lead to improved knowledge of the natural history of each disorder, definition of more exact relationships with atherosclerosis, and use of more rational therapies. A more thorough and more exact classification is obviously in the process of development.

Lipoprotein disorders may also be classified in the classic manner, i.e., as primary or secondary. Primary disorders are genetic (familial) or nongenetic (sporadic). The secondary disorders arise from some cause such as diet, use of alcohol or drugs, or disease of metabolic, hormonal, infectious, or malignant etiology. Note, however, that the diagnostic strategy, when a phenotype pattern of plasma findings is obtained, is first to rule out secondary causes, which are common, and then to explore for primary disorders which are, by comparison to secondary disease, relatively uncommon.

In summary, the first step in diagnosis of hyper- and hypolipoproteinemias is the definition, by chemical analysis of plasma for lipids and lipoproteins, of a phenotype or plasma lipoprotein pattern. The picture acquired will be only a plasma lipoprotein pattern; it will not define a homogeneous disease entity from the clinical, genetic, or pathophysiological point of view. It will, however, provide a basis for additional testing that may lead to a definitive diagnosis, for establishing treatment regimens that ameliorate the symptoms of primary disease and reduce risks of atherogenesis associated with both primary and secondary dyslipoproteinemias, and for monitoring the course of treatment. These applications will be discussed and illustrated in the following sections.

HYPERLIPOPROTEINEMIAS

Disorders are listed and findings are summarized in Table 7-8.

Increased Chylomicrons (Fredrickson Type I Hyperlipoproteinemia)

Clinical presentation. In the majority of cases, this extremely uncommon disorder has been diagnosed before the patient reached 10 years of age. In a significant number of patients, the diagnosis has been made in the first year of life. In those patients whose disorders were diagnosed after the first decade, either the symptoms were mild or the condition was misdiagnosed earlier. The most frequent presentation is recurrent abdominal pain, often accompanied by splenomegaly or hepatomegaly or both, and eruptive xanthomas. The xanthomas usually appear when triglyceride values exceed 2000 mg/dL. They take the form of erythematous papules, commonly appearing on the buttocks and proximal portions of the extremities and lasting several weeks until triglyceride values are reduced. When triglyceride values exceed 2000 mg/dL, lipemia retinalis may be seen and may be an important clue to diagnosis. The real significance of hyperchylomicronemia rests in its ability to produce abdominal pain, which may range from mild and transient to a severe peritonitis associated with acute pancreatitis. Pancreatitis is the most serious complication and has led to death in a number of patients. Serum amylase may be falsely normal due to an unknown inhibitory factor. With dilution of the serum with normal saline, amylase activity may increase to levels consistent with pancreatitis. When this disorder is suspected in the differential diagnosis of an acute abdomen, the diagnosis should be apparent from the thick, creamy plasma which often can be recognized even without routine centrifugation.

Primary versus secondary hyperchylomicronemia. In early life, the disease is usually primary or familial in nature. However, Type I phenotypes have been observed, often transiently, in older patients with acute pancreatitis, uncontrolled diabetes, hypothyroidism, immunologic disorders such as dysglobulinemia and systemic lupus erythematosus, and during oral contraceptive therapy. The exclusion of most of these underlying disorders is by history and physical examination. In some cases, however, detailed laboratory investigations may be necessary.

After elimination of the possibility of secondary hyperchylomicronemia, it is still necessary to confirm the primary disorder. At present, there are no specific genetic markers, although an isotopic immunoassay for circulating lipoprotein lipase appears promising. Reliance must be placed on family screening for the same phenotype and a history of eruptive xanthoma, hepatosplenomegaly, abdominal pain, and/or pancreatitis.

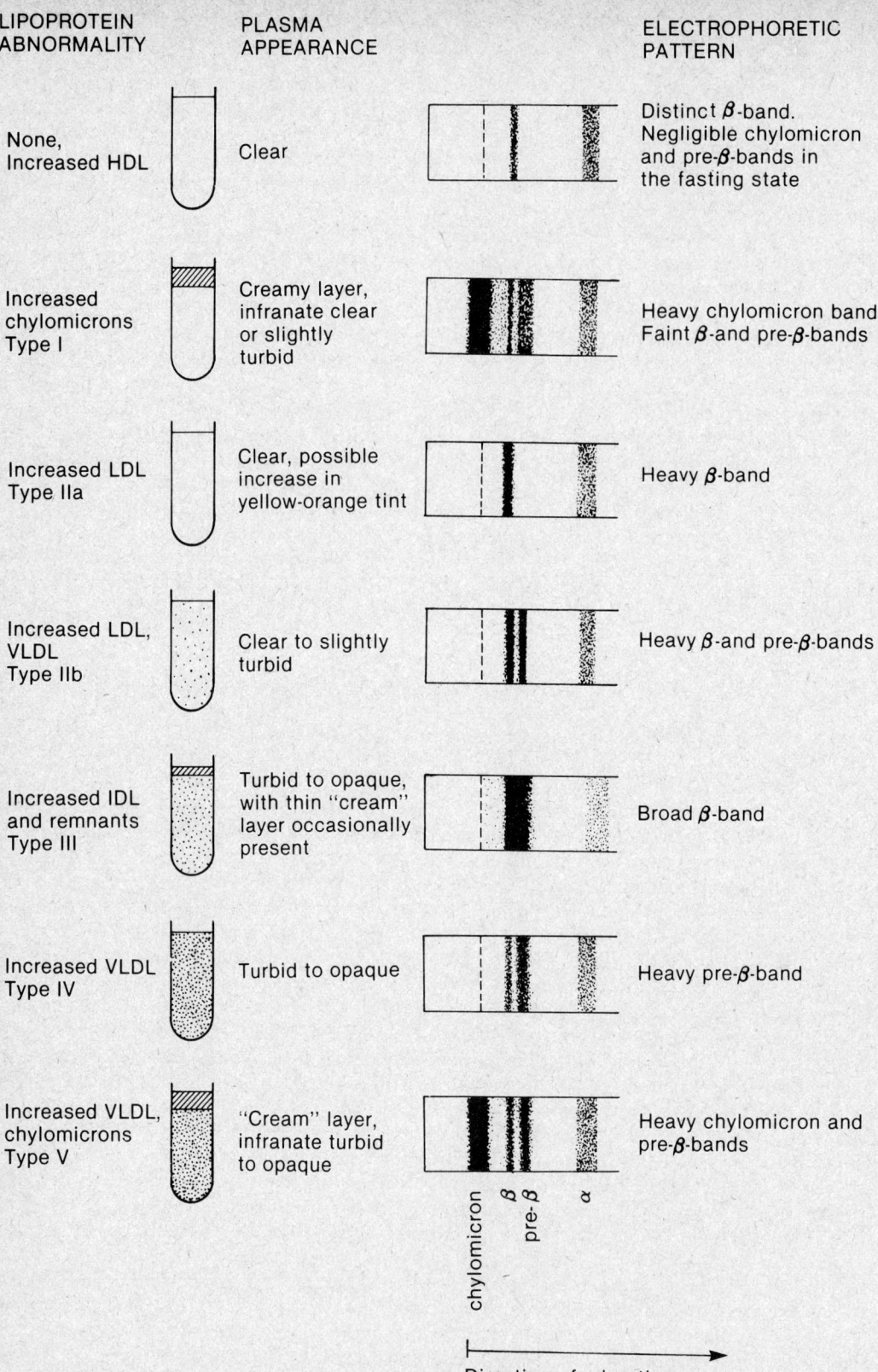

LIPOPROTEIN ABNORMALITY	PLASMA APPEARANCE	ELECTROPHORETIC PATTERN
None, Increased HDL	Clear	Distinct β-band. Negligible chylomicron and pre-β-bands in the fasting state
Increased chylomicrons Type I	Creamy layer, infranate clear or slightly turbid	Heavy chylomicron band. Faint β- and pre-β-bands
Increased LDL Type IIa	Clear, possible increase in yellow-orange tint	Heavy β-band
Increased LDL, VLDL Type IIb	Clear to slightly turbid	Heavy β- and pre-β-bands
Increased IDL and remnants Type III	Turbid to opaque, with thin "cream" layer occasionally present	Broad β-band
Increased VLDL Type IV	Turbid to opaque	Heavy pre-β-band
Increased VLDL, chylomicrons Type V	"Cream" layer, infranate turbid to opaque	Heavy chylomicron and pre-β-bands

chylomicron β pre-β α

Direction of migration

Figure 7-19. Hyperlipoproteinemia. Plasma appearance after 16 h at 4 °C. Electrophoretic migration is from left to right. Dotted line = origin. Relative separations are those typically found on agarose or paper. On polyacrylamide gel, the pre-β-lipoprotein band migrates slower than the β-lipoprotein band. (Modified from Ellefson, R. D., and Caraway, W. T.: Lipids and lipoproteins. *In*: Fundamentals of Clinical Chemistry, 2nd ed. N. W. Tietz, Ed., Philadelphia, W. B. Saunders Co., 1976.)

Biochemical findings. Hyperchylomicronemia is associated with some of the highest triglyceride levels reported and is easily diagnosed on the basis of the creamy layer of chylomicrons present in a fasting sample (Figure 7-19).

Biochemical defect. A deficiency in the activity of the enzyme lipoprotein lipase (LPL) is thought to be the major cause of impaired chylomicron clearance. The enzyme LPL is responsible for initial triglyceride hydrolysis, and its inactivity prevents the removal of ingested fat from the circulation. Hyperchylomicronemia is rarely due to an absence of Apo CII, the co-factor apolipoprotein in chylomicrons responsible for LPL activation. Apo CII absence has been reported as a genetic defect inherited as a recessive disorder and can be detected by quantitating Apo CII in plasma. LPL can be released from tissues following administration of intravenous heparin (100 units/kg), thus allowing its measurement in plasma. In the investigation of hyperchylomicronemia, this test is important. After administration, heparin-released LPL can be assessed by at least two methods. Lipoprotein electrophoresis is first performed on plasma collected before and 15 min after the heparin injection. In subjects with normal LPL activity, the heparin-induced lipolysis of chylomicrons leads to release of free fatty acids (FFA) which then bind to other lipoproteins. This results in smeared pre-β-, β-, and α-bands. In a more recently developed assay, LPL is measured directly using specific antisera and an immunological technique such as RIA.[60a]

Treatment. The major objective is to prevent recurrent abdominal episodes, especially pancreatitis. This requires maintenance of triglyceride values below 2000 mg/dL, a target achieved only by strict adherence to a prescribed diet. Total fat should be limited to < 0.5 mg/kg. Medium chain triglycerides (MCT) which enter the portal venous system directly without chylomicron formation may be consumed ad lib. In an acute attack, a totally fat-free diet should be consumed and supportive intravenous fluid therapy provided for the first 12–24 h. If the Type I phenotype is secondary to an underlying disorder, the treatment is that for the underlying disease.

Increased Low Density Lipoprotein (Fredrickson Type II Hyperlipoproteinemia)

Prevalence. The gradual but definite increase in atherosclerotic risk with increasing LDL-cholesterol concentrations makes a definitive cut-off point for the diagnosis of Type II hyperlipoproteinemia extremely difficult. Many "normal" ranges have been established in the United States and other industrialized countries where CAD is pandemic. Note that these reference ranges are not related in any way to health or disease but are merely statistical parameters. Long-term studies in Framingham suggest that total cholesterol values rising above 180 mg/dL are associated with a progressively escalating risk of CAD.[43] The recent National Institutes of Health Coronary Primary Prevention Trial[52] has for the first time demonstrated, unequivocally, that men with an LDL-cholesterol > 175 mg/dL (85th percentile) will decrease their risk for CAD by reducing their LDL-cholesterol levels. Therefore, it would seem logical to base our acceptable lipid concentrations on "healthy" rather than "statistical" findings. If this were done and populations relatively free of atherosclerosis (e.g., Southeast Asians or black Africans) formed our reference base, a great majority of persons living in the United States would be classified as being hyperlipemic.[3]

There are also a number of disease entities which may be associated with or which may cause Type II hyperlipoproteinemia (Table 7-10). The familial disorder, also known as familial hypercholesterolemia (FH), is inherited as an autosomal dominant gene. The heterozygote gene frequency has been estimated in the United States and Britain at between 0.2 and 0.5%. In white South Africans, especially those of Jewish and Afrikaner background, the gene frequency is as high as 1.3%.[69] Similar frequency has been reported in Lebanon. The homozygous disorder has been estimated to have a prevalence of approximately 1 in 1 000 000 in Britain; however, it is

Table 7-10. SECONDARY CAUSES OF LDL INCREASES

Diet high in cholesterol and total and saturated fat	Diabetes
Hypothyroidism	Chronic renal failure
Nephrotic syndrome	Drugs:
Dys-γ-globulinemia, multiple myeloma	Estrogens
Hepatic obstruction, hepatic disease	Androgens (anabolic steroids)
Porphyria	β-Blockers
Pregnancy	Carbamazepine
Anorexia nervosa	Progestins

nearly ten times more frequent in South America and in Lebanon, where within-family marriage is common.

Clinical presentation. Type II hyperlipoproteinemia is recognized as a metabolic problem of major clinical interest. It is characterized by a striking incidence of premature atherosclerosis, especially among young persons with FH, and has a propensity to cause morbidity and mortality at a young age (Table 7-11).

In subjects homozygous for FH, very marked elevations of cholesterol (usually 800–1200 mg/dL) and major elevations of LDL are observed even with severe reductions in dietary fat and cholesterol. The disorder is usually manifest in infancy, with tuberous and plantar xanthomas, and is characterized by the early appearance of tendon xanthomas and of aortic stenosis secondary to accumulation of atheromas around the aortic ring. Often in childhood, and without exception by early adulthood, clinical symptoms and electrocardiographic signs of CAD develop. Death due to severe atherosclerosis in the untreated homozygote is usual before age 30.

Recent studies by Goldstein and Brown[35] have demonstrated that homozygotes can be subdivided into two groups according to their LDL receptor activity. In subjects in whom no LDL receptor activity is detectable (known as receptor-negative), there is earlier onset of coronary artery disease and a higher frequency of coronary deaths as compared with those homozygotes in whom some residual receptor activity (2–25% of normal) could be detected (receptor-defective). A third receptor abnormality has also been described where one receptor-negative allele is combined with a normal binding receptor; however, this latter receptor lacks the ability to internalize the bound LDL. Each of these mutations in the LDL receptor impairs the ability of the cells to take up and degrade LDL.

In heterozygous FH, diagnosis—especially in childhood—is often a consequence of routine or family plasma lipid screening. However, a comprehensive physical examination is always necessary since the presence of Achilles tendon xanthomas in late adolescence or early adulthood, signified by straightening and nodular irregularity of the tendon, together with increased LDL-cholesterol is almost pathognomonic of FH. Other commonly missed clinical signs include extensor tendon xanthoma, especially of the middle and ring fingers where nodular thickening and irregularities are often easier felt than seen, and corneal arcus. Less common signs are xanthomas of the hands, knees, or elbows and xanthelasma (yellow cholesterol infiltration of the skin around the eye). These are not, however, specific markers for hypercholesterolemia. Very occasionally, acute tendinitis and arthritis may be presenting symptoms.

Primary versus secondary increases in LDL. When taking the initial history, it is important to seek actively other disease or environmental states or influences which can produce LDL elevation (Table 7-10). This entails a detailed medical, social, dietary, alcohol, drug, and family history. To differentiate primary from secondary elevations of LDL, it is necessary to assess other biochemical parameters such as renal, hepatic, and selected endocrine functions. If no primary disease entity is revealed after reasonable investigation, it is still advisable to differentiate the dietary-induced (polygenic hypercholesterolemia or hyper-responders) from the monogenic disorder. This can be done by instituting a suitable diet and ensuring compliance to it. A mild (5–15%) reduction in total cholesterol and LDL-cholesterol is usually observed in familial hypercholesterolemia, while a more marked drop is probably indicative of a dietary-induced disorder. This distinction is clouded in the first few years of life when lipid and lipoprotein levels in children with heterozygous FH are acutely sensitive to dietary manipulation.

Even in the absence of states known to cause secondary hypercholesterolemia, a positive diagnosis of the familial disorder must still be made. A family screen including both parents and other siblings is often necessary. The reports of highly specific characteristics of in vitro cultured fibroblasts in individuals homozygous for familial hyperlipidemia provide a much firmer basis for diagnosis but have not yet proved useful in the detection of heterozygotes.

Biochemical findings. The hallmark of Type II is the elevation of LDL, usually measured

Table 7-11. RISK OF ISCHEMIC HEART DISEASE IN HETEROZYGOUS FAMILIAL HYPERCHOLESTEROLEMIA

Age (y)	Male (%)	Female (%)
< 30	5.4	
< 50	51.1	12.2
< 65	85.4	57.7

indirectly by quantitating or calculating the LDL-cholesterol content. In the Fredrickson classification, division is made into Type IIa, with normal VLDL and triglyceride concentrations, and Type IIb, with elevation of VLDL and triglycerides. Accurate diagnosis of Type IIb requires an appreciation of factors that determine triglyceride levels. Studies in free-living populations in the United States have documented increasing triglyceride values with increasing age. Therefore, one might expect a greater prevalence of Type IIb with increasing age and obesity. In the patient with cholesterol and triglyceride values above 300 mg/dL, it is often necessary to resort to ultracentrifugal and Apo E studies to differentiate between Types IIb and III.

Biochemical defect. In familial hypercholesterolemia, turnover studies have implicated faulty catabolism rather than excess synthesis of LDL. In their studies of cultured human fibroblasts from subjects homozygous for familial hyperlipidemia, Brown and Goldstein provided elegant evidence for the basic pathophysiologic defect.[11] Cells obtained from subjects with homozygous FH have a deficiency in the number of functional LDL receptors and thus are resistant to the biochemical events mediated by LDL, such as suppression of HMG-CoA reductase activity.

With lack of uptake of LDL via the receptor-dependent pathway, LDL levels increase significantly. The receptor-independent pathway, which usually accounts for only 15% removal of the LDL pool daily, continues to function. Part of the receptor-independent clearance of LDL occurs in macrophages and histiocytes of the reticuloendothelial system, which are referred to collectively as scavenger cells. The observation that in FH homozygotes there is a large accumulation of cholesterol in tendon macrophages (xanthomas), splenic macrophages, hepatic Kupffer cells, bone marrow histiocytes, and similar scavenger cells in many organs, confirms that these cells are taking up and degrading large amounts of LDL.

Treatment. Therapy can be divided into three stages. The dietary stage is the foundation of treatment of Type II disease. The response to dietary therapy will depend on a number of factors including age and genetics. In the Type IIb disorder, the first goal is weight reduction. Once ideal body weight has been achieved, dietary adjustment takes the form of three modifications unrelated to total calorie intake: a reduction of calories from total fat to <30% of total calories; an increase in polyunsaturated fat at the expense of saturated fats to a P:S ratio of 1:1; and reduction in dietary cholesterol to <300 mg/d.

The maximum effect of diet is usually seen within 4–5 weeks, and the best response is seen in young persons. Although it is possible to reduce cholesterol levels up to 15% in older patients with heterozygous familial hyperlipidemia, few, if any, achieve normal total or LDL-cholesterol concentrations. The homozygous condition is virtually unchanged by even the most rigid of dietary regimens. The majority of patients with primary or familial hypercholesterolemia will require, in addition to diet, drug therapy to assist in normalization of LDL-C values. Although many hypercholesterolemic agents are currently available, a number of problems are common to them all, making therapy extremely difficult. The most effective group of drugs for treatment of heterozygous FH are the bile acid sequestrants. Included in the group are cholestyramine and cholestipol, both of which bind bile salts in the gastrointestinal tract, thus interrupting the enterohepatic circulation and increasing the degradation of cholesterol to new bile acid production. There is no set dose for the resins, the optimal one being the amount taken regularly that will maintain a total cholesterol and LDL-C below arbitrary levels (in practice <215 mg/dL and <160 mg/dL, respectively). Resin treatment can reduce both total and LDL-C levels some 15–35% below those established by preliminary dietary control; failure to respond to resin treatment is usually due to poor compliance with the drug regimen. Cholestyramine and other resins have the ability to bind other therapeutic agents in the gut, e.g., digitalis and thyroxin.

Several other drugs are used for the treatment of FH, but in general they are either less effective, have undergone less evaluation, or are associated with severe systemic side effects. Clofibrate, the best known and most widely used hypolipemic agent, is no longer considered a rational, safe, or effective "first line" drug for treating Type II disease. Nicotinic acid, although an effective cholesterol-lowering aid, has severe and often intolerable side effects such as gastric irritation, pruritus, flushing, and skin changes, and it is potentially hepatotoxic. Probucol, a drug with few side effects, lowers both HDL-C and LDL-C. The HDL-C reduction is marked and consistent while the LDL-C lowering is about 12% and variable. The effectiveness of probucol in reducing CAD risk is still not clear.

A new group of drugs which suppress HMG-CoA reductase activity appears very promising. In this group, two drugs, Mevinolin and Compactin, have produced large reductions in LDL-C in patients with the Type II disorder.

The third mode of therapy is surgery, but it is usually used only for very severely affected patients. Ileal bypass is a potential therapy. The most common operation is partial ileal bypass, which is purported both to reduce dietary cholesterol absorption and to increase fecal sterol excretion. The operation has been used to treat a number of patients with heterozygous FH, but the results have been variable; some patients have little or no response. Some patients have increased triglycerides and others protracted diarrhea. Nearly all patients require regular intramuscular injection of vitamin B_{12}. A recent advance in the treatment of homozygous FH has been the use of portacaval shunt. An alternative approach has been the use of repeated exchange using a continuous-flow blood-cell separator. The cholesterol- and LDL-containing plasma is replaced every three weeks; initial results with this technique have been promising.[75]

Increased Intermediate Density Lipoproteins (Fredrickson Type III Hyperlipoproteinemia)

Prevalence. The prevalence has varied depending on the techniques and diagnostic criteria employed to define the disease, but it is suggested to be clinically manifest in about 1 in 10 000 subjects. Some studies report that 1% of the population has the phenotype EII/EII associated with dys-β-lipoproteinemia or increased levels of IDL (see section on apoprotein E). With the exception of six reported kindreds, primary Type III hyperlipoproteinemia has rarely been found in childhood.

Clinical presentation. Clinical characteristics vary with a number of factors such as age, sex, obesity, and presence of associated disorders. In population surveys, the majority of patients detected are asymptomatic. When physical signs of the disease are present, the most characteristic is a skin lesion, xanthoma striata palmaris, which ranges from yellowish discoloration to elevation and even obliteration of the palmar creases. Other sites for xanthomas, usually of the tuberous or tubero-eruptive variety, are the buttocks and extensor surfaces of all four limbs. In Type III hyperlipoproteinemia, cerebral and peripheral vascular disease are as common as CAD.

Primary versus secondary hyperlipoproteinemia. Type III hyperlipoproteinemia, or dys-β-lipoproteinemia, is only rarely acquired secondary to hypothyroidism, uncontrollable diabetes mellitus, and systemic lupus erythematosus. Confirmation of primary Type III disorders should be made after excluding these secondary causes.

Biochemical findings. In the last few years, analysis of the isoforms of Apo E has become the definitive test in the diagnosis of Type III or dys-β-lipoproteinemia. However, most laboratories do not, or are currently unable to, do uni- or bidirectional isoelectric focusing combined with ultracentrifugation to study the Apo E isoforms. It is therefore important to consider alternatively the physical and chemical properties of the abnormal lipoproteins that are secondarily associated with the dysfunctional apolipoprotein. Clinical manifestations of dys-β-lipoproteinemia are occasionally seen in persons with phenotypes other than EII/EII. A number of patients with primary elevations of VLDL may show remnant-like particles with an increased cholesterol:trigylceride ratio, which, although different from Type III, is associated with atherosclerosis.

Type III is characterized by elevations of both cholesterol and triglyceride levels (Table 7-8) and an abnormal lipoprotein which floats with the VLDL fraction on preparative ultracentrifugation. This floating β-lipoprotein or β-migrating VLDL often merges with the normal pre-β-band on electrophoresis, giving rise to a broad β-band. For a more reliable diagnosis of Type III, an ultracentrifugal study with measurement of the lipid composition in the separated fractions is necessary. Following isolation of the $d < 1.006$ g/dL fraction ("VLDL" fraction) on the preparative ultracentrifuge, its cholesterol and triglyceride content are measured. Type III lipoproteinemia should be suspected when the VLDL-cholesterol to VLDL-triglyceride ratio exceeds 0.35 or when the VLDL-cholesterol to whole plasma triglyceride ratio exceeds 0.25. These ratios may be particularly valuable in preliminary differentiation of Type III from Type IIb and Type IV hyperlipoproteinemia patients. Once these ratios are detected, confirmation of Type III should be done by studying the Apo E isoforms.

Biochemical defect. In hyperlipidemic patients with dys-β-lipoproteinemia, the hallmark of the disease in terms of plasma lipoproteins is the increase in the serum of two populations of remnant particles, i.e., chylomicron remnants and IDL. These are thought to arise from partially degraded chylomicrons and VLDL. As discussed in an earlier section, the ultimate catabolism of these remnant particles and IDL occurs by hepatic receptors which depend upon specific Apo E isoforms for binding and uptake of the degraded chylomicrons and VLDL. Both lipoprotein particles that occur in Type III appear to be rich in Apo E, a previously used marker termed "arginine-rich apolipoprotein." Despite the fact that there is increased Apo E in the particles, they are metabolized extremely slowly. With detailed studies of Apo E structure and isoforms, it has become obvious that in dys-β-lipoproteinemia, hepatic receptors, which usually have high affinity and specificity for Apo E, do not bind or remove the lipoproteins in a normal manner.

It is now known that there are four isomorphic forms of Apo E termed EI through EIV and named in order of increasing isoelectric point. For normal remnant and IDL binding to receptors, it appears that at least the isoforms EIII or EIV or both are necessary. In ~1% of the population, the EII/EII phenotype is found, and it is this phenotype which is most commonly associated with the clinical manifestations of dys-β-lipoproteinemia. Heterozygosity for one allele specifying functionally abnormal Apo E isoform also influences the expression of primary hyperlipidemia. From a number of studies, it appears that EIV/EII and EIII/EII phenotypes are twice as likely to have severe hypertriglyceridemia as subjects with phenotypes EIV/EIV, EIV/EIII, and EIII/EIII.

Patients with dys-β-lipoproteinemia have characteristically low levels of normal LDL, and kinetic studies have confirmed that the conversion of VLDL to LDL is impaired. Type III lipoproteinemia is another receptor-mediated lipoprotein abnormality in which the normal VLDL catabolic cascade to form LDL is defective.

Treatment. This uncommon disorder is of importance because it currently is the only hyperlipoproteinemic state in which regression of atherosclerotic arterial lesions appears to occur. In addition, treatment results in a striking remission and disappearance of cutaneous-palmar xanthomas, and the size of large tuberous xanthomas is substantially reduced. As with all forms of hyperlipoproteinemia, therapy is initially aimed at underlying or exacerbating disorders, such as hypothyroidism and obesity. Once weight loss has been achieved, maintenance dietary therapy should restrict caloric intake, cholesterol, saturated fats, and alcohol consumption.

The treatment of patients with Type III is usually most gratifying since they uniformly respond to diet or diet plus clofibrate. Failure of clobifrate to normalize or maintain normal lipid levels may indicate poor dietary adherence. Other drugs which are effective in Type III include niacin and gemfibrozil.

Increased Very Low Density Lipoprotein (Fredrickson Type IV Hyperlipidemia)

Prevalence. A high proportion of adults with hyperlipoproteinemia have an increase in VLDL. The familial disorder is less commonly found in childhood than is familial hypercholesterolemia, probably because the genetic disorder is less penetrant in children. Although a number of investigators have found Type IV to be inherited in some families as an autosomal dominant trait, the disorder is generally regarded as recessive. The frequency of the monogenic form of Type IV in the general population has been estimated to be between 0.2 and 0.3%.

Clinical presentation. Glucose intolerance, hyperuricemia, hypertension, premature CAD, and peripheral vascular disease are associated with both primary and secondary Type IV. A significant number of subjects with hypertriglyceridemia are obese. It is rare to observe xanthoma, corneal arcus, or xanthelasma in young patients with primary Type IV; however, the last two signs may be present at a later age. Acquired hypertriglyceridemia is more commonly associated with underlying disease entities such as poorly controlled diabetes, nephrotic syndrome, and obesity.

Primary versus secondary hypertriglyceridemia. VLDL elevation often occurs secondary to a wide variety of diseases, drugs, and dietary habits. Because of the numerous acquired hypertriglyceridemias, an approach to the management of the hypertriglyceridemic patient involves a careful distinction between primary and secondary Type IV. It is extremely important to obtain fasting blood for triglyceride measurement when the patient is on a habitual diet and weight is stable. The causes for secondary Type IV listed in Table 7-12 are common, particularly in hospital populations, and require routine assessment. Useful in making the distinction between secondary and primary Type IV are a history of ethanol intake and of estrogen or steroid administration; results of urinalysis; measurements of fasting or 2-h postprandial blood glucose; and liver, thyroid, and renal function tests.

After exclusion of secondary hypertriglyceridemia, confirmation of a primary or possibly familial Type IV disorder must be made. In familial endogenous hypertriglyceridemia with the Type IV pattern, physical examination usually reveals little of note. A particularly vexing problem is the coexistence of hypertriglyceridemia with juvenile diabetes. When there is poor control of fasting blood glucose (> 250 mg/dL), discrimination between hypertriglyceridemia secondary to poorly controlled diabetes and primary hypertriglyceridemia with conjoint diabetes is difficult. If hypertriglyceridemia is persistent despite adequate control of blood glucose, the triglyceride elevation is probably not due to diabetes.

Table 7-12. CAUSES OF SECONDARY INCREASE IN VLDL AND CHYLOMICRONS

Alcoholism	Drugs:
Excessive alcohol intake without clinical alcoholism	Estrogen-progestin oral contraceptives
Excess caloric intake (especially simple carbohydrates)	Estrogens
	Corticosteroids
Obesity	β-blockers
Poorly controlled diabetes mellitus	Thiazide diuretics
Hypothyroidism	Pancreatitis (usually alcoholic)
Nephrotic syndrome	Glycogen storage diseases
Uremia	Dysproteinemias, systemic lupus erythematosus
Pregnancy	Storage diseases (Gaucher, Niemann-Pick, lecithin cholesterol acyltransferase deficiency)

Biochemical findings. Type IV hyperlipoproteinemia is characterized by an increase in triglyceride and VLDL-cholesterol levels (Table 7-8). Plasma cholesterol ranges from normal to moderately increased. LDL-cholesterol is normal or low, and HDL-cholesterol is often lower than in normotriglyceridemic populations. Chylomicrons are not present in fasting plasma. On electrophoresis, there is increased staining of the pre-β-lipoprotein band, due to elevated VLDL. The accurate biochemical diagnosis of Type IV, however, cannot be made by electrophoresis alone. In a considerable number of individuals with normal triglyceride levels, the Lp(a) antigen (sinking pre-β-lipoprotein) may produce increased staining in the pre-β-lipoprotein region.

Biochemical defect. At present, the exact etiology of primary hypertriglyceridemia remains speculative. Some studies indicate that increased VLDL production rates by the liver are the predominant factors in the development of primary hypertriglyceridemia. Other studies, however, have concluded that reduced VLDL removal is important in the development of hypertriglyceridemia. Still other reports provide evidence for both mechanisms.

A second controversial area implicated in the development of endogenous hypertriglyceridemia has been hyperinsulinemia. Positive correlations between insulin and triglyceride levels have been reported in some studies while others reported no correlations between the two levels. In the assessment of insulin as an etiologic agent in endogenous hypertriglyceridemia, one confounding factor has been the frequent presence of obesity. In obese patients there appears to be no clear relationship between insulin and triglyceride levels.

A third pathophysiologic feature of hypertriglyceridemia is the ability of carbohydrate to induce it. Evidence is again conflicting; carbohydrate inducibility is a variable phenomenon in Type IV patients, but excessive sensitivity to carbohydrate is not an obligatory precursor of Type IV. Obesity, which is common in Type IV hyperlipoproteinemia, is another variable which has been suggested to be etiologic in the development of hypertriglyceridemia. A number of studies, however, found no correlation between any measure of obesity and the magnitude of the response of endogenous hypertriglyceridemia to high carbohydrate feeding, and concluded that obesity was not a necessary feature of the Type IV abnormality.

Treatment. Treatment of patients with hypertriglyceridemia is most successful after differentiation between primary and secondary Type IV. Identifiable underlying disease states, excessive ethanol intake, poorly controlled diabetes, triglyceride-elevating drugs, and the like must be recognized and altered if possible (Table 7-12). Fortunately, most patients with primary and familial hypertriglyceridemia (Type IV) are extremely sensitive to dietary manipulation and caloric restriction.

The use of hypolipidemic agents is rarely, if ever, necessary. The initial therapy is always dietary and this entails (1) a reduction of total caloric intake to achieve ideal body weight; (2) restriction of sucrose-rich foods; (3) moderate decrease of saturated fat and replacement with polyunsaturated fat; and (4) moderate restriction of cholesterol intake. Alcohol, if being consumed, should be reduced or eliminated. There is no single drug of choice for treating patients with Type IV hyperlipoproteinemia, but if triglyceride levels are not normalized by diet alone, the most commonly used group of drugs consists of fibrate analogs. The triglyceride-lowering effects of these drugs are probably achieved by inhibition of hepatic triglyceride formation. In certain patients not responsive to diet plus treatment with clofibrate, nicotinic acid may be effective. However, many patients with Type IV hyperlipoproteinemia have concurrent gout, hyperuricemia, or diabetes mellitus, which may be exacerbated by nicotinic acid treatment. In certain patients unresponsive to diet and drugs, oxandrolone (a synthetic anabolic-androgenic steroid) has proved useful. This compound apparently augments efficiency of triglyceride clearing. Almost no pharmacologic triglyceride-lowering agent is effective for treating hypertriglyceridemic patients in the face of elevated blood glucose levels. It is particularly important to control hyperglycemia as an approach to control of hypertriglyceridemia.

Increased Chylomicrons and Very Low Density Lipoproteins (Fredrickson Type V Hyperlipoproteinemia)

Prevalence. Hypertriglyceridemia is the most common lipid abnormality encountered. The majority of hypertriglyceridemic patients have Type IV hyperlipoproteinemia; Type V is encountered less often. Expression of Type V in childhood or adolescence is uncommon. In addition, the disorder is more likely to be secondary to one of a number of other disease states.

Primary versus secondary Type V. The precise genetic mechanism for familial Type V is not clear, nor is its genetic relationship to familial Type IV well understood. Type V hyperlipoproteinemia may be secondary to a wide variety of diseases, drugs, and dietary habits (Table 7-12).

Due to the numerous forms of acquired Type V hyperlipoproteinemia, a careful distinction between primary and secondary causes must be made. A history of ethanol intake or estrogen or steroid administration; urinalysis; measurement of fasting or 2-h postprandial blood glucose; and liver, thyroid, and renal function tests are useful in the distinction. Poorly controlled diabetes mellitus, excess alcohol, or estrogens or estrogen-containing oral contraceptives in an individual with pre-existing Type IV hyperlipoproteinemia will often produce the Type V pattern.

Biochemical findings. Type V is characterized by elevation of triglyceride levels due to increases in both VLDL and chylomicrons, as shown in Table 7-8. The increase in chylomicrons can be seen as a surface layer (cream) when a plasma specimen is allowed to stand for 16 h at 4 °C. Visual detection of a very thin chylomicron layer over a turbid infranate may be difficult, and electrophoresis may be required for determination. The turbid to lactescent layer below the surface layer is due to the VLDL increase. Plasma cholesterol may be slightly to moderately increased. LDL-cholesterol and HDL-cholesterol are usually normal to low. In some patients, a reduction of lipoprotein lipase has been reported.

Biochemical defect. The presence of circulating chylomicrons and VLDL in fasting plasma indicates that normal clearing mechanisms for triglyceride-rich lipoproteins are inadequate. The usually normal post-heparin lipoprotein lipase activity (PHLA) suggests that the defect is not the same as that found in Type I hyperlipoproteinemia. Conflicting studies have resulted in the proposal of three different hypotheses. The first is that endogenous triglyceride and resultant VLDL secretion are abnormally increased, leading to a saturation of removal pathways which are shared by chylomicrons, thus causing an elevation of both lipoproteins. The second hypothesis contends that triglyceride synthesis is normal and that the defect lies in the clearance of VLDL and chylomicrons. This mechanism is operative for subjects with Apo CII deficiency, where lack of LPL activation results in failure to hydrolyze triglyceride from both chylomicrons and VLDL. Finally, some investigators have shown that Type V may be due to a mixture of both abnormalities.

Treatment. Secondary disease states must be treated and use of aggravating pharmacologic agents discontinued. In common with all hypertriglyceridemias, reduction to and maintenance of ideal body weight may be therapeutic. A fat- and carbohydrate-restricted diet high in protein is useful but expensive and relatively difficult to follow. In contrast to Type IV, treatment with clofibrate is usually of limited effectiveness, and nicotinic acid (1–3 g/d) is considered to be the drug of choice. Very often, however, patients cannot tolerate nicotinic acid on a prolonged basis, and several specialized drugs of research interest are effective. Oxandrolone (Anavar), an anabolic-androgenic agent, is often effective in patients unresponsive to nicotinic acid and clofibrate. Norethindrone acetate, a synthetic steroid with progestational activity, may also be effective. It is also particularly important to maintain reasonably normal blood glucose levels in patients who have both Type V hyperlipoproteinemia and hyperglycemia.

Increased High Density Lipoprotein

Prevalence. A newly recognized genetic lipoprotein abnormality, familial hyper-α-lipoproteinemia, has been described. Statistical analysis of families selected by longevity shows that familial hyper-α-lipoproteinemia appears to be transmitted as an autosomal dominant trait.[31] The condition is fully expressed in children, and may also be diagnosed in neonates. No epidemiologic studies for hyper-α-lipoproteinemia in the general population have been reported. Secondary hyper-α-lipoproteinemia associated with use of estrogenic preparations or alcohol has been described.

Clinical presentation. No clinical features have been associated with hyper-α-lipoproteinemia. As far as is known, elevated HDL-cholesterol levels do not have an adverse effect on health. Longevity analyses demonstrate an 8–12-y prolongation of life expectancy for kindred members. Myocardial infarction morbidity and mortality are reduced. In those children having HDL-cholesterol levels of ≥ 70 mg/dL, the presence of familial hyper-α-lipoproteinemia can be documented by quantitation of HDL-cholesterol levels in first degree relatives. Elevated HDL-cholesterol levels are rarely the result of estrogen, oral contraceptive, or excessive alcohol intake, but these factors must be ruled out.

Biochemical findings. This syndrome is characterized by distinctive primary elevations of HDL-cholesterol levels, slight elevation of total cholesterol, and normal triglyceride concentrations.

Reference values for HDL-cholesterol have been established. In males, there is a sharp decrease of nearly 10 mg/dL, or 20%, in HDL-C during the second decade of life. The HDL-cholesterol remains constant at a mean of 44–46 mg/dL until age 55. During the ensuing ten years (ages 55–65) there is a small increase (10%) in HDL-cholesterol, and levels remain fairly constant at ~50 mg/dL until well into the 90s. The female HDL-cholesterol levels rise from a mean of about 53 mg/dL at age 20 to 65 mg/dL by age 65. Thereafter, they drop slightly to ~60 mg/dL and are maintained at this level well into the ninth and tenth decades and are still about 20% higher than HDL-cholesterol values in age-matched males. Some of this sex difference in HDL-cholesterol may be due to the use of sex hormone preparations in females, especially after age 40. The HDL findings of the Lipid Research Clinic Program[50] showed that, after age 40, females using hormone preparations had slightly higher HDL-cholesterol levels (maximal 10 mg/dL) than did those not taking such therapy. In order to avoid a diagnosis of hypercholesterolemia based on a total cholesterol, HDL-cholesterol levels should routinely be measured whenever patients are seen for evaluation of lipids and lipoproteins.

In view of the fact that hyper-α-lipoproteinemia is at present merely a biochemical state associated with less than average cardiovascular disease risk, extreme caution should be used in its diagnosis and clinical interpretations. This is discussed in more detail later in the section on investigation of lipid disorders.

Biochemical defect. No hypothesis for the increased HDL levels found in familial hyper-α-lipoproteinemia has been made, but it may be associated with enhanced triglyceride-rich lipoprotein clearance and LPL activity.

Treatment. Hyper-α-lipoproteinemia in the presence of normal levels of LDL appears to confer reduced risk for CAD. Thus, no treatment for these subjects is indicated.

HYPOLIPOPROTEINEMIA

Only in the last few years has the clinical significance of hypolipoproteinemia become apparent. (See Table 7-9.) Epidemiological studies have tended to concentrate on the hyperlipoproteinemias, probably because clinically significant hypolipoproteinemias occur less commonly than lipoprotein elevations. Hypolipoproteinemias usually have important consequences in that they involve abnormalities in one or more of the major metabolic roles served by lipoproteins. For example, deficiency of chylomicron formation is known to impair the absorption of triglycerides and cholesterol as well as the fat soluble vitamins. Deficiencies of VLDL and LDL, as noted previously, impede the delivery of endogenous triglycerides and cholesterol synthesized in the liver to the peripheral cells. Abnormalities of HDL or the LCAT system are manifest both in the clearance of cholesterol from the periphery and the impairment of normal metabolism of the triglyceride-rich lipoproteins.

Reduced Low Density Lipoprotein (Hypo-β-lipoproteinemia)

Prevalence. As recently as 1972, only five kindreds with hypo-β-lipoproteinemia had been fully described. However, the studies by Glueck in Cincinnati detected 13 such kindreds in a small metropolitan area and in a population sample of less than 6000 kindreds.[32] Hypo-β-lipoproteinemia is more common than was previously suspected; this is probably due to the limited attention focused on the persons in the lower fifth percentile of plasma total cholesterol and LDL-cholesterol distributions.

Clinical presentation. Familial hypo-β-lipoproteinemia is inherited as an autosomal dominant trait as shown in a study of 61 siblings from 22 hypo-β-lipoproteinemic and normolipoproteinemic matings,[32] where an equal number with low LDL-cholesterol and normal LDL-cholesterol were found. That hypo-β-lipoproteinemia is a Mendelian dominant gene was further supported by a bimodal distribution of LDL-cholesterol in first-degree relatives of the originally detected subjects.

Like increased HDL-cholesterol, reduced plasma LDL-cholesterol is associated with significant increases in life expectancy in both males and females when compared with that expected for the overall U.S. white population. This exceptional longevity is probably related to a relative diminution of morbid and lethal myocardial infarctions. The reduced incidence of myocardial disease in patients with reduced LDL-cholesterol is consistent with population reports which demonstrate a direct relationship to total and LDL-cholesterol. A number of other disorders have been shown to produce low LDL concentrations. (See Table 7-13.)

Biochemical findings. The syndrome is arbitrarily defined as an LDL-cholesterol level below the fifth percentile for a given sex- and age-matched population. The condition should not be confused with a-β-lipoproteinemia, which is a rare disorder characterized by a total lack of Apo B-containing lipoproteins, fat malabsorption, failure to thrive in infancy, and mental retardation. In hypo-β-lipoproteinemia, LDL-cholesterol is always detectable, while the other lipoproteins such as VLDL and HDL may be low, normal, or even increased.

Biochemical defect. Studies of Apo B-LDL turnover in plasma in patients with hypo-β-lipoproteinemia indicate that these patients synthesize LDL at only one half the normal rate but that their fractional catabolic rate for LDL remains normal. Since LDL is the breakdown product of VLDL, patients with hypo-β-lipoproteinemia also have low triglyceride values, indicating decreased VLDL production.

Treatment. In the absence of any underlying illness, hypo-β-lipoproteinemia is an advantageous biochemical finding, and treatment is therefore unnecessary.

Absent Low Density Lipoproteins (A-β-lipoproteinemia)

This rare autosomal recessive disorder has a characteristic absence of lipoprotein species containing Apo B. This is manifest in the total absence of LDL with consequent very low levels of plasma total cholesterol. The heterozygotes have no known clinical or biochemical abnormality but may present as having low LDL levels (hypo-β-lipoproteinemia). Homozygotes have no detectable chylomicrons, VLDL, or LDL, and immunofluorescent studies have failed to detect apolipoprotein B in intestinal mucosa. As a consequence of the absence of Apo B, levels of triglyceride, cholesterol, and phospholipids are extremely low. Total plasma cholesterol concentrations are usually below 50 mg/dL and virtually all of this cholesterol is associated with the HDL fraction. Free fatty acid concentrations are normal or slightly reduced with palmitoleic acid predominating. Levels of HDL in plasma are usually slightly decreased; HDL composition is abnormal, with an increased ratio of free to esterified cholesterol and of sphingomyelin to phosphatidylcholine. The other apolipoproteins, AI, AII, and the C series, are usually all present.

Biochemical defect. The exact underlying biochemical defect is not known but probably involves an abnormality in the synthesis or secretion of the lipoproteins containing Apo B. This results in severe malabsorption of fat due to failure of chylomicron formation and the restricted delivery of endogenously synthesized cholesterol to peripheral tissue via the LDL pathway. Adrenal production of cortisol in response to ACTH is deficient in a-β-lipoproteinemic subjects. This supports the view that there is a functional disturbance in cholesterol transport at the site of synthesis rather than in the receptor-mediated delivery. Studies of Apo E-containing lipoproteins indicate that some cholesterol may be delivered to peripheral tissues by lipoproteins containing apolipoprotein E.

Clinical presentation. Clinically, the features of a-β-lipoproteinemia are variable and are dependent to some extent on its severity. Malabsorption of dietary fat is present in infancy with poor weight gain and failure to thrive. Steatorrhea is present if the individual continues to take

---

Table 7-13. DISEASE STATES ASSOCIATED WITH REDUCED PLASMA LEVELS OF LDL

Malnutrition	Myeloproliferative disorders
Intestinal malabsorption	Chronic anemias
Intravenous hyperalimentation	Severe hepatocellular dysfunction
Dolman's disease	Reye's syndrome
Hyperthyroidism	Acute severe stress (myocardial infarction,
Myeloma	surgery, trauma)

in a normal fat diet. Growth retardation is a common feature. While radiographic findings are nonspecific, jejunal biopsy normally shows villi with extensive vacuolated mucosal cells packed with lipid droplets. The malabsorption of fat is associated with severe malabsorption of fat soluble vitamins.

There is progressive degeneration of the central nervous system, including cerebellar and lateral spinal tract involvement. This is usually evident in the first decade of life with clumsiness, ataxia, dysarthria, and muscle weakness as early symptoms. Most subjects are unable to walk by their mid-twenties. Histological studies have shown degenerative changes in myelin sheaths of the peripheral nerves as well as in the posterior columns, spinocerebellar tracts, and cerebellum. These neurological lesions may, in part, be related to the profound deficiency of vitamin E. Another abnormality associated with a-β-lipoproteinemia, probably due to the lack of carotene and vitamin A absorption and transport, is decreased visual acuity and night blindness. Blood smears from many of these patients show 50–70% of erythrocytes with spiny projections; the abnormal cells are called acanthocytes, i.e., "thorn," or "spur" cells. A prolonged prothrombin time may also occur due to vitamin K malabsorption, but it does not usually cause abnormal bleeding.

Reduced High Density Lipoprotein (Hypo-α-lipoproteinemia)

Prevalence. While data on hypo-α-lipoproteinemia are still extremely limited, studies have indicated a significant increase in atherosclerosis risk in subjects with lower HDL-cholesterol levels as compared with groups with higher HDL-cholesterol concentrations.[62] The differences in HDL-cholesterol levels between various risk groups are small (5–10 mg/dL), and extreme caution should be exercised by both the laboratorian and the clinician when diagnosing HDL deficiency states. In view of the variability of HDL-cholesterol measurements, it is likely that in the future they will be replaced by other measurements of the HDL moiety, such as measurement of Apo AI.

A number of underlying disease states or environmental influences may be associated with reduced levels of HDL. (See Table 7-14.) One of the most common associations is with hypertriglyceridemia. Subjects whose levels of HDL-cholesterol are appreciably below the age-sex specific fifth percentile and who have normal levels of triglyceride may have specific genetic determinants for hypo-α-lipoproteinemia. A number of studies have suggested that familial hypo-α-lipoproteinemia follows Mendelian dominant transmission.[33] At present the number of genes governing the level of HDL in plasma is unknown.

The interaction of environmental influences, such as exercise, obesity, and alcohol, will influence the underlying genetic pattern controlling HDL levels. HDL levels are also known to be reduced by a number of drugs, especially probucol, β-blockers, and some diuretics. Identification of individuals with familial hypo-α-lipoproteinemia is important in order to treat other associated risk factors and perhaps eventually to increase HDL concentrations by a number of drugs currently being investigated.

A number of newly recognized lipoprotein abnormalities have been described in which extremely low levels of HDL-cholesterol and Apo AI have been found. Among these disorders are AI$_{Milano}$ disease, first described in Italy.[26] Patients are generally asymptomatic and clinical signs of atherosclerosis have not been found. The underlying defect appears to be in the primary structure of apolipoprotein AI with mild to moderate impairment of VLDL catabolism. Without Apo AI, HDL does not form, and without HDL, Apo CII cannot be returned to the liver during VLDL catabolism. This results in a relative Apo CII deficiency and increased VLDL levels.

Another disorder, known as fish-eye disease, has been described in which HDL levels are reduced to ~10% of normal, with an increase in triglyceride.[13] Clinically, there is severe corneal opacity. The tonsils and other reticuloendothelial tissues do not appear to contain lipid as is found in Tangier disease. Although the HDL levels are diminished, their apolipoprotein composition appears normal. Atherosclerosis is present

---

Table 7-14. DISEASE STATES ASSOCIATED WITH REDUCED LEVELS OF HDL

Smoking	Diabetes
Acute and chronic hepatocellular disease	Hypo- and hyperthyroidism
Acute stress (myocardial infarction, surgery, trauma)	Chronic anemias
	Myeloproliferative disorders
Intravenous hyperalimentation	Drugs (probucol, androgens, β-blockers)
Severe malnutrition	

in some patients but is usually a late manifestation. Low plasma HDL-cholesterol levels and extremely low levels of Apo AI have also been described in two sisters from Detroit who presented with early severe coronary artery disease.[60] As assays for Apo AI and HDL_2 become more routinely used, further clinical associations with reduced HDL will probably be reported.

Treatment. No current treatment exists for hypo-α-lipoproteinemia. A number of general health measures may be associated with raising HDL. These include weight loss, vigorous exercise, and cessation of smoking. A number of drugs are reported to increase plasma HDL-cholesterol. These include niacin and gemfibrozil. The results of raising HDL-cholesterol to decrease the risk of atherosclerosis is as yet unknown.

Absent High Density Lipoprotein (Tangier Disease)

The best known disorder of HDL metabolism is Tangier disease. It is a rare autosomal recessive disorder characterized by accumulation of cholesterol esters in the reticuloendothelial system and in other tissues. The disorder is associated with the virtual absence of HDL in plasma. The diagnosis is suspected when extremely low HDL-cholesterol levels are detected on conventional precipitation techniques, and confirmed by the absence of an α-lipoprotein band on electrophoresis. The homozygote is found to have undetectable amounts of HDL-cholesterol and extremely low levels of Apo AI and AII. Plasma total and LDL-cholesterol concentrations are usually low, and mild hypertriglyceridemia is often observed. The genetic defect most likely involves altered synthesis or catabolism of AI or AII. Recent studies indicate that there are extremely rapid rates of HDL and Apo AI catabolism.

The clinical findings in Tangier disease reflect the sites of accumulation of the cholesterol esters and most notably include enlarged, lobulated, orange-yellow tonsils and adenoids. Mucosa of the pharynx and rectum may also show orange discoloration. Recurrent peripheral neuropathy, motor weakness including ptosis, ocular motor palsies, muscle atrophy, and depressed tendon reflexes are also found. Splenomegaly and mild thrombocytopenia are common. The course of the disease is extremely variable, and the diagnosis has been made between ages 3 and 69 years. The disorder is relatively benign since the lipid accumulation in the reticulendothelial tissue does not usually affect any organ function. A recent review of subjects with Tangier disease has suggested earlier than usual occurrence of atherosclerosis, mainly between the ages of 35–65 years. Unlike findings in LDL abnormalities, the abnormality in Tangier disease is predominantly cerebrovascular in nature.

INVESTIGATION OF PLASMA LIPOPROTEIN DISORDERS

Factors affecting diagnostic tests. All too often, the diagnosis and classification of disturbances in lipoprotein metabolism are based on a single plasma sample which is flawed or nonrepresentative. The three most common errors leading to incorrect diagnosis are improper sampling, unreliable laboratory determination, and failure to repeat sampling. Table 7-15 lists conditions for proper sampling. The requirement for a fasting sample is relatively unimportant when total cholesterol is the only lipid being assayed. However, total cholesterol alone is a very poor test

Table 7-15. FACTORS IMPORTANT TO EVALUATION OF LIPID, LIPOPROTEIN, AND APOLIPOPROTEIN DISORDERS

Laboratory methods
  Accurate, as judged against the assay used to obtain clinically relevant reference ranges
  Precise
Sample collection
  12–14 h fast, usually overnight—water only
  Habitual diet and activity
  Stable weight over past four weeks
  No acute or recent illness, injury, or surgery (especially myocardial infarction or coronary artery bypass surgery)
  No lipid raising or lowering drugs, e.g., oral contraceptives, thyroid hormones, steroids
Repeat sampling
  Base diagnosis on at least two (preferably three) specimens, one to four weeks apart, while observing sampling
    conditions and laboratory principles above

for lipoprotein assessment. The usual routine screening tests require a fast of at least 12 h (overnight) during which time only water should be consumed; water can be taken ad lib.

Obtaining a blood sample from the patient while he is on his habitual diet is particularly important. The physician must be cautious in scheduling sampling time with the patient's knowledge since subjects may, prior to sampling, consciously or unconsciously alter their life-styles. Decreasing intake of calories, cholesterol, or alcohol may allow lipid and lipoprotein levels to be normal or closer to normal than usual and the physician's interpretation to be more optimistic than is warranted.

Weight loss acutely lowers triglyceride concentrations and may transiently elevate cholesterol and low-density lipoprotein levels. Any acute febrile illness, trauma, or recent surgical procedure may affect plasma lipids, often by elevating triglyceride levels and depressing total cholesterol and LDL-cholesterol concentrations. Chronic and/or debilitating illness may severely decrease LDL and HDL concentrations. Finally, the patient may be taking drugs which either lower lipid levels or substantially affect lipid metabolism. Often, young women do not consider estrogens or estrogen-progestin oral contraceptives as drugs and will not volunteer the information unless specifically asked about these compounds. Similarly, young males participating in contact sports may often illegally take anabolic steroids and may not reveal this even on direct questioning. Ideally, the plasma lipid sampling is best made while the patient is off all drugs; if this is not possible, their use should be noted in order to allow the most accurate interpretation of the lipid data.

As with other metabolic and endocrine disorders, the physician is heavily dependent on laboratory determinations of plasma constituents for the diagnosis and management of lipid disorders. Long-term precision of laboratory methods is clearly necessary in monitoring plasma lipid and lipoprotein concentrations of individuals undergoing either dietary or drug therapy. The physician is handicapped in both achieving a diagnosis and following treatment when uncertainty exists in regard to the reliability of the lipid and lipoprotein measurements available to him. The danger is even further increased when analytical problems exist that the physician is unaware of. The physician should be encouraged to demand of the laboratory performing any lipid analysis detailed, current, and continuous information as to its standardization, quality control program, and performance. Once a suitable laboratory has been selected, the physician should consistently use that laboratory as long as its quality control remains acceptable.

Current medical and public information has ensured that the diagnosis of hyperlipoproteinemias, especially of hypercholesterolemia, will generate extreme anxiety, because of the necessity for lifelong dietary restriction or drug therapy. The diagnosis may also carry with it penalties in terms of health and life insurance. This fact is often overlooked or underestimated by physicians and laboratories who tend to screen healthy individuals for the many biochemical parameters that can be conveniently and inexpensively accommodated on the large multichannel analyzers present in many clinical laboratories. In clinical practice, many errors in diagnosis of and therapy for lipoprotein abnormalities are based on one nonrepresentative specimen. If an abnormal lipid or lipoprotein concentration is detected, it is extremely important that the biochemical finding be confirmed on at least two subsequent occasions, preferably two to four weeks apart, and with consistent observation of the above precautions in sampling and laboratory quality.

Diagnostic Tests for Lipoprotein Disorders

The useful tests in diagnosis of lipoprotein disorders are listed in Table 7-16.

Plasma Appearance

This simple, convenient, and inexpensive test is often overlooked by many clinicians and seldom, if ever, reported by clinical laboratories. Its value can be considerable because the information is occasionally diagnostic. If the plasma is clear, the triglyceride level is most likely to be either normal or near normal (<200 mg/dL). When the triglyceride level increases to ~300 mg/dL, the plasma is usually hazy-turbid in appearance and is not translucent enough to allow clear reading of newsprint through the tube. When the plasma triglyceride level is >600 mg/dL, the plasma is usually opaque and milky (lipemic, lactescent). If chylomicrons are present, a thick homogeneous creamy layer may be observed floating at the plasma surface after several hours at 4 °C. As summarized earlier in Figure 7-19, a uniformly opaque plasma sample

Table 7-16. ASSESSMENT OF LIPOPROTEIN DISORDERS

Routine tests
  Appearance of plasma (overnight at 4 °C)
  Total cholesterol (TC)
  Triglyceride (TG)
  HDL-cholesterol (HDL-C)
  LDL-cholesterol (LDL-C) by calculation*
    LDL-C = TC − (HDL-C) − TG/5
Additional or research tests
  HDL Subfractionation (HDL$_2$ and HDL$_3$)
  IDL and remnant assessment
    Ratio of cholesterol in d < 1.006 fraction (by ultracentrifugation) to total triglyceride
    Isoelectric focusing of Apo E for isomorph pattern
  Apo CII
  Apo AI
  Apo B
  Post-heparin lipolytic activity
  Lipoprotein lipase activity
  Lecithin cholesterol acyl transferase (LCAT) activity

*Provided TG <400 mg/dL.

usually denotes a Type IV pattern. An opaque plasma sample with a creamy layer on top is usually consistent with the Type V pattern. A thick chylomicron creamy layer with generally clear plasma infranate is usually consistent with the Type I pattern. In nonfasting subjects, a chylomicron layer may also be found. It does not constitute an abnormal finding unless at least 12 h of fasting has occurred prior to the sampling of blood. If chylomicrons are suspected but are not clearly discernible, electrophoresis may confirm their presence. In patients with hypercholesterolemia due only to elevated LDL concentrations, the plasma is clear but may have an orange-yellow tint since carotenoids are carried in LDL.

Determination of Total Cholesterol and Triglyceride

After visual observation of the specimen, the next most useful and reliable tests are determinations of triglyceride and total cholesterol concentrations. Provided that the sampling and laboratory conditions outlined previously are observed, a single cholesterol and triglyceride value more than 20% below the suggested age- and sex-adjusted upper reference limits (shown later in Tables 7-19 to 7-22) virtually eliminates a diagnosis of hyperlipoproteinemia; it does not, however, address the equally important *lipoprotein deficiency* states (Table 7-9). Changes in weight, age, and diet can produce significant alterations in plasma lipids, and a single normal test should not be expected to set the pattern for the ensuing decade. If lipid values are close to the suggested upper or lower normal or reference levels, tests should be repeated on at least two subsequent occasions, preferably two or four weeks apart. It is important that no intervention (dietary or drug) be instituted while evaluation is taking place. Not only may such intervention eventually prove inappropriate but also a normal laboratory variation in a downward or upward direction (regression toward the mean) may lead the physician to continuing such therapy indefinitely.

In addition to repetitive sampling for triglyceride and total cholesterol determinations in patients with marginal or elevated levels, specific lipoprotein estimations should be carried out.

Determination of High Density Lipoprotein Cholesterol (HDL-C)

The value of high density lipoprotein cholesterol in reducing the risk of atherosclerotic heart disease has recently gained increased prominence among both the scientific and the lay communities. Many laboratories now routinely offer HDL-cholesterol as part of a lipid profile. Since it is currently extremely difficult and impractical to quantitate HDL directly, most methods depend on the measurement of the plasma content of HDL-cholesterol. The majority of techniques are based on selective precipitation of VLDL and LDL with various polyanions, e.g., heparin and manganese chloride, followed by measurement of the cholesterol concentration in the supernatant containing the HDL. Although the precipitation stage is relatively reproducible, the estimation of the remaining cholesterol is subject to significant variation in accuracy and precision because of the low cholesterol content of the supernatant.

There has been growing use of indexes or ratios, the most common ones being total cholesterol:HDL-C. While some authors believe these ratios can be attached to coronary artery disease risk indexes, the ratios contribute little to the understanding of the underlying disease and may prove misleading. Incorrect diagnostic or "risk" information may also be generated, for example, where the total cholesterol:HDL-C ratio is used without the measurement of triglycerides and without estimation of LDL-cholesterol. The potential for obtaining misleading information when using ratios rather than specific lipoprotein cholesterol levels is illustrated in Table 7-17.

Determination of Low Density Lipoprotein Cholesterol (LDL-C)

The use of LDL-C for diagnosing specific lipoprotein phenotypes is well established. Previously, it was necessary to resort to the ultracentrifuge to obtain isolated LDL upon which to perform the assay. The establishment and validation of a formula by Friedewald, Levy, and Fredrickson in 1972 has led to the use of a calculated LDL-C value.[28] The formula hinges on the assumption that VLDL-cholesterol is present in a concentration equal to one fifth of the triglyceride concentration. This assumption is valid for triglyceride concentrations of <400 mg/dL; thereafter, inconsistencies in the VLDL triglyceride/cholesterol ratio occur, and the formula cannot be used. Thus, by measuring total cholesterol, triglyceride, and HDL-C levels, the LDL-C level may be calculated. (See Table 7-16.) This valuable information may be obtained by a simple calculation if the laboratory does not report the LDL-cholesterol.

The value of fractionating lipoprotein cholesterol determinations (as an indication of circulating specific lipoprotein levels) is shown in Table 7-18. Note the difference in diagnosis and treatment between the stages before and after lipoprotein subfractions have been determined. The diagnosis of hyper-β-lipoproteinemia and hypo-α-lipoproteinemia may be missed in patients X and Z, respectively, by relying only on total cholesterol and triglyceride values. An incorrect diagnosis may also be made in patient Y and a therapy started which could conceivably lower HDL, and perhaps increase the long-term susceptibility to CAD.

Lipoprotein Electrophoresis (LPE)

The value of electrophoresis as part of the routine lipid or lipoprotein profile offered by many clinical laboratories has been in question for a number of years. Clinical and analytical experts now discourage the use of LPE in the primary assessment of lipid or lipoprotein disorders. They recommend instead quantitative assays for triglycerides and for total and HDL-cholesterol, calculation of VLDL- and LDL-cholesterol, inspection of plasma, and follow-up on abnormal findings with ultracentrifugation to establish the phenotype. In this context, LPE remains valuable as a supplemental, qualitative adjunct to preparative ultracentrifugation for characterization of uncommon disorders such as a-β-lipoproteinemia, Tangier disease, and Type III hyperlipoproteinemia (broad or floating β-disease). LPE also continues to be important for assessing postheparin lipolytic activity.

There are two instances in which LPE serves useful purposes by first intent. One is when a patient undergoing evaluation for a lipid disorder is suspected of being nonfasting. A small

Table 7-17. PROBABILITY OF CAD IN TWO MEN, AGED 60, WITH THE SAME TC:HDL-C AND LDL-C:HDL-C RATIOS, BUT DIFFERENT ACTUAL LIPOPROTEIN CONCENTRATIONS[27,62]

	mg/dL	mg/dL
TC	180	360
TG	100	100
HDL-C	30	60
LDL-C	130	280
TC:HDL-C	6:1	6:1
LDL-C:HDL-C	4.3:1	4.6:1
Cumulative probability of CAD	0.1	0.62

Table 7-18. CLINICAL IMPORTANCE OF LIPOPROTEIN SUBFRACTIONATION

	Patient X Male, 50 y	Patient Y Male, 50 y	Patient Z Male, 50 y	Reference range, mg/dL*
Plasma after 16 h at 4 °C	Clear	Clear	Clear	Clear
Total cholesterol, mg/dL	260	280	200	156–274
Triglyceride, mg/dL	50	140	100	63–313
Diagnosis made at this stage	Normal	Hypercholesterolemia	Normal	
Possible treatment	?Reassurance	?Diet/drug therapy	None	
After subfractionation of lipoproteins:				
HDL-C, mg/dL	40	90	25	28–63
LDL-C, mg/dL	210	162	155	89–197
Diagnosis made at this stage	Hyper-β-lipoproteinemia (Type IIa)	Hyper-α-lipoproteinemia	Hypo-α-lipoproteinemia	
Risk of CAD	Increased	Reduced	Increased	
Treatment	Diet/drug therapy	None	??	

*5th–95th percentile, Tables 7–19 to 7–22.

chylomicron band at the origin will alert the physician to retest the patient after proper preparation. The second instance is for the detection of Lp-X, an abnormal lipoprotein which is a marker in obstructive jaundice. (See p. 877.)

Ultracentrifugation

In many research centers dealing with lipoprotein abnormalities, preparative ultracentrifugation is now widely used in conjunction with density adjustments to separate the various lipoprotein classes. The most common method for expressing the concentration of the isolated lipoprotein fractions is then by their cholesterol or apolipoprotein content. It is seldom that the routine clinical laboratory will have need for such isolations, which are best referred to a research center. When cholesterol and triglyceride are both elevated above 300–350 mg/dL, the presence of IDL (Type III) is suspected and should then be confirmed by ultracentrifugation at $d = 1.006$ g/mL. (See page 864.) When triglycerides exceed 400 mg/dL and an accurate LDL-cholesterol is needed, ultracentrifugation at $d = 1.006$ g/mL will allow removal of the VLDL in the supernatant layer and measurement of LDL-cholesterol and HDL-cholesterol in the bottom fraction. Precipitation of the LDL fraction using heparin-Mn(II) allows measurement of HDL-cholesterol. Subtraction of the HDL-cholesterol then gives an accurate assessment of LDL-cholesterol.

Determination of HDL Subfractions

The fact that HDL is heterogeneous has been recognized since 1954. The major fractions isolated by ultracentrifugation are HDL_2 and HDL_3. Epidemiological studies have suggested that subjects with lower levels of HDL_2-cholesterol are those most predisposed to the development of early coronary artery disease. HDL_3-cholesterol is the more stable fraction, and major differences in total HDL-cholesterol are related to variation in the HDL_2 fraction.

Separation and analysis of the HDL subclasses have, for the most part, been carried out by ultracentrifugal techniques. With the advent of polyanion precipitating reagents in 1970, a significant amount of work has been done in order to adapt this technique to HDL subfractionation analysis.[30] The fact that the insoluble lipoprotein-polyanion complex forms more readily when the ratio of protein to lipid in the lipoprotein is low enables HDL_2, which has a lower protein to lipid ratio than HDL_3, to be selectively precipitated. The HDL_3 fraction remains in solution and its cholesterol content can then be measured. HDL_2-cholesterol is then calculated by subtracting HDL_3-cholesterol from total HDL-cholesterol. Preliminary studies on precipitation methods for HDL_2 and HDL_3 measurement appear promising and should eventually allow larger scale epidemiological studies to confirm the clinical value of HDL_2-cholesterol measurement.

Methods for the Determination of Lipids and Lipoproteins

Separation and Isolation of Lipoproteins

Lipoproteins may be separated in at least five different ways depending upon their physical, chemical, and immunological properties. Lipoproteins are relatively light macromolecules and ultracentrifugal techniques can be used to separate various fractions on the basis of differential density. Chylomicrons and VLDL can be isolated by ultracentrifugation of unmodified plasma. The other, denser lipoproteins (LDL and HDL) can be caused to float by selective adjustment of plasma density by the addition of salt solutions. Lipoproteins can also be separated electrophoretically based on their differing surface charges and molecular size. The lipoproteins form insoluble complexes with polyanions and divalent cations and may be precipitated by appropriate choice and concentration of polyanion and metal ion. Lipoproteins may also be separated on the basis of their molecular size by gel or membrane filtration, or isolated with antibodies to apolipoproteins.

Preparative Ultracentrifugation[37]

Separation of plasma lipoproteins by ultracentrifugation is currently the reference method, and preparative ultracentrifugation is the most common technique used. The different densities of the individual lipoproteins permit their separation from other plasma proteins and from each other. Ultracentrifugation usually requires expensive instrumentation and skilled technical input and, therefore, is used mainly in research facilities.

To achieve the separation, an aliquot of plasma is ultracentrifuged for 16 h (overnight) at plasma density $(d) = 1.006$ g/mL, and $10\ 000 \times g$. The floating VLDL is recovered by a tube-slicing technique. The plasma infranate is then adjusted to $d = 1.063$ g/mL by adding a salt solution (KBr) and is subjected to a second ultracentrifugation for 20 h at $10\ 000 \times g$. The HDL fraction is separated from the $d = 1.063$ infranate by further adjustment of the density to 1.210 g/mL with solid KBr and additional ultracentrifugation. Isolated lipoprotein fractions are quantitatively recovered and reconstituted to the original volume of the plasma aliquot. Quantitation of the lipoprotein fractions is accomplished by measurement of their cholesterol, triglyceride, phospholipid, or protein content. For additional methodological details, the reader is referred to the Manual of Laboratory Operations for the Lipid Research Clinic (LRC) program.[56]

The proposed reference method for separation of the HDL fraction for clinical application involves the use of both ultracentrifugation and precipitation techniques.[56] Plasma is ultracentrifuged at $d = 1.006$ g/mL, as described for the removal of VLDL. The recovered and reconstituted infranate is then treated with heparin and manganese chloride (see below) to precipitate LDL. Brief centrifugation in a standard refrigerated centrifuge sediments the insoluble LDL and allows the recovery of the clear HDL-containing fraction.

A relatively inexpensive tabletop air-driven ultracentrifuge (Beckman Airfuge, Beckman Instruments, Inc., Palo Alto, CA 94304) permits the ultracentrifugal separation of the lipoproteins in 175 μL plasma in 2.5 h. Both whole plasma $(d = 1.006)$ and plasma adjusted to $d = 1.063$ g/mL are centrifuged. Cholesterol quantitation of the infranate fractions as well as the uncentrifuged plasma provides sufficient information to calculate the cholesterol content of each of the plasma lipoprotein classes.[10] The samples may be prestained with Sudan Black B before ultracentrifugation to permit visual inspection of the lipoprotein distribution and detection of abnormal patterns.

Lipoprotein Electrophoresis

The major classes of lipoproteins migrate electrophoretically based on their charge, as well as their size, shape, and interaction with the supporting medium. Agarose gel and paper electrophoresis produce similar separations of the lipoproteins, with the agarose gel offering increased resolution and, occasionally, increased separation within classes. Cellulose acetate may be inadequate to detect chylomicrons which co-migrate with VLDL; it is therefore not recommended.

Paper and agarose gel lipoprotein electrophoresis are performed by the same procedures as serum protein electrophoresis except for sample size and staining techniques. Paper electrophoresis is carried out for 16 h at 120 V in a Durrum cell with an albumin-containing barbital buffer.

Albumin is added to the buffer to improve separation and definition of the lipoprotein bands. The paper strips are stained in an alcoholic solution of Oil Red O, rinsed, and air-dried prior to qualitative visual examination. For further details, the reader is referred to the LRC Manual for Laboratory Operations.[56]

Most agarose gel electrophoresis methods are based on the procedure developed by Noble.[59] Several commercially prepared agarose gel media are available; they eliminate the inconvenient, technically demanding preparation of gels and offer adequate resolution for clinical purposes. However, laboratory-prepared gels may produce a much higher degree of resolution within the lipoprotein classes. Agarose gel electrophoresis is performed for about 90 min with a barbital buffer and is followed by fixing, drying, and staining with Fat Red 7B or Sudan Black B. For quantitative estimation of HDL-cholesterol, an enzymatic reagent for visualization of cholesterol has been applied as a stain to agarose gel and cellulose acetate electrophoresis films; these films may then be examined by densitometry. However, a number of studies have found the quantitative procedures to be unreliable.[16,76]

Types of lipoproteins separated by LPE are classified according to the serum protein band of corresponding mobility. On agarose gel and paper, HDL is conventionally associated with the α-globulin region, VLDL with the α_2-globulin region, and LDL with the β-globulin region; chylomicrons remain at the origin. (See Figure 7-19.) Aside from chylomicrons, however, these higher density lipoproteins are poorly resolved in *native serum,* and, in fact, the β-, pre-β-, α_2-, and α_1-bands which are observed on the electrophoretic strip cannot consistently be assigned with certainty to classes of lipoproteins defined by their density characteristics. Nevertheless, LPE of native serum is helpful in several instances. For example, an abnormal serum lipoprotein called lipoprotein-X (Lp-X) can be identified by its peculiar electrophoretic behavior. On most support media, Lp-X migrates with the β-lipoproteins, but in agar gels, where endosmosis is strong, Lp-X moves cathodically behind the origin. Lp-X is characterized by a low protein content and relatively large amounts of phospholipid and cholesterol, and is found in sera of patients with obstructive jaundice.[70] LPE of native serum is also helpful in evaluating post-heparin lipolytic activity and in distinguishing Lp-A, a minor β-lipoprotein that is also called "sinking pre-β-lipoprotein." For the former, blood specimens are obtained before and 15 min after injection of heparin (100 units/kg body weight) and both sera are put to lipoprotein electrophoresis. Evidence of heparin-releasable lipoprotein lipase consists of greater mobility and smearing of bands in the post-heparin sample as compared to the pre-heparin sample.

LPE is most useful when applied to fractions of serum isolated by differential density preparative ultracentrifugation. Recall that chylomicrons have a hydrated density of < 0.95 g/mL and that definition of VLDL is relative to d of 0.95–1.006, of LDL to d of 1.006–1.063, and of HDL to d of 1.063–1.21. So preparative ultracentrifugation isolates infranates and supernatants containing particular classes of lipoprotein. When individual fractions are subjected to LPE, specific types of lipoproteins can be identified with more certainty and their relative amounts estimated with more reliability than by LPE of native, unfractionated serum. In Type III disease, LPE of native serum shows a "broad-β-" band, but LPE of supernatants from ultracentrifugation at $d = 1.006$ will show the β-VLDL or "floating" β-lipoproteins as a distinct β-band. Furthermore, observation with LPE of increased or decreased amounts of particular α- or β-bands in infranate or supernatant fractions isolated at particular densities allows the detection of uncommon disorders. Comparisons of LPE on paper or agarose with LPE on polyacrylamide for the same fraction may also be helpful. For instance, on polyacrylamide gel the $d < 1.006$ g/mL supernatant of Type III samples will not show a discrete β-migrating band; on paper, the band will be present.

Even when indicated in lipid studies, lipoprotein electrophoresis is recommended only as a qualitative supplement to quantitative lipid measurements. Problems of incomplete resolution and differences in staining sensitivity of lipoproteins prevent accurate quantitation of their concentrations on the basis of electrophoresis alone. Calibrations and subsequent calculations derived from densitometric scans of lipoprotein electrophoretograms must correct for the different amounts of stain taken up by different bands; but even with corrections, results have dubious value.

Procedure for the Separation of Lipoproteins by Electrophoresis

Specimen

After a 12-h fast, collect blood using an evacuated blood collection tube containing dry disodium EDTA (1 mg/mL). Store plasma at 4 °C prior to analysis. Whenever possible, lipoprotein electrophoresis should be performed on the same day the specimen is collected. Mix sample well before use in assay.

Principle

Lipoproteins are separated electrophoretically due to their differing surface electrical charges. The rate of migration is also affected by the type of supporting medium, e.g., paper, agarose gel, cellulose acetate, polyacrylamide gel, and starch. The last two support media also provide size separation while the first three do not. The HDL or α-lipoprotein migrates most rapidly toward the anode. With starch and polyacrylamide, the LDL or β-band is the next fastest band followed by VLDL or pre-β-lipoprotein. With paper, cellulose acetate, and agarose, the β- and pre-β-migrations are reversed. On all media, except cellulose acetate, chylomicrons do not migrate well and generally remain at the origin.

A typical agarose gel electrophoresis procedure using commercially available agarose gel film and reagents (Corning Medical, Medfield, MA 02052) is presented in detail.

Reagents

1. Barbital Buffer, 0.05 mol/L. Dissolve 10.6 g of sodium barbital and 0.70 g disodium EDTA in 950 mL deionized water. Adjust to pH 8.6 with HCl, 2 mol/L. Adjust volume to 1000 mL with additional deionized water. Store in a clean, tightly closed, brown glass bottle at 4 °C for a maximum of one month.

2. Agarose Film.

3. Fat Red 7B Stock Stain. Transfer 0.225 g Fat Red 7B into a 1-L volumetric flask which has been prerinsed with reagent grade methanol. Adjust the volume to 1 L using reagent grade methanol. Cover the flask tightly; mix with a magnetic stirrer for a minimum of 6 h until solution is complete. Store in a clean, tightly closed, brown glass bottle at room temperature for a maximum of two months. Allow stain to age for one week prior to use.

4. Fat Red 7B Working Stain. Prepare immediately before use. Add 10 mL of the stock stain to a clean, dry 25-mL flask. Gradually add 2 mL of NaOH, 0.1 mol/L, dropwise with mixing and continue mixing until clear. Precipitation of the stain indicates contaminated glassware, in which case preparation of the working stain must be repeated.

5. Clearing Solution. Add 50 mL deionized H_2O to 12.5 mL reagent grade methanol and mix thoroughly.

6. Control. A fresh plasma specimen with total cholesterol, triglyceride, and HDL-cholesterol levels at about the 50th percentile should be used as a control.

Procedure

1. Prepare the electrophoresis cell base by pouring 95 mL of buffer into each compartment and then plug in the base. Fresh buffer must be used for each film.

2. Carefully peel the agarose film from the protective plastic cover, touching only the edges of the film not coated with agarose.

3. Using a microliter dispenser, fill each of the eight specimen wells. Typically, a control and seven specimens are run per film.

4. Load the film into the cell cover, place the cover on the base, and electrophorese for 35 min at 90 V. Do not exceed 150 V.

5. Remove the cell cover and the film. Blot excess moisture from the back of the film and place in a drying oven at 55 °C for 15–20 min. Remove from oven and allow to cool.

6. Place the dried film in a staining dish and pipet 10 mL of working stain evenly over the film. Stain for 4 min until the stain begins to turn blue and precipitate. Drain the excess stain and remove the film.

7. Place the film into a clean staining dish containing clearing solution and gently agitate for 1 min.

8. Remove the film and blot excess moisture from the back. Oven-dry again at 55 °C for 15–20 min.

Interpretations

Visual interpretation of lipoprotein electrophoresis is preferred (see Tables 7-8 and 7-9), and should be made in conjunction with quantitative lipid and lipoprotein measurements. The control sample should demonstrate distinct α- and β-bands and a faint pre-β-band. Densitometric quantitation is not recommended.

Methods for the Determination of HDL-Cholesterol

Numerous studies[88] have compared the different methods for quantitating HDL-cholesterol. Isolation procedures involving preparative ultracentrifugation are considered reference methods and are generally used only in specialized research centers. Electrophoretic techniques, on the other hand, lack precision and accuracy in the range of 20–40 mg/dL, the range of greatest clinical interest,[76] and are not suitable for clinical or epidemiological studies. Ion-exchange and gel-permeation chromatography are too complex and time consuming for routine analysis but are commonly used for research purposes. Various precipitation techniques are now recommended.

Techniques for preparative isolation of the lipoprotein classes by precipitation have been used for years but have been adopted only recently in routine clinical chemistry for the separation of HDL in small volumes of plasma. The major lipoproteins are precipitated by divalent cations and sulfated polysaccharides or by sodium phosphotungstate. The lipoprotein classes can be selectively precipitated by appropriately chosen polyanions and metal ions in the requisite concentrations. The ability of the reagents to precipitate lipoproteins is also dependent upon plasma ionic strength and the total protein concentration. For clinical purposes, a reagent precipitating system must form an insoluble complex with all the plasma lipoproteins except HDL so that HDL remaining in the supernatant after centrifugation can be quantitated by its cholesterol content. A frequently used and well-studied method incorporates heparin-Mn(II) as the polyanion-divalent cation pair.[85] Other precipitating reagents include dextran sulfate 500 or sodium phosphotungstate with Mg(II),[23,54] heparin with Ca(II),[88] concanavalin A, and polyethylene glycol 6000.[84] HDL-C results by the original LRC procedure using heparin-Mn(II), 46 mmol/L, agree best with results by ultracentrifugation ($d > 1.063$) of serum samples. Results using the LRC modified reagent[88] with a final Mn(II) concentration of 92 mmol/L agree most closely with results by the reference method (ultracentrifugation at $d = 1.063$) on plasma samples. The modified reagent has been shown to remove completely from plasma those lipoproteins that contain Apo B without precipitating HDL. Reagent systems containing sodium phosphotungstate-Mg(II) and dextran sulfate-Mg(II) have been reported as more stable and more compatible with enzymatic assays for cholesterol.[54] Although these reagents also produce a better sediment of VLDL and LDL, especially in lipemic samples, they do precipitate small but significant amounts of HDL.[88] Presence of Mn(II) and its interaction with phosphate buffer in an enzymatic cholesterol assay may cause an overestimation of the cholesterol content of HDL separated by heparin-Mn(II) precipitation.[75] But the interference is totally eliminated by reconstituting the cholesterol reagent with EDTA solution[73] to chelate the Mn(II) or by using a reagent that does not contain phosphate. When lipemic samples are precipitated with heparin-Mn(II), the insoluble complex has a low density and may not sediment completely. Ultrafiltration or high-speed centrifugation[87] may then be used to achieve separation of the precipitate and HDL-containing supernate. The source of heparin used for the lipoprotein precipitations is probably not critical since the heparin-Mn(II) procedure is effective even if the heparin concentration is varied over a wide range.[40]

A detailed description of the heparin-Mn(II) precipitation technique follows.[88]

Procedure for the Separation of High Density Lipoprotein (HDL)

Specimen

After a 12-h fast, collect blood using an evacuated blood collection tube containing dry disodium EDTA (1 mg/mL). Store plasma at 4 °C prior to analysis. HDL is stable for 4 d at 4–6 °C; slight but significant changes occur by 7 d. Specimens frozen at −20 °C show statistically but not clinically significant decreases in HDL measured at 7–14 d. Whenever possible, HDL separation should be performed on the same day as specimen collection.

Principle

The major classes of plasma lipoproteins are isolated by the formation of insoluble lipoprotein-polyanion-divalent cation complexes with lipoproteins. In the presence of Mn(II) and heparin, chylomicrons, VLDL, and LDL are selectively precipitated, leaving only HDL in solution. The precipitated lipoproteins are sedimented by centrifugation and the clear, HDL-containing supernatant is recovered.

Reagents

1. Sodium heparin, 40×10^6 USP units/L (Lipo-Hepin, Riker Laboratories, Northridge, CA, 91324).

2. Manganese chloride, 1.06 mol/L. Dissolve 20.98 g of $MnCl_2 \cdot 4 H_2O$ in distilled water and dilute to 100 mL. Stable for three months at 4 °C.

3. Working precipitation reagent. Add 0.6 mL of the sodium heparin solution to 10 mL $MnCl_2 \cdot 4 H_2O$, 1.06 mol/L. Stable for one month at 4 °C.

Procedure

1. Pipet 1.0 mL aliquots of specimens and controls into disposable 10×75-mm glass tubes.

2. Add 0.1 mL of the working precipitation reagent to each tube and mix thoroughly on a vortex mixer.

3. Allow tubes to stand for 10 min at room temperature.

4. Centrifuge at $1500 \times g$ for 30 min at 4 °C.

5. Remove the clear supernatant carefully to a clean tube for use in cholesterol analysis. (See section on cholesterol methods.)

Calculations

The result of the quantitative analysis for cholesterol in the supernatant is multiplied by 1.1 to account for dilution with the precipitating reagent.

Comments

1. Specimens anticoagulated with liquid EDTA will be somewhat diluted; results of all lipids, including HDL-C, may be low.

2. Lipoproteins precipitated from lipemic samples may not sediment; their low density keeps them suspended. Supernatants remain or become turbid, contain non-HDL lipoproteins, and are not suitable for further analysis unless subjected to one of the following procedures:

a. Ultrafiltration of the turbid supernatant through a 0.22-nm filter in a 25-mm filter holder. Aspirate the turbid supernatant into a syringe which is then used to introduce the fluid into the filter and to exert the pressure necessary for forcing the filtrate through the assembly. Collect the *clear* filtrate in a small, clean, glass disposable tube.[87]

b. Dilution of lipemic samples 1:1 with NaCl, 0.15 mol/L, prior to precipitation. Dilution decreases density of the solution and facilitates sedimentation of the precipitate. However, HDL-C which is usually already low in lipemic samples is also diluted and analytical error is increased.

c. Ultracentrifugation for 18 h at $10 000 \times g$, to remove chylomicrons and VLDL, prior to performing the precipitation.

3. Supernatants should be mixed well before analysis for cholesterol.

4. EDTA plasma is the preferred specimen and is required if the reagent formulation specified above is followed. Serum may be used in the same procedure but only if the Mn(II) concentration is reduced from 92 to 46 mmol/L in the precipitation mixture (by preparing Reagent 2 at half concentration, i.e., 0.53 mol/L). The higher concentration of Mn(II) used with plasma counteracts the EDTA it contains. But if the higher concentration of Mn(II) is used with serum, small amounts of HDL can be precipitated by the excess Mn(II); HDL-cholesterol is subsequently underestimated.

5. To check for completeness of precipitation of VLDL and LDL, lipoprotein electrophoresis or apolipoprotein B quantitation can be carried out. This check should be done when establishing the procedure and to confirm unusual results.

6. Results of other precipitating methods have been compared to the heparin-Mn(II) procedure.

The most widely used precipitating reagent is *phosphotungstate-Mg(II)*. This system has the advantages of reagent stability and improved sedimentation of precipitate from lipemic samples. It also does not require the use of a refrigerated centrifuge to produce a sediment. However, detailed studies by Warnick, Cheung, and Albers[88] have shown the procedure to be less specific than the heparin-Mn(II) method. Small changes in either the phosphotungstate or Mg(II) concentrations resulted in more or less lipoprotein precipitation. Temperature also significantly affected the degree of precipitation. Variability with reagent concentration and temperature accounts for poorer precision with phosphotungstate-Mg(II) precipitation. The most important difference with phosphotungstate-Mg(II) is its significant negative bias in HDL isolation compared to ultracentrifugation ($d > 1.063$ fraction) and heparin-Mn(II), 46 mmol/L, precipitation. Since virtually all the HDL-cholesterol data relating risk to reference ranges have been obtained with the heparin-Mn(II) technique, use of phosphotungstate-Mg(II) procedures is fraught with interpretive problems.

The *Dextran 500-Mg(II)* procedure, when compared to heparin-Mn(II), is associated with a constant negative bias for HDL-cholesterol, due to its precipitation of some HDL. It too is sensitive to small changes in either reagent concentration, but it is relatively unaffected by temperature.

Polyethylene glycol 6000 may produce a constant, almost 20% negative bias in HDL-cholesterol, when compared to ultracentrifugation ($d > 1.063$) and heparin-Mn(II), 46 mmol/L. Modifications of this procedure have, however, been recently published and appear more satisfactory.

7. Combined coefficient of variation for lipoprotein precipitation methods and cholesterol determinations should be <8% for HDL-cholesterol levels between 20 and 40 mg/dL. High precision and accuracy are necessary since differences as small as 5 mg/dL may be interpreted as large changes in risk for atherosclerosis.

Reference Ranges[51]

HDL-cholesterol levels vary considerably with age and sex. Table 7-19 shows the most comprehensive data available; these are based on the extensive work of the Lipid Research Clinics Prevalence Program using the heparin-Mn(II) precipitating method. Systematic differences between different methods of separating HDL are sufficiently great to recommend caution in evaluating results obtained by other methods against LRC reference ranges.

Standing Plasma Test (Plasma Appearance)

Specimen

Collect blood in evacuated tubes containing dry disodium-EDTA (1 mg/mL). Although serum can be used, EDTA plasma is convenient since it can be used for other lipoprotein analyses when the standing plasma test has been completed. The test should be carried out on *fresh* specimens and never on previously frozen plasma.

Principle

Chylomicrons have a density < 1.006 (plasma) and insufficient protein-protein interaction to keep them in solution. After plasma is left standing, chylomicrons will float to the uppermost surface and form a distinct opaque band. VLDL is large enough to cause dispersion of light (turbidity) but will remain uniformly distributed in plasma on standing.

Table 7-19. REFERENCE VALUES FOR HDL-CHOLESTEROL (mg/dL)[51]

	Males								Females						
Age in Years	Percentiles							Age in Years	Percentiles						
	5	10	25	50	75	90	95		5	10	25	50	75	90	95
5–9	38	43	49	55	64	70	75	5–9	36	38	48	52	60	67	73
10–14	37	40	46	55	61	71	74	10–14	37	40	45	52	58	64	70
15–19	30	34	39	46	52	59	63	15–19	35	38	43	51	61	68	74
20–24	30	32	38	45	51	57	63	20–24	33	37	44	51	62	72	79
25–29	31	32	37	44	50	58	63	25–29	37	39	47	55	63	74	83
30–34	28	32	38	45	52	59	63	30–34	36	40	46	55	64	73	77
35–39	29	31	36	43	49	58	62	35–39	34	38	44	53	64	75	82
40–44	27	31	36	43	51	60	67	40–44	34	39	48	56	65	79	88
45–49	30	33	38	45	52	60	64	45–49	34	41	47	58	68	82	87
50–54	28	31	36	44	51	58	63	50–54	37	41	50	62	71	84	92
55–59	28	31	38	46	55	64	71	55–59	37	41	50	60	73	85	91
60–64	30	34	41	49	61	69	74	60–64	38	44	51	61	75	87	92
65–69	30	33	39	49	52	74	75	65–69	35	38	49	62	73	85	96
70+	31	33	40	48	56	70	75	70+	33	38	45	60	71	82	92

Procedure

1. A 500-μL aliquot of fresh, well-mixed plasma is placed in a 6 $\times$ 50-mm disposable glass culture tube, covered with Parafilm and allowed to remain overnight (16–18 h) at 4 °C.

2. The tube is then carefully examined in a strong light against a black or dark background.

Results

Chylomicrons, lactescence, and turbidity are noted if present. (See also Figure 7-19.) Chylomicrons appear as a thick creamy layer over the top of the remainder of the sample which may be clear or turbid. Turbidity is a cloudy or milky appearance. The sample should also be assessed for color; increased concentrations of LDL may be associated with an orange color. Hemolysis and jaundice should also be noted if present.

Comments

1. This procedure must be performed in a consistent, standardized manner.

2. Chylomicrons should not be confused with the white material that appears to float to the top of plasma with increased triglyceride levels and which has been stored for several days in the refrigerator.

3. Samples showing a creamy layer or turbidity indicate a need for dilution or other special handling for analysis of triglycerides and isolation of HDL.

METHODS FOR THE DETERMINATION OF CHOLESTEROL

Cholesterol may be quantitatively determined either by enzymatic or by chemical methods. The methods are either *direct,* utilizing serum or plasma directly for the assay, or *indirect,* treating the sample with solvent extraction or other isolation procedures prior to assay of cholesterol. The direct methods are simple, convenient, and readily adapted to automatic analysis; they include chemical procedures using iron salts (Salkowski reaction) and the Lieberman-Burchard reagent, as well as the more popular coupled-enzymatic reagent systems.

In the chemical methods, cholesterol reacts as a typical alcohol with strong, concentrated acids; the products are colored substances, chiefly cholestapolyenes and cholestapolyene carbonium ions. In virtually all of these procedures, acetic acid and acetic anhydride are used as solvents and dehydrating agents, with sulfuric acid being used as a dehydrating and oxidizing reagent. The reaction is enhanced by addition of various metal ions. The general reaction sequence of the most common chemical procedure, the Lieberman-Burchard reaction, is shown below.

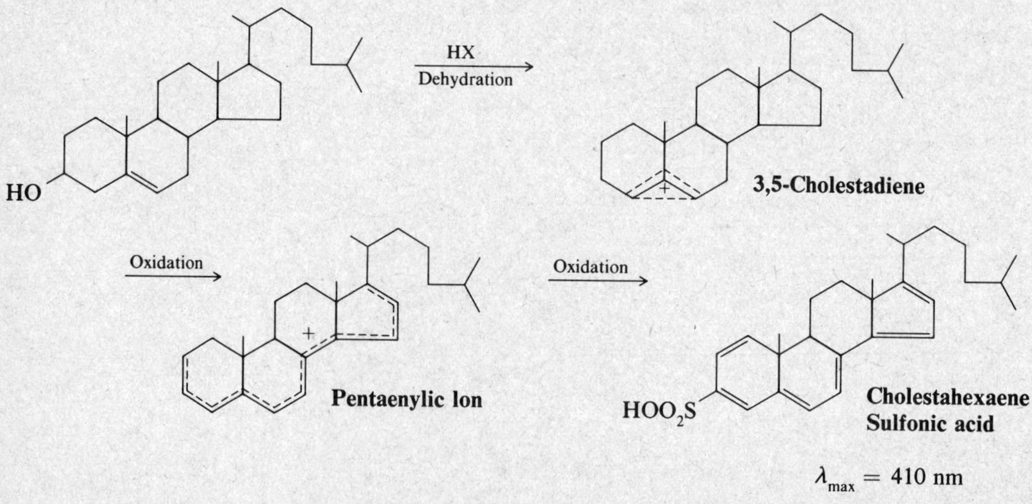

$\lambda_{max} = 410$ nm

Cholesterol is first exposed to strongly acid reagents, generalized as HX, where X might stand for the sulfate ion. Such reagents remove a molecule of water, then oxidize the intermediate to produce 3,5-cholestadiene (two double bonds) or its cation. The oxidizing agent, sulfuric acid, is converted to sulfur dioxide. The cholestadiene reacts further to form cholestapolyene carbonium ions; the stabilities of these cations are dependent on the sulfuric acid concentration. In the Liebermann-Burchard reaction, the polyenes are the main chromophores. Due to the concentration of sulfuric acid and the formation of cholestapolyene sulfonic acid, a green color is formed with a maximal absorption at 410 nm.

Enzymatic procedures have virtually replaced the chemical methods in the clinical laboratory. The initial reaction steps are common to all enzymatic procedures. These are the hydrolysis of cholesteryl esters at C-3 to form free cholesterol and the subsequent oxidation step utilizing O_2 to produce H_2O_2.

$$\text{Cholesteryl esters} \xrightarrow[\text{esterase}]{\text{Cholesterol}} \text{Cholesterol} + \text{fatty acids}$$

$$\text{Cholesterol} + O_2 \xrightarrow[\text{oxidase}]{\text{Cholesterol}} \text{Cholest-4-ene-3-one} + H_2O_2$$

It is during or subsequent to this second reaction that the enzymatic methods differ. In some, the amount of O_2 consumed is measured amperometrically by an oxygen sensing electrode. This measurement can be made as a rate response. The most popular methods involve quantitation of the H_2O_2 by formation of a colored oxidation product or a reduced pyridine nucleotide. In the reaction

$$2 H_2O_2 + \text{4-aminoantipyrine} + \text{phenol} \xrightarrow{\text{Peroxidase}} \text{Quinoneimine dye} + 4 H_2O$$

the quinoneimine product absorbs at 510 nm; p-hydroxybenzenesulfonate may be substituted for phenol. In the reactions with pyridine nucleotide,

$$H_2O_2 + \text{ethanol} \xrightarrow{\text{Catalase}} \text{Acetaldehyde} + 2 H_2O$$

$$\text{Acetaldehyde} + \text{NAD(P)}^+ \xrightarrow[\text{dehydrogenase}]{\text{Aldehyde}} \text{Acetate} + H^+ + \text{NAD(P)H}$$

an increase in absorbance is measured at 340 nm.

In direct methods, interference observed with turbid, lipemic, icteric, or hemolyzed specimens is an important consideration. Bilirubin causes positive interference with the Lieberman-Burchard and iron-salts methods by forming biliverdin which absorbs in the spectral region of the cholesterol reaction product. It causes negative interference in the enzymatic methods that produce a colored chromophore because bilirubin reacts with hydrogen peroxide, thus reducing the amount of peroxide available for formation of the colored complex. Bilirubin can also interfere directly due to its absorbance around 500 nm. Effects of bilirubin interference are minimized in enzymatic assays where oxygen consumption in the cholesterol oxidase reaction is measured electrochemically.[21] Blanking of cholesterol assays is a complex and unresolved issue because of the varied nature of the interferents and their effects. Other sterols with 3β-hydroxyl functions and a double bond in the 4–5 or 5–6 position can also react in the cholesterol oxidase step of a coupled-enzymatic assay.

Problems caused by interferents can be reduced using methods that partially purify the specimen by means of solvent extraction of cholesterol. Typical methods in this category include the candidate reference method—the Abell-Kendall procedure[1]—and the semi-automated method of the Lipid Research Clinics.[56] In the Abell-Kendall method, cholesterol and cholesteryl esters are first extracted from serum. The esterified cholesterol is then enzymatically hydrolyzed or chemically saponified before the Lieberman-Burchard reagent is added for color development. Esters are hydrolyzed because free and esterified cholesterol do not produce equivalent amounts of color in the reaction. The saponification step also reduces interference from nonspecific chromogens. The method of the Lipid Research Clinics differs from the Abell-Kendall procedure only by dispensing with the hydrolysis of esterified cholesterol. The LRC method compensates for the color difference between free and ester cholesteryl by using a plasma calibrator containing a ratio of free and esterified cholesterol similar to that of normal plasma.[53] This calibrator is assigned its value after repeated analysis by the Abell-Kendall method.

An even more complicated method involves extraction, saponification, and isolation of free cholesterol by precipitation as the digitonide. The digitonide precipitate, washed and resaponified, is then subjected to the color reaction. This method of Schoenheimer and Sperry[67] was, prior to development of the Abell-Kendall method, considered a reference method. Cholesterol and cholesteryl esters can also be isolated and quantitated by several types of chromatography. Gas-liquid chromatography,[41] with an internal standard and following solvent extraction and saponification, has been found to correlate well with the Abell-Kendall reference procedure. Since required sample size is of the order of microliters, gas-liquid chromatography is an advantageous technique for analyzing pediatric and animal samples.

Determination of Cholesterol by an Enzymatic Method

Specimen

After a 12-h fast, collect blood using an evacuated tube containing dry disodium-EDTA (1 mg/mL). Separated plasma can be stored at 4 °C for up to 4 d prior to analysis. The choice between plasma and serum is controversial. EDTA-plasma has the advantages that samples kept at 4 °C can be separated immediately and that lipoproteins have enhanced stability during storage. Specimens should preferably be analyzed on the day of collection. Plasma is stable at −20 °C for three months and for many years at −70 °C. Thawed specimens should be well mixed and brought to room temperature prior to analysis.

Principle

All cholesteryl esters present in plasma are hydrolyzed quantitatively into free cholesterol and fatty acids by cholesterol esterase. In the presence of oxygen, free cholesterol is then oxidized by cholesterol oxidase to cholest-4-ene-3-one and hydrogen peroxide (H_2O_2). The hydrogen peroxide reacts with phenol and 4-aminophenazone (4-aminoantipyrine) in the presence of peroxidase (POD) to form an o-quinoneimine dye (Trinder reaction):

$$H_2O_2 + phenol + 4\text{-aminophenazone} \xrightarrow{POD} o\text{-Quinoneimine dye} + 2\ H_2O$$

The intensity of the color formed is proportional to the cholesterol concentration and can be measured photometrically between 480 and 520 nm.

Reagents

Reagents are available in kit form (Boehringer Mannheim Diagnostics, Indianapolis, IN 46250). Concentrations of components in the final reaction mixture for the endpoint and kinetic modes are shown below.

Component	Endpoint Method	Kinetic Method
Tris HCl, pH 7.7	100	100 mmol/L
Magnesium aspartate	50	50 mmol/L
4-Aminophenazone	1	1 mmol/L
Phenol	6	6 mmol/L
3,4-Dichlorophenol	4	4 mmol/L
Sodium cholate	10	10 mmol/L
Detergent (fatty alcohol polyglycolether)	3	3 g/L
Cholesterol esterase	≥ 400	≥ 400 U/L
Cholesterol oxidase (*Nocardia*)	≥ 250	— U/L
Cholesterol oxidase (*Streptomyces*)	—	≥ 1000 U/L
Peroxidase	≥ 200	≥ 200 U/L

Procedure

The assay may be run in endpoint or kinetic modes and is adaptable to automated continuous flow and centrifugal analyzers. The manual endpoint procedure is described here.

1. To 2 mL of substrate mixture, add 200 µL of standard, sample, or control.
2. Incubate at 37 °C for 10 min.
3. Measure absorbance at 500 nm against a blank of substrate mixture set at zero absorbance.

Standardization/Calibration

The method should be standardized using a secondary plasma or serum calibrator previously assigned a target value by an independent reference technique such as the Abell-Kendall procedure. The optimal target value for a total cholesterol calibrator is between 250 and 300 mg/dL. If standardization uses a primary standard containing only unesterified cholesterol, patient values will be underestimated, probably due to incomplete hydrolysis of esters.[6] A second problem with primary standards is slow reaction time due to inhibition of cholesterol oxidase by the detergents used in the system to solubilize free cholesterol; this latter point may be especially important in the kinetic procedure. Recent development of an esterase that achieves total and rapid hydrolysis appears to eliminate the difference between the primary standard and the secondary calibrator.

Comments

1. The method shows good linearity up to 500 mg/dL. Specimens with values above 500 mg/dL should be reanalyzed after 1:1 dilution with NaCl, 0.15 mol/L.

2. The method also demonstrates good linearity at low cholesterol concentrations, an important feature for the measurement of HDL-cholesterol. For measurement of HDL-C, a secondary calibrator with a target value between 45 and 60 mg/dL should be used.

3. For total cholesterol measurements, a normal (170–190 mg/dL) and an elevated (280–320 mg/dL) control should be included in each run.

4. For HDL-cholesterol measurements, at least four controls are required. These will probably have to be prepared as serum pools. Two of them should have low and high *total* cholesterol concentrations of 25–35 and 60–70 mg/dL, respectively; these are used only in the run for quantitating cholesterol. The other two should have low and high HDL-cholesterol concentrations of 25–35 and 60–70 mg/dL, respectively; these should be carried through the HDL separation step *and* the cholesterol quantitation.

5. If a secondary serum calibrator is used, primary cholesterol standards with values of 100, 200, and 400 mg/dL should still be included in the cholesterol assay to check linearity. For the HDL-cholesterol determination, the 100 mg/dL standard and a 25 and 50 mg/dL standard should be used.

6. Commercial lyophilized or stabilized serum-based preparations are not suitable for the calibration of enzymatic cholesterol methods. The turbidity of these products is not cleared by the surfactant present in the reagent as is turbidity of patient samples. Therefore, no blank correction is possible. If possible, a frozen (−70 °C) plasma or serum calibrator should be used.

7. Total cholesterol and HDL-cholesterol measurements should be carried out by the same methodology, particularly if LDL-cholesterol is to be calculated by the Friedwald formula.[28]

8. The endpoint assay is slightly more affected by interfering substances, such as bilirubin, hemoglobin, and ascorbic acid, than is the kinetic procedure.

Reference Ranges[51]

Table 7-20 shows the percentile values obtained for total cholesterol using LRC methodology on a North American population according to age and sex. By using the formula of Friedwald described earlier, it is possible to calculate the LDL-cholesterol concentrations accurately (Table 7-21). It should be emphasized that these percentile values are representative of a population which is known to have one of the highest prevalences of atherosclerosis and cardiovascular disease. These values are, therefore, not ideal or necessarily representative of a healthy state. On the basis of results obtained from epidemiological studies done in populations where atherosclerosis is uncommon and also based on recent clinical trials which have unequivocally demonstrated a reduced risk of atherosclerotic complications with reduced cholesterol levels, it appears prudent to aim at keeping cholesterol levels closer to the 50th percentile. From the previous discussions, it should be clear that significant changes in total, LDL, and HDL cholesterol occur with age and that there are also significant differences between the sexes. It is, therefore, vital that laboratories provide reference values (based on population percentiles) for specific ages and sexes if clinically significant lipid abnormalities are not to be overlooked. Many laboratories, however, provide a single reference value with the upper limit usually between 250 and 300 mg/dL, irrespective of age and sex. Such values may be misleading to both the physician and the patient.

Table 7-20. REFERENCE VALUES FOR TOTAL CHOLESTEROL (mg/dL)[51]

Males								Females							
Age in Years	Percentiles							Age in Years	Percentiles						
	5	10	25	50	75	90	95		5	10	25	50	75	90	95
5–9	125	131	141	153	168	183	189	5–9	131	135	150	164	177	189	197
10–14	124	132	144	161	173	191	204	10–14	125	131	142	159	171	191	205
15–19	118	123	135	152	168	183	191	15–19	119	126	140	157	176	198	208
20–24	118	126	142	159	179	197	212	20–24	121	132	147	165	186	220	237
25–29	130	137	154	176	199	223	234	25–29	130	142	158	178	198	217	231
30–34	142	152	161	190	213	237	258	30–34	133	141	158	178	197	215	227
35–39	147	157	176	195	222	248	267	35–39	139	149	165	186	209	233	249
40–44	150	161	179	204	229	251	260	40–44	146	156	172	193	220	241	259
45–49	163	171	188	210	234	255	275	45–49	148	162	182	204	213	256	268
50–54	156	168	189	211	234	262	274	50–54	163	171	188	214	240	267	281
55–59	161	172	188	214	236	260	280	55–59	167	182	201	229	251	270	294
60–64	163	170	191	215	237	262	287	60–64	172	186	207	226	251	282	300
65–69	166	174	192	213	250	275	288	65–69	167	179	212	233	259	282	291
70+	144	160	185	214	236	253	265	70+	173	181	196	226	249	268	280

METHODS FOR THE DETERMINATION OF FREE AND ESTERIFIED CHOLESTEROL

Separate measurements of free and esterified cholesterol are tedious and complicated. They require some form of chromatographic separation, usually TLC, after extraction of cholesterol into an organic solvent. Following separation of free cholesterol from esterified cholesterol by digitonin precipitation, quantitation of each component is made by colorimetric or enzymatic assay. A simpler approach is to perform a total cholesterol assay by the enzymatic method described above, with and without addition of cholesterol hydrolase to the reaction mixture, and calculate esterified cholesterol as the difference. Cholesteryl esters are customarily expressed as a percentage age of total cholesterol. For reasons discussed above, standardization is difficult. Quality assurance must usually be pursued using selected pooled plasma, but reliable precision data are usually attainable.

METHODS FOR THE QUANTITATION OF TRIGLYCERIDE

Methods for the quantitation of plasma triglycerides include chemical and enzymatic methods.[8,24,86] The chemical methods require solvent extraction of the plasma to solubilize triglyceride and to denature and remove protein. The extract must be treated with an adsorbent material to

Table 7-21. REFERENCE VALUES FOR LDL-CHOLESTEROL (mg/dL)[51]

Males								Females							
Age in Years	Percentiles							Age in Years	Percentiles						
	5	10	25	50	75	90	95		5	10	25	50	75	90	95
5–9	63	69	80	90	103	117	129	5–9	68	73	88	98	115	125	140
10–14	64	73	82	94	109	123	133	10–14	68	73	81	94	110	126	136
15–19	62	68	80	93	109	123	130	15–19	59	73	78	93	110	129	137
20–24	66	73	85	101	118	138	147	20–24	57	65	82	102	118	141	159
25–29	70	75	96	116	138	157	165	25–29	71	75	90	108	126	148	164
30–34	78	88	107	124	144	166	185	30–34	70	77	91	109	129	146	156
35–39	81	92	110	131	154	176	189	35–39	75	81	96	116	139	161	172
40–44	87	98	115	135	157	173	186	40–44	74	84	104	122	146	165	174
45–49	97	106	120	140	163	185	202	45–49	79	89	105	127	150	173	186
50–54	89	102	118	143	162	185	197	50–54	88	94	111	134	160	186	201
55–59	88	103	123	145	168	191	203	55–59	89	97	120	145	168	199	210
60–64	83	107	121	143	165	188	210	60–64	100	105	126	149	168	191	224
65–69	98	104	125	146	170	199	210	65–69	92	99	125	151	184	205	221
70+	88	100	119	142	164	182	186	70+	96	108	126	147	170	189	206

remove phospholipids and interfering substances; isopropanol extracts are treated with a zeolite mixture or with alumina, chloroform extracts with silicic acid. Alternatively, an extraction system which employs nonane and isopropanol may be used to separate triglycerides into the upper, nonane layer; contaminating phospholipids, free glycerol, and glucose are left in the lower layer which is discarded. Once isolated and purified, triglycerides are quantitated by either chemical or enzymatic reactions directed against their glycerol component.

The release of glycerol from triglyceride in the purified extract is effected by saponification with potassium hydroxide or by transesterification with alkoxides. In the chemical methods, the glycerol is then oxidized to formaldehyde by sodium periodate. The formaldehyde either is reacted with a chromotropic–sulfuric acid mixture[14] to form a product which absorbs at 570 nm or is treated with acetylacetone and ammonium acetate to produce a fluorescent condensation product (Hantzsch condensation reaction), which may be quantitated by either fluorometry or colorimetry.[47]

$$H-\overset{\overset{\displaystyle O}{\|}}{C}-H + NH_4^+ + 2\ CH_3-\overset{\overset{\displaystyle O}{\|}}{C}-CH_2-\overset{\overset{\displaystyle O}{\|}}{C}-CH_3 \longrightarrow$$

Formaldehyde

Acetylacetone
(2,4-pentanedione)

$$+ 3\ H_2O + H^+$$

3,5-Diacetyl-1,4-dihydrolutidine

The multiple steps and complexities of the chemical methods complicate their adaptation to automation. The semi-automated procedure used by the Lipid Research Clinics[56] is a widely used chemical method and will be described below in detail.

In the enzymatic methods, hydrolysis may be achieved by preliminary saponification or by lipase (triacylglycerol acylhydrolase).

$$\text{Triglyceride} + 3\ H_2O \xrightarrow{\text{Lipase}} \text{Glycerol} + 3\ \text{free fatty acids}$$

In one commercial method, cholate is used to improve triglyceride solubility and hydrolysis, thus improving linearity of the assay.

Glycerol produced by hydrolysis may be assayed by a variety of coupled-enzyme approaches. In each of these methods, the first step is conversion of glycerol to glycerol-3-phosphate.

$$\text{Glycerol} + \text{ATP} \xrightarrow[\text{kinase}]{\text{Glycerol}} \text{Glycerol-3-phosphate} + \text{ADP}$$

The final product of subsequent coupled reactions, the product that is measured, is proportional to the amount of glycerol, which is, in turn, proportional to the amount of triglycerides.

One method uses the ADP of the first step to generate pyruvate which is consumed by an indicator reaction:

$$\text{ADP} + \text{phosphoenolpyruvate} \xrightarrow[\text{kinase}]{\text{Pyruvate}} \text{ATP} + \text{pyruvate}$$

$$\text{Pyruvate} + \text{NADH} + H^+ \xrightarrow[\text{dehydrogenase}]{\text{Lactate}} \text{Lactate} + \text{NAD}^+$$

Disappearance of NADH is followed spectrophotometrically at 340 nm. Other methods utilize the glycerol-3-phosphate. The reaction

$$\text{Glycerol-3-phosphate} + \text{NAD}^+ \xrightarrow[\text{dehydrogenase}]{\text{Glycerol phosphate}} \text{Dihydroxyacetone phosphate} + \text{NADH} + H^+$$

allows the rate of appearance of NADH to be monitored at 340 nm. An alternative to glycerol-phosphate dehydrogenase is glycerol phosphate oxidase to produce hydrogen peroxide:

$$\text{Glycerol-3-phosphate} + O_2 \xrightarrow[\text{oxidase}]{\text{Glycerol phosphate}} \text{Dihydroxyacetone phosphate} + H_2O_2$$

The H_2O_2 is then reacted by means of peroxidase to give a red quinoneimine dye which can be measured spectrophotometrically at 505 nm.

$$H_2O_2 + \text{4-chlorophenol} + \text{4-aminophenazone} \xrightarrow{\text{Peroxidase}} HCl + 2\ H_2O + \text{4-(}p\text{-benzoquinoneimino)-phenazone}$$

Potassium ferrocyanide　　　　　　　Potassium ferricyanide

This last approach confers a greater degree of specificity by avoiding oxidative/reductive side reactions of NAD. Improved linearity and sensitivity of the peroxidase method has been described when 2-hydroxy-3,5-dichlorobenzenesulfonate is substituted for 4-chlorophenol. The red chromophore produced in this modification is measured at 510 nm.

The presence of *free glycerol* in patient samples should be recognized as a potential source of error in both chemical and enzymatic methods. Increased concentrations of free glycerol may occur in plasma with greatly increased triglyceride concentrations, in specimens which are not fresh, or in plasma from patients taking glycerol-containing medications. Bias due to free glycerol can be eliminated in both colorimetric and enzymatic assays. In the colorimetric procedure, a specimen blank is run in an identical manner except that the saponification step is omitted and the amount of free glycerol is subtracted. In the enzymatic procedure, the first step involving lipase is omitted and free glycerol is subtracted from the value obtained with the hydrolysis reaction. With chemical methods, other interferences can occur since the periodate produces formaldehyde from compounds containing *cis*-hydroxy groups. For the majority of fresh refrigerated specimens with triglycerides below 300 mg/dL, the contribution to the triglyceride measurement caused by free glycerol and other interfering substances is relatively low and constant, and is of minimal clinical importance. When false elevations of triglycerides are suspected, a sample blank should be performed as described.

Chromatography has also been used to quantitate triglycerides. Triglycerides are extracted from plasma, separated from the other lipids by thin-layer chromatography, then eluted, dried, and weighed for quantitation.[91]

Determination of Triglycerides by an Enzymatic Procedure[56a]

Specimen

Specimens should not be obtained for triglyceride determination unless the subject has been fasting for 10–14 h. Collect blood in an evacuated tube containing dry, disodium EDTA (1 mg/mL). Store specimens at 4 °C prior to analysis. Specimens can be stored at 4 °C for seven days, frozen at −20 °C for three months, or frozen at −70 °C for years. Lipemic specimens may require warming to 37 °C and vigorous mixing prior to analysis, especially if they have been frozen.

Principle

Triglycerides are hydrolyzed by microbial lipase to produce glycerol and free fatty acids. The glycerol participates in a series of coupled enzymatic reactions, the last of which results in the formation of a stable red quinoneimine dye that is read at 510 nm. The reactions are as follows:

$$\text{Triglycerides} \xrightarrow{\text{Lipase}} \text{Glycerol} + \text{free fatty acids}$$

$$\text{Glycerol} + ATP \xrightarrow[\text{kinase}]{\text{Glycerol}} \text{Glycerol-3-phosphate} + ADP$$

$$\text{Glycerol-3-phosphate} + O_2 \xrightarrow[\text{oxidase}]{\text{Glycerophosphate}} \text{Dihydroxyacetone phosphate} + H_2O_2$$

$$H_2O_2 + \text{4-aminoantipyrine} + \text{sodium 2-hydroxy-3,5-dichlorobenzenesulfonate} \xrightarrow{\text{Peroxidase}}$$
$$\text{Quinoneimine dye} + H_2O$$

Reagents

1. Lipase (triacylglycerol acylhydrolase, EC 3.1.1.3) and glycerophosphate oxidase (EC 3.1.1.21) from Fermco Biochemics, Elk Grove Village, IL 60007.

2. Glycerokinase (EC 2.7.1.30), peroxidase (EC 1.11.1.7), ATP, Triton X-100, 4-aminoantipyrine (AAP), α-cyclodextrin, and glycerol (Sigma Chemical Co., St. Louis, MO 63178).

3. Sodium 2-hydroxy-3,5-dichlorobenzenesulfonate (Research Organics Inc., Cleveland, OH 44125). The working reagent is prepared in Tris · HCl buffer (50 mmol/L, pH 7.6) containing per liter 0.1 g Triton X-100, 1 mmol of AAP, α-cyclodextrin, 1.5 mmol sodium 2-hydroxy-3,5-dichlorobenzenesulfonate, 5 mmol $MgCl_2$, 0.5 mmol ATP, 10^4 units peroxidase, 0.25×10^3 units glycerol kinase, and 10^5 units of lipase. For a blanking reagent, lipase is left out of the working reagent. The working reagent is stable for 24 h at 4 °C.

Procedure

A 1:100 ratio of specimen to reagent is used in the assay. Dilute lipemic specimens 1:1 with NaCl, 0.15 mmol/L, before analysis. Mix after addition of specimen to reagent. Incubate for 15 min at 37 °C and read absorbance at 510 nm against working reagent set to zero absorbance. Assay using blanking reagent is performed the same way, and the blank absorbance is subtracted from the test absorbance to eliminate the effect of preformed glycerol in the specimen. Specimens with triglyceride values in excess of 700 mg/dL should be diluted with NaCl, 0.15 mmol/L, and the assay repeated on the diluted specimen.

Calibration and Standardization

Aqueous primary standards of glycerol can and have been used. However, an aqueous primary standard of glycerol is not identical to a patient specimen or a plasma calibrator containing triglyceride. Triolein, a triglyceride which has been used as a primary standard in some chemical methods, is not water soluble; when triolein is made up in an organic solvent and used as a standard, the solvent causes interference in enzymatic assays. A plasma- or serum-based secondary standard should therefore be prepared from a frozen pool of plasma with a triglyceride value of about 200 mg/dL. The pool must first be repeatedly assayed for triglyceride by the colorimetric reference method to assign its secondary standard value. Primary standards of glycerol at 100, 200, 400, and 600 mg/dL should be included in each run as a monitor of linearity.

Comments

1. Blood collection tubes whose stoppers have been lubricated with glycerol should not be used for collection of specimens for triglyceride assay. Such tubes, usually for collection of 2-mL specimens, are sometimes found in use for pediatric and micro specimen collections.

2. Plasma specimens should preferably be analyzed on the day of collection.

3. Controls at both normal and abnormal triglyceride levels (90–110 and 190–200 mg/dL, respectively) should be included in each assay run.

4. For specimens with triglyceride levels below 200 mg/dL, preformed glycerol content is almost always low and blanking of the assay is not necessary. Larger amounts of preformed glycerol occur when triglycerides exceed 300 mg/dL but are probably still not clinically significant; elevated glycerol levels are found in subjects with acute or severe illnesses characterized by excessive release of fatty acids from tissue at storage depots. Frozen or lyophilized controls may contain measurable amounts of free glycerol.

5. When enzymatic assays are performed on automated instrumentation, stability of reagents must be carefully monitored. Some reagents have extremely short stability after reconstitution, e.g., one hour.

Reference Ranges

Table 7-22 shows age- and sex-specific percentile ranges for a broad cross section of the North American population. These ranges were derived using the LRC combined procedure for cholesterol and triglycerides (described below). Many laboratories, however, prefer to set an upper limit of normal for triglycerides somewhere between 130 and 200 mg/dL using the 95th percentile value for a given population distribution. Reference ranges based on an "apparently healthy population" are generally higher than the recommended (desirable) levels for adults: 40–160 mg/dL (males) and 35–135 mg/dL (females).

Table 7-22. REFERENCE VALUES FOR TRIGLYCERIDES (mg/dL)[51]

Age in Years	Males							Age in Years	Females						
	Percentiles								*Percentiles*						
	5	*10*	*25*	*50*	*75*	*90*	*95*		*5*	*10*	*25*	*50*	*75*	*90*	*95*
5–9	28	34	39	48	58	70	85	5–9	32	37	45	57	74	103	126
10–14	33	37	46	58	74	94	111	10–14	39	44	53	68	85	104	120
15–19	38	43	53	68	88	125	143	15–19	36	40	52	64	85	112	126
20–24	44	50	61	78	107	146	165	20–24	37	42	60	80	104	135	168
25–29	45	51	67	88	120	141	204	25–29	42	45	57	76	104	137	159
30–34	46	57	76	102	142	214	253	30–34	40	45	55	73	104	140	163
35–39	52	58	80	109	167	250	316	35–39	40	47	61	83	115	170	205
40–44	56	69	59	123	174	252	218	40–44	45	51	66	88	116	161	191
45–49	56	65	88	119	165	218	279	45–49	44	55	71	94	139	180	223
50–54	63	75	94	128	178	244	313	50–54	53	58	75	103	144	190	223
55–59	60	70	85	117	167	210	261	55–59	59	65	80	111	163	229	279
60–64	56	65	84	111	150	193	240	60–64	57	66	78	105	143	210	256
65–69	54	61	78	108	164	227	256	65–69	56	64	86	118	158	221	260
70+	63	71	87	115	152	202	239	70+	60	68	83	110	141	189	289

Simultaneous Determination of Cholesterol and Triglyceride, Using Continuous Flow Analysis (AutoAnalyzer II)[56]

Specimen

After a 12-h fast, collect blood in an evacuated collection tube containing dry Na_2 EDTA (1 mg/mL). Store specimens at 4 °C prior to analysis.

Principle

The specimen is manually extracted with isopropanol, to separate lipids from protein, in the presence of a zeolite mixture (reagent 9) which adsorbs interfering phospholipids. The extract is then aspirated into the cholesterol assay stream containing the Lieberman-Burchard reagent and incubated at 60 °C; absorbance of the color developed is measured in a colorimeter at 630 nm. Simultaneously sampled extract also enters a triglyceride assay stream of 80% isopropanol to which is added a stream of KOH. Following incubation at 50 °C for triglyceride saponification, periodate reagent enters the stream to oxidize the glycerol to formaldehyde, followed by acetylacetone in ammonium acetate to condense with the formaldehyde. Condensation takes place in a second heating bath (50 °C) to form the fluorescent product 3,5-diacetyl-1,4-dihydrolutidine; the fluorescent product is measured in a fluorometer.

Reagents

1. Isopropanol, spectroquality.
2. Lieberman-Burchard color reagent. Prepare fresh the day of use by adding 300 mL acetic anhydride and 150 mL glacial acetic acid to a low actinic Pyrex bottle surrounded with packed, crushed ice in an insulated container resting on a magnetic stirrer. Allow to mix one-half hour until thoroughly chilled. Add 50 mL of chilled (4 °C) concentrated H_2SO_4 dropwise, with stirring, to the acetic anhydride mixture from a buret or separatory funnel. Continue to stir for 20 min. Stopper the bottle tightly and seal with Parafilm to prevent absorption of moisture. Store at −20 °C for a maximum of 24 h. This reagent must be prepared in a fume hood. The reagent is a strong oxidant and *must be handled with care.*
3. Isopropanol in water (60:40).
4. Isopropanol in water (80:20).
5. KOH, 0.8 mol/L. Place 400 mL of deionized water in a 1-L volumetric flask. Add 45.0 g KOH (depending on its water content) and swirl to dissolve completely. Bring to volume with deionized water.
6. Periodate reagent. Add 115 mL glacial acetic acid to 400 mL deionized water in a 1-L volumetric flask and mix. Add 5.4 g sodium metaperiodate ($NaIO_4$) and allow to dissolve. Bring to volume with deionized water. Prepare this reagent in the hood. The reagent is stable for one month at room temperature.
7. Ammonium acetate, 2 mol/L. Dissolve 154 g ammonium acetate in 700 mL deionized water in a beaker. Adjust the pH to 6.0 with HCl, 2 mol/L. Dilute to 1 L with deionized water. The reagent is stable for one month at room temperature. Protect from excessive exposure to air.
8. Working acetylacetone reagent. Mix 1.5 mL of 2,4-pentanedione and 5.0 mL isopropanol in a 200-mL volumetric flask. Dilute to volume with ammonium acetate, 2 mol/L, and transfer to a brown glass bottle. This reagent must be prepared fresh daily.

9. Zeolite mixture. Grind 200 g of zeolite to a fine powder (80–100 microns) in a blender and heat overnight at 110 °C. Cool and add 20 g Lloyd's reagent (sodium alumina silicate), 10 g cupric sulfate pentahydrate, and 20 g calcium hydroxide. Mix thoroughly by shaking in a large jar. Store in a tightly closed clean container. If not used within two weeks, the mixture should be reactivated by heating overnight at 110 °C.

10. Stock triglyceride standard, 1000 mg/dL. Dissolve 1000 mg triolein in 50 mL isopropanol in a 100-mL volumetric flask. Bring to volume with isopropanol. Store in a brown glass bottle in a desiccator over isopropanol for a maximum of one month. Triolein is used as a standard since its oleic fatty acid component approximates closely the *average* length of the numerous fatty acids found in plasma triglycerides.

11. Stock cholesterol standard, 1000 mg/dL. Dissolve 1000 mg cholesterol in 50 mL isopropanol in a 100-mL volumetric flask and bring to volume with isopropanol. Store in a brown glass bottle in a desiccator over isopropanol for a maximum of one month. Storage over isopropanol prevents evaporation of stock standards and avoids the need to bring refrigerated solutions to room temperature before each use.

12. Prepare five working standards as follows:

	Cholesterol stock std. (mL)	Triglyceride stock std. (mL)	Isopropanol to final vol. of (mL)	Concentration of cholesterol (mg/dL)	Concentration of triglyceride (mg/dL)
Standard 1	10	5	100	100	50
Standard 2	20	10	100	200	100
Standard 3	75	50	250	300	200
Standard 4	40	30	100	400	300
Standard 5*	5	0	100	50	0

*This standard is used only when measuring HDL-cholesterol.

13. Serum calibrator (SC). A serum based calibrator with the same ratio of free to esterified cholesterol as typically occurs in patient specimens should be used. This is necessary because most plasma cholesterol is esterified and because esterified cholesterol produces a different amount of color in the Lieberman-Burchard reaction. The calibrator is prepared from a serum pool which is frozen in small aliquots and which has a cholesterol value of ∼300 mg/dL. The calibration value is determined by repeated cholesterol analyses by the Abell-Kendall procedure.[1] The serum calibrator may be stored at −70 °C for a maximum of one year. A serum calibrator is not required for triglycerides by this colorimetric procedure since triglycerides are completely saponified to glycerol and fatty acids. (For enzymatic triglyceride methods, a serum calibrator is probably more valid since triolein standards are difficult to use and triglycerides may not be completely hydrolyzed.)

Procedure

Preparation of Extracts

1. Place ∼0.8 g zeolite mixture into a sufficient number of disposable 13 × 100-mm glass culture tubes for the blanks, controls, calibrators, and patient specimens to be extracted.

2. With an automatic dilutor, aspirate 200 μL of specimen, control, or calibrator and deliver along with 3.8 mL isopropanol into a prepared tube. Close tubes with a Teflon-lined screw cap and mix well on a vortex mixer.

3. Allow to stand for 30 min. During this period, mix the tubes twice.

4. Centrifuge for 30 min at 1200 × g. Decant the clear extract into clean screw-capped tubes for analysis. The procedure can be halted at this point and specimens stored at 4 °C for up to one week before continuing the analysis.

5. To prepare a blank, substitute saline (NaCl, 0.15 mmol/L) for the specimen and proceed as in step #2.

6. To prepare a standard, aspirate 200 μL of standard and deliver along with 1.8 mL of isopropanol into a prepared tube. Aspirate 200 μL of saline and deliver along with 1.8 mL of isopropanol into the same tube. Cap and mix and proceed as in step #3.

Continuous Flow Analysis

1. Refer to the flow diagram (Figure 7-20) for instrument parameters.
2. Set up trays of specimens as follows:

	Tray 1		*Subsequent Trays*
Cups 1–2	Blanks	Cups 1–2	Blanks
Cups 3–4	S-3	Cups 3–4	S-3
Cups 5–6	S-1	Cups 5–39	Specimens, controls
Cups 7–8	S-2		
Cups 9–10	S-3		
Cups 11–12	S-4		
Cups 13–14	S-3		
Cups 15–16	SC		
Cups 17–18	S-3		
Cups 19–39	Specimens, controls		
Cup 40	Blank		

3. Pump reagents until the system is equilibrated. Set up the fluorometer and colorimeter. Adjust the fluorometer reference aperture and light pipe wheel if necessary. Set the fluorometer recorder at zero. Peak the flow cell if necessary and then adjust the reagent baseline to zero. The colorimeter zero and full scale adjustments should be made and the cholesterol reagent baseline set at zero.

4. Begin sampling extracts. When blank peaks appear on the recorder, set the blanks to zero. Adjust the S-3 standard in cups 4, 14, and 18 to 60% of scale (100 units is full scale) for cholesterol (300 mg/dL) and to 50% of scale (100 units is full scale) for the triglyceride (200 mg/dL). The series of standards in cups 5 through 12 serves as a linearity check. At the beginning of each subsequent tray, the blanks are set to zero and the S-3 to its calibration value.

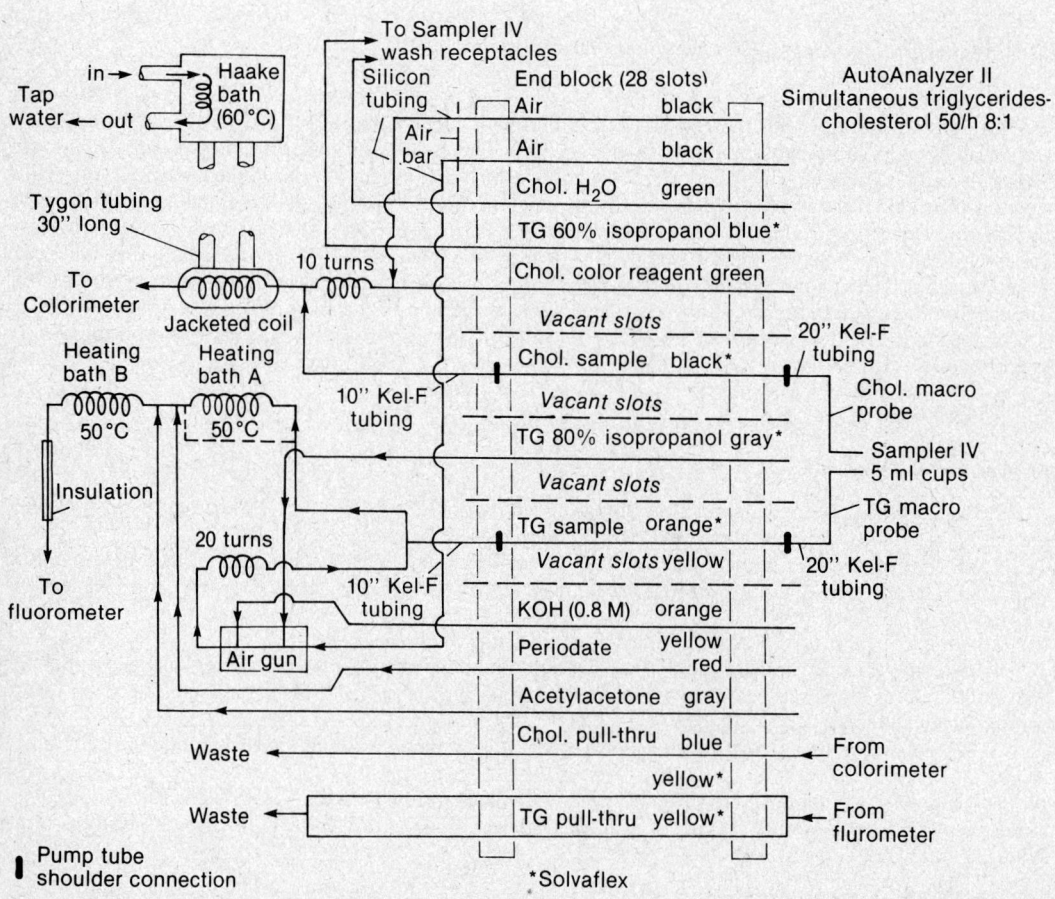

Figure 7-20. Flow sheet for simultaneous determination of total cholesterol and triglyceride concentrations in plasma by AutoAnalyzer II.

5. When the analyses have been completed, the systems should be shut down and cleaned. The system is first allowed to pump air for 15 min until dry. The triglyceride reagent lines are placed in distilled water and the cholesterol lines in NaOH, 1.0 mol/L, and pumped for 15 min. All lines are then allowed to pump air until dry. The cholesterol lines must pump H_2SO_4, 0.125 mol/L, for an additional 10 min and then all lines must pump air until the system is completely dry.

Calculations

1. For cholesterol, the concentration value, SC_{AA-II} in mg/dL, of the serum calibrator relative to the primary cholesterol standard in this run is first determined:

$$SC_{AA-II} = SC \text{ scale reading} \times \frac{300 \text{ mg/dL}}{60 \text{ scale units}}$$

Then the concentration value of the patient serum is determined relative to the primary cholesterol standard and multiplied by a correction factor which is the ratio of its assigned value (from repeated Abell-Kendall assay) to the value obtained in this run.

$$\text{Patient cholesterol, mg/dL} = \text{Patient scale reading} \times \frac{300 \text{ mg/dL}}{60 \text{ scale units}} \times \frac{SC_{Abell-Kendall}}{SC_{AA-II}}$$

2. For triglyceride, the calculation is:

$$\text{Patient triglyceride, mg/dL} = \text{Patient scale reading} \times \frac{200 \text{ mg/dL}}{50 \text{ scale units}}$$

3. Further information will be found in the Lipid Research Clinics Manual of Laboratory Operations.[56]

Comments

1. The reagents for triglycerides are generally available from a number of manufacturers. Commercial preparations of the Lieberman-Burchard reagent are *not recommended*. The stock triglyceride standard (triolein) may be purchased commercially, while the stock cholesterol standard should be prepared from NBS cholesterol (SRM #911A).

2. Specimens with cholesterol concentrations > 400 mg/dL or triglycerides > 300 mg/dL should be appropriately diluted with NaCl, 0.15 mmol/L, prior to extraction.

3. Plasma specimens should be *extracted* the same day they are collected.

4. The specimen tray cover must be used at all times to prevent excessive evaporation of isopropanol. The use of 5-mL specimen cups which contact the cover is recommended.

5. Nonsaponified triglyceride blanks of fresh specimens with normal triglyceride concentrations (< 200 mg/dL) are fairly low and may be ignored. Blanks run for specimens with increased concentrations (> 300 mg/dL) may be increased but are probably not clinically significant. The determination of blank values may be important when analyzing specimens which have been frozen or lyophilized. If desired, blanks can be determined on the same extract after modification of the triglyceride lines to bypass the first heating bath where saponification takes place. (See Figure 7-20.)

6. Controls with normal and high concentrations of analytes should be included in each run.

7. The user is cautioned in regard to the highly corrosive properties of the Lieberman-Burchard reagent which can cause severe burns. The reagent also produces a harmful vapor, is combustible, and reacts violently with water. During analysis, the Lieberman-Burchard reagent should be packed in ice in an insulated container and continually mixed on a magnetic stirrer.

QUALITY CONTROL AND CALIBRATION CONSIDERATIONS IN LIPID ANALYSIS

Measurements of lipids and of lipoproteins are among the more difficult and exacting clinical laboratory tests; for these and other reasons, lipid and lipoprotein separation and quantitation have often left much to be desired. This is an unfortunate situation since initial accuracy and long-term precision are vital for the diagnosis and treatment of the hyperlipoproteinemias. A recent survey of over 2000 laboratories conducted by the College of American Pathologists[15] revealed a considerable variation in both accuracy and precision for the estimation of cholesterol and triglycerides. For triglycerides, the results obtained by enzymatic methods were variable, with coefficients of variation (CV) ranging from 3.8–15.9% at a level of ~283 mg/dL. Cholesterol estimation proved to be a little more dependable, with most laboratories reporting a CV between 3.1% and 9.3% on a 310-mg/dL specimen. However, the results of enzymatic methods generally showed better precision. The Centers for Disease Control Survey of High-Density Lipoprotein Cholesterol Measurement enrolled 130 participants who analyzed five specimens using a variety of lipoprotein precipitation and cholesterol methods.[36] The conclusions of the study were that laboratory accuracy of the HDL-cholesterol at the low, critical concentration of 34.5 mg/dL was generally inadequate. Statements regarding the cause of the bias between precipitating reagents could not be made because of the great degree of variability due to improper calibration.

The exemplary performance of reference and research laboratories can be attributed not only to their tightly standardized methodology but also to the superiority of the frozen calibration and control material available to them. The modified Abell-Kendall method as used by the Centers for Disease Control, Atlanta, Georgia, produced remarkable precision on frozen pools with a CV of 1.0% at a level of 162 mg/dL and 0.8% at 255 mg/dL.[17] In the analysis of cholesterol, twelve Lipid Research Clinics achieved total CVs < 2.0% for a pool with a concentration of 263 mg/dL and total CVs of < 3.0% at a level of 166 mg/dL.[53] For triglycerides, the Lipid Research Clinics have shown total CVs of about 4% for pools with concentration ranges from 36–274 mg/dL.[2]

Commercially available lyophilized quality control materials may be suitable for monitoring quantitative measurements of cholesterol and triglyceride. For total cholesterol, controls must be routinely assessed in the normal (150 mg/dL) and abnormal (250 mg/dL) ranges as well as at the level of HDL-cholesterol (45 mg/dL). Triglyceride quality assurance should be tested at normal (75 mg/dL) and abnormal (250 mg/dL) concentrations. Commercial controls used for triglycerides should be examined for glycerol blanks since some may have had glycerol or formaldehyde-generating substances added during their manufacture.

The integrity of lipoproteins is disturbed by lyophilization or freezing of plasma, so that material preserved in these ways is less than ideal for control of lipoprotein separations. Electrophoretic mobility of components in reconstituted, lyophilized serum is altered and the bands become less discrete. Precipitation behavior is also affected. Examining differently preserved materials for the control of HDL separation by precipitation, workers have reported that Apo B lipoproteins (LDL) are not completely removed when lyophilized, reconstituted serum is treated with precipitating reagent, and Apo AI lipoproteins (HDL) are precipitated when ethylene glycol-stabilized serum is so treated.[38] The same study presented evidence that HDL is not lost in the treatment of frozen plasma or serum and that LDL and VLDL precipitation is complete. These findings would seem to indicate that frozen plasma or serum is a more reliable control material than a lyophilized or otherwise stabilized material, where HDL, LDL, and VLDL separations are concerned.

The overall accuracy of cholesterol quantitation is greatly dependent on correct calibration of nonsaponification methods with a human serum-based calibrator that has nearly the same cholesterol/cholesteryl ester ratio as is found in patient specimens. Unavailability of suitable frozen material analyzed by a reference method and the turbidity problem associated with lyophilized and stabilized material may force the user of enzymatic methods to calibrate with aqueous standards in which the cholesterol is solubilized with the aid of detergents or similar agents.

For the measurement of cholesterol in HDL, linearity and precision are required at very low levels. Increased sample size is often required to provide the necessary sensitivity, and an alternate cholesterol calibrator in the concentration range of HDL-cholesterol is recommended. HDL-cholesterol measurements should be monitored by a control that is carried through the HDL isolation step as well as by a nonprecipitated control with a low total cholesterol concentration. The need to correct HDL-cholesterol values by a factor to account for the dilution by the precipitating reagent is a potential source of error.

TESTS FOR QUANTITATING OTHER LIPIDS

Phospholipid measurements are seldom requested for clinical purposes. The procedure utilizes a plasma extract[24] which is digested and analyzed for inorganic phosphorus.[8] Total lipid phosphorus (in mg/dL) is multiplied by 25 to obtain the total phospholipid concentration expressed as phosphatidyl choline.

Free fatty acids (FFA) or nonesterified fatty acids (NEFA) are measured occasionally. In certain disease states, such as diabetes and acute stress, fatty acids are released from adipose tissue into the bloodstream. In patients with acute myocardial infarction, increased free fatty acids have been postulated as a cause of cardiac arrhythmias. The plasma levels of free fatty acids can be determined by titration, thin-layer or gas-liquid chromatography, or colorimetry.[22,81] In all three methods, fatty acids are extracted from acidified plasma by a mixture of heptane and isopropyl alcohol. In the titration method, the extract is titrated with dilute alkali using an ethanolic solution of thymol blue or thymolphthalein as an indicator. During the titration, the system is protected from exposure to atmospheric carbon dioxide, which would also consume the dilute alkali. This is accomplished by bubbling a stream of alkali-washed nitrogen through the titration tube, which also stirs the system to ensure thorough mixing of the added alkali with the extract. The alcohol promotes the contact of fatty acid with aqueous base since it is a solvent for both.

Colorimetric methods are based on the formation of *cupric soaps* or fatty acids. The soaps are extracted into a mixture of chloroform-heptane-methanol, and the amount of bound cupric ion is determined colorimetrically after reaction of the copper with a suitable reagent such as sodium N,N-diethyldithiocarbamate. The levels of free fatty acids found in plasma from healthy adults range from 0.3–0.9 mmol/L. Concentrations in children and obese adults may be as high as 1.1 mmol/L.

Measurement of Apolipoproteins

Measurement of apolipoproteins is rapidly gaining in popularity as specific antisera and various purified apolipoproteins become more widely available. A number of established immunochemical techniques are used: radial immunodiffusion (RID),[20] electroimmunoassay (EIA) in agar or agarose gel,[20] radioimmunoassay (RIA),[44] immunonephelometry (INA),[89] enzyme-linked immunoassay (ELISA),[38,39,77] and fluorescence immunoassay (FIA).[48] Nonimmunological procedures such as molecular sieve chromatography, isoelectric focusing, and polyacrylamide gel electrophoresis are laborious and imprecise but are still used as reference methods to standardize or validate immunological assays.

Immunochemical assays are more sensitive, specific, reproducible, and potentially more capable of automation than are nonimmunological procedures. There are, however, significant problems in producing antisera and in standardization of the apolipoprotein assays in the United States and elsewhere in the world. Even within a single laboratory that makes its own antisera, continuity of measurement of a specific apoprotein is difficult since even in the same animal, differences in batches of antiserum are commonplace. The problems arise from the genetic polymorphism of apoproteins, masking of the antigenic determinants of apoproteins in the lipoprotein form, and the tendency of apoproteins to self-aggregate when in dilute solution or when the lipoprotein form is delipidated.[29]

Polymorphism is a characteristic of apoproteins AI, B, and E, and an antiserum produced to one of them may not react with all the isoforms of the specific apoprotein. Masking of antigenic sites on the apoprotein while it is residing in the lipoprotein is a source of analytical error. An estimate has been made that 90–95% of the Apo AI in HDL is not antigenically available.[71] Analytical methods using heat or delipidation to expose the Apo AI sites detect significantly more Apo AI than do other methods. Specificity of antiserum produced against an apoprotein is also highly dependent on whether the intact lipoprotein, a delipidated lipoprotein, or a purified apoprotein is used as immunizing agent. The tendency of apoproteins to self-associate, especially when purified and in dilute solutions, poses a problem in their use as standards or labeled tracers in certain immunochemical approaches. Their instability in this respect indicates a need to produce secondary calibrators, either isolated lipoprotein subfractions (e.g., HDL or LDL) or whole plasma or serum.

Problems of apolipoprotein quantitation and standardization are being addressed by the National Institutes of Health (NIH) and the Centers for Disease Control (CDC). In 1982, NHLBI sponsored a workshop on apolipoprotein methodology and the proceedings were published in a detailed monograph the following year.[61] In 1983, an international committee on apolipoprotein standardization was formed under the auspices of a subcommittee of the Standardization Committee of the International Union of Immunological Societies (IUIS) and of the Centers for Disease Control. Initial activity has centered on the definition of a stable reference plasma pool.

The pool was analyzed for apolipoprotein AI and B by 75 laboratories worldwide. RIA was used by 13% of participating laboratories, EIA by 18%, ELISA by 10%, RID by 30%, and INA by 29%. Preliminary results[18] indicate that the European mean of values for Apo AI is 10% lower than the American mean, but there was no significant difference for Apo B. Within-run coefficients of variation for both assays in most laboratories were under 10%. Mean concentration by all methods was 109 mg/dL for Apo AI and 67 mg/dL for Apo B. The 5th–95th percentile range was 76–144 mg/dL for Apo AI and 40–98 mg/dL for Apo B. These efforts are expected to continue and to result eventually in a consensus reference calibrator and reference procedure using a standard antiserum for each apolipoprotein.

Commercial assays for Apo AI and B have recently become available in the USA: RID from Tago Laboratories (Burlingame, CA 94010); and INA from Hyland Diagnostics (Deerfield, IL 60015). Their performance is similar to that described for RID and INA below.

Radial immunodiffusion (RID). RID for apolipoprotein analysis is slow (2 or 3 d are required for the diffusion-precipitation process), and calls for relatively large sample volumes and for pretreatment of sample. Assays are critically dependent on antibody characteristics, antibody-antigen ratio, and the properties of the gel. Chylomicrons and VLDL either will not diffuse or will diffuse at very much slower rates than do LDL and HDL.[72] For example, using 1.5% agarose gel and an 18–24-h incubation period, only LDL-Apo B can be quantitated. Apo AI analysis usually requires dilution of specimen in a tetramethylurea-Tris buffer to delipidate lipoprotein, precipitate Apo B, and expose Apo AI immunoreactive sites. Sensitivity of RID procedures is generally lower than that of EIA and INA, and considerably less than that of FIA, ELISA, and RIA. Between-run coefficients of variation for RID assays of most apoproteins at their normal plasma concentrations are 8–15%.

Electroimmunoassay (EIA). EIA, or "rocket" assay, is a more rapid and more sensitive immunoprecipitation technique than RID although equipment is more expensive. In EIA, rocket formation depends on low intrinsic charge of the agarose support gel, type of buffer, concentration of polyethylene glycol (PEG), and the duration of electrophoresis.

Most EIAs are optimized to produce a rocket height of 1–4 cm over the range of the standard curve. The lowest possible concentrations of antigen and antibody that will produce a clearly visible rocket, and an antiserum that will produce a clearly defined edge for the rocket, are used. Rocket visibility may be enhanced by increasing the concentration of PEG. Need to pretreat samples for EIA varies and depends to a great extent on the properties of the antiserum. Any laboratory setting up apolipoprotein assays and using commercially or inhouse prepared reagents must assess the reactivity of individual lipoprotein fractions and the effects of manipulations such as heating or delipidation on the assay system. With careful technique and proper optimization, EIA can have good precision—CV of 6–10% over the normal range of Apo AI and B levels. CVs for apolipoproteins found in lower concentrations (Apo AII, E, and CII) are usually 10–15%.

Immunonephelometry (INA).[63,79] In INA, the antigen-antibody interaction results in turbidity which is quantitated by measuring changes in light scattering intensity. The procedure can be performed in either endpoint or kinetic mode. In endpoint mode, antigen must be in excess; in kinetic mode, antibody excess is required. INA has been used for quantitating Apo AI, AII, and B. The procedure is subject to interference by triglyceride-rich particles so that a pretreatment step to delipidate or solubilize the lipoproteins becomes necessary. Nonuniformity of HDL particles causes variation in size of the immune complexes formed and introduces an uncontrolled variable in the system. However, the procedure is potentially capable of automation, requires small sample volume, and has high sample throughput.

Enzyme-linked immunoassay (ELISA).[38,39,77] Enzyme-linked immunoassays have been described for the measurement of Apo AI, AII, and B. The procedures differ; some require labeling of the apoprotein with the enzyme and are based on competition of labeled and unlabeled apoprotein for a solid phased antibody. Labeling of apoproteins raises the same problems discussed earlier, namely self-aggregation and long term instability. Other ELISAs are "direct" or sandwich procedures. In these systems, the antibody is solid phased, the standard or unknown apoprotein is added, followed by a second antibody labeled with enzyme. Monoclonal antibodies can also be used in these systems. A variant of competitive ELISA, in which labeling of both apoprotein and the apoprotein antibody has been avoided, has been described by Stein and Pesce.[77] In this prototype method, purified Apo AII or delipidated HDL is solid-phased to wells of a microtiter plate. The diluted specimen and the monoclonal anti-Apo AII antibody are added, the system is incubated for 2 h and is then washed. Quantity of monoclonal antibody bound to the solid-phase apoprotein is detected with a second antibody that has been raised against mouse IgG and has been labeled with alkaline phosphatase. This kind of assay has the advantage of the long term consistency conferred by use of a monoclonal antibody reagent. The selected monoclonal antibody reacts only with a surface epitope, and the need to pretreat the specimen is thereby avoided. Coated wells are stable for more than six months. Labeled mouse IgG antiserum is commercially available and has a long shelf life. It also has potential for use in different apoprotein assays which could be set up by changing only the well coating and the monoclonal antibody. The method utilizes a frozen plasma calibrator and selected frozen plasma pools for quality assurance.

ELISA systems in general require minimal upkeep once they have been developed. Only small amounts of antiserum are required, assay time is short, and assay precision is acceptable. CVs are usually under 10% for plasma apoprotein concentrations normally encountered. Other than INA, only ELISA methods have potential for automation and high throughput. Commercial ELISA methods for apoproteins are likely to become available in the near future and will probably be the method of choice in both routine and research laboratories.

Fluorescent immunoassay (FIA).[48] FIA has been used for the measurement of Apo B. It has the advantages of ELISA in terms of analytical specificity and reagent stability, but has greater sensitivity than ELISA. For plasma apoproteins with concentrations generally > 1 mg/dL, the additional sensitivity is probably not necessary.

Radioimmunoassay (RIA).[46,63,79] RIA is the most traditional approach to apolipoprotein measurement. It is also the technique that has been used for most apolipoproteins—Apo AI, AII, B, CII, CIII, E, and D. Radio-labeling of the apoprotein antigen risks denaturation of the apoprotein and altered immunologic behavior in the reaction system. Self-aggregation again can be a problem of the tracer solution. Because tracer has a short half-life and nonisotopic assays are generally more popular alternatives to RIA, this technique has fallen out of favor in many clinical laboratories. However, for assays of the minor apolipoproteins and research studies, RIA is likely to remain the most commonly used method.

The future of apolipoprotein measurements. Despite extensive efforts and an extensive literature, measurement of apolipoproteins is still time consuming and technically demanding, and has low specimen throughput. Measurements are heavily influenced by variations in antisera, standardization, and the methodology of a particular assay system. These factors have limited the use of apolipoprotein measurement in prospective clinical studies and in prevalence or screening studies. The epidemiological group in the 1982 NIH workshop concluded that present techniques for plasma apolipoprotein assay precluded organization of mass population studies as had been done for HDL- and LDL-cholesterol.[61] The optimal assay for apolipoproteins is one that is reliable and reproducible over time, offers high specimen throughput at reasonable cost, and requires minimal technical skills. Ideally, it should be adaptable to implementation on equipment currently available in routine clinical laboratories.

Reference ranges for plasma apolipoprotein levels are not well established because of the variety of current assays used for measurement. Table 7-6 synthesizes reference ranges that have been found in a number of studies and is a rough guide to levels found in healthy individuals. These reference ranges are likely to change significantly in the coming years since these initial estimates are based on small populations.

References

1. Abell, L. L., Levey, B. B, Brodie, B. B., et al.: A simplified method for the estimation of total cholesterol in serum and demonstration of its specificity. J. Biol. Chem., *195*:357, 1952.
2. Ahmed, S., Lippel, K., Bachorik, P. S., et al.: External quality control survey of triglyceride (triacylglycerol) analyses performed by 12 lipid research clinics. Clin. Chem., *25*:880, 1979.
3. Arteriosclerosis. A Report by the National Heart and Lung Institute Task Force on Arteriosclerosis. Department of Health, Education, and Welfare publication (NIH) 72-137. Washington D.C., National Institutes of Health, 1971, Vol. 1.
4. Avogaro, P., Bittolo, B. G., Cazzolato, G., et al.: Plasma levels of apolipoprotein AI and apolipoprotein B in human atherosclerosis. Artery, *4*:385-394, 1978.
5. Avogaro, P., Cazzolato, G., Bittolobon, G., et al.: Are apolipoproteins better discriminators than lipids for atherosclerosis? Lancet, *1*:901-903, 1979.

6. Bacharik, P. S., and Wood, P. D. S.: Laboratory considerations in the diagnosis and management of hyperlipopro-teinemia. *In*: Hyperlipidemia: Diagnosis and Therapy. B. M. Rifkind and R. I. Levy, Eds. New York, Grune and Stratton, 1977.

7. Barr, D. P., Russ, E. M., and Eder, H. A.: Protein-lipid relationships in human plasma. II. In atherosclerosis and related conditions. Am. J. Med., *11*:480-493, 1951.

8. Bartlett, G. R.: Phosphorus assay in column chromatography. J. Biol. Chem., *234*:466, 1959.

9. Breckinridge, W. C., Little, J. A., Steiner, G., et al.: Hypertriglyceridemia associated with deficiency of apolipoprotein CII. N. Engl. J. Med., *298*:1265-1273. 1978.

10. Bronzert, T. J., and Brewer, H. B.: New micromethod for measuring cholesterol in plasma lipoprotein fractions. Clin. Chem., *23*:2089-2098, 1977.

11. Brown, M. S., and Goldstein, J. L.: Receptor-mediated control of cholesterol metabolism. Science, *191*:150, 1976.

12. Bucolo, G., and David, H.: Quantitative determination of serum triglycerides by use of enzymes. Clin. Chem., *19*:475, 1973.

13. Carlson, L. A., and Philipson, B.: Fish-eye disease: A new familial condition associated with massive corneal opacities and dyslipoproteinemia. Lancet, *1*:921-925, 1979.

14. Carlson, L. A., and Wadström, L. B.: Determination of glycerides in blood serum. Clin. Chim. Acta, *4*:197, 1959.

15. College of American Pathologists: Comprehensive Set C-D Chemical Procedures, Participant Summary. Skokie, Ill., College of American Pathologists, 1979.

16. Conlon, D. R., Blankstein, L. A., Pasakarnis, P. A., et al.: Quantitative determination of high density lipoprotein cholesterol by agarose gel electrophoresis. Clin. Chem., *25*:1965-1969, 1979.

17. Cooper, G. R., Gill, J. B., Biegeleisen, J. J., et al.: CDC point of reference for total cholesterol measurements. Clin. Chem., *26*:966, 1980.

18. Cooper, G. R., and Rosseneu, M.: Report of preliminary results of CDC-NHLBI-IUIS apolipoprotein AI and B survey. Personal correspondence, 1984.

19. Coronary Drug Project Research Group: The Coronary Drug Project: Clofibrate and niacin in coronary heart disease. JAMA, *231*:360-381, 1975.

20. Curry, M. D., Gustafson, A., Alaupovic, P., et al.: Electroimmunoassay and radial immunodiffusion assay evaluated for quantification of human apolipoprotein B. Clin. Chem., *24*:280-286, 1978.

21. Dietschy, J. M., Weeks, L. E., and Delente, J. J.: Enzymatic measurement of free and esterified cholesterol levels in plasma and other biological preparations using the oxygen electrode in a modified glucose analyzer. Clin. Chim. Acta, *73*:407, 1976.

22. Duncombe, W. G.: The colorimetric microdetermination of long-chain fatty acids. Biochem. J., *88*:7, 1963.

23. Finley, P. R., Schifman, R. B., Williams, R. J., et al.: Cholesterol in high-density lipoprotein: Use of Mg^{2+}/dextran sulfate in its enzymic measurement. Clin. Chem., *24*:931, 1978.

24. Folch, J., Lees, M., and Sloane-Stanley, G. H.: A simple method for the isolation and purification of total lipids from animal tissues. J. Biol. Chem., *266*:497, 1957.

25. Franceschini, G., Bondioli, A., Mangero, M., et al.: Increased lipoprotein B in very low density lipoproteins of patients with peripheral vascular disease. Arteriosclerosis, *2*:74-80, 1982.

26. Franceschini, G., Sirtori, C. R., Capurso, A., et al.: AI$_{milano}$ apoprotein decreased high density lipoprotein cholesterol levels with significant lipoprotein modifications and without clinical atherosclerosis in an Italian family. J. Clin. Invest., *66*:892-900, 1980.

27. Fredrickson, D. S., and Levy, R. I.: Familial hyperlipoproteinemias. *In*: The Metabolic Basis of Inherited Disease. 3rd ed. J. B. Stanbury, J. B. Wyngaarden, and D. S. Fredrickson, Eds. New York, McGraw-Hill, 1972, pp. 545-614.

28. Friedwald, W. T., Levy, R. I., and Fredrickson, D. S.: Estimation of the concentration of low-density lipoprotein cholesterol without the use of the preparative ultracentrifuge. Clin. Chem., *18*:499, 1972.

29. Gibson, J. C., and Brown, W. V.: The human plasma apolipoproteins: Assay methods. Part I. Lab. Management, *21*:19-27, 1983.

30. Gidez, L. I., Miller, G. J., Burstein, M., et al.: Separation and quantitation of subclasses of human high density lipoprotein by a simple precipitation procedure. J. Lipid Res., *23*:1206-1223, 1982.

31. Glueck, C. J., Fallat, R. W., Millet, F., et al.: Familial hyperalphalipoproteinemia. Metabolism, *24*:1243, 1975.

32. Glueck, C. J., Gartside, P. O., Steiner, P. M., et al.: Hyperalpha- and hypobeta-lipoproteinemia in octogenarian kindreds. Atherosclerosis, *27*:387, 1977.

33. Glueck, C. J., Mellies, M. J., Tsang, R. C., et al.: Low and high density lipoprotein cholesterol interrelationships in neonates with low density lipoprotein cholesterol ≤ the 10th percentile and in neonates with high density lipoprotein cholesterol ≥ the 90th percentile. Pediatr. Res., *11*:957, 1977.

34. Gofman, J. W., DeLalla, O., Glazier, F., et al.: The serum lipoprotein transport system in health, metabolic disorders, atherosclerosis and coronary heart disease. Plasma, *2*:413-484, 1954.

35. Goldstein, J. L., and Brown, M. S.: The LDL receptor defect in familial hypercholesterolemia. Med. Clin. North Am., *66*:335-362, 1982.

36. Hainline, A., Cooper, G. R., Olansky, A. S., et al.: Report of the CDC survey of high density lipoprotein cholesterol measurement. Atlanta, Center for Disease Control, 1980.

37. Havel, R. J., Eder, H. A., and Bragdon, J. H.: Distribution and chemical composition of ultracentrifugally separated lipoproteins in human serum. J. Clin. Invest., *34*:1345-1354, 1955.

38. Holmquist, L.: Quantitation of human serum apolipoprotein B by enzyme immunoassay. Clin. Chim. Acta, *121*:327-336, 1982.

39. Holmquist, L.: Quantitation of human very low density apolipoproteins Cl, C2, C3 and E by enzyme immunoassay. J. Immunol. Methods, *34*:234-251, 1980.

40. Ishikawa, T. T., Brazier, J. B., Steiner, P. M., et al.: A study of the heparin-manganese chloride method for determination of plasma alpha-lipoprotein cholesterol concentration. Lipids, *11*:628, 1976.

41. Ishikawa, T. T., Brazier, J. B., Stewart, L. E., et al.: Direct quantitation of cholesterol in plasma by gas-liquid chromatography. J. Lab. Clin. Med., *87*:345, 1976.
42. Jackson, R. L.: Lipid transport and atherosclerosis: Role of apolipoproteins in vascular smooth muscle. *In*: Topics in Physiology; Metabolic, Ionic, and Contractile Mechanisms. M. F. Krass III and C. D. Barnes, Eds. New York, Academic Press, 1982, pp. 163-196.
43. Kannel, W. B., Castel, W. P., Gordon, T., et al.: Serum cholesterol, lipoprotein and risk of coronary heart disease: The Framingham study. Ann. Intern. Med., *74*:1, 1971.
44. Karlin, J. B., Juhn, D. J., Goldberg, R., et al.: Recent, progress in the development of radioimmunoassays for human serum lipoproteins. Ann. Clin. Lab. Sci., *8*:142-154, 1978.
45. Kashyap, M. L., and Glueck, C. J.: Hyperlipoproteinemia in diabetes mellitus. *In*: Diabetes and the Heart. R. C. Scott, Ed. Mt. Kisco, N. Y., Futura Publishing Company, 1981.
46. Kashyap, M. L., Srivastava, L. S., Chen, C. Y., et al.: Radioimmunoassay of human apolipoprotein CII: A study of normal and hypertriglyceridemic subjects. J. Clin. Invest., *60*:171-180, 1977.
47. Kessler, G., and Lederer, H.: Fluorometric measurement of triglycerides. *In*: Automation in Analytical Chemistry. L. T. Skeggs, Jr., et al., Eds. Technicon Symposia 1965. New York, Mediad, 1966, p. 341.
48. Lee, J. K., and Carrico, R. T.: Substrate labeled fluorescence immunoassay for the determination of serum protein. *In*: Proceedings of the Workshop on Apolipoprotein Quantification. K. Lippel, Ed. U.S. Department of Health and Human Services, National Institutes of Health, Publication No. 83-1266, 1983, pp. 332-342.
49. Levy, R. I.: Declining mortality in coronary heart disease. Arteriosclerosis, *1*:312-325, 1981.
50. The Lipid Research Clinics Population Studies Data Book, Vol. 1. The prevalence study. U.S. Department of Health and Human Services, National Institutes of Health, Publication No. 80-1527, 1980.
51. Lipid Research Clinics Program Epidemiology Committee: Plasma lipid distributions in selected North American population. The Lipid Research Clinics Program Prevalence Study. Circulation, *60*:427-439, 1979.
52. Lipid Research Clinics Program: The Lipid Research Clinics Coronary Primary Prevention Trial Result. II. The relationship of reduction in incidence of coronary heart disease to cholesterol lowering. JAMA, *251*:365-374, 1984.
53. Lippel, K., Ahmed, S., Albers, J. J., et al.: Analytical performance and comparability of the determination of cholesterol by 12 lipid research clinics. Clin. Chem., *23*:1774, 1977.
54. Lopes-Virella, M. F., Stone, P., Ellis, S., et al.: Cholesterol determinations in high-density lipoproteins separated by three different methods. Clin. Chem., *23*:882, 1977.
55. Maciejko, J. J., Holmes, D. R., Kottke, B. A., et al.: Apolipoprotein A-I as a marker of angiographically assessed coronary artery disease. N. Engl. J. Med., *309*:385-389, 1983.
56. Manual of Laboratory Operations, Lipid Research Clinics Program. Lipid and lipoprotein analysis. U.S. Department of Health, Education and Welfare, National Institutes of Health, Publication No. 75-628, 1974.
56a. McGowan, M. W., Artiss, J. D., Strandbergh, D. R., et al.: A peroxidase-coupled method for the colorimetric determination of serum triglycerides. Clin. Chem., *29*: 538-542, 1983.
57. Nestel, P. J., Fridge, N. H., and Tan, M. H.: Increased lipoprotein-remnant formation in chronic renal failure. N. Engl. J. Med., *307*:329-333, 1982.
58. Nikkila, E.: Studies on the lipid-protein relationships in normal and pathological sera and the effect of heparin on serum lipoproteins. Scand. J. Clin. Lab. Invest., *5*:1-101, 1953.
59. Noble, R. D.: Electrophoretic separation of plasma lipoproteins in agarose gel. J. Lipid Res., *9*:693, 1968.
60. Norum, R. A., Lakier, J. B., Goldstein, S., et al.: Familial deficiency of apolipoproteins A-I and C-III and precocious coronary artery disease. N. Engl. J. Med., *306*:1513-1519, 1982.
60a. Olivecrono, T., and Bengtsson, G.: Immunochemical properties of lipoprotein lipase; development of an immunoassay applicable to several mammalian species. Biochim. Biophys. Acta, *752*:38-45, 1983.
61. Proceedings of the Workshop on Apolipoprotein Quantification. K. Lippel, Ed. U.S. Department of Health and Human Services, National Institutes of Health, Publication No. 83-1266, 1983.
62. Rhoads, G. C., Gulbrandsen, C. L., and Kagan, A.: Serum lipoproteins and coronary heart disease in a population of Hawaii Japanese men. N. Engl. J. Med., *294*:293, 1976.
63. Rosseneu, M., Vercaemst, R., Steinberg, K. K., et al.: Some considerations of methodology and standardization of apolipoprotein B immunoassays. Clin. Chem., *29*:422-433, 1983.
64. Royer, M. E., and Ko, H.: A simplified semiautomated assay of plasma triglycerides. Anal. Biochem., *29*:405, 1969.
65. Schaefer, E. J., Eisenberg, S., and Levy, R. I.: Lipoprotein apoprotein metabolism. J. Lipid Res., *19*:667-687, 1978.
66. Schaefer, E. J., Heaton, W. H., Wetzel, M. G., et al.: Plasma apolipoprotein A-1 absence associated with a marked reduction of high density lipoproteins and premature coronary artery disease. Arteriosclerosis, *2*:16-26, 1982.
67. Schoenheimer, R., and Sperry, W. M.: A micromethod for the determination of free and combined cholesterol. J. Biol. Chem., *106*:745, 1934.
68. Schonfeld, G.: The effect of antibody specificity on the quantitation of Apo B concentrations in plasma. *In*: Proceedings of the Workshop on Apolipoprotein Quantification. K. Lippel, Ed. U.S. Department of Health and Human Services, National Institutes of Health, Publication 83-1266, 1983, pp. 362-372.
69. Seftel, H. C., Baker, S. G., Sandler, M. P., et al.: A host of hypercholesterolaemic homozygotes in South Africa. Br. Med. J., *281*:633-636, 1980.
70. Seidel, D., Alaupovic, P., Furman, R. H., et al.: A lipoprotein characterizing obstructive jaundice. II. Isolation and partial characterization of the protein moieties of low density lipoproteins. J. Clin. Invest., *49*:2396, 1970.
71. Shonfeld, F., and Pfleger, B.: The structure of human high density lipoprotein and levels of apolipoprotein AI in plasma as determined by radioimmunoassay. J. Clin. Invest., *54*:235, 1974.
72. Sniderman, A., Shapiro, S., Marpole, D., et al.: Association of coronary atherosclerosis with hyperapobetalipoproteinemia (increased protein but normal cholesterol levels in human plasma low density lipoproteins). Proc. Natl. Acad. Sci. USA, *77*:604-608, 1980.
73. Steele, B. W., Kochler, D. F., Azar, M. M., et al.: Enzymatic determinations of cholesterol in high-density lipoprotein fractions prepared by a precipitation technique. Clin. Chem., *22*:98, 1976.

74. Stein, E. A., and Glueck, C. J.: Hyperlipoproteinemia: Implications, diagnosis and therapy. *In*: Cardiac Diagnosis and Treatment. 3rd ed. N. O. Fowler, Ed. New York, Harper and Row, 1980, pp. 677-695.

75. Stein, E. A., Glueck, C. J., Wesselman, A., et al.: Repetitive intermittent-flow plasma exchange in patients with severe hypercholesterolemia: Results and complications. Atherosclerosis, *38*:149-164, 1981.

76. Stein, E. A., McNeely, S., and Steiner, P.: Electrophoretic separation of high-density lipoprotein cholesterol evaluated and compared with the modified lipid research clinic procedure. Clin. Chem., *24*:1934, 1979.

77. Stein, E. A., and Pesce, A.: Enzyme linked immunoassays for apolipoproteins: Advantages, problems and prototype assay. *In*: Proceedings of the Workshop on Apolipoprotein Quantification. K. Lippel, Ed. U.S. Department of Health and Human Services, National Institutes of Health, Publication No. 83-1266, 1983, pp. 319-331.

78. Stein, E. A., and Steiner, P. M.: Lipid metabolism and methodology. *In*: Pediatric Clinical Chemistry. J. M. Hicks and R. L. Boeckx, Eds. Philadelphia, W. B. Saunders Co., 1983.

79. Steinberg, K. K., Cooper, G. R., Graiser, S. R., et al.: Some considerations of methodology and standardization of apolipoprotein A-I immunoassays. Clin. Chem., *29*:422-433, 1983.

80. Steiner, P. M., Freidel, J., Bremner, W. F., et al.: Standardization of micromethods for plasma cholesterol, triglyceride and HDL-cholesterol with the Lipid Clinics' methodology. J. Clin. Chem. Clin. Biochem., *19*:850, 1981.

81. Trout, D. L., Estes, E. H., and Friedberg, S. J.: Titration of free fatty acids of plasma: A study of current methods and a new modification. J. Lipid Res., *1*:199, 1960.

82. Turpeinen, O.: Effect of cholesterol lowering diet on mortality from coronary heart disease and other causes. Circulation, *59*:1-7, 1979.

83. Vergani, C., Trovato, G., and Dioguardi, N.: Serum total lipids, lipoproteins, cholesterol and apoproteins A and B in cardiovascular disease. Clin. Chim. Acta, *87*:127-133, 1978.

84. Viikari, J.: Precipitation of plasma lipoproteins by PEG-6000 and its evaluation with electrophoresis and ultracentrifugation. Scand. J. Clin. Lab. Invest., *36*:265, 1976.

85. Warnick, G. R., and Albers, J. J.: A comprehensive evaluation of the heparin-manganese precipitation procedure for estimating high density lipoprotein cholesterol. J. Lipid Res., *19*:65, 1978.

86. Warnick, G. R., and Albers, J. J.: Evaluation of quality control materials for high-density lipoprotein cholesterol separation. Clin. Chem., *26*:998, 1980.

87. Warnick, G. R., and Albers, J. J.: Heparin-MN^{2+} quantitation of high-density lipoprotein cholesterol: An ultrafiltration procedure for lipemic samples. Clin. Chem., *24*:900, 1978.

88. Warnick, G. R., Cheung, M. C., and Albers, J. J.: Comparison of current methods for high-density lipoprotein quantitation. Clin. Chem., *25*:596, 1979.

89. Weinstock, N., Bartholome, M., and Seidel, D.: Determination of apolipoprotein A1 by kinetic nephelometry. Biochim. Biophys. Acta, *663*:279-288, 1981.

90. Whayne, T. F., Alaupovic, P., Curry, M. D., et al.: Plasma apolipoprotein B and VLDL, LDL and HDL-cholesterol as risk factors in the development of coronary artery disease in male patients examined by angiography. Atherosclerosis, *39*:411-424, 1981.

91. Wood, P. D., and Wiezel, A.: Mass determination by direct weighing in estimation of cholesterol specificity activity. Clin. Chim. Acta, *26*:539-546, 1969.

92. Yamamoto, K., Koga, S., and Ibayashi, H.: Apolipoprotein A1 in cholestatic liver diseases. Clin. Chim. Acta, *87*:85-92, 1978.

93. Zak, B.: Cholesterol methodologies: A review. Clin. Chem., *23*:1201, 1977.

94. Zannis, V. I., and Breslow, J. L.: Apolipoprotein E. Mol. Cell. Biochem., *42*:3-20, 1982.

CHAPTER 8

NUTRITION, VITAMINS, AND TRACE ELEMENTS

Nutrition

by Dorice M. Czajka-Narins, Ph.D.

With the progress of science and technology, the cellular role of various nutrients has become more clearly delineated and the importance of nutrition, both as a vital component of health and in minimizing the effects of disease, has become more apparent. Good nutrition is essential to individuals of all ages; it provides materials for growth and development of the young and optimizes the health of the elderly. Nutrition is also important to the critically ill patient debilitated by disease whose diet can frequently be replaced by total parenteral nutrition (TPN).

As the role of nutrition has expanded, so has the need for methods of assessing nutritional status. No longer is there solely an interest in detecting severe nutritional deficiency; there is now a need to determine the degree of depletion and to monitor recovery. Ideally, patients at risk must be identified before they become debilitated. The clinical chemist should have an active interest in all aspects of nutritional assessment, from standardizing existing methods to developing new ones. What follows is an initial step in fostering an improved cooperation between clinical nutritionists and clinical chemists by providing a better understanding of the needs in this area.

DIETARY REQUIREMENTS

We eat to live. When the diet supplies all essential nutrients and sufficient energy, the cells and tissues of the body are capable of synthesizing the thousands of compounds required for life and well-being. When the dietary supply is inadequate, the body makes a series of regulatory adjustments that tend to conserve essential nutrients. Prolonged inadequacy results in progressive deterioration and, ultimately, death.

Approximately forty dietary components and elements are recognized as essential for humans. Additional minerals may be required in such small amounts that their essentiality has not yet been recognized. The amounts of individual nutrients required per day range from grams to picograms. In general, *macronutrients* (proteins, carbohydrates, and lipids) constitute structural and functional components of the cells and sources of energy. *Micronutrients* (vitamins and minerals) are components of enzymes or cofactors needed for metabolic reactions. Some nutrients, such as calcium or zinc, are present in two or more body compartments, from which they may be mobilized at different rates. *Conditionally essential nutrients* are nonessential in the healthy person but are required by certain sick individuals who have lost the ability to synthesize such compounds at an adequate rate.

Determining quantitative requirements for energy and essential nutrients is of practical importance but is also more difficult than generally realized. For animals, the requirements (minimum needs) can be determined in relation to a particular function, such as growth, milk yield, or fleece production. In humans, however, requirements should be sufficient to optimize health and provide reserves for emergency needs; therefore they are more difficult to define. Another complicating factor is the variability of requirements among individuals. Finally, because the requirement is really the amount needed by individual cells to function properly and not the amount consumed, bioavailability (the degree to which a nutrient is absorbed by the body) must be considered when developing recommendations regarding intake of some nutrients. Thus, the true requirements of a single individual are seldom known.

Dietary standards or, in the United States, *Recommended Dietary Allowances (RDA)*, have been established to provide standards that serve as a goal for good nutrition. RDA are not absolute nutritional standards, nor are they recommendations for an ideal diet. As originally defined, the Recommended Dietary Allowances are "goals at which to aim in providing the nutritional needs of groups of people."* Unfortunately, the term connotes to many people recommendations for an ideal diet. Some scientists have suggested the term "Acceptable Nutrient Intake" as more descriptive of the original concept and less open to misinterpretation. Except for energy,† the RDA are set at levels sufficiently above the average physiological requirements in order to cover the variability which arises from differences in genetic makeup and to ensure that the needs of nearly all are met. Most frequently this value is two standard deviations above the mean requirement.

RDA for energy are categorized by age and sex. Energy needs per kilogram of body weight are highest for the young infant; after age 10 they are generally higher for males than females. For young adults, the allowances are set for individuals doing light work; for example, for males 19–22 years of age, the RDA is 2900 $\pm$ 400 kcal. The range emphasizes the variation in energy intake appropriate for any group of people. Allowances for older groups are set to compensate for a 2% decrease in basal metabolic rate (BMR) per decade and for a reduction in activity. (The *basal metabolic rate* is the sum of the energy needed for all involuntary activities.) The energy allowance, as opposed to other nutrients, is set at a level believed to be consistent with good health of average persons. Insufficient intake of total calories, when extreme and prolonged, results in a condition called marasmus.

Proteins, carbohydrates, and lipids are dietary sources of energy. Proteins and carbohydrates yield 4 kcal/g (17 kJ/g), and triglycerides yield 9 kcal/g (38 kJ/g). Alcohol provides 7 kcal/g (29 kJ/g), which may explain in part the inadequate food intake by chronic alcoholics. Foods typically contain all three macronutrients in varying proportions. *Carbohydrates* are the primary source of energy for the body; they come from grain, pulses, and root crops. Consumption of insufficient carbohydrates results in symptoms of general malaise. *Proteins* and *amino acids* are needed as the precursors of structural proteins, enzymes, antibodies, some hormones, and other metabolically active compounds, but they can also be used for energy. Insufficient intake of protein, when extreme and prolonged, leads to a condition called *kwashiorkor*.

The requirement for protein is determined by the requirement for essential amino acids and is highest during periods of growth. Amino acids are also needed to replace losses during protein turnover. Some protein is lost in feces, sweat, sloughed skin, hair, and nails. Amino acids are not stored; those consumed in excess of immediate requirements are degraded. The RDA for protein are corrected for 75% efficiency of utilization of mixed protein in the diet.

Lipids are sources of essential fatty acids and are a concentrated source of energy. The greater the proportion of fat, the greater the number of calories in a given portion of food. Lipids function as components of all cell membranes and as precursors of prostaglandins. Failure to consume essential fatty acids results in poor growth and skin lesions. However, essential fatty acid deficiency is not seen in individuals consuming food by mouth.

RDA have been established for those *vitamins* and *minerals* for which sufficient information on requirements is available. For another group of vitamins and minerals, there is less information available, and for these only "Estimated Safe and Adequate Daily Dietary Intakes" have been developed. Values for specific vitamins and minerals are included in appropriate sections of Chapters 8B and 8C. For the remainder, so little information is available that not even an estimate of a safe and adequate intake can be developed.

RDA are frequently used for a purpose not originally intended, i.e., evaluation of adequacy of intake by individuals. Such evaluations require caution, since two questions remain unanswered: how well does a short-term survey reflect the long-term nutrient intake, and what is the significance of short-term suboptimal intake? Given individual variation and differences in bioavailability, the lack of strong correlation between dietary intake and biochemical data should not be surprising. The lower the intake of an essential nutrient in relation to the RDA and the greater the duration

---

*Food and Nutrition Board, National Academy of Science: The Recommended Dietary Allowances, 9th ed. Washington, D.C., National Research Council, National Academy of Sciences, 1980.

†In the nutritional sciences the word "energy" refers to food energy; the unit is the calorie or, in the SI, the joule.

of decreased consumption, the more likely a deficiency will develop. Final confirmation of nutritional status must come through biochemical tests.

FACTORS ALTERING NUTRITIONAL REQUIREMENTS

Several physiological and pathological conditions affect nutritional requirements. Anabolic processes such as growth increase energy and nutrient needs. While some clinical disorders may not change the physiological requirements of patients compared to those of a healthy person, other conditions, such as those associated with reduced gastrointestinal absorption or increased renal excretion, result in the need to increase dietary intake. In patients with other clinical disorders, such as trauma or burns, the requirements are also increased. Finally, drug-nutrient interactions and nutrient-nutrient interactions can result in deficiency in the face of what appears to be adequate intake.

Physiological Factors

Growth. Physiological conditions characterized by expansion of tissue mass, i.e., growth and pregnancy, result in increased requirements. Requirements, however, are highest in the newborn. As the growth rate slows, the requirements decrease and do not rise again until the adolescent growth spurt. These changes are reflected in the RDA, which are highest in units per kilogram of body weight in infants less than six months of age. Increased requirements for most nutrients (e.g., during pregnancy) can be met with increased caloric intake.

Lactation. Lactation is the most nutritionally demanding physiological state. The energy required for lactation is proportional to the volume of milk produced. The vitamin content of milk appears to be more sensitive to maternal intake than the mineral content.

Sex. Requirements for energy and for several nutrients are different in males and females after onset of puberty. Energy requirements per kg body weight are higher in the male than in the female due to different proportions of body fat and the effects of hormones. Iron requirements are substantially higher in females from menarche to menopause. On the other hand, several vitamins in the B group are required in higher amounts by the male than by the female because of the higher caloric intake by males.

Age. Many older people are less active than younger adults and therefore have reduced needs. Lean body mass and basal metabolic rate, as well as caloric need, decline with age; however, requirements for essential nutrients do not. Information on requirements of elderly individuals is generally inadequate. Most available information relates to young adults.

Nutrient-Nutrient Interactions

Nutrients in food can interact with each other or with non-nutrient components. The most thoroughly studied of these interactions is the enhancement of iron absorption by ascorbic acid. Meat, poultry, or fish proteins exert a similar effect, but the specific compound responsible has not been identified. The presence or absence of these interactions between food constituents in a given meal should be taken into account when evaluating iron status. Another example of a nutrient-nutrient interaction is the decrease in absorption of several trace minerals by diets containing a large proportion of cereals. Phytic acid or fiber present in the cereals has been found to bind these minerals so that they are not absorbed.

A similar interaction is that between calcium and protein. Increased fecal excretion of calcium occurs when intake of some proteins increases. Since protein intake of most Americans is higher than the requirement, the American RDA for calcium are also higher than those recommended for international use.

Drug-Nutrient Interactions

Within the past decade much attention has been focused on the effect of drugs on nutritional status. Some drugs alter food intake, some inhibit synthesis, and others interact to reduce absorption or to alter distribution, transport, utilization, storage, or excretion (Table 8A-1). Antineoplastic drugs may cause nausea or gastrointestinal damage and thus reduce intake and absorption. Insulin, steroids, some psychotropic drugs, and certain antihistamines increase food requirements. Tetracyclines and other broad-spectrum antibiotics inhibit vitamin K synthesis by

Table 8A-1. MAJOR DRUG EFFECTS ON NUTRIENTS

Therapeutic Class	Major Drugs	Effect on Nutritional Status
Alcohols	Ethanol	Vitamin B_1 deficiency; impaired vitamin B_6 activation; folic acid deficiency; magnesium excretion
Antacids*	Those containing Al or Ca	Reduced PO_4 absorption
Anticonvulsants and sedatives	Diphenylhydantoin Phenobarbital Glutethimide	Accelerated vitamin D metabolism; accelerated vitamin K metabolism; folic acid deficiency; vitamin B_6 deficiency
Anti-inflammatory drugs	Aspirin	Iron loss caused by GI bleeding
Antitubercular drugs	Isoniazid	Vitamin B_6 deficiency; niacin deficiency
Corticosteroids	Cortisone Prednisone	Accelerated vitamin D metabolism; accelerated vitamin C excretion; increased vitamin B_6 requirement; increased zinc excretion
Diuretics*	Chlorothiazide Spironolactone	Increased K and Mg excretion / Reduced K excretion
Hypocholesterolemic agents	Cholestyramine	Malabsorption of fat, vitamins A, K, D, and B_{12}
Hypotensive drugs	Hydralazine	Vitamin B_6 depletion
Laxatives*	Mineral oil Phenolphthalein	Malabsorption of vitamins A, D, and K / Malabsorption of vitamin D and calcium
Nonabsorbed antibiotics	Neomycin Kanamycin	Lower absorption of vitamins A, D, K, B_{12}, carotene, protein
Oral contraceptive agents	Mestranol Ethinyl estradiol Conjugated estrogens	Folic acid deficiency; reduced calcium excretion

*Categories of drugs which may cause significant mineral depletion in the elderly.

microorganisms in the gastrointestinal tract. Chloramphenicol decreases protein synthesis. Malabsorption can also be caused by laxatives, hypocholesterolemic drugs, antibiotics, and anticonvulsants.

Chronic anticonvulsant drug therapy results in biochemical signs and symptoms of vitamin D deficiency in children and adults. Some anticonvulsants enhance the breakdown of vitamin D by inducing the synthesis of hepatic microsomal enzymes, which convert vitamin D to polar, inactive metabolites. Prolonged use of phenytoin may result in megaloblastic anemia, which responds to therapeutic doses of folate. These and other observations demonstrate the necessity of including information on use of over-the-counter and prescription medication as part of the diet history.

Adaptation to Altered Nutrient Intake

In the face of decreased intake or increased need for nutrients, physiological adjustments take place, particularly if the change comes about slowly and is not too severe. The following are examples of such adaptations.

Consumption of pharmacological doses of vitamin C by pregnant women can lead to rebound scurvy in their formula-fed infants, since formulas were developed for infants of mothers with an average intake of vitamin C. Once the tissues of the fetus are saturated with vitamin C, the excess is metabolized and excreted. The greater the excess, the more efficient the excretion. When intake of the newborn drops suddenly, the high rate of excretion continues and depletes the body of vitamin C. Apparently, the mechanisms for increased excretion change slowly. Infants fed formula seem also to have greater requirement for vitamin E than infants fed human milk, because the type of fat in the formula differs from that in human milk.

In many countries children and adults grow healthy bones while consuming considerably less calcium and protein than recommended for North Americans. Gastrointestinal absorption may adapt by becoming more efficient in the individual consuming less calcium. Increased absorption is also an early adaptation to a lower iron intake.

Effects of Disease on Nutrient Requirements

Although the RDA are usually preceded by a statement that they are intended only to satisfy the needs of healthy people, they are nevertheless adequate for the increased requirements of minor trauma. However, the RDA are inadequate for patients with serious acute diseases or chronic conditions. Many serious diseases also cause anorexia, thus complicating the nutritional picture. Some data on patients with cancer suggest an increased basal metabolic rate; other data suggest no change. The confusion results from inter-individual variation in basal metabolic rate and from the fact that the basal metabolic rate of the patient with cancer is compared with a "normal" range instead of the precancer value for that particular patient. If the basal metabolic rate is indeed increased, there will be an increase in requirements.

Other examples of diseases which alter nutrient requirements are measles, infection, and burns. Measles affects the wall of the gastrointestinal tract and causes leakage of albumin and an associated drop in serum albumin concentration. Infection causes a decrease in the serum concentrations of zinc and iron, but this is the result of redistribution, not deficiency. The most dramatic increase in requirements occurs in patients with extensive burns. This condition will be discussed in more detail in the section dealing with nutritional support of patients.

DEVELOPMENT OF A NUTRITIONAL PROBLEM

Deficiencies can develop in a few weeks or may require several years, depending on the nutrient in question. The lower the intake of an essential nutrient in relation to need, the more rapidly the deficiency will develop. The longer this intake is below the recommendation, the more likely that a deficiency will develop.

A deficiency develops by degree in a series of steps; there may or may not be laboratory tests available to detect each step of deficiency. The rapidity with which a deficiency develops depends on whether or not there are stores or reserves of that nutrient. Stores are an accumulation of nutrients which can be mobilized in times of need. Decreased stores in no way diminish the ability of the body to do work and to meet strain and stress. Reserves are nutrients which are available for use in an emergency, but use of reserves is accompanied by some impairment of health or physiological function. For those nutrients which are not stored, a prioritization of need apparently takes place to preserve, as well as possible, the body pool of nutrient.

For those nutrients which are stored, the initial response to insufficient intake is utilization of body stores. Iron and vitamin A are both stored, but our ability to monitor the development of deficiency in each case is quite different. For iron, serum ferritin provides an estimate of stores. For vitamin A, there is no measure of stores.

As the deficient intake for any stored nutrient continues, either in severity or duration, changes in serum concentration or altered urinary excretion of the nutrient or its metabolites may occur. For example, at an intermediate stage of iron deficiency, serum iron begins to decrease and erythrocyte protoporphyrin will begin to increase. In vitamin A deficiency, a moderate decrease in the serum concentration reflects depletion of stores. These changes suggest that the risk of frank anemia or vitamin A deficiency is now much greater than at the earlier stages and functional changes may now also appear. Finally, if the depletion is sufficiently severe or continues for a longer time, first general and then more specific physical signs of deficiency will appear.

While malnutrition is frequently used as a synonym for undernutrition, it is actually "poor nutrition" that results from either deficient or excess intake. For many essential nutrients, such as most of the water-soluble vitamins, any excess is merely excreted; for others, such as vitamins A and D, excess intake may result in toxicity. Excessive intake for most nutrients is very difficult to achieve by ordinary means. However, with the availability of high potency vitamin-mineral supplements, cases of excessive intake have been reported. Malnutrition can be classified by cause, type, degree, duration, or outcome. Cause is characterized as primary or secondary; type as excess or deficiency; degree, in general, as mild, moderate, or severe (more specifically as depleted stores, biochemical lesion, functional change, and structural lesion); duration is characterized as acute, subacute, or chronic; and outcome as reversible or irreversible.

NUTRITIONAL ASSESSMENT

A nutritional assessment is made by observing physical signs, making anthropometric measures, collecting dietary information, taking medication and medical histories, measuring metabolic and immunological parameters, and evaluating these data in relation to reference standards. Nutritional assessment can be done at several levels. The tests and measurements used for surveys of large populations are different from those used for individuals during health maintenance or for acutely ill patients. Assessment of hospitalized patients varies from minimal screening of all admitted patients to an extensive evaluation of selected patients by a metabolic support service.

Physical Examination

Physical signs are the last evidence of deficiency to develop and the first to disappear when correction of the deficiency has begun. Physical signs are rarely definitive for, diagnostic of, or specifically associated with deficiency of a single nutrient. Observers frequently fail even to agree on the presence or absence of physical signs. Environmental factors (such as excessive heat or sun), poor personal hygiene, and cultural factors can cause or contribute to physical signs associated with malnutrition. Manifestations may also vary with age. For example, in children one symptom of scurvy is painful, swollen joints due to hemorrhages under the periosteum. In the elderly, ecchymoses appear. Bleeding gums, the classic sign of ascorbic acid deficiency, do not occur in young infants and edentulous adults. Table 8A-2 summarizes the physical signs suggestive of malnutrition.

Anthropometry

Anthropometric measurements can provide information on general nutritional status. Although such measurements require only simple instruments, they are frequently not done. Among

Table 8A-2. PHYSICAL SIGNS SUGGESTIVE OF MALNUTRITION

Organ	Signs Associated with Malnutrition	Deficient Nutrient
Hair	Lack of luster, thin, sparse, easily pluckable, dyspigmentation, flag sign	Energy, protein
Face	Nasolabial dyssebacea	Riboflavin
	Moon face	Energy, protein
Eyes	Papilledema	Vitamin A
	Pale conjunctiva	Iron
	Poor dark adaptation, Bitot's spots, conjunctival xerosis	Vitamin A
	Angular palpebritis	Riboflavin, pyridoxine
	Intraocular hemorrhage	Ascorbic acid
Lips	Angular stomatitis	Riboflavin
	Cheliosis	Niacin and/or riboflavin
Tongue	Sore tongue	Vitamin B_{12}
	Scarlet and raw tongue	Nicotinic acid
	Magenta tongue	Riboflavin
	Glossitis	Folic acid, niacin, pyridoxine
Teeth	Mottled enamel/caries	Fluoride
Gums	Pale	Iron
	Spongy, bleeding gums	Ascorbic acid
Glands	Thyroid enlargement	Iodine
Skin	Purpura	Vitamin K
	Hyperpigmentation	Energy, B_{12}, folic acid, niacin
	Dry rough skin	Hypervitaminosis A
	Yellow discoloration	Carotene
	Desquamatory dermatitis	Biotin, essential fatty acids, zinc, B_6
	Follicular hyperkeratosis	Vitamin A
	Petechial hemorrhages	Ascorbic acid
	Pellagrous dermatosis	Nicotinic acid
	Scrotal and vulval dermatosis	Riboflavin
Nails	Koilonychia	Iron
Subcutaneous tissue	Fat decreased/increased	Energy deficit/surfeit
	Edema	Energy deficit
Muscle and skeletal system	Wasting	Energy
	Craniotabes, frontal bossing, knock knees, beading of ribs, bow legs	Vitamin D
	Musculoskeletal hemorrhage	Ascorbic acid
Cardiac system	Enlargement, tachycardia	Thiamin
Nervous system	Mental confusion, motor weakness, calf tenderness	Thiamin

the large number of anthropometric measurements, four are most useful: height, weight, skinfold measurements, and arm circumference.

In the adult, *height for age* reflects total previous history as well as genetic potential. In the child, length or height for age reflects adequacy of caloric intake and genetic potential. *Velocity of growth or incremental growth* (growth per unit time) is a more sensitive indicator of calorie adequacy than height/age, since the first observable response to undernutrition is frequently a decrease in growth rate.

Changes in *body weight* reflect caloric intake. Insufficient intake results in weight loss, while surfeit leads to weight gain. Comparison of the adult patient's weight to the Metropolitan Life Insurance tables as a measure of appropriate body weight may be misleading; instead, comparison to the usual weight of the patient should be used to assess weight loss (or gain). A current weight which is 85–95% of the usual weight suggests mild malnutrition, 75–84% suggests moderate malnutrition, and less than 75% suggests severe malnutrition. The rapidity of weight loss is as important as the actual amount, but a 10% loss in body weight is always clinically significant.

Energy needs. Calorie balance is achieved when intake equals expenditure. If calorie intake exceeds expenditure, the patient is in positive balance and excess calories are stored as fat. If calorie intake is less than expenditure, the patient is in negative balance and deficient calories are mobilized from stored fat and body protein. Energy needs are the sum of the energy required for basal needs, activity, and specific dynamic action (the thermogenic effect of food). An estimate of basal energy expenditure (BEE) in kcal/d can be made using the Harris-Benedict equations.[7,20] For men,

$$BEE = 66 + (13.7 \times wt \text{ in kg}) + (5 \times ht \text{ in cm}) - (6.8 \times age)$$

For women,

$$BEE = 655 + (9.6 \times wt \text{ in kg}) + (1.7 \times ht \text{ in cm}) - (4.7 \times age)$$

Another but more crude method is to estimate basal needs as 1 kcal/kg wt/h for men and 0.9 kcal/kg wt/h for women. The estimate is increased by 10–30% for multiple fractures, 20–50% for sepsis, and 90–110% for burns. A further increase of 10, 15, or 20% can be added for the patient's physical activity.[45]

To understand the effects of various disease processes on energy requirements, a more accurate method is needed. One proposed technique for predicting energy expenditure is based on heart rate and oxygen consumption.[44] However, for most hospitalized patients activity is limited and the degree of accuracy provided by the mentioned procedures is unnecessary; therefore the Harris-Benedict equations suffice.

The amount of reserve energy stored as fat is also important. Methods are not widely available to estimate reserves, and body weight does not accurately reflect them. Among the experimental techniques to estimate fat stores, ultrasound has the greatest potential for widespread use, while skinfold measurements are presently the most frequently used method.

Body composition. Using weight to assess energy reserves is insufficient, since weight does not provide information on body composition. Below age 6 or 7 years, total body weight leads to underestimation of the amount of fat; during adolescence, weight frequently leads to overestimation of the amount of fat.

For purposes of nutritional assessment the body is divided into six compartments: fat (primary energy stores), skin and skeleton, extracellular mass, plasma protein, visceral protein mass, and skeletal muscle (also referred to as somatic protein mass). Anthropometric measurements are used to estimate energy stores (skinfolds) and skeletal muscle mass (arm circumferences and the calculated value, arm muscle area). Biochemical tests are used to determine depletion of visceral protein as well as some aspects of skeletal protein and will be discussed later.

Estimation of body fat. Clinically, body fat is estimated primarily from skinfold (SF) measurements. Ultrasound, computed tomography, prompt gamma neutron activation, and underwater weighing have been used to provide more accurate estimates of muscle and fat reserves, but at present these are primarily research techniques.

Estimation of body fat from *skinfold measurements* is based on the assumption that approximately one half of the total body fat is deposited as subcutaneous tissue loosely attached to underlying muscles. One to four skinfold measurements are generally used. To make the actual measurements, the skin and underlying fat are lifted up with the fingers into a fold and the thickness of the fold is measured with a caliper. The triceps skinfold thickness is most commonly measured, but a better characterization of individual overall fatness is obtained by measuring both the subscapular and triceps skinfolds and summing the two. Subcutaneous depots change slowly and reflect chronic undernutrition if depleted. In patients with edema, however, the measurements are invalid. Age and sex-specific reference values are available for the sum of the two skinfolds.[18]

Because of the need to evaluate the effects of chronic wasting disease and to select management regimens correctly, there is a need for a simple, rapid, and reliable method to measure lean body mass and total body fat. One proposed method entails the measurement of total body electrical conductivity and is based on the principle that conductivity of lean tissue is far greater than that of fat.[33] More research and comparative studies are needed to evaluate this method completely.

Mid-upper arm circumference (MAC) of the nondominant arm is an objective measurement of thinness and correlates with weight. MAC can also be used to calculate arm muscle area, a value used to evaluate changes in somatic protein mass.

Nutritional History

A nutritional history usually involves an estimate of present intake or an estimate of past intake in terms of foods actually eaten or usually eaten, as well as social, cultural, economic, and medical data. These data can be used to estimate the patient's risk of deficiency and to provide information that can serve as the basis of a care plan.

Two methods of estimating intake are diaries and recalls. *Diaries* of food intake are maintained for varying intervals, most commonly for three to seven days. The patient is asked to record everything eaten or drunk. Serving size may be estimated, or the patient may be asked to use measuring cups and spoons to measure the servings. *Recalls* aim at eliciting information on actual intake, as remembered at the time of the interview. Most common is the 24-h recall, defined as a record of food consumed in the immediate past. The main disadvantage of this approach is the inability of the patient to remember the foods consumed and their correct quantity. Recall data do not correlate with physical and biochemical findings on an individual basis, since the former measure recent intake and the latter long-term intake.

Once the list of foods is obtained, the dietitian codes and enters the data into a computer which contains a database of the nutrient composition of various foods based on USDA Handbook #8 or other tables of nutrient content of foods. The printout provides the dietitian with the estimate of the total amount of each nutrient of interest and the per cent RDA for that nutrient. If a computer is not available, the estimates can be done manually. The most commonly used standards to determine adequacy are the RDA described earlier. These guidelines, like others, contain the statement that they are to be used to estimate population needs and should be used cautiously to evaluate the diet of an individual.

Another method of evaluating adequacy is to categorize the patient's intake by food groups and compare these to the recommended number of servings. The grouping of specific foods is determined by nutrient content. The basic groups are (1) milk and milk products, (2) meat or protein group, (3) fruits and vegetables, (4) grains and pulses, and (5) other, i.e., foods which may be added to meet total energy needs. Adults should consume two servings of groups 1 and 2 and four servings of groups 3 and 4 each day, plus additional servings from these groups to meet their energy needs. Recommendations regarding the number of servings are different for children and for pregnant or lactating women.

Questions regarding frequency of intake of specific foods are used as a crosscheck of data collected by other methods and provide information about usual intake in terms of how often key foods are consumed. Food frequency provides another bit of information suggestive of relative risk and long-term pattern of intake. The nutritional history should include questions about prescription and nonprescription drugs currently in use, recent weight loss or gain, presence of chronic illness, dietary habits, and allergic reaction to food.

Biochemical Data

Biochemical markers of food intake are needed to confirm dietary information. For some nutrients, these markers are available; for example, urinary excretion of vitamin C indicates adequacy of intake. For many other nutrients, these markers are not available.

Biochemical measurements represent the most objective and sensitive assessment of the nutritional status of an individual and reflect dietary intake, degree of storage and saturation of tissues, or functional changes. There is no single master test to determine nutritional status. Multitest systems should be used, but as yet there are few data to define these systems. Biochemical procedures are used to measure (1) nutrient concentrations in blood, (2) urinary excretion rate of nutrients, (3) abnormal metabolic products in blood or urine, (4) changes in blood components or enzyme activities, and (5) changes after a test load or the administration of stable isotopes.

The clinical chemist is of paramount importance in this aspect of nutritional assessment. Clinical chemists should work with nutritionists to develop the biochemical markers needed to confirm food intake, to develop multisystem tests, and generally to be at the forefront of developing new methodology.

Protein-calorie malnutrition (PCM). For the past 25 years, there has been a search for a method to indentify mild to moderate undernutrition in children. Such a method would allow the clinician to estimate the severity of the deficit and the prognosis for recovery. More recently, increased interest in assessing nutritional status of hospitalized adult patients has spurred new research in methodology. Overall change in protein (nitrogen) status is estimated from nitrogen balance data. Of the two main components of protein reserves, skeletal muscle (approximately 60% of the body cell mass) is estimated from anthropometric measurements and visceral protein (20–30% of the body cell mass) is estimated from biochemical measurements.

Nitrogen balance. Clinically, adequacy of protein intake is mainly determined by measurement of nitrogen balance (nitrogen intake minus nitrogen losses via urine, feces, and skin).[1] Nitrogen intake is calculated from protein intake by dividing the intake in g/d by 6.25 (based on the assumption that all protein contains 16% nitrogen). The major route of nitrogen loss is urine (70–90%). The major catabolite of protein in the urine is urea, and the amount of urea nitrogen is closely correlated with total urinary nitrogen and protein intake. Ammonia and creatinine excretion are fairly constant in adults eating either high- or low-protein diets. There are changes, however, during fasting.

For precise determinations of nitrogen balance, fecal and urine nitrogen losses are measured directly. In most hospitals, a factor of 4 g/d is used as an estimate for the sum of fecal losses, dermal desquamation, and urinary nonurea nitrogen.[29] If there are no fecal losses, the factor is taken as 3 g/d.

$$\text{Nitrogen balance} = \frac{\text{protein intake, g/d}}{6.25} - (\text{Urinary urea N, g/d} + 3 \text{ or } 4)$$

(Intake)   *(Excretion)*

A positive nitrogen balance is indicative of net protein synthesis; a negative balance indicates net protein catabolism. In adults, negative nitrogen balance is associated with starvation, fever, and illness. In growing children, negative nitrogen balance can be more serious than in adults because growth can be compromised.

There are several sources of error in estimating nitrogen balance; usually intake is overestimated and output underestimated. Exudation from burns, oozing from wounds, or copious sputum can result in significant additional losses. Knowledge of the patient's previous nutritional status is essential to interpretation of data. Severely malnourished patients receiving diet supplementation can be in strong positive nitrogen balance before their nutritional status returns to normal. While nitrogen balance is therefore useful in formulating nutritional therapy and as an indicator of metabolic stress, it is not an indicator of nutritional status. The goal of therapy for depleted patients is a positive nitrogen balance of 4–6 g/d.

Urinary urea nitrogen can also be used as an indicator of the level of catabolic stress. As the severity of stress increases and catabolism becomes pronounced, the urinary excretion of urea nitrogen increases. Bistrian[5] has proposed a "Stress Index" to estimate the severity of the stress:

$$\text{Stress Index (SI)} = 24\text{-h urinary urea nitrogen} - (\tfrac{1}{2} \text{ nitrogen intake} + 3)$$

In patients with an SI < 1 there is no significant stress. An SI between 1 and 5 denotes moderate stress and > 5, severe stress.

Visceral protein mass. Visceral proteins are depleted as the protein status of the patient worsens. Protein undernutrition (kwashiorkor in children and kwashiorkor-like syndrome in adults) is characterized by depletion of the visceral protein mass. Marasmus, a severe deficiency in energy stores, is characterized by depletion of the skeletal muscle mass with relative preservation of the visceral component.

Concentrations of the various serum transport proteins are used to indicate adequacy of the visceral component, based on the observation that nearly every transport protein evaluated has been shown to decrease as patients lose body mass. The concentration of a serum protein is the result of the rate of synthesis, utilization, excretion, and transport of the protein as well as the hydration of the patient. Administration of whole blood or certain fractions of blood also affects the concentration of serum proteins.

Although serum *albumin* is widely measured to assess protein nutrition, it is a poor indicator of marginal protein depletion since it is depressed only in severe undernutrition. Only a slight drop in serum albumin has been observed in patients consuming 20 g protein/d for 5 weeks. (Serum albumin values are, however, decreased in patients with infections.[15]) Clinical nutritionists differ on the values that are indicative of mild or moderate protein undernutrition, but they generally agree that in the absence of liver or kidney disease, <2.1 g/dL serum albumin is indicative of severe kwashiorkor-like depletion. Patients who have low serum albumin at admission have frequently been shown to stay in the hospital significantly longer than those with normal values.[2] Serum albumin is lower in the elderly than in young adults by a slight but statistically significant amount.[14]

Plasma *transferrin* has been proposed as a more sensitive indicator of protein status than albumin, because of its smaller pool (mean plasma pool = 5.29 g) and shorter half-life (8.8 d). In the first week after injury, transferrin concentration cannot be used, because transferrin is an acute-phase reactant. After the acute-phase reaction has subsided, measurements of transferrin may be of value. Plasma transferrin concentration falls during catabolic stress and rises in patients with iron deficiency anemia, during pregnancy, and after blood loss. Many texts discussing nutritional assessment suggest calculating serum transferrin from total iron binding capacity (TIBC).[7] The most common formula[7] is

$$\text{Transferrin} = 0.8 \text{ TIBC} - 43$$

Others[18] suggest that each hospital develop its own predictive equation since there are a variety of procedures available for measuring both transferrin and TIBC. In the absence of a proven clinical method for defining nutritional status, clinicians are seeking a nutritional test which will accurately predict clinical outcome. On the basis of limited testing, a weighted index of serum albumin and serum transferrin appears to be an effective combination for selecting candidates for nutritional support. Verification of the use of such an index would require further testing with other populations.

Prealbumin and *retinol-binding protein* have also been proposed as indicators of protein status. Due to their short half-lives, prealbumin (2 d) and retinol-binding protein (12 h) are sensitive to changes in protein intake. However, prealbumin concentration also drops whenever there is a sudden increase in protein synthesis. Also, a rapid fall in prealbumin immediately preceeds the final, irreversible collapse of the malnourished patient. Changes in retinol-binding protein occur with minor stress, thus obviating its use in nutrition assessment. Retinol-binding protein is cleared by the kidney, and the blood concentration is therefore affected by kidney disease and possibly by some nephrotoxic antineoplastic agents.

The status of other nutrients also affects serum proteins. Concentrations of albumin, transferrin, and prealbumin are depressed in patients with *zinc deficiency* as compared with controls, and values rise after a short period of zinc supplementation.[3]

Somatic protein mass. Twenty-four–hour urinary creatinine excretion is the most widely used biochemical marker for estimation of body muscle mass. Undernourished patients who have lost muscle mass excrete less creatinine than well-nourished patients; during recovery, urinary creatinine excretion increases. There is a high correlation between values for lean body mass, determined by [40]K counting, and values calculated from urinary creatinine excretion.[12] The relationship is linear over a wide range of creatinine values, and there is no obvious influence of sex. Urinary creatinine excretion is most accurate as an index of lean body mass if 24-h urine specimens can be collected on three consecutive days. Renal function must be normal to obtain valid results.

The creatinine height index (CHI) is the ratio of 24-h creatinine excretion of a patient to that of well-nourished controls of the same height.[41] For well-nourished children, the ratio is close to 1.0. Children with protein-calorie malnutrition have CHI's ranging from 0.25 to 0.75. More recently, the use of CHI has been proposed as a measure of skeletal muscle mass for adults.[6] As reference standards, "expected" 24-h excretion values for various heights have been calculated from the mean creatinine excretion of healthy young adults and from the weight determined from Metropolitan Life Insurance tables, based on medium frame. This method, however, fails to take into account the errors of the standard and changes in creatinine excretion due to age. Therefore, great care must be used in interpretation of the CHI.

Urinary hydroxyproline. Urinary hydroxyproline is derived almost entirely from collagen and therefore reflects the rate of collagen catabolism. Hydroxyproline released during collagen breakdown cannot be reused and is excreted. Hydroxyproline excretion has been suggested as an indicator of protein status, since the amount excreted is decreased when a low protein diet is consumed.[8] Total hydroxyproline excretion is high during periods of rapid bone turnover, such as in patients with Paget's disease of bone.

Urinary 3-methylhistidine (3MH). Measurement of 3MH has also been proposed as a reliable index of protein breakdown in muscles.[42] 3MH is a unique amino acid present in myofibrillar proteins. Methylation of histidine residues occurs after synthesis of actin and myosin peptide chains. Once released from muscle protein by catabolism, the amino acid is neither reutilized nor metabolized further but is excreted in the urine either as the amino acid or its N-acetyl derivative. Dietary 3MH from meat adds to the body pool and causes an increased and variable excretion, which must be taken into consideration when data are evaluated.

Several factors alter 3MH excretion in the healthy individual. Adult women excrete less 3MH than adult men at all ages, probably due to differences in total muscle mass.[4] Elderly men and women excrete significantly less 3MH than younger men and women when excretion is expressed per kilogram of body weight, but there is no significant difference when the excretion is expressed per gram of creatinine.

The presence of disease or other conditions also alters excretion patterns. Children recovering from protein-calorie malnutrition initially have low values that reflect the reduced muscle mass. The values increase during recovery.[43] Skeletal trauma, catabolic steroids, and sepsis also produce increased excretion.[28]

Measuring 3MH serially in patients receiving total parenteral nutrition (TPN) makes it possible to assess nutritional changes. Excretion of 3MH by patients receiving TPN increases gradually during the course of therapy and is nearly twice the pretreatment level by the end of the third week.[23] A urinary 3MH/creatinine ratio has been suggested as a potentially useful parameter to identify clinically stressed premature infants.[37]

Host defenses. Infection is a leading cause of the morbidity and mortality of nutritionally debilitated individuals. Acute infection may also be a factor in precipitating acute nutritional deficiencies.

Mucocutaneous integrity is interrupted by deficiencies of several vitamins and minerals, some of which are commonly consumed in less than adequate amounts by certain segments of the population. Breaks in this tissue barrier can result from deficiencies of vitamin A, niacin, riboflavin, folic acid, vitamin B_{12}, pyridoxine, ascorbic acid, iron, and protein.

A significant impairment of *cell-mediated immunity* occurs in patients with protein-calorie malnutrition. Manifestations of impaired cell-mediated immunity are reduced lymphocyte blastogenesis, lymphopenia, diminished number of circulating T-lymphocytes, tonsils of reduced size, decreased reactivity to delayed skin test antigens, and a less effective response to nonspecific mitogens. Whether the specific cause of these defects is a deficiency of either protein or calories is difficult to ascertain because intake of insufficient amounts of protein and/or calories results in the concurrent intake of insufficient amounts of several vitamins and minerals. Protein-calorie malnutrition is in actuality a multiple deficiency syndrome; therefore, the specific cause might be one or another of the deficiencies which occur concurrently. According to evidence currently available, iron deficiency can account for most of the changes, although zinc deficiency also produces some of the aberrations in cell-mediated immunity. For nutritional assessment, the most common measures of cell-mediated immunity are the total number of lymphocytes and skin testing. Impairment of delayed hypersensitivity can be expected when the serum albumin falls below 3.5 g/dL in patients with kwashiorkor-like syndrome. In patients with marasmus, immune function is typically intact until their weight falls below 85% of the reference standard. With nutritional repletion, the defects are reversed.

Serum immunoglobulin levels are usually normal or elevated in children with various forms of undernutrition. Antibody responses to some antigens, such as typhoid, diphtheria, and influenza, are depressed but are normal or increased for other antigens, such as tetanus toxoid, attenuated poliovirus, and attenuated measles virus.[38] Nutritional supplementation and clinical improvement are associated with rapid recovery of immune responsiveness except in the small-for-gestation-age newborn in whom the defect persists for several months and perhaps years.[10]

Lipids

Essential fatty acid deficiency can be established by determining the concentration of 5,8,11-eicosatrienoic acid in serum or plasma lipids. Deficiency of essential fatty acids has been seen only in patients who are not consuming food by mouth. Serum lipid profiles are frequently performed to measure cholesterol, triglycerides, and apolipoproteins A and B and to identify types of hyperlipidemia, some of which are treated by dietary measures.

Vitamins

Biochemical tests of vitamin status may reflect recent dietary intake or metabolic changes that occur in a deficiency state. For instance, plasma ascorbic acid or plasma vitamin A concentrations reflect dietary intake or nutrients in transit from one organ to another. On the other hand, measurement of erythrocyte transketolase activity or erythrocyte glutatathione reductase activity reflects adequacy or deficiency of thiamin and riboflavin, respectively. Usually determination of vitamin status is performed only during in-depth assessment of individuals and is based on clues provided by the patient's history or by clinical and dietary information. For example, the presence of certain diseases or chronic use of certain drugs might suggest a potential for deficiencies. Roe[35] recommends that the nutritional profile of alcoholics should include assessment of niacin, ascorbic acid, and vitamins B_6 and B_{12}.

When laboratory tests are used in international, national, or local surveys, factors such as difficulties in drawing blood or collecting urine, the transport of samples, differences in technique, and variations in reference range, as well as expected deficiencies, must be considered in selecting the tests. The goal of a nutrition survey (i.e., assessment of a population) is completely different from the goal in assessment of an individual. In the main, deficiencies of vitamins in hospitalized patients are secondary to deficiencies of energy and protein. Use of multivitamins, season of the year, and social class are among the factors which should be taken into account when evaluating results of tests for vitamin status. See Chapter 8B, Vitamins, for further discussion.

Minerals

Assessment of nutritional status of trace minerals is still in the early stages of development, since appropriate instrumentation and techniques were not available until recently. As with vitamins, assessment of nutritional status with regard to particular minerals should be based on indicators such as personal appearance or from histories that indicate high risk in a particular subset of the population. Interpretation of results of individual tests can be complicated by the presence of disease. For example, values for serum iron and zinc are lower in patients with infections. Hair analysis also may have some value in the assessment of minerals (see Hair Analysis).

Of all the essential minerals, only four are usually measured in surveys—calcium, phosphorus, magnesium, and iron. Of these, calcium and iron intake are frequently severely short of the RDA. According to recent data, nearly half of the population consumed 70% or less of the 1980 RDA for calcium; for iron the figure was one third.[29] See Chapter 8C, Trace Elements, for further discussion.

Biopsy Samples

Until iron stores could be estimated from serum ferritin values, staining a bone marrow sample to determine the amount of iron present was the only way to estimate iron stores. It remains the definitive method, but it is not used in nutritional assessment or in the diagnosis of simple iron deficiency. Megaloblastic anemia due to vitamin B_{12} or folate deficiency can be confirmed by bone marrow biopsy, which shows the presence of erythroid hyperplasia and megaloblastic changes in the erythroid precursors.

Adipose tissue. Adipose tissue can be obtained by needle aspiration, and methods have been developed to measure the number and size of adipocytes in the sample. Total body fat can be measured using tritiated water and total adipocyte number by dividing the total body fat by the mean cell size. However, for the immediate future this technique will remain in the armamentarium of the researcher and not the practitioner.

Hair. Many parameters of hair have been studied. Microscopic examination is tedious and time consuming, taking 1–2 h per subject. Many trace minerals can be measured in small samples of hair, but analysis is complex and still in the developmental stage. Hair analysis appears to have promise for assessment of zinc status.[19] Unfortunately, hair analysis has appealed to pseudonutritionists, who perform the test without taking into account its limitations and who extrapolate results far beyond what has been scientifically demonstrated.

NUTRITIONAL ADEQUACY IN NORTH AMERICA

During the past two decades, several nutrition surveys of samples of the total population of North America and selected subsets of the population have been conducted.[24,31,39,40] Before the first survey, some cases of severe protein-calorie malnutrition (PCM) were reported among the American Indian population of the southwest. More recently, isolated cases of mild to moderate malnutrition have been reported in infants whose mothers received improper education on breast feeding or in young children fed bizarre diets. Severe malnutrition has been seen only in hospitalized patients in whom the underlying disease process has contributed substantially to the depletion.

Surveys that indicate many individuals consume an adequate intake on a given day do not provide information on intake of a specific nutrient over an extended period of time. Long-term substandard intake is reflected only in low values of biochemical parameters. Those individuals with substandard intake but adequate biochemical values represent the population at risk for deficiency. Nonhospitalized individuals at risk include infants, adolescents, pregnant women, lactating women, and the elderly. Hospital patients at high risk include those who are grossly underweight, those who are grossly overweight, alcoholics, patients with malabsorption syndromes or extensive burns, or those on renal dialysis. Limited surveys, i.e., those looking at nutritional status of specific population groups, may actually define the problem better than general surveys. For example, one survey revealed that pregnant teenagers frequently had inadequate intake of calcium, iron, vitamin A, and total calories.[24] Another limited survey showed that 26.7% of poor preschoolers had low transferrin saturation.[31] These details might have been lost in a large general survey.

Optimal nutritional status, however, is more than prevention of deficiency disease. Of the ten leading causes of death in the United States, six (heart disease, stroke and hypertension, cancer, diabetes, arteriosclerosis, and cirrhosis of the liver) have been statistically linked to diet. Epidemiological studies provide data which show a positive relationship between cancer of the colon and annual per capita consumption of specific foods such as beef. Cereal and potato intake are highly negatively correlated with the incidence of colon cancer. Changes in the diet of the American population play a role in the development of the diseases mentioned above. Consumption of complex carbohydrates, for example, has decreased. As the total consumption of carbohydrates decreased from 56% of the food energy to 46%, the amount of simple carbohydrates consumed as sugars and syrups has increased. In addition, fat consumption, primarily as fatty meats, has increased to approximately 42% of the total caloric intake.

The general recommendations for improving the diet of the American population include increased consumption of fruits, vegetables, and whole grains; decreased consumption of refined and processed sugars; decreased consumption of fat; decreased total energy intake; increased consumption of leaner meats and fish; and decreased consumption of salt and food high in salt. The percentage of energy from protein should remain at the present level, but when the total caloric intake is reduced, the total amount of protein must also be reduced. Americans as a group consume protein in excess of needs, without any known benefit. A wide variety of foods should be consumed in order to have a high probability of consuming adequate amounts of bioavailable nutrients. Salt should be used in moderation. Fruits, vegetables, and grains add fiber as well as complex carbohydrates to the diet; thus, their intake is encouraged. For the past few years exercise has also been recommended to aid in weight control.

Worldwide, many people live in conditions where the availability and intake of food are inadequate. Protein may be of poor quality as well as insufficient in quantity; iron is consumed in a less bioavailable form. Poverty, war, floods, crop failure, and parasites are only a few of the problems which further contribute to malnutrition. Malnutrition also results in decreased ability to work, think, absorb, and metabolize, all of which affect the person's ability to buy food and can lead to further malnutrition.

The most extensive nutrition problems are protein-energy malnutrition, iron deficiency anemia, and vitamin A deficiency. Supplementation or fortification of the diet can improve physiological measures and in many cases increase productivity. Any program addressing the problems of malnutrition must be tailored to meet the need as defined by the target population.

NUTRITIONAL SUPPORT IN PATIENT MANAGEMENT

There are two basic methods of nutritional support of patients. *Enteral* methods use the gastrointestinal (GI) tract and include supplemental oral feeding, feeding through tubes placed into the GI tract, and feeding via portals into the GI tract. *Parenteral* methods are either supplementary (also called peripheral venous alimentation) or total (TPN, also called central venous alimentation). Typically, patients who need nutritional support have been malnourished for a long period of time, may have a condition which is exacerbated by sepsis, may have biochemical deficiencies of one or several nutrients, and may have requirements substantially above the RDA.

Supplements. Oral supplements should be used with patients who are not consuming adequate calories and protein but are still able to eat and absorb nutrients. Supplements should be high in energy and protein, easily ingested, and palatable. Proprietary preparations are on the market and can be served as hot or cold drinks, ice cream, or puddings. Commercial products are convenient and easily prepared; homemade formulas allow for additional alternatives. Patients with malabsorption problems need formulas which are partially digested or which contain simple nutrients that are readily absorbed. Patients with bowel fistulas, inflammatory bowel disease, or acute pancreatitis may also require semipurified diets.

Enteral feeding. Patients who cannot eat or drink adequate amounts to meet their needs should receive enteral feeding by nasogastric or nasoduodenal tubes. This method of management is used for low-birth-weight infants and for patients with severe illnesses, such as trauma, burns, or surgery, which reduce appetite or make the patient too weak or unable to eat (e.g., unconscious or delirious patients). Special formulas are available for infants with inborn errors of metabolism such as phenylketonuria.

Patients and physicians may resist tube feeding because of unpleasant past experiences. Newer tubing is smaller in diameter and can be left in place for weeks without significant discomfort to the patient or damage to the mucous membranes of the nose and throat. With these tubes in place, the patient can cough, discharge oral secretions, or eat.

There are many commercial formulas available.[11] However, many of the formulas used for enteral nutrition are not palatable orally. Proteins in commercial preparations may be intact or partially digested; carbohydrates may be isolated from foodstuffs or simple sugars, and lipids may be from milk, vegetables, or medium chain triglycerides. Nutritionally complete formulas contain vitamins and minerals; others contain only single nutrients. Patients should be monitored for bloating, nausea, fluid and electrolyte balance, glucosuria, and hyperglycemia, as well as hepatic and hematological status. Many formulas used for tube feeding do not contain sufficient water to allow for the renal excretion of urea and electrolytes; thus there is the possibility that patients can become dehydrated, hyperosmolar, and hyperglycemic. Blood glucose should be monitored several times a week. In patients with liver dysfunction, hyperammonemia and hepatic encephalopathy are possible complications. Edema is a relatively common complication. Formulas should be started at a low concentration and should be gradually increased. Most patients tolerate feedings by slow drip infusion better than periodic bolus feedings.

Metabolic alterations in patients receiving enteral nutrition are generally less severe than in patients receiving TPN. Tube feeding is limited to patients with at least partially functioning GI tracts. Major advantages are the avoidance of possible sepsis from the catheter required for TPN and having the GI tract act as a buffer.

Tube *gastrostomy* or *jejunostomy feedings* are also possible. The general principles are the same as for nasogastric feedings. Intermittent feeding of hyperosmolar solutions is associated with large intraluminal shifts of fluid. Diarrhea can result when the initial infusion rate is too rapid, the formula is hyperosmolar, or the patient is lactose intolerant. Jejunostomy feeding has been recommended for patients during the early postoperative period after gastric stapling or after extensive surgery on the upper GI tract, pancreas, or biliary tract.

Parenteral nutrition. Patients with GI disease can be put into positive nitrogen balance by infusion of a hypertonic nutrient solution through a catheter inserted into the subclavian vein and threaded to the superior vena cava. Hospitals where enteral and parenteral feeding are extensively used have a metabolic support team of clinicians, scientists, dietitians, pharmacists, and nurses to provide the expertise needed to manage the patients successfully on either regimen.

Parenteral nutrition (PN) is often indicated as primary therapy for patients with conditions such as bowel fistula, pancreatitis, acute renal failure, impending hepatic coma, inflammatory bowel disease, and radiation enterocolitis. PN is supportive nutritional therapy for comatose patients; for patients with neoplasms, transient bowel obstruction, or anorexia nervosa; for infants with failure to thrive; and for burn patients. PN is contraindicated when the patient is in good nutritional status and will be able to eat in a few days or is in the terminal stages of illness. The use of PN is also questionable in patients with systemic sepsis or those with fluid overload.

Hypertonic *glucose* can be administered in sufficient quantity to provide caloric needs, but the patient will achieve positive nitrogen balance only when essential amino acids are given concurrently. Therefore glucose is not used alone. *Amino acids* are available either as protein hydrolysates or as crystalline amino acids. Protein hydrolysates are not used as effectively as are pure amino acid preparations and may cause mild hypersensitivity reactions. However, pure amino acids occasionally cause metabolic acidosis. Hyperammonemia can occur in patients receiving either. Both the absolute amounts of essential amino acids and the total amount of amino acids are important for protein metabolism. Arginine, alanine, glutamic acid, and proline are necessary for optimal utilization of all amino acids, although they are nonessential amino acids. Approximately half the total nitrogen should be in the form of essential amino acids to meet the increased needs of protein synthesis.

Several specialized amino acid solutions have been developed for use in patients with clinical conditions in which amino acid metabolism is changed or mechanisms of nitrogen excretion are impaired. Albumin can be added to the solution to increase the serum concentration if it is less than 2.5 mg/dL. Albumin infusion repletes the albumin pool more rapidly than using amino acids alone. Table 8A-3 gives the composition of commonly used amino acid solutions.

In addition to amino acids, the TPN solution must contain an adequate supply of calories, usually as glucose. These solutions have 25–35% glucose, provide 2550–3570 kcal/3-L bag, and are diluted with an amino acid source to give the appropriate calorie to gram nitrogen ratio. Parenteral solutions for peripheral infusion have 10% glucose. The caloric density of these solutions depends on whether or not lipids are also used. Since fat emulsions are now used in the United States, the energy requirements of most patients are easier to meet. Use of the emulsions has increased the flexibility for implementing programs of TPN and eliminates the problem of essential fatty acid deficiency.

Vitamins are either added to the solution or given intramuscularly. There are several preparations available which must be combined or given on alternate days to meet the requirements. Vitamin K is usually omitted from preparations of multivitamins because of incompatibility with the solutions.

Potassium and *phosphate* are needed in larger amounts than usual in TPN solutions to meet the needs for protein synthesis. Hypophosphatemia is averted by adding 10–15 mEq of phosphate/L. *Calcium gluconate* and *magnesium* are added to prevent the neuromuscular irritability, tremor, and seizures seen in chronically malnourished patients. Other *minerals,* such as zinc, chromium, and copper, have been added recently because of reported cases of deficiency. One case of molybdenum deficiency has also been reported. Preterm infants receiving TPN present special problems. If they are maintained on TPN for long periods, they develop pathologic fractures and rickets, although infants given large amounts of calcium develop fewer fractures. A Ca/P ratio of approximately 3.0 has been suggested as a safe compromise to prevent the complications of phosphate deficiency and the chronic problem of fractures and rickets due to calcium deficiency.[25] Ferrous citrate addition has been suggested to increase the concentration of iron.[36] Interactions between various components of TPN solutions are not well studied.

Table 8A-3. COMPOSITION OF FREQUENTLY USED PARENTERAL NUTRITION SOLUTIONS*

	Amigen, 5%	Amigen, 10%	Aminosol, 5%	CPH, 5%	Hyprotogen, 5%	Aminosyn, 7%	Freamine II, 8.5%	Travasol, 5.5%	Travasol, 8.5%	Veinamine, 8.0%
Essential amino acids, g/L										
Lysine	3.1	6.2	4.0	3.5	3.5	5.1	8.7	3.2	4.9	5.4
Tryptophan	0.35	0.7	0.5	0.5	0.4	1.2	1.3	1.0	1.5	0.8
Phenylalanine	2.0	4.0	1.0	2.3	2.0	3.1	4.8	3.4	5.3	4.0
Methionine	1.3	2.6	1.0	2.2	1.6	2.8	4.5	3.2	4.9	4.3
Threonine	1.9	3.8	2.3	1.8	1.9	3.7	3.4	2.3	3.6	1.6
Leucine	4.1	8.2	6.4	4.1	4.1	6.6	7.7	3.4	5.3	3.5
Isoleucine	2.6	5.2	2.2	2.4	2.5	5.1	5.9	2.6	4.1	4.9
Valine	3.1	6.2	1.6	3.0	3.0	5.6	5.6	2.5	3.9	2.5
Nonessential amino acids, g/L										
Histidine	1.3	2.6	1.2	1.2	—	2.1	2.4	2.4	3.7	2.4
Glutamate	13	26	1.4	—	—	—	—	—	—	4.3
Proline	4.5	9.0	3.2	—	—	6.1	9.5	2.3	3.6	1.1
Aspartate	3.5	7.0	0.5	—	—	—	—	—	—	—
Serine	3.0	6.0	3.4	—	—	3.0	5.0	—	—	4.0
Arginine	1.8	3.6	2.9	1.5	—	6.9	3.1	5.7	8.8	7.5
Alanine	1.5	3.0	2.2	—	—	9.0	6.0	11	18	—
Glycine	1.1	2.2	2.1	—	—	9.0	18	11	18	34
Tyrosine	0.6	1.2	5.6	—	—	0.44	0.2	0.22	0.34	—
Cysteine	—	—	0.3	—	—	—	—	—	—	—
Ornithine	—	—	—	—	—	—	—	—	—	—
Hydroxyproline	—	—	—	—	—	—	—	—	—	—
Electrolytes, mEq/L										
Na	35	60	10	39	25	—	10	70	79	49
K	19	31	17	18	18	5.4	—	60	60	30
Cl	20	44	7	14	18	—	—	70	70	50
Mg	2	4	—	2	2	—	—	10	10	6
Ca	5	10	—	6	5	—	—	—	—	—
HPO$_4$	30	60	—	14	25	—	20	60	60	—

*Adapted from Grant, J.: Handbook of Total Parenteral Nutrition.[17]

Insulin is frequently added to the TPN solution or given subcutaneously to maintain serum glucose concentrations within or slightly above normal limits. *Heparin* is added in low concentrations to prevent occlusion of the subclavian catheters used for TPN. Measurement of clotting time has demonstrated no anticoagulant effect.

The continuous administration of nutrients is in contrast to the usual meal-eating pattern, where the mode shifts from storage to utilization and back. Normally, fat, glycogen, and protein are deposited in the fed state and fat and glycogen are mobilized in the fasting state. One study[30] showed that serum albumin, transferrin, and total lymphocytes recover more rapidly in patients on cyclic TPN. Serum insulin and glucose fall during the "off" period. Serum free fatty acids, ketone bodies, alanine, and lactate do not change. In comparison to continuous TPN, cyclic TPN requires equal or slightly fewer calories to maintain nitrogen balance. Cyclic PN is used more for patients on home PN. Whether feeding should take place during the day or night has not been clearly defined, but there are advantages to each approach.

Osmotic diuresis, sepsis, and metabolic changes are the main *complications of TPN* (Table 8A-4). Most problems can be quickly detected with proper monitoring. Hyperosmolar, hyperglycemic, nonketotic dehydration can result if hyperglycemia is not detected soon enough and the infusion rate adjusted. In more severe cases, hypotonic saline, 5% dextrose in water, and insulin should be given to the patient instead of the PN solution. Middle-aged or elderly patients who are obese, physically inactive, and possibly mildly diabetic are more likely to develop this problem.

Hypophosphatemia is a serious problem which results when a semi-starved patient, deriving energy from fat metabolism, which is less dependent on phosphate, is suddenly given glucose as the predominant fuel. Serum glucose concentration rises and causes a release of insulin, which not only promotes uptake of glucose and amino acids by tissues but also favors storage of organic phosphorus in the liver. The clinical changes are complex and can result in cerebral hypoxia and acute cardiopulmonary decompensation and greater susceptibility to infection.

Most complications of parenteral nutrition can be avoided or minimized by monitoring the patient at regular intervals. Table 8A-5 lists those parameters which should be monitored and the suggested frequency of monitoring. The frequency of monitoring depends to some extent on the accepted routine of the hospital. When the patient is an infant or small child, the frequency should also be evaluated in terms of total amount of blood required for the tests.

If a lipid emulsion is not given routinely, there is the possibility of essential fatty acid deficiency, which in infants may result in deposition of abnormal myelin, and in both infants and adults in scaly dermatitis. Vitamin deficiencies can also occur, particularly in infants where the margin of safety between requirements and allowances is much less than in adults.

Table 8A-4. POSSIBLE METABOLIC COMPLICATIONS OF PARENTERAL NUTRITION

Complication	Symptom	Cause
Hyperglycemic, hyperosmolar nonketotic coma	Hyperglycemia; dehydration; increase in serum osmolality and serum sodium; somnolence; seizures	Carbohydrate intolerance; too rapid initiation of PN; infection; latent diabetes mellitus; pancreatic disease
Hypoglycemia	Hypothermia; somnolence; lethargy; peripheral vasoconstriction	Rapid cessation of infusion
Hyperchloremic metabolic acidosis	Decrease in blood pH and serum bicarbonate	Excessive renal or GI losses of base; concentration of cationic amino acids greater than anionic amino acids
	Increase in serum chloride and sodium	Excessive chloride in solution
Hyperammonemia	Elevated blood ammonia; somnolence; coma; lethargy; seizures	Hepatic dysfunction; deficiency in urea cycle amino acids
Hypophosphatemia	Paresthesis; mental confusion; hyperventilation; lethargy; decreased RBC function	Inadequate inorganic phosphate in solution; concentrated glucose infusion may precipitate syndrome; redistribution into cells or bone
Hypokalemia	Muscular weakness; cardiac arrhythmias; altered digitalis sensitivity	Excessive gastrointestinal or urinary losses of potassium
Hyponatremia	Lethargy; confusion	Excess GI or urinary losses of sodium; water intoxication

Table 8A-5. PARAMETERS USUALLY MONITORED DURING PARENTERAL NUTRITION*[22]

Frequency	Parameters
Every six hours	Urinary glucose
	Vital signs (initially)
Daily	Weight
	Input and output (nitrogen balance)
	Vital signs after initial period
Daily until stable, then twice weekly	Serum electrolytes and acid-base status
	Serum glucose
	Blood urea nitrogen
	Serum calcium, phosphorus, magnesium, transferrin, and albumin
Twice a week	Hemoglobin or hematocrit
	WBC
	Platelet count
Weekly	Plasma zinc and copper; prothrombin time
	Blood ammonia
	3-Methylhistidine
	Height in infants and children
	Head circumference in infants
	Liver function tests
Monthly	Serum folate and B_{12}
Four hours after lipid infusion	Serum triglycerides and cholesterol
As needed	Serum iron, reticulocyte count

*If patient is a child, parameters are measured less frequently.

NUTRITIONAL SUPPORT IN SPECIAL CLINICAL SITUATIONS

Most of what we know about nutrition relates to the nutrition of the basically healthy individual. We have some understanding of the effects of specific diseases on nutritional status, but more information is needed to provide the best possible nutrition care.

Ill patients frequently have anorexia or are physically unable to eat so that their nutritional status is altered not only by the disease but also by starvation. In simple starvation, the patient's body adapts through a series of steps in order to best utilize reserves to meet the needs of the body during relative or total energy restriction. Fat, stored as triglycerides, is by far the largest reservoir of energy in the body. Typically, *fat stores* are estimated at 20% of the total body weight, or approximately 130 000 kcal. In obese individuals, the value is higher.

The major body reservoir of protein is muscle tissue, roughly 10 kg or approximately 40 000 kcal. All body proteins are either functional or structural and as such are not reserves in the same sense as adipose tissue. Many scientists believe that protein depletion limits survival in starvation. During the initial response to protein deprivation, somatic protein is lost. As depletion continues and more critical functions are compromised, progressively more visceral protein is lost, which leads eventually to death. The magnitude of protein loss cannot be measured with precision.

In the human body there is limited storage of *carbohydrates*. If fasting continues for more than 18 h, liver glycogen is depleted. The endocrine metabolic response to starvation mainly involves changes in the insulin/glucagon ratios which enable the patient to provide adequate substrates for brain metabolism and to minimize loss of body protein. If normal homeostatic mechanisms fail, the body does not conserve protein stores.

The normal response to starvation is a continuum which can be divided into three stages: postabsorptive (6–12 h after eating), short-term starvation (lasting up to 3–7 d), and prolonged starvation (lasting 2 weeks or longer). During an overnight fast, 75% of the glucose released is derived from glycogen; the rest is from gluconeogenesis. Much of this glucose is utilized by the brain, an organ with a high metabolic rate. In the adult, the liver represents a larger proportion of the body weight and the brain a smaller proportion than in the child, and thus in adults the liver is able to provide sufficient glucose to meet the needs of the brain during starvation. Children, who have relatively larger brains and smaller livers, place more demand on the glycogen stores during a short fast. As fasting continues, the body metabolism shifts to the utilization of lipids, with only 13% of the caloric requirements coming from catabolism of tissue protein. At this stage, the relatively small muscle mass in children limits the amount of amino acids which can be mobilized, and ketosis may develop sooner than in adults.

Physical stress. In patients who have undergone severe physical stress, such as injury, surgery, trauma, burns, or infection, there is extensive loss of tissue protein. *Hypermetabolism* is an adaptive response by which the body maximizes its ability to survive. The requirements of a patient under stress differ quantitatively rather than qualitatively from those of the healthy individual. There is a range of response, depending on the extent and type of trauma. For instance, uncomplicated surgical procedures increase energy requirements the least. Multiple fractures, severe infection (sepsis), and third degree burns produce progressively greater increases in energy requirements. If the patient has sepsis in addition to one of the other stresses, the requirements are increased even further. Patients with extensive thermal injury may require up to 8000 kcal/d.

Skeletal muscle can be depleted of 25–30% of its protein after seven days of stressful illness. Skeletal trauma increases the rate of whole body protein depletion, consistent with the concept that muscle protein metabolism is most seriously affected by severe injury. Protein synthesis is more sensitive to severe stress than is protein breakdown. During moderate to severe stress, urinary excretion of zinc, copper, magnesium, potassium, and calcium increases. Blood levels of vitamin A, ascorbic acid, zinc, and iron are decreased, and sodium and water are retained. Decisions as to whether to manage the stressed patient by enteral or parenteral nutrition must be made on an individual basis.

Diseases That Produce Nutrition Problems

Diseases of the liver and biliary systems. In patients with cholestasis, the quantity of bile salts in the intestine decreases and the absorption of fat diminishes. Dietary fats stimulate gallbladder contraction and relaxation of the sphincter of Oddi. Reduction of dietary fat will result in less stimulation. Therefore, patients with cholecystitis should not be given fatty or fried food. After the attack, the diet initially should be limited to small amounts of carbohydrate foodstuffs. Later, cereals, skim milk, broiled fish, and lean broiled meats can be added.

In patients with liver disease of non-nutritional origin, dietary management is limited to improving the patient's nutritional status and enabling the liver to function as efficiently and easily as possible. During the acute stage of hepatitis, patients may require 10% glucose intravenously. When appetite returns, simple foods, given as frequent small feedings, are tolerated best. The meals should be well balanced and contain all essential nutrients. If food intake is not adequate, feeding by tube may be necessary.

Patients with cirrhosis may develop hidden water retention, frank edema, or ascites associated with nutritional, endocrine, or secondary metabolic disorders. Initial treatment of bed rest and a low-salt diet may be sufficient to cause diuresis. Keeping a diet as low as possible in sodium content while maintaining an adequate protein intake is difficult. Patients must learn to enjoy their food seasoned with herbs and spices which do not contain sodium. In patients with persistent ascites, a low-sodium diet is used in combination with diuretics. Hidden sources of sodium such as softened water and drugs may cause unexpected reaccumulation of fluid.

For patients with hepatic failure, the amount of protein needed to sustain liver function while minimizing protein intolerance is not well defined. Hyperammonemia is seen in patients with severe cirrhosis. Some foods are more likely to cause hyperammonemia because they contain substantial amounts of preformed ammonia. Foods which raise the serum ammonia most in patients with cirrhosis include several types of cheese, chicken, ham, potatoes, peanut butter, and salami. Low ammonia foods include lima beans and egg yolks. Complete lists are available in most books on diet therapy. When hepatic cells are damaged, increased amounts of ammonia enter the systemic blood stream and precipitate hepatic coma. (See also Chapter 13.) Ammonia is also formed by the action of intestinal bacteria on protein foods. Elevated plasma levels of aromatic amino acids and decreased levels of branched-chain amino acids have been noted in patients with liver disease. Plasma aminograms (a listing of the concentrations of all of the amino acids) determined at regular intervals would be helpful to monitor patients, but such tests are rarely done.

Diseases of the GI tract. Dietary indiscretions and nutritional deficiencies have been implicated in the etiology and pathogenesis of chronic gastritis, but the evidence is inconclusive. Chronic nonspecific gastritis is variable in course but may result in a deficiency of vitamin B_{12} that does not progress to pernicious anemia.

Arguments which started in the nineteenth century regarding dietary management of patients with ulcers continue to the present. In the acute phase, there is some justification for traditional dietary modifications. At later stages, the diet should be adequate to meet nutritional needs and should include foods as tolerated.

Edema occurs in patients with protein malnutrition, when the individual fails to absorb amino acids, has high losses of protein from the kidneys or the damaged intestine, or has defective synthesis of albumin by the liver. If the patient does not have liver disease or nephrotic syndrome, as for example in protein malnutrition resulting from anorexia due to cancer, increased protein intake will probably correct these defects.

Nutrition is recognized as an integral part of the therapy of patients with cystic fibrosis or Crohn's disease. Children with cystic fibrosis who are adequately nourished are better able to resist infection and to gain weight. Most of these patients have pancreatic insufficiency and malabsorption; thus, pancreatic enzymes are given with each feeding. Use of medium-chain triglycerides in the diet helps decrease the steatorrhea and increase weight gain. Starches are poorly tolerated; therefore carbohydrate is usually consumed in the form of simple sugars. Protein is usually consumed in amounts well in excess of the RDA, and vitamin supplements are given to satisfy requirements.

During exacerbations, patients with Crohn's disease are usually in negative nitrogen balance, which can be reversed by parenteral nutrition or the use of an elemental diet. At other times, the diet should be low in residue and high in calories, animal protein, vitamins, and minerals.

Food intolerances. Gluten in the diet of individuals with celiac disease leads to a malabsorption syndrome. If gluten is removed from the diet of these patients, improved absorptive function returns as the mucosa of the small intestine recovers. The anemia seen in these patients is caused by an unknown interference with folate absorption. Oral supplements of folate may not alleviate the problem, and intramuscular injections may be necessary.

Disaccharide intolerance, of which lactose intolerance is the most common, occurs with high frequency among Blacks, Arabs, Japanese, and other groups. (See also Chapter 14.) Affected individuals can frequently tolerate small amounts of milk or milk products as sources of calcium. Analysis of breath for hydrogen production after an oral load of lactose appears to be more sensitive for the detection of deficiency than absence of a rise in serum glucose.[27] The definitive test is peroral intestinal mucosal biopsy and demonstration of the disaccharidase involved.

A gastroenteropathy is occasionally seen in younger infants fed cow's milk products and in older infants fed whole cow's milk. In young infants the symptoms are diarrhea, steatorrhea, growth retardation, anemia, hypoproteinemia and edema, respiratory and/or skin allergies, and peripheral eosinophilia. Older infants are most often asymptomatic, and anemia is detected only during routine testing. Bleeding from the gastrointestinal tract is usually not apparent but can be detected by testing the stool for occult blood to distinguish this condition from strictly dietary causes of anemia. In young infants, the cow's-milk–based formula can be replaced by soybean milk, meat-based formulas, or a formula containing an enzymatic digest of casein or goat's milk. Tolerance for milk is regained with maturity. If the introduction of cow's milk into the diet is delayed until the infant is approximately 9 months old, the problem will be minimized, since tolerance is gained with maturity of the GI tract.

Renal disease. Patients with chronic renal failure retain nitrogenous waste products and phosphate, have a low serum calcium, and may develop edema. Restriction of protein intake results in a trend toward normalization of serum urea values. Complicating the nutritional status of the patient with renal failure is the loss during hemodialysis of nutrients (free amino acids, glucose, vitamins, and minerals). Approximately 40% of the amino acids lost during hemodialysis are essential; therefore, the patients need to consume high-quality proteins such as meat. Patients treated with peritoneal dialysis require more protein because of the greater losses of amino acids during this procedure. Circulating levels of folic acid, pyridoxine, and ascorbic acid are often decreased; therefore a supplement of these compounds is advised. Children with renal failure have retarded growth even if treated with hemodialysis.

Most adult patients with renal disease maintain sodium and water balance with an intake of 2–8 g of salt per day and 1.5 to 3 L of water. These patients should avoid diets high in potassium and magnesium and should take calcium supplements. Absorption of calcium is impaired and diets usually include few calcium-rich dairy products because of the need to restrict protein. The metabolic changes in renal failure also stem from the kidney's inability to transform 25-hydroxycholecalciferol to 1,25-dihydroxycholecalciferol; the resulting vitamin deficiency contributes to the deranged calcium metabolism. Response to vitamin D therapy can be determined by the observance of improved bone structure and the return of serum calcium concentrations toward normal. Patients with kidney disease can be managed by enteral or parenteral therapy as well as by regular diet.

Cancer. The debilitation of patients with cancer is due to the effects of treatment as well as to the disease itself. Weight loss may occur in patients with a variety of malignancies. One third of the patients with cancer of the esophagus, stomach, or pancreas will have lost more than 10% of their weight before the institution of treatment. Due to the increased energy and protein requirements, surgery, radiation, and chemotherapy tend to accelerate weight loss.

Tumors of the GI tract cause problems by their very presence, i.e., obstruction. Absorption is altered even before obstruction is complete. Resection of the GI tract for removal of the tumor, together with early and late effects of therapy, may cause destruction of some cells and atrophy of the villi.

Anorexia in patients with cancer is due to anxiety, depression, and metabolic changes. Certain patients have protein aversion and alteration of taste preferences. The lower threshold for bitter taste may result in the patient's avoiding meat and protein products. Consequently both protein and energy malnutrition are seen frequently in patients with cancer. Protocols have been developed both to assess status and as prognostic indices. One protocol[21] uses serum albumin, delayed hypersensitivity, the presence or absence of sepsis, and the presence or absence of cachexia to identify high-risk patients and to evaluate effectiveness of therapy. Another group suggests a weighted index using serum albumin and serum transferrin.[34]

Diabetes mellitus. For the insulin-dependent diabetic patient, the goal of diet therapy is the balance of food intake with exercise and insulin administration. Physicians have different philosophies about diet management, varying from relative freedom to teaching the qualified patient to weigh food and use tables of food composition for the selection of his diet.

The most common method for dietary management uses exchange lists developed jointly by the American Diabetes Association, the U.S. Public Health Service, and the American Dietetic Association. The patient with diabetes limits consumption of foods high in simple sugars and increases the amount of complex carbohydrates. Principal foods are classified into six groups according to composition. Each group contains similar kinds and amounts of food according to the value of carbohydrate, protein, and fat. The term *exchange* is used to denote equivalent amounts. For example, one *bread exchange* is one slice of whole wheat bread or $1/2$ cup of bran flakes, because each of these quantities contains 15 g carbohydrate, 2 g protein, and 70 kcal. One vegetable exchange contains 0.5 g of carbohydrates, 2 g of protein, and 25 kcal and is equivalent to one half cup of asparagus, beets, broccoli, carrots, or several other vegetables. Starchy vegetables are included in the bread exchange because the amount of carbohydrate is high. The number of various exchanges allowed each day is determined by energy needs and carbohydrate allowance. Distribution of the food into meals is determined by the type, dosage, and time of administration of insulin.

Non-insulin-dependent diabetes occurs mainly in adults, particularly in those who are obese. For many of these patients, diet management or weight reduction results in a correction of the diabetes. Frequent urine testing for glucose is a necessary part of the management of all patients with diabetes. Blood glucose is also monitored, but less frequently.

There has been the suggestion that the diabetes of some elderly patients may actually be due to chromium deficiency. Although chromium does have a role in glucose metabolism, its role in diabetes is not yet defined.

The fiber content of the diet has been suggested to be a factor in the treatment of diabetes. While its importance has not been unequivocally demonstrated, patients with non-insulin-dependent diabetes should consume fiber if they follow the suggestion of increasing the intake of complex carbohydrates. Furthermore, the consumption of more complex carbohydrates as fruit or vegetables enables the patient to lose weight which may also improve his clinical status.

Hypertension. While some degree of sodium restriction is recommended for patients with hypertension, the main mode of treatment is medication. If oral diuretics are prescribed, the side effects are hypokalemia, hyperuricemia, and hyperglycemia. Hypokalemia is particularly common and can be treated with diet or a potassium supplement.

Hyperlipidemia. Dietary treatment of patients with hyperlipoproteinemia varies with the phenotype. Type I patients are treated with a diet low in fat; type II, with restricted cholesterol intake and increase in the ratio of polyunsaturated to saturated fats; Type III, with weight monitoring, limited cholesterol intake, controlled carbohydrate intake, and fats with increased ratio of polyunsaturated to saturated fatty acids. Type IV patients are treated by weight monitoring, controlled carbohydrate intake, moderate cholesterol restriction, and increase in the ratio of polyunsaturated to saturated fats, while type V patients should restrict their intake of total fat and to some extent carbohydrate. (See also Chapter 7, Lipids, Lipoproteins, and Apolipoproteins.)

Inborn errors of metabolism. For patients with some *inborn errors of carbohydrate metabolism,* the dietary treatment is avoidance or decreased intake of the offending carbohydrate. Treatment of patients with *inborn errors of amino acid metabolism* involves bypass of the block while providing enough energy for adequate growth and development of the child. This is done by supplying the offending amino acid in minimal amounts and giving all other amino acids as needed. Dietary management of the individual with phenylketonuria is currently the best defined. Lofenalac is a commercially available casein hydrolysate from which 95% of the phenylalanine has been removed. The use of this product minimizes increase in the blood levels of phenylalanine and prevents the resulting derangements of brain metabolism, impaired mental development, aggressive behavior, hyperactivity, neuromuscular instability, seizures, and eczema.

Maple syrup urine disease is more complex to treat, because three essential amino acids, namely, leucine, isoleucine, and valine, must be monitored. Mental retardation is severe if the diet is not restricted. Dietary management of patients with other less common metabolic disorders is also difficult.

Physical handicaps. Nutrition for the handicapped is particularly important during the period of growth and development. Children with motor dysfunction severe enough to prevent ambulation require only 75% of the calories of active children. Those children with less severe motor dysfunction have calorie needs which are unaffected by their handicap. Nonambulatory spastic children are at risk for overweight; therefore energy intake must be carefully monitored. While some children with athetotic motions do not appear to need more energy than normal active children, others do. These children may require drugs which in themselves affect nutritional status.

The handicapped adult compromised by limb injury, progressive neurologic disease, or dysphagia has such symptoms as fatigue, loss of food from the mouth, and greater length of time required for eating. With certain older patients, drug-nutrient interactions complicate dietary management. Other conditions, such as arthritis, may make food preparation and consumption painful. Patients may eat less because they don't want to go to the trouble and endure the pain. Improper fitting or lack of dentures may be another problem that results in failure to eat adequately.

When determining the total effect of a clinical condition on a patient, economics, physical environment, and other factors must be taken into account in the evaluation of the patient and the implementation of dietary change.

Diseases of Eating Behavior

Obesity. Many weight-reduction diets result in weight loss; unfortunately most patients gain the weight back in a few weeks or a few months. Modification of eating behavior, modification of diet composition, and an understanding of factors which are important in the etiology of obesity may help the patient maintain a lower body weight once reduction has been achieved.

A balanced diet is the safest, most effective way to lose weight; however, since weight loss is gradual, more time is required to lose a given amount of weight. Any diet which advocates the consumption of ≤ 900 kcal/d does result in rapid weight loss, but it leads to dangerous losses of potassium and utilization of protein from muscle. All fad diets produce an initial rapid weight loss from the diuresis which occurs when intake is suddenly reduced. Fad diets are innumerable. Fortunately, most individuals cannot tolerate these diets for more than a few days, and thus they avoid potentially dangerous or fatal consequences.

Fad diets can be categorized as high-fat, high-protein, or balanced. High-protein weight reduction diets can result in dehydration due to increased loss of water that is needed to excrete urea, a product of protein catabolism. Most advocates of high-protein diets suggest consumption of a large quantity of water to prevent this dehydration; however, some people need more water than the recommended amount, and some do not adhere to the advice. High-fat weight reduction diets are satisfying but carry the potential increased risk for some types of cancer and heart disease. Any diet which restricts the caloric intake makes it difficult for the patient to consume the requirements for critical nutrients, such as iron, zinc, and some of the water-soluble vitamins. Starvation should not be used as a treatment unless the patient is carefully monitored, especially for potassium losses.

Other treatments of obesity can also result in complications. Jejunoileal bypass surgery may result in several metabolic derangements, such as renal oxalate stones and renal failure and sometimes fatal hepatic disease. Gastroplasty has not yet been shown to be as damaging metabolically. Drug therapies are effective on a short-term basis, but effects may not extend beyond the period of ingestion. Several new drugs are being developed.

Anorexia nervosa. Anorexia nervosa is a psychological disorder of eating behavior. It results in a severe weight loss, which frequently endangers the life of the patient. Nutrition treatment is the same as for other starving patients. There are also some healthy individuals who have trouble gaining weight; for these individuals there is no reason for concern.

Conclusion

Nutritional assessment of populations has been done for a number of years. Nutritional assessment of individuals is a relatively new concept. The role of the clinical chemist in nutritional assessment should be as an expert on the biochemical aspect of assessment with enough of an understanding of the other components to know the contribution of each part. There is a need for the clinical chemist to work with physicians and nutritionists to develop better methods of determining body composition, functional status of many vitamins and minerals, and biochemical markers which will confirm dietary intake data.

Of necessity, the discussion here has been very limited. Not all diseases could be covered and the discussion of nutritional management has provided only an overview. More complete discussions of nutrition are available in Human Nutrition: Clinical and Biochemical Aspects,[13] published by the American Association for Clinical Chemistry; in Food, Nutrition and Diet Therapy;[26] and in Modern Nutrition in Health and Disease.[16] There are also specific texts and review articles listed under Additional Reading.

References

1. Allison, J. B., and Bird, J. W. C.: Nitrogen excretion from the body. *In:* Mammalian Protein Metabolism, Vol. 1. H. N. Munro and J. B. Allison, Eds. New York, Academic Press, 1964.
2. Anderson, C. F., and Wochos, D. N.: The utility of serum albumin values in the nutritional assessment of hospitalized patients. Mayo Clin. Proc., 57:181-184, 1982.
3. Bates, J., and McClain, C. J.: The effect of severe zinc deficiency on serum levels of albumin, transferrin and prealbumin in man. Am. J. Clin. Nutr., 34:1655-1660, 1981.
4. Bilmazes, C., Uauy, R., Haverberg, L. N., et al.: Muscle protein breakdown rates in humans based on N^r-methylhistidine (3-methylhistidine) content of mixed proteins in skeletal muscle and urinary output of N^r-methylhistidine. Metabolism, 27:525-530, 1978.
5. Bistrian, B. R.: A simple technique to estimate severity of stress. Surg. Gynecol. Obstet., 148:675-678, 1979.
6. Bistrian, B. R., Blackburn, G. L., Sherman, M., et al.: Therapeutic index of nutritional depletion in hospitalized patients. Surg. Gynecol. Obstet., 141:512-516, 1975.
7. Blackburn, G. L., Bistrian, B. R., Maini, B. S., et al.: Nutritional and metabolic assessment of the hospitalized patient. J. Parent. Ent. Nutr., 1:11-22, 1977.
8. Cabacungan, N. B., Miles, C. W., Abernathy, R. P., et al.: Hydroxyproline excretion and nutritional status of children. Am. J. Clin. Nutr., 26:173-176, 1973.
9. Chanarin, I., Laidlaw, J., Loughridge, L. W. et al.: Megaloblastic anemia due to phenobarbitone—The convulsant action of therapeutic doses of folic acid. Br. Med. J., 1:1099-1104, 1960.
10. Chandra, R. K.: Fetal malnutrition and postnatal immunocompetence. Am. J. Dis. Child., 129:450-454, 1975.
11. Chernoff, R.: Enteral feedings. Am. J. Hosp. Pharm., 37:65-74, 1980.
12. Forbes, G. B., and Bruining, G. J.: Urinary creatinine excretion and lean body mass. Am. J. Clin. Nutr., 29:1359-1366, 1976.
13. Garry, P. J., Ed.: Human Nutrition: Clinical and Biochemical Aspects. Washington, D.C., American Association for Clinical Chemistry, 1982.
14. Gersovitz, M., Munro, H. N., Udall, J., et al.: Albumin synthesis in young and elderly subjects using a new stable isotope methodology: Response to level of protein intake. Metabolism, 29:1075-1086, 1980.
15. Golden, M. H. N.: Transport proteins as indices of protein status. Am. J. Clin. Nutr., 35:1159-1165, 1982.
16. Goodhart, R. S., and Shils, M. E.: Modern Nutrition in Health and Disease. 6th ed. Philadelphia, Lea and Febiger, 1980.
17. Grant, J. P.: Handbook of Total Parenteral Nutrition. Philadelphia, W. B. Saunders Company, 1980.
18. Grant, J. P., Custer, P. B., and Thurlow, J.: Current techniques of nutritional assessment. Surg. Clin. North Am., 61:437-463, 1981.

19. Hambidge, K. M., Hambidge, C., Jacobs, M., and Baum, J. K.: Low level of zinc in hair, anorexia, poor growth and hypogeusia in children. Pediat. Res., 6:868-874, 1976.
20. Harris, J. A. and Benedict, F. A.: A biometric study of basal metabolism in man. Washington, D. C., Carnegie Institute, Publication No. 120, 1919.
21. Harvey, K. B., Moldawer, L. L., Bistrian, B. R., et al.: Biological measures for the formulation of a hospital prognostic index. Am. J. Clin. Nutr., 34:2013-2022, 1981.
22. Kerner, J. A., Jr., Ed.: Manual of Pediatric Parenteral Nutrition. New York, John Wiley and Sons, 1983.
23. Kim, C. W., Okada, A., Itakura, T., et al.: Urinary excretion of 3-methylhistidine in patients receiving parenteral nutrition. J. Parent. Ent. Nutr., 3:255-257, 1979.
24. King, J. C., Cohenour, S., Calloway, D., et al.: Assessment of nutritional status of teenage pregnant girls. 1. Nutrient intake and pregnancy. Am. J. Clin. Nutr., 25:916-925, 1972.
25. Knight, P., Heer, D., and Abdenour, G.: Ca x P and Ca/P in the parenteral feeding of preterm infants. J. Parent. Ent. Nutr., 7:110-114, 1983.
26. Krause, M., and Mahan, L. K.: Food, Nutrition and Diet Therapy. 7th ed. Philadelphia, W. B. Saunders Company, 1984.
27. Levitt, M. D., and Donaldson, R. M.: Use of respiratory hydrogen (H_2) excretion to detect carbohydrate malabsorption. J. Lab. Clin. Med., 75:937-945, 1970.
28. Long C. L., Birkhahn, R. H., Geiger, J. W. et al.: Urinary excretion of 3-methylhistidine: An assessment of muscle protein catabolism in adult normal subjects and during malnutrition, sepsis and skeletal trauma. Metabolism, 30:765-776, 1981.
29. Mackenzie, T. A., Blackburn, G. L., and Flatt, J. P.: Clinical assessment of nutritional status using nitrogen balance. Fed. Proc., 33:683 (Abstract #2673), 1974.
30. Maini, B., Blackburn, G. L., Bistrian, B. R., et al.: Cyclic hyperalimentation: An optimal technique for preservation of visceral proteins. J. Surg. Res., 20:515-525, 1976.
31. Owen, G. M., Kram, K. M., Garry, P. J., et al.: A study of nutritional status of preschool children in the United States, 1968–1970. Pediatrics, 53(Suppl.):597-646, 1974.
32. Pao, E. M., and Mickle, S. J.: Problem nutrients in the United States. Food Tech., 35(Pt. 2):56-79, 1981.
33. Presta, E., Wang, J., et al.: Measurement of total body electrical conductivity: A new method for estimation of body composition. Am. J. Clin. Nutr., 37:735-739, 1983.
34. Rainey-MacDonald, C. G., Holliday, R. L., Wells, G. A., et al.: Validity of a two-variable nutritional index for use in selecting candidates for nutritional support. J. Parent. Ent. Nutr., 7:15-20, 1983.
35. Roe, D. A.: Nutritional assessment of alcoholics. In: Alcohol and the Diet. Westport, Conn., AVI Publishing Company, 1979.
36. Sayers, M. H., Johnson, D. K., Schumann, L. A., et al.: Supplementation of total parenteral nutrition solutions with ferrous citrate. J. Parent. Ent. Nutr., 7:117-120, 1983.
37. Seashore, J. H., Huszar, G., and Davis, E. M.: Urinary 3-methylhistidine/creatinine ratio as a clinical tool: Correlation between 3-methylhistidine excretion and metabolic and clinical states in healthy and stressed premature infants. Metabolism, 30:959-969, 1981.
38. Stiehm, E. R.: Humoral immunity in malnutrition. Fed. Proc., 39:3093-3097, 1980.
39. U.S. Department of Health, Education, and Welfare: Ten State Nutrition Survey in the United States 1968–1970.HSM 72-8134. Washington, D.C., 1972.
40. U.S. Public Health Service–Health Resources Administration, National Center for Health Statistics: Dietary Intake and Biochemical Findings, Preliminary Findings of First Health and Nutrition Examination Survey, U.S., 1971–72. Rockville, Md., Jan., 1974.
41. Viteri, F. E., and Alvarado, J.: The creatinine height index: Its use in the estimation of the degree of protein depletion and repletion in protein calorie malnourished children. Pediatrics, 46:696-706, 1970.
42. Young, V. R.: The role of skeletal and cardiac muscle in the regulation of protein metabolism. In Mammalian Protein Metabolism. H. N. Munro, Ed. New York, Academic Press, 1970, Vol. 4.
43. Young, V. R., and Munro, H. N.: N$^\tau$-methylhistidine (3-methylhistidine) and muscle protein turnover: An overview. Fed. Proc., 37:2291-2300, 1978.
44. Warnold, I., and Lenner, R. A.: Evaluation of the heart rate method to determine the daily energy expenditure in disease. A study of juvenile diabetics. Am. J. Clin. Nutr., 30:304-315, 1977.
45. Weinsier, R. L., and Butterworth, C. E.: Handbook of Clinical Nutrition. St. Louis, C. V. Mosby, 1981.

Additional Readings

Deitel, M., Ed.: Nutrition in Clinical Surgery, Baltimore, Williams and Wilkins, 1980.
Fischer, J. E.: Hyperalimentation. Med. Clin. North Am., 63:973-985, 1979.
Gross, R. L., and Newberne, P. M.: Role of nutrition in immunologic function. Physiol. Rev., 60:198-256, 1980.
Newell, G. R., and Ellison, N. M.: Cancer and Nutrition: Etiology and Treatment. New York, Raven Press, 1981.
Schneider, H. A., Ed.: Nutritional Support of Medical Practice. 2nd ed. Philadelphia, J.B. Lippincott Company, 1983.
Taylor, K. B., and Anthony, L. E.: Clinical Nutrition. New York, McGraw-Hill Book Company, 1983.

Vitamins

by Donald B. McCormick, Ph.D.

Among those factors that affect the health and development of humans, none is more essential than vitamins. These nutrients occupy the attention of those concerned with the physical well-being of a public made increasingly aware of the need for the quality as well as the quantity of dietary intake. Although deficiency of a single vitamin is relatively uncommon in humans, it can occur as a result of an inborn error of metabolism or as a result of unusual restriction in dietary intake. More frequently encountered are complex deficiencies that arise as a result of food fads; as complications of certain diseases, especially those affecting food absorption; as a result of massive losses of blood or from hemodialysis; and with the use of certain drugs. There are even instances, more common in the Western world, where excessive use of some vitamins is encountered. These and other influences that can cause vitamin imbalance, together with advances in knowledge of what constitutes imbalance, have led to an increased demand for laboratory tests for the evaluation of the vitamin status of individuals. This, in turn, has served to advance the improvement and interpretation of results obtained by newer methods.

DEFINITION OF VITAMINS

Vitamins are organic compounds required in trace amounts (μg to mg quantities per day) in the diet for health, growth, and reproduction. It is commonly understood that vitamins are natural materials that can be isolated from organisms such as plants that have the ability to make most of these compounds, or that can be chemically synthesized. Additionally, there are synthetic analogs and derivatives of vitamins which have been designed to serve as inhibitors, e.g., amethopterin as an antifolate, or others that substitute in part for the natural vitamin, e.g., 8-ethylriboflavin for riboflavin. Only small amounts of vitamins are required for the functional, often catalytic (coenzymatic) roles they serve, in contrast to the relatively large amounts of such macronutrients as protein, lipid, and carbohydrate which constitute the bulk of the ingesta and which serve primarily as sources for energy and reconstitution of body mass. Hence, it is somewhat inappropriate to consider inositol or choline as vitaminic substances, when they are utilized in rather sizable quantities largely for incorporation into such constitutive material as phospholipids; this is especially true when the organism can synthesize these compounds, which become limiting only during periods of rapid growth. Such compounds are not covered in this chapter.

Although a full listing of presently known vitamins includes compounds that fit the definition for one or more species, e.g., lipoic acid for certain bacteria, the human appears able to synthesize adequate quantities of some of these; therefore, they are not properly considered a vitamin for our species. In other cases, several naturally occurring compounds seemingly serve the defined vitaminic roles; following ingestion they are either metabolically interconverted to a common functional form, e.g., the vitamin B_6 group of pyridoxine, pyridoxal, and pyridoxamine to the coenzyme pyridoxal-5'-phosphate, or they directly substitute for each other, e.g., phylloquinones (K_1) and menaquinones (K_2). Such vitamers are reasonably subclassed under the parent common or chemical name for the respective vitamin group.

CLASSIFICATION OF VITAMINS

Historically, vitamin groups such as A, B, and D bear an Arabic subscript number following the letter either to designate structural and functional similarity, e.g., A_1 (retinol) and A_2 (3-dehydroretinol), or to indicate the approximate order in which they are identified as the members of the so-called B-complex, e.g., B_1 (thiamin) and B_2 (riboflavin). Common chemical names, which are receiving greater usage, give a better indication of the types of compounds involved. These often reflect the presence of some specific atom (*thia*min), prime functional group (pyridox*amine*), or even larger portion of the molecular structure (phyllo*quinone*). Parts of some names reflect functional properties (chole*calciferol*).

Another classification pertains to relative solubility of vitamins. The "fat-soluble" group, i.e., A, D, E, and K, are more soluble in organic solvents, whereas the B-complex group and vitamin C are "water-soluble." This general separation on the basis of solubility is useful not just for purposes of noting gross physical properties, but as a reminder that the fat-soluble vitamins are absorbed, transported, and stored for longer times and in a manner generally similar to fats. Most water-soluble vitamins share the fate of other solutes more compatible with an aqueous, physiological medium; this includes a lesser tendency to be retained for long times in the body and a greater loss by way of urinary excretion.

Some vitamins have been designated in earlier or more restricted literature under other names, e.g., vitamin H for biotin; in other cases the term vitamin has been incorrectly applied to material not proven essential and even potentially toxic, e.g., Laetrile or B_{17}.

VITAMINS ESSENTIAL FOR HUMANS

Table 8B-1 provides a current list of those 13 known vitamins and vitameric groups essential to the human. For 10 of these, there are well-discerned levels of requirements that have led to recommended daily dietary allowances (RDA) in the U.S.A.; for three (K, biotin, and pantothenic acid), there is somewhat less information on which to base allowances; thus only "estimated safe and adequate daily dietary intakes" have been published.[35]

VITAMIN A

Chemistry and Sources

The two natural forms of vitamin A, *retinol* (A_1) and *3-dehydroretinol* (A_2), are C_{15}-isoprenoid alcohols that have a substituted β-ionone and 3-dehydro-β-ionone ring, respectively, as shown in Figure 8B-1. These compounds are yellowish oils or low-melting solids (depending on isomeric purity) that are practically insoluble in water but soluble in organic solvents and mineral oil. Vitamin A is sensitive to oxygen and ultraviolet light, which induces a greenish fluorescence; absorbances are at 325 and 351 nm for A_1 and A_2, respectively. Vitamin A_1 predominates, especially as long-chain fatty acid esters, in the liver of mammals and salt-water fish, e.g., cod liver, whereas the biologically less active (by rat assay) vitamin A_2 is found in fresh-water fish oils. Although higher animals are unable to synthesize the β-ionone type ring structure, they can derive the aldehyde retinal from intestinal dioxygenase-catalyzed cleavage of plant-derived caro-

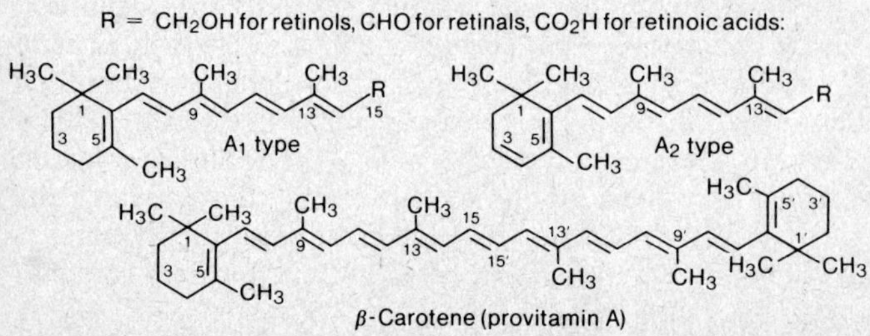

$R = CH_2OH$ for retinols, CHO for retinals, CO_2H for retinoic acids:

Figure 8B-1. Vitaminic forms of A_1, A_2 and β-carotene.

β-Carotene (provitamin A)

Table 8B-1. VITAMINS REQUIRED BY THE HUMAN*

Common Name	Trivial Chemical Name	General Roles	Deficiency Symptoms or Disease	Direct and Indirect Assays
		Fat Soluble		
Vitamin A_1 A_2	Retinol 3-Dehydroretinol	Vision, growth, reproduction	Nyctalopia, xerophthalmia, keratomalacia	Colorimetric, fluorometric, dark adaptation, RIA, HPLC
Vitamin D_2 D_3	Ergocalciferol Cholecalciferol	Modulation of Ca(II) metabolism, calcification of bone and teeth	Rickets (young), osteomalacia (adult)	CPB, HPLC
Vitamin E	Tocopherols, $\alpha, \beta, \gamma, \delta$	Antioxidant for unsaturated lipids	Lipid peroxidation including red cell fragility, hemolytic anemia (premature, newborn)	Colorimetric, HPLC, erythrocyte hemolysis
Vitamin K_1 K_2	Phylloquinones Menaquinones	Blood clotting, osteocalcins	Increased clotting time, hemorrhagic disease (infant)	Colorimetric, HPLC, prothrombin time, RIA (abnormal prothrombin)
		Water Soluble		
Vitamin B_1	Thiamin	Carbohydrate metabolism, nervous function	Beriberi, Wernicke-Korsakoff syndrome	Fluorometric, microbiological, transketolase
Vitamin B_2	Riboflavin	Oxidation-reduction reactions	Angular stomatitis, dermatitis, photophobia	Fluorimetric, microbiological, glutathione reductase
Vitamin B_6	Pyridoxine, pyridoxal, pyridoxamine	Amino acid, phospholipid, and glycogen metabolism	Epileptiform convulsions, dermatitis, hypochromic anemia	Microbiological, tyrosine decarboxylase
Niacin Niacinamide	Nicotinic acid Nicotinamide	Oxidation-reduction reactions	Pellagra	Microbiological, fluorimetric
Folacin	Pteroylglutamic acid	Nucleic acid and amino acid biosynthesis	Megaloblastic anemia	CPB, microbiological
Vitamin B_{12}	Cyanocobalamin	Amino acid and branched-chain keto acid metabolism	Pernicious and megaloblastic anemia, neuropathy	CPB, microbiological
Biotin	—	Carboxylation reactions	Dermatitis	Microbiological, colorimetric, carboxylases
Pantothenic acid	—	General metabolism	Burning feet syndrome	Microbiological
Vitamin C	Ascorbic acid	Connective tissue formation	Scurvy	Colorimetric

*Only one vitamer from a vitamin group, e.g., A_1 for vitamin A or pyridoxine for vitamin B_6, is sufficient for that vitamin.

tenes and cryptoxanthin which serve as provitamins. Retinal is then reversibly reduced by pyridine nucleotide–dependent enzymes to retinol.

The structure for the most common and effective provitamin A, *β-carotene*, is also given in Figure 8B-1. This compound is an orange-to-purple, water-insoluble solid which is air oxidized to inactive products. The other carotenes, cryptoxanthin and β-apocarotenals, are asymmetrical with only one β-ionone ring and yield less vitamin A_1 activity. Although such carotenoid compounds, which constitute the yellow-to-orange pigments of most vegetables and fruits, show considerable variation in their availability to the human, they often constitute the main dietary source of what ultimately becomes vitamin A. On the basis of structure alone, it has been estimated that the number of provitamin A precursors is between 50 and 60 known carotenoid and apocarotenoid compounds.[16]

Requirements and Allowances

Conclusions from studies with humans indicate that 500–600 μg of retinol, or twice as much β-carotene, is a minimum requirement for adults to maintain an adequate blood concentration and to prevent all deficiency symptoms.[35] Greater intakes are necessary to produce significant liver storage. Recommended allowances (RDA) for adult males are 1000 and for females 800 retinol equivalents, where 1 retinol equivalent = 1 μg retinol or 6 μg β-carotene or 12 μg of other provitamin A carotenoids. In terms of international units (IU), 1 retinol equivalent = 3.33 IU of retinol or 10 IU of β-carotene. These equivalencies are based primarily on rat bioassays. The RDA for infants, based on content and volume of human milk, is 420 retinol equivalents until the age of 6 months and 400 retinol equivalents from then until the first year. Higher levels are recommended to satisfy growth needs of children and adolescents. An additional 200 retinol equivalents is recommended for pregnant women and an additional 400 retinol equivalents for lactating women. A comprehensive review of the vitamin A requirements of the human, including the effects of exercise, stress, and genetic defects, has been published.[65]

Absorption, Transport, and Metabolism

The emulsification of vitamin A and pro A forms to the micellar level by bile salts enhances their uptake by mucosal cells of the small intestine and, for retinyl esters, facilitates hydrolysis by pancreatic retinyl ester hydrolase;[24] remaining ester is hydrolyzed by a brush border enzyme. Absorption of the free retinol is followed by re-esterification with long-chain fatty acids, predominantly palmitic and stearic, within the mucosal cell. Those carotenoids capable of being cleaved by the cellular dioxygenase system are converted in part to retinal which is largely reduced to retinol and esterified. Retinyl esters in association with chylomicrons then pass via the lymphatic system to the liver where uptake by parenchymal cells again involves hydrolysis. Cellular retinol-binding proteins bind retinol which, when adequately supplied, can be re-esterified by acyl donors, mainly palmitoyl-CoA, stearyl-CoA, and oleyl-CoA, and stored in a lipoglycoprotein complex. Retinol released from the complex by esterases allows association with plasma retinol-binding protein, which is synthesized by the rough endoplasmic reticulum. Release from hepatocytes via secretory vesicles presents holo-retinol-binding protein (M.W. ~21 000) for further association with circulating prealbumin (M.W. ~55 000) to form a molecular aggregate of sufficient size to avoid loss by glomerular filtration.

In addition to retinal, other products from cleavage of β-carotene include the 8'-, 10'-, and 12'-apo-β-carotenals which are further degraded in intestinal mucosal cells to retinal.[16] Although most of the aldehyde is reversibly reduced to retinol, lesser amounts are oxidized to retinoic acid in several tissues, e.g., intestine, liver, and kidney. Retinoic acid from the intestinal mucosa is transported bound to serum albumin via the portal vein.[24] Retinoic acid cannot be significantly reduced to retinal but is rapidly metabolized in tissue such as liver to yield more polar catabolites, e.g., 5,6-epoxyretinoic acid, or conjugates, such as retinoyl β-glucuronide, which are excreted. A small amount of retinoic acid undergoes enterohepatic circulation after intestinal hydrolysis of the glucuronide excreted in bile.

Functions

Among the physiological functions of vitamin A, the best understood is the participation of retinal in vision. Upon receiving the predominant circulating form of vitamin A, viz., all-*trans-*

retinol, cells of the retina can isomerize this to the 11-*cis* alcohol which is reversibly dehydrogenated to 11-*cis* retinal. This sterically hindered geometrical isomer of the aldehyde combines as a lysyl-linked Schiff base with suitable proteins, e.g., opsin, to generate photosensitive pigments, such as rhodopsin. Illumination of such pigments causes photoisomerization and release of all-*trans*-retinal plus the protein, a process which couples the large conformational change to ion flux and optic nerve transmission. The all-*trans*-retinal can be isomerized to the 11-*cis* isomer, which again combines with the liberated protein to reconstitute the photopigment in a visual cycle shown in Figure 8B-2. The pyridine nucleotide–dependent dehydrogenase (reductase) can also reduce the all-*trans*-retinal to all-*trans*-retinol.

Other broader functions of vitamin A are in reproduction and growth. Systemic effects that reflect an optimal level of vitamin A are the stabilization of cellular and intracellular membranes, the maintenance of the integrity of epithelial tissue, and the synthesis of glycoproteins. All-*trans*-retinyl-1-β-phospho-D-mannose is formed from retinyl phosphate and UDP mannose and is a good donor in vitro of mannose to certain glycoproteins. It has been reported that retinoic acid can maintain normal growth but is unable to replace vitamin A in reproduction or vision.

Vitamin A Deficiency

Clinically, degenerative changes in eyes and skin are commonly observed in vitamin A deficiency.[81] Poor dark adaptation or *night blindness* (nyctalopia) is an early symptom which is followed by degenerative changes in the retina. There is a *xerophthalmic condition* in which the conjunctiva becomes dry and small gray plaques with foamy surfaces (Bitot's spots) develop. These lesions are reversible with vitamin A administration. More serious effects of deficiency are known as *keratomalacia* and cause ulceration and necrosis of the cornea, which lead to perforation, prolapse, endophthalmitis, and blindness. Usually there are associated skin changes which include dryness, roughness, papular eruptions, and follicular hyperkeratosis. The general change is atrophy of certain specialized epithelia followed by metaplastic hyperkeratinization.

Vitamin A Toxicity

Toxic effects occur with hypervitaminosis A as a result of ingestion of excess vitamin or as a side effect of inappropriate therapy.[9,10,81] One of the most important factors is the form of vitamin A administered.[30] Symptoms can appear more rapidly after administering aqueous emulsions rather than oily solutions. Hypervitaminosis A occurs after the liver storage of retinol and its esters exceeds 10 000 IU/g tissue, a level ten times the estimated RDA for adult males, or if plasma vitamin A levels exceed 140 μg/dL. Acute toxicity from a single massive dose presents as abdominal pain, nausea, vomiting, severe headaches, dizziness, sluggishness, and irritability, followed within a few days by desquamation of the skin and recovery. Chronic toxicity from moderately high doses taken for protracted periods is characterized by bone and joint pain, hair loss, dryness and fissures of the lips, anorexia, benign intracranial hypertension, weight loss, and hepatomegaly. Carotenemia results from chronic excessive intake of carotene-rich foods, principally carrots. This condition, in which yellowing of skin is observed, is not injurious, since the excess carotene is deposited rather than converted to vitamin A.

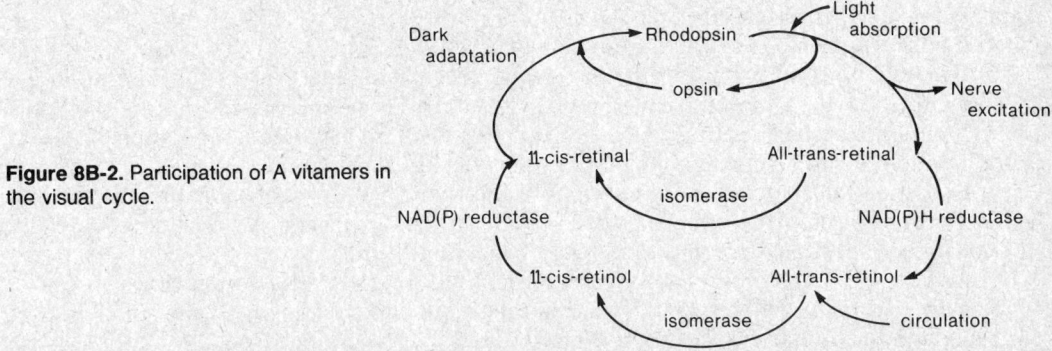

Figure 8B-2. Participation of A vitamers in the visual cycle.

Methods for the Determination of Vitamin A and β-Carotene

Principle

Trifluoroacetic acid is reacted with the conjugated double-bond system of the organic solvent-extracted compounds to produce a blue color (A_{620}) according to the Neeld-Pearson procedure.[38] A correction for the color contribution by carotenes is necessary.

Specimen

Collect blood for the determination of serum carotene and vitamin A with the patient in the fasting state. The specimen should be free from hemolysis and protected from light. Separate serum promptly and freeze at −20 °C if not analyzed immediately; vitamins A and β-carotene are stable for at least two weeks if frozen and protected from light.

Reagents

1. Ethanol, 95% (v/v), reagent grade.
2. Petroleum ether, reagent (pesticide) grade.
3. Chloroform, anhydrous, reagent grade.
4. Trifluoroacetic acid reagent. Mix 1 vol of trifluoroacetic acid (CF_3COOH), reagent grade, with 2 vol of anhydrous chloroform just prior to use. This solution is stable for 4 h at 25 °C.
5. Vitamin A stock standard, 160 μg/mL. Transfer 16.0 mg of all-*trans*-retinyl acetate (Eastman Organic Chemicals, Rochester, N.Y.) to a 100-mL volumetric flask and dilute to volume with anhydrous chloroform.
6. Vitamin A working standard. Pipet 2.5, 5.0, 7.5, and 10.0 mL of the stock standard (160 μg/mL) into 100-mL volumetric flasks and dilute to volume with anhydrous chloroform to obtain working standards with vitamin A concentrations of 4.0, 8.0, 12.0, and 16.0 μg/mL. These standards are stable for one week at 4 to 8 °C if protected from light. Preparation of standards and all analytical operations must be performed in low actinic glassware or under very low intensity light.
7. β-Carotene stock standard, 200 μg/mL. Transfer 20.0 mg of synthetic crystalline β-carotene (Sigma Chemical Company, St. Louis, Mo.) to a 100-mL volumetric flask. Dissolve in approximately 4 mL of chloroform, and then dilute to volume with petroleum ether.
8. β-Carotene working standard. Pipet 10.0 mL of the 200 μg/mL standard into a 100-mL volumetric flask and dilute to volume with petroleum ether to obtain a concentration of 20 μg/mL. Then transfer 2.5, 5.0, 10.0, 15.0, and 20.0 mL of the 20-μg/mL standard to a 100-mL volumetric flask and dilute to volume with petroleum ether to obtain working standards with concentrations of 0.5, 1.0, 2.0, 3.0, and 4.0 μg/mL. These standards are stable for only a few hours at 25 °C and should be made up fresh each time an analysis is performed. All standard solutions must be protected from direct light.

Procedure

1. Pipet 1.0 mL of serum into 15-mL glass-stoppered centrifuge tubes (duplicates).
2. Add 2.0 mL of 95% (v/v) ethanol, stopper, and mix well with a vortex mixer.
3. Add 3.0 mL of petroleum ether and place tubes in a Kraft "Shaker in the Round" (Kraft Apparatus, Inc., Queens, N.Y.) or some other suitable shaker to extract vitamin A and the carotenes into the petroleum ether phase.
4. Centrifuge the stoppered tubes for 10 min at 2500 × g.
5. Carefully pipet off 2.0 mL of the petroleum ether phase (upper layer) and transfer into a dry cuvet (e.g., 10 × 75 mm Coleman cuvet). Read the absorbance at 450 nm in a suitable spectrophotometer against a petroleum ether blank without delay to prevent evaporation of solvent and destruction of carotenoids by light. Mark this reading A_1.
6. Evaporate the contents of the cuvets to dryness in a 50 °C water bath with the aid of a fine stream of nitrogen. After removing the cuvets from the water bath, dry each cuvet carefully with nonabrasive paper (to prevent scratching or marking them).
7. Add 0.1 mL of chloroform to each cuvet and mix briefly with a vortex mixer.
8. Add 1.0 mL of trifluoroacetic acid reagent to the cuvet serving as blank, mix, and set spectrophotometer to 0 absorbance at 620 nm.
9. Add forcefully (to facilitate immediate mixing) 1.0 mL of trifluoroacetic acid reagent to

all other cuvets, and record the reading at 620 nm at exactly 2 s after addition of the reagent. (Add the trifluoroacetic acid reagent to only one cuvet at a time so that readings may be taken as required.) *Caution*: Trifluoroacetic acid is a strong acid, and care should be exercised to prevent spilling or splattering it.

Best results are obtained if the trifluoroacetic acid reagent is added forcefully with an automated pipet. If a recorder is used, the absorbance value can be read at the peak or inflection point after the initial surging peak caused by the introduction of the trifluoroacetic acid reagent. This peak or inflection point occurs about 2 s after the introduction of the color reagent. Mark this reading A_2.

Calculations

Carotene. Determine the amount of carotene per mL (see A_1 reading in step 5 of Procedure) from the carotene standard curve and carry out the following calculations:

$$\mu g \text{ carotene/dL serum} = \mu g \text{ carotene/mL} \times 3.0 \times 100$$

where
  3.0 = volume petroleum ether containing the carotene from 1.0 mL of serum after extraction
  100 = conversion factor from μg/mL to μg/dL.
Vitamin A. For accurate calculation of the vitamin A content, it is necessary to correct for the absorbance contributed by carotenes at 620 nm:

$$A_3 = A_2 - (F \times A_1)$$

where
  A_1 = absorbance of carotene at 450 nm (step 5).
  A_2 = absorbance at 620 nm due to both carotene and vitamin A (step 9).
  A_3 = absorbance at 620 nm of vitamin A (corrected for absorbance contributed by carotene).
  F = factor which converts the carotene absorbance at 450 nm (step 5) into the equivalent absorbance at 620 nm in the color reaction (step 9).
There have been reports indicating considerable variation in the value of this factor. Therefore, each laboratory must determine its own factor (see Calibration, below).

After A_3 has been found, the actual concentration of vitamin A per dL of serum is calculated as follows:

$$\mu g \text{ vitamin A (free alcohol)/dL} = \frac{A_3 \times \mu g \text{ retinyl acetate standard/cuvet}}{A_{620} \text{ retinyl acetate standard}} \times \frac{3}{2} \times 100 \times 0.872$$

or

$$\text{vitamin A, } \mu g/\text{dL} = \frac{A_3 \times \mu g \text{ retinyl acetate standard/cuvet}}{A_{620} \text{ retinyl acetate standard}} \times 130.8$$

where
  3 = volume of the petroleum ether extract from 1.0 mL serum.
  2 = aliquot of the petroleum ether extract used for the assay.
  100 = conversion of μg retinyl acetate/mL to μg retinyl acetate/dL.
  0.872 = ratio of molecular weight of retinol to molecular weight of retinyl acetate. Thus, this factor corrects for the use of retinyl acetate instead of retinol as the standard.

Calibration

β-Carotene. Place β-carotene working standards in concentrations of 0.5, 1.0, 2.0, 3.0, and 4.0 μg/mL into appropriate cuvets (e.g., 10 × 75 mm Coleman cuvets), and read the absorbance at 450 nm against a petroleum ether blank. Plot carotene concentration (μg/mL) against absorbance.

For the purpose of calculating the carotene correction factor F for the vitamin A procedure, 2.0 mL of each of the β-carotene standards is treated as a sample, beginning with step 6 of the vitamin A procedure. The average ratio of absorbance at 620 nm to concentration of β-carotene (in μg/mL) is then calculated and used in the computation of the β-carotene correction factor F.

Vitamin A. Pipet 0.1 mL of each retinyl acetate working standard into cuvets and treat as in vitamin A procedure. Plot μg retinyl acetate per cuvet against the absorbance at 620 nm.

Discussion of Methods

The present *Neeld-Pearson procedure* has greater sensitivity than the classic Carr-Price method using antimony trichloride in chloroform and is not affected by moisture. However, this method is still relatively imprecise, tedious, and nonspecific. Several colorimetric/spectrophotometric and fluorometric methods applied after solvent extractions with and without additional separation techniques, such as high pressure liquid chromatography (HPLC), have been used to quantitate carotenoids and retinoids in numerous biological specimens.[46,49] The HPLC method by Bieri et al.[5] is an example of newer methods which are gaining rapid acceptance. This method is rapid and specific for retinol as well as individual carotenes. It also allows the simultaneous determination of vitamins A and E. The method uses a reverse-phase column and a UV detector with a 280-nm filter. Additionally, methods widely applied for carotenoids in food materials have been detailed by the Association of Official Analytical Chemists.[26]

The application of HPLC to the separation of retinoids has markedly improved vitamin A methodology. Both straight-phase and reverse-phase techniques have been used.[49] In the former, compounds to be separated are adsorbed to microparticulate silica gel and eluted in the order of least polar to most polar. Good separation and quantitative yields of neutral and charged retinoids are obtained. Reverse-phase HPLC is preferable for acid-sensitive compounds such as 5,6-epoxyretinoic acid. The principle of hydrophobic interaction is involved. Solutes are forced out of the mobile phase and are bound with the hydrocarbon ligands of the stationary phase, usually octadecylsilanized silica gel. Increasing polarity of the mobile phase increases retention so that elution order is most polar to least polar.

Reference Ranges

The concentration of *vitamin A* in serum or plasma from a fasting subject is presently considered the only practical biochemical indicator of vitamin A status.[34] The reference ranges for newborns, children, and adults are 35–75, 30–80, and 30–65 μg/dL, respectively. Values above 30 μg/dL are associated with appreciable reserves in the liver and correlate well with vitamin A intake. Within the reference range, levels for healthy subjects increase modestly with age; values for men are generally about 20% higher than those for women. Values above 100 μg/dL indicate toxicity.

The reference range for serum *carotenoids* (mainly β-carotene and xanthophyll) is 60–200 μg/dL in petroleum ether extracts, and 50–300 μg/dL in iso-octane extracts. Elevated levels are found in hypothyroid cases where conversion to vitamin A is decreased and in patients with hyperlipemia associated with diabetes mellitus.

Plasma *retinol-binding protein,* usually measured by RIA, has been used to assess nutritional status of vitamin A; values normally range between 40 and 50 μg/mL.[24] Liver, thyroid, and kidney diseases (with protein loss), cystic fibrosis, and protein-calorie malnutrition cause decreases in serum concentration of this protein. This is particularly the case in liver disease where biosynthesis is impaired. Increased concentrations are seen in renal disease where there is decreased excretion. See also the discussion in Chapters 4 and 8A.

VITAMIN D

Chemistry and Sources

The two forms of vitamin D, *ergocalciferol* (D_2) and *cholecalciferol* (D_3), are steroid-derived compounds. D_2 does not occur naturally but is produced by ultraviolet irradiation of ergosterol, which occurs in molds, yeast, and higher plants; the action of light causes ring cleavage to yield an intermediate pre-ergocalciferol (plus lumisterol and other related congeners), which rearranges under thermal conditions to ergocalciferol. Vitamin D_3 does occur in our bodies, but similarly derives from irradiation of provitamin 7-dehydrocholesterol in skin to produce, via thermal arrangement of pre-cholecalciferol, the natural cholecalciferol. As shown by structures in Figure 8B-3, the only chemical difference between the two vitamin and provitamin forms is in the side chain. There are at least 10 compounds known to yield vitamin D–active compounds upon irradiation. Most differ only in the side chain at C-17 of the sterol nucleus, e.g., 22,23-dihydro-ergocalciferol. Vitamin D forms white, odorless crystals which are soluble in fats and organic solvents. It is generally stable to heat, alkalis, acids, and oxidation. The absorption maximum for both D_2 and D_3 in hexane is at 264.5 nm. (See also Chapter 12.)

Requirements and Daily Allowances

It has been determined that 2.5 μg (100 IU) of vitamin D per day is required to prevent rickets in the young, but 10 μg (400 IU) seems to promote better calcium absorption and allow optimal growth.[35] Therefore, this higher level is recommended as the daily allowance from infancy through 18 years of age. As the rate of skeletal growth and calcium need decreases, the daily allowances are reduced to 7.3 μg between the ages of 19 and 22 years and to 5 μg thereafter. An additional 5 μg/d is recommended for pregnant and lactating women. In such diseases as uremic osteodystrophy and with autosomal recessive vitamin D dependency, requirements are increased to the pharmacologic dose range of 30–50 μg (1200–2000 IU).[71]

Absorption, Transport, and Metabolism

The extent to which ultraviolet irradiation of 7-dehydrocholesterol in skin contributes to production of vitamin D_3 is quite variable since it depends upon seasonal variation in sunlight, amount of clothes worn, degree of skin pigmentation, and other factors. Availability and consumption of such vitamin D_3–containing animal foods as fatty fish, eggs, liver, and butter also are variable. Hence, widespread fortification of food has been adopted to ensure a more secure supply. Because dairy foods in the United States supply about 75% of dietary calcium in a calcium/phosphorus ratio near optimal for bone growth, milk and infant formulas were judged to be desirable vehicles for vitamin D fortification.[1] Though a poor source of vitamin D itself, milk is now routinely marketed to contain 10 μg (400 IU) of added vitamin D per quart. Both ergocalciferol and cholecalciferol are about equally converted in the human to hormonally active dihydroxy forms.

Vitamin D is efficiently absorbed from the gastrointestinal tract unless there is fat malabsorption due to lack of bile salts, pancreatic insufficiency, or a defect of intestinal mucosal function. When absorbed, the vitamin is bound directly to chylomicrons and transported initially via the lymphatics. To exert its biological activity, vitamin D must be metabolically altered.[17,24] Much of the prohormone-like vitamin initially supplied to the liver is hydroxylated at the terminal side-chain position to yield 25-hydroxy-D; this process involves molecular oxygen and a pyridine nucleotide–dependent, microsomal, mixed-function oxidase system. The 25-hydroxycholecalciferol, which represents the major metabolite of D_3 in plasma, circulates bound to a vitamin D–binding α-globulin (M.W. ~52 000). In kidneys, a second hydroxylation at the 1α-position results from the action of a cytochrome P_{450}-mediated, mixed-function oxidase located on the inner mitochondrial membrane. The kidney 1α-hydroxylase is regulated by vitamin D status and by calcium/phosphate levels which are controlled by parathormone (PTH) and calcitonin.

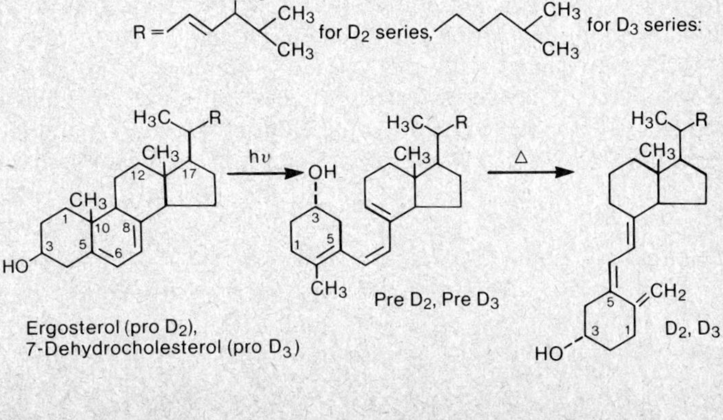

Figure 8B-3. Pro-, pre-, and vitamin D_2 and D_3.

Other aspects of vitamin D metabolism include the ability of the kidney to direct formation of 24,25-dihydroxy-D when calcium and phosphate ion levels are normal. Increased formation of 1α,25-dihydroxy-D occurs when decreased blood Ca (II) leads to parathormone release, cyclic AMP formation, renal loss of phosphate, and stimulation of the 1α-hydroxylase. The 1α,25-dihydroxycholecalciferol formed in the kidney complexes with the vitamin D–binding protein for transfer via blood to target tissues, e.g., small intestine and bone. In the intestinal mucosal cell, the hormonally active dihydroxy-D binds to a cytosolic receptor protein (M.W. ~45 000) before entering the nucleus and associating with a chromatin receptor. A prime effect is on DNA-dependent RNA polymerase II to mediate synthesis of a specific calcium-binding protein.

An outline of the transport and metabolism of vitamin D is given in Figure 8B-4. More detailed treatment of the influence of 1α,25-dihydroxycholecalciferol on calcium and phosphorus metabolism;[31,77] the regulatory interplay of parathormone, calcitonin, and prolactin;[55] and the clinical significance of insufficient and excess vitamin D[9,33,39] and suitable analytical methods for its determination[9,46,49] is found in the indicated references and in Chapter 12.

VITAMIN E

Chemistry and Sources

The form of vitamin E determined to be biologically most active, and upon which units are based, is *d-α-tocopherol*. There are eight related natural compounds in this vitaminic group, which is biosynthesized in plants and is especially abundant in vegetable oils.[29] As shown by structures in Figure 8B-5, all have a 6-chromanol nucleus substituted with methyl groups at positions 2 and 8 and with a branched isoprenoid chain at position 2. The isoprenoid chain is saturated (phytyl) for tocopherols but unsaturated at positions 3', 7', and 11' for tocotrienols. The Greek letter prefixes indicate presence or absence of methylations at positions 5 and 7. Members of the vitamin E group are viscous oils at room temperature. They are soluble in fat solvents and insoluble in water, stable to acid and heat in the absence of oxygen, but labile to oxygen in alkaline solutions and to ultraviolet light; the absorption maximum is 294 nm for α-tocopherol.

Requirements and Allowances

The requirement for vitamin E is related to the polyunsaturated fatty acid content of cellular structures and, therefore, depends upon the nature and quantity of dietary fat which affects such composition. Hence, the minimum adult requirement for vitamin E is not certain but is probably not more than 3–4 mg (4.5–6 IU) of *d-α-tocopherol* per day for those who ingest a diet containing the minimum of essential fatty acids (3% of calories).[35] Since vitamin E activity derives from a series of tocopherols and tocotrienols in usual mixed diets, calculations based on their abundance and activity relative to the biologically most active *d-α-tocopherol* are used. The milligrams of β-tocopherol are multiplied by 0.5, those of γ-tocopherol by 0.1, and those of α-tocotrienol by 0.3. Their sum plus mg of α-tocopherol accounts for the mg of α-tocopherol equivalents. It has been estimated that a range of 7–13 mg of α-tocopherol equivalents (10–20 IU) can be expected in balanced diets supplying 1800–3000 kcal. This intake will maintain plasma concentrations of total tocopherols within the reference range of 0.5–1.2 mg/dL, which also ensures an adequate concentration in all tissues.[3] It has been claimed by some that the ratio of circulating α-tocopherol to total lipids (or triglycerides, or β-lipoproteins) is a more accurate indicator of tissue vitamin E status than circulating α-tocopherol alone.

Recommended daily dietary allowances based on the foregoing considerations are 10 mg α-tocopherol equivalents for adult males and 8 mg for males from 11 to 14 years old and adult females. The increased caloric intake during pregnancy and lactation should be accompanied by increased intakes of 2 and 3 mg, to compensate for amounts deposited in the fetus and secreted

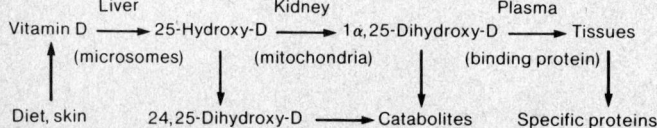

Figure 8B-4. Transport and metabolism of vitamin D.

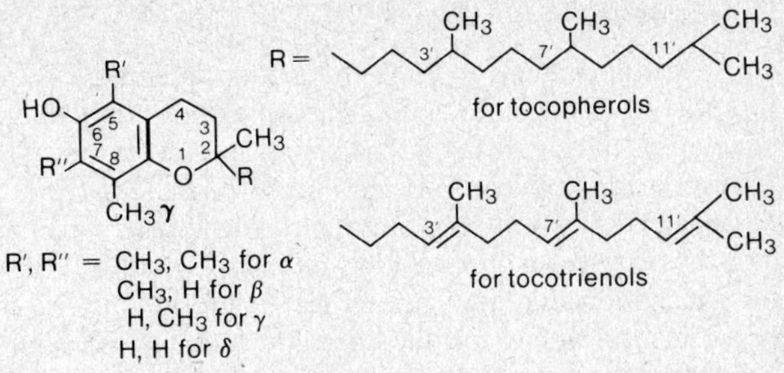

Figure 8B-5. Vitaminic forms of E.

in milk, respectively. It is recommended that infants receive 3 mg during the first half year, since the vitamin E content of human milk is from 1.3–3.3 mg d-α-tocopherol equivalent per liter; yet this is sufficient to raise blood tocopherols to the adult levels in two or three weeks. Intake from 4–8 mg/d is suggested for children.

Absorption, Transport, and Metabolism

Vitamin E is absorbed in the presence of bile and from the small intestine. Most tocopherol enters the blood stream via lymph, where the tocopherol is associated with chylomicrons and very low density lipoproteins. The vitamin is stored in most tissues, with the largest amount stored in adipose tissue. Some of the deposition is in association with lipoproteins in cellular membranes. Rapid exchange of tocopherol occurs between the erythrocyte membranes and plasma lipoproteins. When physiological amounts are administered, only a small fraction of the dose appears in urine; urinary forms are the quinoid tocopheronic acid and the β-glucuronide conjugate of the hydroquinoid form.[73]

Some of the dynamics of metabolism undoubtedly are reflected in the relative biological activity of tocopherols; for example, the less active γ-vitamer is taken up as effectively as the α-vitamer but is turned over more rapidly.[24]

Functions

The best defined role for vitamin E is as an antioxidant for unsaturated fatty acyl moieties of lipids within membranes. Oxidative damage to polyunsaturated fatty acyl parts of membrane phospholipids can occur as a result of hydrogen peroxide production by flavoprotein oxidases. Concurrent free-radical damage can ensue, as has been shown during oxidation of NADPH by liver microsomes in the presence of erythrocyte membranes.[56] Interactions among vitamin E, selenium, and sulfur amino acids have been rationalized on the basis that while E is oxidized in lieu of unsaturated fatty acyl functions, the Se-containing glutathione peroxidase helps reduce such lipid peroxides as are formed, thereby decreasing peroxidative autocatalysis.[24]

In the mammal, it has not been demonstrated that any enzyme reaction is specifically affected by vitamin E, although decreases in liver microsomal drug hydroxylation[11] and increase in net synthesis of xanthine oxidase[13] have been found in E-deficient animals. At present, the only direct evidence for participation of E-like compounds is in the anaerobic rumen bacterium, *Butyrivibrio fibrisolvens*, where *cis*-9, *trans*-11-octadecadienoate is hydrogenated by a system coupled to α-tocopherolquinol to form α-tocopherolquinone and *trans*-11-octadecanoate.[27]

Deficiency

Premature and low-birth-weight infants are particularly susceptible to development of vitamin E deficiency, since placental transfer is poor and they have such limited adipose tissue where much of the vitamin is normally stored.[82] Signs of deficiency include irritability, edema, and a hemolytic anemia. The anemia reflects the shortened life span of erythrocytes with fragile membranes; it does not respond to iron therapy, which may aggravate the condition. Deficiency symptoms rarely occur in children or adults except in cases of severe malabsorption, as for example in cystic fibrosis. Even in those cases where there are low blood levels of the vitamin, symptoms are not corrected with vitamin E therapy alone. In adults with longstanding inability to absorb fat, laboratory studies revealed enhanced fragility of erythrocytes, increased urinary excretion of creatine indicating muscle loss, and deposition of ceroid pigment in the musculature of the small intestine.[6]

Excess

Health claims for large dietary supplements of vitamin E are doubtful at best. There is, however, support from animal experiments to indicate that supplements of tocopherol can protect against chemical oxidants that constitute a portion of atmospheric pollutants, such as ozone and nitrous oxide.[24] There may also be some benefit from vitamin E therapy for humans with intermittent claudication resulting from arteriosclerotic, peripheral vascular disease.[4]

Toxicity from chronic high intake of vitamin E has been noted in animals, where competition for absorption may increase requirements for other fat-soluble vitamins, notably D and K. This problem appears to be the case for humans who already have limited vitamin K because of anticoagulant therapy following coronary infarcts, and who have also ingested large amounts of vitamin E which may additionally suppress vitamin K absorption.[9] Malaise, intestinal distress, depressed prothrombin, and ecchymoses have been reported. In general, however, relatively high doses of vitamin E, e.g., 300 mg/d, appear to be tolerated.

Methods for the Determination of Vitamin E and E Status

Methods[46,70] involve solvent extraction of vitamin E forms after saponification of samples, most commonly blood plasma. Molecular distillation and two-dimensional reverse-phase paper, thin-layer, or column chromatographies have been used for quantitation. More recently *gas chromatography* and *high-pressure liquid chromatography* permit rapid separation of the different tocopherols and tocotrienols. The method by Bieri et al.,[5] which was mentioned in the section on vitamin A, is specific for α- and λ-tocopherol, and is recommended.

Chemical methods are based on an oxidation-reduction reaction. Following specific elution techniques, fractions are commonly subjected to the Emmerie-Engel procedure in which tocopherol is oxidized to tocopheryl quinone by $FeCl_3$, and the Fe (II) in the resultant $FeCl_2$ is complexed with α,α'-dipyridyl to produce a red color. In a modification, nitrous acid reacts to form a yellow color with all of the tocols (2-methyl-2-phytyl-6-chromanols) except the α-forms which are substituted in both positions ortho to the phenolic hydroxyl. The difference between the total tocopherols as measured by the $FeCl_3$ method and the non-α forms as determined by the nitroso reaction gives a value for α-tocopherol plus α-tocotrienol.

Vitamin E status can also be monitored by the degree of *erythrocyte hemolysis* after treatment of cells with H_2O_2 (preferred), dialuric acid, or isotonic saline–phosphate buffer.[70] When feasible, evaluation should be based on the hemolysis test and the measurement of plasma tocopherol. The latter is assessed in a total lipid extract by oxidation with ferric chloride–bipyridyl to form α-tocopherol quinone, which is purified by TLC and quantitated by GLC.

VITAMIN K

Chemistry and Sources

Compounds in the vitamin K series are 2-methyl-1,4-napthoquinones, which are substituted with side chains at carbon 3. The two principal natural classes, as shown in Figure 8B-6, are the *phylloquinones* (K_1 type) synthesized in plants and the *menaquinones* (K_2 type) of bacterial origin. Most commonly, vitamin K_1 bears a saturated, phytyl, 20-carbon side chain derived from 4 isoprenoid units; K_2 shows greater variation, but an all-*trans*-farnesylgeranylgeranyl, 35-carbon chain of 7 isoprenoid units is typical. Several synthetic analogs and derivatives have been used; most relate to or derive from *menadione* (K_3), which lacks a side-chain substituent at position 3. In a sense, the menadione-type compound is a synthetic provitamin, since K_3 is converted to menaquinone, e.g. $K_{2(20)}$, by addition of the side chain in the liver. The K vitamins are insoluble in water but dissolve in organic fat solvents. They are destroyed by alkaline solutions and reducing agents and are also sensitive to ultraviolet light.

Estimated Adequate Intake

No specific RDA are presently given for vitamin K, since intestinal bacteria in normal individuals synthesize the menaquinones, which are partially absorbed. However, because the sufficiency of this source of K is uncertain over long periods, an estimated adequate range of dietary intake has been suggested.[35] The lower levels of vitamin K are based on the assumption

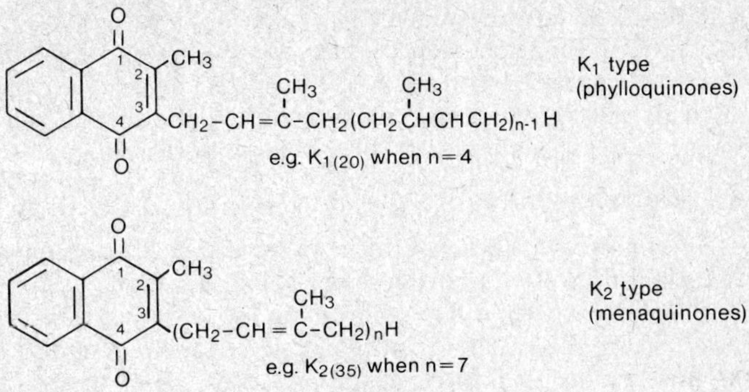

Figure 8B-6. Vitaminic forms of K.

that about half the requirement, estimated to be 2 $\mu g/kg$ of body weight,[53] is contributed by intestinal synthesis and the other half by diet. This is reasonable, since analyses of human liver indicate that half of the vitamin K present is of bacterial menaquinone type and half is of plant-derived phylloquinones.[62] The upper levels assume that all the requirement is of dietary origin. The intake suggested for adults, then, is 70–140 $\mu g/d$, which is easily supplied by an average mixed diet estimated to provide 300–500 μg of vitamin K daily.[53] The intake suggested for infants, 12 $\mu g/d$, assumes no intestinal synthesis but is within the range supplied by mature breast milk.

Absorption, Transport, and Metabolism

The absorption of natural vitamin K from the small intestine is facilitated by bile, as is true for other fat-soluble materials. Efficiency of absorption varies from 15–65% as reflected by recovery in lymph within 24 hours. Vitamins K_1 and K_2 are bound to chylomicrons for transport from mucosal cells to the liver. Menadione (K_3), on the other hand, is more rapidly and completely absorbed from the gut before entering the portal blood. In liver, intracellular distribution is mostly in the microsomal fraction, where phenylation of menadione to form K_2 occurs. Release of vitamin K to the blood stream allows association with circulating β-lipoproteins for transport to other tissues. Significant levels of vitamin K have been noted in spleen and skeletal muscle.

Within metabolically active and vitamin K–utilizing tissue, especially liver, it appears that a microsomal vitamin K cycle exists.[23,54] The vitamin is reduced by an NADH-dependent and thiol-sensitive flavoprotein system to the hydroquinone, which then can couple to the oxygen- and carbon dioxide–utilizing γ-carboxylation of glutamyl residues or specific proteins, e.g., pro-thrombin. The 2,3-epoxide of vitamin K that may be concomitantly formed is reduced to the starting vitamin K quinones, a process that can be antagonized by such vitamin K antagonists as warfarin. Whereas only traces of urinary metabolites of vitamins K_1 and K_2 appear in urine, a considerable portion of vitamin K_3 (menadione) is conjugated at the hydroquinone level to form β-glucuronide and sulfate esters, which are excreted.

Function

Initially, vitamin K was recognized as a dietary antihemorrhagic factor. It has been found to be necessary for liver syntheses of plasma clotting factors II (prothrombin), VII (proconvertin), IX (plasma thromboplastin component), and X (Stuart factor); the last three are also called autoprothrombins I, II, and III, respectively. These and other factors, including Ca(II), are known to initiate a process whereby an aggregate composed of several proteins with prothrombin, calcium ion, and phosphatide react to form thrombin, which then catalyzes the proteolytic conversion of fibrinogen ultimately to a polymerized fibrin clot. It is now known that reduced (hydroquinone) K participates in the oxygen-dependent incorporation of CO_2 into the γ-methylene of specific L-glutamyl residues of prothrombin and other plasma proteins.[76]

The search for γ-carboxyglutamyl (Gla)-containing proteins has led to the realization that the vitamin K–dependent formation of such Ca(II)-binding proteins also includes bone Gla protein or osteocalcin, which may serve a regulatory function in mineralization. The synthesis of this protein[57-59] is regulated by 1,25-dihydroxy-D_3 and acts by virtue of its specific ability to bind to hydroxyapatite.[46,60] The occurrence of Gla-containing, Ca(II)-binding proteins in kidney, urine, and renal stones has also been noted.[23,54]

Bis-4-hydroxycoumarin (dicumarol), the anti-clotting compound from spoiled sweet clover, and synthetic 4-hydroxycoumarins, such as warfarin, have found use as anticoagulants and have expedited the understanding of the metabolic cycling of vitamin K.[23] These anticoagulants interfere with the reductase-catalyzed conversion of epoxide to quinone forms of vitamin K as well as the reduction of the latter to the functional hydroquinone. Hence, 4-hydroxycoumarins are able to suppress the formation of prothrombin and other vitamin K–dependent Ca(II)-binding proteins.

Deficiency

Hemorrhagic disease of the newborn can develop readily, since the menaquinone-synthesizing intestinal flora have not become established within the first week postpartum and early breast milk is low in vitamin K. Prothrombin levels during this period are only about 25% of the adult level. Severe diarrhea and antibiotics used to suppress diarrhea readily exacerbate the situation, so that prothrombin levels can drop below 5% of the adult level and bleeding can occur. Development of vitamin K deficiency in the adult may require both reduction of dietary intake and antibiotic inhibition of intestinal microflora or 4-hydroxycoumarin-type anticoagulant therapy. Deficiency is thus relatively uncommon and found in cases with chronic malabsorption of fats, including fat-soluble vitamins, or during long-term antibiotic or anticoagulant treatments.[81] Defective blood coagulation and demonstration of abnormal noncarboxylated prothrombin are at present the only well-established signs of vitamin K deficiency.

Excess

The use of high doses of naturally occurring vitamin K (K_1 and K_2) appears to have no untoward effect; however, menadione (K_3) treatment can lead to the formation of erythrocyte cytoplasmic inclusions known as Heinz bodies and hemolytic anemia.[9] With severe hemolysis, increased bilirubin formation and undeveloped capacity for its conjugation may produce kernicterus in the newborn.

Methods for the Determination of Vitamin K and K Status

Direct methods for the quantitation of the different forms of vitamin K, following extraction and chromatographic separations,[36] involve spectrophotometry.[49] Colorimetric determinations based on reactivity of the quinoid nucleus of K vitamins include reactions with acidic phenylhydrazine, 2,6-dichloroindophenol, active methylene compounds, alkalis, and piperidine. Also, spectrophotometric determination of chemically reduced vitamin K has been used.

Conventional assessment of vitamin K status relies on the clotting ability of plasma as reflected by the *prothrombin time*. In this procedure, it is usual to add tissue thromboplastin (often in crude form from rabbit brain) to recalcified plasma and measure the time required for clot formation compared with a normal control. When the prothrombin concentration declines below 30% of normal, prothrombin time rises above 30 s. Deficiency of vitamin K can be distinguished from hypoprothrombinemia of liver disease by measurement with RIA of the non-γ-carboxylated prothrombin precursor (i.e., "abnormal prothrombin") that accumulates in plasma when the vitamin is deficient. Alternatively, if a vitamin K deficiency exists, parenteral administration of the vitamin will result in prompt correction of an abnormal prothrombin or clotting time. It has been suggested that routine determination of prothrombin time should be performed before all surgical procedures and deliveries.[81]

THIAMIN

Chemistry and Sources

Thiamin (thiamine, vitamin B_1) is a pyrimidyl-substituted thiazole [3-(4-amino-2-methyl-pyrimidyl-5-methyl)-4-methyl-5-(β-hydroxyethyl)thiazole] (Figure 8B-7). The basic vitamin is isolated or synthesized and handled as a solid thiazolium salt, e.g., thiamin chloride hydrochloride. The principal if not sole coenzyme form is the pyrophosphate ester (TPP) formed at the β-hydroxyethyl substituent and also shown in Figure 8B-7. Monophosphate and triphosphate esters occur naturally. Small amounts of thiamin and its phosphates are present in most plant and animal tissue, but more abundant sources are unrefined cereal grains, liver, heart, kidney, and lean cuts of pork. The enrichment of flour and derived food products has considerably

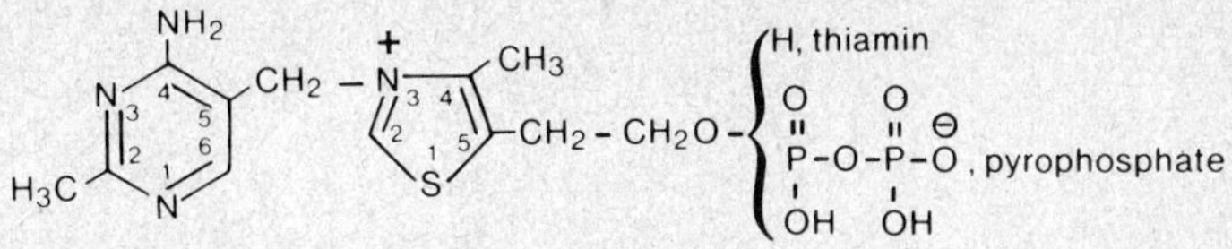

Figure 8B-7. Thiamin and the pyro-phosphate coenzyme.

increased the availability of this vitamin. Thiamin is somewhat heat labile, particularly in alkaline solutions, where base attack at carbon 2 of the thiazolium ring occurs.

Requirements and Allowances

The requirement levels for thiamin vary considerably, since there is a direct correlation of need with amount of metabolizable carbohydrate intake. There is a generally increased requirement under numerous situations where metabolism is heightened, for example, in the normal conditions of increased muscular activity, pregnancy, and lactation or in the abnormal cases of protracted fever and hyperthyroidism.[22] Clinical signs of deficiency in adults can be prevented with intakes of thiamin above 0.15–0.2 mg/1000 kcal, but 0.35–0.4 mg/1000 kcal may be closer to a level necessary to maintain urinary excretion and TPP-dependent erythrocyte transketolase activity within normal ranges.[35] On the basis of these considerations, an adult allowance of 0.5 mg/1000 kcal has been recommended. With further considerations of average caloric intakes and activities in different age groups, the RDA expressed as mg/d range from 1.0 for older women to 1.5 for young men. The requirement for pregnant women increases early in pregnancy and then remains constant; an additional allowance of 0.4 mg/d is recommended. The lactating woman secretes 0.1–0.2 mg of thiamin per day in milk, so an additional 0.5 mg/d allowance is suggested. Based on the thiamin content of human milk and with an increment considered to provide a margin of safety, 0.3 mg/d is the allowance for young infants. Increases above this are suggested for growing children.

Absorption, Transport, and Metabolism

Thiamin is readily absorbed in the small intestine by an active transport process that is probably carrier-mediated so long as intake is less than 5 mg/d; at higher intakes, passive diffusion increasingly contributes to absorption.[24,81] Phosphorylation takes place in the jejunal mucosa to yield thiamin pyrophosphate. Thiamin is carried by the portal blood to the liver. The free vitamin occurs in the plasma, but the coenzyme, TPP, predominates in the cellular components. Approximately 30 mg is stored in the body with 80% as the pyrophosphate, 10% as triphosphate, and the rest as thiamin and its monophosphate. About half of the body stores are found in skeletal muscles, with much of the remainder in heart, liver, kidneys, and nervous tissue including brain, which contains most of the triphosphate. The three tissue enzymes known to participate in formation of the phosphate esters are thiaminokinase (a pyrophosphorylase), which catalyzes formation of TPP and AMP from thiamin and ATP; TPP-ATP phosphoryl-transferase, which forms the triphosphate and ADP from TPP and ATP; and thiamin triphosphatase, which hydrolyzes TPP to the monophosphate. Though thiaminokinase is widespread, the phosphoryl transferase and membrane-associated triphosphatase are mainly in nervous tissue.

Thiamin, as well as several of its catabolites (illustrated in Figure 8B-8), is excreted into the urine by the renal tubules.[12,24]

As a thiamin deficiency develops, there is a rather rapid loss of the vitamin from all tissues except the brain. The decrease of TPP in the erythrocyte roughly parallels the decrease of this coenzyme in other tissues. During this time, the thiamin in urine falls to near zero; the urinary metabolites remain high for some time before decreasing.

Functions

There are two general types of reactions in the human, where TPP functions as the Mg(II)-coordinated coenzyme for so-called "active aldehyde" transfers. These are the oxidative decarboxylation of α-keto acids catalyzed by dehydrogenase complexes and the formation of α-ketols (ketoses) as catalyzed by transketolase.[80]

The multienzymic dehydrogenase complexes that effect decarboxylative conversion of α-keto acids to acyl-CoA derivatives, for example, pyruvate dehydrogenase, are localized in the mitochondria, where efficient utilization in the Krebs tricarboxylic acid (citric acid) cycle follows.

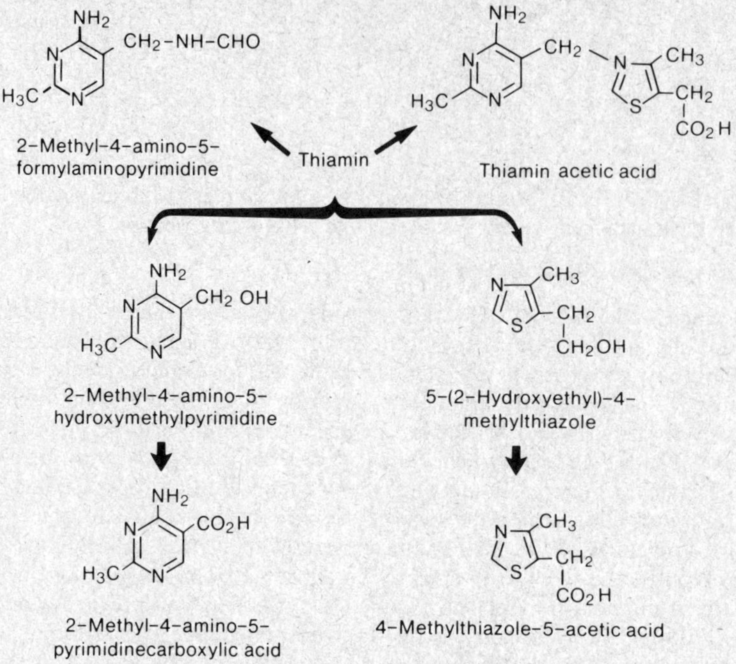

Figure 8B-8. Principal urinary catabolites of thiamin.

Three types of subunit proteins constitute such dehydrogenase complexes: a TPP-dependent decarboxylase, which converts the α-keto acid to an α-hydroxyalkyl-TPP complex; a transacylase core, which contains lipoyl residues that are acylated by the α-hydroxyalkyl-TTP; and an FAD-dependent dihydrolipoyl dehydrogenase, which reoxidizes the reduced lipoyl residues produced after transfer of their acyl functions to reduced CoA. In addition to energy and ultimate ATP supply derived from reactions in the Krebs cycle, the initial pyruvate dehydrogenase–catalyzed step importantly provides acetyl-CoA as a biosynthetic precursor to other essential compounds such as lipids and acetylcholine of the parasympathetic nervous system.

Transketolase is a TPP-dependent enzyme found in the cytosol of many tissues, especially liver and blood cells, where principal carbohydrate pathways exist. In the pentose phosphate pathway, which additionally supplies NADPH needed for biosynthetic reactions, this enzyme catalyzes the reversible transfer of a glycoaldehyde moiety from the first two carbons of a donor ketose phosphate to the aldehyde carbon of an aldose phosphate.

Although thiamin as its pyrophosphate contributes to nervous system composition and function in such essential reactions as energy production and biosynthesis of lipids and acetylcholine, it appears that there is another incompletely understood role, particularly for the triphosphate. Thiamin and its phosphate esters are located in axonal membranes of nerves; electrical stimulation leads to hydrolysis and release of both the di- and triphosphate.[24] As noted before, enzymes involved in formation and cleavage of thiamin triphosphate are in nervous tissue. Moreover, a subacute necrotizing encephalomyelopathy in patients with Leigh's syndrome results from the presence of an inhibitor of TPP-ATP phosphoryl transferase.[24,80]

Deficiency

Beriberi is the disease resulting from thiamin deficiency. The causes for deficiency[24,35] include inadequate intake due to diets largely dependent on milled, nonenriched grains such as rice and wheat, or the ingestion of raw fish containing microbial thiaminases,[19] which hydrolytically destroy the vitamin in the GI tract. Tea may contain antithiamin factors that have been detected in certain other plant extracts. Chronic alcoholism is a common contributor to deficiency in that there is not only a low intake of thiamin (and other B vitamins) but also impaired absorption and storage. There are several thiamin-responsive inborn errors of metabolism;[67,72] these include a megaloblastic anemia of unknown mechanism, lactic acidosis due to low or defective pyruvate decarboxylase, branched-chain ketoaciduria with poor activity of the keto acid dehydrogenase system, and subacute necrotizing encephalomyelopathy where there is lack of thiamin triphosphate in neural tissues. Therapeutic doses of 5–20 mg of thiamin daily have proved beneficial in some cases.[9] Finally, other at-risk patients are those undergoing long-term renal dialysis or intravenous feeding and even those with chronic febrile infections.[35]

Clinical signs primarily involve the nervous and cardiovascular systems.[35] In the adult, symptoms most frequently observed are mental confusion, anorexia, muscular weakness, ataxia, peripheral paralysis, ophthalmoplegia, edema (wet beriberi), muscle wasting (dry beriberi), tachycardia, and an enlarged heart. In infants, symptoms appear suddenly and severely, often involving cardiac failure and cyanosis. Commonly, the distinction between wet (cardiovascular) and dry (neuritic) manifestations of beriberi relate to duration and severity of the deficiency, the degree of physical exertion, and caloric intake.[81] The wet or edematous condition results from severe physical exertion and high carbohydrate intake, whereas the dry or polyneuritic form stems from relative inactivity with caloric restriction during the chronic deficiency. The three major physiologic derangements that involve the cardiovascular system are peripheral vasodilatation leading to a high-output state, biventricular myocardial failure, and retention of sodium and water leading to edema. Nervous system involvement includes peripheral neuropathy, Wernicke's encephalopathy, and the amnesic psychosis of Korsakoff's syndrome.[9]

Methods for the Determination of Thiamin and Thiamin Status

Earlier microbiologic methods and the chemical conversion (by ferricyanide in alkaline solution) of thiamin to fluorometrically determined *thiochrome* were the principal means to determine the vitamin in various biologic fluids. There are now several useful modifications for the chemical determination of the free vitamin as well as its phosphate esters, which can be separated by electrophoresis, ion-exchange, or HPLC techniques.[44,47] Determination of the urinary excretion of thiamin in a 4-h specimen, especially with comparison of excretion before and after a test load, is helpful in differentiating between extremes of thiamin status. But as is the case in most assessments based on amount of water-soluble vitamins in urine, excretion can be influenced considerably by dietary intake, absorption, and other factors. Measurements of certain urinary metabolites, notably thiamin acetic acid, have also been suggested as being reflective of thiamin status.[36]

Biochemical tests based on the functional level of TPP may reflect more adequately the immediate status of an individual. One such diagnostic test is the measurement of the ratio of lactic to pyruvic acids in the blood after administration of glucose. Blood and urinary levels of these metabolites are characteristically elevated in deficiency, as would be expected when the TPP-dependent pyruvate dehydrogenase is decreased. One disadvantage of this test is the potential of provoking a lactic acidosis in a severely deficient patient given a glucose load.[22] Other catabolites elevated as a consequence of thiamin deficiency include α-ketoglutarate, glyoxylate, and urinary methylglyoxal.[24]

The most useful and reliable method presently used for assessing thiamin status is a measurement of whole blood or **erythrocyte transketolase.** This enzyme catalyzes two reactions in the pentose phosphate pathway (see also Figure 8B-9):

Xylulose-5-phosphate + ribose-5-phosphate → sedoheptulose-7-phosphate + glyceraldehyde-3-phosphate
Xylulose-5-phosphate + erythrose-4-phosphate → fructose-6-phosphate + glyceraldehyde-3-phosphate

Transketolase activity in blood is usually measured by determining the rate of disappearance of D-ribose-5-phosphate with the orcinol reagent (orcinol and ferric chloride in concentrated hydrochloric acid). Transketolase activity in blood can also be measured by determining the amount of fructose-6-phosphate formed in the second reaction above, utilizing the anthrone reagent.

The enzyme assay is performed by incubating a sample of heparinized blood (previously hemolyzed in a phosphate buffer containing sodium chloride and magnesium sulfate) with ribose-5-phosphate and removing aliquots of the incubation mixture at three successive time intervals. Trichloroacetic acid is immediately added to the aliquots to stop the reaction. After centrifugation, the supernatant is analyzed for the remaining level of ribose-5-phosphate by the orcinol reaction. The product of the transketolase reaction, sedoheptulose-7-phosphate, also reacts with the orcinol reagent to yield a colored product with an absorption spectrum (λ_{max} = 670 nm) different from but overlapping with the spectrum of the colored product formed when ribose-5-phosphate reacts

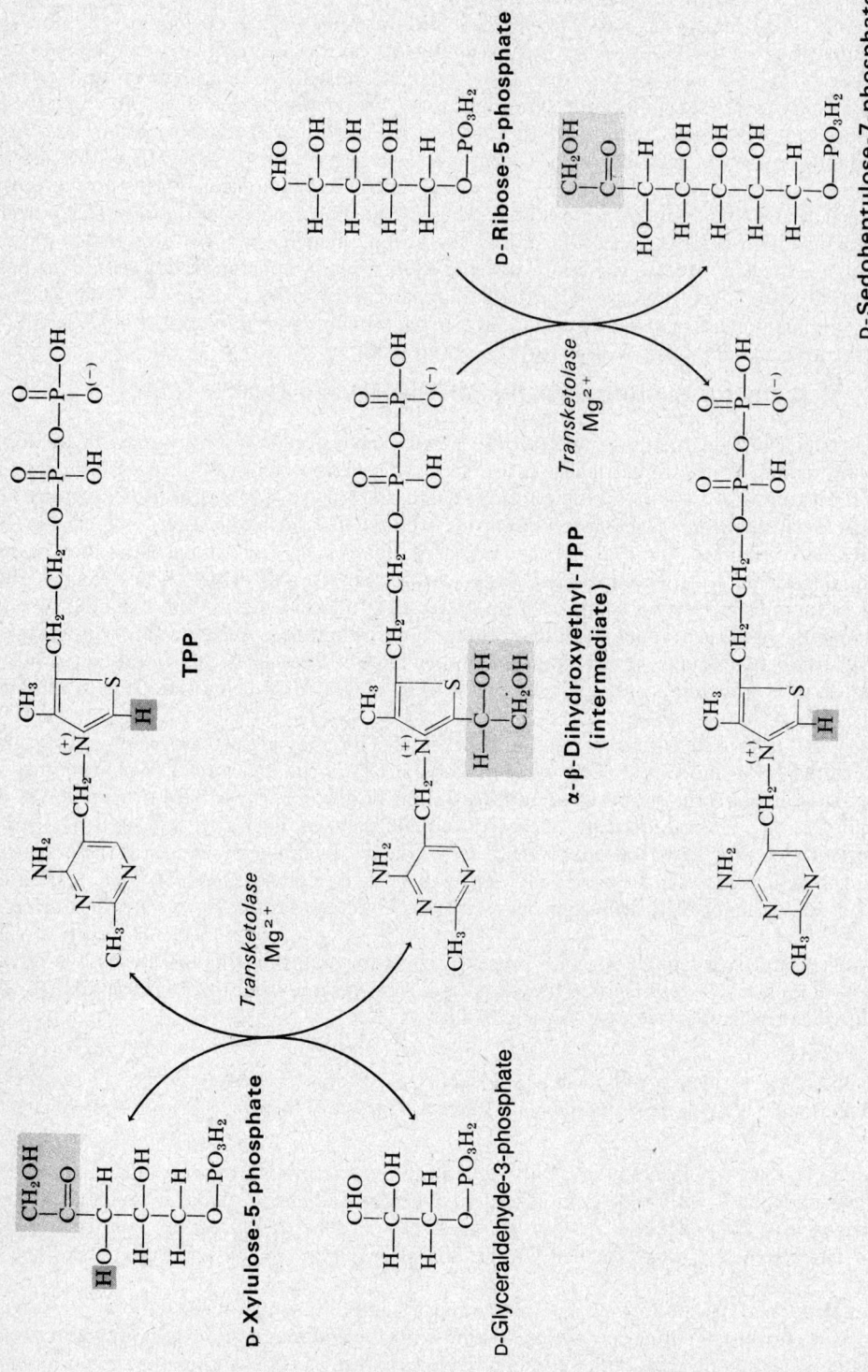

Figure 8B-9. The transketolase reaction. In this reaction a two-carbon fragment (shaded area) is transferred from a donor ketose (D-xylulose-5-phosphate) to thiamin pyrophosphate (TPP), and then finally to an acceptor aldose (D-ribose-5-phosphate). The products of this reaction are D-glyceraldehyde-3-phosphate and D-sedoheptulose-7-phosphate.

with the orcinol reagent ($\lambda_{max} = 580$ nm). By measuring the absorbances of samples at both 670 nm and 580 nm, and solving simultaneous equations, one can measure transketolase activity in terms of both rate of substrate (ribose-5-phosphate) disappearance and rate of product (sedoheptulose-7-phosphate) formation.[79]

Although ribose-5-phosphate is used as the substrate for the transketolase assay, xylulose-5-phosphate is also needed for this reaction. Xylulose-5-phosphate is not included in the assay mixture because the erythrocytes contain nonlimiting amounts of two extremely active enzymes that can convert ribose-5-phosphate to xylulose-5-phosphate. These enzymes are *phosphoribose isomerase*, which forms ribulose-5-phosphate from ribose-5-phosphate, and *pentose phosphate epimerase*, which forms xylulose-5-phosphate from ribulose-5-phosphate. Ribose-5-phosphate added to an incubation mixture containing these enzymes is rapidly converted to an equilibrium mixture of ribose-5-phosphate, ribulose-5-phosphate, and xylulose-5-phosphate in a 1/9/1 ratio. There is a rapid decrease in ribose-5-phosphate during the first 5 min of incubation due to the formation of this equilibration mixture of pentoses. The decrease in ribose-5-phosphate concentration after the 5-min equilibration period is linear with time and is presumed to be due only to transketolase activity. Some previously proposed assays of transketolase activity measure ribose-5-phosphate disappearance before the equilibration of the pentoses takes place. Such methods give falsely high values for transketolase activity.

The sensitivity of the above procedure as a measure of thiamin deficiency can be improved by performing the assay before and after addition of thiamin pyrophosphate (TPP). Presence of TPP results in optimal enzyme activity. This increase in activity is known as the TPP effect. (See below.)

Methods for the Determination of Erythrocyte Transketolase Activity

In the Brin procedure[47] activities of holo- and apo- forms of transketolase in erythrocyte hemolysates are measured before and after addition of TPP, by colorimetric determinations of the amount of ribose-5'-phosphate utilized or hexose-6-phosphate formed. A detailed description of the Brin method has been published.[47] This method is reliable but very time consuming.

In an alternative method by Smeets, Muller and De Wael,[74] the rate of formation of glyceraldehyde-3-P is measured indirectly by a coupled reaction in a system containing excess triosephosphate isomerase (TIM), glycerolphosphate dehydrogenase (GDH), and NADH. Glyceraldehyde-3-P is converted by TIM to dihydroxyacetone-P, which, in the presence of GDH and NADH, is reduced to glycerol-1-P. The rate of NADH oxidation, measured at 340 nm, is proportional to the TK activity.

Specimen

Heparinized blood is drawn (1 mL of blood is sufficient for the semimicro method) before administering any oral or parenteral therapy. The specimen is centrifuged, and the plasma as well as the buffy coat is removed by suction. The packed red cells are transferred to a preweighed vial. A volume of distilled water equal to the weight of packed cells is added, and the cells are resuspended by stirring with a wood applicator stick. The mixture is frozen with the vial tipped to prevent breakage. (Repeated freezing and thawing, however, result in loss of transketolase activity.)

Discussion

Transketolase is an enzyme within the erythrocyte, and as such it is independent of nonspecific changes in the extracellular plasma. As vitamin B_1 deficiency becomes more severe, (1) thiamin becomes limiting in the body cells, (2) the amount of the coenzyme is depleted, and therefore (3) the transketolase activity diminishes. The "TPP effect" measures the extent of depletion of the transketolase enzyme for coenzyme.

Ranges

Most normal individuals have an RBC-TK activity range of 850–1000 μg of hexose formed per mL of hemolysate per hour. Duplicate determinations may vary by 5%, and repeat assays by 10%. No differences have been observed between males and females. The ranges of TPP effect in per cent are normal, 0–15; marginally deficient, 15–25; severely deficient (with clinical signs) ≥ 25.

Hexose activity of less than 800 μg hexose formed/mL of hemolysate per h has been generally associated with a TPP effect greater than 15%.

The TPP effect is generally eliminated within 2–4 h after the parenteral administration of 50 mg of thiamin to the patient. Transketolase activity responds slowly to oral therapy. Occasionally, however, patients with very low transketolase activity may be encountered (i.e., 200–400 μg of hexose/mL of hemolysate per h) in whom the TPP effect is effectively eliminated by the parenteral therapy, but whose total enzyme activity may not return to normal values for 1–14 days. These are generally very malnourished individuals, often with liver involvement, who may have an apotransketolase deficit in addition to a coenzyme depletion. Although this situation may appear to be a deterrent to the use of the assay, the TPP-effect data are still applicable in that they reflect the proportion of the available transketolase apoenzyme which was depleted of thiamin pyrophosphate.

The hexose activity and the TPP effect associated with hexose activity may be more reflective of objective and subjective clinical findings than is the TPP effect associated with pentose activity, but some laboratories have found the pentose data more useful.

RIBOFLAVIN

Chemistry and Sources

Riboflavin [vitamin B$_2$; 7,8-dimethyl-10-(1'-D-ribityl)isoalloxazine] is a yellow fluorescent compound which is widely distributed throughout the plant and animal kingdoms. Numerous naturally occurring flavins are now known to have the vitamin or closely related derivatives as an integral part of their structures.[16] Principal among these are the coenzymes flavin mononucleotide (FMN) and flavin adenine dinucleotide (FAD). The components of the naturally predominant FAD are indicated in the structure given in Figure 8B-10. Such coenzyme forms are rich sources of the vitamin in liver, kidney, and heart. Many vegetables are also good sources, but cereals are rather low in flavin content. Raw milk is a good source of the vitamin, but considerable loss can occur from exposure to light during pasteurization and bottling or as a result of irradiation to increase the vitamin D content. Flavins are stable to heat but are decomposed by light, which causes photodegradation of the D-ribityl side chain at position 10 of the isoalloxazine ring system to yield ultimately lumiflavin (7,8,10-trimethylisoalloxazine) under alkaline conditions and lumichrome (7,8-dimethylalloxazine) at all pH values, especially in neutral-to-acidic solutions. Flavins are chemically and biologically reduced to nearly colorless compounds which rapidly reoxidize upon exposure to air (oxygen).

Requirements and Allowances

Assessment of riboflavin status has been made on the bases of the relationship of dietary intake to overt signs of hyporiboflavinosis, urinary excretion of the vitamin, erythrocyte riboflavin content, and, more recently, erythrocyte glutathione reductase activity.[35] Calculations have been based on protein allowances, energy intakes, and metabolic body size, but these do not differ significantly, since they are interdependent. At least 0.5 mg of riboflavin per 1000 kcal is required by the adult, and 0.6 mg/1000 kcal constitutes the allowance suggested for all ages; this translates to levels of 1.2–1.7 mg/d for middle-aged women and young men. Since elderly people and occasionally others may take in less than 2000 kcal/d, a minimum intake of 1.2 mg riboflavin/d is recommended for safety. Since pregnant women tend to excrete less riboflavin as pregnancy progresses and additionally exhibit FAD stimulation of erythrocyte glutathione reductase activity, recommended allowances call for an additional 0.3 mg/d during pregnancy. During lactation, approximately 40 μg of riboflavin is secreted into 100 mL of human milk. A mean daily secretion of milk, calculated over six months, is 850 mL. This translates into a required additional intake of 0.34 mg riboflavin per day. Assuming 70% utilization of ingested vitamin, an increment of 0.5 mg/d is recommended for the lactating woman.

Absorption, Transport, and Metabolism

The processes by which riboflavin and lesser amounts of natural derivatives are released by digestion of complexes with food proteins, and then absorbed, transported, and metabolically altered, has been reviewed in a fairly comprehensive manner.[16] Salient features are that coenzyme

Riboflavin

Figure 8B-10. Riboflavin and FMN as components of FAD.

FMN

AMP

FAD

forms of the vitamin (mainly FAD and less FMN) are released from noncovalent attachment to proteins as a consequence of gastric acidification. Nonspecific action of pyrophosphatase and phosphatase on the coenzyme forms occurs in the upper gut. Several per cent of 8α-(amino acid)riboflavins originally in covalent attachment to certain enzymes, e.g., mitochondrial succinate dehydrogenase or monoamine oxidase, and traces of other ring and side-chain substituted flavins are also released by these actions following proteolysis. The vitamin is primarily absorbed in the proximal small intestine by a saturable transport system that is rapid and proportional to intake before leveling off at doses near 25 mg riboflavin per day. Bile salts appear to facilitate the uptake, and a modest amount of the vitamin circulates via the enterohepatic system.[24] Active transport at lower levels of intake may be Na^+-dependent and involve phosphorylation. In the human, some of the riboflavin circulating in blood plasma is loosely associated with albumin, though significant amounts complex with other proteins. One subfraction of IgG has been found to bind avidly a small portion of the total free flavin in blood.[51]

Metabolic interconversions of flavins at the cellular level are outlined in Figure 8B-11. Conversion of riboflavin to coenzymes occurs within the cellular cytoplasm of most tissue, but particularly in the small intestine, liver, heart, and kidney.[16,24] The obligatory first step is the ATP-dependent phosphorylation of the vitamin catalyzed by flavokinase. The FMN product can be complexed with specific apoenzymes to form several functional flavoproteins, but the larger quantity is further converted to FAD in a second ATP-dependent reaction catalyzed by FAD synthetase (pyrophosphorylase). It seems likely that the biosynthesis of flavocoenzymes, particularly at the flavokinase step, is tightly regulated. Thyroxine and triiodothyronine stimulate FMN and FAD synthesis in mammalian systems.[63,64] FAD is the predominant flavocoenzyme present in tissues where it is mainly complexed with numerous flavoprotein dehydrogenases and oxidases. Less than 10% of the FAD can also become covalently attached to specific amino acid residues of a few important apoenzymes. Examples include the 8α-N(3)-histidyl FAD within succinate dehydrogenase and 8α-S-cysteinyl FAD within monoamine oxidase, both of mitochondrial localization. Turnover of covalently attached flavocoenzymes requires intracellular proteolysis, and further degradation of the coenzymes involves nonspecific pyrophosphatase cleavage of FAD to FMN and AMP and further action by nonspecific phosphatases on FMN and AMP. Since there

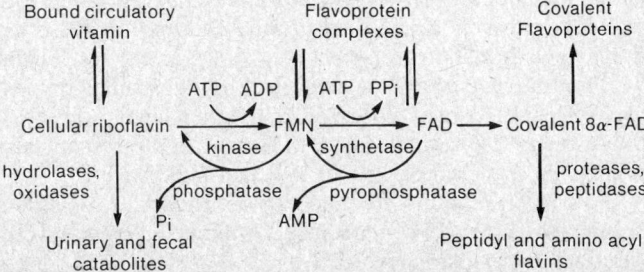

Figure 8B-11. Cellular interconversions of flavins.

is little storage of riboflavin as such, the urinary excretion reflects dietary intake. Milk contains reasonable quantities of the vitamin and lesser amounts of coenzyme, principally FMN. Smaller amounts of sidechain degradation products, such as lumichrome, 10-formylmethylflavin, and 10-(2′-hydroxyethyl)flavin, and ring-altered compounds are also excreted and may largely result from action of intestinal microorganisms.[16] Traces of 8α-flavin peptides and catabolites are found in urine and feces.

Functions

In bound coenzymic form, riboflavin participates in oxidation-reduction reactions in numerous metabolic pathways and in energy production via the respiratory chain. There is a diversity of chemical reactions catalyzed by flavoproteins.[16,24,40,64] Their functions include one-electron transfers during which the oxidized quinone level of flavin is half reduced to the radical semiquinone, two-electron transfer to a full-reduced hydroquinone in dehydrogenations that are both pyridine nucleotide dependent and independent, reactions with sulfur-containing compounds, hydroxylations, oxidative decarboxylations, dioxygenations, and reduction of O_2 to hydrogen peroxide. The intrinsic abilities of flavins to be varyingly potentiated to serve as redox carriers upon differential binding to proteins, to participate in both one- and two-electron transfers, and in reduced (1,5-dihydro) form to react rapidly with oxygen, permit wide scope in their operation.

Deficiency

Although riboflavin has a wide distribution in foodstuffs, many people live for long periods on low intakes, and consequently minor signs of deficiency are common in many parts of the world.[82] Several diseases affect riboflavin status.[37] Moreover, deficiency is encountered almost invariably in combination with deficit of other water-soluble vitamins. Clinical deficiency of riboflavin has been induced by feeding a riboflavin-deficient diet and by the administration of an antagonist such as galactoflavin. The deficiency syndrome is characterized by sore throat, hyperemia and edema of the pharyngeal and oral mucous membranes, cheilosis, angular stomatitis, glossitis (magenta tongue), seborrheic dermatitis, and normochromic, normocytic anemia associated with pure red cell aplasia of the bone marrow.[81] As noted above, though, some of these symptoms, such as glossitis and dermatitis, when encountered in the field may have resulted from other complicating deficiencies. Severe riboflavin deficiency can also affect the conversion of vitamin B_6 to its coenzyme[20] and even curtail conversion of tryptophan to niacin.[80]

Methods for the Determination of Riboflavin and Flavin Status

There are numerous biochemical methods aimed at the separation and quantitation of the diverse natural flavins.[45,48] Among the more sensitive are those that invoke specific binding, e.g., riboflavin with egg white riboflavin-binding protein, FMN with apoflavodoxin, and FAD with apoproteins for D-amino acid oxidase or glucose oxidase. However, nutritional status is commonly assessed by measuring urinary excretion of the vitamin in fasting, random, or 24-h specimens; or by load return tests, measurement of erythrocyte riboflavin concentration, or determination of the erythrocyte glutathione reductase activity coefficient.[9,37]

Urinary riboflavin can be measured by *fluorometric* as well as by microbiological procedures. Under conditions of adequate intake, the amount excreted per day is more than 120 μg or 80 μg/g creatinine. The rate of excretion expressed as μg/g creatinine is greater for children than for adults. Conditions causing negative nitrogen balance and the administration of antibiotics and certain psychotropic drugs (phenothiazine) increase urinary riboflavin as a consequence of tissue depletion and displacement. A load return test augments the reliability in a given case.

Erythrocyte riboflavin can also be determined by either fluorometric or microbiological means. Since changes observed are rather small, there is some problem with sensitivity and interpretation of results. Nevertheless, it is clear that values below 15 μg/dL cells should be considered as reflecting low or deficient status.

Currently, the most commonly used method for assessing riboflavin status utilizes the determination of FAD-dependent glutathione reductase activity in freshly lysed erythrocytes as described by Sauberlich et al.[69]

Method for the Determination of Erythrocyte Glutathione Reductase Activity

Principle

Activities of holo and apo forms of glutathione reductase in erythrocyte hemolysates are measured, with and without addition of FAD, by spectrophotometric determinations of NADP formed:

$$Apoenzyme + FAD \rightarrow Holoenzyme$$

$$G\text{-}S\text{-}S\text{-}G + NADPH + H^+ \xrightarrow{\text{Holoenzyme}} 2G\text{-}SH + NADP^+$$

Specimen

Venous blood is obtained from a fasting subject, using EDTA or heparin as anticoagulant. Washed erythrocytes are obtained by mixing 0.2 mL of blood with 1.0 mL of cold NaCl, 0.15 mol/L, and centrifuging; this is repeated twice more. Prior to the assay, 1.5 mL of double-distilled water is added and the mixture is centrifuged to obtain a dilute hemolysate.

Reagents

1. Potassium phosphate buffer, 0.1 mol/L, pH 7.4.
2. Reduced NADP. Dissolve 16.6 mg of tetrasodium salt in 10 mL of 1% sodium bicarbonate. Prepare daily.
3. Oxidized glutathione, 7.5 mmol/L. Prepare daily by dissolving 46 mg in 10 mL of double-distilled water to which 0.1 mL of sodium hydroxide, 1 mol/L, has been added.
4. FAD, 0.25 mmol/L. Dissolve 2.4 mg of monosodium salt in 10 mL of double-distilled water. Prepare daily.
5. EDTA, 80 mmol/L. Dissolve 1.5 g of dipotassium salt in 50 mL of double-distilled water.

Procedure

1. Pipet 0.1 mL of FAD solution into one cuvet and 0.1 mL of distilled water into another cuvet.
2. Add successively to the FAD-containing cuvet 2.0 mL of phosphate buffer, 0.05 mL of EDTA solution, 0.1 mL of dilute hemolysate, and 0.1 mL of oxidized glutathione solution.
3. Equilibrate cuvet and contents at 37 °C for 8 min.
4. Add 0.1 mL of reduced NADP solution.
5. Measure decrease in absorbance at 340 nm over a 10-min period to obtain ΔA_{340} in the presence of FAD.
6. Repeat steps 2 through 5 with the water-containing cuvet to obtain ΔA_{340} without FAD.

Calculations

Erythrocyte glutathione reductase assays are expressed in terms of "activity coefficients" (AC), which represent the degree of stimulation of apoenzyme resulting from addition in vitro of FAD.

$$AC = \Delta A_{340} \text{ with FAD}/\Delta A_{340} \text{ without FAD}$$

Discussion

The biochemical method of choice for assessing riboflavin status, the erythrocyte glutathione reductase assay, has one drawback.[37] The test cannot be used in persons with glucose-6-phosphatase deficiency because of an increased avidity of the glutathione reductase for FAD in this disease.

Reference Ranges

An AC of 1.0 indicates no stimulation and that only holoenzyme is present as a result of adequate amounts of FAD (and riboflavin) in the original erythrocytes. Suggested guidelines for interpretation of such coefficients are < 1.2, acceptable; 1.2–1.4, low; > 1.4, deficient.

VITAMIN B$_6$

Chemistry and Sources

The vitamin B$_6$ group comprises three natural forms: *pyridoxine* (pyridoxol), *pyridoxamine,* and *pyridoxal,* which are 4-substituted 2-methyl-3-hydroxyl-5-hydroxymethyl pyridines as shown in Figure 8B-12. During metabolic conversions, each vitamer becomes phosphorylated at the 5-hydroxymethyl substituent as also shown in Figure 8B-12. Although both pyridoxamine-5′-phosphate and pyridoxal-5′-phosphate (PLP, P-5′-P) interconvert as coenzyme forms during aminotransferase (transaminase)-catalyzed reactions, PLP is the coenzyme form that participates in the large number of B$_6$-dependent enzyme reactions. Vitamin B$_6$ is widely distributed in animal and plant tissues, where the phosphorylated forms, and particularly PLP, predominate. Meats, poultry, and fish are good sources, as are yeast, certain seeds, and bran; somewhat more limited sources are milk, eggs, and green leafy vegetables.[22,35,81,82] The common commercial form of the vitamin is pyridoxine hydrochloride, which is a water-soluble, white, crystalline solid. Solutions of the B$_6$ vitamers are decomposed by light, especially in the UV region at neutral to alkaline pH. The reactive aldehyde function of PLP leads to significant loss during thermal processing of foods.[21]

Requirements and Allowances

Requirements for vitamin B$_6$ are complicated by differences in protein intake, by the probable provision of a fraction of the needed quantity through bacterial synthesis in the intestinal tract, by the use of alcohol and oral contraceptives, and less frequently by cases in which extra needs are apparent.[22,35,81] Estimates of requirements with some margin of safety have been based on the production and cure of clinical signs of deficiency, but more often on biochemical parameters. The latter include the determination of the urinary excretion of vitamin B$_6$ and 4-pyridoxic acid or xanthurenic acid after a tryptophan load test, the plasma levels of PLP, and red cell transaminase activity.[35] A ratio of 0.02 mg of vitamin B$_6$/g of protein intake has been suggested for normal adults and may be extrapolated to children and adolescents. For adult males ingesting 100 g of protein per day, the RDA is 2.2 mg; for females ingesting 100 g protein per day, it is 2.0 mg. An addition of 0.6 mg B$_6$ per day is suggested for the pregnant woman to match the increased protein allowance during gestation. During lactation, an additional 0.5 mg/d is recommended to accommodate for extra protein intake and provide a level of 0.10–0.25 mg/L of the vitamin in milk, which is adequate for the breast-fed infant.

Absorption, Transport, and Metabolism

The processes by which vitamin B$_6$ is taken in and utilized in numerous biochemical reactions have been treated in a recent volume.[78] The three B$_6$ vitamers, mostly released from their 5′-phosphate esters by intraluminal action of intestinal alkaline phosphatase, are readily absorbed by the mucosal cells, which contain cytoplasmic pyridoxal kinase responsible for catalyzing the ATP-dependent phosphorylation of all three vitamin forms. As shown in Figure 8B-13, it is probable that other cells utilizing vitamin B$_6$ also allow passive diffusion of free vitamin into the cell followed by "metabolic trapping" as phosphates. Most cells contain a cytosolic FMN-dependent, pyridoxine (pyridoxamine)-5′-phosphate oxidase responsible for catalyzing the O$_2$-dependent conversion of pyridoxine phosphate and pyridoxamine phosphate to PLP (and H$_2$O$_2$). The coenzyme can enter directly into subcellular organelles, such as hepatocyte mitochondria.[32] PLP binds for catalytic function with numerous specific apoenzymes throughout the

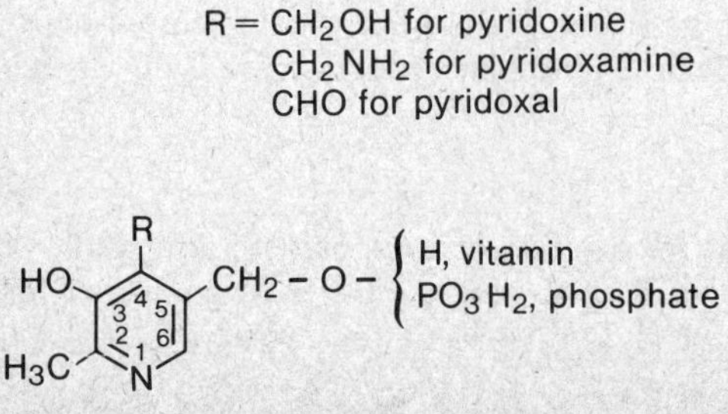

R = CH$_2$OH for pyridoxine
CH$_2$NH$_2$ for pyridoxamine
CHO for pyridoxal

Figure 8B-12. Free and phosphorylated forms of vitamin B$_6$.

HO $-$ ... CH$_2$ $-$ O $-$ $\begin{cases} \text{H, vitamin} \\ \text{PO}_3\text{H}_2\text{, phosphate} \end{cases}$

H$_3$C

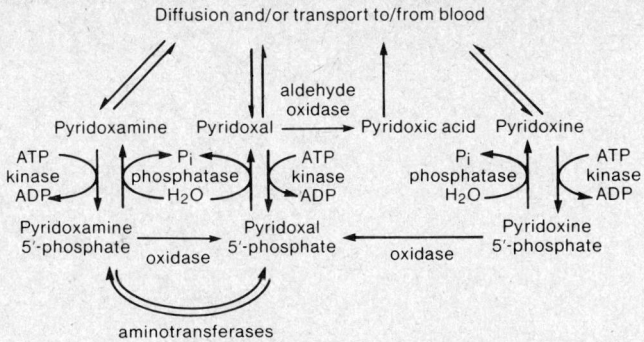

Figure 8B-13. Metabolism of vitamin B_6.

cell. The erythrocyte, in addition, traps PLP as a conjugate Schiff base with hemoglobin.[50] Glycogen phosphorylase contains most of the PLP in skeletal muscle.[7]

Release of free vitamin, mainly pyridoxal, when physiological nonsaturating levels of vitamin are absorbed, occurs when the phosphates are hydrolyzed by nonspecific alkaline phosphatase located in the plasma membrane of cells. However, some PLP is also released into the circulation by the liver.[78] Since the reactive aldehyde is capable of forming Schiff bases with amino groups, PLP in plasma is more tightly complexed to proteins, mostly albumin, than is pyridoxal, which forms an intramolecular hemiacetal between the 4-formyl and 5-hydroxymethyl functions. While PLP is the principal tissue form of vitamin B_6 and pyridoxal constitutes much of the circulating vitamin, the main catabolite excreted in urine is 4-pyridoxic acid, which is formed by the action of the FAD-dependent general liver aldehyde oxidase and especially by NAD-specific aldehyde dehydrogenase, which is found in most tissues.

Functions

As coenzyme PLP, vitamin B_6 functions in numerous reactions that embrace the metabolism of macronutrients, i.e., proteins, carbohydrates, and lipids.[80] Especially diverse are PLP-dependent enzymes that are involved in amino acid metabolism. By virtue of the ability of PLP to condense its 4-formyl substituent with the α-amino group of an amino acid to form an azomethine (Schiff base) linkage, a conjugated double bond system, extending from the α-carbon of the amino acid to the pyridinium nitrogen in PLP, results in reduced electron density about the α-carbon. This configuration potentially weakens each of the bonds from the amino acid α-carbon to the adjoined hydrogen, carboxyl, and side-chain functions. A given apoenzyme then locks in a particular configuration of the coenzyme-substrate compound, such that maximal overlap of the bond to be broken will occur with the resonant, coplanar, electron-withdrawing system of the coenzyme complex. Aminotransferases effect rupture of the α-hydrogen bond with the ultimate formation of an α-keto acid and pyridoxamine-5'-phosphate; this reversible reaction provides an interface between amino acid metabolism and that for ketogenic and glucogenic reactions. Amino acid decarboxylases lead to formation of amines, including several that are functional in nervous tissue, e.g., epinephrine, norephinephrine, serotonin, and γ-aminobutyrate. The biosynthesis of heme depends upon the early formation of δ-aminolevulinate from PLP-dependent condensation of glycine and succinyl-CoA, followed by decarboxylation. There are many examples of enzymes, such as cysteine desulfhydrase and serine hydroxymethyltransferase, which effect the loss or transfer of amino acid side chains. PLP is the essential coenzyme for phosphorylase, which catalyzes phosphorolysis of the α-1,4-linkages of glycogen. An important role in lipid metabolism is the PLP-dependent condensation of L-serine with palmitoyl-CoA to form 3-dehydrosphinganine, a precursor of sphingomyelins.

As with other water-soluble vitamins that function as coenzymes, the relative affinity of the coenzyme for a given apoenzyme and the extent to which a particular holoenzyme-catalyzed reaction is essential are reflected in the progressive symptomatology of deficiency in the vitamin.

Deficiency

Investigations of the consequences of vitamin B_6 deficiency in the human utilize diets deficient in the vitamin and/or diets containing an antagonist, usually 4'-deoxypyridoxine.[14] Otherwise, a deficiency of vitamin B_6 alone is uncommon, and it is more usual to expect the problem to occur in association with deficits in other vitamins of the B-complex. There are, however, instances where chemotherapeutic use or fortuitous ingestion of antagonists has led to hypovitaminosis B_6.[24,81] These antagonists include the tuberculostatic drug isoniazid (isonicotinic acid hydrazide), which can form hydrazones with pyridoxal and PLP. As with other "carbonyl reagents," such compounds not only cause loss by displacement and urinary excretion, but the Schiff bases formed with pyridoxal inhibit pyridoxal kinase,[42,43] and the PLP Schiff bases may additionally inhibit some PLP-dependent enzymes.[78] There are several naturally occurring substituted hydrazines and hydroxylamines which pose such risk.[47] One such compound is D-cycloserine.[15] Penicillamine (β-dimethyl cysteine), used in treatment of patients with Wilson's disease in an attempt to decrease the damaging levels of Cu found in liver, inactivates PLP by forming a thiazolidine derivative.[28] There are several genetic conditions in which abnormalities in the function of vitamin B_6 occur. Pyridoxine-responsive genetic diseases include the following: (1) cases of infantile convulsions where the apoenzyme for glutamate decarboxylase has a poor affinity for the coenzyme, (2) a type of chronic anemia where the number but not morphologic abnormality of erythrocytes is improved by pyridoxine supplementation, (3) xanthurenic aciduria where affinity of the mutant kynureninase for PLP is decreased, (4) primary cystathioninuria due to similarly defective cystathionase, and (5) homocystinuria where there is less of the normal cystathionine synthetase.[52] Usually these inborn errors of metabolism respond to increased levels (5–50 mg/d) of administered vitamin B_6.[67]

Biochemical changes occur early and become more marked as deficiency of B_6 progresses.[22,24,81] Plasma levels of PLP and urinary output of B_6 and 4-pyridoxic acid decrease within a week of removal of vitamin from the diet. There is increased xanthurenic acid in urine, since liver kynureninase activity is decreased. Transaminase activity in serum and red cells also decreases. Electroencephalographic abnormalities appear within three weeks. Epileptiform convulsions are a common finding in young vitamin B_6–deficient subjects. In addition, skin changes include a dermatitis with cheilosis and glossitis. Hematologic manifestations may include a decrease in circulating lymphocytes and possibly a normocytic, microcytic, or sideroblastic anemia.

Methods for the Determinations of Vitamin B_6 and B_6 Status

Several methods have proved useful for assays of vitamin B_6 and PLP, PLP-dependent enzymes, and metabolites of vitamin B_6 and those amino acids which reflect vitamin B_6 status in the human.[9,24,78] There are a number of reliable modifications that have been adapted for diverse biological samples.[44,47]

Direct assessment of all or separate vitaminic forms of B_6 in urine and blood have utilized microbiological assays with specific strains of *Saccharomyces carlsbergensis* (*S. uvarum*) for all three natural vitamers, *Streptococcus faecium* for pyridoxal and pyridoxamine, and *Lactobacillus casei* for pyridoxal. Levels of 20 μg vitamin B_6 per g creatinine in urine are considered indicative of marginal or inadequate dietary intake of the vitamin. Fluorometric assays of *urinary 4'-pyridoxic acid* and blood PLP after conversion of the latter to the cyanide complex or condensation with a fluorophore, such as methyl anthranilate followed by reduction, have also found application. 4'-Pyridoxic acid is now best determined by HPLC. During deficiency, the level of 4-pyridoxic acid will drop well below the normal of at least 0.8 mg/d in urine. The *PLP concentration in plasma,* usually measured by using radioactive tyrosine and the apodecarboxylase, has been judged the most reliable indicator of B_6 status. The *reference range* is 5–23 ng/mL of plasma. This range was supported by a recent study.[61]

Activities of blood transaminases have been frequently used as a reflection of vitamin B_6 status. Though the enzyme activity in serum is depressed in B_6 deficiency, a considerable variability results because release of these enzymes reflects cell death and breakdown in various tissues. Erythrocyte levels of aspartate and alanine aminotransferases provide a better reflection of vitamin B_6 status. Enzymatic assays are best run after and before addition in vitro of PLP, to yield an activity coefficient ratio. Ratios of less than about 1.5 for aspartate aminotransferase and 1.2 for alanine aminotransferase are considered normal, but may depend somewhat on the assay method used.[9]

Measurement of urinary tryptophan metabolites, particularly *xanthurenic acid,* following an oral load (2–5 g) of L-tryptophan, is one of the most common indices used in studies of vitamin B_6 nutriture, because changes can be recognized early, and measurements are relatively easy.

Amounts of xanthurenate well above the normal (near 25 mg/d) are seen in vitamin B_6 deficiency. Levels of other metabolites, such as kynurenic acid and 3-hydroxykynurenine, are also increased. More recently, the methionine load test has also been utilized.[9,24] The ratio of cystathionine to cysteine sulfinic acid is elevated in a 24-h urine sample from vitamin B_6–deficient patients who have received a 3-g methionine load. This method, however, requires an amino acid analyzer.

NIACIN AND NIACINAMIDE

Chemistry and Sources

Though the term *niacin* is chemically synonymous with nicotinic (pyridine-3-carboxylic) acid, it is now used as the generic name for the specific compound, as well as for derivatives exhibiting qualitatively the biological activity of niacinamide (nicotinamide, nicotinic acid amide). Thus "niacin activity" and "niacin deficiency" carry this broader meaning in nutritional literature.[35] A distinction between the two primary vitamin forms needs to be borne in mind, however, when dealing with some aspects of their metabolism, and especially their different pharmacologic actions at high doses. Structures of both vitamers and the two coenzyme forms containing the nicotinamide moiety are given in Figure 8B-14. Nicotinamide adenine dinucleotide (NAD; diphosphopyridine nucleotide, DPN) and nicotinamide adenine dinucleotide phosphate (NADP; triphosphopyridine nucleotide, TPN) represent most of the niacin activity found in good sources, which include yeast, lean meats, liver, and poultry.[22] Milk, canned salmon, and several leafy green vegetables contribute lesser amounts, but still sufficient to prevent deficiency. Additionally, some plant foodstuffs, especially cereals such as corn and wheat, contain niacin bound in forms nutritionally not readily available.[24,35] Among these are niacinogens and niacytin. At least part of such material appears to be constituted by nicotinic acid that is amide-linked to the ϵ-amino lysyl groups of peptides. Protein provides a considerable portion of niacin equivalent because of the tryptophan content. As much as two thirds of niacin required by adults can be derived from tryptophan metabolism via nicotinic acid ribonucleotide to NAD and NADP. It has been found that 60 mg of tryptophan is equivalent to 1 mg of niacin in the adult. Free forms of the vitamin are white, stable solids that are quite soluble in water. The oxidized coenzymes are labile to alkali, whereas the reduced (dihydro) coenzymes are labile to acid. Reduction of the oxidized coenzymes commonly occurs by addition of a hydride ion to the *para* (4) position of the nicotinamide ring, with simultaneous formation of a solvated proton. NADH and NADPH (but not NAD and NADP) absorb light in the near ultraviolet region (339 nm).

Requirements and Allowance

Estimations of niacin requirements take into account the contribution of tryptophan derived from protein. Average diets in the U.S. supply 0.5–1 g of tryptophan plus 8–17 mg of niacin for a total of 16–34 mg of niacin equivalents (NE).[35] Moreover, utilization of niacin within pyridine nucleotide coenzymes can be generally related to energy expenditure. Hence, niacin equivalents have been related to caloric intake. The allowance recommended for adults is 6.6 niacin equivalents per 1000 kcal (1.6 NE/MJ). For females and males during childhood and as adults, the recommended range is then 9–19 NE/d. An increase of 2 NE/d during pregnancy is based on the

Nicotinic acid (niacin)   Nicotinamide (niacinamide)

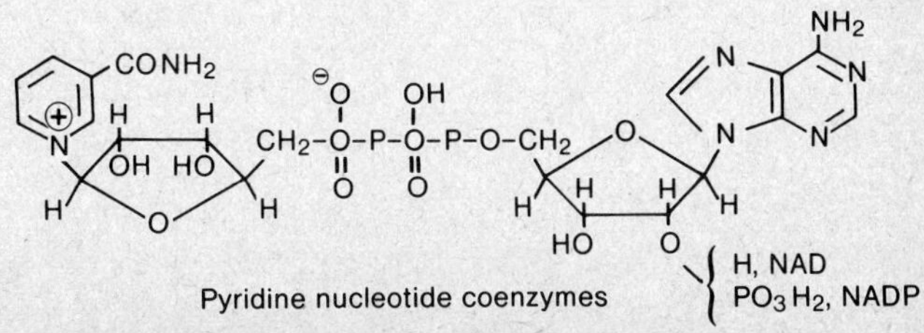

Figure 8B-14. Niacin, niacinamide, and coenzymes.

Pyridine nucleotide coenzymes   H, NAD
                                       PO_3H_2, NADP

recommended daily increase in energy intake of 300 kcal/d. Similarly, 5 NE daily for lactation relates to an increased energy intake of 500 kcal/d and will offset the 1.6 mg of preformed niacin lost in 850 mL of milk. Human milk contains approximately 0.17 mg of niacin and 22 mg of tryptophan/dL or 70 kcal, and these amounts are adequate to meet niacin needs of the infant. RDAs are set at 6 NE/d for infants up to six months and 8 NE/d for the period of six months to a year.

Absorption, Transport, and Metabolism

The coenzymes are hydrolyzed in the intestinal tract, and both the acid and amide forms of the vitamin are readily absorbed. Nicotinic acid and nicotinamide are both present in blood and plasma and move, possibly by facilitated diffusion, between blood and cerebrospinal fluid.[75] Both compounds are converted to the coenzyme forms in blood cells, kidney, brain, and liver. The first step involves the cytosolic pyrophosphorylase (phosphoribosyltransferase)-catalyzed reaction of nicotinate or nicotinamide with 5-phosphoribosyl-1-pyrophosphate to form pyrophosphate and nicotinic acid mononucleotide or nicotinamide mononucleotide, respectively.[80] Additionally in liver, quinolinate from the metabolism of tryptophan is similarly converted with concomitant decarboxylation to nicotinic acid mononucleotide. A nuclear mononucleotide adenylyltransferase catalyzes attachment of the AMP moiety from ATP to form deamido-NAD pyrophosphate from nicotinic acid mononucleotide. The deamido compound subsequently reacts with glutamine and a cytosolic ATP-dependent synthetase step to yield NAD, glutamate, and phosphate. Nicotinamide mononucleotide is directly converted by the adenylyltransferase to NAD. NADP is formed by a kinase-catalyzed phosphorylation of NAD. In the tissues most of the vitamin is present as nicotinamide in NAD and NADP, although liver may contain a significant fraction of the free vitamin. There is little storage of niacin as such.

Although nicotinamide can be converted to nicotinic acid by a rather widespread microsomal deamidase, there is no direct reamidation of nicotinic acid. A considerable number of catabolites of vitaminic forms have been identified from several animal species. Catabolites include conjugates of the acid with glucuronic acid or glycine, and methyl derivatives of nicotinamide with some further oxidation of 1-methylnicotinamide to the 6-pyridone. The human excretes 1-methylnicotinamide and 1-methyl-3-carboxamido-6-pyridone, which are primary urinary metabolites.[22,24,35,80]

Functions

There are hundreds of enzymes that require the nicotinamide moiety within either NAD or NADP. Most of these oxidoreductases function as dehydrogenases and catalyze such diverse reactions as the conversion of alcohols (often sugars and polyols) to aldehydes or ketones, hemiacetals to lactones, aldehydes to acids, and certain amino acids to keto acids. The common mechanism of operation involves the stereospecific abstraction of a hydride ion from substrate, with *para* addition to one or the other side of carbon 4 in the pyridine ring of the nucleotide coenzyme. The second hydrogen of the substrate group oxidized is concomitantly removed as a proton and ultimately exchanges as hydronium ion. Most dehydrogenases utilizing NAD or NADP function reversibly. Glutamate dehydrogenase, for example, favors the oxidative direction, whereas others, such as glutathione reductase, preferentially catalyze reduction. A further generality is that most NAD-dependent enzymes are involved in catabolic reactions, whereas NADP systems are more common to biosynthetic reactions.

Deficiency

Pellagra is a classic deficiency disease of man that has been most often found among those who subsist chiefly on corn.[66] Though the pathogenesis has been attributed to a deficiency of niacin (and tryptophan), other associated complicating factors may be an imbalance of amino acid intake, particularly the ingestion of high levels of leucine,[24,81] and the presence of mycotoxins elaborated by mold infestations, mainly by *Fusarium*.[81] Pellagra is also an occasional secondary manifestation of two disorders which profoundly affect tryptophan metabolism:[81] *carcinoid syndrome*, in which up to 60% of tryptophan is catabolized by what is ordinarily a minor pathway of metabolism, and *Hartnup disease*, an autosomal recessive disorder in which several amino acids, including tryptophan, are poorly absorbed.

The typical presentation of pellagra is that of a chronic wasting disease associated with dermatitis, dementia, and diarrhea. The characteristic erythematous dermatitis is bilateral and symmetrical. It occurs on skin areas exposed to sunlight. Mental changes include fatigue, insomnia, and apathy, which precede an encephalopathy characterized by confusion, disorientation, hallucination, loss of memory, and eventually frank organic psychoses. The diarrhea, when it occurs, reflects a widespread inflammation of the intestinal mucous surfaces; other gastrointestinal manifestations include achlorhydria, glossitis, stomatitis, and vaginitis.

Excess

While relatively large daily intakes of niacin (40–200 mg) may be required in treatment of Hartnup disease and carcinoid syndrome,[81] the use of pharmacologic doses is of doubtful value for other dysfunctions and may even prove harmful.[10,24,35] Massive (g) doses of nicotinic acid (but not the amide) produce vascular dilation or "flushing," with an accompanying sensation of burning or stinging of the face and hands. Pruritus, nausea, vomiting, and diarrhea have been commonly reported, but often abate with continued therapy. Varying degrees of hyperpigmentation and acanthosis nigricans occur in rare cases. Additional effects are abnormal glucose tolerance, hyperuricemia, peptic ulcer, hepatomegaly, jaundice, and increased serum transaminases. Hence, high chronic doses of nicotinic acid appear to be potentially hepatotoxic.

Methods for the Determination of Niacin

Since the urinary excretion of metabolites of niacin and tryptophan is lower than average in patients with generalized malnutrition, measurements of these analytes are not entirely satisfactory as diagnostic indices for niacin deficiency.[24,70,81] However, the measurement of the *excretion of N(1)-methylnicotinamide and N(1)-methyl-3-carboxamide-6-pyridone* (also named N(1)-methyl-2-pyridone-5-carboxamide) has received continued use in the biochemical assessment of niacin nutriture. Normally, adults excrete 20–30% of their niacin in the form of methylnicotinamide and 40–60% as the pyridone.[70] An excretion ratio of pyridone to methylnicotinamide of 1.3–4.0 is acceptable, but latent niacin deficiency is indicated by a value below 1.0. As depletion occurs, the pyridone is absent for weeks, before clinical signs are noted, while the methylnicotinamide excretion falls to a minimum at about the time clinical signs are evident. The pyridone to methylnicotinamide ratio is best determined with HPLC methods.

Although assays for coenzymes derived from niacin have not yet been adapted well to assess nutritional status, there are several efficient means to separate and quantitate both vitaminic and coenzymic forms.[45,48] Some of these rely on the absorbance of reduced pyridine nucleotides at 340 nm in coupled enzyme systems, and others utilize fluorescent properties, especially of addition products, e.g., with methyl ethyl ketone.

FOLIC ACID AND VITAMIN B$_{12}$

Discussion of these vitamins, which interrelate in one-carbon metabolism, is covered in Chapter 15 on Biochemical Aspects of Hematology.

BIOTIN

Chemistry and Sources

Biotin (vitamin H) is *cis*-tetrahydro-2-oxothieno[3,4-*d*]-imidazoline-4-valeric acid as shown in Figure 8B-15. The vitamin in most organisms occurs mainly bound to protein. The ε-amino group of the lysyl side chain of protein is linked via an amide function, involving the carboxyl group of the valeryl side chain of biotin. In addition, some biotin is linked noncovalently as a complex with avidin, a protein in egg white. Good sources of biotin include liver, kidney, pancreas, eggs, yeast, and milk. Cereal grains, fruit, and meat are regarded as poor sources.[35] The ureido ring and the ionizable carboxyl group of biotin allow modest solubility of the white crystalline solid in aqueous solution, especially at an alkaline pH. Oxidizing agents convert the thioether to sulfoxides and sulfones which are not active per se.

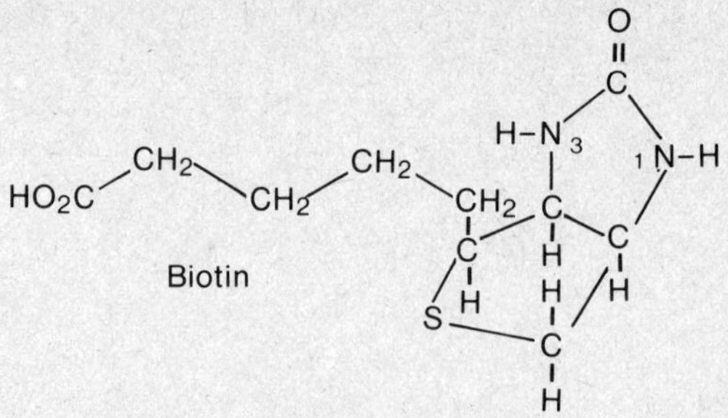

Figure 8B-15. Biotin.

Estimated Adequate Intakes

Intestinal microflora make a significant contribution to the body pool of available biotin, making determination of the dietary requirement difficult.[35] Mean urinary excretion, reflective of dietary intake, ranges from 18–46 $\mu g/d$ for adults who ingest 50–200 $\mu g/d$. It is felt that an intake of 100–200 $\mu g/d$ is quite adequate. Since urinary concentration of biotin in infants after the age of six months is comparable with that of adults, the requirement for older infants and children is probably proportional to body weight and energy consumption. A ratio of 50 $\mu g/$ 1000 kcal has been suggested, and adequate intakes are increased for the half-year old from 50 $\mu g/d$ up to the adult level. The suggested intake of 35 $\mu g/d$ for young infants is more than adequate, when compared with the low biotin content of human milk at 10 $\mu g/1000$ kcal or 0.16 $\mu g/dL$.

Absorption, Transport, and Metabolism

Digestion of dietary proteins containing bound biotin yields considerable biocytin (ϵ-N-biotinyl lysine). Biocytin is resistant to hydrolysis by proteolytic enzymes in the intestinal tract, as is the biotin-avidin complex. Biocytin and biotin are readily absorbed. An enzyme called biocytinase (biotin amidohydrolase) in plasma and erythrocytes catalyzes the hydrolysis of biocytin to yield free biotin. Biotin is cleared from the circulating blood more rapidly in deficient than in normal mammals, is taken up by such tissues as liver, muscle, and kidney, and is localized in cytosolic and mitochondrial carboxylases.[24,41] Covalent attachment of biotin to apoenzymes involves ATP-dependent conversion of the vitamin to biotinyl-5′-adenylate followed by condensation of the biotinyl moiety with ϵ-amino groups of specific lysyl residues in apoenzymes preformed from subunits. The enzymes responsible for catalyzing the formation of the ϵ-N-biotinyl-L-lysyl (biocytinyl) moiety of proteins are holoenzyme synthetases.

Following turnover and during fractional catabolism of biotin, some oxidation of the thioether function occurs and both *d*- and *l*-sulfoxides are formed. Partial cleavage of the valeric acid side chain is effected by β-oxidation. Trace amounts of biotin sulfoxides and bis-norbiotin are excreted in the urine with larger (but still μg) amounts of free vitamin. Careful balance studies in man, where perhaps only 1 mg is the total body content, showed that urinary excretion of biotin often exceeded dietary intake, and that in all cases fecal excretion was as much as three to six times greater than dietary intake because of microfloral biosynthesis.[22,35]

Functions

At present, nine biotin-dependent enzymes are known, six carboxylases, two decarboxylases, and a transcarboxylase; four carboxylases are found in human tissues.[41] These latter are carboxylases for acetyl-CoA, propionyl-CoA, β-methylcrotonyl-CoA, and pyruvate. The biotin-dependent carboxylases operate via a common mechanism, which involves phosphorylation of bicarbonate by ATP to form carbonyl phosphate, followed by transfer of the carboxyl group to the sterically less hindered nitrogen of the biotin moiety. The resulting N(1)-carboxybiotinyl enzyme can then exchange the carboxylate function with a reactive center in a substrate. With cytosolic acetyl-CoA carboxylase, the product is malonyl-CoA utilized for fatty acid biosynthesis. In mitochondria, pyruvate carboxylase catalyzes formation of oxaloacetate, which together with acetyl-CoA forms citrate. The other carboxylases are involved in the metabolism of odd-numbered fatty acids and branched-chain fatty acids.

Deficiency

Dietary deficiency of biotin is infrequently seen and has been produced in the adult usually only after ingestion of diets that have included large amounts of raw egg white (containing avidin).[35,70] Symptoms include anorexia, nausea, vomiting, glossitis, pallor, depression, and a dry scaly dermatitis. A seborrheic dermatitis in infants under six months of age can also be caused by inadequate biotin, but the condition responds promptly to biotin therapy. Significantly lowered urinary excretion or circulating blood levels have also been found in pregnant women, alcoholics, and patients with achlorhydria, as well as among the elderly and some athletes.[8] Finally, there are rather rare genetic enzyme defects such as in holoenzyme synthetase (reflected in inadequate conversion of apo- to holocarboxylases) and propionyl-CoA carboxylase (reflected in a distinguishing acidemia).[41]

Methods for the Determination of Biotin

At the trace concentrations present in biological samples, biotin can be quantitated using microbiological assays.[70] Bound biotin is first liberated by proteolytic digestion of the sample using, for example, papain with whole blood. Then aliquots are added to a biotin-deficient medium inoculated with a test organism such as *Lactobacillus plantarum*. Standard curves are derived from growth in controls containing known amounts of biotin. Other methods include isotopic dilution assays that are generally applicable and even a colorimetric determination with acidic *p-dimethylaminocinnamaldehyde*, which forms a red Schiff base; this is suitable only when quantity and purity of specimens are adequate.[44,47]

PANTOTHENIC ACID

Chemistry and Sources

Pantothenic acid is of ubiquitous occurrence in nature, where it is synthesized by most microorganisms and plants from pantoic acid (D-2,4-dihydroxy-3,3-dimethylbutyric acid) derived from L-valine, and β-alanine derived from L-aspartate. The vitamin is an integral part of 4'-phosphopantetheine, which serves as a covalently attached prosthetic group of acyl carrier proteins, and within the structure of coenzyme A as shown in Figure 8B-16. The vitamin is widely distributed in foods, mostly within CoA-containing compounds, and is particularly abundant in animal sources, legumes, and whole grain cereals.[22,35] Excellent food sources (100–200 μg/g dry weight) include egg yolk, kidney, liver, and yeast. Fair sources (35–100 μg/g) include broccoli, lean beef, skimmed milk, sweet potatoes, and molasses. Over half the pantothenate in wheat may be lost during manufacture of flour, and up to a third is lost during cooking of meat. Pantothenic acid is a hygroscopic, viscous oil that is easily destroyed by heat, especially at extremes of pH. The most common commercial synthetic form is the calcium salt.

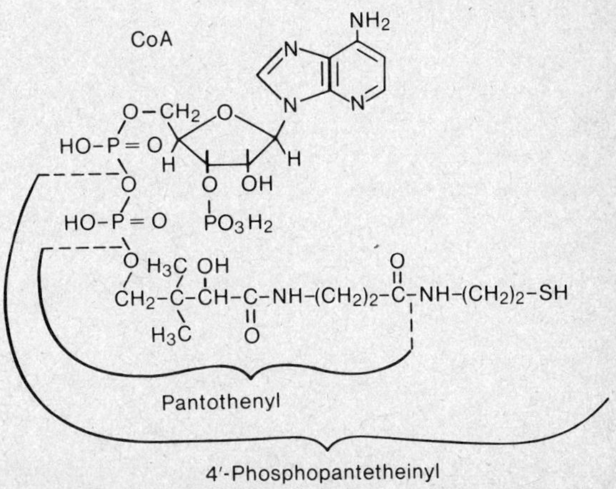

Figure 8B-16. Pantothenate and 4'-phosphopantetheine as components of CoA.

Estimated Adequate Intake

An intake of pantothenic acid of 4–7 mg/d seems sufficient for adults, since consumption averages 7 mg/d when supplied with a range of ordinary foods, and even 4 mg/d was consumed by low-income women who were deemed at least marginally adequate in other vitamins.[35] Urinary excretion, which generally correlates with dietary intake, is 2–7 mg/d in adults consuming 5–7 mg/d; another 1–2 mg/d is lost in feces. A slightly higher intake may be warranted for pregnant or lactating women. Human milk contains approximately 2 mg/L. Adequate intakes for younger age groups are based on proportional energy needs and range from a suggested 2 mg/d for young infants to 4–5 mg/d for adolescents.

Absorption, Transport, and Metabolism

Coenzyme A, the form in which much of the pantothenic acid is ingested, is hydrolyzed by intestinal pyrophosphatase and phosphatase to pantetheine (pantothenyl cysteamine), which together with pantothenate is absorbed into the portal circulation. Within cells resynthesis of coenzyme level compounds occurs by successive conversions of pantothenate to 4'-phosphopantothenate, and ultimately to 4'-phosphopantetheine, diphospho-CoA, and finally CoA (some of which is incorporated as 4'-phosphopantetheine into the acyl carrier protein of the fatty acid synthetase complex).[22,24,80] About 80% of the vitamin in animal tissues is in CoA form, and the rest exists mainly as phosphopantetheine and phosphopantethenate. Cleavage enzymes catalyzing hydrolysis of the phosphate moieties and release of β-mercaptoethylamine from pantothenate operate during turnover and release of the vitamin, which is excreted in the urine. Only a small fraction of pantothenate is secreted into milk and even less into colostrum.

Functions

The myriad acyl thiol esters of CoA, of which pantothenic acid is a constituent, are central to the metabolism of numerous compounds, especially lipids and the ultimate catabolic disposition of carbohydrates and ketogenic amino acids.[22,80] The chemical properties of the thiol ester, which has a high group-transfer potential, permits facile acylations and hydrolysis; the ready formation of enolate ions and the carbanion-like property of the carbon α to the carbonyl facilitate condensation reactions. For example, acetyl-CoA, which derives from the metabolism of carbohydrates, fats, and amino acids, can acetylate compounds such as choline and hexosamines to produce essential biochemicals; it can also condense with other metabolites such as oxaloacetate to supply citrate and cholesterol.

Another essential role of pantothenic acid is in its participation in the 4'-phosphopantetheine moiety of acyl carrier protein (ACP), where the phosphodiester-linked prosthetic group utilizes the sulfhydryl terminus to exchange with malonyl-CoA to form an ACP-S-malonyl thioester, which can chain elongate during fatty acid biosynthesis.

Although the reactive thiol function of CoA and ACP is not an integral part of pantothenate, the steric and chemical properties conferred by the vitamin structure are important for enzymic recognition.

Deficiency

The widespread occurrence of pantothenic acid in foods is commensurate with its many roles, and makes an uncomplicated dietary deficiency of pantothenate unlikely in the human.[24,35] Symptoms have been produced in a few volunteers who have received ω-methylpantothenic acid as an antagonist, and more recently in persons fed semisynthetic diets virtually free of pantothenate. Subjects became irascible and developed postural hypotension and rapid heart rate on exertion, epigastric distress with anorexia and constipation, numbness and tingling of the hands and feet, hyperactive deep tendon reflexes, and weakness of finger extensor muscles. The eosinopenic response to ACTH was impaired. More severe deficiency in animals leads to adrenal cortical failure.

Methods for the Determination of Pantothenic Acid

Since urinary output of pantothenate is directly proportional to dietary intake, present assessment relies mainly on this index.[9,70] Urinary excretion of < 1 mg/d is considered abnormally low. Suspicion of inadequate intake is further supported if whole blood values are < 100 μg/dL.

Functional tests based on acetylation of sulfanilamide by red cells[18] and urinary excretion of acetylated *p*-aminobenzoic acid after a load test of the acid[68] have been suggested, but their value is not yet established.

Pantothenic acid is usually measured by microbiological procedures, often using *Lactobacillus plantarum.*[24,70] *Pediococcus acidilactici* NCIB 6990 has been found to be especially sensitive as an assay organism. A radioimmunoassay recently developed is also being used.[83] Gas chromatographic methods are employed for pharmaceutical preparations. Enzymatic assays can be used for determining the quantity of CoA and ACP.[44,47]

ASCORBIC ACID

Chemistry and Sources

L-*Ascorbic acid (vitamin C)* is the enol form of 2-oxo-L-gulofuranolactone as seen in Figure 8B-17. The acidic enolic hydroxyl on ring carbon 3 has a pK_a of 4.2. The vitamin is a white, crystalline solid which is readily soluble in water. Acidic solutions (below pH 3) show an absorption maximum at 245 nm, whereas solutions of the ionized material (above pH 5) have an absorption peak at 265 nm. Ascorbic acid is a relatively strong reductant with an E_o' (pH 7) of +0.58 volt. It is reversibly oxidized to dehydroascorbic acid (ascorbone), also shown in Figure 8B-17. The dehydro form is more labile than the reduced form to hydrolytic ring opening to yield 2,3-diketo-L-gulonic acid, which is not antiscorbutic when taken orally.[35] Plants and most animals possess the ability to synthesize the vitamin from D-glucose via the lactones of D-glucuronic and L-gulonic acids; however, some mammals, including the human, lack L-gulonolactone oxidase, the enzyme that catalyzes the formation of 2-keto-L-gulonolactone, which spontaneously tautomerizes to L-ascorbic acid. Best sources of the vitamin are citrus fruits, berries, melons, tomatoes, green peppers, raw cabbage, and leafy green vegetables.[22] Losses during processing, especially with heat and aerobic conditions, can be considerable.

Requirements and Allowances

The amount of vitamin C sufficient to alleviate and cure the clinical signs of scurvy is only 10 mg/d, which may be near the requirement level for adults.[35] This amount, however, does not provide for any reserves. An RDA of 60 mg seems adequate to maintain near saturation of tissues in the adult male, who has a body pool near 1.5 g. Only when this total pool falls below about 300 mg do clinical symptoms of deficiency occur.[2,25] To provide for fetal needs, an additional 20 mg/d is recommended for the pregnant woman to offset the decrease in plasma vitamin C level during pregnancy. A lactating woman should receive an additional 40 mg/d, since an average of 25–45 mg may be secreted in 850 mL of milk. A daily recommendation of 35 mg has been made for breast-fed infants. A higher RDA of 100 mg is suggested for premature and newborn babies during the first week to protect against possible tyrosinemia. An RDA of 45 mg is given for children who may have a somewhat higher requirement, on a weight basis, than adults.

Absorption, Transport, and Metabolism

Absorption of vitamin C occurs readily, mostly from the stomach, where some of the ascorbic acid is converted to the dehydro form. At physiological pH, the uncharged dehydroascorbic acid passes across cell membranes faster than the monoanionic L-ascorbate. Passive diffusion of vitamin C may largely account for entry into some cells, such as leukocytes and erythrocytes, but an

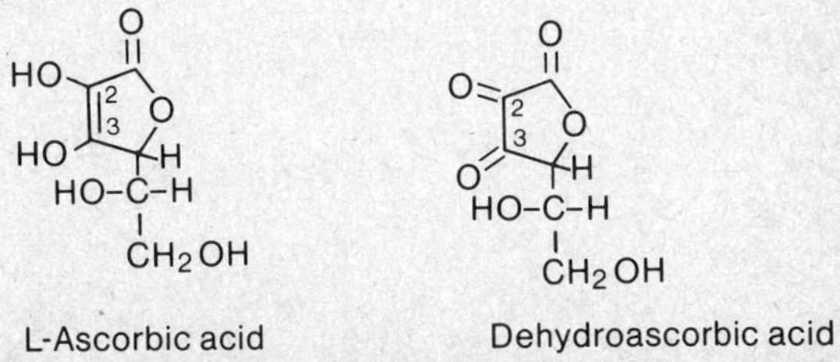

Figure 8B-17. L-Ascorbic and dehydroascorbic acids.

L-Ascorbic acid

Dehydroascorbic acid

active transport mechanism may also operate, especially for platelets, adrenals, and retina. The free diffusion of dehydroascorbic acid into cells, followed by intracellular reduction to the less diffusible ascorbate ion, could explain the occurrence of the higher concentration of ascorbate in leukocytes than in plasma. Vitamin C is found in most tissues, but glandular tissues such as the pituitary, adrenal cortex, corpus luteum, and thymus have the highest amounts,[22] and the retina has 20 to 30 times the plasma concentration. The half-life for vitamin C in the human is only about 16 days. In addition to the presence of ascorbate and dehydroascorbic acid in urine, lesser amounts of a number of catabolites are also present. These include oxalate, derived from the first two carbons of the vitamin, some ascorbate-2-sulfate from tissue (liver) sulfurylation with 3'-phosphoadenosine-5'-phosphosulfate, and the 2-methyl ascorbic acid formed by catechol-O-methyltransferase with S-adenosyl methionine. Some ring-opened and chain-degraded sugar acids, e.g., 2,3-diketo-L-gulonate and L-threonate, xylonate, and lyxonate, may also occur.

Functions

The most clearly established and critical functional role for ascorbic acid is as a cofactor for protocollagen hydroxylase, the enzyme responsible for hydroxylation of prolyl and lysyl residues within nascent peptides in connective tissue proteins. Among these are collagen and related proteins which comprise intercellular material of cartilage, dentin, and bone.[22,80] Vitamin C functions similarly in the hydroxylation of γ-butyrobetaine to carnitine.[80] Vitamin C may also be involved in tyrosine metabolism, microsomal drug metabolism, synthesis of epinephrine and anti-inflammatory steroids by the adrenals, folic acid metabolism, and leukocyte functions.[16] These roles seem to relate to the favorable reductive properties of L-ascorbic acid, especially on Fe(II)-enzyme systems. Iron absorption, as Fe(II), is also enhanced by simultaneous ingestion of the vitamin.

Deficiency

Protracted deficiency of vitamin C leads to the classic disease called *scurvy*. Inability to form adequate intercellular substance in connective tissue is reflected in swollen, tender, and often bleeding or bruised loci at joints and in other areas where structurally weakened tissue cannot withstand stress.[82] Infantile scurvy, also known as Barlow's disease, exhibits a bayonet-rib syndrome. The gums are livid and swollen, particularly in the regions of the papillae between the teeth, and sometimes "scurvy buds" develop, which may project beyond the biting surface. Cutaneous bleeding often begins on the lower thighs as perifollicular hemorrhages, which may then spead to the buttocks, abdomen, legs, and arms. Petechial hemorrhages due to the rupture of capillaries often appear. Thereafter, large spontaneous bruises (ecchymoses) may arise almost anywhere on the body. Ocular hemorrhages, drying of salivary and lacrimal glands, parotid swelling, femoral neuropathy, edema of the lower extremities, and psychological disturbances have also been described. Some scorbutic patients may develop anemia, display radiologic changes characteristic of osteoporosis, or die suddenly from heart failure.

Excess

The use of "megadoses" of vitamin C, as suggested by some for the prevention or amelioration of the common cold, does not appear warranted on the basis of present information. As reviewed elsewhere,[10,22,24,35] large doses of ascorbic acid have generally been considered nontoxic, except for gastrointestinal symptoms, which are experienced by some subjects; however, more serious adverse effects have been observed and suspected as potential hazards. These include reductive destruction of concomitantly ingested vitamin B_{12}, facilitation of too much iron absorption, reduced catabolism leading to higher requirements (dependency), increased production of oxalate favoring deposition of calcium oxalate stones in kidney and bladder (particularly by those who are congenital, familial hyperoxalurics), and uricosuria. Both benefits and hazards from high levels of vitamin C seem minimal for most people, who simply excrete the excess.

Methods for the Determination of Ascorbic Acid

Ascorbic acid can be determined colorimetrically with *2,4-dinitrophenylhydrazine* to form the red *bis*-hydrazone or with *2,4-dichlorophenol-indophenol,* which is reduced to a colorless form. Particulars on these and related methods for clinical biochemical use have been reviewed;[9,70] other methods, including fluorometric and HPLC techniques, have also been developed.[47] Assessment of vitamin C status is generally done by measuring serum (plasma) and leukocyte levels of the vitamin. Urinary excretion and red cell concentrations have not been found to be specific and useful indices of vitamin C status; however, urinary levels of ascorbic acid, especially following a load test, can be helpful in the clinical diagnosis of scurvy.

Principle

Ascorbic acid in plasma is oxidized by Cu(II) to form dehydroascorbic acid, which reacts with acidic 2,4-dinitrophenylhydrazine to form a red *bis*-hydrazone, which is measured at A_{520}.

Specimen

Collect heparinized blood and centrifuge to obtain plasma, which should be analyzed immediately, or not later than 3 h if the specimen is refrigerated.

Reagents

1. Metaphosphoric acid solution, 6.0 g/dL. Dissolve 30.0 g of metaphosphoric acid (HPO_3) in distilled water and bring to a final volume of 500 mL. Prepare immediately before use.

2. Sulfuric acid, 4.5 mol/L. Add slowly 250 mL of concentrated sulfuric acid, reagent grade, to 500 mL of cold water in a 1-L flask and fill to mark with distilled water. *Caution*: Since significant heat is generated when concentrated sulfuric acid is diluted, the flask should be placed in an ice bath. The concentrated acid should be added slowly and the resulting solution mixed constantly.

3. Sulfuric acid, 12 mol/L. Add 650 mL of concentrated sulfuric acid to 300 mL of cold water in a 1-L flask, cool, and fill to mark with distilled water. Refrigerate.

4. 2,4-Dinitrophenylhydrazine reagent, 2.0 g/dL in sulfuric acid, 4.5 mol/L. Dissolve 10 g of 2,4-dinitrophenylhydrazine in sulfuric acid, 4.5 mol/L, and dilute to a final volume of 500 mL. Let stand in the refrigerator overnight, and then filter.

5. Thiourea solution, 5.0 g/dL. Dissolve 5 g of thiourea in glass-distilled water and dilute to a final volume of 100 mL. This reagent is stable for one month at 4 °C.

6. Copper sulfate solution, 0.6 g/dL. Dissolve 0.6 g of anhydrous copper sulfate in glass-distilled water and dilute to a final volume of 100 mL.

7. Dinitrophenylhydrazine–thiourea–copper sulfate (DTCS) reagent. Combine 5 mL of the thiourea solution, 5 mL of the copper sulfate solution, and 100 mL of the 2,4-dinitrophenylhydrazine reagent. Store in a bottle at 4 °C for a maximum of one week.

8. Standards. All ascorbic acid standards should be prepared daily.

   a. Ascorbic acid stock standard, 50.0 mg/dL. Dissolve 50 mg of ascorbic acid in metaphosphoric acid (6.0 g/dL) and bring to a final volume of 100 mL with metaphosphoric acid.

   b. Intermediate ascorbic acid standard, 5.0 mg/dL. Pipet 10.0 mL of stock standard into a 100-mL volumetric flask and dilute to mark with metaphosphoric acid (6.0 g/dL).

   c. Working standards. In a series of 25-mL volumetric flasks, pipet the following amounts of intermediate standard: 0.5, 2.0, 4.0, 6.0, 10.0, 15.0, and 20.0 mL. Bring to a final volume of 25 mL with metaphosphoric acid (6.0 g/dL) to yield working standards of 0.10, 0.40, 0.80, 1.20, 2.00, 3.00, and 4.00 mg/dL.

Procedure

1. Add 0.5 mL of heparinized plasma to 2.0 mL of freshly prepared metaphosphoric acid in a 13 × 10 mm test tube, and mix well on a vortex mixer. Centrifuge the plasma–metaphosphoric acid mixture for 10 min at 2500 × g. Pipet 1.2 mL of the clear supernatant into a 13 × 100 mm Teflon-lined, screw-cap test tube.

2. Add 1.2 mL of each concentration of working standard into 13 × 100 mm screw-cap test tubes. Prepare standards in duplicate. Add 1.2 mL of metaphosphoric acid to two tubes for use as blanks.

3. Add 0.4 mL of DTCS reagent to all tubes. Cap tubes, mix contents, and incubate the tubes in a water bath at 37 °C for 3 h.

4. Remove the tubes from the water bath and chill for 10 min in an ice bath. While mixing, slowly add to all tubes 2.0 mL of cold sulfuric acid, 12 mol/L, cap, and mix with a vortex mixer. (The temperature of the mixture must not exceed room temperature.)

5. Adjust the spectrophotometer with the blank to read zero A at 520 nm, and read the standards and unknowns. Plot the concentration of each working standard versus absorbance values. The standard curve obeys Beer's law up to an ascorbic acid concentration of 2.0 mg/dL.

Calculation

The concentration of the samples is obtained from the standard curve and is multiplied by 5 (to correct for dilution of the plasma by metaphosphoric acid) to give the concentration of ascorbic acid per dL of plasma.

Discussion

Plasma (or serum) levels of ascorbate show a linear increase with dietary intake of the vitamin up to a level of 1.2–1.4 mg/dL, beyond which urinary excretion rapidly increases.[9,22,70] Plasma values above 0.3 mg/dL are considered acceptable, those from 0.2–0.29 are "at risk," and those below 0.2 are an indication of deficiency. Leukocyte ascorbate levels are somewhat more difficult to measure but are more representative of tissue stores. Values ranging from 0–7 mg/dL of leukocytes are suggestive of deficiency, whereas values > 15 indicate satisfactory vitamin C status. Values may also be expressed as 20–50 μg/10^8 WBC.

Reference Ranges

In persons with adequate intake of vitamin C, plasma concentrations of total vitamin (ascorbic acid plus dehydroascorbic acid) are between 0.6 and 2.0 mg/dL. The lower limit value may be seen in some cases with subclinical vitamin C deficiency and in older individuals.

References

1. American Medical Association, Council on Foods and Nutrition: Improvement of the nutritive quality of foods. JAMA, *225*:1116, 1973.
2. Baker, E. M., Hodges, R. E., Hood, J., et al.: Metabolism of [14]C- and ^{3}H-labeled L-ascorbic acid in human scurvy. Am. J. Clin. Nutr., *24*:444, 1971.
3. Bieri, J. G., and Evarts, R. P.: Tocopherols and polyunsaturated fatty acids in human tissues. Am. J. Clin. Nutr., *28*:717, 1975.
4. Bieri, J. G., and Farrell, P. M.: Vitamin E. *In*: Vitamins and Hormones, Vol. 34. P. L. Munson, J. Glover, E. Diczfalusy, et al., Eds. New York, Academic Press, 1976, p. 31.
5. Bieri, J. G., Tolliver, T. J., and Catignani, G. L.: Simultaneous determination of α-tocopherol and retinol in plasma or red cells by high pressure liquid chromatography. Am. J. Clin. Nutr., *32*:2143, 1979.
6. Binder, H. J., Hertig, D. C., Hurst, V., et al.: Tocopherol deficiency in man. N. Engl. J. Med., *273*:1289, 1965.
7. Black, A. L., Guirard, B. M., and Snell, E. E.: The behavior of muscle phosphorylase as a reservoir for vitamin B_6 in the rat. J. Nutr., *108*:670, 1978.
8. Bonjour, J. P.: Biotin in man's nutrition and therapy—a review. Internat. J. Vit. Nutr. Res., *47*:107, 1977.
9. Briggs, M., Ed.: Vitamins in Human Biology and Medicine. Boca Raton, Fla., CRC Press, Inc., 1981.
10. Campbell, T. C., Allison, R. G., and Carr, C. J.: Feasibility of Identifying Adverse Health Effects of Vitamins and Essential Minerals in Man. Bethesda, Md., Life Sciences Research Office, Federation of American Societies for Experimental Biology, 1980.
11. Carpenter, M. P., and Howard, C. N., Jr.: Vitamin E, steroids, and liver microsomal hydroxylations. Am. J. Clin. Nutr., *27*:966, 1974.
12. Catabolites of thiamine from the rat. Nutr. Rev., *29*:120, 1971.
13. Catignani, G. L., Chytil, F., and Darby, W. J.: Vitamin E deficiency: Immunochemical evidence for increased accumulation of liver xanthine oxidase. Proc. Nat. Acad. Sci., *71*:1966, 1974.
14. Coburn, S. P.: The Chemistry and Metabolism of the Vitamin B_6 Antagonist, 4'-Deoxypyridoxine. Boca Raton, Fla., CRC Press, Inc., 1981.
15. Cohen, A. C.: Pyridoxine in the prevention and treatment of convulsions and neurotoxicity due to cycloserine. Ann. N.Y. Acad. Sci., *166*:346, 1969.
16. Darby, W. J., Ed.: Annual Review of Nutrition, Vol. 1. Palo Alto, Cal., Annual Reviews, Inc., 1981.
17. DeLuca, H. F.: Recent advances in the metabolism of vitamin D. *In*: Annual Review of Physiology, Vol. 43. I. S. Edelman and S. G. Schultz, Eds. Palo Alto, Cal., Annual Reviews, Inc., 1981, p. 199.
18. Ellestad, J. J., Nelson, R. A., Adson, M. A., et al.: Pantothenic acid and coenzyme A activity in blood and colonic mucosa from patients with chronic ulcerative colitis. Fed. Proc., *29*:820 Abst., 1970.
19. Evans, W. C.: Thiaminases and their effects on animals. *In*: Vitamins and Hormones, Vol. 33. P. L. Munson, J. Glover, E. Diczfalusy, et al., Eds. New York, Academic Press, 1975, p. 467.
20. Genetic determination of coenzyme synthesis in red cells. Nutr. Rev., *39*:331, 1981.
21. Gregory, J. F., and Kirk, J. R.: Vitamin B_6 in foods: assessment of stability and bioavailability. *In*: Human Vitamin B_6 Requirements. Washington, D.C., National Academy of Sciences, 1978, p. 72.
22. Harper, H. A.: The water-soluble vitamins. *In*: Review of Physiological Chemistry. H. A. Harper, V. A. Rodwell, and P. A. Mayes, Eds. Los Altos, Cal., Lange Medical Publications, 1979, p. 159.

23. Hauschka, P. V., Lian, J. B., and Gallop, P. M.: Vitamin K and mineralization. Trends Biochem. Sci., 3:75, 1978.

24. Hegsted, D. M., Chichester, C. O., Darby, W. J., et al.: Eds.: Present Knowledge in Nutrition. 4th ed. Washington, D.C., The Nutrition Foundation, Inc., 1976.

25. Hodges, R. E., Hood, J., Canham, J. E., et al.: Clinical manifestations of ascorbic acid deficiency in man. Am. J. Clin. Nutr., 24:432, 1971.

26. Horwitz, W., Ed.: Association of Analytical Chemists, Official Methods of Analysis. Washington, D.C.

27. Hughes, P. E., and Tove, S. B.: Identification of an endogenous electron donor for biohydrogenation as α-tocopherolquinol. J. Biol. Chem., 255:4447, 1980.

28. Jaffe, I. A.: The antivitamin B_6 effect of penicillamine: clinical and immunological implications. Adv. Biochem. Psychopharmacol., 4:217, 1972.

29. Janiszowska, W., and Pennock, J. F.: The biochemistry of vitamin E in plants. In: Vitamins and Hormones, Vol. 34. P. L. Munson, J. Glover, E. Diczfalusy, et al., Eds. New York, Academic Press, 1976, p. 77.

30. Körner, W. F., and Völlm, J.: New aspects of the tolerance of retinol in humans. Internat. J. Vit. Nutr. Res., 45:363, 1975.

31. Lawson, D. E. M., and Davie, M.: Aspects of the Metabolism and Function of Vitamin D. In: Vitamins and Hormones, Vol. 37. P. L. Munson, J. Glover, E. Diczfalusy, et al., Eds. New York, Academic Press, 1979, p. 1.

32. Lui, A., Lumeng, L., and Li, T.-K.: Metabolism of vitamin B_6 in rat liver mitochondria. J. Biol. Chem., 256:6041, 1981.

33. National Dairy Council: Recent developments in vitamin D. Dairy Council Digest, 47:13, 1976.

34. National Nutrition Consortium, Inc.: Vitamin-Mineral Safety, Toxicity, and Misuse. Chicago, The American Dietetic Association, 1978.

35. National Research Council, Committee on Dietary Allowances: Recommended Dietary Allowances. 9th revised ed. Washington, D.C., National Academy of Sciences, 1980.

36. Neal, R. A.: Vitamin deficiencies: Thiamin. In: Proc. Workshop on Problems of Assessment and Alleviation of Malnutrition in the U.S. R. G. Hansen and H. N. Munro, Eds. Nashville, GSMHA and National Institutes of Health, 1970, p. 129.

37. Nichoalds, G. E.: Riboflavin. Clin. Laboratory Med., 1:685, 1981.

38. Nino, H. V., and Shaw, W.: Vitamins. In: Fundamentals of Clinical Chemistry. N. W. Tietz, Ed. Philadelphia, W.B. Saunders Co., 1976.

39. Norman, A. W., and Henry, H.: 1,25-Dehydroxy D_3. A hormonally active form of Vitamin D_3. Recent Prog. Horm. Res., 30:431, 1974.

40. Massey, V., and Williams, C. H., Jr., Eds.: Flavins and Flavoproteins. New York, Elsevier Biomedical, 1982.

41. McCormick, D. B., and Olson, A. E.: Biotin. In: Present Knowledge in Nutrition. 5th ed. R. E. Olson et al., Eds. Washington, D.C., The Nutrition Foundation, Inc., 1984.

42. McCormick, D. B., and Snell, E. E.: Pyridoxal kinase of human brain and its inhibition by hydrazine derivatives. Proc. Natl. Acad. Sci., U.S., 45:1371, 1959.

43. McCormick, D. B., and Snell, E. E.: Pyridoxal phosphokinases. II. Effects of inhibitors. J. Biol. Chem., 236:2085, 1961.

44. McCormick, D. B., and Wright, L. D., Eds.: Vitamins and coenzymes. Methods in Enzymology, Vol. 18, part A. New York, Academic Press, 1970.

45. McCormick, D. B., and Wright, L. D., Eds.: Vitamins and coenzymes. Methods in Enzymology, Vol. 18, part B. New York, Academic Press, 1971.

46. McCormick, D. B., and Wright, L. D., Eds.: Vitamins and coenzymes. Methods in Enzymology, Vol. 18, part C. New York, Academic Press, 1971.

47. McCormick, D. B., and Wright, L. D., Eds.: Vitamins and coenzymes. Methods in Enzymology, Vol. 62, part D. New York, Academic Press, 1979.

48. McCormick, D. B., and Wright, L. D., Eds.: Vitamins and coenzymes. Methods in Enzymology, Vol. 66, part E. New York, Academic Press, 1980.

49. McCormick, D. B., and Wright, L. D., Eds.: Vitamins and coenzymes. Methods in Enzymology, Vol. 67, part F. New York, Academic Press, 1980.

50. Mehansho, H., and Henderson, L. M.: Transport and accumulation of pyridoxine and pyridoxal by erythrocytes. J. Biol. Chem., 255:11 901, 1980.

51. Merrill, A. H., Jr., Froehlich, J. A., and McCormick, D. B.: Isolation and identification of alternative riboflavin-binding proteins from human plasma. Biochem. Med., 25:198, 1981.

52. Mudd, S. H.: Pyridoxine-responsive genetic disease. Fed. Proc., 30:970, 1971.

53. Olson, R. E.: Vitamin K. In: Modern Nutrition in Health and Disease. R. S. Goodhart and M. E. Shils, Eds. Philadelphia, Lea and Febiger, 1973, p. 166.

54. Olson, R. E., and Suttie, J. W.: Vitamin K and γ-carboxyglutamate biosynthesis. In Vitamins and Hormones, Vol. 35. P. L. Munson, J. Glover, E. Diczfalusy, et al., Eds. New York, Academic Press, 1977, p. 51.

55. Pahuja, D. N., and DeLuca, H. F.: Stimulation of intestinal calcium transport and bone calcium mobilization by prolactin in vitamin D–deficient rats. Science, 214:1038, 1981.

56. Pfeifer, P. M., and McCay, P. B.: Reduced triphosphopyridine nucleotide oxidase–catalyzed alterations of membrane phospholipids. V. Use of erythrocytes to demonstrate enzyme-dependent production of a component with the properties of a free radical. J. Biol. Chem., 246:6401, 1971.

57. Price, P. A., and Baukol, S. A.: 1,25-Dihydroxy D_3 increases serum levels of the vitamin K–dependent bone protein. Biochem. Biophys. Res. Commun., 99:928, 1981.

58. Price, P. A., and Baukol, S. A.: 1,25-Dihydroxy D_3 increases synthesis of the vitamin K–dependent bone protein by osteosarcoma cells. J. Biol. Chem., 225:11660, 1980.

59. Price, P. A., Baukol, S. A., and Williamson, M. D.: 1,25-Dihydroxyvitamin D_3 regulates the synthesis of the vitamin K–dependent bone protein. Calcif. Tissue Int., 33:341, 1981.

60. Price, P. A., Williamson, M. K., and Lothringer, J. W.: Origin of the vitamin K–dependent bone protein found in plasma and its clearance by kidney and bone. J. Biol. Chem., *256*:12760, 1981.

61. Reynolds, R. D.: Nationwide assay of vitamin B_6 in human plasma by different methods. Fed. Proc., *42*:665 (Abst. No. 2185), 1983.

62. Rietz, P., Gloor, U., and Wiss, O.: Menadione aus menschlicher Leber und Faulschlamn. Internat. J. Vit. Nutr. Res., *40*:351, 1970.

63. Rivlin, R. S.: Hormones, drugs, and riboflavin. Nutr. Rev., *37*:241, 1979.

64. Rivlin, R. S., Ed.: Riboflavin. New York, Plenum Press, 1975.

65. Rodriguez, M. S., and Irvin, M. I.: Vitamin A requirements of man. *In*: Nutritional Requirements of Man. M. I. Irwin, Ed. Washington, D.C., The Nutrition Foundation, Inc., 1980, p. 75.

66. Roe, D. A.: A Plague of Corn. Ithaca, N.Y., Cornell University Press, 1973.

67. Rosenberg, L. E.: Vitamin-responsive inherited diseases affecting the nervous system. *In*: Brain Dysfunction in Metabolic Disorders; Research Publication for the Association for Research in Nervous and Mental Disease, Vol. 53. F. Plum, Ed. New York, Raven Press, 1974, p. 263.

68. Sarma, P. S., Menon, P. S., and Venkatachalam, P. S.: Acetylation in the laboratory diagnosis of "burning feet syndrome" (pantothenic acid deficiency). Curr. Sci., *18*:367, 1949.

69. Sauberlich, H. E., Judd, J. H., Jr., Nichoalds, G. E., et al.: Application of the erythrocyte glutathione reductase assay in evaluating riboflavin nutritional status in a high school student population. Am. J. Clin. Nutr. *25*:756, 1972.

70. Sauberlich, H. E., Skala, J. H., and Dowdy, R. P.: Laboratory Tests for the Assessment of Nutritional Status. Boca Raton, Fla., CRC Press, Inc., 1974.

71. Scriver, C. R.: Diet, genes and (so-called) vitamin D-resistant rickets. Nutrition and the M.D., *5*:(2), Feb. 1979.

72. Scriver, C. R.: Vitamin-responsive inborn errors of metabolism. Metabolism, *22*:1319, 1973.

73. Simoin, E. J., Eisengart, A., Sundheim, L., et al.: The metabolism of vitamin E. II. Purification and characterization of urinary metabolites of α-tocopherol. J. Biol. Chem., *221*:807, 1956.

74. Smeets, E. H. J., Muller, H., and De Wael, J.: A NADH-dependent transketolase assay in erythrocyte hemolysates. Clin. Chim. Acta, *33*:379, 1971.

75. Spector, R.: Niacin and niacinamide transport in the central nervous system. *In vivo* studies. J. Neurochem., *33*:895, 1979.

76. Suttie, J. W., Ed.: Vitamin K Metabolism and Vitamin K–Dependent Proteins. Baltimore, University Park Press, 1979.

77. Symposium on vitamin D and membrane structure and function. Fed. Proc., *41*:60, 1982.

78. Tryfiates, G. P., Ed.: Vitamin B_6 Metabolism and Role in Growth. Westport, Conn., Food and Nutrition Press, Inc., 1980.

79. Warnock, L. G.: A new approach to erythrocyte transketolase measurement. J. Nutr., *100*:1057, 1970.

80. White, A., Handler, P., Smith, E. L., et al.: Principles of Biochemistry. New York, McGraw-Hill Book Company, 1978.

81. Wilson, J. D.: Disorders of vitamins—deficiency, excess and errors of metabolism. *In*: Harrison's Principles of Internal Medicine. 10th ed. R. G. Petersdorf, R. D. Adams, E. Braunwald, et al., Eds. New York, McGraw-Hill Book Company, 1982, p. 461.

82. Winick, M.: Nutrition in Health and Disease. New York, John Wiley and Sons, Inc., 1980, Chapters 8 and 9.

83. Wyse, B. W., Wittmer, C., and Hansen, R. G.: Radioimmunoassay for pantothenic acid in blood and other tissues. Clin. Chem., *25*:108, 1979.

Trace Elements

by Robert A. Jacob, Ph.D.

The analytical chemist's definition of a trace concentration is $< 100\ \mu g/g$ (0.01%). Concentrations of major elements in biological tissues are reported in mg/g amounts, whereas *trace element concentrations* are expressed in $\mu g/g$ or smaller amounts. Correspondingly, human intake requirements for major elements are expressed in grams or fractions of a gram per day, while intakes of trace elements are in mg or $\mu g/d$.

Further improvements in the sensitivities of analytical methods have allowed the study of *ultratrace elements,* those present in ng/g or less and those whose dietary requirements are near 50 ng/g diet.

ESSENTIAL TRACE ELEMENTS

The capability of reliably quantitating trace element levels in biological specimens has led to discoveries of the vital roles that trace elements play in human and animal metabolism. Basically, an *essential element* is one that is uniquely required for growth or for the maintenance of life or health. A deficiency of the element consistently produces a functional impairment which is alleviated by physiological supplementation of only that element. A biochemical basis for the element's essential functions must be demonstrated. For trace metals this is often the identification of a unique metalloenzyme which contains the metal as an integral part or as an enzyme activator. A simple model of the relationship between tissue concentration of an essential nutrient and a dependent biological function is shown in Figure 8C-1. The plateau of the curve reflects the operation of a homeostatic regulation system which ensures optimal health over a range of nutrient levels. At low nutrient levels, dependent biological functions are impaired, thus defining a deficiency state. Very high nutrient levels may result in another impairment, i.e., a toxicity state. Each nutrient has its own characteristic curve, and since pathology due to deficiency is often different from that due to toxicity, such curves may be asymmetric.

Over the past several decades, as analytical detection limits have been lowered, the list of trace elements claimed to be essential for humans or animals has grown steadily. A complete list of essential trace elements is somewhat controversial, partly because of fragmentary data on elements under current study. Also, comprehensive criteria which define trace element essentiality are not universally accepted. Before essentiality is acknowledged, an element should be demonstrated to be essential in several species and by more than one independent investigator. The elements presently accepted as essential in animals or humans are shown in a periodic table format in Figure 8C-2. All macro and trace (shaded) elements are considered essential for humans with the exception of nickel, vanadium, arsenic, and silicon, for which essentiality has been demonstrated only in animals. Limited evidence in animals suggests possible essentiality for lithium and tin. Trace elements which are consistently present in human tissues and which are toxic, but considered nonessential, are aluminum, cadmium, lead, and mercury.

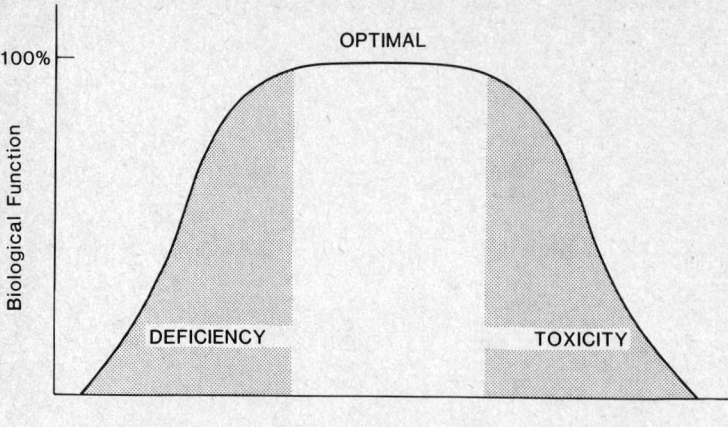

Figure 8C-1. Model of the relationship between tissue concentration or intake of an essential nutrient and dependent biological function.

ULTRATRACE ELEMENTS

Because of ubiquitous contamination, creation of experimental deficiency conditions is much more difficult for ultratrace level elements (ng/g), such as Cr, Ni, V, and As, than for trace level elements (μg/g), such as zinc, copper, and iron. Cases of human ultratrace element intake deficiencies are unlikely; however, this does not preclude the possible occurrence of adverse effects due to suboptimal ultratrace element status. The biological availability of the particular form of an element in the diet and the presence of factors in the diet which affect its availability are important determinants of the ability of a diet to provide the trace element requirement. Conditions associated with iron, zinc, and copper deficiencies may be of potential clinical importance for ultratrace metal nutrition as well. Such conditions include (1) inborn errors of metabolism which affect absorption, retention, or excretion of the element; (2) disruptions to element metabolism as a secondary consequence to malnutrition, disease, injury, or stress; (3) the absence of an element in total parenteral or synthetic formula solutions fed over long periods of time; and (4) marginal deficiencies induced by various dietary manipulations or by antagonistic interactions with other nutrients or drugs. Because marginal ultratrace metal deficiencies are difficult to detect and verify in humans, most current evidence for the essentiality of ultratrace elements is from animal studies.

Figure 8C-2. Chemical elements accepted as essential for human or animal nutrition (trace elements shaded).

CHARACTERISTICS OF TRACE ELEMENT FUNCTIONS

Although many details of trace element functions are not yet understood, some general characteristics are well known. These are *amplification* of trace element action, *specificity, homeostasis,* and *interactions.*[41]

Amplification of Trace Element Action

The action of a very small amount of trace element is necessary for optimal performance of a whole organism. Lack of a small amount of a trace element (e.g., iron) can result in disease (anemia) seemingly disproportionate to the amount of element missing. The basis for this *amplification* of trace element action is that trace elements are constituents of, or interact with, enzymes and hormones that regulate the metabolism of much larger amounts of biochemical substrates. If the substrates are also regulatory, the effect is even further amplified.

Specificity of Trace Elements

Essential trace elements are *specific* for their in vivo functions; they cannot be effectively replaced by chemically similar elements. The essential trace metal or element interacts with electron donor atoms such as nitrogen, sulfur, and oxygen, the types of interaction depending on configurational preferences and bond types. Certain trace metals are stable in more than one valence state (e.g., Fe, Cu, Mo), allowing biological redox function, while others are stable in only a single state [e.g., Zn(II), Ni(II)]. Metals with partially filled *d* electron orbitals (e.g., Fe, Cu, Co) tend to coordinate with a larger number of electron donor atoms than metals with filled *d* orbitals (Zn, Se), and exhibit qualitatively different preferences for the donor atoms. Specificity of trace element function is also promoted by specific carrier and storage proteins, such as transferrin and ferritin for iron, albumin and α_2-macroglobulin for zinc, ceruloplasmin for copper, transmanganin for manganese, and nickeloplasmin for nickel. These carrier proteins recognize and bind specific metals and transport them to, or store them at, specific sites within the organism.

Homeostasis of Trace Elements

Mechanisms which ensure optimal body distribution of an element over a range of intakes constitute a system of *homeostatic regulation* for that element, and include absorption, storage, and excretion. Although many details of trace element absorption processes are still unknown, the rate of absorption of a trace element generally decreases with its increasing concentration in the intestinal lumen. Active transport mechanisms involving absorption by specific metal binders and/or feedback inhibition have been postulated for iron, zinc, and copper. Carrier and storage proteins have some capability to buffer against excess free metal levels, but they generally are not as important for regulation as are absorption control mechanisms. The principal excretory route for elimination of trace metals is via the feces. Fecal excretion reflects dietary intake as well as homeostatic regulatory mechanisms such as gastrointestinal absorption and endogenous metal secretion into the intestine. Relatively small amounts of trace metals are excreted via the urine, but halides (iodide and fluoride) as well as selenium and chromium are effectively eliminated via the urine. The loss of trace elements through other routes such as hair, skin cell desquamation, and sweat is generally of a minor degree. Body surface losses of some trace metals, particularly zinc, copper, selenium, and iron, can be appreciable in hot climates or in certain stressful conditions. Sources for information on sweat or surface losses of essential trace elements are available.[25,29] Menstrual iron loss and seminal zinc loss are also minor but can be significant in some cases.

Interactions of Trace Elements

An overabundance of one trace element can interfere with the metabolic utilization of another element present in normal or marginal concentrations.[37,42] Alternatively, the effect of a toxic trace element may be ameliorated by another "protective" trace element. The addition of large amounts of zinc to a diet interferes with the intestinal copper absorption system, resulting in copper deficiency in spite of an otherwise adequate copper intake. Hence, zinc is said to *antagonize* copper absorption. Copper deficiency, in turn, is known to provoke iron deficiency and anemia. Molybdenum deficiency is much easier to produce in animals when the chemically similar element tungsten is simultaneously administered in large amounts. Interactions which involve toxic elements include the increased retention of cadmium and lead associated with iron deficiency and the protective effect of selenium against cadmium and mercury toxicity.

LABORATORY ASSESSMENT OF TRACE METAL STATUS

No wholly satisfactory laboratory methods for clinical assessment of trace metal status have yet been established. Currently the only definitive test of human trace metal deficiency is the clinical response to therapeutic supplementation with the trace metal in question.

Measurements of metalloenzyme activities have been proposed as useful assessment tests because plasma trace metal levels are often affected by factors unrelated to the whole body metal status. Although hair is relatively easy to analyze for trace metal content, its chief problem as a sample for assessing metal nutriture is its susceptibility to environmental contamination. Nutritionists and toxicologists who wish to assess body metal burden must measure only endogenous (body derived) hair metal.

A simultaneous battery of tests involving body tissue or body fluid metal determinations, metalloenzyme assays, and functional-morphological indices would provide a more reliable assessment of metal nutriture. In the busy clinical diagnostics setting, however, it is most practical to assess trace metal status from analysis of a single blood specimen. Further investigations are needed on the clinical value of whole blood, leukocyte, erythrocyte, saliva, skin, hair, and fingernail analyses as metal nutriture indices.

SAMPLE COLLECTION AND TESTING

Accurate determination of trace elements in biological specimens presents special analytical difficulties and requires special precautions.[26] *Sampling* procedures must be carefully considered since heterogeneity of trace element distributions in tissues is the rule rather than the exception. Analysis of seemingly homogeneous specimens such as blood, saliva, sweat, and hair can be significantly affected by sampling and sample processing practices. For example, levels of zinc can be 5–15% higher in serum than in plasma due to release of zinc from erythrocytes and platelets during clotting. Zinc concentrations in whole uncentrifuged saliva are greater than in the supernatant from centrifuged saliva. Similarly, zinc, copper, and iron levels in uncentrifuged whole sweat are greater than in a cell-free supernatant fraction.

Hair samples should be cut with stainless steel scissors to avoid contamination. The best source for hair is the occipitonuchal region, since this area is low in the back of the head and is less visible, and the hair is probably less contaminated. Different longitudinal sections along the hair shaft represent body metal accumulated at different times. Since hair grows slowly (1–2 cm/month), distal ends reflect body metal burdens many months in the past, while proximal (scalp) ends reflect the most recent metal status. Thus, the proximal hair segments more nearly represent the current body metal status; in addition, they are less likely to be contaminated with exogenous metal. Approximately 0.3 g of hair sample should be taken and the proximal 3 or 4 cm cut for metals analysis. Washing procedures have been developed to remove oils and exogenous surface metal contamination, while leaving all or most of the endogenous metal.[27,40] Hair samples should be weighed dry and wet-ashed with nitric or nitric/perchloric acid mixtures,[4] and the metals determined in the digestates. Hair samples that are thoroughly delipidated with organic solvents in the washing step digest easily and cleanly in strong acids. Comprehensive reviews on hair analysis are available.[9,27]

Contamination of Samples

The chief analytical problem encountered in trace element analysis is external contamination. Many trace metals are prevalent in the laboratory environment in ng and even μg amounts; hence, an appreciable portion of the result of a trace element analysis may be due to contamination. This is also a major reason for the wide variation in reported reference values. A laboratorian contemplating trace metals analysis must be prepared to take precautions through all sampling and analytical procedures to minimize contamination. When careful attention is given to proved practices and pitfalls, the determination of μg/g amounts of trace metals in biological specimens (e.g., iron, zinc, and copper) is straightforward. The accurate determination of ultratrace metals (e.g., Mn, Co, Cr, Ni, V, As), however, presents considerably greater methodological difficulties. Thus, ultratrace metal tests should not be part of the routine test armamentarium of the clinical laboratory.

Common sources of trace metal contamination in analytical work include rubber, wood, paper products, metal surfaces, skin, dandruff, and hair. Plastic, borosilicate glass, and stainless steel equipment are best suited for *trace* metal analysis and, of these, the fluorocarbon, polypropylene, and polyethylene plastics are generally best. Stainless steel knives or scissors are suited for cutting specimens for determination of iron, zinc, and copper, but not for *ultratrace* metal determinations. Water should meet ACS specifications for type I water, with ≥ 14.0 megohm/cm^2 resistance and iron, zinc, and copper content of less than 10 μg/L. Glassware should be cleaned of surface trace metals by soaking overnight in dilute nitric acid or diluted commercial metal-scavenging solutions marketed for radioactive cleanup. The latter solutions are effective and are considerably safer and easier to handle than large volumes of dilute acids. Only evacuated blood collection tubes specified for trace metal analysis should be used for blood collection; other tubes may contaminate the specimen due to the leaching of metals from the rubber stoppers. Plastic disposable syringes with stainless steel needles are also suitable. Other materials involved in sample collection or processing, such as clotting aids or gel serum separators, should be avoided or checked before use for absence of the trace metals of interest. The level of trace metals should be checked in all anticoagulants used, particularly heparin, which is derived from animal tissue sources. Common laboratory practices which can contribute to contamination include the use of wooden applicator sticks to ring blood clots off tube walls, the wiping of pipet tips with tissues, and the mixing or storing of samples in rubber-stoppered tubes. Skin contact with any surface which will be exposed to samples or solutions must be avoided.

ANALYTICAL METHODS

Any analytical method for the determination of trace metals in biological specimens must be sensitive, specific, precise, accurate, and relatively fast. Sensitivity is most important, since concentrations of trace or ultratrace elements in some specimens are in the μg/g or lower range. Without adequate sensitivity in the range of clinical interest, a method cannot provide adequate precision or accuracy. Since the above method requirements are not of the same importance for all metals and for all specimens, the analytical technique of choice is dependent on the sample type and the metal to be determined.

The most popular of currently employed techniques for trace metal analysis include atomic absorption spectrophotometry (AAS), emission spectroscopy (ES), neutron activation analysis (NAA), colorimetry, mass spectrometry (MS), and, to a lesser extent, X-ray fluorescence spectrometry (XRF) and electrochemical techniques, such as anodic stripping voltammetry (ASV). Emission spectroscopy and NAA offer simultaneous multi-element determination. The NAA technique is one of the most sensitive, and the analysis can be uniquely carried out without any sample preparation, destruction, or contamination. The ES technique also requires minimal sample preparation but lacks sensitivity for some ultratrace metal determinations and is susceptible to spectral interferences. These last two problems are still present with inductively coupled plasma emission spectroscopy (ICP-PES). The ICP-PES technique provides time saving due to simultaneous multi-element determinations and an unusually wide analytical range. However, appreciable sample preparation, such as ashing, is usually necessary to convert specimens to the required soluble form. High organic content of sample solutions causes nebulization and interference problems as well as decreased sensitivity for most ultratrace metal determinations. Ashing to remove interfering organic matrix is also necessary for MS techniques; however, MS offers very high sensitivity and multi-element capability. The XRF technique also provides multi-element detection but suffers from lack of sensitivity.

The chief drawbacks to ES, ICP-PES, NAA, MS, and XRF techniques are the complexity and expense of equipment and materials and the high level of operator skill required. For this reason, these techniques are generally not practical for most clinical diagnostic laboratories. Electrochemical or ASV techniques are highly sensitive but have been applied to only a few metals in biological specimens, primarily to zinc, copper, cadmium, and lead. The matrix-sensitivity of these techniques complicates determinations in biological specimens, which reduces sample throughput considerably. This leaves AAS and colorimetry as alternative methods for trace metal determinations. Because nearly all *d*-type transition elements are active in forming colored complexes, the use of a color-forming agent is often not wholly specific for a particular metal. Sample preparation or procedural steps must then be taken to eliminate the interference.

Atomic Absorption Spectrophotometry

Since the demonstration of its utility as a technique for the specific determination of metals in biological matrices in 1955, atomic absorption spectrophotometry (AAS) has supplanted colorimetric techniques as the method of choice for those laboratories without access to advanced instrumental techniques. The AAS technique detects only one element at a time, and some sample preparation work is often necessary. Metal determination, however, is quick and accurate once the specimen is in the proper form. Besides offering good sensitivity and specificity, AAS instruments are readily available and relatively simple to operate. Proved AAS procedures for the determination of most trace metals in biological specimens are available in either the instrument manuals or the general scientific literature. The principles of AAS are detailed in Chapter 1B, Section One, and other sources.[10,13,65,66]

Methods in which diluted serum or plasma is aspirated directly into the AAS flame are effective for the determination of magnesium, zinc, copper, and iron. Discrete nebulization techniques minimize sample aspiration waste and allow metal determinations on microliter volumes of serum.[38,77] Electrothermal or "flameless" AAS micromethods, using as little as 10 μL of serum, have also been reported.[24,75] Serum dilutions of ten-fold or more decrease the problems of clogging nebulizers and burners, but they result in poor sensitivity. Therefore, sample dilutions of five-fold or less are recommended. Differences in aspiration rates due to viscosity differences in aqueous standard solutions versus diluted serum specimens can be overcome by the use of glycerol diluents for the standards.[7,62]

The sensitivities of the AAS techniques for determining various metals depend on the particular metal of interest and on the technique used. The flame AAS mode is simpler and less tedious to perform than the flameless mode. Generally, if analyte concentrations of a specimen are below 50 ng/g, flameless AAS techniques are necessary. Sunderman has written useful reviews on the application of AAS to clinical pathology,[65] and on the use of flameless AAS for the determination of metals in biological specimens.[66]

Atomic fluorescence spectrometry, a technique somewhat analogous to AAS, can be performed on many commercial AAS instruments with certain optional equipment. The technique is considerably more sensitive than AAS for certain elements, zinc and cadmium especially.

QUALITY CONTROL OF TRACE ELEMENT DETERMINATIONS

Because trace element testing is not standardized and is disposed to exogenous contamination problems, effective quality control measures must be incorporated into trace metal analysis schemes. In general, control or reference materials which contain certified levels of trace elements are not available. One commercially available control serum provides analyzed values for magnesium, zinc, and copper, but the levels are above the reference range and thus are of limited value as controls. Other lyophilized commercial serum controls have been found to be zinc contaminated.

The Centers for Disease Control of the U.S. Department of Health and Human Services have made available trace metal reference sera for survey purposes. However, these are not routinely available for quality control monitoring. Some commercially available lyophilized urine controls have assayed values for zinc and copper. A convenient and reliable approach for control of biological fluid assays (blood, urine, saliva) is the use of frozen aliquots of a specimen pool. Aliquots should be stored frozen, in tightly capped polypropylene tubes, with a minimum of headspace. No changes in zinc or copper levels were seen in serum samples refrigerated or frozen for up to 50 days.[17] Frozen-pool aliquots can also be used to assess the day-to-day precision of methods. Thawed specimens should be well mixed before sampling, since analytical errors due to metal concentration gradients in unmixed thawed specimens have been demonstrated. Recovery studies of metal from samples containing known quantities of added analyte are useful initially for assessing method accuracy and linearity, but such studies do not need to be performed daily.

Because standard hair reference material is not available, several laboratories have prepared homogeneous hair "pools" that can be used as controls for hair analysis and for assessing precision and accuracy through standard metal recovery tests. A hair pool can be prepared by cutting several grams of hair as finely as possible with stainless steel scissors, washing successively with acetone-ether-detergent, drying, and storing in a desiccator. The reproducibility (CV) of replicate analysis of such a hair pool was 3% for zinc, 7% for copper, and 13% for chromium. Alder et al. investigated the use of silk and animal hairs as a standard for hair analysis.[3] NBS Standard Reference Materials, such as Bovine Liver (see below) and Orchard Leaves, can be used as standards whose organic matrices are representative of hair samples.

For quality control of trace element determinations in tissue specimens, the NBS Standard Reference Material 1577a, Bovine Liver, is available (U.S. Department of Commerce, National Bureau of Standards, Office of Standard Reference Materials, Washington, D.C. 20234). The material is a lyophilized, homogeneous liver powder which has certified values (based on results obtained by the use of two or more independent reference analytical methods) for magnesium, iron, zinc, copper, manganese, molybdenum, cobalt, selenium, and arsenic. Also available from the NBS is Standard Reference Material 1569, Brewer's Yeast, which has a certified value for chromium and contains a volatile organo-chromium component analogous to that found in some biological specimens.

ESSENTIAL TRACE METALS AND CARDIOVASCULAR DISEASE

Many studies have associated trace metals with human diseases. These studies are usually controversial and thus are not covered in detail here. All of the essential trace metals shown in Figure 8C-2 have in some way been implicated in cardiovascular disease.[34] The associations are based primarily on epidemiologic studies and the effects of the metals on blood lipid (mainly cholesterol) levels. Trace metals have also been shown to affect vascular and myocardial integrity, arterial tension, and hemostasis. The cardiovascular diseases which have been associated with trace metals are atherosclerosis and, to a lesser extent, cerebrovascular, hypertensive, and myocardial diseases. Evidence suggests that trace metal effects on cardiovascular integrity may exist for copper, zinc, and magnesium. Epidemiological evidence showing strong and consistent inverse correlations between heart disease rates and drinking-water hardness (the "water factor") was an initial impetus to studies of the relationship between trace metal intake and heart disease.

Zinc and copper have been associated with cholesterol metabolism and maintenance of cardiovascular integrity. Studies are not consistent as to what the atherogenic effects of differing zinc intakes might be. Evidence implicates copper more strongly than zinc in cardiovascular disease. Hypercholesterolemia due to copper deficiency has been observed in humans and in a variety of animal species. Decreases in the activity of copper-dependent enzymes are cited as causes for neural and cardiac abnormalities and vascular defects such as aneurysm and rupture of aorta and large vessels in copper-deficient animals.

Magnesium ions are essential for maintenance of the functional and structural integrity of the myocardium. Studies have shown that magnesium depletion produces cardiac necrosis and influences coronary flow, hemostasis, and atherogenesis.[60]

Hypocholesterolemia in a man fed a formula diet inadvertently deficient in manganese was attributed to a deficit in manganese-dependent farnesylpyrophosphate synthesis.

MAGNESIUM

Magnesium is not a true trace element. The adult body contains about 24 g of magnesium, and dietary intake is about 0.3 g/d. Along with potassium, magnesium is a major intracellular cation.

Biochemistry

Being an alkaline earth metal, magnesium has chemical properties distinctly different from those of the transition metals. Magnesium atoms contain no low-energy *d*-orbitals available for coordinate-covalent bonding. Compared with the transition metals, magnesium interacts with other chemical species with a much stronger electrostatic bonding component and a relative preference for oxygen over nitrogen atoms. Magnesium is an essential factor in many important enzymatic reactions, either as an integral part of a metalloenzyme or as an intermediary activator. Principal magnesium-dependent pathways are those involving enzyme-catalyzed transfer of phosphate groups, especially those involving ATP. Magnesium is also involved in amino acid activation and protein synthesis and is required for ribosomal integrity, RNA and DNA reactions, and proper neuromuscular function.

Metabolism

The 70-kg adult human contains about 20–28 g of magnesium, mostly in bone (about 55%, which is available for mobilization during deficiency) and muscle (about 27%). Thus, bone magnesium decreases concomitant with a calcium increase in animals made magnesium deficient, and it rapidly increases when the animals are repleted with magnesium.

Magnesium intakes vary appreciably, an approximate range for U.S. and western European populations being 180–480 mg/d. The Recommended Dietary Allowance (RDA) for magnesium is listed in Table 8C-1, and is 300–350 mg/d for adults. The magnesium content of food varies widely. Seafood, meats, nuts, and grains contain appreciable amounts, while oils, fats, and sugars contain very little.

Table 8C-1. RECOMMENDED DIETARY INTAKES OF ESSENTIAL TRACE ELEMENTS*

A. Recommended Dietary Allowances (RDA)†

	Age (Years)	Magnesium (mg)	Iron (mg)	Zinc (mg)	Iodine (μg)
Infants	0.0–0.5	50	10	3	40
	0.5–2.0	70	15	5	50
Children	1–3	150	15	10	70
	4–6	200	10	10	120
	7–10	250	10	10	120
Males	11–14	350	18	15	150
	15–18	400	18	15	150
	19–22	350	10	15	150
	23–50	350	10	15	150
	51+	350	10	15	150
Females	11–14	300	18	15	150
	15–18	300	18	15	150
	19–22	300	18	15	150
	23–50	300	18	15	150
	51+	300	10	15	150
Pregnant		+150	‡	+5	+25
Lactating		+150	‡	+10	+50

B. Estimated Adequate and Safe Daily Dietary Intakes§

	Age (Years)	Copper (mg)	Manganese (mg)	Fluoride (mg)	Chromium (mg)	Selenium (mg)	Molybdenum (mg)
Infants	0–0.5	0.5–0.7	0.1–0.5	0.1–0.5	0.01–0.04	0.01–0.04	0.03–0.06
	0.5–1	0.7–1.0	0.7–1.0	0.2–1.0	0.02–0.06	0.02–0.06	0.04–0.08
Children and	1–3	1.0–1.5	1.0–1.5	0.5–1.5	0.50–1.50	0.02–0.08	0.05–0.10
adolescents	4–6	1.5–2.0	1.5–2.0	1.0–2.5	0.03–0.12	0.03–0.12	0.06–0.15
	7–10	2.0–2.5	2.0–3.0	1.5–2.5	0.05–0.20	0.05–0.20	0.10–0.30
	11+	2.0–3.0	2.5–5.0	1.5–2.5	0.05–0.20	0.05–0.20	0.15–0.50
Adults		2.0–3.0	2.5–5.0	1.5–4.0	0.05–0.20	0.05–0.20	0.15–0.50

*From Food and Nutrition Board, National Research Council; Recommended Dietary Allowances, 9th ed. Washington, D.C., National Academy of Sciences, 1980.

†The RDA are the daily levels of intake of essential nutrients considered to be adequate to meet the known nutritional needs of practically all healthy persons.

‡Supplemental iron (30–60 mg) is recommended during pregnancy, and for 2–3 months after parturition.

§Estimated ranges of adequate but safe intakes of nutrients for which RDA cannot be established on the basis of present knowledge. Since toxic levels for some of the elements are only severalfold more than usual intakes, the upper levels of estimated intakes should not be greatly and habitually exceeded.

Magnesium is absorbed in the upper intestine. The greater proportion of ingested magnesium (60–70%) is not absorbed but is excreted in the feces. Unlike calcium absorption, magnesium absorption is not vitamin D dependent, but can be affected by the amount of calcium, phosphate, protein, lactose, or alcohol ingested.

The intestinal absorption of magnesium has been found to be inversely related to the total magnesium intake. This intestinal absorption mechanism, plus an effective renal regulation of urinary excretion, allows homeostasis of magnesium over a wide range of dietary intakes. The mechanisms of renal regulation of plasma and urinary magnesium levels are not yet established. The renal clearance and plasma levels of magnesium are often related to those of calcium, phosphate, sodium, and potassium. There is evidence of hormonal regulation of the renal clearance of magnesium similar to that for potassium. The major part of magnesium in plasma (about 65–70%) exists as free ions or in the form of various diffusible complexes; the remainder is protein bound, mostly to albumin.

Clinical Significance

The best-defined manifestation of magnesium deficiency is impairment of neuromuscular function, e.g., hyperirritability, tetany, convulsions, and electrocardiographic changes.

In experimentally induced human magnesium deficiency, levels of serum magnesium were found to indicate the depletion much sooner than red cell magnesium levels. Serum magnesium levels do not fall below 0.5 mmol/L until at least 25% of intracellular magnesium has been lost. Symptoms of magnesium deficiency usually do not occur until serum magnesium levels decrease below 0.5 mmol/L. Since renal clearance plays a large role in magnesium homeostasis, the urinary excretion of magnesium is also a useful index of magnesium deficiency or excess. Human magnesium deficiency, as indicated by reduced serum magnesium levels (hypomagnesemia), has been shown to occur with either normal or reduced serum calcium levels. Hypomagnesemia may be a secondary effect in hypocalcemic or calcium-deficiency tetany, yet a hypomagnesemic-normocalcemic tetany has been described which is effectively treated with magnesium supplementation alone. During tetany, serum magnesium levels of 0.15–0.5 mmol/L have been reported, accompanied by normal serum calcium and pH levels. There is evidence that tetany accompanied by hypocalcemia and hypomagnesemia may not be optimally treated with calcium administration alone. Decreased serum potassium concentrations (hypokalemia) have also been found to accompany magnesium depletion, and the occurrence of otherwise unexplained hypokalemia and hypocalcemia should suggest magnesium deficiency.

Conditions which have been associated with *hypomagnesemia* include chronic alcoholism, childhood malnutrition, lactation, malabsorption, acute pancreatitis, hypoparathyroidism, chronic glomerulonephritis, aldosteronism, digitalis intoxication, and prolonged intravenous feeding. Magnesium depletion occurs in conditions which disrupt the normal renal conservation of magnesium, e.g., in patients with renal tubular reabsorption defects and those taking chlorothiazides, ammonium chloride, or mercurial diuretics for congestive heart failure.

Increased serum magnesium concentrations have been observed in dehydration, severe diabetic acidosis, and Addison's disease. Conditions which interfere with glomerular filtration, e.g., uremia, result in retention of magnesium and hence elevation of serum levels. Hypermagnesemia leads to an increase in the atrioventricular conduction time of the electrocardiogram.

DETERMINATION OF MAGNESIUM IN BODY FLUIDS

Specimens

Serum magnesium determinations should be done on samples drawn without venous stasis. Since the magnesium concentration in erythrocytes is substantially greater than that in serum, the specimen should be separated from the erythrocytes as soon as possible, and hemolyzed specimens should not be used for analysis. Magnesium levels in serum are stable for several days if the specimen is stored in the refrigerator, separated from the red cells.

Urine should be acidified to pH 1.0 and mixed well prior to analysis.

METHODS FOR THE DETERMINATION OF MAGNESIUM

Atomic absorption spectrophotometry (AAS) is the preferred technique for the determination of magnesium in biological specimens. Some of the analytical difficulties are similar to those discussed under calcium analysis. La(III) or Sr(III) is added to the sample diluent to bind with phosphates which otherwise would form refractory calcium and magnesium compounds in the AAS flame, resulting in falsely low values. Some methods involve release of magnesium from proteins by treatment with HCl or trichloroacetic acid (TCA) followed by centrifugation. Protein precipitation minimizes imprecision due to nebulizer and burner fouling. An acid precipitation method involves diluting the sample with a diluent containing 6.6 g La(III) and 40 g TCA/L. After centrifugation, the supernatant is analyzed by AAS, and results are calculated from a standard curve.

Direct dilution techniques are faster and simpler and have been found to be precise and accurate, provided proper care is taken to maintain burner and nebulizer performance. Because relatively large sample dilutions (e.g., 1:50) are made for magnesium determinations, the problem of nebulizer or burner clogging is minor. A direct dilution AAS method for determining serum magnesium is given below; urinary magnesium can be similarly determined using a 50-fold dilution with 0.5% La_2O_3 diluent [La(III), 30.8 mmol/L]. Detailed procedures for the determination of magnesium in biological specimens by AAS are also given in AAS instrument manuals.

Many *fluorometric* (8-hydroxyquinoline) and colorimetric methods have been used for determining serum magnesium.[68] For laboratories without an AAS instrument, the Calmagite colorimetric method has been adapted for both manual[1] and automated[18] use. The colored Calmagite-Mg complex is formed immediately upon mixing of Calmagite reagent with 50 µL of plasma and is stable for over 30 min. Plasma magnesium is calculated by comparing the spectral absorbance of the sample at 520 nm with magnesium iodate standards. The method is simple and rapid. Between-batch precision (CV) is 3%, and method results correlate well with results from an atomic absorption procedure.

Determination of Serum Magnesium by Atomic Absorption Spectrophotometry Using Direct Sample Dilution

Principle

Serum specimens are diluted 50-fold with an acidic lanthanum chloride or oxide diluent and aspirated directly into the AAS flame without further treatment. Serum magnesium levels are calculated from a linear calibration curve of standards made up with the same lanthanum diluent and amounts of Na^+ and K^+ equivalent to those present in 50-fold diluted serum.

Reagents

1. Lanthanum oxide diluent, 0.43 g La(III)/dL (0.5 g La_2O_3/dL). Add 10 mL of concentrated HCl to about 800 mL of deionized water, then add 5 g of La_2O_3 and dilute to 1 L. Use only high purity La_2O_3 with low Ca and Mg content. Stir until La_2O_3 dissolves and the whitish suspension is clear. The diluent is stable at room temperature.

2. Calibration standards (0–0.04 mmol Mg(II)/L final concentration). Prepare at least four standards in the range of 0–0.04 mmol Mg/L by diluting commercially available Mg standard solution (1 g/L) with lanthanum oxide diluent containing 164 mg NaCl (2.8 mmol Na^+)/L and 7.5 mg KCl (0.1 mmol K^+)/L. The lanthanum oxide–salt diluent serves as the zero standard. Standards should be stored in polypropylene or polyethylene bottles and are stable at room temperature.

Procedure

1. Dilute serum specimens 50-fold with the salt-free lanthanum oxide diluent, e.g., 100 µL of serum diluted to 5 mL, and vortex.

2. Zero the AAS instrument with the lanthanum-salt diluent. Aspirate standards, then dilute samples, taking readings as prescribed by the instrument manufacturer. Check the baseline periodically, readjusting if necessary.

3. Calculate the concentration of magnesium in the samples from the calibration curve.

Comments

1. An air-acetylene flame and the 285.2-nm wavelength of a magnesium hollow cathode lamp can be employed. Check the instrument manufacturer's recommendations for instrumental settings for magnesium determinations.

2. The sodium and potassium salts are added to the standard solutions and the baseline-zeroing solution to compensate for spectral interferences contributed by the same salts in the diluted serum samples.

3. The above method is primarily that of Bowers and Pybus for determining serum calcium, but the method is equally applicable to serum magnesium. Hence both calcium and magnesium can be determined on the same serum dilution.

Reference Ranges

The reference range for magnesium in *serum* with the atomic absorption method is 0.6–1.1 mmol/L. Serum levels in newborns are lower than in adults (0.5–0.9 mmol/L).[69] Infants older than 5 months, children, and adolescents have levels essentially the same as adults. Levels do not change appreciably throughout the day, but in females the highest serum values are obtained at the time of menses.

Erythrocytes contain about 1.7–2.7 mmol/L. The concentration range in *cerebrospinal fluid* is 1.0–1.4 mmol/L. The urinary excretion is 3.0–5.0 mmol/d; values for males are slightly higher than those for females.

Reference

Bowers, G. N., and Pybus, J.: Total calcium in serum by atomic absorption spectrophotometry. *In*: Standard Methods of Clinical Chemistry, Vol. 7. S. Meites, Ed. New York, Academic Press, 1972, p. 143.

IRON

Iron is the most important of the essential trace metals. An appreciable number of human diseases are related to iron deficiency or disorders of iron metabolism. These topics as well as methods for the determination of iron are covered in Chapter 15, Biochemical Aspects of Hematology. The RDA for iron is listed in Table 8C-1.

ZINC

The essentiality of zinc for the growth and well-being of both plants and animals is well established. Zinc is second only to iron in importance as an essential trace metal. Since zinc deficiency was first described in the early 1960's, research has elucidated many specific metabolic interactions of zinc. Moreover, the discovery of a variety of zinc-related clinical disorders has directly demonstrated the importance of zinc in human nutrition.

Biochemistry and Physiology

Zinc is found in biological systems only in the $+2$ valent state. This is the result of the extra stability associated with a filled d-orbital electron configuration. This also accounts for the biochemical differences between zinc and the other first-row transition elements with unfilled d-orbitals, such as iron, copper, and manganese. Biological redox functions are not possible with zinc.

The *metabolic functions* of zinc are largely based on its presence in zinc metalloenzymes, over 70 of which have been identified in various living systems.[57] Important zinc metalloenzymes in humans include carbonic anhydrase, alkaline phosphatase, RNA and DNA polymerases, thymidine kinase, carboxypeptidases, and alcohol dehydrogenase. The zinc atoms are an integral, firmly bound part of the metalloprotein molecule and often are directly involved in the active catalytic site; they contribute also to the structural stability of metalloenzymes.

Loss of zinc metalloenzyme activity upon zinc depletion varies with different zinc metalloenzymes. Activity loss is dependent on the enzyme turnover rate and the affinity of the enzyme for zinc. Pancreatic carboxypeptidase A, thymidine kinase, and alkaline phosphatase activities are appreciably reduced by zinc depletion, whereas the activities of certain dehydrogenases are not compromised. Carbonic anhydrase activity has been shown to be reduced in the blood, stomach, and intestines of zinc-deficient animals; it is also reduced in the blood of patients with sickle cell anemia, whose erythrocytes have a decreased zinc content. The activity of these enzymes may serve as a marker for zinc deficiency.

The proportion of RNA to DNA in various animal tissues has been shown to be affected by zinc deficiency. Lowered RNA levels and RNA/DNA ratios could result from both increased catabolism and decreased biosynthesis of the polynucleotides as well as from increased plasma ribonuclease activity. The conformation and composition of polynucleotides have also been shown to be affected by zinc deficiency. Thus, altered tissue RNA and DNA content may be the cause of the impaired protein synthesis and growth retardation seen in zinc deficiency.

Zinc has been shown to be an important element in *wound healing*. The mechanism for this effect has not been clearly established; however, several studies have implicated zinc as a necessary factor in the biosynthesis and integrity of connective tissue. For this reason, adequate zinc nutrition is especially important for the postsurgical patient.

There is increasing evidence that zinc deficiency compromises the *immune response*.[59] Zinc deficiency causes deficits in lymphocyte and thymocyte functions that are reversed by zinc supplementation. In renal patients undergoing regular hemodialysis, cellular immunity was found to be impaired in untreated patients compared with those receiving zinc supplementation.

Metabolism

High-protein foods such as meat, fish, and dairy products are good sources of available zinc, containing about 15–60 μg Zn/g wet weight. The *bioavailability* of zinc from vegetables and cereal grains is reduced because phytates (inositol phosphates) inhibit zinc absorption, as do cellulose, hemicellulose, and other dietary fiber. The effect of dietary fiber on mineral absorption has been shown to be a general one, decreasing absorption of the macrominerals calcium and magnesium and of other trace metals such as iron and copper. The availability of dietary zinc is decreased by calcium and phosphate, and is increased by some animal proteins and chelating agents such as EDTA.[63] A low-molecular-weight, zinc-binding ligand facilitating zinc absorption has been cited as the explanation for the improved availability of zinc in human milk over cow milk, and the reported higher plasma zinc levels of breast-fed over bottle-fed infants.[2] Western diets generally supply adults with 10–15 mg Zn/d. The RDA for zinc is 15 mg/d for adults (see Table 8C-1).

Zinc *absorption* occurs mostly in the duodenum and proximal jejunum. The absorption process is active, energy dependent, and apparently mediated by specific zinc transport (binding) ligands. Evidence suggests that the zinc absorption mechanism plays a significant role in homeostatic regulation. Zinc is transported in blood plasma mostly by albumin (60–70%) and by α_2-macroglobulin (30–40%), with a small amount associated with transferrin and free amino acids.

Only a small portion of dietary zinc is absorbed from the intestine, the major fraction passing out through the feces. Pancreatic secretion accounts for about 25% of total excretion. Biliary and urinary losses are small, the latter being less than 0.6 mg/d in an adult consuming 12 mg/d. Sweat losses are generally minor but can be appreciable in tropical climates or under physical stress.[29]

Zinc is second to iron as the most abundant trace element in the body, 1.4–2.3 g being present in a 70-kg adult. Tissues and fluids especially rich in zinc are prostate, semen, liver, kidney, retina, bone, and muscle.[25] Zinc in bone, skin, and muscle (70–80% of total) is apparently not an effective pool of readily mobilized zinc. The zinc concentration of erythrocytes is about ten-fold that of plasma because of their rich content of carbonic anhydrase and other zinc metalloenzymes.

Clinical Significance

Human *zinc deficiency* has been documented in Iranian and Egyptian males as being due to a low-zinc diet in which a high fiber content decreased the availability of zinc for intestinal absorption. Primary clinical features included retardation of growth and skeletal maturation, testicular atrophy, and hepatosplenomegaly. Growth failure, reduced taste acuity, and/or hypogonadism in young adults living in New York and Tennessee and in Colorado school children have been ascribed to zinc deficiency. Other manifestations of human zinc deficiency include susceptibility to infection, impaired healing of wounds, sores and ulcers, scaly dermatitis, and diarrhea. Substantial sections of the U.S. population are at risk of suboptimal zinc nutrition, particularly children, pregnant or lactating women, and those subsisting on low-income diets. Replacement of red meat (a good source of available zinc) in the diet with cheaper cereal grains reduces both the content and availability of dietary zinc.

In most clinical cases reported, the physical sequelae of zinc deficiency are rapidly cleared by administration of dietary zinc sulfate supplements, not to exceed two or three times the 15 mg/d RDA.

The causes of human zinc deficiency are more varied than simple dietary intake deficiency. Besides nutritional factors, many diseases and medical treatments may produce *conditioned zinc deficiency*. Zinc deficiency in hepatic cirrhosis patients (e.g., alcoholism, viral hepatitis) is characterized by low serum and hepatic zinc levels, along with increased urinary excretion of zinc. Lowered zinc status has been documented in patients with gastrointestinal disorders such as ulcers, ulcerative colitis, Crohn's disease, sprue, intestinal bypass, and regional enteritis. The occurrence of zinc deficiency in patients with renal disease has been attributed to loss of protein-zinc complexes in proteinuria, or to decreased tubular reabsorption of zinc. In burn patients, loss of zinc in exudates and increased zinc requirements for healing are likely causes for observed zinc deficiencies. Supplemental zinc sulfate has been shown to be effective in healing pilonidal sinus lesions, bedsores, and leg ulcers, although the findings have not been totally consistent.

Iatrogenic causes of conditioned zinc deficiency include administration of anabolic and metal-chelating drugs such as corticosteroids and penicillamine, and synthetic diet therapies. Synthetic oral diets and total parenteral alimentation fluids are commonly trace metal deficient, and have been shown to produce zinc and other trace metal deficiencies in patients receiving long-term treatment.[58]

Zinc deficiencies in patients with neoplastic and inflammatory diseases (arthritis, lupus erythematosus) have been attributed to anorexia, starvation, loss of zinc from catabolized tissue, and increased urinary excretion of zinc subsequent to its mobilization by *leukocyte endogenous mediator (LEM)*. This polypeptide hormone, released by granulocytes, mediates a redistribution of body zinc during the acute phase reaction, resulting in an increased hepatic zinc sequestration and urinary excretion of zinc.[47] The long-term effect of LEM in cases of chronic infection or injury is toward increased body zinc loss through hyperzincuria. Some parasitic infections also contribute to zinc deficiency due to intestinal blood loss of zinc.

Pregnant women are at higher risk of acquired zinc deficiency due to the uptake of zinc by the fetus and associated tissues. Plasma zinc declines during pregnancy. Evidence suggests that zinc is a crucial nutrient for normal human fetal development. The use of oral contraceptives produces a decrease in plasma zinc with an increase in erythrocyte zinc, i.e., a redistribution of blood zinc.[61]

The most clearly defined genetic disorder of zinc metabolism is *acrodermatitis enteropathica*. Manifestations of the disease are those of zinc deficiency, including retarded growth and hypogonadism, dermatologic and ophthalmic lesions, and gastrointestinal disturbances. The patients exhibit lowered plasma zinc levels, and symptoms are completely cleared with zinc sulfate supplementation. Zinc deficiency is also often associated with sickle cell anemia (SCA), and zinc supplementation may be of benefit in decreasing symptoms and crises of some sickle cell patients.[54]

For the reader seeking more detailed information on the metabolic interactions of zinc in humans and animals, comprehensive reviews are available.[5,23,54,55,72]

LABORATORY ASSESSMENT OF ZINC NUTRITURE

Laboratory tests for assessing zinc nutriture can be classified into two groups: those involving the *determination of zinc in a body tissue or fluid*, and those *testing a zinc-dependent function*. Useful tests in the former group are determinations of zinc in plasma or serum, hair, urine, erythrocytes, and saliva, while functional tests include measurement of zinc metalloenzyme activities and assessment of taste acuity. All of the above indices decrease with zinc deficiency except ribonuclease activity, which increases. Other tests which are either not well investigated or too complex for routine diagnostics include skin, fingernail, and leukocyte zinc; determinations and measurement of macrophage chemotaxis; erythrocyte uptake of ^{65}Zn, buccal mucosal cell morphology; ^{65}Zn retention and turnover; and zinc balance.[64]

While the useful tests mentioned above have been shown to reveal zinc deficiency in animals and humans, *no one test has proved to be a definitive indicator of zinc status.* Plasma zinc determination is the best single choice, while serum alkaline phosphatase or erythrocyte zinc measurements may also be useful. Test results must be interpreted with caution, since they may be confounded by clinical conditions unrelated to the subject's zinc status per se. Methodological problems unique to trace metal analysis have also hampered the routine diagnosis of zinc deficiency. Clinical factors affecting interpretation of zinc assessment tests are considered below, while analytical problems are discussed under Laboratory Methods.

Plasma or serum zinc. While circulating zinc levels in plasma and serum have often been shown to indicate human zinc deficiency, they do not accurately reflect whole body zinc status in all cases. Circulating zinc levels closely correlate with the major carrier protein albumin. Thus, lowered plasma zinc levels observed in hypoalbuminemic conditions such as hepatic cirrhosis and malnutrition may reflect depressed plasma protein binding of zinc. Circulating zinc levels will also be decreased by the action of bacterial endotoxin or LEM and by various forms of steroid regimens (e.g., oral contraceptives).

Hair zinc. Lowered hair zinc concentrations have been documented in zinc-deficient Egyptian dwarfs, in marginally deficient U.S. infants and children, and in conditions associated with zinc deficiency, including acrodermatitis enteropathica, sickle cell anemia, and celiac disease. In some cases of severe zinc deficiency, however, above-normal levels of hair zinc are explained as being due to accumulation of zinc in hair whose growth was stunted as a result of the deficiency. While hair zinc levels have been shown to correlate with zinc levels in serum and bone in some studies, correlations between hair zinc and blood or tissue zinc levels are usually poor.

Lack of correlation of hair zinc with circulating zinc levels likely results from the fact that the hair reflects long-term chronic zinc nutriture while the blood reflects acute zinc status.

Urinary zinc. Decreased urinary zinc excretion usually accompanies human zinc deficiency. However, certain conditions associated with body zinc depletion, such as hepatic cirrhosis, viral hepatitis, sickle cell anemia, postsurgical periods, and total parenteral nutrition, often give rise to increased urinary zinc excretion. The utility of urinary zinc determinations as a measure of zinc status is also limited by the difficulties encountered in collecting a 24-h urine specimen without exogenous zinc contamination.

Erythrocyte zinc. Erythrocyte zinc has been shown to be reduced in human zinc deficiency and in certain conditions associated with zinc deficiency, such as malnutrition and sickle cell anemia. However, in other conditions where plasma zinc levels were significantly depressed (hepatitis, postnecrotic cirrhosis, diabetes, pulmonary tuberculosis), erythrocyte zinc levels remained normal. The normal pattern of progressive rise of erythrocyte zinc levels during childhood limits the usefulness of erythrocyte zinc assays to the adult age groups. The highly lipid and organic composition of erythrocytes presents analytical problems which are not encountered in the analysis of body fluids.

Salivary zinc and taste acuity. Gustin is a zinc-containing protein which can be isolated from saliva and which is thought to mediate taste acuity. Salivary zinc is found to be significantly depressed in patients with idiopathic hypogeusia (taste impairment), a postulated functional zinc-deficient state.

In an initial study, supplementation with zinc sulfate was found to be beneficial to patients with idiopathic hypogeusia; however, a later, double-blind study showed no effect of zinc over placebo in improving taste acuity. Considerable interest in taste acuity as a functional index of zinc nutriture has been generated. A group of Denver school children with decreased hair zinc, poor appetite, and impaired taste thresholds showed improvement in all of the above after zinc supplementation. Mixed results from studies regarding the utility of taste acuity tests may be due in large part to the studied populations being zinc adequate.

Enzyme activities and zinc status. Of the many zinc-dependent enzymes, alkaline phosphatase, carbonic anhydrase, lactate dehydrogenase, and ribonuclease are most consistently useful indices of zinc deficiency.[12] Depression of serum alkaline phosphatase activity has been observed in a number of animal species and in a variety of human zinc-deficient conditions. A study of zinc-deficient patients receiving zinc supplementation showed increases in serum alkaline phosphatase activity that paralleled the degree of zinc repletion. However, alkaline phosphatase levels are nonspecific and are affected by a number of metabolic conditions unrelated to zinc status. Alkaline phosphatase test methods that contain zinc in the reagents are not suitable as tests for zinc status.

METHODS FOR THE DETERMINATION OF ZINC IN PLASMA OR SERUM

The determination of plasma or serum zinc concentrations by atomic absorption spectrophotometry (AAS) is the simplest and analytically most reliable test for the routine assessment of zinc nutriture. Colorimetric methods are available for those laboratories without an AAS instrument. The method using 4-(2-pyridylazo) resorcinol is simple and rapid and gives results comparable to those obtained by AAS.[32] Serum metals are released from protein by guanidine hydrochloride, and cyanide is added to complex all metals except zinc, which is selectively demasked by chloral hydrate. Day-to-day precision (CV) is 2% and 200 μL of serum is required for analysis. Other methods for determining plasma zinc include anodic stripping voltammetry, emission spectroscopy, and neutron activation analysis.

A method utilizing five-fold diluted plasma and standards in 5% glycerol matrix has been published as a proposed selected method.[62] The method requires 0.5 mL of specimen and is recommended for plasma or serum zinc determination.

The Determination of Plasma Zinc by Atomic Absorption Spectrophotometry

Principle

Five-fold diluted plasma or serum samples are aspirated into the atomic absorption flame. The zinc level is determined by comparing the signal from diluted plasma with the signal from aqueous standards which are prepared in a diluted glycerol matrix (5 mL/dL) to simulate the viscosity of the diluted plasma.

Specimens

Precautions must be taken throughout all procedures to avoid zinc contamination during sample collection or from reagents.

Collect at least 3 mL of blood by venipuncture as described previously under Sample Collection and Testing. Anticoagulants may be used, but they must be free from zinc contamination. After centrifugation, use zinc-free pipets to transfer the serum or plasma to polypropylene tubes. Hemolysis must be avoided since erythrocytes contain at least ten times more zinc than plasma. The specimens may be stored in the refrigerator for several days but should be frozen at −20 °C for long-term storage.

Reagents

1. Glycerol diluent. Dilute 50 mL of reagent grade glycerol to 1000 mL with deionized water.
2. Standards
   a. Stock standard. Use commercially available 1 g/L aqueous zinc calibration standard. Alternatively, dissolve 1.0000 g of pure zinc metal in 100 mL of ten-fold diluted nitric acid and dilute to 1 L with deionized water.
   b. Working standards. 100, 200, 300, and 400 μg Zn/L. Deliver 1 mL of 1 g/L zinc standard into a 100-mL volumetric flask and dilute to volume with glycerol diluent. Mix by inverting at least 10 times. Place 1, 2, 3, and 4 mL aliquots of this intermediate stock into four 100-mL volumetric flasks and dilute to volume with the glycerol diluent. The standards, 10, 20, 30, and 40 μg Zn/dL, correspond to plasma zinc concentrations of 50, 100, 150, and 200 μg/dL, respectively. The standards are stable in polypropylene bottles at room temperature and do not need to be freshly prepared with each test run.

Instrumental Conditions

1. *Instrument settings*: Wavelength, 213.8 nm; Readout, absorbance; Slit width, 0.7 nm.
2. *Burner gas mixture*: Air-acetylene. Fuel rich, luminescent flame. See the instrument manual for specific gas pressures and flow rate, sample aspiration rate, lamp current, and optimizing wavelength and burner position to achieve maximum sensitivity.

Procedure

1. Allow plasma or serum to come to room temperature and then mix by gently inverting the tubes.

2. Deliver 0.5 mL of specimen or control into a 16-mm plastic test tube. Add 2.0 mL of deionized water and immediately mix the solutions thoroughly.

3. Establish instrumental and gas-flow settings and aspiration rate to optimize signal and minimize background noise. Check instrument manual for specific instrumental settings. Aspirate glycerol diluent into the flame and set the baseline to read zero absorbance. To correct for baseline drift, aspirate the glycerol diluent before and after each aspiration of standard and specimen and reset the baseline to zero as required.

4. Aspirate the zinc working standards sequentially from most dilute to most concentrated, aspirating until the reading is stable (± 0.002 A). The resulting values are used to establish the working curve, preferably by use of a least-squares regression fit.

5. Aspirate the specimens and the plasma pool control.

6. Calculate specimen concentrations from absorbance readings by interpolation from the working curve. Results for the control should be within 6% of the previously established mean.

Comments

In order to obtain reproducible results it is most important to maintain consistently optimal aspiration and flame conditions. This is especially true when aspirating solutions with high solid concentrations rather than protein-free or digested specimens. Dilute hydrochloric acid should be periodically aspirated to help keep the nebulizer system clean. The system should occasionally be taken apart and cleaned. The burner head should be soaked overnight in a non-acid cleaner and cleaned thoroughly before every run. The inside of the burner orifice should be scraped with a thin metal file to dislodge aggregated solids and maintain smooth orifice surfaces. An extra burner head and nebulizer system should be kept on hand for installation when unexplained variability in results occurs.

Reference Range

The accepted reference range for zinc in plasma is 70–150 μg/dL. An exhaustive compendium of element composition in human tissues and body fluids is available.[25] Serum zinc levels are generally 5–15% higher than plasma levels due to zinc released from platelets and erythrocytes during clotting. Plasma zinc levels exhibit both circadian and postprandial fluctuations. The levels decrease after eating and from morning to evening. Fasting morning values of plasma zinc below 70 μg/dL represent deficient zinc status. Routine quality control of plasma zinc measurements is accomplished by including aliquots of a frozen plasma pool in each analytical batch.

Reference

Smith, J. C., Jr., Butrimovitz, G. P., and Purdy, W. C.: Direct measurement of zinc in plasma by atomic absorption spectroscopy. Clin. Chem., 25:1487, 1979.

Other Methods for Determining Zinc Status

Hair. Acid digestion of 15–20 mg of hair and dilution of the digestate to 10 mL final volume will provide zinc concentrations suitable for flame AAS determinations. The reference range for zinc in hair is 100–280 μg/g. A hair pool can be used for quality control as previously discussed.

Urinary zinc. Urine specimens should be collected acidified (10 mL of concentrated HCl per 24-h collection) to preserve the sample and maintain metal solubility. Urinary zinc can be determined by direct aspiration of the acidified urine into the AAS flame. Because of the high salt content of urine, a sample background correction mode should be used. Samples of commercially available, lyophilized urine controls with assayed values for zinc, or frozen urine pool aliquots, should be included with each analysis batch for quality control. The reference range for urinary zinc excretion is 0.15–1.00 mg/d.

Erythrocyte zinc. The high organic content of erythrocytes complicates their analysis for trace metals. Solutions of diluted hemolysates are too heterogeneous to be aspirated into the AAS flame with uniformity and without clogging. The high organic content causes background and fouling problems when direct injection into graphite atomizers is used. Wet digestion of the erythrocyte organic matrix leaves a "clean" solution which is much more amenable to precise quantitation by AAS.

Salivary zinc. Both mixed saliva and sterile parotid saliva collected directly over Stensen's ducts have been utilized for zinc determinations. Usually saliva flow is stimulated by chewing paraffin or placing drops of citrus juice on the tongue. Since food particles and bacteria in the mouth contribute to exogenous zinc contamination of the saliva, sterile parotid saliva is best for assessing zinc nutriture. Data suggest that the zinc content of a centrifugal supernatant reveals differences in zinc intakes better than the zinc content of whole saliva. Steps involved in collecting saliva and information on saliva-collecting devices have been published.[28] The reference range for the zinc content of sterile parotid saliva is 20–80 mg/L. Levels of zinc in whole saliva may be considerably greater. Since no reference control samples are available for salivary trace metal determinations, samples of a frozen saliva pool aliquot should be included in each analytical batch.

Taste acuity. Taste acuity may be quantitated by measuring taste thresholds. Four tastes can be tested: salt (NaCl), bitter (urea), sour (HCl), and sweet (sucrose). Both detection (a taste is detected) and recognition (identification of which of the four tastes) thresholds can be determined. Basically, the test involves utilizing pure water and various concentrations of taste test solutions to determine the lowest concentration of taste (threshold) that can be detected and correctly identified. Obtaining valid taste acuity data demands considerable time and attention from both the subject and the test administrator.

COPPER

Biochemistry and Physiology

Copper is an integral component of many metalloenzymes, including cytochrome oxidase, superoxide dismutase, uricase, dopamine-β-hydroxylase, lysyl oxidase, ceruloplasmin, and tyrosinase. The important protein metallothionein binds copper as well as other heavy metals.

A variety of pathological conditions have been attributed to loss of cuproenzyme activity. Failure of pigmentation has been attributed to depressed tyrosinase activity, required for the biosynthesis of melanin. A variety of connective tissue cross-linking defects (skeletal, cardiac, and vascular) are believed to be caused by loss of amine oxidase activity, particularly that of lysyl oxidase. A nervous disorder, ataxia, may be a result of depressed cytochrome oxidase activity in motor neurons. Depressed cuproenzyme activities (dopamine-β-hydroxylase and superoxide dismutase) result in abnormal catecholamine conversions.

Copper plays an important role in iron metabolism. Copper deficiency impairs iron absorption, and anemia accompanies severe copper deficiency. *Ceruloplasmin,* the major plasma copper protein, is a ferroxidase which oxidizes ferrous iron to the ferric state prior to its binding by plasma transferrin.

Metabolism

Absorption and transport. Copper absorption in humans occurs primarily in the stomach and duodenum. Evidence suggests two mechanisms for copper absorption. One involves the transport of copper–amino acid complexes across the mucosa. (L-Amino acids facilitate the absorption of copper.) The other mechanism is protein mediated, involving a high molecular weight copper-binding protein on the luminal side of the mucosa and a metallothionein-like protein in the mucosa that serves as a copper storage depot and releases copper to the serosal side of the mucosa. The metallothionein-like protein is rich in sulfhydryl groups and binds copper through the formation of mercaptide bonds. The metallothionein-type storage and binding of copper is believed to play a role in homeostatic regulation of circulating copper levels. Other metal ions, particularly cadmium and zinc, compete with copper for sulfhydryl binding sites, thus explaining the antagonism of zinc and cadmium toward copper absorption.

The amount of orally ingested copper absorbed from the intestine is variable. A recent review of the data suggests 40–60% as a reasonable range.[39] Factors affecting copper absorption include the amount ingested, the chemical form (bioavailability), and certain dietary constituents. The latter include other trace metals, sulfate, ascorbic acid, fiber, and phytates. Copper absorption may be impaired in patients with diffuse small bowel disease such as sprue, lymphosarcoma, and scleroderma.

Absorbed copper is transported by albumin to the liver where it is stored, mostly as metallothionein-like cuproproteins. The liver releases a small portion of copper in the form of amino acid complexes; however, most copper is released as ceruloplasmin, which constitutes 90–95% of circulating plasma copper. Ceruloplasmin is an important copper oxidase enzyme, and its oxidase activity usually correlates well with values obtained with the immunochemical determination of the ceruloplasmin protein. Since nearly all plasma copper is present as ceruloplasmin, levels of the two parameters almost always change proportionately. The major cuproprotein in erythrocytes is erythrocuprein, i.e., superoxide dismutase (SOD).

Distribution and homeostasis. The adult human body contains about one twentieth as much copper (about 75 mg) as zinc. The highest levels are found in the liver (about 10% of total) with relatively high levels also found in brain, heart, and kidneys. Muscle and bone levels are low but constitute about 50% of total body copper because of their large mass.

Copper is excreted primarily in the feces, mostly from unabsorbed dietary copper, and from biliary and gastrointestinal secretions. Biliary copper excretion ranges from 0.5–1.3 mg/d and involves enterohepatic circulation of copper with bile acids. Gastrointestinal sources include copper from gastric juice and from desquamated mucosal cells. A small amount of copper, 0.3–3% of dietary intake, is excreted in the urine. Evidence suggests that loss of copper through sweat may be appreciable.[29] Menstrual copper losses appear to be minor, 0.1–0.8 mg per period.

The homeostatic regulation of copper metabolism is apparently complex. As with zinc, the liver is the key organ, copper being stored there and incorporated and released as ceruloplasmin to maintain blood copper levels. Intestinal absorption mechanisms also appear to play a role in the regulation of copper homeostasis. The level of plasma ceruloplasmin consistently remains at about 93% of total plasma copper.

Dietary intakes. Most populations appear to consume adequate copper, and frank dietary copper deficiency in humans is rare. The copper content of foods is variable and depends on the copper levels in the soil in the geographical area from which the food is obtained, and on the copper loss or contamination throughout processing. Liver, crustaceans, and shellfish contain large amounts of copper. Cow milk has a low copper content; therefore, infants solely on a milk diet may develop copper deficiency and anemia. A list of the content of inorganic elements in foods used in hospital diets[21] and an extensive listing of the copper content of foods[48] have been published. Copper intake data have been reviewed,[76] and considerable worldwide variability is apparent. Western diets generally supply 2–4 mg/d. However, several U.S. studies have revealed intakes between 0.2 and 1.8 mg/d. Estimates of the copper content of regular, vegetarian, and renal diets of a U.S. hospital were 0.90, 1.10, and 0.51 mg/d, respectively. Estimated "adequate and safe" intakes of copper are shown in Table 8C-1. The range for adults is 2.0–3.0 mg/d.

Clinical Significance

Human copper deficiency. A variety of human copper deficiency conditions are recognized. Copper deficiency in infants has been observed in prematurity, malnutrition, malabsorption, chronic diarrhea, hyperalimentation, and prolonged feeding with low-copper, total-milk diets. The susceptibility of premature infants to copper deficiency is related to their lower stores of liver and spleen copper (which accumulate rapidly in the later stage of gestation) and the probability of longer maintenance on exclusively milk formulas. A deficit in protein nutriture sufficient to impair copper transport and storage may also be a causal factor. Symptoms of copper deficiency include (1) neutropenia and hypochromic anemia in the early stages, both of which are responsive to oral copper but not iron; (2) osteoporosis and various bone and joint abnormalities ("scurvy-like" changes) that reflect deficient copper-dependent cross-linking of bone collagen and connective tissue; (3) decreased pigmentation of the skin and general pallor, attributed to loss of tyrosinase required for melanin synthesis; and (4) in the later stages, possible neurological abnormalities (hypotonia, apnea, psychomotor retardation), probably due to a cytochrome *c* oxidase deficit. Copper deficiency symptoms in infants are generally reversed by copper supplementation. The levels of plasma copper and ceruloplasmin directly reflect the degree of copper deficiency and replete states. The monitoring of body copper status and copper intake in the premature infant is warranted.

Long-term hyperalimentation with infusates deficient in trace metals has been shown to produce copper deficiency in both infants and adults. Observed symptoms are those usually seen in copper deficiency; they include lowered plasma copper and ceruloplasmin, anemia, neutropenia, leukopenia, and hypocellular bone marrow. Parenteral infusates have often been found to contain little or no copper. Although the need for trace metals in parenteral fluids is obvious, recommendations on the optimal amounts have not been established. A minimum parenteral intake of copper of 0.3 mg/d for adults is recommended.[30] Human copper deficiency has also been associated with zinc therapy for sickle cell disease and treatment with copper-chelating agents such as penicillamine.

Menkes' syndrome. A genetically determined defect in copper transport and storage underlies the rare *Menkes' steely-hair syndrome.* Clinical manifestations include kinky or steely hair, depigmentation of the skin and hair, hypothermia, seizures, cerebral degeneration, and vascular defects. The formation of steely hair is attributed to loss of copper-catalyzed disulfide bond formation and is also found in sheep grazing on copper-deficient pastures. The affected infants have been shown to be copper deficient, with low serum, hepatic, and cerebral copper levels; low ceruloplasmin; and little or no cytochrome oxidase activity in nerve tissue. Symptoms appear usually by 3 months of age, with death by 5 years. Affected infants show a buildup of copper in the duodenal mucosal tissue resulting from defective copper absorption. Intravenous administration of copper may help raise plasma copper levels; however, urinary copper excretion increases accordingly, and the course of vascular and cerebral degeneration is irreversible. Surprisingly, red cell copper is not decreased, and neutropenia and anemia, two usual signs of copper deficiency, do not appear.

Wilson's disease. Wilson's disease (hepatolenticular degeneration) is a genetically determined copper accumulation disease which usually occurs between 6 and 20 years of age, and with a prevalence of about 1:200 000. (See also Chapter 13.) Copper accumulates in the liver, brain, kidney, and cornea, and urinary excretion of copper is increased. The hepatic synthesis of ceruloplasmin is decreased, resulting in a low serum level of this enzyme (< 20 mg/dL). Although free and albumin-bound copper is increased, total serum copper is generally decreased because of low ceruloplasmin levels. Thus, the simultaneous determinations of serum ceruloplasmin, serum copper, and urinary copper are useful in the diagnosis of this disease. Hepatic copper content, however, was found to be more accurate than serum ceruloplasmin or urinary copper excretion in differentiating between Wilson's disease and chronic active hepatitis in children and adolescents.[49] Disease symptoms include neurological disorders, cirrhosis of the liver, and Kayser-Fleischer rings (green-brown deposits) due to corneal copper deposits. The hemolysis, necrosis, and other cellular damage which occurs are probably due to lipid peroxidation, a known effect of copper toxicity. The biochemical basis for Wilson's disease is not known. Several hypotheses suggest copper-protein binding of unusually high affinity. The disease can be controlled, and in some cases halted, by early and persistent treatment with copper chelators, such as penicillamine and British anti-Lewisite (2,3-dimercaptopropanol).

Copper toxicity, characterized by nausea, vomiting, epigastric burning, and diarrhea, can occur subsequent to ingestion of copper-contaminated solutions, the use of copper-containing IUD's, the use of copper salts in animal feeds, and exposure to copper-containing fungicides. Systemic toxic effects include hemolysis, hepatic necrosis, gastrointestinal bleeding, oliguria, azotemia, hemoglobinuria, hematuria, proteinuria, hypotension, tachycardia, convulsions, coma, and death.

Abnormal Copper Metabolism Associated with Other Conditions

Estrogens increase serum copper levels, probably by increasing hepatic ceruloplasmin synthesis. Serum copper is normally higher in women than in men, and this difference increases further for women taking estrogenic oral contraceptives. Serum copper during *pregnancy* is increased two- to three-fold over normal and is diagnostically useful. A significant fall in serum copper between the third month and term may serve as an indicator of placental insufficiency or spontaneous abortion. Testosterone and progesterone administration has also been shown to increase plasma copper. Abnormal copper levels seen in diseases of the pituitary-adrenal axis include elevated plasma copper and decreased urinary copper in Addison's disease and hypopituitarism and the reverse in Cushing's syndrome and adrenal cortical carcinomas.

During *infections or inflammatory stress,* serum copper levels rise due to the acute-phase action of leukocyte endogenous mediator (LEM), previously discussed under Zinc. The action of LEM may cause the elevated serum copper and decreased serum zinc levels seen during the acute phase of myocardial infarction. Copper seems to be associated with *rheumatic diseases.* Increases of ceruloplasmin have been observed in the serum and synovial fluid of patients with rheumatoid arthritis.

Since the liver is the important organ for copper storage and regulation of homeostasis, abnormal copper metabolism is seen in *liver disease.* Elevated serum copper levels are seen in portal cirrhosis, biliary tract disease, and hepatitis, probably because excess copper, which would otherwise be excreted in the bile, is retained in the circulation. Hypocupremia has been observed in hemolytic jaundice, hemochromatosis, and some types of hepatic cirrhosis because of the inability of the damaged liver to synthesize ceruloplasmin.

Hypocupremic states are found in the genetic disorders Menkes' steely-hair syndrome and Wilson's disease, and in malnutrition, kwashiorkor, and marasmus. In Wilson's disease and malnutrition, hypocupremia is attributed to limited synthesis of apoceruloplasmin rather than to dietary copper deficiency. Hypocupremia has also been associated with gastrointestinal disease (sprue, small bowel, and celiac), cystic fibrosis, and nephrosis. In the last-named case, copper as ceruloplasmin is lost through the urine.

Reviews which further detail the biochemical and clinical aspects of human copper nutrition are available.[11,23,39,46,51,67]

LABORATORY ASSESSMENT OF COPPER NUTRITURE

Measurement of plasma or serum copper levels provides the best routine test for the clinical assessment of copper nutriture. As in the case of zinc, factors not related to copper nutriture can affect plasma copper levels, e.g., the amount of binding protein (apoceruloplasmin), the effect of LEM, and the effect of hormones, both endogenous (pregnancy) and exogenous. Plasma copper levels exhibit diurnal variation, with the highest levels in the morning.

Hair copper, urinary copper, and cuproenzyme activities are also useful for indexing copper deficiency.[64] Only copper in the most proximal hair segments should be analyzed as a test for copper nutriture. Urinary excretion of copper is quite low and somewhat independent of intake. An effect of depleted body copper may be lowered urinary copper excretion, as observed in some patients undergoing intravenous feeding with copper-deficient solutions. Urinary copper excretion in infants with Menkes' syndrome may be normal, however. The best cuproenzymes for assessing copper status are erythrocyte superoxide dismutase and leukocyte cytochrome oxidase.[12]

Other potential indices of copper status, which either have not been well investigated or are too tedious for routine use, include fingernail and erythrocyte copper content, skin morphology, and measurements of copper retention and turnover.

METHODS FOR THE DETERMINATION OF COPPER

Atomic absorption spectrophotometry (AAS) after direct dilution is the method of choice for determining serum copper. For laboratories without an AAS instrument, colorimetric methods are available, such as those using bis-cyclohexanone oxaldihydrazone (Cuprizone)[56] and Bathocuproine[79] as chromogens. The latter chromogen is used in a method adapted for the centrifugal analyzer.[35] Hemolysis is not a great concern for copper determinations, since levels of copper in plasma and red cells are nearly equal. Reference ranges for serum copper have been compiled[25,65,69] and are higher in pregnancy (120–300 μg/dL), in children 6–12 years (80–190 μg/dL), and in adult females (80–155 μg/dL) than in adult males (70–140 μg/dL) and infants (20–70 μg/dL).

Serum ceruloplasmin can be determined by a colorimetric copper oxidase reaction, by immunonephelometry, or by radial immunodiffusion. (See Chapter 4, Amino Acids and Proteins.) The reference range is 15–60 mg/dL.

The Determination of Serum Copper by Atomic Absorption Spectrophotometry

Principle

Serum specimens diluted with an equal volume of deionized water are aspirated directly into the AAS flame, and the copper concentrations are calculated against copper standards with a 10 mL/dL glycerol matrix to approximate the viscosity of the diluted specimens. The procedure is similar to that detailed for plasma zinc determination by AAS; however, because the sensitivity of AAS detection for copper is less than for zinc, a two-fold rather than a five-fold dilution of serum is recommended for copper determination. Aspiration of the higher-solids solution for copper determination requires more frequent cleaning of the burner orifice or use of a high-solids (Boling) burner to avoid clogging and erratic absorbance readings. To match the viscosity characteristics of two-fold diluted serum, copper standards and the blank solution for zeroing the instrument should consist of a 10 mL/dL glycerol solution. At least four working standards in the range of 0–100 μg/dL should be prepared. This range corresponds to serum copper concentrations of 0–200 μg/dL. The standards are stable in polypropylene bottles at room temperature and do not need to be freshly prepared with each test run.

Instrumental Conditions

1. Instrument settings: Wavelength, 324.8 nm; Readout, absorbance; Slit width, 0.7 nm.
2. Burner gas mixture: Air-acetylene, oxidizing (lean), blue.

See instrument manual for specific gas pressures and flow rates, sample aspiration rate, and lamp current, and for optimizing wavelength and burner position to achieve maximum sensitivity.

Comments

Hair copper is determined by careful sample collection and preparation, wet digestion, and AAS determination, as discussed previously for hair zinc. However, because hair copper levels (10–40 μg/g) are about one tenth of those for zinc, copper in hair digests should be determined by the more sensitive flameless AAS techniques.

Despite claims that *urinary* copper can be determined by direct aspiration of acidified (or even diluted) samples into the AAS flame, the sensitivity of flame AAS is not sufficient to differentiate between normal and subnormal urinary copper levels. The high-salt matrix complicates the use of the flameless AAS mode. Therefore, separation of the copper from the salt by ion exchange or extraction with an organic chelator is recommended. Many effective schemes for specific extraction of trace metals from high-solids solutions such as blood, urine, or seawater have been published.[80] Depending on the degree of concentration effected by the extraction scheme, copper can then be determined in the extractant by either flame or flameless AAS. The reference range for urinary copper excretion is 15–50 μg/d.

MANGANESE

Biochemistry

The biochemical basis for manganese essentiality is its role as a cofactor in certain enzymatic reactions. Metal ion competitors of manganese, such as copper, iron, and magnesium, can replace manganese as an activator of enzymes. Such activation can mask manganese deficiency and make it difficult to observe biochemical or clinical effects of manganese deficiency. Manganese-activated enzymes include hydrolases, kinases, decarboxylases, and transferases.

Manganese is the most effective of several metal ions that activate glycosyl transferases, which are necessary for polysaccharide and glycoprotein synthesis. Manganese is involved in cholesterol biosynthesis; hence, manganese deficiency may result in decreased serum cholesterol. Manganese metalloenzymes incorporate manganese in the +2 or +3 valent state and include pyruvate carboxylase, superoxide dismutase, avimanganin, manganin, and concanavalin A.

Metabolism

Manganese is absorbed in the small intestine by an unknown mechanism. Calcium, phosphorus, ferric citrate, and soy protein may interfere with manganese absorption.

The whole body content of manganese in adults is 12–20 mg. Although distributed widely throughout body tissues and fluids, manganese is concentrated in mitochondria. Liver and retina contain relatively high amounts of manganese, and the concentration of manganese in tears is 50-fold or higher than in serum.

Manganese homeostasis appears to be regulated more through adjustments in body excretion than by intestinal absorption. It is known that the liver can adjust to a manganese load by increased biliary excretion.

Average manganese dietary intakes among different populations have been found to vary from 2.7–8.3 mg/d. The estimated adequate and safe intake of manganese is 2.5–5.0 mg/d for adults. (See Table 8C-1.)

Chronic *manganese poisoning* has occurred in miners, foundry workers, welders, and workers manufacturing drugs, pottery, ceramics, glass, varnish, and food additives. The symptoms are schizophrenic-like psychiatric effects and neurological disorders clinically similar to Parkinson's disease. The similarity between symptoms of manganese poisoning and Parkinson's disease led to discovery of a link between manganese and catecholamine metabolism. Decreases in striatal dopamine are seen in both syndromes, and administration of L-dopa (a dopamine precursor) is effective in alleviating several of the common symptoms of Mn poisoning. Treatment of human manganese intoxications may also include various therapies for respiratory symptoms and the administration of metal chelators such as EDTA to reduce body burden.

Clinical Significance

Unlike iron, zinc, and copper, which are considered essential because clinical evidence of deficiency exists, manganese is accepted as essential for humans on the basis of its proven role in manganese-dependent enzymes rather than on direct evidence of human deficiency. Manganese supplementation was found to be necessary to alleviate blood clotting defects and hypocholesterolemia in a manganese-deficient experimental subject. It was postulated that manganese-dependent glycosyl transferase activity may be needed for synthesis of normal prothrombin, a glycoprotein.[14] Serum manganese levels are increased following industrial exposure, acute hepatitis, and myocardial infarction. Increased erythrocyte levels have been found in patients with rheumatoid arthritis.

Methods for the Determination of Manganese

Laboratory tests which reliably assess body manganese status have not been established. The determination of blood manganese levels rather than evaluation of functional indices such as manganese-dependent enzyme activities has been proposed. Widely varying levels of manganese in blood have been reported.[25] Adult reference ranges for manganese are 0.4–2.4 $\mu g/dL$ for whole blood, 0.04–1.4 $\mu g/dL$ for serum, 0.5–4.6 $\mu g/dL$ for erythrocytes, 1–50 $\mu g/L$ for urine, and 0.1–2.1 $\mu g/g$ for hair.[25,69] The large variation in reported manganese levels can be partly attributed to sample contamination during collection or processing, the use of older analytical methods which were relatively insensitive and nonspecific, and the use of different analytical methods.

The most popular analytical techniques for determining manganese in biological specimens are neutron activation analysis (NAA), emission spectroscopy (ES), and atomic absorption spectrophotometry (AAS). Spectrophotometric and electrochemical methods have also been used. For the clinical laboratory, the AAS technique is the method of choice. Because of the low levels of manganese in blood, hair, and urine, the more sensitive flameless AAS mode should generally be used. Sunderman has summarized instrumental parameters and references to published methods for flameless AAS determination of manganese in biological specimens.[66] The usual problems of direct sampling into an electrothermal atomizer are avoided if the sample is ashed or the sample metal selectively extracted before AAS determination. Manganese in urine should be separated from the high salt matrix by ion exchange or selective extraction[80] before AAS determination. Manganese in hair can be determined by flameless AAS after sample digestion with nitric acid or a 4:1 nitric/perchloric acid mixture.

MOLYBDENUM

Biochemistry

The essentiality of molybdenum for animals and humans is based on its incorporation into three metalloenzymes: xanthine oxidase, aldehyde oxidase, and sulfite oxidase. Xanthine oxidase participates in the degradation of purines to uric acid. Aldehyde oxidase catalyzes the oxidation of aldehydes, and sulfite oxidase catalyzes the final stage of sulfur-containing amino acid oxidations.

Metabolism

The majority of ingested molybdenum is absorbed in the intestine. Dietary sulfate and copper decrease molybdenum retention. Normally, molybdenum is excreted largely via the urine; however, bile is also an important route of excretion.

As with other ultratrace metals, the wide variance in reported molybdenum levels for tissues and body fluids may be due largely to contamination and the use of varied instrumental techniques, rather than to true biologic variation. The *average intake* of molybdenum in several populations varies from 70–300 µg/d. Estimated adequate and safe dietary intakes of molybdenum are listed in Table 8C-1, and are 150–500 µg/d for adults.

Molybdenum is relatively nontoxic to humans. Biochemical manifestations of acute molybdenum poisoning in mammals include inhibition of ceruloplasmin and cytochrome oxidase (copper-dependent enzymes), glutaminase, cholinesterase, and sulfite oxidase activities.

Clinical Significance

No well-defined cases of human molybdenum deficiency have been reported. Epidemiologic and animal studies suggest that molybdenum may have a preventive effect on development of dental caries.[22] A possible congenital defect in molybdenum metabolism was suggested for an infant who showed feeding difficulties, mental retardation, skull asymmetry, dislocation of the left lens, and biochemical deficits in xanthine and sulfite oxidase activities.[16]

High activities of xanthine oxidase associated with elevated molybdenum levels may lead to high serum uric acid levels and a predisposition to gout. Though accepted as being an essential element, molybdenum does not appear to be directly related to any clinically observed conditions.

Methods for the Determination of Molybdenum

Current methods for the determination of molybdenum in biological specimens are inadequate. The methods have mostly employed emission spectroscopy, neutron activation analysis, and atomic absorption spectrophotometry (AAS) techniques. Use of a hotter nitrous oxide–acetylene flame has been suggested for molybdenum determinations by AAS, but molybdenum concentrations in biological specimens are so low that preconcentration or prior extraction of the sample is necessary for flame AAS determinations.

Reference Ranges

Reported *reference ranges* for healthy adults are 1–15 ng/mL for whole blood, 18 ng/mL (mean) for erythrocytes, 0.1–6 ng/mL for plasma or serum, 10–16 µg/L for urine excretion, and 0.06–0.20 µg/g in hair.[25]

CHROMIUM

Biochemistry

Chromium is accepted as being essential only because of its relation to the maintenance of normal glucose tolerance. Studies have shown that chromium is a potentiator of insulin action. Insulin resistance may be a consequence of chromium deficiency, and insulin appears to be ineffective as a glucose regulator without chromium.

The agent which potentiates the action of insulin has been isolated from brewer's yeast and kidney powder and has been termed "glucose tolerance factor" (GTF). GTF is an organic low-molecular-weight complex containing trivalent chromium; the exact structure of GTF has not been elucidated. The site and pathway of GTF biosynthesis are unknown.

Metabolism

The site and mechanism of intestinal chromium absorption in humans have not been determined. Animal studies indicate that chromium is absorbed in the upper small intestine.

A small portion of circulating blood chromium is apparently bound to a protein in the β-globulin fraction, possibly transferrin, while the major portion is presumed to be bound to GTF.

The regulation of chromium homeostasis is still obscure, although urinary excretion appears to be the key. The majority of orally absorbed chromium is excreted by the kidneys in an unknown form.

The total content of chromium in foods has been studied but is of limited value since chromium absorption and its biological activity are highly dependent on the chemical form. With this in mind, various food items were rated (in arbitrary units) for biologic GTF activity and were found to range from 44.9 for brewer's yeast to 1.6 for skim milk.[70] Mean chromium intakes of various human populations range from 52–320 μg/d. The estimated adequate and safe dietary intakes of chromium are 50–200 μg/d for adults. (See Table 8C-1.)

Trivalent chromium has a low order of *toxicity*. An apparently wide margin of safety exists between supplementation to cure deficiency and toxicity. Chromium(VI) is considerably more toxic than Cr(III). Occupational exposure to chromates occurs in electroplating, steelmaking, leather tanning, photography, dyeing, and chemical manufacture operations. Acute exposure may produce an allergic reaction, conjunctivitis, nasal septum ulceration, dermatitis, edema, and ulcer. Chronic exposure may produce gastrointestinal symptoms, hepatitis, and lung cancer. Gastric lavage followed by administration of demulcents and close monitoring of fluid and electrolyte balance are useful in the treatment of acute chromium poisoning. For dermatitis, topical aluminum acetate or 10% EDTA ointment may be useful. British anti-Lewisite (BAL) may be used to treat systemic poisoning.[36]

Clinical Significance

The clinical significance of chromium rests primarily on its relation to glucose metabolism. No definitive studies of human chromium deficiency have been carried out, largely because of analytical difficulties in determining ultratrace chromium levels in tissue. Evidence of human chromium deficiency is mostly indirect, based on the improvement of insulin-resistant glucose intolerance after supplementation with physiological amounts of chromium compounds.

In the U.S., simple addition of 150 μg of Cr(III) to the daily diet improved glucose tolerance of maturity-onset diabetics and middle-aged and elderly subjects. Malnourished children in Jordan and Turkey showed normalization of intravenous glucose tolerance tests after supplementation with 250 μg of chromium per day in their diet formula. However, malnourished children in Egypt showed no beneficial response to dietary chromium supplementation.

Deficiencies of chromium have been documented in two cases of total parenteral feeding. Jeejeebhoy described severe chromium deficiency in a female patient fed intravenously with a chromium-deficient (8 μg/d) solution for more than five years.[31] Low blood and hair chromium levels, negative chromium balance, weight loss, glucose intolerance, decreased respiratory quotient, and peripheral neuropathy were all corrected by intravenous administration of 250 μg/d of chromium for two weeks. A second, independent report showed that the glucose intolerance, insulin resistance, and central nervous system disorder of a total parenterally fed patient was corrected after chromium, but not after insulin administration.[19]

The potentiation of insulin action due to dietary supplementation with GTF (as a component of brewer's yeast) has been demonstrated in several human studies.[15,50] In insulin-dependent diabetics, a daily insulin requirement was lowered from 60–133 units/d to 20–45 units/d after a 1- to 2-month supplementation with GTF. Upon supplementation with brewer's yeast, elderly diabetics required less insulin within 24 h, and their need for insulin progressively declined 1–2 units every 2–3 d. The complete elimination of insulin requirements by GTF supplementation has not been achieved, probably because of a deficit in endogenous insulin production. A risk of hypoglycemia during the initial GTF supplementation period has been emphasized.

METHODS FOR THE DETERMINATION OF CHROMIUM STATUS

No laboratory tests which reliably define body chromium status have been established. The determination of chromium levels in human tissues and fluids has proved to be among the most difficult of trace metal determinations. Reported serum and urine chromium levels have often been grossly overestimated in the past. As method sensitivity has improved, the capability of identifying and eliminating exogenous chromium contamination has resulted in decreasing estimates of chromium levels in biological specimens. The reported "normal" levels of serum chromium of 0.08–45 μg/L appear to be decreasing over the years.[74] Rather than reflecting a true chronological decrease in serum chromium levels, the changing values are likely due to use of increasingly sensitive and specific analytical instruments and techniques and to improved specimen collection procedures. Reported reference ranges for urinary chromium excretion have also decreased over the years to presently reported concentrations of less than 1 μg/d. The one biological tissue notably less difficult to analyze for chromium is hair because of its relatively high chromium content. The reference range for adult hair chromium content is 0.1–3.6 $\mu g/g$. Reported mean values of chromium for healthy adults range from 19–66 ng/mL in erythrocytes and 0.7–28 ng/mL in whole blood.[25,69]

Colorimetric determinations of chromium using 1,5-diphenylcarbazide are neither sensitive nor specific, the ligand apparently preferring to chelate Cr(VI) rather than Cr(III). The commonly used instrumental techniques for determining chromium in biological specimens are neutron activation and emission and atomic absorption spectrophotometry (AAS). The AAS technique is recommended as the most practical for clinical and medical research laboratories. The flameless mode should be used because of the very low levels of chromium in biological specimens. Instrumental parameters, as well as references to published methods for AAS determination of chromium in biological specimens, have been summarized.[65,66] Because of the low chromium levels, background absorption corrections by the conventional deuterium lamp technique are inadequate. More powerful background correction techniques such as use of the Zeeman effect or a tungsten-halide lamp are necessary. Flameless AAS methods for serum and urinary chromium using AAS instruments with enhanced background correction capability have been described.[33,73] Nitric acid/peroxide digestion of samples in quartz tubes with threaded Teflon caps provides an organically clean digest well-suited for chromium quantitation in a graphite atomizer. Losses of volatile organo-chromium components of biologic specimens during analytical procedures have been reported, and techniques to avoid loss, such as closed-tube digestions, are recommended. A National Bureau of Standards (NBS) Reference Material (SRM 1659), Brewer's Yeast, contains a volatile organo-chromium component and a certified value for chromium content. This material is recommended as a reference standard for assessing accuracy of methods for chromium determination. The NBS reference material, Bovine Liver (SRM 1577), also has a certified value for chromium concentration.

COBALT

Cobalt is essential for humans only as an integral part of vitamin B_{12}, i.e., cobalamin. No other function for cobalt in the human body is known. Details of vitamin B_{12} biochemistry and function in humans are discussed in Chapter 15, Biochemical Aspects of Hematology. Microflora of the human intestine cannot utilize cobalt to synthesize physiologically active cobalamin. The human vitamin B_{12} requirement must be supplied from the diet. Free (non–vitamin B_{12}) cobalt does not interact with the body vitamin B_{12} pool.

Because cobalt itself is nonessential, mechanisms for homeostatic control of body cobalt levels apparently have not evolved. Reports regarding the degree of intestinal absorption of dietary cobalt do not agree and range from 20% to considerably higher. Body cobalt is turned over rather rapidly, with urinary excretion as the main route of elimination. The major part of absorbed cobalt is lost within 1 week. Total body cobalt in the adult is about 1 mg with a majority residing in bone, although liver and blood can contain up to 0.5 mg. Plasma transport is believed to be primarily by serum albumin and to a lesser extent by α_2-macroglobulin.

Reports of dietary cobalt intakes in the U.S. vary widely, from 0.1–1.0 mg/d. Cobalt in fruits, vegetables (about 0.02 $\mu g/g$), and cereals is mostly in noncobalamin form, while meats (about 0.05 $\mu g/g$) contain significant amounts of the cobalamin form. The recommended dietary allowance for vitamin B_{12} is 3 $\mu g/d$, while a requirement for free cobalt is not recognized.

Reference ranges of cobalt in specimens from healthy adults are 1–10 ng/mL in whole blood, 0.1–12 ng/mL in erythrocytes, 0.2–2 ng/mL in plasma or serum, 0.7–10 $\mu g/d$ excretion in urine, and 0.2–1.0 $\mu g/g$ in hair.[25,69]

Relatively large doses of cobalt can be ingested without appreciable ill effects. However, *toxic effects* due to excess cobalt intake have been noted in renal patients receiving cobalt as an erythropoietic agent and in heavy drinkers of beer containing cobalt as a foam stabilizer. In both cases, cardiomyopathy resulting in death was observed. Possible contributing factors were reduced urinary excretion of cobalt in anemic renal patients and alcohol toxicity in the beer drinkers.

Determination of ultratrace levels of cobalt in human body fluids and tissues has proved to be fraught with analytical difficulty, as evidenced by the variability in the orders of magnitude of reported serum values.[74] Emission spectroscopy, neutron activation analysis (NAA), and flameless AAS provide the required sensitivity and have been the most commonly used analytical techniques. The NAA and flameless AAS techniques are recommended, with the latter being the most practical for clinical and research laboratories.

Further information on the human biochemistry, metabolism, and toxicity of cobalt is available.[8,78]

NICKEL, VANADIUM, SILICON, AND ARSENIC

Nickel, vanadium, silicon, and arsenic are considered to be the "new" ultratrace elements, with tissue concentrations in ng/g amounts, and dietary requirements about 50 ng/g diet or less (excepting silicon). These elements have been shown to be essential for animals; however, in humans no effects of deficiency are known. Several publications review the evidence for essentiality of these elements, variously called "ultratrace," "minor," or "abstruse" trace elements.[43–45]

NICKEL

Consistent symptoms of nickel deficiency in animals are ultrastructural abnormalities in the liver, depressed growth, and iron deficiency. Nickel is involved in facilitating the intestinal absorption of ferric iron. Findings that nickel absorption is significantly increased in iron-deficient rats suggest that nickel ions use the iron transport system located in the proximal part of the small intestine. Nickel is apparently transported across the mucosal epithelium via an active energy-dependent process rather than by simple diffusion. Dietary substances such as phytate and EDTA have been reported to decrease nickel absorption.

Limited studies indicate that *human dietary* intakes of nickel range from 170–700 $\mu g/d$. A human dietary requirement for nickel has been estimated to be 50 $\mu g/kg$.[45] Diets based on foods of animal origin or high in fats are likely to be low in nickel. Rich sources of nickel include chocolate, nuts, peas, beans, and grains.

Reference ranges for nickel in the blood of healthy adults are 1–22 ng/mL in whole blood, 48–106 ng/ml in erythrocytes, and 1–21 ng/mL in serum or plasma. Levels of nickel found in hair of healthy adults are 0.01–1.8 $\mu g/g$, and urinary excretion has been reported as 0.1–20 $\mu g/d$.[25,69]

Like other essential trace metals, nickel salts and metallic nickel are relatively nontoxic, exerting mainly gastrointestinal irritation. Oral nickel, however, can have adverse effects in copper or iron deficiency and in cases of nickel allergy. The most significant human nickel poisoning has been through industrial exposure to nickel carbonyl, a gaseous compound some 100 times more toxic than carbon monoxide.

Clinical Significance

Several human conditions in which nickel nutrition may be important have been reported. Since iron and nickel interact, conditions affecting iron absorption and turnover (e.g., those characterized by pancreatic insufficiency) may also affect nickel absorption. Nickel metabolism may be altered in individuals with excessive tissue levels of iron or copper, such as in hemosiderosis and hemochromatosis. Prescription of a low-nickel diet may be appropriate for individuals with nickel allergy, which occurs most often in women consuming marginal amounts of iron in their diet. Nickel metabolism may be disturbed in individuals with diseases that result in excessive levels of thiols in the body.

Very sensitive techniques are necessary for the *determination* of the ultratrace elements in biological specimens; flameless AAS is the recommended analytical technique. Instrumental parameters and published methods for AAS determinations of nickel in biological samples have been summarized.[65,66]

VANADIUM

In vitro and experimental animal evidence suggests that vanadium is essential for humans. Probable biological roles of vanadium are as an enzyme regulator or cofactor, as a redox couple (V^{+4}/V^{+5}), or as a potentiator of certain endocrine functions. The pentavalent vanadate anion is structurally similar to the phosphate anion, and many reports have appeared on the effects of vanadium on ATPase and related enzymes. Vanadates affect many enzyme reactions that involve phosphate, such as the potent inhibition of human alkaline phosphatase by *o*-vanadate. Evidence indicates that vanadium acts as a cofactor or activator of δ-aminolevulinic acid transaminase, cardiac adenylate cyclase, and monoamine oxidase enzymes.

In humans, most ingested vanadium remains unabsorbed (> 85%) and is excreted in the feces. Urinary excretion of vanadium is quite low compared with dietary intake and fecal levels. An estimated daily human vanadium requirement is 10–25 $\mu g/d$.[45] The vanadium content of most foods is < 1 ng/g, with daily intakes varying from about 10–40 $\mu g/d$.

Reported *vanadium levels* in healthy adults are 0.02–8 ng/mL in whole blood, 0.1–17 ng/mL in erythrocytes, 0.02–10 ng/mL in plasma or serum, 0–10 $\mu g/d$ in urine, and 0.01–2.2 $\mu g/g$ in hair.[25,69]

Current analytical techniques for the determination of vanadium in biological specimens are inadequate. The techniques most commonly employed are NAA and flameless AAS, with the latter being the practical choice for most laboratories.

SILICON

The claim of essentiality for silicon is based on the consistent observation of bone and collagenous disorders in animals made silicon deficient. Silicon is a constituent of certain glycosaminoglycans and polyuronides and may be involved in formation of polysaccharide linkages. Silicon is found as a bound constituent of collagens and is a constituent of elastin. The evidence suggests that silicon, like zinc and copper, plays a fundamental role in collagen and elastin cross-linking and is probably involved in bone calcification.

Silicon is an ubiquitous element and is found in significant amounts in many foods, especially in unrefined grains. Animal-derived foods (excepting skin) are generally low in silicon. A possible human requirement for silicon has not been estimated.

Evidence suggests several possibilities whereby silicon may influence human health. Silicon has an anti-atheromatous effect, and silicon deficiency may be involved in the genesis of atherosclerotic heart disease. Evidence suggests that silicon may be involved in the development of hypertension.

The reported *levels* of silicon in specimens from healthy adults are 1.2–8.9 μg/mL in whole blood, 4.1 μg/mL in erythrocytes, 0.4–10.0 μg/mL in plasma or serum, 4.7–5.2 mg/L in urine, and 20–44 mg/g in hair.[25] Mass spectrometry, emission spectroscopy, and AAS have been used to determine silicon in biological materials, with AAS being the method of choice for most laboratories.

ARSENIC

Since no specific biological function for arsenic has been established, the biochemical basis for the essentiality of arsenic is speculative. Arsenic is apparently associated with the lipid phase of the human brain, and in vitro evidence suggests the possibility of arsenic's being involved in phospholipid analogs. Arsenic can be an activator or inhibitor of enzymes, although arsenic-specific enzyme action has not been identified. Many arsenic deprivation symptoms are similar to those of zinc deficiency, and signs of arsenic deprivation may be influenced by zinc and arginine intake.

Data are insufficient to estimate a possible human arsenic *requirement*, but most diets probably provide adequate amounts of arsenic if hypothetical human requirements are similar to those for animals. Recent food surveys indicate intakes of 20–130 μg/d in the U.S. Fish and seafood contain relatively high amounts of arsenic, while dairy products and certain vegetables and fruits contain much less.

Presently, too little is known of human arsenic interactions to suggest pathologies which might be linked to arsenic metabolism. The poisonous nature of arsenicals is legend, and some reports have linked the element to certain cancer forms. Yet data from animal studies suggest that arsenic is likely to be essential for humans in some small amount.

Reported *levels* of arsenic in specimens from healthy adults are 2–62 ng/mL in whole blood, 3 ng/mL in erythrocytes, 1–20 ng/mL in plasma or serum, 5–50 μg/d excretion in urine, and 0.1–1.1 μg/g in hair.[25,69] Mass spectrometry, neutron activation, emission spectroscopy, and AAS have been commonly utilized to determine arsenic in biological specimens. For most laboratories AAS may be the most practical instrumental technique. Hydride generation–AAS techniques are required for arsenic determinations in some specimens since organo-arsenic forms may be quite volatile. Methods for determining arsenic are discussed in Chapter 18, Analysis of Toxic Substances.

SELENIUM

The human essentiality for selenium is based on its incorporation into the enzyme *glutathione peroxidase* (GSH-Px). Selenium has no other confirmed human function but may be important for the activity of other human enzymes, such as muscle cytochrome.

Biochemistry and Physiology

Glutathione peroxidase, discovered in 1957, catalyzes the breakdown of hydrogen peroxides and lipid hydroperoxides in body tissues and fluids, thereby protecting against oxidative damage to body tissues. Erythrocyte GSH-Px is a selenoprotein containing 4 selenium atoms per enzyme molecule. Selenium is a necessary part of this biochemical mechanism to prevent cellular oxidative damage. The relation of selenium and GSH-Px to human health and disease has been reviewed.[6,20,53] In long-term selenium deficiency, all body tissues show decreased GSH-Px activity.

In erythrocytes, GSH-Px operates to prevent peroxidation of hemoglobin (precipitation of Heinz bodies) and cell membranes (hemolytic anemia). Many dietary, physiological, pathological, and environmental factors have been shown either to increase or to decrease GSH-Px activity. One such dietary constituent is vitamin E, which acts synergistically with selenium to prevent peroxide damage.

High activity of GSH-Px in phagocytic cells, such as leukocytes and macrophages, helps to protect the phagocytic cell itself from peroxides it generates during oxidative destruction of foreign matter. Reduced GSH-Px activity in platelets has been linked to bleeding disorders and reduced activity in plasma to edema due to peroxidation of capillary membranes.

The substantial amount of GSH-Px activity in liver may be more important for detoxifying lipid hydroperoxides than for hydrogen peroxide, since catalase, which decomposes the latter, has no activity toward lipid hydroperoxides. Selenium, vitamin E, glutathione, and sulfur-containing amino acids all provide some protection against hepatic damage due to oxidative stress, such as occurs in carbon tetrachloride poisoning. The sulfur-containing amino acids increase the availability of glutathione, the substrate involved in GSH-Px catalyzed lipid peroxide reduction.

Evidence suggests that selenium and GSH-Px activity may be important for preventing oxidative damage to lens tissue, i.e., inhibiting cataract formation. In humans, selenium concentrations increase about four-fold in normal lenses between birth and age 85, while in cataractous lenses the selenium concentrations are less than one sixth of those of normal lenses for the same age group.

Metabolism and Dietary Intake

Selenium appears to be well absorbed from the gastrointestinal tract (about 40%); however, little is known about the site and mechanism of absorption. Urinary selenium excretion appears to correlate well with dietary intake and plays a major role in regulation of selenium homeostasis. In blood, selenium is preferentially taken up by erythrocytes, while plasma selenium is bound less avidly by protein sulfhydryl groups.

Selenium exists in foods in a number of forms; protein-bound selenomethionine, selenoamino acids, selenothiosulfide, and mercury-bound selenium. Selenium content in foods is directly related to the protein content and the soil selenium content in the area from which the food was produced. Determinations of selenium in foodstuffs have shown about 0.2 $\mu g/g$ in meats, 1 $\mu g/g$ or more in fish products, and 1 $\mu g/g$ or less in grains.

The estimated adequate and safe *dietary intakes* of selenium are listed in Table 8C-1, and are 50–200 $\mu g/d$ for adults. Dietary intakes vary, depending on the selenium status of the geographical area. Selenium intakes were undetectable in Baltimore, Maryland, 31 $\mu g/d$ in a Vermont hospital diet, and 6–70 $\mu g/d$ in New Zealand, all low-selenium areas. Effective selenium nutriture is also dependent on the form, i.e., bioavailability, of the dietary selenium.

Selenium *poisoning* is well known in animals, but few cases of human selenium toxicity have been reported. Studies of higher intakes of selenium in human populations living in high-selenium areas (South Dakota, Nebraska, and Venezuela) showed no associated health hazard. A reported sign of excessive selenium exposure is garlic-like breath due to dimethyl selenide excretion.

Clinical Significance

There are no well-defined cases of human disease due to selenium deficiency except in China, where selenium deficiency has been associated with Keshan's disease, a specific type of cardiomyopathy. A number of clinical disorders have been shown to involve glutathione peroxidase (GSH-Px) deficiency, including hemolytic anemia; Glanzmann's thrombasthenia, a platelet disorder; the Hermansky-Pudlak syndrome, a chronic granulomatous disease; and carcinoma. Selenium has been reported to be both a carcinogen and an anticarcinogenic element. Current evidence does not suggest any role for selenium in causing human cancers. It has been suggested that selenium derivatives of glutathione, such as selenopersulfides, might act as anticarcinogenic agents by helping to detoxify active metabolites of certain carcinogens.

The determinations of urinary and blood selenium are useful measures of human selenium status. Plasma or serum selenium levels may be a more sensitive indicator of selenium status than whole blood levels. In China, hair selenium levels were found to correlate with blood levels and hair selenium was used to assess risk of selenium deficiency. Assay of erythrocyte GSH-Px activity has been shown to correlate with blood selenium up to 100 ng/mL and is a useful functional test of selenium status.

The reported selenium concentrations in specimens from healthy adults range from 57–340 ng/mL in whole blood, 71–340 ng/mL in erythrocytes, 78–320 ng/mL in serum or plasma, 5–100 μg/L excretion in urine, and 0.6–2.6 μg/g in hair.[25,69] Blood and tissue selenium levels vary depending on the selenium status of the particular geographical area. Low blood selenium concentrations have been observed in patients with gastrointestinal cancer, during pregnancy, and in protein-calorie malnutrition, while high values have been reported in patients with reticuloendothelial neoplasia.

Methods for the Determination of Selenium

Neutron activation analysis, spectrofluorometry, and AAS are the most popular techniques for quantitating selenium in biological specimens. Selenium, like arsenic, has the capability of forming covalent organo-compounds. This characteristic impacts on its analytical determination in two ways: (1) the organo-selenium forms are likely to be quite volatile and therefore can be lost in certain sample preparation steps such as high temperature ashing; (2) the facile reduction of sample selenium to the volatile hydride form allows the determination of selenium by the AAS hydride generation technique. Recommended methods for determining selenium in biological specimens are flameless AAS[71] and spectrofluorometry.

FLUORIDE

The first interest in human fluoride status resulted from the discovery in 1931 of chronic, endemic fluoride toxicity in several countries. Earlier, it was observed that children with mottled enamel (a sign of fluoride excess) had very little dental decay. Since then, numerous clinical and epidemiological studies have demonstrated the unique property of fluoride in preventing tooth decay. Although fluoride has been shown to affect enzyme activities and growth in animals, its benefit for human nutrition lies in its anticariogenic action.

Biochemistry and Metabolism

Inorganic fluoride is readily absorbed in the small intestine and distributed almost entirely to bone and teeth. Fluoride replaces hydroxyl groups in tooth apatite. Renal excretion is most important for regulation of body fluoride levels.

The *reported concentrations* of fluoride in specimens from healthy adults are 10–370 ng/mL in serum or plasma, 450 ng/mL in erythrocytes, and 0.2–1.9 mg/L in urine.[25,69] Concentrations of fluoride in these specimens vary widely, depending on whether or not the water in an area is fluoridated. Fluoride intakes exclusive of drinking water are about 1–3.4 mg/d. The estimated safe and adequate dietary intakes of fluoride are listed in Table 8C-1, and are 1.5–4.0 mg/d for adults. Fluoridated drinking water contributes appreciably to total fluoride intake. Seafoods (5–10 μg/g) and tea (100 μg/g) are higher in fluoride than cereal grains (1–3 μg/g) and cow milk (1–2 μg/g dry weight).

Clinical Significance

Some evidence suggests that low levels of fluoride intake are associated with osteoporosis and that sodium fluoride and calcium supplementation may stimulate bone formation.

The numerous studies that document the anticariogenic properties of fluoride have been reviewed.[52] The exact form of fluoride treatment which provides maximal anticariogenic effect is not definitively known. Supplemental fluoride is generally most effective during the period of tooth eruption and immediately thereafter.

The mechanism responsible for the cariostatic effects of fluoride is not precisely known. Elements of the action may include alteration of tooth crown morphology, greater integrity of apatite crystal formation, stimulation of enamel surface processes, decreased enamel solubility, and decreased bacterial enzyme activity. Excessive fluoride intake produces mottled or pitted enamel in growing teeth. Chronic fluorosis may produce osteosclerosis, calcification of ligaments and tendons, and crippling deformities such as kyphosis, stiffness of the spine, and bony exostoses.

Methods for the Determination of Fluoride

Fluoride assay methods are discussed in Chapter 18, Analysis of Toxic Substances.

References

1. Abernethy, M. H., and Fowler, R. T.: Micellar improvement of the calmagite compleximetric measurement of magnesium in plasma. Clin. Chem., 28:520, 1982.
2. Aggett, P. J., and Harries, J. T.: Current status of zinc in health and disease states. Arch. Dis. Child., 54:909,. 1979.
3. Alder, J. F., Pankhurst, C. A., Samuel, A. J., et al.: The use of silk and animal hairs as a standard for hair analysis. Anal. Chim. Acta, 91:407, 1977.
4. Analytical Methods Committee: Methods for the destruction of organic matter. Analyst, 85:643, 1960.
5. Brewer, G. J., and Prasad, A. A.: Zinc Metabolism: Current Aspects in Health and Disease. New York, Alan R. Liss, Inc., 1977.
6. Burk, R. F.: Selenium in man. In: Trace Elements in Human Health and Disease, Vol. 2, Essential and Toxic Elements. A. S. Prasad and D. Oberleas, Eds. New York, Academic Press, 1976, pp. 105–133.
7. Butrimovitz, G. P., and Purdy, W. C.: The determination of zinc in blood plasma by atomic absorption spectrometry. Anal. Chim. Acta, 94:63, 1977.
8. Carson, B. L., and Smith, I. C.: Cobalt. In: Trace Metals in the Environment. Woburn, Mass., Ann Arbor Science/Butterworth Group, 1981.
9. Chittleborough, G.: A chemist's view of the analysis of human hair for trace elements. Sci. Total Environ., 14:53, 1980.
10. Christian, G. D., and Feldman, F. J.: Atomic Absorption Spectroscopy. New York, Wiley-Interscience, 1970.
11. Clayton, B. E.: Clinical Chemistry of Trace Elements. Advances in Clinical Chemistry, Vol. 21. New York, Academic Press, 1980.
12. Danks, D. M.: Diagnosis of trace metal deficiency with emphasis on copper and zinc. Am. J. Clin. Nutr., 34:278, 1981.
13. Dean, J. A., and Rains, T. C.: Flame Emission and Atomic Absorption Spectrometry, Vols. 1–3. New York, Marcel Dekker, Inc., 1971.
14. Doisy, E. A., Jr.: Effects of deficiency in manganese upon plasma levels of clotting proteins and cholesterol in man. In: Trace Element Metabolism in Animals. W. G. Hoekstra et al., Eds. Proceedings of the 2nd International Symposium. Baltimore, University Park Press, 1974, pp. 668–670.
15. Doisy, R. J., Streeten, D. H. P., Freiberg, J. M., et al.: Chromium metabolism in man and biochemical effects. In: Trace Elements in Human Health and Disease, Vol. 11, A. S. Prasad, Ed. New York, Academic Press, 1976, pp. 79–104.
16. Duran, M., Beemer, F. A., Van der Heiden, C., et al.: Combined deficiency of sulphite oxidase and xanthine oxidase: A defect of molybdenum metabolism or transport? Proc., 16th Ann. Meet. Soc. Study Inborn Errors Metab., 1978, p. 165.
17. Fisher, G. L., Davies, L. G., and Rosenblatt, L. S.: The effects of container composition, storage duration, and temperature on serum mineral levels. National Bureau of Standards, TS, p. 58. Symposium on Accuracy in Trace Analysis, Proc. 6th IMR Symposium. P. D. LaFleur, Ed. National Bureau of Standards, Special Publ., 1975, p. 422.
18. Fowler, R. T., Abernethy, M. H., Walmsley, T. A., et al.: Measurement of magnesium by continuous-flow colorimetry. Clin. Chem., 28:523, 1982.
19. Freund, H., Atamian, S., and Fisher, J. E.: Chromium deficiency during total parenteral nutrition. JAMA, 241:496, 1979.
20. Ganther, H. E., Hafeman, D. G., Lawrence, R. A., et al.: Selenium and glutathione peroxidase in health and disease—a review. In: Trace Elements in Human Health and Disease, Vol. 2, Essential and Toxic Elements. A. S. Prasad and D. Oberleas, Eds. New York, Academic Press, 1976, pp. 165–234.
21. Gormican, A.: Inorganic elements in foods used in hospital menus. J. Am. Diet. Assn., 56:397, 1970.
22. Hadjimarkos, D. M.: Effect of trace elements on dental caries. Adv. Oral Biol., 3:253, 1968.
23. Hambidge, K. M., and Nichols, B. L.: Zinc and Copper in Clinical Medicine. New York, Medical and Scientific Books, 1978.
24. Issaqi, H. J.: Effects of matrix on the determination of volatile metals in biological samples by flameless atomic absorption spectrometry with the graphite tube atomizer. Anal. Chem., 51:657, 1979.
25. Iyengar, G. V., Kollmer, W. E., and Bowen, H. J. M.: The Elemental Composition of Human Tissues and Body Fluids. Weinheim, Verlag Chemie, 1978.
26. Iyengar, G. V., and Sansoni, B.: Sampling and sample preparation for trace element analysis. In: Elemental Analysis of Biological Materials, Technical Report Series No. 197. Vienna, International Atomic Energy Agency, 1980.
27. Jacob, R. A.: Hair as a biopsy material. In: Systemic Aspects of Biocompatibility. D. F. Williams, Ed. Boca Raton, Fla., CRC Press, Inc., 1981.
28. Jacob, R. A.: Zinc and copper. Clin. Laboratory Med. 1:743, 1981.
29. Jacob, R. A., Munoz, J. M., Sandstead, H. H., et al.: Whole body surface loss of trace metals in normal males. Am. J. Clin. Nutr., 34:1379, 1981.
30. Jacobson, S., and Wester, P. O.: Balance study of twenty trace elements during total parenteral nutrition in man. Br. J. Nutr., 37:107, 1977.
31. Jeejeebhoy, K. N., Chu, R. C., Marliss, E. B., et al.: Chromium deficiency, glucose intolerance, and neuropathy reversed by chromium supplementation, in a patient receiving long-term total parenteral nutrition. Am. J. Clin. Nutr., 30:531, 1977.
32. Johnson, D. J., Djuh, Y. Y., Bruton, J., et al.: Improved colorimetric determination of serum zinc. Clin. Chem., 23:1321, 1977.
33. Kayne, F. J., Komar, G., Laboda, H., et al.: Atomic absorption spectrophotometry of chromium in serum and urine with a modified Perkin-Elmer 603 atomic absorption spectrophotometer. Clin. Chem., 24:2151, 1978.

34. Klevay, L. M.: The role of copper and zinc in cholesterol metabolism. *In*: Advances in Nutritional Research, Vol 1. H. H. Draper, Ed. New York, Plenum Publishing Corp., 1977, pp. 227–252.

35. Kossman, K. T.: Copper in serum measured with the Cobas-Bio centrifugal analyzer. Clin. Chem. *29*:578, 1983.

36. Langard, S., and Hensten-Pettersen, A.: Chromium toxicology. *In*: Systemic Aspects of Biocompatibility, Vol. 1. Boca Raton, Fla., CRC Press Inc., 1981, pp. 144–157.

37. Levander, O. A., and Cheng, L., Eds.: Micronutrient interactions: Vitamins, minerals, and hazardous elements. Ann. N.Y. Acad. Sci., *355*:1, 1980.

38. Makino, T., and Takahara, K.: Direct determination of plasma copper and zinc in infants by atomic absorption with discrete nebulization. Clin. Chem., *27*:1445, 1981.

39. Mason, K. E.: Copper metabolism and requirements of man. J. Nutr., *109*:1979, 1979.

40. Mattera, V. D., Jr, Arbige, V. A., Jr., and Tomellini, S. A., et al.: Evaluation of wash solution as a preliminary step for copper and zinc determinations in hair. Anal. Chim. Acta, *124*:409, 1981.

41. Mertz, W.: The essential trace elements. Science, *213*:1332, 1981.

42. Mills, C. F.: Interactions between elements in tissues: Studies in animal models. Fed. Proc., *40*:2138, 1981.

43. Nielsen, F. H.: Evidence of the essentiality of arsenic, nickel, and vanadium and their possible nutritional significance. *In*: Advances in Nutritional Research, Vol. 3. H. H. Draper, Ed. New York, Plenum Publishing Corp., 1980.

44. Nielsen, F. H.: Possible functions and medical significance of the abstruse trace metals. *In*: Inorganic Chemistry in Biology and Medicine, ACS Symposium Series, No. 140. A. E. Martell, Ed. Washington, D.C., American Chemical Society, 1980.

45. Nielsen, F. H.: Possible future implications of nickel, arsenic, silicon, vanadium and other ultratrace elements in human nutrition. *In*: Clinical and Public Health Significance of Trace Minerals in the World Population. A. S. Prasad, Ed. New York, Alan R. Liss Inc., 1982.

46. Owen, C. A., Jr.: Copper. *In*: Biology and Medicine Series, Vols. 1–3. Park Ridge, N.J., Noyes Publications, 1981–1982.

47. Pekarek, R. S., Wannenmacher, R. W., and Beisel, W. R.: The effect of leukocyte endogenous mediator (LEM) on the tissue distribution of zinc and iron. Proc. Soc. Exp. Biol. Med., *140*:685, 1972.

48. Pennington, J. T., and Calloway, O. H.: Copper content of foods. J. Am. Diet. Assn., *63*:143, 1974.

49. Perman, J. A., Werlin, S. L., Grand, R. J., et al.: Laboratory measures of copper metabolism in the differentiation of chronic active hepatitis and Wilson's disease in children. J. Pediatr., *94*:564, 1979.

50. Prasad, A. S.: Chromium. *In*: Trace Elements and Iron in Human Metabolism. New York, Plenum Medical Book Company, 1978, pp. 3–15.

51. Prasad, A. S.: Copper. Ibid., pp. 17–54.

52. Prasad, A. S.: Fluoride. Ibid., pp. 55–62.

53. Prasad, A. S.: Selenium. Ibid, pp. 215–250.

54. Prasad, A. S.: Zinc. Ibid., pp. 251–436.

55. Prasad, A. S., and Oberleas, D.: Ibid., pp. 363–438.

56. Rice, E. W.: Principles and Methods of Clinical Chemistry. Springfield, Ill., Charles C Thomas, 1960, pp. 157–159.

57. Riordan, J. F., and Vallee, B. L.: Structure and function of zinc metalloenzymes. *In*: Trace Elements in Human Health and Disease, Vol. 1. A. S. Prasad, Ed. New York, Academic Press, 1976, pp. 227–251.

58. Sandstead, H. H., Vo-Khactu, K. P., and Solomons, N.: Conditioned zinc deficiencies. *In*: Trace Elements in Human Health and Disease, Vol. 1. A. S. Prasad, Ed. New York, Academic Press, 1976, pp. 33–49.

59. Schloen, L. H., Fernandes, G., Grofalo, J. A., et al.: Nutrition immunity and cancer—a review. Part II: Zinc, immune function and cancer. Clin. Bull., *9*:63, 1979.

60. Seelig, M. S.: Magnesium interrelationships in ischemic heart disease: A review. Am. J. Clin. Nutr., *27*:59, 1974.

61. Smith, J. C., Jr., and Brown, E. D.: Effects of oral contraceptive agents on trace element metabolism—a review. *In*: Trace Elements in Human Health and Disease, Vol. 2, Essential and Toxic Elements. A. S. Prasad, and D. Oberleas, Eds. New York, Academic Press, 1976, p. 315.

62. Smith, J. C., Jr., Butrimovitz, G. P., and Purdy, W. C.: Direct measurement of zinc in plasma by atomic absorption spectroscopy. Clin. Chem., *25*:1487, 1979.

63. Solomons, N. W.: Factors affecting the bioavailability of zinc. J. Am. Diet. Assn., *80*:115, 1982.

64. Solomons, N. W.: On the assessment of zinc and copper nutriture in man. Am. J. Clin. Nutr., *32*:856, 1979.

65. Sunderman, F. W., Jr.: Atomic absorption spectrometry of trace metals in clinical pathology. Hum. Pathol., *4*:549, 1973.

66. Sunderman, F. W., Jr.: Electrothermal atomic absorption spectrometry of trace metals in biological fluids. Ann. Clin. Lab. Sci., *5*:421, 1975.

67. Sunderman, F. W., Jr.: Trace elements. *In*: Chemical Diagnosis of Disease. S. S. Brown, F. L. Mitchell, and D. S. Young, Eds. Amsterdam, Elsevier North Holland Biomedical Press, 1979.

68. Thiers, R. E.: Magnesium (fluorometric) and magnesium (titan yellow). *In*: Standard Methods of Clinical Chemistry, Vol 5. S. Meites, Ed. New York, Academic Press, 1965, pp. 131–142.

69. Tietz, N. W., Ed.: Clinical Guide to Laboratory Tests. Philadelphia, W. B. Saunders Company, 1983.

70. Toepfer, E. W., Mertz, W., Roginski, E. E., et al.: Chromium in foods in relation to biological activity. J. Agric. Food Chem., *21*:69, 1973.

71. Tulley, R. T., and Lehmann, H. P.: Flameless atomic absorption spectrophotometry of selenium in whole blood. Clin. Chem., *28*:1448, 1982.

72. Underwood, E. J.: Trace Elements in Human and Animal Nutrition. 4th ed. New York, Academic Press, 1977.

73. Veillon, C., Patterson, K. Y., and Brydeh, N. A.: Chromium in urine as measured by atomic absorption spectrometry. Clin. Chem., *28*:2309, 1982.

74. Versieck, J., Hoste, J., Barbier, F., et al.: Determination of chromium and cobalt in human serum by neutron activation analysis. Clin. Chem., *24*:303, 1978.

75. Vieir, N. E., and Hansen, J. W.: Zinc determined in 10-μL serum or urine samples by flameless atomic absorption spectrometry. Clin. Chem., *27*:73, 1981.

76. Waslien, C. I.: Human intake of trace elements. *In*: Trace Elements in Human Health and Disease, Vol. 2, Essential and Toxic Elements. A. S. Prasad and D. Oberleas, Eds. New York, Academic Press, 1976, p. 347.

77. Weinstock, N., and Uhlemann, M.: Automated determination of copper in undiluted serum by atomic absorption spectroscopy. Clin. Chem., *27*:1438, 1981.

78. Williams, D. F., Ed.: Systemic Aspects of Biocompatibility, Vol. 1. Boca Raton, Fla., CRC Press Inc., 1981.

79. Williams, H. L., Johnson, D. J., and Haut, M. J.: Simultaneous spectrophotometry of Fe^{+2} and Cu^{+2} in serum denatured with guanidine hydrochloride. Clin. Chem., *23*:237, 1977.

80. Wilson, D. L.: Separation and concentration techniques for atomic absorption spectrometry: A guide to the literature. Atomic Absorption Newsletter, *18*:13, 1971.

CHAPTER 9

ENDOCRINOLOGY

by Sati C. Chattoraj, Ph.D., and Nelson B. Watts, M.D.

Endocrinology is a science that deals with the products of a group of glands and their action in maintaining the chemical integrity of cell environment. Endocrine glands are ductless and release their secretory products, termed *hormones* (Greek: *hormon*—exciting, setting in motion), directly into the bloodstream for distribution to distant tissues responsive to hormones.

Historically, a hormone has been defined as any substance normally produced by specialized cells in one part of the body and carried by the bloodstream to another part, on which its effect is produced. For example, adrenocorticotropin (ACTH) is secreted by the pituitary, but it affects the functional activities of the adrenal cortex. Similarly, blood-borne hormones of the adrenal cortex regulate the carbohydrate, fat, protein, and mineral metabolism of the body. However, in the light of recent information, such a restricted characterization of a hormone is no longer appropriate. It is now apparent that a hormone can be transported by mechanisms other than the bloodstream and that it can act both in close proximity to and distant from its site of release. Table 9-1 lists different modes of transport and functional characteristics of hormones, while Figure 9-1 shows the approximate location of the endocrine glands in the body.

Hormones vary widely in chemical composition; nevertheless, most hormones can be grouped into three general classes: (1) steroids, such as cortisol; (2) polypeptides or proteins, such as ACTH and growth hormone; and (3) substances derived from amino acids, such as thyroxine and catecholamines. Steroid hormones are generally hydrophobic. Many steroid hormones circulate in plasma bound to high-affinity plasma proteins such as cortisol-binding globulin (CBG) or sex steroid hormone–binding globulin (SHBG), leaving only a small percentage of the hormone

Table 9-1. TYPES OF HORMONE ACTION

Endocrine:	Hormone synthesized in one location and released into plasma; binds to specific receptor in cells at a distant site to elicit characteristic response. *Example*: The action of TSH from the anterior pituitary on the thyroid.
Neuroendocrine:	Hormone synthesized in nerve ending and released into extracellular space; interacts with receptors of cells at distant site. *Example*: The action on the heart of norepinephrine synthesized in splanchnic nerve ending.
Neurocrine:	Hormone synthesized in neurons and released into extracellular space; binds to receptor in nearby cell and affects its function. *Example*: The action on cardiac muscle cells of norepinephrine synthesized in nerve endings in heart.
Neurotransmission:	Hormone synthesized in neurons and released from nerve endings; crosses synapse and binds to specific receptor in another neuron, affecting its action. *Example*: Release of acetylcholine from preganglionic nerve fibers in sympathetic ganglia and binding to receptor in postganglionic neuron with liberation of norepinephrine.
Paracrine:	Hormone synthesized in endocrine cells and released into extracellular space; binds to specific receptor of nearby cell and affects its function. *Example*: Release of somatostatin from islet D cells and its subsequent action on nearby α- and β-cells in the same pancreatic islet.
Exocrine:	Hormone synthesized in endocrine cells and released into lumen of gut; binds to cells lining the gut at varying distances from endocrine cell, thereby affecting their function. *Example*: The release of gastrin by mucosal cells and its action on the gastric acid secretion by the stomach.

Modified from Williams, R. H., Ed.: Textbook of Endocrinology. 6th ed. Philadelphia, W.B. Saunders Co., 1981.

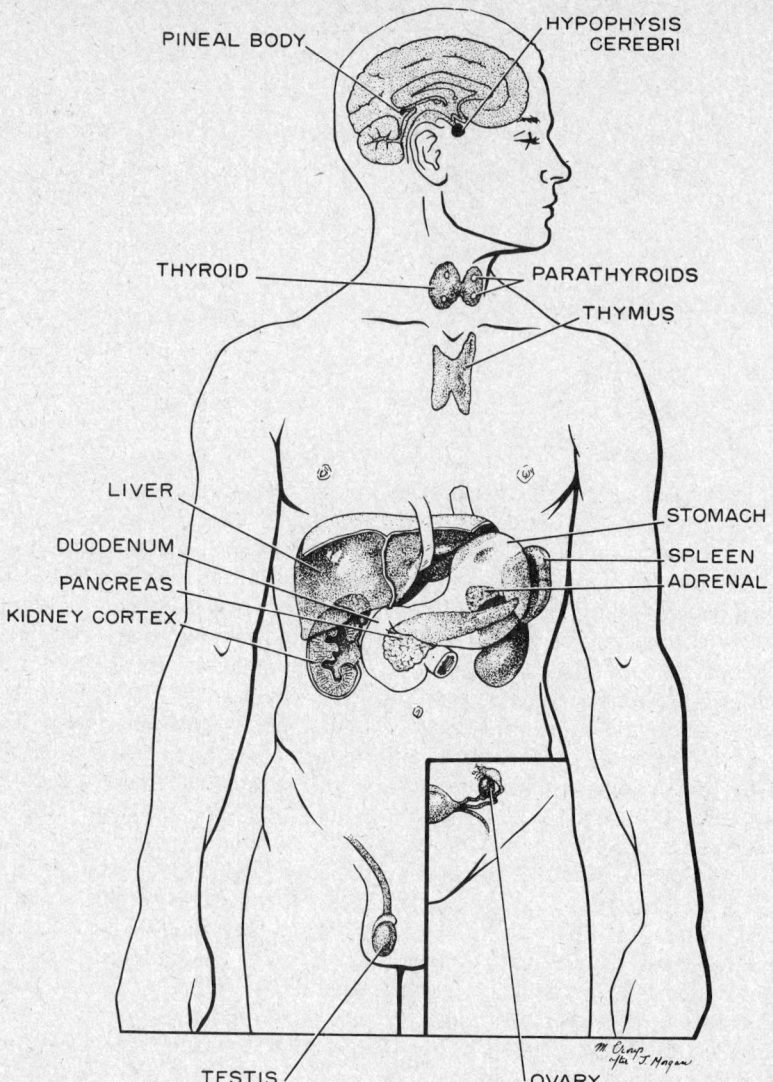

Figure 9-1. Location of the endocrine glands in humans. (From Turner, C. D.: General Endocrinology. 4th ed. Philadelphia, W.B. Saunders Co., 1966.)

free for biologic activity. The half-lives of these hormones vary from 60–100 min. The polypeptide and protein hormones are water soluble, circulate unbound in plasma, and may undergo rapid fluctuations. Their half-lives vary from 5–60 min. Although thyroxine and catecholamines are both water soluble, thyroxine circulates bound to three binding proteins and has a half-life of almost one week. In contrast, catecholamines such as epinephrine are not protein bound; epinephrine has a half-life of <1 min.

Hormones possess a high degree of structural specificity. A slight alteration in the molecular composition of a hormone may bring significant changes in its physiological activity. For example, the structural difference between the female sex hormones estradiol and estriol is only an additional α-hydroxyl group (---OH) at the C-16 position, but estradiol is the most potent estrogen, whereas estriol is almost inert as far as its effect on accessory sex organs is concerned. Similarly, when norepinephrine is N-methylated to produce epinephrine, this minor structural change alters the nature of its biological activity. Classifications of hormones, their source, and a brief description of their action are given in Table 9-2.

Actions of Hormones

The actions of the hormones are complex and diverse. They may, however, be broadly divided according to three general aspects:

Regulatory function. One of the major functions of the endocrine system is to maintain constancy of chemical composition (homeostasis) of extracellular and intracellular fluids for proper and efficient function and growth of the organism. This homeostatic mechanism is maintained through the sensitively regulated metabolism of salt, water, carbohydrate, fat, and protein by secretion of appropriate hormones. When there is derangement in salt and water balance, hormones such as vasopressin and aldosterone come into play. If there is an increased concentration of blood glucose (hyperglycemia, after a carbohydrate-rich meal), insulin is promptly secreted from the pancreas so that the glucose will be utilized at a faster rate, until the glucose concentration decreases to its normal level. The other important regulatory functions of hormones, in addition to the maintenance of normal body environment, are the responses to emergency demands such as starvation, infection, trauma, and psychological stress. Hormones also regulate the process of sexual reproduction, including gametogenesis, sexual behavior, fertilization, nourishment of the fetus, and parturition.

Morphogenesis. Some hormones play an important part in controlling the growth and development of an organism. The development of the male and female sex characteristics under the influence of the respective sex hormones (testosterone and estradiol) is perhaps the best example.

Integrative action. This aspect of hormonal function is the most complex and the least understood. Broadly speaking, each hormone has a specific function. For example, estrogens and progesterone, produced in the ovaries and called female sex hormones, regulate the development of secondary sex characteristics; the adrenal hormone aldosterone controls salt and water balance; the pancreatic hormone insulin regulates carbohydrate metabolism; and so on. However, even though a particular hormone dramatically influences a single biochemical event or changes the morphology and rate of metabolism of a single organ, other hormones produced by different endocrine glands may also be important for regulating a single function. Insulin alone would not be adequate to maintain the balance of carbohydrate metabolism. The concerted action of glucagon (from the pancreas) and of other hormones from glands such as the pituitary (growth hormone), adrenals (glucocorticoids, epinephrine), thyroid (thyroxine), and even gonads (estrogens) is also important. This interrelation is not limited to the endocrine glands but extends to the nervous system as well. While it is true that mineralocorticoids (deoxycorticosterone, aldosterone) have a profound influence on the maintenance of salt and water balance, this control mechanism would fail without the simultaneous adjustment of the rate of blood flow, blood pressure, and vasoconstriction by the autonomic nervous system. There exists an integrative functioning of the endocrine and nervous systems that is reflected by the maintenance of a constant body environment. Derangements of such interdependence give rise to disease states.

Control of Hormone Secretion

There are several mechanisms for maintaining the delicate balance between the production of hormones and the need of the organism for hormones. Detailed discussions of these mechanisms are beyond the scope of this book, and students are referred to other texts.[312,334] The following is a brief outline, covering only the salient features of these mechanisms.

The anterior pituitary occupies a central position in the control of hormone secretion. (See Regulation of Anterior Pituitary Function, p. 1017.) It secretes several *tropic* hormones that in turn stimulate and maintain other target endocrine glands. In the absence of these tropic hormones, the target glands are unable to maintain a normal rate of secretion. The main target organs of the pituitary tropic hormones are the thyroid gland, adrenal cortex, and gonads. The specific tropic hormones are described elsewhere in the text.

Feedback regulation is important in the control of hormone secretion. A feedback system is one where the function of one variable (A) affects another variable (B). If A increases as B increases, the relationship is described as *positive feedback*. If a decrease in A causes an increase in B, this is *negative feedback*. Negative feedback—acting either directly, secondary to change in the concentration of a circulating metabolite, or indirectly, via a tropic hormone—is the principal mechanism for the control of hormone secretion. An example of negative feedback by a metabolite is the suppressive effect of a high blood glucose level on the secretion of insulin. (See Chapter 6.) Indirect negative feedback regulation by tropic hormones is operative in the control of adrenocortical, gonadal, and thyroidal hormone secretions and is discussed in respective sections of the text.

Text continued on page 1004

Table 9-2. HORMONES, THEIR SOURCE, AND THEIR ACTION

Endocrine Gland and Hormone	Nature of Hormone	Site of Action	Principal Actions
Hypothalamus			
Thyrotropin-releasing hormone (TRH)	Peptide (3 aa)*	Anterior pituitary	Release of TSH and PRL
Gonadotropin-releasing hormone (GnRH) or LH-releasing hormone (LHRH)	Peptide (10 aa)	Anterior pituitary	Release of LH and FSH
Corticotropin-releasing hormone (CRH)	Polypeptide (41 aa)	Anterior pituitary	Release of ACTH and β-LPH
Growth hormone–releasing hormone (GHRH)	Polypeptide (40 aa)	Anterior pituitary	Release of GH
Somatostatin†(SS) or growth hormone–inhibiting hormone (GHIH)	Peptide (14 aa)	Anterior pituitary	Suppression of GH and TSH; inhibition of gastrin, VIP, GIP, secretin, motilin, and insulin
Prolactin-releasing factor (PRF)	Peptide?	Anterior pituitary	Release of PRL
Prolactin-inhibiting factor (PIF)	Dopamine?	Anterior pituitary	Suppression of PRL
Anterior pituitary			
Thyrotropin or thyroid-stimulating hormone (TSH)	Glycoprotein‡ (α, 89 aa; β, 112 aa)	Thyroid	Stimulation of thyroid hormone formation and secretion
Follicle-stimulating hormone (FSH)	Glycoprotein‡ (α, 89 aa; β, 115 aa)	Ovary	Growth of follicles and, with LH, secretion of estrogens and ovulation
		Testis	Development of seminiferous tubules, spermatogenesis
Luteinizing hormone (LH)	Glycoprotein‡ (α, 89 aa; β, 115 aa)	Ovary	Ovulation, formation of corpora lutea, secretion of progesterone
		Testis	Stimulation of interstitial tissue; secretion of androgens
Prolactin (PRL)	Protein (198 aa)	Mammary gland	Proliferation of mammary gland; initiation of milk secretion; antagonist of insulin action
Growth hormone (GH) or somatotropin	Protein (191 aa)	Body as a whole	Growth of bone and muscle
β-Lipotropin (β-LPH)	Polypeptide (91 aa)	Unknown	Precursor of β-MSH and the endorphins
Corticotropin or adrenocorticotropin (ACTH)	Polypeptide (39 aa)	Adrenal cortex	Stimulation of adrenocortical steroid formation and secretion
β-Endorphin (β-END)†¶	Polypeptide (31 aa)	Brain	Endogenous opiate; raising of pain threshold and influence on extrapyramidal motor activity

α-Melanocyte-stimulating hormone (α-MSH)	Peptide (13 aa)	Skin	Dispersion of pigment granules, darkening of skin
Leu-enkephalin (LEK)[¶] and met-enkephalin (MEK)[¶]	Peptide (5 aa)	Brain	Same as β-endorphin
Posterior pituitary			
Vasopressin or antidiuretic hormone (ADH)	Peptide (9 aa)	Arterioles / Renal tubules	Elevation of blood pressure / Water reabsorption
Oxytocin	Peptide (9 aa)	Smooth muscles (uterus, mammary gland)	Contraction, action in parturition and in sperm transport, ejection of milk
Pineal			
Serotonin or 5-hydroxytryptamine (5-HT)	Indoleamine	Cardiovascular, respiratory, and gastrointestinal systems, brain	Neurotransmitter; stimulation or inhibition of various smooth muscles and nerves; possible role in mental illness
Melatonin	Indoleamine	Hypothalamus	Suppression of gonadotropin and GH secretion; induction of sleep
Thyroid			
Thyroxine (T$_4$) and triiodothyronine (T$_3$)	Iodoamino acids	General body tissue	Stimulation of oxygen consumption and metabolic rate of tissue
Calcitonin or thyrocalcitonin	Polypeptide (32 aa)	Skeleton	Inhibition of calcium resorption; lowering of plasma calcium and phosphate
Parathyroid			
Parathyroid hormone (PTH) or parathormone	Polypeptide (84 aa)	Skeleton, kidney, gastrointestinal tract	Regulation of calcium and phosphorus metabolism
Adrenal cortex			
Cortisol	Steroid	General body tissue	Metabolism of carbohydrates, proteins, and fats; inflammation, resistance to infection; hypersensitivity
Aldosterone	Steroid	Kidney	Salt and water balance
Adrenal medulla			
Norepinephrine and epinephrine	Aromatic amines	Sympathetic receptors	Stimulation of sympathetic nervous system
Epinephrine		Liver and muscle / Adipose tissue	Glycogenolysis / Lipolysis

Table continued on following page

Table 9-2. HORMONES, THEIR SOURCE, AND THEIR ACTION (*Continued*)

Endocrine Gland and Hormone	Nature of Hormone	Site of Action	Principal Actions
Ovary			
Estrogens	Phenolic steroids	Female accessory sex organs	Development of secondary sex characteristics
Progesterone	Steroid	Female accessory reproductive structure	Preparation of the uterus for ovum implantation, maintenance of pregnancy
Relaxin	Polypeptide	Uterus	Inhibition of myometrial contraction
Inhibin	Polypeptide	Hypothalamus	Suspected role in the control of FSH secretion
Testis			
Testosterone	Steroid	Male accessory sex organs	Development of secondary sex characteristics, maturation, and normal function
Inhibin	See above	See above	See above
Placenta			
Estrogens	See above	See above	See above
Progesterone	See above	See above	See above
Relaxin	See above	See above	See above
Human chorionic gonadotropin (hCG) or choriogonadotropin	Glycoprotein‡ (α, 92 aa; β, 144 aa)	Same as LH	Same as LH; prolongation of corpus luteal function; suspected role in steroidogenesis during fetal life
Human chorionic somatomammotropin (hCS) or human placental lactogen (hPL)	Protein (191 aa)	Same as PRL	Same as PRL
Pancreas			
Insulin	Polypeptide§	Most cells	Regulation of carbohydrate metabolism; lipogenesis
Glucagon	Polypeptide (29 aa)	Liver	Glycogenolysis
Pancreatic polypeptide (PP)	Polypeptide (36 aa)	Gastrointestinal tract	Increased gut motility and gastric emptying; inhibition of gallbladder contraction
Gastrointestinal tract			
Gastrin¶	Peptide (17 aa)	Stomach	Secretion of gastric acid, gastric mucosal growth

Secretin	Polypeptide (27 aa)	Pancreas	Secretion of pancreatic bicarbonate and digestive enzymes
Cholecystokinin-pancreozymin (CCK-PZ)¶	Polypeptide (33 aa)	Gallbladder and pancreas	Stimulation of gallbladder contraction and secretion of pancreatic enzymes
Motilin	Polypeptide (22 aa)	Gastrointestinal tract	Stimulation of gastrointestinal motility
Vasoactive intestinal peptide (VIP)¶	Polypeptide (28 aa)	Gastrointestinal tract	Neurotransmitter; relaxation of smooth muscles of gut and of circulation; increase of release of hormones and secretion of water and electrolytes from pancreas and gut
Gastric inhibitory polypeptide (GIP)	Polypeptide (42 aa)	Gastrointestinal tract	Inhibition of gastric secretion and motility; increase of insulin secretion
Bombesin¶	Peptide (14 aa)	Gastrointestinal tract	Stimulation of release of various hormones and pancreatic enzymes, smooth muscle contractions and hypothermia, changes in cardiovascular and renal function
Neurotensin¶	Peptide (13 aa)	Gastrointestinal tract and hypothalamus (gut and brain)	Uncertain
Substance P (SP)¶	Peptide (11 aa)	Gastrointestinal tract and brain	Sensory neurotransmitter, analgesic; increase in contraction of gastrointestinal smooth muscle; potent vasoactive hormone; promotion of salivation, increased release of histamine
Kidney			
1,25-(OH)$_2$ Vitamin D	Sterol	Intestine	Facilitation of calcium and phosphorus absorption
		Bone	Increase in bone resorption in conjunction with PTH
		Kidney	Increase in reabsorption of filtered calcium
Erythropoietin	Glycoprotein	Bone marrow	Stimulation of red cell formation

*aa = Amino acid residues.
†Also produced by gastrointestinal tract.
‡Glycoprotein hormone composed of two dissimilar peptides. The α-chain is similar in structure or identical; the β-chain differs for each hormone and confers specificity.
§Two chains linked by disulfide bonds: A, 21 aa; B, 30 aa.
¶Also produced in the brain.

As opposed to negative feedback, positive feedback does not operate in isolation but constitutes an integral part of a control system. Evidence for positive feedback regulation is primarily observed in hormonal control of the menstrual cycle. (See the Normal Menstrual Cycle, p. 1109.)

Control by the central nervous system (CNS). Hormones secreted by the neurohypophysis (e.g., oxytocin, vasopressin) and the adrenal medulla (e.g., epinephrine) are controlled primarily by the nervous system.

The importance of the relationship between the nervous system and endocrine system is apparent through various feedback control mechanisms that involve the hypothalamus. Since the hypothalamic nuclei also receive signals from higher centers in the brain, one special feature of the regulation of pituitary hormone release is the ability of the brain to override or fine-tune other control mechanisms. This type of control by the higher centers of the brain is called an *open-loop control system*; the hormonal response to stress is an example of this system in operation.

Hormone-Receptor Interactions

The biological response of a target cell or organ to a particular hormone is initiated by binding of the hormone to target cell receptors. The term "target" is used to refer to the site of action of any hormone, regardless of whether or not the hormone is tropic in nature. For example, the thyroid gland and the uterus are the target organs for thyrotropin and estrogens, respectively. The receptor provides the target cell with a mechanism for recognizing the hormone, and the hormone-receptor complex activates the target cell to begin the chain of biochemical events that become manifest as the biological effect(s) of that hormone. The hormone-receptor complex has several characteristics:

1. It is highly *specific.* The specific nature of the hormone-receptor interaction enables a target tissue to accumulate the biologically active hormone from among the myriad of other substances to which the cells are exposed. The physiological concentrations of circulating hormones range from 10^{-9} to 10^{-12} mol/L, whereas the total concentrations of other proteins and peptides are $\sim 10^{-3}$ mol/L. Consequently, the specificity for recognition of a hormone by a receptor is such that one molecule is picked up out of a million or a billion molecules of other substances.

2. It is an *equilibrium system* that may be expressed by the reaction scheme, H + R $\rightleftharpoons$ HR. This model assumes homogeneity of both the hormone (H) and receptor (R). These two species react reversibly according to the first order mass-action law, forming the hormone-receptor complex (HR); the reaction is assumed to reach equilibrium.

3. It is *saturable.* Since there are a finite number of receptors in cells, there is a maximum hormone-binding capacity. The degree of a biological response of a target tissue is directly proportional to the number of hormone-receptor complexes (i.e., to receptor occupancy); a maximal response is obtained when all sites are filled.

4. It is of high *affinity.* In order for hormone-receptor complexes to form in the presence of very low circulating hormone levels, the affinity constant (K) must be very high. Other terms, such as equilibrium or association constant (K_a), dissociation constant (K_d), and Michaelis constant (K_m), are also used to express affinity. The estimation of equilibrium constants for the enzyme-substrate system has been directly applied to hormone-receptor interactions. Thus, the relationship between the hormone and its receptor can be represented as follows:

$$[H] + [R] \rightleftharpoons [HR]$$

According to the law of mass action,

$$K_a = \frac{[HR]}{[H][R]}$$

where K_a = association constant

[HR] = concentration of hormone complexed with the receptor

[H] = concentration of free hormone

[R] = concentration of free receptors.

The reciprocal of K_a $(1/K_a)$ represents the dissociation constant, K_d. Therefore

$$K_a = 1/K_d \text{ or } K_d = 1/K_a$$

The affinity for most of the hormones, when described as *equilibrium dissociation constant* (K_d), falls approximately in the range of circulating hormone concentrations (i.e., 10^{-9} to 10^{-12} mol/L).

Receptor Regulation

Receptor proteins are not static components of the cell. The number of receptors may increase or decrease in response to various stimuli, and receptor properties may change with varying physiologic conditions. Current evidence suggests that exposure of some responsive cells to high concentrations of hormone will decrease the number of surface membrane receptors. This change in binding sites has been called by various names such as *desensitization, down regulation,* and *tachyphylaxis.* The physiological result is to protect the cell from the intense stimulation by chronically elevated hormone levels. Decreased sensitivity by responsive cells to insulin as a result of a reduced number of insulin receptors has been demonstrated in obesity, which is characterized by high circulating insulin levels. In addition, the binding of a hormone molecule to its receptor influences the affinity of neighboring binding sites. When an increase in receptor occupancy results in a decrease in the affinity of remaining receptors for a hormone, the process is called *negative cooperativity.* Such a mechanism consists in modulating hormone action by providing high cell sensitivity (receptor affinity) at low hormone concentrations and low sensitivity at high hormone concentrations. In the case of insulin, when the plasma concentrations are high (e.g., obesity, glucocorticoid excess), gradual saturation of receptors progressively lowers the affinity of all of the receptors for that hormone. In clinical entities with low normal or subnormal levels of circulating insulin (e.g., anorexia nervosa, glucocorticoid deficiency, growth hormone deficiency), a heightened responsiveness to injected insulin is observed due to elevated concentrations of insulin receptors or increased receptor affinity or both.

A hormone generally controls the activity of its own receptors (homologous effect), but the receptors may also be modulated by other hormones (heterologous effect). Thus, estrogens increase oxytocin binding to receptors of the uterus, FSH and LH increase the concentrations of LH receptors in the ovary, and both glucocorticoids and growth hormone regulate insulin receptors and reduce cell sensitivity to insulin. Progesterone reduces the concentration of its own receptors as well as the concentration of estrogen receptors. Conversely, estrogen and progesterone receptors are both increased by estrogens. Furthermore, one hormone may have a strong affinity for its own receptor and some affinity for the receptor of another hormone (*specificity spillover*). Such a spillover, or cross reaction, generally occurs between hormones with structural similarities. At physiological levels of hormones, the specificity spillover is of no consequence. If one hormone is produced in excess, that hormone not only shows an excessive biological effect of its own but also may elicit the response of a structurally similar hormone through receptor interactions. Thus, glucocorticoids at high concentrations, in addition to excess glucocorticoid effects, can produce effects of mineralocorticoid excess. Similarly, in acromegaly, high levels of growth hormone in blood may act through the prolactin receptors to produce prolactin-like effects (galactorrhea).

Dysfunction of receptors may cause a variety of diseases.[242] Thus, in pseudohypoparathyroidism and nephrogenic diabetes insipidus, parathyroid hormone and vasopressin fail to elicit the increase in cyclic AMP (cAMP, 3′,5′-AMP) in their target organs that normally occurs as a result of postreceptor interaction. The deficiency of nucleotide regulatory protein G may also be a causative factor in some cases of pseudohypoparathyroidism. The roles of the regulatory protein G and cAMP in the mechanism of hormone action are discussed below. Androgen resistance observed in some male pseudohermaphrodites may be due to abnormal receptors or to the absence of normal receptors. When the loss of androgen receptor function is total, the *testicular feminizing syndrome* results. Androgen resistance can also occur in the presence of normal androgen receptors when there are congenital defects in the molecular events that occur after receptor binding (receptor-positive androgen resistance).

MECHANISM OF ACTION OF HORMONES

Two general models of hormone action prevail at present—one for water-soluble peptides, proteins, and catecholamines, and one for lipid-soluble hormones, such as steroids, $1,25-(OH)_2$ vitamin D, and iodothyronines.

Protein Hormones and Catecholamines

Protein hormones and catecholamines are too polar to diffuse passively through lipoprotein membranes; protein hormones are also too large to pass through membrane pores. Instead, these hormones initiate their response by binding to receptors located on or in the cell membrane. For several hormones, combination with their respective receptors activates adenylate cyclase, a membrane protein that catalyzes the formation of cAMP from ATP at the inner membrane surface. Cyclic AMP diffuses throughout the cell and activates a group of closely related enzymes known as cAMP-dependent protein kinases. These kinases phosphorylate intracellular proteins, especially other enzymes, and thereby regulate (activate or inactivate) the activity of these intra-cellular enzymes or proteins. The hormone, or "first messenger," carries the message to the target cell, but the action of the hormone at the cellular level is carried out by cAMP, which is called the "second messenger." Hormones that utilize cAMP as a second messenger are listed in Table 9-3.

The steps from adenylate cyclase activation through cAMP production to protein kinase activation represent a common intracellular pathway that acts through cAMP. The activation of adenylate cyclase by the hormone-receptor interaction depends on the presence of three proteins, all of which are intrinsic proteins of the cell membrane: the specific receptor, the adenylate cyclase, and a nucleotide regulatory subunit, called N or G protein. The functional relationships of these components are shown in Figure 9-2. G protein is a guanine nucleotide–binding protein that can also cleave GTP (guanosine triphosphate) to GDP (guanosine diphosphate) and P_i (inorganic phosphate). When the appropriate hormone binds to the receptor, G protein associated with GTP activates adenylate cyclase. The GDP complex of the G protein inhibits the enzyme. When the receptor is unoccupied, GDP is released very slowly from the nucleotide regulatory protein. Binding of hormones to receptors, however, causes rapid replacement of bound GDP with cytoplasmic GTP, thereby activating adenylate cyclase as well as facilitating the release of hormone from its receptor. Adenylate cyclase activity remains high until GTP is hydrolyzed to GDP and P_i. The cycle repeats with a newly formed hormone-receptor complex.

As mentioned above, cAMP produced by adenylate cyclase activates protein kinases within the cell. cAMP-dependent protein kinases consist of four subunits—two catalytic subunits (C) and two regulatory (dimeric) subunits (R-R)—that can bind two molecules of cAMP (Figure 9-3). When cAMP binds to the regulatory dimer, the two catalytic subunits are released and become fully active as phosphorylating enzyme. With removal of cAMP, the regulatory dimer reassociates with the catalytic subunits, quenching the enzyme activity. Free cAMP, but not the bound moiety, is rapidly hydrolyzed by the enzyme phosphodiesterase. A rise in the intracellular concentration of cAMP favors binding to the regulatory unit and activation of kinase, whereas a fall in the concentration favors the opposite reaction.

The cAMP-dependent protein kinases catalyze the phosphorylation of enzymes or other proteins. Phosphorylation of these proteins generally changes their conformation and, in the case of enzymes, increases or decreases their activity, thus producing changes in metabolism. For

Table 9-3. HORMONES THAT STIMULATE ADENYLATE CYCLASE

Adrenocorticotropic hormone	LH-releasing hormone
Calcitonin	Melanocyte-stimulating hormone
Catecholamines (β-adrenergic)	Nerve growth factor
Chorionic gonadotropin	Parathyroid hormone
Follicle-stimulating hormone	Prostaglandin E_1
Glucagon	Thyroid-stimulating hormone
Lipotropin	Thyrotropin-releasing hormone
Luteinizing hormone (LH)	Vasopressin

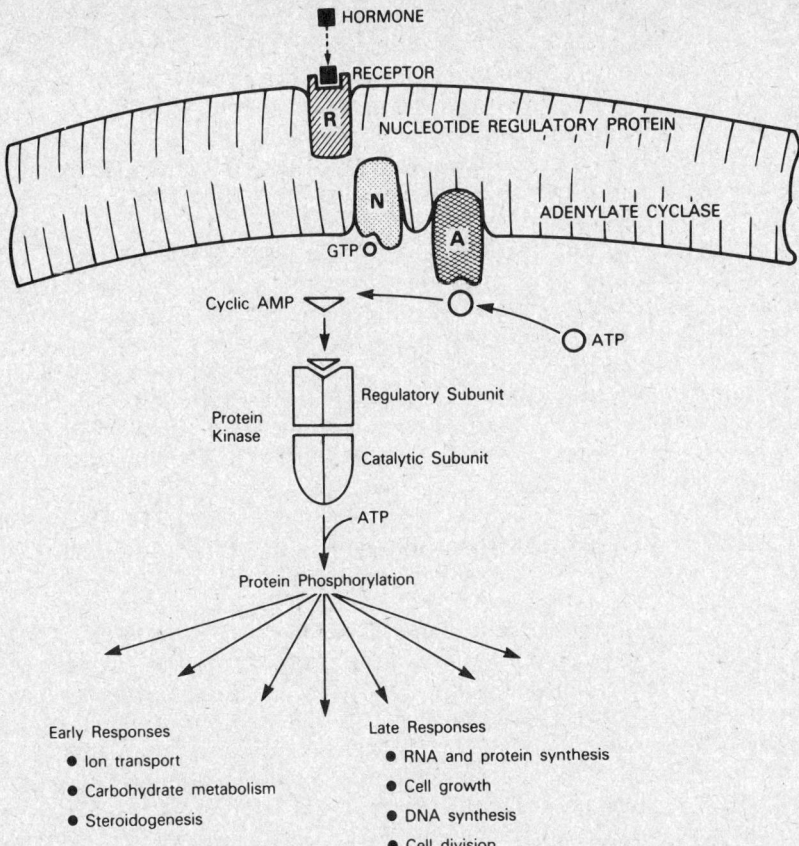

Figure 9-2. Cyclic AMP as a "second messenger" mediating the effects of extracellular chemical messengers such as hormones and neurotransmitters. The ligand (hormone) binds to the receptor protein (R), and this reaction activates adenylate cyclase (A) through the nucleotide regulatory protein (N), which binds GTP. Activation of adenylate cyclase converts ATP to cAMP. The cyclic AMP activates protein kinase, which in turn phosphorylates enzymes and other proteins to produce a variety of effects. (Reproduced with permission from Catt, K. J., et al.: Recent Prog. Horm. Res., *36:*557, 1980.)

example, epinephrine-stimulated phosphorylation of triglyceride lipase in fat cells and of phosphorylase kinase and phosphorylase in hepatic cells causes breakdown of lipids and glycogen, respectively. Conversely, phosphorylation of glycogen synthetase in the epinephrine-stimulated hepatic cell inhibits glycogen synthesis.

The cAMP regulatory system is also affected by intracellular calcium ion concentrations. The interaction of hormone with its receptors leads to increased permeability of plasma membrane to Ca^{2+}. Inside the cell, Ca^{2+} is thought to exert its effect by first binding to a calcium-binding

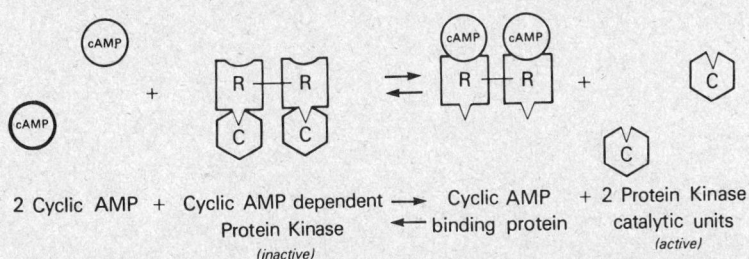

Figure 9-3. Activation of protein kinase by cyclic AMP. See text for explanation. Note that R is widely used as an abbreviation for "receptor" as well as for "regulatory" components, especially for the regulatory component of the cyclic AMP–dependent protein kinases. Also, note that C is widely used as an abbreviation for the catalytic component of an enzyme, especially for the adenylate cyclase and the cAMP-dependent protein kinase. (Reproduced with permission from Roth, J., and Greenfeld, C.: *In:* Textbook of Endocrinology, R. H. Williams, Ed. 2nd ed. Philadelphia, W.B. Saunders Co., 1981.)

protein, calmodulin, present in all nucleated cells. The calmodulin-Ca^{2+} complex is active and binds to calcium-sensitive proteins, altering their activity. The activity of many phosphodiesterases is markedly enhanced by Ca^{2+}-calmodulin. Similarly, phosphorylase kinase, which is activated by cAMP-dependent protein kinase, requires Ca^{2+} for catalytic function.

Several hormones that bind to membrane receptors (e.g., insulin, prolactin, angiotensin) fail to activate adenylate cyclase. For these agents, other effects of hormone-receptor interaction that may mediate a response are currently being actively explored. These effects include changes in Ca^{2+} or K^+ and Na^+ fluxes, variations in the cytoplasmic cyclic GTP levels, and phospholipid turnover. Some hormones (e.g., insulin and epidermal growth factor) have been shown to be internalized after binding to their membrane receptors, and it has been suggested that the internalized hormones or their degradation products may also act within the cell. Recently, a small peptide considered to be a soluble intracellular messenger for insulin has been detected in several cell types, including liver and fat cells. This soluble peptide activates in mitochondria an insulin-sensitive enzyme, pyruvate dehydrogenase, apparently by activating a specific phosphatase that converts the enzyme from its inactive *phospho* to its active *dephospho* form. It has not yet been established, however, if such soluble peptide is the sole mediator of all of the metabolic effects of insulin in all tissues.

Lipid-Soluble Hormones (Steroids, 1,25-(OH)₂ Vitamin D, Thyroid Hormones)

Unlike the preceding hormones, this class of hormones acts by regulating gene expression to induce synthesis of specific proteins or enzymes in their target cells. In some cases the specific proteins have been identified. For example, sex steroids act on the hen oviduct to increase synthesis of ovalbumin, avidin, and other egg proteins; glucocorticoids act to induce specific enzymes (e.g., glucose-6-phosphatase) in liver; and 1,25-(OH)₂ vitamin D stimulates the intestinal mucosa to produce a protein needed for Ca^{2+} absorption. The lipid-soluble hormones passively diffuse across cell membranes. After entry into cells they bind to specific receptor proteins in the cytoplasm, followed by one or more events, as yet unknown, that result in activation of the hormone-receptor complex and its translocation to the nucleus. The activation process may alter the size, shape, conformation, and chemical properties of the receptor. The nuclear receptor complex binds largely to the nuclear chromatin, which is composed of DNA, histone, and nonhistone proteins. The nuclear binding sites associated with chromatin are defined as nuclear acceptors.

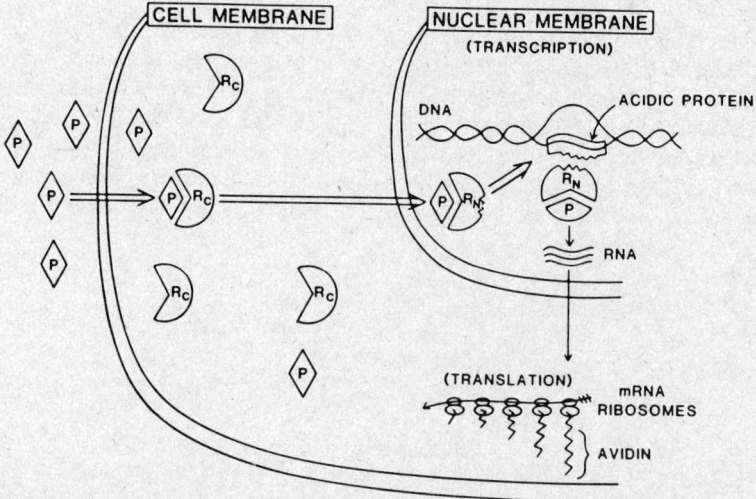

Figure 9-4. General model of steroid hormone action. Binding of hormone, in this case progesterone (P), to the cytoplasmic receptor (R_c), is followed by translocation (arrow) and induction of gene transcription by nuclear receptor (R_N) to produce messenger RNA (mRNA). mRNA is translated on cytoplasmic ribosomes to synthesize new protein, in this case avidin, a protein whose synthesis is induced by progesterone in the chick oviduct. (Reproduced with permission from O'Malley, B. W., et al.: Endocrine Rev., *3:*141, 1982.)

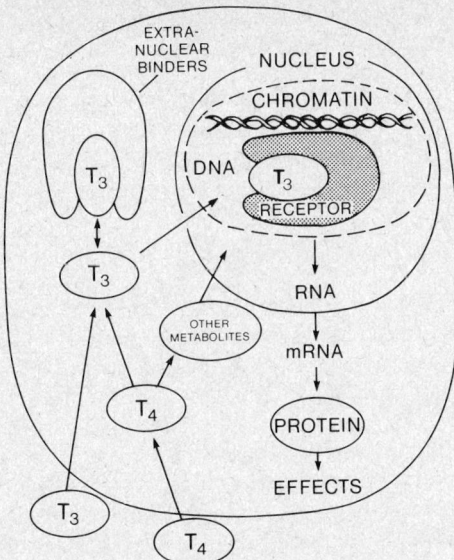

Figure 9-5. Thyroid hormone action. (Reproduced by permission from Baxter, J. D.: N. Engl. J. Med., *301*:1149, 1979.)

The nature of the interaction between hormone-receptor complexes and chromatin, as well as the chemical nature of the so-called nuclear acceptors, is as yet poorly defined. Studies of the progesterone receptor suggest that the receptor has two protein subunits, A and B. Subunit A binds to DNA and subunit B binds to chromatin protein. It is suggested that subunit B provides site specificity for the action of subunit A on DNA. The mechanism of steroid hormone action is illustrated in Figure 9-4.

Thyroid hormones do not fit this conventional model for steroid hormone action. A large body of evidence suggests that thyroid hormones enter cells and bind directly to receptors in nuclear chromatin (Figure 9-5). The nuclear receptor affinity for triiodothyronine (T_3) is approximately ten times that for thyroxine (T_4). Some 85–90% of nuclear thyroid hormone is T_3; the rest is T_4. The nuclear T_3 is derived both from serum T_3 and from intracellular deiodination of T_4 to T_3. The T_3-nuclear receptor complex acts on DNA to increase the synthesis of messenger RNA (mRNA) and ribosomal RNA. The mRNA that is formed dictates the synthesis of proteins by the ribosomes. Some of these induced proteins are enzymes that modify cell function. It is generally assumed that a variety of different proteins are involved, since it is unlikely that a single group of enzymes or proteins could bring about the multiple and varied effects of thyroid hormones. The previous explanation that thyroid hormones increase energy consumption by uncoupling oxidation of substrate from phosphorylation has been found to be inadequate; such uncoupling cannot be demonstrated in hyperthyroid rats, and uncoupling agents such as 2,4-dinitrophenol could not reproduce the metabolic effects of thyroid hormones.

Numerous hypotheses involving interactions of mitochondrial and cell membrane receptors, such as the so-called sodium pump, modulation of adrenergic receptor sensitivity, and neurotransmitters, have been put forward for the mechanism of thyroid hormones. However, there is controversy about these postulations and their significance. For comprehensive discussions of this topic, the reader is referred to a review article.[270]

PROTEIN HORMONES

The hormones secreted by the pituitary, the pancreas, and the parathyroids are proteins. The number and the type of hormones secreted by each gland are enumerated in Table 9-2. These hormones differ from each other in their physiological action and in their chemical structure, ranging from a simple nonapeptide (e.g., oxytocin) to complex glycoprotein molecules (e.g., FSH, LH, TSH). Certain problems arise in assaying protein and peptide hormones. Nonprotein hormones of low molecular weight, such as steroids and catecholamines, have distinctive chemical groups toward which a specific color reaction may be directed as a means of assay (see chemical methods for the Determination of 17-Ketosteroids, p. 1091, and catecholamines, p. 1143). In the case of protein hormones, no such singularity in chemical composition exists. Consequently, the direct chemical determination of these hormones in biological fluids cannot be made. In the absence of chemical assay procedures, measurements of protein hormones depend on other techniques. The basic principles of some of these techniques are discussed in the following paragraphs.

Bioassay Techniques

Bioassay is based on observation of a physiologic response specific for the hormone being measured. A concentrate of hormone is injected into a suitably prepared animal (test) and a standard amount of a reference preparation of hormone into a matched animal (standard). Magnitudes of test and standard responses are compared in order to estimate the amount of hormone administered to the test animal. This technique is fraught with problems; among them the most noteworthy are nonreproducibility due to variability of response in the experimental animals and lack of a sufficiently pure reference standard. Furthermore, biological fluids may contain substances other than the hormone that augment, decrease, or abolish the specific response. The probability of chemical change in the hormone before or after injection introduces uncertainty in evaluation of the results. Bioassay methods have consequently been largely replaced by techniques such as radioimmunoassay. At present, bioassays are primarily employed as reference methods to ascertain biological activity of a purified hormone. Further discussion of bioassay techniques may be found in a monograph.[69]

Radioreceptor Assays

Sensitive and specific radioreceptor assays are now available.[175] These depend on an in vitro interaction of hormone with a preparation of its biological receptor. Radioreceptor assays are radioligand assays based on displacement of trace amounts of radioactively labeled hormone from receptor sites by unlabeled hormone in the sample. Methods for assaying such hormones as ACTH, LH, prolactin, growth hormone, insulin, glucagon, and 1,25-$(OH)_2$ vitamin D have been published. Radioreceptor assays have a distinct advantage in that they measure biologically active hormone. By contrast, radioimmunoassays may measure not only active hormone but also inactive prohormone, hormone polymer, and catabolites which have the same antigenic determinants. Radioreceptor assays nevertheless have some important disadvantages. In some instances, the binding sites provided in the assay, although quite specific for interaction with hormone, are not true receptors (e.g., talcum in insulin assay). Interference with binding by certain substances may lead to spurious results. GTP, for example, has been shown to decrease the binding of glucagon to receptors in hepatocyte membrane; Ca^{2+} has a similar effect on ACTH interaction with adrenal receptors. Additionally, peptidases and proteases in the sample may destroy labeled tracer, or phospholipases may alter the membrane which contains the receptors and thus decrease specific binding.

Immunoassay Techniques

General principles and characteristics of immunochemical procedures have been described in Chapter 1C. Of all the immunological techniques, radioimmunoassay (RIA) is used most often for the quantitation of protein hormones. The sensitivity of RIA has made possible the measurement of most protein and polypeptide hormones without prior extraction of the hormone from the sample. Complete commercial kits as well as individual components for performing these assays are available, and sources for these have been listed in a product guide edition of Clinical Chemistry.[78] Since the use of commercial kits from many different manufacturers has become almost universal in clinical laboratories, no detailed description is given of the RIA procedures for the assay of specific hormones. Instead, a general discussion on the assay of protein hormones follows.

Although RIA procedures for different hormones vary in detail as to the mode of sample preparation and separation of bound from free hormone in the assay, the basic steps are similar. Unlabeled standards or samples are added to predetermined amounts of radiolabeled hormone and antibody. The unlabeled and radiolabeled hormones compete for binding sites on the antibodies during an incubation period. Whenever possible, length of incubation has been minimized in modern procedures because protein hormones are susceptible to damage by radiation, oxidants, and proteolytic enzymes present in the assay system. After incubation, the antibody-bound and free hormones are separated using one of a variety of methods, such as adsorption with dextran-coated charcoal, double-antibody precipitation, or use of primary antibody attached to a solid phase. (See Chapter 1C.) The selection of an appropriate separation method depends on the nature of the hormone and the efficiency of the method for complete and rapid separation of the bound and unbound fractions so that the radioactivity of either fraction can be accurately measured. A calibration curve is constructed by plotting the radioactivity in either fraction (e.g., cpm, bound cpm/total cpm, bound cpm/cpm of zero standard, or free cpm/total cpm) against the standard concentrations.

Circulating concentrations of protein hormones are very low, in the ng/mL or pg/mL range. Therefore, with relatively insensitive techniques such as bioassay, concentration and extraction of hormone from a 24-h urine may be necessary. With higher assay sensitivity, most hormones can be measured in blood, plasma, serum, or urine without these preliminary steps. The selection of specimen type used for assay depends on the sensitivity of the assay, the ease of sample collection, and the physiology of the hormone to be measured. If fluctuations of the hormone in blood are extreme or rapid, measurements in timed urine collections may be more useful, since results would represent an integration over time.

Most protein hormones (e.g., insulin, growth hormone, gonadotropins) can be analyzed in the plasma or serum without preliminary extraction and concentration. Concentration procedures may be necessary, however, if plasma levels of the hormones to be determined are below the sensitivity of the assay available, and if the concentration of degradative enzymes in plasma samples is high enough to cause significant in vitro inactivation of particular hormones (e.g., angiotensin II, ACTH). However, enzyme degradation may be prevented by the use of proteolytic enzyme inhibitors and the collection of samples in tubes chilled in ice. In the past, polypeptide hormones were not thought to circulate bound to specific plasma proteins as do steroid hormones and thyroxine; recently, the existence of specific binding proteins such as neurophysins for oxytocin and vasopressin have been established. Consequently, for the analyses of these hormones, the preliminary treatment of plasma is generally required to inactivate any endogenous binding proteins. Furthermore, unknown substances in plasma may interfere with the assays of some hormones, e.g., vasopressin and angiotensins. Extraction procedures are often used to minimize such interferences. For a discussion of additional aspects of RIA, the section on Determination of Steroid Hormones by Radioimmunoassay, p. 1049, should be consulted.

INSULIN

Insulin is a protein hormone produced by the β-cells of the pancreatic islets of Langerhans. Insulin was the first protein hormone to be sequenced, the first substance to be measured by radioimmunoassay, and the first compound produced by recombinant DNA technology for practical use. Insulin is clinically important in disorders of carbohydrate metabolism (e.g., diabetes mellitus and hypoglycemia). The clinical implications of alterations in insulin production and action are discussed further in Chapter 6, Carbohydrates.

Chemistry

Human insulin (M.W. 6000) consists of 51 amino acids in two chains (A and B) joined by two disulfide bridges. The amino acid sequence of human insulin differs from insulin of other species. The 22–26 amino acids of the β-chain are similar in most animal insulins and appear crucial for the biologic action of insulin. Although insulins from most animal species show immunologic and biologic activity similar to human insulin, insulin from some species, particularly fish, exerts biologic activity in man but reacts quite differently immunologically.[292]

Synthesis, Release, and Degradation

Preproinsulin, a protein of ~100 amino acids, is formed by ribosomes in the rough endoplasmic reticulum of the pancreatic β-cells. Preproinsulin is not detectable under normal conditions because it is rapidly converted to proinsulin. Proinsulin (M.W. 9000) is stored in secretory granules in the Golgi complex of the pancreatic β-cells, where proteolytic cleavage to insulin and

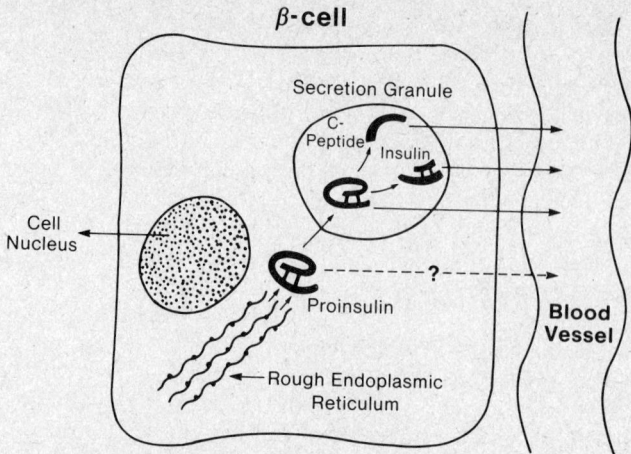

Figure 9-6. Diagram of insulin secretion and release from the pancreatic β-cell. (From Rubenstein, A. H., et al.: Arch. Intern. Med., *137:*625, 1977. Copyright 1977, American Medical Association.)

C-peptide occurs. At the cell membrane, insulin and C-peptide are released into the portal circulation in equimolar amounts. Small amounts of proinsulin and intermediate cleavage forms are also released. This sequence is shown schematically in Figure 9-6.[261]

Insulin release is stimulated by glucose, amino acids, other pancreatic and gastrointestinal hormones (e.g., glucagon, gastrin, secretin, pancreozymin, gastrointestinal polypeptide), and some medications (e.g., sulfonylureas and β-adrenergic stimulators such as isoproterenol). Insulin release is inhibited by hypoglycemia, somatostatin (produced in the pancreatic delta cells), and a variety of drugs (e.g., α-adrenergic stimulators, β-adrenergic blocking agents, diazoxide, phenytoin, phenothiazines, and nicotinic acid).[237]

Fifty to 70% of insulin is extracted by the liver on the first pass through the portal circulation and is degraded there. The kidney is also a site of insulin degradation. The half-life of insulin in the circulation is between 5 and 10 min.[292]

Proinsulin has very little biologic activity but is the major storage form of insulin. Under normal conditions, only small amounts of proinsulin enter the circulation. In the fasting state, the molar ratio of proinsulin to insulin is in the range of 0.4–0.8, with the ratio in peripheral blood being approximately twice that of portal blood.

Two families have been described in whom proinsulin accounts for almost all of the circulating biologic and immunologic insulin activity. These subjects appear to have impaired cleavage of proinsulin. Most of these subjects have had mild hyperglycemia similar to non-insulin-dependent diabetes (NIDDM, formerly type II). This rare condition, called familial hyperproinsulinemia, seems to be inherited in an autosomal dominant fashion. Unusually high concentrations of proinsulin may also be seen in patients with β-cell tumors of the pancreas. Occasionally, proinsulin is present in sufficient quantity to cause hypoglycemia, in spite of its weak biologic activity. Most patients with β-cell tumors will have increases in the concentration of insulin and C-peptide as well as proinsulin, but an occasional case will be seen in which only proinsulin is elevated. Measurement of proinsulin is primarily of clinical usefulness in those patients with fasting hypoglycemia who are suspected of having an islet-cell tumor.[313]

C-peptide (M.W. 3600). Proinsulin is cleaved to form a 31 amino acid connecting ("C") peptide and insulin. C-peptide is devoid of biologic activity but appears necessary to ensure the correct structure of insulin.[237]

Although insulin and C-peptide are secreted into the portal circulation in equimolar amounts, the concentration of C-peptide is greater than that of insulin in both portal and peripheral blood due to the longer half-life of C-peptide (about 20 min). The molar ratio of C-peptide to insulin is > 5 in normal individuals (fasting), and falls to < 2.5 after β-cell stimulation with glucose or glucagon. In contrast to insulin, C-peptide is not taken up by the liver but is removed from the circulation by the kidney and degraded, with a fraction excreted in the urine.[73]

C-peptide *measurements* in blood or urine are helpful in a variety of situations.[115] They are particularly useful in assessing β-cell function in the presence of exogenous insulin or circulating antibodies to insulin. Fasting serum concentrations of C-peptide range from 0.78–1.89 ng/mL in healthy subjects. Values after stimulation with glucose or glucagon range between 2.73 and 5.64 ng/mL (5–6 times the prestimulation value).[115] Measurements of urine C-peptide are useful when a continuous assessment of β-cell function is desired or frequent blood sampling is not practical. The 24-h urine C-peptide content correlates well with fasting serum C-peptide concentration or with the sum of C-peptide concentrations in sequential specimens after a glucose load. Normal urinary C-peptide is 74 $\pm$ 26 μg/d.[119]

Since C-peptide is not significantly metabolized by the liver, its measurement provides a better indication of β-cell function than does the peripheral insulin concentration. C-peptide should be undetectable after a radical pancreatectomy and should increase after a successful pancreas or β-cell transplant. In subjects with diabetes mellitus, those with C-peptide levels >1.8 ng/mL after stimulation with glucagon behave clinically like non-insulin-dependent diabetics, and those with low peak C-peptide levels (<0.5 ng/mL) like patients with insulin-dependent diabetes (IDDM, formerly type I).[116] Measurements of C-peptide after glucagon stimulation may allow identification of some diabetic subjects on insulin treatment who can be managed with diet alone.[310] Also, patients with IDDM who have no C-peptide response are usually more labile with regard to swings of blood glucose than those with some residual β-cell function.

The primary diagnostic use of C-peptide measurement has been in patients with fasting hypoglycemia and suspected of having an insulin-producing β-cell tumor, where concentrations of insulin, C-peptide, and proinsulin are all likely to be high. In some of these subjects, particularly if hyperinsulinism is intermittent, C-peptide concentration may be diagnostically elevated when the insulin concentration is not. Patients previously treated with porcine or bovine insulin are likely to form antibodies to insulin that interfere with insulin immunoassay. C-peptide measurement is indicated if insulinoma is suspected in a patient with insulin antibodies. When hypoglycemia occurs due to surreptitious insulin injection, insulin levels will be high but C-peptide levels will be low; that is because C-peptide is not found in commercial insulin preparations and exogenous insulin will suppress β-cell function.

Insulin receptors and insulin action. Insulin acts on its target tissue cells (primarily liver, muscle, and adipose tissue) by combining with specific *insulin receptors* on the plasma membrane. The insulin-receptor combination initiates changes within the cell (shown in Figure 6-11 and discussed further in Chapter 6). The end result is anabolism—facilitation of glucose entry into cells for storage as glycogen and inhibition of protein breakdown and lipolysis.

The final action of insulin depends on (1) the concentration of insulin that reaches the target cells, (2) the number of insulin receptors on the target cells, (3) the affinity of the insulin receptors, and (4) "postreceptor" intracellular events.

The concentration of insulin reaching the target cells is dependent on the response of the β-cells to stimuli that cause them to secrete the hormone. In healthy humans, insulin response to a standard pulse of glucose can be shown to take place in two phases. The first phase begins 1–2 min after the intravenous injection of the glucose and ends within 10 min. This phase is illustrated by the sharp spike in Figure 9-7,*A* and represents prompt release of stored insulin. The second phase, beginning at the point where the first phase ends, depends on continuing insulin synthesis and release[292] and lasts until normoglycemia has been restored, usually within

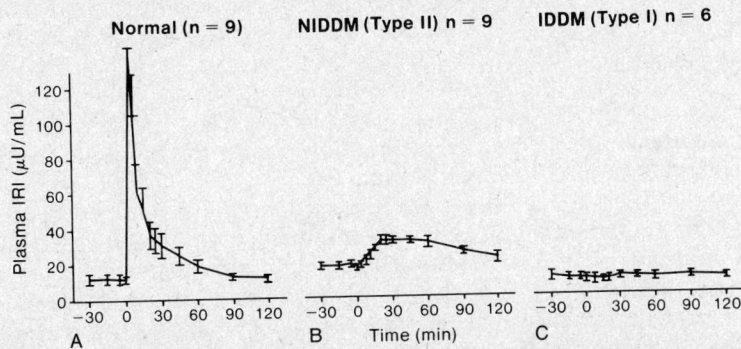

Figure 9-7. Response of plasma insulin to glucose stimulation. A 20-g glucose pulse is given IV at 0 time. *A*, Healthy subjects; *B*, patients with non-insulin-dependent diabetes (NIDDM); *C*, patients with insulin-dependent diabetes (IDDM). IRI=immunoreactive insulin. Values before 0 time represent baseline values. (Used with permission from Pfeifer, M. A., et al.: Am. J. Med., *70*:579-588, 1981.)

60–120 min. With progressive failure of β-cell function, the first-phase insulin response to glucose is lost, although it may remain for other stimuli such as glucagon or amino acids.[237] In most patients with type II diabetes (i.e., non-insulin-dependent diabetes, NIDDM), the first-phase insulin response is lost but the second-phase response is preserved. (See Figure 9-7,*B*.) In type I diabetes (i.e., insulin-dependent diabetes, IDDM), minimal or no insulin response is observed (Figure 9-7,*C*). Decreased levels of plasma insulin occur in type I diabetes mellitus (β-cell failure) or after pancreatectomy. Patients with insulin-producing islet-cell tumors usually present with fasting hypoglycemia and elevated plasma insulin levels.

A reduction in insulin action associated with elevated concentrations of insulin and fasting or postprandial hyperglycemia suggests insulin resistance. (See Chapter 6.) Such resistance is commonly seen in obesity, diabetes mellitus, uremia, and hypercortisolism. There appear to be two mechanisms producing insulin resistance. In one, the change in insulin sensitivity results from a decrease in the number of available insulin receptors.[221] The reduced action of insulin in this case is compensated for by an increase in insulin concentration. The finding that large doses of insulin are relatively ineffective in such conditions as obesity has suggested that post-receptor defects as well as decreased receptor binding are involved in reducing insulin action. In general, the more severe the resistance to insulin, the more prominent the post-receptor defect.[132] The mechanism of the post-receptor defects in most clinical situations has not yet been defined.

Increases in insulin action may be due to an increase in insulin receptors, such as that associated with growth hormone deficiency, glucocorticoid deficiency, or anorexia nervosa. In these subjects, hypoglycemia may be seen even when the insulin concentration is normal. Changes in insulin receptor number, increases or decreases, are usually reversible.

Clinical application of insulin measurement. The primary clinical application of insulin measurement has been in the evaluation of patients with fasting hypoglycemia. This is discussed in detail in Chapter 6. A few investigators have recommended routine measurement of insulin along with glucose during an oral glucose tolerance test as an aid in the diagnosis of early diabetes mellitus.[156] This approach has not been recommended by most diabetologists, and its expense should discourage general application until satisfactory diagnostic criteria are developed and effective treatment is available for diabetes mellitus in its early, prehyperglycemic stages.

Another potential use for insulin determination is to identify those patients with known diabetes mellitus who are at high risk for degenerative complications or to distinguish those patients who will require insulin treatment from those who can be controlled with diet alone. Preliminary studies indicate that those individuals with a peak plasma insulin $> 60\ \mu U/mL$ after a 75-g oral glucose load are likely to remain free from microvascular complications and can usually be controlled with diet alone, while those with peak insulin concentrations $< 40\ \mu U/mL$ will require insulin treatment and are more likely to develop microvascular disease.[311] Others recommend the use of fasting or stimulated (glucose or glucagon) C-peptide measurements to make this type of distinction.[116] Presently there is not sufficient information to recommend the general use of insulin or C-peptide measurements in diabetes mellitus.

Antibodies to insulin. Antibodies to insulin develop in almost all patients who are treated with exogenous bovine or porcine insulin.[8] These antibodies are usually of the IgG class and are rarely of sufficient quantity or affinity to cause clinical problems other than interference in the radioimmunoassay for insulin. Occasionally, however, binding of insulin to circulating high-affinity antibodies will interfere with the action of exogenously administered insulin; erratic fluctuations of plasma glucose may then occur as free insulin levels rise and fall due to erratic release from the antibody-bound form. Beef insulin is more immunogenic than pork insulin. Preliminary studies with human insulin indicate that it is slightly less immunogenic in man than is purified pork insulin when administered subcutaneously in repository form. Rarely, antibodies to insulin may develop in patients who have not received insulin.

Patients with *antibodies to the insulin receptors* have been described.[85] These antibodies fall into two categories—antibodies that occupy the insulin receptor site and block the action of insulin (the rare syndrome of insulin-resistant diabetes with acanthosis nigricans), and antibodies that occupy the receptor site and mimic the action of insulin, resulting in hypoglycemia.[304]

The method for detection of such antibodies is similar to that of RIA procedures for insulin except that the patient's serum is used as the source of antiserum. In the assay, different dilutions of the patient's serum are incubated with labeled porcine or bovine insulin. The dilution at which the binding of the labeled material is equivalent to that of a negative control serum is considered the end point of the titration. The result is expressed as the antibody titer, which is the inverse of the degree of dilution at the end point.[22] Different aspects of endogenous insulin antibodies, including assay procedures, have been extensively discussed elsewhere.[6]

Nonsuppressible insulin-like activity (NSILA). Insulin accounts for only ~7% of the hypo-glycemic effect of compounds in the peripheral blood as determined by bioassay. The remainder of the effect is due to "insulin-like" activity of certain other compounds. Since insulin activity is variable and suppressible—for instance, it can be blocked by antibodies to insulin—and activity of these other compounds is rather constant and nonsuppressible, the term NSILA has come into use. Such NSILA compounds are heterogeneous and include the somatomedins, responsible for 4–8% of total insulin-like activity,[238] and a high-molecular-weight protein called NSILAP that accounts for 85–89% of total insulin-like activity. The somatomedins are also known to have NSILAs where the *s* stands for *soluble* fraction. The significance of these compounds in normal carbohydrate metabolism is not known. They are occasionally produced in excess by extrapan-creatic neoplasms and patients may present with fasting hypoglycemia. NSILA may also be a tumor marker in some patients who do not have hypoglycemia.

METHODS FOR THE DETERMINATION OF INSULIN, PROINSULIN, AND C-PEPTIDE

The clinical significance, methods, biologic activity, metabolism, and structure-function re-lationship of insulin, proinsulin, and C-peptide have been reviewed by Kitabchi.[142]

Radioimmunoassay of Insulin

Because of the limitations of biological methods,[175] RIA has become the technique of choice for measurement of insulin in biological fluids. The subject has been reviewed by Wilson and Miles.[335]

Principle

Various RIA methods and commercial kits for insulin measurement are now available. The principle and general procedures of RIA have been described previously (see p. 1010). The following is a brief outline of the method described by Wilson and Miles.[335]

The antibody is produced against porcine insulin in guinea pigs. (Antisera raised against porcine insulin react identically with porcine and human insulin.) ^{125}I-labeled porcine insulin is used as a radioactive tracer. Samples and serial dilutions of the insulin standard are incubated with antisera for 4 h at 4 °C, followed by addition of ^{125}I-insulin and overnight incubation at 4 °C. Antibody-bound insulin is then precipitated by polyethylene glycol (PEG-6000), 12.5 g/dL. Following centrifugation, the radioactivity of the precipitate is determined. A standard curve is obtained by plotting on logit-log graph paper the per cent of total radioactivity bound (B/T%) against the concentration of the standards. The amount of insulin in the sample is determined by comparing the B/T% of the sample to the standard curve.

Specimen Collection and Storage

No special specimen preparation is required for sample collection. The specimen should be collected from a fasting patient except when otherwise indicated in the protocols for stimulation tests. Measurements of insulin may be carried out on serum or plasma. If, however, the analysis is performed with plasma, the anticoagulant heparin should be avoided because high concentra-tions of this polyanion substance interfere with the antigen-antibody reaction. Plasma is best collected in the presence of 1.5 mg EDTA/mL of whole blood. Hemolysis is more frequent in serum and can cause falsely low insulin values. Insulin is stable in whole blood at room temperature for several hours. Plasma or serum, if stored, should be stored at −20 °C; repeated thawing and freezing should be avoided.

Urine may be assayed for insulin content. Albumin must be added to the collection vessel in a sufficient quantity to ensure a final albumin concentration of 0.2–0.5 g/dL. This level of protein is necessary to minimize denaturation of insulin in the specimen.

Reference Range

After an overnight fast, insulin levels in healthy *normal* subjects range from 2–25 μU/mL. Representative values for insulin levels after glucose are shown in Figure 9-7.

Comments

Various insulin preparations, including human insulin, are used as insulin standards for RIA. For ease of comparison of results between different laboratories, the insulin standard is expressed in terms of International Units (IU). One IU of insulin is equal to 0.04167 mg of the WHO Fourth International Standard, which is a mixture of porcine (48%) and bovine (52%) insulin.

Reliability criteria for an insulin assay are 3% intra-assay coefficient of variation for single replicates and 5% inter-assay coefficient of variation. Minimal detectable amount of insulin, as calculated from the dose response curve, should be 0.2 μU.

Antisera raised against insulin show some degree of cross-reactivity with proinsulin. Under most circumstances, proinsulin is present in blood in relatively small amounts and does not appreciably affect the absolute values of immunoreactive insulin. In certain situations (e.g., islet-cell tumors), however, proinsulin may be the major component of circulating insulin. In such instances, direct assay of plasma may falsely estimate the true insulin concentration. The magnitude of the error will depend on the concentration of proinsulin and on the relative reactivity of the antiserum with insulin and proinsulin.

Radioimmunoassay of Proinsulin

Principle

Direct RIA of proinsulin has not been possible. Because antisera specific for proinsulin are not available, an anti-insulin antibody must be used. However, preliminary separation of insulin from proinsulin prior to immunoassay of proinsulin with anti-insulin antibody provides adequate specificity. Chromatographic separation on a molecular sieve column, either Sephadex G-50 or Biogel P-30, is commonly used. A simpler alternative involves incubation of plasma or serum with a proteolytic enzyme, insulin-specific protease (ISP). ISP degrades the insulin into nonimmunoreactive products, and proinsulin can then be assayed specifically.

The following is a brief description of a method based on the one reported by Starr and his colleagues.[120,291] Further discussions may be found in a review article.[142]

One or two milliliters of serum is applied directly to a column of Biogel P-30 equilibrated with immunoassay buffer (borate buffer, pH 8.0). The immunoreactive insulin-like activity in each fraction is determined by the insulin assay as described above. Two well-separated peaks corresponding to the proinsulin-like components (PLC) and insulin are found. When standards of human proinsulin and insulin are used in RIA, calculations can be made for the absolute amounts of each peptide in the serum; the values obtained in the PLC fractions are read against the proinsulin standard, and those in the insulin region are read against the insulin standard. The proportion of insulin immunoreactivity due to proinsulin is calculated by expressing the sum of the proinsulin fractions measured against the insulin standard as a percentage of the total immunoreactive insulin concentration in all the fractions. The latter procedure is generally used because of a limited supply of human proinsulin.

Reference values for fasting proinsulin are 0.05–0.5 ng/mL; the figures represent 5–48% of the insulin concentration.

Radioimmunoassay of C-Peptide

Principle

C-peptide, in contrast to proinsulin, does not react with anti-insulin antibodies. In order to raise a specific antiserum, C-peptide (which is a poor immunogen because it contains only 33 amino acids) must be conjugated to a larger protein such as albumin. Furthermore, because there are marked variations in sequences of C-peptide among species, homology must be preserved in the assay system. The radiolabeled C-peptide antigen must be tyrosylated before it can be iodinated, since native C-peptide contains no tyrosine. Despite the difficulties, various RIA methods for C-peptide assay have been described.[142] The following is a brief outline of the method described by Horwitz et al.[120]

The antiserum is raised in guinea pigs against a synthetic human C-peptide analog (M.W. 3600) conjugated to rabbit serum albumin. A similar tyrosylated ^{125}I-analog serves as the radio-active ligand. The assay buffer is Tris/HCl, 0.1 mol/L, pH 7.7, containing sodium chloride, 0.05 mol/L, and bovine albumin, 2.5 g/L. The incubation period varies between 16 and 24 h.

(The duration of incubation can be shortened by the use of high-affinity antisera.) The antigen-antibody complex is precipitated by adding a second antibody raised against guinea pig globulin. After centrifugation at 4 °C for 20 min at $800 \times g$, the supernatant fluid is decanted and the precipitate counted in a gamma counter. The calculation is similar to that described for insulin. The standard employed is synthetic human C-peptide analog.

The reliability criteria are comparable to those for the insulin assay. The RIA is sensitive to 75 pg/mL, and insulin does not cross-react in this assay. Proinsulin reacts about 1/15 as well as C-peptide on a weight basis. Since proinsulin is present at only 10–20% of the level of C-peptide in normal serum, it contributes $< 1\%$ to the C-peptide values. Consequently, extraction and gel filtration of samples are not necessary for this method. In samples from insulin-treated patients, however, precipitation of insulin antibodies and insulin antibody–proinsulin complexes with polyethyleneglycol (PEG) prior to assay for C-peptide is often recommended.

The *reference range* for serum from fasting individuals is 0.7–1.9 ng/mL.

ANTERIOR PITUITARY HORMONES (ADENOHYPOPHYSEAL HORMONES)

The pituitary gland (hypophysis) lies in a bony cavity, the sella turcica, located at the base of the skull. The gland is small, ~1 cm in diameter, and weighs ~500 mg. Anatomically, the gland is divided into two lobes, anterior (adenohypophysis) and posterior (neurohypophysis). The intermediate lobe, which is present in the pituitary of most vertebrates and in the human fetus, is rudimentary in the adult human. Blood reaches the pituitary from two sources. Arterial blood is supplied by the superior hypophyseal artery, a branch of the internal carotid; venous blood reaches the pituitary through a venous portal system that originates in the median eminence of the hypothalamus and ends in the sinusoidal capillaries of the anterior lobe of the pituitary. This venous blood system, which comprises primary and secondary capillary plexuses connected by sinusoids, is called the hypothalamo-hypophyseal portal system. These vessels carry neurosecretory materials from the hypothalamus to the adenohypophysis. (See below.) Recent studies indicate that there may be a retrograde flow from the adenohypophysis to the median eminence. If such upstream flow exists, it could then carry pituitary hormones and influence hypothalamic function by a so-called short feedback loop.

Until recently, human anterior pituitary cells have been divided on the basis of their histologic staining reactions into three categories: acidophils (about 40% of the cells), which stain with acidic dyes; basophils (about 10% of the cells), which stain with basic dyes; and chromophobes (the remaining 50% of the cells), which do not stain with either. With more modern techniques of immunocytochemistry and electron microscopy, five types of cells can be distinguished. They are as follows: somatotropes, which secrete growth hormone (GH); mammotropes, which secrete prolactin (PRL); thyrotropes, which secrete thyroid-stimulating hormone (TSH); gonadotrophs, single cells that appear to secrete both luteinizing hormone (LH) and follicle-stimulating hormone (FSH); and corticotropes, which secrete both adrenocorticotropic hormone (ACTH) and β-lipotropin (β-LPH).

The pituitary gland secretes at least eight hormones, all of which are proteins or peptides. (See Table 9-2.) Of the eight hormones, six—GH, PRL, TSH, ACTH, FSH, and LH—are produced in the anterior pituitary. The anterior pituitary also secretes β-LPH and a number of smaller peptides of undetermined significance. The other two hormones, vasopressin (also known as antidiuretic hormone, ADH) and oxytocin, are produced in the hypothalamus and are carried through the neurohypophyseal nerve axons to the posterior pituitary. The posterior lobe of the pituitary is therefore not a discrete endocrine organ but acts as a reservoir for these two hormones.

Two of the six principal hormones from the adenohypophysis, GH and PRL, act primarily on diffuse target tissues; the other tropic hormones—TSH, ACTH, and the gonadotropins, LH and FSH—act primarily on specific target endocrine glands, i.e., thyroid, adrenal cortex, and gonads. These peptides and protein hormones, as well as related hormones of the placenta, can be classified based on their molecular structure and biochemical evolution.

Regulation of Anterior Pituitary Function

The pituitary gland was once considered the "conductor of the endocrine orchestra" or the "master gland." Recent evidence indicates that its activity is actually under the control of the hypothalamus, which manufactures small peptides (neuropeptides) known as *releasing* or *inhibiting factors.* Because these substances qualify as hormones, i.e., are transported in the bloodstream to produce a specific effect on the activities of a target organ, it has been suggested that they should be termed "releasing or inhibiting hormones." At present, however, the term "releasing factor" is applied to hypothalamic substances of unknown chemical nature, while substances with established chemical identity are referred to as *releasing hormones.* The existence of at least seven different and specific hypothalamic substances has been established. They are corticotropin-releasing hormone (CRH), thyrotropin-releasing hormone (TRH), growth hormone–releasing hormone (GHRH), somatostatin (also called growth hormone–inhibiting hormone, GHIH), luteinizing hormone–releasing hormone (LHRH or LRH), prolactin-releasing factor (PRF), and prolactin-inhibiting factor (PIF). LHRH stimulates the secretion of FSH as well as LH; whether or not there is a separate releasing factor for FSH has not yet been established. The proponents of a single gonadotropin-regulating hormone refer to LHRH as GnRH (gonadotropin-releasing hormone). TRH, somatostatin, LHRH, GHRH, and CRH have been isolated, characterized, and synthesized. The role of dopamine as the sole prolactin-inhibiting factor (PIF) still remains controversial. The principal hypothalamic hormones and their chemical nature and action are also summarized in Table 9-2.

The releasing hormones, in contrast to some tropic hormones (e.g., gonadotropins), are not species-specific. Porcine LHRH and TRH are equally effective in the human. The availability of highly purified and synthetic releasing hormones has very important diagnostic and therapeutic usefulness. Until recently no definitive tests were available to distinguish between hypothalamic and pituitary dysfunction. This differentiation is now facilitated by administering individual releasing hormones and measuring the response by the changes in the plasma concentration of the respective tropic hormones. In the case of pituitary dysfunction the response is minimal. If, however, the defect is in the hypothalamus, the level of tropic hormone may increase significantly. The therapeutic use of releasing hormones may be indicated in those conditions in which insufficient secretion of anterior pituitary hormones is due to hypothalamic dysfunction. RIA methods for measuring plasma concentrations of these hormones have been reported.[129]

The anterior pituitary is not innervated. As discussed above, the functional connection between the hypothalamus and the anterior pituitary is due to the intimate contact between the hypothalamic nerve endings and the source of the pituitary blood supply. Neurosecretory materials are released into the primary capillary loops of the hypophyseal portal system and are thus distributed to the anterior pituitary. In this manner, the hypothalamus controls the synthesis and release of the pituitary hormones.

However, the neurons that elaborate hypophysiotropic hormones are themselves influenced by the hypothalamic neurotransmitters such as dopamine, norepinephrine, serotonin, acetylcholine, and endorphins, which modify the secretory activity of the anterior pituitary (Table 9-4). Indeed, basal and episodic secretion, diurnal rhythm, and nocturnal release of pituitary hormones

Table 9-4. NEUROTRANSMITTER EFFECT ON HORMONAL SECRETIONS BY THE ANTERIOR PITUITARY

	Secretion Stimulated by	Secretion Inhibited by
ACTH	Serotonin Acetylcholine Endorphins	γ-Aminobutyric acid (GABA)
TSH	Norepinephrine	Dopamine Serotonin Endorphins
PRL	Norepinephrine Endorphins	Dopamine
GH	Dopamine Norepinephrine Serotonin Endorphins	
Gonadotropins	Norepinephrine Acetylcholine γ-Aminobutyric acid (GABA)	Serotonin Dopamine Endorphins

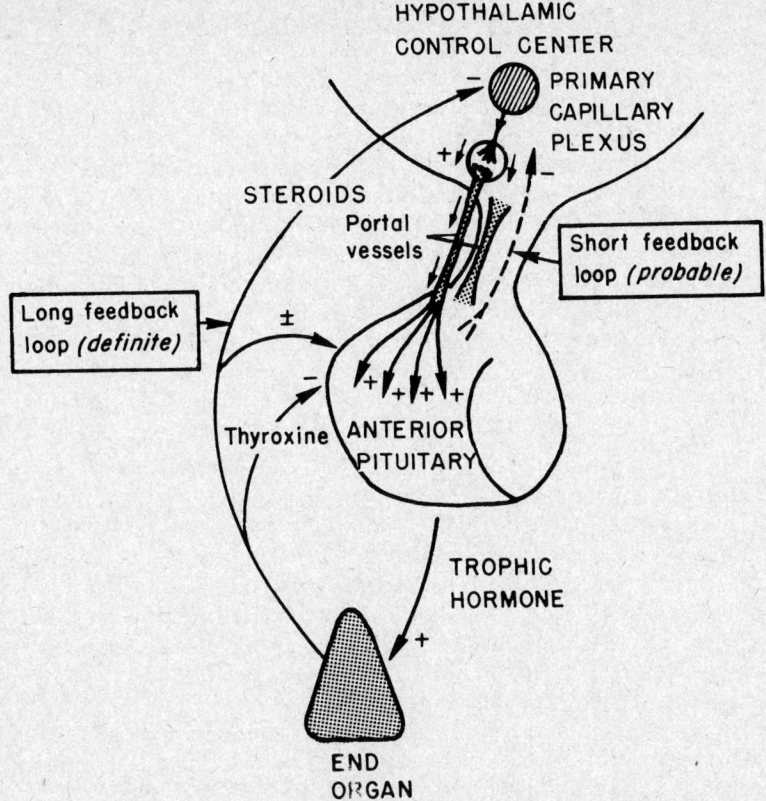

Figure 9-8. Control of anterior pituitary function. (After Catt, K. J.: Lancet, *1*:827, 1970.)

are all considered to be secondary to central nervous system events mediated through hypothalamic hormones.

The functional relationship between the pituitary and its target glands is based on feedback regulation, primarily negative feedback to the pituitary and hypothalamus (Figure 9-8). The effect of negative feedback control is characteristically opposite to the initial stimulus. Thus, an elevated level of blood cortisol (initial stimulus) diminishes the synthesis and release of CRH, resulting in decreased secretion of ACTH and, consequently, of cortisol (final response). Conversely, a diminished level of cortisol in the blood triggers the release of ACTH and hypothalamic CRH, which causes additional release of ACTH; ACTH in turn stimulates the adrenal cortex to increase secretion of cortisol. The importance of such feedback control lies in the maintenance of an optimal concentration of hormones in the blood under a variety of circumstances in order to preserve optimal function of the target tissues. Although hormones such as thyroxine, cortisol, estradiol, progesterone, and testosterone primarily interact with hypothalamic receptors to regulate the release of hypothalamic hormones, recent studies indicate that these secretory products may directly modulate the response of pituitary cells to respective releasing hormones. For example, progesterone and testosterone have been shown to suppress the response of gonadotropin-producing cells to LHRH, while small doses of estradiol potentiate the response. In addition to the long feedback loops comprising pituitary, hypothalamus, and target gland hormones, there exists a *short feedback loop* by which pituitary hormones can act upon the hypothalamus and modify their own secretion. (See Figure 9-8.)

GROWTH HORMONE AND SOMATOMEDINS

Growth hormone (GH, somatotropin, STH), the most abundant hormone in the anterior pituitary, is a single-chain polypeptide of M.W. 21 500 containing 191 amino acids and two intramolecular disulfide bridges. It bears a marked structural resemblance to prolactin and to the placental hormone chorionic somatomammotropin (CS) or placental lactogen, with which it shows overlapping biological activities. Some consider that GH, prolactin, and CS may be phylogenetically descended from a common single primordial peptide.

GH is synthesized in somatotropic (acidophilic) cells of the anterior pituitary and is stored in intracellular granules. During most of the day, the plasma concentration of GH in normal adults remains stable and relatively low (≤ 3 ng/mL, often below the level of detectability of most assays), with one or two sharp spikes occurring 3–4 h after meals. However, adults and children show a marked rise in GH secretion 60–90 min after the onset of sleep; GH concentration reaches a peak value during the period of deepest sleep (stages III and IV) that occurs during the first 2 h. If the onset of deep sleep is delayed, the peak secretion of GH is also delayed. The significance of the sleep-related GH release is not completely understood. This pattern of GH secretion may be important in anabolic and repair processes and in proper skeletal growth.

Regulation of Secretion

The release of GH appears to be under the control of two hypothalamic factors released into the portal blood system from nerve terminals in the median eminence. One is GH-releasing hormone (GHRH), which stimulates GH release, and the other is somatostatin or GHIH, which inhibits GH release. Somatostatin is also found in the delta cells of the pancreatic islets of Langerhans and in other sites in the digestive tract; it has effects on gastrointestinal hormones and causes inhibition of insulin release. The hypothalamic influence on GH release seems to be predominantly stimulatory. This effect is evidenced by the fact that damage to the hypothalamic pituitary connection (e.g., pituitary stalk section, lesions of the median eminence) is followed by dramatic decreases in both basal and induced GH release. The release of the two hypothalamic factors is in turn controlled by the higher centers of the brain. Thus, various stimuli such as exercise; physical and emotional stress; hypoglycemia; increased amino acid levels, particularly arginine; and hormones such as testosterone, estrogens, and thyroxine stimulate GH secretion. In the presence of abnormally high levels of glucocorticoids, GH secretion is suppressed. In addition, the blood level of circulating GH is thought to influence the release of hypothalamic hormones through a short feedback loop.

Physiological Action

The overall action of GH is to promote growth in many soft tissues, in cartilage, and in bone. This action results from the stimulation of protein synthesis that is partly induced by increased amino acid transport through cell membranes. The increased growth of soft and skeletal tissues is accompanied by changes in electrolyte metabolism, including a positive nitrogen and phosphorus balance, a rise in plasma phosphorus, and a fall in the blood urea nitrogen and amino acid levels. Intestinal absorption of calcium is increased; urinary excretion of sodium and potassium is decreased, probably as a result of increased uptake of these ions by the growing tissues. GH has additional effects on intermediary metabolism. It stimulates uptake of nonesterified fatty acids (NEFA) by muscle and accelerates fat mobilization from adipose tissue to the liver. GH also stimulates hepatic glycogenolysis and antagonizes the effect of insulin on glucose uptake by peripheral cells (see Metabolism of Carbohydrates, Chapter 6), so that blood glucose concentrations increase. It is important to note that the roles of GH and insulin are similar in inducing growth, since they both have protein anabolic effects and stimulate the transport of amino acids into peripheral cells. However, their respective effects on the blood glucose level oppose each other. Most GH effects are delayed rather than immediate; the molecular mechanisms by which GH exerts these effects are not established.

The effects of GH on cartilage and on linear growth are now believed to be mediated by other hormonal factors, known as somatomedins, that are synthesized in the liver and possibly in other tissues under the influence of GH.

Clinical Significance

Clinically important states of *growth hormone* excess or deficiency are rare. GH levels are quite variable under normal circumstances, so that reference values often are below the level of assay detectability but may be transiently above the reference range. Measurement of GH under random conditions is generally not of value. A single GH measurement usually cannot adequately separate normal variations of low and high from low or high levels seen with pituitary disease.

GH measurements are best done as part of a dynamic test that involves pharmacologic or physiologic stimuli for release or suppression of GH.[75]

Somatomedin levels are fairly constant; thus, a single measurement of somatomedin can be taken as an accurate reflection of somatomedin production.[238] Somatomedin concentrations are elevated in conditions of growth hormone excess, such as acromegaly, and low in states of growth hormone deficiency. However, somatomedin levels may be elevated or depressed in conditions where GH levels are normal. Correlation of somatomedin levels with high or low GH concentrations is not linear, but measurement of somatomedins in addition to GH may be helpful in selected cases.

Growth hormone excess. Excess growth hormone production is seen in association with eosinophilic or chromophobe adenomas of the pituitary; abnormal X-rays or tomograms of the sella area are present in ～50% of cases; current generation CT scanners will demonstrate pituitary tumors in 80–90% of cases. A prolonged excess of GH activity results in an overgrowth of the skeleton and soft tissues. This most commonly occurs in adults and is known as *acromegaly.*[62] When GH excess is seen before long-bone growth is complete, there is also a striking acceleration of linear growth; this condition is termed *pituitary gigantism.* In advanced cases, there is coarsening of the facial features and enlargement of the hands and feet. In this severe state of the disease, the diagnosis may almost be made on the basis of physical appearance alone. Unfortunately, at this stage the tissue changes are largely irreversible. Untreated acromegaly may cause severe disability or death from cardiac or neurologic sequelae. Early diagnosis and treatment can lessen morbidity and mortality. The most important requirement for diagnosis is demonstration of inappropriate GH secretion.[207] Patients who have pituitary tumors that produce GH can frequently be shown to release GH in response to hypothalamic hormones (TRH and GnRH) that under normal circumstances do not cause release of GH. Occasionally, pituitary tumors are seen that produce excess amounts of both GH and prolactin.

As many as 10% of patients with active acromegaly may have GH levels that fall within the normal range.[207] Essentially all patients with acromegaly can be shown to have an abnormal response to a glucose load. (See Protocol A.) The response in acromegaly may be either no change in the basal level or a paradoxical increase of GH; normal individuals, on the other hand, will suppress GH levels to <5 ng/mL and usually to <1 ng/mL.[74] Higher GH values are quite suggestive of GH hypersecretion. Patients who have been shown to lack normal suppressibility of GH after oral glucose may be studied further by a profile of GH levels throughout a 24-h period and by testing the GH responses to TRH and GnRH, except when the clinical picture and growth hormone abnormality seem clear cut without this additional information.

Somatomedin-C levels have been shown to be elevated in active acromegaly (Figure 9-9). In one study these levels have been shown to correlate better with clinical indexes of severity than glucose-suppressed or basal GH levels.[59] In this study, patients with acromegaly had mean

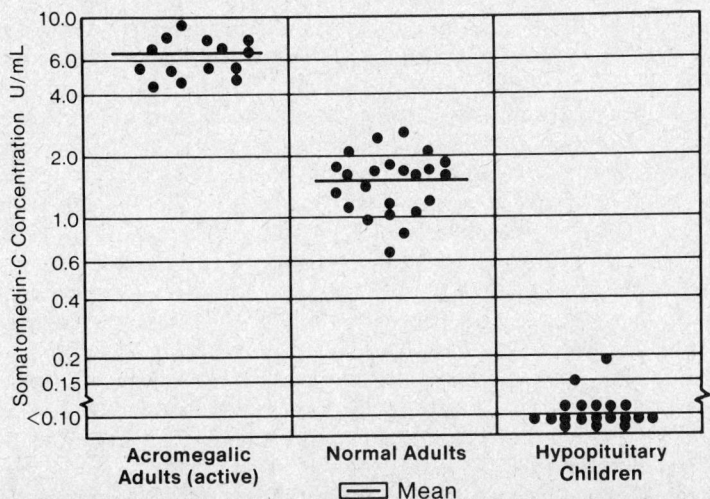

Figure 9-9. Somatomedin levels in growth hormone excess and deficiency states. (From Van Wyk, J. J., and Underwood, L. E.: Hosp. Pract., *13*:57, 1978.)

PROTOCOL A: GLUCOSE SUPPRESSION OF GROWTH HORMONE

Rationale: Normal subjects will show suppression of growth hormone after the oral administration of glucose; subjects with acromegaly will not show this suppression.

Procedure: The test is done after an overnight fast with the patient maintained at bed rest. A baseline sample is collected for the growth hormone measurement. A solution of 100 g glucose is given orally; growth hormone is measured on a blood sample collected 60 min later.

Interpretation: Growth hormone concentration in normal individuals will fall to < 1 ng/mL; subjects with acromegaly will fail to show this suppression and may show a paradoxical rise.

---

somatomedin-C levels of 6.8 U/mL (range, 2.6–21.7 U/mL). Subsequent studies have not found such good correlation.[297] In many cases, elevated GH levels; degree of failure to suppress GH after oral glucose; paradoxical increase of GH after glucose, TRH, or GnRH; or somatomedin levels have not correlated particularly well with symptoms prior to treatment. Also there is often no good correlation between somatomedin values and clinical improvement after treatment.

Growth Hormone Deficiency States and Growth Retardation

Children who lack sufficient GH production or action will not grow normally. GH deficiency may be congenital or acquired, idiopathic or due to anatomic damage to the pituitary or hypothalamus, isolated or associated with deficiencies of other pituitary hormones. There is an interesting reversible GH deficiency state known as *psychosocial dwarfism,* in which environmental stress inhibits pituitary and hypothalamic function, leading to GH suppression and growth retardation. These children may show clinical and chemical evidence of growth deficiency when first evaluated, but a few days of hospital stay will usually correct the abnormalities of pituitary function. GH deficiency is not a common cause of growth retardation, however; for about half of the children evaluated for growth retardation, no organic cause will be found. About 15% of children with growth retardation have endocrine problems as the cause, and perhaps only half of these (around 8% of children with short stature) will be found to have GH deficiency. However, children with growth retardation or dwarfism that has no clear explanation should at least be screened for GH deficiency.

GH deficiency in adults is probably the most common demonstrable abnormality in patients with large pituitary adenomas or after pituitary irradiation, but this is rarely of clinical significance.

There are abnormalities of the GH-somatomedin system that account for growth failure without straightforward diminution of GH levels. An unusual type of dwarfism described by Laron[162] is characterized by high basal levels of GH and low somatomedin levels. Variants of this abnormality may show normal GH levels and low somatomedin levels that do not increase further in response to GH administration or may show normal GH levels and increased levels of somatomedin.[159] In these cases, the defect is thought to be impaired tissue responsiveness to somatomedin.

In evaluating the possibility of GH deficiency, it should be remembered that the basal level of GH is often low in normal individuals and that the half-life of circulating GH is brief (~30 min). Thus, standard assays of GH performed on samples obtained in the basal state may not distinguish between values that are abnormally low and values that fall in the low range of normal. When evaluating GH production, use of provocative tests is essential. While a normal GH response to a provocative test points strongly against GH deficiency, no single test can be considered diagnostic in this situation; as many as 20–30% of subjects with normal GH secretory capacity may be unable to secrete an adequate amount of GH in response to these known provocative stimuli at any given time. For this reason, to diagnose GH deficiency as a cause for growth retardation, it is necessary to demonstrate that the serum concentration of GH remains low after at least two different provocative stimuli.[75,168] A GH value > 10 ng/mL after stimulation is generally considered to be a normal response; values between 5 and 10 ng/mL should be considered borderline, and values < 5 ng/mL subnormal.

There are a number of physiologic and pharmacologic circumstances that stimulate GH release. These are shown in Table 9-5. A simple screening test is to subject the patient to vigorous

Table 9-5. STIMULI FOR GROWTH HORMONE RELEASE

Test	Test Conditions and Dosages	Time of Peak Growth Hormone Release	Reference
Exercise	20 min of vigorous exercise	20 min after exercise has begun	130
Sleep	Sample is drawn 1 h after the onset of deep sleep (stage III or IV); use EEG documentation if possible	Usually 1 h after onset of deep sleep, coincident with onset of stage III or IV sleep	179
Arginine*†	Arginine hydrochloride, 0.5 g/kg body weight, given intravenously over a 30-min period	60–120 min	201, 225
Insulin*†‡	Regular insulin, 0.1–0.15 unit/kg, IV push (severe hypoglycemia may result, and a physician should be in constant attendance)	45–75 min	248
Glucagon†	0.03 mg/kg IM or subcutaneously (not to exceed 1 mg)	120–180 min	267
L-Dopa	0.5 g/1.73 m², usually given with lunch (responses may be improved by giving priming doses of L-Dopa, 0.25 g/1.73 m², for one or more days prior to the test)	30–120 min	76
Clonidine	0.15 mg/m²	90 min	100
Diazepam	0.15 mg/kg, orally	60 min	155
Pentagastrin	IV infusion, 1.5 mg/kg/h	75 min	76

*Sequential administration of arginine and insulin may reduce the need for additional testing.[231]
†Propranolol, 0.75 mg/kg (20–40 mg orally), given 30–60 min before glucagon, insulin, or arginine may enhance a normal response.[226]
‡Insulin, TRH, and GnRH can be given as a "cocktail" to evaluate GH, ACTH, TSH, PRL, LH, and FSH in a single test period.[265]

PROTOCOL B: EXERCISE TEST FOR GROWTH HORMONE

Rationale: Brisk exercise normally causes an increase in growth hormone.

Procedure: The test is best performed in the morning after an overnight fast but may be done at any time. Vigorous physical exercise (running, calisthenics) is performed for 20 min. A venous sample for determination of growth hormone is drawn immediately.

Interpretation: If growth hormone is > 6 ng/mL, growth hormone deficiency is unlikely. An intermediate response (3–6 ng/mL) should be considered suspect; further testing should be done if clinically warranted.

---

exercise for 20 min and promptly draw a blood sample for determination of GH (see Protocol B),[130] a test that can easily be done on an outpatient basis. Taking advantage of the known rise in GH that occurs with Stage III or IV sleep, one may obtain a sample 1 h after the onset of deep sleep.[179] For the interpretation of growth hormone response, the same criteria are used as for postexercise (Protocol B). The obvious limitation of this approach is that the patient needs to be in the hospital for the initial level of testing. Insulin infusion[246] (see Protocol C) or arginine infusion[201,225] (see Protocol D) has been the standard pharmacologic stimulus for GH release; protocols for these are well established and standardized. Other medications known to stimulate GH release are L-dopa and glucagon.

Somatomedin levels are low in patients with growth hormone deficiency and growth failure (Figure 9-9), generally < 0.15 U/mL. Patients with growth failure due to other endocrine disease or to nonendocrine organic disease often have low levels of somatomedin; thus, a low level is not a specific test for GH deficiency.

Somatomedins

The first factor isolated was called *sulfation factor* because of its ability to enhance the incorporation of sulfate into cartilage. However, this substance has other effects on cartilage, including stimulation of synthesis of DNA and RNA and of collagen formation. For this reason, the name of this factor and of other peptides with similar biologic activities that are stimulated by GH has been changed to *somatomedins* (M.W. ranging from 6000–8500). These peptides include somatomedin-A, somatomedin-C, and insulin-like growth factors I and II (IGF I and II), which were formerly known as nonsuppressible insulin-like activity (NSILA, p. 1015).

In addition to the growth-promoting effects on cartilage, somatomedins also show insulin-

---

PROTOCOL C: INSULIN TOLERANCE TEST

Rationale: The stress of insulin-induced hypoglycemia will trigger the release of GH and ACTH in normal subjects. Growth hormone response is measured directly. Cortisol is measured as an indication of ACTH response.

Procedure: The test is done after an overnight fast with the patient at bed rest. An indwelling needle or intravenous line is inserted. Sampling is begun after a 30-min rest period. Baseline samples are drawn for determination of glucose, growth hormone, and cortisol. Regular insulin, 0.1–0.15 unit/kg body weight, is given intravenously. Samples are then obtained at 30, 60, and 90 min for glucose, growth hormone, and cortisol. The patient must become symptomatic (sweating, tremor) or the glucose concentration must fall to < 40 mg/dL for the test to be valid. Additional intravenous insulin may be given if this has not occurred by 30 min, in which case sampling would be prolonged by 30 min. The physician should be in attendance throughout the test, and 50% dextrose for intravenous administration should be kept on hand to be used in the event of a severe hypoglycemic reaction.

Interpretation: Cortisol should increase by > 7 μg/dL, to a peak value > 20 μg/dL. Growth hormone should rise to > 20 ng/mL. Failure to respond may be due to pituitary hormone deficiency or a hypothalamic lesion.

PROTOCOL D: ARGININE STIMULATION TEST FOR GROWTH HORMONE

Rationale: In normal subjects, intravenous administration of arginine hydrochloride stimulates growth hormone release.

Procedure: The test is done after an overnight fast with the patient maintained at bed rest. Arginine hydrochloride, 0.5 g/kg body weight, is infused intravenously over 30 min. Blood samples are drawn for determination of growth hormone before the infusion is started, and 30, 60, and 90 min after the infusion is begun.

Interpretation: Growth hormone should rise at least three-fold above the baseline value. A subnormal response is seen in growth hormone–deficient subjects; a subnormal response should be confirmed by repeating the test on another day or testing growth hormone reserve with another provocative stimulus.

---

like activity in some tissues. They can inhibit lipolysis and increase glucose oxidation in adipose tissues and stimulate glucose and amino acid transport into diaphragm and heart muscle. The insulin-like activity is ascribed in part to the similarity of the structures of insulin and certain somatomedins (e.g., IGF I and II). Unlike most other peptide hormones, somatomedins circulate in blood complexed with a plasma-binding protein. Dissociation occurs prior to passage through capillary membranes and entrance into dense tissues such as cartilage. As previously mentioned, somatomedin-C levels are low in growth hormone deficiency and high with growth hormone excess.

Plasma levels of immunoreactive somatomedin are increased in acromegaly and are reduced in growth hormone deficiency and in many other forms of growth retardation, hypothyroidism, chronic illness, nutritional deficiency, and liver disease.

Methods for the Determination of Growth Hormone and Somatomedins

Principle

The determination of *growth hormone* in blood and urine has been reviewed by Loraine and Bell.[175] (See also Protein Hormones, p. 1009.) Available biological assay methods do not have adequate sensitivity to detect the presence of this hormone in human blood or urine. At present, radioimmunoassays based on the procedure originally reported by Hunter and Greenwood are the methods of choice for the analysis of GH.[123] The basic steps involve (1) incubation of an optimal amount of ^{125}I-labeled growth hormone (antigen) with the test sample and a predetermined amount of anti-GH (antibody) for several days, (2) separation of the free antigen and the antigen-antibody complex using a second antibody, and (3) determination of the radioactivity of the bound fraction. The amount of unlabeled antigen (GH) in the test sample is then calculated by comparison with a simultaneously run standard curve.

A method using a solid-phase system (antibody-coated tube) for the separation of bound and free fractions has been reported.[40] From the practical point of view, this technique may be superior to the previously described methods. Reliable RIA kits for GH are now also available.

Reference Range

The mean plasma levels of GH by this method in 21 ambulatory adult males and 35 females were found to be 4.2 (SE ± 1.2) and 5.1 (SE ± 1.0) ng/mL, respectively. Plasma GH levels are generally higher in children than in adults. A comparison of levels by different methods in normal and abnormal conditions has been reported by Skelley et al.[279]

Somatomedins can be measured in plasma or serum by in vitro bioassay using cartilage, radioreceptor assay, or radioimmunoassay.[238] Although these assays have provided some overall information about the concentrations of somatomedins in health and disease, they have limitations. For example, radioligands and antisera used in radioimmunoassays do not interact equally with all forms of somatomedin; this results in poor precision and specificity. Bioassays lack precision and are affected by the presence in serum of inhibitors of biological activity. However, immunoassay and bioassay techniques may give complementary results when levels of growth factors in the presence of suspected inhibitors are evaluated. The *reference range* for somatomedin-C is 0.4–2.0 U/mL.

PROLACTIN

The structure of prolactin (PRL) is similar to that of growth hormone. PRL contains 198 amino acids with three intramolecular disulfide bridges. Its molecular weight is 22 000. It is secreted by the pituitary mammotropic cells, which are acidophilic. A biologically inactive high-molecular-weight "big prolactin" may be produced by adenomas. The relative number of lactotropic cells with consequent PRL content is increased in fetal pituitaries and during pregnancy. The increase in the number of mammotropic cells is the result of the very high concentration of circulating estrogens during human pregnancy. PRL secretion is pulsatile; values are higher in the afternoon than in the morning and reach peak values during the night.

Regulation of Secretion

PRL secretion is under hypothalamic control. The existence of a PRL-inhibiting factor (PIF) has long been suspected, but so far such a factor remains unidentified. According to many, dopamine is that inhibitory factor. Since thyrotropin-releasing hormone (TRH) stimulates PRL secretion when injected intravenously into human subjects, it has been suggested that TRH is the prolactin-releasing factor (PRF). Yet if this is true, TRH is probably not the only PRF, since hypothalamic extracts cause PRL release even when the TRH content has been inactivated or removed. Furthermore, the regulation of PRL is not altered in experiments with anti-TRH antibody, and suckling, which causes a sharp rise in plasma PRL, is without effect on TSH secretion. However, the chemical nature of the true PRF has not yet been determined.

The physiologic stimulus for PRL release is suckling. Maternal plasma levels increase within minutes after initiation of breast feeding. Following delivery, they remain elevated longer if the newborn is breast fed. Many other stimuli induce PRL release, probably through suppression of dopamine from the hypothalamus; stress is one of the important factors. ACTH, which is also associated with stress, causes elevated PRL levels. Plasma hyperosmolality is another stimulus. Thus, administration of hypertonic saline stimulates PRL release, whereas hypo-osmolality brought about by administration of hypotonic fluids or water ingestion (water loading) reduces the plasma PRL concentration. Whether this effect is stress related or associated with the function of PRL in osmolality control has not been established. Finally, the most important influence on PRL regulation appears to be exerted by PRL itself by way of the short feedback loop. Short feedback loop inhibition by PRL of LHRH secretion has also been suggested as responsible for the gonadotropin inhibition that occurs in women who are nursing and in patients with PRL-secreting adenomas of the pituitary. (See Hyperprolactinemia, below.)

Physiological Action

The initiation and maintenance of lactation are the primary functions of PRL. However, for proper manifestation of PRL action, breast tissue must be primed by estrogens, progestins, corticosteroids, and insulin. PRL induces ductal growth, development of the lobule alveolar system, and synthesis of specific milk proteins including casein and α-lactalbumin. It has also been found to be important in the control of osmolality and in subcutaneous fat metabolism, carbohydrate metabolism, calcium and vitamin D metabolism, fetal lung development, and steroidogenesis. This last function may be related to its antigonadotropin effect.

PRL, unlike other pituitary hormones, has a receptor on the cell membrane of its target organs—the breast, adrenal, ovary, testis, kidney, and liver—where PRL exerts its action through the second messenger, cAMP. (See Mechanism of Action of Hormones, p. 1006.) However, the exact mechanism of PRL action at the molecular level has not been delineated. It has been suggested that PRL may control lipoprotein transport in the adrenal, ovary, and testis, thereby assuring a constant supply of cholesterol for steroidogenesis. In addition, PRL may stimulate the synthesis of enzymes of the adrenal androgen pathway which facilitate the conversion of pregnenolone to dehydroepiandrosterone (DHEA) and DHEA sulfate. (See Adrenocortical Steroids, p. 1057.) PRL may also stimulate in testes 3-β-ol-dehydrogenase activity, resulting in increased testosterone production and ovarian steroidogenesis, particularly of progesterone.

Table 9-6. CAUSES OF PROLACTIN ELEVATION

Chronic renal failure
Pregnancy
Breast stimulation or chest wall trauma
Primary hypothyroidism
Empty sella syndrome
Pituitary adenoma (microadenoma or macroadenoma)
   "Nonsecretory"
   With galactorrhea and amenorrhea or oligospermia
Idiopathic
Drugs
   Dopaminergic-blocking agents: phenothiazines, butyrophenones, benzamides
     (metoclopramide, sulpiride)
   Dopamine-depleting agents: α-methyldopa, reserpine
   Noncatecholamine-dependent agents: thyrotropin-releasing hormone, estrogens
Hypothalamic etiologies

Clinical Significance

Once the assay for PRL became widely available, it became apparent that patients with PRL elevations were being seen frequently.[140,232] There is still lack of agreement as to what constitutes a significant PRL elevation, what (if any) stimuli for PRL release or suppression give helpful information, and what tests help explain the cause of PRL elevations. PRL deficiency does not appear to have clinical importance except as an additional test of pituitary function in a patient already known to have pituitary disease.

Hyperprolactinemia. PRL levels may be elevated in patients who have no clinical manifestations but who have relatively subtle alterations of fertility such as (1) anovulation with or without menstrual irregularity or (2) amenorrhea and galactorrhea or (3) galactorrhea alone. PRL excess in males may present as oligospermia or impotence or both. As many as 30% of patients with clinically *silent pituitary adenomas* will have elevated PRL levels. The magnitude of the PRL elevation often correlates with the size of the tumor, and post-treatment levels correlate with clinical improvement. *Microadenomas* of the pituitary that produce PRL are commonly seen and may present as galactorrhea-amenorrhea or simply as infertility. Many consider a marked increase (> 5 times the upper reference limits) in the basal level of PRL strong evidence for presence of a pituitary tumor.[146] Intermediate elevations of PRL may be seen with pituitary macroadenomas, with microadenomas, or even without demonstrable endocrine pathology. Other causes of PRL elevation are shown in Table 9-6 and should be kept in mind when evaluating patients with PRL elevations. The administration of thyrotropin-releasing hormone (TRH) will stimulate PRL in normal individuals. (See Protocol E.) Failure of TRH to stimulate PRL is seen with pituitary adenomas.[148] This finding is not specific for pituitary adenomas; it may also be seen in women with hyperprolactinemia from other causes. Apparently, failure of TRH to stimulate PRL is simply a nonspecific characteristic of hyperprolactinemia.[148] Other pharmacologic agents have been studied as stimulants or suppressants of PRL (see Table 9-7), but none seem to offer any real advantage over measuring the basal level of PRL.

While basal gonadotropin levels are low in most hyperprolactinemic patients, the response

---

PROTOCOL E: TRH STIMULATION OF PROLACTIN RELEASE

Rationale: TRH normally stimulates release of prolactin from the pituitary gland.

Procedure: A baseline blood sample is drawn for prolactin determination. 500 μg TRH is given intravenously; prolactin is measured on a sample drawn 15 min later.

Interpretation: Prolactin should increase by > 20 ng/mL in males and > 40 ng/mL in females. Subjects with hypopituitarism will have a low baseline and a blunted rise. Patients with hyperprolactinemia will usually have a blunted rise after TRH regardless of the cause of hyperprolactinemia.

Table 9-7. AGENTS FOR PROLACTIN RELEASE AND SUPPRESSION

Stimulus	Dose	Time of Peak	Magnitude of Normal Response
TRH[148]	500 µg, IV	15 min	> 100%
Perfenazine	8 mg, orally	60–120 min	> 100%
Chlorpromazine[145]	25 mg, IM	2–3 h	> 100%
Insulin hypoglycemia[35]	0.1 unit regular insulin, IV	15–40 min	> 100%
Dompyridone[187]	4 mg, IV	30–60 min	> 100%
Metoclopramide[308]	10 mg, IV	1–2 h	> 100%
Nocturnal[181]	N/A	(ratio of 0200–1000 h values)	ratio > 1.2

Suppressants	Dose	Time of Nadir	Magnitude of Normal Response
L-Dopa[43]	500 mg, orally	3 h	Decrease to < 4 ng/mL
Nomifensine[187,133]	200 mg, orally	3 h	> 30% fall

to gonadotropin-releasing hormone (GnRH) is normal or increased, but the response to clomiphene, an agent that normally stimulates release of LH and FSH through release of GnRH, is blunted. This suggests that PRL inhibits the release of GnRH from the hypothalamus, resulting in a state of functional hypogonadotropism. Other pituitary function tests are usually normal. It has been reported that the usual nocturnal increase in PRL is blunted in patients with pituitary adenomas and hyperprolactinemia, but not in normal subjects or patients with the empty sella syndrome. A ratio of night-to-day PRL of > 1/2 is considered normal. The degree of PRL increase in patients with PRL-secreting pituitary adenomas is generally greater than is seen with other causes. However, none of these laboratory approaches is definitive in differentiating hyperprolactinemia due to nonpituitary causes from that due to pituitary disease.

If only a slight elevation of PRL is present in an initial sample, it is advisable to repeat the measurement on at least two other occasions, obtaining the sample with minimal trauma, no breast stimulation, and in the morning; the patient should not take any medication that stimulates PRL release.

A lateral skull X-ray will be sufficient to document a pituitary macroadenoma in most cases, but computed tomography (CT) of the sella is necessary to reveal the radiologic changes of a microadenoma. Microadenomas may be present with no radiologic abnormality. False positive CT scans are not uncommon; that is, an abnormal CT may suggest a microadenoma where no microadenoma is present.[36,43] Thus, there is no specific or definitive test for a prolactin-producing microadenoma.

Methods for the Determination of Prolactin

Principle

Before radioimmunoassays for PRL became available, the hormone was estimated by inconvenient, insensitive, and nonspecific *bioassays*; these assays are based on the action of PRL on the pigeon crop sac or the rat mammary gland, both of which also measure effects of growth hormone and human placental lactogen. The development of *radioimmunoassays* was hampered by extensive cross-reactivity and unavailability of purified prolactin. The use of a heterologous system in which the hormones used for immunization and labeling are derived from different species had been employed to minimize such problems. However, the isolation and purification of human PRL have made possible the development of homologous radioimmunoassays.[91,277] At present, a number of commercial kits are available for radioimmunoassay of PRL.

Like other protein hormones, PRL has also been measured by *radioreceptor assays*.[88] The radioiodinated PRL is allowed to bind to specific membrane receptors for PRL. The receptors are prepared from rabbit or mouse mammary gland or from rabbit liver. Biological lactogenic activity is determined by the ability of samples to inhibit the binding of radioiodinated PRL to these receptors. One of the major drawbacks of the radioreceptor assay is nonspecificity, since this method does not discriminate between PRL and growth hormone or chorionic somatomammotropin (formerly called human placental lactogen, hPL). However, since radioreceptor methods theoretically have a biological relevance to bioassays (see Mechanism of Action of Hormones, p.

1006), they are most useful when results by RIA appear to be at variance with the clinical effects of PRL.

CORTICOTROPIN (ACTH) AND RELATED PEPTIDES (β-LPH, ENDORPHINS, AND ENKEPHALINS)

Corticotropic cells of the anterior pituitary appear to synthesize a large precursor protein of M.W. 31 000, often called *pro-opiocortin* or "31 K precursor." The structure of this precursor is diagrammatically presented in Figure 9-10. The precursor is hydrolyzed to ACTH and β-lipotropin in the corticotropes before these two hormones are secreted. Where an intermediate pituitary lobe is present, ACTH is further hydrolyzed to α- and β-melanotropins (melanocyte-stimulating hormone, MSH), corticotropin-like intermediate-lobe peptide (CLIP), and other fragments. Since the intermediate pituitary lobe is rudimentary in adult humans, neither α- nor β-MSH is synthesized. The plasma values for MSH obtained by RIA are most likely due to cross-reactions of anti-MSH antiserum with β- and γ-LPH and ACTH. The skin pigmentary changes in several endocrine diseases (e.g., adrenal insufficiency) are probably due to the α-MSH activity of excess ACTH.

Beta-LPH contains the amino acid sequences of the endogenous opioid peptides: *α-endorphin* (amino acids 61–76), *γ-endorphin* (61–77), *β-endorphin* (61–91), and *enkephalin* (61–65). Although the sequence of enkephalin is an intrinsic part of the β-LPH, there is evidence to indicate that enkephalin is not the product of β-LPH breakdown but that it arises independently, presumably from another prohormone. These peptides are formed in the brain by neurons composing a distinct peptidergic system related to pain perception. The most active peptide, β-endorphin, is secreted in parallel with ACTH, but the physiologic significance of pituitary endorphin remains to be established. Enkephalins (pentapeptides—Tyr-Gly-Gly-Phe-Met and Tyr-Gly-Gly-Phe-Leu, generally referred to as met-enkephalin and leu-enkephalin, respectively) and endorphins are the endogenous opiates that have so far been isolated and characterized. Their discovery has been one of the most exciting episodes in modern biology.[121] The physiological actions of the endogenous opiates include sedation, an increase in the threshold of pain, changes in extrapyramidal motor activity (e.g., postural control), and possibly changes in behavior. Endorphins and enkephalins may also form a major peptidergic system in the brain involved in modifying endocrine responses to stress.

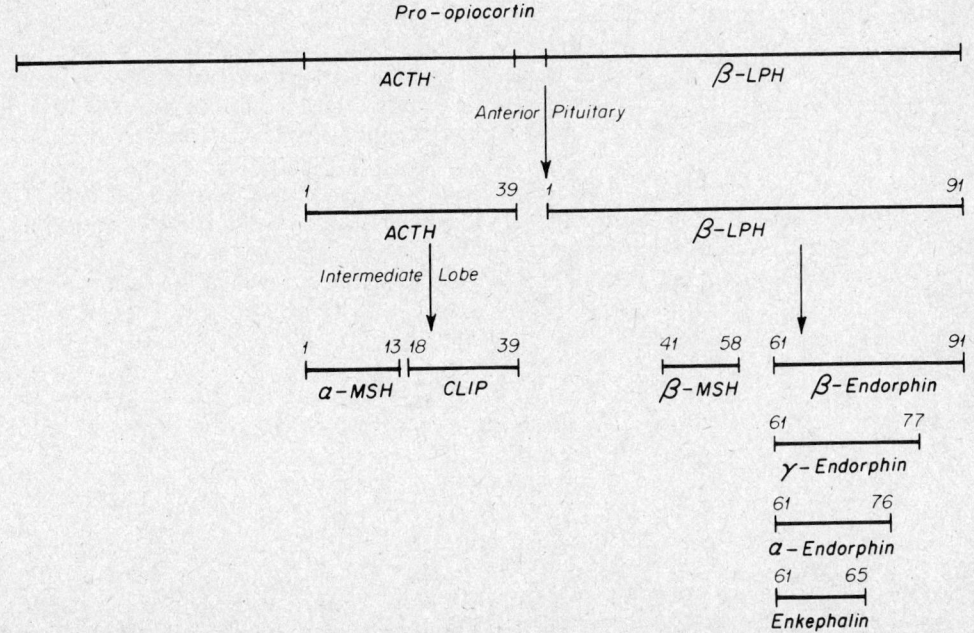

Figure 9-10. Diagrammatic representation of pro-opiocortin and its precursor relationship to ACTH, β-LPH, α- and β-MSH, and the endorphins.

Regulation of secretion, physiological action, and clinical significance of ACTH are described in the section on adrenocortical hormones.

Determination of ACTH by RIA

Principle

ACTH can be measured by bioassay, radioimmunoassay, or receptor assay. However, RIA is at present the method of choice for the analysis of plasma ACTH.[175] Radioimmunoassays have employed porcine or human synthetic ACTH and a variety of antibodies to the whole molecule or fragments thereof, particularly amino acids 1–24. Full biologic activity resides in the first 24 amino acids; the physiologic significance of residues 25–39 is unknown. Commercial reagent kits for the measurement of ACTH are now widely available, and many commercial reference laboratories offer ACTH assays.

The principles, precautions, and procedures for ACTH RIA have been discussed in a recent monograph.[1] Availability of highly specific antibody to ACTH has allowed determination of ACTH in unextracted plasma. The general procedure includes incubation of an aliquot of freshly collected or thawed plasma with ^{125}I-ACTH (tracer) and antibody for 16–24 h at 2–8 °C, followed by separation of bound from free tracer by a double antibody technique. After centrifugation the radioactivity in the precipitate is counted in a gamma scintillation counter. The levels of ACTH in the samples are determined by comparing the per cent bound radioactivity in the samples with that in the standard curves. The sensitivity of most of the methods lies between 10 and 20 pg/mL.

Specimen Collection and Storage

Proper precautions must be taken in the collection, transportation, and storage of samples. ACTH is adsorbed onto glass surfaces, is very unstable in whole blood, and is destroyed by proteolytic plasma enzymes that are activated during freezing and thawing of the sample. Factors that influence plasma ACTH levels, such as prior administration of corticosteroids, time of the day at which the sample is taken (diurnal variation), and stress resulting from a poorly performed venipuncture, must be taken into account.

The following procedure for sample collection is recommended. Blood samples should be drawn by an experienced phlebotomist using a lightly heparinized plastic syringe at a set time. If repeated serial samples are required, these should be obtained via an indwelling venous needle or catheter. High concentrations of heparin may interfere with antigen-antibody binding. If vacutainer EDTA tubes are used, they must be completely filled with blood in order to avoid too high EDTA concentrations; concentrations >1 mmol/L may interfere with the RIA. Blood should be collected into a plastic tube chilled in iced water and then centrifuged for 15 min at 1000–1500 × g at 4 °C. The supernatant should then be carefully aspirated, transferred to another plastic tube, and recentrifuged at 6000 × g at 4 °C for another 10 min. The second centrifugation is very important in reducing incubation damage; it removes additional formed elements that, if frozen and then thawed for assay, apparently rupture and release proteolytic enzymes that damage labeled and unlabeled ACTH. The plasma is decanted from the tube, stored at −20 °C, and transported in a polystyrene foam container filled with dry ice to the laboratory performing the assay.

Reference Ranges

$$pg/mL$$

0800 h: 25–100
1800 h: <25

PITUITARY GLYCOPROTEIN HORMONES (LH, FSH, TSH)

The glycoprotein hormones of the pituitary (LH, FSH, TSH) and of the placenta (hCG; see Chapter 19) are composed of two peptide chains, usually referred to as α- and β-subunits, each with a carbohydrate substituent group attached. The carbohydrate moiety, accounting for 15–31% of the molecular weight, includes fucose, mannose, galactose, glucosamine, galactosamine, and sialic acid. The α-subunits of these hormones are similar to one another and are interchangeable. For example, the hybrid molecule made up of the α-subunit of TSH and the β-subunit of LH possesses LH activity. Isolated α-subunits are devoid of biological activity. The β-subunits, which have greater differences in amino acid sequences between the various hormones, confer hormonal and immunological specificity. Isolated β-subunits may have little intrinsic biologic activity, but full activity is attained when α- and β-subunits are recombined. This suggests that the presence of both α- and β-subunits is important for specific receptor recognition and that the β-subunit is responsible for eliciting the specific biological response. The availability of specific antisera for β-subunits has led to the development of radioimmunoassay techniques for the specific measurements of these structurally similar glycoprotein hormones. (See hCG, Chapter 19, p. 1752.)

GONADOTROPINS (FSH, LH)

The gonadotropic cells of the anterior pituitary secrete two gonadotropins, follicle-stimulating hormone (FSH or follitropin, M.W. 30 000) and the luteinizing hormone (LH or luteotropin, M.W. 32 000). In the male, LH is often called interstitial cell–stimulating hormone (ICSH). Since these two hormones control the functional activity of gonads, they are grouped together under the generic term gonadotropins.

Physiological Action

In the female, FSH stimulates growth of the ovarian follicles and, in the presence of LH, promotes the secretion of estrogens by the maturing follicles. LH in the female causes release of the ovum from the ovarian follicle that has previously been ripened by FSH. Following ovulation, the transformation of the ruptured follicle into the corpus luteum (i.e., luteinization) and secretion of progesterone by the corpus luteum are also believed to be under the influence of LH and FSH. In the male, FSH stimulates spermatogenesis, and LH is responsible for the production of testosterone by the Leydig cells of the testes.

Regulation of LH and FSH secretion and their clinical significance in reproductive endocrinology are discussed in the Reproductive Endocrinology section, p. 1109.

Methods for the Determination of LH and FSH

Radioimmunoassays have largely replaced time-consuming bioassays. Many methods for the determination of FSH and LH in blood and urine have been described[1] and reliable commercial kits are available. Until recently, it has been difficult to obtain specific antibodies to FSH and LH because these two hormones possess immunologic similarities not only to each other but also to hCG and TSH. However, the use of highly purified human FSH and LH has made it possible to raise specific antisera showing as little as 1% cross-reactivity with structurally similar hormones.

Plasma or serum can be utilized for LH and FSH measurements. However, because of the episodic, circadian, and cyclic variations in the secretion of gonadotropins, a meaningful clinical evaluation of these hormones requires determinations in pooled blood samples, in multiple serial blood samples, or in a timed sample of urine.

RIA procedures for FSH and LH are similar to those described for other protein hormones. Samples, standards, and controls are incubated with ^{125}I-labeled hormone and antisera for 16–18 h at 4 °C or at room temperature. The antibody bound hormone is separated from free hormone by precipitation with a second antibody. The precipitate is counted in a gamma scintillation counter. The concentration of gonadotropins in the sample is determined by comparing radioactivity in the sample tubes to the standard curve prepared with known concentrations of gonadotropins.

Hormonal results are best expressed as the mass of the pure hormone per unit volume of sample (e.g., ng/mL) as measured in the RIA system. However, due to variation in the purity of the gonadotropins employed as the standard, most procedures express concentrations in terms of the International Units of the Second International Reference Preparation of Human Menopausal Gonadotropin (IRP-2-hMG). Use of these preparations aids the comparison of interlaboratory data.

Reference Ranges

Published FSH and LH ranges differ because of variations in method and technique. Each laboratory must establish its own normal and abnormal ranges. The following is an example of such ranges:

	FSH mIU/mL	LH mIU/mL
Female,		
Follicular phase:	2–10	0–14
Mid-cycle peak:	9–18	20–70
Luteal phase:	0–9	0–16
Postmenopause:	20–100	20–70
Male:	2–10	0–9

Mean FSH and LH levels in prepubertal boys and girls are approximately half those of adults.

THYROTROPIN

Thyrotropin (thyroid stimulating hormone, TSH) is a glycoprotein with an approximate M.W. of 26 600; it is secreted by the thyrotropic cells of the anterior pituitary. As previously discussed, the α-chain of TSH is similar to the α-chain of LH, FSH, and hCG. TSH has several effects on the anatomy and function of the thyroid gland. It stimulates the vascularity of the thyroid gland and hypertrophy of the follicular cells. Practically every step in thyroid hormone synthesis is stimulated by TSH, including the uptake of iodine, organic binding, coupling of tyrosines, and release of thyroid hormones. TSH may also have extrathyroidal effects such as induction of lipolysis in isolated tissues, but the physiological importance of such actions is unclear.

Regulation of secretion of TSH, its clinical significance, and its relation to thyroid disease are discussed under Thyroid Hormones, p. 1116.

Methods for the Determination of TSH

A great number of *bioassay* methods of varying sensitivity, reliability, and ease have been developed for the measurement of TSH; none of them is suitable for routine clinical application. The methods that appear to be most sensitive are *radioimmunoassays.*[1]

As in the case of LH and FSH, the antisera raised against highly purified human TSH have <1% cross-reactivity with structurally related glycoprotein hormones and are widely employed for RIA. The principle and procedures for TSH RIA are similar to those described for LH and FSH. Reliable RIA kits for TSH are now commercially available. Note, however, that the sensitivity of most of the currently available kits is insufficient to allow distinction of low values seen in primary hyperthyroidism from euthyroid subjects. The methodology, reference ranges, and clinical significance of TSH are discussed in the section on thyroid hormones.

POSTERIOR PITUITARY HORMONES (NEUROHYPOPHYSEAL HORMONES)

The posterior pituitary is part of a neurosecretory system consisting of the supraoptic and paraventricular nuclei in the hypothalamus, the tracts of nerve fibers from the cell bodies in the hypothalamic nuclei, and the posterior lobe of the pituitary. The cell bodies of the supraoptic and paraventricular nuclei synthesize and secrete two hormones, vasopressin and oxytocin. *Vasopressin* (antidiuretic hormone, ADH) is synthesized mainly in the supraoptic nuclei while oxytocin is produced in the paraventricular nuclei. Vasopressin is a nonapeptide of M.W. 1080, consisting of a cyclic hexapeptide and a tripeptide side chain. *Oxytocin* has a similar structure (see Figure 9-11) but with isoleucine instead of phenylalanine at position 3 and leucine instead

```
   ┌ S ─────────────── S ┐
   │                     │
Cys-Tyr-Phe-Gln-Asn-Cys-Pro-Arg-Gly-NH₂
 1   2   3   4   5   6   7   8   9
```

Arginine Vasopressin

```
   ┌ S ─────────────── S ┐
   │                     │
Cys-Tyr-Ile-Gln-Asn-Cys-Pro-Leu-Gly-NH₂
 1   2   3   4   5   6   7   8   9
```

Oxytocin

Figure 9-11. The amino acid sequences of arginine vasopressin and oxytocin.

of arginine at position 8. The structure of vasopressin (arginine vasopressin) in most animal species is identical to that of humans. The only known exceptions are the pig and hippopotamus, in which lysine replaces arginine (lysine vasopressin) in position 8.

Following their synthesis, both oxytocin and ADH become bound to specific proteins called neurophysins, which are synthesized in the same cell bodies. There are separate and specific neurophysins for both ADH and oxytocin. These molecular complexes are incorporated into granules that then migrate down the nerve axons as a result of axoplasmic flow (axonal transport). The nerve endings are intimately associated with the posterior pituitary capillaries, which carry the hormone into the bloodstream. Release of the hormones occurs with the arrival of the action potential at the nerve endings, which results in depolarization of the terminal membranes. The hormone-containing granules are thought to be released by reverse pinocytosis, also known as exocytosis.

Regulation of Secretion

ADH. The principal stimulus for ADH secretion is an increase in plasma osmolality. Hypertonic saline infusion and administration of hyperosmolar carbohydrate solutions stimulate ADH release. Osmoreceptors located in the hypothalamus and probably in the posterior pituitary respond to changes in plasma osmolality. Because the response to sodium-containing solutions is more consistent and greater than that observed with carbohydrate solutions, some believe that the posterior pituitary contains specific sodium receptors. Other important stimuli for ADH release include a reduction in plasma volume or arterial pressure or both; pain; stress; sleep; exercise; and drugs such as nicotine, morphine, and barbiturates. Hypo-osmolality, volume expansion, hypotension, exposure to cold, alcohol, and some drugs (e.g., phenytoin, glucocorticoids) inhibit the release of ADH.

Oxytocin. The primary stimulus for oxytocin release is suckling. Stimulation of tactile receptors located around the nipples of the breasts initiates an action potential that propagates along afferent nerve fibers through the spinal cord and midbrain to the hypothalamus. The cell bodies in the paraventricular nucleus are then stimulated, resulting in the episodic release of oxytocin. Stretch receptors in the uterus and possibly in the vaginal mucosa may also initiate action potentials in afferent nerve fibers, which ultimately stimulate the release of oxytocin from the posterior pituitary. Estrogens enhance the response of oxytocin to these stimuli. The influence of other parts of the brain on the release of oxytocin has been reported; emotional stress, for instance, inhibits lactation.

Physiological Action

ADH. The principal physiological function of ADH is the stimulation of water reabsorption from the renal tubular fluid. The blood level of ADH directly regulates water balance and indirectly regulates the concentration of osmotically active solutes, primarily sodium, in the extracellular fluid. ADH increases the volume of water reabsorbed by the renal tubules; consequently, the urine becomes concentrated and its volume decreases.

A further action of ADH is expressed by its alternative name, vasopressin. ADH, released in sufficient quantity, induces a generalized vasoconstriction that leads to a rise in arterial blood pressure. ADH is believed to play an important role in maintaining arterial blood pressure during significant blood loss.

Oxytocin is present in both males and females but its physiological effects are known only in females. Oxytocin stimulates the contraction of the uterus only in the estrogen-primed uterus. It is a useful therapeutic agent in inducing labor, but the physiological mechanism whereby it induces uterine contractions remains obscure. Oxytocin also stimulates the lactating mammary gland. Progestins are believed to inhibit the action of oxytocin.

Clinical Significance

Polyuric states. Deficient production or action of ADH results in polyuria due to the failure of the renal tubule to reabsorb solute-free water.[14] Under normal circumstances, urine output is largely dependent on fluid intake; thus, an arbitrary limit for normal urine output cannot be defined. When urine output is > 2.5 L/d, further investigation is usually indicated; with complete deficiency of ADH, urine output may approach 1 L/h. If thirst response is normal, increased ingestion of fluid (polydipsia) will follow. If access to water is not restricted, serum osmolality and electrolytes will usually remain normal.

Polyuric states can be divided into three general categories: (1) deficient ADH production (central diabetes insipidus), (2) deficient ADH action on the kidney (nephrogenic diabetes insipidus), and (3) excessive water intake (psychogenic polydipsia).

Central diabetes insipidus can be caused by destructive or infiltrative lesions involving the hypothalamus or pituitary, among them tumor, hemorrhage, infarction, or infection, or by head trauma. Often, central diabetes insipidus occurs without apparent cause (idiopathic); idiopathic diabetes insipidus may be familial. Central diabetes insipidus is due to deficiency of ADH, which may be partial or complete.

Nephrogenic diabetes insipidus may be due to a variety of conditions affecting the kidney; they include amyloidosis, multiple myeloma, chronic electrolyte abnormalities, as well as medications including lithium, demeclocycline, methoxyflurane, propoxyphene, or colchicine. Nephrogenic diabetes insipidus may also be seen in the absence of these factors (idiopathic), and this form may be inherited. Both central and nephrogenic diabetes insipidus may be partial[205] or complete, transient or permanent. Nephrogenic diabetes is due to an inability of the renal tubule to respond to normal or increased concentrations of ADH.

Psychogenic polydipsia may be difficult to distinguish from forms of diabetes insipidus. Patients with psychogenic polydipsia generally have significant psychiatric illness. They may deny that their intake of liquids is excessive. In many cases, it may be difficult to determine whether the primary problem is due to excessive fluid intake or to uncontrolled water loss as a result of deficient ADH action.

Diagnostic studies. Since patients with diabetes insipidus may be extremely sensitive to water restriction, diagnostic studies are best performed in the hospital under close supervision. Prior to hospitalization the patient should collect urine for two separate days on ad lib fluid intake to document polyuria, urine volume > 2.5 L/day. Creatinine in these collections should be measured to provide an index of the adequacy of the collection. Substances such as nicotine, alcohol, and caffeine that interfere with the action of ADH must be avoided.

The traditional approach to the diagnosis of polyuric disorders is the water deprivation test.[205,245] (See Protocol F.) During a period of water deprivation, patients with central or nephrogenic diabetes insipidus lose free water, with the result of weight loss, hypernatremia, and hyperosmolality of plasma. Urine osmolality remains low, often less than that of plasma. There may be a slight increase in urine osmolality compared to baseline as a result of osmotic diuresis, particularly if the patient loses > 3% of body weight during water deprivation. Patients with psychogenic polydipsia may show a persistently hypotonic urine during water deprivation, but they will not develop hypernatremia or hyperosmolality.[256] Water deprivation should be continued until the patient shows a stable urine osmolality (a change of < 30 mOsm/kg on two consecutive hourly samples). If maximum urine osmolality is < 400 mOsm/kg, some form of diabetes insipidus is likely. At this point, 5 units of aqueous vasopressin may be given subcutaneously and urine

PROTOCOL F: OVERNIGHT WATER DEPRIVATION TEST

Rationale: Water deprivation provides a maximal stimulus for ADH release. This can be assessed indirectly by measuring urine osmolality or directly by measuring plasma ADH. If urine remains hypo-osmolal during water deprivation, the administration of ADH can help differentiate central diabetes insipidus from nephrogenic diabetes insipidus.

Procedure: The patient is weighed at 2200 h, at which time plasma and urine samples are collected for the determination of sodium concentrations and osmolality. No oral intake is allowed until the test is terminated. Care is taken to be sure the patient's body weight does not fall by $\geq 10\%$. Beginning at 0600 h, the patient is again weighed and urine collected hourly for measurement of volume and determination of osmolality. When urine osmolality is stable (a change of ≤ 30 mOsm/kg for 2 consecutive hours, which usually takes 8–12 h to occur), samples are collected for plasma osmolality (and plasma ADH, if desired). Five units of aqueous vasopressin (ADH) is given subcutaneously, and urine osmolality is measured 1 h later. The test is then terminated.

Interpretation: Normal individuals will lose $\leq 3\%$ of body weight, will not develop an elevated plasma sodium or osmolality, and will produce a urine with osmolality ≥ 400 mOsm/L, with no further increase in urine osmolality after ADH administration. Patients with *central diabetes insipidus* may show an increase in plasma osmolality and plasma sodium; their urine will be less concentrated than normal (≤ 400 mOsm/kg; but often less than plasma), and urine osmolality will increase $\geq 10\%$ after ADH administration. Patients with *nephrogenic diabetes insipidus* will show plasma and urine osmolalities similar to those seen with central diabetes insipidus during water deprivation; however, they will not show a significant increase in urine osmolality after ADH injection. In patients with *psychogenic polydipsia,* a normal response may be seen, but the period of water deprivation may need to be prolonged. However, some of them may show responses suggestive of diabetes insipidus. Measuring plasma ADH in response to hypertonic saline (see Protocol G) may be needed if the diagnosis remains unclear at this point.

osmolality measured 60 min later. An increase in urine osmolality of $> 10\%$ would indicate central diabetes insipidus, whereas failure to increase urine osmolality after ADH injection would indicate nephrogenic diabetes insipidus. Patients with psychogenic polydipsia will usually produce a normal urine osmolality after water deprivation, but some patients with psychogenic polydipsia will fail to produce a concentrated urine and may show a variable response to ADH injection.

Measurement of serum ADH concentration will help to separate these difficult cases.[343] Water deprivation until urine osmolality has stabilized is felt to be a maximum stimulus for ADH release. Patients with central diabetes insipidus will show low or inappropriately "normal" values for ADH in the presence of hyperosmolality. Patients with nephrogenic diabetes insipidus will have high plasma ADH levels but will continue to produce hypotonic urine. Patients with psychogenic polydipsia will have low or normal levels of plasma ADH, and their plasma will not be hyperosmolal. Figure 9-12 shows the relationships between plasma ADH levels and concurrent plasma osmolality. Since the best discrimination occurs when the patient is hyperosmolal, an infusion of hypertonic saline may be given to produce a hyperosmolal state (Protocol G). A systematic approach to the diagnosis of polyuric disorders is shown in Table 9-8.

Syndrome of inappropriate ADH secretion (SIADH). Excess ADH causes retention of free water by the kidney, which leads to volume expansion and dilutional hyponatremia. SIADH may

Table 9-8. DIAGNOSIS OF DIABETES INSIPIDUS

Document polyuria (urine volume ≥ 2.5 L/d) and exclude glycosuria.

If plasma osmolality is ≥ 295 mOsm/kg or serum sodium ≥ 145 mmol/L, primary polydipsia is unlikely; proceed to Protocol F or Protocol G.

Overnight water deprivation test (Protocol F): If the ratio of urine to plasma osmolality is < 1.5 at the end of the test, primary polydipsia is unlikely. Measure plasma and urine osmolality and plasma ADH at the end of the test; use these relationships to differentiate normal, nephrogenic or central diabetes insipidus, and psychogenic polydipsia. If urine osmolality is < 400 mOsm/kg at end of test, give 5 units of aqueous vasopressin subcutaneously. If urine osmolality increases $\geq 10\%$, central diabetes insipidus is probable; if urine osmolality does not increase, nephrogenic diabetes insipidus is highly probable.

Hypertonic saline infusion (Protocol G): Plot plasma osmolality vs plasma ADH. (See Figure 9-12.)

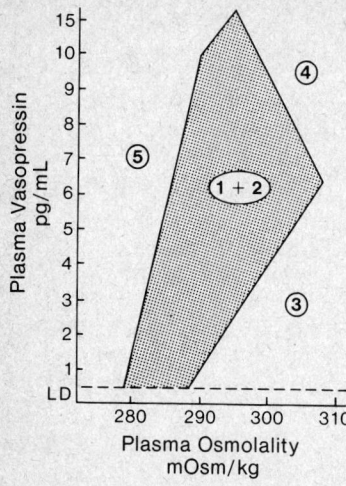

Figure 9-12. Relation of plasma vasopressin to osmolality. Regions 1 + 2 represent the normal range and values seen in subjects with psychogenic polydipsia. Region 3 indicates central diabetes insipidus, 4, nephrogenic diabetes insipidus, and 5, syndrome of inappropriate ADH secretion. LD denotes the lowest detectable plasma vasopressin correlation with this assay. (Modified and abstracted by permission of Zerbe, R. L., and Robertson, G. L.: N. Engl. J. Med., *305:*1539, 1981.)

be caused by a variety of acute and chronic disorders of the central nervous system or by pulmonary disorders, malignancies, and other illnesses. Other conditions that cause dilutional hyponatremia and that must be differentiated from SIADH include congestive heart failure, renal insufficiency, chronic liver disease, glucocorticoid deficiency, and hypothyroidism as well as excessive administration of hypotonic fluids, diuretics, and drugs that stimulate or potentiate ADH (e.g., chlorpropamide, vincristine, clofibrate, carbamazepine, nicotine, phenothiazines, cyclophosphamide).[206] Hyponatremia may also occur from depletion of total body sodium content as a result of mineralocorticoid deficiency, renal sodium wasting, or excessive gastrointestinal losses of sodium in excess of water.

The clinical manifestations of *hyponatremia* are nonspecific. These include weakness and central nervous system changes such as lethargy, confusion, coma, and convulsions. There are no signs or symptoms that are specific for SIADH. History, physical examination, and routine laboratory tests often indicate whether hyponatremia is dilutional or depletional and may indicate the presence of one of the causes of hyponatremia listed above. It may be difficult to differentiate SIADH from subtle presentations of these conditions.

Measurement of sodium and osmolality in plasma and urine will usually permit this differentiation. The typical patient with SIADH will have hypo-osmolal plasma (<275 mOsm/kg), urine osmolality that slightly exceeds plasma (300–400 mOsm/kg), and a urine sodium concentration that is neither high nor low (often 40–80 mmol/kg). Patients with hyponatremia from excess water intake will have hypotonic plasma, urine sodium concentration usually >80 mmol/kg, and urine osmolality less than plasma. Patients with hyponatremia from total body sodium depletion will have hypotonic plasma, a low urine sodium concentration (usually <20 mmol/kg), and urine osmolality greater than serum (often >600 mOsm/kg).

PROTOCOL G: SALINE INFUSION TEST FOR DIABETES INSIPIDUS

Rationale: An increase in plasma osmolality is the strongest stimulus for release of ADH. The administration of hypertonic saline intravenously will produce a hyperosmolal state; plasma ADH can be plotted against plasma osmolality and compared with a nomogram.

Procedure: An intravenous infusion of 3% saline is begun at a minimum rate to establish flow; two baseline blood samples are drawn 15 min apart. The flow rate is then adjusted to 0.1 mL/kg/min and maintained with an infusion pump for 2 h. Blood samples are collected every 15 min during this 2-h period. Each sample is then analyzed for plasma osmolality and ADH.

Interpretation: Normal subjects will show a plasma ADH of >7 pg/mL when plasma osmolality is ≥ 310 mOsm/kg. Abnormal response will fall, depending on probable cause, into regions 3, 4, or 5 shown in Figure 9-12. Figure 9-12 shows typical responses for patients with disorders of ADH secretion or action.

PROTOCOL H: RAPID ACTH (COSYNTROPIN) TEST

Rationale: The administration of ACTH to normal subjects results in a rapid rise in plasma cortisol. Patients with adrenal destruction (Addison's disease) will show no change in plasma cortisol after ACTH. Patients with disuse atrophy of the adrenal cortex due to exogenous glucocorticoids or dysfunction of the pituitary or hypothalamus may show a slight rise in plasma cortisol, but not of normal magnitude.

Procedure: A baseline blood sample is drawn for determination of plasma cortisol. 250 μg cosyntropin (1–24 ACTH) is given IM or IV. Further samples for plasma cortisol determination are drawn 30 and 60 min after injection.

Interpretation: A normal rise in plasma cortisol is >7 mg/dL, and a normal peak plasma cortisol is >20 mg/dL.

Laboratory findings in patients with adrenocortical insufficiency may be similar to those of SIADH, since these patients have some renal sodium wasting. However, patients with adrenocortical insufficiency are volume depleted; volume depletion is an appropriate stimulus for ADH secretion. If adrenal insufficiency is suspected, a specific diagnostic test should be done. (See Protocol H.)

If the cause for hyponatremia remains unclear after determination of sodium concentration and osmolality in plasma and urine, a water loading test (see Protocol I) may be done. Patients with SIADH show a reduced ability to excrete a water load. Measurement of plasma ADH is not usually needed in these patients, but plasma ADH would be expected to be high relative to the patient's hypo-osmolal state. A systematic approach to the diagnosis of hyponatremic disorders is shown in Table 9-9.

Methods for the Determination of ADH and Oxytocin

Numerous radioimmunoassay methods for measuring ADH or oxytocin have been described.[129] However, their routine clinical application has been hampered because of their complexity and the lack of sensitivity and specificity. With most assays, a preliminary extraction procedure is required, not only to concentrate the minute amount of hormone that is present in the sample but also to remove some of the cross-reacting substances. The clinical significance of oxytocin measurements is obscure. Furthermore, measurements of serum and urine osmolality are sufficient in most clinical situations for evaluating patients with ADH problems. If measurement of ADH is important for diagnostic purposes, then the sample should be analyzed in a laboratory that provides a detailed characterization of the method employed. Samples should be collected, processed, and transported according to the particular specifications of the laboratory. The results should be compared with reference ranges established by the same laboratory, because the same reference standard is not used by all.

PROTOCOL I: WATER LOADING TEST FOR SIADH

Rationale: Inappropriate excess of antidiuretic hormone leads to hyponatremia by interfering with the ability of the kidneys to excrete a water load normally.

Procedure: The test is started in the morning 2 h after a light breakfast. Plasma and urine osmolality are measured. The patient is given water to drink (20 mL/kg) over a 15–30 min period; lightly salted crackers may be given with the water if needed. The patient is kept in a recumbent position and samples taken hourly for the next 4 h for plasma and urine osmolality. Total urine output is measured.

Interpretation: Plasma osmolality should decrease by ≥5 mOsm/L and urine osmolality should drop to ≤100 mOsm/kg, with ≥90% of the water load excreted in 4 h. SIADH is characterized by <90% of the water load being excreted and urine osmolality remaining >100 mOsm/kg. Plasma ADH may be measured at 90 or 120 min after the water load to confirm the diagnosis.

Table 9-9. DIAGNOSIS OF SIADH

Document plasma hypo-osmolality (≤ 270 mOsm/kg) and hyponatremia (sodium ≤ 130 mmol/L).
Use history, physical exam, and appropriate tests to exclude cardiac, hepatic, renal, or adrenal failure or hypovolemia. (SIADH cannot be diagnosed unless these factors are corrected.)
Measure urine sodium and osmolality.
　Low urine sodium (often ≤ 20 mmol/L) means total body sodium depletion.
　Urine osmolality less than plasma osmolality means dilutional hyponatremia.
　Urine osmolality greater than plasma osmolality *and* without corresponding low urine sodium (usually > 60 mmol/L) means SIADH is probable.
If further tests seem needed:
　Water loading test (Protocol I). Normal results exclude SIADH.
　Measure plasma ADH and plasma renin. SIADH is characterized by high ADH and low renin. If both are low, the problem lies with a primary defect in renal water excretion.

STEROID HORMONES

　　Steroids are compounds containing the cyclopentanoperhydrophenanthrene ring system (Figure 9-13). The three six-sided rings (A, B, C) constitute the phenanthrene nucleus to which is attached a five-sided ring (D), cyclopentane. The prefix "perhydro" refers to the fact that all the necessary hydrogen atoms have been added to the compound to make it fully saturated. This class of compounds includes such natural products as sterols (e.g., cholesterol), bile acids (e.g., cholanic acid), sex hormones (e.g., estrogens, androgens), corticosteroids, cardiac glycosides (e.g., digitoxigenin, the aglycone of digitoxin), sapogenins (e.g., tigogenin), and some alkaloids (e.g., solasodine). The steroid hormones with which we are concerned contain up to 21 carbon atoms (C_{21} steroids), numbered as shown in Figure 9-13. Each carbon atom of a ring bears two hydrogen atoms, except when it is common to two rings, in which case it bears only one hydrogen atom (i.e., at C-5, C-8, C-9, and C-14). C-17 bears one hydrogen atom, while C-10 and C-13 are bound only to other carbons (i.e., C-19 and C-18, respectively). The carbon atoms composing the rings and the hydrogen atoms attached to them are not usually written into the structure unless it is required to draw special attention to configuration. Furthermore, in all naturally occurring steroid hormones, the projected solid line from the carbon atom at position 10 or 13 usually designates the presence of an angular (—CH$_3$) group, unless otherwise indicated.

　　Steroids consist of tetracyclic rings and are three-dimensional. Thus, the constituent carbon atoms and the hydrogen atoms or their substituents lie in different planes, giving rise to *isomers*. The direction of the hydrogen atoms, the substituents, and the side chain plays a much more important role in the distinction of various isomers of the steroid compounds than does the relative position of the carbon atoms in the rings. Thus, the isomers resulting from fusion of two rings are decided on the basis of the spatial relationship between the hydrogen atoms or the substituents at common carbon atoms. When rings A and B are fused, two isomers are possible, depending upon whether the hydrogen atom at C-5 and the methyl group at C-10 are on the same or the opposite side of the plane of the rings. If the hydrogen atom points in the same direction as the angular methyl group at C-10, the compound is said to be the *cis* or *normal*

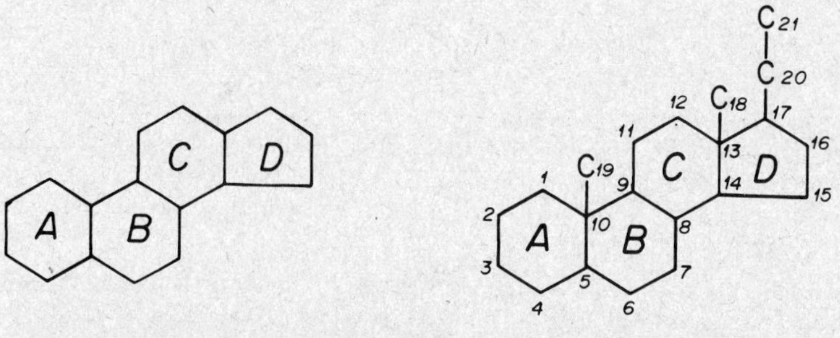

Cyclopentanoperhydrophenanthrene

Figure 9-13. Common features and numbering system of steroids.

Figure 9-14. Fusion of rings A and B in naturally occurring steroids.

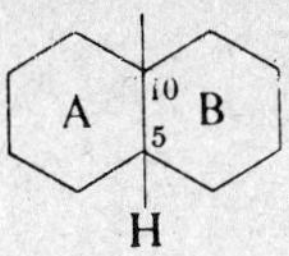

Cis-

Hydrogen atom and methyl group are on the same side (above the plane of the paper —5β-isomer).

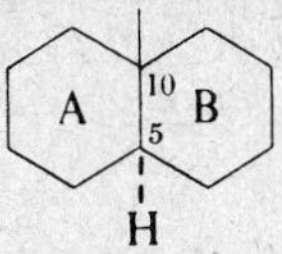

Trans-

Hydrogen atom is on the opposite side (below the plane of the paper—5α— isomer).

form. If, however, they are on the opposite sides, the compound is said to be the *trans* or *allo* form. While the rings A and B may be either *cis* or *trans*, the rings B/C and C/D have *trans* configuration in all naturally occurring steroid hormones.

The two methyl groups attached to C-10 and C-13 lie above the plane of the molecule and are customarily the points of reference for describing the spatial orientation of other substituents in the steroid nucleus. Substituents on the same side as these two methyl groups are said to possess a β-configuration, which is indicated by a solid line (—) joining them to the appropriate carbon atoms in the nucleus. Substituents on the opposite side are attached by a broken line (- - -) to denote an α-configuration. Thus, in the structures shown in Figure 9-14 when the hydrogen substituent at C-5 is *cis*, the isomer is the 5β-isomer, and when it is *trans* the isomer is accordingly the 5α-isomer. Similarly, the substituents at C-3, C-11, C-17, or any other carbon atoms are indicated as either α- or β-configuration, depending on their spatial orientation relative to these methyl groups (C-10 and C-13).

The innumerable steroids containing the cyclopentanoperhydrophenanthrene nucleus differ from one another by the introduction of double bonds between certain pairs of carbon atoms, by the introduction of substituents for the hydrogen atoms, or by the addition of a specific type of side chain. On the basis of such structural characteristics, the steroidal compounds are classified as derivatives of certain parent hydrocarbons, namely *estrane* (for estrogens, Figure 9-15), *androstane* (for androgens, Figure 9-16), and *pregnane* (for corticosteroids and progestins, Figure 9-17).

It should be noted that the parent substance *estrane* lacks one methyl group at C-10, and hence it is a C_{18} compound. Furthermore, estrogens are actually derivatives of the compound estratriene, since the benzenoid ring structure is a common feature of all the naturally occurring estrogens.

The parent substance *androstane* is a C_{19} compound and possesses 5α- or 5β-configuration.

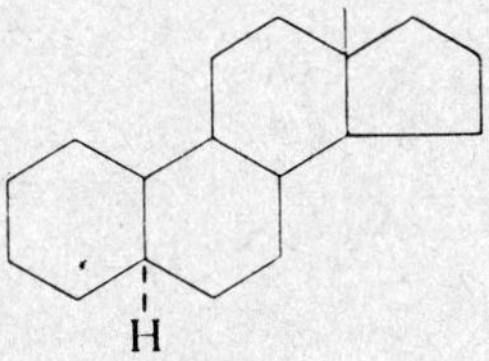

5α-Estrane

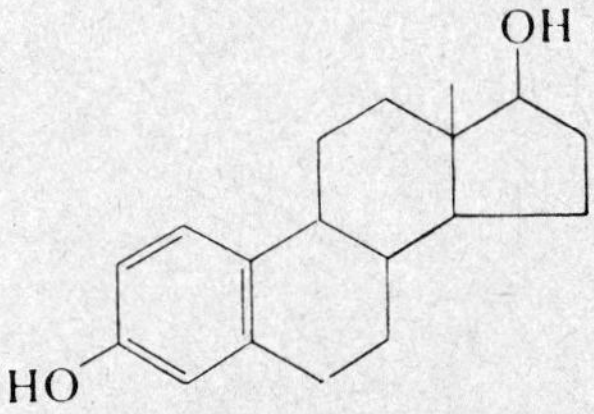

Estra-1,3,5(10)-triene-3,17β-diol (Estradiol-17β)

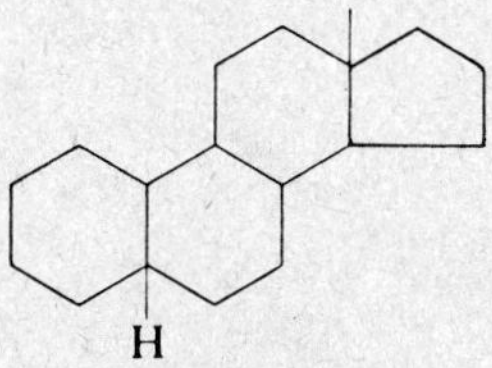

5β-Estrane

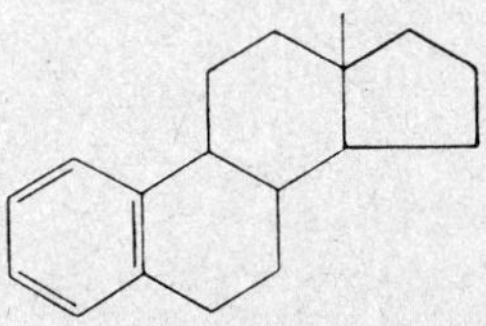

Estra-1,3,5(10)-triene (Estratriene)

Figure 9-15. Parent hydrocarbons of estrogens.

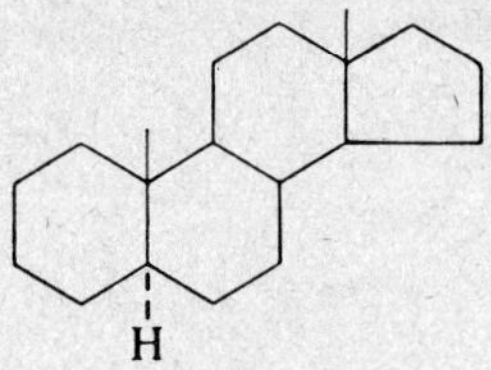

5α-Androstane

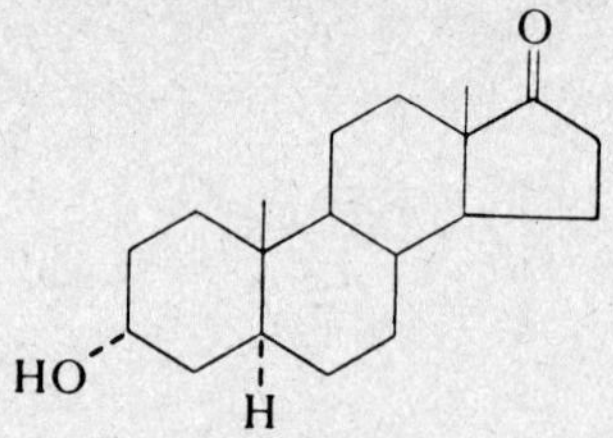

3α-Hydroxy-5α-androstan-17-one
(Androsterone)

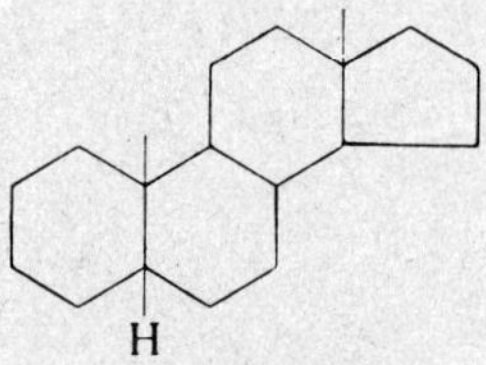

5β-Androstane
(Etiocholane)

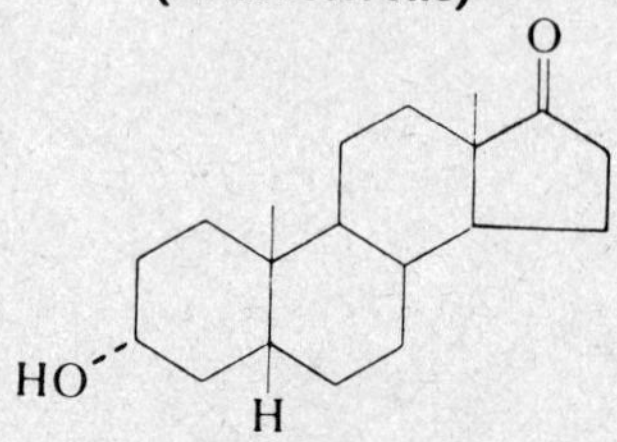

3α-Hydroxy-5β-androstan-17-one
(Etiocholanolone)

Figure 9-16. Parent hydrocarbons of androgens.

Naturally occurring *androsterone* and *etiocholanolone* are the examples of the respective derivatives of these isomers.

The special feature of the hydrocarbon *pregnane* is an ethyl side chain (—CH₂—CH₃) attached to C-17, making it a C_{21} compound. The side chain is in *cis* relationship to the methyl groups at C-10 and C-13 and is therefore β-oriented. In all naturally occurring steroids or pregnane derivatives, the side chain is β-oriented and the hydrogen or its substituent at this position is always α-oriented. Consequently, in recent steroid nomenclature, the spatial orientation of the substituent at this position of the pregnane derivative is no longer specified as α (e.g., cortisol: 1β,17,21-trihydroxypregn-4-ene-3,20-dione). When the configuration at position 20 in the side chain of a pregnane derivative is as depicted in the projection formula (e.g., pregnanediol), substituents shown to the right of C-20 are termed α, and those to the left are termed β. Like

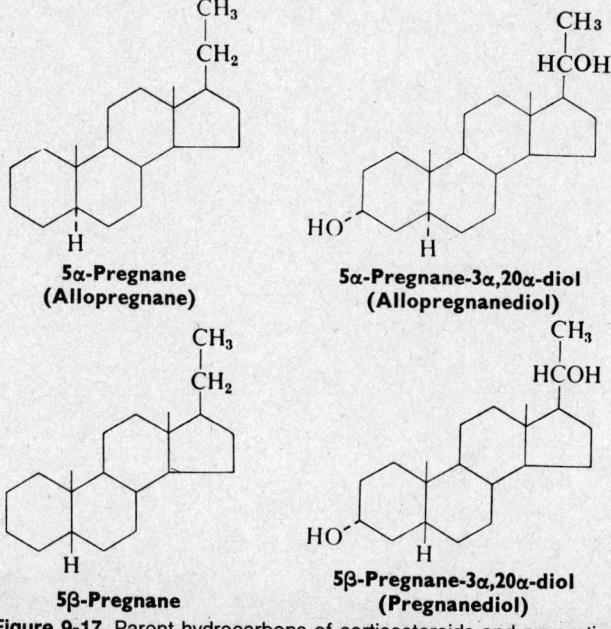

5α-Pregnane
(Allopregnane)

5α-Pregnane-3α,20α-diol
(Allopregnanediol)

5β-Pregnane

5β-Pregnane-3α,20α-diol
(Pregnanediol)

Figure 9-17. Parent hydrocarbons of corticosteroids and progestins.

androstane, this parent substance also has two isomers, 5α- and 5β-pregnane. *Allopregnanediol* and *pregnanediol* are the respective derivatives of these two isomers.

It should be noted that the prefix *allo* refers only to stereoisomerism of the hydrogen atom at C-5. When the configuration differs at any other carbon atom, the prefix *epi* is used; for example, androsterone possesses a 3α-hydroxyl group, and epiandrosterone a 3β-hydroxyl group; testosterone possesses a 17β-hydroxyl group, and epitestosterone a 17α-hydroxyl group.

To describe a compound with chemical nomenclature, a variety of other *suffixes* and *prefixes* are used. The suffix *-ane* indicates a fully saturated compound (e.g., pregn*ane*); *-ene,* the presence of one double bond (e.g., pregn*ene*); *-diene,* two double bonds; and *-triene,* three double bonds; the terminal "e" is omitted before a vowel, e.g., -4-en-3β-ol. The position of the double bond is indicated by the number of the carbon atom from which it originates, and it is understood to terminate at the next higher carbon atom (i.e., 4-ene means that a double bond lies between C-4 and C-5). However, when an alternative is possible (for example, a double bond originating at C-8 in the *estrane* nucleus can terminate at C-9 or C-14), the number of the carbon atom at which the bond ends is written in parentheses. Thus, a double bond at C-8 terminating at C-14 is designated as 8(14)-ene. A formerly used prefix for a double bond is the symbol Δ with a superscript indicating the position of the double bond (e.g., Δ^5). An alcohol (—OH substituent of the nucleus) is indicated by the suffix *-ol* (two alcohol groups as *-diol,* three as *-triol,* etc.), or by the prefix *hydroxy* or *oxy* (dihydroxy for two, trihydroxy for three, etc.). Ketones ($>C=O$) are identified by the suffixes *-one* for one keto group, *-dione* for two keto groups, etc., or by the prefix *oxo-.* (See Table 9-10.)

In naming a compound containing double bonds, hydroxyl groups, and ketones, priorities are given to the use of suffixes and prefixes. Thus, hydroxyl groups are indicated by the prefix followed by the suffixes for other substituents. Accordingly, the systematic name of dehydro-epiandrosterone (Figure 9-18) is written as 3β-hydroxy-androst-5-en-17-one. Note that to denote the bond of unsaturation, the first part of the parent hydrocarbon is followed by the position of the bond and the suffix (i.e., androst-5-en). If the prefix Δ for unsaturation is chosen, then suffixes for both the hydroxyl and the ketone groups are used, e.g., Δ^5-androsten-3β-ol-17-one. When there is only one kind of substituent, the use of a suffix is customary, e.g., pregnanediol (5β-pregnane-3α,20α-diol), androstenedione (androst-4-ene-3,17-dione).

The usefulness of the systematic name of a compound lies in the fact that it gives information about the parent substance, the position of unsaturation, and the nature, position, and orientation of substituents. The trivial name, as the term suggests, conveys little or no information about the chemical origin and characteristics of a compound, e.g., cortisol, progesterone, testosterone. The trivial and systematic names of some of the important steroid hormones are given in Table 9-11.

Table 9-10. COMMON SUFFIXES AND PREFIXES FOR STEROIDS

Suffix or Prefix	Definition
Suffix	
-al	Aldehyde group
-ane	Saturated hydrocarbon
-ene	Unsaturated hydrocarbon
-ol	Hydroxyl group
-one	Ketone group
Prefix	
hydroxy- (oxy-)	Hydroxyl group
keto- (oxo-)	Ketone
deoxy- (desoxy-)	Replacement of hydroxyl group by hydrogen
dehydro-	Loss of two hydrogen atoms from adjacent carbon atoms
dihydro-	Addition of two hydrogen atoms
cis-	Spatial arrangement of two substituents on the same side of the molecule
trans-	Spatial arrangement of two substituents on opposite sides of the molecule
α-	Substituent which is *trans* to the methyl group at C-10
β-	Substituent which is *cis* to the methyl group at C-10
epi-	Isomeric in configuration at any carbon atom except at the junction of two rings
Δ^n-	Position of unsaturated bond

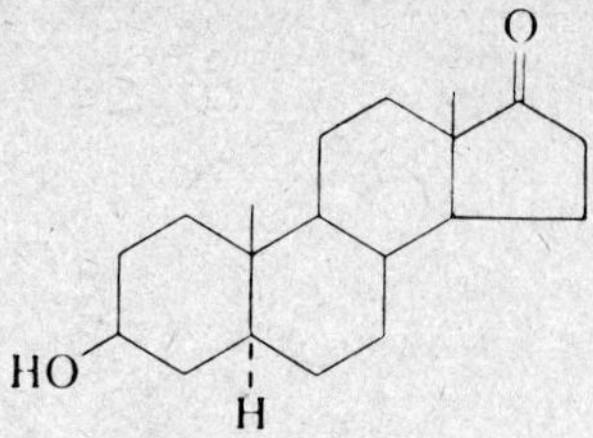

3β-Hydroxy-5α-androstan-17-one
(Epiandrosterone)

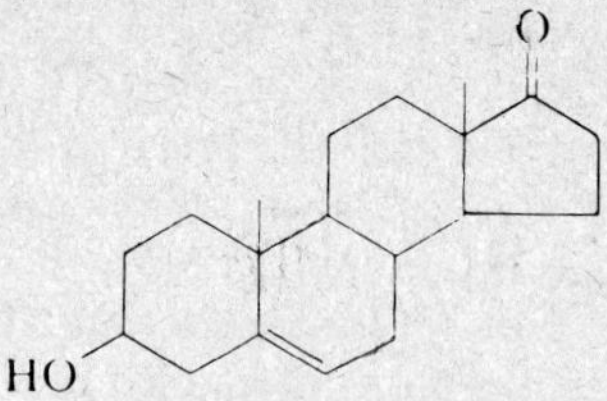

3β-Hydroxyandrost-5-en-17-one
(Dehydroepiandrosterone)

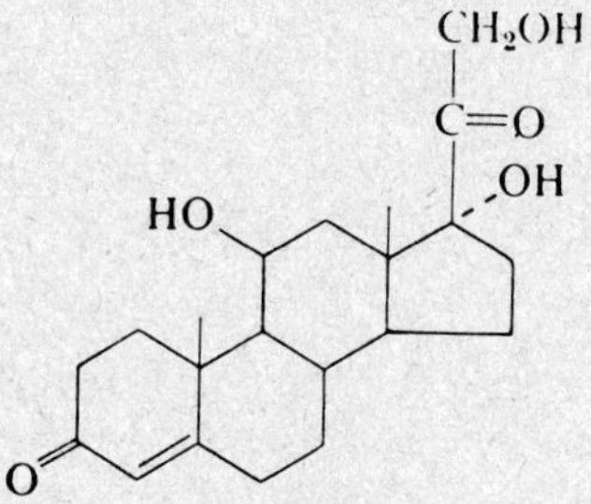

11β,17,21-Trihydroxypregn-4-ene-3,20-dione
(Cortisol)

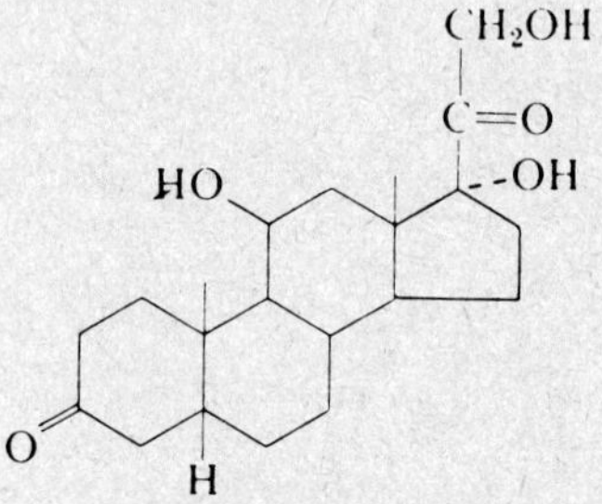

11β,17,21-Trihydroxy-5β-pregnane-
3,20-dione
(Dihydrocortisol)

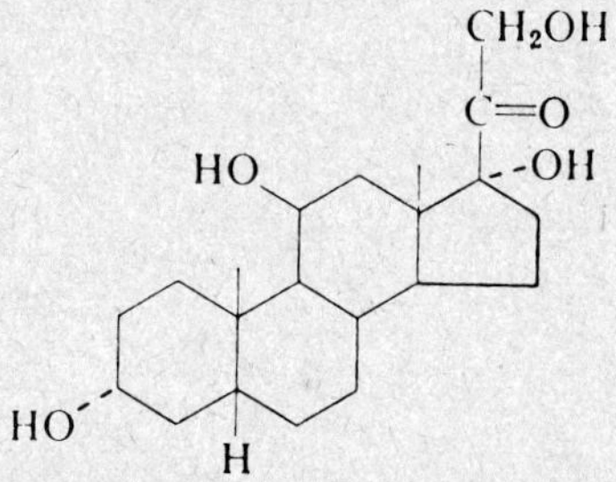

3α,11β,17,21-Tetrahydroxy-5β-pregnan-20- one
(Tetrahydrocortisol, Urocortisol)

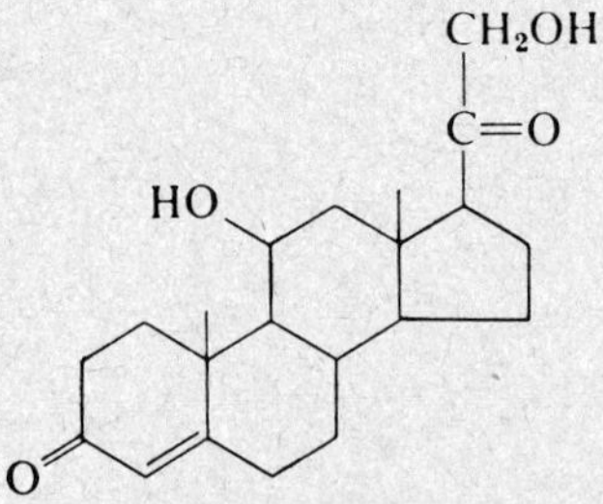

11β,21-Dihydroxypregn-4-ene-3,20-dione
(Corticosterone)

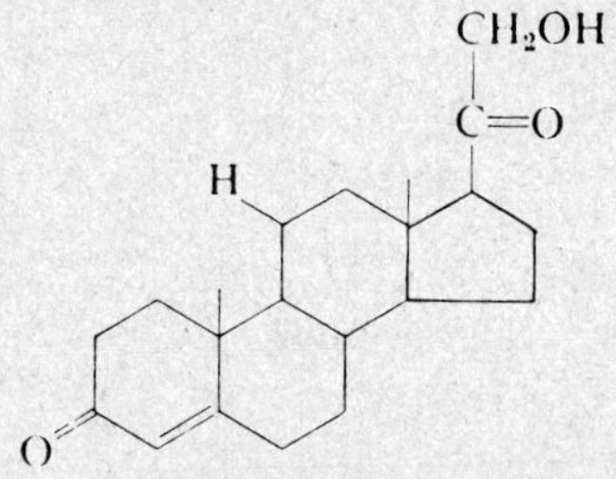

21-Hydroxypregn-4-ene-3,20-dione
(11-Deoxycorticosterone)

Figure 9-18. Illustration of semitrivial names.

In addition to the usual suffixes and prefixes just discussed, there are some special prefixes which are generally used for the semitrivial names of the compounds. Thus, the prefix *dehydro* is used to indicate the loss of two hydrogen atoms from adjacent carbon atoms with the formation of a double bond, e.g., dehydroepiandrosterone (Figure 9-18). The prefix *dihydro-* or *tetrahydro-*

Table 9-11. TRIVIAL AND SYSTEMATIC NAMES OF SOME IMPORTANT STEROID HORMONES

Trivial Name	Systematic Name
Estrone	3-Hydroxyestra-1,3,5(10)-trien-17-one
Estradiol-17β	Estra-1,3,5(10)-triene-3,17β-diol
Estriol	Estra-1,3,5(10)-triene-3,16α,17β-triol
Testosterone	17β-Hydroxyandrost-4-en-3-one
Androsterone	3α-Hydroxy-5α-androstan-17-one
Etiocholanolone	3α-Hydroxy-5β-androstan-17-one
Dehydroepiandrosterone	3β-Hydroxyandrost-5-en-17-one
Adrenosterone	Androst-4-ene-3,11,17-trione
Progesterone	Pregn-4-ene-3,20-dione
Pregnanediol	5β-Pregnane-3α,20α-diol
Cortisol	11β,17,21-Trihydroxypregn-4-ene-3,20-dione
Urocortisol (tetrahydro F)	3α,11β,17,21-Tetrahydroxy-5β-pregnan-20-one
Aldosterone	11β,21-Dihydroxy-3,20-dioxopregn-4-en-18-al

indicates the addition of two or four hydrogen atoms to the molecule, respectively, as in dihydrocortisol and tetrahydrocortisol (Figure 9-18). The replacement of a hydroxyl group by hydrogen (COH → CH) is prefixed by *deoxy-* (or *desoxy-*), for example, 11-deoxycorticosterone.

GENERAL ASPECTS OF BIOSYNTHESIS AND METABOLISM

Availability of radiolabeled compounds has played a very important role in the elucidation of biosynthesis and metabolism of steroid hormones. The use of radioactive acetic acid and cholesterol for the study of steroidogenesis in vivo and in vitro has produced radioactive steroid hormones and led to identification of both compounds as precursors of steroid compounds. Similarly, the administration of radioactive steroid hormones followed by the separation and identification of radioactive metabolites in the urine has helped to delineate the metabolic pathways. Even though acetate and cholesterol are both precursors of steroid hormones, they do not, in all probability, constitute a separate pathway of biosynthesis but follow the general sequence of acetate → cholesterol → steroid hormones. It has been amply documented that acetate is the sole precursor of cholesterol; among the 27 carbon atoms constituting cholesterol, 12 originate from the carboxyl carbon (C) and 15 from the methyl carbon (M) of acetic acid (CH_3COOH), as shown in Figure 9-19. Stepwise degradations of radioactive steroid hormones (e.g., cortisol, corticosterone) synthesized from ^{14}C-labeled acetate indicate that the individual carbon atoms of steroid hormones correspond to those of cholesterol originating from carboxyl- and methyl-carbon atoms of acetic acid. There are, however, 30 separate biochemical reactions involved in the biosynthesis of cholesterol from acetate, and many more enzymatic reactions come into play to convert cholesterol to a variety of steroid hormones. For a detailed discussion of these topics students are referred to a monograph by Dorfman and Ungar.[70] Recent studies indicate that cholesterol may be synthesized by steroidogenic cells or taken up from circulating lipoproteins.[107] (See Chapter 7.) In the adrenal, ovary, placenta, and possibly the testes of many species including humans, lipoprotein cholesterol is the major source of steroidogenic substrate. With the exception of the placenta, lipoprotein cholesterol uptake is hormonally regulated in these tissues. It is coordinated with intracellular cholesterol synthesis and the mobilization of cholesterol esters to

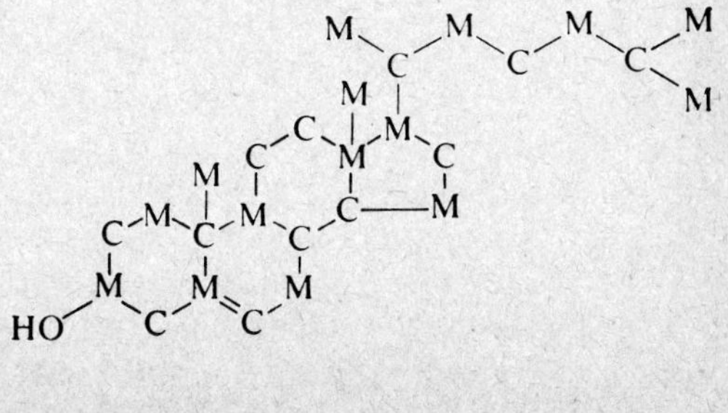

Figure 9-19. Carbon atoms of cholesterol derived from carboxyl carbon (C) and methyl carbon (M) of acetate.

ensure a continuous supply of free cholesterol for steroid synthesis. When the rate of cholesterol uptake exceeds the rate of steroidogenesis, intracellular cholesterol synthesis is suppressed and cholesterol in excess of cellular needs is esterified and stored for future use.

In normal men and women, steroid hormones are produced in the adrenals, ovaries, and testes. All these glands utilize the same precursors, acetate and cholesterol, but the nature and the quantity of steroid hormones produced by each gland are different. The difference is inherent in the degree of activity of certain enzymatic systems. For example, the enzymes 11β-hydroxylase and 21-hydroxylase, present only in the adrenals, synthesize the characteristic hormones of the adrenal glands, the corticosteroids. Similarly, the enzymatic distinction between the ovaries and the testes lies in the fact that, in addition to the enzymes found in the testes, the ovaries contain an active aromatizing enzyme system that converts male sex hormones to female sex hormones, e.g., testosterone to estradiol.

The different enzymes participating in the biosynthesis of steroid hormones may be broadly classified into the following functional groups:

1. Hydroxylases. These enzymes catalyze the substitution of the hydroxyl group (—OH) for hydrogen (—H). For example, 21-hydroxylase introduces a hydroxyl group at C-21. Similarly, 11β-hydroxylase introduces an OH-group at the β-position of C-11. There are numerous examples of other important hydroxylases, such as 20α-hydroxylase, 19-hydroxylase, and 17α-hydroxylase. The required cofactors are NADPH and molecular oxygen; the reaction is irreversible. The hydroxylation is carried out in conjunction with an electron-transport chain that consists of NADPH $\rightarrow$ flavoprotein $\rightarrow$ nonheme iron protein $\rightarrow$ cytochrome $P_{450} \rightarrow O_2$. Cytochrome P_{450} couples with specific hydroxylases for the introduction of an atom of oxygen into the steroid molecule.

2. Desmolases. These enzymes are required for cleaving the side chain. There are two desmolases, 20,22-desmolase and 17,20-desmolase; both are very important in steroidogenesis. The former participates in the conversion of the C_{27} carbon compound, cholesterol, to a C_{21} compound, pregnenolone, whereas the latter transforms C_{21} steroid hormones to C_{19} steroid hormones. The required cofactors are thought to be NADPH and molecular oxygen.

3. Dehydrogenases. This group of enzymes catalyzes the transfer of hydrogen (oxidation and reduction). The reaction is generally reversible. The cofactor is either the oxidized or reduced form of NAD or NADP, depending on the direction of the reaction. Examples include 3β-hydroxysteroid dehydrogenase, 11β-hydroxysteroid dehydrogenase, 17β-hydroxysteroid dehydrogenase, Δ^5-3β-hydroxysteroid dehydrogenase, and 3α-hydroxysteroid dehydrogenase.

4. Isomerases. These enzymes catalyze the migration of a double bond. The most important enzyme of this group that is involved in steroidogenesis is Δ^5-ketosteroid isomerase (Δ^{4-5}-isomerase). The concerted action of Δ^5-3β-hydroxysteroid dehydrogenase and Δ^5-ketosteroid isomerase on pregnenolone produces progesterone through the oxidation of the 3β-hydroxyl group and the migration of a double bond (Δ^5 to Δ^4).

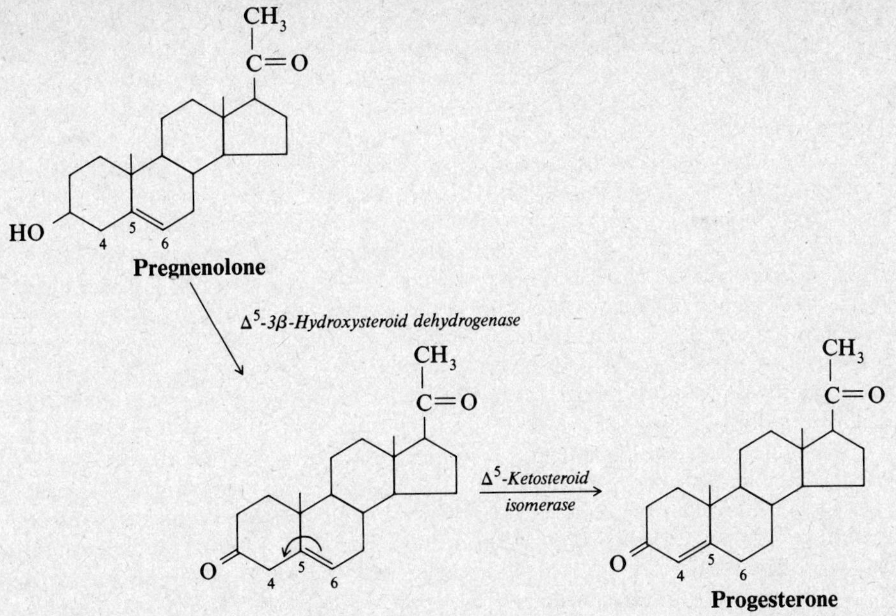

Pregnenolone

Δ^5-3β-Hydroxysteroid dehydrogenase

Δ^5-Ketosteroid isomerase

Progesterone

The liver is the major site of steroid metabolism. There is some evidence, however, that the kidney and the gastrointestinal tract may also carry out some of the metabolic transformation of steroids.

Important biochemical steps for neutralizing biological activity of hormones and for their rapid elimination from the systemic circulation are (1) introduction of a new hydroxyl group, e.g., estradiol to estriol; (2) dehydrogenation, e.g., testosterone to androstenedione; (3) reduction of a double bond, e.g., cortisol to dihydrocortisol; or (4) conjugation of an essential hydroxyl group(s) with a chemical moiety such as glucuronic acid, e.g., testosterone to testosterone glucuronide. The conjugation of these hormones and their metabolites with sulfuric or glucuronic (glucosiduronic) acid is, by far, the most efficient single metabolic process for their excretion in the urine. Almost all steroid metabolites are excreted as water-soluble glucuronides or sulfates.

METHODS FOR THE DETERMINATION OF STEROID HORMONES

Choice of Specimen

Until recently, very few methods with adequate sensitivity for determination of steroids in blood were available for use in a routine clinical laboratory; therefore, urinary measurements were widely used. Although the *urinary excretion* of a hormone, its metabolites, or both, does not account for the total amount of hormone secreted by the gland, it usually represents an approximate proportion of the amount secreted during the period of urine collection. Thus, urinary assays are considered to reflect the secretory activity of the endocrine glands. However, factors such as incompleteness of collection, altered renal function, and contribution by more than one gland to the total excretion of the same hormone(s) (e.g., adrenals and gonads both contribute to the urinary 17-ketosteroids) warrant special precaution in the interpretation of urinary values. Analyses of total urinary metabolites have further shortcomings, since they measure only a fraction of the active steroid hormones that are metabolized through multiple pathways; the quantity and the nature of these metabolites may, in turn, be dependent not only on the pathologic condition but also the intake of drugs and diet. Other arguments against urinary assay of metabolites include the delay and inconvenience encountered in the collection of 24-h urine specimens. In contrast, *plasma* sampling, transport, and processing are so much more convenient for patient, clinician, and laboratory that assays of specific plasma steroids by sensitive radioimmunoassays have displaced many assays of urine whenever plasma levels are equally useful. In addition, in certain clinical investigations where rapid and repeated sample analyses are dictated, such as in dynamic tests, control of ovulation induction, and the monitoring of fetoplacental function, plasma assays are more appropriate than urine measurements. However, plasma estimations may be limited in value because there are rapid fluctuations in hormone levels, and thus a plasma sample is representative only of the concentration that existed at the time of sampling. This is not a limitation of measurements of unconjugated active steroid hormones, such as free cortisol in timed urine collections. Thus, such values may, indeed, be a better index of endocrine function and hormonal activity at the site of tissues because it is the biologically active non-protein-bound (free) fraction of plasma steroids that is excreted in the urine. Furthermore, since the glomerular filtration and tubular reabsorption of steroid hormones are passive and thus concentration dependent, urinary free steroid is directly related to plasma non-protein-bound steroid concentration. Such analyses, therefore, obviate the need for measuring free hormone in plasma. So far, only urinary cortisol estimation for the assessment of adrenal cortical function has found useful application in clinical practice. The importance and applicability of other active urinary steroid hormones such as progesterone, estradiol, and testosterone await further investigation. An excellent discussion on the advantages of plasma and urinary hormone assays can be found elsewhere.[175]

Recently, it has been suggested that measurement of *salivary steroids* provides information similar to that derived from urinary free steroids.[254] Salivary sampling regimens have the advantage of frequent, easy collection by noninvasive, stress-free techniques. Moreover, they obviate difficulties of ensuring completeness of a 24-h urine collection. Patients find little difficulty in salivating directly into disposable tubes, thus providing an adequate volume in ∼10 min. Assays of samples collected at 1–2-h intervals during waking hours provide an accurate assessment of baseline endocrine activity. Since smaller aliquots (500 μL) can be collected at 15- or even 10-min intervals, salivary samples could well be more useful than either plasma or urine samples in short-term dynamic tests. Most steroids of clinical interest have been measured in saliva, e.g., cortisol, dehydroepiandrosterone sulfate, 17-hydroxyprogesterone, and aldosterone, as well as progesterone, estradiol, and testosterone. Although results appear promising, the clinical usefulness of salivary measurements of steroids is not yet fully established.

There is no experimental evidence that steroid hormone levels in serum are different from those in plasma. However, rapid separation of red blood cells in the specimen is important since red cells at room temperature can alter plasma concentrations of active steroid hormones; red cells degrade estradiol to estrone and cortisol to cortisone and adsorb testosterone. Thus, plasma is preferred because it can be more rapidly harvested and put in cold storage than serum.

Hydrolysis, Extraction, and Separation

Although steroid hormones differ greatly in their physiological activities in the body, the assay procedures have many similarities. The following general steps are required: hydrolysis, extraction, purification and separation, and final quantitation. Note, however, that with the advent of newer techniques described below (e.g., HPLC, RIA), steps such as hydrolysis, extraction, purification, and separation are seldom necessary for routine clinical analyses of steroids.

Hydrolysis. Steroid hormones and their metabolites are present in blood and urine as the water-soluble conjugates of glucuronic (glucosiduronic) acid and sulfuric acid. Indeed, the metabolites are predominantly excreted in the urine as these conjugates. When total concentrations (i.e., unconjugated plus conjugated) of a steroid hormone or its metabolites are determined, the hydrolysis of ether and ester linkages is an obligatory step. Two general types of procedures are available, namely, acid hydrolysis and enzymatic hydrolysis. In *acid hydrolysis,* an aliquot of a 24-h urine sample is boiled, generally with reflux, in the presence of a specified concentration of mineral acid for a specified length of time (10–60 min). For *enzymatic hydrolysis,* a portion of a 24-h urine specimen or an aliquot of plasma is adjusted with buffer to the optimal pH for the enzyme employed; after the addition of an adequate amount of the necessary enzymes (β-glucuronidase to hydrolyze glucuronates [glucosiduronates] and sulfatase to hydrolyze sulfate conjugates), the test sample is incubated for 18–72 h at a specified temperature (e.g., 37 °C).

From the technical point of view, acid hydrolysis is always preferred, except for acid-labile hormones, because the process has simplicity and speed, and results in completeness of reaction regardless of the nature of conjugates. Enzyme hydrolysis, on the other hand, requires special attention to optimal concentration and type of enzyme, pH, temperature, and duration of incubation. In addition, the possible presence of enzyme inhibitors, varying in amount and nature with different specimens, casts some doubt on the degree of completeness of hydrolysis. In spite of such drawbacks, enzymatic hydrolysis is used for plasma analysis of steroids that are labile in strong acid solution (e.g., pregnanetriol, corticosteroids) and to avoid interfering substances produced by acid hydrolysis.

Extraction. Following hydrolysis, deconjugated steroids become sparingly soluble in aqueous solution. Thus, when an immiscible organic solvent in which steroids are highly soluble is added to the hydrolyzed sample and is shaken, the vast majority of steroids are extracted into the organic layer. Repeating the extraction process with a fresh volume of the organic solvent increases the recovery of steroids. The same procedure is followed for the assay of unconjugated steroid hormones in plasma or serum. Note that addition of an organic solvent to the plasma denatures the binding proteins. Thus, the extracted steroid represents both protein-bound and unbound fractions.

Selection of the organic solvent is based on the polarity of the steroid hormone to be measured. The nonpolar tetracyclic ring system is common to all steroids, but the polarity increases as the number of oxygen groups (i.e., ketone and hydroxyl groups) and double bonds increases. Steroids with one or two oxygens (e.g., androgens, estrogens) are of low polarity; the best solvent for their extraction is a relatively nonpolar solvent such as diethyl ether or benzene. Similarly, steroids with three or more oxygens (e.g., corticosteroids and their metabolites) are quite polar; for their extraction, polar organic solvents such as chloroform, dichloromethane, or ethyl acetate are most suitable.

The relative solubility of substances in two immiscible solvents is exploited not only for the extraction process but also for the separation and purification of substances (partition chromatography). The ratio of the concentration of a substance in a nonpolar phase to the concentration of the same compound in the polar phase is known as the partition coefficient (K). Substances with high K values will mostly be in the nonpolar phase, whereas substances with low K values will preferentially move into the polar phase. In the extraction process, the solvent system is composed of an organic solvent (a relatively nonpolar phase) and the hydrolyzed urine or plasma (polar phase). Steroids with high partition coefficients will consequently be extracted into the organic layer. A better recovery of polar compounds (e.g., urinary corticosteroids, estriol) from the polar phase can be conveniently achieved by adding ammonium sulfate or sodium chloride prior to extraction. The addition of inorganic salts increases the partition coefficient by decreasing the solubility of the steroid in the aqueous solution.

Purification and separation. Although a proper choice of solvent improves the selectivity of extraction, a large number of closely related steroids, chromogenic substances, and other nonspecific materials are also extracted with the steroids. Removal of such contaminants, especially those which will interfere in the final estimation, is very important.

The *solvent partition* method is the simplest and most suitable method for a clinical laboratory. Thus, it is widely used for preliminary purification and separation of compounds of interest. The basic principle is the same as that for extraction. Steroids and other urinary impurities contained in the organic solvent are treated with weakly basic solutions (e.g., sodium bicarbonate, sodium carbonate). By virtue of their greater solubility, strongly acidic components partition into the basic aqueous layer. The separation of neutral and phenolic steroids can be achieved in a similar way. Because of the acidic nature of the phenolic steroids (estrogens), they can be readily extracted from an organic phase with an aqueous solution of sodium hydroxide. After lowering the pH of the alkaline solution, estrogens are re-extracted with a suitable solvent (diethyl ether) and are processed further for final estimation. Most often the organic extract is washed to neutrality with water to ensure complete removal of alkali which, if allowed to remain, might interfere in subsequent workup of the extract.

The degree of purification and separation needed prior to quantitative measurement will depend on the method used for final quantitation. For example, if the final mode of estimation is a color reaction that is very specific for an individual steroid or for a group of steroids, further purification of the steroid extract may be omitted. In fact, many colorimetric assays of steroid hormones (e.g., urinary 17-ketosteroids) are performed on crude extracts. Similarly, in many radioimmunoassays using fairly specific antisera, the quantitation is carried out without prior purification of organic extracts. Although such determinations may yield adequate information for most clinical purposes, specific measurement of an individual steroid or group of steroids may occasionally necessitate further purification and separation of the extract. Various techniques are available for this purpose. For detailed descriptions of the theory and application of these methods to steroid analysis, the reader is referred elsewhere.[68] It suffices here to mention that these methods are based either on some selective chemical reaction (e.g., digitonide formation to separate 3β-hydroxy and 3α-hydroxy steroids) or on physical techniques such as countercurrent distribution (CCD), paper and column-partition chromatography, column and thin-layer adsorption chromatography, gel filtration using LH-20, gas chromatography (GC), or high performance liquid chromatography (HPLC).

Quantitation

Methods for the quantitative estimation of steroids may be divided into five categories: colorimetric, fluorometric, chromatographic (GC and HPLC), radioisotope methods (double-isotope derivative formation), and saturation analysis (radioimmunoassays). In recent years, the use of methods based on colorimetry, fluorometry, and double-isotope derivative formation has been very limited. Radioimmunoassay is by far the most sensitive and widely applied assay method at present.

Colorimetric estimation. In these methods a certain functional group of a steroid is reacted with a particular chemical reagent to form a specific colored product. For example, steroids containing a keto ($>C=O$) group in position 17 (17-ketosteroids) react with *m*-dinitrobenzene in alcoholic alkali to produce a reddish-purple compound. The intensity of the color is proportional to the concentration of the steroid and is measured with a spectrophotometer or a colorimeter using the wavelength of maximum absorption. The principal source of error in this procedure is the interference from nonspecific chromogens derived from other steroidal and nonsteroidal components of biological extracts. Colorimetric methods that are used in clinical laboratories for estimations of steroid hormones are described in the appropriate sections.

Fluorometric estimation. Many steroids produce a characteristic fluorescence if present in a suitable medium such as sulfuric acid or phosphoric acid. Activation and emission wavelengths under specified experimental conditions are relatively specific for a given substance. Although this technique offers certain advantages in increased sensitivity (e.g., for estrogen measurements) and, at times, in increased specificity, from a practical point of view it has many drawbacks, including nonspecific fluorescence and quenching effects caused by solvent residues and improperly purified reagents. The need for controlled reaction conditions and the instability of fluorescence have further limited the usefulness of this technique in routine clinical laboratories. Despite certain advantages of fluorometric assays, radioimmunoassays have now replaced most of these procedures.

Gas chromatographic estimation. Historically, this technique has been an important addition to steroid methodology. Speed, sensitivity, accuracy, precision, and specificity are important criteria for selection of procedures to measure compounds of clinical interest. Gas chromatographic techniques fulfill these requirements. As a result, gas chromatography has become an important tool for clinical laboratories and has replaced many of the time-consuming and less specific chemical methods.[79,337] It has been found to be quite suitable for routine clinical use (e.g., urinary pregnanediol, pregnanetriol, 17-ketosteroids, estriol). The theory and the instrumentation involved in gas chromatography have been described in Chapter 1B, Section Seven.

High performance liquid chromatography (HPLC). HPLC is the fastest-growing technique in clinical chemistry and offers many advantages over traditional liquid chromatography. These advantages include high resolution that is superior even to GLC; increased speed, since analysis times of <1 h are common; increased sensitivity, since UV absorption detectors can detect nanogram quantities and the detection sensitivity can be increased to picogram quantities by the use of fluorescence and electrochemical detectors; reusable columns; and easy sample recovery, since most detectors are nondestructive. In contrast to GLC, nonderivatized compounds and compounds that are thermally labile or nonvolatile can also be analyzed. The theory of instrumentation and its application are discussed elsewhere in the text. (See p. 1148 and Chapter 1B, Section Eight.)

Both adsorption and reverse-phase partition chromatography using isocratic or gradient elution modes have been employed for the separation and quantitation of steroid hormones. The most widely used detectors are fixed or variable-wavelength ultraviolet photometers. In a typical assay procedure, steroids are co-assayed with an internal standard, a substance of the same chemical class with a similar degree of recovery during sample preparation and with comparable separation and detection characteristics. Under particular experimental conditions, the elution times relative to the internal standard are used for identification of individual steroids. Quantitation is generally carried out by measuring the relative peak heights or areas of respective steroids. Several methods of separation and quantitation of steroid hormones have been reported. For a detailed description, see the monograph by Kautsky.[135] Most of the methods published so far are investigative analytical procedures and are less popular than other routine clinical techniques such as radioimmunoassay. Equipment costs and the experience necessary to obtain reliable results are further limitations.

Double-isotope derivative assay. In the double-isotope derivative assay, a steroid labeled with an isotope of high specific activity and negligible mass (e.g., ${}^{3}H$ or ${}^{14}C$) is added to the specimen to be analyzed. Following the extraction and purification of the sample, a suitable reagent labeled with a second isotope (e.g., ${}^{35}S$) is added to yield a chemically stable derivative. After further purification, the measurement of the isotope ratio of the derivative provides the amount of steroid originally present in the sample. Two isotopes (e.g., ${}^{3}H$ and ${}^{35}S$) that emit particles of different energy spectra are chosen so that the radioactivity of both isotopes can be measured simultaneously in a liquid scintillation counter. The radioactive steroid added prior to processing of the samples serves as an indicator for calculating losses during the extraction and purification, whereas the formation of a derivative with a radiolabeled reagent aids in the specific detection and the quantitation. Double-isotope derivative assays are highly sensitive, specific, and accurate.[68] For these reasons, double-isotope assays are considered reference methods; their complexity, tediousness, and expense generally preclude their routine use.

Determination of Steroid Hormones by Radioimmunoassay

The measurement of steroid hormones by radioimmunoassay is the most significant advance in steroid methodology of the past decade. Because of its speed, sensitivity, specificity, simplicity, and application to simultaneous assay of a large number of samples, the technique has been applied to the determinations of virtually every steroid of physiological importance. The theoretical and practical aspects of the radioimmunoassays of steroid and other hormones have been discussed in several monographs.[2,129] The basic principles of the procedure have also been described in other sections of the text. (See p. 1010 and Chapter 1C.)

Techniques which exploit the fundamental analytical principles of immunoassay have come to be known by a variety of names, including saturation analysis, competitive protein binding assay (CPB), radioligand assay (RLA), radioenzymatic assay, displacement analysis, radioreceptor assay, fluoroimmunoassay, viroimmunoassay, and, more recently, enzyme immunoassay and spin-immunoassay. Some of these assays have been described under a general term, *ligand assays*. The principle of these assays is similar to RIA technique in which unlabeled and labeled forms of the same substance compete for a limited number of sites on a specific binding protein. At equilibrium the amount of labeled materials bound to protein is inversely proportional to the amount of unlabeled material in the reaction mixture. The proportion of label bound to protein is determined by introducing a separation system into the assay. By the addition of known amounts of unlabeled ligand (or standard) to the assay mixture, a dose-response curve is constructed from which the value of an unknown sample is determined. A ligand is a substance that will bind to an appropriate binding reagent; the term usually refers to both labeled and unlabeled molecules of the same species. For a ligand assay, the type of label employed as a marker is usually indicated by the prefix, and the type of binding protein is indicated by the suffix. Thus, a radioimmunoassay is a ligand assay using a radioisotope as a label and antibodies as binding protein. Similarly, an assay using a radiolabeled tracer (a substance used to monitor the degree of binding in a ligand assay, also called marker or label) in conjunction with a tissue receptor as binding protein is referred to as a radioreceptor assay; when an enzyme-labeled tracer and antibody binder are used, the assay is called an enzyme immunoassay.

Steroid hormones are not immunogenic by themselves. Therefore, antibodies against steroids are produced by active immunization of sheep, rabbits, or other animals with steroids covalently linked to carrier proteins (steroid-protein conjugates). The most commonly used carrier protein is bovine serum albumin (BSA), which is available in a highly purified form at low cost. Many other proteins, including polylysine and hemocyanin, have been used, but they seem to have no particular advantage over BSA.

The most important property of an antiserum is its specificity, i.e., the degree of cross-reactivity of the antiserum with other steroids. Cross-reactivity is expressed in terms of the mass of unlabeled cross-reacting steroid required to displace 50% of the antibody-bound labeled steroid derived from the steroid used to produce the antiserum. For example, the antibody being tested is produced against steroid X, and the % cross reaction of steroid Y needs to be determined. Then % cross reaction of steroid Y = $(m/n) \times 100$, where m = mass of steroid X required to displace 50% of bound labeled X, and n = mass of steroid Y required to displace 50% of bound labeled X.

The specificity of a particular antiserum is influenced by the position on the steroid nucleus that is used for conjugation. Several investigators have shown that antisera tend to be specific for that part of the steroid molecule farthest removed from the site of conjugation. Thus, an antiserum is more likely to recognize differences in the D ring if the site of conjugation is at the 3 position in the A ring. In an attempt to produce highly specific antisera, conjugates have been prepared by coupling steroids to protein carriers through the B and C rings. This approach leaves the immunodeterminant groups present in the A and D rings of most naturally occurring steroids available for antibody recognition. Indeed, on the basis of this rationale, highly specific antisera against 6-O-carboxymethyloxime-BSA conjugates of estrone, estradiol, and estriol have been produced. With these types of antisera, no significant cross reactions have been observed among estrogens or with other steroids except with corresponding estrogens substituted at carbon 6.

Recently, the potential use of monoclonal antibodies for steroid immunoassay has received considerable attention.[122] The hybridoma technique provides the feasibility of selecting an antibody with desired specificity and sensitivity from a large number of antibodies produced by numerous clones, each secreting a different antibody. The availability of such standardized antibodies holds a great potential for achieving significant improvement in overall specificity, particularly in nonextraction (direct) assays. Note that the antibodies used in direct assays need to be indifferent to other steroids and their metabolites in the sample which, in conventional procedures, are often removed by extraction procedures. The application of monoclonal antibodies to routine clinical assay of steroids is currently at an exploratory stage.

RIA techniques are most suitable for plasma steroid determination because of the high sensitivity that can be achieved, often in the order of 1 pg (10^{-12} g). Several factors, including the quality of antiserum, the mode of separation of bound and unbound fractions, and inherent experimental errors, affect the overall sensitivity and specificity of determination. However, the quality and nature of the radiolabeled steroid also play an important role in this regard. The radiolabeled steroid is required both for evaluating recoveries of steroid when an extraction or purification step, or both, is carried out prior to assay and for indicating, during RIA, the distribution of protein-bound and unbound steroid. A radioactive steroid of high-molar specific activity (Ci/mmol) is a prerequisite for a sensitive assay system. In general, the low-molar specific activities of ^{14}C-labeled steroids preclude their use in highly sensitive assays. Tritium-labeled steroids, on the other hand, can be prepared at much higher specific activities and are the β-emitting radioligands of choice for RIA of these compounds. A wide range of tritiated steroids with specific activities as high as 200 Ci/mmol is commercially available. Because of the relatively long half-life of tritium (12.26 years), no correction for decay is necessary. Tritiated steroids behave essentially like the unlabeled steroids. Therefore, corrections for losses during pre-assay workup and assumption of identical interactions of labeled and unlabeled steroids with binding proteins are possible. The stability of most tritiated steroids is not normally a problem over at least a six-month period provided they are stored under conditions recommended by the manufacturer. Samples containing tritiated steroids are counted by β-liquid-scintillation counters using one of the commercially available scintillation cocktails.

In spite of wide application of tritiated steroids, β-scintillation counting has several drawbacks, among them the expense of special counting vials and cocktails and the inability to count solid phases that are insoluble in the cocktail. In addition, counting efficiency and accuracy can easily be impaired in liquid scintillation counting because true counts can be quenched or enhanced by the presence of different solvents or chemical impurities in the cocktail. Many workers, particularly in clinical laboratories performing large numbers of steroid assays, prefer to use steroids labeled with γ-emitters. Gamma-emitters usually have shorter half-lives (more dpm/mole per unit of time) than β-emitters and greater potential for high specific activity. This high specific activity results in shorter counting times and greater assay sensitivity. Further, γ-nuclides can be counted in any form (solid, liquid, or gas) without quench or other interferences and without scintillation cocktails or special counting vials. As a consequence, the use of γ-labeled steroids is less expensive per specimen assayed and more convenient than use of β-labeled steroids.

The most widely used γ-emitting isotope is ^{125}I, which has an adequate half-life (60 d compared with 8.1 d for ^{131}I). It is also less hazardous (lower energy of emitted radiation) than the other γ-nuclides. Steroids labeled with ^{125}I can be prepared at significantly higher specific activities (~1000 Ci/mmol) than tritiated steroids. However, with the exception of phenolic estrogens, direct radioiodination of steroid molecules is not possible. Steroids are first conjugated to moieties such as tyramine, histamine, and tyrosine methyl ester that can be easily radioiodinated. There are, however, several drawbacks to using iodinated steroid ligand. High energy of penetrating radiation from γ-emitting isotopes poses more radiation hazard than do β-labels. The short half-life of ^{125}I limits usability of a tracer preparation to <2 months. Incorporation of an iodine atom whose size is approximately the same as that of the A ring may adversely affect binding between antibody and labeled steroid. Iodinated haptens commonly used as tracers may have different binding avidity (usually higher) for antibodies than the native steroid being measured, such that detection sensitivity is reduced. Problems and advantages of using radioiodinated steroids have been further discussed in a monograph by Hunter and Corrie.[122]

In recent years commercial assay kits for almost all steroid hormones have become available. There are no rigid criteria by which a commercial kit can be selected or rejected. However, certain performance characteristics of an assay kit are very important for its selection—reliability and practicality. Reliability of an assay includes high *precision* (within-assay and between-assay reproducibility), low *systematic error* (evaluated by comparing results with those obtained by other established methods), high *sensitivity* range (the area on the dose-response curve of maximum precision and high accuracy in correspondence with the range of values with the greatest clinical significance), high *specificity,* and *ruggedness* (the degree to which changes in results occur in response to changes in assay reagents and procedure). Practicality includes *speed* (turnaround time), *technical simplicity* (minimum intertechnologist error), *cost,* and, finally, *resources* required to perform the test.

General Procedures for Steroid RIA

Despite structural differences among steroid hormones, the radioimmunoassay of these substances is carried out following similar procedures. Variations are generally observed in the sample preparations and the degree of purification prior to assay. The principal steps in a typical steroid radioimmunoassay method are described below.

Reagents

Assay buffer. The binding of steroids to antisera is likely to be influenced by the pH and ionic strength of the incubation medium. Sodium phosphate buffer, ranging in molarity from 0.01 to 0.1 and with pH near neutrality (7.4), is found to be optimal for most steroid assays. Generally, sodium azide is added to the buffer to protect the antiserum as well as radioactive and nonradioactive steroids from bacterial decomposition. The presence of gelatin or other proteins such as albumin or globulin in the buffer prevents adsorption of steroids onto the surfaces of glassware, tubes, or pipets; decreases adsorption of the antibody-bound steroid to charcoal (see below); minimizes nonspecific binding; and stabilizes the overall assay system.

Dextran-coated charcoal (DCC). Charcoal Norit A (neutral) and Dextran T70 are suspended in assay buffer. Stock solution can be stored for 4 weeks at 4 °C. Charcoal alone not only readily adsorbs unbound steroids but also shows some affinity for steroid-antibody complex. Coating the charcoal with Dextran T70 (M.W. 70 000) apparently produces a molecular sieving effect; i.e., unbound steroids of low molecular weight are readily adsorbed while substances of high molecular weight, such as antibody complex, are excluded.

Radioactive tracer. Two kinds of radiolabeled steroids are used: tritium-labeled steroids or ^{125}I-labeled steroid derivatives, such as iodinated tyrosine methyl ester conjugates.

The stock solution of tritiated steroids in the solvents should be stored as suggested by the supplier at a concentration of 25–50 μCi/mL and at a temperature just above the freezing point of the solvent. Decomposition is markedly reduced by using suitable organic solvents, such as toluene, ethanol, benzene, or a benzene:methanol mixture (9:1). Storage temperature also plays an important role in the rate of decomposition of tritiated steroid. Freezing accelerates decomposition. For use in the assay, the tritiated steroid is dissolved in assay buffer at a concentration of 0.01 μCi per 0.1 mL of assay buffer. This solution is stable for up to 4 weeks when stored at

4 °C. Significant decomposition of the labeled steroid usually results in a decrease in the slope of the dose-response curve and in increased assay blanks.

In general, steroids labeled with ^{125}I are stable for 6–8 weeks when stored in ethanol at 2–4 °C. Many ^{125}I-labeled steroids are sold commercially as components of kits and should be stored for the duration and in the manner suggested by the supplier. Note that the stability varies with the nature of the steroid derivatives used for iodination.

Standard solutions. 1. Stock standard, 1 mg steroid per mL ethanol. Dissolve 10 mg of recrystallized pure steroid in 10 mL absolute ethanol.

2. Intermediate solutions of standard, 1 μg/mL. Dilute stock solution 1:10 in absolute ethanol to give a solution of 100 μg/mL. Further dilute 0.1 mL of 100 μg/mL solution to 10 mL with absolute ethanol. Store standard solutions 1 and 2 at −20 °C to minimize evaporation.

3. Working standard solution, 1000 pg/0.1 mL. Dilute 0.1 mL of standard solution, 1 μg/mL, to 10 mL with assay buffer. The steroid is stable for months in this solution stored at 4 °C.

4. Assay standard set (e.g., 20, 40, 100, 200, 500, 1000 pg/0.1 mL) should be prepared fresh by diluting the working standard with assay buffer. Note that for direct (nonextraction) assay all dilutions of working standard should be made with serum or plasma that reflects the biological matrix of the sample, but it must be free of the steroid to be measured.

Reconstitution of steroid antiserum. The procedure for reconstitution is usually prescribed by the supplier. An antiserum is supplied either frozen or lyophilized. When the supply is received and reconstituted, it is often convenient to make whatever dilution is appropriate and then to freeze the diluted antiserum in aliquots suitable for daily use. Repeated freezing and thawing of bulk preparations must be avoided because of the risk of denaturing the antibody protein.

The optimal dilution titer of an antiserum will depend on the particular affinity and concentration of the antibody. The optimal titer is usually defined as that dilution that binds 50% of a standard amount of radiolabeled ligand under specified conditions of incubation time, temperature, and volume and with a specified separation technique. In practice, analysis can be performed with good precision and sensitivity when per cent binding ($\%B_0$) ranges from 30–50%. Note, however, that when $\%B_0$ is low, sensitivity is comparatively high but precision is poor. The reverse is true when $\%B_0$ is high.

Specimen Collection and Storage

Except for catechol estrogens and estetrols, steroids are very stable and no special precaution is necessary during collection of biological samples. Steroid levels can be determined either in plasma (heparin or EDTA) or serum. However, erythrocytes, which metabolize steroid hormones (see p. 1046), should be separated rapidly from serum or plasma by centrifugation for 10 min at $900 \times g$. Consistency in the method of blood collection is important in order to avoid sources of error. Repeated freezing and thawing of plasma, serum, or urine causes hydrolysis of conjugates and yields falsely high values for unconjugated steroids.

For urinary measurements, a 24-h urine specimen is collected in the presence of a preservative such as boric acid (1 g/L) to prevent bacterial growth. Note that bacteria can degrade steroids and hydrolyze steroid conjugates. After recording the total volume, a well-mixed aliquot is saved for analysis. Creatinine determination was once believed to be a fair check for completeness of the specimen collection. However, recent reports cast doubt on the effectiveness of creatinine determinations for this purpose,[61] although creatinine determinations in serial collections from the same individual are of help in this regard.

Plasma, serum, or urine without preservative may be stored overnight at 2–8 °C or for longer periods (up to 2 years) at −20 °C in airtight containers. Storage of steroids in aqueous solution such as buffer or urine should be made with an inert protein (i.e., gelatin) added at a concentration of 0.1–0.5 g/dL. Nonpolar steroids (e.g., progesterone and androstenedione) are very unstable when stored at −20 °C in an aqueous solution without protein; as much as 80% of steroids may be lost after 6 months of storage.

Sample Preparation

The mode and nature of specimen preparation depend primarily on the affinity constant and cross-reactivity of the antiserum. The affinity constant determines the theoretical limit of detection: the higher the affinity constant, the greater the sensitivity. Cross-reactivity dictates whether or not samples need to be purified before assay; when cross-reactivity between the steroid of interest and other steroids is high, purification of samples is necessary.

The volume of plasma required for the analysis depends on the expected amount of steroid in the sample, the sensitivity of the assay system, and the possible loss of analyte in preparation procedures. For the quantitation of estradiol in male plasma (30–80 pg/mL), as much as 500 μL of plasma may be required, while as little as 50 μL may be sufficient for testosterone measurement on the same plasma (3000–8000 pg/mL).

Until recently, extraction of steroids from plasma with an organic solvent was a preliminary step for almost all radioimmunoassays. Extraction serves to separate steroids from plasma binding proteins, such as sex hormone–binding globulin (SHBG) and corticosteroid-binding globulin (CBG), which would otherwise cause additional interaction in the assay system. It is also employed to remove interfering lipids and cross-reacting steroids. However, methods for the measurement of steroids in unextracted serum or plasma are now available.[122] In these methods, extraction is avoided by the use of agents or by providing conditions of low pH that quantitatively displace steroids from plasma binding proteins and enable direct assay of steroids. In addition, the availability of antisera with improved avidity and specificity has reduced the need for extraction or purification of steroids prior to assay. Kits for direct assay of many steroids are commercially available.

In situations where extraction is necessary before assay, an aliquot of the organic extract is evaporated to dryness at 40 °C under a gentle stream of nitrogen to preclude oxidation and decomposition of steroids. The dried residue is dissolved in a volume of assay buffer appropriate for expected levels and heated for 10 min in a 40 °C water bath. Heating ensures dissolution of steroids.

Procedure

Incubation. Appropriate aliquots of unknown samples, controls, and standards are incubated with properly diluted antiserum and radioactive ligand. The tube containing tracer and assay buffer (substituted for antibody) gives information on the nonspecific binding (NSB). The final incubation volume may vary from 0.3–1.5 mL.

For some steroid immunoassays, incubation is carried out at 4 °C for 2–16 h. Although equilibrium is reached faster at room temperature or at 37 °C, the lower incubation temperature is preferred because the binding affinity and the stability of the assay system are greater at 4 °C. For the sake of speed, the incubation period has been shortened to 30–60 min in many commercially available kits for steroid RIA. It should be recognized, however, that in nonequilibrium conditions small inequalities in incubation times reduce precision of the assay. Under these circumstances, each laboratory must determine the best compromise between acceptable precision and the clinical need for shorter turnaround time.

Separation of Free from Antibody-Bound Steroid. After incubation, the free steroid is separated from the bound form by the addition of a cold dextran-charcoal suspension. The tubes are further incubated for 10–15 min at 4 °C and centrifuged. The radioactivity either of the supernatant (bound fraction) or the sediment (free fraction) is measured in a suitable radiation counter. The supernatant is generally counted when tritium-labeled ligand is used, whereas the radioactivity of the sedimented charcoal is customarily determined for ^{125}I-labeled ligand.

An ideal separation technique must completely separate the antibody-bound and free fractions; it should not interfere with the antigen-antibody reaction, and it should be simple, rapid, and inexpensive. Many techniques have been used for separating antibody-bound steroid from free steroid, including dextran-coated charcoal and Florisil adsorption, ammonium sulfate or double antibody precipitation, and solid phase systems.[122] For direct RIA of steroids using ^{125}I-ligand, techniques such as accelerated liquid-phase second antibody (in which the precipitating reagent is a mixture of polyethylene glycol and second antibody) or solid phase second antibody are widely employed.

Dextran-coated charcoal adsorption is the most popular technique in steroid RIA. However, the concentration of protein (or gelatin) and lipid, the concentration of the dextran-charcoal suspension, and the time and temperature of charcoal addition greatly influence results. The effect may be either spuriously positive or negative, i.e., an increase or decrease in the estimated hormone concentration. Therefore, each of these variables should be optimized for every assay method. Time may become a factor when many samples are assayed in the same batch. The tube at the beginning of the experiment may have a time of exposure to the dextran-charcoal different from that of the tubes at the end. Temperature may play a role if samples are allowed to remain on the bench top for variable periods of time while awaiting decanting or aspiration of supernatant. The concentration of dextran-charcoal may vary if the suspension is not kept agitated at all times during its addition.

The more polar the steroid, the less it is adsorbed by the charcoal and the less critical the length of its exposure to charcoal becomes. Also, the higher the affinity of the antibody for ligand, the less the dissociation of antibody-bound fraction and the less critical the time. Dextran-charcoal is not a suitable separation material if the affinity constant of the antiserum is $< 10^9$ L/mol. Whatever the separation technique, it must be judiciously selected, and specified conditions must be strictly observed in order to achieve optimal accuracy, precision, and sensitivity of the assay. Advantages and disadvantages of presently available separation techniques have been discussed elsewhere.[122]

Calculations

Results from RIA are calculated by comparing the values of the unknowns with a standard curve. Therefore, it is imperative that all conditions of the assay be identical for the unknowns and the standards. Several types of plots are currently used. The simplest is to plot radioactivity in cpm of the antibody-bound fraction (B) on the ordinate of linear cross-sectional graph paper relative to concentrations (doses) of standards on the abscissa. Another approach is to convert B to a fraction, or more usually to a percentage, of either total cpm added to each tube (T) or cpm bound in the zero standard tube (B_0). The equations are, respectively, $\%B = B/T \times 100$ and $\%B_0 = B/B_0 \times 100$. In such case, $\%B$ is plotted on the ordinate. Some workers determine both bound (B) and free (F) cpm and plot the B/F ratio. All these plots produce curvilinear relationships that can be partially linearized (to sigmoidal curves) by use of semilogarithmic paper, with the dose plotted on the logarithmic axis. However, first-order linear curves are generally preferred. First-order linearity is most often accomplished by converting B/B_0 to its logit:

$$\text{logit } (B/B_0) = \ln [(B/B_0)/(1 - B/B_0)]$$

The logit is then plotted on the ordinate of semilog paper and the dose is plotted on the abscissa. Another and easier option is the use of logit-log graph paper whose ordinate is already in logit divisions. The logit transform requires subtraction of nonspecific counts (NSB) from each B of unknown and standards and from B_0 before the logit relation of B/B_0 is calculated, e.g.,

$$\text{logit } (B/B_0) = \ln \left[\frac{(B - NSB)/(B_0 - NSB)}{[1 - (B - NSB)]/(B_0 - NSB)} \right]$$

Although the unknown concentrations can be calculated by means of a hand calculator and a graph of the logit-log relation, software is now available for logit-log data reduction by programmable microcomputers or microprocessors integrated to automated counters.

Data reduction by the logit-log transform has the following special advantages: (1) within-assay variations between duplicates with 95% confidence limits are readily obtained from it; (2) between-assay variations may be assessed by comparing standard curves of successive runs; (3) accuracy may be assessed by measuring parallelism between sample dilutions and the standard curve (see below); and (4) the characteristic slope and intercept of each standard curve can be calculated.

The one drawback of the transform is the need for extrapolation at both ends of the curve with corresponding expansions of error. The mathematical background of the logit-log method has been discussed by Rodbard.[258]

Competitive Protein Binding (CPB) and Radioreceptor Assays (RRA)

In addition to RIA procedures, CPB and RRA have been employed for the measurement of steroid hormones. The important difference between these assays is the type of binding protein. For steroid CPB assays the binding agent is a naturally occurring protein found in blood. Until recently, before antibodies to steroids were readily available, CPB assays were the most sensitive methods for steroid determination. Plasma proteins widely used as binding proteins include sex hormone binding globulin (SHBG) for the assays of estradiol, testosterone, and dihydrotestosterone; and corticosteroid binding globulin (CBG, transcortin) for the measurements of cortisol, corticosterone, progesterone, and other corticosteroids. Plasma binding proteins have advantages in that they are readily available and large quantities can be obtained at relatively low cost. However, the major disadvantage of this assay technique is its lack of specificity. For example, as many as eight different corticosteroids can be measured using CBG. Consequently, CPB assays have largely been replaced by specific RIA methods.

Target tissues are characterized by the presence of high-affinity hormone receptors (see Receptor Regulation, p. 1005) and have been used as a source of binding proteins in RRA. Target tissues that have been utilized for steroid hormone determinations include uterus for estrogens; prostate and seminal vesicles for dihydrotestosterone; liver, hepatoma cells, and thymus for glucocorticoids; and kidney and toad bladder for mineralocorticoids. Since receptor binding of an active hormone is a prerequisite for its ultimate biologic effect, the tissue receptor interacts specifically only with biologically active steroid hormones. (In RIA, antibodies bind the immunoreactive component that may or may not possess biological activity.) In addition, tissue receptors have high affinities, a condition that results in highly sensitive assays. However, the unstable nature and complexity of preparation and purification of these receptors have limited their routine use.

Reliability Criteria for Steroid Hormone Assays

For any method to be used, reliability criteria such as accuracy, precision, specificity, and sensitivity must be established. (See Chapter 2D.) Furthermore, these evaluations must be repeated at frequent intervals for the maintenance of good quality control.

Accuracy. This term refers to the degree with which a measured value approaches the true value of the quantity being measured. Accuracy is evaluated for the preparative steps by determining per cent of recovery of steroid added to buffer or to a previously assayed specimen. One adds the steroid as a small amount of labeled internal standard of high specific activity prior to sample preparation. An aliquot of prepared sample is then examined for radioactivity. The calculation is

$$\% \text{ recovery} = (\text{cpm found}/\text{cpm added}) \times 100$$

A range of 90-110% recovery is considered good validation of the preparative step and obviates the need to correct for losses of analyte during preparation.

Another way of evaluating accuracy is to examine parallelism between sample dilutions and the standard curve. This is performed by analyzing unknown samples at multiple dilutions along with the standard. Responses for unknown specimens and the standards are then plotted against log-dose. The most suitable response curve for such evaluation is obtained by using a logit-log plot. (See above.) A response curve for the unknown specimen parallel to that of the standard curve proves the validity of the assay because parallelism will occur only when the specimen behaves as if it were a dilution of the standard preparation.

Accuracy of the assay, however, depends also on the quality of standard (reference) material. For steroid assays, either the standard substance must be very pure or its type and proportion of impurity must be known with a high degree of certainty. The standard must be correctly weighed and prepared in correct solvent volume. It must be protected from chemical degradation until use.

Precision. This term refers to the magnitude of random error and thus to the reproducibility of measurements. Customarily, it is expressed as coefficient of variation (CV) of within- (or intra-) assay error and between- (or inter-) assay error. Intra-assay CV is determined by multiple measurements on the same sample in the same batch. Inter-assay CV is determined by measurement of the same plasma pool (control) in 20 or more successive batches. In RIA, a CV deemed acceptable for one analyte may not be the same as for another, but for steroid hormone RIA a CV of 10% or less is usually adequate. A CV is usually larger for lower hormone levels; intra-assay CV is usually less than inter-assay CV. Establishing inter-assay precision is particularly important since small intra-assay CV's are not necessarily reflected by inter-assay CV. Factors affecting batch-to-batch variation include variable recovery in extraction or purification steps, variation of standard reference material or reagents over time, and differences in technique among workers or inattention to a standardized procedure.

For a particular method there is usually an intrinsic range of concentration, which can be measured with maximum reproducibility (minimum variation between replicate analyses). In the case of RIA, precision is also related to the slope of the assay curve, i.e., the magnitude of difference in response (B/F or %B) values observed for two doses in the standard set. Good precision is generally achieved for an RIA method when large differences in the response parameter occur for small differences in doses over the range of steroid concentrations of maximum clinical interest. The introduction of automated procedures has greatly enhanced precision. However, in automatic pipetting and dispensing systems, contamination of subsequent samples by samples with high concentrations of analyte may occur. The effect of an automated procedure on carry-over should always be checked.

Specificity. This term refers to the exclusive measurement of a compound or compounds for which the method has been designed. In other words, when a method is designed to measure a particular steroid or steroid group, it should measure only those steroids and nothing else. The clinical usefulness of the determination of hormones, their metabolites, or both, in blood or urine lies in the proper assessment of their production in the body. A method without appropriate specificity defeats this purpose.

Generally, the specificity of a steroid assay hinges on a selective feature of the assay method— selective color reaction (e.g., Zimmermann color reaction for 17-ketosteroids), distinctive chromatographic behavior of the compounds (e.g., retention times in gas chromatographic measurement of pregnanediol or estriol), or selective interaction of steroid with binding proteins (e.g., competitive protein-binding assay of cortisol using CBG; radioimmunoassay of estradiol using antibodies). However, the chromatographic property of a compound, the selectivity of the color reaction, or the selective interaction of the binding protein does not necessarily impart specificity. Interference from many drugs, reagents, and other materials in the biological extract may yield spurious results.

For RIA methods, specificity problems occur with antisera because of their interactions with steroids of similar structure. Therefore, the cross-reactivity of each batch of antiserum must be established, and a purification procedure for exclusion of interfering steroids must be included in the method. Nonspecific factors such as ionic strength, pH, and excessive concentrations of heparin, urea, and bilirubin may alter the affinity of binding of an antibody for a specific steroid. Another form of nonspecific interference in RIA is so-called *nonspecific binding* (NSB), which refers to the amount of labeled steroid that remains, after separation, in the bound fraction when antibody is absent. Corrections for NSB are usually accomplished by including assay blanks in each assay. An assay blank tube contains unknown (often as a zero standard or normal pooled plasma) and all reagents except antibody.

Sensitivity. This term is defined as a minimum amount of a substance in biological medium that can be determined with accuracy and precision by a particular method. It applies essentially to the degree of sensitivity of the final phase of quantitation. Generally, the methods involving fluorescence are more sensitive than those based on color reactions. In some cases sensitivity can be further increased by the use of gas chromatographic methods. However, the highest sensitivity seems to be obtained by radioligand-binding assays, including radioimmunoassays.

Although the theoretical limit of detection in RIA is primarily determined by the affinity constant of an antiserum, the actual experimental limit of detection is affected by the tracer-specific activity and, more importantly, by blank values. Indeed, the blank value can cause under- or overestimation of a steroid. The blank value is defined as an *apparent* amount of a given steroid in a sample free of that steroid. The blank value has been ascribed to several sources. Steroidal or nonsteroidal materials present in the plasma extract or unknown materials in the solvent or in the eluate from the chromatograms may either compete for the binding sites or denature part of the binding protein. Binding of the radioactive steroid will be decreased, and the steroid level will be overestimated. If the interfering substances adsorb the steroid or decrease efficiency of a separation adsorbent such as dextran-coated charcoal, an apparent increase in the bound fraction will occur and will result in an underestimation of the steroid. The limit of sensitivity of steroid assays is usually defined as two times the standard deviation of the blank determination.[2] As discussed above, factors such as precision of the standard curve, the conditions of the assay (pH, buffer, temperature), the reaction volume, and the overall accuracy and reproducibility of the method must be considered when assessing assay sensitivity. By judicious choice of appropriate antiserum and experimental parameters, steroid hormone levels as low as 10 pg/tube can be quantitated.

ADRENOCORTICAL STEROIDS

The human adrenal cortex secretes a variety of steroid hormones that are intimately concerned with a wide range of metabolic processes. More than 40 different steroids have been isolated from the adrenals. These include the corticosteroids, which are formed exclusively by the adrenals, as well as androgens, progestogens, and estrogens, which are also secreted by the gonads.

CORTICOSTEROIDS

The corticosteroids are physiologically and quantitatively the most important group of adrenal steroids. The structural formulas of some of the most significant biologically active corticosteroids are shown in Figure 9-20; their trivial and systematic names are listed in Table 9-11. These compounds all possess a Δ^4-3-keto group (unsaturation between carbon atoms 4 and

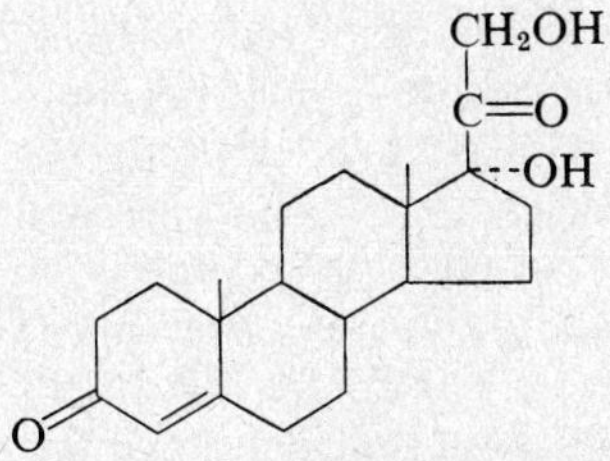

II-Deoxycortisol (Compound S)

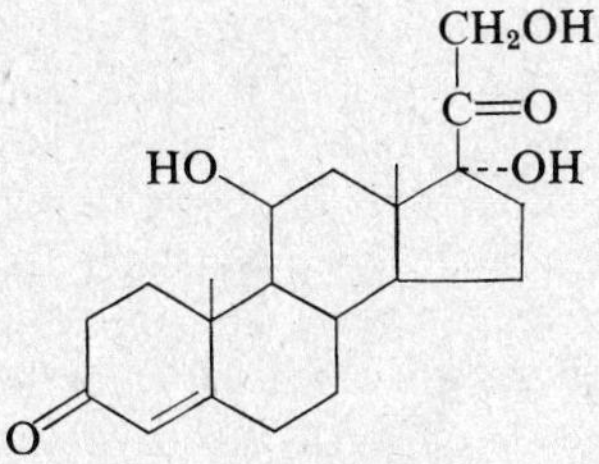

Cortisol (Compound F)

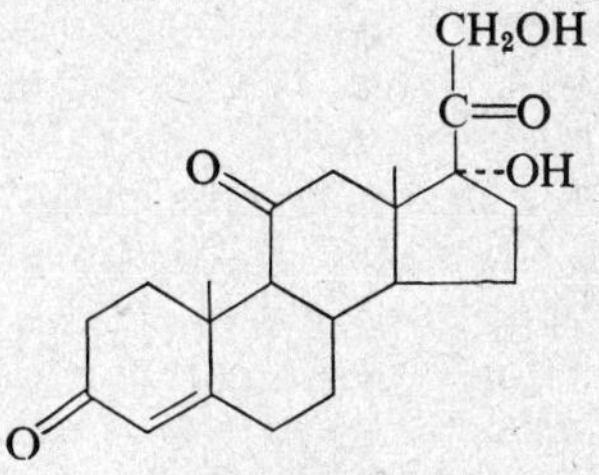

Cortisone (Compound E)

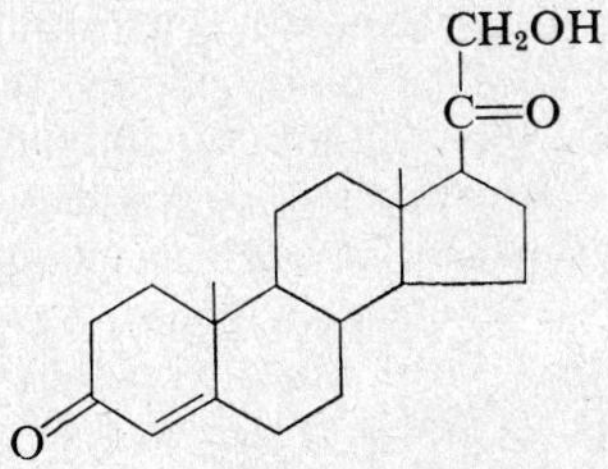

Deoxycorticosterone (DOC)

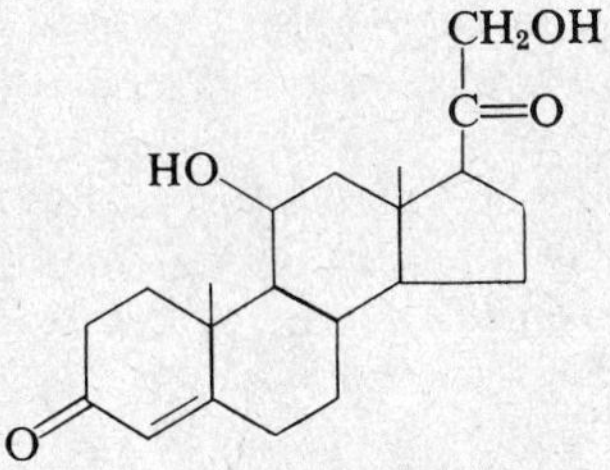

Corticosterone (Compound B)

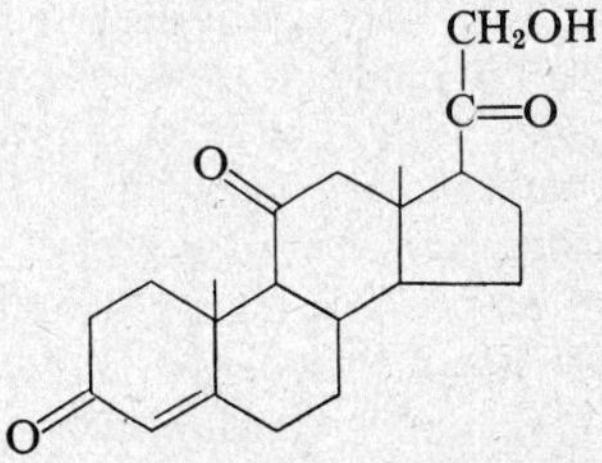

II-Dehydrocorticosterone (Compound A)

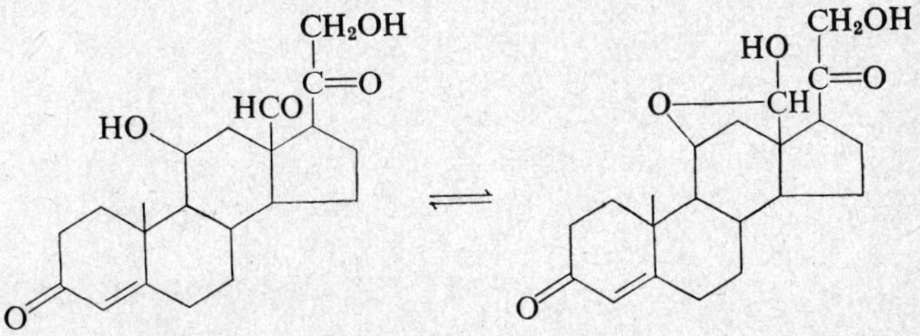

Aldehyde **Hemiacetal**

Aldosterone

Figure 9-20. Structural formulas and trivial names of some biologically active corticosteroids.

Table 9-12. RELATIVE POTENCIES OF CORTICOSTEROIDS

	Glucocorticoid Activity	Mineralocorticoid Activity
Cortisol	1	1
Cortisone	0.7	0.7
Corticosterone	0.2	2
11-Deoxycorticosterone	0	20
Aldosterone	0.1	400
Fludrocortisone*	10	400
Prednisone*	4	0.7
Prednisolone*	4	0.7
Dexamethasone*	30	2
Triamcinolone*	3	0
6α-Methylprednisolone*	5	0.5

*Synthetic.

5 and a keto group at carbon atom 3); a side chain (CH_2—$\overset{O}{\overset{\|}{C}}$—) substituted at C-17 in the β-position (above the plane of the paper); and, with the exception of compound S and deoxycorticosterone, an oxygen function (keto or β-hydroxyl) at C-11. Cortisone and hydrocortisone (cortisol) also have a 17α-hydroxyl group (below the plane of the paper). Corticosteroids show maximum structural specificity. Structural alterations, especially the reduction of the Δ^4-3-keto group, render them biologically inactive. The major corticosteroids—cortisol and aldosterone— are secreted by the adrenals at the rate of approximately 25 mg/d and 200 μg/d, respectively.

Functionally, the adrenal cortical steroids may be subdivided into glucocorticoids and mineralocorticoids. The influence of **glucocorticoids** on carbohydrate metabolism includes promotion of gluconeogenesis, deposition of liver glycogen, and elevation of blood glucose concentration due to decreased carbohydrate utilization. Increased gluconeogenesis is principally due to the stimulation of protein catabolism. In addition, glucocorticoids inhibit amino acid uptake and protein synthesis in peripheral tissues (muscle, skin, bone). These steroids also affect fat metabolism; when present in excess (see Cushing's syndrome, p. 1066), glucocorticoids produce a central distribution of fat in the face, neck, and trunk. Large quantities of circulating glucocorticoids have other clinically important effects, such as anti-inflammatory actions. Glucocorticoids have therefore been used therapeutically in inflammatory conditions such as rheumatoid arthritis. Closely related to the anti-inflammatory actions of glucocorticoids are their immunosuppressive actions. Inhibition of the normal immune response results from gradual destruction of lymphoid tissue, followed by a decrease in antibody production as well as in the numbers of eosinophils, basophils, and lymphocytes. The inhibition of immune response is beneficial during transplant and graft operations, although it renders the patient more susceptible to infection. Excess glucocorticoids also have anti-allergic properties, probably due to inhibition of the intracellular synthesis of histamine in mast cells and basophils. Of the naturally occurring corticosteroids, only cortisol and corticosterone possess glucocorticoid activity. Of these, cortisol is the most active. Cortisone and 11-dehydrocorticosterone lack glucocorticoid activity as such but have potential glucocorticoid activity, since they are converted to cortisol and corticosterone, respectively.

The term **mineralocorticoids** refers to those steroids that regulate salt (sodium conservation and loss of potassium) and water (control of extracellular fluid) metabolism. The most potent naturally occurring mineralocorticoid is aldosterone, followed in potency by deoxycorticosterone (DOC), 18-hydroxy-DOC, corticosterone, and cortisol. A large number of analogs with mineralocorticoid and glucocorticoid activity have been synthesized in recent years; some are more potent than the naturally occurring ones (Table 9-12).

OTHER ADRENAL HORMONES

Besides corticosteroids, the adrenals also secrete androgens, progesterone, and estrogens, all of which are known to be produced by the gonads as well. Androgens include androstenedione, testosterone, dehydroepiandrosterone (DHEA), and 11β-hydroxyandrostenedione (Fig. 9-21).

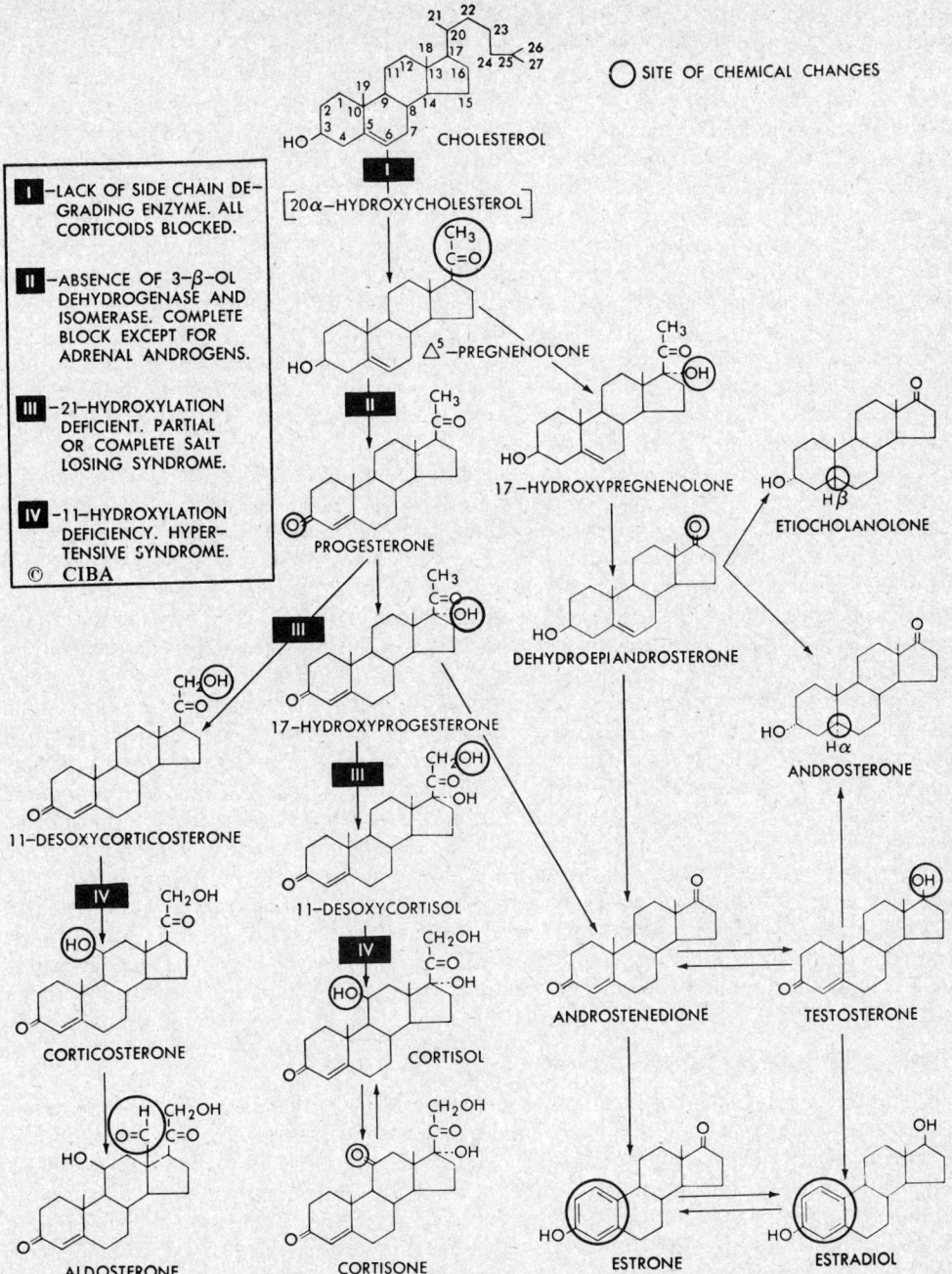

Figure 9-21. Biosynthesis of corticosteroids. I, II, III, and IV indicate sites of major blocks causing adrenogenital syndromes. (© Copyright 1959 CIBA Pharmaceutical Company, Division of CIBA-GEIGY Corporation. Reproduced, with permission, from *The CIBA Collection of Medical Illustrations* by Frank H. Netter, M.D. All rights reserved.)

From the quantitative standpoint, androstenedione and DHEA are the most important adrenal androgens. DHEA is believed to be secreted at the rate of as much as 25 mg/d. It is secreted both as the free steroid and sulfate conjugate (DHEAS); in the circulation the latter predominates. DHEAS is produced almost exclusively by the adrenal glands. Estrogens (e.g., estrone) and progesterone of adrenal origin are quantitatively insignificant.

Biosynthesis of Adrenal Corticosteroids

Investigations with radiolabeled compounds have shown that acetate, cholesterol, pregnenolone, and progesterone are all precursors of corticosteroids. The rate-limiting step in the biosynthesis of steroids is the conversion of cholesterol to pregnenolone. Cholesterol itself can be synthesized within the gland or be taken up from the circulation (see p. 1043). Important and characteristic biochemical events involved in the formation of adrenal steroids include the introduction of hydroxyl groups at C-21, C-17, and C-11, catalyzed by the specific enzyme systems known as hydroxylases. The biosynthesis of aldosterone has not yet been completely elucidated, but it is believed that corticosterone is hydroxylated at C-18 and then dehydrogenated by 18-hydroxysteroid dehydrogenase to produce aldosterone. Androstenedione and testosterone arise from either 17-hydroxyprogesterone, in a reaction catalyzed by the enzyme 17,20-desmolase, or from dehydroepiandrosterone, which is synthesized directly from 17-hydroxypregnenolone following removal of the side chain. Subsequent hydroxylation of androstenedione at C-11 forms 11-β-hydroxyandrostenedione. Aromatization of androstenedione and testosterone produces estrogens.

Under normal circumstances, cortisol, aldosterone, and DHEA are physiologically the most significant secretory products and undergo rapid enzymatic transformation through multiple steps. Impairment of these enzymatic steps results in accumulation and secretion of intermediates in quantities sufficient to be pathophysiologically important.

The adrenal cortex consists of three relatively distinct layers of cells (zones), each having different functions in terms of its secretory products: the outer layer or the zona glomerulosa, the zona fasciculata, and the zona reticularis. The *zona glomerulosa* produces aldosterone. This layer lacks the 11β-hydroxylase enzyme required for cortisol synthesis, but it uniquely contains 18-hydroxysteroid dehydrogenase, an obligatory enzyme for the formation of aldosterone. ACTH stimulates steroidogenesis in the zona glomerulosa from cholesterol to progesterone and corticosterone. Subsequent transformations of corticosterone to aldosterone are controlled by the renin-angiotensin system. (See below.)

The principal secretory products of the *zona fasciculata* in humans are cortisol and, to a lesser extent, corticosterone and 11-deoxycorticosterone. The *zona reticularis* is the principal site of secretion of androgens such as androstenedione and dehydroepiandrosterone, and probably of estrogens. This layer is characterized by the presence of 17,20-desmolase, an enzyme that cleaves the side chain of 17α-hydroxyprogesterone and 17α-hydroxypregnenolone to form androstenedione and dehydroepiandrosterone, respectively. To a very limited extent these androgens may be converted to estrogens by aromatization reactions similar to those occurring in the ovary. However, the major source of adrenal estrogens is the peripheral aromatization of androstenedione. When stimulated by ACTH, the fascicular and reticular zones are able to transform cholesterol to glucocorticoids and androgens. (See Figure 9-21.) In the absence of ACTH, the fascicular and reticular zones atrophy; however, the zona glomerulosa, which is under the control of the renin-angiotensin system, is relatively preserved.

Metabolism of Adrenal Cortical Steroids

Common aspects of steroid metabolism have been discussed before (p. 1043). The major site of corticosteroid metabolism is the liver. Unlike aldosterone, cortisol circulates in blood bound to a specific plasma protein, transcortin (also called corticosteroid-binding globulin, CBG), and consequently cortisol is metabolized more slowly ($t_{1/2}$ ~100 min) than aldosterone ($t_{1/2}$ ~30 min). The metabolism of cortisol is shown in Figure 9-22. Reduction of the double bond between carbons 4 and 5 by Δ^4-5β- or Δ^4-5α-reductase produces 5β-dihydro- or 5α-dihydrocortisol (dihydro F). In humans, the direction of reduction is predominantly toward the 5β-form; dihydro compounds are not found in the urine. Further reduction of the ketone group at C-3 by 3α-hydroxysteroid dehydrogenase forms tetrahydro compounds. While the latter products are the major excretory metabolites, the hydrogenation of ketone groups at C-20 produces some hexahydro compounds (cortols). A small percentage of tetrahydro- and hexahydro-cortisol is converted to 11β-hydroxy- or 11-ketoetiocholanolone by the removal of the side chain (Figure 9-22). The latter compounds are also formed from 11β-hydroxyandrostenedione following the reduction of ring A. The metabolism of other androgens will be considered elsewhere. (See discussion of androgen metabolism, p. 1087.)

Corticosterone, aldosterone, and other C_{21}-steroids such as 11-deoxycorticosterone and 11-deoxycortisol (Compound S) follow the same catabolic sequences as cortisol. However, compounds devoid of 17-hydroxyl groups (e.g., corticosterone, aldosterone, 11-deoxycorticosterone) are not metabolized to C_{19} 17-ketosteroids. Ninety per cent of cortisol metabolites are conjugated

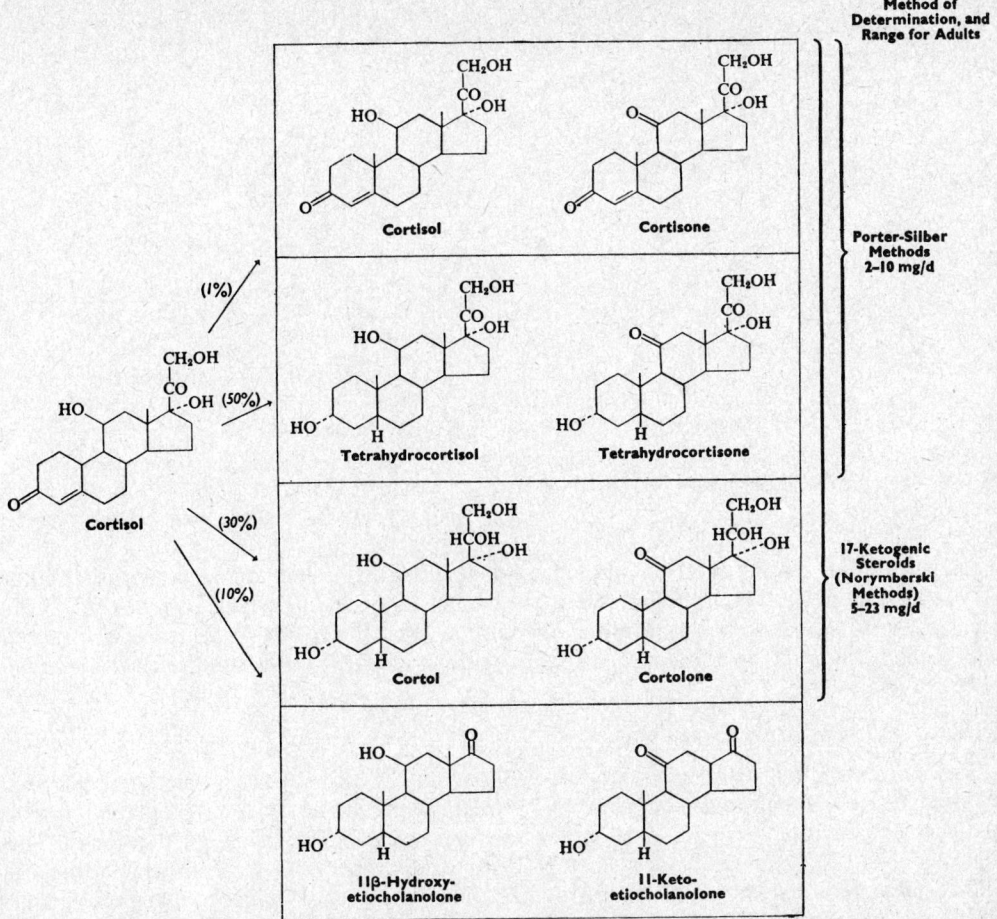

Figure 9-22. Major urinary metabolites of cortisol showing the approximate extent of conversion and the metabolites as determined by different methods. (After James, V. H. T., and Landon, J.: *In:* Recent Advances in Endocrinology. 8th ed. Boston, Little, Brown and Co., 1968.)

at C-3 and excreted as conjugates of glucuronic acid. Under normal circumstances ~1% of the total amount of cortisol secreted appears unchanged in urine.

Regulation of Cortisol Secretion

It is now well established that the synthesis and release of glucocorticoids and androgens in the zona fasciculata and the zona reticularis of the adrenal cortex are under the control of the anterior pituitary hormone, corticotropin (ACTH), through a negative feedback mechanism. (See p. 1019.) Of the products of the human adrenal cortex, only cortisol is involved in the regulation of ACTH secretion. As the level of free (non-protein-bound) cortisol in the blood rises, release of ACTH is inhibited by the negative feedback effect. Conversely, if cortisol levels are subnormal, negative feedback is diminished, ACTH levels rise, and the adrenal cortex secretes cortisol until normal blood levels are restored. The release of ACTH itself is under the control of hypothalamic corticotropin-releasing hormone (CRH); negative feedback systems involving cortisol have been identified at both hypothalamic and pituitary levels.

There is diurnal variation in the secretion of cortisol and ACTH. In the early morning hours, blood levels of ACTH and cortisol start to rise. They both reach a maximum between 0600 and 0800 h, after which there is a gradual decline to the lowest level between early evening and midnight. The level at 2000 h is normally two-thirds the level at 0800 h. The regularity of this rhythm is a function of one's sleep-wake habits and can be altered by consistent revision of the time and duration of the sleep-wake schedule. The normal regulatory system is superseded by stimulatory factors such as trauma and stress (e.g., pyrogens, acute hypoglycemia, injections of histamine, electroconvulsive treatments, and acute anxiety). Regardless of the time of day and

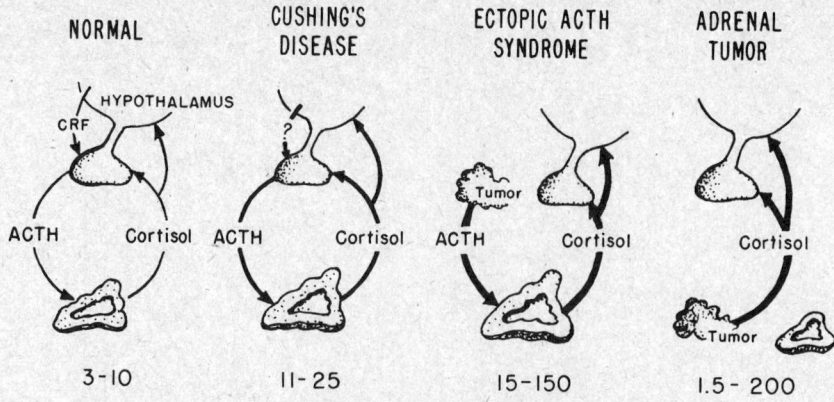

Figure 9-23. Hypothalamic-pituitary-adrenal relations in the normal situation and in various adrenal disorders. (After Lipsett, M. B., et al.: Ann. Intern. Med., *61*:733, 1964.)

independent of the plasma level of cortisol, normal individuals respond to stressful situations with a sharp rise in ACTH secretion and a consequent increase in cortisol secretion. In certain adrenal diseases (e.g., Cushing's syndrome), the diurnal rhythmicity of cortisol secretion is absent. The hypothalamic-pituitary-adrenal relationships in normal and in various adrenal disorders are depicted in Figure 9-23.

Regulation of Aldosterone Secretion

The control mechanism for the secretion of aldosterone involves the renin-angiotensin system. *Renin* is a proteolytic enzyme synthesized and stored in specialized cells, called juxtaglomerular cells, located along the terminal part of the afferent arterioles of the renal glomeruli. These specialized cells constitute part of the juxtaglomerular apparatus (JGA). Following stimulation of the JGA, renin is released into the blood, where it hydrolyzes its substrate, angiotensinogen, to produce a decapeptide called *angiotensin I.* (Angiotensin is a globulin, synthesized in the liver.) Angiotensin I is then rapidly converted to an octapeptide, angiotensin II, by a circulating *angiotensin converting enzyme* that is found in abundance in the lung. *Angiotensin II*, a potent vasoconstrictor, stimulates the cells of the zona glomerulosa to produce aldosterone. Although the precise mechanism has not yet been established, it is thought that angiotensin II stimulates the conversion of cholesterol to pregnenolone, a precursor of aldosterone. (See Figure 9-21.) In addition angiotensin II also has a minor stimulatory effect on later steps in the aldosterone biosynthetic pathway. Consequently, any increase in renin release that results in stimulation of angiotensin II formation will cause increased aldosterone secretion. Primary stimuli for renin release are a decrease in perfusion pressure to the JGA and a negative sodium balance. Stimulation of renin release and aldosterone secretion in various physiologic and pathologic conditions will be further discussed below.

Congenital Adrenal Hyperplasia (CAH; Adrenogenital Syndrome)

The biosynthesis of cortisol and aldosterone from cholesterol requires the action of specific enzymes for the chemical modification and introduction of different functional groups. CAH is characterized by an inherited absence or deficiency of any of the biosynthetic enzymes that lead to cortisol production.[203] As noted in Figure 9-21, defective or deficient enzymes may block corticoid synthesis at four sites. Recall also that low levels of plasma free cortisol reduce inhibition of ACTH secretion, and that ACTH stimulates the conversion of cholesterol to pregnenolone. Enzyme defects therefore cause (1) hyperplasia of the adrenal cortex and (2) accumulation of intermediate compounds behind the block. The accumulated intermediates indicate the particular deficient enzyme. A partial block may cause either marked or subtle clinical manifestation, whereas a complete enzyme block is incompatible with life.

The alternative names at the head of this section reflect the natural history of the enzymatic defect: congenital, present at birth; adrenal hyperplasia, overgrowth of cortical tissue; and ad-

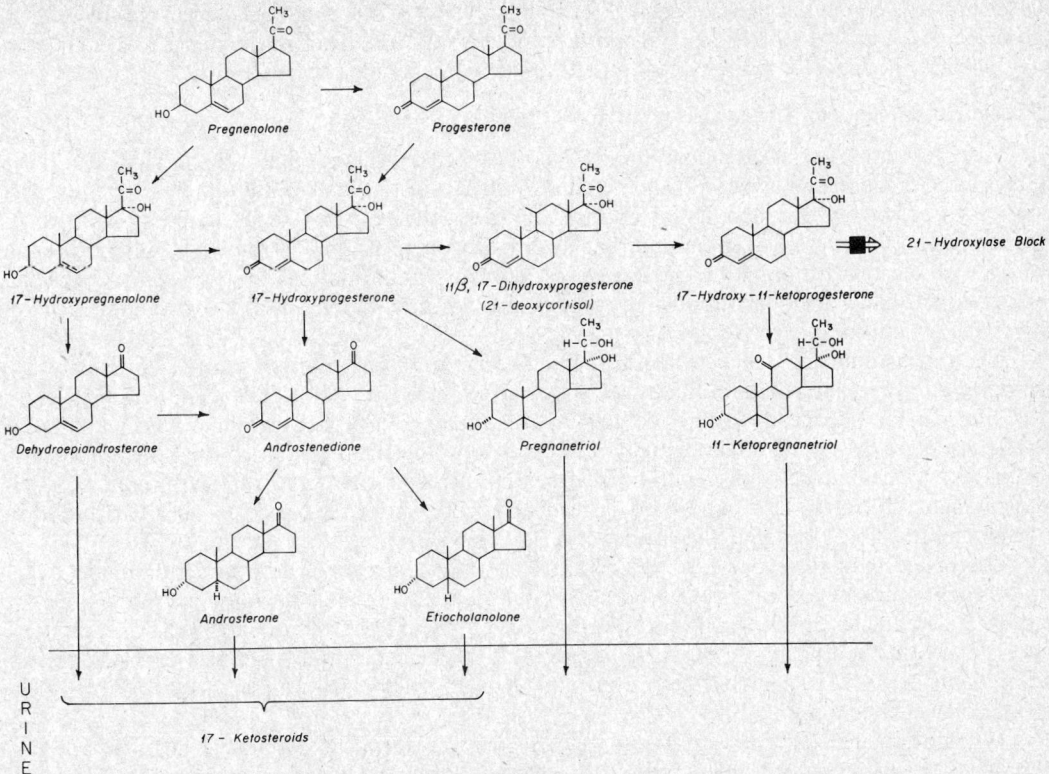

Figure 9-24. Formation of steroids in adrenogenital syndrome (21-hydroxylase deficiency).

renogenital, affecting genitalia or secondary sex characteristics. In females, particularly neonates, the diagnosis is commonly suggested by ambiguous genitalia. In males, the abnormality may not be suspected until there are signs of precocious puberty or accelerated growth. Because aldosterone production may be affected by diversion of intermediates, hypertension and salt-wasting may also be present. Adrenogenital syndrome is now being recognized with increased frequency in adults, often presenting with abnormalities at the time of puberty. In adult women, the clinical presentation may be indistinguishable from the polycystic ovary syndrome or idiopathic hirsutism.[57]

The most frequently encountered enzyme defect causing congenital adrenal hyperplasia is 21-hydroxylase deficiency (Figure 9-24). Other enzyme defects include 11β-hydroxylase and 3β-hydroxysteroid dehydrogenase deficiencies (Figure 9-21). The standard initial diagnostic approach to these disorders has been the measurement of 17-ketosteroids in the urine. For the first two weeks of life, normal infants excrete up to 2.5 mg of 17-ketosteroids per day; the quantity is < 2 mg/d before the age of six, and between 2 and 4 mg/d from age six to puberty. Elevation of urinary pregnanetriol (> 0.2 mg/d), the primary metabolite of 17-hydroxyprogesterone, is diagnostic of 21-hydroxylase deficiency.[203] It is now possible to measure the specific precursor steroids in blood—17-hydroxyprogesterone in 21-hydroxylase deficiency, 11-deoxycortisol in 11β-hydroxylase deficiency, and 17-hydroxypregnenolone in 3β-hydroxysteroid dehydrogenase deficiency. These tests have largely replaced the less specific urinary steroid determinations.[176,196] When congenital adrenal hyperplasia presents in an adolescent or an adult, basal levels of these precursors may not be clearly elevated but will rise to abnormal levels after ACTH stimulation.[57] With 21-hydroxylase deficiency, a 30-min infusion of cosyntropin will cause the level of serum 17-hydroxyprogesterone to rise above 330 ng/dL, the upper limit of a normal response.

The effectiveness of a treatment program for congenital adrenal hyperplasia is judged by normal linear growth, normal sexual development, and suppression of abnormal blood and urine steroid levels into the normal range. With virilizing congenital hyperplasia due either to 21-hydroxylase deficiency or to 11β-hydroxylase deficiency, excess androstenedione from the adrenals is converted to testosterone in peripheral tissues; plasma testosterone values in the adult male range are common in these infants. With effective treatment, suppression of plasma testosterone to <0.2 ng/mL correlates with normal values for 17-ketosteroids and pregnanetriol and may obviate the need for urine collections to monitor the results of treatment.[154]

Clinical Significance of Glucocorticoid Measurements

A large number of steroid compounds are produced by the adrenal cortex. Most can now be measured in blood and urine with specific and sensitive assays. Many of these steroid compounds have little or no biologic activity in themselves. Nevertheless, they serve as precursors for a smaller number of steroid compounds that are important in normal physiology and in disease states. Familiarity with common disease states involving abnormalities of adrenal cortical hormones should allow the appropriate use of laboratory tests to reach an accurate diagnosis or to assess the effectiveness of a treatment program.

The most important glucocorticoid in humans is cortisol. Although plasma cortisol levels are reflected in measurements of urinary 17-hydroxysteroids and 17-ketogenic steroids (i.e., steroids that can be chemically converted to 17-ketosteroids), these nonspecific assays have been largely replaced by specific measurements of cortisol in blood and urine. Cortisol measurements are the key to evaluating suspected abnormalities of glucocorticoid production. A single cortisol measurement under random conditions is rarely of value, however, since cortisol varies under normal circumstances through the course of a day and with stress.

Glucocorticoid deficiency (Addison's disease). Deficient glucocorticoid production may be due to destructive processes that involve both adrenal glands, such as autoimmunity, infection, or neoplasm, or due to deficient pituitary production of ACTH that is secondary to disorders of the pituitary or hypothalamus. Pharmacologic doses of glucocorticoids may also cause sufficient suppression of ACTH such that cessation of treatment will be followed by a period of hypoadrenocortical function.[290]

When adrenal insufficiency presents in early or mild form, symptoms or signs of adrenal hypofunction may not be evident unless the patient is under stress, in which case failure to increase cortisol and aldosterone appropriately may lead to hypotension, hyponatremia, and hyperkalemia. Symptoms of chronic cortisol deficiency include easy fatigue and weakness. With *primary adrenal insufficiency,* deficient mineralocorticoid production may cause postural hypotension and electrolyte imbalances; excess ACTH release, unchecked by negative feedback, may cause darkening of the skin and mucous membranes through a melanocyte-stimulating hormone effect.

In healthy individuals the adrenal gland will increase cortisol production rapidly in response to ACTH stimulation. The biologically active 1–24 amino acid sequence of human ACTH has been synthesized and is available as cosyntropin or Cortrosyn (Organon). This compound is a potent stimulant of cortisol secretion and has a very brief half-life and minimal antigenicity; this is a welcome replacement for bovine ACTH, which is highly antigenic and has been a cause of fatal anaphylaxis. Administration of cosyntropin should be followed by a ≥7μg/dL rise in plasma cortisol, reaching a peak value of ≥20 μg/dL.[288] (See Protocol H.) A normal response establishes

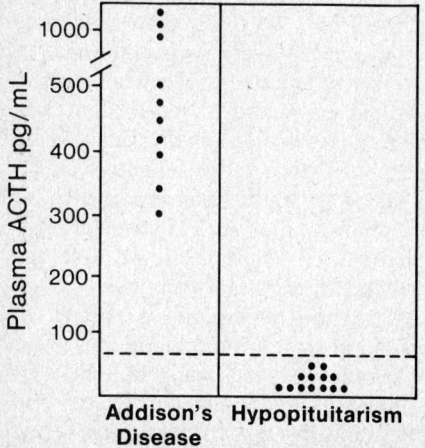

Figure 9-25. Plasma ACTH levels in patients with adrenal insufficiency due to primary adrenal insufficiency (Addison's disease) and hypopituitarism. (From Watts, N. B., and Keffer, J. H.: Practical Endocrine Diagnosis. 3rd ed. Philadelphia, Lea & Febiger, 1982.)

PROTOCOL J: OVERNIGHT METYRAPONE TEST

Rationale: Metyrapone inhibits 11β-hydroxylase, the enzyme that catalyzes the step immediately preceding cortisol synthesis. As the blood level of cortisol falls, the negative feedback effect is diminished, causing release of ACTH from the pituitary. The stimulatory effect of ACTH on the adrenal cortex leads to a rise in 11-deoxycortisol, the compound immediately preceding cortisol in the biosynthetic pathway.

Procedure: Metyrapone (3 g) is given orally at 2300 h with milk or a snack to delay absorption. Blood is drawn at 0700 h the next day for determination of 11-deoxycortisol and cortisol.

Interpretation: Normally, 11-deoxycortisol rises to >7 μg/dL after metyrapone. Failure to respond may be seen in pituitary or hypothalamic disease or with inadequate enzyme blockade (plasma cortisol >3 μg/dL).

---

that the adrenal cortex is capable of releasing cortisol in a normal fashion in response to ACTH and excludes a diagnosis of primary adrenal insufficiency.

Failure to respond to a single dose of cosyntropin strongly suggests adrenal insufficiency, either primary *or* secondary to disease of the pituitary or hypothalamus. Measurement of ACTH in plasma (Figure 9-25) will usually allow the definitive diagnosis of primary adrenal insufficiency in the unstressed patient (elevated ACTH level, low cortisol) or secondary adrenal insufficiency (inappropriately low ACTH level in relation to a low cortisol level).[31]

If secondary adrenal insufficiency is suspected and the diagnosis is still in doubt after these studies,[170] additional tests may be done. Unfortunately, direct measurement of ACTH after various stimuli has not been predictable enough to be helpful.[169] Many clinicians prefer to follow abnormal or equivocal results of the rapid cosyntropin stimulation test (Protocol H) by giving 500 μg of cosyntropin by intravenous infusion over 6 h and measuring plasma cortisol. The normal adrenal gland responds with a prompt increase in cortisol secretion. In primary adrenal failure, no response in cortisol secretion is observed. In patients with secondary adrenal insufficiency, an inadequate or absent response is observed at first, but a delayed or staircase response is seen after successive administration of cosyntropin for 2–3 d. In order to test the integrity of the pituitary-adrenal axis, other studies rely on the adrenal response to maneuvers that stimulate endogenous ACTH release. One must document that the adrenal glands are capable of responding to ACTH stim-

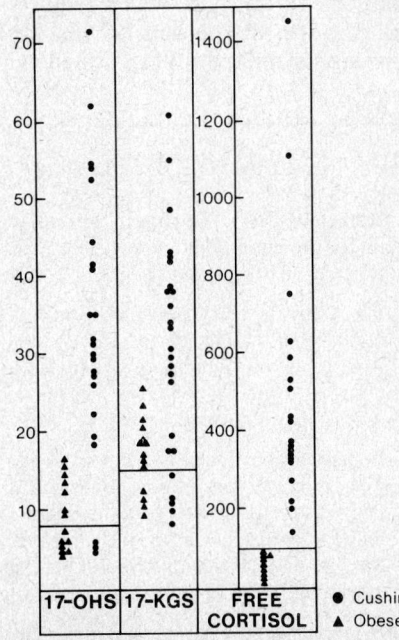

Figure 9-26. Urinary steroid in obese subjects and in patients with Cushing's syndrome, illustrating the value of urinary free cortisol. (From Watts, N. B., and Keffer, J. H.: Practical Endocrine Diagnosis. 3rd ed. Philadelphia, Lea & Febiger, 1982, after data of Eddy et al., 1974.)

● Cushing's
▲ Obese

PROTOCOL K: OVERNIGHT DEXAMETHASONE SUPPRESSION TEST

Rationale: Dexamethasone, a cortisol analog, suppresses cortisol production in normal subjects but not in patients with Cushing's syndrome.

Procedure: 1.0 mg of dexamethasone is given orally at 2300 h, along with a sedative. Blood is drawn for determination of plasma cortisol at 0700 h.

Interpretation: Normally, plasma cortisol will be ≤ 5 $\mu g/dL$ after 1 mg of dexamethasone. Plasma cortisol > 5 $\mu g/dL$ is seen in Cushing's syndrome but may also be seen due to stress, failure to take the dexamethasone, treatment with diphenylhydantoin or phenobarbital (enhancement of dexamethasone metabolism), or endogenous depression.

---

ulation before doing any of these tests. Insulin can be given to stimulate ACTH release through hypoglycemia, and plasma cortisol can be observed for a rise (Protocol C).[158] This test involves risks and should be done only with an experienced physician in attendance. A less risky test of the H-P-A axis is to give metyrapone, an inhibitor of the 11β-hydroxylase enzyme that converts 11-deoxycortisol to cortisol. In a normal person, the resulting fall in plasma cortisol would stimulate pituitary ACTH, and cortisol precursors such as 11-deoxycortisol would accumulate. Several protocols are acceptable for metyrapone testing; one that is simple and relatively safe for outpatient testing is described in Protocol J.[287]

   Glucocorticoid excess (Cushing's syndrome). Excessive cortisol production results in the clinical picture known as *Cushing's syndrome;* truncal obesity, hypertension, hypokalemic metabolic alkalosis, carbohydrate intolerance, disturbance of reproductive function, and neuropsychiatric symptoms are seen. Since these symptoms are also commonly noted in patients with normal adrenal function, a simple and sensitive test to screen for Cushing's syndrome is needed. Although cortisol overproduction is the hallmark of Cushing's syndrome, patients with other varieties of adrenal cortical hyperfunction may produce excessive cortisol and other steroids as well (androgens, inactive steroids, or aldosterone, which will be covered in more detail later). Measurements of cortisol in blood or urine under conditions of suppression and stimulation are needed to evaluate patients suspected of Cushing's syndrome and to provide a differential diagnosis. Cortisol excess may be due to excess ACTH production by a pituitary adenoma, a benign or malignant adrenal cortical neoplasm, or the production of ACTH by a nonendocrine neoplasm.

   Testing for Cushing's syndrome. Urinary free cortisol is high in most patients with Cushing's syndrome.[77] This test, which requires a 24-h urine collection and cortisol assay, appears to be the most specific and sensitive one for initial screening. A normal value is strong evidence against Cushing's syndrome. When a high value is found, repeat testing may be indicated since transient

---

PROTOCOL L: LOW-DOSE DEXAMETHASONE SUPPRESSION TEST

Rationale: Normal subjects will show lowering of plasma and urinary free cortisol concentrations under the conditions of low-dose dexamethasone suppression. Patients with excess cortisol production will usually show persistent elevation of cortisol under these conditions.

Procedure: Twenty-four–hour urine samples are collected daily for four consecutive days. Dexamethasone, 0.5 mg, given orally every 6 h is begun at 0800 h on day 3 and continued for 8 doses. Free cortisol, 17-hydroxycorticosteroids, and creatinine are measured in each 24-h urine sample. Other measurements include plasma cortisol at 0800 h and 2000 h on day 1 or 2 to look for diurnal variation and plasma cortisol at 0800 h on day 5.

Interpretation: Urine free cortisol should be $< 50\%$ of the upper limit of normal on day 4 and urine 17-hydroxycorticosteroids < 4 mg/g of creatinine. Plasma cortisol on day 4 should be ≤ 5 $\mu g/dL$. With normal diurnal variation, plasma cortisol at 2000 h should be two-thirds the value at 0800 h. Patients with Cushing's syndrome will usually not show diurnal rhythmicity and suppression with the low-dose dexamethasone. Those patients taking diphenylhydantoin or phenobarbital, or both, metabolize dexamethasone more rapidly than normal and may not show suppression.

Table 9-13. ANATOMIC TESTS FOR CUSHING'S SYNDROME

Pituitary studies: Tomograms; computed tomography; selective venous sampling for ACTH measurements (jugular to peripheral ratios)
Adrenal studies: Intravenous pyelography; ultrasonography; computed tomography; selective arteriography; selective venography; selective venous sampling for cortisol measurement
Ectopic ACTH production: Conventional radiography, ultrasonography, and computed tomography to look at sites of suspected neoplasms

increases can occur as a result of stress. Urine free cortisol is proportional to the integrated value of free (non-protein-bound) cortisol in blood. Under normal circumstances, only about 1% of secreted cortisol appears in the urine as free cortisol; the remainder is metabolized to other 17-hydroxycorticosteroids or conjugated to sulfates or glucuronides. In Cushing's syndrome, a higher proportion of cortisol in blood is free, and integrated cortisol secretion is increased. These factors contribute to the increase in urine free cortisol in subjects with Cushing's syndrome. In contrast to values for urinary 17-hydroxycorticosteroids and 17-ketogenic steroids, urinary free cortisol values are not increased by obesity (Figure 9-26). Values are decreased, however, in patients with renal disease.

Blood samples may be more easily obtained than timed urine samples. Because plasma cortisol levels vary with time of day, standardized collection times are necessary. The overnight dexamethasone suppression test (Protocol K) utilizes plasma cortisol levels obtained under controlled circumstances. Dexamethasone is a potent glucocorticoid; a dose of 1.0 mg given at bedtime is normally adequate to suppress the nocturnal rise in ACTH level and thus the normal morning rise in cortisol. In normal individuals the cortisol level at 0800 h will be suppressed to <5 $\mu g/dL$, while values >10 $\mu g/dL$ will be found in patients with Cushing's syndrome.[230] Borderline values (0800 h levels between 5–10 $\mu g/dL$) and false positive responses make the dexamethasone suppression test somewhat less specific and less sensitive than urinary free cortisol measurement.

The greater reliability of urinary free cortisol measurement relates to the secretory pattern of cortisol. Cortisol is secreted in a pulsatile fashion in normal individuals and levels at 0800 h and 2000 h are predictable. Diurnal variation, however, is often lost in patients with Cushing's syndrome, whose plasma levels stay relatively constant at higher concentrations regardless of the time of day.[25] Patients with Cushing's syndrome are therefore more readily distinguished from normal individuals by 24-h urinary free cortisol than by plasma cortisol determinations because of the integrating effect of collection over time. Further discussion on the significance of urinary free cortisol will be found below.

An equivocal result on a screening test can be pursued with the low-dose dexamethasone test (Protocol L)[166]. The dose is sufficient to suppress cortisol production into the normal range in persons whose elevation is due to stress but should not suppress hypercortisolism seen in patients with Cushing's syndrome.

Differential diagnosis of Cushing's syndrome. In the patient whose hypercortisolism has been documented, additional testing is needed to identify the cause. Most commonly, Cushing's syndrome is due to overproduction of ACTH by a pituitary adenoma, in which case it is called "pituitary" Cushing's syndrome or *Cushing's disease*. However, the possibilities of a benign adrenal adenoma, adrenocortical carcinoma, or ectopic ACTH syndrome (ACTH production by a non-endocrine malignant tumor) must also be considered. Methods of anatomic localization,[137] listed in Table 9-13, supplement information gained from suppression and stimulation tests,[102] listed in

Table 9-14. DIFFERENTIAL DIAGNOSIS OF CUSHING'S SYNDROME

	Plasma ACTH	Suppression with High-Dose Dexamethasone	Response to Cosyntropin or Metyrapone
Pituitary Cushing's syndrome (bilateral adrenal hyperplasia)	Normal or slight elevation	Yes	Yes
Adrenal adenoma	Low	No	Variable
Adrenal carcinoma	Low	No	No
Ectopic ACTH production	Very high	No	Variable

PROTOCOL M: HIGH-DOSE DEXAMETHASONE SUPPRESSION TEST

Rationale: Patients with Cushing's disease due to an ACTH-producing pituitary adenoma will usually show suppression of cortisol with high-dose dexamethasone. Patients with Cushing's syndrome due to other causes (adrenal cortical adenoma, adrenal cortical carcinoma, ectopic production of ACTH) will usually not demonstrate any change in cortisol under these conditions.

Procedure: Twenty-four–hour urine collections are obtained daily for 4 d for free cortisol and 17-hydroxycorticosteroid determinations. Dexamethasone, 2.0 mg orally every 6 h, is begun at 0800 h on day 3 and continued for 8 doses. Free cortisol and creatinine are measured in each 24-h sample. Other measurements include plasma cortisol at 0800 h and 2000 h on day 1 or 2 to look for diurnal variation, and at 0800 h on day 5.

Interpretation: Patients with Cushing's disease due to an ACTH-secreting pituitary adenoma will usually show (1) suppression of urine free cortisol and 17-hydroxycorticosteroid excretion $\geq 50\%$ of baseline by day 4, (2) lack of diurnal variation in plasma cortisol, and (3) plasma cortisol $\geq 10 \ \mu g/dL$ at 0800 h on day 5.

Table 9-14. Differentiation of pituitary Cushing's syndrome from other causes is possible with the high-dose dexamethasone suppression test (Protocol M), in which suppression of urine and plasma cortisol does not occur except in pituitary Cushing's syndrome. Cosyntropin stimulation and metyrapone testing usually give exaggerated results in pituitary Cushing's syndrome but not with other causes. Measurement of plasma ACTH is also useful (Figure 9-27). Low or undetectable ACTH levels are virtually diagnostic of adrenal Cushing's syndrome. Plasma ACTH values > 200 pg/mL are highly predictive of ectopic Cushing's syndrome. Values between 75 and 200 pg/mL are consistent with, but not highly predictive of, pituitary Cushing's disease. Since none of these tests is totally specific for any of the causes of Cushing's syndrome, a number of chemical and radiologic studies may be needed to make a differential diagnosis with confidence.

Clinical Significance of Measurements of Renin and Aldosterone

The interaction of renin, angiotensin, and aldosterone is important in the regulation of extracellular fluid volume, blood pressure, and the balance of sodium and potassium. A change

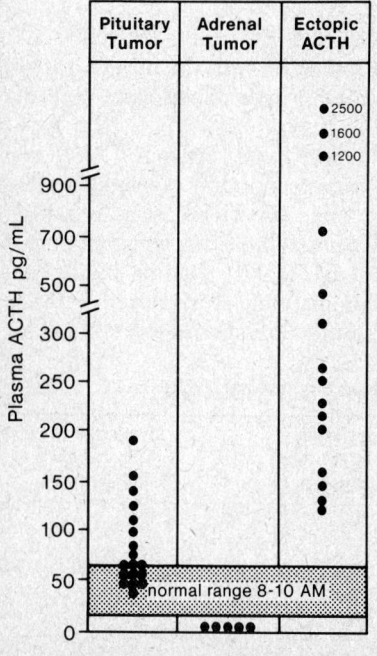

Figure 9-27. Plasma ACTH levels in patients with bilateral adrenal hyperplasia due to excess ACTH secretion from the pituitary are compared with levels in patients with Cushing's syndrome due to adrenal neoplasia or ectopic secretion of ACTH. (After Besser, G. M.: Br. Med. J. *4*:522, 1968; from Watts, N. B., and Keffer, J. H.: Practical Endocrine Diagnosis. 3rd ed., Philadelphia, Lea & Febiger, 1982.)

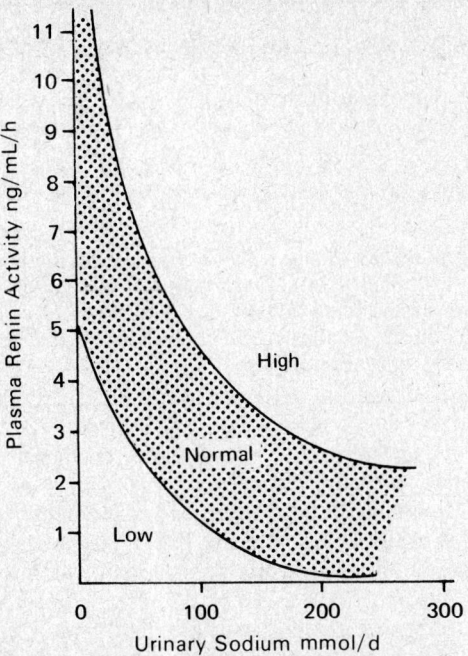

Figure 9-28 Plasma renin compared with urinary sodium excretion. (From Laragh, J. H., et al.: Am. J. Med., *53*:649, 1972.)

in one of these variables leads to changes in the others. These secondary changes are much more common than primary abnormalities. A decrease in effective plasma volume or mean arterial pressure leads to release of renin from juxtaglomerular cells of the kidneys; more angiotensin is formed, and an increased production of aldosterone by the adrenal glands occurs. The result is retention of water and sodium, an increase in extracellular volume, and a decrease in serum potassium. This state of "secondary aldosteronism" is commonly present in congestive heart failure, nephrotic syndrome, cirrhosis of the liver, other hypoproteinemic states, or any condition of chronic depletion of plasma volume. Secondary hyperaldosteronism can be inferred when one of these clinical conditions is observed in patients with volume depletion, edema, and hypokalemic alkalosis. Measurements of renin and aldosterone are seldom needed in these cases. Measurements of renin and aldosterone are invaluable, however, when investigating primary abnormality in the

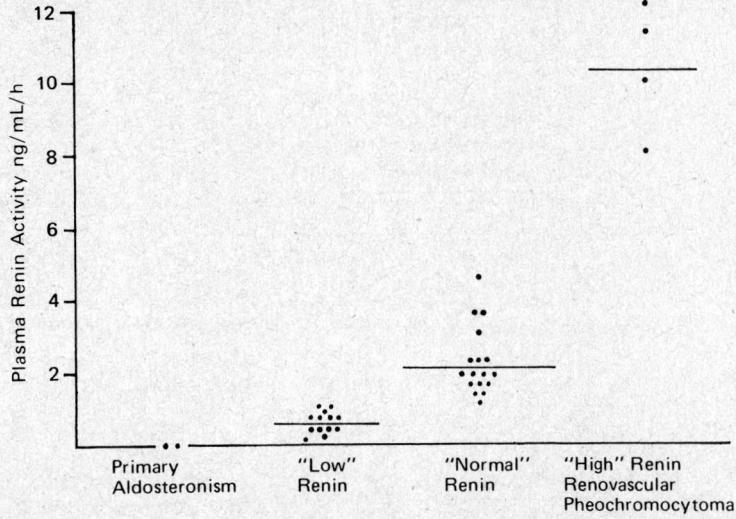

Figure 9-29. Plasma renin 5 h after 60 mg furosemide taken orally. (From Wallach, L., et al: Ann. Intern. Med., *82*:27, 1975.)

PROTOCOL N: FUROSEMIDE STIMULATION TEST

Rationale: Plasma renin activity varies with the state of hydration and sodium intake. The administration of furosemide, a potent diuretic, provides a stimulus to increase plasma renin production.

Procedure: Furosemide, 60 mg, is given orally at 0700 h. The subject is maintained in an upright posture (seated or standing) until blood is drawn at 1200 h for the determination of plasma renin activity.

Interpretation: Responses must be defined for the assay technique employed. Patients with renovascular hypertension will show values approximately five times normal. Patients with primary aldosteronism usually have plasma renin activity below the level of assay sensitivity. Patients with hyporeninemic hypoaldosteronism usually have low levels of plasma renin as well as low aldosterone levels. Figure 9-29 shows typical responses.

---

renin-angiotensin-aldosterone system and when assessing the significance of renal artery stenosis in genesis and maintenance of arterial hypertension.

For patients suspected of primary aldosteronism, measurement of plasma renin activity (PRA) is a good initial screening test. Since plasma renin levels vary with sodium balance, it is helpful to compare an individual's ambulatory plasma renin activity with his sodium excretion (Figure 9-28).[161]

Another approach to screening patients suspected of having primary aldosteronism is the furosemide stimulation test, a test that stimulates renin production.[298] In primary hyperaldosteronism, low renin levels and high aldosterone levels would be expected. However, many factors influence basal secretion of renin and aldosterone, and these factors must be controlled prior to measurement. Diuretic medications, for example, would spuriously raise renin; hypokalemia would lower aldosterone. In combination, these two effects would lead to an erroneous diagnosis. Furosemide is a potent diuretic which will stimulate plasma renin production in the normal individual (Figure 9-29 and Protocol N).[21] As shown in Figure 9-30, patients with a blunted or suppressed PRA response would be highly suspicious for primary aldosteronism and should be studied further. Patients with very high renin values, on the other hand, might be studied for other identifiable causes of hypertension such as renal artery stenosis or pheochromocytoma. Plasma renin measurements will also aid in selecting specific antihypertensive drug therapies in patients who have renin values slightly above or below normal (Figure 9-30).

Primary aldosteronism. This disorder is characterized by hypertension and hypokalemia.[96,203] However, <2% of hypertensive patients have primary aldosteronism as the underlying cause of

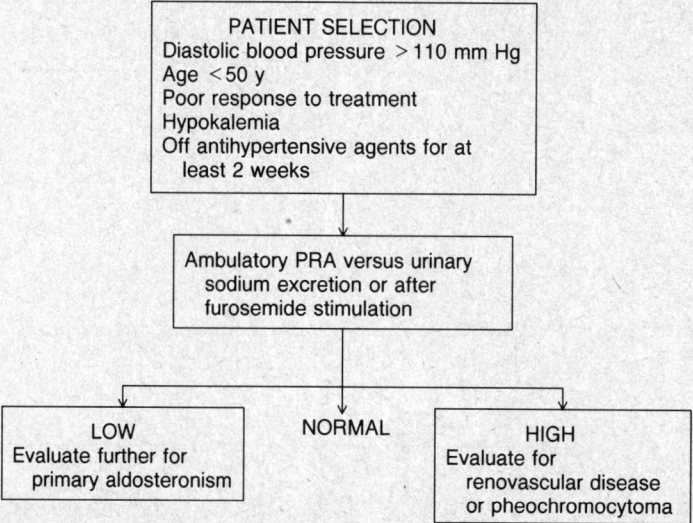

Figure 9-30. Screening with plasma renin activity (PRA) in hypertension. (From Watts, N. B. and Keffer, J. H., Practical Endocrine Diagnosis, 3rd ed., Philadelphia, Lea & Febiger, 1982.)

PROTOCOL O: SALINE SUPPRESSION TEST

Rationale: Rapid volume expansion with intravenous saline should suppress plasma aldosterone in normal subjects but not in patients with primary aldosteronism.

Procedure: Care must be taken to ensure that the subject is not hypokalemic before starting the test. The subject is awakened at 0600 h and kept in an upright posture for 2 h. Blood is drawn for determination of plasma aldosterone at 0800 h. The subject then assumes a supine position, and 2 L of saline, 0.9 g/dL, is infused over 4 h. Blood is drawn for plasma aldosterone at 1200 h.

Interpretation: Normal individuals show plasma aldosterone < 5 ng/dL after saline infusion. Levels > 5 ng/dL are usually seen in patients with primary aldosteronism.

---

their hypertension. The indications that should lead to further evaluation of the hypertensive patient for primary aldosteronism include hypokalemia without an apparent cause, development of profound or refractory hypokalemia upon treatment with diuretics, or lack of response to combination antihypertensive drug treatment.

Most patients with primary aldosteronism will be hypokalemic, but most patients with hypokalemia do not have primary aldosteronism. Primary aldosteronism or other mineralocorticoid excess can be excluded if renal conservation of potassium is normal (random urine potassium < 30 mmol/L). Renal potassium wasting associated with an elevation of plasma renin indicates secondary aldosteronism. Evaluation for primary aldosteronism should be done in the hypertensive patient with hypokalemia, renal potassium wasting, and low plasma renin activity.

Since hypokalemia has a suppressive effect on aldosterone secretion, potassium replacement should be considered before aldosterone measurements are undertaken. The demonstration of an elevated level of aldosterone in blood or urine in a patient with a low plasma renin is presumptive evidence for primary aldosteronism. Demonstration of a lack of suppression of aldosterone after saline infusion (Protocol O)[139] or treatment with a mineralocorticoid such as fludrocortisone (Protocol P)[21] confirms the diagnosis of autonomous aldosterone overproduction.

Primary aldosteronism may be due to an adenoma of one adrenal gland, i.e., an aldosterone-producing adenoma (APA, or Conn's syndrome), or to idiopathic hyperplasia of aldosterone-producing cells in both adrenal glands, i.e., idiopathic adrenal hyperplasia (IAH). This differentiation is important, since most patients with APA respond well to surgical removal of the tumor, whereas patients with IAH do not. X-ray localization[137] can help (CT scans, ultrasonography, adrenal scanning with iodocholesterol,[9,117] and adrenal venography with or without selective adrenal venous sampling for aldosterone measurement), but the tumors may be too small to be detected in this way. Aldosterone secretion and plasma renin suppression are usually greater with APA, and patients with APA will usually show no change or a fall in plasma aldosterone with sodium depletion and upright posture; patients with IAH will show a rise in plasma aldosterone after sodium restriction and upright posture (Figure 9-31).[113]

A small percentage of patients with APA or IAH will respond to dexamethasone treatment with a normalization of blood pressure and correction of the biochemical abnormalities.[95] Dexa-

---

PROTOCOL P: FLUDROCORTISONE SUPPRESSION TEST

Rationale: Fludrocortisone, a potent mineralocorticoid, will suppress aldosterone production in normal subjects but not in subjects with primary aldosteronism.

Procedure: Hypokalemia must be corrected before starting this test, and serum potassium monitored during the test. Fludrocortisone, 0.1 mg every 6 h, is given for 3 d. Twenty-four-hour urine collections for measurement of aldosterone are obtained 1 d before fludrocortisone is started and on day 3.

Interpretation: Normal subjects will have urine aldosterone ≤ 20 μg on day 3. Patients with primary aldosteronism will have urine aldosterone > 20 μg/d.

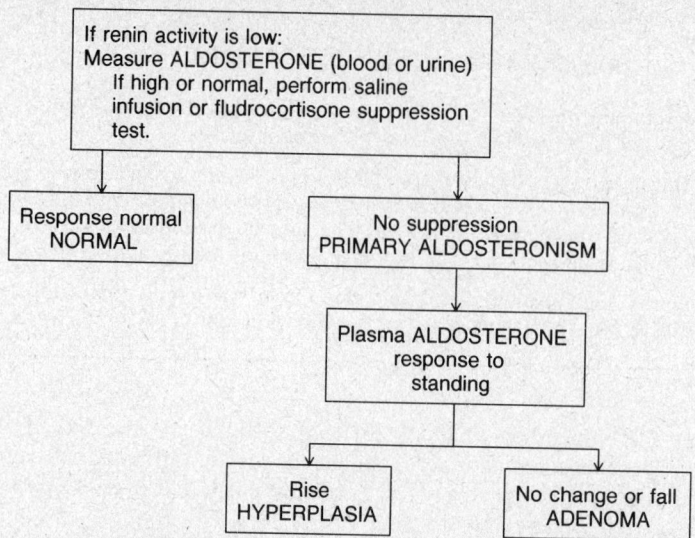

Figure 9-31. Clinical approach to patients with low plasma renin activity. Further testing may be needed to differentiate aldosterone-producing adenoma from idiopathic adrenal hyperplasia. See text. (From Watts, N. B., and Keffer, J. H.: Practical Endocrine Diagnosis. 3rd ed. Philadelphia, Lea & Febiger, 1982.)

methasone treatment should be tried before surgery, particularly in young patients, or when the normal fall in plasma aldosterone after salt depletion and standing does not occur but no unilateral lesion is seen by X-ray. Also, trial treatment using spironolactone, an antagonist of aldosterone, will usually normalize blood pressure in patients who are likely to have a good response to surgery.

States of low renin or low aldosterone. Deficient production of aldosterone may be seen due to (1) enzyme blocks in aldosterone synthesis as in congenital adrenal hyperplasia, (2) destructive lesions of both adrenal glands as in primary adrenal insufficiency with mineralocorticoid deficiency associated with glucocorticoid deficiency, or (3) deficient production of renin by the kidney leading to secondary hypoaldosteronism.[318] The resulting chemical changes are hyponatremia and hyperkalemia, often with hypochloremic acidosis. Mild or moderate volume depletion, often with postural or unprovoked hypotension, may also occur. Hyporeninemic hypoaldosteronism is most commonly seen in patients with diabetes mellitus and mild or moderate renal insufficiency. Finding low plasma renin and aldosterone levels under circumstances that should cause an elevation (furosemide stimulation, upright posture) would be diagnostic. (See Figure 9-29.)

Several unusual conditions present chemical abnormalities suggesting aldosterone excess or deficiency. However, these are primary disorders not connected to the renin-angiotensin-aldosterone system.[269] *Liddle's syndrome* (pseudohyperaldosteronism)[167] resembles primary aldosteronism clinically, but aldosterone production is low and hypertension is absent. *Bartter's syndrome*[329] is a prostaglandin excess that induces renal potassium wasting, but both aldosterone and renin levels are increased. In renal tubular acidosis and pseudohypoaldosteronism, the clinical picture of hypoaldosteronism is seen concurrent with greater than normal levels of aldosterone.

PROTOCOL Q: DIFFERENTIAL RENAL VEIN RENIN

Rationale: In patients with renovascular hypertension, plasma renin activity will be higher in the renal vein on the involved side.

Procedure: Ideally, the patient should be on a low-sodium, high-potassium diet and on a diuretic for 3 d before the procedure. Under fluoroscopic guidance, percutaneous catheterization is performed and blood samples obtained from both renal veins and the inferior vena cava for determination of plasma renin activity.

Interpretation: A ratio of plasma renin activity (affected side to unaffected side) that is > 1.5 indicates functionally significant renovascular disease.

Plasma renin in renovascular hypertension. Used as a screening test, elevated plasma renin activity after furosemide stimulation or related to urinary sodium excretion[161] is suggestive of renal artery stenosis as the cause of the hypertension. (See Figure 9-30). If there is arteriographic evidence for renal artery stenosis, measurement of plasma renin on samples obtained from selective renal vein catheterization can be helpful in predicting the response to correction of the renal vascular lesion or nephrectomy. Lateralization of renal vein renin to the radiographically involved side, especially after sodium depletion, is predictive of a good response to surgery in 90% of cases (Protocol Q).[191] Further distinction can be made by comparing renin activity in the renal artery to that in the renal vein.[317]

METHODS FOR THE DETERMINATION OF CORTISOL IN BLOOD AND URINE

Cortisol represents almost 80% of the total 17-hydroxycorticosteroids in the blood. Most cortisol circulates in its original form along with small amounts of conjugated reduced derivatives. About 90% of the circulating cortisol is bound to plasma proteins, primarily to an α-globulin known as *corticosteroid-binding globulin* (CBG) or *transcortin*; there is also weak binding to albumin. While the precise function of such protein binding is still obscure, this mechanism may assure a ready source of available circulating hormone as well as protection from inactivation and conjugation in the liver or filtration by the kidney. The concentration of CBG in the plasma rises during pregnancy and during estrogen therapy; there is a concomitant increase of total cortisol in plasma. However, since the amount of free, physiologically active hormone remains at normal levels, there are no untoward effects. Procedures involving organic solvent extraction or protein precipitation estimate both free and protein-bound cortisol.

Some authors[46] have suggested that *urinary free cortisol* may be a more reliable index of adrenocortical hyperfunction than plasma cortisol. Normally only ~1% of the total amount of secreted cortisol appears unchanged in urine, since most cortisol is protein bound and, therefore, not filtered through the glomeruli. At plasma cortisol concentrations > 20 μg/dL, transcortin becomes saturated. Consequently, the amount of unbound cortisol, and thus the amount available for filtration, increases rapidly. The result is a relatively large increase of urinary free cortisol. The urinary free cortisol, therefore, represents the excretion of biologically active cortisol responsible for the signs and symptoms of diseases such as *Cushing's syndrome.*

There are three general methods for the estimation of blood cortisol in a routine clinical laboratory. They are based on the Porter-Silber color reaction,[243] the measurements of the sulfuric acid–induced fluorescence,[189] or the ligand assays (e.g., RIA, CPB assay, and enzyme immunoassay).[30] In recent years CPB and RIA methods have been more widely used than colorimetric or fluorometric procedures.

The *fluorometric method* is more sensitive than the colorimetric technique and thus requires relatively small volumes of plasma. Steroids such as prednisone, prednisolone, and dexamethasone interfere in the colorimetric, but not in the fluorometric, method. The same is true of ketones, hexoses, and some commonly used drugs such as propoxyphene. A major drawback of the fluorometric technique is an overestimation of plasma cortisol by ~2.5 μg/dL due to the presence of nonspecific fluorogens.[189] These nonspecific fluorogens are dependent upon the functional status of the adrenal glands and appear to correlate closely with the true plasma cortisol level. Therefore, nonspecific plasma fluorescence does not negate the clinical usefulness of the method. However, the fluorometric procedure has the disadvantage of not measuring 11-deoxycortisol (compound S), which is increased in patients undergoing the metyrapone test.

Until recently, the *CPB assay* was widely used because it is rapid, fairly simple to perform, and adequately sensitive for the measurement of cortisol in blood. As opposed to fluorometric or colorimetric procedures, the results obtained with this method are not affected by the presence of nonspecific substances, drugs, and synthetic analogs (e.g. spironolactone, dexamethasone). Further advantages and disadvantages of the CPB assay and its principle have been discussed previously (see p. 1054). CPB assays[214] generally involve deproteinization of the plasma sample by the addition of ethanol. These assays use human plasma as a source of the binding protein (CBG) and Florisil for the separation of bound and free fractions.

Determination of Cortisol by RIA

A large number of cortisol radioimmunoassays have been developed using antibodies produced in animals immunized with a variety of cortisol derivatives conjugated to protein.[175] Antisera raised against cortisol-21-hemisuccinate and cortisol-3-(carboxymethyl)oxime conjugates have been found to be most suitable and are widely used. The latter antiserum, in particular, shows minimum cross-reactivity with other corticosteroids and allows direct assay for cortisol without extraction.

Principle

Most cortisol assays performed in clinical laboratories are direct assays and require no initial extraction of cortisol from the sample. In these direct assays, cortisol is quantitatively displaced from endogenous binding proteins by protein-binding agents such as ANS (8-anilino-1-naphthalene–sulfonic acid) or salicylate, by low pH, or by heat treatment.[122] However, in those cases where increased sensitivity is desirable and specific antisera are not available, an extraction of cortisol from serum before measurement may be required. This extraction can be carried out in a manner similar to urine sample extractions.[46] (See below.) An important advantage of direct assays is their superior precision as compared with assays involving extraction. In part this may reflect the variability inherent in manual extraction techniques.

^{125}I- instead of ^{3}H-labeled cortisol is used as tracer to overcome the cost and inconvenience of β-scintillation counting. The performance and technical simplicity of direct assay methods make them the techniques of choice for routine clinical application.

Popular commercial cortisol kits use antibody-coated tubes in which the cortisol antibody is immobilized onto the lower inner wall of polyethylene tubes. Unknown samples and standards are first incubated with ^{125}I-labeled cortisol in the antibody-coated tubes. After incubation, the contents of the tubes are then aspirated or decanted, and the radioactivity remaining in the tube (^{125}I-cortisol bound to antibody) is counted. This relatively quick and simple procedure does not require a centrifugation step.

Other commercial kits employ *double-antibody separations* or solid-phase antibody suspensions. In double-antibody systems, a second antibody is used to precipitate the primary antibody, and, after centrifugation and aspiration of the supernatant, the antibody-bound tracer is counted. Solid-phase suspension systems employ antibodies bound to small solid-phase particles that remain suspended in the reaction medium. After antigen-antibody reactions have occurred, the suspended particles are separated by centrifugation, the supernatant is aspirated, and the precipitate is counted to measure bound radioactivity.

A *standard curve* is prepared from serum-based standards provided by plotting either bound counts per minute (cpm) or %B for each standard against its concentration. Concentrations of samples are then determined by interpolation from the standard curve. Other methods of plotting standard curves have been discussed on page 1054.

Most commercial kits can also be used for the measurement of urinary free cortisol after appropriate *extraction* of the urine sample. Preliminary extraction of free cortisol before measurement is required because of the large number of cortisol metabolites and conjugates in urine that cross-react with the cortisol antibody in the assay. Since most interfering substances are more soluble in urine than is cortisol, the cortisol can be extracted with organic solvents such as dichloromethane and ethylacetate. Generally, the sample is thoroughly mixed with the organic solvent in a ratio of $\sim$1:5, the layers are allowed to separate, and the aqueous layer is removed. An aliquot of the organic layer containing the free cortisol is transferred to a separate tube, evaporated to dryness, and redissolved in a buffered solution. This solution can then be used as an unknown sample in the assay described for serum cortisol. Extraction efficiency can be monitored by adding a known cortisol standard to an aliquot of the sample and then extracting the spiked sample in parallel with an aliquot of sample to which no standard has been added. Extraction efficiency of cortisol can be calculated from the difference in measured cortisol between the two samples divided by the known amount of cortisol standard added to one of the samples. The calculation of the amount of cortisol in an extracted sample must take into account the dilution with organic solvent and the extraction efficiency.

Specimen collection and storage have been described on page 1052.

Reagents and Procedures

A variety of commercial kit procedures are available for the measurement of cortisol. Since these kits contain reagents, detailed instructions for performing the assay, and the necessary calculations, no details are given here.

Comments

In direct assays, steroids are quantitatively displaced from binding proteins by use of conditions that supposedly do not interfere in the subsequent immunoassay. However, the efficiency of steroid displacement by protein-binding agents such as ANS may be influenced by the levels of binding protein present in the sample. For example, the amount of ANS that is adequate for plasma obtained from normal men and nonpregnant women may be insufficient to displace cortisol from the elevated level of CBG present in pregnancy. Furthermore, the concentration of ANS required for complete displacement may reduce specific binding of the steroid to the antiserum.

In general, direct assays for cortisol are precise, and results show high correlation with those obtained by reference methods such as gas chromatography–mass spectrometry (GC-MS). However, a lack of accuracy in direct assay methods for cortisol has been reported.[160] This has been ascribed to variations in the plasma or serum matrix of the standards. Accuracy is improved by selecting plasma or serum that reflects the biological matrix of the samples.

Because of low cross-reactivity of the antiserum, corticosterone and 11-deoxycortisol do not normally interfere. However, when extremely high plasma concentrations of 11-deoxycortisol are found (such as in 11β-hydroxylase deficiency or after administration of metyrapone), preliminary extraction with carbon tetrachloride to remove this steroid is important.

Inclusion in the extraction procedure of an appropriate blank (plasma or water) is important. If the blank value is greater than the detection limit of the assay, the value in pg/tube of the blank should be subtracted from the value in pg/tube of unknowns before proceeding with the calculation. If the recovery of added cortisol is 92–95%, a recovery check on each sample may be omitted. The inter-assay and intra-assay coefficients in variation for the extraction method are generally 7.5–10.5% and 6.0–8%, respectively, for plasma concentrations ranging from 10–15 μg/dL.

Reference Range

Plasma cortisol values range from 5–23 μg/dL plasma (average 12 μg) between 0800 and 1000 h and from 2–15 μg/dL plasma (average 8 μg) at 1600 h. Urinary excretion ranges from 10–100 μg/d. In adults, there is no significant dependence of values on age and sex.

METHODS FOR THE DETERMINATION OF ALDOSTERONE IN PLASMA AND URINE

Until recently, estimations of aldosterone in plasma and urine have been beyond the scope of most routine clinical laboratories. Techniques based on double isotope derivative formation[68] or gas chromatography[79] are time consuming, complicated, and expensive; their application has been feasible only in specialized laboratories. Simple and reliable RIA methods, including commercial kits, for measuring minute quantities of aldosterone in plasma and urine[129] are now available. RIA methods differ primarily in the types of aldosterone-protein conjugates of antigens used and in the specificity of the antisera raised against these antigens. Many of the former antisera exhibited significant cross-reactivity with cortisol so that paper chromatography or other separation techniques were required to isolate and purify aldosterone. Newer radioimmunoassays use highly specific antisera and avoid organic extractions and tedious purification procedures.[122]

Determination of Aldosterone by RIA

Principle

Most direct RIA methods for plasma aldosterone employ highly specific antisera generated against an aldosterone-3-mono-oxime-BSA conjugate, ^{125}I-labeled ligand, and ANS at pH 3.6 to displace aldosterone from plasma-binding proteins. Separation of free from antibody-bound aldosterone is achieved using either a solid-phase first antibody, an accelerated liquid-phase second antibody, a solid-phase second antibody, or dextran-coated charcoal. Most assay procedures for urinary total aldosterone are similar to those for plasma except that urine is acid-hydrolyzed prior to assay.[129]

A brief outline of a direct plasma assay method[5] is as follows. An aliquot of the plasma or serum sample is pipetted into an assay tube. ^{125}I-aldosterone-3-mono-oxime-iodohistamine (tracer) and antiserum, both diluted in phosphate/citrate buffer, 0.05 mol/L, pH 3.6, are added to the sample tube. Dilutions of standards made in steroid-free serum (charcoal stripped) and ranging from 0–500 pg/tube are used for the standard curve. The tubes are mixed and incubated at 4 °C for 4 h. Separation of antibody-bound and free fractions is achieved by adding dextran-coated charcoal suspension followed by centrifugation. The supernatant is aspirated and the free fraction (charcoal sediment) counted in a gamma counter. The standard curve is obtained by plotting % free steroid against concentration. Concentrations of samples are calculated by interpolation from the standard curve.

Assay procedures involving preliminary extraction have been reported by Farmer and his colleagues for plasma[82] and urine[81] aldosterone. Antiserum is raised against aldosterone-lactone coupled to BSA, and purification consists of (1) partitioning a dichloromethane extract of plasma or urine between cyclohexane and water to remove nonpolar steroids such as progesterone; (2) periodate oxidation of aldosterone and cortisol to aldosterone-lactone and etienic acids, respectively; and (3) washing with sodium hydroxide solution to remove etienic acids and other interfering substances. The purified organic extract is evaporated to dryness. The residue is then dissolved in buffered solution and subjected to RIA.

Specimen collection and storage. Follow the procedure as described on page 1052.

Reagents and procedures. Commercial kits are available for the measurement of plasma and urine aldosterone. These kits contain instructions for the preparation of all reagents as well as descriptions of procedures. For the general information regarding RIA techniques, the reader is referred to the section on radioimmunoassay of steroids (p. 1051).

Reference Ranges. The reference range for plasma aldosterone for individuals on ad libitum sodium intake varies from 48–148 pg/mL when recumbent and from 120–300 pg/mL after standing for 4 h. Urinary excretion is 6–8 μg/d.

Determination of 11-Deoxycortisol in Plasma

Principle

Estimation of plasma 11-deoxycortisol is required for the metyrapone test. Because metyrapone inhibits 11β-hydroxylase, a large increase in 11-deoxycortisol is expected in patients with adequate pituitary-adrenal reserve. Consequently, immunoassay methods for 11-deoxycortisol need not be highly specific or sensitive. Even though a method for direct analyses of plasma 11-deoxycortisol has been reported,[122] to date very few reliable commercial kits are available. Several RIA methods involving extraction are available.[129] The method of Mahajan et al.[180] appears suitable for clinical use. Plasma samples are extracted with dichloromethane, and the organic phase is dried and redissolved in ethanol. An aliquot of this solution is used for the analyses. The assay utilizes tritium-labeled ligand and antibodies raised against the 21-hemisuccinate derivative of 11-deoxycortisol. Antibody-bound and free steroid are separated by dextran-coated charcoal adsorption.

Reference Ranges

In normal subjects receiving a single dose of metyrapone, values range from 8.0–22.3 μg/dL. In patients with hypopituitarism or Addison's disease, response values are < 1.0 μg/dL.

Determination of 17-Hydroxyprogesterone in Plasma

Principle

Until recently, plasma 17-hydroxyprogesterone was assayed with CPB techniques using CBG as binding reagent; these methods have now been replaced by RIA.[175] The method described by

Sommerville and his colleagues[341] employs an antiserum to 17-hydroxyprogesterone-3-carboxymethyloxime-BSA and appears to be simple, rapid, specific, and suited for use in routine clinical laboratories.

Specimen collection and processing as well as reagents and procedures are generally as described on pages 1052–1053. A special feature is extraction of plasma samples with diethyl ether and removal of progesterone by passage of the extract over a Sephadex LH-20 column. Free radioactivity is removed by dextran-coated charcoal.

The method has been evaluated.[48] Inter-assay CV's on samples taken in the follicular phase ranged from 5–15%. The lower limit of steroid determined with reasonable precision varied from 5–10 pg, depending on volume of plasma extracted, volume of aliquot taken for assay, and magnitude of the blank. The specificity of the method hinges on the quality of the antiserum and the efficiency of the chromatographic separation of progesterone and 17-hydroxyprogesterone. The antiserum employed by Sommerville has minimal (10%) cross-reactivity with steroids of similar structure and chromatographic properties (e.g., 20α-dihydroprogesterone, 17-hydroxy-pregnenolone, deoxycorticosterone, and progesterone). A chromatographic step to remove progesterone is still required, however, because its level is six times that of 17-hydroxyprogesterone in the luteal phase of the menstrual cycle and twenty times in pregnancy. A similar method has been proposed that uses ^{125}I-radioligand and a pH of 4.0 in the assay;[122] its clinical application and reliability remain to be established.

Reference Ranges

	ng/mL
Men (2–40 years):	0.31–3.17
Women (days 1–10 of cycle):	0.20–0.72
(day 18 to end of cycle):	0.35–2.21
Pregnancy (12th week to term):	2.18–12.89

Determination of Plasma Dehydroepiandrosterone Sulfate (DHEAS)

Principle

The clinical value of plasma assays of DHEA or its sulfated conjugate (DHEAS) is associated with the diagnosis of adrenal hyperplasia and the differential diagnosis of hirsutism. Since DHEAS predominates in the circulation and is a more specific product of the adrenals, determination of this steroid is widely used in clinical practice. Various methods involving gas-liquid chromatography, double-isotope derivative techniques, and competitive protein binding assays have been used.[30] These methods have been replaced by RIA in most clinical laboratories, and a variety of commercial kits are now available.

The RIA method by Buster and Abraham[37] is suited for routine analysis of DHEAS. Plasma samples are diluted 1:1000 in buffer and assayed using antiserum generated against DHEA-3-hemisuccinate-HSA. This antiserum completely cross-reacts with unconjugated DHEA but only minimally with androstenedione. Because of the presence of relatively large amounts of DHEAS, this steroid can be measured directly in plasma by diluting out interfering steroids that are present in relatively low concentration. Values of plasma DHEAS obtained by this method are consistent with the range of values reported for other accepted methods. Intra- and inter-assay CV's are 12.6% and 18.8%, respectively, at levels ranging from 0.1–4.56 μg/mL. The range of values that can be measured with confidence is between 0.02–5 μg/mL for plasma.

Reference Ranges

	μg/mL
Men:	1.99–3.34
Women,	
Premenopausal:	0.82–3.38
Term pregnancy:	0.23–1.17
Postmenopausal:	0.11–0.61
Newborns, both sexes:	1.67–3.64

METHODS FOR THE DETERMINATION OF CORTICOSTEROIDS IN URINE

With the introduction of RIA, the use of urinary corticosteroid determinations has been drastically reduced in favor of plasma cortisol determinations. Notwithstanding, many clinicians find urinary estimations of corticosteroid to be valuable in specific circumstances, since blood levels of the hormone fluctuate; thus, the level at the moment of blood withdrawal may not fairly represent the secretory activity of the adrenals. The urinary metabolites derived from cortisol (see Figure 9-22) may be grouped as follows:

1. Tetrahydro metabolites: tetrahydrocortisol (THF), tetrahydrocortisone (THE), and allotetrahydrocortisol.

2. Hexahydro metabolites: α- and β-cortols and cortolones.

3. 11-Oxygenated 17-ketosteroids: 11β-hydroxyetiocholanolone, 11-ketoetiocholanolone, 11β-hydroxyandrosterone, and 11-ketoandrosterone.

In addition to these cortisol metabolites, urinary products of some clinical importance also include tetrahydro-11-deoxycortisol (THS), pregnanetriol, 11-keto- and 11β-hydroxypregnanetriol, and 17-hydroxypregnanolone.

Colorimetric methods based on the Porter-Silber reaction for the steroids containing the dihydroxyacetone side-chain are widely used in routine clinical laboratories. This color reaction, however, does not include all the C_{21} metabolites; α- and β-cortols α- and β-cortolones are not determined. In contrast, the assay based on the oxidation of cortisol and its metabolites to 17-ketosteroids (i.e., 17-ketogenic steroids) measures all the major C_{21} metabolites.

Determination of Urinary 17-Hydroxycorticosteroids by a Modified Porter-Silber Method[300]

Principle

In 1950 Porter and Silber described a color reaction based upon the formation of a yellow pigment (absorption maximum at 410 nm) when certain corticosteroids react with phenylhydrazine in the presence of alcohol and sulfuric acid.[243] They demonstrated that this color reaction occurs primarily with corticosteroids that possess a dihydroxyacetone side chain as illustrated below:

17,21-Dihydroxy-20-ketone   **Yellow pigment**

Corticosteroids with this configuration include cortisol, cortisone, 11-deoxycortisol, and their tetrahydro-derivatives (Figures 9-21 and 9-22). Steroids such as corticosterone, progesterone, pregnanediol, and pregnanetriol lack the dihydroxyacetone side chain configuration and are therefore excluded from the color reaction. In urine, tetrahydrocortisol and cortisone are the major corticosteroids reacting with the Porter-Silber reagent; in certain types of adrenogenital syndromes (e.g., 11β-hydroxylase deficiency) and during the metyrapone test, tetrahydro-11-deoxycortisol constitutes the bulk of the Porter-Silber chromogens.

The basic steps of the procedure include hydrolysis of conjugates by β-glucuronidase; extraction with chloroform; washing the chloroform extract with dilute alkali to remove estrogens, bile acids, and interfering chromogens; and color reaction with alcoholic phenylhydrazine–sulfuric acid reagent.

Reagents

1. Chloroform, HPLC grade.
2. Sodium hydroxide, 0.1 mol/L. Dissolve 4 g of sodium hydroxide pellets in 1 L of distilled water.
3. Ethanol, HPLC grade.
4. Sulfuric acid, 64% (v/v). To 360 mL distilled water slowly add 640 mL of concentrated sulfuric

acid, reagent grade, with constant swirling. Prepare the reagent in a Pyrex 2-L Erlenmeyer flask immersed in an ice water bath.

5. Alcoholic–sulfuric acid reagent (blank reagent). Mix 100 mL 64% sulfuric acid with 50 mL absolute ethanol. The reagent is stable indefinitely.

6. Phenylhydrazine hydrochloride, recrystallized. A commercially available, chemically pure grade of phenylhydrazine hydrochloride is purified further as follows. Add 100 g of phenylhydrazine hydrochloride to 500 mL of warm water at 70 °C. Add 1 g activated charcoal. Heat 1 L ethanol to boiling and add to the dissolved phenylhydrazine in the water. Quickly filter while hot through Whatman No. 2 filter paper. Cool the filtrate in the refrigerator and collect the crystals in a sintered glass filter with medium porosity. Repeat the procedure of recrystallization, dissolving the crystals in proportionally less water. Wash the last collection of crystals with cold ethanol and dry thoroughly. Store in a tightly stoppered brown bottle in a desiccator over anhydrous calcium chloride. The purified material should have a melting point of 240–243 °C.

7. Alcoholic phenylhydrazine–sulfuric acid reagent. Dissolve 50 mg recrystallized phenylhydrazine hydrochloride in 50 mL alcoholic–sulfuric acid reagent. The reagent should be prepared fresh before use.

8. β-Glucuronidase, 1000 U/mL. The optimal pH and buffer to be used will vary with the source of the enzyme. Beef liver β-glucuronidase (Ketodase, Warner-Chilcott Laboratories, Morris Plains, NJ) is incubated in the presence of acetate buffer, 0.1 mol/L, pH 5. Bacterial β-glucuronidase (Sigma Chemical Company, St. Louis, MO) is incubated in phosphate buffer, 0.1 mol/L, pH 6.8. Prepare fresh before use.

9. Buffer solutions. Phosphate buffer, 0.5 mol/L, pH 6.8. To 500 mL solution of KH_2PO_4, 1 mol/L (68.0 g in 500 mL), add NaOH, 1 mol/L, to bring the pH to 6.8. Adjust the solution to a final volume of 1 L with distilled water.

Acetate buffer, 1 mol/L, pH 5. Dissolve 95 g of sodium acetate · $3H_2O$ and 17.2 mL glacial acetic acid in distilled water, and dilute to a volume of 1 L with water.

10. Stock standard, 100 μg/mL. Transfer 25 mg cortisol or tetrahydrocortisone to a 250-mL volumetric flask, and dilute to the mark with absolute ethanol.

11. Working standard, 5 μg/mL. Transfer 5 mL of the stock standard solution to a 100-mL volumetric flask, and dilute to the mark with distilled water.

Specimen Collection

Follow the procedure as described on page 1052.

Procedure

Hydrolysis, Extraction, and Washing

1. Transfer 10 mL of urine specimens and urine controls to 250-mL glass-stoppered cylinders. Adjust the pH of each urine to 6.8 using indicator paper. Add 1 mL β-glucuronidase (1000 U/mL), 2 mL phosphate buffer (0.5 mol/L), and 0.1 mL chloroform.

2. In a similar manner, prepare a water blank and standard using 10 mL of distilled water and 10 mL of working standard, respectively, instead of urine.

3. Mix the samples well and incubate at 37 °C for 18–24 h.

4. Add ~3 g ammonium sulfate to each tube and mix. Then add 100 mL chloroform to each glass-stoppered cylinder and mix by repeated inversion for 30 s. Let the cylinders stand for 5 min in order to separate the aqueous and organic phases.

5. Aspirate and discard the aqueous supernatants.

6. Add 10 mL of NaOH, 0.1 mol/L, to each cylinder and shake for 30 s. Allow to stand for 5 min. Aspirate and discard the aqueous alkali layer.

7. In a similar manner, wash the chloroform extracts twice with 10 mL of distilled water.

Porter-Silber Reaction

1. Transfer 40-mL aliquots of each chloroform extract to properly labeled 50-mL glass-stoppered centrifuge tubes and add 5 mL of alcohol–sulfuric acid reagent (blank reagent) or phenylhydrazine reagent as follows:

Blank-Blank	Phenyl-Blank	Standard-Blank	Standard-Phenyl	Test-Blank	Test-Phenyl
40 mL blank extract	40 mL blank extract	40 mL standard extract	40 mL standard extract	40 mL urine extract	40 mL urine extract
+	+	+	+	+	+
5 mL blank reagent (alcoh. H_2SO_4)	5 mL phenyl-hydrazine reagent	5 mL blank reagent	5 mL phenyl-hydrazine reagent	5 mL blank reagent	5 mL phenyl-hydrazine reagent

2. Tightly stopper all tubes and shake vigorously for 30 s. Allow to stand for 15–20 min. Alternatively, centrifuge the tubes at 900 $\times$ g for 10 min.

3. Transfer ~2.5 mL of the supernatant phase from each tube into labeled cuvets.

4. Incubate the cuvets in a water bath at 60 °C for 30 min, or overnight in the dark at room temperature.

5. Measure the absorbance (A) of each sample with a spectrophotometer at 410 nm as follows: Adjust the photometer to zero absorbance using the blank-blank, and read the standard and test blanks. Similarly, set the phenyl-blank at zero absorbance and read the standard and test-phenyl tubes.

Calculation

The standard sample contains 0.05 mg of cortisol. Incorporating this value and the appropriate dilution factor (10) to calculate the concentration of corticosteroids/dL of urine, the following equation is derived:

$$\text{Corticosteroids, mg/dL} = \frac{A_{\text{test}} - A_{\text{test blank}}}{A_{\text{standard}} - A_{\text{standard blank}}} \times 10 \times 0.05$$

$$\text{Corticosteroids, mg/d} = \frac{\text{conc (mg/dL)} \times \text{urine volume (mL/d)}}{100}$$

Comments

Acid hydrolysis is unsuitable because the unconjugated corticosteroids are labile in a strongly acidic medium. The metabolites of cortisol contain numerous hydroxyl and keto groups, making them relatively hydrophilic. The use of a polar organic solvent such as chloroform ensures quantitative extraction of these steroids from hydrolyzed urine. To remove acidic components and phenols, including estrogens, the solvent extract is washed with dilute alkali. The use of alkali stronger than 0.1 mol/L destroys the corticosteroids. The alkali-washed extract, termed the neutral fraction, contains metabolites of cortisol and other steroids excreted as glucuronides as well as any other neutral lipid-soluble material found in urine. The selectivity of the color reaction toward steroids with dihydroxyacetone side chains obviates the need for further purification. Impurities present in the extract form nonspecific brown chromogens in the presence of sulfuric acid. The use of a urine blank corrects for such background interference.

Various nonsteroidal substances, including acetone, fructose, and dehydroascorbic acid, also form a colored complex with the Porter-Silber reagent. In addition, several drugs and their metabolites, as well as bilirubin, have been reported to cause interference with the colorimetric estimation: iodides, paraldehyde, chloral hydrate, nitrofurantoin, colchicine, coffee, most sulfa drugs, phenothiazines, spironolactone, quinine, and propoxyphene. Administration of these drugs should be withheld for several days prior to determination of 17-hydroxycorticosteroids.

Reference Ranges

Children (up to 1 year):	0.5–1.0 mg/d
Adult male:	3–10 mg/d
Adult female:	2–8 mg/d

Determination of Total 17-Hydroxycorticosteroids (Total 17-Ketogenic Steroids)[284]

Principle

In 1952, Appleby[7] reported that sodim bismuthate oxidizes several groups of 17-hydroxy-corticosteroids to 17-ketosteroids, which can then be measured by the Zimmermann reaction. (See Determination of 17-Ketosteroids in Urine, p. 1091.) He termed these steroids "17-ketogenic steroids." The characteristic side chains that are oxidized by sodium bismuthate are shown at the top of the next page.

Group I includes cortisol, cortisone, their tetrahydro derivatives, 11-deoxycortisol (compound S), and tetrahydro S; group II includes cortols and cortolones: group III constitutes pregnanetriol and its 11-oxygenated derivatives; group IV includes 17-hydroxyprogesterone and 17-hydroxy-pregnenolone.

Note that the first two groups consist of active corticosteroids (cortisol, cortisone) and their metabolites, whereas groups III and IV comprise mainly the metabolites of the precursors of

I. 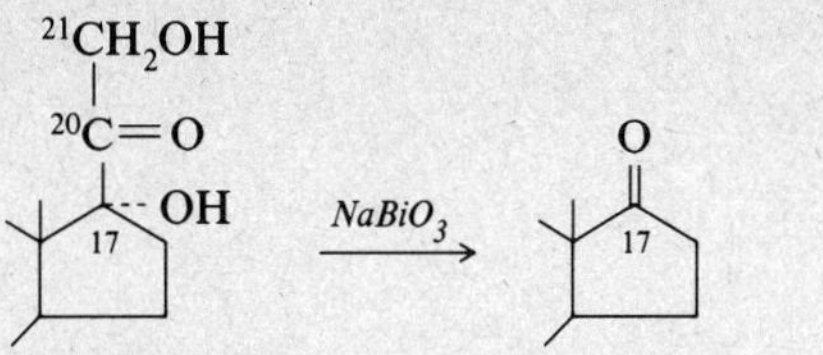 $^{21}CH_2OH$ — $^{20}C{=}O$ — ^{17} ---OH $\xrightarrow{NaBiO_3}$ (17, O)

17,21-Diol-20-one
(dihydroxyacetone)

II. 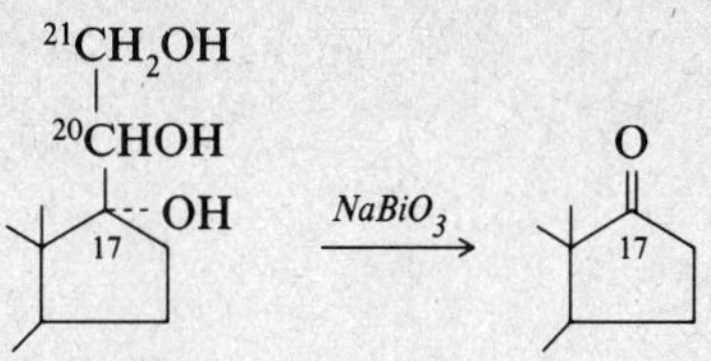 $^{21}CH_2OH$ — $^{20}CHOH$ — ^{17} ---OH $\xrightarrow{NaBiO_3}$ (17, O)

17,20,21-Triol
(glycerol)

III. 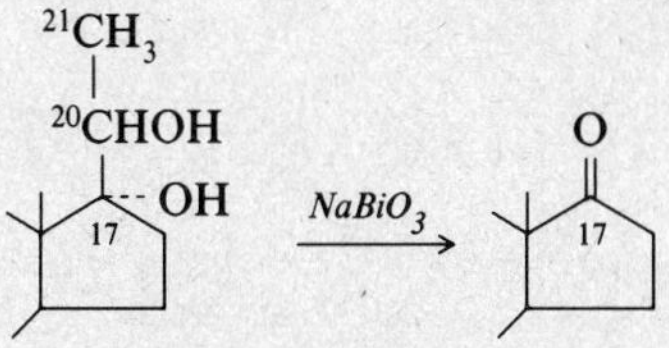 $^{21}CH_3$ — $^{20}CHOH$ — ^{17} ---OH $\xrightarrow{NaBiO_3}$ (17, O)

(17,20-glycol)

IV. 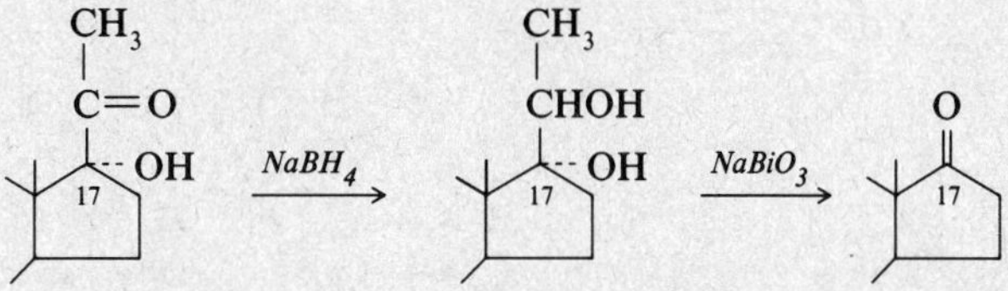 CH_3 — C$=$O — ^{17} ---OH $\xrightarrow{NaBH_4}$ CH_3 — CHOH — ^{17} ---OH $\xrightarrow{NaBiO_3}$ (17, O)

(17,20-ketol)

V. 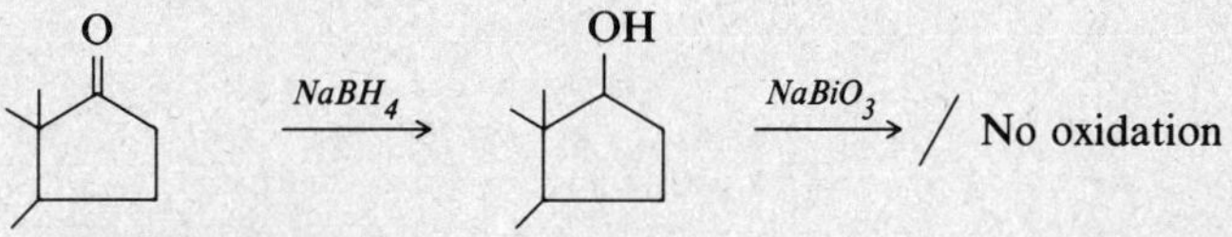 (O at 17) $\xrightarrow{NaBH_4}$ (OH) $\xrightarrow{NaBiO_3}$ / No oxidation

17-Ketosteroids

active corticosteroids (e.g., 17-hydroxyprogesterone). The excretion of the latter is quantitatively very significant in certain forms of the adrenogenital syndrome. Sodium bismuthate does not oxidize 17-hydroxy compounds containing a ketone at C-20 and a methyl group at C-21. In later modifications, a reduction step using sodium borohydride prior to bismuthate oxidation was introduced. This made it possible to measure the metabolites containing a 21-deoxy keto side chain (e.g., 17-hydroxypregnenolone together with the compounds included in groups I, II, and III). Following borohydride reduction, the 17-hydroxy-20-keto-21-deoxy steroids are reduced to 17,20-dihydroxy-21-deoxysteroids, and naturally occurring urinary 17-ketosteroids are reduced to C_{19} 17-hydroxysteroids. Subsequent treatment of the urine with sodium bismuthate produces 17-ketosteroids from all four groups of C_{21} 17-hydroxysteroids. Since sodium bismuthate does not reoxidize the C_{19} 17-hydroxysteroids, the 17-ketosteroids originally present in the urine become negative to the Zimmermann reaction. As a result, a determination of 17-ketosteroids after borohydride reduction and sodium bismuthate oxidation provides a direct measure of all total C_{21} 17-hydroxycorticosteroids.

Instead of sodium bismuthate, the oxidizing agent sodium m-periodate (10 g in 100-mL solution of NaOH, 0.1 mol/L) may be used.[83] The advantage of using this reagent lies in the fact that in addition to oxidizing 17-hydroxycorticosteroids to 17-ketosteroids, it oxidizes glucuronides to the free steroids or to their formates, which are easily hydrolyzed in alkaline solution; thus, the need for acid hydrolysis is eliminated.

Reagents

1. Ethylene dichloride, HPLC grade.
2. Sodium bismuthate, Merck, reagent grade. Since the quality of sodium bismuthate varies from lot to lot, the optimal amount required for procedure step 5 should be established for each lot.

3. Sodium bisulfite solution, 5 g/dL. Prepare fresh before use.

4. Sodium borohydride, 10 g/dL. Dissolve 2.5 g $NaBH_4$ (Metal Hydrides, Inc., Beverly, MA) in distilled water, to a final volume of 25 mL. Prepare fresh daily; use promptly.

5. Tes-Tape (Eli Lilly Co., Indianapolis, IN). Other reagents are as those described for urinary 17-ketosteroids determination.

Apparatus

Special glassware: glass-stoppered, heavy-walled centrifuge tubes of 35- and 50-mL capacity.

Mechanical shaker: Burrell, wrist-action shaker.

Procedure

1. Test urine with pH paper. If alkaline, acidify with glacial acetic acid (to dissolve phosphate precipitate if present).

2. Using Tes-Tape, determine the approximate concentration of glucose in the sample. If the specimen contains <0.5 g glucose/dL, proceed to step 3. If the specimen contains >0.5 g glucose, separate the glucose from the steroids as follows: Transfer 20 mL of urine to a glass-stoppered centrifuge tube, add 10 g ammonium sulfate, and mix to dissolve the salt. Extract three times with 20-mL portions of solvent (ether-ethanol, 3:1). Evaporate the combined extracts to dryness under nitrogen in a water bath at 50 °C. Add 10 mL ethanol to the residue and warm the solution in hot water to dissolve the steroids. (Ignore the insoluble material.) Cool, centrifuge, and transfer two 4-mL aliquots (equivalent to 8 mL urine) of the supernatant fluid (for duplicate analysis) to 50-mL centrifuge tubes. Evaporate the ethanol to dryness. Redissolve the residue in 0.5 mL methanol and dilute to 8 mL with water. Proceed to step 3, beginning with addition of sodium borohydride.

Reduction, Oxidation, Hydrolysis, and Extraction

3. Place 8 mL of urine in a 125-mL Erlenmeyer flask. Add 0.8 mL of sodium borohydride, 10 g/dL. Let stand for 2 h or overnight at room temperature.

4. Add 8 mL glacial acetic acid and allow to stand for 15 min. (The acid decomposes the excess borohydride.)

5. Transfer to a 50-mL centrifuge tube. Add 2 g sodium bismuthate. Stopper and shake mechanically for 30 min away from direct sunlight. (The samples may be covered with a heavy black cloth during the treatment with bismuthate.) Add 2 g fresh sodium bismuthate and shake for an additional 15 min. Leave the samples overnight at room temperature. The following morning shake the tubes for 15 min.

6. Centrifuge for 10 min at $900 \times g$. Transfer 6.0 mL of the supernatant fluid to 35-mL glass-stoppered centrifuge tubes containing 1.5 mL of freshly prepared sodium bisulfite solution. Mix the solution and allow to stand for 5 min.

7. Add 5 mL distilled water and 3.6 mL concentrated hydrochloric acid. Let stand for 15 min.

8. Place in a boiling water bath for 10 min. Remove and cool the samples in a cold water bath.

9. Add 12 mL ethylene dichloride and shake mechanically for 15 min. Centrifuge for 2 min at $900 \times g$.

10. Aspirate and discard the upper aqueous phase as completely as possible without losing any organic solvent.

11. Add 25–30 pellets of sodium hydroxide to the organic extract. Place in a shaking machine for 15 min, centrifuge, and filter through 7-cm Whatman No. 1 filter paper into a test tube.

12. Transfer 4 mL of filtrate (= 1 mL of urine) to a test tube and evaporate to dryness under nitrogen in a water bath at 50–55 °C. (In the case of a 24-h collection of large volume, use 8 mL of filtrate.)

Color Reaction

13. Perform the Zimmermann color reaction and measure the absorbance as described in the method for urinary 17-ketosteroid determination, page 1091.

Calculations

Total 17-ketogenic steroids, mg/d =
(corrected A of sample/corrected A of standard) $\times$ 0.05 $\times$ total urine volume (mL).

Comments

Since the 17-ketosteroids formed from the 17-hydroxycorticosteroids are fairly stable in a hot acid medium, the hydrolysis of steroid conjugates can now be performed with acid as opposed to the enzymatic hydrolysis used in the direct method based on the Porter-Silber reaction.

The presence of glucose in urine interferes with the bismuthate oxidation. All urine specimens should, therefore, be routinely tested with Tes-Tape and the glucose removed before the determination is begun. The most suitable means to rid the sample of the glucose appears to be the procedure outlined in step 2.

Errors due to the presence of glucose may also be avoided, however, by increasing the amount of sodium bismuthate (1 g for each gram of glucose above 0.5 g/dL).

The presence in urine of varying amounts of reducing substances other than glucose makes the use of a large excess of borohydride necessary. Addition of sufficient borohydride is indicated by effervescence on the addition of acetic acid (step 4). The absence of effervescence suggests an insufficient amount of boro-hydride, which may yield misleading results because of incomplete reduction of different ketone groups. Necessary precautions and drug interference in the Zimmermann color reaction will be discussed elsewhere. (See Determination of 17-Ketosteroids in Urine, p. 1091.)

Since a greater number of cortisol metabolites (cortol and cortolone) are estimated by the ketogenic method, urinary excretion values are generally higher than those obtained by the Porter-Silber method. This ketogenic method also yields high values in the adrenogenital syndrome because of the presence of excessive amounts of urinary metabolites such as pregnanetriol, which are not measured by the Porter-Silber procedure.

Reference Ranges

Children, up to 1 year:	< 1 mg/d
1–10 years:	2.3–3.8 mg/d
Adult, male:	5–23 mg/d
female:	3–15 mg/d
	Values decrease with age.

Methods for the Determination of Renin and Angiotensins

Until the introduction of RIA, the estimation of renin activity and renin substrate (angio-tensinogen) were carried out by bioassay methods. For a detailed discussion and description of these methods, the reader is referred to the monograph by Page and McCubbin.[224a] Several commercial RIA kits for the measurement of renin are available. However, none of the kits measure *renin* concentration. Instead, they are designed to measure plasma renin activity (PRA) by assaying angiotensin I.

Although numerous radioimmunoassays for angiotensins have been reported, most do not appear to differ significantly in their essential characteristics. Like steroid hormones, angiotensins are poor immunogens because of their small molecular size. Therefore, they must be conjugated to proteins such as albumin or succinylated polylysine for the production of antisera. The procedure of raising antisera, the special problems encountered in the preparation of plasma samples for RIA of angiotensins, and other related methodological aspects have been discussed by Freedlender and Goodfriend.[89] It should be emphasized, however, that angiotensins are very labile in plasma, and they are generated in vitro in untreated plasma. Therefore, great care must be taken in collection and storage of specimens for angiotensin assay, including special attention to the inactivation of angiotensinases.

A brief description of a procedure for determining PRA and angiotensin I by RIA follows. For detailed information, refer to the article by Freedlender and Goodfriend.[89]

Determination of Plasma Renin Activity (PRA)

Principle

PRA is defined as the rate of angiotensin I produced from angiotensinogen by renin in a patient's plasma. PRA is expressed in ng angiotensin I produced per mL of plasma per hour and is determined by assaying angiotensin I before and after incubation of plasma at 37 °C.

Specimen Collection and Storage

Draw blood into a Vacutainer tube containing EDTA (1 mg/mL), centrifuge the blood quickly at room temperature to sediment cells, then freeze the plasma at $-20\,°C$ or lower. Plasma should be transported frozen to a reference laboratory. Samples with high renin activities may generate considerable amounts of angiotensin I before and during freezing. This will not affect results, since angiotensin I is measured before and after the incubation step. Note, however, that storing chilled or acidified samples may activate prorenin so that an increased, but artifactual, renin activity is present before (or will develop during) assay. Prompt freezing of plasma, as described above, is considered the safest procedure.

Procedure

1. Add the following inhibitors to thawed plasma: phenylmethyl sulfonyl fluoride (PMSF), 8-hydroxyquinoline sulfate, dimercaprol, and neomycin. (See comments below for the explanation of the use of inhibitors.)

2. Add 0.1 mL phosphate buffer, 0.5 mol/L, pH 7.4, per mL of plasma.

3. Divide plasma into four aliquots.

4. Incubate 2 aliquots for 0.5–18 h at 37 °C (see comments below for discussion of choice of period of incubation) and store the other two aliquots at 4 °C.

5. At the end of the incubation period, chill the aliquots to 4 °C.

6. Assay incubated and nonincubated aliquots for angiotensin I as described below. Incubated plasma represents the endpoint of renin activity whereas nonincubated plasma represents renin activity at zero time.

7. Calculate renin activity in terms of ng of angiotensin I produced/mL plasma per hour of incubation, adjusting for the volume of plasma, the time of incubation, and the difference of angiotensin I concentrations in the incubated and nonincubated sample.

Comments

EDTA not only acts as an anticoagulant but also inhibits converting enzyme, stops the renin reaction at angiotensin I, and inhibits other enzymes that can destroy angiotensins. PMSF added to thawed plasma inhibits angiotensinases that can cleave the generated angiotensin I. The other inhibitors help retard proteolysis and bacterial growth.

To obtain the maximum rate of renin's enzymatic activity in the assay, saturating amounts of the angiotensinogen substrate should be present. But in most assay procedures the only substrate provided is that present in the test plasma, and saturation of renin is seldom achieved. According to some investigators, plasma renin activity is best estimated when the test plasma is incubated with an excess of exogenous renin substrate such as sheep angiotensinogen. This type of assay is generally known as a plasma renin concentration (PRC) assay rather than a plasma renin activity (PRA) assay. The method is similar to the one described above except that 50 µL sheep substrate is added to 50 µL plasma and the pH is adjusted to 7.0 with phosphate buffer, 0.1 mol/L.

The duration of incubation will vary with the sensitivity of the angiotensin I antiserum and with the level of renin activity. If the approximate range of renin activity is not known, incubation can be terminated at 3 h, aliquots taken for analysis, and the remainder frozen. The frozen mixture can then be thawed and incubated for longer times if the initial assay reveals very low activity. Prolonged incubation, however, may cause the generation of renin inhibitors or increase the degradation of renin or both. These problems can be minimized by brief incubations that, in turn, will depend on using antisera of high affinity.

Reference Ranges

Reference ranges vary. For patients on a low-sodium (10 mmol/d) diet, a reference range of 5–24 ng angiotensin I/mL/h is frequently observed. Sodium intake, posture, hydration state, and other factors may affect renin concentration. High values may be seen in patients receiving diuretics (volume and sodium depletion) or estrogen-containing medications (stimulation of hepatic synthesis of angiotensinogen). Most antihypertensive drugs alter plasma renin activity. For results to be meaningful, such drugs should be discontinued 1–2 weeks before the test is ordered.

Determination of Angiotensin I by RIA

Principle

The principle of RIA procedures has been described on page 1049.

Specimen Collection and Storage

Collect plasma as described previously for plasma renin activity.

Reagents

1. Buffer. Potassium phosphate, 0.01 mol/L, pH 7.4, containing EDTA, 0.003 mol/L, 8-hydroxyquinoline sulfate, 0.00015 mol/L, neomycin sulfate, 0.02 g/dL, and crystalline bovine serum albumin, 0.02 g/dL. The buffer is heated at 56 °C for 30 min to destroy angiotensinase activity in albumin and then stored at 4 °C. This buffer is used for all dilutions.

2. Stock standard concentration. Synthetic angiotensin I (Beckman Instruments, Palo Alto, CA).

3. Antiserum (Calbiochem-Behring, San Diego, CA).

4. ^{125}I-angiotensin I (New England Nuclear, Boston, MA).

5. Dextran-charcoal suspension. Stock charcoal suspension consists of Norit A neutral charcoal, 2.5 g/dL (Fisher Scientific, Medford, MA), with 0.25% Dextran T-70, 0.25 g/dL (Pharmacia, Uppsala, Sweden), in phosphate buffer, 0.01 mol/L, pH 7.4. The stock suspension is diluted 1:5 with the buffer just before use.

Procedure

All steps are carried out at 4 °C.

1. Add 50 μL of incubated or nonincubated plasma (see PRA above) to each tube, and add buffer to a final volume of 1.0 mL.

2. Prepare standards in duplicate in 1.0-mL volumes. Concentrations generally range from 2 pg to 1 ng/mL.

3. Add 0.1 mL ^{125}I-angiotensin I to give $\sim$10 000 cpm/tube.

4. Add 0.1 mL diluted antiserum to the tubes.

5. Incubate at 4 °C for 2–24 h.

6. Add 1.0 mL of dextran-charcoal suspension to each tube and mix. Centrifuge, and decant the supernatant.

Note: Usually the charcoal suspension must be diluted before use. Dilutions will vary with both different antisera and different samples. For example, more avid antisera and samples containing protein require a more concentrated charcoal suspension. With the use of an avid antiserum and a protein concentration of 2.5 mg/mL, a 1:5 dilution of the stock charcoal suspension has proved satisfactory. To achieve reproducibility, it is important that samples be exposed to the charcoal for an equal and brief period of time ($<$ 10 min), e.g., during centrifugation and decanting.

7. Count supernatant (antibody-bound hormone) and charcoal pellets (free hormone) for radioactivity. Plot standard curve using either antibody-bound or free hormone by any of the methods described on page 1054.

8. Calculate amounts of angiotensin I in each sample from the standard curve.

Comments

The radioimmunoassay of angiotensin I is complicated by factors similar to those encountered in the immunoassay of other substances. Antisera are seldom strictly specific, and samples usually contain cross-reacting decapeptide, heptapeptide, and hexapeptide congeners of angiotensins. In addition, plasma may continue to produce or destroy angiotensin I during the assay and may contain materials that affect the assay by interfering with the antigen-antibody reaction or by binding angiotensin I itself. Isolation of angiotensin I from the sample before assay has been found to minimize these problems. Fuller's earth is considered by many to be the most suitable, convenient, and simple agent for this purpose. Other chromatographic procedures for separation and purification of angiotensin I have also been suggested.[89] It should be noted that problems regarding cross-reactivity and interference can best be eliminated by using sensitive and specific antisera. At present, several antisera are commercially available that can detect 10 pg of angiotensin I.

Reference Ranges

Angiotensin I: 11–88 pg/mL

Conditions or drugs that increase or decrease PRA (see above) will also influence plasma levels of angiotensin I.

ANDROGENS

Androgens are a group of C_{19} steroids. They are required for masculine differentiation of the genital tract and for the development and maintenance of male secondary sex characteristics; they also contribute to muscle bulk, bone mass, sex drive, and sexual performance in males. In normal females, androgens serve as precursors for estrogens but have no other well-defined function.

In males, androgen excess is rarely seen; androgen deficiency results in diminished libido and potency, and in infertility. In females, androgen deficiency has no known clinical significance; mild or moderate androgen excess disturbs the normal menstrual cycle and leads to excessive facial and body hair. Severe androgen excess in females causes virilization. Biologically active androgens include testosterone, dihydrotestosterone (DHT), androstanediol, and androstenediol; plasma concentrations of the latter two compounds are too low to be significant in normal physiology.

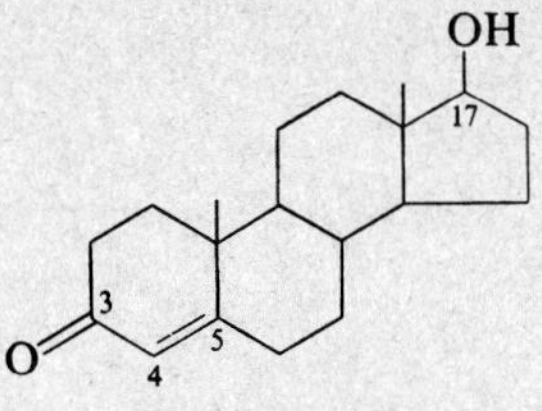

Testosterone
(17β-hydroxyandrost-4-en-3-one)

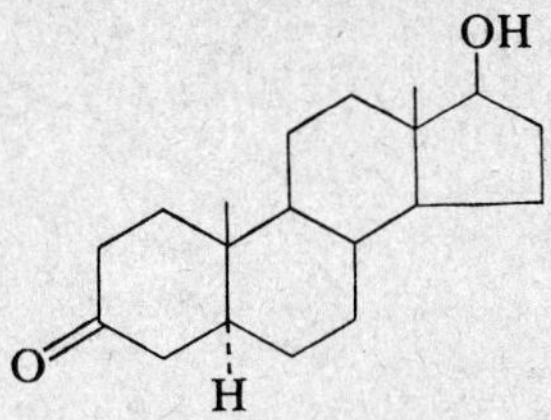

Dihydrotestosterone
(17β-hydroxy-5α-androstan-3-one)

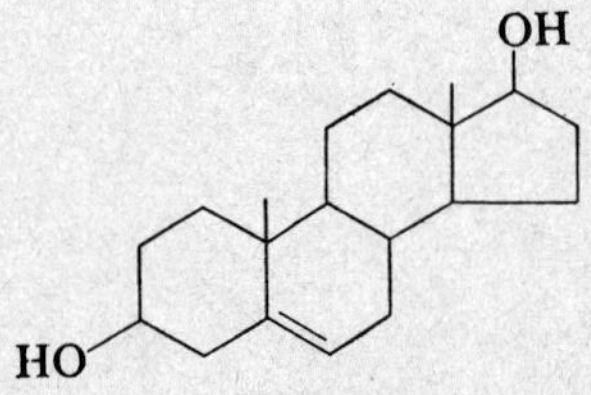

Androstenediol
(3β,17β-dihydroxyandrost-5-ene)

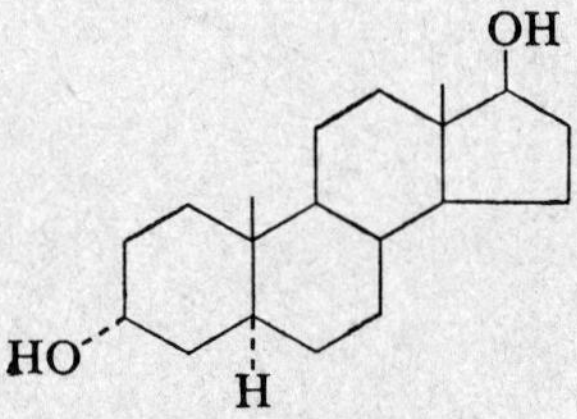

Androstanediol
(3α,17β-dihydroxy-5α-androstane)

Figure 9-32. Chemical structure of androgens.

Testosterone is the most important androgen secreted into the blood. In males, testosterone is secreted primarily by the Leydig cells of the testes; in females, ~50% of circulating testosterone is derived from peripheral conversion of androstenedione, 25% from the ovary, and 25% from the adrenal glands (except at midcycle, when the ovarian contribution increases by 10–15%). The structural characteristics of this steroid include an unsaturated bond between C-4 and C-5, a ketone groupt at C-3 (Δ^4-3-keto), and a hydroxyl group in the β-position at C-17 (Figure 9-32).

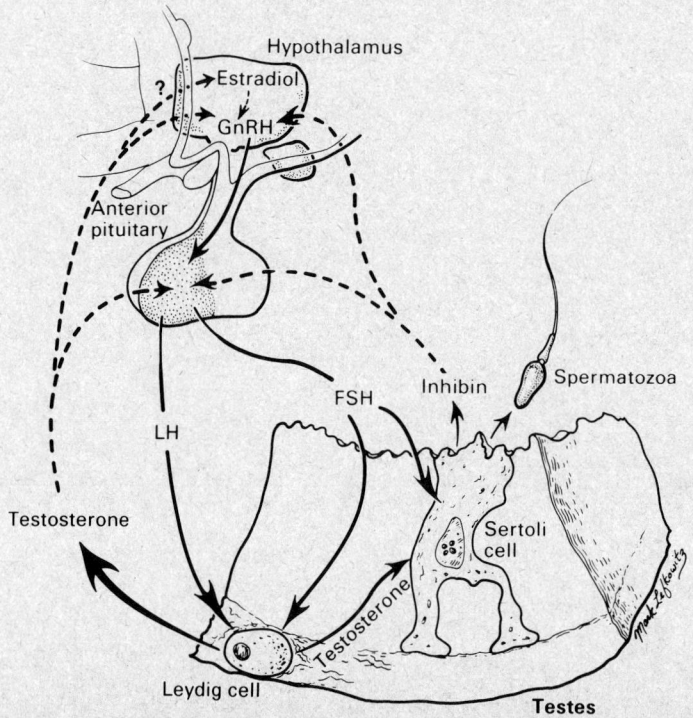

Figure 9-33. Control of testicular function. See text for explanation.

The hydroxyl group in the β-position (above the plane of the paper) is essential for biological activity; epitestosterone (17α-hydroxyl) and androstenedione (17-keto) have little biological activity.

Testosterone is converted to DHT by the addition of two hydrogens (at C-4 and C-5) catalyzed by the enzyme 5α-reductase, found in the cytoplasm of cells in the prostate, skin (particularly genital skin), and seminal vesicles. DHT exerts major effects in the tissues where it is produced and may also enter the blood and produce effects in other tissues. Data from bioassays indicate that DHT is a much more potent androgen than testosterone. The affinities of DHT and testosterone for their common receptor are about the same. Their respective effects on a tissue will therefore depend on the rate of conversion of testosterone to DHT in that tissue.

Androstanediol is produced from testosterone in peripheral tissues and passes into the blood. *Androstenediol* may be secreted by the adrenals or ovaries or be derived from peripheral conversion of plasma dehydroepiandrosterone (DHEA). *Androstenedione* and *DHEA* are weakly androgenic; they are much more important as prehormones of testosterone. A prehormone is defined as a substance with little or no inherent biologic potency that is converted to a more active product in peripheral tissue. Androstenedione is derived in equal amounts from the adrenals and the ovaries. The contribution of DHEA to testosterone levels in blood is negligible in normal women. DHT is derived from extraglandular conversion of androstenedione.

The mean production rates of testosterone and DHT in normal males are 700 and 300 μg/d, respectively, while the corresponding production rates in females are 300 and 60 μg/d. The production rate of androstenedione in healthy men is ~2.5 and in healthy women ~3.5 mg/d.

Control of the Testicular Hormone Secretion

Testicular function is regulated by means of the hypothalamic-pituitary-testicular axis (Figure 9-33). Hypothalamic gonadotropin-releasing hormone (GnRH) mediates pituitary release of LH (also known as ICSH—interstitial cell stimulating hormone) and FSH. The primary effect of LH is to stimulate the synthesis and secretion of testosterone by Leydig cells; this effect of LH is enhanced by FSH, which increases the number of LH receptors on the cells. Increasing levels of free testosterone, in turn, inhibit LH secretion. According to some investigators, the negative feedback of testosterone is carried out through its conversion to estradiol by an aromatase enzyme system within the hypothalamus. In contrast to its effects on LH, testosterone does not suppress FSH at physiologic levels. This phenomenon led to the discovery of *inhibin*, a polypeptide that specifically inhibits FSH. Inhibin is produced by the Sertoli cells of the seminiferous tubules. When the seminiferous tubules fail to function, FSH secretion increases, presumably the result of a lack of inhibin. Recent studies suggest that inhibin reduces FSH secretion both by direct action on the pituitary cell and through inhibition of GnRH release. However, secretion of inhibin is not under direct negative feedback control of FSH but is dependent upon the functional activity of Sertoli cells.

Biosynthesis

There are two pathways for the formation of androgens from pregnenolone (Figure 9-34). The Δ^5-pathway includes the transformation of pregnenolone to 17-hydroxypregnenolone to DHEA and then to androstenedione and testosterone. The Δ^4-pathway route leads from pregnenolone to progesterone and then to androstenedione. Although both pathways are functional in the testes and adrenals, the Δ^5-route is probably more active. The pathway of androgen synthesis in the ovary has been discussed in the section on estrogen biosynthesis. DHEA may also be converted to testosterone via androstenediol without involving androstenedione as an obligatory intermediate. The transformation of androstenedione into testosterone is a reversible reaction. In the ovary, the equilibrium lies far toward the formation of androstenedione, which is secreted. In the testes the forward reaction is more active, and production and secretion of testosterone are favored.

Metabolism

After entering the bloodstream, testosterone is transported bound to plasma proteins. One of the proteins that binds testosterone with high affinity is termed the sex hormone–binding globulin (SHBG), also known as testosterone-estradiol–binding globulin (TeBG) to indicate its capacity for binding estradiol. SHBG is a β-globulin distinct from CBG, the major transport

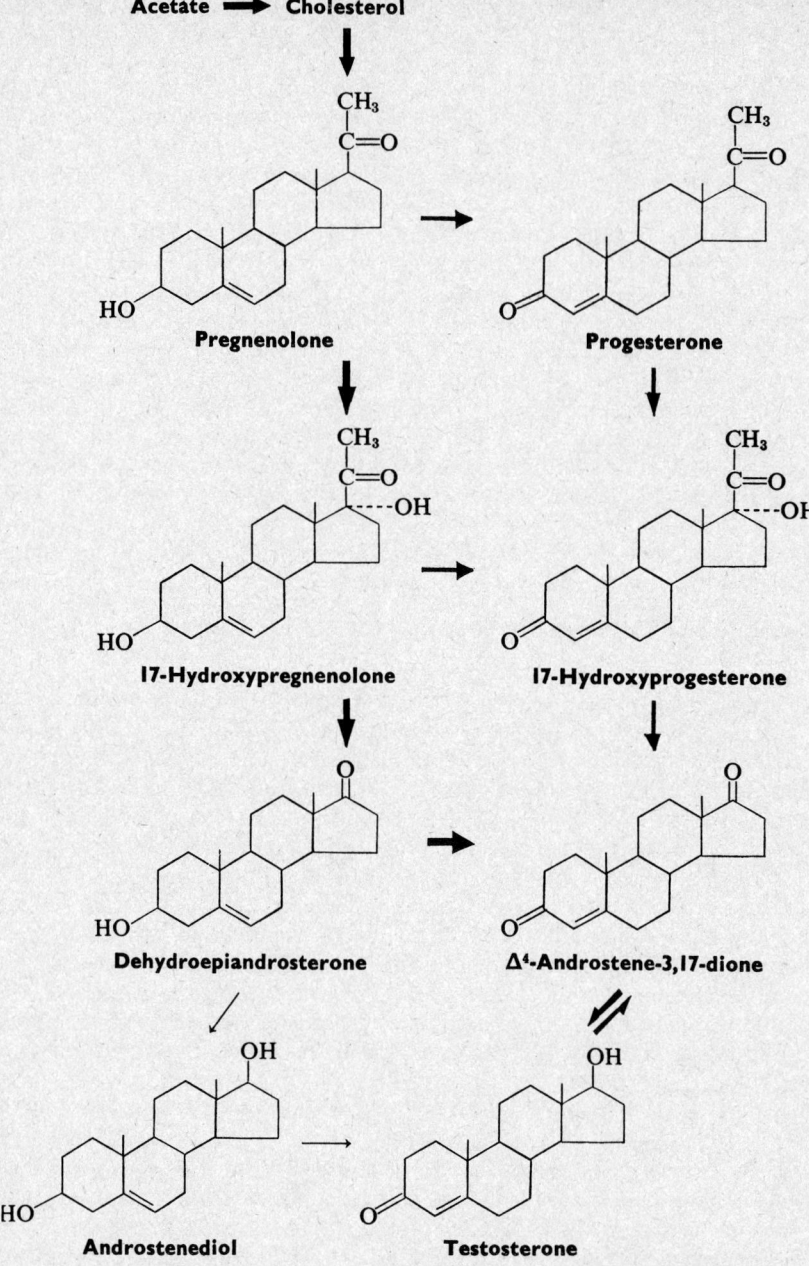

Figure 9-34. Biosynthesis of androgens (adrenal and testis). The heavy arrows indicate the preferred pathway. See text.

protein for cortisol. In women, about 80% of circulating testosterone is bound to SHBG and 19% to albumin, but ~1% is not bound to any plasma protein. It is this unbound or free form that is biologically active. In men, 3% of total testosterone is free; 78% is bound to SHBG and 19% is bound loosely to albumin. Only the unbound fraction is available for metabolism. Increases or decreases in the metabolic clearance rate (MCR) of testosterone are primarily due to changes in the concentration of SHBG. Indeed, the higher MCR's of DHEA, DHEAS, and androstenedione relative to the MCR of testosterone are due to their lack of binding to plasma proteins. The MCR of testosterone is increased in hypothyroidism and virilization, and decreased in hyperthyroidism and aging. Treatment with androgens, medroxyprogesterone, or dexamethasone increases MCR; treatment with estrogens or barbiturates decreases it.

The main metabolites of androstenedione, testosterone, and DHEA are shown in Figure 9-

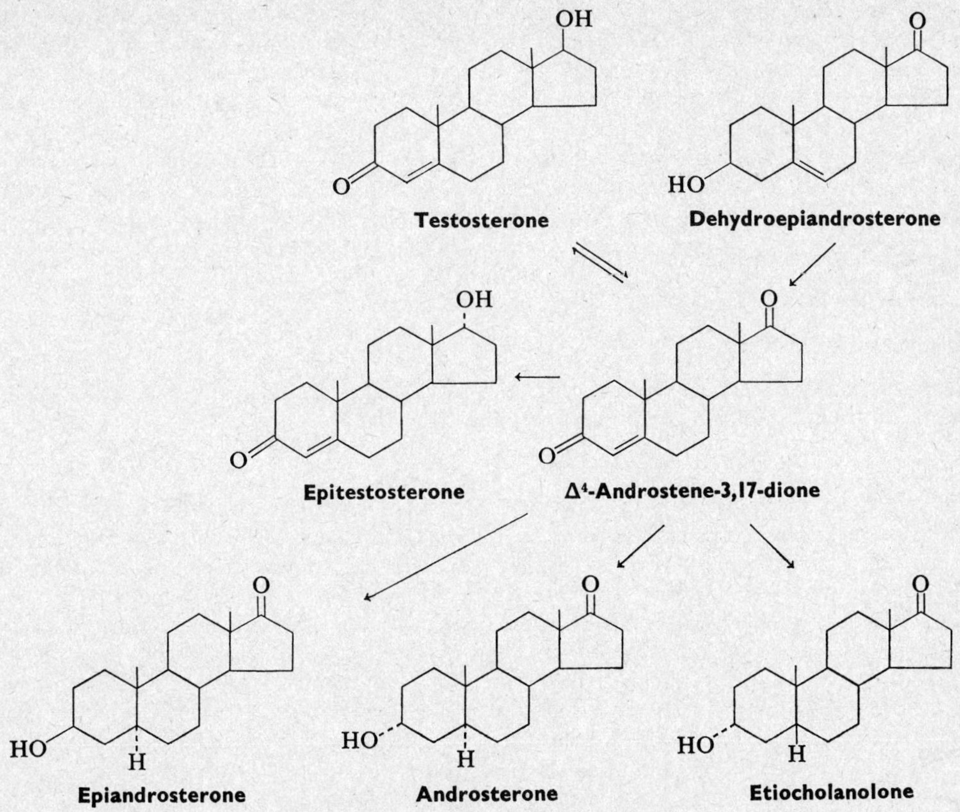

Figure 9-35. Catabolism of $C_{19}O_2$ androgens.

35. As in corticosteroid metabolism, reduction of the double bond between carbon atoms 4 and 5 of androstenedione and testosterone gives rise to two isomers that differ in the spatial configuration of H at C-5. When H is on the same side as the methyl group at C-10 (*cis*), it is a 5β-isomer (etiocholanolone), and when the H is on the opposite side (*trans*), the isomer is a 5α-isomer (androsterone). Similarly, hydrogenation of the ketone group at C-3 produces either 3α-hydroxy (androsterone, etiocholanolone) or 3β-hydroxy (epiandrosterone) compounds. Reduction of the ketone group at C-17 of androstenedione by 17α-hydroxysteroid dehydrogenase produces epitestosterone. Quantitatively, the 3α-hydroxysteroids predominate in urine. The metabolism of DHEA follows the same biochemical sequences as described above except that DHEA is first converted irreversibly to androstenedione. DHEA may also be excreted unchanged as the sulfate conjugate. The metabolism of 11-oxygenated androgens has been described in a previous section. (See Metabolism of Adrenal Cortical Steroids, p. 1060.) Except for epitestosterone, these catabolites constitute a group of steroids known as 17-ketosteroids. Conjugation of these steroids may occur with sulfuric or glucuronic acid. The glucuronide predominates for androsterone, etiocholanolone, and the 11-oxygenated 17-ketosteroids; DHEA, however, is present exclusively as the sulfate conjugate.

Determination of Plasma Testosterone by RIA

Principle

Earlier methods for measuring testosterone by fluorometry, double isotope derivative, electron-capture gas-liquid chromatography, and competitive protein binding assays have been largely replaced by radioimmunoassays using highly specific antisera.[175] RIA methods differ mainly in the mode of sample preparation, in the quality of antisera used, and in the nature of the steroid-protein conjugate used to generate the antiserum.[129] General aspects of radioimmunoassay of steroid hormones are discussed on page 1049.

Direct methods for the determination of plasma testosterone have been reported.[122] Testosterone bound to SHBG is displaced by the addition of excess estradiol in assays employing

[3]H-ligand and charcoal separation.[244] Testosterone can also be displaced from SHBG at pH 4.0. However, testosterone levels are overestimated when displacing conditions of pH 4 are used in a direct assay that employs an antiserum raised against testosterone-3-CMO conjugate and ^{125}I-labeled tracer.[250] Several commercial assay kits for direct assay of testosterone have recently become available. Unfortunately, details of assay reagents and displacing conditions have not been revealed by the manufacturers. The principles of the procedures of these kits are similar to those described for cortisol determination by RIA. (See p. 1074.) Most in-house assays still employ solvent extraction, ^{3}H-radioligand, and charcoal separation as described on page 1049.

Specimen Collection and Storage

Follow the directions as described on page 1052.

Reagents and Procedures

For specific and detailed instructions, refer to the monograph by Jaffe and Behrman[129] or to inserts of the commercial kit chosen for the assay.

Comments

Direct assay methods for total plasma testosterone suffer from lack of accuracy for some samples, despite claims of a high correlation between results obtained by direct and extraction methods. Most testosterone antisera presently available lack the specificity required for direct assays. According to one report,[250] overestimations of testosterone by direct methods are caused by the presence of water-soluble cross-reacting substance(s) in the specimen. Clinical laboratories using direct assay kits must be aware that lack of specificity may occasionally yield spurious results. Experience suggests that the validity of results employing kit methods is best proved by tests for parallelism of the response curves for both standards and randomly selected samples from both males and females. Almost all testosterone antisera show some degree of cross-reactivity with DHT. In most clinical situations, however, estimations of testosterone without prior separation of DHT are permitted, since plasma concentrations of DHT are only 10–20% of those for testosterone. Moreover, testosterone and DHT are the two most important androgens in the circulation; even when a method measures both of them, clinically useful information about the total androgen load is obtained. However, if specific estimations of testosterone are needed, the level of DHT must be separately determined and subtracted from the value of testosterone. An RIA method involving preliminary chromatographic separation of testosterone and DHT is available.[129]

Reference Ranges

	ng/dL
Males,	
Prepubertal:	10–20
Adult:	300–1000
Females,	
Prepubertal:	10–20
Adult:	20–75 (higher at mid cycle peak)
Pregnant:	3–4 times adult level
Postmenopausal:	8–35

In males, serum testosterone levels rise briefly in the twelfth week of gestation and again shortly after birth. A sharp increase occurs at puberty. Subsequently, the level remains relatively constant except for the diurnal variation. (The highest level occurs in the morning and gradually decreases to 25% of the high by evening.) A slow and progressive decrease begins after the fifth decade. In females, there is an increase in plasma testosterone at puberty, and a cyclic pattern of elevation for 1–2 d at midcycle is established.

Determination of Free (Nonprotein Bound) Testosterone in Plasma

Principle

Only a small fraction of unconjugated testosterone exists free, i.e., nonprotein bound, in plasma. None of the conventional assay methods, including RIA, is sufficiently sensitive to quantitate the free steroid directly in a protein-free ultrafiltrate of plasma. Instead, one can estimate free steroid in plasma by adding a known amount of radiolabeled compound to the plasma and allowing labeled compound and unlabeled endogenous compound to reach equilibrium in their competition for the same binding sites on the proteins. Bound and free radioactive labels are then separated, and the ratio of free label to total label determined. At equilibrium, the ratio of free to total label is taken as a measure of the ratio of free to total endogenous testosterone. Most problems with this procedure have involved the separation of bound and free label. Several separation techniques have been used, including equilibrium dialysis, membrane ultrafiltration, and steady-state gel filtration. Their shortcomings include a requirement for a large sample volume, the need for complicated correction of sample volume for changes that occur in it during the separation, and difficulties of collecting and measuring radioactivity in numerous fractions of each sample. Equilibrium dialysis has been used most often in the past, but more recent work[319] has pointed out some of the errors arising from sample dilution required by this method. A recent method of ultrafiltration[109] appears to overcome the problems listed above and to obviate errors due to dilution. The following is a brief outline of a modified method that is rapid and convenient for routine application.[105]

A sample is equilibrated with radioactive steroid. Free steroid is then separated from bound by filtration through an anisotropic, hydrophilic YMT ultrafiltration membrane (MPS-1 Micropartition System, Amicon Corp., Danvers, MA, 01923). The driving force for ultrafiltration is provided by centrifugation at $1000-2000 \times g$. Filtrate containing free steroid collects in the filtrate cup while protein-bound steroid remains in the retentate. Radioactivity in the filtrate is a measure of the free fraction.

The procedure is as follows: a small volume of an ethanolic solution of freshly purified ³H-testosterone is placed in a glass tube and evaporated under nitrogen. The residue is reconstituted with 0.6 mL of the serum or plasma to be analyzed and incubated at 37 °C for 15 min. A 50-μL aliquot is then transferred to a scintillation vial for counting (total counts). The 550 μL remaining is then added to the MPS-1 micro-partition device which has been pre-equilibrated to 37 °C for 1 h. The device is centrifuged at $1000 \times g$ for 5 min in a fixed-angle (35 °) rotor in a centrifuge operated in a controlled-temperature environment at 37 °C. About 100 μL of ultrafiltrate is collected; a 50-μL aliquot of ultrafiltrate is transferred to a scintillation vial for counting. A 50-μL aliquot of the patient's untreated serum is also counted (background counts). The per cent of free testosterone is calculated:

$$\% \text{ free testosterone} = \frac{\text{cpm, ultrafiltrate} - \text{cpm, background}}{\text{cpm, total} - \text{cpm, background}} \times 100$$

Coefficients of variation for two pooled serum controls (one male, one female) are about 7% for inter-assay and about 4% for intra-assay.

Reference Ranges

The mean ($\pm$SD) for free testosterone in healthy, nonpregnant women is $1.09 \pm 0.25\%$ and for healthy men is $2.47 \pm 0.61\%$. These values compare favorably with those reported for membrane ultrafiltration, gel filtration, and equilibrium dialysis methods.

Determination of 17-Ketosteroids in Urine

The 17-ketosteroids are metabolites of precursors secreted by the adrenals, the testes, and to some extent the ovaries. In males, approximately one-third of the total urinary 17-ketosteroids represent metabolites of testosterone secreted by the testes, whereas most of the remaining two-thirds are derived from the steroids produced by the adrenals. In females, who normally excrete smaller quantities than men, the total 17-ketosteroids are derived almost exclusively from the adrenals.

The bulk of the urinary 17-ketosteroids consists of androsterone, epiandrosterone, etiocholanolone, dehydroepiandrosterone (DHEA), 11-keto- and 11β-hydroxyandrosterone, and 11-keto- and 11β-hydroxyetiocholanolone. DHEA and 11-oxygenated 17-ketosteroids are the products of adrenals only, while the others also arise from precursors (androstenedione, testosterone) elaborated by the gonads. Thus, the main purpose of measuring these steroid metabolites is to assess adrenal androgen production.

Clinical Significance

Decreased values of 17-ketosteroids are generally found in males with primary hypogonadism (Klinefelter's syndrome, castration), in secondary hypogonadism (panhypopituitarism), and in women with primary hypoadrenalism (Addison's disease). *Inceased values* are obtained in testicular tumors (interstitial cell tumor, chorioepithelioma), adrenal hyperplasia, and adrenal carcinoma, and in some women with hirsutism.

All 17-ketosteroids are not androgens, and not all androgens are 17-ketosteroids. DHEA and androstenedione are 17-ketosteroids with only weak androgenic properties, and etiocholanolone is a 17-ketosteroid with no androgenic effects. The potent androgens, testosterone and DHT, are not 17-ketosteroids. A specific measurement of testosterone and DHT in blood is required to evaluate the production of these androgens. The measurement of DHEAS in blood serves as a more convenient marker for adrenal androgen production than does urinary 17-ketosteroid excretion.[173] Many clinicians now prefer plasma DHEAS to urinary 17-ketosteroid measurements.

There are a number of chemical methods available for estimating total 17-ketosteroids. Final quantitation in most of these is based on the color reaction originally described by Zimmermann. The method described by Sobel et al.[284] has been shown to be adequate for routine clinical use and is given below.

Principle

The 17-ketosteroids are excreted as water-soluble conjugates of glucuronic and sulfuric acid. Cleavage of these conjugates with acid is followed by extraction, washing with alkali, and finally color development. Estrone, which is an "acidic" 17-ketosteroid, is removed by alkali treatment because of its phenolic nature and thus is eliminated prior to the colorimetric reaction of the remaining "neutral" 17-ketosteroid fraction. The color formation is based on the reaction of 17-ketosteroids with *m*-dinitrobenzene in alcoholic potassium hydroxide to produce a reddish-purple color with maximum absorption at 520 nm. Marlow[183] has demonstrated that the development of color depends on the presence of an active methylene group adjacent to a carbonyl group, most likely giving the following product:

17-Ketosteroid   m-Dinitrobenzene   Purple compounds

When the ketone group is situated at other positions (e.g., Δ^4-3-keto in testosterone, progesterone, cortisol) the color development is less intense and the absorption maxima differ. Quantitation is carried out by comparison of the color density obtained with the sample with that obtained with a known amount of pure standard, such as dehydroepiandrosterone.

Specimen Collection and Storage

See page 1052.

Reagents

1. Ethanol, HPLC grade.
2. Ethanol, 70% (v/v). Dilute 700 mL absolute ethanol to 1 L with distilled water.
3. Ethylene dichloride, HPLC grade.
4. Sodium hydroxide, solid pellets, reagent grade.
5. Potassium hydroxide, reagent grade. Prepare a saturated aqueous solution.
6. Potassium hydroxide–ethanol solution. Add 1 vol of saturated solution of potassium hydroxide to 4 vol of absolute ethanol. Centrifuge and use the supernatant. Prepare this reagent just before use.
7. *m*-Dinitrobenzene, Grade V, Sigma Chemical Co. (St. Louis, MO 63178), 1.16 g/dL. Dissolve 116 mg *m*-dinitrobenzene in 10.0 mL absolute ethanol. Prepare fresh daily.
8. Dehydroepiandrosterone (DHEA) standard, 100 μg/mL. Dissolve 10.0 mg of DHEA in 100.0 mL of absolute ethanol. Store at −20 °C. Stable for 3–6 months.

Procedure

1. Test urine with pH paper. If alkaline, acidify with glacial acetic acid to dissolve any phosphate precipitate.

Hydrolysis, Extraction, and Washing

2. Transfer 8 mL of urine to a 35-mL glass-stoppered centrifuge tube. Add 2 mL of glacial acetic acid and 3 mL concentrated hydrochloric acid.
3. Stopper and place into a 100 °C bath for 10 min. Cool under cold, running tap water.
4. Add 10 mL of ethylene dichloride. Cap the tube(s) and place in a shaking machine for 15 min.
5. Centrifuge for 2 min at 900 × g and aspirate off the top aqueous layer as completely as possible.
6. Add 25–30 pellets of sodium hydroxide to the sample. Cap the tube(s) and place in a shaking machine for 15 min. Alternatively, the extract may be washed with NaOH, 10 g/dL, followed by two water washes. Centrifuge as before and filter the solvent through Whatman No. 1 filter paper.
7. Transfer 2.5 mL of the filtrate (equivalent to 2 mL urine) to a test tube. Also, transfer 0.5 mL of the 100 μg/mL DHEA standard to a test tube.
8. Evaporate all tubes to dryness under nitrogen in a water bath at 50–55 °C. (If very low concentrations are expected, use 5 mL of the filtrate.)

Color Reaction and Spectrophotometric Reading

9. Perform the Zimmermann reaction as follows: (a) to a blank tube, the sample tube, and the standard tube containing 50 μg DHEA, add 0.2 mL of *m*-dinitrobenzene solution. (b) Add 0.2 mL of freshly prepared alcoholic potassium hydroxide solution and mix. (c) Place the tubes in a water bath at 25 °C in the dark for 30 min. (d) Add 5 mL of 70% ethanol to each tube and mix.
10. Measure the absorbance of the standard and the sample in a spectrophotometer or a colorimeter at 480, 520, and 560 nm, setting the instrument at zero absorbance with the blank solution.

Calculation

Calculate the corrected absorbance of the standard and sample using the following formula:

$$\text{Corrected absorbance} = (A_{520}) - \frac{(A_{480} + A_{560})}{2}$$

Calculate 24-h excretion of 17-ketosteroids as follows:

$$\text{mg 17-ketosteroids/d} = \frac{\text{corrected } A \text{ of sample}}{\text{corrected } A \text{ of standard}} \times 0.05 \text{ mg DHEA} \times \frac{\text{24-h urine volume (mL)}}{2}$$

Comments

Most of the urinary 17-ketosteroids are excreted as sulfate and glucuronide conjugates, which are hydrolyzed by strong acid and heat. The duration of hydrolysis is very critical. Less than 10 min will cause incomplete hydrolysis and more than 10 min will lead to gradual destruction of the steroids and formation of an increased amount of nonsteroidal chromogens. Addition of glacial acetic acid helps to minimize the formation of nonspecific chromogens during the hydrolytic procedure, particularly in the case of an alkaline urine specimen. Although solvents such as benzene, carbon tetrachloride, and ether are suitable for extraction, ethylene dichloride is particularly suitable because the steroid hormones are extracted from hydrolyzed urine more quantitatively at a relatively low ratio of the solvent to urine. Thus, this solvent is technically advantageous for a routine laboratory method since it avoids the need for handling and evaporating large quantities of solvent. Also, ethylene dichloride possesses a convenient temperature of evaporation.

According to Drekter et al.,[71] the treatment of the extract with pellets of sodium hydroxide is superior to the customary treatment with aqueous sodium hydroxide because solid NaOH removes estrogenic phenols and other acidic and chromogenic pigments more completely.

Two different alkaline reagents have been in common use for the Zimmermann reaction, namely, aqueous and alcoholic KOH. The former yields colors of much less intensity than the alcoholic reagent and the latter has the disadvantage of being unstable. The saturated solution of KOH, as proposed by Sobel et al.,[284] is stable and yields very low blank absorbance. According to the same authors, the time, temperature, and dilution with 70% ethanol give maximal color development and stability. To avoid deleterious effects in the color reaction, the ethanol must be of highest quality. However, in spite of meticulous care in the preparation of reagents and in color development, the formation of nonspecific background chromophores arising from other ketonic steroids and nonsteroidal ketones cannot be avoided. The reading of the absorbance at three wavelengths and the use of the correction formula (Allen correction) serve to eliminate the effect of such background interference in the estimation. The correction is based on the assumption that the absorbances of nonspecific materials at the three chosen wavelengths lie on a straight line. Alternatively, the preparation of a urine blank and subtraction of its reading from the sample serve the same purpose. The following drugs and their metabolites in urine are examples of those known to result in either under- or overestimation of 17-ketosteroids: ascorbic acid, glutethimide, morphine, meprobamate, and penicillin G.

Reference Ranges

	mg/d
Up to 1 year:	< 1
1–4 years:	< 2
5–8 years:	< 3
9–12 years:	3–10
13–16 years:	5–12
Young adult male:	9–22
Adult male:	8–20
Adult female:	6–15

The excretion values are the same for both sexes throughout childhood. After about age 60, the rate of excretion declines progressively in both sexes.

PROGESTERONE

Progesterone is a female sex hormone; in conjunction with estrogens, it regulates the accessory organs during the menstrual cycle. This hormone is especially important in preparing the uterus for the implantation of the blastocyst and in maintaining pregnancy. In nonpregnant women, progesterone is secreted mainly by the corpus luteum, a yellow glandular mass in the ovary formed by an ovarian follicle following the discharge of its ovum. During pregnancy the placenta becomes the major source. Minor sources are the adrenal cortex in both sexes and the testes in males.

The structural formula of progesterone, a C_{21} compound, is shown in Figure 9-36. Like the corticosteroids and testosterone, progesterone (pregn-4-ene-3,20-dione) contains a keto group (at C-3) and a double bond between C-4 and C-5 (Δ^4); both structural characteristics are essential for progestational activity. The 2-carbon side chain (CH_3—C=O) on C-17 does not seem to be very important for its physiological action. Indeed, the synthetic compound 19-nortestosterone (Figure 9-36) and its derivatives—widely used as oral contraceptives—are more potent progestational agents than progesterone itself.

Biosynthesis and Metabolism

Biosynthesis of progesterone in ovarian tissues is believed to follow the same path from acetate to cholesterol through pregnenolone as it does in the adrenal cortex. In luteal tissue, however, low-density lipoprotein cholesterol (LDL-cholesterol) is thought to serve as the preferred precursor despite the potential of the corpus luteum to synthesize progesterone de novo from

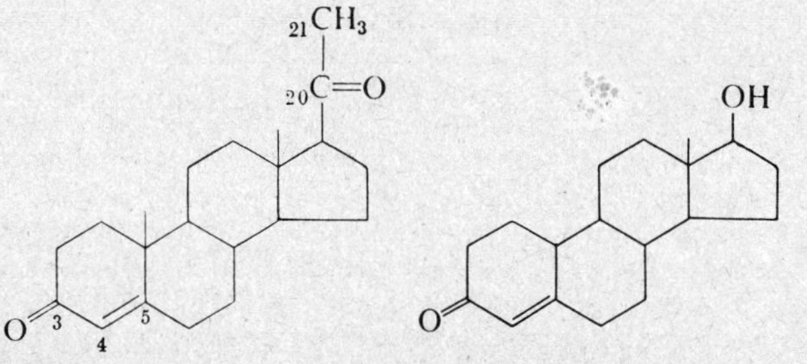

Figure 9-36. Structural formulas of progesterone and 19-nortestosterone.

Progesterone
(Pregn-4-ene-3,20-dione)

19-Nortestosterone
17β-Hydroxy-19-norandrost-4-en-3-one)

acetate.[275] Initiation and control of luteal secretion of progesterone is regulated by LH and FSH. The interrelationship of progesterone and gonadotropins will be discussed under Female Reproductive Endocrinology, page 1109.

Progesterone does not have a specific plasma binding protein but, like cortisol, is bound to CBG. Reported values for plasma free progesterone vary from 2–10% of total concentrations, and the percentage of unbound progesterone remains constant throughout the normal menstrual cycle.[86] The production rate of progesterone during the luteal phase reaches as much as 30 mg/d, whereas the production rate of progesterone by the placenta during the third trimester of pregnancy is ~300 mg/d.

The important metabolic events leading to inactivation of progesterone are reduction and conjugation. An examination of the chemical structure of progesterone shows that three different sites are susceptible to hydrogenation (reduction): the double bond between carbon atoms 4 and 5 and the keto groups at C-3 and C-20. The reduction at each site produces two isomers differing in the spatial orientation of the hydrogen at C-5 or the hydroxyl groups at C-3 and C-20. The formation of an α- or β-isomer is catalyzed by specific enzymes, e.g., Δ^4-5α- or Δ^4-5β-reductase, 3α- or 3β-hydroxysteroid dehydrogenase, or 20α- or 20β-hydroxysteroid dehydrogenase. The main metabolic pathway is outlined in Figure 9-37. Reduced metabolites are eventually conjugated with glucuronic acid and excreted as water-soluble glucuronides.

Metabolites of progesterone may be classified into three groups based on the degree of reduction.

1. *Pregnanediones.* The C4-5 double bond is reduced, producing two compounds: pregnanedione (H at C-5 is in β-orientation, i.e., above the plane of the paper) and allopregnanedione (H at C-5 is in α-orientation, i.e., below the plane of the paper).

2. *Pregnanolones.* The keto group at C-3 is reduced, producing hydroxyl groups in α- or β-orientation. However, most urinary pregnanolones exist in the α-configuration.

3. *Pregnanediols.* The keto group at C-20 is also reduced. As in the previous case, metabolites containing the 20-hydroxyl group in α-orientation are quantitatively more important. In fact, urinary measurement of pregnanediol (5β-pregnane-3α,20α-diol) is used as an index of endogenous production of progesterone, since this metabolite is quantitatively very significant and correlates with a majority of clinical conditions.

Clinical Significance

During the follicular phase of the menstrual cycle, some progesterone is believed to be produced by the adrenal; plasma levels, however, remain consistently low (<1 ng/mL). (See Reproductive Endocrinology, p. 1109.) Following ovulation, production of progesterone by the corpus luteum increases rapidly, reaching a maximal concentration of 10–20 ng/mL in 4–7 d. Progesterone concentrations remain elevated for 4–6 d and then fall abruptly to baseline levels ~24 h before the onset of menstruation. Since the rise and fall of progesterone parallel the activity of the corpus luteum, measurements of plasma progesterone are used clinically to confirm ovulation and normal function of the corpus luteum in nonpregnant women. If ovulation does not occur, the corpus luteum is not formed, and no cyclical rise of plasma progesterone is observed. Abnormal progesterone secretion has been implicated in premenstrual tension, irregular shedding of the endometrium, dysmenorrhea, and luteal insufficiency.

During pregnancy, progesterone is largely produced by the placenta. Plasma progesterone levels rise steadily, reaching values as high as 200 ng/mL at term. Although decreased excretion of the urinary progesterone metabolite pregnanediol has been reported[175] in cases of threatened abortion, toxemia of pregnancy, and intrauterine fetal death involving the placenta, no systematic study of plasma levels of progesterone has yet been made in these conditions.

Progesterone concentrations vary considerably among individuals, and in the same individual from day to day or even hour to hour. Consequently, in gynecological disorders or in abnormal pregnancies, serial rather than single measurements of plasma progesterone are preferred for proper interpretation of results. Abraham et al.[4] found that the functional status of the corpus luteum is normal in females with normal menstrual cycles if the sum of three plasma progesterone values obtained 4–11 d prior to menstruation is ≥15 ng/mL. In females with abnormal cycles, the sum was <15 ng/mL. According to some investigators,[254] measurements of progesterone in saliva may be a better means of evaluating luteal function. The significance of salivary steroids has been discussed on page 1046.

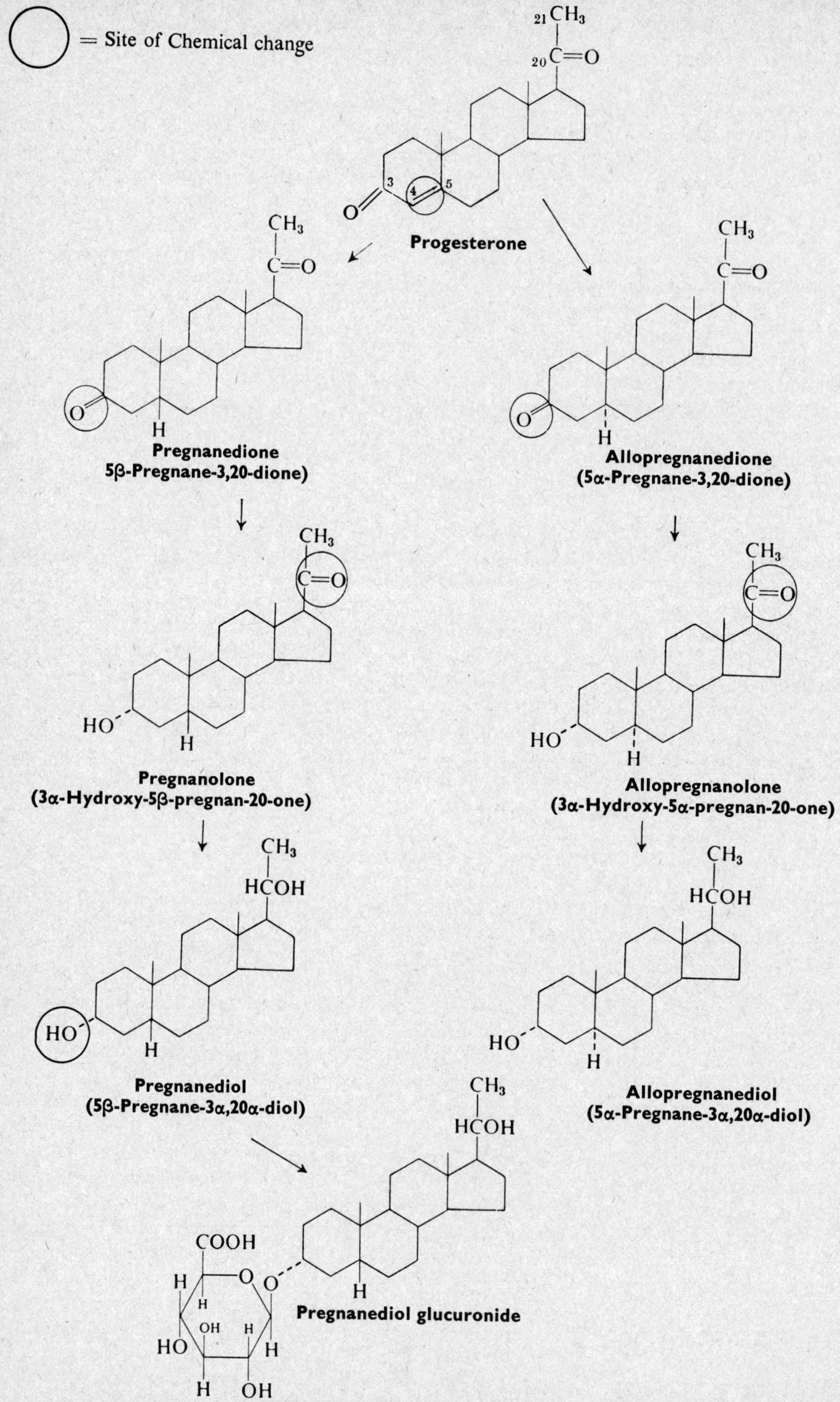

Figure 9-37. Metabolism of progesterone.

Methods for the Determination of Progesterone in Plasma

Direct measurement of progesterone in plasma is considered to be the most reliable way to assess its rate of production. Techniques based on gas-liquid chromatography and competitive protein binding (CPB) have been described,[30] but these assays have now been largely replaced by RIA. Various RIA procedures using antisera generated against BSA conjugates of 11-deoxy-cortisol-21-hemisuccinate or 20-O- or 3-O-carboxymethyloxime-progesterone[175] have been reported. Unfortunately, these antisera cross-react with other steroids, and sample purification is required prior to radioimmunoassay. Recently, a highly specific antiserum has been generated against a progesterone-11α-hemisuccinate-BSA conjugate.[45] With the availability of specific antisera, direct measurement of progesterone in serum or plasma is now feasible. One reported method uses ³H-progesterone as tracer, danazol or cortisol as displacing agents, and charcoal for separation of bound from free.[250] Results agree closely with those obtained by conventional assays involving extraction.

Determination of Progesterone by Radioimmunoassay

Principle

A direct assay method employing ¹²⁵I-labeled progesterone as tracer and danazol or ANS (8-anilino-1-naphthalene sulfonic acid) as displacing agent[251] is summarized below.

Danazol at pH 7.4 or ANS at pH 4.0 is added to serum standards (progesterone added to pooled serum) and specimens. Following addition of tracer (¹²⁵I-labeled progesterone-11α-glucuronyl tyramine) and antiserum generated against the 11α-hemisuccinyl conjugate of progesterone, standards and specimens are incubated at room temperature for 3 h. For the method employing danazol, separation of bound from free is accomplished by adding donkey anti-rabbit γ-globulin. Following overnight incubation at 4 °C, tubes are centrifuged, supernatants aspirated, and the activities in the pellets counted. For the method employing ANS, separation is achieved by using a second antibody coupled to Sepharose (solid phase system) and a sucrose layering technique. This separation procedure consists of adding a second antibody suspension, agitating for 0.5 h at room temperature, and layering a 10% sucrose solution beneath the incubate. In a short time (∼15 min), the solid-phase particles carrying the bound fraction settle to the bottom of the tube, leaving the free fraction in the top layer. During this settling process the particles are effectively washed as they pass through the sucrose layer. The free fraction and most of the sucrose layer are then removed by suction, leaving the bound fraction in the tube ready for counting. For efficient washing and complete separation, this procedure is generally repeated. When this sucrose layering technique is used in conjunction with the solid phase system, centrifugation is avoided and shorter second incubation periods are permitted. Results for either the danazol or the ANS method are interpolated from the standard curve, constructed as a log-logit plot. Numerous commercial kits for direct assay of progesterone are now available; therefore, descriptions of reagents and procedures are omitted. *Specimen collection and storage* have already been discussed (p. 1052).

Comments

According to the authors,[251] direct assays of progesterone using either danazol or ANS as displacing agent give quantitative recovery of progesterone added to human serum; progesterone values for patients' sera were also found to correlate well with results of assays involving serum extraction. As discussed earlier (see Determination of Cortisol by RIA, p. 1074), the accuracy may be affected by the nature of the matrix used in preparing the standards. For progesterone assays, satisfactory accuracy may be achieved by using pooled sera from males who have progesterone concentrations in the range of 0.11–0.26 ng/mL. Precision is similar with each displacing agent but superior to that of extraction assays. Both intra- and inter-assay coefficients of variation are below 10%. Poor sensitivity is observed in homologous systems due to the fact that the common structural features of both the immunogen and tracer result in the tracer having a much higher affinity than the analyte for the antibody.

Antisera produced against C-11 conjugates of progesterone are highly specific and cross-react minimally (< 10%) with other C_{21}-corticosteroids. However, relatively large amounts of corticosteroids are present in plasma and may cause overestimation of progesterone. If these corticosteroids become a problem, a simple extraction procedure prior to RIA may be followed.[45,46] Fortunately, these interfering steroids are much more polar than progesterone. The use of a nonpolar solvent such as petroleum ether or hexane quantitatively extracts progesterone, while the polar corticosteroids remain mostly in the plasma sample.

Reference Ranges

	ng/mL
Male:	0.12–0.3
Female,	
Menstrual cycle:	
Follicular phase:	< 1
Luteal phase:	5–20
Pregnancy	
1st trimester:	20–50
2nd trimester:	50–100
3rd trimester:	100–400

Determination of Pregnanediol in Urine

Clinical Significance

Pregnanediol is a major metabolite of progesterone. Clinically, measurements of urinary pregnanediol are used as an indirect measure of progesterone and to indicate ovulation and normal function of the corpus luteum in nonpregnant women and to evaluate the placental function during pregnancy. In children, little or no pregnanediol can be found in urine. In men, pregnanediol is derived mainly from progesterone and 11-deoxycorticosterone secreted by the adrenals; excretion in men rarely exceeds 1 mg/d. In women with normal menstrual cycles, the excretion pattern of urinary pregnanediol mirrors the secretion of progesterone by the corpus luteum. During the follicular phase, the level of urinary pregnanediol is similar to that found in men. During the luteal phase, which follows ovulation and corpus luteum formation, the excretion of pregnanediol increases gradually to its maximum (luteal peak) between the twenty-first and twenty-fourth days. In the absence of ovulation, no corpus luteum is formed. Consequently, no cyclical rise of urinary pregnanediol is observed in anovulatory menstrual cycles.

During the first three months of pregnancy, the amount of urinary pregnanediol is only slightly higher than that found during the luteal phase of the menstrual cycle. As secretion of progesterone by the placenta increases, there is a steady rise of pregnanediol excretion until about the thirty-second week of gestation, at which time excretion levels off. Within 24 h of delivery, excretion values begin to drop, reaching nonpregnancy levels in 4–5 d. In threatened abortion, toxemia of pregnancy, and intrauterine fetal death involving the placenta, values are lower than those observed at corresponding weeks of normal gestation. For detailed information on the excretion of pregnanediol in abnormal pregnancies and other gynecological disorders, the reader is referred to the monograph by Loraine and Bell.[175]

Principle

Until recently, suitable methods for the direct analysis of progesterone were not available. Consequently, the measurement of urinary pregnanediol, the primary metabolite of progesterone, was used as an index of progesterone secretion. However, with the advent of radioimmunoassay methods for measuring plasma progesterone, assays of urinary pregnanediol are now being discontinued. In the recent past, assays of urinary pregnanediol were mainly colorimetric and gas chromatographic. The main steps in the *colorimetric procedure* by Klopper et al.[149] consist of acid hydrolysis, toluene extraction, permanganate oxidation of the extract to remove steroidal artifacts formed during acid hydrolysis, alumina chromatography, acetylation, a second alumina chromatography step, color reaction with sulfuric acid, and spectrophotometric measurement of the chromogen.

Various *gas chromatographic* methods for urinary pregnanediol have been described in the literature.[175] The basic steps involved in one such method[49] include acid hydrolysis, extraction with toluene, washing with alkali and water, addition of cholesterol propionate as internal standard, acetylation with acetic anhydride and pyridine, and injection into a gas chromatograph containing a 3% SE-30 stationary phase in a glass column, 2 m × 4 mm. Quantitation is carried out by comparing the ratio of peak height or peak area obtained from the known concentrations of authentic standard and internal standard with that of pregnanediol and internal standard in the sample.

Reference Ranges

	mg/d
Male:	0.1–0.7
Female,	
Normal menstrual cycle:	
Proliferative (follicular) phase:	0.10–1.3
Luteal phase:	1.2–9.5
Pregnancy:	
10–12 weeks	5–15
12–18 weeks	5–25
18–24 weeks	13–33
24–28 weeks	20–42
28–32 weeks	27–47

Note: There is much interindividual variation.

ESTROGENS

Estrogens, like progesterone, are female sex hormones. They are responsible for the development and maintenance of the female sex organs and secondary female sex characteristics. They

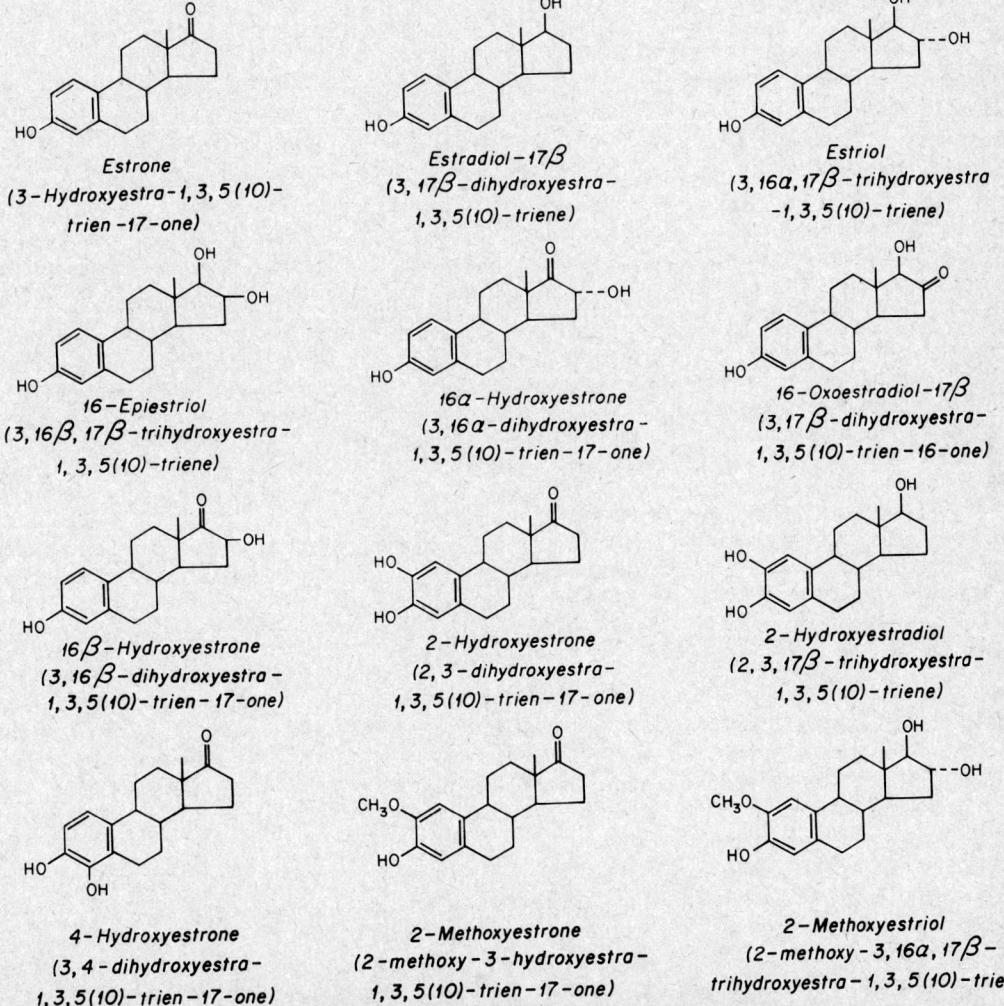

Estrone
(3–Hydroxyestra–1, 3, 5 (10)–
trien –17–one)

Estradiol–17β
(3, 17β–dihydroxyestra–
1, 3, 5(10)–triene)

Estriol
(3, 16α, 17β–trihydroxyestra
–1, 3, 5 (10)–triene)

16–Epiestriol
(3, 16β, 17β–trihydroxyestra–
1, 3, 5(10)–triene)

16α–Hydroxyestrone
(3, 16α–dihydroxyestra–
1, 3, 5 (10)–trien–17–one)

16–Oxoestradiol–17β
(3, 17β–dihydroxyestra–
1, 3, 5(10)–trien–16–one)

16β–Hydroxyestrone
(3, 16β–dihydroxyestra–
1, 3, 5(10)–trien–17–one)

2–Hydroxyestrone
(2, 3–dihydroxyestra–
1, 3, 5(10)–trien–17–one)

2–Hydroxyestradiol
(2, 3, 17β–trihydroxyestra–
1, 3, 5(10)–triene)

4–Hydroxyestrone
(3, 4–dihydroxyestra–
1, 3, 5(10)–trien–17–one)

2–Methoxyestrone
(2–methoxy–3–hydroxyestra–
1, 3, 5(10)–trien–17–one)

2–Methoxyestriol
(2–methoxy–3, 16α, 17β–
trihydroxyestra–1, 3, 5 (10)–triene)

Figure 9-38. Structural formulas of important estrogens.

also participate in the regulation of the menstrual cycle and in the maintenance of pregnancy. In normal women, most estrogens are secreted by the ovarian follicles, the corpus luteum, and, during pregnancy, by the placenta. Adrenals and testes are also believed to secrete estrogens, but only in minute quantities. The names and structural formulas of some of the important estrogens are shown in Figure 9-38. Estradiol is by far the most potent naturally occurring estrogen.

Structurally, estrogens are derivatives of the parent hydrocarbon estrane. They consist of 18 carbon atoms and possess the following characteristic features: (1) an aromatic ring A; (2) a ketone (estrone) or hydroxyl group (estradiol) at C-17 and frequently at C-16 (16-ketoestradiol, estriol); (3) a phenolic hydroxyl group at C-3, which gives the compounds acidic properties; and (4) absence of the methyl group at C-10. The phenolic A ring and the oxygen function at C-17 are essential for biological activity. Substituents at other positions in the molecule diminish feminizing potency. For example, estriol and 2-methoxyestrone, which contain a hydroxyl group at C-16 and a methoxy group at C-2, respectively, possess very little biological activity.

Biosynthesis

The ovary follows the same steroidogenic pathway as other steroid-producing organs. In vivo and in vitro studies indicate that acetate, cholesterol, progesterone, and testosterone can all serve as precursors of estrogens. The normal human ovary produces all three classes of sex steroids: estrogens, progestins, and androgens; estradiol and progesterone, however, are its primary secretory products. Unlike the testes, the ovary possesses a highly active aromatase system that rapidly converts androgens such as testosterone to estrogens. Unlike the adrenal cortex, the normal ovary lacks both the 21-hydroxylase and the 11β-hydroxylase enzymes and therefore cannot produce glucocorticoids and mineralocorticoids.

The first biochemical event in the aromatization of testosterone is hydroxylation of the C-19 methyl group to produce 19-hydroxytestosterone (Figure 9-39). This hydroxylated compound is further oxidized to 19-oxotestosterone. The C-19 carbon atom and the C-1 hydrogen atom of this intermediate are eliminated as formaldehyde. The resulting 3-oxo-androst-1(10),4-diene aromatizes spontaneously to estradiol. The biochemical sequence for aromatization of androstenedione to estrone is the same as that described for testosterone. Even though the reactions of testosterone to androstenedione and estradiol to estrone are reversible, the pathway involving conversion of testosterone to estradiol is quantitatively more significant in the ovary.

There are two theories regarding the synthesis of estrogens during the ovarian cycle: the two-pathway theory and the two-cell theory. According to the two-pathway theory, estrogen production by thecal and interstitial cells of the ovaries is mainly provided by the Δ^5-3β-hydroxy pathway during the proliferative stage (follicular phase) of the cycle whereas, following ovulation and corpus luteum formation, these estrogens are synthesized from progesterone by the Δ^4-pathway. According to the two-cell hypothesis, the thecal cells provide androgens, primarily androstenedione, which diffuse into the granulosa cells where they are converted to estrogens. The production of androgens in the thecal cells is stimulated by LH, whereas the conversion of androstenedione to estradiol (aromatization) is regulated by FSH. The two-cell theory appears to be more logical and is widely accepted. Steroidogenesis and its regulation during the ovarian cycle have been discussed in detail by Channing et al.[44] The role of estrogens in normal and abnormal menstrual cycles will be described in the female reproductive endocrinology section.

Estrogens are also produced by peripheral aromatization of androgens, primarily androstenedione. In normal men and women, $\sim$1% of the secreted androstenedione is converted to estrone. Although the ovaries of menopausal women do not secrete estrogens, these women have significant blood levels of estrone originating from the peripheral conversion of androstenedione secreted by the adrenal. Since a major site of this conversion is adipose tissue, estrone will increase in obese postmenopausal women, sometimes yielding enough estrogen to produce bleeding. The significance of extraglandular estrogen production in health and disease has been discussed at length by Siiteri and MacDonald.[274]

Circulating estradiol is largely unconjugated; it is strongly bound to a specific globulin, sex hormone binding globulin (SHBG), and loosely bound to albumin. Only 2–3% of total estradiol circulates in the biologically active unbound form. In contrast, estrone and estrone sulfate circulate bound almost exclusively to albumin. In the normal nonpregnant female, estradiol is produced at the rate of 100–300 μg/d. The production of estrone ranges from 100–200 μg/d. Diurnal variation of blood estrone levels occurs in postmenopausal women, presumably reflecting the

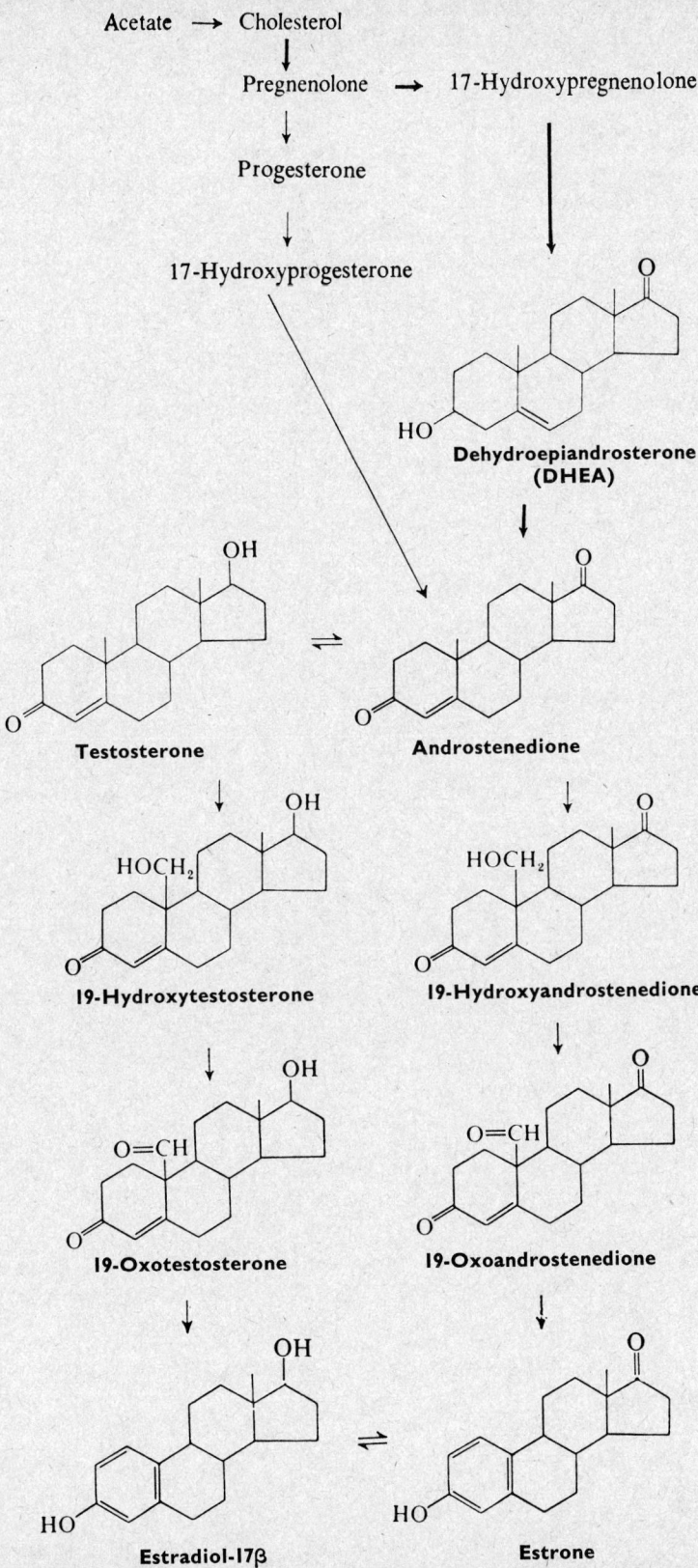

Figure 9-39. Biosynthesis of estrogens. Heavy arrows indicate the Δ^5-3β-hydroxy pathway.

variation in the androstenedione precursor that is originating in the adrenals. However, no such diurnal rhythms have been demonstrated for estradiol.

Biosynthesis of estriol during pregnancy. Research in recent years has shown that biosynthesis of estrogens differs qualitatively and quantitatively during pregnancy from the nonpregnant state. During pregnancy, the major source of estrogens is the placenta, whereas in women who are not pregnant, the ovaries are the main site of synthesis. In contrast to microgram quantities secreted by nonpregnant women, the amount of estrogen excreted during pregnancy increases to milligram amounts. The major estrogen secreted by the ovary is estradiol, whereas the major product secreted by the placenta is estriol. For further discussions and the clinical significance of estriol, see Chapter 19, Biochemical Aspects of Pregnancy.

Metabolism

Estrogen metabolism is far more complex than the simple reaction sequence estradiol $\rightleftharpoons$ estrone $\rightarrow$ estriol. In the normal course of estrogen metabolism, estradiol is believed to form a reversible redox system with estrone. Estrone is then metabolized along two alternative pathways (Figure 9-40): the 2-hydroxylation pathway leading to the formation of catechol estrogens (2-hydroxyestrone, 2-hydroxyestradiol, 2-hydroxyestriol, and their corresponding methoxy-derivatives) and the 16α-hydroxylation pathway leading predominantly to estriol.[112] Recent evidence

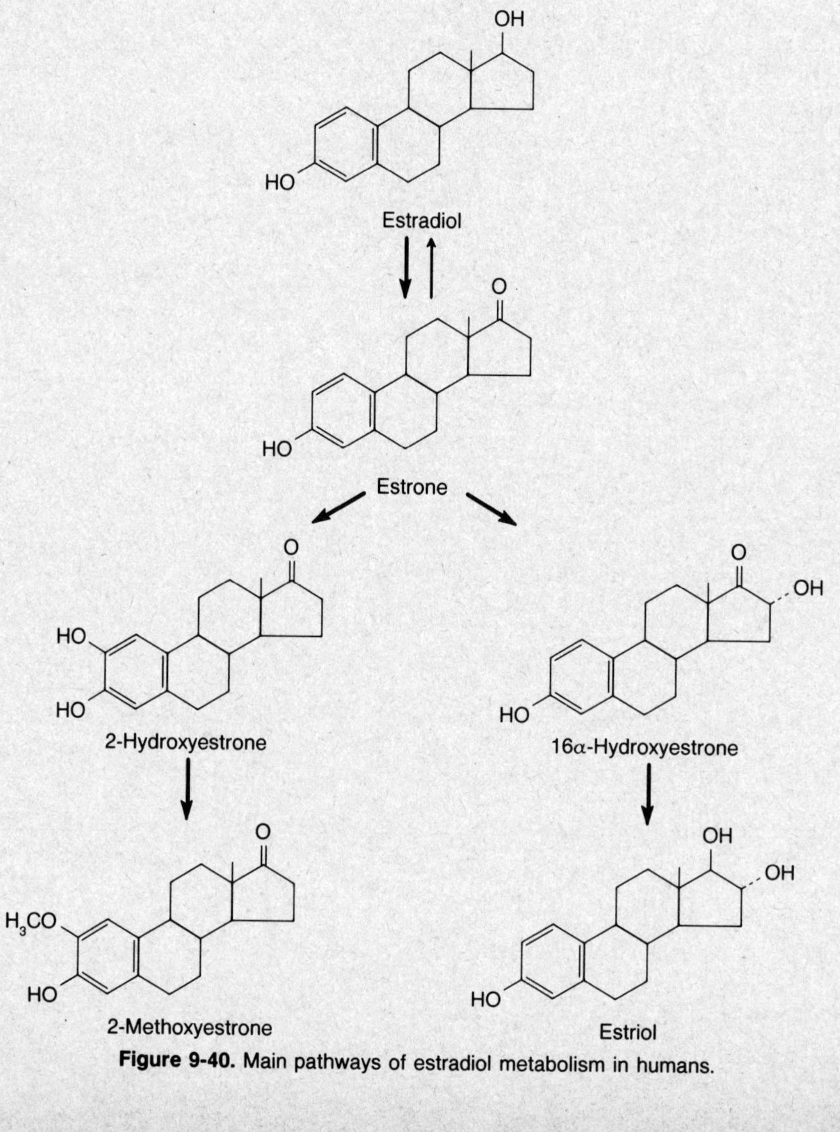

Figure 9-40. Main pathways of estradiol metabolism in humans.

indicates that in nonpregnant women the 2-hydroxylation pathway is quantitatively more important than the 16α-pathway. During the menstrual cycle, 2-hydroxyestrone is excreted to the extent of 60 μg/d and at a level comparable to that of estriol.[47] The direction of estradiol metabolism is dependent upon the pathophysiological state. Thus, hyperthyroidism and anorexia nervosa increase the 2-hydroxylation pathway at the expense of 16-hydroxylation of estradiol. Hypothyroidism, liver damage, and obesity reverse the direction of metabolism.[185] According to some investigators, catechol estrogens may play an anti-estrogen role,[11] a role currently ascribed to estriol.[338] The significance of these estrogen metabolites in health and disease, however, is far from being defined.

As is the case for other steroids, the liver is the primary site for the inactivation of estrogens. The main biochemical reactions are hydroxylation, oxidation, reduction, and methylation. Conjugation with glucuronic or sulfuric acid is the final step in the metabolic process. Conjugation imparts more water solubility to these steroids, allowing them to be eliminated rapidly through the kidney. Catechol estrogens, in addition, can bind covalently with protein through O-quinone or O-semiquinone intermediates.

METHODS FOR THE DETERMINATION OF ESTROGENS IN PLASMA

Most plasma estradiol methods measure both free and bound fractions of the unconjugated steroid. Because plasma estradiol concentrations are very low throughout the menstrual cycle, adequate assay sensitivity can be achieved in conventional chemical assays only by using very large sample volumes. Although double isotope-derivative methods[10] or gas chromatography with electron capture detection[79] are sufficiently specific and sensitive, these methods find limited utility in the clinical laboratory for reasons cited earlier. These methods have now been superseded by radioimmunoassays, which combine advantages of sensitivity, reliability, and practicality.[2]

Clinical Significance

Estradiol is not only the most potent natural estrogen, it is also derived almost exclusively from the ovaries. Consequently its measurement is often considered sufficient to evaluate ovarian function. The secretory pattern of estrogens during the menstrual cycle, as well as their clinical significance in menstrual disorders, is described under Female Reproductive Endocrinology. The clinical value of estrone determinations is limited to diagnosis of postmenopausal bleeding and to menstrual dysfunction due to extraglandular estrone production. Normally, blood estrone levels parallel estradiol levels throughout the menstrual cycle but at one-third to one-half their magnitude.[3] Except during pregnancy, measurements of estriol have little clinical value, since in nonpregnancy estriol is derived almost exclusively from estradiol. Although assays for catechol estrogens are now available,[47] their clinical value remains to be established.

Determination of Estradiol by RIA

Principle

Antisera have been generated against estradiol-6-O-carboxymethyl oxime-BSA, leaving the 3-hydroxyl and 17-hydroxyl groups free. These antisera are highly specific and permit direct quantitation of estradiol in plasma or serum.[131] Estradiol, which is bound with high affinity to SHBG in plasma, is displaced by adding a large excess of testosterone to samples. Standards in buffer containing 5% BSA also receive the same amount of the displacing steroid. Following addition of ^{3}H-estradiol and antibody, the tubes are incubated in a water bath at 37 °C for 1 h. Separation of bound from free is achieved by means of dextran-coated charcoal. Interference of plasma proteins with this separation step is corrected by including sample blanks (i.e., sample plus all reagents except the antiserum) and standard blanks. Bound activity of standards and plasma samples is corrected for its own blank. A standard curve is drawn by plotting the reciprocal of the bound activity on the ordinate versus the concentration of each estradiol standard. The concentrations of unknown samples are then interpolated from the standard curve. Commercial kits employing ^{125}I-labeled tracer for direct assay of estradiol are available. Instructions for specimen collection, reagent preparation, and the assay procedure have been discussed further elsewhere in the text (p. 1051).

Comments

Antisera raised against estradiol-6-O-CMO-BSA do not generally cross-react with any other steroids or estrogens except the 6-oxygenated estradiols. Relative to estradiol, blood levels of the 6-oxygenated compounds are absent or very low, and therefore direct quantitation without extraction is possible. Problems of accuracy in direct assays for plasma or serum estradiol are similar to those described earlier for cortisol and testosterone. Experience suggests that extraction as well as purification using a Sephadex LH-20 column[222] is necessary for accurate measurement of estradiol in men or in prepubertal children. For a specific analysis of estrone, the reader is referred to the article by Dobson and Dean.[67]

Reference Ranges

	pg/mL
Female,	
Prepubertal:	4–12
Early follicular phase:	30–100
Late follicular phase:	100–400
Luteal phase:	50–150
Postmenopausal:	5–18
Male,	
Prepubertal	2–8
Adult:	10–60

METHODS FOR THE DETERMINATION OF ESTROGENS IN URINE

Studies based on the administration of tracer doses of radioactive estrogens have shown that urine is the principal route of estrogen excretion; the urinary determination of three estrogens—estrone, estradiol, and estriol, either individually or together—renders adequate information about the endogenous production of estrogenic hormones. In recent years a variety of chemical methods for urinary estrogen determination have appeared in the literature.[175] In most methods, detection is carried out by either fluorometry or colorimetry. When heated with sulfuric acid, estrogens form an orange-yellow color with intense yellowish-green fluorescence. Such acid-induced fluorescence is sufficiently sensitive to permit detection of as little as 0.005 μg of estrogens. However, nonspecific fluorescence and a number of other variables—such as length of exposure of acid and steroids to elevated temperatures, the amount of water added, and the presence or absence of solvents—are sources of difficulty in utilizing fluorometry for routine analysis. Of the colorimetric methods available, the Kober reaction is best known and most widely accepted for quantitative determination of estrogens. The specific functional groups that are responsible for the development of the pink color in the Kober reaction (absorbance maximum at 520 nm) include the phenolic or phenolic ether groups at C-3 of ring A and an intact ring D oxygenated at C-17. Although the Kober reaction is specific for estrogens, urinary contaminants will produce a yellow-brown color during the reaction sequence. When the brown color is superimposed on the pink color caused by estrogens, a decrease in sensitivity results. For this reason, extensive purification of the urinary extract prior to color development, as well as a background color correction, becomes necessary for reliable measurement of estrogens.

A modification of the Kober reaction[32] is presently the most widely applied method. This method has the distinction of being the first published procedure that is both sensitive and specific enough to measure individually estrone, estradiol, and estriol in the urine of nonpregnant women. The main steps in this method are acid hydrolysis, extraction, separation into neutral and phenolic fractions, conversion of the 3 estrogens into their respective 3-methyl ethers, purification and separation by alumina chromatography, and final quantitation by the Kober color reaction using the Allen correction.

Ittrich[127] has introduced an ingenious method for avoiding the contributions of nonspecific chromogens. The products of the second stage in the Kober reaction are diluted with water to an acid concentration of 20–30% and are shaken in the cold with chloroform containing 2% p-nitrophenol and 1% ethanol. The color resulting from the estrogens is extracted into the chloroform layer, whereas the yellow-brown, nonestrogen color remains in the acid layer. The novelty of this modification lies in the fact that the color complex in chloroform can be measured either colorimetrically or fluorometrically, since it emits an intense yellowish-green fluorescence when excited with visible light. When the complex is measured colorimetrically, an amount equivalent to 0.2 μg of estrogens can be determined; fluorometrically, as little as 0.005 μg of estrogens can be detected. This procedure has been incorporated into various methods for estimating estrogens in urine and blood.[175] Indeed, fluorometric measurement of total estrogens in crude urine extracts, without chromatographic purification but with the Ittrich extraction, has been shown to be fairly specific and highly suitable for routine analysis. A method based on such a principle for urinary determination of total estrogens in nonpregnant women is described below.

Clinical Significance

Either plasma estradiol or urinary estrogen excretion reflects the excretion pattern of estrogens in normally menstruating women. Which of the two is chosen usually depends on the nature of the information sought, the convenience of the patient or analyst, or the availability of skilled technical personnel. Prior to the introduction of RIA, measurements of urinary estrogens enjoyed full clinical confidence; in most instances, plasma estradiol and urinary estrogen concentrations correlate quite well. Nevertheless, one should recall that most estrogens are excreted in urine as metabolites, and therefore changes may represent altered liver and renal metabolism as well as changes in estrogen production. On the other hand, certain clinical entities such as polycystic ovarian disease exhibit significant extraovarian estrogen production. In these cases, measurement of urinary estrogen may be more reliable diagnostically. In the case of induction of ovulation, urinary determinations of estrogens have also been suggested to be useful.[305]

Determination of Urinary Total Estrogens in Nonpregnant Subjects[34]

Principle

Estrogen glucuronides are acid hydrolyzed; free estrogens are then extracted into diethyl ether, washed with carbonate buffer to remove acidic impurities, and separated into phenolic and neutral steroids by partitioning between a sodium hydroxide solution and the organic extract. Following re-extraction of phenolic steroids with diethyl ether, the Kober color is developed, extracted into chloroform containing 2% p-nitrophenol (Ittrich modification), and measured fluorometrically using 530 nm as the wavelength for excitation and 550 nm as the wavelength for emission. Urinary estrogen excretion is calculated by comparing the intensity of fluorescence of the sample with that obtained from standard mixtures (estrone, estradiol, and estriol) of known concentrations.

Reagents

All reagents should be reagent grade.
1. Concentrated hydrochloric acid.
2. Diethyl ether, glass-distilled, HPLC grade.
3. Sodium carbonate buffer, pH 10.5, prepared by mixing 150 mL of sodium hydroxide, 20 g/dL, with 1 L of sodium bicarbonate, 8 g/dL.
4. Light petroleum ether (b.p. 40–60 °C), HPLC grade.
5. Sodium bicarbonate, 8 g/dL.
6. Sodium hydroxide, 1 mol/L. Dissolve 40 g sodium hydroxide in distilled water and dilute to 1 L.
7. Sodium sulfate, anhydrous.
8. Ethanol, HPLC grade.
9. Hydroquinone solution, 4 g/dL ethanol recrystallized from ethanol.
10. p-Nitrophenol solution, recrystallized from benzene; dissolve 2 g p-nitrophenol in 98 mL of chloroform containing 1% (v/v) ethanol.
11. Concentrated sulfuric acid.
12. Chloroform, HPLC grade.
13. Standard solution: Weigh 8 mg each of estrone and estriol and 4 mg of estradiol and dissolve in 100 mL absolute ethanol. Dilute 0.1 mL of this stock solution with absolute ethanol to 100 mL in a volumetric flask. (This working standard contains 0.008 μg of estrone and estriol and 0.004 μg estradiol per 0.1 mL.) The stock solution should be stored at 4 °C.

Procedure

1. Collect a 24-h urine specimen and test the urine for glucose as described for the total 17-ketogenic steroids (p. 1082). If the glucose concentration is > 0.5 g/dL, follow the procedure as described there, using 1% of the total volume of urine. Dilute the glucose-free residue to 25 mL with distilled water and proceed to step 3 below.

Hydrolysis and Extraction

2. Transfer 1% of the total volume of urine into a 250-mL round-bottom flask and dilute to 25 mL with distilled water.

3. Add several glass beads (to prevent bumping) and heat to boiling under a reflux condenser.

4. Add 5 mL of concentrated hydrochloric acid through the condenser and continue boiling for 30 min.

5. Cool the flask rapidly under running tap water and transfer the contents to a separatory funnel.

6. Extract the hydrolyzed urine once with 25 mL of ether and twice with 12.5 mL of ether.

7. Shake the combined ether layers with 10 mL of the sodium carbonate buffer (pH 10.5). Discard the aqueous layer.

Separation of Phenolic Steroid

8. Add 50 mL of petroleum ether to the ether extract in the separatory funnel.

9. Extract the organic solvent mixture two times with 25 mL of NaOH, 1 mol/L, collecting the alkali layer in an Erlenmeyer flask.

10. Partly neutralize the alkaline solution by adding solid $NaHCO_3$ in portions until the pH is 10.

11. Transfer the aqueous solution to a separatory funnel and extract the solution three times with ether— once with 50 mL and twice with 25 mL.

12. Shake the combined ether layers with 20 mL of sodium bicarbonate (8 g/dL). Discard the aqueous layer.

13. Wash the ether extract with 10 mL of distilled water and drain off the water as completely as possible.

14. Transfer the ether extract to an Erlenmeyer flask containing 5 g anhydrous sodium sulfate.

15. Rinse the separatory funnel with a few mL of fresh ether and add to the flask.

16. Filter the ether extract through a Whatman No. 1 filter paper into a 250-mL round-bottom flask.

17. Evaporate the ether to dryness at 30 °C under reduced pressure.

Fluorometry

18. Dissolve the residue in the flask in 5 mL of absolute ethanol and transfer 2-mL aliquots to two glass-stoppered tubes.

19. Prepare standard tubes (in duplicate) containing 0.1 mL of working standard solution (total steroids = 0.02 µg, i.e., sum of estrone = 0.008 µg, estriol = 0.008 µg, and estradiol = 0.004 µg) and a blank tube containing pure ethanol.

20. Add 0.5 mL of hydroquinone solution, 4 g/dL, to each tube.

21. Evaporate each tube to dryness under nitrogen in a water bath at 50 °C.

22. Cool the tubes in an ice-water bath for 3 min.

23. Add 0.4 mL of distilled water and 0.75 mL of concentrated sulfuric acid. Stopper the tubes and heat in a boiling water bath for 40 min with occasional shaking.

24. Cool the tubes in an ice bath for 3 min.

25. Add 1.5 mL of distilled water to each tube and mix thoroughly. Allow the tubes to stand in ice not less than 5 min and not more than 25 min.

26. Add 2.5 mL of p-nitrophenol solution, 2 g/dL.

27. Stopper the tubes and shake vigorously for 30 s.

28. Centrifuge the tubes for 3 min at 900 × g.

29. Aspirate and discard the upper layer.

30. Transfer the lower organic phase into a cuvet and set the instrument to zero absorbance with the blank. Read the fluorescence of standards and samples at 550 nm following excitation at 530 nm.

Calculation

If 1% of the total volume of urine is used, then

$$\mu g \text{ of total estrogens/d} = \frac{\text{reading of sample}}{\text{reading of standard}} \times 0.02 \text{ } \mu g \text{ standard} \times \frac{5}{2} \times 100$$

Comments

Since the bulk of the estrogens is excreted as water-soluble conjugates of glucuronic and sulfuric acids, hydrolysis is necessary to allow extraction of the steroids with an organic solvent such as ether. As a matter of expediency, acid hydrolysis is generally used. However, the presence of excess glucose, hydrochlorothiazide, and a wide variety of formaldehyde-generating drugs (e.g., methenamine mandelate) during acid hydrolysis can seriously decrease the recovery of estrogens, particularly estriol. Interferences are also observed in the presence of phenolphthalein, cascara, senna, and diethylstilbestrol. Acid hydrolysis of urine from patients with liver disease results in negative values for estrogens. In the presence of significant amounts of glucose, the destruction of estrogens may be > 50%, and destruction is > 95% in the presence of sucrose, fructose, and inulin. A change in specific gravity of urine from 1.010–1.020 also causes a linear decrease from 90% to 60% in the recovery of estriol. Measures suggested to eliminate the effects of interfering substances include dilution of the urine, isolation of estrogen conjugates by ammonium sulfate precipitation, solvent extraction, gel filtration, or extraction of conjugates by neutral polystyrene resin.

Washing the specimen with sodium carbonate buffer (pH 10.5) removes strongly acidic components. As described before, estrogens are slightly acidic in nature because of the presence of a phenolic group at C-3. This acidic property has been utilized for the separation of phenolic estrogens and neutral steroids. Shaking the organic solvent with sodium hydroxide solution moves the estrogens into the alkali layer, while the neutral steroids (17-ketosteroids, pregnanediol, corticosteroids, and others) stay in the organic phase. Petroleum ether is added to achieve better recovery of some of the estrogens (estrone, estradiol) that would be left in the ether layer when partitioned with sodium hydroxide solution. Since estrogens are quite soluble at a pH above 11, the adjustment of the pH to <10.5 is very important for quantitative re-extraction of estrogens with ether.

The addition of hydroquinone protects the estrogens from oxidation and facilitates formation of the Kober color complex. Directions for the addition of each reagent necessary for color development, the duration of heating, and the extraction of the color complex with p-nitrophenol solution after the specified length of time should be followed as closely as possible. The tubes should be kept in ice at all times. The intensity of fluorescence decreases with time; thus, the fluorometric reading should be completed within half an hour after extraction of the color complex. Since the relative intensities of fluorescence of the three estrogens—estrone, estradiol, and estriol—are different, determination of the total estrogen content without separation of the individual estrogens may result in some error. The use of a standard solution of three estrogens mixed in a ratio generally excreted in the urine of a normal nonpregnant woman (estrone: estradiol: estriol, 2:1:2) reduces the probability of such errors.

Reference Ranges

The excretion of estrogens in children is generally <1 µg/d. In men, a constant amount of estrogen is excreted that is derived from the adrenals and probably also from the testes. The average value is ~11 µg/d with a range of from 5–18 µg/d.

Brown[32] has extensively studied the excretion of estrone, estradiol, and estriol during the normal menstrual cycle. The following are excretion values of total estrogens computed from his data:

	µg/d	
Onset of menstruation:	4–25	(mean 13)
Ovulation peak:	28–99	(mean 56)
Luteal peak:	22–105	(mean 43)
Menopausal women:	1.4–19.6	(mean 6.4)

Estrogens and Breast Cancer

Association of an etiologic role for estrogens in the development of human breast cancer stems primarily from epidemiologic observations. Breast cancer is predominantly confined to females after puberty. There is a significantly smaller risk of mammary carcinoma in women subjected to oophorectomy prior to their fortieth year. Thus, early menarche and later natural menopause are associated with increased risk of breast cancer. A two-fold mechanism has been postulated: initiation of a precancerous state by ovarian activity during the early reproductive years and continuation of ovarian activity in later years as a promoting influence on already initiated tumor cells. Ovarian estrogen has been assumed to be the causative factor because protective effects of early oophorectomy are negated by administration of estrogen. Moreover, treatment of males with estrogen for prostatic cancer or after transsexual operations is also associated with an increased risk of breast cancer.

Low risk for breast cancer has consistently been connected with high parity. The evidence may be summarized as follows:

1. Pregnancy occurring before age 25–30 has a protective effect; women bearing children for the first time after the age of 30 actually have a higher risk than nulliparous women.

2. Only the first, full-term, early pregnancy conveys protection. Women having their first pregnancies before 18 years of age have approximately one third the risk of women bearing their first child after age 35. Interrupted pregnancies do not afford protection; in fact, some studies indicate that abortions may increase risk.

3. The protective effects of pregnancy have been deduced from other observations. For example, married women without children are at higher risk than those with one or more children; unmarried women and nuns are at higher risk than women who have experienced pregnancy.

Cole and MacMahon[60] advance a popular hypothesis relating estrogen to risk of breast cancer. According to them, the relative levels of individual estrogen fractions (estrone, estradiol, and estriol) produced in the first decade or so after puberty are important determinants of a woman's lifetime risk of breast cancer. In particular, pregnancy at a young age is associated with both favorable estrogen-fraction ratios and decreased risk. The urinary estriol ratio, i.e., the ratio of estriol to sum of estrone and estradiol, is higher in Asian women, who have a lower incidence of breast cancer, than in American women; these differences are particularly marked in the early years of reproductive life. These findings imply a protective role for estriol against breast cancer. However, more recent biological, epidemiological, and metabolic studies have cast serious doubt on this hypothesis. Further discussion of the role of estrogen in the genesis of breast cancer will be found in recent monographs and review articles.[138,172,338]

MEASUREMENTS OF ESTROGEN AND PROGESTERONE RECEPTORS IN BREAST CANCER

The first step in steroid hormone action in general, and for estrogens in particular, is the binding of the hormone to specific cell receptors. An *estrogen receptor* is a specific cytoplasmic protein with high affinity and great specificity for estrogen. The estrogen receptor protein is found in target tissue cells such as the uterus, pituitary, hypothalamus, and breast. Since estrogen stimulates biochemical processes in target cells that normally contain estrogen receptor protein, a reduction in blood estrogen levels would be expected to reduce the biochemical activity of these cells. This has been the rationale for using endocrine therapy in women with breast carcinoma. Approximately one third of women with metastatic breast carcinoma obtain an objective remission following various types of endocrine therapy directed at lowering their estrogen levels. Such therapy includes oophorectomy, hypophysectomy, and adrenalectomy (ablative therapy) as well as administration of antiestrogens and androgens (additive therapy).

Cytoplasmic estrogen receptors are now routinely measured in samples of breast tissue after surgical removal of the tumor. Sixty per cent of patients with carcinoma of the breast have tumors that are estrogen-receptor positive. Approximately three fourths of patients with estrogen receptor-positive tumors respond to endocrine therapy; 95% of the patients with estrogen receptor-negative tumors fail to respond. The greater the estrogen receptor content of the tumor, the higher the response rate to endocrine therapy.

Occasionally, a tumor is defined as estrogen-receptor negative but the patient responds to endocrine therapy (*false negative results* of estrogen receptor assay). Explanations of false negatives include:

1. Incorrect handling and storage of tissue sample that may degrade the thermolabile receptor proteins.

2. Insufficient sampling of malignant tissue or inadvertent biopsy of neighboring nonmalignant tissue.

3. A tumor that is heterogeneous in terms of receptor status so that the biopsied site is nonrepresentative of other tumor deposits.

4. High levels of endogenous estrogens that saturate receptor sites and render them undetectable when standard methodology is used.

False positive results of estrogen receptor assays (estrogen-receptor positive tumor but no response to endocrine therapy) are more common than false negatives. The most frequent explanation is heterogeneity of tumor with biopsy of a site which is not representative of the tumor deposits taken as a whole.

Progesterone receptor assay is a useful adjunct to the assay of estrogen receptors. Since progesterone receptor synthesis appears to be dependent upon estrogen action, measurement of progesterone receptor activity provides confirmation that all the steps of estrogen action are intact. Indeed, metastatic breast cancer patients with estrogen *and* progesterone receptor–positive tumors have a response rate of 75% to endocrine therapy while those with estrogen receptor–positive and progesterone receptor–negative tumors have only a 30% response rate.

Determination of Estrogen and Progesterone Receptors

The most reliable and reproducible methods for cytosol receptor assay appear to be the multiple-point dextran-coated charcoal (DCC) assay analyzed by a Scatchard plot. These conventional assays measure only the receptor in the cytosol that is not occupied by estradiol.

Cytosol of the submitted tissue is prepared as a supernatant of tissue homogenate centrifuged at $100\ 000 \times g$ for 45 min at 2 °C. For the *estrogen receptor assay*, the estrogen receptors are titrated to saturation with increasing concentrations of radioactive estradiol, both in the presence and absence of unlabeled diethylstilbestrol added in excess, under equilibrium conditions (incubation at 0 °C for 16–24 h). Because diethylstilbestrol binds only to receptor and not to SBHG or to other binding proteins, labeled estradiol bound when receptors are saturated with diethylstilbestrol is a measure of nonspecific binding. Separation of bound and free labeled steroid is achieved by adsorption of free steroid to dextran-coated charcoal. Estimates of the quantity and affinity of estrogen receptors are determined from a Scatchard plot of the ratio of bound/free ^{3}H-steroid versus bound ^{3}H-steroid. Generally, linear functions are obtained. The dissociation constant (K_d) is determined from the inverse of the slope; concentration of binding sites (n) is read at the intersect of the line on the abscissa. Concentration of sites is expressed as fmoles ^{3}H-estradiol per milligram of tissue protein. If the concentration is > 10 fmoles/mg of protein, the test is considered positive.

The principle and procedure for *progesterone receptor assay* are similar to those described for the measurement of estrogen receptor except that ^{3}H-progesterone is used as the labeled ligand, and excess unlabeled R 5020 (17,21-dimethyl-19-nor-pregna-4,9-diene-3,20-dione), a synthetic progestin, is substituted for diethylstilbestrol. The test is considered positive if the concentration of receptors exceeds 2 fmoles/mg of protein. Authoritative discussions on the clinical significance and assay of estrogen and progesterone receptors have been compiled in recent monographs.[194,195]

Reproductive Endocrinology

Puberty

During childhood, circulating levels of sex steroids and gonadotropins are low and are similar for both sexes. The transition from sexual immaturity appears to begin with a diminished sensitivity of the pituitary or hypothalamus, or both, to the negative feedback effect of the sex steroids. The mechanism for this change is unclear. Values for LH, FSH, and gonadal steroids rise gradually over several years before stabilizing in the adult range, when full sexual maturity is reached. In girls, precocious puberty is defined as the onset of pubertal development before the age of 8½, and puberty is considered delayed if there has been no development by the age of 13½ or if menarche has not occurred by age 16½. In boys, puberty is considered precocious before the age of 10 and delayed when the onset is later than age 16.[252]

True precocious puberty is more common in girls. The changes are generally identical with those of normal puberty. When isosexual precocity is suspected, careful neurological examination and CT scan should be performed to exclude central nervous system lesions. Sex hormone–producing tumors of the ovaries or adrenal glands or gonadotropin-secreting tumors may occur, but these are unusual. When puberty is delayed, measurement of gonadotropins may be useful; elevated levels would indicate gonadal failure. Screening for chromosome abnormalities is particularly important in a girl with delayed puberty. Pituitary function testing and X-rays may be helpful if gonadotropin levels are not elevated. Often, delayed puberty is not due to specific organic disease but is simply the unusual end of the spectrum of maturation.

Female Reproductive Endocrinology

The normal menstrual cycle. During a normal menstrual cycle, there is a closely coordinated interplay of feedback effects between the hypothalamus, the anterior pituitary gland, and the ovaries, as well as cyclic hormone changes that lead to functional and structural changes in the ovaries (follicle maturation, ovulation, and corpus luteum development), uterus (preparation of the endometrium for possible implantation of the fertilized ovum), cervix (to permit transport of sperm), and vagina.[92] See Figure 9-41.

The menstrual cycle is usually counted as beginning with day 1 on the first day of menstrual bleeding. The follicular phase, which is about the first half of the cycle (14 d of a 28-d cycle) actually begins a few days before. During the *follicular phase*, pituitary gonadotropin production is suppressed by low levels of estrogen produced by the developing follicle. Hypothalamic release of gonadotropin-releasing hormone increases, with concomitant rise of estradiol production by the ovary. During the *ovulatory phase*, a period of about 48 h at midcycle, there is a rise in estradiol, LH, and FSH that is followed by ovulation. The last half of the cycle, the *luteal phase*, is characterized by increasing production of progesterone and estrogen with consequent gradual

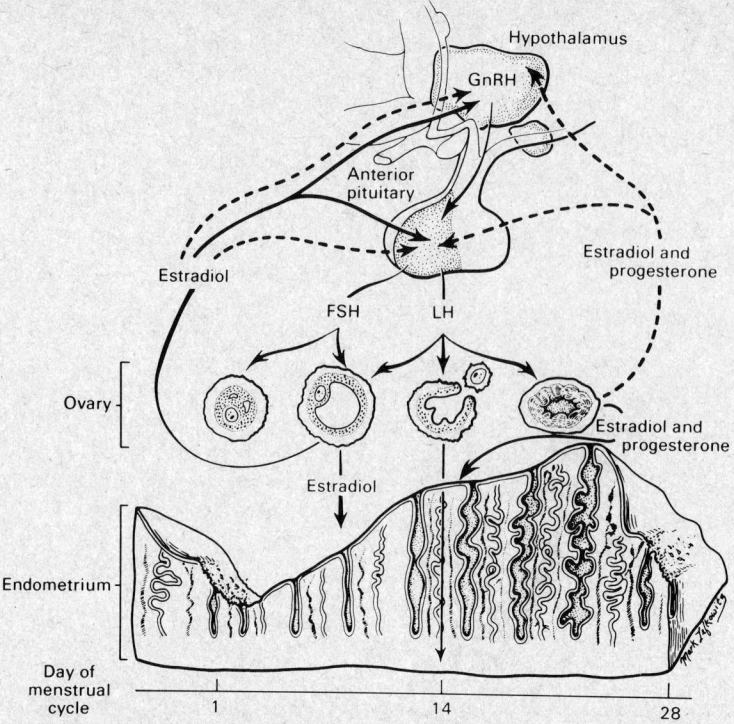

Figure 9-41. Summary of endocrine control and changes in ovaries and endometrium during the menstrual cycle. Dashed arrows indicate inhibitory effects and solid lines stimulatory effects.

lowering of LH and FSH levels. A schematic representation of the hormone changes of a normal menstrual cycle is shown in Figure 9-42. LH and FSH release are pulsatile throughout the cycle; therefore the values shown in the figure represent integrated levels. To explain further the intricacies of the normal menstrual cycle, each important hormone change will be considered separately with regard to control and effects (Figure 9-41).

Gonadotropin-releasing hormone (GnRH, also known as luteinizing hormone–releasing hormone or LHRH) from the hypothalamus acts to trigger the surge of luteinizing hormone that precedes ovulation. There appear to be two separate feedback centers in the hypothalamus—a tonic negative feedback center in the basal medial hypothalamus and a cyclic positive feedback center in the anterior hypothalamus. Low levels of estradiol, such as occur during the follicular phase, affect the negative feedback center, while high levels of estradiol, such as are seen just before the midcycle LH peak, trigger the positive feedback center. Progesterone in combination with estrogen affects the negative feedback center in the luteal phase. GnRH is released in a pulsatile fashion and has a self-priming effect; the first dose potentiates the effect of subsequent doses. The magnitude of the LH response to GnRH increases steadily through the follicular phase and is greatest at the time of the preovulatory surge of LH, after which it declines again.

A few days before day 1 of the cycle, *follicle-stimulating hormone* (FSH) shows a slight but important peak, probably triggered by a fall in estradiol that briefly eliminates the negative feedback effect. This peak of FSH begins the growth and maturation of a receptive ovarian follicle. FSH falls again and remains fairly low through the follicular phase due to negative feedback from estradiol produced by the developing follicle. However, the effect of FSH on the maturing follicle is increased through estradiol-induced changes in FSH receptors. FSH, aided by estradiol, acts on the cells of the follicle to increase responsiveness of LH receptors by the time of the midcycle surge. FSH and LH receptors are increased either in their number or in their affinity for corresponding gonadotropin, or in both. There is a rise in FSH at midcycle that is triggered by GnRH, but the function of this peak is unknown. During the luteal phase FSH is suppressed by negative feedback from estradiol until a lesser FSH peak, occurring near the end of the cycle, starts off the follicular maturation of the next cycle.

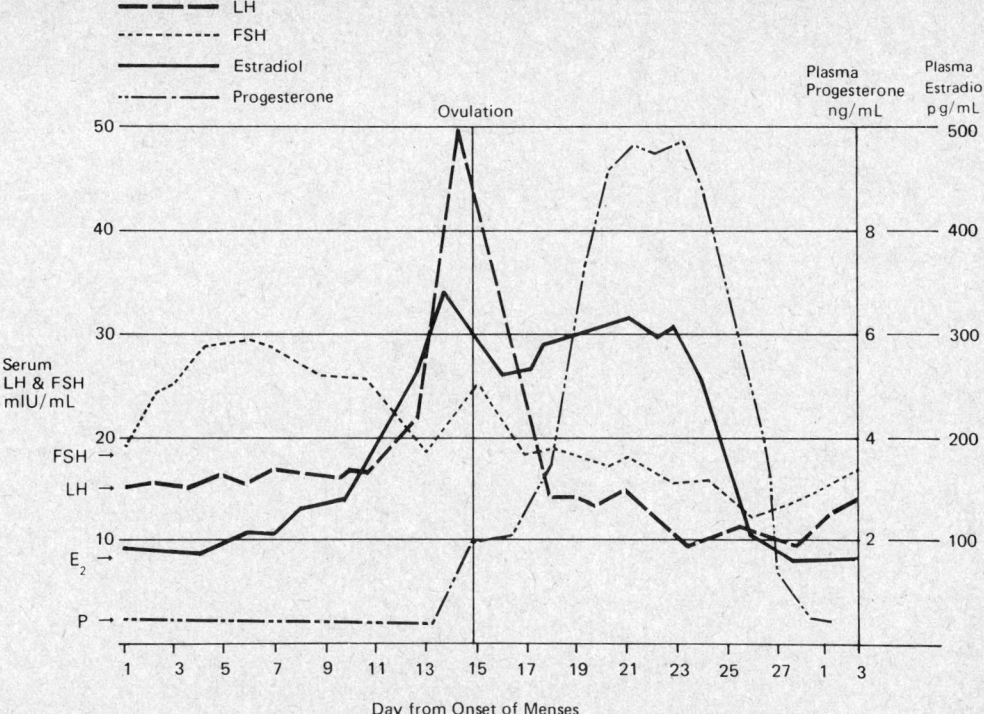

Figure 9-42. Composite of hormone changes during normal menstrual cycle. (From Watts, N. B., and Keffer, J. H.: Practical Endocrine Diagnosis. 3rd ed. Philadelphia, Lea & Febiger, 1982.)

LH secretion is suppressed in the follicular phase by negative feedback from estradiol. As estradiol production by the developing follicle increases, the effect of estradiol on the positive feedback center becomes important. Increasing release of GnRH from the hypothalamus and increasing sensitivity of the anterior pituitary to GnRH result in the midcycle surge of LH. Ovarian follicle receptors for LH, sensitized by FSH and estradiol, transmit the stimulus to enhance differentiation of the theca cell and the production of progesterone by the developing corpus luteum. LH production is suppressed during the luteal phase by negative feedback from progesterone combined with estradiol, but a low level of LH is probably necessary to prolong corpus luteum function.

Estradiol production by the ovary begins to fall near the end of a cycle but begins to increase again under the influence of a late-cycle surge of FSH. Estradiol enhances the FSH effect on a ripening follicle through changes in FSH receptors of the follicular cells but supresses pituitary FSH and LH release during the follicular phase through negative feedback. As estradiol production increases near midcycle, the effect on the positive feedback center takes over, with increased release of GnRH from the hypothalamus leading to increased release of LH and FSH from the anterior pituitary. During the luteal phase, progesterone produced by the corpus luteum, combined with estrogen, exerts a negative effect on the hypothalamus and anterior pututiary. As a result, LH and FSH are suppressed again during the luteal phase. Estradiol is essential for the development of proliferative endometrium, and is synergistic with progesterone for the development of secretory endometrium. The fall of estradiol near the end of the luteal phase causes the vascular changes in the endometrium that initiate shedding, and the decrease in negative feedback from estradiol on the anterior pituitary triggers the FSH surge that begins the development of an ovarian follicle for the next cycle.

Progesterone is not produced in significant amounts until after the midcycle LH surge and ovulation. LH enhances theca cell differentiation and progesterone production, which rises 10- to 20-fold, to a maximum about 8 d after the midcycle peak of LH. Progesterone promotes the growth of secretory endometrium, which is necessary for implantation of the fertilized ovum.

Growth of ovarian follicles appears to be continuous; how an individual follicle is singled out for each menstrual cycle is not known. The late-cycle peak of FSH is important in this process. Once a follicle has been stimulated, estradiol production causes that specific follicle to be more receptive to effects from FSH. The high level of estradiol just before midcycle is responsible for triggering the positive feedback in the hypothalamus that leads to the midcycle LH surge. The precise cause of ovulation is not known, but ovulation occurs 16–24 h after the LH peak. After ovulation, LH is suppressed by progesterone and estradiol, but the effect of LH on the corpus luteum is increased. In the event of successful fertilization and implantation, corpus luteum function is sustained by chorionic gonadotropin produced by the placenta. Otherwise, the declining level of estradiol leads to regression of the corpus luteum and to the late-cycle FSH peak that starts the process again.

There are changes in androgen production during the menstrual cycle; a peak at midcycle may possibly serve to increase libido at the time. Estradiol is not the only estrogen produced; estrone, mainly from peripheral sources, also rises through the cycle. Estrogen and progesterone have visible effects on vaginal cytology and cervical mucus, and progesterone elevates body temperature. These responses may have more clinical value than estrogen and progesterone measurements for detecting ovulation or identifying the luteal phase of the menstrual cycle.[202]

Infertility and irregular menses. In a normal ovulatory menstrual cycle, bleeding occurs 14 or 15 d after the midcycle LH surge. Normal women display a fair amount of variation in cycle length; these differences are due to variations in the length of the follicular phase. In infertile women, the luteal phase may be shortened. Progesterone secretion is low, the corpus luteum regresses early, and menses occur only 5–10 d after the midcycle peak. When basal body temperature is used as an index of ovulation, menses occurring 5–10 d after the temperature rise are suggestive of a short luteal phase. Measurement of serum progesterone 4–7 d after the basal temperature rise will help establish this diagnosis. Normally the level of progesterone should be > 10 ng/mL. The most specific diagnostic information may be gained from an endometrial biopsy[326] obtained on day 26 of a 28-d cycle.

Other abnormalities of the menstrual cycle may occur. The cycle length itself may be normal but anovulatory; a record of basal body temperature or profile of gonadotropin release pattern may elucidate the problem.[208] A diagnosis of *primary amenorrhea* should be considered if menses have not begun by age 16½, and an underlying cause should be sought. Sex chromosome abnormalities that cause gonadal dysgenesis, such as *Turner's syndrome*, constitute the most common cause of primary amenorrhea. This diagnosis, usually suggested by associated physical findings, can be confirmed by chromosome analysis.[316] Typical hormone findings in gonadal failure are high levels of LH and FSH, low levels of estrogens, and low effects of estrogens on target tissues. Less common causes of primary amenorrhea include hypopituitarism, testicular feminization syndrome, imperforate hymen, or congenital absence of the vagina.[193]

Secondary amenorrhea is diagnosed when menses have stopped after months or years of cyclic bleeding. Depending on the clinical findings, the evaluation might simply focus on the exclusion of a serious underlying disease such as hypopituitarism or on detailed evaluation of reproductive hormone function. If fertility is important, a more detailed evaluation is necessary. A scheme proposed by Kletsky et al.[147] (Table 9-15) illustrates a systematic approach to diagnosis in patients with secondary amenorrhea. Endometrial responsiveness can often be demonstrated by the occurrence of bleeding after progesterone withdrawal. If a single course of progesterone fails to cause withdrawal bleeding, several cycles of estrogen and progesterone should be given. Previous overvigorous uterine curettage (Asherman's syndrome) is suspected if cyclic estrogen and progesterone treatment does not produce bleeding. Measuring LH and FSH in blood provides information regarding the pituitary.

In the evaluation of women with menstrual disorders or suspected infertility, measurements

Table 9-15. DIAGNOSTIC STRATEGY FOR EVALUATION OF AMENORRHEA[147]

Bleeding After Progesterone Withdrawal	Serum LH	Serum FSH	Probable Diagnosis
Yes	High	Normal or low	Polycystic ovary syndrome
Yes	Normal or low	Normal or low	Hypothalamic dysfunction (includes psychogenic causes)
No	Low	Low	Hypothalamic, pituitary, or uterine failure
No	High	High	Ovarian failure

of gonadotropins and sex steroids are useful. Since serum gonadotropins vary considerably because of pulse secretion, a single sample may not provide accurate information. The effect of pulse secretion may be countered by obtaining blood samples at intervals (e.g., 3 samples 15 min apart), then pooling equal aliquots for a single assay to give an "average" value.[111]

The physical effects of gonadal steroids, or the lack of them, are easily read from evaluations of vaginal cytology, cervical mucus, basal body temperature, and endometrial biopsy. Laboratory measurements of specific estrogens or total estrogens, though readily made, are relatively insensitive and rarely add much to these other clinical maneuvers.[203] However, progesterone measurement is helpful in the diagnosis of the short luteal phase syndrome. Menstrual disturbance due to androgen excess will be considered later; prolactin excess as a cause of infertility and amenorrhea has been discussed previously. In any case, when infertility is the problem, a complete history and evaluation of both partners are essential.[209,306] Anatomic abnormalities that cause infertility may occur with or without endocrine abnormalities.

Menopause. With advancing age, the ovaries fail to produce adequate amounts of estrogen and progesterone, ovulation ceases, and cycle length shortens. Cycles then become irregular, due in part to irregular maturation of the follicles, which results in anovulatory bleeding. FSH begins to rise, then fluctuates irregularly; both FSH and LH then become tonically elevated as the failure of ovarian estrogen production eliminates the negative feedback effect on the pituitary. Ovarian failure may occur at any age, but menopause prior to age 40 is considered premature.[253] In an unusual variation of gonadal failure, the resistant ovary syndrome,[152] primary or secondary amenorrhea may present associated with the typical hormone pattern of the menopause. Surprisingly, some of the affected women have shown evidence of ovulation and have conceived. The basic defect is not clear but appears to be related to a deficient response of the ovary to gonadotropin stimulation.

Hirsutism. Most women find an excess of facial and body hair (hirsutism) distressing. While hirsutism is sometimes a side effect of various medications or does not appear to have an identifiable underlying cause, unwanted hair growth is often due to an increase in androgen production. Evaluation is appropriate in all women in whom hirsutism is of moderate or severe degree and in women with mild hirsutism if the patient is particularly concerned or when there is associated oligomenorrhea or infertility.

It is important to be sure that hirsutism is not due to androgen production from a neoplasm.[184] Such androgen excess is usually severe and often causes virilism, including temporal recession of the hair line and balding, clitoral enlargement, deepening of the voice, increased muscle bulk, male body habitus, and amenorrhea, in addition to hirsutism. DHEA-S concentrations in plasma of > 700 μg/dL would be strongly suggestive of an adrenal tumor that is secreting androgens. Total serum testosterone concentrations of > 200 ng/dL would suggest a testosterone-producing tumor; these tumors are usually found in the ovaries but may arise in the adrenal glands. Signs of androgen excess associated with signs of glucocorticoid excess would suggest Cushing's syndrome. Appropriate testing for Cushing's syndrome has previously been discussed.

There are a variety of nontumorous causes of androgen excess (Table 9-16). The source of excess androgen may be the adrenal glands, the ovaries, peripheral conversion of steroid precursors to androgens by skin and other tissues, or a combination of these. While the capability exists to measure all androgenic compounds in blood or urine, selective use of hormone measurements is appropriate in evaluating patients with hirsutism for quantitation of the hormone imbalance and to follow the response to treatment.

Table 9-16. CAUSES OF HIRSUTISM

Drugs	Ovarian diseases
Endocrine disorders	Polycystic ovary syndrome
Adrenal diseases	Hyperthecosis, hilus cell or stromal cell
Classic congenital adrenal hyperplasia	hyperplasia
21-Hydroxylase deficiency	Androgen-producing ovarian tumors
11-Hydroxylase deficiency	Familial hirsutism
3β-Hydroxysteroid dehydrogenase	Idiopathic hirsutism
deficiency	
Adult or attenuated adrenal hyperplasia	
Androgen-producing adrenal tumors	
Cushing's syndrome	

Interpretation of basal androgen concentrations in blood can be difficult because these hormones are bound to protein; the degree of binding affinity or changes in the concentration of binding protein will alter total hormone values without changing free hormone concentrations. Diurnal variation concordant with that of cortisol (higher values in the morning than in the evening) occurs with some androgenic compounds, including DHEA, androstenedione, and testosterone. Concentrations of testosterone and androstenedione vary with menstrual cycle phase, higher values being present in the luteal phase than in the follicular phase. Finally, most androgenic compounds are secreted in an episodic fashion, so a single blood sample may not be a true reflection of the integrated hormone concentration over time. No matter how precise or controlled the measurement, determination of basal androgen levels in blood or urine does not identify the source of the androgen production.

Methods to identify the source of androgen excess include (1) catheterization of ovarian and adrenal veins and measurement of androgen levels[326] and (2) dynamic tests, including stimulation (with ACTH or chorionic gonadotropin) and suppression (with glucocorticoids such as dexamethasone or cyclic estrogen and progesterone). Interpretation of dynamic tests is difficult because the agents used for stimulation or suppression are not specific (e.g., dexamethasone may have some effect on suppressing ovarian steroidogenesis in addition to suppressing ACTH and adrenal steroid synthesis). As an additional factor, increases in specific androgenic compounds may alter enzymatic activities in various pathways of steroid synthesis, so that an increase in androgen from one organ may alter androgen production in another. Usually, it is not important to the management of the patient to determine the specific defect.

A simple diagnostic approach for a woman with mild or moderate hirsutism is to collect blood samples in the morning during the early follicular phase of the menstrual cycle, drawing three samples 15 min apart, and then pooling equal aliquots from each sample for the assays. Minimum studies to be done include serum LH, FSH, total testosterone, and DHEA-S. In women with polycystic ovary syndrome[104] serum LH is usually > 20 mIU/mL with a ratio of LH to FSH of > 2.[19] An elevated DHEA-S[153] with a normal total testosterone suggests an adrenal origin for androgen excess, while an elevated total testosterone with a normal DHEA-S indicates ovarian androgen excess. Elevation of both DHEA-S and total testosterone indicates a mixed source of the androgens. Treatment could then be selected based on these results; if elevations are seen in LH or testosterone, or both, initial therapy with oral contraceptives or cyclic estrogen and progesterone would be indicated. If DHEA-S is elevated as an isolated finding, initial treatment should be with dexamethasone, usually 0.5 mg at bedtime. Repeat measurements should be done after three or four weeks of treatment; if a complete response is not seen, dexamethasone can be added to oral contraceptives or vice versa.

Short-term stimulation and suppression tests do not correlate well with the results of long-term suppressive treatment and probably add little to patient management. A number of different protocols have been reported, varying in dose of medication, route of administration, time of study, and other factors. No specific protocol for short-term tests can be recommended for general use. Some authorities recommend the measurement or calculation of free testosterone,[229,260] which takes into account variations in testosterone-binding to sex hormone–binding globulin but probably adds little to the workup and management of most patients.[188] Some recommend the routine measurement of androstenedione, but elevations of this weak androgen are rarely seen as an isolated finding; most patients with elevated androstenedione concentration will have an elevation of either DHEA-S or total testosterone, or both.

Recent studies by Horton and colleagues[174] confirmed work done some years ago by Mauvais-Jarvis and associates[190] that androstanediol and its glucuronide conjugate are markedly elevated in almost all women with hirsutism, regardless of the specific androgen excess involved. Androstanediol is a metabolite of dihydrotestosterone, the most potent androgen, and is derived primarily from the metabolism of dihydrotestosterone in target tissues. While experience with this measurement has been limited, use of this approach seems promising to identify women with androgen excess and thus to select them for further study.

Male Reproductive Endocrinology

Infertility is rarely the sole presentation of primary or secondary hypogonadism in men; other clinical features (particularly impotence) generally point to that diagnosis. In men who have no obvious clinical manifestations or androgen deficiency, azoospermia or oligospermia should always be considered as a possibility. A variety of studies have been suggested for the evaluation of the pituitary-gonadal axis in "idiopathic oligospermia," but no specific trends have been demonstrated. Some studies indicate normal LH and FSH values with a normal GnRH response; others have indicated an increased magnitude of response to GnRH or increased basal levels of LH or FSH. Some studies have found low levels of testosterone; others have accounted for this by the findings of lowered levels of sex hormone–binding globulin. There is no indication of any estrogen abnormality in oligospermic men. Hyperprolactinemia may cause oligospermia along with secondary hypogonadism. Unless evidence of androgen deficiency is present, endocrine profiles in patients with oligospermia are not generally helpful.[50]

Impotence is the persistent inability to develop or maintain a penile erection sufficient for intercourse and ejaculation. A wide variety of organic and psychological abnormalities may cause changes in sexual drive, erection, or ejaculation, and the physician must pursue a careful evaluation of the possible psychological factors, neuropathy, or vascular abnormalities interfering with proper sexual function.[281] If there is no obvious explanation for troublesome impotence, LH and FSH measurements are indicated; elevated values establish a diagnosis of primary hypogonadism. Total and even free testosterone levels may be in the normal range, yet still may be subnormal for a specific individual, as elevated LH or FSH values (or both) would indicate. Hyperprolactinemia is an infrequent cause of impotence but should be considered in unusual situations.

The pituitary-gonadal axis is affected by a number of systemic diseases. Chronic renal failure causes a decrease in testosterone and sperm count, but LH and FSH are not elevated. With chronic cirrhosis, testosterone concentration tends to be low and estrogen levels elevated, but the plasma LH and FSH values are in the normal range. Even in the absence of liver disease, alcohol may have a direct effect on the testes, impairing spermatogenesis. With severe malnutrition, plasma testosterone declines but LH levels are variably affected. Cortisol excess and deficiency states, thyroid hormone excess and deficiency states, or any chronic debilitating illness may affect reproductive hormone levels and sexual function.

Gynecomastia. In adult males, the growth of glandular breast tissue should be viewed with suspicion.[39] Gynecomastia may develop due to excessive use of alcohol or marijuana or due to medication with certain drugs known to cause it—estrogens, phenothiazines, and spironolactone. Hyperthyroidism or hypothyroidism may be associated with gynecomastia. Chromosomal abnormalities such as *Klinefelter's syndrome* are often accompanied by gynecomastia, but hypogonadism is usually obvious in these patients. Germinal-cell or nonendocrine tumors that produce chorionic gonadotropin; estrogen-producing tumors of the adrenals, testes, or liver; and prolactin-producing pituitary adenomas all may cause gynecomastia. These are uncommon disorders, and the cause of gynecomastia may remain unknown in many cases. In cases of striking gynecomastia when history and physical examination point to no specific disorder, measurement of hCG, plasma estradiol or total estrogens, and prolactin is appropriate.

The presence of glandular breast tissue in pubertal males is quite common. This condition is usually mild and almost always resolves spontaneously.

Gynecomastia may be unilateral or bilateral; unilateral gynecomastia has no specific diagnostic significance.

Hypogonadism. There is an excellent review of this subject by Odell and Swerdloff.[219] Generally, the etiology of male hypogonadism can be classified by using hormone measurements. A simplified classification includes (1) abnormalities in hypothalamic or pituitary function (secondary hypogonadism); (2) primary gonadal abnormalities; and (3) chemical or cellular defects in androgen synthesis, androgen action, or both.

Measurement of serum testosterone, LH, and FSH will provide the necessary data to plan a comprehensive study of an individual case. Since testosterone, LH, and FSH are secreted in pulses, three serum samples should be collected 15–20 min apart and hormone measurements performed on a pooled specimen.[111]

Most cases of male hypogonadism will fit a description of either primary or secondary hypogonadism.[302] In primary hypogonadism, LH and FSH are elevated but testosterone is low. In hypogonadism secondary to pituitary or hypothalamic hypofunction, LH, FSH, and testosterone are all low. Less commonly, primary germ cell failure will be found, with elevated FSH but normal LH and testosterone. In syndromes of partial androgen resistance,[106] LH is increased, FSH is normal or high, and testosterone is normal or increased. In specific cases, a stimulation test with gonadotropin-releasing hormone may aid the diagnosis. Chromosome analysis[98] is useful for providing specific differential information.

THYROID FUNCTION

The fully developed thyroid gland in man is composed of two lobes connected by a thin band of tissue, the isthmus, which gives the gland the appearance of a butterfly. The gland is closely attached to the trachea in the anterior aspect of the neck. Each lobe measures ~2.0–2.5 cm in both thickness and width and 4.0 cm in length. The isthmus measures 2.0 cm in both length and width and 0.5 cm in thickness.

The secretory units of the gland are the follicles, which consist of an outer layer of epithelial cells that rest on a basement membrane and enclose an amorphous material called colloid. Colloid is mainly composed of thyroglobulin (see below), an iodinated glycoprotein, and small quantities of iodinated thyroalbumin. The follicles are embedded in stromal tissue, which contains blood vessels and autonomic nerve fibers. Increased activity of the gland is characterized by a decrease in the quantity of colloid, with the subsequent reduction of follicular volume; the lining cells become columnar and may even proliferate into the colloid. During decreased activity, the follicles are enlarged due to accumulation of colloid and the flattening of the follicular cells. The cytoplasm of the follicular cells has a microtubular network, and microvilli extend from the apices of these cells into the colloid. The important reactions of thyroid hormone synthesis, such as iodination and the initial phase of hormone secretions (colloid resorption), are believed to take place at or near this surface of the cell. These follicular cells, like those of other endocrine glands, have a prominent endoplasmic reticulum.

The thyroid also contains another type of cell known as parafollicular or C cells. These cells have been shown to produce the polypeptide hormone *calcitonin.* (See Chapter 12.) These cells never border the follicular lumen but are confined within the follicular basement lamina or exist in clusters in the interfollicular spaces.

THYROID HORMONES

The thyroid gland secretes two hormones, thyroxine (3,5,3′,5′-L-tetraiodothyronine) and triiodothyronine (3,5,3′-L-triiodothyronine), which are commonly known as T_4 and T_3, respectively. In addition, the thyroid secretes minute amounts of biologically inactive 3,3′5′-L-triiodothyronine (reverse T_3 or rT_3). Approximately 35% of the secreted T_4 is deiodinated by the liver and other peripheral tissues to yield T_3, and about 40% to yield rT_3. Therefore, with a normal T_4 production of 90 µg/d, ~26 µg of T_3 and 30 µg of rT_3 would be produced by peripheral deiodinations. From the estimated daily production rates for T_3 (30 µg) and rT_3 (30 µg), it is evident that at least 80% of normal T_3 production and essentially all of rT_3 production can be accounted for by peripheral deiodination of T_4 rather than by direct secretion by the thyroid (Figure 9-43). Until recently, T_4 was considered the principal biologically active hormone and T_3 an adjunct and relatively minor thyroid hormone. However, new, precise methods of measurement of T_3 in biological fluids and new knowledge of peripheral kinetics of T_3 and T_4 have shown that T_3 has a primary role. Most investigators regard T_3 as more important than T_4 because T_3 is biologically 4–5 times more potent than T_4. Some even consider T_4 primarily as a prohormone and T_3 as the biologically significant hormone.

Although thyroid hormones have many actions, the primary one is a calorigenic effect (increased oxygen consumption) on many tissues. However, some tissues—for example, the brain, retina, lungs, spleen, and testes—do not appear to be affected by this action of the hormones. Thyroid hormones are also indispensable for growth, development, and sexual maturation in mammals. Other actions include stimulation of heart contractions, maintenance of body weight, stimulation of protein synthesis and carbohydrate metabolism, increase in the synthesis and degradation of cholesterol and triglycerides, increase in vitamin requirements, and enhancement of sensitivity of the β-receptors to catecholamines. These effects are usually magnified in patients with hyperthyroidism and minimized in patients with hypothyroidism. The physiological action has been described further by Shambaugh[271] and Sterling.[293]

Biosynthesis, Secretion, and Metabolism

The biosynthesis of thyroid hormones involves thyroidal trapping of serum iodide (iodide transport), incorporation of iodine into tyrosine, coupling of iodinated tyrosyl residues of thyroglobulin, and proteolytic cleavage of follicular thyroglobulin to free T_4 and T_3 and to mono-

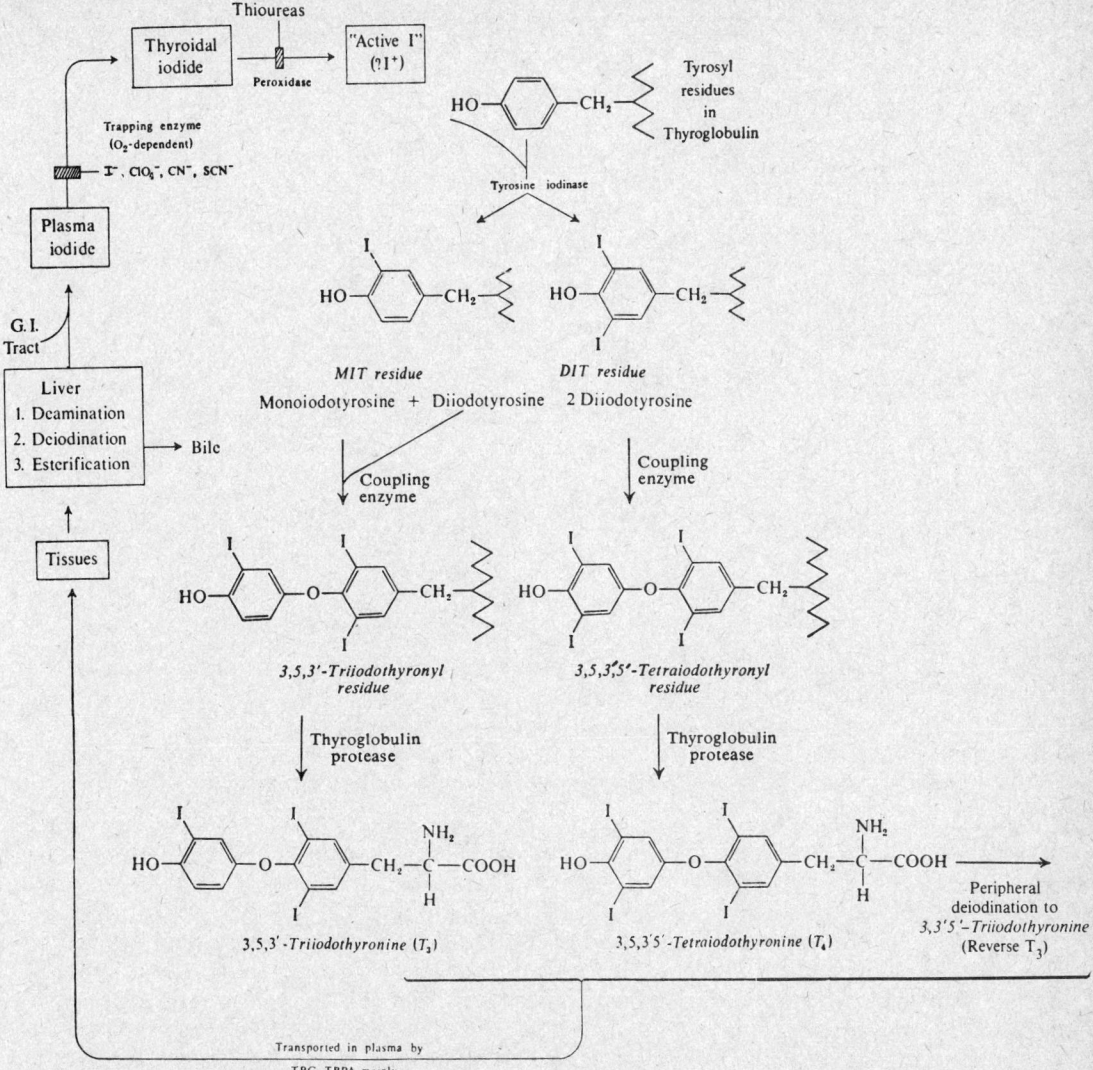

Figure 9-43. The metabolism of iodine, emphasizing formation and secretion of the thyroid hormones. ▨ indicates block in the pathway. Iodine transport is inhibited by monovalent anions such as thiocyanate (SCN⁻), perchlorate (ClO_4^-), and pertechnetate (TcO_4^-). The oxidation and organic binding of iodide to thyroglobulin is blocked by thiourylenes, sulfonamides, and high concentrations of iodide. (Modified from Berger, S., and Quinn, J. L., *In:* Tietz, N. W.: Fundamentals of Clinical Chemistry, 2nd ed. Philadelphia, W. B. Saunders, Co., 1976.)

and diiodotyrosine. Schematic outlines of iodine metabolism and formation and secretion of thyroid hormones are given in Figures 9-43 and 9-44.

The important element involved in the synthesis of thyroid hormones is iodine, which is normally ingested in the form of iodides. Iodide transport to the follicles is the first and rate-limiting step in the synthetic process. The follicular cells concentrate iodide, by means of an energy-dependent pump mechanism, to some 30–40 times the normal plasma levels. Upon stimulation of the gland or blockade by antithyroid drugs such as propylthiouracil or methimazole, the gland iodide concentration may reach 850 times the plasma level. Other anions are accumulated by the thyroid and act as competitive inhibitors of iodide transport. Two clinically useful anions are pertechnetate (TcO_4^-) and perchlorate (ClO_4^-). The ability of perchlorate to inhibit iodide transport allows its use in the so-called perchlorate-discharge test for detection of defects in the thyroid organic-binding mechanism, and uptake of the radioactive anion pertechnetate by the gland makes this agent suitable for thyroid imaging by radioscan. The salivary gland, the gastric mucosa, the placenta, the ciliary body of the eye, the choroid plexus, and the mammary glands

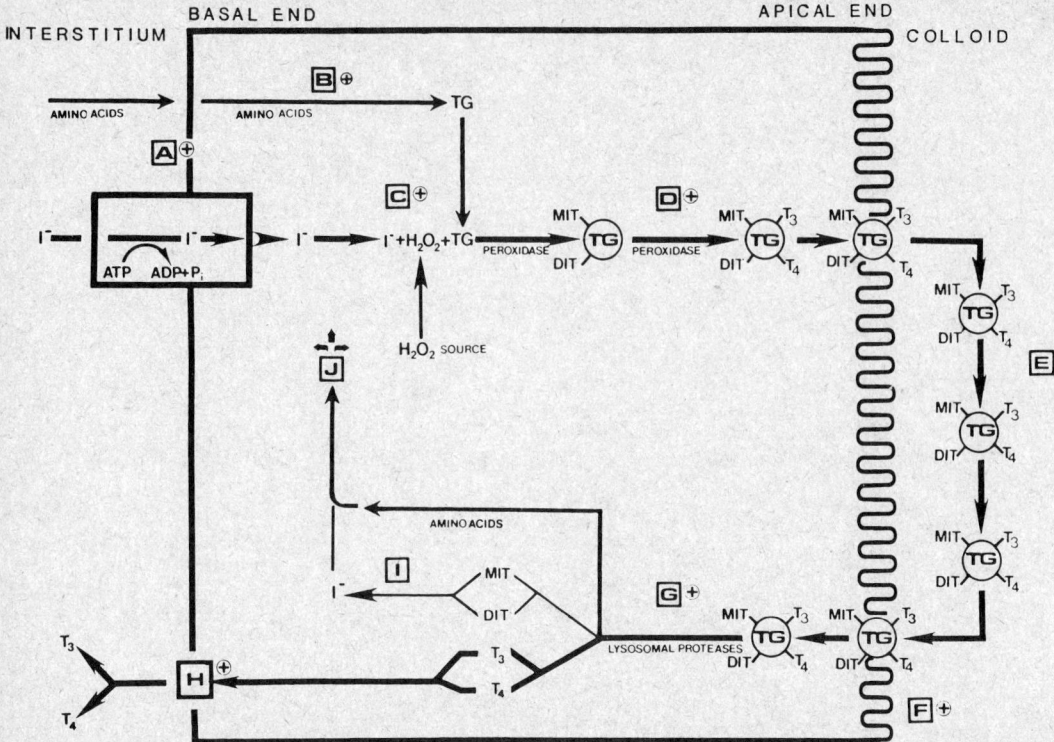

Figure 9-44. Schematic diagram of the thyroid cell, depicting stages of thyroid hormonogenesis and intrathyroidal iodine metabolism. *A*, Iodine transport. *B*, Thyroglobulin (TG) synthesis. *C*, Iodine organification. *D*, Intrathyroglobulin coupling. *E*, Storage. *F*, Endocytosis. *G*, Hydrolysis. *H*, Hormone secretion. *I*, Intrathyroidal deiodination. *J*, Recycling. Steps influenced by the thyroid-stimulating hormone (TSH) are indicated by the symbol +. (Reproduced with permission from Fernandez-Ulloa, M., and Maxon, H. R.: Thyroid. *In:* Clinical Chemistry. L. A. Kaplan and A. J. Pesce, Eds. St. Louis, The C. V. Mosby Co., 1984.)

also transport iodide against a concentration gradient, but their iodine uptake is not affected by TSH. The physiologic significance of extrathyroidal iodide-concentrating mechanisms is obscure.

Iodide is oxidized in the thyroid gland to more reactive iodine, probably as the free radical of iodine, I*; in a matter of seconds it is bound to tyrosine molecules attached to a thyroidal protein called thyroglobulin (M.W. 660 000). Thyroglobulin is also synthesized within the follicular cells before being secreted into the follicular lumen by exocytosis. The enzyme responsible for the oxidation and binding of iodide (organification) is thyroid peroxidase; hydrogen peroxide accepts the electron. Monoiodotyrosine (MIT) and diiodotyrosine (DIT) are formed (Figure 9-44). Two DIT molecules then undergo an oxidative condensation with the release of an alanine residue for the formation of T_4, still in peptide linkage to thyroglobulin. Similarly, T_3 results from coupling of MIT and DIT. A small amount of rT_3 is also formed, probably by condensation of DIT and MIT. The condensation reaction is an aerobic, energy-requiring reaction, and like the oxidation and binding, it is considered to be catalyzed by thyroid peroxidase. Iodide thus bound to tyrosine residues can no longer be discharged by thiocyanate, perchlorate, or other inhibitors of iodide transport.

The synthesis of T_3, T_4, DIT, and MIT in the thyroglobulin molecules occurs mainly at the follicular cell colloid interface but also within the colloid. Thyroglobulin is present in highest concentrations within the colloid, where it is stored. The follicular cells engulf colloid globules by endocytosis; these globules then merge with the lysosomes in the follicular cell. Lysosomal proteases break the peptide bonds between iodinated residues and the thyroglobulin; and T_4, T_3, DIT, and MIT are liberated into the cytoplasm of the follicular cell. T_4 and T_3 diffuse into the systemic circulation. DIT and MIT are deiodinated by an intracellular microsomal iodotyrosine dehalogenase. The freed iodide is reutilized for hormonal synthesis.

Each step involved in the synthesis of thyroid hormones is regulated by pituitary thyrotropin, TSH. This hormone stimulates the "iodide pump," thyroglobulin synthesis, and colloidal uptake by follicular cells. The rate of proteolysis of thyroglobulin for the liberation of T_4 and T_3 is also regulated by TSH. In addition, the hormone induces an increase in size and number of the thyroidal follicular cells. Prolonged TSH stimulation leads to increased vascularity and eventual hypertrophic enlargement (goiter) of the gland.

Metabolism of T_4 to T_3 and transport of both in the blood are discussed in the clinical context below. Degradative metabolism of T_3 and T_4 by oxidative deamination produces pyruvic acid analogs, which are converted to thyroacetates by subsequent decarboxylation. These analogs have some biologic activity, but there is no evidence that they are physiologically significant. In the liver, T_3 and T_4 are conjugated to form sulfates and glucuronides. These conjugates enter the bile and pass into the intestine. The thyroid hormone conjugates are hydrolyzed, and some are reabsorbed (enterohepatic circulation) or are excreted in the stool. The amounts of thyroidal substances in urine are very small and urinary assays have little clinical value. For a thorough discussion of thyroid metabolism, secretion, and synthesis, the reader is referred to the article by Ingbar and Woeber.[125]

Clinical Significance of Thyroid Hormone Measurement

Normal Thyroid Physiology

There is a tightly coordinated feedback relationship between the thyroid gland, the hypothalamus, and the pituitary gland.[164] These systems are closely interrelated and integrated, the net result being maintenance of thyroid hormone levels in blood within the normal range in the face of a wide variety of circumstances.[33] Thyrotropin-releasing hormone (TRH), a tripeptide, is produced in the hypothalamus. TRH acts on the pituitary thyrotropes to cause the synthesis and release of thyroid-stimulating hormone (TSH).[128] A rise in thyroid hormone level inhibits the pituitary response to TRH (negative feedback). A fall in thyroid hormone causes an increase in TRH and TSH secretion.

T_4 is the primary secretory product of the normal thyroid gland. T_4 undergoes peripheral deiodination of the outer ring at the 5' position to yield T_3. This deiodination occurs at a number of tissue sites, but primarily in the liver.[268] Reverse T_3, produced by removal of one iodine from the inner ring, is metabolically inactive and is a metabolic end-product of thyroxine metabolism (Figure 9-43). Peripheral deiodination is a rapidly responsive mechanism of control for thyroid hormone balance. Acute or chronic stress or illness will cause a shift in the direction of this deiodination, favoring formation of reverse T_3 rather than T_3, while the T_4 level remains essentially unchanged. Various medications (see below) will also shift peripheral deiodination toward the inactive product, rT_3.

T_4 and T_3 in the circulation are almost completely but reversibly bound to carrier proteins. These carrier proteins (thyroxine-binding globulin, thyroxine-binding prealbumin, and albumin) bind ~99.97% of T_4, and 99.7% of T_3. This means that only a very small fraction of each of these hormones is free for biologic activity. Because there is a wide variation in the concentration of thyroxine-binding proteins, even under normal circumstances, there is a wide variation in total T_4 levels among euthyroid individuals. Total T_3 concentrations will also vary with alterations in binding proteins, though not quite to the same degree as T_4 levels. Common circumstances in

Table 9-17. ALTERATIONS IN PLASMA
CONCENTRATION OF THYROID HORMONE–BINDING
PROTEINS

Decreased levels in:
Treatment with anabolic steroids, androgens, diphenylhydantoin
Major illness or surgical stress
Nephrotic syndrome
Active acromegaly
Genetic (inherited) deficiency
Increased levels in:
Treatment with estrogens, perphenazine
Pregnant or newborn state
Acute intermittent porphyria
Infectious hepatitis
Genetic (inherited) increase in synthesis

which thyroid hormone–binding proteins are altered are shown in Table 9-17. When relying on thyroid hormone measurements to diagnose thyroid hormone excess or deficiency states, knowledge of possible changes in binding proteins or in free hormone concentration can be extremely important.

Diagnosis of Thyroid Hormone Excess and Deficiency States

Laboratory tests most commonly used to evaluate patients for thyroid hormone dysfunction are listed in Table 9-18. Familiarity with normal physiology and with pathophysiology is important in the proper selection of these tests. However, normal serum thyroid hormone levels do not exclude thyroid disease. Diffuse or nodular thyroid enlargement may be seen in euthyroid patients.

Clinical signs and symptoms of thyroid hormone excess or deficiency are nonspecific. When hypothyroidism or hyperthyroidism is suspected, confirmation with laboratory tests is generally necessary. Depending on the clarity of the clinical presentation, the necessary laboratory tests may be either extremely simple and straightforward or quite complicated.

Patients not suspected of thyroid hormone excess or deficiency states on clinical grounds can be screened using either (1) a total T_4 measurement, (2) a calculation of an FT_4I from total T_4 and T_3 uptake ratio measurements, or (3) one of the newer assays for free thyroxine (FT_4) measurement.[199] T_4 is the major secretory product of the thyroid gland and is a good reflection of thyroid hormone production. Changes in serum thyroid hormone–binding proteins affect total T_4 measurements, and even the most reliable free thyroxine index or free T_4 measurement may not completely correct for extreme variations in binding proteins. Also, because acute changes in T_4 conversion to T_3 and rT_3 may occur with nonthyroidal illnesses, relationships between these hormones may not be totally predictable. Screening a large adult population for thyroid hormone excess or deficiency is not generally advisable because of frequent false positive and false negative results; however, if adult screening is done, a determination of total T_4 with follow-up of high, low, or borderline values with a T_3 uptake test will improve the cost-effectiveness.[136] For individual patient testing, a total T_4, T_3 uptake, and calculated FT_4I provide more reliable information than a total T_4 value alone.

Hypothyroidism. Primary hypothyroidism is a common disease, affecting 1–2% of the population. Primary hypothyroidism results when the thyroid gland is damaged and is unable to produce an adequate amount of T_4, such as in chronic lymphocytic thyroiditis; or in inherited

Table 9-18. ABBREVIATIONS AND NOMENCLATURE FOR THYROID TESTS

Hormone concentration	
T_4(RIA)	Total serum thyroxine (radioimmunoassay)
T_3(RIA)	Total serum triiodothyronine (radioimmunoassay)
FT_4D	Free thyroxine (dialysis method)*
FT_3D	Free triiodothyronine (dialysis method)
rT_3	Reverse 3,3′,5′-triiodothyronine
TSH	Thyroid-stimulating hormone, thyrotropin
Indirect measure of binding sites	
T_3U	T_3 resin uptake†
Indirect estimates of free hormone concentration	
FT_4 index (FT_4I)	Free thyroxine index [T_4(RIA) $\times$ T_3U]
FT_3 index (FT_3I)	Free triiodothyronine index [T_3(RIA) $\times$ T_3U]
Serum binding proteins	
TBG	Thyroxine-binding globulin‡
TBPA	Thyroxine-binding prealbumin

*Any method other than dialysis for measurement of "free" thyroxine must be critically evaluated; results are not always comparable with dialysis techniques.

†The recommended test is the T_3 uptake ratio, where the T_3U of the patient is divided by the T_3U from a pool of normal standard reference serum included in the same assay run. This ratio is highly preferred and should be used in calculating the free thyroxine index. The normal range for T_3U ratio, although centered on 1.0, is not necessarily the same for each method.

‡Thyroxine-binding globulin, the protein that binds thyroxine in serum, is distinct from thyroglobulin, the intracellular carrier protein that is involved in thyroid synthesis. Thyroglobulin concentrations may be elevated in well-differentiated thyroid carcinoma and other conditions.

PROTOCOL R: TRH STIMULATION OF TSH RELEASE

Rationale: The pituitary cells that produce TSH are exquisitely sensitive to changes in thyroid hormones. When thyroid hormones rise above normal, TSH release from the pituitary is reduced or blocked; when thyroid hormones fall below normal, TSH release in response to TRH is exaggerated.

Procedure: No patient preparation is necessary. A baseline sample is collected for TSH determination (and T_4, if not already done). 500 μg TRH is given IV, and repeat samples are drawn for TSH at 30 and 60 min after injection.

Interpretation: A normal response is at least a doubling of the basal TSH concentration, with the peak value at 30 min. Blunted or flat responses are seen in hyperthyroidism (usually with a high T_4 or T_3) or in hypothyroidism secondary to hypopituitarism (usually with a low T_4). An exaggerated rise (peak TSH over 35 μIU/mL) is seen in early primary hypothyroidism. Patients with hypothyroidism secondary to hypothalamic disorders may show a response of normal magnitude but a delayed peak. Illustrative responses are shown in Figure 9-45.

conditions where thyroid hormone synthesis is inefficient, such as in enzyme abnormalities associated with dyshormonogenesis; or after ablation or removal of thyroid tissue with radioactive iodine or surgery. The fall in T_4 and T_3 leads to an increase in TSH, confirming a diagnosis of primary hypothyroidism.[114] Even with a strong clinical presentation of primary hypothyroidism, TSH measurement is helpful in confirming the diagnosis and as a guide for following therapy. In early hypothyroidism, symptoms may be mild; thyroid hormone levels may still be above the lower limit of the reference range, but TSH will be elevated. In mild hypothyroidism, TSH measurement is invaluable in establishing the diagnosis.

Secondary hypothyroidism is due to pituitary or hypothalamic disease. With secondary hypothyroidism, serum thyroid hormone concentrations are low but the TSH value is either within the normal range (and inappropriately low for the low T_4) or only very slightly elevated. When T_4 and TSH are both low, a TRH test is helpful (Protocol R). In patients who have destructive lesions of the pituitary that result in TSH deficiency, no TSH response is expected. In patients with hypothalamic abnormalities affecting TRH and TSH release, the peak TSH response to TRH may be normal, but it is generally delayed until 45 or 60 min after the TRH administration. (See Figure 9-45.)

Congenital hypothyroidism may be due to absence of the thyroid gland (athyreosis) or

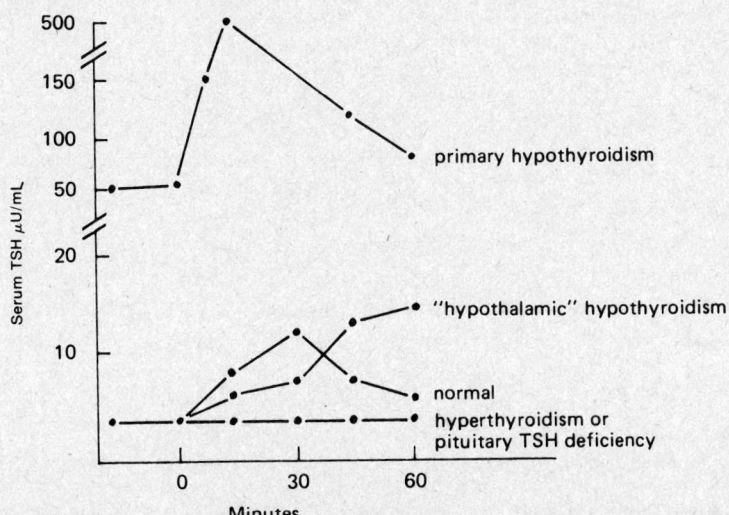

Figure 9-45. Patterns of TSH response to TRH stimulation. (From Watts, N. B., and Keffer, J. H.: Practical Endocrine Diagnosis. 3rd ed. Philadelphia, Lea and Febiger, 1982.)

secondary to defects of thyroid hormone synthesis. Screening programs for *congenital hypothyroidism*[322] have been established in most states. This disorder occurs once in every 3500–4000 live births, and early treatment is critical in preventing mental retardation. For a primary screening test, T_4 is measured on blood eluted from a filter paper spot collected from a heelstick of the newborn infant. When results are outside the lower 2–3 SD of the normal distribution, TSH determination on another sample from the same filter paper spot is indicated. In many instances, the diagnosis of primary hypothyroidism in the neonate can be made at this time; in some cases, further testing is needed on follow-up.

Hyperthyroidism. This disorder can be caused by a number of conditions, including diffuse toxic goiter, toxic multinodular goiter, solitary toxic nodule, and damage to the thyroid gland from thyroiditis.[144] Hyperthyroidism is generally easier to diagnose on clinical grounds than is hypothyroidism. If a patient has diffuse thyroid enlargement, including signs and symptoms of rapid pulse, tremor, nervousness, weight loss, increased sweating, easy fatigue, heat intolerance, and exophthalmos, the diagnosis of hyperthyroidism requires only a confirmatory determination of free thyroxine.

If the clinical picture is not clearly that of hyperthyroidism, or if the T_4 concentration is not clearly elevated, measurement of serum T_3 is helpful. In hyperthyroidism, T_3 is often elevated earlier and to a greater degree than T_4. Since T_3 is bound to protein, though to a lesser degree than T_4, total T_3 may be altered by changes in binding proteins. Since only the free fraction of T_3 is active, the estimation of free T_3 by the use of a total T_3 measurement with a T_3 uptake ratio (FT_3I) may be helpful in adjusting the total T_3(RIA) for abnormalities in binding proteins. A number of medications and acute or chronic illnesses may cause a transient lowering of T_3. In a patient with nonthyroidal illness, the diagnosis of hyperthyroidism using T_4, total T_3, and T_3 uptake may not be possible.

Hyperthyroidism may be seen with normal total T_4 and FT_4, but an elevated T_3. This syndrome of *T_3 toxicosis* is most commonly seen following previous radioiodine treatment or surgery for hyperthyroidism or with thyroid hormone treatment preparations that contain both T_4 and T_3. *T_4 toxicosis* [high T_4 and normal T_3(RIA)] is not a common situation; in many patients with high T_4 levels but low T_3 levels and clinical hyperthyroidism, the low T_3 level is probably the result of the transient influence of drugs or stress. These patients will have high T_3 levels if tested again in a basal state.

Occasionally, increases in T_4 and T_3 may occur due to release of thyroid hormones from damage to the thyroid parenchyma associated with *subacute thyroiditis* or *chronic lymphocytic thyroiditis* (hyperthyroiditis). The increase in T_4 and T_3 may be associated with clinical findings suggestive of hyperthyroidism. The diagnostic dilemma can be solved by finding a low radioactive iodine uptake (per cent of orally administered radioactive iodine taken up by the gland at 6 or 24 h) accompanying the hyperthyroidism of thyroiditis. In most cases, this self-limited condition will resolve without residual thyroid function abnormality.

The most conclusive test for hyperthyroidism is the TRH test. (See Protocol R.) High levels of thyroid hormone will block the release of TSH from the pituitary. In a patient with equivocal but suggestive symptoms of hyperthyroidism and T_4 or T_3 levels slightly above the normal range, demonstration of failure of TSH to rise after TRH establishes the diagnosis of hyperthyroidism. This test, however, is not needed in most patients who have hyperthyroidism.

Drugs and Nonthyroidal Illness Affecting Thyroid Studies

As discussed before, peripheral deiodination of T_4 is the primary source of T_3, the more potent of the two thyroid hormones in blood. This enzymatic conversion is inhibited by a number of medications, by acute and chronic stress, and in a variety of acute and chronic nonthyroidal

Table 9-19. EFFECTS OF ACUTE AND CHRONIC ILLNESS ON THYROID FUNCTION

Reduced peripheral conversion of T_4 to T_3	Decreased degradation of T_4
Increased production of rT_3	Inhibition of TSH or TRH secretion
Reduced production of serum binding proteins	Elevation of serum TSH
Circulating inhibitors of extrathyroidal binding	Reduced cellular T_3 and T_4 receptors
Circulating inhibitors of serum hormone binding	Reduced cellular T_3 action
Increased degradation of T_4	

Table 9-20. EFFECTS OF DRUGS ON THYROID FUNCTION

Altered hypothalamic or pituitary function
Altered biosynthesis or release of thyroid hormones
Displacement of T_4 and T_3 from binding proteins
Reduced peripheral conversion of T_4 to T_3
Inhibition of peripheral hormone activity

illnesses.[55,323] As conversion of T_4 to T_3 is reduced, conversion of T_4 to rT_3 increases, resulting in higher serum rT_3 concentrations.[56] Effects of acute and chronic illness on thyroid function are shown in Table 9-19. Common test abnormalities accompanying nonthyroidal illnesses in euthyroid patients include an increase in rT_3, lowering of T_3(RIA) and FT_3I, and lowering of TBG and other binding protein levels. Less commonly, the T_4 concentration is increased.[97] Free T_4 levels may be high with acute illness or low with severe chronic illness. TSH and TRH response are rarely affected. Thus, in nonthyroidal illness, abnormal thyroid tests are not necessarily suggestive of thyroid disease.[134] Low T_4 levels are commonly seen in sick euthyroid patients. Hyperthyroid patients with acute or chronic nonthyroidal illness may have T_3 levels that are lowered into the normal range, and these patients sometimes may show transient lowering of T_4 levels as well. TSH and TRH responses are usually valid despite nonthyroidal illness, and these tests should be employed if hyper- or hypothyroidism is suspected.

Medications may alter thyroid function and thyroid function tests; general effects are shown in Table 9-20. The most commonly encountered variations are reduced peripheral conversion of T_4 to T_3 or displacement of T_4 and T_3 from binding proteins. Some drugs that affect laboratory findings are shown in Table 9-21.

Familiarity with the effects of nonthyroidal illness and various medications on thyroid function tests is extremely important in interpreting these test results.

Cost-Effective Use of Laboratory Tests in Thyroid Diagnosis

Figure 9-46 outlines a scheme for the effective use of various laboratory tests for the diagnosis of thyroid hormone excess or deficiency states.

As the laboratory detection of thyroid diseases has been refined through wide-scale availability of newer techniques and commercial kits, some confusion in terminology has developed. The inaccurate naming of a test can lead not only to unnecessary expense but also to errors in diagnosis. Table 9-18 is a review of accepted nomenclature for tests of thyroid hormones in serum. The nomenclature and definition of thyroid hormone tests have been further described by Solomon et al.[285] Almost all laboratory tests for thyroid function are now commercially available in kit form. A comparison of methods and kits for assessing thyroid function has been described in a recent publication.[199] The following is a brief description of the tests that are currently considered

Table 9-21. DRUGS THAT AFFECT RESULTS OF THYROID FUNCTION TESTS

Drug	Effect	Cause
Salicylates, phenylbutazone, diphenylhydantoin	Total T_4 and T_3(RIA) reduced, FT_4I low, FT_4D (dialysis) normal	Inhibition of serum protein binding
Propylthiouracil, methimazole, lithium, iodides	Total T_4 and T_3(RIA) reduced	Inhibition of TSH production or release
Propylthiouracil	Total T_3(RIA) reduced, T_4 normal, TSH high or normal	Inhibition of conversion of T_4 to T_3
Propranolol	Total T_3(RIA) reduced, T_4 and TSH normal	Same
Glucocorticoids	Total T_3(RIA) reduced, both T_4 and TSH low or normal	Same
Oral radiographic dyes	Total T_3(RIA) reduced, both T_4 and TSH high	Same
Dopamine, L-dopa, glucocorticoids	Basal TSH and response to TRH reduced	Direct effect to inhibit TSH production in pituitary
Amiodarone, TRH, benzamide (transient), metoclopramide, sulpiride	Basal TSH increased	Increased TSH production

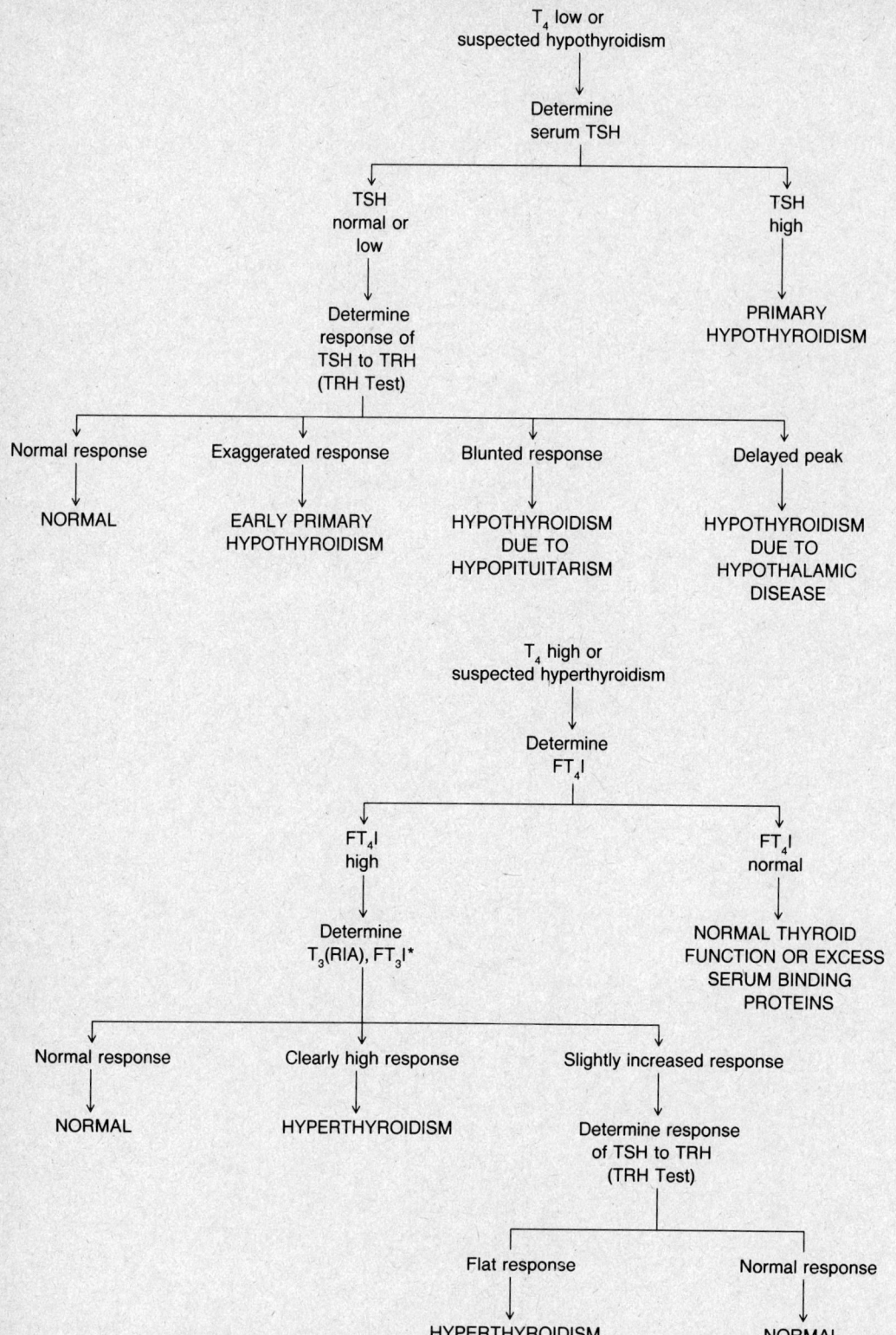

*Suspect possible T_3 toxicosis.

Figure 9-46. Scheme for diagnosis of thyroid dysfunction.

most useful for evaluation of thyroid status. For detailed descriptions of methods, refer to original reports or to commercial kit inserts.

Methods for the Determination of Thyroxine (T₄) in Serum*

Serum T_4 concentrations were initially determined indirectly by methods which measured the amount of iodine in protein precipitates of serum, the protein-bound iodine (PBI).[13] Somewhat more specific measurements were the procedure for BEI (butanol-extractable iodine) and T_4 by column.[186] All of these methods were useful because iodine in T_4 normally accounts for 80–90% of all iodine in serum. Both butanol extraction and ion-exchange column chromatography selectively separate iodothyronines from other organic iodine compounds. These methods are of historical interest only, since they are now wholly obsolete. Occasionally, PBI is used to detect the presence of iodoprotein abnormalities. More sensitive and specific methods for total serum T_4, known collectively as competitive protein binding assays (CPBA), became popular for several years. Several of these methods were reported using TBG obtained from pooled human sera as the binding protein.[213,257] Today these methods have been replaced almost universally by RIA methods.

Determination of T₄ by Radioimmunoassay [T₄(RIA)]

Principle. Principles and general procedures for RIA have been described in Chapter 1C and on page 1049. An extensive discussion of RIA for iodothyronines can be found in Chopra's report.[53] Here we discuss points of particular concern in the utilization of kits for T_4 assay.

Most methods use antibodies produced against a serum albumin-thyroxine conjugate. These antibodies are highly specific and are able to distinguish between molecules differing by only one atom (e.g., T_3 and T_4). Thyroglobulin (Tg) is also used as an immunogen, since this protein contains T_4 (as well as T_3) as part of its structure.

Current RIA methods do not require extraction of T_4 and T_3 from serum prior to assay. But, since 99.7% or more of T_4 and T_3 circulate tightly bound to TBG, albumin, and TBPA in serum, accurate measurement of the hormones requires dissociation of the hormones from the serum proteins prior to assay. Binding of iodothyronines to albumin is not a concern since the association constant of an iodothyronine for antibody (usually 10^9 L/mol or more) is several orders of magnitude higher than that of iodothyronine for albumin ($\sim 1.6 \times 10^6$ L/mol). However, binding constants for iodothyronine with TBG (for T_4, 2×10^{10} L/mol; for T_3, 2×10^9 L/mol) and with TBPA (for T_4, 2×10^8 L/mol; for T_3, 2×10^7 L/mol) are high. Binding of iodothyronine to TBPA is overcome by use of barbital buffers, since barbital ions inhibit this binding. Various agents are used to inhibit binding of iodothyronine to TBG.[53] The fluorescent probe, 8-anilino-1-naphthalene-sulfonic acid (ANS), appears to be the agent of choice, although salicylate, thimerosal, and phenytoin are also used. These blocking agents are effective in displacing T_3 and T_4 from TBG without affecting their binding characteristics with antibody.

A typical method[53] for T_4 assay consists of incubation of 25-μL aliquots of serum and standards with antiserum and ^{125}I-T_4 in ANS-barbital buffer for 1 h at room temperature, followed by chilling for 5 min at 4 °C. Bound and free label are separated by double-antibody technique (addition of goat-antirabbit γ-globulin) and further incubation for 20 h at 4 °C. Radioactivity of the bound label in the precipitate is determined; per cent bound activity of standards is plotted against their T_4 concentrations in μg/dL. Double-antibody separation of bound and free label is convenient and precise. Alternatively, separation with dextran-coated charcoal, ammonium sulfate, or polyethylene glycol is faster and more practical. When this method was so modified, the average recovery of T_4 added at various concentrations to a serum pool was $\sim$100%. Intra- and interassay reproducibilities, expressed as CV, were 4.3 and 7.1%, respectively. The method is sufficiently sensitive to allow use of a sample as little as 2.5 μL. Such sensitivity is highly desirable if T_4 is to be determined on specimens obtained from infants by heel puncture and collected in capillary tubes or on filter paper. Dried blood specimens are stable and easily

---

*The assistance of Joseph D. Musto, M.Sc., Ph.D., Newton-Wellesley Medical Laboratory, Inc., Charles River Medical Center, Wellesley, MA, in the preparation of this section is gratefully acknowledged.

transported and are ordinarily obtained for purposes of screening neonates for congenital hypothyroidism.[165] When this collection technique is used, a $1/8$-inch dot is punched out from the blood-saturated paper and is extracted into buffer prior to T_4 assay.

In recent years, many commercial kits with good analytical reliability have become available. These kit procedures differ primarily in their approach to separation of antibody-bound and unbound fractions. Some utilize a double-antibody technique with polyethylene glycol (PEG) to hasten and enhance separation; some use a solid-phase technique with T_4-specific antibody bound to beads, magnetizable particles, or tubes. The latter two modes, for example, simplify the procedure by eliminating centrifugation for separation of bound and free label. Basic steps of a popular kit method that uses an antibody-coated tube are as follows:

1. Displacement of T_4 from serum proteins by ANS or salicylate.
2. Incubation of sample and standards with ^{125}I-T_4 tracer in antibody-coated tubes for ~1 h.
3. Aspiration or decantation of fluid from tubes.
4. Counting of tubes and preparation of a standard curve for a concentration range of 0.0–20.0 $\mu g/dL$.
5. Determination of values of unknowns by interpolation on the curve.

Kit procedures are generally fast; as a rule, they require only about 1 h incubation either at room temperature or at 37 °C. Overall, a batch of 20 patient samples, 2 or 3 controls, and 6 standards (each in duplicate) will take less than 3 h to process. A typical *reference range* for serum T_4 is 4.5–12.0 $\mu g/dL$. The reference range varies among procedures and reference populations; therefore, each laboratory should establish its own.

Nonisotopic immunoassays are also commercially available. As in RIA, the principle is based on competition of labeled and unlabeled analyte for an antibody and measurement of a property of the labeled analyte that is bound to antibody. The label may be an enzyme or a fluorophore, allowing measurement of an enzyme reaction or of fluorescence, respectively. In an *enzyme-linked immunosorbent assay (ELISA)*, offered by ICL Scientific, Fountain Valley, CA, a serum sample is incubated with peroxidase-labeled T_4 in a test tube coated with T_4-antibody. Endogenous T_4 competes with peroxidase-labeled T_4 for binding sites on the antibody. After incubation, the contents of the tube are aspirated, and the inside of the tube is washed to remove any free labeled T_4. As in an RIA method, activity of bound label (here T_4-conjugated peroxidase activity complexed with antibody on the tube) is inversely proportional to the T_4 concentration in serum. Enzyme activity is determined by photometric measurement of color intensity following incubation with a suitable substrate. Quantitation of analyte in the unknown specimen is made by comparison with a calibration curve of standards treated in the same manner. The *enzyme-multiplied immunoassay technique* (EMIT) for T_4, as developed by Syva, Palo Alto, CA, is a *homogeneous immunoassay system*; i.e., separation of free from antibody-bound T_4 is not required. In this system, malate dehydrogenase, covalently linked to T_4, is used as the enzyme label. T_4 bound to the enzyme blocks the enzyme's active site. In the presence of antibody this inhibition is reversed, thus allowing a suitable substrate access to the catalytic site. Endogenous T_4 competes with the enzyme-labeled T_4 for antibody binding sites. Consequently, as the concentration of T_4 increases, less of the enzyme-labeled T_4 becomes bound to the antibody and enzyme activity decreases. In the case of malate dehydrogenase, the reaction involves reduction of NAD and can be observed photometrically. Since enzyme activity is a property only of antibody-bound enzyme, separation of bound from free enzyme-labeled T_4 is not required. The enzymatic activities observed with unknowns are compared with enzymatic activities observed with standards in order to quantitate T_4 in the sample.

Fluorescent labels can also be used in T_4 immunoassays. In the *homogeneous fluorescence immunoassay* developed by Syva, fluor-labeled T_4 is added to the specimen and competes with endogenous T_4 for binding to a T_4-specific antibody, which is labeled with a fluor-acceptor compound such as rhodamine. When the fluor-labeled T_4 is bound to the antibody, its fluorescence intensity is reduced. This decrease in fluorescence is inversely proportional to the amount of unlabeled T_4 in the sample. Quantitation is carried out on a fluorometer, with interpolation of specimen fluorescence readings from a standard curve. A *heterogeneous fluorescence immunoassay* developed by Bio-Rad Laboratories, Richmond, CA, uses a solid-phase antibody technique. Fluor-labeled T_4 competes with T_4 in the sample for antibody which is covalently linked to polyacrylamide beads. The beads are sedimented and the supernatant discarded. The beads are then resuspended, the fluorescence intensity of the suspension is measured in a fluorometer, and the concentration of unknowns is determined from a standard curve.

In some cases, radioimmunoassays and nonisotopic immunoassays have been partially or fully automated. A technique based on fluorescence polarization immunoassay developed for the Abbott TDx has also been introduced. For a detailed discussion on commercially available, fully automated systems, the reader is referred to the article by Chen.[51]

Specimen collection and storage. The preferred sample is serum; plasma with EDTA or heparin as anticoagulant may also be used. Note, however, that plasma tends to form fibrin clots after freezing and thawing, which may interfere mechanically with the assay, especially in an automated system. T_4 in serum is quite stable; storage of serum samples at room temperature up to 7 d results in no appreciable loss of T_4. However, serum samples should be stored at 2–8 °C if they will not be tested within 24 h. For longer periods of storage, freezing the specimen is recommended. Frozen specimens are stable for at least 30 d. Repeated freezing and thawing of the sample should be avoided. Grossly lipemic specimens should not be used, since fatty acids compete with T_4 for binding sites on TBG. Further, as in the case of all radioassays, if patients have received diagnostic or therapeutic radionuclides within 2 weeks before the T_4 determination, the radioactivity of the serum sample from such patients should be checked in a γ-counter set for ^{125}I to determine whether the radioactivity in the sample could significantly affect RIA results.

Reference Ranges[307]

	$\mu g/dL$
Cord blood:	8.0–13.0
Newborn:	11.5–24.0
(Lower in low-birth-weight infants)	
Newborn screen (filter paper):	6.2–22.0
Neonate:	9.0–18.0
Infant:	7.0–15.0
1–5 years:	7.3–15.0
5–10 years:	6.4–13.3
10–60 years:	5.0–12.0
> 60 years	
Male:	5.0–10.0
Female:	5.5–10.5

Determination of T_3 by Radioimmunoassay [T_3(RIA)]

Unlike T_4 antiserum, high-titer, specific T_3-antiserum is seldom obtained by immunization with naturally occurring thyroglobulin. The reason has been attributed to the lower T_3 content in normal thyroglobulin compared with T_4. Good quality antiserum can be produced by using T_3-enriched thyroglobulin, T_3-human serum albumin (HSA), or T_3-BSA conjugates. The preparation of conjugates, the protocol for raising antisera to T_3, and the RIA procedure have been described by Chopra.[53]

Principle. T_3 measurement is similar to that described for serum T_4(RIA). In brief, the various reagents are added to tubes in the following order: (1) barbital buffer; (2) unknown samples or varying concentrations of standards to provide 10 pg–10 ng T_3 in T_3-free serum for the standard curve; (3) ANS solution (2 mg/mL); (4) T_3-antiserum diluted so as to bind about 30–50% of ^{125}I-T_3 at zero concentration of standard; (5) ^{125}I-T_3 in barbital buffer.

The reagents are mixed and incubated for 24 h at 4 °C. A potent goat antirabbit γ-globulin (second antibody) is then added and the tubes are reincubated to allow precipitation of all antibody-bound T_3. The amount of T_3 present in unknowns is read from the standard curve and the results are expressed as ng/dL.

The reliability criteria of the method are adequate. The mean recovery of T_3 added to hypothyroid and normal human sera in concentrations varying between 20 and 800 ng/dL was 99.5%; the within-assay CV of sera with T_3 values between 27 and 975 ng/dL was 5.8% and the between-assay CV of sera with T_3 values between 25 and 1120 ng/dL was 9.8%. This method permits detection of as low as 15–25 ng/dL of T_3 when 200 μL of serum is assayed.

A similar but more sensitive RIA method for T_3 has been described by Larsen.[163] The method employs 50 μL of unknown serum, specific T_3 antiserum, and sodium salicylate as an inhibitor of T_3 binding to plasma proteins. Bound and free hormone are separated by addition of dextran-coated charcoal, thus making this method less time consuming than the one described above. However, the method of Chopra[53] has found wider applications to physiological and clinical studies.

As with T_4(RIA), numerous commercial RIA kits for measuring serum T_3 are available. The comments and the typical procedure are similar to those described for T_4 except for utilizing a ^{125}I-T_3 tracer and T_3-specific antibody. A standard curve is prepared from five standards ranging from 0.0–800 ng/dL. Values for unknown specimens are obtained from the standard curve by interpolation.

The *specimen collection and storage* are the same as described for T_4; sample volume per test is usually 50–100 μL. The time required for testing 20 samples is equivalent to that for T_4, usually < 3 h. Typical CV's for both intra- and interassays are ~4%; sensitivity is ~15 ng/dL.

Reference ranges. The typical reference range for serum T_3 is 110–190 ng/dL. This range may vary from one procedure and reference population to another. Therefore, each laboratory should establish its own reference range.

Determination of Reverse T_3 by Radioimmunoassay [rT$_3$(RIA)]

Principle. Antisera for rT$_3$ are generally obtained by immunization of rabbits using rT$_3$ conjugates with human or bovine serum albumin.[52] The antisera are highly specific and devoid of cross-reactivity with both T_3 and T_4. A perceptible cross reaction is observed only with 3,3'-diiodothyronine (T_2). However, since the serum concentration of T_2 is < 10 ng/dL, 10% cross reaction has negligible effect on the specificity of rT$_3$ measurement.

Like T_3 and T_4, rT$_3$ also binds to TBG, TBPA, and albumin in serum. The binding of rT$_3$ to serum proteins has been overcome either by the use of an ethanol extract of serum or by the addition of ANS to the serum in a manner similar to that described above for T_3 and T_4(RIA). Numerous RIA methods for rT$_3$ have been published.[54] The following is a brief outline of the method described by Chopra.[52]

The various reagents are added to the test tubes in the following order: (1) barbital buffer containing 1% normal rabbit serum, (2) ethanol into standard tubes and equal volumes of an ethanol extract of test sera into patient specimen tubes, (3) nonradioactive rT$_3$ into standard tubes in varying concentrations ranging from 5 pg to 3 ng, (4) appropriately diluted rT$_3$-antiserum, and (5) ^{125}I-rT$_3$ in barbital buffer. Reagents in tubes are mixed and tubes are incubated at 4 °C for 24 h. An appropriate quantity of goat-antirabbit γ-globulin (second antibody) is then added and the tubes are reincubated to allow separation of bound from free radioactive rT$_3$. The plotting of the standard curve and calculations of rT$_3$ in samples are carried out in a manner similar to that described above for T_4(RIA).

The validity and the reliability of the rT$_3$(RIA) procedure have been established. The data on specificity, recovery of added rT$_3$ to serum, and intra- and interassay variability compare favorably with those described for T_3(RIA) and T_4(RIA).

Commercial kits for direct serum analysis of rT$_3$ are available. However, requests for this test are not as frequent as for T_4 or T_3. Also, these kit procedures are not as convenient as those for T_3. But in comparison with the reference method described above, the kit procedure is relatively fast; it requires 3 h incubation at room temperature, followed by PEG precipitation and centrifugation. The entire procedure requires < 6 h. Typical intra- and interassay CV's are between 2 and 10%; sensitivity is ~3 pg/mL.

Specimen collection and storage. Collection and storage of plasma or serum are similar to those described for T_4(RIA). Amniotic fluid obtained by amniocentesis or by transmembrane aspiration at delivery may also be analyzed for rT$_3$. The amniotic fluid should be clear. To avoid debris and other cellular particles, centrifugation of the fluid for 15 min at 900 $\times$ g is suggested. Specimens are stable for 1 d at 2–8 °C and for up to 3 months frozen. Amniotic fluid contaminated with hemolyzed blood should not be used.

Reference Ranges[307]

	ng/dL
Cord blood:	315 ± 16 SE
Maternal serum:	79 ± 5 SE
1–5 years:	15–71
5–10 years:	17–79
10–15 years:	19–88
Adult:	20–53
Amniotic fluid:	82 ± 25 SE

Methods for the Determination of Thyroid Hormone Binding

Total serum T_4 and T_3 concentrations are dependent on their production rates and on the level of circulating thyroxine-binding globulin (TBG). Thus, tests measuring thyroid hormone binding are performed to determine whether a change in the total concentration of the hormone is due to a change in the level of binding protein or to a change in the hormone production. Furthermore, such tests provide evidence of whether an abnormality in the total concentration of serum T_3 or T_4 is also associated with alterations in the free (non-protein-bound) fraction of these hormones.

Several approaches have been undertaken to assess thyroid hormone binding to proteins. They include direct estimation of the free thyroxine concentration (FT_4), measurement of the concentration of unoccupied binding sites on TBG (T_3U), and the direct determination of TBG concentration by RIA.

Determination of Serum Free Thyroxine (FT_4)

The interaction between serum FT_4 and the unoccupied or available binding sites on its principal carrier protein, TBG, conforms to the law of mass action and can be described by the following equation:

$$[FT_4] \times [TBG^0] = K \times [T_4 \cdot TBG]$$

where $[FT_4]$ = concentration of free T_4 in serum
$[TBG^0]$ = concentration of unoccupied binding sites on serum TBG
K = an association constant
$[T_4 \cdot TBG]$ = concentration of T_4-occupied binding sites on serum TBG

The relationship among these entities has important diagnostic implications. A primary increase in either $[FT_4]$ or $[TBG^0]$ drives this reaction to the right, increasing serum $[T_4 \cdot TBG]$. Since 99.97% of the total thyroxine is bound to TBG, the value of $[T_4 \cdot TBG]$ is considered equivalent to that of total $T_4(RIA)$. Hyperthyroidism produces a primary increase in FT_4, while estrogens and idiopathic or genetic conditions may produce a primary increase in TBG. In both cases there is an increase in $[T_4 \cdot TBG]$, but in the first case, the patient is ill and requires treatment; in the latter case, the patient is euthyroid and requires nothing but an understanding of the equation. Likewise, a low serum $[T_4 \cdot TBG]$ may be due to a primary decrease in $[FT_4]$ or to a primary decrease in $[TBG]$. It is therefore clinically important to differentiate between changes in $[T_4 \cdot TBG]$ that are due to primary changes in $[FT_4]$, such as hyper- or hypothyroidism, and those due to primary changes in $[TBG^0]$.

In normal serum the proportion of FT_4 averages 0.03% of the total. In the case of T_3, the percentage of free hormone is approximately ten times that of T_4 (i.e., 0.3%), reflecting the lesser affinity of TBG for T_3.

Principle. The concentrations of FT_4 and FT_3 in serum are exceedingly low and consequently were not amenable to direct measurement in the past. Recently commercial RIA kits have been developed which make FT_4 and FT_3 determinations practical for the clinical laboratory. Measurements have been performed in the past by equilibrium dialysis,[124,295] ultrafiltration,[286] and chromatography on polyacrylamide gel[192] or Sephadex.[283] However, these earlier techniques, including FT_4 measurement by radioimmunoassay in serum dialysates,[339] are too cumbersome and time consuming for routine clinical use and, except for equilibrium dialysis, have found little application.

The method of Ingbar and his coworkers[124] consists of a *double dialysis procedure.* In the first dialysis, test serum is enriched with a very small quantity of radioiodine-labeled T_4, which quickly distributes between free and bound unlabeled T_4. This mixture is then dialyzed against phosphate buffer (pH 7.4) overnight at 37 °C. The dialysate (i.e., the part that passes through the semipermeable membrane) contains both free radiothyroxine and labeled inorganic iodide as well as other contaminating iodinated compounds generally present in commercial preparations of radioiodine-labeled thyroxine. The second dialysis is employed to remove these contaminants. In this procedure, a mixture of the first dialysate and pooled plasma is dialyzed against phosphate buffer containing anion-exchange resin. The resin adsorbs only the radioactive contaminants and thereby maintains an effective diffusion gradient for impurities until all of them are removed. Because the binding proteins in the pooled plasma possess a high affinity for T_4, little if any radioactive T_4 is available for diffusion and adsorption by the resin. All radioactivity that remains within the dialysis bag is considered to be the free T_4.

The method of Sterling and Brenner[295] simplifies the FT_4 method significantly by replacing the second dialysis with a much simpler precipitation procedure. It involves dialysis of the serum-radioactive T_4 mixture against phosphate buffer as described above and precipitation of the radioactive T_4 in the dialysate with magnesium chloride after addition of carrier T_4. This method is often called the *magnesium precipitation method,* and the method of Ingbar et al.[124] is called the *resin dialysis technique.* Although these two procedures have been found to give identical results, because of its relative technical simplicity the Sterling-Brenner method has found wider acceptance. Values for $\%FT_4$ are calculated according to the formula, $\%FT_4 = (C_0/C_i) \times 100$, where C_0 = cpm of dialysate–magnesium precipitate and C_i = cpm/mL dialysand. To calculate the concentration of FT_4, it is necessary to determine the total T_4 in the serum by an RIA method as described above, and then compute with the equation:

$$FT_4, \text{ng/dL} = \frac{\%FT_4 \times \text{total } T_4 \text{ } (\mu g/dL) \times 1000}{100}$$

Commercial kits are available for the measurement of FT_4. Most kit methods are direct serum procedures and are based on the principle of conventional RIA. The assay is so designed that the original equilibrium between free and protein-bound T_4 in the patient sample remains undisturbed. To this end, it uses as a tracer a chemically modified ^{125}I-labeled thyroxine derivative which does not bind to TBG; in addition, it uses an antiserum which binds both endogenous T_4 and the radioactive T_4 derivative. Since the tracer competes only with FT_4 for the binding sites on the antibody, the amount of ^{125}I-thyroxine derivative bound to the antibody is inversely proportional to the concentration of FT_4 in the test sample. In widely used single-step procedures, the entire assay is carried out under physiological conditions of temperature, pH, and ionic strength in a tube whose lower inner wall is coated with the antibody. Following incubation with the sample or standard, the tube is aspirated or decanted and counted for radioactivity. Values for unknown samples are interpolated from a standard curve. The kit procedures are generally fast with total incubation times between 1 and 2 h at room temperature or at 37 °C. Total analysis time for 20 patient samples, controls, and standards is generally < 4 h. The inter- and intra-assay coefficients of variation are ~5% and ~7%, respectively. The sensitivity lies in the vicinity of 0.01 ng/dL.

The standard method for free T_4 measurement has been the equilibrium technique. A number of commercial kits have become available for FT_4 using a variety of methods. In some of these, results may differ from FT_4 determined by dialysis due to the influence of extremely high or low TBG concentrations on the commercial method. Different methods give slightly different results in the setting of severe nonthyroidal illness.

Specimen collection and storage. The specimen collection and storage should be carried out as described for T_4(RIA).

Reference range. The typical reference range for FT_4 in adults is 0.7–1.9 ng/dL.

Determination of Free Triiodothyronine (FT₃)

FT$_3$ is determined by essentially the same procedure as that employed for the measurement of FT$_4$. Since T$_3$ is less firmly bound by TBG than is T$_4$, the dialyzable fraction of T$_3$ is appreciably greater (almost 10 times) than that of T$_4$. Thus, the *reference ranges* for %FT$_3$ are 0.2–0.4%, and for FT$_3$, 230–660 pg/dL.

Determination of T₃ Uptake (T₃U) by Radioassay

Principle. Determinations of total T$_3$ and T$_4$ concentrations in serum are clinically more meaningful if the functional levels (capacity to bind hormone) of thyroid hormone–binding protein in blood are known. The latter information is important, since variations in total thyroid hormones in blood, as discussed above, can result from changes in the binding protein levels. The T$_3$ uptake (T$_3$U) assay reveals the number of available free binding sites in TBG. Interrelationships between serum TBG, T$_3$U, and other functional tests are illustrated in Table 9-22.

The T$_3$U test is sometimes referred to as the "T$_3$ test," a term that is erroneous, since it is not an estimation of T$_3$. The T$_3$U test evolved from the empirical observation that when trace amounts of radioactive T$_3$ (or T$_4$) were added to a mixture of plasma and red blood cells, the fraction which adhered to the red blood cells was directly proportional to the concentration of thyroid hormone present in the plasma.[110] Methodological difficulties were minimized by replacing red blood cells with an anion exchange resin, either in granular form or impregnated in a sponge.[218,294] Use of such resins is still widespread, in spite of the recent introduction of dextran-coated charcoal, talcum, Sephadex, and high-affinity antiserum as a secondary binder.

An aliquot of test serum is incubated at room temperature for 1 h in the presence of ^{125}I-T$_3$ (or ^{125}I-T$_4$) and resin. At the end of the incubation period, the mixture is centrifuged to separate the resin, and the supernatant is decanted or aspirated. The radioactivity on the resin is then counted, and values are expressed in terms of percentage of the added radioactivity taken up by the resin. The use of labeled T$_3$ is preferred because of its relatively low affinity for TBG compared to T$_4$. The radioactive T$_3$ binds to the unoccupied binding sites on TBG and does not displace any bound T$_4$. In addition, T$_3$ is more stable (less photosensitive) than T$_4$.

Many commercial T$_3$U kits are currently available; they utilize a wide range of secondary binders. A convenient procedure that yields accurate, sensitive, and precise results is based on a solid-phase RIA technique.

Reference controls and samples are incubated with ^{125}I-T$_3$ tracer in tubes coated with T$_3$ antibody. Contents of each tube are aspirated or decanted and the radioactivity in the tube is counted. The T$_3$ tracer binds to the TBG in the serum in proportion to the number of unsaturated binding sites on TBG. Therefore, the amount of T$_3$ tracer binding to the antibody on the tube varies inversely with the number of available binding sites on TBG in the serum sample and directly with the level of endogenous T$_4$. The per cent T$_3$ uptake for each unknown sample is determined by comparison with a euthyroid reference serum of known per cent uptake. The per cent uptake is calculated for each specimen from the equation:

$$\%T_3U = \frac{cpm_x}{cpm_r} \times \%T_3U_r$$

where cpm$_x$ = cpm for each patient tube

cpm$_r$ = cpm for the reference tube

%T$_3$U$_r$ = % uptake of euthyroid reference serum

Specimen collection and storage. Procedures for serum collection and storage are similar to those described for T$_4$(RIA). Repeated freezing and thawing, mixing by vortexing, or vigorous agitation of serum samples should be avoided since such treatment causes denaturation of proteins. The use of hemolyzed or lipemic samples is also not recommended.

Comments. The concentration of total TBG is frequently normal in patients with thyroid dysfunction but may be slightly low in hyperthyroidism and slightly high in hypothyroidism. However, in hyperthyroidism more binding sites, and in hypothyroidism fewer binding sites, on TBG are occupied by FT$_4$. As a result, the uptake of labeled T$_3$ by *resin* is high in hyperthyroidism

Table 9-22. RELATIONSHIPS OF FINDINGS IN THYROID FUNCTION TESTS IN
VARIOUS CLINICAL ENTITIES

Clinical Entity	Thyroid Function Tests					
	T_4	T_3	FT_4	T_3U	FT_4I	TBG
Normal	N	N	N	N	N	N
Hyperthyroidism	↑	↑	↑	↑	↑	N,↓
Hypothyroidism	↓	↓	↓	↓	↓	N,↑
Increased TBG	↑	↑	N	↓	N	↑
Decreased TBG	↓	↓	N	↑	N	↓
Displacement of hormone by drugs	↓	↓	N	↑	N	N

and low in hypothyroidism. In cases of primary increase in TBG in serum, free thyroid hormone levels are normal, but the total T_4 is increased, since the total number of binding sites in TBG is increased. In addition, the increased number of TBG binding sites will bind labeled T_3, and thus less labeled T_3 will remain free to bind with resin; therefore, the T_3U is low. The opposite changes occur in conditions of decreased TBG. Certain drugs, such as phenytoin (Dilantin), dicumarol, heparin, and salicylates, compete with thyroid hormones for the TBG binding sites. This situation is reflected by normal FT_4, low total T_4, and high T_3U (Table 9-22).

The kit procedures are generally rapid, requiring only 10–45 min of incubation time. Analysis of specimens from 20 patients can be completed in <2 h. Typical intra- and interassay CV's are 4 and 5%, respectively. Recently, Witherspoon et al.[336] made an evaluation of eight commercially available T_3 uptake test kits employing various secondary binders. They observed that the uptake values obtained from all eight kits showed varying degrees of dependencies on incubation time and temperature. The kits utilizing charcoal and antibody-coated tubes were particularly sensitive to variation in incubation time. The charcoal kit also exhibited marked temperature-dependent variations in the uptake values. Results of the study on the relationships between T_3U and TBG concentrations in serum showed a pronounced change in the T_3U value at TBG concentrations <10 mg/L and very little change in T_3U values at TBG concentrations >40 mg/L. The latter finding indicates that the T_3U test has no significance at high TBG concentrations.

Semiautomated or fully automated uptake tests using nonisotopic labels are commercially available. However, they have not been widely accepted for routine clinical use.

Reference ranges. Since reference ranges vary considerably, each laboratory must establish its own. The variations in values stem from the fact that the tracer uptake depends on such conditions of incubation as time, temperature, pH, quantity of serum added, and nature and quantity of the secondary binder. A typical reference range is 25–35%.

Free Thyroxine Index (FT$_4$I)

Total T_4 levels in serum are not wholly indicative of thyroid function since it is the level of *free* hormone that determines clinical status. As mentioned previously, a number of methods are available for estimation of FT_4 in blood. Alternatively, an *indirect* assessment of free hormone can be obtained by computing a *free thyroxine index* (FT$_4$I). The FT$_4$I originally suggested by Clark and Horn[58] correlates closely with FT_4; it is calculated by multiplying the value of total T_4 by the value of T_3U.

Principle. The calculation of FT$_4$I is based on the equilibrium relationship that exists between bound [TBG · T_4] and free T_4 [FT$_4$]:

$$[FT_4] + [TBG°] \rightleftarrows [TBG \cdot T_4] \tag{1}$$

where [TBG°] represents unoccupied binding sites on TBG. If K is an equilibrium constant, then

$$K = \frac{[TBG \cdot T_4]}{[FT_4] \times [TBG°]} \tag{2}$$

or

$$K \, [FT_4] = \frac{[TBG \cdot T_4]}{[TBG^\circ]} \tag{3}$$

Since 99.97% of the total thyroxine is bound to TBG, the term $[TBG \cdot T_4]$ can be essentially replaced by the total thyroxine concentration. And since $[TBG^\circ]$, the concentration of unoccupied TBG binding sites, is inversely proportional to the T_3U, then

$$K \times [FT_4] = [T_4] \times \frac{1}{[TBG^\circ]} = [T_4] \times T_3U \tag{4}$$

If FT_4I is defined as $K \times [FT_4]$, then

$$FT_4I = [T_4] \times \%T_3U \tag{5}$$

T_4 is expressed in $\mu g/dL$ and T_3U is expressed as % of a euthyroid control. Thus, FT_4I has units of $\mu g/dL$ like T_4. (Alternative units for expressing FT_4 have been discussed in the literature.[285]) Therefore the *reference range* for FT_4I will vary with reference ranges established for T_4 and T_3U. One can calculate the lower limit of the FT_4I range by multiplying the lower limit of the T_3U range times the lower limit of the T_4 range; the upper limit of the FT_4I range can be obtained by multiplying the upper limit of the T_3U range times the upper limit of the T_4 range.

Comments. In recent years, a number of pseudonyms for the FT_4I have appeared in the literature and in commercial kits, leading to some confusion. Effective thyroxine ratio (ETR), T-7, and T-12 are only a few of the examples. These tests differ either in the mode of calculation of results or in the mode of performance of the tests. The underlying principle of all these tests is to evaluate the FT_4 indirectly. The Committee on Nomenclature of the American Thyroid Association[285] has recommended that all tests designed to estimate FT_4 indirectly be designated as FT_4I.

There is no general agreement regarding the clinical value of FT_4I versus FT_4 and no agreement regarding the method of choice for measurement of FT_4. In most situations, using the FT_4I is less expensive than determining the FT_4, and FT_4I provides two pieces of potentially useful information (T_4 and T_3U). In acute medical illness, FT_4I appears to be a better index of thyroid status than FT_4. With chronic nonthyroidal illness, FT_4 seems to be better than FT_4I in resolving problem cases. In the final analysis, neither FT_4I nor FT_4 can be counted on to indicate thyroid status in all circumstances; the method of choice depends on the clinical situation, and either test may require supplementary data (T_3, TSH, TRH stimulation test, or even rT_3 measurement) for precise evaluation of a particular case.

Determination of FT_3 Index (FT_3I)

The FT_3I is calculated like the FT_4I, as $T_3 \times T_3U$, and used in the same way as an estimate for FT_3. FT_3I may be helpful in some patients to exclude a diagnosis of T_3 toxicosis. In general, however, the FT_3I offers no advantage over the FT_4I and is utilized less frequently in clinical practice.

Determination of Thyroxine Binding Globulin (TBG)

Tests described above are quite adequate in most clinical situations for the assessment of TBG or its effect. However, direct measurement of TBG is occasionally required to confirm the results of the free hormone assay in serum. For example, in patients with familial deficiency or absence of TBG, low T_4 values and high T_3U ratio are conveniently clarified by direct measurement of TBG. Similarly, during pregnancy, elevated T_4 levels accompanied by a less than normal T_3U are confirmed by the observation of an elevated TBG.

Principle. TBG can be measured indirectly in terms of its T_4-binding capacity or directly by RIA.[41] Tests based on *binding capacity* are usually performed by adding to the test serum a mixture of radiolabeled and unlabeled T_4 in a concentration sufficient to saturate all the binding sites on the TBG. Total T_4 is determined on this enriched serum, which is then subjected to electrophoresis. Radioactivity bound to individual protein bands (albumin, TBG, TBPA, and others) is counted, and the fraction associated with each band is calculated. The fractional radioactivity for TBG or TBPA is then multiplied by the total T_4 concentration in the T_4-enriched serum to give T_4-binding capacity (units: μg T_4/dL) for the particular protein. Binding capacity of TBG in normal serum generally ranges from 15 to 23 μg T_4/dL, for TBPA from 150 to 300 μg T_4/dL.

Direct measurement of TBG by RIA[99] is more popular because it is technically less complex. The antiserum is generated against purified TBG, which is also used as standard and for preparation of the radioiodinated ligand. In the test procedure, appropriately diluted serum is incubated overnight at 40 °C with antiserum and ^{125}I-TBG; bound and free fractions are separated by a double-antibody technique using goat-antirabbit antibody. Quantitation is made against a standard curve. The reference range for adults is 2.5–4.8 mg/dL using this procedure and differs significantly from the range obtained with commercially available TBG procedures.

Several kits are commercially available for RIA of TBG. Although the assay is not as often requested as those for T_4, T_3U, and FT_4, results can be diagnostically useful in the special clinical situations previously discussed.

In one commercial TBG assay, a two-site immunoradiometric assay (IRMA) or "sandwich" assay is used (Corning Medical, Medfield, MA 02052). TBG antibody bound to glass particles is added in excess to the patient sample and binds all endogenous TBG. ^{125}I-T_4 is then added and binds to the TBG, forming an anti-TBG–TBG–^{125}I-T_4 sandwich. After a 15-min incubation period the precipitate is collected by centrifugation and its radioactivity is counted. Patient values are interpreted from a standard curve run simultaneously with the patient specimens. The adult *reference range* for this procedure is 15–34 μg/mL. For *specimen collection and storage*, see requirements listed for T_4(RIA).

T_4/TBG and T_3/TBG Ratios

These ratios are indices derived to correct for the hormonal variation that is a function of TBG concentration or binding capacity. They give essentially the same information as the FT_4I and FT_3I. It has been debated in the literature whether FT_4I and FT_3I or T_4/TBG and T_3/TBG ratios correlate better with clinical status;[332] the calculation of choice seems to depend on the particular clinical situation. The reference range for T_4/TBG ratios is 0.35–0.42 when normal values for T_4 and TBG are 4.5–12.5 μg/dL and 13–30 μg/mL, respectively.

Determination of Thyrotropin (TSH) by RIA

The reference methods for the measurement of TSH have been described in the section on pituitary hormones. Several commercial kits are currently available.

Principle. TSH kit methods are based on the conventional RIA or the immunoradiometric assay (IRMA) procedure. Described below is a typical example of an IRMA TSH procedure that is based upon the general principles of a two-site immunoradiometric assay.

The patient sample is incubated first with ^{125}I-labeled antibody, followed by addition of an immobilized antibody. After a second incubation, saline is added to the tubes to dilute the tracer, which minimizes assay variation due to variable decanting and reduces nonspecific binding. The tubes are then centrifuged and decanted, and the pellets that contain the "sandwiches" are counted in a gamma-counter. A standard curve is constructed, and concentrations of TSH in serum are obtained by interpolation from the curve. Unlike the typical RIA procedures, the radioactivity in an IRMA method is directly proportional to the endogenous TSH concentration.

Specimen collection and storage. The preferred sample is serum, free of hemolysis and lipemia. Serum specimens are stable for 7 d at 2–8 °C and for at least one month when frozen.

Comments. Kit procedures vary widely in incubation time from approximately 2–18 h at room temperature or at 37 °C. The above procedure requires two 2-h incubations at 37 °C. The total assay time for 20 patients is ~6 h. Typical CV's for intra- and interassays within the reference range are ~5% and ~8%, respectively. The sensitivity is generally <0.5 μIU/mL and the upper reference range is 5.0 μIU/mL.

Many of the commercially available kits suffer from poor performance in the low normal range due to matrix effects, as well as from a lack of analytical sensitivity and specificity. The poor quality of TSH data in this range can be misleading to the requesting physician. The evaluation of commercial TSH kits should be carried out not only using the conventional approaches but also including the integration of basic thyroid physiology with the analytical results. Musto et al.[217] have proposed a comprehensive approach to TSH test evaluation with this concept in mind. According to these investigators, a thyrotropin kit that consistently yields undetectable thyrotropin results (within the sensitivity of the method) for serum samples from a statistically significant number of hyperthyroid patients will have minimal or no matrix effect and will provide reliable results.

Determination of Thyroglobulin (Tg) by RIA

Principle. Measurement of serum Tg by RIA was first introduced by Van Herle et al.[315] Before Tg is measured, the presence of anti-Tg in the serum should be determined. If such antibodies are detected, further analysis of Tg in the serum cannot be performed, since anti-Tg autoantibodies will interfere with the radioimmunoassay. The RIA method involves the incubation of serum with ^{125}I-Tg and Tg-specific antibody overnight at 4 °C. The separation of bound from free Tg is accomplished by precipitation with a second antibody. The Tg content is determined by comparing the per cent bound radioactivity of the test serum with the standard curve.

There are two types of commercial kits currently available. One is a competitive double antibody RIA technique and the other is a solid phase, two-site immunoradiometric assay (IRMA).

Reference ranges. Tg is present in 75% of normal sera at a concentration of ∼5 ng/mL. Tg has also been detected in normal pregnant women and in cord blood. Tg is elevated in thyroid carcinoma, thyrotoxicosis, and subacute thyroiditis. Measurement of Tg is most useful in detecting recurrence of thyroid carcinoma following surgical resection or radioactive iodide ablation.

Other Tests Related to Thyroid Hormone Function

Human Thyroid-Stimulating Immunoglobulins (hTSI)

Principle. The involvement of the immune system in thyroid disease has been discussed by Werner.[328] The indication that autoantibodies might play a role in the pathogenesis of Graves' disease came with the discovery of *long-acting thyroid stimulator* (LATS) in the serum of some patients. Until recently, the estimation of LATS by bioassay[94] in the diagnosis and treatment of Graves' disease was considered important, but the poor correlation between LATS and the severity or fluctuation of the disease has led investigators to regard LATS as merely an epiphenomenon, irrelevant to the hyperthyroidism of Graves' disease. For this reason, its use in clinical medicine has become more and more limited.

Sera obtained from LATS-positive and LATS-negative patients contain an immunoglobulin that inhibits LATS-binding to thyroid subcellular fractions. This substance, designated as LATS-protector (LATS-p), is highly species specific and has no action on thyroid fractions obtained from mouse, guinea pig, or monkey. It acts on the human thyroid in vivo to cause release of radioiodine and in vitro to increase colloid droplets and cAMP formation. The LATS-p is also known as human thyroid-stimulating immunoglobulins (hTSI). However, the generic term for thyroid stimulators, including hTSI, is *thyroid-stimulating immunoglobulins* (TSI). TSI have been shown to displace TSH from receptor sites and to block their binding to these sites on human thyroid cell membranes that are used in the cell membrane receptor assay for TSH. These immunoglobulins are postulated to be antibodies to the TSH receptor that interact with the receptor to activate the adenylate cyclase system and cause thyroid stimulation similar to that caused by TSH. TSI in general demonstrate substantial heterogeneity. Some interfere with binding of TSH to its receptor, some are agonists of the adenylate cyclase system, and some stimulate the thyroid of other species (e.g., mouse).

The clinical significance and assay methods of TSI have been discussed by Zakarija et al.[342] TSI can be detected by a receptor assay.[197] The method is based on the fact that TSI interfere in a dose-dependent manner with the binding of radioiodinated TSH to a crude preparation of human thyroid membranes. The method is rapid, simple, and capable of handling a large number of specimens. For further discussion on other cytochemical assay methods, the reader is referred to the report by Zakarija et al.[342]

TSI are undetectable in normal human serum but are present in 83–93% of patients with hyperthyroidism due to Graves' disease and in patients who are unresponsive to long-term antithyroid drug therapy. The presence of high levels of TSI in pregnant women is predictive of neonatal hyperthyroidism.

Determination of Antithyroglobulin and Antimicrosomal Antibodies

Antibodies are found in a variety of thyroid disorders that are directed against thyroglobulin and thyroid cell microsomes. Various methods for detection of these antibodies have been employed. They include the precipitation reaction, complement fixation, Ouchterlony diffusion plate analysis, competitive protein binding (CPB) radioassay,[211] tanned erythrocyte agglutination,[26] and the indirect Coombs immunofluorescence method.[126] Only the CPB assay, the tanned erythrocyte agglutination method, and the technique involving indirect immunofluorescence have so far found clinical application. The most widely applied method is based on hemagglutination.

Principle. In the tanned erythrocyte hemagglutination method of antithyroglobulin antibodies, an aliquot of patient's serum is mixed with erythrocytes that have been treated with tannic acid and then coated with thyroglobulin. When antibodies in the patient's serum combine with antigen (i.e., thyroglobulin) bound to erythryocytes, agglutination of the erythrocytes occurs. The use of thyroglobulin-coated erythrocytes makes the agglutination reaction much more sensitive than a simple antigen-antibody reaction. Serial dilutions of the patient's serum are employed to establish the thyroglobulin antibody titer. A similar procedure is followed for the measurement of *antimicrosomal antibodies,* where erythrocytes are used that have been coated with a microsomal antigen derived from human hyperplastic thyroid glands.

These antibodies are detected in 5–10% of normal patients. However, in patients with primary myxedema and Graves' disease, and most strikingly in Hashimoto's thyroiditis, reactivity occurs much more frequently and to a higher titer. An agglutination test positive at a dilution of 1:2500 is equivalent to histologic proof for Hashimoto's disease. However, in patients who have multinodular goiter, thyroid adenomas, and thyroid carcinomas, antibody titers are not found with any greater frequency than in normal control subjects.[64]

Commercial kits for both antithyroglobulin and antimicrosomal antibodies are available. The one most widely used is based upon the hemagglutination method of Boyden.[26] The preferred *specimen* for testing is serum deactivated at 56 °C for 30 min. Heating is important for inactivation of complement and TBG which otherwise would interfere with the assay. Serum specimens should be kept frozen if the test is not performed the day blood is drawn.

CATECHOLAMINES AND METABOLITES

Catecholamines are compounds with amines attached to a benzene ring bearing two hydroxyl groups (catechol). The most important endogenously produced compounds of this group are epinephrine (adrenalin), norepinephrine (noradrenalin), and dopamine [(3,4-dihydroxyphenyl)ethyl amine]. The structural formulas and the numbering system of these compounds are shown in Figure 9-47. Norepinephrine differs from dopamine in that it possesses an hydroxyl group on the β-carbon atom; epinephrine is distinguished from norepinephrine by the presence of a methyl group on the nitrogen of the terminal amino group.

Dopamine and norepinephrine are primary amines; epinephrine, on the other hand, is a secondary amine because of the substitution of a methyl group for an amino hydrogen atom. Epinephrine and norepinephrine are dihydroxylated phenyl-β-ethanolamines and therefore exhibit the chemical properties of phenols, alcohols, and amines. Chemically, epinephrine is rapidly oxidized in neutral and alkaline solutions; norepinephrine is much more resistant to oxidation. These differences in properties of the two compounds have been utilized for their individual estimations in biological fluids. For further information on the chemical properties of catecholamines, see the review article by von Euler.[321]

The main *sites of production* of the catecholamines are the brain, chromaffin cells of the adrenal medulla, and the sympathetic neurons. The chromaffin cells are so designated because of their distinctive brown or black color on staining with chromic acid. Epinephrine is quantitatively the most important substance produced by the adrenal medulla. Norepinephrine is the

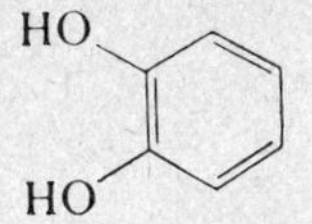

Catechol (Dihydroxybenzene)

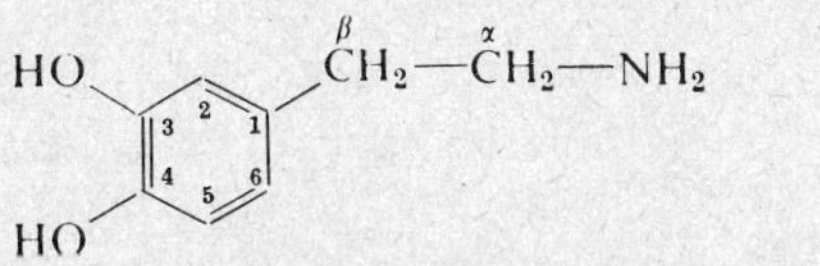

Dopamine [β(3,4-dihydroxyphenyl)ethyl amine]

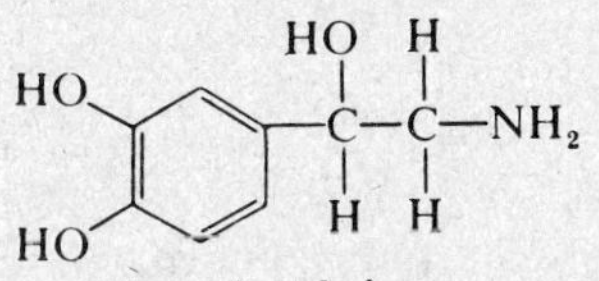

Norepinephrine (Noradrenaline)

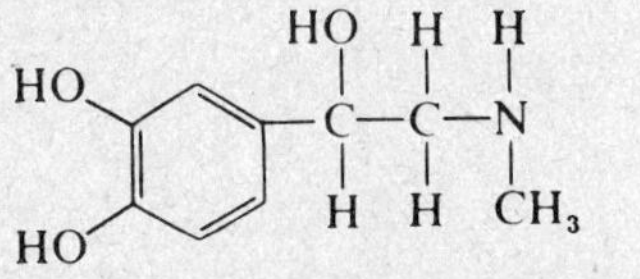

Epinephrine (Adrenaline)

Figure 9-47. Structure and numbering system of catecholamines.

major substance liberated by the postganglionic sympathetic nerves. In addition, norepinephrine is the predominant catecholamine secreted in mammalian fetal life by the organ of Zuckerkandl.

Dopamine and norepinephrine are important neurotransmitters in the central nervous system (the brain and the spinal cord). The brain contains three main dopamine neurotransmitter systems. These consist of the extrapyramidal system involved in the coordination and integration of fine muscular movement (e.g., picking up small objects), the mesolimbic system involved in the control of memory and emotion, and the hypothalamic-pituitary axis involved in such endocrine-related activity as the release of prolactin. Norepinephrine-containing neurons are present in the cerebellum and cerebral cortex, the parts of the brain that control the fine coordination of body movement and balance as well as alertness and emotion. Another part of the brain containing norepinephrine is the hypothalamus, which controls many of the body's functions, such as hunger, thirst, temperature regulation, blood pressure, reproduction, and behavior. Manipulation of the brain's norepinephrine levels can change many of the functions of the hypothalamus. Norepinephrine-producing neurons in the brain appear also to be involved in sleep, mood elevation, and depression. Catecholamines and other substances that serve as neurotransmitters have been reviewed by Bleich and Moore.[23]

Each of the three catecholamines—dopamine, norepinephrine, and epinephrine—has characteristic *physiological functions* and *pharmacologic actions*. Detailed consideration of these topics may be found in a textbook of endocrinology by Williams.[334] Dopamine and norepinephrine have, in general, a marked influence on the vascular system, whereas epinephrine, which is considered to be the true adrenal medullary hormone, influences metabolic processes, especially carbohydrate metabolism. Furthermore, there are certain tissues in which epinephrine and norepinephrine

Table 9-23. PHYSIOLOGIC RESPONSES INDUCED BY
α- AND β-RECEPTORS

Site of Effect	Alpha Response	Beta Response
Vascular beds	Vasoconstriction	Vasodilatation
Intestinal smooth muscle	Relaxation	Relaxation
Bronchial smooth muscle	None	Relaxation
Cardiac contraction	None	Increase
Heart rate	None	Increase
Pupils	Dilatation	None
Piloerection	Stimulation	None
Insulin release	Inhibition	Stimulation
Blood glucose	Increase	Decrease

produce divergent effects. The existence of two types of adrenergic receptors, α and β, with different sensitivities for the various catecholamines, accounts for the varying responses. The α-receptors interact with both epinephrine and norepinephrine; the β-receptors respond to epinephrine but are relatively insensitive to norepinephrine. A summary of these responses is presented in Table 9-23.

Biosynthesis, Storage, and Release

The aromatic L-amino acid tyrosine has been shown by in vivo and in vitro studies to be the precursor of the catecholamines (Figure 9-48). Tyrosine is derived from the diet or synthesized from phenylalanine. The first and rate-limiting step in the biosynthetic path of catecholamines is hydroxylation of tyrosine in the mitochondria by tyrosine hydroxylase to produce dihydroxyphenylalanine (dopa). Dopa is converted in the cytoplasm to dopamine through the action of aromatic L-amino acid decarboxylase. Dopamine is then transported by an energy-dependent mechanism into storage granules (vesicles) which are present both in the sympathetic nerve endings and in the adrenal medulla. These granules contain dopamine-β-oxidase in association with a macromolecule called chromogranin A. Chromogranin A itself is a water-insoluble substance, probably composed of subunits of dopamine-β-hydroxylase. When in contact with ATP, chromogranin A is solubilized and the enzyme becomes activated. Dopamine-β-oxidase is responsible for final conversion of dopamine to norepinephrine, which is then stored in the granule. In the adrenal medulla, norepinephrine released from storage granules becomes the substrate for another enzyme, phenylethanolamine-N-methyl transferase, which transfers a methyl group from S-adenosylmethionine to norepinephrine, thus forming epinephrine.

Biosynthesis and release of catecholamines are regulated both by intracellular compartmentalization and by feedback inhibition on rate-limiting steps by the products. The vesicles sequester the catecholamines, thereby preventing premature degradation by cytoplasmic monoamine oxidase (MAO) until neurogenic stimulation commands their release. When norepinephrine is released, it inhibits tyrosine hydroxylase activity and transport of dopamine into the vesicles. Epinephrine is a potent noncompetitive inhibitor of phenylethanolamine-N-methyl transferase.

Pharmacologic manipulation of catecholamine synthesis is facilitated by the nonspecificity

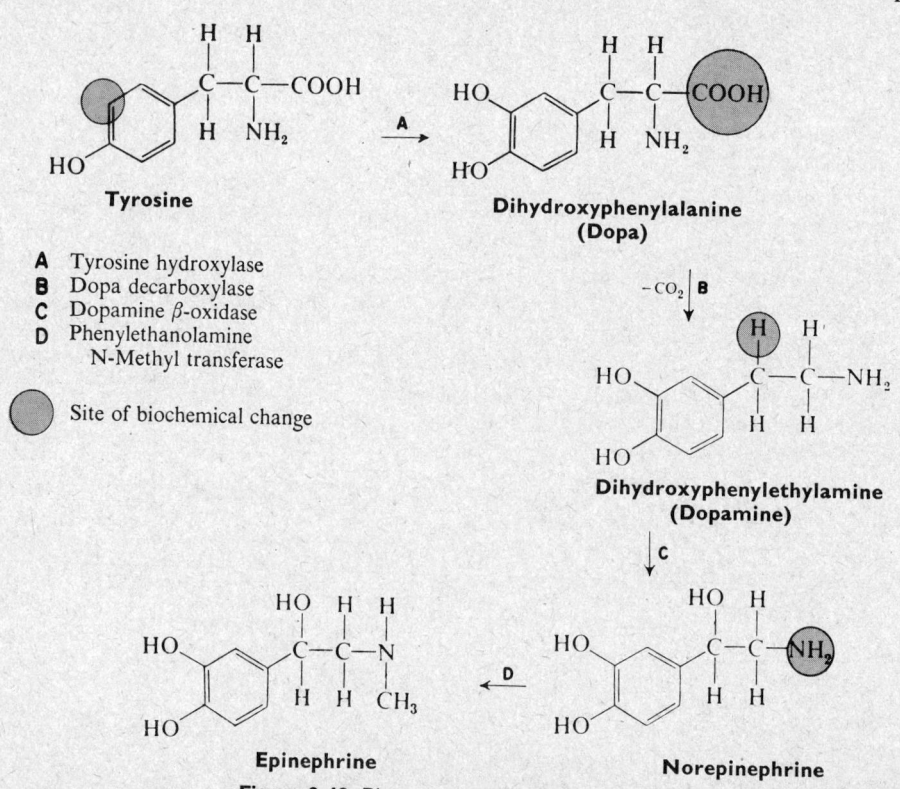

A Tyrosine hydroxylase
B Dopa decarboxylase
C Dopamine β-oxidase
D Phenylethanolamine
    N-Methyl transferase

Site of biochemical change

Figure 9-48. Biosynthesis of catecholamines.

of some of the enzymes in the pathway and of the storage and release mechanism. The cytoplasmic aromatic L-amino acid decarboxylase has low substrate specificity and can decarboxylate most aromatic amino acids. Dopamine-β-oxidase has a similar low specificity, since it will hydroxylate many substances that have an aromatic ring bearing a side chain of two or three carbon atoms and terminating with an amino group (e.g., tyramine, phenylalkylamines). For these reasons, synthetic congeners of amine intermediates can be administered in clinical treatments to generate substitute or false neurotransmitters. For instance, α-methyldopa, known to stimulate α_2-adrenergic activity centrally, can be decarboxylated and subsequently β-hydroxylated by the same pathway to form α-methylnorepinephrine. This compound can displace norepinephrine, can be stored in the granules, and can be released upon nerve stimulation. The ultimate result of the substitution is to diminish the physiological effects mediated by norepinephrine. Artificial intermediates such as α-methyl hydrazine or α-methyl metatyrosine can also be introduced to block the pathway, or the pathway can be supplemented with an exogenous source of the natural intermediate L-dopa, as is done in the treatment of Parkinson's disease.

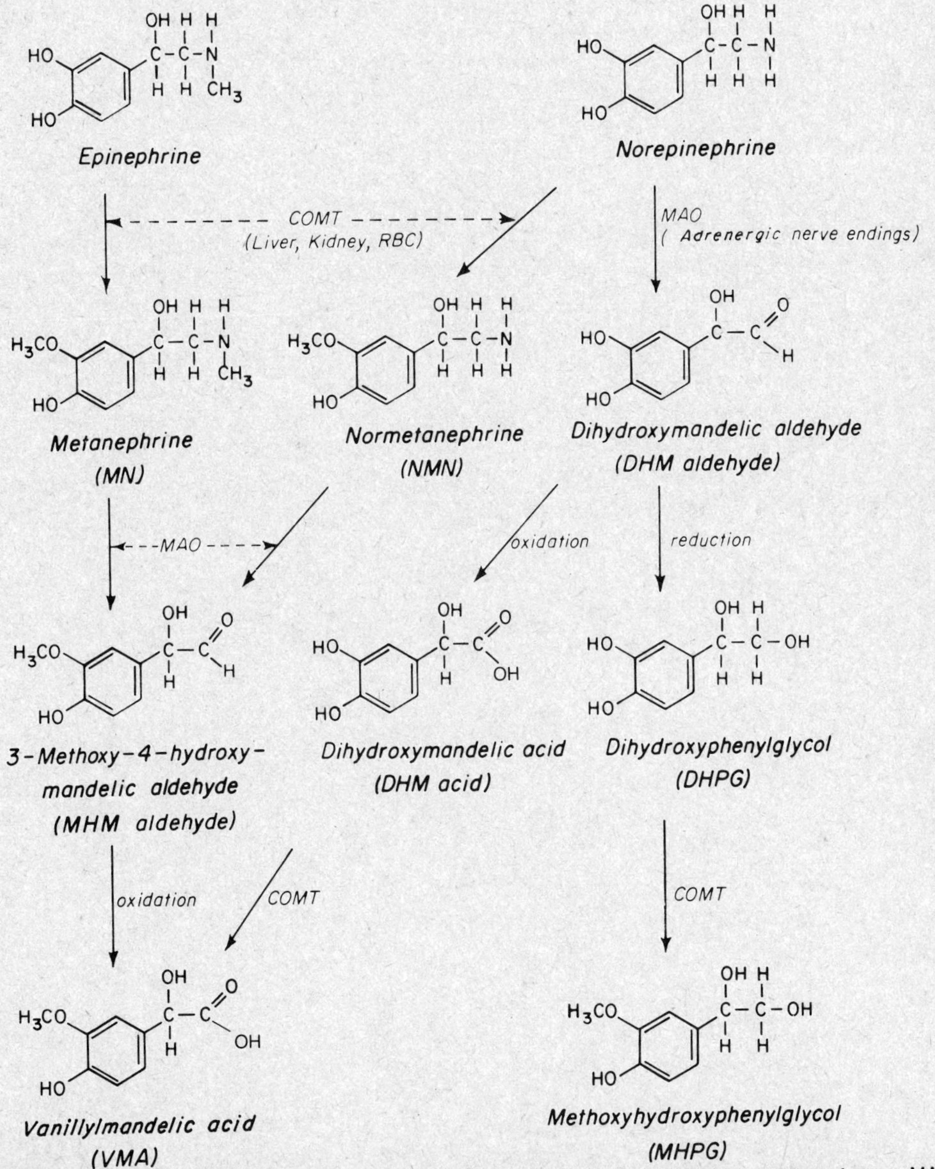

Figure 9-49. Metabolism of epinephrine and norepinephrine. COMT = Catechol-O-methyltransferase; MAO = Monoamine oxidase.

Nerve stimulation is necessary for the *release of catecholamines*. The chemical mediation of catecholamine release is thought to be linked to the presence of acetylcholine or to the influx of calcium ions, or both. The process is energy dependent; it requires glycolysis or oxidative metabolism to produce ATP. The catecholamines are generally thought to be released from their vesicles by the process of exocytosis, aided by contraction of the microfibrils of the cells. The involvement of microfibrils in this process is suggested by the evidence that the release of catecholamines is inhibited by colchicine, vinblastine, and vincristine, each of which interferes with microtubular function. For further information on the storage and release of catecholamines, refer to the review article by Smith.[282]

Metabolism

Upon release from storage vesicles, catecholamines act on effector sites. They are then rapidly inactivated through reuptake by storage granules, conversion to metabolites, or excretion. Catecholamines are transported in blood unbound to proteins and have a very short plasma half-life of ~2 min. Two important enzymatic events for the inactivation of catecholamines in the body involve catechol-O-methylation and oxidative deamination. Catechol-O-methyl transferase (COMT) is present in most tissues, and is especially abundant in liver, kidney, and red blood cells. This cytoplasmic enzyme methylates the C-3 hydroxyl group of norepinephrine and epinephrine, resulting in the formation of normetanephrine and metanephrine, respectively (Figure 9-49). A certain portion of these methylated products is excreted either in the free form or as conjugates of sulfuric or glucuronic acid. However, the majority of these 3-methoxy derivatives undergo deamination by the ubiquitous mitochondrial monoamine oxidase to 3-methoxy-4-hydroxymandelic aldehyde (MHM aldehyde). This latter intermediate is either oxidized to vanillylmandelic acid (VMA) or reduced to methoxyhydroxyphenylglycol (MHPG). These end products are subsequently excreted. When oxidative deamination of catecholamines occurs first, as in adrenergic nerve endings, dihydroxymandelic aldehyde (DHM aldehyde) results. Oxidation and reduction of DHM aldehyde and subsequent O-methylation by COMT yield the excretory products VMA and MHPG, respectively (Figure 9-49). Thus, regardless of the nature of the initial enzymatic attack, the principal end products of both norepinephrine and epinephrine are VMA and, to a lesser degree, MHPG. For this reason urinary measurements of these metabolites, specifically VMA, reflect the total rather than the differential production of norepinephrine and epinephrine in the body. The final metabolite of dopamine following oxidative deamination and O-methylation is homovanillic acid, HVA (Figure 9-50).

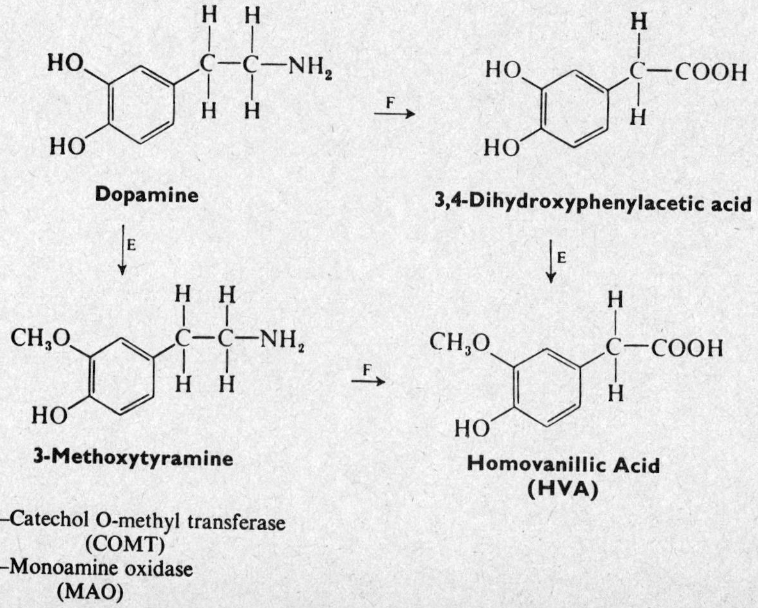

E—Catechol O-methyl transferase
    (COMT)
F—Monoamine oxidase
    (MAO)

Figure 9-50. Metabolism of dopamine.

Clinical Significance of Measurement of Catecholamines—Epinephrine and Norepinephrine—and Their Metabolites

Catecholamines play an important role in health and disease. For example, excess catecholamines are associated with stress, a fall in blood pressure or blood volume, thyroid hormone deficiency, congestive heart failure, and arrhythmias; deficient catecholamines are seen in idiopathic postural hypotension. However, laboratory measurements of epinephrine, norepinephrine, or their metabolites are primarily useful for the diagnosis of catecholamine-secreting neurochromaffin tumors (pheochromocytomas, paragangliomas, or neuroblastomas).

Benign or malignant neoplasms may arise from neurochromaffin cell tissue in the autonomic nervous system or the adrenal medulla. These tumors may produce excessive amounts of catecholamines or catecholamine metabolites. Such tumors are rare, occurring in only 0.1–0.3% of the population; up to 75% are found only at autopsy, having been unsuspected premortem.[301] However, when the classic picture of catecholamine excess occurs (i.e., sustained or paroxysmal hypertension, weight loss, "spells" of sweating, headache, palpitations, and anxiety), the clinical picture is quite striking. Diagnosis requires a high degree of clinical suspicion. Although life-threatening, this condition is usually curable with surgery.

In adults, 90% of these rare tumors occur in the adrenal medulla (pheochromocytoma). In adults, ~10% of these adrenal tumors may be bilateral. The 10% that occur in extra-adrenal sites are known as paragangliomas. The clinical picture of pheochromocytoma may be seen with adrenal medullary hyperplasia in the absence of tumor.[263] Pheochromocytoma occurs at any age, but is most common in the fourth and fifth decades. There is a slight female predominance. Nine out of ten pheochromocytomas in adults are benign; hormonally "silent" tumors are more likely to be malignant. Malignant tumors and bilateral tumors occur more frequently in children.[90]

Although hypertension is a common symptom of patients with pheochromocytoma, this catecholamine-secreting tumor is not a common cause of hypertension; it accounts for <0.5% of all hypertensive cases. Pheochromocytomas are familial in about 10% of cases, usually in kindreds with multiple endocrine neoplasia syndromes. Screening or testing all hypertensive patients for pheochromocytoma is not cost effective. However, patients with clinical features that suggest adrenergic excess should be tested, particularly those with severe hypertension. Patients in kindreds known to have multiple endocrine neoplasia syndromes, types IIa or IIb, should also be tested, even when they are asymptomatic. Despite selectivity in testing, false positive test results are frequent and are much more common than true positive tests.

A number of aspects of normal physiology make it difficult for one test of a single aspect of catecholamine production or action to be satisfactory for screening or for definitively diagnosing pheochromocytoma. Norepinephrine levels in plasma usually reflect only a small fraction of the concentration of norepinephrine at the synaptic nerve ending. The half-life of catecholamines in plasma is brief. Rapid and striking increases in plasma catecholamine concentration may be induced by stress, upright posture, exercise, hypoglycemia, hypovolemia, cold, hypoxia, hypercapnia, or the mental states of anxiety or anger. Also, there is a diurnal variation of catecholamines, with the zenith in the morning and the nadir at night. Norepinephrine levels are higher than normal in subjects with mitral valve prolapse or essential hypertension.[63] In addition, medications may change catecholamine concentrations through their usual mechanism of action. Alpha- and β-adrenergic blockers, vasodilators, and theophylline increase norepinephrine levels; adrenergic neuron blockers, clonidine, α-methyldopa, bromocriptine, and phenothiazines decrease the levels of norepinephrine. Other factors that further complicate the selection of these tests include drug interference with specific assay procedures and the conditions of sample collection. In patients with pheochromocytoma, the release of catecholamines by the tumor may be intermittent, or the uptake and degradation of catecholamines by the tumor may be so great that normal or near-normal concentrations of catecholamines will be seen in blood and urine when striking increases in metanephrines and VMA are observable in urine.

Most authorities recommend measurement of urinary metanephrine as the initial screening test for pheochromocytoma.[101] Although recent studies suggest some false negatives may occur with this test, most large studies have found measurement of urinary metanephrines to produce almost no false negatives (nearly 100% sensitivity). False positive tests are often seen in time of severe stress, but results are likely to be normal when the test is repeated. While a 24-h urine collection is desirable and provides an "integrated" picture of catecholamine production, a 12-h overnight sample can be satisfactory, provided that urinary metanephrine is expressed per gram of creatinine; a normal metanephrine/creatinine ratio is <2.2 μg metanephrine/g creatinine.[101,299]

Measurement of VMA or catecholamines or both in urine is best applied to follow-up testing in patients with an elevated urinary metanephrine/creatinine ratio or in those subjects highly suspect for pheochromocytoma with normal urinary metanephrine.[101] Urinary VMA and catecholamine determinations may be more specific (fewer false positives) but are less sensitive (more false negatives) and are more subject to drug interferences than are urinary metanephrine measurements.

Some clinicians advocate measurement of plasma catecholamines as an alternative initial test to measuring urinary metanephrines. Cases of pheochromocytoma with normal levels of catecholamines, metanephrine, and VMA in urine but with definite elevations of norepinephrine or epinephrine in plasma have been reported.[29] Conditions of sampling, however, are most important; blood should be drawn in the morning after 30 min of rest in the recumbent position and, ideally, from an indwelling needle placed 20–30 min beforehand. Venipuncture may cause stress and increase plasma levels of catecholamines. Plasma must be promptly separated from blood. In subjects with a borderline elevation of urinary metanephrines or where the suspicion for pheochromocytoma is high but urine results are normal, plasma catecholamines should be measured.

If measurements of basal catecholamines and metabolites fail to establish or exclude the diagnosis of pheochromocytoma, pharmacologic tests to suppress or stimulate catecholamines may be considered. Stimulation tests are somewhat risky and are best avoided. Agents such as glucagon, histamine, or tyramine that cause release of catecholamines may also precipitate a dangerous rise in blood pressure. Drugs that suppress catecholamine release can be used with greater safety, though serious hypotension may result. Clonidine inhibits neurogenic catecholamine release and will cause a fall of plasma norepinephrine into the reference range in hypertensive subjects without pheochromocytoma.[28] (See Protocol S.) However, intermittent norepinephrine secretion by a pheochromocytoma may sometimes give the false impression that clonidine-induced suppression had occurred. A recent article summarizes the tests needed in the diagnosis, localization, and management of pheochromocytoma.[27a]

If there is biochemical evidence of catecholamine excess, studies to locate the tumor are appropriate. The value of CT scanning, arteriography, and selective venous sampling for measurement of epinephrine and norepinephrine is well established. Scanning with ^{131}I-benzylguanidine appears to be quite sensitive in locating small tumors or in identifying adrenal medullary hyperplasia.

The clinical significance of determining norepinephrine and its metabolites, specifically 3-methoxy-4-hydroxy phenylglycol (MHPG) in depressed patients stems from the hypothesis that a norepinephrine deficiency at critical CNS synapses underlies the cyclic emergence of depression in many patients with affective disorders.[178] While urinary norepinephrine, normetanephrine, metanephrine, and perhaps VMA originate in catecholamine pools external to the CNS, there is evidence that a significant fraction of the MHPG excreted in the urine has its origin in the metabolism of brain norepinephrine. Several investigators have demonstrated that patients with low urinary excretion of MHPG have a favorable therapeutic response to antidepressants such as desipramine and maprotiline, known to inhibit selectively the reuptake of norepinephrine from the synaptic cleft into the presynaptic neuron. The inhibition of reuptake makes more norepinephrine available in the synaptic cleft for interaction with the postsynaptic receptor. Such augmented activity of norepinephrine may underlie the clinical response of the antidepressant drugs.

PROTOCOL S: CLONIDINE SUPPRESSION TEST

Rationale: Clonidine is a drug that inhibits catecholamine release in normal subjects but not in patients with pheochromocytoma.

Procedure: The test is best done in the morning after an overnight fast. The patient is to remain recumbent through the entire procedure. Thirty minutes after the insertion of an indwelling needle for repeated sampling, blood is drawn for plasma catecholamine determination. Clonidine, 4.3 μg/kg body weight, is given orally, and a repeat sample for plasma catecholamines is drawn 3 h later.

Interpretation: Plasma norepinephrine should be within the established reference range at 3 h. Plasma norepinephrine above normal at 3 h is highly suggestive of pheochromocytoma.

Methods for the Determination of Catecholamines and Metabolites

Numerous methods have been used for the determination of catecholamines and their metabolites in biological fluids. They include fluorometric, spectrophotometric, gas and liquid chromatographic, and radioenzymatic techniques. The current state of the art in the analysis of catecholamines has been reviewed by Krstulovic[157] and Holly and Makin.[118] Among the methods, fluorometric, spectrophotometric, radioenzymatic, and HPLC assays with electrochemical detection are widely used in clinical laboratories. The following sections describe their principles and applications to the analyses of catecholamines and their metabolites in blood and urine.

Determination of Plasma Catecholamines

The concentrations of epinephrine and norepinephrine in plasma are normally very low, necessitating the use of highly sensitive assay techniques. Of the many techniques applied to the quantitation of plasma catecholamines, HPLC methods are the latest and possibly most promising.[118] The specificity of liquid chromatography, combined with the sensitivity and additional specificity of electrochemical detection, makes HPLC a very powerful technique. However, extremely low plasma levels of epinephrine and dopamine require that HPLC systems operate close to their limit of detectability, resulting in occasional unreliable estimations. Until further improvements in methodology are achieved, routine application of HPLC is largely limited to analysis of urinary catecholamines. For many years, the only sufficiently sensitive assay methods available were the ethylenediamine (EDA) and the trihydroxyindole (THI) fluorometric methods. More recently, however, a radioenzymatic assay, originally reported by Engelman et al.,[80] has found increasing application in the clinical field. A general outline of this technique and the fluorometric method are given below.

Fluorometric Determination

Principle. Fluorometric methods for the determination of epinephrine and norepinephrine in plasma depend upon the conversion of the amines into derivatives that emit a characteristic fluorescence greater than that of the parent amine. The methods employed most extensively are modifications of the EDA method and the THI method. A comparative study of these two fluorometric procedures for the analyses of plasma catecholamines has been made by Manger et al.[182] Although both methods are very accurate in quantitating epinephrine and norepinephrine in plasma, the EDA method has been considered to be more practical for use in a routine clinical laboratory.[182]

In the EDA method, epinephrine is first oxidized in dilute acetic acid to adrenochrome; this compound then condenses with one mole of ethylenediamine with elimination of 2 H_2O and 2 H.

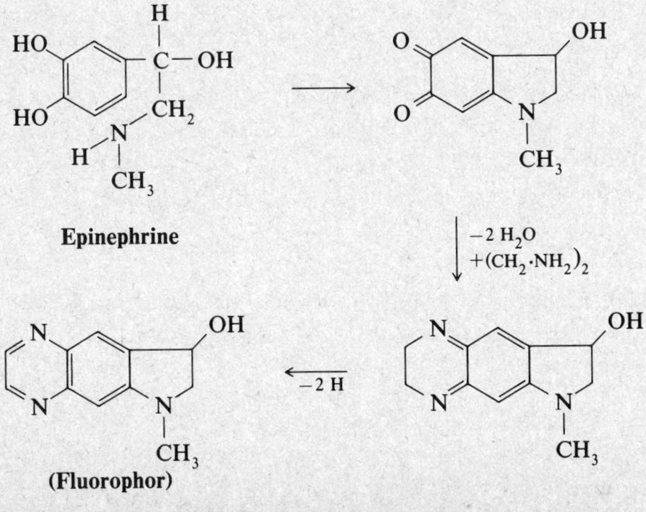

Epinephrine

$-2\ H_2O$
$+(CH_2 \cdot NH_2)_2$

$-2\ H$

(Fluorophor)

The reaction of norepinephrine with ethylenediamine is more complex and not yet entirely resolved. Epinephrine gives but one fluorescent product, and norepinephrine gives two major products and a third minor product. According to the present concept, norepinephrine is oxidized to noradrenochrome; EDA then condenses with this compound, resulting in ring formation and the elimination of the side chain of norepinephrine and 2 H. The compound then condenses with another molecule of EDA to produce the compound shown below as one of two principal fluorescent products formed; the second product has not been identified.

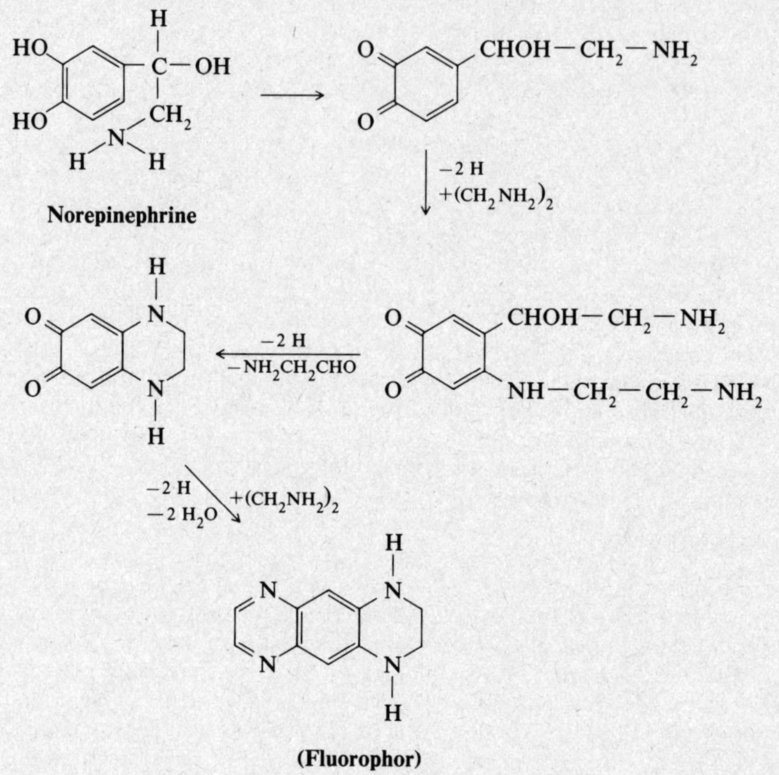

(Fluorophor)

Note that norepinephrine reacts with 2 mol of EDA and loses its side chain, whereas epinephrine reacts with 1 mol of EDA and retains its side chain as part of the final product. The fluorescent products of epinephrine and norepinephrine have different emission spectra, allowing simultaneous measurement of both hormones without prior separation.

The most satisfactory EDA method is that described by Weil-Malherbe.[325] A general outline of the method is given below.

Specimen collection and storage. Venous blood is collected in a tube containing 1 g EDTA and 2 g sodium thiosulfite/dL at pH 7.4. Heparin can also be used as anticoagulant. Specimens should be chilled immediately in ice water. Plasma should be separated in a refrigerated centrifuge and frozen immediately at $-70\,°C$ in plastic vials. For further discussions on specimen collection and storage, see under Comments below.

Procedure[118,182]

1. Adjust a measured volume of plasma to pH 8.4 with Na_2CO_3, 0.5 mol/L, and pass through a column (5 mm I.D.) prepared with acid-washed, activated alumina.

2. Wash column with water and elute catechols with acetic acid, 0.2 mol/L, followed by water.

3. Mix the eluate with EDTA (1 g/dL) and adjust to pH 6.0.

4. Transfer the mixture onto a column containing Amberlite CG-50, wash column with water, and elute with acetic acid, 1 mol/L.

Note: Perform condensation reaction in red glassware to exclude the effects of light upon the fluorophors, since the product formed from norepinephrine is unstable in light.

5. Mix the eluate with redistilled ethylenediamine (EDA), shake vigorously, and keep in a water bath at 55 °C.

6. Cool mixture, saturate with sodium chloride, and extract with isobutanol.

7. Prepare standard solutions containing epinephrine and norepinephrine together with two blanks; one blank (column blank) is processed through the entire procedure, and the other (reagent blank) through the condensation reaction only. Measure fluorescence at emission wavelengths of 510 nm (reading *b*) and 580 nm (reading *y*) using the same activation wavelength of 420 nm. Correct the readings for column blank and the standards for reagent blank.

The fluorescence of the product formed from norepinephrine at 510 nm is twice as great as that from epinephrine; at 580 nm the ratio is reversed. The content of epinephrine and norepinephrine is calculated using the following equations:

$$E + \frac{N}{m} = y$$

$$E + \frac{N}{n} = b$$

where   m = ratio of fluorescence of E/N standards at 580 nm
  n = ratio of fluorescence of E/N standards at 510 nm
  E = amount of epinephrine in sample
  N = amount of norepinephrine in sample
  y = amount of apparent epinephrine in sample indicated by the reading at 580 nm
  b = amount of apparent epinephrine in sample indicated by the reading at 510 nm

Hence, $N = mn(b - y)/(M - n)$ and $E = y - N/m$.

Comments. Blood samples should be taken as quickly as possible and the plasma separated and frozen immediately; small delays at this stage cause significant losses in catecholamine content. Prolonged storage of frozen plasma samples and repeated thawing and freezing also result in losses. Venipuncture can also raise blood catecholamine content. When multiple estimations are necessary, samples should be taken through an indwelling venous catheter. Certain common drugs (e.g., ampicillin, methyldopa, promethazine, protamine, sulfonamides, vitamin B complex) and beverages (e.g., cocoa, coffee, tea) interfere with this fluorometric method for catecholamine estimation. Failure to standardize rigorously such factors as posture and avoidance of interfering substances may invalidate the most careful biochemical analysis and confound interpretation of the data.

Reference ranges. With the procedure outlined above, the ranges in the plasma from 37 healthy subjects for plasma epinephrine and norepinephrine were 140–300 pg/mL and 360–800 pg/mL, respectively.

Radioenzymatic Assay

Principle. The principle of this technique is based upon the use of the enzyme catechol-O-methyltransferase (COMT) to transfer a radioactive methyl group from the tritiated reagent, S-adenosyl-methionine (^{3}H-SAM), to catecholamines present in the sample, thereby forming their corresponding labeled methylated derivatives. A variation of this assay for the estimation of norepinephrine has been the use of the enzyme phenylethanolamine-N-methyltransferase (PNMT) to transfer a radioactive methyl group from SAM to the amine moiety of the norepinephrine molecule. The following is a brief description of the simultaneous single-isotope radioenzymatic assay of plasma norepinephrine (NE), epinephrine (E), and dopamine (DA) reported by Peuler and his coworkers.[227,236] A commercial kit based on this technique is also available (CAT-A-KIT, Upjohn Co., Kalamazoo, MI 49001).

The procedure consists of simultaneous conversion of E, NE, and DA in the sample (plasma or cerebrospinal fluid) to their corresponding ^{3}H-methoxy derivatives by the use of partially purified COMT in the presence of ^{3}H-SAM.

Dopamine → ^{3}H-3-Methoxytyramine

Norepinephrine → ^{3}H-Normetanephrine

Epinephrine → ^{3}H-Metanephrine

A duplicate incubation mixture containing a second aliquot of the sample and known amounts of NE, E, and DA as internal standards is run in parallel with the sample.

Following the formation of O-methylated derivatives, the products are extracted and separated by thin-layer chromatography. The direct measurement of the radioactivity of the ^{3}H-3-methoxytyramine fraction provides quantitation for dopamine. The separated ^{3}H-normetanephrine and ^{3}H-metanephrine are further converted by periodate oxidation to ^{3}H-vanillin, which is directly proportional to the amount of NE or E.

R = H ^{3}H-Normetanephrine
R = CH$_3$ ^{3}H-Metanephrine

^{3}H-Vanillin

Estimates of individual endogenous catecholamines in a plasma sample are determined by comparison with the radioactivity produced by corresponding internal standards in the duplicate sample.

Total plasma catecholamines are measured using a similar assay procedure except for omitting the separation step by thin layer chromatography. The assay system is sensitive to 1 pg (20 pg/mL of plasma) for NE and E and 6 pg (120 pg/mL) for DA. With this procedure, the plasma concentrations in supine normotensive adults were 111–603 pg/mL for norepinephrine, 0–62 pg/mL for epinephrine, and 0–83 pg/mL for dopamine.

Determination of Catecholamines and Their Metabolites In Urine

A large number of methods involving colorimetry,[68] fluorometry,[325] gas chromatography,[38] and HPLC[141] for measuring catecholamines and their metabolites in urine have been proposed. Colorimetric methods for catecholamines are not sufficiently sensitive, and the current methods of choice are those based on fluorometry and HPLC. However, for the analyses of such metabolites as metanephrines and VMA, the colorimetric procedures are widely used in clinical laboratories. Numerous HPLC methods have now become available.[20,93,210,223,273] These methods are replacing the cumbersome and less specific fluorometric and colorimetric techniques for the analysis of catecholamines and their metabolites.

Metanephrine and Normetanephrine

The colorimetric method described by Pisano[240] has found wide application in clinical laboratories. After acid hydrolysis, metanephrines are adsorbed on Amberlite CG-50 and eluted with NH$_4$OH, 3 mol/L. The eluted compounds are then converted to vanillin by periodate oxidation. (See VMA determination, p. 1152.) The resulting compound is then assayed spectrophotometrically at 360 nm. This method does not distinguish between metanephrine and normetanephrine but does serve as a rapid, easy, and reliable method for screening urine for pheochromocytoma. The reference range for total metanephrines is reported to be 0.6 mg/d ±0.3 (SD).

The gas-chromatographic method[38] allows estimation of metanephrines and VMA in the same urine sample and appears to be simple, accurate, and suitable for clinical application. Acidified urine (pH 1.0) is extracted with ethyl acetate; the organic extract is saved for VMA estimation. The aqueous phase is subjected to hydrolysis by hydrochloric acid, followed by oxidation of the metanephrines to vanillin with periodate at pH 10–11. The vanillin is extracted with methylene dichloride containing the internal standard, p-hydroxybenzoic ethyl ester, and is then quantitated as a trimethylsilyl (TMS) derivative by gas chromatography using a 3% OV-1 column.

The ethyl acetate phase containing VMA is extracted with acetate buffer, pH 6.2, and VMA is oxidized to vanillin by sodium periodate at pH 11–12. After extraction into methylene dichloride, vanillin is estimated by gas chromatography as above. Results obtained by gas chromatography have been compared with those of Pisano's spectrophotometric procedure.[38] While the values for VMA by both methods were very similar, the quantities of metanephrines measured by GLC were substantially lower than those obtained by chemical methods. Such a disparity has prompted the authors to conclude that Pisano's procedure overestimates metanephrines because of the presence of interfering contaminants.

Several methods based on HPLC with electrochemical detection have been reported.[20,223,273] Most of these procedures are based on modifications of the method described by Shoup and Kissinger[273] and measure not only metanephrines but also homovanillic acid (3-methoxytyramine). In principle, these methods consist of isolating the urinary metabolites of interest by passing the hydrolyzed urine through small ion-exchange columns. Samples are then concentrated by solvent extractions and injected onto the liquid chromatographic column. Concentration and instrument response are linear up to 2.0 mg/L for all three compounds; practical detection limits are ~20 μg/L. The combination of the extraction procedure, HPLC separation, and electrochemical detection makes the analyses highly selective. Within-assay and between-assay coefficients of variation for the three compounds are ~5 and 7%. The *reference ranges* are 105–354 μg/d for normetanephrine, 74–297 μg/d for metanephrine, and 66–222 μg/d for homovanillic acid.

3-Methoxy-4-Hydroxy Phenylglycol (MHPG)

As discussed above, MHPG is an important metabolite of norepinephrine and may have particular relevance to the disposition of norepinephrine in brain. In humans, MHPG is found in the CSF, urine, and plasma. Because of methodological considerations and the ease of collection of specimens, urinary measurements of MHPG have found wide clinical applications. Although MHPG may be assayed by different methods, including HPLC,[272a] the GLC method of Dekirmenjian and Maas[65] is commonly used. This procedure consists of enzyme hydrolysis of urinary conjugates, followed by organic solvent extraction, acetylation, and injection of the acetylated derivative into a gas chromatograph equipped with a glass column packed with 6% QF-1 coated on Anakrom ABS. The detector is an electron capture system with 150 mCi of tritiated foil. According to the authors, the procedure provides a high degree of recovery, reproducibility, and specificity.

Reference range. The mean 24-h MHPG urinary excretion for 5 normal males was 1600 ± 380 μg (±SD) and for 6 normal females 1320 ± 360 μg (±SD).

Homovanillic Acid (HVA)

HVA has been identified as the principal urinary metabolite of dopa and dopamine. The lack of suitable methods for measuring dopamine in blood and urine has aroused interest in the estimation of urinary HVA, which is excreted in free form in relatively large amounts. The application of gas chromatography[333] and HPLC[93,273] allows simultaneous determination of both HVA and VMA. HVA is the major terminal metabolite of dopamine, and VMA is the major terminal metabolite of norepinephrine. Since dopamine or norepinephrine may be elevated in patients with neuroblastoma, it is diagnostically important to determine both HVA and VMA.

In addition to the HPLC method[273] already discussed, the general outline of the GLC method described by Williams and Greer[333] is given below.

A volume of urine containing 10 mg of creatinine is acidified to pH 1–2, saturated with NaCl, and extracted five times with equal volumes of diethyl ether. The ether extracts are pooled and evaporated to dryness under a stream of nitrogen. The residue is acetylated with acetic anhydride and trifluoroacetic acid at 75 °C for 30 min. After evaporation of the acetylating mixture to dryness, the residue is then methylated with ethereal diazomethane to form methylacetoxy esters of HVA and VMA. The sample (200 μg creatinine equivalent) is injected into a gas chromatograph containing a nonpolar, 3% OV-1 column. Either argon or flame-ionization detectors can be used. If a large peak is observed at the exact retention time of either the HVA standard or the VMA standard, the identity of the peak is further checked by repeating the analytical run using a polar column (10% EGA). Quantitation is carried out by comparing the peak area of the standard with that of the sample. The specificity of measurements has also been confirmed by analyzing the sample in a combined gas chromatograph–mass spectrometer.

Reference ranges are 1–40 μg HVA/mg creatinine, and 1.5–7.1 μg VMA/mg creatinine. Urinary excretion is expressed in terms of μg/mg creatinine to allow random specimens of urine to be used when complete collections are difficult to obtain. In addition, variations due to sex and age are minimized.

Determination of Urinary Free Catecholamines (Fractionated) by HPLC-EC

Until recently, measurements of urinary total catecholamines[324] and VMA[241] were common routine clinical procedures. With the advent of HPLC with electrochemical detection, rapid and reliable determinations of individual catecholamines are possible. The following is a description of a method for urinary free catecholamines based on high-performance liquid chromatography with electrochemical detection (HPLC-EC) originally reported by Kissinger and his coworkers.[141,212] The colorimetric procedure of Pisano et al.[241] for the quantitation of VMA is also given in detail.

Principle

Catecholamines are excreted in the urine as free amines, and as glucuronide and sulfate conjugates. In order to measure the microgram quantities of the free (unconjugated) catecholamines, preliminary removal of interfering substances is necessary. Twenty-four–hour urinary excretion of total catecholamines (conjugated and unconjugated forms) may be measured by hydrolyzing the urine prior to assay. However, the determination of unconjugated amines is clinically more useful because they closely reflect the rate of their endogenous production.

Following protein precipitation with perchloric acid, an aliquot of a 24-h urine collection (preserved in acid) is first applied to a weak-acid cation exchange resin. Unconjugated catecholamines are selectively adsorbed at pH 6.5 and then eluted with dilute boric acid, pH 4.0. An intermediate water wash removes interfering urine impurities. Subsequent resolution of the individual catecholamines is achieved by reverse-phase, paired-ion high-performance liquid chromatography (HPLC) under optimized isocratic conditions. Alkyl-bonded silica is used as the nonpolar stationary phase, and an organic/aqueous buffer mixture (pH 2.8) is used as the polar mobile phase. To enhance the affinity of the polar catecholamines for the hydrophobic stationary phase, an ion of opposite charge (octyl sodium sulfonate) is also included in the mobile phase. This "counter-ion" is capable of forming uncharged ion pair conjugates with catecholamine cations prior to partitioning into the lipophilic stationary phase:

$$\underset{\substack{\text{Catecholamine}\\\text{cation}}}{RNH_3^+} \quad + \quad \underset{\substack{\text{Sulfonate}\\\text{counter-ion}}}{CH_3(CH_2)_7SO_3^-} \longrightarrow \underset{\substack{\text{ion pair}}}{\text{Neutral}} \left(\substack{\text{Partition}\\\text{chromatography}} \right)$$

This separation mechanism may also be described by postulating that the counter-ion itself partitions into the stationary phase with its ionic groups oriented at the surfaces:

(Ion-exchange chromatography)

The reverse-phase column then has the physical characteristics of a conventional ion-exchange resin.

A thin-layer glassy carbon or carbon-paste working electrode, in conjunction with an Ag/AgCl reference electrode and a stainless steel auxiliary electrode, are utilized as the amperometric detection system. Each catecholamine passing through the detector cell undergoes a rapid two-electron oxidation at a fixed potential to form an o-quinone:

$R = -\ CH_2CH_2NH_2$, dopamine

$= -\ \underset{\underset{OH}{|}}{CH}CH_2NH_2$, norepinephrine

$= -\ \underset{\underset{OH}{|}}{CH}CH_2\underset{\underset{CH_3}{|}}{NH}$, epinephrine

The current resulting from this reaction is converted to a voltage signal and monitored as a function of time. At a constant temperature and flow rate, this oxidation current is directly proportional to the concentration of the analyte.

Catecholamine standards, previously checked for purity, are used to calibrate the system on the basis of peak heights and retention times. To calculate sample concentrations, peak height ratios relative to the internal standard, dihydroxybenzylamine, for unknowns are compared to those of standards.

Specimen Collection and Storage

All antihypertensive medications should be withheld from the patient for at least 2 d prior to and during specimen collection. If the patient cannot be completely removed from a drug regimen, the test may still be performed with appropriate judgment applied to the results based on the expected physiological response to the drug(s).

A complete 24-h urine is collected in a gallon jug containing 10 mL HCl as a preservative. The specimen should be refrigerated during collection. Upon receipt of the specimen, adjust pH between 2 and 5 by adding HCl, 6 mol/L. The total urine volume is measured and recorded; a 100-mL aliquot is stored in the refrigerator until the test is performed or may be frozen and stored indefinitely.

Reagents

1. Perchloric acid $(HClO)_4$, 72%. Analytical grade.
2. Phosphoric acid (H_3PO_4), 85%. Analytical grade.
3. Sodium phosphate, 10 mmol/L. Dissolve 1.42 g anhydrous Na_2HPO_4 in distilled water to final volume of 1 L. Stable at 4 °C for 3 months.
4. Potassium phosphate, 10 mmol/L. Dissolve 1.36 g KH_2PO_4 in distilled water to a final volume of 1 L. Stable at 4 °C for 3 months.
5. Phosphate buffer, 10 mmol/L, pH 6.5. Mix 32 mL Na_2HPO_4 (10 mmol/L), and 68 mL KH_2PO_4 (10 mmol/L), to give 100 mL of a phosphate buffer, pH 6.5. Stable at 4 °C for 3 months.

6. Disodium ethylenediaminetetraacetate, 2.7 mmol/L. Dissolve 1 g Na_2EDTA in 1 L distilled water. Stable at room temperature for 3 months.

7. Sodium hydroxide, 0.5 mol/L. Dissolve 20.0 g NaOH in distilled water and dilute to 1 L. Store tightly capped. Prepare monthly.

8. Boric acid, 0.65 mol/L. Dissolve 40 g H_3BO_3 in 800 mL distilled water. Warm in a 50 °C water bath to aid dissolution. Allow to cool and dilute to 1 L with water. Store at room temperature and prepare every two weeks.

9. Catecholamine stock standard. Weigh and transfer the following catecholamines to a 500-mL volumetric flask: 30.7 mg norepinephrine bitartrate · monohydrate (M.W. 346.3); 9.1 mg epinephrine bitartrate (M.W. 333.3); 55.8 mg dopamine HCl (M.W. 189.6). Dissolve and dilute to the mark with perchloric acid, 0.1 mol/L, to give the following free base concentrations: norepinephrine, 30 mg/L; epinephrine, 10 mg/L; dopamine, 90 mg/L. Store the stock standard at 4 °C. Stable for at least 6 months.

Note: Free base concentrations are calculated by multiplying the mass of the catecholamine salt by the molecular weight ratio of free base to salt. For norepinephrine,

$$\left(\frac{30.7 \text{ mg norephinephrine bitartrate}}{0.5 \text{ L perchloric acid}}\right) \times \left(\frac{\text{M.W. free base} = 169.2}{\text{M.W. salt} = 346.3}\right) = 30 \text{ mg/L (free base concentration)}$$

10. Internal standard, 3,4-dihydroxybenzylamine, 10 mg/L. Dissolve 8 mg 3,4-dihydroxybenzylamine HBr (Sigma D-7012, M.W. 220.1) in perchloric acid, 0.1 mol/L. Dilute to 500 mL in a volumetric flask to give a free base concentration of 10 mg/L. Store at 4 °C. Stable for at least 6 months.

11. Catecholamine calibration standard. Add 50 μL of the catecholamine stock standard and 50 μL of the internal standard solution to 7.0 mL boric acid, 0.65 mol/L. Prepare fresh daily.

12. Methanol, spectrograde.

13. Sodium octyl sulfate.

14. Phosphoric acid, 0.1 mol/L. Dilute 6.9 mL H_3PO_4, 85%, to 1 L with distilled water. Store at room temperature.

15. Mobile phase, methanol/phosphate buffer (10/90). Dissolve 21.5 g KH_2PO_4, 74 mg Na_2EDTA, and 100 mg sodium octyl sulfate in 1 L distilled water. Add 3.3 mL phosphoric acid (85%) and dilute to 2 L with distilled water. Buffer is phosphate, 0.1 mol/L, pH 2.8; octyl sodium sulfate, 50 mg/L; and disodium EDTA, 0.1 mmol/L. Filter the buffer through a Millipore filtration apparatus using a 2-μm membrane filter. De-gas for 1 h with a vacuum and then add methanol (100 mL/L buffer). Stir the mixture for 10 min at 40 °C. Further de-gas the mixture for 10 min by slowly bubbling nitrogen gas through the mixture. Store tightly capped at room temperature.

Note: Changes in the composition of the mobile phase will be necessary as the HPLC column ages and catecholamine retention times decrease. To maintain peak integrity (i.e., to increase retention time), concentration of methanol is decreased and that of the ion pair reagent increased over the lifetime of the column (5–6 months). With a new column, begin with mobile phase containing 50 mg/L octyl sulfate and 10% methanol. Adjust the methanol concentration to sharpen peaks as well as optimize retention times. Allow 1–2 h between adjustments for the column to equilibrate with the mobile phase. Full equilibration will require 3–4 h. Overnight equilibration may be more convenient.

16. BioRex-70 exchange resin (50–100 mesh, BioRad Laboratories). Prepacked columns may be obtained from BioRad. Alternatively, the resin may be washed and packed into plastic columns as follows:

a. With gentle manual agitation, wash the resin with successive volumes of hydrochloric acid, 3 mol/L; sodium hydroxide, 3 mol/L; acetic acid, 3 mol/L; ammonium acetate, 1.0 mol/L (pH 6.5); and ammonium acetate, 0.1 mol/L (pH 6.5).

b. Prepare a slurry of about 3 g resin and 15 mL ammonium acetate buffer, 0.1 mol/L (pH 6.5).

c. Pipet the resin slurry into a plastic chromatography column, 1 × 8 cm.

d. Allow the resin to pack in each column to a height of about 2.5 cm.

e. Store packed columns at 4 °C after carefully sealing both tip and cap with Parafilm. Stable for at least 1 year.

17. Nitrogen gas, prepurified, 99.995% minimum purity.

Equipment. Commercially available HPLC system, equipped with 25 cm × 0.4 cm (I.D.) reverse phase ODS C_{18} (average particle diameter 5 or 10 μm) column and an electrochemical detector, is suitable.

Controls. Three urine controls are included in every run. For example, Ortho urine controls I and II (Ortho Diagnostics, Inc., Raritan, NJ 08869) are reconstituted each day and processed for immediate use. A 24-h urine collection from a healthy individual is also used as a control pool. This urine is adjusted to pH 3 with hydrochloric acid, 6 mol/L, and spiked with the catecholamine stock standard, 10 mL/L urine. Six-mL aliquots of the spiked urine are frozen in polypropylene vials and stored for future test runs.

Procedure

Sample Preparation

1. Into separate 50-mL centrifuge tubes, pipet 25.0 mL of each urine and urine control.
2. Add 1.0 mL concentrated perchloric acid. Cap and vortex for 15 s. Allow to stand for 10 min and then centrifuge for 5 min at 900 × g.
3. Pipet 5.0 mL of each supernatant into 50-mL plastic beakers. Store the remaining supernatants in the freezer.
4. Transfer 5.0 mL of phosphate buffer, 10 mmol/L (pH 6.5), to a 50-mL beaker and add 50 μL of the catecholamine stock solution.
5. To all beakers, add 50 μL of the internal standard solution (3,4-dihydroxybenzylamine, 10 mg/L) and 15 mL of Na$_2$EDTA, 2.7 mmol/L. Adjust the pH to 6.5 ± 0.1 with NaOH, 0.5 mol/L, or phosphoric acid, 0.1 mol/L.

Ion-Exchange Chromatography

6. Prepare one BioRex-70 ion-exchange column for each urine, urine control, and standard to be run. Insert the column into a suitable support rack and allow to drain into a drainage tray.
7. Quantitatively transfer the entire contents of each beaker onto separate columns. Then rinse each beaker with ~5 mL water. Apply the rinse solutions to the respective columns and allow to drain completely.
8. Wash the columns with two 10-mL portions of water to remove urine contaminants.
9. When completely drained, add 7.0 mL boric acid, 0.65 mol/L, to each column and collect the eluate in clean polypropylene vials, 17 × 100 mm.
10. Cap and mix the vials by inversion.
Note: At this time, samples may be stored at 4 °C for up to 48 h before injection into the HPLC system.

High-Performance Liquid Chromatography

11. Establish operating conditions of the chromatographic and detection systems as instructed in the manufacturer's manual.
12. Prior to the introduction of specimens onto the reverse-phase HPLC column, adjust the pH of each sample to 2.8 ± 0.1 with phosphoric acid. Follow the injection procedure as described in the manufacturer's manual.

Calculations

1. Identify catecholamine and internal standard peaks by relative retention times (minutes). Measure peak heights (namp) directly from the chromatographic tracings.

Example:

Calibration Standard	Retention Time (min)	Peak Heights ht(namp)
Norepinephrine	6.0	56
Epinephrine	8.8	14
Internal Standard	11	18
Dopamine	17	108

2. Calculate ratio of peak heights (pk ht of catecholamine/pk ht of internal standard) for each urine, urine control, and standard.
3. Calculate concentrations of unknown catecholamines:

$$CAT = \frac{U}{S} \times C \times 0.0104$$

where   CAT = unknown catecholamine concentration (μg/mL)
      U = peak height ratio for unknown catecholamine
      S = peak height ratio for corresponding catecholamine stock standard applied to ion-exchange resin
      C = free base concentration of catecholamine stock standard solution (mg/L = μg/mL)
Factor 0.0104 is derived as follows:

$$\frac{0.05}{5.0} \times \frac{26.0}{25.0} = 0.0104$$

where   0.05 = volume (mL) of catecholamine stock standard applied to ion-exchange column
         5.0 = volume (mL) of deproteinized urine applied to exchange column
        25.0 = 24-h urine aliquot (mL)
        26.0 = total volume (mL) of deproteinized mixture (urine and perchloric acid)
    4. Calculate 24-h catecholamine excretion:

$$CAT/d = CAT \times TV$$

where CAT, $\mu g/d$ = catecholamine excretion per day
      CAT, $\mu g/mL$ = unknown catecholamine concentration
      TV, mL   = 24-h urine volume

Comments

For the quantitation of catecholamines in urine, a preliminary purification is needed. For this purpose, cation-exchange resin as described above and adsorption chromatography on aluminum oxide are extensively used. However, according to some investigators,[289] such methods are not very specific and do not yield complete recoveries. The use of affinity chromatography on immobilized boric acid for the isolation of catecholamines from urine is claimed to be reproducible, specific, and easy to perform.

Until recently, nearly all electrochemical detectors for catecholamine analysis by HPLC consisted of carbon-paste electrodes. Carbon paste is composed of very finely divided graphite and a binder, mixed thoroughly to the consistency of a semidry paste. The binder merely holds the graphite together as an integral unit and should be inert, both chemically and physically, to maintain sensitive electrode performance with minimal noise. Paraffin oil and silicone grease have proved most satisfactory and popular in this regard. However, the choice of the binder depends on its solubility in the mobile phase; acetonitrile, a popular constituent of many mobile phases, dissolves paraffin oil; carbon paste using paraffin oil as binder would be obviously unsuitable under this condition. More recently, glassy carbon has come into use as a popular electrode for HPLC-EC. Glassy carbon is a hard, brittle, solvent-impervious electrode material. Since it is 100% carbon, the chemical and physical resistance of glassy carbon to all mobile phases is unequaled. Detector characteristics such as linearity and electron transfer property are comparable for both the electrodes. However, the carbon-paste electrode under optimal operating conditions provides higher sensitivity than that obtained by the glassy carbon electrode.

Reference Ranges[212]

	Age	Urinary Excretion Rate ($\mu g/d$)		
		Norepinephrine	Epinephrine	Dopamine
Adult:	>15 years	14–80	0.5–20	65–400
Children:	0–1 years	0–10	0–2.5	0–85
	1–2 years	0–17	0–3.5	10–140
	1–4 years	4–29	0–6.0	40–260
	4–7 years	8–45	0.2–10	65–400
	7–10 years	13–65	0.5–14	65–400
	10–15 years	15–80	0.5–20	65–400

Determination of VMA (Vanillylmandelic Acid, 3-Methoxy-4-Hydroxymandelic Acid) in Urine[24]

Principle

VMA and other phenolic acids are extracted from acidified urine with ethyl acetate and then re-extracted into aqueous potassium carbonate solution. Addition of sodium metaperiodate to the carbonate extract, oxidizes VMA to vanillin.

Vanillin is then separated from contaminating phenolic acids by selective extraction into toluene and back-extraction into carbonate. The vanillin concentration is determined spectrophotometrically at a wavelength of 360 nm.

Specimen Collection and Storage

The intake of chocolate, coffee, bananas, foods containing vanilla, citrus fruits, and drugs such as aspirin and antihypertensive agents (e.g., Aldomet) may cause falsely high values for urinary VMA with some methods. No dietary restrictions during urine collection are necessary if the VMA method based on the oxidation of VMA to vanillin is used. However, for methods that employ a reaction of the phenolic acids with diazotized p-nitroaniline, rigid control of diet and drugs is still necessary.

The pH of the urine should be kept at ~2 during the collection by placing 10 mL of HCl, 6 mol/L, into a suitable container (dark-brown bottle). After measurement of the total volume, 100-mL aliquots may be stored at 4 °C for subsequent analysis. The specimen so preserved is stable for several weeks.

Reagents

All reagents should be of reagent grade purity.

1. Hydrochloric acid, 6 mol/L. Slowly add 500 mL concentrated HCl to a 1-L volumetric flask containing ~300 mL distilled water and dilute to mark with water.

2. Sodium chloride.

3. Ethyl acetate.

4. Potassium carbonate, 1 mol/L. Dissolve 138 g potassium carbonate in 1 L of distilled water. Store at room temperature and discard after one month.

5. Sodium metaperiodate, 2 g/dL, in distilled water. Make fresh daily.

6. Sodium metabisulfite, 10 g/dL, in distilled water. Prepare fresh daily.

7. Acetic acid, 5 mol/L. Dilute 286 mL glacial acetic acid with distilled water to 1 L.

8. Phosphate buffer, 1 mol/L, pH 7.5. Solution A: dissolve 268 g disodium phosphate ($Na_2HPO_4 \cdot 7 H_2O$) in distilled water and dilute to 1 L. Store in a refrigerator. Solution B: dissolve 27.22 g potassium dihydrogen phosphate (KH_2PO_4) in 200 mL distilled water. Mix 168.2 mL of solution A with 31.8 mL of solution B. Check pH on pH meter and make any necessary adjustment to obtain a pH of 7.5. Store at room temperature and discard after three months.

9. Hydrochloric acid, 0.01 mol/L. Dilute 0.83 mL of concentrated HCl to 1 L with distilled water.

10. Standard solutions.

Stock solution of VMA, 1 mg/mL. Accurately weigh 100 mg of VMA and dissolve in 100 mL of HCl, 0.01 mol/L, in a volumetric flask. The solution is stable approximately three months under refrigeration.

Working solution, 10 μg/mL. Dilute 1 mL of the stock solution to 100 mL with HCl, 0.01 mol/L. Prepare fresh before use.

Procedure

1. Pipet 0.2% of the 24-h volume into duplicate 50-mL glass-stoppered or Teflon-lined, screw-capped centrifuge tubes marked as "tests," "internal standards," and "unoxidized blanks." To the internal standard tubes add 1 mL of the working standard.

2. Dilute the contents of all tubes to 5.5 mL with distilled water, and further acidify with 0.5 mL of HCl, 6 mol/L.

3. Add a saturating amount of sodium chloride (~3 g), mix, and extract with 30 mL of ethyl acetate by shaking on a mechanical shaker for 30 min. Centrifuge for 5 min.

4. Transfer 25 mL of the organic extract (upper layer) to a second glass-stoppered centrifuge

tube containing 1.5 mL of potassium carbonate, 1 mol/L. Shake mechanically for 3 min and centrifuge for 5 min. Aspirate and discard the upper organic phase.

5. Pipet 1 mL of the carbonate phase (lower layer) to a third glass-stoppered centrifuge tube.

6. To the test and standard tubes, add 0.1 mL of sodium metaperiodate, 2 g/dL, mix, and stopper loosely; place all tubes, including the tubes marked "unoxidized blank" (metaperiodate solution is omitted at this stage), into a water bath at 50 °C for 30 min.

7. At the end of the incubation period, remove the tubes and cool to room temperature.

8. To the "unoxidized blank" tubes, add 0.1 mL of sodium metaperiodate and mix.

9. Without delay, add to all tubes 0.1 mL of metabisulfite solution to reduce residual periodate.

10. Add 0.3 mL of acetic acid, 5 mol/L, to neutralize all solutions. Allow to stand for 10 min.

11. Add 0.6 mL of phosphate buffer, 1 mol/L, pH 7.5. (The pH can be checked at this point by adding one drop of aqueous cresol red, 0.04 g/dL. The solution should be yellow, indicating a pH of < 8.8.)

12. Shake mechanically for 3 min with 20 mL of toluene to extract vanillin, the oxidized product of VMA.

13. Centrifuge for 5 min, and transfer 15 mL of the toluene extract into a fourth glass-stoppered centrifuge tube containing 4.0 mL of potassium carbonate, 1 mol/L. Shake mechanically for 3 min and centrifuge for 5 min.

14. Transfer the carbonate layer containing vanillin into a microcuvet, and determine the absorbance at 360 nm against a water blank.

Calculations

$$\text{mg VMA/d} = \frac{A_t - A_b}{A_{st} - A_t} \times \frac{10}{1000} \times \frac{100}{0.2} = \frac{A_t - A_b}{A_{st} - A_t} \times 5$$

where A_b = absorbance of "unoxidized" urine blank
A_t = absorbance of test
A_{st} = absorbance of internal standard (standard + test)
10 = concentration of standard
1000 = converts standard from μg to mg
0.2 = urine aliquot used

Comments

Care in the collection and preservation of the urine is very important. Diets and drugs contributing to the excretion of related phenoxy acids that may be oxidized to vanillin may yield falsely elevated results. However, as a precautionary measure, an unoxidized blank is prepared for every sample to correct for the presence of vanillin in urine, even when the dietary restrictions prior to and during collection of the specimen have been followed. The absorbance may be measured against the unoxidized blank instead of the water blank, and in that case the need for subtraction of the absorbance of the urine blank from the absorbance of the test samples is obviated.

The internal standard compensates for procedural losses, for decomposition of vanillin, and for the relative inhibition of its formation by unknown urinary factors. Indeed, at room temperature the oxidation of VMA by periodate proceeds smoothly in pure solutions, whereas an elevated temperature (50 °C) is required for the oxidation of VMA in urinary extracts. In occasional urine samples, the oxidation may be strongly inhibited even at 50 °C.[324]

The oxidation of VMA to vanillin is also sensitive to hydrogen ion concentration. In neutral and acidic solutions, oxidation results in the formation of a yellow pigment; strongly alkaline solutions, on the other hand, delay the formation of vanillin and cause its decomposition. Optimal conditions are obtained in the presence of 1–15 g sodium or potassium carbonate solution/dL urine at an approximate pH of 11.[241] The maximum absorption of vanillin occurs at 348 nm. However, at this wavelength there is considerable absorbance of the oxidation product (*p*-hydroxybenzaldehyde) of *p*-hydroxymandelic acid, a normal constituent of urine. This necessitates measurement at 360 nm, where the absorbance of vanillin is 80% of its peak value and interference is minimal. The absorbance of vanillin drops sharply between 350 and 380 nm, so the wavelength setting must remain exactly at 360 nm.[241]

Reference Ranges[307]

	mg/d
Newborn:	< 1.0
Neonate:	< 1.0
Infant:	< 2.0
Child:	1–5
Adolescent:	1–5
Thereafter:	2–7 (or 1.5–7.0 μg/mg creatinine)

SEROTONIN AND 5-HYDROXYINDOLEACETIC ACID (5-HIAA)

Biosynthesis and Metabolism

Serotonin (5-hydroxytryptamine, 5-HT), a powerful smooth-muscle stimulant and vasoconstrictor, is a derivative of the amino acid tryptophan. It is transported in the blood by the platelets. The formation and breakdown of serotonin is depicted in Figure 9-51. The essential amino acid tryptophan is hydroxylated to form 5-hydroxytryptophan (5-HTP). Approximately 1–3% of dietary tryptophan is normally metabolized by this pathway. The 5-HTP is decarboxylated to serotonin (5-hydroxytryptamine, 5-HT). The enzymatic decarboxylation is very active in carcinoid tumors.

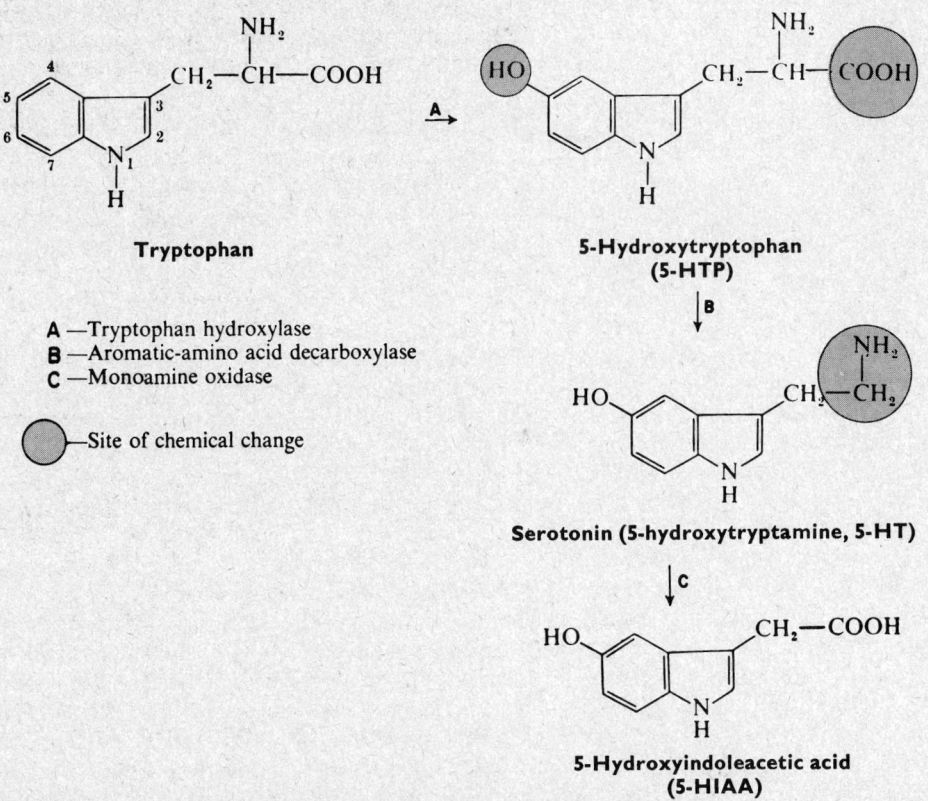

A —Tryptophan hydroxylase
B —Aromatic-amino acid decarboxylase
C —Monoamine oxidase

—Site of chemical change

Figure 9-51. Biosynthesis and metabolism of serotonin.

Pharmacologically, 5-HT is the most active indole amine; however, its biological activity is apparently lost when it is bound to tissues or platelets. It may rapidly undergo oxidative deamination in a tumor or in the blood after release from a tumor. The oxidative deamination of serotonin by the enzyme monoamine oxidase (MAO) leads to the formation of 5-hydroxyindoleacetic acid (5-HIAA), which is quantitatively the most significant metabolite of the 5-hydroxyindole pathway. The majority of the 5-HIAA is excreted in the free form, although a small amount may be conjugated as the O-sulfate ester before excretion.

Clinical Significance of Measurements of Serotonin and Metabolites

Depression. The most important physiological role of serotonin is that of a transmitter in neurons (serotoninergic neurons) within the brain. In humans, serotonin has been implicated in a variety of behavioral patterns including sleep, perception of pain, social behavior, and mental depression.[24] However, the role of serotonin in depression is, at present, a subject of active investigation. Evidence for this relationship comes from numerous reports of diminished levels of 5-HIAA in the cerebrospinal fluid (CSF) of depressed patients and from the observation that 5-HTP, a precursor of 5-HT, is an effective antidepressant only in depressed patients with decreased CSF 5-HIAA. Furthermore, in such serotonin-deficient patients, treatment with antidepressant drugs such as amitriptyline and trazodone gives a favorable clinical response.[178] These pharmacological agents selectively inhibit the reuptake of 5-HT by the presynaptic neurons, thereby increasing the concentration of 5-HT in serotoninergic synapses.

Carcinoid Tumors. Serotonin is produced in excess by carcinoid tumors.[233] The syndrome associated with the humoral manifestations of these tumors is quite striking but is seen only rarely. In a large series, carcinoid tumors were found in 1% of autopsies, but 90% of these tumors had not been suspected premortem.[18] Carcinoid tumors usually arise in the small intestine, appendix, or rectum; patients usually present with bleeding, obstruction, or metastases, without humoral manifestations.

Carcinoid tumors develop from enterochromaffin cells, which are widely distributed throughout the gastrointestinal tract, biliary tract and gallbladder, pancreatic ducts, and bronchial tree. These cells are also found in the thymus, thyroid, ovary, uterus, and salivary glands. Carcinoid tumors may occur in any of these sites;[228] the tumors may be malignant and often metastasize early.[200]

The production and metabolism of serotonin vary relative to the tissue of origin of the tumor. Tumors from mid-gut cells, such as ileal carcinoid, usually contain and release large quantities of 5-HT. These amounts may not be fully reflected in the amount of the metabolite (5-HIAA) in urine, because little is metabolized. Tumors derived from fore-gut cells (bronchial, pancreatic, duodenal, or biliary carcinoid) produce large amounts of 5-HT, which is oxidized within the tumor to 5-HIAA. With these tumors, urinary excretion of 5-HIAA is often much higher than would be expected from the clinical presentation. Tumors derived from hind-gut cells (rectal carcinoid) only rarely produce excess 5-HT or 5-HIAA.

The classic clinical presentation of carcinoid syndrome includes pronounced flushing, bronchial constriction, diarrhea, and cardiac valvular lesions, often associated with right-sided heart failure.[200] Not all of these manifestations can be explained by serotonin excess, but carcinoid tumors often produce excesses of other substances as well, among them histamine, catecholamines, prostaglandins, vasoactive peptides, kallikrein, ACTH, growth hormone, and insulin. Carcinoid tumors may also be associated with multiple endocrine neoplasia (MEN-1: hyperparathyroidism and pituitary and pancreatic adenomas).

Patients with functioning carcinoid tumors usually have striking increases in urinary excretion of 5-HIAA (> 25 mg/d, when normal is < 6 mg/d). If a borderline elevation of 5-HIAA is found (6–15 mg/d), repeat collections should be made and care taken to avoid food and medications that might elevate 5-HIAA, e.g., pineapples, avocados, bananas, walnuts, chocolate, guaifenesin, and reserpine. Nontropical sprue may cause a slight increase in urinary 5-HIAA.[150] 5-HIAA levels are lowered by phenothiazines, in renal insufficiency, and after small bowel resection.

When a patient strongly suspect for carcinoid syndrome shows normal or only borderline increases of 5-HIAA, two possibilities should be considered: one, that large amounts of serotonin produced are not being metabolized, in which case blood levels of 5-HTP or 5-HT are needed to document the diagnosis; the other, that secretion of 5-HIAA by the tumor is intermittent, in which case repeat specimen collections are needed to demonstrate the abnormality.

METHODS FOR DETERMINATION OF SEROTONIN AND ITS METABOLITE

5-HT and its metabolite in body fluids and tissues have been measured by spectrophotometry, fluorometry, and gas chromatography. The procedures and principles have been reviewed by Lovenberg and Engelman.[177] A specific and sensitive radioenzymatic method for the determination of 5-HT has been reported by Saavedra et al.[264] HPLC with electrochemical detection has also been applied to the analysis of serotonin and 5-HIAA in human body fluids, including cerebrospinal fluid.[151,235] See the original articles for detailed description of these methods.

Determination of 5-Hydroxyindoleacetic Acid (5-HIAA) in Urine

Estimation of the parent hormone, 5-HT, in blood and urine has been severely limited in the clinical laboratory because of its very low concentration and because of methodological complications. As a result, the urinary determination of 5-HIAA continues to be the most useful means for the diagnosis of carcinoid tumors. In such cases, this serotonin metabolite is excreted in very large amounts, often exceeding 350 mg/d, and a positive result is obtained on simple qualitative (screening) tests. However, for early diagnosis when tumors are small and have not metastasized, and in some carcinoid tumors where the excretion values barely exceed 8 mg, the more sensitive and specific quantitative test is required. Both qualitative and quantitative procedures based on the methods reported by Udenfriend and his associates[314] are described below.

Principle of Screening Test [278]

This test is based on the development of a purple color, specific for 5-hydroxyindoles, on the addition of 1-nitroso-2-naphthol and nitrous acid. Other interfering chromogens are extracted into ethylene dichloride.

Specimen Collection and Storage

A random specimen is usually suitable for the screening test. For quantitative analysis, a 24-h urine specimen is collected without preservatives. The specimen should be refrigerated during collection. Upon receipt in the laboratory, the urine specimen is thoroughly mixed and the total volume measured and recorded. If desired, aliquots may be removed at this time for determination of acid-labile substances (e.g., MHPG). The pH of the urine is then adjusted between 2 and 3 by addition of HCl, 6 mol/L. The acidified urine can be kept at 4 °C for 2 weeks and for longer periods of time at −20 °C. False negative results may occur in patients taking phenothiazine drugs. The ingestion of bananas, avocados, red plums, eggplants, tomatoes, or cough medications containing glycerol guaiacolate may produce false positive results. Therefore, these drugs and diets should be restricted 3 to 4 d prior to and during the collection.

Reagents

1. 1-Nitroso-2-naphthol, 0.1 g/dL, in 95% ethanol.
2. Sulfuric acid, 1 mol/L.
3. Sodium nitrite, 2.5 g/dL in water. Prepare fresh at frequent intervals. Refrigerate.
4. Nitrous acid reagent. Prepare fresh before use by adding 0.2 mL sodium nitrite, 2.5 g/dL, to 5 mL of sulfuric acid, 1 mol/L.
5. Ethylene dichloride, redistilled.

Procedure

1. Pipet into a test tube 0.2 mL of urine, 0.8 mL of distilled water, and 0.5 mL of 1-nitroso-2-naphthol. Mix. Prepare another tube in the same way with normal urine to serve as a negative control.
2. Add 0.5 mL of freshly prepared nitrous acid reagent to both tubes and mix again. Let the tubes stand at room temperature for 10 min.
3. Add 5 mL of ethylene dichloride and shake. If turbidity results, centrifuge. A positive test shows a purple color in the top aqueous layer. The negative control with normal urine produces a slight yellow color.

Comments

Dietary and drug restrictions as outlined are important if false negative or positive results are to be avoided. The substance p-hydroxyacetanilide derived from acetanilide or related drugs reacts similarly and adds to the color. Color formation may be inhibited in specimens from patients who excrete large amounts of keto acids.

A purple color (positive test) will be seen at levels of 5-HIAA excretion as low as 40 mg/d. At higher levels, the color is more intense and is almost black at levels > 300 mg/d. A positive result should be verified with a quantitative method.

Quantitative Determination of Urinary 5-Hydroxyindoleacetic Acid[103,314]

Principle

5-Hydroxyindoleacetic acid (5-HIAA), other phenolic acids, and drug metabolites are extracted into diethyl ether from acidified urine; a saturating amount of sodium chloride is added to promote quantitative transfer into the ether phase. The 5-HIAA is then back-extracted into a phosphate buffer (0.1 mol/L, pH 7) and reacted with nitroso-naphthol and nitrous acid at 37 °C to form a violet color. Phosphate buffer at pH 7 is chosen for efficient extraction, since 5-HIAA becomes progressively more unstable at higher pH values. Urinary phenols lacking an acid group are not ionized at pH 7 and remain behind in the ether layer, thereby providing a relatively clean extract. An intense blue chromophore is rapidly formed upon subsequent addition of 2-mercaptoethanol:

5-HIAA   Nitrosonaphthol   Nitrous acid   Mercaptoethanol

Extraneous colors, caused by reactive phenols and indoleacetic acid, are removed by treatment with the mercaptoethanol and extraction into ethyl acetate. The absorbance maximum of the remaining blue solution occurs at 645 nm, but measurements are generally made at 590 nm, where Beer's law is obeyed through a suitable range of 5-HIAA concentration.

Specimen Collection and Storage

Follow the procedure as described on page 1157.

Reagents

All reagents should be of analytical reagent quality unless otherwise stated.

1. Hydrochloric acid, 1 mol/L. Dilute 83 mL concentrated HCl in distilled water to a final volume of 1 L.

2. Hydrochloric acid, 1 mmol/L. Dilute 1 mL HCl, 1 mol/L, in distilled water to a final volume of 1 L.

3. Sodium chloride.

4. Ethyl ether, peroxide free, distilled in all-glass apparatus. Protect from light, heat, and air. Test for peroxides using Quantofix Peroxide Test Strips (Macherey Nagel, Duren, Germany).

5. Phosphate buffer, 0.1 mol/L, pH 7.0. Dissolve 5.18 g monobasic potassium phosphate, KH_2PO_4, and 8.80 g anhydrous dibasic sodium phosphate, Na_2HPO_4, in distilled water and dilute to a final volume of 1 L. Store at 4 °C and discard after 3 months.

6. Absolute ethanol, USP-NF reagent quality.

7. 1-Nitroso-2-naphthol, 2 g/L. Dissolve 0.2 g nitroso-naphthol in absolute ethanol to a final volume of 100 mL. Store in a dark brown bottle at 4 °C.

8. Sodium nitrite, 25 g/L. Dissolve 0.25 g $NaNO_2$ in distilled water and dilute to a final volume of 10 mL. Prepare fresh daily.

9. Nitrous acid solution. Mix 1 mL $NaNO_2$, 25 g/L, with 25 mL HCl, 1 mol/L. Prepare fresh daily.

10. Mercaptoethanol, 250 mL/L. In a fume hood, dilute 12.5 mL of 2-mercaptoethanol in distilled water to a final volume of 50 mL. Store in the hood in a dark brown bottle at room temperature. Stable indefinitely.

11. Ethyl acetate, glass distilled. Store in a dark brown glass bottle.

12. Hydroxyindoleacetic acid, stock standard, 250 mg/L. Dissolve 25 mg 5-HIAA in 100 mL HCl, 1 mmol/L. Store in a dark brown glass bottle at 4 °C. Discard after 3 months.

13. Thiourea, 1 g/L. Dissolve 1 g thiourea in distilled water and dilute to a final volume of 1 L.

Controls

Two urine controls are included in every run. Ortho urine controls I and II (Ortho Diagnostics, Inc.) are reconstituted each day with distilled water. Aliquots are adjusted to pH 3 with hydrochloric acid, 6 mol/L, prior to use.

Procedure

Separation of 5-HIAA from Urine

1. Transfer 5.0 mL of each urine specimen and urine control into 50-mL glass-stoppered centrifuge tubes. Include a 5.0-mL water blank.
2. Also transfer 0.1-, 0.2-, and 0.3-mL aliquots of the 5-HIAA standard, 250 mg/L, into labeled centrifuge tubes. (Use precision pipets.) Dilute with 4.9, 4.8, and 4.7 mL of thiourea, 1 g/L, respectively.
3. To all tubes, add 5.0 mL of hydrochloric acid, 1 mol/L, 25.0 mL ethyl ether, and a saturating amount of NaCl (about 4 g). Cap, check for leaks, and shake mechanically in a horizontal position for 5 min. Centrifuge for 3 min at $2000 \times g$.
4. Pipet 20.0 mL of the upper ether phase into disposable, screw-capped glass test tubes, 20×150 mm.
5. Add 4.0 mL of phosphate buffer, 0.1 mol/L, pH 7. Shake mechanically for 5 min and then centrifuge for 3 min. Aspirate and discard the upper ether phase.
6. Transfer 2.0 mL aliquots of the lower aqueous phase to 10×150 mm disposable, screw-capped test tubes.

Note: Do not discard the remaining aqueous solutions until the analysis is completed.

Color Development

7. Add, with mixing, 0.5 mL of nitroso-naphthol, 2 g/L to each tube.
8. Add 1 mL of nitrous acid, 25 g/L, and remix. Incubate in a water bath at 37 °C for 5 min.
9. Add 0.2 mL of mercaptoethanol, 250 mL/L, to each tube. Mix and reincubate for 20 min at 37 °C.

Note: Formation of a deep red color in step 8 may suggest the presence of glycerol guaiacolate or related drugs. If incubation with mercaptoethanol produces a dark, olive-green color comparable in intensity to the 50 μg standard, place the tube in a hot water bath (85–100 °C) for 5 min. The heat treatment eliminates the drug interference but does not affect the 5-HIAA chromophore, which is thermostable. Cool and then proceed with Step 10.

10. At the end of the incubation period, add 5.0 mL of ethyl acetate to each tube. Vortex vigorously for 30 s and allow layers to separate. Aspirate and discard the upper organic layer.
11. Transfer the lower aqueous phase to 12×75 mm disposable glass test tubes.
12. Immediately determine the absorbance of each sample at 590 nm in the spectrophotometer against the water blank.

Note: If the absorbance of the urine sample is more than 3 times greater than the absorbance of the 50-μg standard, repeat the color development using 0.2 mL of the phosphate extract in Step 6. Add 1.8 mL of phosphate buffer to the sample and proceed with steps 7–12. Multiply the calculated 5-HIAA concentration (mg/L) by 10.

13. Standard curve: Plot absorbance on the ordinate versus amount of 5-HIAA in each standard tube (25, 50, 75 μg) on the abscissa.

Calculations

$$5\text{-HIAA, mg/L} = \frac{U}{S} \times 10$$

where U = absorbance of urine sample
S = absorbance of 50 μg standard

Factor 10 and its implied units are:

0.05 mg (mass of 5-HIAA in standard) $\times$ 1/0.005 L (urine sample volume) = 10 mg/L

Calculate daily 5-HIAA excretion by multiplying 5-HIAA concentration (mg/L) by 24-h urine volume (L).

Reference Range

Adults: 1.8–6.0 mg/d.

Comments

Acetic acid should not be used as a urine preservative since it decreases the extraction efficiency of 5-HIAA into the phosphate buffer. Stability of 5-HIAA in 24-h specimens is not reduced in the absence of preservatives, provided the urine is kept under refrigeration during the collection. In the 5-HIAA method of Udenfriend et al.,[314] urine is subjected to a 2-step cleanup prior to extraction of the 5-HIAA. Preliminary treatment of urine was found not to be necessary using the present procedure. The reaction of 5-HIAA with nitrosonaphthol to form a violet chromophore is claimed to be very specific. Serotonin and 7-hydroxy-indoles do not react.[314]

Interferences[307]

Apparent Increase in 5-HIAA	Apparent Decrease in 5-HIAA

Foods containing serotonin:
  Avocados, bananas, red plums, walnuts,
  pineapples, eggplant or tomatoes
Drugs:
  Glycerol guaiacolate, mephenesin, phenacetin,
  acetaminophen
	Drugs and chemicals:
	  Methenamine, phenothiazine tranquilizers,
	  homogentisic acid, gentisic acid, acetic acid,
	  levodopa

ECTOPIC HORMONE AND PARANEOPLASTIC SYNDROMES

Malignant neoplasms may produce a variety of ectopic products,[220] exhibiting striking clinical effects (Table 9-24). These products are termed ectopic because they are not normally derived from the tissue affected by the tumor. The ability of neoplasms to produce hormones ectopically is due in part to the frequent occurrence of tumor cells derived from neural ectoderm. These are small cells that have amine precursor uptake and decarboxylase activity (APUD cells).[15,323a] Although not all hormone-producing malignancies are derived from APUD cells, probably all cancers can produce and secrete one or more proteins or peptides. Not all of the products are biologically active, and none is specific for a particular cell type. As tumor markers, however, these substances continue to attract great interest.

Hormones are among the markers produced by neoplasms (Table 9-25). In many instances, these markers lack biologic activity, but their measurement may be useful in the diagnosis of certain malignancies or in following the response of these malignancies to treatment.[143] Hormonal tumor markers are not specific, and elevations of these hormones may occur in non-neoplastic states. Some neoplasms may exhibit these markers inconsistently or not at all, or variably with the natural progression of the tumor.

Most frequently, these ectopically produced substances are immunochemically similar to the naturally occurring hormones and are therefore detectable with standard radioimmunoassay techniques. Usually, sufficient immunochemical differences do not exist between the ectopic and natural hormone for most assays to distinguish normal from ectopic production.

Concurrence of a neoplasm with an endocrine syndrome or with excess hormone levels is

---

Table 9-24. SUBSTANCES AND SYNDROMES
ASSOCIATED WITH MALIGNANCIES

Substances
  Hormones (including intact biologically active hormones,
    fragments and/or precursors of active hormones, and immu-
    nologically detectable but biologically inactive hormones)
  Fetal proteins (carcinoembryonic antigen, α-fetoprotein)
  Enzymes (alkaline phosphatase)
Syndromes
  Hormone excess states
  Central nervous system degenerative conditions
  Myopathies, myasthenia
  Dermatologic syndromes
  Arthropathies, digital clubbing
  Hematologic abnormalities (erythrocytosis, leukemoid reaction,
    thrombophlebitis, coagulopathies)
  Fever

Table 9-25. ECTOPIC HORMONES PRODUCED BY MALIGNANT TUMORS

ACTH, proACTH	Corticotropin-releasing hormone (CRH)
Lipotropin	Calcitonin
Chorionic gonadotropin (intact and β-subunit)	Pituitary glycoprotein hormones (α-subunit)
Somatomedins	Antidiuretic hormone (vasopressin)
Parathyroid hormone	Prostaglandins
Prolactin, "big" prolactin	Growth hormone
Gastrin	Secretin
Glucagon	Somatostatin
Chorionic somatomammotropin	Vasoactive intestinal polypeptide
Erythropoietin	

not sufficient to establish the diagnosis of ectopic hormone secretion. Evidence for ectopic hormone production should be sought only after exclusion of other causes for the elevated hormone levels. The evidence should include (1) demonstration that the abnormal hormone does not respond to usual agents for stimulation or suppression; (2) demonstration of an arteriovenous gradient for the hormone across the tumor capillary bed; (3) decrease of the hormone level after successful treatment of the tumor; (4) demonstration of elevated quantities of the hormone in tumor tissue; and (5) evidence of in vitro synthesis of the hormone by tumor cells.[15,259]

Common Endocrine Syndromes Associated with Malignancies[224]

Ectopic ACTH (or CRH) production. The production of biologically active ACTH or corti- cotropin-releasing hormone (CRH) by a malignant tumor leads to the overproduction of cortisol and other steroids by the adrenal glands. Differentiation of ectopic ACTH production from ectopic production of CRH requires techniques that are not presently available outside research laboratories.[102] Often, the typical picture of cortisol excess (Cushing's syndrome) is overshadowed by the wasting associated with the tumor. Common manifestations of the ectopic ACTH syndrome are glucose intolerance, hypokalemia, hypertension, and edema. The diagnosis is suggested in a patient with a diagnosed malignant tumor by demonstrating high concentrations of cortisol in blood or urine, or both, with a concomitant high blood level of immunoreactive ACTH. In mild or equivocal cases, selective venous catheterization and measurement of ACTH to demonstrate a gradient outside the pituitary circulation may be necessary to differentiate pituitary Cushing's syndrome from ectopic ACTH production.[84] Tumors most frequently associated with ectopic ACTH production are oat-cell carcinoma of the lung,[249] bronchial and other carcinoid tumors, and medullary carcinoma of the thyroid.[276]

Ectopic parathyroid hormone (PTH) production. Hypercalcemia occurs in association with a number of malignancies, but there are various causes.[272] The clinical presentation is the same as with other causes of hypercalcemia—no symptoms; or gastrointestinal difficulties such as con- stipation, anorexia, or vomiting; or central nervous system dysfunction such as confusion or coma—and is not dependent on the severity of the hypercalcemia. Metastasis of the tumor to bone with osteolysis is probably the most common cause of tumor-associated hypercalcemia. A variety of humoral factors such as prostaglandins,[66] vitamin D sterols, and osteoclast-activating factor are being implicated as causative, in addition to tumor production of PTH. Measurement of nephrogenous cAMP, the mediator of PTH action on the kidney, is not generally helpful in this setting, since elevations of nephrogenous cAMP are seen despite low PTH levels with hypercalcemia and malignancy.[262,296] An elevated PTH level does not specifically differentiate the hypercalcemia of malignancy from primary hyperparathyroidism. Primary hyperparathyroidism and cancer may coexist by chance.[72] Results differ with different assays for PTH,[108] but comparing the PTH value to the serum calcium will usually show some separation, because patients with primary hyperparathyroidism[17,108] have higher PTH values relative to the degree of hypercal- cemia.[108] (See Figure 12-5, Chapter 12.) Tumors commonly associated with hypercalcemia include renal adenocarcinoma, squamous cell[16] and other carcinomas of the lung,[249] and carcinomas of the ovary.

Ectopic production of antidiuretic hormone. Tumor production of ADH and the syndrome of inappropriate ADH secretion (SIADH) are usually associated with oat-cell carcinoma of the lung and other lung tumors,[249] carcinoma of the colon, carcinoid tumors, and lymphomas.

Other ectopic hormones. There are other striking clinical syndromes of ectopic hormone production that occur less frequently than those previously discussed. These include (1) ectopic production of gonadotropins[27] from various lung, liver, and gonadal tumors presenting as isosexual precocious puberty or gynecomastia; (2) ectopic production of growth hormone or growth hormone-releasing hormone associated with bronchial carcinoid or pancreatic tumors that present as acromegaly; (3) ectopic production of vasoactive intestinal polypeptide associated with pancreatic islet-cell carcinomas, bronchogenic carcinomas, pheochromocytomas, or neuroblastomas and presenting as watery diarrhea, hypochlorhydria, and hypokalemia (Verner-Morrison syndrome);[320] (4) prolactin production[309] by lung or renal carcinoma associated with galactorrhea; and (5) various tumor mechanisms causing hypoglycemia,[42] hypokalemia, or hypophosphatemia. Multiple hormones may be produced by the same tumor.

References

1. Abraham, G. E., Ed.: Handbook of Radioimmunoassay. New York, Marcel Dekker, Inc., 1977.
2. Abraham, G. E., Ed.: Radioassay Systems in Clinical Endocrinology. New York, Marcel Dekker, Inc., 1981.
3. Abraham, G. E., and Chakmajian, Z. H.: Serum steroid levels during the menstrual cycle in bilaterally adrenalectomized women. J. Clin. Endocrinol. Metab., 37:581-587, 1973.
4. Abraham, G. E., Maroulis, G. B., and Marshall, J. R.: Evaluation of ovulation and corpus luteum function using measurement of plasma progesterone. Obstet. Gynecol., 44:522-525, 1974.
5. Al-Dujaili, E. A. S., Williams, B. C., and Edwards, C. R. W.: The development and application of a direct radioimmunoassay for corticosterone. Steroids, 37:157-176, 1981.
6. Anderson, O. O., Deckert, T., and Nerup, J., Eds.: Immunological aspects of diabetes mellitus. Acta Endocrinol., Suppl. 205, 1976.
7. Appleby, J. E., Gibson, G., Norymberski, J. K., et al.: Indirect analysis of corticosteroids. 1. The determination of 17-hydroxycorticosteroids. Biochem. J. 60:453-467, 1955.
8. Asplin, C. M., Hollander, P., Pecoraro, R. E., et al.: Insulin, pancreatic polypeptide, and glucagon antibodies in insulin-dependent diabetes mellitus. Diabetes Care, 4:337-342, 1981.
9. Baer, L., Sommers, S. C., Krakoff, L. R., et al.: Pseudoprimary aldosteronism. Circ. Res., 27:203-216 (Suppl. 1), 1970.
10. Baird, D. T.: A method of the measurement of estrone and estradiol-17β in peripheral human blood and other biological fluids using 35-S pipsyl chloride. J. Clin. Endocrinol. Metab., 28:244-258, 1968.
11. Ball, P., and Knuppen, R.: Catecholoestrogens (2- and 4-hydroxyoestrogens). Acta Endocrinol., Suppl. 232, 1980.
12. Bardin, C. W., and Mahoudeau, J. A.: Dynamics of androgen metabolism in women with hirsutism. Ann. Clin. Res., 2:251-262, 1970.
13. Barker, S. B., Humphrey, M. J., and Solely, M. H.: The clinical determination of protein-bound iodine. J. Clin. Invest., 30:55-62, 1951.
14. Bartter, F. C., and Delea, S. C.: Diabetes insipidus—its nature and diagnosis. Lab. Management, 20:23, 1982.
15. Baylin, S. B., and Mendelsohn, G.: Ectopic (inappropriate) hormone production by tumors: Mechanisms involved and the biological and clinical implications. Endocr. Rev., 1:45-77, 1980.
16. Bender, R. A., and Hansen, H.: Hypercalcemia in bronchogenic carcinoma. A prospective study of 200 patients. Ann. Intern. Med., 80:205-208, 1974.
17. Benson, R. C., Riggs, B. L., Pickard, B. M., et al.: Radioimmunoassay of parathyroid hormone in hypercalcemic patients with malignant diseases. Am. J. Med., 56:821-826, 1974.
18. Berge, T., and Linell, F.: Carcinoid tumors: Frequency in a defined population during a 12-year period. Acta Pathol. Microbiol. Scand. [A], 84:322-330, 1976.
19. Berger, M. J., Taymore, M. L., and Patton, W. C.: Gonadotropin levels and secretory patterns in patients with typical and atypical polycystic ovarian disease. Fertil. Steril., 26:619-626, 1975.
20. Bertani-Dziedzic, L. M., Krstulovic, A. M., Dziedzic, S. W., et al.: Analysis of urinary metanephrines by reversed-phase high performance liquid chromatography and electrochemical detection. Clin. Chim. Acta, 110:1-8, 1981.
21. Biglieri, E. G., and Stockigt, J. R.: A preliminary evaluation for primary aldosteronism. Arch. Intern. Med., 126:1004-1007, 1970.
22. Bio-Science Laboratories: The Bio-Science Handbook. 12th ed. Van Nuys, California, 1979, p. 25.
23. Bleich, H. L., and Moore, M. J.: The evolutionary origins of hormones, neurotransmitters and other extracellular chemical messengers. N. Engl. J. Med., 306:523-527, 1982.
24. Boullin, D. J., Ed.: Serotonin in Mental Abnormalities. New York, John Wiley & Sons, 1978.
25. Boyar, R. M., Witkin, M., Carruth, A., et al.: Circadian cortisol secretory rhythms in Cushing's disease. J. Clin. Endocrinol. Metab., 48:760-765, 1979.
26. Boyden, S. V.: Adsorption of proteins on erythrocytes treated with tannic acid and subsequent hemagglutination by antiprotein sera. J. Exp. Med., 93:107-120, 1951.
27. Braunstein, G. D., Vaitukaitis, J. L., Carbone, P. P., et al.: Ectopic production of human chorionic gonadotropin by neoplasms. Ann. Intern. Med., 78:39-45, 1973.
27a. Bravo, E. L., and Gifford, R. W., Jr.: Pheochromocytoma: Diagnosis, localization and management. N. Engl. J. Med., 311:1298-1303, 1984.
28. Bravo, E. L., Tarazi, R. C., Found, F. M., et al.: Clonidine-suppression test: A useful aid in the diagnosis of pheochromocytoma. N. Engl. J. Med., 305:623-626, 1981.
29. Bravo, E. L., Tarazi, R. C., Gifford, R. W., et al.: Circulating and urinary catecholamines in pheochromocytoma. N. Engl. J. Med., 301:682-686, 1979.
30. Breuer, H., Hamel, D., and Krüskemper, H., Eds.: Methods of Hormone Analysis. New York, John Wiley & Sons, 1976.

31. Broughton, A.: Application of adrenocorticotropin assays in a routine clinical laboratory. Am. J. Clin. Pathol., 64:618-624, 1975.

32. Brown, J. B.: A chemical method for the determination of oestriol, oestrone and oestradiol in human urine. Biochem. J., 60:185-193, 1955.

33. Brown, J. B., Chopra, I. J., Cornell, J. S., et al.: Thyroid physiology in health and disease. Ann. Intern. Med., 81:68-81, 1974.

34. Brown, J. B., MacNaughton, C., Smith, M. A., et al.: Further observations on the Kober colour and Ittrich fluorescence reactions in the measurement of oestriol, oestrone, and oestradiol. J. Endocrinol., 40:175-188, 1968.

35. Buckman, M. T., and Peake, G. T.: Concordance of insulin-induced hypoglycemia and the phenothiazine-induced prolactin secretion in man. J. Clin. Endocrinol. Metab., 48:213-216, 1979.

36. Burrow, G. N., Wortzman, G., Rewcastle, N. B., et al.: Microadenomas of the pituitary and abnormal sella tomograms in an unselected autopsy series. N. Engl. J. Med., 304:156-158, 1981.

37. Buster, J. E., and Abraham, G. E.: Radioimmunoassay of plasma dehydroepiandrosterone sulfate. Anal. Letters, 5:543-551, 1972.

38. Calseyde, J. F. van de, Scholtis, R. J., Schmidt, N. A., et al.: Gas chromatography in the estimation of urinary metanephrines and VMA. Clin. Chim. Acta, 32:361-366, 1971.

39. Carlson, H. E.: Gynecomastia. N. Engl. J. Med., 303:795-799, 1980.

40. Catt, K. J., Tregear, G. W., Burger, H. G., et al.: Antibody-coated tube method for radioimmunoassay of human growth hormone. Clin. Chim. Acta, 27:267-279, 1970.

41. Cavalieri, R. R., and Ingbar, S. H.: Methods for measuring the thyroxine-binding proteins and free thyroid hormone concentration in serum. Methods Enzymol., 36:126-132, 1975.

42. Chandalia, H. B., and Boshell, B. R.: Hypoglycemia associated with extrapancreatic tumors. Arch. Intern. Med., 129:447-456, 1972.

43. Chang, R. J., Keye, W. R., Monroe, S. E., et al.: Prolactin-secreting pituitary adenomas in women. IV. Pituitary function in amenorrhea associated with normal or abnormal serum prolactin and sellar polytomography. J. Clin. Endocrinol. Metab., 51:830-835, 1980.

44. Channing, C. P., Schaerf, F. W., Anderson, L. D., et al.: Ovarian follicular and luteal physiology. In: Reproductive Physiology III. R. O. Greep, Ed. Baltimore, University Park Press, 1980, p. 117.

45. Chattoraj, S. C., Rankin, J. L., Turner, A. K., et al.: Urinary progesterone as an index of ovulation and corpus luteal function. J. Clin. Endocrinol. Metab., 43:1402-1405, 1976.

46. Chattoraj, S. C., Turner, A. K., Pinkus, J. L., et al.: The significance of urinary free cortisol and progesterone in normal and anencephalic pregnancy. Am. J. Obstet. Gynecol., 124:848-854, 1976.

47. Chattoraj, S. C., Fanous, A., Cecchini, D., et al.: A radioimmunoassay method for urinary catechol estrogens. Steroids, 31:375-391, 1978.

48. Chattoraj, S. C.: Unpublished observation, 1984.

49. Chattoraj, S. C., and Wotiz, H. H.: A routine method for the gas chromatographic determination of pregnanediol. Fertil. Steril., 18:342-352, 1967.

50. Chehval, M. J., and Mehan, D. J.: The appropriateness of gonadotropin determination in the work-up of the infertile male. Fertil. Steril., 32:233-234, 1979.

51. Chen, I.-W.: Commercially available fully automated sytems for radioligand assays. Part I. Overview. Ligand Rev., 2(2):46-50, 1980. Part II. Performance characteristics. Ligand Rev., 2(3):46-48, 1980.

52. Chopra, I. J.: A radioimmunoassay for measurement of 3,3',5'-triiodothyronine (reverse T₃). J. Clin. Invest., 54:583-592, 1974.

53. Chopra, I. J.: Radioimmunoassay of iodothyronines. In: Handbook of Radioimmunoassay. G. E. Abraham, Ed. New York, Marcel Dekker, Inc., 1977, p. 679.

54. Chopra, I. J.: Thyroid Function Tests: Thyroxine and Reverse Triiodothyronine Measurements. In: Radioassay Systems in Clinical Endocrinology. G. E. Abraham, Ed. New York, Marcel Dekker, Inc., 1981, pp. 101-115.

55. Chopra, I. J., Hershman, J. H., Pardridge, W. M., et al.: Thyroid function in nonthyroidal illnesses. Ann. Intern. Med., 98:946-957, 1983.

56. Chopra, I. J., Solomon, D. H., Hepner, G. W., et al.: Misleading low free thyroxine index and usefulness of reverse triiodothyronine measurement in nonthyroidal illness. Ann. Intern. Med., 90:905-912, 1979.

57. Chrousos, G. P., Loriaux, D. L., Mann, D. L., et al.: Late onset 21-hydroxylase deficiency mimicking idiopathic hirsutism or polycystic ovarian disease. Ann. Intern. Med., 96:143-148, 1982.

58. Clark, F., and Horn, D. B.: Assessment of thyroid function by the combined use of the serum protein-bound iodine and resin uptake of 131-I-triiodothyronine. J. Clin. Endocrinol. Metab., 25:39-45, 1965.

59. Clemons, D. R., Van Wyk, J. J., Ridgway, E. C., et al.: Evaluation of acromegaly by radioimmunoassay of somatomedin-C. N. Engl. J. Med., 301:1138-1142, 1979.

60. Cole, P., and MacMahon, B.: Oestrogen fractions during early reproductive life in the aetiology of breast cancer. Lancet, 1:604-606, 1969.

61. Curtis, G., and Fogel, J.: Creatinine excretion: Diurnal variation and variability of whole and part-day measurements. Psychosom. Med., 32:337-350, 1970.

62. Daughaday, W. H., and Cryer, P. E.: Growth hormone hypersecretion and acromegaly. Hosp. Pract., 13(8):75-80, 1978.

63. DeChamplain, J. D., Farley, L., Cousineau, D., et al.: Circulating catecholamine levels in human and experimental hypertension. Circ. Res., 38:109-114, 1976.

64. DeGroot, L. J., and Stanbury, J. B.: The Thyroid and Its Diseases. New York, John Wiley & Sons, 1975.

65. Dekirmenjian, H., and Maas, J. W.: An improved procedure of 3-methoxy-4-hydroxyphenyl-ethylene glycol determination by gas-liquid chromatography. Anal. Biochem., 35:113-122, 1970.

66. Demers, L. M.: Plasma prostaglandins in hypercalcemic patients with neoplastic disease. Cancer, 39:1559-1562, 1977.

67. Dobson, H., and Dean, P. D. G.: Radioimmunoassay of oestrone, oestradiol-17α and -17β in bovine plasma during the oestrous cycle and last stages of pregnancy. J. Endocrinol., *61*:479-486, 1974.

68. Dorfman, R. I., Ed.: Methods in Hormone Research, Vol. I. 2nd ed. New York, Academic Press, 1968.

69. Dorfman, R. I., Ed.: Methods in Hormone Research, Vol. IIA. 2nd ed. New York, Academic Press, 1969.

70. Dorfman, R. I., and Ungar, F.: Metabolism of Steroid Hormones. New York, Academic Press, 1965.

71. Drekter, I. J., Heisler, A., Scism, G. R., et al.: The determination of urinary steroids. I. The preparation of pigment-free extracts and a simplified procedure for the estimation of total 17-ketosteroids. J. Clin. Endocrinol. Metab., *12*:55-65, 1952.

72. Drezner, M. K., and Lebovitz, H. E.: Primary hyperparathyroidism in paraneoplastic hypercalcaemia. Lancet, *1*:1004-1006, 1978.

73. Duckworth, W. C., and Kitabchi, A. E.: Insulin metabolism and degradation. Endocr. Rev., *2*:210-233, 1981.

74. Earll, J. M., Sparks, L. L., and Forsham, P. H.: Glucose suppression of serum growth hormone in the diagnosis of acromegaly. JAMA, *201*:628-630, 1967.

75. Eddy, R. L., Gilliland, P. F., Ibarra, J. D., et al.: Human growth hormone release: Comparison of provocative test procedures. Am. J. Med., *56*:179-185, 1974.

76. Eddy, R. L., Jones, A. L., Chakmakjian, Z. H., et al.: Effects of levodopa (L-dopa) on human hypophyseal tropic hormone release. J. Clin. Endocrinol. Metab., *33*:709-712, 1971.

77. Eddy, R. L., Jones, A. L., and Gilliland, P. F.: Cushing's syndrome: A prospective study of diagnostic methods. Am. J. Med., *55*:621-630, 1973.

78. Edwards, L., Ed.: Product guide for radioassay and nonisotopic assays. Clin. Chem., *29*:889-986, 1983.

79. Eik-Nes, K. B., and Horning, E. C.: Gas Phase Chromatography of Steroids. New York, Springer-Verlag, 1968.

80. Engelman, K., Portnoy, B., and Lovenberg, W.: A sensitive and specific double-isotope derivative method for the determination of catecholamines in biological specimens. Am. J. Med. Sci., *255*:259-268, 1968.

81. Farmer, R. W., Roup, W. G., Pellizzari, E. D., et al.: A rapid aldosterone radioimmunoassay. J. Clin. Endocrinol. Metab., *34*:18-22, 1972.

82. Farmer, R. W., Brown, D. H., Howard, P. Y., et al.: A radioimmunoassay for plasma aldosterone without chromatography. J. Clin. Endocrinol. Metab., *36*:461-465, 1973.

83. Few, J. D.: A method for the analysis of urinary 17-hydroxycorticosteroids. J. Endocrinol., *22*:31-46, 1961.

84. Findling, J. W., Aron, D. C., Tyrrell, J. B., et al.: Selective venous sampling for ACTH in Cushing's syndrome. Differentiation between Cushing's disease and ectopic ACTH syndrome. Ann. Intern. Med., *94*:647-652, 1981.

85. Flier, J. S., Kahn, C. R., Roth, J., et al.: Antibodies that impair insulin receptor binding in an unusual diabetic syndrome with severe insulin resistance. Science, *190*:63-68, 1976.

86. Fotherby, K.: Progesterone. II. Clinical aspects. *In*: Hormones in Blood. 3rd ed. C. H. Gray and V. H. T. James, Eds. London, Academic Press Ltd., 1979, pp. 439-491.

87. Franchimont, P.: Comments made during discussion of paper presented by G. M. Besser (Effects of TRH on and interaction with other pituitary hormones). *In*: Thyrotropin Releasing Hormone (Frontiers of Hormone Research), Vol. I. R. Hall, Ed. Basel, S. Karger, 1972, pp. 139-140.

88. Frantz, W. L., MacIndoe, J. H., and Turkington, R. W.: Prolactin receptors: Characteristics of the particulate fraction binding activity. J. Endocrinol., *60*:485-497, 1974.

89. Freedlender, A. E., and Goodfriend, T. L.: Renin and the angiotensins. *In*: Methods of Hormone Radioimmunoassay. B. M. Jaffe and H. R. Behrman, Eds. New York, Academic Press, 1979, pp. 889-907.

90. Freier, D. T., Tank, E. S., and Harrison, T. S.: Pediatric and adult pheochromocytoma. Arch. Surg., *107*:252-255, 1973.

91. Friesen, H., Webster, B. R., Hwang, P., et al.: Prolactin synthesis and secretion in a patient with the Forbes-Albright syndrome. J. Clin. Endocrinol. Metab., *34*:192-199, 1972.

92. Fritz, M. A., and Speroff, L.: The endocrinology of the menstrual cycle. The interaction of folliculogenesis and neuroendocrine mechanisms. Fertil. Steril., *38*:509-529, 1982.

93. Fujita, K., Maruta, K., Ito, S., et al.: Urinary 4-hydroxy-3-methoxymandelic (vanillylmandelic) acid, 4-hydroxy-3-methoxy-phenylacetic (homovanillic) acid, and 5-hydroxy-3-indoleacetic acid determination by liquid chromatography with electrochemical detection. Clin. Chem., *29*:876-878, 1983.

94. Furth, E. D., Rathbun, M., and Postillico, J.: A modified bioassay for the long-acting thyroid stimulator (LATS). Endocrinology, *85*:592-593, 1969.

95. Ganguly, A.: New insights and questions about glucocorticoid-suppressible hyperaldosteronism. Am. J. Med., *72*:851-854, 1982.

96. Ganguly, A., Grim, C. E., and Weinberger, M. H.: Primary aldosteronism. The etiologic spectrum of disorders and their clinical differentiation. Arch. Intern. Med., *142*:813-815, 1982.

97. Gavin, L. A., Rosenthal, M., and Cavalieri, R. R.: The diagnostic dilemma of isolated hyperthyroxinemia in acute illness. JAMA, *242*:251-253, 1979.

98. Gerald, P. S.: Sex chromosome disorders. N. Engl. J. Med., *294*:706-708, 1976.

99. Gershengorn, M. C., Larsen, P. R., and Robbins, J.: Radioimmunoassay for serum thyroxine-binding globulin: Results in normal subjects and in patients with hepatocellular carcinoma. J. Clin. Endocrinol. Metab., *42*:907-911, 1976.

100. Gid-Ad, I., Topper, E., and Laron, Z.: Oral clonidine as a growth hormone stimulation test. Lancet, *2*:278-280, 1979.

101. Gitlow, S. E., Mendlowitz, M., and Bertrani, I. M.: The biochemical techniques for detecting and establishing the presence of a pheochromocytoma. Am. J. Cardiol., *26*:270-279, 1970.

102. Gold, E. M.: The Cushing syndromes: Changing views of diagnosis and treatment. Ann. Intern. Med., *90*:829-844, 1979.

103. Goldenberg, H.: Specific photometric determination of 5-hydroxyindoleacetic acid in urine. Clin. Chem., *19*:38-44, 1973.

104. Goldhzhier, J. W.: Polycystic ovary disease. Fertil. Steril., *35*:371-394, 1981.
105. Green, P. J., and Yucis, M. J.: Free testosterone determination by ultrafiltration and comparison with dialysis. Clin. Chem., *28*:1237-1238, 1982.
106. Griffin, J. E., and Wilson, J. D.: The syndromes of androgen resistance. N. Engl. J. Med., *302*:198-209, 1980.
107. Gwynne, J. T., and Strauss, J. F., III: The role of lipoproteins in steroidogenesis and cholesterol metabolism in steroidogenic glands. Endocr. Rev., *3*:299-329, 1982.
108. Habener, J. F., and Segre, G. V.: Parathyroid hormone radioimmunoassay. Ann. Intern. Med., *91*:782-785, 1979.
109. Hammond, G. L., Nisker, J. A., Jones, J. A., et al.: Estimation of the percentage of free steroid in undiluted serum by centrifugal ultrafiltration dialysis. J. Biol. Chem., *255*:5023-5026, 1980.
110. Hamolsky, M. W., Golodetz, A., and Freedberg, A. S.: The plasma protein-thyroid hormone complex in man. III. Further studies on the use of the in vitro red blood cell uptake of ^{131}I-l-triiodothyronine as a diagnostic test of thyroid function. J. Clin. Endocrinol. Metab., *19*:103-116, 1959.
111. Hansen, J. W., and Ross, G. T.: A new method simplifying collection of serial specimens for gonadotropin determinations. J. Clin. Endocrinol. Metab., *41*:241-244, 1975.
112. Hellman, L., Bradlow, H. L., and Zumoff, B.: Recent advances in human steroid metabolism. Adv. Clin. Chem., *13*:1-35, 1970.
113. Herf, S. M., Teates, D. C., Tegtmeyer, C. J., et al.: Identification and differentiation of surgically correctible hypertension due to primary aldosteronism. Am. J. Med., *67*:397-402, 1979.
114. Hershman, J. M.: Important role of serum TSH in the diagnosis of hypothyroidism. Thyroid Clinics, *1*:1, 1981.
115. Hoekstra, J. B. L., van Rijn, H. J. M., Erkelens, D. W., et al.: C-peptide. Diabetes Care, *5*:438-446, 1982.
116. Hoekstra, J. B. L., van Rijn, H. J. M., Thijssen, J. H. H., et al.: C-peptide reactivity as a measure of insulin dependency in obese diabetic patients treated with insulin. Diabetes Care, *5*:585-591, 1982.
117. Hogan, M. J., McRae, J., Schambelan, M., et al.: Location of aldosterone-producing adenomas with 131-I-19-iodocholesterol. N. Engl. J. Med., *294*:410-414, 1976.
118. Holly, J. M. P., and Makin, H. L. J.: The estimation of catecholamines in human plasma. Anal. Biochem., *128*:257-274, 1983.
119. Hoogwerf, B. J., and Goetz, F. C.: Urinary C-peptide—a simple measure of integrated insulin production with emphasis on the effects of body size, diet and corticosteroids. J. Clin. Endocrinol. Metab., *56*:60-67, 1982.
120. Horwitz, D. L., Starr, J. E., Mako, M. E., et al.: Proinsulin, insulin and C-peptide concentrations in human portal and peripheral blood. J. Clin. Invest., *55*:1278-1283, 1975.
121. Hughes, J., Ed.: Opioid peptides. Br. Med. Bull., *39*:1-106, 1983.
122. Hunter, W. M., and Corrie, J. E. T., Eds.: Immunoassays for Clinical Chemistry. 2nd ed. Edinburgh, Churchill Livingstone, 1983.
123. Hunter, W. M., and Greenwood, F. C.: A radio-immunoelectrophoretic assay for human growth hormone. Biochem. J., *91*:43-56, 1964.
124. Ingbar, S. H., Braverman, L. E., Dawber, N. A., et al.: A new method for measuring the free thyroid hormone in human serum and an analysis of the factors that influence its concentration. J. Clin. Invest., *44*:1679-1689, 1965.
125. Ingbar, S. H., and Woeber, K. A.: The thyroid gland. *In*: Textbook of Endocrinology. 6th ed. R. H. Williams, Ed. Philadelphia, W. B. Saunders Co., 1981, pp. 117-247.
126. Irvine, W. J.: Automated determinations of thyroid and gastric complement fixing antibody: Comparison with the fluorescent antibody and manual complement fixation methods. Clin. Exp. Immunol., *1*:341-354, 1966.
127. Ittrich, G.: Eine neue Methode zur chemischen Bestimmung der oestrogenen Hormone in Harn. Hoppe-Seylers Z. Physiol. Chem., *312*:1-14, 1958.
128. Jackson, I. M. D.: Thyrotropin-releasing hormone. N. Engl. J. Med., *306*:145-155, 1982.
129. Jaffe, B. M., and Behrman, H. R., Eds.: Methods of Hormone Radioimmunoassay. New York, Academic Press, 1979.
130. Johnsonbaugh, R. E., Bybee, D. E., and Georges, L. P.: Exercise tolerance test: Single-sample screening technique to rule out growth hormone deficiency. JAMA, *240*:664-666, 1978.
131. Jurjens, H., Pratt, J. J., and Woldring, M. G.: Radioimmunoassay of plasma estradiol without extraction and chromatography. J. Clin. Endocrinol. Metab., *40*:19-25, 1975.
132. Kahn, C. R.: Insulin receptors and syndromes of insulin resistance. Diabetes Care, *5*:98-101 (Suppl. 1), 1982.
133. Kamoi, K., Tchuchida, L., Sato, H., et al.: Comparison of the responses in the nomifensine test with hyperprolactinemia due to prolactin-secreting pituitary tumors and nonprolactin-secreting hypothalamic tumors. J. Clin. Endocrinol. Metab., *53*:1285-1287, 1981.
134. Kaplan, M. M., Larsen, P. R., Crantz, F. R., et al.: Prevalence of abnormal thyroid function test results in patients with acute medical illness. Am. J. Med., *72*:9-16, 1982.
135. Kautsky, M. P., Ed.: Steroid Analysis by HPLC. New York, Marcel Dekker, Inc., 1981.
136. Keffer, J. H.: Thyroid diagnosis and the progressive thyroid profile. Lab. Med., *6*(10):23-26, 1975.
137. Kehlet, H., Blichert-Toft, M., Hancke, S., et al.: Comparative study of ultrasound, 131-I-19-iodocholesterol scintigraphy, and aortography in localizing adrenal lesions. Br. Med. J., *2*:665-667, 1976.
138. Kelsey, J. L.: A review of the epidemiology of human breast cancer. Epidemiol. Rev., *1*:74-109, 1979.
139. Kem, D. C., Weinberger, M. H., Mayes, D. M., et al.: Saline suppression of plasma aldosterone in hypertension. Arch. Intern. Med., *128*:380-386, 1971.
140. Keye, W. R., Chang, R. J., Wilson, C. B., et al.: Prolactin-secreting pituitary adenomas: III. Frequency and diagnosis in amenorrhea-galactorrhea. JAMA, *244*:1329-1332, 1980.
141. Kissinger, P. T., Riggin, R. M., and Alcorn, R. L.: Estimation of catecholamines in urine by high performance liquid chromatography with electrochemical detection. Biochem. Med., *13*:299-306, 1975.
142. Kitabchi, A. E.: Proinsulin and C-peptide: A review. Metabolism, *26*:547-587, 1977.
143. Klee, G. G., and Go, V. L. W.: Serum tumor markers. Mayo Clin. Prac., *57*:129-132, 1982.
144. Klein, I., and Levey, G. S.: Silent thyrotoxic thyroiditis. Ann. Intern. Med., *96*:242-244, 1982.

145. Kleinberg, D. L., Noel, G. L., and Frantz, A. G.: Chlorpromazine stimulation and L-dopa suppression of plasma prolactin in man. J. Clin. Endocrinol. Metab., 33:873-876, 1971.
146. Kleinberg, D. L., Noel, G. L., and Frantz, A. G.: Galactorrhea: A study of 235 cases, including 48 with pituitary tumors. N. Engl. J. Med., 296:589-600, 1977.
147. Kletzky, O. A., Davajan, V., Nakamura, R. M., et al.: Classification of secondary amenorrhea based on distinct hormone patterns. J. Clin. Endocrinol. Metab., 41:660-668, 1975.
148. Klijn, J. G. M., Lamberts, S. W. J., De Jong, F. H., et al.: The value of the thyrotropin-releasing hormone test in patients with prolactin-secreting pituitary tumors and suprasellar non-pituitary tumors. Fertil. Steril., 35:155-161, 1981.
149. Klopper, A., Michie, E. A., and Brown, J. B.: A method for the determination of urinary pregnanediol. J. Endocrinol., 12:209-219, 1955.
150. Kowlessar, O. D., Williams, R. C., Law, D. H., et al.: Urinary excretion of 5-hydroxyindoleacetic acid in diarrheal states, with special reference to nontropical sprue. N. Engl. J. Med., 259:340-341, 1958.
151. Koch, D. D., and Kissinger, P. T.: Determination of serotonin in serum and plasma by liquid chromatography with precolumn sample enrichment and electrochemical detection. Anal. Chem., 52:27-29, 1980.
152. Koninckx, P. R., and Brosens, I. A.: The "gonadotropin-resistant ovary" syndrome as a cause of secondary amenorrhea and infertility. Fertil. Steril., 28:926-931, 1977.
153. Korth-Schultz, S., Levine, L. S., and New, M. E.: DHEA-S levels. A rapid test for abnormal adrenal androgen secretion. J. Clin. Endocrinol. Metab., 42:1005-1013, 1976.
154. Korth-Schultz, S., Virdis, R., Saenger, P., et al.: Serum androgens as a continuing index of adequacy of treatment of congenital adrenal hyperplasia. J. Clin. Endocrinol. Metab., 46:452-458, 1978.
155. Koulu, M., Lammentausta, R., Kangas, L., et al.: The effect of methysergide, pimozide and sodium valproate on the diazepam-stimulated growth hormone secretion in man. J. Clin. Endocrinol. Metab., 48:119-122, 1979.
156. Kraft, J. R.: Detection of diabetes mellitus in situ (occult diabetes). Lab. Med., 6(2):10-22, 1975.
157. Krstulovic, A. M.: The current state of the art in the analysis of catecholamines. Adv. Chromatogr., 17:279-309, 1979.
158. Landon, J., Wynn, V., and James, V. H. T.: The adrenocortical responses to insulin-induced hypoglycaemia. J. Endocrinol., 27:183-192, 1963.
159. Lanes, R., Plotnick, L. P., Spencer, E. M., et al.: Dwarfism associated with normal serum growth hormone and increased bioassayable, receptor assayable and immunoassayable somatomedin. J. Clin. Endocinrol. Metab., 50:485-488, 1980.
160. Lanto, O., Bjorkhem, I., Blomstrand, R., et al.: Interlaboratory evaluation of four kits for determination of plasma cortisol, with special reference to accuracy: Influence of matrix in calibration standards. Clin. Chem., 26:1899-1902, 1980.
161. Laragh, J. H., Sealey, J., and Brunner, H. R.: The control of aldosterone secretion in normal and hypertensive man: Abnormal renin-aldosterone patterns in low renin hypertension. Am. J. Med., 53:649-663, 1972.
162. Laron, Z., Pertzelan, A., and Karp, M.: Administration of growth hormone to patients with familial dwarfism with high plasma immunoreactive growth hormone. J. Clin. Endocrinol. Metab., 33:332-342, 1971.
163. Larsen, P. R.: Radioimmunoassay of thyroxine triiodothyronine and thyrotropin in human serum. In: Manual of Clinical Immunology. N. R. Rose and H. Friedman, Eds. Washington, D.C., American Society for Microbiology, 1976, pp. 222-230.
164. Larsen, P. R.: Thyroid-pituitary interaction. N. Engl. J. Med., 306:23-32, 1982.
165. Larsen, P. R., and Broskin, K.: Thyroxine immunoassay using filter paper blood samples for screening neonates for hypothyroidism. Pediatr. Res., 9:604-609, 1975.
166. Liddle, G. W.: Tests of pituitary-adrenal suppressibility in the diagnosis of Cushing's syndrome. J. Clin. Endocrinol. Metab., 20:1539-1560, 1960.
167. Liddle, G. W., Bledsoe, T., and Coppage, W. S.: A familial renal disorder simulating primary aldosteronism but with negligible aldosterone secretion. Trans. Assoc. Am. Physicians, 76:199-213, 1963.
168. Lin, T., and Tucci, J. R.: Provocative tests of growth hormone release. A comparison of results with seven stimuli. Ann. Intern. Med., 80:464-469, 1974.
169. Lindholm, J., Hummer, L., Kehlet, H., et al.: Plasma corticotropin and cortisol in patients with pituitary adenomas. J. Clin. Endocrinol. Metab., 48:348-351, 1979.
170. Lindholm, J., Kehlet, H., Blickert-Toft, M., et al: Reliability of the 30-minute ACTH test in assessing hypothalamic-pituitary-adrenal function. J. Clin. Endocrinol. Metab., 47:272-279, 1978.
171. Lindner, H. R., Perel, E., Friedlander, A., et al.: Specificity of antibodies to ovarian hormones in relation to the site of attachment of the steroid hapten to the peptide carrier. Steroids, 19:357-375, 1972.
172. Lipsett, M. B., and Lippman, M. E.: Endocrine responsive cancers of man. In: Textbook of Endocrinology. 6th ed. R. H. Williams, Ed. Philadelphia, W. B. Saunders Co., 1981, pp. 1213-1227.
173. Lobo, R. A., Paul, W. L., and Goebelsmann, U.: Dehydroepiandrosterone sulfate as an indicator of adrenal androgen function. Obstet. Gynecol., 57:69-73, 1981.
174. Lobo, R. A., Goebelsmann, U., and Horton, R.: Evidence for the importance of peripheral tissue events in the development of the polycystic ovary syndrome. J. Clin. Endocrinol. Metab., 57:393-397, 1983.
175. Loraine, J. A., and Bell, E. T., Eds.: Hormone Assays and Their Clinical Application. 4th ed. Edinburgh, Churchill Livingstone, 1976.
176. Loriaux, D. L., Ruder, H. J., and Lipsett, M. D.: Plasma steroids in congenital adrenal hyperplasia. J. Clin. Endocrinol. Metab., 39:627-630, 1974.
177. Lovenberg, W., and Engelman, K.: Assay of serotonin, related metabolites and enzymes. In: Methods Biochemical Analysis. (Suppl. volume on biogenic amines.) D. Glick, Ed. New York, Interscience Publishers, 1971, pp. 1-34.
178. Maas, J. W.: Biogenic amines and depression. Arch. Gen. Psychiatry, 32:1357-1361, 1975.

179. Mace, J. C., Gotlin, R. W., and Beck, P.: Sleep-related human growth hormone release, a test of physiologic growth hormone secretion in children. J. Clin. Endocrinol. Metab., *34*:339-341, 1972.

180. Mahajan, D. K., Wahlen, J. D., Tyler, F. H., et al.: Plasma 11-deoxycortisol radioimmunoassay for metyrapone tests. Steroids, *20*:609-620, 1972.

181. Malarkey, W. B., Goodenow, T. J., and Lanese, R. R.: Diurnal variation of prolactin secretion differentiates pituitary tumors from the primary empty sella syndrome. Am. J. Med., *69*:886-890, 1980.

182. Manger, W. M., Steinsland, O. S., Nahas, G. G., et al.: Comparison of improved fluorometric methods used to quantitate plasma catecholamines. Clin. Chem., *15*:1101-1123, 1969.

183. Marlow, H. W.: Groups involved in the Zimmermann and Kober reactions. J. Biol. Chem., *183*:167-171, 1950.

184. Maroulis, G. B.: Evaluation of hirsutism and hyperandrogenemia. Fertil. Steril., *36*:273-305, 1981.

185. Martucci, C., and Fishman, J.: Direction of estradiol metabolism as a control of its hormonal action—uterotrophic activity of estradiol metabolites. Endocrinology, *101*:1709-1715, 1977.

186. Masen, J. M.: A simplified procedure for serum butanol extractable iodine. Am. J. Clin. Pathol., *48*:561-567, 1967.

187. Massara, F., Genazzi, A. R., Camanni, F., et al.: Tests of prolactin secretion in the diagnosis of hyperprolactinemic states: Nomifensine and domperidine. Fertil. Steril., *35*:149-154, 1981.

188. Mathur, R. S., Moody, L. O., Landgrebe, S., et al.: Plasma androgens and sex hormone–binding globulin in the evaluation of hirsute females. Fertil. Steril., *35*:29-35, 1981.

189. Mattingly, D.: A simple fluorimetric method for the estimation of free 11-hydroxycorticoids in human plasma. J. Clin. Pathol., *15*:374-379, 1962.

190. Mauvais-Jarvis, P., Charransol, G., and Bobas-Masson, F.: Simultaneous determination of urinary androstanediol and testosterone as an evaluation of human androgenicity. J. Clin. Endocrinol. Metab., *36*:452-459, 1974.

191. Maxwell, M. H., Marks, L. S., Lupu, A. N., et al.: Predictive value of renin determinations in renal artery stenosis. JAMA, *238*:2617-2620, 1977.

192. McDonald, L. J., Robin, N. I., and Siegel, L.: Free thyroxine in serum as estimated by polyacrylamide gel filtration. Clin. Chem., *24*:652-656, 1978.

193. McDonough, P. B.: Amenorrhea—etiologic approach to diagnosis. Fertil. Steril., *30*:1-15, 1978.

194. McGuire, W. L., Carbone, P. P., and Vollmer, E. P., Eds.: Estrogen Receptors in Human Breast Cancer. New York, Raven Press, 1975.

195. McGuire, W. L., Raynaud, J-P., and Baulieu, E. E., Eds.: Progesterone Receptors in Normal and Neoplastic Tissues. New York, Raven Press, 1977.

196. McKenna, T. J., Jennings, A. S., Liddle, G. W., et al.: Pregnenolone, 17-OH pregnenolone, and testosterone of patients with congenital adrenal hyperplasia. J. Clin. Endocrinol. Metab., *42*:918-925, 1976.

197. Mehdi, S. Q., and Nussey, S. S.: A radioligand receptor assay for the long-acting thyroid stimulator. Inhibition by the long-acting thyroid stimulator of the binding of radioiodinated thyroid stimulating hormone to human thyroid membranes. Biochem. J., *145*:105-111, 1975.

198. Melani, F., Rubenstein, A. H., Oyer, P. E., et al.: Identification of proinsulin and C-peptide in human serum by a specific immunoassay. Proc. Natl. Acad. Sci., USA, *67*:148-155, 1970.

199. Melmed, S., Geola, F. L., Reed, A. W., et al.: A comparison of methods for assessing thyroid function in nonthyroidal illness. J. Clin. Endocrinol. Metab., *54*:300-306, 1982.

200. Melmon, K. L.: The endocrinologic function of selected autacoids: Catecholamines, acetylcholine, serotonin, and histamine. *In*: Textbook of Endocrinology. 6th ed. R. H. Williams, Ed. Philadelphia, W. B. Saunders Co., 1981, pp. 515-588.

201. Merimee, T. J., Rabinowitz, D., and Fineberg, S. E.: Arginine-initiated release of human growth hormone: Factors modifying the response in normal man. N. Engl. J. Med., *280*:1434-1438, 1969.

202. Migeon, C.: Diagnosis and management of congenital adrenal hyperplasia. Hosp. Pract. *12*(3):75-82, 1977.

203. Mikhail, G.: Hormone assays and the gynecologist. Fertil. Steril., *27*:229-237, 1976.

204. Miles, L. E. M.: Immunoradiometric assay (IRMA) and two-site IRMA systems (assay of soluble antigens using labeled antibodies). *In*: Handbook of Radioimmunoassay. G. E. Abraham, Ed. New York, Marcel Dekker, 1977, p. 131.

205. Miller, M., Dalakos, T., Moses, A. M., et al.: Recognition of partial defects in antidiuretic hormone secretion. Ann. Intern. Med., *73*:721-729, 1970.

206. Miller, M., and Moses, A.: Drug-induced states of impaired water excretion. Kidney Int. *10*:96-103, 1976.

207. Mimbs, R. B., and Bethune, J. E.: Acromegaly with normal fasting growth hormone concentration but abnormal growth hormone regulation. Ann. Intern. Med., *81*:781-784, 1974.

208. Moghissi, K. S.: Prediction and detection of ovulation. Fertil. Steril., *34*:89-98, 1980.

209. Moghissi, K. S., and Wallach, E. E.: Unexplained infertility. Fertil. Steril., *39*:5-21, 1983.

210. Moleman, P., and Borstrok, J. J. M.: Determination of urinary vanillylmandelic acid by liquid chromatography with electrochemical detection. Clin. Chem., *29*:878-881, 1983.

211. Mori, T., and Kriss, J. P.: Measurements by competitive binding radioassay of serum antimicrosomal and antithyroglobulin antibodies in Graves' disease and other thyroid disorders. J. Clin. Endocrinol. Metab., *33*:688-698, 1971.

212. Moyer, T. P., Jiang, N.-S., Tyce, G. M., et al.: Analysis for urinary catecholamines by liquid chromatography with amperometric detection: Methodology and clinical interpretation of results. Clin. Chem., *25*:256-263, 1979.

213. Murphy, B. E. P.: The determination of thyroxine by competitive protein-binding analysis, employing an anion-exchange resin and radiothyroxine. J. Lab. Clin. Med., *66*:161-167, 1965.

214. Murphy, B. E. P.: Some studies of the protein-binding of steroids and their application to the routine micro and ultramicro measurement of various steroids in body fluids by competitive protein-binding radioassay. J. Clin. Endocrinol. Metab., *27*:973-990, 1967.

215. Murphy, B. E. P.: Clinical evaluation of urinary cortisol determinations by competitive protein-binding radioassay. J. Clin. Endocrinol. Metab., *28*:343-348, 1968.

216. Murphy, B. E. P., and Pattee, C. J.: Determination of thyroxine utilizing the property of protein binding. J. Clin. Endocrinol. Metab., *24*:187-196, 1964.

217. Musto, J. D., Pizzolante, J. M., and Chesarone, V. P.: A comment on thyrotropin measurement and evaluation. Clin. Chem., *30*:329-330, 1984.

218. Nava, M., and De Groot, L. J.: Resin uptake of ^{125}I-labeled tri-iodothyronine as a test of thyroid function. N. Engl. J. Med., *266*:1307-1310, 1962.

219. Odell, W. D., and Swerdloff, R. S.: Abnormalities of gonadal function in men. Clin. Endocrinol., *8*:149-180, 1978.

220. Odell, W. D., and Wolfsen, R. A.: Hormones from tumors: Are they ubiquitous? Am. J. Med., *68*:317-318, 1980.

221. Olefsky, J. M.: Insulin resistance and insulin action—an in vitro and in vivo perspective. Diabetes, *30*:148-162, 1981.

222. Orczyk, G. P., Caldwell, B. V., and Behrman, H. R.: Plasma estradiol, estrone, estriol and urinary estriol glucuronide. *In*: Methods of Hormone Assay. B. M. Jaffe and H. R. Behrman, Eds. New York, Academic Press, 1979, p. 675.

223. Orsulak, P. J., Kizuka, P., Grab, E., et al.: Determination of urinary normetanephrine and metanephrine by radial compression liquid chromatography and electrochemical detection. Clin. Chem., *29*:305-309, 1983.

224. Orth, D.: Ectopic hormone production. *In*: Endocrinology and Metabolism. P. Felig, J. D. Baxter, A. E. Broadus, et al., Eds. New York, McGraw-Hill, Inc., 1981, pp. 1191-1217.

224a. Page, I. H., and McCubbin, J. W.: Renal Hypertension. Chicago, Year Book Medical Publishers, 1968.

225. Parker, M. L., Hammond, J. M., and Daughaday, W. H.: The arginine provocative test: An aid in the diagnosis of hyposomatotropism. J. Clin. Endocrinol. Metab., *27*:1129-1136, 1967.

226. Parks, J. S., Amrhein, J. A., Vaidya, V., et al.: Growth hormone response to propranolol-glucagon stimulation. A comparison with other tests of growth hormone reserve. J. Clin. Endocrinol. Metab., *37*:85-92, 1973.

227. Passon, P. G., and Peuler, J. D.: A simplified radiometric assay for plasma norepinephrine and epinephrine. Anal. Biochem., *51*:618-631, 1973.

228. Patchefsky, A. S., Solit, R., Phillips, L. D., et al.: Hydroxyindole-producing tumors of the pancreas. Carcinoid–islet cell tumor and oat cell carcinoma. Ann. Intern. Med., *77*:53-61, 1972.

229. Paulson, J. D., Keller, D. W., Wiest, W. G., et al.: Free testosterone concentration in serum: Elevation is the hallmark of hirsutism. Am J. Obstet. Gynecol., *128*:851-857, 1977.

230. Pavlatos, F., Smilo, R. P., and Forsham, P. H.: A rapid screening test for Cushing's syndrome. JAMA, *193*:720-723, 1965.

231. Penney, R., Blizzard, R. M., and Davis, W. T.: Sequential arginine and insulin tolerance tests. J. Clin. Endocrinol. Metab., *29*:1499-1501, 1969.

232. Pepperell, R. J.: Prolactin and reproduction. Fertil. Steril., *35*:267-274, 1981.

233. Pernow, B., and Waldenström, J.: Determination of 5-hydroxytryptamine, 5-hydroxyindole acetic acid and histamine in thirty-three cases of carcinoid tumor (argentaffinoma). Am. J. Med., *23*:16-25, 1957.

234. Peterson, R. E.: Measurement of plasma or serum cortisol. *In*: Lipids and Steroid Hormones in Clinical Medicine, F. W. Sunderman, Jr., Ed. Philadelphia, J. B. Lippincott Co., 1960, pp. 164-167.

235. Petruccelli, B., Bakris, G., Miller, T., et al.: A liquid chromatographic assay for 5-hydroxytryptophan, serotonin and 5-hydroxyindoleacetic acid in human body fluids. Acta Pharmacol. Toxicol., *51*:421-427, 1982.

236. Peuler, J. D., and Johnson, G. A.: Simultaneous single isotope radioenzymatic assay of plasma norepinephrine, epinephrine and dopamine. Life Sci., *21*:625-636, 1977.

237. Pfeifer, M. A., Halter, J. B., and Porte, D.: Insulin secretion in diabetes mellitus. Am. J. Med., *70*:579-588, 1981.

238. Phillips, L. S., and Vassilopoulou-Sellin, R.: Somatomedins (in two parts). N. Engl. J. Med., *302*:371-380, 438-446, 1980.

239. Pileggi, V. J., Lee, N. D., Golub, O. J., et al.: Determination of iodine compounds in serum. I. Serum thyroxine in the presence of some iodine contaminants. J. Clin. Endocrinol. Metab., *21*:1272-1279, 1961.

240. Pisano, J. J.: A simple analysis for normetanephrine and metanephrine in urine. Clin. Chim. Acta, *5*:406-414, 1960.

241. Pisano, J. J., Crout, R. J., and Abraham, D.: Determination of 3-methoxy-4-hydroxymandelic acid in urine. Clin. Chim. Acta, *7*:285-291, 1962.

242. Pollet, R. J., and Levey, G. S.: Principles of membrane receptor physiology and their application to clinical medicine. Ann. Intern. Med., *92*:663-680, 1980.

243. Porter, C. C., and Silber, R. H.: A quantitative color reaction for cortisone and related 17,21-dihydroxy-20-ketosteroids. J. Biol. Chem., *185*:201-207, 1950.

244. Pratt, J. J., Wiegman, T., Lappohn, R. E., et al.: Estimation of plasma testosterone without extraction and chromatography. Clin. Chim. Acta, *59*:337-346, 1975.

245. Price, J. D. E., and Lauener, R. W.: Serum and urine osmolality in the differential diagnosis of polyuric states. J. Clin. Endocrinol. Metab., *26*:143-148, 1966.

246. Rabkin, M. T., and Frantz, A. G.: Hypopituitarism: A study of growth hormone and other endocrine functions. Ann. Intern. Med., *64*:1197-1207, 1966.

247. Raisz, L. H., Yajnik, C. H., and Bockman, R. S.: Comparison of commercially available parathyroid hormone immunoassays in the differential diagnosis of hypercalcemia due to primary hyperparathyroidism or malignancy. Ann. Intern. Med., *91*:739-741, 1979.

248. Rakoff, A. E.: Ovulatory failure: Clinical aspects. Fertil. Steril., *27*:473-492, 1976.

249. Rassam, J. W., and Anderson, G.: Incidence of paramalignant disorders in bronchogenic carcinoma. Thorax, *30*:86-90, 1975.

250. Ratcliffe, W. A.: Direct (non-extraction) serum assays for steroids. *In*: Immunoassays for Clinical Chemistry. W. M. Hunter and J. E. T. Corrie, Eds. Edinburgh, Churchill Livingstone, 1983, pp. 401-409.

251. Ratcliffe, W. A., Corrie, J. E. T., Dalziel, A. H., et al.: Direct ^{125}I-radioligand assays for serum progesterone compared with assays involving extraction of serum. Clin. Chem., *28*:1314-1318, 1982.

252. Rayner, P. H. W.: Puberty: Precocious and delayed. Br. Med. J., *1*:1385-1387, 1976.

253. Rebar, R. W., Erickson, G. F., and Yen, S. S. C.: Idiopathic premature ovarian failure: Clinical and endocrine characteristics. Fertil. Steril., 37:35-41, 1982.

254. Riad-Fahmy, D., Read, G. F., Walker, R. F., et al.: Steroids in saliva for assessing endocrine function. Endocr. Rev., 3:367-395, 1982.

255. Riggin, R. M., and Kissinger, P. T.: Determination of catecholamines in urine by reverse-phase liquid chromatography with electrochemical detection. Anal. Chem., 49:2109-2111, 1977.

256. Robertson, G. L.: Psychogenic polydipsia and inappropriate diuresis. Arch. Intern. Med., 140:1574-1575, 1980.

257. Robin, N. J., Hagen, S. R., and Collaco, F.: Serum tests for measurement of thyroid function. Hormones, 2:266, 1971.

258. Rodbard, D.: Statistical quality control and routine data processing for radioimmunoassays and immunoradiometric assays. Clin. Chem., 20:1255-1270, 1974.

259. Rosen, S. W., and Weintraub, B. D.: Humours, tumors and caveats. Ann. Intern. Med., 82:274-276, 1975.

260. Rosenfield, R. L.: Plasma free androgen patterns in hirsute women and their diagnostic implications. Am. J. Med., 66:417-421, 1979.

261. Rubenstein, A. H., Kuzuya, H., and Horwitz, D. L.: Clinical significance of circulating C-peptide in diabetes mellitus and hypoglycemic disorders. Arch. Intern. Med., 137:625-632, 1977.

262. Rude, R. K., Sharp, C. F., Fredericks, R. S., et al.: Urinary and nephrogenous adenosine 3',5'-monophosphate in the hypercalcemia of malignancy. J. Clin. Endocrinol. Metab., 52:765-771, 1981.

263. Rudy, F. R., Bates, R. D., Cimorelli, A. J., et al.: Adrenal medullary hyperplasia: A clinicopathologic study of four cases. Hum. Pathol., 11:650-657, 1980.

264. Saavedra, J. M., Brownstein, M., and Axelrod, J.: A specific and sensitive enzymatic-isotopic microassay for serotonin in tissues. J. Pharmacol. Exp. Ther., 186:508-515, 1973.

265. Sato, T., Inoe, M., Masuyama, T., et al.: Simultaneous evaluation of the pituitary reserve of GH, TSH, ACTH, and LH in children. J. Clin. Endocrinol. Metab., 39:595-599, 1974.

266. Sawin, C. T., Chopra, D., Albano, J., et al.: The free triiodothyronine (T_3) index. Ann. Intern. Med., 88:474-477, 1978.

267. Sawin, C. T., and Mitchell, M. L.: A comparison of response of serum growth hormone to glucagon and vasopressin. Metabolism, 19:898-903, 1970.

268. Schimmel, M., and Utiger, R. D.: Thyroidal and peripheral production of thyroid hormones: Review of recent findings and their clinical implications. Ann. Intern. Med., 87:760-768, 1977.

269. Sebastian, A., Hulter, H. N., Kurtz, I., et al.: Disorders of distal nephron function. Am. J. Med., 72:289-307, 1982.

270. Segal, J., and Ingbar, S. H.: Plasma membrane–medicated effects of thyroid hormones. In: Endocrinology. I. A. Cumming, J. W. Funder, and F. A. O. Mendelsohn, Eds. Canberra, Australian Academy of Science, 1980.

271. Shambaugh, G. E., III: Biologic and cellular effects. In: The Thyroid. 4th ed. S. C. Werner and S. H. Ingbar, Eds. Hagerstown, Md., Harper & Row, 1978, pp. 115-124.

272. Sherood, I. M.: The multiple causes of hypercalcemia in malignant disease. N. Engl. J. Med., 303:1412-1413, 1980.

272a. Shipe, J. R., Savory, J., and Wills, M. R.: Improved liquid-chromatographic determination of 3-methoxy-4-hydroxyphenylethyleneglycol in urine with electrochemical detection. Clin. Chem., 30:140-143, 1984.

273. Shoup, R. E., and Kissinger, P. T.: Determination of urinary normetanephrine, metanephrine and 3-methoxytyramine by liquid chromatography with amperometric detection. Clin. Chem., 23:1268-1274, 1977.

274. Siiteri, P. K., and MacDonald, P. C.: Role of extraglandular estrogens in human endocrinology. In: Handbook of Physiology. S. R. Geiger, E. B. Astwood, and R. O. Greep, Eds. Washington, D.C., American Physiol. Society, 1973, p. 615.

275. Simpson, E. R.: In: Initiation of Parturition, Prevention of Prematurity. Report of the fourth Ross Conference on Obstetric Research. P. C. MacDonald and E. C. Hasselmeyer, Eds. Columbus, Ohio, Ross Laboratories, 1983, p. 94.

276. Singer, W., Kovacs, K., Ryan, N., et al.: Ectopic ACTH syndrome: Clinicopathological correlations. J. Clin. Pathol., 31:591-598, 1978.

277. Sinha, Y. N., Selby, F. W., Lewis, U. J., et al.: A homologous radioimmunoassay for human prolactin. J. Clin. Endocrinol. Metab., 36:509-516, 1973.

278. Sjoerdsma, A., Weissbach, H., and Udenfriend, S.: Simple test for diagnosis of metastatic carcinoid (argentaffinoma). JAMA, 159:397, 1955.

279. Skelley, D. S., Brown, L. P., and Besch, P. K.: Radioimmunoassay. Clin. Chem., 19:146-186, 1973.

280. Slag, M. F., Morley, J. E., Elson, M. K., et al.: Free thyroxine levels in critically ill patients: A comparison of currently available assays. JAMA, 246:2702-2706, 1981.

281. Slag, M. F., Morley, J. E., Elson, M. K., et al.: Impotence in medical clinic outpatients. JAMA, 249:1736-1740, 1983.

282. Smith, A. D.: Mechanisms involved in the release of noradrenaline from sympathetic nerves. Br. Med. Bull., 29:123-129, 1973.

283. Snyder, S. M., Cavalieri, R. R., and Ingbar, S. H.: Simultaneous measurement of percentage free thyroxine and triiodothyronine: Comparison of equilibrium dialysis and Sephadex chromatography. J. Nucl. Med., 17:660-664, 1976.

284. Sobel, C. S., Golub, O. J., Henry, R. J., et al.: Study of the Norymberski methods for determination of 17-ketogenic steroids(17-hydroxycorticosteroids) in urine. J. Clin. Endocrinol. Metab., 18:208-221, 1958.

285. Solomon, D. H., Benotti, J., DeGroot, L. J., et al.: Revised nomenclature for tests of thyroid hormones in serum. J. Clin. Endocrinol. Metab., 42:595-598, 1976.

286. Sophianopoulos, J. A., Jerkunica, I., and Sgoutas, D. S.: A rapid ultrafiltration method for the measurement of free thyroxine (FT_4) or free triiodothyronine (FT_3) in serum. Clin. Chem., 25:1080-1081, 1979.

287. Spark, R. F.: Simplified assessment of pituitary-adrenal reserve. Measurement of serum 11-deoxycortisol and cortisol after metyrapone. Ann. Intern. Med., 75:717-723, 1971.

288. Speckart, P. F., Nicoloff, J. T., and Bethune, J. E.: Screening for adrenocortical insufficiency with cosyntropin (synthetic ACTH). Arch. Intern. Med., *128*:761-763, 1971.

289. Speek, A. J., Odink, J., Schrijver, J., et al.: High-performance liquid chromatographic determination of urinary free catecholamines with electrochemical detection after prepurification on immobilized boric acid. Clin. Chim. Acta, *128*:103-113, 1983.

290. Spiegel, R. J., Oliff, A. I., Vigersky, R. A., et al.: Adrenal suppression after short-term corticosteroid treatment. Lancet, *1*:630-633, 1979.

291. Starr, J. I., Horwitz, D. L., Rubenstein, A. H., et al.: Insulin, proinsulin and C-peptide. *In*: Methods of Hormone Radioimmunoassay. B. M. Jaffe and H. R. Behrman, Eds. New York, Academic Press, 1974, pp. 289-311.

292. Steiner, D. E.: Insulin today. Diabetes, *26*:322-340, 1977.

293. Sterling, K.: Thyroid hormone action at the cell level (in two parts). N. Engl. J. Med., *300*:117-123, 173-177, 1979.

294. Sterling, K., and Tabachnik, M.: Resin uptake of I^{131}-triiodothyronine as a test of thyroid function. J. Clin. Endocrinol. Metab., *21*:456-464, 1961.

295. Sterling, K., and Brenner, M. A.: Free thyroxine in human serum: Simplified measurement with the aid of magnesium precipitation. J. Clin. Invest., *45*:153-163, 1966.

296. Steward, A. F., Horst, R., Deftos, L. J., et al.: Biochemical evaluation of patients with cancer-associated hypercalcemia. N. Engl. J. Med., *303*:1377-1383, 1980.

297. Stonesifer, I. D., Jordan, R. M., and Kohler, P. O.: Somatomedin C in treated acromegaly: Poor correlation with growth hormone and clinical response. J. Clin. Endocrinol. Metab., *53*:931-934, 1981.

298. Streeten, D. H. P., Tomycz, N., and Anderson, G. H.: Reliability of screening methods for the diagnosis of primary aldosteronism. Am. J. Med., *67*:403-413, 1979.

299. Sullivan, J. M., and Solomon, H. S.: The diagnosis of pheochromocytoma: Overnight excretion of catecholamine metabolites. JAMA, *231*:618-619, 1975.

300. Sunderman, F. W., Jr.: Measurement of urinary corticosteroids; modified Porter-Silber method. *In*: Lipids and Steroid Hormones in Clinical Medicine. F. W. Sunderman and F. W. Sunderman, Jr., Eds. Philadelphia, J. B. Lippincott Co., 1960, pp. 162-164.

301. Sutton, M. G., Sheps, S. G., and Lie, J. T.: Prevalence of clinically unsuspected pheochromocytoma. Mayo Clin. Proc., *56*:354-360, 1981.

302. Swerdloff, R. S., and Boyers, S. P.: Evaluation of the male partner of an infertile couple—an algorithmic approach. JAMA, *247*:2418-2422, 1982.

303. Szpunar, W. E., Stoffer, S. S., Bednarz, M. N., et al.: Screening for minimal hypothyroidism. How sensitive is the free thyroxine assay? Postgrad. Med., *74*:259-262, 1983.

304. Taylor, S. I., Grunberger, G., Marcus-Samuels, B., et al.: Hypoglycemia associated with antibodies to the insulin receptor. N. Engl. J. Med., *307*:1422-1426, 1982.

305. Taymor, M. L., Yussman, M. A., and Gminski, D.: Estrogen monitoring in ovulation induction. Fertil. Steril., *21*:759-762, 1970.

306. Thomas, A. K., and Forrest, M. S.: Infertility: A review of 291 infertile couples over eight years. Fertil. Steril., *34*:106-111, 1980.

307. Tietz, N. W., Ed.: Clinical Guide to Laboratory Tests. Philadelphia, W. B. Saunders Co., 1983.

308. Tucker, H. St. G., Lankford, H. V., Gardner, D. F., et al.: Persistent defect in regulation of prolactin secretion after successful pituitary tumor removal in women with the galactorrhea-amenorrhea syndrome. J. Clin. Endocrinol. Metab., *51*:968-971, 1980.

309. Turkington, R. W.: Ectopic production of prolactin. New Engl. J. Med., *285*:1455-1458, 1971.

310. Turkington, R. W., Estowski, A., and Link, M.: Secretion of insulin or connecting peptide. A predictor of insulin dependence of obese "diabetics." Arch. Intern. Med., *142*:1102-1105, 1982.

311. Turkington, R. W., and Weindling, H. D.: Insulin secretion in the diagnosis of adult-onset diabetes mellitus. JAMA, *240*:833-836, 1978.

312. Turner, C. D., and Bagnara, J. T.: General Endocrinology. 6th ed. Philadelphia, W. B. Saunders Co., 1976.

313. Turner, R. C., and Heding, L. G.: Plasma proinsulin, C-peptide and insulin in diagnostic suppression tests for insulinomas. Diabetologia, *13*:571-577, 1977.

314. Udenfriend, S., Titus, E., and Weissbach, H.: The identification of 5-hydroxy-3-indoleacetic acid in normal urine and a method for its assay. J. Biol. Chem., *216*:499-505, 1955.

315. Van Herle, A. J., Uller, R. P., Matthews, N. I., et al.: Radioimmunoassay for measurement of thyroglobulin in human serum. J. Clin. Invest., *52*:1320-1327, 1973.

316. Van Niekerk, W. A.: Chromosomes and the gynecologist. Am. J. Obstet. Gynecol., *130*:862-875, 1978.

317. Vaughan, E. D., Buhler, F. R., and Laragh, J. H.: Renovascular hypertension: Renin measurements to indicate hypersecretion and contralateral suppression, estimate renal plasma flow, and score for surgical curability. Am. J. Med., *55*:402-414, 1973.

318. Veldhuis, J. D., and Melby, J. C.: Isolated aldosterone deficiency in man: Acquired and inborn errors in the biosynthesis or action of aldosterone. Endocr. Rev., *2*:495-517, 1981.

319. Vermeulen, A., Stoica, T., and Verdonck, L.: The apparent free testosterone concentration, an index of androgenicity. J. Clin. Endocrinol. Metab., *33*:759-767, 1971.

320. Verner, J. V., and Morrison, A. B.: Endocrine pancreatic islet disease with diarrhea: Report of a case due to diffuse hyperplasia of nonbeta islet tissue with a review of 54 additional cases. Arch. Intern. Med., *133*:492-500, 1974.

321. Von Euler, U. S.: The catecholamines. Adrenaline; noradrenaline. *In*: Hormones in Blood. C. H. Gray and A. L. Bacharach, Eds. New York, Academic Press, 1961, p. 515-582.

322. Walfish, P. G.: The best way to screen for neonatal hypothyroidism. Diag. Med., *7*(2):67-75, 1984.

323. Wartofsky, L., and Burman, K. D.: Alterations in thyroid function in patients with systemic illness: The "euthyroid sick syndrome." Endocr. Rev., *3*:164-217, 1982.

323a. Weichert, R. F.: The neural ectodermal origin of the peptide-secreting endocrine glands. Am. J. Med., *49*:232-241, 1970.

324. Weil-Malherbe, H.: The estimation of total (free and conjugated) catecholamines and some catecholamine metabolites in human urine. *In:* Methods Biochem. Anal., D. Glick, Ed. New York, Interscience Publishers, 1968. Vol. 16, pp. 293-326.

325. Weil-Malherbe, H.: The chemical estimation of catecholamines and their metabolites in body fluids and tissue extracts. *In:* Methods Biochem. Anal. (Supplemental volume on biogenic amines), D. Glick, Ed., New York, Interscience Publishers, 1971, pp. 119-152.

326. Wentz, A. C.: Diagnosing luteal phase inadequacy. Fertil. Steril., *37*:334-335, 1982.

327. Wentz, A. C., White, R. I., Jr., Migeon, C. J., et al.: Failure of ovarian and adrenal vein catheterization to aid in the evaluation of hirsutism. Fertil. Steril., *27*:218, 1976.

328. Werner, S. C.: Immune System, III. Role in thyroid disease. *In:* The Thyroid. 4th ed. S. C. Werner and S. H. Ingbar, Eds. Hagerstown, Md., Harper and Row, 1978, pp. 615-623.

329. White, M. G.: Bartter's syndrome: A manifestation of renal tubular defects. Arch. Intern. Med., *129*:41-47, 1972.

330. No entry.

331. Wilke, T. J.: Five kits for estimating free thyroxine concentration in serum evaluated and correlated with other indices to thyroid status. Clin. Chem., *28*:2051-2056, 1982.

332. Wilke, T. J.: Free thyroid hormone index, thyroid hormone/thyroxine-binding globulin ratio, triiodothyronine uptake and thyroxine-binding globulin compared for diagnostic value regarding thyroid function. Clin. Chem., *29*:74-79, 1983.

333. Williams, C. M., and Greer, M.: Estimation by gas chromatography of urinary homovanillic acid and vanilmandelic acid in neuroblastoma. Methods Med. Res., *12*:106-114, 1970.

334. Williams, R. H., Ed.: Textbook of Endocrinology. 6th ed. Philadelphia, W. B. Saunders Co., 1981.

335. Wilson, M. A., and Miles, L. E.: Radioimmunoassay of insulin. *In*: Handbook of Radioimmunoassay. G. E. Abraham, Ed. New York, Marcel Dekker, Inc., 1977, p. 275.

336. Witherspoon, L. R., Shuler, S. E., and Garcia, M. M.: The triiodothyronine uptake test: An assessment of methods. Clin. Chem., *27*:1272-1276, 1981.

337. Wotiz, H. H., and Chattoraj, S. C.: The role of gas-liquid chromatography in steroid hormone analysis. J. Chromatogr. Sci., *11*:167-174, 1973.

338. Wotiz, H. H., Chattoraj, S. C., Kudisch, M., et al.: Impeding estrogens and the etiology of breast cancer. Cancer Res., *38*:4012-4020, 1978.

339. Yeo, P. P. B., Lewis, M., and Evered, D. C.: Radioimmunoassay of free thyroid hormone concentrations in the investigation of thyroid disease. Clin. Endocrinol., *6*:159-165, 1977.

340. Young, R. L., Fuchs, R. J., Woltjen, M. J., et al.: Glucose-insulin response to oral glucose in a healthy obese population. Diabetes, *28*:208-212, 1979.

341. Youssefnejadian, E., Florensa, E., Collins, W. P., et al.: Radioimmunoassay of 17-hydroxyprogesterone. Steroids, *20*:773-788, 1972.

342. Zakarija, M., McKenzie, J. M., and Banovac, K.: Clinical significance of assay of thyroid-stimulating antibody in Graves' disease. Ann. Intern. Med., *93*:28-32, 1980.

343. Zerbe, R. L., and Robertson, G. L.: A comparison of plasma vasopressin measurements with a standard indirect test in the differential diagnosis of polyuria. N. Engl. J. Med., *305*:1539-1546, 1981.

ELECTROLYTES, BLOOD GASES, AND ACID-BASE BALANCE

Section One

Electrolytes

by Norbert W. Tietz, Ph.D., Elizabeth L. Pruden, Ph.D., and Ole Siggaard-Andersen, M.D., Ph.D.

Electrolytes are described as anions or cations, depending on whether they move in an electrical field toward the anode or toward the cathode, that is, whether they have a positive or negative charge. They are essential components of all living matter and include the major electrolytes Na^+, K^+, Ca^{2+}, Mg^{2+}, Cl^-, HCO_3^-, $HPO_4^{-,2-}$, SO_4^{2-}, and lactate, as well as the trace elements. Although amino acids and proteins in solution also carry an electrical charge, in clinical chemistry they are usually considered separately from electrolytes. The major electrolytes occur primarily as free ions. The trace elements occur primarily in combination with proteins and thus are also frequently considered separately. (See Chapter 8C.)

The dietary requirement for electrolytes varies widely; most need to be consumed only in small amounts or at rare intervals, and are retained when in short supply. Some, like calcium and phosphorus, are continuously excreted and must be ingested regularly in order to prevent deficiency. Excessive intake leads to correspondingly increased excretion, mainly in the urine. Abnormal loss of electrolytes as a result of profuse perspiration, vomiting, or diarrhea is readily assessed by laboratory tests and can be corrected by administration of oral or parenteral salt solutions.

The role of electrolytes in the human body is manifold. There are almost no metabolic processes which are not dependent on or affected by electrolytes. Among other functions of the electrolytes are maintenance of osmotic pressure and water distribution in the various body fluid compartments, maintenance of the proper pH, regulation of the proper function of the heart and other muscles, involvement in oxidation-reduction (electron transfer) reactions, and participation in catalysis as cofactors for enzymes. Thus, it becomes quite apparent that abnormal levels of these electrolytes and trace elements may be either the cause or the consequence of a variety of disorders, and that determination of electrolytes is one of the most important functions of the clinical laboratory.

In this section, we discuss sodium, potassium, chloride, and bicarbonate in body fluids as they are commonly grouped in the familiar test order for an "electrolyte profile."

Other substances which are also electrolytes, but which have special functions in particular contexts, are discussed elsewhere in the book—calcium and phosphorus in Chapter 12, iron in Chapter 16, magnesium and the trace elements in Chapter 8C, and amino acids and proteins in Chapter 4.

Specimens for Electrolyte Determinations

Blood, collected by venipuncture into an evacuated tube, is the usual specimen employed for assay of sodium, potassium, chloride, and bicarbonate. Capillary blood, collected in micro sample tubes or in capillary tubes, or heparinized arterial specimens obtained in conjunction with blood gases and pH determinations may also be presented for analysis. (Collection procedures are described in detail in Chapter 3.) Either serum or plasma is appropriate for assay and with ion-selective electrodes whole blood may be used. Differences for values of these analytes, between arterial and venous samples and between serum and plasma, have been documented (Tables 3-3 and 3-5) and must be taken into consideration, mainly when reporting and interpreting results in particular situations. Heparin, either the lithium or ammonium salt, is required if plasma is assayed. Use of plasma has the advantage of expediting the assay and shortening turnaround time in emergency situations. Specimen tubes should be centrifuged unopened and the serum or plasma should be separated promptly. Since grossly lipemic blood can be a source of analytical error, ultracentrifugation of serum or plasma prior to analysis may be necessary.

Urine collection for sodium, potassium, or chloride assay should be made without addition of preservatives. Although a physician, on occasion, may opt for a random specimen, timed collections are recommended so that results can be compared to established reference ranges. Feces as well as aspirates and drainages from different portions of the gastrointestinal tract may also be submitted for analysis.

Note that specific concerns of specimen collection and handling will be addressed in the following pages under individual analytes. Collection of sweat is described in Chapter 14.

SODIUM

Sodium is the major cation of extracellular fluid. Because it represents ∼90% of the 154 mmol of inorganic cations per liter of plasma, sodium is responsible for almost half the osmolality of the plasma. It therefore plays a central role in maintaining the normal distribution of water and the osmotic pressure in the extracellular fluid compartment. The normal daily diet contains 8–15 g (130–260 mmol) of sodium chloride, which is nearly completely absorbed from the gastrointestinal tract. Since the body requirement is only 1–2 mmol/d, the excess is excreted by the kidneys, which are therefore the ultimate regulators of the amount of sodium and potassium in the body. Sodium is initially filtered by the glomeruli, but 60–70% of the filtered load is reabsorbed in the proximal tubules along with bicarbonate and water. Another 25–30% is reabsorbed in the loop of Henle along with chloride and more water. In the distal tubules, interaction of the adrenocortical hormone aldosterone with the coupled systems of Na^+-K^+ and Na^+—H^+ exchange directly modulates the reabsorption of sodium, and indirectly of chloride, from the remaining 5–15% of the filtered load. It is the management of this fraction of the filtered sodium that sets the renal threshold* for sodium at 110–130 mmol/L and determines the amount of sodium excreted in the urine. The mechanisms of these processes are discussed in detail in Chapter 11.

Hyponatremia (decreased plasma sodium concentration) is a predictable consequence of decreased intake of sodium, particularly that precipitated or complicated by unusual losses of sodium from the gastrointestinal tract, kidneys, or sweat glands. *Depletional* hyponatremia may be the result of prolonged vomiting, persistent diarrhea, or salt-losing enteropathies, i.e., conditions where sodium loss is greater than water loss. Renal loss due to diminished tubular reabsorption may be caused by inappropriate choice, dose, or use of diuretics; by primary or secondary deficiency of aldosterone and other mineralocorticoids; or by severe polyuria. Hyponatremia is common in metabolic acidosis (e.g., diabetic ketoacidosis) where cations are lost by co-excretion with large amounts of organic anions, and in renal tubular acidosis where either impaired reabsorption or defect in Na^+-H^+ exchange may be the cause. Any form of alkalosis or any condition associated with alkalinized urine is also a factor contributing to increased renal loss of sodium. The *dilutional* type of hyponatremia occurs with edema, ascites in chronic cardiac failure, uncontrolled diabetes, hepatic cirrhosis, nephrotic syndrome, and malnutrition. This type of hyponatremia is secondary to excessive retention of water; total body sodium may be normal or increased. In SIADH (syndrome of inappropriate antidiuretic hormone secretion, Chapter 9, p. 1035), hyponatremia is seen because of the expansion of extracellular fluid in conjunction with diminished Na^+ reabsorption in the distal tubules.

---

*The renal threshold of a substance is the plasma concentration at which the rate of glomerular filtration of the substance begins to exceed the rate at which the tubules can reabsorb it. See also Chapter 11.

Hypernatremia (increased plasma sodium concentration) is often attributable to excessive loss of sodium-poor body fluids, i.e., to water loss disproportionate to sodium loss, as may occur in profuse sweating, prolonged hyperpnea, vomiting or diarrhea, or polyuria. Other causes include decreased production of antidiuretic hormone or decreased tubular sensitivity to the hormone, osmotic diuresis, inappropriate forms of parenteral therapy with saline solutions, or high salt intake without corresponding intake of water. Hypernatremia may also occur in hyperaldosteronism; in hyperadrenocorticism (Cushing's syndrome), in which increased production of mineralocorticoids causes increased tubular reabsorption of sodium; in certain types of brain injury; and in response to insulin treatment of uncontrolled diabetes. In the last instance, falling plasma concentration of glucose is believed to cause transfer of extracellular sodium into intracellular fluid in order to equalize osmotic pressure in both compartments. Also, decrease in plasma glucose, by decreasing plasma osmolality, causes contraction of the extracellular fluid volume. Hypernatremia is often associated with hypercalcemia and hypokalemia.

Hypernatriuria (increased urinary sodium excretion) is obviously a frequent corollary to initial stages of development of hyponatremia; it is observed in hypoaldosteronism, adrenal failure, salt-losing nephritis, diuretic therapy, and SIADH. Physiologic causes are increased dietary intake and postmenstrual diuresis. Conversely, *hyponatriuria* is associated with low dietary intake of sodium and premenstrual retention of sodium and water. It occurs pathologically in adrenocortical hyperfunction, hyperaldosteronism, with decreased glomerular filtration rates as in congestive heart failure, and in acute oliguria and prerenal azotemia; in these instances, hyponatriuria is a corollary to sodium and water retention, i.e., to expansion of extracellular fluid volume.

Specimens. Serum, heparinized plasma, sweat, urine, feces, or gastrointestinal fluids may be assayed. Timed collections of urine, feces, or gastrointestinal fluids are needed to allow comparison of values with reference ranges or for the determination of rates of electrolyte loss. Serum, plasma, and urine may be stored at 2–4 °C or frozen for delayed analysis. Erythrocytes contain only one tenth the sodium present in plasma; thus hemolysis of blood does not cause a significant decrease in observed sodium values for serum or plasma unless hemolysis is severe, in which case there is a dilutional effect.

Fecal and gastrointestinal fluid specimens will require preparation prior to assay. Only those stools which are liquid justify the trouble of analysis since it is only when liquid feces are passed that fecal losses of electrolytes are significant. Liquid stool specimens should be clarified of particulate matter by filtration through gauze or filter paper or by centrifugation. Because even filtered or centrifuged liquid samples can be expected to contain bacteria, the risk of contaminating sampling systems of automated instrumentation is high; either special cleaning and flushing procedures must follow the assay, or the liquid specimen, in a loosely capped tube, should be boiled prior to assay. Semisolid stool can be prepared either by dry ashing or by homogenation of 30 g of well-mixed specimen in 100 mL of distilled water, followed by clarification.[27] The chief difficulty with these specimens is their unpredictable concentration which may necessitate a number of different dilutions to bring the concentration within the analytical range of the method to be employed. Feces and gastrointestinal fluids may be stored frozen to avoid unwarranted microbial growth.

Determination of Sodium in Body Fluids

Sodium may be determined by atomic absorption spectrophotometry (AAS), flame emission spectrophotometry (FES), or electrochemically with a sodium ion-selective electrode (ISE). Of these methods, FES and ISE are currently most often used. Excellent accuracy and coefficients of variation of $<1.5\%$ are readily achieved with modern equipment, reliable standards, and a good quality control program. Sodium and potassium are routinely assayed together and flame emission and ion-selective electrode approaches to their analysis are described under the heading of Methods for the Determination of Sodium and Potassium.

Reference Ranges.[33] The range for *serum* sodium is 136–146 mmol/L* from infancy throughout life. The lower limit of the range for premature newborns at 48 h and for cord blood from full term newborns is ∼127 mmol/L. *Urinary sodium* excretion varies with dietary intake; for persons on an average diet, a range of 40–220 mmol/d is typical. The rate of sodium excretion during the night is only 20% of the peak rate during the day, an indication of large diurnal

---

*In many laboratories, mEq/L remains the unit for expressing electrolyte concentrations; mmol/L is the SI unit.

variation. The sodium concentration of cerebrospinal fluid is 138–150 mmol/L, and that of sweat is 10–40 mmol/L; a value > 70 mmol/L is considered highly suggestive of cystic fibrosis. Values >40 mmol/L may be seen in Addison's disease and in a variety of familial disorders.[33] Fecal sodium excretion (for eight healthy subjects) has been reported as 7.8 ± 2.0 mmol/d (SEM),[4] but is cited elsewhere as <10 mmol/d.[33]

POTASSIUM

Potassium is the major intracellular cation. In tissue cells, its average concentration is 150 mmol/L. In erythrocytes the concentration is 105 mmol/L, $\sim$23 times the concentration in plasma. High intracellular concentrations are maintained because K^+ diffuses only slowly outward through the cell membrane while the Na^+,K^+-ATPase pump, which is fueled by oxidative energy, continually transports K^+ into the cell against the concentration gradient. The pump is a critical factor in maintaining and adjusting the ionic gradients upon which nerve impulse transmission and contractility of cardiac and skeletal muscle depend. When the efficiency of the pump decreases because of depletion of metabolic substrates for ATP production or because of competition for ATP between the pump and other energy-consuming activities of the cell, diffusion (mass transfer) of K^+ down the concentration gradient, out of the cell into the plasma, exceeds pump-mediated K^+ uptake. Human blood banked for transfusion is fortified with glucose (dextrose) to forestall ATP depletion in the living blood cells; refrigeration, however, slows cellular metabolism so that K^+ leakage into the plasma still occurs. Skeletal muscle activity, because it draws its energy from anaerobic glycolysis (which has limited capacity for ATP production), causes K^+ efflux from muscle cells into plasma.

The body requirement for potassium is satisfied by a dietary intake of 50–150 mmol/d. Potassium absorbed from the gastrointestinal tract is rapidly distributed; a small amount is taken up by cells but the majority is excreted by the kidneys. Potassium filtered through the glomeruli is almost completely reabsorbed in the proximal tubules and is then secreted in the distal tubules. The amount of potassium excreted in urine varies relative to intake. The secretory system responds almost immediately to K^+ loading with an increase in K^+ output, and urine collected relative to a period of high intake may have a potassium concentration as high as 100 mmol/L.[8] But tubular secretory response to conserve K^+ in instances of total body depletion is very slow in the initial stages of depletion; the normal kidney continues to excrete potassium at a rate of 20–30 mmol/d. Unlike the prompt response of the tubules to conserve Na^+ in deficit states, 1–2 weeks may be required for the tubules to reduce K^+ excretion to 5–10 mmol/d, even when there is a total body deficit of K^+. Fortunately, in deficit states intracellular stores of K^+ remain adequate to maintain the K^+ concentration in the extracellular compartment at a normal level. Factors that regulate distal tubular secretion of potassium are intake of Na^+ and K^+, flow rate in the distal tubules, plasma level of mineralocorticoids, and acid-base balance. Because the conservation mechanism is slow to respond, potassium depletion is an early consequence of restricted K^+ intake or of unusual losses of K^+ by extrarenal routes. Diminished glomerular filtration rate is typical of renal failure, and the consequent decrease in distal tubular flow rate is an important factor in the retention of potassium. The mineralocorticoid aldosterone enhances K^+ secretion in portions of the distal tubules and collecting ducts; the effect is interrelated with the control of Na^+ reabsorption by aldosterone and with Na^+-K^+ exchange. Renal tubular acidosis, as well as metabolic and respiratory acidosis and alkalosis, affects renal regulation of potassium excretion. General aspects of these topics are discussed in Section Three of the present chapter and in Chapter 11.

Disturbance of potassium homeostasis has serious consequences. For example, decrease of extracellular K^+ is characterized by muscle weakness, irritability, and paralysis; fast heart rate and specific conduction effects apparent by electrocardiographic examination; and eventual cardiac arrest. Because extracellular potassium concentrations are maintained at the expense of intracellular supply, plasma potassium and neuromuscular function can be normal even with substantial total body deficit of K^+; plasma potassium <3.0 mmol/L is associated with marked neuromuscular symptoms and is evidence of a critical degree of intracellular depletion. Abnormally high extracellular potassium produces symptoms of mental confusion; weakness, numbness, and tingling of the extremities; weakness of the respiratory muscles; flaccid paralysis of the extremities; slowed heart rate, typical conduction defects on the electrocardiogram, and eventually peripheral vascular collapse and cardiac arrest. The symptoms are apparent at potassium levels > 7.5 mmol/L; levels > 10.0 mmol/L have been, in most cases, fatal.

Causes of *hypokalemia* (decreased plasma potassium concentration) can be grouped into decreased intake, redistribution of extracellular K^+ into intracellular fluid, and increased loss of potassium-rich body fluids. Situations of decreased intake include chronic starvation and postoperative therapy with potassium-poor fluids. Redistribution is illustrated by the fall in plasma potassium when insulin therapy of diabetic hyperglycemia is instituted; cellular uptake of glucose is accompanied by uptake of potassium. Hypokalemia is a feature also of alkalosis, where K^+ moves from the extracellular fluid into the cell as H^+ moves in the opposite sense; thus, all other things being equal, alkalosis of itself causes hypokalemia. On the other hand, intracellular potassium depletion may be a cause of alkalosis, as will be explained in Section Three of this chapter. Gastrointestinal loss of potassium occurs with fluids lost because of vomiting, diarrhea, or intestinal fistulas; renal losses may be the result of renal tubular acidosis, primary or secondary aldosteronism, or Cushing's or Bartter's syndromes. Aldosterone acts on renal tubular excretion to decrease potassium reabsorption; this effect is in contrast to aldosterone's effect to enhance sodium reabsorption.

Hyperkalemia (increased plasma potassium concentration) may be precipitated by intravenous infusion of K^+ at a rate in excess of 20 mmol/h or by treatment using K^+ solutions with concentrations > 40 mmol/L or doses of > 80 mmol/d. Overtreatment is unlikely to produce hyperkalemia so long as renal function is normal because excess K^+ is readily excreted in the urine. Transfer of intracellular K^+ into extracellular fluid may occur in dehydration and shock with tissue hypoxia, diabetic ketoacidosis, massive intravascular or extracorporeal hemolysis, severe burns, and with violent muscular activity such as that in status epilepticus. Decreased excretion of potassium in acute renal failure or endstage renal failure with oliguria or anuria and acidosis is a common cause of hyperkalemia. In renal failure due to shock and in renal tubular acidosis, interference with Na^+-H^+ exchange in the tubules leads to retention of potassium. The hyperkalemia in acidosis is the result of K^+ moving from intracellular fluids into the plasma as H^+ moves into the cells from extracellular fluid. Hyperkalemia occurs along with sodium depletion in adrenocortical insufficiency because, in the absence of adequate amounts of aldosterone and other mineralocorticoids, diminished Na^+ reabsorption and Na^+-K^+ exchange and decreased K^+ secretion lead to retention of K^+.

Hyperkaliuria, increased urinary excretion of potassium, occurs at the onset of starvation, in primary and secondary aldosteronism, in primary renal diseases, in renal tubular syndromes, and during the recovery phases of acute tubular necrosis, metabolic acidosis, and metabolic alkalosis. Although *hypokaliuria* is eventually a mark of body depletion of potassium, its occurrence is far less worthy of note than the hypokalemia and symptoms of depletion which appear much earlier in a disease process.

Specimens. Comments made earlier on specimens for sodium analysis are generally applicable to those for potassium analysis. Some additional points should, however, be considered.

Specimens for serum or plasma assay of potassium must be collected in such a way as to minimize hemolysis; release of potassium from as few as 0.5% of the erythrocytes can increase the serum level by 0.5 mmol/L. With greater use of extracorporeal circulation in cardiac surgery and renal dialysis, however, the clinical laboratory now often analyzes plasma from heparinized patients in which case hemolysis is the result of the procedure, and the amount of potassium added to the plasma by the procedure is important clinical information.

Potassium levels in plasma have been shown to be 0.1–0.7 mmol/L lower than in serum. The difference is due to release of potassium from platelets ruptured in the coagulation process. These considerations now make plasma the specimen of choice, and emphasize the necessity of noting on reports whether serum or plasma was assayed and whether the specimen was visibly hemolyzed.

Plasma potassium levels may be increased 10–20% as a result of muscle activity if the patient opens and closes his fist repeatedly prior to venipuncture. Increases can also occur as a result of K^+ leakage from erythrocytes when plasma or serum is not promptly separated from cells after collection, or if the whole blood specimen is chilled prior to separation. The increase in serum is of the order of 0.2 mmol/L in 1.5 h at 25 °C; at 4 °C, the increase is 1.5 times greater.[16] Extreme thrombocytosis or leukocytosis can also affect measured potassium levels. The effect of thrombocytes is prominent when serum is taken for assay; the apparent potassium concentration may be twice the actual concentration. In such cases, when levels from serum and a paired platelet-poor plasma have been compared, normal potassium levels were observed for the plasma.[19] The effect of leukocytosis is biphasic—in the first 30–60 min to decrease plasma potassium as the metabolically active leukocytes consume glucose, and subsequently to increase potassium when glucose substrate is exhausted and K^+ leakage begins.[2] When the leukocyte count is > 100 000/μL and hypokalemia is already a characteristic of the disease as it is in acute myeloid leukemia, glycolysis at room temperature may cause the potassium deficit indicated by the assay to seem more profound than it really is. The current recommendation for the most reliable potassium determinations is to collect blood with heparin, to maintain it at 37 °C, and to separate the plasma within minutes by high-speed centrifugation and without cooling.[18] In practical terms, separation within 15–60 min at room temperature is unlikely to introduce great error, in a vast majority of instances.

Determination of Potassium in Body Fluids

Potassium may be determined by atomic absorption spectrophotometry, flame emission spectrophotometry, or electrochemically with a potassium ion-selective electrode. The last two approaches are currently the most popular. In the physiological range, it should be possible with modern equipment and good quality control to achieve coefficients of variation of <2% for plasma. Methods are described below.

Reference Ranges.[33] The reference range for serum of adults is 3.5–5.0 mmol/L; for plasma the range is 3.5–4.5 mmol/L. Serum values for newborns are higher than for adults, 3.7–5.9 mmol/L. Cerebrospinal fluid levels are ~70% of those determined simultaneously in serum. The range for sweat is 5–17 mmol/L; values > 60 mmol/L are seen in cystic fibrosis. Urinary excretion of potassium varies with dietary intake but a typical range observed for persons on an average diet is 25–125 mmol/d. Gastric juice contains potassium at ~10 mmol/L. Fecal excretion has been reported as 18.2 $\pm$ 2.5 mmol/d (SEM) in one instance[4] and ~2 mmol/d in another;[33] with severe diarrhea, fecal loss may be as much as 60 mmol/d.

Methods for the Determination of Sodium and Potassium

Although atomic absorption spectrophotometry may be used for sodium and potassium analysis, the majority of laboratories employ either flame emission spectrophotometry or ion-selective electrode methods. The principles of each of these approaches (which are discussed in detail in Chapter 1B, Sections One and Four) are the same whether the instrumentation is dedicated or integrated into a multichannel system.

Flame Emission Spectrophotometry

Principle. Sample is diluted in a lithium diluent (or cesium diluent, if lithium is an analyte) and aspirated into a propane-air flame. Na^+, K^+, Li^+, and Cs^+, when excited, emit spectra with sharp, bright lines at 589, 768, 671, and 852 nm, respectively. Light emitted from the thermally excited ions is directed through separate interference filters to corresponding photodetectors. The lithium or cesium emission signal is taken as a reference (internal standardization) against which sodium and potassium (or lithium) signals are individually compared. The system is standardized relative to low and high concentrations of each analyte, and the relation of signal to concentration is defined by an associated microprocessor. Upon aspiration of controls and patient samples into the standardized instrument, results are presented in mmol/L on a digital readout device. (See also Chapter 1B, Section One.)

Reagents. Commercially prepared standards of NaCl and KCl are convenient and are widely used. Many are specifically provided by the instrument manufacturer, as is diluent, in the form of either a concentrate or ready-for-use solution. In diluting concentrated standards in diluent for direct aspiration, care should be taken that unknowns assayed against the dilute standards are diluted in the same batch or lot of diluent. If standards or diluent, or both, are prepared in-house, NaCl and KCl salts should be dried overnight at 110 °C. Standard reference materials (SRMs; see Table 1A-4) are available from the National Bureau of Standards (Washington, DC 20334), but for routine purposes analytical reagent grade chemicals from established sources are satisfactory. Lithium salts (nitrate, sulfate, or carbonate) should be carefully chosen for high purity; the lithium concentration in the diluent is usually 15 mmol/L. The cesium diluent is prepared at a concentration of 1.5 mmol/L. For some analytical systems, addition of Acationox, Sterox SE, or Brij is specified; these nonionic surface active agents facilitate uniform aspiration and atomization of diluent and diluted sample. Depending upon the specifications for standards for a particular analytical system, the standard solutions may contain only K^+ or only Na^+, or both. The concentration for a high Na^+ standard may be 160 mmol/L and for K^+ 8.0 or 10.0 mmol/L; the low concentration standard may be 100 mmol/L for Na^+ or 2.0 mmol/L for K^+. For urine analysis, low "standard" for both Na^+ and K^+ may be diluent only (zero concentration). Standard solutions must be kept tightly closed to avoid evaporation and are preferably stored in polyethylene bottles to avoid the leaching of Na^+ and K^+ that may occur from glass. The polyethylene container should be kept relatively full, since this type of bottle "breathes" and allows water vapor to escape and the concentration of the standard to rise slowly over time.

Procedure. The dilution of sample is usually fixed at 100- or 200-fold by a manual dilution procedure, by a diluting accessory on the instrument, or, as in continuous flow analysis, by dialysis of prediluted sample into a recipient stream. The dilution ratio is chosen to optimize linearity of detector response and to bring the signal within the range of sensitivity of the detector. By decreasing the protein concentration and viscosity, dilution promotes uniform flow of sample and decreases or eliminates positive or negative interferences by other sample constituents on the emissions of analyte ions.

Method control. Reliable assay of unknown samples is dependent on meticulous maintenance of instrumentation to keep it within the manufacturer's operating specifications, and on close adherence to operating and maintenance instructions provided by the manufacturer. Careful cleaning of the atomizer-burner and, in some instruments of the receptacle into which the diluted sample is delivered, is important to maintain uniform aspiration. Cleanliness is promoted by frequent flushing of the sampling and aspirating systems with fresh diluent or pure water. The optical quality of the photometer must be preserved by appropriate and frequent cleaning, particularly of the glass chimney that characteristically surrounds an atomizer-burner. The correct fuel and oxidant must be provided at adequate pressures and ratio. Of special importance is proper adjustment of flame size and aspiration rate of the atomizer-burner; geometry of the flame with respect to the detectors and uniformity of entry of atomized, diluted sample into the flame are critical to both accuracy and precision of analysis.

After ignition of the flame, diluent should be aspirated for a few minutes. A warm-up period is necessary because the initial evaporation of water in the flame decreases the temperature of the burner body and the air in the burner chamber. Thermal equilibrium must be established with the flame lighted in order for the flame to stabilize and readings to become reproducible and reliable.

Safety

Propane, commercially supplied in small, easily replaceable tanks, is currently an almost universal fuel for flame photometers in the USA. This circumstance has been largely dictated by OSHA (Occupational Safety and Health Administration) regulations for safety in the work place and by policies adopted by instrument manufacturers' associations. Somewhat larger tanks of propane, similar to those for portable home barbecue units, are also available but are commonly used only on multichannel instruments for high-volume analysis (e.g., SMA and SMAII, Technicon Instruments, Tarrytown, NY 10591). Sizes supplied by instrument manufacturers are often usable to exhaustion of their contents without any particular effect on the flame quality. With the larger tank, however, flame quality generally begins to deteriorate before the tank is empty. The reason is that such tanks are ordinarily filled with liquefied propane gas (LPG) and appreciable quantities of other hydrocarbons (butane, for example) may be present in the liquid. As the tank empties, the relative proportions of the other hydrocarbons increase in the fuel and flame characteristics change for the worse.

Most contemporary flame photometers are fitted with valves which open to allow fuel and oxidant to flow only when the igniter is energized and that close promptly if ignition does not occur at once. Some instruments emit audible warnings when the fuel valve is open but the flame is not lit. Every flame photometer should be lighted according to prescribed, safe procedures. Leakage of propane from tanks, instrument fittings, or from automatic or manually operated valves is always possible, but may be detected with the aid of soap solution (e.g., SNOOP).

Because propane is heavier than air, it readily collects at low spots in the laboratory, thus posing an explosion hazard. Work areas should therefore be well ventilated. Safety regulations should be enforced to prevent storage of numerous small tanks or more than one large tank in the same location of the laboratory. Accreditation standards of the College of American Pathologists specify that no more than a two-day supply be stored. In some hospital laboratories, safety rules require storage of all explosive or flammable compressed gases at a site isolated from the building.

Applications

Various types of specimens can be assayed for Na^+ and K^+ if available instrumentation is flexible enough to permit it. Almost all equipment is tailored primarily for the analysis of serum or fluids with similar analyte concentration. Some equipment is provided with electronic modes that expand or contract analytical ranges and allow analysis at concentrations more common in urine or sweat. Some instruments which are fitted with automated, accessory dilutors set to fixed dilution ratios appropriate to serum can be disconnected from the dilutor. The aspirator can then be fed special standards and specimens diluted by a manual procedure to an analyte concentration compatible with the analytical range of the instrument. Note, however, that internally standardized instrumentation is not adaptable to analysis of specimen dilutions that do not contain the reference ion at the proper concentration. Diluting the sample with standard in diluent and determining sample concentration by the method of addition may be necessary.[42]

Reference Methods

Reference methods for serum have been proposed for both sodium[37] and potassium,[36] although sodium is the only method that has been evaluated in a multi-laboratory study.[29] Using the same strict operating protocol, seven laboratories analyzed material containing sodium at concentrations from 114–158 mmol/L, as previously established by the definitive method of ion-exchange/gravimetry. Instruments included IL Models 143, 343, and 443 (Instrumentation Laboratory, Inc., Lexington, MA 02173) and the KLiNa (Beckman Instruments, Inc., Brea, CA 92621). The mean values determined showed ≤ 1.0 mmol/L bias relative to values set by the definitive method, and standard deviations for the equivalent of day-to-day replication were < 0.9 mmol/L. Bias and precision acceptable for a reference method had been defined a priori as ≤ 2.0 and ≤ 1.5 mmol/L, respectively. The method was judged satisfactory in meeting these as well as other general criteria.

Methods Using Ion-Selective Electrodes (ISE)[25]

Analyzers fitted with ISEs usually contain Na^+ electrodes with glass membranes and K^+ electrodes with liquid ion-exchange membranes that incorporate valinomycin. (Typical electrodes are described on pages 116–121.) High selectivity and Nernstian response are common features of contemporary ISEs. The principle of potentiometry can be simply stated as determination of change in electromotive force (E, potential) in the potential-measuring circuit between a measurement electrode (the ISE) and a reference electrode, as the selected ion interacts with the membrane of the ISE. In instrumental applications, the measuring system is calibrated by introduction of one (high concentration) or two (low and high concentration) calibrator solutions containing Na^+ and K^+. The potential of the standard is determined and the $\Delta E / \Delta \log$ concentration is stored in microprocessor memory as a factor for the calculation of the unknown concentration when E of the unknown is measured. The concentration of the unknown is then presented in a digital output. Frequent standardization, initiated either by keyboard command or by an automatic, microprocessor-controlled sample uptake from a reservoir of calibrator, is characteristic of most systems. A number of instruments have both a serum and a urine mode, selected from the keyboard and implemented by electronic adjustment of the analytical range of the instrument.

Two types of ISE methods can be distinguished. In the *indirect methods*, sample is introduced into the measurement chamber mixed with a rather large volume of diluent of high ionic strength; the ASTRA systems (Beckman Instruments, Brea, CA 92621) provide a good example of this approach. In the *direct methods*, sample is presented to the electrodes without dilution; the NOVA systems (NOVA Biomedical, Newton, MA 02164) exemplify this approach. In some direct method applications, either serum, plasma, or whole blood may be entered, in a manner analogous to sample admission to blood gas instrumentation. Single-use, thin-film ion-selective electrodes for

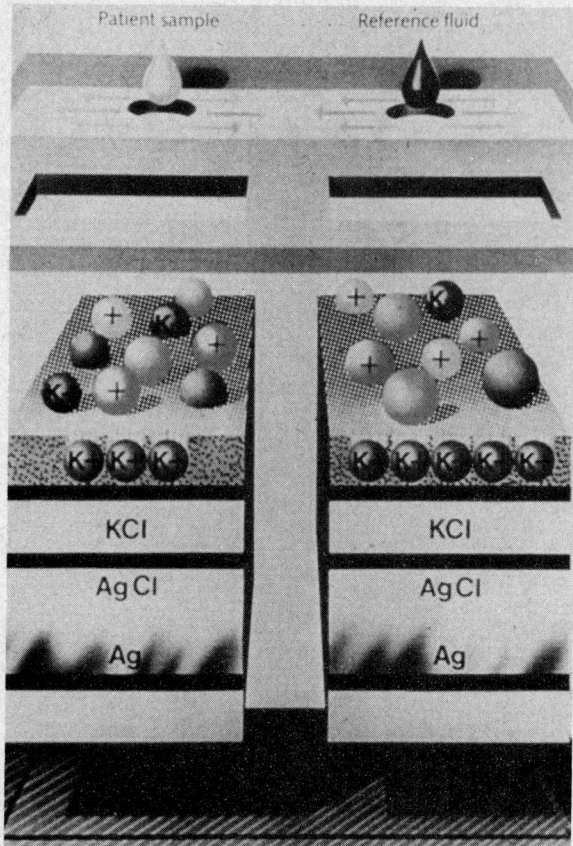

Figure 10-1. A thin-film ion-selective electrode slide (Kodak EKTACHEM) for the determination of potassium. (Courtesy of Eastman Kodak Co., Rochester, NY 14650).

Na^+, K^+, Cl^-, and CO_2 determinations are unique applications of a direct method (EKTACHEM, Eastman Kodak, Rochester, NY 14650). Figure 10-1 illustrates a potassium slide; slides for the other analytes differ chiefly in terms of the ion-selectivity of the membrane upon which sample and reference solution are dispensed. The analyte ions selected at the membrane diffuse to the silver/silver chloride electrode and generate a potential. The difference between potentials of the reference and sample electrodes is used as a measure of ion concentration in the sample.

The indirect method is historically the earlier one and was developed at a stage of ISE technology prior to miniaturization of electrodes; dilution was (and remains) a way to present a small sample in a volume large enough to cover adequately a large electrode's surface. The logic of dilution was perhaps also influenced by the fact of competition at the time between ISEs and flame photometry, because dilution is routine in flame photometry and flame photometry was the usual comparison method against which ISE methods were tested. The direct method, particularly as applied to whole blood, gives the advantage of more rapid determinations; the analytical approach is also nondestructive of the sample, i.e., sample can be recovered, essentially unchanged by the direct ISE measurement process, for additional testing. There are important differences in the applications of potentiometry between the direct and indirect methods that cause significant differences in analytical results. These aspects are discussed in the section on Electrolyte Exclusion Effect.

Errors observed in the use of *ISEs* fall into two categories. One includes the obvious errors due to lack of selectivity and those due to protein coating of the ion-sensitive membrane, or to contamination of the membrane by ions that compete or react with the selected ion and thus alter electrode response to the selected ion. The second category, the "electrolyte exclusion effects," apply chiefly to indirect methods and are mainly due to the solvent displacing effect of lipid and protein in the sample. (This effect is frequently referred to as the exclusion error.) Electrolyte exclusion effects are not confined to indirect ISE methods, or to sodium and potassium determinations, even though their importance has been highlighted chiefly by comparisons of analytical

Table 10-1, *A.* METHODS MEASURING THE CONCENTRATION IN THE WHOLE SAMPLE VOLUME AND THUS SUBJECT TO ELECTROLYTE EXCLUSION EFFECTS

Method	Analytes
Flame photometry	Na^+, K^+, Li^+
Atomic absorption	Ca^{2+}, Mg^{2+}, and others
Amperometry/coulometry	Cl^-
Indirect potentiometry	Na^+, K^+, Ca^{2+}, Cl^-

results for sodium as obtained by direct and indirect ISE methods.[11] Table 10-1 summarizes methods and analytes that are subject to electrolyte exclusion effects.

Electrolyte Exclusion Effect

The cause of this effect is the exclusion of electrolytes in plasma from that fraction of the total plasma volume which is occupied by solids. The volume fraction of total solids of an aliquot of plasma is ordinarily ~8% and the bulk of total solids is protein and lipid; 92% of the plasma is water. The main electrolytes are almost entirely associated with plasma water, i.e., they are not evenly distributed throughout the whole volume of plasma but are instead essentially confined to the water phase. For this reason, a concentration of sodium determined by flame photometry or by an indirect ISE method as 145 mmol/L in *plasma* containing 92% water is, in fact, a concentration of 145 $\times$ (100/92) or 158 mmol/L in plasma *water*. This negative "error" in plasma electrolyte analysis has been recognized, though not often emphasized, for many years; one of the earliest correction formulas to deal with the error was used by McLean and Hastings[26] in 1935 when determining ionized calcium in ultrafiltrates (protein-free aqueous fractions) of plasma. As determinations of electrolytes in plasma by flame photometry became routine after the 1950's, however, a tacit assumption became entrenched that the volume fraction of water in plasma and the difference between the concentrations of Na^+ and K^+ in plasma and plasma water were sufficiently constant that this difference could be ignored. The assumption was extended to Li^+ determinations, to atomic absorption spectrophotometry for Ca^{2+} and Mg^{2+}, and to the amperometric/coulometric assay for Cl^-. In fact, virtually all the concentrations we measure are related to the total sample volume rather than to the water phase. Development of direct ISE methods for Na^+ and K^+ left the assumption largely unchallenged.

Indirect ISE methods, because they dilute sample in a diluent of fixed high ionic strength, control the activity coefficient of individual ions and set it to a virtual constant; that for sodium approaches a value of 1. Under these circumstances, measurement of activity, a (where $a = \gamma \times$ concentration, and γ is the activity coefficient) is tantamount to measurement of concentration. Flame photometry measures emission of a specific ion in dilute solutions of high reference ion concentration where measurement of specific emission is tantamount to measurement of concentration of the specific ion. Since indirect ISE method and flame photometry are both based on measurement of ionic behaviors under conditions where they are locked into a strong, relatively invariant relation with concentration, plasma samples with widely divergent mass fractions of plasma water give essentially identical results by both methods. But the reason for the identity is that both methods are subject to the electrolyte exclusion effect to approximately the same degree. In certain clinical situations, notably diabetic ketoacidosis with severe hyperlipidemia,[15] and multiple myeloma,[20] the negative exclusion effect may be so large that laboratory results lead attending physicians to believe that electrolyte concentrations are normal or low when, in fact,

Table 10-1, *B.* METHODS MEASURING THE ACTIVITY, MOLALITY, OR CONCENTRATION IN THE WATER PHASE AND THUS NOT SUBJECT TO ELECTROLYTE EXCLUSION EFFECTS

Method	Analytes
Ion-selective electrode with *undiluted* sample	H^+ (pH), Na^+, K^+, Ca^{2+}, Cl^-, Li^+
Gas electrodes	CO_2 (pCO_2), O_2 (pO_2)
	HCO_3^- (calculated from pH and pCO_2)
Freezing point depression	H_2O (osmolality)

the concentration in the water phase may be dangerously high, or, in less extreme cases, to believe them low when concentrations in plasma water are actually normal.

In the *direct ISE methods*, sample is not diluted. The concentration is still determined relative to activity, but the activity is directly proportional to the concentration in the water phase, not the concentration in the total volume. ($a = \gamma \times$ concentration), but γ is not a constant, it is about 0.7 in normal plasma, and varies slightly with ionic strength, I, which is in turn a function of all the ions in the plasma water. The concentration in total plasma equals the concentration in plasma water times the volume fraction of water in total plasma. The latter may vary widely as a result of variation of quantities of lipid or protein; but as long as the activity of the specific ion is constant, the concentration of the ion in the water phase becomes independent of the relative proportions of water and total solids. Since direct ISE methods are based on activity measurement in the undiluted sample, they become free of electrolyte exclusion effects. Furthermore, the values determined by direct methods, because they are directly proportional to the activity, define electrolyte concentrations in a physiologically and physicochemically true sense, i.e., in the sense in which electrolytes interact with physiologic and pathologic processes.

Some clinical chemists and physicians have reached a conclusion that *direct* ISE methods for electrolyte analysis are the methods of choice.[12] They base their conclusion on the fact that great changes in plasma lipid, protein, and other solids can be expected in relatively common clinical conditions and in increasingly common therapies such as parenteral alimentation with lipid emulsions. They emphasize further that even in the absence of great changes in the volume fraction of solids, results by direct methods most realistically reflect clinical status and are therefore more effectively employed in diagnosis and management than are results obtained by methods subject to electrolyte exclusion effect. Should the conclusion become universal that concentrations should refer to the water phase, it would pose an important problem of "retooling" for clinical laboratories with currently significant investments in analyzers committed to methods measuring the concentration in the total sample volume. Levy[24] has explored some of the options that could make results for sodium (and by inference, for other electrolytes) obtained by these methods interchangeable with or comparable with results obtained by direct ISE methods; his discussion of the practicality and of the advantages and disadvantages of these options is illuminating but discouraging to anyone in search of simplistic solutions.

The most common causes of the electrolyte exclusion effect are hyperlipidemia[6] and hypo- and hyperproteinemia.[24] Some approaches have been proposed for qualifying reported results of electrolyte values determined by methods subject to the exclusion effect. For simply hypertriglyceridemic sera, centrifugation at 100 000 $\times$ g and analysis on the chylomicron-poor infranate have been suggested.[3] Alternatively, triglyceride concentration can be determined and electrolyte values found in conjunction with triglyceride > 1.5 g/L can be reported along with an estimate of error (% underestimation) calculated by an empirical equation,[32]

$$\% \text{ negative error} = (2.1 \times \text{triglycerides, g/dL}) - 0.6$$

The error, however, is <3% for triglyceride concentrations of <1500 mg/dL and <5% for concentrations of <2500 mg/dL.

In situations where both lipid and protein content are altered, presenting plasma electrolyte values along with concurrent estimates of plasma water has been suggested.[6] Use of Waugh's empirical equation[38] would require determination of serum total protein and total lipids and the calculation,

$$\text{Plasma water, g/dL} = 99.1 - 0.73(P_s) - 1.03(L_s)$$

where P_s is serum total protein in g/dL and L_s is serum total lipid in g/dL. Osmometry by freezing point depression can be used to estimate plasma water independently of total protein and total lipid measurements.[3] The procedure requires determination of the osmolality of the native serum, then repetition of the osmolality measurement on 2 mL of serum which has been added to dry NaCl (18.19 mg = 322 μmol $\rightarrow$ 161 mmol/L). Serum water is calculated as

$$\text{Serum water, L/kg} = 289.3/(\text{Osmolality}_2 - \text{Osmolality}_1)$$

where Osmolality$_2$ is the determined osmolality of the NaCl/serum mixture (in mmol/kg), Osmolality$_1$ is the determined osmolality of native serum, and 289.3 is the net effective osmolality of the added NaCl assuming an osmotic coefficient of 0.90. Estimate of serum water is also possible, rather simply, by gravimetry; a measured aliquot of serum is weighed, evaporated to dryness, and reweighed, the decrement in weight then being the measure of the aqueous fraction.

None of these possibilities, however, appear practical for routine application to all the specimens analyzed in a clinical laboratory that performs significant numbers of electrolyte assays. Analysts must certainly be aware of the electrolyte exclusion effects and the probability of their adverse effect on electrolyte results obtained on particular sera by certain methods. The analyst should as well be sensitive and responsive to the need to invoke special procedures in special cases to mitigate these errors.

CHLORIDE

Chloride is the major extracellular anion; at a concentration of ~103 mmol/L, it represents the largest fraction of the total inorganic anion concentration of ~154 mmol/L. Sodium and chloride together represent the majority of the osmotically active constituents of plasma. Chloride is therefore significantly involved in maintenance of water distribution, osmotic pressure, and anion-cation balance in the extracellular fluid compartment. The concentration of chloride in intracellular fluid of erythrocytes is 45–54 mmol/L, and in intracellular fluid of tissue cells ~1 mmol/L. In both gastric and small and large intestinal secretions, chloride is the most abundant anion.

Chloride ions in food are almost completely absorbed from the intestinal tract. They are filtered from plasma at the glomeruli and passively reabsorbed, along with Na$^+$, in the proximal tubules. In the thick ascending limb of the loop of Henle, Cl$^-$ is actively reabsorbed by the so-called chloride pump whose action promotes reabsorption of Na$^+$ as well. Loop diuretics such as furosemide and ethacrynic acid inhibit the chloride pump. Surplus chloride is excreted in the urine and is also lost in the sweat. Excessive losses in sweat, as can occur in hot weather, are normally minimized by the action of aldosterone, which is secreted by the adrenal cortex in response to decrease in plasma sodium and chloride. Aldosterone acts to cause the sweat glands to reabsorb more Na$^+$ and Cl$^-$, producing sweat with less than the usual sodium and chloride content, so that secretion of a more dilute sweat conserves body chloride. The conservation mechanism is successful, providing that loss of body chloride by other routes is not excessive.

Hypochloremia (decreased plasma chloride concentration) is observed in salt-losing nephritis as associated with chronic pyelonephritis. The loss is probably due to defective tubular reabsorption despite a body deficit of chloride. In Addison's disease, chloride levels are usually maintained close to normal except in Addisonian crisis, when chloride as well as sodium levels may drop significantly. Hypochloremia may also be seen in those types of metabolic acidoses that are caused by increased production or diminished excretion of organic acids (e.g., diabetic ketoacidosis or renal failure). In such cases, the fraction of total anion concentration represented by chloride is diminished because the complementary fraction of β-hydroxybutyrate, acetoacetate, lactate, and phosphate is increased. Prolonged vomiting, whatever the cause, results in significant loss of chloride, and ultimately in hypochloremia and depletion of total body chloride. In metabolic alkalosis, plasma chloride tends to fall while bicarbonate increases.

Hyperchloremia (increased plasma chloride concentration) occurs with dehydration, renal tubular acidosis, acute renal failure, metabolic acidosis associated with prolonged diarrhea and loss of NaHCO$_3$, and in salicylate intoxication. A slight rise in chloride is seen in respiratory alkalosis. Hyperchloremic acidosis may be a sign of severe renal tubular pathology. Extremely high dietary intake of salt and overtreatment with saline solutions are also causes of hyperchloremia.

In a study of individuals with hypercalcemia due either to primary hyperparathyroidism or to other causes, plasma chloride concentrations were 106 ± 5 mmol/L for cases of primary hyperparathyroidism compared to 103 ± 3 mmol/L for the other group.[41] The difference in chloride levels was believed to be due to the effect of parathyroid hormone on distal tubular function.

Urinary excretion of chloride normally approximates dietary intake; physiologic increase occurs with postmenstrual diuresis and decrease with premenstrual salt and water retention, in parallel with increase and decrease of urinary sodium. Massive diuresis of any cause is accompanied by increased chloride excretion, as is potassium depletion and adrenocortical insufficiency. Urinary excretion of chloride decreases when losses by other routes are increased, as well as in adrenocortical hyperfunction and in postoperative stress syndrome.

Methods for Determination of Chloride in Body Fluids

Specimens. Chloride is most often measured in serum or plasma, urine, and sweat. Under aerobic conditions, separation of cells from plasma should be prompt since loss of gaseous CO_2 alters distribution of Cl^- between cells and plasma. With anaerobic separation, chloride in venous samples is found to be 3–4 mmol/L less than in arterial samples. Measurement of chloride lost in gastric aspirates or intestinal drainages is an adjunct to parenteral replacement therapy. Fecal chloride determination is clinically useful for the diagnosis of congenital hypochloremic alkalosis with hyperchlororrhea (increased excretion of chloride in stool). In this condition, concentration of Cl^- in feces may reach 180 mmol/L, but almost no Cl^- is found in urine. A filter paper test has been described for rapid semi-quantitative estimate of fecal chloride.[35]

Mercurimetric Titration (Schales and Schales, Modified)

Principle. A tungstic acid protein-free filtrate of specimen is titrated with mercuric nitrate solution in the presence of diphenylcarbazone added as indicator. Free mercuric ions combine with chloride ion to form soluble but essentially un-ionized mercuric chloride:

$$2\ Cl^- + Hg(NO_3)_2 \rightarrow HgCl_2 + 2\ NO_3^-$$

Excess Hg^{2+} reacts with diphenylcarbazone to form a blue-violet color complex. The first appearance of this color is considered the endpoint of the titration.

Reagents

1. Sulfuric acid, 0.36 mol/L. To ~700 mL of reagent quality water in a 1-L volumetric flask, add 20 mL of concentrated sulfuric acid, reagent grade. Cool and dilute to volume with water. Stable at room temperature.

2. Sodium tungstate, 10 g/dL. Dissolve 112 g sodium tungstate ($Na_2WO_4 \cdot 2\ H_2O$) in ~700 mL of water in a 1-L volumetric flask and dilute to volume with water. Let stand for several days; decant or filter if a precipitate has formed. Reagent is stable at room temperature.

3. Mercuric nitrate, 5 mmol/L. Place 1.0833 g of HgO (red) into a small beaker or flask. Add 3 mL concentrated nitric acid and 20 mL water. Stir until dissolved and transfer with rinsing to a 1-L volumetric flask. Dilute to volume with water. Stable at room temperature. Standardized preparations of this reagent are also available from commercial sources (e.g., Fisher Scientific Company, Pittsburgh, PA 15219).

4. Diphenylcarbazone, 250 mg/dL. Dissolve 250 mg *s*-diphenylcarbazone in 100 mL of absolute methanol or ethanol. The solution should have an orange-red color. Store in a brown glass bottle in the refrigerator. Do not allow the reagent to come in contact with rubber. Degradation of the reagent is indicated by a color change to dark cherry red or to yellow. Make fresh every two months, or sooner if color changes or control values deteriorate.

5. Chloride standard, 100 mmol/L. Place 5.845 g of NaCl, previously dried overnight at 110 °C, in a 1-L volumetric flask. Add 3 mL of concentrated nitric acid and ~100 mL water to dissolve. Dilute to volume with water. Stable at room temperature in a tightly closed container.

Procedure

1. Add 500 μL of serum, standard, or control to a 16 × 100 mm test tube. Add in succession 3.5 mL water, 500 μL sulfuric acid, 0.36 mol/L, and 500 μL sodium tungstate solution. Mix and let stand for 5 min. Centrifuge at 1000 × g. The procedure provides a modified Folin-Wu protein-free filtrate of sample. See Comments below for the handling of body fluids other than plasma or serum.

2. Transfer 2.0 mL clear supernatant into a suitable titration vessel and add 100 μL diphenylcarbazone indicator. Titrate with mercuric nitrate, 5 mmol/L, using a microburet (see Comment 3, below).

3. When approaching the endpoint (first appearance of faint blue-violet color), add mercuric nitrate one drop at a time. Approximately 2 mL of reagent will be required to titrate the 100 mmol/L standard or a serum sample with similar concentration.

Calculation

$$\text{Chloride, mmol/L} = \frac{\text{Titration of unknown, mL}}{\text{Titration of standard, mL}} \times 0.02 \times \frac{1000}{0.2}$$

$$= \frac{\text{Titration of unknown, mL}}{\text{Titration of standard, mL}} \times 100$$

where 0.02 = mmol chloride/mL of standard
   0.2 = mL sample in titration
   1000 = mL/L

Comments

1. If *sweat* collected on gauze or filter paper is to be analyzed, first extract the gauze or filter paper with a known volume (3.0 or 5.0 mL) of pure water. Some workers prefer to use water acidified with nitric acid (2.5 mL of ∼0.07 mol/L HNO_3 in 100 mL of water, or 500 μL of concentrated HNO_3 in 100 mL of water; see also Comment 2, below). Transfer 2.5 mL of extract to the titration vessel, add 100 μL of indicator, and proceed with the titration as above. Set up 200 μL of standard, 100 mmol/L, with 2.3 mL of water (or dilute acid) and treat in the same way. Calculate with the first equation above, but replace 0.2 with the quotient of

$$\frac{2.5 \times \text{sweat collected, g}}{\text{Volume added for extraction, mL} + \text{sweat collected, g}}$$

where "sweat collected, g" represents the increment in weight (assumed to be equal to the volume of sweat in mL) of gauze or filter paper used in the sweat test to absorb the specimen.

For the analysis of *spinal fluid or urine* of low protein content, place 2.0 mL of water in the titration vessel and add 200 μL of specimen and 100 μL of indicator. Then add 50 μL of HNO_3, ∼0.07 mol/L, to provide an acid pH (the bluish color of the indicator in the alkaline mixture should change after acidification). Set up 200 μL of standard and an appropriate control in the same way. Proceed with the titration as in Step 2, above. Calculate by the first or second equation, as for serum.

The authors have no information whether mercurimetric titration is applicable to analysis of chloride in feces or gastrointestinal fluids.

2. Color at the titration endpoint is sensitive to pH. Most workers use pH 3–4.5 as is provided by the protein-free filtrate described or by additions of dilute HNO_3 in other situations. It is important to assure that standards and unknowns are titrated in the same pH range and in similar volumes of fluid.

3. The microburet used for titration should be calibrated in intervals of 0.05 mL and should deliver drops no larger than 0.02 mL. The delivery tip should be fine-drawn glass or platinum since mercuric nitrate reacts with other metals. A 5-mL glass beaker, containing a Teflon-coated mini-stirring bar and placed on a white filter paper on a magnetic stirrer, is suitable for the titration. Placing the delivery tip of the microburet so that the motion of the fluid in the beaker will pull off the drop makes titration easier. Titration in an appropriate light, such as that from a blue-white fluorescent lamp (like a desk lamp), improves detection of the endpoint and long-term precision. The light should be close and directed to minimize glare.

4. Titration of serum or plasma without preparation of a protein free filtrate is fast but has an inherent positive error of ∼2% because of reaction of Hg^{2+} with sulfhydryl groups of protein. Hemoglobin, bilirubin, or other pigments in the sample may also mask the endpoint; some authors have reported positive errors for some specimens as high as 15 mmol/L.

5. Mercuric nitrate reacts with any halide ion (Br^-, I^-, Cl^-), as well as with CN^- and SCN^-, and —SH groups. In bromide poisoning, because Br^- replaces some Cl^-, mercurimetric titration will measure the sum of the two ions and thus overestimate the true Cl^- concentration. Despite this limitation, mercurimetric titration is simple and sufficiently fast to serve as backup for automated analysis and as a useful alternative for unusual samples with chloride concentrations outside the analytical range of an automated procedure.

6. Precision of the method is largely dependent on the quality of the microburet and the titrating skills of the analyst. In the authors' experience, a long-term coefficient of variation of < 3% has been possible for a control with chloride concentration of 50 mmol/L when several skilled and experienced technologists were involved. This procedure is one of the few in the modern clinical chemistry laboratory where a defect in color perception may disqualify a worker as an analyst.

Reference

Schales, O., and Schales, S. S.: A simple and accurate method for the determination of chloride in biological fluids. J. Biol. Chem., *140*:879–884, 1941.

Spectrophotometric Methods

Spectrophotometric methods based on the reaction of chloride with *mercuric thiocyanate* have been implemented on a number of automated analyzers. The principle is illustrated in the following equations:

$$Hg(SCN)_2 + 2\ Cl^- \rightarrow HgCl_2 + 2\ SCN^-$$

$$3\ (SCN)^- + Fe^{3+} \rightarrow Fe(SCN)_3$$

Chloride ions react with undissociated mercuric thiocyanate to form undissociated mercuric chloride and free thiocyanate ions. The thiocyanate ions react with ferric ions to form the highly colored, reddish complex of ferric thiocyanate with an absorption peak at 480 nm. Perchloric acid increases the intensity of the red color. High levels of globulins in the serum interfere in methods where serum is added directly to the reagent, because turbidity develops.[21] The method has been recently adapted to a benchtop multi-channel analyzer (RA-1000, Technicon Instrument Corp., Tarrytown, NY 10591).[13]

The useful analytical range of the method, as developed, is limited to chloride concentrations of 80–120 mmol/L, but even in the stated range the reaction is not linear. Addition of a constant amount of $Hg(NO_3)_2$, equivalent to 60 mmol Cl^-/L, has improved the linearity.[23] The free mercuric ions bind a constant amount of Cl^- and only those chloride ions that are in excess are left to react with the mercuric thiocyanate and to produce color.

Mercurimetric automated methods applied to high-volume testing present the problem of disposal of reagent waste containing a significant amount of toxic mercury. Use of *TPTZ*, mercuric-2,4,6-tri(2-pyridyl)-5-triazine, in a spectrophotometric method applied directly to diluted serum, reduces the amount of mercury in wastes by half.[14] TPTZ reacts with Cl^- to produce an intense blue color that is stable for 5–30 min and can be measured at 560–600 nm. The method can be used manually and has been automated on SMAC (Technicon Instruments, Tarrytown, NY 10591) and Parallel (American Monitor Corp., Indianapolis, IN 46268) analyzers. It is sensitive and specific.[7]

The *ferric perchlorate* method, described by West and Coll[39] and recently automated on a centrifugal analyzer,[22] eliminates mercurial reagent entirely. Ferric perchlorate and Cl^- react in dilute perchloric acid to form a color complex with absorption maxima at 344 and 562 nm. The absorbance/concentration relationship is linear and the reaction is more specific for Cl^- than are most others.

The analytical range of spectrophotometric automated methods is, more often than not, limited to concentrations of chloride that can be expected in serum, plasma, or spinal fluid. The methods are therefore often not applicable to chloride analysis of other body fluids in which chloride concentrations may be much more or much less than 80–120 mmol/L.

Coulometric-Amperometric Titration

The reactions in coulometric-amperometric determinations of chloride depend on the generation of Ag^+ from a silver electrode at a constant rate and the reaction of silver ion with chloride ion in the sample to form insoluble AgCl:

$$Ag^+ + Cl^- \rightarrow AgCl$$

After the stoichiometric point is reached, excess Ag^+ in the mixture triggers shutdown of the silver ion generation system. A timing device records the elapsed time between start and stop of silver ion generation. (A more detailed description of this principle can be found in Chapter 1B, Section Four, pages 127–128.) Since the time interval is proportional to the amount of chloride ion present in the sample, the concentration of chloride can be calculated:

$$\text{Chloride, mmol/L} = \frac{\text{Time}_{unknown} - \text{time}_{blank}}{\text{Time}_{standard} - \text{time}_{blank}} \times C_s$$

Alternatively, a factor may be calculated from the concentration of standard (C_s) and the corrected time for the standard and then be used to multiply the corrected time of the unknown. Contemporary instruments furnished with microprocessors usually follow the second approach and results are presented on a digital readout in mmol/L.

The liquid reagent into which sample is delivered is commonly an acid solution (acetic or nitric or a mixture) to which gelatin or polyvinyl alcohol (PVA) has been added. Strong acid provides good electrolytic conductivity; weak acid depolarizes the medium and reduces solubility of silver chloride and thus provides a sharper endpoint. The presence of hydrogen ions prevents reduction of silver chloride at the cathode. Gelatin or PVA enchances reproducibility of the titration because they are adsorbed preferentially to high spots on the silver anode and the reaction rate is equalized over the entire electrode surface. Some error is introduced in the direct analysis of protein-rich samples since silver ions react with sulfhydryl groups of proteins.

A typical dedicated instrument is the CMT10 Chloride Titrator (Radiometer America, Inc., Westlake, OH 44145), which has an analytical range of 0–999 mmol/L for samples of 5–25 μL. The ASTRA instruments (Beckman Instruments, Brea, CA 92621), fitted with the appropriate module, utilize the principle in modified form for automated chloride analysis.[10] The modification is called proportional control coulometry. Cl^- in the sample is titrated with silver ions generated from the anode. As the concentration of Cl^- decreases, an electronic detector circuit recognizes slowing of the reaction rate and responds by slowing the generation of silver ions. When the reaction rate passes through the zero chloride threshold, silver ion generation stops entirely; total silver ion generated is measured, and this measure is proportional to chloride in the sample. Two modes are provided, one for serum and the other for urine analysis. For details the reader is referred to manuals available from instrument manufacturers.

Applications of the coulometric-amperometric principle are the most accurate methods for chloride over the entire range of concentrations displayed in body fluids. Surveys of the College of American Pathologists have shown coefficients of variation <2.5% over the concentration range 100–125 mmol/L. Standard concentrations should, however, be chosen to correlate with the concentration range of unknowns, e.g., a standard of 25 or 50 mmol/L should be used for sweat analysis rather than the 100 mmol/L standard used for serum assays. The method is subject to interferences by other halide ions, by CN^- and SCN^- ions, by sulfhydryl groups, and by heavy metal contamination. Maintenance of the systems is crucial to proper operation; electrodes and reaction vials or chambers must be kept scrupulously clean. Maintaining proper shape and size of silver-generating electrodes is particularly important.

Reference

Cotlove, E.: Determination of chloride in biological materials. *In:* Methods of Biochemical Analysis, Vol. 12. D. Glick, Ed. New York, Interscience Publishers Inc., 1964.

Ion-Selective Electrode Methods

Recent advances in ion-selective electrode (ISE) technology have made available chloride-selective electrodes for analysis of sweat and serum. Overall, applications to sweat analysis have had mixed reviews,[9,28] and are not accepted by the Cystic Fibrosis Foundation. The STAT/ION electrolyte analyzer (Technicon Instrument Corp., Tarrytown, NY 10591), and NOVA3 (NOVA Biomedical, Newton, MA 02164) utilize ISEs for measurement of Na^+, K^+, and Cl^-; the STAT/ION uses serum,[30] the NOVA3 whole blood.[31,40] Commentary on differences between direct and indirect ISE methods and on the electrolyte exclusion effect can be expected in the literature as direct ISE methods become more common. The Ektachem analyzer (Eastman Kodak, Rochester, NY 14650)[5] and Hitachi-705 (Boehringer Mannheim Diagnostics, Indianapolis, IN 46250)[1] have also been fitted with ISEs for serum analysis.

Reference Ranges.[33] Values for chloride in *serum or plasma* are 98–106 mmol/L. For full term and premature neonates, the upper limit of the range is extended to 110 mmol/L. Serum values vary little during the day, although there is a slight decrease after meals due to the diversion of Cl^- to the production of gastric juice. The range for *spinal fluid* is 118–132 mmol/L. *Urinary excretion* of chloride varies with dietary intake but a range of 110–250 mmol/d is typical. Sweat chloride is clearly normal at <30 mmol/L; in 98% of the subjects with cystic fibrosis, sweat chloride concentration exceeds 60 mmol/L. Values between 30 and 60 mmol/L may be encountered in adrenocortical insufficiency, nephrogenic diabetes insipidus, and certain familial diseases.[33] If these disorders can be ruled out, values in the range of 30–60 should be considered suspicious for cystic fibrosis and the test repeated at a later date. *Fecal excretion* of chloride (for eight healthy subjects) has been reported as 3.2 ± 0.7 mmol/d (SEM)[4] and elsewhere as ~ 2 mmol/d.[33]

Bicarbonate (Total Carbon Dioxide)

Total carbon dioxide is used in the immediate context to describe the quantity that is measured most often in automated analyzers by acidification of a serum or plasma sample and measurement of the carbon dioxide released by the process. Under certain conditions of collection and specimen handling, total CO_2 values determined in this manner may be almost identical with values for the calculated concentration of total carbon dioxide obtained in blood gas analysis.

Clinical Significance. Total carbon dioxide of plasma consists of CO_2 in physical solution, or loosely bound to amine groups in proteins (carbamino compounds), or present as HCO_3^- or CO_3^{2-} ions or as carbonic acid ($HHCO_3$). Bicarbonate ions make up all but ~ 2 mmol/L of the total carbon dioxide of plasma;[17] about 1 mmol/L of the bicarbonate exists as undissociated $NaHCO_3$. Measurement of the total carbon dioxide concentration (or CO_2, as we can conveniently call it henceforth) as part of an electrolyte profile is useful chiefly to evaluate bicarbonate concentration. Thus, in the following paragraphs we discuss (total) CO_2 and bicarbonate interchangeably.

Alterations of bicarbonate and CO_2 dissolved in plasma are characteristic of acid-base imbalance. The nature of the imbalance cannot, however, be inferred from the CO_2 value by itself and determination of CO_2 is rarely ordered alone. Its value has significance in the context of other electrolytes determined with it and in screening for electrolyte imbalance. When acid-base imbalance is suspected in association with electrolyte imbalance, evaluation of blood gases and pH is required to provide a definitive picture of the overall pattern of imbalances. The full clinical significance of the determination of total carbon dioxide will become apparent in Section Three. Table 10-2 summarizes alterations generally associated with some disorders.

Specimens. Either serum or heparinized plasma may be assayed. The usual specimen is venous blood drawn into an evacuated tube, although capillary blood taken in micro tubes or capillary tubes may also be presented for analysis. Given a specimen in a vacuum-draw tube, the concentration of total carbon dioxide is most accurately determined when the assay is done immediately after opening the tube and as promptly as possible after collection and centrifugation of the blood in the unopened tube. Ambient air contains far less CO_2 than does plasma, and gaseous dissolved CO_2 will escape from the specimen into the air, with a consequent decrease in CO_2 value of up to 6 mmol/L in the course of an hour.[34] In practical terms, the logistics of high-volume processing and automated analysis of specimens almost assures that most CO_2 measurements are done on specimens that have lost some dissolved, gaseous carbon dioxide, simply because preservation of anaerobic conditions between the time plasma is placed on a presentation device and the time it is sampled is not practical; for this reason the term bicarbonate may be preferable to total CO_2. On the other hand, it is probable that the result of a stat specimen, promptly introduced by an interrupt mode into an automated analyzer, has a significantly smaller error. The concerns an analyst might have for these errors are, however, almost obviated by lack of concern of physicians who employ both low-error and large-error results indiscriminately in their clinical practice without grave effects.

An electrolyte profile is sometimes requested on an arterial sample collected anaerobically for blood gas analysis. Aerobic handling to separate plasma is likely to have some effect on what kinds of values are obtained in this situation. But with special handling to preserve anaerobic

Table 10-2. CONDITIONS ASSOCIATED WITH ALTERED BICARBONATE CONCENTRATION OF PLASMA OR BLOOD

Increase	Decrease
Metabolic alkalosis due to severe vomiting in pyloric stenosis, hypokalemic states, or excessive intake of alkali	Metabolic acidosis in renal, glomerular, or tubular failure
	Renal tubular acidosis with coexisting hyperchloremia (normal anion gap)
	Diarrhea or intestinal fistula
	Ureteral transplant to ileum or colon
	Hypotension, dehydration, or other state of poor tissue perfusion
Respiratory acidosis (especially when compensated)	Respiratory alkalosis (especially when compensated)

character until the time of sampling, both venous and arterial plasma will display their characteristic differences in the concentration of total CO_2. (See Reference Ranges.)

Methods for Determination of Serum or Plasma Total CO_2

The first step in automated methods is the acidification of the sample to convert the various forms of CO_2 in plasma to gaseous CO_2. The step is ordinarily accomplished by dilution with an acid buffer.

In *continuous flow analysis* (SMAC, Technicon Instruments Corp., Tarrytown, NY 10591) the gaseous CO_2 is allowed to diffuse across a silicone membrane into a recipient solution buffered at pH 9.2 and containing the indicator phenolphthalein. As CO_2 is absorbed by the alkaline medium to HCO_3^-, the increased acidity of the recipient stream intensifies the red color. The increase in color over the baseline is determined spectrophotometrically and is proportional to the amount of CO_2 released from the sample. Baseline drift is the most frequent evidence of trouble with the method. The fault can usually be traced either to the color reagent, whose pH must be maintained within 0.2 unit, or to exhaustion of the CO_2 trap. Frequent preparation or frequent adjustment and restandardization of the reagent is necessary to keep the method in control. Air lines that supply bubbles to streams are fitted with a CO_2-absorbing trap; the trap often has a shorter effective life than the analyst expects.

In the *ASTRA method* (Beckman Instruments, Brea, CA 92621), the released gaseous CO_2 is determined by a pCO_2 electrode (of the type illustrated in Figure 1B-39, page 122) that is set in the reaction chamber of the CO_2 module. The *rate* of change of pH of the buffer inside the membrane of the measuring electrode is taken as the measure of total CO_2 present in the sample.

An *enzymatic method* for CO_2 is illustrated by the method used in the *aca* (DuPont, Wilmington, DE 19898). The specimen is first alkalinized to convert all forms of CO_2 to HCO_3^-. The enzymatic reactions are as follows (PEPC is phosphoenolpyruvate carboxylase and MDH is malate dehydrogenase):

$$HCO_3^- + \text{phosphoenolpyruvate} \xrightarrow{PEPC} \text{Oxaloacetate} + P_i$$

$$\text{Oxaloacetate} + NADH + H^+ \xrightarrow{MDH} \text{Malate} + NAD^+$$

Decrease in absorbance of NADH at 340 nm is proportional to the total carbon dioxide content.

The *manometric method* for total CO_2 content, using the Natelson microgasometer, has been almost entirely supplanted in clinical laboratories by automated methods. The method is described in some detail in an older text.[34] Measurement of CO_2 is based on the difference between pressure of mixed gases released from plasma in a closed system by acid and vacuum and the pressure of remaining gases after gaseous CO_2 has been converted with alkali to CO_3^{2-}. The disadvantage of the manometric method lies in the exposure to metallic mercury contained in the equipment and used for seals during sampling and reagent admission. An advantage, at a time before electrochemical methods were available, was application of the principles and equipment to accurate and reliable measurement of both total CO_2 in plasma and other body fluids and total O_2 in whole blood in micro specimens.
O_2 in whole blood in micro specimens.

Reference Ranges. Reference ranges generally are instrument dependent, and manufacturers' manuals should be consulted in specific cases. The following are reference ranges recently published in a compendium:[33]

Anaerobic Samples	mmol/L
Whole blood, venous:	22–26
Whole blood, arterial:	19–24
Plasma/serum, venous:	22–29
Plasma, capillary	
Premature, 1 week:	14–27
Newborn:	13–22
Infant, child:	20–28
Adult:	22–28
Cord blood:	14–22

Quality Control of Concurrent Electrolyte Measurements

An estimator called the anion gap can be calculated from concentrations of Na^+, Cl^-, and total CO_2 concentration determined on the same plasma or serum. The value of the anion gap can be used clinically to assess unmeasured anions and in the laboratory as a quality control estimate. Both aspects of anion gap are discussed in Section Three under the heading Estimates Derived from Electrolyte Measurements.

References

1. Adam, A., Boulanger, J., Ers, P., et al.: Analytical performance of the ion-selective electrodes used in the Hitachi 705. Clin. Chem., *30:*1720, 1984.
2. Adams, P. C., Woodhouse, K. W., Adela, M., et al.: Exaggerated hypokalaemia in acute myeloid leukaemia. Br. Med. J., *282:*1034-1035, 1962.
3. Albrink, M. J., Hald, P. M., Man, E. B., et al.: Displacement of serum water by the lipids of hyperlipemic serum. A new method for the rapid determination of serum water. J. Clin. Invest., *34:*1483-1488, 1955.
4. Caprilli, R., Sopranzi, N., Colaneri, O., et al.: Salt-losing diarrhoea in idiopathic proctocolitis. Scand. J. Gastroenterol., *13:*331-335, 1978.
5. Costello, P., Kubasik, N. P., Brody, H. E., et al.: Multilayer film analysis: Evaluation of ion-selective electrode slides. Clin. Chem., *29:*129-132, 1983.
6. Creer, M. H., and Ladenson, J.: Analytical errors due to lipemia. Lab. Med., *14:*351-355, 1983.
7. Feldkamp, C. S., Palmer, D. J., Salancy, J. A., et al.: Interference by other halides in the automation of chloride. Contributions to the general methodology of continuous flow systems. Z. Klin. Chem. Klin. Biochem., *12:*146-150, 1974.
8. Fernandez, P., and Cox, M.: Basic concepts of renal physiology. Int. Anesthesiol. Clin., *22:*1-34, 1984.
9. Finley, P. R., Dye, J. A., Lichti, D. A., et al.: A modified ion-selective electrode method for measurement of chloride in sweat. Am. J. Clin. Pathol., *69:*615-618, 1978.
10. Finley, P. R., Williams, J. R., Lichti, D. A., et al.: Evaluation of a new multichannel analyzer, "Astra-8." Clin. Chem., *24:*2125-2131, 1978.
11. Fogh-Anderson, N., Wimberley, P. D., Thode, J., et al.: Determination of sodium and potassium with ion-selective electrodes. Clin. Chem., *30:*433-436, 1984.
12. Forest, A. R. W., and Shenkin, A.: Dangerous pseudohypernatraemia. Lancet, *2:*1256, 1980.
13. Frey, M. J.: A quantitative colorimetric method for determination of serum chloride using the Technicon RA-1000 system. Clin. Chem., *29:*1255, 1983.
14. Fried, Von R., Hoeflmayr, J., and Velösy, G.: Eine neue, hochempfindliche Methode zur Bestimmung des Chlorids in Körperflüssigkeiten ohne Eiweissfällung. Z. Klin. Chem. Klin. Biochem., *10:*280, 1972.
15. Frier, B. M., Steer, C. R., Baird, J. D., et al.: Misleading plasma electrolytes in children with severe hyperlipidaemia. Arch. Dis. Child., *55:*771-775, 1980.
16. Goodman, J. T., Vincent, J., and Rosen, I.: Serum potassium changes in blood clots. Am. J. Clin. Pathol., *24:*111-113, 1954.
17. Hall, W. D.: Serum total CO_2 content. *In:* Clinical Methods: The History, Physical, and Laboratory Examinations. H. K. Walter, W. D., Hall, and J. W. Hurst, Eds. Boston, Butterworth Publisher, 1980.
18. Hultman, E., and Bergström, J.: Plasma potassium determination. Scand. J. Clin. Lab. Invest. Vol. *14* Supplementum, *64:*87-93, 1962.
19. Ingram, R. H., Jr., and Seki, M.: Pseudohyperkalemia with thrombocytosis. N. Engl. J. Med., *267:*895-900, 1962.
20. Ladenson, J. H., Apple, F. S., Aguanno, J. J., et al.: Sodium measurements in multiple myeloma: Two techniques compared. Clin. Chem., *28:*2383-2386, 1982.
21. Landowne, R. A.: Immunoglobulin interference with phosphorus and chloride determinations with the Coulter Chemistry. Clin. Chem., *25:*1189-1190, 1979.
22. Law, W. T., and Ertingshausen, G.: A novel reagent and method for determination of chloride in serum with a centrifugal analyzer. Clin. Chem., *26:*1874-1877, 1980.
23. Levinson, S. S.: Direct determination of serum chloride with a semiautomated discrete analyzer. Clin. Chem., *22:*273-274, 1976.
24. Levy, G. B.: Determination of sodium with ion-selective electrodes. Clin. Chem., *27:*1435-1438, 1981.
25. Maas, A. H. R., Kofstad, J., Siggaard-Andersen, O., et al., Eds.: Ionized Calcium, Sodium, and Potassium by Ion Selective Electrodes. Vol. 5. Proceedings of the first meeting of the European working group on ion selective electrodes. IFCC Workshop, Oslo, 1983. Copenhagen, Private Press, 1984.
26. McLean, F. C., and Hastings, A. B.: The state of calcium in the fluids of the body. J. Biol. Chem., *108:*285-322, 1935.
27. Peter, J. B.: Usefulness of fecal electrolytes. Diag. Med. Nov/Dec, 1983, pp. 19-20.
28. Price, C. P., and Spencer, K.: Problems associated with measuring sweat chloride concentration with an ion-specific electrode. Ann. Clin. Biochem., *14:*171-178, 1977.
29. Schaffer, R., Velapoldi, R. A., Paule, R. C., et al.: A multilaboratory evaluated reference method for the determination of serum sodium. Clin. Chem., *27:*1824-1828, 1981.
30. Slaunwhite, D., Clements, J. C., and Reynoso, G.: Clinical evaluation of the Technicon Stat/Ion system. Clin. Biochem., *10:*44-46, 1977.
31. St. Andre, J., and Young, C. C.: A new instrument for clinical determination of chloride and total CO_2. Clin. Chem., *26:*991, 1980.

32. Steffes, M. W., and Freier, E. F.: A simple and precise method of determining sodium, potassium, and chloride concentrations in hyperlipemia. J. Lab. Clin. Med., *88:*683-688, 1976.

33. Tietz, N. W., Ed.: Clinical Guide to Laboratory Tests. Philadelphia, W. B. Saunders Co., 1983.

34. Tietz, N. W., Ed.: Fundamentals of Clinical Chemistry, 2nd ed. Philadelphia, W. B. Saunders Co., 1976.

35. Tomaszewski, L., and Konarska, L.: A simple, rapid, and accurate screening test for hyperchlororhoea. J. Clin. Chem. Clin. Biochem., *15:*385-388, 1977.

36. Velapoldi, R. A., Paule, R. C., Schaffer, R., et al.: A reference method for the determination of potassium in serum. NBS Special Publication 260-263, U.S. Department of Commerce, National Bureau of Standards. Washington, D.C., U.S. Government Printing Office, 1978.

37. Velapoldi, R. A., Paule, R. C., Schaffer, R., et al.: A reference method for the determination of sodium in serum. NBS Special Publication 260-60, U.S. Department of Commerce, National Bureau of Standards. Washington, D.C., U.S. Government Printing Office, 1978.

38. Waugh, W. H.: Utility of expressing serum sodium per unit of water in assessing hyponatremia. Metabolism, *18:*706-712, 1969.

39. West, P. W., and Coll, H.: Direct spectrophotometric determination of small amounts of chloride. Anal. Chem., *28:*1834-1838, 1956.

40. Willis, J. P., Young, C. C., Olson-Mank, L., et al.: Clinical evaluation of a liquid membrane chloride ion selective electrode. Clin. Chem., *29:*1193, 1983.

41. Wills, M. R.: Value of plasma chloride concentration and acid-base status in the differential diagnosis of hyperparathyroidism from other causes of hypercalcaemia. J. Clin. Pathol., *24:*219-227, 1971.

42. Winter, S. D.: Measurement of urine electrolytes: Clinical significance and methods. CRC Crit. Rev. Clin. Lab. Sci., June, 1981, pp. 163-187.

Additional Reading

Maas, A. H. J., Siggaard-Andersen, O., Weisberg, H. F., et al.: Ion-selective electrodes for sodium and potassium: A new problem of what is measured and what should be reported. Clin. Chem., *31:*482-485, 1985.

Section Two

BLOOD GASES AND pH

by Elizabeth L. Pruden, Ph.D., Ole Siggaard-Andersen, M.D., Ph.D., and Norbert W. Tietz, Ph.D.

Clinical management of respiratory and metabolic disorders and surgical intervention to correct cardiopulmonary damage have been greatly improved over the last decade by the development of instrumentation for rapid, accurate measurements of oxygen and carbon dioxide in blood and expired air. Vigorous measures to preserve life in premature neonates and to support life in other patients with cardiopulmonary impairment depend largely on assisted ventilation using mixtures of gases tailored, in response to laboratory findings, to serve individual needs. Determination of blood gases also plays an important part in the detection of acid-base imbalance and in following the effect of therapy. Modern instrumentation for blood gas determinations, although highly sophisticated, is nevertheless easy to operate, and with meticulous maintenance and quality control it is capable of quick turnaround of very reliable laboratory data. The responsibility for blood gas analysis in some institutions is undertaken by the clinical laboratory, in others by a respiratory therapy department, and in still others by a special pulmonary laboratory. Supervision of the service may fall under a physician, a respiratory therapist, a clinical chemist, or a clinical chemistry technologist. The discussion here is particularly directed to the student of clinical chemistry.

Nomenclature in this rapidly developing aspect of analysis has not yet been fully systematized. Table 10-3 addresses nomenclature used in the following discussion and alternative terminology which may be encountered in practice.[8]

Respiration

Respiration is responsible for supplying oxygen needed by tissue cells for their metabolic processes and for removing the carbon dioxide produced by those processes.

External Respiration

Exchange of O_2 and CO_2 in the lungs between alveolar air and blood is called external respiration, in contrast to internal respiration occurring at the tissue level. At inspiration, con-

Table 10-3. CONVERSION FACTORS, PREFIXES, SYMBOLS, AND DESCRIPTORS USED IN DISCUSSIONS OF GASES MEASURED IN BLOOD AND EXPIRED AIR*

Conversion factors
1 mm Hg = 0.133 kPa
1 kPa = 7.5 mm Hg
kPa: 1 kilopascal = 1000 pascal. The pascal is the SI derived unit of pressure; it equals 1 Newton/m². See also Chapter 1A.

General prefixes
p: partial pressure or tension
  Usage: pO_2, pCO_2, pH_2O
  Alternative: PO_2
s: saturation fraction
  Usage: sO_2
  Alternative: SO_2
c: substance concentration (trivial, "content," especially when referring to total amounts)
  Usage: ctO_2 for concentration of total oxygen
        $ctCO_2$ for concentration of total carbon dioxide
        $cHCO_3^-$ for concentration of bicarbonate
  Alternative: C. Note, however, the risk of confusion when used as CO_2 to mean oxygen concentration. Not recommended.
d: dissolved gas, used with substance concentration (c)
t: total, used with substance concentration (c), thus $ctCO_2 = cHCO_3^- + cdCO_2$

Symbols for gases, ions, hydrogen ion concentration
Standard chemical forms, e.g., O_2, CO_2, HCO_3^-, pH

Specimen origin is indicated by lower case letters. Whole blood and plasma are distinguished by capitals.
  a: arterial   B: blood
  v: venous   P: plasma
  c: capillary

Usage: $pO_2(aB)$, for partial pressure of oxygen in arterial blood
A bar over the lower case letter means "mixed." A prime after the lower case letter means "ends of" so that $pO_2(c'B)$ would mean partial pressure of O_2 in blood at the end of a capillary proximal to a venule, and $pCO_2(\bar{v}B)$ would mean partial pressure of CO_2 in sample taken from a peripheral vein, rather than from a central vein or from a vein draining a particular organ.

Prefixes associated with external respiration
  V: volume of air or blood (unit, L)
  $\dot{V}$: volume rate (unit, L/min)
  x: substance fraction, also called mole fraction
  E: expired air
  I: inspired air
  A: alveolar air
Usage: $\dot{V}(A)$ means alveolar ventilation; $\dot{V}(B)$ cardiac output; $xO_2(I)$ fraction of O_2 in inspired air; $pO_2(A)$ partial pressure of O_2 in alveolar air; and $pCO_2(E)$ partial pressure of CO_2 in expired air.
A bar over the letter means "mixed," i.e., $\bar{E}$ means mixed expired air. In the physiological literature, the corresponding terminology is $\dot{V}_A$; $\dot{Q}$; FIO_2; P_AO_2; and P_ECO_2.

Other descriptors
  BTPS: *B*ody *T*emperature (37 °C or 310.16 K) and ambient *P*ressure, fully *S*aturated (pH_2O = 47 mm Hg or 6.25 kPa)
  STPD: *S*tandard *T*emperature (0 °C or 273.16 K) and standard *P*ressure (760 mm Hg or 101.08 kPa) of *D*ry gas
  Atm: atmospheric
    B: barometric (atmospheric)
      Usage: $p(atm)$, $p(Atm)$
      Alternative: P(B). Note risk of confusion with B for blood. Not recommended.
  SVP: *S*aturated *V*apor *P*ressure, the vapor pressure of water. SVP_T means SVP at a specified temperature, e.g., $SVP_{37 °C}$ = 47 mm Hg; pH_2O(saturated)
  ATPS: *A*mbient *T*emperature and *P*ressure, *S*aturated with water vapor

*This list is not complete and not all the terms and quantities shown are the forms recommended by the IFCC.[16] The list is presented to facilitate interpretation of terms used in the text and to illustrate a variety of unofficial and semi-official alternative forms that may be encountered in the literature.

traction of the diaphragm and thoracic musculature expands intrathoracic volume and creates a negative intrapulmonary pressure. Atmospheric air is drawn through the nasopharynx and trachea and into the bronchial tree, which consists of successively smaller passages, the bronchioles, terminating at the alveoli. Alveoli are small sac-like chambers with very thin walls in close approximation to pulmonary capillaries. The exchange of gases between alveolar air and pulmonary blood occurs across the alveolar and capillary membranes. Expiration takes place passively by recoil as the elastic tissues of the lung and chest wall rebound and the intrathoracic volume is decreased. Loss of elasticity of the lung and destruction of the alveolar membranes are basic causes of pulmonary distress. Paralysis of thoracic musculature with curare or similar drugs stops breathing altogether.

In a resting state, the respiration rate is normally 12–15/min. For an average-sized adult with a tidal volume of about 0.5 L, 6–8 L of air is moved per minute; CO_2 is exchanged at 0.2 L/min and O_2 at 0.25 L/min. Physical activity increases ventilation (respiratory rate $\times$ tidal volume), i.e., the amount of air turned over per minute. Voluntary efforts can increase the rate of ventilation 20–30 times over the resting level and increase exchange of CO_2 to a maximum of 8.0 L/min and exchange of O_2 to a maximum of 4.0 L/min, but only briefly. Involuntary increases in rate and depth of respiration are regulated by the medullary respiratory center in the brain stem, which in turn is stimulated by central chemoreceptors located on the anterior surface of the medulla oblongata and by peripheral chemoreceptors, including the carotid bodies and aortic bodies. Peripheral chemoreceptors are stimulated by a fall in arterial blood pH due to accumulation of CO_2, or, more precisely, by a fall in the pH of the extracellular fluid surrounding the sensitive cells. The central chemoreceptors are stimulated by a decrease in pH of the cerebrospinal fluid, which is in close contact with extracellular fluid bathing the sensitive cells in the brain.

A decrease in arterial pO_2 stimulates the peripheral, but not the central, chemoreceptors. Several other factors also influence the respiratory center. Apparently, reflexes from the working muscles are responsible for the initial hyperventilation during exercise. A rapid fall in skin temperature causes an inhibition of respiration, while increased temperatures accelerate the respiratory rate.

This regulatory system is sometimes called the *basic chemical control mechanism*. Its importance will be explored in a later discussion of acid-base regulation by the lungs. *Assisted ventilation* utilizes a mechanical device to provide intermittently gas mixtures via an endotracheal tube inserted through the mouth or through a tracheostomy. Appropriate ventilation conditions can be provided by adjusting the cycle and pressure of the respirator, thus assisting or, in some life support situations, replacing normal ventilation. The selection of conditions for ventilation carries risks of hyperventilation and excessive loss of CO_2, or of hypoventilation and excessive retention of CO_2 and reduced uptake of O_2. Gas mixtures containing different fractional compositions of O_2 and CO_2 may be administered in conjunction with assisted ventilation. The physician's adjustments of the conditions of assisted ventilation, or his choice of gas composition, or both, depend on the results of blood gas and pH determinations. In effect, the patient's ventilation is directed by laboratory results and the clinical condition of the patient and not by the normal physiologic chemical control mechanism.

Peripheral venous blood reaches the pulmonary circulation from the right ventricle of the heart and is arterialized in the capillaries of the lung by uptake of O_2 and loss of CO_2. Pulmonary venous blood then returns by way of the left atrium to the left ventricle and is directed through the aorta to the peripheral tissues. In the peripheral tissue capillaries, the arterial blood gives up oxygen to the tissue cells and takes up carbon dioxide. With return of blood to the lungs, the cycle is completed.

Exchange of Gases in the Lungs and Peripheral Tissues

Movement of oxygen and carbon dioxide across alveolar and cell membranes is governed by gradients in the partial pressure of each gas. In air, the partial pressure of each gas is determined by its fractional composition. Blood pO_2 and pCO_2 values actually represent the partial pressures that would exist in a hypothetical gaseous phase in equilibrium with the blood. In Figure 10-2

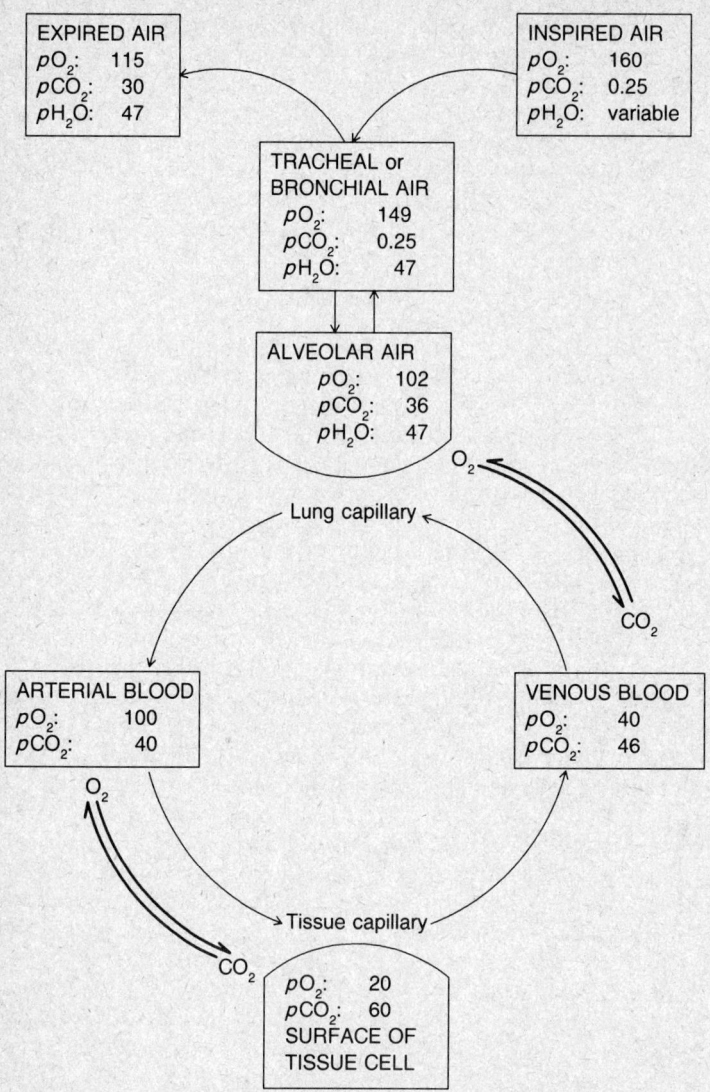

Figure 10-2. Partial pressures of oxygen and carbon dioxide in air, blood, and tissue. Values shown are approximations in mm Hg. Heavy arrows show directions of gradients. (Modified from Tietz, N. W.: Fundamentals of Clinical Chemistry. 1st ed. Philadelphia, W. B. Saunders, Co., 1970.)

and the following discussion, gradients of the pulmonary and tissue exchanges are described as gradients in partial pressure. Dry air inspired at a pressure of 1 atmosphere (760 mm Hg) consists of 20.95% O_2 (pO_2 ~159 mm Hg), 0.03% CO_2 (pCO_2 ~0.25 mm Hg), 78.1% nitrogen, and ~0.1% other inert gases. As inspired air passes over the moist mucous membranes of the upper respiratory tract, it is warmed to 37 °C, becomes saturated with water vapor, and mixes with air in the respiratory tree, resulting in partial pressures of 149 mm Hg for oxygen, 0.3 mm Hg for carbon dioxide, 47 mm Hg for water, and 564 mm Hg for nitrogen. Further mixing with alveolar air results in partial pressures, at the alveolar membrane, of 102 mm Hg for oxygen, 36 mm Hg for carbon dioxide, and 47 mm Hg for water vapor. Venous blood on the opposite side of the alveolar membrane contains oxygen at a partial pressure of ~40 mm Hg and carbon dioxide at ~46 mm Hg; the gradient for oxygen is therefore inward, to the blood, and for carbon dioxide it is outward, from the blood. Carbon dioxide removal is so efficient that the pCO_2 in expired air is > 100 times the pCO_2 in inspired air. At the arterial end of capillaries of peripheral tissues, partial pressures of oxygen and carbon dioxide in blood are about the same as in alveolar air, but that of oxygen at ~100 mm Hg is substantially higher than the average pO_2 at the surface of the tissue cells (20 mm Hg), and the partial pressure of carbon dioxide at 35–46 mm Hg is substantially lower than that in the cells (50–70 mm Hg). In the tissue capillary, the gradient for oxygen is therefore inward, to the cell, and for carbon dioxide is outward, from the cells.

The arteriovenous difference in partial pressures is ~60 mm Hg for O_2 and ≤6 mm Hg for CO_2; the arteriovenous difference for oxygen is an indicator of the efficiency of oxygen extraction in the passage of blood through the capillaries.

Mechanism of Exchange of O_2 and CO_2 at the Lungs and Tissue Cells

Internal respiration of tissue cells consumes oxygen and produces carbon dioxide and organic acid metabolites of biochemial substrates. As will be discussed at length in Section Three, the acid products of metabolism are buffered or removed by various mechanisms so that close control (homeostasis) of hydrogen ion concentration (pH) is maintained in the extracellular fluids of the body. The major buffer system of the extracellular fluid is the bicarbonate system, which interacts with CO_2 physically dissolved (dCO_2) in the blood.

Carbon dioxide is transported in the blood in several forms; the three most abundant and important are bicarbonate ions (HCO_3^-) in the erythrocytes and plasma, carbaminohemoglobin ($HbCO_2$) in erythrocytes, and dCO_2 in fluids of erythrocytes and plasma. (See Table 10-4.) These forms are not simply transport forms; they constitute a dynamic system that buffers pH change in the extracellular fluid and participates in concurrent and integrated delivery of O_2 and removal of CO_2 at the tissue cell, and uptake of O_2 and discharge of CO_2 at the alveolar membrane. *Oxygen is transported* to the tissues in two forms: one reversibly bound to hemoglobin in the erythrocytes (HbO_2); the other, dO_2, physically dissolved in fluids of the erythrocytes and plasma. Hemoglobin in the erythrocytes is not simply an oxygen carrier; it is also a dynamic buffer system that interacts with the bicarbonate system of erythrocytes and plasma during the processes of O_2 delivery and uptake.

The mechanism of O_2 delivery and uptake and CO_2 removal and discharge is called the *isohydric and chloride shift*. This mechanism is illustrated and discussed in detail in Section Three. Here we will only call attention to the forms of CO_2 that participate and the equilibria among the forms that affect oxygen exchange. An understanding of these forms is basic to understanding terminology used for acid-base parameters derived from measurements of pCO_2 and pH, and to understanding corrections based on pCO_2 and pH determinations and applied to measurements of pO_2.

Gaseous CO_2 exposed to *plasma* is dissolved in proportion to the partial pressure, pCO_2. A small portion of dissolved CO_2 (dCO_2), reacting with plasma water, forms carbonic acid (Figure 10-3) and another small amount, reacting with amino groups of plasma proteins, forms carbamino compounds. At a blood pH of 7.4 (cH^+ ~40 nmol/L), the equilibrium of the hydration reaction to form carbonic acid lies far to the left; only one part in 700–1000 parts of dCO_2 is converted to carbonic acid. The equilibrium of the carbonic acid → bicarbonate dissociation lies far to the right while that of the bicarbonate → carbonate dissociation lies far to the left. As a result, the carbonic acid concentration is only about 1.7 μmol/L and the CO_3^{2-} concentration is even less. The weak acid, carbonic acid, and its salt, bicarbonate, together constitute the bicarbonate buffer

Table 10-4. AVERAGE DISTRIBUTION OF CARBON DIOXIDE IN ONE LITER OF NORMAL BLOOD, ASSUMING A HEMATOCRIT OF 40%

Carbon Dioxide	Arterial		Venous		Difference*	
	mmol/L	*% of Total*	*mmol/L*	*% of Total*	*mmol/L*	*% of Total*
In plasma (600 mL)						
As dissolved CO_2:	0.72	3.32	0.81	3.47	0.09	5.38
As HCO_3^-†:	15.27	70.56	16.26	69.75	0.99	59.28
In erythrocytes						
As dissolved CO_2:	0.36	1.66	0.40	1.71	0.04	2.39
As carbamino-CO_2:	0.98	4.52	1.39	5.96	0.41	24.55
As HCO_3^-:	4.31	19.91	4.45	19.09	0.14	8.38
Total	21.64		23.31		1.67	

*The difference between arterial and venous blood is considered to be that amount of carbon dioxide which is disposed of by the lungs.

†Plasma contains a small amount of carbamino-CO_2 (about 0.2 mmol/L), which is traditionally included in the HCO_3^- fraction. A small amount of bicarbonate exists as undissociated $NaHCO_3$.

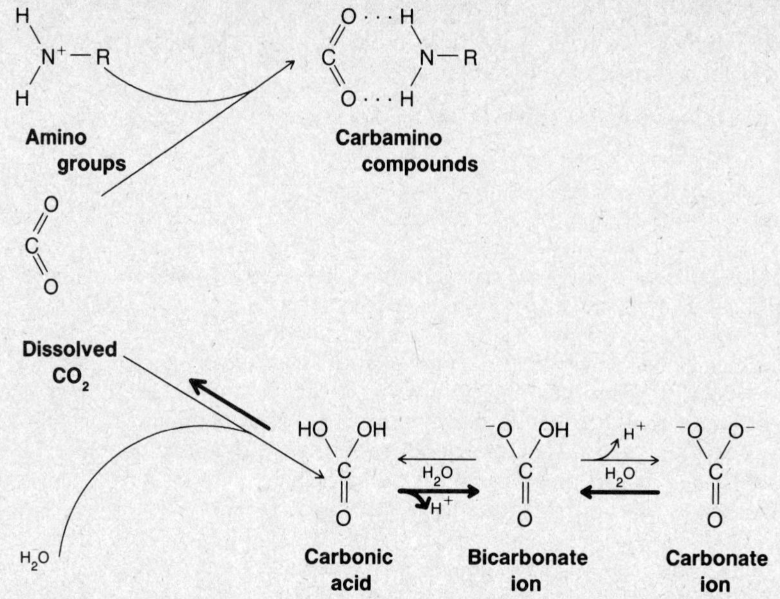

Figure 10-3. Reactions of carbon dioxide with water and amino groups. Hydrogen bonding is denoted as . . . .

pair, the major buffer system of the extracellular fluid. The dCO_2 of the plasma therefore is part of a complex equilibrium of several forms of CO_2, although the primary one is:

$$\text{Dissolved } CO_2 \rightleftharpoons \text{Carbonic acid} \rightleftharpoons \text{Bicarbonate ion} + H^+$$

The topic of buffering is discussed in Section Three of this chapter, as is the relationship of pCO_2, dCO_2, HCO_3^-, and pH and its role in acid-base balance. The mathematical aspects of the relationship are expressed in the Henderson-Hasselbalch equation as it is applied to plasma. (See also page 1200.)

In *erythrocytes*, the same reactions occur between dCO_2 and intracellular water and between dCO_2 and amino groups of hemoglobin and other proteins. Here, however, the equilibria are dependent (see Figure 10-9) on the presence and action of carbonic anhydrase to interconvert dCO_2 and HCO_3^-; the interaction of hemoglobin with H^+, dO_2, and dCO_2; flux of HCO_3^- across the erythrocyte membrane; *and* the relationship of HCO_3^-, dCO_2, and pH in the plasma. The buffer systems are therefore more complex than in plasma, and dCO_2 (as pCO_2), plasma pH, and factors such as 2,3-diphosphoglycerate concentration and the structure of hemoglobin protein in the erythrocyte together affect the association and disassociation of hemoglobin and oxygen. pCO_2 and plasma pH therefore affect oxygen transport in the blood and the distribution of oxygen in the forms of HbO_2 and dO_2.

pCO_2 of *whole blood* then reflects equilibria and concentrations of forms of CO_2 that are present in both erythrocytes and plasma. In order to evaluate the many facets of acid-base balance from measurements of pCO_2 and pH, a number of derived quantities have been defined by means of calculation and experimental determinations. They include buffer base, standard bicarbonate, actual bicarbonate, base deficit or base excess, and others; these terms are explained in Section Three of this chapter. We note them here since some are calculated by microprocessors of blood gas instruments and are presented in the results display. The basic terms that are of concern in acid-base relationships described by the Henderson-Hasselbalch equation are concentration of total CO_2 ($ctCO_2$), bicarbonate concentration ($cHCO_3^-$), concentration of dissolved CO_2 ($cdCO_2$), and pH.

Physical Principles

Behavior of Gases

Determination of gas pressures in expired air or blood depends on the application of certain physical principles (Table 10-5) with which the student of clinical chemistry is already familiar.

Table 10-5. PHYSICAL PRINCIPLES APPLIED IN BLOOD GAS MEASUREMENTS

Boyle's law	
The volume of an ideal gas at a constant temperature varies inversely with the pressure exerted to contain it.	$V \propto 1/P$
Charles' (Gay-Lussac's) law	
The volume of an ideal gas at a constant pressure varies directly with its absolute temperature.	$V \propto T$
Avogadro's hypothesis	
Equal volumes of different ideal gases at the same temperature and pressure contain the same number of molecules.	$n_i/V_i = n_j/V_j$
Dalton's law	
The total pressure exerted by a mixture of ideal gases is the sum of the partial pressures of each of the gases in the mixture.	$P = \Sigma p_i$
Henry's law	
The amount of a sparingly soluble gas dissolved in a liquid is proportional to the partial pressure of the gas over the liquid.	$c = \alpha \times p$

Many of the applications and assumptions, however, are quite different from those encountered in basic physics or chemistry courses, and they are discussed here in order to put them in the context of the present topic.

Gas laws are defined in terms of closed systems. In the immediate context, however, they are applied to open systems, i.e., spaces whose components can move into other spaces. For this reason, the amount of a gas dissolved in blood is conceptualized as the pressure of the gas in the liquid phase of the blood when it has reached equilibrium with an imaginary gas phase in contact with the liquid phase. Furthermore, assumptions are made that the partial pressure of the gas is the same in the liquid phases of erythrocytes and plasma and that the partial pressure of the gas in whole blood is equal to its partial pressure in plasma. The term "equilibrium" means an even balance of forces that results in a stable condition. "Equilibration" of blood with a gas mixture of known composition is a procedure of exposing the blood to the mixture such that each gas in it partitions to an equilibrium state between liquid and gaseous phases, i.e., the partial pressure of each gas-in-liquid is brought to equilibrium with its partial pressure in the gas phase. Equilibration is a routine procedure known in the laboratory as tonometry (see also Tonometry section below). In surgery, equilibration refers to the procedure whereby a heart-lung machine is used to saturate the patient's blood with oxygen and bypass the cardiopulmonary system of the patient during an operation.

A variety of spaces where gases are present are involved in the following discussion: the room, the tracheobronchial tree and alveoli of the patient, and the measuring chamber of an instrument. In all these spaces, atmospheric (barometric) pressure, $p(\text{Atm})$, is the prevailing pressure and partial pressures of each of the gases present in these spaces add up to the value of $p(\text{Atm})$, which is a variable dependent on geographic location. Scientific convention reduces measurements made at $p(\text{Atm})$ to *STPD*, standard temperature (0 °C or 273.16 K) and pressure (760 mm Hg or 101.08 kPa) for dry gas, and thus makes experimental data transferable, i.e., independent of the location where measurements are made. The student should realize that in blood gas work STPD is only one of the standard conditions; when it is used, it is used selectively and ordinarily only for expression of constants and similar transferable data.

BTPS is another set of standard conditions used in blood gas work (see Table 10-3). It specifies that measurements of partial pressure in biological specimens are always made at a temperature of 37 °C, in the presence of a partial pressure of water vapor of 47 mm Hg, and at the local atmospheric pressure. Use of this convention has the following practical effects:

1. It relates laboratory data for blood gases strictly to the geographic location of the patient, so that reference ranges become altitude-dependent.

2. It assumes that the standard temperature of a normal human is exactly 37 °C and that the measuring device also holds the sample of blood at exactly 37 °C. This assumption requires special concern for thermal stability of the instrument (a matter of quality control). Just as

importantly, it implies that in circumstances such as imposed hypothermia, when a patient's temperature is not 37 °C, his blood gas values determined at 37 °C must be corrected to his actual temperature if a reasonable estimate of partial pressures of the gases in the patient's body is to be obtained. These points will be addressed in greater detail later on.

3. It recognizes that the partial pressures of measured gases in the blood coexist with a partial pressure (SVP) of water vapor which is a constant and is therefore identical for both the calibration conditions of the instrument and the measurement conditions of the blood sample.

Boyle's and Charles' laws and Avogadro's hypothesis are combined in what is called the general gas equation:

$$P = (nRT)/V$$

where P = pressure in units of mm Hg or kilopascals (kPa),
   V = volume in L in which an ideal gas is contained,
   T = temperature in K (0 °C = 273.16 K),
   n = number of moles of gas, also called amount of substance of gas, and
   R = gas constant.

The SI units of P are Pa (or multiples thereof); the unit of mm Hg (also called torr) has, however, continued to be more popular (see Table 10-3 for conversion factors). Use of the SI unit has an advantage in that one atmosphere almost equals 100 kPa (1 atm = 101.325 kPa). Pressures expressed in kPa therefore are about equal to the percentage of atmospheric pressure, i.e., partial pressures in kPa for gases in a mixture of 1 atm are quite close estimates of percentages of the gases in the mixture. Numerical value and units of R differ depending on the units used for P. R = 62.36 mm Hg $\times$ L^{-1} $\times$ K^{-1} $\times$ mol^{-1} or 8.31 kPa $\times$ L^{-1} $\times$ K^{-1} $\times$ mol^{-1}. A trial calculation, after rearranging terms and evaluating n as 1 mol and P as 760 mm Hg, shows that the volume of 1 mol of an ideal gas at STPD is 22.4 L. Not all gases display ideal behavior; water vapor and oxygen approach the ideal mole volume of 22.4 L, but carbon dioxide has a mole volume of 22.2 L.

The general gas equation is the justification for accepting partial pressures of gases in blood as estimators of their concentrations, since a partial pressure (p) as well as P, pressure, is directly proportional to n/V, a concentration term in units of mol/L. This equation was, in the era of manometric gas analysis, central to the definition of the concentration of gases in blood. In current concepts of blood gas analysis, however, pO_2 is related only to the concentration of O_2 physically dissolved in the blood and pCO_2 only to the concentration of CO_2 physically dissolved in the blood (see Henry's law, p. 1199, and definitions of cdO_2 and $cdCO_2$). Concentration is given in mmol/L. The total concentration of oxygen in blood is the sum of concentrations of dissolved O_2 and of O_2 bound to hemoglobin. The total concentration of carbon dioxide is defined operationally as the sum of concentrations of dissolved CO_2, carbonic aid, bicarbonate ion, undissociated bicarbonate, and carbonate ion. These definitions of total concentration again take their origin from measurements made by manometric and volumetric techniques. In the past, content was usually expressed as mL of gas per dL of blood or plasma, a volume/volume relationship. That convention has some value in describing the content of oxygen in blood, since the amount of O_2 is chiefly related to hemoglobin concentration, which is expressed as g/dL or g/L of blood. References to normal oxygen content of ~20 mL/dL (or ~20 vol %) still occur, but such units have been entirely discontinued for carbon dioxide content.

Pressure, P or p, may mean either total pressure, as in the expression p(Atm) for the mixture of gases in room atmospheric air, or partial pressure in blood as in pO_2(aB).

Dalton's law (see Table 10-4) may be written for room air as:

$$p(\text{Atm}) = pO_2 + pCO_2 + pN_2 + pH_2O + pX$$

where pH_2O is the partial pressure of water vapor and pX is that of any other gas in the air sample. For venous blood, in a measuring device, the equation could be written

$$p(\text{vB}) = pO_2 + pCO_2 + pX'$$

by making the assumption that

$$pX' = pH_2O + pN_2 + pX$$

is constant or, if variable, has no clinical significance. In manometric and volumetric methods for blood gas determinations, Dalton's law and this assumption have central importance. But current determination of pO_2 and pCO_2 by specific electrochemical techniques is direct and essentially free of effect by any other gases that might be present. Dalton's law of partial pressures remains important, however, for standardization and control of the measuring devices, as later sections will illustrate.

Consider the example of a calibrator gas certified to contain 15% O_2 (L/L or mol/mol) and 5% CO_2, the remainder being N_2. The mole fractions (or F) of the gases in the dry mixture are 0.15, 0.05, and 0.80, respectively. This mixture, after saturation with water vapor at 37 °C (to mimic a patient's blood or alveolar air), is introduced into the instrument's measuring chamber (held at 37 °C to mimic a patient's body temperature) for the purpose of standardizing the instrument for subsequent measurements of gases in patients' samples. If the local barometric pressure, $p(Atm)$, on this occasion is 747 mm Hg, then the humidified calibrator gas is present in the chamber at ambient, barometric pressure, such that

$$p(Atm) = 747 \text{ mm Hg} = pO_2 + pCO_2 + pN_2 + pH_2O$$

To set the instrument to the pO_2 and pCO_2 of the calibrator gas, we must, first of all, account for pH_2O at 37 °C, which we know is equal to SVP (saturated vapor pressure) of water, 47 mm Hg. Therefore,

$$p(Atm) - pH_2O = pO_2 + pCO_2 + pN_2$$

$$747 - 47 = 700 \text{ mm Hg}$$

If $p(Atm)$ corrected for pH_2O represents the sum of partial pressures for the dry gases whose mole fractions we know, we can calculate the exact pO_2 and pCO_2 values for the calibrator gas, under the circumstances of measurement, and then enter these standard values into the instrument.

$$pO_2 = 700 \times 0.15 = 105 \text{ mm Hg}$$

$$pCO_2 = 700 \times 0.05 = 35 \text{ mm Hg}$$

The law of partial pressure is also applied in defining gas mixtures used to determine $pO_2(0.5)$ and other derived quantities, and to control instrumentation with tonometered samples.

Henry's law predicts the amount of dissolved gas in a liquid in contact with a gaseous phase (see Table 10-5). The concentration (c, mol/L) of free dissolved gas (dG) in blood (B) is calculated:

$$cdG(B) = \alpha G(B) \times pG(B)$$

where $\alpha G(B)$ is the coefficient of solubility for the gas in blood at 37 °C.

The coefficient for oxygen, $\alpha O_2(B)$, is 0.00140 mol/L per mm Hg (the corresponding coefficient for the volume/volume relationship at STPD is 31 μL/L per mm Hg). For plasma the coefficient is slightly less, 0.00126 mmol/L per mm Hg. When $pO_2(aB)$ is normal, $\sim$98 mm Hg, $cdO_2(aB)$ will therefore be 0.00140×98, or 0.137 mmol/L—a very small proportion of ctO_2. Increasing the O_2 fraction of inspired air to 100%, or increasing the pressure of inspired air as can be done in hyperbaric chambers, forces more O_2 into solution. In therapy with pure O_2, when $pO_2(aB)$ may rise to 640 mm Hg, $cdO_2(aB)$ could be as high as 0.9 mmol/L. In hyperbaric treatment, a $pO_2(aB)$ of 2500 mm Hg ($\sim$3.2 atmospheres), is equivalent to $cdO_2(aB)$ of 3.5 mmol/L. The prediction of levels of free dissolved O_2 in special therapies is extremely useful since tissue oxygenation by dissolved O_2 becomes increasingly important when hemoglobin-mediated oxygen delivery is impaired.

The concentration of dissolved CO_2 can be calculated in the same way. $\alpha CO_2(P)$ at 37 °C is 0.0306 mmol/L per mm Hg. At a pCO_2 of 40 mm Hg therefore, $cdCO_2$ is $40 \times 0.0306 =$

1.224 mmol/L. In the determination of blood gases, $p\mathrm{CO_2}$ is determined along with blood pH. As will be subsequently explained, these two parameters in conjunction with the Henderson-Hasselbalch equation permit the calculation of $c\mathrm{HCO_3^-}$, using the relation:

$$\log c\mathrm{HCO_3^-} = \mathrm{pH} - \mathrm{p}K' - \log\,[\,p\mathrm{CO_2} \times \alpha\mathrm{CO_2(P)}\,]$$

The antilog is then taken to derive $c\mathrm{HCO_3^-}$. Modern blood gas instrumentation, furnished with algorithms for this calculation by its microprocessor, not only calculates bicarbonate concentration but also several other derived values useful in evaluation of acid-base balance. Among these are base excess or base deficit, and standard bicarbonate.

The Interrelationship Between $ct\mathrm{CO_2}$, $c\mathrm{HCO_3^-}$, $cd\mathrm{CO_2}$, and pH

The Henderson-Hasselbalch Equation

Carbon dioxide and water react to form carbonic acid, which in turn dissociates to hydrogen ions and bicarbonate ions:

$$\mathrm{CO_2} + \mathrm{H_2O} \underset{}{\overset{K_{\text{hydration}}}{\rightleftharpoons}} \mathrm{H_2CO_3} \underset{}{\overset{K_{\text{dissociation}}}{\rightleftharpoons}} \mathrm{H^+} + \mathrm{HCO_3^-}$$

According to the law of mass action,

$$K_{\text{hydration}} = \frac{a\mathrm{H_2CO_3}}{a\mathrm{CO_2} \times a\mathrm{H_2O}} = 0.00229\ (\mathrm{p}K = 2.64 \text{ at } 37\ ^\circ\mathrm{C})$$

and

$$K_{\text{dissociation}} = \frac{a\mathrm{H^+} \times a\mathrm{HCO_3^-}}{a\mathrm{H_2CO_3}} = 2.04 \times 10^{-4}\ (\mathrm{p}K = 3.69 \text{ at } 37\ ^\circ\mathrm{C})$$

These can be combined to give

$$K_{\text{combined}} = \frac{a\mathrm{H_2CO_3}}{a\mathrm{CO_2} \times a\mathrm{H_2O}} \times \frac{a\mathrm{H^+} \times a\mathrm{HCO_3^-}}{a\mathrm{H_2CO_3}} = \frac{a\mathrm{H^+} \times a\mathrm{HCO_3^-}}{a\mathrm{CO_2} \times a\mathrm{H_2O}} = 4.68 \times 10^{-7}\ (\mathrm{p}K = 6.33)$$

In the classical formulation, Henderson (1908) used concentrations (c) rather than activities (a) for bicarbonate, $\mathrm{CO_2}$, and $\mathrm{H^+}$; the concentration of water was assumed to be constant and was therefore incorporated into the constant K'.

$$K' = \frac{c\mathrm{H^+} \times c\mathrm{HCO_3^-}}{cd\mathrm{CO_2}}$$

The symbol $cd\mathrm{CO_2}$ stands for the concentration of dissolved $\mathrm{CO_2}$ *including* the small amount of undissociated (dissolved) $\mathrm{H_2CO_3}$. It can be expressed as $cd\mathrm{CO_2} = \alpha \times p\mathrm{CO_2}$, where α is the solubility coefficient for $\mathrm{CO_2}$. The symbol $c\mathrm{HCO_3^-}$ represents the concentration of total $\mathrm{CO_2}$ ($ct\mathrm{CO_2}$) minus the concentration of dissolved $\mathrm{CO_2}$ ($cd\mathrm{CO_2}$), which includes $\mathrm{H_2CO_3}$.

$$c\mathrm{HCO_3^-} = ct\mathrm{CO_2} - (\alpha \times p\mathrm{CO_2})$$

The "bicarbonate" concentration by definition includes undissociated $\mathrm{NaHCO_3}$, carbonate ($\mathrm{CO_3^{2-}}$), and carbamate (carbamino-$\mathrm{CO_2}$; $\mathrm{RCNHCOO^-}$), which are present only in small amounts in plasma.

If the Henderson equation is rearranged, and $cd\mathrm{CO_2}$ is replaced by $\alpha \times p\mathrm{CO_2}$, the following equation results:

$$c\mathrm{H^+} = K' \times \frac{\alpha \times p\mathrm{CO_2}}{c\mathrm{HCO_3^-}}$$

Hasselbalch (1916) showed that a logarithmic transformation of the equation was a more useful form and employed the symbols pH ($= -\log c\mathrm{H}^+$) and pK' ($= -\log K'$). Currently, the pH is defined as the negative log of the activity of H^+ ($a\mathrm{H}^+$), which is the entity actually measured with pH meters. The resulting Henderson-Hasselbalch equation becomes:

$$\mathrm{pH} = pK' + \log\frac{c\mathrm{HCO_3^-}}{\alpha \times p\mathrm{CO_2}} = pK' + \log\frac{ct\mathrm{CO_2} - (\alpha \times p\mathrm{CO_2})}{\alpha \times p\mathrm{CO_2}}$$

K' is the first, apparent, overall (combined) dissociation constant for carbonic acid; apparent, because concentrations are employed rather than activities and because it includes both the hydration and dissociation constants; overall, because both the concentration of dissolved CO_2 and the true concentration of H_2CO_3 are used. K' depends not only on the temperature but also on the ionic strength of the solution. For an aqueous sodium bicarbonate solution at 37 °C the following approximate relationship exists between K' and ionic strength (I) measured in mol/kg H_2O:

$$pK' = 6.33 - 0.5\ \sqrt{(I)}$$

For blood plasma at 37 °C the normal mean value is $pK'(\mathrm{P}) = 6.103$ with a normal biological standard deviation of about ± 0.0015, mainly due to normal variations in ionic strength. (The normal average pK' at 38 °C is 6.100). In pathological cases with markedly deviant ionic strength, the standard deviation for pK' may be significantly greater. Changes in ionic strength of $\pm 20\%$ cause changes in pK' between 6.08 and 6.12. The variation of pK' of plasma with temperature can be expressed approximately as $\Delta pK'(\mathrm{P})/\Delta T = -0.0026 \times \mathrm{K}^{-1}$ or a decrease of 0.0026/°C increase in temperature. Due to the inclusion of carbonate and carbamate in the bicarbonate concentration ($c\mathrm{HCO_3^-}$), $pK'(\mathrm{P})$ appears to vary with pH, decreasing slightly with increasing pH. For most clinical purposes this variation of pK' with pH change can be ignored.

α, the solubility coefficient for CO_2 gas (including that present in its hydrated form, H_2CO_3), varies with temperature and composition of the solution. For pure water at 37 °C the solubility coefficient is $\alpha = 0.0329$ mmol/L $\times$ mm Hg ($= 0.0329$ mmol $\times$ L^{-1} $\times$ mm Hg^{-1}). The presence of salts or proteins in the solution decreases the solubility coefficient, while lipids increase it. The mean value for normal plasma at 37 °C is 0.0306 mmol $\times$ L^{-1} $\times$ mm Hg^{-1} with a biological SD of about ± 0.0003 mmol $\times$ L^{-1} $\times$ mm Hg^{-1}. In lipemic plasma the value of α may be 0.033 or even higher. The temperature variation, expressed as $\Delta\log \alpha\mathrm{CO_2}(\mathrm{P})/\Delta T$, is approximately $-0.0092 \times \mathrm{K}^{-1}$.

Inserting pK' and α for normal plasma at 37 °C, the Henderson-Hasselbalch equation takes the following form:

$$\mathrm{pH} = 6.103 + \log\frac{c\mathrm{HCO_3^-}}{0.0306 \times p\mathrm{CO_2}}$$

or

$$\mathrm{pH} = 6.103 + \log\frac{ct\mathrm{CO_2} - 0.0306 \times p\mathrm{CO_2}}{0.0306 \times p\mathrm{CO_2}}$$

where $p\mathrm{CO_2}$ is measured in mm Hg, and $c\mathrm{HCO_3}$ and $ct\mathrm{CO_2}$ are measured in mmol/L.

Taking the antilogarithm and combining constants, the equation becomes:

$$a\mathrm{H}^+ = 24.1 \times \frac{p\mathrm{CO_2}}{c\mathrm{HCO_3^-}} \times 10^{-9}$$

or, taking the activity coefficient of hydrogen ions in plasma to be 1.00 (the true value is probably about 0.80), the equation can be written:

$$c\mathrm{H}^+ = 24.1\frac{p\mathrm{CO_2}}{c\mathrm{HCO_3^-}}$$

where $c\text{H}^+$ is now given in nmol/L, $p\text{CO}_2$ in mm Hg, and $c\text{HCO}_3^-$ in mmol/L. If normal values are substituted in the equation,

$$c\text{H}^+ = 24.1 \times \frac{40}{25.4} \text{ nmol/L} = 38.0 \text{ nmol/L}$$

Application of the Henderson-Hasselbalch Equation in Blood Gas Measurements

Clearly, by measuring any two of the four parameters, $p\text{CO}_2$ or $c\text{dCO}_2$, pH, $c\text{tCO}_2$, $c\text{HCO}_3^-$, and using the Henderson-Hasselbalch equation with appropriate values for pK' and α, the other two parameters may be calculated. The values chosen for pK' and α are 6.103 and 0.0306, respectively. Although used as constants, these values should be recognized as means and susceptible to biological variation (see discussion of effects of altered ionic strength and lipid concentrations of plasma, page 1201). Parameters calculated on the assumption that pK' and α are invariant may therefore have significant error under certain pathological circumstances. The Henderson-Hasselbalch equation was originally used to calculate blood pH from alveolar $p\text{CO}_2$ and plasma $c\text{tCO}_2$, both measured manometrically. As electrochemical measurement of pH became feasible, plasma $p\text{CO}_2$ was calculated from measured pH and $c\text{tCO}_2$. Now that both pH and $p\text{CO}_2$ are measured electrochemically, $c\text{HCO}_3^-$, $c\text{dCO}_2$, and $c\text{tCO}_2$ are usually calculated by algorithms in the microprocessor of state-of-the-art blood gas instrumentation. Tables, nomograms, and specially designed slide rules are also available for these estimates.

Example

$$\text{pH} = pK' + \log \frac{c\text{HCO}_3^-}{(\alpha \times p\text{CO}_2)}$$

Since pK' is 6.103 and $\alpha = 0.0306$ mmol/L per mm Hg, with measured blood pH = 7.42 and $p\text{CO}_2 = 37.5$ mm Hg, $c\text{HCO}_3^-$ can be calculated:

$$7.42 = 6.103 - \log c\text{HCO}_3^- - \log (0.0306 \times 37.5)$$

Rearranging,

$$\log c\text{HCO}_3^- = 7.42 - 6.103 + \log 1.1475$$

$$= 7.42 - 6.103 + 0.0598 = 1.3768$$

Taking the antilog of 1.3768,

$$c\text{HCO}_3^- = 23.97 \text{ mmol/L}$$

Dissolved CO_2 and total CO_2 concentrations are:

$$c\text{dCO}_2 = \alpha \times p\text{CO}_2 = 0.0306 \times 37.5 = 1.1 \text{ mmol/L},$$

$$c\text{tCO}_2 = c\text{dCO}_2 + c\text{HCO}_3^- = 1.1 + 23.97 = 25.1 \text{ mmol/L}.$$

Oxygen in Blood

Most of the oxygen in blood is transported as HbO_2; when $c\text{tO}_2$ of blood is $\sim$9 mmol/L, $c\text{dO}_2$ is only $\sim$0.14 mmol/L. The oxygen in HbO_2 is reversibly bound to Fe(II) in the heme group of hemoglobin contained in erythrocytes. Each mole of hemoglobin-Fe(II) binds one mole of O_2 at STPD; one mole of hemoglobin tetramer (M.W. 64 456) therefore binds four moles of O_2. Theoretically, hemoglobin in 1 L of blood with a hemoglobin concentration ($c\text{Hb}$) of 15 g/dL (150 g/L), when all the Hb is in the form of HbO_2, carries 9.2 mmol of O_2 at STPD.

$$\frac{150 \text{ g Hb/L blood} \times 4 \text{ mmol } O_2/\text{mmol Hb}}{64.456 \text{ g/mmol Hb}} = 9.2 \text{ mmol/L}$$

Conversely, 1 g of fully functional hemoglobin protein is capable of binding 1.39 mL (0.062 mmol) of O_2 at STPD. The amount of O_2 bound to hemoglobin must, of necessity, be expressed relative to hemoglobin concentration of the blood. "Functional" is defined in terms of hemoglobin A protein (the normal adult gene product) capable of reversibly binding O_2 at the active site (the heme moiety of the molecule) and optimal binding of biologic effectors at the allosteric sites of the molecule. Some hemoglobins—abnormal because of genetically determined, variant globin chains—are not normally functional because allosteric binding is altered; oxygen affinity may be either reduced or (more rarely) increased. Methemoglobin, carbon monoxide hemoglobin (carboxyhemoglobin), sulfhemoglobin, and cyanmethemolgobin are examples of forms of hemoglobin made nonfunctional because the heme moiety is altered. See also Chapter 15.

Uptake of O_2 by the blood in the lungs is governed primarily by the pO_2 of alveolar air and by the ability of O_2 to diffuse freely across the alveolar membrane into the blood, and—to a lesser extent—by the affinity of deoxyhemoglobin (HHb) in the erythrocytes for O_2. At the pO_2 normally present in alveolar air, and with a normal membrane and normally functional hemoglobin protein, $> 95\%$ of the hemoglobin is bound to oxygen. At a pO_2 above 100–120 mm Hg, 100% of the hemoglobin is bound to oxygen, i.e., is *saturated* with O_2. When all hemoglobin capable of binding O_2 is saturated, further increase in the pO_2 of alveolar air simply increases cdO_2 of arterial blood. Delivery of O_2 by the blood to the tissues is governed by the large gradient between pO_2 of the blood and that of the tissue cells, by the working of the isohydric and chloride shift, and by the dissociation behavior of oxyhemoglobin (HbO_2) in the erythrocytes at the lower pO_2 of the blood-tissue cell interface. Hemoglobin oxygen dissociation behavior is an alternative description for affinity of hemoglobin for O_2. Under all circumstances, however, the total oxygen content of a blood sample is the sum of concentrations of hemoglobin-bound and physically dissolved oxygen.

Oxygen Saturation

Traditionally, sO_2, the oxygen saturation of blood, has been defined as

$$sO_2, \% = \frac{\text{Oxygen content}}{\text{Oxygen capacity}} \times 100\%$$

This equation was shaped by determination of O_2 by manometric methods. Oxygen *content*, i.e., the total concentration of O_2 in the sample, was determined on blood collected anaerobically, and then treated with ferricyanide reagent to liberate O_2 from hemoglobin and with negative pressure to release all free oxygen from solution. Oxygen *capacity*, i.e., the total concentration of O_2 in the fully saturated sample, was the amount of O_2 in the same sample and determined in the same way, but after the blood had been equilibrated with air or a gas mixture sufficiently rich in O_2 to saturate the blood to capacity with O_2. Now that spectrophotometric methods (described in Chapter 15) are used to determine oxyhemoglobin and reduced hemoglobin (along with carboxyhemoglobin and methemoglobin) specifically, sO_2 is calculated according to the equation

$$sO_2, \% = \frac{c\,HbO_2, \text{g/dL}}{c\,Hb, \text{g/dL}} \times 100\%$$

where $c\,HbO_2$ is the concentration of oxyhemoglobin, and $c\,Hb$ is the concentration of total functional hemoglobin (sum of oxy- and deoxyhemoglobin).* sO_2 is usually expressed in per cent in the USA but may also be expressed as a decimal fraction of 1.00. Oxygen saturation can also be approximated mathematically from measured pO_2 with the use of empirical equations;[19]

---

*sO_2 is sometimes calculated as

$$sO_2, \% = \frac{c\,HbO_2}{ctHb} \times 100\%$$

where $ctHb$ includes carboxy-, met-, and sulfhemoglobin. With this equation, the value of sO_2 can never reach 100% if any of the nonfunctional hemoglobin derivatives is present. For example, should carboxyhemoglobin amount to 10% of $ctHb$, sO_2 could be no more than 90% in fully saturated blood. This finding would be misleading, indicating increased shunting of blood in the lungs. Therefore, this calculation for sO_2 should be abandoned.

it is sometimes calculated by microprocessors of blood gas instruments. Such calculated values should, however, be interpreted with reservations since the algorithmic approach assumes normal oxygen affinity of the hemoglobin.

Oxygen saturation of arterial blood is normally > 95%; decrease in oxygen saturation below the critical level necessary for adequate oxygenation of tissue is a grave clinical situation. Such decrease indicates either a low pO_2 or seriously impaired ability of hemoglobin to serve as an oxygen carrier. The amount of oxygen that the blood can carry is determined by three major factors: the amount of functional hemoglobin available within erythrocytes; the pO_2 that determines how much oxygen will dissolve in the blood; and the affinity of the available hemoglobin for oxygen. Decreases in the amount of hemoglobin can come about as a result of a decreased number of erythrocytes that contain a normal concentration of hemoglobin (normochromic anemia), or a decreased concentration of hemoglobin in the erythrocytes (hypochromic anemia). Decreased amounts of functional hemoglobin can occur as a result of poisonings that convert large quantities of hemoglobin into nonfunctional carboxyhemoglobin, methemoglobin, sulfhemoglobin, or cyanmethemoglobin. Impairment of diffusion of oxygen across the alveolar membrane, as well as inadequate circulation due to cardiac insufficiency, results in decrease of pO_2, and may affect oxygen delivery to tissues in varying degrees. Compensatory factors that govern affinity of hemoglobin for oxygen may intervene to keep delivery at an adequate level. Clinically, it is important to distinguish between hypoxia (decreased pO_2 and decreased sO_2 because of decreased availability of O_2) and cyanosis (decreased sO_2) because of abnormally high concentrations of reduced hemoglobin or chemically altered hemoglobin incapable of carrying O_2.

Hemoglobin-Oxygen Dissociation

The degree of association or dissociation of oxygen with hemoglobin is determined by pO_2 and the affinity of hemoglobin for oxygen. When the saturation of blood, sO_2, is determined over a range of pO_2 and plotted against pO_2, a sigmoidal curve called the oxygen dissociation curve is obtained; this curve could also be described accurately as an association curve. The *shape* of the curve arises from the increasing efficiency with which deoxyhemoglobin molecules bind more oxygen once some oxygen has been bound (see also Chapter 15, p. 1519, and Figure 15-7). The *location* of the curve relative to the pO_2 required to achieve a particular degree of saturation (association of O_2 with hemoglobin) in the blood is a function of the affinity of the hemoglobin in the blood for oxygen.

Affinity of hemoglobin for oxygen is dependent on five factors: temperature, pH, pCO_2, concentration of 2,3-diphosphoglycerate (DPG), and the type of hemoglobin. Dissociation of oxygen is discussed in terms of DPG concentration and abnormal hemoglobin proteins in Chapter 15. Here we note the effects of temperature, pH, and pCO_2 on dissociation behavior, introduce the correction coefficients that can be applied to measured pO_2, and describe the $pO_2(0.5)$ or P_{50} (in alternative terminology) as an indicator of alteration of hemoglobin affinity.

Figures 15-7 illustrates the effect of different concentrations of DPG on the sO_2/pO_2 relationship. In Figure 10-4, the graph on the left illustrates the effect of plasma pH on this relationship. If a similar graph were made for variation of pCO_2 (with pH at 7.4, $cDPG$ at 4.8 mmol/L, and temperature at 37 °C), and another for variation of temperature (with pH at 7.4, pCO_2 at 40 mm Hg, and $cDPG$ at 4.8 mmol/L), the shifting locations of the curves could be described as in the chart in the legend of Figure 10-4. The shifts are obviously considered in terms of deviations of the parameters from some reference point—temperature from 37 °C; pCO_2 from 40 mm Hg; pH(P) from 7.40; and $cDPG/cHb$ from some value defined as "normal." In the third column of the chart in the legend of Figure 10-4, the shifts are related to their effects on affinity of hemoglobin. The sigmoidal curves shown on the left of Figure 10-4, however, are difficult to use when quantitative estimates of the amount of shift are desired. The Hill transform (illustrated as the Hill plot, on the right side of Figure 10-4) converts the curvilinear dissociation function into a linear function. The slope of the linear function, called the Hill slope or n_{Hill}, is essentially constant at 2.7 for sO_2 of 30–70%. The transform of sO_2 is to logit sO_2 (ordinate of the Hill plot), where

$$\text{logit } sO_2 = \log [sO_2/(1 - sO_2)] \qquad 0 \leq sO_2 \leq 1$$

pO_2 is transformed to log pO_2 on the abscissa. (Students of clinical chemistry will recognize the Hill transform as an application of the logit-log relations so frequently utilized in reduction of immunoassay data.) The linear transforms of the dissociation curves have allowed linear coeffi-

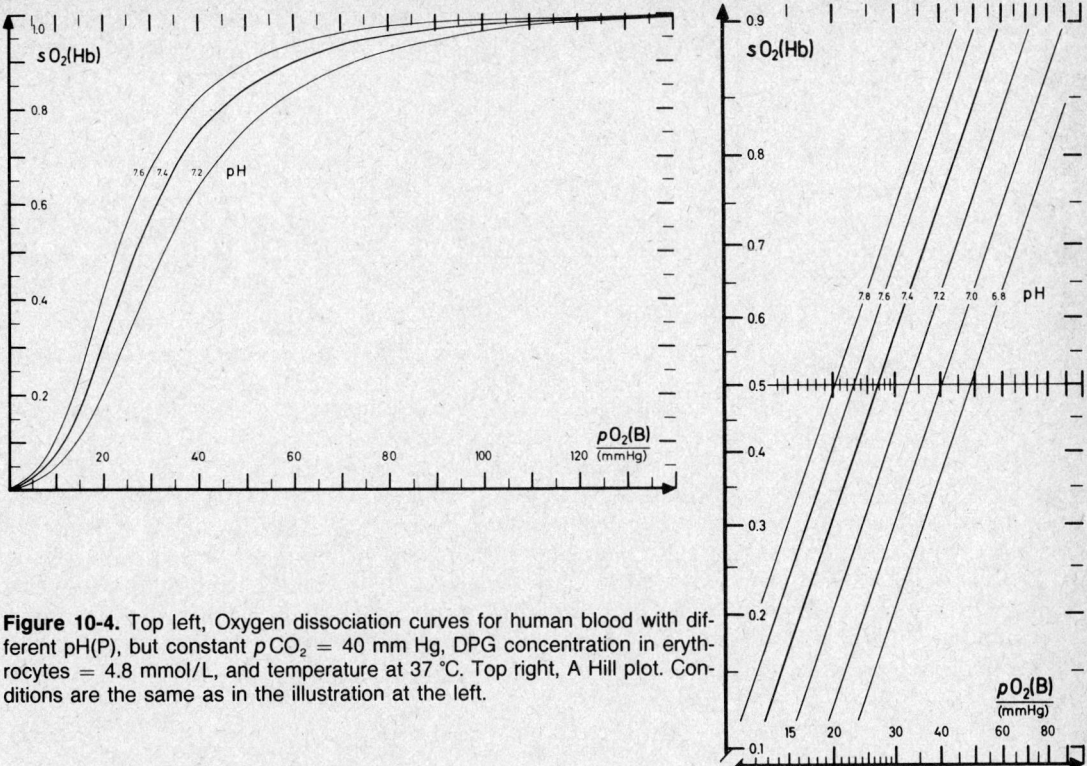

Figure 10-4. Top left, Oxygen dissociation curves for human blood with different pH(P), but constant $pCO_2 = 40$ mm Hg, DPG concentration in erythrocytes = 4.8 mmol/L, and temperature at 37 °C. Top right, A Hill plot. Conditions are the same as in the illustration at the left.

Deviation from Standard Conditions	Shift in Dissociation Curve	Affinity of Hemoglobin for Oxygen	Coefficient of Change*
pH(P) > 7.4	←	↑	$\dfrac{\Delta \log pO_2}{\Delta pH(P)} = -0.38$
pH(P) < 7.4	→	↓	
Temperature			
> 37 °C	→	↓	$\dfrac{\Delta \log pO_2}{\Delta T} = +0.024\ K^{-1}$
< 37 °C	←	↑	
pCO_2 > 40 mm Hg	→	↓	$\dfrac{\Delta \log pO_2}{\Delta \log pCO_2} = +0.10$
pCO_2 < 40 mm Hg	←	↑	
DPG/Hb ratio			
> normal	→	↓	$\dfrac{\Delta \log pO_2}{\Delta(cDPG/cHb)} = +0.60$
< normal	←	↑	

*The coefficients given in this chart are the basis for the correction of measured pO_2; their application to $pO_2(0.5)$ correction is discussed in the text. The effect of pH(P) to shift the dissociation curve is called the Bohr effect; the coefficient above for $\Delta \log pO_2 / \Delta pH(P)$ applies to conditions when the pCO_2 is 40 mm Hg and changes in pH(P) are due to changes in concentrations of noncarbonic acids and bases. If, however, the changes in pH(P) are being caused by changes in pCO_2, then the absolute value of the coefficient is greater, i.e., $\Delta \log pO_2 / \Delta pH(P) = -0.53$. The coefficients for the Bohr effect are specified for pO_2 of whole blood but utilize the pH of plasma, pH(P). The coefficient for DPG effect utilizes the common form of expression of DPG concentration, ratio of DPG concentration to hemoglobin concentration.

cients of change (fourth column of the chart in the legend of Figure 10-4) to be determined for each of the factors that shift dissociation curves, and hence the lines on the Hill plot. These coefficients find application, singly or in combination, for correcting measured pO_2 or calculated $pO_2(0.5)$ for temperature, pCO_2, pH(P), and DPG effects.

$pO_2(0.5)$ or P_{50}

$pO_2(0.5)$ is defined as the pO_2 for a given blood at which hemoglobin of the blood is half-saturated with O_2. Note that on the Hill plot a horizontal line has been placed where logit $sO_2 = 0$

and sO_2 is equal to 0.5. Note also that the dissociation lines shown on the plot for different pH(P) values intersect the horizontal line. If a perpendicular is dropped from such an intersection to the abscissa, it will cut the log scale of the abscissa at some pO_2. That particular log pO_2 is called $pO_2(0.5)$, or in alternative terminology P_{50}. The $pO_2(0.5)$ represents what the pO_2 of the blood *would* be if it were half-saturated with O_2, i.e., if its sO_2 were equal to 0.5. The measured value of $pO_2(0.5)$ differs from the standard value of pO_2 (0.5) by some amount determined by the effect that factors different from $pCO_2 = 40$, T = 37 °C, DPG = 4.8 mmol/L have to shift the dissociation line away from the reference line when T $\neq$ 37 °C (310.16 K), pH $\neq$ 7.40, pCO_2 $\neq$ 40 mm Hg, or $cDPG \neq 4.8$ mmol/L (the reference conditions specified for Figure 10-4). The value of $pO_2(0.5)$ therefore becomes a measure of change imposed on hemoglobin affinity by the factors that affect it, as they vary individually or together.

Determination of $pO_2(0.5)$

The simplest method for determining $pO_2(0.5)$ consists of measuring pO_2 potentiometrically and sO_2 spectrophotometrically on an anaerobically collected venous whole blood sample. Alternatives are measurement of sO_2 on the whole blood equilibrated by tonometry against a gas mixture of known pO_2, or measurement of pO_2 on a 1:1 mixture (equivalent to 50% saturated) of the fully oxygenated and completely deoxygenated whole blood. With any of the methods pH and pCO_2 should be measured along with the other parameters. The calculating equation for $pO_2(0.5)$ is shown in equation (1). (With the first method, both terms on the right side would be evaluated with measured values; with the same equation and the second method, log pO_2 would be evaluated with the pO_2 of the equilibrating gas; with the third method and the same equation, sO_2 would be evaluated as 0.5.)

$$\log pO_2(0.5) = \log pO_2 - \frac{\text{logit } sO_2}{2.7} \tag{1}$$

where logit $sO_2 = \log [sO_2/(1 - sO_2)]$ and 2.7 is the Hill slope, which is assumed to be independent of the other variables. Taking the antilog, the value for $pO_2(0.5)$ as found is uncorrected, i.e., it simply defines $pO_2(0.5)$ for the patient's blood at 37 °C, the temperature of measurement, without reference to deviation of the patient's pH(P) and pCO_2, or $cDPG/cHb$ ratio or deviations of his actual body temperature from reference conditions. To convert log $pO_2(0.5)_{\text{uncorrected}}$ to a value relative to pH(P) 7.4, the term in which reference ranges are expressed, further calculation is necessary:

$$\log pO_2(0.5, 7.4) = \log pO_2(0.5)_{\text{uncorr}} - 0.38[\text{pH(P)} - 7.40]$$

where the first term of the right hand side is the value obtained from equation (1) and the second term is a correction factor utilizing the measured pH(P) of the patient's sample and the coefficient for ΔpH(P) that is shown beneath Figure 10-4. Further corrections can be made on the same principle, so that $pO_2(0.5, 7.4, 40$ mm Hg, 30 °C) would be the antilog of the number obtained by the following equation set up for a hypothermic, acidotic patient with pH(P) 7.20, pCO_2 80 mm Hg, and body temperature of 30 °C.

$$\log pO_2(0.5, 7.4, 40 \text{ mm Hg, } 30 °C) = \log pO_2(0.5)_{\text{uncorr}} - 0.38(7.20 - 7.40)$$
$$+ 0.10(\log 80 - \log 40) + 0.024 (30 - 37)$$

Using the same equation for a patient whose body temperature is 37 °C, but leaving out the correction term for temperature, the antilog of log $pO_2(0.5, 7.4, 40$ mm Hg, 37 °C) would define the effect of an altered ratio of $cDPG/cHb$ and thus allow the measurement of $pO_2(0.5)$ to become an indirect measurement of 2,3-diphosphoglycerate. The individual coefficients can therefore be used singly or in combination to obtain different definitions of $pO_2(0.5)$ in different circumstances. A nomogram for obtaining $pO_2(0.5)$ from the measurements of pO_2, sO_2, pH, and $cDPG/cHb$ has been devised.[14]

Sources of Error. Besides the errors that may arise in measurements of sO_2, pO_2, pH, and pCO_2, additional error may enter from the use of coefficients to correct $pO_2(0.5)$. Assuming that the random error (CV) of a coefficient for a change is at best 1%, the random error of $pO_2(0.5)$ corrected to pH 7.4 would be at least 1.5%. Extrapolation to $sO_2(0.5)$ along the Hill slope is another source of error because the value of the slope is subject to biological variation; for this reason, sO_2 is best measured as close as possible to 50% and the method of mixing deoxygenated and oxygenated blood in equal proportions carries less risk for this type of error. Large corrections due to very deviant pH from 7.40 and pCO_2 from 40 mm Hg can cause significant errors because of the variation of the coefficients with changes in pH, pCO_2, and DPG concentrations.

Reference Ranges.[18] For adults, the 95% limits for $pO_2(0.5)$, measured at 37 °C and corrected to pH(P) of 7.4, are 25–29 mm Hg. For newborn infants, the limits are 18–24 mm Hg.

Clinical Significance. *Increased values* for $pO_2(0.5)$ indicate displacement of the oxygen dissociation curve to the right, i.e., a decreased affinity of the hemoglobin for oxygen. The chief causes are hyperthermia, acidemia, hypercapnia, high concentrations of 2,3-diphosphoglycerate, or presence of an abnormal hemoglobin with decreased oxygen affinity. DPG concentrations tend to be increased in chronic alkalemia and in anemia. An example of hemoglobins with decreased O_2 affinity is hemoglobin Seattle.

Physiological effects of decreased affinity of hemoglobin are small. Generally, the affinity is still sufficient to allow the hemoglobin to bind adequate amounts of oxygen in the lungs. Low affinity merely facilitates dissociation of HbO_2 at the peripheral tissue cell; in anemia, low affinity (as a result of increases in 2,3-DPG) is a desirable compensatory mechanism. Patients with hemoglobin Seattle have $pO_2(0.5)$ of ~41 mm Hg and low cHb, but are otherwise unaffected.

Low values for $pO_2(0.5)$ signify displacement of the oxygen dissociation curve to the left, i.e., increased affinity of hemoglobin. The main causes are hypothermia, acute alkalemia, hypocapnia, low concentration of 2,3-diphosphoglycerate, or an abnormal hemoglobin. Decreases of DPG are commonly observed in acidemic states which have persisted for more than a few hours; the initial increase in $pO_2(0.5)$ caused by the acidemia is gradually compensated by decrease in DPG so that $pO_2(0.5)$ then falls to lower than normal values. Banked human blood for transfusion develops decreased affinity of the hemoglobin if DPG concentrations are not protected by the addition of inosine, pyruvate, or inorganic phosphate. The increase in $pO_2(0.5)$ in the presence of carboxyhemoglobin is due to the Haldane-Smith effect, which occurs because of a slight increase in the affinity of the remaining unaffected heme groups for oxygen. Methemoglobin has a similar effect. Both effects are slight and without clinical significance. An example of an abnormal hemoglobin with increased oxygen affinity is hemoglobin Yakima. Fetal and neonatal blood have a higher affinity because of the presence of hemoglobin F, which has a higher affinity for O_2 than hemoglobin A.

The physiological consequence of increased affinity of hemoglobin for oxygen is less efficient dissociation of HbO_2 at the peripheral tissues and lower tissue pO_2. Patients with hemoglobin Yakima have $pO_2(0.5)$ of ~12 mm Hg; but they are compensated by an increase in cHb and are relatively unaffected, except for less than normal stamina upon strenuous muscular exercise.

Determination of pCO_2, pO_2, and pH

Specimens

Blood is the most likely specimen that a clinical laboratory will receive for gas analysis; in some facilities, expired gases may also be analyzed.

Arterial or venous blood specimens may be obtained from any site accessible to contemporary procedures of vascular catheterization or entry; these sites include chambers of the heart, the great vessels of the chest, and vessels of the extremities. Analysts should recognize that specimens that are difficult to obtain are precious and therefore should be handled with special care. Furthermore, the volume of specimen collected for an analysis, especially from premature neonates being monitored by frequent and repetitive testing, needs to be as small as is possible without sacrifice of specimen quality and accuracy.

Quality control of blood analysis for gases and pH is as much (perhaps more) dependent on control of preanalytical error, i.e., on proper collection and handling of specimen, as it is on control of the analytical instrument and process. Laboratory personnel do not always control collection of arterial or venous specimens; the clinical chemist must therefore work closely and cooperatively with those physicians, nurses, respiratory therapists, and other personnel who obtain them and must have taken an active part in developing interdepartmental mechanisms to guarantee their quality.

Arterial puncture carries a medical risk; under no circumstances should it be undertaken by anyone who has not been properly trained to perform it. In some institutions, only physicians are considered qualified to perform arterial punctures; in others, nonphysician personnel with special training, among them selected medical technologists or laboratory phlebotomists, may be allowed to perform them. Arterial puncture is always done with syringe and needle. No tourniquet is used and no pull applied to the plunger of the syringe; the arterial blood pressure pushes the blood into the syringe.

Venous blood for blood gases and pH is best collected with a needle and syringe, although some workers also accept specimens drawn to a complete fill of an evacuated tube containing a dry heparin salt. In the collection of venous blood from an arm vein, the specimen should be obtained in the first few seconds after application of the tourniquet; the patient should not be allowed to flex the fingers or clench the fist. Prolonged application of the tourniquet or muscular activity decreases venous pO_2 and allows accumulation of acid metabolites. Placement of indwelling cannulas with heparin locks for short and long term therapies is now fairly common; the cannula serves often as a port for specimen collection. The lock must be thoroughly flushed with blood before a specimen for blood gases and pH (or any other assay) is drawn. Failure to flush the lock properly has unpredictable effects on measured quantities and frequently is indicated by bizarre results.

Arterial or venous specimens are best collected anaerobically with liquid heparin anticoagulant in all-glass, sterile syringes with capacities of 1–5 mL.* A syringe with a Luer-Lok hub is best for collecting specimens by arterial puncture and may also be used for venous puncture. Glass is preferred to plastic to avoid loss of gases through the syringe wall, and in arterial collections to avoid friction on the plunger as arterial blood pressure is allowed to fill the syringe. The size of the syringe, the concentration of heparin, and the maximum volume of blood drawn into the syringe are important. For syringes of 1–5 mL capacity, dead space varies roughly from 0.1–0.2 mL. If a sterile heparin solution of 500 U/mL (5 mg/mL) is used, the dead space will hold 0.5–1.0 mg of heparin. Less than 0.2 mg is needed to anticoagulate 5 mL of blood. Adequate anticoagulation is achieved by drawing enough sterile heparin solution into the syringe in such a way as to wet the interior of the barrel over the maximum capacity of the syringe, and then ejecting air and excess heparin in such a way as to leave the dead space of the syringe filled with heparin. Blood drawn to the mark in a 1-mL syringe may be diluted with up to 10% heparin solution; drawn to the mark in a 5-mL syringe, the blood will contain ≤4% heparin solution. Increasing the dead space volume of the syringe or filling to less than full capacity further increases the proportion of heparin solution, the amount of heparin in the specimen, and the dilution of

---

*Packaged, disposable sets for collection of blood gases are commercially available, as are cold-packs for their subsequent transport to the analyst. The expense of the sets usually prevents their utilization in laboratories that do many blood gas determinations. The reader interested in such materials is referred to sales representatives of manufacturers of blood gas instrumentation.

Table 10-6. CHANGES IN pH, pCO_2, BASE EXCESS, pO_2, AND ACTUAL BICARBONATE WITH DIFFERENT VOLUMES OF HEPARIN SOLUTION IN VENOUS SAMPLES

	Percentage Volume of Heparin Solution in Blood			
	2	*4*	*10*	*20*
pH	7.37	7.37	7.37	7.36
pCO_2, mm Hg	51.0	48.8*	44.2*	39.0**
Base excess, mmol/L	+3	+3	0	−3**
pO_2, mm Hg	32.2	35.2	36.8	38.2
Actual HCO_3^-, mmol/L	29	28***	25	22**

Percentage volume of 2 for heparin in blood represents 0.2 mg heparin in 0.1 mL of solution in 5 mL of blood. Significance of difference from percentage volume of 2: *P < 0.01; **P < 0.001; ***P < 0.05.

Modified from Hutchison, A. S., Ralston, S. H., Dryburgh, F. J., et al.: Too much heparin: Possible source of error in blood gas analysis. Br. Med. J., *287*:1131-1132, 1983.

the specimen. Certain lots of heparin solution, commercially prepared for parenteral injection and recommended as anticoagulant, are acidic. An increasing ratio of heparin to blood (see Table 10-6) can have an increasingly marked effect on measured pCO_2 and parameters calculated from it.[7] Each lot of heparin should be tested before its use in patient specimens and rejected if found unacceptable. A bead or other solid object in the syringe when blood is admitted is not necessary for mixing of the blood and heparin, and may, if it enlarges the dead space to be filled with heparin, actually be a source of error. For subsequent mixing to ensure homogeneous distribution of erythrocytes throughout the specimen, a bead may be added prior to analysis; however, with prompt transit of the specimen to the analyst and immediate analysis, mixing by simple but vigorous rolling of the syringe between the palms is usually adequate to establish the desired homogencity and avoids breaking the closure on the specimen until it is actually sampled.

Anaerobic technique for collection has as its guiding principle either no or minimal exposure of blood to atmospheric air. The pCO_2 of dry air is about 0.25 mm Hg and therefore much less than that of blood (about 40 mm Hg); the CO_2 content and pCO_2 of blood exposed to air will decrease and blood pH, which is a function of pCO_2, will rise. The pO_2 of atmospheric air is ~60 mm Hg higher than that of arterial blood and ~120 mm Hg higher than that of venous blood. Blood exposed to atmospheric air therefore gains O_2; the gain is much less for arterial blood, which is ordinarily > 95% saturated with O_2, than it is for venous blood. Blood with pO_2 >150 mm Hg, as might be obtained from patients undergoing oxygen therapy, will lose O_2 on exposure to air. In experimental and reference range studies exposure to air must be avoided. In routine arterial or venous puncture, depending on whether the needle is full of heparin or empty, a single small bubble of air may appear in the blood in the syringe. Error will be minimal if the bubble is ejected immediately upon removing the needle from the puncture site and before capping the syringe tip. An air bubble equal to 10% of the specimen volume causes a 15% increase in measured pO_2 when actual pO_2 is ~30 mm Hg; pCO_2 decreases. Numerous small air bubbles have similar effects. When a full syringe is disconnected from a needle or cannula, it should be held tip up, a drop of blood ejected from it, and a tightly fitting cap placed on the tip of the hub. The practice of leaving or putting a needle in place and pushing the needle into a rubber stopper is, in the authors' experience, fraught with risk of leakage of air in, or blood out, at some stage of handling of the syringe. Suitable caps are inexpensive and readily available from laboratory supply houses.

Arterialized capillary blood is sometimes an acceptable alternative to arterial blood when blood losses must be minimized, when an arterial cannula is not available, or when repeated arterial puncture must be avoided. Arterialized capillary blood is not acceptable, however, in cases of decreased cardiac output when systolic blood pressure is < 95 mm Hg; in cases of vasoconstriction; in patients on oxygen therapy; in newborns in the first few hours or days after birth; or in newborns with respiratory distress syndrome. In these situations, there is a particular risk of admixture with blood from the venules that causes erroneously low pO_2 values. Freely flowing cutaneous blood originates in the arterioles and corresponds to arterial blood in composition. Capillary puncture must be preceded by warming the selected skin puncture site for 10 min to achieve vasodilation and good blood flow through local capillary vessels. For collection from the finger of a child or adult or from an infant's heel, warming may be accomplished by immersing the arm or leg in water warmed to 45 °C. The limb should be rosy and warm to the touch before the skin is punctured. The first blood drop to appear should be wiped away, and subsequent free-forming drops taken up in a capillary collection tube containing dried heparin. Only free-flowing blood provides a satisfactory sample; taking up the drops as they form minimizes exposure of specimen to air. Appropriate capillary tubes are 80–250 mm long; depending on their diameter, they hold 70–300 μL of blood. The larger diameter tubes require mixing. They are allowed to fill to about two-thirds their length; then immediately, a flea (a slender wire ~5 mm in length) is inserted into the tube, a cap fitted to the filled end, and a magnet used to move the flea back and forth in the blood to mix it with the heparin coating on the tube walls.

Transport and analysis of specimens should be prompt, even when there is no emergency in returning test results. Physicians who employ blood gas and pH measurements in acute care management often require a turnaround time of < 15 min between specimen acquisition and reporting of results; many will not tolerate elapsed time of > 30 min. Ideally, specimens should never be stored, although analysis of properly preserved specimens may be delayed up to 1 h.

The pH of freshly drawn blood decreases on standing at a rate of 0.04–0.08 pH unit/h at

37 °C, 0.03/h at 25 °C, and 0.008/h at 4 °C. pCO_2 increases at $\sim$5 mm Hg/h at 37 °C but at 2–4 °C only one-tenth as fast. The decrease in pH is accompanied by corresponding decrease in glucose and an equivalent increase in lactate. The primary cause of these changes is thought to be glycolysis by leukocytes, thrombocytes, and reticulocytes. In leukemia, the pH drop during the first 30 min after specimen collection may be as high as 0.6 pH unit. In freshly drawn blood with normal pO_2, kept anaerobic, cell respiration causes pO_2 to decrease at a rate of 2–6 mm Hg/h at room temperature and twice as rapidly at 37 °C; with pO_2 $\sim$400 mm Hg, the rate of decrease is 2 mm Hg/min. Adverse effects of glycolysis and respiration on pH, $ctCO_2$, pO_2, and pCO_2 of blood can best be avoided by analysis within 30 min after collection. If analysis must be delayed, or if circumstances create a risk of delay, the syringe or tube containing the blood should be immersed in a mixture of ice and water until analysis is possible. Under these conditions, changes in blood pH are negligible. With state-of-the-art blood gas instrumentation, introduction of a chilled sample carries little risk of low temperature effect on measurements; thermal equilibration of sample to 37 °C is rapid and complete before the measurement signal on the sample is acquired.

Thorough *mixing*, to ensure homogeneity of the whole blood specimen, is essential immediately before sampling for measurement of blood gases and pH. With a capillary specimen, mixing is accomplished by repeating the magnet manipulation to move the flea. With a venous specimen in a stoppered tube, gentle inversion is appropriate. With syringe specimens, rolling the syringe between the palms is usually adequate.

Anaerobic separation of plasma is seldom required, although instances may arise when determination of plasma pH is specifically desired. Separation of arterial or venous blood collected in a syringe is described in Chapter 12, page 1345. A venous specimen collected in an evacuated tube must be warmed to 37 °C before centrifugation, then centrifuged without opening, and the plasma sampled immediately upon removal of the stopper. Failure to warm either arterial or venous blood to 37 °C before separation by centrifugation causes falsely high values for measured pH. Specimens in capillary tubes closed with simple, plastic stopper caps must be centrifuged with special care to avoid leakage of sample from the bottom cap, to avoid breakage in swinging bucket rotors, and to avoid excessive heating in the centrifuge. Good practice requires sampling, whenever possible, from the center of the plasma phase, as distant as possible from any surface exposed to air and from the erythrocytes.

Instrumentation

A schematic diagram characteristic of contemporary instrumentation is shown in Figure 10-5. Electrochemical principles and structural features of electrodes are discussed in Chapter 1B, Section Four. Prominent names among manufacturers of blood gas equipment include Corning Medical and Scientific, Medfield, MA 02173; Instrumentation Laboratories, Lexington, MA 02173; and Radiometer America, Westlake, OH 44145. The reader is referred to these manufacturers for details and operational features of the various instruments.

The operation of a blood gas instrument begins with the operator presenting a blood specimen at the sample probe and initiating its uptake by a keyboard command. The pump device then draws the sample through the probe and into the measuring chamber (C), bringing the sample into contact with the electrodes. The pump then pauses to allow the sample to reside in the chamber for temperature equilibration prior to measurement. Upon completion of measurement, the pump pushes the sample to waste, while digital output is being made available on a display, printed tape, or a printed multi-copy requisition/report ticket. The instrument (see Figure 10-5) is designed so that a manually or electronically actuated valve (V) admits calibrator gases, standard buffers, or sample to a small chamber (C) holding 60–150 μL and maintained by a thermostated fluid or metal bath (B) to a constant temperature of 37 $\pm$ 0.1 °C. Measuring and reference electrodes (E) protrude into the chamber. In the pH calibration phase of the instrument, high pH standard buffer and low pH standard buffer are alternately admitted into the chamber and electronic responses of the upper and lower limits of a linear pH curve are established. In the gas calibration phase, gas mixtures with high and low fractional concentrations of O_2 and CO_2 are alternately admitted into the chamber and electronic responses of the upper and lower limits of linear pO_2 and pCO_2 curves are set. In the measurement phase, either an anaerobically collected blood sample or expired air is admitted; in the case of expired air, only pCO_2 and pO_2 are measured since no fluid is present in the chamber; in the case of blood, pH, pCO_2, and pO_2 are measured in the fluid sample.

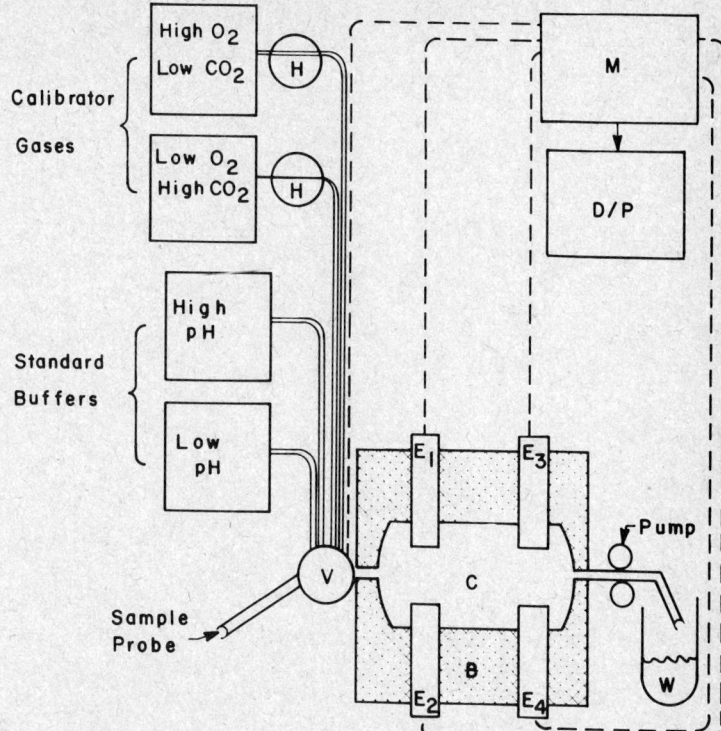

Figure 10-5. Diagram of blood gas instrumentation. H, humidification device; V, valve; C, chamber; B, constant temperature bath at 37 °C; W, waste; M, microprocessor; D/P, display/printer. E (electrodes) where E_1 is pO_2, E_2 pCO_2, E_3 pH, and E_4 reference for pH. A combination electrode is possible for pH measurement but is not often used.

A majority of instruments are designed to be self-calibrating. Under the command of the microprocessor (M), calibrator gases and buffers are cycled at short intervals through the chamber and electronic responses are continually monitored and reset to the constants initially entered for high and low pCO_2 and pO_2 and high and low pH of the calibrator materials. A keyboard-entered command to sample a specimen interrupts the cycling and initiates uptake of sample through the probe. The peristaltic pump (P in the diagram) loads the chamber by drawing sample in and subsequently emptying it by pushing its contents to waste. This pump is also under microprocessor control since it must pause after admission of sample in order to let the sample reside in the chamber long enough to allow measurement to be completed. Some instruments are designed for multiple, sequential readings at short intervals on the contained sample; in these designs, the pump is not actuated to empty the chamber until the signal from the slowest electrode has reached a maximum plateau. Other instruments are designed to empty the chamber at a fixed time after filling is complete.

Electrodes. Electrodes are typically constructed with threaded collars or bayonet locks for a leak-proof fit when they are installed in the chamber. They are positioned to bring their tips into contact with the contained sample. The tip of the pH measuring electrode is made of H^+-sensitive glass. The pH reference electrode is a calomel or silver/silver chloride electrode filled with saturated KCl which contacts the sample through a port, thus forming a salt bridge. Aside from specialized construction for application in the blood-gas chamber, most pH measuring and reference electrodes differ little from those of free-standing pH meters. The potentiometer is somewhat more sensitive since it is calibrated to such a narrow range.

Construction of pO_2 and pCO_2 electrodes is shown in Figures 1B-39 and 1B-41. The gas electrodes are fitted at the tip with gas-permeable membranes held in place with O-rings. Inside these membranes are electrolyte solutions in contact with the measuring and reference elements of the electrode. The membrane of the pCO_2 electrode consists usually of Teflon or silicone rubber ~ 25 μm thick. The electrolyte solution is a thin film containing $NaHCO_3$ at 0.005 mol/L and NaCl at 0.1 mol/L and saturated with AgCl. A spacer of nylon net or cellophane lies between the solution and the H^+-sensitive glass of the measuring element proper. The electrolyte also contacts the silver/silver chloride reference electrode. As CO_2 diffuses from sample into electrolyte, the slight rise in $[H^+]$ from its hydration reaction is measured as ΔpH by an especially sensitive potentiometer, and transformed to $\Delta\log pCO_2$ electronically. Sensitivity of a pCO_2 electrode ($-\Delta pH/\Delta\log pCO_2$) is ordinarily 0.96–1.00 over the range 5–200 mm Hg.

The membrane of a pO_2 electrode is usually polypropylene $\sim20\ \mu m$ thick. The electrolyte solution is a thin film of phosphate buffer saturated with AgCl and also containing KCl, and is in contact with the polarized platinum cathode and the silver/silver chloride anode. As O_2 diffuses into the electrolyte, it reacts with the cathode to cause current to flow; the generated current is measured. The sensitivity ($\Delta I/\Delta pO_2$, change in current per mm Hg) of a pO_2 electrode with a polypropylene membrane is ~20 pA/mm Hg. At a pO_2 of ~100 mm Hg, 20 pA corresponds to a reduction of oxygen of 5.2 fmol/s. The sensitivity is almost constant up to pO_2 of 500 mm Hg. At zero pO_2 the zero current is 1–50 pA.

Significant Figures of Results

Contemporary instruments commonly display pH readings to three decimal places, and pCO_2 and pO_2 readings to one decimal place. Calculated parameters may be displayed to 1–3 decimal places. pH should be reported to two decimals, and pCO_2 and pO_2 as integers. The number of significant digits reported for calculated data take precision into account. Values with two significant digits are generally adequate for clinical purposes; often the imprecision and inaccuracy lie in the second digit.

Tonometry

Tonometry is the process of exposing a liquid to an ambient gas phase in such a way that each gas in the gaseous phase partitions to an equilibrium between the liquid and gas phases. This equilibration, in effect, imparts the pCO_2 and pO_2 of an equilibrating gas to blood that is exposed to it. Equilibration by tonometry employs gases of known fractional composition, humidified at 37 °C to give a saturated water vapor pressure of 47 mm Hg. The pCO_2 or pO_2 of such gases is calculated according to Dalton's law (see p. 1198 for an example).

Tonometry is used to treat blood samples for various purposes. For $pO_2(0.5)$ determination, the patient's sample is equilibrated with a gas mixture with known pO_2 of about 27 mm Hg and pCO_2 of about 40 mm Hg, prior to measurements of sO_2 from which $pO_2(0.5)$ is calculated. For determination of standard bicarbonate, patient's sample is equilibrated with gas of pCO_2 40 mm Hg and $pO_2 > 100$ mm Hg (i.e., pO_2 sufficient to guarantee $sO_2 > 0.90$), prior to measurement of pH and pCO_2. Other applications include preparation of a blood sample for quality control of blood gas measurements, for determining liquid-gas difference for a pO_2 electrode, and for verifying accuracy and linearity of pO_2 and pCO_2 electrodes.

The principles which should be observed in tonometric equilibration of blood are rapid equilibration, preferably within 3 min, to minimize glycolysis in the blood sample; temperature control at 37 ± 0.1 °C; humidity and temperature control to prevent evaporation of water from the blood and condensation of water into the blood; and a mode of exposure of gas to blood that causes no more hemolysis than 50 μmol Hb/L of blood plasma.

Tonometry may be achieved by simple, homemade assemblies or by commercially available equipment. One widely used tonometer (IL) consists of a glass or plastic cup fitted on a shaft and enclosed in a humidified chamber whose temperature is maintained at 37 °C. A few mL of blood is placed in the cup and gas flow is initiated. The chamber is continuously flushed with the humidified equilibrating gas directed to play over the inside surface of the cup. A controller unit causes the cup to rotate rapidly and periodically, in short bursts, so that blood in the cup is thrown in a thin layer over the inside walls of the cup. Setting a timer on the controller unit defines the length of time for the treatment. Another type of tonometer is a syringe that is specially constructed to allow gas to be introduced and humidified through the plunger. A blood sample is loaded into the syringe to fill the space between the bottom of the plunger and orifice of the syringe body. The gas line is then connected to the plunger and gas flow is initiated. From the plunger, humidified gas bubbles into the blood in the syringe, through it, and out into the atmosphere. During tonometry, the syringe is laid in a thermostatically controlled aluminum heat block. An antifoaming agent added to the blood prevents the formation of tenacious bubbles.

Removal of tonometered blood from a treatment container should be made with a syringe that has been thoroughly flushed with the equilibrating gas. The orifice through which the syringe is filled should be as large as possible and the blood should be drawn up with care to avoid creating a vacuum ahead of the plunger. For most purposes, the sample should be analyzed immediately; otherwise, it should be tightly sealed for delayed analysis. *Note* that equilibrium between gas with high pO_2 and blood is extremely difficult to achieve, due to the continuous consumption of O_2 by the blood.

Calibration

pH Measurement. The pH measurement system is calibrated against primary standard buffers admitted either manually or automatically into the sample chamber. The buffers are phosphate solutions with a temperature coefficient, $\Delta pH / \Delta T$, of $-0.0015 \times K^{-1}$ (or $°C^{-1}$). The buffers should meet NBS specifications (National Bureau of Standards, Washington, D.C.). Such buffers are prepared from Standard Reference Materials, SRM #186Ic (potassium dihydrogen phosphate) and SRM #186IIc (disodium monohydrogen phosphate), according to directions furnished with the materials, to give two buffer solutions: one at ionic strength of 0.1 mol/kg and pH 7.386 at 37 °C; the other at ionic strength of 0.1 mol/kg and pH 6.841 at 37 °C. A Sørensen buffer of ionic strength of ∼0.17 mol/kg and pH 7.383 at 37 °C is sometimes used.

Calibration buffers meeting NBS specifications are available from the manufacturer of an instrument, usually in containers of appropriate size and shape for mounting as a reservoir on the instrument. Container openings are threaded to accept special caps that are connected to hydraulic uptake lines. It is preferable to purchase calibrator buffers rather than to prepare them in-house; the responsibility for their quality thus becomes the burden of the manufacturer and the laboratory's responsibility becomes one of preserving quality. The pH values of the low and high calibrator buffers are set by the manufacturer but always lie close to 6.8 and 7.4 at 37 °C; tolerances for the specified values appear on the label. The tolerance should be less than or equal to an SD of ± 0.003 in order to achieve SD's of ± 0.005 to ± 0.01 in measuring blood pH. Another independent buffer of as good or better quality should always be kept on hand and sampled as part of routine quality control in order to assure that buffers mounted on the instrument have not degraded because of heat or contamination with dust and bacteria.

Unopened containers should be stored at room temperature. When visual observation or an instrument display warns of low levels in the reservoir, recommended practice is to replace the reservoir with a newly opened container rather than to replenish fluid in the current one. Pooling tag ends from several almost empty containers is not recommended. The containers should always be mounted and connected correctly and completely in order to assure protection of their contents while they are in use.

pCO_2 and pO_2 Measurement. Calibration of the gas measurement systems is made against gases of known O_2 and CO_2 composition, admitted into the sample chamber. Compressed gases, with a certificate of analysis provided by the manufacturer, are employed as primary standards. Pure O_2, CO_2, and N_2 may be obtained in individual tanks and mixed as desired with a precision gas mixer in the blood gas laboratory; gas mixtures of various fractional compositions may also be purchased. Many laboratories find the latter approach more convenient. The "low gas" mixture for calibration usually has a fractional composition of 5% CO_2, 0% O_2, and 95% N_2; the "high gas" mixture has fractional composition of 10% CO_2, 20% O_2, and 70% N_2. These compositions correspond roughly to a calibration range of 38–76 mm Hg for pCO_2 and 0–152 mm Hg for pO_2. When numerous specimens with expected pO_2 of > 200 mm Hg must be analyzed with high accuracy, 99.8% pure (medical) oxygen should be available for establishing calibration over the interval from 150 to ∼700 mm Hg.

The mode of calibration is determined by the design of the instrument. Some instruments contain a barometer or a transducer responsive to ambient atmospheric pressure so that barometric pressure is always known to the microprocessor. With such instruments, only a keyboard entry of the fractional composition of O_2 and CO_2 in low and high calibrator gas mixtures needs to be made; the microprocessor then calculates the values for pO_2 and pCO_2, according to Dalton's law, for gases saturated with water vapor at 37 °C. For other instruments, the operator must read a free-standing barometer, manually calculate the set points for the calibrator gases, and then enter them into the instrument.

Liquid-Gas Difference of a pO_2 Electrode. When calibration gases are used to standardize an instrumental system for measurement of gases in blood samples, a particular property of the pO_2 electrode needs to be considered. This property is called the liquid-gas or blood-gas difference.

pO_2 and pCO_2 electrodes are alike in that gas diffusing from liquid or gaseous sample passes through a gas-permeable membrane into the enclosed electrolyte solution in contact with measuring and reference elements. In both, the rate of diffusion of a gas through their membranes is slower from a liquid phase than from a gaseous phase. But pCO_2 and pO_2 electrodes use different principles of measurement (see Chapter 1B, Section Four). CO_2 entering bicarbonate-electrolyte solution enters into a reversible equilibrium relation in which the change of H^+ activity (pH) in the electrolyte becomes a measure of pCO_2. The CO_2 is not consumed; therefore the rate of its diffusion through the membrane is not limiting on the measurement. The pCO_2 electrode senses pCO_2 to the same extent whether the sample is liquid and diffusion is relatively slow or the sample is gaseous and diffusion relatively fast.

By contrast, O_2 entering the electrolyte solution enters an irreversible reaction at a polarized cathode set to a fixed voltage so that current generated by the reaction is proportional to the amount of O_2 consumed at the cathode. The amount of O_2 available for consumption is dependent on the rate of diffusion of O_2 through the membrane; operation of a pO_2 electrode is therefore diffusion-limited. The electrode responds to a greater degree to O_2 diffusing from a gaseous phase than from a liquid phase.

The liquid-gas difference for a pO_2 electrode calibrated on gas and measuring pO_2 in expired air, or for one calibrated on liquid equilibrated with O_2 and measuring pO_2 in liquid, is negligible. It becomes significant, however, when the electrode calibrated on gas is used to measure pO_2 of blood. The difference is usually expressed as a ratio of pO_2 (gas sample) to pO_2 (liquid sample). For contemporary electrodes, the ratio is commonly 1.02–1.06. For routine clinical work, a ratio of 1.04 is frequently assumed rather than determined. But for highest accuracy of blood pO_2 analysis, the ratio should be determined and applied as a correction factor to measured pO_2.

The factor is determined by measuring pO_2 of a gas with a known content of O_2, and pO_2 of a normal blood sample equilibrated against the same gas, and by calculating

$$\text{Factor} = \frac{pO_2, \text{ equilibrating gas}}{pO_2, \text{ equilibrated blood}}$$

The factor is then applied to correct measured blood pO_2, thus:

$$\text{Blood } pO_2, \text{ corrected} = \text{Blood } pO_2, \text{ measured} \times \text{factor}$$

Quality Control

Maintenance of Instrumentation. Sophistication of contemporary equipment and availability of high quality calibrator materials have made reliable and accurate determination of blood pH and gases primarily a matter of meticulous maintenance and control of the equipment and meticulous control of specimens admitted to it. In many instances, monitor programs of the microprocessor display warning and diagnostic routines that alert the operator and then assist in the troubleshooting process. Nevertheless, regular maintenance of all aspects of the instrumentation, with close adherence to the manufacturer's recommended procedures, is essential to satisfactory operation. The frequency with which maintenance should be scheduled is in direct proportion to the volume of blood gas analyses performed in the laboratory; in very active laboratories, certain maintenance procedures may have to be performed more than once in a 24-h period, but in any laboratory some maintenance will be required every 24 h. The manufacturer's suggested schedule should be considered a guideline; often, experience will indicate necessity to speed up the schedule.

Cleanliness of the sample chamber and path is especially important. Automatic flushing to cleanse the sample chamber and path after each blood sample measurement is a feature of many instruments. When it is not, then manual modes of flushing recommended by the manufacturer should be faithfully practiced. Despite proper flushing, however, complete or partial clogging of chamber or path or both may occur. Frequency of clogging is related to the number of heparinized capillary blood samples that are analyzed. Fibrin threads and small clots may be present in the specimen or may form while the sample resides in the warm chamber. If allowed to remain, they will affect subsequent measurements or calibrations by interfering with contact of blood, buffers, or gases with electrode membranes or with KCl from the pH reference electrode, or by causing segmentation in the stream as the next blood sample is admitted. Visibility of the path through the heat sink is helpful for detecting clogs and dirt. Cleaning should be performed by procedures recommended in the instrument's manual; the procedure may be as simple as passing a horsehair through the sample path or as drastic as dismantling the chamber and its accessory structures.

Prompt and reliable service by the manufacturer or in-house by a biomedical engineer is essential for the laboratory performing many analyses per day, even when backup instruments are available. Also important is ready availabilty, from the manufacturer or from laboratory supply houses, of calibrator materials (pH buffers and gases of certified quality) and replacement membranes and small parts for maintenance of electrodes. Almost every manufacturer provides operator training upon installation of an instrument. In-service training and retraining should rely more heavily on instrument operating manuals provided by the manufacturer than on simple word-of-mouth. Numerous publications by NCCLS (C21-P, C12-T, I6-T, EP2-T, EP3-T, EP4-T, and EP5-P, available from the National Committee on Clinical Laboratory Standards, Villanova, PA 19085) describe manufacturing standards and basic operational requirements for blood gas instrumentation.

Control Materials

Quality control materials can be classed as tonometered whole blood for pCO_2 and O_2 control; independent standard buffers for pH control; and commercial blood-base or blood-base-like fluids and aqueous fluids for pH, pCO_2, and pO_2 control. Pure and mixed gases are also needed for some control procedures.

Tonometered whole blood. A fresh blood specimen with a normal hemoglobin is preferred. When equilibrated against gas mixtures of primary standard quality, a tonometered sample measured on a properly calibrated and properly operating instrument should give measured values for pCO_2 and pO_2 that are within 1–4% of those calculated for the equilibrating gas. A typical gas mixture for one level of control by tonometry might contain 10% O_2, 5% CO_2, and 85% N_2. At a p(Atm) of 760 mm Hg and after saturation with water vapor at 37 °C, the pO_2 of the gas would be calculated (according to Dalton's law) as

$$pO_2, \text{ mm Hg} = (760 - 47) \times 0.10 = 71$$

and pCO_2 as

$$pCO_2, \text{ mm Hg} = (760 - 47) \times 0.05 = 36$$

Workers using citrate-dextrose-phosphate preserved, banked human blood have shown at pCO_2 levels of 20, 47, and 65 mm Hg, SD's less than 1, 2, and 2.5 mm Hg, respectively; and at pO_2 levels of 41, 100, and 195 mm Hg, SD's less than 2, 2.5, and 6 mm Hg, respectively.[3] Quality control of pO_2 and pCO_2 by tonometered blood is considered by many workers the method of choice since the control material most nearly approximates patients' samples in its interaction with gas electrodes. Tonometered blood is acknowledged to have greater sensitivity for detecting deterioration of the gas-permeable membranes of electrodes than do aqueous fluid controls. The disadvantages are the time required for tonometry, particularly if two or three levels of control are desired; difficulties with availability of fresh, normal blood samples when needed often or at inconvenient times; necessity for repeated calculation (and risk of miscalculation) of pCO_2 and pO_2 of equilibrating gases; need to keep special gas mixtures for equilibration; and inapplicability to direct control of pH measurement.

Standard buffers. The buffer or buffers used for control of pH measurement should be of a quality equal to or greater than that of the calibrator buffers. Buffer salts with larger temperature coefficients than phosphate calibrator buffers—e.g., TES (2-[tris(hydroxymethyl)methylamino]-1-ethanesulfonic acid) or HEPES (N-2-hydroxyethylpiperazine-N'-2-ethanesulfonic acid)—may be used to detect incipient or frank failure of temperature control of the sample chamber.[3] Tris(hydroxymethyl)aminomethane and its hydrochloride are available as SRM's #922 and #923 from NBS. The temperature coefficient of a Tris buffer is $-0.026 \times K^{-1}$ (or °C^{-1}), almost 20 times that of phosphate buffers and almost 200 times that of whole blood. Desirable precision of pH measurements on control material is of the order of ± 0.003 pH unit.

Blood-base (or blood-base-like) control material is a relatively recent development. One type of material (Quantra Plus, American Dade, Miami, FL 33152) consists of aldehyde-tanned human erythrocytes suspended in buffered medium and sealed in vials with a gas mixture of known O_2

and CO_2 content. Another (Prime, Fisher Scientific, Pittsburgh, PA 15219) contains human hemoglobin in a similar medium and package. A nonblood fluorocarbon material, with oxygen-carrying properties similar to those of blood,[4] is also available (abc, Instrumentation Laboratories, Lexington, MA 02173). These products are made at three levels of pH, pCO_2, and pO_2. Prior to sampling, Quantra Plus must be incubated in the vials at 37 °C; all of them must be shaken vigorously for a prescribed length of time in order to equilibrate the gas and fluid phases. Unopened, these types of control material have the advantages of long shelf-life in the refrigerator, for the tanned erythrocytes for 20–28 d and for the others even longer. The buffered medium allows for control of pH along with that of pO_2 and pCO_2. Within-day and day-to-day consistency within the lifetime of a given lot have been reported to be excellent; within-day CV's over the range of levels were 1–2% for pH, pCO_2, and pO_2.[6] In the authors' experience, Quantra Plus and Prime have been useful adjuncts to twice-a-day control by tonometry since they are conveniently and immediately applicable for spot checking at any hour of the day. Quantra Plus is the more technique-dependent of the two.

Aqueous fluid control materials consist of a buffered medium sealed in vials with gas mixtures; the fluid is equilibrated with the gas by vigorous shaking by hand for a prescribed length of time immediately before the vial is opened and sample is admitted to the instrument. Examples of these are CONFIRM (Corning Medical and Scientific, Medfield, MA 02052), GAS (General Diagnostics, Morris Plains, NJ 07950), and ContrIL (Instrumentation Laboratories, Lexington, MA 02173). For CONFIRM, CV's of 0.1% for pH, 2.5% for pCO_2, and 3.2% for pO_2 have been reported.[9]

Gases: pure and mixed. Keeping a supply of gases on hand is necessarily a part of certain quality control procedures such as tonometry control or of certain analyses such as determination of $pO_2(0.5)$. The quality of pure or mixed gases for these purposes should be as good as that of gases used for calibration. The variety of pure gases or mixtures kept depends on applications. Some laboratories invest in a precision gas mixer; an example of such a device is the microprocessor-controlled, two-channel mixer, Model 192, available from Corning Medical. Keeping only compressed CO_2, O_2, and N_2 of greater than 99.6% purity, they mix gases in-house to any desired composition for any desired application. Advantages of this approach include avoiding storage and handling of numerous tanks of compressed gas, and ability to redefine continually the proportions of mixing in terms of current $p(Atm)$, and thus to utilize mixtures with essentially constant pO_2 and pCO_2 over long periods of time.[3] Laboratories that do not have a precision mixer can purchase in small tanks a variety of commonly used mixtures with certified composition or can have special mixtures prepared to order.

Data obtained from assays of the various control materials may be handled in the same way as data from other clinical chemistry determinations, i.e., by calculation of mean, SD, and CV, and of control and confidence limits for construction of Levy-Jennings plots. Equally important features of quality control in an active blood gas service are the sixth sense of practiced operators for detecting subtle manifestations of deterioration of instrument performance and the suspicion of trouble expressed by physicians closely supervising care of individual patients being monitored with frequent blood gas analysis. These intangible aspects of blood gas quality control are not to be despised.

Linearity of Electrodes

Linearity of new electrodes should be verified upon their installation. For the pCO_2 electrode, linearity can be tested after calibration with 5% and 10% CO_2, by determination of pCO_2 on a gas containing 7% CO_2. A similar test of an intermediate point of the calibration range for a pO_2 electrode calibrated with O and 20% O_2 might use 10% O_2, or for a pO_2 electrode calibrated in the extended range, 50% O_2. The gases used in the initial verification can subsequently be used for periodic checks in the quality control program. In some laboratories, checking linearity with tonometered whole blood samples is preferred to using gases.

Barometer

A barometer, whether built into the instrument or free-standing, should be checked for accuracy on a periodic basis. Since the classic mercury-filled barometer has become a rarity in clinical laboratories, the most reliable reference is local atmospheric pressure as recorded by the nearest official meteorological station and obtained by a phone call.

Temperature Control

Because an exact temperature of 37 °C is essential to accurate measurement of blood gases and pH, state-of-the-art instrumentation is furnished with thermal sensors embedded in the heat sink around the measuring chamber and communicating to the microprocessor. Audible or visible alarms signal deviation of temperature outside of preset tolerances (usually 37 $\pm$ 0.1 °C). Older instruments depend on visual inspection of a thermometer by the operator. As noted earlier, inclusion of temperature-sensitive buffers, HEPES or TES, in pH quality control procedures is helpful in monitoring temperature control.

There is sufficient variation in the design of these heat sinks that only reference to an instrument's manual can illustrate the kinds of problems that can occur and the kinds of solutions that should be applied. Suffice it to say here that failure of temperature control must be detected early and corrected promptly.

Sources of Error

With good control of analytical error, the most probable source of error is pre-analytical, i.e., in the collection and handling of a specimen. General causes of analytical error are calibration of the instrument with incorrect set points for pH buffers or calibrator gases, failure of temperature control of the measurement chamber, a dirty sample chamber or path, and inadequate resuspension of cellular components of blood specimens prior to sampling. Incorrect calibration may arise from wrong entries made for buffer or gas values to the microprocessor memory, from incorrect manual calculations of pCO_2 and pO_2 values by Dalton's law for calibrator gases, or from using gases that are dry because the humidification device is not working properly. Another cause can be the use of calibrator gases that have no certificate of analysis or that have been obtained from a manufacturer whose control of quality is poor.

Failure of temperature control may be due to hydraulic causes in instruments whose heat sinks are maintained from a circulating fluid system. Delivery tubings may become pinched or kinked; with a decrease in the flow of warmed fluid, temperature equilibration of the sample in the chamber may be delayed or the chamber temperature may fall slightly. Dirt or clogs in sample spaces may also lead to slow equilibration of temperature for the sample, but are more likely to interfere with the contact of sample with electrodes. Segmenting of the sample stream upon admission may place air gaps or void spaces opposite the electrodes. The junction potential of electrodes is sensitive to the presence of erythrocytes, and thus, whole blood specimens need to be homogeneous when admitted.

A source of error for the pO_2 electrode is using it in a range above $\sim$500 mm Hg when it has not been calibrated for use in this range. pO_2 measurements are particularly sensitive to temperature error; in order to keep systematic error to 1–2%, the temperature control at 37 °C must be within $\pm$0.1 °C. Failure to account for the liquid-gas difference of a pO_2 electrode can introduce a 2–6% error. The error will, of course, be less if a general liquid-gas correction factor is incorporated into the algorithm that generates results. If the specific correction factor for a particular electrode is used, the error will be further minimized. Some instruments are provided with an automatic carry-over correction which may apply to blood but not to aqueous solutions.

Gases other than O_2, present in a blood sample, may affect performance of the pO_2 electrode. CO_2 exerts its effect by altering pH of the electrolyte; buffering the electrolyte to keep its pH constant minimizes CO_2 effect. The anesthetic gases halothane and nitrous oxide have a direct effect since both can be reduced at the polarized cathode in competition with O_2. Under most circumstances, however, these effects are small and ordinarily ignored.

Temperature Correction of Measured pH and pCO_2

In the Henderson-Hasselbalch equation, pK' and α are used as constants for a temperature of 37 °C. The temperature-controlled sample chamber of an instrument is specified to be 37 $\pm$ 0.1 °C, and it is at that temperature that all measurements of pH and partial pressure of gases are made. But the body temperature of a patient may, because of fever, be elevated to 40–41 °C, while the temperature of a patient made hypothermic for cardiopulmonary bypass surgery may be taken as low as 26 °C. Although temperature correction of pH and pCO_2 measured at 37 °C is the subject of considerable clinical controversy, most blood gas instruments, upon keyboard entry of the patient's actual temperature, can calculate and present temperature-corrected pH and pCO_2, as well as values derived from the temperature-corrected primary data. Algorithms

Table 10-7. TEMPERATURE-CORRECTION FORMULAS USED BY BLOOD-GAS ANALYZERS

pH

1. $pH = pH_m + [-0.0146 + 0.0065(7.4 - pH_m)] (t - 37)$
2. $pH = pH_m - 0.015(t - 37)$
3. $pH = pH_m + [-0.0147 + 0.0065(7.4 - pH_m)] (t - 37)$
4. $pH = pH_m - 0.0146(t - 37)$

pCO_2

1. $pCO_2 = pCO_{2,m}\ 10^{0.021(t-37)}$
2. $pCO_2 = pCO_{2,m}\ 10^{0.019(t-37)}$
3. $pCO_2 = pCO_{2,m}\ 10^{0.019(t-37)}$
4. $pCO_2 = pCO_{2,m}\ 10^{0.019(t-37)}$

pO_2

1. $pO_2 = pO_{2,m}\ 10^{\left[\left(\frac{0.0252}{[0.243(pO_2/100)^{3.88}]+1}\right) + 0.00564\right](t-37)}$
2. $pO_2 = pO_{2,m}\ 10^{(0.0052 + 0.27[1 - 10^{-0.13(100-Sat)}])\ (t-37)}$
3. $pO_2 = pO_{2,m}\ 10^{\left[\frac{(5.49\times10^{-11}\ pO_2^{3.88})+0.071}{(9.72\times10^{-9}\ pO_2^{3.88})+2.30}\right](t-37)}$
4. Sat $\leq 95\%$

   $pO_2 = pO_{2,m}\ 10^{0.31(t-37)}$

   Sat $> 95\%$

   $pO_2 = pO_{2,m}\ 10^{[0.032 - 0.0268e^{(0.3\ Sat-30)}]\ (t-37)}$

Formulas are for (1) Radiometer ABL-3; (2) Corning models 178, 170, 168, and 158; (3) Instrumentation Laboratory models 1301 and 1303; and (4) Instrumentation Laboratory model 813. An improved equation for correction of pO_2 has recently been published by Siggaard-Andersen:[17]

$$pO_{2,t} = pO_{2,37} \cdot \exp\left[\frac{0.012 \cdot (pO_{2,37}/714) + sO_2(1 - sO_2)(cHb/0.6) \cdot 0.073}{(pO_{2,37}/714) + sO_2(1 - sO_2)(cHb/0.6)} \cdot (t-37)\right]$$

where cHb is in g/dL, pO_2 in mm Hg, sO_2 is decimal fraction, t is temperature (Celsius), and m = measured.

Used with permission from Ashwood, E. R., Kost, G., and Kenny, M.: Temperature correction of blood gas and pH measurements. Clin. Chem., 29:1877-1885, 1983.

---

used in some contemporary instruments have been listed in a recent publication[2] and are shown in Table 10-7.

Algorithms shown in the table represent a selection from the equations developed over many years of study of the change of pH or pCO_2 caused by change of temperature in vitro in closed anaerobic systems. Although the constants used in the equations remain a subject of discussion and investigation, $\Delta pH/\Delta T = 0.0147 \times K^{-1}$ (or °C^{-1}) and $\Delta\log pCO_2/\Delta T = 0.021$ (or 0.019) $\times K^{-1}$ (or °C^{-1}) appear to be reasonable estimates over the range of 25–42 °C.[2] These constants are experimentally validated descriptions of ΔpH and $\Delta\log pCO_2$ when temperature of anaerobic blood in a test chamber is varied; their validity for calculating corrected values is not controversial.

There is a general clinical consensus that routine correction of pH and pCO_2 values in hyperthermic states is unnecessary. The magnitude of correction to 40 °C (104 °F) for pH would be +0.045 and for pCO_2 would be +13%. Furthermore, clinical management of hyperthermic states seldom is accompanied by adjustment of acid-base status by CO_2 content of gas mixtures supplied for respiration and by assisted ventilation. The disagreement is in respect to hypothermic states. Anesthetists, by choosing ventilation conditions and gas mixtures, can manage the hypothermic patient's pH at 7.40 and pCO_2 at 40 mm Hg, i.e., they can, guided by values corrected to the patient's actual body temperature, *force* the pH and pCO_2 at 26–30 °C to the same values that are considered ideal at 37 °C. The real question is whether pH and pCO_2 values that are ideal at 37 °C are ideal at 26–30 °C, and whether there is risk of cardiac arrhythmias when they are imposed on the hypothermic patient. The physiology of acid-base balance in hypothermia has been addressed by Rahn and his colleagues,[10,11,13] and pros and cons of management of acid-base balance of the hypothermic patient have been noted by other authors.[2,5,12] The issue, however, remains unresolved. Prudent policy for the laboratory might be to generate and report temperature-corrected results for pH and pCO_2 only upon specific request of the physician. However, temperature-corrected results should never be reported unaccompanied by the original results measured at 37 °C.

Temperature Correction of Measured pO_2

Values of pO_2 measured at 37 °C should be corrected if the patient's body temperature is different by more than 1 °C from 37 °C. The equations shown in Table 10-7 illustrate the complexity of the calculations. Complexity is unavoidable because at pO_2 <100 mm Hg ($sO_2 \leq 0.95$) the hemoglobin-oxygen dissociation curve is shifted to the left by decrease in temperature as well as by the concomitant rise in pH (see Figure 10-4). The combined effect can be expressed by $\Delta\log pO_2/\Delta T = +0.031 \times K^{-1}$ (or $°C^{-1}$). At pO_2 >100 mm Hg, the hemoglobin is fully saturated ($sO_2 = 1.00$) whatever the temperature, and $\Delta\log pO_2$ is determined by the solubility coefficient for O_2 relative to temperature, i.e., $\Delta\log \alpha/\Delta T = +0.005 \times K^{-1}$ (or $°C^{-1}$). For temperature corrections for pO_2 100–400 mm Hg, accurate formulas become even more complicated.

A recently recommended[2] formula for calculation is that of Severinghaus[15] (Number 1 in Table 10-7). No analysis of algorithms given in Table 10-7 will be made here. However, an analysis of the general algorithmic approach to correction can be found in the monograph of Adams and Hahn.[1] There is no clinical controversy with respect to the desirability of correcting measured pO_2 to the patient's body temperature. But because routine specification of the patient's body temperature is unlikely to accompany most requests for blood gas analysis, request for correction is improbable except in instances of imposed hypothermia.

Clinical Significance of pO_2 Levels in Blood

Increased arterial oxygen tension, $pO_2(aB)$, occurs upon breathing oxygen-enriched air; with administration of 100% O_2, the pO_2 may be as high as 640 mm Hg. Exercise, in both healthy subjects and cardiac patients, results in an increase over the resting level.

Decreased arterial oxygen, *hypoxemia*, may occur in certain situations: at high altitudes, with exposure to carbon monoxide, with improper anesthesia, or because of near-drowning. General causes of hypoxemia can be classified as:

1. Decreased capacity for pulmonary diffusion, as for example due to respiratory distress syndrome in adults or newborns, lymphangitic carcinomatosis, pulmonary adenomatosis, sarcoidosis, Hamman-Rich syndrome, berylliosis, or pulmonary hemosiderosis secondary to mitral stenosis.

2. Decreased surface area of alveolocapillary membranes as the result of resection or compression of the lung.

3. Inequalities of ventilation/perfusion in the cardiopulmonary system because of bronchitis, asthma, emphysema, bronchiectasis, atelectasis, pneumoconiosis, granulomas, neoplasms, pulmonary infarction, pneumonia, mucoviscidosis, or obstruction of the airway by a neoplasm, scar tissue from resection, a foreign body, or secretions (e.g., croup).

4. Generalized alveolar hypoventilation of *peripheral origin*, i.e., as a result of suffocation, submersion, skeletal abnormalities or trauma to the chest that prevent full expansion; or neuromuscular abnormalities that affect respiration, as for example phrenic nerve paralysis, tetanus, or acute poliomyelitis; or the Pickwickian syndrome; or of *central origin* because of depression of the respiratory center by drugs such as barbiturates or morphine. Additional causes are right-to-left shunt as in congenital heart disease or patent ductus arteriosus, pneumonia, atelectasis, pulmonary edema, or shock. Often more than one of these mechanisms are operating simultaneously.

Clinical Significance of pCO_2 Levels in Blood

Increased blood pCO_2, *hypercapnia*, causes respiratory acidosis. pCO_2 rises with decreased alveolar ventilation due to diseases of the lungs or bronchial tree, to obstructions of the airway, or to breathing CO_2-enriched air. Depression of the respiratory center by drugs or impairment of the neuromuscular respiratory apparatus, by decreasing rate and depth of respiration, may lead to retention of CO_2. In metabolic alkalosis, compensatory hypoventilation is responsible for CO_2 retention and increased pCO_2.

Decreased blood pCO_2, *hypocapnia*, is the cause of respiratory alkalosis; hypocapnia occurs upon increased alveolar ventilation as a result of too rapid rates of mechanically assisted respiration or of stimulation of the respiratory center to increase rate and depth of respiration.

Specific causes of respiratory acidosis and alkalosis are listed in Tables 10-11 and 10-12; the effect of altered CO_2 content of the blood on acid-base status is discussed in Section Three.

Clinical Significance of pH Levels in Blood

The pH range of blood compatible with life is 6.80–7.80. The normal arteriovenous difference is 0.01–0.03 but may be greater in cases of congestive heart failure or shock.

Increase in blood or plasma pH is called *alkalemia*. *Metabolic alkalosis* is due to increased plasma bicarbonate. It may be the result of excessive alkali administration, or of loss of Cl^- as gastric HCl by prolonged vomiting or aspiration (hypochloremic alkalosis), or of potassium depletion.

Alkalemia may also be a feature of *respiratory alkalosis* which is due to increased elimination of carbonic acid as CO_2. Hyperventilation in hysteria and stimulation of the respiratory center are the usual causes. With normal lung function and normal alveolar diffusion, stimulation of the rate and depth of respiration "blows off" an excessive amount of CO_2 and thus alters the relationship of pCO_2 and HCO_3^- that maintains blood pH around 7.40.

Decrease in blood or plasma pH is called *acidemia*. *Metabolic acidosis* may occur because of increased formation of organic acids, as in diabetic acidosis or in lactic acidosis because of cellular hypoxia. Acidemia may also arise from decreased excretion of H^+ in renal disease because of acquired or inherited prerenal, renal, or postrenal disorders. Increased acid intake can be a factor; examples are salicylate poisoning or NH_4Cl administration. Increased loss of alkaline body fluids can also decrease blood pH. Hyperkalemia may be another cause of acidemia.

Acute *respiratory acidosis* may be a result of decreased alveolar ventilation (increased pCO_2) because of pulmonary edema, bronchoconstriction or airway obstruction, or drugs that depress the respiratory center. Chronic respiratory acidosis is characterized by increased plasma bicarbonate and arises from chronic obstructive or restrictive respiratory disorders.

Alterations of blood pH are discussed in conjunction with acid-base disorders in Section Three.

Reference Ranges[18]

Reference ranges for arterial blood pO_2, sO_2, pCO_2, and pH are extensively described in Chapter 20. Arterial blood pO_2, low at birth, rises to an adult level of 83–108 mm Hg. An empirical relationship of age to $pO_2(aB)$ is given by the equation:

$$pO_2(aB), \text{ mm Hg} \simeq (-0.27 \times \text{age, y}) + 104$$

Values decrease with altitude.

Saturation fraction, $sO_2(aB)$, may be as low as 0.40 at birth, but thereafter is 0.95–0.98. $pO_2(0.5)$ corrected to pH 7.40 is 18–24 mm Hg for newborns and 25–29 mm Hg for adults.

Arterial blood pCO_2 ranges, at sea level, are somewhat lower for infants than for adults; the range for adult males is 35–48 mm Hg, for adult females, 32–45 mm Hg. Values decrease with altitude above sea level at a rate of 3 mm Hg/km (5 mm Hg/mile). A physiologic change occurs with change in posture; $pCO_2(aB)$ is 2–4 mm Hg higher for a sitting or standing subject than for one in the supine position. During pregnancy, pCO_2 falls gradually to a mean of $\sim$28 mm Hg just prior to term.

Arterial blood pH, in the first few hours of life, may vary normally over a range of 7.09–7.50 but thereafter is 7.35–7.45. pH measured on arterial *plasma*, separated anaerobically at 37 °C, is 0.01–0.03 unit greater than that of a corresponding whole blood sample but the difference is not physiological. It arises instead from the effect of erythrocytes on the junction potential of the pH electrodes.

References

1. Adams, A. P., and Hahn, C. E. W.: Principles and Practice of Blood Gas Analysis. London, Franklin Scientific Projects, Ltd., 1979.
2. Ashwood, E. R., Kost, G., and Kenny, M.: Temperature correction of blood gas and pH measurements. Clin. Chem., 29:1877-1885, 1983.
3. Elser, R. C., Stiler, J., and Garver, C.: A flexible and versatile program for blood-gas quality control. Am. J. Clin. Pathol., 78:471-478, 1982.
4. Feil, M. C., Legg, K. D., and Cormier, A. D.: A blood gas control with improved oxygen buffering. Clin. Chem., 27:1111, 1981.
5. Hansen, J. E., and Sue, D. Y.: Should blood gas measurements be corrected for patient's temperature? N. Engl. J. Med., 303:341, 1980.

6. Hansen, J. L., and Forleo, D. J.: Precision and long-term stability of newer controls for blood pH and blood gases. Clin. Chem., 30:366, 1984.

7. Hutchison, A. S., Ralston, S. H., Dryburgh, F. J., et al.: Too much heparin: Possible source of error in blood gas analysis. Br. Med. J., 287:1131-1132, 1983.

8. Joint Committee on Pulmonary Nomenclature, American College of Chest Physicians and American Thoracic Society: Pulmonary terms and symbols. A report of the AACP-ATS. Chest, 67:583-593, 1967.

9. Moran, R. F., Draper, J. P., and Lesica, J. P.: Blood gas analyzer goals—tonometry, buffer and a multi-sample control material. Clin. Chem., 27:1111, 1981.

10. Rahn, H.: Why are pH of 7.4 and P_{CO_2} of 40 normal values for man? Bull. Eur. Physiopathol. Respir., 12:5-13, 1976.

11. Rahn, H., Reeves, R. B., and Howell, B. J.: Hydrogen ion regulation, temperature, and evolution. Am. Rev. Respir. Dis., 112:165-172, 1975.

12. Ream, A. K., Reitz, B. A., and Silverberg, G.: Temperature correction of P_{CO_2} and pH in estimating acid-base status: An example of the emperor's new clothes? Anesthesiology, 56:41-44, 1982.

13. Reeves, R. B., and Rahn, H.: Patterns in vertebrate acid-base regulation. In: Lung Biology in Health and Disease, Vol. 13. Evolution of Respiratory Processes; A Comparative Approach. S. C. Wood and C. Lefant, Eds. New York, Marcel Dekker, 1979.

14. Samaja, M., Mosca, A., Luzzana, M., et al.: Equations and nomograms for the relationship of human blood p_{50} to 2,3-diphosphoglycerate, CO_2, and H^+. Clin. Chem., 27:1856-1861, 1981.

15. Severinghaus, J. W.: Simple accurate equations for human blood O_2 dissociation computations. J. Appl. Physiol., 46:599-602, 1979.

16. Siggaard-Andersen, O. S., Durst, R. A., and Maas, A. H. J.: Physico-chemical quantities and units in clinical chemistry. International Federation of Clinical Chemistry and International Union of Pure and Applied Chemistry. Pure Appl. Chem., 56:567-594, 1984.

17. Siggaard-Andersen, O. S., Wimberley, P. D., Göthgen, I., et al.: A mathematical model of the hemoglobin-oxygen dissociation curve of human blood and of the oxygen partial pressure as a function of temperature. Clin. Chem., 30:1646-1651, 1984.

18. Tietz, N. W., Ed.: Clinical Guide to Laboratory Tests. Philadelphia, W. B. Saunders Co., 1983.

19. Thomas, L. J.: Algorithms for selected blood acid-base and blood gas calculations. J. Appl. Physiol., 33:154-158, 1972.

Additional Reading

Maas, A. H. J., Kofstad, J., Siggaard-Andersen, O., et al., Eds.: Physiology and methodology of blood gases and pH, Vol. 4. Proceedings of the 9th meeting of the IFCC expert panel on pH and blood gases, IFCC Workshop, Oslo, 1983. Copenhagen, Private Press, 1984.

Section Three

ACID-BASE BALANCE AND ACID-BASE DISORDERS

by Norbert W. Tietz, Ph.D., Ole Siggaard-Andersen, M.D., Ph.D., and Elizabeth L. Pruden, Ph.D.

The normal human diet is almost neutral and contains very little titratable acid. However, metabolic processes in the body result in the production of relatively large amounts of carbonic, sulfuric, phosphoric, and other acids. Lactic acid and β-hydroxybutyric acid are intermediary products which are normally metabolized to carbon dioxide and water before excretion. However, in some abnormal conditions, such as diabetes mellitus, they may accumulate to a significant extent.

A person weighing 70 kg disposes daily of about 20 moles of carbon dioxide through the lungs and about 70–100 millimoles of titratable, nonvolatile acids, mainly sulfuric and phosphoric acids, through the kidneys. These products of metabolism are transported to the excretory organs (lungs and kidneys) via the extracellular fluid and the blood without producing any appreciable change in the plasma pH and with only minimal pH difference between arterial and venous blood plasma. This is accomplished by the combined functions of the *buffer systems* of the blood, the *respiratory system*, and the *renal mechanisms*.

Table 10-8. CONCENTRATIONS OF CATION AND ANION CHARGES IN SERUM (EXPRESSED IN mmol/L).

Cation charges		Anion charges	
Na^+	142	Cl^-	103
K^+	4	HCO_3^-	27
Ca^{2+}	5	HPO_4^{2-}	2
Mg^{2+}	2	SO_4^{2-}	1
Others (trace elements)	1	Organic acids$^-$	5
	$\overline{154}$	Protein$^-$	16*
			$\overline{154}$

*The ion charge, mmol/L, for protein in serum of a healthy individual can be calculated by multiplying the total protein, g/dL, by the factor 2.41 (Van Slyke, D. D., Hastings, A. B., Hiller, A., et al: Studies of gas and electrolyte equilibrium in blood. XIV. The amounts of alkali bound by serum albumin and globulin. J. Biol. Chem., *79*:769-780, 1928). Van Leeuwen (Acta Med. Scand. Suppl., *422*, 1-212, 1964), however, has found the following relationship: net negative charge (mEq/100 g protein) = 6.9 (pH − 4.96)

Acid-Base Balance. In physiology, a component is *in balance* if the rates of input and output of this component are equal for a given time interval. A *positive balance* indicates that the mean rate of gain of the component in the body is positive (net gain), while a *negative balance* indicates that the mean rate of gain of the component in the body is negative (net loss). Imbalance results in either an increase or a decrease in the amount of the component in the body. A *balance account* is a detailed account of all input (intake plus production) and all output (excretion plus metabolic conversion) over a given time interval. A description of the *acid-base balance* involves a description of the CO_2 balance as well as the balance of noncarbonic acid and base. The *acid-base status* of the body fluids (i.e., the pH, the pCO_2, the concentration of titratable acid or base, and other acid-base variables) is the resultant of the input and output of acids and bases in the period preceding the sampling.

Whole blood, plasma, and serum are the main specimens used to evaluate the acid-base and electrolyte status of the body, since these specimens are readily available. It must be kept in mind, however, that the composition of the various body fluid compartments is different (see p. 1223) and that the movement of ions and other solutes between these compartments is not necessarily free. Furthermore, the vascular compartment contains only 2.6–3.5 L water, and the total extracellular compartment contains 16–18 L. This compares to ∼43 L of total body water. Thus, results obtained on plasma or serum reflect mainly the composition of extracellular fluid. Nevertheless, when properly interpreted, values obtained on plasma can be of great help to the physician.

The pH of the plasma may be considered to be a function of two independent variables: the pCO_2, which is regulated by the respiratory mechanism, and the concentration of titratable base (base excess or deficit), which is regulated by the renal mechanism. For a pure bicarbonate solution, changes in the concentration of titratable base equal the changes in the bicarbonate concentration. Thus, the plasma bicarbonate concentration is often taken as a measure of the base excess or deficit in plasma and extracellular fluid, although it is recognized that conditions in plasma differ greatly from those present in pure bicarbonate solutions.

The H^+-balance is closely associated with the balance of other electrolytes because H^+ cannot be introduced without a concomitant anion (e.g, Cl^-) or in exchange for a cation (e.g., Na^+). Table 10-8 gives the concentrations of the cation and anion charges for normal plasma. Note that there is an exact equality of the concentrations of total anion charge and of total cation charge; i.e., the sum of negative and positive charges is zero. This law of electrical neutrality* can be expressed mathematically as:

$$\sum_{i=1}^{n} z \times cI_i = 0 \qquad (1)$$

*The term electrical neutrality is not to be confused with acid-base neutrality (pH = 7.0, where the activity of H^+ equals the activity of OH^-), nor does it indicate the pH is normal (7.4); it indicates only that there is an equality between the concentrations of total anion charge and total cation charge.

where z is the number of charges of the ion I_i (positive for cations, negative for anions) and c is the substance concentration. Any increase in the concentration of one anion is accompanied either by a corresponding decrease of other anions or by an increase of one or more cations, or both, so that total electrical neutrality is invariably maintained. Similarly, any decrease in the concentration of anions involves either a corresponding increase in other anions, or a decrease in cations, or both.

According to Brønsted's concept, an acid is a substance that can donate protons (H^+) and a base is a substance that can accept protons (H^+).

$$acid \rightleftharpoons H^+ + conjugate\ base$$
$$HCl \rightleftharpoons H^+ + Cl^-$$
$$NH_4^+ \rightleftharpoons H^+ + NH_3 \tag{2}$$
$$glycine \rightleftharpoons H^+ + glycinate^-$$
$$glycinium^+ \rightleftharpoons H^+ + glycine\ (electrically\ neutral)$$

Thus, HCl, NH_4^+, and amine ions ($glycinium^+$) are acids, and NH_3, $glycinate^-$, and free (electrically neutral) amines are bases. Some hydrogen-containing anions, such as HCO_3^- and HPO_4^{2-}, as well as all amino acids and proteins, can act as acid or base:

$$H_2PO_4^- \rightleftharpoons H^+ + \underset{(conjugate\ base)}{HPO_4^{2-}}$$
$$\underset{(acid)}{HPO_4^{2-}} \rightleftharpoons H^+ + PO_4^{3-} \tag{3}$$

Concentrations of electrolytes in body fluids are commonly expressed in units of millimoles per liter (mmol/L).* In the case of polyvalent ions it is important to distinguish between the (substance) concentration of the ion (cI) and the (substance) concentration of the ion charge ($cI_{1/z}$). Thus, the concentration (c) of the total calcium ions in normal plasma is $cCa^{2+}(P) = 2.5$ mmol/L, while the concentration of the total calcium ion charge is $cCa_{1/2}^{2+} = 5.0$ mmol/L (formerly called 5 mEq/L).

Electrolyte Composition of Body Fluids

The body has two main fluid compartments, the *intracellular* and the *extracellular*, the latter being further divided into blood plasma (vascular compartment) and interstitial fluid. For simplicity, we will adhere to this division, although it should be realized that these compartments can be divided further into subcompartments with varying composition in regard to water, electrolytes, and other components. The masses and volumes of the different body compartments are summarized in Table 10-9.

Plasma, which is of main interest in our discussion, generally has a volume of 1300–1800 mL/m² of body surface and constitutes approximately 5% of the body volume. Generally,

---

*It is still the custom in the United States to express the electrolyte concentrations in terms of mEq/L (in case of calcium, also in mg/100 mL). The use of these units is now discouraged.

Table 10-9. MASS (m) AND VOLUME (V) OF VARIOUS BODY
COMPARTMENTS IN A 70-kg ADULT

	m (kg)	V (L)	mH_2O (kg)	m Bone (kg)	m Fat (kg)
Interstitial fluid, cerebrospinal fluid, and bone	19	17	14	4.5	
Plasma	3.4	3.2	3.0		
Erythrocyte fluid	2.3	2.1	1.5		
Intracellular fluid	45.5	44	24.5		12
Total	70	66	43		

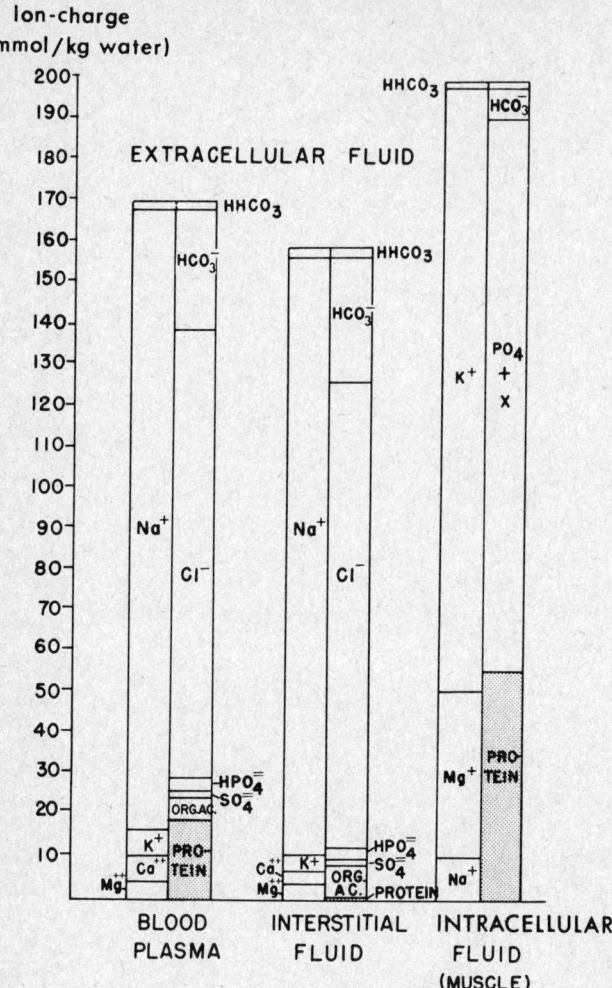

Figure 10-6. Electrolyte composition of blood plasma, interstitial fluid, and intracellular fluid. Note that the solutes are expressed in molality (i.e., amount of substance of solute divided by the mass of solvent). This form of expression is most meaningful when comparing concentrations of solute in two aqueous solutions of widely differing water content, since it is only in solvent that solutes are free to move and be osmotically active. By using molality, the mass of solvent is kept constant at 1 kg. (After Gamble, J.: Chemical Anatomy, Physiology, and Pathology of Extracellular Fluid. Cambridge, Harvard University Press, 1950.)

body volume is derived from body mass by using a body density of 1.0 kg/L. (The true value is ~1.06 kg/L). The electrolyte composition of plasma is summarized in Table 10-9 and Figure 10-6.

Interstitial fluid is essentially an ultrafiltrate of blood plasma. Its volume makes up approximately 15% of the total body volume; however, when all extracellular spaces except plasma are included, the volume accounts for about 26% (17/66) of the total body volume. (See Table 10-9.) Plasma is separated from the interstitial fluid by the endothelial lining of the capillaries, which acts as a semipermeable membrane and allows passage of water and diffusible solutes, but not of compounds of large molecular mass such as proteins. However, this "impermeability" is not absolute, as is demonstrated by the varying (although low) concentration of protein in interstitial fluids.

The exact composition of *intracellular fluid* is extremely hard to measure because of the relative unavailability of cells free of contamination. Although erythrocytes are easily accessible, it would be incorrect to make any generalizations based on the composition of these highly specialized cells. Data on cell composition (Figure 10-6), therefore, are considered only approximations. The volume of intracellular fluids contributes approximately 66% to the total body volume. (See Table 10-9.)

The mass concentration of water in normal plasma is about 0.933 kg/L,* depending upon the protein and lipid content. Thus, a concentration of sodium in the plasma of 140 mmol/L

---

*The total weight (mass) of 1 liter of plasma is about 1.026 kg; thus, the total weight of solute in 1 liter of plasma is about 1.026 − 0.933 = 0.093 kg.

would correspond to a molality of sodium in plasma water of 150 mmol/kg H_2O, i.e., 140 mmol $\times L^{-1}/(0.933$ kg $\times L^{-1})$.

Gibbs-Donnan Equilibrium. Two solutions separated by a semipermeable membrane will establish an equilibrium in such a way that all ions are equally distributed in both compartments, provided that the solution contains only solutes that can freely move through the membrane. At the state of equilibrium the total ion concentration and, therefore, the total concentration of osmotically active particles (osmolutes) are the same on both sides of the membrane.

If the solution on one side of a membrane contains ions that cannot freely move through the membrane (e.g., proteins), distribution of the diffusible ions at the steady state is unequal.

Before steady state		*After steady state*	
Compartment I	Compartment II	Compartment I	Compartment II
$cNa^+ = c_I$ $cCl^- = c_I$	$cNa^+ = c_{II}$ $cR^- = c_{II}$	$cNa^+ = c_I - \Delta c$ $cCl^- = c_I - \Delta c$	$cNa^+ = c_{II} + \Delta c$ $cCl^- = \Delta c$ $cR^- = c_{II}$ (4)

In our example, compartment I initially contains only Na^+ and Cl^- and compartment II initially contains only Na^+ and R^-, where R^- is a nondiffusible ion such as protein. The system is of fixed size and the membrane is nonelastic. The initial concentrations are symbolized by c_I and c_{II}, respectively. The change in the concentrations of Na^+ and Cl^- on the two sides of the membrane is symbolized by Δc. In order to preserve electrical neutrality, Na^+ and Cl^- diffuse in pairs across the membrane until a steady state is established.

The unequal distribution of the diffusible ions (Na^+ and Cl^-) generates a membrane potential (ΔU), the magnitude of which is given by the Nernst equation. (See Chapter 1B, Section Four.)

Applied to equation (4) this becomes

$$\Delta U(I \mid II) = U(II) - U(I) = -\frac{R \times T}{(+1) \times F} \times \ln\frac{aNa^+(II)}{aNa^+(I)}$$

$$\simeq -\frac{R \times T}{(+1) \times F} \times \ln\frac{cNa^+(II)}{cNa^+(I)} \qquad (5)$$

The symbol $\simeq$ is used because the expression in terms of ion concentrations (or molalities) is accurate only if the activity coefficients of the ions are identical for compartments I and II.

Similarly:

$$\Delta U(I \mid II) = -\frac{R \times T}{(-1) \times F} \times \ln\frac{aCl^-(II)}{aCl^-(I)}$$

$$\simeq -\frac{R \times T}{(+1) \times F} \times \ln\frac{cCl^-(I)}{cCl^-(II)}$$

Since $\Delta U(I \mid II)$ is identical for the Na^+ and Cl^-, therefore

$$\frac{R \times T}{(+1) \times F} \times \ln\frac{cNa^+(II)}{cNa^+(I)} = \frac{R \times T}{(+1) \times F} \times \ln\frac{cCl^-(I)}{cCl^-(II)}$$

This reduces to:

$$\frac{cNa^+(II)}{cNa^+(I)} = \frac{cCl^-(I)}{cCl^-(II)} \qquad (6)$$

or:

$$cNa^+(I) \times cCl^-(I) = cNa^+(II) \times cCl^-(II)$$

Consequently: *The product of the concentrations of the diffusible ions in one compartment is equal to the product of the concentrations of the diffusible ions in the other compartment* (Gibbs-Donnan law).

Application of the data from equation (4) to equation (6) allows calculation of Δc in terms of c_I and c_{II}:

$$\frac{c_{II} + \Delta c}{c_I - \Delta c} = \frac{c_I - \Delta c}{\Delta c} \qquad (7)$$

and rearrrangement gives:

$$(c_{II} + \Delta c)\,\Delta c = (c_I - \Delta c)(c_I - \Delta c)$$

$$(c_{II} \times \Delta c) + \Delta c^2 = c_I^2 - 2(c_I \times \Delta c) + \Delta c^2$$

$$c_{II} \times \Delta c = c_I^2 - 2(c_I \times \Delta c)$$

$$(c_{II} \times \Delta c) + 2(c_I \times \Delta c) = c_I^2$$

$$\Delta c \times (c_{II} + 2c_I) = c_I^2$$

$$\Delta c = \frac{c_I^2}{c_{II} + 2c_I}$$

If, for example, we assume the same initial concentration for NaR(II) and NaCl(I) of 3.0 mmol/L, then

$$\Delta c = c\mathrm{Cl}^-(\mathrm{II}) = \frac{3^2}{3 + (2 \times 3)} = 9/9 = 1 \text{ mmol/L}$$

If we enter this figure into Equation (4), the following ion distribution results:

Before steady state		After steady state	
Compartment I	Compartment II	Compartment I	Compartment II
$c\mathrm{Na}^+ = 3$ mmol/L	$c\mathrm{Na}^+ = 3$ mmol/L	$c\mathrm{Na}^+ = 2$ mmol/L	$c\mathrm{Na}^+ = 4$ mmol/L
$c\mathrm{Cl}^- = 3$ mmol/L	$c\mathrm{R}^- = 3$ mmol/L	$c\mathrm{Cl}^- = 2$ mmol/L	$c\mathrm{Cl}^- = 1$ mmol/L
			$c\mathrm{R}^- = 3$ mmol/L

Note that after steady state has been established, the law of electrical neutrality is obeyed for both compartments.

The membrane potential according to the Na^+ distribution is

$$\Delta U(\mathrm{I} \mid \mathrm{II}) = -\frac{R \times T}{1 \times F} \times \ln\frac{4}{2}$$

or according to the Cl^- distribution:

$$\Delta U(\mathrm{I} \mid \mathrm{II}) = -\frac{R \times T}{-1 \times F} \times \ln\frac{1}{2}$$

which gives

$$\Delta U(\mathrm{I} \mid \mathrm{II}) = -18.5 \text{ mV}$$

when $T = 310.15$ K ($= 37°C$); R(gas constant) $= 8.314$ Joules (J) $\times$ mol^{-1} $\times$ K^{-1}; and F (Faraday constant) $= 96487$ Coulombs (C) $\times$ mol^{-1}.

Note that the concentration of total particles in compartment II is greater than that of compartment I (compartment II: 8 mmol/L; compartment I: 4 mmol/L). By osmosis, water therefore tends to diffuse from compartment I into compartment II, resulting in an increased pressure in compartment II (since the volume in compartment II is fixed). This increase in pressure is due to the presence of the nondiffusible ions (colloids) in compartment II and is called the *colloidal osmotic pressure*. This colloidal osmotic (oncotic) pressure due to proteins is especially important in biological systems because it aids in the proper distribution of water in the various fluid compartments. Decrease in the colloidal osmotic (oncotic) pressure of plasma (e.g., owing to loss of protein in renal disease) results in a loss of water from the vascular compartment to the interstitial fluid, causing the clinical condition known as edema.

The body cells that contain nondiffusible protein anions can withstand only a limited and temporary difference in osmotic pressure across the cell membrane. The osmotic pressure or the concentration of osmolutes is normally identical inside and outside the cells, since the cell membrane can correct concentration differences by excluding some small ions by active transport processes. If these processes cease, the cells gradually swell and eventually burst.

Distribution of Ions by Active and Passive Transport

Examination of Figure 10-6 reveals that the electrolyte compositions of blood plasma and interstitial fluid (both extracellular fluids) are similar, but their compositions differ markedly from that of intracellular fluid. The major extracellular ions are Na^+, Cl^-, and HCO_3^-; in intracellular fluids the main ions are K^+, Mg^{2+}, organic phosphates, and protein. The mechanism for this unequal distribution of most electrolytes (or other constituents) between intracellular and extracellular fluids has so far not been adequately explained.

The Na^+ distribution is greatly different from the distribution predicted by the membrane potential, according to which the concentration of sodium should be higher inside the red cells than outside. It is now established that this distribution is due to an active transport of Na^+ from inside to outside the cell against an electrochemical potential gradient. This process requires energy supplied by the metabolic processes in the cell (e.g., glycolysis). An active sodium pump deriving its energy from ATP is now thought to be present in most cell membranes, frequently coupled with a transport of K^+ in the opposite direction. The actual chemical nature of the Na^+ pump is still unknown.

Other ions also appear to be transported actively, for instance H^+ from the muscle cells. The pH of the intracellular fluid of muscle cells has been measured to be about 6.9, while the pH of the interstitial fluid may be about 7.3. Therefore, on the basis of calculations using the Nernst equation, H^+ ions are thought to be actively pumped out of the muscle cells, possibly by the same mechanism which pumps Na^+ out of the cells.

The Henderson-Hasselbalch Equation

The Henderson-Hasselbalch equation aids in the understanding and explanation of pH control of body fluids; this will become clearer in the later discussions of the compensatory mechanisms of the body. The equation derived on page 1200 can also be written as follows:

$$pH = pK' + \log \frac{cHCO_3^-}{cdCO_2} \tag{8}$$

where $cdCO_2$ is the concentration of dissolved gas which by definition includes also the low concentration of undissociated carbonic acid. It is equal to $\alpha \times pCO_2$, where α is the solubility coefficient for CO_2. Note that dissolved CO_2 itself does not change the pH. However, dissolved CO_2 participates in the following equilibrium:

$$CO_2 + H_2O \rightleftharpoons H_2CO_3 \rightleftharpoons H^+ + HCO_3^-$$

The average normal ratio of the concentrations of bicarbonate and dissolved carbon dioxide in plasma is 25 (mmol/L)/1.25 (mmol/L) = 20/1; the $\log_{10}$ of 20 is 1.3. Formula (8) applied to normal plasma can then be written:

$$pH = 6.1 + 1.3 = 7.4$$

It follows that any change in the concentration of either bicarbonate or dissolved CO_2 and, therefore, in the ratio of $cHCO_3^-/cdCO_2$ must be accompanied by a change in pH. Such changes in the ratio can occur through a change either in the numerator ($cHCO_3^-$) or in the denominator ($cdCO_2$), as will be discussed later.

Clinical conditions characterized as metabolic disturbances of acid-base balance are classified as primary disturbances in HCO_3^- concentration. Those characterized as respiratory disturbances

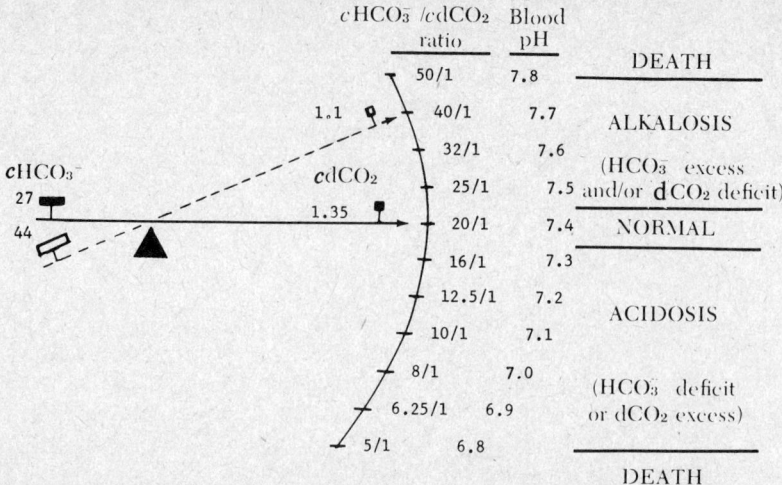

Figure 10-7. Scheme demonstrating the relation between pH and ratio of bicarbonate concentration to the concentration of dissolved CO_2. If the ratio in blood is 20/1 ($cHCO_3^- = 27$ mmol/ $cdCO_2 = 1.35$ mmol/L), the resultant pH will be 7.4 as demonstrated by the solid beam. The dotted line shows a case of uncompensated alkalosis (bicarbonate excess) with a bicarbonate concentration of 44 mmol/L and a dCO_2 concentration of 1.1 mmol/L. The ratio therefore is 40/1 and the resultant pH is 7.7. In a case of uncompensated acidosis, the pointer of the balance would point to a pH between 6.8 and 7.35, depending on the $cHCO_3^-/cdCO_2$ ratio. pH values below 6.8 or above 7.8 are incompatible with life. (After Weisberg, H. F.: Surg. Clin. N. Amer., *39*:93, 1959; Snively, W.D., and Wessner, M.: J. Ind. State Med. Assn., *47*:957, 1954.)

are classified as primary disturbances in $cdCO_2$. Various compensatory mechanisms attempt to re-establish the normal ratio of $cHCO_3^-/cdCO_2$ and hence normal pH. Here again, the compensatory mechanisms may result in changes in the bicarbonate and/or dissolved CO_2 concentration. The application of the Henderson-Hasselbalch equation can be illustrated by the lever-fulcrum (teeter-totter) diagram (Figure 10-7).

Acid-Base Parameters—Definitions and Abbreviations

Bicarbonate. Bicarbonate is the second largest fraction of the anions of the plasma ($\sim$25 mmol/L). Conventionally, it is defined to include plasma bicarbonate, carbonate, and CO_2 bound in plasma carbamino compounds. At the pH of blood, the plasma carbonate concentration is $\sim$25 μmol/L, which is $\sim$1/700 to 1/1000 of the bicarbonate concentration. CO_2 bound in plasma carbamino compounds is 0.2 mmol/L; in erythrocytes, it is 1.5 mmol/L. Bicarbonate concentration of blood is expressed as $cHCO_3^-$, in units of mmol/L. About 1 mmol/L exists as undissociated $NaHCO_3$.

Dissolved CO_2 (dCO_2). The dissolved CO_2 fraction is defined to include both the undissociated carbonic acid and physically dissolved, free CO_2. At the pH of the blood, the amount of dissolved CO_2 is 700–1000 times greater than the amount of carbonic acid and therefore $cdCO_2$ is the term used to express their combined concentration. $cdCO_2$ is calculated from the solubility coefficient of CO_2 in blood at 37 °C ($\alpha = 0.0306 \times$ mmol $\times L^{-1} \times$ mm Hg^{-1}) as noted earlier. At a pCO_2 of 40 mm Hg, $cdCO_2$ is 1.224 mmol/L.

pCO_2, the Partial Pressure of CO_2. The abbreviation PCO_2 is also frequently used to designate the partial pressure of CO_2, mostly in the physiological literature. In the USA, pCO_2 is generally expressed in mm Hg, but the IFCC recommends the use of kPa (kilopascal). 1 kPa = 7.5 mm Hg.

Total CO_2. The concentration of total CO_2 (formerly CO_2 content) of blood consists of the sum of bicarbonate and dissolved CO_2 concentrations, as each of them is defined above. In equation form,

$$ctCO_2, \text{mmol/L} = cHCO_3^- + cdCO_2$$

CO_2 Combining Power (CO_2 Capacity). This now obsolete term refers to the $ctCO_2$ that can be bound by blood, serum, or plasma when pCO_2 is 40 mm Hg and the temperature is 25 °C.

Standard Bicarbonate of Blood. This term has units of mmol/L and describes plasma $c\text{HCO}_3^-$ in blood that has been equilibrated with a gas mixture having $p\text{CO}_2 = 40$ mm Hg and $p\text{O}_2 >100$ mm Hg at 37 °C. It is a reference value to which bicarbonate calculated from actual $p\text{CO}_2$, found in conjunction with any $p\text{O}_2$, can be compared.

Buffer Base. Buffer base is the sum of concentrations of all the buffer anions in the blood: hemoglobin, bicarbonate, proteins, and phosphate. It may therefore be defined as the concentration of titratable base when titrating to the apparent isoelectric pH (pI) of the proteins at a $p\text{CO}_2$ of zero. Hence, the fundamental difference between the "buffer base" and the "base excess" (noted below) is the choice of endpoint of titration (pI of proteins vs pH 7.40). *Whole blood buffer base* refers to buffer base of fully oxygenated blood.

Normal Buffer Base (NBB) is the sum of concentrations of all the buffer anions in the blood when $p\text{CO}_2$ is 40 mm Hg and pH is 7.4, as corrected for hemoglobin concentration. It is calculated:

$$\text{NBB, mmol/L} = 41.7 + (0.42 \times \text{hemoglobin, g/dL})$$

where 41.7 is the buffer base value in mmol/L of anions other than hemoglobin and a relative constant, and 0.42 is a conversion factor for hemoglobin in g/dL to hemoglobin anion in mmol/L.

Base Excess. The base excess concentration is defined as the concentration of titratable base when titrating the blood or plasma with a strong acid or base to a plasma pH of 7.40 at a $p\text{CO}_2$ of 40 mm Hg at 37 °C. Positive values (base excess) indicate a relative deficit of noncarbonic acid and suggest presence of a metabolic alkalosis, and negative values (base deficit) indicate a relative excess of noncarbonic acids in the blood or plasma and suggest presence of a metabolic acidosis. The numeric value of the base excess or deficit indicates the *theoretical* amount of acid or base, in mmol/L, that should be given to correct the blood pH. (In reality, less acid or base is given to avoid any rapid changes in pH and because body regulatory mechanisms will generally aid in restoring normal acid-base status.) The base excess value can be obtained from a nomogram or, more commonly, is calculated by algorithms programmed in the microprocessor of a blood gas instrument.

pH. The pH is the negative logarithm of the hydrogen ion activity ($\text{pH} = -\log a\text{H}^+$). Thus, the average pH of blood (7.40) corresponds to a hydrogen ion concentration of 4×10^{-8} mol/L = 40 nmol/L, assuming that the activity coefficient of H^+ is 1.

Potentiometric determinations of blood pH measure the H^+ activity ($a\text{H}^+$) and not the H^+ concentration, and $c\text{H}^+$ is assumed to equal $a\text{H}^+$. It has been recommended that the acidity of blood in terms of its hydrogen ion concentration be expressed in nmol/L. This form of expression has the advantage that one does not have to deal with logarithmic figures when performing acid-base calculations. (See below.) The relationship between hydrogen ion activity and pH is illustrated in Figure 10-8,*A* in the form of a bar graph and in Figure 10-8,*B* in the form of a double scale. It can readily be seen that the relationship is inverse and nonlinear. Some practicing physicians use a simple way to relate pH and $c\text{H}^+$ by considering a change of 0.01 unit in pH equal to a change of ~1 nmol/L in $c\text{H}^+$, in the pH range of 7.20–7.60. However, even within this narrow range, errors of >7% can be introduced. A better approximation is obtained by multiplying sequentially 40 nmol/L by 1.25 for each 0.10 pH decrease and by multiplying 40 nmol/L sequentially by 0.8 for each 0.10 pH increase. For example, the $c\text{H}^+$ of blood with a pH of 7.20 can be approximated by multiplying $40 \times 1.25 = 50$ and again $50 \times 1.25 = 62.5$ nmol/L.

To approximate $c\text{H}^+$ in blood, the following equation has been derived from the Henderson-Hasselbalch equation (p. 1201):

$$c\text{H}^+ = 24 \times \frac{p\text{CO}_2}{c\text{HCO}_3^-}$$

This equation not only approximates $c\text{H}^+$, it also relates $p\text{CO}_2$ and $c\text{HCO}_3^-$ to $c\text{H}^+$ and readily allows for calculation of the $c\text{H}^+$ without resorting to the cumbersome Henderson-Hasselbalch equation. Despite the clinical utility of the equation to estimate the $c\text{H}^+$ and to identify acid-base disorders, calculated estimates are not a substitute for the actual determination of the blood acid-base parameters involved.

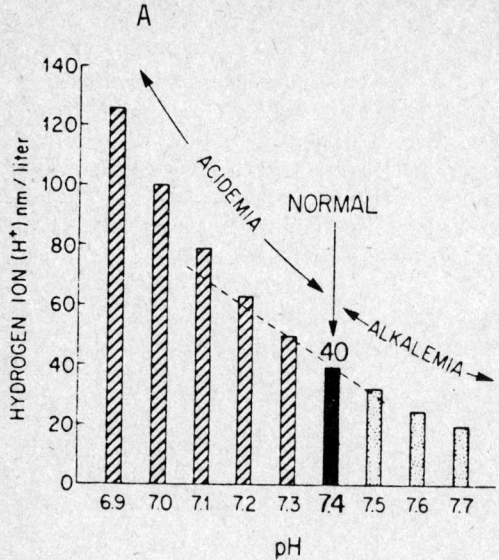

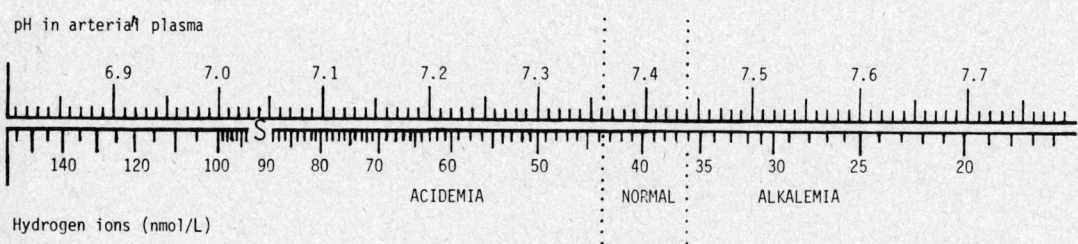

Figure 10-8. Relationship of pH to hydrogen ion concentration. *A,* Broken line is drawn to emphasize the (approximately) linear relationship between hydrogen ion concentration and pH over the pH range of 7.2–7.5. (Used by permission from Narins, R. G., and Emmett, M.: Simple and mixed acid-base disorders; a practical approach. Medicine, *59:*161–187, 1980.) *B,* Scale for the interconversion of pH and hydrogen ion concentration.

Buffer Systems and Their Role in Regulating the pH of Body Fluids

The action of buffers in the regulation of body pH can be explained by using the bicarbonate buffer system as an example. (For definition of buffers, see Chapter 1A.) If we add a strong acid to a solution containing HCO_3^- and H_2CO_3, the H^+ will react with HCO_3^- to form more $HHCO_3$ and subsequently CO_2. The hydrogen ions are thereby bound, and the increase in the H^+ concentration will be minimal:

$$HCO_3^- + H^+ \rightarrow H_2CO_3 \rightarrow CO_2 + H_2O$$

On the other hand, if we add a base to the same buffer solution, the base will react with H_2CO_3 directly [equation (9)] or OH^- will react with the H^+ of $H_2PO_4^-$ and H_2CO_3 to form HPO_4^{2-} or HCO_3^- and water [equations (10a) and (10b)]. The pH change, therefore, will be small:

$$HPO_4^{2-} + H_2CO_3 \rightleftharpoons H_2PO_4^- + HCO_3^- \tag{9}$$

$$OH^- + H_2PO_4^- \rightleftharpoons HPO_4^{2-} + H_2O \tag{10a}$$

$$OH^- + H_2CO_3 \rightleftharpoons H_2O + HCO_3^- \tag{10b}$$

The buffer systems of most physiological interest in connection with regulation of the pH of body fluids are those of plasma and erythrocytes. A discussion of the most important buffers follows.

The Bicarbonate/Carbonic Acid Buffer System

The most important buffer of plasma is the bicarbonate/carbonic acid pair; it is also present in red cells, but at a lesser concentration. The effectiveness of the bicarbonate buffer is based on its high concentration and on the fact that CO_2 can readily be disposed of or retained in the lungs. In addition, the renal tubules can increase or decrease the rate of reclamation of bicarbonate from the glomerular filtrate. (See p. 1238.) The bicarbonate/carbonic acid buffer system obviously buffers only noncarbonic acid or base.

The buffer value (β) is defined as the amount of base required to cause a change in pH of one unit. Therefore the buffer value of the bicarbonate buffer in plasma is by definition

$$\beta HCO_3^-(P) = \frac{\Delta c HCO_3^-(P)}{\Delta pH(P)} \tag{11}$$

This coefficient β can be derived by taking partial differentials of the Henderson-Hasselbalch equation. For a *closed system* where the concentration of total CO_2 ($ctCO_2$) is constant, the result is:

$$\beta HCO_3^-(P, \text{closed}) = \left(\frac{\partial c HCO_3^-(P)}{\partial pH(P)}\right)_{ctCO_2} = 2.303 \times ctCO_2(P) \times \frac{K' \times aH^+}{(K' + aH^+)^2} \tag{12}$$

For an *open system* where the pCO_2 is constant, as in circulating blood, the result is:

$$\beta HCO_3^-(P, \text{open}) = \left(\frac{\partial c HCO_3^-(P)}{\partial pH(P)}\right)_{pCO_2} = 2.303 \times c HCO_3^-(P) \tag{13}$$

Inserting $pK' = 6.10$, $pH = 7.40 \leftrightarrow aH^+ = 40 \times 10^{-9}$, $ctCO_2 = 25.7$ mmol/L, and $cHCO_3^- = 24.5$ mmol/L gives the buffer values:

$$\beta HCO_3^-(P, \text{closed}) = 2.7 \text{ mmol/L}$$

$$\beta HCO_3^-(P, \text{open}) = 56.6 \text{ mmol/L}$$

The above calculation illustrates the insignificance of the bicarbonate buffer in a closed system at pH 7.4 as against its great importance in an open system of constant pCO_2, which pertains to the situation in the living organism.

The Phosphate Buffer System

At a plasma pH of 7.4, the ratio $cHPO_4^{2-}/cH_2PO_4^-$ is 80/20 ($pK' = 6.8$). The total concentration of this buffer in both erythrocytes and plasma is less than that of other major buffer systems. Inorganic phosphate accounts for only about 5% of the nonbicarbonate buffer value of plasma. Organic phosphate, however, in the form of 2,3-diphosphoglycerate (present in erythrocytes in a concentration of about 4.5 mmol/L), accounts for about 16% of the nonbicarbonate buffer value of erythrocyte fluid.

The phosphate buffer reacts with acids and with bases as follows:

$$HPO_4^{2-} + H^+ \rightarrow H_2PO_4^-$$

$$H_2PO_4^- + OH^- \rightarrow HPO_4^{2-} + H_2O$$

This system is important in the excretion of acids in the urine, as will be explained in the section on renal mechanism.

The Plasma Protein Buffer System

The buffer value of the nonbicarbonate buffers of plasma is about 7.7 mmol/L at pH 7.40 for a normal plasma protein concentration of 72 g/L (7.2 g/dL). The value varies slightly in the physiological pH range with a maximum at pH $\sim$7.3. The proteins, especially albumin, account for the greatest portion (95%) of the nonbicarbonate buffer value of the plasma. The most important buffer groups of proteins in the physiological pH range are the imidazole groups of histidines ($pK \sim 7.3$), of which 16 are present for each albumin molecule:

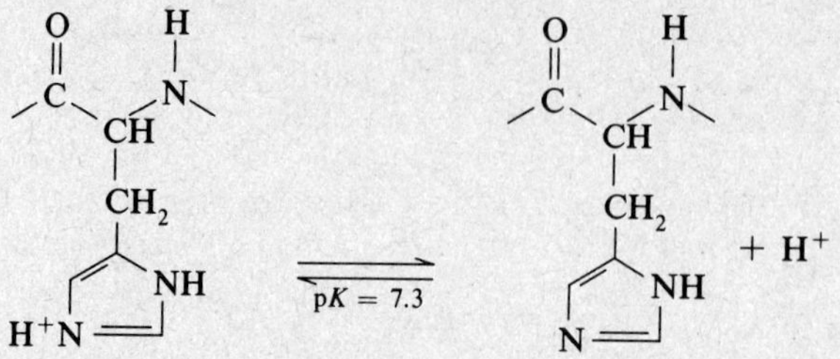

The significance of the nonbicarbonate buffers of the plasma can be illustrated by plotting the CO_2 equilibrium curve of the plasma in a diagram of pH versus $cHCO_3^-$. The CO_2 equilibrium curve is obtained by equilibrating the plasma with gas mixtures of varying pCO_2. The slope of the CO_2 equilibrium curve in a pH-$cHCO_3^-$ diagram is equal to the buffer value of the nonbicarbonate buffers (with opposite sign). This is apparent when considering the chemical reactions during CO_2 equilibration:

$$CO_2 + H_2O \rightarrow H_2CO_3 \rightarrow H^+ + HCO_3^-$$

$$HPr \leftarrow H^+ + Pr^-$$

where the HPr/Pr^- system represents all nonbicarbonate buffers. For each molecule of HCO_3^- which is generated, one molecule of nonbicarbonate buffer base disappears because the concentration of H^+ remains virtually constant compared to the changes in $cHCO_3^-$ and cPr^-. Mathematically this can be expressed as follows:

$$-\left(\frac{\partial cHCO_3^-}{\partial pH}\right)_{Base} \sim \left(\frac{\partial cPr^-}{\partial pH}\right)_{Base} = \beta Pr^- = 7.7 \text{ mol/L} \tag{14}$$

which says that the negative value of the slope of the CO_2 equilibrium curve in a pH-$cHCO_3^-$ diagram equals the buffer value of the nonbicarbonate buffers, which is approximately equal to the buffer value of the plasma proteins.

The equation for the CO_2 equilibrium curve of plasma can therefore be written:

$$\Delta cHCO_3^-(P) = [-7.7 \text{ (mmol/L)} \times \Delta pH(P)] + \Delta cB'(P) \tag{15}$$

where $\Delta cHCO_3^-(P) = cHCO_3^-(P) - 24.5$ mmol/L, $\Delta pH(P) = pH(P) - 7.40$, and $\Delta cB'(P)$ is the concentration of titratable base when titrating the plasma with strong acid or base to $pH(P) = 7.40$ at $pCO_2 = 40$ mm Hg and 37 °C.

The Hemoglobin Buffer System

The buffer value of the nonbicarbonate buffers of erythrocyte fluid is about 63 mmol/L at pH 7.20 for an erythrocyte hemoglobin (Fe) concentration of 21 mmol/L (33.8 g/100 dL). Hemoglobin accounts for the major part (53 mmol/L), the remainder being due mainly to 2,3-diphosphoglycerate. The imidazole groups of hemoglobin are quantitatively the most important buffer groups.

The slope of the CO_2 equilibrium curve of whole blood depends on the buffer value of nonbicarbonate buffers, i.e., mainly on the hemoglobin concentration of the blood. It is possible to derive an approximate equation for the CO_2 equilibrium curve of whole blood:

$$\Delta cHCO_3^-(P) = -\beta \times \Delta pH(P) + \zeta^{-1} \times \Delta cB'(B) \tag{16}$$

where $\Delta cHCO_3^-(P) = cHCO_3^-(P) - 24.5$ mmol/L
$\Delta pH(P) = pH(P) - 7.40$
$\Delta cB'(B)$ = concentration of titratable base (B) − concentration of titratable acid (B), titrating the blood with strong acid or strong base to $pH(P) = 7.40$ at $pCO_2 = 40$ mm Hg and 37 °C

$$\beta = 2.3 \times c\text{Hb(B)}, \text{mmol/L} + 7.7 \text{ mmol/L}$$

$$\zeta = 1 - c\text{Hb(B)}, \text{mmol/L} \times 0.023 \times \text{L} \times \text{mmol}^{-1}$$

This equation, together with the Henderson-Hasselbalch equation, provides the simplest algorithm for calculation of the various acid-base variables by means of an electronic calculator. The derivation of equation (16) is complicated by the fact that blood is a two-phase system and is therefore not given here. When $c\text{Hb(B)} = 0$ mmol/L, equations (15) and (16) are identical.

The buffer value of deoxyhemoglobin is slightly lower than that for oxyhemoglobin at pH $\sim$ 6.5 but higher at pH $\sim$ 7.8. This is due to a decrease in the pK-value of the so-called oxygen-linked acid-base groups when deoxyhemoglobin is oxygenated. This also causes a liberation of H^+ upon oxygenation of hemoglobin, a phenomenon called the *Haldane effect*. In a hemoglobin solution a close relationship exists between the Bohr effect and the Haldane effect. The linkage equation

$$\left(\frac{\partial \log p\text{O}_2}{\partial \text{pH}}\right)_{c\text{HbO}_2} = \left(\frac{\partial c\text{Hb-H}^+}{\partial c\text{HbO}_2}\right)_{\text{pH}} \tag{17}$$

relates the coefficients of the Bohr and Haldane effects. The first coefficient is the Bohr coefficient, i.e., the change in log $p\text{O}_2$ with pH at constant $c\text{HbO}_2$ (usually at a constant oxygen saturation of 0.5). The second coefficient is the Haldane coefficient, i.e., the degree of change in the concentration of hemoglobin-bound hydrogen ion with changing concentration of hemoglobin-bound oxygen at constant pH. The value of the coefficients varies with pH, $p\text{CO}_2$, and the concentration of 2,3-diphosphoglycerate. For erythrocyte fluid with pH $= 7.20$, $p\text{CO}_2 = 40$ mm Hg, and a normal concentration of 2,3-diphosphoglycerate (4.5 mmol/L), the mean value of the Haldane coefficient, when the oxygen saturation increases from zero to one, is about -0.47. The value varies slightly with the oxygen saturation, with a maximum value at half-saturation. For whole blood, the heterogeneity of the blood must be taken into account and the "Haldane coefficient for whole blood," i.e., the rise in the concentration of titratable base of the whole blood [$\Delta c\text{B}'\text{(B)}$], is only 0.3 mmol/L when oxyhemoglobin of a concentration of 1 mmol/L is converted to deoxyhemoglobin.

The oxygen-linked acid-base groups are the imino groups of the C-terminal histidines of the two β-chains, the amino groups of the N-terminal valines of the two α-chains, and possibly also the imino group of a histidine of the α-chains. In deoxyhemoglobin these groups participate in the formation of salt bridges, and the pK values are thereby increased. In oxyhemoglobin the salt bridges are ruptured by changes in the tertiary and quaternary structures of the molecule, and the pK values thereby fall. The situation is considerably complicated by the fact that CO_2 and 2,3-diphosphoglycerate are bound to some of the oxygen-linked acid-base groups. CO_2 is bound as carbamino-CO_2 to the four terminal amino groups (valines) of the four peptide chains with a greater affinity to deoxyhemoglobin than to oxyhemoglobin. 2,3-DPG is bound in the central cavity of the deoxyhemoglobin molecule by salt bridges to the aminium groups of the two N-terminal valines and the iminium groups of the two histidines. In oxyhemoglobin the central cavity narrows and 2,3-DPG is expelled. (See also Chapter 15.)

The Isohydric and Chloride Shift

Because of the continuous production of carbon dioxide within the tissue cells, there is a concentration gradient for carbon dioxide from these cells to the plasma and the erythrocytes. This gradient causes a shift of physically dissolved carbon dioxide from the tissue cells into the plasma and the erythrocytes. A small portion of the carbon dioxide entering the plasma stays as dissolved carbon dioxide and another small portion reacts with water to form carbonic acid. The increased amount of H^+ is buffered by the plasma buffers, including the proteins (Figure 10-9, reaction 1). Another small portion combines with the amino groups of proteins and forms carbamino compounds ($PrCO_2$; Figure 10-9, reaction 2). The normal concentration of carbamino compounds in the plasma is about 0.2 mmol/L, and the arteriovenous difference is negligible. Most of the carbon dioxide enters the erythrocytes and reacts with water to form carbonic acid. This reaction is catalyzed by the enzyme carbonic anhydrase (C.A.) and therefore proceeds at a relatively high speed (Figure 10-9, reaction 3). Some CO_2 remains as dissolved CO_2 and some combines with Hb to form $HbCO_2$ (Figure 10-9, reaction 4).

The carbonic acid formed in reaction 3 contributes to the H^+-concentration. The pH change, however, is fully or partially compensated by the release of oxygen from oxyhemoglobin, which involves the conversion of the stronger acid ($H^+HbO_2^-$) into a weaker acid (HHb); the oxygen-linked acid-base groups will accept the H^+, as outlined in the discussion of the hemoglobin buffer. For each mole of O_2 given off, the hemoglobin binds about 0.5 mol of H^+. Furthermore,

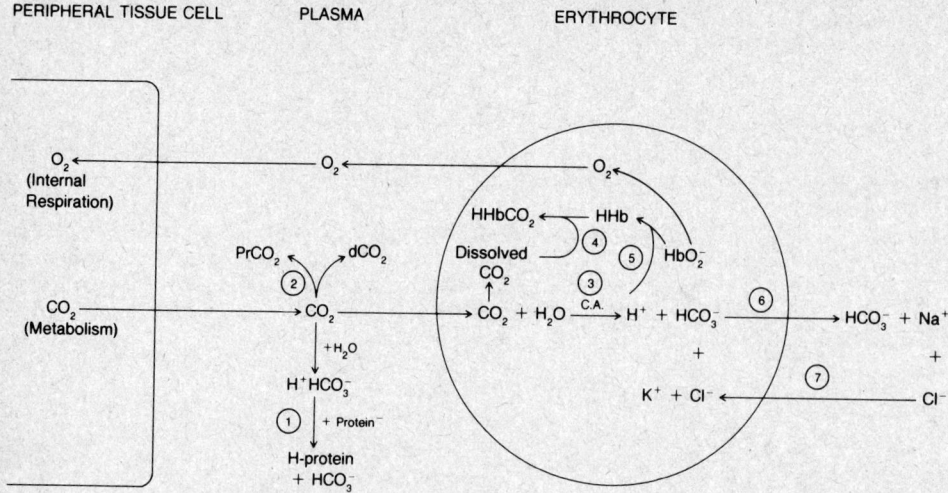

Figure 10-9. Scheme demonstrating the isohydric and chloride shift. The encircled numbers refer to the reactions described in the text. For details, see text.

the deoxyhemoglobin binds significantly more CO_2 in the form of carbamino-CO_2 than does oxyhemoglobin, and thus a significant fraction of CO_2 is transported in this form (Table 10-4). The oxygen released from HbO_2^- moves from the erythrocytes through the plasma into the peripheral tissue cells. (See Figure 10-9.)

The remainder of the hydrogen ions formed in reaction 3 are buffered by the nonbicarbonate buffers of the erythrocyte fluid, while the concentration of HCO_3^- increases to the same extent that the concentration of Hb anions falls. The transformations described so far (Figure 10-9, reactions 1–5) are referred to as the *isohydric shift* (= a shift in which the hydrogen ion concentration remains essentially unchanged).

The equilibrium between plasma and red cells has been disturbed by the reactions described so far. The concentration of HCO_3^- has increased relatively more in the erythrocytes than in the plasma, the pH of plasma has fallen relatively more than the pH of the erythrocytes, and, most important, the nondiffusible ion concentration in the erythrocytes has fallen due to the increase in protonation of proteins and hemoglobin. The membrane potential of the erythrocytes therefore falls numerically (becomes less negative) and the distribution of all diffusible ions must change in accordance with the new membrane potential. The ion shifts that occur rapidly are a movement of HCO_3^- out of the erythrocytes and a movement of Cl^- into the erythrocytes to provide electrochemical balance. The Cl^- ratio $[mCl^-(E)/mCl^-(P) =$ molal concentration of Cl^- in erythrocytes/molal concentration of Cl^- in plasma] thereby increases in accordance with the change in the membrane potential. This shift of chloride ions is referred to as the *chloride shift*. (See also Figure 10-9, reactions 6 and 7.) Although protonation of macromolecular ions decreases the number of osmotically active particles, retention of HCO_3^- (30% of the amount formed) causes a net increase of osmolutes inside the erythrocyte; therefore, a small amount of H_2O will pass from the plasma into the erythrocytes. As a result of these ion and water fluxes, the concentration of chloride in the venous plasma is about 1 mmol/L lower than that in the arterial plasma, and the mean volume of an erythrocyte becomes about 1 femtoliter (fL) greater in the venous blood than in the arterial blood. This increases the packed cell volume of arterial blood by ∼0.4, but the hematocrit of venous blood will increase by 2–3% due to loss of plasma water. The permeability of the erythrocyte membrane for Na^+ and K^+ is so low that the movements of these ions in and out of the erythrocytes during the circulation of the blood can be ignored; i.e., during a short time interval the erythrocyte membrane is practically impermeable to these ions.

The HCO_3^-, the carbamino compounds, and the dissolved carbon dioxide are transported in venous blood to the pulmonary capillaries and alveoli. The comparatively low pCO_2 in the alveoli will cause a shift of carbon dioxide from the erythrocytes and the plasma into the alveoli.

On the other hand, the high pO_2 in the alveoli causes a shift of oxygen into the plasma and the erythrocytes. This exchange causes a reversal of reactions 1–7 in Figure 10-9. The removal of carbon dioxide from the blood and the oxygenation of the blood are the major reactions that convert venous blood into arterial blood.

The Respiratory Mechanism in the Regulation of Acid-Base Balance

The respiratory mechanism contributes to the maintenance or restoration of normal body pH through retention of CO_2 in metabolic alkalosis and increased elimination of CO_2 in metabolic acidosis. These processes also support the bicarbonate buffer system, as was explained earlier. The regulation of the respiratory system has been discussed in detail in Section Two of this chapter.

The respiratory system responds immediately to a change in acid-base status, but about 3–6 h are required for the response to become maximal. (Of course, most metabolic acid-base disorders develop slowly—within hours in diabetic ketoacidosis and months or even years in chronic renal disease.) The maximum response is not attained until both the central and peripheral chemoreceptors are fully stimulated. For example, in the early stages of metabolic acidosis plasma pH decreases, but because ions equilibrate rather slowly across the blood brain barrier, the cerebrospinal fluid pH remains nearly normal. The peripheral chemoreceptors are stimulated by the decrease in plasma pH, hyperventilation occurs, and plasma pCO_2 decreases. pCO_2 of the extracellular fluid of the brain decreases immediately because CO_2 equilibrates rapidly across the blood brain barrier. Therefore, initially the pH of the extracellular fluid of the brain tends to rise and the central chemoreceptors are inhibited. But as plasma bicarbonate gradually falls, bicarbonate concentration in extracellular fluid of the brain also falls over the subsequent 3–6 h, and its pH returns to normal or slightly below normal. It is at this point that stimulation of respiration becomes maximal.

The reverse is true when a patient with metabolic acidosis is treated with HCO_3^-. When the pH in plasma increases as the result of HCO_3^- administration, stimulation of the peripheral chemoreceptors returns to normal. However, due to the slow equilibration of HCO_3^- between plasma and extracellular fluid of the brain, the central chemoreceptors continue to be stimulated and the patient continues to hyperventilate, even when the blood pH has returned to normal. Respiration does not return to normal until normal acid-base balance in the extracellular fluid of the brain is restored.

Examples of the action of the respiratory mechanism will be discussed in the sections on metabolic and respiratory disorders.

Renal Excretion of Acid, Ammonia Formation, and Reabsorption of Bicarbonate—The Renal Compensatory Mechanism

The average pH of plasma, as well as that of the glomerular filtrate, is $\sim$7.4; the average urinary pH of fasting individuals is $\sim$6.0. This drop in pH is brought about by the kidneys, which excretes the nonvolatile acids produced by metabolic processes. The various functions of the renal mechanism respond to specific requirements: in the case of acidosis, there is an increased excretion of acids and a conservation of base; in alkalosis, there is an increased excretion of base and conservation of acids. The pH of the urine changes correspondingly and may vary in *random* specimens from pH 4.5 to 8.2 (reference range from 4.8 to 7.8). This ability to excrete variable amounts of acid or base is of utmost importance and makes the kidney the final defense mechanism against any changes in body pH or cation-anion composition.

The various acids produced during metabolic processes are buffered in the extracellular fluid at the expense of HCO_3^-. (See Bicarbonate/Carbonic Acid Buffer System, p. 1231.) Renal excretion of acid and conservation of HCO_3^- occur through several mechanisms: (1) the Na^+-H^+ exchange, (2) production of ammonia and excretion of NH_4^+, and (3) reclamation of bicarbonate.

Excretion of Acids

Strong acids such as sulfuric, hydrochloric, and phosphoric acids are fully ionized at the pH of urine and are excreted only after the H^+ derived from these acids reacts with a buffer base

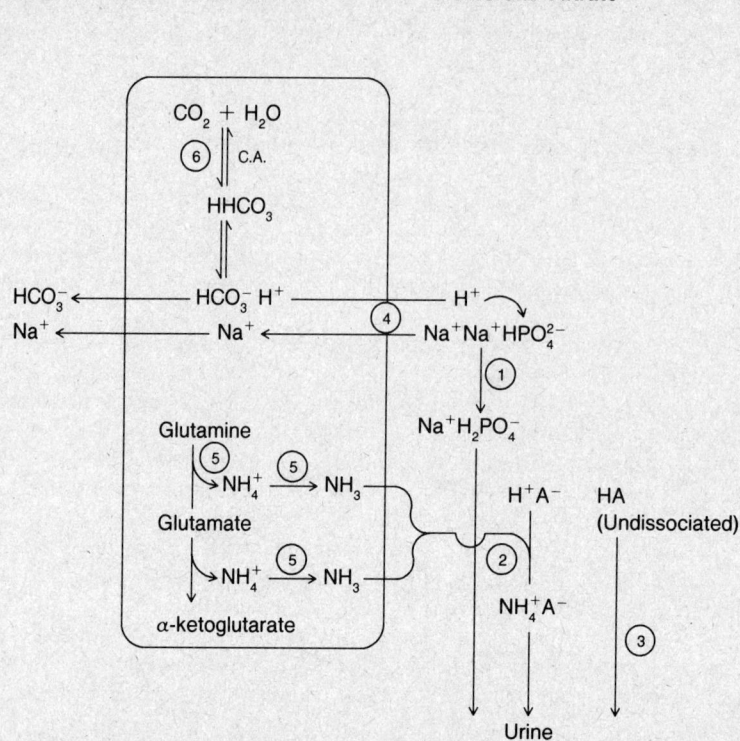

Figure 10-10. Hydrogen ion excretion, sodium hydrogen ion exchange, and ammonia production in the renal tubules. 1, Conversion HPO_4^- to $H_2PO_4^-$; 2, reaction of hydrogen ions with NH_3; 3, excretion of undissociated acids; 4, Na^+H^+ exchange; 5, NH_3 production; and 6, synthesis of carbonic acid from CO_2.

(e.g., HPO_4^{2-}, Figure 10-10,1; or NH_3, Figure 10-10,2), as will be explained below. Excretion of the anions of these acids is accompanied by the simultaneous removal of an equal number of cations, such as Na^+, K^+, or NH_4^+, to provide electrochemical balance. Some acids, such as acetoacetic acid ($pK = 3.58$) and β-hydroxybutyric acid ($pK = 4.7$), are present in blood almost entirely in ionized form, but at the acid pH frequently prevailing in urine some are nondissociated and thus may be excreted partially as the nondissociated acid (Figure 10-10,3). For example, 50% of β-hydroxybutyric acid at pH 4.7 is nonionized, and thus 50% of its hydrogen can be excreted as nondissociated acid.

Na^+-H^+ Exchange

In the renal tubules, predominantly the proximal but also the distal tubules, hydrogen ions from the tubular cells can be exchanged for sodium from the tubular fluid—an energy-dependent process involving an ATPase (Figure 10-10,4). This process is enhanced in acidosis and decreased in alkalosis. The proximal tubules, however, cannot maintain a H^+-gradient of more than ~ 1 pH unit and the distal tubules cannot maintain one of more than ~ 3 pH units. Thus, maximum urine acidity is reached at $\sim$pH 4.4. Hydrogen ions of the glomerular filtrate or those that have been transported through the Na^+-H^+ exchange into the tubular fluid may react with NH_3 or HPO_4^{2-}, as shown in Figure 10-10,1 and 2. In some forms of renal tubular acidosis, this exchange process is defective and may lead to a decrease in blood pH and an increase in urinary pH, as will be explained later.

Potassium ions compete with hydrogen ions in the Na^+-H^+ exchange. If the intracellular K^+ level of renal tubular cells is high, more K^+ and less H^+ are exchanged for Na^+; therefore, the urine becomes less acid and the acidity of body fluids increases. If there is K^+ depletion, more H^+ ions are exchanged for Na^+, and the urine becomes more acid and the body fluids more alkaline. Since the body's compensatory mechanism against metabolic alkalosis is relatively ineffective, K^+ depletion frequently results in a metabolic alkalosis. (See Metabolic Alkalosis, p. 1244.) Remember that serum K^+ concentrations are not always an accurate indicator of the intracellular K^+ concentrations; in potassium depletion and in acidosis there is a tendency for K^+ to move from the cells into the serum. This movement will, at least temporarily, maintain normal serum K^+ concentrations despite intracellular K^+ depletion.

Renal Production of Ammonia and Excretion of Ammonium Ions

The renal tubular cells are able to generate ammonia from glutamine and other amino acids derived from muscle and liver cells according to the following reaction:[9]

$$\text{Glutamine} \xrightarrow[\text{H}_2\text{O} \quad \text{NH}_4^+]{\textit{Glutaminase}} \text{glutamate} \xrightarrow[\text{NAD}^+ \quad \text{NADH} \quad \text{NH}_4^+]{\substack{\textit{Glutamate} \\ \textit{dehydrogenase}}} \alpha\text{-ketoglutarate}$$

$$\text{NH}_4^+ \rightleftharpoons \text{NH}_3 + \text{H}^+$$

The ammonium ion produced dissociates into ammonia and hydrogen ions to a degree dependent on the pH (Figure 10-10,5). At normal blood pH, the ratio of NH_3 to NH_4^+ is about 1 to 100. Ammonia, being a gas, diffuses readily across the cell membrane into the tubular lumen, where it combines with hydrogen ions to form ammonium ions (Figure 10-10,2). The NH_4^+ formed in the tubular lumen cannot easily cross cell membranes and thus is trapped in the glomerular filtrate and excreted with anions such as phosphate, chloride, or sulfate. In normal individuals, NH_4^+ production in the tubular lumen accounts for the excretion of ~60% (30–60 mmol) of the hydrogen ions associated with nonvolatile acids.

The amount of H^+ excreted bound to NH_3 can be measured as NH_4^+. The H^+ required for NH_4^+ formation may be present in the glomerular filtrate or may be generated within the tubular cell through the synthesis of carbonic acid from CO_2, catalyzed by carbonic anhydrase (Figure 10-10,6). These hydrogen ions are secreted into the tubular lumen through the Na^+-H^+ exchange (Figure 10-10,4). In systemic acidosis, glutamine is released from muscle; glutamine uptake by the liver is decreased and glutamine release is increased. More glutamine is therefore present in the circulation, and renal production of ammonia increases. In fact, under these circumstances, NH_4^+ excretion accounts by far for the greatest net excretion of H^+ by the kidneys. However, the maximum rate of glutamine release and therefore of NH_3 production is not achieved until acidosis has persisted for 3 d. At the acid pH of urine, the equilibrium between NH_4^+ and NH_3 shifts to the left and reaches a ratio of 10 000 to 1 at a urine pH of 5.3. However, in patients with chronic renal insufficiency, the kidneys are unable to generate sufficient NH_3 to buffer the nonvolatile acids produced, and the defect contributes significantly to the acidosis in such patients. Potassium depletion increases ammonia genesis; K^+ overload decreases it. This reciprocal relationship between availability of K^+ and NH_3 production is of great clinical importance in those acid-base disturbances that are associated with mineralocorticoid deficiency or excess and in the pathogenesis of some forms of renal tubular acidosis. (See Chapter 11.)

Excretion of H^+ as $H_2PO_4^-$

H^+ secreted into the tubular lumen by the Na^+-H^+ exchange mechanism may react with HPO_4^{2-} to form $H_2PO_4^-$. (See Figure 10-10,1.) This process depends on the amount of phosphate filtered by the glomeruli and the pH of urine. Under normal physiological conditions, ~30 mmol of H^+ is excreted per day as $H_2PO_4^-$, and this amount accounts for ~90% of the titratable acidity of urine. In case of high-protein intake, phosphate production and filtration may be increased; after low-protein intake it may be decreased. Acidemia increases phosphate excretion and thus provides additional buffer for reaction with H^+. A decrease in the glomerular filtration rate, as observed in renal disease, may result in a decrease of $H_2PO_4^-$ excretion.

The secretion of H^+ and the subsequent reaction with HPO_4^{2-} allow the removal of one H^+ without any significant decrease in urinary pH. As a result, the pH gradient is not greatly affected, and more H^+ can be secreted into the tubules and more Na^+ reabsorbed and preserved. The HCO_3^- formed in the tubular cells as the result of H^+ secretion reclaims the HCO_3^- lost in buffering the body burden of ingested or produced acid.

**Plasma and
Interstitial Fluid**   **Tubular Cell**   **Glomerular Filtrate**

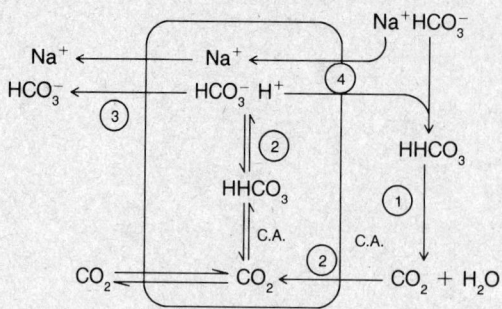

Figure 10-11. 1, Formation of CO_2 from bicarbonate in the tubular fluid; 2, formation of H^+ and HCO_3^- from CO_2 in the tubular cell; 3, new generation of HCO_3^-; and 4, NA^+–H^+ exchange.

Reclamation of Filtered Bicarbonate

The unmodified glomerular filtrate has the same concentration of HCO_3^- as does plasma; however, with increasing acidification of the proximal tubular urine, the urine pCO_2 increases and the HCO_3^- concentration decreases. It is believed that these changes are triggered by the excretion of H^+ by the Na^+-H^+ exchange mechanism, which results in a decrease in urinary pH. The H^+ thus excreted reacts with HCO_3^- (catalyzed by carbonic anhydrase, C.A., in the brush border of the proximal tubular cells) to form H_2CO_3 and subsequently CO_2 and H_2O (Figure 10-11,1).

$$HCO_3^- + H^+ \rightleftharpoons HHCO_3 \rightleftharpoons CO_2 + H_2O$$

The increase in urinary pCO_2 causes carbon dioxide to diffuse across the tubular wall into the tubular cell, where it reacts with H_2O in the presence of carbonic anhydrase in the tubular cells to form H_2CO_3 and subsequently H^+ and HCO_3^-. (See Figure 10-11,2.) Thus, reclamation of bicarbonate is in fact diffusion of CO_2 into tubular cells and its subsequent conversion to HCO_3^-. (The term "bicarbonate reabsorption" that is often used for this process is therefore inappropriate.) The increase in HCO_3^- helps to reclaim the filtered HCO_3^- and therefore to maintain or normal pH in restore a normal pH in the general circulation. Normally, nearly 90% of the HCO_3^- (or about 4500 mmol/d) is reclaimed in the proximal tubule and the extent of HCO_3^- reclamation parallels the Na^+ reabsorption. Thus, for each H^+ secreted into the tubular fluid, one each Na^+ and HCO_3^- enters the tubular cell and returns to the general circulation. In addition, about 50 mmol HCO_3^- per day is newly formed, mainly in the distal tubules (Figure 10-11).

The reclamation of HCO_3^- in the proximal tubule is a high-capacity, low-gradient process. Probably due to the fast dehydration of carbonic acid, the pH of tubular fluid at the end of the proximal tubule is only about pH 6.7. When plasma HCO_3^- concentration increases above 26 mmol/L, the capacity of the proximal and distal tubules to reclaim HCO_3^- is exceeded and HCO_3^- is excreted in the urine. The process of bicarbonate reclamation is enhanced in acidosis (and is decreased in alkalosis), most likely as a result of increased Na^+-H^+ exchange. In this way the kidneys in acidosis or alkalosis support the other compensatory mechanisms to restore the normal $cHCO_3^-/cdCO_2$ ratio. (See discussions on compensatory mechanisms.)

CONDITIONS ASSOCIATED WITH ABNORMAL ACID-BASE STATUS AND ABNORMAL ANION-CATION COMPOSITION OF THE BLOOD[2,10,11]

Many pathological conditions are accompanied by or caused by disturbances of the acid-base balance and electrolyte composition of the blood. These changes are usually reflected in the acid-base pattern and anion-cation pattern of extracellular fluid. Because blood is the only sample of extracellular fluid that can be readily obtained, acid-base status is evaluated from measurements of blood pH and pCO_2 and from the determination of $cHCO_3^-$ (for blood by calculation or for plasma by measurement of $ctCO_2^-$). It is important to note, however, that results obtained on blood or plasma may not always be a true indication of acid-base status of *intra*cellular fluid.

Abnormalities of acid-base balance of the blood are usually accompanied by characteristic changes in electrolyte concentrations in the plasma, especially in the case of the metabolic acid-base disorders. Hydrogen ions cannot accumulate without concomitant accumulation of anions such as Cl^-, SO_4^{2-}, or lactate, without exchange for cations such as K^+ or Na^+. For this reason, electrolyte composition of blood serum or plasma is often determined along with measurements of blood gases and pH and the acid-base parameters derived from them. Additional reasons for electrolyte imbalances in plasma and, in some instances, the mechanisms by which such imbalances develop will be noted below in the discussion of acid-base disturbances. Acid-base disturbances are classified in one of the following four groups: metabolic acidosis, metabolic alkalosis, respiratory acidosis, or respiratory alkalosis.

Metabolic Acidosis (Primary Bicarbonate Deficit)

Metabolic acidosis is usually caused by one or more of the following processes:

1. Production of organic acids that exceeds the rate of elimination (e.g., production of acetoacetic acid and β-hydroxybutyric acid in diabetic acidosis and of lactic acid in lactic acid acidosis). Acidosis may be accompanied by loss of cations that are excreted with the anions, as was explained earlier.

2. Reduced excretion of acids (e.g., renal failure, tubular acidosis).

3. Excessive loss of bicarbonate (base) because of excessive loss of duodenal fluid (as in diarrhea). Plasma $cHCO_3^-$ falls and is replaced by inorganic anions such as chloride, phosphate, sulfate, or organic anions.

When any of these conditions exists, the ratio of $cHCO_3^-/cdCO_2$ is decreased because of the primary decrease in bicarbonate, and compensatory mechanisms act to restore the normal pH. If this restoration is complete (or nearly complete so that the pH remains between 7.35 and 7.45), the condition is called *fully compensated metabolic acidosis*. If, in spite of the compensatory mechanisms, the pH stays below 7.35, the condition is called *partially compensated metabolic acidosis*.

Classification and Selected Examples of Metabolic Acidoses

The various forms of metabolic acidosis can be classified as normal anion gap acidosis, which is characterized by hyperchloremia, and increased anion gap acidosis, in which the chloride is normal (normochloremic metabolic acidosis) or at times low.

Normal Anion Gap Acidosis

Normal anion gap acidosis can be divided into hypokalemic and normokalemic acidosis. *Hypokalemic acidosis* includes renal tubular acidosis caused either by proximal or distal tubular dysfunction or by buffer deficiency; acidosis caused by gastrointestinal loss of Na^+, K^+, and HCO_3^-; acidosis due to diuretics that are carbonic anhydrase inhibitors, such as acetazolamide and mafenide; and acidosis caused by ureteral diversions such as ureterosigmoidostomy, or obstructed ileal bladder or ureter. *Normokalemic acidosis* is seen in early renal failure; in hydronephrosis; after treatment with NH_4Cl, arginine·HCl, or similar therapeutic agents; and in hypoaldosteronism or after administration of aldosterone antagonists such as spironolactone, triamterene, or amiloride. In the latter cases a hyperkalemia is often present.

Diarrhea may cause acidosis as a result of loss of Na^+, K^+, and HCO_3^-; the resulting hyperchloremia is due to the replacement of lost bicarbonate with Cl^- to maintain electrical balance.

Renal tubular acidosis is predominantly characterized by loss of bicarbonate due to decreased tubular secretion of H^+. (See Chapter 11.) Lost HCO_3^- is replaced by chloride, resulting in hyperchloremic acidosis.

Carbonic anhydrase inhibitors cause wasting of Na^+, K^+, and HCO_3^-. Posthypocapnic acidosis may develop, but it rarely lasts more than 6–12 h and rarely has severe consequences.

Hyperkalemic acidosis inhibits renal ammonia formation and therefore elimination of H^+. If associated with increased extracellular fluid (ECF) volume, HCO_3^- reclamation in the tubules may be depressed. Failure of the kidneys to synthesize renin, failure of the adrenal cortex to secrete aldosterone, and renal tubular resistance to aldosterone are the most common causes of this type of acidosis.

Increased Anion Gap Acidoses

Renal failure may result in loss of functional renal tubular mass and consequently in a decrease in ammonia formation, Na^+-H^+ exchange, and decreased glomerular filtration rate. All result in decreased acid excretion. (See Chapter 11.) Acidosis usually develops if GFR falls below 20 mL/min and when urea nitrogen increases to >40 mg/dL and creatinine to >4 mg/dL.

Ketoacidosis. The pathogenesis of this disorder has been discussed in Chapter 6. Recall that the accumulation of ketone bodies causes a decrease in HCO_3^-, a normal or low serum chloride, and a high anion gap.

Salicylate intoxication generally occurs with blood salicylate concentrations above 30 mg/dL. Initially the drug stimulates the respiratory center to increase rate and depth of respiration, resulting in a low pCO_2, low HCO_3^-, and respiratory alkalosis. Salicylates, however, also alter peripheral metabolism, causing production of various organic acids, without dominance of any specific acid. The processes eventually lead to a metabolic acidosis with a high anion gap. In adults, mixed respiratory alkalosis and metabolic acidosis are more common, while in children metabolic acidosis predominates.

Lactic acidosis. Lactic acid, present in *blood* entirely as lactate ion ($pK_1 = 3.86$), is an intermediate of carbohydrate metabolism and is derived mainly from muscle cells and erythrocytes. (See also Chapter 6.) It is normally metabolized by the liver. The blood lactate concentration is, therefore, affected by the rate of production as well as the rate of metabolism. Any increase in the concentration of lactate and the associated H^+ to >2 mmol/L after resting leads to a condition called *lactic acidosis.* During exercise, lactate levels may increase significantly, from an average normal level of ∼0.9 mmol/L to ∼12 mmol/L. However, pyruvate levels increase under these conditions as well, and the ratio of lactate/pyruvate remains ∼6/1 or 7/1. (Higher normal ratios of 9/1 or even higher have been reported but are probably due to variations in technique and sample collection.)

Lactic acidosis caused by severe tissue hypoxia is called *type A lactic acidosis.* It is seen in severe anemia, shock, cardiac decompensation, and pulmonary insufficiency. Severe oxygen deprivation of tissues blocks aerobic oxidation of pyruvic acid in the tricarboxylic acid cycle and results in the reduction of pyruvate to lactate. Lactic acidosis is associated with a significant increase in the lactate/pyruvate ratio in blood. Such extreme findings signal deterioration of the cellular oxidative process and are associated with marked hyperpnea, weakness, fatigue, stupor, and finally coma. Conditions at these stages are frequently irreversible, even when treatment for acidosis and hypoxia is instituted; examples are irreversible stage of shock, diabetic coma without ketosis, and a variety of illnesses in the terminal stage.

Type B lactic acidosis is caused by drugs and toxins, such as ethanol, methanol, biguanides, and streptozotocin; by acquired and hereditary defects in enzymes involved in gluconeogenesis; by disorders such as severe acidosis, uremia, liver failure, tumors, and seizures; and during anesthesia.

Alcohol taken in excess tends to prevent gluconeogenesis from lactate in the liver because oxidation of ethanol to acetaldehyde competes for the NAD^+ that is necessary for the conversion of lactate to pyruvate. Fructose-1,6-diphosphatase deficiency is an example of congenital absence of enzymes involved in gluconeogenesis. Infants with this disorder may have blood lactate concentrations >20 mmol/L. Severe acidosis, such as diabetic acidosis, may suppress lactate utilization and cause a shift in the lactate-pyruvate equilibrium with the accumulation of H^+:

$$\text{Lactate} + NAD^+ \rightleftharpoons \text{Pyruvate} + NADH + H^+$$

This shift may in part be responsible for the lactic acidosis seen in diabetics.

Hyperventilation in lactic acidosis is more intense than in other forms of metabolic acidosis. It is believed that this is due to the participation of the respiratory center in lactic acid production and the resulting greater acidification of the respiratory center. The pCO_2 as the result of the more intense hyperventilation is $>1.5(HCO_3^-) + 8$.

Simultaneous determination of lactate in *arterial* and *coronary sinus blood* after fast arterial pacing is sometimes requested to facilitate the diagnosis of doubtful angina pectoris. In such cases, high accuracy and precision in the analytic method are required, since differences in the lactate concentration in the two samples may be very small.

Lactate in *spinal fluid* normally parallels blood levels. In case of biochemical alterations in the central nervous system, however, CSF (Csf) lactate values change independently of blood values. Increased CSF levels are seen in intracranial hemorrhage, bacterial meningitis, epilepsy, and other CNS disorders.

Toxins and other chemical agents can be the cause of metabolic acidosis, as seen in the following examples. *Methanol* is metabolized to formaldehyde and formic acid. Accumulation of this acid leads to metabolic acidosis with a high anion gap and to clinical symptoms of optic papillitis, retinal edema, blindness due to optic nerve damage and atrophy, neurologic defects, and ultimately coma. Methanol itself is not toxic. High serum and urine amylase is often associated with methanol intoxication. *Ethylene glycol,* if ingested, is metabolized to glycolic and oxalic as well as other acidic metabolites. Its metabolism leads to an acidosis with a high anion gap. Accumulation of toxic metabolites may lead to lactic acid production and further contribute to the acidosis. Precipitation of calcium oxalate and hippurate crystals in the urinary tract may lead to acute renal failure. Clinically, patients develop a variety of neurologic symptoms that may lead to coma. Some patients may develop bronchial pneumonia, pulmonary edema, congestive heart failure, hypertension, or cardiopulmonary arrest, or a combination of these symptoms. The minimal lethal dose of ethylene glycol is $\sim$100 mL. *Paraldehyde* toxicity may develop after chronic paraldehyde ingestion. The pathogenesis is ill defined, but the acidosis may actually be a ketosis with (nitroprusside negative) β-hydroxybutyric acid as the main acidic product. Patients with paraldehyde toxicity have a pungent odor.

Compensatory Mechanisms in Metabolic Acidosis

Buffer systems. The buffer systems of the blood, mainly the bicarbonate/carbonic acid buffer, minimize changes in pH. The bicarbonate concentration decreases to give a ratio of $c\text{HCO}_3^-$/$cd\text{CO}_2$ of < 20/1. The respiratory mechanism responds to correct the ratio with increased rate and depth of respiration to eliminate HHCO_3 as CO_2.

Respiratory mechanism. The decrease in pH in metabolic acidosis stimulates the respiratory mechanism and produces hyperventilation (Kussmaul respiration), which results in the elimination of carbonic acid as CO_2, a decrease in $p\text{CO}_2$ (hypocapnia), and consequently a decrease in $cd\text{CO}_2$. There is also a decrease in $c\text{HCO}_3^-$ that is smaller than that of $cd\text{CO}_2$. For example, the ratio of $c\text{HCO}_3^-$/$cd\text{CO}_2$ (e.g., before compensation 16/1.28 or 12.5/1 $\Rightarrow$ pH 7.2) changes to 14.5/0.9 or 16.1 $\Rightarrow$ pH 7.30 after compensation. The $cd\text{CO}_2$ diminishes and the ratio of $c\text{HCO}_3^-$/$cd\text{CO}_2$ approaches its normal value of 20/1 (e.g., HCO_3^-/$d\text{CO}_2$ before compensation = 15/1.2 or 12.5/1 $\Rightarrow$ pH 7.2; after compensation = 14.5/0.86 or 16/1 $\Rightarrow$ pH 7.3).

Renal mechanism. The kidneys attempt to restore the normal pH by increased excretion of acid and preservation of base (increased rate of Na^+-H^+ exchange, increased ammonia formation, and increased reabsorption of bicarbonate). The total amount of H^+ excreted may be as much as 500 mmol/d. As a result, $c\text{HCO}_3^-$ will increase, e.g., to 22/1.1 or 20/1 $\Rightarrow$ pH 7.40. This is a fully compensated metabolic acidosis—fully compensated because the pH has returned to normal; however, acidosis still exists because a primary $c\text{HCO}_3^-$ deficit persists.

Laboratory Findings in Metabolic Acidosis

Plasma bicarbonate, total carbon dioxide content, and blood $p\text{CO}_2$ are decreased in metabolic acidosis. In uncompensated cases, blood pH is decreased to a degree depending on the ratio of $c\text{HCO}_3^-$/$cd\text{CO}_2$. Clinicians sometimes use the bicarbonate concentration to arrive at empirical estimates of pH and $p\text{CO}_2$. For pH, they add 15 to the determined bicarbonate concentration to obtain an estimate of the decimal digits of pH. For example, for a patient who has been acidotic for at least 12–24 h and has a bicarbonate of 10 mmol/L, 10 + 15 gives an estimated pH of 7.25. For estimating $p\text{CO}_2$, the formula

$$p\text{CO}_2, \pm 2 = 1.5(c\text{HCO}_3^-) + 8$$

is used.[1] For our example the $p\text{CO}_2$ estimate would be 1.5(10) + 8, or 23 $\pm$ 2. (See also Table 10-10 and Figure 10-12.) Use of this equation may have important clinical implications, since it

Table 10-10. CLASSIFICATION AND CHARACTERISTICS OF SIMPLE ACID-BASE DISORDERS

	Primary Change	Compensatory Response	Expected Compensation
METABOLIC			
Acidosis	$\downarrow\downarrow\downarrow cHCO_3^-$	$\downarrow\downarrow pCO_2$	$pCO_2 = 1.5\ (cHCO_3^-) + 8 \pm 2$ pCO_2 falls by 1–1.3 mm Hg for each mmol/L fall in $cHCO_3^-$ Last 2 digits of pH $= pCO_2$ (e.g., if $pCO_2 = 28$, pH $= 7.28$) $cHCO_3^- + 15 =$ last 2 digits of pH ($cHCO_3^- = 15$, pH $= 7.30$)
Alkalosis	$\uparrow\uparrow\uparrow cHCO_3^-$	$\uparrow\uparrow pCO_2$	pCO_2 increases 6 mm Hg for each 10 mmol/L rise in $cHCO_3^-$ $cHCO_3^- + 15 =$ last 2 digits of pH ($cHCO_3^- = 35$, pH $= 7.50$)
RESPIRATORY			
Acidosis			
Acute	$\uparrow\uparrow\uparrow pCO_2$	$\uparrow cHCO_3^-$	$cHCO_3^-$ increases by 1 mmol/L for each 10 mm Hg rise in pCO_2
Chronic	$\uparrow\uparrow\uparrow pCO_2$	$\uparrow\uparrow cHCO_3^-$	$cHCO_3^-$ increases by 3.5 mmol/L for each 10 mm Hg rise in pCO_2
Alkalosis			
Acute	$\downarrow\downarrow\downarrow pCO_2$	$\downarrow cHCO_3^-$	$cHCO_3^-$ falls by 2 mmol/L for each 10 mm Hg fall in pCO_2
Chronic	$\downarrow\downarrow\downarrow pCO_2$	$\downarrow\downarrow cHCO_3^-$	$cHCO_3^-$ falls by 5 mmol/L for each 10 mm Hg fall in pCO_2

Modified from Narins, R. G., and Gardner, L. B.: Simple acid-base disturbances. Med. Clin. North Am., 65:321–346, 1981.

indicates whether a given pCO_2 value is appropriate for a given degree of metabolic acidosis or if one deals with a mixed acid-base disorder. If, for example, a respiratory acidosis is superimposed on a pre-existing metabolic acidosis, the pCO_2 would be higher than expected.

Electrolytes are also altered in various ways, depending on the cause of metabolic acidosis. In diabetic ketoacidosis the increase in the fraction of organic acids caused by increased ketone body production is reflected by a decrease in plasma bicarbonate and sometimes in chloride. Plasma Na^+ and K^+ are also decreased because of the associated polyuria and coexcretion of these cations with acetoacetate and β-hydroxybutyrate. (See Figure 10-13,B.) Furthermore, there is a dilutional effect as a result of the osmotically induced increase in the vascular volume. When glucose concentration decreases as a result of insulin treatment, water leaves the vascular compartment and Na^+ concentration increases. Serum potassium levels, however, may be normal or even high, despite severe total body depletion of K^+; the serum level represents a balance struck between the amount of K^+ lost in the urine, the amount of K^+ shifted from cells into extracellular fluid, and the degree of dehydration. In renal failure, organic acids, phosphate, and sulfate are increased because of retention (Figure 10-13,C). Chronic metabolic acidosis enhances mobilization of calcium from bone; the decrease in plasma pH increases dissociation of plasma protein–bound calcium so that more Ca^{2+} is filtered through the glomerulus and less is reabsorbed in the tubule. (See also Chapter 11.) Other types of electrolyte changes are described in the discussions of specific acidoses and in the section on renal tubular acidosis in Chapter 11.

So long as the renal compensating mechanisms are functioning, urinary acidity and urinary ammonia are increased.

Metabolic Alkalosis (Primary Bicarbonate Excess)

Metabolic alkalosis is most frequently caused by one of the following causes of bicarbonate excess:

1. Administration of excess alkali, such as $NaHCO_3$, citrate (in transfused blood), or some antacids.

2. Excessive loss of hydrochloric acid from the stomach, as seen, for example, after prolonged vomiting, in pyloric or high intestinal obstruction, and after gastric suction.

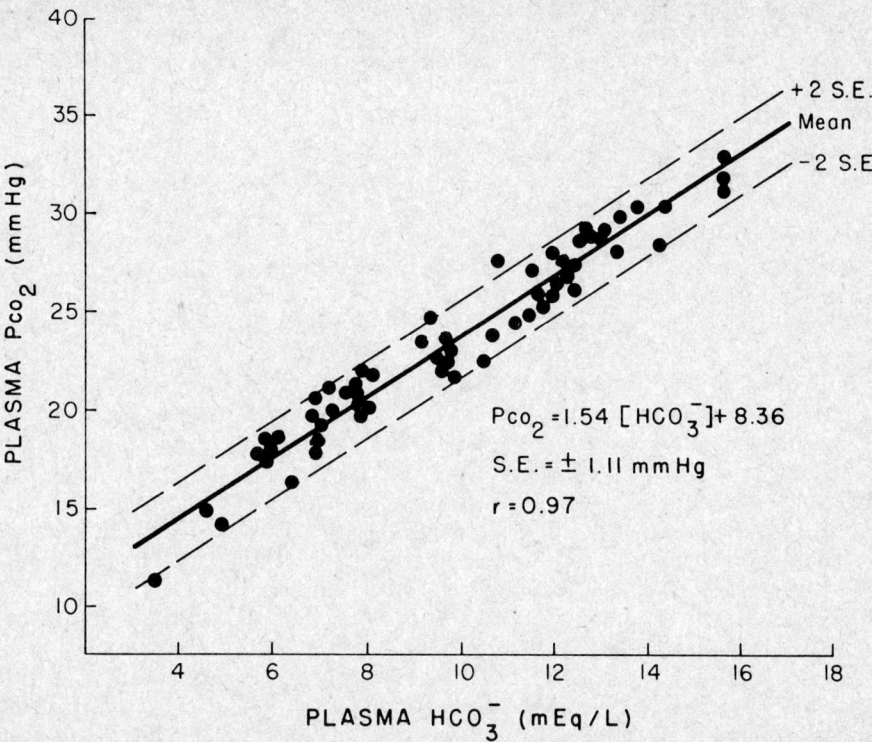

Figure 10-12. Relationship of pCO_2 to serum bicarbonate in chronic metabolic acidosis. (From Albert, M. S., et al.: Quantitative displacement of acid-base equilibrium in metabolic acidosis. Ann. Intern. Med., *66*:312, 1968.)

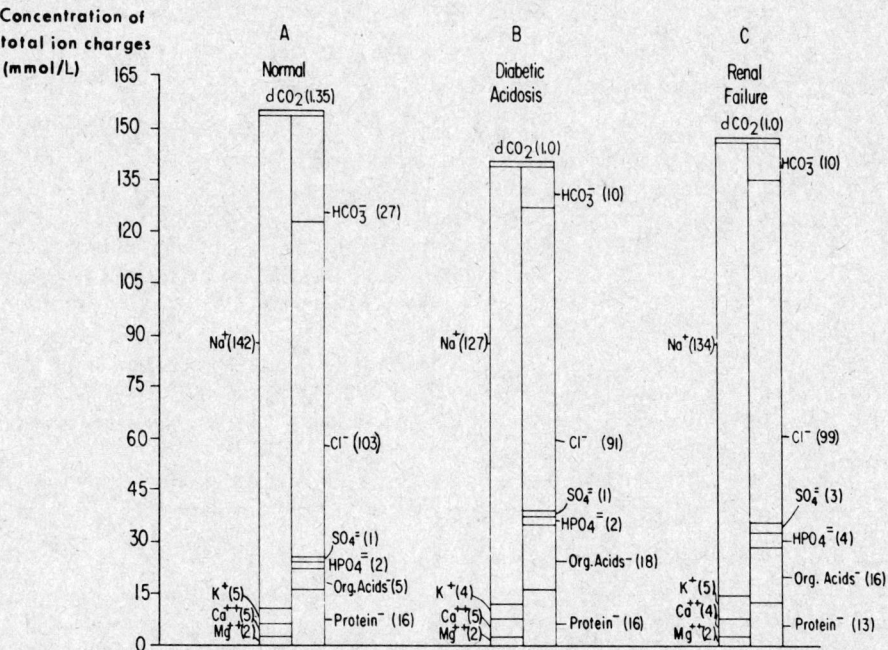

Figure 10-13. *A,* Gamblegram illustrating normal electrolyte composition of plasma. *B,* Example of anion-cation pattern as may be found in diabetic acidosis. Na^+ and Cl^- are decreased because of polyuria and osmotically induced hypervolemia due to increased plasma glucose. The organic acid fraction is increased because of excessive formation of ketone bodies. The ratio of HCO_3^-/dCO_2 in this example is 10/1; thus, the plasma pH must be 7.10. *C,* Example of anion-cation pattern as may be seen in renal failure. Organic acids as well as phosphates and sulfates are retained due to decreased renal function. (From Tietz, N. W.: The Chicago Medical School Quarterly, *22*:156, 1962.)

3. Potassium depletion, as seen, for example, in Cushing's syndrome, after administration of ACTH or adrenocortical hormones, in hyperaldosteronism, after intake of licorice and carbenoxolone, or because of low K^+ intake.

In these conditions, renal Na^+ reabsorption and H^+ secretion are enhanced secondary to K^+ depletion. Decreased K^+ also stimulates NH_3 production and therefore H^+ excretion as NH_4^+. Enhanced secretion of H^+ increases HCO_3^- reclamation. Potassium plays a less critical role in non-mineralocorticoid–induced alkalosis. Hypokalemia suppresses aldosterone secretion and thus removes its stimulatory effect on H^+ secretion. At the same time, hypokalemia stimulates ammonia genesis and chloruresis, both of which are alkalinizing. The various opposing forces tend to minimize the effect of K^+.

4. Renal bicarbonate retention, as seen in hypovolemic-induced aldosterone stimulation, results in increased reabsorption of Na^+ together with HCO_3^- or Cl^-.

5. Prolonged administration of certain diuretics. Diuretics causing alkalosis include those acting on the proximal tubule (organomercurials); those acting on the proximal convoluted tubule as well as on the ascending portion of the distal convoluted tubule (carbonic anhydrase inhibitors such as sulfanilamide, acetazolamide [Diamox], and dichlorphenamide); and those acting on the ascending limb of the loop of Henle (furosemide [Lasix], ethacrynic acid). All block sodium absorption. The resulting increase in Na^+ concentration in the tubular fluid reaching the distal convoluted tubule stimulates aldosterone secretion and therefore secretion and loss of K^+. The loss of K^+ with furosemide is much less than with thiazides.

6. After administration of laxatives, carbenoxolone, and potassium-poor intravenous fluids.

In these conditions, the ratio of HCO_3^-/dCO_2 becomes $> 20/1$ because of a primary increase in bicarbonate, e.g., $48/1.5 = 32/1$ (pH of 7.6). The compensatory mechanisms of the body then act to restore the normal plasma pH. If compensation is complete, we have a state of fully compensated metabolic alkalosis with a pH value within the normal range. With progression of the disturbance, the compensatory mechanisms are not effective enough, and the pH will increase.

If the increase in pH is great enough, increased neuromuscular activity may be seen, and above pH 7.55 tetany may develop, even in the presence of a normal serum total calcium concentration. The cause is a decreased concentration of ionized calcium due to increased binding of calcium ions by protein and other anions.

Compensatory Mechanisms in Metabolic Alkalosis

Buffer systems. As a result of loss of acid (e.g., HCl), excess base reacts with $HHCO_3$ of the HCO_3^- buffer system to form an increased amount of HCO_3^-, thereby minimizing pH change.

Respiratory mechanism. The increase in pH depresses the respiratory center, causing a retention of carbon dioxide (hypercapnia), which in turn causes an increase in $cHHCO_3$ and $cdCO_2$. Thus, the ratio of $cHCO_3^-/cdCO_2$, which was originally increased (see earlier section), approaches its normal value although the actual levels of both $cHCO_3^-$ and $cdCO_2$ remain increased (e.g., the ratio of $cHCO_3^-/cdCO_2$ before compensation was $48/1.5$ or $32/1 \Rightarrow$ pH 7.6; the ratio after partial compensation is $49.5/1.86$ or $26.7/1 \Rightarrow$ pH 7.53 and the total CO_2 is $49.5 + 1.86 = 51.4$ mmol/L). The respiratory response to metabolic alkalosis is erratic, and increases in pCO_2 are variable. Therefore, the empirical formula $pCO_2 = 0.9 (cHCO_3^-) + 15.6$ is of only limited value and may often underestimate the patient's pCO_2. After respiratory compensation, the formula $pCO_2 = 0.9 (HCO_3^-) + 9$ may be used.

Renal mechanism. The kidneys respond to the state of alkalosis by decreased Na^+-H^+ exchange, decreased formation of ammonia, and decreased reabsorption of bicarbonate.

Laboratory Findings in Metabolic Alkalosis

Blood plasma values for $cHCO_3^-$, $cdCO_2$, and pCO_2 and therefore the plasma total CO_2 concentration are increased and the ratio of $cHCO_3^-/cdCO_2$ is high. In uncomplicated metabolic alkalosis, the pCO_2 increases by ~ 6 mm Hg for each 10 mmol/L rise in $cHCO_3^-$. A higher than expected pCO_2 may indicate superimposed respiratory acidosis. The pH in uncompensated metabolic acidosis is increased. The extent of increase can be estimated by adding 15 to the $cHCO_3^-$ to give the last two digits of the pH. If the $cHCO_3^-$ is 35 mmol/L, the estimated pH $= 35 + 15 = 7.50$. Other electrolyte ions vary, depending on the condition. In cases of prolonged vomiting, Cl^- and possibly K^+ levels are low because of the loss of these ions through the vomitus (Figure 10-14). Protein values may be increased owing to dehydration; if food intake

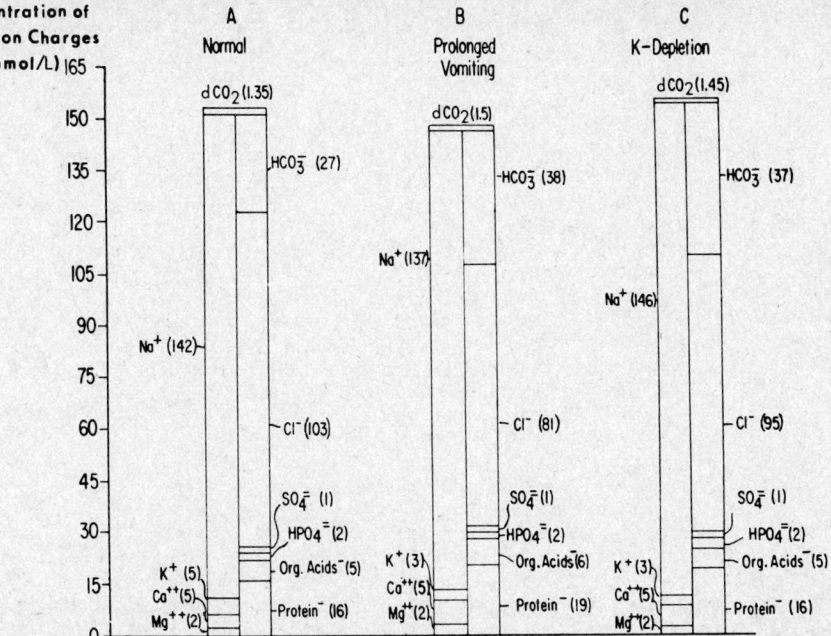

Figure 10-14. *A,* Electrolyte composition of normal plasma. *B,* Example of plasma electrolyte composition after prolonged vomiting, showing the decrease in K^+ and Cl^- and the increase in protein$^-$, organic acids$^-$, HCO_3^-, and dCO_2. The ratio of the last two is $38/1.5 = 25.3$. The pH is therefore 7.50. *C,* Example of typical plasma electrolyte composition in a patient with intracellular K^+ depletion. There is a decrease in K^+ and Cl^- and an increase in HCO_3^- and dCO_2, resulting in a ratio of $37/1.45 = 25.4$ (pH $= 7.50$). (From Tietz, N. W.: The Chicago Medical Quarterly, *22*:156, 1962.)

is inadequate, formation of ketone bodies may increase the organic acid fraction. In cases of excessive administration of $NaHCO_3$, Na^+ levels are increased. In K^+ depletion, decreased concentrations of Cl^- are common (Figure 10-14,*C*). Serum K^+ concentrations are generally but not necessarily low, since a total intracellular K^+ loss of about 100–200 mmol is required before serum K^+ concentrations decrease below normal.

Urinary pH values are usually increased because of the decreased excretion of acid and increased excretion of bicarbonate. Urinary ammonia values are decreased because of decreased formation of ammonia in the tubules. In K^+ depletion, H^+ is preferentially exchanged for Na^+ and the pH of the urine may be low in spite of a metabolic alkalosis; this is called paradoxical aciduria.

Respiratory Acidosis

Any condition that decreases elimination of carbon dioxide through the lungs results in an increase in pCO_2 (hypercapnia) and a primary dCO_2 excess (respiratory acidosis). Causes of decreased CO_2 elimination (see Table 10-11) can be classified as acute or chronic, or conditions may be separated into those caused by factors that directly depress the respiratory center (such as centrally acting drugs, CNS trauma, and infections) and those that affect the respiratory apparatus or cause mechanical obstruction of the airways. Cardiac disease may also cause respiratory acidosis, although generally it causes a slight respiratory alkalosis because the hypoxemia stimulates hyperventilation. Rebreathing, or breathing of air high in CO_2 content, may also cause high pCO_2. Increase in pCO_2 results in an increase of $cdCO_2$, which in turn causes a decrease in the $cHCO_3^-/cdCO_2$ ratio (e.g., the ratio may be $28/1.7 \doteq 16$, resulting in a pH of ~7.30).

Compensatory Mechanisms in Respiratory Acidosis

Buffer system. Carbonic acid present in blood in excess is to a great extent buffered by the hemoglobin and protein buffer systems. The buffering of CO_2 causes a slight rise in $cHCO_3^-(P)$.

Respiratory mechanism. The increase in pCO_2 stimulates the respiratory center and results in increased pulmonary rate and depth of respiration, provided that the primary defect is not in

Table 10-11. CONDITIONS LEADING TO RESPIRATORY ACIDOSIS

Factors that directly depress the respiratory center
  Drugs such as narcotics and barbiturates
  Central nervous system trauma, tumors, and degenerative disorders
  Infections of the CNS such as encephalitis and meningitis
  Comatose states such as CVA due to intracranial hemorrhage
  Primary central hypoventilation
Conditions that affect the respiratory apparatus
  Chronic obstructive pulmonary disease, COPD (most common cause)
  Pulmonary fibrosis
  Status asthmaticus (severe)
  Diseases of the upper airways such as laryngospasm or tumor
  Pulmonary infections (severe)
  Impaired lung motion due to pleural effusion or pneumothorax
  Adult respiratory distress syndrome
  Chest wall diseases and chest wall deformities
  Neurologic disorders affecting the muscles of respiration
Others
  Abdominal distention, as in peritonitis and ascites
  Extreme obesity (pickwickian syndrome)
  Sleep disorders such as sleep apnea

the respiratory center. The elimination of carbon dioxide through the lungs results in a decrease in $c\mathrm{d}CO_2$, the ratio of $c\mathrm{HCO_3^-}/c\mathrm{d}CO_2$ approaches the normal value, and pH changes toward normal.

The respiratory response is proportional to the degree of acidosis and is therefore fairly predictable. The expected pCO_2 can be calculated with the empirical equations noted earlier:

$$pCO_2, \text{mm Hg}, \pm 2 = 1.5\,(\text{measured } c\mathrm{HCO_3^-}, \text{mmol/L}) + 8$$

Renal mechanism. The kidneys respond to respiratory acidosis in the same way that they do to metabolic acidosis, namely, with increased Na^+-H^+ exchange, increased ammonia formation, and increased reclamation of bicarbonate. In a chronic respiratory acidosis in steady state, the normal renal compensation returns the plasma pH about halfway toward normal as compared with the acute (uncompensated) situation. A so-called fully compensated chronic respiratory acidosis, in which the pCO_2 is high but the pH is within the reference range, is actually not the result of a true compensation but is generally the result of a chronic respiratory acidosis with a superimposed metabolic alkalosis arising from some cause, such as prolonged administration of diuretics. Renal compensation is not effective before 6–12 h and is not optimal before 2–3 d.

Laboratory Findings in Respiratory Acidosis

Plasma $c\mathrm{d}CO_2$, pCO_2, $c\mathrm{HCO_3^-}$ and, therefore, $c\mathrm{t}CO_2$ are elevated. Due to an increase in $c\mathrm{d}CO_2$, the ratio of $c\mathrm{HCO_3^-}/c\mathrm{d}CO_2$ is decreased, resulting in a decreased pH. In the acute phase, $c\mathrm{HCO_3^-}$ will increase about 1 mmol/L for each 10 mm Hg rise in pCO_2. If respiratory acidosis persists, the change will be 3.5 mmol/L, mainly as a result of renal compensation. The pH, responding to an increase in pCO_2, changes in the *acute* phase by $\sim$0.10 pH unit, and in chronic conditions by slightly less than 0.05 pH unit for every 15 mm Hg increase in pCO_2. For example, if the pCO_2 increases by 30 mm Hg, the pH in the acute phase changes to $\sim$7.20 and in the chronic condition to $\sim$7.31. The $c\mathrm{H}^+$ may be calculated by the following empirical formula:

$$\Delta\mathrm{H}^+ = 0.8\,(\Delta pCO_2)$$

or:

$$c\mathrm{H}^+ = 24\,\frac{pCO_2}{c\mathrm{HCO_3^-}}$$

If the $pCO_2 = 70$ mm Hg, and the $c\mathrm{HCO_3^-} = 27$ mmol/L, $c\mathrm{H}^+ = 24 \times \dfrac{70}{27} = 64$ nmol/L

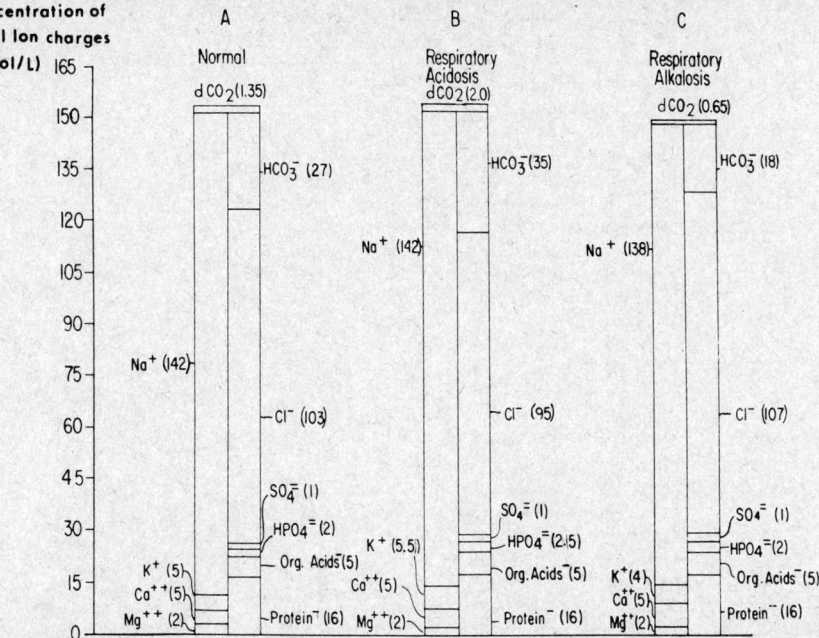

Figure 10-15. *A,* Electrolyte composition of normal plasma. *B,* Example of possible electrolyte pattern in a patient with respiratory acidosis. Note the increase in HCO_3^- and dCO_2. The increase in the latter fraction is more pronounced than that of HCO_3^-. The ratio therefore is decreased, and the chloride fraction shows a decrease. *C,* Theoretical electrolyte pattern of patient in respiratory alkalosis. There is a decrease in the HCO_3^- and especially dCO_2 fraction. Therefore, the ratio of HCO_3^-/dCO_2 and the pH are increased. The Na^+ is at the lower limit of normal. (From Tietz, N. W.: The Chicago Medical School Quarterly, *22:*156, 1962.)

(pH $\simeq 7.2$). These approximations can yield important clinical information. Taking our example, an increase in $cHCO_3^-$ of 3.0 mmol/L and an increase in pCO_2 of 30 mm Hg when the pH is 7.20 suggest an acute respiratory acidosis, while a pH of 7.31 in the presence of a pCO_2 increase of 33 mm Hg suggests a chronic condition.

Uncomplicated acute respiratory acidosis will generally not result in increases of $cHCO_3^-$ > 32 mmol/L or cause decreases in pH below 7.20. Greater decreases are generally due to a mixed acidosis.

The plasma chloride decreases as plasma bicarbonate increases. Hyperkalemia may occur but is not as predictable as in some forms of metabolic acidosis. For every 0.1 unit decrease in pH, there is generally an inverse change of 0.6 mmol/L in K^+. This increase in K^+ is mainly due to the movement of K^+ from the cells into the plasma in exchange for the movement of H^+ into the cells. Urinary acidity and ammonia content are increased. (See Figure 10-15,*B*.)

Respiratory Alkalosis

A decrease in pCO_2 (hypocapnia) and the resulting primary deficit in dCO_2 (respiratory alkalosis) are caused by an increased rate or depth of respiration, or both. Excessive elimination of carbon dioxide reduces the pCO_2 and causes an increase in the $cHCO_3/cdCO_2$ ratio (due to decrease in $cdCO_2$). The latter shifts the normal equilibrium of the bicarbonate/carbonic acid buffer system, reducing the hydrogen ion concentration and increasing the pH. This shift also results in a decrease in $cHCO_3^-$, which somewhat ameliorates the change in pH. In respiratory alkalosis the compensation is very efficient and returns the pH almost to the original value.

Analogous to causes of respiratory acidosis, causes of respiratory alkalosis can be classified as those with a direct stimulatory effect on the respiratory center and those due to effects on the pulmonary mechanism. These and some additional causes are listed in Table 10-12.

Compensatory Mechanisms in Respiratory Alkalosis

The compensatory mechanisms respond to respiratory alkalosis in two stages. In the first stage, erythrocyte and tissue buffers provide H^+ ions that consume a small amount of HCO_3^-.

Table 10-12. FACTORS CAUSING RESPIRATORY
ALKALOSIS

Nonpulmonary stimulation of respiratory center
  Anxiety, hysteria
  Febrile states
  Gram-negative septicemia
  Metabolic encephalopathy, e.g., due to liver disease
  CNS infections such as meningitis, encephalitis
  Cerebrovascular acccidents
  Intracranial surgery
  Hypoxia, e.g., severe anemia, high altitudes (acute)
  Drugs and agents such as salicylates, catecholamines, and
    progesterone
  Pregnancy, mainly third trimester (↑ progesterone?)
  Hyperthyroidism
Pulmonary mechanism *
  Pneumonia
  Asthma
  Pulmonary emboli
  Interstitial lung disease
  Large right to left shunt (pO_2 < 50 mm Hg)
  Congestive heart failure
  Respiratory compensation after correction of metabolic acidosis
Others
  Ventilator-induced hyperventilation

*Some of these disorders, in severe stages, may be associated with respiratory acidosis if elimination of CO_2 is severely impaired.

The second stage becomes operational in prolonged respiratory alkalosis and depends on the renal compensation as outlined for metabolic alkalosis.

Laboratory Findings in Respiratory Alkalosis

In this condition, the $cdCO_2$, pCO_2, $cHCO_3^-$, and, thus, the total CO_2 concentration are decreased. The ratio of $cHCO_3^-/cdCO_2$ is increased and causes an increase in pH which, however, rarely exceeds 7.60. Greater increases are usually due to a mixed alkalosis.

During the *acute* phase, $cHCO_3^-$ falls by 2.0 mmol/L for each 10 mm Hg decrease in pCO_2; i.e., if the pCO_2 falls by 20 mm Hg, $cHCO_3^-$ decreases by 4 mmol/L. For the same fall in pCO_2, the cH^+ will decrease by 16 nmol/L.

$$\Delta H^+ = 0.8\,(\Delta pCO_2)$$

$$\Delta H^+ = 0.8 \times 20 = 16$$

If the original cH^+ was 40 nmol/L, it would now be 24 nmol/L (40 − 16 = 24), which corresponds to a pH of 7.61.

In *chronic* respiratory alkalosis, the $cHCO_3^-$ falls 5 mmol/L for every 10 mm Hg decrease in pCO_2 and the pH will return to near normal.

$$\Delta H^+ = 0.17\,(\Delta pCO_2)$$

$$\Delta H^+ = 0.17 \times 20 = 3.40$$

$$cH^+ = 40 - 3.40 = 36.6, \text{ which corresponds to a pH of } \sim7.43.$$

Individuals living at high altitudes chronically hyperventilate due to hypoxia and have pCO_2 values lower than those seen at sea level. In one study, the average pCO_2 was 31 mm Hg, but the blood pH was within reference range as a result of the compensatory mechanism.

Mild hypokalemia is caused by translocation of K^+ into cells in exchange for H^+ that move into plasma and to mild K^+ wasting by the kidneys. Chloride is retained to replace the decreasing HCO_3^-. There may be a small increase in the anion gap due to enhanced glycolysis and lactate formation. Lactate levels may increase 2–4 mmol/L, probably due to decrease in hepatic blood flow. In prolonged severe alkalosis there may be an increase of ketone bodies due to decreased carbohydrate utilization. Phosphate levels may be significantly decreased.

Estimates Derived from Electrolyte Measurements

Electrolyte analysis is usually requested as an electrolyte panel or profile, defined as concurrent determinations of Na^+, K^+, Cl^-, and total CO_2. Multichannel instruments perform such panels quickly and inexpensively, and the set of results returned is clinically more informative than each result by itself. Supplemental data derived from the set of results can have additional clinical value. Calculated anion gap and calculated osmolality are examples of such data. In calculations of this kind, the concentration of total CO_2 is taken as the concentration of HCO_3^- in mmol or mOsmol per liter of plasma.

Anion Gap

Anion gap is a mathematical construct that is most effectively applied in detecting altered concentrations of anions other than Cl^- and HCO_3^-. The law of electroneutrality demands that the number of positive charges (Na^+, K^+, Ca^{2+}, Mg^{2+}, and traces of other cations) in plasma exactly equal the number of negative charges (Cl^-, HCO_3^-, SO_4^{2-}, $HPO_4^{-,2-}$, protein anions, organic anions, and traces of other anions). Therefore, the term "anion gap" is a misnomer and refers to the "analytical gap"; under no circumstances is there ever a deficit of anions relative to cations. The anion gap, in reality, is the difference between unmeasured anions (cUA^-) and unmeasured cations (cUC^+) as also shown in the nonshaded area in Table 10-13.

$$\text{Anion gap, mmol ion charge/L} = cUA^- - cUC^+$$
$$= (cHPO_4^{2-} + cSO_4^{2-} + cOrg^- + cPr^-)$$
$$- (cK^+ + cCa^{2+} + cMg^{2+} + c\,\text{others}^+)$$
$$= (2 + 1 + 5 + 16) - (4 + 5 + 2 + 1)$$
$$= 24 - 12$$
$$= 12 \text{ (mean)}$$

In a serum with normal composition, $cUA^- - cUC^+$ is equal to the concentration of the measured cations (e.g., Na = 142) minus the concentration of measured anions (e.g., Cl = 103 plus HCO_3^- = 27). In practical terms, the anion gap is defined in a way that avoids dealing with ion charges of unmeasured ions and capitalizes on the ease of measurement of Na^+, Cl^-, and HCO_3^-, thus:

$$\text{Anion gap, mmol ion charge/L} = cNa^+ - (cCl^- + cHCO_3^-) \tag{18}$$

or alternatively,

$$\text{Anion gap, mmol ion charge/L} = (cNa^+ + cK^+) - (cCl^- + cHCO_3^-) \tag{19}$$

Equation (18) represents the conventional and more commonly used definition, but both definitions are acceptable and one or the other is used by automated electrolyte analyzers that display a

Table 10-13. ILLUSTRATION OF GAP BETWEEN UNMEASURED ANIONS (UA^- = 24 MMOL ION CHARGE/L) AND UNMEASURED CATIONS (UC^- = 12 MMOL ION CHARGE/L)

Na^+	142		Cl^-	103	
K^+	4		HCO_3^-	27	
Ca^{2+}	5	12	HPO_4^{2-}	2	
Mg^{2+}	2		SO_4^{2-}	1	24
Others	1		Org. anions	5	
			Protein	16	
	154			154	

Shaded area indicates measured anion and cations.

calculated anion gap as part of results output for an electrolyte panel. Note, however, that the reference ranges for the two equations are different.

Clinical Significance.[6] In terms of the definition of anion gap given by equation (18), K^+, Ca^{2+}, and Mg^{2+} are classified as unmeasured cations, and protein anions (Pr^-), organic anions (Org^-), SO_4^{2-}, and $HPO_4^{.2-}$ are classified as unmeasured anions. It is important to realize that the value of the calculated anion gap may be affected in five situations:[8]

1. Anion gap values may be *increased* due to real increase in the concentration of one or more of the unmeasured anions. Such real increases include increased β-hydroxybutyrate and acetoacetate in diabetic, alcoholic, or starvation ketoacidosis; increased lactate in lactic acidosis; increased anions such as sulfate and phosphate that have been retained because of renal failure; the presence of salicylate, or formate accumulating from the metabolism of methanol, or metabolites such as oxalate, hippurate, and glycolate accumulating from the metabolism of glycols in poisonings; increased negative charge of protein anions as a result of increased blood pH in respiratory or metabolic alkalosis; or increased number of protein anions because of dehydration (hemoconcentration).

2. Anion gap values may also be slightly *increased* due to an *apparent* increase in unmeasured anions that is actually a real decrease in unmeasured cations (hypocalcemic, hypomagnesemic, or hypokalemic states).

3. Anion gap values may be *decreased* due to a real decrease in the number of negatively charged protein anions (hypoalbuminemia), or to a real increase of special proteins that carry relatively few negative charges (hypergammaglobulinemia), or to an increase in plasma water (hemodilution).

4. Anion gap values may also be *decreased* due to an *apparent* decrease in unmeasured anions that is actually a real increase in unmeasured cations (hypercalcemic, hypermagnesemic, or hyperkalemic states, or in lithium toxicity) or in the presence of special proteins (e.g., myeloma proteins), carrying a net positive charge.

5. Anion gap values can be affected by analytical error in estimation of Na^+, Cl^-, or HCO_3^-. Analytical errors due to technical problems that overestimate the concentration of total CO_2 (bicarbonate) decrease the value; those that overestimate Na^+ or underestimate bicarbonate increase it. Analytical error may also develop when a particular sample contains an interferent that reacts with the method available for the assay. Low values for anion gap will be observed if Cl^- is overestimated because the sample contains Br^- or some similar positive interferent for the method used.

In and of itself, the finding of an abnormal anion gap is nonspecific. In the absence of laboratory errors, however, a high value, taken together with patient history, physical examination, and other laboratory data, has been found very useful for the detection of organic acidoses in which real increases occur in anions that are difficult or impossible to measure, especially on a "stat" basis.[6,7] Low and high values are considered less useful in detection of altered levels of unmeasured cations or proteins.[7]

High and Normal Anion Gap Acidosis

Use of anion gap values to screen for and follow acidoses of various etiologies has prompted classification of metabolic acidoses into high anion gap and normal anion gap types. Typical causes of *high anion gap* acidosis are shown in Table 10-14. Among the causes of metabolic acidosis characterized by *normal anion gap* are excessive loss of bicarbonate from the gastrointestinal tract, ureteral transplant to the ileum or colon, certain drugs (e.g., acetazolamide, cholestyramine), treatment with NH_4Cl, rapid intravenous hydration, hyperalimentation, and renal tubular acidosis.

Use of Anion Gap in Quality Control of Electrolyte Analysis

Technologists in clinical laboratories use equation (18), above, informally to check a set of electrolyte results for gross analytical error. On the assumption that unmeasured ion concentrations lie within their reference ranges, an anion gap calculated with equation (18) that exceeds 12–20 mmol/L indicates a need for review of calculations and procedures for Na^+, Cl^-, and HCO_3^-. If results are confirmed, reasons for a high anion gap may be sought; e.g., a high glucose or a high urea nitrogen would suggest accumulation of organic anions above normal concentrations (ketoacidosis and renal acidosis, respectively).

Formal use of a calculated average of patient anion gaps has also been proposed[4] and

Table 10-14. METABOLIC ACIDOSES WITH HIGH ANION GAP

Etiology	Acids Retained	Anion Gap, mmol/L	Osmolal Gap, mOsm/kg	Other Laboratory Findings
Renal failure	Sulfuric, phosphoric, organic	15–20 (usually)	< 10	Increased plasma urea nitrogen, creatinine
Ketoacidosis				
Diabetic	Acetoacetate,	> 25 (often)	< 10	Urinary ketones
Alcoholic	β-hydroxybutyrate			positive
Starvation				Increased plasma uric acid and urinary ketone bodies
Poisoning				
Aspirin	Salicylate	> 25 (often)	< 10	Respiratory alkalosis
Methanol	Formate		> 15	
Ethanol			> 15	
Ethylene glycol	Hippurate, glycolate, oxalate		> 15	Urine oxalate crystals
Lactic acidosis	Lactate	> 25 (often)	< 10	Normal arterial pO_2 (for type B lactic acidosis)

Modified from Narins, R. G., and Gardner, L. B.: Simple acid-base disturbances. Med. Clin. N. Am., *65*:321-346, 1981.

evaluated[3] as a quality control measure for the detection of systematic error in electrolyte analysis. In the application, values for 8 consecutive patient anion gaps (excluding those < 0 and > 22 mmol/L) are averaged; the mean is then tested against 99% confidence limits around an established, long-term mean. A mean of 8 patient values that falls outside the confidence limits is taken as a signal for possible systematic error in Na, Cl, or CO_2 determinations and as justification for repeating assays after an analytical problem has been found and corrected. When evaluated,[3] the approach was found to be sensitive (having a high probability of error detection) and specific (having a low probability of unnecessary repetition) as an indicator of systematic analytical error, particularly in Na and Cl estimations.

Reference Ranges. For anion gap calculated without the inclusion of K^+ [equation (18)], the reference range is 8–16 (mean 12) mmol/L. For anion gap calculated with inclusion of K^+ [equation (19)], the range is 10–20 (mean 16) mmol/L. Low values are characteristic of hypo- and hyperproteinemia (cationic type). The highest values are associated with forms of acidosis summarized in Table 10-14. When using electrolyte analysis and anion gap calculated without K^+ to screen for acid-base imbalance, most physicians react to an anion gap of < 5 or > 19 mmol/L, since there is a 2–4 mmol/L random variation at the upper and lower limits of the reference range.[6]

Osmolal Gap

The osmolal gap is a mathematical construct similar to anion gap, but it applies to osmolute (osmotically active solute) concentrations rather than to concentrations of ion charges. The value of the gap is:

Osmolal gap, mOsm/kg* = measured osmolality, mOsm/kg — calculated osmolality, mOsm/kg

---

*The IFCC does not recognize milliosmoles (mOsm) as a unit, and osmolality (unit: mol/kg) is defined as $-(\ln a_{H_2O})/M_{H_2O}$ where a_{H_2O} is the activity of water and M_{H_2O} is molar mass of water (= 0.018 kg/mol). Osmolarity (unit: mol/L) is defined similarly as $-(\ln a_{H_2O})/V_{H_2O}$, where V_{H_2O} is the molar volume of pure water (~0.018 L/mol). The numerical difference between the two is the density of pure water, which is very nearly 1 kg/L; i.e., numerically osmolality and osmolarity are identical.

Osmolality of a biological fluid is generally determined by measuring the freezing point depression (ΔT_{fus}) and dividing by the molal freezing point depression constant (1.855 kg · K · mol^{-1}). Measuring vapor pressure (p_{H_2O}), the osmolality is calculated as $[-\ln (p_{H_2O}/p_{H_2O}*)]/M_{H_2O}$, where $p_{H_2O}*$ is the vapor pressure of pure water.

For the calculation of the osmolality, the concentration (mmol/L) of solutes (osmolutes) must be converted into molality (amount of substance divided by mass of solvent, mmol/kg) by dividing by the mass concentration of water in plasma. The average value for mass concentration of water in normal plasma is 0.933 kg/L. The molality of total solutes must then be converted into osmolality by multiplying with the osmotic coefficient for plasma ($\sim$0.92 for normal plasma).

Medical literature uses the terms osmolal and osmolar interchangeably to describe the gap, although the term osmolal is now recommended. (See also Chapter 1B, Section Five.)

Calculation of the osmolal gap requires measurement of plasma or serum osmolality by freezing point depression osmometry and measurement of the plasma or serum concentrations of Na^+, glucose, and urea (nitrogen). Osmolality measured by freezing point depression, as opposed to vapor pressure measurements, reflects the total concentration of osmolutes in the specimen, including volatiles such as alcohols or acetone. Since total osmolute concentration is essentially determined by the five major osmolutes—Na^+, Cl^-, HCO_3^-, urea, and glucose—and since each Na^+ ion can be assumed to be accompanied by an anion, only sodium, glucose, and urea need to be measured in order to calculate the osmolality of total solutes in plasma.

The best documented equation[5] for calculating plasma osmolality is:

$$\text{Calculated osmolality, mOsm/kg} = 1.86\ (Na^+, \text{mmol/L}) + \text{glucose, mmol/L} + \text{urea, mmol/L} + 9$$

or

$$= 1.86\ (Na^+, \text{mmol/L}) + \frac{\text{glucose, mg/dL}}{18} + \frac{\text{urea N, mg/dL}}{2.8} + 9$$

or

$$= 1.86\ (Na^+, \text{mmol/L}) + 0.056\ (\text{glucose, mg/dL}) + 0.36\ (\text{urea N, mg/dL}) + 9$$

The equation is empirical. It was developed by regression analysis to solve for the factor 1.86 (slope for the relation of Na^+ concentration in mmol/L to the calculated osmolality) and the composite intercept represented by the last three terms in the equation. The equation was validated by comparison of calculated osmolality with measured osmolality; the mean difference was approximately zero, and the standard deviation of the differences $\sim$6 mOsm/kg. The factors applied to glucose and urea N convert concentration expressed in mg/dL to mmol/L, and for urea N to urea,

for glucose,

$$0.056 = \frac{1}{18} = \frac{1\ \text{mg/dL} \times 10\ \text{dL/L}}{180\ \text{mg/mmol glucose}}$$

for urea N,

$$0.36 = \frac{1}{2.8} = \frac{1\ \text{mg/dL} \times 10\ \text{dL/L}}{28\ \text{mg N/mmol urea}}$$

A number of other equations can be used;[5] the above equation, as well as any of the others, may be applied on electrolyte analyzers that display the calculated osmolality together with the results of a test panel that includes sodium, urea nitrogen, and glucose. The test of suitability of an equation used to calculate osmolality is whether calculated osmolality of almost all specimens containing a normal amount of plasma water compares closely to measured osmolality on the same specimens.

The ability of contemporary analyzers to calculate and display a result for calculated osmolality is a mixed blessing. On the one hand, the calculated osmolality may lead some physicians and laboratorians to believe that whatever equation has been used for the algorithm in instrument software is the only or the best equation for calculating osmolality, or to believe that calculated osmolality is a uniformly valid substitute for measured osmolality. On the other hand, it provides informed laboratorians and physicians with the opportunity to calculate the osmolal gap more conveniently.

Clinical Utility of Osmolal Gap. The mean for the plasma osmolal gap calculated as described above is approximately zero, with a standard deviation of $\sim$6 mOsm/kg. The value of the gap is increased in hyperosmolal states due to increase in osmolutes other than the five major ones.

Values up to 10 (i.e., very moderate increases) are associated with ketoacidosis, renal acidosis, and lactic acidosis. Values frequently fall within the reference range and thus are not very helpful in these clinical situations. However, the osmolar gap is most effectively used to detect and follow poisonings with nonelectrolytes such as acetone, ethanol, methanol, isopropanol, ethylene glycol, diethyl ether, paraldehyde, or trichloroethane. High concentrations of both parent compounds and metabolites cause a hyperosmolality and a large increase in the osmolal gap that are often correlated with the severity of the intoxication. Such increase in osmolality is not detected by calculating the osmolality on the basis of the Na^+, glucose, and urea concentrations. The osmolal gap therefore becomes valuable in detecting and following these particular hyperosmolal states without recourse to difficult or specific assays for individual, and perhaps unknown, analytes. Osmolal gaps expected in some clinical conditions are shown in Table 10-14.

References

1. Albert, M. D., Dell, R. B., and Winters, R. W.: Quantitative displacement of acid-base equilibrium in metabolic acidosis. Ann. Intern. Med., 66:312, 1967.
2. Bakerman, S.: Clinical Chemistry Review. East Carolina University Department of Pathology, Greenville, N.C.
3. Bockelman, H. W., Cembrowski, G. S., Kurtycz, D. F. I., et al.: Quality control of electrolyte analyzers: Evaluation of the anion gap average. Am. J. Clin. Pathol., 81:219-223, 1984.
4. Cembrowski, G. S., Westgard, J. O., and Kurtycz, D. F. I.: Use of anion gap for the quality control of electrolyte analyzers. Am. J. Clin. Pathol., 79:688-696, 1983.
5. Dorwart, W. V., and Chalmers, L.: Comparison of methods for calculating serum osmolality from chemical concentrations, and the prognostic value of such calculations. Clin. Chem., 21:190-194, 1975.
6. Emmett, M., and Narins, R. G.: Clinical use of the anion gap. Medicine, 56:38-54, 1977.
7. Gabow, P. A., Kaehny, W. D., Fennessey, W. D., et al.: Diagnostic importance of an increased serum anion gap. N. Engl. J. Med., 303:854-858, 1980.
8. Gambino, R.: The anion gap in '82. Lab Report for Physicians, 4:28-30, 1982.
9. McGilvery, R. W.: Biochemistry. Philadelphia, W. B. Saunders Co., 1983, pp. 778-782.
10. Narins, R. G., and Emmett, M.: Simple and mixed acid-base disorders: A practical approach. Medicine, 59:161-187, 1980.
11. Narins, R. G., and Gardner, L. B.: Simple acid-base disturbances. Med. Clin. N. Am., 65:321-346, 1981.

CHAPTER 11

NITROGEN METABOLITES AND RENAL FUNCTION

by Robert C. Rock, M.D., W. Gordon Walker, M.D., and C. Darrell Jennings, M.D.

The Kidneys

The kidneys are a paired organ system located in the retroperitoneal space. They extend from the level of the lower part of the eleventh thoracic vertebra to the upper portion of the third lumbar vertebra, with the right kidney situated slightly lower than the left (Figure 11-1,*A*). Each kidney weighs about 150 g in the adult human; their collective weight represents about 0.4% of total body weight. Each kidney is bean shaped with a central hilus on the medial side where vessels, lymphatics, and the renal pelvis join (Figure 11-1,*B*). The kidney parenchyma is divided into lobes that consist of an outer cortex which covers a pyramidal medulla. This cortex is composed mostly of glomeruli and the proximal and distal convoluted tubules. The medulla consists mostly of loops of Henle, vasa recta (elongated vessels accompanying the loops of Henle), and collecting ducts. The collecting ducts drain through the rounded apex of the pyramid into branches of the renal pelvis called calyces. From here urine collects in the pelvis and flows through the ureteropelvic junction into the ureter. The ureters carry urine from each kidney into the bladder, where it is stored until voided through the urethra. The renal blood supply represents ~25% of the cardiac output. The renal artery divides into posterior and anterior portions with subsequent divisions into interlobar, arcuate, and, finally, interlobular arteries, which ultimately terminate in the afferent arterioles, each supplying a single glomerulus.

The short course of the renal artery and rapid branching leading into the afferent arterioles are very important for renal function. This vascular arrangement allows transmittance of adequate hydrostatic pressure to the glomerular capillary bed. The hydrostatic pressure gradient across the glomerular capillary membrane provides the driving force for glomerular ultrafiltration. The oncotic pressure difference between the capillary lumen and Bowman's space opposes filtration. Since most proteins are not normally filtered, the oncotic pressure in the glomerular capillary progressively rises until filtration ceases. This limits the amount of filtrate that can be obtained from a given volume of plasma. Thus, maintenance of glomerular filtration rate is dependent upon an adequate renal blood flow rate as well as adequate perfusion pressure. The normal renal blood supply is well designed to meet these needs.

Renal tubular function is also dependent upon the anatomy of the vessels providing the blood supply. Efferent arterioles form capillary plexuses as well as elongated vessels accompanying the loops of Henle (vasa recta), which provide the rich, intimate perfusion necessary for the highly active tubular cells to reabsorb various urine constituents either in large quantities (in the proximal tubule) or against significant concentration gradients (in the distal tubules). Further, the flow through the vasa recta is vital for the steady state operation of the countercurrent exchange mechanism that allows formation of the medullary osmotic gradient that is so important in renal handling of water.

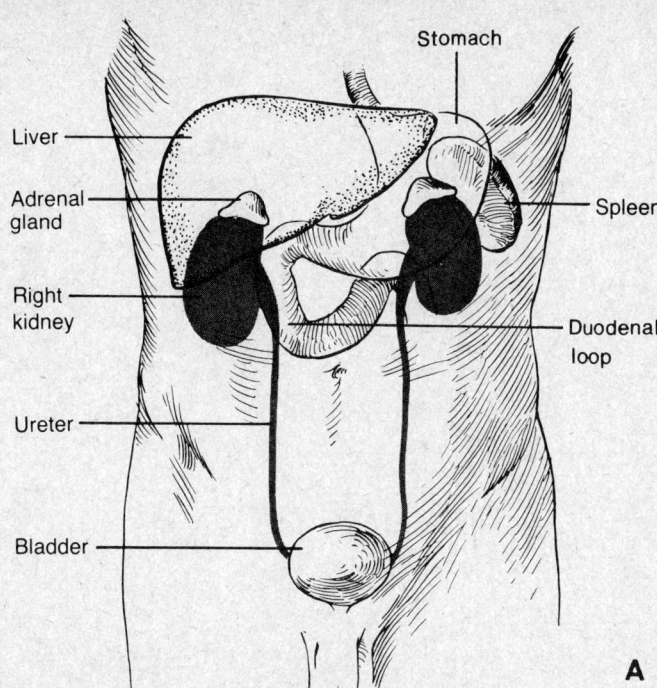

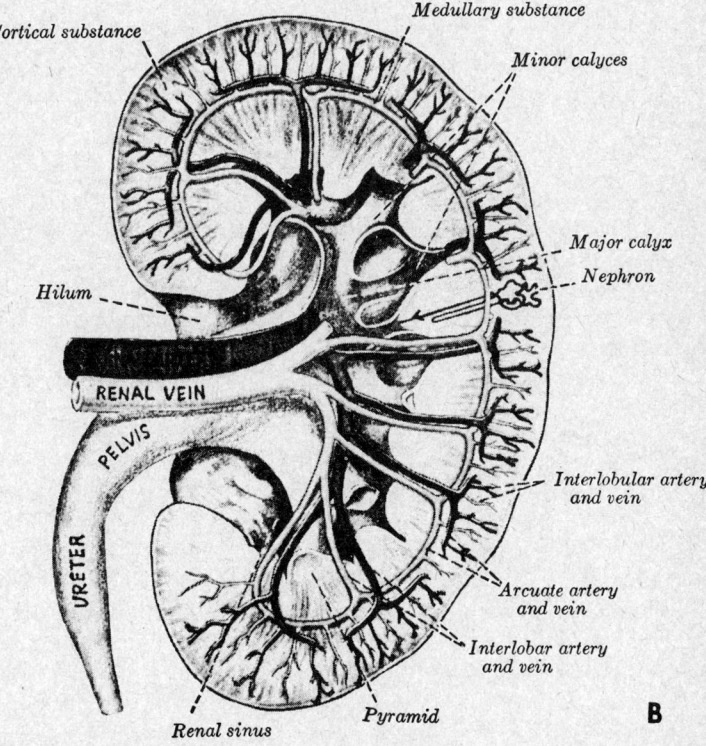

Figure 11-1. *A,* Schematic drawing of right and left kidneys with respect to other abdominal viscera. *B,* Diagram of vertical section through the kidney. Nephron and blood vessels are greatly enlarged. (From Goss, C. M., Ed.: Gray's Anatomy. 28th ed. Philadelphia, Lea & Febiger, 1966.)

The nephron. The nephron is the functional renal unit. It consists structurally of (1) the glomerulus, a spherical epithelial space invaginated by a capillary tuft that connects the afferent and efferent arterioles, and (2) a tubule of epithelial cells, continuous with the glomerular epithelial space and ultimately leading to collecting ducts that empty into the renal pelvis. Figure 11-2 schematically depicts the functional regions of the nephron. Note that the efferent arteriole is located in close association with the exterior surface of the nephron.

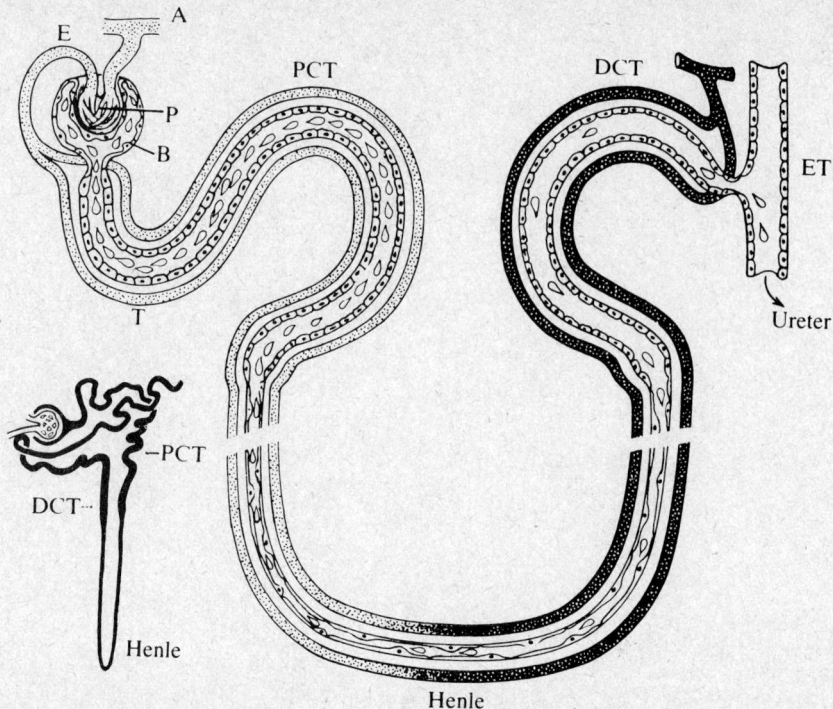

Figure 11-2. Schematic drawing of the glomerulus and the tubular system of the nephron. A: afferent arteriole; E: efferent arteriole; P: plexus of capillaries (glomerular tuft); B: Bowman's capsule; T: tubular blood supply; PCT: proximal convoluted tubule; Henle: loop of Henle; DCT: distal convoluted tubule; ET: excretory tubule or duct. The blood capillaries shown along the tubular system (T) gradually change to venous capillaries as they pass down the tubular system.

Formation of urine. The first step in urine formation is filtration of plasma water at the glomerulus. A net filtration pressure of about 15 mm Hg in the capillary bed of the tuft drives the filtrate through the glomerular membrane. The filtrate is called an ultrafiltrate because its composition is essentially the same as that of plasma, with the notable absence of molecules of molecular weight above ~15 000. Since each nephron produces about 100 μL of ultrafiltrate per day and each kidney contains about one million nephrons, ~170–200 L of ultrafiltrate pass through the glomeruli in 24 h. In the passage of ultrafiltrate through the tubules, reabsorption of solutes and water in various regions of the tubules reduces the total volume, which ranges between 0.4 and 2 L of urine per day.

In the tubules, the solute composition of the ultrafiltrate is altered by the processes of reabsorption and secretion, so that the urine excreted may have a very different composition from that of the original filtered fluid. Different regions of the tubule have been shown to specialize in certain functions. See Table 11-1 for a summary of the functions and characteristics of different regions of the tubule. In the *proximal tubule*, 60–80% of the ultrafiltrate is reabsorbed in an obligatory fashion along with sodium, chloride, bicarbonate, calcium, phosphate, and other ions. The threshold substance,* glucose, is also reabsorbed in the proximal tubule. Certain nonbiological compounds, such as phenolsulfonphthalein (PSP), *p*-aminohippurate (PAH), and iodopyracetate (Diodrast), are secreted by the proximal tubule and have been used for the evaluation of renal

---

*Certain substances, almost completely reabsorbed by the tubule when their concentrations in plasma are relatively low, appear in the urine when their plasma levels are above a certain set point or "threshold" level. Creatinine, which is reabsorbed only slightly or not at all, is a "low-threshold" substance. "High-threshold" substances, such as glucose and amino acids (whose conservation is important to the body economy), are almost completely reabsorbed by means of specific transport systems in the tubular cells. The appearance of a high-threshold substance in the urine is evidence that the filtered load of the substance is exceeding the maximal reabsorption rate of its transport system. This may occur when the filtered load is excessive due to elevated plasma concentration or to rapid urine flow; alternatively, the defect may be in the transport system and related to chemical inhibitors, injury to tubular epithelial cells, or genetic defects in the system. The "renal threshold" of a filterable compound is therefore that plasma concentration of the substance at which the tubular reabsorptive rate, operating at maximum, is not sufficiently fast to remove all of the substance from the plasma ultrafiltrate so that the substance "spills" into the urine.

Table 11-1. RENAL TUBULAR FUNCTIONS

Segment of Renal Tubule	Characteristic	Functions
Proximal tubule	High volume, low gradient system; reabsorbs major fraction of filtrate; location of several specific transport systems	Reabsorption of ~85% of filtered bicarbonate; reabsorption of 60–75% of filtered Na^+, K^+ and H_2O in isosmolar fashion; reabsorption of virtually all glucose, most amino acids, and a major fraction of divalent ions
Loop of Henle	Lower volume, higher gradient system; generates gradient for urinary dilution and concentration; site of action of loop diuretics such as Lasix	Reabsorption of 20–25% of filtered Na^+ and K^+ without H_2O reabsorption in the ascending limb, generating the corticopapillary osmotic gradient; active transport into tubular cells of Cl^- with accompanying Na^+ to generate a dilute urine
Distal tubule and collecting ducts	Low volume, high gradient system; regulates sodium and water balance as well as acid-base balance; site of action of thiazide diuretics and aldosterone antagonists	Reabsorption of ~10% of filtered Na^+, mostly coupled to H^+ and K^+ secretion; aldosterone controlled Na^+ reabsorption coupled to H^+ and K^+ secretion; regulates Na^+ and extracellular fluid volume; secretion of H^+ and generation of NH_4^+ to regulate acid-base status; ADH regulated reabsorption of H_2O or further urinary dilution

tubular secretory capacity. When blood levels of creatinine increase above normal, creatinine is secreted in this region of the nephron. In the *loop of Henle*, chloride and more sodium without water are reabsorbed, generating a dilute urine. Water reabsorption in the more distal tubules and collecting ducts is then regulated by the antidiuretic hormone (ADH). In the *distal tubule*, secretion is the prominent activity; organic ions, potassium ions, and hydrogen ions are transported from the blood in the efferent arteriole into the tubular fluid. It is also this region that secretes H^+ and reabsorbs Na^+ and HCO_3^- in aid of acid-base regulation. (See Renal Excretion of Acid, p. 1235.)

Homeostasis. The obvious role of glomerular filtration and tubular secretory functions is to eliminate waste products of metabolism. Another equally important role is to maintain the constant optimal chemical composition of the blood and thereby to maintain the constant optimal chemical composition of the interstitial and intracellular fluids throughout the body. Homeostasis, the maintenance of the internal milieu, largely depends on the efficient operation of the reabsorptive and secretory mechanisms of the kidney. The kidney must therefore be viewed not only as an excretory organ but also as a regulatory organ.

FUNCTIONS OF THE KIDNEY

The functions of the kidney may be characterized as excretory, regulatory, and endocrine. The *excretory function* serves to rid the body of most of the undesirable end products of metabolism, as well as any excess of inorganic substances ingested in the diet. Waste products include the nonprotein nitrogenous compounds urea, creatinine, and uric acid; a number of other organic acids, including amino acids, are excreted in small quantities. Diet contains a variable and usually excessive supply of sodium, potassium, chloride, calcium, phosphate, magnesium, sulfate, and bicarbonate. The dramatic efficiency of the excretory function is illustrated by sodium; body sodium content is maintained essentially constant, regardless of whether daily sodium intake is 1 or 150 mmol or more. Daily intake of water is also variable and may, on occasion, greatly exceed the minimal requirements of the body. Under such circumstances, water becomes additional waste material requiring excretion. In order to achieve the excretion of the metabolic wastes and the ingested surpluses without disrupting homeostasis, the kidney must exercise both its excretory and regulatory functions.

Mechanisms for the excretion of excess electrolytes, nitrogenous wastes, and organic acids are similar although not identical. For all except potassium and hydrogen ions and a few organic acids, the maximal excretory rate is limited by or established by their plasma concentrations and

Table 11-2. COMPARISON OF RENAL FUNCTION BETWEEN THE NORMAL AND SEVERELY IMPAIRED KIDNEY

	Normal	Impaired*
Glomerular filtration rate (GFR), L/d	180	5
Renal blood flow, L/d	1720	52
Maximum urine volume, L/d	25	1.4
Minimum urine volume, L/d	0.5	1.0
Maximum solute concentration, mOsm/kg H_2O	1200	300
Maximum free water clearance, L/d	23	0.4
Maximum free water reabsorption, L/d	8	<0.2
Maximum total hydrogen ion secretion,[†] mmol/d	5000	100
Titratable acidity,[‡] mmol/d	150	10
Ammonium production,[‡] mmol/d	350	20
Average solute excretion,[§] mOsm/d	600	350
Daily solute excretion rate/GFR, mOsm/L	3.3	70
Maximum Na^+ excretion, mmol/d	>450	~60
Minimum Na^+ excretion, mmol/d	<1	~15
Maximum K^+ excretion, mmol/d	>400	40
Minimum K^+ excretion,[¶] mmol/d	4	10
Phosphate excretion, mmol/d	110	10

*Values are arbitrarily chosen to represent usual limits exhibited when GFR is reduced to 3.5 mL/min.

[†]Represents sum of bicarbonate reabsorption + NH_4^+ production + titratable acid.
[‡]Represents maximal values developed during acidosis.
[§]Determined principally by intake.
[¶]These values are achieved only after development of a severe K^+ deficit.

the rate of their filtration through the glomeruli. Bulk transfer of substances from blood to urine determines the initial mass upon which the nephron must operate to produce and excrete urine. Thus, the maximal amount of substance excreted in urine does not exceed the amount transferred through the glomeruli by ultrafiltration except in the case of those substances capable of being secreted by tubular cells. Depending upon the activity of the renal tubular epithelial cells and their several reabsorptive capacities, excreted amounts of urinary constituents will be, in general, variably less than the amounts filtered. Because of this general behavior, for many substances one may obtain a general estimate of the excretory capacity of the kidney by measuring either the glomerular filtration rate or some variable that is closely related to it.

Among the data presented in Table 11-2 are comparisons of maximal excretory capacities exhibited by the kidney for some common substances in health and in advanced renal failure. The primary objective in evaluation of renal excretory function is to detect quantitatively the degradation of normal capacities or the improvement of impaired ones.

The *regulatory function* of the kidney plays a major role in homeostasis. The mechanisms of differential reabsorption and secretion, located in the tubule of the nephron, are the effectors of regulation. The mechanisms operate under a complex system of control in which both extrarenal and intrarenal humoral factors participate. Further discussion of the mechanisms and factors appears in a later section of this chapter.

The *proximal convoluted tubule* (PCT) is generally and predominantly concerned with reabsorption. Here about 75% of the sodium, chloride, and water of the ultrafiltrate is reabsorbed as is most of the bicarbonate, phosphate, calcium, and potassium. Water reabsorption in the PCT is termed "obligatory," since its volume is related to the heavy load of solutes being returned to the blood in the efferent arteriole. The amount of bicarbonate reabsorption is related to the glomerular filtration rate and the hydrogen ion secretory rate; the amount of phosphate reabsorption is controlled in part by plasma calcium concentration and in part by the effect of parathormone on the tubular cells. Normally, the high-threshold substances, glucose and, to a great extent, amino acids, are reabsorbed here by means of specific, intracellular, active transport systems. Uric acid may be either reabsorbed or secreted in the PCT by a two-way, carrier-mediated process.

In the ascending *loop of Henle*, 20–25% of filtered sodium is reabsorbed without concomitant reabsorption of water. This process generates a dilute urine with an osmolality of 100–150 mOsm/kg H_2O and helps establish the corticomedullary osmotic gradient. The resulting hypertonicity of the interstitium is important in the pathogenesis of renal infections, since the hypertonic environment interferes with leukocyte function. Subsequent water reabsorption is regulated by antidiuretic hormone (ADH). Although the reabsorption of Na^+ in the loop of Henle is complex and incompletely understood, at least one mechanism consists of an active Cl^- pump with subsequent reabsorption of Na^+ along on electrochemical gradient. This mechanism is apparently the one inhibited by the powerful "loop" diuretics.

The *distal tubule* (DT) is functionally the most active region of the nephron for the homeostatic regulation of plasma electrolytes and plasma acid-base levels. Here a combination of secretion and reabsorption, which can be described as exchange, takes place among Na^+, K^+, and H^+. Although excess plasma hydrogen ions are secreted all along the tubule, it is in the distal tubule that exchange of H^+ for Na^+ (which is reabsorbed) fine-tunes the balance between H^+ loss and retention. (See section on H^+ excretion, Chapter 10.) Potassium ions are also secreted in the DT. Aldosterone is a potent modulator of Na^+ reabsorption in the DT, particularly when need arises to conserve Na^+. Production of aldosterone in the adrenal cortex is stimulated by the renin-angiotensin system and by high plasma potassium concentration. Renal secretion of renin is complex but is at least partly regulated by renal perfusion and plasma sodium concentration. Inadequate perfusion and a low concentration of plasma sodium both stimulate renin secretion. Organic anions such as acetoacetate and β-hydroxybutyrate also consume H^+ as they are eliminated, in part, in their nondissociated acid form. When H^+ must be conserved in order to maintain plasma pH, DT cells reduce the secretion of H^+, reduce NH_4^+ generation, reduce Na^+-H^+ exchange, and increase bicarbonate excretion. The net effect is a reduction in plasma bicarbonate and restoration of normal plasma pH.

The *endocrine functions* of the kidney may be regarded either as primary, because the kidney is an endocrine organ producing hormones, or as secondary, because the kidney is a site where hormones produced elsewhere are activated. In its *primary endocrine function*, the kidney produces renin, prostaglandins, and erythropoietin. Erythropoietin acts on the bone marrow to stimulate proliferation and maturation of erythrocyte precursors. Anemia, therefore, is a prominent feature of renal insufficiency, and its severity parallels deterioration of renal function as renal failure progresses. In the anephric state, the hematocrit can be expected to fall as low as 10–12% before stabilizing.[66] Renin and prostaglandins affect the vascular system. The renin-angiotensin system is an important regulator of blood pressure; it is responsive to plasma sodium and potassium concentrations and regulates the aldosterone effect on K^+-Na^+ exchange in the distal tubule. (See also Chapter 9.)

Renal kallikrein is most likely produced in the cortex, but it is unclear whether the quantity is sufficient to have demonstrable extrarenal effects. In the kidney, it acts on a kininogen substrate to produce kallidin, a potent vasodilator with natriuretic effects. Urinary excretion of kallikrein is affected by aldosterone levels, sodium intake, arterial pressure, and angiotensin II. The role of the renal kallikrein system is not entirely clear, but it may be related to local regulation of renal blood flow and sodium excretion. It does appear to have significant interrelationships with the renin-angiotensin system.

Another interrelated system that is important in regulation of renal blood flow in states of underperfusion is the group of prostaglandins. The prostaglandins PGA_2, PGE_2, and PGF_2, made in the renal medulla, act on arteries to cause vasodilation and to lower blood pressure. The renal prostaglandins appear to work by increasing renal blood flow and sodium loss, while inhibiting ADH effect on water reabsorption, with little effect upon filtration. The kidney increases prostaglandin synthesis, especially that of PGE_2 and PGF_2, in response to ischemia and decreased perfusion. PGE_2 is a potent vasodilator that decreases renal vascular resistance. These prostaglandins have complex effects on renal production of renin and thus may affect renal sodium handling and regulation of blood pressure. Much work remains to be done in the elucidation of the complex interrelationships of these several systems.

In the *secondary* but no less important aspect of its *endocrine function*, the kidney is a site of degradation of insulin, glucagon, and aldosterone. Furthermore, the kidney is the location of the important 1-hydroxylation of 25-(OH)-D_3 that produces 1,25-(OH)$_2$-D_3, the most active known form of vitamin D. Progressive, chronic renal failure is regularly associated with a complex metabolic disorder of bone, renal osteodystrophy. This disorder is a combination of osteomalacia (inadequate mineralization of bone), due to altered vitamin D metabolism, and osteitis fibrosa cystica, caused by secondary elevation of PTH levels. Recent evidence suggests aluminum may also play a significant role in the bone disease of patients with chronic renal failure. Consumption of excess dietary aluminum salts or the administration of phosphate binding gels [e.g., $AL(OH)_3$] and failure to excrete the absorbed aluminum results in high blood and tissue levels of aluminum, which is incorporated into the osteoid seams and which is associated with the neurologic disorders of dysarthria and apraxia.

Renal Dialysis

The development of technologies for hemodialysis (treatment with the "artificial kidney") and peritoneal dialysis* has changed end-stage renal disease from an illness that once terminated fatally within a few months to a chronic illness with an average survival time of 10–15 years. Hemodialysis is generally more efficient than peritoneal dialysis in removal of waste compounds in uremic patients, although peritoneal dialysis is of value in situations when arteriovenous fistulas ("shunts") fail. For diabetic patients with end-stage renal disease, chronic ambulatory peritoneal dialysis has been found to be most useful.

Hemodialysis applies ultrafiltration to achieve clearance, i.e., removal of filterable substances from the blood. (Clearance is discussed in more detail on p. 1262). The concept of "dialysance" is analogous to that of clearance and is defined by the relation:

$$D_B = M/(C_{bi} - C_{di})$$

where D_B　= whole-blood dialysance (the rate, in terms of volume of inlet blood, at which blood is completely cleared of solute)

　　　　M　= the mass transfer rate of the solute in moles per unit of time

　　　　C_{bi}　= the molar concentration of solute in inlet blood

　　　　C_{di}　= the molar concentration of solute in inlet dialysate

D_B, dialysance, is a quantitative measure of the efficiency of the hemodialysis procedure to remove particular solutes from a given patient's blood during the treatment. Manipulation of the hydrostatic pressure within the dialyzer controls the volume of fluid removed from the patient.

Complications of hemodialysis include mechanical difficulties such as air embolism, hemorrhage, or obstruction of the shunt, as well as toxicity caused by the dialysate itself. The syndrome of "dialysis dementia" in chronic renal dialysis patients has, for example, been attributed to high levels of aluminum in dialysate.[40] Fluid and electrolyte shifts during dialysis produce symptoms at the time and in the immediate postdialysis period. Rapid removal of urea, sodium, and other osmotically active substances causes "dialysis disequilibrium," with symptoms ranging from lethargy and headache to stupor and seizures. Rapid lowering of potassium during dialysis may lead to myocardial toxicity. Vitamins such as folate and pyridoxine are removed by dialysis and must be replaced. Patients on hemodialysis will sometimes have a transient leukopenia at the initiation of the procedure.

Although dialysis converts end-stage renal disease to a stable, chronic illness, it does not completely correct for lost renal function. Some of the complications of uremia persist. Patients on long-term hemodialysis remain more susceptible to infections, and both anemia and atherosclerosis contribute to their morbidity. Viral hepatitis (B and non-A, non-B) is a particular problem for dialysis services. Renal osteodystrophy, with secondary hyperparathyroidism and vitamin D deficiency resulting in osteomalacia, persists despite hemodialysis. Control of hyperparathyroidism occasionally requires subtotal parathyroidectomy.

Laboratory tests routinely used to monitor patients on hemodialysis[24] are listed in Table 11-3. Most renal diseases that produce chronic renal failure and a need for dialysis result in a decrease in the number of functioning nephrons. The laboratory picture in these individuals is remarkably consistent. Effectiveness of dialysis is usually monitored by measurement of creatinine;

---

*Hemodialysis, the general term, refers to a procedure where blood solutes are allowed to diffuse down a concentration gradient, across a semipermeable membrane, into a recipient fluid, the dialysate. The artificial kidney is an application of extracorporeal dialysis, since an extracorporeal circulation is set up via an implanted arteriovenous cannula to lead the patient's blood to the membrane cartridge immersed in the dialyzer bath. Peritoneal dialysis is intracorporeal, so to speak; the patient's abdominal cavity is perfused with (or filled with and then emptied of) the recipient fluid, and dialysis occurs across the peritoneal membrane. Clearance or dialysance achievable by peritoneal dialysis is less than 15–20% of that achieved by hemodialysis. It is especially slow at removing K^+. Hence, when hyperkalemia is the prime indication for dialysis, hemodialysis is the preferred and more efficient means of treatment unless contraindicated.

Table 11-3. BIOCHEMICAL MONITORING OF HEMODIALYSIS

Disorder	Tests* (Plasma or Serum)
Fluid and electrolyte imbalance	Sodium
	Potassium
	Chloride
	Bicarbonate
	Osmolality
Nitrogen imbalance	Total protein
	Albumin
	Urea
	Creatinine
	Uric acid
Osteodystrophy	Calcium
	Phosphorus
	Alkaline phosphatase
	Magnesium
	Parathyroid hormone
Hepatitis	Hepatitis B antigen
	Bilirubin
	Aspartate aminotransferase
	Alanine aminotransferase

*None of these tests provide direct data bearing on balance but must be interpreted in relation to changes in body weight, fluid intake, and food intake as well as changes attributable directly to dialysis.

serum potassium and urea nitrogen are usually determined to assess dietary compliance; measurement of body weight, blood pressure, and serum sodium assess regulation of body hydration and sodium balance; measurement of bicarbonate and arterial pH indicates acid-base status and buffering capacity; and serum phosphate, calcium, and periodically PTH (and possibly hand X-rays) assess the effects upon bone metabolism. The most reliable parameters for following residual renal function are urine volume, which represents a significantly greater fraction of the entire filtered load than in healthy kidneys, and a reliably measured glomerular filtration rate (GFR). In the patient on dialysis, GFR is much more responsive to sodium balance and blood pressure than in normals. Thus, the patient should have blood pressure controlled and be in relative sodium balance before the GFR is measured. Furthermore, in the 24 h following dialysis there is a transient decrease in GFR, limiting the usefulness of values from this time interval. The method for determining GFR is also problematic. When plasma creatinine levels are high, creatinine is secreted by tubules so that creatinine clearance will overestimate GFR. Urea clearance underestimates GFR due to tubular back-diffusion of urea. Inulin clearance is reliable but cumbersome. ^{125}I-iothalamate clearance has been advocated by several investigators because this compound (1) can be administered subcutaneously, (2) is easily measured in serum and urine by scintillation counting without chemical analysis, (3) requires a short period of urine collection, and (4) allows detection of residual urine in the bladder by scanning.[54] Several therapeutic drugs, especially the nephrotoxic aminoglycosides, should be monitored by measuring serum levels in dialysis patients, since the effects of poor renal function and dialysis are not always predictable. Table 11-3 makes clear the full range of concern that the laboratory data must address in order to support the total care of the end-stage renal disease patient undergoing dialysis.

Transplantation

Transplantation of kidneys into immunologically suitable recipients corrects many of the abnormalities of chronic renal failure that hemodialysis can only partially correct. But immunosuppression, which is required to forestall rejection of the transplanted kidney, causes new problems for the recipient. The most immediate danger is infection by opportunistic organisms such as fungi, protozoa, certain bacteria, and certain viruses. Long-range problems relate to the development of malignancies in immunosuppressed recipients of renal allografts.[45]

Biochemical monitoring of renal transplant patients is designed to detect deterioration of function in the transplanted kidney, especially as it relates to delayed rejection of the graft. Serial monitoring of plasma creatinine, with a reliable and accurate method, is most useful in the early recognition of decline of glomerular filtration. Detection of tubular dysfunction, by following urinary excretion of markers such as β_2-microglobulin (low molecular weight) or N-acetylglucosaminidase (high molecular weight),[49] may allow earlier detection of graft rejection. β_2-Microglobulin assay may also be helpful in the oliguric patient with a recent transplant. Here it may help differentiate acute tubular necrosis from early acute rejection. The patient with acute renal failure will have higher urine values of this protein than the patient with acute humoral rejection. (See also Chapter 4.)

TESTS OF RENAL FUNCTION

The kidney's excretory, regulatory, and endocrine roles display complex interactions. Not surprisingly, the composition of urine, as well as plasma, reflects not only functional disorders of the nephrons but also various systemic disorders. The practical evaluation of the status of the kidney in renal disease is more or less limited to examining the nephron's functions of glomerular filtration, secretory capacity for particular endogenous and exogenous compounds, and reabsorptive capacity for water, i.e., concentrating ability.

Quantitation of overall renal function is based on a simplifying assumption, the *intact nephron hypothesis.* The intact nephron hypothesis holds that each individual and intact nephron is normally functional and that decline in renal function is due to, and quantitatively related to, loss of functional nephrons. Thus, in nearly all types of diffuse renal disease impaired function of the kidney is attributed to a diminished number of functioning nephrons rather than to the diminished function of individual nephrons. Since glomerular filtration is the initiating phase of all nephron functions, measurement of glomerular filtration rate (or of some variable that bears a constant relationship to it) provides the most useful general index for the physician in his assessment of the severity and progress of renal damage.

Specific defects in particular functions of the nephron can also be identified and evaluated. For instance, measurement of *titratable acidity* of urine (*net acid secretion*) can identify defective function of the distal tubule with respect to hydrogen ion secretion; assessment of the *maximal concentrating capacity* of the kidney gives an estimate of ADH-controlled reabsorption of solute-free water in the distal portion of the tubule. "Pinpoint defects," such as aminoacidurias due to genetically determined deficiency of specific tubular transport systems or toxic and pharmacologic lesions of such transport systems, may be evaluated by *determining quantities of endogenous marker compounds* appearing in urine. (See Chapter 4.) *Secretory capacity* (*tubular secretory mass*) may be estimated by the *p*-aminohippurate (PAH) load test, which measures the T_m (transport maximum) for PAH. T_m is the maximal secretory capacity of the tubule for the compound under investigation. The blood level of PAH is brought to 50 mg/dL by injection of PAH to force the secretory carriers to operate at maximal capacity. Urine is then collected and the amount of PAH excreted is determined. The difference between PAH excreted and PAH filtered at the glomerulus (determined from GFR) represents the PAH secreted; the normal maximum is about 80 mg/min/1.73 m². GFR must be measured by a different technique, such as inulin clearance, so that it may be used to calculate the expected filtered load of PAH. Iodopyracet (Diodrast) may be used in a similar way, but it is less satisfactory. Diminished T_m's can be used to gauge the extent of damage to the tubular secretory apparatus. Most of the tests that measure localized functions of the nephron are employed in a narrow diagnostic context; since these tests are not commonly performed, the reader is referred to the literature.[9,49,61,75]

Of all the possible tests that may be used for evaluating renal function, however, those discussed here in some detail are by far the most practical on a routine basis for diagnosis of impaired renal function and for following the course of progressive, chronic renal disease. They are clearance of various compounds (creatinine, in particular) to estimate glomerular filtration rate; assessment of glomerular permeability by establishing the types of proteins appearing in urine; measurement of plasma creatinine; and measurement of the concentrating ability of the tubules.

Renal Clearance and Glomerular Filtration Rate

Most clinical laboratory information used to assess kidney function is derived from or related to measurement of "clearance" of some substance by the kidney. Clearance is defined as the (hypothetical) quantity of blood or plasma completely cleared of a substance per unit of time and is expressed mathematically as:

$$C_s = (U_s \times V)/P_s$$

where U_s is the urinary concentration of the substance, s; P_s is its plasma concentration; V is urine flow rate in mL/min; and C_s, the *clearance* in units of mL of plasma cleared of the substance per min. C_s is proportional to the rate of solute mass excretion ($U_s \times V$).

The clearance definition is a special application of the Fick principle, which has been widely employed in the evaluation of cardiac output. Simply stated, the principle requires that the amount of a substance entering an organ (here, the kidney) must exactly equal the amount leaving it. Amount in and amount out are functions of arterial and venous concentrations (A_s, V_s) of the substance and of renal arterial and venous plasma flow rates (RAPF, RVPF). When a substance is *not* lost to the urine during its transit of the kidney, at steady state the rate of input equals the rate of output and thus:

$$A_s \times RAPF = V_s \times RVPF$$

or, since both RAPF and RVPF may be represented as RPF without introducing significant error,

$$A_s \times RPF = V_s \times RPF$$

When substance *is* lost to urine as it passes through the kidney carried by the blood, the amount lost must be accounted for by including in the equation the excretion rate, $U_s \times V$, thus:

$$A_s \times RPF = V_s \times RPF + (U_s \times V)$$

This equation, rearranged, becomes

$$(A_s - V_s)RPF = U_sV$$

If *all* of the substance is lost to the urine,

$$V_s = 0 \text{ and}$$
$$A_sRPF = U_sV$$

Since A_s is the same as P_s, the clearance equation above, $C_s = (U_sV)/P_s$, is seen to be equivalent to $RPF = (U_sV)/A_s$, i.e., the volume of plasma cleared of substance per unit of time.

Para-aminohippurate (PAH) is a substance almost completely cleared from the blood by the kidney tubules, and thus its clearance is a measure of tubular excretory capacity.* The small amount ($\sim8\%$) that is not removed is in blood, perfusing nonfunctional regions of the organ such as the pelvis and peripelvic fat. The PAH clearance test is also a useful means of determining total renal plasma flow if tubular function is known to be normal. Renal *plasma* flow determined by PAH clearance is ~574 mL/min per 1.73 m^2.

Clearance of substances that are filtered exclusively or predominantly by the glomerulus but neither reabsorbed nor secreted by other regions of the nephron can be used to measure the glomerular filtration rate. Inulin (a polyfructan) is such a substance, and thus the GFR may be determined by the inulin clearance:

$$GFR = (U_{Inulin} \times V)/P_{Inulin}$$

Furthermore, when a substance is freely filterable through the glomerulus, it is possible to assess quantity reabsorbed from or secreted into the tubule by multiplying its plasma concentration, P_s, times GFR. If the product (quantity filtered per min) is greater than $U_s \times V$ (excretion rate), then the substance must be secreted by the tubule. The magnitude of the difference between filtration and excretion rates is a quantitative measure of tubular reabsorption or secretion of the substance. The inulin clearance measurement, however, is cumbersome and thus has been replaced by some clinicians with the iothalamate clearance, as discussed earlier.

---

*The PAH clearance test differs in both procedure and purpose from the PAH load test noted earlier. For PAH clearance, the dose of PAH injected is calculated to give a blood concentration of ~2 mg/dL.

The clearance technique, with appropriate choice of substance for the measurements, thus allows a quantitative estimation of the glomerular filtration rate, of renal plasma flow, and of tubular reabsorption or secretion. The primary or reference substance for measuring GFR is inulin, and p-aminohippurate is the reference substance for measurement of renal plasma flow. Both substances have been extensively used in clinical and experimental studies, and their clearances are standards against which other proposed clinical tests are evaluated. The necessary plasma and urine assays for inulin and PAH are too difficult and too time consuming to be practical in routine clinical laboratories; thus, creatinine clearance is almost universally used for the clinical assessment of GFR. Plasma and urine creatinine are easily and conveniently measured by either manual or automated methods whose precision and reliability are well documented. These methods will be discussed below. The reader should keep in mind that specificity and precision of any method for any substance whose clearance is measured are critical factors in the clinical application of clearance values.

Glomerular Permeability. The glomerular capillary wall is a special membrane uniquely designed to support rapid ultrafiltration and to retain nearly all proteins within the plasma. This wall (see Figure 11-3) is composed of three structural elements: (1) a layer of endothelial cells that have typical fenestrae; (2) the glomerular basement membrane that separates the endothelial cells from the next layer; and (3) the outer epithelial cell layer that is attached to the basement membrane by discrete cytoplasmic extensions (pedicels or foot processes). Its structure allows for ultrafiltration at a rate that removes ~20% of the plasma water (with small-molecular-weight solutes) during transit of plasma through the glomerular capillary but that restricts passage to protein molecules over a range of molecular weights beginning at ~15 000. Figure 11-4 shows the characteristic relationship between molecular size (as effective solute radius) and fractional clearance of large molecules of increasing size. In addition, molecular shape affects glomerular filtration. Fractional clearance of globular proteins is less than that of similar-sized but linear neutral dextrans. This better clearance presumably results from the flexibility of the dextrans and the ability to turn end on during passage through the filtration barrier.[58] The difference in fractional clearance between neutral dextran and dextran sulfates indicates that not only is size an important determinant of permeance but also molecular charge. Thus, molecules that are highly negatively charged, as are those of the dextran sulfates, exhibit much lower glomerular permeance than do neutral molecules of the same size. This is strong direct evidence that the filtering membrane has

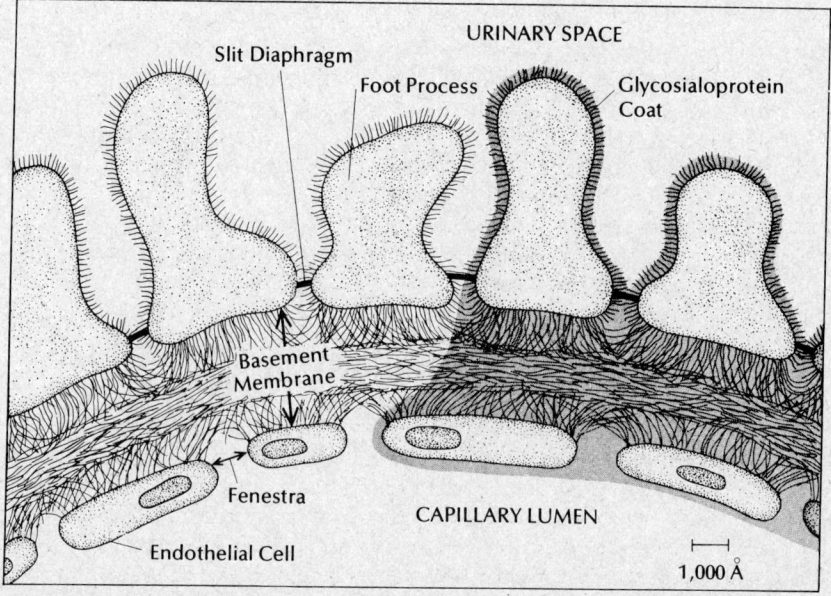

Figure 11-3. Glomerular capillary wall. In glomerular filtration, filtered fluid is believed to traverse the capillary wall via an extracellular route, i.e., through endothelial fenestrae, basement membrane, and slit diaphragms. Circulating polyanions (e.g., albumin) are thought to be retarded by the rich distribution in inner barriers of negatively charged glycosialoproteins (shaded area in schematic diagram). (Used with permission from Brenner, B. M., and Beeuwks, R., III: The kidney in health and disease. III. The renal circulations. Hosp. Prac., *13* [July]:35-46, 1978.)

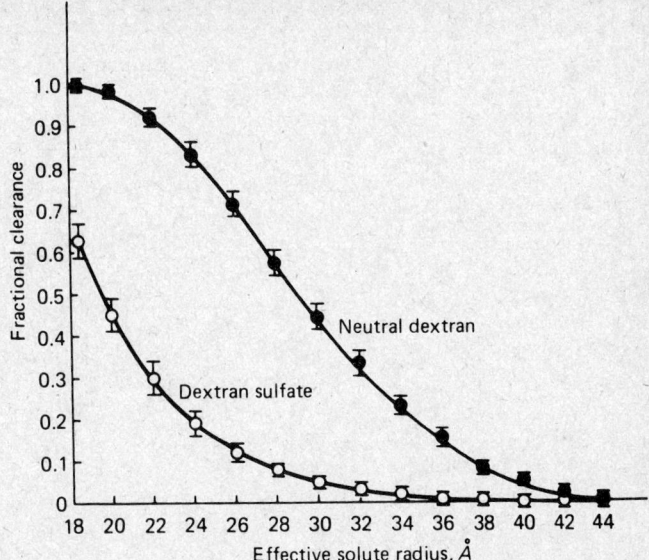

Figure 11-4. Fractional clearances (ratio of dextran to inulin clearance) of tritiated neutral dextrans and dextran sulfates as a function of their effective molecular radii. Decreased filtration of dextran sulfates is consistent with a charge limitation to filtration. For orientation, the effective molecular radius of albumin is approximately 36 Å, of lactoperoxidase 38 Å, of catalase 52 Å, and of ferritin 61 Å. (From Brenner, B. M., Bohrer, M. D., Baylis, C., et al.: Kidney Int., *12*:229, 1977. Reprinted from Kidney International with permission.)

a negative charge that tends to repel negatively charged molecules and thus decreases the effective radius of "pores" in the membrane as molecules with like charge approach it. (See Figure 11-3.) Studies suggest that a major form of injury to the glomerulus in human renal disease may well be loss of glomerular polyanionic charge, perhaps because of decrease or alteration of sialoprotein content of the basement membrane.[6]

The glomerular filtration apparatus may become grossly distorted during the course of numerous diffuse renal diseases, with the primary derangement being loss of the ability to retain selectively the large protein molecules. The result is loss of large quantities of low- and high-molecular-weight proteins in the urine. Three types of measurements provide diagnostic and prognostic information when proteinuria is a feature of renal disease; they are (1) measurement of excretion rate of total urinary protein; (2) electrophoresis of urinary proteins; and (3) measurement of selective clearance of proteins of different molecular sizes.

Measurement of the excretion rate of total urinary protein is the standard means to quantitate the severity of proteinuria. Table 11-4 illustrates the rates of protein excretion associated with several categories of renal disease. The magnitude of protein loss generally directs the physician toward selection of an appropriate diagnostic category. Some diseases rarely have protein excretion rates of more than 1 g/d; others may have rates of more than 3 g/d. The latter category includes patients who have persistent or intermittent clinical manifestations of the nephrotic syndrome. A characteristic means of evaluating the effectiveness of therapy (e.g., when corticosteroids are used to treat some forms of the nephrotic syndrome) is to follow the protein excretion rate sequentially. The excretion rate is ordinarily determined on 24-h specimens since it tends to fluctuate during the day with changes of activity and posture.

Qualitative evaluation of urine proteins can be valuable in the differential diagnosis of diseases producing proteinuria. The relatively benign cause of nephrotic syndrome, minimal change disease (lipoid nephrosis), produces little alteration in glomerular structure other than loss of glomerular basement membrane charge characteristics, and hence produces a selective proteinuria consisting almost entirely of albumin. More severe glomerulonephritis causes significant glomerular injury with resulting loss of larger-molecular-weight globulins into the urine and hence an increased globulin/albumin or IgG/albumin ratio that can readily be measured. (See also Chapter 4.) Tubular diseases result in decreased reabsorption of those small proteins that are normally filtered freely. Thus, one can detect increased urinary levels of lysozyme, β_2-microglobulin, and immunoglobulin light chains in such conditions as acute tubular necrosis (shock kidney), heavy metal poisoning, chronic transport rejection, and nephrotoxicity after intake of certain drugs.

Electrophoresis and immunoelectrophoresis of urinary proteins are valuable in the diagnosis of Bence Jones proteinuria (light chain disease). On occasion, the presence of a monoclonal urinary protein of a M.W. of ~22 000 can be seen in the urine for months or years before any evidence of myeloma or other lymphoproliferative disorder becomes manifest.

Table 11–4. RANGES OF PROTEIN EXCRETION RATES IN RENAL DISORDERS*

	Range, mg/d				
	50	100	1000	3000	>3000
Exercise proteinuria	··········				
Febrile proteinuria	········				
Postural proteinuria	············	··········	··········		
Bence Jones proteinuria	··········	··········	··········	··········	··········
Arteriosclerotic renovascular diseases	··········	····			
Hypertension, arterial	··········	····			
Congestive heart failure	··········	··········			
Interstitial nephritis	··········	··········			
Pyelonephritis	··········	········			
Polycystic kidney disease	········	··········	··········		
Medullary cystic disease	··········				
Acute glomerulonephritis†	··········	··········	··········	··········	··········
Membranous glomerulonephritis			··········	··········	··········
Membranoproliferative glomerulo-nephritis			··········	··········	··········
Nephrotic syndrome‡				··········	··········
Kimmelstiehl-Wilson syndrome (diabetic glomerulopathy)			··········	··········	··········
Systemic lupus erythematosus	··········	··········	··········	··········	··········
Polyarteritis nodosa	··········	··········	··········	··········	··········
Goodpasture's syndrome	··········	··········	··········	··········	··········
Systemic sclerosis	··········	··········	········		
Multiple myeloma§	··········	··········	··········	··········	··········
Nephrotoxins (e.g., Hg, CCl₄)	··········	··········	··········		

* The divisions are to be considered approximations rather than the absolute limits of protein excretion rates. Some of the disease categories in which rates of > 3000 mg/d are reached may in fact have excretion rates that exceed 15 000 mg/d. Large dots indicate common, small dots rare (< 10%) occurrences. Note that scale is logarithmic.

† Approximately 10% of cases show rates greater than 3000 mg/d.

‡ By definition > 3 g/d.

§ Refers to generalized disease rather than to light chain (Bence Jones) disease.

NONPROTEIN NITROGEN-CONTAINING COMPOUNDS

Catabolism of proteins and nucleic acids results in formation of so-called NPN compounds (*nonprotein nitrogenous* compounds). The principal nonprotein nitrogenous substances, their biochemical origins, and situations in which their assay has significant clinical utility are summarized in Table 11-5.

Several of these metabolic products are sequentially derived from the catabolism of either exogenous (dietary) or endogenous (tissue) proteins:

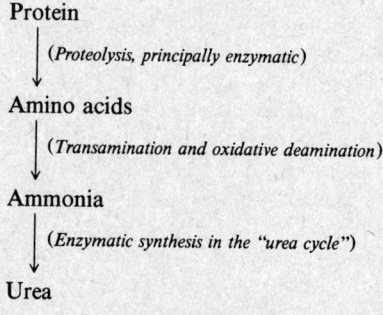

Protein

    ↓ (*Proteolysis, principally enzymatic*)

Amino acids

    ↓ (*Transamination and oxidative deamination*)

Ammonia

    ↓ (*Enzymatic synthesis in the "urea cycle"*)

Urea

UREA

Biochemistry and Physiology

Urea is the major nitrogen-containing metabolic product of protein catabolism in man, comprising over 75% of the nonprotein nitrogen eventually excreted. The biosynthesis of urea

Table 11-5. URINARY NITROGENOUS METABOLITES*

Metabolite	Biochemical Origin	Clinical Utility of Measurement	Per cent of Urine Nitrogen
Amino acids	Proteins, endogenous and exogenous	Liver disease; inborn errors of metabolism; tubular disorders	< 1
Ammonia†	Amino acids	Liver disease; renal disease, congenital and acquired; inborn errors of metabolism	10–20
Urea	Ammonia	Liver disease; renal disease	55–90
Creatinine	Creatine	Renal function	2–3
Uric acid	Purine nucleotides	"Marker" for cell turnover; disorders of purine synthesis	1–1.5

*These compounds account for > 90% of urinary nonprotein nitrogen.
†Present in urine as NH_4^+.

from amino nitrogen-derived ammonia is carried out exclusively by hepatic enzymes of the urea cycle shown in Figure 11-5. (See also Chapter 13.)

Over 90% of urea is excreted through the kidney, with losses through the gastrointestinal tract and skin accounting for most of the remaining minor fraction. Urea is neither actively reabsorbed nor secreted by the tubules but is filtered freely by the glomerulus. In the normal kidney, 40–70% of the highly diffusible urea moves passively out of the renal tubule and into the interstitium, ultimately to re-enter plasma. The back-diffusion of urea is also dependent on urine flow rate, with more entering the interstitium in slow flow states. Consequently, urea clearance underestimates GFR. More importantly, urea production is too dependent upon several

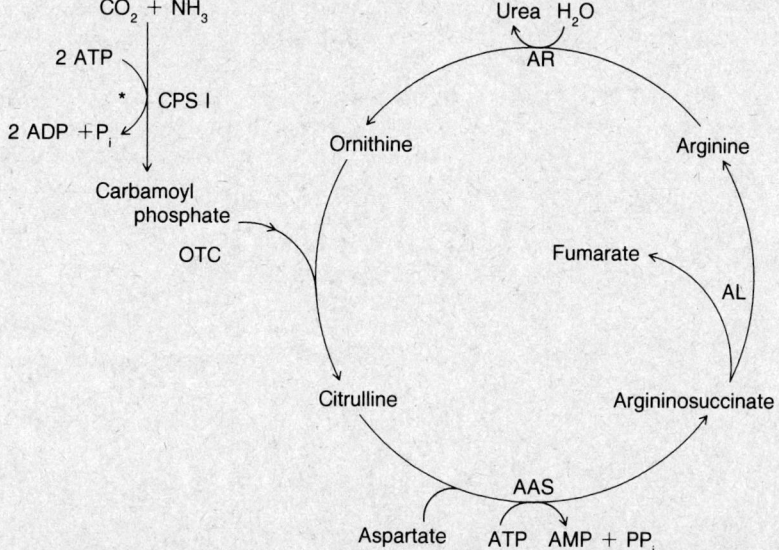

Figure 11-5. Biosynthesis of urea. CPS I: carbamoyl phosphate synthetase I; *N-acetylglutamate as positive allosteric effector; OTC: ornithine transcarbamoylase; AAS: argininosuccinate synthetase; AL: argininosuccinate lyase; AR: arginase.

nonrenal variables such as diet and hepatic synthesis to make it useful as a measure of GFR. At present, utility of urine urea measurement is limited to partition of urinary NPN among the various nitrogen-containing constituents and to the measurement of the urea production rate. In chronic renal failure the osmotic diuresis in the remaining functional nephrons limits the back-diffusion of urea so that urea clearance approaches inulin clearance. Measurement of plasma or serum urea, however, may provide useful clinical metabolic information in particular circumstances.

Clinical Utility of Plasma Urea Measurement

A wide variety of renal diseases with different permutations of glomerular, tubular, interstitial, or vascular damage can cause an increase in plasma urea concentration. Unfortunately, the usefulness of urea as an independent indicator of renal function is limited by the variability of its blood levels as a result of nonrenal factors. Mild dehydration, high-protein diet, increased protein catabolism, muscle wasting as in starvation, reabsorption of blood proteins after a GI hemorrhage, treatment with cortisol or its synthetic analogs, and decreased perfusion of the kidneys may cause an azotemia (increased blood urea), which is called *prerenal azotemia*. Impaired perfusion may be due to decreased cardiac output or shock due to blood loss or other causes. The key to identifying the azotemia as prerenal is documentation of an increase of plasma urea without concomitant increase of plasma creatinine. *Postrenal azotemia* has its cause in conditions that obstruct urine outflow through the ureters, bladder, or urethra; examples of these conditions are nephrolithiasis, prostatism, and tumors of the genitourinary tract. With obstruction, plasma urea and creatinine both rise, but there is a disproportionately greater rise of urea than of creatinine, because the obstruction of urine flow causes a back-pressure on the tubule and back-diffusion of urea into the blood from the tubule.

Thus, the principal clinical utility of plasma urea determination lies in its measurement in conjunction with the measurement of plasma creatinine, and in discrimination between prerenal and postrenal azotemia. Clinicians frequently calculate a convenient relationship, the *urea nitrogen/creatinine ratio*:

$$\frac{\text{Serum urea nitrogen, mg/dL}}{\text{Serum creatinine, mg/dL}}$$

For a normal individual on a normal diet, the ratio ranges between 12 and 20, with values for most individuals lying between 12 and 16. Significantly lower ratios usually denote acute tubular necrosis, low protein intake, starvation, or severe liver disease (decreased urea synthesis). High ratios with *normal* creatinine levels may be seen with catabolic states of tissue breakdown; prerenal azotemia; high protein intake, especially in uremic patients; and after gastrointestinal hemorrhage. High ratios associated with *elevated* creatinine concentrations may denote either postrenal obstruction or prerenal azotemia superimposed on renal disease. The ratio may also be affected by the degree of specificity of the method used for urea and creatinine determination. Thus, the urea nitrogen/creatinine ratio can show great variability and is only a rough guide to the nature of the underlying abnormality. Its magnitude is not tightly regulated in health or disease and should not be considered a precise quantity.

Methods for the Determination of Urea

Indirect methods are based on preliminary hydrolysis of urea with urease (urea amidohydrolase, EC 3.5.1.5), followed by some process that quantitates the ammonium ion.[18]

$$\begin{matrix} H_2N \\ \searrow \\   C=O \ + \ H_2O \xrightarrow{\text{Urease}} 2\,NH_3 + CO_2 \\ H_2N \nearrow \end{matrix}$$

$$\downarrow H_2O$$

Urea $2\,NH_4^+ + CO_3^{2-}$

Spectrophotometric approaches to ammonium quantitation include the *Berthelot reaction* (described in the manual method given below) and the *enzymatic assay* with glutamate dehydrogenase [L-glutamate:NAD(P) oxidoreductase (deaminating), EC 1.4.1.3].

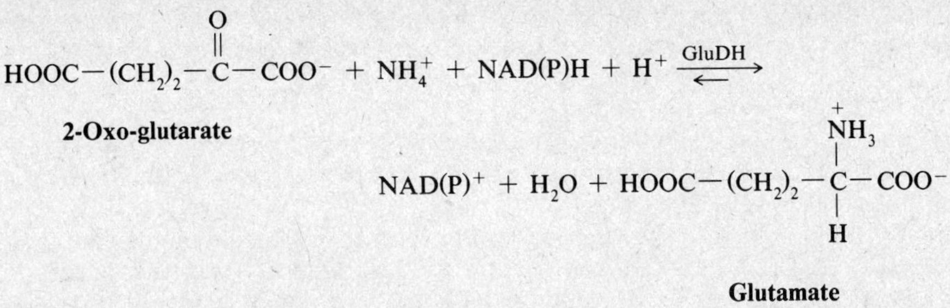

$$HOOC-(CH_2)_2-\overset{\overset{\textstyle O}{\|}}{C}-COO^- + NH_4^+ + NAD(P)H + H^+ \xrightleftharpoons{GluDH}$$

2-Oxo-glutarate

$$NAD(P)^+ + H_2O + HOOC-(CH_2)_2-\overset{\overset{\textstyle +}{\underset{\textstyle |}{NH_3}}}{\underset{\underset{\textstyle H}{|}}{C}}-COO^-$$

Glutamate

For serum or plasma assays, the reaction system is usually formulated with urease so that the addition of sample containing urea starts the reaction. Decrease in absorbance, resulting from the glutamate dehydrogenase reaction, is monitored at 340 nm in either an endpoint or kinetic mode.[60,71] The coupled-enzyme system has been automated on the DuPont *aca* (DuPont Co., Wilmington, DE 19898). *Electrochemical* approaches to quantitation of ammonium involve conductimetry and potentiometry. The conductimetric method has been automated on the ASTRA instruments (Beckman Instruments, Brea, CA 92621), where a sample and a urease-containing reagent are introduced into a conductivity cell. The rate of change of the conductivity of the mixture, as nonionized urea is transformed to an ionic species, is measured.[1] Potentiometry using an ammonium ion-selective electrode fitted with a membrane on which urease has been immobilized,[25] although promising, has not yet come into widespread use; electrode response tends to be undesirably slow for routine application.

Indirect methods may be *standardized* with either urea or ammonium salt solutions, but urea is much to be preferred. Any of these methods are at risk of ammonia contamination of reagents or glassware. Some caution must be observed when selecting an indirect method for assay of urine, where significant amounts of preformed ammonia may be present in specimens. Either the preformed ammonia must be removed with Permutit prior to assay (as in the method described below) or the method must prescribe a step that compensates for it before urease is added. Use of the enzymatic assay would require blanking by running sample in the reaction system both before and after urease addition. The automated conductimetric method noted above compensates for preformed ammonia by making a measurement of the rate of change in conductivity due to ammonia production.

Modern *direct methods* for urea are usually variations on the Fearon reaction, the condensation of diacetyl with urea to form the chromogen diazine.[72] Since diacetyl is unstable, it is usually generated in the reaction system from diacetyl monoxime and acid.

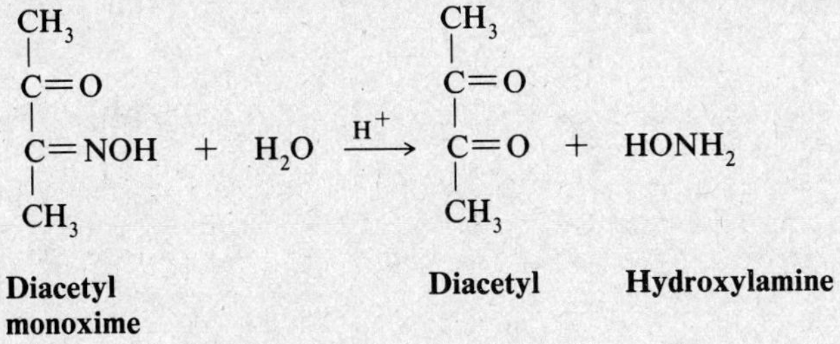

Diacetyl
monoxime   **Diacetyl**   **Hydroxylamine**

The reaction of diacetyl and urea gives diazine

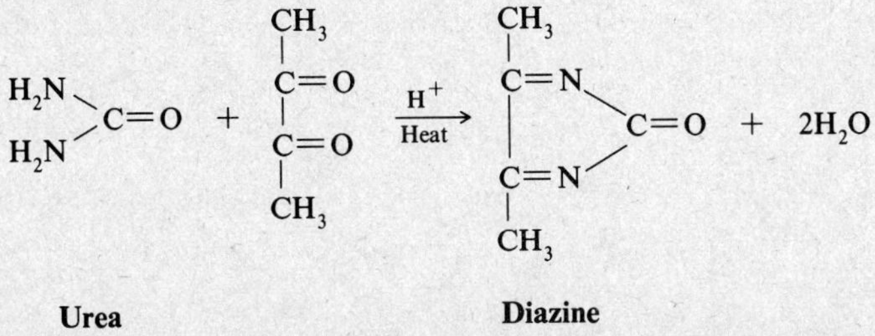

Urea   **Diazine**

which absorbs strongly at 540 nm. Thiosemicarbazide and Fe(III) are added to the system to enhance and stabilize the color. The method may be used manually and has been automated on various systems, among them the Technicon continuous flow systems.[43] Urea must be used as standard for a direct method. The Fearon reaction is applicable to both plasma and urine assays.

Comments

Contemporary methods for serum or plasma, whether direct or indirect, rarely call for deproteinization of sample. The need has been obviated by selection of small sample volumes and by optimization of reagent systems.

Although BUN (*b*lood *u*rea *n*itrogen) continues to be the terminology for ordering the plasma or serum urea nitrogen test, analysis of *blood* for urea has essentially been discontinued.

The long-established habit of reporting and expressing results of a urea assay in units of urea nitrogen appears to be strongly entrenched in the USA, although the SI system recommends use of urea, expressed in mmol/L. Thus, it behooves the student of clinical chemistry to have in mind the conversion factors for urea to urea nitrogen. Since 60 g (one gram molecular weight) of urea contains 28 g (two gram atomic weights) of nitrogen, the factor is 0.467 for converting urea mass units to those of urea N, and 2.14 for converting urea nitrogen mass units to those of urea. The factor for converting urea nitrogen in mg/dL to urea in mmol/L or mOsmol/L is 0.357.

Method for the Determination of Urea (Berthelot Reaction)

Specimen. Either serum or urine is suitable for the assay. Plasma is also suitable, provided that the anticoagulant used contains neither ammonium nor fluoride salts. Urea in serum is stable for up to 24 h at room temperature, for at least several days at 4–6 °C, and for at least 2–3 months when frozen.

Urea is stable in urine at pH < 5.0 for several days at 4–6 °C. However, contamination of urine with bacteria that hydrolyze urea is common; rapid growth of such organisms will cause loss of urea and formation of ammonia.

Principle. Urea is hydrolyzed with urease, and the ammonium ion formed is reacted with phenol and hypochlorite in alkaline medium to form indophenol.

Nitroprusside is used to catalyze the reaction. Absorbance of dissociated indophenol, a blue chromogen, is measured at 560 nm.

Reagents

1. Ammonia-free water. Pass distilled water through a mixed cation-anion exchange resin bed and store the product in a glass-stoppered bottle.

2. Phenol-nitroprusside solution. Place 10.0 g phenol and 0.05 g of sodium nitroferricyanide dihydrate (reagent grade) in 1.0 L of ammonia-free water. Store at 4–6 °C for up to 60 d.

3. Alkaline hypochlorite solution. Dissolve 5.0 g sodium hydroxide in ~500 mL of ammonia-free water, cool, and add 0.42 g sodium hypochlorite; dilute to 1.0 L with ammonia-free water. Store in an amber bottle, refrigerated, for up to 60 d.

4. Urease. Prepare a stock suspension by mixing 0.2 g urease (activity 3000–4000 units/g) with 10.0 mL ammonia-free water and 10.0 mL glycerol (stable at 4–6 °C for 6 months). Prepare a working reagent by diluting 1.0 mL of stock suspension to 100 mL with disodium EDTA solution (10.0 g/L, pH 6.5, in ammonia-free water). Working reagent is stable for up to 1 month if refrigerated.

5. Urea standard. Dissolve 1.0717 g anhydrous urea, reagent grade, in 100 mL ammonia-free water, adding 0.1 g sodium azide as preservative. The stock standard represents 500 mg urea *nitrogen* per dL and is stable for up to 6 months at 4–6 °C.

6. Working standards (10, 25, and 50 mg/dL as urea nitrogen). Dilute 2.0, 5.0, and 10.0 mL stock standard to 100 mL with ammonia-free water.

7. Permutit, according to Folin, 40- to 60-mesh.

Procedure (Serum or Plasma)

1. Label 16 × 125 mm tubes for each unknown and control sample, for each working standard, and for a blank.

2. Add 1.0 mL working urease solution to each tube.

3. Add 10 μL of samples and standards to appropriate tubes.

4. Mix and incubate all tubes at 37 °C for 15 min.

5. Rapidly add 5.0 mL phenol-nitroprusside solution and mix; then add 5.0 mL of alkaline hypochlorite solution and mix.

6. Incubate all tubes at 37 °C for 20 min.

7. Read absorbance for each tube at 560 nm against reagent blank set to zero absorbance.

8. Prepare a calibration curve of absorbance versus concentration (in mg/dL urea nitrogen) and read values for unknowns and controls from the curve.

Procedure (Urine)

1. Treat the urine to remove preformed ammonia by mixing 1.0 mL urine with 0.5 g of washed Permutit in a total volume of 25 mL made up with ammonia-free water.

2. Perform steps 1–7 of procedure for serum or plasma and read urea nitrogen concentration from the curve. Multiply the value by 25 to account for the initial dilution. The result is urea nitrogen in mg/dL.

3. To calculate urea nitrogen excretion per day, use the formula:

$$\text{Urea N, g/d} = (\text{Urea N, mg/dL})/100 \times \text{24-h urine volume in mL}$$

Comments

1. The method was simplified by Chaney and Marbach,[12] who introduced the combined reagents, phenol-nitroprusside and alkaline hypochlorite.

2. The sources of error include ammonia contamination of reagents or glassware from ammonia vapors in the air or of specimen with ammonium salt anticoagulant. Fluoride present in plasma in high concentrations will cause falsely low results by inhibiting urease.

Reference Ranges

The reference range for plasma of healthy ambulatory adults is 7–18 mg/dL as urea nitrogen (15–39 mg/dL or 2.5–6.4 mmol/L expressed as urea). Levels in the neonatal period are 3–6 mg/dL lower than in adults; in adults over 60 years of age, levels are 1–4 mg/dL higher than in younger adults.[70] Plasma concentrations also tend to be slightly higher in males than in females. High-protein diet causes significant increases in plasma urea concentrations and urinary excretion. On an average protein diet, urinary excretion expressed as urea N is 17–20 g/d.

References

Chaney, A. L., and Marbach, E. P.: Modifed reagents for determination of urea and ammonia. Clin. Chem., 8:130-132, 1962.

Kaplan, A.: Urea nitrogen and urinary ammonia. In: Standard Methods of Clinical Chemistry, Vol. 5. S. Meites, Ed. New York, Academic Press, 1965, pp. 245-256.

CREATININE AND CREATINE

Biochemistry and Physiology. Creatine is synthesized in the kidney, liver, and pancreas by two enzymatically mediated reactions. In the first, transamidation of arginine and glycine forms guanidinoacetic acid; the second, methylation of guanidinoacetic acid, occurs with S-adenosyl methionine as methyl donor.[51] (See p. 1272.)

Creatine is then transported in blood to other organs such as muscle and brain, where it is phosphorylated to phosphocreatine, a high-energy compound. Interconversion of phosphocreatine and creatine is a particular feature of metabolic processes of muscle contraction; some of the free creatine in muscle spontaneously converts to creatinine, its anhydride. (See p. 1272.) Between 1 and 2% of muscle creatine is converted to creatinine daily. Because the amount of endogenous creatinine produced is proportional to muscle mass, the production varies with age and sex; nonobese adult males excrete ∼1.5 g/d, females 1.2 g/d. Daily excretion of creatinine can be 10–30% greater as a result of dietary intake of creatine and creatinine in meats. On the whole, however, dietary fluctuations of creatinine intake cause only minor variation in daily creatinine excretion of the same individual. The excretion rate in any one individual, in the absence of renal disease, is relatively constant and parallels endogenous production. Most of the interindividual variations of creatinine excretion in healthy subjects are attributable in the main to age, sex, and lean body mass, and intraindividual variation tends to be <15% from day to day.

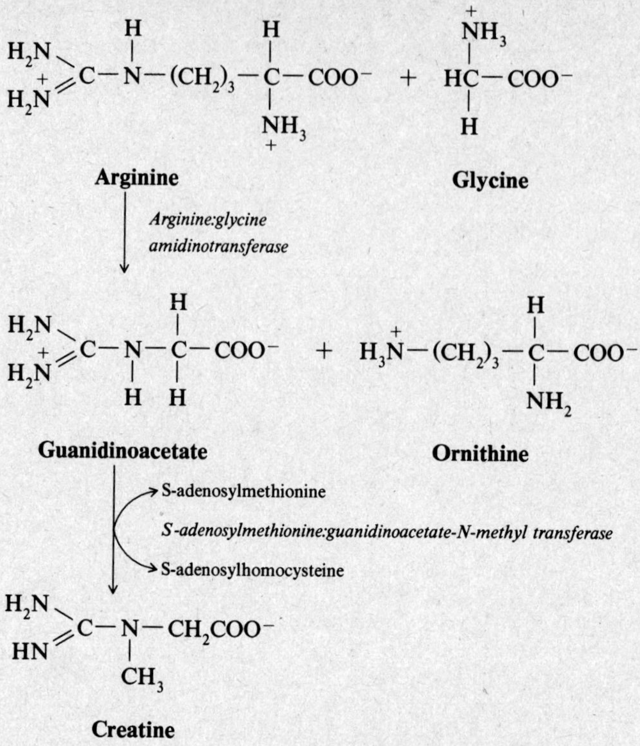

Arginine   **Glycine**

Arginine:glycine amidinotransferase

Guanidinoacetate   **Ornithine**

→ S-adenosylmethionine

S-adenosylmethionine:guanidinoacetate-N-methyl transferase

→ S-adenosylhomocysteine

Creatine

Clinical Utility of Creatinine Clearance* and Determination of Plasma Creatinine

Constancy of endogenous creatinine production and its release into the body fluids at a constant rate, and constancy of plasma levels of creatinine over the 24 h of a day, make creatinine a useful endogenous substance whose clearance may be measured as an indicator of glomerular filtration rate. However, a small quantity of creatinine is reabsorbed by the tubules and a small

---

*The creatinine clearance is performed by obtaining a 4-, 12-, or 24-h urine specimen and also a blood specimen sometime within the period of urine collection. The volume of the urine is measured, urine flow rate is calculated (mL/min), and the assay for creatinine is performed on plasma and urine to obtain the concentration in mg/dL or mg/mL. The procedure is fully described in the method section below.

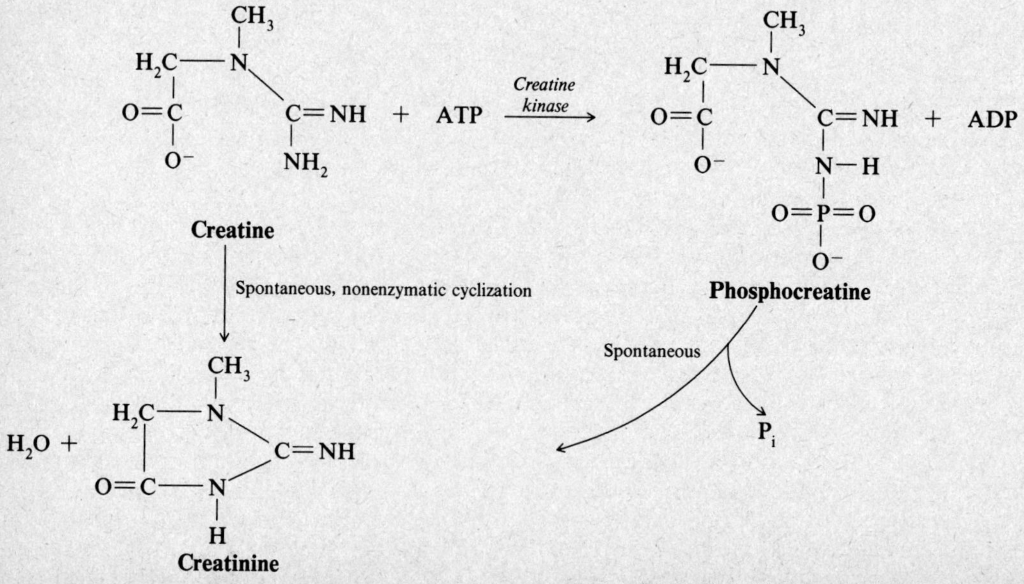

Creatine $+$ ATP $\xrightarrow{\text{Creatine kinase}}$ **Phosphocreatine** $+$ ADP

Spontaneous, nonenzymatic cyclization

Spontaneous

P_i

Creatinine

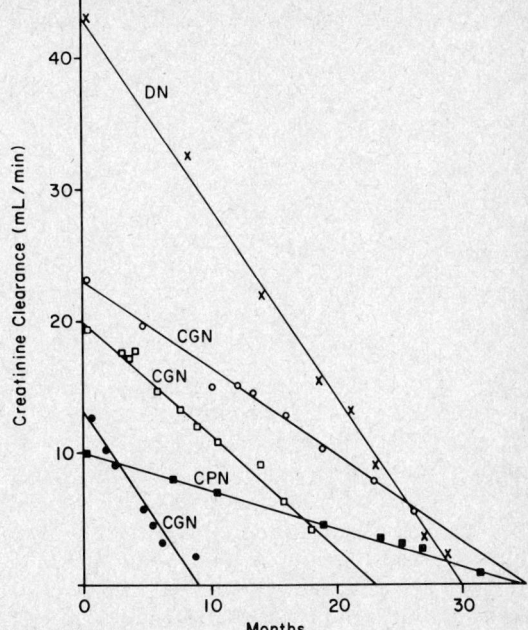

Figure 11-6. Changes in creatinine clearance with time for some diffuse renal diseases. CGN: chronic glomerulonephritis; DN: diabetic nephropathy; CPN: chronic pyelonephritis. Sequential determinations of creatinine clearance over periods ranging from 10 to 30 months illustrate clearly the linear decrease in renal function in progressive renal disease.

quantity of creatinine appearing in the urine (7–10%) is due to tubular secretion. As a result, creatinine clearance (if creatinine is measured with a specific method) is ~7% greater than inulin clearance.[46] Most methods for creatinine used in clinical laboratories, however, are nonspecific, and thus this difference is often smaller. Variability of individual clearance results still remains. (See later section, Endogenous Creatinine Clearance.) An occasional patient with congestive heart failure may have a creatinine clearance less than the inulin clearance, thus raising the possibility of tubular reabsorption.

Two factors influence measurement of creatinine clearance and thus its correct interpretation. First, the most common methods for measuring creatinine employ the nonspecific alkaline picrate reaction, and thus noncreatinine chromogens in plasma increase the apparent plasma concentration by as much as 30% if serum values are < 1.0 mg/dL and by ~10% if values are > 1.0 mg/dL. The per cent increase is progressively less with higher creatinine concentrations. (Urine contains considerably less noncreatinine chromogens.) This overestimate of plasma creatinine concentration

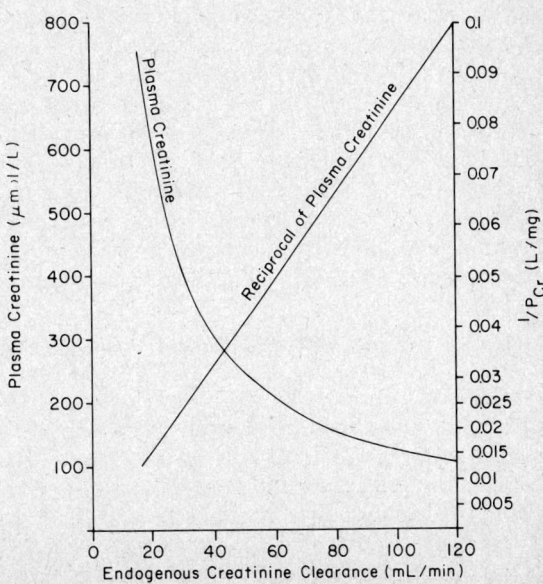

Figure 11-7. Relation between creatinine clearance, creatinine excretion rate, and reciprocal of plasma creatinine concentration. Note that reciprocal of plasma creatinine concentration exhibits a linear relationship with clearance and is thus a more convenient means of following change than is the hyperbolic relationship between plasma creatinine concentration and clearance.

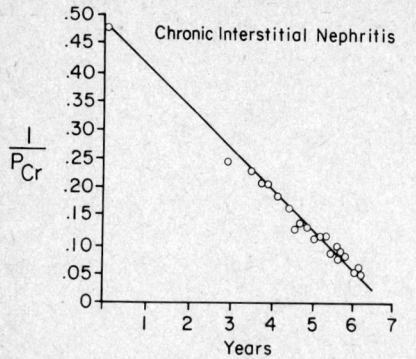
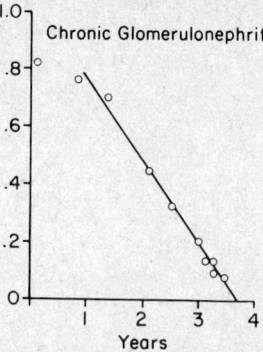

Figure 11-8. Changes in the reciprocal of plasma creatinine concentration as a function of time in cases of progressive renal disease. The straight line relationship allows a quantitative estimate of the rate of loss of renal function in an individual case.

results in an underestimate of creatinine clearance and partially offsets the apparent high clearance of creatinine that is due to tubular secretion. As a result, the endogenous creatinine clearance agrees closely with the inulin clearance over a substantial range of clearances.[46] If highly specific methods are used for assay of plasma creatinine, the GFR estimated by creatinine clearance may not be as well correlated with the GFR estimated by inulin clearance. Second, GFR measured by creatinine clearance and GFR measured by inulin clearance in the same patient progressively diverge as renal failure progresses and plasma creatinine rises. The greater apparent GFR found by creatinine clearance may be due to an increase in tubular secretory activity for creatinine when plasma levels increase much above normal and to the relatively smaller contribution of noncreatinine chromogens in a nonspecific assay of plasma creatinine.

The creatinine clearance as a measure of GFR has several unique advantages over the clearance of inulin or other exogenous substances. Not only are plasma and urine creatinine easily measured, but extensive data are available for reference ranges of both plasma creatinine levels and creatinine clearances in individuals of all age groups. Presently a majority of physicians feel that the advantages of creatinine clearance outweigh its acknowledged failings. Pending accumulation of reference data for creatinine clearance that are based on "true" plasma creatinine values or the widespread availability of alternative and less ambiguous clearance tests, the creatinine clearance—flawed as it is—remains the most practical and popular method for estimating GFR. (See page 1280.)

The linear decrease of creatinine clearance over time, as renal function fails, has been thoroughly documented for several forms of chronic glomerulonephritis (including chronic membranous glomerulonephritis), diabetic nephropathy, chronic interstitial nephritis, and at least some cases of chronic pyelonephritis. Moreover, graphic representation of creatinine clearance as a function of time (see Figure 11-6) has some prognostic value since it may allow the physician to estimate when the severity of renal failure may make dialysis mandatory. Another parameter of creatinine metabolism, the reciprocal of plasma creatinine concentration, has also been used as a guide in following and documenting the rate of loss of renal function in progressive, diffuse renal disease. As may be seen in the left side of Figure 11-7, P_{Cr} decreases hyperbolically as C_{Cr} increases linearly. The reciprocal, however, $1/P_{Cr}$ (right-hand side of Figure 11-7), has a first-order linear relationship versus C_{Cr}. As the creatinine clearance of a patient increases (or decreases), the reciprocal of his plasma creatinine concentration increases (or decreases) linearly. Thus, sequential plots of the reciprocal of P_{Cr} (as shown in Figure 11-8) are an alternative to plots of creatinine clearance versus time when following the clinical course of the patient with progressive renal impairment.

Estimation of Creatinine Clearance from Plasma Creatinine Level

Clinical pharmacologists, pharmacokineticists, and clinicians, when supervising the administration of potentially toxic drugs cleared by the kidney, commonly use creatinine clearance as an adjunct to dose management. Among the drugs of concern are the nephrotoxic aminoglycosides and also digoxin, guanethidine, sulfonamides, methotrexate, and 5-fluorouracil. Because plasma creatinine determinations are quicker and easier than the creatinine clearance procedure, a one-time plasma creatinine level is sometimes used, together with the patient's body weight, age, and sex, to arrive at an estimate of clearance.[13] The estimate may be based on a calculation[5,14,37] or derived from a nomogram.[62] Examples of a nomogram are shown in Figure 11-9.

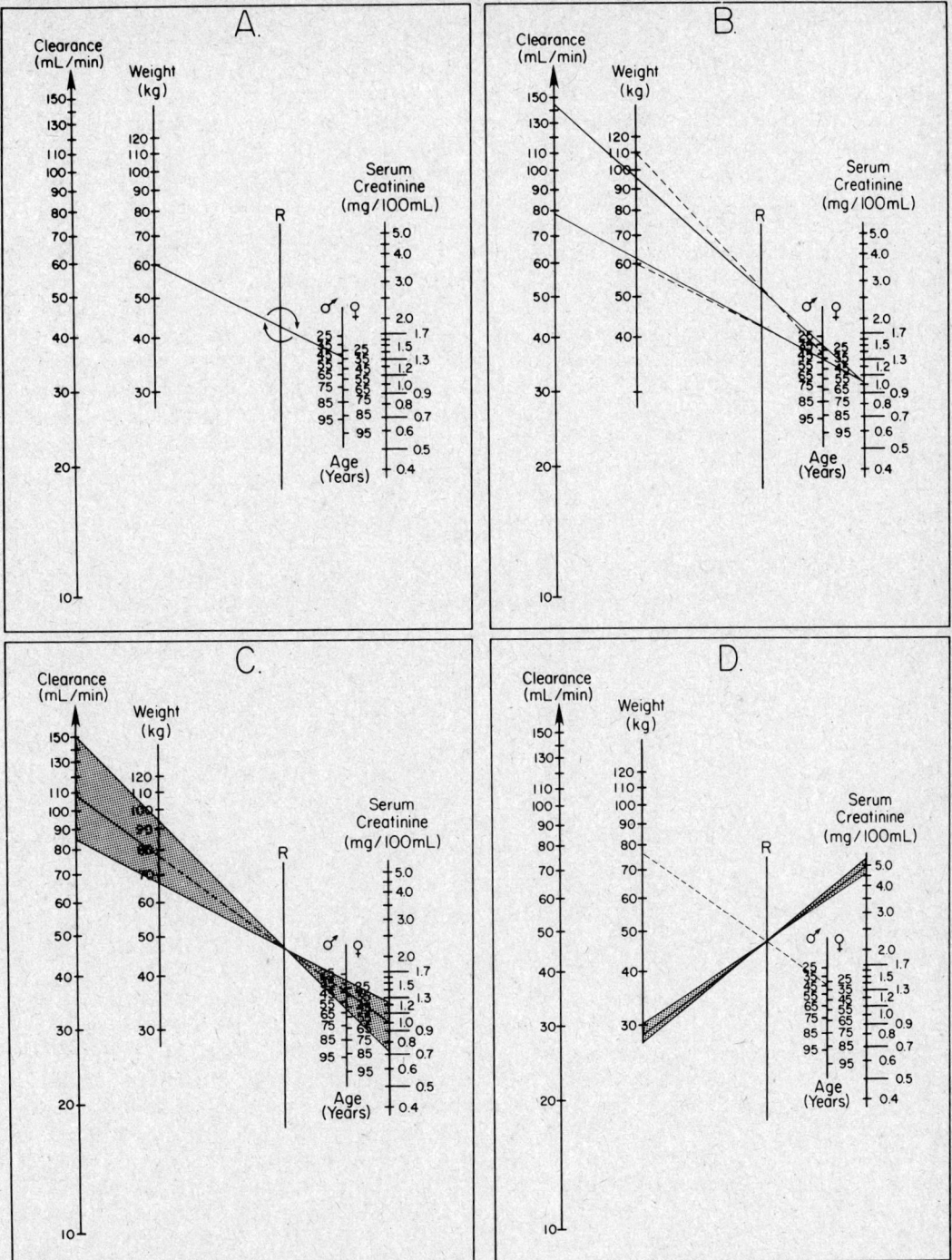

Figure 11-9. The Siersbaek-Nielsen nomogram for estimation of endogenous creatinine clearance. The first step in estimation (part *A*) is the identification of the axis along the reference line "R" around which the relation between plasma creatinine and creatinine clearance rotates. The point on the line "R" is defined as a function of the age, weight, and sex of the patient. The nomogram is based on the assumption that an increase in body weight represents mainly increase in lean body mass. *B*, Potential obesity error. *C*, 95% confidence limits of creatinine clearance estimate from serum creatinine in the normal individual. *D*, 95% confidence limits of creatinine clearance estimate in advanced renal failure. (Drawn with permission from Siersbaek-Nielsen, K., Molholm-Hansen, J., Kampmann, J., et al.: Rapid evaluation of creatinine clearance. Lancet *1*: 1133-1134, 1971.)

Estimations of clearance made by these techniques must be interpreted with great caution! Note that in Figure 11-9,*A* a point is first defined on the "R" line by the patient's body weight, age, and sex. The point is then lined up with the observed plasma creatinine (right-hand scale) and the estimate of clearance is read from the left hand scale. Obesity of the patient is a source of potentially large error in the clearance estimate even if correction for weight is made. Note in Figure 11-9,*B* that 60 kg of lean body mass, observed as 100 kg of body weight in the obese patient, would estimate clearance at about half the estimate obtained if a body weight of 100 kg had been used. The error contributed by obesity can lead to the belief that clearance is normal when, in fact, it is low. This error can be reduced by correcting obese body weight to estimated lean body mass by one or another of the algorithms used for this purpose. In general practice, use of the nomogram for obese patients is discouraged.

The random error of the plasma creatinine determination is another source of error in estimating clearance. Note that in Figure 11-9,*C*, a single creatinine value of 1.0 mg/dL with 95% ($\pm$2s) confidence limits (between 0.72 and 1.28) may estimate clearance anywhere from 85–155 mL/min. Because precision of plasma creatinine determinations increases as the concentration rises, the error in estimate of clearance decreases, as is evident in Figure 11-9,*D*. As a general guide, an estimated clearance should not be considered reliable when plasma creatinine is < 2.5 mg/dL but might be acceptable when creatinine is 2.5–5.0 mg/dL. When plasma creatinine exceeds 5.0 mg/dL, the clinical situation is sufficiently critical that a preliminary, estimated clearance should be followed as soon as possible by a full, standard creatinine clearance measurement.

Determination of Creatinine

The methods most widely used today are based on the *Jaffe reaction,* first described in 1886.[34] The reaction occurs between creatinine and the picrate ion formed in alkaline medium; a red-orange adduct develops, but in spite of extensive study its structure remains uncertain. A postulated structure for the adduct is a 1:1 Janovsky complex.[11]

Picrate-creatinine complex

Concentration of alkali in the reagent system is critical. The observed rate constant of adduct formation is a linear function of the hydroxyl ion concentration over a broad range of picric acid concentrations.[11] Not only does hydroxyl ion concentration determine the rate of reaction, it also specifies to a large extent the behavior of the spectral absorbance curve of the adduct over the wavelengths 485–520 nm.[51] Modern methods usually employ a sodium hydroxide concentration below 1.0 mol/L, in most instances below 0.5 mol/L, and picric acid in excess of stoichiometric amounts. In most methods, wavelengths of 505–520 nm are specified for spectrophotometry and a narrow-bandpass spectrophotometer is required to provide appropriate analytical sensitivity and specificity.

The Jaffe reaction has long been known to be nonspecific when applied to the measurement of creatinine in plasma. Among the noncreatinine Jaffe-reacting chromogens are protein, glucose, ascorbic acid, guanidine, acetone, cephalosporins and α-ketoacids such as acetoacetate and pyruvate.[63,64] Depending on the method, these "quasi-creatinine" compounds cause an overestimate of plasma "true" creatinine by 0.2–0.4 mg/dL. The numerous modifications of Jaffe-reaction methods have had as their chief purpose improvement of specificity for creatinine in plasma; urine has a relatively small content of the same interferents. Protein effect has been eliminated in automated continuous flow systems by dialysis, in manual methods by a protein precipitation step (as in the method described below), or by adding sodium dodecyl sulfate to the reagent in order to form nonreactive protein complexes. Glucose and ascorbic acid have also been complexed, these with borate in a buffered system.[30]

Another approach to eliminate interferences in the Jaffe reaction has been acid-blanking, i.e., addition of acid to the reaction mixture after color development and subtraction of this value from the absorbance value obtained without acid addition. Treating a protein-free filtrate with Lloyd's reagent (aluminum silicate)[28,55] or with ion-exchange resin[57] to adsorb and remove non-creatinine chromogens, although generally successful in improving specificity, is considered an excessively time-consuming procedure and tends to increase method imprecision.[51]

Manual methods have traditionally been endpoint methods, with 10–15 min allowed for color development at room temperature. *Kinetic assays, i.e.,* those measuring the rate of the pseudo first-order reaction that forms adduct, were developed in a quest both for specificity and for faster and automatable analyses. Importance of temperature control to assure reproducibility of rate measurements was clearly recognized early in their development.[8,15,42] In the early studies of interferences in the kinetic methods, two kinds of noncreatinine chromogens were identified—those whose rates of adduct formation were very rapid only in the first 20 s after mixing reagent and sample, and those whose rates did not become rapid until 80–100 s after mixing.[21] The "window" between 20 and 80 s therefore was a period in which the rate signal being observed could be attributed predominantly to the creatinine-picrate reaction. (Some investigators found 60 s as the upper limit of this window.) Thus, improvement of specificity gained in the kinetic assays was achieved by selecting times for rate measurements 25–60 s after initiation of the reaction (mixing). This approach has been implemented on a variety of automated microprocessor-controlled instruments.[21,32] A critical evaluation of kinetic methods in 1980 pointed out that many of the kinetic methods are still subject to positive interference by α-keto compounds and to negative interference by bilirubin and its metabolites.[8,64]

Enzymatic or partially enzymatic assays for creatinine have had a long history. Two enzymes are of particular interest: (1) creatininase (EC 3.5.4.21), also called creatinine *imino*hydrolase or deiminase, catalyzes creatinine degradation to *N*-methylhydantoin and ammonium ion; and (2) creatinine hydrolase (EC 3.5.2.10), also called creatinine *amido*hydrolase, catalyzes the hydrolysis of creatinine to creatine. Partially enzymatic methods, both early manual and recent automated ones, have used these two enzymes, either singly or in combination, to achieve specificity of analysis. The enzymes have been added to destroy creatinine, so measurement of the difference of total Jaffe-reacting chromogens (before enzyme addition) and of residual, noncreatinine Jaffe-reacting chromogens (after enzyme addition) represents the true creatinine. Despite the gain in specificity, other and basic problems with an alkaline picrate system remain. Creatininase has also been used in conjunction with an ammonium ion-selective electrode[69] or with indicator dyes for quantitating ammonium. An example of the latter approach is the method on the Ektachem analyzer (Eastman Kodak Co., Rochester, NY 14650) where ammonium ion released by creatininase reacts with bromphenol blue and the color is determined by reflectance spectrophotometry. Methods using creatininase are subject to risks of high levels of endogenous ammonia and contamination of the reaction mixture with environmental ammonia.

In one completely enzymatic assay, creatinine is hydrolyzed to form creatine, which can be measured using creatine kinase. The indicator reaction in this coupled-enzyme system is monitored spectrophotometrically at 340 nm in either an endpoint or a kinetic mode. The reactions are:

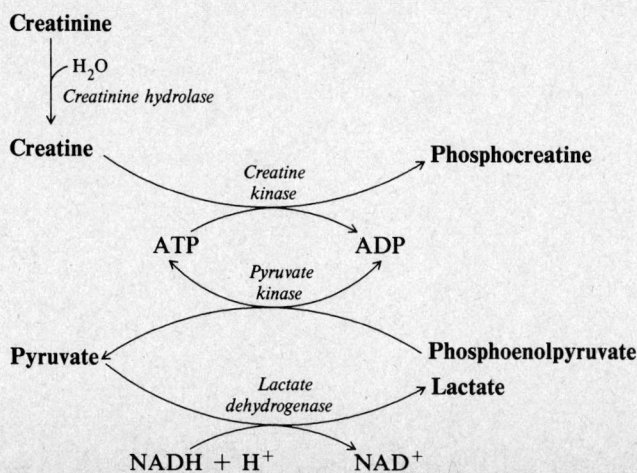

The method is highly specific and has been adapted for both manual and automated use.[35,50]

Isolation of creatinine by high performance liquid chromatography (HPLC) using paired-ion technique on a reverse-phase column followed by detection at 200 nm has also been described for creatinine assay.[65] Although the HPLC assay is impractical for routine analysis of creatinine, its specificity recommends it as a reference method and for validating other methods.

Importance of Specificity

Currently popular methods for assaying plasma creatinine have varying degrees of specificity. Laboratorians who serve active renal disease centers must recognize how the specificity of the method used in their facility affects the clinical application of results. Reference ranges for creatinine clearance, the reciprocal of plasma creatinine concentration, and plasma urea nitrogen/creatinine ratio—because each of these is a calculated value—will vary depending on how "true" is the creatinine value used in the calculation, i.e., how specific is the method used to obtain the result. The more a method overestimates true creatinine, the greater the underestimate of C_{Cr}, $1/P_{Cr}$, and urea N/creatinine ratio.

As methods of more uniform and greater specificity are developed and utilized, reference intervals for plasma creatinine levels probably will become less method dependent and will be 0.1–0.3 mg/dL (10–25 μmol/L) lower than many of those currently quoted.

Determination of Creatinine Using the Jaffe Reaction[18,55,68]

Specimen. Either plasma or serum is suitable. Creatinine in the sample is stable for at least 7 d in the refrigerator and indefinitely when frozen. Significant hemolysis of a blood specimen may cause spurious elevations of creatinine values for serum or plasma since rupture of erythrocytes increases the concentration of noncreatinine chromogens.

Creatinine in urine is stable for 2–3 d at room temperature and for at least 5 d refrigerated. If urine must be collected with a preservative for other analytes, only thymol or toluene may be used.

Principle. Creatinine in a protein-free supernatant of plasma or serum is reacted with alkaline picrate to form a color complex whose intensity is measured at 510 nm.[67] (See page 1276.)

Instrumentation. A spectrophotometer with a bandwidth of < 20 nm is necessary in order to achieve optimal analytical sensitivity for the method.

Reagents

1. Picric acid, 0.036 mol/L. Dissolve 9.16 g picric acid, reagent grade, in about 500 mL of pure water at 80 °C. Cool and dilute to 1.0 L with water. The solution is stable when stored protected from light.

2. Sodium hydroxide, 1.4 mol/L. Dissolve 54 g sodium hydroxide, reagent grade, in about 500 mL of water. Cool and dilute to 1.0 L with water and store in a plastic bottle. Stable for at least 12 months at room temperature.

3. Tungstic acid, 0.035 mol/L. Dissolve 1 g polyvinyl alcohol with heat (do not boil) in 100 mL water, then cool and transfer to a 1-L volumetric flask in which 11.1 g sodium tungstate dihydrate, reagent grade, has been dissolved in ~300 mL of water. In a separate container, mix 2.1 mL concentrated sulfuric acid with 300 mL of water. Then mix the sodium tungstate and sulfuric acid solutions together and dilute the mixture to 1.0 L with water. This solution is stable at room temperature for at least 12 months. Do not refrigerate.

4. Creatinine stock standard, 20 mmol/L. Dissolve 0.226 g creatinine, reagent grade, in 100 mL of HCl, 0.1 mol/L. This solution is stable at 4–6 °C for 12 months. Prepare working standards volumetrically as follows:

Stock standard (mL) diluted to 100 mL with HCl, 0.1 mol/L	Equivalent plasma concentration of creatinine	
	μmol/L	mg/dL
0.5	100	1.1
1.0	200	2.3
2.0	400	4.5
3.0	600	6.8
4.0	800	9.0
6.0	1200	13.6
8.0	1600	18.1

Working standards are stable at room temperature for at least one month.

Procedure

1. Deproteinize serum or plasma unknowns and controls by placing 4.5 mL tungstic acid in a 16×125 mm tube and adding 0.5 mL sample. Mix thoroughly for 10 s and centrifuge at $1500 \times g$ for 10 min.

2. Dilute urine samples 1:200 with water.

3. Into a series of appropriately labeled tubes pipet 3.0 mL deproteinized supernatant, diluted urine, or standards. Set up also a reagent blank of 3.0 mL water.

4. Add 1.0 mL picric acid to each tube and mix thoroughly.

5. At accurately timed 30-s intervals add 0.5 mL of NaOH to each tube and mix thoroughly.

6. Exactly 15 min after adding NaOH, read the absorbance of each tube at 500 nm against the reagent blank set to zero absorbance.

7. Plot absorbance versus concentration for standards and read concentrations of unknowns and controls from the curve. Multiply concentrations found for urine samples by 20; the factor takes into account that the concentration on the abscissa of the calibration graph is already multiplied by 10, so that multiplying by 20 corrects for the original dilution of 1:200.

8. If urinary excretion of creatinine is desired based on a specimen collected over a timed period, the calculation is:

Urine creatinine, mg excreted per period

$$= \text{Urine creatinine, mg/dL} \times 10 \text{ dL/L} \times \text{Urine volume (L) per period}$$

Comments

1. Pretreatment of deproteinized supernatant with aluminum silicate (Lloyd's reagent) improves specificity by removing noncreatinine chromogens and produces results that are 20–35 μmol/L (0.2–0.4 mg/dL) lower. Use of Lloyd's reagent, however, is not essential to clinical validity. Urine contains very little noncreatinine chromogenic material and the method is therefore inherently more specific for urine than it is for serum.

2. The random error of manual methods for creatinine may be significantly greater than that of automated methods. Recommended practice is to determine reproducibility for whatever method is being used, particularly for plasma or serum, and to inform clinicians who use plasma creatinine values for supervision of their renal patient of the magnitude of random error. Standard deviation for day-to-day reproducibility should be of the order of 0.10–0.15 mg/dL so that the physician can rely on a change of 0.2–0.3 mg/dL as signifying real change in his patient's plasma creatinine level.

3. *Creatine* in urine may be determined by the same method by diluting the urine 1:200 in HCl, 0.1 mol/L, in a tightly closed screw-capped tube, and placing the tube in a boiling water bath for 1 h to form the condensation product creatinine. Creatinine is then determined by the method above on heated and unheated dilutions, and each of the two values obtained is multiplied by 20 as noted above. The value of the unheated tube (preformed creatinine) is subtracted from the value of the heated tube (total creatinine), and the difference is multiplied by 1.16 to correct for the difference in molecular masses of creatinine (M.W. 113) and creatine (M.W. 131). It is not uncommon to obtain negative values when the creatine concentration is very low.

Reference Ranges[70]

Reference ranges for serum or plasma creatinine, measured by methods similar to the one described above, are 0.9–1.5 mg/dL (80–133 μmol/L) in men and 0.7–1.3 mg/dL (62–115 μmol/L) in women.

Table 11-6. REFERENCE RANGES FOR "TRUE" CREATININE

| | | True Plasma Creatinine | | | |
| | | mg/dL | | μmol/L | |
Age	Height, cm	*Mean*	*Range**	*Mean*	*Range**
Cord blood		0.75	0.51–0.99	66.3	45.1–87.5
0–2 weeks	50	0.50	0.34–0.66	44.2	30.0–58.3
2–26 weeks	60	0.39	0.23–0.55	34.5	20.3–48.6
26–52 weeks	70	0.32	0.18–0.46	28.3	15.9–40.7
2 years	87	0.32	0.20–0.44	28.3	17.7–38.9
4 years	101	0.37	0.25–0.49	32.7	22.1–43.3
6 years	114	0.43	0.27–0.59	38.0	23.9–52.2
8 years	126	0.48	0.31–0.65	42.4	27.4–57.4
10 years	137	0.52	0.34–0.70	46.0	30.1–61.9
12 years	147	0.59	0.41–0.78	52.2	36.2–69.0
Adult, male	174	0.97	0.72–1.22	85.7	63.6–108
Adult, female	163	0.77	0.53–1.01	68.1	46.8–89.3

*±2 SD.

Urinary creatinine excretion is 14–26 mg/kg/d (124–230 μmol/kg/d) in adult men and 11–20 mg/kg/d (97–177 μmol/kg/d) in adult women. There is some intraindividual variability in renal excretion of creatinine that can be assigned to the amount of dietary (exogenous) creatinine ingested in meats. This variability may account for the 15–20% between-day variation in calculated creatinine clearances for a given individual. Creatinine excretion decreases with age.

Reference ranges for serum or plasma creatinine, as determined by highly specific methods, are given in Table 11-6. Values in elderly individuals increase only slightly, since the decreased excretion of creatinine is partially compensated for by the decreased production of creatinine as a result of decreased muscle mass.

Endogenous Creatinine Clearance

Principle. The principle is discussed on pages 1271 to 1274.

Procedure

1. Hydrate the patient with at least 600 mL of water. Note that special orders may have been given by the attending physician of a patient with documented renal disease; these special orders may override recommendations made here with respect to water intake or withholding of medication.

2. Withhold tea, coffee, and drugs on the day of the test.

3. Have the patient void and discard the specimen. Note time, and from then on collect all urine passed for 4, 12, or 24 h. Keep patient well hydrated during collection period in order to assure a urine flow rate of ≥ 2 mL/min.

4. Collect a blood specimen. Because plasma creatinine concentration is relatively constant, the collection can be made at any time during the urine collection period. For the sake of consistency, however, blood collection at midpoint of the urine collection period is recommended.

5. In the laboratory, measure the volume of the total urine collected and record both volume and minutes (h $\times$ 60) of the period in which it was collected. Perform the assays of plasma (serum) and urine creatinine.

6. Calculate clearance:

$$\text{mL plasma cleared/min per standard surface area} = \frac{U_{Cr} \times V}{P_{Cr}} \times \frac{1.73}{A}$$

where U_{Cr} = concentration of creatinine in urine
P_{Cr} = concentration of creatinine in plasma in the same units as for urine
V = volume of urine flow in mL/min
A = body surface area in square meters

The factor 1.73/A normalizes clearance for average body surface, correcting as it does for variation in creatinine excretion relative to lean body (muscle) mass. Nomograms for determining body surface area from height and weight are found in the Appendix (Tables 20-18 and 20-19).

Sources of Error

1. Error in recording timing of collection period or loss of a portion of the urine during collection is the most common source of error.

2. Vigorous exercise during the urine collection period may alter clearance.

3. Proper hydration of the patient to assure urine flow rate of ≥ 2 mL/min improves the accuracy of the measurement of filtration rate and tends to eliminate retention of urine in the bladder as a source of negative error.

Reference Ranges[70]

With the creatinine method described in this chapter, clearance for males is 105 $\pm$ 20 mL/min; for females it is 95 $\pm$ 20 mL/min. When a more specific method for creatinine is used, clearance is 117 $\pm$ 20 mL/min for males and 108 $\pm$ 20 mL/min for females. Clearance rates decrease with age. (See Table 11-7.)

Intraindividual variability in creatinine clearance is approximately $\pm$15%; sequential deter-

Table 11-7. CREATININE CLEARANCE
(mL/min/1.73 m²)

Age (Years)	Males	$\bar{x}$	Females	$\bar{x}$
20–30	88–146	117	81–134	107
30–40	82–140	110	75–128	102
40–50	75–133	104	69–122	96
50–60	68–126	97	64–116	90
60–70	61–120	90	58–110	84
70–80	55–113	84	52–105	78

Calculated from a nomogram for ascertaining age-adjusted percentile rank in creatinine clearance.[62]

minations of creatinine clearance and averaging of values are required to reduce this variation appreciably.

Creatine in Biological Fluids

Creatine comprises a small fraction of the total nonprotein nitrogen of plasma and urine. It is unstable at both alkaline and acidic pH's and rapidly undergoes conversion to creatinine. Creatine in urine is usually measured as the difference between preformed creatinine in the sample and the total "creatinine chromogens" found after the specimen has been subjected to acid condensation. The methods employed are the same as those used for creatinine determination. Determination of creatine in plasma or urine has little clinical utility for evaluating renal disease. Urine creatine levels, once used in diagnosis of muscle dystrophic and atrophic diseases, have been superseded for that purpose by assays for activity of the muscle enzymes, particularly creatine kinase.

URIC ACID

Biochemistry and Physiology. In man, uric acid is the major product of the catabolism of the purine nucleosides, adenosine and guanosine. Purines from catabolism of dietary nucleic acid are converted to uric acid directly. However, the bulk of purines ultimately excreted as uric acid in the urine arises from degradation of endogenous nucleic acids. The daily synthesis rate of uric acid is approximately 400 mg; dietary sources contribute another 300 mg. In adult males on a purine-free diet, the total body pool or exchangeable urate is estimated at 1200 mg, 600 mg in adult women. By contrast, patients with gouty arthritis and tissue deposition of urate may have urate pools as large as 18 000–30 000 mg.[7]

Overproduction of uric acid may result from increased synthesis of purine precursors. Synthesis and metabolism of the major precursors are illustrated in outline in Figure 11-10. The second enzymatic step in the synthetic pathway (Figure 11-10,A), formation of 5′-phosphoribosylamine, is the first irreversibly committed step in purine biosynthesis. The intracellular concentration of the substrate phosphoribosylpyrophosphate (PRPP) regulates de novo purine synthesis. The enzyme PRPP-amidotransferase is controlled through feedback inhibition by the purine nucleotides that are the final products of the biosynthetic pathway. The first purine nucleotide formed by ring closure is inosine monophosphate (IMP); adenosine and guanosine monophosphates are derived from IMP through enzymatically mediated interconversions. Adenine and guanine nucleotides may then be utilized as precursors to the corresponding nucleosides that are the building blocks of DNA and RNA, or, when further phosphorylated, these nucleotides become carriers of high-energy bonds in the forms of ATP and GTP.

Catabolism of the nucleotides (Figure 11-10,B) begins with removal of their ribose-linked phosphate, a process catalyzed by purine 5′-nucleotidase. Removal of the ribose moiety of inosine and guanosine by action of purine nucleoside phosphorylase forms hypoxanthine and guanine, both of which are converted to xanthine. Xanthine is converted to uric acid through the action of xanthine oxidase.

Reutilization of the major purine bases, adenine, hypoxanthine, and guanine, is achieved through "salvage" pathways (Figure 11-10,C), in which phosphoribosylation of the free bases causes resynthesis of the respective nucleotide monophophates. Adenine is converted to AMP through the action of adenine phosphoribosyl transferase (APRT), hypoxanthine, and guanine

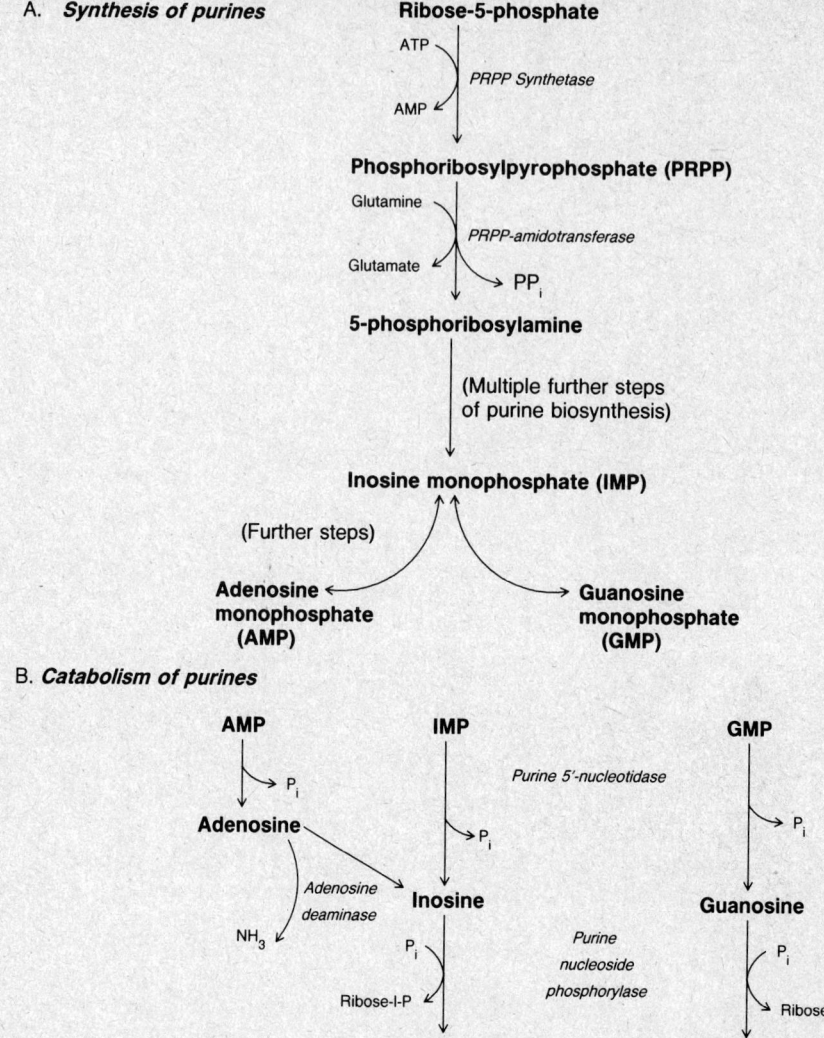

A. *Synthesis of purines*

B. *Catabolism of purines*

C. *Salvage pathways of purines*

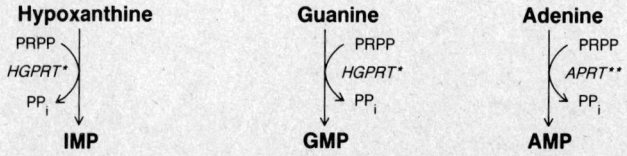

*Hypoxanthine-guanine phosphoribosyl transferase
**Adenine phosphoribosyl transferase

Figure 11-10. Metabolism of purines

to their monophosphates through hypoxanthine-guanine phosphoribosyl transferase (HGPRT). The HGPRT pathway is quantitatively more important than the APRT pathway.[74]

Lower primates and mammals other than man carry purine metabolism one step further with the formation of allantoin from uric acid, a step mediated by uricase (urate:oxygen oxido-reductase, EC 1.7.3.3). Approximately 75% of uric acid excreted is lost in the urine; most of the remainder is secreted into the gastrointestinal tract, where it is degraded to allantoin and other compounds by bacterial enzymes.

Renal handling of uric acid is complex and involves four sequential steps: (1) glomerular filtration of virtually all the uric acid in capillary plasma entering the glomerulus; (2) reabsorption in the proximal convoluted tubule of about 98–100% of filtered uric acid; (3) subsequent secretion of uric acid into the lumen in the distal portion of the proximal tubule; and (4) further reabsorption in the distal tubule. The net urinary excretion of uric acid is 6–12% of the amount filtered.

The physicochemical properties of uric acid are important in considering uric acid concentrations in the circulation, in tissues, and in the kidney. The first pK_a of uric acid is 5.57; above this pH, uric acid exists chiefly as urate ion, which is more soluble than uric acid.[74] At urine pH below 5.75, uric acid is the predominant form.

Clinical Significance. *Hyperuricemia* is most commonly defined by serum or plasma uric acid concentrations > 7.0 mg/dL (0.42 mmol/L) in men, or > 6.0 mg/dL (0.36 mmol/L) in women (if specific methods are used to measure uric acid). The major causes of hyperuricemia are summarized in Table 11-8. Asymptomatic hyperuricemia is frequently detected through biochemical screening; long-term follow-up of asymptomatic hyperuricemic subjects is undertaken because many are at risk for renal disease that may develop as a result of hyperuricemia and hyperuricosuria; few of these patients go on to develop the clinical syndrome of gout.

Gout occurs when monosodium urate precipitates from supersaturated body fluids; the deposits of urate are responsible for the clinical signs and symptoms. Gouty arthritis may be associated with urate crystals in joint fluid as well as with deposits of crystals (tophi) in tissues surrounding the joint. The deposits may occur in other soft tissues as well, and wherever they occur they elicit an intense inflammatory response consisting of polymorphonuclear leukocytes and macrophages. Renal disease associated with hyperuricemia may take one or more of several forms: (1) gouty nephropathy with urate deposition in renal parenchyma, (2) acute intratubular deposition of urate crystals, and (3) urate nephrolithiasis.[7] The justification for medical treatment of sustained asymptomatic hyperuricemia is avoidance of urate-induced renal damage.

Gout may be classified as primary or secondary. *Primary gout* is associated with "essential" hyperuricemia, i.e., that due to metabolic overproduction of purines or to underexcretion of uric acid. *Secondary gout* is a result of hyperuricemia attributable to several identifiable causes. Renal retention of uric acid may occur in acute or chronic renal disease of any type or as a consequence of adminstration of drugs; diuretics, in particular, are implicated in the latter instance. Organic acidemia due to increased acetoacetic acid in diabetic ketoacidosis or to lactic acidosis may interfere with tubular secretion of urate. Increased nucleic acid turnover and consequent increase in catabolism of purines may be seen in rapid proliferation of tumor cells as well as in massive destruction of tumor cells upon therapy with certain chemotherapeutic agents.

Table 11-8. CAUSES OF HYPERURICEMIA

"Essential" hyperuricemia
    Overproduction (associated with hyperuricaciduria)
    Underexcretion (associated with normal or decreased renal excretion of uric acid)
Renal retention
    Renal failure
    Drug therapy: diuretics, salicylates, pyrazinamide, ethambutol
    Poisons: lead, alcohol
    Organic aciduria: acetoacetate, lactate
    Endocrinopathies: hypothyroidism, hyperparathyroidism
Increased turnover of nucleic acids
    Myeloproliferative syndromes
    Chemotherapy of malignant tumors, especially leukemias and lymphomas
Specific enzyme defects
    Deficiency of hypoxanthine-guanine phosphoribosyl transferase
        Complete (Lesch-Nyhan syndrome)
        Partial
    Abnormal phosphoribosyl pyrophosphate synthetase

Hyperuricemia is also attributable to primary defects of enzymes in the pathways of purine metabolism. The *Lesch-Nyhan* syndrome is characterized by complete deficiency of hypoxanthine-guanine phosphoribosyltransferase (HGPRT), the major enzyme of the purine salvage pathways. This sex-linked genetic disorder is manifested clinically by mental retardation, abnormal muscle movements, and behavioral problems (self-mutilation and pathological aggressiveness); and biochemically by hyperuricemia, hyperuricaciduria, and markedly decreased levels of HGPRT in erythrocytes, fibroblasts, and other cells. Intracellular levels of PRPP and rates of purine synthesis are increased. Neurological symptoms of this syndrome may be related to decreased availability of purines to the developing brain, which has limited capacity for de novo purine synthesis and therefore relies upon the purine salvage pathways to supply it with most of the purine nucleotides it requires.[74] Less severe deficiency of HGPRT displays a clinical spectrum of mild to moderate neurologic defects. Affected fetuses can be identified by HGPRT assays on cultured fibroblasts obtained by amniocentesis; unaffected heterozygous female carriers of the defective gene can be identified by observing HGPRT mosaicism in cultured fibroblasts or in individual hair follicles. Increased levels of intracellular PRPP production with consequent increased uric acid can also occur due to mutations in PRPP synthetase, which are inherited as X-linked recessive traits. Glucose-6-phosphatase deficiency also leads to hyperuricemia due to both overproduction and underexcretion of uric acid.

Quantitation of urinary uric acid excretion is an aid in selecting appropriate treatment for asymptomatic hyperuricemia. Hyperuricemic patients excreting < 600 mg uric acid daily are candidates for treatment with uricosuric drugs such as probenecid or sulfinpyrazone; uricosuric drugs enhance renal excretion of uric acid by blocking the carriers in the tubular cells that mediate reabsorption. Patients excreting > 600 mg/d are candidates for treatment with allopurinol, a drug that decreases intracellular concentrations of PRPP and inhibits xanthine oxidase activity,[7] thereby suppressing purine synthesis and degradation of hypoxanthine to uric acid.

About one in five patients with clinical gout will also have urinary tract uric acid stones. Although formation of urinary tract stones is a complex process, in about 50% of patients uric acid stones are relatable to either hyperuricosuria or excretion of a persistently acid urine, or both. Undissociated uric acid (pK_a 5.57) is relatively insoluble whereas urate at pH 7.0 is more than ten times as soluble. Thus, in patients with urinary pH persistently < 6.0, relatively small amounts of uric acid in urine may produce supersaturation. Measurement of both urine pH and uric acid excretion are important in the investigation of uric acid urolithiasis. In any patient with urolithiasis, identification of crystals present in urine may provide a significant clue as to the nature of stones present. Pure uric acid stones are radiolucent—unlike many of the calcium-containing stones—and represent 5–10% of all urinary tract stones.

Hypouricemia, often defined as serum urate concentrations below 2.0 mg/dL (0.12 mmol/L), is much less common than hyperuricemia. It may be secondary to any one of a number of underlying conditions. Severe hepatocellular disease with reduced purine synthesis or xanthine oxidase activity is one possibility. Another is defective renal tubular reabsorption of uric acid. Defective reabsorption may be congenital, as in generalized Fanconi's syndrome, or acquired. The reabsorption defect may be acquired acutely because of injection of radiopaque contrast media or chronically because of exposure to toxic agents. Overtreatment of hyperuricemia with allopurinol or uricosuric drugs, or cancer chemotherapy with 6-mercaptopurine or azathioprine (inhibitors of de novo purine synthesis) may also cause hypouricemia. Hypouricemia in combination with xanthinuria is rarely seen and suggests deficiency of xanthine oxidase.

Methods for the Determination of Uric Acid

Methods in current use for assay of uric acid fall into two groups: phosphotungstic acid (PTA) methods and uricase methods.

Phosphotungstic acid methods rely on the development of a blue color ("tungsten blue") as PTA is reduced by urate in alkaline medium; the color is read by spectrophotometry at wavelengths of 650–700 nm. The presence of protein during the color development causes both turbidity and an unpredictable quenching of the absorbance. In methods for plasma uric acid, protein removal is therefore an obligatory step. Removal has been accomplished in automated continuous flow methods by dialysis and in manual methods by preparation of a $ZnSO_4/Ba(OH)_2$ protein-free filtrate or supernatant. These approaches avoid pH values of < 6.0 and consequent negative error due to co-precipitation of urate as poorly soluble uric acid. Another approach uses PTA in excess as the precipitant in the protein removal step; the color reaction is then initiated by adding base to the supernatant that contains the PTA-urate complex.

Even in the absence of protein, the color system may develop turbidity. Sodium cyanide has been used in some methods both to suppress turbidity and to provide the alkaline pH for the reaction; manual versions of this method have fallen out of use, but at least one adaptation is still used on a popular sequential multichannel automated analyzer. Other manual[3] or automated[52] PTA methods employ sodium hydroxide as alkali and, for control of turbidity, incorporate glycerine, sodium silicate, or surfactants such as polyanethole sulfonate (Liquoid) into reagents. PTA methods are subject to many interferences, including *endogenous* compounds such as glucose and ascorbic acid in plasma or urine and glutathione, ergothionine, and cysteine spilled into plasma from hemolyzed erythrocytes; and *exogenous* compounds such as acetaminophen, acetylsalicylic acid, gentisic acid (a salicylate metabolite), and the substituted purines such as caffeine, theobromine, and theophylline. All of these compounds also reduce phosphotungstic acid and thus introduce a positive error. Efforts to modify PTA methods have had little success in improving their specificity.

Uricase methods are inherently more specific because they have, either as a single step or as the initial step, urate oxidation catalyzed by the enzyme uricase [(urate:oxygen) oxidoreductase, EC 1.7.3.3]. Uricase methods became feasible and popular as a result of availability of high-quality, low-cost preparations of the bacterial enzyme. Preliminary precipitation of protein is not required. In a majority of uricase methods only guanine, xanthine, and a few other structural analogs of uric acid interfere, and then only at concentrations improbable in biologic fluids.[19]

The reaction common to all enzymatic methods is

Urate   **Allantoin**

Some uricase methods are based on this reaction after the plasma or urine is diluted with a borate or Tris buffer. At an alkaline pH, gaseous CO_2 is converted to bicarbonate, which prevents bubbles from interfering with spectrophotometry. The reaction can be observed in either the kinetic or the endpoint mode. The decrease of absorbance as urate is converted may be monitored spectrophotometrically at 282–292 nm. The spectrophotometric applications have been made on the *aca* (DuPont Co., Wilmington, DE 19898) and on the SMAC (Technicon, Tarrytown, NY 10591); on the SMAC, uricase is immobilized on the inner surface of a coil in the flowing stream.

Coupled uricase methods as a rule utilize the hydrogen peroxide produced by an initial uricase reaction as substrate for a chemical indicator reaction involving peroxidase or catalase.[73] An example of an automated uricase/catalase system[26] with a chemical indicator reaction is the following:

$$H_2O_2 \ + \ CH_3OH \ \xrightarrow{Catalase} \ H_2C=O \ + \ 2\,H_2O$$

Methanol   **Formaldehyde**

$$H_2C=O \ + \ 3\,H_3C-\overset{\overset{O}{\|}}{C}-CH_2\overset{\overset{O}{\|}}{C}-CH_3 + NH_3 \longrightarrow 3\,H_2O + \textbf{3,5-Diacetyl-1,4-dihydrolutidine}$$

Acetylacetone

Other compounds such as the redox indicators o-dianisidine and o-toluidine are feasible but less satisfactory alternatives to MBTH-DMA in this system:[36]

$$H_2O_2 \quad + \quad \text{[benzothiazolinone structure]} \quad + \quad \text{[dimethylaniline structure]}$$

3-Methyl-2-benzothiazolinone   **N,N-Dimethylaniline (DMA)**
hydrazone (MBTH)

$$\longrightarrow \quad \text{[indamine dye structure]} \quad + \quad 2\,H_2O$$

Indamine dye, absorbs at 600 nm

The chromophore absorbs at 410 nm.

The method given below illustrates in detail an enzymatic indicator system coupled to uricase and catalase:[29]

$$H_2O_2 \quad + \quad CH_3CH_2OH \quad \xrightarrow{\text{Catalase}} \quad CH_3CHO \quad + \quad 2\,H_2O$$

Ethanol   **Acetaldehyde**

$$CH_3CHO \quad + \quad NAD^+ \quad \xrightarrow[\substack{H_2O}]{\substack{\text{Aldehyde} \\ \text{dehydrogenase}}} \quad CH_3COO^- \quad + \quad NADH \quad + \quad 2\,H^+$$

Acetate

Increase in absorbance is followed spectrophotometrically at 340 nm.

High performance liquid chromatographic methods on reversed-phase columns have also been presented. An ion-pair technique can determine uric acid, hypoxanthine, and xanthine simultaneously and with high sensitivity at 280 nm, using neutralized acid extracts of biological fluids.[10] In another method, plasma spiked with internal standard is injected directly into the HPLC, and uric acid is determined by spectrophotometry at 254 nm.[53] The methods are specific and fast; in both, mobile phases are simple and retention time for uric acid is < 6 min, reasonable conditions that recommend them for either reference or routine use.

Enzymatic Method for the Determination of Uric Acid

Specimen. The method is suitable for serum, heparinized plasma, or urine. Uric acid in serum is stable for 48–72 h at room temperature, for 3–7 d at 4–6 °C, and for 6–12 months when frozen. Uric acid in urine is stable for several days at room temperature; specimens should nevertheless be refrigerated to avoid bacterial growth and an in vivo effect of microbial uricase.

Principle. Uric acid is oxidized in the presence of uricase to allantoin and hydrogen peroxide. The reaction of hydrogen peroxide with ethanol is catalyzed by catalase and produces acetaldehyde. Acetaldehyde then reacts with NAD$^+$ in the presence of aldehyde dehydrogenase to produce acetate and NADH. Increase in absorbance at 340 nm is measured for samples and standard solutions of uric acid, and the concentration of uric acid in the sample is determined from a calibration curve. Refer to the original publication of the method[29] for important additional details and variations on approach.

Instrumentation. A spectrophotometer with bandwidth < 10 nm at 340 nm is required.

Reagents

1. Uric acid standards.
   a. Stock standard, 100 mg/dL (5.95 mmol/L). Use NBS Certified Uric Acid, Standard Reference

Material (SRM) #913 (National Bureau of Standards, US Department of Commerce, Washington, DC). Dissolve 1.000 g uric acid and 0.750 g lithium carbonate in sterile distilled water, and dilute to a final volume of 1 L. Aliquots of the solution, stored frozen at -20 °C, are stable for at least 3 months. A 1:100 dilution of stock standard should have $A_{293\,nm} = 0.750 \pm 0.010$.

b. Working standards, 2.0–16.0 mg/dL (0.12–0.94 mmol/L). Dilute 2.0, 4.0, 8.0, 12.0, and 16.0 mL of stock standard to 100 mL with sterile distilled water. Store at 4–6 °C; working standards are stable 5–7 d if protected from bacterial contamination.

2. KCl-phosphate buffer, pH 8.5. Make a glycerol solution of 260 g/L in pure water. Dissolve 3.75 g KCl and 22.25 g tetrasodium pyrophosphate decahydrate ($Na_4P_2O_7 \cdot 10\ H_2O$) in ~800 mL of glycerol solution. Adjust pH to 8.5 with HCl and bring to a final volume of 1 L with glycerol solution.

3. NAD$^+$, 10 g/L. Dissolve 100 mg NAD$^+$ in 10.0 mL pure water. Stable at 4–6 °C up to 48 h.

4. Ethanol, absolute. Reagent grade, denaturant-free.

5. Enzymes (available from Sigma Chemical Co., St Louis, MO 63178)

a. Uricase (urate:oxygen oxidoreductase, EC 1.7.3.3)

b. Catalase (hydrogen peroxide:hydrogen peroxide oxidoreductase, EC 1.11.1.6)

c. Aldehyde dehydrogenase (Aldehyde:NAD(P) oxidoreductase, EC 1.2.1.5)

6. Working reagent (sufficient for 100 determinations). Mix 50 mL KCl-phosphate buffer, 5.0 mL ethanol, and 5.0 mL NAD$^+$ solution. The final mixture will have the following concentrations:

$Na_4P_2O_7 \cdot 10\ H_2O$	45 mmol/L
KCl	45 mmol/L
NAD$^+$	1.35 mmol/L
Glycerol	~2.5 mmol/L
Ethanol	1.54 mmol/L

Add enzymes to give the following concentrations in the final mixture.

Catalase	900	kU/L
Uricase	162	U/L
Aldehyde dehydrogenase	500	U/L

The reaction mixture is stable up to 6 h at room temperature.

Procedure

1. Pipet 500 μL of working reagent into cuvets labeled for each standard, unknown, and control. Set up one cuvet labeled B.

2. Dilute urine 1:10 with pure water.

3. Add 50 μL of water to cuvet labeled B, and 50 μL of appropriate sample to other cuvets. Mix.

4. Let stand at room temperature for 20 min.

5. Read absorbance (A) of each cuvet against B cuvet set at zero absorbance at 340 nm.

6. Plot absorbance of standards versus concentration to obtain a calibration curve and read concentration of unknowns from curve. Multiply concentration found for a urine specimen by 10 to correct for dilution. Calculate urinary excretion of uric acid:

$$\text{Uric acid, mg/d} = \text{mg/dL} \times 10\ \text{dL/L} \times \text{urine volume, L/d}$$

Comments

1. Add enzymes to working reagent in a minimum volume of fluid. Reaction velocity may be increased by increasing the amount of uricase added.

2. This method has been made available in commercial kits and has been automated.

3. The method has been shown to be linear up to 32.0 mg/dL (2.00 mmol/L); recovery of uric acid was >97%.

4. Bilirubin, hemoglobin, ascorbic acid, and many drugs known to introduce positive errors in reduction methods do not interfere. Potential interferences from homogentisic acid, endogenous alcohol dehydrogenase, or inhibitors of aldehyde dehydrogenase such as disulfiram (Antabuse) are unlikely because concentrations high enough to interfere are improbable in biological fluids.

Reference Ranges[70]

Consensus in a number of studies of serum or plasma uric acid levels in healthy subjects would indicate reference intervals of 3.5–7.2 mg/dL (0.21–0.42 mmol/L) for males and 2.6–6.0 mg/dL (0.15–0.35 mmol/L) for females. However, reference intervals are method dependent; studies using PTA methods show higher values than those using enzymatic assays. For PTA methods, ranges are 4.5–8.2 mg/dL (0.27–0.48 mmol/L) for males and 3.0–6.5 mg/dL (0.18–0.38 mmol/L) for females.

An alternative approach to interpretation of serum uric acid levels is to consider the degree of hyperuricemia in relation to the risk of developing gout; men with plasma uric acid concentrations > 9.0 mg/dL (0.54 mmol/L) are ~150 times more likely to have coexisting gouty arthritis than are men with uric acid concentrations < 6.0 mg/dL (0.36 mmol/L).[74]

Urinary uric acid excretion in individuals on a diet containing purines is 250–750 mg/d (15–45 mmol/d). Excretion may decrease by 20–25% on a purine-free diet to < 400 mg/d.

The Renal Compensatory Mechanism in Renal Disease

In renal disease, a decreased glomerular filtration rate may result in the retention of metabolic acids with resultant acidosis and accumulation of anions such as phosphates, sulfates, keto acids, amino acids, and so on. The decreased filtration of phosphates reduces the ability of the body to remove H^+ by formation of $H_2PO_4^-$. The decreased ability of NH_3 formation results in the decreased formation NH_4^+ and the associated decrease in removal of H^+. There may also be an impairment of the Na^+-H^+ exchange, especially in renal tubular acidosis. For a more detailed discussion on H^+ secretion, Na^+-H^+ exchange, NH_3 formation, and HCO_3^- reclamation, see Chapter 10 and standard textbooks of renal physiology.

The kidneys provide the only mechanism for the body to remove nonvolatile acids. However, in terms of absolute removal of acids, this amounts to only 3% of the amount of acid excreted by the lungs as CO_2. Hence, disruption of the kidneys' ability to secrete an acid urine and regenerate bicarbonate leads to the slow development of acidosis over a period of days. On the other hand, disruption of CO_2 excretion by the lungs may produce life-threatening acidosis over a span of minutes.

Renal Excretion of Amino Acids

Amino acids in plasma are filtered by the glomeruli and appear in the glomerular filtrate in the same proportions as they do in plasma. A great portion of amino acids is reabsorbed by the renal tubular cells through a process of active transport mediated by membrane-bound carrier systems. Thus, the normal urinary excretion of amino acids is only a small fraction of the filtered load and is about 50–200 mg/d when measured as α-amino acid nitrogen. In some congenital disorders there is a defect in the reabsorption of amino acids that results in aminoaciduria. An example of such a condition is cystinuria, in which there is a failure to reabsorb dibasic amino acids (cystine, lysine, arginine, and ornithine). In Fanconi's syndrome there is a failure to reabsorb a wide variety of amino acids. These and other forms of aminoaciduria are discussed in more detail in Chapter 4.

In conditions associated with decreased glomerular filtration, amino acids are retained, together with other nonprotein nitrogenous compounds, and contribute to the acidity and the organic anion fraction of plasma.

Renal Tubular Acidosis (RTA)

RTA is characterized by hyperchloremia, a normal anion gap, and urinary HCO_3^- or H^+ excretion inappropriate for the plasma pH. Hyperchloremia is caused by enhanced Cl^- reabsorption stimulated by contraction of the extracellular volume (ECV). Other causes for hyperchloremic acidosis with normal anion gap must be excluded, but this can generally be done on clinical grounds. RTA is the result of loss of or decreased reabsorption of bicarbonate by the proximal renal tubules, or it results from insufficient acidification of the distal tubular fluid due to different etiologies. According to the specific defect, tubular renal acidosis is classified as proximal renal tubular acidosis, distal renal tubular acidosis, or a combination of both. The etiology is not always well established. A detailed discussion of these disorders is beyond the scope of this textbook, and thus only the main features of each type will be mentioned here. The interested reader is referred to more detailed discussions of the topic.[4,16,22,79] A summary of the most important laboratory findings in RTA is given in Table 11-9.

Proximal (Type II) Renal Tubular Acidosis

Primary or idiopathic proximal RTA (PRTA) can be either sporadic or genetically transmitted. It occurs chiefly in male infants and is commonly associated with growth retardation. It is less frequently seen in adults. Proximal RTA may also occur secondary to inherited systemic diseases, such as cystinosis, tyrosinosis, hereditary fructose intolerance, Wilson's disease, and Lowe's syndrome. Other diseases associated with proximal RTA are multiple myeloma, Sjögren's syndrome, amyloidosis, nephrotic syndrome, renal transplantation, hypervitaminosis D, and sec-

Table 11-9. CHARACTERIZATION OF RENAL ACIDIFICATION DEFECTS

	Minimal Urine pH	Renal NH_4^+ Excretion	Reabsorption of Filtered HCO_3^-	Plasma Cl^-	Plasma K^+	Other Frequent Laboratory Findings
End-stage renal disease	<4.8	Markedly reduced	Reduced	Normal or low	Normal or increased	Azotemia, hyperuricemia
Proximal renal tubular acidosis	>5.5, but <5.5 during acidosis	Low at urine pH >6.0; normal at urine pH <5.5	Severely impaired; near normal at low plasma HCO_3^-; fractional HCO_3^- excretion >15%	Increased	Low or normal	Hypocalcemia, hypophosphatemia, hypouricemia; volume depletion, increased aldosterone; aminoaciduria, glycosuria; increased phosphate and uric acid clearance; lysozyme and immunoglobulin light chains in urine; Na and K wasting
Distal renal tubular acidosis	>5.5	Reduced	Slightly reduced; fractional HCO_3^- excretion 3–5%	Increased	Low	Hypophosphatemia, hypokalemia; increased PTH if hypovolemia present; increased excretion of Ca, Na, K, uric acid; decreased titratable acidity; no increase in urine pCO_2 during alkaline diuresis
Hyperkalemic DRTA	>5.5	Reduced	Reduced	Increased	Increased	K excretion not lowered in response to Na_2SO_4
Selective aldosterone deficiency (SAD; DRTA IV)	>5.5	May be reduced due to hyperkalemia		Increased in 75%	Increased (modestly)	Urine pCO_2 normal during alkaline diuresis

ondary hyperparathyroidism. In PRTA, contrary to DRTA, nephrocalcinosis and nephrolithiasis are rarely observed. Drugs or toxic agents known to produce PRTA are carbonic anhydrase inhibitors, heavy metals, outdated tetracyclines, acetazolamide, and streptozotocin. (Note that several of these disorders and agents can also cause distal renal tubular acidosis.) PRTA is usually not an isolated defect but is accompanied by other tubular dysfunctions that result in hyper-aminoaciduria, glycosuria, phosphaturia, and uricosuria (Fanconi's syndrome). There may also be a decrease in reabsorption of filtered low-molecular-weight proteins, resulting in increased excretion of lysozyme, β_2-microglobulin, and immunoglobulin light chains.

Diagnosis. Patients with PRTA develop metabolic acidosis that is associated with hyper-chloremia; urinary pH values may be either > 5.5 or < 5.5, depending on the severity of the acidemia and therefore the plasma bicarbonate concentration. At normal or near normal plasma bicarbonate concentrations ($\sim$26 mmol/L), significant bicarbonate loss in urine occurs, and the fractional excretion of HCO_3^- may exceed 15%. This loss of bicarbonate could be due to a defect in the Na^+-H^+ exchange mechanism, or to decreased luminal or intracellular carbonic anhydrase activity, or to other causes. The fractional bicarbonate excretion can be calculated as follows:

$$\text{Fractional bicarbonate excretion, \%} = \frac{\dfrac{\text{Urine } HCO_3^-}{\text{Plasma } HCO_3^-}}{\dfrac{\text{Urine creatinine}}{\text{Plasma creatinine}}} \times 100\%$$

However, if plasma bicarbonate concentrations go below a certain threshold level (which is lower than in normal individuals), the remaining proximal tubular capacity to reabsorb bicarbonate, together with the normally functioning distal tubular reabsorption process, is adequate to acidify the urine sufficiently and to assure reabsorption of most of the bicarbonate. As a result, the urine pH may be < 5.4. This low pH occurs mainly in severe forms of acidosis in which plasma HCO_3^- is low. For this reason, proximal renal tubular acidosis is termed a "quantity-limited" variety of RTA. In contrast, patients with plasma bicarbonate levels above the modified threshold observed in this disease, or patients treated with bicarbonate, cannot absorb enough bicarbonate in the proximal tubules, and thus the capacity of the normally functioning distal tubules is exceeded. The result is significant loss of bicarbonate and a urine pH > 5.5.

Other laboratory findings include acidemia (except in incomplete RTA), normo- or hypo-kalemia, hypocalcemia, hypophosphatemia, hypouricemia, glucosuria, aminoaciduria, increased phosphate and uric acid clearance, and sodium as well as potassium wasting. Renal potassium wasting is a result of the increased delivery of bicarbonate to the distal tubule and thus will be worse if bicarbonate treatment (usually $\sim$10 mmol/kg per day) is initiated. Therefore, potassium supplementation may have to be increased in this form of therapy. The renal loss of ions and compounds mentioned above may lead to chronic volume depletion with secondary hyperaldo-steronism, malnutrition, failure to thrive, and osteomalacia. If severe bone disease occurs, vitamin D therapy may be required. Titratable acidity and ammonium excretion rate may be reduced. Urine bicarbonate and pH values are as outlined above. Nephrocalcinosis and urinary stone formation are rare in PRTA, possibly due to the normal excretion rate of citrate or other organic acids in these patients. Infusion of HCO_3^- to reach a normal plasma concentration of $\sim$26 mmol/L will result in a loss of > 15% of the filtered bicarbonate.

Treatment of the underlying disorder should be carried out when possible; otherwise it consists mainly of administration of base to correct the acidosis. If normal pH is restored, *maintenance* HCO_3^- administration of 6–10 mmol/kg per day in adults and 10–15 mmol/kg per day in children is generally adequate. If hypokalemia is present, potassium must be restored to avoid clinical signs of muscle weakness, respiratory paralysis, or cardiac arrhythmias.[59] Such conditions could occur if the administered HCO_3^- results in alkalosis that would cause further lowering of K^+ due to a shift into the cells. If the plasma pH is corrected before normocalcemia is restored, tetany and seizures may be precipitated. Vitamin D treatment is rarely needed.

Distal Renal Tubular Acidosis (DRTA)

Classic or Type I DRTA occurs most often in infants (sometimes transient) or young children, but it may also be seen in adults. In fact, DRTA is much more common in adults than is PRTA.

The condition may occur:

1. As an autosomal dominant condition with higher incidence in females.

2. As a sporadic, nonfamilial disease.

3. As the result of drug intake such as amphotericin B, gentamicin, phenacetin, and lithium or after inhaling toluene (e.g., glue sniffing).

4. As a consequence of a variety of autoimmune disorders and other diseases such as systemic lupus erythematosus, Sjögren's syndrome, idiopathic hypergammaglobulinemia, primary biliary cirrhosis, and thyroiditis. If autoimmune and hyperglobulinemic disorders are also the cause of PRTA, it is felt that tubular dysfunction is the result of immunological injury.

5. In low renin/low aldosterone states, resulting in low H^+ and low K^+ excretion as in diabetes (Type IV RTA).

Other conditions associated with DRTA include hyperparathyroidism, hyperthyroidism, and vitamin D intoxication as well as idiopathic hypercalciuria, obstructive uropathy, renal transplantation, renal medullary cystic disease, pyelonephritis, and sickle cell anemia. The pathogenesis of nephrocalcinosis and nephrolithiasis may be the result of decreased urinary citrate excretion as a result of cellular acidosis. (Citrate complexes calcium and thus prevents precipitation.) The condition may also be caused by the mobilization of carbonate (and calcium) from the bones to neutralize the excess of H^+ in DRTA. This process also produces progressive decalcification of bones. In children with DRTA, the observed characteristic bone changes are termed *renal rickets*.

Diagnosis. The condition presents frequently with growth retardation; rickets and osteomalacia resulting from increased excretion of phosphate and hypophosphatemia; hypercalciuria often associated with nephrocalcinosis, renal calculi, renal colic, and pyelonephritis; and muscle weakness often caused by hypokalemia.

The acidosis is of the hyperchloremic type and is characterized by a normal anion gap. Other laboratory findings include hypokalemia associated with urinary potassium excretion of > 20 mmol/d; urinary sodium excretion of > 20 mmol/d even if volume depletion is evident (in volume depletion renin and aldosterone may be increased); possible urinary calcium excretion of > 300 mg/d, which causes an increase in plasma PTH; acidemia; a urine pH > 5.5 (a urine pH below 5.5 excludes DRTA, except the low renin/low aldosterone variety); and increased fractional excretion of bicarbonate of 3–5% at normal plasma HCO_3^- concentration. In contrast, the fractional bicarbonate excretion in proximal RTA at normal bicarbonate concentrations is $> 15\%$. Strong support for distal tubular defect in acidification is a urinary pCO_2 that fails to increase above that of plasma during alkaline diuresis. In normal individuals the urinary pCO_2 increases by > 30 mm Hg above that of plasma when the urine pH is > 7.5.

The distal nephron is the site where urine pH reaches its lowest value; therefore, an inability to lower urine pH is a standard finding in distal tubular acidosis. If serum bicarbonate levels are raised to normal by bicarbonate administration, neither urinary excretion of bicarbonate nor urinary pH increases greatly. This suggests that the proximal tubular reabsorption mechanism is functional and suggests an inability of the distal nephron to acidify the urine properly.

The diagnosis of distal RTA can be confirmed by an *acid loading test*[78] which requires administration of 0.1 g NH_4Cl/kg body weight. A urine pH below 5.2 after 2–8 h and a steady blood bicarbonate concentration are observed in normal individuals and thus exclude DRTA. Patients with DRTA have a urine pH > 5.5, often > 6. Other screening tests would be a low urinary citrate excretion (< 200 mg/d) or a low urine/blood pCO_2 after bicarbonate loading.

Treatment. Treatment generally requires only administration of modest amounts of bicarbonate (or sodium citrate) sufficient to neutralize the daily unexcreted load of nonvolatile acids (1–2 mmol/kg in adults and 4–6 mmol/kg in children). Correction of acidosis generally also corrects the accompanying hyperparathyroidism and potassium wasting.

Classification of DRTA. The mechanism for the acidification defect is not fully understood. Several theories have been advanced and have led to the definition of several subclasses of DRTA.

The cause of DRTA may be a true impairment of H^+ secretion in the distal tubules, and this condition is called *secretory DRTA*. If the distal nephron has an increased permeability for hydrogen ions (or carbonic acid), the ions may diffuse back into the tubules, and thus proper acidification of urine cannot occur. These patients also cannot increase the urine pCO_2 during alkalinization. The condition is referred to as *back-leak DRTA*. A third type is a defect in sodium reabsorption, limiting H^+ secretion or exchange, as in lithium toxicity or aldosterone deficiency.

Some patients acidify urine at a submaximal rate which, however, is generally sufficient to maintain acid-base balance. Ammonia production in the distal tubules is generally high in these patients. Potassium wasting and hypokalemia occur only in acidosis. Hyperchloremia is generally not present and the fractional excretion of HCO_3^- is < 1%. However, when the patient is stressed or when he is given an acid load, his ability to excrete acid and lower urine pH is suboptimal and urine pH values > 5.5 may be seen. The condition is called *incomplete distal RTA.*

Hyperkalemic DRTA is a rather common clinical condition characterized by a hyperchloremic acidosis with hyperkalemia instead of hypokalemia. It is frequently associated with mild to moderate chronic renal insufficiency. As in other forms of DRTA, patients cannot properly acidify urine because of a decrease in H^+ excretion in the distal tubules. The urinary pH is frequently > 5.5 in the presence of systemic acidosis. Plasma aldosterone is normal or increased unless selective aldosterone deficiency (SAD) is also present. Hyperkalemia is caused in part by the decreased filtration rate (due to chronic renal disease), but there is also a decreased potassium excretion relative to the filtered load. Contrary to classic DRTA, potassium excretion cannot be increased in response to Na_2SO_4 administration; this suggests an isolated defect in H^+ secretion. However, since H^+ secretion is dependent on Na^+ reabsorption in the distal tubules, the primary defect could also be an impairment of distal tubular Na^+ reabsorption. Although the NH_4^+ generating mechanism is intact, the hyperkalemia may cause decreased NH_4^+ production.

Selective Aldosterone Deficiency

In this form of RTA (also called RTA IV) hyperchloremia is present in ∼75% of cases and hyperkalemia, although mild, is a usual manifestation. The hyperchloremic, hyperkalemic acidosis is primarily caused by a decreased glomerular filtration rate; it does not appear to be the result of impaired proximal bicarbonate reabsorption.

The urine pH is lowered as in normal individuals and the urine pCO_2 is normal after bicarbonate loading. The mechanism for NH_4^+ production is intact, but ammonia formation may be reduced by hyperkalemia. The symptoms of this disorder are similar to those seen in *aldosterone resistance.*

Combined Proximal and Distal RTA

Certain features of both proximal and distal RTA have been reported in the same individual. The clinical and laboratory findings depend on the specific combination of defects that are present. The condition is also called RTA III.

Water Homeostasis

The kidney is the principal organ for regulating both the total water content and the total solute content of the body. It maintains a precise balance between water and solute, and it controls plasma osmolality within an extremely close tolerance (283 ± 3 mOsm/kg).[41] In man, the kidney can vary the rate of excretion of water and solute independently to achieve urine osmolalities as low as 50 mOsm/kg and as high as 1200 mOsm/kg or greater.

Urine volume. In health, water homeostasis is determined by several interrelated processes: (1) water intake and water formed through oxidation of foodstuffs; (2) extrarenal water loss through the feces, insensible water loss, and sweating; (3) a solute load to be excreted that is derived from ingested minerals and nitrogenous substances; and (4) the ability of the kidney to produce a concentrated or dilute urine. Other factors such as vomiting and diarrhea become important in various disease states; loss of ability to produce a concentrated urine is a hallmark of virtually all types of chronic renal diseases.[31]

To maintain water homeostasis, the kidney must produce urine in a volume that precisely balances water intake and production on the one hand and water loss through extrarenal routes on the other. Minimum urine volume is determined by the solute load to be excreted and may be as low as 400 mL/d, while maximum urine volume is determined by the amount of excess water that must be excreted and may be as high as 20–25 L/d. The kidney is more proficient in excreting excess water than in conserving water; a urine osmolality of 50 mOsm/kg represents a six-fold dilution from the normal plasma osmolality of 283 mOsm/kg, while a urine osmolality of 1200 mOsm/kg represents more than a four-fold concentration. Fluid intake that is inadequate to provide the volume necessary to excrete the solute load will result in a net loss of body water and will produce prerenal azotemia. This condition is frequently seen in elderly patients with compromised renal concentrating ability and can be treated simply by increasing fluid intake.

Solute load. The solute load that must be excreted by the kidney is determined by the mineral and nitrogenous content of the diet or, in the fasting state, by the amounts of these materials produced by breakdown of body tissues. Carbohydrates and triglycerides produce no metabolic end-products that require renal excretion; however, metabolism of proteins and nucleic acids results in formation of urea, creatinine, uric acid, and various inorganic ions that contribute to the solute load. Thus, the solute load can be reduced by providing a high-carbohydrate, high-fat, low-protein, low-salt diet, which is usually prescribed for patients in acute or chronic renal failure. In certain disease states the oxidation of carbohydrates and triglycerides results in the production of large quantities of organic acids (e.g., lactate, acetoacetate, β-hydroxybutyrate), and the excretion of these compounds by the kidneys requires both water and Na^+.

The usual American diet produces a solute load of 1000–1200 mOsm/d, although the load will be higher in those who consume large meat meals and who use excessive amounts of salt. A fasting person will have a solute load of 600–800 mOsm/d, but provision of carbohydrate sufficient to minimize breakdown of body tissues will result in a solute load as low as 200 mOsm/d.

The highest rate of solute excretion encountered in disease (in the absence of administration of exogenous solute such as mannitol) is in diabetes mellitus, where solute excretion rates may approach 5000 mOsm/d as the plasma glucose approaches 800 mg/dL. This rate of solute excretion can lead very rapidly to profound water deficits, particularly in elderly diabetics, in whom it produces a syndrome termed hyperosmolar nonketotic coma.

Control of Osmolality[38]

All cell membranes are freely permeable to water, and osmolality is the same in all fluid compartments of the body (intracellular, extracellular, intravascular). Measurement of plasma osmolality therefore provides an indication of the total body water osmolality. An increase in plasma osmolality initiates two responses: (1) thirst, which stimulates an increase in water intake, and (2) secretion of antidiuretic hormone (ADH), which stimulates renal tubular reabsorption of water. The threshold for thirst is higher than for ADH secretion, and body osmolality is controlled largely by the effects of ADH on renal water conservation and excretion.

Antidiuretic hormone is synthesized as a high-molecular-weight precursor in cells of the supraoptic and paraventricular nuclei of the hypothalamus. ADH is then transported complexed with a carrier protein, neurohypophysin, in the neural tracts connecting the hypothalamus with the posterior portion of the pituitary (the neurohypophysis). Release of ADH from its storage site in the posterior pituitary is controlled by osmoreceptors located in the hypothalamus. Decreased body water leading to plasma hyperosmolality stimulates release of ADH. Contraction of extracellular fluid volume can also stimulate secretion of ADH and can apparently override the stimulus from hyperosmolality; however, this does not occur until there has been a 10% loss in fluid volume.

ADH acts on the cortical collecting tubules and the collecting ducts of the nephrons to make them permeable to water. The degree of permeability is determined by the rate of ADH secretion; in the absence of ADH the tubules are impermeable to water. The mechanism of action of ADH consists of activation of adenylate cyclase at the basal-lateral cell surface with the production of cAMP, transport of cAMP to the luminal side of the cell membrane with the activation of a protein kinase, and increased phosphorylation of key proteins that increase the permeability to water. The processes through which permeability or impermeability to water affects urine osmolality are discussed in the following section.

The Countercurrent Multiplier System

The countercurrent multiplier system is the mechanism through which a large osmolal gradient is established in the interstitial tissue between the corticomedullary junction and the tips of the renal papillae (Figure 11-11).[38,41] The medullary collecting ducts course through this hyperosmolal interstitium. In the presence of ADH, which renders the collecting ducts permeable to water, fluid is reabsorbed until the osmolality within the collecting ducts is equal to that of the interstitium, thus producing a hyperosmolal urine. In the absence of ADH, the collecting ducts are impermeable to water and the dilute tubular fluid traverses the medulla unchanged, resulting in a hyposmolal urine.

The urine concentrating/diluting system involves the loops of Henle, the distal convoluted tubules, the

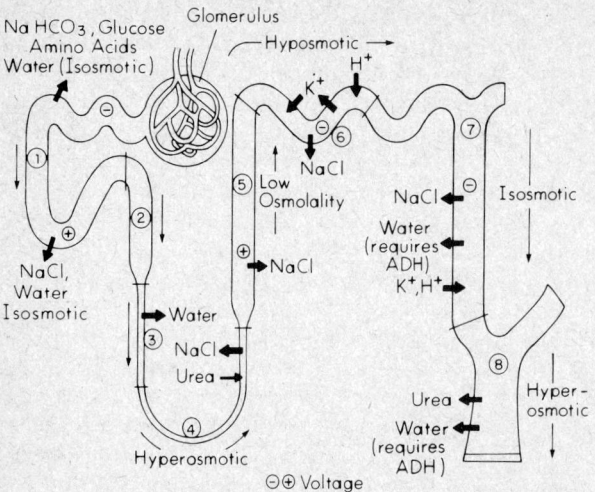

Figure 11-11. Schematic representation of the principal processes of transport in the nephron. In the convoluted portion of the proximal tubule (1) salt and water are reabsorbed at high rates in isotonic proportions. Bulk reabsorption of most of the filtrate (65–70%) and virtually complete reabsorption of glucose, amino acids, and bicarbonate take place in this segment. In the pars recta (2) organic acids are secreted and continuous reabsorption of sodium chloride takes place. The loop of Henle comprises three segments: The thin descending (3) and ascending (4) limbs and the thick ascending limb (5). The fluid becomes hyperosmotic, because of water abstraction, as it flows toward the bend of the loop and hyposmotic, because of sodium chloride reabsorption, as it flows toward the distal convoluted tubule (6). Active sodium reabsorption occurs in the distal convoluted tubule and in the cortical collecting tubule (7). This latter segment is water-impermeable in the absence of ADH, and the reabsorption of sodium in this segment is increased by aldosterone. The collecting duct (8) allows equilibration of water with the hyperosmotic interstitium when ADH is present. For further details see text. (Adapted from a figure by Iselin, A., In: Burg, M. B.: Hosp. Pract., 13 (Oct.): 100, 1978.)

cortical collecting tubules, the collecting ducts, and the associated capillary vasa recta. It is important to recognize that in man there is no physiological mechanism for active transport of water. Solutes, primarily sodium and chloride, can be actively "pumped" from the tubular lumen into the interstitium, and water will follow passively if the wall of the tubule is permeable to water.

The loop of Henle is a hairpin-shaped segment of the tubule that descends into the renal papilla and that connects the proximal convoluted tubule and the distal convoluted tubule. The descending and ascending limbs of the loop are in close proximity, and fluid flow is in opposite directions, i.e., countercurrent. The descending limb is impermeable to sodium and chloride, quite permeable to urea, and freely permeable to water. As fluid descends into the loop, water diffuses into the increasingly hypertonic interstitium, and the tubular fluid becomes progressively more concentrated as it approaches the end of the loop. In contrast to the descending limb, the ascending limb is impermeable to water but is permeable to sodium and chloride and moderately permeable to urea. As fluid ascends into the region of lower interstitial osmolality, sodium, chloride, and urea diffuse out of the hyperosmolal tubular fluid into the less hyperosmolal interstitial fluid. In the final portion of the ascending limb of the loop—the thick segment—chloride is actively transported from the tubules into the interstitium accompanied by a passive transfer of sodium, further reducing the osmolality of the tubular fluid to as low as 100 mOsm/kg and increasing the osmolality of the interstitium. The effect of this active, energy-requiring transport of chloride and sodium is "multiplied" by the countercurrent flow in the two limbs of the loop of Henle. Water reabsorbed from the descending limb of the loop is offset by sodium chloride transported into the interstitium from the ascending limb. The concentration gradient between the tubular fluid and the interstitial fluid at any level is small (<200 mOsm/kg), but the total gain in osmolality of the interstitial fluid from the corticomedullary junction to the tip of the papilla is large (from about 300 to 1200 mOsm/kg).

The concentration of urea in the interstitium also contributes substantially (about half of the total) to the osmotic gradient. Urea concentration in the interstitium of the medulla is maintained at a high level by a cyclic trapping process. Urea diffuses from the interstitium into the loops of Henle and is carried through the distal convoluted tubules to the cortical collecting tubules and thence to the medullary collecting ducts. In the presence of ADH, the cortical collecting tubules become permeable to water but not to urea, and diffusion of water raises the urea content of the fluid within the ducts. The medullary collecting ducts respond to ADH by becoming permeable to both water and urea. Urea can therefore diffuse out of the collecting ducts into the medullary interstitium, from which it enters the loops of Henle and begins the cycle once more.

Urea apparently plays an important role in extending the countercurrent multiplier's effect to the tip of the papilla below the site of active chloride transport in the thick segment of the ascending limb of the loop of Henle. This process occurs in the following manner: fluid entering the descending limb of the loop has a high concentration of sodium chloride but low urea content. During transit down the loop, water is passively absorbed from the tubular fluid into the increasingly hypertonic interstitium. At the tip of the papilla, the tubular fluid has a very high concentration of sodium chloride and a much lower concentration

Table 11-10. DISEASES ASSOCIATED WITH
DISTURBANCES IN THE RENAL CONCENTRATING
MECHANISM

Marked polyuria and hypotonic urine after water deprivation
    Diabetes insipidus
    Hereditary nephrogenic diabetes insipidus
    Chronic lithium toxicity
    Sickle cell nephropathy
    Hypokalemia (rarely)
Moderate polyuria and inability to produce hypertonic urine
    Hypercalcemia
    Hypokalemia
    Chronic pyelonephritis
    End-stage renal disease
    Amyloidosis
    Interstitial nephritis

of urea, in contrast to the interstitial fluid in the region, which has high concentrations of both sodium chloride and urea. The concentration gradients of sodium and chloride cause diffusion of these ions out of the thin segment of the ascending limb of the loop, since it is freely permeable to them. Some diffusion of urea in the opposite direction, into the ascending limb, also takes place. Since the ascending limb of the loop is impermeable to water, the net effect is transfer of sodium chloride into the interstitium at a rate proportional to the concentration gradient between the tubular fluid and the interstitial fluid.

Maintenance of the osmotic gradient in the medulla requires that water absorbed from the tubules be removed while solute remains in the interstitium. Since capillaries are permeable to both water and solute, a unique arrangement of the blood vessels supplying the medulla, the vasa recta, makes this possible. The vasa recta form a looped countercurrent exchange system. Solute removed from the interstitium by the descending limb is returned to the interstitium by the ascending limb. Thus osmolality in both limbs of the vasa recta becomes progressively higher toward the tip of the papilla, yet blood leaving the medulla has approximately the same osmolality as blood entering it.

Failure of ADH production results in marked polyuria (diabetes insipidus), which stimulates thirst and greatly increases water intake. Lack of response of the kidney to ADH has a similar effect (nephrogenic diabetes insipidus). A number of other conditions are also characterized by diminished ability to produce a concentrated urine (Table 11-10).

Urine Production

Consideration of the volumes and osmolalities of fluid during the sequential steps of urine production will illustrate the function of the concentrating/diluting mechanism of the kidney (Table 11-11). Renal blood flow represents about one fifth of the total cardiac output and results in a renal plasma flow of 600 mL/min. One fifth, or 120 mL/min, of the renal plasma flow is filtered through the glomeruli as isosmotic fluid of very low protein content. About two thirds of the glomerular filtrate is reabsorbed isosmotically in the proximal convoluted tubules, and thus 40 mL/min of isosmotic fluid enters the loops of Henle. Reabsorption of solute and water does not occur isosmotically in the loops of Henle; about 25% of the sodium chloride and 15% of the water filtered by the glomeruli are reabsorbed. The higher solute reabsorption, of course,

Table 11-11. FLUID FLOW RATES AND OSMOLALITIES

	Flow Rate, mL/min	Flow Rate, L/day	Osmolality, mOsm/kg
Renal blood flow	1100	1600	283
Renal plasma flow	600	860	283
Glomerular filtrate	120	180	283
End of proximal tubule	40	60	283
End of loop of Henle	18	25	100
End of distal tubule	18	25	50
Urine	0.4–18	0.6–25	50–1200

Figures given are approximate and for illustrative purposes. Clinical measurements can be made or inferred for volumes and osmolalities in renal blood flow, renal plasma flow, glomerular filtrate, and urine.

is necessary to maintain the hyperosmolal condition in the medullary interstitium and results in a hyposmolar fluid (100–150 mOsm/kg) entering the distal convoluted tubules. Additional solute is reabsorbed in the distal convoluted tubules and in the cortical collecting tubules and further lowers the osmolality of the fluid (50–100 mOsm/kg). It is only on those structures beyond the distal convoluted tubules that ADH exerts its effect.

It is important to note that the major portion of water reabsorption by the tubules is not influenced by the rate of ADH secretion. Two thirds of glomerular filtrate is reabsorbed by the proximal convoluted tubules, which are unaffected by ADH. More than half of the fluid entering the loops of Henle is reabsorbed, also without influence from ADH. It is only on the fluid leaving the distal convoluted tubules that ADH exerts its effect.

In the absence of ADH the cortical collecting tubules and the collecting ducts remain impermeable to water, and fluid flowing from the distal convoluted tubules is excreted as a large volume of dilute urine.

In the presence of ADH the cortical collecting tubules and the medullary collecting ducts become permeable to water, and rapid osmolal equilibration takes place between the tubular fluid and the interstitial fluid. Water is reabsorbed in the cortex until the osmolality of the tubular fluid approximates that of plasma (~300 mOsm/kg), and this process accounts for the major part of water conservation during antidiuresis. It is in the medullary collecting ducts that the final concentration of tubular fluid takes place, increasing the osmolality of the tubular fluid from that of the cortical interstitium to the osmolality of the interstitium of the papillae.

Quantitation of Water Excretion and Conservation

The quantitative description of water conservation or excretion by the kidney can be derived from clearance concepts previously presented. The kidneys are required to excrete a given amount of solute each day as determined by the daily intake of minerals and nitrogenous substances. The solute excretion rate expressed in mOsm/min can be calculated from the solute concentration in the urine (U_{Osm}) expressed in mOsm/kg and the urine flow rate (V) expressed in mL/min. (The calculations assume that 1 kg = 1 L of urine.)

$$\text{Solute excretion rate} = U_{Osm} \times V \times 1000$$

Water excretion with varying solute loads can be expressed as the osmolal clearance (C_{Osm}), which is the hypothetical volume of plasma completely cleared of solute per unit time expressed in mL/min.

$$C_{Osm} = \frac{U_{Osm} \times V}{P_{Osm}}$$

The osmolal clearance also represents the volume of urine that would be required for isosmolal clearance of the solute load. During renal water conservation, formation of hypertonic urine results in a urine flow rate that is less than that which would be required if isosmolal urine were produced, i.e., less than C_{Osm}. Conversely, during excretion of excess water by means of a hypotonic urine, the actual flow will exceed C_{Osm} by a variable degree. These differences between actual urine flow rates and flow rates calculated for isosmotic excretion provide important measures of the kidney's ability to concentrate and dilute urine. Two useful quantitative terms that represent these differences have been termed "free water clearance" (C_{H_2O}) and "negative free water clearance" (TC_{H_2O}).

Free water clearance is defined as the hypothetical rate at which water is excreted during production of hypotonic urine in excess of that required for excretion of the solute load in isosmolal urine. It is calculated as follows:

$$C_{H_2O} = V - C_{Osm}$$

Negative free water clearance is a quantitative expression of the rate at which water is reabsorbed from tubular fluid during the production of a hypertonic urine. It is calculated as follows:

$$TC_{H_2O} = C_{Osm} - V$$

Figure 11-12. Histogram representing the relation between osmolal clearance and urine flow. C_{osm}: osmolar clearance (mL/min); CH_2O: free water clearance (mL/min); TcH_2O: negative free water clearance (mL/min). In each case the total solute excretion and the osmolal clearance are the same. When the urine is excreted as an isotonic solution, the urine flow and osmolal clearance are identical. When water is removed by the distal concentrating process within the kidney, the urine becomes hypertonic and urine flow is less than osmolal clearance. The amount of water removed represents that amount of water which would be required to return the urine to the isotonic state and is depicted here by the dashed bar. Production of a hypotonic urine in the presence of water excess leads to excretion of additional quantities of water, and thus urine flow exceeds osmolal clearance by the amount of solute-free water produced in the distal nephron. These relationships between urine flow, osmolal clearance, and solute-free water excretion or conservation vary over a substantial range under different physiological conditions. This range is illustrated in Figure 11-13. (Reprinted with permission from Walker, W. G., and Burrow, C. R.: Pathophysiology of renal insufficiency, pathogenesis of uremia, and clinical evaluation of renal function. *In:* Principles and Practice of Medicine. 21st ed. A. M. Harvey, R. J. Johns, V. A. McKusick, et al., Eds. Norwalk, Conn., Appleton-Century-Crofts, 1984.)

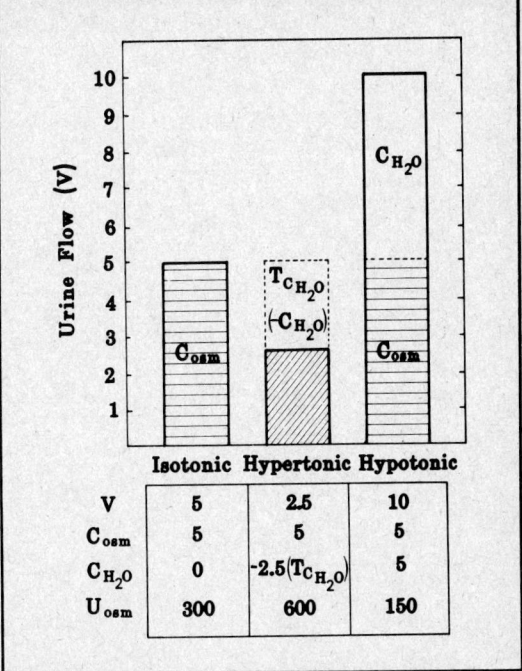

	Isotonic	Hypertonic	Hypotonic
V	5	2.5	10
C_{osm}	5	5	5
C_{H_2O}	0	$-2.5(Tc_{H_2O})$	5
U_{osm}	300	600	150

Negative free water clearance is the converse of free water clearance; it is the hypothetical volume of water that is conserved when compared to the volume necessary to excrete the solute load in isosmolal urine.

These three parameters—osmolal clearance, free water clearance, and negative free water clearance—constitute the three primary values that allow definition of the limits of water conservation and excretion in health and disease. Figure 11-12 illustrates these concepts. The limits which the healthy kidney can achieve and the limitations produced by advanced renal disease are illustrated in Figure 11-13.

At a low rate of solute excretion, the ability of the kidney to reabsorb solute-free water is limited by its capacity to increase the osmolality of urine compared to plasma, the U:P osmolal ratio. This ratio varies between 3.5 and 4.0 in normal individuals. As the solute load increases, there is a corresponding increase in reabsorption of solute-free water (TC_{H_2O}); however, this process is limited by the maximal rate of water diffusion. Under these circumstances, negative free water clearance (TC_{H_2O}) becomes maximal at high rates of osmolar clearance. This causes the osmolality of urine to fall progressively toward that of plasma. At high values of osmolal clearance, maximal water conservation may be accompanied by a urinary osmolality of 400 mOsm/kg or less. This accounts, in part, for the large water losses during the osmotic diuresis of diabetic acidosis. In far advanced renal failure, the osmotic load may be so great that the kidney is incapable of producing a significantly hypertonic urine and water conservation is seriously impaired.

Similarly, free water clearance (C_{H_2O}) at low solute excretion rates is limited by the minimal osmolal concentration that can be produced by the kidney. With increasing solute load, the rate of increase in free water clearance slows remarkably, and urine osmolality approaches plasma osmolality during water diuresis just as it does during hydropenia.

Assessment of Renal Concentrating Ability

The capacity of the kidney to conserve water can be assessed simply by demonstrating that the solute concentration of the urine approaches the maximal range that can be achieved in health. The usual test consists of withholding fluids overnight, obtaining the first voided specimen in the morning, and measuring osmolality. Maximal urine concentration requires fluid deprivation for 36–48 h; however, after fluid deprivation of 18 h or more a urine osmolality that exceeds 850 mOsm/kg is considered to reflect a normal renal concentrating mechanism.[38]

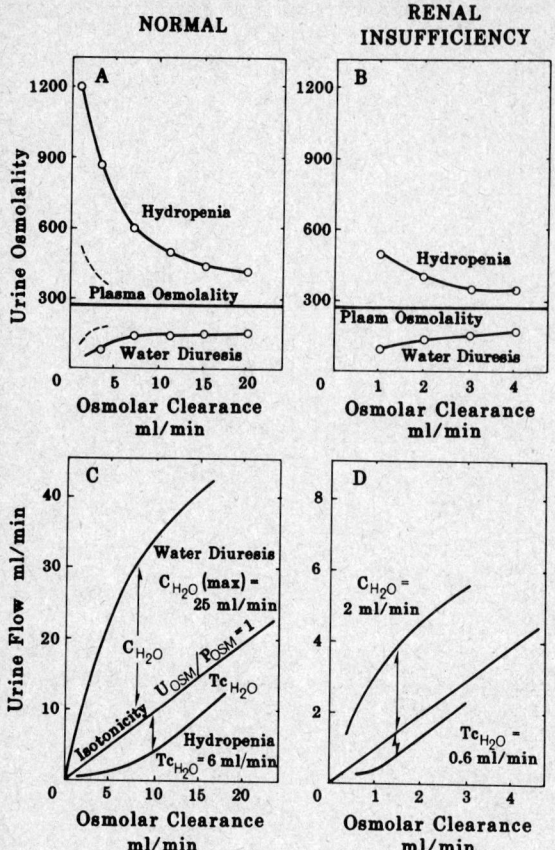

Figure 11-13. Relationship between urine osmolality and osmolar clearance during hydropenia and water diuresis in health (upper left) and disease (upper right). As osmolar clearance increases, urine osmolality approaches that of plasma. The lower figures show the relationship between urine volumes and osmolar clearance during hydropenia and water diuresis in health (lower left) and disease (lower right). It is evident that the kidney has a much greater capacity to excrete solute-free water than to extract solute-free water from hypertonic urine. Water diuresis under some circumstances can be associated with excretion of as much as 25 mL/min of solute-free water. Although this rate of water excretion is rarely encountered in the healthy individual, it can occur in subjects who are ADH deficient and who are given large solute loads. The two right-hand figures depicting renal insufficiency illustrate how markedly restricted this capacity to excrete and conserve water becomes as progressive renal failure occurs. (Reprinted with permission from Walker, W. G., and Burrow, C. R.: Pathophysiology of renal insufficiency, pathogenesis of uremia, and clinical evaluation of renal function. *In*: Principles and Practice of Medicine. 21st ed. A. M. Harvey, R. J. Johns, V. A. McKusick, et al., Eds. Norwalk, Conn., Appleton-Century-Crofts, 1984.)

In patients with either hypothalamic or pituitary disorders causing complete ADH deficiency or in patients who lack the normal renal response to ADH (nephrogenic diabetes insipidus), urine osmolality rarely exceeds 300 mOsm/kg. However, partial defects in ADH secretion or renal response to ADH will result in urine osmolalities between 300 and 800 mOsm/kg. The assessment of partial deficiencies of ADH secretion or deficient renal response to ADH requires measurement

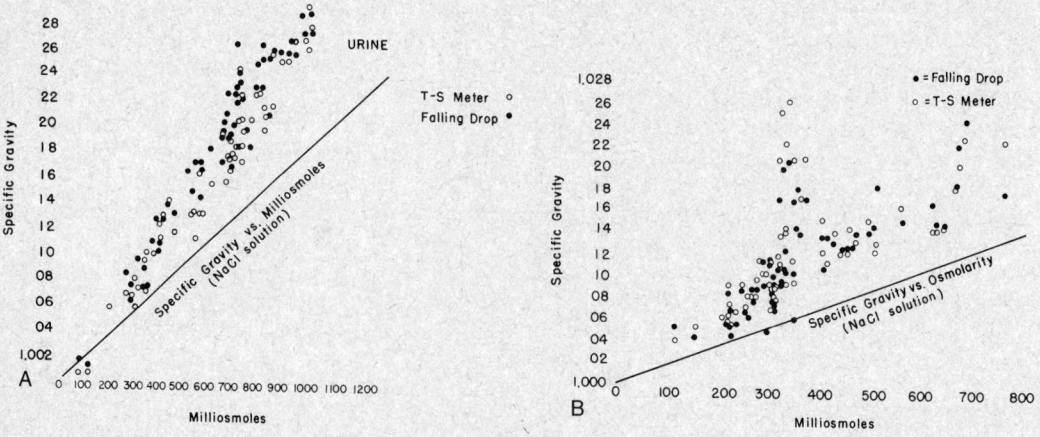

Figure 11-14. *A,* Comparison of the urinary specific gravity and urinary osmolality in a series of urines obtained from healthy medical students. The straight line represents comparative readings on various concentrations of sodium chloride solutions. *B,* Comparison of the urinary specific gravity and urinary osmolality in a series of unselected urines obtained from patients on the renal service. (From Holmes, J. H.: Workshop on Urinalysis and Renal Function Studies. Chicago, American Society of Clinical Pathologists, © 1962. Used by permission.)

of urinary osmolality or measurement of plasma ADH levels or both during specific manipulations of plasma osmolality through dehydration of the patient or intravenous osmotic loading.[2,47]

A useful guide to the adequacy of the renal concentrating mechanism is the measurement of urine specific gravity, as long as the limitations of this procedure are recognized. This simple test measures the density of urine relative to the density of water, and in most circumstances density bears a constant relationship to osmolality. This relationship is illustrated in Figure 11-14. In some conditions there is a remarkable divergence between osmolality and specific gravity. Urine collected after intravenous administration of iodine-containing radiopaque compounds for radiological studies may give extraordinarily high values. Glucose and protein may also contribute substantial increments to the density of urine, and semiquantitative determination of these substances is necessary for valid interpretation or correction of urine specific gravity measurements. Diabetic patients with uncontrolled hyperglycemia and glucosuria may have a high urine specific gravity even when the normal renal concentrating function is seriously impaired.

Determination of Specific Gravity with the Urinometer

Specimen. Specific gravity is most often determined as a part of a routine urinalysis on a random urine specimen. Less frequently, however, specific gravity is measured on timed specimens after water restriction, in which case more exact information is derived.

Reference Range. Specific gravity values observed in healthy individuals vary greatly with fluid intake and the state of hydration. Thus, the normal values for a 24-h specimen are usually considered to be from 1.015–1.025. Values after fluid restriction are > 1.025.

Principle. The urinometer is a hydrometer designed for the measurement of urinary specific gravity. When placed in the specimen contained in a cylinder, it sinks to the level characteristic of the specific gravity of the specimen. The value may then be read directly from the calibrations on the stem.

Procedure

1. Pour the specimen into the urinometer tube until it is about three-fourths full. Sufficient space should be allowed so that it will not overflow when the urinometer is floated in the sample.

2. Place the urinometer in the specimen with a slight twisting motion so that it will spin and have less tendency to stick to the sides of the tube. It is important that the urinometer float freely without sticking to the walls of the container.

3. Read the scale on the stem where it is intersected by the lowest line of the meniscus.

4. For greatest accuracy, measure the temperature and make the following correction: Add 0.001 to the specific gravity for each 3 °C or 5.4 °F that the temperature is above the urinometer calibration temperature. Subtract 0.001 for each 3 °C or 5.4 °F that it is below the calibration.

Notes

1. The urinometer should be calibrated against distilled water and should read 1.000 at its calibration temperature. It should also be calibrated at a high value. This may be done by testing in a mixture of 75 mL of xylene and 28 mL of bromobenzene, which has a specific gravity of 1.030.

2. If the urine contains significantly large quantities of protein, a correction should be applied to compensate for this factor. Subtract 0.003 from the reading for each 1 g of protein/dL of urine.

3. Subtract 0.004 for each 1 g of glucose/dL of urine.

Sources of Error

Readings may be in error if:

1. The urinometer is not allowed to float freely. It must not adhere to the sides of the tube, and no bubbles should cling to the stem.

2. There is failure to compensate for temperature or for gross proteinuria and glycosuria.

Determination of Specific Gravity by Refractometry

Specimen. See under Determination of Specific Gravity with the Urinometer.

Principle. The refractive index and the specific gravity of a urine specimen are both related functions of the quantity and type of dissolved substance in the specimen. Each substance contributes differently to the refractive index and also to the specific gravity; however, because various urine specimens are likely to contain dissolved substances of similar types and proportions, the refractive index and the specific gravity may be correlated. Increased amounts of abnormal substances such as glucose and protein may partially invalidate the correlation and give specific values that are misleading.

The instrument most commonly used for this purpose is the TS (total solids) Meter (American Optical Corp., Scientific Instrument Div., Buffalo, NY 14240). This is a hand refractometer with two temperature-compensated scales, which allows direct determination of total solids of serum or the specific gravity of urine. Other refractometers have two scales, one calibrated in refractive index and the other in total serum protein concentration.

Procedure

1. Place a small drop of sample on the lower glass plane surface of the TS Meter. Then bring the upper hinged surface down firmly on the drop so that the two glass planes are parallel.

2. Hold the meter toward a source of light so that the beam passes through the sample and the prisms.

3. Read the specific gravity from the proper scale at the sharp line of contrasting light and dark areas that falls across the scales.

Osmolality of Serum and Urine

Urine Osmolality Measurements

Measurements of the osmotic concentration of urine are considered more valid than specific gravity measurements in assessing the concentrating ability of the kidney, since the regulation of water excretion is, in part, determined by the osmolality of the fluid compartments of the body. Consequently, measurement of the urine osmolality, especially as part of a concentration test, is preferred. Holmes[33] performed a comparative study of specific gravity versus osmolality values in normal individuals and in unselected patients with renal disease. Although in the first group

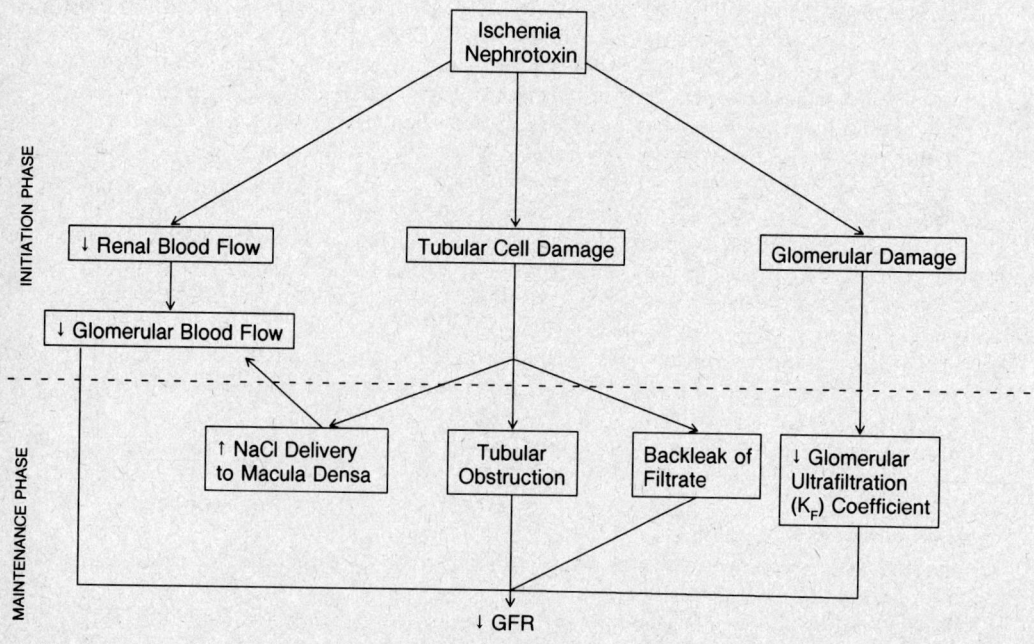

Figure 11-15. Pathogenesis of acute renal failure. (Used with permission from Martin, K.: Pathophysiology of acute renal failure. *In:* The Kidney and Body Fluids in Health and Disease. S. Klahr, Ed. New York, Plenum Medical Book Co., 1983.)

Table 11-12. VALUES FOR OSMOLALITY OBSERVED UNDER VARIOUS CONDITIONS

Specimen	mOsm/kg H_2O
Urine	
Osmotic limits of renal dilution and concentration	50–1200
Normal random specimen (average fluid intake)	300–900
Normal range during maximum urine concentration (fluid restriction for 12 h)	850–1200
Serum	
Osmotic limits observed	230–490
Reference range	278–298 (mean: 283)
Ratio $\dfrac{\text{Urine osmolality}}{\text{Serum osmolality}}$	
Random urine (average fluid intake)	1.0–3.0
Specimen after 12-h fluid restriction	3.0–4.7
Ratio $\dfrac{\text{Serum Na}}{\text{Serum osmolality}}$	0.43–0.50

there appears to be a reasonably close relationship between specific gravity and osmolality, the relationship is maintained to a much lesser degree in patients with renal disease. (See Figure 11-15.) This lack of correlation can be explained at least partially by the fact that the presence of heavy molecules, such as protein, glucose, or iodine-containing compounds, affects the specific gravity of urine substantially more than its osmolality.

The urine osmolality of normal individuals may vary widely depending on the state of hydration. After excessive intake of fluids, for example, the osmotic concentration may fall as low as 50 mOsm/kg, while in individuals with severely restricted fluid intake concentrations of up to 1200 mOsm/kg can be observed. In individuals on an average fluid intake, values of 300–900 mOsm/kg are most frequently seen.

If a *random urine specimen* of a patient has an osmolality of 600 mOsm/kg H_2O or higher (or > 850 mOsm/kg H_2O after 12-h fluid restriction), it can be assumed that the renal concentrating ability is normal.

In *chronic progressive renal failure*, the concentrating ability of the tubules is diminished, which may be readily and reliably shown by means of urine osmolality measurements. In acute tubular necrosis, the urine osmolality, if there is urine output at all, approaches the osmolality of the glomerular filtrate.

In *polyuria due to diabetes insipidus*, the urine osmolality is extremely low, and this may serve as a means to differentiate this condition from polyuria due to neurogenic origin. (See also Ratio of urine osmolality to serum osmolality.) Also, if the fluid intake of patients with diabetes insipidus is restricted, only slight changes in osmolality are noted, while in the latter type of disease, urine will be concentrated to above 850 mOsm/kg. (See Table 11-12.)

Ratio of Serum Sodium to Serum Osmolality

The solutes most often seen in excess and responsible for the hyperosmolality of serum are glucose and urea (e.g., in diabetes mellitus and uremia). In diabetes, ketone bodies may also accumulate and contribute as much as 10 mOsm/kg serum H_2O to the osmotic concentration.

Abnormally *low ratios* occur in such diverse conditions as *lymphomas, cancer, liver failure, shock, acute infections*, and *myocardial infarctions*, and indicate, in general, unfavorable prognosis. In special cases of *acute toxicity* due to overdose of drugs (e.g., salicylate poisoning) the finding of normal serum sodium and high serum osmolality may often be considered as an indication for dialysis.

In *dehydration*, where water loss exceeds salt loss, the osmolality and the serum sodium increase but the ratio stays normal. Examples are *prolonged diarrhea* and *peritonitis* or *intestinal obstruction* in which fluid has been sequestered outside the vascular system.

Ratio of Urine Osmolality to Serum Osmolality

In normal individuals on an average fluid intake, the ratio is most frequently between 1.0 and 3.0. After fluid restriction, as in the case of the concentration tests, the ratio is 3.0 or above and may rise in extremes to 4.7. In patients with *renal tubular deficiency*, the ratio will be below that observed in normal individuals. In polyuria of *diabetes insipidus*, the ratio will be between 0.2 and 0.7 even after fluid restriction. This allows for differentiation of this type of polyuria from that seen in other conditions such as diabetes mellitus. In polyuria of neurogenic origin, the ratio may be normal without fluid restriction and increases after fluid restriction.

Determination of Osmolality in Serum and Urine

Specimen. Blood should be collected by venipuncture with a minimum of stasis, and the serum should be separated by centrifugation soon after collection. In order to lessen the possible presence of particulate matter, a second centrifugation is recommended. If the serum is not to be analyzed soon after centifuging, it should be refrigerated or frozen. Heparinized plasma is also satisfactory, but oxalated plasma is not.

Urine should be collected in clean, dry containers without preservatives and centrifuged at sufficiently high speed to remove all gross particulate material. If the analysis cannot be carried out soon after collection, the specimen should be refrigerated. Before analysis, refrigerated specimens should be warmed to aid the complete solution of any precipitated substances.

Procedure. Several different osmometers are commercially available, and each one must be operated in a manner appropriate to the particular instrument, as described in its corresponding manual. However, the following procedure is general and may be applied to any instrument.

1. Centrifuge the specimen twice to eliminate any gross particulate matter.
2. Place the proper volume of specimen in a sample tube and position the tube in the instrument.
3. Supercool the specimen (about $-7\ °C$).
4. Initiate the freezing process and allow the temperature equilibrium to take place.
5. Measure the freezing point and read the value in milliosmoles from the instrument.

Standardization. Osmometers are generally standardized in accordance with the procedures supplied by the manufacturer, utilizing standard sodium chloride solutions.

Comments and Sources of Error. It has been reported that a reproducibility of ± 1 mOsm/kg H_2O with an equal accuracy can be obtained if the instrument is standardized correctly; however, under routine conditions an accuracy and precision of ± 2 mOsm/kg H_2O is quite acceptable.

Faulty standardization, faulty use of the osmometer, and the presence of particulate matter in the specimen are the most common sources of inaccurate results.

Renal Disease and the Role of the Laboratory

There are few renal diseases for which we know the entirety of etiology, pathophysiology, structural change, and clinical presentation. Therefore, there is presently no generally accepted consistent, systematic nomenclature of renal disease, and many renal diseases are defined in terms of their clinical presentation and structural change alone. Consequently, many names for renal diseases actually represent clinical syndromes and their structural basis as revealed by renal biopsy. An additional difficulty in classifying renal diseases is the tendency of distinct primary diseases to merge with time into a single clinical entity, the end-stage kidney, regardless of their initial etiology or anatomic region of injury. This is primarily due to the intimate relationship between the anatomy and function of renal structures and their blood supply, as discussed below. This delicately balanced system is vulnerable to disease processes at many points. A diminution of the kidney's share of the normal cardiac output, about one fifth of the total, may impair all the functions of the nephron and produce the clinical syndrome of prerenal azotemia. If severe, it may lead to ischemic injury of renal tubules and the syndrome of acute renal failure. Sustained renal ischemia triggers renin release from the kidney with resulting hypertension. Hypertension, in fact, is a complication of many renal diseases of varied etiologies. Additionally, severe hypertension itself can damage renal arterioles, further compromising nephron function. The complex interdependence of various nephron functions and their interrelated blood supply makes loss of entire individual nephrons, together with hypertrophy of those remaining, the common denominator of most chronic renal diseases. Whether the initial insult is immunologic injury of the

Table 11-13. STAGES OF CHRONIC PROGRESSIVE RENAL DISEASE

Stage	Renal Function Remaining (%)	Serum Creatinine (mg/dL)	Serum Urea N (mg/dL)
1. Decreased renal reserve	50–75	1.0–2.5	15–30
2. Renal insufficiency	25–50	2.5–6.0	25–60
3. Renal failure	10–25	5.5–11.0	55–110
4. Uremic syndrome	0–10	> 8.0	> 80

glomerulus, infection of the interstitium, or toxic injury to tubular cells, the ultimate result is loss of the complete nephron. The loss of > 90% of nephrons, resulting in the end-stage kidney, produces a remarkably consistent clinical and laboratory picture. Once chronic renal disease has reached this point, it may be impossible to determine the site and nature of the initial injury.

Our discussion of renal disease begins with the end-stage kidney. We will then discuss acute renal failure and differentiate underperfusion of the kidney from acute reversible loss of nephron function, the syndrome of acute tubular necrosis. This will be followed by discussions of diseases that initially involve one component of the nephron. These discussions will begin with diseases affecting glomeruli and proceed to conditions affecting tubules and the interstitium, and finally to obstruction of the collection system.

End-stage Renal Disease and the Pathophysiology of the Uremic Syndrome. The uremic syndrome, the terminal clinical manifestation of kidney failure, is the constellation of symptoms, physical signs, and abnormal diagnostic studies that results from the failure of the kidneys to maintain adequate excretory, regulatory, and endocrine function. Classic signs of uremia include progressive weakness and easy fatigability, loss of appetite followed by nausea and vomiting, muscle wasting, tremors, abnormal mental function, frequent but shallow respirations, and metabolic acidosis. The syndrome progresses to produce stupor, coma, and ultimately death unless support is provided by hemodialysis or successful renal transplantation. Regulation of body fluids is impaired in uremia patients because of failure to excrete excess ingested fluid or to cope with fluid losses due to vomiting or diarrhea. Patients also have difficulty excreting a salt load or retaining sodium when intake is low or vascular volume inadequate. Acid excretion is impaired, as is the ability to excrete nitrogenous metabolites from dietary sources. Consequently, the composition of plasma is abnormally labile in response to such factors as diet, state of hydration, gastrointestinal bleeding, vomiting, diarrhea, and intake of therapeutic drugs. It is helpful to characterize progressive renal disease as occurring in four stages, defined by the percentage of renal function remaining and by the plasma concentrations of creatinine and urea nitrogen (Table 11-13), although there are no well-defined clinical syndromes that necessarily characterize these four stages.

The most characteristic *laboratory findings* are increased concentrations of nitrogenous compounds in plasma, e.g., urea nitrogen and creatinine (azotemia), as a result of reduced glomerular filtration rate and decreased tubular function. Retention of these compounds and of metabolic acids is followed by progressive hyperphosphatemia, hypocalcemia, and potentially dangerous hyperkalemia. Although most patients eventually become acidemic, respiratory compensation by elimination of CO_2 is extremely important. In addition, there is reduced endocrine function manifested by inadequate synthesis of erythropoietin and vitamin D with resulting anemia and osteomalacia. Disordered regulation of blood pressure generally leads to hypertension. Biochemical characteristics of the uremic syndrome are summarized in Table 11-14.

End-stage renal disease and the resulting uremic syndrome can be the outcome of any of a wide variety of renal diseases, including chronic glomerulonephritis, chronic pyelonephritis, immunological diseases with renal involvement (e.g., systemic lupus erythematosus), hypertension, acute obstruction of the lower urinary tract, and toxic or ischemic damage to the kidney. Whatever the cause of end-stage renal disease, the eventual pathophysiology is the result of reduced GFR and tubular function due to loss of individual nephrons.

In addition to the consequences of reduced excretory, regulatory, and endocrine function of the kidney, there are several systemic manifestations of the uremic syndrome—among them pericarditis, pleuritis, disordered platelet and granulocyte function, and encephalopathy—that

Table 11-14. BIOCHEMICAL CHARACTERISTICS OF THE UREMIC SYNDROME

Retained nitrogenous metabolites	*Abnormal lipid metabolism*
Urea	Hypertriglyceridemia
Cyanate	Decreased HDL-cholesterol
Creatinine	Hyperlipoproteinemia
Guanidine compounds	*Altered endocrine function* [39]
"Middle molecules"	Secondary hyperparathyroidism
Uric acid	Osteomalacia (secondary to abnormal vita-
Fluid, acid-base, and electrolyte disturbances	min D metabolism)
Fixed urine osmolality	Hyperreninemia and hyperaldosteronism
Metabolic acidosis (decreased blood pH,	Hyporeninemia
bicarbonate)	Hypoaldosteronism
Hypo- or hypernatremia	Decreased erythropoietin production
Hypo- or hyperkalemia	Altered thyroxine metabolism
Hyperchloremia	Gonadal dysfunction (increased prolactin
Hypocalcemia	and luteinizing hormone, decreased tes-
Hyperphosphatemia	tosterone)
Hypermagnesemia	
Carbohydrate intolerance	
Insulin resistance	
Plasma insulin normal or increased	
Delayed response to carbohydrate load-	
ing	
Hyperglucagonemia	

have been difficult to explain. Many retained metabolites have been implicated in the systemic toxicity of the uremic syndrome. The major nitrogenous metabolites are listed in Table 11-5. Although urea was the first of these metabolites to be identified as being increased in uremia, it does not appear to be responsible for the systemic manifestations of uremia.[77] Cyanate, produced endogenously from urea, may produce toxic effects by irreversible carbamylation of proteins.[39] Guanidine compounds, especially methylguanidine, have been implicated in toxicity of experimental renal failure, but their significance in human uremia remains to be proved.[27] Unidentified compounds with molecular weights between 300 and 1500 may play a role in the pathogenesis of the clinical syndrome of uremia. However, it is more likely that the syndrome is a result of the cumulative effect of many retained compounds, which may act as toxins and which may also have an effect on metabolism in general, for example through enzyme inhibition or derangement in membrane transport. The decreased ability of the kidney to degrade or eliminate hormones may also play a role.[39,76]

Acute Renal Failure. A significant proportion of patients with renal disease manifest the clinical picture of acute renal failure. This clinical syndrome occurs when there has been a rapid and severe reduction of GFR to <10 mL/min. Most of these patients are oliguric with urine outputs of <400 mL/d after fluid challenge. During this oliguric phase, several physiological and biochemical alterations become evident; one of the most significant is the inability to excrete sodium and water. This leads to expansion of extracellular fluid with resulting peripheral edema, hypertension, and congestive heart failure. In many patients, water is retained in excess of salt, producing hyponatremia. If the hyponatremia is severe, central nervous system (CNS) symptoms may result, beginning with somnolence and leading to seizures and coma. As in chronic renal failure, the inability to excrete potassium may result in life-threatening hyperkalemia manifested by cardiac arrhythmias. This situation may be aggravated by metabolic acidosis due to the failure to excrete endogenous acid. Retention of phosphate leads to hyperphosphatemia, hypocalcemia, and later to secondary hyperparathyroidism. Hypermagnesemia may contribute to CNS depression and coma, as well as to suppression of cardiac pacemakers. As in chronic renal failure, there is an accumulation of the metabolic products creatinine and urea nitrogen. When acute renal failure is due to underperfusion of the kidney without intrinsic renal disease (prerenal azotemia), urea nitrogen will accumulate in excess of creatinine (high urea/creatinine ratio). This phenomenon is especially marked in slow-flow states (decreased GFR). The distinction of acute renal failure secondary to underperfusion (prerenal azotemia) from that due to intrinsic renal disease (renal azotemia) is discussed more fully below. The rise in serum creatinine and, less often, the increase in urea nitrogen are used to monitor the degree of renal impairment. As discussed previously, creatinine clearance will overestimate GFR when plasma creatinine is increased.

Patients with acute renal failure may develop all the symptoms of the uremic syndrome. In general, however, the anemia and metabolic bone disease seen in chronic renal failure are not as prominent in acute renal failure. Patients with acute renal failure are particularly at risk for gastrointestinal hemorrhage and systemic infection, which are the major causes of death.

Laboratory monitoring is crucial for assessing the severity of renal impairment and for the management of the fluid and electrolyte disorders that accompany acute renal failure.[17] Patients with reversible acute renal failure generally remain oliguric for about two weeks. Following this period, they enter the diuretic phase in which there is gradually increasing urine output as the GFR begins to rise. Small changes in GFR may give rise to significant changes in urine volume so that urine output may rise while urea nitrogen and creatinine are still being retained. During the diuretic phase, recovery of tubular function may lag behind the recovery of GFR so that regulation of sodium, potassium, and water reabsorption may be inadequate, with resulting hypovolemia and hypokalemia. Serial monitoring of creatinine, GFR, serum electrolytes, and urine volume may show more rapidly changing values than in chronic renal failure which may require prompt attention.

A complete discussion of the *causes of acute renal failure* is beyond the scope of this chapter. In the majority of cases the initial event is either ischemic or toxic injury to the kidney. Inadequate perfusion of the kidney results in reduced GFR and may mimic acute renal failure. As long as injury of the kidney does not occur, tubular function will continue and will produce a distinctive biochemical picture discussed below. (See prerenal azotemia.) Further confusion arises in that prolonged or severe underperfusion may result in ischemic injury to the kidney and actual acute renal failure. The pathophysiology of the initial insult to the kidney and the subsequent maintenance of reduced GFR have been summarized well by Martin[44] (Figure 11-15). Judicious use of the chemistry laboratory is essential in recognizing and monitoring acute renal failure and in distinguishing it from inadequate perfusion of an intrinsically normal kidney. Poor perfusion of a normal kidney results in prerenal azotemia with urea nitrogen retained in excess of creatinine due to reabsorption of urea but not of creatinine. With normal glomerular and tubular function, the kidney responds to the underperfusion with avid reabsorption of sodium and water to produce a highly concentrated urine containing little sodium. Following ischemic or toxic damage to renal tubules, the ability to concentrate urine as well as maximally reabsorb sodium is lost. Thus, there is frequently an isosmotic urine with higher urine sodium values.

One measure of urine concentration is the ratio of urine creatinine concentration to plasma creatinine concentration. As urine becomes more concentrated, this ratio will rise such that it is typically > 30:1 in underperfusion states, and < 15:1 with tubular injury. Urine sodium is usually < 15 mmol/L in underperfursion and > 20 mmol/L with tubular injury, although the amount is dependent on dietary intake. Urine sodium < 10 mmol/L is highly suggestive of inadequate renal perfusion. Urine to plasma osmolality ratios may also serve as a diagnostic guide; a ratio of 1:1 or less (fixed osmolality) is characteristic of tubular injury, while a ratio of 1.5:1 or more suggests prerenal azotemia.

Other investigators have found that the fractional excretion of sodium calculated as (U/P) Na$/(U/P)$ creatinine is the best diagnostic test for distinguishing underperfusion from tubular injury.[20,48] The ratio is almost always < 1 in prerenal azotemia (underperfusion) and > 1.5 in acute tubular necrosis. This distinction is very important, since prerenal azotemia would be expected to respond to volume expansion or improved cardiac function, whereas acute tubular necrosis would require careful fluid and electrolyte management until renal function returns. Many conditions that initially result in underperfusion and prerenal azotemia may eventually produce acute tubular necrosis. It is of obvious therapeutic importance to know if and when this has occurred. In addition to the urine chemistry tests discussed above, examination of the urinary sediment can also be of great value. Moderate numbers of hyaline and finely granular casts are typical of prerenal azotemia. In contrast, acute tubular injury produces free renal epithelial cells as well as epithelial cell casts and fatty casts. The presence of erythrocyte casts strongly suggests acute glomerular injury, while leukocyte casts are most often seen in pyelonephritis.

Although the preceding tests are very useful in distinguishing underperfusion of the kidney from acute tubular injury, they may be misleading in certain patients. In the early stages, acute glomerular injury may resemble prerenal azotemia. Furthermore, prior treatment with diuretics may make interpretation of urine chemistry values impossible. Finally, acute oliguric renal failure due to urinary tract obstruction does not have a characteristic picture and may require the use of other techniques such as radiographic studies for proper diagnosis.

Glomerular Diseases. Diseases that cause direct injury to the renal glomerulus while preserving tubular function may initially resemble prerenal azotemia. Eventually chronic progressive glomerular disease will affect the blood supply of tubules and result in the loss of complete nephrons to produce chronic renal disease and the uremic syndrome. In many patients significant primary glomerular disease may be recognized by the presence of distinctive clinical syndromes that result from glomerular injury. Among the more important are acute nephritis, rapidly progressive glomerulonephritis, chronic glomerulonephritis, and the nephrotic syndrome. Again, a discussion of all the causes of these syndromes is beyond the scope of this book and only examples of each will be discussed to illustrate the use of the laboratory in their evaluation.

The *acute nephritic syndrome* (acute glomerulonephritis) is characterized by the rapid onset of hematuria, proteinuria, reduced GFR, and sodium and water retention with resulting hypertension and sometimes localized peripheral edema. Congestive heart failure and oliguria may also be seen. Renal biopsy shows enlarged, inflamed glomeruli with narrowed capillary lumina.

In a number of patients with the acute nephritic syndrome, the pathologic process is related to recent group A beta-hemolytic streptococcal infection of the pharynx or, less commonly, the skin. Only certain strains of streptococci are capable of inducing acute nephritis. A latent period averaging about two weeks exists between the streptococcal infection and clinical evidence of nephritis; it is slightly longer for cutaneous than for pharyngeal-associated disease. The initial presentation usually consists of oliguria, hematuria, hypertension, and facial and hand edema with some evidence of vascular congestion. The presentation may vary considerably from patient to patient. Percutaneous renal biopsy of patients with poststreptococcal glomerulonephritis reveals enlarged hypercellular glomeruli infiltrated by polymorphonuclear leukocytes and monocytes, usually within the mesangium or capillary lumina. There is also proliferation of mesangial and endothelial cells. Involvement of the kidney is diffuse. Electron microscopy reveals deposits, presumably immune complexes, on the epithelial side of the basement membrane. Glomerular injury with damage of the glomerular basement membrane leading to reduction of GFR is thought to be due to activation of the inflammatory response by immune complexes. Abnormal *laboratory results* are usually present early in the course of acute nephritis. Hematuria, gross or microscopic, and proteinuria, usually < 3 g/d, are almost always present. Red cell casts are highly suggestive of glomerulonephritis. They are commonly present in urine but will be observed only if the specimen is fresh and acidic, centrifugation is light, and sediment (after decantation) is resuspended gently. Large numbers of hyaline and granular casts are also common; waxy casts suggest a chronic process and should raise a question of acute exacerbation of a pre-existing disease. The proteinuria is progressively more nonselective, with larger-molecular-weight species having greatly increased clearances. (See also Chapter 4.) To some extent the degree to which selectivity is lost is indicative of the severity of damage to the glomerular basement membrane. The urine protein may also have greatly increased amounts of fibrinopeptides (cleavage products of fibrinogen), although these may be moderately elevated in renal insufficiency of many causes.

In patients suspected of acute poststreptococcal glomerulonephritis, evidence of recent infection may be found in increased titers of antibodies to streptococcal extracellular products, i.e., antistreptolysin O (ASO), antihyaluronidase (AHase), and antideoxyribonuclease-B (ADNase-B). ASO titers of > 200 units are present in the serum of a majority of patients with pharyngeal infection; in cutaneous infection, AHase and ADNase-B titers are elevated in over 90% of cases. Serial measurements that document rising antibody titers against streptococcal antigens are stronger evidence of recent infection than a single determination.

Most patients will have moderate reductions in total hemolytic complement activity (CH_{50}) and in the C3 component of the complement cascade. Persistent and severe depression of C3 levels should suggest membranoproliferative glomerulonephritis, SLE, endocarditis, or other forms of sepsis. Although depressed levels of complement imply disease activity, they are not useful for grading the severity or determining the prognosis of the illness.

Other causes of acute nephritis are reactions to drugs, acute infection of the kidney, systemic diseases with immune complexes such as SLE, bacterial endocarditis, and, finally, disease where the antigen is unknown but possibly related to antecedent viral infections.

Rapidly progressive glomerulonephritis (RPGN) refers to a heterogeneous group of disorders characterized by a fulminant clinical course, leading to renal failure in weeks to a few months. This group of diseases is characterized by a consistent histologic picture of glomerular crescent formation. The glomerular epithelial crescent is a proliferation of blood-derived macrophages and parietal epithelial cells lining Bowman's capsule. The proliferating epithelial cells and macrophages eventually compress the glomerulus as well as obstruct the proximal convoluted tubule, thus severely compromising nephron function. The proliferation of epithelial cells and macrophages is in response to fibrinogen and fibrin polymers present in Bowman's space. This proliferation is a nonspecific reaction to severe glomerular injury where there is leakage of fibrinogen, activated enzymes, and macrophages out of the capillaries into Bowman's space. These crescents may develop rapidly, and thus the parameters of declining renal function show changes in terms of days. RPGN can be divided into three broad categories: postinfectious RPGN, RPGN that is associated with systemic diseases, and isolated idiopathic RPGN. Since choice of treatment depends on knowing the etiology of the RPGN, full investigation becomes very important. The clinical laboratory is a useful adjunct in this process. Some of the *infectious diseases* that may be associated with RPGN are streptococcal pharyngitis or impetigo, bacterial endocarditis, septicemia, and hepatitis B. Blood cultures; hematologic studies; and other studies such as ASO titers, HB_sAg, HB_sAg antibodies, and antibody titers for several other viruses may be quite helpful. In postinfectious RPGN there is a tendency to reduced levels of serum complement manifested as low CH_{50} and C3. Acute poststreptococcal glomerulonephritis may present with an appearance of RPGN; this presentation, more common in adults than children, is a poor prognostic sign.

In *RPGN associated with systemic disease,* diagnostic studies are frequently aimed at identifying the underlying disease. Examples of these studies include ANA, anti-double-stranded DNA, antibodies to Sm antigen, and complement studies in patients with SLE; radioimmunoassay for antiglomerular basement membrane antibodies in patients with Goodpasture's syndrome; and assays for cryoglobulins in patients with essential cryoglobulinemia. The final category, *idiopathic RPGN,* is relatively rare. Urinalysis shows the usual hallmarks of glomerular injury: hematuria, proteinuria, and casts, usually including RBC casts. About 25% of these patients will also have antiglomerular basement membrane antibodies but not the pulmonary involvement characteristic of Goodpasture's syndrome. Most patients will have normal levels of complement; cryoimmunoglobulins, antinuclear antibodies, and immune complexes are usually absent. The diagnosis is based on the rapid downhill course, crescentic glomerulonephritis on renal biopsy, and the exclusion of infectious or multisystem diseases.

Chronic glomerulonephritis refers to the clinical syndrome that may result from a number of glomerular diseases that have a prolonged downhill course with progressive loss of nephron mass. Many of these diseases may remain entirely asymptomatic except for mild hematuria, proteinuria, and slightly reduced renal function. In some cases, the first indication of disease is the gradual onset of the uremic syndrome. In the late stages of chronic glomerulonephritis, hypertension is frequently a complication and end-stage renal disease becomes apparent, regardless of the initial etiology. Many of the glomerular diseases discussed under other headings may lead to chronic glomerulonephritis, or chronic glomerulonephritis may occur as an idiopathic process. The kidneys are small due to loss of mass but not scarred and distorted as in chronic pyelonephritis. The remaining functional nephrons may hypertrophy, giving the kidney surface a granular appearance. Specific etiological diagnosis is only occasionally possible. Differentiation from nonglomerular diseases is not always possible, but heavy proteinuria or red cell casts suggest a glomerular process. Ultimately, chronic glomerulonephritis becomes indistinguishable from chronic tubulointerstitial nephritis and chronic vascular diseases such as hypertension.

The *nephrotic syndrome* consists of massive proteinuria of ≥ 3.5 g/d, hypoalbuminemia of < 3 g/dL, generalized edema, and hyperlipidemia. This syndrome results from glomerular injury that produces excessive permeability to plasma proteins and massive proteinuria. GFR is usually depressed as vascular volume becomes inadequate. In rare patients this may lead to superimposed acute renal failure. The loss of albumin in urine overwhelms hepatic synthetic capacity and hypoalbuminemia results. Loss of intravascular oncotic pressure results in edema, which is enhanced by sodium and water retention. This retention is secondary to aldosterone secretion stimulated by intravascular hypovolemia and by the presence of unidentified intrarenal factors that promote sodium reabsorption. The genesis of the hyperlipidemia is not clear, but experimental evidence points to a close relationship with decreased plasma oncotic pressure, which appears to stimulate increased hepatic synthesis of lipoproteins. There is also decreased peripheral catabolism of the lipids. Leakage of lipoproteins into the urine results in lipiduria with oval fat bodies (degenerated tubular cells that have reabsorbed lipoproteins).

The nephrotic syndrome can be seen with several systemic diseases that involve the kidney, such as diabetes mellitus, amyloidosis, and SLE. Systemic disease accounts for 40% of instances of the nephrotic syndrome in adults but only 5% in children. The remainder of cases of nephrotic syndrome are due to primary glomerular disease. In children between 2 and 6 years of age, *minimal change disease* (a synonym for lipoid nephrosis) accounts for over 90% of cases of primary nephrotic glomerular disease. The name minimal change disease is used because of the lack of light microscopic structural alteration on renal biopsy. There is experimental evidence suggesting an altered filtration barrier possibly related to abnormal electric charge. This alteration allows smaller-molecular-weight proteins such as albumin to be filtered. Electron microscopy shows fusion of epithelial foot processes, but this may be a nonspecific reaction to the proteinuria. The exact etiology is unknown, but the disease frequently follows an acute viral syndrome. Other causes of primary nephrotic glomerular disease are focal segmental glomerulosclerosis and membranoproliferative glomerulonephritis. These two diseases are defined on the basis of their renal biopsy appearance. Although presumably of immunologic nature, the exact antigens involved are unknown, and in fact these two categories may represent a common reaction to injury by several agents. In adults, lipoid nephrosis accounts for only 15% of cases of primary nephrotic glomerular disease. Membranous glomerulonephritis accounts for 40% and is most common, while focal segmental glomerulosclerosis and membranoproliferative glomerulonephritis are less common. Membranous glomerulonephritis refers to the renal biopsy appearance, is presumably immunologic in nature, and is multifactorial in etiology. Except in the child between 2 and 6 years of age where a nephrotic primary glomerular disease is almost certainly minimal change disease, the diagnosis of a particular cause of the nephrotic syndrome usually requires percutaneous renal biopsy. The clinical laboratory is most useful in detecting and confirming the presence of the nephrotic syndrome and in helping to eliminate certain systemic diseases, such as diabetes mellitus and SLE, as possible causes.

In addition to massive proteinuria, hypoalbuminemia, and hyperlipidemia, there are other important *laboratory changes* in the nephrotic syndrome as well. Serum levels of α_2- and β-globulins are usually increased while α_1- globulins are normal to slightly decreased. (See Figure 4-12.) IgG levels may be dramatically decreased while other immunoglobulins are usually normal. Factor B of the alternative pathway of complement activation may also be significantly decreased and, along with depressed IgG, may give rise to the significant risk of infection in these patients. Other complement components are not significantly decreased. There is an increase in the levels of fibrinogen and several other coagulation factors along with a decrease in antithrombin III. These changes, together with increased platelet aggregation, hypovolemia, and edema, contribute to the occurrence of thromboembolic problems in the nephrotic patient. Renal vein thrombosis is seen in 30–50% of nephrotic patients with membranous glomerulonephritis. There is significant alteration of several important binding proteins including depression of serum levels of transferrin, transcortin, and thyroxine binding globulin. Thus, tests such as total iron binding capacity, total serum iron, total T_4, and T_3 resin uptake will be affected. Total cholesterol and triglycerides will be elevated secondary to changes in lipoproteins. Finally, patients may have reduced serum concentrations of vitamin D, copper, and zinc due to loss of binding proteins.

Tubular Diseases. Two important types of tubular disease, ischemic and toxic tubular necrosis, were discussed previously under acute renal failure because this is their usual clinical presentation. Another group of diseases in which primary tubular dysfunction affects acid-base balance are discussed separately under renal tubular acidosis. Here we limit discussion to diseases affecting predominantly the tubules and interstitium and, in particular, those that present clinically as interstitial nephritis rather than acute renal failure. Note that tubulointerstitial disease is exceeded in frequency only by chronic glomerular disease as a cause of chronic renal failure. This group of diseases is characterized in the acute stage by leukocytic infiltration of the interstitium and tubules with varying mixtures of lymphocytes and by polymorphonuclear leukocytes including eosinophils, interstitial edema, and scattered foci of tubular necrosis. In the chronic stage, the infiltrate is mostly mononuclear with accompanying interstitial fibrosis and prominent tubular atrophy. The causes of the initial inflammatory response can be quite varied and include bacterial, fungal, or viral infection; hypersensitivity reactions to drugs such as methicillin; toxic effects of analgesic drugs; radiation toxicity; and immunologic rejection of a transplanted kidney.

Hallmarks of glomerular disease such as severe proteinuria, nephrotic syndrome, and red cell casts are usually missing in the tubulointerstitial nephritides. Clinical presentation is marked by reduced GFR associated with tubular dysfunction and loss of concentrating ability, reduced acid excretion, and inadequate regulation of sodium balance. In general, the degree of tubular functional impairment correlates well with the extent of the disease process.

The most clear-cut instance of acute tubulointerstitial disease is acute pyelonephritis, which is easily recognized by the accompanying clinical manifestations, including acute febrile episodes associated with more or less severe flank pain and tenderness, and usually associated with urinary symptoms of frequency and urgency. Characteristically, the urine contains numerous leukocytes and frequent leukocyte casts. The diagnosis is nearly always readily made by culture. The search for predisposing factors such as diabetes mellitus, lower urinary tract obstruction, nephrolithiasis, uric acid nephropathy, nephrosclerosis, and oxalate nephropathy may require further investigations, including several biochemical tests. Confirmation of chronic pyelonephritis may be sought by radiologic demonstration of cortical scarring, retraction of papillae, and caliceal abnormalities that include loss of the normal delicate cupping associated with progressive distention and the deformation of these structures with repeated infection.

In contrast to the characteristic clinical features of acute or recurrent pyelonephritis, non-infectious tubulointerstitial disease is virtually never symptomatic until the symptoms of uremia appear. In the majority of instances, the findings on urinalysis are quite benign, though on occasion modest proteinuria with moderate numbers of leukocytes in the urine is seen. The great majority of these cases can be attributed to drug toxicity. This includes conditions such as analgesic nephropathy; penicillin nephropathy; and reaction to nonsteroidal anti-inflammatory agents and, rarely, to diuretics, allopurinol, anticonvulsive agents, and a long list of other drugs. Careful examination may reveal impaired acidification and other urinary abnormalities such as the presence of glucose and amino acids in the urine; more commonly, reduced concentrating ability and slight proteinuria are the only manifestations. Aminoglycoside toxicity initially causes reduced concentrating ability. If the drug is not discontinued, this reduced concentrating ability may progress to nonoliguric acute renal failure with rising urea nitrogen and creatinine. Therapeutic drug monitoring may help prevent toxicity in some patients. On rare occasions, acute tubulointerstitial disease may be associated with polyuria and hematuria, but this is rarely seen except in the instances of acute hypersensitivity reactions involving the kidney.

Urinary Tract Obstruction. Urinary tract obstruction may cause disease, either indirectly by predisposing to urinary tract infection or directly by raising tubular pressures that destroy entire nephrons and lead to chronic renal failure. Obstruction may be at any level in the urinary tract from the renal tubule to the distal urethra. Causes range from precipitation of myeloma proteins in renal tubules, to ureteral obstruction by stones or tumors, to bladder outlet obstruction by prostatic hypertrophy, to urethral obstruction by congenital urethral valves. Whatever the location, obstruction is frequently complicated by urinary tract infection; laboratory findings may then reflect both the underlying and superimposed problems. In the progression of obstructive disease of one or both kidneys, the earliest impairment is loss of concentrating ability followed by reduced acid excretion, and ultimately by reduction of GFR and renal blood flow. The most useful *laboratory tests* are routine urinalysis, urine culture, CBC, serum urea nitrogen, and creatinine. Diagnosis is generally based on radiological evaluation. The laboratory can also be of aid in evaluating possible systemic conditions and diseases predisposing to obstruction, such as multiple myeloma, gout, hypercalcemia, and tuberculosis.

Two very important systemic diseases, diabetes mellitus and hypertension, account for a sizable portion of the chronic renal disease in the western world. Both produce renal disease primarily by means of a vascular mechanism that causes progressive renal ischemia and consequently gradual loss of nephrons. Diabetes mellitus has a propensity to produce a nephrotic picture at times. Both diseases and their laboratory findings are discussed more fully elsewhere.

Use of the Laboratory for the Patient with Renal Disease

The diagnostic process in nephrology is heavily dependent upon the clinical laboratory because a majority of the diseases of the kidney produce few or no clinical manifestations until

50–75% of the kidney is destroyed. Laboratory tests often provide initial recognition of renal impairment and indicate the direction for further productive laboratory investigations.

The initial step of diagnosis is to determine biochemical or physiological parameters that characterize, in a general manner, a relatively small number of subgroups of the myriad diseases possible in the kidney. The next step is to determine specific parameters that focus on particular diseases within an indicated subgroup. The first step of the process is illustrated by Table 11-15, which shows laboratory findings in conjunction with clinical syndromes; in Table 11-16 are shown particular causes that associate closely with one or another clinical syndrome. The classifications shown in these tables are those of Frederick L. Coe,[14] who has proposed that ten distinct clinical syndromes of nephrology may be distinguished with a limited, easily obtained base of laboratory data. Although not all nephrologists adopt this approach or concur in the organization, systematic classifications such as these help to put laboratory studies in a manageable perspective. The initial laboratory data base needed to identify one of the clinical syndromes

Table 11-15. LABORATORY FINDINGS IN TEN CLINICAL SYNDROMES
(AFTER COE,[14] MODIFIED)

Clinical Syndrome	Characteristic Laboratory Findings	Other Possible Laboratory Findings
Acute nephritis	Hematuria or RBC casts and documentation of one or more of following: decreased GFR, oliguria, edema, or hypertension	S: ↑ Creatinine, urea N, K^+ S: ↓ Na^+, CO_2, CH_{50}, C3, proteinuria usually < 3 g/d For specific serological studies, see p. 1306
Nephrotic syndrome	Proteinuria (> 3.5 g/d)/1.75 m^2	S: ↑ Triglyceride, cholesterol, α_2- and β-globulin S: ↓ Albumin, IgG U: oval fat bodies, free fat, ↓ Na^+ excretion For secondary laboratory changes, see p. 1308
Asymptomatic urinary abnormalities	Hematuria or proteinuria without the presence of one of the other nine syndromes	
Acute renal failure	Oliguria, anuria, or reduced GFR and documentation of one of the following: normal GFR < 3 months previously, GFR falling in days to weeks, creatinine rising daily, or the absence of anemia	S: ↑ Urea N, creatinine, K^+, PO_4^{2-}, Mg^{2+} S: ↓ Na^+, Ca^{2+}, CO_2, pH U: epithelial cells and epithelial casts For differential diagnosis, see p. 1305
Chronic renal failure	Reduced GFR stable for > 3 months, uremic symptoms for > 3 months, renal osteodystrophy, or demonstration of small kidneys	See Table 11-3
Urinary tract infection	Bacteriuria > 10^5 colonies/mL or documentation of other infectious agent	B: leukocytosis U: pyuria, leukocyte casts, positive nitrite (some organisms); positive urine leukocyte esterase (dipstick)
Urinary tract obstruction	Obstruction visualized directly or seen on X-ray, or documentation of urinary retention or residual volume	S: ↑ Urea N, creatinine U: ↓ GFR, osmolality following fluid restriction, acid excretion Findings of urinary tract infection
Renal tubule defect	Documentation of actual tubular defects such as depressed secretion or reabsorption, impaired concentrating ability, renal tubular acidosis	S: ↑ Urea N, creatinine, K^+ U: ↓ GFR, osmolality after fluid restriction Proteinuria, glucosuria, phosphaturia, aminoaciduria, uricosuria, natriuria, renal tubular acidosis
Hypertension	Increased blood pressure, documented on three separate occasions, of > 145/95	Renal vascular disease S: ↑ renin, aldosterone, Na^+, renal vein renin differential, S: ↓ K^+ U: ↑ K^+ For hyperaldosteronism, see Chapter 9 For pheochromocytoma, see Chapter 9
Nephrolithiasis	Stone passed, removed, or seen by X-ray	Depends on underlying metabolic defect, e.g.: S: ↑ uric acid in gout, ↑ PTH and calcium in primary hyperparathyroidism

S: serum; U: urine; B: blood; ↑↓ increase or decrease.

Table 11-16. SELECTED CAUSES OF TEN CLINICAL SYNDROMES[14]

ACUTE NEPHRITIS
Post-infections: streptococcal, pneumococcal, staphylococcal, toxoplasmosis, hepatitis B, mumps, varicella, infectious mononucleosis, *Falciparum malariae*, and echovirus. Henoch-Schönlein purpura, acute tubulo-interstitial nephritis, systemic lupus erythematosus (SLE), bacterial endocarditis, thrombotic thrombocytopenic purpura (TTP), cryoglobulinemia.

NEPHROTIC SYNDROME
Lipoid nephrosis, membranous glomerulonephritis, amyloidosis, diabetes mellitus, membranoproliferative glomerulonephritis, focal sclerosing glomerulonephritis, SLE, renal vein thrombosis.

ASYMPTOMATIC URINARY ABNORMALITIES
Proteinuria only: Diabetes mellitus, hypertension, Waldenström's macroglobulinemia, Fabry's disease, toxemia of pregnancy, multiple myeloma, amyloidosis, Sjögren's syndrome.
Proteinuria with hematuria: Chronic glomerulonephritis, chronic pyelonephritis, bacterial endocarditis, Alport's syndrome, cryoglobulinemia, TTP, SLE, renal vein thrombosis, Henoch-Schönlein purpura.

ACUTE RENAL FAILURE
Rapidly progressive glomerulonephritis, polyarteritis nodosa, Wegener's granulomatosis, SLE, malignant hypertension, acute cortical necrosis, TTP, hemolytic-uremic syndrome, acute tubulointerstitial nephritis or acute pyelonephritis, hyperuricemia or oxalate nephropathy, hypercalcemic nephritis, prerenal azotemia or acute tubular necrosis due to heavy metals or other toxins, shock, hypovolemia, cardiac failure, vascular obstruction, or liver failure.

CHRONIC RENAL FAILURE
Any of the chronic diseases involving the kidney.

URINARY TRACT INFECTION
Most commonly with *E. coli*, followed by other organisms. Important to rule out remediable factors such as stones, obstruction, or vesicoureteral reflux.

URINARY TRACT OBSTRUCTION
Benign prostatic hypertrophy, carcinoma of the prostate, bladder, uterine cervix, endometrium, or ovaries; lymphoma, ureteropelvic junction stricture, renal stones, urinary tuberculosis, pelvic irradiation, urethral strictures.

RENAL TUBULE DEFECTS
Renal cysts, medullary sponge kidneys, medullary cystic disease, polycystic disease, renal tubular acidosis, Fanconi's syndrome, Bartter's syndrome, cystinuria.

HYPERTENSION
(INCLUDES NONRENAL CAUSES)
Essential hypertension, renal artery stenosis, hyperaldosteronism, Cushing's syndrome, pheochromocytoma, coarctation of the aorta, hyperthyroidism, primary renal disease, central nervous system disease.

NEPHROLITHIASIS
Calcium stones due to primary hyperparathyroidism, idiopathic hypercalciuria, renal tubular acidosis, hyperoxaluria, uric acid stones due to gout, hyperuricosuria, leukemia, struvite stones due to infection; cystine stones due to cystinuria.

consists of a complete urinalysis, complete blood count, and the following serum chemistries: urea nitrogen, creatinine, sodium, potassium, chloride, CO_2, phosphorus, calcium, and uric acid. Supplemental data that aid in establishing the identification are 24-h protein excretion, urine protein electrophoresis, urine culture, and serial measurements of serum creatinine, of GFR, and of 24-h urine volume.

Presumptive diagnosis is therefore made by history and physical examination in conjunction with the data base provided by the laboratory. The diagnosis is refined by selective testing, which is also used to follow the progress of the disease. Although biochemical studies are important in the initial detection process, their use is overshadowed by the history and physical findings. But in the subsequent confirmation of a particular syndrome, in the differentiation of particular causes of diseases within the syndrome, and in monitoring the disease process, the role of biochemical studies dominates that of clinical observations.

RENAL CALCULI

The majority of kidney stones are composed of one or more of the following substances: calcium oxalate, calcium phosphate, uric acid, cystine, or a mixture of these with magnesium ammonium phosphate (struvite). These substances crystallize within an organic matrix, the nature of which is not well understood. All of the substances listed above are poorly soluble, and, for some, solubility is influenced to a major degree by the urinary pH. The mechanisms responsible for the multiple recurrences of renal stones in only certain individuals are not completely understood but probably involve a multitude of factors including (1) urine flow (fluid intake), (2) excretion of excess quantities of the relatively insoluble substances listed above, and perhaps (3)

Table 11-17. THE COMMON VARIETIES OF
RENAL STONES

Type of Stone	Urine pH	Cause(s)

Calcium oxalate	Variable	Concentrated urine
		Hypercalciuria
		    Vitamin D intoxication
		    Hyperparathyroidism
		    Sarcoidosis
		    Milk-alkali syndrome
		    Osteoporosis
		    Renal tubular acidosis
		    Idiopathic hypercalciuria
		Hyperoxaluria

Calcium phosphate	Alkaline	Renal tubular acidosis Alkali ingestion Infection with urea-splitters (produces mixed stones)
Magnesium ammo- nium phosphate	Alkaline	Infection

Uric acid	Acid	Hyperuricaciduria
		    Hyperuricemia
		    Gout
		    High purine diet
		Urinary hyperacidity

Cystine stone	Acid	Cystinuria

the absence, in individuals who form stones, of a substance or substances in the urine that, under normal circumstances, inhibit precipitation of some of these nearly insoluble agents.[12] Certain types of stones are recognized as being associated with specific disorders and these are listed in Table 11-17.

Calcium oxalate stones are perhaps the most common stones that are encountered, and they occur in a worldwide stone belt that is mainly confined to the tropical and subtropical regions of the globe. They may be associated with either a persistently concentrated urine or a consistently increased excretion of urinary calcium or oxalate. Most commonly this type of stone is not composed of pure calcium oxalate, but calcium oxalate predominates, with small quantities of calcium phosphate and uric acid also present. In many instances it is not possible to find any abnormality in the stone-forming individual beyond a persistently small urine volume; however, it is important to obtain data on calcium and oxalate output in the urine and to ascertain that the mechanism of urinary acidification is normal.

Hypercalciuria, which can be documented in instances of perhaps half of the calcium oxalate stones, can be categorized by appropriate studies into *absorptive, resorptive,* or *renal hypercalciuria* (Table 11-18). As the name implies, *absorptive hypercalciuria* results from increased gastrointestinal absorption of calcium; it is documented by demonstrating that oral calcium administration leads to an increase in urinary calcium excretion. Serum calcium and phosphorus are normal and dietary reduction of calcium reduces the rate of urinary calcium excretion. *Resorptive hypercalciuria* represents excretion of calcium that has been resorbed from bone. Most commonly, this resorption is caused by hyperparathyroidism or neoplastic bony metastases. Most of these patients have elevated serum calcium and their conditions are thus easily distinguished from the other forms of hypercalciuria. Patients with either primary or secondary hyperparathyroidism exhibit elevated levels of urinary cyclic AMP and elevations in circulating parathormone. The diagnosis of met-

Table 11-18. DIFFERENTIATION OF TYPES OF HYPERCALCIURIA

Type	Serum [Ca]	Serum [P]	Bone Density	Serum PTH	Urinary cAMP
Absorptive	Normal	Normal	Normal	↓	↓
Resorptive	↑ or Normal	Normal or ↓	↓	↑	↑
Renal	Normal	Normal	↓	↑	↑

Table 11-19. FREQUENCY, CHARACTERISTICS, AND ETIOLOGIES OF KIDNEY STONES

Chemical Type	Approximate Frequency[14]	Characteristics	Etiologies[14,23,56]
Calcium	65%	Hard, small to medium-sized, often multiple	Idiopathic lithiasis (30%)
	Oxalate 30% Phosphate 10% Mixed 25% (oxalate-phosphate)	X-ray: radiopaque; may have diffuse renal calcium deposits (nephrocalcinosis)	Normocalcemic hypercalciuria (30%) Hyperuricosuria (20%) Primary hyperparathyroidism (7%) Other causes, e.g., hyperoxaluria (5%), renal tubular acidosis (3%), medullary sponge kidney, sarcoidosis, immobilization
Magnesium ammonium phosphate (Mg NH$_4$ PO$_4$)	25%	May be large ("staghorn") X-ray: radiopaque	Urinary tract infection (with hydrolysis of urea by organisms such as *Proteus*)
Uric Acid	5%	Yellow, friable, may be large ("staghorn") X-ray: radiolucent	Idiopathic hyperuricosuria Primary or secondary gout (e.g., malignancies)
Cystine	2%	X-ray: radiolucent May be large (staghorn"); brownish color ("maple sugar")	Primary cystinuria

astatic bone disease with hypercalciuria is almost always clinically obvious. Renal hypercalciuria represents persistent loss of calcium via the urine as a result of a renal leak, as in renal tubular acidosis. Details of other clinical entities associated with stones are found in Table 11-19. Other laboratory procedures which may be useful in evaluating patients who form kidney stones are indicated in Table 11-20.

Laboratory Analysis of Kidney Stones

Both qualitative and quantitative analysis of the chemical constituents of kidney stones may be useful in establishing the etiology and in planning rational therapy. Analysis is not needed for every urinary calculus, especially when other laboratory and clinical findings are diagnostic. In some patients, however, two or more possible etiologies exist, and analysis of the calculus may help distinguish between different diagnoses. For example, patients with hypercalciuria and calcium oxalate stones may develop renal infections with deposition of magnesium ammonium

Table 11-20. QUANTITATIVE LABORATORY EVALUATION OF THE PATIENT WITH KIDNEY STONES[14]

Procedure	Serum	24-h Urine	Comments
Calcium	X	X	Collect urine in acid preservative (HCl, 50% v/v, 10 mL/L of urine)
Oxalate		X	
Cystine		X	
Creatinine	X	X	
Uric acid	X	X	
Sodium	X		
Potassium	X		
Chloride	X		
Total CO$_2$ content	X		
Phosphate	X	X	

phosphate upon the calcium oxalate stones, to form "mixed stones." Chemical analysis of these stones is needed to confirm this sequence of events.

Specialized physical techniques such as infrared spectroscopy and X-ray diffraction are gradually replacing less specific qualitative chemical methods for stone analysis.

References

1. Acord, W., and Feldman, F. J.: Design considerations for a fully automated rate electrochemical glucose and urea nitrogen analyzer. (Abst.) Clin. Chem., 20:903, 1974.
2. Alsever, R. N., and Gotlin, R. W.: A Handbook of Endocrine Tests in Adults and Children. Chicago, Year Book Medical Publishers, 1978.
3. Archibald, R. M.: Colorimetric measurement of uric acid. Clin. Chem., 3:102-105, 1957.
4. Batlle, D., and Kurtzman, N. A.: Acid-base physiology and pathophysiology. In: Contemporary Nephrology, Vol. 2. S. Klahr and S. G. Massry, Eds. New York, Plenum Publishing Corp., 1983.
5. Bjornsson, T. D.: Use of serum creatinine concentrations to determine renal function. Clin. Pharmacokinet., 4:200-222, 1979.
6. Blau, E. B., and Haas, J. E.: Glomerular sialic acid and proteinuria in human renal disease. Lab. Invest., 28:477-481, 1973.
7. Boss, G. R., and Seegmiller, J. E.: Hyperuricemia and gout. Classification, complications and management. N. Engl. J. Med., 300:1459-1468, 1969.
8. Bowers, L. D., and Wong, E. T.: Kinetic serum creatinine assays. II. A critical evaluation and review. Clin. Chem., 26:555-561, 1980.
9. Brenner, B. M., and Rector, F. C., Jr.: The Kidney. 2nd ed. Philadelphia, W. B. Saunders Co., 1981.
10. Brown, N. D., Kintzios, J. A., and Koetitz, S. E.: Determination of hypoxanthine, xanthine, and uric acid in biological fluids by ion-pair high-performance liquid chromatography. J. Chromatogr., 177:170-173, 1979.
11. Butler, A. R.: The Jaffe reaction. Identification of the coloured species. Clin. Chim. Acta, 59:227-232, 1976.
12. Chaney, A. L., and Marbach, E. P.: Modified reagents for determination of urea and ammonia. Clin. Chem., 8:130-132, 1962.
13. Cockcroft, D. W., and Gault, M. H.: Prediction of creatinine clearance from serum creatinine. Nephron, 16:31-41, 1976.
14. Coe, F. L.: Clinical and laboratory assessment of the patient with renal disease. In: The Kidney. Vol. 1 2nd ed. B. M. Brenner and F. C. Rector, Jr., Eds. Philadelphia, W. B. Saunders Co., 1981.
15. Cook, J. G. H.: Factors influencing the assay of creatinine. Ann. Clin. Biochem., 12:219-232, 1975.
16. DeFronzo, R. A., and Thier, S. O.: Inherited disorders of renal tubular function. In: The Kidney, Vol. 2. B. M. Brenner and F. C. Rector, Jr., Eds. Philadelphia, W. B. Saunders Co., 1981.
17. Del Greco, F., and Krumlovsky, F. A.: Role of the laboratory in management of acute and chronic renal failure. Ann. Clin. Lab. Sci., 11:283-291, 1981.
18. DiGiorgio, J.: Nonprotein nitrogenous constituents. In: Clinical Chemistry: Principles and Technics. 2nd ed. R. J. Henry, D. C. Cannon, and J. W. Winkelman, Eds. New York, Harper and Row, 1974.
19. Duncan, P. H., Gochman, N., Cooper, T., et al.: A candidate reference method for uric acid in serum. I. Optimization and evaluation. Clin. Chem., 28:284-290, 1982.
20. Espinel, C. H., and Gregory, A. W.: Differential diagnosis of acute renal failure. Clin. Nephrol., 13:73-77, 1980.
21. Fabiny, D. L., and Ertingshausen, G.: Automated reaction-rate method for determination of serum creatinine with the CentrifiChem. Clin. Chem., 17:696-700, 1971.
22. Fernandez, P., and Cox, M.: Basic concepts of renal physiology. In: The Kidney in Anesthesia, Vol. 22. Boston, Little, Brown and Co., 1984.
23. Fiereck, E. A.: Analysis of calculi. In: Fundamentals of Clinical Chemistry. 2nd ed. N. W. Tietz, Ed. Philadelphia, W. B. Saunders Co., 1976.
24. Gardner, M. D., and Henderson, A. R.: Test-request patterns for clinical chemistry in a British and a Canadian dialysis unit. Clin. Chem., 29:1546-1548, 1983.
25. Georges, J.: Determination of ammonia and urea in urine and of urea in blood by use of an ammonia-selective electrode. Clin. Chem., 25:1888-1890, 1979.
26. Gochman, N., and Schmitz, J. M.: Automated determination of uric acid, with use of a uricase-peroxidase system. Clin. Chem., 17:1154-1159, 1971.
27. Gotch, F. A.: Progress in hemodialysis. Clin. Nephrol., 9:144-155, 1978.
28. Haeckel, R.: Assay of creatinine in serum, with use of fuller's earth to remove interferents. Clin. Chem., 27:179-183, 1981.
29. Haeckel, R.: The use of aldehyde dehydrogenase to determine H_2O_2-producing reactions. I. The determination of the uric acid concentration. J. Clin. Chem. Clin. Biochem., 14:101-107, 1976.
30. Heinegård, D., and Tiderström, G.: Determination of serum creatinine by a direct colorimetric method. Clin. Chim. Acta, 43:305-310, 1973.
31. Heptinstall, R. H., Ed.: Pathology of the Kidney. 3rd ed. Boston, Little, Brown, and Co., 1983.
32. Hicks, J. M., Iosefson, M., and Lewis, S. A.: Evaluation of the Beckman creatinine analyzer. Clin. Chem., 25:1005-1008, 1979.
33. Holmes, J. H.: Measurement of osmolality in serum, urine, and other biological fluids by the freezing point determination. In: Workshop on Urinalysis and Renal Function Studies. Commission on Continuing Education. American Society of Clinical Pathologists, 1962.

34. Jaffe, M: Über den Niederschlag welchen Pikrinsäure in normalem Harn erzeugt and über eine neue Reaktion des Kreatinins. Z. Physiol. Chem., *10*:391, 1886.
35. Jaynes, P. K., Feld, R. D., and Johnson, G. F.: An enzymic, reaction-rate assay for serum creatinine with a centrifugal analyzer. Clin. Chem., *28*:114-117, 1982.
36. Kageyama, N.: A direct colorimetric determination of uric acid in serum and urine with uricase-catalase system. Clin. Chim. Acta, *31*:421-426, 1971.
37. Kampmann, J., Siersbaek-Nielsen, K., Kristensen, M., et al.: Rapid evaluation of creatinine clearance. Acta. Med. Scand., *196*:517-520, 1974.
38. Klahr, S.: Structure and function of the kidney. *In*: Cecil Textbook of Medicine, 17th ed. J. B. Wyngaarden and L. H. Smith, Jr., Eds. Philadelphia, W. B. Saunders Co., 1985.
39. Knochel, J. P., and Seldin, D. W.: The pathophysiology of uremia. *In*: The Kidney. 2nd ed. B. M. Brenner and F. C. Rector, Jr., Eds. Philadelphia, W. B. Saunders Co., 1981.
40. Lazarus, J. M.: Complications in hemodialysis: An overview. Kidney Internat., *18*:783-796, 1980.
41. Leaf, A., and Cotran, R. S.: Renal Pathophysiology. 2nd ed. New York, Oxford University Press, 1980.
42. Lustgarten, J. A., and Wenk, R. E.: Simple, rapid kinetic method for serum creatinine measurement. Clin. Chem., *18*:1419-1422, 1972.
43. Marsh, W. H., Fingerhut, B., and Miller, H.: Automated and manual direct methods for the determination of blood urea. Clin. Chem., *11*:624-627, 1965.
44. Martin, K.: Pathophysiology of acute renal failure. *In*: The Kidney and Body Fluids. S. Klahr, Ed. New York, Plenum Medical Book Co., 1983.
45. Merrill, J. P.: Dialysis versus transplantation in the treatment of end-stage renal disease. Ann. Rev. Med., *29*:343-358, 1978.
46. Miller, B. F., and Winkler, A. W.: The renal excretion of endogenous creatinine in man. Comparison with exogenous creatinine and inulin. J. Clin. Invest., *17*:31-40, 1938.
47. Miller, M., Moses, A. M., and Streeten, D. H.: Recognition of partial defects in antidiuretic hormone secretion. Ann. Intern. Med., *73*:721-727, 1970.
48. Miller, T. R., Anderson, R. J., Linas, S. L., et al.: Urinary diagnostic indices in acute renal failure: A prospective study. Ann. Intern. Med., *89*:47, 1978.
49. Morgan, D. B.: Assessment of renal tubular function and damage and their clinical significance. Ann. Clin. Biochem., *19*:307-313, 1982.
50. Moss, G. A., Bondar, R. J. L., and Buzzelli, D. M.: Kinetic enzymatic method for determining serum creatinine. Clin. Chem., *21*:1422-1426, 1975.
51. Narayanan, S., and Appleton, H. D.: Creatinine: A review. Clin. Chem., *26*:1119-1126, 1980.
52. Nishi, H. H.: Determination of uric acid. An adaptation of the Archibald method on the Auto Analyzer. Clin. Chem., *13*:12-18, 1967.
53. Nissen, P.: Simultaneous determination of allopurinol, oxipurinol, and uric acid in human plasma by high performance liquid chromatography. J. Chromatogr., *228*:382-386, 1982.
54. Ott, N. T., and Wilson, D. M.: A simple technique for estimating glomerular filtration rate with subcutaneous injection of [^{125}I] iothalamate. Mayo Clinic Proc., *50*:664-668, 1975.
55. Owen, J. A., Iggo, B., Scandrett, F. J., et al.: The determination of creatinine in plasma or serum, and in urine; a critical examination. Biochem. J., *58*:426-437, 1954.
56. Peacock, M., and Robertson, W. G.: The biochemical aetiology of renal lithiasis. *In*: Urinary Calculous Disease. J. E. A. Wickham, Ed. London, Churchill Livingstone, 1979.
57. Polar, E., and Metcoff, J.: "True" creatinine chromogen determination in serum and urine by semi-automated analysis. Clin. Chem., *11*:763-770, 1965.
58. Rennke, H. G., and Venkatachalam, M. A.: Glomerular permeability of macromolecules: Effect of molecular configuration on the fractional clearance of uncharged dextran and neutral horseradish peroxidase in the rat. J. Clin. Invest., *63*:713-717, 1979.
59. Rodriguez, S. J.: Renal tubular acidosis. *In*: Pediatric Kidney Disease. C. M. Edelmann, Ed. Boston, Little, Brown and Co., 1978.
60. Sampson, E. J., Baird, M. A., Burtis, C. A., et al.: A coupled-enzyme equilibrium method for measuring urea in serum: Optimization and evaluation of the AACC Study Group on urea candidate reference method. Clin. Chem., *26*:816-826, 1980.
61. Scriver, C. R., Clow, C. L., and Lamm, P.: On the screening, diagnosis and investigation of hereditary aminoacidopathies. Clin. Biochem., *6*:142-188, 1973.
62. Siersbaek-Nielsen, K., Molholm-Hansen, J., Kampmann, J., et al.: Rapid evaluation of creatinine clearance. Lancet, *1*(Part 2):1133-1134, 1971.
63. Smith, C. H., Landt, M., Steelman, M., et al.: The Kodak Ektachem 400 Analyzer evaluated for automated enzymic determination of plasma creatinine. Clin. Chem., *29*:1422-1425, 1983.
64. Soldin, S. J., Henderson, L., and Hill, J. G.: The effect of bilirubin and ketones on reaction rate methods for the measurement of creatinine. Clin. Biochem., *11*:82-86, 1978.
65. Soldin, S. J., and Hill, J. G.: Micromethod for determination of creatinine in biological fluids by high-performance liquid chromatography. Clin. Chem., *24*:747-750, 1978.
66. Stenzel, K. H., Cheigh, J. S., Sullivan, J. F., et al.: Clinical effects of bilateral nephrectomy. Am. J. Med., *58*:69-75, 1975.
67. Taussky, H. H.: Creatinine and creatine in urine and serum. *In*: Standard Methods of Clinical Chemistry, Vol. 3. D. Seligson, Ed. New York, Academic Press, 1961, pp. 99-113.
68. Taussky, H. H., and Kurzmann, G.: A microcolorimetric determination of creatine in urine by the Jaffe reaction. J. Biol. Chem., *208*:853-861, 1954.

69. Thompson, H., and Rechnitz, G. A.: Ion electrode based enzymatic analysis of creatinine. Anal. Chem., *46*:246-249, 1974.
70. Tietz, N. W., Ed.: Clinical Guide to Laboratory Tests. Philadelphia, W. B. Saunders Co., 1983.
71. Tiffany, T. O., Jansen, J. M., Burtis, C. A., et al.: Enzymatic kinetic rate and end-point analyses of substrate, by use of a GeMSAEC fast analyzer. Clin. Chem., *18*:829-837, 1972.
72. Veniamin, M. P., and Vakirtzi-Lemonias, C.: Chemical basis of the carbamidodiacetyl micromethod for estimation of urea, citrulline, and carbamyl derivatives. Clin. Chem., *16*:3-6, 1970.
73. Watts, R. W. E.: Determination of uric acid in blood and urine. Ann. Clin. Biochem., *11*:103-111, 1974.
74. Watts, R. W. E.: Purines and nucleotides. *In*: Chemical Diagnosis of Disease. S. S. Brown, F. L. Mitchell, and D. S. Young, Eds. Amsterdam, Elsevier/North Holland, 1979.
75. Wesson, L. G., Jr.: Physiology of the Human Kidney. New York, Grune and Stratton, 1969.
76. Wills, M. R.: Uremic toxins, and their effect on intermediary metabolism. Clin. Chem., *31*:5-13, 1985.
77. Wills, M. R., and Savory, J.: Biochemistry of renal failure. Ann. Clin. Lab. Sci., *11*:292-299, 1981.
78. Wrong, O., and Davies, H. E. F.: The excretion of acid in renal disease. Q. J. Med., *28*:259-313, 1959.
79. Wyngaarden, J. B., and Smith, L. H., Jr., Eds.: Cecil Textbook of Medicine. 16th ed. Philadelphia, W. B. Saunders Co., 1982.

Additional Reading

Brenner, B. M., and Rector, F. C., Jr., Eds.: The Kidney. 2nd ed. Philadelphia, W. B. Saunders Co., 1981.
Ganong, W. F.: Review of Medical Physiology. 10th ed. Los Altos, Cal., Lange Med. Pub., 1981.
Klahr, S., Ed.: The Kidney and Body Fluids in Health and Disease. New York, Plenum Medical Book Co., 1983.

CHAPTER 12

CALCIUM AND PHOSPHATE METABOLISM

by Donald Fraser, M.D., Ph.D., Glenville Jones, Ph.D., Sang Whay Kooh,
M.D., Ph.D., F.R.C.P.(C), and Ingeborg C. Radde, M.D., Ph.D., F.R.C.P.(C)

In this textbook, as in most, calcium and phosphate are considered in the same chapter. The justification for this lies in the important physiological interrelation the inorganic forms of the elements have to one another. Calcium and inorganic phosphate co-exist in relatively fixed proportions in the mineral phase of the hard tissues—bone, dentin, and enamel. Bone calcium and inorganic phosphate exchange with those in the tissue fluids. The concentrations of calcium and inorganic phosphate in plasma are dependent in part on the balance between bone mineral deposition and bone resorption; at the same time, the concentration of each element has an important bearing on the mineralization process itself. From these observations, it is not surprising that the three main regulators of calcium and inorganic phosphate homeostasis—parathyroid hormone, vitamin D–hormone, and calcitonin—each exert independent effects on both ions.

It has often been stated that a reciprocal relationship exists between the plasma calcium and phosphate concentrations. This concept is an oversimplification, as will be discussed below. Nevertheless, it is important to pay close attention to the interrelation of these plasma constituents in health and disease.

CALCIUM

Distribution

Calcium is the fifth most common element in the body.[11] It is a divalent cation, atomic weight 40.08, consisting mainly of ^{40}Ca, with trace amounts of the naturally occurring stable isotopes ^{46}Ca and ^{48}Ca. Practically all of the ~25 mol (1 kg) of calcium in the human adult is extracellular; 99% of it is in the hard tissues, where it exists as extracellular crystalline hydroxyapatite of the general formula $Ca_{10}(PO_4)_6(OH)_2$. The extracellular fluid contains ~25 mmol (1 g) calcium; blood contains 7.5 mmol (300 mg), virtually all of which is in the plasma. Approximately 125 mmol (5 g) of calcium is in the exchangeable calcium pool, located mainly on bone-forming surfaces and in soft tissues, where the calcium is readily accessible for physiological functions.

Calcium exists in plasma in three physicochemical states: protein-bound (~45%); complexed with small diffusible ligands such as citrate, lactate, phosphate, and bicarbonate (~10%); and ionized (~45%). Under normal conditions, the concentration of total calcium (Ca_T) in human plasma (or serum) is ~2.5 mmol/L (10 mg/dL), and that of the ionized fraction is ~1.125 mmol/L (4.5 mg/dL). Eighty per cent of the protein-bound fraction is bound to albumin, 20% to globulins. The ionized fraction (Ca_I) is the physiologically active form. This fraction is in dynamic equilibrium with the other forms, its magnitude depending largely upon pH and the concentration of the plasma proteins. Acidosis favors dissociation and raises the proportion existing as Ca_I, while alkalosis has the opposite effect.

In contrast to the large quantity of extracellular calcium, only 1 mmol (40 mg) is present in the cytosol. Yet, despite the smallness of this amount, intracellular calcium serves a vital role in controlling the function of all cells.

Physiological Functions

The functions of calcium are diverse.[11] Intracellularly, calcium is the prime inorganic messenger for regulation of cell functions. It determines activity of enzymes, notably adenylate cyclase and phosphodiesterase, through reversible combination with calmodulin.[16] Calmodulin is a small protein, M.W. 16 790, found in all nucleated cells. Its primary structure is virtually identical in all organisms and all cell types so far studied. It is involved in the regulation of a multitude of functions within cells, including fertilization, mitosis, cell motility, ciliary action, and the numerous complex functions of the various organ systems. Each molecule of calmodulin contains four binding sites for calcium, and the activities of enzymatic processes are modulated through reversible formation of Ca-calmodulin-enzyme complexes. In striated muscle, calcium activates contraction of the myosin fibril through combination with troponin, a calcium-binding protein similar, if not identical, to calmodulin in structure and function.

Calcium also functions at the plasma membrane itself. It serves to regulate membrane permeability, to cause neurotransmitter release, and to diminish neuromuscular excitability. Secretions of several endocrine glands are controlled by the extracellular Ca_I concentration at the cell surface. Examples are the parathyroid glands, thyroid C-cells, and pancreatic β-cells. Calcium is involved in cell-to-cell adhesion and possibly communication. The ion plays a vital role in blood coagulation. As a principal constituent of hydroxyapatite, it serves to confer rigidity to the extracellular protein matrices of bones and teeth.

Calcium is constantly being lost from the body through excretion in feces, urine, and, to a small extent, sweat. The calcium requirement of the human depends upon age, growth, pregnancy, and lactation. Dairy products contribute 70% of the calcium in the North American diet. The U.S. Recommended Dietary Allowance (RDA) for calcium is 800 mg/d for adults and children above one year of age. An additional 400 mg/d calcium intake is recommended during adolescence, pregnancy, and lactation. Net absorption of ingested calcium is about 25–30%.

PHOSPHATE

Distribution

The body of a human adult contains $\sim$20 mol (620 g) of phosphorus, entirely in the form of phosphate.[11] Phosphate is distributed fairly equally between extracellular and intracellular compartments, in which regard it contrasts with the preponderantly extracellular location of calcium. Intracellularly, phosphate occurs mainly as an integral component of structural organic macro-molecules such as phospholipids and phosphoproteins. This fraction is termed "organic phosphate." A small, but extremely important fraction exists intracellularly as "inorganic phosphate" which participates in high-energy transfer reactions.

About 85% of extracellular phosphate occurs in inorganic form as hydroxyapatite. In plasma (or serum), most phosphate exists in the inorganic form (P_i); this is present as the mono- and dihydrogen forms, the relative proportions varying with the pH. At pH 7.40, the HPO_4^{2-}:$H_2PO_4^-$ ratio is close to 80:20; it increases with increasing pH and vice versa. Approximately 15% of plasma P_i is protein bound; the remainder exists in complexed and free forms, and is ultrafiltrable. In urine, the HPO_4^{2-}:$H_2PO_4^-$ ratio varies over a wide range, from 1:100 at pH 4.5 to 90:10 at pH 8.0. On the other hand, the total plasma and total urinary P_i concentrations are not altered by pH changes.

Physiological Functions

Phosphate in the form of phospholipid is a structural component of cell membranes. As phosphoprotein it participates in many important cell functions, including energy metabolism (high energy nucleotides) and gene replication (nucleic acids). Chronic deficiency of intracellular inorganic phosphate can affect cell metabolism at several levels. Although the intracellular concentration of inorganic phosphate is low, it participates in many reactions concerned with generation of metabolic energy. P_i is a substrate for oxidative phosphorylation, the mitochondrial process that produces ATP from ADP and serves as a primary source of metabolic energy. Glycogen breakdown in liver and kidney requires P_i to produce free glucose and in muscle to provide glycolytic intermediates. In red cells, production of 2,3-diphosphoglycerate from glyceraldehyde-3-phosphate consumes P_i to generate the reducing equivalents required to maintain integrity of the cell membrane. Illustrations of metabolic pathways in standard biochemistry textbooks provide a clear indication where and when P_i is essential to the intracellular energy economy.

Inorganic phosphate is a major component of hydroxyapatite; the concentration of P_i in plasma and extracellular fluid is an important determinant of bone mineral turnover. Chronic hypophosphatemia is associated with rickets and osteomalacia, partly through failure to provide an adequate concentration of this major constituent of bone crystal and partly because inorganic phosphate probably plays a role in mineralization by initiating formation of matrix rich in phosphorylated proteins which form a nucleus upon which hydroxyapatite can crystallize.

Phosphorus is present in most foods; the most important sources in North America are meat and milk. The Recommended Dietary Allowance for phosphorus (in mg/d) is the same as that for calcium.

HORMONES THAT REGULATE CALCIUM AND INORGANIC PHOSPHATE TRANSPORT[3,4,5,17,25,49,93]

The three primary hormonal regulators of calcium and inorganic phosphate are parathyroid hormone (PTH), calcitonin (CT), and $1\alpha,25$-dihydroxyvitamin D ($1,25$-$(OH)_2D$), the hormonal form of the vitamin. To understand the manner in which these hormones interact to maintain calcium and inorganic phosphate homeostasis requires a brief description of each.

PARATHYROID HORMONE

PTH is synthesized by the parathyroid glands. In man there are usually four glands, situated close to or on the posterior surface of the thyroid gland. However, additional parathyroid glands may be located elsewhere in the neck or within the thymus in the superior mediastinum. There are two types of cells in the gland: the chief (principal, clear) cells and the oxyphil cells. The chief cells are responsible for the production, storage, and secretion of PTH. Their cytoplasm contains numerous small vesicles believed to contain the synthesized hormone. The oxyphil cells appear in the gland at about the time of puberty and increase in number with age; their function is unknown.

Parathyroid hormone (PTH) is a single-chain polypeptide of 84 amino acids, M.W. 9500.[17,49,103] The precursor of PTH, pre-proparathyroid hormone, a 115-amino acid polypeptide, is synthesized on the ribosomes within the chief cells. It is rapidly converted by proteolysis to pro-PTH (90 amino acids, M.W. 10 000). In turn, pro-PTH is converted to intact PTH in the Golgi region of the chief cells by a second proteolytic reaction. Intact PTH 1–84 is the form stored in the glands and the principal form secreted into the bloodstream. PTH 1–84 is metabolized in the peripheral tissues, principally the liver and kidney, to an amino-terminal (N-terminal) fragment of at least 34 amino acids, M.W. $\sim$2000, and to several carboxy-terminal (C-terminal) and mid-molecule fragments (e.g., amino acids 53–84, 44–68, 35–64 and sequences that overlap these*).

The biological activity of PTH resides in the first 30 or so amino acids of the N-terminal portion of the molecule. Thus, intact PTH 1–84 and PTH 1–34 possess full biological activity. The C-terminal and mid-molecule fragments are biologically inert. Intact PTH and the N-terminal fragment are removed from the plasma by attachment to hormone-specific receptors on the plasma membranes of target cells, by renal excretion, and by degradation in the liver, kidney, and possibly bone. The inactive mid-molecule and C-terminal fragments are removed exclusively by renal excretion. The plasma $t_{1/2}$ of PTH 1–84 is 15–30 min, and the $t_{1/2}$ of the mid- and C-terminal fragments are 2–4 h; estimates of $t_{1/2}$ of PTH 1–34 vary.

Because of these differences in metabolism and excretion, the plasma concentrations and relative distributions of PTH-derived peptides cover a wide range and vary according to the physiological or pathological status. The concentration of biologically active hormones is very low in normal plasma (about 10 pmol/L, or 0.1 ng/mL when determined by radioimmunoassay),

---

*Bovine and human PTH (bPTH, hPTH) and many of their fragments have been synthesized by genetic engineering and chemical techniques. Synthetic bPTH 1–34 and hPTH 1–34 are available commercially.

whereas that of the inert mid-region and C-terminal fragments is 5–10 times greater. Recent studies show that intact PTH is the dominant biologically active form in plasma; little PTH 1–34 is present except in chronic renal failure when the two forms occur in approximately equal proportions.

Functions of PTH and Regulation of PTH Synthesis and Secretion

The primary role of PTH is regulation of the extracellular fluid (ECF) calcium concentration. This is accomplished mainly by stimulating osteoclasts, osteocytes, and the calcium-reabsorbing cells of the renal tubules (mainly the distal tubules). Synthesis and secretion of PTH are under precise feedback regulation by the concentration of ionized calcium (Ca_I) in the circulation. Ionized hypocalcemia stimulates PTH synthesis and secretion within a few minutes; ionized hypercalcemia suppresses secretion. Although PTH also affects the concentration of plasma P_i, the plasma P_i concentration does not control PTH secretion directly. However, hyperphosphatemia stimulates PTH secretion indirectly, causing movement of both calcium and phosphate into bone, decreasing resorption of mineral from bone, and increasing formation of undissociated calcium-phosphate complexes in plasma. As a result, there is a decrease of plasma Ca_I, thus stimulating PTH secretion. Phosphate retention occurs in chronic renal failure and elevation of the plasma PTH concentration is an early feature. Severe hypermagnesemia is associated with suppression of PTH secretion.

Mode of Action of PTH

The mode of action of the hormone is depicted in Figure 12-1. PTH-responsive cells possess a PTH-specific receptor protein in the plasma membrane. The hormone-receptor complex activates adenylate cyclase in the inner part of the plasma membrane. The activity of the hormone-receptor–adenylate cyclase complex is stimulated or inhibited by a recently identified regulatory factor, N-protein (or G-protein), which itself is activated by interaction with the nucleotide, guanosine triphosphate (GTP).[116] Activated adenylate cyclase stimulates conversion of ATP to 3'5'-cyclic adenosine monophosphate (cAMP); the latter, through stimulation of protein kinase, elicits target cell responses. For example, in PTH-responsive proximal renal tubular cells, increased intracellular cAMP causes inhibition of P_i transport and increased phosphaturia results. Increased cAMP production within the tubular cells is reflected in increased excretion of this nucleotide in the urine. A genetic defect in N-protein activity is implicated in the pathogenesis of one form of pseudohypoparathyroidism (see below).

Actions of PTH

Parathyroid hormone:
1. Stimulates osteoclasts and probably osteocytes to mobilize calcium from bone. This action requires the synergistic action of 1,25-$(OH)_2D$. Mobilization of calcium, if persistent, leads to demineralization of bone and to release of phosphate and hydroxyproline into the plasma.
2. Increases renal tubular reabsorption of calcium.
3. Inhibits proximal renal tubular reabsorption of inorganic phosphate.
4. Inhibits proximal tubular reabsorption of HCO_3^-, Na^+, K^+, H_2O, and amino acids.

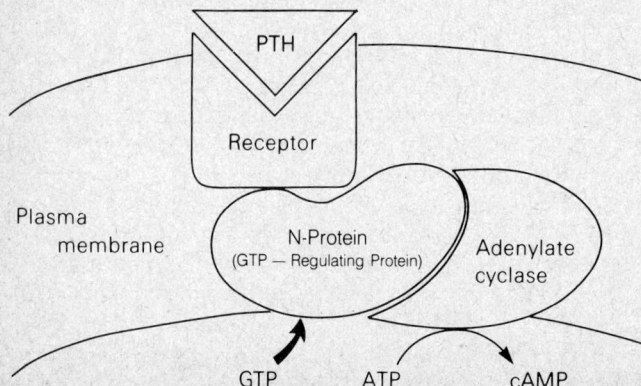

Figure 12-1. Postulated mechanism of cAMP-mediated PTH response. (From Daneman, D., Kooh, S. W., and Fraser, D.: Clin. Endocrinol. Metab., *11*:211–231, 1982.)

5. Stimulates renal hydroxylation of 25-(OH)D to 1,25-(OH)$_2$D, by increasing the synthesis of renal l-hydroxylase.

6. Indirectly induces an increase in intestinal calcium absorption by stimulating 1,25-(OH)$_2$D synthesis. PTH has no direct action on intestinal calcium absorption.

7. Promotes bone remodeling; low concentrations of PTH are thought to stimulate bone formation.

PTH affects plasma and urine composition by:

1. Increasing plasma Ca$_T$ and Ca$_I$.

2. Decreasing plasma P$_i$.

3. Usually increasing urinary calcium excretion rate, because the increase in plasma Ca and hence in the filtered load of calcium tends to override the effect of PTH to increase tubular calcium reabsorption.

4. Increasing renal phosphate clearance.

5. Increasing urinary cAMP excretion rate.

6. Causing generalized aminoaciduria and metabolic acidosis, though these effects are of minor physiological importance.

CALCITONIN

Calcitonin (CT) is produced by specialized, histologically distinctive cells (C-cells) distributed throughout the thyroid gland.[6] Low concentrations of a CT-like peptide have been identified in thymus, adrenal medulla, parathyroid glands, and elsewhere. Whether its presence in these sites is due to synthesis in situ or to transfer from thyroid C-cells remains controversial.

CT consists of a 32 amino acid polypeptide chain, M.W. 3500, with a disulfide bridge between the N-terminal and the 7th amino acid in the chain, and a prolinamide residue at the C-terminus. It is derived from a large precursor polypeptide chain, M.W. 15 000.

CT has been demonstrated in fish, reptiles, birds, and mammals. The amino acid sequences of human, rat, cow, pig, sheep, and salmon CT differ from one another. The structural differences affect potency; in the human, porcine CT is one tenth as active as human CT, whereas salmon CT is 10 times more potent than human CT. The differences in amino acid sequence also confer immunospecificity. Patients treated with salmon or porcine CT may develop antibodies to the agent.

Actions of CT and Regulation of CT Secretion

Calcitonin binds to specific membrane receptors of target cells and stimulates cAMP production. The effect of small doses of CT is to decrease resorption of calcium and inorganic phosphate from bone by inhibiting the action of osteoclasts and probably osteocytes, leading to lower plasma Ca$_T$, Ca$_I$, and P$_i$ concentrations.

Secretion of CT is regulated by the level of Ca$_I$ in the circulation. Healthy individuals normally have low plasma concentrations of CT. Hypercalcemia stimulates CT release, while hypocalcemia suppresses secretion. Whether other physiological factors regulate C-cell function in the normal individual is not established; however, biogenic amines, somatostatin, and estrogens have been implicated. When administered parenterally, gastrin, pentagastrin, glucagon, and cholecystokinin are secretagogues, but it is not yet settled whether gastrointestinal hormones have a role in regulating CT secretion in vivo.

Pharmacological doses of CT (1) decrease renal tubular reabsorption of calcium, phosphate, sodium, potassium, and magnesium; (2) decrease gastrin secretion; and (3) increase secretion of sodium, potassium, chloride, and water by the small intestine.

Physiological Functions of CT

Since the discovery of CT in 1962, it has been known that an action of the hormone is to lower the plasma Ca$_T$ and Ca$_I$ concentrations. At first, CT was considered to be a major plasma calcium-regulating hormone which functioned along with PTH and vitamin D to maintain the constancy of plasma Ca in the normal individual. However, the importance of CT as a plasma calcium regulator in humans is now in considerable doubt because thyroidectomized persons with no detectable CT do not have hypercalcemia, and patients with marked hypercalcitoninemia due to medullary thyroid carcinoma (see below) are rarely hypocalcemic. Other hypotheses—

that it functions in long-term maintenance of skeletal mass, that it plays a role in preventing bone resorption during short-term dietary calcium deficiency, that it helps to control secretion of gastrin—are attractive but still controversial. Secretion of CT in response to gastric hormones has been proposed as a mechanism to prevent postprandial hypercalcemia, but the rise in serum gastrin that follows eating is not associated with an increase in immunoreactive CT. CT has its greatest physiological effect in the fetus and the young, probably because bone calcium turnover is very rapid at this stage of development. Plasma CT levels are higher in neonates and in pregnant and lactating women than in normal adults. CT may have a function in promoting deposition of calcium in the skeleton of the human fetus, and may be a causative factor in neonatal hypocalcemia.

VITAMIN D AND METABOLITES

Chemistry and Distribution

There are two families of vitamin D, both of which are sterols.[25] Vitamin D_3 (cholecalciferol) is the natural form produced in the skin from the action of sunlight. Vitamin D_2 (ergocalciferol) is manufactured commercially from precursors of plant origin. The chemical difference lies in the functional groups attached at carbons 22, 23, and 24 of the side chain.

In nature, 7-dehydrocholesterol, a normal secretion of skin, is converted by certain ultraviolet wavelengths (290–310 nm) of sunlight to previtamin D_3 and thence by the heat of the body to vitamin D_3. Skin exposure to sunlight is a very efficient means of acquiring vitamin D.[1] Deep pigmentation of the skin reduces the efficiency in the temperate zones. Vitamin D_3 is transported from the skin as a complex with vitamin D–binding protein (DBP), a specific plasma protein that migrates in the α_1-globulin region of an electrophoretic separation.

Until half a century ago vitamin D was acquired mainly through synthesis of vitamin D_3 in the skin, since only a few natural foods—fish, fish-liver oils, egg yolk—contain significant amounts of vitamin D. In developed countries today, a considerable fraction of the individual's vitamin D is acquired by ingestion of pre-formed vitamin, either taken as a proprietary vitamin preparation or consumed in commercially enriched foods, mainly milk. In North America, but not in Europe, vitamin D_2 is the only form used in prophylactic and therapeutic vitamin preparations; both vitamins D_2 and D_3 are employed for food enrichment. As a result, the vitamin D body stores of the North American contain both the vitamin D_2 and the vitamin D_3 series of metabolites, with the vitamin D_3 series predominating in the average healthy individual. These points must be considered when designing vitamin D assay techniques. Further, the ability of a specific assay to discriminate between the vitamin D_2 and D_3 series must be considered in interpreting assay results. (See section on Methods.)

The Recommended Dietary Allowance for ingested vitamin D from all sources is 10 $\mu g/d$ (400 IU/d) during childhood and 5 $\mu g/d$ for adults. Human milk is a poor source of vitamin D; on the other hand, virtually all cow's milk sold commercially in North America is enriched with either vitamin D_2 or D_3 (10 $\mu g/L$, 400 IU/L). Except for breast-fed infants and individuals consuming strict vegetarian diets without eggs or milk, the vitamin D requirement of the healthy individual is ordinarily met at all ages through consumption of usual North American foods and exposure of only small areas of skin to sunlight.

Metabolism of Vitamin D

Neither vitamin D_2 nor vitamin D_3 is active as such.[25] Before the vitamins D can exert physiological action, each must undergo two hydroxylations, both of which are catalyzed by specific enzymes (Figure 12-2). It should be noted that most of the recent research on vitamin D metabolism has been carried out with the D_3 series. It is probable that the principal metabolites of vitamins D_2 and D_3 exist in the plasma in proportion to the amounts of the parent vitamins D_2 and D_3 acquired by the individual. Vitamins D_2 and D_3 appear to have equal potency in man.

The first metabolic step, which takes place in the liver, is conversion of vitamin D_3 to 25-hydroxyvitamin D_3 (25-(OH)D_3). This reaction is catalyzed by specific microsomal and mitochondrial enzymes, collectively designated vitamin D_3-25-hydroxylase (25-(OH)ase). The 25-(OH)D_3 is transported by vitamin D binding protein (DBP) to the kidney where it undergoes further hydroxylations. The important metabolites produced by the kidney are 1α,25-dihydroxyvitamin D_3 (1,25-(OH)$_2$D$_3$) and 24,25-dihydroxyvitamin D_3 (24,25-(OH)$_2$D$_3$). Synthesis of the respective metabolites is catalyzed by the specific mitochondrial enzymes, 25-hydroxyvitamin

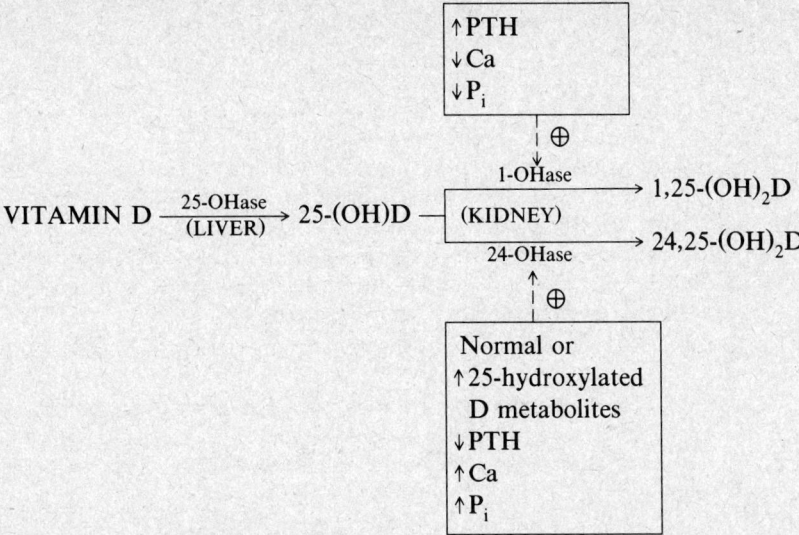

Figure 12-2. Simplified diagram to illustrate the pathways of vitamin D metabolism and the main factors that stimulate 1,25-(OH)$_2$D and 24,25-(OH)$_2$D synthesis.

D_3-1-hydroxylase (1-(OH)ase) and 25-hydroxyvitamin D_3-24 hydroxylase (24-(OH)ase). 1-Hydroxylation occurs mainly in kidney mitochondria but also may occur in bone and placenta. 24-Hydroxylation occurs in the kidney and, in small amounts, probably also in skeletal cartilage and elsewhere. 1,25-(OH)$_2$D$_3$ disappears quickly from the plasma ($t_{1/2}$, 0.5 h) and its usual concentration in the plasma is low. 24,25-(OH)$_2$D$_3$ disappears more slowly ($t_{1/2}$, 10 h) and its concentration is considerably greater, approximating one third the concentration of 25-(OH)D$_3$, its substrate. Other metabolites are synthesized from 25-(OH)D$_3$ in the kidney, e.g., 1,24,25-(OH)$_3$D$_3$, 23,25-(OH)$_2$D$_3$, 25,26-(OH)$_2$D$_3$, 25-(OH)D$_3$-26,23-lactone, and 24-oxo-23,25-(OH)$_2$D$_3$. These compounds appear to have little physiological importance; however, some can interfere in certain assay procedures.

Control of Vitamin D Metabolism

1,25-(OH)$_2$D is by far the most potent of the known metabolites. It has rapid and powerful stimulatory action on membrane transport of calcium. The pattern of vitamin D metabolism corresponds to the physiological state of the individual. When calcium mobilization is required, for example in states of hypocalcemia, vitamin D deficiency, hyperparathyroidism, and hypophosphatemia, 1-OHase activity is stimulated and synthesis of 1,25-(OH)$_2$D takes place. Other factors that stimulate 1-OHase activity are estrogens, progestins, testosterone, growth hormone, and prolactin.

When calcium mobilization is unnecessary or would be detrimental to the individual, for example when plasma Ca is normal or elevated or when plasma P$_i$ is normal, 1-hydroxylation ceases and 25-(OH)D is converted to 24,25-(OH)$_2$D. 24,25-(OH)$_2$D has no activity in the mobilization of bone calcium. It has been shown in vitro that 24-(OH)ase activity is stimulated in the presence of 25-hydroxylated metabolites of vitamin D, e.g., 25-(OH)D, and 1,25-(OH)$_2$D.[109]

In the healthy individual on a diet containing adequate calcium, the plasma contains a small but detectable amount of 1,25-(OH)$_2$D and relatively large amounts of 24,25-(OH)$_2$D. With calcium deprivation or impaired calcium transport, plasma 1,25-(OH)$_2$D increases and the plasma 24,25-(OH)$_2$D concentration decreases.

Biologic Actions of Vitamin D Metabolites

1,25-(OH)$_2$D has the following biological actions:

1. Stimulation of calcium absorption by the small intestine.

2. Stimulation of calcium resorption from bone. In this action, PTH acts synergistically.

3. Increase of calcium reabsorption by the distal renal tubules of the kidney. An earlier suggestion that 1,25-(OH)$_2$D increases renal tubular reabsorption of phosphate has not been proved.

4. Direct stimulatory action on intestinal absorption of phosphate independent of its action on intestinal calcium absorption.

5. Differentiation of a variety of cells. It is suggested that 1,25-(OH)$_2$D stimulates transformation of progenitor cells into osteoclasts. Stimulation of cell differentiation and metabolism may prove to be fundamental actions of vitamin D metabolites.

1,25-(OH)$_2$D-receptors have been identified not only in intestinal mucosa, bone, and renal tubules but also in a large variety of organs, for example, parathyroid glands, brain, and muscle. In the activated T-lymphocyte, which also possesses 1,25-(OH)$_2$D receptors, 1,25-(OH)$_2$D$_3$ suppresses production of the growth-promoting lymphokine interleukin-2. However, a physiological action of 1,25-(OH)$_2$D has not been proved at sites other than intestine, bone, and kidney. 1,25-(OH)$_2$D does not stimulate mineralization of bone directly.

1,25-(OH)$_2$D has the following net effects on plasma and urinary calcium and inorganic phosphate:

1. Increase of plasma Ca$_T$ and Ca$_I$ concentrations and of plasma P$_i$ concentration.

2. Increase of urinary calcium excretion rate because the effect of 1,25-(OH)$_2$D to increase plasma calcium, and hence the filtered load of calcium, is greater than its effect to increase tubular reabsorption of calcium.

Biological Effects of Other Vitamin D Metabolites

An inordinately high concentration of 25-(OH)D in plasma has been found experimentally to increase plasma Ca in the absence of 1,25-(OH)$_2$D, i.e., in anephric animals and man. In this circumstance, 25-(OH)D is thought to possess all the actions of 1,25-(OH)$_2$D. Although some evidence of biological activity of 24,25-(OH)$_2$D has been presented, it remains controversial whether this metabolite has a significant role in calcium homeostasis. The significance of other hydroxylated forms of vitamin D is unknown; they too may represent stages of vitamin D degradation.

Plasma Ca and P$_i$ Regulation

The concentrations of Ca and P$_i$ in plasma are determined by the balance between absorption from the intestine and excretion by the kidneys, intestinal tract, and, to a small extent, the skin; and by bi-directional exchanges among plasma, extracellular fluid (ECF), and tissues including bone. These dynamic processes are determined by many factors, principal among which are PTH, CT, and vitamin D–hormone. Functioning together in a coordinated fashion, the actions of these three hormones result in remarkably precise control of the plasma Ca$_I$ concentration in the healthy individual. The main transport routes that determine the concentration of calcium in plasma and extracellular fluid, and the manner in which PTH, CT, and 1,25-(OH)$_2$D influence these processes, are shown diagrammatically in Figure 12-3. The plasma P$_i$ level similarly represents a balance

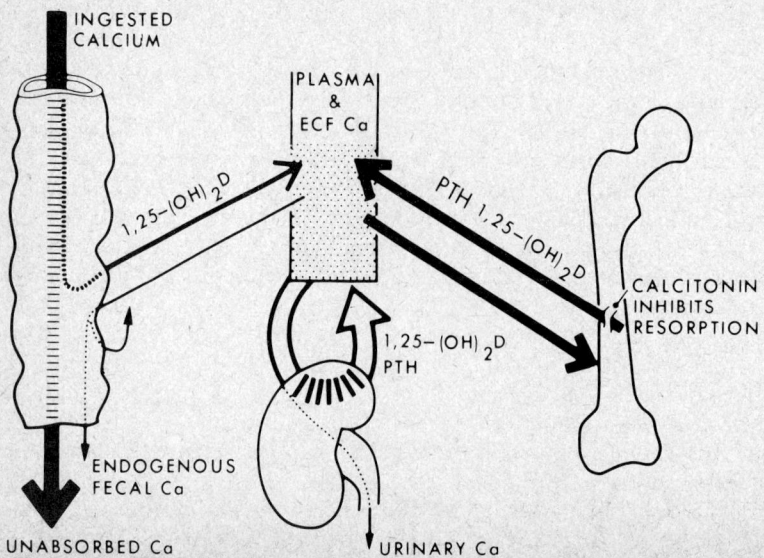

Figure 12-3. Diagram showing the role of 1,25-(OH)$_2$D, PTH, and CT in maintaining calcium homeostasis. The relative magnitudes of the various homeostatic processes are indicated by the thickness of the arrows. In the normal adult, rates of calcium deposition and resorption are equal. The processes stimulated by 1,25-(OH)$_2$D, PTH, and CT are indicated. (From Fraser, D., and Kooh, S. W., *In*: Textbook of Paediatrics. 3rd ed. J. O. Forfar and G. C. Arneil, Eds. Edinburgh, Churchill Livingstone, 1984, Vol. 2, pp. 1157–1172.)

between absorption from the intestine, excretion in the feces and urine, and movement into and out of extracellular fluids, cells, and hard tissues. However, in the case of plasma P_i regulation, the kidneys play a particularly important role and PTH exerts an important effect by influencing phosphate transport by the kidney tubules. These aspects are depicted in Figure 12-4.

A number of other factors affect calcium and phosphate metabolism. There is evidence that prostaglandin E_2, osteoclast activating factor (OAF), and other local bone resorbing factors affect the plasma Ca concentration. It is also clear that alterations in secretions of other endocrine glands, for example, the adrenal cortex, the gonads, and the anterior pituitary gland, can influence calcium homeostasis secondarily.

SIGNIFICANCE OF PLASMA LEVELS OF Ca_T, Ca_I, P_i, PTH, CT, AND VITAMIN D METABOLITES

DISEASES ASSOCIATED WITH HYPERCALCEMIC STATES

Diseases associated with hypercalcemia and their important clinical findings are listed in Table 12-1. Primary hyperparathyroidism and hypercalcemia of malignancy are discussed in detail because of their frequent occurrence and varied clinical and biochemical manifestations.

PRIMARY HYPERPARATHYROIDISM

Primary hyperparathyroidism is a disorder caused by persistent semi-autonomous hypersecretion of parathyroid hormone.[3,103] The diagnosis is made much more frequently today because hypercalcemia may be identified in asymptomatic individuals by plasma calcium determinations during the conduct of routine plasma screening. Adenoma of a single parathyroid gland is found in about 80% of patients, and hyperplasia involving two or more glands in about 20% of patients with primary hyperparathyroidism. Parathyroid carcinoma occurs in 2% of patients. Familial forms of hyperparathyroidism, such as that associated with multiple endocrine neoplasia (MEN) syndromes and familial hypocalciuric hypercalcemia (FHH), are usually caused by hyperplasia of all parathyroid glands.

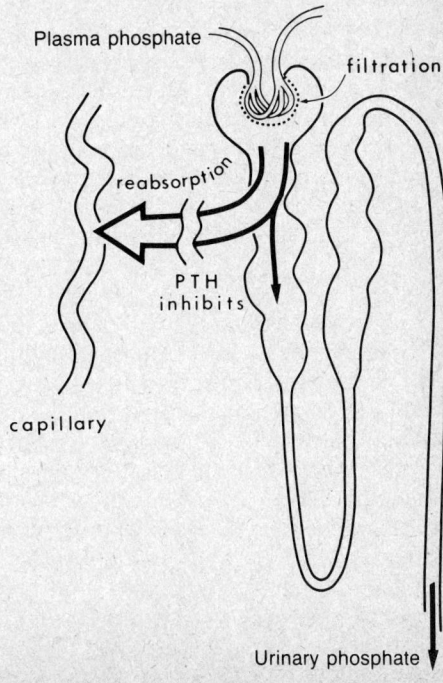

Figure 12-4. Diagram showing the effect of PTH on the handling of P_i by the nephron. (From Fraser, D., and Kooh, S. W., *In*: Texbook of Paediatrics, Vol. 2. 3rd. ed. J. O. Forfar and G. C. Arneil, Eds. Edinburgh, Churchill Livingstone, 1984, pp. 1157–1172.)

Table 12-1. PLASMA CHEMISTRY FINDINGS IN CONDITIONS
ASSOCIATED WITH HYPERCALCEMIA

	P_i	ALP	PTH	25-(OH)D	1,25-(OH)$_2$D
Primary hyperparathyroidism	↓ or N	↑ or N	↑	N	↑
Malignancy					
PTH-producing tumor	↓ or N	↑ or N	↑ or N		
Prostaglandin-producing tumor	↓ or N	↓ or N	↓ or N		
Skeletal metastasis of tumor	↑ or N	↑ or N	↓ or N		
Multiple myeloma, lymphoma, leukemia	↑ or N	↑ or N	↓ or N		
Local osteolytic metastasis	N	N	↓ or N		
Vitamin D toxicity	↑ or N	↓ or N	↓ or N	↑	N
Drug-induced hypercalcemia					
Thiazides, Ca-containing antacids, parenteral Ca	↓ or N	N	↓ or N		
Sarcoidosis	↓ or N	↑ or N	↓ or N	N	↑
Thyrotoxicosis	↑ or N	↑ or N	↓ or N		
Chronic renal failure	↑	↑ or N	↑	N	N or ↓
Adrenal insufficiency	N	N	↓ or N		
Immobilization	N	N	↓ or N		
Familial hypocalciuric hypercalcemia	↓ or N	N	↑ or N	N	N or ↑
Infantile hypercalcemia syndrome	↑ or N	↓ or N	↓	N	↓

Abbreviations: N, normal range; ↓, significantly decreased or undetectable; ↑, significantly increased; ALP, alkaline phosphatase.

Clinical Manifestations

Most patients with primary hyperparathyroidism either are asymptomatic or have vague, nonspecific symptoms such as weakness and fatigue. Manifestations are related to three causes:

1. Hypercalcemia. Hypercalcemia may cause nocturia, polyuria, and thirst as early signs, and anorexia, nausea, and vomiting if hypercalcemia is protracted and severe. Patients may have peptic ulcer, pancreatitis, hypertension, or symptoms related to calcium deposits in the eyes (band keratopathy), joints, and skin. Arthralgia may be due to deposits of calcium pyrophosphate crystals in the joint (chondrocalcinosis). When hyperparathyroidism is a manifestation of MEN (multiple endocrine neoplasia) Type I, in which tumors of pituitary and pancreas are also present, the patient may have Zollinger-Ellison syndrome (excessive gastric acid secretion, gastrointestinal ulceration, and pancreatic islet cell tumors). When hyperparathyroidism is part of MEN Type IIa,* it is associated with medullary thyroid carcinoma (MTC) and pheochromocytoma. In MEN Type IIb,* multiple neuromas of tongue, lips, and intestine are present in addition to hyperparathyroidism, MTC, and pheochromocytoma.

2. Kidney lesions. Apart from polyuria and polydipsia, stone formation and nephrocalcinosis are the most common renal manifestations. Renal colic occurs in about one third of patients with primary hyperparathyroidism. Chronic renal failure is now an uncommon complication.

3. Bone lesions. In present practice, the diagnosis of primary hyperparathyroidism is usually made before bone lesions become apparent. Patients with longstanding hyperparathyroidism may occasionally have bone pain or pathological fractures. Skeletal X-rays may show subperiosteal erosion, generalized demineralization, and cyst formation (brown tumors).

Laboratory Findings

Hypercalcemia is the hallmark of hyperparathyroidism. Those patients with so-called normocalcemic hyperparathyroidism usually have increased Ca_I concentrations[82] or hypercalcemia may be discovered intermittently during the day, as a result of episodic PTH hypersecretion or from consumption of a calcium-containing meal. For these reasons, the importance of conducting multiple calcium determinations and determining plasma Ca_I cannot be overemphasized. True normocalcemic primary hyperparathyroidism can occur, however, if the patient has co-existing vitamin D deficiency or some other condition in which the bone action of PTH is blunted. The best discrimination between hyperparathyroid and euparathyroid individuals is achieved by measuring Ca_I.[10,31,73,76,82] However, the "corrected Ca_T" value,[98] calculated by taking into account total

---

*Some investigators classify MEN Type IIa and Type IIb as MEN Type II and MEN Type III, respectively.

protein concentration,* somewhat improves discrimination between hyperparathyroid and eu-parathyroid individuals.

Hypophosphatemia is present in about 50% of patients with hyperparathyroidism. Plasma P_i may be normal, particularly in patients with diminished renal function; it may be elevated when severe renal failure supervenes. *Hyperchloremic acidosis* occurs in about one third of patients. It is due to the inhibitory action of PTH on renal tubular reabsorption of bicarbonate. In contrast, patients with hypercalcemia from other causes tend to develop metabolic alkalosis and a low plasma P_i concentration. The plasma urate concentration may be elevated. The plasma *alkaline phosphatase* (ALP) activity is usually normal. However, if bone lesions have developed, ALP tends to be increased roughly in proportion to the degree of skeletal involvement. In some patients with bone disease, ALP may be normal.

Urinary excretion of calcium is usually increased. This is due to an increased filtered load of calcium. The magnitude of hypercalciuria is less in hyperparathyroidism than in hypercalcemia from other causes because PTH enhances tubular reabsorption of calcium. If urinary calcium excretion is low, (< 2.5 mmol/d, < 100 mg/d) in the presence of hypercalcemia, the diagnosis of familial hypocalciuric hypercalcemia (FHH) should be considered. Renal phosphate clearance, as assessed by tubular reabsorption of phosphate (% TRP), is usually decreased in hyperparathyroidism. About 80% of hyperparathyroid patients excrete increased amounts of cAMP in the urine. The plasma $1,25\text{-}(OH)_2D$ concentration tends to be elevated but levels may overlap with normal values.

The plasma PTH concentration is raised in primary hyperparathyroidism and its determination is extremely important in diagnosis. However, many factors must be considered in interpreting a given plasma PTH radioimmunoassay value. This aspect is dealt with separately under Clinical Significance of Parathyroid Hormone Determinations (see page 1336).

Differential Diagnosis

The differential diagnosis of primary hyperparathyroidism includes conditions associated with hypercalcemia, hypophosphatemia, skeletal demineralization, and renal calculi. In making the diagnosis of primary hyperparathyroidism, other causes of hypercalcemia should be systematically eliminated. (See Table 12-1.)

Treatment

Patients in whom hyperparathyroidism has had adverse effects on the kidney, bone, or soft tissues should be treated by parathyroidectomy. However, if the patient is asymptomatic, there is no consensus whether or not he should be subjected to parathyroidectomy. Successful removal of an adenomatous gland is curative. Plasma Ca usually starts to fall within 12 h of surgery, reaches the nadir in about 5 d, and then rises fairly promptly to normal. About 5% of patients become permanently hypoparathyroid.

HYPERCALCEMIA OF MALIGNANCY

Hypercalcemia is a common complication in patients with malignant disease.[4,12,50,91] Estimates indicate that 10–20% of patients with malignancy have increased plasma Ca concentrations. Malignant disease and primary hyperparathyroidism account, in approximately equal numbers, for the vast majority of hypercalcemic individuals in the general population.

Malignancies associated with hypercalcemia are usually classified into three categories:

1. Solid tumor without metastasis (10% of patients). The most usual malignancies associated with this category are squamous cell carcinomas of the lung, as well as tumors of the head, neck, kidney, and ovary. The majority of patients in this group have increased urinary excretion of nephrogenous cAMP, suppressed plasma PTH, low plasma $1,25\text{-}(OH)_2D$ concentration, and decreased TRP. Complete removal of the tumor corrects the hypercalcemia. These findings indicate that such tumors secrete a humoral factor which activates PTH receptors; indeed, this particular

---

$$*\text{Ca (corrected), mg/dL} = \frac{\text{Ca(observed), mg/dL}}{0.6 + \dfrac{\text{total protein, g/dL}}{18.5}}$$

This relationship does not hold, however, for conditions discussed on page 1332.

condition is referred to as *humoral hypercalcemia of malignancy*. There is evidence that the humoral factor is a macromolecular protein that can mobilize calcium and stimulate renal adenylate cyclase in vitro. Biological activity is detected in the sensitive cytochemical bioassay referred to below, but the protein does not cross-react with various antisera used in the PTH radioimmunoassay.

2. Solid tumor with metastasis (70% of patients). In this, the most common category of hypercalcemia of malignancy, the tumor has metastasized to the skeleton and has caused bone destruction. Bronchogenic and mammary carcinoma are the most common types in this category.

3. Hematologic malignancies (20% of patients). Multiple myeloma is the most common neoplasm in this group.

The cause of hypercalcemia in categories (2) and (3) is controversial. Resorption of bone mineral by the invasive tumor tissue is a suggested mechanism. The macromolecular humoral factor mentioned above is not demonstrable. Several other factors have been suggested as potential mediators: ectopic PTH, prostaglandins of the E series, osteoclast activating factor (OAF), and the vitamin D metabolites. The role of these factors in humans is controversial.

Clinical Features

As with other types of hypercalcemia, the clinical symptoms are mainly gastrointestinal, renal, and neurological. However, hypercalcemia of malignancy runs a more rapidly progressive course and the levels of plasma Ca_T and Ca_I are higher than in primary hyperparathyroidism. Patients tend to show symptoms of hypercalcemia at lower levels of plasma Ca, probably because of the more rapid onset of hypercalcemia in malignancy than in primary hyperparathyroidism. On the whole, patients with hypercalcemia of malignancy are sicker and lose more weight than patients with primary hyperparathyroidism, but they are less likely to have renal calculi.

Laboratory Findings

No single laboratory feature distinguishes hypercalcemia of malignancy from hypercalcemia due to other causes including primary hyperparathyroidism. However, some features suggest hypercalcemia of malignancy. In the early stages of multiple myeloma, the plasma Ca_T may be elevated, as a reflection of an increase in abnormal plasma protein, without any increase in the plasma Ca_I concentration. In other forms of malignancy, plasma Ca_T and Ca_I concentrations tend to be more elevated and more variable than in primary hyperparathyroidism. The plasma P_i concentration in hypercalcemia of malignancy may be elevated, normal, or decreased; it is likely to be decreased in primary hyperparathyroidism. ALP is often elevated in hypercalcemia of malignancy, but is usually normal in primary hyperparathyroidism without bone lesions. The plasma chloride concentration is often < 103 mmol/L in hypercalcemia of malignancy, while this value is > 103 mmol/L in primary hyperparathyroidism. In consequence, the ratio of Cl (mmol/L) to P_i (mmol/L) is > 90 in over half the patients with hypercalcemia of malignancy, whereas the ratio is likely to be < 90 in primary hyperparathyroidism. The plasma HCO_3^- content is elevated or normal in hypercalcemia of malignancy; it is normal or decreased in hyperparathyroidism.

Typically, plasma PTH concentrations are elevated in primary hyperparathyroidism and low in hypercalcemia of malignancy. However, as indicated below, PTH immunoassays vary in their ability to differentiate the two conditions. The discrimination is improved when the PTH value is correlated with the corresponding serum Ca value. (See Figure 12-5.) The importance of employing a PTH immunoassay with the demonstrated ability to differentiate primary hyperparathyroidism and hypercalcemia of malignancy cannot be overemphasized (see page 1336).

Measurements of the tubular reabsorption of phosphate and urinary cAMP excretion are usually not helpful in the diagnosis of hypercalcemia of malignancy, although a % TRP value > 90 would make the diagnosis of primary hyperparathyroidism unlikely.

Plasma 25-(OH)D levels are normal. The 1,25-$(OH)_2$D levels tend to be low in hypercalcemia of malignancy; this is in contrast to elevated levels observed in primary hyperparathyroidism.

Treatment

Although elimination of hypercalcemia by medical or surgical management of the underlying tumor is the objective, it is rarely achieved. Attention must therefore be directed toward symptomatic management of hypercalcemia. Treatment includes fluid therapy, diuretics, or administration of phosphate, glucocorticoids, calcitonin, diphosphonates, and mithramycin.

Figure 12-5. Graph showing correlations of serum PTH and serum Ca concentrations in patients with primary hyperparathyroidism (1° HPT) (●), cancer (□), hypoparathyroidism (hypo) (○), chronic renal failure (▲), parathyroid carcinoma (*), and hypernephroma (H). (From Kao, P. C., Jiang, N., Klee, G. G., et al.: Clin. Chem., 28:69–74, 1982.)

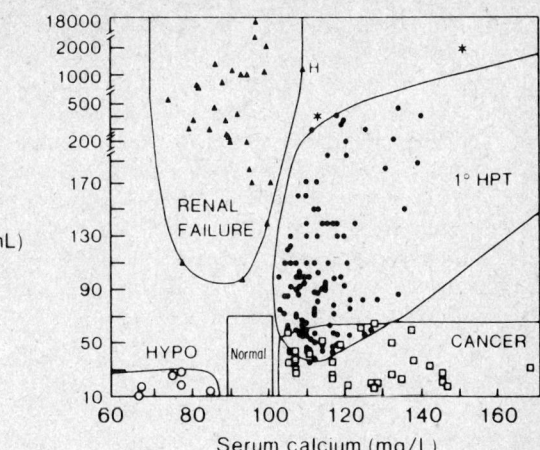

DISEASES ASSOCIATED WITH HYPOCALCEMIC STATES

The main diseases associated with hypocalcemia and the salient plasma chemistry findings are listed in Table 12-2. Some of the more important conditions are discussed in greater detail below.

HYPOPARATHYROIDISM AND PSEUDOHYPOPARATHYROIDISM

Hypoparathyroidism results from deficient secretion of PTH and pseudohypoparathyroidism from impaired response of the target organs to the hormone.[21,41] In both conditions, hypocalcemia and hyperphosphatemia are present. Plasma Ca_T is usually < 1.75 mmol/L (7.0 mg/dL), Ca_I < 0.8 mmol/L (3.2 mg/dL). In children with hypoparathyroidism, plasma P_i levels are usually > 2.0 mmol/L (6.2 mg/dL); the degree of plasma P_i elevation is less marked in pseudohypoparathyroidism, and in adults with hypoparathyroidism.

The plasma PTH concentration is low or undetectable in hypoparathyroidism. The concentration is often found to be elevated in pseudohypoparathyroidism, but values that fall within the reference range of the particular RIA are not uncommon. In a few patients, differentiation between hypoparathyroidism and pseudohypoparathyroidism is impossible by clinical examination and standard biochemical tests including PTH radioimmunoassay. The PTH response test is required in this situation. However, in many instances, the history, clinical examination, and standard laboratory tests permit the differentiation to be made. Most patients with idiopathic hypoparathyroidism have, or will eventually develop, additional endocrine deficiencies or other manifestations of autoimmune states. Among these manifestations are Addison's disease, alopecia, ovarian failure, and mucocutaneous candidiasis (MEDAC syndrome), the presence of which help to differentiate these patients from individuals with pseudohypoparathyroidism.

Table 12-2. PLASMA CHEMISTRY FINDINGS IN CONDITIONS
ASSOCIATED WITH HYPOCALCEMIA

	P_i	ALP	PTH	25-(OH)D	1,25-(OH)$_2$D
Hypoparathyroidism	↑	↓ or N	↓	N	↓
Pseudohypoparathyroidism					
Type 1	↑	N	↑ or N	N	↓
Type 2	↑	N	↑ or N	N	
Neonatal hypocalcemia	↑	N	Variable	N	
Hypomagnesemia	N or ↑	N	↓	N	
Vitamin D deficiency, and malabsorption syndromes	↓	↑	↑	↓	N or ↓
Renal failure	↑	↑ or N	↑	N	N or ↓
Hereditary vitamin D dependency rickets					
Type I	↓	↑	↑	N	↓
Type II	↓	↑	↑	N	↑

Abbreviations: N, normal range; ↓, significantly decreased or undetectable; ↑, significantly increased.

The pathogenesis of hypoparathyroidism includes autoimmune destruction of parathyroid glands, inadvertent removal or damage during thyroidectomy, and occasionally congenital aplasia.

Pseudohypoparathyroidism is a hereditary disorder characterized by symptoms and biochemical signs of hypoparathyroidism, although production of PTH is normal and administration of exogenous PTH does not correct the abnormalities. The condition is caused by a failure of the target tissues to respond to PTH. The patient often has characteristic somatic and mental abnormalities (so-called Albright's hereditary osteodystrophy or AHO), the manifestations of which include short stature, short metacarpals, round face, soft tissue calcifications, primary ovarian dysgenesis, and mental defect. Pseudohypoparathyroid patients with AHO have a 50% reduction in N-protein activity in red blood cell membranes or cultured skin fibroblasts; those without AHO have normal N-protein values.[35,116]

The defect in target tissue response to PTH can be demonstrated by the *PTH response test.*[21,78] Synthetic bPTH 1–34, 1.0 μg/kg body weight (~7 USP units/kg) to a maximum dose of 40 μg, is given intravenously over a 15-min period. Urinary excretion rates of cAMP and P_i (the latter measured as % TRP) are determined for two 1-h periods before the infusion of PTH, and for two half-hour and three 1-h periods after the infusion. Typical data are shown in Figure 12-6.

In *hypoparathyroidism,* as in normal individuals, PTH administration causes an immediate marked (50-fold) increase in the rate of urinary cAMP excretion; in *pseudohypoparathyroidism Type 1,* PTH causes very little or no increase in urinary cAMP excretion. Thus, urinary cAMP excretion in response to PTH challenge reliably differentiates hypoparathyroidism from pseudohypoparathyroidism. Additionally, the phosphaturic effect tends to be blunted in pseudohypoparathyroidism Type 1 in comparison to that observed in hypoparathyroidism. However, measurement of the phosphaturic response to PTH is not a reliable means of differentiation because individual responses frequently overlap in the two conditions.

A rare and poorly characterized sub-group of pseudohypoparathyroidism has been reported and named Type 2 pseudohypoparathyroidism. In the Type 2 condition, administration of PTH causes normal nephrogenous cAMP production but a markedly diminished phosphaturic response. A marked increase in urinary cAMP excretion in conjunction with little or no phosphaturic response in a patient with hypocalcemia, hyperphosphatemia, and normal or elevated plasma PTH is consistent with a diagnosis of Type 2 pseudohypoparathyroidism.

Pseudopseudohypoparathyroidism is a term that has been applied to individuals with the somatic and mental defects of pseudohypoparathyroidism (i.e., AHO), but without hypocalcemia or hyperphosphatemia. As far as we are aware, however, all such patients have failed to show a rise in urinary cAMP excretion with the PTH response test, and all patients have shown a 50% reduction in erythrocyte N-protein activity. In our opinion, the term pseudopseudohypoparathyroidism should no longer be used; such patients should be considered to have normocalcemic pseudohypoparathyroidism.

As noted previously, N-protein has an integral role in regulating the activation of adenylate cyclase by polypeptide hormones. It is therefore not surprising that pseudohypoparathyroid patients with AHO characteristically have other endocrine deficiencies such as hypothyroidism and ovarian failure.

HYPOPARATHYROIDISM   PSEUDOHYPOPARATHYROIDISM

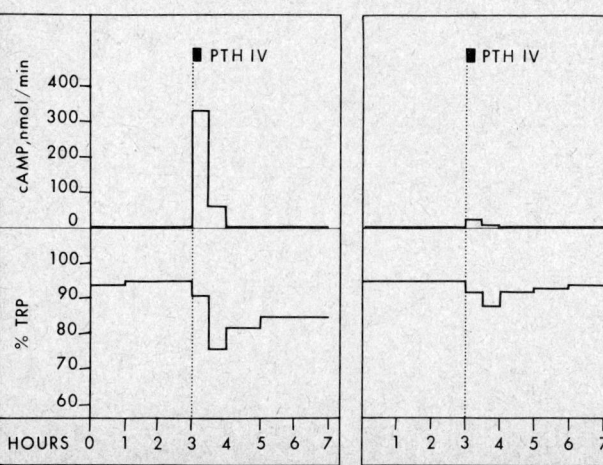

Figure 12-6. Typical results of PTH response test in a patient with untreated hypoparathyroidism (left-hand panel) and a patient with untreated pseudohypoparathyroidism (right-hand panel). The increase in urinary cAMP excretion (upper part of panel) and decrease in renal tubular reabsorption of phosphate, % TRP (lower part of panel), following a single IV dose of PTH are compared for the two conditions. (From Fraser, D., and Kooh, S. W., *In:* Textbook of Paediatrics, Vol. 2. 3rd ed. J. O. Forfar and G. C. Arneil, Eds. Edinburgh, Churchill Livingstone, 1984, pp. 1157–1172.)

VITAMIN D DEFICIENCY

Vitamin D deficiency, which was once commonplace, still occurs occasionally in North America. It is caused by insufficient skin exposure to sunlight and deficient oral intake of the vitamin. Neither human milk nor nonsupplemented cow's milk contains a significant amount of vitamin D. Therefore, breast-fed infants and individuals consuming strictly vegetarian diets and not exposed to sunlight are at risk of developing vitamin D deficiency. Dairy milk and infant formulas in North America are enriched with vitamin D and are effective sources of this essential vitamin.

Three stages of vitamin D deficiency can be distinguished.[42] *Stage I* deficiency involves impaired intestinal calcium absorption and diminished calcium mobilization from bone, with resultant hypocalcemia. Plasma P_i is not usually affected at this stage, and skeletal lesions on radiographs are minimal. In *Stage II* deficiency, hypocalcemia stimulates PTH secretion which returns the plasma Ca_T and Ca_I to normal or nearly normal, but the secondary hyperparathyroidism results in hyperphosphaturia, hypophosphatemia, and aminoaciduria. The diminished concentrations of calcium and phosphate contribute to the classical rachitic and osteomalacic lesions. In *Stage III* deficiency, vitamin D stores have become severely depleted; the consequent widespread rachitic lesions and absence of vitamin D make the patient less sensitive to the plasma-calcium raising actions of PTH. Hypocalcemia is again present in this advanced stage of rickets, while PTH-stimulated hypophosphatemia and aminoaciduria persist.

The plasma 25-(OH)D level, which is the most practical measure of the state of vitamin D nutrition, is low or undetectable in all stages of vitamin D deficiency. In response to the stimulus of secondary hyperparathyroidism, plasma 1,25-$(OH)_2D$ concentrations are often within the normal range or even elevated during the early stages of vitamin D deficiency; the level becomes low or undetectable as vitamin D deficiency advances.

HEREDITARY VITAMIN D DEPENDENCY RICKETS, TYPE I[118]
(PSEUDO-VITAMIN D DEFICIENCY RICKETS)

Hereditary vitamin D dependency rickets Type I is characterized by failure to convert 25-(OH)D to 1,25-$(OH)_2D$, because of a recessively inherited defect in 25-(OH)D-1-hydroxylase.[118] The disease resembles Stage III vitamin D deficiency and is characterized by hypocalcemia, hypophosphatemia, increased serum alkaline phosphatase activity, secondary hyperparathyroidism, generalized aminoaciduria, and severe rachitic bone lesions. The plasma 25-(OH)D concentration is normal but the 1,25-$(OH)_2D$ concentration is extremely low.

An essential feature of the disorder is complete healing of the biochemical and skeletal lesions with massive doses of vitamin D or 25-$(OH)D_3$, or with very small amounts of synthetic 1α,25-$(OH)_2D_3$ (trivial name, calcitriol*) or 1α-hydroxyvitamin D_3.†

HEREDITARY VITAMIN D DEPENDENCY RICKETS, TYPE II

Hereditary vitamin D dependency Type II is a rare autosomal recessive form of rickets.[32] As with Type I, patients have hypocalcemia, hypophosphatemia, aminoaciduria, increased serum alkaline phosphatase activity, raised plasma PTH, and severe rickets. The plasma 25-(OH)D level is normal, but in contrast to Type I, the plasma 1,25-$(OH)_2D$ concentration is raised, sometimes to a level 5–10 times the upper limit of reference values. The pathogenesis is refractoriness of end organs to 1,25-$(OH)_2D$, due to deficiency of cytosolic receptors for 1,25-$(OH)_2D$.

MALABSORPTION OF VITAMIN D

Rickets and osteomalacia sometimes complicate chronic hepatobiliary and gastrointestinal diseases, because bile salts and intact intestinal mucosa are essential for efficient intestinal absorption of vitamin D. The severe bone disease that occurs in longstanding liver disease appears to be caused by malabsorption of vitamin D more than by defective hepatic conversion of vitamin D to 25-(OH)D.[71]

---

*Rocaltrol, Hoffmann-La Roche, Ltd.

†One-Alpha, Leo Laboratories, Ltd.

NEONATAL HYPOCALCEMIA

The plasma Ca_T level in cord blood is $\sim$10% higher than in maternal plasma (1.4 mmol/L vs 0.95 mmol/L) and the cord blood Ca_I is appreciably higher than the maternal value (5.6 mg/dL vs 3.8 mg/dL). Plasma Ca_T level decreases considerably during the first 24–72 h after birth and reaches a value of 1.9–2.1 mmol/L (7.6–8.4 mg/dL); plasma Ca_I decreases proportionately.[107] The level then gradually increases to the normal infant range by 5 or 6 d of age. The plasma P_i concentration of the term infant is much above that in childhood and may rise to 2.25 mmol/L (7 mg/dL).

During the first days of life, about 1% of full term infants have significant, usually symptomatic, hypocalcemia (plasma Ca_T < 1.75 mmol/L, < 7.0 mg/dL).[39] The incidence of more marked hypocalcemia is much greater in premature infants and infants with perinatal complications. Twitching, hyperirritability, and convulsions are the symptoms. Symptomatic hypocalcemia usually starts between 12 h and 4 d after birth (early neonatal hypocalcemia). In the less common "late" form, onset is between 4 and 28 d of age.

Neonatal hypocalcemia may be the first sign of a more serious form of parathyroid gland hypofunction such as that associated with DiGeorge syndrome, primary hypomagnesemia, permanent isolated hypoparathyroidism, or transient hypoparathyroidism due to maternal hyperparathyroidism.

HYPOMAGNESEMIA

Hypomagnesemia, either primary or secondary, can cause hypocalcemia. There is evidence in support of two suggested mechanisms for the hypocalcemia—defective secretion of PTH or end organ unresponsiveness to the hormone. With a plasma Mg concentration of < 0.3 mmol/L (0.7 mg/dL) hypocalcemia may often develop. The plasma P_i concentration in hypomagnesemia varies; both high and low levels have been reported. Plasma PTH levels are low in hypocalcemic hypomagnesemia.[122] This type of hypocalcemia can be corrected only by magnesium supplementation.

CONDITIONS IN WHICH DETERMINATION OF PLASMA Ca_I IS ESSENTIAL

It is the ionized calcium which has physiologic effects. Ordinarily, Ca_I is a fairly constant fraction of Ca_T of plasma and measurement of the plasma Ca_T is adequate to define the physiologically effective calcium concentration in the blood. However, the proportion of Ca_I:Ca_T depends on the degree to which calcium is bound by proteins and other blood constituents; the degree of binding depends on concentrations of ligands and on the blood pH. Many clinical situations occur in which pH and ligand concentrations are sufficiently altered that Ca_I is abnormal, despite normal plasma Ca_T levels.

In the newborn infant, particularly the premature, plasma protein concentrations are low and pH aberrations are common; hence Ca_T concentrations may not be indicative of the physiologically important ionized calcium fraction.[24,108] Transfusions with large volumes of blood or plasma containing calcium-binding anticoagulants, as with exchange transfusions and cardiopulmonary bypass surgery, are often associated with sudden marked depression of the Ca_I concentration without a corresponding decrease in Ca_T.[22,65,96] ACD (acid-citrate-dextrose) and CPD (citrate-phosphate-dextrose) blood transfusions can cause ionized hypocalcemia. As a result, acute myocardial depression has been documented during rapid blood transfusions.[96] Patients who are treated with β-adrenergic blocking agents are particularly sensitive to cardiac depression by hypocalcemia. On the other hand, the increase in Ca_I that occurs after hemodialysis with a high-calcium dialysate is associated with improved cardiac contractility.[53] Hemodialysis may produce acute depression of plasma Ca_I unless the dialysate contains an adequate calcium concentration.

Evaluation of the Ca_I concentration may be made either by direct measurement by electrode or by calculating an estimate based on protein concentrations and pH, i.e., Ca_T (corr). Although formulas have been suggested for the calculation, there are many situations in which the formulas do not apply.[18] For example, in the growing child, plasma Ca_I is higher relative to the Ca_T. Ca_T

(corr) does not always correctly diagnose hypercalcemia in hyperparathyroidism,[73] chronic renal failure,[18] and liver disease.[87] Furthermore, the commonly used plasma protein methods tend to have greater error than most calcium methods.

DISEASES ASSOCIATED WITH HYPERPHOSPHATEMIA

The main factors that influence the plasma P_i level are intake, distribution within the body compartments, glomerular filtration rate, and reabsorption by the renal tubules. In deciding whether a given plasma P_i result is normal, elevated, or diminished, it is essential to take into account the significant influences of age and sex. Conditions associated with hyperphosphatemia and hypophosphatemia are listed in Table 12-3. Hyperphosphatemia per se does not result in any definable disorder. The abnormalities attributable to hyperphosphatemia are secondary to changes in calcium homeostasis. They include secondary hyperparathyroidism, soft tissue calcification, and possibly tumoral calcinosis.

RENAL OSTEODYSTROPHY

The term "renal osteodystrophy" (ROD) refers to the skeletal lesions that occur in chronic renal failure. Rickets, osteomalacia, hyperparathyroid bone changes (osteitis fibrosa), osteosclerosis, and osteoporosis are present in varying proportions.[17] ROD is described here because hyperphosphatemia plays a pivotal role in its pathogenesis.

Pathogenesis

In chronic renal failure, increased plasma PTH is one of the earliest and most consistent findings; it tends to occur even with relatively mild failure (e.g., GFR 50 mL/min per 1.73 m^2). Hyperparathyroidism is secondary; it is caused by a reduction in the concentration of plasma Ca_I. A number of factors contribute to this hypocalcemia. Phosphate retention due to reduction in the filtered load of phosphate is an important cause. Intestinal absorption of calcium is decreased because of impaired synthesis of 1,25-$(OH)_2D$. The bones become resistant to the calcium-mobilizing action of PTH.

The rachitic and osteomalacic components of chronic renal failure are not completely understood. Impaired synthesis of 1,25-$(OH)_2D$ alone cannot explain defective mineralization since there is no proof that 1,25-$(OH)_2D$ directly stimulates bone mineralization. Other factors such as metabolic acidosis and retention of pyrophosphate may contribute to the lesions. Pyrophosphate retention interferes with bone mineral formation.

Laboratory Findings

Plasma calcium. Plasma Ca_T and Ca_I concentrations are usually within the normal range because of secondary hyperparathyroidism. However, normocalcemia is achieved at the expense of excessive bone resorption. When hypocalcemia occurs it is usually mild and in inverse relation to the level of plasma creatinine. Severe hypocalcemia (<1.5 mmol/L, <6 mg/dL) is seen occasionally. Hypercalcemia is rarely observed in uncomplicated renal failure; when present, it may be caused by severe secondary hyperparathyroidism (tertiary hyperparathyroidism) or by aggressive therapy with vitamin D or calcium. Occasionally, it may signal the presence of pre-existing primary hyperparathyroidism or sarcoidosis.

Plasma magnesium. Plasma Mg concentrations are within the normal range in mild renal failure. However, with advancing disease (GFR <30 mL/min per 1.73 m^2) the concentration is likely to be increased. Hypomagnesemia is rare, but may be seen if the patient has a very poor food intake, suffers from intestinal malabsorption, or has a magnesium-wasting lesion of the kidney.

Plasma inorganic phosphate. In the early stages of chronic renal failure, the plasma P_i concentration is usually maintained within the normal range by the increased levels of PTH; occasionally, it is low. As renal failure advances, the plasma P_i concentration is usually increased.

Alkaline phosphatase. Increased plasma ALP is not common in chronic renal failure; many patients with clear-cut skeletal involvement show normal alkaline phosphatase activities. Elevated levels are seen in 10–15% of patients, occur late in the disease, and tend to be related to the severity of the skeletal lesions.

Table 12-3. CONDITIONS ASSOCIATED WITH HYPO- AND HYPERPHOSPHATEMIA

Hyperphosphatemia	Hypophosphatemia
	Intake
Poisoning by phosphate-containing enema or laxative	Deficiency of dietary phosphate
	Phosphate malabsorption in dialyzed patients, alcoholics
	Overuse of phosphate-binding agents
	Total parenteral alimentation (TPN)
	Redistribution
Respiratory acidosis	Glucose infusion
Lactic acidosis	Treatment of diabetes mellitus
Diphosphonate therapy	Respiratory alkalosis
	β-Adrenergic agents
	Increased skeletal uptake in healing phase of rickets
	Osteoblastic metastases of cancer
	Nutritional recovery syndrome
	Diuretic phase of severe burns
Renal Causation	
Acute and chronic reduction in glomerular filtration rate	Specific phosphate transport defect:
Hypoparathyroid state	X-linked dominant hypophosphatemia
Primary hypoparathyroidism	Autosomal dominant hypophosphatemia
Pseudohypoparathyroidism	Multiple renal tubular transport defect
Suppression of PTH secretion from any hypercalcemic	Idiopathic Fanconi syndrome
condition	Cystinosis
	Hereditary fructose intolerance
	Hereditary tyrosinemia
	Galactosemia
	Wilson's disease
	Oculocerebrorenal (Lowe's) syndrome
	Phosphaturia due to primary or secondary hyperparathyroidism
	Primary hyperparathyroidism
	Secondary hyperparathyroidism due to hereditary vitamin D
	dependency, Types I and II
	Hypocalcemia from any cause, provided parathyroid glands
	are intact
	Miscellaneous Causes
Vitamin D intoxication	Tumor-induced hypophosphatemia
Acromegaly	Post-transplantation hypophosphatemia
Tumoral calcinosis	Hypercalciuric nephrolithiasis
Cortical hyperostosis	

PTH. The plasma PTH concentration is invariably increased in chronic renal failure, even early in the disease. The increase is caused by secondary hyperparathyroidism and may also reflect decreased renal clearance of C-terminal fragments of PTH. With assays directed toward the N-terminal fragments, elevation in the plasma PTH level is less marked.

DISEASES ASSOCIATED WITH HYPOPHOSPHATEMIA

Severe phosphate depletion can cause neuromuscular, skeletal, and other abnormalities. Muscle abnormalities include weakness, fatigue, and cardiomyopathy. Neurological manifestations include paresthesia, confusion, ataxia, convulsions, coma, decreased nerve conduction velocity, and electroencephalographic abnormalities. Phosphate deficiency may cause reduction in red blood cell 2,3-diphosphoglycerate and probably accounts for the hemolytic anemia occasionally observed. Decreased ATP in leukocytes and platelets may explain the diminished chemotaxis and clot retraction sometimes seen in hypophosphatemic individuals. Chronic hypophosphatemia from any cause will eventually result in rickets and osteomalacia.

FAMILIAL HYPOPHOSPHATEMIC RICKETS AND OSTEOMALACIA

The most common examples of chronic rickets and osteomalacia from this cause are X-linked hypophosphatemia and autosomal dominant hypophosphatemia.[118]

X-linked hypophosphatemia results from a selective inborn defect in transepithelial phosphate transport, located in the proximal renal tubule cells. The patient usually has skeletal deformities, short stature, and radiographic and histological evidence of longstanding rickets; some females with X-linked hypophosphatemia have no physical or radiographic abnormalities.

Autosomal dominant hypophosphatemia is a less common condition in which the pathology of the defect in renal phosphate conservation is less well understood, and on the whole the severity of rickets is less marked.

The main chemical abnormality in the two conditions is hypophosphatemia. Plasma Ca_T and Ca_I are normal. Raised plasma PTH and generalized aminoaciduria, hallmarks of hereditary vitamin D dependence Type I, are absent, suggesting that hypophosphatemia is a primary abnormality. Plasma ALP is frequently but not always increased.

The diagnosis of familial hypophosphatemia depends upon demonstrating hypophosphatemia. The plasma P_i concentration of the normal individual is age- and sex-dependent, and this dependence must be taken into account in assessing the significance of the plasma P_i concentration. Since food intake also affects P_i concentration, a fasting blood sample should be taken. Defective renal tubular reabsorption of P_i is indicated by abnormally high urinary P_i clearance, diminished tubular reabsorption of P_i (low % TRP), and low TmP_i/GFR (ratio of transport maximum for P_i vs glomerular filtration rate), even at low plasma P_i concentrations. Measurement of these parameters is described by Bijvoet.[7] The plasma 25-(OH)D concentration is normal, ruling out vitamin D deficiency. The plasma 1,25-$(OH)_2D$ concentration is normal. The significance of this observation is discussed on page 1339.

Currently, the recommended treatment is neutral phosphate in divided oral doses, combined with synthetic 1,25-$(OH)_2D_3$ or, instead, vitamin D in a supraphysiological dosage.

DISEASES ASSOCIATED WITH NORMOCALCEMIC STATES

Conditions discussed under this heading have either normal or variable plasma Ca concentrations.

OSTEOPOROSIS

Osteoporosis is a general term used to describe a group of diseases of diverse causation characterized by reduction of bone mass below that required for normal mechanical function. The ratio of mineral to organic matrix in bone is normal. Osteoporosis can be secondary to well-defined conditions such as Cushing's disease and to glucocorticoid administration and chronic immobilization. More often it is not associated with any other identifiable disease and is then called primary osteoporosis.

Laboratory Findings

Characteristically, the usual laboratory tests, including plasma Ca and P_i, are normal in osteoporosis. Hypercalcemia may occur transiently when the patient is immobilized. Increased ALP may occur in patients who have sustained fractures. Decreased levels of plasma 1,25-$(OH)_2D$ have been reported in some osteoporotic individuals.

PAGET'S DISEASE OF BONE

Paget's disease is among the most common of the nonmetabolic bone diseases. It is characterized by regions of excessive bone resorption and repair, increased turnover of bone minerals and matrix, and formation of abnormal bone. Evidence for a viral etiology has recently been presented.[86a] The condition is often discovered incidentally during radiographic examination of the pelvis or spine for unrelated disease or by the finding of increased serum ALP upon biochemical screening. Deep bone pain and headache are usually the first symptoms. Deformity, recurrent fractures, and swelling of a long bone or enlargement of the head may occur.

Laboratory Findings

The plasma Ca and P_i concentrations are usually normal. However, hypercalcemia develops occasionally; when present, it is related to immobilization or fracture. A higher than usual incidence of primary hyperparathyroidism and hypercalcemia has been noted. Serum ALP is usually markedly increased. Its elevation correlates with the extent and activity of the disease, and also with urinary hydroxyproline excretion. The urinary calcium excretion rate varies; when bone resorption predominates, it tends to be increased, and this may account for the high incidence of urolithiasis in the disease.

MEDULLARY THYROID CARCINOMA

Medullary thyroid carcinoma (MTC) is a malignancy of the calcitonin-secreting cells (C-cells).[6] It accounts for about one third of malignant thyroid tumors. Occurrence may be sporadic or familial (autosomal dominant). One third of the familial cases have primary parathyroid hyperplasia and/or pheochromocytoma (MEN Type IIa). Some families may in addition manifest multiple neuromas of buccal mucous membranes and intestine (MEN Type IIb). Despite excessive secretion of CT (plasma levels may be increased 1000-fold), the plasma Ca concentration is usually normal, probably because of down-regulation of the response to CT. If hypocalcemia does occur, it is related to diarrhea, a commonly associated feature. The plasma P_i concentration is usually normal.

CLINICAL SIGNIFICANCE OF PLASMA PTH DETERMINATIONS[3,4,52,85,103,127]

Small amounts of biologically active PTH are continually present in the plasma of healthy individuals. The rate of secretion of PTH may be increased or decreased because of primary pathological changes within the glands or because of changes in glandular function secondary to alterations in the body's calcium metabolism. Thus, a significant elevation in the circulating PTH concentration is consistent either with autonomous hypersecretion by disordered glands, i.e., primary hyperparathyroidism, or with overstimulation of functionally normal parathyroid glands, i.e., secondary hyperparathyroidism. Conversely, a PTH concentration below the normal range can be attributed either to diminished PTH secretion because of inherent hypofunction of the parathyroid glands (transient or permanent) or to secondary suppression of the secretory activity of normal glands in response to chronic hypercalcemia. Conditions that result in increased and decreased concentrations of plasma PTH are shown in Table 12-4.

Measurements of PTH should clearly distinguish patients with hyperparathyroidism and hypoparathyroidism from normal individuals and PTH-induced hypercalcemia from hypercalcemia due to other causes.[3,4,51,67,103,127] Unfortunately, a fully satisfactory means of achieving this

Table 12-4. CAUSES OF INCREASED AND DECREASED PLASMA PTH CONCENTRATIONS

Conditions in which plasma PTH concentration is elevated:
 Primary hyperparathyroidism
  Hyperplasia
  Adenoma
  Carcinoma
  Familial hypocalciuric hypercalcemia, familial neonatal primary hyperparathyroidism
  Medullary thyroid carcinoma
  MEN: Types I, IIa, and IIb
 Secondary hyperparathyroidism (in response to hypocalcemia of various causes)
  Vitamin D deficiency
  Hereditary vitamin D dependency rickets: Types I and II
  Chronic renal failure
  Pseudohypoparathyroidism
  Dietary calcium deficiency

Conditions in which plasma PTH concentration is diminished or undetectable
 Hypoparathyroidism
  Isolated
  Autoimmune (MEDAC syndrome)
  DiGeorge syndrome (sometimes temporary)
  Injury of parathyroid glands (hemochromatosis, thyroid or parathyroid surgery)
  Secondary to severe hypomagnesemia
  Neonatal
   Transient neonatal hypocalcemia
   Maternal hyperparathyroidism
  Secondary to hypercalcemia of various causes:
   Hypercalcemia of malignancy
   Infantile hypercalcemia syndrome
   Hypervitaminosis D
   Sarcoidosis
   Sudden immobilization
   Hyperthyroidism
  Hypomagnesemia

goal is not yet at hand. Radioimmunoassays are the only widely available procedures for measuring plasma PTH, but these have many shortcomings, and cannot be relied upon always to provide unequivocal information. There is optimism that recent sensitive bioassay procedures may provide a practical means of diagnosis, but such techniques are not yet widely available. Fortunately, data gained from selected RIA procedures, when combined with plasma Ca and other information about the patient, have proved of great help in establishing the diagnosis.

Numerous factors must be considered in interpreting RIA results. As described earlier, the relative concentrations of intact PTH and various PTH fragments vary from condition to condition and patient to patient. Furthermore, PTH assays vary markedly in their capacity to measure these various fragments depending upon assay sensitivity, and the sequence specificities of the particular antiserum used.[67,83,92] All immunoassays recognize the intact hormone, but intact hormone constitutes only a small proportion of the PTH-derived peptides in the plasma of the normal individual, and the concentration may be below the detection limit of many assays. A mid-molecule assay will react with fragments in the region of amino acids 35–64, and it will usually recognize long C-terminal fragments (e.g., 34–84); however, it does not crossreact with fragments 1–34 or 64–84. An N-terminal assay recognizes intact PTH and PTH 1–34, but does not react with mid-molecule and C-terminal fragments that make up most of the circulating PTH protein. C-terminal assays mainly recognize C-terminal fragments; however, many C-terminal assays cross-react also with mid-regional fragments and therefore measure multiple inert PTH fragments in the blood.

For any assay, regardless of its immunochemical specificity, the most important criterion of usefulness for diagnosis is a compilation of PTH values correlated with plasma Ca determinations in large numbers of patients with various disorders, such as hyper- and hypoparathyroidism and various other hypercalcemic and hypocalcemic disorders not due to parathyroid dysfunction. Only then is it possible to state whether a given radioimmunoassay has the ability to discriminate different clinical entities. Figure 12-5 presents such data for an assay with good discriminating capability.[67] It illustrates that the assay usually differentiates hypercalcemia due to primary hyperparathyroidism from that caused by malignancy, and almost always from that caused by renal failure.

The result of a PTH immunoassay should never be interpreted in isolation, but should be considered in relation to the corresponding plasma Ca_T and Ca_I concentrations. The findings should be interpreted in the light of the patient's history, physical findings, and other data. Indeed, there is a general tendency for clinicians to request the determination of PTH too frequently when, in fact, other evidence might better have resolved the question of diagnosis.

The choice of the radioimmunoassay procedure depends partially on the patient's tentative diagnosis. One might suppose that N-terminal radioimmunoassays would be the most effective for determining parathyroid function. Actually, PTH assays that measure mid- and C-terminal fragments tend to have greater diagnostic usefulness.[80,83,85] The poor correlation between N-terminal assays and parathyroid function may be because of the low circulating concentrations of intact PTH and PTH 1–34 which must be measured in the presence of substances that react nonspecifically with the antiserum. Hyperparathyroidism must be severe before the concentration of intact PTH rises above the "background noise."

N-terminal assays are mainly useful for samples in which the concentration of intact hormone is relatively high and that of inert fragments low (e.g., in selective jugular vein catheterization procedures for localization of parathyroid adenomas),[103] and to measure rapid changes in parathyroid gland function (e.g., suppression by intravenous calcium infusion or stimulation by intravenous EDTA administration—maneuvers performed in certain research applications to establish the relationship between plasma Ca_I and PTH.) N-terminal assays are also preferred by some physicians for evaluating patients for renal dialysis and those with early renal failure.

Assays directed toward the C-terminal and mid-molecule fragments are preferred for the differential diagnosis of hypercalcemia; when combined with plasma calcium determinations, as illustrated in Figure 12-5, selected assay systems show good discrimination between patients with primary hyperparathyroidism and those with hypercalcemia due to malignancy.[67] Although, in the example shown, a minor zone of overlap remains, the problem of diagnosis may not be as great as would first appear. Most patients with hypercalcemia caused by malignancy have grossly evident disease; it is very rare that hypercalcemia is caused by truly occult cancer.

For most PTH immunoassays, an overlap is seen between the PTH values of normal individuals and of those with primary hyperparathyroidism. Most assays show an elevated concentration of PTH in only 80–90% of patients with surgically proved primary hyperparathyroidism; the remaining 10–20% of affected individuals have PTH concentrations that fall within the normal range. Fortunately, other biochemical findings help in determining the diagnosis. Approximately 80–90% of patients with primary hyperparathyroidism can be shown to have elevated plasma Ca_T values on occasion, and 95% or more of patients have elevated plasma Ca_I values when plasma is tested repeatedly.[76,82] Furthermore, in those 10–20% of hyperparathyroid patients with normal immunoreactive PTH concentrations, the PTH value tends to be inappropriately high in relation to the corresponding plasma Ca concentration. In addition to measurement of plasma Ca_T, Ca_I, and PTH, a careful history, physical examination, and other tests appropriate to the clinical setting are important in establishing the diagnosis of primary hyperparathyroidism. It is generally conceded that other secondary manifestations of hyperparathyroidism (e.g., increased renal phosphate clearance and urinary cAMP excretion) are of no more than ancillary value for diagnosis.

There is disagreement as to the best assay to use in patients with chronic renal disease. Biologically inert mid-region and C-terminal fragments are retained to a greater extent than intact PTH and the N-terminal fragment; hence, C-terminal and mid-molecule assays are more likely to exaggerate the apparent degree of hyperparathyroidism than are N-terminal assays. On the other hand, it has been shown that mid-molecule and C-terminal assays correlate well with the bone histomorphometric criteria of secondary hyperparathyroidism, and thus are useful for monitoring the effectiveness of therapy in renal osteodystrophy.

Any assay of sufficient sensitivity, regardless of its immunologic characteristics, can be used when hypoparathyroidism is the tentative diagnosis. However, assay sensitivity is often a limiting factor in the diagnosis of hypoparathyroidism since many radioimmunoassays are not sufficiently sensitive to measure plasma PTH in all euparathyroid individuals.

CLINICAL SIGNIFICANCE OF ELEVATED LEVELS OF CALCITONIN IN PLASMA

The only clinical indications for measuring plasma CT are the various conditions associated with medullary thyroid carcinoma (MTC).[4,6] Measurement of the plasma CT concentration is particularly useful in screening asymptomatic relatives of patients with autosomal dominant MTC and in identifying postoperative recurrence of MTC at an early stage. The basal plasma CT concentration is usually increased in individuals who have MTC, sometimes as much as 1000-fold, but it may be normal or only intermittently increased. In such circumstances, hypercalcitoninemia can usually be established by conducting one, or preferably more than one, of the established provocative tests. Calcium, pentagastrin, or glucagon, administered by intravenous injection, is employed to stimulate secretion, and plasma CT is determined at prescribed times. (See Measurement of Calcitonin in Plasma, p. 1356.) Although these secretagogues cause a rise in plasma CT in normal individuals, the response of individuals with MTC is usually significantly greater. Because of differences in assays and variability of responses to the secretagogues, any laboratory that performs these tests must establish its own reference ranges.

Although a significantly increased plasma CT level is strongly suggestive of the presence of MTC, elevated CT can also occur in patients with chronic renal failure, pancreatic islet cell tumors, breast cancer, and small cell carcinoma of the lung.

CLINICAL SIGNIFICANCE OF CHANGES IN VITAMIN D METABOLITES IN PLASMA

Before vitamin D can exert an effect, several steps must occur: (1) synthesis of the parent vitamin in the skin or absorption of ingested vitamin D by the intestine, or both; (2) metabolism of vitamin D in the liver and kidney to produce $1,25\text{-}(OH)_2D$; (3) transport of vitamin D and its metabolites from one site to another by the plasma; and finally (4) response of the target cells to $1,25\text{-}(OH)_2D$.

A defect in any of these steps can lead to deficiency or excess of $1,25\text{-}(OH)_2D$ or its precursors, and both states can lead to disease. The plasma concentrations of vitamin D and its two main biologically active metabolites, $25\text{-}(OH)D$ and $1,25\text{-}(OH)_2D$, may be decreased; such a circumstance implies deficiency of the parent vitamin as the primary defect. Alternatively, their levels can change reciprocally, implying more subtle metabolic or regulatory alterations. Excessive amounts of vitamin D from excessive exposure to sunlight or ultraviolet light or from excessive ingestion of high potency vitamin preparations or enriched food can cause significantly increased plasma levels of vitamin D, $25\text{-}(OH)D$, and $24,25\text{-}(OH)_2D$, but not of $1,25\text{-}(OH)_2D$. The last observation is indicative of precise feedback control of $1,25\text{-}(OH)_2D$ synthesis, and accounts for the wide range of vitamin D intakes an individual is capable of tolerating. Mild to moderate elevation of vitamin D and $25\text{-}(OH)D$ in plasma is usually not accompanied by any evidence of pathological response. However, marked elevation of $25\text{-}(OH)D$ may cause vitamin D intoxication or, more correctly, $25\text{-}(OH)D$ intoxication. The intoxication is explained by the fact that a markedly elevated concentration of $25\text{-}(OH)D$ is known to exert a biological action on target cells similar to that of an elevated concentration of $1,25\text{-}(OH)_2D$. Measurements of vitamin D metabolites in plasma are useful to assess patient compliance in patients treated with vitamin D, $1,25\text{-}(OH)_2D_3$, or $1\alpha\text{-}(OH)D_3$, and to measure vitamin D absorption and its half-life.

Not all low or high vitamin D metabolite concentrations are indicative of an abnormality. Certain physiological states cause significant elevation of the plasma $1,25\text{-}(OH)_2D$ concentration above adult *reference values* (41–142 pmol/L; 17–59 pg/mL). In pregnancy, for example, the $1,25\text{-}(OH)_2D$ level is markedly elevated to a mean of 190 pmol/L, range of 140–300 pmol/L (80 pg/mL, range 60–120 pg/mL). It is higher in children and lower in the elderly than the adult reference value, and it is increased by low dietary calcium or phosphate, or both.

SIGNIFICANCE OF REDUCED LEVELS OF VITAMIN D METABOLITES IN PLASMA

In the presence of rickets, a low level of vitamin D, $25\text{-}(OH)D$, and $1,25\text{-}(OH)_2D$ implies a deficient supply of vitamin D which in turn signifies deficient dietary intake and inadequate exposure to sunlight. Alternatively, the low level may be due to intestinal malabsorption of vitamin D, as well as inadequate exposure to sunlight. There is no known example of defective cutaneous synthesis of vitamin D_3, nor do plasma protein defects result in inadequate transport of vitamin D to the liver. In states of mild to moderate vitamin D deficiency, in which the plasma 25-(OH)D concentration is decreased, the plasma 1,25-$(OH)_2$D may be normal or even increased. This situation arises because vitamin D deficiency results in a decrease in plasma Ca and P_i, and an increase in PTH, all of which cause increased efficiency of 1,25-$(OH)_2$D synthesis in the face of a gradually decreasing pool of 25-(OH)D. In the presence of a normal concentration of vitamin D in plasma, a low level of 25-(OH)D might suggest defective 25-hydroxylation in the liver. However, the liver possesses great reserve capacity for 25-hydroxylation;[71] furthermore, a genetic absence of 25-hydroxylase has not been described. The nephrotic syndrome is a reported cause of a low 25-(OH)D level. In this condition, there is excessive urinary excretion of DBP to which 25-(OH)D is bound. In such a patient, the 1,25-$(OH)_2$D level may be normal or low depending upon whether the patient does or does not have sufficient vitamin D intake to compensate for excessive losses. In patients receiving long-term phenytoin or phenobarbital therapy, plasma levels of 25-(OH)D may be low and bone disease may be present. It is speculated that anticonvulsants cause induction of hepatic microsomal enzymes that divert metabolism of 25-(OH)D to inactive metabolites. The 1,25-$(OH)_2$D level is often normal in such patients.

A low 1,25-$(OH)_2$D level in the face of a normal plasma 25-(OH)D level indicates a defect in 1-hydroxylation by the kidney. Chronic kidney disease is the most common explanation. Other causes are a genetic defect of the renal 1-hydroxylase enzyme, as in vitamin D dependency rickets (VDDR) Type I, or inadequate stimulation of renal 1-hydroxylase by PTH, as in hypoparathyroidism or pseudohypoparathyroidism.

Patients with familial hypophosphatemic rickets have 1,25-$(OH)_2$D levels that fall within the reference range; these "normal" levels of 1,25-$(OH)_2$D have been interpreted as inappropriately low, on the rationale that one would expect the hypophosphatemia of this condition to stimulate 1-hydroxylase and cause higher than normal levels of 1,25-$(OH)_2$D.

Other conditions in which low 1,25-$(OH)_2$D levels have been reported in the presence of adequate vitamin D intake are osteoporosis, tumor-induced hypophosphatemia, infantile hypercalcemia syndrome, and osteomalacia associated with total parenteral nutrition.

SIGNIFICANCE OF ELEVATED LEVELS OF VITAMIN D METABOLITES

Because of efficient feedback regulation of 1,25-$(OH)_2$D synthesis, elevated levels of 1,25-$(OH)_2$D do not usually occur, even in hypervitaminosis D. Elevated levels of vitamin D, 25-(OH)D, and 24,25-$(OH)_2$D without an elevated level of 1,25-$(OH)_2$D suggest excessive ingestion or excessive synthesis of vitamin D in the skin.

The finding of a high 1,25-$(OH)_2$D level without elevation of the 25-(OH)D level can usually be explained by one of the physiological adaptations discussed earlier. Alternatively, the patient may be receiving 1,25-$(OH)_2D_3$ therapy. Hereditary vitamin D dependency, Type II, is a very rare cause of a high 1,25-$(OH)_2$D level in plasma.[32] In this disease, compensatory elevation of 1,25-$(OH)_2D_3$ occurs because of defective 1,25-$(OH)_2$D-receptors in target cells; vitamin D and 25-(OH)D levels are normal unless the patient is receiving high dosage vitamin D treatment.

Mild elevations of 1,25-$(OH)_2$D can occur in hyperparathyroidism. Elevated levels are seen in some patients with renal calculus because of excessive renal loss of calcium. Sarcoidosis is accompanied by an elevation of the 1,25-$(OH)_2$D level; synthesis of additional 1,25-$(OH)_2$D by the sarcoid tissue is the explanation.

CLINICAL SIGNIFICANCE OF HYPERCALCIURIA AND HYPOCALCIURIA

In the steady state, the rate of urinary calcium excretion is determined by calcium transport functions in the intestine, bones, and kidneys. Thus, depending on the cause, hypercalciuria may be termed *absorptive,* due to increased intestinal calcium absorption; *resorptive,* due to increased mobilization of skeletal calcium; or *renal,* due to decreased tubular calcium reabsorption. The normal limits of urinary calcium excretion depend on the sex and age of the individual and on dietary calcium intake. Men excrete more calcium than women. Urinary calcium excretion decreases after age 60 in men and after age 50 in women, probably because of decreased intake of calcium and decreased bone mass. Dietary calcium intake influences urinary calcium excretion in normal individuals and the effect is exaggerated in patients with "absorptive" hypercalciuria.

Definition of "Hypercalciuria"

There is no universally accepted reference value for the 24-h urinary calcium excretion rate because of the numerous factors that influence excretion in healthy individuals.[7,55,95,115] Ninety per cent of healthy women excrete less than 6.25 mmol (250 mg) per day, and 90% of healthy men excrete less than 7.50 mmol (300 mg) per day. In children, this figure is equivalent to 0.1 mmol (4 mg)/kg body weight per day. Recent evidence suggests that the upper limit may be somewhat higher. In a voided urine sample, the ratio of the concentration of calcium to creatinine (Ca:Cr) provides a convenient and fairly reliable index of the rate of calcium excretion but only when muscle mass is constant (i.e., not during growth). In normal individuals, urinary Ca (mmol/L):Cr (mmol/L) is usually < 0.40; (Ca [mg/dL]:Cr [mg/dL] < 0.14). A value > 0.57 (mmoL/L units) or > 0.20 (mg/dL units) signifies hypercalciuria.[97] The lower limit of normal calcium excretion is about 1.8 mmol (75 mg)/d. It must be stressed that 24-h calcium excretion rates span wide ranges both in healthy and in sick individuals.

IDIOPATHIC HYPERCALCIURIA

It has been estimated that approximately 70% of kidney stones are caused by idiopathic hypercalciuria. Two main types have been defined. In *absorptive hypercalciuria,* the primary abnormality is considered to be overabsorption of calcium from the gut. In this group of patients, it is proposed that plasma Ca increases significantly after meals, with the result that PTH secretion is suppressed. The urinary calcium excretion rate is increased postprandially because of (1) an increased filtered load of calcium and (2) the absence of PTH action to increase tubular calcium reabsorption. Fasting urinary calcium is normal. In *renal hypercalciuria,* the primary abnormality is considered to be a selective defect in the ability of the renal tubules to reabsorb calcium. Increased urinary calcium excretion is relatively constant both postprandially and during fasting. The urinary loss of calcium elicits secondary hyperparathyroidism, which in turn stimulates synthesis of 1,25-$(OH)_2$D to maintain normocalcemia. In establishing the diagnosis of idiopathic hypercalciuria, the other causes of hypercalciuria should be systematically excluded, especially conditions in which hypercalcemia is a salient feature.

The initial investigation of kidney stone disease should include the following determinations: plasma Ca_T, Ca_I, P_i, creatinine, uric acid, electrolytes, blood pH, pCO_2, and HCO_3^-. The 24-h urinary excretion rates of calcium, creatinine, oxalate, and uric acid should be measured. A fresh urine sample should be examined for specific gravity and pH. Microscopy and a bacterial culture should be carried out to rule out urinary tract infection. The plasma PTH concentration and the urinary excretion of nephrogenous cAMP should be measured if hyperparathyroidism is suspected. Assessment of the urinary amino acid pattern and a qualitative test for cystine should be routinely carried out on a voided sample to exclude the diagnosis of cystinuria.

Thiazide diuretics and oral phosphate are employed to reduce urinary calcium excretion in both absorptive and renal hypercalciuria.

FAMILIAL HYPOCALCIURIC HYPERCALCEMIA (FHH)

FHH, also called "familial benign hypercalcemia," is inherited as an autosomal dominant trait.[86] The distinctive feature is hypocalciuria (< 2.5 mmol/d, < 100 mg/d) in the presence of mild to moderate hypercalcemia; indeed, urinary calcium is often < 1.25 mmol/d (< 50 mg/d). PTH levels are either normal or slightly elevated. Parathyroidectomy does not correct hypercalcemia or hypocalciuria, and therefore is contraindicated. The disorder is thought to be due to decreased sensitivity of the parathyroid glands to plasma Ca_I, i.e., due to an abnormally high "set-point" for regulation of secretion of the parathyroid glands.

ANALYTICAL METHODS

ANALYSIS OF CALCIUM IN BIOLOGICAL FLUIDS

Calcium in biological fluids has traditionally been difficult to measure accurately and precisely. Of the many methods that have been introduced, atomic absorption spectrophotometry (AAS) and spectrophotometry of calcium-dye complexes are most often used for measuring Ca_T. Ca_I measurement, although long known to be clinically valuable, is not yet a common procedure. However, development of calcium ion–selective electrodes (ISE) and sensitive potentiometers now makes Ca_I measurement feasible on a routine basis.

Determination of Ionized Calcium (Ca_I)

Several analytical systems are available for determination of physiologically active, "free" or "ionized" calcium. The term "free" calcium indicates that the calcium fraction is not bound to small or large molecular weight ligands. However, the term ionized calcium (Ca_I) is readily understood and more widely used and thus will be used in this chapter. The potentiometric determination of Ca_I, strictly speaking, does not determine Ca-ion concentration but rather the activity of the ionic species in the sample. The term "concentration" will subsequently be used in this qualified sense.

Principle

A sensitive potentiometer measures the voltage differential (E, EMF) developed between a reference electrode and a calcium ion-selective electrode when both are exposed to a fluid containing the calcium ion. The equilibrium potential across the membrane of the ISE is defined by the Nernst equation:

$$E = E_0^1 + \frac{RT \times 2.303}{z_i \times F} \times \log_n \times aCa^{2+}$$

where

E = the potential of the Ca^{2+} electrode
E_0^1 = the standard potential of the reference half-cell
R = the gas constant ($8.3143\ J \times K^{-1} \times mol^{-1}$)
T = the absolute temperature, °K
F = the Faraday constant ($96\ 487$ coulombs $\times\ mol^{-1}$)
z_i = the number of electrons exchanged ($+2$ for Ca^{2+})
aCa^{2+} = the calcium ion activity.

The potential established by the unknown solution is compared with the potential developed by a standard solution of known ionic strength and ion composition and containing calcium ions in exactly known concentration.

Instruments

The several commercial instruments available for measuring Ca_I each consist of a sampler pump, a calcium ion–selective electrode, a reference electrode, a potentiometer, and electronic components to compute and display results. Table 12-5 summarizes features of current models offered by different manufacturers. Note that the instruments differ in sample volume requirements, length of time for measurement, whether the electrode dips into the sample or the sample flows through the electrode, and the type and frequency of standardization required. A vertical flow path has the advantage of easier elimination of inadvertently introduced air bubbles and of less frequent clogging with small fibrin clots and thrombi. Automatic self-standardization simplifies the operation during long sample runs and allows the instrument to be used at any time without delay. Ability to accept small sample volumes is of particular importance for pediatric applications. Such considerations will largely govern the choice of an instrument.

Table 12-5. COMPARISON OF ISE SYSTEMS FOR MEASURING IONIZED CALCIUM

	Manufacturer			
	Orion, Cambridge, MA	*Applied Medical Technology, Menlo Park, CA*	*Nova Biomedical, Newton, MA*	*Radiometer, Copenhagen, Denmark*
Model	SS-20	Auto-Electrion	Nova 2 (Nova 6, Nova 7)*	ICA-1
Sample introduction	Syringe/internal peristaltic pump	Sample cup	Suction probe/internal peristaltic pump	Suction probe/internal peristaltic pump
Calcium electrode	Solid state/neutral carrier	Solid state/organophosphate	Solid state/organophosphate	Solid state/neutral carrier/PVC membrane
Liquid junction	Flowing	Static	Flowing	Flowing
Reference electrode	Ag/AgCl	Ag/AgCl	Ag/AgCl	Hg/HgCl$_2$
Standardization	Automatic, 1 standard; external calibration	Manual, 2 standards; pooled sera	Automatic, 2 standards; external calibration	Automatic, 2 standards; external calibration
Measurement time (min)	3	6	1.3	3.7
Sample volume (mL)	1.0	1.3	0.45	0.15
Selectivity of Ca ISE	K_{Na} 2.5 × 10^{-3} K_{Mg} 5.3 × 10^{-2}	K_{Na} 3.0 × 10^{-3} K_{Mg} 4.4 × 10^{-2}	K_{Na} 2.3 × 10^{-3} K_{Mg} 6.0 × 10^{-2}	K_{Na} 9.5 × 10^{-3} K_{Mg} 6.0 × 10^{-3}

*Nova 6 and Nova 7 models combine the same Ca$_I$ electrode with additional electrodes (pH, Na, K).

Calcium Ion-Selective Electrodes

The instruments shown in Table 12-5 use calcium-selective electrodes designed around organophosphate or other organic ion exchanger compounds that have been modified for selective interaction with calcium and that are called the ion-exchanger. The ion-exchanger may carry a charge or may be neutral, i.e., uncharged. It may be a liquid or a solid dispersed in a mediator solvent. The liquid ion-exchangers are usually supported by an inert compound dissolved in a water-immiscible solvent; this system functions as a liquid membrane between the exterior, aqueous solution of the analyte and the interior, aqueous solution in contact with the metal element of the electrode. The reaction of ionic calcium that generates the potential occurs with the ion-exchanger of the membrane; the process is analogous to the interaction of hydrogen ions with the substance of the glass electrode.

Selectivity is a property of an electrode and is defined in terms of relative electrode response to calcium ions and competing ions. Selectivities of different electrodes are compared in Table 12-5.[112,113] The smaller the selectivity value shown opposite the K of the competing ion, the greater is the selective response of the electrode to calcium ion. By this criterion, the Radiometer ICA 1 electrode is the most calcium selective relative to Mg^{2+} of the four electrodes shown.

Because of less-than-perfect selectivity, cation interference may occur in the measurement of Ca$_I$ by ISE. A slight positive bias has been observed, for instance, in Ca$_I$ values obtained with serum Mg concentrations > 4 mmol/L, as might occur in eclampsia or renal failure. Lithium does not interfere at concentrations as high as 8 mmol/L. The earlier electrodes were subject to hydrogen ion interference, but with calcium-selective electrodes on the market since 1978, such interference is small over the pH range 4–9.[68,113] Electrodes made with neutral ion exchangers are particularly H$^+$ resistant; indeed such electrodes have been used in instruments for potentiometric determination of both Ca$_I$ and Ca$_T$, the latter after the sample has been acidified to convert to the protein-bound calcium to the ionized species.[2] The selectivity of any ISE may change with use and over its lifetime.

Anion interference may also be a problem. Interference in the Ca$_I$ measurement may be due to a reaction of the anion with Ca^{2+} to form nonionized complexes, to interaction of the anion with the electrode itself, or to both causes. Many physiological anions (phosphate, citrate, lactate, sulfate, and oxalate) form complexes with calcium ions. At concentrations of these anions encountered clinically, they affect the circulating Ca$_I$ concentration by formation of complexes but not by direct interference with the performance of the Ca-ISE. Chloride ions do not affect the measurement of Ca by either mechanism. Because of the formation of nonionized calcium complexes, citrate and oxalate cannot be used as anticoagulants in samples intended for Ca_I determination.

Older electrodes displayed a *protein effect*; readings drifted slightly after the first protein-containing fluid was introduced into the electrode, making it necessary to add protein to standard solutions to compensate for this effect. A recently introduced electrode with a PVC membrane at the junctional interface is said to eliminate the protein effect;[3,17,25] in the other commercially available instruments, the effect on the calcium concentration due to the changing from protein-free standards to plasma or serum is less than 0.02 mmol/L, and is therefore of little clinical importance.

Temperature has an effect both on electrode response (see Nernst equation) and on the calcium binding equilibrium with plasma proteins and small molecule ligands. Instruments for measuring Ca_I therefore are designed to control and maintain temperature constant at 37 °C.

Chemical interference with the operation of the ISE may be caused by contamination of the specimen with surfactants having anionic properties or by added or intrinsic chemicals in the sample.[20] Ethanol, when present at increased blood levels, may interfere directly with the Ca_I measurement by affecting the ion-exchanger liquid membrane. EGTA, EDTA, and other strong calcium chelators do not interfere directly with the operation of the newer electrodes. However, they chelate calcium in the plasma sample, and thus lower the Ca_I value; they should not be used as anticoagulants in specimens intended for Ca_I determination.

Heparin is the only suitable anticoagulant for whole blood or plasma samples intended for Ca_I determination. However, it causes a 0.01 mmol/L decrease in the Ca_I value for each unit of heparin added per mL blood. The effect is attributable to both chelation of Ca_I and interaction with the Ca-ISE in vitro. For this reason it must be stressed that, in practice, the amount of added heparin should be kept constant and at the lowest concentration that prevents clotting (see section on Preparation of Heparinized Syringe, below). Intravenous injection of 10 000 units of heparin has been shown, in adults, to cause a significant decrease in Ca_I of about 0.03 mmol/L.[8]

Specimen

Venous, arterial, and capillary blood specimens are suitable for Ca_I determination. Whole blood, plasma, or serum can be used. However, because pH has a great effect on the concentration of ionized calcium in plasma, it is necessary that the pH of the specimen at the time of measurement be the same as that of the patient's blood at the time of sampling. This is most readily achieved by collecting and processing the specimen anaerobically. Exposing the specimen to air leads to loss of CO_2, increase in pH, and shift of the Ca_I-protein binding equilibrium to give artifactually low Ca_I values. Techniques for handling whole blood and plasma anaerobically are relatively straightforward. Also, techniques are described below by which serum can be handled anaerobically. The type of specimen used by a given laboratory will depend on technical and instrumental considerations, and on the preference of its director or the client physicians.

When the specimen is not collected anaerobically, its pH should be adjusted to the patient's blood pH at the time of sampling—a complicated and time-consuming procedure. We do not advocate the practice of adjusting the pH of specimens to pH 7.4 because values of Ca_I obtained in this manner do not reflect the in vivo situation. For example, a severely acidotic patient (pH 7.15) with an ionized calcium of 1.15 mmol/L is "hypocalcemic" when the Ca_I value is corrected to 7.40.

We recommend that measurement of Ca_I be carried out on heparinized whole blood, collected anaerobically. Whole blood has the following advantages over plasma: (1) minimal handling; the specimen can be introduced directly into the electrode; (2) requirement of smaller volume of blood per sample (important in pediatrics); (3) greater ease of maintaining anaerobic conditions; and (4) rapid acquisition of test results, i.e., within a few minutes of drawing the blood. If the specimen cannot be tested within 30 min, it must be stored at 4 °C, in which case the concentration of Ca_I remains constant for 48 h. The technique for whole blood collection is analogous to that for obtaining specimens for the determination of blood gases.

A drawback of using whole blood is the effect of erythrocytes on the junction potential of the ISE, an effect that cannot be compensated for by standards. Ca_I determination in whole blood with a normal hematocrit gives a value 1.6% higher than plasma or serum of the same specimen. Differences are proportional to the hematocrit reading. Indeed, in vitro studies have demonstrated that a 38% suspension of washed erythrocytes in Tris-buffered calcium standard increased the Ca_I readings by 7% over the value for cell-free fluid.[5]

When plasma is used for analysis, the heparinized blood specimen must be centrifuged at $1000 \times g$ at 4 °C in the sampling syringe in such a way that the cells become packed against the plunger. The plasma fraction is then sampled directly from the inlet of the syringe within 30 min, using the suction probe of the instrument. Alternatively, the plasma fraction can be drawn anaerobically from the tip of the sampling syringe into another syringe for immediate analysis or storage. Plasma (or serum) in a tightly sealed syringe can be stored at 4 °C for several days or at −20 °C for six months.

To obtain an anaerobic sample of serum, the specimen is collected from the patient into a vacuum tube filled to the top without permitting air to leak into the tube. The vacuum tube is then centrifuged with the stopper in place, and the serum harvested anaerobically into a syringe which is then sealed.[117] With some instruments, the serum can be sampled directly from the centrifuged vacuum tube. Disadvantages of using serum are the risk of exposure to air and the possibility of contamination from rubber stoppers, which frequently contain calcium introduced during their manufacture.[104]

Preparation of Heparinized Syringe

Although unnecessarily large amounts of heparin may cause an appreciable decrease in the Ca_I measurement, the effect is insignificant when sampling syringes are prepared according to either of the following procedures that ensure a final heparin concentration of about 2 U/mL blood. This concentration of heparin causes a decrease in the Ca_I value of not more than 0.02 mmol/L.

Method 1. Mix 100 mg (15 000 units) of sodium heparin with 10.0 g soluble starch in a pharmaceutical mill. This mixture is stable at room temperature for one year or more.

Place portions of the mixture into syringes in amounts appropriate to the blood volume to be collected, so that about 1.5 mg of mixture (1.5 units sodium heparin) will be present for each mL of blood. The syringe may be autoclaved without loss of heparin activity.

Method 2. Dilute 0.1 mL of heparin (1000 units/mL) with 1.9 mL NaCl, 150 mmol/L. Mix well.

Draw up this solution in a 3.0 mL syringe, wetting the internal surfaces, and then expel; this will leave fluid only in the dead space of the syringe. The final heparin concentration will be about 2 units/mL blood.

Note: Method 1 is preferred because the heparin concentration in the sample is more reproducible, and sample dilution is avoided.

Patient Preparation

Concentrations of Ca and protein in plasma and serum are known to be affected by posture of the patient and by venous stasis at the time of blood sampling.[111] Stasis causes efflux of water from the vascular compartment; this increases both Ca_T and protein levels. Ca_I concentrations are less affected than are Ca_T concentrations by stasis and posture.[111] However, Ca_I in individuals confined to bed is 0.10–0.15 mmol/L lower than in ambulatory individuals.[113] Exercise or muscular activity increases plasma lactate and protein, and decreases blood pH.[111] Ca_T is usually elevated during exercise but the net effect on Ca_I is unpredictable. Whether food ingestion affects Ca_I concentration is controversial; increased, decreased, and unchanged postprandial levels have been described.

The recommended collection procedure is to keep the patient recumbent and at rest for 30 min before drawing the blood and to obtain the specimen without the use of a tourniquet.

Reagents

Ready-prepared standards are available from instrument manufacturers. However, standards may also be prepared in house.

1. Calcium stock standard, 100 mmol/L (4 g/L). Dry ~12 g of anhydrous $CaCl_2$, ACS grade, to constant weight at 110 °C and allow to cool in a desiccator. Weigh out 11.099 g of dried $CaCl_2$, place in a 1-L volumetric flask, dissolve, and dilute to mark with deionized water. Verify concentration by appropriate dilution and analysis by AAS.

2. Calcium working standards, 0.5, 1.0, and 2.0 mmol/L (2.0, 4.0, 8.0 mg/dL). Add 5.0, 10.0, or 20.0 mL of stock standard to a 1-L volumetric flask. Add 150 mL of NaCl, 1 mol/L, to each flask. Dilute each to mark with deionized water.

In some instruments, the 1.0 and 2.0 mmol/L solutions are used for continuous self-standardization. For standardization of some systems, KCl (4 mmol/L) is also incorporated into the standard solutions; in that circumstance, the NaCl concentration is reduced to 146 mmol/L.

Procedure

The procedure for the standardization and introduction of the sample into the analysis system depends on the design and the operational directions for a particular instrument. The microprocessor in most modern instruments computes the value of the unknown sample from standard readings and slope of the concentration-response relationship, and the instrument displays the value of the unknown in concentration units of mmol/L or mg/dL. Some instruments also display Ca_I computed to a standard pH of 7.40. Our reservations about the validity of this procedure are discussed on page 1344. Millivolt (potential) values are accessible on some instruments. On a plot of mV versus log Ca_I of standards, the calibration relation can be extrapolated and Ca_I concentrations lying outside the range of standards can be determined.

Comments

1. *Sources of errors* include instrument or electrode malfunction and incorrect specimen collection and handling. Other causes of erroneous Ca_I values are interference by drug or intravenous treatment of the patient, and interference from chemicals contaminating the specimen or the analytical system.

2. *Carryover* from one sample to another has negligible effects, except when a specimen with a value of Ca_I less than 0.5 mmol/L follows one with normal or high Ca_I concentration; in this situation up to 8% positive error may occur. Improving the wash-out procedure can reduce this carryover.

3. *Abnormalities of ionic strength* of a patient's blood sample are a potential source of error. ISE's respond to the activity of the selected ion. Ion activity is a function not only of the concentration of the selected ion but also of the ionic strength of the solution in which it exists. Standards for Ca_I are therefore prepared with a "normalized" ionic strength approximating that of NaCl at 150 mmol/L. As long as the sum of the patient's sodium and potassium concentration lies between 120 and 160 mmol/L, differences in the ionic strengths of the unknowns and standards will have no clinically significant effect. If the sum of the Na and K concentration is < 120 or > 160 mmol/L, accurate determination of Ca_I by ISE requires a comparison of the unknown with standards containing NaCl concentrations that mimic the ionic strength of the unknown sample. Such corrective action may be necessary when measuring Ca_I in blood or plasma from sick infants or patients undergoing cardiopulmonary bypass surgery. Measuring Ca_I in dialysis fluids or in urine also requires such standardization.

4. A difference in *pH of the sample* at the time of measurement compared to the patient's pH at the time of sampling is a further source of error. The consequences of aerobic blood collection and the change of specimen pH on Ca_I measurement have been discussed above. (See section on Specimen.) Metabolism of erythrocytes or contaminating bacteria, as may occur if the specimen is allowed to stand at room temperature, can result in production of acid metabolites causing the pH to decrease and the measured Ca_I to increase. Precautions in sample collection and handling are essential to prevent spurious elevations or reductions in Ca_I concentration. Overnight storage, weekend storage, or transport of the specimen to a remote laboratory for assay are frequent occurrences in clinical situations. When samples are collected anaerobically, as recommended above, whole blood samples can be stored at 4 °C for 48 h before effects of erythrocyte metabolism become significant. Plasma or serum, in a tightly sealed syringe, can be stored at 4 °C for several days or at −20 °C for as long as six months.[101]

5. No *reference method* is presently available for Ca_I measurement in whole blood, plasma, or serum. Quality control materials are available from one of the instrument manufacturers. Frozen, pooled serum may be used, provided that special attention is paid to stability.

Bowers has recently proposed procedures for evaluating instrumental precision.[9]

Reference Ranges

Age-related variations in Ca_I occur during the neonatal and pubertal periods. Ca_I is 0.4 mmol/L higher in cord blood than in maternal blood.[23,106] In healthy neonates, we have observed a rapid decrease in plasma Ca_I from a mean value of 1.50 to a mean value of 1.10 mmol/L during the first 36 h of life, followed during the subsequent 72 h by an increase to a value slightly below that of cord blood.[108] Thereafter, through infancy and childhood, the reference range for whole blood Ca_I is 1.10–1.50 mmol/L (4.40–6.00 mg/dL); the value for plasma Ca_I is 0.05 mmol/L lower (0.20 mg/dL). The concentration decreases at puberty. The adult reference range for whole blood Ca_I is 1.12–1.32 mmol/L (4.48–5.28 mg/dL); the range for plasma Ca_I is 1.03–1.23 mmol/L (4.12–4.92 mg/dL). After age 60, another decline in Ca_I concentration is observed.

Differences in the reported reference ranges[28,128] are attributable largely to differences in sample preparation and in the selectivity characteristics of the electrodes.

DETERMINATION OF TOTAL CALCIUM (Ca_T)

The definitive method for Ca_T is isotope dilution followed by mass spectrometry (ID-MS method).[88] It was designed to measure absolute calcium concentrations and to serve as reference for atomic absorption spectrometry (AAS), the secondary reference method.[13] The instrumental and technical demands of the definitive method preclude its use for routine analysis in the clinical laboratory. Although atomic absorption spectrophotometry also requires special instrumentation and skills, the technique is well within the scope of many laboratories.

Accuracy and reproducibility of the AAS methods are usually comparable to the simpler automated methods such as the cresolphthalein complexone (CPC) method.[19,114] However, in four of eight comparisons, the differences between AAS and ID-MS were > 2%.[19] The observed lack of precision in the case of values lying outside the physiological range has been attributed to lack of standardization at the extreme ends of the calcium range.[114]

We describe here the AAS method applied to a variety of biological specimens. We also describe the CPC method, a simple spectrophotometric procedure based on calcium-dye binding and used in both manual and automated modes.

Determination of Ca_T in Plasma, Serum, or Urine by Atomic Absorption Spectrophotometry[105]

Principle

Calcium, free or bound, introduced into a flame of appropriate temperature forms a small amount of free calcium atoms, a majority of which are in the ground state. The atoms at ground state absorb incident light emanating from a calcium hollow cathode lamp at the resonance wavelength of 422.7 nm. The degree of absorption is concentration dependent. (See Chapter 1B, Section One.)

Lanthanum, usually in the form of $LaCl_3$, is added to the diluent for samples and standards for several reasons: (1) it eliminates the effect of anions (phosphate, sulfate), which depress the light absorption by atomic calcium (see Comment 3);[129] (2) it serves to liberate calcium from its ligands;[129] and (3) it decreases the interferences from protein and variations in Na and K concentrations in the sample. Deproteinization of serum or plasma and the addition of Na, K, and Mg salts to standards also help to minimize interference effects. The use of strontium as an internal reference element decreases analytical variability caused by changes in sample atomization and flame stability.

Specimen

Serum, heparinized plasma, and urine are suitable specimens. Specimens may be stored at 4 °C for weeks or months; falsely low values may result from adsorption of calcium to glass and plastic containers or, in the case of plasma, from coprecipitation of calcium with fibrin (see Comments, below). Urine specimens, preferably timed (24 h), should be collected in acid-washed bottles containing 5 mL of HCl, 6 mol/L. If the urine specimen is collected without acid, the pH should be adjusted to 3–4 with no more than 10 mL of HCl, 6 mol/L, at the time of receipt in the laboratory and before an aliquot is taken for assay or storage.

Reagents

1. Trichloroacetic acid (TCA), 500 g/kg. To 500 g trichloroacetic acid add 500 g deionized water and mix. The solution is stable indefinitely at 4 °C.

2. Sample diluent for plasma and serum samples.* Lanthanum chloride, 10 mmol/L; strontium chloride, 0.12 mmol/L. Add 1.63 g La_2O_3 (Spectrograde), 0.0306 g $SrCl_2 \cdot 6 H_2O$ (Spectrograde), and 6.7 mL concentrated HCl to a 1-L volumetric flask. Mix until dissolved. Add 200 mL deionized water and 80 mL TCA, 500 g/kg, and dilute to mark with deionized water. Mix thoroughly. Stable indefinitely.

3. Calcium stock standard, 50 mmol/L. Dry about 6 g $CaCO_3$ (SRM #915, National Bureau of Standards, Gaithersburg, MD) at 140 °C for 16 h. Allow to cool in a desiccator. Weigh 5.0044 g of dried $CaCO_3$ into a 1-L volumetric flask. Add concentrated HCl dropwise (~5 mL) until $CaCO_3$ is dissolved. Dilute to mark with deionized water. Mix thoroughly. Stable indefinitely.

4. Stock electrolyte solutions.

    a. NaCl, 1.4 mol/L (81.76 g/L) in water.

    b. KCl, 1.0 mol/L (74.6 g/L) in water.

    c. $MgSO_4$, 100 mmol/L. Dry ~15 g $MgSO_4$ (anhydrous) at 140 °C overnight. Weigh 12.039 g of dry $MgSO_4$ into a 1-L volumetric flask, dissolve, and dilute to mark with deionized water.

Solutions are stable indefinitely.

5. Calcium working standards, 0, 1.0, 2.0, and 3.0 mmol/L (0–12.0 mg/dL). Into 200-mL volumetric flasks, add 0.0, 4.0, 8.0, or 12.0 mL calcium stock standard, 50 mmol/L. Add to each flask 20.0 mL stock NaCl, 1.4 mol/L; 1.0 mL stock KCl, 1.0 mol/L; and 2.0 mL stock $MgSO_4$, 100 mmol/L. Dilute to volume with deionized water and mix thoroughly. Solutions are stable indefinitely.

Procedure

1. To 0.1 mL of standard, unknown sample, or control sample in a test tube, add 4.9 mL of sample diluent. Zero standard serves as blank.

2. Cover each tube with Parafilm and let stand for 10 min. Mix, let stand another 5 min, and centrifuge at $1000 \times g$ for 10 min.

3. While aspirating an appropriate calcium solution, adjust flame to produce a peak response of the instrument at 422.7 nm. This is usually achieved with a reducing (bluish) flame. In general, follow the operating instructions provided by the manufacturer of the equipment.

Set instrument read-out or concentration display to zero (baseline) while aspirating the zero standard. Aspirate and read standards and supernatants of control(s) and unknowns.

Comments

1. Good *analytical technique* includes the following precautions. Aspirate deionized water frequently, preferably between successive samples. Keep the burner clean and replace it when corrosion appears at the edges of the slot or when the signal-to-noise ratio decreases. Bracket unknowns between standards and repeat aspiration of blank and standards often, readjusting the instrument as often as necessary. Use of an assayed control material is required.

2. *Pre-analytical errors* may arise from unreliable standards, contamination of specimen or reagents with calcium, or loss of calcium from stored specimens or standard solutions. If standards are purchased, they should be "certified" for AAS or checked against a primary standard such as the NBS material described in the method. Calcium contamination of glassware can be eliminated or minimized by acid-wash (HCl, 0.1–0.3 mol/L), followed by copious rinse with deionized water; such an acid-wash is particularly desirable if a high-phosphate detergent is

---

*For use in urine assay, modify the diluent by substituting 8.15 g La_2O_3 to give a lanthanum chloride concentration of 50 mmol/L. The modified "urine diluent" is required because the concentration of phosphate (anionic interferent) is greater and more variable in urine than in plasma. The "urine diluent" should also be used for assay of digests of feces, tissue, and diet, prepared as described on page 1349.

routinely used in a mechanical glassware washer. Corks, rubber stoppers including those used with evacuated tubes, and reused plastic bottles or pipet tips are sources of calcium contamination and should be avoided.

Calcium loss from stock and working standards can be minimized by storage in borosilicate glass. Containers should be kept tightly closed to avoid evaporation; Pyrex vials with Teflon-lined screw caps are recommended for specimen storage. Plastic containers should be avoided because calcium may be lost by adsorption to the container material; the loss is particularly serious for specimens of small volume.[90] Use of acid-etched glassware may lead either to contamination with calcium because the etched areas cannot be cleaned thoroughly or to calcium loss because of adsorption of calcium onto the damaged surface. The risk of calcium loss is high when heparinized plasma is stored, since calcium coprecipitates with fibrin during storage and is removed by centrifugation.

3. *Analytical interferences* have been minimized in the method described. *Anionic interference* can arise from the phosphate, sulfate, and citrate which are coconstituents with calcium in biological materials. Phosphate is the most serious interferent; not only does it form a complex with calcium that is not readily dissociated in an air-acetylene flame, but it also directly depresses absorption of light by the ground state calcium atoms in the flame. In plasma, phosphate concentrations may range from 0.3–4.0 mmol/L; in urine, the highest concentrations may differ from the lowest up to 100-fold. La^{3+} and Sr^{2+} incorporated into the sample diluent dissociate calcium from anion complexes and reduce the depressant effect of anions on light absorption by atomic calcium.[129] *Cationic interference* of Na^+, K^+, Mg^{2+}, Fe^{2+}, and Zn^{2+} is less troublesome than anionic interference. The effect of the main cations, Na^+, K^+, and Mg^{2+}, can be obviated by their inclusion in standards at an ionic strength similar to that of plasma. These cations exert no direct spectral interference because their resonance lines lie well away from the 422.7 nm calcium line. *Protein interference* may occur due to calcium-protein complex formation and the mechanical effect of viscosity on aspiration and atomization of the sample.[36] Addition of TCA to the sample diluent acidifies the sample, breaks the calcium-protein complex, and precipitates the protein. Viscosity, rate of aspiration, and efficiency of atomization are made constant; the burner stays clean longer, the stability of the baseline readings is improved, and a favorable signal-to-noise ratio is maintained.

Determination of Calcium in Feces, Tissues, or Diet

Calcium assay of feces, tissue, or diet is achieved by taking weighed aliquots of homogenized specimens, ashing them, and dissolving the ash in dilute nitric acid. The acid digest may then be analyzed for calcium or magnesium by AAS and for phosphate, Na, and K by other methods.

Reagents

1. Sulfuric acid, 1.80 mol/L. Place ~500 mL deionized water into a 1-L volumetric flask. Place the flask into an ice-water bath, and add slowly 100 mL concentrated H_2SO_4, analytical grade. Allow solution to cool. Fill to mark with deionized water. The solution is stable indefinitely.

2. Nitric acid, 1.58 mol/L. Place 500 mL deionized water in a 1-L volumetric flask. Add slowly, in a fume hood, 100 mL concentrated HNO_3, analytical grade. Allow solution to cool and fill to mark with deionized water. The solution is stable indefinitely.

Procedure

1. Weigh into porcelain crucibles triplicate aliquots (0.5–1.0 g) of homogenized specimen. Add 1.0 mL H_2SO_4, 1.80 mol/L, and dry at 70 °C overnight.

2. Heat the crucibles on a hotplate, increasing the heat gradually over a period of 1–2 h to avoid splattering. This step is complete when the sample is thoroughly dry and no longer gives off fumes.

3. Place crucibles in a muffle furnace, gradually increasing the temperature to 500 °C. Keep crucible and specimen at this temperature for 24–72 h. This step is complete when the ash is gray or white.

4. Cool, and dissolve ash in 2.0 mL of HNO_3, 1.58 mol/L, by scraping the ash and crucible with a transfer pipet. Transfer the solution into a 10-mL volumetric flask.

5. Rinse the crucible twice with 2.0 mL of deionized water, adding the rinses to the flask. Fill the flask to mark with deionized water. Mix thoroughly on a vortex mixer or by inversion.

6. Dilute 1.0-mL aliquots of sample solution five- to ten-fold with sample diluent that contains 500 mmol $LaCl_3$/L (see footnote, p. 1348), and proceed as for determination of calcium in plasma, serum, or urine.

Reference Ranges

Mean *plasma* or *serum* Ca_T concentrations reported for adults are in most cases close to 2.35 mmol/L (9.4 mg/dL) with 95% confidence ranges from 2.20–2.55 mmol/L (8.8–10.2mg/dL). Slightly lower means have been reported for persons over 50 y;[69] the mean values are close to 2.25 mmol/dL (9.0 mg/dL), and the 95% confidence limits are between 2.15 and 2.45 mmol/L (8.6–9.8 mg/dL). The 95% range for Ca_T concentrations in cord blood plasma is 2.46–2.94 mmol/L (9.8–11.8 mg/dL);[23,107,108] the level of Ca_T in cord blood is higher than in maternal blood at the time of delivery. The 95% confidence range for Ca_T in children, aged 4–20 y, is 2.30–2.75 mmol/L (9.2–11.0 mg/dL).[14,15,46]

Urinary Ca_T is usually determined and expressed as a 24-h excretion rate; the excretion depends on many variables but particularly on dietary intake of calcium. For infants and children, the upper limit of the reference range is 0.15 mmol/kg/d (6.0 mg/kg/d).[70] For adults on a high calcium diet, the upper limit is 7.5 mmol/d (300 mg/d), but decreases to 3.75 mmol/d (150 mg/d) if the calcium intake is restricted. Further details on urinary calcium excretion and conditions that affect urinary calcium excretion are discussed on page 1340.

Fecal calcium. Assuming an average intake of 17.5 mmol calcium per 24 h and net absorption of 20–40%, the fecal calcium excretion rate in healthy adults ranges from 10.5–14 mmol/24 h (420–560 mg/d).

Spectrophotometric Determination of Ca_T in Plasma or Serum with ortho-Cresolphthalein Complexone (CPC)[81]

Variations of this method have been widely applied to automated analysis. The method described here is a manual method.

Principle

The metal-complexing dye *o*-cresolphthalein complexone (CPC) forms a chromophore with calcium in alkaline solution; the color produced is measured spectrophotometrically at 578 nm. The chromogen reagent contains 8-hydroxyquinoline to mask interfering cations (mainly Mg), urea to diminish turbidity of lipemic specimens and to enhance metal-dye complex formation, and ethanol to inhibit color development in the blank. After measuring the developed color, addition of an excess of the calcium chelator EDTA corrects for hemolysis. This treatment dissociates the calcium-chromogen complex but the hemoglobin-chromogen complex remains undissociated; the absorbance due to this hemoglobin-chromogen complex is subtracted from the total absorbance to give Ca_T absorbance.

Reagents

1. Buffer. Diethanolamine, 2 mol/L, pH 11.7; urea, 5 mol/L. Dissolve 210 g of diethanolamine and 300 g of urea in about 500 mL of water. Adjust pH to 11.7 with acetic acid, and dilute to 1 L. Stable for two months.

2. Chromogen. *o*-Cresolphthalein complexone, 100 μmol/L; 8-hydroxyquinoline, 8 mmol/L; urea, 5 mol/L; acetic acid, 43 mmol/L; and ethanol, 25% (v/v). Dissolve 64 mg of CPC, 1.16 g of 8-hydroxyquinoline, and 2.5 mL acetic acid in 250 mL ethanol in a 1-L volumetric flask. Add 300 g urea and dilute to mark with deionized water. Stable for three months.

3. Color reagent. Mix equal volumes of buffer and chromogen reagent. Use within one day.

4. Calcium standard, 2.5 mmol/L. Weigh out 125.1 mg $CaCO_3$, reagent grade, and dissolve in 500 mL of trichloroacetic acid, 0.1 mol/L.

5. Ethylenediaminetetraacetic acid (EDTA), 150 mmol/L. Dissolve 558 mg of $Na_2EDTA \cdot 2\ H_2O$ in 10 mL of deionized water.

Procedure

1. Add 1.0 mL of color reagent to test tubes labeled blank, unknown, and standard.

2. To blank tube add 20 μL of water, to standard tube 20 μL of standard, and to unknown tube 20 μL of serum or heparinized plasma.

3. Mix and let stand for 5–60 min.

4. Read at 578 nm and record absorbance of standard (A_{std}) and unknowns (A_{tot}) against blank set to zero absorbance.

5. To blank and unknowns, add 10 μL of EDTA. Mix.

6. Read and record absorbance of unknown (A_{hem}) with blank set to zero absorbance at 578 nm as in step 4.

Calculations

$$A_{unknown} = A_{tot} - A_{hem}$$

$$Ca_T, mmol/L = \frac{A_{unknown}}{A_{std}} \times 2.5$$

Comments

1. The method has been adapted to automated analysis on the SMAC and other Technicon systems (Technicon Instruments Corporation, Tarrytown, NY, 10591); *aca* (E.I. DuPont Instrument Products, Clinical Systems Division, Wilmington, DE, 19898), Cobas-Bio (Roche Analytical Instruments Inc., Nutley, NJ, 07110), the Diagnostic M Discrete Analyzer (Boehringer Mannheim Diagnostics, Houston, TX, 77063), and others. Details of particular applications may be obtained from the manufacturers.

2. The intensity of color developed by the Ca-CPC complex increases proportionately with an increase in pH and is also dependent on the buffer species employed. Diethanolamine, 1 mol/L in the final mixture, enhances the color intensity; the concentration is adequate to maintain the pH at 11.7 without increasing the viscosity or absorbance excessively. 2-Amino-2-methyl-1-propanol (AMP) is used as an alternative buffer in some adaptations.[89]

3. 8-Hydroxyquinoline in the color reagent almost completely eliminates interference from Mg at concentrations encountered clinically, although it decreases the color intensity by 25–40%.

4. Bilirubin in high concentrations introduces a negative error in Ca_T results.[44] The error becomes significant for Ca_T determinations on specimens from severely icteric newborns. Acetaminophen and the antihypertensive agent hydralazine cause a positive interference. A positive bias from hemolysis has been reported for some automated methods (5% at a hemoglobin value of 2.5 g/L),[102] but addition of EDTA or EGTA in the manual method compensates for this error.[48] Overall, the CPC-method has been found to be accurate and reproducible.[114]

Reference Ranges

Reference ranges are identical to those obtained by AAS.[114]

Reference

Lorentz, K.: Improved determination of serum calcium with 2-cresolphthalein complexone. Clin. Chim. Acta, *126*:327-334, 1982.

Ektachem Slide Method for Ca_T

Principle

A sample (0.01 mL) is deposited automatically on multilayered slides (Kodak, Rochester, NY). The calcium in the sample reacts with an indicator dye (Arsenazo III) at pH 7.8 to form a colored complex, the amount of which is proportional to the calcium present in the sample. Intensity of color is measured by reflectance spectrophotometry at 680 nm.

METHODS FOR THE DETERMINATION OF INORGANIC PHOSPHATE*

Since the method of Fiske and SubbaRow was introduced in 1925,[37] almost all methods for assay of inorganic phosphate (P_i) have relied on the formation of a complex of phosphate ion

---

*Standard methods applied to plasma and serum measure only the inorganic phosphate fraction. However, by tradition, the result is expressed as phosphorus; hence, the term "phosphorus" is still frequently employed, though its use is unjustified from the physiological standpoint. When results are expressed in mmol/L rather than mg/dL, the need to specify "phosphate" or "phosphorus" is obviated.

with a molybdate compound. The colorless hexavalent complex can be extracted into an organic solvent, concentrated (if need be) by evaporation, reconstituted in aqueous or organic media, and quantitated by UV spectrophotometry or as molybdenum by AAS. In the more commonly used methods, however, the colorless hexavalent molybdenum phosphate complex, exemplified as $(NH_4)_3[PO_4(MoO_3)_{12}]$, is reduced to the pentavalent form (colored) and then measured spectrophotometrically at wavelengths above 400 nm, most often near 700 nm.

Many reducing agents are suitable for producing the colored pentavalent molybdenum-phosphate complex. However, some have disadvantages worthy of note. *Aminonaphtholsulfonic acid* (ANSA) is relatively unstable and the precipitate that forms in the reagent often necessitates its filtration before use. The color intensity of the reduced molybdate increases over several hours so that strict timing of standard and sample readings is required. *Stannous chloride* is an excellent reducing agent and gives greater color intensity than ANSA. However, the color is also unstable, its formation is very sensitive to small differences in acid concentrations, and it deviates from Beer's law at concentrations observed in clinical measurement. Mixtures of *stannous chloride* and *hydrazine*, used in continuous flow systems, tend to precipitate and to cause hydraulic problems. *Methyl-p-aminophenol sulfate*,[47] because it has a greater acid tolerance than ANSA, allows the use of a one-component acid-molybdate reagent while still maintaining sufficient color development. *Ascorbic acid* and *ferrous sulfate* have also been used, more often in research than in clinical methods. They have the advantages of less breakdown of the labile phosphate esters commonly present in biological fluids and of producing a more stable color. The HCl salt of *N-phenyl-p-phenylenediamine (semidine)* provides good sensitivity for phosphate and good color stability.[29] This method has been published as a "Selected Method"[43] by the American Association for Clinical Chemistry and will be described in detail below.

Certain other methods, although involving a phosphomolybdate complex, do not employ reduction to produce measurable color. A *molybdivanadophosphate complex* with a stable yellow color is formed by addition of excess molybdate to an acid solution of vanadate and orthophosphate.[123] The chief disadvantage of this method is acid hydrolysis of labile phosphate esters and hence overestimation of P_i in biological specimens. Complex formation of certain *basic dyes* with *phosphomolybdate* will cause a marked shift in the absorbance maximum of the uncomplexed dye. Itaya and Ui,[59] in a study of these dyes, found the method based on the triphenylmethane dye Malachite Green to be 30 times more sensitive than the Fiske-SubbaRow procedure.

The method selected for description in this book is the semidine method employed in the manual mode.[43] Several automated versions have been published.

Principle

Serum or plasma proteins are precipitated with liberation of inorganic phosphate from ligands by trichloroacetic acid. The protein-free filtrate is mixed with ammonium molybdate in acid solution to form ammonium phosphomolybdate.

$$7\,H_3PO_4 + 12\,(NH_4)_6Mo_7O_{24} \cdot 4\,H_2O \rightarrow 7\,(NH_4)_3[PO_4(MoO_3)_{12}] + 51\,NH_4^+ + 51\,OH^- + 33\,H_2O$$

Semidine HCl is then added to reduce the hexavalent molybdenum-phosphate complex to the pentavalent, blue-colored complex. The intensity of the blue color is measured spectrophotometrically.

Specimens

Serum or heparinized plasma is suitable. Common anticoagulants such as citrate, oxalate, or EDTA should not be used because they interfere with the formation of the phosphomolybdate complex. Serum or plasma should be separated from cells within 1 h after collection of the specimen. Hemolysis must be avoided because the P_i concentration of erythrocytes is seven times that of plasma and because hemoglobin interferes with the color reaction. Plasma or serum may be stored at 4 °C for several days or in the frozen state for several months.

Ideally, specimens should be obtained without the use of a tourniquet from a recumbent fasting patient. Venous stasis, because it causes hemoconcentration, affects plasma P_i in a manner analogous to its effect on Ca_T, with the added effect of hypoxia which increases the breakdown of phosphate esters. Food ingestion causes a transient decrease in plasma P_i because (1) the increase in blood pH after meals (alkaline tide) enhances in vivo complex formation of calcium with phosphate and its subsequent deposition in the bone mineral phase; and (2) glycogen deposition in muscle and liver draws on the P_i pool for formation of glucose-phosphate intermediates.

Urine for P_i determination should be collected in acid-washed, detergent-free bottles. If urinary calcium is to be determined on the same specimen, the bottle should contain 5 mL of HCl, 6 mol/L.

Reagents

1. Trichloroacetic acid, 100 g/L (0.61 mol/L). Add to 100 g of trichloroacetic acid (ACS grade) from a freshly opened bottle, 800–850 mL of deionized water, and transfer into a 1-L volumetric flask. When the solution has reached room temperature, fill to mark and mix. Solution is stable at room temperature for one month, and at 4 °C for six months.

2. Ammonium molybdate, 8 mmol/L. Dissolve 9.887 g of ammonium molybdate, $(NH_4)_6Mo_7O_{24} \cdot 4 H_2O$, ACS grade, in 700–750 mL of deionized water in a 1-L volumetric flask. Stir vigorously for 15 min and fill to mark. Store solution in a polyethylene bottle at room temperature. Stable for one month.

3. Sodium bisulfite, 96 mmol/L. Dissolve 10.00 g of sodium bisulfite ($NaHSO_3$), ACS grade, with deionized water in a 1-L volumetric flask. Mix well. Adjust volume to mark. Use this reagent for preparation of semidine hydrochloride solution; discard remainder.

4. Semidine hydrochloride, 2.76 mmol/L. Weigh out 250.0 mg of solid semidine HCl (N-phenyl-p-phenylenediamine HCl) in a polystyrene weighing boat and add 0.5 mL of absolute ethanol to wet the semidine HCl. Wash semidine with sodium bisulfite solution into a 500-mL volumetric flask. Mix well. Adjust to mark with the bisulfite solution. Filter solution through a glass wool plug into a dark borosilicate bottle. Store in the dark at room temperature. Discard if solution becomes discolored.

5. Phosphorus stock standard, 10 mmol/L (31 mg/dL). Dry ~1.5 g anhydrous KH_2PO_4, reagent grade, to constant weight. Weigh out 1.360 g and dissolve in deionized water in a 1-L flask. Dilute to mark. This solution is stable indefinitely at 4 °C.

6. Phosphorus working standards, 0.01–0.20 mmol/L (0.031–0.62 mg/dL). Into appropriately labeled 1-L volumetric flasks place 1, 2, 4, 8, 12, 16, or 20 mL of phosphorus stock standard. These concentrations are equivalent to plasma, serum, or urine concentrations of 0.1–2.0 mmol/L. Fill to mark with deionized water. Standards are stable indefinitely at room temperature.

Procedure

1. Label screw-capped tubes with Teflon-lined caps for standards, controls, and unknowns (serum, heparinized plasma, appropriately diluted urine).

2. Deliver 1.8 mL of trichloroacetic acid into each tube with an automatic pipet.

3. Add 0.2 mL of sample.

4. Let mixture stand for 45 min with occasional gentle mixing. Do not vortex-mix or shake vigorously.

5. Centrifuge controls and unknowns at 1200 × g for 10 min at room temperature; after repeated gentle mixing of tube contents, let stand for 10 min, then centrifuge again at 1200 × g for 10 min.

6. Decant supernatant into correspondingly marked tubes. Recentrifuge if particulate matter is observed.

7. For each standard, control, and unknown, pipet 1.00 mL of supernatant into another set of correspondingly labeled screw-capped tubes.

8. Dispense 2.00 mL of the semidine-HCl reagent into each tube and vortex-mix.

9. Add 0.20 mL of ammonium molybdate solution into each tube and vortex-mix.

10. Let tubes stand for at least 10 min.

11. Read absorbance against water set to zero absorbance. The optimum wavelength is 740 nm. If the spectrophotometer cannot be set at that wavelength, a setting between 650 and 700 nm can be used.

12. Plot a calibration curve for absorbance versus concentration of standards and read values of unknowns from the curve.

Comments

1. The order of addition of reagents—semidine before molybdate—must be observed; the excess semidine-HCl immediately and preferentially reduces the complex formed of phosphate

with endogenous ascorbic acid before other side reactions with the complex can occur. Reversing the order can lead to 8% overestimation of P_i.

2. The pH range for maximal color development is much wider than for methods using stannous chloride or ANSA as reducing agents. Color intensity reaches a plateau about 10 min after the addition of molybdate except for specimens with very high P_i concentrations.

3. Interference by lipemia is not found with the manual or the SMAC methods because protein separation, by precipitation in one case and dialysis in the other, obviates such an effect. However, "creamy" serum caused a positive error of 0.435 mmol/L (1.35 mg/dL) with the DuPont *aca* method. Interference by hemoglobin at a concentration of 5 g/L caused a negative error of 0.548 mmol/L (1.70 mg/dL) with the *aca* method, and a positive error of 0.22 mmol/L (0.69 mg/dL) with the manual and SMAC methods.[43] Neither ascorbic acid (up to 25 mg/dL) nor bilirubin interferes in any of the methods.[43] Mannitol, widely used to reduce intracranial pressure and also in cardiopulmonary bypass surgery, interferes by decreasing the rate of formation of the phosphomolybdate complex.[26] Citrate and oxalate prevent full color development but do not interfere with the determination at concentrations usually found in plasma and urine.[126] Neither citrate nor oxalate should be used as anticoagulants.

4. The manual method described above has been adapted to automatic analysis on the SMAC (Technicon Instruments, Tarrytown, NY 10591) and *aca* (E.I. DuPont Instrument Products, Clinical Systems Division, Wilmington, DE 19898).

5. Because urinary P_i excretion reflects to a great extent the dietary intake of phosphate, the concentration of P_i in urine varies widely. Urine specimens will often require dilution with deionized water to bring the P_i concentration within the range of the assay standards. Testing will be expedited if, on the first try at assay, one uses urine undiluted, diluted 1:10, and diluted 1:100 in the same run. The value that falls within the standard curve should be multiplied by the appropriate dilution factor to obtain the concentration in mmol/L or mg/dL.

6. The method is also applicable to assay of phosphate in feces, diet, or tissue. The specimen should be prepared as for assay of Ca_T by AAS (p. 1349); however, separate aliquots must be taken, since H_2SO_4 must be omitted in the ashing procedure.

7. Day-to-day variation of the method, expressed as CV, is ~1% at both normal (1.2 mmol/L) and elevated (2.5 mmol/L) concentrations of P_i.[43]

Reference Ranges

The reference ranges for fasting plasma or serum P_i are shown in Table 12-6. Values vary greatly with the age of the patient. In infancy, normal plasma P_i concentrations are 50% higher

Table 12-6. REFERENCE RANGES FOR FASTING PLASMA INORGANIC PHOSPHATE ACCORDING TO AGE AND SEX IN mmol/L (mg/dL)

Age	Male	Male and Female	Female	Reference
0–2		1.20–2.23 (3.7–6.9)*		14
2–5		0.97–2.13 (3.0–6.6)*		14
4–9		1.23–1.81 (3.8–5.6)*		15
9–13			1.07–1.71 (3.3–5.3)*	15
10–15	1.03–1.74 (3.2–5.4)*			
14–20			0.84–1.49 (2.6–4.6)*	15
16–20	0.78–1.49 (2.4–4.6)*			15
20	0.81–1.45 (2.5–4.5)†		0.94–1.52 (2.9–4.7)†	69
40	0.78–1.36 (2.4–4.2)†		0.84–1.39 (2.6–4.3)†	69
70	0.65–1.26 (2.0–3.9)†		0.94–1.55 (2.9–4.8)†	69

*±2 Sd calculated from published raw data.
†Estimated 2.5 and 97.5 percentiles.

than in adults and they decline gradually through childhood.[15,46] No sex difference is observed before puberty. The reference values decrease significantly between 12 and 14 years in females, and between 14 and 16 years in males,[46] to reach concentrations characteristic of young adults. The adult levels diminish slightly in both sexes until middle age. According to several surveys, values then increase in elderly females, and tend to decrease slightly in elderly males.[69] In general, values for adult females are approximately 10% higher than for males of corresponding age.

Reference ranges for urinary excretion of phosphate in children have been difficult to develop because of the great variability of urinary phosphate excretion with growth, renal function, muscle mass, dietary phosphate intake, and hormonal activity. The use of the phosphate excretion index (PEI) of Nordin and Fraser[94] has been found helpful in the diagnosis of hyperphosphaturia.[124]

Reference

Garber, C. C., and Miller, R. C.: Revisions of the 1963 semidine HCl standard method for inorganic phosphorus. Clin. Chem., *29*:184-188, 1983.

MEASUREMENT OF PTH IN PLASMA

The PTH molecule and its fragments can be detected by biological assay procedures and by radioimmunoassays. Most biological assays require extraction of PTH from plasma or serum and are relatively cumbersome, slow, and insensitive. Radioimmunoassays are presently the only widely available procedures for measuring PTH. A very sensitive cytochemical bioassay has recently been described that promises to have important clinical applications in the future.[45,91]

The measurement of PTH by RIA is complicated by several factors. As discussed earlier, plasma contains not only intact PTH 1-84 but also N-terminal, C-terminal, and mid-molecule fragments, the relative concentrations of which vary from condition to condition and from patient to patient. Furthermore, radioimmunoassays differ in their ability to detect the various PTH fragments. Many assays are not sufficiently sensitive to measure PTH in plasma of all euparathyroid individuals, and are therefore unsatisfactory for diagnosis of hypoparathyroidism. Some assays are ineffective in differentiating primary hyperparathyroidism from hypercalcemia of malignancy. Presently, no one radioimmunoassay is considered ideal in all disorders, and there is lack of agreement about the best method to use in various clinical situations.

In employing a PTH radioimmunoassay system, it is essential that the characteristics of the antibody for recognition of the various fragments be accurately established and that the reference range of the particular assay be known. The choice of the particular radioimmunoassay system depends partly on the tentative diagnosis of the patient. Antisera have been developed with recognition sites for the N-terminal, the C-terminal, and the mid-region amino acid sequences; a few antisera recognize amino acid sequences along the entire length of the molecule.[92] Assays directed toward the C-terminal and mid-molecule fragments presently appear to have the greatest general support.[83,85] Mid-molecule procedures tend to have a shorter assay time (< 8 h) than do C-terminal procedures ($\sim$24 h).

Procedure for Radioimmunoassay of PTH in Plasma (Serum)

Regardless of the specificity of the different antisera, the same principles of methodology apply in most PTH radioimmunoassays. In the mid-molecule assay selected,[83] specific antiserum is added to individual plasma (serum) specimens or standards, and the mixtures are incubated for 15 min. ^{125}I-PTH is then added, and the reaction mixture is incubated for 2 h to allow ^{125}I-PTH to compete with unlabeled PTH for binding sites on the antiserum. After incubation, a second antibody directed specifically against the first antibody is added to precipitate the antibody-bound PTH. The mixture is centrifuged, the supernatant discarded, and the radioactivity of the precipitate determined in a γ-scintillation counter to measure bound ^{125}I-PTH. The concentration of PTH in samples is determined by interpolation using a standard curve.

Reference Ranges

Reference ranges common to all methods cannot be presented. This is because of differences in sequence specificity of the antisera employed, differences in the proportions of the various

PTH fragments in the plasma of individuals under investigation, and different modes of expressing results [pmol/L or ng/mL of the PTH standard used (e.g., bPTH 1–84); or microequivalents/mL of a standardized hyperparathyroid serum]. The reference range for a typical radioimmunoassay with recognition sites for both N- and C-terminal fragments is 2–25 pmol/L (0.02–0.25 ng/mL).[92] PTH values of infants, children, and pregnant women tend to be somewhat higher than those of other healthy subjects.

Cytochemical Bioassay of PTH

In the recently described cytochemical bioassay of Goltzman et al.,[45] PTH-stimulated glucose-6-phosphate dehydrogenase activity of cultured distal convoluted tubules of rat kidney is measured by a micro-colorimetric procedure. The method detects only intact PTH and the biologically active N-terminal fragment. It is much more sensitive than any radioimmunoassay yet described. Values reported appear to differentiate clearly specimens from hypoparathyroid (<10 fmol/L; <1 pg/mL), normal (29–290 fmol/L; 2.9–29 pg/mL), and primary or secondary hyperparathyroid (0.34–110 pmol/L; 34–11 000 pg/mL) patients.

This bioassay procedure is not yet widely applied. It appears, however, that it will not only complement radioimmunoassay data by determining the concentration of biologically active PTH in plasma of normal and diseased individuals, but will also be particularly helpful when RIA results are discordant with the clinical findings.

Measurement of Calcitonin in Plasma

Several techniques have been described for measuring CT in human serum or plasma: bioassays based on the hypocalcemic effect of test serum injected into young rats, competitive displacement radioreceptor assays, receptor assays linked to generation of cAMP, and classical radioimmunoassays. Only homologous radioimmunoassays using high titer antisera raised against synthetic human CT (hCT) are sufficiently sensitive to measure the low concentrations (<29.3 pmol/L, <100 pg/mL) of hormone present in plasma of normal adults.[51,54,85]

Such methods follow the established principles for radioimmunoassay of peptide hormones. Suitable dilutions of hCT standard or 50–200 μL of patient's serum or plasma are preincubated with antibody for 3–7 d at 4 °C. ^{125}I-hCT or ^{131}I-hCT is added and the mixture incubated for an additional 3–6 d at 4 °C. Antibody-bound and free fractions are separated by dioxane precipitation or adsorption on dextran-coated charcoal and the radioactivities of the bound and free fractions are determined. The concentration of CT in the patient's sample is determined by interpolating from a standard curve.

Comment

It should be pointed out that to establish a calcitonin radioimmunoassay in the laboratory involves practical problems similar to those mentioned in the section on PTH radioimmunoassay.

Reference Ranges

Basal (i.e., unstimulated) CT values reported for healthy adults vary from laboratory to laboratory depending upon such factors as sequence specificity of the antiserum and heterogeneity of the secreted hormone. Thus, results of assays must be interpreted in the light of reference data of the same laboratory. Reference ranges for three sensitive radioimmunoassays are given in Table 12-7. The plasma CT level in newborn infants is higher than the basal value in adults and decreases gradually in the elderly. The plasma CT level is also somewhat higher during pregnancy and lactation than for the nonpregnant adult female. Plasma CT does not appear to be strongly influenced by meals. Values of CT in serum are somewhat higher than in plasma. Hemolysis causes artifactually high values.

Despite the high sensitivity of recently described radioimmunoassays, the basal plasma CT level is less than the detection limit in an appreciable proportion of normal subjects tested. Provocative secretion tests usually cause a significant increase of plasma CT and the hormone becomes readily detectable in nearly all normal individuals. The most commonly used secretagogues for provocative tests are pentagastrin, glucagon, and calcium salts, each of which is

Table 12-7. BASAL REFERENCE RANGES OF PLASMA CT

Mean, Range (pmol/L)	Mean, Range (pg/mL)	Individuals with Undetectable Basal CT (%)	Reference Number
14, <7–21 (♂)	49, <25–73	10	52
9, <7–15 (♀)	31, <25–51	50	52
7, <3–22	24, <10–75	10	99
5, <2–23	19, < 8–74	50	54

administered intravenously.[51,99] (See Table 12-8.) The tests are of particular value in the diagnosis of medullary thyroid carcinoma (MTC), to identify asymptomatic affected relatives of patients with MTC and multiple endocrine neoplasia (MEN, Types IIa and IIb), and to monitor surgically treated MTC patients for possible recurrence of the tumor.

METHODS FOR THE DETECTION OF VITAMIN D AND ITS METABOLITES

Vitamin D and its metabolites are a family of lipid soluble, nonionic compounds that differ principally in the number of their hydroxyl groups. They can be resolved from the bulk of the plasma components by straightforward solvent extraction techniques. All vitamin D compounds isolated to date possess the distinctive secosteroid nucleus and contain a cis-triene bond system (Figure 12-7). This conjugated double-bond system gives rise to an unusually intense UV-absorption with characteristic $\lambda_{max} = 265$ nm, $\lambda_{min} = 228$ nm, and $\epsilon_{molar} = 18\ 300$. These features can be utilized in designing methodology for separation and quantitation.

A number of natural binding proteins and receptors for vitamin D and its metabolites can be utilized in radioligand assays. Vitamin D–binding protein (DBP) has strong binding affinities for 25-(OH)D and 24,25-(OH)$_2$D and is a basic component of the most popular form of 25-(OH)D assay, despite species differences in binding affinities. The same protein can be used, though less easily, for measuring vitamin D. The cytosolic binding protein of chick intestine, the probable intestinal receptor for 1α,25-(OH)$_2$D in vivo, facilitates quantitation of extremely small amounts of 1,25-(OH)$_2$D. Recently, specific antibodies have been generated for measuring individual vitamin D metabolites; however, these antibodies are not widely available and much work must still be done to optimize their sensitivity and specificity.

Vitamin D$_2$ and D$_3$ appear to have equivalent biological activity in man. Rates of hydroxylation of vitamin D$_2$ to 25-(OH)D$_2$, 24R,25-(OH)$_2$D$_2$, and 1,25-(OH)$_2$D$_2$ are generally assumed to equal those of their vitamin D$_3$ counterparts. Thus in an individual adult receiving dietary vitamin D$_2$ supplements and exposure to sunlight, the plasma will contain vitamin D$_2$ and D$_3$ metabolites in proportion to the amounts of the parent vitamins the individual has acquired. Methods designed to measure metabolites of vitamins D$_2$ and D$_3$ must have equal sensitivity for both types.

Numerous assays, either for single or for multiple metabolites, have been reported, often with appreciable differences in procedural details. Most procedures involve three basic steps: (1) extraction, (2) purification, and (3) radioligand or UV assay. As yet, no assay is sufficiently specific or sensitive to eliminate the need for steps (1) and (2), but future radioimmunoassays may be sufficiently specific to do so. Since most assays require several steps, internal standards must be used to assess losses of vitamin D compounds during the purification process. In most

Table 12-8. PROVOCATIVE TESTS FOR CT SECRETION

Secretagogue	Dose	Times for Plasma CT Determination	Reference Number
Calcium gluconate	15 mg elemental Ca/kg, IV over 4 h	0, 1, 2, 3, 4 h	5
Calcium chloride	3 mg elemental Ca/kg, IV over 10 min	0, 10, 20, 30, 40 min	99
Pentagastrin (Pentavlon, Ayerst)	0.5 μg/kg, IV over 10 sec	0, 2, 5, 10, 20, 30 min	52
Glucagon (Parke-Davis)	0.01 mg/kg rapid IV, plus 0.01 mg/kg per h for 3 h	0, 1, 2, 3, 4 h	99

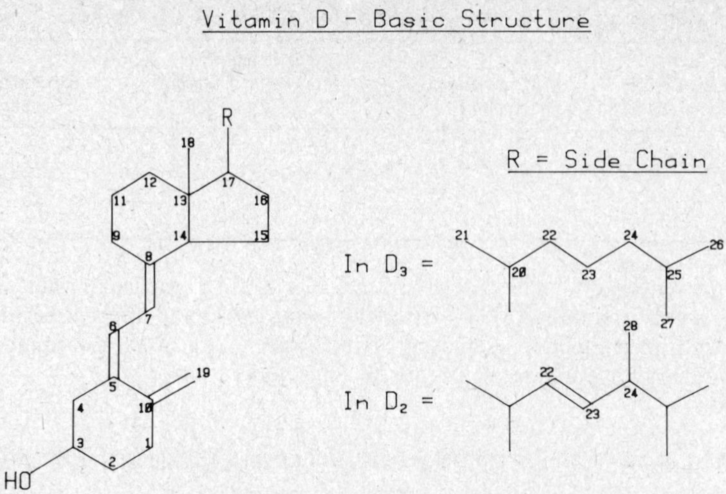

Vitamin D - Basic Structure

Figure 12-7. Basic structure of vitamin D.

cases, the corresponding radioactive vitamin D metabolite is used as an internal standard. Radioactive [³H] preparations with high specific activity (10–200 Ci/mmol) currently available include vitamin D_3; 25-(OH)D_3; 24R,25-(OH)$_2D_3$; 1α,25-(OH)$_2D_3$; and 1α,24R,25-(OH)$_3D_3$. Unfortunately, radioactive vitamin D_2 metabolites are not commercially available; many workers assumed, in some instances erroneously, that recoveries of vitamins D_2 and D_3 are identical through the extraction and chromatography procedures.

Numerous attempts to measure simultaneously several vitamin D metabolites, including 25-(OH)D, 24,25-(OH)$_2$D, and 1,25-(OH)$_2$D, are reported in the literature.[56,58,74,100,119] Most of the procedures are extremely complex; in some there are technical problems alluded to in later sections, and in almost all there is a difficulty with conducting the assay on plasma containing D_2 metabolites. Furthermore, the clinical usefulness of assaying some of the minor metabolites has yet to be established. Therefore, multi-metabolite procedures are not recommended. Rather, we suggest methodology designed to measure metabolites individually. We describe methods to measure 24,25-(OH)$_2D_3$ and total amounts of 25-(OH)D, 1,25-(OH)$_2$D, and vitamin D.*

Determination of Total 25-(OH)D in Plasma by Competitive Binding Assay

Principle

25-(OH)D along with other plasma lipids is extracted from heparinized plasma or serum, partially purified on Sephadex LH-20 or by high performance liquid chromotography (HPLC), and assayed by virtue of its strong binding affinity for rat vitamin D-binding protein (DBP). The ability of nonradioactive 25-(OH)D from the plasma extract to compete with [26,27-³H]25-(OH)D_3 for DBP binding is compared with the competition by known amounts of pure 25-(OH)D_3 standard.

The competitive binding assay outlined below is an updated hybrid version of a published procedure.[61]

Reagents

1. DBP reagent. Dilute rat plasma, preferably from vitamin D–deficient animals, 1:10 000 in Tris-HCl buffer (0.25 mol/L, pH 8.6) containing 1% bovine serum albumin (BSA) (Pentex, Miles Laboratories, Elkhart, IN 46515).

2. [26,27-methyl-³H] 25-(OH)D_3, ~20 Ci/mmol (Amersham, Arlington Heights, IL). Dissolve compound in ethanol to give an activity of 5000 cpm/50 μL.

3. 25-(OH)D_3 standard. Dissolve 25-(OH)D_3, crystalline (Hoffmann-LaRoche, Nutley, NJ 07110), in ethanol to give a series of standard solutions of 25, 50, 100, 200, 300, 400, 500, 750, 1000 or 10 000 pg/50 μL concentration.

4. Ethanol, methanol, methylene chloride, and hexane, HPLC grade (e.g., distilled-in-glass solvents, Burdick & Jackson Laboratories, Muskegon, MI 49442).

---

*Total 25-(OH)D = 25-(OH)D_2 plus 25-(OH)D_3; total 1,25-(OH)$_2$D = 1,25-(OH)D$_2$ plus 1,25-(OH$_2$)D_3; and total vitamin D = vitamin D_2 plus vitamin D_3.

5. Potassium chloride and sodium chloride, reagent grade.

6. Sephadex LH-20 (Pharmacia, Uppsala, Sweden); or microparticulate silica (5–10 μ particles) in 25 cm $\times$ 6.2 mm HPLC columns (e.g., Zorbax-SIL, DuPont Instruments, Wilmington, DE 19898).

7. 20-mL test tubes with Teflon-lined screw-caps.

8. 5-mL conical-based Reactivials with Teflon-lined screw-caps. (Pierce Chemical Co., Rockford, IL 61105).

9. Neutral Charcoal (Norit A, Fisher, Toronto, Canada) 1%, and Dextran T70 (Pharmacia, Uppsala, Sweden), 0.1% as a suspension in Tris-HCl buffer (0.25 mol/L, pH 8.6).

Instrumentation

HPLC is performed on a standard liquid chromatograph fitted with an automated or manual sample injector, pump(s) capable of generating a flow rate of 2 mL/min at a back-pressure of up to 2000 psi, and a fixed wavelength detector set at 254 nm or variable wavelength detector set at 265 nm. The system is operated in an isocratic mode and effluent fractions are collected using manual or automatic (fraction collector) techniques.

Procedure

Sample Extraction and Purification

Perform steps 1–7, then choose either Steps 8–13 for purification with Sephadex LH-20 or Steps 14–17 for purification by HPLC on microparticulate silica, and then continue with Step 18.

1. Dilute between 100 μL and 2 mL plasma or serum sample to 2.0 mL volume with 0.9% saline in 20 mL test tubes, with Teflon-lined screw-caps, and add 25 μL of Reagent 2 (1000–5000 cpm).

2. After equilibration for 30 min, extract lipids by adding 7.5 mL of methanol:methylene chloride (2:1, v/v). Vortex the monophasic mixture and allow to stand for 5 min.

3. Add an additional 2.5 mL of methylene chloride and 2.5 mL of saturated potassium chloride and vortex.

4. Centrifuge the resultant milky suspension at 500 $\times$ g for 10 min in a standard centrifuge to produce two phases, an upper aqueous and a lower organic. Take care to avoid disturbing the compact, proteinaceous interface of precipitated plasma proteins during the next two steps.

5. Aspirate the upper aqueous phase, containing water-soluble plasma components and phospholipids, with the aid of a vacuum into a trap attached to a water aspirator.

6. Decant the lower organic phase, containing vitamin D metabolites into a Reactivial, after carefully tapping the test tube to displace the protein interface. The protein interface will adhere as a disk to the sides of the test tube.

7. Evaporate the organic solvent, consisting mainly of methylene chloride, using heat ($\sim$35 °C) and a stream of prepurified N_2. To avoid lipid peroxidation, redissolve the lipid extract as described in step 8 or step 14 soon after it reaches dryness.

For Purification Using Sephadex LH-20

8. Dissolve the lipid extract in 500 μL hexane:methylene chloride:methanol, 9:1:1.

9. Pour suspensions of Sephadex LH-20 in hexane:methylene chloride:ethanol, 9:1:1, into glass columns (internal diameter 0.6 cm) to a depth of 15 cm. Before use, wash the Sephadex LH-20 successively with 10 volumes hexane, methylene chloride, and methanol and air-dry to minimize contaminants.

10. Transfer the lipid extract onto the column with two 100-μL washes of hexane:methylene chloride:methanol solvent.

11. Discard the first 5-mL effluent which contains neutral and highly nonpolar lipids.

12. Collect the next 5-mL effluent containing the 25-(OH)D.

13. Evaporate and redissolve the residue in 2–10 mL ethanol, depending on the concentration of 25-(OH)D in the sample. One should attempt to achieve a sample dilution around the B_{50} of the binding assay ($\sim$200 pg/50 μL).

Go to Step 18.

For Purification Using HPLC on Microparticulate Silica

14. Dissolve the lipid extract (Step 7) in 200 μL of hexane:isopropanol:methanol, 96:3:1. Centrifuge to remove any precipitant lipid or protein as necessary.

15. Perform HPLC on microparticulate silica using, for example, a 25 cm $\times$ 6.2 mm column of Zorbax-SIL (6 μ microparticulate silica, DuPont Instruments, Wilmington, DE 19898). Use hexane:isopropanol:methanol, 96:3:1 as solvent with a flow rate of 1.5 mL/min at 600 psi, and elute 25-(OH)D$_2$ and 25-(OH)D$_3$ at 10 and 12 min, respectively.

16. Collect the two forms of 25-(OH)D as a single fraction in a Reactivial.

17. Evaporate and redissolve the sample in ethanol as described in Step 13 for LH-20 procedure.

18. At this point in the assay, count the radioactivity of an aliquot of the purified ethanolic 25-(OH)D fraction and compare with radioactivity added in Step 1, in order to assess recovery through the extraction and purification steps.

Assay Procedure

1. Mix 500 μL of DBP reagent (Reagent 1), plus 50 μL of [³H]25-(OH)D₃ solution (5000 cpm) in ethanol (Reagent 2), plus either 50 μL of the various 25-(OH)D₃ standards (Reagent 3) or 50-μL plasma extract.

2. Incubate the mixture at 0–4 °C for 4–20 h.

3. Add 200 μL of a suspension containing 1% neutral charcoal coated with 0.1% Dextran T70 in Tris-HCl buffer, 0.25 mol/L, pH 8.6. Allow the mixture to stand for an additional 1 h.

4. Separate free and protein-bound [³H]25-(OH)D₃ by centrifugation at 3000 × g for 15 min. Using 600 μL of supernatant fluid, measure the protein-bound radioactive ligand by liquid-scintillation counting using a water-based cocktail.

5. Compare counts of bound ligand in the sample with those in the standard curve (Figure 12-8).

6. Calculate the concentration of 25-(OH)D after correction for sample volume, various dilution factors, and the fraction of 25-(OH)D recovered through the purification steps. The result is usually expressed in nmol/L. See Table 12-9 for reference ranges.

Comments

1. The method described will give accurate results provided that the chromatographic procedures are adhered to. Falsely low levels stem from the presence of large amounts of neutral lipid which can arise from inadequate chromatographic separation and cause interference with binding of the free tracer to charcoal-Dextran. Falsely high results could occur from incomplete separation of 25-(OH)D₃-26,23-lactone, 24,25-(OH)₂D₃, or 25,26-(OH)₂D₃ from the 25-(OH)D fraction in patients treated with large doses of vitamin D.

2. Chromatographic steps prior to HPLC used to isolate 25-(OH)D from impurities sometimes resolve 25-(OH)D₂ and 25-(OH)D₃ and the final fraction does not contain all 25-(OH)D₂. Under such conditions tracer [³H]25-(OH)D₃ gives an apparent recovery of 25-(OH)D₂ that is falsely high and this provides falsely low results for total 25-(OH)D in plasma of vitamin D₂–treated patients.

3. Solvent systems described here must be finely tuned to local conditions in the reader's laboratory. Variations in chromatographic separation due to column packing, particle manufacturer, and environment can be expected. Similar retention times to those reported here can be reproduced by altering the solvent strength slightly. For example, to increase solvent strength, change solvent composition from hexane:isopropanol:methanol, 96:3:1 to 94:5:1.

4. The purity of BSA preparations and the activity and pH of various charcoal preparations vary greatly from manufacturer to manufacturer and even between batches. Before setting up a standard curve for 25-(OH)D₃ the optimum concentrations of BSA, charcoal-Dextran, and DBP reagent should be determined, in that order. In this way local differences in these critical factors will be avoided and a standard curve of the type shown in Figure 12-8 will be obtained.

5. The attractiveness of the procedure using microparticulate silica is its ease of automation for analysis

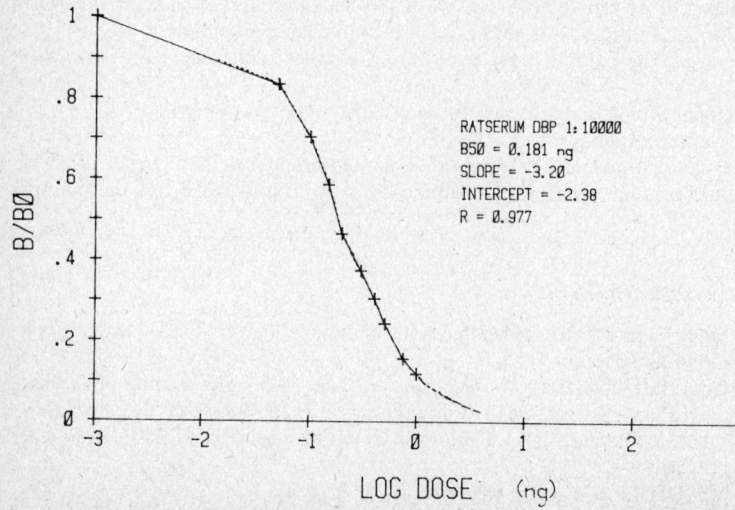

RATSERUM DBP 1:10000
B50 = 0.181 ng
SLOPE = -3.20
INTERCEPT = -2.38
R = 0.977

Figure 12-8. Standard curve for competitive protein binding assay of 25-(OH)D, constructed with rat serum diluted 1:10 000 in buffer.

Table 12-9. REFERENCE RANGES FOR VITAMIN D AND
METABOLITES IN SERUM

	Mean ($\pm$ SD)	Range
Vitamin D[61]	5.7 $\pm$ 2.9 nmol/L	2.1–12.2 nmol/L
25-(OH)D		
Summer:[30]	55.3 $\pm$ 17.8 nmol/L	20.0–82.5 nmol/L
Winter:[61]	40.0 $\pm$ 9.8 nmol/L	22.8–59.8 nmol/L
1,25-(OH)$_2$D[60]	88.9 $\pm$ 30.5 pmol/L	40.9–141.8 pmol/L
24,25-(OH)$_2$D$_3$[27]	5.8 $\pm$ 2.6 nmol/L	1.9–10.3 nmol/L

of large numbers of samples. Automatic injectors (e.g., WISP, Waters Associates, Milford, MA 01757) permit as much as 200 μL of a 220 μL sample of hexane:isopropanol:methanol extract to be loaded on the HPLC, thereby maximizing yields through this purification stage. Newer programmable fraction collectors (e.g., LKB Superac, Bromma, Sweden) can be set to select specific, timed fractions from HPLC runs, thus eliminating the need for an operator during preparative HPLC of multiple samples.

Determination of Total 25-(OH)D in Plasma by Ultraviolet Detection

Principle

25-(OH)D$_2$ and 25-(OH)D$_3$ along with plasma lipids are extracted from the plasma, separated from each other and the bulk of other UV-absorbing impurities by two HPLC steps, and measured independently by their strong UV-absorption at 265 nm or 254 nm.

Reagents and Apparatus

5–10 μ octadecasilane microparticulate column for HPLC (e.g., 25 cm $\times$ 6.2 mm Zorbax-ODS, DuPont, Wilmington, DE 19898). For other reagents see section describing reagents for determination of 25-(OH)D in plasma by competitive-binding assay, page 1358.

Procedure

Perform Step 1, then choose either Step 2a or 2b, and then continue with Steps 3–5.

1. Extract 25-(OH)D from plasma as described in steps 1–7 of procedure for competitive binding assay of 25-(OH)D.

2a. Dissolve the lipid extract of 25-(OH)D in 500 μL hexane:methylene chloride:methanol, 9:1:1. Purify on Sephadex LH-20 as described in Steps 8–13 of the competitive protein-binding assay of 25-(OH)D. Then further purify the effluent from HPLC on a microparticulate silica column, e.g., μ-Porasil[34] or Zorbax-SIL,[64] using hexane:isopropanol:methanol, 96:3:1, as solvent, at a flow rate of 1.5 mL/min.

2b. Dissolve the lipid extract of 25-(OH)D in 200 μL of hexane:isopropanol:methanol, 96:3:1. Purify by HPLC on a Zorbax-SIL microparticulate silica column using hexane:isopropanol:methanol, 96:3:1 as solvent, in the manner described in Steps 14–16 of the procedure for competitive protein-binding assay of 25-(OH)D. Redissolve the evaporated 25-(OH)D fraction in 200 μL methanol:water, 91:9, and perform HPLC on a microparticulate reverse phase packing such as Zorbax-ODS. (See scheme shown in Figure 12-9. Note that this scheme also permits measurement of total vitamin D in the same plasma sample.)

3. Measure UV-absorption of effluent peaks at 265 nm or 254 nm. Integrate the areas of peaks with retention times corresponding to 25-(OH)D$_2$ and 25-(OH)D$_3$. Figure 12-10 shows the results obtained when the method described in Step 2b and Figure 12-9 is applied to plasma from normal individuals and from patients treated with pharmacological doses of vitamin D$_2$ or D$_3$.

4. To assess recovery, collect the HPLC effluent fractions containing [^{3}H]25-(OH)D$_3$ and measure radioactivity by liquid scintillation counting.

5. Calculate concentrations of 25-(OH)D$_2$ and 25-(OH)D$_3$ based upon plasma volumes, dilution factors, and the recovery of [^{3}H]25-(OH)D$_3$ added to the original plasma sample.

Comments

1. The procedure separates 25-(OH)D$_2$ and 25-(OH)D$_3$, permitting an estimate of relative amounts of vitamins D$_2$ and D$_3$ acquired by patients.

2. The method is faster and more accurate than the competitive protein-binding assay, is more readily automated, and permits analysis of 25-(OH)D over a concentration range of 12–2500 nmol/L (5–1000 ng/mL). To encompass such a range with the competitive protein-binding assay requires initial dilutions of the plasma extract.

3. The ultraviolet detection method has a lower sensitivity than the competitive protein-binding assay. Larger volumes must be used for plasma samples containing < 10 ng 25-(OH)D/mL.

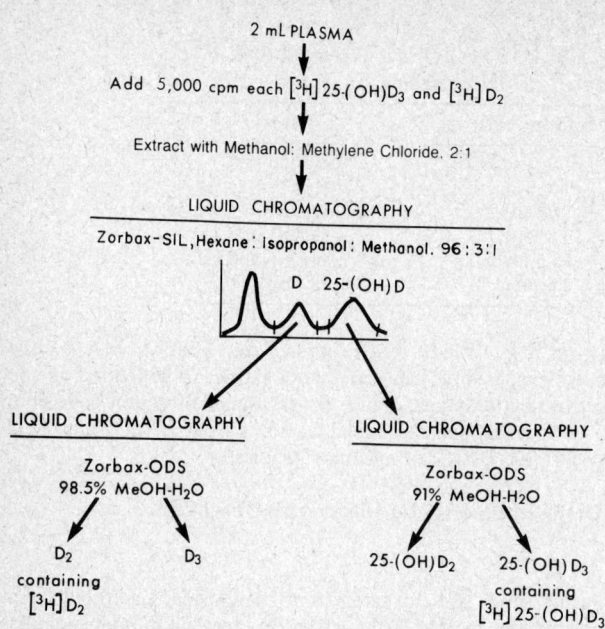

Figure 12-9. Scheme for the assay of vitamins D₂ and D₃, 25-(OH)D₂, and 25-(OH)D₃ in human plasma. (Adapted from Jones, G.: Clin. Chem., *24*:287–298, 1978.)

4. The major criterion for identification of 25-(OH)D₂ and 25-(OH)D₃ using UV-detection is retention time. Overloading of HPLC column with plasma lipids influences retention times, thus removing the major criterion for identification. The lipid interference is overcome in this method by using the Sephadex LH-20 purification step in the procedure or alternatively by employing a semi-preparative, 25 cm × 6.2 mm HPLC column. This column is also better able to handle the large sample volumes (200 μL) needed to minimize losses during the HPLC stage of the procedure.

5. There is occasional interference from peaks at 254 nm or 265 nm, probably due to aromatic compounds. Such peaks have similar retention times to 25-(OH)D and vary somewhat in amount from sample to sample. They make quantitation less precise; when retention times are identical to those for 25-(OH)D, quantitation becomes impossible.

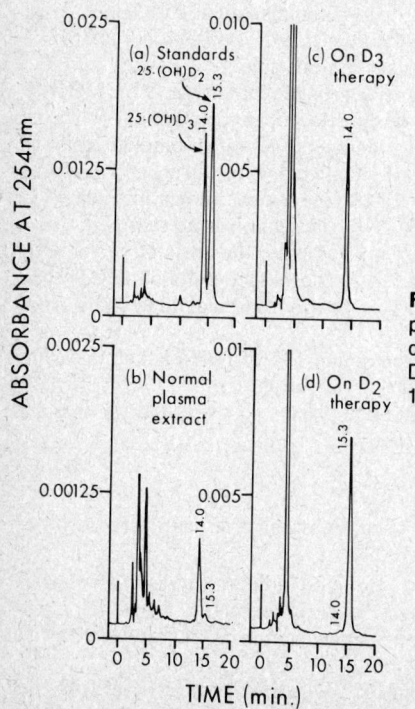

Figure 12-10. Ultraviolet detection of 25-(OH)D₂ and 25-(OH)D₃ in plasma extracts chromatographed on Zorbax-ODS. (a) Reference standards; (b) normal individuals; (c) patients on D₃ therapy; (d) patients on D₂ therapy. (Reproduced from Jones, G.: Clin. Chem., *24*:287–298, 1978.)

References

Eisman, J. A., Shepard, R. M., and DeLuca, H. F.: Determination of 25-hydroxyvitamin D_2 and 25-hydroxyvitamin D_3 in human plasma using HPLC. Anal. Biochem., *80*:298-305, 1977.

Jones, G.: Assay of vitamin D_2 and D_3 in human plasma by high performance liquid chromatography. Clin. Chem., *24*:287-298, 1978.

Determination of Total 1,25-(OH)$_2$D in Plasma by Competitive Protein Binding Assay

Principle

1,25-(OH)$_2$D is selectively extracted from plasma by use of Extrelut adsorbent (Kieselguhr) at alkaline pH 10, partially purified on Sephadex LH-20 and HPLC, and assayed by virtue of its strong binding affinity for cytosolic vitamin D–binding receptors in homogenates of chick intestinal mucosa. The extracted 1,25-(OH)$_2$D is compared with a standard curve of pure 1,25-(OH)$_2$D$_3$ with respect to its ability to compete with [^{3}H]1,25-(OH)$_2$D$_3$ for intestinal cytosol binding sites.

Reagents

1. Chick intestinal cytosol suspension. Cytosol is derived from intestinal mucosa of vitamin D–deficient chicks by the method of Eisman et al.[33] This receptor is stable for about one month at 4 °C. Suspend the chick intestinal cytosol in potassium phosphate buffer, 0.05 mol/L, pH 7.4, containing KCl, 0.2 mol/L, and dithiothreitol, 10 mmol/L, to achieve a concentration of 0.2–1.0 mg cytosol protein/mL. Alternatively, chick intestinal cytosol may be purchased from Yamasa Shoyu Company, Choshi, Japan.

2. [26,27-^{3}H]1α,25-(OH)$_2$D$_3$, 100–200 Ci/mmol (Amersham, Arlington Heights, IL). Dissolve compound in ethanol at a concentration of 5000 cpm/50 μL.

3. 1α,25-(OH)$_2$D$_3$ Standard. Dissolve 1α,25-(OH)$_2$D$_3$, crystalline (Hoffmann-La Roche, Nutley, NJ) in ethanol, to give a series of standard solutions at concentrations of 2, 5, 10, 20, 30, 40, 50, 75, 100, 200, or 10 000 pg/50 μL.

4. Extrelut adsorbent (e.g., Kieselguhr; Merck, Darmstadt, West Germany).

5. Sephadex LH-20 (Pharmacia, Uppsala, Sweden).

6. Microparticulate silica column (5–10 μ particles); 25 cm × 6.2 mm for HPLC (e.g., Zorbax-SIL, DuPont Instruments, Wilmington, DE 19898).

7. 20 mL test tubes with Teflon-lined screw-caps.

8. 5 mL conical base Reactivials with Teflon-lined screw-caps (Pierce Chemical Co., Rockford, IL 61105).

9. Neutral charcoal (0.5%) and Dextran T70 (0.05%), as a suspension in potassium phosphate buffer, 0.05 mol/L, pH 7.4, containing KCl, 0.1 mol/L, and dithiothreitol, 10 mmol/L.

Procedure

Sample Extraction and Purification

1. To 5 mL plasma, add 2000 cpm of [26,27-methyl-^{3}H] 1α,25-(OH)$_2$D$_3$ (~160 Ci/mmol).

2. After equilibration for 30 min, dilute the plasma with 15 mL of sodium hydroxide, 0.1 mol/L, and pour the mixture into an Extrelut column.

3. Elute 1,25-(OH)$_2$D from column by successive additions of 40-mL portions of hexane and of methylene chloride:methanol, 80:20.

4. Evaporate pooled extract to dryness in rotary evaporator; dissolve in hexane:chloroform:methanol, 90:10:10, and transfer to a 5-mL Reactivial and evaporate with N_2.

5. Redissolve residue in 500 μL hexane:chloroform:methanol, 90:10:10, and chromatograph on a 15.5 cm × 0.6 cm column of Sephadex LH-20 using the same solvent.

6. Discard the first 13 mL of effluent.

7. Elute 1,25-(OH)$_2$D with the next 6 mL of solvent. The exact characteristics of this elution step should be determined in the particular laboratory, using [^{3}H]1,25-(OH)$_2$D$_3$.

8. Evaporate the 1,25-(OH)$_2$D fraction to dryness and redissolve in 200 μL of hexane:isopropanol:methanol, 87:10:3.

9. Perform HPLC of the 1,25-(OH)$_2$D fraction on a 6.2 cm × 2.5 cm column of Zorbax-SIL, using hexane:isopropanol:methanol, 87:10:3, as solvent at a flow rate of 1.5 mL/min. Collect 1,25-(OH)$_2$D between 20 and 24 min as shown in Figure 12-11. The selected fraction of 1,25-(OH)$_2$D is devoid of 25-(OH)D$_3$, 24,25-(OH)$_2$D$_3$, and 25,26-(OH)$_2$D$_3$.

10. Evaporate 1,25-(OH)$_2$D fraction to dryness in a Reactivial and redissolve immediately in 210 μL ethanol.

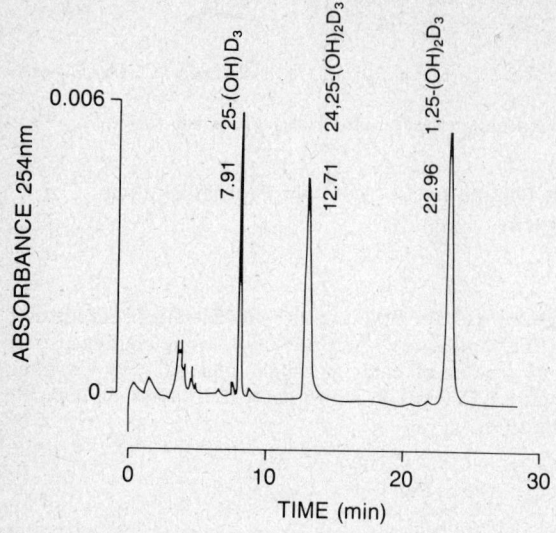

Figure 12-11. Typical elution pattern of vitamin D metabolites on HPLC. Performed with a Zorbax-SIL column (6.2 mm × 2.5 cm); solvent hexane: isopropanol:methanol, 87:10:3; flow rate 1.5 mL/min. (Redrawn from Jones, G.: J. Chromatogr. *221*:27–37, 1980.)

11. Assess recoveries of [³H]1,25-(OH)₂D₃ internal standard by measuring radioactivity in 50 μL of sample. Use remaining 160 μL for competitive binding assay.

Competitive Binding Assay

1. Mix 500 μL of chick intestinal cytosol suspension (Reagent 1) with 50 μL of [³H]1,25-(OH)₂D₃ solution (5000 cpm in ethanol). Add 50 μL of standard 1,25-(OH)₂D₃ or 50 μL plasma extract from Step 11. Perform assay in triplicate.
2. Incubate for 1 h at 25 °C in a shaking water-bath.
3. Plunge tubes into an ice bath and simultaneously add 100 μL of charcoal-Dextran suspension.
4. Centrifuge at 3000 × g for 15 min to separate supernatant (bound ligand) from pellet (free ligand).
5. Measure radioactivity in a 500-μL aliquot of the supernatant using a water-based cocktail in a liquid scintillation counter.
6. Compare bound ligand in the sample to the standard curve (Figure 12-12).
7. Calculate the concentration of 1,25-(OH)₂D after correction for sample volume, various dilution factors, and the fraction of 1,25-(OH)₂D recovered through the purification steps. The result is usually expressed in pmol/L. See Table 12-9 for reference range.

Comments on the Procedure, Modifications, and Other Procedures

1. Because chick intestinal cytosol has a lower binding affinity for 1,25-(OH)₂D₂ than for 1,25-(OH)₂D₃, the assay provides a slight but measurable underestimation of 1,25-(OH)₂D₂ in plasma specimens. This may result in low 1,25-(OH)₂D values if the patient is treated with large doses of vitamin D₂.
2. Reinhardt et al.[110] have described an extraction procedure for 1,25-(OH)₂D which utilizes alkaline extraction of plasma on a C₁₈-SEP PAK (Waters Associates, Milford, MA 01757) as an alternative to the

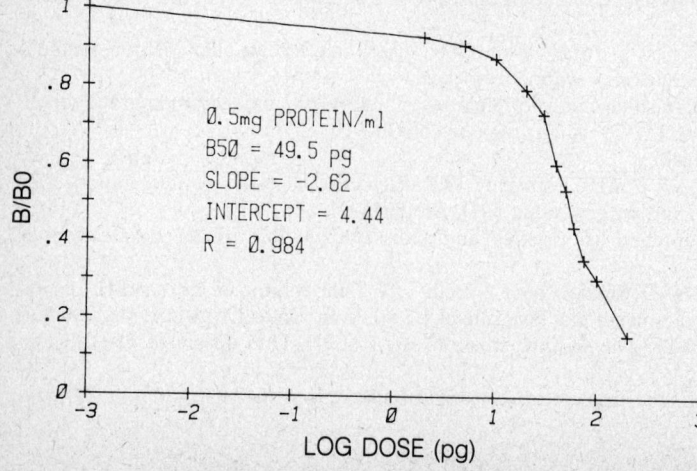

Figure 12-12. Competitive protein binding assay of 1,25-(OH)₂D. Standard curve constructed with chick intestinal cytosol at a protein content of 0.5 mg/mL.

Kieselguhr column (Extrelut) (Steps 2 and 3). Their method also substitutes a purification on a SIL-SEP PAK (Waters, Milford, MA) for chromatography on Sephadex LH-20 and HPLC on Zorbax-SIL. A mammalian 1,25-$(OH)_2D_3$ receptor derived from bovine thymus cytosol is used in the binding assay. Kao and Heser[66] report a similar procedure for measurement of 25-(OH)D and 1,25-$(OH)_2D_3$ based upon Bond Elut C_{18} and NH_2 cartridges (Analytichem International, Harbor City, CA 90710) and a step of HPLC prior to the binding assay.

3. Several antibodies[40] have now been prepared against modified vitamin D molecules and have been used with varying degrees of success to measure total 1,25-$(OH)_2D$ or 1,25-$(OH)_2D_3$. Newer antibodies have decreased cross-reactivity against vitamin D_3 metabolites other than 1,25-$(OH)_2D_3$, and improved cross-reactivity against 1,25-$(OH)_2D_2$. Sample purification prior to radioimmunoassay is identical to that described for the competitive binding assay (Steps 1–11). The conditions for radioimmunoassay differ slightly in the use of an overnight incubation step at 0–4 °C for binding and the use of 1% BSA in the incubation buffer. (This necessitates the use of 1% charcoal coated with 0.1% dextran in a later step.) The main advantage of the radioimmunoassay over the competitive binding assay for 1,25-$(OH)_2D$ is that the antibody is much more stable than the chick cytosolic receptor and the results are more consistent from assay to assay.

4. An ingenious method of Manolagas and Deftos[84] reduces the degree of purification of 1,25-$(OH)_2D$ necessary prior to radioreceptor assay by incubating the 1,25-$(OH)_2D$ fraction from the plasma specimen with intact cultured rat bone cells containing 1,25-$(OH)_2D_3$ receptors. The cell membrane serves the dual purpose of reducing the entry of interfering vitamin D metabolites into the cell for binding to the 1,25-$(OH)_2D_3$ receptor and of helping in the separation of bound from free ligand. The use of a mammalian 1,25-$(OH)_2D_3$ receptor in this assay and that of Reinhardt et al.[110] overcomes the slight underestimation of 1,25-$(OH)_2D_2$ observed in the competitive binding assay using chick intestinal cytosol.

5. Another variation of the measurement of 1,25-$(OH)_2D$ is the use of a bone resorption bioassay[121] in conjunction with the purification and chromatography described above; the purified 1,25-$(OH)_2D$ fractions are tested for their ability to release deep-seated ^{45}Ca from cultured long bones of fetal rats. The amount of ^{45}Ca released into the culture medium is a measure of bone (^{45}Ca-labeled hydroxyapatite) resorption. This method offers a sensitivity limit of 2.5 fmol (1 pg) of 1,25-$(OH)_2D$.

Reference

The method is based upon the published procedure of Eisman, J. A., Hamstra, A. J., Kream, B. E., et al.: A sensitive, precise and convenient method for determination of 1,25-dihydroxyvitamin D in human plasma. Arch. Biophys. Biochem., *176*:235-243, 1976.

Determination of 24,25-$(OH)_2D_3$ in Plasma by Competitive Protein Binding Assay*

Principle

24,25-$(OH)_2D_3$ is extracted from plasma along with total lipids including 25-(OH)D; the metabolite is then separated from 25-(OH)D by chromatography on HPLC, and assayed by virtue of its strong binding affinity for DBP. The affinity of the 24,25-$(OH)_2D_3$ for DBP is equal to that of 25-$(OH)D_3$. Per cent bound 24,25-$(OH)_2D_3$ is compared with a standard curve of pure 25-$(OH)D_3$ in respect to its ability to compete with [^3H]25-$(OH)D_3$ for DBP binding. Need for a standard curve of 24,25-$(OH)_2D_3$ is thus avoided. The method described is based upon a procedure of Jones.[62]

Reagents

As in the determination of total 25-$(OH)D_3$ in plasma by CPB assay, with the following exceptions:
1. Internal standard. [26,27-^3H]24,25-$(OH)_2D_3$, 100–200 Ci/mmol, dissolved in ethanol, 200 cpm/50 µL.
2. 5–10 µ cyano-bonded microparticulate column (e.g., 25 cm × 4.6 mm Zorbax-CN, DuPont) is used in place of microparticulate silica.

Procedure

1. To 250–500 µL of plasma, add 50 µL internal standard and dilute to 2 mL with saline, 0.15 mol/L.
2. After 30 min equilibration, carry out lipid extraction, as in the Determination of Total 25-(OH)D, Sample Extraction and Purification, steps 2–7, page 1363.
3. Dissolve the resultant lipid extract in 220 µL hexane:isopropanol:methanol, 94:5:1.
4. Perform HPLC by injecting 200 µL of this solution onto the cyano-bonded microparticulate column using hexane:isopropanol:methanol, 94:5:1, as solvent at a flow rate of 1.3 mL/min.
5. Collect the 24,25-$(OH)_2D_3$ in a Reactivial between 10.5 and 13 min (the retention time of 24,25-

*This assay is not suitable if the patient is receiving large doses of vitamin D_2. (See comments on Methods.)

$(OH)_2D_3$ is 11 min, Table 12-10). Discard 25-(OH)D_3 and 25-(OH)D_3-26,23-lactone, which elute at around 7 and 20 min, respectively.

6. Evaporate the plasma 24,25-(OH)$_2D_3$ fraction to dryness under a stream of N_2 and redissolve in 250 μL ethanol.

7. Measure radioactivity in a 50-μL aliquot of the plasma extract by liquid scintillation counting. Assess recovery by comparison of radioactivity with that added to original plasma.

8. Perform CPB assay in triplicate on 50-μL aliquots of the plasma extract as described under Determination of Total 25-(OH)D Assay Procedure, steps 1–5, using [26,27-³H]25-(OH)D_3 as the radioactive ligand and a standard curve prepared with 25-(OH)D_3.

9. Calculate the concentration of 24,25-(OH)$_2D_3$ after correction for sample volume, various dilution factors and the fraction of 24,25-(OH)$_2D_3$ recovered through the extraction and purification steps. The result is usually expressed in nmol/L. See Table 12-9 for reference range.

Comments

1. Because extraction and purification steps common to both 25-(OH)D and 24,25-(OH)$_2D_3$ are used, both can be measured simultaneously in the same aliquot of serum by performing competitive protein binding assays on the 25-(OH)D and 24,25-(OH)$_2D_3$ fractions against a single standard curve of 25-(OH)D_3. Use in the assay of expensive [26,27-³H]24,25-(OH)$_2D_3$ is limited to its role as an internal standard.

2. The detection limit of the assay is ~1 ng/mL (2.5 nmol/L). To detect concentrations of 24,25-(OH)$_2D_3$ below this value, a more sensitive assay using a 25 cm $\times$ 6.2 mm column and a plasma volume of 2 mL could be employed.

3. Dreyer and Goodman[27] described a method for measurement of 24,25-(OH)$_2D_3$ based upon two steps of HPLC and ultraviolet detection of the metabolite at 254 nm. Their method utilizes Zorbax-SIL HPLC with methylene chloride:methanol, 97:3, as solvent to separate 24,25-(OH)$_2D_3$ and 25-(OH)D_3-26,23-lactone.

4. This method should not be used in individuals receiving large doses of vitamin D_2.

Reference

This procedure is modified from Jones, G.: Chromatographic separation of 24R,25-dihydroxyvitamin D_3 and 25-dihydroxyvitamin D_3-26,23-lactone using a cyano-bonded phase packing. J. Chromatogr. Biomed. Appl., 276:69-75, 1983.

Factors Complicating Current Procedures for Determining Total 24,25-(OH)$_2$D

1. Many of the present methods, including that presented here, employ DBP in the assay. This protein binds 24,25-(OH)$_2D_2$ with less affinity than 24,25-(OH)$_2D_3$, therefore resulting in an underestimate of 24,25-(OH)$_2D_2$. Thus, if an individual receives large amounts of vitamin D_2, the form employed in prophylactic vitamin D medications and in some fortified foods in North America, total 24,25-(OH)$_2$D in plasma is underestimated.

2. Most chromatographic methods separate 24,25-(OH)$_2D_2$ and 24,25-(OH)$_2D_3$, and since 24,25-(OH)$_2D_3$ is used as the indicator of recovery of 24,25-(OH)$_2$D, great care must be taken to ensure that there is no unrecognized greater loss of 24,25-(OH)$_2D_2$ during the sample preparation stage that precedes the actual assay.

Table 12-10. RETENTION TIMES OF VITAMIN D METABOLITES ON ZORBAX-CN AND ZORBAX-SIL[62]

Metabolite	Zorbax-CN 25 cm $\times$ 4.6 mm Hexane:Isopropanol: Methanol, 94:5:1 1.3 mL/min	Zorbax-SIL 25 cm $\times$ 4.6 mm Hexane:Isopropanol: Methanol, 91:7:2 1.5 mL/min
Vitamin D_3	3.87	3.60
25-(OH)D_2	6.66	6.28
25-(OH)D_3	7.11	6.78
24-Keto-25-(OH)D_3	9.78	7.54
24R,25-(OH)$_2D_2$	11.30	10.39
24R,25-(OH)$_2D_3$	11.43	11.08
24-Keto-23,25-(OH)$_2D_3$	14.82	11.32
25S,26-(OH)$_2D_3$	14.09	14.49
1α-25-(OH)$_2D_3$	18.66	21.90
25-(OH)D_3-26,23-lactone	20.71	11.71

Note that incomplete resolution of 24R,25-(OH)$_2D_3$ and 25-(OH)D_3-26,23-lactone occurs on Zorbax-SIL.

3. In patients with hypoparathyroidism and vitamin D refractory syndromes treated with pharmacological doses of vitamin D_2, the form present in all high potency vitamin D preparations in North America, significant amounts of $25,26\text{-}(OH)_2D_2$ are present in plasma. This metabolite co-migrates with $24,25\text{-}(OH)_2D_3$ on standard liquid chromatograms resulting in a falsely high value with many $24,25\text{-}(OH)_2D$ assays.

4. $25\text{-}(OH)D_3\text{-}26,23\text{-lactone}$ has at least three-fold greater affinity for DBP than has $24R,25\text{-}(OH)_2D_3$. The two compounds co-migrate on most LC systems used for purification of $24,25\text{-}(OH)_2D$ prior to the competitive protein binding assay. Thus, inadequate resolution of $25\text{-}(OH)D_3\text{-}26,23\text{-lactone}$ and $24,25\text{-}(OH)_2D_3$ could contribute to falsely high values of total $24,25\text{-}(OH)_2D$.

Determination of Vitamin D in Plasma by Ultraviolet Detection[61]

Principle

Vitamin D is extracted from plasma along with total lipids, separated from the bulk of the lipids and the vitamin D metabolites by HPLC on microparticulate silica, then further purified by a second HPLC step on octadecasilane-bonded silica prior to quantitation by ultraviolet detection. This method is suitable for measurement of vitamin D in plasma of individuals receiving therapeutic doses of vitamin D_2 or D_3.

Reagents

As described in the determination of total $25\text{-}(OH)D$ in plasma by CPB assay, with the following exception: $[1,2\text{-}^3H]$ vitamin D_3, 10–20 Ci/mmol (Amersham, Arlington Heights, IL), dissolved in ethanol, 5000 cpm/50 μL, is used in place of $[^3H]25\text{-}(OH)D_3$.

Procedure

1. To a volume of plasma ranging from 250 μL to 2 mL, add 50 μL of the ethanolic solution of $[1,2\text{-}^3H]$ vitamin D_3 (5000 cpm). Dilute to 2 mL with saline, 0.15 mol/L.

2. After 30 min equilibration, extract as described in the determination of total $25\text{-}(OH)D$ in plasma by CPB assay, steps 2–7, and purify by HPLC on microparticulate silica, steps 14–15.

3. Collect and pool fractions at 6–8 min from the Zorbax-SIL column eluted with hexane:isopropanol:methanol, 96:3:1, corresponding to the elution time of vitamin D (Figure 12-9). These fractions contain a mixture of vitamins D_2 and D_3.

4. Evaporate under a stream of N_2 and redissolve the residue in 220 μL methanol:water, 98.5:1.5.

5. Using 200 μL of sample, perform a second step of chromatography on Zorbax-ODS using methanol:water, 98.5:1.5, as solvent.

Figure 12-13. Ultraviolet detection of vitamins D_2 and D_3 in plasma extracts, chromatographed on Zorbax-ODS. (a) Reference standards; (b) normal individual; (c) patients on vitamin D_3 therapy; (d) patients on vitamin D_2 therapy. (Reproduced from Jones, G.: Clin. Chem., 24:287–298, 1978.)

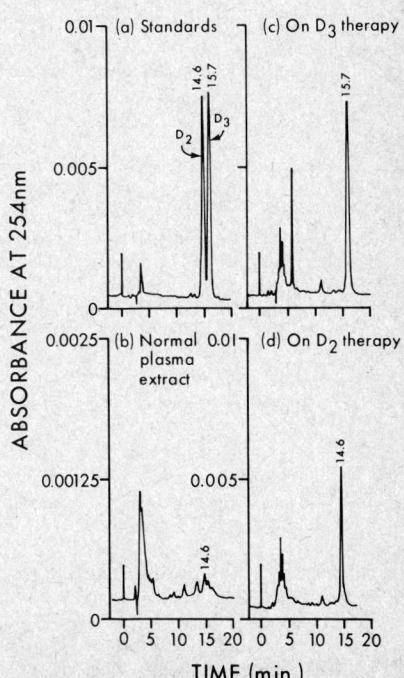

6. Assess recovery of $[1,2-{}^3H]$ vitamin D_3 internal standard by collecting the eluting vitamin D peak in a scintillation vial. Evaporate to dryness and measure radioactivity in a scintillation counter.

7. Measure UV-absorption of eluting peaks at 254 or 265 nm. Integrate the areas of HPLC peaks corresponding to vitamins D_2 and D_3, 14.6 min and 15.7 min, respectively (Figure 12-13).

8. Calculate concentrations of vitamins D_2 and D_3 in plasma after correction for sample volume and the fraction of vitamin D recovered through the purification steps. The result is expressed in nmol/L.

Reference

This procedure is based on a method published by Jones, G.: Assay of vitamin D_2 and D_3 in human plasma by high performance liquid chromatography. Clin. Chem., *24*:287-298, 1978.

Comment

The detection limit for vitamin D in plasma by the UV detection method described above is approximately 50 nmol/L. This sensitivity is ample for assaying plasma of patients treated with large doses of vitamin D, but is not entirely satisfactory in individuals under natural conditions of vitamin D intake and exposure to sunlight.

A more sensitive procedure using sheep DBP has been reported by Horst et al.[57] (detection limit approximately 1.25 nmol/L). However, this method has several drawbacks. It is less convenient, extensive purification of samples is required, binding of vitamin D to DBP is slow, and the binding affinities for vitamins D_2 and D_3 may differ. The method using sheep DBP is recommended only in the relatively unusual circumstance when knowledge of the load of vitamin D is required in individuals under natural conditions of vitamin exposure.

References

1. Adams, J. S., Clemens, T. L., Parrish, J. A., et al.: Vitamin D synthesis and metabolism after ultraviolet irradiation of normal and vitamin D–deficient subjects. N. Engl. J. Med., *306*:722-725, 1982.
2. Anker, P., Wieland, E., Ammann, D., et al.: Neutral carrier based ion-selective electrode for the determination of total calcium in blood serum. Anal. Chem., *53*:1970-1974, 1981.
3. Arnaud, C. D.: The parathyroid glands. *In*: Cecil Textbook of Medicine. 16th ed. J. B. Wyngaarden and L. H. Smith, Jr., Eds. Philadelphia, W. B. Saunders Co., 1982, pp. 1286-1302.
4. Aurbach, G. D., Marx, S. J., and Spiegel, A. M.: Parathyroid hormone, calcitonin and the calciferols. *In*: Textbook of Endocrinology. 6th ed. R. H. Williams, Ed. Philadelphia, W. B. Saunders Co., 1981, pp. 922-1031.
5. Aurbach, G. D., and Phang, J. M.: Vitamin D, parathyroid hormone, and calcitonin. *In*: Medical Physiology, Vol. 2. V. B. Mountcastle, Ed. St. Louis, C. V. Mosby Co., 1980, pp. 1519-1557.
6. Austin, L. A., and Heath, H., III: Calcitonin. Physiology and pathophysiology. N. Engl. J. Med., *304*:269-278, 1981.
7. Bijvoet, O. L. M.: Kidney function in calcium and phosphate metabolism. *In*: Metabolic Bone Disease, Vol. 1. L. V. Avioli and S. M. Krane, Eds. New York, Academic Press, 1977, pp. 49-140.
8. Biswas, C. K., Ramos, J. M., and Kerr, D. N. S.: Heparin effect on ionized calcium concentration. Clin. Chim. Acta, *116*:343-347, 1981.
9. Bowers, G. N., Jr.: A reference system for ionized calcium. Scand. J. Clin. Lab. Invest., *43* (Suppl. 165):49-56, 1983.
10. Boyd, J. C., Lewis, J. W., Slatopolsky, E., et al.: Parathyrin measured concurrently with free or total calcium in the differential diagnosis of hypercalcemia. Clin. Chem., *27*:574-579, 1981.
11. Brinkhurst, F. R., and Potts, J. T., Jr.: Calcium and phosphate distribution, turnover, and metabolic actions. *In*: Endocrinology, Vol. 2. L. J. DeGroot et al., Eds. New York, Grune & Stratton, 1979, pp. 551-585.
12. Broadus, A. E., and Stewart, A. F.: Humoral hypercalcemia of malignancy. *In*: Clinical Disorders of Bone and Mineral Metabolism. B. Frame and J. T. Potts, Jr., Eds. Amsterdam, Excerpta Medica, 1983, pp. 284-287.
13. Cali, J. P., Bowers, G. N., Jr., and Young, D. S.: A reference method for the determination of total calcium in serum. Clin. Chem., *19*:1208-1213, 1973.
14. Cheng, M. H., Lipsey, A. I., Blanco, V., et al.: Microchemical analysis for 13 constituents of plasma from healthy children. Clin. Chem., *25*:692-698, 1979.
15. Cherian, A. G., and Hill, J. G.: Percentile estimates of reference values for fourteen chemical constituents in sera of children and adolescents. Am. J. Clin. Pathol., *69*:24-31, 1978.
16. Cheung, W. Y.: Calmodulin. Sci. Am., *246*:62-70, 1982.
17. Coburn, J. W., and Slatopolsky, E.: Vitamin D, parathyroid hormone, and renal osteodystrophy. *In*: The Kidney, Vol. 2. B. M. Brenner and F. C. Rector, Jr., Eds. Philadelphia, W. B. Saunders Co., 1981, pp. 2213-2305.
18. Conceicao, S. C., Weightman, D., Smith, P. A., et al.: Serum ionized calcium concentration: Measurement versus calculation. Br. Med. J., *1*:1103-1105, 1978.
19. Copeland, B. E., Grisley, D. W., Casella, J., et al.: Comparison of serum calcium measurements with respect to five models of atomic absorption spectrometers using NBS-AACC calcium reference method and isotope-dilution mass spectrometry as the definitive method. Am. J. Clin. Pathol., *66*:619-633, 1976.

20. Craggs, A., Moody, G. J., Thomas, J. D. R., et al.: Effect of anionic surfactants on calcium ion–selective electrodes. Analyst, *105*:426-431, 1980.

21. Daneman, D., Kooh, S. W., and Fraser, D.: Hypoparathyroidism and pseudohypoparathyroidism in childhood. Clin. Endocrinol. Metab., *11*:211-231, 1982.

22. Das, J. B., Eraklis, A. J., Filler, R. M., et al.: Serum ionic calcium: Changes with large volume blood transfusion in the infant. J. Pediatr. Surg., *6*:333-338, 1971.

23. David, L., and Anast, C. S.: Calcium metabolism in newborn infants: The interrelationship of parathyroid function and calcium, magnesium, and phosphorus metabolism in normal, sick, and hypocalcemic newborns. J. Clin. Invest., *54*:287-296, 1974.

24. Day, G. M., Chance, G. W., Radde, I. C., et al.: Growth and mineral metabolism in very low birthweight infants. II. Effects of calcium supplementation on growth and divalent cations. Pediatr. Res., *9*:568-575, 1975.

25. Deluca, H. F., and Schnoes, H. K.: Vitamin D: Recent advances. Ann. Rev. Biochem., *52*:411-439, 1983.

26. Donhowe, J. M., Freier, E. F., Wong, E. T., et al.: Factitious hypophosphatemia related to mannitol therapy. Clin. Chem., *27*:1765-1769, 1981.

27. Dreyer, B. E., and Goodman, D. B. P.: A simple direct spectrophotometric assay for 24,25-dihydroxyvitamin D_3. Anal. Biochem., *114*:37-41, 1981.

28. Drop, L. J., Tochka, L. N., and Misiano, D. R.: Comparative evaluation of two calcium ion–selective electrode systems and their utility for monitoring steady-state changes in [Ca^{2+}]. Clin. Chem., *28*:129-133, 1982.

29. Dryer, R. L., and Routh, J. I.: Determination of serum inorganic phosphorus. Stand. Methods Clin. Chem., *4*:191-195, 1963.

30. Duthie, D., and Fraser, D.: Unpublished observations.

31. Edmondson, J. W., and Li, T-K.: The relationship of serum ionized calcium and total calcium in primary hyperparathyroidism. J. Lab. Clin. Med., *87*:624-629, 1976.

32. Eil, C., Libermann, U. A., Rosen, J. F., et al.: A cellular defect in hereditary vitamin D–dependent rickets Type II: Defective nuclear uptake of 1,25-dihydroxyvitamin D in cultured skin fibroblasts. N. Engl. J. Med., *304*:1588-1591, 1981.

33. Eisman, J. A., Hamstra, A. J., Kream, B. E., et al.: A sensitive, precise and convenient method for determination of 1,25-dihydroxyvitamin D in human plasma. Arch. Biochem. Biophys., *176*:235-243, 1976.

34. Eisman, J. A., Shepard, R. M., and DeLuca, H. F.: Determination of 25-hydroxyvitamin D_2 and 25-hydroxyvitamin D_3 in human plasma using HPLC. Anal. Biochem., *80*:298-305, 1977.

35. Farfel, Z., Brickman, A. S., Kaslow, H. R., et al.: Defect of receptor-cyclase coupling protein in pseudohypoparathyroidism. N. Engl. J. Med., *303*:237-242, 1980.

36. Fassel, V. A., and Becker, D. A.: Chemical or solute vaporization interference in flame atomic emission and absorption spectrometry. Anal. Chem., *41*:1522-1526, 1969.

37. Fiske, C. H., and SubbaRow, Y.: The colorimetric determination of phosphorus. J. Biol. Chem., *66*:375-400, 1925.

38. Fogh-Anderson, N., Christiansen, T. F., Komarmy, L., et al.: Measurement of free calcium ion in capillary blood and serum. Clin. Chem., *24*:1545-1552, 1978.

39. Forfar, J. O.: Normal and abnormal calcium, phosphorus, and magnesium metabolism in the perinatal period. Clin. Endocrinol. Metab., *5*:123-134, 1976.

40. Fraher, L. J., Adami, S., Clemens, T. L., et al.: Radioimmunoassay of 1,25-dihydroxyvitamin D_2: Studies on the metabolism of vitamin D_2 in man. Clin. Endocrinol., *18*:151-165, 1983.

41. Fraser, D., and Kooh, S. W.: Disorders of the endocrine glands: Disturbances of parathyroid hormone and calcitonin. *In*: Textbook of Paediatrics, Vol. 2. 3rd ed. J. O. Forfar and G. C. Arneil, Eds. Edinburgh, Churchill Livingstone, 1984, pp. 1157-1172.

42. Fraser, D., Kooh, S. W., and Scriver, C. R.: Hyperparathyroidism as the cause of hyperaminoaciduria and phosphaturia in human vitamin D deficiency. Pediatr. Res., *1*:425-435, 1967.

43. Garber, C. C., and Miller, R. C.: Revisions of the 1963 semidine HCl standard method for inorganic phosphorus. Clin. Chem., *29*:184-188, 1983.

44. Gitelman, H. J.: An improved automated procedure for the determination of calcium in biological specimens. Anal. Biochem., *18*:521-530, 1967.

45. Goltzman, D., Henderson B., and Loveridge, N.: Cytochemical bioassay of parathyroid hormone. J. Clin. Invest., *65*:1309-1317, 1980.

46. Gomez, P., Coca, C., Vargas, C., et al.: Normal reference-intervals for 20 biochemical variables in healthy infants, children, and adolescents. Clin. Chem., *30*:407-412, 1984.

47. Gomori, G.: A modification of the colorimetric phosphorus determination for use with a photoelectric colorimeter. J. Lab. Clin. Med., *27*:955-960, 1942.

48. Haas, R. G., and Mushel, S.: Modified DuPont *aca* calcium method for hemolyzed specimens. Am. J. Clin. Pathol., *77*:216-219, 1982.

49. Haefner, J. F., and Potts, J. T., Jr.: Biosynthesis of parathyroid hormone. N. Engl. J. Med., *299*:580–585, 635-644, 1978.

50. Heath, D. A.: Hypercalcemia of malignancy. *In*: Calcium Disorders. D. A. Heath and S. J. Marx, Eds. London, Butterworth, 1982, pp. 233-247.

51. Heath, H., III: Tests of parathyroid function: Utility and limitations. Endocrinol. and Metab., *2*:1-10, 1984.

52. Heath, H., III, and Sizemore, G. W.: Plasma calcitonin in normal man: Differences between men and women. J. Clin. Invest., *60*:1135-1140, 1977.

53. Henrich, W. L., Hunt, J. M., and Nixon, J. V.: Increased ionized calcium and left ventricular contractility during hemodialysis. N. Engl. J. Med., *310*:19-23, 1984.

54. Hillyard, C. J., Cooke, T. J. C., Coombes, R. C., et al.: Normal plasma calcitonin: Circadian variation and response to stimuli. Clin. Endocrinol. (Oxf), *6*:291-298, 1977.

55. Hodgkinson, A., and Pyrah, L. N.: The urinary excretion of calcium and inorganic phosphate in 344 patients with calcium stone of renal origin. Br. J. Surg., 46:10-18, 1958.
56. Horst, R. L., Littledike, E. T., Riley, J. L., et al.: Quantitation of vitamin D and its metabolites and their plasma concentrations in five species of animals. Anal. Biochem., 116:189-203, 1981.
57. Horst, R. L., Reinhardt, T. A., Beitz, D. C., et al.: A sensitive competitive protein binding assay for vitamin D in plasma. Steroids, 27:581-591, 1981.
58. Imawari, M., Kozawa, K., Yoshida, T., et al.: A simple and sensitive assay for 25-hydroxyvitamin D, 24,25-dihydroxyvitamin D and 1,25-dihydroxyvitamin D in human serum. Clin. Chim. Acta, 124:63-73, 1982.
59. Itaya, K., and Ui, M.: A new micromethod for the colorimetric determination of inorganic phosphate. Clin. Chim. Acta, 14:361-366, 1966.
60. Jones, G.: Unpublished observations.
61. Jones, G.: Assay of vitamin D_2 and D_3 in human plasma by high performance liquid chromatography. Clin. Chem., 24:287-298, 1978.
62. Jones, G.: Chromatographic separation of 24R,25-dihydroxyvitamin D_3 and 25-hydroxyvitamin D_3-26,23-lactone using a cyano-bonded phase packing. J. Chromatogr., 276:69-75, 1983.
63. Jones, G.: Ternary solvent mixtures for improved resolution of hydroxylated metabolites of vitamin D_2 and vitamin D_3 during high performance liquid chromatography. J. Chromatogr., 221:27-37, 1980.
64. Jones, G., and DeLuca, H. F.: High pressure liquid chromatography: Separation of the metabolites of vitamin D_2 and D_3 on small-particle silica columns. Lipid Res., 16:448-453, 1975.
65. Kan, R. C., Jascott, D., Carlon, G. C., et al.: Massive blood replacement: Correlation of ionized calcium, citrate and hydrogen ion concentration. Anesth. Analg., 58:274-278, 1979.
66. Kao, P. C., and Heser, D. W.: Simultaneous determination of 25-hydroxy- and 1,25-dihydroxyvitamin D from a single sample by dual-cartridge extraction. Clin. Chem., 30:56-61, 1984.
67. Kao, P. C., Jiang, N., Klee, G. G., et al.: Development and validation of a new radioimmunoassay for parathyrin. Clin. Chem., 28:69-74, 1982.
68. Kaufman, R. A., and Tietz, N. W.: Ion effects in measurement of ionized calcium with a calcium-selective electrode. Clin. Chem., 26:640-644, 1980.
69. Keating, F. R., Jones, J. D., Elveback, L. R., et al.: The relation of age and sex to distribution of values in healthy adults of serum calcium, inorganic phosphorus, magnesium, alkaline phosphatase, total proteins, albumin, and blood urea. J. Lab. Clin. Med., 73:825-834, 1969.
70. Knapp, E. L.: Factors influencing the urinary excretion of calcium. I. In normal persons. J. Clin. Invest., 28:182-196, 1947.
71. Kooh, S. W., Jones G., Reilly, B. J., et al.: Pathogenesis of rickets in chronic hepatobiliary disease in children. J. Pediatr., 94:870-874, 1979.
72. Ladenson, J. H., and Bowers, G. N., Jr.: Free calcium in serum. I. Determination with the ion-specific electrode and factors affecting the results. Clin. Chem., 19:565-574, 1973.
73. Ladenson, J. H., Lewis, J. W., McDonald, J. M., et al.: Relationship of free and total calcium in hypercalcemic conditions. J. Clin. Endocrinol. Metab., 48:393-397, 1978.
74. Lambert, P. W., DeOreo, P. B., Hollis, B. W., et al.: Concurrent measurement of plasma levels of vitamin D and five of its metabolites in normal humans, chronic renal failure patients and anephric subjects. J. Lab. Clin. Med., 98:536-548, 1981.
75. Larsson, L., Finnström, O., Nilsson, B., et al.: Evaluation of Radiometer ICA 1 as a routine instrument for serum ionized calcium and its application for whole blood capillary samples from newborn infants. Scand. J. Clin. Lab. Invest., 43 (Suppl. 165):21-26, 1983.
76. Larsson, L., and Öhman, S.: Serum ionized calcium and corrected total calcium in border-line hyperparathyroidism. Clin. Chem., 24:1962, 1978.
77. Larsson, L., and Öhman, S.: Serum calcium ion activity. Some aspects on methodological differences and intraindividual variation. Clin. Biochem., 12:138, 1979.
78. Law, W. M., Jr., Nissenson, R. A., Klee, G. G., et al.: Preparation of synthetic bovine parathyroid hormone fragment 1–34 for parenteral use in human studies. J. Clin. Endocrinol. Metab., 56:1335-1337, 1983.
79. Levine, M. A., Downs, R. W., Jr., Moses, A. M., et al.: Resistance to multiple hormones in patients with pseudohypoparathyroidism. Am. J. Med., 74:545-555, 1983.
80. Lindall, A. W., Elting, J., Ellis, J., et al.: Estimation of biologically active intact parathyroid hormone in normal and hyperparathyroid sera by sequential N-terminal immunoextraction and midregion radioimmunoassay. J. Clin. Endocrinol. Metab., 57:1007-1014, 1983.
81. Lorentz, K.: Improved determination of serum calcium with 2-cresolphthalein complexone. Clin. Chim. Acta, 126:327-334, 1982.
82. Low, J. C., Schaaf, M., Earl, J. M., et al.: Ionic calcium determination in primary hyperparathyroidism. JAMA, 223:152-155, 1973.
83. Mallette, L. E., Tuma, S. N., Berger, R. E., et al.: Radioimmunoassay for the middle region of human parathyroid hormone using a homologous antiserum with a carboxyl-terminal fragment of bovine parathyroid hormone as radioligand. J. Clin. Endocrinol. Metab., 54:1017-1024, 1982.
84. Manolagas, S. C., and Deftos, L. J.: Cytoreceptor assay for 1,25-dihydroxyvitamin D_3: A novel radiometric method based on binding of the hormone to intracellular receptors in vitro. Lancet, 2:401-402, 1980.
85. Martin, K. J., Hruska, K., Freitag, J., et al.: Clinical utility of radioimmunoassays for parathyroid hormone. Min. Electr. Metab., 3:283-290, 1980.
86. Marx, S. J., Spiegel, A. M., Brown, E. M., et al.: Divalent calcium metabolism: Familial hypocalciuric hypercalcemia versus typical primary hyperparathyroidism. Am. J. Med., 65:235-242, 1978.
86a. Mills, B. G., Singer, F. R., Weiner, L. P., et al.: Immunohistological demonstration of respiratory syncytial virus antigens in Paget's disease of bone. Proc. Natl. Acad. Sci. USA, 78:1209-1213, 1981.

87. Moore, E. W.: Studies with ion-exchange calcium electrodes. III. The state of serum calcium in patients with cirrhosis. Gastroenterology, *60*:43-54, 1971.

88. Moore, L. J., and Machlan, L. A.: High accuracy determination of calcium in blood serum by isotope dilution mass spectrometry. Anal. Chem., *44*:2291-2296, 1972.

89. Moorehead, W. R., and Biggs, H. G.: 2-Amino-2-methyl-1-propanol as the alkalizing agent in an improved continuous-flow cresolphthalein complexone procedure for calcium in serum. Clin. Chem., *20*:1458-1460, 1974.

90. Muhlert, M., Julita, M., and Quamme, G.: Disappearance of calcium and other electrolytes from microvolume samples. Am. J. Physiol., *242*:F202-F206, 1982.

91. Mundy, G. R., and Martin, J.: The hypercalcemia of malignancy: Pathogenesis and management. Metabolism, *31*:1247-1277, 1982.

92. Murray, T. M., and Keutmann, H. T.: The immunochemical specificity of antisera to bovine parathyroid hormone: An approach to region-specific radioimmunoassay. J. Endocrinol., *56*:493-501, 1973.

93. Neer, R. M.: Calcium and inorganic phosphate homeostasis. *In*: Endocrinology, Vol. 2. L. Groot et al., Eds. New York, Grune and Stratton, Publishers, 1979, pp. 669-692.

94. Nordin, B. E. C., and Fraser, R.: Assessment of phosphate excretion. Lancet *1*:947-951, 1960.

95. Nordin, B. E. C., Peacock, M., and Wilkinson, R.: Hypercalciuria and calcium stone disease. Clin. Endocrinol. Metab., *1*:169-183, 1972.

96. Olinger, G. N., Hottenrott, C., Mulder, D. G., et al.: Acute clinical hypocalcemic myocardial depression during rapid blood transfusion and postoperative hemodialysis. J. Thorac. Cardiovasc. Surg., *72*:503-511, 1976.

97. Pak, C. Y. C., Britton, I., Peterson, R., et al.: Ambulatory evaluation of nephrolithiasis. Am. J. Med., *69*:19-30, 1980.

98. Parfitt, A. M.: Chlorothiazide-induced hypercalcemia in juvenile osteoporosis and hyperparathyroidism. N. Engl. J. Med., *281*:55-59, 1969.

99. Parthemore, J. G., and Deftos, L. H.: Calcitonin secretion in normal human subjects. J. Clin. Endocrinol. Metab., *47*:184-188, 1978.

100. Parvainen, M. T., Savolainen, K. E., Korhonen, P. H., et al.: An improved method for routine determination of vitamin D and its hydroxylated metabolites in serum from children and adults. Clin. Chim. Acta, *114*:233-247, 1981.

101. Plant, S. B., McCarron, D. A.: Effects of sample freezing on ion-selective electrode determinations of serum calcium. Clin. Chem., *28*:1362-1363, 1982.

102. Porter, W. H., Carroll, J. R., and Roberts, R. E.: Hemoglobin interference with the DuPont Automatic Clinical Analyzer procedure for calcium. Clin. Chem., *23*:2145-2147, 1977.

103. Potts, J. T., Jr.: Disorders of parathyroid glands. *In*: Harrison's Principles of Internal Medicine. 10th ed. R. G. Peterdorf, R. D. Adams, E. Braunwald, et al., Eds. New York, McGraw-Hill Book Co., 1983, pp. 1929-1943.

104. Pragay, D. A., Howard, S. F., and Chilcote, M. E.: Inorganic ion contamination in vacutainer tubes and micropipets used for blood collection. Clin. Chem., *17*:350-352, 1971.

105. Pybus, J., Feldman, F. J., and Bowers, G. N., Jr.: Measurement of total calcium in serum by atomic absorption spectrophotometry with use of a strontium internal reference. Clin. Chem., *16*:998-1007, 1970.

106. Radde, I. C., Hoffken, B., Parkinson, D. K., et al.: Practical aspects of a measurement technique for calcium ion activity in plasma. Clin. Chem., *17*:1002-1006, 1971.

107. Radde, I. C., Parkinson, D. K., and Höffken, M.: Ionized calcium levels in the healthy and diseased neonate. Proc. XIII International Congress of Paediatrics, Vol. 7, Metabolism. Vienna, Verlag der Wiener Medizinischen Akademie, 1971, pp. 363-368.

108. Radde, I. C., Parkinson, D. K., Höffken, B., et al.: Calcium ion activity in the sick neonate: Effect of bicarbonate administration and exchange transfusion. Pediatr. Res., *6*:43-49, 1972.

109. Reddy, G. S., Jones, G., Kooh, S. W., et al.: Stimulation of 24R,25-dihydroxyvitamin D_3 synthesis by metabolites of vitamin D_3. Am J. Physiol., *245*:E359-E364, 1983.

110. Reinhardt, T. A., Horst, R. L., Orf, J., et al.: A microassay for 1,25-dihydroxyvitamin D which does not require HPLC. J. Lab. Clin. Med., *58*:91-98, 1984.

111. Renoe, B. W., McDonald, J. M., and Ladenson, J. H.: The effects of stasis with and without exercise on free calcium, various cations, and related parameters. Clin. Chim. Acta, *103*:91-100, 1980.

112. Robertson, W. G.: Measurement of ionized calcium in body fluids—a review. Ann. Clin. Biochem., *13*:540-548, 1976.

113. Robertson, W. G., and Marshall, R. W.: Ionized calcium in body fluids. CRC Crit. Rev. Clin. Lab. Sci., *15*:83-125, 1981.

114. Robertson, W. G., and Marshall, R. W.: Calcium measurements in serum and plasma—total and ionized. CRC Crit. Rev. Clin. Lab. Sci., *13*:271-304, 1979.

115. Robertson, W. G., and Morgan, D. B.: The distribution of urinary calcium excretion in normal persons and stone formers. Clin. Chim. Acta, *37*:503-508, 1972.

116. Rodbell, M. L.: The role of hormone receptors and GTP-regulatory protein in membrane transduction. Nature, *284*:17-22, 1980.

117. Schwartz, H. D.: New techniques of ion-selective measurements of ionized calcium in serum after pH adjustment of aerobically handled sera. Clin. Chem., *22*:461-467, 1976.

118. Scriver, C. R., Fraser, D., and Kooh, S. W.: Hereditary rickets. *In*: Clinical Endocrinology 2, Calcium Disorders. D. A. Heath and S. J. Marx, Eds. London, Butterworths' Intl. Med. Rev. 1982, pp.1-46.

119. Shepard, R. M., Horst, R. L., Hamstra, A. J., et al.: Determination of vitamin D and its metabolites in plasma from normal and anephric man. Biochem. J., *182*:55-69, 1979.

120. Smith, S. C. H., Buckley, B. M., Wedge, G., et al.: An evaluation of the ICA 1 ionized calcium analyzer in a clinical chemistry laboratory. Scand. J. Clin. Lab. Invest., *43* (Suppl. 165):33-37, 1983.

121. Stern, P. H., Hamstra, A. J., DeLuca, H. F., et al.: A bioassay capable of measuring 1 picogram of 1,25-dihydroxyvitamin D_3. J. Clin. Endocrinol. Metab., *46*:891-896, 1978.
122. Suh, S. M., Tashjian, A. J., Matsuo, N., et al.: Pathogenesis of hypocalcemia in primary hypomagnesemia: Normal end-organ responsiveness to parathyroid hormone, impaired parathyroid gland function. J. Clin. Invest., *52*:153-160, 1973.
123. Taussky, H. H., and Shorr, E.: A microcolorimetric method for the determination of inorganic phosphorus. J. Biol. Chem., *202*:675-685, 1953.
124. Thalassinos, N. C., Leese, B., Latham, S. C., et al.: Urinary excretion of phosphate in normal children. Arch. Dis. Child., *45*:269-272, 1970.
125. Thode, J., Wandrup, J., Aas, F., et al.: Evaluation of a new-semiautomatic electrode system for simultaneous measurement of ionized calcium and pH. Scand. J. Clin. Lab. Invest., *42*:407-415, 1982.
126. Vreman, H. J., and Jöbsis, F. F.: Interference by mannitol and other compounds with phosphate determinations. Anal. Biochem., *17*:108-118, 1966.
127. Woodhead, J. S., and Walker, D. A.: Assay of parathyroid hormone in human serum and its uses. Ann. Clin. Biochem., *13*:549-554, 1976.
128. Worth, G. K., Retallack, R. W., Devlin, R. D., et al.: Ionized calcium: Aspects of technique and some factors affecting results. Clin. Chim. Acta, *114*:283-286, 1981.
129. Zettner, A., and Seligson, D.: Application of atomic absorption spectrophotometry in the determination of calcium in serum. Clin. Chem., *10*:869-890, 1964.

CHAPTER 13

LIVER FUNCTION

By William F. Balistreri, M.D., and Leslie M. Shaw, Ph.D.

The human liver contains complex parenchymal cells which perform multiple diverse functions that are essential for life. Hepatocytes have a unique regenerative capability as well as a marked capacity to respond to increased metabolic demands of the organism. The liver directly receives, processes, and stores materials absorbed from the digestive tract such as amino acids, carbohydrates, fatty acids, and cholesterol, as well as vitamins, and is capable of releasing metabolites of these compounds on demand. The liver synthesizes multiple plasma proteins including albumin, α- and β-globulins, clotting factors, and transport proteins. These factors influence homeostasis, since binding proteins modulate the circulating total concentrations of calcium, magnesium, and many drugs, while albumin concentrations regulate the plasma oncotic pressure and thus influence the fluid dynamics between the blood and the tissues. The liver is the main organ of detoxification and is the site of metabolic conversion of endogenous and exogenous compounds. Another major function of the liver is to synthesize bile acids from cholesterol and to secrete these compounds from the hepatocyte into the intestine, thereby generating bile flow and facilitating dietary fat emulsification and absorption. Because the liver is a major site of catabolism of thyroid, steroid, and other hormones, it participates in regulation of plasma hormone levels. The liver responds to multiple hormonal and neural stimuli to regulate the blood glucose concentration and contributes to the body's immune system.

Assessment of Liver Function

Tests of liver function have been used for many years, and have been of great clinical utility in the assessment of hepatic dysfunction (Table 13-1). Numerous diagnostic modalities have been devised in order to meet several purposes: (1) *screening* for abnormalities in liver function (sensitive detection of suspected dysfunction), (2) *documenting* an abnormality, (3) specifying the *type* (e.g., cholestasis versus hepatocellular disease) and *site* (i.e., intrahepatic versus extrahepatic) of injury, as well as (4) *prognostication* and follow-up of patients with hepatic disease. Many aspects of the myriad functions of the liver have been quantitated. Because a single laboratory test will not satisfy all of the postulated purposes, tests must be used in combination. It is, therefore, important to indicate the value of each test in regard to (1) diagnostic *sensitivity* in screening for hepatic dysfunction, (2) *specificity* for liver disease, and (3) *selectivity* in differentiating these disorders. Rapid emergence and refinement of radiographic, endoscopic, ultrasonic, and scintigraphic techniques have provided sophisticated second-line diagnostic modalities. However, laboratory assessment of hepatic dysfunction is a very valuable initial discriminator, which complements historical findings and physical examination prior to selection of more definitive or invasive tests in the patient with suspected or proven hepatic disease.

Table 13-1. TESTS OF HEPATIC FUNCTION

Tests based on substances *produced* or metabolized by the liver
  Substances *synthesized* by the liver (albumin, coagulation factors)
  Substances *metabolized* by the liver (drugs, xenobiotics, bilirubin, cholesterol, triglycerides)
Tests based on substances *released* from damaged tissue
  Endogenous compounds released by damaged *hepatocyte* (enzymes such as AST, ALT)
  Endogenous compounds *synthesized* at an increased rate or *released* by canalicular membrane, *bile duct* epithelium,
    and endothelium of central and periportal veins (alkaline phosphatase, γ-glutamyltransferase, 5′-nucleotidase)
Tests based on substances *cleared* by the liver
  Endogenous metabolites (bile acids, bilirubin, ammonia)
  Exogenous compounds (indocyanine green, caffeine, aminopyrine)
Tests measuring enterohepatic circulation
  Endogenous metabolites (bile acids)

ANATOMY OF THE LIVER

Gross Anatomy

The liver is the largest organ in the human body, weighing ~1.2–1.5 kg in the human adult; it accounts for ~5% of body weight in early life and ~2% in the adult. The liver is located in the right upper quadrant of the abdomen, beneath and attached by its ligaments to the diaphragm. In an adult the liver can be percussed from approximately the right fifth intercostal space in the midclavicular line down to or slightly below the right costal margin.

Grossly, the liver appears divided into a left and right half by the falciform ligament, an extension of the peritoneal folds that connect the liver to the diaphragm and anterior abdominal wall (Figure 13-1). The liver is actually divided into four lobes (left, right, quadrate, and caudate), which are supplied by the left and right branches of the portal vein and the hepatic artery. The biliary drainage is into the left and right hepatic ducts. A thin connective tissue capsule (Glisson's capsule) covers the entire liver surface. Connective tissue that provides an internal supporting framework for the liver parenchyma arborizes to ensheath vessels and nerves and subdivides the parenchyma into lobules.

The *circulatory system* of the liver is characterized by a dual blood supply. The portal vein carries blood from the capillary bed of the alimentary tract; this blood is rich in nutrients absorbed from the gastrointestinal tract. The portal vein supplies ~75% of the afferent blood volume to the liver. The other source is the hepatic artery, which is a branch of the celiac trunk; this vessel carries well-oxygenated blood to the liver. Both vessels enter the liver at the porta hepatis and arborize throughout the hepatic parenchyma. The venous drainage from the liver is via the right and left hepatic veins, which enter into the inferior vena cava near the right atrium. In addition, there are numerous lymphatic channels within Glisson's capsule and stroma.

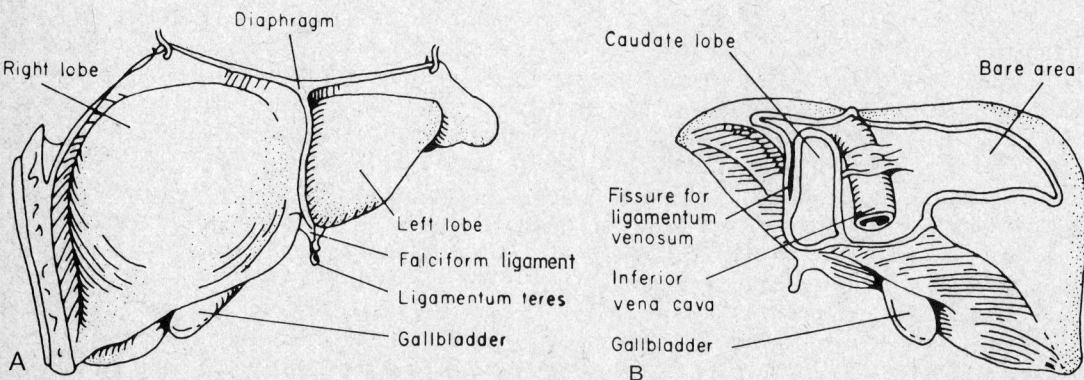

Figure 13-1. *A,* Anterior view of the liver. *B,* Posterior view of the liver. (From Sherlock, S.: Diseases of the Liver and Biliary System. 6th ed. London, Blackwell Scientific Publications, 1981.)

The *portal triad* consists of bile ducts that accompany the hepatic artery and portal vein. Bile flows from the hepatocyte into the bile canaliculi and ductules, to the larger intrahepatic bile ducts, and finally to the left and right hepatic bile ducts, which emerge from the liver at the porta hepatis and form the common hepatic duct. The hepatic duct is joined by the cystic duct from the gallbladder, to form the common bile duct. The common bile duct and pancreatic duct enter the duodenum at the ampulla of Vater.

Innervation of the liver is provided by the hepatic nerve plexus, which enters the liver at the porta hepatis. Parasympathetic innervation is derived from the preganglionic fibers of the vagus; sympathetic innervation is derived from preganglionic fibers originating from cell bodies at spinal level T-5 to T-9. The principal modulation of the nervous system is exerted upon the biliary tract and blood vessels.

The *gallbladder* is divided into three segments: the fundus, body, and neck; the latter is characterized by a sacculation called Hartmann's pouch. The gallbladder occupies a fossa that extends from the inferior border of the liver to the right margin of the porta hepatis. The fundus (or broad end) is directed forward and may be palpable on abdominal examination if enlarged. The gallbladder itself, in the adult, is 9–10 cm in length and has a capacity of ~50 mL. The gallbladder wall is made up of musculoelastic tissue but has no submucosa or muscularis mucosa. The Rokitansky-Aschoff sinuses are invaginations that branch from the lumen into the mucosa and muscularis. The musculature is most prominent in the neck and fundus. The interior of the gallbladder is marked by deep indentations of the mucosa (crypts of Luschka) that penetrate to the muscular layer. The gallbladder receives blood from the cystic artery, a large branch of the hepatic artery. The venous drainage is via the cystic vein into the portal venous system.

Microscopic Anatomy

The liver is composed of a large number of functional units or *lobules*. The classic liver lobule is a polyhedral prism of tissue demarcated by connective tissue septa and the vascular and biliary vessels (Figure 13-2). The central vein (terminal hepatic venule) is in the center of

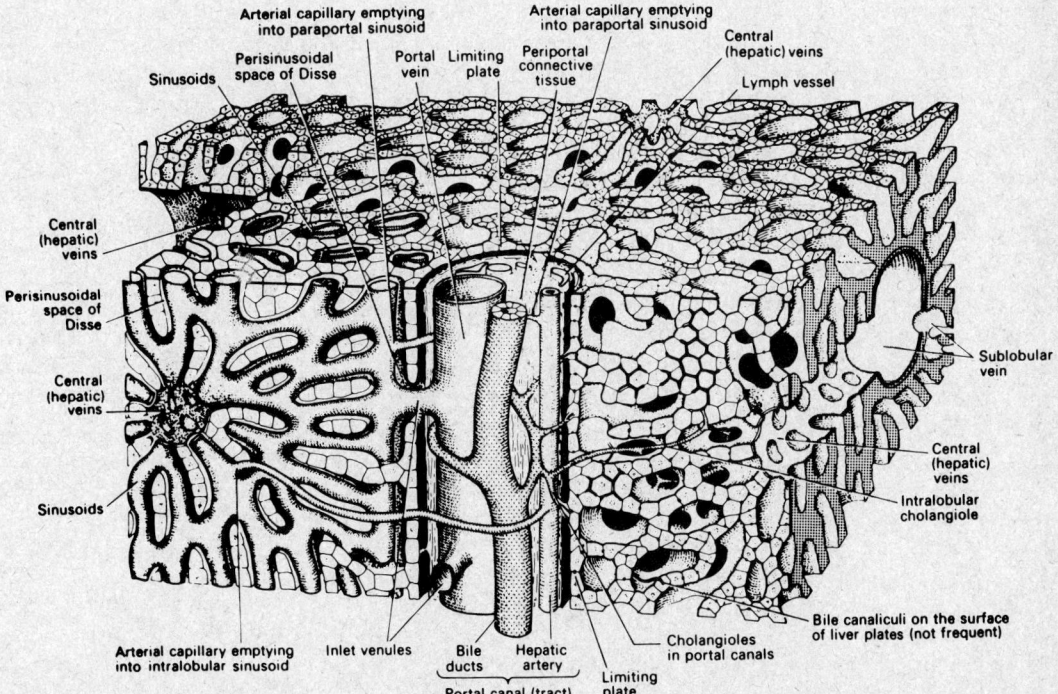

Figure 13-2. The classic liver lobule. (From Sherlock, S.: Diseases of the Liver and Biliary System. 6th ed. London, Blackwell Scientific Publications, 1981.)

the lobule (Figure 13-3), and single cell plates of parenchymal epithelial cells radiate from the central vein. Rappaport has conceptualized the liver lobule as an acinus in which parenchymal cells are grouped into concentric zones surrounding the terminal afferent vessels. Therefore, zone 1 cells are the hepatocytes that receive incoming portal blood. Those cells in regions more distal to the afferent blood vessels (i.e., in zones 2 and 3) receive less-enriched blood, since the oxygen and nutrient levels decrease from zone 1 through zone 3.

The *sinusoids* are blood-carrying vascular channels on either side of the liver cell plates; they form a rich, intralobular vascular network that converges toward the central vein. The sinusoidal lining cells are either endothelial cells or Kupffer cells. The latter constitute the largest group of fixed macrophages in the human and have phagocytic activity. They accumulate ferritin and hemosiderin and thus play a role in iron metabolism. The tissue space between the endothelial cells and hepatocytes (i.e., the space of Disse) contains interstitial fluid, across which the transfer of nutrients and waste products from the blood and liver cells occurs.

Hepatocytes, which constitute ∼60% of the liver mass, are polygonal cells ∼30 µm in diameter. The average life span of hepatocytes in experimental animals is 150 d. The surface of the cell is specialized such that three aspects are defined: (1) the sinusoidal surface, which faces the incoming blood (sinusoid and space of Disse); (2) the intercellular surface, contiguous with the sinusoidal surface; and (3) the canalicular surface.

Ultrastructure

There have been major refinements of our understanding of the ultrastructure of hepatocytes. Hepatic parenchymal cells contain an extremely well-developed organellar substructure (Figures 13-4 and 13-5) that includes:

1. Numerous *mitochondria*, which constitute ∼18% of the liver cell volume and which participate in energy generation (oxidative phosphorylation and fatty acid oxidation).

2. *Lysosomes*, which have specific degradative functions (hydrolytic enzymes).

3. The *endoplasmic reticulum*, which is a highly developed system of membrane-lined intracellular canals that are the site of many functions, including hepatic drug metabolism and bile acid synthesis. The *smooth* endoplasmic reticulum assumes the form of tubules and vesicles and is the site of bilirubin conjugation, drug detoxification, and cholesterol synthesis. The *rough* endoplasmic reticulum forms lamellar profiles lined with ribosomes, which are the site of specific protein synthesis, such as albumin, coagulation factors, and various enzymes.

4. The multifunctional *Golgi complex*, composed of closely packed, smooth-surfaced cisternae that produce lipoproteins (VLDL) and that are involved in glycoprotein synthesis and albumin secretion (a "packaging" process).

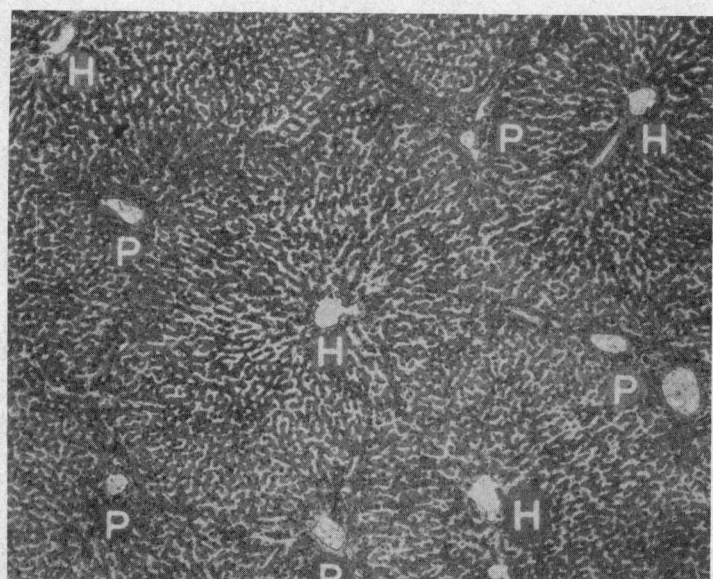

Figure 13-3. Normal hepatic histology. H = terminal hepatic venule; P = portal tract (×60). (From Sherlock, S.: Diseases of the Liver and Biliary System. 6th ed. London, Blackwell Scientific Publications, 1981.)

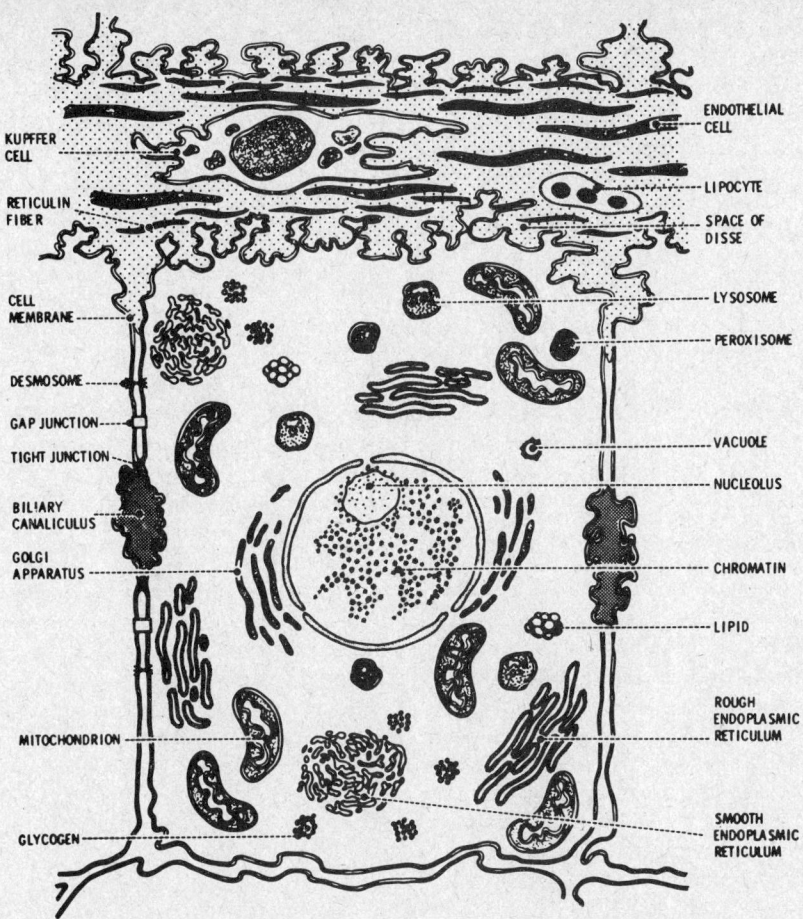

Figure 13-4. Hepatic ultrastructure. (From Sherlock, S.: Diseases of the Liver and Biliary System. 6th ed. London, Blackwell Scientific Publications, 1981.)

Figure 13-5. Electron microscopic appearance of a normal human liver cell: the zone around the bile canaliculus × 20 000 (original magnification). N = nucleus; M = mitochondria; B = bile canaliculus, with abundant microvilli, enclosed by tight junction (t) between adjacent hepatocytes; G = Golgi zone (lamellar stacks of smooth membrane); S = smooth endoplasmic reticulum; R = rough endoplasmic reticulum; C = cell membrane; g = glycogen particles (stellate α-particles). (Courtesy of Dr. Cynthia Daugherty, Pathology Department, EM Lab, Children's Hospital Medical Center, Cincinnati, Ohio.)

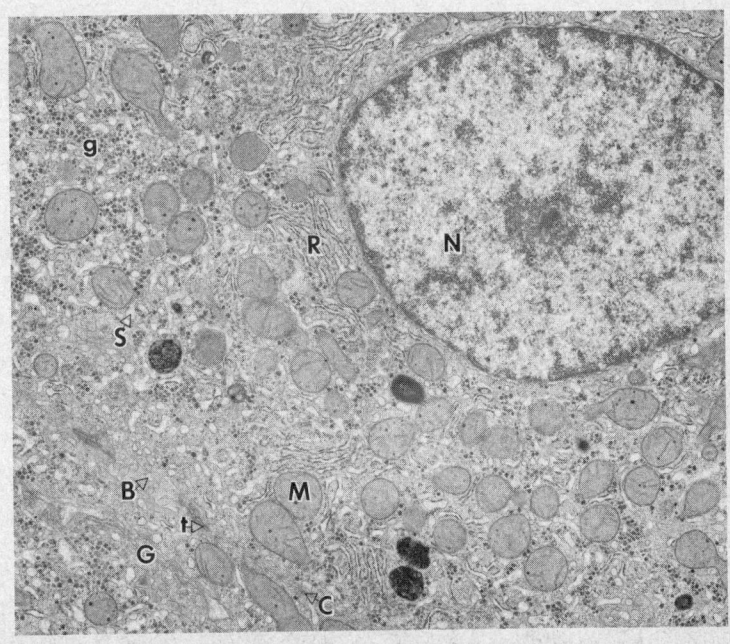

5. *Microtubules* and *microfilaments,* which maintain cell shape and provide contractile force.

In addition to the highly developed intracellular structure, the *cell membrane* itself is complex and specialized. For example, the bile canaliculus is made up of a highly adapted area on the surface of adjacent hepatic parenchymal cells and is isolated from the remainder of the intercellular space (i.e., the sinusoidal space) by the junctional complexes or tight junction.

IMPLICATIONS OF LIVER STRUCTURE FOR CLINICAL CHEMISTRY

Structural Heterogeneity of Liver Cells

Histochemical studies and differential staining techniques have demonstrated a considerable heterogeneity of hepatocytes; the cells in the portal zone of the liver lobule differ structurally and functionally from hepatocytes in the central lobular area.[24] For example, the centrilobular cells contain larger amounts of lysosomes and smooth endoplasmic reticulum than do periportal cells. The metabolic differences noted between hepatocytes in different zones either may be intrinsic to the cell or may be due to their location within the lobule and their exposure to different levels of oxygen tension, substrate concentration gradients, and other factors.

Reserve Capacity of the Hepatocytes

The unique responsiveness of this organ to various stimuli is illustrated by studies performed on laboratory animals. For example, many exogenous, lipid-soluble compounds are metabolized by the hepatic microsomal mixed function oxidase system that includes a chain of enzymes such as cytochrome P_{450} and NADPH cytochrome c reductase. The administration of such compounds as phenobarbital or halogenated hydrocarbon insecticides will induce hypertrophy of specific organelles (the smooth endoplasmic reticulum) and increase the associated enzymatic activity (enzyme induction). It is also of interest that in situations in which an enhanced excretory function is required, e.g., in selective biliary obstruction in the rat, compensation for the obstructed portion by the remaining liver tissue occurs. Previously inactive cells in the *central* portion of the lobule undergo hypertrophy, suggesting that the reserve capacity of these cells may be "recruited" during periods of enhanced bile secretory activity.

CLINICAL MANIFESTATIONS OF LIVER DISEASE

Jaundice

The classic clinical manifestation of hepatic disease is yellow discoloration of the plasma, skin, and mucous membranes, caused by bilirubin accumulation and staining. Jaundice, or icterus, may be the first, and often the only, manifestation of liver disease. This clinical finding is most easily seen in the peripheral portions of the conjunctivae and in the mucous membranes of the lips or palate. The yellow discoloration is clinically apparent when the serum total bilirubin concentration reaches 2–3 mg/dL. This feature may be highlighted by compressing the tissue with a glass slide. There may be other signs of disturbances in bilirubin metabolism, e.g., the passage of dark urine or acholic (white, tan) stools. The clinical finding of jaundice is not specific and may indicate a variety of diseases. Multiple classification schemata of the jaundiced patient have been devised; however, initial evaluation should focus upon whether the hyperbilirubinemia arises from the predominance in serum of the unconjugated form of bilirubin or whether there is a significant increase of the conjugated form. (See also later section on bilirubin.) A classification of jaundice based on the site of altered bilirubin metabolism (see below) is shown in Table 13-2.

Portal Hypertension

The blood entering the liver via the portal vein represents the influx from the splanchnic area (i.e., from the abdominal portion of the digestive tract, pancreas, and spleen) into the hepatic sinusoids. Under normal conditions, the portal venous system differs from other venous blood systems in that the pressure is slightly higher in order to overcome the resistance of the liver sinusoidal system. In addition, the oxygen content is low; however, multiple nutrients and bacterial toxins are present. In the face of sinusoidal infiltration, scarring, or hepatic vein (outlet) obstruc-

Table 13-2. CLASSIFICATION OF JAUNDICE (BASED ON TYPE AND SITE)

Unconjugated hyperbilirubinemia
  *Increased **production** of unconjugated bilirubin from heme*
    Hemolysis
      Hereditary
      Acquired
    Ineffective erythropoiesis
    Rapid turnover of increased red cell mass (neonate)
  *Decreased **delivery** of unconjugated bilirubin (in plasma) to hepatocyte*
    Right-sided congestive heart failure
    Portacaval shunt
  *Decreased **uptake** of unconjugated bilirubin across hepatocyte membrane*
    Competitive inhibition
      Drugs
      Others?
    ? Gilbert's syndrome
    ? Sepsis, fasting
  *Decreased **storage** of unconjugated bilirubin in cytosol (↓ Y and Z proteins)*
    Competitive inhibition
    Fever
  *Decreased **biotransformation** (conjugation)*
    Neonatal jaundice (physiologic)
    Inhibition (drugs)
    Hereditary (Crigler-Najjar)
      Type I (Complete enzyme deficiency)
      Type II (Partial deficiency)
    Heptatocellular dysfunction
    ? Gilbert's syndrome
Conjugated hyperbilirubinemia (cholestasis)
  *Decreased **secretion** of conjugated bilirubin into canaliculi*
    Hepatocellular disease
      Hepatitis
      Cholestasis (intrahepatic)
    Dubin-Johnson and Rotor syndromes
    Drugs (estradiol)
  *Decreased **drainage***
    Extrahepatic obstruction
      Stones
      Carcinoma
      Stricture
      Atresia
      Primary sclerosing cholangitis
    "Intrahepatic" obstruction
      Drugs
      Granulomas
      Primary biliary cirrhosis
      Tumors

tion, the normal portal pressure of ~5–10 mm Hg will be increased to varying degrees, causing portal hypertension, operationally defined as > 20 mm Hg. The resistance to the flow of portal venous blood is further increased and there is dilatation of the venous blood system behind the obstruction with enlargement of the spleen. With time, there is an increase in collateral flow around the liver and the development of variceal vessels in the esophagus and at other locations that are potential sites of bleeding.

The site of obstruction causing portal hypertension has classically been of three major types: (1) extrahepatic (presinusoidal), (2) intrahepatic (parenchymal), and (3) suprahepatic (outflow). *Presinusoidal portal hypertension* is most commonly due to portal vein thrombosis or cavernous transformation of the portal vein. Various *parenchymal diseases* (cirrhosis, fatty liver, hepatitis, infiltration) may cause portal hypertension. The most common cause of *outflow obstruction* is hepatic vein occlusion due to either thrombosis or fibrous obliteration (Budd-Chiari syndrome). Other causes of obstruction include tumors, abscesses, membranous obstruction of the vena cava, veno-occlusive disease (as may be seen in patients following bone marrow transplant), or cardiac disease (heart failure or constrictive pericarditis).

In addition to hepatosplenomegaly and the presence of collateral vessels (varices), ascites and the hepatorenal syndrome are major complications of cirrhosis and portal hypertension. *Ascites* represents the accumulation of free fluid within the peritoneal cavity and is a consequence of sinusoidal hypertension, excess lymph formation, and hypoalbuminemia. Medical management of portal hypertension and of its attendant complications includes control of variceal bleeding and judicious management of fluid and electrolyte balance (e.g., sodium restriction). Surgical management of portal hypertension has frequently been attempted. The two major approaches that have been used for the emergency treatment of bleeding varices are obliteration of the varix and emergency portacaval shunt. In a stable patient, elective surgery for portal hypertension involves the creation of a portal-systemic shunt, which has been carried out with varying degrees of success. The theoretical benefit of the operation is to alter the hemodynamics and decompress the elevated vascular resistance. The use of selective shunts, i.e., decompression of the venous collaterals in the upper stomach and lower esophagus with preservation of blood flow to the liver, has achieved good success. Despite multiple variations in the types of shunts, serious consequences such as encephalopathy may result.

Hepatic Failure and Encephalopathy

Neuropsychiatric disturbances may result from the metabolic abnormalities that occur in the presence of chronic liver disease. The resultant syndrome of hepatic encephalopathy leads to altered mentation and neuromuscular dysfunction, as well as altered consciousness and coma; it is a well-described complication of various forms and degrees of liver disease.[16] The episodes may be recurrent and may occur either spontaneously or when induced by an inciting factor such as intercurrent illness, drug administration, bleeding, or electrolyte and acid-base imbalance.

As mentioned above, the surgical creation of a portal-systemic shunt may precipitate encephalopathy. In many respects, hepatic encephalopathy can be attributed to the shunting of blood from the intestine to the systemic circulation, bypassing the liver, and therefore allowing substances which are normally metabolized by the liver to accumulate and to exert a cerebrotoxic effect. The mediators of hepatic encephalopathy have not been definitely identified but may be ammonia, mercaptans, or other endogenous or exogenous compounds.

Multiple metabolic abnormalities are seen in the patient with liver failure and hepatic coma. Some are primary, and others represent secondary compensatory or decompensatory mechanisms. The major findings are increased ammonia concentrations in blood and the accumulation of ammonia, glutamine, and α-ketoglutarate in brain and CSF. There is an increase in short and medium-chain fatty acids in the plasma; mercaptans are also increased. The latter may account for the characteristic odor known as "fetor hepaticus" in these patients. There is also hyperaminoacidemia as well as an increased concentration of amino acids in the brain, cerebrospinal fluid, and urine. The most marked increases involve methionine, phenylalanine, tyrosine, aspartate, and glutamate, while the branched-chain amino acids (leucine, isoleucine, and valine) are decreased. Neurotransmitters are generally decreased in brain and muscle, and there is an accumulation of so-called false neurotransmitters such as octopamine.

Altered Drug Metabolism

Another manifestation of hepatic dysfunction is altered drug metabolism. The liver plays a major role in enzymatic transformation and disposition of therapeutic drugs.[9] Therefore, during the course of diseases causing hepatic dysfunction, many drugs will undergo limited or aberrant transformation, will accumulate, and will exert unwanted effects upon the host. The type and severity of liver disease will determine both the fate of the administered drug as well as the adverse effect. For example, drug disposition may be severely altered due to (1) decreased enzymatic activity as a result of liver cell damage, (2) altered blood flow with shunting of the drug past metabolic sites, or (3) an absence of "carriers" or drug receptors. A clear understanding of pharmacokinetic principles may also allow the use of *drug clearance* as an investigative tool to assess hepatic function.

Endocrine Abnormalities

Hormone imbalances are another manifestation of hepatic disease, since various aspects of hepatic metabolism are regulated by hormones brought to the liver by the portal and systemic circulation. In addition to serving as a target tissue for the action of these compounds, the liver plays a key role in the uptake and metabolism of hormones. Therefore, in the presence of failing liver function or impairment by various disease processes, endocrine alterations may be found, and hormonal balance markedly altered.

Among the common manifestations of chronic liver disease in men is feminization, i.e., indication of increased estrogen effects. Clinical findings may include testicular atrophy, gynecomastia, and spider angiomata. Biochemical alterations such as reduced circulating testosterone and elevated estrogen levels are presumably due to aberrant estrogen metabolism by the liver, although the exact defects are not understood. These are, of course, merely some aspects of the altered hormonal milieu and deranged endocrine function that occur in cirrhosis or chronic liver disease; alterations in pituitary, adrenal corticosteroid, thyroid, and peptide hormone responses have also been noted. Alterations of the specific hepatocyte receptors for many of these hormones may occur in chronic liver disease.

Nutritional and Metabolic Abnormalities

Because intake and disposition of nutrients in patients with chronic liver disease are altered, these patients are subject to nutritional imbalance. There may be severe metabolic and nutritional derangements in the cirrhotic patient. Examples are alterations in glucose metabolism due to insulin resistance and hypokalemia, secondary to excess losses of potassium as a result of vomiting and diarrhea. Secondary hyperaldosteronism may occur with hyponatremia. Hypoalbuminemia is frequently present due to decreased synthesis. In the presence of cholestasis, inefficient delivery of bile acid to the duodenum impairs absorption of fats and fat-soluble vitamins and therefore leads to chronic deficiencies of vitamins A, D, E, and K and to malnutrition and steatorrhea.

Immunoglobulin Abnormalities

There are complex immunologic alterations present in various forms of liver disease, and polyclonal gammopathy is frequently present in chronic liver disease. It is difficult to assess whether or not the immunologic disturbance represents a primary pathogenetic process or a secondary supervening process that perpetuates the liver disease. For example, in chronic active hepatitis, altered host responsiveness and the role of membrane-specific antibodies are subjects of much controversy. In addition, primary biliary cirrhosis appears to be a chronic, progressive, immunopathic disorder affecting the intrahepatic bile ducts.

A number of observations implicate the liver in immunological responses; for example, the liver efficiently sequesters dimeric IgA and secretes it into the bile. In liver disease, however, there are increased serum concentrations of IgA due to defects in the uptake process.

Disordered Hemostasis in Liver Disease

It has been recognized for over 150 years that chronic liver disease produces alterations in hemostasis and a generalized hemorrhagic tendency. Since the liver plays a central role in the synthesis of coagulation factors and in the regulation of hemostasis, the pathogenesis of bleeding is complex. For example, in liver disease there may be decreased synthesis of specific clotting factors and production of qualitatively abnormal proteins. There may be alterations in platelet number and function due either to hepatic disease or to associated hypersplenism. In addition, endogenous anticoagulant substances may be present, fibrinolytic systems may be aberrant, and there may be intravascular coagulation. These abnormalities make management of patients with variceal bleeding due to chronic liver disease a difficult proposition.

BIOCHEMICAL ASSESSMENT OF LIVER FUNCTION

Hepatic Excretory Function

Organic anions of both endogenous and exogenous origin are removed from the sinusoidal blood and excreted into the bile. Assessment of this excretory function of the liver can provide valuable clinical information. The most frequently used means for assessment are measurement of serum concentrations of endogenously produced compounds such as bilirubin and bile acids and the rate of clearance of an administered exogenous dye such as indocyanine green.

Bilirubin Metabolism

Bilirubin IXα, the orange-yellow bile pigment, is produced from protoporphyrin IX by microsomal heme oxygenase. The tetrapyrrolic product of the ring opening at the α-methene bridge is the green pigment biliverdin, which is subsequently hydrogenated to bilirubin by the cytosolic enzyme biliverdin reductase, an NADPH-dependent enzyme (Figure 13-6). For each mole of heme catabolized by this pathway, 1 mole each of carbon monoxide, bilirubin, and ferric iron is produced. Daily bilirubin production in man from all sources averages 250–300 mg. Approximately 85% of the total bilirubin produced is derived from the heme moiety of the hemoglobin released from senescent erythrocytes that are destroyed in the reticuloendothelial cells of the liver, spleen, and bone marrow. The remaining 15% of bilirubin is produced from red cell precursors destroyed in the bone marrow (so-called "ineffective erythropoiesis") and from the catabolism of other heme-containing proteins such as myoglobin, cytochromes, and peroxidases, which are distributed throughout the body.

After production in peripheral tissues, bilirubin is transported to the liver in association with albumin ($K_D \sim 10^{-8}$ mol/L). Bilirubin is then rapidly taken up by hepatocytes by what is presumed to be a carrier-mediated active-transport process across the sinusoidal membrane (Figure 13-7). Once inside the liver cells, bilirubin is tightly, though reversibly, bound to soluble proteins. There are two cytosolic binding proteins, ligandin and Z protein, that account for most of the bilirubin binding. Ligandin, which constitutes $\sim$5% of the total protein of human liver cytosol, binds bilirubin more tightly than does Z protein. Ligandin also binds a variety of other compounds such as steroids, bromsulfophthalein, indocyanine green, cholecystographic substances, and some carcinogens. Ligandin may play a significant role in the processing of these compounds; it may increase the net efficiency of uptake by retarding the reflux of these substances back to plasma.

Inside the hepatocytes, bilirubin is rapidly conjugated with glucuronic acid to produce bilirubin mono- and diglucuronide, which are then excreted into bile (Figure 13-7). The microsomal enzyme bilirubin UDP–glucuronyl transferase catalyzes the formation of bilirubin monoglucu-

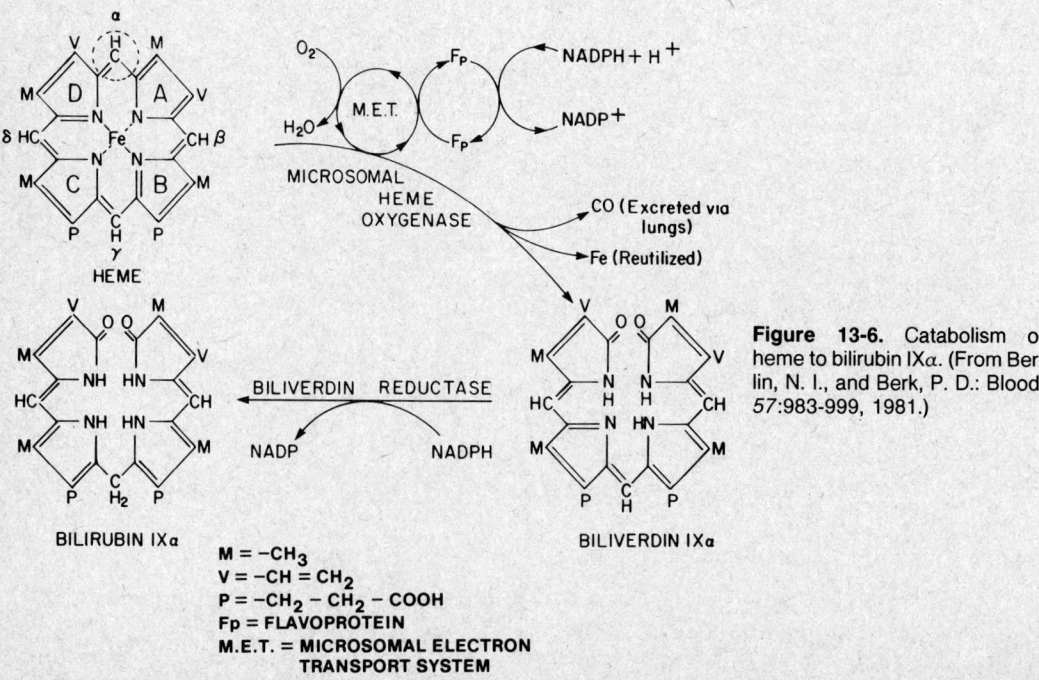

Figure 13-6. Catabolism of heme to bilirubin IXα. (From Berlin, N. I., and Berk, P. D.: Blood, *57*:983-999, 1981.)

M = −CH₃
V = −CH = CH₂
P = −CH₂ − CH₂ − COOH
Fp = FLAVOPROTEIN
M.E.T. = MICROSOMAL ELECTRON TRANSPORT SYSTEM

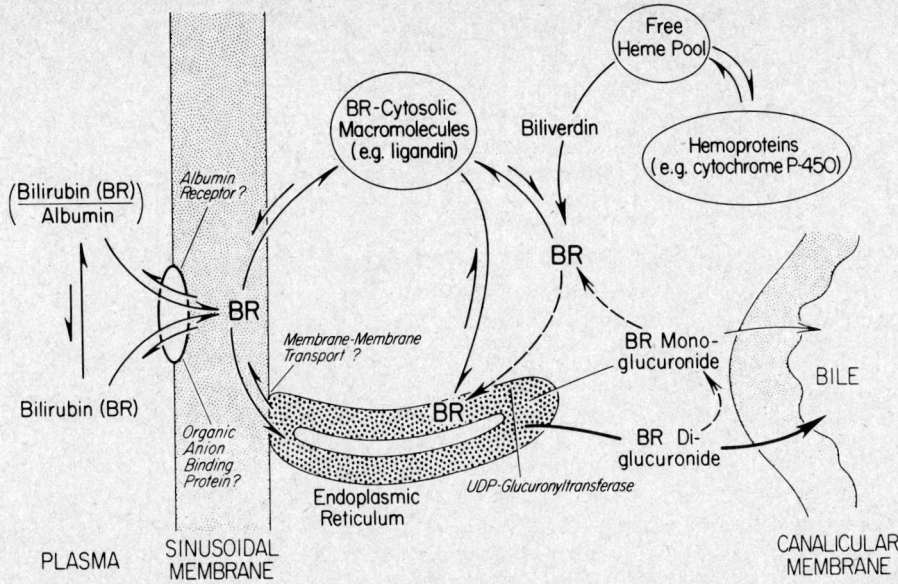

Figure 13-7. Bilirubin uptake, metabolism, and transport in the hepatocyte. (From Gollan, J. L., and Schmid, R.: Progress in Liver Diseases, Vol. VII. Chap. 15, 1982. By permission.)

ronide. It is not certain whether the conversion of the monoglucuronide to bilirubin diglucuronide is catalyzed by the same enzyme or by another enzyme located in or near the canaliculus. The excretion of conjugated bilirubin into bile against a significant concentration gradient is thought to be an energy-dependent, active-transport process.

In human adults, virtually all bilirubin excreted in bile is in the form of glycosidic conjugates; glucuronides account for ~95% of them, glucosides and xylosides for the remainder. Of the glucuronides, the diglucuronide is the major fraction (~90%) and monoglucuronide is the minor fraction (~10%).

Once secreted into the upper small intestine, bilirubin glucuronides are not reabsorbed to any great extent. In the intestinal tract, bilirubin glucuronides are hydrolyzed to the unconjugated pigment in the alkaline pH of the upper small intestine by the catalytic action of β-glucuronidase from liver, intestinal epithelial cells, and intestinal bacteria. The unconjugated bilirubin is then reduced by the anaerobic intestinal microbial flora to form a group of three colorless tetrapyrroles collectively called *urobilinogens*. In each of these three bilirubin reduction products, all bridge carbons are in the saturated (methylene) form. The urobilinogens differ from one another in the degree of hydrogenation of the vinyl side chains as well as the two end pyrrole rings; urobilinogens contain 6, 8, or 12 more hydrogen atoms than bilirubin and are named *stercobilinogen, mesobilinogen,* and *urobilinogen,* respectively. Up to 20% of the urobilinogens produced daily are reabsorbed from the intestine and enter the enterohepatic circulation. The majority of the reabsorbed urobilinogens is taken up by the liver and is re-excreted in the bile; only 2–5% enter the general circulation and appear in urine. In the lower intestinal tract the three urobilinogens spontaneously oxidize at the middle methylene bridge to produce the corresponding bile pigments *stercobilin, mesobilin,* and *urobilin,* which are orange-brown and are the major pigments of stool. Approximately 50% of the conjugated bilirubin excreted in bile is metabolized to products other than the urobilinogens. The detailed structure of these metabolites has not been elucidated.

Disturbances of Bilirubin Metabolism

A number of inherited and acquired diseases affect one or more of the steps involved in the production, uptake, storage, metabolism, and excretion of bilirubin. Bilirubinemia is frequently a direct result of these disturbances. Depending on the disorder, unconjugated bilirubin or conjugated bilirubin or both are major contributors to the hyperbilirubinemia.

Unconjugated Hyperbilirubinemia

The most commonly occurring form of unconjugated hyperbilirubinemia is that seen in newborns and referred to as *physiological jaundice*. All newborns have serum unconjugated bilirubin concentrations greater than values obtained in the healthy adult population, and about 50% of the newborn population is clinically jaundiced during the first 5 d of life. In the normal full-term neonate, unconjugated bilirubin values rise to 40–50 mg/L and, in a small percentage of cases, to as high as 100 mg/L by 48 h, with a decrease to normal values by 7–10 d. This temporary increase is caused by an increased bilirubin production as a result of hemolysis of erythrocytes and incomplete maturation of several steps involved in bilirubin metabolism and excretion. In ∼5% of neonates unconjugated bilirubin values of > 150 mg/L are seen.

The increased production of bilirubin that accompanies the premature *breakdown of eryth-rocytes* and *ineffective erythropoiesis* results in hyperbilirubinemia in the absence of any liver abnormality. Total bilirubin concentrations are usually not higher than 40 mg/L in these conditions; unconjugated bilirubin is the predominant form since the rate of bilirubin production is usually greater than the bilirubin excretory capacity.

A common feature of the rare genetic diseases Crigler-Najjar Type I and Type II syndromes is a lower than normal or absent activity of hepatic bilirubin UDP–glucuronyltransferase. A more common inherited disorder, Gilbert's syndrome, is a heterogeneous condition that is usually ascribed to decreased UDP-glucuronyltransferase activity but may also be caused by defects in membrane transport. In the case of Crigler-Najjar Type I syndrome, bilirubin UDP–glucuronyl-transferase activity is totally absent; reduced but measurable bilirubin UDP–glucuronyltransferase activity is present in liver tissue of patients with Crigler-Najjar Type II syndrome. Since bilirubin cannot be conjugated at the normal rate in patients with these disorders, the rate of excretion of bilirubin is significantly reduced and the serum concentration of unconjugated bilirubin rises. Serum total bilirubin concentrations of 200–500 mg/L are commonly seen in Crigler-Najjar Type I syndrome. Patients with this disease usually die in infancy due to the development of kernicterus (bilirubin staining of the basal ganglia of the brain). Total bilirubin concentrations of < 200 mg/L and of < 30 mg/L are the usual ranges in patients with Crigler-Najjar Type II and Gilbert's syndrome, respectively. These disorders have a benign prognosis.

Conjugated Hyperbilirubinemia

Acquired abnormalities of bilirubin excretion into bile are common and are discussed below. In hepatobiliary diseases of various causes bilirubin uptake, storage, and excretion are impaired to varying degrees. Thus, both conjugated and unconjugated bilirubin are retained in these disorders, and a wide range of abnormal serum concentrations of each form of bilirubin may be observed.

When any portion of the biliary tree becomes blocked or abnormally permeable, biliary passage of bilirubin and of all other components of bile is retarded and these substances are retained. As a result, plasma concentrations of conjugated bilirubin, cholesterol, γ-glutamyl-transferase, alkaline phosphatase, and bile acids increase to abnormal values. In addition, obstruction of the biliary tree also leads to an increased synthesis of enzymes such as γ-glutamyl-transferase and alkaline phosphatase and increases blood levels further. Drug- and steroid hormone–induced cholestasis and, occasionally, alcoholic hepatitis and acute viral hepatitis are examples of diseases that may cause *intrahepatic cholestasis*. Bile canaliculi and bile ductules are injured by the disease process but larger biliary channels are normal.

Inflammation and fibrosis of the interlobular bile ducts in the portal triads produce the blockage of bilirubin drainage seen in sclerosing cholangitis and primary biliary cirrhosis. Blockage of interlobular and larger intrahepatic bile ducts is probably the major cause of jaundice in the space-occupying lesions of the liver that are produced in patients with sarcoidosis, other granulomatous diseases, or primary and metastatic hepatic carcinomas. Mechanical obstruction of the biliary tree produces the increased concentrations of serum conjugated bilirubin observed in such disorders as carcinomas of the head of the pancreas, common bile duct, or ampulla of Vater; choledocholithiasis; fibrosis of the head of the pancreas; and common-duct strictures secondary to ductal injury from surgery.

The *Dubin-Johnson syndrome* is a rare and benign genetic disorder of bilirubin excretion. Kinetic studies on the disappearance from plasma of bilirubin, indocyanine green, and other anionic dyes in patients with this disorder have shown that hepatic uptake and storage are normal but bilirubin excretion into bile is markedly impaired. In these patients serum total bilirubin concentrations are usually in the range of 20–50 mg/L and the predominant form is conjugated bilirubin.

Disturbances of Urobilinogen Metabolism

The formation of urobilinogen is decreased in all conditions in which liver secretion or the biliary drainage of bilirubin is impaired. Thus, in these conditions there is a decreased urobilinogen load returning to the liver and a greater proportion of the reabsorbed urobilinogen is diverted into plasma and urine. As biliary obstruction becomes more complete, delivery of bilirubin to the gut becomes limiting and urine as well as stool urobilinogen excretion decreases to very low concentrations. The clay-colored or chalky white stool in obstructive jaundice reflects this impairment of bile pigment metabolism.

Structure of Bilirubins

In 1942 Fisher and Plieninger succeeded in synthesizing bilirubin IXα and represented the structure as shown in Figure 13-8. This linear tetrapyrrolic structure for the bilirubin molecule was accepted for more than 30 years. An important chemical property of the bilirubin molecule is its insolubility in water and its ready solubility in a variety of nonpolar solvents. However, lipid solubility of bilirubin is not predicted from the linear tetrapyrrole structure illustrated in Figure 13-8 because the two propionic acid side chains would be expected to make the bilirubin molecule highly polar and therefore water soluble. Recent X-ray crystallographic studies of crystalline bilirubin confirmed the overall chemical structure assigned to bilirubin by Fisher and colleagues. However, two additional important structural features were cited: (1) a so-called "Z-Z" (trans) conformation for the double bonds between carbons 4 and 5 as well as 15 and 16, and (2) an involuted hydrogen-bonded structure in which the propionic acid carboxylic acid groups are hydrogen bonded to the pyrrole ring nitrogens (Figure 13-9). When exposed to light, bilirubin in the "Z-Z" configuration is converted to the "E-E" (cis) conformation as well as other combinations, namely, 4E-15Z and 4Z-15E. The "E-E" conformation and other E-containing isomers do not permit the degree of internal hydrogen bonding that occurs in the "Z-Z" conformation and are therefore more water soluble than is bilirubin in the "Z-Z" conformation.[49] Thus, exposure to light produces forms of bilirubin that are more water soluble than the normal "Z-Z" isomer and therefore are more readily excretable. This is the rationale for exposing to light neonates who have clinically significant jaundice in order to reduce plasma unconjugated bilirubin concentrations.

The bilirubin molecule in the crystalline state is not a linear tetrapyrrole but takes the form of a ridge tile, the ridge being along the line C8-C10-C12. In this configuration, rings A and B lie in one plane and rings C and D in another with the angle between the two rings being 98°. The preferred conformation of bilirubin in aqueous solution at pH 7.4 is not known, but the occurrence of a hydrogen-bonded structure in aqueous solution would explain some of the unique chemical properties of bilirubin IXα. For example, the addition of hydrogen bond–breaking chemicals such as caffeine, methanol, ethanol, or urea, 6 mol/L, is required for unconjugated bilirubin to react with the diazo reagent in the van den Bergh reaction. These reagents probably act by breaking internal hydrogen bonds of the bilirubin molecule to allow it to react with diazotized sulfanilic acid. On the other hand, bilirubin IXα diglucuronide is quite soluble in water and readily reacts with the diazo reagent. Due to the presence of the bulky glucuronic acid moiety it is unlikely that conjugated bilirubin can undergo internal hydrogen bond formation.

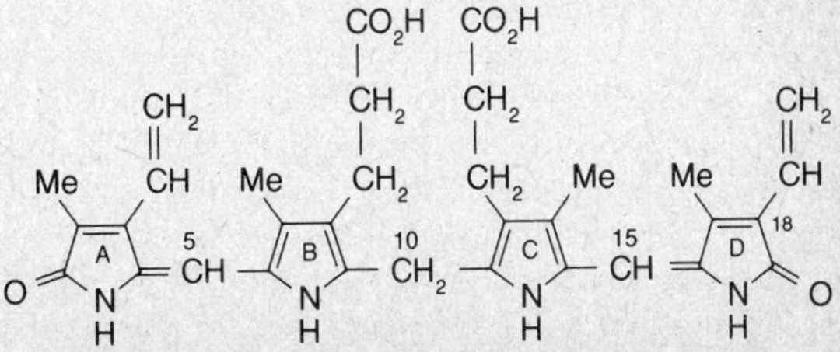

Figure 13-8. Structure of bilirubin IXα according to Fisher and Plieninger.[17]

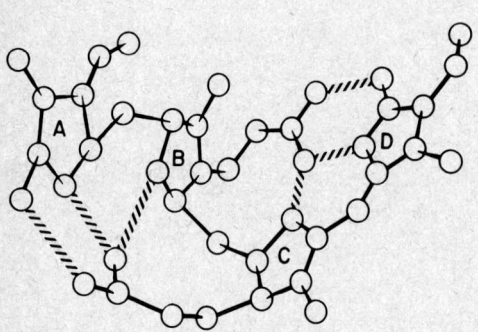

Figure 13-9. Bilirubin IXα structure. *Top,* The unfolded or linear tetrapyrrole structure showing the "Z" configuration at the C-4 to C-5 and C-15 to C-16 double bonds. *Bottom,* The folded conformation showing extensive internal hydrogen bonding. (Reprinted by permission of Elsevier Science Publishing Co., Inc., from Schmid, R.: Gastroenterology, 74:1307-1312, 1978. Copyright 1978 by The American Gastroenterological Association.)

The conformational change conferred by esterification with glucuronic acid may be one of the reasons why bilirubin IXα glucuronides are readily excreted in bile and urine whereas unconjugated bilirubin is not.

METHODS FOR THE DETERMINATION OF BILIRUBIN

Numerous separation methods such as paper and thin-layer chromatography have been used to study the distribution and clinical significance of bilirubin and its metabolites in serum. These methods were severely limited by the fact that instability of the pigments made the identification and quantitation of the separated bilirubin species unreliable. Kuenzle and colleagues[28] were the first to use successfully an open column chromatography technique that avoided the use of a deproteinization step. Four bilirubin fractions were obtained, namely, unconjugated bilirubin (α), monoconjugated bilirubin (β), diconjugated bilirubin (γ), and a fraction that was irreversibly bound to protein (δ). The last fraction was clearly distinct from the albumin-unconjugated bilirubin complex that exists in serum and in solutions of purified albumin to which purified bilirubin is added. This technique was limited by the large serum sample required and by problems in the quantitation of the separated fractions. Recently, HPLC methods have been developed that overcome the problems of the instability of bilirubin and its metabolites and that make use of purified bilirubin conjugates for standardization of the analysis of the respective metabolites. In the method of Blanckaert,[7] bilirubin conjugates, but not unconjugated bilirubin, are converted to the corresponding bilirubin methyl esters by base-catalyzed transesterification in methanol followed by a chloroform extraction step. With this procedure virtually all of the α-, β- and γ-bilirubin fractions are recoverable but the δ-fraction remains in the denatured protein pellet that the chloroform extraction step produces. In the HPLC method of Lauff et al.,[29] all four bilirubin fractions remain in solution after a step that involves salting-out globulins (but not albumin) with sodium sulfate. Both methods require the use of dim incandescent or yellow light to minimize photodegradation of the various bilirubin species. Current evidence, primarily from the studies of Lauff and coworkers,[30] implies that the δ-bilirubin fraction consists of one or more bilirubin species that are tightly, probably covalently, linked to albumin. Existence of covalent linkage is supported by the fact that the associated bilirubin species are not released from the albumin fraction by treatment with strong acid or base or a variety of strong denaturing agents, by hydrolysis with proteolytic enzymes, or by boiling in methanol. The δ-bilirubin fraction has been shown to react directly with diazotized sulfanilic acid.

From HPLC analyses of a large number of sera from patients with various liver disorders, Wu and colleagues[59] found the following proportions of bilirubin in the 4 fractions: 27% (8–55%) unconjugated bilirubin, 23.9% (8–37%) monoconjugated bilirubin, 13.1% (7–21%) diconjugated bilirubin, and 36.8% (10–77%) protein-bonded bilirubin. Using the alkaline methanolysis HPLC method, Scharschmidt et al.[48] found in 5 urine specimens from 4 patients with liver disease, bilirubin fraction values of 8 ± 6% (mean ± SD) unconjugated bilirubin, 30 ± 9% monoconjugated bilirubin, and 62 ± 14% diconjugated bilirubin.

The HPLC method of Lauff et al.,[29] which uses unconjugated, monoconjugated, and diconjugated bilirubin standards, has provided clinical chemists with a contemporary, rigorous reference method against which to compare routine laboratory bilirubin assays for accuracy. HPLC methods, however, are presently not recommended for measurements of bilirubin and its metabolites in the routine laboratory since they are labor intensive and require specialized equipment and highly specialized skill.

The most widely used methods for bilirubin measurement are those based on the diazo reaction, which was first described in 1883 by Ehrlich. In this reaction diazotized sulfanilic acid (the diazo reagent) reacts with bilirubin to produce 2 azodipyrroles (Figure 13-10), which are reddish purple at neutral pH and blue at low or high pH values. In 1916 van den Bergh and Muller applied this reaction to the quantitation of bilirubin in serum and established that alcohol

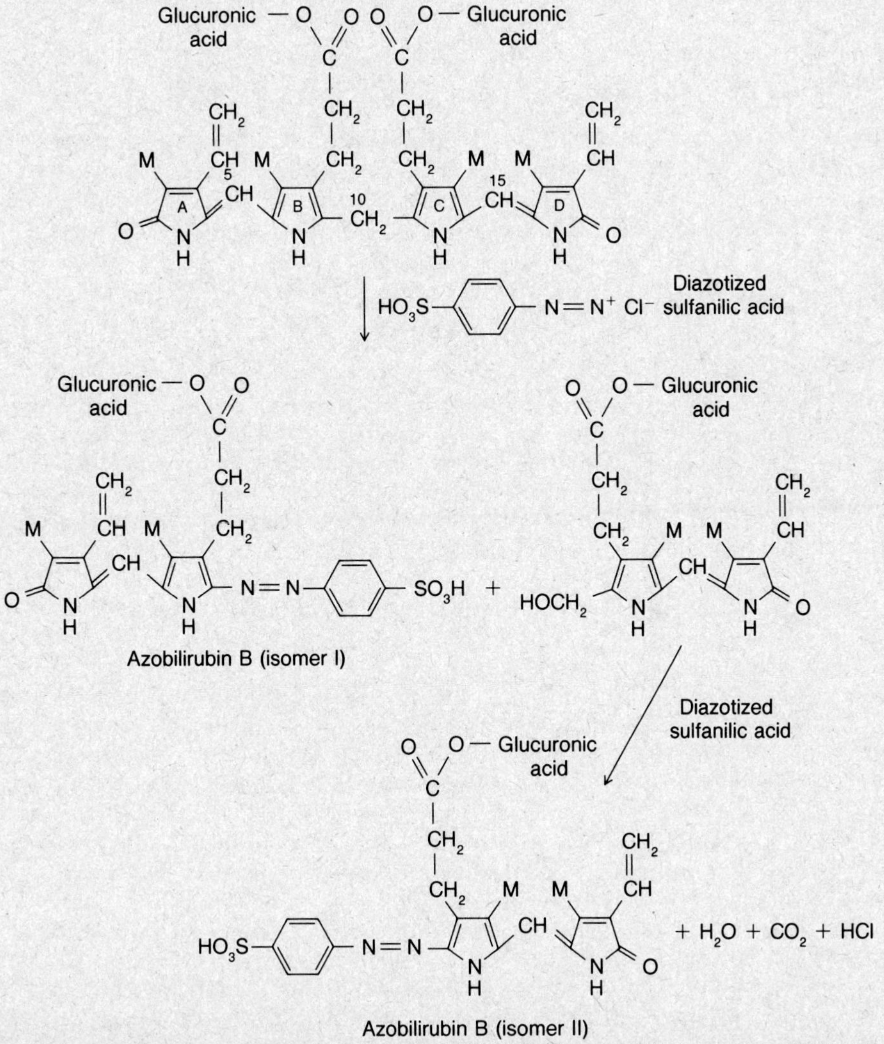

Figure 13-10. The reaction of bilirubin glucuronide with diazotized sulfanilic acid to produce isomers I and II of azobilirubin B. Unconjugated bilirubin reacts in the same way to produce isomers I and II of azobilirubin A.

accelerated the diazotization reaction of unconjugated bilirubin. These investigators described the fraction of bilirubin that reacted with the diazo reagent in the absence of alcohol as the "direct" bilirubin fraction. They used the term "indirect" bilirubin for the difference between total bilirubin, found after the addition of alcohol to the reaction mixture, and the so-called direct bilirubin fraction. Numerous variations of the van den Bergh procedure have been developed. All utilized one of a variety of "accelerators" that, analogous to the alcohol used by van den Bergh, facilitate the reaction of unconjugated (indirect) bilirubin with the diazo reagent. It has long been thought that direct-reacting bilirubin is primarily conjugated bilirubin, that total bilirubin consists of both the conjugated and unconjugated species, and that indirect bilirubin is the unconjugated species. In light of the recent work described above, we now know that there are at least 4 bilirubin species in serum. Direct reacting bilirubin consists of mono- and diconjugated bilirubin and the δ-fraction, which is bilirubin tightly bound to albumin. Unconjugated bilirubin, which is water insoluble and is associated with albumin by nonconvalent bonding, only reacts with the diazo reagent after the addition of an "accelerator" such as alcohol or caffeine.

The diazo method originally described by Jendrassik and Grof in 1938 and later modified by Doumas and colleagues in 1973[14] gives results for serum bilirubin that are reliable judged by the HPLC method of Lauff et al. In this procedure an aqueous solution of caffeine and sodium benzoate serves as the accelerator. The precise mechanism by which the caffeine-benzoate solution facilitates the reaction of unconjugated bilirubin with the diazo reagent has never been formally proved, but this reagent probably works by displacing unconjugated bilirubin from its association sites on albumin and by making bilirubin more water soluble, possibly by complex formation and disruption of internal hydrogen bonds. In careful, detailed studies using samples prepared by the addition of unconjugated bilirubin and authentic human diconjugated bilirubin to low-bilirubin pooled sera, Lo and Wu[33] have shown that the modified Jendrassik-Grof total bilirubin assay detects unconjugated and diconjugated bilirubin quantitatively (as unconjugated bilirubin equivalents). This method has acceptable transferability as recently demonstrated by the Study Group on Bilirubin of the Standards Committee of the American Association for Clinical Chemistry, and is currently the method of choice.[42]

Determination of Conjugated and Total Bilirubin in Serum
(Jendrassik and Grof Method)

Specimen

Serum or plasma may be used as the specimen. A morning fasting specimen is preferred in order to avoid lipemia. Hemolysis should be avoided, since it produces falsely low values with diazo methods. Because both conjugated and unconjugated bilirubin are photo-oxidized when exposed to white or UV light, specimens should be protected from direct exposure to either artificial light or sunlight as soon as they are drawn. The sensitivity to light is temperature dependent; for optimal stability, storage of specimens in the dark and at low temperatures is essential. When specimens are stored in the refrigerator, stability is maintained for 3 d. Samples are stable for three months when stored frozen at −70 °C in the dark.

Principle

For the measurement of *conjugated bilirubin* the serum or plasma is acidified with dilute HCl and then mixed with diazotized sulfanilic acid to produce azobilirubin. Only the conjugated forms of bilirubin will react with the diazo reagent in the absence of the accelerator caffeine-benzoate. The reaction is stopped by the addition of an ascorbic acid solution. Then an alkaline tartrate solution is added to the reaction mixture, followed by the addition of an aliquot of the caffeine reagent. The latter shifts the absorbance peak of azobilirubin from 585 to 600 nm, at which the absorbance is measured, while the tartrate reagent provides an alkaline pH to produce the blue and more intense color of azobilirubin.

The measurement of *total bilirubin* in serum or plasma is achieved by adding caffeine reagent (accelerator) to the specimen followed by the addition of diazotized sulfanilic acid. During the incubation period both conjugated and unconjugated bilirubin react with the diazo reagent to produce azobilirubin. Ten minutes after the addition of diazotized sulfanilic acid, solutions of ascorbic acid, alkaline tartrate, and dilute hydrochloric acid are added to the reaction mixture. The absorbance of the resulting blue azobilirubin solution is measured at 600 nm.

Reagents

1. Caffeine-benzoate reagent. Dissolve 56 g anhydrous sodium acetate, 56 g sodium benzoate, 1 g disodium EDTA, and 37 g caffeine in 700 mL of reagent grade water. Dilute to 1 L. This reagent is stable for at least six months at room temperature.

2. Hydrochloric acid, 0.05 mol/L.

3. Sulfanilic acid, 5 g/L. Add 5 g of sulfanilic acid to 700 mL of reagent grade water. Add 15 mL of concentrated hydrochloric acid. Dilute to 1 L. Store at room temperature.

4. Sodium nitrite, 5 g/L. Dissolve 0.5 g of sodium nitrite in 70 mL of reagent grade water and dilute to 100 mL. Store at 4 °C. Prepare fresh every two weeks.

5. Diazotized sulfanilic acid. Mix 20 mL of the stock sulfanilic acid solution (reagent 3) with 0.5 mL of a stock solution of sodium nitrite (reagent 4). Prepare fresh daily and store in a refrigerator.

6. Ascorbic acid solution, 40 g/L. Dissolve 200 mg ascorbic acid in 5 mL water. Prepare fresh daily and store in a refrigerator.

7. Alkaline tartrate solution. Dissolve 75 g sodium hydroxide and 320 g of sodium potassium tartrate in 700 mL of reagent grade water. After the solution has cooled, dilute to 1 L. This reagent is stable for at least six months at room temperature.

Procedure

Conjugated Bilirubin

1. Add 1.0 mL of HCl, 0.05 mol/L, to each of two glass test tubes, one labeled C (conjugated) and the other B (blank). Add 2.0 mL of caffeine-benzoate reagent to tube B only.

2. Add 0.2 mL of sample to each tube.

3. Add 0.5 mL of sulfanilic acid solution to tube B only and mix.

4. Add 0.5 mL of freshly prepared, diazotized sulfanilic acid reagent to tube C and mix.

5. Exactly 10 min later, add 0.1 mL ascorbic acid solution to tubes B and C followed immediately by the addition of 1.5 mL of alkaline tartrate solution to tubes B and C, and mix.

6. Add 2.0 mL of caffeine reagent to tube C and mix.

7. Read the absorbance of tube C at 600 nm against tube B set at zero absorbance.

Total Bilirubin

1. Add 2.0 mL caffeine reagent to a glass test tube labeled T (Total).

2. Add 0.2 mL of sample to the tube.

3. Add 0.5 mL of freshly prepared diazotized sulfanilic acid reagent to the tube and mix.

4. Exactly 10 min after the addition of diazotized sulfanilic acid in step 3, add, in succession, 0.1 mL of ascorbic acid solution, 1.5 mL of alkaline tartrate, and 1.0 mL of HCl, 0.05 mol/L, to the tube and mix.

5. Read the absorbance of the solution in tube T at 600 nm against that of tube B (from step 5 in the conjugated bilirubin method) set at zero absorbance.

Calibration

Only unconjugated bilirubin of acceptable purity should be used for establishing calibration curves. Such preparations of bilirubin may be obtained from the National Bureau of Standards (Washington, DC 20234) or Pfanstiehl Laboratories (reference grade bilirubin, Waukegan, IL 60085) and should have a molar absorptivity at 453 nm of 60 700 ± 1600 in chloroform at 25 °C. Extensive multireference laboratory studies have shown that pure bilirubin, as azobilirubin, has a molar absorptivity at 600 nm of 75 080 ± 1520.[42] An acceptable diluent for the unconjugated bilirubin is bovine serum albumin (BSA) Cohn Fraction V, 40 g/L. The bilirubin standard to be used for the preparation of the calibration curve should be prepared as follows:

1. Weigh out 20.0 mg of bilirubin of acceptable purity, transfer to a 100-mL volumetric flask, and dissolve by adding 1.0 mL of dimethyl sulfoxide and 2.0 mL of Na_2CO_3, 0.1 mol/L. Weigh bilirubin on a plastic weighing dish, not glassine paper. When transferring the bilirubin to the 100-mL volumetric flask, use the dimethyl sulfoxide for quantitative transfer of any bilirubin that sticks to the weighing dish. Dilute the solution to 100 mL with the bovine serum albumin diluent solution (40 g/L) previously adjusted to pH 7.4. Since bilirubin is light sensitive, the flask should immediately be wrapped with aluminum foil to protect the solution from light as

much as possible. Stability of the standard is dependent on the storage temperature. Deterioration is about 1.5% per month at -20 °C and about 1% in six months at -70 °C.

2. Standard solutions for the preparation of the calibration curve are prepared as follows:

Total Bilirubin (mg/L)	40 g/L BSA (mL)	200 mg/L Standard (mL)
0	4.0	0
20	9.0	1.0
50	3.0	1.0
100	2.0	2.0
150	1.0	3.0
200	0	4.0

Note: Minimize exposure to light.

3. Determine the absorbance values of azobilirubin produced by each of the standard solutions using the procedure for the analysis of serum bilirubin. The procedure is linear to 200 mg/L.

Reference Ranges

Reference ranges for serum total bilirubin of newborns, infants, and adults are summarized below:

Age	Premature mg/L	Premature µmol/L	Full-Term Newborn mg/L	Full-Term Newborn µmol/L
Up to 24 h	10–80	17–137	20–60	34–103
Up to 48 h	60–120	103–205	60–100	103–171
Days 3–5	100–140	171–239	40–80	68–137
Infants after 1 month and adults:				
Conjugated:	0–2.0	0–3.4		
Total:	2.0–10.0	3.4–17		

Comments

1. It is important to determine the molar absorptivity of the bilirubin standard solution after color development; it should fall within the range of 75 080 $\pm$ 1520.

2. Check the acceptability of the bilirubin stock standard solution by centrifuging 5–10 mL of the standard in a table top centrifuge at 1500 $\times$ g for 10 min and inspecting the tube. If there is any orange sediment on the wall or the bottom of the tube, the solution is unacceptable. Prepare it again and be sure to allow bilirubin to dissolve completely in the dimethyl sulfoxide–sodium carbonate solution before adding the bovine serum albumin solution.

Reference

Perry, B. W., Doumas, B. T., Bayse, D. D., et al.: A candidate reference method for determination of bilirubin in serum. Test for transferability. Clin. Chem., *29*:297-301, 1983.

Determination of Bilirubin in Infants by Direct Spectrophotometry

Specimen

Draw a capillary specimen from the heel or fingertip. Avoid hemolysis.

Principle

The absorbance of bilirubin in serum at 454 nm is proportional to its concentration. The serum of newborn infants does not contain carotene and other pigments that increase the absorbance at 454 nm. However, these pigments may be present in serum from older children and adults, and thus use of the direct spectrophotometric method should be restricted to newborns. The contribution of hemoglobin to the absorbance at 454 nm is corrected by subtracting the absorbance at 540 nm. (See calculations.)

Reagent

Phosphate buffer, pH 7.4. Weigh out 7.65 g of $Na_2HPO_4 \cdot 7H_2O$ and 1.74 g of anhydrous KH_2PO_4. Dissolve in water and dilute to 1 L in a volumetric flask. Check the pH with a pH meter and adjust if necessary.

Procedure

1. With a micropipet, add 20 μL of serum to a microcuvet (10 mm light path), containing 1 mL of phosphate buffer. Rinse the pipet several times with the buffer.

2. Add 1 mL of phosphate buffer to another microcuvet and, using this as a blank, set the spectrophotometer at zero absorbance; read the absorbance of the diluted serum at 454 and 540 nm. To obtain the accurate absorbance readings required in this method, a spectrophotometer with a bandpass of ≤ 10 nm should be used.

Calculations

Absorbance values observed at 454 and 540 nm represent the sum of absorbances at each wavelength of bilirubin and hemoglobin present in the sample. Since absorbance contributed by hemoglobin is essentially the same at both wavelengths, the difference in absorbance between the two wavelengths ($A_{454}-A_{540}$) represents absorbance of bilirubin only; i.e., hemoglobin absorbance is canceled out.

Determine the absorption constant of bilirubin on the spectrophotometer to be used for assay of unknown sera. A set of bilirubin standards from 0 to 200 mg/L is prepared as described for the diazo method, *except* that hemoglobin-free, pooled human serum with a bilirubin concentration < 3 mg/L is used instead of BSA. Each standard is put through the procedure above and $A_{454}-A_{540}$ (ΔA_2) recorded for each. ΔA_2 of each standard is then divided by its corresponding concentration in mg/L, and values are averaged. The average ΔA_2 value should be $0.084 \cdot mg^{-1} \cdot L$; its reciprocal is 11.9 mg/L. The calculating equation for unknowns is therefore

$$\text{Bilirubin, mg/L} = (A_{454} - A_{540})_{unknown} \times 11.9 \times 50$$

where 50 is the dilution factor.

Determination of Bilirubin with Direct-Reading Bilirubinometers

The principle of the direct spectrophotometric method for measuring total bilirubin in serum has been applied in the design of direct-reading bilirubinometers. These instruments are in essence differential spectrophotometers, either with two light paths passing at right angles to each other through the same cuvet or with a single light path that is split into two beams after passing through the cuvet. The resulting light beams are passed through two individual narrow-bandpass filters to separate photodetection cells. One filter has a wavelength of 454 (or 461) nm, which is the absorption peak wavelength for bilirubin. The other filter wavelength is 540 (or 561) nm, chosen as the point at which oxyhemoglobin has the same absorbance as at 454 (or 461) nm.

When bilirubin is present in the sample, the absorbance at 454 (or 461) nm increases and unbalances the photodetection system. The bilirubin concentration in the sample is read directly on a meter, or the null point needle is returned to zero and the concentration is read on a scale calibrated in mg/dL or mg/L. If oxyhemoglobin is present in the sample, no interference will occur, because the absorbances of oxyhemoglobin at 540 (or 561) nm and at 454 (or 461) nm are equal and will therefore cancel.

The instruments can be standardized with bilirubin solutions of known concentrations or with secondary standards, such as a methyl orange solution at pH 7.4 or with a multilayered colored glass standard.

These instruments are available commercially from Advanced Instruments, Inc. (Newton Highlands, MA), and American Optical Corp. (Buffalo, NY).

Determination of Bilirubin by Reflectance Spectrophotometry

A new and very promising thin-film technique for the spectrophotometric determination of total bilirubin has been developed by Eastman Kodak Co. (Rochester, NY) for the Kodak Ektachem Analyzer, which utilizes reflectance spectrophotometry.[59] The thin film for total bilirubin determination has three main components. The top layer, or spreading layer, contains caffeine and sodium benzoate to dissociate bilirubin from albumin; the second layer, the screen layer, contains gelatin, which traps and holds serum proteins in place; the third layer, the reaction layer, contains a cationic polymer, called a mordant, that binds the bilirubin. The interaction of both unconjugated and conjugated bilirubin with the mordant produces spectral shifts to lower wavelengths and increases molar absorptivity values. The reflectance spectra of the mordant-diconjugated bilirubin and mordant-monoconjugated bilirubin complexes are similar, but they differ from that of the unconjugated bilirubin-mordant complex. However, the reflectance densities of all three bilirubin species are comparable at 400 nm, which is the measuring wavelength for this method. Substances such as hemoglobin, chylomicrons, lipoproteins, and pigmented proteins that interfere with other direct spectrophotometric methods are trapped and retained in the screen layer. This reflectance technique measures only bilirubin or other species that absorb light at 400 nm and that are located in the reaction layer. The δ-bilirubin fraction is also retained in the screen layer and thus will not be detected by the Ektachem pediatric bilirubin technique. This is not a significant problem for pediatric bilirubin determinations, since the δ-bilirubin fraction from neonates up to 14 d of age is < 5% of total bilirubin. In contrast, adult specimens can contain up to 90% of total bilirubin as the δ-fraction.

The Kodak system requires a total of about 35 μL of serum, 10 μL of which is dispensed onto the spreading layer from a specimen cup. For neonatal sera the method agrees very well with the modified Jendrassik-Grof diazo method,[14] and experience in the General Chemistry Laboratory of the Hospital of the University of Pennsylvania has been excellent.[27]

There is now available from Eastman Kodak Co. a bilirubin procedure that uses thin-film technology and reflectance spectrophotometry for the determination of total bilirubin and of the individual bilirubin fractions—unconjugated bilirubin, conjugated bilirubin (mono- plus dicon-jugates), and δ-bilirubin.[56a] The procedure for the determination of the unconjugated and conjugated bilirubin fractions makes use of the same thin-film slide described above. This procedure takes advantage of the difference in reflectance spectra of unconjugated bilirubin and that of monoconjugated and diconjugated bilirubin. Reflectance densities are measured at two wavelengths, 400 nm and 460 nm, and the concentrations of unconjugated and conjugated bilirubin are then derived from these measurements. A separate determination of total bilirubin is made with a different thin-film slide. In this thin-film slide all of the bilirubin fractions react in the top layer—the spreading layer—with a diazonium salt in the presence of an accelerator, dyphylline, and a surfactant, Triton X-100, to produce the corresponding azo pigments. The reflectance densities of the azo derivatives of all bilirubin fractions are comparable at 540 nm, the measuring wavelength for this method. The concentration of the δ-bilirubin is derived by subtraction of the unconjugated and conjugated bilirubin values from the total bilirubin result. Investigations of the clinical significance of determining serum concentrations of these bilirubin fractions are currently under way.

Methods for the Determination of Urine Bilirubin

The qualitative detection of bilirubin in urine has proved to be a useful test in the differential diagnosis of jaundice. Because conjugated, but not unconjugated, bilirubin is excreted in urine, the urine test can demonstrate whether a patient with hyperbilirubinemia has increased concentrations of conjugated bilirubin. The most commonly used test method for detecting bilirubin in urine, besides visual examination, involves the use of a "dipstick" impregnated with a diazo reagent. Dipstick methods can detect concentrations of 5 mg/L or higher.

Specimens

A fresh urine specimen is required since bilirubin is very unstable when exposed to light and room temperature. If the test must be delayed, protect the specimen from light and store in a refrigerator at 2–8 °C for not more than 24 h.

Procedure

Dip the reagent strip into the urine specimen for no longer than 1 s. Allow 60 s for the bilirubin to react with 2,6-dichlorobenzene-diazonium-tetrafluoroborate (Chemstrip, Bio-Dynamics/bmc, Indianapolis, IN) at an acidic pH. A pink to red-violet color is produced, the intensity of which is proportional to the bilirubin concentration. The reaction mechanism for urinary conjugated bilirubin is the same as that described in Figure 13-10 with diazotized sulfanilic acid except that the diazo reagent is 2,6-dichlorobenzene-diazonium-tetrafluoroborate. Another commonly used dipstick is the Ictotest (Ames Co., Div. Miles Laboratories, Elkhart, IN) procedure. In this test procedure p-nitrobenzenediazonium p-toluenesulfonate is the diazo reagent.

Sources of Error

The Chemstrip test for bilirubin in urine is a highly specific test. It has a low incidence of false positive results. Medications that color the urine red or that give a red color in an acid medium, such as phenazopyridine, can produce a false-positive reading. Large quantities of ascorbic acid or of nitrite lower the sensitivity of the test.

Urobilinogen in Urine and Feces

Any pathological process which leads to increased concentrations of urobilinogen in the gastrointestinal tract results in an increase in the quantity of urobilinogen excreted in urine. Examples of such conditions are excess hemolysis, liver damage produced by hypoxia or exposure to various toxic agents, and various forms of liver disease. As noted in the discussion of disturbances of urobilinogen metabolism, when biliary obstruction becomes more complete, urinary excretion of urobilinogen decreases due to the very limited delivery of bilirubin to the gut and the extremely low rate of urobilinogen production. The clay-colored or chalky-white stool in patients with obstructive jaundice results from the very low quantities of the bile pigment metabolites of urobilinogen produced in this condition. These disturbances of urobilinogen excretion are the basis for the use of urobilinogen analysis in urine and feces as an index of liver disease. It is the opinion of the authors that urobilinogen measurement is of little help in the evaluation of liver disease since these measurements do not add to the diagnostic information obtained from the commonly measured tests of liver disease described below in the section on liver diseases.

Since the analysis of urobilinogen in urine and feces is a procedure that is still of interest to some laboratories, a brief discussion of the methodology for this analyte follows.

Urobilinogen in freshly collected urine is most commonly measured by reaction with Ehrlich's reagent (p-dimethylaminobenzaldehyde in concentrated HCl) to produce a red-colored product.

In 1964, Henry critically evaluated this measurement principle for urobilinogen and developed a rapid semiquantitative technique that gives reliable values. In this method urobilinogen is determined spectrophotometrically after reaction of an aliquot with p-dimethylaminobenzaldehyde in concentrated HCl. In order to maintain urobilinogen in a reduced state and prevent re-formation of urobilin, ascorbic acid is added as a reducing agent. Sodium acetate is added to reduce the acidity after reaction of urobilinogen with Ehrlich's reagent. Another attribute of sodium acetate is that it inhibits color formation from indole and skatole and enhances the color produced by the reaction of urobilinogen with Ehrlich's reagent. The principle of the method for the determination of fecal urobilinogen is the same as that described for urinary urobilinogen except that an aqueous extract of fresh feces is first treated with alkaline ferrous hydroxide (to reduce urobilin to urobilinogen) prior to the addition of Ehrlich's reagent.

References

Henry, R. J., Fernandez, A. A., and Berkman, S.: Studies on the determination of bile pigments. VI. Urobilinogen in urine as urobilinogen-aldehyde. Clin. Chem., *10*:440-446, 1964.
Routh, J. I.: Liver function. *In*: Fundamentals of Clinical Chemistry. N. W. Tietz, Ed. Philadelphia, W.B. Saunders Co., 1976.

BILE ACIDS

Bile acids are a class of endogenous organic anions whose chemistry and physiology have been extensively studied for many years.[12] However, it has only been in recent years that definition of the unique aspects of bile acid physiology and alterations in bile acid metabolism that occur with disease have been applied to clinical medicine. The major impetus for this renewed interest has been the recognition that surgical and pharmacologic *manipulation* of bile acid metabolism may have important implications in cholesterol gallstone disease and in cholestatic liver disease. In addition, derangements in intestinal bile acid absorption may lead to a severe secretory diarrhea (cholerheic enteropathy). A major reason for our improved understanding of bile acid metabolism and physiology has been the recent development of valid, practical methods of analyses, including radioimmunoassay and immunoenzymatic assays that are specific and accurate for measuring concentrations of bile acids in various body fluids. Therefore, knowledge of the clinical utility of these tests is changing rapidly.

Since the regulation of bile acid metabolism is a major function of the liver, detectable derangements in bile acid metabolism are a reflection of liver dysfunction. Through an orderly process of bile acid synthesis, conjugation, and secretion, the liver serves to maintain cholesterol balance and to provide surface-active detergent molecules to facilitate both hepatic excretion of cholesterol and intestinal absorption of lipid. The proper functioning of this system involves an efficient ileal conservation mechanism through which bile acid molecules are rescued from fecal loss. Therefore, alterations in hepatic bile acid synthesis, intracellular metabolism, excretion, intestinal absorption, or plasma extraction are reflected in derangements in bile acid metabolism.

Bile Acid Chemistry and Metabolism

Cholesterol is continually synthesized by all tissues, but primarily by the liver and small intestine. It must be eliminated from body pools in order to prevent accumulation with the attendant risk of atherosclerosis. A portion of cholesterol is converted in the liver to highly polar bile acids, which are subsequently secreted in bile. This transformation and the ability of bile acids to solubilize additional cholesterol in bile are the major mechanisms of cholesterol elimination from the body.

The products of cholesterol metabolism are cholic acid and chenodeoxycholic acid, which are termed "primary" bile acids because of their hepatic origin. The sequence of reactions involved in the synthesis of bile acids from cholesterol is shown in Figure 13-11. Comparison of the structure of cholesterol and cholic acid demonstrates that a number of alterations of the cholesterol molecule are necessary to complete this conversion: (1) addition of a hydroxyl group in the 7-α-position, with subsequent further hydroxylation at the 12-α-position; (2) inversion of the hydroxyl group at position 3 from the 3-β-position (above the plane of the molecule) to the 3-α-position (below this plane); (3) saturation of the double bond at position 5-6; (4) loss of a 3-carbon group from the cholesterol side chain giving rise to a 24-carbon molecule; and (5) oxidation of the C-24 carbon to a carboxyl group. The other primary bile acid formed in the liver, chenodeoxycholic acid, differs from cholic acid only by the absence of a hydroxyl group at position 12. The initial step in the synthesis of both bile acids, 7-α-hydroxylation of cholesterol, occurs under the influence of the rate-limiting enzyme for bile acid synthesis, 7-α-hydroxylase. The reaction is affected by bile acids returning to the liver, via the portal vein, to exert feedback inhibition to control their own synthesis rate.

Prior to secretion into the bile canaliculi, both of the primary bile acids are conjugated at the carboxylic acid carbon with an amino acid (Figure 13-12). This combination with either glycine or taurine increases polarity, and hence water solubility. Through this mechanism of conjugation, four primary bile acids (cholyltaurine, cholylglycine, chenodeoxycholyltaurine, and chenodeoxycholylglycine) are formed. Conjugation decreases the pK_a values of bile acids from ~6 to 4 for glycine conjugates and 2 for taurine conjugates. The conjugated bile acids are present in the intestinal lumen in the ionized form because their pK_a values are low compared with the pH of the intestinal lumen, which is much higher. In health, the glycine conjugates predominate in a ratio of approximately 3:1 to 4:1. Unconjugated (free) bile acids are not present in bile.

During passage through the small intestine and colon, both cholic acid and chenodeoxycholic acid are subject to alterations by enzymes produced by the indigenous bacterial flora. Bacterial 7-α-dehydroxylase gives rise to the secondary bile acids deoxycholic acid and lithocholic acid (Figure 13-13). After variable reabsorption, secondary bile acids are also conjugated in the liver with glycine or taurine and join the primary bile acids as components of bile. The average bile acid composition of normal bile in the human adult is ~38% cholate conjugates, 34% chenodeoxycholate conjugates, 28% deoxycholate conjugates, and 1–2% lithocholate conjugates.

The Role of Bile Acids in Hepatic Bile Formation

The carrier-mediated active transport of bile acids into the bile canaliculi generates osmotic water flow and is a major factor regulating bile formation and secretion. Transport of these organic anions also influences secretion of the remainder of the major components of bile, such as bilirubin, cholesterol, and phospholipids. There is no secretion of the latter two compounds

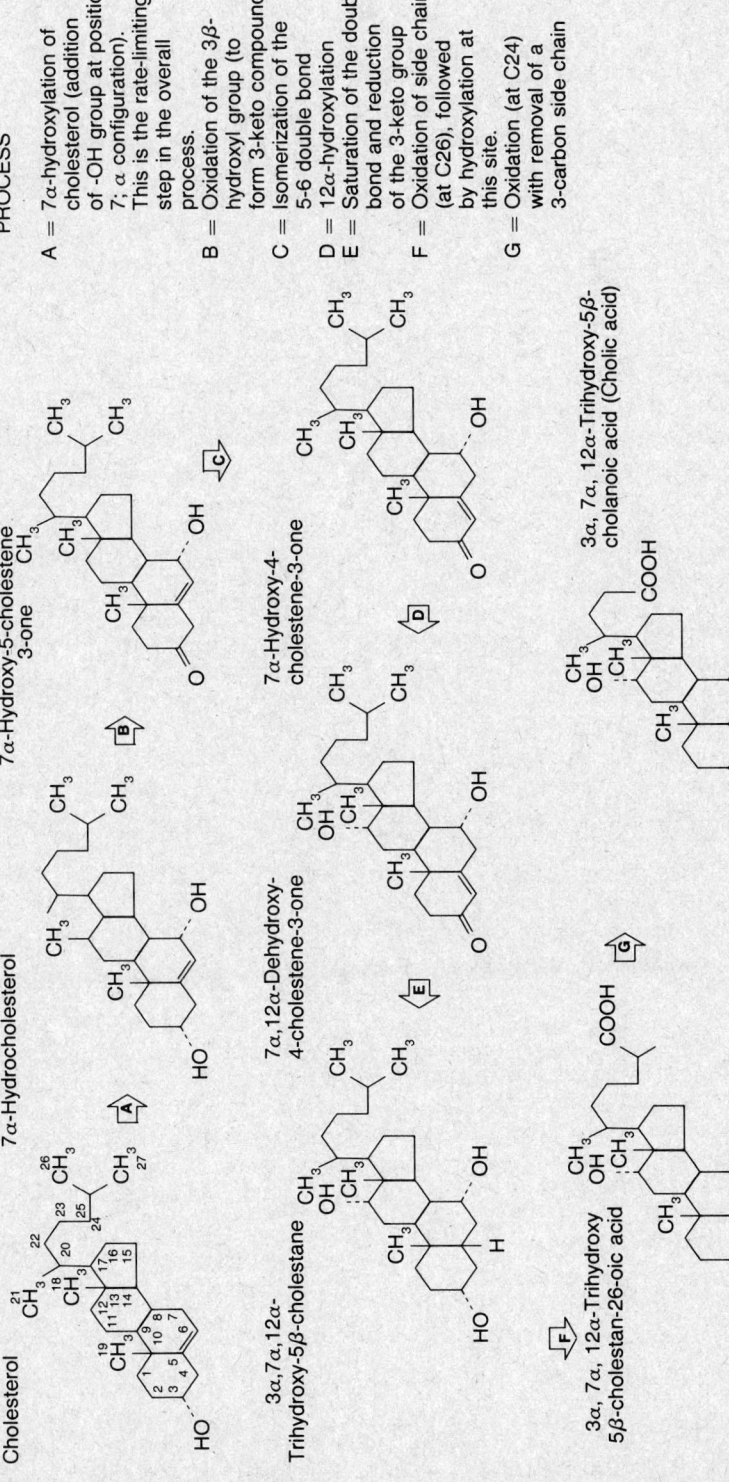

Figure 13-11. Synthesis of cholic acid, a primary bile acid, from cholesterol. (From Balistreri, W. F., and Soloway, R. D.: Clinical Guide to Bile Acid Physiology and Alterations in Disease States. Reproduced with the permission of Abbott Laboratories, Diagnostics Division, North Chicago, Ill. © 1979 by Abbott Laboratories.)

PROCESS

A = 7α-hydroxylation of cholesterol (addition of -OH group at position 7; α configuration). This is the rate-limiting step in the overall process.

B = Oxidation of the 3β-hydroxyl group (to form 3-keto compound)

C = Isomerization of the 5-6 double bond

D = 12α-hydroxylation

E = Saturation of the double bond and reduction of the 3-keto group

F = Oxidation of side chain (at C26), followed by hydroxylation at this site.

G = Oxidation (at C24) with removal of a 3-carbon side chain

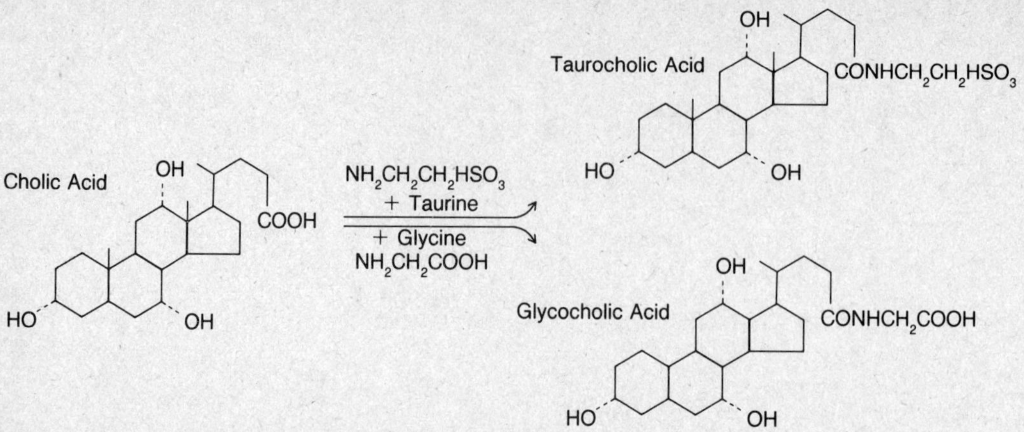

Figure 13-12. Conjugation of cholic acid with either taurine or glycine. (From Balistreri, W. F., and Soloway, R. D.: Clinical Guide to Bile Acid Physiology and Alterations in Disease States. Reproduced with the permission of Abbott Laboratories, Diagnostics Division, North Chicago, Ill. © 1979 by Abbott Laboratories.)

in the absence of bile acid secretion. The influence of bile acid secretion on biliary lipid excretion is, in large part, due to the ability of bile acids to solubilize cholesterol and phospholipids in an aqueous medium in mixed micelles.

Bile formation occurs by processes that are not fully defined and takes place in the bile canaliculi, which are specialized modifications of the hepatocyte membrane and which ultimately unite to form bile ductules. Hepatic bile contains 5–15% total solids, the major components of which are bile acids. Bile acids have a major role in regulating bile flow, and they have been shown to be potent choleretic agents. The increase in water and small solute (electrolyte) excretion in bile due to the osmotic effect of actively secreted bile acid solute represents the bile acid–dependent fraction of bile flow. However, even with total depletion of the circulating bile acid pool, as is seen with bile duct diversion, some bile flow continues. It has been postulated that active transport of sodium into the canaliculi is responsible for an additional portion of bile flow. This sodium-induced fraction, independent of bile acid excretion, may account for as much as 50% of total bile flow. It has also been shown that hormones such as secretin increase bile flow without concomitant increase in bile acid output. This indicates an effect at sites distal to the canaliculi, such as the bile ductules. The relative contribution of the bile ductules or major bile ducts to normal bile flow is unknown.

The ability of bile acids to solubilize biliary lipids is due to their amphophilic nature, i.e., they possess polar and nonpolar regions in the same molecule (Figure 13-14). Such molecules align at water-lipid interfaces, reduce surface tension, and therefore act as detergents. In an aqueous solution, bile acids aggregate to form small polymolecular aggregates called micelles, ~5 nm in diameter, which are capable of incorporating cholesterol and phospholipids (Figure 13-15). Co-micellization with these water-insoluble constituents will maintain cholesterol in so-

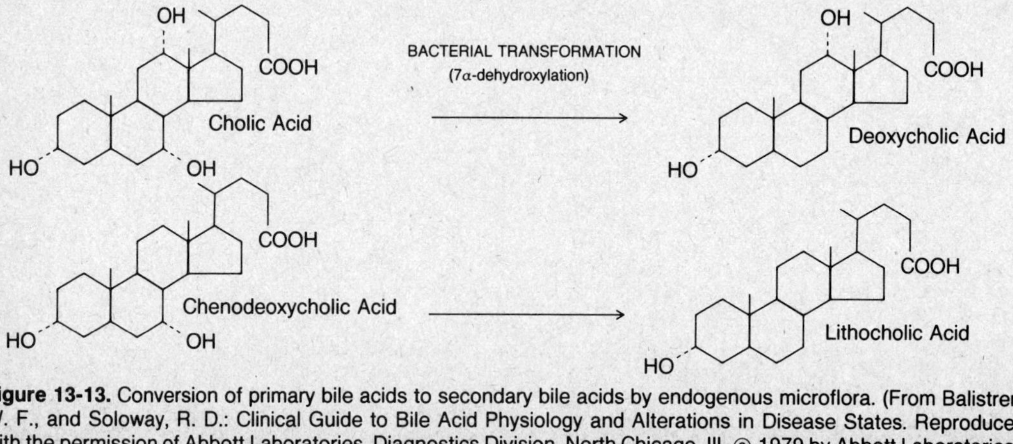

Figure 13-13. Conversion of primary bile acids to secondary bile acids by endogenous microflora. (From Balistreri, W. F., and Soloway, R. D.: Clinical Guide to Bile Acid Physiology and Alterations in Disease States. Reproduced with the permission of Abbott Laboratories, Diagnostics Division, North Chicago, Ill. © 1979 by Abbott Laboratories.)

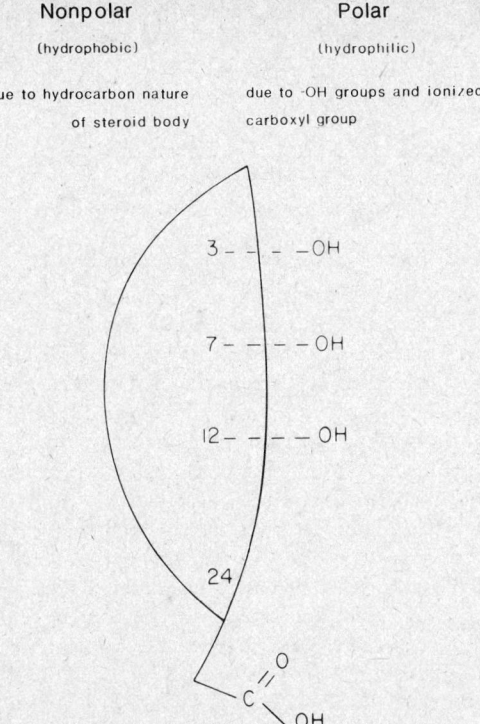

Figure 13-14. Schematic diagram of bile acid molecule, which demonstrates alignment of —OH and —COOH groups that assure amphophilic properties. (From Balistreri, W. F., and Soloway, R. D.: Clinical Guide to Bile Acid Physiology and Alterations in Disease States. Reproduced with the permission of Abbott Laboratories, Diagnostics Division, North Chicago, Ill. © 1979 by Abbott Laboratories.)

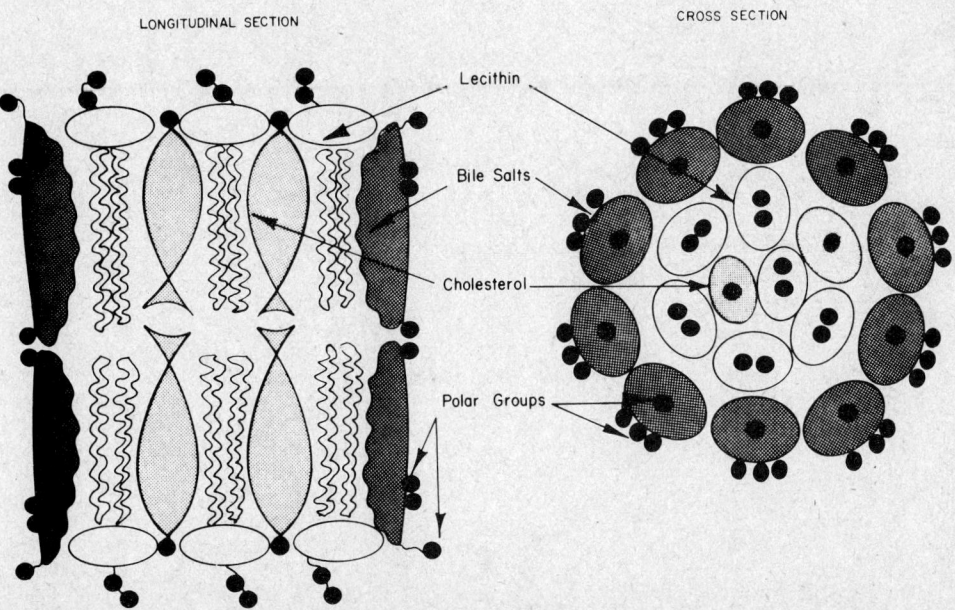

Figure 13-15. Structure of mixed micelles in bile. (From Balistreri, W. F., and Soloway, R. D.: Clinical Guide to Bile Acid Physiology and Alterations in Disease States. Reproduced with the permission of Abbott Laboratories, Diagnostics Division, North Chicago, Ill. © 1979 by Abbott Laboratories.)

lution. Formation of mixed micelles of bile salts and phospholipids will enhance the aqueous solubility of cholesterol, a weakly polar compound, and permit cholesterol excretion in bile, an aqueous polar medium.

The Enterohepatic Circulation of Bile Acids

The body conserves the bile acid pool through an effective conservation and recirculating system, the enterohepatic circulation. The anatomical components of the enterohepatic circulation are the liver, biliary tract, intestine, and portal venous system. The restriction of the bulk of the bile acid pool to this compartment is made possible by efficient intestinal reabsorption and hepatic extraction of these molecules, allowing a very efficient recycling mechanism.

During fasting, bile acids pass down the biliary tree to enter the gallbladder (Figure 13-16). Here they undergo a ten-fold increase in concentration due to reabsorption of water and electrolytes. After an overnight fast, 95% of the bile acid pool may be sequestered in the gallbladder. This sequestration results in low levels of bile acids in the intestine, portal vein, and liver and, in fasting patients, a very low concentration of bile acids in plasma.

In response to a meal, hormones such as cholecystokinin that have been released from the intestinal wall cause relaxation of the sphincter of Oddi and contraction of the gallbladder. This allows a concentrated solution of mixed micelles of bile salts and the phospholipid lecithin to enter the intestine, carrying ∼1 g of cholesterol daily. In the intestinal lumen, endogenous (biliary) phospholipid and cholesterol leave the micelles and are replaced by dietary cholesterol and the products of triglyceride digestion, predominantly free fatty acids and monoglycerides. Micelles serve to facilitate fat absorption in the jejunum by accelerating the hydrolytic action of pancreatic lipase on triglyceride and by solubilizing the hydrolytic products. In addition, bile acid micelles assure delivery of the lipolytic products to the mucosal surface. In order to carry out these functions, a minimal concentration of bile acids, ∼2 mmol/L, is necessary; below this critical concentration, micelles are no longer formed and fat digestion and absorption are impaired.

Bile acids are subsequently reabsorbed in the intestine by two mechanisms. The predominant means is by active transport of all types of bile acids; transport occurs exclusively in the distal portion of the ileum. The alternative mechanism, nonionic diffusion, is dependent upon the variable ability of each bile acid to remain dissociated at the pH of jejunal contents; only unconjugated bile acids and the glycine conjugates of the dihydroxy bile acids will be in a nonionized form and will be reabsorbed by means of this mechanism. Selective reabsorption of these less polar, freely diffusible, nonionized compounds takes place by passive transport in the jejunum and

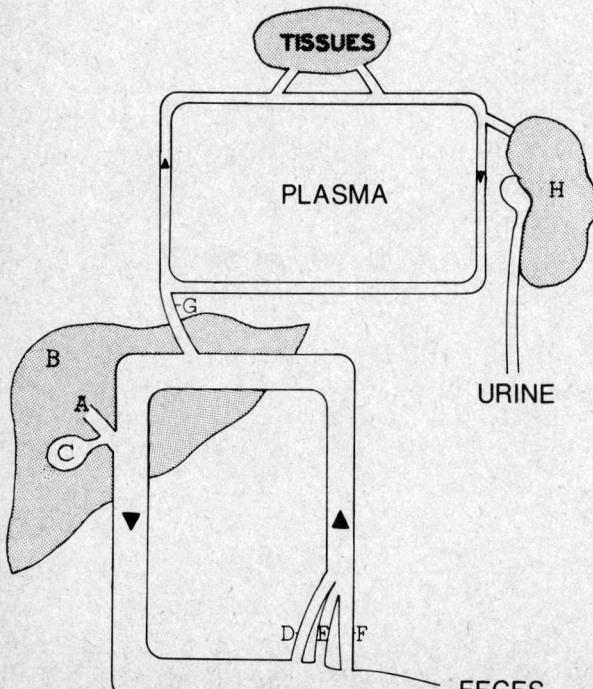

Figure 13-16. Fasting bile salt kinetics depicted as in a normal adult with estimated degree of compartmentation in the various organs involved in the enterohepatic circulation. **A** = Hepatic synthesis is followed by delivery from the liver (**B**) to the gallbladder (**C**); intestinal reabsorption occurs in several sites (**D** = Jejunal *passive* transport, **E** = Ileal *active* transport; **F** = colonic *passive* transport); **G** = Portal-systemic shunting; **H** = Kidney.

Kinetic data: Pool size = 2–4 g; jejunal concentration = 2–3 mmol/L; synthesis rate = 0.3 g/d; fecal loss = 0.3 g/d; peripheral plasma level = 3–14 μmol/L. With ingestion of a *meal*, jejunal concentration will increase to 5–10 mmol/L due to gallbladder contraction. An increase in portal venous levels will follow due to ileal reabsorption; there will be "spillover" into peripheral plasma of this increased load. (Modified from Balistreri, W. F., and Soloway, R. D.: Clinical Guide to Bile Acid Physiology and Alterations in Disease States. Reproduced in part with the permission of Abbott Laboratories, Diagnostics Division, North Chicago, Ill. © 1979 by Abbott Laboratories.)

colon. This difference in reabsorption may account for the observation that fasting serum concentrations of chenodeoxycholic acid are higher than those of cholic acid. Furthermore, serum chenodeoxycholic acid levels rise more rapidly following a meal than do cholic acid levels.

Bile acids are carried in blood tightly bound to protein. The concentration of bile acids in the portal vein is high; however, because of the efficiency of hepatic extraction, systemic blood levels remain low. The efficiency of extraction of cholic acid is ~80% in a single pass through the liver and ~60% for chenodeoxycholic acid. Hepatic extraction efficiency is unchanged over a broad range of portal bile acid concentrations. In response to the increased portal vein concentration of bile acids after a meal, systemic plasma levels rise due to spillover, and the plasma pool of bile acids is increased.

Measurement of bile acid concentrations in serum should be a key determination in the evaluation of a patient with suspected liver disease. Increased concentrations as a result of decreased liver uptake or altered metabolism are sensitive indicators of liver dysfunction. The clinical applicability of detecting alterations in this physiologic sequence is indicated by the fact that in many patients with liver disease, the 2-h postprandial bile acid level increase is the only detectable abnormality.

Kinetics of Bile Acid Metabolism

The size of the bile acid pool and its rate of turnover have been measured by the isotope dilution technique, in which a small amount of radiolabeled bile acid is administered either orally or intravenously, followed by serial sampling of bile (duodenal fluid) over 3–5 d. The extent of initial dilution of the labeled bile acid by the patient's own circulating bile acids is a measure of pool size. The rate at which this dilution increases with time reflects the rate of turnover of the pool (i.e., the bile acid synthetic rate). Using this technique, it has been found that in normal adults the bile acid pool averages 2–4 g. When hepatic synthesis and fecal loss are in balance, a steady state has been achieved. In health, the magnitude of each process is 0.3–0.8 g/d. There are 4–10 enterohepatic cycles per day, and these occur predominantly with meals. Because of this recycling mechanism, the jejunal concentration of bile acids is maintained at ~5–10 mmol/L during the postprandial state; this is much higher than the critical micellar concentration of ~2 mmol/L, the level needed for effective micellar solubilization of dietary fats. Between meals, with decreased entry of bile acids into the intestine, the intraluminal concentration decreases. Plasma bile acid levels depend on input from the enterohepatic circulation; therefore, levels of total plasma bile acids will increase by 50% above fasting values 90–120 min following a meal.

The Importance of Secondary Bile Acids

Lithocholic acid is a secondary monohydroxy bile acid formed from chenodeoxycholic acid by the intestinal microflora. This compound differs significantly in physicochemical properties from the other bile acids. It is insoluble in water and will precipitate out of solution in the colonic lumen. Lithocholic acid has been shown to be toxic to a number of cell membranes, including those of hepatocytes and erythrocytes. Cirrhosis has been produced in rabbits by intragastric instillation of lithocholic acid. Degeneration of hepatocytes as well as bile duct and ductular cell inflammation has been produced in a wide variety of reptiles, birds, and mammals by lithocholic acid administration. These anatomical changes have been associated with inhibition of bile flow. Intravenous lithocholate administration causes an immediate cessation of bile flow; therefore, in contrast to cholic acid, which is a potent choleretic agent, lithocholic acid may cause cholestasis. Small doses of lithocholic acid injected intramuscularly can produce fever, headache, nausea, and malaise, in addition to a local inflammatory reaction.

Despite limited intestinal absorption, lithocholic acid is a normal, albeit minor, constituent of human bile and serum. Therefore, certain protective mechanisms must exist in healthy subjects to prevent toxicity. This purpose is served by sulfation of lithocholate at the 3-α-position, a process which is 60% complete in one circulation through the liver (Figure 13-17). Sulfated bile acid esters of lithocholic acid are less well absorbed from the intestine and are more rapidly excreted in feces than in the nonsulfated form. The sulfate group, which is a fully ionized polar portion of the molecule, enhances water solubility and thus renal excretion.

Despite this "safety valve" mechanism, which acts to reduce the hepatotoxicity of lithocholic acid, elevated levels of this compound have been reported in patients with hepatic disease. Increased concentrations of lithocholate have been noted in livers obtained at autopsy of patients with acute hepatic necrosis and other hepatic diseases. Enhanced renal excretion of lithocholate,

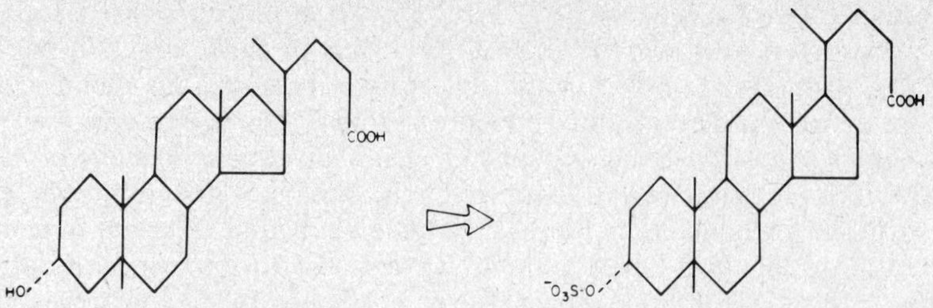

Lithocholic Acid
(3α -OH)

Sulfolithocholic Acid
(3α -OSO₃)

Figure 13-17. Lithocholic acid is converted to the more polar sulfated derivative, a process mediated by lithocholate sulfotransferase. (From Balistreri, W. F., and Soloway, R. D.: Clinical Guide to Bile Acid Physiology and Alterations in Disease States. Reproduced with the permission of Abbott Laboratories, Diagnostics Division, North Chicago, Ill. © 1979 by Abbott Laboratories.)

predominantly in the sulfated form, has been noted in alcoholic cirrhosis, although in normal subjects, both sulfated and nonsulfated bile acids appear in the urine in only small amounts.

Abnormalities of Bile Acid Metabolism

In view of the multiple processes involved in bile acid synthesis, conjugation, and excretion, as well as hepatic and intestinal uptake, there are several potential sites for primary or secondary disturbances (Table 13-3).

Defective bile acid synthesis. Specific defects in bile acid synthesis have been noted in two unusual conditions. One is *cerebrotendinous xanthomatosis,* a condition in which there exists an altered bile acid metabolism and a diminution of daily total bile acid production. In this disease, chenodeoxycholic acid is affected more than cholic acid, and unusual metabolites such as bile alcohols are present in bile and feces. Decreased activity of a specific enzyme causes a decrease in side-chain degradation, which leads to the eventual excretion of an accumulated intermediate of the bile acid synthetic pathway. The other is an unusual form of *intrahepatic cholestasis,* a disease first noted in infancy and childhood. Although the majority of patients with abnormalities of the intrahepatic bile duct system do not have identifiable *primary* abnormalities of bile acid metabolism, many cases have been found to have high concentrations of abnormal bile acids such as trihydroxycoprostanoic acid (THCA) in bile and urine. This compound, a 27-carbon bile acid, is a precursor of cholic and chenodeoxycholic acids in man. A metabolic block in the conversion of THCA to cholic acid is responsible for accumulation of this compound and may be, in part, responsible for the initiation or perpetuation of the liver injury.

More frequently encountered are *acquired* defects in bile acid synthesis that have been noted in liver diseases such as hepatitis and cirrhosis. In acute hepatitis, there are alterations of bile acid synthesis and conjugation due to hepatic parenchymal cell disease. Investigation of cirrhotic patients has documented a marked reduction in cholic acid synthesis with a very low concentration of biliary deoxycholic acid. These abnormalities are due to changes in hepatocyte function and the loss of liver cell mass as well as the presence of portosystemic shunting. The severity of cirrhosis correlates closely with the loss of cholic acid synthetic capacity. The decreased ratio of the trihydroxy bile acid (cholic acid) to the dihydroxy bile acid (chenodeoxycholic acid) in serum has therefore been used as a diagnostic test of liver cell dysfunction.

Abnormalities of bile acid delivery to the bowel. Delayed emptying or mechanical obstruction of the gallbladder or common bile duct can prevent bile acids from reaching the duodenal lumen. Extrahepatic bile duct obstruction due to biliary atresia, stricture, stone, or carcinoma will result in bile acid retention and regurgitation from the liver cell into plasma, as well as decreased delivery to the intestine.

Interruption of the enterohepatic circulation of bile acids. Approximately 95% of the bile acids that are secreted during a single enterohepatic cycle are recirculating bile acids. Therefore, a significant interruption of this cycle will lead to a decrease in hepatic bile acid secretion. The increased bile acid synthesis caused by this negative feedback, however, can only partially compensate for the losses. The active transport system for bile acid absorption, located in the terminal ileum, has a major input to the enterohepatic circulation. Therefore, resection, inflammation, or bypass of the ileum will be associated with specific clinical symptoms and disturbances of bile

Table 13-3. DISTURBANCES IN BILE ACID METABOLISM

Defective bile acid *synthesis*
  Congenital impairment of hepatics synthesis
  Specific defects in bile acid synthesis as seen in:
    Cerebrotendinous xanthomatosis
    Intrahepatic cholestasis
      Qualitative
      Quantitative
  Acquired defects in bile acid synthesis, as observed in liver diseases such as hepatitis and cirrhosis
Abnormalities of bile acid *delivery* **to the bowel**
  Celiac sprue
  Extrahepatic bile duct obstruction due to:
    Congenital biliary atresia
    Stricture
    Stone
    Carcinoma
Interruption **of the enterohepatic circulation of bile acids**
  An external bile fistula
  Ileojejunal exclusion for exogenous obesity or hypercholesterolemia
  Cystic fibrosis
  Contaminated small bowel syndrome (with bile acid precipitation, increased jejunal absorption, and "short
    circuiting")
  Entrapment of bile acids in intestinal lumen by:
    Cholestyramine
    Trivalent cations
    Fiber
Bile acid *malabsorption*
  Primary bile acid malabsorption (absent or inefficient ileal active transport)
    Intractable diarrhea (infancy)
    Irritable bowel (adults)
  Secondary bile acid malabsorption
    Ileal disease or resection
      Crohn's disease
      Ileal resection
      Ileal bypass
      Radiation enteritis
      Postinfectious enteritis
    Exogenous bile acid administration (e.g., gallstone dissolution)
    Cystic fibrosis
  Tertiary bile acid malabsorption
  Postcholecystectomy
  Renal failure
  Drugs
Defective *uptake* **or altered intracellular metabolism**
  Parenchymal disease (acute hepatitis, cirrhosis) associated with regurgitation from cells or portosystemic shunting
  Cholestasis

acid metabolism (Figure 13-18). These abnormalities are most closely reflected in the prototype of ileal resection (Figure 13-19).[4] Depending upon the length of the small intestine that has been removed, these patients will have moderate to severe watery diarrhea with or without steatorrhea. The amount of bile acid return to the liver is reduced, and there is a failure of feedback inhibition and an accelerated hepatic synthesis of bile acid. The concentration of serum cholesterol is reduced, since an increased proportion of this compound is utilized for bile acid synthesis. Levels of serum bile acids can, therefore, be an accurate reflection of ileal dysfunction, since the expected postprandial rise (due to ileal absorption of bile acids) will not be present (Figures 13-20 and 13-21).

Disturbances of bile acid metabolism in cholestasis. Cholestasis implies a disturbance in hepatic excretory function and a resultant decrease in bile flow; intrahepatic disease or extrahepatic processes may cause this condition. The causes of bile secretory failure can be classified according to whether there exists a functional disturbance or a mechanical obstruction to bile flow. In cholestasis of any cause, biliary excretion of bile acids is decreased, hepatic accumulation occurs, and there is a shift of the bile acid pool to nonintestinal compartments such as plasma and peripheral tissues. Elevated peripheral serum bile acid levels are present.[3]

Disturbances of bile acid metabolism in hepatocellular disease. The decrease in bile acid secretion that is associated with cirrhosis and hepatitis has been previously discussed. The mechanisms responsible for increased fasting serum levels in the presence of liver disease are regur-

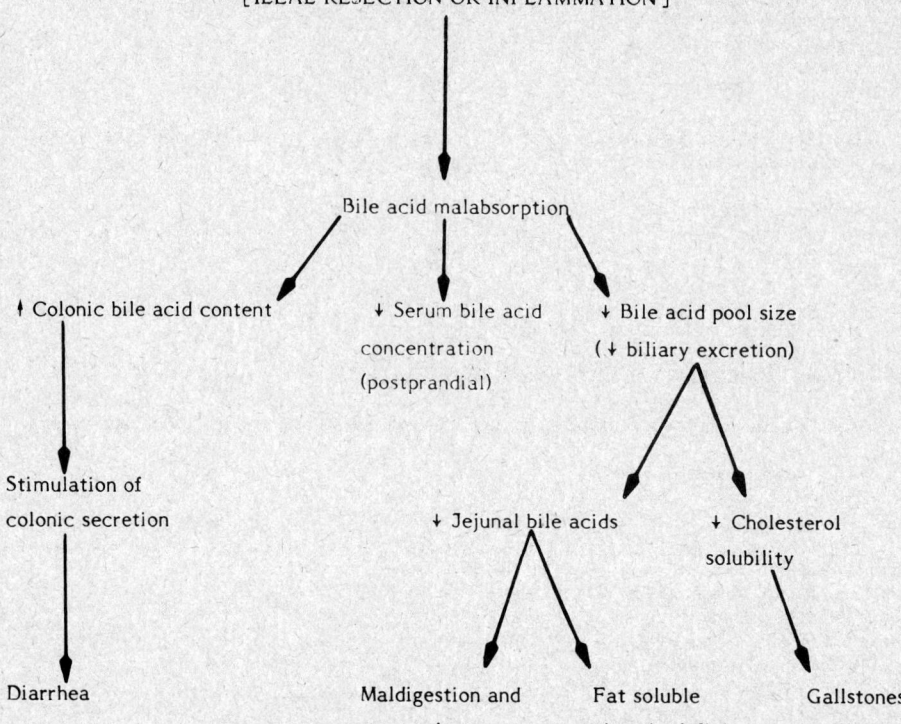

- PRIMARY BILE ACID TRANSPORT DEFECT
- SECONDARY (ACQUIRED) BILE ACID TRANSPORT DEFECT
  [ILEAL RESECTION OR INFLAMMATION]

Bile acid malabsorption

↑ Colonic bile acid content   ↓ Serum bile acid   ↓ Bile acid pool size
concentration   (↓ biliary excretion)
(postprandial)

Stimulation of
colonic secretion   ↓ Jejunal bile acids   ↓ Cholesterol
solubility

Diarrhea   Maldigestion and   Fat soluble   Gallstones
steatorrhea   vitamin deficiency

Figure 13-18. Theoretical consequences of ileal dysfunction leading to bile acid malabsorption.

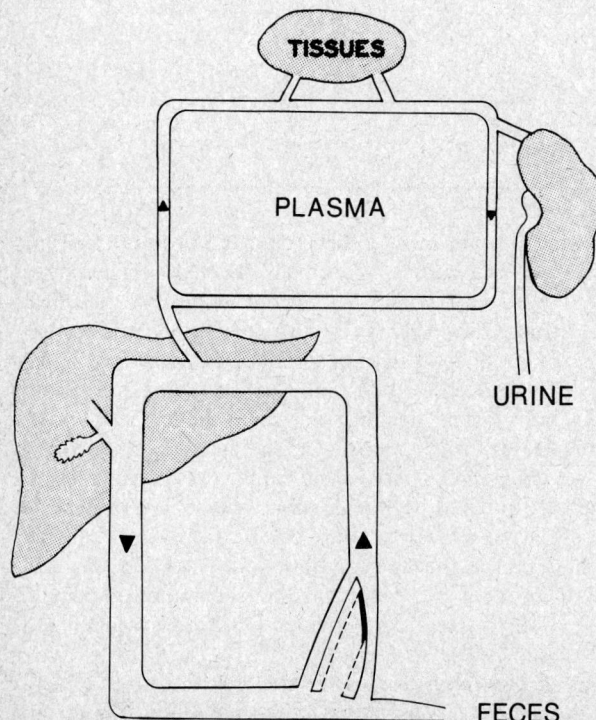

TISSUES

PLASMA

URINE

FECES

Figure 13-19. Bile acid kinetics after major ileal resection (absence of ileum denoted by dotted lines). Compared to a normal adult, pool size is markedly decreased, synthesis rate is increased maximally in an attempt to maintain pool size in the face of a markedly increased fecal loss. Serum levels are decreased in the fasting state; there is no change with feeding. The jejunal concentrations are also diminished. (Modified from Balistreri, W. F., and Soloway, R. D.: Clinical Guide to Bile Acid Physiology and Alterations in Disease States. Reproduced in part with the permission of Abbott Laboratories, Diagnostic Division, North Chicago, Ill. © 1979 by Abbott Laboratories.)

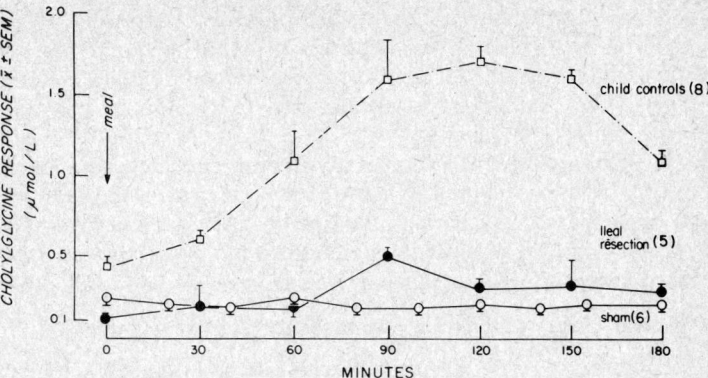

Figure 13-20. The mean response in serum cholylglycine concentration that occurred following meal ingestion in eight normal children (□) and in five patients who had undergone ileal resection (●) is plotted versus time. For comparison, the response obtained in six normal adults following ingestion of the sham feeding (○) is also plotted. Bars represent mean ± standard error of the mean. (Reproduced with permission, from Balistreri, W. F., et al.: J. Pediatr., 96:582-589, 1980.)

gitation of bile acids from cholestatic hepatocytes and portosystemic shunting through hepatic fibrous tissue or past ischemic or obstructed regenerative nodules. These defects will allow the serum levels to rise proportionately much higher than normal following meals, suggesting that the postprandial rise in serum bile acids may be a sensitive test for the detection of liver disease.

Methods for the Measurement of Bile Acids in Serum

Several assays have been used to quantitate either total or individual bile acids in biological fluids.[51,56] The methods that have been used specifically to analyze *serum* bile acids are gas liquid chromatography (GLC), high performance liquid chromatography (HPLC), enzymatic assays, and immunoassays (RIA and ELISA).

GLC. The composition of the steroid moiety of individual bile acids in serum can be determined by GLC. This requires extensive preparation of each sample, i.e., removal of the glycine or taurine moiety by means of alkaline or enzymatic deconjugation, desulfation, and preparation of a volatile derivative. This method is time consuming and tedious, and the sensitivity is variable; therefore, the use of GLC in routine clinical practice is not appropriate and should be reserved for qualitative evaluation of individual bile acids in serum.

HPLC. Several HPLC methods have been developed for analysis of duodenal bile acids; these techniques permit quantitation with minimal derivatization. The major problem in the assay of serum is limited sensitivity of currently available detectors. However, as modified methods and detectors are developed, the application of this technique to the determination of serum bile acids may be feasible.

Enzymatic method. An enzyme produced by *Pseudomonas testosteroni*, 3-α-hydroxy steroid dehydrogenase, catalyzes the oxidation of the 3-α-hydroxyl group of all bile acids, producing a 3-keto group in the presence of NAD^+. The assay is usually carried out as an endpoint determination by measuring the change in absorbance at 340 nm as NADH is formed. Sensitivity is increased by measuring the NADH fluorometrically or by coupling the reaction to the diaphorase-catalyzed oxidation of resazurine, a fluorescent dye. A recent modification using luciferase to

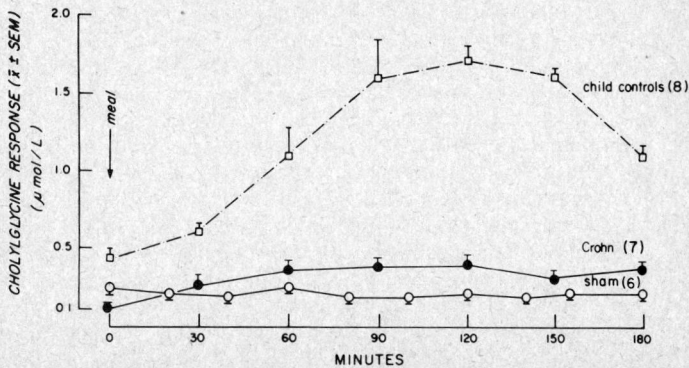

Figure 13-21. The response in serum cholylglycine concentration occurring after ingestion of the standard meal is plotted for seven patients with Crohn's disease (●) and contrasted with the normal children (□) and sham-fed group (○). (Reproduced with permission, from Balistreri, W. F., et al.: J. Pediatr., 96:582-589, 1980.)

produce chemiluminescence further increases the ease and sensitivity. If individual bile acid quantitation is needed, the various species can be separated by thin layer chromatography into classes and quantitated by the enzymatic method. The major disadvantages at present are the cost of the commercial assay kits and the limitation on the number of samples that can be processed in a single day.

Immunoassays. Antibodies have been prepared that have high specificity for the steroid moiety of bile acids (and to some extent the conjugated moiety), permitting individual bile acids to be measured by radioimmunoassay or enzyme-linked immunoassay. Selectivity depends upon antibody affinity; sensitivity is determined by the specific activity of the tracer. Several standardized kits for measuring individual bile acids in serum are available and the assays are simple.

Clinical Utility of Fasting Serum Bile Acid Measurements

Increased serum bile acid concentrations in the fasting state suggest impaired hepatic uptake or secretion, or portal-systemic shunting. Thus, such measurements may be used as a sensitive endogenous clearance test. However, a diagnosis suggested by an increase in serum bile acid concentrations should be confirmed by standard liver function tests. In a similar manner, abnormal standard liver function tests can be confirmed as indicative of hepatic dysfunction by concomitant measurement of serum bile acids. Serum bile acid measurements may also be used serially to monitor patients with suspected or proven hepatic disease.

Bile Acid Concentrations in Serum Following a Meal or Exogenous Load

The 2-h postprandial serum bile acid determination may be a more sensitive test of liver function than a test performed in the fasting state. The administration of bile acids, either endogenously by stimulation of gallbladder contraction or exogenously by administration of a bile acid in a capsule, followed by bile acid determination in serum, may serve as a "stress test" and as a clearance test to provide further diagnostic information.

Serum bile acid measurement in the detection of ileal disease. Since cholate conjugates are absorbed exclusively in the ileum, the postprandial increase in the serum bile acid level may be an index of ileal absorption. In this instance, ileal disease or dysfunction would be manifest by a lack of a rise in the postprandial serum bile acid levels or in a diminished area under the curve. Clinical utility of this test remains to be proved; however, it is suggested that this test may be useful in screening patients with insidious abdominal complaints that may originate in diseases of the terminal ileum as in Crohn's disease. The test may also be utilized in the serial assessment of patients receiving various treatment regimens in the management of inflammatory bowel disease.

Dye Excretion Tests

It has been recognized for almost 60 years that a number of lipophilic synthetic dyes are cleared by the liver and are excreted into bile either as the intact parent compound or as both parent compound and its conjugates. Among these dyes are bromsulfophthalein, indocyanine green, and rose bengal. The clearance of these dyes by the liver is normally very rapid, and it is believed that uptake by hepatocytes is a carrier-mediated active transport process. Little, if any, dye is cleared by other tissues. Excretion into bile is slow. The elimination of these dyes from the bloodstream, therefore, depends on hepatic blood flow, patency of the biliary tree, and hepatic parenchymal function.

In the past, dye excretion tests have been found to be sensitive indicators of the presence of liver disease, but it must be emphasized that they do not distinguish between intra- and extrahepatic cholestasis. Thus these tests are usually reserved for evaluation of patients who are not jaundiced but who are suspected of having a liver disorder. With the development of more sensitive and liver-specific indicators of liver disease such as γ-glutamyltransferase and with the development of the 2-h postprandial serum bile acid test, the dye excretion tests have become obsolete and, for the most part, unnecessary. Of the dye excretion tests, the bromsulfophthalein (BSP) clearance test was, until the 1970's, most frequently used. But because of reports of fatalities resulting from hypersensitivity reactions to BSP, as well as rare adverse reactions in a small per cent of patients (nausea, syncope, headache, chills, and thrombophlebitis at the site of injection), this test has been discontinued. Indocyanine green (ICG) can be used instead of BSP. ICG has the advantage over BSP of having a much lower incidence of side effects. Currently, the ICG excretion test is used for investigation of hepatic blood flow and for predicting clearance rates of drugs such as lidocaine that undergo first-pass clearance by the liver.[61] Typical ICG clearance values in healthy subjects range from 6.5–14 mL/min/kg.

HEPATIC SYNTHETIC FUNCTION

The liver has an extensive synthetic function and plays a major role in the regulation of carbohydrate, lipid, and protein metabolism. There is a bidirectional flux of precursors and products such as glucose, amino acids, free fatty acids, and other nutrients across the hepatocyte membrane. Normal blood glucose concentrations are maintained during short fasts by the breakdown of hepatic glycogen stored after the last dietary carbohydrate intake, and during prolonged fasts by hepatic gluconeogenesis. The primary source of carbon atoms for gluconeogenesis is amino acids that are derived from muscle proteins. To a lesser extent, lactate, produced in skeletal muscle and erythrocytes, and glycerol, obtained from hydrolysis of triglycerides, also serve as substrates for gluconeogenesis. In humans, even-numbered fatty acids (e.g., C-16) are not converted to glucose because of the lack of enzymes that convert acetyl CoA to pyruvate. However, the oxidation of odd-numbered fatty acids yields one propionyl-CoA residue per fatty acid molecule, which can be converted to glucose. But the formation of glucose in this manner is not quantitatively significant. The main source of energy from metabolic processes, adenosine triphosphate, arises from hepatic metabolism of glucose by means of the citric acid and Embden-Meyerhof pathways, and fatty acid oxidation. Protein, triglyceride, fatty acid, cholesterol, and bile acid synthesis also occur within the liver cell.

PROTEIN SYNTHESIS

The liver is the primary site for synthesis of plasma proteins. Its reserve synthetic capacity is very high and not readily exhausted. Proteins are synthesized on the rough endoplasmic reticulum of the liver cells and released from there into the hepatic sinusoids.

Qualitative and quantitative disturbance of protein synthesis is a consequence of impaired hepatic function. *Decreased synthesis of proteins* produced under normal circumstances—albumin, α_1-antitrypsin, fibrinogen, ceruloplasmin, haptoglobin, transferrin, and the coagulation proteins— will lead to a decrease in plasma concentrations of all of them. However, significant decreases in most of them will not become apparent except in severe or longstanding hepatic disease.

The liver responds to injury or inflammation with *increased synthesis of certain proteins*, the acute-phase reactants (haptoglobin, α_1-antitrypsin, orosomucoid, C-reactive protein, C3, and ceruloplasmin). The pattern of plasma protein alterations depends on the type, severity, and duration of liver injury or disease. For example, in acute hepatic dysfunction, there is usually very little change in the plasma protein profile or the total plasma protein concentration. In chronic liver disease, however, serum albumin levels decrease and γ-globulin levels increase. Specific information can be obtained by quantitation of individual protein bands on serum protein electrophoresis or immunoelectrophoresis. Serial determination of plasma proteins may also provide prognostic information; for instance, continued normality of serum albumin and γ-globulin concentrations during acute hepatitis suggests a benign outcome.

Additional information on specific proteins can be found in Chapter 4.

Albumin is synthesized only in the liver; it is secreted across the sinusoidal surface of the hepatocyte into plasma from the Golgi apparatus. In man, absolute rates of synthesis of 10 mg/d per g liver have been found; the half-life is 18–20 d. The synthesis rate varies depending upon the hormonal environment, the nutritional status, age, and other local factors.

Hypoalbuminemia is noted in various liver disorders. The mechanism remains conjectural; however, it is thought to be a consequence of decreased hepatic synthesis of albumin. This is not the only effect, however, since some patients with cirrhosis are capable of synthesizing normal amounts of albumin. The latter observation suggests that albumin distribution is altered, with albumin distributed to the peritoneal cavity rather than being confined to the vascular system. This redistribution may contribute to the production of ascites through leakage of fluid into the peritoneal cavity.

Recent studies have more clearly characterized albumin mRNA and the mechanism by which synthesis of albumin is directed; there are, however, less data regarding the modulation of albumin secretion. This process, which appears to be energy-dependent, is not linked to the rate of albumin synthesis.

Very little albumin is stored in the liver, so factors modulating albumin synthesis and homeostasis exert their effects at the synthesis step and not on the secretion process. The key role of nutritional status and the delivery of amino acids to the liver cell has been shown by studies in which fasted animals (or animals fed low-protein diets) have a 50% decrease in albumin synthesis. A decrease in hepatic protein and albumin mRNA results in a significantly reduced albumin output. The absolute and fractional catabolic rates and the synthesis rate of albumin are increased in animals made hyperthyroid; conversely, the synthetic and catabolic rates are decreased in hypothyroidism. Cortisol and growth hormone are also capable of modulating albumin degradation and synthesis rates.

The feedback of albumin on its synthesis rate is attributed to one of its key functions, the maintenance of normal oncotic pressure by serum albumin. Albumin production rates correlate with changes in the colloid content of the extracellular space. For example, if solutions of low osmotic pressure are perfused into livers, the albumin production rate will increase in direct relationship to the decrease in the osmotic activity. The converse has also been demonstrated, i.e., a decrease in albumin synthesis with perfusates that are hyperosmolar.

Prealbumin is another protein synthesized by the liver. Because it has an extremely short half-life of ~1.9 d, the concentration of prealbumin is likely to serve as a very sensitive index of alterations in hepatic synthetic or catabolic function. Prealbumin concentrations have been measured in sera of patients with various liver diseases; these studies have documented the fact that prealbumin concentration is significantly decreased in liver dysfunction.[25] Therefore, measurement of the concentration of prealbumin may serve as an index of liver function.

Serum levels of **immunoglobulins** indirectly reflect impairment of hepatocyte uptake and that of the reticuloendothelial system of the hepatic sinusoid in its function to filter incoming intestinal tract antigens from the portal venous blood. In cirrhosis, the mechanism of hyperglobulinemia may be due in part to shunting of antigens past the liver. The antigens stimulate the extrahepatic reticuloendothelial system and elicit the immune response, i.e., increased antibody production. In acute viral hepatitis there is often an increase in serum γ-globulin concentrations; there may be specific patterns noted, but these changes are transient. Persistence of hypergammaglobulinemia suggests the presence of chronic active liver disease.

Ceruloplasmin levels will increase in liver diseases such as chronic active liver disease, biliary cirrhosis, and hemochromatosis. Of specific interest are the decreased levels that have been noted in Wilson's disease. (See Chapters 4 and 8C.)

Alpha₁-antitrypsin, a glycoprotein synthesized and secreted by the liver, is the major protease inhibitor in serum. One of its multiple allelic variants, PiZZ, is characteristic of a homozygote deficiency state, in which serum α_1-antitrypsin concentrations are significantly reduced. Children of the PiZZ genotype are predisposed to liver disease; homozygous and heterozygous α_1-antitrypsin-deficient adults (e.g., MZ) are prone to develop chronic obstructive lung disease. The finding of low serum α_1-antitrypsin levels, in conjunction with the accumulation of periodic acid–Schiff positive deposits in periportal hepatocytes, is therefore useful in subcategorizing patients presenting with liver disease.

The liver is also the primary site of **haptoglobin** synthesis, and low levels of haptoglobin will be seen in liver disease.

Serum levels of **β_2-microglobulin** can be increased in hepatobiliary disease complicated by renal disease.[44] This plasma protein has a M.W. of 11 800; therefore, its serum level is affected by the integrity of the glomerular membrane. In chronic active hepatitis and alcohol-induced liver cirrhosis without renal disease, an increase in β_2-microglobulin levels has been noted. These increases appear to be independent of simultaneous changes in serum transaminase, alkaline phosphatase, or γ-glutamyltransferase levels. These preliminary studies need further confirmation.

Estimation of levels of **serum cholinesterase** (acylcholine acylhydrolase, SChE, pseudocholinesterase), an enzyme synthesized by hepatocytes, has been postulated as a useful measurement of hepatic function and status. However, because of inconsistent results, this test has not been widely used, either for the differential diagnosis of jaundice or in serially monitoring patients with hepatic disease. Patients with liver disease may have low serum cholinesterase levels. The finding is not specific for a particular disease because low levels have been found in patients with viral hepatitis, cirrhosis, tumors, and liver disease secondary to cardiac disease. SChE levels return to normal in the recovery phase of acute viral hepatitis, and in this case its assay may have prognostic value. Conversely, persistent depression of SChE, for example in chronic liver disease or in cirrhosis, is a poor prognostic sign. In patients with extrahepatic obstruction, SChE levels are normal.

Ligandin

Ligandin is one of a class of multifunctional proteins known as glutathione-S-transferases; its name is derived from its ability to bind a wide range of ligands (small molecules). Ligandin has a M.W. of 46 000 and is abundant in mammalian liver, constituting ~5% of the total protein in human liver cytosol. Some of the roles postulated for ligandin are (1) detoxification by binding certain toxic chemicals or, through its glutathione-S-transferase activity, by expediting conjugation of a large number of electrophilic compounds; (2) storage or transport of organic ions; and (3) enzymatic degradation of some compounds. For example, it has been suggested that detoxification of carcinogens or mutagens is possible since ligandin serves as a scavenger. In addition, ligandin may play a role in hepatic uptake and secretion of organic anions such as bilirubin. The exact role of ligandin in mediating the latter process remains to be defined; it is suggested that ligandin may facilitate uptake either by providing a site of intracellular storage or by mediating intracellular transport.

Ligandin is present in low concentrations in serum, and the recent development of a sensitive and specific radioimmunoassay has raised the question of the clinical significance of ligandin assay. The intracellular abundance of ligandin suggests that serum levels would be markedly elevated in acute hepatitis. Several reports have indeed shown that in experimental hepatocellular necrosis (e.g., in carbon tetrachloride–treated or bile duct–ligated rats) plasma ligandin levels rise at a greater rate and to higher levels than transaminases, but return to normal very rapidly.[1,54] Elevated levels of ligandin have been found in patients with primary hepatocellular carcinoma. Therefore, measurement of ligandin may prove useful in the diagnosis of these conditions.

Coagulation Proteins

As Table 13-4 shows, most coagulation proteins are synthesized in the liver. These proteins interact in a cascade fashion to produce the fibrin clot. Two fibrinolytic system proteins, plasminogen and α_2-antiplasmin, and the anticoagulant protein antithrombin III are also synthesized in the liver. Current evidence suggests that Factor VIII is synthesized in the vascular endothelium and possibly also in the liver. Some of the coagulation factors (II, VII, IX, and X) require vitamin K for post-translational carboxylation within the hepatocyte. Protein C is also synthesized in the liver by a vitamin K–dependent plasma zymogen. Activated protein C in plasma, in contrast to other vitamin K–dependent clotting factors, *inhibits* coagulation by inactivating Factors V and VIII. It also facilitates fibrinolysis in vivo by elevating circulating plasminogen activator levels.[16a,50a] Parenchymal liver disease of sufficient severity to impair protein synthesis or obstructive disease sufficient to impair intestinal absorption of the fat-soluble vitamin K is therefore a potential cause of bleeding disorders.[11] Because of the great functional reserve of the liver, failure of hemostasis may not be a complication of every liver disease or may not become a complication except in liver disease of long standing or of great severity. Thus, testing for a coagulation defect is not a screening procedure but rather a means of following the progress of the disease or of assessing risk of bleeding before undertaking a traumatic diagnostic procedure such as biopsy.

The prothrombin time (PT) is abnormal in coagulation defects due to liver disease because it is affected by deficiencies of more than one factor. Partial thromboplastin time (PTT) or assays of individual factors do not ordinarily give any additional diagnostic information. Thus, a prolonged PT, in conjunction with other evidence of liver disease, is an important finding.

Prolonged obstructive jaundice can cause vitamin K deficiency as a result of interruption of normal bile flow, consequent failure of fat emulsification, and poor absorption of fat-soluble vitamins. In this case, the coagulation abnormality will be corrected within 24 h by parenteral injection of 5–10 mg of vitamin K. In hepatocellular disease, fat and vitamin K absorption are not usually significantly affected and administration of vitamin K is not likely to correct a coagulation abnormality.

PT is usually normal or only slightly prolonged in the majority of patients with acute infectious or toxic hepatitis; its determination has limited value also in patients with chronic liver disease or with obstructive

Table 13-4. BLOOD COAGULATION FACTORS

Number	Name	Comment
I	**Fibrinogen**	Also an acute-phase-reaction protein
II	**Prothrombin**	Synthesis requires vitamin K
III	Tissue factor	
IV	Ca^{2+}	
V	**Proaccelerin**	
VI	—	
VII	**Proconvertin**	Synthesis requires vitamin K
VIII	Antihemophilic factor	
IX	**Christmas factor**	Synthesis requires vitamin K
X	**Stuart-Prower factor**	Synthesis requires vitamin K
XI	**Plasma thromboplastin antecedent**	
XII	**Hageman factor**	
XIII	**Fibrin-stabilizing factor**	
PK	**Prekallikrein (Fletcher factor)**	
HMWK	**High-molecular-weight kininogen**	

Note: Names are usually used in referring to Factors I through IV. (Proteins synthesized in liver appear in boldface type.)

disease due to biliary calculus or neoplasm. However, in patients with fulminant hepatitis of infectious or toxic etiology, serious hemostatic abnormalities and hemorrhagic symptoms often occur during the acute stage of the disease, and PT will be significantly prolonged. A good prognostic sign is a return to a normal PT; a bad prognostic sign is a rapid increase in the PT values of serial blood samples. Some patients with fulminant hepatitis develop a marked hypofibrinogenemia and increased levels of fibrinogen-fibrin degradation products in serum, which are probably the result of disseminated intravascular coagulation.

In the presence of normal or even elevated levels of plasma fibrinogen, prolongation of PT persists in some patients with hepatic dysfunction. This abnormality has been attributed to increased plasma antithrombin activity due to circulating fibrinogen-fibrin degradation products whose hepatic clearance has been decreased. An alternative hypothesis is that the fibrinogen molecule is qualitatively abnormal in patients with liver disease. Palascak and Martinez have studied fibrinogen in five patients with documented liver disease.[40] They found the plasma prothrombin time to be prolonged by at least 40%; there was no clinical or laboratory evidence of disseminated intravascular coagulation or fibrinolysis. Their studies suggest that some patients with liver disease and prolonged plasma thrombin times have a dysfibrinogenemia functionally characterized by an abnormality of fibrin monomer polymerization. Subsequent studies by these investigators of purified fibrinogen from 12 patients with liver disease demonstrated a biochemical alteration of the functionally abnormal fibrinogen and indicated that excess sialic acid played an important role in the qualitative defect of this protein.[34]

Liebman et al.[31] have detected high circulating levels of abnormal prothrombin (des-γ-carboxyl) in patients with hepatocellular carcinoma. This abnormal prothrombin was not detected in the blood of normal individuals. Determination of serum levels of this product of aberrant protein synthesis may be a tumor marker useful in screening and in serial monitoring of at-risk patients.

An important use of coagulation tests in liver disease patients is in the evaluation of the clotting process before liver biopsy. A typical battery of coagulation tests for this purpose is a PT and platelet count. A prolonged PT or an abnormally low platelet count would be cause for postponing the biopsy until these parameters return to normal values. The policy in effect at both the Hospital of the University of Pennsylvania and the University of Cincinnati Medical Center allows a liver biopsy to be performed only when the PT is no more than 2 s above the reference range of 10–12 s and the platelet count exceeds 50 000 μL.

LIPID AND LIPOPROTEIN SYNTHESIS

The liver is the major source of plasma lipoprotein production and metabolism. (See Chapter 7.) In acute hepatocellular injury, levels of hepatic enzymes such as lecithin cholesterol acyltransferase (LCAT) and triglyceride lipase (H-TGL: triacylglycerol lipase) are decreased.[20] Patients with acute liver disease, therefore, have increased levels of plasma triglycerides, a decreased percentage of cholesterol esters, and abnormal lipoprotein electrophoretic patterns. In acute viral or alcoholic hepatitis, there is an absence of α- and pre-β-bands and presence of a wide, densely staining β-band. The hypertriglyceridemia characteristic of acute hepatocellular injury is mild (150–300 mg/dL); there is, however, accumulation of a triglyceride-rich low-density lipoprotein (LDL), which migrates in the β-region and gives rise to the broad β-bands seen on electrophoresis. Deficiency of hepatic TGL may be responsible for the increased LDL triglyceride seen in acute hepatitis.

Lipoprotein abnormalities are common in chronic cholestasis of either intra- or extrahepatic origin. There are marked elevations in the plasma levels of cholesterol and phospholipids. In addition, in patients with obstructive jaundice, an abnormal lipoprotein (LP-X) is present. Hypercholesterolemia, hyperphospholipidemia, and the presence of LP-X may be useful in differentiating cholestasis from hepatic parenchymal disease; however, the test has no value in differentiating intra- from extrahepatic cholestasis. LP-X is also found in sera of patients with familial deficiency of the enzyme LCAT.

Another potential contribution of the liver in modulating lipid metabolism is its synthesis of hepatic triglyceride lipase (H-TGL). H-TGL differs from lipoprotein lipase, which is released from extrahepatic sites such as adipose tissue, in that it does not require activation by apoproteins and is not inhibited by protamine sulfate. However, the function of H-TGL is not clearly understood. It has been suggested that H-TGL functions in the degradation of the remnant lipoproteins produced by the catabolism of VLDL and chylomicrons. In patients with liver disease such as alcohol-induced hepatitis, a triglyceride-rich LDL is present that is deficient in cholesterol esters. The deficiency in these esters is presumed to be secondary to a decrease in LCAT activity; however, the accumulation in plasma of a lipoprotein similar in its triglyceride composition to remnant lipoproteins suggests that impaired lipoprotein lipolysis due to decreased H-TGL may be the cause.

METABOLIC FUNCTION

A recurring theme is the central importance of the liver in metabolic and regulatory pathways. The functional expression of the complex, integrated organelle structure includes the metabolism of drugs (activation and detoxification) as well as the disposal of exogenous and endogenous metabolites such as galactose and ammonia. As a corollary, it is a truism that multiple metabolic abnormalities due to specific, inherited enzyme deficiencies can primarily or secondarily affect the liver. A classic example is galactosemia; in this condition, the congenital absence of the galactose-1-phosphate uridyl transferase enzyme allows accumulation of the toxic metabolite galactose-1-phosphate, which causes injury to the liver, brain, and kidneys.

Urea Synthesis

Patients with end-stage cirrhosis may have low levels of urea in plasma after ingestion of a high-protein diet, and the rate of urea excretion in urine will be lower than in healthy individuals. In addition, plasma levels of the urea precursors ammonia and amino acids are elevated, and the specific activities of the enzymes involved in urea synthesis are low. All these findings suggest an impairment in the ability of patients with liver disease to metabolize protein nitrogen and to synthesize urea.

The rate of hepatic urea synthesis is dependent upon exogenous intake of nitrogen as well as endogenous protein catabolism. Newly synthesized urea will equilibrate throughout the total body water compartment and, to a large extent, will be excreted in the urine. A small portion will diffuse into the intestine, where bacterial enzymes will hydrolyze urea to produce NH_3. Carbon dioxide is excreted, while NH_3 is reabsorbed and must be recycled through the liver. Thus, the rate of urea synthesis can be calculated through analysis of urinary urea excretion and the blood urea nitrogen (BUN) if the total body water and the extent of gastrointestinal hydrolysis of urea can be estimated. In cirrhotic patients, the Krebs-Henseleit pathway is unable to incorporate excess ammonia nitrogen into urea. Recent studies have defined a simplified method of measurement of the maximum rate of urea synthesis (MRUS). In normal subjects, MRUS appears to have no upper limit. However, in patients with cirrhosis, the MRUS is quickly reached.[45b] Results of this test could also be predictive of a need for dietary adjustment to prevent hepatic encephalopathy or in the preoperative identification of patients with the greatest risk for developing encephalopathy following a shunt procedure.

AMMONIA

A number of animal and human studies have shown that excess ammonia exerts toxic effects on the central nervous system. The major source of circulating ammonia is the gastrointestinal tract. Portal vein plasma ammonia concentration is typically five- to ten-fold higher than that in the general circulation and is derived from the action of bacterial proteases, ureases, and amine oxidases on the contents of the colon as well as from the hydrolysis of glutamine in both the small and large intestines. Under normal circumstances most of the portal vein ammonia load is metabolized to urea in hepatocytes in the Krebs-Henseleit urea cycle during the first pass through the liver; this process includes intramitochondrial and cytosolic enzyme-catalyzed steps (Figure 13-22).

There are several causes of hyperammonemia; inherited deficiencies of urea cycle enzymes are the major cause of hyperammonemia in infants. Two other types of inherited metabolic disorders produce hyperammonemia, namely, those involving the metabolism of the dibasic amino acids lysine and ornithine and those involving the metabolism of organic acids such as propionic acid, methylmalonic acid, isovaleric acid, and others (Figure 13-22). Several cases of transient neonatal hyperammonemia have been described with no clearly established mechanism.[18]

Of the possible acquired causes of hyperammonemia the most commonly encountered is advanced liver disease. Severe liver failure, whether acute, as in toxic or fulminant viral hepatitis or Reye's syndrome, or chronic, as in cirrhosis, can lead to a significant impairment of normal ammonia metabolism. Hepatic encephalopathy, which may develop in the cirrhotic patient, can be precipitated by an episode of gastrointestinal bleeding that enhances ammonia production by bacterial metabolism of the blood proteins in the colon and subsequently increases blood ammonia levels. Other precipitating causes of encephalopathy include excess dietary protein, constipation, infections, drugs, or electrolyte and acid-base imbalance. When cirrhosis is accompanied by impaired venous drainage from the intestine into the liver via the portal vein, venous anastomoses develop. These collaterals shunt ammonia of intestinal origin away from the liver and into the

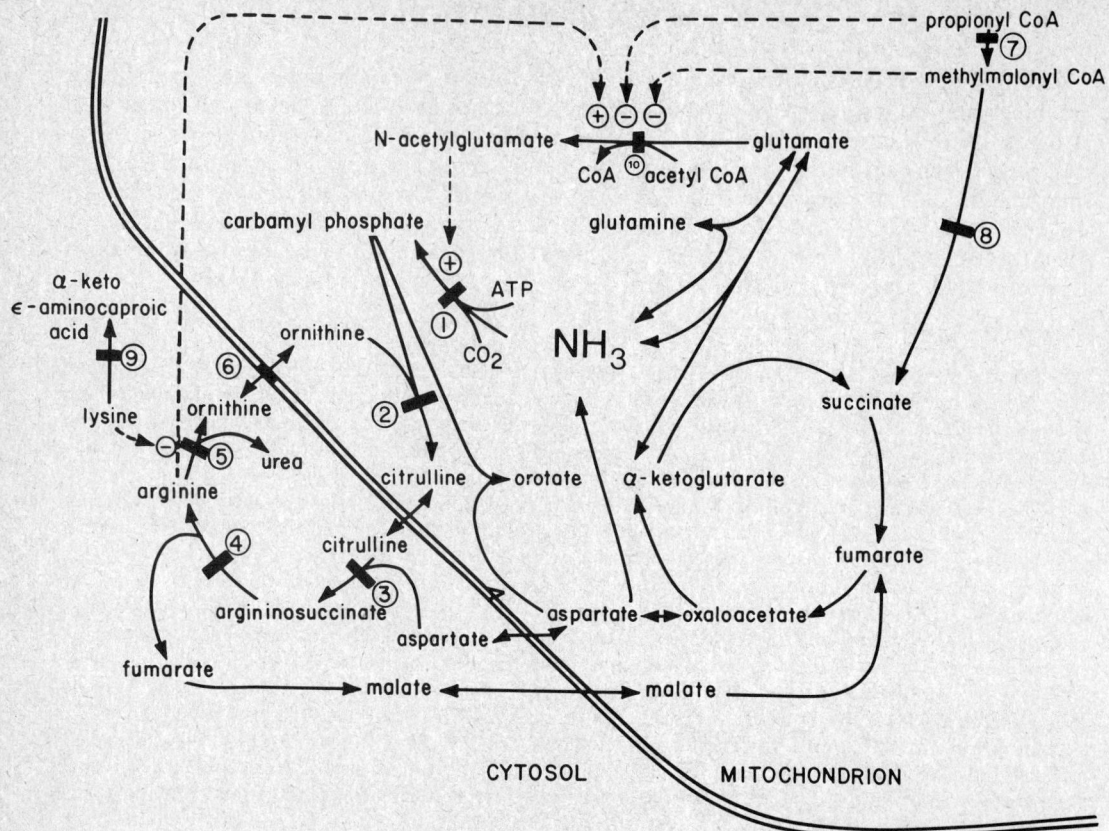

Figure 13-22. The major metabolic pathways for the utilization of ammonia by the hepatocyte. The *solid bars* indicate the sites of primary enzyme defects in various inherited metabolic disorders associated with hyper-ammonemia: (1) carbamyl phosphate synthetase I; (2) ornithine transcarbamylase; (3) argininosuccinate synthetase; (4) argininosuccinate lyase; (5) arginase; (6) mitochondrial ornithine transport; (7) propionyl CoA carboxylase; (8) methylmalonyl CoA mutase; (9) L-lysine dehydrogenase; (10) N-acetylglutamate synthetase. The *dotted lines* indicate the site of pathway activation (⊕) or inhibition (⊖). (From Flannery, D. B., Hsia, Y. E., and Wolf, B.: Current status of hyperammonemic syndromes. Hepatology, *2*:495-506, 1982.)

general circulation and cause increased blood ammonia levels. Impaired renal function may also accompany severe liver disease. As urine output decreases, blood urea concentration increases, leading to increased excretion of urea into the intestine, where it is converted to ammonia.

Ammonia enters central nervous tissue by passive diffusion. The rate of entry increases in proportion to the plasma concentration and is dependent on pH. As pH increases, the rate of entry of ammonia into the central nervous tissue increases. This is thought to be due to the fact that increases in pH produce a shift to the right in the equilibrium

$$NH_4^+ + H_2O \rightleftharpoons NH_3 + H_3O^+$$

that results in increased ammonia base concentration. NH_3 crosses blood-brain barrier membranes more readily than NH_4^+. At 37 °C the pK_a of NH_4^+ is 8.9, i.e., at the normal physiological pH of 7.4, ~3% of blood ammonia is NH_3. A pH rise to a value of 7.6 produces an increase in NH_3 to ~5% of total blood ammonia, a 67% increase in concentration.

During a recent evaluation of 73 cirrhotic patients using 8 conventional laboratory test procedures and 20 measures of nitrogen metabolism (including plasma levels of 16 amino acids), fasting venous plasma ammonia determinations correctly identified 15 of 17 encephalopathic patients and 55 of 56 who did not have encephalopathy.[2]

The fasting venous plasma ammonia concentration can be useful in the differential diagnosis of hepatic encephalopathy, and the ammonia tolerance test has been used in evaluating the patency of portacaval shunts. In the latter procedure ammonia levels are determined before and after ingestion of a standard dose of ammonium chloride.

Methods for the Determination of Ammonia

Methods for measuring ammonia can be divided into two major classes: two-stage procedures and one-stage (direct) procedures. The Conway technique is a two-stage procedure. The sample is alkalinized in a closed system to convert all ammonium ions to ammonia, which then is trapped in an acid medium in a second chamber of the diffusion cell. Trapped ammonia may then be analyzed spectrophotometrically or back-titrated with standard acid. Another, newer two-stage procedure is flow-injection analysis. Here the specimen is injected into a flowing stream of alkaline solution from which the ammonia diffuses through a membrane into a pH-sensitive indicator stream. The method is fast, requires minimal sample, and carries minimal risk of ambient ammonia contamination. It works well in routine practice[15] but does require dedicated equipment. A cation-exchange resin is used in another two-stage assay to remove NH_4^+ from its matrix.[19] The ammonium ion is then eluted from the resin and the eluate reacted with Nessler's reagent or Berthelot (phenol hypochlorite) reagent prior to spectrophotometric quantitation.

One-stage methods do not separate NH_4^+ from the specimen before the analytic step. The enzymatic assay used in these procedures is most frequently based on the reaction,

$$2\text{-Oxoglutarate} + NH_4^+ + NADPH \underset{\text{Dehydrogenase}}{\overset{\text{Glutamate}}{\rightleftharpoons}} \text{Glutamate} + NADP^+ + H_2O$$
$$\text{pH 7.4}$$

as described by Mondzac and coworkers[38] in 1965 and later modified by Van Anken and Schiphorst.[57] Change in absorbance at 340 nm is measured, as NADPH is transformed to $NADP^+$. The method is fast, requires no special equipment as a manual procedure, and is readily automated (for example, on the DuPont *aca*).[13] The DuPont *aca* method is advantageous by virtue of the closed analytical system and its freedom from interference by ammonia in the atmosphere.[26] A direct method using an ammonia-selective electrode is promising in principle, but so far it is not sufficiently tested for its performance to be judged relative to other methods.

Reliable ammonia levels can be achieved only by meticulous precautions to avoid false elevations. Four concerns are of paramount importance:

1. *Smoking* is a source of ammonia contamination both of the patient and of the specimen. The patient must not smoke after midnight before the morning when the fasting blood sample is drawn. Gerron et al. have found that one cigarette smoked 1 h before venipuncture increased fasting venous blood ammonia by 100–200 $\mu g/L$.[23] If the patient is a heavy smoker, he should shower before the test. All patients should put on fresh pajamas before the blood specimen is taken. The technologist performing the test must be a nonsmoker.

2. *Laboratory atmosphere* is a source of ammonia contamination for the specimen and for the assay method. In order to minimize contamination of specimens and glassware by ammonia in the laboratory atmosphere, blood collection for the ammonia analysis as well as the performance of the analysis itself should ideally be done in a special laboratory. Traffic in and out of this laboratory should be restricted at all times and should be prohibited when the test is being performed. Glassware must be chemically clean. It should be soaked in a solution of hypochlorite, 52.5 g/L, and rinsed thoroughly with deionized water the day before use.

3. *Poor venipuncture* technique may result in increased ammonia levels. Use of a heparin lock, probing for a vein, drawing blood into a syringe and transferring it to an anticoagulated tube, or partial fill of the evacuated tube allowing subsequent entry of air may cause elevation of ammonia levels by 100–200 $\mu g/L$. EDTA and heparin have been reported to be acceptable anticoagulants. Ammonia values in serum are significantly but variably higher than the corresponding plasma values.

4. *Metabolism of nitrogenous constituents* in the specimen is a source of ammonia contamination. Production of ammonia by deamination of amino acids such as glutamine in the blood sample may occur once the specimen has been drawn. The specimen must be put on ice immediately and centrifuged without delay, and the analysis performed immediately.

The key to reliable values from ammonia assays is control of the preanalytic sources of error described above.

Reference Ranges

Conway diffusion assays often give falsely high ammonia values (400–1100 $\mu g/L$ as ammonia nitrogen) and are no longer recommended. For the enzymatic method the reference range is 140–490 $\mu g/L$ as ammonia nitrogen or 11–35 $\mu mol/L$. The reference range for the resin method is 220–470 $\mu g/L$ as ammonia nitrogen or 16–33 $\mu mol/L$. Reference ranges should be determined in the laboratory performing the assay. Should values in normal subjects be significantly higher than those noted for the enzymatic method or the resin method, consideration should be given to the existence and correction of sources of error identified above.

CARBOHYDRATE METABOLISM

Because the liver is a major processor of dietary and endogenous carbohydrates, liver disease can be expected to affect carbohydrate metabolism in a variety of ways. (See Chapter 6 and biochemistry textbooks.) Only a few of the conventional and classic modes of evaluating overall carbohydrate metabolism have value in the diagnosis of liver disease. However, many aspects of the metabolism of galactose are unique to the liver, and thus the determination of the rate of galactose elimination is a helpful, but not widely used, measure of hepatic metabolic capacity.[32] Galactose elimination is independent of both hepatic excretory function and metabolism by other organs. Hepatic metabolism begins with specific carrier-mediated entry of D-galactose into the liver cell followed by phosphorylation to galactose-1-phosphate (Gal-1-P) by galactokinase. Since Gal-1-P does not pass through cell membranes, phosphorylation prevents return of the galactose moiety to the circulation; this allows further metabolism of Gal-1-P into UDP-galactose and then to UDP-glucose by galactose-1-phosphate uridyl transferase and epimerase. The transferase enzyme is absent in patients with galactosemia; in healthy individuals, the rate-limiting enzyme appears to be galactokinase.

In the face of relatively low galactose concentrations (<30 mg/dL), hepatic extraction averages ~90%. Therefore, measurement of galactose removal has been used as a basis for estimating hepatic blood flow during infusion studies. When the blood galactose concentration is >30 mg/dL, a constant maximal rate of hepatic galactose sequestration is noted.

Galactose Elimination

The rate of elimination of test compounds that are metabolized by the liver can be a useful functional parameter, provided the elimination of the selected substance is directly determined by the amount of functional hepatic mass or total hepatic metabolic activity.

The intravenous galactose tolerance test can be performed by (1) determination of a single blood galactose concentration; (2) calculation of the galactose elimination coefficient; or (3) calculation of the galactose elimination capacity. The reproducibility of the intravenous galactose tolerance test has been investigated by Bohmer et al.[8] by performing the tests repeatedly under identical conditions in individual patients. The results showed that the interindividual scattering was much greater than the intraindividual variation. The authors suggest that the intravenous galactose tolerance test is more suitable for the longitudinal evaluation of patients than for the detection of impaired liver function. In 25 patients with fulminant hepatic failure, the galactose elimination capacity was significantly higher in the five patients who survived than in the 20 patients who died. That none of the other liver function tests was significantly different emphasizes the potential prognostic value of this procedure.

Hepatic Storage Function

There is a complex interrelationship and interdependency in the regulation of energy metabolism in the organs of healthy human subjects. Since individual cells are unable to store a sufficient supply of all the materials needed to maintain energy and to replete their structure, storage forms of substrates and storage sites are needed to supply them on a continuous basis. The liver serves as the major site for carbohydrate storage. For example, hepatic storage of glycogen allows the release of glucose to other tissues when the need exists, i.e., when plasma levels of glucose diminish. Other organs such as muscle and adipose tissue store proteins and triglycerides respectively and are capable of adaptation; depending upon the availability of oxidizable fuels, these organs can also switch from a storage mode to that of synthesis or release during periods of decreased intake.

MECHANISMS OF ENZYME RELEASE FROM DISEASED LIVER TISSUE

Subcellular Distribution

In various forms of liver disease, serum levels of numerous cytosolic, mitochondrial, and membrane-associated enzymes are increased; the degree of elevation varies with the type of disease. We will discuss the subcellular distribution and mechanisms of release from liver tissue of the enzymes currently regarded as most useful in liver disease evaluation. The preanalytical variables, methodology, and reference values for these enzymes, as well as for other enzyme tests used in the evaluation of liver disease, are described in detail in Chapter 5. Alanine and aspartate aminotransferases, alkaline phosphatase, and γ-glutamyltransferase are the enzymes most often measured for evaluation of liver disease. Although alanine aminotransferase is present in mitochondria and the cytosol, the mitochondrial form is low in concentration and very unstable. Aspartate aminotransferase occurs in the hepatocytes in mitochondrial and cytosolic isoenzyme forms, which are genetically distinct and different in their amino acid composition, kinetic behavior, electrophoretic mobilities, and immunochemical properties.[45] The possible clinical significance of the measurement of the cytosolic and mitochondrial forms of aspartate aminotransferase is currently being investigated.[41]

Alkaline phosphatase and γ-glutamyltransferase are membrane-bound glycoprotein enzymes. Alkaline phosphatase has been demonstrated with histochemical techniques to be present in highest concentration in the sinusoids and in the endothelium of the central and periportal veins; smaller concentrations occur in the biliary canaliculi.[35] In the normal human liver lobule, γ-glutamyltransferase activity is most intense in bile canaliculi and in the bile duct epithelium of the periportal zone. Activity has also been demonstrated in the lipocytes that are located in the space of Disse.[52] A common finding in pathological liver tissue is a pronounced increase in the γ-glutamyltransferase of the canalicular membranes and bile duct epithelium.

Cytosolic and mitochondrial enzymes. The detailed mechanism by which enzymes are released from the cytosol and mitochondria of hepatocytes into the bloodstream is not known. Clinical observations and experimental model studies using perfused rat liver have made it clear that subtle membrane changes are sufficient to allow passage of intracellular enzymes to the extracellular space.[21] A considerable concentration gradient, dependent on normal cellular metabolism for its maintenance, exists for enzymes between the hepatocyte and the sinusoidal space. Upon damage to the process of cellular energy production, as in infections such as viral hepatitis, permeability of the hepatocyte membrane increases and cytosolic isoenzymes of the aminotransferases spill into the sinusoids and from there into the peripheral blood.[21] Permeability of mitochondrial membranes may also increase and mitochondrial isoenzymes will then be released as well.

Membrane-bound enzymes. How membrane-bound enzymes such as γ-glutamyltransferase, alkaline phosphatase, and 5'-nucleotidase reach the peripheral blood is not clearly understood. The increased activity of γ-glutamyltransferase and alkaline phosphatase in diseased human liver is believed to be due to enhanced synthesis; but how enhanced synthesis of tissue-bound enzymes translates into increased activity in serum is not at all clear. However, fragments of hepatocyte membrane rich in γ-glutamyltransferase and alkaline phosphatase activities have been detected in serum of patients with cholestasis. The finding suggests that elevation of serum activity may be due to the release of cell membrane fragments into the circulation, a process that may be a result of membrane fragmentation by bile acids. Particularly in cholestatic disease, the stasis and prolonged contact of bile acids (which are detergents) with canalicular and bile duct epithelia could solubilize and release γ-glutamyltransferase and alkaline phosphatase from plasma membranes. In vitro studies of membranes treated with bile acids show that the possibility exists.

DYNAMICS OF LIVER ENZYME CHANGES IN LIVER DISEASE

The degree of increase of serum enzyme activities in liver disease and the pattern of changes can vary markedly throughout the clinical course of the disease. (See Chapter 5, Enzymes.) Among the factors that may affect the degree of elevation and the pattern of changes are (1) specific activity (i.e., the units of catalytic activity per gram of tissue) of the enzymes in the liver cell at the time they are released into the bloodstream; (2) the pattern of relative specific activities of the enzymes in the liver at the time serum activities are determined; and (3) the relative rates of catabolic clearance of the enzymes from the bloodstream.

The amount of an enzyme in a diseased liver cell can be significantly different from that in a normal liver cell. Analyses of 144 liver biopsy specimens obtained from normal subjects and subjects with liver disease have shown that the average specific activity (U/g liver protein) of GGT was only about 20% higher than normal in liver specimens from patients with acute hepatitis, but was three-fold higher than normal in alcoholic hepatitis and biliary obstruction, and four-fold higher than normal in metastatic liver disease.[50] The last categories of liver disease are typically associated with much higher serum levels of GGT than is acute hepatitis. Another illustration of such changes can be found for the aminotransferases when chronic liver disease is compared with acute hepatitis. In chronic liver disease, the specific activity of alanine amino-transferase (ALT) in liver tissue decreases to a significantly greater degree than does that of aspartate aminotransferase (AST). On the other hand, in acute hepatitis the specific activities of both aminotransferases decrease significantly. These changes in aminotransferase patterns in tissue are reflected by the pattern of enzyme activities observed in serum. It is now believed that changes in the aminotransferase pattern in tissue explain why patients with cirrhosis of the liver have a higher serum activity of AST than of ALT and why in acute hepatitis ALT is usually higher than AST.

Clearance of liver enzymes from plasma occurs at greatly different rates and by a mechanism not yet identified. The average half-life of alanine aminotransferase has been found to be 47 h, of cytosolic aspartate aminotransferase 17 h, of alkaline phosphatase 9.7 d, and of γ-glutamyl-transferase 4.1 d.

DISEASES OF THE LIVER

Mechanisms and Patterns of Liver Cell Injury

Hepatic injury can occur by means of various mechanisms and may be either diffuse or localized. The different patterns of liver cell injury that can be seen can be correlated with abnormalities in structure and function of specific hepatocyte organelles. Sherlock has suggested that there are three specific patterns of liver cell injury: (1) direct, (2) immunologic, and (3) cholestatic.[53]

In the *direct pattern of liver injury,* individual cell death occurs in specific areas such as the centrilobular zone. Damaged cells may contain fat or they may be lysed; Kupffer cells proliferate. Localization of injury may be due to the lobular vascular gradient; for example, with hypoxia the central location of necrosis can be attributed to a lower oxygen content of sinusoidal blood in the central areas.

Certain toxins will cause injury to specific hepatocyte organelles. For example, following chronic alcohol ingestion there may be predominantly mitochondrial damage eventuating in cell death; this sequence can be attributed to oxidation products of ethanol, e.g., acetaldehyde. The necrotic cells may show evidence of degenerated cytoplasmic constituents, i.e., the eosinophilic hyaline body of Mallory, with damage to the rough endoplasmic reticulum and cytoskeletal elements.

Metabolic conversion may occur in the hepatocyte; this may activate drugs or produce potentially toxic agents.[37] Certain toxic metabolites, such as those formed from acetaminophen, may bind covalently to liver cell macromolecules which are essential to hepatocyte viability; therefore, liver cell death may occur. Once again, the necrosis will be variable throughout the lobule but will occur predominantly at the site of maximum activity of drug-metabolizing enzymes, e.g., the centrilobular zone.

Immunologically mediated hepatocyte injury may be more complex and enigmatic. It appears that cytotoxic lymphocytes are capable of attacking liver cell membrane antigens. Morphologically, there is an inflammatory infiltrate that is most prominent in an expanded portal area and consists predominantly of lymphocytes and plasma cells. There may also be extension into the neighboring parenchyma, i.e., piecemeal necrosis. The classic example of immunological hepatic injury is chronic active hepatitis. In this type of liver disease, it is postulated that a hepatocyte membrane protein serves as an antigen to which immunologically mediated activity is directed. It is also possible that drugs may serve as haptens and may combine with normal liver cell membrane antigens, thus providing a focus for immunologically directed hepatocyte injury. Primary biliary cirrhosis has also been cited as an example of immunologically mediated liver disease. In this situation, there is a marked infiltration of lymphocytes and plasma cells with injury to the bile ducts. The role of histocompatibility antigens in determining susceptibility to injury remains to be more fully defined.

Cholestatic liver injury can be due to either intrahepatic or extrahepatic obstruction. Extrahepatic obstruction is most frequently due to a mechanical cause such as atresia, stenosis, or obstruction of the bile ducts secondary to carcinoma or gallstones. Intrahepatic cholestatic liver injury is less fully defined; it may be due to functional disturbances in the bile secretory mechanism. The perpetuating effect of retained toxic compounds, such as bile acids and trace metals normally excreted in the bile, also remains to be defined. Morphologic evidence of cholestatic liver injury includes an accumulation of bile in hepatocytes and in the bile canaliculi. Ultrastructural alterations such as blunting and distortion of canalicular microvilli can also be seen. Liver biopsy may provide definite information regarding the type of injury, course, and management of hepatic dysfunction.

Acute Viral Hepatitis

Viral hepatitis is a primary infection of the liver most commonly caused by hepatitis A virus (HAV) or hepatitis B virus (HBV). However, other viruses, tentatively grouped under the name "non-A, non-B," can produce clinical pictures similar to those produced by the classic HAV or HBV. Of less importance as agents of viral hepatitis are cytomegalovirus, herpes simplex, varicella-zoster, Epstein-Barr, rubella, and Coxsackie viruses. Clinical features of acute viral hepatitis caused by HAV and HBV are summarized in Table 13-5. Detection of viral hepatitis is greatly facilitated by availability of a number of tests based on morphology of particular viruses and serologic response of the host.

Antigens related to HAV infection are not detectable in serum, although a viral particle, hepatitis A antigen, has been identified in stools of patients during early hepatitis A infection. The humoral immune response to HAV produces, in the initial phase of the infection, IgM type anti-HAV and, in the later phase, IgG type anti-HAV (Figure 13-23). The latter antibodies persist for a prolonged period in patients, possibly indefinitely.

HBV is a double-shelled structure originally called the Dane particle. The outer coat, called the HBV surface antigen (HB$_s$Ag), surrounds the HBV core antigen (HB$_c$Ag), double-stranded DNA, and the enzyme DNA polymerase. A soluble antigen (HB$_e$Ag) is closely associated with HB$_s$Ag.

HB$_s$Ag appears in serum before clinical features of hepatitis B infection appear and persists throughout the clinical course of the disease (Figure 13-24). The humoral immune response to HBV infection produces antibodies to the surface antigen (anti-HB$_s$) that first appear during the resolution and convalescent phases of the disease. Anti-HB$_s$ are considered essential for the

Table 13-5. CLINICAL FEATURES OF ACUTE VIRAL HEPATITIS

	Hepatitis A	**Hepatitis B**
Incubation period	2–6 weeks	6 weeks–6 months
Onset	Usually acute	Usually insidious
Mode of transmission		
Oral (fecal)	Usual	Infrequent oral; not fecal
Parenteral	Infrequent	Usual; intimate contact
Epidemiology		
Sporadic cases	Mainly children	Males predominate
Outbreaks	Food or water borne	Contaminated blood products
Polyarthralgia	Rare	Common
Rashes	Occasional	Common
Duration	Brief (14–21 d)	Prolonged
Sequelae		
Carrier state	No	10% of cases
Chronic hepatitis	Rare	Frequent (~10% of cases)

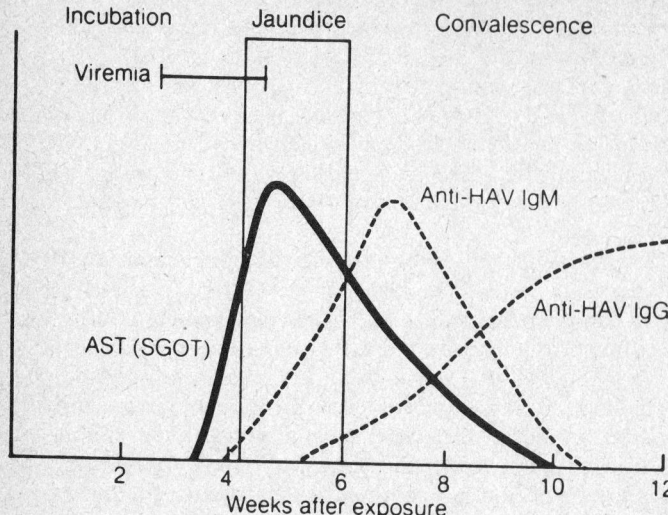

Figure 13-23. In *hepatitis A* infection, jaundice occurs 14–35 d (median 32 d) after exposure. AST elevations precede the development of jaundice and usually remain abnormal after the serum bilirubin returns to normal values. Viremia and fecal excretion of virus usually occur during the preicteric stage of the disease. The limited period of viremia makes hepatitis A virus a very uncommon cause of post-transfusion hepatitis. The appearance and disappearance of serologic markers to hepatitis infection (anti-HAV) are depicted by dotted lines. (From Balistreri, W. F.: Consultant, *24*:131-153, 1984.)

resolution of HBV infection. Anti-HB$_s$ are also believed to provide prolonged immunity to patients in whom they persist. A newly discovered agent, the delta antigen, may co-infect patients who are HB$_s$Ag positive. This may be responsible for worsening of a previously stable clinical condition, e.g., exacerbation of chronic hepatitis.

Those patients who do not develop anti-HB$_s$ become chronic carriers of HBV infection typified by the prolonged presence in serum of HB$_s$Ag and antibodies to HB$_c$Ag (anti-HB$_c$) only. This occurs in about 10% of patients infected with HBV. Chronic carriers whose serum is also persistently positive for another antigen derived from HBV, hepatitis B$_e$ antigen (HB$_e$Ag), are at an especially high risk for the development of chronic active liver disease. Patients who develop antibodies to HB$_e$Ag (anti-HB$_e$) usually have a normal liver biopsy and normal biochemical test values. Assessment of HB$_e$Ag and its antibodies may be an important predictor of which chronic carriers of HBV will develop chronic active liver disease. The blood products of chronic carriers of HBV, that is, patients whose serum is persistently positive for HB$_s$Ag and anti-HB$_c$, should not, of course, be used in transfusions. The screening of donated blood for the chronic carriage of HBV is now an essential laboratory activity to minimize the risk of developing HBV infection in patients undergoing blood transfusion.

The *epidemiology* of HAV and HBV is quite different. Hepatitis A is transmitted by the fecal-oral route or via contaminated food, water, or shellfish; parenteral transmission is uncommon.

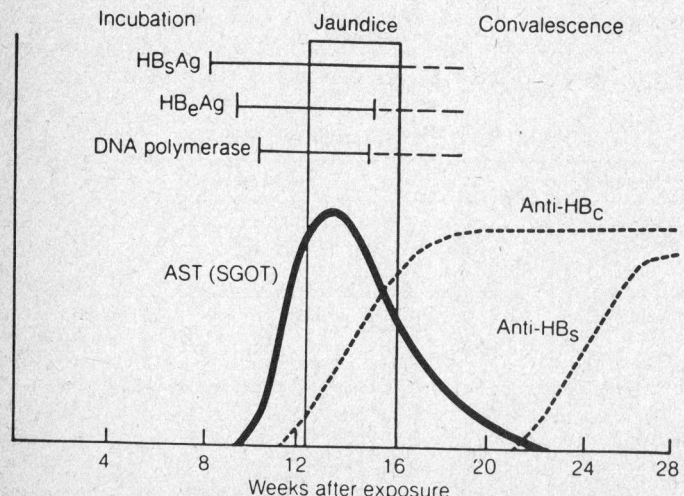

Figure 13-24. Course of *acute type B hepatitis* with recovery. Illustration indicates (1) onset of hepatitis with jaundice three months after exposure; (2) detection of hepatitis B surface antigen (HB$_s$Ag) ~2–8 weeks after exposure, followed by appearance of its antibody (anti-HB$_s$) ~2–4 weeks after HB$_s$Ag is no longer detectable; (3) detection of hepatitis B$_e$ antigen (HB$_e$Ag) shortly after appearance of HB$_s$Ag and its disappearance shortly after HB$_s$Ag disappears; this is usually followed by the appearance of antibody to HB$_e$Ag (anti-HB$_e$) that persists; and (4) detection of hepatitis B core antibody (anti-HB$_c$) at the time of onset of disease two to three months after exposure; anti-HBc IgM will be detectable in high levels for ~5 months. (From Balistreri, W. F.: Consultant, *24*:131-153, 1984.)

In contrast, hepatitis B is transmitted most commonly by parenteral injection or by the exchange of body secretions. There is no known carrier state for HAV; however, as discussed above, chronic carriage of HBV is associated with significant implications for both the individual and his contacts.

Attack rates for hepatitis A appear highest for children in age groups attending nursery schools, primary schools, or day care centers, where hygienic practices may be lax. Incidence appears to be cyclic, with epidemic surges occurring after sufficient numbers of susceptible individuals have accumulated; areas of endemic infection occur worldwide. In contrast, high risk for hepatitis B correlates with exposure to blood and body fluids, as is frequent with health care personnel such as dentists, physicians, and laboratory or blood bank personnel. Certain lifestyles, such as male homosexuality and drug addiction, are also associated with high risk for hepatitis B. Epidemiology of non-A, non-B hepatitis is less clear, but transmission modes and groups at risk seem to parallel those of hepatitis B. With the institution of testing for HB_sAg among blood donors, non-A, non-B hepatitis has become the most frequent transfusion-transmitted form of viral hepatitis.

The spectrum of *pathologic changes* caused by these viruses may vary from a mild bout of acute hepatitis that resolves without further consequence to more severe disease or even fulminant hepatic failure and death. Alternatively, there may be resolution with scarring or chronic liver disease, either of a persistent (benign) nature or of an aggressive nature. Cirrhosis and subsequent development of hepatocellular carcinoma are known pathological outcomes of chronic hepatitis B infection. Fulminant hepatic failure complicates acute viral hepatitis of any etiology in a small percentage of patients.

Treatment of the infected patient is supportive. A major goal is to interrupt spread of infection among the patients' intimates either by passive immunization with immune globulin or, in the case of hepatitis B, with high-titer specific immune globulin. Active immunization against HBV infection is now possible with a highly effective and safe vaccine.

The *laboratory features* of acute hepatitis A and hepatitis B differ. The duration of abnormal laboratory values is generally much shorter in hepatitis A patients; as discussed above, this is also true for the duration of the clinical features of *hepatitis type A* compared with type B. The ranges of the peak values of ALT, AST, conjugated and unconjugated bilirubin, ALP, and GGT among patients with a diagnosis of hepatitis A or B are similar. Peak values for ALT and AST range from ten to one hundred times normal, with ALT higher than AST activity at the peak and throughout the course of abnormal values. (See also Chapter 5, Enzymes.) Peak total bilirubin concentrations vary widely in acute viral hepatitis. In the anicteric form of the disease, values range from normal to slightly above normal, whereas in icteric viral hepatitis, values range from about 50–200 mg/L. In ∼10% of the patients with icteric viral hepatitis, intrahepatic cholestasis is a significant feature. The ability of the liver to excrete bilirubin into the gastrointestinal tract is more severely impaired in such patients than in the majority of patients with icteric viral hepatitis. The intrahepatic cholestasis in patients with cholestatic viral hepatitis may last for only a few days or for as long as 4–5 weeks. Total bilirubin values are higher in the cholestatic form of viral hepatitis than in the noncholestatic form of the disease, with peak values of up to ∼300 mg/L. In the absence of histologically significant intrahepatic cholestasis, acute viral hepatitis is associated with peak alkaline phosphatase values of up to about two times the upper limit of their reference range and GGT values that are up to about five times the upper limit of the reference range. In contrast, in the presence of histologically defined intrahepatic cholestasis, both enzyme activities are significantly higher. Peak values for alkaline phosphatase of about five times normal and for GGT of about ten times normal are a characteristic finding.

The sequence of events associated with *hepatitis A* infection is shown in Figure 13-23. Within a week after onset of clinical illness, alanine and aspartate aminotransferase levels rise to a peak of more than ten-fold normal; they may fall to normal as early as 1–2 weeks later. Serum bilirubin levels rise later, if at all, but return to normal before transaminases; levels of both unconjugated and conjugated bilirubin may be elevated to varying degrees. Viremia and fecal excretion of the virus usually occur in the preicteric stage of the disease. Because the period of viremia is short, HAV is rarely a cause of post-transfusion hepatitis. Serologic response is characterized by early appearance of IgM anti-HAV, followed by the more long-lasting IgG anti-HAV.

The course of events following infection with *hepatitis B* is depicted in Figure 13-24. Approximately 8–10 weeks after infection, there is a gradual rise in aminotransferases, followed by the onset of clinical icterus. The peak increase in serum aminotransferase levels occurs at ∼14–

15 weeks after infection; the jaundice abates before the aminotransferases become normal. The typical sequence of viral markers is depicted; antibody to the hepatitis B core antigen (anti-HB$_c$) appears early, followed by appearance of antibody to the surface antigen (anti-HB$_s$); the latter is protective. In the chronic carrier state, HB$_s$Ag remains in the patients' serum, and there is no *detectable* anti-HB$_s$ rise (Figure 13-25).

Chronic Hepatitis

The clinical presentation of patients with chronic active liver disease is highly variable and may not be heralded by a clinically apparent acute episode; the onset is often insidious. The histology, which forms the basis for classification, does not correlate with the symptomatology. The operational definition is hepatic inflammation, with histologic or biochemical documentation, that persists for > 6 months. The etiology may be viral, autoimmune, or drug induced. Regardless of the initiating event, the classification, prognosis, and treatment are based on the histologic features. At present there are three classifications: (1) chronic persistent hepatitis (periportal inflammation); (2) chronic lobular hepatitis; and (3) chronic active (aggressive) hepatitis (piecemeal necrosis with extension of portal inflammatory cells into the adjacent parenchyma). The two major *etiologic* subclassifications are (1) hepatitis B–related in which HB$_s$Ag is positive (Figure 13-25), and (2) autoimmune. The latter group of patients may have a variety of associated findings, such as positive antinuclear antibody (ANA), LE preparation, and smooth muscle antibody (SMA), as well as increased γ-globulin.

Alcoholic Liver Disease

Depending upon the degree and duration of alcohol consumption, significant abnormalities in liver structure and function can occur.[6] These may range from mild changes, such as fatty liver, to moderate disease (alcoholic hepatitis), or to full-blown hepatic fibrosis or alcoholic cirrhosis. The mechanism of induction of hepatic disease by alcohol or metabolites such as acetaldehyde is complex. A multitude of metabolic abnormalities are present in the patient with alcohol-induced liver disease. For example, the ingestion of alcohol and subsequent metabolism by the liver result in the formation of acetaldehyde and an increased NADH:NAD$^+$ ratio. There are certain metabolic abnormalities, such as hyperlipidemia and ketosis, that may be exacerbated by the presence of underlying abnormalities such as Type IV hyperlipoproteinemia. Alterations in glucose metabolism may be a consequence of the associated pancreatitis and enhanced gluconeogenesis.

The chronic consumption of alcohol will induce a rise in serum γ-glutamyltransferase. However, levels of the enzyme will return to normal if the patient abstains from alcohol. Per-

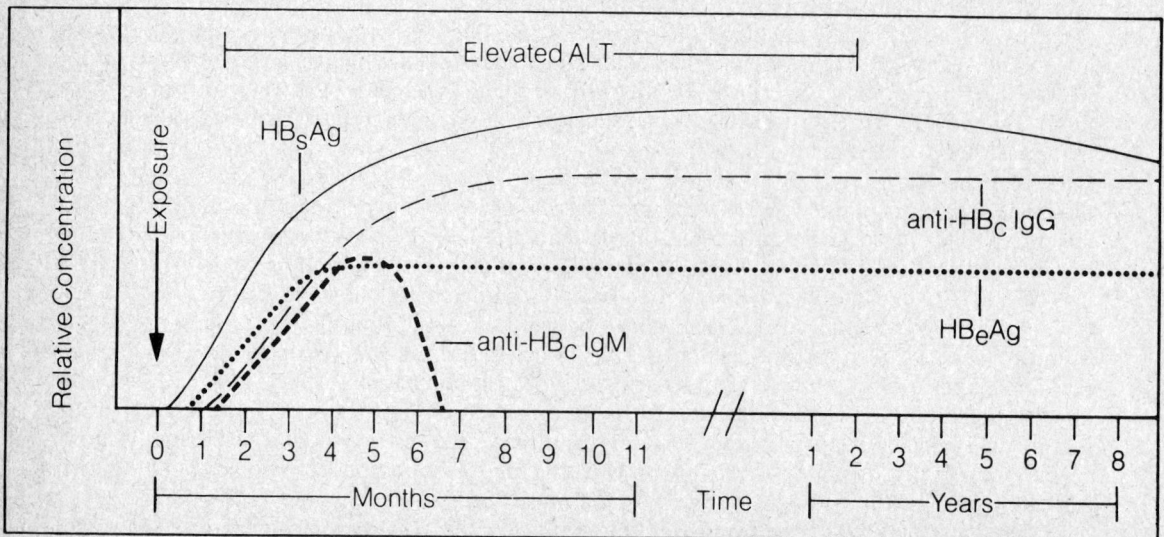

Figure 13-25. *Chronic active type B hepatitis.* Illustration shows (1) prolonged period of abnormal serum alanine aminotransferase (ALT) levels; (2) persistence of HB$_s$Ag and HB$_e$Ag; (3) high levels of anti-HB$_c$ IgM for about five months; and (4) no detectable anti-HB$_e$ or anti-HB$_s$. (Courtesy of Abbott Laboratories, North Chicago, Ill.)

sistently abnormal values in the absence of continuing exposure suggest the presence of underlying structural liver disease. In the absence of liver disease, the serum γ-glutamyltransferase is useful as an index to monitor alcohol consumption or abstention by the subject. Hyperbilirubinemia and other manifestations of cholestasis may complicate severe alcoholic liver disease, and there may be bile duct obstruction secondary to pancreatitis. Depending upon the severity of the liver disease and the nutritional status, hypoalbuminemia or hypergammaglobulinemia or both may be present. In some cases of alcoholic liver disease, as in cirrhosis and alcohol-induced fatty liver, the syndrome described by Zieve is a classic finding—acute hemolysis accompanied by marked elevation of serum lipids.[60]

The increase in GGT activity is a more sensitive and specific index for detecting liver injury in alcoholics than are increases in aminotransferases or alkaline phosphatase. In chronic alcoholics serum aminotransferases, especially AST, are frequently elevated, but such findings are not specific indicators of liver injury since AST elevations can also result from the skeletal muscle injury (alcoholic myopathy, usually asymptomatic) that occurs in ∼80% of alcoholics. Abnormal AST values may also be due to alcoholic cardiomyopathy.

Monitoring serum GGT activity is helpful in distinguishing chronic alcoholics with liver disease from those without liver disease.[39,46] In the chronic alcoholic, liver disease is assumed if GGT activity initially is 8–10 times normal and if the elevation persists after 6–8 weeks of abstention from alcohol. On the other hand, if initial GGT levels are only 2–3 times normal and return to normal after abstention, the patient is assumed to be free of liver disease. In the latter case, increase of GGT is considered to be due to the inductive effect of ethanol on the synthesis of enzymes in the membranes of the liver; the degree of induction does not correlate with daily quantity of alcohol consumed or with the length of time over which consumption occurred.[5,6,36]

The utility of laboratory parameters in the diagnosis and classification of alcoholic hepatitis was examined by Mendenhall et al. (Veterans Administration Hospital Cooperative Study Group).[36] Patients were grouped into those with mild disease (Group I: "compromised liver disease" but serum bilirubin of 5 mg/dL or less); moderate disease (Group II: liver decompensation, serum bilirubin > 5 mg/dL, prothrombin time not prolonged); and severe disease (Group III: prothrombin time prolonged). Despite the presence of biopsy-proven alcoholic hepatitis, the biochemical markers usually associated with cellular injury, inflammation, and necrosis were only mildly abnormal, even in severe cases. In this study, the mean AST activity in Group I was 84 U/L and in Group III 99 U/L. Overall, 13% of cases presented with normal AST activity and no case had a value > 500 U/L. The ALT activity was abnormal even less frequently; mean values were 56 U/L in Group I, 57 U/L in Group III, and 75 U/L in fatal cases; in 36% of cases, ALT levels were normal. Other abnormalities included a variable degree of anemia in over 80% of the patients; macrocytosis was present in 80%, and the MCV increased with increasing severity. Alkaline phosphatase activity was elevated in 60% of the Group I, 100% of the Group II, and 88% of the Group III patients. Elevation of serum bile acids (cholylglycine) was the most frequently encountered abnormality in the Group I patients, occurring in 85% of them; in 100% of all other patients, serum bile acids were abnormal. This study also demonstrated that laboratory parameters were useful in predicting a fatal outcome, which could be anticipated in 90% of patients over 55 years of age when the bilirubin exceeded 25 mg/dL and the prothrombin time was 5 s above control.

Drug-Induced Liver Diseases

The ingestion of any exogenous compound has the potential to induce an adverse reaction. Certain drugs or their metabolites are able to produce a direct, predictable injury to liver cells and to elicit a dose-related hepatotoxic reaction. Other drugs are less predictable in their hepatotoxic potential; they may produce a pattern of injury similar to that seen in either viral hepatitis or cholestatic liver injury. The mechanism in these cases may be an immune hypersensitivity reaction.

The mechanism of cell injury of the predictable hepatotoxins is not totally clear; however, there may be specific organelle damage such as alteration in the smooth endoplasmic reticulum with bile secretory failure. The presentation, symptomatology, clinical features, and laboratory abnormalities of drug-induced liver damage are highly variable; therefore, all patients presenting with suspected hepatocellular disease should be carefully questioned about alcohol or drug ingestion and exposure to environmental toxins.

Two classic examples of drug-induced hepatotoxicity are those of acetaminophen and isoniazid. Acetaminophen is metabolized by the hepatic microsomal P_{450} system to produce an electrophilic, potentially poisonous metabolite that is, in low concentrations, inactivated by sulfation and glutathione-S-transferase-catalyzed conjugation with glutathione. Ingestion of acetaminophen in toxic doses, however, overwhelms the capacity of the normal elimination mechanisms. In the face of glutathione depletion, the toxic metabolite accumulates, binds to and arylates hepatocyte proteins, and causes necrosis. Clinical manifestations include protracted vomiting and nausea, followed by an apparent recovery phase; marked jaundice and hepatic tenderness ensue and deterioration of liver function is rapid. There may be extreme increases in transaminase activity; prothrombin times are grossly abnormal. Early diagnosis of acetaminophen-induced hepatotoxicity is imperative since initiation of therapy within 16 h of ingestion with drugs such as N-acetylcysteine lessens the potential for hepatic injury and decreases the mortality rate. Measurement of serum acetaminophen concentrations and clearance rate may be quite helpful in both diagnosis and prognosis.

Administration of isoniazid for treatment of tuberculosis has been associated with transient aminotransferase elevations in up to 20% of patients. This elevation is ∼5–10 times normal and usually occurs within the first 2–3 months of therapy; it does not appear to be associated with clinical hepatitis and apparently does not dictate drug withdrawal. However, this response must be differentiated from true isoniazid-induced liver injury, which occurs in ∼1% of all isoniazid recipients and generally develops during the second and third months of treatment. In addition to elevated serum aminotransferase levels, there may be an increase in serum alkaline phosphatase and bilirubin and in prothrombin time. However, the injury may have its onset much later in the course of drug treatment. Hepatotoxic reactions are especially prevalent in elderly patients and females. The mechanism of isoniazid toxicity is related to genetic variations in its metabolism. Orientals and black females may be more susceptible to isoniazid-induced hepatotoxicity because of a unique metabolic variation that allows for rapid acetylation of isoniazid. A high percentage of these patients produce potent acylating agents and excrete large amounts of acetylated derivatives such as acetylhydrazine. The latter compound can bind covalently to cytosolic proteins and thereby induce hepatic necrosis.

Cirrhosis

The International Association for the Study of the Liver defines cirrhosis as the existence of nodular parenchyma and diffuse fibrosis. The pattern may be classified as either micronodular or macronodular. Cirrhosis can be the cause of biochemical abnormalities that reflect specific alterations in metabolism, i.e., decreased albumin synthesis, increased prothrombin time, decreased cholesterol synthesis, and insulin resistance. Cholestasis may also be present as evidenced by a mild increase in alkaline phosphatase and GGT, as well as total and conjugated bilirubin. The mild degree of liver cell injury that is usually present in cirrhosis may be responsible for a slight increase in AST and ALT as well as LDH. Any type of hepatic cirrhosis may cause portal hypertension—an increase in blood pressure within the portal vein and its tributaries. This portal hypertension leads to development of portal-systemic collateral venous drainage, splenomegaly, and ascites. Collaterals developing in the mucosa of the gastric fundus and esophagus (esophageal varices) are a serious clinical problem since they may bleed profusely, which frequently results in the patient's death.

Primary Biliary Cirrhosis

Primary biliary cirrhosis (PBC) is characterized by progressive destruction of the intrahepatic bile ducts with eventual development of biliary cirrhosis. The etiology of PBC is poorly defined; however, accumulating evidence suggests an autoimmune basis. The disease primarily affects middle-aged women; the incidence in females is 9 times higher than in males. PBC is classically associated with other immunopathic diseases such as Sjögren's syndrome, scleroderma, and rheumatoid arthritis. The clinical picture is characterized by intractable itching, which precedes onset of jaundice; persistent pruritus may lead to excoriated, hyperpigmented skin. Due to elevated serum triglycerides and cholesterol, xanthomas and xanthelasma develop. The progressive degree of cholestasis is associated with a decrease in bile excretion; therefore, bile acids do not reach the duodenum, and fat malabsorption and fat-soluble vitamin deficiency can occur.

The classic biochemical alteration early in PBC is a marked (two- to ten-fold) elevation in the alkaline phosphatase level; transaminases are only modestly elevated, and bilirubin elevations are variable. The values may rise as the disease progresses. Eighty-five to 90% of patients with PBC have a positive antimitochondrial antibody (AMA) that is neither species nor organ specific and may be present in other liver diseases. Antinuclear (ANA) and smooth muscle (SMA) autoantibodies may also be present.

Hepatic Tumors

Benign and malignant lesions in the liver include primary and secondary neoplasms, cysts, and abscesses. Histological features and clinical data may often suggest the presence of a specific type of lesion. Metastatic lesions are much more common (20:1) than primary liver tumors, since the liver is the second most common site of spread. Factors such as pre-existing cirrhosis or chronic infection (carrier state) of hepatitis B virus are frequently associated with primary liver cell carcinoma. Worldwide, the incidence of hepatocellular carcinoma correlates with the prevalence of HB_sAg carrier states. The suspicion of a tumor should be high when abnormalities in serum biochemical tests occur without an obvious reason. Suspicion is particularly well founded if biochemical abnormalities occur that are not usually seen in cirrhosis. Among these abnormalities are hypoglycemia, hypercalcemia, dysfibrinogenemia, two- to ten-fold elevations of LDH activity, up to 20-fold increases in GGT activity, increases in alkaline phosphatase activity, and elevated serum vitamin B_{12} levels (the last due to increased production by the tumor of B_{12}-binding protein).

Hepatic cancer may develop in a patient with pre-existing cirrhosis. In such patients, a chronic inflammatory process appears to be present prior to the detection of the tumor; this inflammation may be reflective of parenchymal damage as well as altered blood flow through the liver. There may be a gradual or abrupt change in the previously stable biochemical patterns. The transaminases may rise slowly initially, with more marked increases as the mass spreads throughout the parenchyma.

Assay of serum alkaline phosphatase activity, which is markedly elevated in cases of tumor, is widely used as a biochemical screening test. Serum alkaline phosphatase levels are frequently elevated when hepatic parenchyma is displaced by tumor (either primary or metastatic), cysts, or granulomas. This elevation may occur even when the serum bilirubin level remains normal. 5'-Nucleotidase, leucine aminopeptidase, and γ-glutamyltransferase activity may also be elevated. A marked rise in GGT activity may be noted in patients with metastatic disease with masses that are small and not easily detected. A more than two-fold increase in serum total LDH activity has been noted in the presence of both primary and secondary hepatic tumors. There has been recent validation of the use of serum α_1-fetoprotein (AFP) concentrations as an indicator of hepatic cell carcinoma. AFP is a normal fetal plasma protein that disappears from serum after birth. Therefore, a resurgence of serum AFP levels (> 400 ng/mL) in later life is highly suggestive of the presence of primary liver cancer, and serum AFP is a useful screening test in patients with cirrhosis or chronic hepatitis B infection. Diagnosis may be aided by the concomitant assay of another oncofetal protein such as carcinoembryonic antigen (CEA). These markers are useful in serial monitoring of the course of the disease and its treatment by means of chemotherapy, radiation therapy, or surgical intervention.

METABOLIC LIVER DISEASES

Iron Storage Diseases

The liver plays an important role in iron metabolism. It synthesizes the plasma protein transferrin, to which much of circulating iron is bound. Like many other tissues, the liver also synthesizes and contains the protein ferritin, which binds and stores iron. Serum transferrin may be measured directly by immunochemical methods or by estimate of serum total iron binding capacity (TIBC). Serum ferritin is believed to reflect reliably the concentration of cellular ferritin; it is determined by RIA. In severe liver disease, impaired hepatic synthesis of transferrin leads to decreased serum concentrations and low values for TIBC. In acute hepatic disease with liver cell necrosis, both serum iron and serum ferritin concentrations increase since the damaged hepatocyte cannot take up iron or retain ferritin.

Pathologic iron overload (hemochromatosis) may be either primary, i.e., idiopathic or familial, or secondary (acquired) as a consequence of alcoholism, thalassemia, or repeated transfusion. Idiopathic familial primary hemochromatosis is a genetic disorder in which iron absorption is inadequately regulated; therefore, progressive iron overload occurs with deposition of iron in the hepatic, cardiac, and pancreatic parenchymal cells. The disease is thought to be inherited as an autosomal recessive trait, although there is variable phenotypic expression. This appears to be an HLA-related disorder; thus, it may be possible to detect at-risk patients before extensive tissue damage has developed. The disease is less frequent in females than in males, possibly due to net iron loss in menses.

Secondary hemochromatosis cannot be differentiated from the primary form by laboratory findings alone. The hepatic histopathology in both is that of increased iron stores within the hepatocytes and reticuloendothelial cells, in association with variable fibrosis and macronodular cirrhosis. The pattern of liver cell damage is the same regardless of the cause of the iron overload, but the degree of fibrosis is proportional to the load. The actual mechanism of hepatocyte damage is unknown but is thought to be due either to lipid peroxidation with attendant membrane damage or to lysosomal rupture with cell damage. Clinical symptoms are due to massive iron deposition in the heart, pancreas, and skin, as well as in the liver; total body iron may be as high as 50–60 g. Abuse of ethanol may exacerbate the liver damage or accelerate iron accumulation. The eventual hepatic lesion may be fibrosis and cirrhosis leading to portal hypertension. There may be an attendant cardiomyopathy, onset of diabetes, and various joint and skin manifestations.

The diagnosis can be made by documenting elevation of serum ferritin levels in conjunction with increased serum iron and per cent of transferrin saturation, together with decreased transferrin and TIBC. The laboratory findings cannot, however, distinguish whether the excess of stored iron is in parenchymal cells or in RE cells. Storage in RE cells is the more benign condition. Computed tomography and nuclear magnetic resonance (NMR) are promising new techniques for quantitating hepatic density, a direct correlate of hepatic iron content. Once the diagnosis is made, treatment consists of repeated phlebotomy for net removal of iron from the body pool.

Wilson's Disease

Wilson's disease, or hepatolenticular degeneration, is a disorder of copper metabolism. (See also Chapters 4 and 8C.) It is inherited in the autosomal dominant mode and is characterized by copper deposition in brain, liver, kidney, and cornea. Clinical manifestations include neurologic degeneration, cirrhosis, and typical corneal deposits called Kayser-Fleischer rings. In younger patients liver disease is predominant, while in patients over 20 years of age neurologic symptoms predominate. The hepatic manifestations may vary from hepatomegaly to subacute or chronic hepatitis, cirrhosis, and fulminant failure.

The effect on the liver and other tissues is essentially that of copper toxicity. Copper acts as a potent inhibitor of enzymatic processes dependent on sulfhydryl groups in the active center of involved enzymes; membrane ATPase is inhibited and the ATP and potassium content of tissue are decreased. The glycolytic pathway and microsomal membrane enzymes are also inhibited.

The defect in Wilson's disease is poorly characterized; however, it appears that there is an alteration of a single gene product consequent to a single mutant gene; this alteration leads to a block in the excretory pathways of ceruloplasmin-bound copper. Impaired biliary excretion of copper is primarily due to an alteration in carriers or receptors that either incorporate copper into ceruloplasmin or mediate transport into bile. Impaired biliary excretion of copper causes the metal to accumulate in the liver with a diffuse distribution throughout the cytosol. Copper will then be incorporated into lysosomes, leading to destruction of their structure and function. Copper is subsequently released into the circulation in a nonceruloplasmin-bound form and can exert its toxic effects on red cells, causing an acute hemolysis, or can deposit in tissues such as the eye, manifest as the classic Kayser-Fleischer ring. It will also cause degenerative change in the brain and kidneys.

The disease is fatal if untreated, but treatment by copper chelation (e.g., D-penicillamine) is specific and effective. The manifestations are reversible. Because of the genetic implications, the presentation of an index case should lead to screening of the whole family. Institution of appropriate treatment in asymptomatic cases can prevent development of hepatic and other manifestations. It is also important to note that the diagnosis of Wilson's disease and the appropriate differentiation of non-Wilsonian liver disease (such as chronic active hepatitis) can allow the early institution of appropriate therapy for the latter.

In addition to standard laboratory tests for liver dysfunction, certain specific tests are valuable in the diagnosis. The best screen is serum ceruloplasmin, since 96% of patients with Wilson's disease have levels < 30 mg/dL. Total serum copper is high in early Wilson's disease and is present in serum unbound to ceruloplasmin. Urinary copper excretion exceeds 100 μg/d and will increase markedly following a test dose of D-penicillamine. An extremely reliable diagnostic finding is an increased copper content of liver tissue.

Alpha₁-Antitrypsin Deficiency

The hepatic alterations found in patients with the homozygous deficient state of α_1-antitrypsin have been discussed on page 1406 in the section on Protein Synthesis.

CHOLESTASIS

Cholestasis implies functional and morphologic disturbances in hepatic excretory processes.[43] The all-encompassing term "obstruction" is inappropriate since, in many instances, there is no demonstrable blockage in the liver or biliary tract. Intrahepatic cholestasis may be due to physicochemical or structural alterations of the bile secretory apparatus, bile constituents, or transport mechanisms. Extrahepatic obstruction can be secondary to a variety of lesions causing obstruction of the biliary tract.

Clinical consequences of prolonged cholestasis of any nature are due to (1) failure of bile to reach the duodenum and consequent malabsorption of fat and the fat-soluble vitamins A, D, E, and K, and (2) accumulation in the liver of biliary constituents—bile acids, bilirubin, cholesterol, and trace elements normally excreted in bile—and their regurgitation into serum. The accumulation of cholesterol is associated with the development of hypercholesterolemia and xanthomas. There may also be deposits in the skin of bile acids or of other undefined compounds normally excreted in bile, with the eventual onset of annoying pruritus. Cholestasis may also result in progressive liver disease with biliary cirrhosis.

Cholestasis can result from interference with bile excretion at any level from the hepatocyte to the ampulla of Vater. The causes of cholestasis have traditionally been divided into intrahepatic and extrahepatic causes. *Intrahepatic cholestasis* implies a defect in hepatic excretory function due to defective liver cell function; it can be due to cell damage or to altered metabolism. Diseases such as viral hepatitis can be associated with cholestasis, as can alcoholic hepatitis; the pattern of hepatotoxicity induced by certain drugs may also be cholestatic. In addition, there are several syndromes, with or without bile duct paucity, that are responsible for intrahepatic cholestasis. In *extrahepatic cholestasis*, there is an apparent mechanical blockade in the biliary tree, as may develop from biliary atresia, common duct stone, strictures, or carcinoma of the biliary tract or head of the pancreas.

Neonatal Cholestasis

Cholestasis or impairment in hepatic excretory function in the neonatal period can be associated with a wide variety of infectious, metabolic, or endocrine disorders. Jaundice is also a frequent concomitant of various syndromes or inborn diseases such as cystic fibrosis, α_1-antitrypsin deficiency, and Down's syndrome.

Two major decisions must be made in evaluating the infant with jaundice. The first is to differentiate quickly neonatal cholestasis from its more benign counterpart, "physiologic hyperbilirubinemia." The latter is characterized by an increase in unconjugated bilirubin (UCB) because of immaturity of processes involved in bilirubin metabolism. In the infant responding to exogenous stress (hypoxia, acidosis, hemolysis, sepsis) or in prematurely born infants of very low birth weight, the UCB can rise to high levels with the attendant risk of the development of kernicterus (bilirubin encephalopathy). In most cases, however, the jaundice will resolve as the liver matures. On the other hand, jaundice in infants with an increased *conjugated* (direct-reacting) bilirubin fraction is never benign, since some form of hepatobiliary disease is implied.

The second major decision is to differentiate *neonatal hepatitis* from *biliary atresia*. Children affected by biliary atresia may be aided by early surgical intervention and creation of bile drainage, but neonatal hepatitis requires medical management for the consequences of cholestasis. A variety of tests have been proposed to differentiate these conditions. The most reliable and simplest test is examination of stool color for the presence of bile pigments. Consistent absence of bilirubin or bile pigments from the stool suggests bile duct atresia; conversely, pigmented stools are strong evidence against this disease. Patients with neonatal hepatitis may have intermittent acholic or normal stools. Other laboratory tests proposed, such as the determination and fractionation of serum bile acids and determination of 5'- nucleotidase and various enzyme activity ratios, have not proved consistently helpful in the differential diagnosis.[3] The ultimate discriminator is liver biopsy. In biliary atresia, bile duct proliferation with lobular fibrosis is characteristic, even early in the disease, whereas in neonatal hepatitis there is marked lobular disarray, hepatocellular necrosis, and inflammation, as well as giant cell transformation.

Reye's Syndrome

Acute encephalopathy in combination with fatty degeneration of the viscera was initially described by Reye et al. in Australia in 1963,[45a] with nearly simultaneous case descriptions by Johnson et al. in the United States.[26a] In the majority of these early cases, the disease was fatal. There were only sporadic case descriptions until 1974, when 379 cases were reported to the CDC. The mortality rate in this series was 40%. Since that time, a steady incidence level (0.3–1.0 case per 100 000 persons under 18 years of age) has been reported; however, the case fatality rate has dropped remarkably with improved methods of diagnosis and management.

The syndrome is characterized by specific clinical features; there is an abrupt onset of protracted vomiting, which usually follows within a week of a prodromal, febrile viral illness caused by either varicella or influenza B. The vomiting is associated with neurologic changes, such as lethargy and confusion, which may deteriorate rapidly into stupor and coma. At the same time, the liver enlarges and marked abnormalities in liver function become evident. This disease most frequently strikes children in the age range of 6–11 years; it appears to be a disease of rural rather than urban children.

The etiology of Reye's syndrome is unknown; however, it is apparent that there is a generalized mitochondrial dysfunction due either to specific disturbances engendered in the hepatocyte by a virus or a toxin or by viral potentiation of chemical toxins. The role of host susceptibility or the effect of exogenous agents such as aspirin needs to be more fully evaluated.

Laboratory features of Reye's syndrome have been characterized as an "explosive release of various enzymes." There is a marked elevation in ALT and in AST from liver or muscle. Plasma ammonia concentrations are increased and the degree of increase may have prognostic significance. The total bilirubin concentration remains normal. There exist hyperaminoacidemia, increase in plasma fatty acids, hyperuricemia, hypoprothrombinemia unresponsive to vitamin K, and, in infants, a tendency toward hypoglycemia.

The liver biopsy is characteristic; on gross examination, the color is striking—yellow to white. By light microscopy, a uniform foaminess of the liver cell cytoplasm is seen; microvesicular fatty accumulation is demonstrated on frozen sections. Triglyceride is distributed throughout the lobule in a monotonous, small-droplet form. There is no inflammation and cell death is uncommon. There is a progression of ultrastructural change, demonstrable by electron microscopy, highlighted by mitochondrial pleomorphism and proliferation of the smooth endoplasmic reticulum.

Management consists of judicious treatment of the increased intracranial pressure secondary to cerebral edema with osmotic diuretics such as mannitol. Cerebral edema is the major factor contributing to the high mortality rate.

Biliary Tract Diseases

The most common biliary tract diseases are cholecystitis, carcinoma of gallbladder and bile ducts, and gallstone disease.

Cholecystitis. Acute cholecystitis may develop in association with gallstones in the gallbladder or with acute obstruction of the cystic duct by a gallstone. Less commonly, cholecystitis can develop without coincident gallstones. Acalculous cholecystitis may be a complication of bacteremia, vascular disease such as polyarteritis, treatment with drugs such as steroids, or trauma.

Approximately 95% of the episodes of acute cholecystitis are due to gallstone-induced obstruction and a subsequent rise in biliary pressure, which in turn causes direct damage to the gallbladder wall and mucosa. The damage is partially due to bile salts and lipids; in addition, increased pressure and distention of the gallbladder wall will compromise the blood flow, causing ischemia and infarction (acute gangrenous cholecystitis). Regurgitation of pancreatic enzymes into the gallbladder may induce further injury. In the presence of stasis and damaged mucosal integrity, bacterial invasion and further inflammation occur.

The clinical features are variable depending upon the degree of inflammation and ischemia. The most susceptible group is obese women over 40 years of age; however, the disease has been described in all age groups. The most common symptom is right upper quadrant abdominal pain, which is referred to the right shoulder or angle of the right scapula. Attacks may be initiated by the ingestion of heavy or fatty food–containing meals, and usually occur late at night. Spasms may be 30–60 min in duration. During the episode there may be an increase in temperature associated with bacterial infection of the gallbladder wall. Other signs that are occasionally present are abdominal rigidity, a tender mass palpable in the gallbladder area, elevated white count, and jaundice.

The most common type of gallbladder disease is *chronic cholecystitis*. This disease also is secondary to gallstone formation; therefore, predisposing factors are those discussed below as being etiologic in gallstone formation. Chronic inflammation may follow a bout of acute cholecystitis, but in most cases, it develops insidiously. The clinical features are ill defined, although obesity, a family history of gallstones, or a previous attack of cholecystitis or jaundice should be reason to suspect the disease. Abdominal distention and epigastric pain, especially postprandially, along with belching and flatus may be present. Techniques such as ultrasound and endoscopic visualization (retrograde cholangiography) are very helpful in documenting the presence and pathogenesis of biliary obstruction.

Gallstones. The most common type of gallstone is composed predominantly of cholesterol. There has been significant effort devoted to defining the chain of physical-chemical events that occur in the gallbladder to allow cholesterol to precipitate out of solution to form gallstones. The role of bile acids in promoting formation of micelles, which incorporate cholesterol and prevent precipitation, is well known. Lithogenic bile contains excess cholesterol relative to the other components of the mixed micelles (i.e., bile salts and phospholipids) due either to increased secretion of cholesterol or to a decrease in bile acid pool size. In susceptible patients, supersaturated (lithogenic) bile is produced by the liver; however, the gallstones are formed in the gallbladder, possibly due to intravesicular factors such as impaired gallbladder contractility and stasis, with or without infection, which favor cholesterol crystallization.

Factors predisposing to gallstone disease include advancing age, obesity, and underlying liver disease such as cirrhosis. Gallstones are twice as common in women as in men, and the incidence is higher in multiparous than in nulliparous women. Relationship of multiple pregnancies to increased incidence is possibly due to incomplete gallbladder emptying during pregnancy. There are certain geographic areas where the incidence of gallstones is higher. American Indians have the highest known prevalence due to an apparent tendency to the production of lithogenic bile.

Pigment gallstones are much less frequent than cholesterol gallstones. These are characterized by a cholesterol content of < 25% and a black or dark brown color; they are composed of calcium bilirubinate, phosphate, and carbonate. The cause of their formation is not known, but they are most commonly found in association with hemolysis or cirrhosis. Gallstones are a frequent complication in sickle cell anemia.

Tumors of the gallbladder and bile ducts. Benign lesions such as papillomas or adenomas may be seen as an incidental finding at cholecystectomy; malignant disease of the gallbladder is uncommon. Cholelithiasis may be an etiological factor in carcinoma of the gallbladder; 85% of gallbladder carcinomas occur in patients with gallstones, but < 1% of patients with gallstones develop carcinoma. It has been suggested that a calcified gallbladder is especially prone to malignant transformation. There are various pathologic forms, including papillary adenocarcinoma, squamous cell carcinoma, and anaplastic tumors. These tumors usually arise in the neck of the gallbladder and rapidly spread, causing obstruction and cholestasis. Physical examination will reveal a hard, tender mass in the gallbladder fossa. These lesions are particularly difficult to deal with, and the majority of cases are inoperable at the time of diagnosis.

Cholangiocarcinoma, or primary carcinoma of the bile ducts, can arise at any point in the biliary tree from the small intrahepatic bile duct radicles to the common bile duct. This lesion is often associated with underlying liver disease, such as sclerosing cholangitis associated with ulcerative colitis, congenital cystic lesions of the bile ducts, or chronic infestation such as *Clonorchis sinensis*. The clinical picture is that of cholestasis; therefore, differentiation from primary liver disease such as chronic active hepatitis or primary biliary cirrhosis should be made. There is a progressive increase in serum bilirubin levels; however, biochemical markers of hepatic dysfunction are not of specific differential diagnostic aid. Diagnosis is usually made by cholangiography or operative search.

SYSTEMIC CONDITIONS THAT PRODUCE LIVER DISEASE

Heart Disease

Liver damage is a common complication of acute or chronic heart failure. Cardiogenic shock can damage the liver by causing an increase in right atrial pressure that is directly transmitted as hepatic venous pressure, thereby producing central venous as well as sinusoidal stasis. Hepatocytes are especially vulnerable to lowered oxygen tensions; therefore, in the face of heart failure, diminished blood pressure, and reduced hepatic blood flow, there is congestion of centrilobular areas with hemorrhage, focal necrosis, sinusoidal dilatation, and impairment in bile secretion. The centrilobular zone cells normally receive less oxygenated blood and thus are most susceptible to anoxia.

Mild changes in serum ALT, AST, total bilirubin, and alkaline phosphatase occur in chronic right ventricular failure. The values of these tests are usually near or slightly above the upper limit of their reference ranges. Gamma-glutamyltransferase values are usually elevated to about twice the upper limit of the reference range. On the other hand, when acute right ventricular failure occurs, hypoxic damage to the hepatocyte results in a rapid release of enzymes into the bloodstream. Thus, marked elevations of serum ALT and AST are produced. A marked rise in activity of the liver isoenzyme of LDH is also commonly associated with this clinical situation. Mildly abnormal unconjugated and conjugated bilirubin and alkaline phosphatase values are obtained; mild to moderate increases in GGT activities are also typical in acute right heart failure. In patients with constrictive pericarditis, a picture similar to the Budd-Chiari syndrome or hepatic vein outlet obstruction is common.

Renal Disease

The so-called hepatorenal syndrome describes alterations in kidney function that may accompany hepatocellular disease. Renal failure in patients with liver failure may also be due to primary alterations in renal function or to acute tubular necrosis resulting from hemorrhage or infection. However, the most common situation is progressive uremia and oliguria. These renal function aberrations are in response to marked alterations in blood volume or compartmental shifts of fluid within the body; therefore, the renal histology is normal. Renal failure appears to be related to a reduction of the effective renal circulation. The serum urea nitrogen and creatinine concentrations increase and serum sodium levels decrease; urinary sodium excretion is low.

Gastrointestinal Disease

Hepatobiliary disease is a frequent complication of chronic inflammatory intestinal diseases such as Crohn's disease and ulcerative colitis. The spectrum may range from primary sclerosing cholangitis to chronic active hepatitis or macronodular cirrhosis. In addition, the incidence of gallstones is increased in patients with terminal ileal dysfunction as in Crohn's disease. A rare complication of ulcerative colitis is bile duct carcinoma.

Effect of Pregnancy on Liver Function

The altered hormonal milieu present during pregnancy may be associated with alterations in hepatic function made manifest by a rise in serum alkaline phosphatase late in gestation. It is important to note that the source of marked elevations in alkaline phosphatase may be the placenta, not the liver. Therefore, evaluation of the pregnant patient for liver disease may be more reliably carried out by assessing serum total and conjugated bilirubin, GGT, and aminotransferase concentrations. There is an impairment of dye elimination late in gestation, but its significance is not known. An unusual complication of pregnancy is cholestasis, which occurs in ~1 out of 1500 pregnancies. In some cases, this is associated with viral hepatitis; however, 20–25% of all cases are due to intrahepatic cholestasis of pregnancy. Other complications of the gravid state are acute fatty liver and toxemia.

The Liver in Diabetes Mellitus

Diminished tissue utilization of glucose and overproduction of glucose by the liver occur in diabetes mellitus. The liver plays a major role in insulin metabolism; therefore, alterations in blood glucose levels may be seen in liver disease due to a failure of hepatic degradation. A picture resembling diabetes results. Histologic alteration such as fatty infiltration of the liver is common in the obese insulin-resistant diabetic.[22] Fatty infiltration and an increased glycogen content may be seen in the young patient with diabetes. Episodes of ketoacidosis may be associated with hyperglobulinemia and a mild increase in serum bilirubin levels. In patients with well-controlled diabetes mellitus, there are no detectable abnormalities in liver function.

USE OF THE LABORATORY IN THE DIAGNOSIS AND MANAGEMENT OF LIVER DISEASE

There have been significant recent advances in our understanding of hepatic physiology and pathophysiology, and these advances have been associated with the introduction of new diagnostic techniques. However, the phenomenon of "many are called, but few are chosen" remains. In fact, with the increased number of tests and diagnostic procedures that are available, there is a concomitant increase in the degree of confusion among clinicians and laboratorians as to their rationale and interpretation. It must be remembered that most of the standard "so-called" liver function tests do not measure any specific hepatic function. They reflect either liver cell injury, made manifest, for example, by a rise in enzyme (AST, ALT) levels, or an immunologic response to hepatic cell injury, evidenced for example by an increase in immunoglobulin levels. The result of any single biochemical assay, found at a point in time (static test), may provide only a limited degree of information. In the future, with wide availability and clinical utility of dynamic tests of hepatic function, such as specific clearance tests or breath tests, quantitative tests of hepatic function may be helpful in defining the severity and in monitoring the rate of progression or regression of liver disease. Currently these procedures are not widely utilized on a routine basis because of complexity, expense, and hardship to the patient. For the time being, we can describe the course of liver disease in only general terms (Figure 13-26).

At present, a prudent course is for the clinician to become familiar with the terminology, rationale, implications, and pitfalls of a group of *selected* laboratory tests and diagnostic procedures. It is rarely productive to "check all the boxes" or to order a standard liver battery, as if embarking on a hunting expedition in hopes of a serendipitous reward. This approach encourages stereotypic responses and discourages careful, thoughtful analysis. Rational and judicious use of a carefully selected group of tests should then allow the clinician to answer the specific questions that arise when confronted with a patient with suspected liver disease. These questions are: (1) Is liver disease present? (2) What type of disturbance is present? (3) How *severe* is the liver disease? (4) How can the disease be *monitored* during treatment? and (5) What is the *prognosis* for this individual patient? None of the currently available tests or procedures can, by itself, answer these questions, and both the clinician and laboratorian must appreciate the context in which testing and use of test combinations are productive. In the following discussion we explore these questions.

Figure 13-26. A dynamic view of the course of chronic liver diseases. The period of compensated liver disease with nonspecific symptoms lasts much longer than the stage of decompensation. The goal in management of patients with progressive disease is to induce a remission before progression to the stage of decompensation. (From Bircher, 1983, Seminars in Liver Disease, Vol. 3, No. 4.)

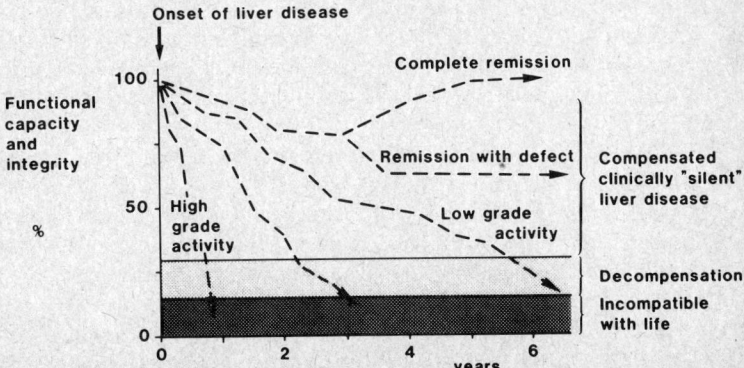

Is Liver Disease Present?

The presence of liver disease may be suspected on the basis of a clinical evaluation that includes a careful history and physical examination. In other cases, routine chemical screening during health maintenance examinations may identify a biochemical abnormality.[10,55] The combination of laboratory tests most frequently used in confirming this suspicion includes total serum bilirubin, AST, and ALT, as well as alkaline phosphatase and γ-glutamyltransferase. (See also Chapter 5, Enzymes.) These tests complement each other and may reflect hepatic excretory function and provide an index of synthetic function, as well as suggest specific abnormalities such as cell damage and cholestasis. Normal values on all of these screening tests can, with substantial certainty, rule out significant liver or biliary tract disease. There are important exceptions, however, such as asymptomatic cholelithiasis, compensated inactive cirrhosis, or small space-occupying lesions. Further workup will be based on the results of these tests if any are abnormal. In their interpretation, however, physiologic variables must be kept in mind. For example, alkaline phosphatase activity is age-related.

Differentiation of potential causes for elevated serum enzyme activities may call for determination of isoenzyme patterns and the use of other indices of hepatic damage. One example of this is the finding of an elevated alkaline phosphatase level in an asymptomatic patient with no history of hepatobiliary disease and a normal physical examination. Since an increase in γ-glutamyltransferase activity in serum is a sensitive indicator of liver disease, but does not occur in bone disease, an increase in alkaline phosphatase with normal γ-glutamyltransferase activity would rule out the liver as the origin of increased alkaline phosphatase activity and would lead the clinician to evaluate the patient for possible nonhepatic disease. On the other hand, should the γ-glutamyltransferase activity be abnormal, other indices of hepatic damage would be required. An algorithm summarizing an approach to this clinical problem is given in Figure 13-27. A normal prothrombin time, when other indices point to liver disease, can be a good prognostic sign; on the other hand, a prolonged prothrombin time in the presence of hepatic dysfunction may be a sign of severe liver damage, but is *not by itself* evidence of liver disease; this test is affected by many nonhepatic diseases.

The recent development of methods for measurement of serum bile acids may help to provide useful information regarding hepatic synthetic and excretory function. *Endogenous* organic anion clearance tests, such as the fasting and 2-h postprandial serum bile acid measurement, are already informative and promise to become more so in the future.

What Type of Disturbance Is Present?

When the presence of liver disease has been suggested by physical and laboratory findings, then specific information regarding the nature of the disease process must be sought.[47] The common questions that arise in clinical practice are: (1) Does the patient have hepatic *parenchymal* disease or *cholestasis*? (2) If cholestasis is present, is the *site* of the obstruction intrahepatic or extrahepatic? (3) What is the nature of the liver disease in an alcoholic? (Is this nonalcoholic liver disease or alcohol-related hepatitis?)

Acute liver cell injury (*parenchymal disease*) can be seen in viral hepatitis, alcoholic liver disease, drug- or toxin-induced liver disease, hypoxemia, shock, or metabolic liver diseases. In most of these instances, there is parenchymal cell necrosis. Reye's syndrome is an exception, as we have noted earlier. Chronic liver cell injury is seen in active liver disease, chronic alcohol ingestion, primary biliary cirrhosis, and also in various chronic metabolic diseases involving the liver. The cell injury in all of these conditions except alcoholic hepatitis is reflected by a marked increase in aminotransferase levels. The highest values correlate with the more severe degrees of hepatocyte injury.

Cholestasis is reflected by an accumulation in serum of the components of bile. Obstruction at any level is accompanied by regurgitation of biliary constituents into serum; the total and conjugated serum bilirubin levels are elevated to some degree in all forms of cholestasis except unilateral or *focal duct obstruction*, as occurs with tumors, cystic masses, or granulomas. If the contralateral duct remains patent, bile excretion may remain functional and serum bilirubin levels will not be elevated. Alkaline phosphatase and, to a greater degree, γ-glutamyltransferase activities are increased in most forms of cholestasis, even in the presence of focal or localized obstruction as noted above or as seen in infiltrative diseases. Therefore, the disparity of a normal serum bilirubin level with increased alkaline phosphatase and γ-glutamyltransferase activities may be

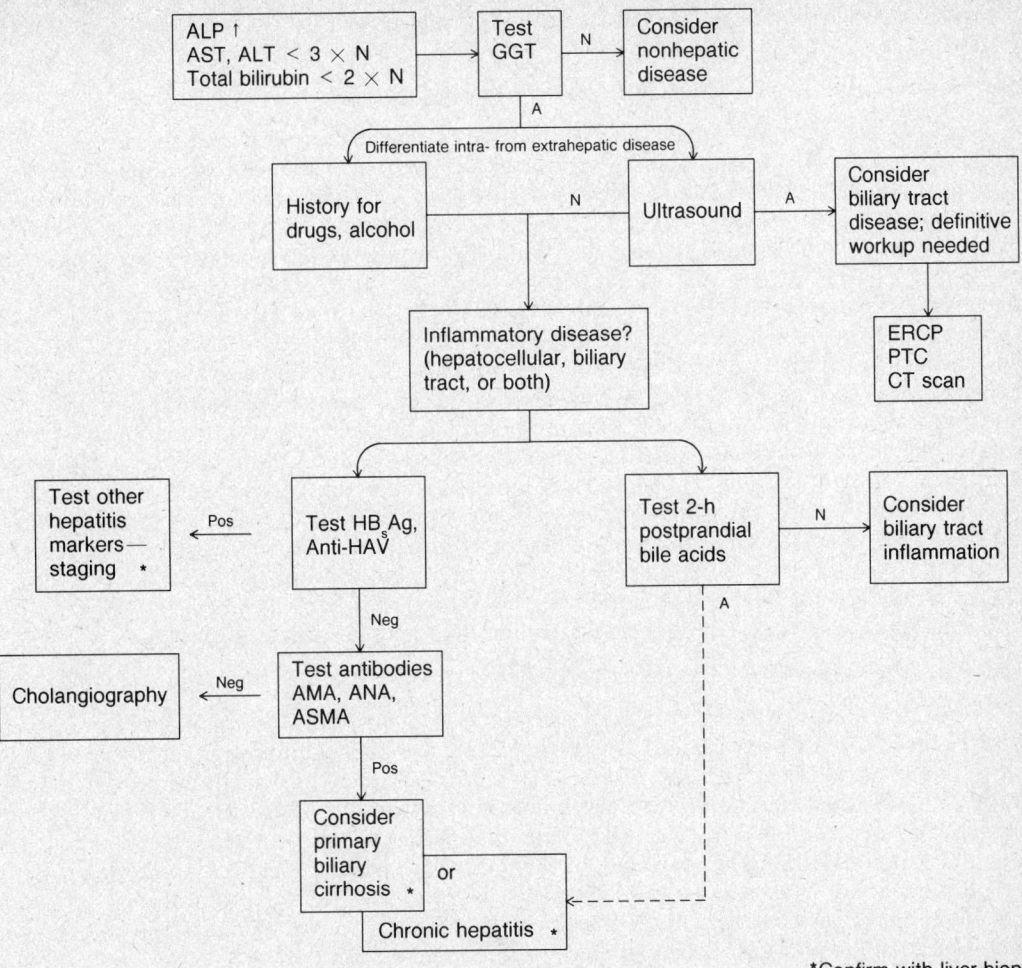

Figure 13-27. Possible investigational pattern for the asymptomatic patient with no history of hepatobiliary disease, a normal physical examination but an elevated alkaline phosphatase activity. A = abnormal results, N = normal results, NEG = negative, POS = positive, ↑ = increase. (Modified from Javitt, N. B., et al.: Materia Diagnostica— A Manual of Liver and Kidney Tests for the Medical Profession. Puerto Rico, Searle and Co., 1977.)

seen in patients with this type of lesion. Precise demonstration and localization of an intrahepatic obstruction are made through the use of ultrasonography, percutaneous transhepatic cholangiography (PTC), endoscopic retrograde cholangiopancreatography (ERCP) or computed tomography, or by operative intervention.

The biochemical feature of acute *alcoholic hepatitis* is the presence of a markedly elevated GGT; this is not true of acute viral hepatitis. Alcoholic hepatitis has also been characterized by an AST/ALT ratio of >1.0 when the AST is <300 U/L. It has been further suggested that if the ratio is <1.0, the presence of liver disease unrelated to alcohol is more likely. Overt liver damage is unlikely to be present in an alcoholic in whom the GGT is within normal limits.[58]

Among other diagnoses that will require consideration are primary biliary cirrhosis and other forms of cirrhosis. A positive test for antimitochondrial antibody (AMA) in the presence of elevated alkaline phosphatase, cholesterol, and serum IgM strongly supports the diagnosis of primary biliary cirrhosis instead of extrahepatic obstruction. These tests can then be supplemented by ancillary studies such as liver biopsy, PTC, ERCP, and other radiographic or scintigraphic techniques. Progression to cirrhosis is a potential consequence of chronic liver disease. Fibrosis is attended by an increased synthesis of two types of collagen (Type III followed later by Type I). Recently developed radioimmunoassays specifically measure peptides cleaved during the formation of collagen from procollagen, allowing the detection of antecedent peptides in serum and ascitic fluid. There is an increased concentration of procollagen III peptide in various liver diseases;

in chronic hepatitis, there is a correlation between serum levels of this peptide and the degree of fibrosis.

How Severe Is the Liver Disease and What Is the Prognosis?

Specific parameters of hepatic function can be of benefit in determining the severity of the liver disease in an individual patient, in monitoring trends, or in making the diagnosis of *fulminant hepatitis*. Fulminant hepatic failure, which carries a very high mortality rate, is most commonly seen with hepatitis B infection but can occur as a complication of any type of acute or chronic hepatic injury. The mechanisms underlying fulminant hepatic failure are poorly understood, but massive hepatic necrosis is seen at autopsy. A harbinger of fulminant failure is the onset of encephalopathy during the course of acute hepatitis. The liver may become nonpalpable as the hepatic parenchymal mass decreases due to massive necrosis. There may be ascites, hypoglycemia, hypokalemia, hyponatremia, and hypoalbuminemia, as well as anemia secondary to gastrointestinal bleeding. The most informative biochemical indicators are serum bilirubin, prothrombin time, and serum albumin. A uniformly poor prognosis has been associated with continued elevation of bilirubin, marked decrease in serum albumin, and prothrombin time prolonged more than 10 s beyond the time of the control and unresponsive to parenteral vitamin K administration. The previously elevated aminotransferases may decrease as hepatic parenchymal mass shrinks and biosynthetic rate is reduced. Blood ammonia levels may also become elevated. A rise in α_1-fetoprotein (AFP) levels is a *favorable* index of recovery. Management is difficult since there is no specific therapy for fulminant hepatic failure. Supportive therapy includes close management of the encephalopathy and bleeding as well as the fluid and electrolyte balance.

How Can Liver Disease Be Monitored?

Serial determination of serum aminotransferase (ALT and AST) levels should be sufficient to chart the course of a patient with acute viral hepatitis. This will allow either documentation of resolution or persistence of the disease. ALT and AST activity usually returns to normal shortly (2–3 months) after clinical recovery, indicating that the liver has healed and is histologically normal. However, abnormal transaminase values that persist for six months or more are indicative of chronic hepatitis; this finding warrants further evaluation of the patient. In monitoring patients with chronic active hepatitis, serum bile acid levels have provided an early index of resolution or relapse. A rise in the serum albumin concentrations, with a concomitant fall in the globulin level, is a good index of response to therapy with corticosteroids.

Monitoring the course of cirrhosis can be accomplished through use of several liver function tests. Measurement of ALT and AST is usually not helpful since their values are typically normal or only slightly abnormal; however, onset of hepatic decompensation is usually indicated by a falling albumin level, hyperbilirubinemia, and increasing PT values.

Systemic Diseases Causing Abnormalities in Biochemical Indices of Hepatic Function

One should remember that laboratory tests used to diagnose hepatic dysfunction may be positive when liver disease is absent or may be negative when liver disease is present. Drugs being administered to the patient may interfere to cause either false positive or false negative results; these interferences will not be discussed here. Other, nonhepatic diseases may affect levels of bilirubin and "liver" enzymes.

In acute pancreatitis, aminotransferase elevations up to 3–5 times normal are not uncommon. However, the clinical picture should aid in the differential diagnosis, and further clinical assessment of pancreatic disease should be considered. In addition, systemic illnesses, such as sepsis due to either specific bacteria or overwhelming fungal infections, can be associated with an elevation of serum bilirubin and alkaline phosphatase activity. There may also be a mild rise in the aminotransferase levels. Septic shock due to gram-negative bacteremia or other causes has classically been stated to cause an extreme rise of the aminotransferases to levels as high as several thousand units per liter and a mild rise in the serum bilirubin. Congestive heart failure of varying degrees will cause hepatic enlargement and abnormalities of hepatic function. Aminotransferase levels will rapidly return to normal as the cardiac status improves.

A diagnostic dilemma often arises in the patient with sickle cell anemia in whom a combination of factors such as hypoperfusion, intrahepatic sickling, and hemolysis may cause an extreme rise in the serum bilirubin level; in addition, the aminotransferase values may also increase. A confounding factor is the tendency to pigment gallstone formation that may be associated with hepatobiliary disease. A number of systemic diseases that have no apparent hepatic manifestations may cause an abnormality in the standard indices of liver function. A situation that commonly arises is an elevated alkaline phosphatase value in diseases such as Hodgkin's disease, myeloid metaplasia, or other malignancies.

A final caution to consider is the fact that significant liver disease can exist in the presence of *normality* of a multitude of tests of hepatic structural and functional integrity. For example, in small space-occupying lesions such as tumors or in congenital hepatic fibrosis, there may be no detectable abnormality in the standard laboratory tests. Similarly, in compensated inactive cirrhosis biochemical tests may be normal.

References

1. Adachi, Y., Horii, K., Takahashi Y., et al.: Serum glutathione-S transferase activity in liver disease. Clin. Chim. Acta, *106*:243-255, 1980.
2. Ansley, J. D., Isaacs, J. W., Rikkers, L. F., et al.: Quantitative tests of nitrogen metabolism in cirrhosis: Relation to other manifestations of liver disease. Gastroenterology, *75*:570-579, 1978.
3. Balistreri, W. F., Suchy, F. J., Farrell, M. K., et al.: Pathologic versus physiologic cholestasis: Elevated serum concentration of a secondary bile acid in the presence of hepatobiliary disease. J. Pediatr., *98*:399-402, 1981.
4. Balistreri, W. F., Suchy, F. J., and Heubi, J. E.: Serum bile acid response to a test meal stimulus: A sensitive test of ileal function. J. Pediatr., *96*:582-589, 1980.
5. Baraona, E., Leo, M. A., Borowsky, S. A., et al.: Pathogenesis of alcohol-induced accumulation of protein in the liver. J. Clin. Invest., *60*:545-554, 1977.
6. Belfrage, P., Berg, B., Cronholm, T., et al.: Prolonged administration of ethanol to young, healthy volunteers: Effects on biochemical, morphological and neurophysiological parameters. Acta Med. Scand. [Suppl.], *552*:5-55, 1973.
7. Blanckaert, M.: Analysis of bilirubin and bilirubin mono- and di-conjugates. Biochem. J., *185*:115-128, 1980.
8. Bohmer, R., Torok, M., and Rommel, K.: Reproducibility of the intravenous galactose tolerance test. Acta Hepatogastroenterol., *25*:271-274, 1978.
9. Branch, R. A.: Drugs as indicators of hepatic function. Hepatology, *2*:97-105, 1982.
10. Corless, J. K., and Middleton, H. M., III: Normal liver function: A basis for understanding hepatic disease. Arch. Intern. Med., *143*:2291-2294, 1983.
11. Corrigan, J. J., Jeter, M., and Earnest, D. L.: Prothrombin antigen and coagulant activity in patients with liver disease. JAMA, *248*:1736-1739, 1982.
12. Danielsson, H., and Sjovall, J.: Bile acid metabolism. Ann. Rev. Biochem., *44*:233-253, 1975.
13. Doumas, B. T., Hause, L. L., Sciacca, R. D., et al.: Performance of the DuPont *aca* ammonia method. Clin. Chem., *25*:175-178, 1979.
14. Doumas, B. T., Perry, B. W., Sasse, E. A., et al.: Standardization in bilirubin assays: Evaluation of selected methods and stability of bilirubin solutions. Clin. Chem., *19*:984-993, 1973.
15. Drott, H.: Personal communication.
16. Duffy, T. E., and Plum, F.: Hepatic encephalopathy. *In*: The Liver, Biology and Pathobiology. I. Arias, H. Popper, D. Schachter, et al., Eds. New York, Raven Press, 1982.
16a. Esmon, C. T.: Protein-C: Biochemistry, physiology, and clinical implications. Blood, *62*:1155-1158, 1983.
17. Fisher, H., and Plieninger, H.: Synthese des Biliverdins (Uteroverdins) und Bilirubins, der Biliverdine XIIIα and IIIα sowie Vinylneoxanthosaure. Hoppe-Seylers Z. Physiol. Chem., *274*:231-260, 1942.
18. Flannery, D. B., Hsia, Y. E., and Wolf, B.: Current status of hyperammonemic syndromes. Hepatology, *2*:495-506, 1982.
19. Forman, D. T.: Rapid determination of plasma ammonia by an ion-exchange technique. Clin. Chem., *11*:1-9, 1964.
20. Freeman, M., Kuiken, L., and Ragland, J. B.: Hepatic triglyceride lipase deficiency in liver disease. Lipids, *12*:443-445, 1977.
21. Friedel, R., Diederichs, F., and Lindena, J.: Release and extracellular turnover of cellular enzymes. *In*: Advances in Clinical Enzymology. E. Schmidt, F. W. Schmidt, I. Trauschold, et al., Eds. Munich, S. Karger, 1979.
22. Galambos, J. T., and Wills, C. E.: Relationship between 505 paired liver tests and biopsies in 242 obese patients. Gastroenterology, *74*:1191-1195, 1978.
23. Gerron, G. G., Ansley, J. D., Isaacs, J. W., et al.: Technical pitfalls in measurement of venous plasma NH_3 concentration. Clin. Chem., *22*:663-666, 1976.
24. Gumucio, J. J., and Miller, D. L.: Functional implications of liver-cell heterogeneity. Gastroenterology, *80*:393-403, 1981.
25. Hutchinson, D. R., Halliwell, R. P., Smith, M. G., et al.: Serum "prealbumin" as an index of liver function in human hepatobiliary disease. Clin. Chim. Acta, *114*:69-74, 1981.
26. Ijpma, S. T., Blijenberg, B. G., and Leijnse, B.: Evaluation of the DuPont *aca* ammonia procedure. Clin. Chem., *24*:489-492, 1978.
26a. Johnson, G. M., Scurletis, T. D., and Carroll, N. D.: A study of sixteen fatal cases of encephalitis-like disease in North Carolina children. N. Carolina Med. J., *24*:464-473, 1963.
27. Kahn, S.: Personal communication.

28. Kuenzle, C. C., Sommerhalder, M., Ruttner, J. R., et al.: Separation and quantitative estimation of four bilirubin fractions from serum and three bilirubin fractions from bile. J. Lab. Clin. Med., 67:282-293, 1966.

29. Lauff, J. J., Kasper, M. E., and Ambrose, R. T.: Separation of bilirubin species in serum and bile by high performance reversed-phase liquid chromatography. J. Chromatog., 226:391-402, 1981.

30. Lauff, J. J., Kasper, M. E., Wu, T-W., et al.: Isolation and preliminary characterization of a fraction of bilirubin that is firmly bound to protein. Clin. Chem., 28:629-637, 1982.

31. Liebman, H. A., Furie, R. C., Tong, M. J., et al.: Des-γ-carboxy (abnormal) prothrombin as a serum marker of primary hepatocellular carcinoma. N. Engl. J. Med., 310:1427-1431, 1984.

32. Lindskov, J.: The quantitative liver function as measured by the galactose elimination capacity. Acta Med. Scand., 212:295-302, 1982.

33. Lo, D. H., and Wu, T.-W.: Assessment of the fundamental accuracy of the Jendrassik-Grof total and direct bilirubin assays. Clin. Chem., 29:31-36, 1983.

34. Martinez, J., Palascak, J. E., and Kwasniak, D.: Abnormal sialic acid content of the dysfibrinogenemia associated with liver disease. J. Clin. Invest., 61:535-538, 1978.

35. McComb, R. B., Bowers, G. N., and Posen, S.: Alkaline Phosphatase. New York, Plenum Press, 1979.

36. Mendenhall, C. L.: Alcoholic hepatitis. Clin. Gastroenterol., 10:417-441, 1981.

37. Mitchell, J. R., Nelson, S. D., Thorgeirsson, S. S., et al.: Metabolic activation: Biochemical basis for many drug-induced liver injuries. Prog. Liver Dis., 5:259-279, 1976.

38. Mondzac, A., Ehrlich, G. E., and Seegmiller, J. E.: An enzymatic determination of ammonia in biological fluids. J. Lab. Clin. Med., 66:526-531, 1965.

39. Moussavian, S. N., Becker, R. C., Piepmeyer, J. L., et al.: Serum gamma-glutamyl transpeptidase and chronic alcoholism: Influence of alcohol ingestion and liver disease. Dig. Dis. Sci., 1985, in press.

40. Palascak, J. E., and Martinez, J.: Dysfibrinogenemia associated with liver disease. J. Clin. Invest., 60:89-95, 1977.

41. Panteghini, M., Malchiodi, A., Calarco, M., et al.: Clinical and diagnostic significance of aspartate aminotransferase isoenzymes in sera of patients with liver diseases. J. Clin. Chem. Clin. Biochem., 22:153-158, 1984.

42. Perry, B. W., Doumas, B. T., Bayse, D. D., et al.: A candidate reference method for determination of bilirubin in serum. Test for transferability. Clin. Chem., 29:297-301, 1983.

43. Popper, H.: Cholestasis: The future of a past and present riddle. Hepatology, 1:187-191, 1981.

44. Rashid, S. A., Axon, A. T. R., Bullen, A. W., et al.: Serum β2-microglobulin in hepato-biliary diseases. Clin. Chim. Acta, 114:83-91, 1981.

45. Rej, R., and Horder, M.: Aspartate aminotransferase. In: Methods of Enzymatic Analysis. H. U. Bergmeyer, J. Bergmeyer, and M. Grassl, Eds. Weinheim, Verlag Chemie, 1983.

45a. Reye, R. D. K., Morgan, G., and Baral, J.: Encephalopathy and fatty degeneration of the viscera: A disease entity in childhood. Lancet, 2:249-252, 1963.

45b. Rudman, B., DiFolco, T. J., Galambos, J. T., et al.: Maximal rate of urea excretion and synthesis in normal and cirrhotic subjects. J. Clin. Invest., 52:2241, 1973.

46. Ryback, R. S., Eckardt, M. J., Felsher, B., et al.: Biochemical and hematologic correlates of alcoholism and liver disease. JAMA, 248:2261-2265, 1982.

47. Ryback, R. S., Eckardt, M. J., Rawlings, R. R., et al.: Quadratic discriminant analysis as an aid to interpretive reporting of clinical laboratory tests. JAMA, 248:2342-2345, 1982.

48. Scharschmidt, B. F., Blanckaert, N., Farina, F. A., et al.: Measurement of serum bilirubin and its mono- and diconjugates: Application to patients with hepatobiliary disease. Gut, 23:643-649, 1982.

49. Schmid, R.: Bilirubin metabolism: State of the art. Gastroenterology, 74:1307-1312, 1978.

50. Schmidt, F. W.: Rationale for the use of enzyme determinations in the diagnosis of liver disease. In: Evaluation of Liver Function, A Multifaceted Approach to Clinical Diagnosis. L. M. Demers and L. M. Shaw, Eds. Baltimore, Urban and Schwarzenberg, 1978.

50a. Seligsohn, U., Berger, A., Abend, M., et al.: Homozygous protein C deficiency manifested by massive venous thrombosis in the newborn. N. Engl. J. Med., 310:559-562, 1984.

51. Setchell, K. D. R., and Matsui, A.: Serum bile acid analysis—the application of liquid-gel chromatographic techniques and capillary column gas chromatography and mass spectrometry. Clin. Chim. Acta, 127:1-17, 1983.

52. Shaw, L. M.: The measurement of gamma-glutamyltransferase. In: Clinical and Analytical Concepts in Enzymology. H. Homburger, Ed. Skokie, Ill., College of American Pathologists, 1983.

53. Sherlock, S.: Patterns of hepatocyte injury in man. Lancet, 1:782-786, 1982.

54. Sherman, M., Bass, N. M., Campbell, J. A. H., et al.: Radioimmunoassay of human ligandin. Hepatology, 3:162-169, 1983.

55. Skrede, S., Solberg, H. E., Ritland, S., et al.: Diagnostic and prognostic value of laboratory tests assessed in a follow-up study of 200 patients with liver disease. Clin. Chem., 28:1177-1181, 1982.

56. Street, J. M., Trafford, D. J. H., and Makin, H. L. J.: The quantitative estimation of bile acids and their conjugates in human biological fluids. J. Lipid Res., 24:491-511, 1983.

56a. Sundberg, M. W., Lauff, J. J., Weiss, J. S., et al.: Estimation of unconjugated, conjugated, and "delta" bilirubin fractions in serum by use of two coated thin films. Clin. Chem., 30:1314-1315, 1984.

57. Van Anken, H. C., and Schiphorst, M. E.: A kinetic determination of ammonia in plasma. Clin. Chim. Acta, 56:151-157, 1974.

58. Weill, J., Schellenberg, F., LeGoff, A.-M., et al.: The predictive value of gamma-glutamyltransferase and other peripheral markers in the screening of alcohol abuse. In: Gammaglutamyltransferases: Advances in Biochemical Pharmacology. G. Siest and C. Heusghems, Eds. Paris, Masson, 1982.

59. Wu, T.-W., Dappen G. M., Powers, D. M., et al.: The Kodak Ektachem clinical chemistry slide for measurement of bilirubin in newborns: Principles and performance. Clin. Chem., 28:2366-2372, 1982.

60. Zieve, L.: Jaundice, hyperlipemia and hemolytic anemia, a heretofore unrecognized syndrome associated with alcoholic fatty liver and cirrhosis. Ann. Intern. Med., *48*:471, 1958.
61. Zito, R. A., and Reid, P. R.: Lidocaine kinetics predicted by indocyanine green clearance. N. Engl. J. Med., *298*:1160-1163, 1978.

Additional Reading

Balistreri, W. F.: Viral hepatitis: Unique aspects of infection during childhood. Consultant, *24*:131-153, 1984.
Balistreri, W. F., and Schubert, W. K.: Liver disease in infancy and childhood. *In*: Diseases of the Liver. L. Schiff and E. Schiff, Eds. Philadelphia, J.B. Lippincott Co., 1982.
Balistreri, W. F., and Soloway, R. D.: Clinical Guide to Bile Acid Physiology and Alterations in Disease States. North Chicago, Ill., Abbott Laboratories, July, 1978.
Demers, L. M., and Shaw L. M.: Evaluation of liver function: A multifaceted approach to clinical diagnosis. Baltimore, Urban and Schwarzenberg, 1978.
Koff, R. S.: Viral Hepatitis. New York, John Wiley and Sons, 1978.
Wallnofer, H., Schmidt, E., and Schmidt, F. W.: Diagnosis of Liver Diseases. Stuttgart, Georg Thieme Publishers, 1977.

CHAPTER 14

GASTRIC, PANCREATIC, AND INTESTINAL FUNCTION

by Norbert W. Tietz, Ph.D., Alan D. Rinker, M.S.,
and A. Ralph Henderson, M.B., Ch.B., Ph.D.

It has long been assumed that, in addition to their anatomical relationship, there is a functional relationship between the stomach, the intestinal tract, and the pancreas. The introduction of sensitive radioimmunoassays for measuring many of the hormones regulating gastrointestinal functions[20] and other chemical parameters has supported this assumption. From those studies it is clear that there is a very delicate and complex interplay between the various gastrointestinal functions, making it appropriate to discuss gastric, pancreatic, and intestinal physiology in one chapter.

The **human stomach** consists of three major zones: the cardiac zone, the body, and the pyloric zone (Figure 14-1). The upper *cardiac zone* contains mucus-secreting *surface epithelial cells*. The *body of the stomach* contains cells or cell groups of four different types: (1) the surface epithelial cells, which secrete mucus; (2) the *parietal cells*, which are the main and possibly only source of hydrochloric acid; (3) the *chief* or *peptic cells*, which secrete a considerable amount of pepsinogen; and, finally, (4) the *neck chief cells* or mucus cells, which secrete mucus and pepsinogen. The third portion of the stomach, the *pyloric zone*, is subdivided into the *antrum*, the *pyloric canal*, and the *sphincter*. Its cells secrete mucus, some pepsinogen, and gastrin but no HCl.

There are three phases of the digestive process, the cephalic, gastric, and intestinal phases. The *cephalic phase* is initiated by the intake of food into the mouth; the sight, smell, and taste of food stimulate the cerebral cortex and subsequently the vagal nuclei. The process may be chemically mediated by acetylcholine from postganglionic parasympathetic nerve endings, which acts upon gastric parietal cells.[96] The vagus also stimulates gastric chief and parietal (oxyntic) cells to secrete pepsinogen and HCl. (Section of the vagus nerve [vagotomy] results in a decrease in volume and acidity of gastric secretion. See p. 1458.) The mechanism of acid secretion is still widely debated; there is, however, significant agreement that acetylcholine, histamine, and gastrin act through their respective neurocrine, paracrine, and endocrine pathways to stimulate the parietal cells and that specific parietal cell receptors to these transmitters exist. Also, there are potentiating interactions between the mentioned secretagogues that probably occur at the parietal cell itself.[58] Histamine may play a role as either a mediator or a potentiator of the actions of other secretagogues. The evidence that administration of histamine markedly increases the secretion provoked by pentagastrin or by cholinergic agonists strongly favors this hypothesis. The important role of histamine in acid secretion was also confirmed by the findings that histamine H_2 receptor antagonists (cimetidine, Tagamet) inhibit acid secretion provoked by most types of stimulation.[79] The secretion of H^+ against a million-fold concentration gradient requires energy from the cell, and this process is coupled to K^+,H^+-ATPase. There is now strong evidence that the histamine-stimulated acid secretion involves cAMP; however, cholinergic stimulation is not associated with rises in intracellular level of cAMP in isolated parietal cells, and this process may be mediated by changes in permeability of Ca^{2+}. The latter activates a chain of events which alters the cell function in various ways.

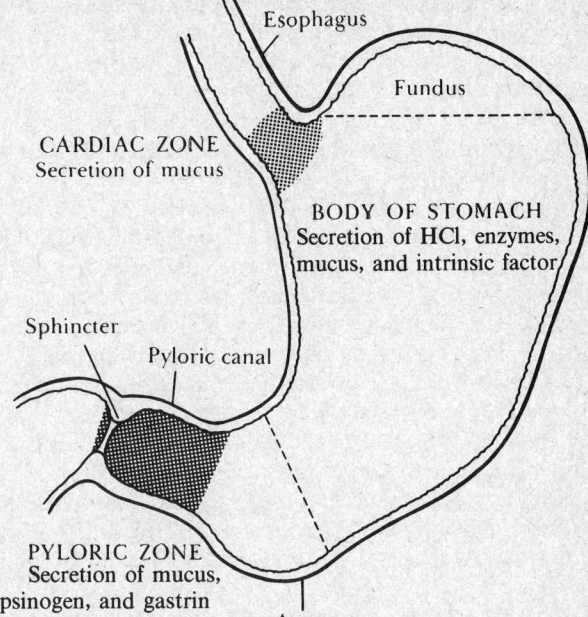

Figure 14-1. Schematic drawing of the stomach, with major zones.

In addition to the above-mentioned mechanisms, vagal fibers to the pyloric glandular mucosa cause the release of gastrin, which also stimulates HCl and pepsinogen secretion.

Agents *inhibiting* acid secretion include somatostatin, prostaglandins, gastric inhibitory polypeptide (GIP), secretin, glucagon, vasoactive intestinal polypeptide (VIP), dopamine, and serotonin. Acid secretion is decreased in the presence of decreased amounts of circulating pituitary, adrenal, thyroid, and parathyroid hormones. Pituitary hormones are apparently essential for the maintenance of the structural integrity of the gastric mucosa and necessary for secretory function; hGH may be necessary for the growth of the gastric mucosa.

As soon as food enters the stomach, the resulting distention initiates the *gastric phase* of digestion, which is mediated by local and vagal reflexes. HCl release is caused by (1) direct stimulation of the parietal cells by the vagus nerve; (2) local distention of the antrum and stimulation of antral cells by the vagus nerve to secrete gastrin, which in turn causes HCl release from parietal cells; and (3) release of gastrin, stimulated by the near neutralization (pH 5–7) of gastric HCl by ingested food entering the pyloric zone. Gastrin also stimulates antral motility, secretion of pepsinogen and of pancreatic fluid rich in enzymes, and release of a number of gastrointestinal hormones (secretin, insulin, acetylcholine, somatostatin, and pancreatic polypeptide). As a result of the acid environment, pepsinogen is rapidly converted to the active proteolytic enzyme pepsin. The contractions of the stomach mix and the chemical secretions of the stomach partially degrade the food into a mucus-containing mixture called *chyme*, which is then moved through the pylorus into the duodenum. The pylorus plays a role in the emptying of food into the duodenum by virtue of its strong musculature.

The *intestinal phase* of digestion begins when the weakly acidic digestive products of proteins and lipids enter the duodenum. Several gastrointestinal hormones are released by both neural and local stimulation and act on various regions of the GI tract to regulate digestion and absorption. The action of gastrin, which has been released by the above-mentioned mechanisms, is potentiated by the secretion of cholecystokinin. (See below.) Additional gastrin is released as the upper duodenal mucosa comes in contact with partially digested proteins and lipids, as well as gastric HCl. (For more information on this and other hormones, see Gastrointestinal Hormones, p. 1437.) *Cholecystokinin* (CCK) is released in the duodenum in response to the presence of fat, protein, and HCl. Its principal actions are stimulation of gallbladder contraction; secretion of enzymes, bicarbonate, insulin, and glucagon from the pancreas; and stimulation of intestinal motility and stomach contraction. *Secretin* is released by gastric acid in the duodenum; it augments the effect of CCK on gallbladder contraction and pancreatic secretions, stimulates pepsinogen secretion by the stomach, inhibits gastrin and gastric acid secretion, and, in contrast to CCK, reduces gastric and duodenal motility. *Gastric inhibitory polypeptide* is secreted by the duodenum and jejunum. It inhibits gastric acid, gastrin, and pepsin secretion; reduces intestinal motility; and increases insulin secretion in the presence of hyperglycemia. *Vasoactive intestinal polypeptide* (VIP), present throughout the gut and in nerve fibers, is a potent vasodilator and aids in the relaxation of smooth muscle. It has a large number of physiologic actions, some of which are shared with secretin and GIP and are described in greater detail later. *Somatostatin* is secreted to inhibit most gastrointestinal secretory and motor functions, thus preventing excessive reactions. *Neurotensin, substance P, VIP, enkephalins,* and *endorphins,* present in both the gut and CNS, act also as neurotransmitters to control the digestive process.

Four of the many hormonal polypeptides and peptides are recognized as legitimate hormones—namely, gastrin, cholecystokinin, secretin, and gastric inhibitory peptide. Others are considered candidate hormones, and their exact role is not completely understood. In addition to the hormones mentioned above, other hormones are released in response to the presence of nutrients in the gastrointestinal tract; however, the detailed physiological functions of some of these are currently not well defined. Digestion, absorption, and storage functions are both stimulated and inhibited by different hormones, creating an intricate hormonal control system that provides for secretion of bile acids, bicarbonate, and numerous enzymes for the digestion of food and that regulates the action of intestinal hormones. Secretin and VIP, for example, inhibit gastrin release and decrease the secretion of HCl and pepsinogen; CCK binds to gastrin receptors and thus also decreases HCl secretion. Somatostatin inhibits gastrin, secretin, CCK, and other hormones.

The *pancreas* lies across the posterior wall of the abdomen. The head is located in the duodenal curve (loop) and the body and tail are directed toward the left, extending to the spleen (Figure 14-2). Pancreatic digestive enzymes, in a bicarbonate-rich juice, enter the duodenum through the ampulla of Vater and the sphincter of Oddi and mix with the food bolus that is moved through the duodenum into the jejunum and ileum (= small intestine) and finally into the cecum, colon, and rectum (= large intestine). During the passage through the small intestine, carbohydrates are broken down by amylase and oligo- and disaccharidases into monosaccharides, which are then actively absorbed into the bloodstream. Protein is further degraded in the duodenum by trypsin, chymotrypsin, and carboxypeptidase from the pancreas and aminopeptidases from the small intestine. The resulting dipeptides and amino acids are absorbed in the jejunum and ileum by specialized absorptive mechanisms in the mucosal surface. Dietary fats are emulsified in the duodenum by the action of bile; are hydrolyzed by lipase (aided by colipase) to individual fatty acids, monoacylglycerols (monoglycerides), and glycerol; and are then absorbed in the remainder of the small intestine. Most nutrients, including vitamins and minerals, have been

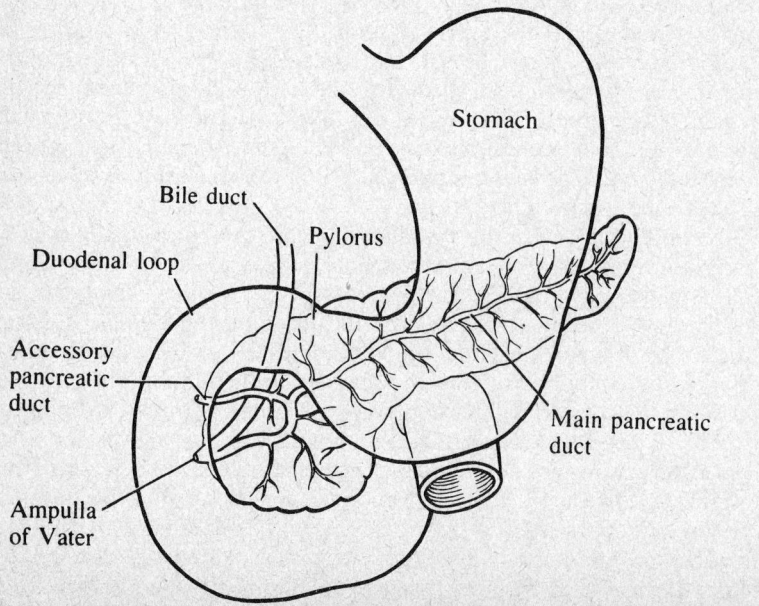

Figure 14-2. Cross section through the pancreas.

absorbed by the time the food passes from the jejunum and ileum into the large bowel. In the large intestine, water is actively absorbed, electrolyte balance is regulated, and bacterial actions take place. These processes result in the formation of feces.

GASTROINTESTINAL HORMONES

The gastrointestinal tract is both a major endocrine organ and a major target for many hormones, released locally and at a distance. GI hormones are released by endocrine cells found throughout the gut mucosa and are so numerous that the gut is recognized as the largest endocrine organ in the body. Collectively, the GI hormones influence motility, secretion, digestion, and absorption in the gut. They regulate bile flow as well as secretion of pancreatic hormones and affect tonicity of vascular walls, blood pressure, and cardiac output. Many of these hormones are present both in the gut and in the central nervous system, making them important in the neuroendocrine control of the gut. In addition, they may have a role as neurotransmitters when released from nerve synapses. The following is a description of the major gastrointestinal hormones as well as some of the candidate hormones. Based on their structural similarity, many of the GI hormones fall into one of two families: the gastrin-cholecystokinin family composed of gastrin and cholecystokinin; and the secretin-glucagon family composed of secretin, glucagon, gastric inhibitory peptide, and vasoactive intestinal peptide. Some of the peptides may not belong to either of the classes, e.g., somatostatin. A listing of gastrointestinal hormones is given in Table 14-1. (See also Figure 14-3.)

GASTRIN

Chemistry and Physiology

Three molecular forms of gastrin that stimulate gastric acid secretion are known to exist in blood and tissues: *big gastrin*, a linear polypeptide of 34 amino acids designated as G-34; *little gastrin* or G-17; and *mini gastrin* or G-14. Each of these polypeptides circulates in nonsulfated (G-17 I) or sulfated (G-17 II) forms as determined by the presence or absence of —SO_3H on the tyrosine residue in position 12 of G-17 (Figure 14-3). This sulfate group does not appear to affect the biologic activity of the peptides.

Other *immunoreactive* forms of gastrin have been found that have unknown structures and little or no biological activities. The largest, called *big-big gastrin*, was identified as an additional component in serum that has been passed through columns of Sephadex G50. Originally thought to be a pre-pro form of gastrin, big-big gastrin is now believed to be an artifact that occurs on gel filtration as a result of nonspecific binding of gastrin peptides to plasma proteins, a phenomenon observed frequently with other peptide hormones. Another form of gastrin, a fragment of G-17 that is possibly the amino-terminal tridecapeptide (G-13), has been identified in tumor extracts and sera from patients with gastrin-secreting tumors.

The smallest peptide sequence of gastrin possessing biological activity is the carboxy-terminal tetrapeptide (G-4, tetrin). However, this tetrapeptide is only one sixth to one tenth as potent as G-17 on a molar basis. Derivatives of gastrin, containing the same terminal tetrapeptide residue, are physiologically active, but the potency of each derivative differs. Synthetic *pentagastrin*, used for maximal stimulation of HCl secretion in gastric function testing, is an example of such a derivative (Figure 14-3). Any deamidization of the tetrapeptide portion or amino acid substitutions either reduces or abolishes physiologic activity.

Gastrin is produced and stored mainly by special endocrine cells (G cells) of the antral mucosa of the stomach, and to a lesser extent by G cells of the proximal duodenum and delta cells of the pancreatic islets. After secretion, gastrin is transported by the blood through the liver to the parietal cells of the fundus of the stomach, where it stimulates the secretion of gastric acid. Gastrin also stimulates secretion of gastric pepsinogen and intrinsic factor by the gastric mucosa, release of secretin by the small intestinal mucosa, and secretion of pancreatic HCO_3^- and enzymes as well as hepatic bile; it also increases gastric and intestinal motility, mucosal growth, and blood flow to the stomach. Gastrin is secreted in response to antral distention, meals, and the presence of partially digested protein products (peptides and polypeptides) in the stomach. Any free amino acids present also stimulate gastrin secretion; of those, glycine, tryptophan, and phenylalanine are the most potent stimulators. Carbohydrates and fats have little

Table 14-1. CHARACTERISTICS OF PRINCIPAL GASTROINTESTINAL HORMONES

Hormone	Molecular Weight	Number of Amino Acids	Tissues (Principal Tissue[s] are listed first)	Half-Life in Blood	Serum Reference Range	Principal Actions
Little gastrin	2098 (I) 2178 (II)	17	Stomach antrum; also vagal fibers, duodenum	~5 min	<100 pg/mL*	Stimulates secretion of gastric acid, pepsinogen, intrinsic factor, secretin, pancreatic enzymes and HCO_3^-, and hepatic bile; increases gastric and intestinal motility and mucosal growth
Big gastrin	3839 (I) 3919 (II)	34	Duodenum, also stomach antrum	~42 min	<100 pg/mL*	
Secretin	3055	27	Duodenum	~4 min	0–500 pg/mL	Increases pancreatic secretion of HCO_3^-, enzymes, and insulin; promotes pancreatic growth; increases HCO_3^- and water secretion from liver and Brunner's glands; stimulates gallbladder contractions; reduces gastric and duodenal motility; inhibits gastrin release and gastric acid secretion
Cholecystokinin-pancreozymin (CCK-PZ)	3884	33	Duodenum and jejunum; also CNS	~3 min	5–800 pg/mL	Regulates gallbladder contraction; increases motility of small intestine; stimulates secretion of pancreatic enzymes, insulin, glucagon, pancreatic polypeptide, and HCO_3^-; stimulates pancreatic growth and secretion from Brunner's glands; slightly stimulates gastric HCl and pepsinogen
Vasoactive intestinal polypeptide (VIP)	3326	28	Nervous system; also all areas of GI tract	~1 min	0–100 pg/mL; about 10-fold higher in CSF	Causes relaxation of smooth muscles of circulatory system, gut, and genitourinary system; increases water and electrolyte secretion from pancreas and gut; releases hormones from pancreas, gut, and hypothalamus; stimulates lipolysis, glycolysis, and bile flow; inhibits gastrin and gastric acid secretion
Gastric inhibitory polypeptide (GIP)	5104	43	Duodenum and jejunum	~15 min	15–100 pmol/L	Stimulates insulin release in hyperglycemia; inhibits gastric acid, pepsin, and gastric secretion; reduces gastric and intestinal motility; increases fluid and electrolyte secretion from small intestine

*This value represents gastrin immunoreactivity as measured in "G-17 equivalents" because gastrin assays are standardized by using synthetic human G-17. Other forms of gastrin (e.g., G-34, G-14) have variable cross-reactivities with G-17 antibodies used in these assays.

Gastrin-Cholecystokinin Family			Secretin-Glucagon Family			
Gastrin-17	Synthetic Pentagastrin	Cholecystokinin	Glucagon	Secretin	GIP	VIP
		Lys	*His*	*His*	Tyr	*His*
		Ala	*Ser*	*Ser*	Ala	*Ser*
		Pro	*Gln*	*Asp*	Glu	*Asp*
		Ser	*Gly*	*Gly*	*Gly*	Ala
		Gly	*Thr*	*Thr*	*Thr*	Val
		Arg	*Phe*	*Phe*	*Phe*	*Phe*
		Val	*Thr*	*Thr*	Ile	*Thr*
		Ser	*Ser*	*Ser*	*Ser*	Asp
		Met	*Asp*	Glu	*Asp*	Asn
		Ile	*Tyr*	Leu	*Tyr*	*Tyr*
		Lys	*Ser*	*Ser*	*Ser*	Thr
		Ala	Lys	*Arg*	Ile	*Arg*
		Leu	*Tyr*	*Leu*	Ala	*Leu*
		Glu	Leu	*Arg*	Met	*Arg*
		Ser	*Asp*	*Asp*	*Asp*	Lys
		Leu	*Ser*	*Ser*	Lys	Gln
Glu		Asp	*Arg*	Ala	Ile	*Met*
Gly		Pro	*Arg*	*Arg*	*Arg*	Ala
Pro		Ser	Ala	Leu	Gln	Val
Tyr		His	*Gln*	*Gln*	*Gln*	Lys
Met		Arg	*Asp*	Arg	*Asp*	Lys
Glu		Ile	*Phe*	Leu	*Phe*	Tyr
Glu		Ser	*Val*	Leu	*Val*	*Leu*
Glu		Asp	*Gln*	*Gln*	Asn	*Asn*
Glu		Arg	*Trp*	Gly	*Trp*	Ser
Glu		Asp	*Leu*	*Leu*	*Leu*	Ile
Ala	N-t-butyloxy-	Tyr-SO₃H	Met	Val	Ala	Asn
Tyr-SO₃H	carbonyl	Met	Asn	NH₂*	Ala	NH₂*
Gly	β-Ala	*Gly*	Thr		Gln	
Trp	*Trp*	*Trp*			Gln	
Met	*Met*	*Met*			Lys	
Asp	*Asp*	*Asp*			Gly	
Phe	*Phe*	*Phe*			Lys	
NH₂*	NH₂*	NH₂*			Lys	
					Ser	
					Asp	
					Trp	
					Lys	
					His	
					Asn	
					Ile	
					Thr	
					Gln	

*Indicates amidation of C-terminal amino acid residue, which occurs with some gut hormones.

Figure 14-3. Amino acid sequences of principal gastrointestinal hormones. Italicized amino acids indicate those occurring in the same positions in different hormones.

effect on gastrin release. Other stimuli of gastrin include alcohol, caffeine, insulin-induced hypoglycemia (see p. 1458), ingestion or intravenous infusion of calcium (see p. 1440), and vagal stimulation initiated by smelling, tasting, chewing, and swallowing of food.

Maximal secretion of gastrin from the fundus occurs at an antral pH of 5–7. This secretion is reduced by ~80% at pH 2.5, and maximal suppression occurs at pH 1.0. Gastrin secretion by antral G cells is inhibited by the direct action of acid on the G cells. This negative feedback regulation appears to be a safeguard against overacidification by any and all stimulants.

The principal circulating form of gastrin is G-34 in healthy individuals and in patients with hypergastrinemia. Trypsin cleaves G-34 into two fragments, one of which is identical to G-17.

On a molar basis, G-17 is six to eight times more potent than G-34 as a stimulant for gastric acid secretion. In the fasting state, the ratio of G-34 to G-17 is about 2:1. After meals, the concentration of G-34 doubles, but that of G-17 increases four times, so that the ratio of these two forms in the circulation becomes 1:1. The half-lives of endogenous human G-17 and G-34 in the circulation are ~5 and 42 min, respectively; this difference probably accounts for the higher concentration of G-34 in peripheral blood of individuals in the fasting state. Removal of circulating G-17 most likely occurs at multiple sites, possibly in all capillary beds, including the kidneys and small intestine. The liver plays a minor role in the metabolism of G-17 and G-34, although fragments with eight or fewer amino acid residues are more than 90% inactivated by the liver.

G-17 is secreted predominantly by the antral mucosa, while G-34 appears to be derived from the duodenum; after a subtotal gastrectomy, G-17 is practically absent from the peripheral blood and does not rise at all following a meal. This decrease in gastrin secretion explains the rationale for subtotal gastrectomy as treatment for peptic ulcer disease.

G-14 represents ~5% of the total gastrin concentration in both fasting and postprandial states. The half-life of G-14 is similar to that of G-17.

Clinical Significance

Zollinger-Ellison syndrome. Knowledge of blood gastrin levels is helpful in the diagnosis of gastrinomas (duodenal or pancreatic endocrine tumors) that produce and secrete large quantities of gastrin. In 1954, Zollinger and Ellison described a syndrome consisting of fulminant peptic ulcers, massive gastric hypersecretion, and non-β-islet cell tumors of the pancreas. Subsequent studies have documented that hypergastrinemia, diarrhea, steatorrhea, and other endocrinopathies are also frequent characteristics of this syndrome, which subsequently has been termed *Zollinger-Ellison (Z-E) syndrome.* This syndrome has been found in patients ranging in age from 7–90, although most cases occur in patients between 30 and 50. It is more common in men (60% of cases) than in women. In Z-E patients, fasting gastrin levels are usually markedly elevated, ranging from 2–2000 times the normal level. Gastrin concentrations > 1000 pg/mL with gastric acid hypersecretion are virtually diagnostic of gastrinoma. However, there is no correlation between the severity of the symptoms and the degree of elevation of the circulating gastrin concentration. Unfortunately, there is substantial overlap in gastrin levels between patients with Z-E syndrome and patients with common peptic ulceration and a variety of other diseases, which will be covered later. Some patients with Z-E syndrome have gastrin levels that are only slightly elevated, and additional testing must be considered.

Management of the Z-E patient requires surgical removal of the tumor, so it is necessary to distinguish hypergastrinemia caused by gastrinoma from that caused by other conditions. This differential diagnosis is aided by the gastrin response to three provocative tests: secretin infusion, calcium infusion, and a standard meal.

The *secretin challenge* is the most useful provocative test for Z-E syndrome. A bolus dose of pure porcine secretin, usually 2–3 U/kg, is injected intravenously over a 30-s interval. Serum gastrin is measured before the injection and at 5-min intervals for 30 min following the injection. A positive test, consistent with the diagnosis of gastrinoma, is indicated by an increase in gastrin concentration of at least 100 pg/mL over the basal level. In normal individuals and in patients with ordinary peptic ulceration, achlorhydria, or isolated retained antrum, secretin causes a slight decrease in serum gastrin. The reason for this paradoxical effect of secretin on gastrinomas is unknown.

A *calcium challenge* is carried out by intravenously infusing 5 mg Ca(II) gluconate per kg of body weight over a 3-h period and collecting blood for gastrin determinations at 30-min intervals during the infusion. Z-E patients demonstrate a 2 times or greater increase in serum gastrin during the third hour and a significant increase in gastric acid output. This exceeds significantly the smaller increases shown by normal patients or patients with duodenal ulcer disease.

It has been reported[104] that secretin stimulations causing an increase of > 100 pg/mL over basal levels identify about 90% of gastrinoma patients who have fasting serum gastrins in the borderline range (100–400 pg/mL) and result in very few false positive responses. Although the gastrin responses to secretin and calcium are similar in gastrinoma patients, calcium infusion appears to be less discriminatory since a substantial proportion of nongastrinoma patients have gastrin responses > 100 pg/mL after calcium administration.

A *standard test meal* (e.g., Lundh meal—5% protein, 6% fat, 15% carbohydrate, and 74% nonnutrient fiber) produces no postprandial rise in gastrin from gastrinomas. Hyperchlorhydria resulting from gastrin hypersecretion by the tumor most likely suppresses any gastrin secretion by normal gastric antral cells.

Miscellaneous conditions. Conditions other than Z-E syndrome may cause hypergastrinemia, although the increases in gastrin concentration are usually not as pronounced as in the Z-E patient. Increased gastrin levels, a markedly exaggerated serum gastrin response to high-protein meals, peptic ulcers, and acid hypersecretion are symptoms of antral *G cell hyperplasia* or hyperfunction, a condition rarely encountered. However, in these conditions, secretin and calcium infusion do not stimulate gastrin secretion as they do in Z-E syndrome.

Other non-gastrinoma patients with hypergastrinemia are either normochlorhydric or achlorhydric. *Pernicious anemia* and *parietal cell antibody–positive chronic atrophic gastritis* are commonly associated with gastrin concentrations that overlap with those found in Z-E syndrome. In these patients, atrophic gastritis results in low gastric secretion of hydrochloric acid (and intrinsic factor) and thus in stimulation of gastrin release through the negative feedback mechanism.

Hypergastrinemia is also observed in patients with *pyloric obstruction* and is further stimulated by test meals. *Chronic renal failure* may cause increased serum gastrin release that appears to be related to the severity of the renal failure. A direct correlation has been found between G cell density and parathyroid function in patients with chronic renal failure, suggesting that secondary hyperparathyroidism may play a role in gastrin elevation.[21] Surgical *resection* or *diseases of the kidneys or small intestine* can cause hypergastrinemia, possibly because these are important sites of gastrin degradation. Fasting gastrin is also increased after *vagotomy*, perhaps because of the removal of cholinergic inhibitory fibers and the negative feedback caused by decreased gastric HCl secretion. Patients with *carcinomas* of the body of the stomach that are associated with achlorhydria or hypochlorhydria usually have increased serum gastrin, while gastrin is normal in patients with carcinoma of the antrum and with normal acid secretion. Hypergastrinemia reported in rheumatoid arthritis[86] and in cirrhosis of the liver[55] is most likely associated with acid hyposecretion.

Increased fasting serum gastrin concentrations are often associated with increasing *age*, especially in patients > 60 years of age who do not demonstrate symptomatic gastrointestinal diseases. These older patients may represent cases of unrecognized gastric mucosal atrophy or may have a reduced rate of gastric acid secretion or both.

Patients with *gastric ulcer disease* usually demonstrate elevated basal and postprandial serum gastrin levels. However, this elevation is most likely due to the lower than normal basal and peak acid output found in these patients, a reduction that results in a decreased inhibition of gastrin secretion.

Fasting serum gastrin levels of patients with *duodenal ulcer disease* do not differ from those of normal subjects, but postprandial serum gastrin responses are usually greater than normal. This increased response occurs with relatively normal as well as with high peak acid outputs. A defect in the autoregulation of gastrin release by acid inhibition may be responsible for this condition in these patients.

Increased basal gastrin concentrations may be classified as "appropriate" or "inappropriate" according to their association with decreased or increased gastric acid secretion. In patients with very low or absent acid secretion and a functionally intact gastric antrum, an increase in serum gastrin is physiologically appropriate and is expected. The increase is due to hyperplasia of antral G cells as observed in *atrophic gastritis, pernicious anemia*, previous *vagotomy*, and *renal failure*. Inappropriate hypergastrinemia may be caused by *gastrinoma, isolated retained antrum* after gastric surgery, or primary *gastrin cell hyperfunction*.

Method for the Determination of Gastrin in Serum

Specimen. Draw a venous blood sample, without additives, after the patient has fasted 12 h or more to avoid postprandial increase in gastrin levels. Plasma is unsuited for analysis.

Gastrin is unstable in serum, even at refrigerator temperatures. Serum specimens and controls may lose up to 50% of their immunoreactivity during a 48-h period at 4 °C, due mainly to the action of proteolytic enzymes in serum. Even when stored at −8 to −20 °C in a self-defrosting freezer, specimens will lose significant immunoreactivity over a period of weeks. For long-term storage, specimens should be kept at −70 °C in a freezer without a self-defrost cycle. Specimens should be centrifuged promptly and then divided immediately after separation from the clot and kept frozen prior to assay. Should a second analysis be required, the duplicate specimen can then be assayed with minimal loss of immunoreactivity. Specimens must be assayed immediately after thawing; repeated freezing and thawing must be avoided.

Principle. An aliquot of serum or standard is incubated with gastrin antiserum and ^{125}I-labeled gastrin in polystyrene tubes. During the incubation period both labeled and unlabeled gastrin compete for a constant and limited number of antibody binding sites in the reaction mixture. The bound and unbound fractions are then separated by addition of a second antibody, which combines with the gastrin-specific antibody to form an insoluble complex. After centrifugation, the radioactivity in either the precipitate or the supernatant is counted, a standard curve is prepared from the counts obtained with the gastrin standards, and specimen results are obtained from the standard curve by interpolation.

Several commercial kit procedures are available for the measurement of gastrin; a sample size of either 100 or 200 μL of serum is required. Most of these procedures utilize a double-antibody separation technique and can be completed within 4 h.

Standards in most available gastrin assays consist of synthetic human G-17. Since G-34 is the predominant form of gastrin in gastrinoma patients and because pure G-34 is difficult to obtain, measurement of serum gastrin in these patients should be interpreted in terms of "G-17 equivalents." If G-17 standards are used, the measured gastrin levels in these patients will be meaningful when clearly elevated, as in most cases of Z-E syndrome, or when more than one determination is performed to confirm trends, or if used as part of stimulation tests.

Reference Ranges.[101] Reference values for serum gastrin in fasting patients range up to 100 pg/mL. Values may be higher in elderly individuals; approximately 15% of people > 60 years of age have gastrin values between 100 and 800 pg/mL. Concentrations may fluctuate throughout the day, according to a circadian rhythm (lowest at 0300–0700 h, highest during daytime), or physiologically in relation to meals.

In healthy individuals, stimulation tests using calcium result in little or no increase in serum gastrin, while use of secretin produces no change or a slight decrease in gastrin levels. A protein meal stimulation should produce a rise of ∼40 pg/mL over baseline levels after 30–60 min with a gradual decline thereafter.

CHOLECYSTOKININ-PANCREOZYMIN (CCK-PZ; CCK)

Chemistry and Physiology. CCK is an intestinal hormone that activates gallbladder contraction and stimulates secretion of pancreatic enzymes. This dual action was discovered separately, and thus it was first thought that two different substances were responsible for these actions. Eventually, it was shown that CCK and PZ are the same substance. Since gallbladder contraction was the first action described for this hormone, the hormone is now usually called cholecystokinin, although CCK-PZ is also used.

CCK is a polypeptide containing 33 amino acids, of which the five C-terminal amino acids are identical to those of gastrin (Figure 14-3). A larger form of CCK with 39 amino acids (CCK-39), as well as a smaller C-terminal octapeptide (CCK-8), has been identified, indicating that CCK, like gastrin, exists in multiple molecular forms. CCK has a sulfated tyrosyl residue similar to that of gastrin II. However, unlike gastrin, this sulfated form is the only one occurring naturally and is essential for the biological activity of CCK. The carboxyl-terminal portion of CCK is necessary for physiologic activity; the C-terminal octapeptide is at least 10 times more potent than the intact hormone.

CCK is found in the I cells of the upper small intestinal mucosa, mainly in the duodenum. Mixtures of polypeptides and amino acids (especially methionine, valine, and phenylalanine) from partially digested protein stimulate CCK secretion, while pure undigested protein does not elicit such a response. Secretion is also stimulated by gastric hydrochloric acid entering the duodenum and by fatty acids with nine-carbon chains or longer, especially in the form of micelles. Therefore, circulating levels of CCK are increased following the ingestion of a mixed meal. CCK is rapidly cleared from plasma, having a $t_{1/2}$ of < 3 min. The kidney appears to be the major organ for removal of CCK from the blood. CCK secretion is completely inhibited after somatostatin infusion.

CCK regulates the contraction of the gallbladder and increases the motility of the duodenum and the small intestine. Because it possesses the same terminal amino acid tetrapeptide as gastrin, it also stimulates to a slight degree gastric HCl and pepsinogen secretion, antral motility, and pancreatic HCO_3^- secretion. Since CCK competes with gastrin for the receptor sites on the HCl-secreting cells, secretion of the less potent CCK hormone results in a decreased output of HCl and may contribute to the termination of gastric secretion after a meal. On the other hand, gastrin and CCK activity are additive in their stimulation of the pancreas, and both increase the effect of secretin on pancreatic function. In addition, CCK stimulates pancreatic growth, relaxes the sphincter of Oddi, and stimulates secretions from Brunner's (duodenal) glands.

CCK has also been found in the brain; however, its function in the central nervous system is not yet clear. There is some evidence that the release of CCK after eating may act as an appetite regulator to induce satiety. In this case, CCK may function not only as a hormone but also as a neurotransmitter. CCK is widely distributed throughout both the central and peripheral nervous systems.

Clinical Significance. Little information is available on plasma CCK concentrations in various disorders. Basal concentrations are increased (up to 8500 pg/mL) in patients with pancreatic exocrine insufficiency and celiac disease. CCK may also be increased in fatty food intolerance, gastric ulcer, postgastrectomy states, and irritable bowel syndrome. In patients with diabetes or duodenal ulcers CCK response to test meals is more rapid than in healthy individuals.

Methods for the Determination of CCK

No reliable and convenient methods are currently available for the measurement of CCK. In vitro bioassays using isolated rabbit gallbladder and in vivo tests for measuring the effect of CCK on pancreatic enzyme secretion have been used to study CCK physiology. Various radioimmunoassays have been developed but have encountered specific problems. These problems include (1) low immunogenicity of CCK, leading to difficulty in antiserum production, (2) nonavailability of pure human or synthetic CCK and lack of a reference preparation for use as a standard, and (3) difficulties in labeling (iodination) of CCK because of the unavailability of a free tyrosine residue. As a result, there is considerable disagreement over circulating CCK values obtained with different radioimmunoassays.

Reference Range. Reference values for fasting serum from healthy individuals range from 5–800 pg/mL and are highly dependent on the method in use.[71] Variation between methods is large and each laboratory must establish its own reference range for its specific method. Estimates of CCK release after stimulation with a test meal are also highly variable; increases of CCK up to 300 pg/mL have been reported.[78]

SECRETIN

Chemistry and Physiology. In 1902, Bayliss and Starling observed that HCl introduced into the small intestine stimulated the flow of pancreatic juice. They concluded that an unidentified chemical messenger was secreted from the duodenum into the bloodstream to act on the pancreas and called it "secretin." However, it was not until 65 years later that secretin was isolated in pure form and its structure elucidated. Secretin is a polypeptide containing 27 amino acids and has structural similarities to glucagon, vasoactive intestinal peptide (VIP), and gastric inhibitory polypeptide (GIP). The positions of 14 amino acid residues within the molecule are identical with those found in glucagon; 10 are the same as in GIP, and 7 are the same as in VIP (Figure 14-3). The intact secretin molecule is required for biological activity, and there is no minimum active fragment comparable to that of gastrin.[71,108]

Secretin is secreted by the mucosal granular S cells located in greatest concentration in the duodenum but present throughout the length of the small intestine. It is released primarily on contact of the S cells with gastric HCl; however, as pancreatic juice flows into the duodenum, it neutralizes gastric acid and thereby removes one stimulus for its own secretion. Secretin is not released until the pH is lowered to at least 4.5. Below this pH, the amount of secretin released is proportional to the amount of acid entering the duodenum. However, conditions of pH < 4.5 are normally obtained only in the first few centimeters of the duodenum, causing little or no increase in plasma secretin after a normal meal. Thus, even though secretin is released after exposure of S cells to HCl, this may not be an important physiological stimulus. On the other hand, plasma secretin levels that are too low to be measured may stimulate the pancreas in the presence of physiological concentrations of cholecystokinin, which is known to potentiate strongly the action of secretin. Undigested fat does not stimulate secretin release; however, fatty acids with chains of 10 or more carbons are weak stimulants. Alcohol appears to increase secretin release by stimulation of gastric acid secretion with subsequent lowering of duodenal pH rather than by a direct stimulatory effect. The half-life of secretin in man is ∼4 min. The kidney is the major site of its degradation (40%). The only known inhibitor of secretin release is somatostatin.

The primary physiological role of secretin appears to be the stimulation of the pancreas to secrete an increased amount of juice with a high bicarbonate content. Other actions include stimulation of bicarbonate and water secretion from the liver and from Brunner's glands; augmentation of gallbladder contraction and increased hepatic bile flow; weak stimulation of secretion of insulin (a pharmacologic effect); stimulation of PTH release and of pancreatic enzymes, as well as of pepsinogen by the chief cells of the stomach; reduction of gastric and duodenal motility; reduction of the lower esophageal sphincter pressure; and promotion of pancreatic growth. Secretin inhibits gastrin release (except in Z-E syndrome) and therefore gastric acid secretion.

Clinical Significance. There is currently little knowledge of abnormal secretion of secretin in pathological conditions. Release of secretin by a tumor has never been described; therefore clinical and pathological changes associated with excessive amounts of secretin are unknown. The only well-known condition characterized by increased secretin concentrations is gastric acid hypersecretion, such as that found in gastrinomas. However, secretin may also be increased in cases of prolonged starvation and in diabetes mellitus. Decreased release of secretin may occur in patients with celiac disease.

Methods for the Determination of Secretin

Specimen. Secretin is unstable in plasma or serum because of the presence of proteases. Fasting blood for secretin measurements should be collected into ice cold, plastic, heparinized tubes, separated immediately, and the plasma stored at −20 °C. Glass containers must be avoided because secretin binds to glass surfaces. Serum is also suitable for analysis.

Method. Radioimmunoassays that are sufficiently sensitive and specific to measure circulating levels of secretin have been described in the literature.[71] Unfortunately, there are large variations in results between methods. No secretin RIA is commercially available at this time. Secretin is an immunogenic hormone, and antisera against both natural and synthetic secretin can be produced without difficulty. Synthetic secretin is generally used for RIA standards.

Reference Ranges. Fasting levels in humans have been reported to range from undetectable to ∼500 pg/mL, depending on the measurement method used.[71] Blood concentrations rise rapidly within 3 min after stimulation by duodenal acidification, and then decline to basal levels in about 60 min.

VASOACTIVE INTESTINAL POLYPEPTIDE (VIP)

Chemistry and Physiology. VIP is a linear polypeptide consisting of 28 amino acids; it has structural similarities to secretin, GIP, and glucagon (Figure 14-3). VIP is present throughout the body and is found in highest concentrations in the nervous system and gut. Unlike other GI hormones, VIP is distributed along the whole length of the gut from esophagus to colon but is more concentrated in the jejunum, ileum, and colon. In the GI tract, special endocrine cells (H cells) and nerve fibers contain VIP; concentrations in the gut are much higher than those of secretin, and VIP is by far the most abundant of all the peptide hormones found in nerves.

Little is known about the conditions that cause VIP to be released into the circulation. There is no evidence that VIP is released during digestion. However, vagal stimulation can cause its release. Its plasma half-life is ∼1 min, and most of the hormone is inactivated by a single passage through the liver.

VIP has a large number of ill-defined physiologic actions, some of which are shared with other similar polypeptide hormones (secretin, GIP). It acts as a neurotransmitter in the central and autonomic nervous systems and causes vasodilation and relaxation of the smooth muscles of the circulatory and genitourinary systems and the gut. Other actions of VIP include increase of water and electrolyte secretion from the pancreas and gut; release of hormones from the pancreas, gut, and hypothalamus; stimulation of lipolysis, glycolysis, and bile flow; and inhibition of gastrin and gastric acid secretion. Most of the actions of VIP tend to be of short duration because of its rapid degradation.

Clinical Significance. *Increased* plasma concentrations of VIP are found in patients with *Verner-Morrison syndrome* (also called pancreatic cholera). This syndrome is characterized by watery diarrhea, hypokalemia and achlorhydria, hypotension, and cutaneous flushing (vasodilation); it is usually associated with a pancreatic tumor. Because overproduction of VIP by the tumor is responsible for these symptoms, these tumors are called *VIPomas*. However, increased amounts of VIP have been found in cases of *extrapancreatic tumors* that are also associated with diarrhea, such as medullary thyroid carcinomas and ganglioneuroblastomas. Measurement of VIP concentrations is a very useful screening test for the diagnosis of VIP-secreting tumors as a cause of intractable diarrhea and is an effective tumor marker for detecting occult metastases. Its measurement can also be used to evaluate the effect of surgery or chemotherapy in patients with VIP-secreting tumors. Other conditions unassociated with intractable diarrhea may also demonstrate increased levels of VIP; patients with *hepatic cirrhosis* have markedly elevated VIP concentrations because of failure of the liver to clear VIP from the blood. In addition, increased tissue concentrations and bowel content of VIP have been found in *Crohn's disease*.

Methods. Radioimmunoassays have been developed to measure circulating plasma concentrations in pathological conditions in which VIP is elevated; however, many RIA methods are not sensitive enough to measure VIP levels accurately in normal individuals.[71] Highly purified synthetic and porcine VIP are available and have been used as standards.

Reference Ranges. Although variations in values between measurement methods are large, reported values for normal individuals are generally < 100 pg/mL.[2,71,101] Concentrations of VIP in CSF are ~10 times higher than in plasma. Patients with VIPomas have demonstrated plasma VIP concentrations ranging from 600–9000 pg/mL.[71]

GASTRIC INHIBITORY POLYPEPTIDE (GIP)

Chemistry and Physiology. Gastric inhibitory polypeptide is a linear peptide consisting of 43 amino acids. Its N-terminal end has a close resemblance to glucagon and secretin, but the C-terminal amino acid sequence of 17 residues is not common to any other known intestinal hormone (Figure 14-3). Most of the biologic activity seems to reside between amino acids 15 and 38.[108] GIP exists in more than one molecular form.

GIP is synthesized and released by K cells located in the duodenal and jejunal mucosa. Plasma GIP is increased by oral administration of glucose, triacylglycerols, or intraduodenal infusions of solutions containing a mixture of amino acids; however, none of these compounds cause changes in GIP concentrations when given intravenously. Protein ingestion has not produced any significant elevation of GIP. In order for food components to stimulate GIP release, they must be absorbed by the intestinal mucosa.

The biological actions of GIP include (1) stimulation of insulin secretion in the presence of hyperglycemia, (2) reduction of intestinal motility with stimulation of small intestinal fluid and electrolyte secretion, and (3) in supraphysiological concentrations, inhibition of gastric acid, pepsin, and gastrin secretion. The insulinotropic action of GIP appears to be the most important of its biological actions and, as a result, this hormone has more recently been called *glucose-dependent insulinotropic peptide* (GIP) as a more accurate description of its physiological action.

Clinical Significance. Fasting baseline levels of GIP are increased in starvation, prolonged fasting, type IV hyperlipoproteinemia, renal failure, and some cases of diabetes such as untreated ketotic juvenile diabetes, where insulin levels are low. Exaggerated GIP responses to mixed test meals are observed in patients with severe adult-onset diabetes and in nondiabetic obese patients. Patients with cystic fibrosis or pancreatitis show an increased response of GIP to glucose and a lower than normal response to triglycerides. This is probably due to a lipase deficiency in the gut that results in decreased triglyceride hydrolysis and therefore decreased stimulation of GIP release. In *duodenal ulcer disease*, GIP shows an increased response to glucose, most likely due to rapid gastric emptying. Surgical procedures and other conditions associated with an accelerated rate of transfer of glucose from the stomach into the small intestine may produce an excessive rise in GIP, glucose, and insulin that leads to hypoglycemia. At this time, there are no documented cases of gastrointestinal disease resulting from the overproduction of GIP. Lowered basal GIP concentrations and decreased GIP response to glucose and triglycerides are seen in patients with untreated *celiac disease* and *malabsorption*. Currently, the diagnostic or therapeutic value of GIP has yet to be established.

Methods and Reference Ranges. GIP is measured by radioimmunoassay, and reported levels of GIP in fasting human plasma range from 15–100 pmol/L.[3] Reference ranges are dependent on the method used and should be determined in the user's laboratory.

OTHER HORMONES OF THE GUT

Somatostatin is a 14-amino acid peptide, with amino acid residues 7–10 being essential for biologic activity. It is found in the hypothalamus, antrum of the stomach, upper small intestine, and pancreas. It is one of the most potent known inhibitors of endocrine secretions. Somatostatin inhibits the release of growth hormone, thyroid-stimulating hormone, insulin, glucagon, gastrin, CCK, secretin, VIP, GIP, motilin, pancreatic polypeptide, enteroglucagon, neurotensin, substance P, and others; it also inhibits the effect of these hormones on their target tissues. This broad inhibition has many biological effects, some of which are inhibition of pepsin secretion and gastric emptying, inhibition of gallbladder contraction, secretion of bile and pancreatic enzymes, and hypotension. Therapeutically, long-acting somatostatin analogs have been used to inhibit hormone secretion and reduce clinical symptoms in patients with gastrinomas, glucagonomas, and VIPomas.

Bombesin is a 14-amino acid peptide and is present in nerves and in special endocrine cells in the gut, with highest concentrations found in the stomach, antrum, and duodenum. Bombesin stimulates release of gastrin as well as CCK and inhibits VIP secretion. In contrast to other hormones that stimulate gastrin, bombesin is not inhibited by antral acidification. Biological actions include increase in secretion of gastric acid, pancreatic enzymes, insulin, glucagon, and pancreatic polypeptide, as well as stimulation of contraction of smooth muscles of the uterus, urinary tract, bronchi, and gallbladder. Bombesin appears to function as a local releasing factor for some gastrointestinal hormones.

Motilin, a 22-amino acid peptide, is present predominantly in the duodenum and jejunum; significant amounts are also found in the pineal and pituitary glands. It is a strong stimulant for contraction of smooth muscles of the upper gastrointestinal tract, and it increases the motility of the fundus, antrum, and duodenum as well as contractions of the lower esophageal sphincter. Experimental motilin infusions in man have shown that motilin plasma concentrations of less than two times the basal concentration increase the rate of gastric emptying; however, supraphysiological concentrations produce a delay in gastric emptying. Ingestion of a mixed meal causes a small rise in motilin that peaks between 15 and 30 min, followed by a gentle decline to initial, or slightly below basal, concentrations after 2–3 h. Fat ingestion stimulates, while oral glucose inhibits, motilin release, and protein ingestion has no significant effect. Motilin is unique in that its actions are generally restricted to the fasting state. The mechanism of physiological control of motilin secretion is still unknown. Motilin concentrations in normal individuals range up to 300 pmol/L.[11,12] *Increases* in motilin concentrations have been observed in conditions involving acute diarrhea, such as *Crohn's disease, acute intestinal infection, irritable bowel syndrome, tropical sprue,* and *ulcerative colitis.*[12]

Pancreatic polypeptide (PP) consists of 36 amino acids, and the pancreas is the major source. Triglycerides, glucose, protein, or a mixed meal stimulates PP secretion; the rise in plasma PP after glucose or triglyceride ingestion is relatively small and of brief duration, whereas that following protein is marked and may last for several hours. Changes in plasma concentrations of amino acids, fat, or glucose do not directly affect PP release. Hypoglycemia caused by insulin results in a large rise in PP concentration. PP release appears to be dependent on vagal nerve stimulation.

The exact physiologic role of PP is not understood. PP has a biphasic effect; it initially increases and then inhibits the secretion of pancreatic enzymes, water, and electrolytes, thus opposing the stimulation effects of secretin and CCK. PP also increases gut motility and gastric emptying, as well as relaxation of the pyloric and ileocecal sphincters, colon, and gallbladder. PP does not affect the secretion of insulin or glucagon or the plasma glucose level.

PP levels are *increased* in a large percentage of patients with *VIPomas* (77%), *glucagonomas* (50%), *gastrinomas* (26%), and *insulinomas* (22%).[108] According to these findings, increased PP concentrations could be used as biochemical markers for pancreatic endocrine tumors. Increased basal levels of PP are found in patients with *duodenal ulcers* and in patients with *juvenile-onset diabetes* of long duration. Some patients with *pancreatic insufficiency* have increased plasma PP, while patients with chronic pancreatitis have decreased basal and meal-stimulated levels.

Reference values for individuals progressively increase from the third to seventh decades of life, possibly because of alterations in vagal tone. Mean values in the third and seventh decades, as determined by radioimmunoassay, have been reported to be 12.2 ± 1.0 (SE) and 49.3 ± 6.7 (SE) pmol/L, respectively.[101]

Glucagon-like immunoreactivity (GLI) or *enteroglucagon* consists of a group of substances that have biological actions similar to those of pancreatic glucagon but are chemically different. GLI is found in all parts of the small intestine and colon. It is released after ingestion of a mixed meal, after administration of fat or CCK, and in insulin hypoglycemia. It is thought to be the growth hormone of the gut; *low concentrations* are found after *starvation,* and *high levels* are associated with *overeating, partial gut resection, jejunoileal bypass,* and in states of *mucosal damage* associated with malabsorption (celiac disease). Its specific physiologic role has not yet been well defined.

The presence of many additional peptides in the gastrointestinal tract has been hypothesized and in some cases these have been identified. Their exact physiological functions in the gut are often speculative or unknown at this time. These substances include *urogastrone, chymodenin, antral chalone, bulbogastrone, gastrone, villikinin, vagogastrone, entero-oxyntin, coherin,* and *enterocrinin.* A variety of peptides, in addition to those described previously, have also been found in both the gut and nervous system, including *neurotensin, substance P,* and endogenous opiates consisting of *enkephalins* and *endorphins.*

ENZYMES OF THE GASTROINTESTINAL TRACT

ENZYMES IN GASTRIC CONTENT

The chief and peptic cells, the neck chief cells or mucus cells, and, to a lesser degree, the cells of the pyloric zone secrete *pepsinogen,* which is activated to pepsin at a strongly acid pH provided by gastric hydrochloric acid.

Some *lipase* is secreted by the stomach and additional lipase enters the stomach through saliva. Gastric content also contains some *amylase,* which is derived from swallowed saliva. Despite the multitude of enzymes present in gastric cells, only pepsinogen has so far been routinely measured for diagnostic purposes.

PEPSIN AND PEPSINOGEN

Pepsin and pepsinogen are general names for several hydrolases (pepsin A, B, C; EC 3.4.23.1, 2, 3) and their precursors found in the gastric mucosa of many species. Pepsins catalyze the hydrolysis of proteins to a mixture of polypeptides. The human form, pepsinogen A (also known as pepsinogen I, but not to be confused with the pepsinogen group I, described below), is present in the fundic as well as pyloric and proximal duodenal mucosa. It is a proteolytic enzyme precursor (proenzyme) that, after secretion, is converted by acid to the active enzyme, *pepsin A* (EC 3.4.23.1). Once pepsin is activated, it is capable of converting more pepsinogen to pepsin (autocatalysis). Pepsin A is an endopeptidase that preferentially splits protein peptide bonds adjacent to phenylalanine, tyrosine, tryptophan, or leucine to form smaller polypeptide molecules; it effects partial digestion of proteinaceous foods during their passage through the stomach.

Of the pepsinogen released from the gastric mucosa, about 99% is secreted into the stomach to become part of the gastric fluid. The remaining 1% diffuses into the interstitial fluid surrounding the chief cells and eventually reaches the blood. Any active pepsin that may get into the blood is rapidly inactivated at the neutral pH of blood. Pepsinogen, however, is stable and circulates in blood as such. That portion which reaches the kidneys passes through the glomeruli and is excreted in the urine. At the mildly acid pH of urine, part or all of the proenzyme, now called *uropepsinogen,* may be converted (activated) to *uropepsin.* Daily excretion is fairly constant for any one individual, although there is a diurnal variation in urine output that parallels the diurnal variation in adrenal activity. The quantity excreted is independent of urine pH, volume, and specific gravity. If the pH of the urine specimen is kept between 5.0 and 6.5, the urine enzyme is stable at room temperature for 2–3 d and in the refrigerator for two weeks. No inhibitor of the enzyme is present in the urine.

At least seven different fractions of pepsinogen in blood have been separated by electrophoresis.[94] The five fractions that migrate toward the anode most rapidly are identical immunologically and are called group I pepsinogens (PG I). They are found only in chief and mucus neck cells of the fundic mucosa. Two other fractions migrate behind the group I pepsinogens; they are immunologically similar to each other but different from the group I pepsinogens. These are called group II pepsinogens (PG II) and are found not only in the oxyntic mucosa but also in stomach, cardiac, and pyloric glands as well as in duodenal Brunner's glands. Both groups of pepsinogens are activated at acid pH and are destroyed at alkaline pH. The pH optima of various pepsin isoenzymes range from 1.8–3.5, depending on substrate concentrations, ionic strength, and pepsin species. Both groups of pepsinogens can be detected in the blood; however, only group I pepsinogens are present in urine and only group II pepsinogens are present in semen.

Pepsinogen secretion is stimulated by the vagus nerve and by some GI hormones (gastrin, secretin, CCK). Its secretion is inhibited by gastric inhibitory polypeptide (GIP), anticholinergics, histamine H_2-receptor antagonists, and vagotomy.

Serum concentrations of pepsinogen I reflect the parietal cell mass and correlate well with the maximum acid secreting capacity. *Increased* pepsinogen as well as increased activity may be observed in diseases associated with increased gastric output and related increased parietal mass; in patients with gastrinomas, such as Z-E syndrome; in 30–50% of duodenal ulcer patients;[94] and in acute gastritis. Because increased pepsinogen concentrations may be observed in any of these diseases, the clinical usefulness of its measurement is limited. *Decreased* levels are seen in diseases associated with decreased chief cell mass; with atrophic gastritis; gastric carcinoma; and in patients with myxedema, Addison's disease, and hypopituitarism. Absence of pepsinogen is observed in patients with achlorhydria, as is seen in pernicious anemia.

In general, for various clinical conditions uropepsinogen concentrations in the urine parallel those of pepsinogen in the blood. However, there is considerable overlap between the values seen in healthy persons and those observed in disease.

Measurement of Pepsinogen and Pepsin

Pepsin, as well as pepsinogen after activation to pepsin, has been measured in mucosal cells, in gastric content, in serum, and in urine. Differential determinations of pepsin and pepsinogen can be performed by measurement of the proteolytic activity at pH 5.3 before and after activation of pepsinogen by exposure of the sample to highly acid pH. Previous methods have been based on the proteolytic cleavage of various protein substrates, such as hemoglobin, casein, or pepsin-specific synthetic substrates, with subsequent determination of the products using chemical, fluorometric, or colorimetric techniques.[32,33,60]

In one approach, tyrosine liberated by proteolysis of hemoglobin is measured with the Folin-Ciocalteu reagent. In analogous procedures using edestin, serum proteins, or radioiodinated serum albumin for substrate, pepsin activity is measured in terms of unhydrolyzed protein remaining; trichloroacetic acid is used to precipitate the protein, which is then separated and assayed with biuret reagent or by measuring radioactivity. Alternatively, the amount of tyrosine liberated is determined by measuring the increase in A_{280nm} of the supernatant.

In another approach, the enzyme is quantitated by measuring the rate at which it clots milk casein under standard conditions (acetate buffer, 0.75 mol/L, pH 4.9; homogenized milk; 37 °C).[107] Pepsin possesses some rennin-like activity; i.e., it attacks soluble casein and modifies it to a form that is insoluble and that clots or precipitates. Although this procedure is probably not as precise as the hydrolysis procedures described above, it is simple to perform, and the results are adequate for most clinical purposes.

A third approach utilizes synthetic substrates of low molecular mass that contain bonds sensitive to peptic action.[83] One such substrate is N-acetyl-L-phenylalanyl-L-3,5-diiodotyrosine, which is rapidly hydrolyzed by (uro)pepsin to 3,5-diiodotyrosine; the latter can be measured colorimetrically after treating it with ninhydrin. Inasmuch as the substrate has only one peptide bond which can be hydrolyzed, activity can be expressed in U/L.

Newer techniques employ radioimmunoassays that are specific for serum pepsinogen group I. In one commercially available RIA,[75] ^{125}I-labeled pepsinogen I competes with authentic pepsinogen I in the specimen for binding sites on rabbit antibodies specific for human pepsinogen I. After a reaction period, the antibody-bound pepsinogen I is precipitated with a second antibody specific for the rabbit antibody. The bound radioactivity in the precipitate is measured with a gamma scintillation counter, and concentrations of pepsinogen I in the specimen are calculated from a standard curve. Reference ranges depend on the method, and ranges of 23–60 ng/mL[75] and 50–100 ng/mL[17] for pepsinogen I in adults have been reported. Clinical correlations indicate that measurement of the pepsinogen I fraction compares well with measurements of total serum pepsinogen.[75]

A simple, sensitive, and relatively fast agarose diffusion method using protein substrate linked to Remazol Brilliant Blue dye (RBB-Hide powder) has been developed for measuring pepsin activity at pH 1.8.[87] The turbid RBB-Hide powder in the agarose is hydrolyzed by pepsin during its diffusion into the agarose gel at 37°C, forming water-soluble and diffusible split products. Clear radial diffusion zones are formed with a sharp boundary against the brilliant blue surrounding; the diameters of the zones are a measure of the enzyme activity. The disadvantage of this method is that it measures all acid protease activity in the specimen and not just that of pepsin.

ENZYMES DERIVED FROM THE PANCREAS

The normal pancreas secretes a number of enzymes that pass almost entirely into the duodenum. Only a small fraction of these enzymes reach the blood where they can be measured. Pancreatic enzymes are synthesized by the acinar cells and are stored there in the form of zymogen granules. (See Figure 14-4.) Enzymes of most clinical interest are *amylase, lipase,* and a group of *proteolytic enzymes.* The relative proportions of lipase, amylase, and proteolytic enzymes in pancreatic juice change if the diet is predominantly of one type for long periods of time; e.g., proteolytic enzymes are secreted in larger amounts if there is a prolonged high-protein intake. Other enzymes present in pancreatic juice (some in the form of precursors and others, like ribonucleases, in the active form) include trypsin, chymotrypsin, carboxypeptidases A and B, phospholipase A, ribonucleases, elastase, and collagenase.

Upon stimulation by CCK and, to a lesser extent, by gastrin and secretin, these enzymes are released from the acinar cells into the acinar lumen in a fluid containing HCO_3^-. From the acinar lumen they pass through the ductules into the main pancreatic duct, which empties into the duodenum through the ampulla of Vater. The bile duct joins the pancreatic duct just proximal to or at the ampulla of Vater. (See Figure 14-2.) The pancreas may have a second or accessory duct (duct of Santorini), which empties separately or jointly with the main duct into the duodenum. In the duodenum, the pancreatic juice mixes with the food material coming from the stomach. The combination of enzymes secreted by the pancreas, at the proper pH provided by the bicarbonate, can digest virtually any food material.

In some disorders, such as pancreatitis or obstruction of the pancreatic duct due to stones or carcinoma, the flow of enzymes and bicarbonate into the duodenum is impeded (Figure 14-5). This results in a decreased secretion of pancreatic juice into the duodenum, as can be demonstrated by direct examination of the duodenal content. (See Secretin-CCK Test.) At the same time, however, an increased amount of pancreatic enzymes reaches the bloodstream (back

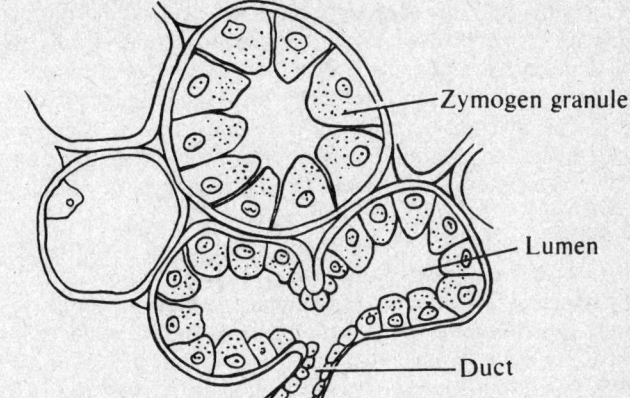

Figure 14-4. Section through pancreatic acinar cells.

flooding). The mechanism by which the enzymes enter the circulation is not exactly known, but it is thought to be due to changes in pressure in the pancreatic duct and ductules, changes in permeability of acinar cells, or disruption of the acinar-limiting membrane.

PANCREATIC ENZYMES IN SERUM

The pancreatic enzymes most commonly measured in serum are *amylase, lipase,* and *trypsin.* The determination of these enzymes and their role in the diagnosis of pancreatic diseases are discussed in more detail in Chapter 5. Here we comment briefly on amylase, lipase, and trypsin and discuss elastase in more detail.

Amylase

Amylase in serum. Measurements are requested chiefly for the diagnosis of acute pancreatitis where serum amylase activity is increased 2–12 h after an acute attack and reaches a peak 12–72 h after the attack. The test is simple, widely available, and therefore widely used. However, elevations of serum amylase are relatively nonspecific since they are observed in a variety of nonpancreatic diseases as well (Table 5-10). Determination of S- and P-isoenzymes increases specificity and aids in the differentiation of hyperamylasemia of pancreatic origin from that due to other causes. (See also Chapter 5.)

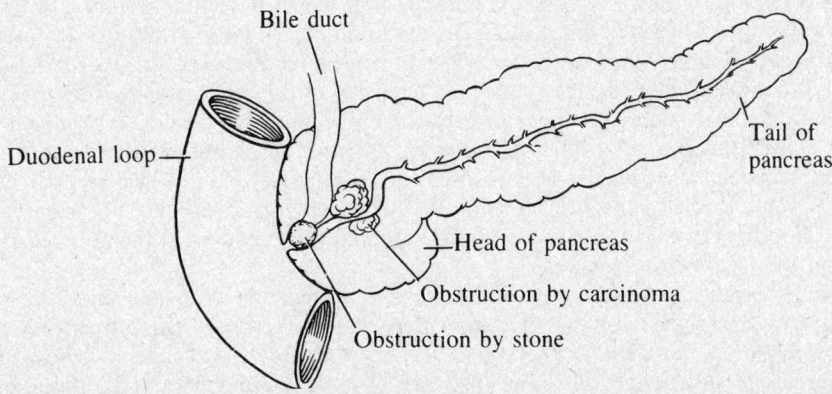

Figure 14-5. Cross section through pancreas with obstruction of the pancreatic duct by stone or carcinoma.

Amylase in urine. Amylase is filtered by the glomeruli and excreted in the urine. In pancreatitis, an increased urinary excretion rate accounts for the earlier, more frequent, and more prolonged elevation of the enzyme in urine, as compared with blood. Increased clearance rate is not a specific response to pancreatitis but instead is due to competition for absorption of low-molecular-weight proteins by the renal tubules. Increases in amylase clearance and in the ratio of amylase to creatinine clearance are observed in a number of nonpancreatic disorders. (See p. 727.) As a result, their measurements have been discouraged.

Lipase

Lipase in serum. Serum assays for lipase are used chiefly for diagnosis of acute pancreatitis, in which levels change in a manner similar to those of amylase. The elevations are, however, more pronounced, more prolonged, and more specific for the disease. For diagnostic purposes, lipase and amylase evaluations complement one another; one study reports clinical sensitivity of 98% for the diagnosis of acute pancreatitis when the two are employed together.[44] (See p. 736.)

Lipase in urine. Lipase in urine is undetectable with most assay methodologies. Although a recent sensitive immunochemical technique[68] can detect traces in urine, urine assays for lipase have not found clinical application.

Trypsin

Use of recently developed immunochemical techniques for the determination of trypsin, trypsinogen, and the trypsin-α_1-antitrypsin complex implies that elevation of serum trypsin is a highly sensitive and relatively specific indicator of pancreatic disease.[30] Very high levels in acute pancreatitis contrast sharply with low or normal levels in chronic pancreatitis (Figure 14-6) and with essentially normal levels in hepatic jaundice. Levels in carcinoma of the pancreas depend on the degree of obstruction involved and the amount of remaining acinar tissue. Since trypsin is excreted by the kidney, levels are elevated in renal disease. Recently, immunoreactive trypsin determinations have been suggested as screening tests for cystic fibrosis in newborns.[22] A more extensive discussion of clinical applications of serum trypsin assay will be found in Chapter 5 (pp. 742–744).

Chymotrypsin

Although chymotrypsin has been determined in the serum of patients with various forms of pancreatic disease, its application for diagnostic use is limited. (See Chapter 5, pp. 745–746.) Direct and indirect chymotrypsin determination on duodenal contents and feces, however, have proved to be of diagnostic value. (See also Trypsin and Chymotrypsin in Stool.)

Elastase

Elastase (EC 3.4.21.11) is an endopeptidase that is found in pancreatic juice and, to a small extent, in serum. Unlike other endopeptidases (trypsin, chymotrypsin, and pepsin) elastase rapidly hydrolyzes elastin, the yellow scleroprotein that is the basic ingredient of yellow elastic connective tissue. Two types of elastase are known to exist: elastase 1, which is anionic and has a molecular weight of 30 000; and elastase 2, which is cationic and has a molecular weight of 25 000.

In the past, assay of this enzyme, due to methodological difficulties and a limited understanding of its clinical utility, has rarely been used in clinical laboratories. Recent reports, however, point out that elastase is increased in acute and relapsing chronic pancreatitis to a greater degree than is serum amylase activity.[91] (See Figure 14-7.) Elevations also persist for a longer time and are said to reflect the clinical course better than amylase activities. Furthermore, no or only minor elevations in elastase are observed in cases of hyperamylasemia due to nonpancreatic origin. Thus there is some evidence that this enzyme is considerably more specific for pancreatitis than is amylase. Elevations of elastase have also been observed in carcinoma of the pancreas, especially carcinoma of the head of the pancreas.

Procedure for the Determination of Elastase. Elastase has been measured by enzymatic determination of elastolytic activity on elastin derivatives. However, these measurements have been hampered by the instability of elastase activity due to the possible presence of elastase inhibitors in serum and pancreatic juice. The recent development of a radioimmunoassay for elastase 1 has generated new clinical interest in this enzyme and avoids the problems associated with measurement of enzymatic activity.[91] The patient specimen is combined with ^{125}I-labeled

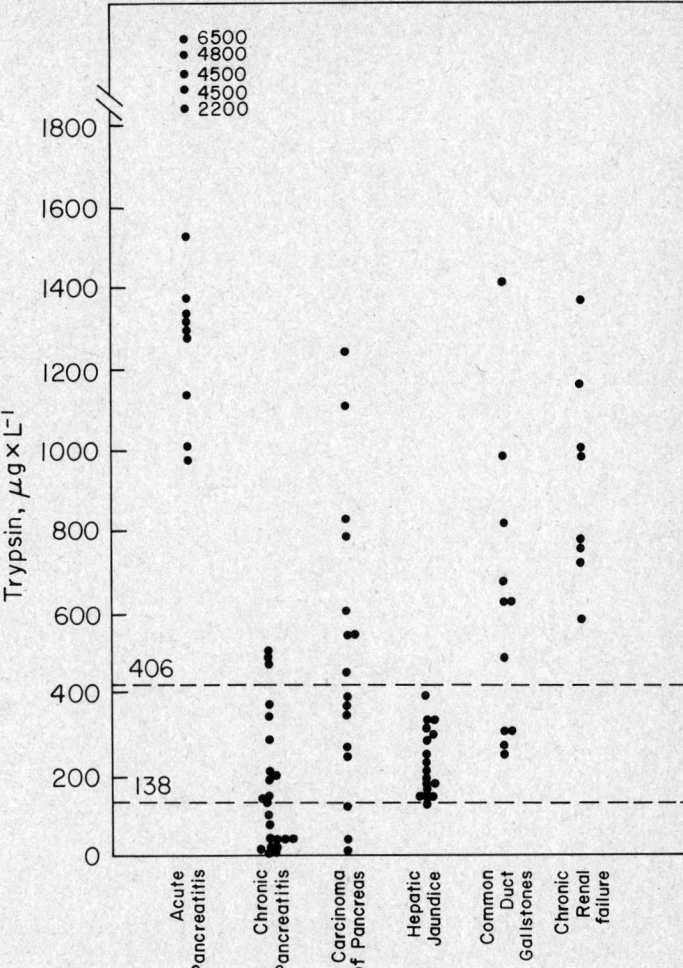

Figure 14-6. Serum immunoreactive trypsin concentrations in various groups of patients. Dotted lines indicate reference ranges. (Reproduced with permission from Elias, E., Wood, T., and Redshaw, M.: Diagnostic importance of changes in circulating concentrations of immunoreactive trypsin. Lancet, 2(1): 66–68, 1977.)

elastase 1 and elastase 1 antibody and is incubated for 3 h at 37 °C. After the incubation period, a second antibody specific for the elastase 1 antibody is added to precipitate the antibody-bound elastase. After centrifugation and decanting, the radioactivity in the precipitate is counted using a gamma scintillation counter, and specimen concentrations of elastase 1 are calculated from a standard curve. This assay has a between-run coefficient of variation of 6.4% and minimal cross-reactivity with elastase 2 (0.5%). The *reference range* is 80–320 ng/dL, with no difference between males and females.

Enzymes in the Intestinal Mucosa

Although the intestinal mucosa contains a wide variety of enzymes, most of these are not measured for routine diagnostic purposes. The importance of disaccharidases for the absorption of dissacharides will be discussed in a later section of this chapter.

GASTRIC FUNCTION TESTS

The functions of the stomach include accepting, mixing, storing, and discharging food into the duodenum as well as secreting enzymes, intrinsic factor, and especially hydrochloric acid. This has been discussed in more detail in the preceding sections. The various motor and secretory

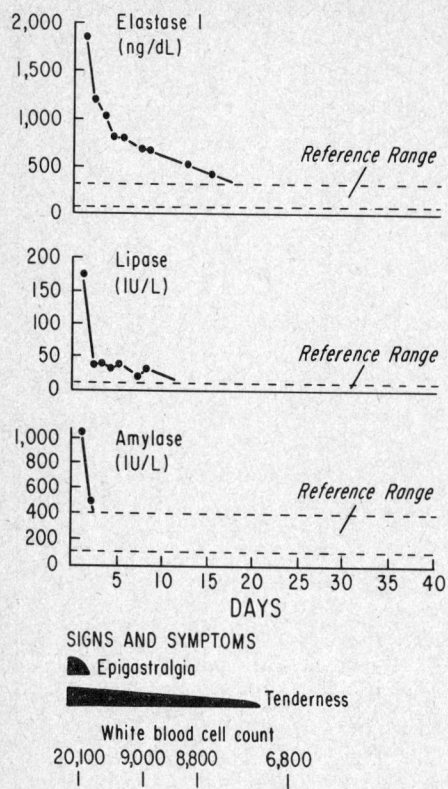

Figure 14-7. Elastase, lipase, and amylase values in patients with chronic relapsing pancreatitis. (Reproduced with permission from Satake, K.: Pancreatic disease and the case for serum elastase 1 RIA. Diagn. Med., Sept./Oct., 1983, pp. 47–55.)

activities of the stomach can be evaluated by a number of clinical laboratory tests, among which are the analysis of gastric residue (the content of the stomach after fasting), the determinations of the secretion rate in the basal state and after stimulation with appropriate stimuli, and the determination of intrinsic factor and pepsinogen. These tests, however, are more helpful in evaluating specific functions of the stomach than in providing specific diagnostic data.[15,16] Detection of compounds not normally seen in gastric contents, such as blood or lactic acid, is also of clinical significance.

Gastric Residue and Its Major Constituents

Gastric residue is defined as the content of the stomach after a fast of ∼12 h. The specimen is obtained by aspiration through a gastric tube. The *normal total volume* is between 20 and 100 mL, but usually below 50 mL; volumes above 100 mL may be considered abnormal. Among the causes for *increased volume* are delayed emptying, as in pyloric obstructions; increase in gastric secretion, as in duodenal ulcer or Zollinger-Ellison syndrome; and admixture of material regurgitated from the duodenum. In this last case, the gastric residue generally contains bile, which can be confirmed by chemical tests for bilirubin (e.g., Ictotest). The *consistency* of gastric residue is rather fluid but may be viscous in the presence of excessive amounts of mucus. The normal *odor* of gastric juice is sharply sour; a foul-smelling gastric juice generally indicates putrefaction or fermentation. The residue is usually colorless, but in the presence of regurgitated bile its color may be slightly yellow or green. This is the case in ∼25% of normal individuals and in the majority of patients after partial gastrectomy or gastroenterostomy. A red or brown color in the gastric residue is usually due to blood, which can readily be confirmed by chemical tests, such as the Gastroccult test (SmithKline Diagnostics, Inc., Sunnyvale, CA 94086).[95]

Free hydrochloric acid. Hydrochloric acid is secreted by the parietal cells at a concentration of ∼155 mmol/L. The pH of parietal secretion is ∼0.9, 6.5 pH units less than in blood. As the secretion mixes with other gastric constituents such as mucus, saliva, regurgitated material, and ingested food, the final concentration of hydrogen ions decreases to 40 mmol/L and the pH increases to 1.5–3.5. It is believed that the variations in the concentration of hydrochloric acid and in the pH of gastric residue are due to changes in the proportions of hydrochloric acid relative to other stomach contents. The secretion of hydrochloric acid by the mucosa is continuous, but the volume fluctuates considerably depending upon the degree and type of stimulation. As previously discussed, the flow is regulated by a neural mechanism mediated through the vagus (*psychic or cephalic phase*) and is stimulated by gastrin.

The *concentration* of free acid in gastric residue of normal individuals varies from 0–40 mmol/L (without stimulation). Approximately 4% of young healthy individuals may have no free hydrochloric acid in the fasting stomach. This percentage increases with increase in age and is about 25% in individuals at age 60. Absence of free hydrochloric acid in gastric residue is considered abnormal only if the condition persists after maximal stimulation with pentagastrin. It follows that patients without free hydrochloric acid should always be subjected to gastric stimulation before a diagnosis of *achlorhydria* is made. *False achlorhydria* is a term used for those cases in which free hydrochloric acid is secreted, but in which the hydrogen ions are subsequently partially or fully neutralized by either saliva, food, or regurgitated materials. A determination of chloride might help avoid misinterpretation in these cases (normal: 45–155 mmol/L).

The determination of the concentration of free hydrochloric acid is generally done by titration. Measurement of the pH of the gastric residue will also give an indication of the amount of hydrogen ions present and is the preferred method according to some investigators.[69] (See Method.)

The hydrochloric acid secreted into the lumen of the stomach is derived from the blood. The mechanism of the process is still not understood but is believed to involve active transport of hydrogen and chloride ions across the membrane of the parietal cells. A $Mg(II)$-dependent ATPase sensitive to K^+ has been implicated in the process.[89] Hydrogen ion concentration, 4×10^{-5} mmol/L in plasma, is raised to 155 mmol/L in parietal secretions, a more than 1 million-fold increase. Chloride concentration is raised from 103 mmol/L to ~170 mmol/L. Rapid secretion of gastric hydrochloric acid after a meal is accompanied by a decrease in blood chloride and an increase in blood pH and HCO_3^-, the so-called *alkaline tide*.

Total acidity. The total acidity in gastric residue includes hydrogen ions occurring as (1) free HCl, (2) mucoprotein, (3) acid salts, and (4) organic acids such as lactic and butyric acid. Lactate and butyrate are normally not found in gastric juice. The total acid concentration is usually between 10 and 50 mmol/L, but may occasionally be slightly higher.

Combined acidity. The difference between free and total acidity is due to acids other than free hydrochloric acid and amounts to about 10–20 mmol/L. The old term "combined acidity" stems from the fact that some of the acid is bound to or reacted with other material, especially proteins (acids or bases with a pK of less than 1×10^{-4}). Moore[69] has recommended that the term "combined acidity" be replaced by the term "nonionized acids." There is considerable doubt whether the determination of the "combined acidity" has any clinical significance, and many investigators suggest that this test be discontinued.

Organic acids. Lactate and butyrate are formed in gastric content by bacterial action when food is retained in the stomach for long periods (> 6 h) at neutral or slightly alkaline pH. Such a condition is usually associated with carcinoma of the stomach or pyloric stenosis and may be due to either decreased secretion of HCl or neutralization of HCl by gastric content. If free hydrochloric acid is present, these organic acids are absent.

Enzymes. Gastric residue contains a number of enzymes such as pepsin, lipase, and salivary amylase. The most important of these is pepsin, which has been discussed previously.

Mucus. Mucus is produced by the surface epithelial cells and the neck chief cells of the stomach. It is present in only small amounts in normal gastric contents; however, increased amounts are sometimes found in gastric carcinoma, in gastritis, and in cases of mechanical irritation due to passage of a stomach tube. Mucus has a pH of 7.4–8.2 and is chemically composed of mucopolysaccharides and protein moieties. There is generally no clinical interest in analyzing gastric content for mucus.

Blood. In some pathologic conditions, such as carcinoma of the stomach, peptic ulcer, gastritis, or bleeding gums, blood may be present in gastric residue. Its appearance will vary with the pH of the gastric content. At the strongly acid pH in the stomach, acid hematin is formed; it has a brownish appearance, resembling coffee grounds. Fresh red blood may be due to accidental trauma from the gastric tube; however, in some instances, underlying lesions, such as ulcer or carcinoma, may be responsible for its presence.

Food. Normal gastric residue does not contain any appreciable amount of food. Excessive accumulation of food particles indicates decreased motor activity of the stomach or pyloric obstruction.

Miscellaneous materials. Gastric residue may contain mucus cells, chief cells, parietal cells, and regurgitated fluids such as pancreatic juice, bile, and duodenal secretions. It also contains the intrinsic factor (IF), which combines in some way with vitamin B_{12} and makes absorption of this vitamin possible. Lack of this factor causes vitamin B_{12} deficiency with resultant arrest in the development of red cells. It is now believed that the intrinsic factor is a mucoprotein with a terminal structure similar to that of specific blood group carbohydrates. It is secreted by the cells of the fundus and behaves electrophoretically as a β-globulin. The determination of the intrinsic factor is now feasible, but it is rarely done except indirectly by the Schilling test, which will be described in Chapter 15.

Stimuli of Gastric Secretion

Test meals, caffeine, and alcohol. In the past, stimulation of gastric secretion was performed by administration of meals such as toast with water or tea (Ewald meal). Caffeine sodium benzoate (500 mg in 200 mL of water) admitted into the stomach through a gastric tube or in the form of tablets or ethanol (3.5 mL ethanol in 50 mL water) has also been used. These stimuli have the advantage that they give essentially no undesirable side reactions; however, the degree of stimulation is relatively weak. Thus, other stimuli capable of eliciting maximum response, such as pentagastrin, are now generally preferred.

Histalog. Histalog (3-β-aminoethylpyrazol dihydrochloride, Eli Lilly & Co., Indianapolis, IN 46285) is a potent stimulus for gastric HCl secretion, with few side effects. (Acute erosive gastritis and gastric bleeding have been reported, in addition to anaphylactic shock.) Nevertheless, all patients should be questioned about any history of allergic reactions, the drug should not be administered if the systolic blood pressure is below 110 mm Hg, and the drug should always be administered by a physician. Administration of an antihistamine 30 min before the Histalog injection is recommended by some and will not affect hydrochloric acid secretion. The most commonly used dose for Histalog is 1.5 mg/kg of body weight. Maximum stimulation is obtained with 1.7–2.0 mg/kg;[57] however, such high levels may not be advisable for routine use because of the higher incidence of side reactions. The peak of hydrochloric acid secretion occurs 90–120 min after Histalog stimulation.

Gastrin and pentagastrin. Gastrin is the natural and most powerful stimulus for gastric HCl secretion (see p. 1437). Gastrin in pure form is administered by a single subcutaneous dose of 2 μg of gastrin/kg of body weight or 1 μg/kg intramuscularly. Gastrin may also be administered by continuous intravenous infusion (67 μg/h) or by a single intravenous injection (50 μg). The extent of stimulation is nearly 20% greater than that produced by the maximal dose of histamine; on a molar basis, gastrin is 500 times as potent as histamine. Although this agent is not yet being used routinely, it is conceivable that it will become the stimulus of choice in the future.

The synthetic product *pentagastrin* contains the same terminal sequence of four amino acids that is responsible for the physiological action of gastrin. This compound is also a potent stimulus for gastric HCl secretion, although a higher dose is required for maximal stimulation (6 μg subcutaneously/kg body weight). The response to the stimulus is highly reproducible, and side reactions are extremely rare and usually limited to epigastric discomfort. Pentagastrin is now commercially available (e.g., Peptavlon, Ayerst Laboratories, Division of American Home Products Corp., New York, NY 10017) and has become the *stimulus of choice.*

Insulin. Hypoglycemia in response to insulin administration stimulates the vagal medullary nuclei and causes increased gastrin release and HCl secretion. The amount of insulin administered is chosen to decrease blood glucose to <45 mg/dL; the effect is usually obtained with a dose of 20 units. (See Hollander Test, below.)

Analysis of Gastric Residue

Preparation of the Patients. The patient should be in the fasting state (no food or liquids for 12 h). The patient is not permitted to smoke the morning of or during the test and should avoid any form of exercise.

Collection of Gastric Content. Insert a gastric (Levin) tube orally, or use nasal intubation if the patient has a hyperactive gag reflex. Confirm by X-ray or fluoroscopy that the tip of the radiopaque tube is in the lowest portion of the stomach. Ten or 15 min after the patient has become calm and adjusted to the presence of the tube, position him with the trunk upright and inclined slightly to the left (to avoid outflow of gastric juice into the duodenum). Then aspirate gastric juice and place the specimen in an appropriate container. Measure the pH or place a drop of Toepfer's reagent into the gastric residue. A pH below 3.0 or a red color after the addition of Toepfer's reagent indicates the presence of free acid. A pH above 3.0 or a yellow color of Toepfer's reagent indicates absence of free acid. For some purposes (e.g., exclusion of pernicious anemia), the test may be discontinued at this point, as presence of HCl excludes this diagnosis. If no HCl is found, a gastric stimulus is administered. (See Stimuli of Gastric Secretion.)

Currently, many clinicians and scientists feel strongly that analysis of gastric residue gives inadequate information and that this test should be replaced by tests measuring the *secretion rate* of HCl. Directions for determinations of the basal secretion rate and maximum or peak acid output after stimulation are given in a later section.

Determination of Free Hydrochloric Acid in Gastric Residue

Principle. A known amount of gastric residue is titrated with NaOH, 0.10 mol/L, to a pH of 3.5 using a pH meter, or until Toepfer's reagent (0.5 g diethylaminoazobenzene/100 mL ethanol) added to gastric juice takes on a salmon color.

Procedure

1. Pipet a convenient volume of gastric juice (e.g., 5.0 mL, but not more than 10.0 mL) into a clean titration vessel. If the gastric juice contains food particles or mucus, centrifuge the sample or filter it through gauze.

2. Determine the pH of the gastric specimen with a pH meter. If the pH is above 3.5, no free acid is present. Such a specimen need not be titrated, and the patient should be administered a gastric stimulus.

3. Titrate the sample with NaOH, 0.10 mol/L, to a pH of 3.5, using a pH meter. If a pH meter is not available, add two drops of Toepfer's reagent and titrate to a salmon color.

Calculation

$$\text{Free HCl, mmol/L} = \frac{\text{mL of NaOH, 0.10 mol/L} \times 0.1 \times 1000}{\text{mL, gastric specimen titrated}}$$

where $0.1 = \text{mmol NaOH/mL titrant}$.

If 5 mL of gastric specimen is titrated, the calculation becomes

$$\text{Free HCl, mmol/L} = \text{mL of NaOH, 0.10 mol/L} \times 20$$

Remarks and Sources of Error

Titration to a pH of 3–3.5 detects essentially all free hydrochloric acid, as can be shown by titration curves established by Michaelis[66] and Lubran.[62] (Since HCl is the only strongly ionized acid in gastric content, this test is essentially a test for free HCl.) Titration beyond a pH of 3.5, as recommended by some, will overestimate the HCl concentration to varying degrees, depending on the composition of the gastric residue. On the other hand, titration to pH 3.5 may underestimate the amount of free H^+ secreted if some of these H^+ ions are bound to or have reacted with other constituents of gastric content. Thus, no fully satisfactory procedure is available to measure accurately the true total amount of free acid secreted by the gastric mucosa.

Moore[69] maintains that titration to a preselected pH is an erroneous concept since the activity coefficients for hydrogen ions in hydrochloric acid and in gastric juice are different. The author recommends pH measurements and calculation of the H^+ concentration using an apparent activity coefficient. It is doubtful whether it is possible to convert pH accurately to H^+ concentration in a material as complex as mucus-containing gastric content, and whether such an approach can yield results of greater accuracy than that of sample titration. It is also questionable whether such an additional effort is warranted in view of the wide overlap of results in various clinical conditions.

Normal gastric content is colorless and has a sour odor; its total volume should not exceed 100 mL. If these criteria are not met, the appropriate observation should be noted on the report. The appearance of the specimen may suggest performance of additional tests such as the following: a test for blood (e.g., Gastroccult test) in case of brownish gastric content; a test for bile (e.g., Ictotest) if the gastric content appears yellowish or green; a test for lactic acid if a foul smell is observed.

Intragastric electrodes, mainly glass electrodes, represent the most recent approach to pH measurement of gastric juice. Technical difficulties, such as problems in proper placement of the electrode and pH fluctuations due to swallowed air, limit this technique mainly to research work and make it impractical for routine work.

Determinations of pH on aspirated gastric juice are best done with a pH meter. Use of pH papers may introduce significant errors.

There is no agreement about whether the gastric juice should be centrifuged before performance of the test. Many investigators recommend this technique, but others feel that some H^+ may be adsorbed to the sediment and thus may not be detected. Clear gastric content with little mucus may be analyzed without centrifugation.

Determination of Total Titratable Acidity in Gastric Residue

Many investigators have recommended the determination of total acidity in gastric residue (or other gastric specimens such as basal secretions or secretions after stimulation) by titration to a pH of 7.0 using a pH meter or phenol red as indicator. Others have titrated gastric juice using phenolphthalein as indicator (pH 8.4).

It must be realized that titration to either pH 7.0 or 8.4 is *not* a measure of free hydrochloric acid and that values thus obtained are less consistent owing to the variable admixture of other constituents of gastric content.

Calculation of total acidity is done in the same way as the calculation of free hydrochloric acid shown previously.

The *reference range* for *total acidity* of gastric residue (without stimulation) is 10–60 mmol/L.

Determination of Basal Acid Output (BAO) and Peak Acid Output (PAO) of Gastric Acid[9]

Determination of the Basal Secretion Rate of Gastric HCl

A basal condition, in the context of gastric analysis, is one in which the patient is at complete rest and not exposed to any visual, auditory, or olfactory stimuli. Such a condition is maintained during sleep. It is for this reason that many clinicians prefer a 12-h nocturnal collection for the determination of the basal acid output. Such an approach has the disadvantage, however, that the patient is exposed to significant discomfort since he has to retain the tube overnight, and in addition he must sit upright and slightly turned to the left to avoid a loss of gastric content into the duodenum. In order to assure that all these conditions are met, the patient needs close supervision throughout the entire night.

A reasonably satisfactory alternative is the collection of a specimen for a shorter period of time (e.g., 30 min) after the patient has had a satisfactory night's sleep in a quiet separate room. This is the procedure most frequently followed.

In either case, the tube has to be placed and the position confirmed as outlined under Collection of Gastric Content. The entire residue is then discarded and the specimen collected as outlined in the following section.

Twelve-hour nocturnal secretion rate. On the day of the test withhold medications that influence gastric secretion (antacids, anticholinergic or adrenergic drugs, reserpine, corticosteroids) after 0800 h. Give the patient a clear-liquid evening meal; thereafter allow no food or drink until completion of the collection. Intubate, aspirate, and discard gastric residue at ~2000 h. Connect the gastric tube to a continuous-operating pump (GOMCO, low suction) fitted with a 1000-mL specimen reservoir. At the end of 12 h, disconnect tube from pump and remove tube from patient. Immediately send the specimen to the laboratory for analysis.

Procedure

Measure and record volume (mL) of specimen. Free acid in the specimen is determined by the procedure outlined on page 1455.

Calculation

$$\text{Free acid, mmol/L} = \frac{\text{mL of NaOH, 0.10 mol/L} \times 0.1 \times 1000}{\text{mL of sample titrated}}$$

$$\text{Free acid, mmol/h} = \frac{\text{mmol free acid/L} \times \text{total volume of specimen (mL)}}{1000}$$

$$\times \frac{60}{\text{collection period of specimen (min)}}$$

The basal acid output, regardless of the collection period, is generally reported as acid output/h.

Thirty-minute basal secretion rate test. Prepare the patient as outlined in the preceding section. Request the patient to expectorate all saliva during the collection period. Then collect gastric fluid for a 30-min period and place specimen into a properly labeled container fitted with a cover. Send the specimen to the laboratory and analyze for free HCl as outlined under Determination of Free HCl in Gastric Residue. Calculate the BAO/h by the formula given under Twelve-Hour Nocturnal Secretion Rate.

Peak Acid Output and Maximum Acid Output (MAO)

After completing basal specimen collection (12 h or 30 min), leave the tube in place and give the patient a gastric stimulant, e.g., pentagastrin, 6 µg/kg. Subsequently collect 6 specimens at 15-min intervals into separate, time-labeled containers. Collection of six 15-min specimens is preferred, since PAO is dependent on maximal stimulation, which in some individuals may start after 30 min and continue beyond the 1-h period.

Determine the amount of HCl in each specimen by the method described on page 1455.

Calculation

Maximum acid output. Calculate the free acid output of four 15-min post-stimulation specimens in mmol/h, using the formula given under Twelve-Hour Nocturnal Secretion Rate. Average the acid output values for all four post-stimulation specimens. This gives the MAO/h.

Peak acid output. Calculate the acid output for six 15-min post-stimulation specimens as described for the MAO/h. Select the two specimens with the highest acid output. Add these two values and divide by two. This gives the peak acid output (PAO/h).

Some investigators prefer the PAO/h instead of the MAO/h value. Blackman[9] has shown that the data for PAO and MAO have a correlation coefficient of 0.998.

BAO/PAO ratio. Additional diagnostic information may be obtained by calculating the ratio of basal acid output to peak acid output as follows: BAO/PAO × 100 = % BAO/PAO. The relationship of BAO to PAO may also be expressed as a fraction.

Interpretation and Reference Ranges[101]

Gastric residue (after 12-h fast)

Volume:	20–100 mL (generally below 50 mL)
pH:	1.5–3.5
Free acid (without stimulation):	0–40 mmol/L
Free acid (after stimulation with pentagastrin):	10–130 mmol/L

12-h gastric secretion

Volume:	150–1000 mL
pH:	1.5–3.5

Gastric acid secretion rate

1. *Basal Acid Output (BAO)*

Normal; gastric ulcer:	0–5 mmol/h
Possible duodenal ulcer:	5–15 mmol/h
Zollinger-Ellison (Z-E) syndrome:	>20 mmol/h (at times >60 mmol/h)

2. *Peak Acid Output (PAO)*

Normal; gastric ulcer:	5–20 mmol/h
High normal; duodenal ulcer; possible Z-E syndrome:	20–60 mmol/h
Z-E syndrome:	>60 mmol/h (generally not more than twice basal acid output)

3. *BAO/PAO*

	%	Fraction
Normal; gastric ulcer or gastric carcinoma:	<20	0.20
Gastric or duodenal ulcer:	20–40	0.20–0.40
Possible duodenal ulcer or Z-E syndrome:	40–60	0.40–0.60
Z-E syndrome:	>60	>0.60

Note: Data related to the acid output are not diagnostic by themselves. A high acid output is compatible with duodenal and prepyloric ulcer or Z-E syndrome. Achlorhydria is compatible with pernicious anemia. Due to wide overlap, all other intermediate values are of little diagnostic significance.

Confirmation of Complete Vagotomy (Hollander Test)[14,31,52]

Hypoglycemia of 45 mg/dL or less, produced by insulin administration (0.15–0.2 unit/kg body weight), strongly stimulates gastric HCl and pepsinogen secretion. It is believed that the hypoglycemia is sensed by the cells in the hypothalamus[23] with subsequent stimulation of the vagal medullary nuclei. Such a response to hypoglycemia is not observed in *completely* vagotomized subjects. The insulin (Hollander) test is therefore performed to confirm the completeness of the nerve division (vagotomy).

The optimal time for performance of the Hollander test is between 3 and 6 months after the surgical procedure. The test should not be performed in the first 2 weeks immediately after surgery.

Procedure

Patient preparation and placement of the gastric tube are the same as outlined under Collection of Gastric Content. A glucose solution for IV injection should be kept at hand in case of severe hypoglycemic reactions. The test is contraindicated in any patient with a condition that already predisposes him to hypoglycemia.

1. Aspirate the gastric residue completely and discard the specimen.

2. Collect a 30-min basal gastric secretion.

3. Draw blood samples for glucose determination before and 30, 60, and 90 min after insulin injection.

4. Administer insulin intravenously (0.15–0.2 unit/kg body weight) immediately after the collection of the basal gastric sample. (This must be done by a physician.) Observe the patient for any possible side reactions.

5. Collect eight 15-min post-insulin gastric specimens (total of 120 min). Place specimens in an appropriate container and send them together with the basal secretions to the laboratory for analysis.

Interpretation. Vagotomy is considered incomplete when (1) acid concentration increases >20 mmol/L (>10 mmol/L if basal concentration was not acid); or (2) total acid output is >2 mmol/h. Test results are invalid if serum glucose has not fallen below 45 mg/dL.

Detection of Lactic Acid in Gastric Contents

Lactic, butyric, and acetic acids in gastric contents are most likely the products of bacterial fermentation as a result of food stagnation, absence of free hydrochloric acid, or both. Testing for lactic acid has relatively little clinical value, but it is occasionally requested because the presence of lactic acid, together with gastric retention and hypochlorhydria, is a common finding in carcinoma of the stomach.

Principle. A portion of gastric content is extracted with ether. An aliquot of this ether extract is treated with ferric chloride, which gives a slight yellow-greenish color with low concentrations of lactic acid and an intense yellow-green color with high concentrations.

Method

1. Place 5 mL of strained or centrifuged stomach contents into a separatory funnel.

2. Add 20 mL of ether, reagent grade, and extract the lactic acid by vigorous shaking for 1 min. Permit the ether phase to separate.

3. Transfer 5 mL of the upper ether layer to another separatory funnel and add 20 mL of distilled water and 2 drops of a ferric chloride solution (10 g/dL). Mix gently.

Interpretation. A slight yellow-greenish color is observed if lactic acid is present in concentrations of more than 50 mg/dL of gastric content. Concentrations of more than 100 mg/dL give an intense yellow-greenish color. Lack of color indicates absence of lactic acid.

Tubeless Gastric Analysis

Passage of a stomach tube is an unpleasant experience for any patient and in some conditions is even contraindicated. At one time a tubeless test for measuring gastric secretory activity, Diagnex Blue, was available, but limitations of the test led the manufacturer to discontinue the product.

TESTS MEASURING THE EXOCRINE FUNCTION OF THE PANCREAS

The predominant exocrine function of the pancreas is the production and secretion of pancreatic juice that is rich in enzymes and bicarbonate. Normal pancreatic juice is colorless and odorless; it has a pH of 8.0–8.3 and a specific gravity of 1.007–1.042. The total 24-h secretion volume is 800–3000 mL.

A number of laboratory tests are available to measure exocrine functions either directly (secretin and CCK tests, trypsin and chymotrypsin in stool, and others) or indirectly (fat absorption tests, β-carotene or vitamin A absorption tests, PABA test, fecal lipids). These tests should be viewed as measurements of pancreatic function or intestinal absorption and not as tests designed to diagnose a specific disorder. When interpreting results one must realize that great overlap occurs between results observed in normal individuals and those found in patients with pancreatic disorders. The overlap is partially due to the large functional reserve of the pancreas. An estimate has been made that pancreatic insufficiency cannot clearly be demonstrated until at least 50% of the acinar cells have been destroyed. Clinical signs of pancreatic insufficiency often do not appear until loss of 90% of acinar tissue has occurred.[4]

The concentration and total output of pancreatic juice, bicarbonate, and enzymes are measured either directly on duodenal content or on pancreatic juice obtained by cannulation and aided by endoscopic retrograde cholangiopancreatography.[84] The enzyme most commonly measured is trypsin, but amylase and lipase measurements, as well as measurements of other enzymes, have been performed. In infants, amylase measurements are unsuitable, since this enzyme is synthesized in only small amounts until the age of one year. Measurement of enzymes, especially chymotrypsin, in feces is simpler and noninvasive and has therefore been performed more frequently than measurements on pancreatic or duodenal content. (See later sections.)

Measurements of the total volume of pancreatic juice, amount or concentration of bicarbonate, or rates of enzyme secretion are generally preceded by stimulation of the pancreas by one of several test meals or stimulants. Examples of those most commonly employed are listed below.

Lundh meal. The meal consists of 5% protein, 6% fat, 15% carbohydrate, and 74% non-nutrient fiber. Advantages of the Lundh meal are that it provides a physiological stimulus of the pancreas and is simple to administer. However, it provides inadequate or no stimulation in the presence of mucosal diseases such as celiac disease, where hormone release from the duodenal mucosa is impaired. Invalid results are also obtained in patients after vagotomy and subtotal gastrectomy with the Billroth II procedure.* Furthermore, results are less sensitive than those obtained with secretin-CCK stimulation. Because of these limitations, the Lundh meal is not advocated by all investigators.[4]

Secretin test. A direct measure of pancreatic function is the determination of the secretion rate and concentration of bicarbonate as well as the secretion rate of pancreatic juice in response to stimulation with secretin. These tests are based on the principle that maximum secretion of pancreatic juice and bicarbonate output are related to the functional mass of pancreatic tissue.

Stimulation is most often accomplished by administration of 1 Crick-Harper-Raper (CHR) unit of secretin/kg body weight intramuscularly or 2 CHR units of secretin/kg per hour by continuous intravenous infusion in sodium chloride, 0.15 mol/L. Secretin preparations differ greatly in their potency, which accounts in part for the significant overlap in results obtained by different investigators. The preparation for which the above doses have been defined can be obtained from the Karolinska Institute, Stockholm, Sweden, and from Pharmacia Laboratories, Piscataway, NJ; it has been found most reliable. Another preparation, made in England by Boots Pharmaceutical Co. and distributed in the United States by Adria Laboratories (Columbus, OH 43215, formerly Warren-Teed Pharmaceuticals, Inc.) is considerably less potent.

Preparation of the Patient and Collection of Specimens

The patient should be in the fasting state. A double-lumen gastroduodenal tube is introduced under fluoroscopic guidance in such a way that the proximal opening of the tube lies in the stomach and the distal opening lies in the duodenum beyond the ampulla of Vater. Constant suction is then applied and aspiration is continued until the duodenal content becomes clear and is not contaminated with gastric juice. At this time, secretin is administered and three to eight consecutive specimens are collected at 10-min intervals, usually between 30 and 80 min. The pH of each specimen is measured. A sudden increase in pH of the gastric content indicates contamination by duodenal fluid; a sudden decrease in pH of the duodenal content indicates contamination by gastric juice.

At the end of the total collection period, all duodenal specimens are pooled, the total volume is measured, and the pooled specimen is tested for blood and assayed for bicarbonate. The bicarbonate secretion rate (mmol HCO_3^-/collection period) is then calculated. Detection of occult or overt blood in any one of the specimens may be helpful in pinpointing gastrointestinal bleedings.

Interpretation. In healthy individuals the flow of pancreatic juice after stimulation is 2.0 mL/kg, with a peak bicarbonate concentration of 80–90 mmol/L during the first 80 min after stimulation. The bicarbonate secretion rate in healthy individuals is > 15 mmol/30 min and values of < 10 mmol/30 min are considered abnormal. In general, the measurement of bicarbonate output is clinically more discriminatory than the measurement of the bicarbonate concentration.

Abnormally *low concentrations* of constituents of pancreatic juice after secretin stimulations have been reported in cases of chronic pancreatitis, pancreatic cysts, cystic fibrosis, and calcification, carcinoma, and edema of the pancreas. Findings in these conditions are very similar and are of relatively little value in the differential diagnosis. The greatest value of the secretin test is probably in excluding pancreatic dysfunction, as seen in cystic fibrosis, and in differentiating steatorrhea of pancreatic origin from that caused by sprue or celiac disease.

In the presence of pancreatic obstruction by carcinoma or stone or due to drugs (morphine sulfate, methacholine chloride, or bethanechol chloride), the outflow of pancreatic juice into the duodenum is diminished and significant decreases in the volume and bicarbonate concentration of duodenal content are observed. However, serum amylase and lipase will be increased. (See Chapter 5.) Patients with substantially decreased amounts of functioning acinar tissue, e.g., patients with cystic fibrosis or chronic pancreatitis, show significantly smaller increases in serum enzyme levels.

Secretin-CCK Test

Stimulation by secretin alone allows only evaluation of the *hydrokinetic* function of the pancreas, i.e., volume and bicarbonate secretion; secretion of pancreatic enzymes is inconsistent. But stimulation with secretin followed by stimulation with CCK will also allow the assessment of the *secretion of enzymes* (ecbolic function). This combination of stimuli is most satisfactory for measuring pancreatic secretory capacity and thus for separating patients with normal from those with impaired pancreatic function. In one study involving over 2000 patients, false abnormal results were reported in only 8% of cases and false normal results in only 6% of cases.[74] Various combinations of secretin and CCK have been used by different investigators. Secretin is given in amounts between 0.25 and 1.0 CHR unit/kg per hour by continuous intravenous infusion in sodium chloride, 0.15 mol/L, followed by 1 Ivy-Hunde unit CCK/kg (Karolinska Institute, Stockholm,

---

*In the Billroth II procedure, the lower two thirds of the stomach is removed and continuity re-established by gastrojejunostomy. The upper end of the duodenum is closed off, creating a blind segment of the small bowel. This "blind loop" may give cause to the blind loop syndrome.

Sweden). For the measurements of enzymes it is important that this lower dose of secretin be given, since the high dose (as given in the simple secretin test) may inhibit the response of pancreatic enzyme secretion to CCK. On the other hand, the smaller dose may not provide the maximum stimulation for bicarbonate secretion.

In his investigations, Otte[4,74] injected 1 Ivy-Hunde unit CCK/kg plus 1 CHR unit secretin (Boots)/kg, then administered 6 g ox bile by tube into the duodenum, followed after 5 min by a second dose of secretin and CCK. Administration of bile prevents false abnormal results caused by inadequate contraction of the gallbladder. The augmented CCK-secretin administration is claimed to result in maximum stimulation and greatest discrimination. Figure 14-8 illustrates findings in control subjects and individuals with various pancreatic and other abdominal diseases. Using this technique, Otte found that the bicarbonate concentration is decreased in 76% of patients with pancreatic insufficiency. If the total bicarbonate output is also considered as a criterion, the detection rate increases to 81%. Abnormally low trypsin results were found in 88% of patients with pancreatic insufficiency. (Trypsin is the most frequently measured enzyme, since it is less sensitive to pH changes and more discriminating than are lipase and phospholipase.) Otte maintains that the greatest diagnostic sensitivity can be obtained by multivariate analysis using more than two parameters.

Regan et al.[80] used a synthetic C-terminal CCK-OP (Sincalide, E.R. Squibb and Sons, Inc.). These investigators administered 40 ng CCK-OP/kg per h plus secretin, 0.2 CHR unit/kg per h.

Cerulein has been used in place of CCK, because it is less expensive and is available in pure form. Cerulein is a decapeptide and its C-terminal octapeptide sequence is identical to the C-terminal octapeptide sequence of CCK except for a single substitution of threonine for methionine. Cerulein is present in the amphibian GI tract.

Comments

There is presently no pancreatic function test that can detect the early stages of pancreatic disease. The tests above, although among the most sensitive, are no exception. Despite some encouraging results, reports by various investigators using the Lundh meal, secretin stimulation, or a combination of secretin and cholecystokinin show considerable disagreement. This may be due to the differences in potencies of secretin and cholecystokinin as well as to use of improper stimulants as explained above, e.g., use of the Lundh meal in individuals with intestinal mucosal disease.

No reference ranges for enzymes after stimulation with either secretin or CCK are given here, since no standardized techniques have been used by the different investigators and thus no direct comparison is possible. The Pancreatic Club of Europe is presently conducting a study involving a number of medical centers in different countries. It is hoped that this effort and the recent advent of new and more reliable techniques for trypsin or immunoreactive trypsin, lipase, and amylase will stimulate the adaptation of standardized techniques.

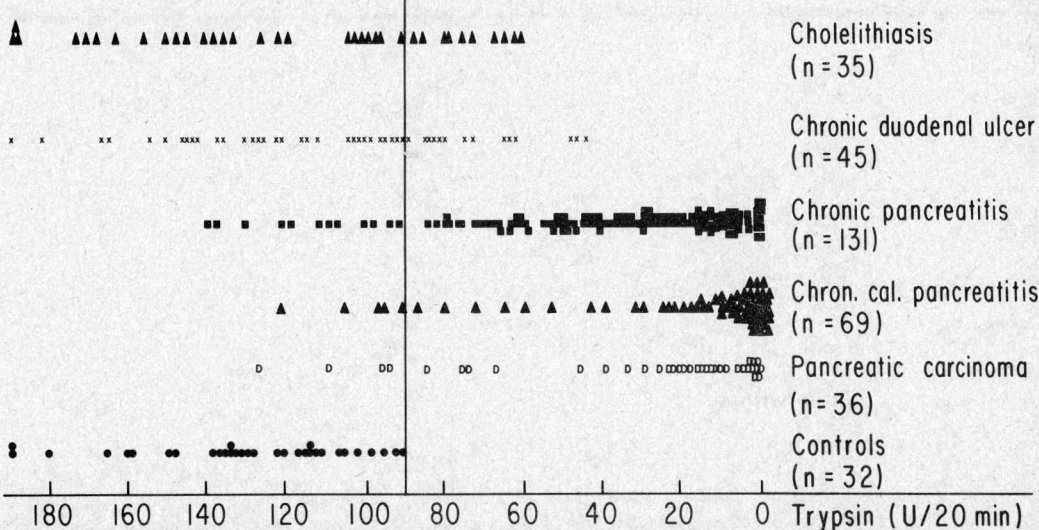

Figure 14-8. Pancreatic secretion of trypsin after secretin, CCK, and bile stimulation. (Reproduced with permission from Otte, M.: Pankreasfunktions-diagnostik. Internist, 20:331–340, 1979. Heidelberg, Springer-Verlag.)

The interested reader is referred to a series of excellent review articles by Otte,[74] Lankisch,[56] and Arvanitakis and Cooke,[4] as well as the article by Rinderknecht et al.[84]

Trypsin and Chymotrypsin in Stool

Trypsin and chymotrypsin secreted by the pancreas and present in duodenal contents mix with food material and, except for the fraction digested in the intestinal tract, are excreted in the feces. Measurement of activity of these enzymes in stool has therefore been used for diagnosis of obstruction of the pancreatic duct or diagnosis of pancreatic insufficiency, as, for example, in cystic fibrosis.

For many years, a *screening test for trypsin* has been used in which a drop of a dilution of emulsified stool is placed on an X-ray film. This procedure is described in Chapter 5. Digestion of the gelatin layer of the film by an appropriate dilution of stool suggests presence of normal amounts of tryptic activity. Although very fast and simple, the test is at best only a screening procedure. It is subject to interferences from proteolytic enzymes present in intestinal bacteria or in contaminating urine. The test is applicable only to children, since lower intestinal motility in adults results in partial autodigestion of the enzyme and falsely abnormal (negative) results.

Limitations of the X-ray film test led first to development of quantitative methods for trypsin in duodenal contents or feces and subsequently to methods for chymotrypsin in such specimens. Several synthetic low-molecular-weight substrates are now available for estimating trypsin or chymotrypsin activity. Of these two enzymes, fecal trypsin estimation has lower sensitivity, whereas chymotrypsin is significantly more stable,[29,90] has a much higher sensitivity, and is therefore a better index of pancreatic function.[90] Its output in stool correlates well with chymotrypsin secretion into duodenal contents when both are measured after stimulation with secretin-CCK. The incidence of falsely low results (falsely positive for diminished pancreatic function) is ~10% and may be due to voluminous stools (> 300 g/d) and thus less enzyme per gram of feces; or to inadequate food intake, partial gastrectomy, or sprue, which cause inadequate stimulation of pancreatic secretion; or to obstruction of the bile duct. Falsely normal results in patients with mild pancreatic insufficiency may be as high as 25%.[74]

Until now, limited diagnostic use of the measurement of stool chymotrypsin could be attributed to the cumbersome procedures available for its measurement. A recently introduced procedure is simple and fast and eliminates inconsistency of results caused by particle-bound enzyme.[48] In this procedure, sample is extracted with a mixture of salts and a detergent, lauryltrimethylammonium chloride, which fully dissociates the enzyme from particles in stool. The extract is mixed with a synthetic pentapeptide of 4-nitroaniline (4-NA); the substrate is hydrolyzed by chymotrypsin to produce free 4-NA.

$$\text{Succinyl—Ala—Ala—Pro—Phe—NH}\!-\!\!\bigcirc\!\!-\!\text{NO}_2 \xrightarrow{\textit{Chymotrypsin}}$$

$$\text{Succinyl—Ala—Ala—Pro—Phe} \;+\; \text{H}_2\text{N}\!-\!\!\bigcirc\!\!-\!\text{NO}_2$$

4-Nitroaniline

Continuous release of 4-NA is measured photometrically at 405 nm. A detailed procedure has been published[48] and is provided with the reagent kit presently available in Europe only. Stool specimens are conveniently prepared for the assay with a system commercially available in Europe from Boehringer Mannheim G.m.b.H. Diagnostica, West Germany (Catalog #718211). Activity of chymotrypsin in stool remains virtually constant at room temperature for up to 10 d.

Bz-Ty-PABA Test (PABA Test, BTP Test) for Chymotrypsin

Tests for the measurement of pancreatic enzymes in duodenal content involve stimulation of pancreatic secretion with subsequent collection and analysis of either pancreatic or duodenal juice. These are lengthy, invasive procedures that are associated with a fair amount of discomfort to the patient. Therefore, indirect tests have been sought, and one example is the PABA test in which the synthetic peptide *N*-benzoyl-L-tyrosyl-*p*-aminobenzoic acid (BTP) is administered orally together with a test meal to stimulate pancreatic secretion. BTP is specifically hydrolyzed by chymotrypsin in the duodenum to release *p*-aminobenzoic acid (PABA). PABA is subsequently absorbed in the intestinal tract and metabolized in the liver to hippurate and to PABA glucuronide and PABA acetylate. These arylamines are then excreted by the kidney into the urine (Figure 14-9). In the presence of low chymotrypsin levels, as found in pancreatic insufficiency, less peptide is hydrolyzed and therefore less chromogen is excreted in the urine. Thus, the amount of PABA (in the form of arylamines) found in the urine is an indirect measure of chymotrypsin activity in duodenal content.[26] This test measures all urinary arylamines; however, the urinary concentrations of arylamines are normally insignificant compared with PABA arylamines following administration of BTP, provided interfering substances have been discontinued prior to dosing.

The test is affected by various medications, especially antibiotics, sulfonamides, and diuretics. Foods such as prunes and cranberries contain hippurate precursors that interfere with the chemical determination of aromatic amines, and they should be discontinued for 2–3 d. Furthermore, test results are affected by the rate of intestinal absorption and the degree of renal excretion. To detect or compensate for these errors, some investigators have administered radiolabeled PABA simultaneously with the substrate.[67] Low recovery of radiolabeled PABA in the urine indicates probable decreased intestinal absorption or decreased renal excretion.

This relatively simple test does not require duodenal intubation or analysis of aspirates to obtain an estimate of chymotrypsin secretion in response to a meal. However, it is subject to the interferences just mentioned and appears to be less sensitive in detecting *slight* or *moderate* forms of exocrine pancreatic insufficiency. Various reports indicate it has a clinical sensitivity between 80 and 100% when compared with the secretin-cholecystokinin test as a reference.[56] The sensitivity of the test in the detection of chronic pancreatitis has been established as being ~86%. The specificity of the test, i.e., the number of true normal results in patients with nonpancreatic disease, has been found to be 93%. There is a high correlation between this test and the direct test for chymotrypsin in stool.

Procedure. The peptide substrate BTP is available commercially (Adria Laboratories, Inc., Columbus, OH 43215) for administration to patients. Detailed instructions are included for oral

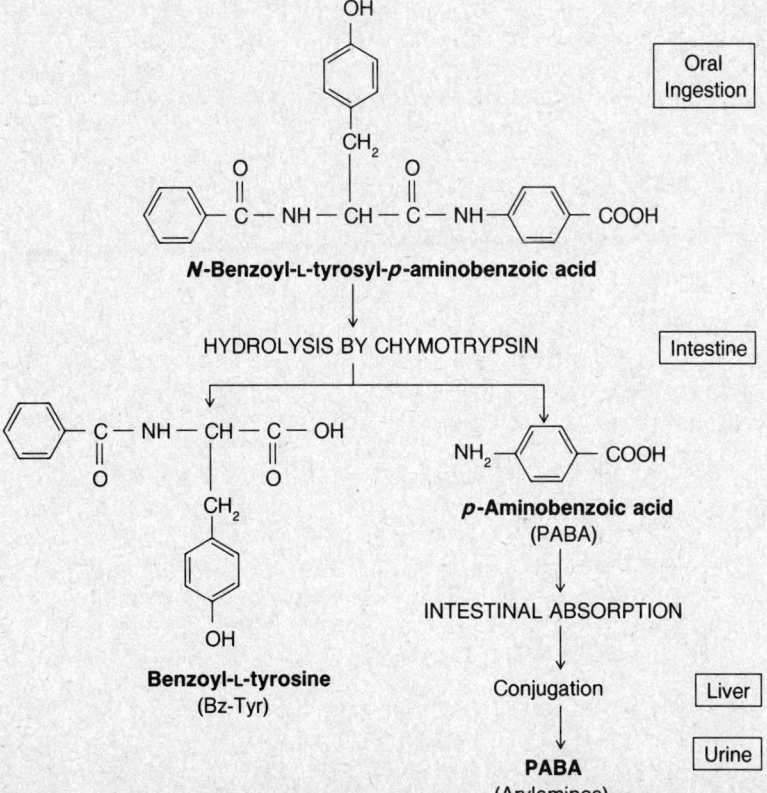

Figure 14-9. Schematic diagram of the digestion of BTP and excretion of PABA into the urine.

administration of BTP, measurement of PABA in urine, calculations of percentage of PABA excreted, and the interpretation of results. Some workers recommend stimulation of the pancreas by either the Lundh meal or an amino acid mixture.[74]

INDIRECT TESTS OF PANCREATIC FUNCTION

This group of tests is based on the fact that proper pancreatic function is essential for normal intestinal absorption of certain substances. Since the pancreas is the major source of amylase, lipase, and proteolytic enzymes, a significant decrease in pancreatic function causes decreased absorption of starch, fats, and protein and a parallel increase in excretion of these food materials in the stool. In pancreatic deficiency, microscopic examination of stool demonstrates large amounts of undigested cell nuclei and meat fibers (creatorrhea), increased fat (steatorrhea), and increased starch (amylorrhea). Demonstration of these conditions is suggestive of impaired absorption. In severe cases, the feces are generally pale, bulky, and unusually foul smelling. In steatorrhea of pancreatic origin, fat droplets usually appear on the surface of feces on standing.

Unfortunately, these and many quantitative tests have no diagnostic value in mild cases of pancreatic disease and in those cases in which the acute phase has subsided. In addition, some patients with pancreatic insufficiency may have almost normal digestion because of the functional reserve of the pancreas and the activity of microorganisms and intestinal enzymes. On the other hand, stools of patients with celiac disease or sprue may be similar to those from patients with pancreatic insufficiency because increased intestinal motility decreases the time that food remains in the intestinal tract and during which digestive enzymes can act. The greatest value of microscopic and other indirect tests of pancreatic function seems to be in the exclusion of pancreatic pathology, especially cystic fibrosis.

FECAL LIPIDS

Total fecal lipids are composed of mono-, di-, and triacylglycerols; fatty acids and fatty acid salts (soaps); phospholipids and glycolipids; and sterols and cholesteryl esters. Most fecal lipids are nonabsorbed dietary fats, but a significant amount ($\sim$2 g) is derived from intestinal bacteria and from epithelial cells that have sloughed normally from the intestinal mucosa.

Fecal lipid determinations are frequently performed as part of metabolic or fat balance studies and in the diagnosis of malabsorption due to pancreatic or intestinal disorders. The determination of fecal fat is the most definitive test for determining the presence of steatorrhea, but not for determining its cause. (See Malabsorption.)

Determination of Total Fecal Lipids

Principle

In the *gravimetric* method for the determination of total fecal lipids, a preweighed, emulsified stool specimen is acidified to decrease the ionization of fatty acids (present as free acids, or soaps). The lipids, including the less polar nonesterified fatty acids, are then extracted from the stool specimen with an organic solvent, the supernatant is evaporated, and the residue quantitated by gravimetry.

In the *titrimetric* method, a weighed stool specimen is subjected to alkaline hydrolysis to convert esterified lipids to fatty acids and their corresponding alcohol (e.g., glycerol or sterol). The mixture is subsequently acidified to facilitate the transfer of fatty acids into an organic solvent. The solvent is then evaporated and the residue dissolved in ethanol and titrated with a standard solution of sodium hydroxide to quantitate the total amount of fatty acids present, which represents about 80% of total fecal lipids.

The solvent systems described in the method below provide significantly better recoveries than does the method by van de Kamer, which is widely used for fecal lipid analysis.[46]

Specimen

The patient should be on a diet containing 50–150 g of fat per day for 3 d prior to this study and during the 72-h collection of the stool specimen. The patient should not have had castor oil as a laxative prior to the specimen collection.

Stool specimens are conveniently collected in a one gallon paint can and the container should be stored refrigerated during the collection period. Collection of 24- or 48-h specimens is not recommended, since results on such specimens are subject to more variability. Any obvious foreign matter should be removed from the specimen prior to processing. Contamination of feces with urine should be avoided.

Reagents

1. n-Heptane, analytical or ACS grade.
2. Ethyl ether, peroxide-free and glass-distilled. Protect from light, heat, and atmospheric air. Test for peroxides by adding 0.5 mL KI solution, 0.5 g/mL, to 10 mL redistilled ether and mixing for 30 s. A brown or yellow color indicates the presence of peroxides. Alternatively, a peroxide test strip may be used.
3. Ethanol, absolute. USP-NF reagent quality.
4. Ethanol, 95%. Dilute 950 mL absolute ethanol with distilled water to a final volume of 1 L.
5. Solvent system A. Mix equal volumes of n-heptane, ethyl ether, and ethanol, 95%. Prepare fresh daily.
6. Solvent system B. Mix equal volumes of n-heptane, ethyl ether, water, and ethanol, 95%. Use the upper phase for the extractions and prepare reagent fresh daily.

Procedure

Mixing of Fecal Specimen

1. Weigh specimen container (1-gal paint can) in advance of specimen collection and record the weight on the container, using an indelible marking pen.
2. When the specimen arrives in the laboratory, weigh the container with specimen and record the weight. Subtract the weight of the empty can to obtain the net weight of the fecal specimen (W).
3. Open the can in a well-vented hood while wearing disposable gloves. If the specimen is firm, add a measured amount of water (e.g., twice the wet weight of feces) to the container. The amount of water should be sufficient to allow blending of the specimen into a homogenate of uniform consistency. Use a smaller volume of water if the stool specimen is less firm. Do not add water to a watery stool. Record the volume of water added (as weight, in grams).
4. Close the can tightly with a hammer and wrap the can in a plastic garbage bag to avoid accidental spillage. Place the can on a commercial paint shaker and agitate vigorously for 10 min.
5. Open the can in a fume hood without delay. Transfer 3.0 g of the fecal emulsion into a tared 50-mL screw-capped centrifuge tube. If necessary, this aliquot may be stored frozen.

Extraction

6. Acidify the specimen by addition of four drops of concentrated HCl and vortex.
7. Add 20 mL of solvent system A to the acidified fecal homogenate. Cap, check for leaks, and vortex for 1 min.
8. Separate the phases by centrifugation for 10 min at ~3000 × g. Transfer the supernatant to a preweighed 100-mL disposable beaker.
9. Re-extract the infranatant two more times with 20-mL portions of solvent system B, adding each extract to the same beaker.
10. Evaporate the combined extracts to dryness under a gentle stream of nitrogen.

Gravimetric Measurement

11. Place the beaker with the residue into a desiccator and dry overnight.
12. Weigh the beaker and subtract the weight of the empty beaker to obtain the weight of the lipid residue (W_r).

Calculations

Calculate the total fecal lipid content as follows:

$$\text{Fecal lipid, g/specimen} = \frac{W_r}{3} \times F \times W.$$

where W_r = weight of lipid residue in grams

    3 = weight of fecal homogenate used for extraction

$$F = \text{dilution factor} = \frac{\text{feces, g} + \text{added } H_2O, \text{ g}}{\text{feces, g}}$$

W = wet weight of entire feces specimen in grams

$$\text{Lipid excretion/d} = \text{fecal lipid, g/specimen} \times \frac{24 \text{ h}}{\text{total collection time (h)}}$$

Reference Ranges

Children up to six years, < 2 g/d. Thereafter, 2–6 g/d. Steatorrhea: > 7 g/d.

Comments

It has been suggested that fractionation of total fecal lipids into neutral fat and nonesterified fatty acids might be of aid in the differential diagnosis of malabsorption due to pancreatic insufficiency from malabsorption due to other causes. Significant overlap in results as well as enzymatic and spontaneous hydrolysis of fats during the collection contributes to the unreliability of this approach.

In metabolic balance studies in infants and children, it is customary to determine the *coefficient of fat retention.* This coefficient can be obtained by expressing the amount of lipid excreted in feces over a specified period as a percentage of the amount of lipid ingested in the same period. For reliable results, a 72-h specimen collection is required. The *reference range* for formula-fed infants is $> 83\%$, for breast-fed infants $> 93\%$, and for children > 1 year $\geq 95\%$.

In order to improve the accuracy of lipid excretion measurements, it has also been suggested that the patient be given a daily oral dose of 240 mg of cuprous thiocyanate in three 80-mg doses. Cuprous thiocyanate is not absorbed and is totally excreted in feces. At the end of the collection period the copper content of the fecal specimen is determined by atomic absorption and the lipid content by a chemical method. If the stool specimen contains only a fraction of the administered dose of cuprous thiocyanate (e.g., 120 mg), the stool specimen is assumed to represent only a portion (in our example, 1/2) of the 24-h excretion period; a correction can then be applied for the 24-h lipid excretion rate.

If it is not possible to collect a 72-h fecal specimen, the amount of fat excreted may also be expressed as fat in per cent of fecal *dry weight.* Dry weight can be determined by taking a 3-g aliquot of the stool homogenate (above) into a preweighed beaker, drying the aliquot overnight at 110 °C, and reweighing the beaker.

$$\text{Fat, \% of dry weight} = \frac{W_r, \text{ g}}{\text{Weight of dried homogenate, g}} \times 100\%$$

The reference range for lipids in this case is 15–25% of dry weight of stool.

Cases of steatorrhea can also be confirmed by performing a microscopic examination of a stool specimen for fat droplets. In one study, investigators found in 17 normal individuals an average of 2.5 fat droplets per high power field and an average of 26.6 fat droplets in 24 individuals with steatorrhea. Patients with steatorrhea have, on the average, 10 times as many fat droplets per standard volume of stool as healthy controls.[92a] Bin et al[7] also confirm that microscopic examination of stool for fat globules (stained with Oil Red O or Sudan III) is a relatively sensitive method for detecting steatorrhea.

Fat Absorption Tests

Normal absorption of fat from the intestinal tract can take place only in the presence of an adequate amount of pancreatic lipase, colipase, and bile. Consequently, a number of fat absorption tests have been devised in which the patient is given a fat meal or iodine-labeled triolein or fatty acids. These tests have largely been discontinued because of their low sensitivity and high variability.

Serum Carotene, Vitamin A, and Vitamin A Tolerance Tests

These tests are based on the fact that carotene and vitamin A are fat soluble and thus can be absorbed to any significant extent only if fats are first hydrolyzed by pancreatic lipase. These tests are subject to the same limitations as the lipid absorption tests mentioned earlier. Details regarding the determination of these analytes are given in Chapter 8B, Vitamins.

$^{14}CO_2$ Breath Tests

In order to avoid the difficulties in obtaining 72-h stool collections for fecal fat, some investigators have administered orally triglycerides containing ^{14}C-fatty acids to determine the degree to which fat was absorbed. The test is based on the assumption that the oxidation of ^{14}C-fatty acids to $^{14}CO_2$ is proportional to the rate of absorption of fatty acids. An extension of this test is the comparison of results obtained after administration of labeled free fatty acids and labeled triglycerides. It was hoped that this approach would distinguish between disorders of fat digestion and fat absorption.

This test appears to be an easy approach to establishing fat malabsorption; however, the necessity of performing this test over a 6-h period and the need to administer isotopes are clear disadvantages of this procedure and may be responsible for its limited use. In addition, there is uncertainty about the effect of a variety of disorders of fat metabolism such as hyperlipemia or obesity on test results.[88]

D-Xylose Absorption Tests

Xylose is a pentose with a sweet taste not normally present in significant amounts in blood. When given orally, it is passively absorbed in the proximal small (duodenojejunal) intestine to about 60%, and most is subsequently excreted by the kidneys. The amount of xylose recovered in the urine or blood in a specified time interval after administration of a measured dose is used to evaluate mucosal absorption ability. Low absorption of xylose is observed in intestinal malabsorption. However, low values are also observed in celiac disease, tropical sprue, Crohn's disease, immunoglobulin deficiency, pellagra, ascariasis, blind loop syndrome, radiation enteritis, and surgical bowel resection, as well as after vomiting, in delayed gastric emptying, inadequate hydration, decreased circulation, intrinsic renal disease, thyroid disease, sequestration in body fluids as occurs in pregnancy and ascites, and after incomplete urine collection. In addition, excretory values decrease with age as a reflection of decreased kidney function. In malabsorption due to pancreatic insufficiency, the absorption of xylose will be essentially normal provided that there is no significant increase in intestinal motility; on the other hand, > 80% of patients with malabsorption due to jejunal malabsorption show low values. Therefore, the test is of some help in distinguishing between these two types of malabsorption and in evaluating the response to therapy.

Principle

A 25-g dose of xylose (or 5 g if so specified by the physician) is given to the patient, and all of the urine voided over the next 5 h is collected. A blood specimen is taken ~2 h after the xylose is given (in children after 1 h). The xylose concentrations in the 5-h urine specimen and in the blood specimen are determined by treating the diluted urine and a protein-free filtrate of the blood with p-bromoaniline in an acid medium. When heated with acid, xylose is dehydrated to furfural, which in turn reacts with p-bromoaniline. A pink color is produced in this reaction. Thiourea minimizes formation of interfering colored compounds. The concentrations of xylose in blood and urine are calculated on the basis of absorbance readings obtained with a standard xylose solution.

$$\text{D(+)-Xylose} \xrightarrow[\text{Acid and heat}]{-3\ H_2O} \text{Furfural} \xrightarrow{p\text{-Bromoaniline}} \text{Pink complex } (A_{max} = 520\ \text{nm})$$

Sample Collection

Adults

1. Keep patient fasting overnight and during test period.

2. In the morning (e.g., at 0700 h) have the patient empty his bladder completely and discard this urine. *Note:* Some investigators recommend checking a fasting urine and blood sample for nonspecific chromogens that react with *p*-bromoaniline; such interferences have been reported in some cases. Correction may be made by subtracting the base value from that found in the test specimen.

3. Give 25 g of D(+)-xylose (or 5 g if so specified by the physician) dissolved in 250 mL of water, followed immediately by an additional 250 mL water to ensure a urine flow of > 60 mL/h. Note starting time of urine collection. D(+)-xylose is available as XYLO · PHAN, Adria Laboratories, Inc., Columbus, OH 45215.

4. Collect and pool all urine specimens voided during the next 5-h period, including the 5-h specimen. Preserve by refrigeration or addition of toluene.

5. After ∼2 h, collect a blood sample with heparin or EDTA as anticoagulant, preferably in a tube that also contains sodium fluoride. Deproteinize and freeze blood specimens if not analyzed the same day.

Children

1. Keep patient fasting overnight and during test period; infants are fasted for 4 h.

2. In the morning (e.g., at 0700 h) have the patient empty his bladder completely.

3. Give 0.5 g of D(+)-xylose/lb of body weight, up to 25 g, reducing the amount of water according to the weight of the patient.

4. Collect and pool all urine specimens during the next 5-h period.

5. Collect a blood sample ∼60 min after xylose administration. The time of blood collection is different from that in adults because the peak of absorption in children is reached sooner.

Reagents

1. Zinc sulfate ($ZnSO_4$·7 H_2O), 5 g/dL in water.

2. Barium hydroxide, 0.15 mol/L. Dissolve 23.7 g of $Ba(OH)_2$·8 H_2O in water and dilute to 500 mL with water. Boil for a few minutes, stopper, cool, and filter. Avoid unnecessary exposure to air and standardize weekly by titration.

3. *p*-Bromoaniline reagent, 2 g/dL. Prepare a saturated solution of thiourea in glacial acetic acid by shaking about 5 g of thiourea in 100 mL of glacial acetic acid; decant the supernatant prior to use. Dissolve 2 g of *p*-bromoaniline in 100 mL of the saturated thiourea solution. Prepare fresh.

4. Stock standard, 200 mg of xylose/dL. Place 200 mg D(+)-xylose into a 100-mL volumetric flask. Dissolve in a saturated solution of benzoic acid, 0.3 g/dL in water, and fill to volume.

5. Working standard, 10 mg/dL. Dilute stock standard 20-fold with saturated benzoic acid.

6. Working standard, 20 mg/dL. Dilute stock standard 10-fold with saturated benzoic acid.

Procedure

1. *Urine.* If 5–15 g of xylose is given, dilute the urine to 500 mL; if 15 g or more of xylose is given, dilute the urine to 1000 mL. (Freeze a part of the first diluted urine for possible repeat.) Further dilute a convenient portion of the urine 40-fold.

2. *Blood.* Deproteinize the blood by using 1 vol blood, 7 vol H_2O, 1 vol $ZnSO_4$ solution (5 g/dL), and 1 vol $Ba(OH)_2$ (0.15 mol/L). Mix after each addition and centrifuge or filter to give clear solution.

3. Add the following amounts (in mL) of blood filtrate, urine, standard solutions, and *p*-bromoaniline to a set of test tubes as indicated:

	Blood		Urine (Unknown and Control)		Standard (0.1 mg/mL)		Standard (0.2 mg/mL)	
	Test	Blank	Test	Blank	Test	Blank	Test	Blank
Supernatant	0.5	0.5	—	—	—	—	—	—
Diluted urine	—	—	0.5	0.5	—	—	—	—

(*Continued at top of page 1469*)

	Blood		Urine (Unknown and Control)		Standard (0.1 mg/mL)		Standard (0.2 mg/mL)	
	Test	Blank	Test	Blank	Test	Blank	Test	Blank
0.1 mg/mL Xylose	—	—	—	—	0.5	0.5	—	—
0.2 mg/mL Xylose	—	—	—	—	—	—	0.5	0.5
p-Bromoaniline	2.5	2.5	2.5	2.5	2.5	2.5	2.5	2.5

4. Place all tests into a 70 °C water bath for 10 min and all blanks in the dark at room temperature.

5. Cool the incubated tubes rapidly and place all test tubes in the dark.

6. After 70 min, read the test against the blank, using a spectrophotometer at a wavelength of 520 nm. Alternatively, all tests and blanks may be read against a water blank.

Note: If the reading of the unknown is > 0.4 absorbance unit, dilute with the reagent blank and repeat the measurement. Readings should be made within 30 min since color fades after this time.

Calculation

Blood

$$\text{Xylose, mg/dL} = \frac{A_{(\text{test})}}{A_{(\text{standard, 0.1 mg/mL})}} \times 0.1 \times 100 \times 10$$

$$= \frac{A_{(\text{test})}}{A_{(\text{standard, 0.1 mg/mL})}} \times 100$$

If all test and blank tubes are read against water:

$$\text{Xylose, mg/dL} = \frac{A_{(\text{test})} - A_{(\text{blank})}}{A_{(\text{standard})} - A_{(\text{blank})}} \times 0.1 \times 100 \times 10$$

$$= \frac{A_{(\text{test})} - A_{(\text{blank})}}{A_{(\text{standard})} - A_{(\text{blank})}} \times 100$$

where 0.1 is the concentration of the standard in mg/mL, 10 is the dilution factor, and 100 converts mg/mL to mg/dL.

Urine

If urine was diluted to 1000 mL:

$$\text{Xylose excreted, g/5 h} = \frac{A_{(\text{test})}}{A_{(\text{standard, 20 mg/dL})}} \times \frac{20 \times 40}{100} \times 1000 \times \frac{1}{1000}$$

$$= \frac{A_{(\text{test})}}{A_{(\text{standard, 20 mg/dL})}} \times 8$$

If urine was diluted to 500 mL:

$$\text{Xylose excreted, g/5 h} = \frac{A_{(\text{test})}}{A_{(\text{standard, 20 mg/dL})}} \times \frac{20 \times 40}{100} \times 500 \times \frac{1}{1000}$$

$$= \frac{A_{(\text{test})}}{A_{(\text{standard, 20 mg/dL})}} \times 4$$

where 20 is the concentration of standard in mg/dL; 40, 500, and 1000 are dilution factors; 1/1000 converts mg to g; and 100 converts mL to dL.

If all test and blank tubes were read against water, the absorbance of the blank must be subtracted from the absorbance of the test, as indicated under blood.

Reference Ranges

	mg/dL	mmol/L
Blood		
Child, 1 h		
(dose, 0.5 g/lb):	> 30	> 2.00
Adult,		
2 h (dose, 5 g):	> 20	> 1.33
2 h (dose, 25 g):	> 25	> 1.67
Range (dose, 25 g)		
1 h:	21–57	1.40–3.80
2 h:	32–58	2.13–3.86
3 h:	19–42	1.27–2.80
4 h:	11–29	0.73–1.93
5 h:	6–18	0.40–1.20

Urine (5-h collection)

Child:	16–33% of ingested dose	Fraction of ingested dose: 0.16–0.33

	g/5 h	mmol/5 h
Adult,		
dose, 5 g:	> 1.2	> 8.00
dose, 25 g:	> 4.0	> 26.64
> 65 y:	> 3.5	> 23.31

Some investigators feel that a 2-h urine collection is more sensitive. The reference range for a 2-h collection is 14.2–29% of dose (mean 22.9%).[1]

Comments on Methods and Sources of Error

The accuracy of the procedure depends not only on the rate of absorption but also on the rate of excretion of xylose by the kidneys. Thus, patients with renal insufficiency will excrete a decreased amount of xylose. In order to eliminate misinterpretations as a result of renal retention, a blood determination of xylose is carried out along with the determination of xylose in urine. A normal blood xylose level in the presence of decreased urine xylose excretion would suggest renal retention. False positive (low) results may also occur in the presence of vomiting, gastric stasis, dehydration, myxedema, massive ascites, edema, or massive bacterial overgrowth.

Slight abdominal discomfort or diarrhea may be observed in some patients. These symptoms can be minimized by use of the 5-g xylose dose.

Kendall et al.[49] recommend the 5-g dose because it is easier to take and much less likely to cause abdominal distention, nausea, and diarrhea. These authors also suggest reducing the urine collection period to 2 h to minimize any possible metabolism of xylose by microorganisms.

Haeney et al.[43] performed a study on 216 subjects with ages ranging from 14–92 and found that administration of 5 g D-xylose, followed by measurement of blood xylose concentrations 1 h after oral administration and after correction of results for body surface, shows maximum discrimination between normal and abnormal subjects. The incidence of false negative results with that technique is only 4.8% and the incidence of false positive results only 2.2%. The blood test circumvents many of the conditions that lead to low urinary excretion, and values are not affected by age. The stated reference range is 0.65–1.33 mmol/L per 1.73 m² of body surface. Measurements of blood xylose concentrations are useful in pediatric patients in whom timed urine specimens are difficult to obtain. Given reference ranges are: 2-h urine, 6.5–31%; 5-h urine, 20.6–49.6% of dose.

Effects of Common Drugs on Procedure[111]

Apparent decrease in urinary xylose excretion (impaired absorption):

Aminosalicylic acid	Ethionamide	Nonfasting state
Arsenicals	Gold	Metformin

(*Continued at top of page 1471*)

Aspirin (impaired renal excretion)	Indomethacin	Nalidixic acid
Colchicine	Isocarboxazid	Neomycin
Digitalis	Kanamycin	Opium álkaloids
	MAO inhibitors	Phenformin

Apparent increase in urinary xylose excretion (procedural interference) occurs only in the presence of galactose > 2 g/dL, glucose > 2 g/dL, and phenazopyridine.

Reference

Reiner, M., and Cheung, H. L.: Xylose. *In:* Standard Methods of Clinical Chemistry. S. Meites, Ed. New York, Academic Press, 1965, Vol. 5, p. 257.

Alternative Methods for the Determination of D-Xylose

Gas chromatography. Mix a 100–200-μL sample with acetonitrile, add internal standard (2-deoxy-D-ribose), and vortex. After centrifugation, decant the supernatant and evaporate under nitrogen. Reconstitute the residue with methanol and again evaporate. This is followed by formation of the methyloxime and trimethylsilyl derivative. Xylose concentrations in plasma and urine are obtained from a standard curve of the area ratios (xylose/internal standard) versus xylose concentration.[47]

Enzymatic procedure. D-Xylose in the sample is oxidized by $NADP^+$ in the presence of D-xylose:$NADP^+$ 1-oxidoreductase (EC. 1.1.1.179). The products of the reaction are NADPH, H^+, and D-xylonolactone. At pH 8.0, the latter hydrolyzes to D-xylonic acid, which contributes to the irreversibility of the reaction. The increase in NADPH concentration can be measured by an increase in absorbance at 339 nm or fluorometrically at > 420 nm using an excitation wavelength of 339 nm.[109]

The Sweat Test

The determination of the chloride concentration of sweat is considered the most reliable single test in the diagnosis of cystic fibrosis. (See also p. 1481.) Sodium and chloride elevations in sweat are seen even in the absence of gastrointestinal or respiratory symptoms or when pancreatic insufficiency cannot be demonstrated by other tests. In most affected infants, the test becomes positive between 3 and 5 weeks of age. Only 1–2% of cystic fibrosis patients have sweat chloride values below the value of 60 mmol/L and only 1 in 1000 has a value < 50 mmol/L.[25] These exceptions were observed predominantly in cystic fibrosis patients who did not have pancreatic involvement.

Collection of Sweat by Iontophoresis

1. With the use of forceps, place two prewashed 2-inch-square gauzes (or two Whatman No. 42 filter papers, diameter 5.5 cm) into a weighing bottle (or Petri dish), stopper, and determine the combined weight accurately, using an analytical balance. Handle weighing bottle with tissue or gauze to avoid direct contact with the fingers.

2. Wash and dry the skin of the anterior (inside) surface of the forearm and place two 2-inch-square gauzes (not the preweighed gauze) over the washed area. Moisten gauzes well with a freshly prepared solution of pilocarpine nitrate, 4 mg/dL. (Pilocarpine is a drug that, when introduced into the skin, induces sweating.) Place two gauzes saturated with saline on the posterior (outside) surface of the arm. The positive electrode of a direct-current power supply designed for iontophoresis is placed over the gauze with pilocarpine, and the negative electrode is placed on the saline-soaked gauze. Ensure good contact and secure the electrodes with rubber strips or similar means. If the arm is too small to secure the electrodes, as in small children, use the thigh or the interscapular area of the back of the patient.

3. Apply a current of 0.16 mAmp/cm^2 for 5 min.[35] The current will tend to increase during this time interval and should be maintained at the above setting. After 5 min take off the electrodes, clean the skin area well with distilled water, and dry the area. Complete kits for iontophoretic stimulation (either battery-operated or A.C.-operated versions) are available commercially from Farrell Instrument Company, P.O. Box 1037, Grand Island, NE 68801.

If the patient complains about discomfort, discontinue the test. A tickling sensation at the site of the electrode is a common finding and should be disregarded. After the test, the skin may be somewhat reddish, but color will disappear within a few hours.

4. With the aid of forceps, place the preweighed gauze or filter paper over the skin area that was exposed to pilocarpine. Place an ~4-inch-square plastic sheet over this area and seal airtight with surgical tape. Allow the sweat to accumulate on the gauze or filter paper. This usually takes ~20–30 min, but the time of sweat collection may be extended as long as necessary. In general, the appearance of droplets on the plastic sheet indicates that enough sweat has accumulated. The droplets must be included in the collection.

5. Remove the gauze or filter paper with forceps, place it immediately into the weighing bottle, and stopper. Send the weighing bottle to the laboratory, wrapped in gauze to avoid direct contact with fingers.

Recovery of Sweat

Accurately (within 1 mg) determine the weight of the bottle or Petri dish containing the gauze or filter paper and calculate the amount of sweat by difference. One gram of sweat is assumed to be 1 mL of sweat. Generally 0.2–0.5 mL of sweat is obtained. A minimum amount of sweat required for reliable quantitation is 100 mg.

Many methods have been suggested for recovery of sweat from the gauzes or filter paper. The method of choice will depend on the particular laboratory setup. The gauzes, for example, may be placed in a funnel and washed off with deionized water. The weighing container should also be rinsed with water and the washings should be combined and brought up to a predetermined volume (e.g., 5.0 mL). The sweat may also be recovered as described below for the determination of sweat chloride. If only chloride is determined, and if a chloridometer is used (coulometric-amperometric technique), the sweat may be recovered by elution with acetic acid–nitric acid reagent.

Determination of Sweat Chloride

An example of sweat recovery for a chloride determination by the Schales and Schales method follows. For the determination of chloride by the coulometric-amperometric technique, consult the manufacturer's instructions supplied with the chloridometer.

1. Add 5.0 mL deionized water to the weighing container with the gauze. Stir well to elute the chloride from the paper or gauze.

2. Place 2.5 mL of the extract into a suitable titration vessel and determine the chloride content as outlined in Chapter 10.

Calculation

$$\text{Chloride, mmol/L} = \frac{\text{Reading of unknown} \times 0.2 \times 100 \times (5.0 + \text{weight of sweat collected})}{\text{Reading of standard} \times \text{total amount of sweat (mL)} \times 2.5}$$

Note: The number 0.2 in the preceding formula indicates the amount of sample for which the chloride procedure has been set up; 100 is the factor for the concentration of the standard; 5.0 is the volume of water added to the weighing container; 2.5 is the amount of extract used in the titration.

Reference Ranges

In children, ranges for sweat sodium and chloride are:

	Healthy Children	Cystic Fibrosis Homozygotes
Na$^+$, mmol/L	10–40	70–190
Cl$^-$, mmol/L	5–35	60–160

Mean values of Na$^+$ and Cl$^-$ for normal children are 27 mmol/L; in normal adults means are slightly higher: Na$^+$ 33, Cl$^-$ 39. Chloride values >70 mmol/L are suggestive of cystic fibrosis. In adults with cystic fibrosis, mean values are Na$^+$ 110, Cl$^-$ 118, and K$^+$ 21 mmol/L.

Remarks and Sources of Error

False high results can be obtained if the skin of the child is not cleaned properly or if contaminated sponges are used. Thus, noncontaminated gauzes should be used (e.g., Curity brand), and a blank determination should be done to assure that the gauze is low in sodium and chloride content.

False normal results can be obtained in cystic fibrosis patients with salt depletion, which is common in these patients during hot weather periods. Additional electrolyte studies on serum help to avoid misinterpretation of results in such instances. Elevated levels of Na, Cl, or K in sweat have also been reported in meconium ileus, in adrenal insufficiency ($\uparrow$K), hypothyroidism ($\uparrow$Cl), ectodermal dysplasia, mucopolysaccharidosis, fucosidosis, hereditary nephrogenic diabetes, glucose-6-phosphate deficiency, papillatonia hyporeflexia, familial cholestasis, alcoholic pancreatitis, and in some cases of renal disease.[92] Most of these conditions are rare and clinically distinct from cystic fibrosis. Also, elevations in these conditions are not consistent.

The volume of sweat may be markedly reduced in premature, malnourished, and severely dehydrated infants. If a small yield is expected, sweating should be induced in two different locations and the sweat collected should be combined.[92] Alternatively, the patient may be tested at a subsequent appointment. Very low sweat rates may lead to falsely low results in cystic fibrosis patients.

It has been stated by some that determination of the chloride concentration is the most reliable index for confirming cystic fibrosis; therefore, many laboratories use only this test. However, no well-documented data are presently available that indicate greater clinical usefulness of chloride over sodium values.[92]

The diagnosis of cystic fibrosis is very critical and difficult. (See also clinical discussion of cystic fibrosis in this chapter.) It is therefore of utmost importance that the laboratory follow a procedure that conforms to the guidelines given by the Cystic Fibrosis Foundation.[35] If such a procedure cannot be made available, patients should be referred for testing to one of the 120 designated centers supported by the Cystic Fibrosis Foundation.

Measurement of Sweat Composition by Ion Specific Electrodes. The use of ion-specific electrodes for Na^+ and Cl^- measurement in sweat has also been recommended.[51] In this method, the electrode is applied directly to the skin, which avoids the necessity for collecting and weighing the sweat. Such electrodes measure the ion activity rather than the ion concentration, but a good correlation between these two measured quantities has been observed. The principle of ion-specific electrodes has been discussed in Chapter 1B, Section Four, Electrochemistry. The technique, in our hands, has been found reliable if results of three consecutive measurements are in close agreement. However, failure to adhere to all details of the method may easily lead to errors. The test is not accepted by the Cystic Fibrosis Foundation.

Agar Plate Test. The palm of the hand of the patient is brought in contact with an agar plate or a commercially available paper prepared with silver salts. An increased amount of chloride in the sweat results in a white precipitate of silver chloride, which can be graded semiquantitatively. Sweating is induced by placing the hand into a plastic bag. Such tests should be used only for screening purposes, since results are not as reliable as those obtained from direct measurements of sweat electrolyte concentrations. However, among the screening tests, this test appears to be most reliable. Shwachman and Mahmoodian,[92] in a series of 29 referral patients, found only one false negative result when using the agar plate test.

GASTRIC, PANCREATIC, AND INTESTINAL DISEASES

Gastric Diseases

Peptic Ulcer Disease. This is the collective name given to chronic duodenal ulcer and chronic gastric (benign) ulcer (as opposed to the acute ulcer disease that is described under erosive gastritis). Although these diseases have many features in common, they differ with respect to incidence, pathogenesis, natural history, outcome, and management. About 500 000 new cases of duodenal ulcer disease occur each year in the United States;[98] the highest incidence occurs during the fifth and sixth decades of life. Duodenal ulcer disease is much more common than gastric ulcer disease.

Chronic duodenal ulcer. A duodenal ulcer is a round, sharply punched-out defect, usually less than 3 cm in diameter, in the mucosa of the duodenum. The defect may be superficial or may penetrate through the serosa. About 90% of duodenal ulcers occur in the first portion of the duodenum (Figure 14-1). Ulcers occurring in the distal antrum and pyloric channel of the stomach are also considered duodenal, since they produce similar clinical syndromes and have similar responses to treatment. In time, chronic duodenal ulcers heal by the formation of fibrous scar tissue and the inward growth of the duodenal mucosal epithelium. Contraction of the fibrous tissue causes the characteristic puckering of the overlying mucosa, which may be detectable on radiological examination or endoscopy.

Apart from the morbidity due to the ulcer itself, duodenal ulcer has four important, although uncommon, complications. *Hemorrhage* due to the ulcer's eroding a blood vessel is a life-threatening emergency; the blood may be vomited (*hematemesis*) or it may be passed as *melena* (dark red or black feces due to partial digestion of blood). *Perforation* through the anterior wall of the duodenal cap causes peritonitis with very severe abdominal pain and is an acute emergency. *Penetration* of the ulcer into adjacent structures such as pancreas, biliary tract, liver, colon, abdominal wall, and even lung have been reported; these cases are characterized by intractable pain. *Luminal obstruction* (pyloric stenosis) is caused by the gradual contraction of a duodenal ulcer scar which reduces the duodenal aperture; the treatment is surgical.

Clinical presentation. Typically 2 h after a meal the patient complains of a deep aching pain, located high in the abdomen (epigastrium), that is relieved by the ingestion of food or antacids. The pain recurs several times during the day, lasting for 15 min to 1 h.

Treatment is directed initially to heal the ulcer by careful dietary habits, the use of antacids, the possible use of H_2-receptor blockers such as cimetidine, and mild sedation. If pain continues or complications develop, then surgical treatment may be necessary.

Pathogenesis.[98] A number of functional abnormalities have been identified in duodenal ulcer patients; these may have some bearing on causation, although further studies are needed. They include increased parietal cell and chief cell mass, increased basal secretory drive, increased postprandial secretory response, rapid gastric emptying, low duodenal pH, increased pepsin activity, and impaired mucosal defense. A number of chronic conditions have been reported to occur more frequently with duodenal ulcer disease than might be expected by chance. These include chronic pulmonary disease (including α_1-antitrypsin deficiency and cystic fibrosis), cirrhosis of the liver, renal failure, renal transplantation, renal stone disease, and, possibly, hyperparathyroidism.

Diagnosis is made primarily by barium radiography (upper GI study); flexible fiberoptic endoscopy and biopsy may subsequently be indicated. Serum levels of pepsinogen I may be of value since they reflect the gastric chief cell mass and correlate with the peak acid secretory capacity (PAO). One third to one half of patients with duodenal ulcer have PAO about 30 mmol/h.[5] (See also p. 1457.) The most important diseases to exclude during the diagnostic process are gastric ulcer and gastric carcinoma.

Chronic benign gastric ulcer. Gastric ulcers occur anywhere in the stomach, although most are found along the lesser curvature. Most gastric ulcers are between 1 and 2 cm in diameter. A single ulcer is most common, although two or more may be found. Some patients with a gastric ulcer will have a coexistent duodenal ulcer. Gastric ulcers heal with the formation of fibrous tissue and the inward growth of the gastric epithelium. In time the fibrous tissue contracts, leaving a characteristic puckering of the epithelial surface. As with duodenal ulcers, gastric ulcers may cause hemorrhage or perforate. Chronic benign ulcers must be differentiated from acute gastric erosion and ulceration. (See Gastritis, p. 1475.) In the United States over 100 000 patients are hospitalized annually for gastric ulcer.[81] Peak incidence is from 55–65 years of age; men and women are equally affected.

Clinical presentation. The rhythm of the pain is typically different from the duodenal ulcer pattern. Epigastric pain occurs 30–90 min after meals and persists until the stomach is empty some 60–90 min later.

Pathogenesis.[81] Gastric ulcer may be due to a number of different factors such as duodenogastric reflux, changes in the gastric mucosal barrier, changes in acid and pepsin secretion, defects in mucosal blood flow, effect of analgesics, effects of smoking, geographic or socioeconomic factors, and heredity.

Diagnosis. It is important to establish that the ulcer is indeed benign; radiological characteristics such as size are important, since ulcers 4 cm or larger are more likely to be malignant. The fact that subsequent therapy causes the ulcer to heal is also good evidence of nonmalignancy. The PAO of men and women with gastric ulcer disease is not significantly different from that of a healthy group of controls.[5] (See also p. 1458.)

Zollinger-Ellison syndrome. The pathophysiology of the Zollinger-Ellison syndrome has already been described (p. 1440). This syndrome is estimated to occur in between 0.1 and 1%

of all patients with duodenal ulcer.[64] *Gastrinomas* (gastrin secreting tumors) are most frequently found in the head or tail of the pancreas, but may be in the wall of the duodenum, and more rarely at other sites. They vary in size from 0.2–20 cm or more. About half of all gastrinomas are multiple and about two thirds are malignant. One fourth of all gastrinomas are part of the multiple endocrine neoplasm syndrome type I (MEN I), with associated tumors in pancreatic islets and in parathyroid and pituitary glands.

Clinical presentation. Symptoms are like those of duodenal ulcer but are more progressive and less responsive to therapy. Nearly all patients with gastrinomas develop recurrent ulcers; the majority of ulcers occur in the first part of the duodenum (like the more common duodenal ulcer disease), but they also occur in the esophagus, stomach, other parts of the duodenum, and even in the jejunum. The ulcers are usually < 1 cm in diameter and are single (but may be multiple). Diarrhea often precedes ulcer symptoms and occurs in about one third of all patients with the syndrome. The hypersecretion of acid causes the diarrhea but the high blood levels of gastrin may also suppress jejunal absorption of sodium and water. Steatorrhea also occurs due to the lowering of the small intestinal pH, but it is less common than diarrhea. Lipase is inactivated at low pH, bile salts are rendered insoluble, and vitamin B_{12} is not absorbed by the distal ileum; gastric secretion of intrinsic factor is unaffected.

Treatment ideally consists of removal of the primary gastrinoma. However, these may be difficult to locate or may have spread to regional lymph nodes, liver, spleen, bone, and other sites by the time the diagnosis is established. Thus, treatment is often directed to reducing gastric acid secretion by means of cimetidine. Alternatively, removal of the target organ (total gastrectomy) has been suggested, although the patient then will be subjected to the severe metabolic complications of total gastrectomy. Since gastrinomas are usually malignant and slowly progressive, treatment has to be modified as symptomatology changes with time.

Diagnosis. The history of the complaint and its progressive nature usually give the clue to the diagnosis. Gastric acid hypersecretion is common although not universal; there is a substantial overlap in gastric acid production between patients with the Z-E syndrome and those with duodenal ulcer disease. The trophic effect of gastrin on the gastric mucosa often leads to prominent gastric mucosal folds that can be identified by radiological examination. Serum gastrin levels are usually markedly increased in the Z-E syndrome but sometimes only marginal elevations are found; additional testing such as secretin or calcium challenge is then necessary. The 12-h overnight secretion of HCl is > 100 mmol and the basal HCl excretion is > 100 mmol/L. (See also p. 1457.)

Gastritis[106] is a term used to denote mucosal inflammation. It can be classified as erosive, nonerosive, and specific; specific gastritis is very rare.

Erosive gastritis (synonyms: stress lesions, stress erosions, or acute gastritis) occurs in four well-defined clinical situations:[85] following severe trauma in which sepsis or serious illness intervenes; following severe burns (Curling's ulcer); in patients with intracranial disease (Cushing's ulcer) and following craniotomy or traumatic head injuries; and in patients who chronically ingest drugs such as corticosteroids, aspirin, or ethanol. Erosive gastritis is characterized by massive upper gastrointestinal bleeding occurring some days after the initial insult. There is a 50% mortality among patients who hemorrhage. Endoscopy is usually the definitive technique to establish the diagnosis and the site of bleeding; the lesions are histologically quite distinct from chronic gastric or duodenal ulcer.

Nonerosive gastritis (synonym: chronic gastritis) is seen associated with gastric ulcer disease or gastric carcinoma, following partial gastrectomy, in pernicious anemia, and in the healthy elderly person. Serum gastrin levels are elevated in achlorhydric patients because there is no negative feedback (acid inhibition of gastrin release).

Gastric cancer (adenocarcinoma) arises from the mucus cells of the gastric mucosa. It is the third most common gastrointestinal cancer in the United States; about 24 000 new cases are diagnosed each year. It is about three times more common in males than females. There is a wide geographic variation in incidence. Hereditary factors, environmental factors, and a preceding disease such as pernicious anemia, atrophic gastritis, or postgastrectomy state, appear to be important.

Presentation.[24] The disease is often asymptomatic, but there may be pain, vague discomfort, or a feeling of epigastric fullness and nausea. In advanced stages there may be abdominal pain, weight loss, obstructive symptoms, or bleeding (hematemesis or melena). The patient may be anemic due to occult bleeding. There may be evidence of metastatic spread, such as abdominal swelling from ascites or a tumor mass, shortness of breath from pleural effusions, or enlarged lymph nodes due to metastases.

Treatment depends on the stage of the tumor. Surgery is "curative" (five-year survival) in only about 10% of patients.

Diagnosis.[5] Diagnosis is made by biopsy. The majority of patients will have fasting achlorhydria, but less than a quarter of all cases have achlorhydria following maximal stimulation. The volume of the gastric residue may be increased, and in some cases blood may be found in the stomach. Lactate is often present, probably due to the presence of lactobacilli.

Postgastrectomy syndrome.[65] Annually, more than 50 000 Americans are treated for peptic ulcer disease by gastric resection or denervation (vagotomy) of the stomach. Of these, about 5000 will suffer some functional incapacity to which the name postgastrectomy dysfunction or syndrome has been given. This term includes the dumping syndrome, diarrhea, weight loss, anemia, and bone disease.

The dumping syndrome occurs because the stomach, after gastrectomy, empties more rapidly than normal. The patient experiences abdominal pain and diarrhea together with vagomotor phenomena such as flushing, lightheadedness, and cardiac palpitations 30–60 min after eating. Because the proximal jejunum contains a hypertonic solution of partially digested food, rapid osmotic equilibration with blood results in a slight but rapid decrease in the blood volume, with concomitant intestinal distention. Between 90 and 180 min after a meal cardiac palpitations, lassitude, mental confusion, or even loss of consciousness is experienced and may be due to reactive hypoglycemia. Diarrhea may occur usually 1 or 2 h after eating as a result of rapid filling of the jejunum after meals.

Weight loss is not uncommon after gastric surgery and there may also be specific nutritional deficiencies. Anemia develops as a result of decreased absorption of iron, vitamin B_{12}, and folate. Maldigestion may occur due to rapid gastric emptying (resulting in less gastric dispersion of solid food and reduced digestion time), small intestinal "hurry," and reduced pancreatic secretion. There is also asynchrony between the arrival of pancreatic and biliary secretions and the entry of food into the small intestine.

There is an increased prevalence of metabolic bone disease following ulcer surgery. Osteoporosis (loss of bone matrix and of bone) occurs more commonly than osteomalacia, which is characterized by bone pain, loss of bone, pseudofractures, low serum calcium, and increased serum alkaline phosphatase activity. These manifestations can be reversed by administration of vitamin D. About half of the patients with osteoporosis also have osteomalacia. The prevalence of metabolic bone disease is estimated to be 5–15% of patients who have had ulcer surgery.

Pancreatic and Intestinal Diseases

The pancreas is inaccessible to direct examination, and thus disease processes may be far advanced before they become manifest. The pancreas lies in close proximity to a number of important structures such as the duodenum, the ampulla of Vater, stomach, colon, liver, common bile duct, arterial blood supply to the gut and spleen, and the portal vein. (See Figure 14-2.) Inflammation or neoplastic processes within the pancreas may therefore cause secondary involvement of adjacent structures. Carcinoma of the pancreas, for example, may be discovered only when it invades the vertebral column and causes pain, or when it occludes the biliary tree and causes jaundice, or when it affects the function of stomach or colon and causes anorexia, nausea, vomiting, and loss of weight.

In clinical practice pancreatic disease other than diabetes mellitus is considerably less common than liver disease. Nonetheless, diseases such as acute and chronic pancreatitis, pancreatic carcinoma, and cystic fibrosis are responsible for a small but significant morbidity and mortality. In the following section, a brief outline is given of those pancreatic diseases that often require the services of the clinical chemistry laboratory.

Acute pancreatitis. Acute pancreatitis is the name given to a discrete, sudden episode of more or less diffuse enzymatic destruction of the pancreatic substance as a result of the escape of active pancreatic enzymes into the glandular parenchyma. It is distinguished from chronic pancreatitis in that, upon elimination of the causative factor, normal endocrine and exocrine functions are restored and there is no permanent functional damage.[97] Chronic pancreatitis is associated with persistence of pain after the acute episode or of symptoms due to impairment of pancreatic function. Relapsing pancreatitis is a term used to characterize recurrent occasions of either acute or chronic pancreatitis. The milder form of acute pancreatitis is edematous; the more severe form is hemorrhagic. Mortality in the edematous form, though significant, is much less than in the hemorrhagic form, where it has been reported to be 20–50%.

Acute pancreatitis has been estimated to occur once in about every 500 acute admissions to a general hospital and occurs most frequently in the age group 50 years and above. It is frequently associated with biliary tract disease, alcoholism, or both.

Apart from the morbidity due to acute pancreatitis itself, there are four important local complications—

pseudocyst formation, ascites, pleural effusion, and pancreatic or peripancreatic abscess formation. *Pseudocysts* may form in areas of necrosis; they contain tissue debris, pancreatic juice, blood, and fat droplets walled off by fibrous tissue. About half of all patients with severe pancreatitis will develop pseudocysts; many of these masses will be palpable on physical examination. About half of all acute pseudocysts will resolve spontaneously over several weeks following formation; the others may require surgical removal or drainage. *Ascites* results from a persistent leak of pancreatic juice from a pseudocyst or, less frequently, from a disrupted pancreatic duct. The ascites may resolve spontaneously, or it may persist and cause a painless enlargement of the abdomen. *Pleural effusion* is caused by spread, via the diaphragmatic lymphatics, of the inflammatory exudate into the pleural (usually left-sided) space. The volume may be sufficient to cause impaired ventilation. *Abscess formation* is rare but very serious. Necrotic pancreatic tissues become infected with coliform bacteria from the colon.

Clinical presentation. The major symptom of acute pancreatitis is abdominal pain. It is often described as steady, dull, or boring in nature. It is located in the epigastrium and periumbilical region and often radiates to the back, flanks, and lower abdomen. Nausea, vomiting and abdominal distention are also usually present. Fever is usually noted. Painless acute pancreatitis is rare; when it occurs the prognosis is grave. A patient with severe pancreatitis may present in coma or shock or with toxic psychosis and is usually hypotensive due to hypovolemia. There are systemic effects due to the release of proteolytic and lipolytic enzymes. Hypocalcemic tetany may be observed, although this is now rare. Mild jaundice is present in 40% of cases due to compression of the common bile duct by the inflamed pancreas. Pulmonary manifestations are evident in about 20% of all cases because of spread of the inflammatory exudate through the diaphragmatic lymphatics.

Treatment. Two aspects must be addressed—general supportive care by reducing pain, correcting hypotension and pulmonary or renal insufficiency, infusing calcium and magnesium, and administering insulin if required; and specific care of the pancreas by resting the pancreas by initially giving nothing by mouth followed by small carbohydrate feedings and by removing gastric secretions by continuous gastric suction.

Pathogenesis. [97] The two leading conditions associated with acute pancreatitis are biliary tract disease, especially cholelithiasis, and excessive alcohol intake. Gallstones are present in about 50% of all cases, and about 5% of patients with gallstones develop acute pancreatitis. Most alcoholic patients with pancreatitis progress to irreversible functional and structural damage to the pancreas, i.e., chronic pancreatitis. It has been estimated that up to 25% of all cases of acute pancreatitis are not associated with known predisposing factors (see later) and are therefore labeled as idiopathic pancreatitis, i.e., of unknown etiology or not secondary to any other disease. But because many of these cases have the characteristics of gallstone-associated pancreatitis, it would seem that undetected gallstones are probable factors. Other known causes of acute pancreatitis include drugs (particularly antimetabolites and sulfonamide derivatives); abdominal operations on or near the pancreas; retrograde injections of radiologic media into the pancreatic duct for visualization purposes (ERCP); physical trauma; hypercalcemia; obstruction to the outflow of the pancreatic duct; inflammation or spasm due to a variety of anatomical defects or coexisting diseases; renal transplantation; and the presence of hyperlipidemia. Patients on estrogen therapy who develop pancreatitis have a type IV or V hyperlipidemia. Up to 20% of alcoholic patients with pancreatitis have hypertriglyceridemia (type V pattern). A similar association occurs in some patients with idiopathic pancreatitis. Finally, patients with type I, IV, or V hyperlipoproteinemias frequently develop acute pancreatitis that often progresses to chronic pancreatitis. A possible mechanism is the lipolysis of triglycerides that produces high concentrations of free fatty acids in the vicinity of the pancreas. Free fatty acids may be toxic to acinar cells, and they also complex Ca^{2+}.

The morphology of the pancreas in acute pancreatitis indicates that proteolysis, lipolysis, and hemorrhage are the principal features. Pancreatic enzymes responsible for this tissue damage are converted from their proenzyme form to the active form (e.g., trypsinogen to trypsin) by one of a number of mechanisms that include bile reflux, hypersecretion-obstruction, alcohol-induced changes, and duodenal reflux.

Bile reflux, if the bile is under increased pressure, might injure the pancreatic ducts and cause leakage and activation of trypsin, which in turn could activate elastase and phospholipase A. Elastase cleaves the elastic fibers of blood vessels (causing hemorrhage), and phospholipase A in the presence of bile produces lysolecithin and lysocephalin, both of which are strongly cytotoxic agents. The presence of trypsin activates kallikrein, which then produces the kinins bradykinin and kallidin from freely circulating kininogens. These kinins are known to produce pain and leukocyte accumulation and to promote shock by causing vasodilatation and increased vascular permeability.

The hypersecretion-obstruction theory suggests that rupture of ducts occurs by pancreatic hypersecretion potentiated by partial ductal obstruction. The concept of partial obstruction is important because complete pancreatic obstruction merely shuts down the exocrine pancreas and does not result in acute pancreatitis, whereas partial obstruction (caused by ductal metaplasia or small calculi) might cause the necessary rupture of the ducts. The theory, like that of bile reflux, lacks experimental substantiation.

Alcohol-induced changes have an important role in the genesis of acute pancreatitis. Alcohol increases pancreatic secretion; it may also cause partial ductal obstruction or cause hyperlipemia in alcoholics after a fatty meal. Duodenal reflux into the pancreatic duct may occur if the sphincter of Oddi has been damaged by biliary disease and thus allows duodenal contents to enter the pancreatic duct and activate pancreatic enzymes.

Whatever the mechanism that activates the pancreatic enzymes, one should recall that enzymes released by the necrotic process can spread not only throughout but also beyond the pancreas. Thus, in 10% of autopsy studies, distant fat necrosis is observed in the retroperitoneum, bones, joints, and subcutaneous tissues.

Diagnosis.[70,97] The diagnosis of acute pancreatitis is principally a diagnosis of exclusion. Alternative causes of the abdominal pain that must also be considered include perforated or strangulated viscus, intestinal obstruction, acute cholecystitis or biliary colic, mesenteric infarction, ureteral or renal colic, ectopic pregnancy, and peptic ulcer. The physician, by consideration of the history and careful physical examination, will use a variety of diagnostic tools in the attempt to establish a diagnosis.[70] Abdominal and chest radiographs, upper gastrointestinal contrast series, ultrasonography, and computed axial tomography (CAT) may be of value, particularly for the detection of complications of pancreatitis such as pseudocyst, abscess, and localized hematoma. There are reports, however, that an acutely inflamed pancreas shows nonspecific changes by ultrasonography and that it appears normal by CAT scanning in 70% of patients with acute pancreatitis.[97] Ultrasonography is useful in detecting gallstones, which may be a cause of pancreatitis.

The *laboratory* is of value in providing serum and urine amylase, serum P-type isoenzyme of amylase, serum lipase, serum immunoreactive trypsin, amylase/creatinine clearance ratio, and a number of general chemical and hematologic assays. The enzyme findings in acute pancreatitis are discussed in Chapter 5 (pp. 724, 726, 736, and 742). Simultaneous determinations of serum amylase and lipase are recommended. The assay of additional enzymes has been proposed, but their clinical value is yet to be conclusively documented.[70] The severity of the attack does not correlate with the elevation of serum amylase activity. However, an assessment of severity can be made by noting the clinical features. Many techniques have been proposed for objectively grading the severity of the disease. For example, experience at the Massachusetts General Hospital suggests that the four most valuable indicators of severe pancreatitis are serum albumin < 30 g/L, hematocrit $< 30\%$, serum creatinine > 2 mg/dL, and serum calcium ≤ 8 mg/dL.[70]

Chronic pancreatitis. Chronic pancreatitis is defined as progressive functional damage to the pancreas that persists even when the causative factor is eliminated.[41] Patients with chronic pancreatitis may suffer episodes of exacerbation indistinguishable from acute pancreatitis.

Clinical presentation.[41] About half of all patients with chronic pancreatitis suffer attacks of acute pancreatitis in the presence of an irreversibly damaged pancreas. One third are made aware of the disease by the insidious onset of pain, which varies in intensity and may be continuous or intermittent in nature. Another, less common presentation is the sudden appearance of diabetes mellitus, malabsorption, jaundice, or upper gastrointestinal bleeding. During the course of chronic pancreatitis, 90% of patients will suffer "boring," dull or sharp and steady epigastric or lower abdominal pain that often radiates to the back. The onset of pain may be gradual and usually lasts for several days to weeks.

Treatment is directed to reducing alcohol intake (when alcohol is the cause), relieving pain, and providing digestive enzymes and nutritional support when necessary.

Etiology and pathogenesis[41] are largely unknown, but associated factors that may be causative are well recognized. Ethanol abuse is clearly associated with the development of chronic pancreatitis. A substantial proportion of these patients may also develop hepatic cirrhosis, alcoholic hepatitis, or fatty liver. Increase of protein content of pancreatic secretions by alcohol ingestion may lead to "plugging" of the pancreatic ducts by protein precipitation. The CCK-stimulated human pancreas has been shown to produce a higher protein secretion in patients with a history of alcoholic abuse than in healthy controls. In alcoholics, an increase in trypsinogen synthesis that is not paralleled by increased synthesis of trypsin inhibitor may allow intraductal activation of trypsinogen; pancreatic stone protein (a constituent of pancreatic stones that has a high affinity for calcium) may be increased; and pancreatic ductal deformation might predispose to blocked ducts. Other causes of chronic pancreatitis are hyperparathyroidism, other hypercalcemic states, and hyperlipidemia.

Diagnosis. In the known alcoholic, a recurring attack of pancreatitis or the presence of pancreatic calcifications suggests the diagnosis. Acute pseudocysts, malabsorption, or diabetes mellitus is a frequent development in these patients. The diagnostic challenge of chronic pancreatitis occurs in patients with painless disease or in those in whom differentiation between acute relapsing and chronic relapsing pancreatitis is difficult. Serum and urine pancreatic enzyme levels are of limited assistance because the acinar cell mass may be markedly decreased or mild inflammation may not elevate serum enzyme activities. It is essential, in view of the close association of chronic pancreatitis with alcoholism, to check liver and biliary function by determination of serum bilirubin and alkaline phosphatase levels. Pancreatic exocrine function should be examined by stimulation with secretin-CCK (p. 1460) or the Lundh meal (p. 1459) followed by duodenal enzyme or HCO_3^- determinations, or with the p-aminobenzoic acid test (p. 1462) and fecal fat estimations (p. 1464).

Radiological investigations and other physical and invasive techniques (see diagnosis of acute pancreatitis) are of considerable value in confirming suspected chronic pancreatitis.

Tumors of the Pancreas. Carcinoma of the pancreas is the major tumor in this organ; in 90% of the cases it is an adenocarcinoma of the ductal epithelium but in 5% of cases it arises from pancreatic islet tissue. In the latter instance the tumor may secrete a hormone so that the clinical presentation can be quite different from that of the ductal adenocarcinoma.

Adenocarcinoma of pancreatic ductal epithelium.[18] Carcinoma of the pancreas is diagnosed in about 24 000 persons every year in the United States. In males, the cancer is the fourth most common after those of lung, colon, and prostate; in females it is the fifth most common after those of breast, colon, lung, and ovary/uterus. The male:female ratio is 2:1; the mean age of onset is in the seventh and eighth decades of life. The tumor is more common in North American blacks than whites. Tumors arise in a fairly consistent distribution within the pancreas: head of pancreas, 60%; body of pancreas, 15–20%; and tail of pancreas, 5% of cases.

Treatment. Curative surgical resection is achieved in only a minority of patients because the disease is so advanced at the time of diagnosis. The majority of patients will therefore receive chemotherapy or radiation therapy or both, with possibly palliative surgical treatment. The prospects of survival, however, are grim.

Pathogenesis. The incidence of carcinoma of the pancreas has increased three-fold in the last 40 years. Cigarette smoking, coffee drinking, dietary agents, diabetes mellitus, chronic pancreatitis, and exposure to certain carcinogens as well as β-naphthylamine and benzidine have been suggested as risk factors. It has been estimated that the risk of pancreatic cancer among heavy smokers is over twice that of nonsmokers. An increased ingestion of fats and protein has been noted in populations in which the incidence of pancreatic cancer has increased. Juvenile-onset diabetes may be associated with a doubling of the incidence of pancreatic cancer.

Clinical presentation and diagnosis. Carcinoma of the pancreas develops insidiously, progresses relentlessly, and is invariably fatal.[18] There are no characteristic clinical features which lead to diagnosis. The appearance of a vague, dull, midepigastric abdominal pain and slow weight loss with possibly diarrhea, vomiting (due to invasion of the stomach or duodenum), and weakness may be features. About half of all patients will have jaundice due to a tumor in the head of the pancreas obstructing or encircling the distal common bile duct. Jaundice usually does not occur until the tumor mass is unresectable. When the tumor is in the body or tail of the pancreas, jaundice is a late manifestation. About a quarter of all patients may have a palpable abdominal mass. A small number (5%) may present with severe back pain, migratory thrombophlebitis, acute pancreatitis, psychiatric disturbance, or diabetes mellitus.

Noninvasive imaging techniques such as an upper gastrointestinal series, ultrasonography, and computed axial tomography are used to investigate the possible presence of pancreatic neoplasms. The diagnostic sensitivity and specificity of the last two techniques exceed 80%; both the tumor mass and enlargement of bile or pancreatic ducts can be readily visualized. Only the smallest tumors that are at the potentially resectable stage are likely to be missed. Invasive techniques[18] include angiography, which has high diagnostic sensitivity for nonresectable tumors but not for potentially resectable tumors; secretin or cholecystokinin stimulation of the pancreas with collection of pancreatic juice, a technique that has had mixed reviews;[18,27] transhepatic cholangiography, which may often be combined with percutaneous aspirate cytology; and ERCP, which has a diagnostic sensitivity and specificity exceeding 90%.[27] Biopsy techniques such as intraoperative transduodenal biopsy or fine needle aspiration biopsy are also used to obtain evidence of tumor.

A wide range of *laboratory tests* for carcinoma of the pancreas have been proposed.[63] Unfortunately, there are only a few studies that rigorously examine the value of these tests. One such study[76] compared the serum markers galactosyltransferase isoenzyme II, carcinoembryonic antigen, α-fetoprotein, ferritin, and ribonuclease with definitive diagnostic results obtained by invasive imaging techniques. The highest diagnostic sensitivity (67%) was obtained using the galactosyltransferase isoenzyme. When this determination was combined with an imaging technique, the diagnostic sensitivity markedly improved: with computed axial tomography to 88%; with ultrasound to 92%; and with ERCP to 100%. This experience should, of course, be tempered by the finding that the galactosyltransferase isoenzyme was positive in 71% of patients with other gastrointestinal tract cancers and in 21% of patients with other malignancies.[76] The reported diagnostic performance of these tests varies between different studies. For example, serum ribonuclease had a diagnostic sensitivity of 30% in this study; other workers have reported values at 70–90% in carcinoma of the pancreas. These examples illustrate well the difficulty of selecting suitable tests for detecting carcinoma of the pancreas. In some patients results of conventional tests for serum amylase, lipase, trypsin, or trypsin/creatinine clearance may be elevated, but the false positive rate is high. Therefore at present reliance must be placed on clinical suspicion and on invasive imaging techniques. A summary of some test findings in carcinoma of the pancreas follows: In carcinoma of the head of the pancreas with obstruction, the secretin-CCK test shows only negligible response. Serum lipase and amylase values may be elevated if a sufficient amount of acinar tissue remains. Alkaline phosphatase elevation is also commonly noted due to either bile duct obstruction that will also raise serum bilirubin levels or to multiple hepatic metastases. Results of fat absorption tests are abnormally low. In carcinoma of the body or tail of the pancreas, amylase and lipase activity in serum is most often normal. The secretin-CCK test shows a decrease in total volume of pancreatic juice; the composition is relatively normal although enzyme concentrations are somewhat more impaired than bicarbonate output. Absorption tests are abnormal. The sweat test is normal.

   Tumors of the pancreatic islet tissues[110] include insulinoma; glucagonoma; gastrinoma (Zollinger-Ellison syndrome, see p. 1440 and 1474); VIPoma (Verner-Morrison syndrome, or watery diarrhea, hypokalemia and achlorhydria syndrome, see p. 1444); somatostatinoma; and multiple endocrine neoplasia (MEN) syndromes.

   Insulinoma.[110] These β-cell islet tumors of the pancreas are usually benign; about 80% are single, some are multiple, and up to 15% are malignant. Diffuse islet cell hypertrophy can also occur. The clinical presentation is highly variable because of the diverse manner in which hypoglycemia can be expressed. The common symptoms of hypoglycemia include sweating, pallor, and tachycardia. However, fasting hypoglycemia produced by insulinomas proceeds more gradually; there may be mild personality changes, confusion, seizures, and even coma. Minor symptoms such as lightheadedness, blurred vision, paresthesias, hunger, fatigue, nausea, and vomiting may be features; the worst symptoms are observed in the morning after an overnight fast.

   The majority of the tumors are benign and solitary, and surgical resection is the treatment of choice. However, localization of these lesions may be difficult because they are only 1 cm in diameter; noninvasive techniques such as computed axial tomography or ultrasound are not sufficiently discriminatory to detect tumors of this size. Arteriography may be useful. Fortunately, these masses can generally be detected at laparotomy by palpating the pancreas. Failure to detect the tumor may require selective portal and splenic vein catheterization to map the source(s) of elevated insulin. If the tumor cannot be detected, a blind, stepwise, distal pancreatectomy may be invoked in the hope that the tumor is located in the region removed. If the tumor cannot be removed, then treatment with diazoxide (a drug which inhibits insulin release from the β-cells) will be necessary. If the tumor is malignant, a combination chemotherapy of streptozotocin and S-fluorouracil has been used.[110]

   The *diagnosis* is normally established by finding a low serum glucose (male < 55, female < 35 mg/dL). About 80% of patients with insulinomas will show the critical level after a 24-h fast followed by excercise, and > 95% after a 48-h fast with exercise. The diagnosis can be confirmed by finding an inappropriately elevated plasma insulin level > 15 mU/L and a plasma proinsulin level > 40 fmol/mL. However, these findings may be mimicked by the surreptitious use of oral sulfonylureas, and screening the serum for these agents may be necessary to exclude this possibility.

   Glucagonoma. The α-cell islet tumors of the pancreas are mostly malignant and demonstrate progressive metastatic growth. The tumor occurs in all age groups. At the time of diagnosis > 50% of patients will have metastatic spread of the tumor beyond the pancreas. Most of the primary tumors will be in the body and tail of the pancreas with a few in the head, and the remainder will be too dispersed to be located. The patient with a glucagonoma presents with the characteristic clinical findings of a dermatitis called migratory necrolytic erythema that occurs on the face and extremities and in areas associated with friction, such as the groin and buttocks. Patients also have substantial weight loss—even when the tumor is small and nonmalignant—and a mild, easily controlled diabetes mellitus. Psychiatric and neurological disturbances may occur in a minority of patients.

   The *diagnosis* is established by the appearance of clinical symptoms and by the finding of a plasma glucagon level exceeding 500 pg/mL. Intravenous glucose administration decreases plasma glucagon in patients with glucagonoma but oral carbohydrates paradoxically increase it, although this finding may be due to cross reaction of enteroglucagon in a glucagon assay. Usually, the tumor can be detected by ultrasound or computed axial tomography since its diameter usually exceeds 3 cm at the time of diagnosis. Surgical resection is the treatment of choice unless there has been metastatic spread. In this latter case, chemotherapy with streptozotocin and 5-fluorouracil is instituted.

Somatostatinoma. Within a few years of the discovery of somatostatin (see p. 1445) pancreatic tumors containing somatostatin-like material were found in association with high circulating levels of the hormone in the blood. Somatostatin may also be secreted ectopically by a variety of neoplasms including medullary carcinoma of the thyroid, oat-cell carcinoma of the lung, and thymic and bronchial carcinoids. The pancreatic tumor may also secrete adrenocorticotropic hormone and calcitonin.

The symptomatology can be related to the actions of somatostatin. Thus, a mild diabetic syndrome is usually present because insulin and glucagon release are inhibited and gallbladder contraction is inhibited, which, together with inhibition of pancreatic juice secretion, can lead to diarrhea and steatorrhea. Delayed gastric emptying and inhibition of acid secretion produce dyspepsia and achlorhydria.

The *diagnosis* of somatostatinoma requires a very high index of suspicion and a knowledge of the characteristic symptomatology. Unfortunately, the patient may present with metastases to liver, skin, or bone already evident. If so, the only treatment is palliative chemotherapy; with earlier presentation, surgical resection of the primary tumor may suffice.

Multiple endocrine neoplasia (MEN) syndrome. A number of syndromes are associated with hyperplasia or tumors in two or more endocrine organs. (See also Chapters 9 and 12.) These have been described as the multiple endocrine neoplasia syndromes. One of these syndromes (MEN I, Wermer's syndrome) consists of a highly variable combination of hyperplasia or tumors of the non-beta islet cells of the pancreas, the parathyroids, and the anterior pituitary, with variable degrees of hyperfunction of these systems. In one large series of cases of MEN I, 87% had involvement of the parathyroids, 81% had disorders of the endocrine pancreas, and 65% had lesions in the pituitary. Some of these patients had evidence of disease of the adrenals (38%) and thyroid gland (19%); the occasional patient had carcinoid tumors. Patients seldom present with clinical signs of more than one endocrinopathy although multiple organs are diseased. The tumors may also be "nonfunctioning" in the sense that insufficient hormone is secreted to produce symptoms; alternatively, even overproduction of some hormones such as prolactin in adult males or children, or pancreatic polypeptide, may not produce symptoms or illness.

The most important relationship between pancreatic tumor and MEN I is in gastrinoma, where up to a quarter of all gastrinoma patients will have MEN I. Other pancreatic tumors (e.g., insulinoma, VIPoma, and glucagonoma) have a much lower incidence of MEN I.

Once a diagnosis is made of the existence of a secretory endocrine tumor, the possibility of other endocrinopathies must be considered. This assessment should include a family history, since the kin of patients with MEN have a 50% chance of being affected.

Cystic fibrosis.[38] This disease is the most common lethal genetic defect of Caucasian populations. It is inherited in an autosomal recessive fashion, and its incidence in Caucasian populations ranges from 1:1900–1:15 000 with an average of 1:2000 live births. These data suggest a heterozygote carrier incidence of about 1:20. Unfortunately, at present heterozygote carriers cannot be identified. Cystic fibrosis is a major cause of malabsorption in infants and children and of chronic pulmonary disease in childhood. All cases of cystic fibrosis will eventually develop chronic pulmonary disease and, indeed, the respiratory disorder is responsible for much of the morbidity and nearly all of the mortality associated with this disease. Cystic fibrosis is a systemic disease affecting many types of exocrine glands. It is associated with a unique sweat gland defect, with abnormalities of mucus secretion, and with abnormalities of the male genital tract. The nature of the primary defect is not known; the disease cannot be reproduced in animals and is therefore difficult to study. Given appropriate long-term treatment, over 80% of patients with cystic fibrosis survive into their twenties; male patients survive slightly longer than females.

Clinical presentation. The disease should be suspected in a child with chronic or recurrent respiratory tract infections. But there are many other presentations, among them pancreatic exocrine and endocrine dysfunctions, meconium ileus in the newborn, distal intestinal obstruction syndrome, intussusception (invagination of a part of the bowel into the following part), rectal prolapse (a protrusion of the rectal mucous membrane through the anus), hepatomegaly, obstructive jaundice, and genital abnormalities in the male. The initial lesion in the lung is bronchial obstruction that leads to chronic infections and bronchopneumonia, emphysema, atelectasis, and abscesses. There can eventually be massive lung destruction. Usually, the upper respiratory tract is also infected and chronic sinus infection is very common. Therefore, cough is an early symptom and can become chronic and so frequent that vomiting may occur. By the age of two years more than three quarters of children with cystic fibrosis have affected lungs. The patient may have finger clubbing and be cyanotic. Serious complications such as pneumothorax, hemoptysis, and cor pulmonale may intervene. Respiratory failure may be, and frequently is, the cause of death.

Nearly 90% of patients with cystic fibrosis present with exocrine pancreatic dysfunction. The pancreatic lesions are caused by obstruction of the small ducts by secretions that eventually cause necrosis of the acinar and ductal cells. Fibrosis replaces necrosis in the pancreatic lobules. The islets of Langerhans are usually spared until late in the process. The resulting malabsorption and maldigestion will cause bulky and frequent stools, described as light, oily, and foul-smelling. The odor of the stools is particularly pungent. In an infant, there may be diarrhea and failure to gain weight. There may also be pancreatic endocrine dysfunction; about 2% of pediatric patients with cystic fibrosis and 13% of patients over 25 years of age have diabetes. Many more have glucose intolerance.

Meconium ileus is the presenting symptom in about 10% of infants with cystic fibrosis. In cystic fibrosis the meconium (the first material discharged from the bowels of a newborn infant) is extremely viscous and blocks the distal ileum. The condition presents with signs of intestinal obstruction within 48 h of birth; it is usually treated with enemas containing N-acetylcysteine, to reduce the viscosity of the meconium by cleaving disulfide bonds, and Tween 80, whose detergent action allows fluid to pass into the inspissated meconium.

Hepatic abnormalities occur in up to 50% of patients with cystic fibrosis. Plugging of the biliary tracts with mucus progresses sometimes to periportal inflammation, biliary fibrosis, and, in <5% of cases, to cirrhosis. Cholestasis is therefore a presentation of cystic fibrosis, and gallstones are an expected complication.

In the male there are developmental abnormalities of the epididymis, the vas deferens, and the seminal vesicles. It has been estimated that about 97% of males are sterile as a result of these abnormalities.

Treatment is largely symptomatic. It must be lifelong and directed to ensuring adequate pancreatic and pulmonary function and to preventing infections. Pancreatic enzyme replacement and vitamin and mineral supplements must be provided.

Pathogenesis.[38,45] Although the specific abnormality causing cystic fibrosis has not been identified, the mechanisms of tissue dysfunction include humoral factors and disorders of electrolyte secretion, of glycoproteins and mucus substances, and of cellular cholecystokinin proteolysis. Up to 90% of patients with cystic fibrosis, many heterozygote carriers, and a smaller number of healthy controls possess serum factors that produce abnormal movement of beating cilia, such as those on cells lining the respiratory tract, in various tissues. These serum factors have molecular weights between 75 000 and 180 000 and have been isolated also from the sweat and saliva of patients with cystic fibrosis. When perfused into normal human sweat glands, the factors cause inhibition of sodium reabsorption.

Abnormal electrolyte secretion is a characteristic and constant finding in cystic fibrosis. The sodium and chloride and less often the potassium content of sweat are elevated. Micropuncture studies of the sweat glands show that the osmolality and sodium content of the fluid elaborated by the secondary coil are normal. The defect appears to be in the reabsorption of sodium and chloride in the sweat ducts. Abnormal sodium and chloride content has also been noted in submaxillary gland saliva and, less often, in parotid gland saliva. Duodenal fluid and tears are not affected, but stimulated pancreatic juice shows an abnormal electrolyte pattern.

In almost all organs there is histological evidence of excessive accumulation of mucus with normal composition. However, glycoprotein fractions obtained from secretions of patients with cystic fibrosis contain more fucose in relation to sialic acid than those obtained from normal subjects. The calcium content of submaxillary saliva may be increased, and as the disease progresses calcium may precipitate the glycoproteins in the duct.

The basic defect in cystic fibrosis may lie in the deficiency of an enzyme that cleaves the Arg-Asp peptide bond in cholecystokinin and produces the active octapeptide CCK-8, which normally stimulates exocrine secretion in the intestine, gallbladder, and pancreas.[37] CCK-8 also potentiates the action of other gastrointestinal hormones. The anomaly of the male genital tract may possibly also be due to the nonactivation of a hormone dependent on the presence of CCK-8.

Diagnosis.[13,45,105] There must always be a suspicion of the disease in unexplained chronic pulmonary disease, chronic hepatobiliary disease, hypoproteinemia, edema, and failure to thrive. Most children with pancreatic insufficiency or meconium ileus and 30% with meconium peritonitis have cystic fibrosis.

The diagnosis rests primarily upon the findings with the sweat test (see p. 1471), although these results must be integrated with the clinical findings and family history. Recently, assay for immunoreactive trypsin in dried blood spots obtained at birth has been suggested as a practical, large-scale screening method for cystic fibrosis;[13] however, more studies will be required to assess this proposal completely.

Diarrhea[8,103] is defined as an increase in fecal water excretion. Normally, the colon absorbs more than 90% of the fluid entering it from the small gut. The absorptive capacity of the colon is about 5 L per day; a fluid load exceeding this volume may therefore result in diarrhea. There are a number of mechanisms responsible for diarrhea—active ion secretion, decreased solute absorption, increased luminal osmolality, and increased hydrostatic pressure.

Stimulated *active ion secretion* (secretory diarrhea) can occur in either the small gut or the colon. The major cation of gut secretions is sodium and the calculated fecal osmolality [$2 \times ([Na^+] + [K^+])$ mmol/kg] approximates the measured osmolality.[103] This type of diarrhea is caused by bacterial toxins (*Vibrio cholerae, Escherichia coli, Shigella dysenteriae* I, *Staphylococcus aureus*), hormones (Zollinger-Ellison syndrome, see p. 1440; VIPoma, see p. 1444, respectively), detergents that alter intestinal fluid and electrolyte movement (bile acids, fatty acids, some over-the-counter laxatives), and damage to the mucosa (regional enteritis, ulcerative colitis).

Decreased solute absorption results from the loss of mucosal transport functions secondary to altered mucosal morphology (celiac disease) or to isolated transport defects (monosaccharide malabsorption, see p. 1484).

Increased luminal osmolality results from the accumulation of osmotically active molecules. These species draw water into the gut, and subsequent absorption of sodium and water by the colon leaves a colonic fluid with a low sodium content. Osmotic diarrhea can be differentiated from secretory diarrhea by the large osmolal gap of the diarrheal fluid. Osmotic diarrhea is caused by laxatives, maldigestion (e.g., due to lactase and lipase deficiencies), the malabsorption of nonelectrolytes such as glucose and other sugars that occurs in mucosal disease, and transport defects of glucose and other sugars.

Increased hydrostatic pressure may occur when there is increased intraluminal pressure and associated luminal distention that stimulates fluid and electrolyte secretion.

The loss of diarrhea fluid results in water depletion (hypovolemia), Na^+ and K^+ depletion, and HCO_3^- loss that may cause a hyperchloremic metabolic acidosis with a normal anion gap.

Maldigestion, Malabsorption, and Related Disorders

The total quantity of fluid absorbed each day by the gut is estimated to be ~9 L, of which 1.5 L is ingested liquids and 7.5 L is gastrointestinal secretions.[1] The majority of secretions enter the gut in the upper portion of the tract—from the mouth, stomach, pancreas, and bile duct. More than 90% of the fluid is absorbed in the small intestine. The efficiency of absorption is due to unique features of the epithelial cell lining and its relation to the underlying rich vascular plexus. The intestinal mucosa lies in folds from which the epithelial lining projects in finger-like villi ~1 mm long. The absorptive face (brush border) of each epithelial cell consists of some 600 microvilli projecting ~1 μm from the cell. Folds, villi, and microvilli together present an absorptive surface ~600 times greater than would be inferred from the length and diameter of this portion of the gut. Guyton[42] estimates the absorptive surface area of the small intestine to be about 250 m^2, which is the surface area of a tennis court!

Several hundred grams of carbohydrates, over 100 g of fat, and 50–100 g of amino acids are absorbed daily in the small gut, but maximal absorptive capacity is believed to be at least 10 times greater. The maximal absorptive capacity for fluid is probably at least 20 L. This considerable reserve of capacity may compensate for mild to moderate degrees of dysfunction induced by disease processes, at least in the early phases.

Different segments of the gut are more or less specialized or adapted to particular stages of digestion and absorption. For convenience, these stages are called the luminal and small intestinal stages. The luminal stage is further divided into the secretory (pancreatic) and the biliary stages. The small intestinal stage is divided into the surface (brush border) and the cellular (delivery, transport) stages. Defects of digestion or absorption may occur at one or multiple stages and in one or more of the mechanisms of any stage. Defects in digestion and absorption are also related to the nature of the foodstuff that is processed.

It is necessary to distinguish between the terms maldigestion and malabsorption. *Maldigestion* is a dysfunction of the *digestive* process that can occur at a number of sites in the gastrointestinal tract. For example, hypoacidity in the stomach will reduce peptic digestion of protein; hyperacidity of the duodenum, due to the Zollinger-Ellison syndrome, can inactivate pancreatic enzymes; loss of brush border enzymes in the small intestine, due to a variety of processes, can prevent oligosaccharides from being further hydrolyzed; pancreatic insufficiency will reduce intraluminal enzyme activity in the small gut and will cause maldigestion of fats and proteins. On the other hand, *malabsorption* is a dysfunction of the *absorptive* process by the small gut; it is due to loss of absorptive epithelial cells caused by gluten, inflammation, infection, surgical resection, infiltrations, and so on.

Carbohydrates

Normal Digestion in the Small Intestine. After the action of salivary and pancreatic α-amylase on dietary starch and glycogen, carbohydrate content in the gut consists of newly formed maltose; ingested monosaccharides; dietary disaccharides such as lactose, sucrose, maltose, and trehalose; oligosaccharides such as dextrins and maltotriose; and indigestible polysaccharides such as cellulose, agar, and heteropolysaccharide food fibers.[73] This mixture of large and small molecules is osmotically active and draws water from the intravascular fluid into the gut. The osmotic load is increased as hydrolases of the brush border (Table 14-2) split di- and oligosaccharides into their constituent monosaccharides but is decreased as the transport mechanisms for the monosaccharide products move them into the epithelial cells (absorption by active transport).[73,77] However, for most oligosaccharides (lactose excepted), hydrolytic rate is greater than transport rate, i.e., transport is the rate-limiting step in reducing the concentration of monosaccharides and the osmotic load in the gut. When the transport system is operating at its maximum rate but monosaccharide concentration is still high, a compensating mechanism—inhibition of hydrolases by their monosaccharide products—slows down hydrolytic activity, keeps monosaccharide concentration relatively constant, and thereby keeps osmotic load and water concentration in the gut under control. The importance of this control can be illustrated by the consequences of intestinal disorders where ingested disaccharide is not split and absorbed. (See p. 1485.) Presence of undigested disaccharide causes increased fluid secretion into the gut and increased intestinal motility. The sugars are fermented by enteric bacteria, and hydrogen, CO_2, and organic acids are produced; abdominal discomfort, such as bloating, distention, and cramping, occurs, and absorption of fermentation products may lead to metabolic acidosis. In the large bowel, the presence of CO_2 and organic acids decreases pH and keeps the osmolality high so that water reabsorption in this segment is less efficient than it should be. The result is an acidic, liquid stool (diarrhea).

For lactose, the rate-limiting step is hydrolysis. Normally, however, accumulation of monosaccharide products does not occur, since the transport system is sufficiently fast to remove them. Mucosal lactase concentration is the lowest of all the oligosaccharidases and is not increased (induced) by feeding large amounts of lactose, as is the case for maltase and sucrase. Lactase, maltase, and sucrase all show diurnal variation in their activities.

Carbohydrate digestion appears to be complete in the midjejunum. The ileum represents some additional absorptive capacity although the hydrolytic activity of its surface is more limited than that of the jejunum. Increased volumes of fluid caused by undigested oligosaccharides can be reduced in the colon since that organ can absorb up to four times the normal colonic water load; for this reason, diarrhea may not always be a consequence of oligosaccharide malabsorption.

Maldigestion and malabsorption of carbohydrates. Generalized impairment of carbohydrate absorption occurs following a number of diseases that cause mucosal damage or dysfunction, e.g., celiac disease, tropical sprue, and acute gastroenteritis.[93] The hydrolase content or the transport mechanism, or both, may be affected. Symptoms are due to effects of undigested oligosaccharides in the intestine; the severity of the symptoms is related to the degree of damage or dysfunction caused by the causative disease.

Absorption defects can be caused by a deficiency in a single or all brush border oligosaccharidases.[73,77] The most common defect is lactose intolerance (lactase deficiency), of which there are two distinct types. *Congenital lactase deficiency* is a rare disorder in which lactase levels in

Table 14-2. BRUSH BORDER OLIGOSACCHARIDASES[73]

Enzyme	Substrate	Product
Lactase	Lactose	Glucose and galactose
Sucrase*	Sucrose	Fructose and glucose
	Maltose	Glucose
Trehalase	Trehalose	Glucose
α-Glucosidase	Short linear oligosaccharides	Glucose
α-Dextrinase* (isomaltase)	α-Dextrins containing some α-1,6 linkages	Glucose

*Sucrase and isomaltase occur as a single complex in the brush border.

the mucosa are low or absent at birth. Symptoms occur as soon as milk is taken; stools have a low pH and contain glucose produced by bacterial action on undigested lactose. Absent or low intestinal lactase in the neonate, however, cannot be taken as proof of *congenital* deficiency, since lactase is normally the slowest of the oligosaccharidases to reach normal levels in the newborn's gut. A definitive diagnosis must be deferred until after maturation of the lactase synthesis system can be expected to have occurred. In the interim, relief is dependent on adjustments of dietary composition that appear to reduce severity of symptoms. An abnormal oral lactose tolerance test obtained a few months after birth could also be due to congenital glucose-galactose intolerance (see below); the differential diagnosis requires performance of an oral glucose tolerance test in conjunction with the lactose tolerance test.

Physiological change is responsible for *acquired lactase deficiency*. Lactase activity declines as a child ages, but the age at which decline begins is genetically determined in an autosomal recessive fashion and differs among ethnic groups. Thai children, for example, have low lactase activity by the age of two, while white American children do not begin to show decline until after the age of five. Prevalence of lactose intolerance in adults also varies among various ethnic groups. In adults of Northern European ancestry, prevalence is < 20%; in those of African or Asian ancestry, prevalence is > 65%. If symptoms of flatulence, abdominal discomfort, bloating, or diarrhea occur after consumption of one or two glasses of milk or of a large portion of ice cream or yogurt, lactose intolerance should be suspected. Suspicion would be increased if the subject were of an ethnic group with high prevalence of lactose intolerance—Blacks, Orientals, Jews, Arabs or other Mediterranean peoples, or Indians of North, South, or Central America. Lactose intolerance aggravates bowel symptoms of diseases such as ulcerative colitis.

Sucrase-isomaltase deficiency is inherited as an autosomal recessive trait. The two enzymes occur together in the mucosa as a complex so both activities are low or absent, although sucrase activity is always more impaired than is isomaltase activity. The deficiency is rare in North Americans but more common in Eskimo tribes. Symptoms of sugar malabsorption occur when sucrose-sweetened foods are consumed. *Trehalase deficiency* is the cause of symptoms produced by ingestion of mushrooms that contain large amounts of trehalose, a disaccharide of two glucose units in 1,2 linkage.

Malabsorption of monosaccharides can also cause intestinal symptoms more commonly attributed to maldigestion of oligosaccharides. *Glucose-galactose malabsorption* is also inherited as an autosomal recessive trait. Symptoms occur in the affected neonate as soon as milk (lactose) is taken but also follow ingestion of glucose- or galactose-containing foods. Symptoms due to *fructose malabsorption* occur upon ingestion of fruit. This dietary intolerance is quite a different disorder from hereditary fructose intolerance in which the hepatic enzyme aldolase is defective.

Tests for the Diagnosis of Disaccharidase Deficiency. The most direct diagnostic test for an intestinal enzyme deficiency is histochemical examination of the brush border of the intestinal epithelium. Biopsy is, however, difficult and undesirably invasive and so is rarely employed. A more practical approach is an oral challenge of the patient with one disaccharide at a time followed by measurement of serum glucose levels. The challenge dose is usually 50 g of the selected sugar per m^2 of body surface area or 1 g/kg body weight. If the appropriate disaccharidase is present, the disaccharide will be hydrolyzed at the brush border and its component monosaccharides absorbed. Since glucose is consistently a component of the common oligosaccharides, hydrolysis and absorption will affect blood glucose levels. Increase of > 30 mg/dL over the fasting glucose level is taken to indicate adequate disaccharidase activity; an increase of < 20 mg/dL is considered to indicate deficiency. Appearance of the monosaccharide in the blood obviously depends not only on the presence of the appropriate hydrolytic enzyme for the disaccharide but also on the ability of the transport mechanism to internalize the monosaccharide product. Verification of absorptive ability is made by performing a glucose tolerance test prior to the disaccharide challenge test. A normal rise from the fasting glucose level (see Chapter 6, Carbohydrates) indicates absence of a transport defect.

Because undigested oligosaccharides undergo bacterial degradation in the gut, measurement of hydrogen excretion in the breath by either gas chromatography or polarography has been used to confirm malabsorption. Positive correlation between hydrogen excretion in expired air and hydrogen production in the intestinal lumen has been claimed; the test may be the test of choice for detection of lactase deficiency.[77] Test of pH on a filtrate or clear supernatant of liquid stool or on an aqueous extract of formed stool is also employed.

Lipids

Normal digestion in the small intestine. The average daily dietary intake in North America contains over 100 g of fat.[73] Less than 5% of this amount is recovered in the feces, indicating the overall efficiency of the normal processes of fat digestion and absorption. Most dietary fat is in the form of triglyceride. Although pancreatic lipase is quantitatively the most important hydrolytic enzyme, a lipase secreted by the tongue and soft palate and by the gastric mucosa can hydrolyze up to 30% of fats in the stomach.[36,73] This nonpancreatic lipase may have a significant role in lipid digestion when pancreatic function is impaired. The lingual and gastric lipases do not require bile salts to function and they are active at gastric pH. Liberation of fatty acids in the stomach serves to stabilize the surface of the triglyceride emulsion and promotes the binding of pancreatic colipase; in addition, the liberated fatty acids stimulate release of cholecystokinin from the duodenal mucosa.

In the duodenum and jejunum, pancreatic lipase in the presence of bile salts, colipase, and the triglyceride emulsion produces fatty acids and monoglycerides. These products form micelles with bile salts and lysophosphoglycerols; the micelles convey the nonpolar lipid molecules from the lumen to the epithelial cell surface and dissociate there to produce a high concentration of monoglycerides, lysophosphoglycerols, and fatty acids that are absorbed into the mucosal cell. This absorption is facilitated by the presence of a fatty acid binding protein, in the cytosol of the cell, that has a high affinity for fatty acids. Within the cell, fatty acids are immediately incorporated into triglycerides. The triglycerides, together with phospholipids, cholesterol and its esters, and a specific apolipoprotein, are formed into spherical chylomicrons that are then released by exocytosis into the lymphatic system of the small gut. From the lymphatics, chylomicrons enter the bloodstream and are distributed to the liver, adipose tissue, and other organs.[73] Table 14-3 lists conditions that cause failure in any stage of the overall digestive/absorptive process and that compromise its efficiency.

Amino Acids and Proteins

The average daily dietary intake of protein in North America is ~100 g. Another 70 g of so-called endogenous protein enters the intestinal contents daily as protein of digestive juices and shed mucosal cells. Of the total protein load, only ~10 g is lost in the feces.[73]

Normal digestion and absorption. Protein digestion is initiated in the stomach by the action of pepsin in a highly acid medium. The acidity also denatures the protein, unfolding the polypeptide chains for better access by the gastric, pancreatic, and intestinal proteolytic enzymes. Additionally, the polypeptides and amino acids produced in the stomach by action of pepsin are potent secretagogues for the hormones that stimulate the pancreas and intestine. Stimulated pancreatic secretion contains the proenzyme forms of trypsin, chymotrypsin, elastase, exopeptidases, carboxypeptidases, and other enzymes. Stimulation of the intestine by gastrointestinal hormones

Table 14-3. CONDITIONS ASSOCIATED WITH MALDIGESTION
AND MALABSORPTION OF LIPIDS

Pancreatic stage
  Isolated pancreatic lipase or colipase deficiency (inherited)
  Pancreatic insufficiency
    Cystic fibrosis
    Chronic pancreatitis
    Obstruction of pancreatic duct
Biliary stage
  Decreased synthesis of bile salts in severe hepatic insufficiency
  Decreased delivery of bile salts in obstruction of biliary tract or cholestatic biliary disease
  Decreased concentration of conjugated bile salts because of increased acidity, drugs affecting micelle formation, intestinal stasis, or bacterial overgrowth
  Increased intestinal loss of bile salts because of distal ileal resection, surgical ileal bypass, or diseased terminal ileum
Cellular and delivery stage in small intestine
  Rapid transit, dumping syndrome
  Improper emulsification, following certain types of gastrectomy
  Altered duodenal pH as in Zollinger-Ellison syndrome
  Diseases of small intestinal lymphatics
Defects at multiple stages of digestion and absorption
  Decreased CCK release due to severe mucosal destruction, as in sprue or regional enteritis

---

liberates several proteolytic enzymes from the brush border. One of them, an enteropeptidase called enterokinase, selectively cleaves a hexapeptide from the N-terminus of trypsinogen to form trypsin. Trypsin then activates more trypsin (autocatalysis) and other pancreatic proenzymes as well. Action of the pancreatic enzymes on the partly digested proteins within the lumen produces peptides, 2–6 amino acid residues in length, and also single amino acids. The peptides are hydrolyzed to single amino acids by the aminopeptidases and other hydrolytic enzymes of the brush border. At least seven carrier systems with overlapping specificities for different amino acids (Table 14-4) are involved in transport of amino acids into the cells. Absorption of amino acids by these transport systems is faster in the jejunum than in the ileum. Another transport system, the γ-glutamyl cycle in which membrane-bound γ-glutamyl transferase plays a central role, has also been suggested.[73] Small quantities both of whole proteins and of peptides are probably absorbed as such into the cell, where they may be hydrolyzed into their constituent amino acids or from which they may be secreted intact into the bloodstream. The major portion of digested protein, however, is taken up by the cell in the form of amino acids and passed into the underlying rich vascular plexus. The plexus is drained by the portal circulation, and it is by this route that absorbed amino acids reach the liver and then the systemic circulation.

Maldigestion and malabsorption of proteins. The foregoing account clearly predicts that pancreatic and small intestinal diseases will be major causes of protein maldigestion and malabsorption. Any of the diseases listed under the pancreatic stage in Table 14-3 lead as well to impaired assimilation of protein. Note, however, that the reserve capacity of the pancreas is very great, and fecal loss of protein may not become significant in pancreatic insufficiency states until trypsin has fallen to ~10% of normal.[34] Two rare disabling disorders, trypsin deficiency and enteropeptidase deficiency, have been reported. As would be expected from the important role each of these plays in the activation of proteolytic proenzymes, deficiency of either of them has far-reaching effects on efficiency of protein digestion.

Mucosal diseases may affect protein assimilation by a number of mechanisms. Reduction in number of mucosal cells decreases peptidase activity in the intestine as well as intestinal absorptive capability for amino acids. Disease may increase the turnover of intestinal cells and the rate of desquamation; this cell loss, together with increased losses of plasma proteins from the damaged intestinal surface, can cause a negative nitrogen balance. *Surgical resection* of the intestine not only reduces the total intestinal absorptive surface but may remove a segment of the gut that is specialized for absorption of certain nutrients; an example is resection of the distal ileum, which removes the active transport system for the vitamin B_{12}–intrinsic factor complex. Resection may also alter intestinal motility and thus lead to stasis and bacterial overgrowth that can intensify a negative nitrogen balance. Another cause for reduced protein assimilation is *protein-losing enteropathy.* Inflamed or ulcerated mucosa can exude considerable quantities of plasma proteins which, if not subsequently digested and reabsorbed, represent large fecal losses of nitrogen; these losses will cause hypoproteinemia.

Clinical Presentation and Diagnosis of Malabsorption and Maldigestion

Clinical presentation[40,61] of the patient suffering from a disorder of digestion or absorption classically includes the following features:

1. Evidence of *underlying disease,* such as the facial flush and enlarged liver of the carcinoid syndrome, or signs of thyroid disease.

Table 14-4. AMINO ACID TRANSPORT SYSTEMS IN CELL MEMBRANES[73]

Amino Acid Carrier System	Amino Acids Transported
A (neutral amino acids with short side chains)	Alanine, glycine, proline, serine, methionine
ASCP (higher specificity than system A)	Alanine, serine, cystine, proline
L (neutral amino acids with branched or aromatic side chains)	Leucine, isoleucine, valine, phenylalanine, methionine, tyrosine, tryptophan
Ly (basic amino acids)	Lysine, arginine, ornithine, histidine
Dicarboxylate (low activity, not very specific)	Glutamate, aspartate
β	Taurine, β-alanine
N (amides)	Glutamine, asparagine, histidine

2. Evidence of *general ill health.* Anorexia, weight loss, fatigue following minor effort, and dyspnea may be seen. Edema due to hypoalbuminemia or weakness, tetany, and dehydration due to electrolyte imbalance and water loss may be present. In pancreatic exocrine insufficiency, however, hyperphagia is the rule; patients often report a very high (5000 kcal/d) food intake.

3. *Isolated nutritional deficiencies.* Iron, folate, or vitamin B_{12} deficiency may manifest as anemia, vitamin K deficiency as a bleeding tendency, and vitamin D deficiency as bone disease. These deficiencies are particularly probable with malabsorption in the small intestine. They are reflected by a variety of signs and symptoms—glossitis, pallor, dermatitis, petechiae, bruising, hematuria, muscle or bone pain, or neurological abnormalities.

4. *Abdominal symptoms* such as discomfort, distention, flatulence, and borborygmi (rumbling and gurgling sounds due to movement of flatus in the intestine).

5. *Diarrhea* and possibly *steatorrhea.* In severe cases of steatorrhea (excess fat in feces), the stool is typically loose, bulky, offensive, greasy, light colored, and difficult to flush away. Alternatively, the stools may appear normal but be more bulky or be passed with greater frequency. Steatorrhea may be absent in cases of small intestinal disease.

Early presentation of maldigestion/malabsorption syndrome will, however, be more subtle than this listing would indicate. There may be only a slight alteration in volume or consistency of the stool and only mild symptoms of anorexia, fatigue, and lack of interest in daily activities. It is in these cases that the physician who suspects malabsorption on clinical grounds must rely on the laboratory to rule in or to rule out the diagnosis.

Causes of the maldigestion/malabsorption syndrome are listed in Table 14-5. Some of the diseases are described here in more detail.

Celiac disease.[102] Synonyms are celiac sprue, idiopathic steatorrhea, nontropical sprue, adult celiac disease, and gluten-induced enteropathy. Celiac disease is characterized by malabsorption, a specific lesion of the small intestinal mucosa, and an improvement when gluten-containing foods are withdrawn from the diet. The mucosal lesion consists of flattening of the mucosal surface, absence of villi, and disappearance of the microvilli of the mucosal epithelial cell. Since the absorptive surface is markedly reduced, the resulting maldigestion and malabsorption are severe. The histological changes are due to interaction of gluten, a water-insoluble protein of cereal grains, with the epithelium.

Clinical features of celiac disease are those of the classic malabsorption syndrome. Iron, folate, and vitamin B_{12} deficiency anemias occur; calcium and vitamin D deficiencies cause osteomalacia and osteoporosis. Neurologic symptoms may be due to multiple vitamin deficiency; secondary hyperparathyroidism due to severely impaired calcium absorption; adrenocortical insufficiency due to electrolyte deficiencies; and panhypopituitarism due to profound malnutrition.

Tropical sprue.[50] This disease is a chronic acquired disorder characterized by abnormalities of small bowel structure and function that become progressively more severe and lead to nutritional deficiency. The disorder is usually curable by treatment with tetracycline and folic acid. It ordinarily begins as an intestinal infection that resolves of itself but then evolves into a stage of milk intolerance as lactase deficiency develops. Eventually, and with a concomitant change in jejunal morphology, a frank syndrome of malabsorption is expressed. At this last stage, anorexia, weight loss, megaloblastic anemia, and edema are evident.

Ulcerative colitis.[19] The etiology of this chronic inflammatory disorder of the intestine is unknown. The mucosae of the rectum and left colon are the sites most commonly affected, but any part of the alimentary canal may be involved. Histologic changes are not characteristic; the appearance of the inflamed mucosa is similar to that of bacterial infection or other inflammatory conditions. Although the disease process is usually confined to the mucosa, in very serious cases it may extend into the muscular wall of the colon to become what is called toxic megacolon. In ulcerative colitis there are intermittent attacks of diarrhea and rectal bleeding; the diarrhea is often voluminous and may last for several weeks, culminating in water and electrolyte imbalance. The bleeding leads to an iron deficiency anemia; hypoalbuminemia occurs due either to low protein intake because of the anorexia or to exudative loss of protein from the ulcerated mucosa, or to both.

In toxic megacolon, there may be shock and fluid and electrolyte depletion that can lead to renal failure and cardiac arrhythmias. Additional complications include perforation or cancer of the colon, massive hemorrhage, liver disease, and renal calculi.

Regional enteritis (Crohn's disease).[28] Regional enteritis is a chronic inflammation of the intestine of unknown etiology. The distal ileum and colon are the sites most often affected, but any part of the gut can be involved. The disease has a variable course, several complications are possible, and the disease may recur after surgical resection. The inflammatory process affects the entire intestinal wall; it can progress to ulceration, to fibrosis that leads to contraction and obstruction, and to formation of fistulae between loops of the gut or between the gut and other organs. The complications further include development of malignant intestinal neoplasms, arthritis, mild abnormalities of liver function, and renal involvement with increased

Table 14-5. CLASSIFICATION OF DISEASES THAT CAUSE
MALDIGESTION OR MALABSORPTION

Intraluminal stage
  Defect at secretory (pancreatic) stage
    Absence of trypsin, lipase, or colipase (inherited)
    Cystic fibrosis*
    Pancreatectomy
    Chronic pancreatitis
    Carcinoma of the pancreas*
    Defective stimulation due to intestinal disease or gastric surgery
    Obstruction of pancreatic duct
    Zollinger-Ellison syndrome
    Malnutrition
  Defect at biliary stage
    Parenchymal liver disease*
    Biliary obstruction*
    Terminal ileal disease*
    Resection of terminal ileum*
    Administration of cholestyramine
    Bacterial action as a result of stasis or bacterial overgrowth
Small intestinal stage
  Defect at surface stage
    Enterokinase deficiency (inherited)
    Disaccharidase deficiencies (inherited and acquired)
  Defect at cellular and delivery stage
    Amino acid transport defects
    Primary vitamin B_{12} malabsorption
    Massive resection
    Radiation enteritis
    Intestinal ischemia
    Celiac sprue
    Tropical sprue*
    Ulcerative colitis
    Regional enteritis
    Whipple's disease
    Primary intestinal lymphoma
    Hypogammaglobulinemia
    Food allergy
    Amyloidosis
    Parasitized states
Multiple stage defects
  Post-gastrectomy
  Diabetes mellitus
  Endocrinopathies
  Collagen disease
  Administration of neomycin

*Most common in occurrence.

risk of renal stone formation. The diffuse involvement of the jejunum and ileum affects digestive and absorptive function throughout the gut and leads to numerous problems—disaccharidase deficiency, protein-losing enteropathy, chronic blood loss, lymphatic obstruction, vitamin B_{12} and folate deficiency, stasis of intestinal contents, and generalized inadequate caloric and protein malnutrition. Decrease of absorptive surface by fistulae or palliative resection and overgrowth of bacterial flora due to stasis exacerbate the nutritional deficiencies.

Diagnosis of Pancreatic and Intestinal Diseases

After the physician has taken the patient's history and made the physical examination, the laboratory plays an important part in detecting or ruling out malabsorption and in elucidating its cause. Table 14-6 illustrates the many diagnostic modalities that are available.

The diagnostic process should begin with the determination of total fecal lipids (p. 1464).[40,100] If steatorrhea is not indicated, hydrogen breath tests and selected disaccharide tolerance tests are utilized to evaluate for brush border oligosaccharidase deficiencies. If steatorrhea is present, the total fecal lipid value is a measure of the degree of malabsorption of lipids. The xylose absorption test is then carried out, along with intestinal X-rays; if xylose absorption is

Table 14-6. SELECTED TESTS USED IN DIAGNOSIS OF PANCREATIC AND INTESTINAL DISEASES

Pancreatitis, pseudocyst
  Amylase, serum
  Amylase, urine
  Isoamylase, serum
  Amylase/creatinine ratio, urine
  Lipase, serum
  Immunoreactive trypsin, serum
  Miscellaneous tests: calcium, serum; glucose, serum; creatinine, serum; albumin, serum; hematocrit
  Visualization techniques*
Pancreatic exocrine insufficiency
  Pancreatic juice or duodenal content after stimulation with Lundh meal, secretin, or secretin-CCK for volume, concentration of HCO_3^-, trypsin, chymotrypsin, lipase, and amylase
  Bz-Ty-PABA test after stimulation with Lundh meal
  Stool for trypsin or chymotrypsin, with or without Lundh meal
  Stool for fat
Malignant disease
  Carcinoembryonic antigen, serum
  Pancreatic oncofetal antigen, serum
  α-Fetoprotein, serum
  Special enzyme tests, pancreatic juice or duodenal content
  Pancreatic exocrine function tests, with cytologic examination
  Visualization techniques*
Malabsorption
  Hydrogen test, breath
  Disaccharide tolerance tests
  Xylose absorption test
  Fat, stool
  Tests for pancreatic exocrine insufficiency (above)
  Intestinal biopsy, histochemical study for intestinal enzymes
  Iron, folate, vitamin B_{12} in serum
  Visualization techniques*

*Visualization techniques include angiography, computed axial tomography (CAT scan), endoscopic retrograde cholangiopancreatography (ERCP), ultrasound scan, and radiologic examination of intestine with contrast media.

normal, as is usual in pancreatic exocrine insufficiency, tests of pancreatic function are required. Jejunal biopsy for histochemical and morphological examination is helpful to confirm congenital oligosaccharidase deficiencies and, in other conditions, is *necessary* to evaluate the nature and degree of damage to the intestinal mucosa. X-rays and other visualization techniques are required to demonstrate abnormal motility or anatomic features of the gut. A wide variety of serum and urine tests (Table 14-6) are employed in specific contexts to refine or to confirm possible diagnoses or to assign the cause for the presenting malabsorption syndrome.

References

1. Adria Laboratories: Xylo-pfan (D-xylose) kit insert. Adria Laboratories, Inc., Columbus, OH 43215.
2. American Bio-Science Laboratories, Van Nuys, CA 91405, 1984.
3. Anderson, D., Elahi, D., Brown, J. C., et al.: Oral glucose augmentation of insulin secretion. J. Clin. Invest., 62:152-161, 1978.
4. Arvanitakis, C., and Cooke, A. R.: Diagnostic tests of exocrine pancreatic function and disease. Gastroenterology, 74:932-948, 1978.
5. Baron, J. H.: Clinical Tests of Gastric Secretion: History, Methodology, and Interpretation. London, Macmillan Press, 1978.
6. Bartlett, K., Dobson, J. V., and Eastham, E.: A new method for the detection of hydrogen in breath and its application to acquired and inborn sugar malabsorption. Clin. Chim. Acta, 108:189-194, 1980.
7. Bin, T. L., Stopard, M., Anderson, S., et al.: Assessment of fat malabsorption. J. Clin. Pathol., 36:1362-1366, 1983.
8. Binder, H. J.: Absorption and secretion of water and electrolytes by small and large intestine. In: Gastrointestinal Disease: Pathophysiology, Diagnosis, Management. 3rd ed. M. H. Sleisenger and J. S. Fordtran, Eds. Philadelphia, W. B. Saunders Co., 1983.
9. Blackman, A. H., Lambert, D. L., Thayer, W. R., et al.: Computed normal values for peak acid output based on age, sex, and body weight. Am. J. Digest. Dis., 15:783-789, 1970.
10. Blankenhorn, D. H., and Ahrens, E. H.: Extraction, isolation and identification of hydrolytic products of triglyceride digestion in man. J. Biol. Chem., 212:69-81, 1955.

11. Bloom, S. R., Mitznegg, P., and Bryant, M. G.: Measurement of human plasma motilin. Scand. J. Gastroenterol., *11* (Suppl. 39):47-56, 1976.

12. Bloom, S. R., and Polak, J. M.: Establishing the physiology of gastrointestinal hormones. *In*: Polypeptide Hormones. R. F. Beers, Jr., and E. G. Bassett, Eds. New York, Raven Press, 1980.

13. Blyth, S. A., and Farrell, P. M.: Advances in the diagnosis and management of cystic fibrosis. Clin. Biochem., *17*:277-283, 1984.

14. Bouchier, I. A. D.: Gastric and pancreatic function tests. Ann. Clin. Biochem., *7*:122-125, 1970.

15. Brooks, F. P.: Clinical usefulness of gastric acid secretory tests. Postgrad Med., *51*:189-193, 1972.

16. Brooks, F. P., and O'Neill, F.: Gastric analysis for hydrochloric acid. *In*: Practice of Medicine, Vol. 2. Hagerstown, Md., Harper & Row, 1973.

17. Broughton, A.: Pepsinogen-I and gastrin radioimmunoassay in the diagnosis of gastric pathology. Lab. Man., April, 1980, pp. 46-48.

18. Cello, J. P.: Carcinoma of the pancreas. *In*: Gastrointestinal Disease: Pathophysiology, Diagnosis, Management. 3rd ed. M. H. Sleisenger and J. S., Fordtran, Eds. Philadelphia, W. B. Saunders Co., 1983.

19. Cello, J. P. Ulcerative colitis. *In*: Gastrointestinal Disease: Pathophysiology, Diagnosis, Management. 3rd ed. M. H. Sleisenger and J. S. Fordtran, Eds. Philadelphia, W. B. Saunders Co., 1983.

20. Chang, T.-M., and Chey, W. Y.: Radioimmunoassay of gastrointestinal peptides in normal and abnormal states. *In*: Progress in Gastroenterology, Vol. 4. G. B. J. Glass and P. Sherlock, Eds. Orlando, Grune & Stratton, 1983.

21. Crivelli, O., Pera, A., Lombardo, L., et al.: Antral G- and D-cell counts in chronic renal failure. Scand. J. Gastroenterol., *14*:327-331, 1979.

22. Crossley, J. R., Smith, P. A., Edgar, B. W., et al.: Neonatal screening for cystic fibrosis, using immunoreactive trypsin assay in dried blood spots. Clin. Chim. Acta, *113*:111-121, 1981.

23. Davenport, H. W.: Control of secretion. *In*: Physiology of the Digestive Tract. 3rd ed. H. W. Davenport, Ed. Chicago, Year Book Medical Publishers, Inc., 1971.

24. Davis, G. R.: Neoplasms of the stomach. *In*: Gastrointestinal Disease: Pathophysiology, Diagnosis, Management. 3rd ed. M. H. Sleisenger and J. S. Fordtran, Eds. Philadelphia, W. B. Saunders Co., 1983.

25. Davis, P. B., Hubbard, V. S., and di Sant'Agnese, P. A.: Low sweat electrolytes in a patient with cystic fibrosis. Am. J. Med., *69*:643-646, 1980.

26. deBenneville, P. L., Godfrey, W. J., Sims, H. J., et al.: New substrates for a pancreatic exocrine function test. J. Med. Chem., *15*:1098-1100, 1972.

27. DiMagno, E. P., Malagelada, J.-R., Taylor, W. F., et al.: A prospective comparison of current diagnostic tests for pancreatic cancer. N. Engl. J. Med., *297*:737-742, 1977.

28. Donaldson, R. M.: Crohn's disease. *In*: Gastrointestinal Disease: Pathophysiology, Diagnosis, Management. 3rd ed. M. H. Sleisenger and J. S. Fordtran, Eds. Philadelphia, W. B. Saunders Co., 1983.

29. Dürr, H. K., Schneider, R., Bode, C., et al.: Fecal chymotrypsin: Study on some characteristics of the enzyme. Digestion, *17*:396-403, 1978.

30. Elias, E., Wood, T., and Redshaw, M.: Diagnostic importance of changes in circulating concentrations of immunoreactive trypsin. Lancet, *2*:66-68, 1977.

31. Farrar, G. E., and Bower, R. J.: Gastric juice and secretion: Physiology and variations in disease. Ann. Rev. Physiol., *29*:141-168, 1967.

32. Folin, A., and Ciocalteu, V.: On tyrosine and tryptophane determination in proteins. J. Biol. Chem., *73*:627-654, 1927.

33. Ford, T. F., Hermon-Taylor, J., and Grant, D. A. W.: A sensitive fluorometric assay for the simultaneous estimation of pepsin and pepsinogen in gastric mucosa. Clin. Chim. Acta, *126*:17-23, 1982.

34. Freeman, H. J., Sleisenger, M. H., and Kim, Y. S.: Human protein digestion and absorption: Normal mechanisms and protein-energy malnutrition. Clin. Gastroenterol., *12*:357-378, 1983.

35. Gibson, L. E., di Sant'Agnese, P. A., and Shwachman, H.: Procedure for the quantitative iontophoretic sweat test for cystic fibrosis. Cystic Fibrosis Foundation, 6000 Executive Blvd., Suite 510, Rockville, MD 20852, 1985.

36. Glickman, R. M.: Fat absorption and malabsorption. Clin. Gastroenterol., *12*:323-334, 1983.

37. Gosden, C. M., and Gosden, J. R.: Fetal abnormalities in cystic fibrosis suggest a deficiency in proteolysis of cholecystokinin. Lancet, *2*:541-546, 1984.

38. Grand, R. J., Park, R. W., and Stafford, R. J.: Pancreatic disorders in childhood. *In*: Gastrointestinal Disease: Pathophysiology, Disease, Management. 3rd ed. M. H. Sleisenger and J. S. Fordtran, Eds. Philadelphia, W. B. Saunders Co., 1983.

39. Gray, G. M.: Intestinal disaccharidase deficiencies and glucose-galactose malabsorption. *In*: The Metabolic Basis of Inherited Disease. 5th ed. J. B. Stanbury, J. B. Wyngaarden, and D. S. Fredrickson, Eds. New York, McGraw-Hill, 1983.

40. Gray, G. M.: Maldigestion and malabsorption: Clinical manifestations and specific diagnosis. *In*: Gastrointestinal Disease: Pathophysiology, Diagnosis, Management. 3rd ed. M. H. Sleisenger and J. S. Fordtran, Eds. Philadelphia, W. B. Saunders Co., 1983.

41. Grendell, J. H., and Cello, J. P.: Chronic pancreatitis. *In*: Gastrointestinal Disease: Pathophysiology, Diagnosis, Management. 3rd ed. M. H. Sleisenger and J. S. Fordtran, Eds. Philadelphia, W. B. Saunders Co., 1983.

42. Guyton, A. C.: Textbook of Medical Physiology. 6th ed. Philadelphia, W. B. Saunders Co., 1981.

43. Haeney, M. R., Culank, L. S., Montgomery, R. D., et al.: Evaluation of xylose absorption as measured in blood and urine: A one-hour blood xylose screening test in malabsorption. Gastroenterology, *75*:393-400, 1978.

44. Hansen, W.: Akute Pankreatitis. Der Informierte Artz, *8*:62-69, 1980.

45. Heeley, A. F., and Watson, D.: Cystic fibrosis—its biochemical detection. Clin. Chem., *29*:2011-2018, 1983.

46. Jeejeebhoy, K. N., Ahmad, S., and Kozak, C.: Determination of fecal fats containing both medium and long chain triglycerides and fatty acids. Clin. Biochem., *3*:157-163, 1970.

47. Johnson, S. L., and Mayersohn, M.: Quantitation of xylose from plasma and urine by capillary column gas chromatography. Clin. Chim. Acta, *137*:13-20, 1984.

48. Kasper, P., Möller, G., and Wahlefeld, A.: A new photometric assay for chymotrypsin in stool. Clin. Chem., *30*:1753-1757, 1984.

49. Kendall, M. J., Nutter, S., and Hawkins, C. F.: Bacteria and the xylose test. Lancet, *1*:1017-1018, 1972.

50. Klipstein, F. A.: Tropical sprue. *In*: Gastrointestinal Disease: Pathophysiology, Diagnosis, Management. 3rd ed. M. H. Sleisenger and J. S. Fordtran, Eds. Philadelphia, W. B. Saunders Co., 1983.

51. Kopito, L., and Shwachman, H.: Studies in cystic fibrosis: Determination of sweat electrolytes in situ with direct reading electrodes. Pediatrics, *43*:794-798, 1969.

52. Korman, M. G., Soveny, C., and Hansky, J.: Radioimmunoassay of gastrin. The response of serum gastrin to insulin hypoglycaemia. Scand. J. Gastroenterol., *6*:71-75, 1971.

53. Kretcher, N.: Lactose and lactase. Sci. Am., *227*:70-78, 1972.

54. Lake-Bakaar, G., McKavanagh, S., Rubio, C. E., et al.: Measurement of trypsin in duodenal juice by radioimmunoassay. Gut, *21*:402-407, 1980.

55. Lam, S. K.: Hypergastrinaemia in cirrhosis of liver. Gut, *17*:700-708, 1976.

56. Lankisch, P. G.: Exocrine pancreatic function tests. Gut, *23*:777-798, 1982.

57. Laudano, O. M., and Roncoroni, E. C.: Determination of the dose of Histalog that provokes maximal gastric secretory response. Gastroenterology, *49*:372-374, 1965.

58. Levine, R. A.: Regulation of gastric secretory functions: A critical overview. *In*: Advances in Ulcer Disease, Proceedings of a Symposium on the Pathogenesis and Therapy of Ulcer Disease, Munich, 1980. K. H. Holtermüller and J. R. Malagelada, Eds. Amsterdam, Elsevier/North Holland Bio-Medical Press, 1980.

59. Lifshitz, F., and Holman, G. H.: Disaccharidase deficiencies with steatorrhea. J. Pediatr., *64*:34-44, 1964.

60. Lombarts, A. J., and Peters, M. J.: Routine determinations of serum pepsinogens. Clin. Chim. Acta, *36*:195-200, 1972.

61. Losowsky, M. S.: Malabsorption. *In*: Oxford Textbook of Medicine. D. J. Weatherall, J. G. G. Ledingham, and D. A. Warrell, Eds. Oxford, Oxford University Press, 1983.

62. Lubran, M.: Measurement of gastric acidity. Lancet, *2*:1070-1071, 1966.

63. Malagelada, J. R.: Pancreatic cancer: An overview of epidemiology, clinical presentation, and diagnosis. Mayo Clin. Proc., *54*:459-467, 1979.

64. McGuigan, J. E.: The Zollinger-Ellison syndrome. *In*: Gastrointestinal Disease: Pathophysiology, Diagnosis, Management. 3rd ed. M. H. Sleisenger and J. S. Fordtran, Eds. Philadelphia, W. B. Saunders Co., 1983.

65. Meyer, J. H.: Chronic morbidity after ulcer surgery. *In*: Gastrointestinal Disease: Pathophysiology, Diagnosis, Management. 3rd ed. M. H. Sleisenger and J. S. Fordtran, Eds. Philadelphia, W. B. Saunders Co., 1983.

66. Michaelis, L.: Harvey Lectures, 1926-1927. New York, Academic Press, 1928.

67. Mitchell, J., Field, H. P., Simpson, F. G., et al.: Preliminary evaluation of a single-day tubeless test of pancreatic function. Br. Med. J., *282*:1751-1753, 1981.

68. Moller-Peterson, J., and Dati, F.: Renal handling of pancreatic lipase. Clin. Chem., *30*:343-344, 1984.

69. Moore, E. W.: The terminology and measurement of gastric acidity. Ann. N.Y. Acad. Sci., *140*:866-874, 1967.

70. Moosa, A. R.: Current concepts: Diagnostic tests and procedures in acute pancreatitis. N. Engl. J. Med., *311*:639-643, 1984.

71. Morgan, L. M., and Marks, V.: The gastrointestinal hormones. *In*: Hormone Analysis: Methodology and Clinical Interpretation. G. W. Pennington and S. Naik, Eds. Boca Raton, Fla., CRC Press, 1981, Vol. 2.

72. Newcomer, A. D.: Disaccharidase deficiencies. Mayo Clin. Proc., *48*:648-652, 1973.

73. Newsholme, E. A., and Leech, A. R.: Biochemistry for the Medical Sciences. New York, John Wiley & Sons, 1983.

74. Otte, M.: Pankreasfunktionsdiagnostik. Der Internist, *20*:331-340, 1979.

75. Plebani, M., DiMario, F., Vianello, F., et al.: Pepsinogen group I radioimmunoassay and total serum pepsinogen colorimetric determination: A comparative study in normal subjects and in peptic ulcer patients. Clin. Biochem., *16*:20-22, 1983.

76. Podolosky, D. K., McPhee, M. S., Alpert, E., et al.: Galactosyltransferase isoenzyme II in the detection of pancreatic cancer: Comparison with radiologic, endoscopic, and serologic tests. N. Engl. J. Med., *304*:1313-1318, 1981.

77. Ravich, W. J., and Bayless, T. M.: Carbohydrate absorption and malabsorption. Clin. Gastroenterol., *12*:335-356, 1983.

78. Rayford, P. L., Miller, T. A., and Thompson, J. C.: Secretin, cholecystokinin, and newer gastrointestinal hormones. N. Engl. J. Med., *294*:1093, 1976.

79. Rees, W. D. W., and Turnberg, L. A.: Biochemical aspects of gastric secretion. Clin. Gastroenterol., *10*:521-554, 1981.

80. Regan, P. T., Go, V. L. W., and DiMagno, E. P.: Comparison of the effects of cholecystokinin and cholecystokinin octapeptide on pancreatic secretion, gallbladder contraction, and plasma pancreatic polypeptide in man. J. Lab. Clin. Med., *96*:743-748, 1980.

81. Richardson, C. T.: Gastric ulcer. *In*: Gastrointestinal Disease: Pathophysiology, Diagnosis, Management. 3rd ed. M. H. Sleisenger and J. S. Fordtran, Eds. Philadelphia, W. B. Saunders Co., 1983.

82. Rick, W., Agnew, J. E., Bär, U., et al.: Diagnostische Verfahren. *In*: Pankreas, Vol. 3 of Handbuch der Inneren Medizin. M. M. Forell, Ed. Berlin, Springer-Verlag, 1976.

83. Rick, W., and Fritsch, W. P.: Pepsin. *In*: Methods of Enzymatic Analysis. 2nd ed. H. U. Bergmeyer, Ed. New York, Academic Press, 1974.

84. Rinderknecht, H., Renner, I. G., Douglas, A. P., et al.: Profiles of pure pancreatic secretions obtained by direct pancreatic duct cannulation in normal healthy human subjects. Gastroenterology, *75*:1083-1089, 1978.

85. Robert, A., and Kauffman, G. L.: Stress ulcers. *In*: Gastrointestinal Disease: Pathophysiology, Diagnosis, Management. 3rd ed. M. H. Sleisenger and J. S. Fordtran, Eds. Philadelphia, W. B. Saunders Co., 1983.

86. Rowden, D. R., Taylor, I. L., Richter, J. A., et al.: Is hypergastrinaemia associated with rheumatoid arthritis? Gut, *19*:1064-1067, 1978.

87. Rudolph, C., and Tauschel, H. D.: Fast determination of pepsin activity in a colored radial diffusion test at pH optimum. Anal. Biochem., *119*:49-54, 1982.

88. Ryan, M. E., and Olsen, W. A.: A diagnostic approach to malabsorption syndromes, a pathophysiological approach. Clin. Gastroenterol., 12:533-550, 1983.

89. Saccomani, G., Chang, H. H., Mihas, A. A., et al.: An acid transporting enzyme in human gastric mucosa. J. Clin. Invest., 64:627-635, 1979.

90. Sale, J. K., Goldberg, D. M., Thjodleifsson, B., et al.: Trypsin and chymotrypsin in duodenal aspirate and faeces in response to secretin and cholecystokinin-pancreozymin. Gut, 15:132-138, 1974.

91. Satake, K.: Pancreatic disease and the case for serum elastase 1 RIA. Diagn. Med., September/October, 1983, pp. 47-55.

92. Shwachman, H., and Mahmoodian, A.: The sweat test and cystic fibrosis. Diagn. Med., June, 1982, pp. 61-77.

92a. Sinko, V.: Fecal fat microscopy. Am. J. Gastroenterol., 75:204-208, 1981.

93. Sleisenger, M. H., Ed.: Malabsorption and nutritional support. Clin. Gastroenterol., 12:1-610, 1983.

94. Sleisenger, M. H., and Fordtran, J. S., Eds.: Gastrointestinal Disease: Pathophysiology, Diagnosis, Management. 3rd ed. Philadelphia, W. B. Saunders Co., 1983.

95. SmithKline Diagnostics: Package insert for Gastroccult test. SmithKline Diagnostics, Inc., P.O. Box 61947, Sunnyvale, CA 94086.

96. Sodeman, W. A., Jr., and Sodeman, T. M., Eds.: Sodeman's Pathologic Physiology, Mechanisms of Disease. 6th ed. Philadelphia, W. B. Saunders Co., 1979.

97. Soergel, K. H.: Acute pancreatitis. In: Gastrointestinal Disease: Pathophysiology, Diagnosis, Management. 3rd ed. M. H. Sleisenger and J. S. Fordtran, Eds. Philadelphia, W. B. Saunders Co., 1983.

98. Soll, A. H., and Isenberg, J. I.: Duodenal ulcer diseases. In: Gastrointestinal Disease: Pathophysiology, Diagnosis, Management. 3rd ed. M. H. Sleisenger and J. S. Fordtran, Eds. Philadelphia, W. B. Saunders Co., 1983.

99. Thompson, J. C.: Gastrin and gastric secretion. Ann. Rev. Med., 20:291-314, 1969.

100. Theodossi, A., and Gazzard, B. G.: Have chemical tests a role in diagnosing malabsorption? Ann. Clin. Biochem., 21:153-165, 1984.

101. Tietz, N. W., Ed.: Clinical Guide to Laboratory Tests. Philadelphia, W. B. Saunders Co., 1983.

102. Trier, J. S.: Celiac sprue. In: Gastrointestinal Disease: Pathophysiology, Diagnosis, Management. 3rd ed. M. H. Sleisenger and J. S. Fordtran, Eds. Philadelphia, W. B. Saunders Co., 1983.

103. Walmsely, R. N., and Guerin, M. D.: Digestive tract disorders. In: Disorders of Fluid and Electrolyte Balance. Bristol, J. Wright, 1984.

104. Walsh, J. H., and Lam, S. K.: Physiology and pathology of gastrin. Clin. Gastroenterol., 9:567-591, 1980.

105. Webster, H. L.: Laboratory diagnosis of cystic fibrosis. CRC Crit. Rev. Clin. Lab. Sci., 18:313-338, 1982.

106. Weinsten, W. M.: Gastritis. In: Gastrointestinal Disease: Pathophysiology, Diagnosis, Management. 3rd ed. M. H. Sleisenger and J. S. Fordtran, Eds. Philadelphia, W. B. Saunders Co., 1983.

107. West, P. M., Ellis, F. W., and Scott, B. L.: A simplified method for determining the excretion rate of uropepsin. J. Lab. Clin. Med., 39:159-162, 1952.

108. Williams, R. H.: Gastrointestinal hormones. In: Textbook of Endocrinology. 6th ed. R. H. Williams, Ed. Philadelphia, W. B. Saunders Co., 1981.

109. Wissler, J. H., and Logemann, E.: D-Xylose. In: Methods of Enzymatic Analysis, Vol. VI. H. U. Bergmeyer, Ed.-in-Chief. Weinheim, Verlag Chemie, 1984.

110. Yamada, T.: Secretory tumors of the pancreas. In: Gastrointestinal Disease: Pathophysiology, Diagnosis, Management. 3rd ed. M. H. Sleisenger and J. S. Fordtran, Eds. Philadelphia, W. B. Saunders Co., 1983.

111. Young, D. S., Pestaner, L. C., and Gibberman, V.: Effects of drugs on clinical laboratory tests. Clin. Chem., 21:1D-432D, 1975.

CHAPTER 15

BIOCHEMICAL ASPECTS OF HEMATOLOGY

by Virgil F. Fairbanks, M.D., and George G. Klee, M.D., Ph.D.

Blood cells differ in biological functions and in their metabolic characteristics. White blood cells, or leukocytes, contain nuclei, mitochondria, ribosomes, and lysosomes. Consequently, they can synthesize proteins and lipids, and their energy requirement is relatively high and is met principally by the tricarboxylic acid (Krebs) cycle. By contrast, mature erythrocytes lack nuclei, mitochondria, and ribosomes. They are incapable of biosynthesis. The metabolic machinery with which they begin their four-month journey, after leaving the bone marrow, cannot be replaced. They lack the Krebs cycle and depend instead principally on anaerobic glycolysis for their energy needs. Unique to mammalian erythrocytes is the Rapoport-Luebering cycle that regulates the oxygen affinity of hemoglobin.

Genetically determined abnormalities of erythrocyte proteins result in several important diseases. Among such abnormalities are those that affect structural proteins of the erythrocyte membrane; those affecting hemoglobin structure, function, or stability; and those affecting important erythrocyte enzymes. Deficiencies of vitamin B_{12}, folic acid, or iron impair the ability of the bone marrow to form erythrocytes and thus cause anemia.

The clinical consequences of abnormalities of erythrocyte structure, function, and metabolism are quite varied. They include *anemia; polycythemia*, or greater than normal number of erythrocytes in the blood; *sickle cell disease*, which results from polymerization of molecules of hemoglobin S; *cyanosis*, a bluish discoloration of skin and mucous membranes that results from insufficient oxygen bound to hemoglobin; and *hemolysis*, or more rapid than normal destruction of blood. The causes of these conditions and the laboratory methods required for their investigation are the subject of this chapter.

DISORDERS OF ERYTHROCYTE METABOLISM

PRINCIPAL PATHWAYS OF ERYTHROCYTE METABOLISM

At about the time that a newly formed erythrocyte (reticulocyte) leaves the bone marrow, it loses its nucleus and mitochondria and therefore the enzymes of the oxidative pathway, or Krebs cycle. Thus, it loses the capacity to regenerate the proteins it contains, as these become denatured or otherwise lose their function due to aging of the erythrocyte. Reticulocytes do contain ribosomes and are thus capable of some protein synthesis. But these very early forms of circulating erythrocytes lose their reticulum (i.e., their ribosomes) in the course of ~48 hours. Thereafter, erythrocytes have no capacity to synthesize proteins.

Disorders of erythrocyte metabolism have been described in the glycolytic pathways, in the glutathione pathways, in enzymes of purine and pyrimidine catabolism, in the *adenosine triphosphate–dependent cation pump* of the erythrocyte membrane, in two enzymes that affect the phospholipid composition of the erythrocyte membrane, and in methemoglobin reductase, which converts methemoglobin to hemoglobin. There are many other heritable enzyme deficiencies that may be demonstrated in erythrocytes. However, this chapter will be limited to a presentation of metabolic disorders that have their principal expression in blood or in the hematopoietic system. The disorder of heme synthesis that causes hemolytic anemia in congenital erythropoietic porphyria will be discussed in Chapter 16 (Porphyrins and Disorders of Porphyrin Metabolism).

HEMOLYTIC DISORDERS RESULTING FROM ERYTHROCYTIC ENZYME ABNORMALITIES

The major pathways of erythrocyte glycolysis are shown in Figure 15-1. Except when the erythrocyte is stressed by redox compounds ("oxidant drugs"), ~90% of its glucose passes through the Embden-Meyerhof pathway (EMP) to pyruvate or lactate (Figures 15-1 and 15-2). During this process, there is a net gain of two moles of ATP and a little less than one mole of reduced nicotine adenine dinucleotide (NADH) per mole of glucose catabolized. Other pathways are closely linked with the Embden-Meyerhof pathway. These are the hexose monophosphate pathway (or pentose-phosphate shunt), which provides both reduced NADPH and pentose phosphates and is in turn linked with the glutathione cycle (Figure 15-2), and the *Rapoport-Luebering cycle* (RLC). The RLC is the principal regulator of O_2 affinity of hemoglobin by increasing or decreasing the amount of 2,3-diphosphoglycerate (2,3-DPG). This will be further addressed below in the section on hemoglobin. The RLC also regulates the relative amounts of NAD and NADH and the relative quantities of ADP and ATP in the erythrocyte.

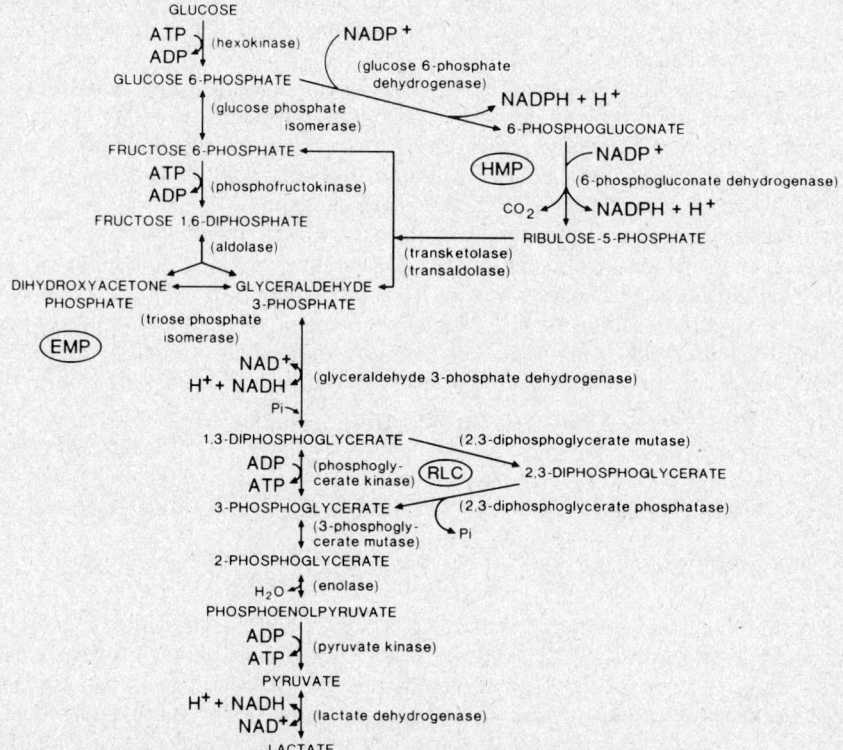

Figure 15-1. Major glycolytic pathways of the erythrocyte. EMP = the Embden-Meyerhof pathway; HMP = hexose monophosphate pathway or pentose shunt; RLC = the Rapoport-Luebering cycle.

Substrates are in upper case type, enzymes in parentheses. ADP = adenosine diphosphate; ATP = adenosine triphosphate; NAD$^+$ = nicotinamide–adenine dinucleotide; NADH = reduced nicotinamide–adenine dinucleotide; NADP$^+$ = nicotinamide–adenine dinucleotide phosphate; NADPH = reduced nicotinamide–adenine dinucleotide phosphate. The step from ribulose-5-phosphate that is shown as being catalyzed by transketolase and transaldolase is an abbreviation of this portion of the HMP.

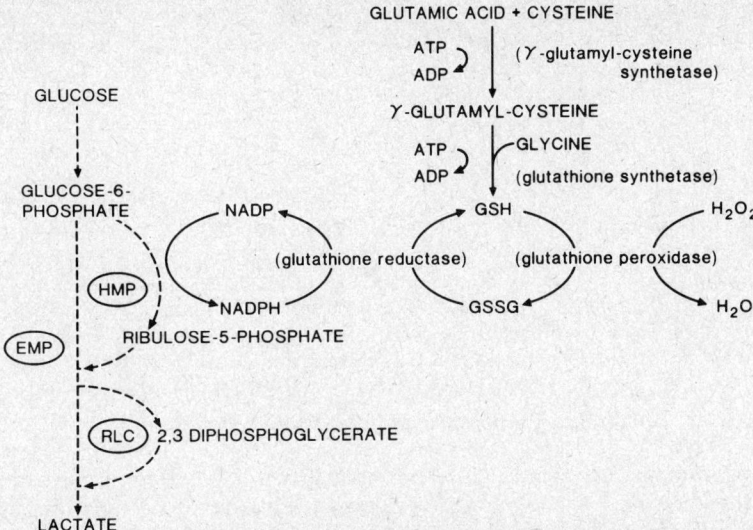

Figure 15-2. Interrelationship of hexose monophosphate and glutathione pathways. GSH = reduced glutathione; GSSG = oxidized glutathione; other abbreviations as in Figure 15-1.

Enzymes of the Embden-Meyerhof Pathway of Anaerobic Glycolysis

Hexokinase (HK, EC 2.7.1.1) effects the phosphorylation of glucose in the first step of glycolysis. In human erythrocytes, there are two isoenzymes: HK-1, which predominates, and HK-3.[47] Deficiency of either isoenzyme may be associated with hemolytic disease. Hexokinase activity in erythrocytes is very low and may be considered rate limiting. The activity of HK shows a marked decline as erythrocytes age. Assay of this enzyme must be related to the reticulocyte count, because in hemolytic disorders erythrocytes have a shorter than normal life span.[100] Several families with HK deficiency resulting in mild to moderate anemia have been identified. Both electrophoretically and kinetically abnormal HK variants have been described in these families.[9]

Glucosephosphate isomerase (GPI, EC 5.3.1.9) represents a pivot point in glycolysis. The reaction it catalyzes is bidirectional between glucose-6-phosphate (G-6-P) and fructose-6-phosphate (F-6-P). Because of it, products of the hexose monophosphate pathway (HMP or pentose shunt) can be recycled to form G-6-P. Glucosephosphate isomerase also catalyzes the formation of the β-anomers of G-6-P and F-6-P. GPI appears to be controlled by a single gene, even though electrophoresis demonstrates three bands. The enzyme is a dimer of identical subunits, each having a M.W. of 60 000.

More than 30 families have been reported in which GPI deficiency was associated with hemolytic anemia of moderate severity.[2,108] GPI deficiency has also been associated with hemolytic disease of the newborn.[2,108] In cases of GPI deficiency, GPI variants have been found with either electrophoretic or kinetic abnormalities or both. Several GPI variants have also been shown to be thermolabile.

Although actually quite an uncommon cause of hemolysis, GPI deficiency is the third most frequently encountered erythrocyte enzymopathy associated with hemolysis (after pyruvate kinase and G-6-PD deficiencies), since all the other erythrocyte enzymopathies are very rare. Except in consanguineous marriages, GPI deficiency has represented not a homozygous state but the inheritance of two different mutant alleles for structural variants of GPI.[108] This is analogous to the inheritance by a person with S-C hemoglobinopathy of an allele for Hb S from one parent and for Hb C from the other parent.

Phosphofructokinase (PFK, EC 2.7.1.11) deficiency is extremely rare. It has been associated with myopathy (in type VII glycogen storage disease) or in a few cases with mild hemolysis with or without myopathy.[106]

Triosephosphate isomerase (TPI, EC 5.3.1.1), following the aldolase-mediated cleavage of hexose, effects an interconversion of the two 3-C fragments. Only a small number of families have been reported in which TPI deficiency was associated with hemolysis. Besides moderate hemolytic anemia, patients with TPI deficiency exhibit mental and neurologic deterioration during the first year of life.[88] Because the gene for TPI deficiency is very rare, this disorder is not likely to be seen except in offspring of consanguineous marriages.

Phosphoglycerate kinase (PGK, EC 2.7.2.3) catalyzes a step in glycolysis in which ATP is produced.

Deficiency of PGK should, therefore, be of serious consequence. Only a few cases have been reported, and these were associated with moderate to severe hemolysis, as well as mental and neurologic·abnormalities.[54]

Pyruvate kinase (PK, EC 2.7.1.40) catalyzes the second energy-yielding step of glycolysis, and defects here have serious consequences. Several distinct forms of PK normally occur in mammalian tissues. These have been designated according to the tissues in which they were recognized, as L (for liver), R (for red cell), and M_1 (for muscle). The pyruvate kinase from each of these sources is a tetramer with a M.W. of about 60 000.[48]

Like hemoglobin, PK is an allosteric protein. Activation of PK requires a monovalent cation such as K^+ and is inhibited by the Mg^{2+}-ATP complex. *Fructose-1,6-diphosphate* is an important regulator of PK activity.

More than 300 cases of PK deficiency have been described, all associated with moderate to severe hemolytic anemia.[10,71,73] PK deficiency may also be a cause of hemolytic disease of the newborn.[10,73] Of all the enzymopathies of the erythrocyte that cause chronic hemolytic disease, this is by far the most common. Many variants of PK with either electrophoretic or kinetic abnormalities or both have been described.[10,71,73]

Hexose Monophosphate (HMP) Pathway

Normally, approximately 10% of glucose is catabolized through the hexose monophosphate pathway, but this fraction may be markedly increased when there is oxidative stress, as in the presence of redox compounds such as methylene blue or acetylphenylhydrazine. The principal function of the HMP is to reduce NADP to NADPH, which in turn maintains glutathione in the reduced state. The HMP also generates five-, four-, three-, and seven-carbon sugar phosphates, but whether these have an important role in the metabolism of mature erythrocytes is unclear. These intermediates may be recycled into the main pathway of glycolysis at the fructose-6-phosphate or the glyceraldehyde-3-phosphate steps, as shown in Figure 15-1. Of some interest is the fact that the activity of transketolase, an enzyme of the HMP, is decreased in thiamin deficiency, and therefore assay of this enzyme may be used as an index of thiamin deficiency. However, of the enzymes of the HMP, only G-6-PD relates to primary blood disorders, and therefore only this enzyme of the HMP will be discussed in this chapter.

The active enzyme **glucose-6-phosphate dehydrogenase** (G-6-PD, EC 1.1.1.49) is a dimer of identical subunits, each with a M.W. of 60 000. A tetramer also occurs and is enzymatically active. Amino acid sequences have been determined for all the tryptic peptides of the normal human erythrocyte enzyme. Partial amino acid sequencing of two G-6-PD variants, types A and Hektoen, reveals that they differ by a single amino acid from the normal (or B type) G-6-PD.[111]

Deficiency of G-6-PD in erythrocytes is widely prevalent throughout the world; approximately 200 million people have this condition. It is particularly prevalent among the people of Southeast Asia, in blacks, in people of the Mediterranean littoral, and in people of India.[7] Twenty-five per cent of male Southeast Asian refugees in North America have erythrocytic G-6-PD deficiency. Since the disorder is X-chromosome–linked, 35% of Southeast Asian women refugees are carriers of the gene for G-6-PD deficiency; the gene is silent in some, partially expressed in others, and fully expressed in about 10% of these women. The gene is also fully expressed in all males who inherit the gene for G-6-PD deficiency. Approximately 10% of black American males have erythrocyte G-6-PD deficiency. Among Mediterraneans, the prevalence varies widely, from a few per cent of males in the northwestern Mediterranean to as high as 50% of males in the eastern Mediterranean area. There is a high prevalence in Iraq and in people of Iraqi ancestry.

In all of these groups, G-6-PD deficiency is accompanied by electrophoretic or kinetic anomalies or both, reflecting molecular variants of G-6-PD. Nearly 200 distinct variants have been characterized electrophoretically and kinetically.[7] Black males may have *G-6-PD B* (the normal or common type), *A*, or *A-*. *G-6-PD A* is a variant that occurs in about 20% of black males. It has more rapid anodal mobility on electrophoresis at pH 8.0 than does G-6-PD B, but A does not have significantly reduced activity. *G-6-PD A-* is found in erythrocytes of black males who are G-6-PD deficient. It has electrophoretic mobility like that of A, but its activity is only about 10% of that of type A or B. The G-6-PD variant that occurs in erythrocytes of some people of the Mediterranean area is called *G-6-PD Mediterranean.* Its electrophoretic mobility is identical with G-6-PD B, but its activity is markedly diminished to less than 5% that of G-6-PD B.[53] Several G-6-PD variants have been described in Southeast Asians, but the most common are the variants *Mahidol* and *Union*.[74,112] These are associated with severe reduction in activity and in this sense resemble the Mediterranean variant. Besides the variants discussed so far, which affect many millions of people but are not accompanied by chronic hemolytic disease, there are many very rare G-6-PD variants. Some of these are of little consequence. Others have been detected because they cause lifelong hemolytic anemia that often is quite severe. A representative of this group is *G-6-PD Chicago.*[52] Many and perhaps most of these variants are thermolabile, unstable molecules that denature at an accelerated rate in vitro and probably in vivo as well.

The G-6-PD variants with severely reduced activity (e.g., Mediterranean and Mahidol) may cause hemolytic disease of the newborn. A small proportion of persons with G-6-PD Mediterranean have mild chronic hemolysis. Persons with G-6-PD Mediterranean are also subject to *favism,* an acute and very severe hemolytic crisis, often with fatal outcome, that is precipitated by ingestion of fava beans (*Vicia fava*). These individuals are also susceptible to severe hemolysis following ingestion of *antimalarial drugs* of the 8-aminoquinoline type or of the older sulfonamides (such as sulfanilamide) or as a consequence of infections. However, hemolysis in blacks with G-6-PD deficiency is rarely severe or life-threatening.

To try to bring some order to the nomenclature of G-6-PD variants, Beutler has divided them into five classes on the basis of severity of enzyme deficiency:[7]

Class I   Deficiency associated with chronic hemolytic anemia (e.g., G-6-PD Chicago)
Class II   Severe deficiency, usually without hemolytic anemia (e.g., G-6-PD Mediterranean, Mahidol, Union)
Class III   Moderate to mild deficiency (e.g., G-6-PD A-)
Class IV   Very mild or no deficiency (e.g., G-6-PD A)
Class V   Increased activity (only one such variant has been described, G-6-PD Hektoen)

A secondary classification of the variants may then be made according to the electrophoretic mobility of G-6-PD variants.

Glutathione Pathway

Glutathione (GSH) protects hemoglobin and other critical erythrocyte proteins from peroxidative injury. As shown in Figure 15-2, its metabolic pathway includes several enzymes. Glutathione reductase links the pathway to the HMP through the reversible oxidation and reduction of NADP. Glutathione peroxidase (GSH-Px, EC 1.11.1.9) effects the conversion of H_2O_2 to water, thus reducing the likelihood of peroxidative denaturation of hemoglobin and other proteins. Of course, erythrocytes also have high catalase activity that converts H_2O_2 to water and O_2; thus GSH-Px is redundant. In fact, deficiencies of either catalase or GSH-Px are without serious consequences for the erythrocytes because the presence of either of these two enzymes can compensate for deficiency of the other. Glutathione must be continually synthesized in the erythrocyte. This requires two enzymes; the first of these, γ-glutamyl cysteine synthetase, forms a dipeptide from glutamic acid and cysteine. Glutathione synthetase (GSH-S) then adds glycine to form the tripeptide γ-glutamyl cysteinyl glycine, or glutathione. Deficiency of each of these erythrocyte enzymes has been found in a few cases, in association with congenital hemolytic anemia.[83]

A few cases have been described in which mild chronic hemolytic anemia is associated with deficiency of **glutathione synthetase** (GSH-S, EC 6.3.2.3).[83] Such persons may also have more severe acute hemolytic crises when exposed to redox substances such as 8-aminoquinolines or other "oxidant" drugs.

There are, in fact, two distinct types of GSH-S deficiency, both associated with mild chronic hemolysis; in one type, hemolysis is the only clinical manifestation.[83] In the other, the major clinical features are mental retardation, severe generalized muscular weakness, tremors, incoordination, and metabolic acidosis. This second and much more severe type of GSH-S deficiency is also known as *5-oxoprolinuria* or *pyroglutamic aciduria.*[91] The difference in severity of these disorders reflects the fact that, in the mild form, GSH-S deficiency is confined to the erythrocyte because in this disorder the GSH-S is unstable. GSH-S activity is present in adequate quantity in young erythrocytes, but it rapidly declines as the cells age, since the cells are unable to synthesize new molecules of GHS-S. Other cells of the body that have nuclei and ribosomes can compensate for accelerated denaturation of GSH-S by synthesizing more. On the other hand, in the severe systemic form of GSH-S deficiency, all cells of the body have low activities of GSH-S because they cannot form this enzyme in adequate amounts. In both types of GSH-S deficiency, erythrocytes exhibit marked reduction in glutathione concentration.

One family has been described in which two siblings had mild to moderate chronic hemolytic anemia associated with deficiency of erythrocyte **γ-glutamyl cysteine synthetase** (γ-GC-S, EC 6.3.2.2).[55] Glutathione concentration was low in their erythrocytes, and they were unable to synthesize GSH or γ-glutamyl cysteine from glutamic acid, cysteine, and glycine.

The Rapoport-Luebering Cycle (RLC)

This cycle, which is unique to mammalian erythrocytes, is the principal physiologic means for regulation of the O_2 affinity of hemoglobin.[5,6,97] From the point of view of O_2 transport, one might assert that the Embden-Meyerhof pathway of the erythrocyte serves principally to generate *2,3-diphosphoglycerate* (2,3-DPG), since quantitatively this substance is the principal glycolytic intermediate of the red cell; the concentration of 2,3-DPG is about equal to the sum of all other glycolytic intermediates. 2,3-DPG content of the erythrocyte is increased in all anemias; in all but one, the increase is inversely proportional to the hemoglobin concentration.[5] The single exception is pyruvate kinase (PK) deficiency, in which 2,3-DPG is increased disproportionately to the hemoglobin deficit as a result of the "bottleneck" at the PK step and of a retrograde accumulation of products of glycolysis. The increase in 2,3-DPG in anemias results in a decrease in O_2 affinity of hemoglobin so that O_2 is more readily transferred to tissues. The O_2 affinity of hemoglobin is also influenced by slight changes in pH of the blood; a corresponding sensitivity to pH exists in the RLC, which again permits change in 2,3-DPG concentration to "fine-tune" the O_2 affinity of hemoglobin.[5,6,97]

Two enzymes involved in the RLC are *2,3-diphosphoglycerate mutase* (2,3-DPGM) and *2,3-diphosphoglycerate phosphatase.** There has been only one well-documented case of deficiency of either enzyme.[86] This was in a case of erythrocytosis with very low erythrocyte concentration of 2,3-DPG, a corresponding increase in O_2 affinity of hemoglobin, and very little activity of 2,3-DPGM. The condition appears to be inherited as an autosomal recessive disorder.

Purine-Pyrimidine Metabolism

Three enzymes of purine or pyrimidine metabolism may be involved in the pathogenesis of hemolytic disease.

Adenosine triphosphatase (EC 3.6.1.3) converts ATP to ADP with the loss of a high-energy bond. The enzyme is involved in the Na^+–K^+ cation pump of the erythrocyte membrane. Rare cases of congenital hemolytic disease have been attributed to deficiency of ATPase. However, within the same family, siblings and mother with equally low ATPase activity had no hemolysis or anemia.[43] Thus, it is debatable whether erythrocyte ATPase deficiency was responsible for, or was only coincidental to, hemolysis in the index case.

Adenosine deaminase (AD, EC 3.5.4.4) converts AMP, ADP, and ATP to inosine following dephosphorylation. An increase in activity of this enzyme would be expected to accelerate depletion of all three adenosine phosphates from the erythrocyte. Only one family has been reported in which some members had mild chronic hemolytic anemia associated with a 30–70-fold *increase* in activity of erythrocyte AD.[102] This is the only abnormality of erythrocyte enzymes in which hemolysis is due to an *increase* in activity. Furthermore, it is the only erythrocyte enzymopathy that is inherited as an autosomal dominant disorder. *Combined immunodeficiency* is associated with a *decrease* in erythrocyte AD activity, without hemolysis.

Pyrimidine-5′-nucleotidase† (P-5′-N) dephosphorylates pyrimidine phosphates (as in RNA); the pyrimidine bases diffuse out of the erythrocyte, and the phosphate is retained. Pyrimidine phosphates are present on ribosomes of erythroblasts and reticulocytes, but there are normally no pyrimidines in the mature erythrocyte. P-5′-N is the principal mechanism for catabolism of the ribosomal RNA. When P-5′-N activity is markedly diminished, basophilic stippling of erythrocytes (from aggregated RNA) becomes pronounced, and total nucleotide content of erythrocytes is increased and is associated with a mild to moderate hemolytic anemia. Numerous cases of hereditary P-5′-N deficiency have now been recognized.[99] In the past, several of these were described as "high ATP syndrome" before it was recognized that cytidylic and uridylic acids were being measured erroneously as ATP. Also, activity of ribosephosphate pyrophosphokinase (EC 2.7.6.1) is diminished, and glutathione concentration is increased. These appear to be secondary changes.

The morphologic appearance of erythrocytes in P-5′-N deficiency is very similar to that in *lead poisoning*; in both conditions basophilic stippling of erythrocytes is pronounced. In lead poisoning the stippling is due to an acquired and transitory functional deficiency of P-5′-N, since lead markedly inhibits P-5′-N.[101]

---

*These enzymes, unique to the mammalian erythrocyte, have not been assigned EC designations.

†P-5′-N has not been assigned an EC designation.

Enzymes of Phospholipid Metabolism

Two enzymes that play a role in maintaining the normal composition of the erythrocyte membrane lipids have been reported as being deficient in certain erythrocytic disorders. One of these, *lecithin:cholesterol acyl transferase* (LCAT, EC 2.3.1.43), is a plasma enzyme that is not found in hemolysates. It transfers an acyl group from lecithin to cholesterol and thus helps maintain the proper ratio of lysolecithin to lecithin in the outer layer of the erythrocyte membrane. Deficiency of plasma LCAT has been described in several families, mostly of northern Europe, in association with morphologic abnormalities of erythrocytes (target cells) and disturbances in composition of plasma lipids. Patients with this disorder also suffer from early atherosclerosis, corneal opacification, hyperlipidemia, hematuria, proteinuria, and mild hemolytic anemia.[36] *Lysolecithin acyl transferase* (LAT, EC 2.3.1.23) is an enzyme of the erythrocyte cytosol that converts lysolecithin to lecithin. A single family has been reported from Austria, in which erythrocyte deficiency of LAT was inherited as an autosomal dominant disorder accompanied by mild hemolytic anemia.[80]

PEROXIDATIVE HEMOLYSIS IN THE NEWBORN

It has long been known that neonates, particularly premature newborns, are susceptible to peroxidative injury to erythrocytes that results in hemolysis. Erythrocyte glutathione concentration in neonates is not readily maintained when such erythrocytes are subjected to oxidant stress (as with vitamin K or acetyl-phenylhydrazine). A factor that may contribute to peroxide sensitivity of neonate erythrocytes is the low selenium content of infant formulas derived from bovine milk. *Glutathione peroxidase* contains selenium; thus, depletion of selenium, required in trace amounts, may result in diminution of activity of gluthathione peroxidase (GSH-Px), which functions to convert H_2O_2 to H_2O. The hemolysis that occurs is due to loss of phosphatidyl ethanolamine from the erythrocyte membrane. The defect can be corrected by tocopherol (vitamin E) supplementation, which appears to counteract the effect of decreased GSH-Px activity, or by increasing the trace selenium content of infant formulas derived from bovine milk.[40] Selenium is a highly toxic element that is required physiologically only in trace amounts. Its only known physiological role is as a component of GSH-Px.

METHEMOGLOBIN REDUCTASE*

For transport of O_2, the iron component of hemoglobin must be in the divalent state. *NADH-dependent methemoglobin reductase,* formerly called "diaphorase," effects the conversion of met-hemoglobin to hemoglobin. Congenital deficiency of this enzyme is very uncommon. There are several causes of congenital methemoglobinemia; among these rare conditions, deficiency of methemoglobin reductase is by far the most frequently encountered. Congenital methemoglobin reductase deficiency is associated with minimal to moderate cyanosis and increased venous blood methemoglobin concentrations of 10–20% of total hemoglobin. This condition is usually of no serious importance; the persistent cyanosis of face and lips may cause distress to patients or their parents, and treatment may be desired if only for cosmetic purposes. Oral administration of methylene blue in small doses three times daily will resolve the cyanosis. Methylene blue, in this situation, serves as an electron donor to the Fe(III) of methemoglobin. The methylene blue in turn is reduced by NADPH, and the utilization of glucose by HMP is accelerated. In the past it was thought that an "NADPH-dependent methemoglobin reductase" was involved in the methylene blue–mediated reduction of hemoglobin. However, this idea has been generally discarded.

METHODS FOR THE DETERMINATION OF ERYTHROCYTE ENZYMES

Very few clinical laboratories need to measure the activity of more than a few of the erythrocyte enzymes. Only procedures for the assays of erythrocyte glucose-6-phosphate dehydrogenase, pyruvate kinase, and glucose phosphate isomerase will be presented, since only these enzymes are commonly encountered as causes of hemolysis. Methods will also be given for the

---

*There is no EC designation for methemoglobin reductase.

determinations of glutathione and 2,3-diphosphoglycerate. A simple, practical screening test will be described that permits identification of pyrimidine-5'-nucleotidase deficiency. Readers who seek additional assay procedures should consult the book by Beutler.[8]

With few exceptions, venous blood specimens for these assays should be anticoagulated either with disodium EDTA or with acid-citrate-dextrose (ACD) solution. The ACD anticoagulant is superior because glutathione concentrations are well sustained for more than a week, whereas if EDTA is the anticoagulant, glutathione concentration should be assayed promptly. With use of either anticoagulant, enzymes are stable at 4 °C for 3 weeks. Plasma should not be removed until shortly before the assay. Frozen specimens cannot be used. Special conditions in specimen collecting and handling are required only for the assay of 2,3-diphosphoglycerate.

Because leukocytes have very much greater enzyme activity than do erythrocytes, assays for G-6-PD, PK, and GPI are performed using saline-washed, packed erythrocytes that are essentially leukocyte-free. The same hemolysate is used for each of these assays.

A UV spectrophotometer is required, and it is convenient to use a recorder that automatically plots absorbance against time, although absorbance readings may be recorded manually. The spectrophotometer cuvet chamber must be temperature controlled at 37 °C unless otherwise stated. An incubation device such as a water bath or a Temp-blok heater is required to bring reagents to 37 °C prior to initiation of the enzyme reaction.

Preparation of Hemolysate

Note: This hemolysate may be used for the *quantitative* determinations of glucose-6-phosphate dehydrogenase, pyruvate kinase, and glucosephosphate isomerase activity.

Reagents and Supplies

1. NaCl, 0.155 mol/L; keep at 4 °C.
2. Hemolyzing solution. Place 10 mL of Na_2EDTA, 0.27 mol/L, pH 8.0, and 0.05 mL of 2-mercaptoethanol in a 1-L volumetric flask and bring to volume with reagent grade water. Store at 4 °C.

Procedure

1. Centrifuge 2 mL of whole blood at 1500 × g for 10 min. Remove and discard plasma and "buffy coat" (the topmost cream-colored layer of cells).
2. Resuspend erythrocytes in 10 mL of cold NaCl, 0.155 mol/L. Mix thoroughly.
3. Centrifuge at 1500 × g for 10 min at 4 °C.
4. Remove and discard supernatant and topmost layer of cells, including any residual buffy coat. Do not disturb uppermost layer of erythrocytes.
5. Repeat the suspension of cells in cold NaCl, 0.155 mol/L, centrifugation, and discarding of supernatant and uppermost layer of cells (to a depth of about 0.5 mm) a total of three times.
6. Following the third saline wash, again remove and discard the supernatant and uppermost layer of cells. Then resuspend the remaining cells in an equal volume of cold NaCl, 0.155 mol/L.
7. Pour about 150 mL of acetone into a 250-mL beaker.
8. Using tongs or insulated gloves, place a small block of dry ice into the beaker. While waiting for frothing of the dry ice-acetone mixture to subside, add 0.2 mL of the erythrocyte-NaCl suspension to 1.8 mL of hemolyzing solution in a glass tube (10 × 150 mm). Cap tube.
9. Place tube into the dry ice–acetone mixture until frozen, usually a few minutes.
10. Thaw at room temperature. (Do not use hot water to thaw; some enzymes are very thermolabile.)
11. Examine for turbidity. Hemolysate should be clear to visual inspection. If turbid, centrifuge at 2250 × g for 30 min at 4 °C.

Determination of Glucose-6-Phosphate Dehydrogenase (G-6-PD) Activity

Principle

The reaction catalyzed by G-6-PD, as shown in Figure 15-1, is:

$$\text{Glucose-6-phosphate} + NADP^+ \xrightarrow{\text{G-6-PD}} \text{6-phosphogluconate} + NADPH + H^+$$

The enzyme activity is determined by measurement of the rate of increase in NADPH concentration. Whereas NADP$^+$ is transparent to ultraviolet (UV) light, NADPH strongly absorbs UV light. Therefore, the rate of increase in absorbance at 340 nm is the measure of enzyme activity.

Reagents

1. Tris-HCl buffer, 1.0 mol/L, pH 8.0. Place 121.14 g of tris(hydroxymethyl)methylamine (Tris, THAM, Trizma) in a 2-L beaker and add ~800 mL reagent grade water. Adjust pH to 8.0 by adding, drop-by-drop, HCl, 1 mol/L, while constantly stirring. After pH adjustment, transfer to a 1-L volumetric flask and add reagent grade water to bring to volume.

2. $MgCl_2$, 0.1 mol/L, aqueous

3. NADP$^+$, 2 mmol/L, aqueous (Sigma Chemical Co., St. Louis, MO)

4. Stock reagent mixture. The above reagents may be premixed in the following proportions, according to the number of specimens to be assayed in cuvets of 1 mL critical volume. (All volumes are given in mL.)

Reagent	Number of specimens to be assayed		
	1	2	4
Tris-HCl buffer	0.50	0.70	1.00
$MgCl_2$ solution	0.50	0.70	1.00
NADP$^+$ solution	0.50	0.70	1.00
H_2O	2.90	4.06	5.80

The volumes specified are sufficient for the number of specimens indicated, the corresponding blank, and the normal control.

The reagent mixture may be prepared in advance and stored frozen for as long as 1 month. If numerous assays are to be done, a 10-fold multiple of one of the above reagent mixtures may be prepared, subdivided into several plastic vials, and stored frozen for up to one month.

5. Glucose-6-phosphate, 6 mmol/L, aqueous. Prepare on the day of assay, in minimum volume of 0.3 mL for one assay and control.

Procedure

1. For each specimen, prepare blank and assay cuvets as follows. Include at least one normal control with each set of specimens.

	Blank Cuvet (mL)	Assay Cuvet (mL)
Stock reagent mixture	0.88	0.88
H_2O	0.10	—
Hemolysate (see p. 1502)	0.02	0.02

2. Cap with Parafilm and invert twice to mix.

3. Incubate at 37 °C for 10 min.

4. Add 0.1 mL of glucose-6-phosphate solution to assay cuvets only.

5. Cap with Parafilm.

6. Rapidly invert twice to mix, then place in spectrophotometer and measure the increase in absorbance at 340 nm for 10 min, at 37 °C.

7. While absorbance is being recorded, transfer 0.2 mL of hemolysate to another cuvet containing 9.8 mL of Drabkin's solution. Mix. Measure the absorbance at 540 nm and calculate hemolysate hemoglobin concentration. (N.B. The dilution of Drabkin's solution is 1:50 instead of the usual 1:502. The factor for calculating the hemolysate hemoglobin concentration is therefore proportionately smaller.)

8. Determine

$$\Delta A_{340}/\text{min} = \frac{A_{10 \text{ min}} - A_{0 \text{ min}}}{10}$$

Calculation:

$$\text{G-6-PD activity, U/g Hb} = \frac{804 \times \Delta A_{340}/\text{min}}{A_{540} \times F_{Hb}}$$

Example:

A_{340}, initial $= 0.230$	$\Delta A_{340}/\text{min for blank}$ $= 0.001$
A_{340} at 10 min $= 0.500$	Corrected $\Delta A_{340}/\text{min for specimen} = 0.026$
$\Delta A_{340}/10$ min $= 0.270$	Hemolysate Hb A_{540} $= 0.269$
$\Delta A_{340}/\text{min} = 0.027$	Factor for cyanmethemoglobin $= 5.48$

$$\text{G-6-PD, U/g Hb} = \frac{804 \times 0.026}{0.269 \times 5.48} = 14.2$$

The derivation of the equation can be found in the method reference listed below.

Reference Values

8–18 U/g Hb (mean $\pm$ 2 SD); values greater than 18 U/g Hb are encountered in any condition associated with younger than normal erythrocytes (as in hemolytic anemias not due to G-6-PD deficiency), but are of no clinical significance.

Comments and Precautions

The G-6-PD assay outlined above reflects NADPH generated in both the G-6-PD and the 6-phosphogluconate dehydrogenase (6-PGD) catalyzed reactions. Correction for the 6-PGD generation of NADPH may be made by use of two additional cuvets: 6-phosphogluconate is added to one cuvet and to the other both 6-phosphogluconate and glucose-6-phosphate are added. This "Glock and McLean assay" then provides both the true G-6-PD activity and an assay for 6-PGD. However, for practical purposes it does not seem warranted to incorporate these additional measurements and calculations. Deficiency of erythrocyte 6-PGD has not yet been reported as a cause of hemolytic disease. Nor will normal 6-PGD activity obscure G-6-PD deficiency. Thus, the simpler procedure is adequate for diagnostic purposes.

Performance of this assay at 37 °C departs from the World Health Organization recommendation that the assay be done at 25 °C. Since the reaction rate is quite temperature sensitive, thermal control is essential, and it cannot be assumed that ambient temperature in all laboratories is 25 °C. Thus, there is no inherent advantage in doing the assay at 25 °C. However, to avoid ambiguity, the assay values calculated may be expressed in International Units (U) (WHO; 37 °C).

It is convenient to express activity in units per gram hemoglobin (U/g Hb). However, it is somewhat more logical to express activity in terms of the number of erythrocytes represented in the hemolysate. To do this requires an independent erythrocyte count of a suspension of the washed packed red cells. From a practical standpoint, either expression is satisfactory. U/g Hb may be converted to $U/10^{10}$ erythrocytes by the relationship

$$U/10^{10} \text{ RBC} = 0.29 \times U/\text{g Hb}$$

References

Betke, K., Brewer, G. J., Kirkman, H. N. et al.: World Health Organization Technical Report Series No. 366. Geneva, World Health Organization, 1967.

Beutler, E.: Red Cell Metabolism. A Manual of Biochemical Methods. 2nd ed. New York, Grune & Stratton, 1975, pp. 66–69.

Beutler, E., Blume, K. G., Kaplan, J. C., et al.: Br. J. Haematol., *35*:331–340, 1977.

Screening Test for Erythrocyte Glucose-6-Phosphate Dehydrogenase (G-6-PD) Deficiency

Principle

$NADP^+$ is reduced to NADPH in both the glucose-6-phosphate dehydrogenase and the 6-phosphogluconate dehydrogenase reactions of the hexose monophosphate pathway. (See Figure 15–1.) If G-6-PD activity is low, only a small quantity of NADPH is formed. This is reoxidized to NADP by glutathione reductase present in erythrocytes, and no fluorescence is observed. In the presence of adequate amounts of G-6-PD, the rate of reduction of $NADP^+$ substantially exceeds the rate of oxidation of NADPH by glutathione reductase, and enough NADPH accumulates in erythrocytes to be demonstrable by its fluorescence.

Specimen

Blood is anticoagulated with EDTA, ACD, or CPD. The sample is stable for one week when stored at 4 °C. Minimum volume required is 0.2 mL, and the test can be done on freshly drawn heparinized blood obtained in a capillary tube, as from finger or heel prick.

Stock Reagent Mixture

To a 20-mL flask or plastic bottle add:

Glucose-6-phosphate, 0.1 mol/L	1 mL
$NADP^+$, 7.5 mmol/L	1 mL
Saponin, 1 g/dL	2 mL
Tris-HCl buffer, 0.7 mol/L, pH 7.8	3 mL
Use 0.75 mole (90.85 g/L) of Tris and adjust to pH 7.8.	
Oxidized glutathione, 8 mmol/L	1 mL
H_2O	2 mL

Mix. Stable for several months when frozen at −20 °C.

Procedure

1. Allow 5–10 min for reagent solution to thaw. Label three glass tubes (12 × 75 mm) and pencil a "tick-tack-toe" grid on a piece of filter paper (Whatman No. 1) for patient, normal control, and abnormal control.

2. Pipet 100 μL of reagent solution into each tube and pipet 10 μL of specimen and controls into the respective tubes containing reaction mixture. Mix.

3. Immediately transfer to filter paper a drop from each tube and start a timer.

4. At 5 min and 10 min repeat step 3.

5. Allow spots to dry completely and examine spots with a longwave ultraviolet light. (Chromato-Vue Model CC-20, Ultraviolet Products, Inc., San Gabriel, CA, is convenient, but a hand-held UV light source is satisfactory.)

Interpretation

Normal specimens show no fluorescence of the initial spot on the filter paper, but progressive increase in fluorescence at 5 and 10 min. Abnormal specimens show no fluorescence at times 0, 5, and 10 min.

Comments and Precautions

The screening test is simple and inexpensive; it can be performed in field surveys or in laboratories that lack UV spectrophotometers. In experienced hands this semiquantitative procedure can give satisfactory results equivalent to those obtained with costly instruments. However, the screen has some disadvantages. It should be used only when testing male subjects for G-6-PD deficiency. Women heterozygotes are often not detected with this method. Black males who are G-6-PD deficient *and who have had recent hemolysis* will commonly have normal results. Although for the sake of simplicity, this procedure was originally devised for use with whole blood rather than washed erythrocytes, occasionally leukocytosis (e.g., WBC count of 12 × 10⁹ cells/L) will cause spuriously normal results because leukocytes have relatively much greater

G-6-PD activity than do erythrocytes. For this reason, it is advantageous to use saline-washed erythrocytes suspended in an equal volume of isotonic saline.

Reference

Beutler, E.: Red Cell Metabolism. A Manual of Biochemical Methods. 2nd ed. New York, Grune & Stratton, 1975.

Determination of Pyruvate Kinase (PK) Activity

Principle

In this assay, the PK reaction is linked with the lactate dehydrogenase (LDH) reaction in which NADH is oxidized to NAD^+.

$$\text{Phosphoenolpyruvate} + \text{ADP} \xrightarrow{\text{PK}} \text{pyruvate} + \text{ATP}$$

$$\text{Pyruvate} + \text{NADH} \xrightarrow{\text{LDH}} \text{lactate} + NAD^+ + H^+$$

Since lactate dehydrogenase is present in excess, the rate of NADH oxidation is limited by the activity of PK. The reaction rate is measured by the rate of decrease in absorbance at 340 nm. Assays are performed at low substrate concentration with and without addition of fructose-1,6-diphosphate (FDP), because some PK variants associated with hemolysis have atypical reaction kinetics (and thus may exhibit "normal" activity at high substrate concentrations but lower than normal activity at lower substrate concentrations), or may show absence of enhancement by FDP, the allosteric activator of PK.

Reagents

1. Tris-HCl buffer, 1 mol/L, pH 8.0. (See G-6-PD assay.)
2. $MgCl_2$, 0.1 mol/L, aqueous.
3. NADH, 0.002 mol/L.
4. Adenosine diphosphate (ADP), 0.03 mol/L.
5. Phosphoenolpyruvate (PEP), 0.05 mol/L.

Small volumes of NADH, ADP, and PEP solution may be prepared in advance, and stored frozen for as long as one month prior to assay.

6. KCl, 1 mol/L.
7. Stock reagent. Prepare a stock reagent solution as follows, according to the number of specimens to be assayed. (All volumes are given in mL.)

Reagent	Number of specimens to be assayed		
	1	2	4
Tris buffer	0.9	1.3	2.1
$MgCl_2$ solution	0.9	1.3	2.1
KCl solution	0.9	1.3	2.1
H_2O	2.97	4.29	6.93

The stock reagent solution mixture may be stored frozen until required in the assay and is stable for at least one month. If a large number of assays are to be performed, a larger volume of stock solution may be prepared as a 10-fold multiple of one of the above proportions, divided, and transferred to plastic vials before freezing. Volumes indicated are sufficient for blank, specimen, and control assays at two substrate concentrations.

8. Fructose-1,6-diphosphate (FDP), 0.01 mol/L (aqueous)
9. Lactate dehydrogenase (LDH), 60 U/mL. Prepare from a slurry of crystalline porcine heart LDH type XVIII, Sigma Chemical Company, St. Louis, MO. This preparation of LDH is free of PK. Its LDH activity is approximately 5000 U/mL in saturated $(NH_4)_2SO_4$ solution; it must be diluted approximately 83-fold with water.

Procedure

1. On the day of assay thaw the frozen reagent mixture vials. Prepare four cuvets, of 1-mL critical volume, for each specimen and control as follows (cuvet 1 is specimen blank, cuvet 2 is

assay with high substrate concentrations, cuvet 3 is assay with low substrate concentrations, and cuvet 4 is assay with low substrate concentrations and fructose-1,6-diphosphate):

Reagents (mL)	Cuvet No.			
	1	2	3	4
Stock reagent	0.630	0.630	0.630	0.630
NADH	0.100	0.100	0.100	0.100
ADP	—	0.050	0.020	0.020
LDH	0.100	0.100	0.100	0.100
FDP	—	—	—	0.050
Hemolysate (see p. 1502)	0.020	0.020	0.020	0.020
H_2O	0.050	—	0.125	0.075

2. Cap and invert twice to mix; incubate for 10 min at 37 °C.

3. Add 0.1 mL of PEP solution to cuvets 1 and 2 and 0.005 mL of PEP solution to cuvets 3 and 4; cap and invert twice to mix.

4. Place in spectrophotometer chamber and set the blank absorbance at 0.5.

5. Record the decrease in absorbance at 340 nm of cuvets 2, 3, and 4 for 10 min, at 37 °C.

6. Assay hemolysate hemoglobin concentration as in G-6-PD assay procedure, p. 1503.

7. Determine ΔA_{340}/min of specimen, after correcting for ΔA_{340}/min of blank.

8. Calculate pyruvate kinase activity for high substrate assay, low substrate assay, and low substrate assay + FDP.

$$PK, U/g\ Hb = \frac{804 \times \Delta A_{340}/min}{A_{540} \times F_{Hb}}$$

Reference Ranges

High substrate assay:	11–19 U/g Hb (mean $\pm$ 2 SD)
Low substrate assay:	1.4–3.0
Low substrate assay + FDP:	4.7–7.3

Comments and Precautions

Precautions that need to be taken with this assay are much the same as those for the G-6-PD assay. A normal control must be run concurrently with every specimen assayed.

References

Beutler, E.: Red Cell Metabolism. A Manual of Biochemical Methods. 2nd ed. New York, Grune & Stratton, 1975, pp. 60-63.

Beutler, E., Blume, K. G., Kaplan, J. C., et al.: Br. J. Haematol., 35:331-340, 1977.

Paglia, D. E., Valentine, W. N., Baughan, M. A., et al.: J. Clin. Invest., 47:1929-1946, 1968.

Tanaka, K. R., Valentine, W. N., and Miwa, S.: Blood, 19:267-295, 1962.

Determination of Glucosephosphate Isomerase Activity

Clinical Significance

Deficiency of glucosephosphate isomerase (GPI) has been associated with chronic hemolytic disease in a few dozen reported cases, and may also cause hemolytic disease of the newborn. Although the third most frequently reported enzymopathy causing hemolytic disease, GPI deficiency is, nonetheless, an infrequent cause of hemolysis.

Principle

As shown in Figure 15-1, GPI catalyzes the interconversion of glucose-6-phosphate and fructose-6-phosphate (F-6-P):

$$\text{Fructose-6-phosphate} \xrightleftharpoons{\text{GPI}} \text{glucose-6-phosphate}$$

$$\text{NADP}^+ + \text{glucose-6-phosphate} \xrightleftharpoons{\text{G-6-PD}} \text{6-phosphogluconate} + \text{NADPH} + \text{H}^+$$

In this assay, F-6-P is used as the substrate for GPI. The GPI reaction is linked to the G-6-PD reaction. The rate of reduction of NADP^+, as indicated by the increase in absorbance at 340 nm, is the measure of GPI activity.

Reagents

The stock reagent solution for the G-6-PD assay is also used in assaying GPI activity.
1. Stock reagent solution (see G-6-PD assay)
2. Fructose-6-phosphate, 0.02 mol/L, aqueous
3. G-6-PD (type XI, crystalline from Sigma Chemical Company, St. Louis, MO. Dilute in EDTA-mercaptoethanol "hemolyzing solution," p. 1502).

Procedure

1. Thaw the stock reagent solution prepared for G-6-PD assays, or prepare fresh stock reagent solution.
2. Prepare the following 1-mL cuvets (of which No. 1 is the blank for a normal control, No. 2 is the normal control assay, No. 3 is the specimen blank, and No. 4 is the specimen assay).

| | Cuvet No. | | | |
Reagents (mL)	1	2	3	4
Stock reagent solution	0.880	0.880	0.880	0.880
H_2O	0.105	0.005	0.105	0.005
F-6-P, 0.02 mol/L	—	0.100	—	0.100

3. Add to each cuvet 0.01 mL of a 10 U/mL suspension of glucose-6-phosphate dehydrogenase.
4. Cap and invert twice to mix.
5. Incubate at 37 °C for 1 h; then add to each cuvet 0.005 mL of hemolysate, prepared as described on p. 1502.
6. Cap and invert twice to mix. Then place in spectrophotometer chamber and measure increase in absorbance at 340 nm for at least 10 min at 37 °C. Determine ΔA_{340}/min, corrected for ΔA_{340}/min of blank.
7. While preceding step is being carried out, measure hemolysate hemoglobin concentration.
8. Calculate erythrocyte GPI activity in U/g Hb:

$$\text{GPI, U/g Hb} = \frac{804 \times A_{340}/\text{min}}{A_{540} \times F_{Hb}}$$

Reference Range

38.8–82.8 U/g Hb (mean $\pm$ 2 SD)

Comments and Precautions

The same precautions discussed under G-6-PD assay apply to the GPI assay. There may be an advantage in also assaying at low substrate concentration (0.005 mL instead of 0.1 mL F-6-P solution in each cuvet) to identify GPI variants with atypical kinetic curves.

References

Baughn, M. A., Valentine, W. N., Paglia, D. E., et al.: Blood, 32:236-248, 1968.
Beutler, E.: Red Cell Metabolism. A Manual of Biochemical Methods. 2nd ed. New York, Grune & Stratton, 1975, pp. 40-42.
Beutler, E., Blume, K. G., Kaplan, J. C., et al.: Br. J. Haematol., 35:331-340, 1977.

Determination of Erythrocyte Glutathione (GSH) Concentration

Clinical Significance

Erythrocyte glutathione concentration is diminished in many persons who have defects in the hexose monophosphate or glutathione synthesis pathways. The GSH stability test, originally devised to permit identification of persons susceptible to hemolysis from primaquine (later shown to be the result of G-6-PD deficiency), still remains a useful "stress test" of the intactness of these closely linked pathways. Since deficiencies of GSH synthetase and γ-glutamyl cysteine synthetase are rare disorders, it is not practical for clinical laboratories to contemplate assays for these enzymes unless the easily performed GSH stability test is abnormal.

Principle

Virtually all of the nonprotein sulfhydryl groups of erythrocytes are in the form of reduced glutathione. 5,5′-Dithiobis(2-nitrobenzoic acid), or DTNB, is a disulfide chromogen which is readily reduced by sulfhydryl compounds to an intensely yellow compound. The absorbance of the reduced chromogen is measured at 412 nm and is directly proportional to the GSH concentration.

Specimen

Whole blood, 3 mL, anticoagulated with ACD. GSH concentration declines in blood anticoagulated with heparin or EDTA; for such specimens the assay should be carried out on the same day. Include at least one normal control specimen with each assay.

Reagents

1. Precipitating solution. Place in a 100-mL volumetric flask 1.67 g of glacial metaphosphoric acid, 0.20 g of disodium or dipotassium ethylene-diaminetetraacetic acid (EDTA), and 30.00 g of NaCl, and bring to volume with distilled water. This solution is stable for three weeks at 4 °C. A fine precipitate may form due to EDTA, but this does not interfere with the test.
2. Phosphate solution, 0.3 mol/L. Place in a 1-L volumetric flask 42.59 g of Na_2HPO_4, and bring to volume with distilled water. Solution is stable indefinitely at 4 °C. If crystals form, dissolve by heating.
3. DTNB reagent. Place in a 100-mL volumetric flask 40.0 mg of 5,5′-dithiobis(2-nitrobenzoic acid) (Sigma Chemical Co., St. Louis, MO). Bring to volume with a solution of sodium citrate, 1 g/dL. This solution is stable for at least 13 weeks at 4 °C.
4. GSH standards. Place 100 mg GSH (Sigma Chemical Co., St. Louis, MO) in a 100-mL volumetric flask and bring to volume with reagent grade water. Invert repeatedly until GSH is completely dissolved. Prepare 50 mg/dL and 10 mg/dL standards by diluting 5 mL of the 100 mg/dL standard with 5 mL and 45 mL, respectively, of reagent grade water. The GSH standards are not stable and must be freshly prepared on the day of the assay.

Procedure

1. Place 0.2 mL of whole blood into a 10-mL test tube and add 1.8 mL distilled water. Mix to hemolyze.
2. Promptly add 3.0 mL of precipitating solution. Mix.
3. Allow to stand 5 min at room temperature and then filter through coarse-grade filter paper.
4. Prepare cuvets as follows:

Reagent (mL)	Blank	Assay
Filtrate	—	2.0
Precipitating reagent	1.2	—
H_2O	0.8	—
Phosphate buffer	8.0	8.0
DTNB solution	1.0	1.0

5. Cap cuvets and invert three times to mix.
6. Read absorbance at 412 nm within 4 min of preparing cuvets.
7. Obtain hematocrit of original whole blood specimen.
8. Assay GSH standards, omitting filtration step.
9. Graph the standard curve. Determine GSH concentration of blood specimen from the graph.
10. Calculate GSH concentration:

$$\text{GSH, mg/dL of erythrocytes} = \frac{\text{GSH conc. (from standard curve)}}{\text{hematocrit}}$$

Reference Range

47–100 mg/dL erythrocytes

Comments and Precautions

The ability of the erythrocyte to maintain a normal GSH concentration when stressed by an oxidant substance may be tested by incubation of blood with 1-acetyl 2-phenylhydrazine

(APH) and glucose for 2 h at 37 °C. In this GSH stability test, GSH concentration is measured before and after incubation of blood with APH. Erythrocytes deficient in G-6-PD, GSH-S, or γ-GC-S exhibit marked decrease in GSH concentration following such stress. Details of this test are given in the first reference cited below.

References

Beutler, E.: J. Lab. Clin. Med., *49*:84-95, 1957.
Beutler, E., Duron, O., and Kelly, B. M.: J. Lab. Clin. Med., *61*:882-888, 1963.

Screening Test for Pyrimidine-5′-Nucleotidase (P-5′-N) Deficiency

Principle

Pyrimidine-5′-nucleotidase effects the release of phosphate from cytidine-5′-monophosphate and uridine-5′-monophosphate and thus is an enzyme involved in the catabolism of RNA. RNA is a normal constituent of reticulocytes, but not of mature erythrocytes. In the absence (or marked deficiency) of P-5′-N, RNA persists in the erythrocyte and is manifested by striking basophilic stippling and mild to moderate hemolytic anemia.

Pyrimidine nucleotides, extracted with acid from the RNA remaining in P-5′-N-deficient erythrocytes, have an absorption peak at 270 nm. This peak is quite distinct from the peak at 257 nm that is due to purine nucleotides (such as ATP) which are normally present. When P-5′-N is deficient and RNA persists, the P-5′-N deficiency can be inferred by observing a pyrimidine nucleotide peak at 270 nm in an acid extract of erythrocytes.

Specimen

A minimum of 1 mL of whole blood anticoagulated with EDTA or ACD is required. Include at least one normal specimen as a control. A solution of cytidine-5′-monophosphate, 3 mmol/L, may be used as a simulated abnormal control. (Because of the rarity of this disorder, true abnormal controls are usually not available.)

Reagents

1. Perchloric acid (PCA), 0.6 mol/L
2. Cytidine-5′-monophosphoric acid (Sigma Chemical Company, St. Louis, MO), 3.0 mmol/L (aqueous)

Procedure

1. Centrifuge blood at 1500 × *g* for 10 min; remove and discard supernatant plasma.
2. Resuspend erythrocytes in NaCl, 0.155 mol/L. Centrifuge 1500 × *g* for 10 min.
3. Remove and discard supernatant.
4. Repeat steps 2 and 3.
5. To 0.5 mL of washed, packed RBC's add 0.5 mL of PCA, 0.6 mol/L. Mix thoroughly.
6. Centrifuge at 1500 × *g* for 20 min.
7. Carefully transfer supernatant to another tube. From this, pipet 0.3 mL to a UV cuvet of 4-mL maximum volume. Add 2.7 mL of PCA, 0.6 mol/L to this (sample) cuvet and 3 mL of PCA, 0.6 mol/L, to a second (blank) cuvet. Use 0.3 mL of reagent 2, plus 2.7 mL of PCA, 0.6 mol/L, for abnormal control.
8. Determine absorbance in the wavelength range of 240–290 nm, at 5-nm intervals and at 257 nm. Plot the absorbance curve, or alternatively scan with a recording spectrophotometer.

Reference Ranges

In normal specimens, or in those of any disorder but P-5′-N deficiency, there is an absorbance curve of low amplitude with a peak at about 257 nm, this being due to the presence of purines such as ATP and ADP. No peak is seen at 270 nm. In P-5′-N deficiency, a markedly higher amplitude curve is observed, with a peak at 270 nm. The absorbance maximum for normal erythrocytes is 257 nm and the observed absorbance value is ~0.3 per 10^8 RBC's/mL. In P-5′-N deficiency, the peak absorbance is at 270 nm, with absorbance values of ~0.8 per 10^8 erythrocytes/mL having been observed. No interfering substances are known.

Comments and Precautions

So far as is known at present, all cases of P-5′-N deficiency have given positive results by this test. False positive results have not been observed. This is not a reliable test for lead poisoning.

Reference

Valentine, W. N., et al.: J. Clin. Invest., *54*:866-878, 1974.

Determination of 2,3-DPG in Erythrocytes

Clinical Significance

2,3-Diphosphoglycerate (2,3-DPG) is the predominant glycolytic intermediate of the erythrocyte and plays a vital role in modulating the affinity of hemoglobin for oxygen. (See p. 1518.)

Principle

The assay of 2,3-DPG is based on an equilibrium system set up among 3-phosphoglycerate, 2-phosphoglycerate, and phosphoenolpyruvate by addition of monophosphoglycerate mutase (MPGM) and enolase.

$$3\text{-Phosphoglycerate} \underset{2,3\text{-DPG}}{\overset{\text{MPGM}}{\rightleftharpoons}} 2\text{-phosphoglycerate} \overset{\text{enolase}}{\rightleftharpoons} \text{phosphoenolpyruvate}$$

The addition of 2,3-DPG (in the sample) shifts the equilibrium so that more 2-phosphoglycerate is converted to 3-phosphoglycerate. This in turn results in the conversion of PEP to 2-phosphoglycerate. The rate of decrease in PEP concentration is measured by the decrease in absorbance at 240 nm, which is proportional to the amount of 2,3-DPG in the specimen.

Specimen

Add two drops of whole blood to 2 mL of ice cold distilled water immediately from syringe tip. Mix to hemolyze. Specimen must arrive in laboratory packed in ice (0–4 °C). Since 2,3-DPG concentration declines rapidly once blood is drawn, it is critical that blood for this assay be lysed and chilled immediately. This is best done by drawing blood with a syringe, and transferring the specimen immediately from the syringe tip to water at 4 °C.

Reagents

1. Stock reagents
   a. Tris-HCl buffer, 2 mol/L, pH 7.4. Prepare as on p. 1503, except for the change in concentration of Tris and pH.
   b. MgCl$_2$, 0.5 mol/L
   c. EDTA, disodium salt, 0.27 mol/L, pH 7.0. Place 0.27 mol (90.78 g/L) of Na$_2$ EDTA in a 2-L beaker. Add ∼800 mL H$_2$O. Add NaOH, 0.1 mol/L, drop by drop, until pH is 7.0. Then transfer to a 1-L volumetric flask and bring to volume with distilled water.
   d. Bovine serum albumin, 35 g/dL
   e. Monophosphoglycerate mutase (MPGM) suspension (may be purchased as phosphoglycerate mutase, crystalline suspension, 5000 U/mL, from Sigma Chemical Co., St. Louis, MO)
   f. Enolase suspension (Sigma Chemical Company, St. Louis, MO)
   g. 2,3-DPG standard solution, 1 mmol/L in distilled water. (Stable for several months if stored at −20 °C.)
2. Working reagent solution (prepare daily and store at 4 °C until used).
   a. Dilute bovine serum albumin to 0.5 g/dL. (Add 1 mL to 69 mL of distilled water.)
   b. Dilute enolase in distilled water to ∼32 U/mL.
   c. Dilute MPGM suspension in distilled water to ∼1600 U/mL.
   d. Prepare PEP solution, 0.25 mol/L.
   e. Mix the following to prepare working reagent solution:

Tris-HCl buffer	0.4 mL
MgCl$_2$ solution	0.2 mL
PEP solution	0.6 mL
EDTA solution	0.2 mL
MPGM (diluted)	0.2 mL
Enolase (diluted)	0.2 mL
Bovine serum albumin (diluted)	1.8 mL
Deionized water	14.4 mL
Total volume	18.0 mL

3. Dilute Tris-HCl buffer. On day of assay, add 1 mL of stock Tris-HCl buffer to 99 mL of distilled water. Mix.

4. 2,3-DPG standards, 1.0 μmol/L. Thaw stock 2,3-DPG standard and dilute 1000-fold by adding 0.1 mL to 99.9 mL of dilute Tris-HCl buffer. Make further dilutions as shown (all volumes are given in mL):

Final Standard Concentration 2,3-DPG $\mu mol/L$	2,3-DPG Std, 1 $\mu mol/L$ (mL)	Tris-HCl (mL)
0	0.0	1.0
0.25	0.4	1.2
0.50	0.8	0.8
0.75	1.2	0.4
1.00	1.0	0.0

Procedure

1. Transfer 0.02 mL of hemolysate to 2 mL of dilute Tris-HCl buffer, pH 7.4.

2. Add 0.9 mL of working reagent solution to cuvets with 1 mL critical volume, preparing one cuvet as a blank for the standard, and samples. Allow to warm to room temperature for at least 10 min.

3. Add to blank cuvet 0.1 mL of distilled water, and add 0.1 mL of 0, 0.25, and 0.50 μmol/L standard solutions to the respective cuvets. Mix by inversion and measure absorbance change at 240 nm for 10 min at 25 °C.

4. Add 0.1 mL of 75 and 100 μmol/L standard solutions to respective cuvets and 0.1 mL of diluted hemolysate to sample cuvets. Mix by inversion and measure absorbance change at 240 nm for a 10-min interval at 25 °C. Plot ΔA_{240}/min against concentration on a graph for standard curve.

5. Using the ΔA_{240}/min calculated for each specimen, find on the standard curve the corresponding concentration of 2,3-DPG. Designate this concentration C_s.

6. Measure the hemoglobin concentration of each hemolysate. Add 0.5 mL of the original hemolysate to 2.5 mL of Drabkin's solution. Mix. Determine $A_{540\ nm}$ of the cyanmethemoglobin standard (80 mg/dL) and of each hemolysate.

Calculations

$$C_{Hb},\ mg/dL = \frac{6 \times 80 \times A_{540}\ (\text{hemolysate})}{A_{540}\ (\text{standard})}$$

$$C_{Hb},\ g/L = C_{Hb},\ mg/dL \times \frac{10}{1000}$$

$$C_{2,3\ DPG},\ \mu mol/gHb = \frac{100 \times C_s}{C_{Hb},\ g/L}$$

where C_{Hb} = hemoglobin concentration of the hemolysate; 6 = dilution of the hemolysate; 80 = concentration of the cyanmethemoglobin standard in mg/dL; 100 = dilution of the hemolysate; 10 = conversion from dL to L; 1000 = conversion of mg to g; C_s = concentration of 2,3 DPG, in μmol/L, read from the standard curve.

Reference Range

13.4–17.4 μmol/g Hb (mean $\pm$ 2 SD)

Comments and Precautions

If the assay is not to be done on the same day, measure the hemolysate hemoglobin concentration of a small aliquot and heat the remainder for 10 min in a boiling water bath to inactivate phosphatases. Freeze the boiled hemolysate and keep frozen until ready to assay.

Reference

Beutler, E.: Red Cell Metabolism. A Manual of Biochemical Methods. 2nd ed. New York, Grune & Stratton, 1975.

DISORDERS OF THE ERYTHROCYTE MEMBRANE CYTOSKELETON

Normal Cytoskeleton

Like other cell membranes, that of the erythrocyte is basically a *phospholipid bilayer.* The lipoproteins are so organized that a net negative charge predominates on the outer layer, and a net positive charge predominates on the inner layer. This arrangement ensures that erythrocytes will tend to repel each other electrostatically. A number of glycoproteins occur within the erythrocyte membrane, like islands in the phospholipid sea. These are the blood group antigens. Some glycoproteins traverse the membrane from the exterior to the inner surface. This is true of *glycophorin* and of another protein simply called "band 3," on the basis of its position after sodium dodecyl sulfate polyacrylamide gel electrophoresis (SDS-PAGE).[60-62] Band 3 is believed to function as an anion channel through which Cl^-, HCO_3^-, and other anions move across the erythrocyte membrane.

On the inner surface of the membrane is a meshwork of filamentous proteins called the *cytoskeleton.* This confers shape and flexibility to the erythrocyte. Several distinct proteins constitute the cytoskeleton: *spectrin, ankyrin, band 3, actin, band 4.1,* and perhaps other as yet unidentified proteins. The manner in which these filamentous proteins are associated in the cytoskeleton is illustrated in Figure 15-3. Spectrin may be isolated as a dimer with a M.W. of 250 000 and as a tetramer.[60-62] The dimer appears to be the functional state of spectrin in the intact cytoskeleton. Its overall amino acid composition has been determined, but the amino acid sequence is as yet unknown. Analysis of the structure of the other components of the membrane cytoskeleton has only begun.

Two types of glycophorin are designated A and B. The complete amino acid sequence of glycophorin A is known; it consists of a chain of 131 amino acids. The function of glycophorins is unknown; glycophorin B contains some of the erythrocyte blood group antigens.

The Na^+-K^+ transmembrane pump is represented in Na^+-K^+ ATPase, a protein that traverses the erythrocyte membrane. Na^+-K^+ ATPase consists of α- and β-subunits with M.W. of 95 000 and 55 000, respectively. The molecular structure and amino acid sequence of Na^+-K^+ ATPase have been partially elucidated, and its complete structure should be known soon.

Besides these intrinsic proteins of the erythrocyte membrane, more than 30 enzymes have been found to be intimately associated with the membrane, since they remain membrane-bound

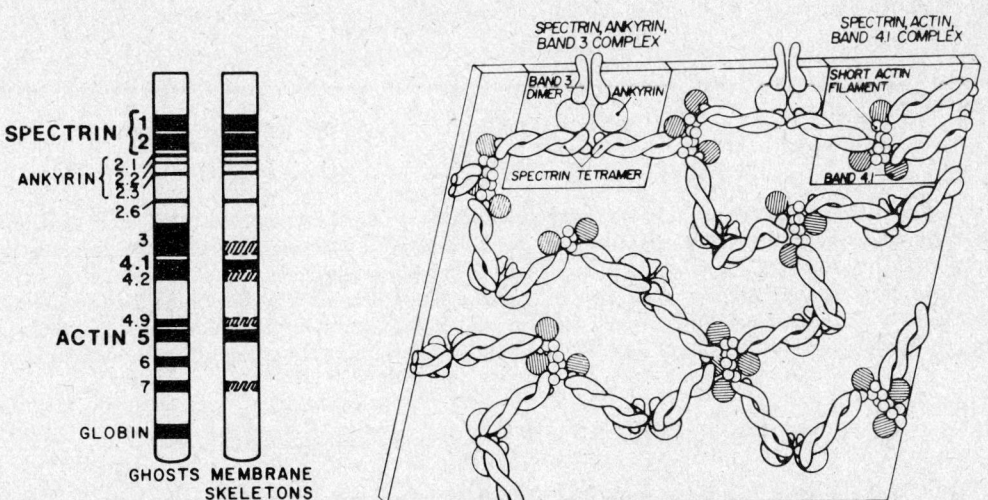

Figure 15-3. Cytoskeleton and SDS gel pattern and components of the erythrocyte membrane. On the left is a representation of the sodium dodecyl sulfate (SDS) polyacrylamide gel electrophoretograms obtained from erythrocyte membranes (ghosts) and from the lipid-extracted membrane cytoskeleton. Principal components are identified as spectrin (2 bands), ankyrin (3 bands), band 4.1, and actin. Several other bands are also seen, but their significance is as yet unknown. On the right is a scheme for the organization of the membrane cytoskeleton viewed as though from the interior of the cell. (Reprinted by permission from Lux, S. E.: Dissecting the red cell membrane skeleton. Nature, *281*:426-429, 1979. Copyright © 1979, Macmillan Journals Limited.)

after hemolysis and repeated washing of erythrocyte ghosts with water or isotonic saline. The reason for this relationship, in most cases, is uncertain.

HEMOLYTIC DISORDERS RESULTING FROM DEFECTS IN THE CYTOSKELETON

Although it has long been thought likely that disorders such as *hereditary spherocytosis* are due to erythrocyte membrane defects, only during recent years have such defects begun to be defined. These will be reviewed briefly. One of the technical problems in identifying defects in membrane proteins is that the method commonly used for distinguishing the proteins of the cytoskeleton, SDS-PAGE, separates them only on the basis of molecular weight. Only the most gross structural defects could be identified by this technique. Substances which differ in a single amino acid only, such as those that characterize hemoglobinopathies, are not detected with this support medium.

Hereditary elliptocytosis. There are two types of hereditary elliptocytosis (HE), a very mild form without hemolysis and a form with mild hemolytic anemia. In the hemolytic type of HE, isolated spectrin appears to be thermolabile in vitro.[66,95] Presumably this characteristic reflects a structurally abnormal spectrin, but further characterization of this defect is yet to be done.

Hereditary pyropoikilocytosis. As of 1983, only 13 cases of hereditary pyropoikilocytosis had been recognized. All cases thus far have been in blacks. The condition causes mild to moderate hemolytic anemia with extreme distortion of erythrocytes and marked microcytosis.[113] Severe hemolytic disease of the newborn may occur. Bidirectional mapping ("fingerprints") of tryptic peptides of spectrin from some cases of pyropoikilocytosis has shown an anomalous peptide fragment, thus implying an amino acid substitution in this "spectrinopathy."[58,62]

Hereditary spherocytosis. The most frequently encountered congenital hemolytic disorder in humans is hereditary spherocytosis (HS). Spectrin from some human HS erythrocytes has a defect in binding of the band 4.1 membrane protein, and a structural abnormality in a spectrin polypeptide fragment is likely.[37,62]

Hydrocytosis and xerocytosis. Two related disorders of the erythrocyte membrane have been called hydrocytosis (stomatocytosis), in which erythrocytes appear osmotically swollen, and xerocytosis (desiccocytosis), in which they appear to be osmotically shrunken.[35] Both conditions are inherited as autosomal dominant traits, and both are usually accompanied by very mild hemolysis, usually without anemia. In both conditions there is an increase in the erythrocyte Na^+ concentration and a reciprocal decrease in K^+ concentration. Total monovalent cation concentration is usually reduced. In hydrocytosis, erythrocyte concentration of 2,3-diphosphoglycerate may be diminished. Both conditions have shown fewer than normal sites of ouabain-inhibitable Na^+-K^+ pumps on erythrocyte membranes.[109] Whether this implies a diminution in the amount of ATPase in the membrane or a functional abnormality of ATPase is as yet unclear. These seem to be rare disorders; fewer than a dozen families have been reported with hereditary hydrocytosis, and only two with hereditary xerocytosis. Yet, these conditions are probably not as rare as the paucity of these reports might seem to imply. Much more commonly encountered is hydrocytosis in alcoholism. The mechanism is unknown.

OTHER DISORDERS DUE TO ABNORMALITIES IN THE ERYTHROCYTIC MEMBRANE

Hereditary abetalipoproteinemia is associated with acanthocytosis of erythrocytes. The cause is extrinsic to the erythrocyte, but the morphologic changes in erythrocytes are presumably due to alterations in the phospholipid composition of the membrane, secondary to the plasma deficiency of β-lipoprotein.

Rh-null phenotype and *McLeod phenotype* are rare hereditary chronic mild hemolytic disorders due to diminished or absent expression of Rh or Kell antigens, respectively, on the erythrocyte membrane. McLeod phenotype is accompanied by persistent elevation in serum creatine kinase, but apparently without myopathy.

HEMOGLOBINOPATHIES AND THALASSEMIAS

NORMAL HEMOGLOBIN; HEMOGLOBIN SYNTHESIS AND DEGRADATION

Hemoglobin is a spheroidal protein with a M.W. of 64 456. It consists of four subunits arranged so as to resemble a thick-walled shell with a small central cavity. The four subunits are two pairs of identical polypeptide chains, each with a cleft or pocket on the exterior of the shell. The cleft contains a heme (iron-porphyrin) group that is the site of O_2 uptake and release.[75]

Hemoglobin generally contains as subunits a pair of α- or α-like globin chains, and a pair of identical non-α chains that may be β, γ, δ, or ϵ. The two α and two non-α chains are arranged around the central cavity as shown in Figure 15-4. Four points of contact between subunits are designated (in Hb A) $\alpha_1\beta_2$, and $\alpha_1\beta_1$. Of these, the $\alpha_1\beta_2$ contacts are narrower, and these function as hinges that permit the hemoglobin tetramer, somewhat like a clamshell, to open and close slightly as O_2 is taken up or released. The details of configurational changes of the hemoglobin tetramer, on oxygenation or deoxygenation, will be discussed further below.

The Structure of Normal Hemoglobins

In normal human adults, at least 96% of the hemoglobin is Hb A, which is composed of two normal α- and two normal β-chains and may thus be represented as $\alpha_2^A\beta_2^A$, or more simply as $\alpha_2\beta_2$. *Hb A_2* is usually about 2.5 or 3.0% of total hemoglobin; it contains two δ-chains, and may thus be designated $\alpha_2\delta_2$. *Fetal hemoglobin (Hb F)* predominates during fetal life but rapidly diminishes during the first year of postnatal life. In normal adults < 1% (actually, when accurately measured, about 0.1%) of hemoglobin is Hb F. It consists of two α- and two γ-chains, and thus is $\alpha_2\gamma_2$. Three embryonic hemoglobins are hemoglobins Portland, Gower-1, and Gower-2. *Hb Portland* is $\zeta_2\gamma_2$, *Hb Gower-1* is $\zeta_2\epsilon_2$, and *Hb Gower-2* is $\alpha_2\epsilon_2$. These three normal embryonic hemoglobins are of little import in hematology or clinical chemistry, as they are not encountered during postnatal life. *Hb H (β_4)* or *Hb Barts (γ_4)* may occur in α-thalassemia.

Each of the globin chains is a long strand of amino acids, 141 in each α-chain and 146 in each β-, δ-, and γ-chain. Each chain contains helical segments separated by short noncoiled segments (Figure 15-5). The helical segments of the β-, γ-, and δ-chains are designated, reading from the amino end of the chain, A, B, C, D, E, F, G, and H. Nonhelical segments are NA, AB, CD, EF, GH, and HC, wherein N- and -C represent amino and carboxyl ends, respectively. The α-chains lack D segments, and the interconnecting nonhelical segment then is not CD, but CE. The β-, γ-, and δ-chains are closely homologous; β- and δ-chains differ in only 10 amino acids.

Each globin chain is looped about itself so as to form a pocket or cleft in which the heme group nestles. Normally, this heme pocket is formed entirely by nonpolar (hydrophobic) amino acids. The heme moiety is suspended within this pocket by an attachment of its iron atom to the imidazole group of the proximal histidine (position 92 of the β-chain [β92] or position 87

a	α30
b	α48
c	α47
d	β1
e	β6
f	β80
g	β83
h	β82
i	β81
j	β143
k	β146
l	β145
m	β90

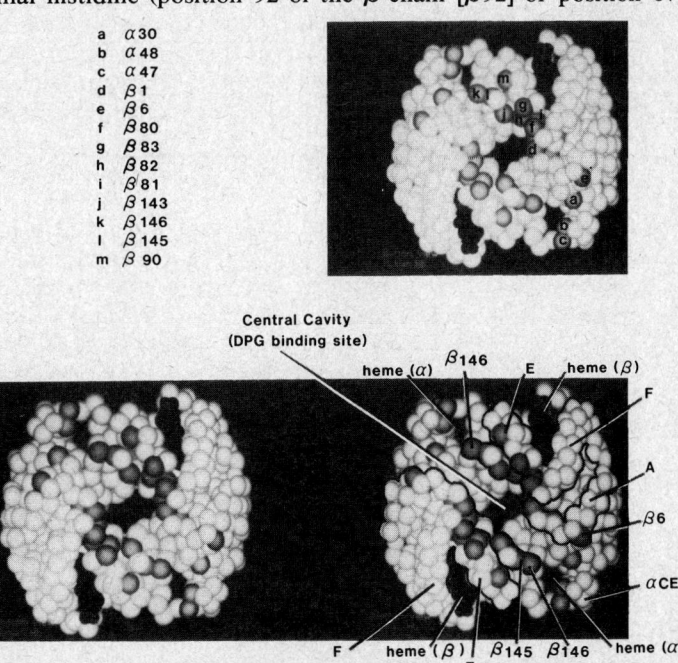

Figure 15-4. Model of the hemoglobin tetramer with the β-chain subunits facing the reader. Amino acids are represented as spherules. Those that are shaded affect hemoglobin mobility in citrate agar gel. Spherules labeled d, j, k, and l in the upper panel constitute the 2,3-DPG binding site and are important in regulation of the oxygen affinity of hemoglobin. The lower panel is a stereo pair for 3-dimensional viewing. To obtain the 3-dimensional effect, hold the illustration at arm's length. The β-globin chains appear to be between the reader and the more distant α-chains. On the right, some of the segments of the β-globin chain have been outlined, and segments and critical components have been marked. β146 is the COO$^-$ end of the β-chain; β6 is the site of amino acid substitutions responsible for hemoglobins S and C. (Reproduced by permission from Winter, W. P., and Yodh, J.: Interaction of human hemoglobin and its variants with agar. Science, *221*:175-178, 1983. Copyright 1983 by the Am. Assoc. for the Adv. of Science.)

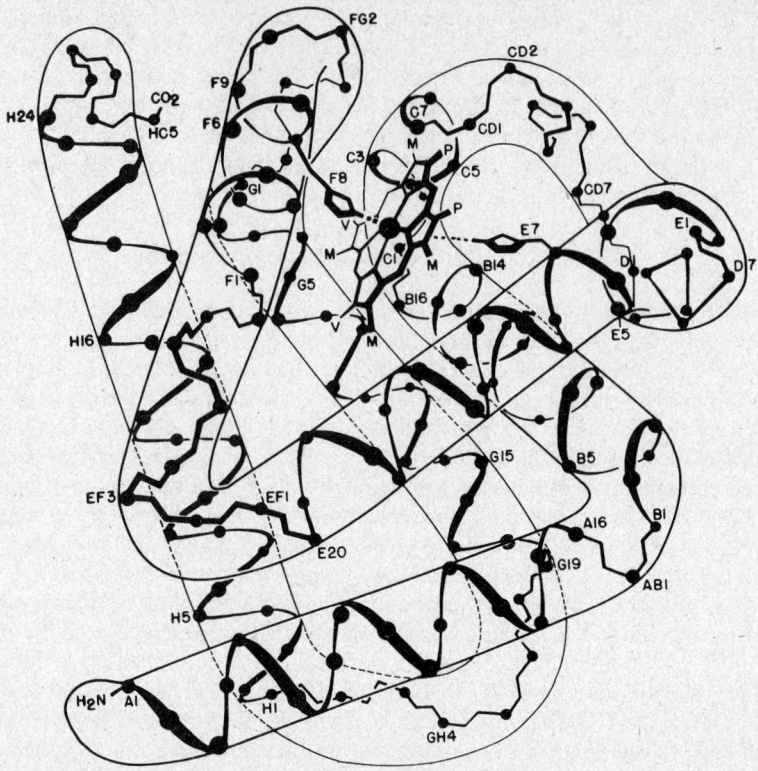

Figure 15-5. Structure of the hemoglobin subunit. Chains of amino acids in spiral or helical segments are linked by short nonhelical segments. The helical segments are designated A through H. In this illustration, amino acids are designated in accordance with the helical or nonhelical segment in which they occur. (From Dickerson, R. E.: X-ray analysis and protein structure. *In:* The Proteins: Composition, Structure and Function, Vol. 2. 2nd ed. H. Neurath, Ed. New York, Academic Press, 1964, pp. 603-778. Reproduced by permission.)

of the α-chain [$\alpha87$]). The imidazole group of the distal histidine ($\beta63$ or $\alpha58$) is also in contiguity with the iron of heme, but it appears to swing in and out of this position to permit the ingress and egress of O_2. The four iron atoms are in the divalent state, whether hemoglobin is oxygenated or deoxygenated. *Methemoglobin* is a nonfunctional derivative with iron in the trivalent state.

Structure of the Globin Gene Loci

The gene loci that determine the structure of the α- and ζ-globin chains are on chromosome 16. These are arranged linearly, together with a pseudo ($\psi\alpha1$) α-gene as follows:[56,69]

$$5' \ldots \zeta'2 \ldots \zeta1 \ldots \psi\alpha1 \ldots \alpha1 \ldots \alpha2 \ldots 3'$$

In this sequence, reading from the 5' ("upstream") end to the 3' ("downstream") end of a segment of DNA, there are two gene loci for the embryonic ζ-globin chains, a pseudo α-gene, and two loci that determine the structure of the α-chains of hemoglobin. Very likely both ζ-genes are expressed in the embryo. Deletions of both ζ-loci are probably incompatible with embryonic life and lead to very early spontaneous abortion. Deletions of one or more α-loci are responsible for various types of α-thalassemias.[24,49,79]

The β-, γ-, and δ-genes are clustered closely together in a region of chromosome 11 that may be called the $\gamma\delta\beta$-*genome complex.* The gene sequence is now known to be:

$$5' \ldots \psi\beta2 \ldots \epsilon \ldots {}^{G}\gamma \ldots {}^{A}\gamma \ldots \psi\beta1 \ldots \delta \ldots \beta \ldots 3'$$

As shown, there are normally two gene loci that determine γ-chain structure, one that determines δ-chain and one that determines β-chain structure. The gene loci are designated ${}^{G}\gamma$ and ${}^{A}\gamma$ because their codons and the γ-chains they specify are not identical. Gamma chains designated [A]γ contain alanine at amino acid position 136 and threonine at position 75. Gamma chains designated [G]γ contain glycine at amino acid position 136 and isoleucine at position 75. The roles of the pseudo ψ-genes are, as yet, unknown. There is reason to believe that a "switch" mechanism of some sort is located between the [A]γ- and δ-gene loci. This switch permits expression of the γ-loci during fetal life, but in the last few weeks of fetal life turns down the expression of the γ-loci and turns on the expression of the β-locus, and to a lesser extent that of the δ-locus also.

Studies with recombinant DNA and with restriction endonucleases have enormously expanded our knowledge of the structure of the globin gene loci.[64,98] The precise sequence of nucleotides has been determined for the regions of DNA that code for the structure of the α-, β-, and γ-chains. Furthermore, the nucleotide sequences of contiguous or flanking DNA regions have also been determined. What is presently known of this for the β-globin gene will be summarized here; the structure of the α-gene regions is analogous.

Although the genetic triplet code requires only 444 nucleotides to code for the amino acid sequence of the β-chain, the β-gene locus is 1900 nucleotides in length (Figure 15-6). On the 5' side of the coding region is a long flanking region. The coding region itself is interrupted by two long intervening segments, or *introns,* one that is 100 nucleotides long, and one that is 700 nucleotides long. Again on the 3' side of the coding region is a long flanking strand of DNA. Although δ- and β-gene loci are in relatively close proximity, there are, nonetheless, 7000 nucleotides (7 kilobases, or 7 kb) between them.

Within the nucleus, the DNA is transcribed to *pre-messenger RNA.* This pre-*m*RNA is then somehow edited by removal of the two introns, and the remaining RNA is spliced together. It is capped with a noncoding 54-nucleotide segment of RNA on its 5' end and with a 135-nucleotide segment, consisting mostly of polyadenylic acid, on its 3' end. It is then released to the cytoplasm, where ribosomes move along the *m*RNA, from the 5' end to the 3' end, translating the message and assembling the globin chains. The sequence of globin chain assembly is such that the 5' end of *m*RNA corresponds to the amino end of globin and the 3' end of *m*RNA to the carboxyl end of globin.

Restriction endonucleases are bacteria-derived enzymes that cleave DNA at specific recognition sites, that is, wherever a specific sequence of nucleotides occurs. For example, the endonuclease *BAM* HI cleaves DNA wherever the sequence 5' . . . GGATCC . . . 3' occurs, cleaving between the two guanyl residues (GG). Two restriction endonucleases cleave DNA at the site of the Hb S mutation, and permit an unequivocal prenatal diagnosis of homozygous Hb S disease. This powerful new methodology is being used for prenatal diagnosis of hemoglobinopathies and thalassemias, and ultimately will be applied for all genetic disorders.

Globin Chain Biosynthesis

As ribosomes move along the strand of mRNA, they join the specified amino acids. Transport of amino acids to the ribosome is effected by several species of transfer RNA, each a carrier for

Figure 15-6. An abbreviated scheme of transcription, processing, and translation of messenger RNA (*m*RNA). This complicated process involves excision of long intervening segments (introns) of "heterogeneous" RNA, with a stepwise reduction in length of the transcribed RNA to the 444 nucleotides (nt) of the final *m*RNA. The *m*RNA is capped at both ends to stabilize it, then released into cytosol, where ribosomes read the code and assemble the globin chains.

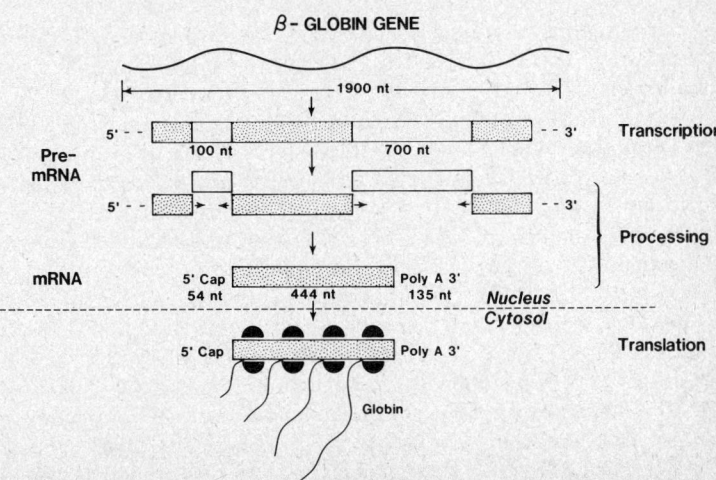

β- GLOBIN GENE

a specific amino acid. When the ribosome comes to a *"nonsense" (terminator) codon* such as UAA, this signals completion of translation, and the polypeptide chain is released from the ribosome. Synthesis of α- and β-chains is somehow coordinated so that a deficiency of either causes a reduction in rate of synthesis of the other. Similarly, a reduction in heme synthesis (as in iron deficiency) results in reduction in synthesis of both α- and β-globin chains. The means by which α- and β-globin chain synthesis and heme synthesis are coordinated is not as yet understood. As will be perceived from the above, globin synthesis takes place in nucleated erythrocyte precursors in bone marrow and in reticulocytes, but not in mature erythrocytes.

Hemoglobin Degradation

Hemoglobin degradation normally occurs within phagocytic cells of the reticuloendothelial system. Aged and metabolically run-down erythrocytes can no longer provide the energy to sustain the Na^+-K^+ cation pump and therefore leak cations at an increasing rate. The sequence of extravascular erythrocyte destruction is (1) phagocytosis of aged cells, (2) opening of the heme ring to form *verdohemoglobin*, and (3) removal of iron and protein from the tetrapyrrole chain. The tetrapyrrole chain is further metabolized to bilirubin. (The details of hemoglobin catabolism are presented in Chapter 13.) The iron is reutilized in the formation of new hemoglobin. The protein is digested, and its component amino acids enter the general amino acid pool. It is not certain whether methemoglobin is an obligatory intermediate in this sequence. When there is accelerated extravascular erythrocyte destruction, as in most autoimmune hemolytic disorders and hypersplenic states, hemoglobin degradation is accelerated.

Under certain circumstances (for example, mechanical hemolysis due to faulty intracardiac valvular prosthesis or *paroxysmal nocturnal hemoglobinuria*), hemolysis may be predominantly intravascular. In this case, the free plasma hemoglobin concentration is bound by the plasma protein *haptoglobin*. The hemoglobin-haptoglobin complex formed is rapidly removed by parenchymal cells of the liver. In hemolytic disorders the unbound haptoglobin concentration is therefore usually decreased or absent. Free hemoglobin, in excess of the binding capacity of haptoglobin, is excreted in the urine, to which it imparts a chocolate-brown color. *Hemoglobinuria* is identified by the occult blood test described later in this chapter. The plasma haptoglobin concentration returns to normal after 2–3 d unless the hemolysis persists. Some of the heme that dissociates from hemoglobin in plasma may be taken up by *hemopexin*, a heme-binding protein present in normal plasma. Some is taken up by serum albumin to form an albumin-heme complex called methemalbumin.

MECHANISMS OF HEMOGLOBIN FUNCTION

Changes in Conformational States of Hemoglobin with Uptake and Release of O_2

Hemoglobin is uniquely adapted to take up and release O_2 in a relatively narrow range of O_2 pressure. This is the result of subunit-to-subunit interaction whereby an alteration in spatial relationships between portions of the molecule facilitates the uptake or release of O_2. The O_2 dissociation curve of normal blood hemoglobin is sigmoidal (Figure 15-7). The shape of this curve implies that the uptake of the first molecule of O_2 by the hemoglobin molecule facilitates the uptake of the second, third, and fourth molecules of O_2 with only a modest increment in oxygen pressure. Conversely, a molecule of hemoglobin with all four heme sites oxygenated readily gives up O_2 at three of these sites. This highly coordinated response to slight changes in oxygen tension, which is absolutely vital for effective O_2 transport, appears to depend in part on (1) reversible binding of certain amino acids across the $\alpha_1\beta_2$ contact point, permitting the $\alpha\beta$-dimers to snap back and forth as the molecule is oxygenated and deoxygenated, and (2) the movement of DPG in and out of the central cavity. (2,3-DPG stabilizes the deoxygenated configuration.) Solutions of pure hemoglobin do not manifest sigmoidal O_2 dissociation curves, but the addition of DPG in approximately equimolar concentration completely restores the normal sigmoidal dissociation curve. In vivo increases in erythrocytic DPG concentration markedly decrease the O_2 affinity of hemoglobin and increase O_2 delivery to tissues. This is an important compensatory

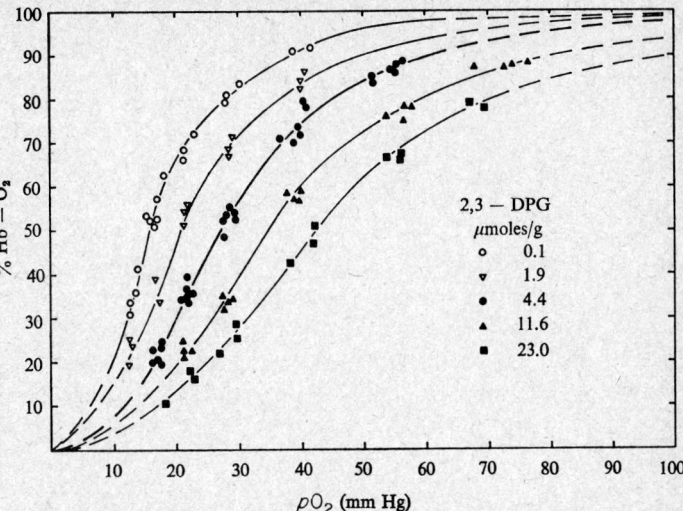

Figure 15-7. Normal oxygen dissociation curve of hemoglobin. Changes in 2,3-diphosphoglycerate (DPG) concentration in the erythrocyte markedly influence the position of the curve. As the concentration of 2,3-DPG increases, the curve shifts to the right. (Reproduced with permission from Duhm, J.: The effect of 2,3-DPG and other organic phosphates on the Donnan equilibrium and the oxygen affinity of human blood. *In*: Oxygen Affinity of Hemoglobin and Red Cell Acid Base Status [Alfred Benzon Symposium, IV]. M. Rorth and P. Astrup, Eds. Copenhagen, Denmark, Alfred Benzon Foundation, 1972.)

mechanism in anemia, and it has been shown that the erythrocytic DPG concentration varies reciprocally with the blood hemoglobin concentration.[5] Conversely, low DPG concentration is associated with a shift of the O_2 dissociation curve to the left, indicating increased O_2 affinity of hemoglobin and decreased O_2 delivery to tissues. This is a problem in blood banking, because blood preserved with ACD (acid-citrate-dextrose) solution or with CPD (citrate-phosphate-dextrose) solution has a rapid progressive decrease in DPG concentration. Thus, transfusion of large amounts of this blood may compromise a patient's ability to oxygenate his tissues for several hours until more DPG is synthesized.

Changes in pH of blood also influence the affinity of hemoglobin for oxygen. A decrease in blood pH or an increase in $p CO_2$ causes the dissociation curve to shift to the right and hemoglobin to become less avid for O_2. This favors increased delivery of O_2 to tissues. Conversely, an increase in blood pH or a decrease in $p CO_2$ causes the curve to shift to the left and hemoglobin to become more avid for O_2. This favors increased uptake of O_2 by hemoglobin, and less delivery of O_2 to tissues. These changes in O_2 affinity of hemoglobin that are caused by changes in pH in the range of 7.2–7.6 are known as the *alkaline Bohr effect*.

The spatial conformational states that occur in the hemoglobin tetramer with oxygenation and deoxygenation have been called the T (for tense) and R (for relaxed) states.[76,78] The T state is the result of the presence of numerous ionic or salt bonds between hemoglobin subunits, and within subunits as well. The glycolytic intermediate, 2,3-DPG, occupies the central cavity in the T state and is bound principally to histidines in the 143rd amino acid positions in the β-chains, but also has bonds to the carboxy ends of all four chains. Also in the T state the penultimate tyrosine of each globin chain is wedged into a crevice between the FG and HC segments, and the last amino acid (at the carboxyl end of the chain) is bound to the FG segment by a H^+ bond between its carboxyl group and the FG amino acid. Thus, in the T conformation, the hemoglobin tetramer may be considered to be tightly laced together by numerous salt and hydrogen bonds. Perhaps because α-chains lack D helices, the heme pockets of the α-chains are a little more open than are those of the β-chains. In the T state, the iron of all four hemes is in a high state of spin, and thus is of larger diameter than that of the R state. For this reason, iron is slightly displaced from the plane of the porphyrin ring (Figure 15-8).[77] The sequence of oxygenation may be conceived as being about as follows: A molecule of O_2 enters the heme pocket of an α-chain. O_2 binds to the Fe(II) of the heme group; the iron atom changes to low spin and drops into the plane of the porphyrin ring, a displacement of 0.28 nm. This slight displacement pulls on the proximal histidine of the F helix, a movement that is transmitted to the FG segment. Then, there is a change in positions of the contiguous C and G helices of the adjacent β-chain that comprise part of the $\alpha_1\beta_2$-contact. This is a sliding movement along the plane of the $\alpha_1\beta_2$-contact. In the process of these conformational changes, many other salt bonds between and within subunits are broken; the bonds to 2,3-DPG are also broken (Figure 15-9). The central cavity becomes smaller, and 2,3-DPG is expelled.[76,77] The sequence of conformational

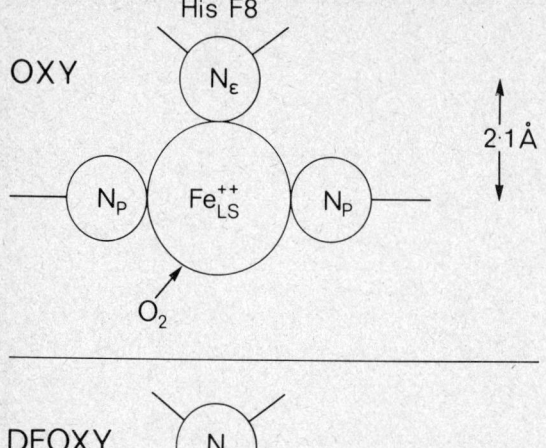

Figure 15-8. Displacement of Fe from heme plane with oxygenation. The "trigger" for the conformational changes accompanying oxygenation of hemoglobin is movement of the central Fe(II) of heme into the plane of the porphyrin ring. This movement, which occurs first in the α-chains, is due to a change in the spin of Fe(II). When in a high-spin state (in the deoxy state), Fe(II) cannot be accommodated in the plane of the porphyrin ring and is consequently displaced. In the lower spin state accompanying oxygenation, Fe(II) slips into the plane of the porphyrin ring, a movement of a few hundredths of a nanometer. This slight movement of the iron atom tugs on the "proximal histidine," setting off the entire sequence of configurational changes. N_ϵ = histidine nitrogen, N_p = nitrogen of porphyrin ring, LS = low spin, HS = high spin. (Illustration provided by Dr. M. F. Perutz, 1984.)

changes here described is for αβ-dimers, which are the functional units for uptake and release of O_2. In the tetramer, when salt bonds are present and 2,3-DPG is in the central cavity, O_2 is expelled and the conformational changes of the R state are reversed. It may also be seen from this presentation that the T conformation will be favored if there is an increase in H^+ or of 2,3-DPG in the environment of hemoglobin. This correlates with the effects of both lowering of pH and elevation of 2,3-DPG concentration in reducing the O_2 affinity of hemoglobin.

THALASSEMIAS AND RELATED CONDITIONS

Thalassemias are conditions in which there is a reduction in the rate of synthesis of one or another type of globin chain. Thus, if there is a reduction in the α-chain synthesis, the condition resulting is an α-thalassemia;[21] if reduction is in β-chain synthesis, the resulting condition is a β-thalassemia, and so forth. Not many years ago it was generally thought that thalassemia was primarily a disorder of people whose ancestors dwelt on the Mediterranean littoral. Now it is known that thalassemias are worldwide in distribution, affecting persons of most ethnic groups; however, thalassemias occur with highest frequency in persons of Asian or African origin. *Alpha-thalassemia* appears to be the most common genetic disorder of man. Approximately 30% of black Americans have a gene for α-thalassemia;[24] a similarly high prevalence of α-thalassemia is observed in persons of Southeast Asian origin. Alpha-thalassemia also occurs in persons of Mediterranean origin, although at a lower prevalence than in the aforementioned groups. *Beta-thalassemia* is the disorder historically associated with Mediterranean peoples. However, the prevalence of β-thalassemia appears to be just as high in Asians, since about 4–6% of Indians, Chinese, and Southeast Asians have β-thalassemia trait. Clinical expressions of thalassemia genes range from none (in the mildest type of α-thalassemia trait) to very severe and often lethal anemia, growth retardation, and bone malformations in homozygous β-thalassemia (classic thalassemia major) or Hb E/β-thalassemia. The most severe forms of thalassemia are incompatible with life; for example, the severest form of α-thalassemia results in fetal death in utero due to extreme anemia and heart failure, resulting in hydrops fetalis. In this chapter the terms *thalassemia minor* or *major* will be used to indicate the severity of the clinical disorder, and the term *thalassemia trait* will signify the *heterozygous* state for a thalassemia gene. Thalassemia minor is characterized by microcytic erythrocytes, by hypochromia, sometimes by mild anemia, and commonly by increased peripheral blood erythrocyte counts. Thalassemia major is characterized by severe

$$\left(\alpha_1^D \alpha_2^D \beta_1^D \beta_2^D\right)^D$$

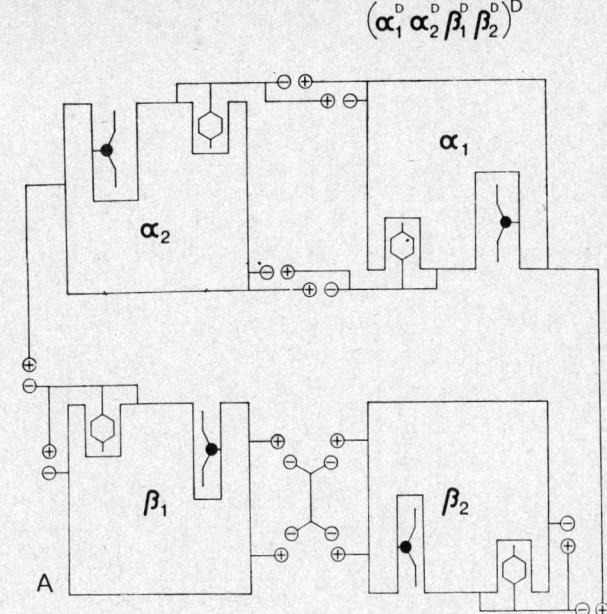

Figure 15-9. A scheme showing in more detail the changes which occur with deoxygenation (A) and oxygenation (B). Relationship of free amino (⊕) and carboxyl (⊖) groups to each other and to 2,3-DPG is seen in A. Ionic bonds (⊕ ⊖) hold the tetramer together in the deoxy configuration. These bonds are broken with oxygenation. The configurational changes widen the heme pockets, permitting the entry of molecules of O_2. (Illustration provided by Dr. M. F. Perutz, 1984.)

$$\left(\alpha_1^o \alpha_2^o \beta_1^o \beta_2^o\right)^O$$

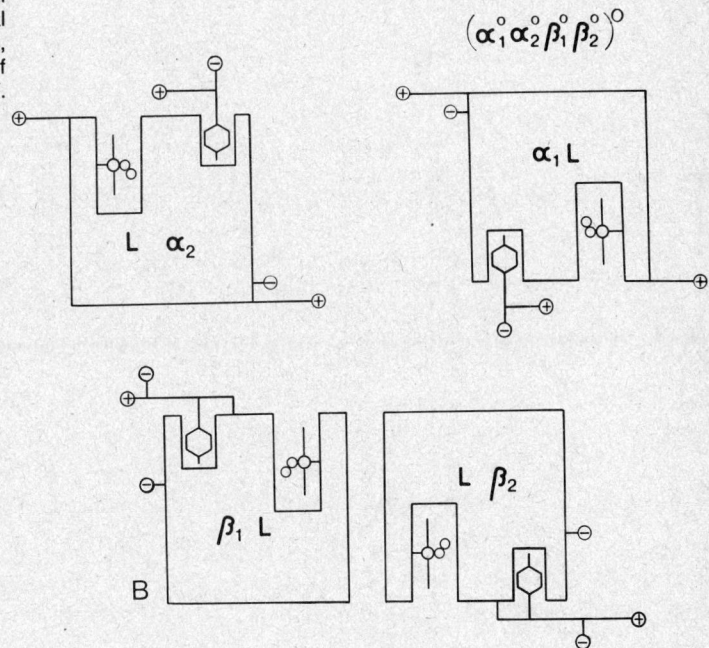

anemia, hypochromia, microcytic erythrocytes, signs of accelerated hemolysis and regeneration (hyperbilirubinemia, reticulocytosis, basophilic stippling), and marked enlargement of liver and spleen. Growth retardation and bony abnormalities are also typical features of thalassemia major. Thalassemias are summarized in Table 15-1.

The α-thalassemias. Most of the α-thalassemias are the result of deletions of one or more α-gene loci.[24,56,81] The mildest form of α-thalassemia is the result of a single α-gene deletion, leaving three functioning α-genes. Persons who have this condition are said to have α-thalassemia-2

Table 15-1. THE THALASSEMIAS AND ASSOCIATED HEMOGLOBIN ABNORMALITIES

Abnormality	Target Cells	Deoxyhemoglobin Solubility	Microcytosis	Anemia	Hemoglobin F	Hemoglobin A$_2$	Hemoglobin H	Cellular Distribution of Hemoglobin F	Other Features
α-Thalassemia									
1 α-chain deleted	0	Normal	0	0	Normal	Normal	0	Not applicable	Small quantities (~2%) hemoglobin Barts (γ₄) detectable at birth
2 α-chains deleted	0	Normal	0	0	Normal	Normal	0	Not applicable	Hemoglobin Barts (~5%) detectable at birth
3 α-chains deleted	+	Normal	+	Mild to severe	May be increased	Normal	10-15%	Nonuniform	Unstable hemoglobin hemolytic anemia (hemoglobin H)
4 α-chains deleted	—	Abnormal	—	Severe	0*	0*	0*	—	Fetal death with severe pallor and edema (hydrops fetalis)
β-Thalassemia and δβ-Thalassemia									
Heterozygous									
β	+	Normal	+	+	Normal	4-10%	0	Not applicable	—
δβ	+	Normal	+	+	5-15%	Decreased	0	Nonuniform	—

Homozygous β	+	Normal	Marked	Marked	70–90%	Up to 20%	0	Uniform	Skeletal abnormalities, cutaneous ulcers, growth retardation
δβ	+	Normal	Moderate	±	100%	0	0	Uniform	—
Doubly heterozygous Hb S-β-thal.	Present	Abnormal	+	Moderate to marked	Increased	Increased	0	Nonuniform	May exhibit all typical features of sickle cell disease
Hb C-β-thal.	Present	Normal	+	Moderate	Increased	Increased	0	Nonuniform	Often marked splenic enlargement†
Hb E-β-thal.	Present	Normal	+	Moderate to severe	Increased	Increased	0	Nonuniform	†
Hereditary Persistence of Fetal Hemoglobin									
Heterozygous	0	Normal	0	0	10–20%	Decreased	0	Uniform	Asymptomatic
Homozygous	0	Normal	0	0	100%	0	0	Uniform	Asymptomatic
Hb S-HPFH	0	Abnormal	0	0	10–20%	Decreased	0	Uniform	Usually asymptomatic

*Hemoglobin Barts, 100%.
†May be present (if β^+ or β^{++}-thalassemia, not in β^0-thalassemia).

trait, a harmless condition that in the adult is usually without any clinical or hematologic expression. Two α-genes may be deleted on the same chromosome ("in cis"). Persons who are heterozygous for this condition are said to have α-thalassemia-1 trait.[24,56] Alpha-thalassemia-1 trait causes mild microcytosis of peripheral blood erythrocytes and no other clinical or hematologic effects. The equivalent condition may be seen in homozygous α-thalassemia-2, in which each of the two chromosomes has one functional α-gene and one deleted α-gene (the deleted α-genes are said to be "in trans"). Persons with three deleted α-genes have a moderately severe thalassemia, with blood hemoglobin concentration of the order of 8–10 g/dL, signs of hemolysis, and splenomegaly. They have Hb H (β_4) and often traces of Hb Barts (γ_4) in their erythrocytes. This condition, Hb H disease, arises as a result of receiving from one parent a chromosome with both α-genes deleted and from the other a chromosome with one α-gene deleted. Homozygosity for the doubly deleted α-thalassemia-1 results in inability of the fetus to make any but embryonic hemoglobins or Hb Barts (γ_4); extreme anemia results, followed by fetal death in utero. In such aborted fetuses, Barts hemoglobin often comprises 100% of the hemoglobin; there may also be traces of embryonic hemoglobins.

Besides the deletional types of α-thalassemia, a "dysfunctional" α-thalassemia gene has been described in persons of the Mediterranean area. A dysfunctional α-thalassemia gene results in reduction in the rate of mRNA formation and a α-globin chain synthesis.[82] Some persons with Hb H disease have both deleted and dysfunctional α-globin genes as well as a single normal α-globin gene.

The β-thalassemias. The β-thalassemias are classified as β^0, β^+ and β^{++}, depending on the degree of deficiency of β-chain synthesis. In *β^0-thalassemia, β*-chain synthesis is absent, and no Hb A is demonstrable; in *β^+-thalassemia* there is slight synthesis, and in *β^{++}-thalassemia* there is moderately reduced β-chain synthesis; the amount of Hb A demonstrable in β^+- and β^{++}-thalassemia varies accordingly. In addition, a rare *$\beta^{Ferrara}$-thalassemia* has been described, in which some β-chain synthesis can be induced in vitro. Heterozygotes for each of these types of β-thalassemia have typical thalassemia minor with increased Hb A_2, often without any degree of anemia and rarely with any clinical consequence. Homozygotes or double heterozygotes (for example, those resulting from inheritance of a gene for β^0-thalassemia and one for β^+-thalassemia) have severe thalassemia major.

Molecular mechanisms. The β-thalassemias result from several abnormalities of DNA that have, as their common expression, reduced β-globin chain synthesis. Approximately two dozen mutations are known as causes of β-thalassemia; more will become recognized in the future.[3,19,88] Two mutations in the coding regions of DNA (the exons) result in premature termination of β-globin chain synthesis, with consequent thalassemia. These are at the DNA codons that normally specify addition of lysine at amino acid position #17 and of glutamine at position #39. In both, the mutation is to a nonsense codon (UAG) that signals termination of chain synthesis. The nonsense mutation at codon 17 is responsible for some cases of β^0-thalassemia in Chinese.[19] The nonsense mutation at codon 39 is the cause of thalassemia in about 18% of cases of β^0-thalassemia in people of Mediterranean ancestry.

Almost all of the other β-thalassemia mutations are in the noncoding introns of the β-gene DNA. One of the effects of an intron mutation is to give a false signal for cutting and splicing, so that the final messenger RNA has completely altered sequences of nucleotides. This is the mechanism for two commonly encountered mutations in intron-1. Approximately 70% of cases of β-thalassemia in India are the result of a G→C mutation, near the 5' end of intron-1, that causes a false splice signal; approximately 50% of cases of β-thalassemias in Mediterranean people are the result of a G→A mutation, near the 3' end of intron-1, that also causes a false splice signal. Other intron mutations include nucleotide insertions or deletions that result in "frame-shifts," i.e., the succeeding DNA codons are all changed, thus scrambling splice signals in DNA and codons for the amino acid sequence of globin. Insertions and deletions are the result of errors in the normal process of crossing over and exchange of DNA fragments between chromosomes during meiosis. A partial listing of the more common mechanisms of β-thalassemia mutations is given in Table 15-2.

In view of the numerous molecular causes of β-thalassemia, the concept of β-thalassemia major as being homozygous β-thalassemia must be revised. Most apparent β-thalassemia homozygotes, in fact, have inherited two quite different mutations, each expressed as a β-thalassemia gene.

Table 15-2. SOME MUTATIONS RESPONSIBLE FOR β-THALASSEMIA

Site and Nature	Type of Mutation	Effect of Mutation	Type of β-Thal	Prevalence* and Ethnic Group
Exon-1, codon 17, A→T	Nonsense codon	Premature termination of globin synthesis	β^0	Chinese, prevalence unknown
Exon-2, codon 39, C→T	Nonsense codon	Premature termination of globin synthesis	β^0	Mediterranean (common)
Exon-1, codons 8 and 9, insertion of G	Frameshift	Premature termination of globin synthesis	β^0	India (common)
Intron-1, nucleotide 143, G→A	5' splice site; wrong signal	Nonfunctional mRNA	β^0	Mediterranean (common)
Intron-1, nucleotide 147, G→C	5' splice site; wrong signal	Nonfunctional mRNA	β^+	India (most common)
Intron-1, nucleotide 148, T→C	5' splice site; wrong signal	Nonfunctional mRNA	β^+	Mediterranean (common)
Intron-1, nucleotide 252, G→A	New (erroneous) splice site near 3' end	Poorly functional mRNA	β^+	Mediterranean (most common)
Exon-2, codons 41 and 42	Deletion, causing frameshift	Nonfunctional mRNA	β^0	India (common)
Exon-2, codon 44, nucleotide 315	Deletion, causing frameshift	Nonfunctional mRNA	β^0	Kurdish Jews (most common)
Intron-2, nucleotide 496, G→A	5' splice site; wrong signal	Nonfunctional mRNA	β^0	Mediterranean (common)
Intron-2, nucleotide 1240, C→G	New (erroneous) splice site near 3' end	Nonfunctional mRNA	β^+	Mediterranean (common)

*Common = 5–20% of cases of β-thalassemia. Most common = 40–70% of cases of β-thalassemia.

Some hemoglobinopathies typically exhibit reduced rates of sythesis of the affected globin chains. This is especially true of Hb E, but it is also true to a lesser degree for Hb S. The explanation for Hb E is that the β^E mutation, which is very near the 3' end of exon 1, causes an ambiguous splice signal, so that about half the messenger RNA molecules contain erroneous nucleotide sequences beyond this codon and are, therefore, nonfunctional; the remainder of RNA molecules result from cutting and splicing at the correct splice signal, and result in functional messenger RNA that contains the codon for Hb E.[70]

Delta-beta thalassemia. Considerably less common than β-thalassemia is $\delta\beta$-*thalassemia trait*, a condition that exhibits the same clinical and hematologic features as β-thalassemia trait, but is not associated with increase in Hb A$_2$. Hemoglobin F concentration is characteristically increased to 5–20% of total hemoglobin. *Homozygous $\delta\beta$-thalassemia* causes, surprisingly, a fairly mild anemia without any severe manifestations.[93] The defect in $\delta\beta$-thalassemia is deletion of most of the $\delta\beta$-gene complex.[72]

Gamma-delta-beta thalassemia trait. A few cases have been observed of severe neonatal thalassemia (erythroblastosis, anemia, and microcytosis) that largely remitted after a few months.[50,103] Older children or adults who have $\gamma\delta\beta$-*thalassemia trait* exhibit a blood picture indistinguishable from that of β-thalassemia trait. However, they have increase neither in Hb A$_2$ nor in Hb F. The neonatal thalassemia is due to inability to produce the requisite quantity of γ-globin chains, and the microcytosis and failure to exhibit increased A$_2$ in later life is, of course, the result of diminished δ- and β-chain synthesis. These few cases appear to be heterozygotes; homozygotes have not been reported. DNA mapping has shown deletion of the $^A\gamma$- and $^G\gamma$-, δ- and β-genes.[33a,103] As yet, only three such cases have been observed, but the condition may be more common than is presently recognized.

Lepore, anti-Lepore, and Kenya hemoglobins. Hemoglobin Lepore is the prototype of a large number of hemoglobinopathies and thalassemias that have arisen as a consequence of nonhomologous meiotic crossing over. This resulted in deletions of portions of the gene loci or in gene fusions. All of the gene deletions (α-thalassemia, $\delta\beta$-thalassemia, and $\gamma\delta\beta$-thalassemia) have arisen in this manner. Indeed, the α-thalassemia-2 gene, which is common in American blacks, is now known to be the result of fusion of the two α-gene loci, with deletion of a portion of each locus, just as in the case of the Hb Lepore fusion gene.[56] Because of the importance of this mechanism, the nonhomologous crossing-over event is illustrated in Figure 15-10. Note that the chromosome bearing the Lepore $\delta\beta$-fusion gene lacks intact δ- and β-genes. The gene product

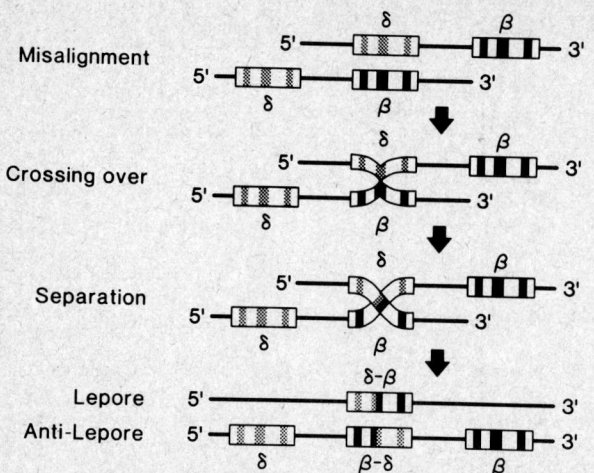

Figure 15-10. Illustration of the concept of nonhomologous meiotic crossing over in the genesis of Lepore and anti-Lepore hemoglobin variants.

is Hb Lepore, which has δ-chain amino acid sequences from helices A through F and β-chain sequences thereafter. Three Lepore hemoglobins have been described, but the only one commonly encountered is *Hb Lepore-Boston*.[27] As shown in Figure 15-10, an anti-Lepore chromosome contains the missing portions of the Lepore chromosome, and thus has intact δ- and β-genes and the δβ-anti-Lepore fusion gene. Several anti-Lepore hemoglobins have been described, such as hemoglobins *Miyada, P-Congo,* and *P-Nilotic*. They are of no clinical consequence and are without associated hematologic abnormalities. On the other hand, Hb Lepore trait causes a mild thalassemia minor blood picture without anemia. Once thought rare, Hb Lepore trait is now known to be a relatively common cause of thalassemia minor. It occurs most commonly in persons of Mediterranean ancestry. In southern Italy, 1 out of every 25 cases of thalassemia minor is due to Hb Lepore trait;[84] in more than 2000 American cases of thalassemia minor ascertained in our laboratory, 2.5% were due to Hb Lepore trait. Hemoglobin Kenya is a kind of Lepore hemoglobin in which the deleted segment of DNA contains most of the δβ-gene locus, resulting in absence of the δ-gene and fusion of the ^Aγ- and β-genes. Because a functional ^Gγ-gene remains, neonatal thalassemia does not occur. Hemoglobin F concentration persists at a high level in adults with Kenya hemoglobinopathy. Homozygotes for this condition have only hemoglobins F and Kenya and exhibit mild erythrocytosis, but are otherwise well.

Hereditary Persistence of High Fetal Hemoglobin

Several types of *hereditary persistence of high fetal hemoglobin* (HPFH) have been defined (Table 15-3). Of these, the African type is most commonly encountered in North America. This condition is the result of deletion of the entire δβ-gene region (Figure 15-11). HPFH trait is harmless and generally considered unassociated with clinical or hematologic features. In children or adults with HPFH (African type), Hb F concentration is usually 15–30% of total. In some cases there may be very slight microcytosis, but anemia is not observed. The molecular mechanism for this condition is very similar to that of δβ-thalassemia, and it is surprising that HPFH has such minimal effect. The explanation may be that in HPFH a switching mechanism is included in the deletion, so that γ-chain synthesis is never turned off, whereas in δβ-thalassemia the switch mechanism may have remained undeleted. Gamma-chain synthesis is turned down postnatally, but then some erythrocyte precursors are able to resume γ-chain synthesis to compensate for inadequate amounts of β-chains. This hypothesis accords with the observation that in HPFH (and also in Kenya hemoglobinopathy) the Hb F is quite uniformly distributed in erythrocytes, while in δβ-thalassemia trait, the distribution of Hb F is nonuniform.

Combinations of Thalassemias

The clinical and hematologic manifestations of homozygous thalassemias have been presented above. Double heterozygotes for α- and β-thalassemias (those who have inherited both an α- and a β-thalassemia gene) have very mild anemia and microcytosis. The same appears to be true

Table 15-3. HEREDITARY PERSISTENCE OF FETAL HEMOGLOBIN (HPFH) AND RELATED CONDITIONS

HPFH Type	% Hb F	$^{G}\gamma$:$^{A}\gamma$	Cellular Distribution
African			
Heterozygous	15–38	1:0	Uniform
		0:1	
		3:2	
		2:3	
Homozygous	100	Variable	Uniform
Greek			
Heterozygous	10–19	1:0	Uniform
Swiss			
Heterozygous	1–3	—	Nonuniform
British			
Heterozygous	4–13	9:1	Nonuniform
Homozygous	20	9:1	Nonuniform
Georgian (US)	4–7	1:0	Nonuniform
Hb Kenya trait	6.5–13	1:0	Uniform
$\delta\beta$-Thalassemia minor	5–15	1:0	Nonuniform
		(Blacks)	
		2:3	
		(Mediterraneans)	
Normal newborn	50–80	3:1	Nonuniform
Normal adult	<2	2:3	Nonuniform

Reprinted with permission from: Hemoglobinopathies and Thalassemias. V. F. Fairbanks, Ed. Thieme-Stratton, Inc., 1980.

of double heterozygotes for β- and $\delta\beta$-thalassemia. A more severe thalassemic disorder characterizes those who are doubly heterozygous for β- or $\delta\beta$-thalassemia and Hb Lepore.[27] Such patients are moderately to severely anemic, have severe microcytosis and hemolysis, and are found to have little or no Hb A and 70–80% Hb F, the rest being Hb Lepore. Since the hemoglobin β^E gene is also expressed as a mild thalassemia,[30,32] it is pertinent to note here the effect of inheritance of both a β-thalassemia gene and a gene for Hb E. Such double heterozygotes (*E/β-thalassemia*) have severe thalassemia that may completely mimic classic β-thalassemia major: severe anemia, microcytosis, hemolysis, growth retardation, skeletal malformations, extramedullary hematopoietic tumors that may cause manifestations of spinal cord compression or brain tumor, and death in childhood. This is the most frequently encountered severe thalassemic disorder in Southeast Asia and in the more than 600 000 Southeast Asian refugees now in the United States. In contrast, a double heterozygote for α-thalassemia-1 and Hb E has moderate microcytosis but little or no anemia. The proportion of Hb E in such cases is usually reduced to 15–20% of total hemoglobin (in contrast with 30–35% in the usual simple Hb E trait). Persons

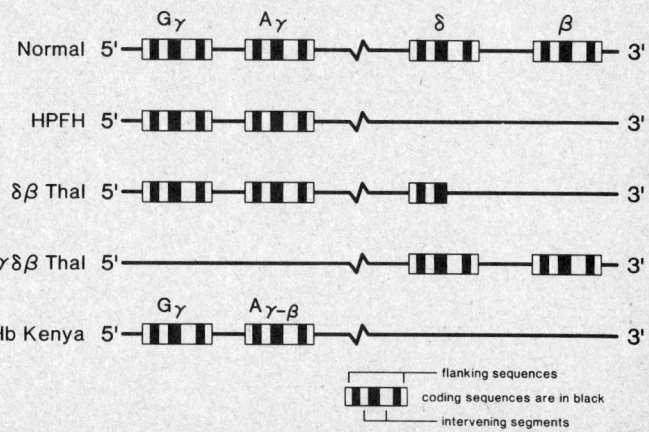

Figure 15-11. The normal $\gamma\delta\beta$-globin gene complex and the deletions affecting this region. The normal $\gamma\delta\beta$-gene region has two similar, but not identical, γ-loci designated $^{G}\gamma$ and $^{A}\gamma$ and a single δ- and a single β-locus. In hereditary persistence of high fetal hemoglobin (HPFH) both the δ- and β-loci are deleted. In $\delta\beta$-thalassemia the β-locus is deleted, as is part of the δ-locus. In $\gamma\delta\beta$-thalassemia, the entire $\gamma\delta\beta$-genome is deleted. In Hb Kenya the δ, β, and $^{A}\gamma$ are deleted, but a $^{A}\gamma$-β fusion gene exists in the place of the normal $^{A}\gamma$ gene. The fusion gene in Hb Kenya is analogous to the δ-β fusion gene of Hb Lepore. All of these deletional disorders appear to be the result of nonhomologous crossing over in meiosis.

who have inherited the α-thalassemia-1 gene (2 α-loci deleted), the α-thalassemia-2 gene (1 α-locus deleted), and the β^E-gene have, clinically and genetically, Hb H disease with moderate hemolytic anemia and microcytosis. They have 10–15% Hb E but, paradoxically, they may show no Hb H upon electrophoresis. This is because the deficit in β-chain synthesis that results from the presence of a β^E-gene balances the deficit in α-chain synthesis; there are not enough surplus β-chains to form Hb H (β_4-tetramers) in large quantities.

HEMOGLOBINOPATHIES

Disorders of hemoglobin that result from structural abnormalities of one or another globin chain are called hemoglobinopathies. Approximately 400 structural variants have been described. By far the majority of these are entirely unattended by clinical or hematologic manifestations and are discovered only incidentally. Some of the abnormal hemoglobins are listed in Table 15-4, together with their clinical and laboratory manifestations. Only a selected number of representative hemoglobin variants are listed. In fact, more than 100 unstable hemoglobin variants have been described, and more than three dozen high-oxygen-affinity variants that are associated with erythrocytosis. (See Additional Reading.)

Table 15-4. CLINICALLY IMPORTANT HEMOGLOBINOPATHIES*

Clinical Manifestation	Hemoglobin Designation	Substitution	Comments
Sickling disorders[†]	S	β(A3)Glu → Val	Target cells
	C	β(A3)Glu → Lys	
	D-Punjab	β(GH4)121 Glu → Gln	
	O-Arab	β(GH4)121 Glu → Lys	
Unstable hemoglobin	Köln	β(FG5)98 Val → Met	Compensated hemolytic anemia
Hemolytic anemia	Zürich	β(E7)63 His → Arg	Sulfonamide-induced hemolysis
	Hasharon	α(CE5)47 Asp → His	No hemolysis or mild hemolysis
Cyanosis	M-Iwate	α(F8)87 His → Tyr	This group of rare abnormal hemoglobins is characterized by abnormal absorption spectra and by normal methemoglobin values when tested by the usual spectrophotometric measurement of methemoglobin (see methemoglobin assay)
	M-Boston	α(E7)58 His → Tyr	
	M-Milwaukee	β(E11)67 Val → Glu	
	M-Hyde Park	β(F8)92 His → Tyr	
	M-Saskatoon	β(E7)63 His → Tyr	
	Kansas	β(G4)102 Asn → Thr	Rightward displacement of O_2 dissociation curve
Erythrocytosis	Chesapeake	α(FG4)92 Arg → Leu	This group of abnormal hemoglobins is characterized by marked leftward displacement of the O_2 dissociation curve
	J-Camptown	α(FG4)92 Arg → Gln	
	Malmö	β(FG4)97 His → Gln	
	Kempsey	β(G1)99 Asp → Asn	
	Ypsilanti	β(G1)99 Asp → Tyr	
	Yakima	β(G1)99 Asp → His	
	Rainier	β(HC2)145 Tyr → Cys	
	Bethesda	β(HC2)145 Tyr → His	
Thalassemic blood picture	E	β(B8)26 Glu → Lys	In Southeast Asians
	Lepore	$\delta\beta$ Fusion	In Mediterraneans
	H	β_4	In Asians
"Physiologic anemia"	Yoshizuka	β(G10)108 Asn → Asp	Decreased O_2 affinity facilitates release of O_2 to tissues

*This is a selected list of abnormal hemoglobins with significant clinical effects.

[†]Hb S and variants which, when inherited together with Hb S (in double heterozygotes), cause sickling disorders.

Abbreviations for amino acids are Glu, glutamic acid; Val, valine; Lys, lysine; Met, methionine; His, histidine; Arg, arginine; Tyr, tyrosine; Gln, glutamine; Asn, asparagine; Leu, leucine; Thr, threonine; Asp, aspartic acid; Cys, cysteine.

Molecular (DNA) Mechanisms Responsible for Hemoglobinopathies

Point mutations in DNA code. By far the largest number of hemoglobin variants are due to point mutations in the code for globin chains. For example, change from GAG to GTG at a certain point in the β-gene results in the substitution of valine for glutamic acid at the sixth amino acid position of the β-globin chain, a mutation responsible for sickling of erythrocytes. There are other possible effects of such point mutations. If the codon for lysine at the 17th position of the β-chain changes from AAG to UAG, chain termination is specified, and no recognizable globin chains are synthesized. This is the cause of one type of β^0-thalassemia. Conversely, if the normal termination codon UAA changes to CAA or GAA, globin chains become extended. For example, the chain hemoglobin variant Constant Spring is the result of mutation from UAA to CAA, which specifies glutamine instead of chain termination immediately after the addition of the 141st amino acid to the α-chain. More amino acids are added until the ribosome encounters a termination codon, so that the *Hb Constant Spring* α-chains are 172 amino acids in length. Other extended chain variants of the α-chain that have arisen in analogous manner are hemoglobins *Koya Dora, Seal Rock,* and *Icaria.* An analogous extended chain variant of the β-chain is *Hb Tak.*

Nonhomologous meiotic crossing over. The mechanism that gave rise to the Lepore hemoglobins has been discussed above in the section on thalassemias. *Nonhomologous crossing over* may result in gross deletions of extensive portions of the globin genomes or in the plucking out or insertion of single nucleotides. Thus, there are hemoglobin variants that result from deletion of one or a few amino acids (nucleotides were deleted in multiples of three), and variants that have resulted from a frame-shift mutation due to insertion or deletion of one or two nucleotides. For example, in *Hb Wayne* a single nucleotide was deleted near the 3' end of the coding region of the β-gene. This altered the codons for at least three amino acids of the β-chain as well as the terminator codon, resulting in an extension of the chain until another terminator codon occurred. This would be comparable to the following changes that illustrate a frame shift with a new termination signal occurring three triplet codons (words) later.

SHE SAW HIM RUN <u>END</u> LET THE PEN DRY NOW . . .

↑
delete N

SHE SAW HIM RUE NDL ETT HEP <u>END</u> RYN . . .

Hemoglobin variants that result from insertions or deletions of small portions of the globin genes are quite rare. Clearly, such events would usually create gibberish in the messenger RNA and would not allow synthesis of molecules recognizable as hemoglobin. A few frame shift mutations are recognized as causes of β-thalassemias.

Clinical Effects of Hemoglobinopathies

Hemoglobinopathies are associated with an astonishing variety of clinical and hematologic manifestations. However, most hemoglobinopathies are clinically and hematologically benign and are detected only incidentally as a result of screening programs.

Sickling disorders. Hemoglobin S forms long rope-like polymers when deoxygenated. Aggregates of such polymers are called tactoids. They distort the shape of the erythrocyte to form sickle cells. The formation of irreversibly sickled erythrocytes is the cause of the severe manifestations that accompany homozygous Hb S disease, such as anemia, joint pains, bone necrosis, and infarcts of various organs. In contrast, *Hb S trait,* the heterozygous state, is usually innocuous. *Hemoglobins C, D-Punjab,* or *O-Arab* copolymerize (interact) with Hb S when both are present, causing a sickling disorder resembling homozygous Hb S disease. On the other hand, hemoglobins A, F, and most hemoglobin variants do not copolymerize with Hb S, and thus when they are present together with Hb S, they prevent severe sickling disorders.

Thalassemias also have important interactions with Hb S trait. When Hb S trait is inherited together with β^0-thalassemia trait, a severe sickling disorder results, since most of the hemoglobin is S. Somewhat milder sickling disorders occur in those who have inherited both Hb S trait and

β^+- or β^{++}-thalassemia. In such cases Hb A is 10–25% of the total, most of the rest being Hb S. Those who inherit Hb S trait together with HPFH trait have no sickling disorder and are not anemic although Hb S is about 70% of the total (the remainder being Hb F). Alpha-thalassemias are protective against severe sickling,[65] so that persons who have homozygous Hb S disease and also α-thalassemia trait have less severe sickling disorders than those who are simple Hb S homozygotes. This phenomenon may account both for the high frequency of α-thalassemia trait in blacks and for many persons with "mild homozygous sickle cell disease."

Unstable hemoglobin hemolytic anemia. In more than 100 hemoglobin variants there is instability of either the α- or β-globin chain. This is usually the result of substitution of a polar (or hydrophilic) amino acid for a nonpolar (or hydrophobic) amino acid that lines the "pocket" in which the heme group is located. Amino acids of the heme pocket are normally all nonpolar to repel water. Leucine, for example, is nonpolar, with a "waxy" hydrocarbon side chain. Hydrophilic side chains are those that contain $COOH$, NH_2, OH, or SH groups (for example, glutamic, aspartic, arginine, lysine, threonine, or cysteine). The instability conferred by these substitutions results in accelerated denaturation of hemoglobin and in a hemolytic disorder that may be mild and fully compensated (no anemia, as in *Hb Köln* disease), or in severe hemolysis (as in *Hb Santa Ana disease*). In some of these disorders (e.g., *Zürich hemoglobinopathy*), hemolysis may be induced by ingestion of sulfonamides. Most of these conditions are quite rare. However, there are probably more than 100 families worldwide with Köln hemoglobinopathy. They are mostly of German or Dutch origin. Köln hemoglobinopathy may be more common than is presently recognized. *Hasharon hemoglobinopathy* has been encountered mostly in Ashkenazic Jewish persons, usually as an incidental finding. It has also been found in more than 100 persons in the Ferrara region of Italy. Hemoglobin E, which occurs in about 30 000 000 people worldwide (mostly Southeast Asians) is unstable in vitro, but seems never to result in hemolytic disease either in those with Hb E trait or homozygous Hb E.[30,32] Hemoglobin H is also very unstable. Since the principal manifestations of hemoglobins E and H are thalassemic, they are not listed in Table 15-4 as causing unstable hemoglobin hemolytic anemias.

Thalassemic blood picture. Hemoglobins H, E, and Lepore characteristically are associated with thalassemic blood pictures, i.e., microcytosis, hypochromia, and target erythrocytes. Hemoglobin E trait, homozygous Hb E, and Hb Lepore trait do not cause anemia; when anemia is present in such persons, it is due to something else. Hemoglobin H disease is associated with moderate anemia (blood hemoglobin concentration 8–9 g/dL) and signs of hemolysis. The combination Hb E/β-thalassemia causes severe anemia.

Erythrocytosis. In the thalassemia minor disorders (β-thalassemia trait, α-thalassemia-1 trait, homozygous α-thalassemia-2, Hb E trait, homozygous Hb E, Hb Lepore trait) the erythrocyte count of venous blood is usually higher than normal, but hemoglobin concentration and hematocrit are usually normal. In contrast, the hemoglobin variants with high oxygen affinity often cause moderate to marked increase in hemoglobin concentration and hematocrit as well as in erythrocyte count. This is because such hemoglobin variants do not readily release O_2 to tissues; many of these hemoglobin variants are functionally inert because they will not release O_2 except under artificial laboratory conditions of markedly reduced oxygen tension. Hemoglobin Malmö may be taken as the prototype for this group of hemoglobin variants. It has been found in four large and apparently unrelated kindreds of Europe and North America. However, each of the more than three dozen known high-O_2-affinity hemoglobins is quite rare.

Cyanosis. Lifelong cyanosis may be the result of a hemoglobin variant in which the iron atom of the heme cannot be reduced by physiological means from the trivalent state. Several of these *"M-hemoglobin" methemoglobinemias* have been defined. Most of them are the results of substitutions of one of the heme-binding histidines—the proximal or distal histidines (F8 or E7, respectively) of the α- or β-chain by tyrosine. The M-hemoglobin methemoglobinemias are extremely rare. Furthermore, they are not identified by the usual chemical test for methemoglobin, since they do not exhibit the 635-nm absorbance peak of methemoglobin obtained from Hb A, and the optical spectra may change little or not at all upon addition of CN^- to a hemolysate in which they are present.[90] An exception to this rule is Hb M-Milwaukee, which has an absorbance spectrum very little different from methemoglobin of Hb A, and which does readily convert to cyanmethemoglobin in the presence of CN^-.

Hemoglobin Kansas causes cyanosis because it deoxygenates more readily than does Hb A. Thus, at the normal pO_2 of arterial blood, a greater than normal proportion of Hb Kansas is in the deoxygenated form. Some other hemoglobin variants result in cyanosis because they convert to the methemoglobin form more readily than normal. Some of these variants are also unstable hemoglobins (e.g., *Hb St. Louis, Hb Freiburg*). They exhibit the 635-nm absorbance peak of methemoglobin.

"Physiological anemia." Some hemoglobin variants (e.g., hemoglobins *Yoshizuka, Presbyterian,* and *Hope*) have slightly diminished O_2 affinity that results in more efficient delivery of O_2 to tissues. Because of this, there is a compensatory reduction in total erythrocyte mass; hemoglobin concentration, hematocrit, and erythrocyte counts are less than normal, a condition termed "physiological anemia." These hemoglobin variants are quite rare.

Classification of Hemoglobin Variants

As should be clear from the preceding discussion, hemoglobin variants are classified according to the globin chains affected and according to the sites of substitution in the globin chains. They are further classified as to whether they represent point mutations (single amino acid substitutions), deletion, fusion, or chain extension. A few hemoglobin variants are classified as having two amino acid substitutions. The best known of these is *Hb C-Georgetown (C-Harlem)*. In this variant the β-chain has the same substitution as in Hb S [β(A3)6 Glu→Val], but also the substitution β(E17)73 Asp→Asn. Although Hb C-Georgetown behaves electrophoretically like Hb C at alkaline pH, in every other respect it behaves like Hb S, and the person who has Hb C-Georgetown trait really has Hb S trait.

Tables of hemoglobin variants, classified in the manner described, are available from the International Hemoglobin Information Center, Comprehensive Sickle Cell Center, Augusta, GA 30912; these are updated periodically.

GENETICS

Inheritance of structurally abnormal hemoglobinopathies follows simple Mendelian laws. Most abnormal hemoglobins cause little or no clinical manifestations in the heterozygote and thus are autosomal recessive traits. These include hemoglobins S, C, and E. On the other hand, a person who is heterozygous for an unstable hemoglobin (such as Hb Köln), for a Hb M, or for a high-O_2-affinity hemoglobin (such as Hb Malmö) has the clinical manifestations indicated in Table 15-4. Therefore, these are transmitted as autosomal dominant traits.

Since the structure of α-chains is controlled by four gene loci, two on each of a pair of chromosomes, abnormal hemoglobins that have substitutions in the α-chain generally make up about 25% of the total hemoglobin (one mutant α-gene, three normal α-genes). Hemoglobinopathies due to α-chain abnormalities are uncommon in comparison with those caused by substitution in the β-chain. The amino acid sequence of β-chains is determined by two gene loci located on a single pair of homologous chromosomes; thus, if one of the two β-chain genes contains an altered sequence of DNA (a mutation), approximately 50% of the β-chains formed will be abnormal. In the formation of the complete tetrameric hemoglobin molecule, there is a preferential matching of like chains. For example, if β^A-chains (normal) and β^S-chains are both being formed at the same rate, Hb A ($\alpha_1\beta_2^A$) and Hb S ($\alpha_2\beta_2^S$) both will be synthesized in nearly equal amounts.* The presence of both Hb A and Hb S in nearly equal quantities is called either Hb S trait or sickle cell trait.

In fact, β^S-chains are not as rapidly produced as β^A-chains, and the proportion of Hb S in a person with sickle cell trait is usually about 40%. Similarly, in Hb C trait, there is about 40% Hb C. Hemoglobin E is about 30% of total hemoglobin in Hb E trait, and Hb O-Arab is also about 30–35% of total hemoglobin. If there is a mutation in only one of a pair of genes controlling β-chain structure such that an unstable hemoglobin is formed (for example, Hb Köln), the proportion of unstable hemoglobin in the blood to Hb A is often about 20%. This may be due in part to accelerated denaturation and preferential removal of circulating erythrocytes that contain a significant amount of unstable hemoglobin.

---

*Hemoglobin notation. The following symbols for Hb S are equivalent: $\alpha_2^A\beta_2^S$, $\alpha_2\beta_2^{6glu\rightarrow val}$, and $\alpha_2\beta_2^{6val}$. However, the currently preferred notation for a hemoglobin variant such as Hb S includes both the helical and the amino acid sequence position of the substitution and does not require indication of the unaffected chain or indicate the tetrameric structure. Thus, Hb S now is properly designated β (A3) 6 Glu→Val.

The inheritance of an identical mutation in both of the genes that determine β-chain structure results in a homozygous hemoglobinopathy. Thus, if both genes direct the synthesis of β^S-chains, only Hb S ($\alpha_2\beta_2^S$) can be formed, and no Hb A will be found. This gives rise to homozygous Hb S disease or sickle cell disease, a very serious disorder that commonly leads to death before puberty. The simultaneous inheritance of different mutations in both of the β-genes may also lead to serious disease, as exemplified by Hb S-C disease. In this disorder, which exhibits all of the features of sickle cell disease, both β^C- and β^S-chains are formed, leading to production of Hb S ($\alpha_2\beta_2^S$) and Hb C ($\alpha_2\beta_2^C$). Again, no Hb A is formed, inasmuch as there are no normal β^A-genes. There would be no genetic basis to explain the simultaneous presence in a hemolysate of hemoglobins A, S, and C; when three major bands are observed in these positions, a different explanation must be sought.

Hemoglobin Derivatives

The term "hemoglobin" implies that the iron atom is in the reduced, divalent (ferrous) state. This is true whether hemoglobin is in the *oxygenated form (oxyhemoglobin)* or *deoxygenated form (deoxyhemoglobin)*. The terms *reduced hemoglobin* and *ferrohemoglobin,* which have been used as synonyms for deoxyhemoglobin, are redundant (since the iron in hemoglobin is, by definition, in the reduced or "ferro" state) and should be avoided. When the iron is trivalent, the brown pigment is *methemoglobin,* also (rarely) called *hemiglobin.*

Carboxyhemoglobin is a hemoglobin–carbon monoxide complex. Carbon monoxide is normally generated in small quantities from the catabolism of heme, although the amount of carboxyhemoglobin formed is too small to be readily measured. (Measurement of carboxyhemoglobin is reviewed in Chapter 18.)

In strongly basic or strongly acidic solution, hemoglobin is denatured to alkaline hematin or acid hematin, respectively. *Sulfhemoglobin, verdohemoglobin,* and *choleglobin* are degradation products of hemoglobin.

Spectral absorbance maxima for the more important hemoglobin derivatives are given in Table 15-5.

Hemoglobin functions physiologically not only in the transport of oxygen but also in the transport of carbon dioxide from tissues to the lungs. (See also Chapter 10.) The hemoglobin-CO_2 complex is designated *carbonyl hemoglobin* (or carbaminoyl hemoglobin, $RNHCOO^-$).

Measurement of Hemoglobin Concentration in Whole Blood

The measurement of hemoglobin concentration in venous or capillary blood is one of the most frequently performed clinical laboratory tests. The principle of the method given here has been almost universally adopted for both manual and automated procedures.

Clinical Significance

A sufficient concentration of hemoglobin in blood is essential for adequate transport of O_2 and CO_2 between lungs and other tissues. Blood hemoglobin concentration may be diminished as a consequence of hemorrhage or hemolysis or as a result of impaired blood formation in the bone marrow. Conversely, blood hemoglobin concentration may be increased when there is impaired gas exchange through the lungs, or in a variety of other disorders. Measurement of the blood hemoglobin concentration is important as an initial step in the detection of anemia (diminished hemoglobin concentration) or erythrocytosis (increased erythrocyte count and hemoglobin concentration).

Table 15-5. ABSORBANCE MAXIMA OF HEMOGLOBIN AND HEMOGLOBIN DERIVATIVES

Compound	Absorbance Maxima (nm)				
Deoxyhemoglobin	428–430		555		
Oxyhemoglobin	412–415		541	576–578	
Carboxyhemoglobin	417–418		537	568–572	
Methemoglobin (neutral pH)	404–407	500	540	578	630
Cyanmethemoglobin	413–418	480	541	580–590	

Principle

The Fe(II) of hemoglobin is oxidized to the Fe(III) of methemoglobin by ferricyanide, and the methemoglobin is converted into stable cyanmethemoglobin by addition of KCN:

$$HbFe(II) + Fe(III)(CN)_6^{3-} \rightarrow HbFe(III) + Fe(II)(CN)_6^{4-}$$
$$HbFe(III) + CN^- \rightarrow HbFe(III)CN$$

where HbFe(II) represents a hemoglobin monomer, HbFe(III) a methemoglobin monomer, and HbFe(III)CN a monomer of cyanmethemoglobin; nonreactive ions are omitted. The absorbance of cyanmethemoglobin is measured at 540 nm, where it exhibits a broad absorbance peak (Figure 15-12).

Specimen

The procedure requires a minimum of 0.02 mL of whole blood. Blood may be anticoagulated with $Na_2(EDTA)$ or may be taken directly from a finger (or heel) puncture without use of an anticoagulant.

Reagents

1. Drabkin's solution. Dissolve, in succession, 0.20 g of $K_3Fe(CN)_6$, 0.05 g of KCN, and 1.0 g of $NaHCO_3$ in distilled H_2O and dilute to 1000 mL. Store the reagents in a dark bottle at 4 °C. Stable for at least 4 months.
2. Cyanmethemoglobin standard (usually 80 mg/dL). The concentration of the standard preparation is certified by an appropriate certifying agency.

Procedure

1. Set spectrophotometer wavelength to 540 nm.
2. Set up a series of 16-mm cuvets labeled blank, standard, control, sample 1, sample 2, sample 3, and so forth.
3. Pipet 6.0 mL of standard into the standard cuvet and 6.0 mL of Drabkin's solution into all other cuvets.
4. Thoroughly mix all blood specimens to be tested by repeated inversion immediately before testing.
5. Transfer 0.020 mL of whole blood from each sample to its respective cuvet. Blow the blood out of the pipet, and rinse pipet with solution in the cuvet.

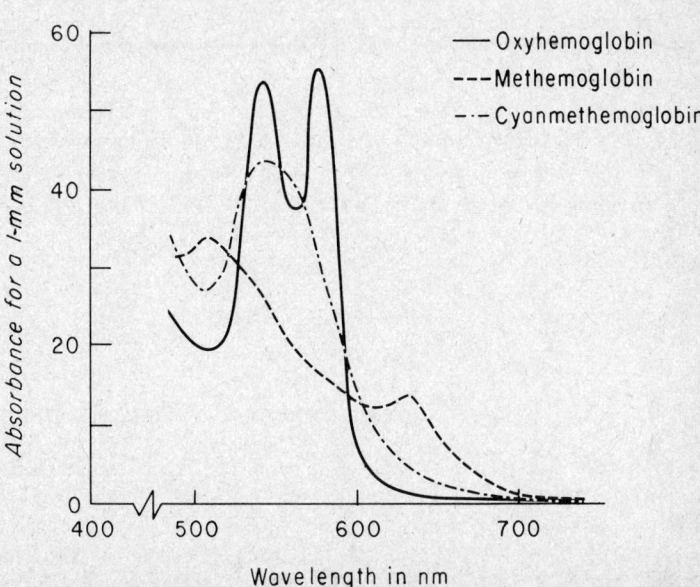

Figure 15-12. Spectrophotometric absorption curves for oxyhemoglobin, methemoglobin, and cyanmethemoglobin (authors' data). Oxyhemoglobin and cyanmethemoglobin are used in measuring the hemoglobin concentration. The peak at 630 nm, which is distinctive for methemoglobin, is abolished by addition of cyanide, and the resultant decrease in absorbance is directly proportional to the methemoglobin concentration. All heme proteins exhibit their maximal absorbance in the Sorét band region of 400–440 nm. Because the absorbance of hemoglobin in the Sorét region is approximately 10 times the absorbance at 540 nm, the Sorét peaks have been omitted from this diagram. The absorbance curve for methemoglobin is markedly influenced by small changes in pH. The curve given here is at pH 6.6.

6. Mix each sample cuvet and then let stand for ~5 min to ensure complete cell lysis and conversion of hemoglobin to cyanmethemoglobin.

7. Place the blank in the spectrophotometer and adjust the meter to read 100% transmittance (0 absorbance). Then replace the blank cuvet with the standard cuvet and read its absorbance (A_s).

8. Measure the absorbance (A_u) of each of the sample cuvets.

Calculations

The dilution of blood in Drabkin's solution is 1:301 (from $\dfrac{0.02}{6.00 + 0.02}$), and the hemoglobin concentration of each sample is calculated:

$$\text{Hb, g/dL} = 301 \left(\frac{A_u \times C_s}{A_s} \right) \times \frac{1}{1000}$$

$$= 0.301 \left(\frac{A_u \times C_s}{A_s} \right)$$

where C_s = concentration of the standard in mg/dL, and 1000 converts mg to g.

Reference Ranges

For persons residing at or near sea level, the normal values of hemoglobin concentration are 14–18 g/dL in adult males and 12–16 g/dL in adult females. In healthy pregnant women, hemoglobin concentration may be as low as 11.0 g/dL. In children, hemoglobin concentration may normally be as low as 10.5 g/dL. In adult males of age ≥ 60 years, hemoglobin concentration may normally be as low as 12.5 g/dL.

Comments and Precautions

The cyanmethemoglobin method has been adopted internationally as the approved reference method for hemoglobin measurement. Although cyanide is a lethal chemical, its concentration in Drabkin's solution is so low that it does not constitute any significant hazard to personnel. Nonetheless, reasonable care must be exercised in handling the solution. The preparation of Drabkin's solution from KCN powder must be performed with much caution, and the disposal of any solutions containing KCN into a sink should be preceded and followed by copious flushing of the sink with water to avoid the potential for generation of lethal HCN gas.

Convenient packets of dry Drabkin's reagent mix are commercially available; the contents of the packet are added to 1000 mL of distilled water.

The method outlined here permits rapid manual measurement of hemoglobin concentration. However, automated methods, based on the same principle, may be more convenient when large numbers of specimens are being processed. In the Coulter automated hematological apparatus, a precalibrated whole blood standard is provided by the manufacturer and is used at least once daily to check the calibration of the instrument for hemoglobin concentration, mean corpuscular volume, erythrocyte count, and leukocyte count. A separate manual method is then required to correlate the manufacturer's hemoglobin standard with the reference method.

References

Eilers, R. J.: Am. J. Clin. Pathol., *47*:212-214, 1967.
van Kampen, E. J., and Zijlstra, W. G.: Adv. Clin. Chem., *8*:141-187, 1965.

Measurement of Plasma Hemoglobin

Virtually all of the hemoglobin in blood is contained within the erythrocytes. A minute quantity of hemoglobin is normally released into plasma by the destruction of the erythrocytes, and this is promptly bound by haptoglobin. The haptoglobin-hemoglobin complex is rapidly removed by parenchymal cells of the liver. Thus, the normal plasma hemoglobin concentration is close to zero. Formerly, benzidine was used as chromogen for plasma hemoglobin measurement, but in view of the carcinogenic potential, benzidine and its derivatives can no longer be used for this purpose. The following procedure has proved sufficiently sensitive and reliable for clinical purposes.

Clinical Significance

An increase in plasma hemoglobin concentration is indicative of acute destruction of erythrocytes (hemolysis) within the vascular system. Therefore, the practical value of the measurement is virtually limited to circumstances in which acute intravascular hemolysis is believed to have occurred, as in hemolytic transfusion reactions, or in evaluation of the degree of hemolysis occurring in extracorporeal treatment of blood. *The measurement of free hemoglobin in plasma is of no practical value in the diagnosis of chronic hemolytic disorders.* Since some degree of hemolysis always occurs when blood clots, the measurement of serum hemoglobin has no diagnostic value whatever, and serum specimens should not be accepted for this purpose.

Principle

The method is based on the measurement of oxyhemoglobin at 415 nm, the Soret band of maximal absorbance of oxyhemoglobin. Measurements at 380 and 450 nm are also required for correction for turbidity and other nonspecific absorbance. A spectrophotometer with 1-nm spectral resolution is required.

Specimen

Measurement of plasma hemoglobin requires a minimum volume of 1 mL of plasma. Blood should be freshly drawn by the following technique, which is designed to minimize in vitro hemolysis from mechanical injury to erythrocytes.

1. Attach a 30-cm length of polyvinyl tubing to the hub of a sterile 18-gauge needle or use an 18-gauge needle with attached tubing from an infusion set. Clamp the other end with a hemostat.

2. Prepare two plastic (or silicone-coated glass) centrifuge tubes, labeled 1 and 2, for collection of blood from the free end of the tubing.

3. Lightly coat the interior of tube 2 by spraying with a fine mist of sodium heparin solution, 1000 units/mL. Only a small amount (0.05 mL) of heparin solution should be applied. Any large droplets or puddles should be wiped away.

4. Place a tourniquet lightly around the upper arm. With minimal delay, perform a clean puncture of the antecubital vein. Release the tourniquet, and permit approximately 5 mL of blood to flow freely into tube 1; then collect 4 mL of blood in tube 2. Be careful that droplets of blood do not adhere to the test tube above the meniscus. Discard the blood in tube 1.

5. Centrifuge tube 2 at 1000 × g for 10 min; carefully draw off the supernatant plasma with a transfer pipet and deliver it into a second centrifuge tube.

6. Recentrifuge plasma at 1600 × g for 20 min; carefully draw off the supernatant plasma and transfer it into a third tube.

7. Proceed promptly to hemoglobin measurement or freeze at −20 °C until a determination is to be completed.

Reagents

1. Na_2CO_3 stock solution, 1 g/dL
2. Na_2CO_3 working solution. Dilute 1 mL of stock solution to 100 mL with distilled water.

Procedure

1. Prepare a blank by adding 4 mL of Na_2CO_3 working solution to a cuvet of 3-mL critical volume and 1-cm light path.

2. To an additional cuvet add 3 mL of Na_2CO_3 working solution and 0.3 mL of specimen. Cover with Parafilm and invert three times to mix.

3. Measure absorbance at 380, 415, and 450 nm.

Calculations

$$\text{Plasma hemoglobin, mg/dL} = \frac{2A_{415} - (A_{380} + A_{450})}{1.655} \times \frac{1000 \, D}{E}$$

(corrected for irrelevant absorption)

where 1000 converts g/dL to mg/dL,

D is the dilution of specimen in working solution, and

E is the absorptivity at 415 nm of a 1 g/dL solution of oxyhemoglobin.

Upon substitution of 79.46 (the figure obtained empirically by Harboe) for E and reducing the terms, the following simpler formula is obtained for calculation of plasma hemoglobin concentration.

$$\text{Plasma hemoglobin, mg/dL} = 83.6 \ (2A_{415} - A_{380} - A_{450})$$

Reference Ranges

Reference values are difficult to define, since the concentration of hemoglobin depends mostly on the care with which samples are obtained and processed. With casually collected blood samples, plasma hemoglobin concentration is usually < 5 mg/dL. With meticulous care in obtaining plasma, normal plasma hemoglobin should be < 1.0 mg/dL.[105] Even this is principally the result of in vitro hemolysis. Very lipemic plasma may give spuriously high results despite measurements at 380 and 450 nm. The method is not reliable when serum bilirubin concentration is increased.

Reference

Harboe, M.: Scand. J. Clin. Lab. Invest., *11*:66–70, 1959.

Determination of Methemoglobin and Sulfhemoglobin

When the iron in hemoglobin is oxidized to the trivalent state, the resulting brownish pigment is *methemoglobin.* The nature of *sulfhemoglobin* has been elusive. Whether sulfhemoglobin is a distinct substance or rather a mixture of "green pigments" has long been debated. Sulfhemoglobin can be formed in vitro by incubation of hemoglobin with an oxidizing agent (such as acetylphenylhydrazine) and a sulfhydryl donor (such as H_2S). The pigment formed is believed to contain sulfur bonded to pyrroles of the porphyrin ring. Such sulfhemoglobin may exist in the Fe(II) or Fe(III) state (i.e., as sulfhemoglobin or metsulfhemoglobin), and sulfhemoglobin may bind O_2 or CO reversibly. In this section it is assumed that the "sulfhemoglobin" formed in vivo corresponds to that which can be prepared in vitro, and that it is the Fe(III) or metsulfhemoglobin form.

Clinical Significance

The normal concentration of *methemoglobin* is < 1.5% of the total hemoglobin. In congenital methemoglobinemia that is due to methemoglobin reductase deficiency, the methemoglobin concentration in blood is increased to 10–30% of the total heme pigment, although a few cases have been reported in which the methemoglobin concentration of peripheral blood has been as high as 40–55% of the total heme pigments. Methemoglobin concentrations greater than 60% are usually lethal regardless of cause. Methemoglobinemia also occurs in persons who have ingested nitrites, nitrates, or certain drugs or chemicals such as sulfones or aniline dyes. *Sulfhemoglobin* is not a normal constituent of blood. It is sometimes found in concentrations of 1–10% of the total hemoglobin after excessive use of certain drugs such as phenacetin or dapsone.

Principle

The absorbance spectrum of methemoglobin exhibits a small, characteristic peak at 630–635 nm (Figure 15-12). Addition of cyanide eliminates this peak by converting methemoglobin to cyanmethemoglobin. The decrease in absorbance is proportional to the methemoglobin concentration.

The normal absorbance spectrum of oxyhemoglobin shows very little absorbance above 600 nm. However, if sulfhemoglobin is present in a hemolysate, there is a broad increase in the absorption curve in the range of 600–620 nm. This sulfhemoglobin plateau is not affected by treatment with cyanide.

Specimen

Blood should be fresh and may be anticoagulated with heparin, EDTA, or ACD (acid-citrate-dextrose) solution. No fluid or food restriction is needed.

Reagents

1. Potassium ferricyanide. Dissolve 2.0 g of $K_3Fe(CN)_6$ in distilled H_2O and dilute to 10.0 mL. If stored in a brown bottle at 4 °C, this solution is stable for at least one year.

2. Potassium cyanide solution (CAUTION: lethal poison). Dissolve 500 mg of KCN in distilled H_2O and dilute to 10 mL. Label "POISON." Stable at 20 °C for at least 4 months.

3. Potassium phosphate buffer, 0.15 mol/L, pH 6.6 (20 °C). Dissolve 17.1 g of K_2HPO_4. 3 H_2O (or 13.2 g anhydrous) in 500 mL of distilled H_2O and 10.2 g of KH_2PO_4 in 500 mL of distilled water. Transfer KH_2PO_4 solution to a 200 mL beaker and add an equal volume of K_2HPO_4 solution. Place a pH electrode in the beaker. Then slowly add more of the K_2HPO_4 solution, with constant stirring, until the mixture has a pH of 6.6. Store at 4 °C. Discard whenever the solution appears turbid. New buffer should be prepared at least once every three months.

Procedure (Methemoglobin Only)

1. Prepare a blank cuvet containing 1.5 mL of phosphate buffer and 1.5 mL of H_2O. Designate this cuvet C_1.

2. Pipet 0.1 mL of whole blood into a test tube containing 3.9 mL of distilled H_2O; swirl to mix.

3. Add 4.0 mL of potassium phosphate buffer and mix thoroughly.

4. Transfer 3 mL of hemolysate to each of two cuvets; designate these C_2 and C_3.

5. To cuvet C_3, add 0.1 mL of $K_3Fe(CN)_6$ solution. Cover with Parafilm, mix by inverting three times, and measure its absorbance at 2 min.

6. Measure the absorbance at 630 nm for cuvets C_2 and C_3, using C_1 as blank. Record as A_{2a} and A_{3a}.

7. Add 0.1 mL of KCN to all cuvets. (Use a safety pipet or add 2 drops from a transfer pipet fitted with a rubber bulb.) Mix by inverting 3 times, and allow to stand for 5 min.

8. Measure absorbance at 630 nm for cuvets C_2 and C_3 with C_1 as blank. Record as A_{2b} and A_{3b}.

Calculations

$$\text{Methemoglobin (per cent of total pigment)} = 100\left(\frac{A_{2a} - A_{2b}}{A_{3a} - A_{3b}}\right)$$

Reference Range

0–1%

Comments and Precautions

The hemoglobin solution is slightly turbid; however, because the turbidity does not change with addition of the $K_3Fe(CN)_6$ or KCN, the absorbance it causes is the same for both readings with each cuvet and therefore is compensated for in these calculations. This simple method is satisfactory for methemoglobin assay if a sulfhemoglobin assay is not also needed.

Reference

Betke, K., Stein, H., and Tönz, O.: German Medical Monthly, 7:217–219, 1962.

Procedure (Methemoglobin and Sulfhemoglobin)

1. Hemolyze 0.1 mL of whole blood with 3.9 mL of distilled H_2O and add 4.0 mL of potassium phosphate buffer as in the preceding section.

2. Centrifuge at 1600 $\times$ g for 30 min to remove stroma.

3. Set up a blank cuvet containing 1.5 mL of phosphate buffer and 1.5 mL of H_2O.

4. Transfer 3.0 mL of clear supernatant (from step 2) to a second cuvet, designated C_2, and 1.0 mL of clear supernatant to a third cuvet, designated C_3.

5. Add 2.0 mL of a mixture of equal volumes of distilled water and phosphate buffer to cuvet C_3.

6. Measure the absorbance of cuvet C_2 at 630 nm. Designate this value A_1.

7. Add 0.1 mL of KCN (use safety pipet or dropper) to cuvets C_2 and C_3. Cover with Parafilm, and invert 3 times to mix thoroughly. Allow to stand 5 min.

8. Measure absorbance of cuvet C_2 at 630 nm and 620 nm against the blank cuvet C_1. Designate these values A_2 and A_3, respectively. Further readings will not be performed on cuvet C_2, and this may be set aside.

9. Add 0.1 mL of $K_3Fe(CN)_6$ solution to cuvets C_1 and C_3. Cover with Parafilm and invert 3 times to mix. Allow to stand for 2 min.

10. Set spectrophotometer to 540 nm and adjust to zero absorbance with cuvet C_1. Measure absorbance of C_3. Designate this value A_4.

Calculations

$$\text{Total hemoglobin, g/dL} = F_1 A_4$$

$$\text{Methemoglobin, g/dL} = F_2(A_1 - A_2)$$

$$\text{Sulfhemoglobin, g/dL} = 80 F_3[A_3 - F_4(A_1 - A_2) - 3F_5 A_4]$$

$$\text{Methemoglobin, per cent of total} = \frac{\text{methemoglobin (g/dL)} \times 100}{\text{total hemoglobin (g/dL)}}$$

$$\text{Sulfhemoglobin, per cent of total} = \frac{\text{sulfhemoglobin (g/dL)} \times 100}{\text{total hemoglobin (g/dL)}}$$

Determination of Factors Used in Calculations

In these calculations, absorptivity constants (F_1, F_2, and so forth) are often taken directly from the data of Drabkin and Austin[25] or van Kampen and Zijlstra.[104] Absorptivity constants for methemoglobin calculation may be obtained independently in the following manner.

Transfer 3 mL of commercial cyanmethemoglobin standard (80 mg/dL or 0.08 g/dL) to a cuvet with a 1.0-cm light path and measure the absorbance at 540 nm. Designate this reading A_s. With a properly calibrated spectrophotometer of narrow bandpass (≤ 1 nm), an absorbance of 0.546 should be obtained for A_s. (A reading between 0.540 and 0.550 is acceptable.) Thus, a cyanmethemoglobin concentration of 1.0 g/dL would have an absorbance of 6.83 (from the ratio $\frac{0.546}{0.080}$). Because the whole blood specimen was diluted 80-fold in the hemolyzing solution and 3-fold again in the cuvet,

$$F_1 = \frac{3 \times 80}{A_s} \times \text{cyanmethemoglobin concentration of standard}$$

$$F_1 = \frac{240}{6.83}$$

$$F_1 = 35.14$$

Prepare a 100% methemoglobin standard as follows. Hemolyze normal blood and centrifuge to remove stroma as previously described. Also prepare a blank as in the procedure for methemoglobin. Transfer 3.0 mL of clear hemolysate to a second and third cuvet and 1.0 mL to a fourth cuvet. To the second cuvet add 0.1 mL and to the fourth cuvet 2.0 mL of a mixture of equal volumes of distilled H_2O and phosphate buffer. Add 0.1 mL of $K_3Fe(CN)_6$ solution to third and fourth cuvets. Cover with Parafilm and invert 3 times to mix. This converts all the hemoglobin to methemoglobin in cuvets 3 and 4. Set the spectrophotometer to 630 nm and use the blank (cuvet 1) to set zero absorbance. Measure the absorbance of the third cuvet and designate this $A_{1(s)}$. Add 0.1 mL (2 drops) of KCN solution to each cuvet. This converts methemoglobin to cyanmethemoglobin, but does not alter hemoglobin or sulfhemoglobin. Cover with Parafilm and invert 3 times to mix. Allow to stand 2 min. Then read absorbance of the third cuvet (630 nm) and designate this value $A_{2(s)}$. Set the spectrophotometer to 620 nm and measure absorbance of cuvets 2 and 3. Designate these $A_{3(s)}$ and $A_{4(s)}$, respectively. Set the spectrophotometer to 540 nm and measure absorbance of the fourth cuvet. Designate this $A_{5(s)}$. The constant F_2 is the reciprocal of the ratio change in absorbance at 630 nm upon addition of CN^- to the absorbance at 540 nm of a cyanmethemoglobin solution corrected to 1 g/dL. Thus, in this procedure,

$$F_2 = \frac{3A_{5(s)} \times F_1}{A_{1(s)} - A_{2(s)}}$$

The constants F_3, F_4, and F_5 are employed in the calculation of sulfhemoglobin concentration. F_3 cannot readily be determined directly. Several estimates of F_3 have been made previously. The early data of Drabkin and Austin suggested an absorptivity of 6.5 for a hypothetical 1 g/dL solution of pure sulfhemoglobin. This would be equivalent to an F_3 value of 0.154. Later estimates have provided higher coefficients of molar absorptivity. Carrico et al.[13] estimated that a pure sulfhemoglobin solution of 1 g/dL concentration would have an absorptivity of 15.7, more than double that proposed by Drabkin and Austin. In the formula that follows, the coefficient 0.0637 is $\frac{1}{15.7}$. Use of this value for F_3 will result in calculated values of sulfhemoglobin that are approximately 1/3 to 1/2 of those reported earlier. Constants F_4 and F_5 are derived from absorptivity measurements $A_{1(s)}$, $A_{2(s)}$, $A_{3(s)}$, $A_{4(s)}$, and $A_{5(s)}$ as follows:

F_4 corrects for the contribution to absorbance at 620 nm from the presence of methemoglobin in the hemolysate.

$$F_4 = \frac{A_{4(s)}}{A_{1(s)} - A_{2(s)}}$$

F_5 corrects for the contribution to absorbance at 620 nm of oxyhemoglobin in the hemolysate. Although the absorptivity of oxyhemoglobin at 620 nm is very small, the quantity of this pigment is very large, thus making this correction essential.

$$F_5 = \frac{A_{3(s)}}{3A_{5(s)}}$$

$$3F_5 = \frac{A_{3(s)}}{A_{5(s)}}$$

[Note: Because, in this procedure, cuvets 3 and 4 differ only in concentration of total hemoglobin pigments, measurement $A_{5(s)}$ might be made on either. Cuvet 4 is specified only to ensure that reading $A_{5(s)}$ is less than 1.0. The coefficient 3 corrects for the dilution of pigments in cuvet 4 relative to other cuvets. Since $3F_5$ is the factor actually used in the formula for SHb, it is a convenient simplification to calculate $3F_5$ rather than F_5. For spectrophotometers with accurate range expansion, cuvet 4, as here specified, may be omitted, and $A_{5(s)}$ may be measured at 340 nm using cuvet 3. Then, F_2 would be $= \frac{A_{5(s)} \times F_1}{A_{1(s)} - A_{2(s)}}$, F_5 would be $= \frac{A_{3(s)}}{A_{5(s)}}$, and F_5 should be used in the sulfhemoglobin formula rather than $3F_5$. A problem in determining F_5 is the precise measurement of $A_{3(s)}$. Because this value is very small it must be measured with much care.]

Coefficients for calculation of total hemoglobin, methemoglobin, and sulfhemoglobin are:

$$F_1 = 35.1$$

$$F_2 = 33.6$$

$$F_3 = 0.0637 \text{ (from Carrico et al.)}[13]$$

$$F_4 = 0.30$$

$$3F_5 = 0.067$$

Using these coefficients, the following formulas may be derived:

$$\text{Total hemoglobin, g/dL} = 35.2A_4$$

$$\text{Methemoglobin, g/dL} = 33.6(A_1 - A_2)$$

$$\text{Sulfhemoglobin, g/dL} = 5.10A_3 - 1.53(A_1 - A_2) - 0.34A_4$$

Reference Ranges

The percentages of methemoglobin and sulfhemoglobin (of total hemoglobin) concentration measured in specimens from 30 healthy persons were:

	Methemoglobin	Sulfhemoglobin
Mean	0.78	0.20
Standard deviation	0.37	0.33
Median	0.85	0.20
95th percentile	1.28	0.70

A methemoglobin concentration $> 1.5\%$ or a sulfhemoglobin concentration $> 1.0\%$ should be regarded as abnormal. The sulfhemoglobin concentration in normal blood is actually very nearly zero; however, due to the imprecision inherent in the measurement of several very small absorptivities, calculated values for sulfhemoglobin may be < 0. All such negative values reflect the absence of sulfhemoglobin in blood and should be reported as 0% sulfhemoglobin.

Use of normal hemolysate to calculate coefficients for the sulfhemoglobin measurements assumes that normal hemolysates contain no methemoglobin or sulfhemoglobin. While this is not entirely correct, the adjustments in the factors that might be made to correct for presence of minute amounts of these pigments are negligible.

Reference

Evelyn, K. A., and Malloy, H. T.: J. Biol. Chem., *126*:655–662, 1938.

Solubility Test for Hemoglobin S

Clinical Significance

The procedure described below is a simple and rapid method for detection of the presence of any sickling hemoglobin, such as *Hb S* or *Hb C-Georgetown*, in blood.[39,59,67] It does not quantitate the amount of abnormal hemoglobin, and therefore does not differentiate Hb S trait from homozygous Hb S disease. Furthermore, other hemoglobin variants may be present in the blood whether results are positive or negative.

Principle

Hemoglobin S, when deoxygenated, is insoluble in concentrated phosphate buffer and produces a visible turbidity. Almost all other hemoglobins, including hemoglobins A, F, C, E, and D, are soluble in such solutions. Thus, this test quickly identifies specimens of blood that contain Hb S. A reducing substance, sodium hydrosulfite ($Na_2S_2O_4$, sodium dithionite), is used to deoxygenate the hemoglobin, and saponin is used to lyse the erythrocytes.

Specimen

The test requires 0.05 mL of whole blood. The blood may be anticoagulated with heparin, EDTA, or ACD (acid-citrate-dextrose). Specimens that have been kept for as long as one to two weeks at 4–5 °C are satisfactory. No preliminary restriction of food or fluid is required.

Reagents

1. Potassium phosphate–saponin buffer, 2.3 mol/L, pH 7.0 (20 °C). Dissolve 303.74 g of $K_2HPO_4 \cdot 3H_2O$ (or 231.9 g anhydrous K_2HPO_4) in 200 mL distilled H_2O. Separately, dissolve 131.86 g KH_2PO_4 (crystalline) in 700 mL distilled H_2O; warm if necessary to hasten solution. When solution is complete, mix both solutions, make up to 1 L with distilled H_2O, and add 0.143 g of saponin. The buffer-saponin solution is stable at room temperature for three months.
2. Sodium hydrosulfite ("dithionite"), dry powder.

Procedure

1. On the day of the test, mix the saponin-buffer solution thoroughly, warming it if necessary to ensure that all crystals are dissolved. Add 100 mg sodium hydrosulfite for each 10 mL of phosphate buffer–saponin solution. Shake thoroughly until the sodium hydrosulfite has dissolved.
2. Transfer 3 mL of this solution to each of several appropriately labeled tubes of 8 mm I.D.; there should be one tube for a normal control, one tube for an abnormal control known to contain hemoglobin S, and one tube for each of the unknown specimens.

3. Centrifuge blood specimens at $1500 \times g$ for 10 min. Carefully remove and discard supernatants. Transfer 0.02 mL of the packed erythrocytes from each centrifuge tube to corresponding reagent tubes. Cover each with Parafilm and invert twice.

4. Wait for at least 5 min, then examine the tubes for turbidity by holding them against a page of print. Good illumination is necessary.

Interpretation

The test is positive if there is sufficient turbidity to prevent reading of print (Figure 15-13). Normal specimens will show only a faint haziness; many will be completely transparent. A positive test is presumptive evidence for Hb S. However, every positive solubility test should be confirmed by hemoglobin electrophoresis, since the rare sickling hemoglobins *C-Harlem* (*C-Georgetown*), *S-Travis,* and *C-Ziguinchor* will also give positive solubility tests, as will *Hb Barts*, if present in the blood of a fetus with hydrops fetalis due to α-thalassemia. In patients with unstable hemoglobins who have had splenectomy, the test may also be positive because of numerous insoluble erythrocyte inclusions.

Comments and Precautions

Use of this procedure for screening for Hb S has been properly criticized on the basis that it does not identify other deleterious hemoglobin variants or thalassemias. Thus, the solubility test is best used not in screening programs but instead in determining whether a hemoglobin variant that appears to be hemoglobin S or D or G by electrophoresis is, in fact, hemoglobin S.

False negative results are very rare but may occur when Hb S concentration is $<15\%$ of total (as in specimens from patients who have been transfused with blood from donors with Hb S trait). If a minor band is observed in the Hb S position by electrophoresis, then doubling or trebling the volume of erythrocytes added in the test will result in a positive test if Hb S is present.

Reagent kits for Hb S screening are commercially available from several sources. In a carefully controlled comparison of seven commercially available kits, Schmidt and Wilson found both false negative and false positive results to be very frequent. They commented, "accuracy of commercial solubility test kits varies greatly and frequently incorrect diagnosis can be expected."[87] This conclusion contrasts markedly with a very favorable experience using reagents prepared in our laboratory. The solubility test has also been adapted for automated screening for hemoglobin S.[68]

References

Itano, H. A.: Arch. Biochem. Biophys., *47*:148–159, 1953.
Nalbandian, R. M., Nichols, B. M., Camp, F. R., et al.: Clin. Chem., *17*:1028–1032, 1971.

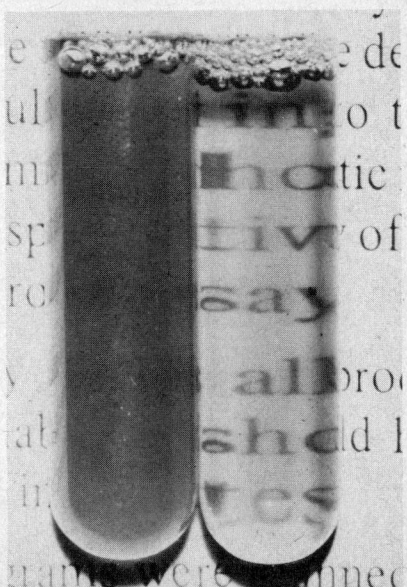

Figure 15-13. Solubility test for Hb S. Deoxyhemoglobin S (left tube) is insoluble in 2.3 mol/L phosphate buffer. By contrast, normal hemolysate (right tube) is sufficiently transparent that print can easily be read through it.

Electrophoretic Separation of Hemoglobins on Cellulose Acetate

Principle

At pH 8.6 hemoglobins are negatively charged ions and move toward the anode (positive electrode) in an electrical field. However, hemoglobins differ in the amount of charge per molecule. Hemoglobin S has two more and Hb C has four more positive charges per molecule than does Hb A. Therefore, Hb S moves toward the anode more slowly than does Hb A, while Hb C moves even more slowly. Conversely, hemoglobins H, I, and J, which have additional negative charges, all move toward the anode more rapidly than Hb A.

Various supporting media, such as potato starch, gel, paper, and polyacrylamide gel, have been used in electrophoresis. *Cellulose acetate* has come into general use because it is convenient, provides sharp resolution of hemoglobin bands in a relatively short time, and may be permanently stored.

Specimen

This procedure requires a minimum of 0.5 mL of whole blood. Any anticoagulant may be used. Fasting and fluid restrictions are unnecessary. A specimen may be stored at 4 °C for one week if the anticoagulant is heparin, and for three weeks if collected with EDTA or ACD (acid-citrate-dextrose) solution. However, if an unstable hemoglobin (for example, *H* or *Köln*) is suspected, electrophoresis should be carried out within 24 hours of collection.

Reagents

1. Tris-EDTA-borate (TEB) buffer, pH 8.6 (25 °C). Dissolve 12.0 g of Tris(hydroxymethyl)aminomethane (Tris), 1.56 g of disodium ethylenediamine-tetraacetate (EDTA), and 0.92 g of boric acid in distilled H_2O and dilute to 1000 mL. Store at 4 °C. This solution is stable indefinitely.

2. Stain. Dissolve 20 g of Ponceau S, 30 g of trichloroacetic acid, and 30 g of sulfosalicylic acid in distilled H_2O and dilute to 1000 mL. Make a new solution at least once every two months.

3. NaCl, 0.155 mol/L.

Procedure

1. If the blood has sedimented, remove the supernatant plasma before proceeding.

2. Place 0.5–1 mL of whole blood or, if plasma has been removed, 0.2–0.5 mL of sedimented cells into a centrifuge tube. Suspend the cells in 10–15 mL of cold NaCl solution, 0.155 mol/L.

3. Centrifuge at 5 °C for 10 min at 900 $\times$ *g*.

4. Aspirate and discard the supernatant saline.

5. Add 9 vol of cold distilled H_2O per 1 vol of packed cells.

6. Add 0.2 mL of xylene or toluene. Resuspend the packed cells with a wooden applicator stick or a glass stirring rod, and then thoroughly mix with a vortex mixer. Hemoglobin H is denatured and removed by this step. If an unstable hemoglobin is suspected, omit step 6.

7. Centrifuge at 1600 $\times$ *g* for 15 min.

8. Aspirate and discard the supernatant xylene or toluene.

9. The clear hemolysate lying between the supernatant and the stromal precipitate is removed with a transfer pipet for immediate use in electrophoresis, or it may be transferred to another tube and stored for short intervals at 4 °C for use the same day or at −20 °C for use in subsequent studies.

10. Prepare a cellulose acetate membrane as described for the methodology in use. Apply samples at a template position near the cathode in alternating positions with hemoglobin type. Place TEB buffer in both buffer chambers, making sure that the ends of the membrane are in contact with buffer and that the buffer level is the same in each chamber.

11. Perform electrophoresis at room temperature for 30 min at 450 V.

12. Stain with Ponceau S; then destain and clear the background.

13. Pick the membrane up on a glass plate of appropriate size. Drain off excess solution. Place the glass plate and membrane for 15 min in an oven, preheated to 100 °C, remove, and allow to cool to room temperature; alternatively, allow the membrane to air dry for approximately 1/2 hour. With a razor blade, carefully peel the membrane from the glass plate. Place the membrane in a plastic envelope and trim off the ends.

Interpretation

The order of electrophoretic mobility of the more commonly encountered hemoglobin variants is shown in Figures 15-14 and 15-15. Usually, Hb A is preceded by a fainter band that merges with it. This has been called A_3; it is a mixture of glycosylated hemoglobins. Hemoglobins H and I have identical mobilities, substantially more anodal than Hb A. Hemoglobins J, K, N, and Barts have mobilities intermediate between those of H and A. Hemoglobin Lepore migrates minimally more toward the anode than does Hb S.

Hemoglobins A_2, C, O-Arab, and E cannot be differentiated on the basis of electrophoretic mobility in this medium, nor can hemoglobins D, G, and Lepore be distinguished from Hb S. Hemoglobin F may be poorly resolved from Hb A. Hemoglobins H and I are indistinguishable. However, the relative amounts found are useful in distinguishing between some of the hemoglobins that have the same mobility. For example, Hb A_2 is never as high as 10% of the total hemoglobin. Thus, if a very slowly moving band constitutes 10% or more of the total hemoglobin, it may be presumed to be hemoglobin C, O-Arab, or E. Hemoglobins C and O-Arab are virtually limited to persons of Central African ancestry, whereas Hb E is virtually limited to those of Southeast Asian ancestry. Hemoglobin Lepore constitutes only 10–15% of the total hemoglobin, whereas hemoglobin S, D, or G constitutes 25–45%. Similarly, Hb H constitutes approximately 5–15% of the total hemoglobin, whereas Hb I (an α-chain variant) is usually about 25%. Very rarely, Hb H may constitute 30–40% of total hemoglobin in acquired Hb H disease associated with erythroleukemia. This condition appears to be the result of a somatic mutation of the erythrocyte precursors, in which there is an acquired severe failure of α-globin chain synthesis. Of the few such cases reported, most have been Caucasians of middle age and have exhibited the features typical of Hb H disease such as anemia, jaundice, splenomegaly, and erythrocytic microcytosis.[107] With these rare exceptions, Hb H occurs most commonly in persons from Southeast Asia and rarely exceeds 15% of the total hemoglobin.

Thus, by measuring the proportions of these hemoglobin bands densitometrically, and by knowing the ethnic origin of a patient, one may deduce the probable identity of an abnormal hemoglobin. However, additional confirmatory tests are often necessary before the hemoglobin type can be established. A hemoglobin solubility test will establish the presence of Hb S; a thermostability test will establish the presence of an unstable hemoglobin (e.g., Köln or H). Agar gel electrophoresis at pH 6.2 differentiates Hb C from E and O-Arab or Hb S from D and G.

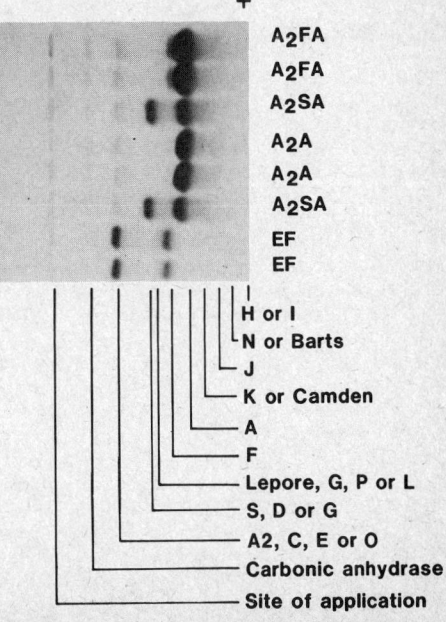

Figure 15-14. Hemoglobin electrophoresis on cellulose acetate at pH 8.6 following staining with Ponceau-S. In the first two tracks, beginning at the top, is a specimen from a patient with β-thalassemia minor. Note prominent bands for Hb A_2 and F. The third and sixth tracks contain a specimen from a patient with Hb S trait. Hemoglobins A and S are present in nearly equal proportion. The fourth and fifth tracks contain specimen from a normal adult. The two lowermost tracks contain specimens from a patient with Hb E/β^0-thalassemia. Note the absence of Hb A, and that Hb E and Hb F are present in proportion of about 2 or 3 to 1, respectively. Hemoglobin C/β^0-thalassemia or O/β^0-thalassemia would result in the same pattern as in the two lowermost tracks.

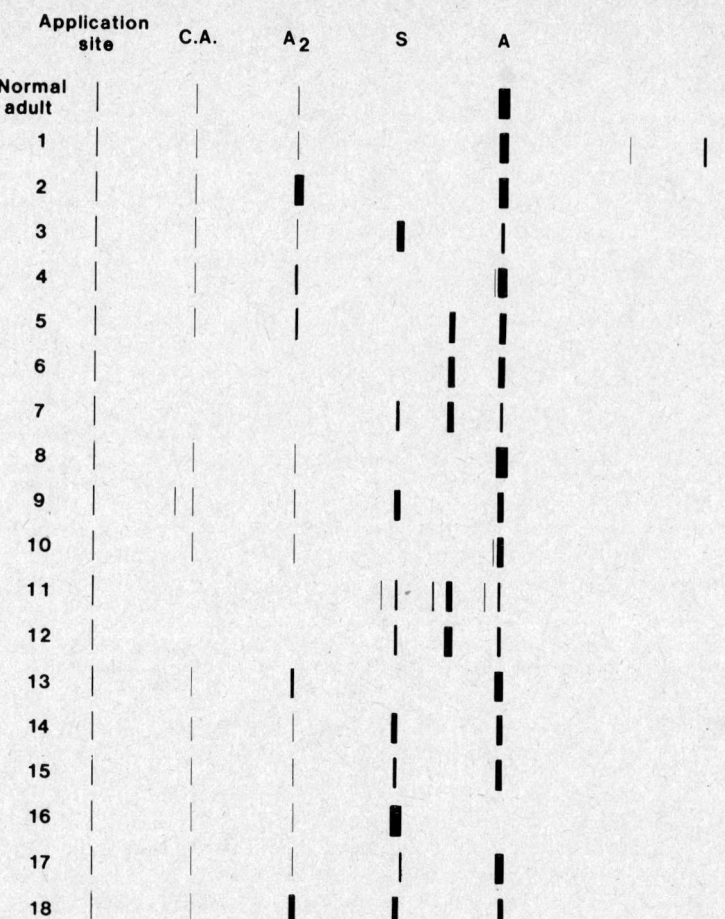

Figure 15-15. Hemoglobin electrophoresis on cellulose acetate, pH 8.6. Some abnormal patterns that are commonly encountered in clinical laboratories are shown in tracks 1–18. For each example, the application site is shown cathodal to protein bands; C.A. = carbonic anhydrase band; A_2, S, and A indicate relative positions of these hemoglobins. Relative proportions of each band are indicated by width. For example, in the "normal adult" track, Hb A_2 is about 2.5% and A is about 97.5%. The following paragraphs explain results displayed for each track.

Track 1. Two minor bands are seen anodal to Hb A. These are Hb H (most anodal) and Hb Barts. H is 10%, Barts 2% of total hemoglobin. Pattern is typical of Hb H disease, although in this condition Hb Barts is often not seen. Hb Barts is seen typically in umbilical cord blood of neonates with α-thalassemia; Hb H is not seen in neonates. Hb N has the same position as Hb Barts; Hb J is slightly cathodal to Hb Barts.

Track 2. Typical pattern for Hb C trait. Hemoglobins O and E cannot be distinguished from Hb C in this medium. However, in Hb C trait, Hb C is about 40% of total hemoglobin, whereas in Hb E trait or O trait, the abnormal band is usually 25–30% of total hemoglobin.

Track 3. Typical pattern for Hb S/β^+-thalassemia. Whenever the proportion of Hb S exceeds that of Hb A, concurrent β-thalassemia is likely. The same pattern would characterize a specimen from a patient with homozygous Hb S disease who had been transfused with blood that contains Hb A. Information about recent transfusion is critical to interpretation of hemoglobin electrophoresis results.

Track 4. Pattern commonly observed in β-thalassemia minor, with increase in the proportion of Hb A_2 to 3.5–9% of total hemoglobin. (Rarely, if ever, does Hb A_2 exceed 9% of hemoglobin even in β-thalassemia. If more than 9% of the hemoglobin is in this position, it is most likely Hb, C, E, or O.) Hemoglobin F (just cathodal to Hb A) is 2–5% of total hemoglobin in about half the cases of β-thalassemia minor and is not increased in the rest. When Hb F is in small proportion it migrates just cathodal to Hb A, but when in high proportion it migrates nearly midway between Hb A and Hb S, as in tracks 5–7.

Track 5. Homozygous β^+-thalassemia or homozygous β^0-thalassemia following transfusion of normal blood. Hb F is markedly increased in all homozygous β-thalassemias, and increase in Hb A_2 may also be observed.

Track 6. Umbilical cord blood from a normal newborn. Characteristic are high proportion of Hb F (50–70% of total) and absence of Hb A_2 and carbonic anhydrase.

Track 7. Umbilical cord blood from a neonate who has Hb S trait. Note approximately equal proportions of hemoglobins A and S, preponderance of Hb F, and absence of Hb A_2 and carbonic anhydrase.

Track 8. Normal adult or child > 1 year of age. However, several abnormal hemoglobins co-migrate with Hb A; a normal pattern therefore does not exclude the presence of an abnormal hemoglobin. An example is Hb Malmö

Comments and Precautions

A circuit breaker must be an integral part of the electrophoresis system and no attempt must be made to bypass this safety feature. Electrocution is a potential hazard of electrophoresis, and death has resulted from failure to observe safety precautions.

References

Bartlett, R. C.: Clin. Chem., 9:325–329, 1963.
Brier, R. O., Golias, T., and Batsakis, J. G.: Am. J. Clin. Pathol., 44:695–701, 1965.
Graham, J. L., and Grunbaum, B. W.: Am. J. Clin. Pathol., 39:567–578, 1963.

Electrophoretic Separation of Hemoglobins in Agar Gel at pH 6.2

Principle

Purified agar contains *agarose* and the negatively charged sulfated polysaccharide *agaropectin*. In an agar gel, agarose is polymerized to constitute an immobile matrix; agaropectin does not polymerize and can bind reversibly with a small number of amino acids of hemoglobin that are on the external surfaces, either in the heme pockets or at the 2,3-DPG binding sites of the β-globin chains. The hemoglobin-agaropectin complex migrates electrophoretically through the immobile agarose matrix toward the anode, whereas noncomplexed hemoglobin is carried toward the cathode by electroendosmotic flow of the citrate buffer.[110] Thus, citrate agar electrophoresis is actually affinity gel electrochromatography. Results of agar gel electrophoresis cannot be predicted just on the basis of differences in charged groups of hemoglobin. Thus, Hb S, which has a substitution near the external surface, has a high affinity for agaropectin, while hemoglobins D-Punjab and G-Philadelphia, with similar substitutions deeper in the molecule, have low affinity for agaropectin. Similarly, Hb C has a high affinity for agaropectin, but Hb E, with like charge change, does not. This simple, useful technique thus permits differentiation of Hb S from D and G, and Hb C from E and O.

Gels prepared with agarose, a refined derivative of agar, lack sulfated agaropectin; thus such gels cannot be used in place of those prepared from agar.

---

that causes erythrocytosis. This pattern would be observed in Hb Malmö trait. This pattern is also typical for β-thalassemia minor or Hb H disease when either is associated with iron deficiency, since iron deficiency reduces the proportion of Hb A_2 and may result in marked diminution in Hb H. The correct diagnosis is made by repeating electrophoresis following treatment of iron deficiency.

Track 9. Hemoglobin S trait together with Hb A_2' trait. Hb A_2' is seen cathodal to carbonic anhydrase. It is found in 1% of blacks and is harmless. The proportion of Hb A_2' is equal to that of Hb A_2.

Track 10. Slight increase in Hb F, as typically seen in many chronic anemias such as hereditary spherocytosis or myelofibrosis.

Track 11. Improperly handled umbilical cord blood from a neonate with Hb S trait. Failure to maintain the specimen at 4 °C during several days that preceded electrophoresis resulted in formation of methemoglobins A, F, and S, shown as minor bands.

Track 12. Same specimen as in Track 11 following addition of KCN to convert methemoglobins to cyanmethemoglobins.

Track 13. Hemoglobin E trait in a patient who also has either iron deficiency or α-thalassemia. In either circumstance, the proportion of Hb E may be diminished to 10–20%, and Hb E may be mistaken for Hb A_2.

Track 14. Hemoglobin S trait, Hb D trait, or Hb G trait. These conditions cannot be distinguished by electrophoresis in this medium.

Track 15. Typical pattern for Hb G-Philadelphia trait. Hb G-Philadelphia is seen in the Hb S position in this medium, and is usually about 25–35% of total hemoglobin.

Track 16. Typical pattern for homozygous Hb S disease or Hb S/Hb D disease. Hemoglobin F proportion may also be increased. Note absence of Hb A. Cannot be distinguished by this technique alone from Hb S/β^0-thalassemia. Family studies are usually needed.

Track 17. Typical pattern for Hb Lepore trait. Hemoglobin Lepore migrates very slightly anodal to Hb S, and is about 10–15% of total hemoglobin. It is a cause of microcytosis.

Track 18. Hemoglobins A, S, and C in a blood specimen? Not genetically possible because β^A, β^C, and β^S are alleles for the same chromosomal locus; it is not possible to have 3 alleles on 2 chromosomes. Three plausible explanations: (1) The patient has Hb S trait together with Hb G-Philadelphia trait, resulting in the combinations $\alpha_2^A\beta_2^A$ or Hb A, $\alpha_2^G\beta_2^A$ or Hb G-Philadelphia (in the Hb S position), $\alpha_2^A\beta_2^S$ or Hb S, and Hb $\alpha_2^G\beta_2^S$ or the hybrid Hb S-G-Philadelphia that migrates in the Hb C position. (2) The patient has Hb S trait and was transfused with blood from a donor with Hb C trait (or has Hb S/C disease and was transfused with blood from a normal donor). (3) Specimens from patients with Hb S trait and Hb C trait were inadvertently mixed in the laboratory. In our experience, the first of these explanations is usually the correct one, and the third is the least likely.

Reagents

1. Citrate buffer, stock solution, 0.5 mol/L, pH 5.8 (25 °C). Dissolve 147 g of trisodium citrate ($Na_3C_6H_5O_7 \cdot 2H_2O$) in distilled H_2O and dilute to 800 mL. Dissolve 10.5 g of citric acid ($H_3C_6H_5O_7 \cdot H_2O$) in distilled H_2O and dilute to 100 mL.

Add the citric acid solution (approximately 68 mL) to the trisodium citrate solution to adjust the mixture to pH 5.8. Add H_2O to a final volume of 1000 mL. Store at 4 °C. The solution is stable for approximately one year.

2. Citrate buffer, working solution, 0.05 mol/L, pH 6.2 (25 °C). Prepare by diluting 1 vol of stock solution with 9 vol of H_2O. On dilution the pH increases to 6.2.

3. Purified agar (Difco Laboratories Detroit, MI).

4. Naphthol Blue-Black stain. Mix 0.6 g of Naphthol Blue-Black dye (Amido Black 10B), 0.6 g of trichloroacetic acid, 50.0 mL of 95% ethanol, and 0.75 mL of glacial acetic acid; dilute with distilled H_2O to 2000 mL. Prepare a fresh solution once a month.

5. Acetic acid rinse, 2%. Dilute 10 mL of glacial acetic acid with distilled H_2O to 500 mL.

Procedure

Preparation of Gel

Satisfactory pre-cast gels of the dimensions specified may be either purchased commercially (Helena Laboratories, Beaumont, TX) or prepared in the following manner, using as supports thin glass plates 8.2 cm in width × 10.0 cm in length × ~0.1-cm thickness.

To 50 mL of working buffer in a 250-mL Erlenmeyer flask, add 0.5 g of agar and suspend it by vigorous swirling. Loosely cap the Erlenmeyer flask (for example, with a 100-mL beaker) to prevent inadvertent contamination and to decrease evaporation. Place the flask in a water bath at 90–100 °C, containing enough water to cover the lower two-thirds of it. At approximately 30-min intervals, remove the flask from the water bath and swirl it to resuspend the agar. After 2–3 h of heating, the agar suspension will begin to clarify. Inspection against a strong light will reveal an extremely fine granularity. Return it to the water bath and continue to heat. In another 30–60 min, the agar will be completely transparent and free of granularity when inspected against a strong light. (It may be necessary to bring the agar momentarily to the boiling point.) Remove the flask from the hot water bath and allow it to cool to 50–60 °C. The cooling may be accelerated by holding the flask at an angle under running cold water while constantly rotating it.

Pouring of Agar

Place the glass supports on a perfectly level surface. (Use of a spirit level will help determine what surfaces are suitable.) Fill a prewarmed 10-mL Mohr pipet to the 0-mL mark with molten agar. Place the tip on the glass plate near one corner and, with a back-and-forth movement of the pipet, allow the agar to run out on the plate. An even coating, 1 mm deep, will result.

If any bubbles are seen on the surface of the agar, they may be eliminated by aspirating them into a Pasteur pipet. Allow the agar to cool undisturbed until solid; this usually requires 15–20 min. Cover with a sheet of polyethylene film or Saran Wrap or place into a tightly sealed container to decrease evaporation and store at 4 °C for at least 24 h prior to use.

Preparation of Hemolysates

1. Centrifuge 1 mL of whole blood at 700 × g for 10 min.
2. Remove and discard plasma, and then resuspend erythrocytes in 10 mL of 0.9% NaCl solution.
3. Centrifuge at 700 × g for 10 min. Remove and discard supernatant.
4. Lyse erythrocytes by alternately freezing and thawing twice.
5. Measure the hemoglobin concentration of the hemolysate and adjust it to 10 g/dL by adding working buffer.

Electrophoresis Procedure

1. Remove cover from agar gel. Electrophoresis is to be performed along the long axis of the gel. Select which end is to be anodal (positive).
2. Cut a tiny notch in one corner of the anodal end of the gel as a marker; then lay a straightedge on top of the gel to define a line 4 cm from the notched end and perpendicular to

the long axis of the gel. Using a razor blade or other very sharp edge, make a series of incisions ~1 cm long and separated from each other by ~0.5 cm of unbroken gel.

3. Moisten the tip of a spatula with hemolysate and insert it into the slot, leaving a very fine line of hemoglobin. Blot any excess buffer or hemolysate from the surface. Note the order of specimens in reference to the notched corner of the gel.

4. Position the gel plate above the buffer chambers, which contain working buffer. Establish the circuit by a bridge composed of several layers of filter paper moistened with working buffer. This filter paper should overlap approximately ~1 cm of gel at each end.

5. Cover the gel with polyethylene film or Saran Wrap, or spray with Krylon. Attach electrodes so that the positive electrode (anode) is in electrical contact with the notched end of the gel. Carry out electrophoresis in a cold room or refrigerator at 4 °C; apply 50 mA and ~100 V (10 V/cm of gel) for 90 min.

Staining

1. Stain gel immediately after termination of electrophoresis by immersing the gel for 20 min in the stain solution listed under Reagents.

2. Wash briefly in distilled water and then destain with 2% aqueous acetic acid. Repeated changes of destaining solution are required for Naphthol Blue-Black, and destaining may not be complete for 24 h. Destaining should be continued until the background is virtually colorless.

3. After completion of staining and destaining, the gel may be dried to a transparent film on the glass plate by leaving the plate overnight on the laboratory bench.

Interpretation

The relative order of hemoglobin migration in this medium is shown in Figure 15-16. Hemoglobins A, D, E, and G co-migrate slightly toward the cathode. Hemoglobin F shows the most marked cathodal movement of the common hemoglobins. Conversely, Hb C shows anodal migration. Hemoglobin S remains close to the site of application or may be anodal. Hemoglobins D and G migrate with A and thus separate well from Hb S. Hemoglobin E also travels with A and thus is easily differentiated from C. Hemoglobin O-Arab has a slightly slower cathodal migration than A, and this serves to distinguish it from C as well as from A, D, G, and E. Hemoglobin A_2 has mobility like that of Hb A. Hemoglobin Lepore also exhibits an agar gel mobility identical to that of Hb A, thus differentiating it from Hb S.

Comments and Precautions

The general precautions described for cellulose acetate electrophoresis should be observed. Known standards must be included with every group of unknowns because apparently minor variations in technique may cause quite variable results. The specified voltage and current will give a maximal rate of electrophoretic mobility without overheating, provided that electrophoresis is carried out at 4 °C and with gels of specified dimensions. Electrophoresis may be performed at room temperature for 16 h with approximately 25 mA and a voltage drop not in excess of 2 V/cm of gel. However, more dependable separations are attained by using shorter runs at higher voltage with adequate cooling.

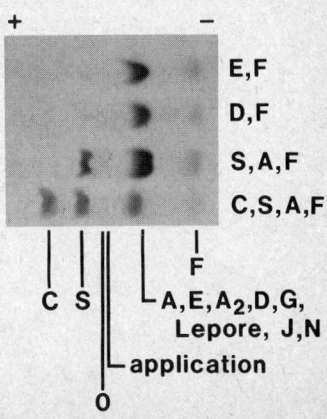

Figure 15-16. Electrophoresis of hemoglobin on citrate agar gel, pH 6.2. Cathode is to left, anode to right. Characteristic hemoglobin positions are indicated below the gel. Letters on the right margin indicate the hemoglobins present in four specimens. The specimen in the uppermost track was from a patient with homozygous Hb E and contains 98% Hb E and 2% Hb F. The second track was from a patient with Hb D-Punjab trait. It contains 51% Hb A and 49% Hb D (comigrating) and 1% Hb F. The third track was a specimen from a patient who has Hb S trait and α-thalassemia trait. It contains hemoglobins S, A, and F. Hb S was 20% of total hemoglobin, a low proportion that indicates concurrent α-thalassemia, consistent with a low mean corpuscular volume of 68 fL. The lowermost track is the control, prepared by mixing blood of hemoglobins C trait and S trait. It contains hemoglobins C, S, A, and F.

The procedure just outlined gives satisfactory results with the agar specified. Agar preparations vary markedly and in poorly understood ways. If a different type of agar or a more highly purified agar is used, altogether different results may be obtained, even though the buffer and pH are identical. This is presumed to be due to differences in reactive groups of the agar. Agarose is unsatisfactory for this purpose.

The concentration of hemoglobin applied to the gel affects mobility. Thus, if there is a larger quantity of Hb A, the apparent electrophoretic mobility will be increased. For this reason, care is required in adjusting the hemoglobin concentration of each sample.

Acid citrate agar gel electrophoresis is a simple procedure that should be performed routinely in laboratories doing hemoglobin electrophoresis for identification of variants. It should not supplant electrophoresis at pH 8.6, since with acid citrate agar electrophoresis such common and clinically significant variants as hemoglobins H, E, and Lepore would be missed, as would also the common variants D-Punjab, G-Philadelphia, and all the other G, D, and J hemoglobins, since none of these differs in mobility, in this medium, from Hb A. Acid agar electrophoresis is an important supplement, not a substitute for alkaline electrophoresis.

References

Marder, V. J., and Conley, C. L.: Bull. Johns Hopkins Hosp., *105*:77-88, 1959.
Robinson, A. R., Robson, M., Harrison, A. P., et al.: J. Lab. Clin. Med., *50*:745-752, 1957.

Globin Chain Electrophoresis

Clinical Significance

Globin chain electrophoresis in urea, 8 mol/L, permits identification of many hemoglobin variants that cannot be distinguished by other electrophoretic techniques. This method is particularly useful in distinguishing hemoglobins D-Punjab, G-Philadelphia, and Hasharon.

Principle

The hemoglobin tetramer is dissociated into heme and globin moieties upon treatment with urea, 8 mol/L, and mercaptoethanol. Mercaptoethanol removes heme. Globin chains are dissociated from each other by urea and are then separated on the basis of charge differences by electrophoresis.

Specimen

Same as for hemoglobin electrophoresis.

Reagents

1. Stock Tris-ethylenediaminetetraacetic acid–borate (TEB) buffer, pH 8.6. Place into a 2-L volumetric flask 20.4 g Tris(hydroxymethyl)aminomethane, 1.2 g EDTA, and 6.4 g boric acid; bring to volume with reagent grade H_2O. Store at 4 °C; stable indefinitely.
2. Stain. See Ponceau S stain for routine hemoglobin electrophoresis on cellulose acetate.
3. Urea, purified (Sigma Chemical Co., St. Louis, MO).
4. 2-Mercaptoethanol. Store in tightly stoppered brown vials at 4 °C.
5. Citric acid, 300 g/L.
6. Working alkaline-urea buffer, pH 8.9. To 263 mL of stock TEB buffer add 135 g urea and 1 mL 2-mercaptoethanol. Stir until the urea dissolves.
7. Working acid-urea buffer, pH 6.3. Prepare as for alkaline buffer, and adjust to pH 6.0 by addition of citric acid.
8. Acetone, analytical reagent grade.
9. Acetone-HCl solution. Prepare by adding 1 mL concentrated HCl dropwise to 100 mL acetone previously chilled to −20 °C.

Procedure

1. Separation of globin
   a. Prepare a 1:10 hemolysate by adding 0.2 mL of saline-washed, packed erythrocytes to 2 mL of water at 4 °C. Mix, then centrifuge for 30 min at 4 °C at 2250 × g to remove stroma, and carefully transfer clear supernatant to a test tube.
   b. Add approximately 0.2 mL of hemolysate dropwise to 2 mL of acetone-HCl at 20 °C. A precipitate will form. Place this suspension in the freezer at −20 °C for 15 min. During this time adjust the temperature of the centrifuge to −20 °C.

c. Centrifuge the precipitate of the previous step for 10 min at 2500 × g. Decant the supernatant and resuspend the precipitate in 10–15 mL of acetone at 20 °C. Do not use acetone-HCl for washing in this step.

d. Repeat centrifugation and washing with cold acetone twice. At the end of the second wash, decant the supernatant. Invert the centrifuge tube over a sheet of filter paper to drain.

e. Dry the globin powder either with a gentle stream of nitrogen gas or by removing it from the centrifuge tube onto filter paper, where it will quickly air dry.

2. Preparation of cellulose acetate membranes

It is convenient to begin preparation of the cellulose acetate membrane before the preparation of globin, since the membranes must first be equilibrated with working buffer solution for at least 1 h. Super Sepraphore cellulose acetate membranes (Gelman Corporation, Ann Arbor, MI) have given very satisfactory results in this method. At the completion of the 1-h equilibration, drain and blot the membrane.

3. Application of sample

Dissolve approximately 1–2 mg of globin from each sample in 0.1 mL of each working buffer. Then apply the globin solutions in a line on the center of the cellulose acetate membrane. If electrophoresis is to be done sequentially in alkaline and acid media, do not apply globin to the second membrane until electrophoresis of the first membrane has been completed. Use of commercial wells and sample applicators is practical. Place working buffer of corresponding pH in electrophoresis chamber wells.

4. Electrophoresis and staining

Carry out electrophoresis at 250 volts for 60 min at room temperature with alkaline pH buffer and at 300 volts for 90 min at room temperature with acid buffer. At the end of these time intervals, remove the membranes and place them in the Ponceau S stain for approximately 4 min. Destain by repeated washing in acetic acid, 0.5 mL/dL. Air-dry the membranes.

Results

Expected results are shown in Figure 15-17. With either acid or alkaline globin chain electrophoresis, α-chains move toward the cathode, β-chains toward the anode. Electrophoresis at pH 8.9 does not distinguish

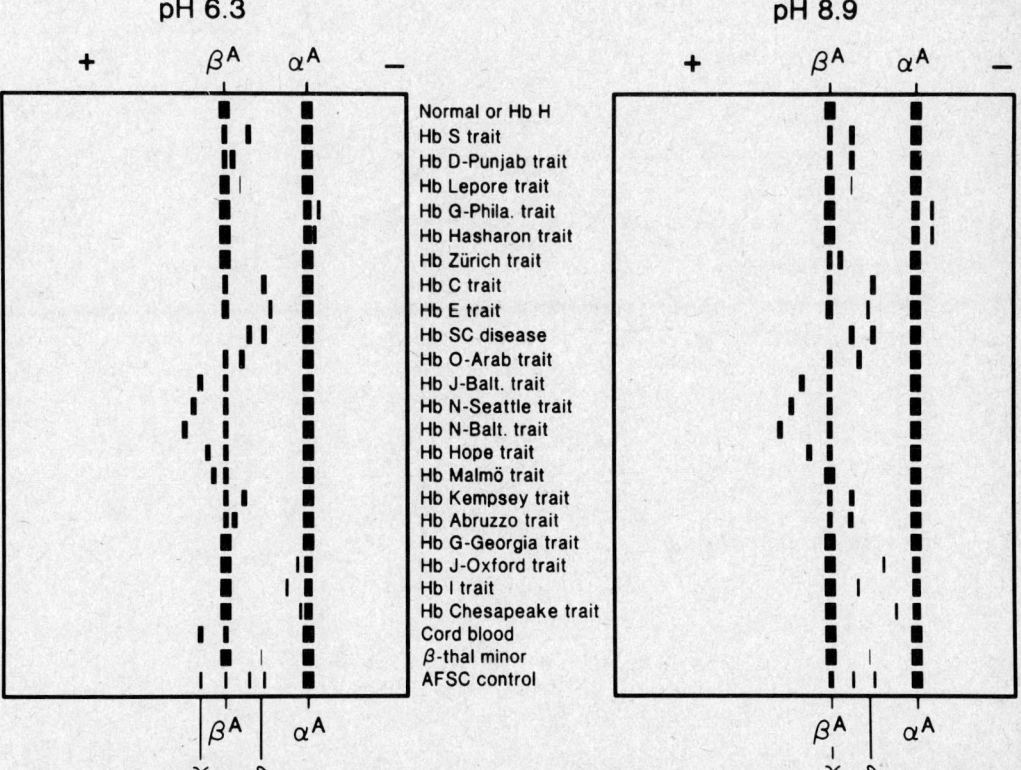

Figure 15-17. Comparison of electrophoretic mobilities of globin chains derived from hemoglobin variants in urea buffers, 8 mol/L. Left panel shows separation at pH 6.3, right panel at pH 8.9. Results are shown for all common and some uncommon hemoglobin variants. Because most of the variants are in specimens from persons who are heterozygotes, the normal β^A- and α^A-globin chains are also present and serve as position markers.

the globin chains β^D from β^S, nor α^G from $\alpha^{Hasharon}$. However, electrophoresis at pH 6.3 easily distinguishes these variants, and they can thus be identified with a high level of confidence.

Comments and Precautions

Schneider has recommended a modification of this procedure in which dissociation of heme and globin is done immediately prior to electrophoresis; the heme separates rapidly during electrophoresis. Our experience with the modification has been less satisfactory than with the procedure outlined above.

References

Schneider, R. G.: Clin. Chem., 20:1111-1115, 1974.
Schneider, R. G.: Crit. Rev. Clin. Lab. Sci., 9:243-271, 1978.
Schneider, R. G., and Barwick, R. C.: Hemoglobin, 2:417-435, 1978.
Ueda, S., and Schneider, R. G.: Blood, 34:230, 1969.

Isoelectric Focusing (IEF) in Polyacrylamide Gel

The principle of *isoelectric focusing* has been presented in Chapter 1B, Section Three. For isoelectric focusing of hemoglobin variants, polyacrylamide gel is a satisfactory medium. A mixture of ampholytes should be used that permits focusing in the pH range 6–8. A thin layer (1 mm) of gel is convenient. Focusing can be completed in 10–15 min, and the gel may then be stained with a general protein stain.

The relative order of hemoglobins observed after IEF is, beginning at the acidic-anodal end, hemoglobins A, S, and C. It has been asserted that with the resolution attained by IEF, at least 70 hemoglobin variants can be distinguished by their relative positions. Because IEF was formerly a relatively time-consuming procedure that required additional equipment and expensive ampholytes, the method has not been widely used for identification of hemoglobin variants, and experience with IEF is still limited. An inexpensive apparatus for IEF and pre-cast thin-layer gels containing ampholytes are now commercially available (Isolab Inc., Akron, OH), and IEF may attain more general acceptance in the future. Our experience is that IEF is useful primarily for identifying Hb Malmö.

References

Basset, P., Beuzard, Y., Garel, M. C., et al.: Blood, 51:971–982, 1978.
Giuliani, A., Marinucci, M., Cappabianca, M. P., et al.: Clin. Chim. Acta, 90:19–28, 1978.
Hedlund, B.: Hemoglobinopathies and Thalassemias. V. F. Fairbanks, Ed. New York, Brian C. Decker, 1980.

Tests for Unstable Hemoglobins

Clinical Significance

More than 100 unstable hemoglobin variants have been described, and most of these are associated with hemolytic anemia. Although heat tests have been widely used in the past, chemical stress tests are more practical for the demonstration of hemoglobin instability. Only rarely is a heat test necessary to determine the presence of an unstable hemoglobin.

Principle

If an unstable hemoglobin is present in a hemolysate, incubation in dilute isopropanol solution at 37 °C results in marked turbidity within 5–10 min, whereas hemolysates containing normal hemoglobins remain transparent much longer.

Specimen

This procedure requires 1 mL of blood if venous hemoglobin concentration is nearly normal; 2 or 3 mL may be required depending upon the severity of anemia. Blood may be collected with any common anticoagulant. Specimens that have been refrigerated at 4 °C for up to one week are satisfactory. No prior restriction of food or fluid is required.

Reagents

Tris-HCl buffer, 0.1 mol/L, pH 7.4. Dissolve 12.1 g of Tris(hydroxymethyl)aminomethane in 800 mL of distilled H_2O. Adjust to pH 7.4 by addition of HCl, 1 mol/L, with constant stirring. Bring volume up to 1000 mL with distilled H_2O.

Procedure

1. Remove plasma by centrifugation at 5 °C at 1500 × g for 10 min and discard supernatant plasma. If the specimen has been stored in a refrigerator, the plasma may simply be aspirated and discarded.

2. Transfer 0.5 mL of packed erythrocytes to a graduated centrifuge tube containing 10–15 mL of NaCl, 0.155 mol/L, at 5 °C. Centrifuge at 5 °C at 1800 × g for 15 min. Remove and discard supernatant.

3. To the washed packed erythrocytes add an equal volume of distilled water, 5 °C. Add 0.3 mL of carbon tetrachloride.

4. Close tube with a stopper and mix thoroughly for 2 min.

5. Centrifuge at 5 °C, at 2200 × g for 30 min.

6. While hemolysate is being centrifuged, prepare buffered isopropanol solution by adding 17 mL of isopropanol to 83 mL of Tris-HCl buffer. Place 2 mL of buffered isopropanol in tubes of 10-mm diameter, one tube for each hemolysate being tested and additional tubes for controls. Place tubes in water bath at 37 °C to bring them to temperature prior to adding hemolysate.

7. Transfer 0.2 mL of clear hemolysate from each specimen to corresponding tubes containing buffered isopropanol. Cover and mix once by inversion. Return to the 37 °C water bath.

8. Examine each tube for turbidity after initial inversion and at 5-min intervals for 30 min.

Interpretation

Hemolysates containing normal hemoglobins usually show no turbidity at 30 min, although occasionally a very faint haziness is observed. Presence of an unstable hemoglobin results in fine turbidity at 5 min and a flocculent precipitate by 20 min.

Comments and Precautions

This procedure has uniformly given positive results in specimens that contain hemoglobins H, E, Köln, Olmsted, Hasharon, Hammersmith, Santa Ana, or Zürich. It has consistently given negative results in specimens that contain hemoglobins S, C, D-Punjab, or other hemoglobins generally considered stable. Specimens that contain 5% or more of Hb F give false positive results, and improperly preserved specimens (e.g., left at room temperature for three or more days) give false positive results because of methemoglobin formation.

An alternative chemical stress procedure has been reported that uses zinc acetate solution, 0.003 mol/L, in place of isopropanol.[15] The zinc acetate test for hemoglobin stability is said to be satisfactory for specimens that have been kept at 4 °C as long as one week and that contain up to 10% Hb F. It appears to identify the same unstable hemoglobins as does the isopropanol test. The procedure and the reagents are identical to the above except that zinc acetate solution is substituted for isopropanol. The zinc acetate solution, 0.003 mol/L, is buffered with the Tris-HCl solution by mixing in a ratio of 0.35 mL to 1.45 mL.

Reference

Carrell, R. W., and Kay, R.: Br. J. Haematol., 23:615–619, 1972.

Determination of Fetal Hemoglobin (Alkali-Resistant Hemoglobin)

Clinical Significance

In δβ-thalassemia minor, the fetal hemoglobin concentration is usually in the range of 5–15% of the total hemoglobin, while in β-thalassemia major, the fetal hemoglobin may be 70–90% or more of the total hemoglobin. Slight increases of Hb F concentration are found in a variety of unrelated hematological disorders such as aplastic anemia and acute leukemia. In homozygous sickle cell disease, the Hb F concentration is often slightly increased. Higher concentrations of Hb F occur in Hb S-thalassemia and in patients who are doubly heterozygous for the Hb S gene and for a gene for hereditary persistence of fetal hemoglobin (HPFH).

Principle

Most human hemoglobins denature readily at alkaline pH and can then be precipitated by addition of ammonium sulfate to 40% saturation. Fetal hemoglobin is resistant to denaturation by alkali and remains soluble. This difference permits a rapid and relatively simple quantitative measurement of the amount of fetal hemoglobin present in human blood. In the following method all hemoglobins are first converted to cyanmethemoglobin by addition of $Fe(CN)_6^{3-}$ and CN^- before addition of alkali (NaOH). Addition of $(NH_4)_2SO_4$ precipitates denatured hemoglobin.

Specimen

Fresh blood anticoagulated with any standard anticoagulant may be used; 1 mL is required. No prior fasting or fluid restriction is needed.

Reagents

1. Cyanide-ferricyanide solution. Dissolve 0.2 g KCN and 0.2 g $K_3Fe(CN)_6$ in distilled H_2O, and dilute to 1000 mL.

2. Saturated solution of ammonium sulfate. Add 750 g of $(NH_4)_2SO_4$ to 1000 mL distilled H_2O. Warm and agitate to dissolve; the final volume of this solution is well in excess of 1000 mL. A few crystals may appear and settle to the bottom of the flask as the solution cools to room temperature.

3. Sodium hydroxide solution, 1.2 mol/L. Dissolve 48 g of NaOH, analytical reagent grade (carbonate-free), in distilled H_2O, and dilute to 1000 mL.

Procedure

1. Place approximately 1 mL of whole blood in a centrifuge tube and add cold isotonic saline to a volume of approximately 10 mL.

2. Centrifuge at 5 °C at $700 \times g$ for 15 min.

3. Aspirate and discard the supernatant.

4. Mix the packed cells with a wooden applicator stick or glass rod. Then pipet 0.2 mL of packed cells into a tube containing 4.0 mL of cyanide-ferricyanide solution.

5. Mix the cells in this solution with a glass stirring rod.

6. Using a safety pipet, transfer 2.8 mL of the hemoglobin-cyanide-ferricyanide mixture to a test tube. Blow in 0.2 mL of NaOH, 1.2 mol/L, and rapidly mix with a glass stirring rod. Begin timing with a stopwatch or other accurate timer at the moment that NaOH solution is blown in.

7. Exactly 2 min after the addition of sodium hydroxide, blow in 2 mL of saturated ammonium sulfate solution and mix well with a stirring rod. Allow to stand for 5 min.

8. Filter through a fine filter paper. (Whatman No. 42 is recommended.) If the filtrate is not clear, refilter through the same paper. Read absorbance of the filtrate at 540 nm, using a cuvet with a 1-cm light path. (If the absorbance exceeds 1.0, dilute 2-fold or 3-fold with a mixture of ammonium sulfate and sodium hydroxide solutions in proportions of 10:1 and reread.) The absorbance of the filtrate is designated A_f.

9. Using a safety pipet, transfer 0.4 mL of the original blood-cyanide-ferricyanide solution mixture (not treated with sodium hydroxide) to another tube and add 6.75 mL of distilled water. Agitate to mix, and read the absorbance of this in the same manner as above. Designate the absorbance A_b.

Calculation

$$\text{Per cent alkali-resistant hemoglobin} = \frac{A_f \times 100}{A_b \times 10}$$

where A_f is the absorbance of the filtrate, and A_b is the absorbance of the diluted blood-cyanide-ferricyanide mixture. The factor 100 converts from a decimal value to percentage of the total hemoglobin. The factor 10 corrects for dilution; there is a 375-fold final dilution of the original washed, packed erythrocytes (blank), while the final dilution of the NaOH-treated, washed, packed erythrocytes (i.e., the filtrate from steps 4–8) is 37.5-fold. Thus, the ratio of dilutions is 10:1, and A_b is multiplied by 10.

Reference Ranges

In normal adults or children more than 1 year old, Hb F by this method is usually <1% and never >2% of the total hemoglobin. In 30 normal subjects we found 29 specimens with values in the range of 0.2–0.7, and only one >1.1%. These results correspond to those reported by Betke et al. and by Pembrey et al.

Comments and Precautions

The rare hemoglobin variants Rainier and Bethesda also exhibit resistance to alkali denaturation and thus may be mistaken for Hb F.

There are several published variations of the alkali denaturation test. The procedure presented is the most sensitive method for measurement of concentrations of fetal hemoglobin < 10% of total hemoglobin. The cyanide-ferricyanide solution has ~4 times the cyanide concentration of Drabkin's solution. It should be handled with appropriate care, including the use of safety pipets. *Do not* pipet by mouth. Other precautions noted under Measurement of Hemoglobin Concentration in Whole Blood (p. 1532) apply to the handling of this reagent. (Drabkin's solution may be directly substituted for the cyanide-ferricyanide solution specified above.)

This procedure and other rapid tests for fetal hemoglobin give erroneously low results when Hb F makes up 30% or more of the total hemoglobin in a specimen. When it is known, or seems likely, that Hb F is present in high concentration, the cumbersome but more accurate Hb F assay of Jonxis and Visser should be used.[45]

In order to obtain reproducible results, the final cyanmethemoglobin concentration should not be < 480 mg/mL. The procedure as outlined here ensures that cyanmethemoglobin concentrations are ~900 mg/dL.

References

Betke, K., Marti, H.R., and Schlicht, I.: Nature, *184*:1877–1878, 1959.
Pembrey, M. E., McWade, P., and Weatherall, D. J.: J. Clin. Pathol., *25*:738–740, 1972.

Determination of Hemoglobin A$_2$

Principle

Diethylaminoethyl (DEAE) cellulose is an ether of cellulose and a substituted, strongly basic amine:

$$\begin{array}{c} C_2H_5 \\ \diagdown \\ \diagup \\ C_2H_5 \end{array} \overset{+}{N} - CH_2 - CH_2 - O - R$$

in which R is the fourth carbon of a glucose in the cellulose chain. Proteins may be bound to the DEAE cellulose and then eluted by changes in pH or salt concentration. In the procedure which follows, Hb A$_2$ is eluted from DEAE cellulose by a solution of glycine-KCN. Then, hemoglobins A and F are eluted with a solution of glycine, KCN, and NaCl.[1]

Specimen

The procedure requires 1 mL of whole blood; any anticoagulant is satisfactory. Since the specimen is prepared as for electrophoresis on cellulose acetate (see p. 1542), both procedures can be performed on the same 1-mL specimen. Fasting or fluid restriction is not necessary. Store specimens at 4 °C until the procedure is started.

Reagents

Columns and reagents can be purchased at reasonable cost from commercial sources (Isolab Inc., Akron, OH; Helena Laboratories, Beaumont, TX; and others). These kits include:

1. Chromatographic "mini-columns" of DEAE cellulose, pre-equilibrated with pH adjusted glycine-KCN solution and prepoured.

2. Hemoglobin elution solutions:

Solution I (for elution of Hb A$_2$), containing glycine, 0.2 mol/L, and KCN, 1.54 mmol/L, adjusted to pH ~7.5 depending on the batch of DEAE cellulose.

Solution II (for elution of hemoglobins A and F), containing glycine, 0.2 mol/L, KCN, 1.54 mmol/L, and NaCl, 0.2 mol/L.

Procedure

Sample Preparation

1. Bring reagents, columns, and samples to room temperature before performing the test. Resuspend sedimented erythrocytes by repeated inversion of specimen tubes.

2. To prepare hemolysate, mix 50 μL of whole blood with 200 μL of distilled water. Shake and let stand for 10 min.

Column Preparation

1. Remove the cap on the top of the column and then the plug at the bottom. NOTE: This order of opening the column is important; otherwise air will enter the column tip and reduce the flow rate.

2. Allow each column to drain (flow rate should be about 3 drops/min) until the meniscus reaches the top disk, when flow will virtually cease.

3. If any column contains air or if the flow rate is slow, see instructions under Comments and Precautions.

4. Add 50 μL of specimen hemolysates to the tops of corresponding columns. Allow each column to drain until the meniscus reaches the upper disk.

5. Rinse the sides of each column with 200 μL of solution I.

6. Allow each column to drain until the meniscus reaches the upper disk. Discard the column effluents collected up to this point.

Collection of the Hemoglobin Fractions

1. Place clean 10-mL sample collection tubes under each column.

2. Add 4 mL of solution I to each column and collect the entire volume of effluents. These are the Hb A_2 fractions. (Collection of these fractions should take less than 30 min.) Elution will stop in each column when the meniscus reaches the top disk. *Do not dilute* these (Hb A_2) fractions before making absorbance measurement.

3. Place a 25-mL flask under each column and add 4 mL of solution II to each column.

4. Allow each column to drain until the meniscus reaches the top disk. The effluent collected is the "Other Hb" fraction. Dilute this fraction with 16 mL of distilled water so that the absorbance reading will be <1.0.

5. Read the absorbance of both fractions at 415 nm in a spectrophotometer with a bandpass of ≤ 8 nm.

Calculations

$$\% \text{ Hb } A_2 = \frac{A}{A + 5H} \times 100$$

where A = absorbance of the Hb A_2 fraction,
           H = absorbance of the "Other Hb" fraction.
           The factor 5 corrects for the 5-fold dilution of the "Other Hb" fraction.

Reference Range

Results in normal adults are in the range 1.5–3%, median 2.2%. In β-thalassemia minor, Hb A_2 values range from 3.5–8%. A very small proportion of specimens yields values for Hb A_2 in the indeterminate range between 3.0 and 3.5%. In our experience, this procedure has a within-day CV of ~2%, while others have reported a CV closer to 5%.

Comments and Precautions

In contrast to other hemoglobins, Hb A_2 must be measured with considerable precision because the diagnosis of β-thalassemia minor may hinge on a Hb A_2 value only 1–2% above the reference range.

Several alternative methods have been used, such as densitometric scanning of cellulose acetate strips or elution from segments cut from starch gel or cellulose acetate and followed by spectrophotometric measurement. These alternative procedures suffer from considerable imprecision.[42]

Hemoglobins A_2, C, E, and O co-elute in this procedure. Thus, in the presence of hemoglobins C, E, or O, Hb A_2 concentration cannot be quantitated by this technique. Any measured values for A_2 of $\geq 10\%$ should be assumed to indicate the presence of hemoglobin C, E, or O.

References

Abraham, E. C., Reese, A., Stallings, M., et al.: Hemoglobin, *1*:27–44, 1976–1977.
Hamilton, S. R., Miller, M. E., Jessop, M., et al.: Am. J. Clin. Pathol., *71*:388–396, 1979.

Test for Fetal Hemoglobin Distribution by Acid Elution

Principle

Prepare blood smears on glass slides, fix in 80% methanol, and then immerse for 5 min at 37 °C in a pH 3.3 buffer (citric acid, 0.075 mol/L, and disodium phosphate, 0.05 mol/L). During this period, agitate the slide gently. Under these conditions, Hb A elutes relatively rapidly while Hb F elutes relatively slowly. The elution time and the pH must be carefully controlled.[89] Rinse the slide with water, then stain it with Ehrlich's acid hematoxylin (Harleco Laboratories, Philadelphia, PA) and counterstain with erythrosin (0.2 g/dL). Examine the stained slide microscopically at 100–400× magnification.

Interpretation and Comments

Normal adult erythrocytes show complete elution of hemoglobin. Erythrocytes containing Hb F stain darkly (Figure 15-18). When fetal erythrocytes are admixed with maternal erythrocytes (as commonly occurs during labor), the mother's blood shows occasional darkly stained erythrocytes containing Hb F. In the genetic trait, hereditary persistence of fetal hemoglobin (HPFH), all erythrocytes show nearly uniform partial retention of hemoglobin. In thalassemias with increased Hb F, there is nonuniform distribution of retained hemoglobin from cell to cell.

When this test is used to determine the magnitude of fetal-to-maternal transfusion that occurs during labor, the per cent of darkly staining cells (F-cells) must be determined. Two per cent F-cells corresponds approximately to a 15-mL fetal-to-maternal transfusion, 4% to 30 mL, 6% to 45 mL, 8% to 60 mL, and so on. Dosage of Rh_0 immune globulin required to prevent Rh immunization of the mother is based on the percent of F-cells.

References

Kleihauer, E., Braun, H., and Betke, K.: Klin. Wochenschr., *35*:637-638, 1957.
Shepard, M. K., Weatherall, D. J., and Conley, C. L.: Bull. Johns Hopkins Hosp., *110*:293-310, 1962.

Measurement of the Oxygen Saturation of Hemoglobin

Clinical Significance

Under normal circumstances arterial blood hemoglobin is at least 95% oxyhemoglobin. During passage of the blood through the capillary network, oxygen is delivered to tissues so that the hemoglobin of venous blood is normally ~75% oxyhemoglobin. Measurement of oxygen saturation of blood may be indicated when a patient has either cyanosis (a bluish discoloration of skin, lips, eyelids, and nail beds) or erythrocytosis. Among the possible causes of cyanosis or erythrocytosis are severe pulmonary disease resulting in diminished oxygenation of blood, and admixture of venous and arterial blood (for example, from an arteriovenous shunt, usually a developmental abnormality of the heart or great vessels). In either of these conditions, cyanosis and erythrocytosis may coexist. Measurement of oxygen saturation of the hemoglobin of arterial blood does not differentiate between these disorders.

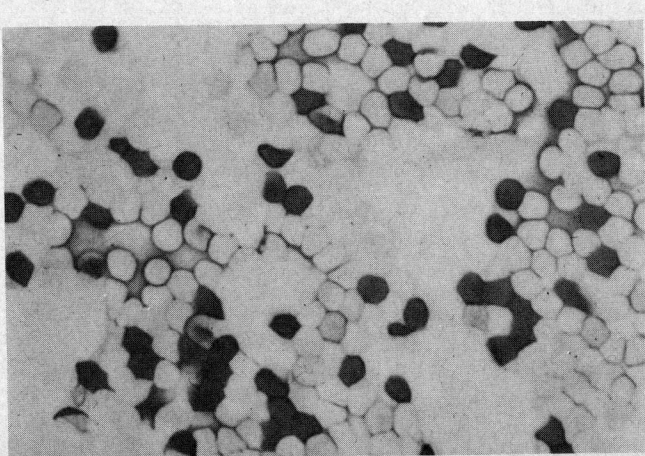

Figure 15-18. Acid elution test for cellular distribution of fetal hemoglobin. Hemoglobin F is more slowly eluted from erythrocytes than is Hb A when exposed to a solution of citric acid. The dark cells in this illustration contain Hb F which has not been eluted. The clear cells contain only Hb A. (Counterstained with hematoxylin and erythrosin, ×300.)

Principle

The oxygen saturation of hemoglobin may be measured by using either of two basic principles.

1. The oxygen contained in a sample of blood is released and measured volumetrically in a Van Slyke apparatus. The total oxygen binding capacity of a blood specimen that has been equilibrated with pure O_2 is measured in the same manner. The oxygen saturation is the ratio of the actual oxygen content to the total oxygen binding capacity of the specimen, expressed as a per cent.

2. The ratio of oxyhemoglobin to deoxyhemoglobin is measured on the basis of differences in spectral characteristics between the two pigments. Spectrophotometers or reflectance meters are suitable for this purpose, spectrophotometers being more generally used. Among applications of the spectrophotometric method are (1) measurement of oxygen saturation in vitro as described below, and (2) in vivo transcutaneous measurement by cuvet oximeters that may be applied to the pinna of the ear for measurement of the absorbance of light by blood in the capillary bed, or (3) by passage of arterial blood through cuvet oximeters to permit continuous monitoring of oxygen saturation. The spectrophotometric method presented below has the advantages that it is at least as accurate as other methods, is relatively simple to perform, requires no unusual or costly equipment, and is the basis of a widely used semi-automated method (CO-Oximeter, Instrumentation Laboratories, Lexington, MA).

The absorbances of any solution of a pure pigment at any two wavelengths have a constant ratio regardless of the concentration of the pigment. (For example, the ratio $A_{577\ nm}:A_{548\ nm}$ is always 1.2:1 for a pure solution of oxyhemoglobin.) However, if another pigment with a different spectral absorption curve is also present in the solution, the absorbance ratio is different in proportion to the concentration of the other pigment. By measuring the ratio of absorbances at two wavelengths for a solution containing two hemoglobin derivatives (oxygenated hemoglobin and deoxygenated hemoglobin or oxygenated hemoglobin and carboxyhemoglobin), the relative proportion of each can be precisely ascertained.

Because the relationship of absorbance ratios to concentration ratios is not a straight-line arithmetic function, it is necessary to construct tables or graphs of the absorbance ratios for every concentration ratio or pigment. Then, for any specimen of blood known to contain only two hemoglobin pigments, the relative proportions can be determined by making absorbance measurements at two wavelengths and using the ratio of absorbances to find the concentration ratio.

The method described below is based on the absorbance ratio method. It differs only in that one of the absorbance measurements is made at an *isosbestic point* of the two pigments. An isosbestic point is a wavelength at which the spectral absorptivity curves for the two pigments intersect (i.e., identical concentrations of the two pigments exhibit identical absorbance values). The absorbance at the isosbestic point, then, is proportional to the total concentration of the two pigments. The other absorbance measurement is made at a wavelength at which the two pigments have markedly different absorbances. For a mixture of oxyhemoglobin and deoxyhemoglobin, a convenient isosbestic point is 548 nm (Figure 15-19). The spectral absorptivity curves of these pigments differ markedly at 577 nm, and this wavelength is suitable for the second absorbance measurement. A major advantage of this method is that the ratio of concentrations of the two pigments is a straight-line arithmetic function of the absorbance ratio. Therefore, a graph or table need not be constructed; the concentration ratios can be calculated by a relatively simple equation. Furthermore, the constants that are required in the calculations can be verified independently.

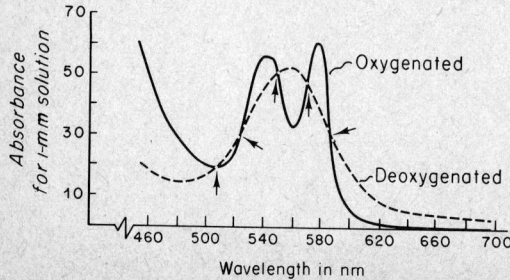

Figure 15-19. Spectral absorbance curves for oxyhemoglobin and deoxyhemoglobin. The data were obtained by the author with whole blood hemolyzed with Triton X-100 in a Nahas cuvet with 0.1-mm light path, as described in the text. Arrows indicate isosbestic points.

The derivation of the formula for oxygen saturation is as follows: C_t represents the total concentration of hemoglobin (oxyhemoglobin plus deoxyhemoglobin), C_o is the concentration of oxyhemoglobin, and C_r is the concentration of deoxygenated (reduced) hemoglobin. A_{548} and A_{577} are the absorbances at 548 and 577 nm, respectively. K_1, K_2, and K_3 are the absorptivity constants for oxyhemoglobin and deoxyhemoglobin at 548 nm, oxyhemoglobin at 577 nm, and deoxyhemoglobin at 577 nm, respectively. Then, when only oxyhemoglobin or deoxyhemoglobin is present

$$A_{548} = K_1C_t = K_1C_o = K_1C_r$$

$$A_{577} = K_2C_o$$

$$A_{577} = K_3C_r$$

Let S = fractional oxygen saturation of hemoglobin. Then

$$S = \frac{C_o}{C_t}$$

or

$$C_o = SC_t$$

For any mixture of oxyhemoglobin and deoxyhemoglobin,

$$A_{577} = K_2C_o + K_3C_r$$

or

$$A_{577} = K_2SC_t + K_3C_r$$

However, $C_r = C_t - C_o$ (assuming presence of only two hemoglobin pigments). Therefore,

$$A_{577} = K_2SC_t + K_3(C_t - C_o)$$

$$= K_2SC_t + K_3(C_t - SC_t)$$

$$= C_tK_2S + C_tK_3 - C_tK_3S$$

$$= C_t(K_2S + K_3 - K_3S)$$

$$= C_t[K_3 + S(K_2 - K_3)]$$

$$\frac{A_{577}}{A_{548}} = \frac{C_t[K_3 + S(K_2 - K_3)]}{C_tK_1}$$

$$= \frac{K_3}{K_1} + \frac{S(K_2 - K_3)}{K_1}$$

$$\frac{A_{577}}{A_{548}} - \frac{K_3}{K_1} = \frac{S(K_2 - K_3)}{K_1}$$

$$S(K_2 - K_3) = K_1\left[\frac{A_{577}}{A_{548}} - \frac{K_3}{K_1}\right]$$

$$S = \left[\frac{K_1}{K_2 - K_3}\right]\left[\frac{A_{577}}{A_{548}} - \frac{K_3}{K_1}\right]$$

$$S = \left[\frac{K_1}{K_2 - K_3}\right]\left[\frac{A_{577}}{A_{548}}\right] - \left[\frac{K_3}{K_2 - K_3}\right]$$

$$\text{Per cent saturation} = 100\left(\left[\frac{K_1}{K_2 - K_3}\right]\left[\frac{A_{577}}{A_{548}}\right] - \left[\frac{K_3}{K_2 - K_3}\right]\right)$$

Since the fractions $\dfrac{100K_1}{K_2 - K_3}$ and $\dfrac{100K_3}{K_2 - K_3}$ are also constants, designate the former K_4 and the latter K_5.

Then,

$$\text{Per cent saturation} = K_4 \frac{A_{577}}{A_{548}} - K_5$$

This is an equation for a straight line, in which per cent saturation represents the ordinate, $\frac{A_{577}}{A_{548}}$ the abscissa, K_4 the slope, and K_5 the ordinate-intercept. When only deoxyhemoglobin is present, $S = 0$, and therefore

$$\frac{A_{577}}{A_{548}} = \frac{K_3}{K_1}$$

Similarly, when only oxyhemoglobin is present, $S = 1.0$, and

$$\frac{A_{577}}{A_{548}} = \frac{K_2}{K_1}$$

Thus, the absorbance ratios $\frac{A_{577}}{A_{548}}$ for pure solutions of either oxyhemoglobin or deoxyhemoglobin are constants, irrespective of concentration. If the ratio $\frac{A_{577}}{A_{548}}$ for oxyhemoglobin is designated R_o and the ratio $\frac{A_{577}}{A_{548}}$ for deoxyhemoglobin is designated R_r, then

$$K_4 = \frac{100}{R_o - R_r}$$

and

$$K_5 = \frac{100 R_r}{R_o - R_r} = R_r K_4$$

Thus, K_4 and K_5 can be calculated from measured values of the ratios R_o and R_r. The constants thus obtained permit subsequent calculation of oxygen saturation in any specimen containing mixtures of oxyhemoglobin and deoxyhemoglobin, by measurement of absorbance at 548 and 577 nm.

The same basic equations are valid with other wavelengths. Nahas used the isosbestic point at 805 nm and the oxyhemoglobin absorbance measurement at 600 nm (where absorbance is very low). Similarly, van Kampen and Zijlstra used absorbance values at 560 and 506 nm, the value at 506 nm being isosbestic. The values of the constants K_4 and K_5 will vary depending on the wavelengths used.

Absorbance measurements at four wavelengths permit determination of oxygen saturation, carboxyhemoglobin, and methemoglobin concentrations. The commercially available and now widely used CO-Oximeter performs simultaneous measurements at 535, 585.2, 594.5, and 626.6 nm. From these data are derived hemoglobin concentration in g/dL, oxygen saturation of hemoglobin, and carboxyhemoglobin and methemoglobin concentrations. Oxygen saturation, carboxyhemoglobin, and methemoglobin are expressed as percentages of the total hemoglobin. The computations are performed automatically, based on the principles and formulas outlined above, though different wavelengths are used.

In interpreting results from the CO-Oximeter one must remember that absorbance measurements are required at an additional wavelength for every additional pigment present: 2 for oxygenated and deoxygenated hemoglobin, 3 if, in addition, carboxyhemoglobin is present, and 4 if methemoglobin is also present; if sulfhemoglobin is present, a fifth wavelength must be selected. Methemoglobinemia and sulfhemoglobinemia commonly coexist, for example in persons being treated with phenacetin or with sulfones such as dapsone. The current model for the CO-Oximeter does not correct for sulfhemoglobin; thus, somewhat spurious results may be obtained in this situation. Further, since whole blood is used, high serum concentration of bilirubin or other pigments may also lead to spurious results. These possible sources of error also apply to the manual procedure given below.

Special Equipment

Nahas cuvets with 0.1-mm light path and 0.01-mL volume are required (NAC-10 cuvet, Waters Corp., Rochester, MN). Because these cuvets do not fit many standard spectrophotometers, the sides may need to be ground to produce a 10×10 mm cross section. The short light path of the Nahas cuvet makes it possible to obtain a satisfactory absorbance reading from minimally diluted hemolysates at the specified wavelengths. Other cuvets with special adapters may be used.

The Determination of Constants in the Formula for Calculation of the Oxygen Saturation

Reagents

1. Hemolyzing reagent. Dilute 1.0 mL of Triton X-100 (Rohm and Haas Co., Philadelphia, PA) with distilled water to 100 mL.
2. Blank reagent. Dilute 1 mL of hemolyzing reagent with 1 mL of distilled water.
3. Sodium hydrosulfite ($Na_2S_2O_4$, dithionite)
4. Heparin sodium. Draw 0.5 mL of solution containing 1000 units/mL into a sterile syringe and eject into a sterile vial containing 10 mL of isotonic saline. The final concentration of heparin is 48 units/mL.

Procedure

1. Obtain approximately 5 mL of fresh heparinized blood from a fasting healthy subject who is a nonsmoker.
2. Equilibrate in a tonometer for 15 min with room air. Air in the tonometer may be either continually or intermittently replaced. Aspirate 1 mL of blood from the tonometer into a syringe containing a glass bead (steel ball or mixing ring) and 1 mL of hemolyzing reagent. Invert several times until clear hemolysate is obtained.
3. Fill one Nahas cuvet (blank) with distilled water and one with hemolysate from step 2.
4. Set a spectrophotometer with a bandpass ≤ 1 nm to 548 nm and determine absorbance of hemolysate. Designate this $A_{548(o)}$.
5. Reset spectrophotometer to 577 nm and determine absorbance of hemolysate. Designate this $A_{577(o)}$.
6. Add 6 mg of $Na_2S_2O_4$ (sodium hydrosulfite, dithionite) to remaining hemolysate; mix.
7. Measure absorbance of treated hemolysate at 577 and 548 nm. Designate these readings $A_{577(r)}$ and $A_{548(r)}$, respectively. $A_{548(r)}$ should be identical (within the replicate specification of the spectrophotometer) with $A_{548(o)}$.
8. Calculate ratios R_o and R_r:

$$R_o = \frac{A_{577(o)}}{A_{548(o)}}; \qquad R_r = \frac{A_{577(r)}}{A_{548(r)}}$$

9. Calculate constants K_4 and K_5:

$$K_4 = \frac{100}{R_o - R_r}; \qquad K_5 = \frac{100R_r}{R_o - R_r}$$

Measurement of Oxygen Saturation

1. Prepare a 5-mL glass syringe by placing a 1-mm glass bead (or a 1-mm steel ball or mixing ring) inside the syringe to facilitate mixing and then insert the plunger as far as possible. Label and autoclave.
2. Prior to drawing a blood sample, aspirate heparin solution (48 units/mL) into syringe. Allow entire area of contact between barrel and plunger to become moistened as a seal. Point syringe upward and eject excess heparin solution so that residual heparin solution fills the small amount of "dead space" and the needle.
3. Draw 5 mL of arterial blood specimen directly from the patient into the heparinized syringe. Seal syringe if it is sent to the laboratory; otherwise, proceed with step 4.
4. Immediately insert needle of syringe into a vial containing a few mL of hemolyzing solution. Eject approximately 0.1 mL of blood into hemolyzing solution to clear needle of any blood exposed to ambient air. Then aspirate 5 mL of hemolyzing solution into the syringe.* Seal needle with plasticine or syringe tip with a steel cap. Mix by repeated inversion. Contents of syringe will become darker and clearer as blood hemolyzes.

---

*When the blood specimen has a hemoglobin concentration of > 15 g/dL, the absorbance at 577 nm may be too high to be read accurately. In this case, dilute 1 mL of blood with 2 mL of hemolyzing solution.

5. Carefully and rapidly replace the needle (or cap) with a 25-gauge needle. Eject 2 drops of blood to eliminate hemoglobin exposed to air in the syringe nipple and in the needle. Then, inject the solution into a Nahas cuvet until cuvet is full and a drop of blood has emerged from the outlet at the bottom. Wipe off the effluent blood.

6. Place the blank cuvet with blank reagent into the spectrophotometer and set to zero A at 548 nm. Then read A_{548} of cuvet containing hemolysate. Reset blank to zero A at 577 nm and determine A_{577} for the cuvet containing the hemolysate.

Calculations

$$\text{Per cent saturation} = K_4 \times \frac{A_{577}}{A_{548}} - K_5$$

Reference Range

95–99% saturation

Comments and Precautions

In the determination of the constants K_4 and K_5, inadvertent exposure of the blood to ambient air will not influence the results. On the other hand, great care must be exerted to ensure that the arterial blood specimen (unknown) is not inadvertently exposed to room air or air bubbles. Furthermore, studies of the arterial blood specimen must be performed promptly so that leukocyte metabolism does not artifactually decrease the oxygen content. Use of a spectrophotometer with narrow bandpass and a narrow slit width (0.01 mm) is essential.

The validity of this technique is based on the assumption that only the two normal forms of hemoglobin (oxyhemoglobin and deoxyhemoglobin) are present in blood in appreciable quantities. Methemoglobin ($> 5\%$), sulfhemoglobin, carboxyhemoglobin, high concentration of bilirubin, and lipemia introduce errors. The studies should be undertaken only in subjects with clear, nonlipemic plasma. (A correction can be made for lipemia by measuring absorbance ratios in plasma, as described by Nahas.)

Cyanosis without erythrocytosis may be due to the presence of an abnormal hemoglobin (for example, methemoglobinemia due to the presence of Hb M); ingestion of nitrates or various drugs; or sulfhemoglobinemia, usually due to drug ingestion. When there is cyanosis due to methemoglobinemia or sulfhemoglobinemia, measurement of arterial oxygen saturation may result in either a normal value or a lower than normal value. Erythrocytosis occurs without cyanosis in polycythemia vera, in the presence of certain neoplastic diseases, as a result of abnormalities of the kidney, or as a result of the presence of a hemoglobin with high oxygen affinity. In these disorders, arterial oxygen saturation is also normal.

Blood concentrations of carboxyhemoglobin in excess of 7% of the total hemoglobin significantly decrease the accuracy of the measurement of oxygen saturation by the method described here. Persons who smoke more than one package of cigarettes daily may have carboxyhemoglobin concentrations as high as 15%. Higher carboxyhemoglobin concentrations may occur in those who have occupational or accidental exposure to carbon monoxide. Whenever there is reason to suspect carbon monoxide exposure, the carboxyhemoglobin concentration should be measured (see Chapter 18); if it is increased, a correction may be made in the oxygen saturation measurement in the following manner. (Instruments such as the CO-Oximeter include this correction automatically.)

For any mixture containing only oxyhemoglobin, deoxyhemoglobin, and carboxyhemoglobin, $A_x = \{a_{x(o)}C_o + a_{x(c)}C_c + a_{x(r)}C_r\}d$ in which A_x is the total absorbance of the solution at any given wavelength (x); C_o is the concentration of deoxyhemoglobin; d is the length of the light path through the cuvet; and $a_{x(o)}$, $a_{x(c)}$, and $a_{x(r)}$ are the absorbances, at wavelength x, of pure solutions of oxyhemoglobin, carboxyhemoglobin, and deoxyhemoglobin, respectively, in equimolar concentrations. Then, at 577 and 548 nm

$$A_{577} = [a_{577(o)}C_o + a_{577(c)}C_c + a_{577(r)}C_r]d$$

$$A_{548} = [a_{548(o)}C_o + a_{548(c)}C_c + a_{548(r)}C_r]d$$

Furthermore, at 568 nm (an isosbestic point of carboxyhemoglobin and deoxyhemoglobin)

$$A_{568} = [a_{568(o)}C_o + a_{568(c)}C_c + a_{568(r)}C_r]d$$

Therefore if the absorbance is measured at the three wavelengths, 577, 548, and 568 nm, solution of these three simultaneous equations will give the concentrations of oxyhemoglobin, deoxyhemoglobin, and carboxyhemoglobin.

$$\text{Oxygen saturation} = \frac{C_o}{C_o + C_c + C_r} \times 100$$

The same results may be achieved manually by measurement of absorbance at two wavelengths (for example, 577 and 548 nm) if the concentration of carboxyhemoglobin has been determined separately (see p. 1699). Thus, in such cases, only the first two of these equations are required. The equations are rewritten and solved for oxygen saturation (per cent oxyhemoglobin) as follows, wherein S_o = per cent oxyhemoglobin, S_r = per cent deoxyhemoglobin, and S_c = per cent carboxyhemoglobin:

$$\frac{A_{577}}{A_{548}} = \frac{S_o a_{577(o)} + S_r a_{577(r)} + S_c a_{577(c)}}{S_o a_{548(o)} + S_r a_{548(r)} + S_c a_{548(c)}}$$

However, $S_r = 100 - S_c - S_o$. Therefore

$$\frac{A_{577}}{A_{548}} = \frac{S_o a_{577(o)} + (100 - S_c - S_o)a_{577(r)} + S_c a_{577(c)}}{S_o a_{548(o)} + (100 - S_c - S_o)a_{548(r)} + S_c a_{548(c)}}$$

$$S_o = \frac{\dfrac{A_{577}}{A_{548}}[100a_{548(r)} - S_c a_{548(r)} + S_c a_{548(c)}] - S_c a_{577(c)} - 100a_{577(r)} + S_c a_{577(r)}}{\dfrac{A_{577}}{A_{548}}[a_{548(r)} - a_{548(o)}] + a_{577(o)} - a_{577(r)}}$$

Since 548 nm is an isosbestic point for oxyhemoglobin and deoxyhemoglobin, $a_{548(o)} = a_{548(r)}$; therefore,

$$S_o = \frac{\dfrac{A_{577}}{A_{548}}[100a_{548(r)} + S_c(a_{548(c)} - a_{548(r)})] - S_c(a_{577(c)} - a_{577(r)}) - 100a_{577(r)}}{a_{577(o)} - a_{577(r)}}$$

The absorptivity constants required for solution of this equation are

$$a_{548(r)} = 49.84; \ a_{548(c)} = 48.40; \ a_{577(c)} = 40.4; \ a_{577(r)} = 38.0; \ a_{577(o)} = 61.48^*$$

By substituting these absorptivity constants in the equation, one obtains the following:

$$S_o = \frac{\dfrac{A_{577}}{A_{548}}[4984 - 1.44 S_c] - 2.4 S_c - 3800}{23.48}$$

$$S_o = \frac{A_{577}}{A_{548}}[212.26 - 0.061 S_c] - 0.102 S_c - 161.8$$

References

Nahas, G. G.: J. Appl. Physiol., *13*:147–152, 1958.
van Kampen, E. J., and Zijlstra, W. G.: Adv. Clin. Chem., *8*:141–187, 1965.

*The constants given are for 1.0 mmol/L concentrations of these hemoglobin pigments. Absorbance values obtained at any other concentrations are equally valid, provided only that the concentrations be identical for all three pigments.

Detection of Hemoglobin in Sputum, Urine, and Feces

Clinical Significance

Hemoglobin is not normally present in sputum, urine, or feces except in minute amounts; the demonstration of the presence of hemoglobin in these materials is an important diagnostic clue that usually implies bleeding within the respiratory, urinary, or gastrointestinal tracts, respectively.

Principle

The tests for the detection of hemoglobin depend on the fact that heme proteins act as peroxidases, catalyzing the reduction of hydrogen peroxide to water. This reaction requires a hydrogen donor. If guaiac or o-tolidine (a benzidine derivative) is used as a hydrogen donor, the oxidation of the donor chromogen results in a blue color, and the intensity of the final color is primarily a function of the hemoglobin concentration:

$$2H_2O_2 + \text{o-tolidine (reduced, colorless)} \xrightarrow{\text{Heme}} 2H_2O + \text{o-tolidine (oxidized, blue)}$$

For details of the procedure, consult the kit insert of various commercially available products.

Comments and Precautions

A distinction must be made between hematuria (the presence of intact erythrocytes in the urine) and hemoglobinuria (the presence of free hemoglobin in the urine). Hematuria implies a bleeding lesion in the urinary tract. Hemoglobinuria, in the absence of hematuria, usually implies severe intravascular hemolysis. Microscopic examination of the urine is the appropriate and most sensitive test for hematuria.

Benzidine derivatives are presumed to be carcinogenic. However, o-tolidine is used in a commercially available dipstick method (Hemastix, Ames Company, Elkhart, IN) for testing for hemoglobin or myoglobin in urine. Use of the Hemastix test of urine or Hematest (also Ames Co.) for feces involves no reagent preparation on the part of laboratory personnel, and the quantity of o-tolidine is very small. There appears to be no carcinogenic hazard from this source.

The o-tolidine tests (Hemastix and Hematest) are sensitive to $\sim$2 mg of hemoglobin/dL of urine and to $\sim$100 mg of hemoglobin/100 g of feces, the equivalent of $\sim$10 erythrocytes/μL of urine or 1 mL of blood/100 g of feces, respectively. Thus, these tests are very sensitive. A negative urine test excludes significant hemoglobinuria or hematuria in the specimen examined. False positive urine tests may occur if the urine container has been contaminated with hypochlorite solution, or if peroxidase-containing bacteria are present in the urine. Positive results occur if there is hemoglobinuria, hematuria (presence of intact erythrocytes in urine), or myoglobinuria.

The o-tolidine based test (Hematest) is not recommended for fecal occult blood testing because it is so sensitive that false positive results are very common.

A guaiac-based test for fecal occult blood is the Hemoccult test (Smith, Kline and French Laboratories, Philadelphia, PA), which is much less sensitive than the o-tolidine-based test; false positive results are therefore very infrequent. Furthermore, false positive results are not observed in stools of patients who are taking medicinal iron orally. On the other hand, false negative reactions are quite common unless the rate of blood loss in feces exceeds $\sim$10 mL/d. Of fecal specimens that contain at least 30 mL of blood in a 24-h collection of feces, 93% are positive by the Hemoccult test.[94] If specimens are not examined on the same day they are obtained, the frequency of positive results diminishes. In screening for occult fecal blood, the Hemoccult test is useful when results are positive because this finding almost always indicates substantial gastrointestinal bleeding. However, negative results do not exclude significant gastrointestinal blood loss.

For further details consult manufacturer's instructions.

MYOGLOBIN

Myoglobin is the oxygen-binding protein of striated (cardiac and skeletal) muscle. It resembles the α-, β-, γ-, and δ-chains so closely that it has served as the prototype in studies of the molecular structure of hemoglobin. Unlike hemoglobin, myoglobin exists only as a monomer, and thus its molecular weight is approximately one-fourth that of hemoglobin. Also, unlike hemoglobin, it is unable to release oxygen except at extremely low oxygen tensions. Its physiological role in man is uncertain. It may serve as an oxygen reservoir of last resort, accessible only under circumstances of extreme hypoxia. Such conditions conceivably may exist when there is strenuous physical exertion, as in athletic events. In contrast to the small amounts of myoglobin contained in the muscle of terrestrial mammals, aquatic mammals have relatively large amounts of myoglobin in their muscles, probably as an adaptation to the need for oxygen during protracted periods of submersion.

MYOGLOBINEMIA

Increase in serum myoglobin concentration occurs following trauma to either skeletal or cardiac muscle, as in crush injury or myocardial infarction, respectively. Tests for myoglobin formerly used urine principally, because even following crush injuries, the serum concentration of myoglobin would not be sufficiently increased to permit detection by the relatively crude techniques then available. A radioimmunoassay for serum myoglobin has now been developed, and a kit for this purpose is available (Nuclear Medical Systems, Inc., Newport Beach, CA).

Measurement of serum myoglobin concentration has been advocated for the diagnosis of myocardial infarction. However, the radioimmunoassay appears to be less sensitive than creatine kinase assay for ascertainment of myocardial injury. False positive results complicate the use of serum myoglobin assay in the diagnosis of myocardial infarction. Furthermore, seemingly minor injury to skeletal muscle cannot be differentiated from myocardial infarction by this method.

Reference Ranges

Ranges for serum myoglobin vary according to age, sex, and ethnic group. Adult males generally have a serum myoglobin concentration $< 80 \mu g/L$ and adult women $< 60 \mu g/L$; children have very low values. Black men have higher serum myoglobin concentrations than do Caucasian men. Serum myoglobin concentration also increases with age: older men have higher values than do younger men.[20] Serum myoglobin concentration following myocardial infarction may be as high as 500 $\mu g/L$.[46]

MYOGLOBINURIA

Clinical Significance

Following injury to skeletal or cardiac muscle, myoglobin may be released and is then excreted in the urine (*myoglobinuria*). There are numerous causes for this condition. In many, the degree of myoglobinuria is so slight that it is not perceived by either the patient or the physician or by routine examination of the urine. More pronounced myoglobinuria, sufficient to cause obvious discoloration of urine, may occur after violent physical exercise or as the result of a severe accident in which there is crushing of muscle. With crush injuries, copious amounts of myoglobin may appear in the urine and may contribute to impairment of renal function in badly injured patients; postmortem studies show myoglobin apparently occluding renal tubules in some instances.

When present in large quantity, myoglobin imparts to the urine a color similar to that of a cola drink or black coffee. Myoglobinuria cannot be distinguished from other causes of dark urine simply on the basis of visual inspection since essentially the same color may be observed in hemoglobinuria. Patients with hemolytic anemia due to unstable hemoglobin also may have coffee-colored urine, but this color is due to a nonprotein, heme-free degradation product of hemoglobin that has not yet been chemically defined.

The possibility of myoglobinuria may be considered when a urine specimen gives both a positive test for protein by the sulfosalicylic acid test and a positive occult blood test. It then becomes necessary to determine whether the heme protein in the urine is hemoglobin or myoglobin. The heat and acetic acid tests may be negative when there is hemoglobin or myoglobin in the urine.

Principle

Several physicochemical properties permit differentiation of hemoglobin and myoglobin: (1) hemoglobin is less soluble than myoglobin in ammonium sulfate solutions; (2) the absorption spectra are different for analogous derivatives of hemoglobin and myoglobin; (3) myoglobin is more electropositive than Hb A and can therefore be differentiated electrophoretically; and (4) the molecular sizes of hemoglobin and myoglobin are different, allowing for differentiation by ultrafiltration or gel filtration. Hemoglobin and myoglobin also can be differentiated by immunodiffusion or immunoprecipitation. These are extremely sensitive techniques capable of measuring minute quantities of myoglobin.

Monoclonal antibodies to human myoglobin are now commercially available. Using these, one may prepare plates for identification of myoglobin in urine by *radial immunodiffusion.* A serious drawback of this procedure is that 24–48 h are required before the plates can be read, and in patients with suspected crush injury clinicians need results right away. The antibodies can also be used in a *nephelometric procedure* for a specific quantitative assay of urinary myoglobin.[63] If a nephelometer and antimyoglobin are available, this is clearly the procedure of choice.

Ultrafiltration of clear, centrifuged urine provides a means of differentiating hemoglobin from myoglobin on the basis of molecular size. Hemoglobin does not pass through a Millipore filter 8–12 nm in pore diameter, whereas myoglobin does. Thus, the presence of heme protein in the ultrafiltrate is indicative of myoglobin. This procedure appears to be satisfactory based on reported studies.[96]

Myoglobin is readily distinguished from hemoglobins by *isoelectric focusing* in polyacrylamide gel. In this medium, myoglobin focuses farthest to the acidic-anode end of the gel, whereas hemoglobins S and C focus at the basic-cathode end of the gel. This method may be of value in patients whose blood contains hemoglobin S or C and in whom hemoglobinuria would be difficult to differentiate from myoglobinuria on the basis of electrophoresis.

Electrophoresis of concentrated "desalted" urine at pH 8.6 on a cellulose acetate membrane (Figure 15-20) permits distinction of myoglobin from hemoglobin, since the former has an electrophoretic mobility like that of hemoglobin C or E. (If this procedure is used for specimens from persons of black or Southeast Asian ethnic groups, a specimen of blood should be examined electrophoretically at the same time to avoid confusion of myoglobin with hemoglobin C or E.) The electrophoretic method was described in *Fundamentals of Clinical Chemistry,* 2nd edition.

A *spectrophotometric method* has also been proposed, based on differences in absorptivity at 600 and 580 nm of methemoglobin and metmyoglobin. This simple, convenient method was also described in *Fundamentals of Clinical Chemistry,* 2nd edition. Our experience with this spectrophotometric test indicates many instances of misleading results.

A *solubility test* for myoglobin is based on the fact that hemoglobin is completely precipitated in 80% saturated ammonium sulfate solution, whereas myoglobin often remains in solution. To perform this test, add 2.8 g of crystalline ammonium sulfate, $(NH_4)_2SO_4$, to 5 mL of urine in a test tube. Mix gently until no crystals are seen at the bottom of the tube. Filter. Test the filtrate for heme protein by the benzidine or *o*-tolidine method. (See preceding section; Hemastix is

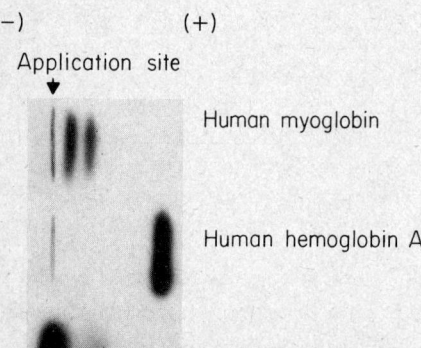

(−)   (+)

Application site

Human myoglobin

Human hemoglobin A

Horse myoglobin

Figure 15-20. Electrophoresis of myoglobin on cellulose acetate at pH 8.6. In the top channel are human myoglobin and metmyoglobin (between origin and myoglobin) from the urine of a patient who sustained a fatal crush injury. The band in the bottom channel is commercially available (Sigma) horse myoglobin in the metmyoglobin form; it has nearly the same electrophoretic mobility as human metmyoglobin. (Myoglobins from other species have quite dissimilar electrophoretic mobilities.)

convenient for this purpose.) If the test for heme protein is positive, this is presumptive evidence for the presence of myoglobin. Unfortunately, this simple and widely used test frequently gives erroneous results.

References

Kagen, L. J.: CRC Crit. Rev. Clin. Lab. Sci., *9*:273-302, 1978.
Kubasik, N. P., Guiney, W., Warren, K., et al.: Clin. Chem., *23*:2047–2049, 1978.
Rosano, T. G., and Kenny, M. A.: Clin. Chem., *23*:69–75, 1977.

VITAMIN B$_{12}$

STRUCTURE AND TERMINOLOGY

The vitamin B$_{12}$ group of substances are physiologically active cobalamins. They are composed of tetrapyrrole rings surrounding central cobalt atoms, with nucleotide side chains attached to the cobalt. The cobalamin tetrapyrrole ring, exclusive of cobalt and other side chains, is called a *corrin*. The cobalt-corrin complex is termed *cobamide*. In cobalamins, 5,6-dimethylbenzimidazole riboside is bound to the cobalt atom by one of its imidazole nitrogens, and its 2'-ribose carbon is linked with an ester of aminoisopropanol and propionic acid to the corrin ring (Figure 15-21). Cobalamins differ in the nature of additional side groups bound to cobalt; examples are methyl (methylcobalamin), 5'-deoxyadenosine (deoxyadenosylcobalamin or coenzyme B$_{12}$), hydroxyl (hydroxocobalamin or vitamin B$_{12a}$), H$_2$O (aquocobalamin or vitamin B$_{12b}$), and cyanide (cyano-cobalamin or vitamin B$_{12}$). *Cyanocobalamin* is a stable compound that forms dark red needle-like crystals; it is the reference compound for measuring serum cobalamin concentration. Less stable serum cobalamins may be converted to this compound for quantitation. The predominant physiological form of cobalamin in serum is *methylcobalamin*, while that in cytosols is *5'-deoxy-adenosylcobalamin*.

Chemistry

Cyanocobalamin has a M.W. of 1355 and a solubility of 1.2 g/dL in water at 20 °C. It is also soluble in lower alcohols and aliphatic acids but is insoluble in acetone, ether, and chloroform. It is gradually destroyed on exposure to light. Aqueous solutions of cyanocobalamin exhibit a distinctive absorptivity spectrum with maxima at 278, 361, and 550 nm, with absorptivity coefficients $\left(\epsilon \dfrac{1\%}{1cm} \right)$ of 115, 204, and 65, respectively.[15] The spectrum is independent of pH but changes when cyanocobalamin binds to intrinsic factor. Because of its stability in aqueous solutions and

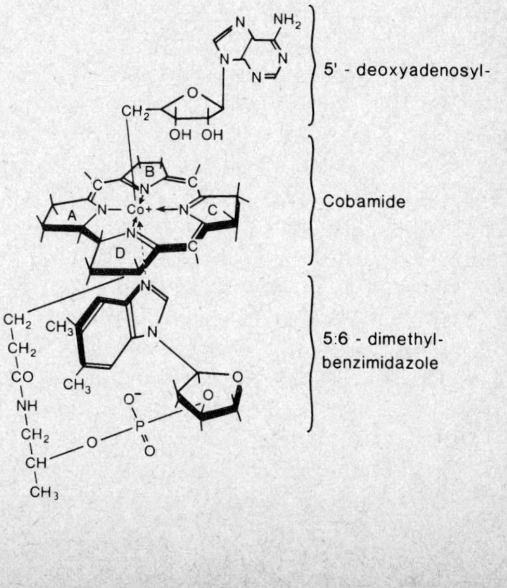

Figure 15-21. The structure of 5'-deoxyadenosyl cobalamin, a physiologically important form of vitamin B$_{12}$. All of the active cobalamins contain a corrin (a porphyrin ring like that in heme, but with cobalt rather than iron at its center), here labeled cobamide, a purine nucleotide (5:6-dimethylbenzim-idazole), and an organic ligand, in this case 5'-deoxyaden-osine. The various cobalamins differ principally in the moiety linked to the cobalt above the plane of the corrin nucleus in this illustration. Methyl cobalamin is the simplest of these, with a methyl group bound to the cobalt atom. Cyanoco-balamin, commercial vitamin B$_{12}$, has CN$^-$ at this same position. In hydroxocobalamin and aquocobalamin, OH$^-$ and H$_2$O groups respectively are linked to Co at this position. (Reprinted from Chanarin, I: The Megaloblastic Anaemias. 2nd ed. Oxford, Blackwell Scientific, 1979; with permission.)

its distinct absorption spectrum, accurate concentrations of cyanocobalamin can be prepared and used as standards for the measurement of serum cobalamin.

Biochemical Function

Vitamin B_{12} is a coenzyme for two physiologically important functions in man: (1) the synthesis of methionine, and (2) the conversion of methylmalonic acid to succinic acid. The synthesis of methionine requires methylcobalamin, while the conversion of methylmalonyl CoA requires deoxyadenosylcobalamin.

(1) tetrahydrofolate   ⟍↗ methylcobalamin   ⟍↗ homocysteine
     N^5-methyltetrahydrofolate ⟋⟍ cobalamin (reduced) ⟋⟍ methionine

(2) methylmalonyl CoA $\xrightarrow[\text{CoA mutase}]{\text{deoxyadenosylcobalamin}}$ succinyl CoA

In reaction (1), methylcobalamin serves as an intermediate in the transfer of a methyl group from N^5-methyltetrahydrofolate to homocysteine for the formation of methionine. Methionine is required for the synthesis of formate and S-adenosylmethionine. Formate is essential for purine synthesis, the conversion of deoxyuridylate to thymidylate, and the production of the active folate coenzyme.[15-18,23] S-adenosylmethionine is thought to be important in neuronal metabolism. Reaction (2) plays an important role in the metabolism of fatty acids and aliphatic amino acids, and deficiency of vitamin B_{12} causes accumulation of abnormal lipids.[34,85] Neurological effects of vitamin B_{12} deficiency may be due to accumulation of these abnormal lipids in the nervous system.

Deficiency of vitamin B_{12} in man is associated with megaloblastic anemia and neuropathy. The metabolic lesion in these disorders is not fully understood, but studies of children with inborn errors of methylmalonyl CoA metabolism indicate that impairment in methionine synthesis (equation 1) is most directly related to these clinical manifestations.[11] Patients with inborn errors which produce methylmalonic acidemia (mut^0, mut$^-$, cbl A, cbl B) do not manifest megaloblastic anemia or neurologic dysfunction, while patients with combined methylmalonic acidemia and homocystinuria exhibit both megaloblastic anemia and neurologic defects.[4] The theory that neurologic dysfunction in pernicious anemia may be caused by aberrant incorporation of odd-numbered branched fatty acyl moieties into myelin, due to blocks in the propionate pathway, is inconsistent with the observations made in children with inborn errors of methylmalonyl CoA metabolism. Therefore, reduced activity of cobalamin-dependent methyltransferase is the most likely cause of both the megaloblastic and neurologic defects.

Metabolism

In humans, the daily vitamin B_{12} requirement is approximately 0.5 μg. Vitamin B_{12} is closely associated with animal proteins; the only significant dietary sources of vitamin B_{12} are meats, milk or milk products, and eggs. In the stomach, vitamin B_{12} forms a complex with intrinsic factor (IF), a glycoprotein of M.W. ∼50 000.[38] When the vitamin B_{12}–IF complex reaches the distal ileum, it is bound by receptors on the surface of mucosal epithelial cells, and then enters the cells. Within the mucosal epithelial cells, the vitamin B_{12}–IF complex is dissociated, and the vitamin then passes into the plasma of the mucosal capillaries and thence to the portal vein. Almost all of the vitamin B_{12} is taken up by hepatocytes as the portal vein blood passes through the liver. B_{12} is stored within the liver, and released into plasma to meet physiological demands. If the quantity of vitamin B_{12} exceeds the capacity of hepatocyte receptors, most of the excess is excreted by the kidneys. Normally there is ∼1 mg of vitamin B_{12} stored in the liver, a quantity equivalent to the daily metabolic requirement for 2000 d. It is for this reason that, when the dietary supply of vitamin B_{12} is interrupted or mechanisms of absorption are impaired, vitamin B_{12} deficiency does not become evident for 5 years or more.

Three classes of vitamin B_{12} transport proteins are found in man: (1) *intrinsic factor,* which occurs only in gastric juice, (2) *transcobalamin II,* which occurs in plasma and other body fluids, and (3) *cobalophilin or "R-proteins,"* which are ubiquitous. An extensive review of these proteins has been published by Jacob, Baker, and Herbert.[44]

Intrinsic factor (IF) is a glycoprotein with a M.W. of ∼40 000–60 000. It is secreted by the parietal cells of the stomach and is required for the intestinal absorption of vitamin B_{12} in the distal ileum. Many other substances can bind vitamin B_{12}, but no other known substance has the property of transporting it across the intestinal wall. One molecule of IF binds one molecule of vitamin B_{12}. The binding reaction is highly specific and has an association constant of approximately 4×10^9 mol/L. At room temperature, equilibrium of this reaction is established within minutes and is not affected by acidic conditions down to pH 3. The gastric secretion of IF is stimulated by food, histamine, and gastrin; it is inhibited by vagal blockade. Pernicious anemia, the most important disorder of vitamin B_{12} metabolism, is due to a defect in the secretion of IF. Antibodies to IF can be demonstrated in a large number of patients, suggesting that pernicious anemia is an autoimmune disease. The ileal receptor for the IF–vitamin B_{12} complex has an association constant of approximately 5×10^9 mol/L between pH 6.4 and 8.4. The binding does not appear to be specific for the configuration of the vitamin B_{12} molecule, since complexes of intrinsic factor with various analogs of vitamin B_{12} bind equally well to the ileal receptors.[57]

The most important vitamin B_{12} transport protein in plasma is *transcobalamin II (TC II)*, a β-globulin. It is synthesized mainly in the liver but also in other tissues. TC II is a polypeptide with a M.W. of $\sim$38 000 with a single vitamin B_{12} binding site per molecule. TC II is less specific for vitamin B_{12} than is IF; it also binds cobalamins that are physiologically inactive. TC II transports vitamin B_{12} to receptors on cell membranes throughout the body. Binding is very rapid; if TC II–vitamin B_{12} is injected intravenously, it is almost completely cleared in one passage through tissues, mostly by the liver. The TC II–vitamin B_{12} complex enters the cell by pinocytosis. Lysosomal proteolysis degrades TC II and releases the vitamin B_{12}. Unbound vitamin B_{12} can also enter the tissue cells, but the process is much less efficient.[41]

Two types of vitamin B_{12} binders occur in human gastric juice, one with slow and one with rapid mobility in zone electrophoresis.[38] The slow component is *intrinsic factor* and the rapid component is designated *"R-proteins"* (R for rapid). Immunologically identical R-proteins occur in plasma, amniotic fluid, milk, saliva, ascitic fluid, and granulocytes. However, the granulocyte-derived protein can be differentiated from the other R-proteins electrophoretically. It is called transcobalamin III, whereas the R-protein from other sources is designated transcobalamin I. Collectively, these two binders are called *cobalophilins*. They are glycoproteins with M.W. between 60 000 and 150 000. Heterogeneity of R-proteins may be due to variations in the carbohydrate moieties (sialic acid residues) rather than in the apoproteins. They have one binding site per molecule and bind vitamin B_{12} analogs to some extent. In gastric juice at pH 2, the cobalophilins have much greater affinity than IF and bind almost all vitamin B_{12}. It has been postulated that cobalophilins aid in host defense against bacteria by depriving them of access to vitamin B_{12}, but no definite physiological function for these proteins is known.

Clinical Significance

Deficiency of vitamin B_{12} causes serious and often irreversible neurologic disorders such as burning pain or loss of sensation in the extremities, weakness, spasticity and paralysis, confusion, disorientation, and dementia. This condition has been given the name "*subacute combined degeneration* of the spinal cord." Neurologic symptoms may occur without any discernible hematologic changes in the blood. However, when hematologic changes are observed, they are usually accompanied by some neurologic changes such as accentuated reflexes or sensory changes. The classic morphologic changes in the blood, in approximate order of appearance, are hypersegmentation of neutrophils, macrocytosis, anemia, leukopenia, and thrombocytopenia, with *megaloblastic* changes in bone marrow accompanying the peripheral blood changes.

Decreased concentrations of serum vitamin B_{12} are frequently, but not always, associated with increased *mean corpuscular volume* (MCV) of erythrocytes. Routine application of electronic cell counters has made possible precise cell measurements on large numbers of patients. However, $\sim$20% of patients with vitamin B_{12} deficiency and 37% of those with megaloblastic bone marrows have normal or low MCV's. Also, $\sim$28% of patients with neurological disorders and low serum vitamin B_{12} concentration have normal MCV's. Therefore, one should not rule out vitamin B_{12} deficiency on the basis of a low or normal MCV.

Vitamin B_{12} deficiency may result from a variety of mechanisms. Most common are (1) lack of intrinsic factor secretion by the stomach and (2) intestinal malabsorption due to other causes. In the absence of intrinsic factor *(pernicious anemia)*, vitamin B_{12} cannot be absorbed from food. This is most common in middle-aged to elderly patients and affects $\sim$1% of such persons. However, cases of pernicious anemia have been observed in children, due either to failure of IF secretion or to secretion of biologically inactive IF. Identification of this type of vitamin B_{12} deficiency can be improved by the measurement of antibodies to IF or by the *Schilling test,* which measures vitamin B_{12} absorption with and without exogenous IF (see p. 1574). Absence of IF also results from total gastrectomy (usually done for gastric carcinoma) but is usually not seen after subtotal gastrectomy (usually done for peptic ulcer disease).

Intestinal malabsorption of vitamin B_{12} may be caused by gastrectomy or ileal resection, with an inverse relationship between the length of ileum resected and the absorption of vitamin B_{12}. Other causes of malabsorption are tropical sprue, inflammatory disease of the small intestine, and intestinal stasis with overgrowth of colonic bacteria which consume the vitamin B_{12} ingested by the host. In very rare cases, infestation of the small intestine by the fish tapeworm *Diphyllobothrium latum* can cause decreased absorption and thus vitamin B_{12} deficiency. This infestation still occurs frequently in Finland where the tapeworm is acquired from eating uncooked freshwater fish. Since ample quantities of vitamin B_{12} are present in all types of meat, milk, cheese, eggs, and other sources of animal protein, only the most extreme form of vegetarianism (*veganism*) can lead to vitamin B_{12} deficiency, and then only after several years of nutritional impoverishment.

FOLIC ACID

Structure and Terminology

Folate and folic acid are general terms for a family of compounds related to pteroic acid. Pteroic acid (Pte) is composed of a pteridine ring joined to a *p*-aminobenzoic acid residue (Figure 15-22). In basic solution, this substance has absorption maxima at 256, 282, and 365 nm and is fluorescent. When pteroic acid is conjugated with one molecule of L-glutamic acid, pteroylglutamic

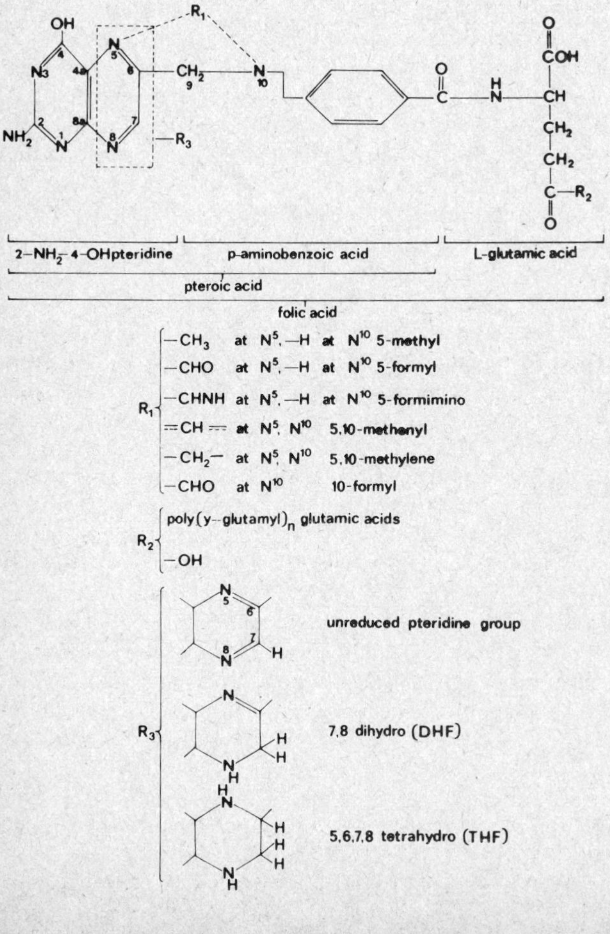

Figure 15-22. Structure and relationships of folic acid and its derivatives. The three basic subunits of folic acid are indicated. Biologically active derivatives have substituent groups at positions R_1, R_2, or R_3. For example, the physiologically active compound N^5N^{10}-methylene tetrahydrofolate heptaglutamate has a methylene bridge between N^5 and N^{10}; four additional protons at positions 5, 6, 7, and 8 in the pteridine ring (R_3); and six additional glutamate residues attached at position R_2 in the glutamic acid side chain. (Reprinted from Row, P.B.: Inherited disorders of folate metabolism. *In*: The Metabolic Basis of Inherited Disease. 5th ed. J. B. Stanbury, J. B. Wyngaarden, and D. S. Fredrickson, Eds. New York, McGraw-Hill, 1983; with permission.)

acid is formed (PteGlu). It can be reduced to dihydrofolic acid (H_2PteGlu or FH_2) with hydrogens in positions 7 and 8 or to tetrahydrofolate (H_4PteGlu or FH_4) with hydrogens in positions 5, 6, 7, and 8. Only the reduced forms are biologically active. Although various forms of folic acid are normally present in human serum and other body fluids, the principal form is N^5-methyl-tetrahydrofolate. It is slowly oxidized in alkaline solution, but this can be reversed by adding ascorbic acid. It is relatively stable in acid solutions but unstable when exposed to light.

Other folate derivatives have multiple glutamic acid residues (H_4PteGlu$_n$), where n may be 1–7. Biochemically, these *polyglutamates* are similar to monoglutamates, but the former function as the natural coenzymes. Multiple forms of folic acid occur with substitutions of functional groups such as methyl, formyl, methylene, or hydroxymethyl at various nitrogen atoms in the pteroic acid residue. Separation of the various forms can be done by column chromatography or high performance liquid chromatography.

Biochemical Functions

Folate coenzymes are essential for the transfer of single carbon units. Five of the major reactions are conversion of serine to glycine, catabolism of histidine, and synthesis of thymidylate, methionine, and purine. These reactions are illustrated in Figure 15-23. Different folates are involved in these reactions depending on the chemical state of the single carbon fragments transferred, as follows:

Reaction	Group Transferred		Folic Acid Derivative
Serine/glycine metabolism	Methylene	($—CH_2—$)	N^5, N^{10}-methylene FH_4
Histidine catabolism	Formimino	($—CHNH$)	N^5-formimino FH_4
Thymidylate synthesis	Methylene	($—CH_2—$)	N^5, N^{10}-methylene FH_4
Methionine synthesis	Methyl	($—CH_3$)	N^5-methyl FH_4
Purine synthesis	Methenyl	($=CH—$)	N^5, N^{10}-methenyl FH_4
	Formyl	($—CHO$)	N^{10}-formyl FH_4

The interconversion of these forms of folic acid takes place through various electron transfer reactions facilitated by specific enzyme systems and coenzymes such as $FADH_2$ and NADPH. (See Figure 15-24.) The conversion between the N^5, N^{10}-methylene form and N^{10}-formyl forms is readily reversible, but the reduction of methylene to methyl and reduction of free tetrahydrofolate to formyltetrahydrofolate are irreversible or nearly so. The conversion of N^5-methyltetrahydrofolate back to free tetrahydrofolate may require cobalamin.

Metabolism

Folate is absorbed from dietary sources, such as green, leafy vegetables. Polyglutamate forms of folate present in food are first converted to monoglutamates by *pteroylpolyglutamate hydrolase* in the intestinal mucosa. Following cellular uptake, most of the folate is reduced and methylated, and enters the circulation as N^5-*methyltetrahydrofolate*. Dietary sources provide ~500 μg of folate/day. The human daily requirement for folate is ~50 μg. Body stores of folate amount to ~5 mg. Interruption of absorption of folate may, therefore, result in deficiency within a few months.

Folic acid and vitamin B_{12} metabolism are linked by the reaction which transfers a methyl group from N^5-methyltetrahydrofolate to cobalamin. In cases of cobalamin deficiency, folate is "trapped" as N^5-methyltetrahydrofolate. It is "metabolically dead" in the absence of vitamin B_{12} and cannot be recycled as tetrahydrofolate back into the folate pool to serve as the main 1-carbon unit acceptor for many biochemical reactions. This deficiency eventually leads to depletion of methylenetetrahydrofolate, causing a reduction in thymidylic acid synthesis, which results in megaloblastic anemia and neuropathies. This concept is supported by the fact that tetrahydrofolate will correct the megaloblastic anemia in patients with congenital methylmalonic aciduria and homocystinuria, whereas methyltetrahydrofolate will not. However, some investigators have suggested that vitamin B_{12} is required for the conversion of folic acid to the formyl form and that formyltetrahydrofolates are the natural substrates for forming folate polyglutamates. They have shown that inactivation of methionine synthetase can be corrected in laboratory rats with the formyl, methenyl, or formative derivatives but not with tetrahydrofolate or methyltetrahydrofolate.

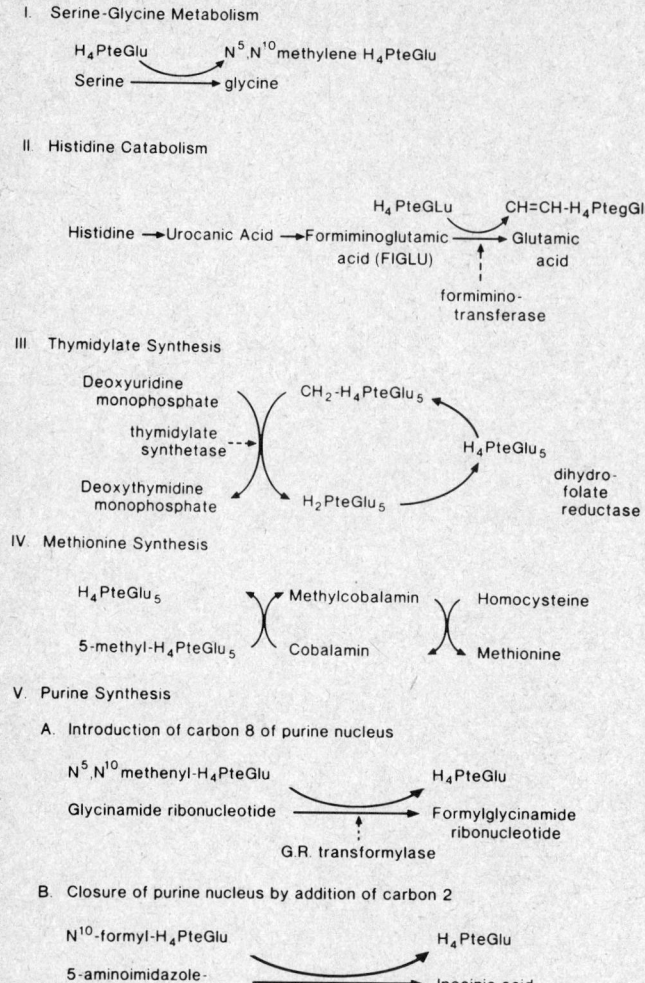

Figure 15-23. The five major metabolic functions of folate in human cells. See text for details.

Both high- and low-affinity folate binders are found in plasma. The high-affinity binding protein has a M.W. of ~35 000 and an affinity constant of 10^{10} mol/L. Another specific binder has a M.W. >200 000 and migrates electrophoretically as an α_2-macroglobulin. However, these specific binders carry <0.2 ng of folate/mL of plasma. About half of the serum folate is bound to a nonspecific low-affinity binder, presumably albumin. Specific binders have been found in the serum of only 15% of normal subjects, although ~30% of normal subjects have folate binders

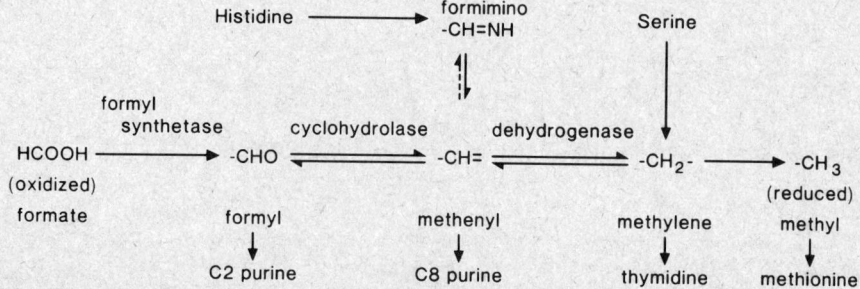

Figure 15-24. The metabolism and interconversion of various forms of folate. Formate is the most oxidized form, and the formyl, methenyl, methylene, and methyl represent progressive reduction states. These various forms are directly related to the metabolic functions illustrated in Figure 15-23. The formyl form closes the purine nucleus by addition of carbon 2. The methenyl form adds carbon 8 to the purine nucleus. The methylene form is a coenzyme in thymidylate synthesis and the methyl forms are coenzymes in methionine synthesis. (Adapted from Chanarin, I., et al.: Lancet 2:505, 1980.)

in their granulocytes. The functional role of these binders is unknown, but they may serve to reduce folate loss from renal filtration, and they possibly could have a role in bacterial defense. Serum folate binding capacity increases in chronic granulocytic leukemia, acute hepatitis, cirrhosis, uremia, and pregnancy. Unlike vitamin B_{12}, no clinical disorders have been reported which are related to the absence of absorption or transport binding proteins for folic acid.

Clinical Disorders

Deficiency of folate may result from the absence of intestinal microorganisms (gut sterilization), poor intestinal absorption (e.g., after surgical resection or in celiac disease or sprue), insufficient dietary intake, excessive utilization (as in pregnancy, liver disease, and malignancies), and administration of antifolate drugs (e.g., methotrexate).

Megaloblastic anemia is the major clinical manifestation of folate deficiency, although sensory loss and neuropsychiatric changes also occur. The combined subacute degeneration of the posterior spinal column does not occur with folate deficiency. Combined deficiency of both folate and iron commonly occurs in malnourished people. In such cases, macrocytosis of erythrocytes, otherwise typical of folic acid deficiency, is not observed.

Determination of Vitamin B_{12} and Folic Acid

Principle

Two types of methods used for the quantitation of vitamin B_{12} and folic acid are microbiological assays and competitive protein binding (CPB) radioassays. The latter have largely replaced the less convenient microbiological assays.

The microbiological assays are reference methods for the determination of biologically active vitamin B_{12} and folate. Microorganisms most widely used for assaying vitamin B_{12} are *Euglena gracilis, Lactobacillus leichmannii,* or a mutant of *Escherichia coli.* Microbiologic assays for folic acid utilize *Lactobacillus casei, Streptococcus faecalis,* or *Pediococcus cerevisiae.* Unfortunately, each of these organisms is susceptible to growth inhibition by antibiotics or other drugs such as methotrexate that may be in a patient's serum. Furthermore, these assays require at least 24 h to establish adequate growth of the microorganism.

Microbiological assays measure the amount of growth of a microorganism which is dependent on the amount of vitamin B_{12} or folic acid in the growth medium. The turbidity caused by the growth of the organism in the presence of the patient's specimen, compared to the growth observed in media containing known standard concentrations of vitamin B_{12} or folic acid, can be used as a measure of the concentration of vitamin B_{12} or folic acid in the patient sample. In addition to the technical considerations, imprecision in bioassays is considerably greater than in CPB methods.

Several commercial kits are available for the *competitive protein binding assays* of vitamin B_{12} and folic acid. Most provide for the simultaneous measurement of both substances in the same reaction tube. The binder used in the folate assay is a protein that occurs naturally in milk, called β-lactoglobulin or milk folate binder. The vitamin B_{12} binder is nonhuman intrinsic factor (IF), usually obtained from hog stomach. If the IF is not highly purified, it may contain R-proteins which will bind vitamin B_{12} as well as related metabolically inactive compounds (vitamin B_{12} analogs) that may be present in the sample, thereby causing artificially elevated vitamin B_{12} results. To measure only metabolically active vitamin B_{12} ("true B_{12}") in CPB assays, either the IF must be highly purified or *cobinamide* (B_{12} analog) must be added to the IF to saturate all binding sites on the R-proteins. Cobinamide is not bound by IF since IF is highly specific for true B_{12}.* This elimination of R-protein binding sites prevents interferences in the assay caused by B_{12} analogs which may be present in the sample or by other binders which may be present in the IF preparation.

In the most widely used CPB assays for simultaneous determination of folate and vitamin B_{12} levels, folate (principally 5-methyltetrahydrofolate) competes with ^{125}I-labeled folate for a limited number of binding sites on the β-lactoglobulin binder, and B_{12} (cobalamin) competes

---

*The National Committee for Clinical Laboratory Standards recommends that methods using purified IF should read < 75 pg/mL when measuring B_{12} analogs at a concentration of 10 000 pg/mL. Also, binding should be inhibited > 95% when specific anti-IF antibodies are added. Methods using cobinamide to block R-proteins also should meet these criteria. If binders other than IF are used, they cannot be blocked with IF antibodies, and it is necessary for the manufacturer to demonstrate directly that they will not measure B_{12} analogs in human serum.

with ^{57}Co-labeled cobalamin for a limited number of binding sites on IF. As a preliminary step, the specimen containing folate and vitamin B_{12} is boiled in a buffered solution containing dithiothreitol, potassium cyanide, and the ^{125}I- and ^{57}Co-labeled tracers. Boiling releases folate and vitamin B_{12} from endogenous binding proteins and inactivates these proteins to prevent their interference in the reaction. Also, folate is reduced and stabilized by dithiothreitol, and serum vitamin B_{12} is converted by reaction with potassium cyanide to the more stable cyanocobalamin. Alternatively, some procedures irreversibly denature endogenous binding proteins by increasing the pH to 12–13 and then readjusting the pH to 9.3 before the binding reagent is added, thereby avoiding a boiling step to release folate and vitamin B_{12}. Next, buffered binder reagent is added and the reaction mixture is incubated at pH 9.3 to allow the competitive reactions to take place. Subsequent separation of bound and free folate and vitamin B_{12} is achieved by contact with dextran-coated charcoal which adsorbs the free (unbound) molecules, leaving protein bound folate and vitamin B_{12} in the solution. Some procedures eliminate charcoal separation by attaching the folate binder and IF to solid-phase particles such as glass or polyacrylamide beads. The free folate and vitamin B_{12} then remain in the supernatant and the bound analytes become part of the solid-phase suspension. Both charcoal and solid-phase particles can be separated from the reaction mixture by centrifugation. For methods employing charcoal, radioactivity of the supernatants is measured, while in solid-phase radiosorbent assays, the radioactivity of the pellet is measured. For the performance of the simultaneous folate/vitamin B_{12} measurement, a γ-scintillation counter must be used which discriminates between the disintegrations produced by the ^{57}Co and ^{125}I tracers, since both are present in the samples to be counted.

Maximum binding (highest counts per minute) of radiolabeled tracer occurs in the standard solution containing no unlabeled analyte (zero standard) because the tracer has no competition for the protein binding sites. Thus, higher standards will give lower bound counts per minute (cpm). The per cent of maximum binding (%B) can be calculated for each sample by dividing sample cpm by the zero standard cpm. A standard curve is prepared by plotting either bound cpm or %B for each standard against its concentration. Concentrations of assayed samples are determined by interpolation from the standard curve.

Specimen Collection and Storage

Serum samples are recommended for this procedure; however, plasma samples collected in EDTA may also be used. Heparin should not be used as an anticoagulant since it has a modest vitamin B_{12}–binding ability. Also, the lithium salt of heparin may stimulate a sharp increase in release of vitamin B_{12}–binding proteins by granulocytes, which may interfere with the vitamin B_{12} assay. Samples should be collected from fasting individuals because recent food intake may appreciably increase the folic acid level.

Collect blood and allow to clot, then centrifuge for 15 min at approximately $900 \times g$ to obtain at least 300 μL of unhemolyzed cell-free serum. Protect specimens from light to prevent folate deterioration and store at 2–8 °C up to 4 h or store frozen (-20 °C) for longer periods.

For the erythrocyte folate assay, collect the blood in a tube containing EDTA as anticoagulant. Determine and record the hematocrit value. Add 100 μL of well-suspended blood from the tube to 2 mL of ascorbic acid solution, 0.2 g/dL. (Note: dilution factor = 21.) Mix several times by gentle inversion; avoid foaming. Allow the sample to stand at room temperature (20–27 °C) protected from light for 60–90 min for complete hemolysis. Hemolyzed specimens may be stored at 2–8 °C up to 3 h or stored frozen (-20 °C) for longer periods.

Reagents and Procedure

A variety of CPB procedures for the measurements of vitamin B_{12} and folate are available commercially in kit form. These kits contain all reagents and detailed instructions for their use in performing the assays; thus, no details are given here.

Calculations

Serum or Plasma Vitamin B_{12} and Folate

Calculation of results is ideally performed using computer-assisted programs, such as those based on log-logit, spline, or other functions. When such equipment is not available, manual plotting of data on log-logit paper is satisfactory. For this, the manufacturer's instructions may be followed.

Red Blood Cell Folate

1. Obtain the concentration of folate (using hemolysate instead of serum) calculated by one of the above methods and multiply it by the dilution factor of 21. This gives the folate concentration ($\mu g/L$) in the whole blood specimen. (Note: if a simultaneous assay for vitamin B_{12} and folate is used, ignore the vitamin B_{12} result obtained on the hemolyzed specimen.)

2. Divide the folate concentration of whole blood by the hematocrit (expressed as a fraction). This gives the folate concentration ($\mu g/L$) of the packed red blood cells. The contribution of serum folate to the total folate measured is ignored since it is insignificant compared to RBC folate.

$$\text{Folate, } \mu g/L \text{ of packed red cells} = \frac{(\mu g/L \text{ of hemolysate}) \times 21}{\text{hematocrit}}$$

Reference Ranges

Depending on the laboratory and the procedure used, much variability in the reference range is expected. Representative reference ranges for adults in the fasting state are: serum folate, 1.8–9.0 $\mu g/L$; RBC folate, 150–450 $\mu g/L$; serum vitamin B_{12}, 200–1100 ng/L. Serum vitamin B_{12} concentration declines significantly as a function of age in healthy adult males, so interpretation of serum vitamin B_{12} concentration must be based on age-matched reference ranges. The following table indicates reference ranges for healthy adult males by decade, using one of the commercial kits.[29] Since different kits have substantially different reference ranges, the ranges shown here cannot be used for all assays. However, the need to determine reference values by age is stressed.

Ages in years (Men)	Serum vitamin B_{12}, ng/L (Central 95% Interval)
20–29	281–1079
30–39	248–965
40–49	218–863
50–59	191–770
60–69	168–687
70–79	152–630

Each laboratory should establish reference ranges using its own selected procedures for measurement of folate and vitamin B_{12}.

Comments and Precautions

Persons with untreated pernicious anemia usually have vitamin B_{12} concentrations of < 100 ng/L. Those with folate deficiency usually have fasting serum folate levels of 1.0 $\mu g/L$ or less. Indeterminate results (B_{12} of 100–150 ng/L or serum folate of 1.0–2.0 $\mu g/L$) are not uncommon. Because these deficiency disorders develop gradually, patients with results in the indeterminate range should have their serum reassayed in a few months, except in cases in which a diagnosis can be clearly established by other means.

Low serum vitamin B_{12} concentration is not necessarily indicative of vitamin B_{12} deficiency. Most of the vitamin B_{12} in serum is bound to transcobalamin-I (TCI), which is released by granulocytes and has no functional role in transport of vitamin B_{12} to cells. Low serum vitamin B_{12} concentration may be due to reduction in TCI as a consequence of low total granulocyte mass. This condition has been observed in benign neutropenia, multiple myeloma, and leukemic reticuloendotheliosis, and may be expected in other conditions in which bone marrow is hypoplastic, aplastic, or replaced by malignant cells. In the future, it may be preferable to measure the TCII-bound serum vitamin B_{12} rather than the total vitamin B_{12}. Methods are being developed for this purpose.

OTHER LABORATORY TESTS FOR DETECTION OF VITAMIN B_{12} DEFICIENCY

The measurement of urinary methylmalonic acid (MMA), the deoxyuridine (dU) suppression test, and cytochemical staining of erythrocyte precursors are ancillary methods for assessing vitamin B_{12} status.

Methylmalonic Acid Excretion

Principle

Since vitamin B_{12} is needed for the conversion of MMA to succinic acid, patients deficient in vitamin B_{12} excrete excess amounts of MMA in their urine. To standardize the procedure, patients should be given a loading dose of valine (a precursor of MMA) prior to the collection of a 24-h urine specimen. The test is of very limited diagnostic value since approximately 30% of patients with vitamin B_{12} deficiency have normal MMA excretion.

Reference

Chanarin, I., England, J. M., Mollin, C., et al.: Br. J. Haematol. 25:45-53, 1973.

dU Suppression Test

Principle

The dU suppression test measures the effect of prior addition of deoxyuridine (dU) on the uptake of radiolabeled thymidine into the DNA of cultured bone marrow cells. Normal cells that contain vitamin B_{12} can convert dU to thymidine and, therefore, do not take up as much thymidine. Bone marrow cells of patients who are deficient in vitamin B_{12} show less suppression than do normal patients. Peripheral blood lymphocytes also can be used for this test. Because it is relatively time-consuming, the dU suppression test has not become widely available as a diagnostic test.

Reference

Ganeshaguru, K., and Hoffbrand, A. V.: Br. J. Haematol., 40:29–41, 1978.

Cytochemical Stain for Homocysteine

Principle

Patients deficient in vitamin B_{12} cannot metabolize homocysteine to methione. Homocysteine accumulates and can be visualized as a brown color in erythrocytes by a specific cytochemical staining reaction with nickel chloride. Wright's stained blood smears can be destained and used for this reaction. As yet, experience with this procedure is very limited; since only four patients with pernicious anemia have been tested we cannot judge its reliability for detection of vitamin B_{12} deficiency.

Reference

Omran, N., and Neumann, E.: Blut, 39:359–363, 1979.

Vitamin B_{12} Absorption (Schilling) Test

Principle

The vitamin B_{12} absorption test permits differentiation of pernicious anemia or intestinal malabsorption as causes of vitamin B_{12} deficiency. The proportion absorbed from orally administered ^{57}Co- or ^{58}Co-labeled vitamin B_{12} can be measured by determining the radioactivity in feces, urine, or serum, or by externally scanning the liver. The usual procedure is to measure radioactivity in a 24-h urine sample, which is collected after oral administration of 0.5 μg of radioactive cobalt-labeled vitamin B_{12} following an overnight fast. The patient is also injected intramuscularly with a "flushing dose" of 1000 μg unlabeled vitamin B_{12} to saturate the binding capacity of plasma and liver for vitamin B_{12} and to cause urinary excretion of the absorbed vitamin B_{12}. When the oral test dose of vitamin B_{12} is 0.5 μg, normal individuals excrete in the urine $\geq 8\%$ of the dose administered, whereas in persons with pernicious anemia, $< 7\%$ (0–3%) is excreted. A confirmatory test for lack of IF requires ingestion of vitamin B_{12} *and* IF and repeat of the flushing dose one or two days after the initial vitamin B_{12} absorption test. Oral administration of IF simultaneously with vitamin B_{12} increases the absorption of vitamin B_{12} by the intestinal tract and thus the urinary excretion to $\geq 8\%$ of the dose administered. In patients with malabsorption, oral administration of IF does not increase the per cent of the dose excreted. However, approximately 25% of patients with pernicious anemia also have intestinal malabsorption that is secondary to vitamin B_{12} deficiency and is reversible following weeks or months of vitamin B_{12} therapy. Thus, a low value for vitamin B_{12} absorption when administered with IF does not exclude pernicious anemia.

Reference

Schilling, R. F.: J. Lab. Clin. Med., *42*:860, 1953.

Tests for Intrinsic Factor–Blocking Antibodies

Principle

Intrinsic factor–blocking antibodies have been observed in the serum of 50–60% of patients with pernicious anemia, but rarely in any other condition.[15,31] The demonstration of these antibodies in the serum is valuable for confirmation of pernicious anemia; negative results do not exclude pernicious anemia. A kit for detection of intrinsic factor–blocking antibody in serum is available from Corning Medical, Medfield, MA 02052. The manufacturer's procedure should be followed.

Reference

Irvine, W. J.: Clin. Exp. Immunol., *1*:99-117, 1966.

IRON METABOLISM AND IRON COMPARTMENTS

Despite the abundance of iron in the environment and in our erythrocytes, the metabolism of iron resembles in many respects that of a trace element. Normally, very small quantities are present in most cells of the body, in plasma, or in other extracellular fluids, and the body rigorously conserves its hoard of iron, so that less than a thousandth part of the body iron content is lost daily. Figure 15-25 illustrates the pathways of iron metabolism and the daily rates of exchange between various iron compartments.

IRON COMPARTMENTS

Hemoglobin

Approximately 2.5 g of iron is in hemoglobin, an iron-protein that contains 0.34% iron by weight. Normally, virtually all hemoglobin iron is contained within erythrocytes or their precursors in the bone marrow.

Transport

Transport of iron from one organ to another is accomplished by a plasma iron-transport protein called *apotransferrin*. This is a β_1-globulin of M.W. 75 000 that has two iron-binding

Figure 15-25. Principal pathways of iron metabolism in humans. Figures indicate approximate number of milligrams that enter or leave the various compartments daily. Omitted from this diagram for simplicity are the very small "tissue iron" compartment, which is also in equilibrium with the plasma iron pool, and the hypothetical "labile iron pool." These are described in the text.

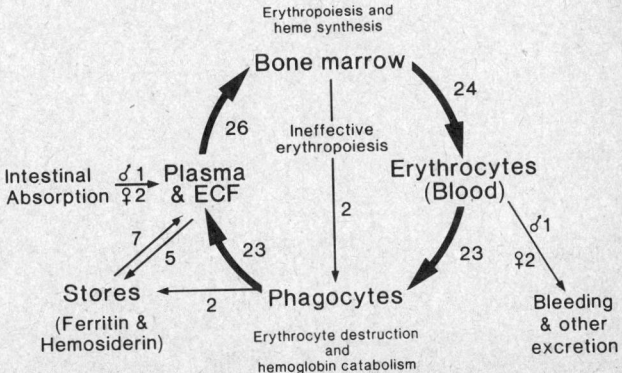

sites per molecule. Each of these sites may bind one Fe(III) ion together with one ion of HCO_3^-. The apotransferrin-Fe(III) complex is called *transferrin*. Normally there is a total of ~2.5 mg of iron in plasma. Transferrin also occurs within the cytosol of many cells and is believed to serve there also as an intracellular iron-transport protein.

Storage

Ferritin is the major iron-storage compound. It is a spherical molecule consisting of an *apoferritin* shell and an interior *ferric oxyhydroxide* crystalline core. The apoferritin shell is composed of 24 subunits or monomers. It is about 13 nm in diameter, and its interior cavity is about 7 nm in diameter. Six pores of approximately 0.7–1.0 nm diameter permit ingress and egress of molecules such as Fe(II), ascorbic acid, and flavin mononucleotide. The edges of the pores serve as enzymatic binding sites for iron. As two Fe(II) ions enter the pore, they are oxidized to FeOOH and released to be added to the surface of the core crystal. The FeOOH core crystal may contain as many as 4000 iron atoms, but usually 2000 or fewer. Release of iron from ferritin is probably nonenzymatic and may involve reduction by reduced flavin mononucleotide or other reducing substances. The resultant Fe(II) leaves the crystal and diffuses out through a pore of the apoferritin shell. The oxidation or reduction of iron takes place rapidly. Thus, ferritin is both a very efficient iron trap and a readily available source of iron for metabolic requirements.

Ferritin occurs in nearly all cells of the body. In hepatocytes of the liver and in the macrophage system of the bone marrow and other organs, ferritin provides a reserve of iron readily available for formation of hemoglobin and other heme proteins. In adult men the total body storage iron, mostly as ferritin, is ~800 mg; in healthy adult women storage iron ranges from 0–200 mg. Minute quantities of ferritin are also present in serum in concentrations proportional to total body storage iron. Liver injury results in release of relatively large amounts of ferritin into plasma.

Hemosiderin, the other form of storage iron, is aggregated, partially deproteinized ferritin. In contrast to ferritin, hemosiderin is insoluble in aqueous solutions, a difference that has traditionally been used to distinguish these two iron-storage compounds. Iron is only slowly released from hemosiderin, probably because it occurs in relatively large aggregates and therefore has a much smaller surface:volume ratio. Like ferritin, hemosiderin normally occurs predominantly in cells of the liver, spleen, and bone marrow.

"Tissue Iron"

Numerous cellular enzymes and coenzymes require iron, either as an integral part of the molecule or as a cofactor. Notable are the *peroxidases* and *cytochromes*, all of which are, like hemoglobin, heme proteins. Nearly half the enzymes of the *Krebs cycle* require iron. These enzymes and coenzymes, which occur in all nucleated cells of the body, are referred to collectively as the "tissue iron compartment." Tissue iron normally amounts to ~8 mg. Although a small compartment, it is metabolically critical; tissue iron diminishes early in the course of iron deficiency.

Myoglobin

Myoglobin very closely resembles a subunit of hemoglobin. In contrast to hemoglobin, myoglobin does not form tetramers and is not influenced by the concentration of 2,3-DPG. Total myoglobin iron content is normally ~130 mg.

"Labile Pool"

There is normally ~80 mg of iron in the "labile pool." This compartment has no clear anatomic location; it is a concept derived from kinetic measurements with radio-labeled iron. It is thought possibly to be either iron that is in the lymphatic circulation or a cytoplasmic iron storage compound other than ferritin, present in most cells, that rapidly takes up or releases iron.

ABSORPTION, TRANSPORT, EXCRETION

The average American diet provides 10–15 mg of iron daily, mostly in the form of the heme proteins, hemoglobin and myoglobin, in meat. In the past a significant component of iron in the diet came from inorganic iron leached from iron utensils. Normally ~1 mg of iron is absorbed each day. Absorption occurs principally in the duodenum. Heme is absorbed directly as such. To be absorbed, inorganic iron must be in the ferrous state.

Both ferritin and transferrin are present in the absorptive cells of the intestinal mucosa and are believed together to regulate iron absorption. When body iron stores are high, the ferritin content of mucosal epithelium is also high, and the transferrin content is low. Iron that enters mucosal cells is trapped in ferritin and lost when the mucosal cell is sloughed into the intestinal lumen. This mechanism reduces iron absorption when body stores of iron are already increased. Conversely, when there is iron deficiency, mucosal cell content of apoferritin is diminished, transferrin (and apotransferrin) content is increased, and iron absorption is accelerated.

The major pathway of iron metabolism (Figure 15-25) is a virtually closed cycle in which iron passes from plasma transferrin to the erythrocyte precursors in the bone marrow, where it is incorporated into hemoglobin; these cells then enter the circulation as mature erythrocytes, where they remain nearly four months before they become metabolically "worn out" and are engulfed by phagocytes; the iron is released from hemoglobin and returns to plasma transferrin, thus completing one cycle and beginning another. From this iron cycle, small quantities of iron are diverted for use in other iron compartments; there is also a slow continual exchange between the storage and transport compartments.

Each day about 1–2 mg of iron absorbed from the intestinal tract enters this cycle to compensate for the 1–2 mg of iron lost each day from the body. Most of the iron loss is due to minute quantities of iron present in epithelial cells and erythrocytes in urine and feces. With each menstrual cycle young women lose ∼40–80 mL of blood, which is equivalent to 20–40 mg of iron. This loss must be made up by increased absorption of iron by the intestinal mucosa. Similarly, about 600–900 mg of iron is lost as a consequence of each pregnancy. Compensation for iron loss from pregnancy and menstruation is often difficult, particularly since many women consume iron-poor diets. Thus, iron deficiency is very common in women, even those of affluence whose diets seem adequate.

DISORDERS OF IRON METABOLISM

IRON DEFICIENCY

Iron deficiency is one of the most prevalent disorders of humans. It is particularly a disease of children, of young women, and of older persons, but it occurs in persons of all ages and all social strata. In children it is frequently due to dietary deficiency, because milk has a low iron content. In adults it almost always indicates chronic blood loss.

Measurements of *serum iron concentration* and *total iron binding capacity* have been widely used as aids in the diagnosis of iron deficiency. However, assay of *serum ferritin concentration* is a much more sensitive and reliable means of demonstration of this disorder. *Free erythrocyte protoporphyrin* (FEP) concentration is increased in patients with iron deficiency or lead poisoning, whereas it is normal in thalassemia minor, a condition often confused with iron deficiency. Measurement of FEP is used as a diagnostic aid and can be performed rapidly on small samples of blood. The most reliable method for diagnosis of iron deficiency is cytochemical staining of bone marrow aspirate by the Prussian blue reaction, which demonstrates presence or absence of hemosiderin. The cytochemical stain for hemosiderin is evaluated microscopically.

IRON OVERLOAD

Hemosiderosis is a term that implies iron overload without associated tissue injury. *Hemochromatosis* implies iron overload with injury to involved organs, as manifested by cellular degeneration and fibrosis. Iron overload most commonly results from chronic excessive absorption of iron from a normal diet. It is rarely the result of protracted (many years) ingestion of iron medication or inappropriate injections of iron.

Hereditary hemochromatosis is the classic disorder of iron overload. It is due to an inborn error of iron absorption; the precise mechanism remains unknown. Approximately 10% of Caucasians of Northern European origin are carriers of a gene conferring susceptibility to hereditary hemochromatosis; 2 or 3 per thousand are homozygous, although only a portion of these will develop the clinical disease. This condition, of lifelong duration, is usually manifested by hyper-pigmentation of skin, diabetes mellitus, cirrhosis of the liver, pituitary insufficiency, hypogonadism, arrhythmias of the heart or congestive heart failure, and arthritis. Some, but usually not all, of these features commonly appear after the age of 20 years in males or after menopause in women. It is important to establish the diagnosis of this condition early, since most patients, when untreated, die within a few years. Treatment is by removal of 500 mL of blood once or twice weekly, usually until a total of ~50 000 mL of blood has been removed. If such a phlebotomy program is instituted early enough, some of the serious complications of hemochromatosis can be avoided, and the prognosis is substantially improved.

Sideroblastic anemia is a group of iron-loading disorders of unknown cause. Abnormal heme synthesis by mitochondria is postulated. According to this theory, iron accumulates in mito-chondria because of this metabolic bottleneck.

Ascertainment of iron-loading disorders is best made by measurement of serum iron con-centration and total iron binding capacity. Usually the serum iron concentration is > 200 μg/dL (35.8 μmol/L) in persons with iron overload. Transferrin saturation of > 55% usually indicates iron overload. In advanced iron-overload states the transferrin saturation often exceeds 90%. Compared with measurement of transferrin saturation, the serum ferritin assay is less sensitive for ascertainment of early iron overload, and the free erythrocyte protoporphyrin test has no role at all in the diagnosis of iron overload.

The Determination of Serum Iron and Iron Binding Capacity

Clinical Significance

Serum iron concentration connotes principally the Fe(III) bound to serum transferrin and does not include the iron contained in serum as free hemoglobin. Serum iron concentration is decreased in many, but not all, patients with *iron deficiency anemia* and in *chronic inflammatory disorders* such as *acute infection, immunization,* and *myocardial infarction* (Table 15-6). Serum iron concentration diminishes markedly in patients who are beginning to respond to specific hematinic therapy for anemias of other causes, for example, *treatment of pernicious anemia* with vitamin B_{12}. Acute or recent hemorrhage, including that due to *blood donation,* results in low serum iron concentration. Serum iron concentration drops at time of *menstruation.* Use of hor-monal contraceptives raises serum iron concentration, but on cessation of contraceptive hormone intake, serum iron concentration decreases as much as 30% concurrent with uterine bleeding.

Greater than normal concentrations of serum iron occur in iron loading disorders such as *hemochromatosis,* in *acute iron poisoning* in children, and following *oral ingestion of iron medication* or *parenteral iron administration* or *acute hepatitis.* For example, one 0.3-g tablet of ferrous sulfate ingested by an adult may raise the serum iron concentration by 300–500 μg/dL (50–90 μmol/L).

Since normally only about one third of the iron-binding sites of transferrin are occupied by Fe(III), serum transferrin has considerable reserve iron binding capacity. This is called the serum unsaturated iron binding capacity (UIBC). The total iron binding capacity (TIBC) is a mea-surement of the maximum concentration of iron that serum proteins, principally transferrin, can bind. The serum TIBC varies in disorders of iron metabolism. It is often increased in iron deficiency and decreased in chronic inflammatory disorders or malignancies.

Principle

Serum Iron Assay

Iron is released from transferrin by reduction in pH of the serum. Proteins and apotransferrin are removed by precipitation and centrifugation. The Fe(III) of the supernatant is reduced to Fe(II) with thioglycolic acid. Fe(II) is complexed with a chromogen that contains the reactive group —N=C—C=N—. (See Figure 15-26.) The iron-chromogen complex has very high ab-sorptivity and is proportional to iron concentration absorbance. Table 15-7 lists some of the chromogens that may be used for serum iron assay. The procedure that follows employs *batho-phenanthroline* (4,7-diphenyl-1,10-phenanthroline). Whereas bathophenanthroline is poorly solu-ble in water, the sodium disulfonate of bathophenanthroline is quite water soluble. A buffer is added to adjust pH to 4.5, ensuring maximum absorptivity of the Fe(II)-bathophenanthroline complex. The reactions involved are:

Transferrin-Fe(III) + HCl-trichloroacetic acid → Fe(III) + transferrin ↓

Fe(III) + thioglycolic acid → Fe(II)

Fe(II) + bathophenanthroline (colorless) → bathophenanthroline-Fe(II) (colored)

Table 15-6. SERUM IRON, TIBC, AND TRANSFERRIN SATURATION IN VARIOUS CONDITIONS

	Serum Iron	TIBC	Transferrin Saturation
*Ingestion of medicinal iron**	↑	N	↑
Parenteral iron administration†	↑	↑	Var
Iron deficiency states			
Chronic blood loss	↓N	↑N	↓N
Acute blood loss	N↓	N	N↓
Post–subtotal gastrectomy	N↓	↑N	↓N
Dietary deficiency (infants and children)	N↓	↑N	↓N
Polycythemia vera	↓N	↑N	↓N
Kwashiorkor	↓	↓	Var
Pregnancy (without Fe supplement)	↓N	↑N	↓N
Iron overload states			
Acute iron poisoning	↑	N	↑
Chronic iron overload	↑	N↓	↑
(Hemosiderosis, hemochromatosis, thalassemia major, sideroblastic anemias, other chronic anemias following numerous transfusions)			
Chronic disorders			
Chronic infections	↓N	↓N	N↓
Rheumatoid arthritis	↓N	↓N	N↓
Malignancies	↓N	↓N	N↓
Laennec's cirrhosis	↓N	↓N	N↓
Acute disorders			
Acute infections	↓	N	↓
Myocardial infarction	↓	N	↓
Hepatitis or hepatic necrosis‡	↑	↑N	Var
Physiological variations			
Diurnal rhythm			
Morning	N	N	N
Evening	↓	N	↓
Menstruation	↓	N	↓
Premenstrual interval	↑	N	N
Contraceptives (except when bleeding)	↑	↑	N
Pregnancy (with Fe supplement)	↑	↑	N
Miscellaneous			
Pernicious anemia			
Untreated	↑	N↓	↑
Early treatment (first week)	↓	N↓	↓N
Treated	N	N	N
Congenital atransferrinemia (*very* rare)	↓	↓	↑
Acquired atransferrinemia (from nephrotic syndrome)	↓	↓	↑
Hemolytic anemias	Var	Var	Var

For each entry in this table, the most characteristic change is indicated first. ↓ = decreased concentration. ↑ = increased concentration. N = normal results. Var = variable results.

*The ingestion of a single tablet or capsule of an iron-containing medicinal can raise the serum iron concentration by 300–500 μg/dL, an effect that often masks iron deficiency. Peak serum iron concentration occurs a few hours after ingestion; the effect is most pronounced in iron-deficient subjects.

†Increased serum iron concentration may be observed for weeks following an injection of iron dextran. The mechanism of this has not been determined; most likely it reflects iron dextran in plasma. The increase in TIBC characteristically observed with automated methods probably does not reflect increased transferrin but a methodological artifact (additional release of iron from iron dextran as a result of the procedure?), since the unsaturated iron binding capacity is normal in this situation.

‡Increased serum iron concentration is the result of extreme hyperferritinemia consequent to hepatocellular injury. The increase in TIBC characteristically observed with automated methods probably does not reflect increased transferrin but a methodological artifact (additional iron release from ferritin as a result of the procedure?).

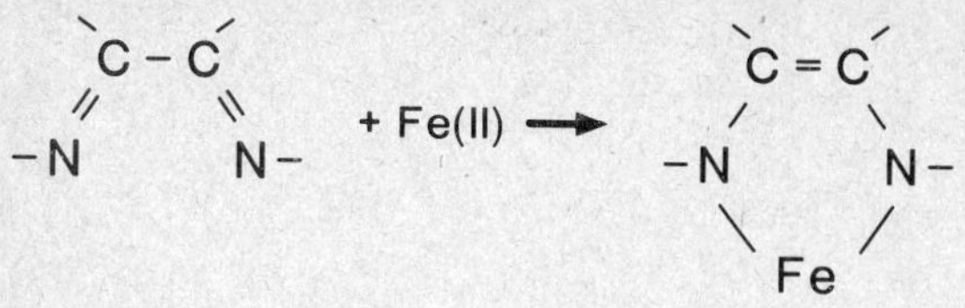

Figure 15-26. General formula for the reaction of Fe(II) with chromogens used in iron assays. The —N=C—C=N— structure chelates Fe(II), forming a 5-membered ring structure that produces an intense pink-to-lavender color with very high absorptivity.

Total Iron Binding Capacity Assay

Ferric ammonium citrate is added to serum in excess to saturate all iron-binding sites of transferrin. The serum–ferric ammonium citrate mixture is incubated to permit maximum Fe(III) binding by transferrin. The unbound Fe(III) is removed by addition of $MgCO_3$ powder and a buffer. The mixture is centrifuged and the iron content of the supernatant is assayed, as described under serum assay.

Specimen

Sufficient blood to provide 3.5 mL serum or heparinized plasma is required. Plasma specimens collected with EDTA, oxalate, or citrate are unsatisfactory, since they bind iron, preventing its reaction with the chromogen. Specimens should be collected in the morning. Markedly hemolyzed specimens should be rejected.

Reagents

Reagents must be of highest purity and virtually iron-free. All glassware must be rendered iron free by immersion overnight in HCl, 6 mol/L, followed by rinsing six times in iron-free water. Water must be free of measurable iron. For this purpose, distilled or de-ionized water may be redistilled in a quartz still.

1. Trichloroacetic acid (Fisher Scientific, Fair Lawn, NJ). Crystals are deliquescent; store over a desiccator. ACS-certified trichloroacetic acid may contain 200 μg Fe/g, an amount sufficient to raise Fe concentration of standards and specimens by 10 μg/dL. Consult the reference for instructions on redistillation to eliminate Fe impurity, or use iron-free trichloroacetic acid solution, 6.1 mol/L (Sigma Chemical Co., St. Louis, MO).

2. Protein precipitant solution. Place in a 1-L volumetric flask 33.3 mL of thioglycolic acid (mercaptoacetic acid), 98 g of trichloroacetic acid, or 98.4 mL of trichloroacetic acid solution (6.1 mol/L) and 400 mL of water. Swirl to dissolve. Slowly add 2 mol of HCl (e.g., 192 mL of concentrated HCl, if assayed at 38% HCl). Add H_2O to volume. Store in a dark brown bottle. Stable for 2 months.

3. Chromogen solution. Place in a 1-L volumetric flask 250 mg of disodium bathophenanthroline disulfonic acid (ICN Pharmaceuticals, Inc., K & K Labs Division, Plainview, NY) and 84.04 g of sodium acetate. Add H_2O to volume.

4. Iron standard, 100 and 200 μg/dL. Commercial standard may be purchased, e.g., from Hach Chemical Co., or a standard may be prepared by dissolving pure iron wire in concentrated HCl.

Table 15-7. CHARACTERISTICS OF SOME CHROMOGENS USED IN IRON ASSAYS*

Chromogen		Absorptivity Maximum of Fe(II) Complex (nm)	Molar Absorptivity of Fe(II) Complex
Common Name	*Chemical Name*		
Bathophenanthroline disulfonate, sodium	4,7-bis(4-phenyl sulfonic acid)-1,10-phenanthroline, sodium salt	534	22.14×10^3
Tripyridyl triazine	2,4,6-tripyridyl-s-triazine	593	22.6×10^3
Ferrozine	3-(2-pyridyl)-5,6-bis(4-phenyl sulfonic acid) 1,2,4 triazine, sodium salt	562	28.0×10^3
Terosite	2,6-bis(4-phenyl-2,2-pyridyl)-4-phenyl pyridine	583	30.2×10^3

*Modified from Carter P., Analyt. Biochem., *40*:450–458, 1971.

5. Ferric ammonium citrate, ∼1.25 mmol Fe(III)/L. Place 400 mg ferric ammonium citrate in a 1-L volumetric flask and add ∼950 mL H_2O. Add a few crystals of citric acid and shake to dissolve. Adjust pH to 7.0 with dilute NH_4OH and dilute to volume with water. Keep at 4 °C. Avoid prolonged exposure to light. Solution should be a light yellow-green and free of rust-colored precipitate. Prepare fresh every 3 months.

6. Barbital-NaCl buffer, 40 mmol barbital/L, pH 7.5. Place 12 g of diethylbarbituric acid, 4.5 g of sodium diethylbarbiturate, and 12.8 g of NaCl in a 2-L beaker. Add 1950 mL of H_2O and stir to dissolve. Add NaOH, 1 mol/L, in 1-mL increments until all reagents have dissolved. Place a pH electrode in the solution and adjust pH to 7.5 by addition of HCl, 0.1 mol/L. Then transfer to a 2-L volumetric flask and bring to volume with H_2O.

Procedure

Serum Iron

1. To 15-mL centrifuge tubes add 2 mL of each serum specimen and 2 mL of protein precipitant solution. Vortex and allow to stand 5 min.

2. Centrifuge at 1500 × g for 10 min, and transfer 2 mL of the clear supernatants to cuvets that have been marked to correspond to the sample tubes.

3. To cuvet marked blank, add 1 mL of water and 1 mL of protein precipitant solution. Mix and allow to stand 5 min.

4. To another cuvet marked standard, add 1 mL of standard solution and 1 mL of protein precipitant solution. Mix and allow to stand 5 min.

5. Add 2 mL of chromogen solution to all cuvets and mix. Allow to stand 5 min.

6. Measure absorbances of specimen, blank, and standard cuvets.

Calculation

$$\text{Serum iron, } \mu g/dL = \frac{(A_u - A_b)}{(A_s - A_b)} \times C_s$$

where A_u is the absorbance of the specimen, A_b is the absorbance of the blank, A_s is the absorbance of the standard, and C_s is the concentration of the standard.

Total Iron Binding Capacity

1. Place 1 mL of serum in a tube of 10-mL minimum volume. Add 0.1 mL of ferric ammonium citrate solution. Mix and allow to stand 10 min. Then add 2 mL of barbital–NaCl buffer and 0.4–0.5 g of magnesium carbonate, light powder. Cap with Parafilm.

2. Mix on rotary mixer for at least 15 min.

3. Centrifuge for 10 min at 1500 × g.

4. Transfer 2 mL of supernatant to another tube and proceed to protein precipitation as for serum iron determination.

Calculation

$$\text{TIBC, } \mu g/dL = 3.1 \times \frac{(A_u - A_b) \times C_s}{(A_s - A_b)}$$

The factor 3.1 corrects for dilution of the specimen with buffer and ferric ammonium citrate solutions.

Transferrin Saturation

$$\text{Transferrin saturation, } \% = \frac{100 \times \text{serum iron concentration}}{\text{TIBC}}$$

Reference Ranges

Serum Iron

Fasting morning specimens from healthy adults range from 70–180 μg/dL (12.5–32.2 μmol/L) in men and from 60–180 μg/dL (10.7–32.2 μmol/L) in women. Serum iron concentrations in normal neonates range from 95–225 μg/dL (17–40.3 μmol/L). After one month of age the range closely approximates that of adults unless there is iron deficiency.

TIBC

250–450 μg/dL (44.8–80.6 μmol/L)

Transferrin Saturation

Men, 20–50%
Women, 15–50%

Comments and Precautions

Because free hemoglobin is precipitated with other serum proteins, hemolysis has very little effect on the serum iron assay results. However, when serum specimens show marked hemolysis, a small amount of iron may be liberated from hemoglobin. Such sera should be rejected.

Many factors influence serum iron concentration and TIBC. Changes that may be observed in various physiologic or pathologic conditions are listed in Table 15-6. Day-to-day variation is quite marked in healthy persons. A distinct diurnal variation results in serum iron concentrations being lower in the afternoon than morning and quite low in the evening (as low as 10–20 μg/dL in healthy persons). Because of the numerous causes of low serum iron concentration, interpretation of results must be made with caution. Furthermore, many persons with iron deficiency have normal values for serum iron concentration and TIBC.

Because of the great quantities of iron in the environment, scrupulous care is needed to ensure that glassware, water, and reagents do not become contaminated with iron.

The ferric ammonium citrate solution used in the TIBC assay has an iron concentration of 7000 μg/dL. Since one volume is added to 10 volumes of serum, the addition of this reagent is sufficient to increase the iron concentration in serum by 636 μg/dL. This concentration also sets an upper limit to the TIBC, which cannot exceed the serum iron concentration + 0.091 × the iron concentration of the ferric ammonium citrate solution. Yet occasionally specimens are encountered with TIBC values very much greater than this upper limit. In such cases, the high TIBC is accompanied by a very high serum iron concentration. There are two common explanations for this paradox: (1) there is marked ferritinemia from hepatocellular injury, as in viral hepatitis; (2) the results reflect recent injection of the patient with iron dextran. In both of these circumstances, serum iron concentration and TIBC may both exceed 1000 μg/dL (179 μmol/L). However, in neither situation does the measured high TIBC reflect elevated serum transferrin.

Serum transferrin concentration may be estimated from the TIBC by the following relationship:

$$\text{Serum transferrin, mg/dL} = 0.70 \times \text{TIBC}$$

The relationship, however, is not entirely linear, as suggested in this formula, since a small portion of iron in serum is bound to other proteins. Furthermore, the magnesium carbonate does not remove quite all of the non-protein-bound iron. Therefore, the calculated TIBC values are a few μg/dL higher than the amount of transferrin-bound iron. These small differences are of no practical consequence. Methods are available for assay of serum transferrin concentration using antitransferrin immunoglobulin. Results of the immunologic measurement of transferrin concentration correlate with those of the TIBC assay. A slight advantage for the immunologic assay of transferrin is that the required volume of specimens is much smaller.

References

Fielding, J.: Methods in Hematology 1, Iron. J. D. Cook, Ed. New York, Churchill Livingstone, 1980, pp. 15-43.
International Committee for Standardization in Haematology: Br. J. Haematol., 38:281-290, 291-294, 1978.
Rice, E. W., and Fenner, H. E.: Clin. Chim. Acta, 53:391-393, 1974.

Other Methods for the Determination of Iron and TIBC

The procedures outlined above are the reference methods approved by the International Committee on Standardization in Hematology (ICSH). However, several other methods are in use for serum iron and TIBC assays. Most of these use chromogens that are structurally closely related to bathophenanthroline (Table 15-7). In one manual method, ascorbic acid is added to serum to denature transferrin, release Fe(III), and reduce it to Fe(II). Absorbance is measured, then chromogen is added, and absorbance is measured again. The difference in absorbance measurements is proportional to the iron concentration. The serum–ascorbic acid mixture is turbid due to protein denaturation, but the absorbance due to turbidity is assumed to be the same following chromogen addition, and so is corrected by subtracting the blank absorbance measurements. In this manner the specimen also serves as its own blank. Either this method or the ICSH reference method given here is readily adaptable to automation, and in most clinical chemistry laboratories that process a large number of specimens each day, automated methods are used.

Atomic absorption spectrophotometry has also been used to measure serum iron concentration. There is no inherent advantage in atomic absorption compared with chromogenic spectrophotometric assays, and the latter are more accurate. If atomic absorption is used for serum iron assay, *every specimen must first be deproteinized* to remove free hemoglobin that is invariably present in serum specimens. The protein precipitating solution described above in the serum iron procedure is satisfactory for this purpose. Failure to deproteinize every specimen results in spurious values that may exceed the true serum iron concentration by 100 μg/dL or more. For example, a serum hemoglobin concentration of 30 mg/dL is barely discernible as hemolysis upon visual inspection of a specimen, yet it represents 102 μg/dL iron.

Some other methods for serum iron assay still in use give spuriously lower assay values than are obtained with the ICSH or other modern procedures. These lower results are usually due to entrapment of iron in the protein coagulum as transferrin is precipitated before the Fe(III) is released. Incorporation of HCl in the precipitating reagent avoids this problem.

SERUM FERRITIN

Clinical Significance

Ferritin is present in the blood in very low concentration. Normally, ~1% of the plasma iron is contained in ferritin. The plasma ferritin is in equilibrium with body stores, and variations in the quantity of iron in the storage compartment are reflected in plasma ferritin concentration. The plasma ferritin concentration declines very early in the development of iron deficiency, long before changes are observed in blood hemoglobin concentration, in erythrocyte size, or in serum iron concentration. Thus, measurement of serum ferritin concentration is a very sensitive indicator of iron deficiency that is uncomplicated by other concurrent disease. On the other hand, a large number of chronic diseases result in increased serum ferritin concentration. These diseases include chronic infections; chronic inflammatory disorders, such as rheumatoid arthritis or renal disease; and numerous malignancies, especially lymphomas, leukemias, breast cancer, and neuroblastoma. In patients who have any of these chronic disorders together with iron deficiency, serum ferritin concentration is very often normal. Increase in plasma ferritin concentration occurs also in viral hepatitis or following toxic liver injury as a result of release of ferritin from damaged liver cells. Plasma ferritin concentration is also increased in patients with hemosiderosis or hemochromatosis. However, as a screening test for detection of early iron overload, measurement of serum ferritin concentration appears to be less sensitive than is measurement of serum iron concentration, TIBC, and per cent transferrin saturation.

Principle

Serum ferritin assay may be performed by any of several methods, including immunoradiometric assay (IRMA) and enzyme-linked immunosorbent assay (ELISA). A widely used procedure is the two-stage IRMA that is the basis for the Ramco kit (Ramco, Inc., Houston, TX). The kit includes plastic beads coated with antihuman ferritin. In the first stage, human serum is incubated with the antiferritin-coated beads, and ferritin in the serum is thus adsorbed to the beads. Following incubation, the beads are washed, and then, in a second stage, they are incubated with antiferritin labeled with [125]I. Following the second incubation, the beads are washed again. The radioactivity remaining on the beads is counted in a gamma spectrometer. The ferritin concentration in each specimen is directly related to the count rate. Standards are processed in the same manner, and a standard curve is prepared that relates count rate to ferritin concentration. The ferritin concentration of serum specimens is read from the curve.

Procedure

In most laboratories, serum ferritin assays are performed using commercially available kits. Follow the manufacturer's instructions.

Reference Ranges

Values are based on the 95 percentile limits.

Adult males	20–300 μg/L
Adult females	10–120 μg/L
Newborn infants	25–200 μg/L
1 month	200–600 μg/L
2 to 5 months	50–200 μg/L
6 months to 15 years	7–142 μg/L

For specimens from adult males, values between 10 and 20 μg/L should be considered indeterminate.

Comments and Precautions

The reference ranges given are appropriate for the Ramco kit. Considerable variation in reference values has been observed with different methods for serum ferritin assay. *Reference values must be determined for each laboratory* to avoid serious misinterpretations.

The two-stage IRMA procedure is reliable. Replicate same-day assays on the same specimen should have a CV of 4% for ferritin concentrations of 100–300 μg/L and of 10% for ferritin concentrations of 10–20 μg/L.

Because precision decreases at very high serum ferritin concentration, all specimens with results > 800 μg/L must be diluted with diluting sera to the 200–400 μg/L range and repeated in the next assay. Any specimen with > 7% difference between duplicate assay results should be re-assayed.

A "high-dose hook effect" that is similar to a prozone phenomenon complicates interpretation of serum ferritin assays; specimens with actual ferritin concentration > 1000 μg/L may exhibit count rates that would be obtained at normal serum ferritin concentration. While it is not practical to do every ferritin assay at two dilutions, this should be done in all patients suspected of having hemosiderosis or hemochromatosis.

FREE ERYTHROCYTE PORPHYRINS (FEP)

Clinical Significance

The free erythrocyte porphyrins (FEP) are the porphyrins that are readily extracted from erythrocytes with organic solvents such as ethyl acetate. The principal free porphyrin is *protoporphyrin IX*. FEP concentration is moderately increased in *iron deficiency* or lead poisoning, since both conditions prevent the final step of heme synthesis, the insertion of iron into the porphyrin ring, and consequently porphyrin precursors accumulate in the erythrocyte. Very marked increase in FEP concentration also occurs in *lead poisoning,* in the very rare hereditary disorder *congenital erythropoietic porphyria,* and in *erythropoietic protoporphyria.* On the other hand, thalassemic disorders, which have many hematologic features in common with iron deficiency, are not associated with increased FEP concentration. The FEP assay is used as a screening test, especially in pediatric practice, for iron deficiency and lead poisoning. Whenever elevated concentrations of FEP are observed, more specific tests are required: measurement of serum iron or serum ferritin concentration and measurement of erythrocyte lead or whole blood lead (not serum lead) concentration. A description of the quantitative determination of erythrocyte protoporphyrin is given in Chapter 16.

References

1. Abraham, E. C., Reese, A., Stallings, M., et al.: Separation of human hemoglobins by DEAE cellulose chromatography using glycine-KCN-NaCL developers. Hemoglobin, *1*:27-44, 1976-1977.
2. Arnold, H.: Inherited glucosephosphate isomerase deficiency. Blut, *39*:405-417, 1979.
3. Baird, M., Driscoll, C., Schreiner, H., et al.: A nucleotide change at a splice junction in the human β-globin gene is associated with β°-thalassemia. Proc. Natl. Acad. Sci. USA, *78*:4218-4221, 1981.
4. Baumgartner, E. R., Wick, H., Maurer, R., et al.: Congenital defect in intracellular cobalamin metabolism resulting in homocystinuria and methylmalonic aciduria. Helv. Paed. Acta, *34*:465-482, 1979.
5. Bellingham, A. J., and Huehns, E. R.: Compensation mechanisms in haemolytic anaemias. Proc. Roy. Soc. Med., *61*:1315-1316, 1968.
6. Benesch, R. E., Benesch, R., and Yu, C. I.: The oxygenation of hemoglobin in the presence of 2,3-diphosphoglycerate. Effect of temperature, pH, ionic strength, and hemoglobin concentration. Biochemistry, *8*:2567-2571, 1969.
7. Beutler, E.: Glucose 6-phosphate dehydrogenase deficiency. *In*: The Metabolic Basis of Inherited Disease. 5th ed. J. B. Stanbury, J. B. Wyngaarden, and D. S. Fredrickson, Eds. New York, McGraw-Hill, 1983, pp. 1629-1653.
8. Beutler, E.: Red Cell Metabolism. A Manual of Biochemical Procedures. 2nd ed. New York, Grune & Stratton, 1975.
9. Beutler, E., Dyment, P. G., and Matsumoto, F.: Hereditary nonspherocytic hemolytic anemia and hexokinase deficiency. Blood, *51*:935-940, 1978.
10. Black, J. A., Rittenberg, M. B., Bigley, R. H., et al.: Hemolytic anemia due to pyruvate kinase deficiency: Characterization of the enzymatic activity from eight patients. Am. J. Hum. Genet., *31*:300-310, 1979.
11. Carmel, R., Bedros, A. A., Mace, J. W., et al.: Congenital methylmalonic aciduria–homocystinuria with megaloblastic anemia: Observations on response to hydroxocobalamin and on the effect of homocysteine and methionine on the deoxyuridine suppression test. Blood, *55*:570-579, 1980.
12. Carrell, R. W., and Lehmann, H.: Zinc acetate as a precipitant of unstable haemoglobins. J. Clin. Pathol., *34*:796-799, 1981.
13. Carrico, R. J., Peisach, J., and Alben, J. O.: The preparation and some physical properties of sulfhemoglobin. J. Biol. Chem., *253*:2386-2391, 1978.
14. Carter P.: Spectrophotometric determination of serum iron at the submicrogram level with a new reagent (ferrozine). Analyt. Biochem., *40*:450-458, 1971.
15. Chanarin, I.: The Megaloblastic Anaemias. 2nd ed. London, Blackwell Scientific Publications, 1979.
16. Chanarin, I.: Cobalamins and nitrous oxide: A review. J. Clin. Pathol., *33*:909-916, 1980.
17. Chanarin, I., Deacon, R., Lumb, M., et al.: Vitamin B_{12} regulates folate metabolism by the supply of formate. Lancet, *2*:505-508, 1980.
18. Chanarin, I., Deacon, R., Perry, J., et al.: Annotation—how vitamin B_{12} works. Br. J. Haematol., *47*:487-491, 1981.
19. Chang, J. C., Temple, G. F., Trecarti, R. F., et al.: β°-Thalassemia: A nonsense mutation in man. Proc. Natl. Acad. Sci. USA, *76*:2886-2889, 1979.
20. Chen, I. W., David, R., Maxon, H. R., et al.: Age-, sex-, and race-related differences in myoglobin concentrations in the serum of healthy persons. Clin. Chem., *26*:1864-1968, 1980.
21. Clegg, J. B., and Weatherall, D. J.: Haemoglobin synthesis in α-thalassemia (haemoglobin H disease). Nature, *215*:1241-1243, 1967.
22. Dickerson, R. E.: X-ray Analysis and Protein Structure in the Proteins, Vol. II. H Neurath, Ed. New York, Academic Press, 1969.
23. Dinn, J. J., Weir, D. G., McCann, S., et al.: Methyl group deficiency in nerve tissue: A hypothesis to explain the lesion of subacute combined degeneration. Irish J. Med. Sci., *149*:104, 1980.
24. Dozy, A. M., Kan, Y. W., Embury, S. H., et al.: Alpha globin gene organization in blacks precludes the severe form of α-thalassemia. Nature, *280*:605-607, 1979.
25. Drabkin, D. L., and Austin, J. H.: Spectrophotometric studies. II. Preparation from washed blood cells: Nitric oxide hemoglobin and sulfhemoglobin. J. Biol. Chem., *112*:51-65, 1935.
26. Duhm, J.: The effect of 2,3-DPG and other organic phosphates on the Donnan equilibrium and the oxygen affinity of human blood. *In*: Oxygen Affinity of Hemoglobin and Red Cell Acid Base Status (Alfred Benzon Symposium, IV). M. Rørth and P. Astrup, Eds. New York, Academic Press, 1972, pp. 583-594.
27. Duma, H., Efremov, G., Sadikario, A., et al.: Study of nine families with haemoglobin-Lepore. Br. J. Haematol., *15*:161-172, 1968.
28. Evelyn, K. A., and Malloy, H. T.: Microdetermination of oxyhemoglobin, methemoglobin and sulfhemoglobin in a single sample of blood. J. Biol. Chem., *126*:655-662, 1938.
29. Fairbanks, V. F., and Elveback, L. R.: Tests for pernicious anemia: Serum vitamin B_{12} assay. Mayo Clin. Proc., *58*:135-137, 1983.
30. Fairbanks, V. F., Gilchrist, G. S., Brimhall, B., et al.: Hemoglobin E trait re-examined: A cause of microcytosis and erythrocytosis. Blood, *53*:109-115, 1979.
31. Fairbanks, V. F., Lennon, V. A., Kokmen, E., et al.: Tests for pernicious anemia: Serum intrinsic factor blocking antibody. Mayo Clin. Proc., *58*:203-204, 1983.
32. Fairbanks, V. F., Oliveros, R., Brandabur, J. H., et al.: Homozygous hemoglobin E mimics β-thalassemia minor without anemia or hemolysis: Hematologic, functional, and biosynthetic studies of first North American cases. Am. J. Hematol., *8*:109-121, 1980.
33. Fairbanks, V. F., Wahner, H. W., and Phyliky, R. L.: Tests for pernicious anemia: The "Schilling test." Mayo Clin. Proc., *58*:541-544, 1983.
33a. Fearon, E. R., Kazazian, H. H., Waber, P. G., et al.: The entire β-globin gene cluster is deleted in a form of gamma delta beta-thalassemia. Blood, *61*:1269-1274, 1983.

34. Frenkel, E. P.: Abnormal fatty acid metabolism in peripheral nerves of patients with pernicious anemia. J. Clin. Invest., *52*:1237-1245, 1973.
35. Glader, B. E., and Sullivan, D. W.: Erythrocyte disorders leading to K loss and dehydration. Prog. Clin. Biol. Res., *30*:503-513, 1979.
36. Glomset, J. A., Norum, K. P., and Gjone, E.: Familial lecithin: Cholesterol acyltransferase deficiency. *In*: The Metabolic Basis of Inherited Disease. 5th ed. J. B. Stanbury, J. B. Wyngaarden, and D. S. Fredrickson, Eds. New York, McGraw-Hill, 1983, pp. 643-654.
37. Goodman, S. R., Shiffer, K. A., Casoria, L. A., et al.: Identification of the molecular defect in the erythrocyte membrane skeleton of some kindreds with hereditary spherocytosis. Blood, *60*:772-784, 1982.
38. Gräsbeck, R.: Intrinsic factor and other vitamin B_{12} transport proteins. Prog. Hematol., *6*:233-260, 1969.
39. Greenberg, M. S., Harvey, H. A., and Morgan, C.: A simple and inexpensive screening test for sickle hemoglobin. N. Engl. J. Med., *286*:1143-1144, 1972.
40. Gross, S.: Antioxidant relationship between selenium-dependent glutathione peroxidase and tocopherol. Am. J. Pediatr. Hematol. Oncol., *1*:61-69, 1979.
41. Hall, C. A.: The transport of vitamin B_{12} from food to use within the cells. J. Lab. Clin. Med., *94*:811-815, 979.
42. Hamilton, S. R., Miller, M. E., Jessop, M., et al.: Comparison of microchromatography and electrophoresis with elution for hemoglobin A_2 (Hb A_2) quantitation. Am. J. Clin. Pathol., *71*:388-396, 1979.
43. Hanel, H. K., and Cohn, J.: Adenosine triphosphate deficiency in a family with non-spherocytic haemolytic anaemia. Scand. J. Haematol., *9*:28-35, 1971.
44. Jacob, E., Baker, S. J., and Herbert, V.: Vitamin B_{12}–binding proteins. Physiol. Res., *60*:918-960, 1980.
45. Jonxis, S. H. P., and Visser, H. K. A.: Determination of low percentage of fetal hemoglobin in the blood of normal children. Am. J. Dis. Child., *92*:588, 1956.
46. Kagen, L., Scheidt, S., Roberts, L., et al.: Myoglobinemia following acute myocardial infarction. Am. J. Med., *58*:177-182, 1975.
47. Kahn, A., Kaplan, J.-C., and Dreyfus, J.-C.: Advances in hereditary red cell enzyme anomalies. Hum. Genet., *50*:1-27, 1979.
48. Kahn, A., Marie, J., Garreau, H., et al.: The genetic system of the L-type pyruvate kinase forms in man. Biochim. Biophys. Acta, *523*:59-74, 1978.
49. Kan, Y. W., Dozy, A. M., Trecartin, R., et al.: Identification of nondeletion defect in α-thalassemia. N. Engl. J. Med., *297*:1081-1084, 1977.
50. Kan, Y. W., Forget, B. G., and Nathan, D. G.: Gamma-beta thalassemia: A cause of hemolytic disease of the newborn. N. Engl. J. Med., *286*:129-134, 1972.
51. Kazazian, H. H., Fearon, E. R., Waber, P. G., et al.: γδβ-Thalassemia: Deletion of the entire β-globin gene cluster. (Abstract.) Blood (Suppl. 1), *60*:54a, 1982.
52. Kirkman, H. N., Rosenthal, I. M., Simon, E. R., et al.: "Chicago I" variant of glucose-6-phosphate dehydrogenase in congenital hemolytic disease. J. Lab. Clin. Med., *63*:715-725, 1964.
53. Kirkman, H. N., Schettini, F., and Pickard, B. M.: Mediterranean variant of glucose-6-phosphate dehydrogenase variant in Thailand. J. Med. Assoc. Thai., *55*:576-585, 1972.
54. Konrad, P. N., McCarthy, D. E., Mauer, A. M., et al.: Erythrocyte and leukocyte phosphoglycerate kinase deficiency with neurologic disease. J. Pediatr., *82*:456-460, 1973.
55. Konrad, P., Richards, F., II, Valentine, W. N., et al.: γ-Glutamyl-cysteine synthetase deficiency. A cause of hereditary hemolytic anemia. N. Engl. J. Med., *286*:557-561, 1972.
56. Lauer, J., Shen, C. J., and Maniatis, T.: The chromosomal arrangement of human α-like globin genes: Sequence homology and α-globin gene deletions. Cell, *20*:119-130, 1980.
57. Lindenbaum, J.: Aspects of vitamin B_{12} and folate metabolism in malabsorption syndromes. Am. J. Med., *67*:1037-1048, 1979.
58. Liu, S.-C., Palek, J., Prchal, J., et al.: Altered spectrin dimer-dimer association and instability of erythrocyte membrane skeletons in hereditary pyropoikilocytosis. J. Clin. Invest., *68*:597-605, 1981.
59. Loh, W.-P.: Evaluation of a rapid test tube turbidity test for the detection of sickle cell hemoglobin. Am. J. Clin. Pathol., *55*:55-57, 1971.
60. Lux, S. E.: Dissecting the red cell membrane skeleton. Nature, *281*:426-429, 1979.
61. Lux, S. E.: Spectrin-actin membrane skeleton of normal and abnormal red blood cells. Semin. Hematol., *16*:21-51, 1979.
62. Marchesi, V. T.: The red cell membrane skeleton: Recent progress. Blood, *61*:1-11, 1983.
63. Markowitz, H., and Wobig, G. H.: Quantitative method for estimating myoglobin in urine. Clin. Chem., *23*:1689-1693, 1977.
64. Marotta, C. A., Forget, B. G., Cohen-Solal, M., et al.: Human β-globin messenger RNA. II. Nucleotide sequences derived from ^{125}I-labeled globin messenger RNA. J. Biol. Chem., *252*:5032-5037, 1977.
65. Mears, J. G., Lachman, H., Labie, D., et al.: Alpha-thalassemia is related to prolonged survival in sickle cell anemia. Blood, *62*:286-290, 1983.
66. Mohandas, N., Clark, M. R., Heath, B. P., et al.: A technique to detect reduced mechanical stability of red cell membranes: Relevance to elliptocytic disorders. Blood, *59*:768-774, 1982.
67. Nalbandian, R. M., Nichols, B. M., Camp, F. R., Jr., et al.: Dithionite tube test: A rapid, inexpensive technique for the detection of hemoglobin S and non-S sickling hemoglobin. Clin. Chem., *17*:1028-1032, 1971.
68. Nalbandian, R. M., Nichols, B. M., Heustis, A. E., et al.: An automated mass screening program for sickle cell disease. JAMA, *218*:1680-1682, 1971.
69. Orkin, S. H., and Goff, S. C.: The duplicated human α-globin genes: Their relative expression as measured by RNA analysis. Cell, *24*:345-351, 1981.
70. Orkin, S. H., Kazazian, H. H., Antonarakis, S. E., et al.: Abnormal RNA processing due to the exon coding mutation of $β^E$-globin gene. Nature, *300*:768-769, 1983.

71. Oski, F. A., and Bowman, H.: A low K_m phosphoenolypyruvate mutant in the Amish with red cell pyruvate kinase deficiency. Br. J. Haematol., *17*:289-297, 1969.
72. Ottolenghi, S., Comi, P., Giglioni, B., et al.: δβ-Thalassemia is due to a gene deletion. Cell, *9*:71-80, 1976.
73. Paglia, D. E., Valentine, W. N., Baughan, M. A., et al.: An inherited molecular lesion of erythrocyte pyruvate kinase: Identification of a kinetically aberrant isozyme associated with premature hemolysis. J. Clin. Invest., *47*:1929-1946, 1968.
74. Panich, V., Sungnate, T., Wasi, P., et al.: G-6-PD Mahidol: The most common glucose-6-phosphate dehydrogenase variant in Thailand. J. Med. Assoc. Thai., *55*:576-585, 1972.
75. Perutz, M. F.: The hemoglobin molecule. Sci. Am., *211*:644-679, 1964.
76. Perutz, M. F.: Hemoglobin structure and respiratory transport. Sci. Am., *293*:92-133, 1978.
77. Perutz, M. F.: Regulation of oxygen affinity of hemoglobin: Influences of structure of the globin on the heme iron. Ann. Rev. Biochem., *48*:327-386, 1979.
78. Perutz, M. F., and Lehmann, H.: Molecular pathology of human haemoglobin. Nature (London), *219*:902-909, 1968.
79. Phillips, J. A., III, Vik, T. A., Scott, A. F., et al.: Unequal crossing-over: A common basis of single α-globin genes in Asians and American blacks with hemoglobin-H disease. Blood, *55*:1066-1069, 1980.
80. Pohl, A., Gugajer-Gleitman, H. E., Lachmann, D., et al.: Glutathionreduktasemangel mit Membrandefekt bei hereditärer hämolytischer Anämie. Acta Haematol., *56*:47-57, 1976.
81. Pressley, L., Higgs, D. R., Clegg, J. B., et al.: Gene deletions in α-thalassemia prove that the 5' α locus is functional. Proc. Natl. Acad. Sci. USA, *77*:3586-3589, 1980.
82. Pressley, L., Higgs, D. R., Clegg, J. B., et al.: A new genetic basis for hemoglobin H disease. N. Engl. J. Med., *303*:1383-1388, 1980.
83. Prins, H. K., Oort, M., Loos, J., et al.: Congenital nonspherocytic hemolytic anemia, associated with glutathione deficiency of the erythrocytes. Blood, *27*:145-166, 1966.
84. Quattrin, N., and Ventruto, V.: Hemoglobin Lepore: Its significance for thalassemia and clinical manifestations. Blut, *28*:327-336, 1974.
85. Ramsey, R. B., Scott, T., and Banik, N. L.: Fatty acid composition of myelin isolated from the brain of a patient with cellular deficiency of coenzyme forms of vitamin B_{12}. J. Neurol. Sci., *34*:221-232, 1977.
86. Rosa, R., Prehu, M.-O., Beuzard, Y., et al.: The first case of a complete deficiency of diphosphoglycerate mutase in human erythrocytes. J. Clin. Invest., *62*:907-915, 1978.
87. Schmidt, R. M., and Wilson, S. M.: Standardization in detection of abnormal hemoglobins: Solubility tests for hemoglobin S. JAMA, *225*:1225-1230, 1973.
88. Schneider, A. S., Valentine, W. N., Hattori, M., et al.: Hereditary hemolytic anemia with triosephosphate isomerase deficiency. N. Engl. J. Med., *272*:229-235, 1965.
89. Shepard, M. K., Weatherall, D. J., and Conley, C. L.: Semi-quantitative estimation of the distribution of fetal hemoglobin in red cell populations. Bull. Johns Hopkins Hosp. *110*:293-310, 1962.
90. Smith, M. H.: Spectral properties of the M haemoglobins. *In*: Haemoglobin-Colloquium (Wien, 1961). H. Lehmann and K. Betke, Eds. Stuttgart, Georg Thieme Verlag, 1962, pp. 49-52.
91. Spielberg, S. P., Garrick, M. D., Corash, L. M., et al.: Biochemical heterogeneity in glutathione synthetase deficiency. J. Clin. Invest., *61*:1417-1420, 1978.
92. Spritz, R. A., Jagadeeswaran, P., Choudary, P. V., et al.: Base substitution in an intervening sequence of a β⁺-thalassemic human globin gene. Proc. Natl. Acad. Sci. USA, *78*:2455-2459, 1981.
93. Stamatoyannopoulos, G., Fessas, P., and Papayannopoulous, T.: F-thalassemia: A study of thirty-one families with simple heterozygotes and combination of F-thalassemia with A₂-thalassemia. Am. J. Med., *47*:194-208, 1969.
94. Stroehlein, J. R., Fairbanks, V. F., McGill, D. B., et al.: Hemoccult detection of fecal occult blood quantitated by radioassay. Am. J. Digest. Dis., *21*:841-844, 1976.
95. Tchernia, G., Mohandas, N., and Shohet, S. B.: Deficiency of skeletal protein band 4.1 in homozygous elliptocytosis: Implications for membrane stability. J. Clin. Invest., *68*:454-450, 1981.
96. Theil, G. B.: Separation and identification of myoglobin and hemoglobin. Am. J. Clin. Pathol., *49*:190-195, 1968.
97. Torrance, J., Jacobs, P., Restrepo, A., et al.: Intraerythrocytic adaptation to anemia. N. Engl. J. Med., *283*:165-169, 1970.
98. Tuan, D., Biro, P. A., de Riel, J. K., Lazarus, H. et al.: Restriction endonuclease mapping of the human γ globin gene loci. Nucl. Acids Res., *6*:2519-2544, 1977.
99. Valentine, W. N., Fink, K., Paglia, D. E., et al.: Hereditary hemolytic anemia with human erythrocyte pyrimidine 5'-nucleotidase deficiency. J. Clin. Invest., *54*:866-879, 1974.
100. Valentine, W. N., Oski, F. A., Paglia, D. E., et al.: Hereditary hemolytic anemia with hexokinase deficiency: Role of hexokinase in erythrocyte aging. N. Engl. J. Med., *276*:1-11, 1967.
101. Valentine, W. N., Paglia, D. E., Fink, K., et al.: Lead poisoning: Association with hemolytic anemia, basophilic stippling, erythrocyte pyrimidine 5'-nucleotidase deficiency, and intraerythrocytic accumulation of pyrimidines. J. Clin. Invest., *58*:926-932, 1976.
102. Valentine, W. N., Paglia, D. E., Tartaglia, A. P., et al.: Hereditary hemolytic anemia with increased red cell adenosine deaminase (45- to 70-fold) and decreased adenosine triphosphate. Science, *195*:783-785, 1977.
103. van der Ploeg, L. H. T., Konings, M., Oort, D., et al.: γ-β-Thalassemia studies showing that deletion of the γ- and δ-genes influences β-globin expression in man. Nature, *283*:637-642, 1980.
104. van Kampen, E. J., and Zijlstra, W. G.: Determination of hemoglobin and its derivatives. Adv. Clin. Chem., *8*:141-187, 1965.
105. Vanzetti, G., and Valente, D.: A sensitive method for the determination of hemoglobin in plasma. Clin. Chim. Acta, *11*:442-446, 1965.
106. Waterbury, L., and Frenkel, E. P.: Hereditary nonspherocytic hemolysis with erythrocyte phosphofructokinase deficiency. Blood, *39*:415-425, 1972.

107. Weatherall, D. J., Old, J., Longley, J., et al.: Acquired haemoglobin H disease in leukaemia: Pathophysiology and molecular basis. Br. J. Haematol., *38*:305-322, 1978.

108. Whitelaw, A. G. L., Rogers, P. A., Hopkinson, D. A., et al.: Congenital haemolytic anaemia resulting from glucose phosphate isomerase deficiency: Genetics, clinical picture, and prenatal diagnosis. J. Med. Genet., *16*:189-196, 1979.

109. Wiley, J. S., Cooper, R. A., Adachi, K., et al.: Hereditary stomatocytosis: Association of low 2,3-diphosphoglycerate with increased cation pumping by the red cell. Br. J. Haematol., *41*:133-141, 1979.

110. Winter, W. P., and Yodh, J.: Interaction of human hemoglobin and its variants with agar. Science, *221*:175-177, 1983.

111. Yoshida, A.: Glucose-6-phosphate dehydrogenase abnormality and hemolysis. Acta Biol. Med. Germ., *36*:689-701, 1977.

112. Yoshida, A., Baur, E. W., and Motulsky, A. G.: A Philippino glucose-6-phosphate dehydrogenase in congenital hemolytic disease. J. Lab. Clin. Med., *63*:715-725, 1964.

113. Zarkowsky, H. S., Mohandas, N., Speaker, C. B., et al.: A congenital haemolytic anaemia with thermal sensitivity of the erythrocyte membrane. Br. J. Haematol., *29*:537-543, 1975.

Additional Reading

Bunn, H. F., Forget, B. G., and Ranney, H. M.: Human Hemoglobins. 2nd ed. Philadelphia, W.B. Saunders Co., in press.

Fairbanks, V. F., Ed.: Current Hematology I. New York, John Wiley & Sons, 1981.

Fairbanks, V. F., Ed.: Current Hematology II. New York, John Wiley & Sons, 1983.

Fairbanks, V. F., and Klee, G. G.: Ferritin. *In* Progress in Clinical Pathology, Vol. 8. New York, Grune and Stratton, 1981.

Jacobs, A., and Worwood, M., Eds.: Iron in Biochemistry and Medicine, II. London, Academic Press, 1980.

Jaffe, E. R.: Hereditary methemoglobinemias associated with abnormalities in the metabolism of erythrocytes. Am. J. Med., *41*:786-798, 1966.

Kagen, L. J.: Myoglobin: Methods and diagnostic uses. CRC Crit. Rev. Clin. Lab. Sci., *9*:273-302, 1978.

Lessin, L. S., and Jensen, W. N.: Sickle cell anemia, 1910-1973. Arch. Intern. Med., *133*:529-537, 1974.

Rucknagel, D. L.: The genetics of sickle cell anemia and related syndromes. Arch. Intern. Med., *133*:595-605, 1974.

Stanbury, J. B., Wyngaarden, J. B., Fredrickson, D. S., et al., Eds.: The Metabolic Basis of Inherited Disease. 5th ed. New York, McGraw-Hill, 1983.

Weatherall, D. J., and Clegg, J. B.: The Thalassaemia Syndromes. 3rd ed. St. Louis, C. V. Mosby, 1981.

Williams, W. J., Beutler, E., Erslev, A. J., et al.: Hematology. 3rd ed. New York, McGraw-Hill Book Company, 1983.

CHAPTER 16

PORPHYRINS AND DISORDERS OF PORPHYRIN METABOLISM

by Robert F. Labbe, Ph.D., and Joel M. Lamon, M.D.

While the primary (inherited) disorders in porphyrin metabolism are relatively uncommon, several secondary (induced) disorders are very common in some groups of the population. All of these disorders are generally considered as rather complex metabolic diseases with overlapping clinical and biochemical features; yet their diagnosis is often attempted using only the most simplistic laboratory procedures. With management and therapy becoming increasingly effective, early and accurate diagnosis assumes added importance. Systematic approaches to diagnoses are now possible utilizing combinations of recently developed methods for measuring metabolite concentrations and enzyme activities.

CHEMISTRY OF PORPHYRINS

Porphyrins are derivatives of porphin, a macrocyclic, highly unsaturated structure composed of four pyrrole rings bonded by four methene bridges ($-CH=$). Porphin itself has no side chain substituent on the pyrrole rings. Porphyrins comprise certain substituted porphins; they are differentiated on the basis of the kind and order of substituents occupying the eight peripheral positions on the four pyrrole rings of porphin (Figure 16-1). Four basic isomer series are possible, but only the type III isomers occur in normal metabolism and heme synthesis. Protoporphyrin IX is derived from type III precursor porphyrins.* Certain type I porphyrins form insignificant amounts as byproducts of a rare biochemical defect, but they have no metabolic function. The II and IV isomers do not occur naturally.

Many kinds of porphyrins are known; however, very few are found in nature and only three of these, *uroporphyrin, coproporphyrin,* and *protoporphyrin,* have established clinical significance. These metal-free porphyrins have no biological function in man; porphyrins are metabolically active only in the form of chelates. The iron chelates of porphyrins are termed *hemes; protoheme* is by far the most common and, in a quantitative sense, the most significant (Figure 16-2). Heme always functions as a prosthetic group of a protein. In mammals, hemoproteins participate in a variety of biochemical processes, all of which are associated with some aspect of oxidative metabolism such as oxygen transport (hemoglobin), cellular respiration (cytochromes), or hydrogen peroxide utilization (catalase). Except in hemoglobin, myoglobin, and catalase, the iron atom undergoes reversible oxidation and reduction during the process of the hemoprotein function. A cobalt chelate, cobalamin or vitamin B_{12}, and a magnesium chelate, chlorophyll, are other forms of naturally occurring tetrapyrroles, although greatly modified in structure from the porphyrin from which they are derived. Zinc protoporphyrin also occurs in trace amounts in normal metabolism; it increases markedly with impaired iron utilization. Some porphyrins occur in both

---

*There are 15 possible isomers of protoporphyrin. The naturally occurring isomer has been designated IX, although it is now known to be derived biochemically from precursors of the type III isomer.

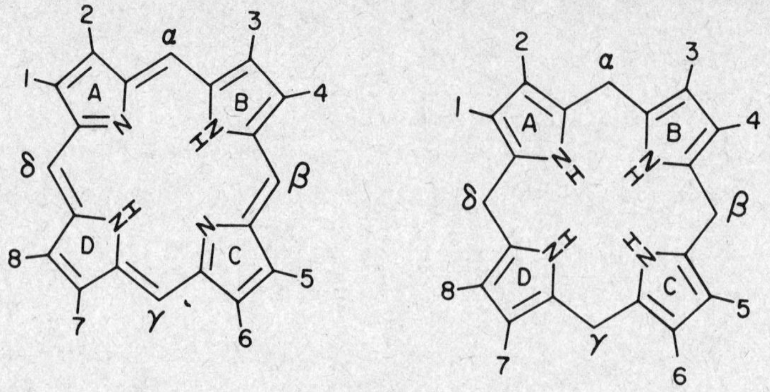

SUBSTITUENTS IDENTIFYING DIFFERENT PORPHYRINS

Ring	Position	URO	COPRO	PROTO
		Porphyrin III or Porphyrinogen III		
A	1	Acetate	Methyl	Methyl
	2	Propionate	Propionate	Vinyl
B	3	Acetate	Methyl	Methyl
	4	Propionate	Propionate	Vinyl
C	5	Acetate	Methyl	Methyl
	6	Propionate	Propionate	Propionate
D	7	Propionate	Propionate	Propionate
	8	Acetate	Methyl	Methyl
		Porphyrin I or Porphyrinogen I		
D	7	Acetate	Methyl	Does not occur in nature
	8	Propionate	Propionate	

Figure 16-1. Structures of porphyrin and porphyrinogen and a listing of the substituents that identify the different porphyrins.

free and chelated forms as pigments in nonmammalian species, e.g., egg shells, bird feathers, and worm integument.

Crystalline porphyrins and their concentrated solutions are very dark red or purple in color. In acid solution porphyrins are known for their intense orange-red fluorescence (620–630 nm) on exposure to long-wavelength ultraviolet light of about 400 nm, a property that is used in most porphyrin analyses. The intense color and fluorescence of porphyrins are due to the high degree of conjugated unsaturation or resonance in the tetrapyrrole ring. In general, free porphyrins tend to be more stable in acid solution and in the dark. Dilute hydrochloric acid solutions of copro-porphyrin are very stable and are most often used as a standard in porphyrin assays by fluorometric techniques.

The degree of polarity of the porphyrins, and thus their water solubility, is influenced by the number of carboxyl groups in the pyrrole substituents. **Uroporphyrin,** having eight carboxyl

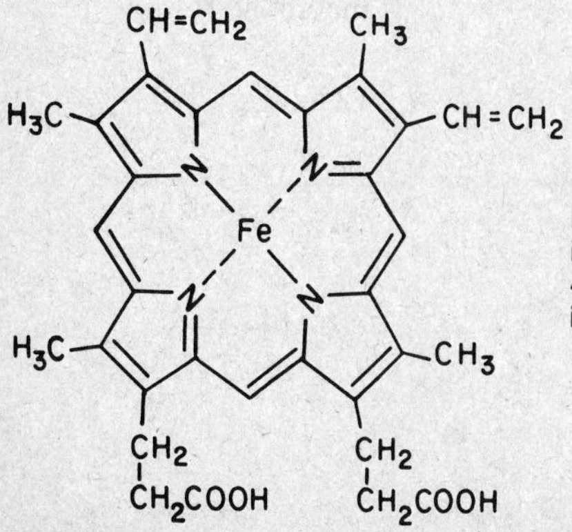

Figure 16-2. The structural formula of protoheme, a chelate of ferrous iron and protoporphyrin IX. The vinyl groups occur in positions 2 and 4 of rings A and B, and the propionic acid (carboxyethyl) groups occur in positions 6 and 7 of rings C and D. (Compare with Figure 16-1.)

groups, is the most soluble porphyrin in aqueous media at physiological pH. **Protoporphyrin,** having only two carboxyl groups, is quite insoluble in aqueous media at this pH, but is very soluble in lipid solvents. **Coproporphyrin,** with its four carboxyl groups, has intermediate solubility. These differing solubility properties and associated degrees of ionization form the basis for the separation and assay of the individual porphyrins. Using suitable analytical procedures, one can also identify and quantitate intermediate porphyrins having seven, six, or five carboxyl groups. Uroporphyrin is excreted for all practical purposes exclusively in the urine, protoporphyrin exclusively in the feces, and coproporphyrin by either route depending upon its rate of formation and the pH of the urine, with alkalinity favoring coproporphyrin excretion in the urine.

Porphyrinogens are reduced forms of porphyrins containing six additional hydrogen atoms, one at each of the four methene bridge carbons, and one at each of the two nonhydrogenated pyrrole nitrogens (Figure 16-1). The porphyrinogens, but not the porphyrins, undergo biochemical alterations in their side-chain substituents. Hence, they are functional precursors of heme. Oxidation of a porphyrinogen to the corresponding porphyrin irreversibly removes that molecule from the heme biosynthetic pathway. Protoporphyrin IX, in which side-chain alteration is complete, is the only nonporphyrinogen intermediate in heme biosynthesis.

The porphyrinogens are colorless and nonfluorescent. Porphyrinogens are also highly unstable, especially in an acid medium, and rapidly oxidize to porphyrins, making their analysis somewhat impractical for laboratory diagnosis. Therefore, porphyrins are clinically the most significant compounds in the diagnosis of disorders of porphyrin metabolism, as these are measurable in various biologic fluids and correlations in disease have been established.

PHARMACOLOGICAL PROPERTIES OF PORPHYRINS AND PORPHYRIN PRECURSORS

The known pharmacological effects of porphyrins are very limited. They cause subtle, if any, direct alterations in metabolism. On the other hand, porphyrins deposited in skin and then exposed to long-wavelength ultraviolet radiation can cause considerable skin damage. The different lesions that are caused by uro-, copro-, or protoporphyrin probably relate to their respective solubilities. The more lipophilic protoporphyrin accumulates predominantly in cell membranes, while the other porphyrins are confined largely to inter- and intracellular fluids. The clinical symptoms that are characterized by porphyrin deposition in the skin are related to the photochemical properties, the stabilities, and the solubilities of the porphyrins. The clinical correlation of these chemical differences is reflected in the observation that uroporphyrin and coproporphyrin typically cause delayed bullous lesions, while protoporphyrin causes an almost immediate burning sensation and inflammatory reaction in sun-exposed areas of the skin.

The porphyrin precursors, δ-aminolevulinic acid and porphobilinogen, have very low renal thresholds which contribute to low blood concentrations. They are considered not to produce pharmacological effects, at least after administration to animals or human subjects. On the other hand, the expression of neurologic symptoms is always accompanied by the excretion of excessive amounts of porphyrin precursors. Several lines of argument suggest that this precursor excess is a significant factor in the pathogenesis of neurologic abnormalities. These include comparison of biochemical and clinical features of lead intoxication with acute intermittent porphyria and the neurotransmitter properties of δ-aminolevulinic acid. Despite these associations, the correlation between the level of precursor excretion and the severity of neurologic symptoms is poor.

BIOSYNTHESIS OF PORPHYRINS AND HEME

Porphyrin and heme biosynthetic activity is quantitatively most prominent in bone marrow and liver; however, porphyrins and heme are synthesized in all mammalian cells. The series of reactions leading to heme (Figure 16-3) begins with the condensation of succinyl coenzyme A and glycine. Pyridoxal phosphate, a cofactor in this condensation, forms a Schiff base with glycine as an initial step. Following the condensation and in the same reaction sequence, the glycine moiety is decarboxylated, forming δ-aminolevulinate. The intramitochondrial enzyme catalyzing the reaction is δ-aminolevulinate synthase. Two molecules of δ-aminolevulinate are then condensed

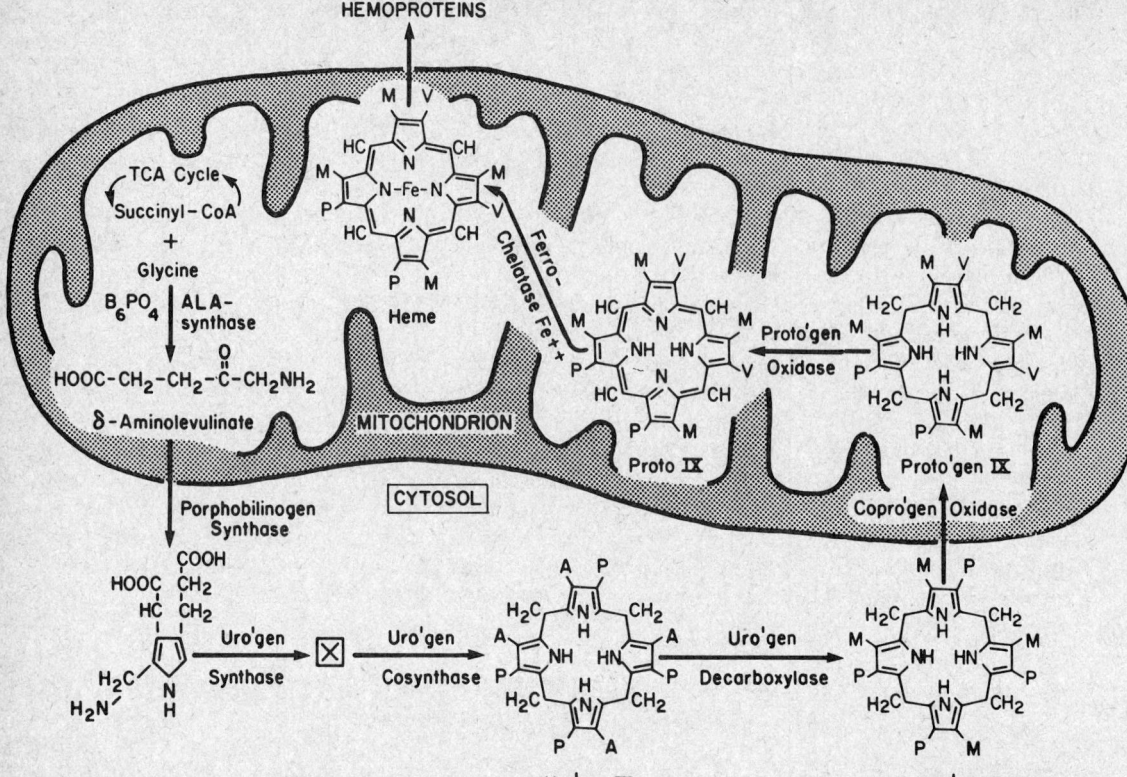

Figure 16-3. The heme biosynthetic pathway showing the distribution of enzymes between the mitochondria and cytoplasm. Intermediates between uroporphyrinogen and coproporphyrinogen, designated by [×], remain unidentified. B_6PO_4 = pyridoxal phosphate.

and cyclized through the action of porphobilinogen synthase to form the monopyrrole porphobilinogen. This synthase is a cytosol enzyme that requires zinc. Porphobilinogen is condensed and cyclized through the concerted action of two enzymes, uroporphyrinogen I synthase and uroporphyrinogen III cosynthase. The former enzyme catalyzes a deamination of the porphobilinogen side chain accompanied by condensation and cyclization of four molecules that would form uroporphyrinogen I if this enzyme acted alone. But with the concerted action of cosynthase, there occurs an intramolecular rearrangement that leads to the formation of uroporphyrinogen III. Any possible intermediates remain speculative. Subsequently, the four acetate side-chains of uroporphyrinogen III are decarboxylated in sequence through the action of uroporphyrinogen decarboxylase to form coproporphyrinogen III. Next, coproporphyrinogen oxidase, a mitochondrial enzyme, decarboxylates and dehydrogenates the propionic acid side chains in positions two and four, converting these to vinyl groups and yielding protoporphyrinogen IX. A less well characterized enzyme, protoporphyrinogen oxidase, then oxidizes protoporphyrinogen to protoporphyrin. Finally, ferrochelatase, which is located on the inner mitochondrial membrane, catalyzes the chelation of a ferrous ion by protoporphyrin to form heme. The reactions between δ-aminolevulinate and coproporphyrinogen that occur in the cellular cytosol have no established cofactors, and only porphobilinogen synthase requires a metal activator, namely, zinc. These four extramitochondrial enzymes are retained in erythrocytes during maturation and after loss of the mitochondria.

Control of this biosynthetic pathway occurs principally via modulation of the activity of the first and rate-limiting enzyme, δ-aminolevulinate synthase. In vitro studies have shown that heme has two effects on this enzyme. First, heme can act as a repressor to control the rate of formation of δ-aminolevulinate synthase. Second, excess cellular heme can act by feedback inhibition to control the activity of formed δ-aminolevulinate synthase. Thus, any metabolic change that affects the rate of either formation or degradation of heme might also affect its biosynthesis. The exact nature of the intracellular localization and chemical nature of the heme(s) which constitutes this "repressor," "regulator," or "free" heme pool is not known. Feedback inhibitions of the other enzymes in the pathway by heme and porphyrins have been demonstrated through in vitro studies with various tissue preparations; however, there is yet little evidence to ascribe these inhibitions to normal physiological control processes. Since δ-aminolevulinate can be converted not only to porphyrins but also back to succinyl-CoA via the succinate-glycine cycle, some physiologic control of porphobilinogen synthase can be assumed, although this has received little study.

Free porphyrins that are found in body fluids and tissues arise as nonenzymatic oxidation byproducts of porphyrinogens from the heme biosynthetic pathway. Normally, only trace amounts of porphyrins escape the biosynthetic processes, because control mechanisms maintain the pathway in a delicate state of balance to meet the requirements of each cell. In erythrocytes, for example, about one molecule of excess protoporphyrin accumulates for each 30 000 molecules of heme that are formed in the process of hemoglobin synthesis, and losses of other porphyrins are at least an order of magnitude less than this. However, a number of pathologic conditions are associated with stimulated or inhibited heme biosynthesis leading to abnormal tissue levels or excretion rates of the porphyrins and their precursors. From a diagnostic point of view, abnormalities in porphyrin metabolism are considered only in terms of *increased* levels of porphyrins and their precursors. Impaired heme synthesis at the point of δ-aminolevulinate synthase or at subsequent metabolic steps may result in levels of porphyrins or precursors below the established reference ranges, but such low concentrations are also outside the capabilities of most analytical procedures. Moreover, no known clinical significance can be ascribed to the occurrence of these low concentrations.

PRIMARY DISORDERS OF HEME BIOSYNTHESIS

An in-depth discussion of the clinical correlations of porphyrin abnormalities is beyond the intended scope of this chapter; however, the topic has been reviewed (see Additional Readings). The metabolic abnormalities observed in the porphyrias result from inherited deficiencies of specific enzymes in the heme biosynthetic pathway (Table 16-1). The disorders which produce a secondary impairment of porphyrin metabolism will be discussed later. Most classifications of the porphyrias have utilized categorization of the diseases into hepatic and erythropoietic groups. This is a historical concept based upon the presumed major sites of excess porphyrin and porphyrin precursor production. Because of the variety of clinical disorders that share the term "porphyria," as well as their overlapping clinical and biochemical features, it is often difficult to know what specific laboratory studies are indicated when the question of porphyria is raised. Therefore, in clinical practice, a classification relating to symptomatology may be more useful (Table 16-2). In outlining a general approach for clinical laboratory participation in the diagnostic assessment of a patient suspected to have a porphyria, one must consider two clinical situations. First is the presence of a presumptive diagnosis of porphyria, and second is the confirmation and assignment of a specific type of porphyria to the patient in whom a porphyrin abnormality has been clinically identified. "Porphyrins," "porphyrin screen," "urine porphyrins," or "porphyria screen" are typical laboratory requests by physicians. These orders reflect the awareness that a disorder of porphyrin metabolism may be present, but they uncover the misconception that porphyrins are specific for "porphyria." The fact that porphyria is more than one disease should not be overlooked. The proposal to describe and classify the porphyrias primarily into broad clinical categories, i.e., neurologic and cutaneous, is an effort to simplify the clinical and chemical correlations in these disorders with the aim of improving the efficacy of the diagnostic evaluations of porphyria. Although this approach has some limitations, it simplifies the ordering and interpretation of laboratory tests for physicians and laboratories.

Neurologic (Acute Attack) Forms of Porphyria

Excess excretion of the porphyrin precursors δ-aminolevulinic acid and porphobilinogen are the characteristic chemical abnormalities in acute intermittent porphyria during symptomatic episodes. This correlation is usually not appreciated. In fact, the acute attack or neurologic forms of porphyria may be considered, for purposes of a simplified discussion, as disorders of porphyrin precursors. Each of the diseases that comprise this group reflects a specific enzyme defect that is inherited as an autosomal dominant trait. They include *acute intermittent porphyria, variegate porphyria,* and *coproporphyria.* The acute attack manifestations of each are identical. Symptomatic illness can be latent for indeterminant periods, and those individuals who experience symptomatic

Table 16-1. ENZYME DEFECTS IN PORPHYRINOPATHIES AND THEIR RELATIONSHIPS TO LABORATORY FINDINGS

Specific Enzyme Defect	Porphyrinopathy	Diagnostic Laboratory Findings
δ-Aminolevulinate synthase	None known; probably lethal lesions if primary	May increase secondary to impaired heme synthesis in some disorders
Porphobilinogen synthase	Symptoms not established	Limited to decreased enzyme activity
Uroporphyrinogen I synthase	Acute intermittent porphyria	↓ Uroporphyrinogen-I-synthase ↑ Urinary δ-aminolevulinic acid* ↑ Urinary porphobilinogen* ↑ Urinary uroporphyrin*
Uroporphyrinogen III cosynthase	Congenital erythropoietic porphyria	↑ Urinary uroporphyrin ↑ Urinary coproporphyrin ↑ Blood porphyrins ↑ Fecal porphyrins
Uroporphyrinogen decarboxylase	Porphyria cutanea tarda	↑ Urinary uroporphyrin ↑ Urinary 7-COOH porphyrin ↑ Fecal isocoproporphyrin
Coproporphyrinogen oxidase	Coproporphyria†	↑ Urinary δ-aminolevulinic acid* ↑ Urinary porphobilinogen* ↑ Urinary coproporphyrin ↑ Fecal coproporphyrin
Protoporphyrinogen oxidase	Porphyria variegata	↑ Urinary δ-aminolevulinic acid* ↑ Urinary porphobilinogen* ↑ Urinary coproporphyrin ↑ Fecal protoporphyrin/coproporphyrin
Ferrochelatase	Protoporphyria	↑ Blood protoporphyrin ↑ Fecal protoporphyrin

*Typical finding during acute neuropathy, may be normal in remission.
†The name "hereditary coproporphyria" is commonly used, but it is redundant because all porphyrias by definition are hereditary.

disease may experience one or multiple symptomatic episodes. During the acute phase, δ-aminolevulinic acid and porphobilinogen are excreted in the urine in excess, but during asymptomatic intervals these chemical abnormalities may resolve. Thus, although measurement of δ-aminolevulinic acid and porphobilinogen may be diagnostic at any particular time, normal results during quiescent periods do not exclude a diagnosis. Although each of these disorders has a unique enzyme abnormality, urine and fecal porphyrin excretion patterns are generally used in the clinical laboratory for establishing the specific form of porphyria (Table 16-1).

Symptoms of an acute attack may include abdominal pain, back pain, nausea, paresthesia, weakness, inability to think clearly, and self-destructive thoughts. Signs of acute attacks often include hypertension and tachycardia; constipation or diarrhea may occur, with the former more common; nausea is occasionally accompanied by vomiting. Paralysis, blindness (may be transient), and seizures are less common. Diaphoresis without fever or infection is part of the autonomic neuropathy. Acute attack episodes generally have a relatively brisk onset of hours to days, but characterization of a typical prodrome is impossible. The attack may last several days to several weeks. The spectrum of symptoms and signs as well as the duration and intensity of symptoms is quite variable among patients. Those individuals with recurrent attacks often experience similar symptomatology with each recurrence. Unusually severe illness with profound neurologic impairment cannot be anticipated, nor can the inciting agent or degree of exposure be correlated with the intensity of clinical disease. An acute attack may be precipitated by drugs and other factors, including sedatives and anticonvulsants, some steroid hormones, alcohol, and starvation by inducing ALA synthetase. Therefore, prevention is a primary mode of patient management. This, of course, requires recognition of the illness. For previously unrecognized patients, the diagnosis is often made after a protracted illness. As a result, neurologic disease may develop. The clinical course of an acute attack may include significant morbidity and occasional mortality.

Table 16-2. CLASSIFICATION OF PORPHYRINOPATHIES

Primary (inherited):
  Neurologic (acute attack)
    Acute intermittent porphyria (uroporphyrinogen I synthase deficiency)
  Cutaneous (photosensitive)
    Congenital erythropoietic porphyria (uroporphyrinogen cosynthase deficiency)
    Porphyria cutanea tarda (uroporphyrinogen decarboxylase deficiency)
    Protoporphyria (ferrochelatase deficiency)
  Mixed (neurologic and cutaneous)
    Coproporphyria (coproporphyrinogen oxidase deficiency)
    Porphyria variegata (protoporphyrinogen oxidase deficiency)
  Asymptomatic (?)
    Porphobilinogen synthase deficiency
Secondary (induced):
  Coproporphyrinuria
    (Examples: tyrosinemia, lead poisoning, alcoholism)
  Protoporphyrinemia
    (Examples: iron deficiency, lead poisoning, inflammation)

The mortality figures for the acute forms of porphyria, especially acute intermittent porphyria, have decreased over the years. This trend is often attributed to improved physician awareness of porphyria as a diagnostic possibility, but it is also due to the avoidance of contraindicated medication or potential precipitating factors in individuals known to have the disease. Considering the importance of prevention and the realization that the most severe clinical problems typically arise in individuals in whom the diagnosis is unsuspected, genetic counseling with appropriate diagnostic studies to identify family members with latent states of porphyria is essential.

Cutaneous (Photosensitive) Forms of Porphyria

The common feature linking this group of diseases is the characteristic excess porphyrin production and excretion (Table 16-1). A bullous dermatosis occurs in **porphyria cutanea tarda**, which is a cutaneous disorder only. Lesions are confined to the dorsum of the hands, face, and ears. Porphyria cutanea tarda is similar to the mixed porphyrin diseases, *variegate porphyria* and *coproporphyria*, in having a distribution of cutaneous lesions related to photosensitivity and a temporal relationship between exacerbation of disease and the sunnier months of the year. However, the direct relationship of sun exposure to the development of skin lesions is frequently not appreciated by affected individuals. Skin fragility and the resulting lesions are probably the most characteristic symptoms in *porphyria cutanea tarda*. Pigment changes, scarring, and milia accompany chronic skin disease. Despite the presence of this disease, the skin on the dorsum of the hands may be remarkably thin and delicate in appearance. Hypertrichosis is frequent. Acute attacks do not occur, and neither neurologic nor psychotic symptoms are present.

Porphyria cutanea tarda has been considered an acquired disease more often than other forms of porphyria. Recently, however, uroporphyrinogen decarboxylase deficiency has been identified in familial clusters with this disorder. These pedigrees have suggested Mendelian dominant inheritance. Further family studies using the measurement of uroporphyrinogen decarboxylase activity have confirmed autosomal dominant transmission of a deficiency of this enzyme in both affected and nonaffected family members. The toxic or acquired cases may result from inhibition of uroporphyrinogen decarboxylase, or from irreversible oxidation of porphyrinogens to porphyrins in the cytosol of hepatocytes with resultant accumulation and excretion of porphyrins. Excessive alcohol consumption is commonly associated with this disorder. However, iron accumulation has been shown to be the major etiologic factor due to an inhibitory effect of iron on uroporphyrinogen decarboxylase. Estrogen exposure in older males being treated for prostate carcinoma and in younger women using oral contraceptives is another etiologic setting. Porphyria cutanea tarda is probably the most common inherited porphyrin disorder in North America.

Congenital erythropoietic porphyria is the rarest form of inherited porphyrin disorders. It is unique among the porphyrias for having an autosomal recessive inheritance. The basic defect is a deficiency of uroporphyrinogen cosynthase in the bone marrow and accumulation of series I isomers. Dark, red urine and severe photosensitivity are typically observed in the neonatal period, but patients with onset in adulthood have been reported. Porphyrins stain the bones and teeth (erythrodontia) of these individuals, which results in a bright fluorescence of the stained areas on exposure to long-wavelength ultraviolet light. Porphyrin deposition in these tissues is a result of the avidity of water-soluble porphyrins for calcium-containing structures. This phenomenon also suggests that excess porphyrin production was occurring in fetal life. Ineffective erythropoiesis, hemolytic anemia, and splenomegaly complicate the clinical course.

Protoporphyria is the result of a ferrochelatase deficiency that leads to protoporphyrin accumulation (Table 16-1), which is associated with severe photosensitivity. The cutaneous reaction occurs within 15–20 min of sun exposure, and this correlation of symptoms with exposure is readily recognized by affected people. The skin reaction begins with burning and itching, followed by eventual painful swelling and erythema, which may last for 48 h or longer. The photosensitivity in protoporphyria can become quite disabling, and photoprotection is important. A therapeutic breakthrough has been the observation that β-carotene affords photosensitivity protection. Of course, carotene is available in carrots and other foods, but consumption of protective amounts of the pigment in its biologic form can be very difficult. A preparation of carotene in capsule form has now been marketed (Solatene) and clinical trials have demonstrated its efficacy. Use of carotene must be planned, however, since several months may be needed to build up adequate tissue levels.

A discussion of the clinical features of protoporphyria would be incomplete without noting the occurrence of hepatobiliary disease in a subgroup of these patients. Liver decompensation may occur abruptly with jaundice and evidence of portal hypertension (ascites, esophageal varices). The major route of protoporphyrin excretion, regardless of erythrocyte or hepatic source, is via the liver. Impaired liver function and bile excretion in this group of patients connote hepatic injury secondary to protoporphyrin accumulation in the liver parenchyma. The large enterohepatic circulation of protoporphyrin can be a very ominous development, since further hepatic protoporphyrin accumulation may then occur. Often, cholelithiasis occurs in protoporphyria at an unusually early age. In some instances, protoporphyrin deposition has been documented in stones that were removed surgically.

Porphobilinogen synthase deficiency is a recently described disorder with an autosomal dominant inheritance. Individuals can be totally asymptomatic, but identification and study of further cases are necessary in order to elucidate the true clinical significance of this enzyme deficiency.

SECONDARY DISORDERS OF HEME BIOSYNTHESIS

A variety of clinical conditions besides porphyrias are accompanied by excess accumulation and excretion of porphyrins or porphyrin precursors. Moreover, symptoms in these conditions may be indistinguishable from those of the porphyrias. In such cases, the disturbance of porphyrin metabolism is a result of a superimposed disorder or toxin and is usually not a result of an inherited defect in the heme biosynthetic pathway. It is interesting to note that those disorders which produce the clinical picture of an acute attack of porphyria exhibit increased amounts of only δ-aminolevulinic acid in the urine, while porphobilinogen excretion is not increased. This demonstrates the specificity of porphobilinogen for acute intermittent porphyria, and may offer a clue with regard to an important etiologic factor in the clinical illness. The two diseases in this category are *lead poisoning* and *hereditary tyrosinemia*. Lead reversibly inhibits both the activity of porphobilinogen synthase and the incorporation of iron into heme. In addition to increased urine excretion of δ-aminolevulinic acid, erythrocyte protoporphyrin concentration (as a zinc chelate) is increased. Urine coproporphyrin is elevated as a delayed response; in fact, chronic lead poisoning is one of the likely causes of significant coproporphyrinuria. Despite the excess coproporphyrin in the urine and protoporphyrin in erythrocytes, individuals with lead poisoning do not experience photosensitivity. Hereditary tyrosinemia also produces an acute porphyria-like illness. One of the metabolites accumulating in excess (succinyl acetone) is a potent inhibitor of porphobilinogen synthase. Interestingly, coproporphyrinuria occurs in this disorder as well. *Heavy metal intoxication* may produce a skin change unrelated to secondary coproporphyrinuria. However, metals such as mercury, bismuth, copper, gold, silver, and arsenic do not cause elevation of urinary δ-aminolevulinic acid or erythrocyte protoporphyrin.

All conditions which produce an imbalance between protoporphyrin formation and iron availability can lead to accumulation of erythrocyte protoporphyrin or its zinc chelate. This abnormal metabolite is best known for its marked increase in chronic lead exposure. Zinc protoporphyrin also increases in iron deficiency, but this has received considerably less attention. This abnormal porphyrin chelate circulates in the erythrocyte attached to a heme site on globin, where it serves no biological function. The mechanism of formation, whether enzymatic or nonenzymatic, is unknown. Sideroblastic anemia, in which iron is not properly utilized; hemolytic anemia, in which erythropoiesis is greatly accentuated; secondary polycythemia, in which there is a stimulus to erythropoiesis extrinsic to the marrow; excessive erythrocyte destruction; and the commonly seen inflammatory block are all associated with elevated erythrocyte protoporphyrin or its zinc chelate. Anemia accompanying chronic disease has also been stated to cause elevated erythrocyte protoporphyrin. The conditions of heme biosynthesis associated with secondary protoporphyrinemia typically have protoporphyrin concentrations below 400 μg/dL of erythrocytes, while erythropoietic protoporphyria typically produces erythrocyte protoporphyrin concentrations in considerable excess of those occurring in these secondary disorders.

Coproporphyrin is by far the most common porphyrin excreted in secondary porphyrinuria. Hexachlorobenzene, alcohol, sedatives, and hypnotics such as chloral hydrate, morphine, ether, and nitrous oxide as well as lead can all cause coproporphyrinuria. Also, neoplasia, liver disease, myocardial infarction, and thalassemia have on occasion been associated with coproporphyrinuria.

DIAGNOSIS OF PORPHYRIN DISORDERS

Porphyrins and Porphyrin Precursors

Essentially, the quantitation of three porphyrins comprises the basis for the laboratory diagnosis of most disorders affecting porphyrin metabolism (Table 16-1). These are uroporphyrin, with eight carboxyl substituents; coproporphyrin, a four-carboxyl molecule; and protoporphyrin, which has two carboxyl groups on the tetrapyrrole ring (Figure 16-1). Certain other porphyrins can also be identified in the blood, urine, or feces, although they have less diagnostic importance. For example, the heptacarboxyl porphyrin is recognized as a major excretion product in porphyria cutanea tarda. Demonstration of this porphyrin in the urine may be helpful but is usually not essential for the diagnosis of this porphyria. Isocoproporphyrin, which has an ethyl group substituent in the 2 position, is unique to porphyria cutanea tarda; it is identified by chromatography of fecal porphyrins. Although isocoproporphyrin may be helpful in differentiating porphyria cutanea tarda from the mild photosensitive dermatoses of variegate and coproporphyria, isocoproporphyrin is not widely measured. Similarly, the determination of isomers of uroporphyrin and coproporphyrin may be necessary for the study of possible defects in heme biosynthesis, but isomer separation is not necessary to diagnose any specific form of porphyria. Recent association of partial deficiencies of specific enzymes in the heme biosynthetic pathway with unique types of porphyria may eventually replace comparative chemical studies of porphyrin excretion and blood levels for the diagnosis of specific porphyrin disorders. Nonetheless, initial laboratory needs, when a diagnosis of porphyria is suspected, will continue to include assessment of the excretion of porphyrins (uroporphyrin, coproporphyrin, or protoporphyrin) or porphyrin precursors (δ-aminolevulinic acid and/or porphobilinogen).

Selection and Interpretation of Laboratory Tests

If an acute neurologic porphyria is suspected, the primary screening procedure is a porphobilinogen determination. Increased urinary porphobilinogen is the characteristic chemical abnormality of acute intermittent porphyria, variegate porphyria, and coproporphyria during an acute attack. Since porphobilinogen concentration in the urine may or may not be increased at nonsymptomatic times, screening procedures should be reserved for the specific question of whether a patient's acute symptomatology is a result of a neuropathic porphyria. False negatives are very rare and false positives are uncovered by assay of porphobilinogen in a 24-h urine sample. A negative screening test is usually reliable in a symptomatic patient. When porphyria is strongly suspected on clinical grounds and quantitative urine studies are performed, it is worthwhile to obtain both δ-aminolevulinic acid and porphobilinogen concentrations. The two measurements allow a differentiation of lead poisoning and tyrosinemia, two disorders that resemble acute porphyria clinically, but typically exhibit normal urinary porphobilinogen and elevated urinary δ-aminolevulinic acid levels. After a porphyria diagnosis has been confirmed, screening tests are not appropriate for following a patient's course.

Since δ-aminolevulinic acid is most stable at acid pH and porphobilinogen is most stable at alkaline pH, organizing a proper urine collection procedure is difficult. This problem is compounded when uroporphyrin and coproporphyrin measurements are also requested. A proven, practical approach, albeit a compromise, is the collection of a 24-h urine specimen that is kept in the cold, protected from light, and with sodium bicarbonate added to maintain pH near neutrality. This provides maximum stability of both precursors as well as the porphyrins in the same specimen.

Urine screening for total porphyrins has only limited specificity for the diagnosis of porphyrinopathies, because there are several causes of porphyrinuria. Moreover, urine porphyrin composition can vary greatly, as discussed previously. Nevertheless, a total porphyrin screen is often requested to identify a suspected neuropathic porphyria. Since excess porphyrins in addition to porphobilinogen can be present in these porphyrias (Table 16-1), a misleading screening result may compound itself with inappropriate quantitative studies, i.e., obtaining quantitative porphyrin data while ignoring porphobilinogen.

Uroporphyrinogen I synthase deficiency, unlike the level of urine porphobilinogen excretion, is a constant finding in acute intermittent porphyria. Also, the deficient state conforms to the expected distribution of an autosomal dominant trait. Latent cases, i.e., without clinical or urine abnormalities, can be identified on the basis of decreased uroporphyrinogen I synthase activity. This is often very useful for counseling family members and in attempting to prevent morbidity and potential mortality from the disease. The uroporphyrinogen I synthase deficiency is specific for acute intermittent porphyria. The two other acute attack forms of porphyria may have an identical clinical and chemical picture during the acute phase, but the uroporphyrinogen I synthase activity will be normal in these porphyrin disorders.

The photosensitive porphyrias are best characterized chemically as disorders of porphyrin excess, as opposed to porphyrin precursor excess. In addition to urine porphyrins, erythrocyte protoporphyrin assay is essential to diagnose photosensitive porphyria; protoporphyrin is not excreted in the urine. Fecal porphyrin assays can be helpful in diagnosing coproporphyria, and these are often considered essential to confirm a diagnosis of porphyria variegata.

Other enzyme assays. Specific enzyme deficiencies along the entire heme biosynthetic pathway have now been described (Figure 16-3, Table 16-1). Demonstration of enzyme deficiencies in erythrocytes, liver, cultured skin fibroblasts, and peripheral blood leukocytes confirms the distribution of enzyme deficiencies among tissues in affected individuals. These new biochemical insights are leading to precise diagnostic capabilities not previously available. However, the feasibility of routine enzyme analysis depends on the accessibility of a suitable tissue to be tested, the stability of the enzyme, and finally the substrate requirements and techniques for assay. Porphyrin fluorescence is a useful property for assay design, but as evident in the pathway (Figure 16-3), a porphyrin is both substrate and product in some reactions. Those enzymes that are present in the cytosol of cells permit easily accessible tissue, such as erythrocytes, to be used for assays. By contrast, mitochondrial enzymes must be measured in fibroblasts, leukocytes, or specific organ tissues. For these reasons, only the measurement of uroporphyrinogen I synthase is readily available at this time in service laboratories; tests for the remainder of the heme biosynthetic enzymes remain research laboratory tools.

ANALYSIS OF PORPHYRIN PRECURSORS

Two porphyrin precursors, δ-aminolevulinic acid and porphobilinogen, accumulate or are overproduced in the neuropathic porphyrin disorders. When used in clinical diagnosis, porphyrin precursors are almost exclusively measured in urine. Both screening tests and quantitative assays are used for porphobilinogen, but no screening test exists for δ-aminolevulinic acid. Serum concentrations can provide clinically useful information, but the much lower concentrations of the porphyrin precursors make their analysis somewhat more difficult; hence, serum assays are restricted to research laboratories.

Determination of δ-Aminolevulinic Acid

Principle

δ-Aminolevulinic acid is condensed with ethyl acetoacetate to form a pyrrole. This derivative is purified by a simple extraction into ethyl acetate, which obviates the need for commonly recommended ion-exchange chromatography. The extracted pyrrole derivative in ethyl acetate is then reacted with Ehrlich's reagent to give a cherry-red colored compound that is measured spectrophotometrically.

Specimen

A 24-h urine collection is obtained and the total volume recorded. The urine collection container should be refrigerated and should contain 2 g barbituric acid to preserve the δ-aminolevulinic acid. If porphyrins or porphobilinogen is also to be assayed, substitute 4–5 g of sodium bicarbonate for barbituric acid to ensure a near neutral pH.

Reagents

1. Acetate buffer, pH 4.6. To 700 mL of water, add 57 mL of glacial acetic acid and 136 g of sodium acetate trihydrate. Dilute to 1 liter with water.
2. Ethyl acetoacetate, reagent grade.
3. Ethyl acetate, reagent grade.
4. Ehrlich's reagent.* To about 30 mL of glacial acetic acid in a 50-mL cylinder, add 1.0 g of p-dimethylaminobenzaldehyde, 5 mL of 70% perchloric acid, and 5 mL water. Dilute to 50 mL with glacial acetic acid.
5. δ-Aminolevulinic acid standard. Stock δ-aminolevulinic acid standard is prepared by dissolving 0.128 g of the commercially available hydrochloride in 100 mL of water to obtain 1000 μg δ-aminolevulinic acid/mL. Dilute 0.10 mL of this stock solution to 10 mL with water to obtain 10 μg/mL. A standard curve is prepared by further diluting this solution as follows: 0.10-mL stock solution to 1.0 mL with water, 0.20-mL stock solution to 1.0 mL with water, . . . continuing in 0.10-mL increments to a maximum of 1.0 mL undiluted stock solution. This gives a standard curve ranging from 1.0–10 μg/mL. The stock solution is stable for 2 months at 4 °C.

Procedure

1. Place 1.0 mL urine in glass-stoppered centrifuge tubes, 3 tubes being required for each urine specimen (duplicate specimens plus 1 blank).
2. Add 1.0 mL acetate buffer to each tube.
3. Add 0.20 mL ethyl acetoacetate to each specimen tube and to a water blank but NOT to the patient specimen blank. Mix by vortexing, stopper, and place in a boiling water bath for 10 min.
4. Allow the tubes to cool, or hasten cooling by placing tubes into ice or cold water.
5. Add 3.0 mL ethyl acetate to each tube, stopper, and shake vigorously 50 times.
6. Remove the glass stoppers and centrifuge lightly (about $1000 \times g$) to separate the phases. A short (1 min) centrifugation will not cause significant evaporation.
7. Using a volumetric pipet, withdraw 2.0 mL of the upper ethyl acetate phases and transfer to separate tubes.
8. Add 2.0 mL Ehrlich's reagent to each tube. Vortex-mix the reactants and allow to stand 10 min.
9. Measure the absorbance immediately at 553 nm, using water to zero the spectrophotometer.

Calculation

Subtract the blank reading from each specimen reading, then determine the μg δ-aminolevulinic acid/mL urine from the standard curve. Calculate the 24-h excretion as follows:

$$\text{mg } \delta\text{-aminolevulinic acid/d} = \frac{\mu\text{g } \delta\text{-aminolevulinic acid/mL}}{1000} \times TV$$

where

$$1000 = \text{conversion of } \mu\text{g to mg}$$

$$TV = \text{total volume (mL) of urine collected in 24 h}$$

Reference Range

δ-Aminolevulinic acid: 1.5–7.5 mg/d

---

*Several modifications of Ehrlich's reagent have been used, each varying in the kind and concentration of acid solvent and the concentration of p-dimethylaminobenzaldehyde. This is the result of attempts to obtain a stabilized chromophore and to enhance its absorbance. The particular Ehrlich's reagent indicated is that described in the original method.

Reference

Tomokumi, K., and Ogata, M.: Clin. Chem., *18*:1534, 1972.

METHODS FOR THE DETECTION OF PORPHOBILINOGEN

Principle

Porphobilinogen condenses with *p*-dimethylaminobenzaldehyde in acid solution (Ehrlich's aldehyde reagent) to form a magenta-colored product (Figure 16-4). Since this reaction can occur with other urinary constituents, the two screening tests use pH adjustment and solvent extractions to remove interfering substances, thus making the tests reasonably specific. For quantitative analysis, porphobilinogen is purified by adsorption to an ion-exchange resin. Color-producing interfering substances, such as urobilinogen, methyldopa, or chlorpromazine, as well as indole and related compounds which interfere by reacting with the chromophore to produce colorless derivatives, may be removed by repeated washings of the column with water prior to the elution of porphobilinogen with acetic acid.

Specimen

Screening tests for porphobilinogen are carried out preferably on a fresh morning specimen. Quantitative analyses should be performed on 24-h urine collections. If the pH of the urine is adjusted to near neutrality (pH 6–8) with sodium bicarbonate, the specimen can be stored for periods up to two weeks frozen, although assays should be performed as soon as possible.

Reagents

1. Ehrlich's reagent, qualitative (Watson-Schwartz test).* Dissolve 0.7 g *p*-dimethylaminobenzaldehyde in 150 mL of concentrated HCl and add 100 mL water. Store in a brown bottle. Although the reagent becomes increasingly yellow with storage, it can be used for a period of several months.

2. Ehrlich's reagent, qualitative (Hoesch test).* Dissolve 2.0 g of *p*-dimethylaminobenzaldehyde in 100 mL HCl, 6 mol/L. The reagent can be stored without loss of activity for at least nine months.

3. Ehrlich's reagent, quantitative.* Dissolve 2.0 g of *p*-dimethylaminobenzaldehyde in 25 mL of concentrated HCl plus 75 mL of glacial acetic acid. The reagent becomes increasingly yellow on prolonged storage. These changes can be retarded by refrigeration in the dark.

4. Sodium acetate, saturated. An excess of sodium acetate · $3H_2O$ (approximately 1 g/mL) is added to water and the reagent is stored at room temperature.

5. Chloroform, reagent grade.

6. *n*-Butanol, reagent grade.

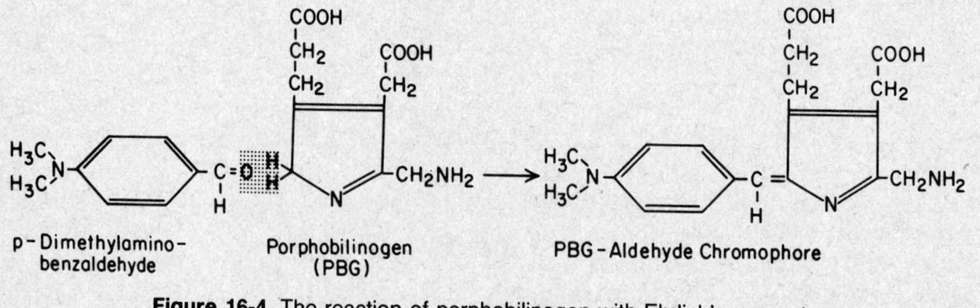

Figure 16-4. The reaction of porphobilinogen with Ehrlich's reagent.

*Several modifications of Ehrlich's reagent have been used, each varying in the kind and concentration of acid solvent and the concentration of *p*-dimethylaminobenzaldehyde. This is the result of attempts to obtain a stabilized chromophore and to enhance its absorbance. The particular Ehrlich's reagent indicated is that described in the original method.

7. Acetic acid, 1.0 mol/L.

8. Ammonia, reagent grade, 28 g NH_3/100 g.

9. Dowex 2-X8 resin (200–400 mesh), acetate form. Suspend 100–200 g of Dowex 2-X8 resin in about 4 volumes of sodium acetate solution (10 g/dL). Mix the suspension well and allow it to settle by gravity until the supernatant solution can be easily decanted. Wash the resin 8–10 times each with about 4 volumes of deionized water, decanting each time to remove both the fines and the excess sodium acetate. A stock suspension of resin is prepared by adding an equal volume of water to the washed, packed resin. Immediately before use, the suspension should be shaken thoroughly, since the resin settles rapidly. This resin preparation is stable for several months.

Screening Tests for Porphobilinogen in Urine

Procedure (Watson-Schwartz Test)

1. To 2.0 mL of urine in a test tube, add 2.0 mL of Ehrlich's reagent, qualitative (Watson-Schwartz), and mix.

2. Add 4.0 mL of saturated sodium acetate and mix again. Check with pH indicator paper to confirm that the solution is in the range of pH 4–5.

3. Add 5 mL of chloroform, stopper, and shake vigorously for 1 min. Permit the phases to separate; centrifuge if necessary.

4. Transfer the upper (aqueous) phase with a pipet to another tube, to which is added 2 mL of n-butanol. Stopper and shake vigorously for 1 min. Allow the phases to separate; centrifuge if necessary.

5. Examine the lower (aqueous) layer for a magenta color. If this color is present, the test is positive for porphobilinogen, indicating a concentration several times normal. A magenta color in the upper (butanol) layer indicates increased urobilinogen. If the interpretation is ambiguous, proceed to the Hoesch test.

Procedure (Hoesch Test)

1. Place 2 mL of Ehrlich's reagent, qualitative (Hoesch), into a test tube, then add 3 drops of urine.

2. If increased porphobilinogen is present, an immediate cherry-red color will develop at the point of mixing.

Comments (Watson-Schwartz and Hoesch Tests)

The best-known qualitative procedure for the detection of porphobilinogen is the Watson-Schwartz test; a modification, the Hoesch test, has been introduced more recently. Despite modifications to improve the specificity of these tests, an inexperienced person can have difficulties with interpretation of some results with either the Watson-Schwartz or Hoesch test. Each of these two screening methods has its unique features and yields useful information, but their clinical value is often enhanced by familiarity with the procedure. An effective solution to the uncertainties of interpretation is the combined use of the Watson-Schwartz and the Hoesch tests, which can be performed simultaneously using similar reagents. Importantly, the Hoesch test allows no reaction with urobilinogen, thereby serving primarily as an added means of eliminating this interference and confirming the results of the Watson-Schwartz test.

In the Watson-Schwartz test, porphobilinogen and the chromogen that it forms always remain in the aqueous phase. The extractions with chloroform and butanol are essential for removing frequently occurring substances that interfere with the test. If no magenta color is observed in the upper (aqueous) phase following the chloroform extraction, the n-butanol extraction may be omitted and the test considered negative. The most common interfering substance is urobilinogen (chloroform soluble), which produces a color with Ehrlich's reagent similar to porphobilinogen. Other substances sometimes present in urine can give a variety of colors, including yellow, orange, and red. All of these tend to make a positive identification of porphobilinogen difficult. Under these circumstances, the quantitative procedure for porphobilinogen should be performed, because this method removes the most common interfering substances, as well as the decolorizing indoles, to permit positive identification of porphobilinogen.

References

Watson-Schwartz Test:

Schwartz, S., Berg, M. H., Bossenmaier, I., et al.: *In*: Methods of Biochemical Analysis, Vol. VIII. D. Glick, Ed. New York, Interscience Publishers, Inc., 1960, p. 249.

Hoesch Test:
Lamon, J., With, T. K., and Redeker, A. G.: Clin. Chem., *20*:1438, 1974.

Quantitative Determination of Porphobilinogen in Urine

Procedure

1. Pipet 4 mL of well-mixed Dowex 2-X8 resin suspension into a 12-mL centrifuge tube and centrifuge about 1 min; discard the supernatant solution. Pipet 1 mL of urine onto the packed resin, followed by 0.1 mL of concentrated ammonium hydroxide (28 g NH_3/100 g). Stir the mixture thoroughly and centrifuge about 1 min; discard the supernatant solution.

2. Wash the resin, to which the porphobilinogen has been adsorbed, 4 times with 5-mL aliquots of deionized water to remove interfering substances. Centrifuge about 1 min after each wash.

3. Elute the porphobilinogen from the washed resin with four 2-mL aliquots of acetic acid, 1 mol/L. After each acetic acid addition, centrifuge the resin for 1 min and decant the supernatant solution into a 12-mL graduated centrifuge tube, adjusting the final volume to exactly 10.0 mL with acetic acid, 1 mol/L. Mix the solution and centrifuge briefly to remove any traces of resin which may have been carried over during decanting.

4. Treat 2 mL of the acetic acid eluate with 2 mL of Ehrlich's reagent (quantitative). Prepare a reagent blank, substituting 2 mL of acetic acid, 1 mol/L, for the eluate. Thoroughly mix the reactants. After allowing the unknowns to stand 6–8 min, read them against the reagent blank at 555 nm and 525 nm in a spectrophotometer having a resolution of 1–2 nm or better.

Calculation

Porphobilinogen excretion is calculated as follows:

$$\text{mg of porphobilinogen/d} = \frac{A_{555} \times 10 \times \text{TV}}{0.114 \times 1000}$$

where

0.114 = absorbance of the reaction product in a 10-mm light path when the porphobilinogen concentration is 1 μg/mL

10 = dilution of the original urine aliquot with acetic acid

TV = total volume (mL) of urine collected in 24 h

1000 = μg to mg conversion factor.

Comments

The factor 0.114 has been determined empirically according to the conditions outlined and using δ-aminolevulinic acid as a standard. δ-Aminolevulinic acid was condensed by heating with excess acetylacetone to give a pyrrole which reacts like porphobilinogen in this test. (Alternatively, porphobilinogen can now be obtained commercially, but it is somewhat more expensive and unstable.)

If porphobilinogen is present in significant amounts, the color developed by Ehrlich's reagent should be rose to crimson, and the ratio of A_{525}/A_{555} should be near 0.83. A ratio of A_{525}/A_{555} > 1.00 is rare and indicates that interfering substances are still present and that the result should not be interpreted as an abnormal concentration of porphobilinogen. Alternatively, an Allen correction can be applied by measuring the absorbance at 535, 555, and 575 nm, and then utilizing the formula $A_{corrected} = 2A_{555} - (A_{535} + A_{575})$. Calculating with this corrected absorbance helps to eliminate the effects of interfering chromogens, but a chomatographic purification remains essential. With a porphobilinogen concentration two to three times normal, which is clinically significant, no difficulty is likely to be encountered in its quantitation.

The reaction of porphobilinogen with Ehrlich's reagent yields a product whose absorbance follows Beer's law from the lower limits of detection through an absorbance of at least 0.750. If the absorbance is too high, an appropriate dilution of the acetic acid eluate should be made, or preferably less of the original sample should be used and a correction made in the dilution factor of the calculation. Upon addition of Ehrlich's reagent, maximum color develops within 6 min. The color remains stable for 2–3 min and then begins to fade slowly. Within 20 min absorbance decreases by about 10%.

A seemingly large amount of resin is used for the amount of porphobilinogen to be adsorbed; however, the optimum ratio for the batch process has been found to be 1 volume of urine to 2 volumes of packed resin. The concentration of acetic acid required to elute porphobilinogen from the resin is not critical, but four elutions are essential to obtain quantitative recovery of the porphobilinogen. The assay has a CV of 7.0%. As an alternative to the batch chromatographic purification, commercially available columns can be substituted, with the remainder of the procedure unchanged.*

Reference Range

The normal urinary excretion rate of porphobilinogen is < 1 mg/d; however, caution should be exercised in interpretation of only slightly elevated values as diagnostic of porphyria.

Reference

Moore, D. J., and Labbe, R. F.: Clin. Chem., *10*:1105, 1964.

METHODS FOR PORPHYRIN ANALYSES

Principle

Virtually all porphyrin analyses are based on the isolation of the porphyrin(s) from the specimen, separation of the individual prophyrins, and observation or measurement of the porphyrin(s) by fluorometry or spectrophotometry. Porphyrins are usually isolated from body excreta or tissues by extraction into an acidified organic solvent. For screening purposes, minimal purification is required. For quantitation, the individual porphyrins are separated by selective solvent extraction or by chromatography. Their characteristic orange-red fluorescence (620–630 nm) on irradiation with long-wavelength ultraviolet light (398–408 nm) allows porphyrins in acid solutions to be detected fluorometrically at concentrations below 10^{-8} mol/L. Alternatively, when the concentrations of porphyrins are sufficiently high, they can be measured spectrophotometrically.

Specimens

Whole blood specimens are collected using any common anticoagulant.

Urine for screening purposes should preferably be a morning specimen, but random specimens can be used. Since the assay of a single urine specimen provides less meaningful data with no reference range, quantitative analyses should be performed on 24-h collections. In this case, the urine should be collected in a container to which 4–5 g of sodium bicarbonate has been added to maintain the specimen at pH 6–8.

If *feces* are used only for qualitative tests, a small specimen (1 g) is adequate.

When porphyrin analyses cannot be performed soon after collection of the specimens, they should be stored in the dark at 4 °C, or they should be kept frozen if more than 1–2 d elapse before analysis.

SCREENING TESTS FOR PORPHYRINS

All screening tests are performed in glass apparatus. Plastic tubes must be avoided because fluorescence quenching by many synthetic materials will interfere with observance of the porphyrin fluorescence.

---

*Prepared chromatographic columns for porphobilinogen assay are supplied by Bio-Rad Laboratories, Richmond, CA 94804.

Reagents

1. Hydrochloric acid, 3 mol/L: Dilute 25 mL of concentrated HCl, reagent grade, to 100 mL with water.
2. Ethyl acetate:glacial acetic acid (4:1): Mix reagent grade solvents in the indicated volume ratio.

Equipment

For all porphyrin screening tests (blood, urine, feces), use a long-wavelength ultraviolet or Wood's lamp of adequate intensity; a 100-W mercury spot bulb (7000 μW/cm^2 at 15 inches) fitted with a red-purple filter is particularly useful. These are available through many laboratory suppliers.

Qualitative Determination of Urinary Porphyrins

Procedure

Place 5 mL of urine plus 3 mL of ethyl acetate:acetic acid into a centrifuge tube and shake thoroughly. Allow the phases to separate by standing, or hasten the process by a brief centrifugation. Irradiate the upper (organic) layer with ultraviolet light and observe for porphyrin fluorescence. This layer fluoresces lavender with moderately elevated concentrations of porphyrins, then pink, and finally red with markedly increased porphyrin concentration. Since drugs, abnormal metabolites, or unknown materials will occasionally impart a fluorescence or color to the organic layer that can make interpretation difficult, this upper layer should be removed carefully with a pipet and transferred to a second tube to which is added 0.5 mL of HCl, 3 mol/L. With mixing, the porphyrins are extracted into the lower acid phase, while most interfering substances remain in the organic phase. In an acid medium, porphyrin fluorescence is greatly intensified; moreover, the bluish tint of common urine components is removed by the extraction to give the orange-red fluorescence typical of porphyrins. Urine specimens from healthy subjects can be used as a guide for normal levels of porphyrins.

Qualitative Determination of Fecal Porphyrins

Procedure

Transfer a small portion of feces (about 50 mg or a volume about the size of a large drop of water) from the end of a glass rod into a 12-mL centrifuge tube. Add 3 mL of ethyl acetate:acetic acid and stir vigorously. Centrifuge briefly and decant the supernatant solution into a second centrifuge tube containing 0.5 mL of HCl, 3 mol/L. Shake or vortex the solutions well for 1 min and allow to separate, or centrifuge briefly. Observe the lower (aqueous) phase in ultraviolet light for porphyrin fluorescence. Normally, this acid solution shows little or no fluorescence; porphyrins, when present in increased concentration, will impart a clearly distinguishable orange-red fluorescence.

The occasional presence of dietary chlorophyll in fecal matter may cause confusion, because it also has a red fluorescence that is not easily distinguishable from that of porphyrins. Even though chlorophyll is insoluble in aqueous solution and remains in the upper organic layer, it can exhibit a deep red fluorescence that might cause uncertainty in interpretation of the test. When high levels of chlorophyll appear to be present, the upper organic layer can be drawn off and discarded and a fresh 3-mL aliquot of ethyl acetate:acetic acid used to extract the remaining HCl layer. This procedure will remove the chlorophyll interference, and any remaining orange-red fluorescence in the aqueous phase can be considered due to porphyrins.

Qualitative Determination of Blood Porphyrins

Procedure

Place 1 mL of whole blood into 3 mL of ethyl acetate:acetic acid in a centrifuge tube. Stir the mixture thoroughly, centrifuge, and decant the supernatant solution into a second centrifuge tube containing 0.5 mL of HCl, 3 mol/L. Shake or vortex the two phases thoroughly and allow to separate. Illuminate the HCl layer with an ultraviolet light. An orange-red fluorescence indicates porphyrins are present. With these volumes of solutions, normal blood will usually show just a trace of porphyrin fluorescence. This will be equivalent to about 60 μg of porphyrin/dL erythrocytes, or near the upper end of the normal range. A distinct porphyrin fluorescence is seen at concentrations of two to three times normal, becoming deep orange-red as the concentration increases further.

Virtually all of the porphyrins found in blood are located within the erythrocytes; the plasma contains only traces of porphyrins. Therefore, an analysis of whole blood actually reflects the porphyrin content of the erythrocytes. Increased cellular porphyrin concentrations can also be easily observed by fluorescence microscopy of an unfixed, unstained blood smear. However, the intense ultraviolet irradiation focused on the slide quickly destroys the protoporphyrin, causing its fluorescence to fade rapidly in the field of view. By contrast, uroporphyrin and coproporphyrin are quite stable under these conditions.

Reference

Haining, R. G., Hulse, T., and Labbe, R. F.: Clin. Chem., *15*:460, 1969.

Identification of Porphyrins

When abnormal levels of porphyrins are seen or suspected in a screening test, identification of the specific porphyrin or porphyrins that are elevated can aid in diagnosis. Identification can be accomplished very simply with thin layer chromatography. Cut Gelman Instant TLC Sheets Type SA (Gelman Sciences, Inc., Ann Arbor, MI) with a sharp knife on a clean flat surface to approximately 1.1 × 6.6 cm; handle the strips with forceps to avoid contamination. Store the strips in a clean test tube or Petri dish. To separate the porphyrins, transfer about 30 μL from the ethyl acetate layer to the chromatographic strip, placing the spot ~1.5 cm from one end. (*Note:* Depending upon the estimated fluorescence, as little as 10 μL or as much as 60 μL can be spotted.) Apply the solutions slowly and dry with a hair dryer or similar device during the course of addition to minimize dispersion of the spot. Place the TLC strip into a 10-mL screw-topped vial (21 × 70 mm) containing about 1.0 mL of developing solvent. (Prepare by mixing 1.2 mL chloroform, 1.4 mL methanol, and 0.3 mL ammonium hydroxide.) Allow the solvent to ascend to near the top of the strip (5–10 min). Remove the strip from the vial and allow to dry briefly. While supporting the strip on a clean firm surface (white filter paper serves well), examine the fluorescence under ultraviolet light and note the pattern. Uroporphyrin is normally visible as a faint round spot at the origin. Coproporphyrin will appear as a sharp band at 1.5–2.0 cm from the origin. Other fluorescence usually moves near the front. Significantly elevated uroporphyrin and coproporphyrin are easy to identify. A unique feature of porphyria cutanea tarda is a sharp band of 7-carboxyl porphyrin appearing slightly above the uroporphyrin spot. Other intermediate porphyrins of five and six carboxyl groups are rarely seen and need not be utilized for diagnosing porphyrinopathies. (See Table 16-1.)

Quantitative Determination of Urinary Porphyrins by HPLC

Reagents

1. Methanol, glass distilled (Burdick and Jackson Laboratories, Inc. Muskegon, MI 49442).
2. Porphyrin acid markers, 100 μg per vial (Porphyrin Products, Logan, UT 84321).
3. Coproporphyrin standards, 5 μg (Sigma Chemical Co., St. Louis, MO 63178). These standards are desiccated for reconstitution.
4. Sodium phosphate, 10 mmol/L, pH 3.0. Dissolve 1.38 g of $NaH_2PO_4 \cdot H_2O$ in 950 mL of deionized water; adjust to pH 3.0 with H_3PO_4, 1 mol/L, and dilute to 1 liter.
5. Sodium phosphate, 10 mmol/L, pH 7.5. Dissolve 1.38 g of $NaH_2PO_4 \cdot H_2O$ in 950 mL of deionized water. Adjust to pH 7.5 with NaOH, 1 mol/L, and dilute to 1 liter.
6. Sodium phosphate, 100 mmol/L, pH 3.5. Dissolve 13.80 g of $NaH_2PO_4 \cdot H_2O$ in 950 mL of deionized water. Adjust to pH 3.5 with H_3PO_4, 1 mol/L, and dilute to 1 liter.
7. Methanol, 650 mL/L, pH 3.0. To prepare 100 mL, mix 65 mL methanol with 35 mL sodium phosphate, 10 mmol/L, pH 3.0. Filter through a 0.45-μm filter membrane under reduced pressure.

8. Methanol, 850 mL/L, pH 3.0. To prepare 100 mL, mix 85 mL methanol with 15 mL sodium phosphate, 10 mmol/L, pH 3.0. Filter as above.

9. Methanol, 800 mL/L, pH 7.5. To prepare 100 mL, mix 80 mL methanol with 20 mL of sodium phosphate, 10 mmol/L, pH 7.5. Filter as above.

10. DEAE-cellulose, DE 52 (Whatman Chemical Separation, Clifton, NJ 07014). Place 25 g of dry DEAE-cellulose in a glass beaker and add 80 mL of sodium phosphate, 100 mmol/L, pH 3.5. Adjust the stirred slurry to pH 5.0 with HCl, 6 mol/L. Pipet 2.2 mL of the slurry into 10-mL Econocolumns (Bio-Rad Laboratories, Richmond, CA 94804) to give a packed volume of 1 mL.

11. Porphyrin standards. Dissolve 100 μg of porphyrin acid markers (containing approximately equal amounts of porphyrins with 2 and 4 through 8 carboxyl groups) in 5 mL of aqueous methanol, 800 mL/L, pH 7.5. Similarly, dissolve 5 μg of coproporphyrin standard in another 5 mL of the same solvent. Do not use HCl for a solvent, as directed in the package insert. Stored well-sealed in the dark at 4 °C, these porphyrin solutions are stable for at least one month. To determine accurately the concentration of the coproporphyrin fluorescence standard, measure absorbance of the methanol solution in a cuvet having a 10-mm light path. At the Soret peak (392 nm), the molar absorptivity is 1.88×10^5. Note that near 392 nm all of these porphyrins have a very sharp spectral peak of high molar absorptivity, which means that slight variations in instrument adjustments can have a relatively large influence on the reading. Hence, it is essential that the concentration be calculated from the amount of absorbance at the Soret peak, whether this occurs above or below 392 nm with any particular instrument.

No internal standard is used, since dicarboxylic acid porphyrins do not have the same recovery from the preparation steps as the urine porphyrins. Measuring a coproporphyrin standard prior to injecting specimens is quite adequate to ensure satisfactory performance.

Specimen Preparation

Urine should be collected for 24 h in a dark brown container to which 5 g of sodium bicarbonate has been added to keep the specimen near neutral (pH 6–8). Specimens can be stored at 4 °C for a day but should be frozen for longer storage.

A 10-mL aliquot of urine is allowed to pass through the DEAE-cellulose column by gravity. Alternatively, the process can be hastened by collecting the eluate in a container with the aid of a vacuum aspirator. Urinary porphyrins are retained by the column. Wash the ion-exchange resin with 10 mL of 100 mmol/L sodium phosphate, pH 3.5, again by aspiration. Add 5 mL of HCl, 0.1 mol/L, to the column and allow to flow by gravity, collecting the eluate which contains the porphyrins. Observation of the eluate under a 100-W long-wavelength ultraviolet light may reveal the orange-red fluorescence of porphyrins.

Using a plastic syringe, pass 10 mL of the 800 mL/L methanol solvent, pH 7.5, through a SEP-PAK column, followed by 10 mL of sodium phosphate, 10 mmol/L, pH 3.5. With a different syringe, inject the entire eluate from the DEAE-cellulose column onto the SEP-PAK column. Wash the SEP-PAK column with another 10 mL of the sodium phosphate buffer. Elute the porphyrins by slowly passing 2 mL of the 800 mL/L methanol solvent, pH 7.5, through the column; wait about 1 min, then complete the elution with an additional 3 mL of the same solvent.

Chromatography System

The HPLC system consists of an M6000 pump, U6K injector, and RCM-100 radial compression module containing a 10 μm μBondapak C_{18} radial compression column, 5.8 mm ID $\times$ 8 cm (all from Waters Associates, Milford, MA 01757). Place between the pump and the analytical column a 3.9 mm ID $\times$ 23 mm guard column packed with C_{18}/Corasil, 37–50 μm particle size. The detector is a Schoeffel FS 970 L.C. fluorometer coupled to a 10-mV recorder. Set the continuously variable excitation wavelength at 405 nm and insert a secondary filter with a cutoff at 580 nm.

Procedure

To equilibrate the HPLC system, aspirate the 650 mL/L methanol solvent, pH 3.0, at a rate of 1.5 mL/min until a steady baseline is achieved (about 10 min). Then inject the porphyrin-containing solution (marker, standard, or specimen). At 7 min post-injection, change the eluting solvent to the 850 mL/L methanol solvent, pH 3.0. After all porphyrins have been eluted (about

20 min), re-equilibrate the system by changing the eluting solvent back to the 650 mL/L methanol solvent, pH 3.0. After 10 min, inject another sample. After the last sample has been analyzed, switch to 100% methanol and allow it to flow for at least 20 min. This cleans the column of any residual material.

Typically, 4–5 μL of a porphyrin marker solution is injected first to establish retention times for each of these porphyrins (Figure 16-5). The detector and recorder are adjusted to provide a full-scale deflection corresponding to about 30 ng of porphyrin. After the markers have eluted, inject 20 μL of the coproporphyrin standard to obtain a peak height equivalent to 20 ng of porphyrin. Finally, inject individual specimens; 100-μL injections are recommended, but this may be increased or decreased as necessary.

Calculations

For each specimen, identify the urinary porphyrins by their retention times. Measure the peak heights of coproporphyrin, uroporphyrin, and any other porphyrins that elute. For each peak, calculate as follows:

$$\mu g \text{ of porphyrin}/d = \frac{\text{Peak (unkn)}}{\text{Peak (std)}} \times 0.02 \times \frac{1000}{100} \times \frac{5}{10} \times TV$$

where
  Peak (unkn) = peak height (mm) of specimen (unknown)
  Peak (std) = peak height (mm) of 20 ng (0.02 μg) of coproporphyrin standard
  1000 = conversion of μL to mL
  100 = μL of sample injected for analysis
  TV = total volume of 24-h urine specimen (mL)
  5/10 = concentration factor (10 mL of urine reduced to 5 mL by preparation)

Comments

The importance of urine preparation to good analyses must not be overlooked. Before processing, urine has much fluorescent material that can obscure the uroporphyrin peak, interfere with interpretation of other peaks, and possibly quench porphyrin fluorescence. The preparative steps outlined remove these interfering substances and leave the baseline virtually flat. The rare, very minor peaks that may appear and that are not due to porphyrins are readily distinguished by their different retention times. Coproporphyrin recovery is quantitative and the CV ranges

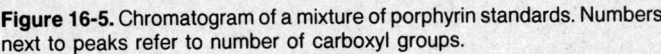

Figure 16-5. Chromatogram of a mixture of porphyrin standards. Numbers next to peaks refer to number of carboxyl groups.

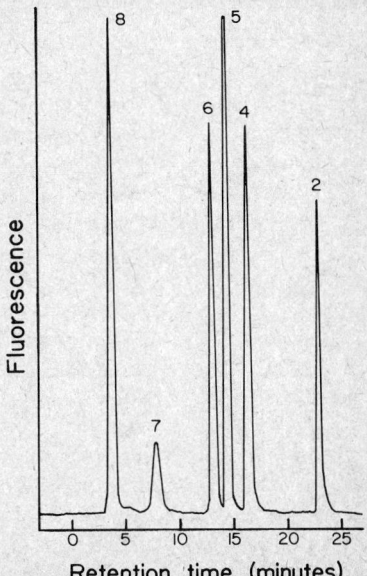

from 9–12%. For uroporphyrin, recovery is about 80% and the CV ranges from 20–38%. Recovery and precision performances improve greatly with the higher levels of porphyrins in pathologic specimens.

The buffer exchange is important for two reasons: (1) the reverse phase columns do not tolerate well solutions of pH < 2.0, and (2) this step allows concentration of the acid eluate to a predetermined volume. The buffer exchange results in an initial 10 mL of urine being decreased to 5 mL of an injectable methanol:water solution; this leads to a two-fold increase in sensitivity.

Unlike conventional clinical laboratory methods, HPLC separates all porphyrins based on the number of carboxyl groups, not simply uro- and coproporphyrin fractions. Since many techniques in common use measure more than one compound as "uroporphyrin" or "coproporphyrin" due to cross contamination as well as the presence of minor components, the HPLC method gives a more accurate account of each porphyrin present. A continuous gradient elution system has also been described.[5]

Reference Range

Uroporphyrin: 4–20 μg/d
Coproporphyrin: 13–179 μg/d

Reference

Schreiber, W. E., Raisys, V. A., and Labbe, R.F.: Clin. Chem., 29:527, 1983.

Quantitative Determination of Erythrocyte Protoporphyrin

Principle

Several simplified, rapid micro-methods for erythrocyte protoporphyrin measurement have been developed. Typically, all red cell porphyrins are first effectively removed from the blood by adsorption or extraction. In this instance Celite is the adsorbant. The described method does not discriminate among uro-, copro-, or protoporphyrin, but this has little clinical consequence, since the predominant extractable porphyrin is protoporphyrin. The adsorbed porphyrin is further purified by eluting it into acidified ethyl acetate. It is then measured fluorometrically, using coproporphyrin as a standard.

Specimen

Whole blood is anticoagulated with EDTA or heparin. The specimen is stable for several weeks when stored at 4 °C. It should not be frozen, since this can markedly affect porphyrin extractability.

Reagents

1. Celite (No. C211, Fisher Scientific Co., Pittsburgh, PA 15219). Suspend 5 g in 100 mL sodium chloride, 0.156 mol/L. Shake well immediately before removing an aliquot. The suspension is stable for at least two months.

2. Ethyl acetate:acetic acid (4:1). Mix reagent-grade solvents in the indicated volume ratio. The solution is stable indefinitely.

3. Hydrochloric acid, 1.5 mol/L. Dilute 12.5 mL of concentrated HCl to 100 mL.

4. Coproporphyrin I standard, 5 μg (Sigma Chemical Co., St. Louis, MO 63178). Dissolve the coproporphyrin standard in 5 mL of HCl, 1.5 mol/L, by heating in a hot water bath according to the package insert. This standard should be stable for many months if stored well sealed in the dark at 4 °C. To determine accurately the concentration of this coproporphyrin reference standard, measure absorbance of the HCl solution in a cuvet with a 10-mm light path. At the Soret peak (399 nm), the molar absorptivity is 4.70×10^5. Note that near 399 nm all porphyrins have a very sharp spectral peak and high molar absorptivity, which means that slight variations in instrument adjustments can have a relatively large influence on the reading. Hence, it is essential that the concentration be calculated from the amount of absorbance at the Soret peak, whether this occurs slightly above or below 399 nm. For a standard curve, prepare concentrations of 10 ng/mL to 50 ng/mL. This concentration range should give a linear response through zero. Coproporphyrins I and III have the same spectral properties; thus, either can be used as a standard.

Procedure

1. Place 200 μL of 5% Celite suspension into a 12-mL centrifuge tube. Add 50 μL of whole blood and mix by vortexing for 10 s.
2. Add 2 mL of ethyl acetate:acetic acid and vortex for 10 s to solubilize the porphyrins.
3. Centrifuge for 5 min at about $1000 \times g$ to pack the Celite-protein precipitate.
4. Decant the supernatant solution into another tube and add 2 mL of HCl, 1.5 mol/L. Vortex mix for 30 s to extract the porphyrins. Centrifuge briefly to separate the layers, the lower layer being the porphyrin-containing HCl fraction.
5. A simple fluorometer fitted with an excitation filter having a peak transmission at 405 nm and emission filter with a cutoff below 600 nm is blanked with HCl, 1.5 mol/L. The fluorescence of the HCl layer is then measured. Calculate the concentration using the standard curve. When the fluorescence reading is too high, the porphyrin solution may be diluted in HCl, 1.5 mol/L, a neutral density filter may be placed behind the secondary filter, or the assay may be repeated using less specimen.

Calculation

Calculation of erythrocyte protoporphyrin must take into account the fact that a coproporphyrin standard is used and that the result is typically normalized to 100 mL of erythrocyte. (Calculation per gram of hemoglobin can also be used.)

$$\mu g \text{ protoporphyrin/dL erythrocytes} = \frac{\text{fluor (unkn)} \times \text{conc (std)} \times 2.6 \times 100}{\text{fluor (std)} \times 0.05 \times 1.11 \times \text{hematocrit}}$$

where:

  2.6 = final volume (mL) of porphyrins in HCl
  0.05 = volume (mL) of blood used for assay
  1.11 = correction for use of a coproporphyrin fluorescence standard to measure protoporphyrin fluorescence
  100 = conversion of hematocrit from per cent to a decimal fraction.

Comments

This measurement has utility well beyond the diagnosis of the porphyrias. Both chronic lead poisoning and iron deficiency cause distinct, albeit more moderate for the latter, elevation of erythrocyte protoporphyrin concentration. These secondary causes of protoporphyrin elevation, which include most predominantly lead poisoning and iron deficiency, are not photosensitive conditions as is protoporphyria. This chemical/clinical discrepancy has been reconciled with the observation of the zinc-protoporphyrin chelate in erythrocytes of the nonphotosensitive conditions, while the metal-free porphyrin occurs in protoporphyria. In this assay, the zinc chelate is destroyed by acid, liberating metal-free porphyrin. (See also Chapter 15.) When the assay is used for diagnosis, any plasma porphyrins are clinically insignificant.

Reference Range

17–77 μg protoporphyrin/dL erythrocytes
These values are highly method dependent. The units and values may also differ whether porphyrin content is expressed per deciliter whole blood, deciliter erythrocytes, gram hemoglobin, or moles heme.

Reference

Piomelli, S.: J. Lab. Clin. Med., *81*:932, 1973; Clin. Chem., *23*:264, 1977.

Determination of Zinc Protoporphyrin by Hematofluorometer

Principle

This method is based on the direct fluorometric measurement of zinc protoporphyrin in blood using front-surface fluorometry as originally described by Blumberg et al. In this case, the specimen must absorb essentially all of the incident light within a thin layer of the specimen surface while allowing all of the emitted light to be efficiently detected. In the hematofluorometer the incident light strikes the bottom of a drop of blood on a glass slide at an angle of 37 degrees. Emitted light is detected in the vertical direction, i.e., at an acute angle to the incident beam. Using this design,

$$I_{em} = KI_{ex} \, \theta \, \frac{A}{A_{tot}} \, C$$

in which A and A_{tot} are the absorbances of the emitting species (zinc protoporphyrin) and the total specimen; I_{em} and I_{ex} are the emission and excitation intensities; K is a constant dependent on the geometry of the instrument; and θ is the quantum yield of the emitting species.

When A is a small fraction of A_{tot}, I_{em} will be proportional to the concentration of the emitting species, C. The optimum excitation wavelength for zinc protoporphyrin is 424 nm. At that wavelength, its absorbance (A_z) will be

$$A_z = \frac{10^{-4} \, C_z \, E_z}{M_z}$$

where C_z is the concentration of zinc protoporphyrin ($\mu g/L$), M_z is its molecular weight, and E_z is its molar absorptivity. The other major absorbing species present is hemoglobin, and its absorbance, A_{Hb}, is

$$A_{Hb} = \frac{C_{Hb} \, E_{Hb}}{M_{Hb}}$$

where C_{Hb} is the hemoglobin concentration (g/L), E_{Hb} is its molar absorptivity at 424 nm, and M_{Hb} is its molecular weight. Since C_{Hb} is usually about 150 g/L while C_z is usually less than 1000 $\mu g/L$, A_z will always be only a small fraction of A_{Hb}. If one assumes that zinc protoporphyrin will be the only emitting species in blood at this wavelength, the original relationship can be modified:

$$I_{em} = \left(10^{-4} \, KI_{ex} \, \theta \, \frac{M_{Hb} \, E_z}{M_z \, E_{Hb}}\right) \frac{C_z}{C_{Hb}}$$

In any specific instrument, the terms enclosed by parentheses are constant and are known. Then the emission intensity is related to the ratio of zinc protoporphyrin concentration to hemoglobin concentration. If C_{Hb} is known, the zinc protoporphyrin concentration can be determined directly from I_{em}.

The instrument originally described by Blumberg et al. uses a 50-W 12-V tungsten halogen lamp and a series of filters transmitting between 420 and 430 nm for excitation. The emitted light is collimated and filtered to select emission in the range of 580–680 nm. A red-sensitive photomultiplier is used for detection.

Specimen

The test requires one drop (approximately 50 μL) of anticoagulated blood obtained by either venipuncture or skin puncture. The zinc protoporphyrin is stable in a refrigerated specimen for at least one week; however, assay by direct fluorometry must be performed before the specimen becomes hemolyzed.

Equipment

The AVIV ZPP meter (AVIV Biomedical, Inc., Lakewood, NJ 08701) is satisfactory. The instrument is available with two electronic modules to allow the expression of results in different units. To allow use of the results with the risk classification system described by the National Centers for Disease Control, the instrument is designed to express the zinc protoporphyrin results in protoporphyrin equivalents as compared with the extraction method of Piomelli (described above). Two modules are available. One, for use with adult specimens, assumes a hematocrit of 42%, while the other, for use with child specimens, assumes a hematocrit of 35%.

Glass microscope cover slips (25 $\times$ 25 mm, 0.15 mm thick; Corning Glass Works, Medfield, MA 02052) having low background fluorescence are used.

Procedure

The instrument calibration should be verified each day using the factory-calibrated checking slides provided with the instrument. These are coated with rhodamine B, a stable dye with an emission spectrum similar to zinc protoporphyrin.

With the instrument on and the calibration verified, place a cover slip in the holder on the front of the instrument. Using a Pasteur pipet or any other pipetting device, place a drop of at least 20 μL of whole, thoroughly mixed and oxygenated blood in the center of the slide. Mix the blood gently and spread it to cover the target area visible below the slide.

Press the MEAS button. The slide holder will be drawn into the instrument and read. The result will be displayed directly.

Comments

Zinc protoporphyrin results using the hematofluorometer have been shown to be linear from normal (300 μg/L) to grossly elevated levels (11 000 μg/L) and to correlate well with results obtained using extraction methods. Precision is $\pm 2\%$.

Because the spectral properties of hemoglobin are altered by oxygen, complete oxygenation of the blood is an essential requirement for this assay. When the blood is partially deoxygenated, erroneously low results will be obtained. Since the results are not influenced by the hematocrit, equal dilution of the blood with isotonic saline can aid in mixing and oxygenation.

Interference due to abnormally high bilirubin can pose a problem. However, designers of the original Bell Laboratories hematofluorometer prototype showed that this bilirubin factor is highly dependent on the choice of interference filters used in the system. Their hematofluorometer and the commercial unit identified were significantly less sensitive to bilirubin, showing a rise of only 5 or 6 μg/L of apparent zinc protoporphyrin for each mg/L rise in plasma bilirubin. At the upper end of the normal range for plasma bilirubin (12 mg/L), this would represent a maximum rise in apparent zinc protoporphyrin concentration of 72 μg/L.

Reference Range

Results from the hematofluorometer and from assays involving extraction procedures can correlate, but confusion about interpretation is common because porphyrin content may be expressed in terms of erythrocyte concentration, whole blood concentration, or hemoglobin (or heme) present. Numerous methodologic variations also affect comparisons and interpretations. Since reference ranges can be greatly influenced by both instrumental and methodological factors, each laboratory should establish its own range. Low concentrations of protoporphyrin or zinc protoporphyrin have no known clinical significance; therefore, a reference range is important only in terms of the upper limit found in persons having normal iron status and lead exposure. However, a protoporphyrin concentration range of 17–77 μg/dL erythrocytes or its equivalent, as noted in the erythrocyte protoporphyrin assay, is a reasonable expectation for most procedures. See also Chapters 15 and 18.

Reference

Blumberg, W. E., Eisinger, J., Lamola, A. A., et al.: J. Lab. Clin. Med., *89*:712–723, 1977.

MEASUREMENT OF HEME BIOSYNTHETIC ENZYMES

The symptomatic classification presented earlier is a useful approach for the initial evaluation of a patient with a suspected porphyrinopathy. Since a unique enzyme deficiency in heme biosynthesis can now be associated with each primary disorder or porphyria (Table 16-1), it is reasonable to expect that a specific diagnosis pertinent to each porphyria of man will eventually be established on the basis of enzyme data alone. Unfortunately, the assay of each enzyme poses unique problems because of specific biochemical features such as substrate instability, the fluorescence of both substrate and product, and distribution of the enzymes between the cytoplasm and mitochondria.

Assays for all of the enzymes of the heme biosynthetic pathway have been described. However, only uroporphyrinogen I synthase activity is being measured regularly in the clinical laboratory. This assay is easily performed because of the ease of obtaining an appropriate patient specimen, the ready availability of substrate, and the requirement for only a simple instrument.

Determination of Uroporphyrinogen I Synthase Activity

Principle

Two features of this enzymatic reaction have allowed for easy adaptability of the assay to the clinical laboratory. First, the enzyme is cytoplasmic and therefore is retained within mature erythrocytes which provide an excellent specimen for assay. Second, the substrate for the reaction is nonfluorescent, while the product (uroporphyrin) is highly fluorescent, which allows for its measurement in picomolar quantities. The enzyme catalyzes the formation of uroporphyrinogen from the monopyrrole precursor, porphobilinogen. Under assay conditions, the reaction forms uroporphyrinogen, which is rapidly oxidized to uroporphyrin during acid deproteinization of the incubation mixture. Uroporphyrin fluorescence is then measured directly without further processing.

Specimen

A whole blood specimen is collected with either heparin or EDTA. The specimen can be stored at 4 °C for 1 week without significant loss of activity.

Reagents

1. Tris, 0.05 mol/L, pH 8.2. Dissolve 600 mg tris (hydroxymethyl) aminomethane in 90 mL water. Adjust to pH 8.2 with HCl, 6 mol/L, and bring volume to 100 mL.
2. Porphobilinogen, 2 mmol/L. Dissolve 4.5 mg in 10.0 mL deionized water. Stable for 1 month if stored at -20 °C.
3. Trichloroacetic acid (TCA), 25% (25 g/dL). Dissolve 25 g trichloroacetic acid in 50 mL water and bring volume to 100 mL.
4. Coproporphyrin standards (5 ng/mL to 50 ng/mL). Prepare as described for quantitative determination of erythrocyte protoporphyrin except that dilutions should be in 12.5% trichloroacetic acid.

Equipment

A simple fluorometer is fitted with an excitation filter with a peak transmission at 405 nm and an emission filter with a cutoff below 606 nm.

Procedure

1. Prepare an erythrocyte hemolysate by adding 50 μL of whole blood to 950 μL of Tris buffer. Leave this solution on ice for 10 min before proceeding.
2. Measure the hemoglobin concentration of the hemolysate (step 1) for calculation of activity on a per gram of hemoglobin basis. The cyanmethemoglobin (Drabkin) or any comparable method is suitable.
3. Combine in two separate tubes (reaction and blank) 0.9 mL of Tris buffer and 0.5 mL of porphobilinogen substrate. Add 5 μL of the hemolysate. Stop the blank reaction immediately by adding 1 mL of trichloroacetic acid.
4. Incubate the reaction tubes in the dark with gentle agitation at 37 °C. After 60 min, stop the reaction by adding 1.0 mL of 25% trichloroacetic acid.
5. Centrifuge the reaction tubes at $1000 \times g$ for 10 min. Measure the fluorescence of the supernatant after blanking the instrument with 12.5% trichloroacetic acid.

Calculations

The rate of uroporphyrin production is calculated and the results are expressed in enzyme units of activity per unit of hemoglobin. One unit of uroporphyrinogen synthase activity is equivalent to the formation of 1 μmol of uroporphyrin/min.

$$ mU/g\ Hb = \frac{fluor\ (unkn)}{fluor\ (std)} \times std\ conc \times \frac{1.09}{0.05} \times \frac{1}{Hb} \times \frac{1}{60} $$

where
std conc = concentration of coproporphyrin standard in nmol/mL
0.05   = mL of blood added (1:20 hemolysate)

Hb	= hemoglobin concentration in the 1:20 hemolysate (g/mL)
1.09	= correction factor to express porphyrin formation as uroporphyrin after measurement against a coporporphyrin standard
60	= minutes of incubation

Comments

The erythrocyte uroporphyrinogen I synthase activity varies with cell age. Therefore, a shift in the erythrocyte population toward younger cells might lead to an enzyme activity that does not accurately represent the true condition. An elevated reticulocyte percentage would indicate that such a situation could be present.

A modification of this procedure has been described in which porphobilinogen is generated from its precursor, δ-aminolevulinic acid, in the hemolysate during the incubation.[6]

Reference Range

1.27–2.01 mU/g hemoglobin

Reference

Magnussen, C. R., Levine, J. B., Doherty, J. M., et al.: Blood, *44*:857–868, 1974.

Uroporphyrinogen Cosynthase

The formation of uroporphyrinogen III requires concerted action of both uroporphyrinogen I synthase and cosynthase. The synthase is essential for tetrapyrrole formation while the cosynthase directs the isomer III synthesis in preference to the nonfunctional isomer I. Assay of uroporphyrinogen III cosynthase requires that isomer production, independent of total porphyrin synthesis, be measured. Although erythrocytes are an accessible source of uroporphyrinogen cosynthase, the technical difficulties of assaying enzyme activity and the relative ease of diagnosis by other means tend to obviate the need for an assay suitable for the clinical laboratory.[7]

Uroporphyrinogen Decarboxylase

The sequential decarboxylation of the eight carboxyl uroporphyrinogen to the four carboxyl copro-porphyrinogen is catalyzed by uroporphyrinogen decarboxylase. A partial deficiency of this enzyme has been associated with porphyria cutanea tarda (Table 16-1). Considering reported experiences with assays of this activity using both uroporphyrinogen III and pentacarboxylic porphyrinogen III, one may assume that this enzyme catalyzes the entire cytosolic decarboxylation sequence from uroporphyrinogen to copro-porphyrinogen. The substrate must be in the reduced or "-ogen" form and, of course, the measurement of the reaction product must allow discrimination between substrate and product porphyrins.[4]

Coproporphyrinogen Oxidase

This enzyme catalyzes the decarboxylation and dehydrogenation of the propionic acid side chains at positions 2 and 4 of coproporphyrinogen to vinyl groups, resulting in the formation of protoporphyrinogen. Since this is a mitochondrial enzyme, erythrocytes cannot be used for the assay. Substrate selection for the enzyme is particularly important. If unlabeled coproporphyrinogen is used, then the protoporphyrinogen product can be difficult to differentiate from the substrate as they have similar chemical and fluorescence properties. A radioisotope approach uses ^{14}C-coproporphyrinogen with the label in the carboxyl carbons of the propionic side chains. The decarboxylation can then be followed by $^{14}CO_2$ release.[3]

Protoporphyrinogen Oxidase

The oxidation of protoporphyrinogen to protoporphyrin occurs in cellular mitochondria. In measuring this oxidase activity, the assay is simplified by the easy discrimination between substrate and product. That is, the substrate, protoporphyrinogen, is nonfluorescent while the protoporphyrin produced is highly fluorescent. Unfortunately, the protoporphyrinogen substrate is unstable and will readily oxidize. Thus, the substrate must be prepared fresh from protoporphyrin. A partial deficiency of this enzyme in cultured fibroblasts from patients with variegate porphyria has recently been described.[2]

Ferrochelatase

This enzyme catalyzes the chelation of a ferrous ion by protoporphyrin with release of two protons. The measurement of ^{59}Fe incorporation into heme, the decrease in porphyrin fluorescence by iron chelation, and the direct quantitation of heme formed have all been used for the assay.[1]

References

1. Bloomer, J. R.: Characterization of deficient heme synthase activity in protoporphyria with cultured skin fibroblasts. J. Clin. Invest., *65*:321, 1980.
2. Brenner, D. A., and Bloomer, J. R.: A fluorometric assay for measurement of protoporphyrinogen oxidase activity in mammalian tissue. Clin. Chim. Acta, *100*:259, 1980.
3. Elder, G. H., and Evans, J. O.: A radiochemical method for the measurement of coproporphyrinogen oxidase and the utilization of substrates other than coproporphyrinogen III by the enzyme from rat liver. Biochem. J., *169*:205, 1978.
4. Elder, G. H., and Tovey, J. A.: Uroporphyrinogen decarboxylase activity of human tissues. Biochem. Soc. Trans., *5*:1470, 1977.
5. Ford, R. D., Ou, C.-N., and Ellefsen, R. F.: Liquid-chromatographic analysis for urinary porphyrins. Clin. Chem., *27*:397, 1981.
6. Piepkorn, M. W., Hamernyik, P., and Labbe, R. F.: Modified erythrocyte uroporphyrinogen I synthase assay and its clinical interpretation. Clin. Chem., *24*:1751, 1978.
7. Romeo, G., Kaback, M. M., and Levin, E. Y.: Uroporphyrinogen III cosynthetase activity in fibroblasts from patients with congenital erythropoietic porphyria. Biochem. Genet., *4*:659, 1970.

Additional Readings

Bishop, D. F., and Desnick, R. J., Eds.: Assays of the heme biosynthetic enzymes. *In*: Enzyme, Vol. 28, 2–3. New York, F. Karger Publishers, Inc., 1982.
Elder, G. H.: Haem Synthesis and Breakdown. *In*: Iron in Biochemistry and Medicine II. A. Jacobs and M. Worwood, Eds. New York, Academic Press, 1980, pp. 245–292.
Falk, J. E.: Porphyrins and Metalloporphyrins. New York, Elsevier, 1964.
Goldberg, A., and Moore, M. R., Eds.: The Porphyrias. *In*: Clin. Haematol., *9*:225–444, 1980.
Kappas, A., Sassa, S., and Anderson, K. E.: The Porphyrias. *In*: The Metabolic Basis of Inherited Disease. 5th ed. J. B. Stanbury, J. B. Wyngaarden, and D. S. Fredrickson, Eds. New York, McGraw-Hill, 1983, pp. 1301–1384.
Meyer, U. A.: Porphyrias. *In*: Harrison's Principles of Internal Medicine. 9th ed. K. J. Isselbacher, R. D. Adams, E. Braunwald, et al., Eds. New York, McGraw-Hill, 1980, pp. 494–500.
Romeo, G.: Enzymatic defects of hereditary porphyrias: An explanation of dominance at the molecular level. Human Genet., *39*:261–276, 1977.
With, T. K.: A short history of porphyrins and the porphyrias. Int. J. Biochem., *11*:189–200, 1980.

CHAPTER 17

THERAPEUTIC DRUG MONITORING

by Thomas P. Moyer, Ph.D., Charles E. Pippenger, Ph.D.,
Robert V. Blanke, Ph.D., and Robert A. Blouin, Pharm.D.

Therapeutic drug monitoring (TDM) is a rapidly developing aspect of clinical laboratory activities. To be effective, TDM requires the acquisition of valid specimens followed by routine, timely, and reliable determinations of drug concentrations in these specimens. Results may be reported individually or collated so that, in conjunction with the dosing schedule, they may be interpreted in a pharmacokinetic context. Clinical pharmacokinetics combines the application of determined mathematical relationships of drugs in biological fluids with the prudent utilization of drug blood concentrations for the development of a safe and effective dosing regimen. It is important, therefore, for the clinical chemist to understand not only how to measure drugs in biological specimens but also how these results are used to promote effective drug therapy. TDM calls for a multidisciplinary approach since it relies on cooperative efforts of the physician, nurse, pharmacologist, pharmacist, and clinical chemist.

TDM has had a long association with clinical chemistry. Wuth[86] in 1927 demonstrated the value of monitoring serum bromide levels to differentiate bromide-induced from intrinsically psychotic behavior. The mathematical description of the time course of drugs in biologic systems is also an established part of pharmacologic science.[69,83,85] The recent rapid evolution of TDM, however, stems from a fortuitous conjunction of several factors:

1. Development of more numerous and more effective drugs with therapeutic potential and application.

2. Development of analytical technologies that are accurate, precise, sensitive, and specific for measurement of drugs in biological fluids.

3. Utilization of computers in analytical technology, in pharmacokinetic data analysis, in dose regimen design, and in the collection and storage of patient data.

Limitations of empirical drug dosing, i.e., standard or fixed-dose regimens, have long been recognized by clinicians, who have responded with all of their clinical skills and their knowledge of basic pharmacology by individualizing each patient's drug dosage. TDM offers a scientific approach to selecting a drug regimen to achieve a targeted blood concentration and thus to optimize therapy. Therefore, blood drug analyses in the clinical laboratory are a critical adjunct to optimal therapeutic drug utilization.

The *pharmacological effect*, i.e., the toxic or therapeutic response to a drug, may be elicited by a direct interaction of the drug with the *receptor* controlling a specific function or by a drug-mediated alteration of the physiologic process regulating the function. For most drugs, the intensity and duration of the observed pharmacologic effect is proportional to the concentration of the drug at the receptor. Within a given tissue, the site at which a drug acts to initiate events leading to a specific biological effect is arbitrarily called the *site of action* of the drug.

The *mechanism of action* of a drug refers to the actual biochemical or physical process that occurs at the site to elicit the biological response. The drug action is generally mediated through a receptor. Cellular proteins such as enzymes, as well as structural or transport proteins, are important examples of drug receptors. Nonprotein macromolecules may also bind drugs, resulting in altered cellular function such as membrane permeability. Some drugs are chemically similar to important, natural, endogenous substances and may compete with these for binding sites. In addition, some drugs may block formation, release, uptake, or transport of essential substances. Others may produce an effect by interacting with relatively small molecules wherever they may be encountered. These and other examples of receptors are more completely discussed in pharmacology texts. Although the exact molecular interactions remain obscure for most drugs, theoretical models have been developed to explain them. One concept postulates that a drug binds to intracellular macromolecular receptors through ionic and hydrogen bonds and van der Waals forces. The theoretical model further postulates that if the drug-receptor complex is sufficiently stable and able to modify the target system, an observable pharmacologic response will occur. As Figure 17-1 illustrates, the response is dose dependent until a maximal effect is reached. A plateau at this point may be due to saturation at the receptor or to a variety of other effects.

The utility of monitoring drug concentration is based on the premise that prediction of a pharmacologic response is best achieved when a targeted concentration of the drug has been established at the site of action, that is, at the receptor. Clearly, measurement of the concentration at the receptor site in a patient is technically impractical, if not impossible. However, studies have shown that, for many drugs, a strong correlation exists between the serum drug concentration and the observed pharmacologic effect. In addition, years of monitoring and relating blood concentrations to drug effects have demonstrated the clinical utility of drug concentration information. One must nevertheless always keep in mind that a serum drug concentration does not necessarily equal the concentration at the receptor; it may merely reflect it.

That said, let us assume that changes in serum drug concentration versus time mirror changes in local concentrations at the receptor site or in body tissues. This assumption is sometimes called the *property of kinetic homogeneity* and is applicable to all pharmacokinetic models in postabsorptive and postdistributive phases of the time course. Figure 17-2 illustrates that property for a hypothetical compound. Parallel concentrations (log C) are expected in blood, at the receptor, and in the kidney and heart as time passes.

The property of kinetic homogeneity is an important assumption in TDM because it is the basis upon which all therapeutic and toxic concentration ranges are established. The ranges collectively define the *therapeutic window*, which is illustrated in Figure 17-3 as a relationship between minimum effective concentration (MEC) and minimum toxic concentration (MTC). Ideally, the *trough blood concentration* (the lowest concentration achieved just before the next dose) within each dosing cycle should not fall below the MEC, and the *peak blood concentration* (the highest concentration achieved within the dosing cycle) should not rise higher than the MTC. Multiple dosing regimens should achieve steady-state serum drug concentrations consistently above the MEC and below the MTC, i.e., within the therapeutic range. Blood concentrations above the MTC put the patient at risk for toxicity; concentrations below the MEC have him at risk for the disorder the drug is supposed to treat. MTC and MEC are useful guidelines in therapy. Dosage must be planned to achieve therapeutic concentrations and these must be monitored to allow adjustment of dosage if necessary. The smaller the difference between MEC and MTC, the more likely TDM will be necessary.

Pharmacology comprises that body of knowledge surrounding chemical agents and their effects on living processes. This is a broad field, and traditionally it has been confined to those drugs that are useful in the prevention, diagnosis, and treatment of disease. *Pharmacotherapeutics* is that part of pharmacology concerned primarily with the application or administration of drugs to patients for the purpose of prevention and treatment of disease. In order for this aspect of

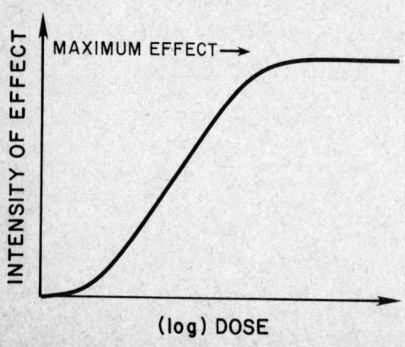

Figure 17-1. The log dose-effect relationship. The plateau (maximum effect) may be due to saturation at the receptor or to a variety of other effects.

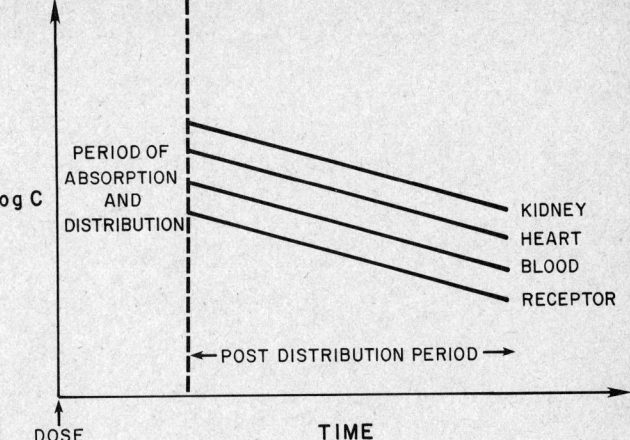

Figure 17-2. Property of kinetic homogeneity. Change in blood concentration of drug occurs in parallel with change in tissues and at receptor site.

medical practice to be effective, the pharmacodynamic and pharmacokinetic properties of drugs should be understood.

Pharmacodynamics may be defined as "the study of the biochemical and physiological effects of drugs and their mechanisms of action."[34] For many drugs, the ultimate effect or mechanism of action at the molecular level is understood poorly if at all. However, effects at the cellular or organ system level or in the whole body can be related to dosage of the drug.

Pharmacokinetics is the study of the rate processes of absorption, distribution, biotransformation, and excretion of drugs or chemicals, generally of exogenous origin. It is this aspect of pharmacology that most strongly influences the interpretation of TDM results and that is dealt with in more detail in this chapter.

Toxicology can be considered an aspect of pharmacology concerned with adverse effects of chemicals on living systems. Toxic effects and mechanisms of action may be different from therapeutic effects and mechanisms for the same drug. Similarly, at the high dosage of drugs at which toxic effects may be produced, rate processes are frequently altered in comparison to those at therapeutic dosages. For these reasons, the terms *toxicodynamics* and *toxicokinetics* are now applied to these special situations.

Figure 17-4 illustrates the conceptual relationship between pharmacodynamics and pharmacokinetics. The former relates drug concentration at the site of action to the observed magnitude of the effect. Pharmacokinetics, on the other hand, relates dose, dosing interval, and route of administration (regimen) to drug concentration in the blood. For more complete discussions of these basic concepts the reader is encouraged to review standard textbooks.[6–8,34,47,49,51,53,54,58,61,77]

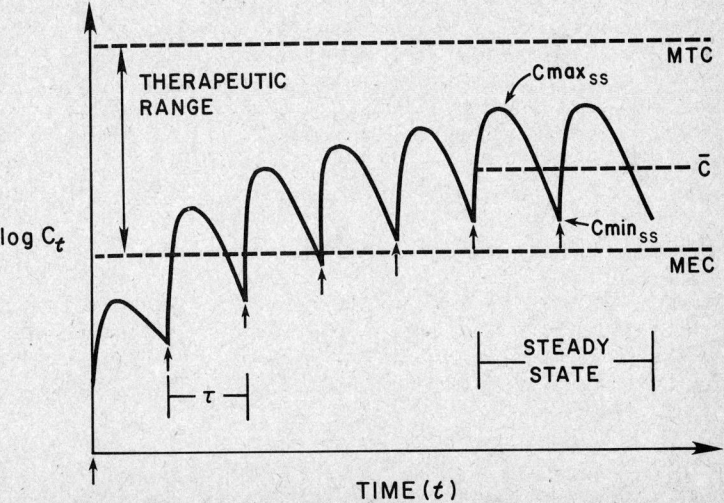

Figure 17-3. Sequence of drug concentration changes with multiple identical doses. Note that at steady state, peak and trough concentrations lie within the therapeutic range (or therapeutic window), and that 5–7 half-lives are required to reach steady state. $C_{max_{ss}}$ and $C_{min_{ss}}$ = maximum and minimum steady-state concentrations; $\overline{C}$ = average steady-state concentration; τ = dosing interval; $\uparrow$ = dose; MTC = minimum toxic concentration; MEC = minimum effective concentration. (Modified from Gilman, A. G., Goodman, L., and Gilman, A., Eds.: The Pharmacological Basis of Therapeutics. 6th ed. New York, Macmillan 1980.)

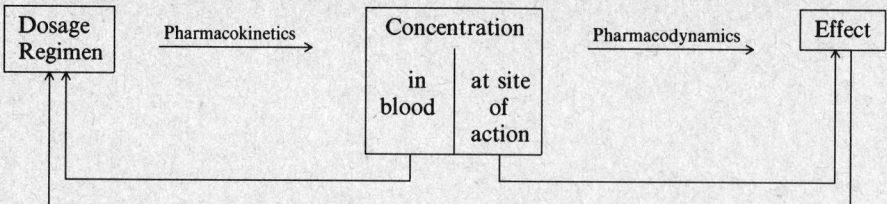

Figure 17-4. Conceptual relationship between pharmacodynamics and pharmacokinetics.

PHARMACOKINETICS

Pharmacokinetics is the study of the rate processes of absorption, distribution, biotransformation, and excretion of drugs, poisons, and chemicals of both exogenous and endogenous origin. *Clinical pharmacokinetics* is the discipline that applies the principles of pharmacokinetics to safe and effective therapeutic management of an individual patient. Figure 17-5 depicts in detail the many factors affecting drug concentration and pharmacologic response.

In pharmacokinetics, various mathematical approaches are utilized to predict or describe certain events, usually for calculating a dosing regimen or predicting the serum drug concentration following a given drug dose. The mathematical tools most often employed in clinical pharmacokinetics are *compartmental models* and *model-independent relationships*. Both of these approaches will be discussed.

Compartmental models have been a mainstay of pharmacokinetic data analysis; however, the model-independent relationships are rapidly being adopted as the more desirable method of analysis. The compartmental approach will be presented here primarily to facilitate defining important pharmacokinetic parameters.

Compartmental models are deterministic in nature; i.e., the blood drug concentration and time data determine or define the model. The number and values of compartments assigned to the model have no true physiologic meaning or anatomic reality. Usually the intravascular fluid is the anatomic reference compartment. The advantage of intravascular fluid as the reference compartment is the ease with which it may be sampled in order to provide a definitive profile of blood concentration of drug versus time. The number of compartments can be quite extensive; however, one- and two-compartment models are most often employed.

In a *one-compartment model*, the body is considered as a single compartment. It is assumed that after introduction of a drug, the substance is rapidly and uniformly distributed throughout the body. In contrast, the *two-compartment model* accounts for an initial decline of drug concentration in the reference compartment due to distribution from the plasma into tissues (the second compartment). Figure 17-6 depicts a one-compartment model, Figure 17-7 a two-compartment model, following an intravenous, bolus administration of a drug. In these figures, X_0 represents the drug dose given and therefore the amount of drug in the system at zero time, X_1 the amount of drug in the central or reference compartment, and X_2 the amount of drug in the peripheral compartment in the case of the two-compartment model. The K's are first order rate constants. K and K_{10} are *elimination* rate constants; i.e., they describe the rates at which the drug leaves the reference compartment and is lost from the system. K_{12} and K_{21} are *transfer* rate constants describing, for the two-compartment model only, rates at which the drug is exchanged between compartments within the system. Figure 17-8 illustrates graphically the relationship between log of concentration and time for the one-compartment model. The curve in Figure 17-8 is described by the equation:

$$C_t = C_0 e^{-Kt} \qquad (1)$$

where C_0 is the initial plasma concentration of the drug; C_t is the blood concentration of the drug as a function of time; and K is again the first order elimination rate constant. C_0 is estimated by extrapolating the line to zero time. From knowledge of C_0 and K, one can theoretically predict C at any time (C_t), or one can determine K and C_0 and C_t either graphically or mathematically. As will be shown later, most drugs are administered in repetitive doses rather than in a single bolus.

Figure 17-9 illustrates the considerably more complex kinetics demonstrated by the two-compartment model. The curve is described by the equation:

$$C_t = Ae^{-\alpha t} + Be^{-\beta t}$$

where the rate constant α is the slope of the curve during the phase in which the drug is being distributed.

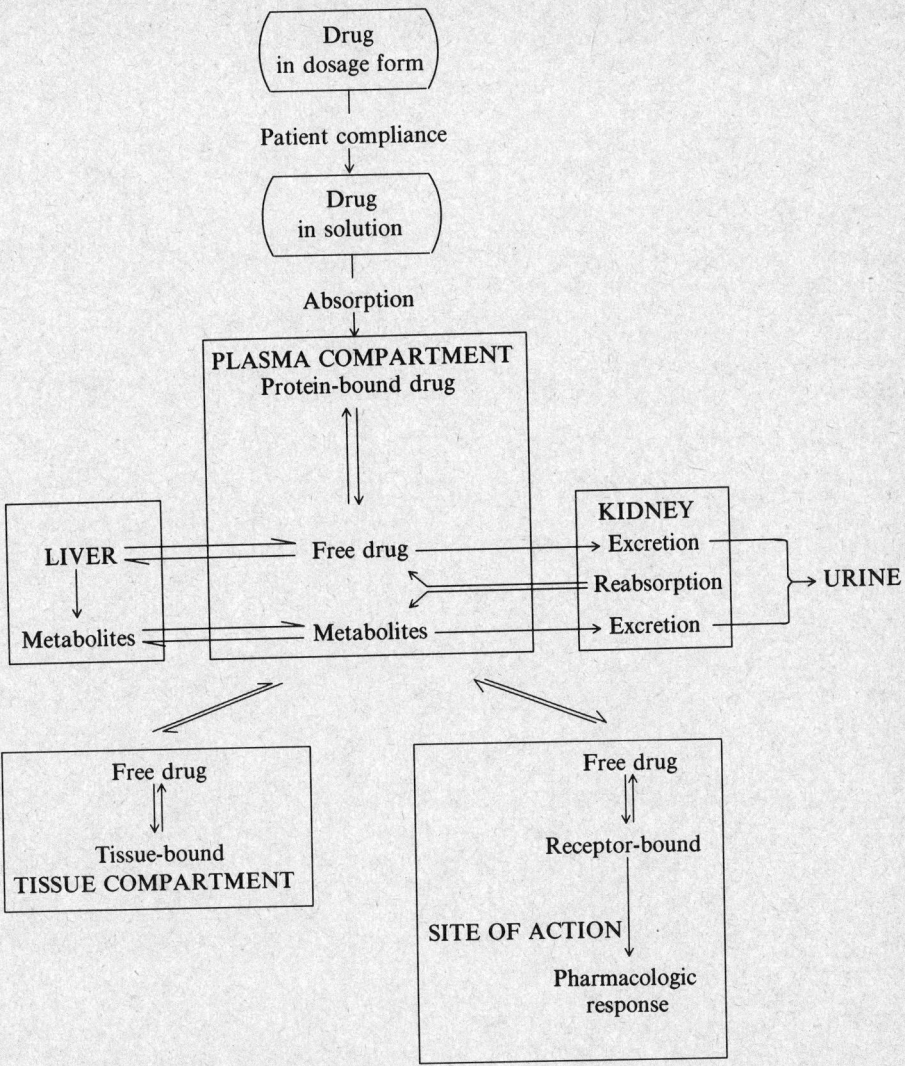

Figure 17-5. Factors affecting plasma drug concentration.

Absorption. Drug must be formulated to assure bioavailability for absorption from the GI tract or other administration site.

Metabolism. Drug is converted to a more soluble compound (metabolite) that may be pharmacologically active or inactive. Metabolism may occur in tissues other than the liver.

Excretion. The more water-soluble drugs and their metabolites are usually excreted in urine. Excretion may also occur via bile, feces, saliva, and expired air.

Tissue storage. The drug may be stored in tissues that exhibit no pharmacologic response to it; side effects may occur from drug interaction with a specific physiological system.

Site of action. Free drug binds to the receptor and produces pharmacologic response. The number and type of receptors to which the drug is bound determine the intensity and duration of the observed response.
(Adapted from Pippenger, C.E.: TDM: Principles of drug utilization. Syva Monitor, Nov., 1978, pp. 1, 3–5. Syva Co., Palo Alto, CA.)

Beta is the slope of the curve during the phase in which the drug is being eliminated by metabolism and excretion (assuming that distribution is complete) and is derived by extrapolating (dashed line) the elimination phase of the curve to time = 0 that would have existed if distribution had been immediate and complete. A is an estimate, using the method of residuals, of the theoretical plasma concentration at time = 0, immediately after intravenous injection of a bolus of drug. B is derived by extrapolation of the terminal slope line to time = 0. The method of residuals merely corrects the measured plasma level for the amount of drug distributed into tissues (as estimated on the extrapolated slope).

For simplicity, the intravenous bolus, one-compartment model will be used for illustrative purposes. Again, note that the discussion is designed only to define parameters and introduce important concepts to the reader.

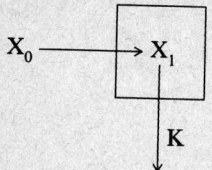

Figure 17-6. A one-compartment pharmacokinetic model. See text for explanation of details.

In the one-compartment model, following introduction of an intravenous bolus dose (X_0) into the single compartment, the drug is assumed to distribute instantaneously through all fluids and tissues of the body. The property of kinetic homogeneity is also assumed.

For purposes of TDM, it is preferable to work with *concentration* of drug in this homogeneous compartment, so X_1 (amount of drug in the compartment) is divided by a volume term. This volume term is called *apparent volume of distribution* or V_d. It is not a real volume in the physiologic sense but instead a proportionality constant to translate the absolute amount of drug present in the compartment into its concentration relative to a volume.

V_d can be determined easily for the one-compartment model. When $t = 0$ (time the drug bolus is introduced), X_0, the amount of drug given, is equal to the amount of drug in the compartment, and the blood concentration (C_0) is:

$$C_0 = \frac{X_0}{V_d} = \frac{X_1}{V_d} \tag{2}$$

then

$$V_d = \frac{X_0}{C_0} \tag{3}$$

The units of V_d are usually liters (L) or liters per kilogram of body weight (L/kg). Although V_d is a mathematical term and not a real physiological parameter, it is useful for contrasting degrees to which different types of drugs distribute. For instance, the polar, hydrophilic drug acetylsalicylic acid has $V_d = 0.15$ L/kg, while the nonpolar, lipophilic drug desipramine has $V_d = 34$ L/kg.

Using the same assumptions, the first-order elimination rate constant can be determined. Referring to Figure 17-8, note that the relation between C_t and time is a natural logarithmic function where

$$\ln C_t = \ln C_0 - Kt \tag{4}$$

Given a zero time blood drug concentration (C_0), a nonzero time concentration (C_t), and a defined time (t), then K can be readily determined either algebraically or graphically. (For example, in a plot of ln C_t versus t, the slope of the linear relationship is $-K$.) The elimination rate constant K represents the fraction of drug removed per unit time and has units of reciprocal time (min^{-1}, h^{-1}, day^{-1}). K is the overall elimination rate constant and includes the renal constant (K_R), the biliary constant (K_B), the metabolic constant (K_M), and others such that $K = K_R + K_B + K_M + \ldots$.

Literature values for K are generally stated for healthy, young adults. Obviously, drugs are generally given to the ill who may not be metabolizing or excreting the drug in a normal fashion. Pharmacokinetic constants are now being derived for special patient populations such as geriatric or pediatric patients, as well as others.

The elimination rate constant can be related to another parameter, *half-life* ($t_{1/2}$), by the following equation:

$$t_{1/2} = \frac{0.693}{K} \tag{5}$$

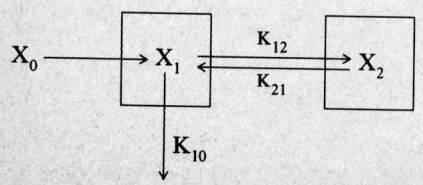

Figure 17-7. A two-compartment pharmacokinetic model. See text for explanation of details.

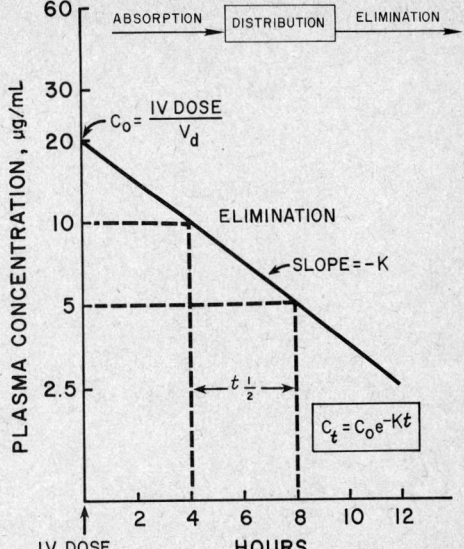

Figure 17-8. Drug concentration in plasma following intravenous administration of a single dose. The data show monoexponential decline that can be described by a one-compartment model.

Half-life can be simply defined as the time required for the amount of drug in the system or for the blood drug concentration to decline to one-half of a measured value. Figure 17-8 demonstrates how the half-life can be rapidly determined from a semilog plot of drug concentration versus time. Two or more successive concentrations are required for the estimate of K used in the calculation of $t_{1/2}$.

$$K = \frac{\ln C_{t_1} - \ln C_{t_2}}{t_2 - t_1}$$

If $t_{1/2}$ differs markedly from reported values for the drug in question, a component of elimination may be the cause, i.e., decreased K_R in renal impairment or decreased K_B or K_M in liver dysfunction.

Total body clearance (Cl_T) is defined as the theoretical total volume of blood, serum, or plasma completely cleared of drug per unit of time. It is usually expressed in units of mL/min, L/h, mL/min/kg, or L/h/kg. Like the elimination rate constant, Cl_T is the sum total of all the clearances contributed by each elimination route, i.e., $Cl_T = Cl_R + Cl_B + Cl_M + \ldots$. Clearance is a most important parameter since it provides a better picture than does K of the body's ability to eliminate a drug. In addition, Cl_T has more physiologic meaning and is readily used to relate the dosing rate to steady-state concentration.

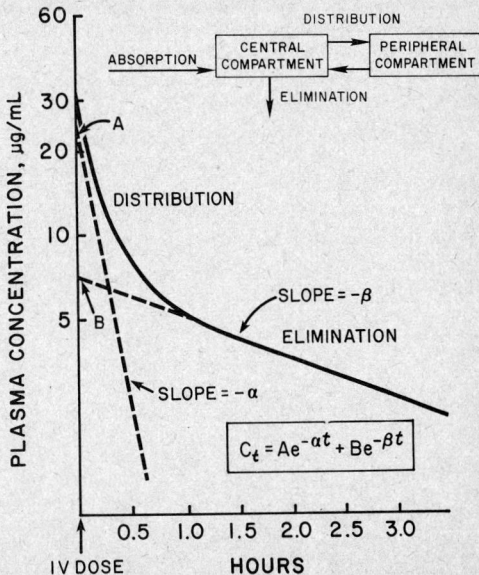

Figure 17-9. Decline of plasma concentration versus time after intravenous administration for a drug best characterized by a two-compartment model.

Clearance is mathematically related to V_d and K as

$$Cl_T = V_d K \tag{6}$$

A different and preferred way to express Cl_T is

$$Cl_T = \frac{X_0}{AUC} \tag{7}$$

where AUC is the area under the curve described by the plasma concentration versus time and integrated over time from zero to infinity. The reader will see later the particular advantage of equation (7).

In therapeutic drug monitoring, we are rarely concerned with a drug administered as a single, one-time intravenous bolus. Drugs administered chronically pose the usual therapeutic situations. Figure 17-3 shows that a drug, repetitively administered at a fixed dosing interval, will accumulate in the body until a steady-state condition exists. Note that the symbol τ stands for dosing interval. *Steady state* can be defined as that point in a dosing scheme when the amount entering the circulation (governed by dosing rate) equals the amount eliminated (governed by elimination rate).

Assuming first-order elimination, 5–7 half-lives of a drug are required to reach 97–99% of a steady-state value. At steady state, assuming τ is constant, the profiles of concentration versus time should be superimposable from dose to dose.

Steady-state drug concentration ($C_{t_{ss}}$) at any time during a dosing interval can be described by equation (8):

$$C_{t_{ss}} = \frac{X_0 e^{-Kt}}{V_d (1 - e^{-K\tau})} \tag{8}$$

In therapeutic drug monitoring, the maximum ($C_{max_{ss}}$) and minimum ($C_{min_{ss}}$) steady-state concentrations are often of primary interest. (See Figure 17-3.) Hence the following equations are more useful.

$$C_{max_{ss}} = \frac{X_0}{V_d (1 - e^{-K\tau})} \tag{9}$$

$$C_{min_{ss}} = \frac{X_0 e^{-K\tau}}{V_d (1 - e^{-K\tau})} \tag{10}$$

Again, recall that equations (8) through (10) apply only to dosing by repeated intravenous bolus. They are useful only when X_0, V_d, K, and τ are known.

Model-independent relationships are becoming increasingly popular in clinical pharmacokinetics. As implied earlier, anything beyond a one-compartment model is impractical for clinical application. The fact that most drugs actually possess multicompartmental characteristics is reason enough to consider noncompartmental alternatives. The main advantages of model-independent relationships are fewer relationships to remember, fewer restrictive assumptions, a more general insight into elimination mechanisms, and easier computations. However, model-independent relationships are not without their disadvantages; conceptualization of compartments or physiologic spaces may be lost, specific information which may be clinically relevant or pertinent to mechanisms of distribution or elimination can be lost, and difficulty in constructing profiles of concentration versus time can be increased.

The most frequently used model-independent approach utilizes the concept of ($\overline{C}$) (see Figure 17-3) and is described by:

$$\overline{C} = \frac{X_0}{Cl_T \times \tau} \tag{11}$$

or a variation of it, considering the extent of absorption,

$$\overline{C} = \frac{X_0 F}{Cl_T \times \tau} \tag{12}$$

where F is the bioavailability (fraction of dose available to reach the systemic circulation) of the drug. Either of these equations may be employed for any drug. The equations are most useful when the half-life of a drug is much greater than τ, the dosing interval.

PRINCIPLES OF DRUG DISPOSITION

Absorption. Most drugs administered chronically to patients are administered extravascularly. Although intramuscular and subcutaneous routes are utilized, the oral route accounts for most of the extravascular doses administered. The absorption process depends on the drug's dissociating from its dosage form, dissolving in gastrointestinal fluids, and then diffusing across biological membrane barriers into the bloodstream. The rate and extent of drug absorption may vary considerably depending on the nature of the drug itself (solubility, pK_a), on the matrix in which it is present, and on the physiologic environment (pH, gastrointestinal motility, vascularity). For example, as seen in Figure 17-10 a drug with the property of a weak acid exists predominantly in the nonionized form in the acid environment of gastric juice. Its absorption in the stomach would be favored over a drug that is a weak base and for which the ionized, poorly absorbed form exists in acidic gastric juice.

The fraction of a drug that is absorbed into the systemic circulation is referred to as its *bioavailability*. The bioavailability (F) of a given drug is usually calculated by comparing, in the same subjects, the area under the plasma concentration time curve (AUC) of an equivalent dose of the intravenous dosage form and oral form.

$$F = \frac{AUC_{oral}}{AUC_{IV}} \tag{13}$$

The bioavailability of a particular drug, if the drug is to be useful, must be great enough so that the active component can pass in sufficient amount and in a desirable time from the gut into the systemic circulation. An exception would be the case where the lumen of the gastrointestinal tract is the site of drug action, e.g., antibiotics used to sterilize the gut. Then low bioavailability would be considered advantageous.

Some drugs that are rapidly and completely absorbed nevertheless have low bioavailability to the systemic circulation. This is true of drugs with a high hepatic extraction rate. After oral administration, drugs that are absorbed in the lumen of the small intestine are carried via the portal vein directly to the liver. A drug with a high hepatic extraction rate may be extensively metabolized by the liver before it reaches the systemic circulation. This phenomenon is called the *first pass effect*.

In addition to the *extent* of absorption, the *rate* of absorption is important. The absorption of a drug is generally considered a first-order process, and the absorption rate constant of a drug is usually much greater than its elimination rate constant. Efforts are now being made in the pharmaceutical industry to decrease the apparent rate of absorption of many drugs by manipulating their dosage formulations (e.g., theophylline, quinidine). Dosage formulations that provide

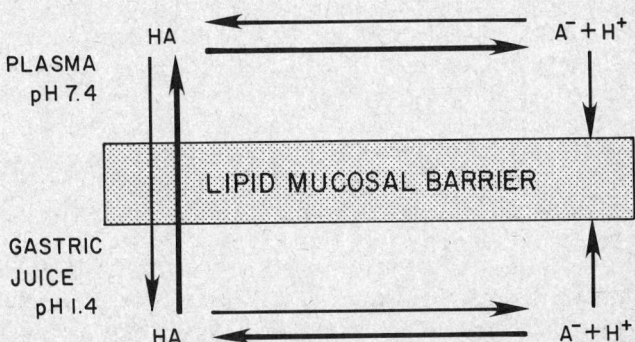

Figure 17-10. Influence of pH on the ionization and the resulting distribution of a weak acid between plasma and gastric juice. In gastric juice, ionization is minimized resulting in enhanced diffusion of the uncharged drug across the lipid barrier. In the plasma compartment, more drug is ionized. The charged species are less likely to diffuse back into the stomach. (Modified from Gilman, A. G., Goodman, L., and Gilman, A., Eds.: The Pharmacological Basis of Therapeutics. 6th ed. New York, Macmillan, 1980.)

sustained release permit drugs taken orally to be taken at less frequent intervals. Conditions that may influence the extent or rate of drug absorption include abnormal gastrointestinal motility; diseases of the stomach as well as of the small and large intestine; gastrointestinal infections; radiation; food; and interaction with other substances in the GI tract. One should be particularly aware of drugs that directly affect gut absorption, such as antacids, kaolin, sucralfate, cholestyramine, and antiulcer medications.

Distribution. Following the entrance of a drug into the vascular compartment, the drug will interact with various blood constituents and be carried by various transport processes to different body organs and tissues. The overall process is referred to as *distribution*. The factors determining the distribution pattern of a drug are binding of the drug to circulating blood components, binding to fixed receptors, ability of the drug to pass through membrane barriers, and the ability to dissolve in structural or storage lipids. Molecular weight, pK_a, and other physical and chemical properties of the drug are also influential.

Once a drug enters the systemic circulation it will rapidly distribute and come to equilibrium with many of the blood components. Probably the most clinically significant of these are *plasma proteins*. Figure 17-5 shows the equilibrium that exists between *free* and protein-*bound* drug. It is generally believed that only the drug unbound to protein (free) in the plasma is available for distribution and elimination. Additionally, only the free drug is available to cross cellular membranes or to interact with the drug receptor to elicit a biologic response. Therefore, changes in the protein-binding characteristics of a drug can have a profound influence on the distribution and elimination of a drug as well as on the manner in which steady-state concentrations are interpreted. Each drug has its own characteristic protein-binding pattern that is dependent on its physical and chemical properties. As a general rule, however, acidic drugs are bound primarily to albumin, and basic drugs primarily to globulins, particularly α_1-acid glycoprotein. Some drugs can bind to both albumin and globulins.

Depending on its affinity for plasma proteins, a drug may be either tightly or loosely bound. A weakly bound drug can be displaced from its protein sites by a drug with a greater affinity for the plasma protein-binding site. Protein binding of a drug is also dependent on the physical characteristics of the plasma proteins, and on the presence or absence of fatty acids or other drugs in the blood. Fatty acids can displace a drug from its protein-binding sites; tightly bound drugs will not be displaced, but a weakly bound drug can be displaced quite rapidly by free fatty acids present in increased concentrations. It is important to recognize that even though the total drug concentration may remain unchanged, displacement of a drug from its plasma protein-binding sites elevates free-drug concentrations and can result in clinical toxicity.

Anything that alters the concentration of *free* drug in the plasma will ultimately alter the amount of drug available to enter the tissues and interact with specific receptor systems. Different disease states can alter free drug concentrations. For example, in uremia, the composition of plasma is altered by an increase in nonprotein nitrogen compounds, by acid-base and electrolyte imbalances, and often by a decrease in albumin. There may even be structural changes of some proteins that alter their binding capacities. In uremic patients, then, free drug concentrations are frequently elevated. If one monitors *total* plasma drug concentration in these patients, it is quite possible that marked alterations of free drug concentration will not be detected since the total drug concentration may not be dramatically different from that observed in healthy patients. However, the patient may be experiencing adverse effects that are a direct consequence of the increased free drug concentrations. For example, phenytoin is 90% bound and 10% free in healthy subjects. In uremic patients, 20–30% of the total plasma concentration of phenytoin may be free. If one considers the clinical situation where a healthy patient has a total plasma phenytoin concentration of 15 μg/mL, the free phenytoin concentration will be 1.5 μg/mL. If a uremic patient has a total concentration of 15 μg/mL, the free drug concentration may be 4.5 μg/mL. A free phenytoin concentration of 4.5 μg/mL may be sufficient to precipitate severe phenytoin side effects including lethargy and increased seizure frequency. In uremic patients, it is advisable to quantitate free phenytoin concentrations and adjust the drug dosage to maintain free phenytoin concentration at ~2.0 μg/mL.

Alteration of protein concentrations in response to acute stress also can alter free drug concentrations. For example, following myocardial infarction, there is a rapid rise in α_1-acid glycoprotein (AAG) concentration. Lidocaine is a drug of choice for control of arrhythmias due to the infarction, but lidocaine is a basic drug that is highly bound to AAG. Doses of lidocaine adequate to control arrhythmia immediately after infarct are likely to become ineffective 48–72 h later because the higher concentration of AAG diminishes the amount of free drug available to tissue. The arrhythmia reappears and because the total lidocaine plasma concentration necessary to control the arrhythmia appears to be in the toxic range, the lidocaine dose is decreased when in reality it should be increased to maintain optimal free concentrations.

Some drugs exhibit saturation of the available plasma protein binding sites at optimal total drug concentrations. For example, disopyramide binding is concentration-dependent and varies widely from patient to patient. Thus the free concentration and the observed clinical response vary markedly from patient to patient. Valproic acid is also a drug that shows saturation at concentrations $> 100~\mu g/mL$. Thus an increase of total plasma valproate concentration from 100 to 125 $\mu g/mL$ represents a significant increase in the free valproate concentration.

Any change in normal physiological status can alter free drug concentrations and thus change the distribution of drugs between plasma and tissue. Geriatric patients often exhibit hypoalbuminemia. Thus, there is a marked decrease in protein binding sites for drugs. In the elderly, the classical signs of drug intoxication are not usually apparent; instead, the clinical symptoms of drug intoxication are manifested as impaired cognitive function—particularly confusion. An elderly patient may often be considered senile, when in reality an increased free drug concentration is affecting his or her cognitive ability. Reduction of drug dosage to decrease the free drug concentrations can result in dramatic improvements in these patients' personalities. At present, total blood drug concentrations (the sum of protein bound and free drug) are routinely analyzed in the therapeutic drug monitoring laboratory. New ultrafiltration and dialysis methods for routine quantitation of free drug are presently being evaluated.

In *equilibrium dialysis*, serum or plasma is placed on one side of a dialysis membrane and an isotonic buffer solution on the other side. The pore size of the membrane is selected to retain binding proteins on the specimen side of the membrane (retentate) while unbound drug passes through the membrane into the buffer solution (dialysate). At equilibrium, the free drug concentrations in the retentate and dialysate are equal.

In practice a two-piece cell, which permits a dialysis membrane to be sandwiched between chambers of equal volume, is used. After filling the chambers with buffer and sample respectively, the two parts are securely fitted together, placed in a water bath at a controlled temperature, and shaken gently until equilibration occurs. Depending on the drug, equilibration may take from 1–24 h. Free drug concentration in the dialysate may be measured by any routine method, provided that sensitivity is adequate. Equilibrium dialysis for estimation of free drug concentration is not sufficiently fast or convenient for use in the clinical laboratory. Furthermore, control of technical variables is difficult, and for drugs that display saturable binding, free drug concentrations determined by dialysis do not reliably reflect in vivo binding relationships.

Estimation of free drug concentration by *ultrafiltration* is more practical for the clinical laboratory. In this procedure, a cup in which the ultrafiltrate is collected is connected to a reservoir tube containing the serum or plasma. A semipermeable membrane is placed between the two compartments and the entire device is centrifuged at about $3000 \times g$ in an angle-head rotor. The free drug concentration is measured in the water-clear ultrafiltrate by a sensitive procedure. As serum or plasma water is driven through the membrane together with any free drug, protein-bound drug and free protein are concentrated at the same rate above the membrane. Under ideal conditions the free drug concentrations in the filtrate will remain constant throughout the filtration. However, each drug and filtration system used should be evaluated as to nonspecific binding of the drug to the membrane or apparatus, protein leakage of the membrane, and other sources of error. Commercially available devices appear to be satisfactory for some drugs but not for all. In addition, variables that affect the drug-protein binding equilibrium constant, particularly temperature, must be controlled.

Estimation of the free drug concentration will continue to be of interest to TDM. Ultrafiltration techniques, after careful evaluation, promise to be useful in satisfying this need. However, it should be remembered that laboratory measurements presently only *estimate* the free drug concentration in circulating blood. Artifacts introduced in drawing, processing, and storing blood can modify dissociation equilibria for some drugs. Despite these drawbacks, free drug estimations by ultrafiltration are superior to estimations of free drug concentration based on measurements in saliva. Few drugs show a strong correlation between salivary concentration and free drug concentration in plasma. In addition, collection of saliva from acutely ill patients is difficult.

Equation (14) reveals the relationship between the apparent volume of distribution at steady state $(V_{d_{ss}})$, blood volume (V_B), extravascular volume (V_T), free fraction of the drug in blood (f_B), and free fraction of the drug in the extravascular space (f_T).

$$V_{d_{ss}} = V_B + \frac{f_B}{f_T} \times V_T \tag{14}$$

It is evident from this equation that increases in $V_{d_{ss}}$ can occur as a consequence of increases in V_B, f_B, V_T or decreases in f_T. On the other hand, decreases in $V_{d_{ss}}$ can occur with decreases in V_B, f_B, V_T, or increases in f_T. In therapeutic drug monitoring, altered physiologic states or drug interactions have been shown to influence significantly all of these variables. Intuitively this is to be expected, of course. Varying states of hydration or high blood concentrations of drugs or other substances such as bilirubin that compete for protein-binding sites are but two examples of situations that lead to altered $V_{d_{ss}}$.

Metabolism. The rate of the enzymatic process to metabolize a drug is usually characterized by the Michaelis-Menten equation:

$$\text{Rate of metabolism:} \quad \frac{dC}{dt} = \frac{V_{max}C}{K_m + C} \tag{15}$$

where V_{max} is the maximum velocity of the reaction; K_m, the Michaelis-Menten constant, is the drug concentration at which the rate of metabolism is one-half of the maximum; and C is the blood drug concentration.

Usually, drugs are administered to achieve concentrations in the blood well below the K_m of a particular drug. Therefore if $K_m >> C$, equation (15) can be simplified to:

$$\frac{dC}{dt} = \frac{V_{max}C}{K_m} \tag{16}$$

and $\dfrac{V_{max}}{K_m}$ can be written as the constant, K, such that

$$\frac{dC}{dt} = KC \tag{17}$$

where K is a simple *first-order rate constant* for the metabolic elimination. In other words, the rate of metabolism is proportional to the concentration of drug. First-order kinetics are characteristic of the metabolism of numerous drugs.

In the event that the rate of metabolism approaches V_{max}, the constant maximum velocity, the rate becomes independent of concentration and thus descriptive of a *zero-order* process. Few drugs in clinical application follow true zero-order kinetics in their metabolism.

There are several drugs, notably phenytoin, salicylates, and ethanol, that cannot be characterized by simple first-order kinetics. Instead, the rate of metabolism of these compounds is said to be *capacity-limited* and would be described by equation (15). The term capacity-limited or *nonlinear* pharmacokinetics means that clearance or the apparent $t_{1/2}$ changes with changes in concentration. Figure 17-11 (Curve B) shows how a dose-response curve may be linear (first order) until a capacity limit of some pathway is reached and then may become nonlinear. Equation (7) has shown the relationship between plasma concentration, total body clearance, half-life, and dose. It should be obvious, therefore, that important clinical considerations arise when a patient is treated with a drug that displays nonlinear kinetics. First, changes in dosing result in disproportionate changes in steady-state drug concentrations so that titration to appropriate serum levels must be approached conservatively. Second, since both clearance and apparent half-life of the drug change with increasing drug concentration, the length of time required to reach a new steady-state concentration will be prolonged.

All of the equations previously described for predicting dose or concentration assume linear kinetic systems; they are therefore not adaptable to treatment with drugs that display nonlinear kinetics. Methods for predicting phenytoin dose and concentration and using a linearized Michaelis-Menten equation have been developed and applied to individualize drug dosing regimens.[61]

Hepatic clearance. Drugs predominantly metabolized by the liver can be divided into two general categories: high and low extracted drugs. (See Table 17-1.) For those drugs dependent solely on hepatic elimination, total body clearance (Cl_T) equals hepatic clearance (Cl_H). When the liver is considered from a purely physiologic perspective, then the hepatic clearance will be determined by the *hepatic blood flow* (Q) and the hepatic extraction fraction (E).

Figure 17-11. Dose-response curves. Line A illustrates the linear relationship between serum drug concentration and total daily dose of a drug that displays first-order kinetics. Line B illustrates the dose-response relationship for a drug that displays capacity-limited kinetics because of a saturable enzyme or transport mechanism; in this situation, serum concentration becomes independent of total daily dose, and the relationship of drug concentration to dose becomes nonlinear. (Adapted from Pippenger, C. E.: Practical pharmacokinetic applications. Syva Monitor, Jan., 1979, pp. 1–4. Syva Co., Palo Alto, CA.)

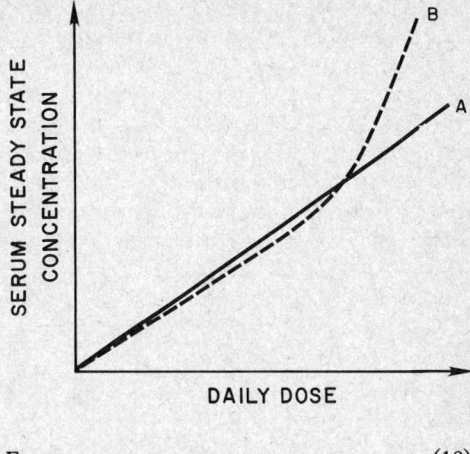

$$Cl_H = Q \times E \qquad (18)$$

The hepatic extraction fraction (E) of a drug reflects the affinity of a particular drug for hepatic microsomal enzymes; E can be found experimentally or calculated by equation (19)

$$E = \frac{C_a - C_e}{C_a} \qquad (19)$$

where C_a is the concentration of the drug in blood entering the organ and C_e is the concentration of the drug in the hepatic venous effluent. For drugs that possess a high extraction fraction, hepatic clearance will approach hepatic blood flow (Q). Drugs with a low extraction fraction will have relatively low hepatic clearances compared to hepatic blood flow. The total body clearance of highly extracted drugs will be primarily dependent upon hepatic blood flow for their elimination. These drugs usually have low bioavailability because of the first-pass effect described earlier. The clearance of low-extracted drugs is less dependent on blood flow and more dependent on the quantity and quality of the hepatic microsomal enzymes. Total body clearance of these drugs will be affected by hepatic function, enzyme inducers and inhibitors, and changes in free drug concentration. The reader should recognize that this is a superficial view of a complex process. Several excellent reviews on this subject are currently available.[69,83,85]

Biotransformation. The liver is the organ principally responsible for drug metabolism. Its role is to convert lipophilic nonpolar molecules to more polar, water-soluble forms. The drug molecule can be modified by Phase I reactions that alter chemical structure by oxidation, reduction, or hydrolysis; or by Phase II reactions that conjugate the drug (glucuronidation, sulfation) to water-soluble forms; or by both Phase I and Phase II reactions. Most drug metabolism takes place in the microsomal fraction of the hepatocytes where many environmental chemicals and endogenous biochemicals are also processed and by the same mechanisms.

The enzymes of the hepatic microsomal system can be induced or inhibited. Enzyme induction and inhibition have greatest significance in instances of drugs with low to moderate hepatic extraction fractions.

Microsomal *enzyme induction* leads to an increase in the amount of enzyme present, most commonly the amounts of the oxidizing enzymes. Numerous *isoenzymes* of *cytochrome* P_{450} are

Table 17-1. EXAMPLES OF SUBSTANCES ELIMINATED BY THE LIVER

High Extraction	Low Extraction
Indocyanine green	Antipyrine
Lidocaine	Diazepam
Meperidine	Phenytoin
Morphine	Prednisone
Propranolol	Theophylline
Verapamil	Warfarin

now known to exist and to be affected differently by different enzyme-inducing drugs. Two classic and clinically relevant enzyme inducers can be contrasted. Phenobarbital represents the type of enzyme inducer with broad induction effects. After a latency period, cytochrome P_{450}, cytochrome P_{450} reductase, and other enzymes are induced. Liver weight, hepatic blood flow, bile flow, and production of hepatic proteins all increase. This induction apparently increases the P_{450} isoenzyme for which antipyrine is a substrate because the hepatic clearance of antipyrine is increased following phenobarbital administration. However, phenobarbital administration induction has little effect on theophylline clearance, suggesting a different isoenzyme for theophylline metabolism. In contrast, polycyclic hydrocarbons in tobacco smoke (3-methylcholanthrene) induce a different P_{450} isoenzyme. No change in P_{450} reductase occurs and a different terminal oxidase appears. After this type of induction, the clearance of theophylline is increased but not that of antipyrine. These substances have served as prototypes for the classification of enzyme inducers. Obviously, when a patient is on a drug with a narrow therapeutic index, his or her dosage regimen would need to be adjusted should a known enzyme-inducing drug be added to or deleted from his therapy.

Since the drug-metabolizing enzymes of the liver are nonspecific and interact with a wide variety of endogenous and exogenous substances, it is not very surprising to witness the *metabolic inhibition* of one drug by another. Several general mechanisms have been proposed to describe these events. They include substrate competition, competitive or noncompetitive inhibition, product inhibition, and repression (where the amount of enzyme is reduced either by decreased synthesis or increased degradation). Most drug-drug interactions probably fall into the categories of substrate competition or competitive or noncompetitive inhibition. Examples of drugs that have been shown to significantly inhibit drug metabolism include chloramphenicol, cimetidine, valproic acid, allopurinol, and erythromycin. As with enzyme inducers, the addition or deletion of an inhibitory drug to a patient's drug therapy requires appropriate therapeutic drug monitoring and probable dosage adjustment.

The value of therapeutic monitoring is nowhere more evident than for those drugs principally metabolized by the liver. There are two cogent reasons for this statement. First, wide variability in the rate of metabolism of any given drug exists not only in different patients in the general population but also in the same patient at different times and in different circumstances. This variability is due to factors such as age, weight, gender, genetics, exposure to environmental substances, diet, co-administered drugs, and disease. Second, there is no good endogenous biochemical marker by which hepatic function, and consequently hepatic capability for drug clearance, can be routinely assessed before drug therapy is initiated.

The biotransformation of drugs may produce *metabolites* that are pharmacologically active. In such instances, the metabolite should also be measured since it is contributing to the effect of the drug on the patient. If the metabolite is inactive, it need not be measured, but steps should be taken to ensure that it does not interfere in the analytical process.

Excretion of drugs or chemicals from the body can occur via biliary, intestinal, pulmonary, or renal routes. Although each of these represents a possible mechanism of drug elimination, *renal excretion* is a major pathway for the elimination of drugs and their metabolites and is the most important in therapeutic drug monitoring. Therefore, alterations in renal function may have a profound effect on the clearance and apparent half-life of the parent compound or its active metabolite(s); decreased renal function will cause elevated serum drug concentrations and increase the pharmacologic response. Three processes are involved in the elimination of the drug via the kidney: *glomerular filtration* (GF), *tubular secretion*, and *tubular reabsorption*.

Kidney function, in contrast to liver function, is readily and reliably evaluated by estimation of *creatinine clearance*. Creatinine is a metabolic product of muscle metabolism and is produced at a constant rate by the body. It is primarily eliminated from the body by the kidney via the glomerular filtration mechanism. Renal clearance of creatinine at 120 mL/min approximates the glomerular filtration rate of 90–130 mL/min. (See also Chapter 11.) Therefore, availability of creatinine clearance measurements on a routine basis provides an effective tool to evaluate kidney function. A strong correlation has been shown to exist between creatinine clearance and the total body clearance or elimination rate constant of those drugs primarily dependent upon the kidney for their elimination. Examples of drugs whose therapeutic use is adjusted to account for changes in creatinine clearance include gentamicin, tobramycin, amikacin, digoxin, vancomycin, and cyclosporine.

Table 17-2. PATHOPHYSIOLOGIC STATES THAT USUALLY DECREASE TOTAL BODY CLEARANCE OF PARTICULAR DRUGS

Renal disease	Amikacin, gentamicin, tobramycin, kanamycin, netilmycin, digoxin, procainamide, *N*-acetylprocainamide, methotrexate, lithium, vancomycin, cimetidine, chloramphenicol*
Liver disease	Phenytoin,* phenobarbital,* valproic acid,* lidocaine, quinidine, verapamil, nifedipine, propranolol, theophylline, chloramphenicol*
Cardiovascular disease	Verapamil, nifedipine, disopyramide, propranolol
Congestive heart failure	Lidocaine, quinidine, theophylline
Hypothyroidism	Digoxin

*Decrease or no change.

Summary of Factors Influencing Drug Disposition

A large number of factors are now recognized to have a profound influence on the pharmacokinetics of drugs and consequently on a patient's pharmacologic response. Consideration of the patient's history, with particular emphasis on his or her pathophysiologic state and adjunct drug therapy, is essential at the initiation of drug therapy and therapeutic drug monitoring. It is beyond the scope of this chapter to detail all of these factors that may influence the disposition of a particular drug. Table 17-2 summarizes the influence various altered pathophysiologic states have on clearance mechanisms of several drugs frequently monitored by therapeutic drug monitoring programs. Factors that influence drug disposition are identified in Table 17-3.

Computer Applications in Pharmacokinetics

The routine application of computer technology has contributed significantly to the development of pharmacokinetics and therapeutic drug monitoring and is continually expanding. Computer programs are

Table 17-3. FACTORS THAT INFLUENCE DRUG DISPOSITION IN MAN

Demographic factors
  Age category (premature infant, neonate, infant, pre- and postpubescent child, adult, elderly adult)
  Weight
  Gender
  Race
  Genetic constitution
Disease-related factors
  Liver disease (cirrhosis, hepatitis, cholestasis)
  Kidney disease
  Thyroid disorders (hypo- or hyperthyroidism)
  Cardiovascular disease (arrhythmias, congestive heart failure)
  Gastrointestinal disease or disorder (sprue or other malabsorption syndromes, peptic ulcer, colitis)
  Cancer
  Surgery
  Burns
  Nutritional status (cachectic or anorexic states)
Extracorporeal factors
  Hemodialysis
  Peritoneal dialysis
  Cardiopulmonary bypass
  Hypo- or hyperthermia
Chemical and environmental factors influencing:
  Absorption of drug
    Food or coadministered drug affecting extent and rate of absorption
  Distribution of drug
    Coadministered drug affecting binding to plasma proteins or tissue receptors
  Metabolism of drug
    Food intake (carbohydrates, proteins, lipids) competing for metabolizing systems
    Coadministration of drug that induces metabolizing enzymes (e.g., phenobarbital)
    Coadministration of drug that inhibits metabolizing enzymes (e.g., cimetidine)
  Excretion of drug
    Coadministration of drug that competes for renal tubular secretory paths (e.g., probenecid or penicillin)
    Changes in urinary flow rate
    Coadministration of compounds that enhance tubular reabsorption (e.g., sodium bicarbonate or phenobarbital)

presently available to meet a number of different needs. These needs can be divided into two main areas of interest: *analysis of pharmacokinetic data* and *calculation of drug dosage regimens*. It is critical to emphasize, particularly with the latter area, that a user of such programs should have a strong understanding of computer programs as well as of pharmacokinetic and clinical pharmacokinetic principles.

Pharmacokinetic data analysis usually involves evaluation of blood drug concentration versus time. Computer programs have been written to arrive at the best fit of data for the concentration versus time relationship, to determine the most appropriate compartmental model which would describe the data, and to calculate the appropriate parameters (Cl_T, V_d, $t_{1/2}$) of the model. These techniques are still employed in research, and the parameters generated remain useful in the therapeutic drug monitoring process. The ability to generate model-independent relationships, particularly clearance, utilizing simpler mathematical techniques, will minimize the value of model fits in the future. These computer programs also provide the user with several alternative approaches. First, given demographics, disease state(s), and drug history of a patient, the program will calculate and present a dosage regimen appropriate for achieving the desired blood drug concentration. The main assumption in this approach is that the patient's condition conforms to those in the population pool used to generate initial pharmacokinetic parameter estimates. Second, drug concentrations and time data can be analyzed by least-squares regression and used to obtain measured estimates of the pharmacokinetic parameters. The value of this approach is often limited by the number of data points one can obtain. With either approach, the user must understand the assumptions of the data base and the models employed.

A relatively new technique has been developed that essentially combines both of the approaches mentioned above. This approach is often called the *Bayesian* method. It is a powerful statistical method that enables the users to incorporate simultaneously pharmacokinetic parameters derived from patient drug concentration and time data with pooled population parameters. The data manipulation starts from population-pooled parameters for estimating dose of blood drug concentration. Then as data begin to accumulate from assays made on the patient's serum, the values are used to derive the patient's own pharmacokinetic parameters in combination with the population parameters. As the number of patient observations increases, the program places more and more weight on the pharmacokinetic parameters derived from the patient's own drug concentration data. The main advantages of this method lie in minimizing the chance of misinterpreting any one drug concentration-time data point and in diminishing the large number of drug concentration-time data points needed to arrive at meaningful interpretations. Such a program is available from Data Med, Inc., Minneapolis, MN.

Clinical Utility of TDM

Therapeutic drug monitoring is most valuable when the drug in question has a narrow therapeutic index and is used chronically. A number of advantages can be realized by a TDM program.

1. *Noncompliance* can be recognized. Many patients, especially those with chronic disease, require prolonged drug therapy. The problem of patient compliance is particularly evident with patients who are characteristically free of pain or unusual discomfort. Among such conditions are epilepsy, asthma, hypertension, and mild heart disease. The end result of noncompliance is exacerbation of the existing disorder and treatment failure. Drug concentration values provide positive feedback to the physician regarding complying and noncomplying patients.

2. Patients with *unusual drug disposition* characteristics can be recognized and documented. Occasionally, the disposition pattern of drugs in particular individuals may deviate considerably from the average patient population parameters. Aberrant behavior may be attributable to an effect of a drug or disease state not previously recognized. The clinical condition and individual metabolism of patients not only may differ from one to another but may change in the same patient during treatment. Both the pharmacokinetic disposition and the pharmacodynamic response to a given drug dose in individual patients may vary widely as a direct consequence of genetic influences. Studies of these genetic factors (pharmacogenetics) have clearly demonstrated that all aspects of pharmacokinetics and pharmacodynamics are under genetic control.

3. Therapeutic drug *regimens can be adjusted* during periods of continuous physiologic change. Normal alterations in physiologic state, as in pregnancy or aging, or continuous pathophysiologic or hemodynamic changes as consequences of disease, surgical treatment of disease, and the healing process complicate assessment of drug dosage needs.

4. *Baseline concentrations* associated with an optimal therapeutic regimen can be identified. After the patient has undergone a strenuous workup to define an appropriate therapeutic regimen, the physician can establish a baseline drug concentration at which the patient responds well.

Should the patient's response alter significantly in the future, the physician can rapidly document whether the patient has been compliant or whether a new disease state may be altering response to the drug.

5. The most *appropriate drug dosing* regimens can be initiated and maintained for a particular patient.

Blood Sample Collection and Processing

The value and quality of a TDM service are only as good as the data produced. Routine analyses for drug concentrations are subject to a large number of problems that could cause gross error or misinterpretation. Issues of continuing concern include the following:

1. *Assay methodology* used for TDM should be accurate and reproducible. All clinical laboratories with a TDM service should be actively involved in an internal quality control and external proficiency testing program. Additionally, sample volume and assay turnaround time should be considered in selecting the most appropriate analytical method.

2. Each laboratory should *inform the medical staff* as to therapeutic and toxic concentration ranges, analytical method (when appropriate), action values, required sample volume, and collection tube specifications.

3. Guidelines should be available for *ideal sample schedules* for each individual drug monitored. Usually, steady-state trough concentrations are most desirable; however, other sample schedules are often invoked, depending on the properties of the drug or the individual needs of the patient.

4. The *time and date of collection* of the drug sample and of the *last dose* should be noted. To assess steady-state conditions, the length of time the patient has been on a particular regimen should be known. In vitro conditions affecting stability of the drug in a sample (example: penicillin or heparin and aminoglycoside antibiotics) or the assay specificity (e.g., presence of hemolysis) should be considered for sample handling procedures.

5. A successful therapeutic drug monitoring program requires a multidisciplinary effort and necessitates *interdepartmental dialogue* and cooperation.

6. Since *laboratory reports* become part of a patient's chart, it is useful to devise a reporting form that incorporates all of the data necessary for interpretation (dosage form, frequency and amount, plasma concentration, trough or peak value, time of dosage, time of draw, and other drugs coadministered).

Drug Assay Technique—A Historical Perspective

The evolution of therapeutic drug monitoring has been primarily dependent on the development of rapid, sensitive, and specific analytical techniques. These technical advances fueled the growth of the fields of pharmacology, biopharmaceutics, and pharmacokinetics.

Gas-liquid chromatography (GLC) represented a major breakthrough over early spectrophotometric methods of drug analysis. It permitted separation of parent drug from metabolite(s) and differentiation from coadministered drugs and endogenous compounds. The ability to separate and quantitate several drugs within a given class of drugs is appreciated when determining the anticonvulsant drugs. Major disadvantages of GLC include the complexity of the instrumentation, which requires a highly trained and skilled analyst, and the need for a relatively large volume of sample. However, more recent advances in the development of the nitrogen-phosphorus detector and capillary columns have increased the sensitivity of the instruments to such an extent that drug analysis can be performed routinely on microliter volumes of plasma.

During the late 1960's and early 1970's *high-performance liquid chromatography* (HPLC) techniques were introduced. This approach offered greater versatility with less sample preparation. Its specificity and sensitivity have made it an extremely effective research tool. Additionally, the relatively small sample requirements and the ease of operation make HPLC an appealing alternative to GLC. HPLC can also be adapted to the simultaneous quantitation of a large variety of drugs as well as their metabolites.

Radioimmunoassay techniques have permitted quantitation of drug concentration in microliter volumes of serum at ng/mL concentrations. But complexity of this technique, long turnaround time, and lack of radioimmunoassays for a wide variety of drugs have prevented its widespread adoption for routine drug assays.

Making TDM available to all laboratories and physicians required simple technology that could be performed by a technologist without extensive training on special instrumentation. This was achieved with the development of the *nonisotopic homogeneous immunoassays* in which analyses are possible on < 40 μL of serum. The major advantages of these systems are their microcapability and specificity as well as their rapidity and ease of performance.

SPECIFIC DRUG GROUPS

Drugs that are routinely monitored are conveniently classified by the kind of therapy they support, i.e., control of epilepsy, management of respiratory or cardiac function. An analytical method for one drug in a grouping is often applicable to other drugs in the same grouping. The following discussion is organized in accordance with classifications commonly recognized. Note that some drugs such as salicylate and nitroprusside (as thiocyanate) are discussed in Chapter 18, Analysis of Toxic Substances.

ANTIEPILEPTIC DRUGS

Phenobarbital, 5-ethyl-5-phenyl-2,4,6(1H,3H,5H)-pyrimidinetrione, known by a wide variety of proprietary names and found in combination with many other drugs, is used in the treatment of all seizures except absence seizures (petit mal). It is particularly useful in the case of tonic-clonic, focal motor, temporal lobe, and febrile seizures. It is also known to reduce synaptic transmission resulting in decreased excitability of the entire nerve cell. The end result is an increase in seizure threshold and inhibition of the spread of discharges from the epileptic foci.

Absorption of oral phenobarbital is slow but complete. The time at which *peak plasma concentrations* are reached is widely variable and ranges from 4–10 h after the dose. Phenobarbital is 40–60% bound to plasma proteins. The *elimination half-life* is from 70–100 h and is age dependent (children average 70 h, geriatric patients 100 h). Since hepatic metabolism is one of the prime routes of elimination, reduced liver function results in prolonged half-life.

The optimally effective *therapeutic concentration* of phenobarbital is between 15 and 40 μg/mL. The predominant side effect observed in adults at blood concentrations > 40 μg/mL is sedation, although with chronic therapy tolerance to this effect develops.

Phenobarbital is metabolized in the liver to *p*-hydroxyphenobarbital, which is largely excreted as the glucuronide or sulfate ester. When renal and hepatic function are decreased, patients experience decreased clearance of the drug. Elimination of phenobarbital may be decreased in the presence of valproic acid and salicylate if a reduction in urinary pH occurs. During chronic administration of either valproate or salicylate, the concentration of phenobarbital may increase 10–20%, and a dosage adjustment may be necessary to avoid intoxication. Phenobarbital induces mixed function oxidative enzymes, resulting in increased metabolism of other xenobiotics after ∼1–2 weeks of therapy.

Because of the long elimination half-life of phenobarbital the blood concentration does not change rapidly. Therefore, a *serum specimen* collected late in the dose interval (trough) is representative of the overall effect. Results from specimens collected 2–4 h after the dose can be misleading, because they may be construed to be the peak concentration when in actuality they are not. Table 17-4 summarizes pharmacokinetic data of the anticonvulsant drugs.

Phenytoin (diphenylhydantoin), 5,5-diphenyl-2,4-imidazolidinedione, most commonly available as *Dilantin* but also available in generic form, is used in the treatment of primary or secondary generalized tonic-clonic seizures, elementary-partial or complex-partial seizures, and status epilepticus. The drug is not effective for absence seizures. The mechanism of action of phenytoin is by the modulation of the enzyme *Na,K-ATPase,* which results in blockage of neurotransmitter release. The physiologic effect of this action is reduction in central synaptic transmission, aiding in control of abnormal neuronal excitability.

Phenytoin is not readily soluble in aqueous solutions. When administered by intramuscular injection, most of the dose precipitates at the site of injection and is then slowly absorbed. *Absorption* of oral phenytoin is slow and sometimes incomplete. Variations in the drug preparation have been blamed for low bioavailability. Once absorbed, the drug is highly protein bound (90–95%). As with all drugs, the pharmacologic effect of phenytoin is directly related to the amount present in the free (unbound) state. Only the *free phenytoin* is available to cross biological

Table 17-4. PHARMACOKINETIC PARAMETERS OF THE ANTICONVULSANT DRUGS

Antiepileptic Drugs	Therapeutic Range (per mL of Plasma)	Half-Life (h)	Volume of Distribution (L/kg)	Bioavailability, Oral (%)	Protein Bound (%)
Bromide	1000–1500 µg	290	*		0
Carbamazepine	8–12 µg	27 ± 4	1.4 ± 0.2	>70	82 ± 5
Clonazepam	10–50 ng	39 ± 12	3.2 ± 1.1	70–100	47
Ethosuximide	40–100 µg	33 ± 6	0.7 ± 0.1		0
Mephobarbital	1–5 µg	30 ± 10			50
Methsuximide	20–40 ng	3			
(Normethsuximide)	(20–40 µg)	(30)			
Phenobarbital	15–40 µg	86 ± 7	0.9 ± 0.3	90	50
Phenytoin	10–20 µg	†	0.6 ± 0.1	90–100	89 ± 23
Primidone	5–10 µg	8 ± 5	0.6 ± 0.5	92 ± 18	20
Valproic acid	50–100 µg	16 ± 3	0.1 ± 0.05	100 ± 10	93 ± 4‡

*Where data are missing, they are either not applicable or not available.

†Phenytoin does not obey first order kinetics at therapeutic concentrations, therefore a true half-life cannot be calculated. In adults, the apparent half-life is 18–30 h; in children it is 12–22 h.

‡Concentration dependent.

membranes and interact at biologically important binding sites. The degree of protein binding can be reduced by the presence of other drugs, anemia, and hypoalbuminemia which can occur in the elderly. In these conditions, an increased effect is observed at the same total drug concentration as in plasma from normal patients.

The optimal *therapeutic concentration* for seizure control without side effects is 10–20 µg/mL. In a large population study, Buchthal et al.[18] found a 50% response rate in patients with plasma concentrations >10 µg/mL and an 86% suppression of seizure activity at concentrations >15 µg/mL. These concentrations also serve as reasonable guidelines when the drug is used as a cardiac *antiarrhythmic* agent. Free phenytoin concentrations of 1–2 µg/mL are optimal. Total phenytoin concentrations in excess of 20 µg/mL do not usually enhance seizure control and are often associated with nystagmus and ataxia. Total phenytoin plasma concentrations in excess of 35 µg/mL have been shown actually to precipitate seizure activity. A side effect of phenytoin, although not related to plasma concentration, is development of gingival hyperplasia.

Phenytoin is metabolized by hepatic microsomal hydroxylating enzymes. The principal *metabolite* is 5-(*p*-hydroxyphenyl)-5-phenylhydantoin (HPPH), which is excreted principally as a glucuronide ester. Other minor metabolites are of minimal clinical importance. Hepatic metabolism of phenytoin may become saturated within the therapeutic range. Once metabolism is saturated, small dosage increments result in large changes in blood concentration; this phenomenon partially explains the wide variation in dosage among patients that is required to accomplish a therapeutic effect. Because of this saturation phenomenon, first-order kinetics do not apply to phenytoin at blood concentrations in excess of 5 µg/mL.

A recent study[41] demonstrated that ~50% of patients receiving the standard therapeutic dose of phenytoin (300 mg/d) achieved steady-state plasma concentrations which were less than ideal. Modification of dosage regimens, to take into account individual drug utilization patterns as well as body weight, have resulted in a significant increase in the number of patients who are maintained within the optimal therapeutic range, with the desired pharmacological effect.

The time to collect the *specimen* is dictated by the reason for monitoring. If the patient displays any symptoms of intoxication, then the peak blood concentration is of interest. This specimen is collected 4–5 h after the dose, although the peak level may be delayed up to 8 h if the drug is given in conjunction with substances that increase stomach acidity. If the principal question at hand is adequate therapy, then the trough concentration is more useful, and the specimen is collected just before the next dose is given.

A number of *drug interactions* result in alteration of the disposition of phenytoin. Alcohol, barbiturates, and carbamazepine induce oxidative enzymes; this induction results in increased metabolism of phenytoin, *reduced* serum concentration of both total and free phenytoin, and reduced pharmacologic effect. Drugs such as chloramphenicol, disulfiram, isoniazid, and dicumarol compete with phenytoin metabolism, resulting in *increase* of both total and free phenytoin concentrations and enhancement of the pharmacologic effect. Salicylate, valproic acid, phenylbutazone, and sulfonylureas compete with phenytoin for serum protein-binding sites. The end result is diminished total serum concentration of phenytoin while the free phenytoin concentration and pharmacologic effect remain approximately the same. The interest in monitoring the free phenytoin concentration is in response to these altered disposition states.

Primidone, 5-ethyldihydro-5-phenyl-4,6(1H,5H)-pyrimidinedione, also known as *Mysoline* and *Primaclone*, is effective in the treatment of tonic-clonic and complex partial seizures. The mechanism of action of this drug is similar to that described for phenobarbital, and the therapeutic effect is due partially to the accumulation of its major metabolite, *phenobarbital.* A second metabolite of primidone, *phenylethylmalonamide* (PEMA), also has some antiepileptic activity.

Primidone is rapidly and completely absorbed after oral administration. Once absorbed, it is not highly protein bound and it has a $t_{1/2}$ of ~8 h. Disposition of the drug is not known to be significantly altered by other disease states or other drugs.

The optimal *therapeutic concentration* of primidone has been established as 5–12 μg/mL. Since phenobarbital is an active metabolite of primidone, concurrent analysis of phenobarbital is required for complete result interpretation. The previously defined therapeutic range for phenobarbital applies to adequate primidone therapy. The phenobarbital concentrations rise gradually over a period of 1–2 weeks after therapy is initiated. *Toxicity* due to accumulation of primidone occurs at serum concentrations > 15 μg/mL and is usually associated with symptoms of sedation, nausea, vomiting, diplopia, dizziness, ataxia, and a phenobarbital concentration > 40 μg/mL. Specimen collection is dictated by the same rules that apply for phenobarbital; the trough concentration is most useful.

Coadministration of acetazolamide with primidone will result in decreased gastrointestinal absorption of primidone and subsequent diminished plasma concentrations. Primidone administered in association with phenytoin will produce a modest elevation of the phenobarbital/primidone ratio since phenytoin competes with the hepatic hydroxylating enzymes associated with phenobarbital's metabolism. Coadministration of valproic acid, for the same reasons outlined for phenobarbital, will cause a modest increase in both primidone and phenobarbital serum concentrations.

Carbamazepine, 5H-dibenz(b,f)azepine-5-carboxyamide, proprietary name *Tegretol*, is used in the treatment of generalized tonic-clonic, partial, and partial-complex seizures. It is also used for the treatment of pain associated with trigeminal neuralgia. While the mechanism of action of carbamazepine is unclear, its anticholinergic, muscle relaxant, antiarrhythmic, antidiuretic, and neuromuscular transmission-inhibitory actions may be explained by blockage of neurotransmitter reuptake.

After oral administration, carbamazepine is rapidly absorbed but with wide individual variability. The drug is highly protein bound (80%). The *elimination half-life* early in therapy is ~1 d. With chronic therapy, the enzymes responsible for metabolism are induced, and the elimination half-life is reduced to 15–20 h. Since hepatic metabolism is the principal means by which plasma levels are reduced, any reduction in liver function will result in drug accumulation.

The *therapeutic concentration range* for optimal pharmacologic effect of carbamazepine is from 8–12 μg/mL. *Toxicity* associated with excessive carbamazepine ingestion occurs at plasma concentrations > 15 μg/mL, and is characterized by symptoms of blurred vision, paresthesia, nystagmus, ataxia, drowsiness, and diplopia. Side effects unrelated to plasma concentration include development of an urticarial rash which usually disappears upon discontinuation of the drug, and hematologic depression (leukopenia, thrombocytopenia, and aplastic anemia).

The active metabolite of carbamazepine is carbamazepine-10,11-epoxide. This metabolite has been found in children to accumulate to concentrations equivalent to carbamazepine. It may contribute to symptoms of intoxication in children who have a therapeutic plasma concentration of the parent drug. Because carbamazepine is metabolized via the hepatic oxidative enzyme system, drugs that induce this system (phenytoin, phenobarbital) increase the rate of clearance of carbamazepine.

Because of carbamazepine's relatively long half-life, the *specimen* yielding the most useful information is the one representing the trough concentration, although in the case of suspected mild intoxication, the peak value of the plasma concentration correlates better with toxicity. The peak specimen should be collected 2–4 h after the oral dose.

Ethosuximide, 3-ethyl-3-methyl-2,5-pyrrolidinedione, proprietary name *Zarontin*, has been used for the treatment of absence seizures (petit mal) characterized by brief loss of consciousness.

A proposed mechanism of action of this drug involves the inhibition of Na,K-ATPase, but in a different manner from that exhibited by phenytoin.

Ethosuximide is readily absorbed from the gastrointestinal tract. Its $t_{1/2}$ is ~33 h, although this may be prolonged in adults. The drug is cleared mainly by metabolism as either the hydroxyethyl compound or the glucuronide ester of the hydroxyethyl metabolite. The *trough specimen* yields the most useful information regarding therapeutic efficacy. The optimal *therapeutic concentration* of ethosuximide is 40–100 μg/mL. (See Table 17-4.) *Toxicity* related to an excessive blood concentration of ethosuximide is rare. Symptoms of gastrointestinal distress, lethargy, dizziness, and euphoria may be encountered early in therapy, but patients usually become tolerant to these symptoms.

Methods for the Determination of Common Antiepileptic Drugs

Phenobarbital, phenytoin, primidone, carbamazepine, and ethosuximide can be coanalyzed by either GLC or high-performance liquid chromatography (HPLC), or individually analyzed by immunoassay. The advantage of HPLC or GLC is that simultaneous analysis can be accomplished. Immunoassay procedures are less labor intensive and are usually quicker than HPLC or GLC for a single analyte; however, multiple drugs or drug plus metabolite requires additional assays. Foerster et al.[31] reviewed GLC procedures, and Tracy et al.[82] reviewed immunoassay procedures. An HPLC method is presented here as a reliable method which simultaneously measures several common anticonvulsants.

Principle

The commonly monitored antiepileptic drugs are separated from proteins by precipitation with acetonitrile, which also serves as solvent. The extract is then chromatographed into its individual components. The drugs are quantitated from the resulting liquid chromatogram (Figure 17-12) based on their ultraviolet energy absorbance, expressed as either peak height or peak area. Inclusion of an internal standard allows for correction of extraction losses and chromatographic variability.

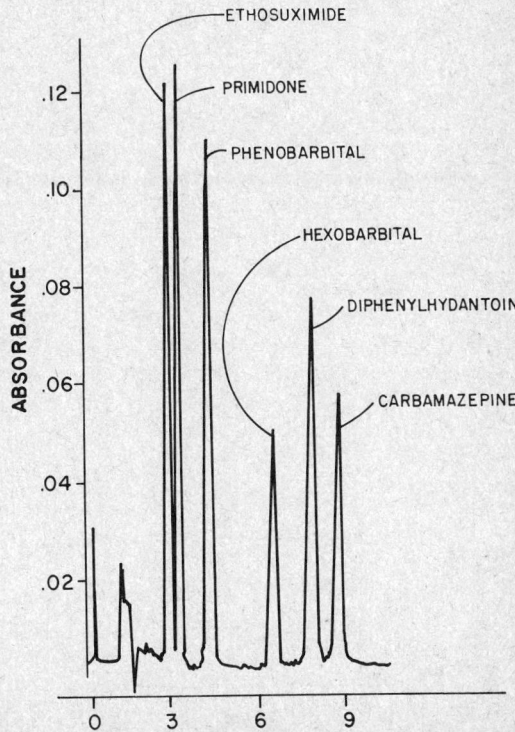

Figure 17-12. High-performance liquid chromatogram of the common antiepileptic drugs.

Reagents

1. Drug standards.

a. Low working standard. 15 μg/mL phenobarbital, 5 μg/mL phenytoin, 3 μg/mL carbamazepine, 30 μg/mL ethosuximide in water. Prepare these from 1 mg/mL stock solutions of each drug in methanol. Stock solutions are stable for periods longer than two years if stored in tightly sealed containers at -20 °C. The working solution is stable for three months if stored in a sealed container at 4 °C.

b. High working standard. 50 μg/mL phenobarbital, 25 μg/mL phenytoin, 15 μg/mL carbamazepine, 100 μg/mL ethosuximide in water. Prepare and store as for the low working standard.

2. Internal standard. Dissolve 20 μg/mL hexobarbital in 100% acetonitrile. Prepare from a stock solution of 1 mg/mL hexobarbital in methanol stored at -20 °C.

3. Solvents.

a. Acetonitrile and methanol. Double glass-distilled, absorbance <0.1 A at 200 nm.

b. KH$_2$PO$_4$, 0.005 mol/L, pH 4.4. Dissolve 680 mg reagent grade KH$_2$PO$_4$ in <1000 mL H$_2$O and titrate to pH 4.4 with KOH, 0.1 mol/L. Dilute to total volume of 1000 mL.

c. Mobile phase. Mix 280 parts KH$_2$PO$_4$, 0.005 mol/L, pH 4.4, 44 parts methanol, and 76 parts acetonitrile; filter through a 0.5 μm filter and de-gas by applying gentle heat and vacuum.

4. Liquid chromatograph.

a. Column. Octadecyl-silane bonded to silica, 10 μm or 5 μm.

b. Mobile phase. Methanol:acetonitrile:K$_2$HPO$_4$, 0.005 mol/L, pH 4.4 (11:19:70), at a flow rate of 2 mL/min.

c. Detector. 15 μL or smaller flow cell, variable wavelength, ultraviolet HPLC detector set at 200 nm, 0.1 Absorbance Units Full Scale (AUFS). Use 0.4 AUFS for ethosuximide.

Procedure

1. Pipet 250 μL of standard, control pool, or specimen into appropriately labeled 12 $\times$ 75-mm glass tubes.

2. Add 250 μL of internal standard in acetonitrile.

3. Vortex vigorously for 15 s.

4. Centrifuge to separate precipitated proteins at 2000 $\times$ g for 3 min.

5. Chromatograph a 20-μL aliquot of the supernatant from step 4. The amount chromatographed can be increased or decreased to alter sensitivity as desired.

6. Identify the drugs by retention time (Figure 17-12) and quantitate by either peak height or peak area (peak height is preferred).

7. Establish for each drug a ratio of peak height or area to the peak height or area of the internal standard.

$$R = \frac{\text{Peak height}_{\text{standard (or sample)}}}{\text{Peak height}_{\text{internal standard}}}$$

8. Using the low and high standards to establish a standard plot (concentration being the independent variable, R the dependent variable), calculate the results for the specimen.

9. Analysis of carbamazepine may be improved by monitoring the chromatographic effluent at 254 nm, 0.005 AUFS.

Reference

Gerson, B., and Anhalt, J. P.: High Pressure Liquid Chromatography and Therapeutic Drug Monitoring. Chicago, ASCP Press, 1980, p. 172.

VALPROIC ACID

Valproic acid (di-n-propylacetic acid), brand name *Depakene*, is used for treatment of absence seizures. It has also been shown to be useful against tonic-clonic and partial seizures when used in conjunction with other antiepileptic agents. The drug inhibits the enzyme γ-aminobutyric acid (GABA) transaminase, resulting in an increase in the concentration of GABA in the brain. GABA is a potent inhibitor of pre- and postsynaptic discharges in the central nervous system.

Valproic acid is rapidly and almost completely absorbed after oral administration. Peak concentrations occur 1–3 h after an oral dose. The principal metabolite, 2-n-propyl-3-ketopentanoic acid, has anticonvulsant activity comparable to that of valproic acid, although this metabolite does not accumulate in plasma. The single dose $t_{1/2}$ is 16 h in healthy adults but this reduces to 12 h on chronic therapy, and the $t_{1/2}$ may be as short as 8 h in children. In neonates and in hepatic disease, when metabolism is reduced, the $t_{1/2}$ becomes prolonged. Valproic acid is highly protein bound (93%). In circumstances when competition for protein binding increases, such as in uremia, cirrhosis, or concurrent drug therapy, the per cent of free valproic acid increases.

The minimum effective *therapeutic concentration* of valproic acid is 40 μg/mL. Concentrations in excess of 100 μg/mL have been associated with hepatic *toxicity* and acute toxic encephalopathy. Glycine has been observed to accumulate in patients on valproic acid therapy.

Clearance of valproic acid is rapid, which presents a dosing dilemma. Dosage must be adequate to provide a plasma concentration >40 μg/mL while avoiding concentrations >100 μg/mL. The ideal *specimen* for monitoring the blood concentration is that drawn just prior to the next dose, usually early in the morning, to confirm that an adequate dose has been prescribed prior to bedtime. Dosage is particularly problematic in young children who might sleep for more than one complete half-life of the drug.

Valproic acid modulates the action of a variety of other common antiepileptic drugs. It inhibits the nonrenal clearance of phenobarbital, resulting in elevated phenobarbital levels. It competes with phenytoin for protein binding sites. The free phenytoin concentration remains approximately the same but the total phenytoin in the plasma decreases. Because the free phenytoin concentration remains unchanged, the pharmacologic effect is retained. Other common antiepileptic drugs which induce hepatic oxidative enzymes result in increased valproic acid clearance; this increased clearance rate requires a higher dose to maintain effective therapeutic levels.

Method for the Determination of Valproic Acid

Analysis of valproic acid can be accomplished by either GLC or immunoassay. GLC analysis is simple and trouble free and is presented below.

Principle

Valproic acid (VPA) is extracted from acidified serum into an organic solvent. A chemical analog, cyclohexane carboxylic acid, is used as an internal standard. The two acids are then extracted into an aqueous alkaline medium (NaOH, 0.5 mol/L). The aqueous phase is again acidified and the two acids are re-extracted into an organic phase (chloroform). A small aliquot of the chloroform extract is subjected to GLC analysis and quantitated using a flame ionization detector.

Reagents

1. Stock drug standard. Dissolve 11.9 μL of di-*n*-propylacetic acid in 10 mL of NaOH, 0.01 mol/L, to yield a solution of 1 mg VPA/mL. If stored in a sealed plastic container at 4 °C, this stock standard is stable for more than six months.

2. Working drug standard.
  a. 25 μg/mL VPA. Dissolve 250 μL of stock standard in a final volume of 10 mL NaOH, 0.01 mol/L.
  b. 100 μg/mL VPA. Dissolve 1.0 mL stock standard in a final volume of 10 mL NaOH, 0.01 mol/L.

3. Working internal standard. Dissolve 25 mg cyclohexane carboxylic acid in NaOH, 0.01 mol/L.

4. Hydrochloric acid, 6 mol/L and 1 mol/L.
  a. HCl, 6 mol/L. Slowly add 49.5 mL concentrated HCl to 40 mL H_2O with stirring. Allow to cool to room temperature, then add H_2O to a final volume of 100 mL.
  b. HCl, 1 mol/L. Pipet 16.6 mL of HCl, 6 mol/L, into 70 mL of H_2O. Allow to cool to room temperature. Adjust final volume to 100 mL.

5. Sodium hydroxide, 0.5 mol/L. Dissolve 2 g NaOH in a total volume of 100 mL H_2O.

6. Extraction solvent. 5 parts isopropanol in 95 parts methylene chloride.

7. Instrument. Gas-liquid chromatograph equipped with flame-ionization detector. The column is 10% Carbowax 20 M on Chromosorb WHP (80/100 μm) packed in a 4 ft × 3-mm I.D. glass column, with nitrogen as the carrier gas flowing at 30 mL/min.

Procedure

1. Into appropriately labeled 16 × 125-mm glass tubes, pipet 200 μL of low or high standard, control pool, or specimen.

2. Add 50 μL of internal standard to all tubes.

3. Acidify the sample with 200 μL of HCl, 1 mol/L, and then add 6 mL of extraction solvent.

4. Vortex for 30 s, and then centrifuge at 2000 $\times$ g for 5 min.

5. Transfer each organic phase to a clean 16 $\times$ 125-mm tube and add 3 mL of NaOH, 0.5 mol/L, to each.

6. Vortex 30 s, and then centrifuge at 2000 $\times$ g for 5 min.

7. Transfer the aqueous phase to a clean 12-mL conical-tip tube.

8. Acidify the same by adding 0.5 mL of HCl, 6 mol/L.

9. Add 200 μL of chloroform and vortex 15 s.

10. Insert a microsyringe through the aqueous phase into the organic phase and withdraw a 2-μL aliquot of the organic phase. Wipe the external surface of the microsyringe dry.

11. Inject the aliquot into the GLC and allow the chromatogram to develop.

12. Calculate results as described in steps 7 and 8 of the antiepileptic drug procedure.

Reference

Dusci, L. J., and Hackett, L. P.: Gas chromatographic determination of valproic acid in human serum. J. Chromatogr., *132*:145-147, 1977.

BROMIDE

Because of the fortuitous observation by Locock[46] in 1857 that potassium bromide proved useful in management of epilepsy, bromide has been widely used for seizure control. The advent of new, more directed therapies has significantly reduced the common use of bromide, although it is still used occasionally in combination therapy. Bromide intoxication, which occurs easily, is achieved accidentally because of its slow excretion. The $t_{1/2}$ in the blood is ~12 d. Consequently, bromide accumulates if taken daily, and over a period of weeks a toxic level may be attained. In *toxicity*, delirium, delusions, hallucinations, mania, lethargy, or coma may occur; EEG changes accompany the intoxication. Neurologic disturbances are manifested as tremors, fixed speech, motor incoordination, decreased superficial reflexes, and positive Babinski great toe sign. Central nervous system pressure may be raised, and in cases of severe intoxication, papilledema may be present. In normal therapy, *serum concentrations* of 1000–1500 μg/mL are attained.

Methods for estimating the blood concentration of bromide have been based on the spectrophotometric properties of a gold bromide complex. Bromide, after deproteinization of plasma by trichloroacetic acid, yields a colored complex with gold chloride that has an absorption maximum at 440 nm. The method was developed by Wuth[86] and modified by Gray and Moore.[37] Gold bromide methods are subject to interference by other blood constituents. Gray and Moore concluded that the nonspecificity in the gold bromide color reaction contributes minimally to the test sensitivity. They suggest that results < 400 μg/mL may be artifactual.

MISCELLANEOUS ANTIEPILEPTIC DRUGS

Mephobarbital owes the greatest portion of its antiepileptic activity to its principal metabolite, phenobarbital. Metabolism is via hepatic demethylating mixed-function oxidases that are induced by phenobarbital. Thus, long-term therapy results in multiphasic elimination profiles of the drug. Early in therapy, high concentrations (> 5 μg/mL) of mephobarbital and low concentrations (< 10 μg/mL) of phenobarbital may be observed. After enzymatic induction, the pattern shifts to one where phenobarbital predominates (20–40 μg/mL) and mephobarbital is a minor constituent (1–3 μg/mL) at *equilibrium*. Mephobarbital concentration is usually determined by chromatography as described for phenobarbital methods. It should be noted that mephobarbital in most immunoassays cross-reacts to some degree with phenobarbital.

Three succinimides have proved useful in the control of absence seizures: *ethosuximide* (*Zarontin*), *methsuximide* (*Celontin*), and *phensuximide* (*Milontin*). Ethosuximide is widely used and is reviewed elsewhere in this chapter. Methsuximide and phensuximide are less commonly used and monitored.

Both methsuximide and phensuximide are active in control of absence seizures and seizures of temporal lobe origin, but phensuximide has been found to be less effective due to its short half-life. Methsuximide is effective because it is metabolized to an active metabolite that is stable and has a long half-life. Steady-state *serum concentrations* of methsuximide and normethsuximide on a standard adult dose of 900 mg/d range from 0.036–0.040 µg methsuximide/mL and 20–40 µg normethsuximide/mL. Methods used are primarily chromatographic.[55]

Clonazepam (*Clonopin*) is a benzodiazepine with chemical structure closely related to diazepam (Valium). The drug appears to limit the spread of EEG discharge from focal lesions without suppressing the primary focus. It is currently approved for use in absence seizures, infantile spasms, akinetic seizures, and Lennox-Gastaut syndrome. *Plasma concentrations* associated with maximal effectiveness of the drug range from 10–50 ng/mL. At concentrations higher than 80 ng/mL, no additional seizure protection is observed, and *toxicity* (drowsiness, ataxia) ensues. The most suitable methods adaptable to routine analysis are based on gas-liquid chromatography with electron capture detection.[30]

CARDIOACTIVE DRUGS

Digoxin, one of a group of cardiac glycosides obtained from digitalis plants (e.g., *Digitalis lanata*), proprietary name *Lanoxin*, restores the force of cardiac contraction in congestive heart failure. It is also used in the management of supraventricular tachycardias. The drug complexes with membrane-bound Na,K-ATPase inhibiting both cellular Na^+ efflux and K^+ influx in myocardial cells. This reduces the sodium/potassium gradient in the Purkinje fibers of the atrial, junctional, and ventricular myocardium, resulting in a decreased transmembrane potential. Inhibition of Na,K-ATPase is postulated to enhance movement of Ca(II) ions into the cell, increasing Ca(II) availability and improving cardiac contractility.

At low concentrations, digoxin causes the atrium to be less electrically excitable. Moderate concentrations of digoxin are required to reduce the rate of depolarization in the spontaneously depolarizing conductive fibers (Purkinje fibers), and high toxic concentrations of digoxin are necessary to diminish depolarization of the ventricular myocardium. Disagreement over the clinical value of digoxin measurements and the failure of the digoxin concentration to correlate with clinical toxicity are usually related to aberrations in serum and tissue concentrations of sodium, potassium, magnesium, and calcium. An increased sensitivity to digoxin can be noted in states of hypokalemia, hypomagnesemia, and hypercalcemia which make establishment of the true therapeutic concentration of digoxin difficult.

Absorption of digoxin is variable and dependent upon the dosage form. The U.S. Pharmacopeia requires more than 65% of digoxin in tablet form to dissolve in 60 min. In plasma, digoxin is 25% protein bound. Digoxin is concentrated in tissues, and at steady state the concentration of digoxin in cardiac tissue is 15–30 times that of plasma. Accumulation of digoxin in tissue lags behind the plasma concentration, i.e., while the *peak plasma concentration* is reached 2–3 h after the oral dose, the *peak tissue concentration* occurs 6–10 h after an oral dose. Although pharmacologic effects and toxicity correlate with tissue concentration, rather than plasma concentration, the effective and safe *therapeutic plasma concentration* of digoxin ranges from 0.5–2.0 ng/mL. This effective range is not determined at the peak plasma concentration, but rather at the time of peak tissue concentration.[17] Thus, to ensure a correlation between plasma concentration and tissue concentration, the appropriate time to collect the *specimen* is 8 h or more after the dose. Results from specimens collected earlier than 8 h after the dose are inappropriate and often misleading.

Digoxin *toxicity* is characterized by nonspecific symptoms of nausea, vomiting, anorexia, and predominance of green/yellow visual distortion. Cardiac symptoms of intoxication include multiform premature ventricular contractions, ventricular bigeminy, ventricular tachycardia, and ventricular fibrillation. Combinations of decreased conduction and increased automaticity may result in paroxysmal atrial tachycardia with atrioventricular node block and nonparoxysmal junction tachycardia.

Elimination of digoxin follows first order kinetics; 50–70% is excreted, unchanged or in the form of digoxigenin mono- or disaccharides in the urine. A small amount is metabolized to dihydrodigoxin and also excreted by the kidney. The remainder is found in the stool as digoxigenin and its saccharides. As a result, digoxin toxicity develops more frequently and lasts longer in patients with renal impairment. Dosage requirements are decreased in patients with renal disease. Bresnahan and Vlietstra[17] present a simple method for calculating dose which is based on creatinine clearance and utilizes the concepts outlined in equation (12) of this chapter.

Decreased gastrointestinal absorption occurs with sprue and small intestinal resections, high-fiber diets, hyperthyroidism, and situations of increased gastrointestinal motility. A more dangerous situation develops due to the interaction of quinidine and digoxin,[44] resulting in an increase in the digoxin concentration. The actual mechanism of this interaction has not been clearly defined, although it has been observed that quinidine decreases the rate of clearance and volume of distribution of digoxin.

Pharmacokinetic parameters of digoxin and other cardioactive drugs are summarized in Table 17-5.

Method for the Determination of Digoxin

The only method with proven clinical utility is the immunoassay. All physical-chemical methods require such large specimen volumes for adequate sensitivity that they are impractical. Recently, a fluorescence-polarization immunoassay has been demonstrated to be clinically useful.[57]

Principle of Immunoassay

Radioimmunoassays have proved to be clinically useful for the measurement and monitoring of digoxin levels and are most commonly used. Accurate results can be obtained with small sample sizes (e.g., 100 μL serum) and relatively short incubation times (e.g., 30 min). In most radioimmunoassays, patient samples, controls, and calibration standards containing unlabeled digoxin are placed in assay tubes with a constant amount of ^{125}I-labeled digoxin and antibody specific for digoxin. During the incubation period, the unlabeled digoxin competes with the ^{125}I-labeled digoxin for binding sites on the specific antibodies. The bound digoxin-antibody complexes are then separated from the unbound fraction by one of a variety of methods, and the radioactivity in the bound fraction is determined with a γ-counter. The amount of radioactivity in the antibody complexes will decrease with increasing amounts of unlabeled digoxin. Levels of digoxin in the samples are then determined by interpolation from a suitable calibration curve constructed with results obtained from the standards.[77a] For more detailed instructions on this technique, the protocol for the specific procedure used should be consulted.

To confirm the standardization of a digoxin assay, independent standards can be prepared by dissolving an appropriate amount of pure digoxin in ethanol. This stock solution is then diluted with phosphate buffered saline and a final dilution is made with digoxin free human serum. For example, dissolve 5.0 mg of digoxin in 50 mL of ethanol (= 100 μg/mL). Dilute this stock solution 100-fold (= 1 μg/mL) and then 50-fold (= 20 ng/mL) with phosphate buffered saline. Prepare the final standard concentrations (e.g., 0.5, 1.0, 2.0, 4.0 ng/mL) by diluting with digoxin-free human serum.

Note: Digitoxin is infrequently prescribed today; however, serum levels should be evaluated in patients suspected of digitalis intoxication with nondetectable digoxin levels. Digoxin RIA procedures cross-react only minimally with digitoxin.

Table 17-5. PHARMACOKINETIC PARAMETERS OF THE CARDIOACTIVE DRUGS

Cardioactive Drugs	Therapeutic Range (per mL of Plasma)	Half-Life (h)	Volume of Distribution (L/kg)	Bioavailability, Oral (%)	Protein Bound (%)
Amiodarone	1.0–2.0 μg	45	12 $\pm$ 3	60–90	90
Digoxin	0.5–2.0 ng	42 $\pm$ 19	5–12	75	25
Disopyramide	2.0–4.5 μg	8 $\pm$ 2	0.8 $\pm$ 0.3	83 $\pm$ 11	*
Lidocaine	2–5 μg	1.8 $\pm$ 0.4	1.1 $\pm$ 0.4	35 $\pm$ 11	51 $\pm$ 8
Mexiletine	0.75–2.0 μg	11–26	†		
Procainamide (plus NAPA)	4–8 μg (30 μg)	3 $\pm$ 0.6 (8)	1.9 $\pm$ 0.3	83 $\pm$ 16	16 $\pm$ 5
Propranolol	50–100 ng‡	3.9 $\pm$ 0.4	3.9 $\pm$ 0.6	36 $\pm$ 10	94 $\pm$ 2
Quinidine	2–5 μg	6.2 $\pm$ 1.8	2.7 $\pm$ 1.2	80 $\pm$ 15	71 $\pm$ 11
Tocainide	8–12 μg	12–16	1.4 $\pm$ 1.6	100	50
Verapamil	50–200 ng	3.5–7	5	20	90

*Blood-concentration dependent; see text.

†Where data are missing, they are not applicable or not available.

‡This range provides for maximal reduction of exercise-induced tachycardia. No therapeutic range exists for antihypertensive use. Concentrations as high as 1000 ng/mL have been observed without toxic effects.

Lidocaine, 2-(diethylamino)-*N*-(2,6-dimethylphenyl) acetamide, proprietary name *Xylocaine*, is the drug of choice for the initial therapy of premature ventricular contractions and the prevention of ventricular arrhythmias. Lidocaine is contraindicated when bradycardias and severe atrioventricular node block appear after myocardial infarction. Lidocaine therapy has also been proposed as a prophylactic measure to avoid life-threatening ventricular fibrillation occurring after myocardial infarction. The mechanism of action of lidocaine is specific for Purkinje fibers. Lidocaine shortens the action potential refractory period in these fibers, and does so at concentrations less than those required to exert pharmacological effects at other sites such as the ventricular myocardium.

Because lidocaine undergoes nearly complete first-pass hepatic metabolism when administered orally, it is administered only as an intravenous or intramuscular injection. Once in the blood, it is 50% bound to protein, mainly to α_1-acid glycoprotein and albumin. Clearance of lidocaine is very rapid. Two half-lives are commonly referenced; the *distribution half-life* is ~0.5 h and the *elimination half-life* is 1–1.5 h. Reduced hepatic function will impair clearance and will cause prolonged elimination and accumulation of the drug. This is due both to reduced blood flow to the liver (seen in heart failure) and to decreased metabolism of lidocaine. The end effect is lidocaine intoxication.

The relationship between optimal *blood concentration* of lidocaine and its *clinical effect* was stated by Rodman,[67] who suggested that the blood concentration of lidocaine is best interpreted in light of the greatest likelihood of therapeutic success, therapeutic failure, or toxicity for selective concentration increments. Blood concentrations < 1.5 µg/mL are rarely effective. Concentrations ranging from 1.5–4.0 µg/mL are usually effective and are rarely associated with any form of central nervous system or cardiovascular *toxicity*. Concentrations ranging from 4–6 µg/mL may be needed for suppressing arrhythmias but may be associated with mild central nervous system depression and slight QRS widening on the electrocardiogram. Concentrations ranging from 6–8 µg/mL are acceptable only if alternative therapy is not possible, since these levels have been associated with significant central nervous system depression and atrioventricular node blockage. Concentrations > 8 µg/mL are commonly associated with seizure activity, significant hypotension, and decreased cardiac output.

Lidocaine has two *metabolites* commonly found in plasma, monoethylglycinexylidide (MEGX) and glycinexylidide (GX). MEGX and lidocaine have nearly identical toxic equivalency, and the sum total of lidocaine and MEGX concentration averaged 18.7 µg/mL (ranging from 17.9–28.0 µg/mL) in patients experiencing lidocaine-induced convulsions. Substitution of MEGX for lidocaine resulted in the same mean concentration for the equivalent convulsive activity. The metabolite GX was shown to be unimportant, because it does not accumulate to significant concentrations.[12]

Since lidocaine is most commonly administered as a constant infusion following a loading dose, the time to collect the *specimen* is determined by the reason for monitoring. If the blood concentration is intended to document an adequate concentration early in therapy, the specimen should be collected 30 min after the loading dose, or 5–7 h after therapy is initiated if no loading dose is given (five half-lives after start of therapy). If the patient shows diminished mental status, QRS widening, or other toxic symptoms, the specimen should be collected as close to the episode as possible and analysis performed immediately since these symptoms present a potentially life-threatening situation (onset of severe lidocaine intoxication).

The total plasma concentration of lidocaine is a result of clearance of the drug and is modulated by hepatic function. There is little impact on clearance in renal disease. In situations of decreased organ perfusion, clearance is reduced and increased blood concentrations of lidocaine should be expected. The principal binding protein of lidocaine, α_1-acid glycoprotein, has been demonstrated to accumulate after myocardial infarction. The result of accumulation of this protein is reduction of free lidocaine, which reduces the pharmacologic effect of the drug.[68] Another interaction of interest is the decreased clearance of lidocaine associated with concomitant dosage of propranolol. Accumulation of lidocaine in this situation could result in *toxicity*. For the discussion of methods used to determine lidocaine, see page 1644.

Quinidine, 6'-methoxycinchonan-9-ol, available as either quinidine sulfate or quinidine gluconate, is used in the treatment of atrial premature contraction, paroxysmal supraventricular tachycardia, supraventricular tachyarrhythmia associated with Wolff-Parkinson-White syndrome, premature ventricular contractions, ventricular tachycardia, and in the prophylactic treatment of postmyocardial infarction patients. It is also used with care in the treatment of atrial fibrillation and atrial flutter, although this treatment is commonly accompanied by the administration of either digoxin or a β-blocker (propranolol) to provide atrioventricular node blockade. These pharmacological responses appear to be based on a mechanism of action involving blockade of cholinergic neurotransmission, which results in depression or abolition of ectopic impulse generation, an increase in the duration of action potential of sinoatrial node cells, and most importantly, prolongation of the effective refractory period of atrial, ventricular, and Purkinje fibers.

Absorption of quinidine is complete and rapid. *Peak serum concentrations* are reached in 1.5–2 h after oral intake, unless the slow release preparation (quinidine gluconate) is used. Peak plasma concentrations are accomplished 4–5 h after quinidine gluconate administration, and the trough concentration occurs 1–2 h after the next administration. Once absorbed, quinidine is 80% protein bound. Clearance of quinidine is dependent on both adequate hepatic and renal function. Reduction of either of these two functions will result in accumulation of the drug. Renal clearance is a function of urine pH. If the urine is alkaline, or if the patient has renal tubular acidosis, clearance will be reduced.

A strong correlation between blood concentration and optimal pharmacologic response has been documented by many authors (reviewed by Moyer[54]). The optimal *therapeutic concentration* defined in these studies is *method dependent* because of interference in some tests by quinidine metabolites and the presence of dihydroquinidine, a naturally occurring analog of quinidine found in all pharmaceutical preparations. Single-extraction fluorescence methods display interference due to coadministration of triamterene (Diazide) and the accumulation of inactive quinidine metabolites in patients with impaired renal function. When a single extraction fluorescence technique is used, serum concentrations in patients responding well to quinidine display blood concentrations in the range of 3–8 μg/mL, and toxicity is normally associated with blood concentrations > 12 μg/mL. Coadministration of triamterene may interfere, yielding apparent blood concentrations as high as 15 μg/mL in a patient with a normal quinidine concentration. Double-extraction fluorescence techniques are more specific for quinidine and a few of its active metabolites [(3s)-3-hydroxyquinidine and quinidine-N-oxide]. The antibody used in the commercially available EMIT procedure for quinidine also appears to cross-react with these compounds; double extraction fluorescence procedures and the EMIT assay yield similar results corresponding to an optimal *therapeutic concentration* of 2–5 μg/mL.[59] Liquid chromatography is a more specific procedure for analysis of quinidine; it is capable of separating dihydroquinidine from quinidine. Because liquid chromatographic procedures can separate this and other metabolites from quinidine, the therapeutic range by liquid chromatography is 1.5–4.5 μg/mL. Quinidine *toxicity* is usually observed at concentrations > 8 μg/mL, and is commonly associated with symptoms of cinchonism, tinnitus, lightheadedness, giddiness, and cardiovascular toxicity, including premature ventricular contractions and atrioventricular node block. The most predominant toxic effect is gastrointestinal in nature and includes nausea, vomiting, anorexia, and abdominal discomfort. Hypersensitivity reactions associated with quinidine are not related to blood concentration.

Routine monitoring of quinidine most commonly is an effort to document adequate dosage. In this case the *trough specimen* is appropriate. When symptoms of intoxication are observed, then a peak specimen may be more useful. When a patient shows alternating signs of toxicity and then loss of therapeutic effect, a half-life determination should be done which may delineate that patient as a fast metabolizer. These rare patients require more frequent dosage schedules or a different drug.

Clearance of quinidine is dependent upon an active mixed-function oxidase system in the liver. Induction of this system, for example by barbiturates, will lead to enhanced clearance of quinidine. Diminished organ perfusion results in decreased clearance. Quinidine itself has been reported to dilate peripheral blood vessels, resulting in mild to moderate hypotension and reduced clearance over the short term.[75] A discussion of methods for quinidine determination is given on page 1644.

Procainamide, 4-amino-N-[2-(diethylamino)ethyl] benzamide, proprietary name *Pronestyl,* is used for therapy of premature ventricular contractions, ventricular tachycardia, atrial fibrillation, and paroxysmal atrial tachycardia. Its mechanism of action is similar to that of quinidine in that it increases the threshold membrane potential, reducing excitability and contraction velocity in Purkinje fibers and ventricular muscle.

Absorption of procainamide is rapid and complete. *Peak plasma concentrations* after oral administration are reached within 0.75–1.5 h if the drug is given in capsule form, or within 1–3 h if given in tablet form. Once absorbed, procainamide is about 20% bound to plasma proteins. Excretion of procainamide is dependent upon hepatic metabolism and renal clearance; therefore, alteration in either organ function will lead to accumulation of procainamide and its metabolites. The $t_{1/2}$ is 3–4 h in healthy adults.

The *concentration* at which procainamide will block premature ventricular contractions and inhibit ventricular tachycardia ranges from 4–8 μg/mL,[42] although most patients can tolerate concentrations higher than this. Studies by Meyerburg et al.[50] suggest that patients experiencing chronic premature ventricular contractions (PVC's) may require blood concentrations as high as 12 μg/mL to reduce PVC's to a reasonable number. Minimum plasma concentrations of 8 μg/mL were required for protection against sustained ventricular tachycardia, and a small number of patients required blood concentrations as high as 15 μg/mL to reduce PVC's to an acceptable level.

The issue of the *ideal therapeutic concentration* for procainamide is complicated by the fact that one of its metabolites, *N*-acetylprocainamide (NAPA), has antiarrhythmic activity similar to procainamide. The compound has been shown to accumulate in patients with impaired renal function and fast acetylators. The optimal therapeutic concentration for NAPA is not well defined. The drug is used in Europe, where the maximum tolerable concentration of NAPA in the absence of procainamide is 30 μg/mL.[78] Coanalysis of NAPA is necessary to provide a complete assessment of therapy or define metabolic status. Fast acetylators will have concentrations of NAPA equal to or exceeding those of procainamide in a specimen collected 3 h after administration, whereas slow acetylators will have procainamide present at greater than twice the NAPA concentration in a specimen collected during the same time interval. Because the effects of procainamide and NAPA are cumulative, peak plasma concentrations of procainamide should be limited to 8–12 μg/mL, and peak concentrations of procainamide plus NAPA should not exceed 30 μg/mL. Interpretation of results requires knowledge of the patient's cardiac status; given concentrations may be intolerable in some patients, while others may require higher levels for control of PVC's.

Symptoms of *intoxication* include bradycardia, prolongation of the QRS interval, AV block, and induced arrhythmias. These symptoms occur at blood concentrations of procainamide and NAPA exceeding 30 μg/mL. Hypotension sometimes encountered in procainamide therapy is not related to excessive plasma concentration. The development of systemic lupus erythematosus (SLE) associated with procainamide therapy is not related to plasma concentration but is associated with the acetylator status of the patient; slow acetylators predominate in the group that develops SLE. A discussion of the methods for the determination of procainamide and NAPA is given on page 1644.

Disopyramide, α-[2-{bis(1-methylethyl)amino}-ethyl]-α-phenyl-2-pyridineacetamide, proprietary name *Norpace,* is used for suppression of unifocal, multifocal, and premature ventricular contractions, ventricular tachycardia, ventricular and supraventricular arrhythmias, and in the treatment of Wolff-Parkinson-White syndrome. The mechanism of action of disopyramide is similar to that of quinidine, and the drug is commonly used as replacement therapy for quinidine when quinidine side effects are intolerable. The chief cardiac effects are prolongation of the effective refractory period of the atrium and the ventricle. The drug is a specific inhibitor of Na,K-ATPase, causing a negative inotropic effect that is dose- and plasma-concentration dependent.

Disopyramide is nearly completely absorbed and a small fraction undergoes first-pass hepatic metabolism. In the blood, disopyramide binds to plasma proteins. Binding is highly variable among individuals, and binding is also dependent upon the concentration of disopyramide. The *elimination half-life* is 4–8 h. Elimination is via renal clearance and hepatic metabolism. Metabolism plays a minor role in elimination, but with renal insufficiency, clearance is prolonged, causing accumulation of disopyramide.

Optimal antiarrhythmic effect is accomplished at plasma concentrations of 2.8–3.2 μg/mL and 3.3–7.5 μg/mL for atrial and ventricular effects, respectively. The relationship between clinical response and optimal therapeutic concentration is confused by a decrease in protein binding that occurs as plasma concentration of disopyramide rises.[28] Disopyramide binds variably to serum proteins; binding ranges from 45–70% at 2 μg/mL total serum concentration and 30–45% at 5 μg/mL. The case can thus be made for monitoring free disopyramide blood concentrations, which should be a better index of therapeutic response. Unfortunately, equilibrium dialysis using radiolabeled disopyramide is the only accurate method available at this time for assessing the fraction of the disopyramide bound to protein. Most ultrafiltration membranes have been shown to bind disopyramide nonspecifically. Several manufacturers expect to have a filtration membrane available for free disopyramide analysis at some time in the future.

Metabolism of disopyramide is by dealkylation. The principal metabolite is nordisopyramide (monodealkylated), a compound reported to have antiarrhythmic activity $\sim$25% that of disopyramide.[36] Under normal circumstances, nordisopyramide accumulates to concentrations ranging from 0.2–1.0 μg/mL, and the compound does not accumulate out of proportion to disopyramide in situations of reduced hepatic or renal function. Therefore, little additional therapeutic information is gained by monitoring nordisopyramide.

The predominant side effects of disopyramide are anticholinergic, i.e., dry mouth, urinary hesitancy, and constipation. These symptoms occur at plasma concentrations > 4.5 μg/mL. Cardiac *toxicity* is usually associated with blood concentrations > 10 μg/mL and is characterized by atrioventricular node blockage, bradycardia, and asystole. Because of the wide degree of variability of protein binding, interindividual variability of the blood concentration at which these symptoms develop is great.

Method for the Determination of Antiarrhythmic Drugs

Lidocaine, procainamide, NAPA, quinidine, and disopyramide can be analyzed by HPLC, GLC, or immunoassay. Quinidine can also be accurately analyzed by double-extraction direct-fluorescence analysis.[59] Because of the versatility of HPLC, it is the method presented here.

Principle

The antiarrhythmic drugs are extracted from alkalinized serum or plasma into a polar organic solvent. After solvent evaporation, the residue is dissolved in a mobile phase and chromatographed. The method requires a liquid chromatograph and a variable-wavelength ultraviolet detector. This permits detection of the analyte of interest with a minimum of interference from other drugs and metabolites.

Reagents

1. Stock standards, 1 mg/mL. Prepare stock standards for each drug by separately weighing 10 mg of drug into a 10-mL volumetric flask and adding methanol to a volume of 10 mL. Each standard is soluble in methanol and stable indefinitely if stored at −20 °C in a sealed container away from fluorescent lights.
   2. Working standards.
      a. Low working standards. Dilute stock standards with HCl, 0.05 mol/L, to give concentrations of 2 μg/mL lidocaine, 4 μg/mL procainamide, 10 μg/mL NAPA, 2 μg/mL quinidine, and 2 μg/mL disopyramide. This standard is stable for three months if stored at 4 °C in a sealed container.
      b. High working standard. Prepare as above to give concentrations of 7 μg/mL lidocaine, 10 μg/mL procainamide, 30 μg/mL NAPA, 7 μg/mL quinidine, and 6 μg/mL disopyramide. This standard is stable for three months if stored as above. Procainamide and disopyramide are unstable if stored longer.
   3. Internal standard.
      a. Stock internal standard. Place *N*-propionyl procainamide (Aldrich Chemical Co.), cinchonidine (Sigma Chemical Co.), or chlorodisopyramide (Searle Laboratories) into separate flasks at concentrations of 1 mg/mL in methanol. These standards are stable for periods up to six months if stored at −20 °C in a sealed container.
      b. Working internal standard. Prepare an internal standard containing 20 μg/mL *N*-propionyl procainamide, 20 μg/mL cinchonidine, or 20 μg/mL chlorodisopyramide in HCl, 0.05 mol/L. These internal standards are stable as described above.
   4. Solvents.
      a. Sodium carbonate, 0.25 mol/L, pH 11.0. Dissolve 31 g $Na_2CO_3 \cdot H_2O$ in 900 mL of H_2O and adjust to pH 11.0 by titration with dilute NaOH. Adjust to a final volume of 1000 mL.
      b. Diethyl ether, ACS grade. Use without further purification. Store in the original tin container.
      c. Acetonitrile, double glass-distilled.
      d. Sodium phosphate, 0.1 mol/L, pH 4.0. Dissolve 15.6 g $NaH_2PO_4 \cdot 2 H_2O$ in 900 mL of H_2O and adjust to pH 4.0 by titration with dilute NaOH. Adjust to a final volume of 1000 mL and filter through 0.5 μm membrane.
   5. Liquid chromatograph.
      a. See instrument description in the antiepileptic drug procedure.

b. Mobile phase and UV settings.

(1) Lidocaine and quinidine. 27 parts acetonitrile and 73 parts NaH_2PO_4, 0.1 mol/L, pH 4.0, flowing at 3.0 mL/min. Monitor absorbance at 230 nm at 0.02 AUFS.

(2) Procainamide and NAPA. 15 parts acetonitrile and 85 parts NaH_2PO_4, 0.1 mol/L, pH 4.0, flowing at 3.0 mL/min. Monitor absorbance at 270 nm at 0.05 AUFS.

(3) Disopyramide. 35 parts acetonitrile and 65 parts NaH_2PO_4, 0.1 mol/L, pH 4.0, flowing at 3.0 mL/min. Monitor absorbance at 270 nm at 0.05 AUFS.

Procedure

1. Into appropriately labeled 16 × 150-mm glass tubes, pipet 1.0 mL of low or high standard, control pool, or specimen.

2. Add 250 μL of working internal standard. Use N-propionyl procainamide for procainamide, cinchonidine for lidocaine or quinidine, or chlorodisopyramide for disopyramide analysis.

3. Add 1 mL of Na_2CO_3, 0.25 mol/L, pH 11.0, and mix thoroughly.

4. Add 8 mL of diethyl ether and mix vigorously for 1 min.

5. Separate the phases, using centrifugation if necessary, and transfer the organic phase to a glass conical-tip tube.

6. Evaporate the ether in a ventilated fume hood.

7. Redissolve the residue in 50 μL of mobile phase and chromatograph it under the conditions listed above.

8. Identify each drug by retention time and quantitate by either peak height or peak area.

9. Establish, for each drug, a ratio of the peak area or height to the peak area or height of the appropriate internal standard.

10. Using the low and high standards to establish a standard plot (concentration being the independent variable), calculate specimen results as described in the antiepileptic drug procedure.

References

Adams, R. F., Vandermark, F. L., and Schmidt, G.: The simultaneous determination of lidocaine and procainamide in serum by use of high pressure liquid chromatography. Clin. Chim. Acta, 69:515, 1976.

Lagerstrom, P., and Persson, B.: Liquid chromatography in the monitoring of plasma levels of antiarrhythmic drugs. J. Chromatogr., 149:331-334, 1978.

Meffin, P. J. Sandera, R. H., and Harrison, D. C.: High pressure liquid chromatographic analysis of drugs in biological fluids. III. Disopyramide. J. Chromatogr., 132:503-510, 1977.

PROPRANOLOL

Propranolol is used in the treatment of arrhythmias of atrial and ventricular origin, angina pectoris, myocardial infarction, and hypertension. It is a nonselective β-blocker with action on cardiac receptors (β_1), and on vascular and bronchial smooth muscle receptors (β_2). Its principal effect is to reduce the heart rate, thus relieving angina, and to slow conduction at the AV node which reduces the ventricular rate in patients with atrial fibrillation.

While propranolol is well absorbed, it undergoes such a high degree of first-pass hepatic metabolism that its final bioavailability is low (20–40%), widely variable between individuals, and dose dependent (the higher the dose, the greater the bioavailability). In the plasma, propranolol is highly protein bound (95%) to albumin and α_1-acid glycoprotein. *Elimination half-life* is 3–4 h. Elimination is predominantly by metabolism; therefore, reduced hepatic function or reduced blood flow to the liver will cause accumulation of propranolol.

There is a close relationship between slowing of heart rate and blood concentration.[60] In *specimens* collected 2 h after dosing for 2 d with 640 mg of drug four times daily, complete β-adrenergic blockade (equated with suppression of the adrenergic component of exercise-induced tachycardia) correlated with blood concentration of 100 ng/mL. Blood concentrations in specimens collected 8 h after the dose, associated with approximately 60% β-adrenergic blockade, ranged from 10–40 ng/mL. The same study also documented a close correlation between the blood concentration and the degree of heart rate slowing, suggesting that observation of the heart rate serves as an adequate means to monitor the therapeutic efficacy, precluding blood measurements in most patients. There is virtually no relationship between plasma concentration of propranolol and the hypotensive effect of the drug.

Method for the Determination of Propranolol

Propranolol as well as other β-blockers such as metoprolol, aprindolol, nadolol, and others can be analyzed by HPLC using the native fluorescence of the molecule for detection.

Principle

Propranolol and an internal standard (pronetholol) are extracted from alkalinized serum into an organic solvent (butanol/hexane). Following transfer of the organic phase to a clean conical-tip tube, the drugs can be back-extracted into an acidified medium and subjected to reverse-phase chromatography. Fluorescence detection allows for quantitation by peak height determination.

Reagents

1. Stock standard, 1 mg/mL. Propranolol or pronetholol at a concentration of 1 mg/mL is stable if stored at −20 °C in methanol.
2. Working standards.
   a. Low standard. 25 ng/mL propranolol in HCl, 0.05 mol/L.
   b. High standard. 100 ng/mL propranolol in HCl, 0.05 mol/L.
3. Working internal standard. 100 ng/mL pronetholol in HCl, 0.05 mol/L.
4. Solvents.
   a. Na_2CO_3, 0.25 mol/L, pH 11.0 (see antiarrhythmic procedure).
   b. Extraction solvent. 20 parts butanol and 80 parts hexane, both ACS grade.
   c. Back-extraction medium. H_2SO_4, pH 2. Add 2 drops of concentrated H_2SO_4 to 100 mL H_2O. This reagent is stable for approximately one month.
5. The liquid chromatograph is as described in the antiepileptic drug procedure, fitted with a fluorescence detector set to an excitation wavelength of 225 nm and emission at 320 nm.

Procedure

1. To appropriately labeled tubes, add 1.0 mL of working standard, control pool, or specimen.
2. Add 200 μL of internal standard.
3. Add 1 mL of Na_2CO_3, 0.25 mol/L, pH 11, and mix thoroughly.
4. Add 7 mL of extraction solvent and shake vigorously for 1 min. Separate phases, using a centrifuge if necessary.
5. Transfer the organic phase to a clean conical-tip tube.
6. Add 200 μL of H_2SO_4, pH 2, and shake vigorously for 30 s.
7. Insert a microsyringe through the organic phase, withdraw an aliquot of the aqueous phase, inject it into the HPLC, and allow the chromatogram to develop.
8. Identify the drugs by retention time and quantitate by peak height or area.
9. Calculate the specimen concentration as described for the antiepileptic drugs.

References

Jatlow, P., Bush, W., and Hochster, H.: Improved liquid chromatographic determination of propranolol in plasma with fluorescence detection. Clin. Chem., 25:777-779, 1979.
Nation, R. L., Peng, G. W., and Chiou, W.: High-pressure liquid chromatographic method for the simultaneous quantitative analysis of propranolol and 4-hydroxypropranolol in plasma. J. Chromatogr., 145:429-436, 1978.

MISCELLANEOUS ANTIARRHYTHMIC DRUGS

Tocainide, 2-amino-N-(2,6 dimethylphenyl) propanamide, is an oral antiarrhythmic analog of lidocaine that is effective in the short-term management of ventricular ectopic beats and ventricular tachycardia. Its mechanism of action is similar to that of lidocaine. The drug has the advantage over lidocaine in that it can be taken orally and it has a relatively long half-life (13–16 h). The *therapeutic concentration* ranges from 8–12 μg/mL.[3] Toxicity is expressed as mild paresthesias, changes in visual perception, decreased mental alertness, and confusion, which occur at concentrations in excess of 12 μg/mL. Analysis of tocainide is described by Reece and Stanley.[65]

Verapamil, 5[(3,4-dimethoxyphenethyl)methylamino]-2-(3,4-dimethoxyphenyl)-2-isopropyl-valeronitrile, is a calcium channel blocker that is effective in the treatment of a variety of cardiovascular disorders including angina (classic and variant), arrhythmias (PSVT), atrial flutter, atrial fibrillation, hypertrophic cardiomyopathy (IHSS), hypertension, congestive heart failure, Raynaud's phenomenon, and preservation of ischemic myocardium. Other future uses may include migraine headaches, intestinal ischemia, allergic disorders, and extrinsic asthma.

McAllister et al.[48] demonstrated a relationship between cardiac responsiveness and blood concentration. The *effective blood concentration* of verapamil ranges from 50–250 ng/mL. Toxic symptoms were characterized by 3:2 Wenckebach-type AV block occurring at blood concentrations >250 ng/mL. All subjects studied developed AV block at concentrations >450 ng/mL. An HPLC method appears most feasible for the quantitation of this drug.[81]

Mexiletine, 1-(1,6-dimethoxyphenoxy)-2-aminopropane, is a class I antiarrhythmic used for control of ventricular dysrhythmias. There is significant interest in mexiletine because of its extended half-life (11–26 h) in comparison with other antiarrhythmic agents; in addition, it undergoes minimal first-pass hepatic metabolism. The *therapeutic concentration* has been identified as ranging from 0.75–2.0 μg/mL.[20] Toxic manifestations include tremor, dizziness, ataxia, dysarthria, diplopia, nystagmus, confusion, and hypotension, which occur at blood concentrations in excess of 2 μg/mL.

Amiodarone, 2-butyl-3-(3,5-diiodo-4-β-diethylaminoethoxybenzoyl) benzofuran, is used to control supraventricular and ventricular tachyarrhythmias and to lengthen the effective refractory period of the accessory pathway in patients with Wolff-Parkinson-White syndrome. The drug is of interest as a substitute for other Group I antiarrhythmics (such as procainamide or quinidine) because it has a very long *elimination half-life* (45 h) and few dose-related side effects. The *effective serum concentration* of the drug, measured 24 h after a single daily dose, ranges from 1.0–2.0 μg/mL.[66]

BRONCHODILATORS

Theophylline, 3,7-dihydo-1,3-dimethyl-1H-purine-2,6-dione, available under many proprietary names, relaxes bronchial smooth muscles to relieve or prevent asthma. Prior to 1980, the mechanism of action of theophylline was thought to be inhibition of cyclic-nucleotide phosphodiesterase. This theory, however, only explains the toxic effects of theophylline. The therapeutic effects would be contrary to the action of increased cAMP concentrations in smooth muscle tissues. At this time, the mechanism of action of theophylline must be defined as unknown or at least not attributable to only one mechanism.

Theophylline is readily absorbed after oral, rectal, or parenteral administration. If the drug is taken orally without food, the blood concentration *peaks* within 2 h. If it is administered with food or as the slow release formula, peak concentrations occur 3–5 h after the dose. Once absorbed, it is 50% protein bound. The drug is rapidly cleared in children and in adults who smoke. In these people, the $t_{1/2}$ ranges from 3–4 h. Nonsmoking adults in good health have an elimination half-life of about 9 h. The $t_{1/2}$ in neonates and in adults with congestive heart failure can be prolonged to 20–30 h depending upon the degree of liver immaturity or loss of liver function. Co-administration of the antibiotics erythromycin and troleandromycin also has been shown to reduce significantly the clearance of theophylline. Cimetidine reduces hepatic metabolism of theophylline, resulting in increased serum concentrations.

The relationship between serum concentration and prevention of symptoms of chronic asthma has been well documented.[38] There is a proportional relationship between forced expiratory volume and theophylline concentration, with the *optimum therapeutic effect* occurring at concentrations ranging from 5–20 μg/mL. Suppression of exercise-induced bronchospasm in the asthmatic patient occurs at concentrations >10 μg/mL and is optimal at 15 μg/mL. Neonatal apnea treated with theophylline responds to slightly lower concentrations, ranging from 5–10 μg/mL.[29]

Theophylline *clearance* is a function of a metabolic process that is dose dependent. At serum concentrations >20 μg/mL, small dosage increases lead to disproportionately large increases in serum concentration and intoxication. Symptoms of theophylline *toxicity* include nausea, vomiting, headache, diarrhea, irritability, and insomnia. Transient CNS stimulation occurring at the initiation of dosage is not directly related to blood concentration. This effect diminishes with chronic use. Serious toxicity characterized by cardiac arrhythmias and seizures is usually associated with serum concentrations in excess of 30 μg/mL. Once seizure activity begins, the final prognosis is very poor. Morbidity is reported in nearly all patients, and mortality can be as high as 50%.[90]

A minor metabolite of theophylline in adults, *caffeine*, 3,7-dihydro-1,3,7-trimethyl-1H-purine-2-6-dione, has been shown to accumulate to significant concentrations in neonates.[15] Caffeine itself is an effective inhibitor of apnea,[5] which may explain the lower therapeutic concentration required for control of neonatal apnea. The case has been made for concurrent determination of both theophylline and caffeine concentrations in neonates treated with theophylline.[79] Therapy with caffeine has also been demonstrated as effective in the treatment of neonatal apnea. It is gaining popularity because of caffeine's long half-life in the neonate (>30 h). The optimal *therapeutic concentration* of caffeine in this situation ranges from 8–14 μg/mL.

Pharmacokinetic details of bronchodilators are summarized in Table 17-6.

Method for the Determination of Bronchodilators

Theophylline, caffeine, and dyphylline can be measured by HPLC. A variety of immunoassays are available for the determination of theophylline[21] and a colorimetric procedure,[72] while not recommended, has also been used to quantitate theophylline. Because of its versatility, the HPLC procedure is reviewed here.

Principle

Theophylline, caffeine, and an internal standard, β-hydroxyethyltheophylline, are extracted into ethyl acetate. Following solvent evaporation, the redissolved residue of the extract can be separated by HPLC and the concentration determined from the resultant chromatogram.

Reagents

1. Stock drug standards, 1 mg/mL. Dissolve 10 mg of theophylline, caffeine, dyphylline, or β-hydroxyethyltheophylline (Sigma Chemical Co.) in 5 mL of water, then add methanol to a final volume of 10 mL. These reagents are stable for more than one year if stored at -20 °C.

2. Working standards.

   a. Low drug standard. Dilute theophylline in water to a final concentration of 5 μg/mL. This standard is stable for six months if stored at 4 °C in a sealed container. Prepare a standard for caffeine or dyphylline in a similar manner.

   b. High drug standard. Dilute theophylline in water to a final concentration of 25 μg/mL. This standard is stable for six months if stored at 4 °C in a sealed container.

   c. Internal standard. Dilute β-hydroxyethyltheophylline in acetonitrile to a final concentration of 15 μg/mL and store in a sealed container at 4 °C.

3. Solvents.

   a. KH_2PO_4, 0.2 mol/L, pH 3.6. Dissolve 2.7 g of KH_2PO_4 in 800 mL water and titrate to pH 3.6 with dilute phosphoric acid. Add water to volume and filter through 0.5-μm membrane.

   b. Ethyl acetate, ACS grade.

   c. Acetonitrile, double glass-distilled.

   d. Mobile phase. 9 parts acetonitrile and 91 parts KH_2PO_4 buffer, 0.2 mol/L, pH 3.6.

4. Liquid chromatograph. See the description of the apparatus in the antiepileptic drug method. Monitor absorbance at 273 nm.

Procedure

1. Pipet 100 μL of low or high standard, control pool, and specimen into appropriately labeled 12 × 75-mm glass tubes.

Table 17-6. PHARMACOKINETIC PARAMETERS OF THE BRONCHODILATOR DRUGS

Bronchodilators	Therapeutic Range (per mL Plasma)	Half-Life (h)	Volume of Distribution (L/kg)	Bioavailability, Oral (%)	Protein Bound (%)
Caffeine	8–14 μg*	30*	†		
Theophylline	10–20 μg	6–8	0.5 ± 0.16	96 ± 8	56 ± 4
	5–10 μg*				

*Neonatal apnea.
†Where data are missing, they are either not applicable or not available.

2. Add 100 μL of internal standard to each tube.

3. Mix each tube for 10 s.

4. Add 1 mL of ethyl acetate and vigorously mix the tubes for 1 min. Centrifuge to separate the phases.

5. Transfer the organic phase into a clean 12 × 75-mm glass tube and evaporate the solvent.

6. Redissolve the residue in 100 μL of mobile phase and chromatograph a 10-μL aliquot.

7. Identify the peaks by retention time and determine peak area or peak height.

8. Determine specimen concentration as described in the antiepileptic drug method.

Reference

Kabra, P., and Marton, L. J.: Liquid chromatographic analysis for serum theophylline in less than 70 seconds. Clin. Chem., *28*:687-689, 1982.

ANTIBIOTICS

The *aminoglycosides* are bactericidal agents that act by specifically binding to the 30S ribosomal subunit of bacteria, thereby inhibiting protein synthesis. They are inactive under anaerobic conditions, because an oxygen-dependent active transport mechanism is involved in the transfer of aminoglycosides across the bacterial cell wall. The aminoglycoside *class* of drugs includes *streptomycin, neomycin, netilmycin, sisomycin, gentamicin, tobramycin, kanamycin,* and *amikacin.* Structures of the common aminoglycosides are shown in Figure 17-13.

The aminoglycosides are a very polar group of compounds and are thus poorly absorbed from the intestinal tract. They are routinely administered intravenously or intramuscularly to bypass this problem. When administered directly into the blood, they rapidly distribute to the extracellular fluid but do not cross cell membranes, nor do they bind to plasma proteins. This behavior is confirmed by their unusually low volume of distribution. Most tissues and secretions contain very small concentrations of aminoglycosides, the exceptions being renal cortex, where the drug is concentrated, and bile, due to active hepatic secretion. The drugs are mainly excreted by glomerular filtration. *Elimination half-lives* are short, ranging from 2–3 h. Since clearance is nearly completely dependent on renal function, any impairment of glomerular filtration will cause accumulation of these drugs.

Therapy with antimicrobial agents differs from the approach used for other drugs. The goal is to accomplish a minimal inhibitory concentration (MIC) in plasma such that the bacterial organism is killed while the host remains undamaged. Because the organisms treated are variable and can become resistant to certain drugs, treatment with specific aminoglycoside agents should always be directed by susceptibility testing. Effective minimal inhibitory concentrations of these drugs are listed in Table 17-7.

Numerous studies, summarized by Schentag[73] and Zaske,[87] recommend a limit to the blood concentration of aminoglycosides, although there is considerable variability reported regarding the relationship of blood concentration to later onset of toxicity, mainly renal tubular necrosis and degeneration of the auditory nerve. The *trough specimen* has been suggested as the best specimen to monitor toxicity. Table 17-8 identifies target maximum, trough serum concentrations; in this mode of monitoring the intent of therapy should be dosing the patient in such a manner that the trough concentration does not exceed these limits. An alternative and equally convincing argument can be made for utilizing the peak serum concentrations (Table 17-8) as a therapeutic index. In a large surgical-patient survey, where dosing was carried out under controlled conditions, limited nephrotoxicity was experienced when the peak serum concentration of gentamicin was maintained < 8 μg/mL.[88] Using similar guidelines, Keys et al.[40] reported a 40% incidence of mild nephrotoxicity using a more sensitive index of renal clearance (iothalamate clearance).

Dosage corrections must be made in patients with compromised renal function, because these patients have prolonged *half-life* and slower elimination. The guidelines prepared by Sarubbi and Hull[70] (Table 17-9) allow application of creatinine clearance to estimate adequate initial dosage. This should then be followed up by the quantitation of the blood concentration and dosage adjustment after the method of Sawchuck et al.[71]

Toxicity associated with aminoglycosides shows up as delayed-onset vestibular and cochlear sensory-cell destruction and acute renal tubular necrosis. The degree and severity of cell damage are variable among the various drugs, but they all will cause cell damage if the concentrations

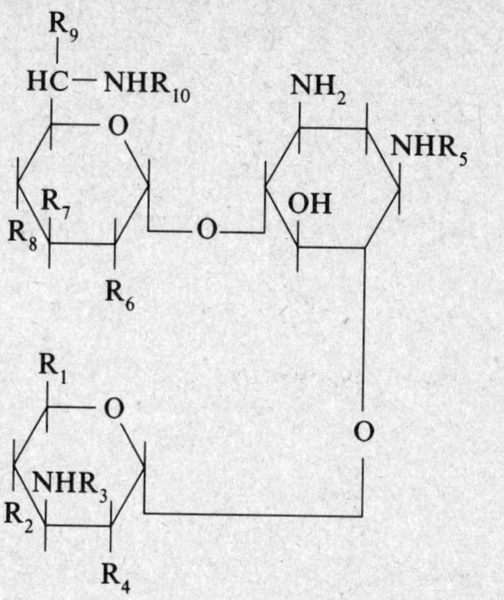

	Amikacin	Gentamicin* (C_1, C_2, C_{1a})	Kanamycin† (A, B)	Tobramycin
R_1	$-CH_2OH$	$-H$	$-CH_2OH$	$-CH_2OH$
R_2	$-OH$	$-CH_3$	$-OH$	$-OH$
R_3	$-H$	$-CH_3$	$-H$	$-H$
R_4	$-OH$	$-OH$	$-OH$	$-OH$
R_5	$-\overset{O}{\overset{\|}{C}}-\underset{OH}{CH}-CH_2-\underset{NH_2}{CH_2}$	$-H$	$-H$	$-H$
R_6	$(-OH)$	$-NH_2$	$(-OH, -NH_2)$	$-NH_2$
R_7	$-OH$	$-H$	$-OH$	$-H$
R_8	$-OH$	$-H$	$-OH$	$-OH$
R_9	$-H$	$(-CH_3, -CH_3, -H)$	$-H$	$-H$
R_{10}	$-H$	$(-CH_3, -H, -H)$	$-H$	$-H$

*Gentamicin is a mixture of isomers, the major constituents being C_1, C_2, and C_{1a}, which differ at R_9 and R_{10}.
†Kanamycin is a mixture of two isomers, A and B, which differ at R_4.

Figure 17-13. Aminoglycoside structures.

Table 17-7. MINIMAL INHIBITORY CONCENTRATIONS* OF ANTIBIOTICS[4]

Antibiotic	Very Susceptible	Moderately Susceptible	Very Resistant	Moderately Resistant
Amikacin	≤ 2	4–16	> 64	32–64
Chloramphenicol	≤ 1	2–8	≥ 16	
Gentamicin	≤ 0.5	1–4	> 64	8–64
Kanamycin	≤ 2	4–16	> 64	32–64
Tobramycin	≤ 0.5	1–4	> 64	8–64
Vancomycin	≤ 0.5	1–4	> 4	

*Concentrations in $\mu g/mL$.
Data taken from Standard Methods for Dilution Antimicrobial Susceptibility Tests for Bacteria which Grow Aerobically, NCCLS Proposed Standard, PSM-7, National Committee for Clinical Laboratory Standards, Villanova, Pa., 1980.

Table 17-8. NORMAL EFFECTIVE CONCENTRATIONS OF ANTIBIOTICS

	μg/mL		
	Peak	*Trough*	*Toxic*
Amikacin	20–25	5–10	> 35
Chloramphenicol	20–30	10	> 50
Gentamicin	5–8	1–2	> 10
Kanamycin	20–25	5–10	> 35
Sulfonamides	125	> 75	> 300
Tobramycin	5–8	1–2	> 10
Trimethoprim	5–8	—	> 50
Vancomycin	20–30	10	> 50

exceed the limits identified in Table 17-8. Unfortunately, the reasonable guidelines identified in Table 17-8 do not guarantee the avoidance of toxicity; some patients will develop it regardless of the concentration, although these patients are rare. Fortunately, most patients reverse the toxic effects without direct intervention if the toxicity is associated with reasonable blood concentrations (Table 17-8). Irreparable loss of vestibular, cochlear, or renal function usually correlates with grossly elevated blood concentrations of any one of the aminoglycosides.

Table 17-9. USEFUL GUIDELINES FOR GENTAMICIN DOSE CALCULATED FROM LEAN BODY WEIGHT AND CREATININE CLEARANCE

1. Select loading dose in mg/kg (lean weight) to provide peak serum level desired. Approximate peak levels from commonly used loading doses are indicated below.

Loading Dose	Expected Peak Serum Level Based Upon 30-Min IV Infusion
2.0 mg/kg	6– 8 μg/mL
1.75 mg/kg*	5–7 μg/mL
1.5 mg/kg	4–6 μg/mL
1.25 mg/kg	3–5 μg/mL
1.0 mg/kg	2–4 μg/mL

*Recommended for most moderate to severe systemic infections.

2. Select maintenance dose (as percentage of chosen loading dose) to continue peak serum levels indicated above according to patient's creatinine clearance (C_{cr}) and desired dosing interval.

Percentage of Loading Dose Required for Dosage Interval Selected[†]			
C_{cr}	8 h	12 h	24 h
90	90%	—	—
80	88	—	—
70	84	—	—
60	79	91%	—
50	74	87	—
40	66	80	—
30	57	72	92%
25	51	66	88
20	45	59	83
15	37	50	75
10	29	40	64
7	24	33	55
5	20	28	48
2	14	20	35
0	9	13	25

†Shaded areas indicate suggested dosage intervals.
From Sarubbi, F. A., and Hull, J. H.: Ann. Intern. Med., *85*:188, 1976, with permission.

Heparin has been implicated as a deactivator of gentamicin by formation of an inactive complex.[74] This complex, while biologically inactive, retains some structural resemblance to the initial aminoglycoside and will cross-react with antibodies to the specific aminoglycoside. Heparin concentrations encountered in therapeutic antithrombotic therapy are < 3 units/mL, making an in vivo complication unlikely. However, specimen collection tubes containing heparin (1000 units/mL) may lead to complex formation, a phenomenon which could interfere with many immunoassay procedures.

Pharmacokinetic details of antibiotics are summarized in Table 17-10.

Method for the Determination of Aminoglycosides

The aminoglycoside antibiotics have been analyzed for years by the bioassay technique. This method, however, is variable and subject to interference by numerous drugs. Recently, both liquid chromatographic and immunochemical methods have become available.[33] Because enzyme immunoassay reagents are generally available and use of these reagents does not require specialized equipment, that technique will be reviewed. These agents are used for treating serious infections due to aerobic and facultatively anaerobic gram-negative bacilli.

Principle

The EMIT assays used for the measurement of drugs are homogeneous enzyme-labeled immunoassays. The products of the reaction are measured directly in the reaction mixture, eliminating the requirement for separation of bound and free label. The reagent system consists of (1) the drug bonded covalently to an appropriate enzyme, (2) an antibody specific for the drug being assayed, and (3) reaction substrates and cofactors for the enzyme used as label. The specimen is mixed with these three reagents and allowed to react during an incubation period. In the absence of drug in the specimen, the enzyme-labeled drug binds to the antibody and forms a complex that alters the conformation or blocks the active site of the enzyme, thus preventing the enzyme reaction from occurring. If the drug is present in the specimen, it will displace enzyme-labeled drug from the antibody; the free enzyme-labeled drug can then react with enzyme substrate to produce a measurable product. The amount of product resulting from the enzymatic activity increases as the concentration of unlabeled drug in the specimen increases. The concentration of the drug can be calculated by interpolation from a standard curve.

EMIT methods for gentamicin, amikacin, and tobramycin use glucose-6-phosphate dehydrogenase (G-6-PD, EC 1.1.1.49) as the enzyme label along with glucose-6-phosphate (G-6-P) and $NADP^+$ as substrate. The type of enzyme used for labeling and its specific substrate may differ in EMIT assay systems for other drugs; however, the principles of the reaction remain the same. The free G-6-PD-labeled drug catalyzes the following reaction:

$$\text{G-6-P} + NAD^+ \longrightarrow \text{6-Phosphogluconolactone} + NADH + H^+$$

Increase in absorbance at 340 nm is measured with a spectrophotometer capable of accurate measurements in the range of 0.02–0.5 A, of accurate temperature control at 30 ± 0.1 °C, and of recording of absorbance versus time. Instruments appropriate for EMIT assays have been reviewed elsewhere.[82] The standard curve is nonlinear; therefore specialized curve-fitting data

Table 17-10. PHARMACOKINETIC PARAMETERS OF THE ANTIBIOTIC DRUGS

Antibiotics	Therapeutic Range (per mL of Plasma)	Half-Life (h)	Volume of Distribution (L/kg)	Bioavailability, Oral (%)	Protein Bound (%)
Amikacin	10–25 μg	2 ± 0.6	0.2 ± 0.08	*	
Chloramphenicol	10–20 μg	2.7 ± 0.8	0.9	75–90	53 ± 5
Gentamicin	2–8 μg	2–3	0.25		< 10
Kanamycin	10–25 μg	2.1 ± 0.2	0.26 ± 0.05		0
Sulfonamides, all	75–125 μg	6–9	0.1–0.2	~100	60–90
Tobramycin	2–8 μg	2.2 ± 0.1	0.26 ± 0.09		< 10
Trimethoprim	5–8 μg	11 ± 1.4	1.8 ± 0.2	~100	70 ± 5
Vancomycin	10–30 μg	5–6	0.4		< 10

*Where data are missing, they are either not applicable or not available.

reduction methods must be used. These are described by Lafara[43] and in EMIT kit instructional inserts.

CHLORAMPHENICOL

Chloramphenicol, D(–)-threo-2,2-dichloro-N-[β-hydroxy-α-(hydroxymethyl)-p-nitrophenyl-ethyl]acetamide, proprietary name *Chloromycetin* and others, is used as a bactericidal agent. It acts by binding to the 50S ribosomal subunit of bacteria and inhibits protein synthesis in prokaryotic organisms. Use of this drug is dependent upon its relative toxicity against the microorganism versus the host. The drug is used against gram-negative bacteria such as *H. influenzae, N. meningitidis, N. gonorrhoeae, S. typhi,* all *Brucella* species, *B. pertussis, V. cholerae,* and *Shigella.* These organisms are all susceptible to a *serum concentration* of 6 μg/mL. Organisms that are susceptible to a concentration of 12 μg/mL are *E. coli, K. pneumoniae, P. pseudomallei, Chlamydia,* and *Mycoplasma.*

Chloramphenicol is rapidly absorbed in the gastrointestinal tract. *Peak serum concentrations* occur 1–2 h after the oral dose. In the plasma, chloramphenicol is ~50% protein bound and is cleared with a $t_{1/2}$ of 2–3 h. Peak serum concentrations after administration of chloramphenicol palmitate or succinate occur 4–6 h after the dose. Chloramphenicol distributes to all tissues, and it concentrates in the CSF. The drug is actively metabolized by the liver by N-acetylation and glucuronidation. Thus, chloramphenicol will accumulate in cases of hepatic disease. Renal disease does not dramatically reduce clearance.

Host *toxicity* displayed after chloramphenicol therapy includes blood dyscrasias and cardiovascular collapse; both show a modest relationship to blood concentration. Other blood concentration-related toxicities include anemia, characterized by maturation arrest in the marrow; cytoplasmic vacuolation of early erythroid and myeloid cells; reticulocytopenia; and increases in both serum iron and serum iron-binding capacity. These symptoms are all associated with serum concentrations in excess of 25 μg/mL.[62] Development of aplastic anemia is not related to dose or blood concentration. Cardiovascular collapse, which occurs primarily in newborns, has been related to a total serum chloramphenicol concentration in excess of 50 μg/mL.[45] An oral dose of 1 g results in a peak serum concentration of 10–20 μg/mL in a healthy adult.

Method for the Determination of Chloramphenicol

Procedures for the determination of chloramphenicol concentrations in blood serum include bioassay, gas chromatography, and HPLC. The technique of HPLC is presented here. Methods for chloramphenicol determination must be able to differentiate between the pro-drug forms, chloramphenicol palmitate or succinate, and their active metabolite, chloramphenicol.

Principle

Chloramphenicol and an internal standard, mephenesin, are extracted from alkalinized serum, plasma, or other biological fluids into diethyl ether. The extract is transferred to a clean tube and the solvent evaporated. Quantitation is accomplished by separating the reconstituted extract by HPLC, using an ultraviolet detector.

Reagents

1. Stock chloramphenicol standard. Chloramphenicol is obtained as either a diagnostic reagent or from alkaline hydrolysis of chloramphenicol succinate. Prepare a solution of 0.5 mg/mL in methanol. Stable for six months if stored in a sealed container at 4 °C.

2. Working standards. Dilute 1.0 mL of the stock standard to a final volume of 5 mL with methanol. Set up from working standards as shown below. These diluted stock standards are stable for one month if stored in sealed containers.

Final concentration (μg/mL)	Diluted stock (mL)	Methanol (mL)
10	0.1	0.9
20	0.2	0.8
40	0.4	0.6
80	0.8	0.2

3. Internal standard. Dissolve 25 mg of mephenesin (Sigma Chemical Co.) in 100 mL methanol. This solution is stable for six months if stored in a sealed container at 4 °C.

4. Tris 0.8 mol/L, pH 10.4. Dissolve 9.69 g of tris(hydroxymethyl)aminomethane (Tris) in water and dilute to a final volume of 10 mL. This solution is stable for one year in a sealed container.

5. Diethyl ether, reagent grade. Store in metal container. Discard by evaporation after can has been open for one month.

Instrumentation

The liquid chromatograph described for antiepileptic drug analysis can be used for this procedure. The mobile phase is a mixture of 35 parts methanol and 65 parts water pumped at 2.0 mL/min. The UV detector is set to monitor 278 nm at 0.02 AUFS.

Procedure

1. Into appropriately labeled 16 × 125-mm screw-capped culture tubes, add 100 μL of each diluted stock standard, and evaporate to dryness at 40 °C using a gentle stream of nitrogen to remove evaporated solvent.

2. Add 100 μL of drug-free serum to each standard tube and mix.

3. Add 100 μL of specimen or control pool to an appropriately labeled tube.

4. To all tubes from steps 2 and 3, add 0.8 mL of Tris 100 μL of internal standard, and 10 mL of diethyl ether.

5. Cap all tubes and gently tumble at a rate of about 20 inversions per min for 10 min.

6. Separate the phases by centrifugation and transfer the ether layer (without carrying over any of the interface emulsion) to a 12-mL disposable polypropylene centrifuge tube.

7. Evaporate the ether with a gentle stream of nitrogen until a wet residue remains. Evaporation to complete dryness is not necessary and too time consuming.

8. Redissolve the residue in 100 μL of methanol and chromatograph.

9. Quantitation is based on the same principles described for the antiepileptic drugs.

Reference

Gerson, B., and Anhalt, J. P.: High Pressure Liquid Chromatography and Therapeutic Drug Monitoring, Chicago, ASCP Press, 1980, pp. 117-119.

SULFONAMIDES

Sulfonamides act by competitively antagonizing bacterial use of *p*-aminobenzoic acid, which is important in the synthesis of folic acid. Therefore, organisms dependent upon self-synthesized folic acid for growth are susceptible to sulfonamides. Sulfonamides are used against common urinary tract pathogens such as *E. coli, K. enterobacter, P. mirabilis*, and indole-positive *Proteus* species. They are not used for infections due to *P. aeruginosa*. The sulfonamides are active against *H. influenzae, S. pneumoniae, S. flexneri*, and *S. sonnei* isolated from the middle ear or bronchial secretions. Chemical structures of the various sulfonamides are shown in Figure 17-14.

The sulfonamides are nearly completely absorbed from the gastrointestinal tract. Once absorbed they are bound to protein (60–90%), mainly to albumin, and are distributed to all tissues. Metabolism is via *N*-acetylation, the products having no antimicrobial activity.

Blood dyscrasias associated with sulfonamide use are not related to dose or blood concentration of the drug. The predominant *toxicity* associated with sulfonamide use is the formation and deposition of crystalline aggregates in the kidneys, ureters, and bladder. The safe and *effective concentration* of these drugs in serum (75–125 μg/mL) is well separated from the serum concentration at which crystallization in the urinary tract occurs. Serum concentrations in excess of 300 μg/mL have been associated with such crystal formation.

Method for the Determination of Sulfonamides

The method of Bratton and Marshall is the standard method used by most laboratories for assessment of sulfonamide blood concentration and is presented here. This method is somewhat nonspecific in that inactive metabolites of the sulfonamides not substituted at the N^4-position are measured in the procedure. The technique is not appropriate for urine monitoring where these

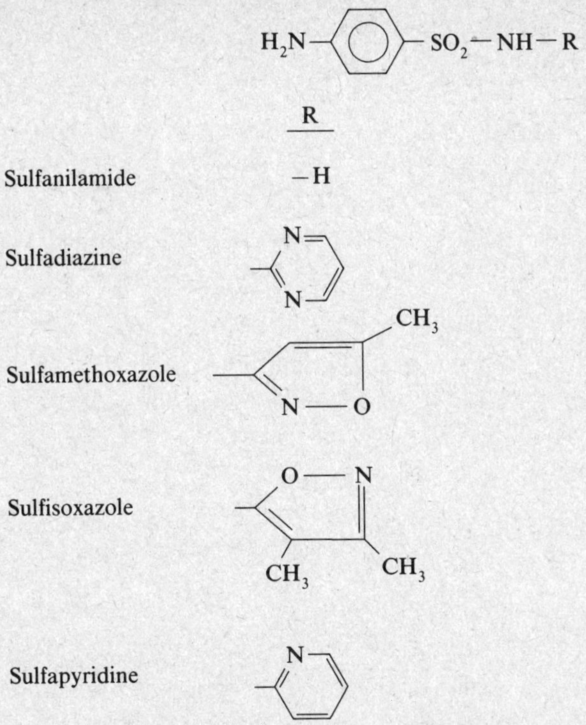

$$H_2N-\text{⟨○⟩}-SO_2-NH-R$$

	R
Sulfanilamide	$-H$

Sulfadiazine

Sulfamethoxazole

Sulfisoxazole

Sulfapyridine

Figure 17-14. Structures of sulfonamides.

metabolites predominate. More specific HPLC methods have been developed and should be used when gross interference is suspected, such as in patients with major renal deficiency.[33]

Principle

Deproteinized blood, serum, or plasma is treated with nitrous acid to diazotize the free, nonacetylated sulfonamide. Excess nitrous acid is destroyed by addition of ammonium sulfamate. Diazotized sulfonamide is then coupled with N-(1-naphthyl)-ethylenediamine dihydrochloride to form a stable complex with a characteristic red color which can be quantitated spectrophotometrically. The degree of color formation is proportional to the concentration of free sulfonamide in the specimen.

Reagents

1. Trichloroacetic acid, 30 g/dL. Add 30 g of trichloroacetic acid to water and dilute to a final volume of 100 mL. This reagent is stable indefinitely if evaporation is avoided.
2. Saponin, 0.05 g/dL. Add 50 mg of saponin (Fisher Chemical Co.) to water and dilute to final volume of 100 mL. This reagent is stable indefinitely if evaporation is avoided.
3. Sodium nitrite, 0.1 g/dL. Add 100 mg of sodium nitrite to water and dilute to a final volume of 100 mL. This reagent is stable for not more than one week.
4. Ammonium sulfamate, 0.5 g/dL. Add 500 mg of ammonium sulfamate to water and dilute to a final volume of 100 mL. This reagent is stable indefinitely if evaporation is avoided.
5. Complex reagent, 0.1 g/dL. Add 100 mg of N-(1-napthyl)-ethylenediamine dihydrochloride to water and dilute to a final volume of 100 mL. Store in a brown bottle at 4 °C. This reagent is stable up to one year.
6. Stock standard, 0.2 mg/mL. Add 20 mg of sulfapyridine to water containing 1 mL of NaOH, 0.1 mol/L, and dilute to final volume of 100 mL. This standard is stable for up to one year if stored in a sealed container at 4 °C.
7. Dilute standard, 6 µg/mL. Add 3 mL of stock standard to water and dilute to a final volume of 100 mL. This standard is stable for one week.

Procedure

1. Place a 1-mL aliquot of blood, serum, or plasma into a 125-mL Erlenmeyer flask.

2. Add 30 mL of saponin, mix thoroughly, and allow to stand at room temperature for 5 min. Saponin is a detergent and cell-lysing agent.

3. Add 4 mL of trichloroacetic acid to the flask slowly with vigorous mixing. Let this mixture stand at room temperature for 10 min.

4. Filter the mixture through Whatman #1 paper into a clean Erlenmeyer flask.

5. Pipet 10 mL of the filtrate into a 50-mL Erlenmeyer flask and treat as in steps 8–12.

6. Prepare a reagent blank by adding 9 mL water and 1 mL trichloroacetic acid to a 50-mL Erlenmeyer flask.

7. Prepare the standard by adding 7 mL of water, 2 mL of dilute standard, and 1 mL of trichloroacetic acid to a 50-mL Erlenmeyer flask.

8. To each of the flasks from steps 5–7, add, in this order:
    a. 1 mL sodium nitrite, mix well, and let stand 4 min.
    b. 1 mL ammonium sulfamate, mix well, and let stand 3 min.
    c. 1 mL complex reagent, mix well, and let stand 5 min.

9. Set the spectrophotometer at 550 nm to zero absorbance using the reagent blank and measure the absorbance of each standard or sample.

Calculation

$$\frac{A_{unk}}{A_{std}} \times 12 \times 3.5 = \mu g \text{ sulfonamide/mL}$$

where 12 is the concentration of the standard in $\mu g/mL$ and 3.5 is the dilution factor.

The colored product gives nearly equivalent absorbance determinations for sulfapyridine, sulfisoxazole, sulfadiazine, and sulfamethoxazole. The colored product for sulfanilamide has a slightly modified absorption spectrum. The calculated result, when multiplied by a correction factor of 1.07, will yield an estimation of sulfanilamide concentration.

Reference

Bratton, A. C., and Marshall, E. K.: A new coupling component for sulfanilamide determination. J. Biol. Chem., *128*:537–550, 1939.

MISCELLANEOUS ANTIBIOTIC DRUGS

Trimethoprim is administered as a combination drug with sulfamethoxazole. The drug is excreted in proportion to sulfamethoxazole; it is not toxic and therefore is not commonly monitored. Sulfa determination may be used to establish compliance. Methods for trimethoprim determinations are available.[33]

Adverse reactions to *β-lactams* (penicillins and cephalosporins) are not related to dose or blood concentrations. They are commonly administered in doses 10–100 times greater than necessary to inhibit organism growth, and appropriate dosing may be guided by susceptibility testing. Patient compliance and evaluation of patients with renal impairment are the most likely reasons to monitor the blood concentration. Methods for specific analysis of most β-lactams have been developed.[33]

Tetracyclines are broad-spectrum antibiotics. They are not highly toxic and therefore, like the β-lactams, the need to monitor their serum concentrations is limited to monitoring patient compliance and adjusting dosage in patients with impaired renal function. Methods have been described by Gerson and Anhalt.[33]

Vancomycin is a glycopeptide which is bactericidal against gram-positive bacteria and some gram-negative cocci. Recently, vancomycin has received attention because of its reported activity against methicillin-resistant staphylococci and corynebacteria. It has thus become popular for treatment of endocarditis and sepsis caused by these organisms. While the drug is poorly absorbed when given orally, a 1-g dose given intravenously every 12 h accomplishes a *peak blood concentration* of 20–30 $\mu g/mL$ and a *trough concentration* of 5–10 $\mu g/mL$. It has an average elimination $t_{1/2}$ of 5–6 h. Blood concentration–dependent *toxicity* involves the auditory nerve. Concentrations < 30 $\mu g/mL$ are rarely associated with this development.[32] Toxicities not related to dose or blood concentration include fever, phlebitis, and pain at the infusion site. These symptoms have been related to impurities present in vancomycin preparations. In patients with impaired renal function, the serum concentration may increase to toxic levels due to reduced clearance. Methods to monitor the serum concentration have been detailed[33] and are useful in these patients so that dosage can be appropriately adjusted. Vancomycin can also be analyzed by commercially available RIA kits.

ANTIPSYCHOTIC DRUGS

Drugs used in psychiatric care and commonly monitored include lithium, antidepressants, and neuroleptics. Pharmacologic parameters of lithium and antidepressants are shown in Table 17-11.

LITHIUM

Lithium, administered as lithium carbonate, proprietary name *Eskalith, Lithane, Lithonate* and others, is used for treatment of the manic phase of affective disorders, mania, and manic-depressive illness. It acts by enhancing reuptake of neurotransmitters, thereby reducing their concentration in the neuronal junction. This produces a sedating effect on the CNS.

Absorption of lithium from the gastrointestinal tract is complete, with *peak plasma concentration* reached in 2–4 h after an oral dose. This cation does not bind to protein. Lithium clearance is biphasic; during the first phase 30–40% of the dose of lithium is cleared, showing a $t_{1/2}$ of 24 h. During the second phase, the remainder of lithium incorporated into the cellular ion pool is cleared, showing a $t_{1/2}$ of 48–72 h. Clearance is predominantly a function of the kidney, where active reabsorption occurs. Reduced renal function will cause prolonged clearance times.

The *optimal therapeutic response* to lithium has not been related to a specific serum concentration; however, *toxicity* is related to serum concentration. Serum lithium concentrations are monitored to assure patient compliance and to avoid intoxication. Amdisen,[1] after a thorough review of lithium pharmacokinetics, recommended that a standardized, 12-h post-dose serum lithium concentration be used to assess adequate therapy. The range of 1.0–1.2 mmol/L was identified as the optimal *trough* therapeutic concentration. Concentrations of 1.2–1.5 mmol/L signified a warning range, and a concentration in excess of 1.5 mmol/L in a *specimen* drawn 12 h after the dose indicated an increased risk of intoxication. Early symptoms of intoxication include apathy, sluggishness, drowsiness, lethargy, speech difficulties, irregular tremors, myoclonic twitchings, muscle weakness, and ataxia. These symptoms, though not life threatening, are uncomfortable for the patient and indicate that the onset of life-threatening seizures is imminent.

Lithium excretion parallels that of sodium. It readily passes the glomerular membrane and is reabsorbed in the proximal convoluted tubules. In situations where the patient is vulnerable

Table 17-11. PHARMACOKINETIC PARAMETERS OF THE ANTIPSYCHOTIC DRUGS

Antipsychotic Drugs	Therapeutic Range (per mL of Plasma)	Half-Life (h)	Volume of Distribution (L/kg)	Bioavailability, Oral (%)	Protein Bound (%)
Amitriptyline	70–225 ng (Ami + Nor)[†]	15 ± 5	8.3 ± 2	*	97 ± 0.8
Amoxapine	200–500 ng (Amox + 8-OH Amox)[†]	8 (8-OH Amox = 30)			
Desipramine	150–300 ng	18 ± 5	35 ± 8	68	92 ± 1
Doxepin	150–250 ng (Dox + Nordox)[†]	17 ± 6	20 ± 8	27 ± 10	
Imipramine	150–250 ng (Imi + Desip)	13 ± 3	15 ± 6	47 ± 21	89–94
Lithium	0.8–1.2 mmol/L	22 ± 8	0.79 ± 0.34	100	0
Maprotiline	200–600 ng	27–58			
Nortriptyline	50–150 ng	31 ± 13	18 ± 4	51 ± 5	94 ± 0.6
Trazodone	0.8–1.6 μg	13			

*Where data are missing, they are either not applicable or not available.
†Indicates where therapeutic range is the sum of the concentration of the parent drug and its metabolite.

to dehydration (fever, watery stools, vomiting, loss of appetite, hot weather) the potential for lithium intoxication is increased. In dehydration, the proximal tubular response to reabsorption of sodium (and lithium) is reduction of clearance. Increased reabsorption of lithium leads to an increased blood concentration of lithium. Severe intoxication, characterized by muscle rigidity, hyperactive deep tendon reflexes, and epileptic seizures, is usually associated with lithium concentrations in excess of 2.5 mmol/L.

Method for the Determination of Lithium

The concentration of lithium in serum, plasma, urine, or other body fluids can be determined by either flame emission photometry or atomic absorption spectrometry. Since most laboratories have a flame photometer, the method using flame photometry is presented here.

Principle

Following dilution of the specimen with a solution containing sodium and potassium (internal standard), the sample is aspirated into the flame of an emission photometer. (In some procedures, cesium is used as an internal standard.) Thermal energy of the flame raises the energy level of both lithium and potassium to an unstable excited state. The molecules revert to ground state by emitting energy in the visible region of the energy spectrum. The energy lines most commonly monitored are 671 nm and 768 nm for lithium and potassium, respectively. The emission intensities of lithium and potassium from the sample are compared as a ratio, and quantitation is performed by comparing that ratio to the ratio observed with standard reference solutions. The standard and sample are diluted with a solution containing a constant amount of sodium because sodium suppresses lithium emission, and a consistent analysis medium in standard and sample reduces this factor as a variable.

Reagents

1. Stock lithium standard, 1 mol/L. Weigh 3.695 g Li_2CO_3 and place in a 100-mL volumetric flask. Add HCl, 6 mol/L, until the lithium carbonate dissolves. (Add no more HCl than necessary to accomplish this.) Allow excess CO_2 to be liberated. Dilute to a final volume of 100 mL with water.
2. Working standards.
   a. Lithium standard, 1 mmol/L. Dilute 1.0 mL of the stock lithium standard to a final volume of 1 L and store in a tightly sealed plastic bottle. This reagent is stable indefinitely as long as evaporation is avoided.
   b. Lithium standard, 2 mmol/L. Dilute 2.0 mL of the stock lithium standard to a final volume of 1 L and store in a tightly sealed plastic bottle. This reagent is stable indefinitely as long as evaporation is avoided.
3. Stock internal standard, potassium, 150 mmol/L. Weigh 11.184 g KCl and place in a 1-L volumetric flask and dilute to a final volume of 1 L.
4. Stock diluent, NaCl, 140 mmol/L, KCl, 5 mmol/L. Weigh 8.182 g NaCl and 0.3728 g KCl and place in a 1-L volumetric flask and dilute to a final volume of 1 L.
5. Diluent. To a 1-L flask, add 10 mL of stock internal standard and 20 mL of stock diluent. Add water to a final volume of 1 L.

Procedure

1. The instrument used may require a modification to allow measurement at the lithium energy line. Check with the instrument manufacturer for appropriate modifications.
2. Turn on flame according to instrument manufacturer's instructions.
3. Aspirate the diluent and set the lithium response to zero.
4. Dilute 0.2 mL of standard, control, or sample with 9.8 mL of diluent and mix thoroughly.
5. Aspirate standards and either record the response for lithium, or, if the instrument is capable of data reduction, set the slope and intercept functions to quantitate the two standards accurately.
6. Aspirate the diluted control and samples and quantitate from lithium response or with data-reduction mode.

Reference

Amdisen, A.: Serum lithium determinations for clinical use. Scand. J. Clin. Lab. Invest., *20*:104-108, 1967.

ANTIDEPRESSANTS

The *tricyclic antidepressants* (and the newer tetracyclic antidepressants) are used to treat endogenous depression characterized by depressed mood, feelings of guilt, appetite suppression, insomnia, weight change, diminished ability to concentrate, loss of interest or pleasure in usual activities, and decreased sexual drive. In more severe cases, depersonalized behavior, paranoid behavior, obsessive-compulsive behavior, and suicidal tendencies are obvious. Endogenous depression implies that there is no apparent organic or societal cause for these behavior changes. Treatment with the tricyclic (amitriptyline, desipramine, doxepin, imipramine, nortriptyline) or tetracyclic (maprotiline) antidepressants results in pharmacologic activity via inhibition of the reuptake of biogenic neurotransmitters in the central nervous system. The end result is a positive effect on mood.

The antidepressants are nearly completely absorbed from the gastrointestinal tract but undergo first-pass hepatic metabolism, so their ultimate bioavailability is moderate and variable. Because these drugs slow gastrointestinal activity and gastric emptying, their absorption may be delayed. Once absorbed, they are highly protein and tissue bound, resulting in large apparent volumes of distribution. Peak plasma concentrations are reached from 2–12 h after the oral dose. Metabolism is via *N*-demethylation and aromatic-ring hydroxylation, followed by conjugation with glucuronic acid. When the drug administered is the tertiary amine (amitriptyline, doxepin, imipramine), metabolism causes accumulation of the respective secondary amine (nortriptyline, nordoxepin, desipramine). These substances have generally equal pharmacologic activity and accumulate to concentrations approximately (but variably) equal to the parent drug. The pharmacologic activities of the hydroxylated metabolites have not been proved, but some activity has been postulated. Taking these factors into consideration (i.e., variable bioavailability, high volume of distribution, variable metabolic activity, and generation of pharmacologically active metabolites), it is not surprising that patient response to these drugs is widely variable. Determining the serum concentration gives the physician the assurance that the patient has been properly dosed.

Drugs such as cimetidine, chloramphenicol, haloperidol, methylphenidate, and phenothiazines inhibit hepatic oxidative enzymes. Inhibition of end-product metabolism of the tertiary tricyclic antidepressants results in a greater accumulation of the secondary amine metabolite (amitriptyline is metabolized to nortriptyline, doxepin to nordoxepin, imipramine to desipramine), since conversion to the aromatic-ring hydroxylated metabolites is blocked. Coadministration of perphenazine with a tricyclic antidepressant will cause accumulation of the secondary amine to concentrations 2–4 times more than normal,[35] with onset of toxicity occurring at the expected blood concentrations.

A relationship between therapeutic response and serum concentration has been established by many researchers.[2] There is a linear relationship between clinical improvement and serum concentration for most of these drugs, the exception being nortriptyline, which has a specific therapeutic window. A serum concentration of nortriptyline below or above the concentration range of 50–150 ng/mL correlates with worsening of mood.[6] The other antidepressants do not display this effect; the upper limit of the optimum blood concentration for these other antidepressants is limited by the onset of mild *toxicity*. Toxicity is expressed as dry mouth and perspiration, signs which may occur also with depression. Thus, it is difficult to differentiate between mild toxicity due to the drug and the disease which is being treated. More serious toxicity is expressed as atrioventricular node block, characterized by a widening of the electrocardiographic QRS interval. Onset occurs at serum concentrations ranging from 800–1200 ng/mL, and the severity of intoxication is related to the serum concentration.[11] The relationship between serum concentration and cardiac toxicity diminishes with time after intoxication as the drug is absorbed into tissues. The *optimal therapeutic concentrations* of antidepressants are listed in Table 17-11.

Method for the Determination of Tricyclic (TCA) and Tetracyclic (TTA) Antidepressants

Numerous methods have been published for the analysis of TCA's and TTA's. These drugs present a variety of problems to the clinical laboratory: (1) the therapeutic serum concentration is 10–100 times lower than that of other commonly monitored drugs, and thus, to be clinically useful, the method must be sensitive to concentrations as low as 25 ng/mL; (2) these drugs have metabolites that must also be measured; and (3) they are structurally similar to mild sleep inducers, antihistamines, and many over-the-counter medications used for appetite suppression which are potential interferences. Of the many hundreds of methods published, only a few have satisfactorily overcome these obstacles.

Analysis by gas-liquid chromatography–mass spectrometry, using selected ion monitoring (GLC/MS-SIM), is the reference method, using either the electron impact mode[10] or the chemical ionization mode.[22] Chemical ionization is the method of choice because of its high sensitivity and selectivity. Gas chromatography with a nitrogen/phosphorus detector (GC-N/P) is the most commonly used technique in clinical laboratories because it is well characterized and has adequate sensitivity. HPLC has been adapted to TCA and TTA analysis, although separation and selectivity continue to be problems with this technique.[80] Because HPLC has yet to prove itself in this area and GC-MS is not routinely available, the GC method is given here. Methods for trazodone[19] and amoxapine[25] are also available.

Principle

Antidepressants, along with appropriate internal standards, are extracted from alkalinized plasma into a polar organic solvent (isoamyl alcohol/heptane). The TCA's and TTA's are then back-extracted into an acidic aqueous phase (dilute HCl) to further purify the sample. Finally, after alkalinization, the TCA's and TTA's are again extracted into an organic phase (ethyl ether), which is then evaporated. An aliquot of the reconstituted residue is then chromatographed on a 3% SP2250 column. The effluent, which passes into a nitrogen/phosphorus detector, causes a response with characteristic retention time for each antidepressant. The intensity of that response, when compared to the response of the internal standard, can be used to quantitate the TCA or TTA in the sample.

Reagents

1. Stock standards. The TCA's and TTA's are available as hydrochloride salts. Thus, preparation of standard solutions requires a weight correction for the hydrochloride portion (~16% of each drug). Prepare individual stock standards of amitriptyline, nortriptyline, doxepin, nordoxepin, imipramine, desipramine, or maprotiline at a final concentration of 1 mg free drug/mL methanol. These stock standards are stable for at least two years if stored at −20 °C in a tightly sealed flask.

2. Stock internal standards. Prepare solutions of promazine and protriptyline at a concentration of 1 mg free drug/mL methanol. These standards are stable for longer than two years if stored at −20 °C in a tightly sealed flask.

3. Diluted standard. Dilute 300 μL of stock standard with 97.7 mL of HCl, 0.05 mol/L.

4. Working standards. Prepare a working standard of each tertiary and secondary TCA pair (i.e., amitriptyline and nortriptyline, doxepin and nordoxepin, imipramine and desipramine) by combining the following:

Final concentration	Diluted std. (Tertiary TCA pair)	Diluted std. (Secondary TCA pair)	Bovine plasma
30 ng/mL	0.1 mL	0.1 mL	9.8 L
100 ng/mL	0.33 mL	0.33 mL	9.34 mL
300 ng/mL	1.0 mL	1.0 mL	8.0 mL

These standards are stable for two months at 4 °C or one year at −20 °C.

5. Working internal standard, 1 μg/mL. In separate 100-mL volumetric flasks, add either 0.3 mL of protriptyline stock or 0.3 mL of promazine stock and dilute to a final volume of 100 mL with HCl, 0.05 mol/L. These reagents are stable for six months if stored at 4 °C in a sealed container.

6. Extraction solvent #1, 3% isoamyl alcohol in heptane. Add 3 parts of isoamyl alcohol, ACS grade, to 97 parts of n-heptane, ACS-grade.

7. Extraction solvent #2, ethyl ether, ACS-grade. Store in metal tin. Discard all used ether by evaporation in an evacuated hood.

8. Ethyl acetate, chromatography grade.

9. Sodium hydroxide, 1.0 mol/L. Add 100 mL of NaOH, 10 mol/L, or 40 g of dry, solid NaOH pellets to 700 mL water. Stir to mix completely and allow to cool to room temperature. Dilute with water to a final volume of 1 L. Store in a plastic bottle.

10. Sodium hydroxide, 0.5 mol/L. Add 500 mL of NaOH, 1.0 mol/L, to a 1-L volumetric flask and dilute to 1 L with water. Store in a plastic bottle.

11. Hydrochloric acid, 1.0 mol/L. Add 82.5 mL of concentrated (12 mol/L) HCl to 500 mL of water, mix thoroughly and allow to cool to room temperature. Dilute to a final volume of 1 L.

12. Hydrochloric acid, 0.1 mol/L. Dilute 100 mL of HCl, 1.0 mol/L, to a final volume of 1 L.

13. Hydrochloric acid, 0.05 mol/L. Dilute 50 mL of HCl, 1.0 mol/L, to a final volume of 1 L.

14. Instrument. This procedure requires a gas-liquid chromatograph equipped with a nitrogen-phosphorus flame-ionization detector. The column used is 6 ft × 2 mm I.D. packed with GP 3% SP2250 on 80/100 Supelcoport (Supelco, Inc., Bellefonte, PA). Instrument conditions are: oven temperature, 255–260 °C; injection port temperature, 300 °C; detector temperature, 300 °C; carrier gas flowrate, 45 mL/min. Detector sensitivity: set to full scale sensitivity on the high standard.

Procedure

1. To 16 × 125-mm screw-capped culture tubes with Teflon-lined caps, add 2.0 mL of working standard, control, or serum specimen; 0.2 mL internal standard (protriptyline for amitriptyline/nortriptyline assay, promazine for all others); 12 mL of isoamyl alcohol-heptane; and 0.4 mL of NaOH, 0.5 mol/L.

2. Shake this mixture gently but thoroughly for 10 min.

3. Centrifuge 5 min at 2000 × g to separate the phases.

4. Transfer the organic phase to a clean 16 × 125-mm screw-capped culture tube and add 0.2 mL of HCl, 0.1 mol/L. Shake for 10 min.

5. Centrifuge 5 min at 2000 × g to separate phases.

6. Aspirate the organic phase to waste. Transfer the aqueous phase to a glass-stoppered 12-mL tube and add 0.2 mL of NaOH, 1 mol/L, and 7 mL ethyl ether. Cap and shake gently for 10 min, and then let stand to allow the phases to separate.

7. Transfer the organic phase to a silanized 15-mL conical tip centrifuge tube.

8. Add an additional 7 mL ethyl ether to each tube, shake 10 min, let the phases separate, and combine the extracts with those collected in step 7.

9. Evaporate the ethyl ether under a gentle stream of nitrogen. Use a water bath to maintain ~25 °C to the tubes, which tend to become chilled otherwise.

10. Redissolve the residue in 50 µL of ethyl acetate and inject a 2-µL aliquot into the gas chromatograph for analysis.

11. Identify the respective drugs in the standard by retention time and compare the peak area (integration) of each drug to the internal standard.

12. Establish a ratio for each standard and sample drug peak area to the peak area of the internal standard.

$$R = \frac{\text{Peak area}_{\text{standard (or sample)}}}{\text{Peak area}_{\text{internal standard}}}$$

13. Establish a standard plot using the three standards, concentration being the independent variable, R the dependent variable.

14. From the standard plot estimate the specimen results.

References

Dawling, S., and Braithwaite, R. A.: Simplified method for monitoring tricyclic antidepressant therapy using gas-liquid chromatography with nitrogen detection. J. Chromatogr., 146:449-456, 1978.

Orsulak, P. J., and Lipchus, S.: Procedure Manual, Psychiatric Chemistry Laboratory, New England Deaconess Hospital, Boston, 1980.

NEUROLEPTIC DRUGS

The psychotic patient is most often treated with one of the *phenothiazines, haloperidol,* or *lithium,* or a combination of these drugs. Because response to these drugs is unpredictable and the patients are difficult to control, monitoring serum concentration should assist in more appropriate treatment. Numerous methods to measure the serum concentration of the various neuroleptic agents have been reported.[24]

The procedures for phenothiazine determination suffer from one drawback; they are too specific. The major problem encountered with monitoring the phenothiazines is the number of active metabolites of each. Cohen et al.[23] reported numerous metabolites for the common phenothiazines, most of which have some pharmacologic activity. Thus, therapeutic monitoring of these drugs requires coanalysis of numerous metabolites, an option which is analytically problematic, and standards for most of the metabolites are not available.

Neuroleptic activity can also be monitored using a competitive protein binding assay. Haloperidol and phenothiazines compete with dopamine binding to the dopamine receptor found in CNS tissue. Using a radiolabeled ligand, Creese and Snyder[27] demonstrated that these drugs could displace that ligand from the receptor, and that the degree of displacement was proportional to the concentration of the active neuroleptic drug. Furthermore, the many active metabolites of the phenothiazines also displace the radiolabeled ligand. Thus, it may be possible, using this one procedure, to assess the degree of dopamine displacement accomplished by a particular drug regimen in a quantitative fashion. The procedure is still under investigation, and therefore is not reviewed here in detail. It may prove to be clinically useful in the future.

ANTINEOPLASTIC DRUGS

METHOTREXATE

Methotrexate has proved useful in the management of acute lymphoblastic leukemia in children, choriocarcinoma and related trophoblastic tumors in women, and carcinomas of the breast, tongue, pharynx, and testes. High-dose methotrexate administration followed by leucovorin rescue may result in regression of carcinoma of the lung and osteogenic sarcoma.

Methotrexate inhibits DNA synthesis by decreasing availability of pyrimidine nucleotides. Methotrexate competitively inhibits the enzyme dihydrofolate reductase, thus decreasing the concentrations of the tetrahydrofolate essential to the methylation of the pyrimidine nucleotides, and consequently the rate of pyrimidine nucleotide synthesis. Leucovorin, a folate analog, is used to rescue host cells from methotrexate inhibition; as a synthetic substrate for dihydrofolate reductase, leucovorin administration allows resumption of tetrahydrofolate-dependent synthesis of pyrimidines and re-initiation of DNA synthesis. Methotrexate is a nonspecific cytotoxin and prolongation of blood levels appropriate to killing tumor cells may lead to severe, unwanted cytotoxic effects such as myelosuppression, gastrointestinal mucositis, and hepatic cirrhosis.

Serum concentrations of methotrexate are commonly monitored during high-dose therapy (>50 mg/m²) in order to identify the time at which active intervention by leucovorin rescue should be initiated. Criteria for blood concentrations indicative of a potential for *toxicity* after single-bolus, high-dose therapy are:

1. Methotrexate concentration > 10 µmol/L, 24 h after dose.
2. Methotrexate concentration > 1 µmol/L, 48 h after dose.
3. Methotrexate concentration > 0.1 µmol/L, 72 h after dose.

Characteristically, blood concentrations are monitored at 24, 48, and 72 h after the single dose and leucovorin is administered when methotrexate levels are inappropriately high for a post-dose phase. The route of elimination for methotrexate is primarily renal excretion. During the period of high blood levels, particular attention must be paid to maintaining output of a large volume of alkaline urine. The pK_a of methotrexate is 5.5; thus, small decreases in urine pH result in significant reduction in its solubility. Keeping urinary pH alkaline diminishes the risks of intratubular precipitation of the drug and obstructive nephropathy during the treatment period. Monitoring blood levels therefore provides the basis for decisions for timing of initiation and continuance of leucovorin treatment and for managing urinary pH. Table 17-12 lists pharmacokinetic parameters for methotrexate.

Method for the Determination of Methotrexate

Methotrexate (MTX) has been measured in biological specimens using a wide variety of techniques. The radioimmunoassay and the folate reductase inhibition techniques are most commonly used, although nonisotopic homogeneous immunoassays are becoming popular. Liquid chromatographic procedures have also been developed which allow for coanalysis of the drug and its metabolites. Since the HPLC method was developed, however, the role of metabolite concentrations has been shown to be of minor importance in toxicity. Because enzyme immunoassay reagents are generally available and use of these reagents does not require specialized equipment, that technique is commonly utilized by clinical laboratories.

Table 17-12. PHARMACOKINETIC PARAMETERS
OF METHOTREXATE

Volume of distribution	0.4 L/kg
Bioavailability, oral	65%
Protein bound	45 $\pm$ 14%
Half-life, at concentrations:	
$> 100 \ \mu mol/L$	1.8 h
$< 100 \ \mu mol/L$	8.4 h
$< 0.1 \ \mu mol/L$	> 10 h
Potentially toxic concentrations:	
24 h after dose (single-bolus)	$> 10 \ \mu mol/L$
48 h after dose	$> 1 \ \mu mol/L$
72 h after dose	$> 0.1 \ \mu mol/L$

Principle

The principle of the MTX assay by EMIT technique is identical to that given for amino-glycoside antibiotics on page 1652. The difference in the assays consists only in the antibody and labeled drug present in the reagent system.[43]

Reference

Buice, R., Evans, W. E., Karos, J., et al.: Evaluation of enzyme immunoassay, radioassay, and radioimmunoassay of serum methotrexate, as compared with liquid chromatography. Clin. Chem., *26*:1902-1904, 1980.

MISCELLANEOUS ANTINEOPLASTIC DRUGS

For various reasons, other antineoplastic drugs are not routinely monitored outside of clinical trials. *Cytosine arabinoside* and *5-fluorouracil*, which are antimetabolites like methotrexate, have been extensively studied. Analytical methods have also been developed but in neither case has the relationship between circulating blood level and therapeutic efficacy been sufficiently firm to justify routine monitoring. Alkylating agents such as cyclophosphamide are metabolically converted to active compounds with life-spans of only seconds before they interact with tissue and are destroyed. Measurement of active metabolite would be extremely useful, but is obviously impractical. *Actinomycin* and *doxorubicin* have toxic effects (bone marrow suppression and dermatitis) that are both immediate and long-acting and that appear to relate not to a circulating blood concentration but to dosage mass and length of exposure. Definition of specific dosage regimens for these drugs is currently of more concern than is control of circulating concentration. *Cisplatin,* easily measurable by platinum analysis, has a renal toxicity that may be related to both blood levels and length of exposure; although monitoring is not now common, a case may be building for it in the future.[64]

IMMUNOSUPPRESSANTS

CYCLOSPORINE

Cyclosporine, proprietary name Sandimmune (formerly Cyclosporin A), is a cyclic peptide composed of 11 amino acids, some of novel structure, isolated from the fungus *Trichoderma polysporum.* The compound has been shown effective in suppressing host versus graft rejection in heterotopic organ transplants. Cyclosporine is approved for use in renal, cardiac, hepatic, and pancreatic organ transplants; its use in bone marrow transplant and suppression of autoimmune disease is under investigation, but appears to be promising.

Cyclosporine acts by inhibiting proliferation of lymphocytes. It interferes with either secretion or action of interleuken-2, a monokine that is a growth factor for T-lymphocytes, resulting in suppression of both humoral and cellular immunity.[13,14] There is also a component related to enhancement of suppressor cell activity or suppression of helper cells as demonstrated by decrease in the helper/suppressor cell ratio after cyclosporine treatment.[39]

Because of variable bioavailability, there is a very poor relationship between dose and blood concentration, but the relationship between whole blood concentration, immunosuppression, and toxicity is predictable.[9] Immunosuppression requires *trough whole blood* concentrations of 100 ng/mL as measured by HPLC. Kahan et al.[39] found that trough whole blood concentrations > 600 ng/mL were associated with hepatic, renal, neurologic, and infective complications. One large research hospital involved in clinical trials of cyclosporine on a large number of patients uses more conservative guidelines for dosage: *therapeutic* trough blood concentrations for renal transplants are 100–200 ng/mL, while 200–300 ng/mL are used as target concentrations for cardiac, hepatic, and pancreatic transplants.[16]

Because cyclosporine is soluble only in an oil-based medium, it is given orally as a suspension. *Bioavailability* is variable, but averages 27%. The drug is rapidly absorbed, and *peak* concentrations are reached in 1–2 h. Cyclosporine is 90% protein bound and concentrated in erythrocytes. The degree of concentration in erythrocytes is temperature dependent in vitro; thus measurement of plasma concentration requires strict attention to specimen temperature if reproducible results are to be obtained.[84] Because of this effect, the best *specimen* for analysis is whole blood. The average $t_{1/2}$ of cyclosporine is 18 h, although there are reports of $t_{1/2}$ as short as 5 h and as long as 35 h in some patients. The volume of distribution is 17 L/kg. Most *metabolites* of cyclosporine are inactive. One of the major metabolites, hydroxylated at the number 1 amino acid, retains ~10% of the immunosuppressive activity of the parent compound.

Several drugs are reported to alter the disposition of cyclosporine. Ketoconazole, melphalan, amphotericin B, and aminoglycoside antibiotics all prolong metabolism of cyclosporine sufficiently to increase the risk of nephrotoxicity.[52] At the time of this writing the drug is very new to the market; therefore more comprehensive reports of such drug interaction should be expected.

Method for the Determination of Cyclosporine

The first procedure available for analysis of cyclosporine was a radioimmunoassay developed by Sandoz Pharmaceuticals, the producer of cyclosporine. The immunoassay exhibits ~30% cross-reactivity with inactive metabolites. Recently, an HPLC method has been developed and has become the method used by most laboratories, including Sandoz Pharmaceuticals. The method differs from most HPLC procedures because it requires heating of the column. Cyclosporine is a peptide and like other peptides interacts with the packing material used in the chromatographic column. The end effect, at room temperature, is a very broadened peak. The phenomenon is found with all brands of chromatography columns. Peak-broadening can be eliminated by keeping the column at a temperature of 75 °C.

Principle

Whole blood, spiked with Cyclosporin D as an internal standard, is acidified and subjected to ether extraction. The cyclosporines, soluble in the organic phase, are transferred to a clean tube and the solvent is evaporated. The residue is redissolved in dilute acid and methanol and washed with hexane to remove interfering substances (hexane is discarded). The aqueous medium is then made alkaline and the cyclosporines are again extracted into ether. The organic phase is transferred to a clean tube and the ether is evaporated. The remaining residue is reconstituted and chromatographed. Quantitation is performed by peak area analysis.

Reagents

1. Standards.
   a. Stock cyclosporine, 1 mg cyclosporine/mL in methanol (Cyclosporin, Sandoz Pharmaceuticals, East Hanover, NJ). This solution is stable for more than six months if stored in a sealed container at −20 °C.
   b. Working cyclosporine. Prepare working standards at three concentrations: 1000, 500, and 50 ng cyclosporine/mL in bovine serum. These standards are divided into 2.0-mL aliquots and stored in 16 × 125-mm glass tubes washed according to the procedure instructions. The standards are capped and refrigerated, and under these conditions are stable as long as significant evaporation and bacterial growth can be avoided.
2. Internal standard, 5 μg/mL. Dilute 0.5 mg of Cyclosporin D (Sandoz Pharmaceuticals) in 100 mL of methanol. This reagent is stable for more than six months if stored in a sealed container at −20 °C.

3. Extraction and chromatography solvents.

    a. Diethyl ether, ACS grade. Store in the original tin container.

    b. Hexane, double glass-distilled.

    c. Acetonitrile, double glass-distilled.

    d. Methanol, double glass-distilled.

    e. Water, double glass-distilled.

4. Hydrochloric acid, 0.18 mol/L. Add 15 mL of concentrated hydrochloric acid to 900 mL of water and dilute to a final volume of 1 L.

5. Hydrochloric acid, 0.01 mol/L in methanol/water. Add 11 mL of HCl, 0.18 mol/L, and 22 mL of water to a 100-mL volumetric flask and dilute to 100 mL with methanol.

6. Sodium hydroxide, 0.1 mol/L. Dissolve 4 g of NaOH pellets in water and dilute to a final volume of 1 L. Store in a plastic bottle and replace every two months.

7. Reconstitution solution: 1 part acetonitrile, 1 part methanol, 3 parts water. Prepare fresh daily.

8. Chromatography system.

    a. Instrumentation. The chromatography system consists of an injector, a high quality solvent pump, a 15-cm Supelcosil LC-18 5-μm column (Supelco, Bellefonte, PA), a column heater capable of maintaining a constant temperature of 75 °C, a fixed wavelength UV-detector set to monitor 214 nm, and a recorder integrator.

    b. Mobile phase. 49 parts acetonitrile, 20 parts methanol, 31 parts water. De-gas this solvent prior to use. This solvent flows through the column at 2.5 mL/min.

    c. Sensitivity. Set instrument sensitivity setting such that the 50 ng/mL standard gives a minimum signal to noise ratio of 20:1.

Procedure

1. Rinse all glassware to be used in this procedure with ether, drain, and air dry under an evacuated hood.

2. Remove standards from refrigerator and warm to room temperature.

3. Transfer 2.0 mL of serum-based control pool and whole blood samples to appropriately labeled 16 × 125-mm tubes.

4. Add 1.0 mL of internal standard and 3.0 mL of hydrochloric acid, 0.18 mol/L, to each tube and mix well.

5. Transfer the contents of each tube to appropriately labeled, glass-stoppered 50-mL centrifuge tubes and add 14 mL of ether to each.

6. Shake vigorously for 15 min.

7. Separate the phases by centrifugation at 1000 × g in a refrigerated centrifuge for 5 min, at 5 °C.

8. Transfer each organic phase to 16 × 150-mm test tubes and evaporate the ether.

9. Redissolve each residue in hydrochloric acid, 0.01 mol/L in methanol/water, add 7.0 mL of hexane, and mix vigorously for 1 min.

10. Aspirate the hexane layer to waste.

11. Pour each aqueous phase into 50-mL glass-stoppered centrifuge tubes and add 3.0 mL sodium hydroxide, 0.1 mol/L, and 14 mL of ether to each.

12. Mix vigorously for 15 min.

13. Transfer the organic phase to conical-tip centrifuge tubes and evaporate the ether from each.

14. Reconstitute each residue in 100 μL of reconstitution solution.

15. Chromatograph a 50-μL aliquot of each reconstituted extract.

16. Identify cyclosporine and Cyclosporin D by retention time and quantitate from peak areas as described in the antiepileptic procedure.

Reference

Faynor, S. M., Sterioff, S., et al.: Therapeutic drug monitoring of cyclosporine. Mayo Clin. Proc., *59*:571-572, 1984.

SUMMARY

    This chapter has reviewed the basic concepts of pharmacokinetics as they are applied to therapeutic drug monitoring, the mechanism of action of the drugs commonly monitored, and therapeutic and toxic blood concentration guidelines for these drugs, and has presented valid methods for their analysis in biological specimens. To accomplish the advantage offered by an individualized pharmacokinetic approach, the methods employed to quantitate blood concentrations must be accurate and precise. Techniques with analytical variability as great as the therapeutic range are no longer acceptable. State-of-the-art analytic techniques can achieve variability of 5% or less for most drugs, a situation which is essential if a pharmacokinetic approach is to be useful.

Advances in methodology continue to be reported. It is likely therefore that improved sensitivity and specificity will continue to permit new or altered techniques to be applied to therapeutic drug monitoring. The methods described in this chapter were selected not only for their reliability but for simplicity. For example, chromatographic methods do not require prior derivatization. Nevertheless, it is anticipated that advances already published will be applied more readily in the future to therapeutic drug monitoring. Capillary gas chromatography can improve sensitivity and specificity dramatically.[89] Derivatization followed by gas chromatography using the electron capture detector is capable of increasing sensitivity by several orders of magnitude.[26] Mass spectrometer detectors for gas chromatographs are now economically feasible and are likely to become readily available, permitting the use of deuterated internal standards and simultaneous analysis of many drug analogs and their metabolites.[56]

There are always special interests in therapeutic drug monitoring. Some rheumatologists monitor salicylate during *aspirin* therapy. (See Chapter 18.) Thiocyanate may be monitored to follow *nitroprusside* therapy. *Anticoagulants* may be of importance to some clinicians. Medical practice often applies traditional drugs in new contexts of therapy. New, more potent drugs whose potential for toxicity has not been fully defined are continually being developed and are coming into use. The clinical chemist faces the constant challenge of keeping current with growth in pharmacological knowledge, medical applications, and analytical technology that is serviceable for therapeutic drug monitoring.

Therapeutic drug monitoring has a demonstrated utility for identifying the noncompliant patient, for assisting in dosage modification in patients with altered metabolic status, and as an early warning of the onset of toxicity. Published optimal therapeutic serum concentrations serve as guidelines only; individual patient results must be interpreted in light of each patient's physical condition and response to the drug. For complete result interpretation and the maximum benefit from the therapeutic drug monitoring exercise, a thorough knowledge of the analytical capability (*and limitations*) and an understanding of both the pharmacology and pharmacokinetics of the drug in question are essential.

References

1. Amdisen, A.: Lithium. *In*: Applied Therapeutics: Principles of Therapeutic Drug Monitoring. W. S. Evans, J. J. Schentag, and W. J. Jusko, Eds. San Francisco, Applied Therapeutics Inc., 1980, pp. 586-617.
2. Amsterdam, J., Brunswick, D., and Mendels, J.: The clinical application of tricyclic antidepressant pharmacokinetics and plasma levels. Am. J. Psychiatry, *137*:653-662, 1980.
3. Anderson, J. L., Mason, J. W., Winkle, R. A., et al.: Clinical electrophysiologic effects of tocainide. Circulation, *57*:685-691, 1978.
4. Anhalt, J. P.: Interpretation of antimicrobial concentrations in serum. American Association for Clinical Chemistry Therapeutic Drug Monitoring Laboratory Improvement Program. Washington, D.C., AACC, 1981.
5. Aranda, J. V., Gorman, W., Bergsteinsson, H., et al.: Efficacy of caffeine in treatment of apnea in the low-birth-weight infant. J. Pediatr., *90*:467-472, 1977.
6. Asberg, M., Cronholm, B., Sjoquist, F., et al.: Relationship between plasma level and therapeutic effect of nor-triptyline. Br. Med. J., *3*:331-334, 1971.
7. Avery, G. S., Ed.: Drug Treatment: Principles and Practice of Clinical Pharmacology and Therapeutics. 2nd ed. Sydney, ADIS, 1980.
8. Baer, D. M., and Dito, W. R., Eds.: Interpretations in Therapeutic Drug Monitoring. Chicago, American Society of Clinical Pathologists, 1981.
9. Beveridge, T., Gratwohl, A., Michot, F., et al.: Cyclosporin A: Pharmacokinetics after a single dose in man and serum levels after multiple dosing in recipients of allogenic bone-marrow grafts. Current multiple dosing in recipients of allogenic bone-marrow grafts. Curr. Ther. Res., *30*:5-18, 1981.
10. Biggs, J. T., Holland, W. H., Chang, S., et al.: Electron beam ionization mass fragmentographic analysis of tricyclic antidepressants in human plasma. J. Pharm. Sci., *65*:261-268, 1976.
11. Biggs, J. T., Spiker, D. G., Petit, J. M., et al.: Tricyclic antidepressant overdose: Incidence of symptoms. JAMA, *238*:135-138, 1977.
12. Blumer, J., Strong, J. M., and Atkinson, A. J.: The convulsant potency of lidocaine and its N-dealkylated metabolites. J. Pharmacol. Exp. Ther., *186*:3136, 1973.
13. Borel, J. F., Feurer, C., Gubler, H. U., et al.: Biological effects of cyclosporin A: A new antilymphocytic agent. Agents Action, *6*:468-475, 1976.
14. Borel, J. F., Feurer, C., Magnee, C., et al.: Effects of the new antilymphocytic peptide cyclosporin A in animals. Immunology, *32*:1017-1025, 1977.

15. Bory, C., Baltassat, P., Porthault, M., et al.: Metabolism of theophylline to caffeine in premature newborn infants. J. Pediatr., *94*:988-993, 1979.
16. Bowers, L.: University of Minnesota Hospital, Minneapolis, personal communication.
17. Bresnahan, J. F., and Vlietstra, R. E.: Digitalis glycosides. Mayo Clin. Proc., *54*:675-684, 1979.
18. Buchthal, F., Svensmark, O., and Schiller, P. J.: Clinical and electroencephalographic correlations with serum levels of diphenylhydantoin. Arch. Neurol. (Chicago), *2*:624-630, 1960.
19. Caccia, S., Ballabio, M., Farelli, R., et al.: Determination of plasma and brain concentrations of trazodone and its metabolite, 1-m-chlorophenyl by gas-liquid chromatography. J. Chromatogr., *210*:311-318, 1981.
20. Campbell, N. P., Kelly, J. G., Adgey, A. A., et al.: The clinical pharmacology of mexiletine. Br. J. Clin. Pharmacol., *6*:103-108, 1978.
21. Chang, J., Gotcher, S., and Gushaw, J. B.: Homogeneous enzyme immunoassay for theophylline in serum and plasma. Clin. Chem., *28*:361-367, 1982.
22. Chinn, D. M., Jennison, T. A., Crouch, D. J., et al.: Quantitative analysis for tricyclic antidepressant drugs in plasma or serum by gas-chromatography-chemical-ionization mass spectrometry. Clin. Chem., *26*:1201-1204, 1980.
23. Cohen, B. M., Herschel, M., and Aoba, A.: Neuroleptic, antimuscarinic, and anticholinergic activity of chlorpromazine, thioridazine and their metabolites. Psychiatry Res., *1*:199-208, 1979.
24. Cooper, T. B.: Plasma level monitoring of antipsychotic drugs. Clin. Pharmacokinet., *3*:14-38, 1978.
25. Cooper, T. B., and Kelly, R. G.: GLC analyses of loxapine, amoxapine, and their metabolites in serum and urine. J. Pharm. Sci., *68*:216-219, 1979.
26. Corkill, J. A., Joppich, M., Kuttub, S. H., et al.: Attogram-level detection and relative response of strong electrophores by gas chromatography with electron capture detection. Anal. Chem., *54*:481-485, 1982.
27. Creese, I., and Snyder, S. H.: A simple and sensitive radioreceptor assay for antischizophrenic drugs in blood. Nature, *270*:180-182, 1977.
28. David, B. M., Madsen, B. W., and Ilett, K. F.: Plasma binding of disopyramide. Br. J. Clin. Pharmacol., *9*:614-618, 1980.
29. Dietrich, B. A., Krauss, A. M., Reidenberg, M., et al.: Alterations in state in apneic pre-term infants receiving theophylline. Clin. Pharmacol. Ther., *24*:474-478, 1978.
30. Edelbroeck, P. M., and De Wolff, F. A.: Improved micromethod for determination of underivatized clonazepam in serum by gas chromatography. Clin. Chem., *24*:1774-1777, 1978.
31. Foerster, E. H., Dempsey, J., and Garriott, J. C.: A gas chromatographic procedure for acid and neutral drugs in blood. J. Anal. Toxicol., *3*:87-91, 1979.
32. Geraci, J. E.: Vancomycin. Mayo Clin. Proc., *52*:631-634, 1977.
33. Gerson, B., and Anhalt, J. P.: High-Pressure Liquid Chromatography and Therapeutic Drug Monitoring. Chicago, Am. Soc. Clin. Path., 1980.
34. Gilman, A. G., Goodman, L. S., and Gilman, A.: The Pharmacological Basis of Therapeutics. 6th ed. New York, Macmillan, 1980.
35. Gram, L. F., and Overo, K. F.: Drug interaction inhibitory effect of neuroleptics on metabolism of tricyclic antidepressants in man. Br. Med. J., *1*:463-465, 1972.
36. Grant, A. M., Marshall, R. J., and Ankier, S. I.: Some effects of disopyramide and its N-dealkylated metabolite on isolated nerve and cardiac muscle. Eur. J. Pharmacol., *49*:389-394, 1978.
37. Gray, M. G., and Moore, M.: Blood bromide determinations: Their use and interpretation. J. Lab. Clin. Med., *27*:680-686, 1941.
38. Hendeles, L., and Weinberger, M.: Theophylline: Therapeutic use and serum concentration monitoring. *In*: Individualizing Drug Therapy. W. J. Taylor, and A. L. Finn, Eds. New York, Gross, Townsend, and Frank, 1981, pp. 32-65.
39. Kahan, B. D., Van Buren, C. T., Lin, S. N., et al.: Immunopharmacological monitoring of cyclosporin A–treated recipients of cadaveric kidney allografts. Transplantation, *34*:36-45, 1982.
40. Keys, T. F., Kurtz, S. B., Jones, J. D., et al.: Renal toxicity during therapy with gentamicin or tobramycin. Mayo Clin. Proc., *56*:556-559, 1981.
41. Koch-Weser, J.: Serum drug concentrations and therapeutic guides. N. Engl. J. Med., *287*:227-231, 1972.
42. Koch-Weser, J.: Serum procainamide levels as therapeutic guides. Clin. Pharmacokinet., *26*:389-402, 1977.
43. Lafara, R. L.: Computer Methods for Science and Engineering. Rochelle Park, N.J., Hayden Book Co., 1973, pp. 153-156.
44. Leahey, E. B., Reiffel, J. A., Drusin, R. E., et al.: Interaction between quinidine and digoxin. JAMA, *240*:533-534, 1978.
45. Lietman, P. S.: Chloramphenicol and the neonate—1979 view. Clin. Perinatol., *6*:151-162, 1979.
46. Locock, C.: Analysis of 52 cases of epilepsy observed by author. Lancet, *1*:527, 1857.
47. Mayer, S. E., Melmon, K. L., and Gilman, A. G.: Introduction, the dynamics of drug absorption, distribution, and elimination. *In*: The Pharmacological Basis of Therapeutics. 6th ed. A. G. Gilman, L. S. Goodman, and A. Gilman, Eds. New York, Macmillan, 1980, pp 1-27.
48. McAllister, R. G., Bourne, D. W. A., and Dittert, L. W.: The pharmacology of verapamil. I. Elimination kinetics in dogs and correlation of plasma levels with effect on the electrocardiogram. J. Pharmacol. Exp. Ther., *202*:38-44, 1977.
49. Melmon, K. L., and Morelli, H. I., Eds.: Clinical Pharmacology: Basic Principles in Therapeutics. 2nd ed. New York, Macmillan, 1978.
50. Meyerburg, R. J., Kessler, K. M., Kiem, I., et al.: Relationship between plasma levels of procainamide suppression of premature ventricular complexes and prevention of recurrent ventricular tachycardia. Circulation, *64*:280-290, 1981.

51. Mirkin, B. L., Ed.: Perinatal Pharmacology and Therapeutics. New York, Academic Press, 1976.
52. Morgenstern, G. R., Powles, R., Robinson, B., et al.: Cyclosporin interaction with ketoconazole and melphalan. Lancet, 2:1342, 1982.
53. Morselli, P. L., Ed.: Drug Disposition During Development. New York, Spectrum Publ., 1977.
54. Moyer, T. P.: Practical therapeutic drug monitoring. In: Clinical Laboratory Annual, Vol. 2. H. A. Homburger and J. G. Batsakis, Eds. Norwalk, Conn., Appleton-Century-Crofts, 1982, pp. 279-322.
55. Moyer, T. P.: Uncommonly used antiepileptic drugs. In: Applied Therapeutic Drug Monitoring, Vol. 1. T. P. Moyer and R. C. Boeckx, Eds. Washington, D.C., American Association for Clinical Chemistry, 1982, pp. 161-166.
56. Narasimhachari, N., Friedel, R., and Saady, J. J.: Improved tricyclic antidepressant assay by GC/MS. Res. Commun. Psychol. Psychia. Behavior, 4:477-490, 1979.
57. Nelson, K., and Bowers, L.: Evaluation and service experience with a fluorescence polarization immunoassay for digoxin. Clin. Chem., 29:1175, 1983.
58. O'Malley, K., Judge, T. G., and Crooks, J., Geriatric clinical pharmacology and therapeutics. In: Drug Treatment: Principles and Practice of Clinical Pharmacology and Therapeutics. G. S. Avery, Ed. Sydney, ADIS, 1980, pp. 158-181.
59. Pape, B. E.: Enzyme immunoassay and two fluorometric methods compared for the determination of quinidine in serum. Ther. Drug Monit., 3:357-363, 1981.
60. Pine, M., Favrot, L., Smith, S., et al.: Correlation of plasma propranolol concentration with therapeutic response in patients with angina pectoris. Circulation, 52:886-889, 1975.
61. Pippenger, C. E.: Pediatric clinical pharmacology of antiepileptic drugs: A special consideration. In: Antiepileptic Drugs, Quantitative Analysis and Interpretation. C. E. Pippenger, J. K. Penry, and H. Kutt, Eds. New York, Raven Press, 1977, pp. 315-319.
62. Polin, H. B., and Plant, M. E.: Chloramphenicol. NY State J Med., 77:378-381, 1977.
63. Powis, G.: Effect of human renal and hepatic disease on the pharmacokinetics of anticancer drugs. Cancer Treat. Rev., 9:85-124, 1982.
64. Powis, G., Ames, M., and Kovach, J.: Dose-dependent pharmacokinetics and cancer chemotherapy. Cancer Chemother. Pharmacol., 6:1-9, 1981.
65. Reece, P. A., and Stanley, P. E.: High-performance liquid chromatographic assay for tocainide in human plasma: Comparison with gas-liquid chromatographic assay. J. Chromatogr., 183:109-114, 1980.
66. Riva, E., Gerna, M., Latini, R., et al.: Pharmacokinetics of amiodarone in man. J. Cardiovasc. Pharmacol., 4:264-269, 1982.
67. Rodman, J. H.: Lidocaine. In: Applied Pharmacokinetics: Principles of Therapeutic Drug Monitoring. W. E. Evans, J. J. Schentag, and W. J. Jusko, Eds. San Francisco, Applied Therapeutics, Inc., 1980, pp. 358-359.
68. Routledge, P. A., Stangel, W. W., Wagner, G. S., et al.: Alpha-1-acid glycoprotein (AAG) as a determinant of lidocaine (L) disposition in myocardial infarction (MI). Clin. Res., 27:776A, 1979.
69. Rowland, M., and Tozer, T. N.: Clinical Pharmacokinetics: Concepts and Application. Philadelphia, Lea and Febiger, 1980.
70. Sarubbi, F. A., and Hull, J. H.: Amikacin serum concentrations: Prediction of levels and dosage guidelines. Ann. Intern. Med., 89:612-618, 1978, and Gentamicin serum concentrations: Pharmacokinetic predictions. Ann. Intern. Med., 85:183-189, 1976.
71. Sawchuck, R. J., Zaske, D. E., Cipolle, R. J., et al.: Kinetic model for gentamicin dosing with the use of individual patient parameters. Clin. Pharmacol. Ther., 21:363-369, 1977.
72. Schack, J. A., and Waxler, S. H.: An ultraviolet spectrophotometric method for the determination of theophylline and theobromine in blood and tissues. J. Pharmacol. Exp. Ther., 97:283-291, 1949.
73. Schentag, J. J.: Aminoglycosides. In: Applied Pharmacokinetics. W. E. Evans, J. J. Schentag, and W. J. Jusko, Eds. San Francisco, Applied Therapeutics, Inc., 1980, pp. 174-209.
74. Schmelter, R. F., Dirksen, J. W., Stalp, J. T., et al.: Effect of heparin on radioimmunoassay of gentamicin. Am. J. Hosp. Pharm., 38:534-537, 1981.
75. Schwartz, M. L., Webb, N. C., Covina, B. G., et al.: Comparative antiarrhythmic effects of intravenously administered lidocaine and procainamide and orally administered quinidine. Am. J. Cardiol., 26:520-523, 1970.
76. Sherwin, A. L.: Clinical Pharmacology of Ethosuximide. In: Antiepileptic Drugs: Quantitative Analysis and Interpretation. C. E. Pippenger, J. K. Penry, and H. Kutt, Eds. New York, Raven Press, 1978, pp. 283-295.
77. Sjoqvest, F., Borga, O., and Orme, M. L. E.: Fundamentals of Clinical Pharmacology. In: Drug Treatment. 2nd ed. G. S. Avery, Ed. Sydney, ADIS, 1980, pp. 1-61.
77a. Smith, T. W., Butler, V. P., and Haber, E.: Determination of therapeutic and toxic serum digoxin concentrations by radioimmunoassay. N. Engl. J. Med., 281:1212-1216, 1969.
78. Soonhag, C., and Karlsson, E.: Comparative antiarrhythmic efficacy of intravenous N-acetylprocainamide and procainamide. Eur. J. Clin. Pharmacol., 15:311-317, 1979.
79. Standefer, J. C., and Callaway, S.: Biotransformation of theophylline in neonates and infants. Clin. Chem., 27:1086, 1981.
80. Thoma, J. J., Bondo, P. B., and Kozak, C. M.: Tricyclic antidepressants in serum by a clin-elut column extraction and high pressure liquid chromatography analysis. Ther. Drug Monit., 1:335-358, 1979.
81. Todd, G. D., Bourne, D. W. A., and McAllister, R. G.: Measurement of verapamil concentration in plasma by gas chromatography and high pressure liquid chromatography. Ther. Drug Monit., 2:411-416, 1980.
82. Tracy, R. P., Ebnet, L. E., and Moyer, T. P.: Use of decreased reagent volumes in enzyme immunoassay. Clin. Chem., 25:1868-1869, 1979.
83. Wagner, J. G.: Fundamentals of Clinical Pharmacokinetics. Hamilton, Ill. Drug Intelligence Publications, 1975.
84. Wenk, M., Follath, F., and Abisch, E.: Temperature dependency of apparent cyclosporin A concentrations in plasma. Clin. Chem., 29:1865, 1983.

85. Winter, M. E.: Basic Clinical Pharmacokinetics. San Francisco, Applied Therapeutics, Inc., 1980.
86. Wuth, O.: Rational bromide treatment. JAMA, *88*:2013-2017, 1927.
87. Zaske, D. E.: Aminoglycosides: Counterpoint discussion. *In*: Applied Pharmacokinetics. W. E. Evans, J. J. Schentag, and W. J. Jusko, Eds. San Francisco, Applied Therapeutics, Inc., 1980, pp. 210-239.
88. Zaske, D. E., Cipolle, R. J., and Strate, R. G.: Gentamicin dosage requirements: Wide interpatient variations in 242 surgery patients with normal renal function. Surgery, *87*:164-169, 1980.
89. Zlatkis, A., Wang, F. S., and Shanfield, H.: Trace gas chromatographic analysis by use of large sample on-column injection with bonded phase capillary columns. Anal. Chem., *54*:2406-2409, 1982.
90. Zwillich, C. W., Sutton, F. D., Neff, T. A., et al.: Theophylline-induced seizures in adults: Correlation with serum concentrations. Ann. Intern. Med., *82*:784-787, 1975.

CHAPTER 18

ANALYSIS OF TOXIC SUBSTANCES

by Robert V. Blanke, Ph.D., and Walter J. Decker, Ph.D.

Clinical toxicology is the branch of toxicology that deals with the diagnosis and treatment of chronic and acute poisoning incidents in humans. Analytical toxicology, effectively performed, is an important and frequently essential part of the sequence of events leading from diagnosis to treatment. The movement toward establishment of clinical laboratories devoted primarily to the measurement of drugs and other toxic substances is increasing constantly. Changes which have occurred over the past 25 years suggest several explanations of this phenomenon.

In the past, reluctance of clinical laboratories to enter the area of clinical toxicology was rationalized on the basis of one or more of the following reasons:

1. Lack of trained personnel for conducting toxicological tests.

2. Difficulty in justifying equipment, personnel, and space costs to meet a clinical need which occurred intermittently in a completely random manner.

3. The fear of becoming involved in medical-legal problems.

The third reason is invalid. Although clinical pathologists and clinical laboratory directors may become involved in medical-legal problems related to toxicology, this infrequent occurrence must not inhibit the provision of quality health care. The first two reasons, however, are valid. Lack of trained personnel is a current problem. However, the recognized need for staff in toxicology laboratories, together with the challenges offered by analytical toxicology, have begun to attract individuals trained in a variety of disciplines such as pharmaceutical chemistry, pharmacology, biochemistry, and analytical chemistry, to name only a few. Some schools of medical technology have begun to offer training in clinical toxicology, but this trend must increase to supply the necessary personnel in the future.

Establishing hard data in order to convince administrators that laboratory budgets should include toxicological activities has also been a difficult problem. Events in the past two decades, however, have begun to supply the necessary justification for this activity. From the middle to late 1960's, our society began to react to the problem of drug abuse. It was not a new problem, of course, but a series of political and social changes dramatically brought it to the attention of the American public. Laboratories competing for contracts to conduct urine drug screening sprang up overnight. It soon became evident that the lack of trained personnel, coupled with a lack of quality assurance, led to a generally poor performance by these laboratories.

At the same time that the social problem of drug abuse was recognized, an additional change was occurring. This was a technological change which resulted in rapid design and development of reliable and reasonably economical instrumentation and analytical techniques which could be applied to drug assays. These technological advances made the implementation of therapeutic drug monitoring practical.

The concept of therapeutic monitoring is not new. However, the technology which makes the concept both practical and effective is new. Improvements and advancements in the general field of drug analysis are so rapid that it is difficult to keep current with new developments. These technological improvements, together with an increased understanding of drug pharmacokinetics and drug interactions, focused interest on the measurement of drug concentrations in serum or blood as an aid to the effective management of patients on prolonged maintenance therapy. Although therapeutic drug monitoring is discussed separately (Chapter 17), there is a considerable amount of overlapping of functions which should be obvious.

Any laboratory equipped to conduct toxicological analyses in drug overdose cases can expand into the area of therapeutic monitoring. Conversely, the techniques used in therapeutic monitoring can obviously be applied as an aid to identifying drug overdose cases. Thus, these related activities should be carried out by utilizing the same facilities and personnel.[6] With the support that has developed from the clinician interested in monitoring drug levels in his patients, the toxicology laboratory is now in a position to approach budget administrators with hard data justifying the needs of the toxicology laboratory in carrying out these requests. It would seem that most of the old arguments used to explain the lack of adequate clinical toxicology facilities are slowly slipping away.

In order to fulfill its role in dealing with the emergency situation and to be of maximal use to the clinician, the toxicology laboratory must provide 24-h service and utilize reliable, rapid methods resulting in a short turnaround time. The monitoring of chemical substances in patients can usually be achieved within a normal workday using dedicated procedures on a relatively continuous basis. Ideally, of course, with unlimited facilities, personnel, and support, these goals can be achieved relatively easily. Since most laboratories can never hope to enjoy this luxury, a more pragmatic approach which involves compromise between the ideal and the practical must be taken.

EMERGENCY TOXICOLOGY

The difficulty and complexity of the diagnosis and treatment of toxic episodes are not obvious to the uninitiated. There are two primary causes for the diagnostic problems. First, there are thousands of chemical substances used as drugs, household products, and in occupational, industrial, and farming applications that are toxic and are possible etiological agents in both acute and chronic poisoning incidents. Recently the extent of the problem was graphically described by Yodaiken:[72]

"The chemical industry has grown exponentially since the 1930's. Every 7 years the production of synthetic chemicals has doubled reaching about 175 billion pounds per year. . . . Approximately 3.5 million to 4.3 million chemicals are known to be in existence, and the number is believed to be growing by about 10% per year. Approximately 25 000 to 50 000 chemicals are in production in the United States alone, and about 50 000 to 65 000 chemicals (not including pesticides, pharmaceuticals and food additives) are believed to be in everyday use.

"In the United States alone, as many as 3000 new industrial chemicals are introduced annually into economic use. . . . In the past decade production of synthetic organic chemicals in the United States has expanded 255%. Some 2500 chemicals or mixtures are reported in current use by the plastics industry alone. The EPA (Environmental Protection Agency) estimates that as many as 1500 active ingredients are found in pesticides, while the Food and Drug Administration estimates that about 4000 active ingredients are in drugs. . . . A 1975 report of the National Academy of Science stated that 1 billion pounds of toxic matter was being introduced yearly for pest control."

Although those chemical substances that are encountered relatively frequently in acute poisoning incidents are diagnosed successfully in most situations, there are still great gaps in information relating to the toxic effects and recommended treatment procedures for both acute and chronic poisoning by the vast majority of chemical substances.

Second, it is unlikely that a single individual exists with the necessary knowledge, training, and skills adequate to respond effectively to all types of poisoning incidents. As a result, the response to emergency toxicology must be a team effort in order to be effective.

There are at least three components involved in this effort (Figure 18-1). First, the patient must undergo a clinical assessment. This means that the signs of illness must be accurately observed and symptoms must be elicited by skillful and sympathetic interrogation. Critical and life-threatening effects must be aggressively and promptly treated, generally in a symptomatic fashion. Treatment is tempered by the judgment of the clinician, depending upon the extent of the information available to him and his past experience in treating patients with the severity of

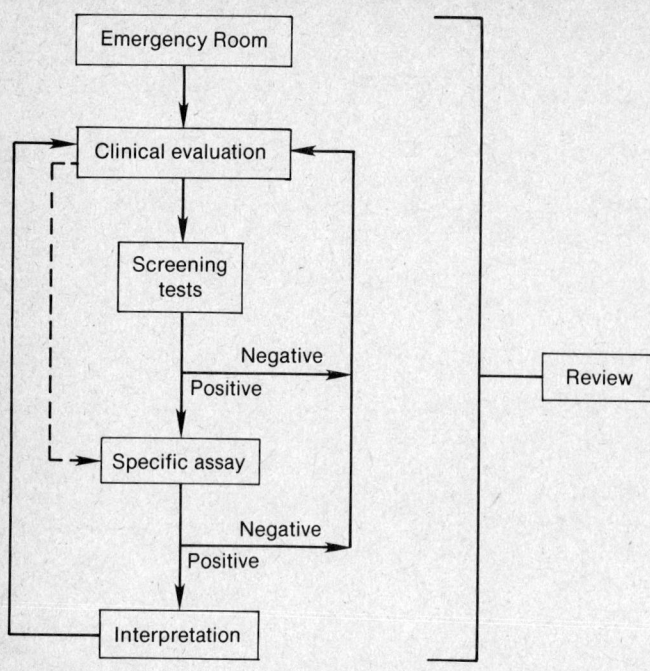

Figure 18-1. A schematic representation of the response to a toxicological emergency which emphasizes the need for maintaining lines of communication and for periodic review.

the presenting condition. Further exposure of the patient to the toxic substance must be prevented or minimized. Depending upon the route of absorption, this may mean (1) removing the patient from the airborne source of a toxic inhalant, (2) initiating emesis or gastric lavage, (3) minimizing further absorption by administering a binding agent such as activated charcoal, and (4) removing or neutralizing chemicals in contact with the skin or clothing.

Subsequently, or even simultaneously with the above procedures, a complete history must be obtained. The completeness or extent of such a history depends, of course, on the age and mental state of the patient, an accompanying witness, or relatives. The history is one of the most important parts of the examination for diagnostic purposes. In addition to a general medical history, a complete occupational and social history should be obtained. This should include not only those drugs or chemicals with which the patient may have been directly in contact but also those substances which might be available in the home, in the workplace, or in the environment from which the patient came. Such a complete history is not easy to document and may require inquiries of employers or manufacturers of products or even the assistance of trained investigators revisiting the scene of the incident. With the facts available, the clinician is generally able to make a presumptive diagnosis or to develop a plan by which the multiple problems of the patient may be resolved. At this juncture a second component of the team effort should be brought into play.

Many toxic substances can be ruled out of, or into, a differential diagnosis by demonstration of the presence or absence of a particular toxic substance in specimens removed from the patient or found in solids, liquids, or gases available to the patient. Although the history, together with the clinical state of the patient, may indicate a specific substance as a causative agent, screening tests employed at this point are best for developing a rapid response. The results of the laboratory analyses may confirm a previous impression of the clinician or may raise the possibility of a toxic substance not previously suspected.

Depending upon the clinical state of the patient and the presumptive report of the laboratory, the decision may then be made to attempt to quantitate the toxic substance present in specimens from the patient which, in turn, may help the clinician to gauge the severity of the poisoning incident. In some cases, this may determine the type of treatment procedure to be instituted. An additional role of the laboratory can include follow-up determinations of the toxic substance to evaluate the effectiveness of the treatment procedure and the prognosis of the case.

These two components of the team effort, clinician and laboratory, might seem sufficient to resolve most cases encountered in emergency toxicology. Yet in the past 20 years, most efforts to support the clinician treating the patient by analytical toxicology services do not appear to have been effective.[11,62] In fact, even in recent years the role of the toxicology laboratory in patient care is still being debated.[38,71] Apparently then, effective response to the toxicological emergency requires something more than a skillful clinician and a well-equipped analytical laboratory.

An essential ingredient for the smooth and effective response to the poisoning incident is a toxicology data coordinator. This third component of the clinical toxicology team provides for an efficient two-way communication between clinician and laboratory. As stated earlier, no single individual can satisfactorily fill all three roles of the team effort. The number of toxic substances encountered and the sophistication and rapid advances in analytical technology, as well as the innovations and modifications of treatment procedures, represent too vast an area for effective assimilation by clinician or laboratorian alone.

The data coordinator serves by correlating a specific patient's history and clinical state with information already available on toxic substances and their known effects. This permits the coordinator to set priorities for the laboratory in order to enable testing procedures to be initiated that have the greatest likelihood of providing useful analytical results. Once the results are available, it is necessary to interpret them in a meaningful way and to communicate this information to the clinician. Without this type of informed communication, the laboratory may waste valuable time in meaningless manipulations or the clinician may be tempted to treat the laboratory result rather than the patient.

Information resources are many and varied. Poison Information Centers have already assembled information on common toxic chemicals. The Toxicology Data Bank at the National Library of Medicine (available through MEDLARS at most libraries) permits rapid retrieval of specific informaton by means of a terminal connected through a telephone line to the computer at the National Library of Medicine. Commercial information resources are also available.

In some successful emergency toxicology applications, the coordinator is a trained and experienced toxicologist. In others, this role may be filled in part by a clinician with a unique and special interest, or by a laboratorian who may have acquired this capability through training and experience. However, without the third component, the coordinator, the effectiveness of the clinical toxicological response to poisoning incidents is frequently nullified.

CHRONIC POISONING: INDUSTRIAL AND ENVIRONMENTAL MONITORING

The problem of chronic poisoning has always been present but only in recent years has it received attention. Long-term exposure to chemical substances may or may not have an adverse effect on human health. Most of the data upon which decisions regarding action levels by regulatory agencies are based, draw on experimental animal studies or retrospective epidemiological studies in which subjects exposed to various concentrations of chemicals developed signs and symptoms of poisoning. Little evidence exists correlating *concentrations* of these chemicals and their metabolites in plasma or target organs with the observed effect.

The nature of toxic substances that give rise to chronic poisoning varies widely, ranging from elements, particularly metals, through complex organic and inorganic compounds. These substances may be encountered as drugs, pesticides, industrial chemicals in the workplace, industrial and economic chemicals in the environment, food additives (either intentional or unintentional), and environmental pollutants (either artificial or natural). Thus, they constitute a spectrum of substances present in a variety of states in a multiplicity of matrices at vanishingly low concentrations. This is an enormous challenge to the analytical toxicologist and an increasingly important challenge to the clinical toxicologist.

Action levels are those concentrations of toxic chemicals in food, air, or water which may be expected to exert adverse effects after prolonged exposure in humans. Little information is available on the actual quantity of toxic substances present in humans when they have been exposed to action levels of these substances in the environment. Techniques are rarely available for measuring extremely low levels of toxic substances in blood, serum, urine, or other specimens. Stewart[60] has found that breath samples can be collected reliably and provide a valid method by which volatile substances present in the environment at the workplace may be reflected by a given concentration of the material in breath from the worker. Surprisingly, some volatile substances continue to appear in the breath many hours or days after exposure. If timed breath samples are collected and analyzed, normal decay curves can be established which permit extrapolating back to a particular concentration of the volatile substances in the air to which the worker was exposed.

Similarly, a variety of nonpolar organic chemicals, particularly pesticides and organochlorine compounds, may be stored in fat depots for extended time periods. Since fat is a convenient specimen which can be collected by aspiration of subcutaneous depots,[28] analysis of this specimen may also provide a simple means by which the exposure of a patient to environmental toxic substances can be evaluated.

There is much that remains to be done in the area of chronic poisoning before laboratory procedures will acquire practical importance. Nevertheless, the technology and the means are now available for acquiring the necessary data and enabling the clinical toxicology laboratory to be an important factor in assessing chronic poisoning by many agents.

RESPONSIBILITY OF THE CLINICIAN

The clinician stands alone in the all-important position of dealing with the patient. This is a demanding, challenging, and often frustrating role to fill. In the case of an acutely poisoned patient, all the resources available to contemporary medicine may be brought to bear on the patient in a life-threatening situation. Frequently rapid decisions must be made while exercising the best judgment possible when all the appropriate facts may not be readily available. It is evident that one of the foremost requirements of the clinician must be extensive experience in emergency medicine built on a solid base of pharmacology and physiology. As with all health-related professions, the clinician dealing with poisoning cases must make every effort to be aware of and familiar with advances in the field of clinical toxicology.

Most attention in the past has been directed toward acute poisonings. Only recently has the importance of chronic poisonings, particularly as they relate to industrial and environmental exposures, become apparent. In both situations, and particularly in the case of chronic poisonings, a complete medical history is of great importance. In order to assure completeness, it is useful to devise a standard approach to the collection of this information. Since the history is frequently obtained in a busy emergency room, it is convenient to design a form by which important information can be collected with a minimum amount of writing. Such a form is shown in Figure 18-2. Included in the information gathered would be those drugs or chemicals to which the patient may be exposed directly or indirectly by means of occupational hazards, household chemicals, drugs both prescribed and over-the-counter (OTC), as well as unusual odors, dusts, and other evidence of possible exposures.

For suspected chronic poisonings, the history must go back in time to the point where a patient may have first experienced an adverse effect. Frequently this experience is associated in the mind of the patient with some event such as the use of a new chemical substance in his daily occupation. Although this association may be valid, it may also be misleading without careful investigation into the nature of all the components of a particular commercial product, and without knowing the interactions of chemicals with drugs and other chemical substances that can also lead to adverse effects.

In addition to becoming part of the patient's chart, the medical and social history must be made available to the coordinator between laboratory and clinician. This is particularly necessary when the possibility exists that a variety of substances may have been encountered leading to a toxic episode. A decision should be made involving the clinician, laboratory analyst, and coordinator as to the priority of tests that should be performed for verification of the presence or absence of particular toxic substances. These priorities are based upon the signs and symptoms presented by the patient, history of exposure to a toxic chemical, and availability of laboratory facilities for the appropriate tests.

It is also important that the clinician exercise some initiative to become familiar with routine problems experienced by most clinical laboratories. These problems include personnel qualifications and restrictions, availability of equipment, and times of peak laboratory loads. Further, the clinician should be familiar with the clearly stated operating procedures of the laboratory, defined in terms of in-house standards as well as those imposed by state and federal laws and regulatory agencies. The procedures include requirements for specimen collection and identifi-

TOXICOLOGY INFORMATION FORM

DATE: _____ TIME: _____ am pm

NAME: _____ SEX: _____ AGE: _____

ID#: _____ LOCATION: _____

Clinical Toxicology Laboratory
Telephone: (804) 786-0339
Division of Clinical Pathology
Medical College of Virginia
Richmond, Virginia 23298

CLINICAL SUMMARY

Behavior

Normal ☐	Disoriented ☐	Coma (Grade) 0 ☐ I ☐ II ☐ III ☐ IV ☐	
Intoxicated ☐	Lethargic ☐	Convulsions ☐ _____ Type	
Drowsy ☐	Hyperactive ☐	Paralysis ☐ _____ Type	

Blood Pressure	**Temperature**	**G.I.**	**Pupils**
Normal ☐	Normal ☐	Vomiting ☐	Normal ☐
Elevated ☐	Elevated ☐	Diarrhea ☐	Dilated ☐
Depressed ☐	Depressed ☐	Abdominal Pain ☐	Constricted ☐

Heart Rate: Normal ☐ Other _____

Respiration: Normal ☐ Other _____

Skin		**Mucous Membranes**	**Secretions**
Normal ☐	Pallor ☐	Foreign Material ☐	Normal ☐
Flushed ☐	Cyanotic ☐	Chemical Burn ☐	Dry ☐
Chemical Burn ☐	Irritation ☐	Inflammation ☐	Excessive ☐
Needle Marks:			
Recent ☐ Old ☐			

Gastric Contents (Submit Sample if Available)

Foreign Material   Acidic ☐ Basic ☐ Neutral ☐ Blood ☐ Unusual Odor _____ Color _____

Urine (Submit 50 mL if available)

Polyuria ☐   Oliguria ☐   Oxalate Crystals ☐   Unusual Odor _____   Color _____

Additional Information and Comments: (Consider onset of illness, vocation and avocation, drug availablity, etc.)

_____

_____

_____

Physician _____ Telephone No. _____

Figure 18-2. A suggested form for easily and rapidly recording clinical information in a toxicological emergency.

cation, timing of specimen withdrawal, and any other aspect affecting the ability of the laboratory to respond to a given request. The STAT test request is frequently abused and should be limited to only those tests whose results may conceivably alter the immediate treatment of the patient. Even then, the clinician must recognize the fact that some toxicological tests, even when run on a STAT basis, may require several hours to complete.

Although the history elicited from the patient may suggest that a particular toxic substance or drug is the causative agent, a confused recollection by the patient or garbled information by a hysterical parent, together with other anecdotal information available, might be completely misleading. In such cases, the history may not match the clinical state of the patient or the laboratory results which are ultimately obtained. In addition, drug interactions are difficult to predict. A therapeutic dose of a drug combined with an evening of social drinking may precipitate a more severe adverse reaction than would be expected when considering the drug alone. Further, an occasional laboratory error or the mislabeling of a specimen can result in laboratory data which are not appropriate for a patient under treatment. For these and many other reasons, it has become axiomatic to encourage the clinician to "treat the patient and not the poison."[13] This is to say, if the clinical state is either more severe or less severe than would be indicated by the history or the laboratory results, the treatment is geared to the clinical state rather than to the other, possibly misleading factors. This rather obvious injunction is frequently overlooked, particularly by the enthusiastic but relatively inexperienced clinician who may be less inclined to rely on his own judgment in these situations.

Finally, the clinician should interact regularly with the other members of the clinical toxicology team. A formal, structured interaction may be achieved through scheduled meetings, seminars, and journal clubs or other means. Every attempt must be made to consider the approach to the toxicological emergency as a team effort, and part of this effort is reflected by the willingness of the clinician to welcome comments and criticisms as well as to communicate his own concerns and needs to the other members of the team.

RESPONSIBILITIES OF THE CLINICAL TOXICOLOGY LABORATORY

Another member of the clinical toxicology team, the analytical toxicologist, also faces a demanding and difficult challenge. Many of the tools available to the analytical chemist must be applied to the assay of a biological specimen of rather limited sample size. In contrast to forensic toxicology, clinical toxicology demands not only reliable and sensitive measurements but also analyses which can be completed in a reasonable time frame to be of practical use to the clinician. Some representative methods for analysis of specific toxic substances are given later in this chapter, but many are available in a variety of sources.[2,3,12,63,65] Depending on the nature of the analyte and the laboratory facilities available, the turnaround time for clinical assays can range from minutes to hours. Except in unusual circumstances, and in the case of chronic exposures to toxic substances, a turnaround time greater than 24 h is of little use to the clinician.

Some practical guidelines which clinical toxicology laboratories can follow should be emphasized. Since tens of thousands of chemicals are available and can conceivably be encountered by potential patients, it is important to identify those substances for which it is practical to devise tests. Some drugs and chemicals are encountered rather commonly by almost all individuals. Thus, ethanol, carbon monoxide, sedatives, analgesics, or tranquilizers would be substances encountered by most clinical toxicology laboratories. On the other hand, depending upon the geographic location and the nature of industries located in the area, some chemical substances may be expected to appear more frequently in an area served by one laboratory than in that served by another.

There are a number of approaches that a clinical toxicology laboratory may consider in determining the procedures to be included in its repertoire of tests. Consultation with the hospital pharmacist or other knowledgeable pharmacists can indicate the most frequently prescribed drugs currently used in a given community. Such an estimate should also include over-the-counter drugs and their specified component ingredients, as well as street drugs currently abused in the community served. The common household products likely to be encountered can be named by the nearest Poison Control Center. These products include a variety of volatile solvents, cleaning agents, pesticides, cosmetics, and toilet articles. Finally, a review of the chemical and other industrial operations conducted in the community served by the clinical toxicology laboratory can identify chemicals which may be encountered either in an industrial accident or in chronic exposures to industrial and environmental pollutants.

Once the nature of the toxic chemicals likely to be encountered in the community is identified, it is necessary to decide which of these substances can reasonably be tested for within the constraints imposed upon the laboratory by equipment limitations, personnel qualifications, and anticipated workload. It is better for a clinical toxicology laboratory to offer a relatively small number of tests by reliable methods with reasonable turnaround times than to attempt to handle all types of situations which might arise. Once having agreed upon a basic nucleus of tests, published procedures can be evaluated and adapted.

IMPLEMENTATION OF A CLINICAL TOXICOLOGY LABORATORY

LABORATORY FACILITIES[7]

Since it is impossible to anticipate every special laboratory need, only general guidelines can be considered here. The *laboratory space* should be designed for "wet" chemical activities. This means that utilities such as water, distilled (deionized) water, gas, compressed air, and electricity must be available in plenteous supply. Laboratory areas previously used for hematology, microbiology, or automated chemistry are generally not appropriate. Utility outlets should be available every 6–8 feet of bench space, and separate electrical circuits should be provided for each important laboratory instrument.

Solvent extractions, evaporations, and thin-layer chromatographic spraying operations require an efficient fume hood. If tissue digestions prior to metal assays are done, a second fume hood should be available. These operations are preferably carried out in a "wet" chemistry section away from expensive instruments. An explosion-proof centrifuge is desirable for centrifuging flammable solvent extracts. Refrigerators designed for storing flammable materials should be available as well as safety cabinets for storing solvents.

Laboratory benches should provide for both "stand up" and "sit down" work. Most instruments can be operated more comfortably and thin-layer plates can be spotted easily at a low bench. Higher benches are generally more convenient for general, wet chemistry operations, such as making up reagents and solutions.

Since drugs must be kept available for references and standards, the laboratory director should acquire state and federal licenses permitting the use of scheduled drugs for analytical purposes. This is a simple and inexpensive procedure but is contingent on space for storing drugs securely and on keeping good records.

Finally, space should be available for filing literature reprints and references, as well as shelving for books. The staff of an active clinical toxicology laboratory will find these resources extremely valuable and will supplement them as the scope of services increases. Sharing of these resources with all three components of the clinical toxicology team is not only essential but also a convenient means by which clinical interaction can be maintained between them.

Instrument requirements for a clinical toxicology laboratory vary widely depending on workload and specific needs. Minimally, the laboratory should have the capability of conducting simple screening tests including thin-layer chromatography. In addition, an ultraviolet spectrophotometer capable of recording continuous spectra between 200 and 700 nm should be available. It is desirable to acquire at least one good gas chromatograph. If budget limitations restrict the laboratory to a single instrument and the number of assays is not great but may be varied, a dual channel, temperature-programmed instrument with flame ionization and electron capture detectors would be appropriate. As with all instruments, the type and complexity to be considered depend upon the skills of the operator and the anticipated workload. For example, a large number of assays for a single substance can best be handled by low cost, dedicated instruments. If a large number of assays are to be performed and a number of technicians with limited skills will be running tests, a microprocessor-controlled instrument equipped with automatic sampler, electronic integrator-calculator, and other automated features is recommended. Additional instrumentation can then be added as the needs develop.

A well-equipped clinical toxicology laboratory should also have available a fluorometer, an atomic absorption spectrophotometer or anodic stripping voltammeter for metal assays, and high performance liquid chromatography and immunoassay facilities. Acquisition of a gas chromatograph–mass spectrometer is desirable only if adequately trained personnel are available.

Purchasers of instruments should be mindful that all instruments require servicing. Many well-designed instruments are available from reliable companies. However, not all companies offer rapid and reliable service to all geographic areas. Quality of service and response time should be a prime consideration and stated as a required specification in purchasing agreements. Additionally, service contracts are well worth the investment if adequately trained individuals are not available on the laboratory staff for troubleshooting and minor repair. Finally, an instrument maintenance program must be established and sustained. Preventive maintenance and monitoring of temperature, gas flow, wavelength calibration, and other factors ensure that instrumental results are valid, and that developing problems are generally identified before they become serious. This is particularly valuable in laboratories with multiple users of instruments.

PERSONNEL REQUIREMENTS

As with any clinical laboratory, the effectiveness of the clinical toxicology component depends greatly on the quality of the personnel. Laboratory personnel must frequently advise the clinician not only on what the laboratory can do but on what it cannot do. Thus, toxicology laboratory personnel are in continuous dialogue with clinicians. In order to communicate effectively, personnel must have training not only in analytical methodology but also in broad areas of toxicology and pharmacology. This will enable the laboratorian to interpret results, plan an analytical approach based on clinical information available, and refer the clinician to appropriate sources of information relating to toxic effects and treatment.

Lack of trained personnel has continued to be a problem. Since demand for training of analytical toxicologists continues to increase, this shortage should be alleviated. Frequently, however, the clinical toxicology laboratory director must recruit young, bright, and trainable individuals from a variety of backgrounds and rely on a tutorial style of teaching and training while they gain experience on the job.

REPORTING RESULTS

The most elegant or difficult of analytical procedures is of little value if the results do not reach the clinician or if they are misinterpreted. Laboratory results should never be interpreted solely on the magnitude of a number generated by the test. The laboratory result has significance only when interpreted in the context of the clinical state of the patient at the time the specimen was drawn as well as a great deal of other information which may not be readily available to laboratory personnel. Certain unusual results should be promptly reported to the physician, but it is extremely hazardous to state unequivocally that a given concentration is "toxic" or "lethal" without knowing the context in which the report should be interpreted. Such factors as age, sex, state of health (particularly relative to diseases of the kidney or liver), degree of tolerance, and the presence of other drugs can all be factors which should be considered in interpreting test results.

The starting point for such interpretations is the laboratory result; however, other factors relating to the analytical procedure can influence interpretation. The sensitivity of the procedure is one such factor. For example, many procedures will detect overdose quantities of drugs in serum but may not be reliable for measuring therapeutic levels. The sensitivity limits of the test should also appear on the report to indicate to the clinician what a negative result might mean. For example, if a particular method used is known to detect 1 μg substance/mL of specimen but only 0.5 mL of specimen were available for this particular test, a concentration of 1 μg/mL may well have been overlooked.

Screening tests may also lead to some confusion. By their very nature, screening tests are designed to rule in or rule out substances in a qualitative manner. Frequently, screening tests are not highly specific; a "positive" result may be the result of interferences. On the other hand, a negative result may effectively rule out a toxic agent or group of substances that no longer need to be considered an etiological factor. If a drug screen of urine is reported as negative, one must be aware that the clinician must know what drugs were tested for in order to appreciate what drugs were absent in that particular specimen. It is misleading to imply in the report that the specimen analyzed was negative for all possible drugs when only a limited number were actually tested for. If a standard array of drugs is routinely included in the urine drug screen, those drugs must be indicated on the report so that the clinician understands which drugs have been excluded by the test used.

Drug nomenclature adds to the confusion. Trade names are used so commonly that a laboratory requisition form may use the trade name of the drug to be tested for and thereby appear to be asking the laboratory to identify not only the drug but also the manufacturer.

Generic names should be used as much as possible unless, for clarity, a trade name may be appropriate.

The problem of adopting a standard unit of measurement for drug levels is far from resolved. It would appear that currently the most popular units of measurement of drugs are in $\mu g/mL$ or mg/L. The latter is particularly useful in therapeutic drug monitoring or in other applications in which pharmacokinetic calculations are required. Some laboratories express the results in parts per million (ppm), but technically this is a weight/weight measurement, that is, weight of drug per weight of specimen. However, any report form can incorporate conversion factors of various commonly used mass/volume units so that users of laboratory results can make the necessary conversions for their own purposes.

SCREENING PROCEDURES

Prime requirements for screening tests are selectivity, sensitivity, and rapid turnaround time. Tests described in this section require the minimum of sophisticated equipment and can usually give at least presumptive evidence of exposure to toxic substances. Two points must always be borne in mind. First, false negative results are not tolerable; they could lead to detrimental or suboptimal treatment of the poisoned patient. Second, and for the same reason, positive results of screening tests are generally presumptive only and must be supported by the use of specific (and ideally independent) confirmatory tests.

The clinician is initially very preoccupied with maintaining or restoring the patient's vital functions, that is to say, maintenance of respiration, cardiac output, and renal function. Upon stabilization of these functions, the clinician turns his undivided attention toward further measures to counteract the poisoning. The antonymic pair, poison/antidote, seems to be etched indelibly into man's memory as something analogous to the one reaction, one enzyme concept; for every poison, says the maxim, there is a specific antidote. However, this is no longer true. In fact, the current armamentarium of specific antidotes is very meager; fewer than a dozen are employed.[16] Clearly, the benefit to the patient can be great if a clinician can be quickly informed, preferably by laboratory confirmation, that a substance is involved which can be treated with a specific antidote. Rapid and reliable screening procedures serve, in part, as initial steps that bioanalysts can take to aid in those situations in which rapid diagnosis is essential.

Many simple reliable tests for commonly encountered drugs and toxic chemicals can be performed readily by colorimetry.[15] If the laboratory has even limited ultraviolet spectrophotometric capability, the range of toxicological tests that can be performed is greatly expanded.[57] Thin-layer chromatography (TLC) is cheap; less than $200 is required to set up this technique. TLC can be performed by minimally trained personnel, and an enormous amount of literature on this subject has developed.[48] Rapid noninstrumental tests, commonly referred to as "spot tests," are very inexpensive. Simplicity is the spot test's big advantage, since requisite technician training is minimal. Spot tests can also be performed at or near the bedside.[17]

Certainly, the most available and least expensive tests are those using the analyst's eyes and nose as detectors.[41] Stains of the skin, such as the bleached white coloration in phenol poisoning, the cherry pink color of carbon monoxide intoxication, and the characteristic yellows of Atabrine and picric acid toxicity, are frequently discernible. Characteristic odors include the almond odor of cyanides, the pear-like smell of chloral hydrate, and the garlic odor of organophosphate intoxication. The foregoing considerations indicate that, in terms of both personnel and equipment, availability and cost are largely determined by the budget of the medical treatment facility. Even with limited resources, however, some degree of laboratory support can be furnished for the clinician.[47]

SPOT TESTS

Spot tests are analytical procedures that are simple and rapid and make use of the analyst's eye as the instrument of detection. In the early days of analytical toxicology, these were the only procedures available. Even today, when highly sophisticated analytical instrumentation is available, judicious use of spot tests can be highly useful in screening for toxic substances. Although most spot tests suffer from lack of sensitivity and/or specificity, they can often be of considerable aid in the laboratory diagnosis of poisoning, by virtue of their rapidity.

A number of chemical spot tests are applicable to bedside or near bedside rapid screening for poisoning. Some of these are quantitative, some are semiquantitative, and others are strictly qualitative. When used with discretion, they can be very helpful, particularly when adequate diagnostic laboratory facilities are not immediately available.

Decker and Treuting[17] reviewed spot tests available in the literature, and selected a number of these which were found to be reliable when used in their laboratories. Some have been slightly modified by these authors; they are listed here in alphabetic order.

Acetaminophen. See *p-Aminophenol.*

Acetone (for detecting uncontrolled diabetes or the metabolite of isopropyl alcohol poisoning). Place 1 drop of urine or serum on an Acetest tablet (Ames Laboratories). Thirty seconds later, scrape the top of the tablet; a purple color indicates the presence of acetone. Reagent test strips (e.g., Ketostix) sensitive to ketones may be more convenient. A violet color indicates acetone. The test result can be made semiquantitative by using the color chart provided for comparison. A commercial urine or serum control containing an increased concentration of acetone should be used in parallel with this test.

Alcohols. See *Volatile reducing substances.*

Aldehydes. See *Volatile reducing substances.*

p-Aminophenol (metabolite of phenacetin or acetaminophen). Add 3 drops HCl, 1.0 mol/L, to 1 mL urine; place in ice bath. When cold, add 3 drops freshly prepared sodium nitrite solution, 0.14 mol/L, and 3 drops α-naphthol reagent (100 mg α-napthol in 5 mL NaOH, 2.0 mol/L). A red color (semiquantitative) is indicative of *p*-aminophenol.

Amphetamine. See *Sympathomimetic amines.*

Antimony. See *Heavy metals.*

Arsenic. See *Heavy metals.*

Barbiturates, glutethimide, ethchlorvynol. Acidify 1 mL serum, urine, or filtered stomach contents or washings with HCl, 1.0 mol/L. Extract with 5 mL chloroform. Evaporate the chloroform layer in a 100 °C water bath. Take up the residue in a few drops of chloroform and spot on a piece of filter paper. Allow the chloroform to evaporate and add a few drops of aqueous mercurous nitrate, 0.02 mol/L. A dark blue to purple color is indicative of a number of acid hypnotic drugs, including barbiturates, glutethimide, and ethchlorvynol.

Bromate. See *Oxidizing agents.*

Carbamates. See *Meprobamate.*

Carbon monoxide. Add 20 mL distilled water to 0.5 mL whole blood and add 1 mL NaOH, 1.0 mol/L, and mix. Blood containing over 20% carbon monoxide hemoglobin (HbCO) will have a light cherry-red color; normal blood will turn brown.

Carbon tetrachloride. See *Halogenated hydrocarbons.*

Chloral hydrate. See *Halogenated hydrocarbons* and *Volatile reducing substances.*

Chlorate. See *Oxidizing agents.*

Chloroform. See *Halogenated hydrocarbons.*

Cholinesterase (activity is depressed by organophosphorus insecticides). Make up a fresh aqueous solution of bromthymol blue, 0.008 mol/L, containing 100 mg/mL acetylcholine bromide. Dip a piece of filter paper in this solution and air dry. Apply two drops of any pH 5 buffer (control) to one piece of impregnated paper; to another apply two drops of cell-free plasma (unknown). The color of the control paper will change from blue to yellow in ~5 min at room temperature. If the unknown paper takes significantly longer to change color (10 min or more), the cholinesterase activity is depressed; the degree of depression is roughly proportional to time.

Codeine. See *Morphine.*

Desipramine. See *Imipramine.*

Ethchlorvynol. See *Barbiturates.*

Ethylene glycol. See *Oxalic acid.*

Glucose (to rule out drug-induced symptomatology or to implicate insulin misadventure). Add 1 drop of fresh whole blood to a Dextrostix reagent strip (Ames Company). After exactly 60 s, wash blood completely off the strip. Compare the developed color with the color chart provided to estimate the glucose concentration.

Glutethimide. See *Barbiturates.*

Halogenated hydrocarbons (chloroform, chloral hydrate, carbon tetrachloride). Extract 1 mL blood with 2 mL ether. Pipet off 1 mL of the ether layer into a test tube; add to it 1 mL pyridine and 1 mL NaOH, 5.0 mol/L. Place the tube in a boiling water bath for 1 min. A red color in the pyridine is positive for halogenated hydrocarbons. Urine (1 mL) may be used without extraction. A blank should also be run.

Note: Not all halogenated hydrocarbons will react positively with this test.

Heavy metals (Reinsch test for antimony, arsenic, bismuth, and mercury). Clean a 10-cm length of heavy-gauge copper wire with sandpaper or emery cloth. Without touching the wire with bare fingers, coil two thirds of it in a loose spiral. To 5–10 mL stomach contents, or urine, add an equal volume of HCl, 2.0 mol/L, and place the copper spiral into the solution. Place in a boiling water bath for 10 min; examine for a stain on the copper. If no stain appears in 1 h, the test may be considered negative. Sensitivity is ∼5 μg.

> Antimony: blue or purple black
> Arsenic: dull black
> Bismuth: shiny black
> Mercury: silver gray

Heroin. See *Morphine.*

Imipramine and desipramine. Add 1 mL urine to 1 mL of reagent consisting of potassium dichromate, 0.007 mol/L; H_2SO_4, 5.4 mol/L; perchloric acid, 2.0 mol/L; and nitric acid, 5.0 mol/L, at a ratio of 1:1:1:1. A green color indicates imipramine; blue indicates desipramine.

Insulin. See *Glucose.*

Iron. To 1 mL filtered stomach contents or washings, add 1 drop concentrated HCl and 0.5 mL potassium ferricyanide, 0.03 mol/L. A blue color is indicative of ferrous iron. To another 1-mL aliquot, add 1 drop concentrated HCl and 0.5 mL potassium ferrocyanide, 0.027 mol/L. A blue color is indicative of ferric ion.

Isopropyl alcohol. See *Acetone.*

Meprobamate. Extract 2 mL serum with 5 mL ether. Pipet off 3 mL of the ether layer and evaporate in a 50-mL beaker. Add 3 mL of a reagent containing 2 g hydroquinone in 100 mL of H_2SO_4:H_2O at a ratio of 15:85. Heat for 25 min at 100 °C. Under ultraviolet illumination (preferably under a Wood's light, 365 nm), a rose fluorescence is positive for meprobamate.

Extract another sample and evaporate the extract as above. Take up residue in 3 drops ether; place on a filter paper. Add 1 drop 10% furfural in ethanol. After the ethanol evaporates, expose the paper to fumes from concentrated HCl. A deep purple color is positive for meprobamate.

Other carbamates, including insecticides such as *N*-methyl carbamate (Carbaryl, Sevin), will also react if present in sufficiently high concentration. Filtered stomach contents or washings are preferred over serum, when these compounds are suspected.

Mercury. See *Heavy metals.*

Methamphetamine. See *Sympathomimetic amines.*

Methanol. 1 mL protein-free filtrate of serum is added to 0.1 mL $KMnO_4$ solution (5 g in 100 mL H_2O). Swirl the test tube gently, and after 5 min add sufficient powdered sodium bisulfite to decolorize the permanganate. Add 0.2 mL freshly prepared chromotropic acid solution (0.5 g in 100 mL H_2O) and 6 mL concentrated H_2SO_4. Mix and heat in a boiling water bath for 5 min. A red-violet color is positive and specific for methanol. Quantitation can be accomplished by colorimetry at 570 nm.

Morphine. Adjust 20 mL urine to pH 9.0–9.5 with NH_4OH, and extract with 10 mL chloroform:ethyl acetate (1:4). Evaporate half of the extract to dryness, add 0.5 mL concentrated HCl and 1 drop concentrated H_2SO_4, and place test tube in a boiling water bath. A reddish-purple color is indicative of morphine. After all HCl is evaporated, add 0.5 mL fresh HCl and neutralize with solid Na_2CO_3. Morphine is indicated by a violet color that changes to green upon dropwise addition of a 57% HI solution to which 0.5 g I_2 in 100 mL H_2O has been added.

Evaporate the other half of the chloroform:ethyl acetate extract to dryness; acidify with 0.5 mL HCl, 2.0 mol/L. Add 0.5 mL Lugol's solution diluted 1:100 with water. The slow formation of reddish-brown crystals is indicative of morphine.

Note: Several other opium alkaloids and synthetic opioids will also react positively. Also, the presence of morphine in urine results from the biotransformation of codeine or heroin.

Nitrate. See *Oxidizing agents.*

Organophosphorus insecticides. See *Cholinesterase.*

Oxalic acid, oxalates (metabolite of ethylene glycol). Acidify urine or filtered stomach contents or washings to pH 1.5 with concentrated HCl, filter, adjust the pH of a 5-mL aliquot to approximately 8 with NH$_4$OH, and acidify with acetic acid. Add saturated CaCl$_2$ solution dropwise. A precipitate is indicative of oxalate. To confirm, add concentrated HCl dropwise. If the precipitate dissolves, oxalate is very probably present. Warm the mixture and add KMnO$_4$ solution (0.1 g in 100 mL H$_2$O) dropwise; the violet color will disappear if oxalate is present.

Oxidizing agents (bromate, chlorate, nitrate, nitrite). To 1 mL filtered stomach contents or washings, add 2 mL of a solution of diphenylamine in concentrated H$_2$SO$_4$ (0.5 g in 100 mL). A blue color is indicative of bromate, chlorate, or nitrate. The latter ion may be a result of ingestion of nitrite. To 2 mL whole blood, add 1 mL HCl, 2.0 mol/L, centrifuge, and decant. Add 0.1 mL Evans blue solution (0.01 g in 100 mL H$_2$O) to the supernatant; decolorization of the dye indicates the presence of one or more of these ions.

Paraquat. To 5 mL filtered stomach contents or urine, add 0.1 g sodium bicarbonate and 0.1 g sodium dithionite. The resulting blue color can be quantitated by colorimetry at 625 nm.

Phenacetin. See *p-Aminophenol.*

Phenols (cresols, naphthols, thymols). Extract 2 mL stomach contents with 10 mL ether. Evaporate the ether layer at room temperature in a small white crucible. Add 1 drop freshly prepared sodium nitrite, 0.14 mol/L, in concentrated H$_2$SO$_4$, 1 drop water, and a NaOH or KOH pellet in that order. Positive reactions are as follows:

Cresols:	dark brown
Naphthols:	green
Phenols:	reddish-green to blue
Thymols:	green to purple

Phenothiazine antipsychotic drugs. See also *Salicylates.* Add 1 mL urine to 1 mL FPN reagent (ferric chloride solution, 5 g in 100 mL H$_2$O; perchloric acid, 2.0 mol/L; nitric acid, 5.0 mol/L, 1:9:10 mixture). Pink to violet color indicates a phenothiazine; the color depends upon dose and metabolites (the latter are predominantly pink). Shake 5 mL urine with 10 mL chloroform; a red metabolite (if present) is extracted. To the chloroform layer, add 5 mL HCl, 2.0 mol/L, and a small piece of tin; a colorless solution will then result. Filter the chloroform layer and add 1 mL 30% H$_2$O$_2$; a red color indicates a positive reaction.

Salicylates (and phenothiazines). Place 1 drop urine or serum on a Phenistix reagent stick (Ames Company). A violet color indicates a salicylate. The test is semiquantitative using the color chart provided.

Note: Phenothiazine antipsychotic drugs also give a violet color that will persist after H$_2$SO$_4$:H$_2$O (1:1) is added; the color due to salicylates will disappear.

Add 1 mL urine or serum to 1 mL ferric chloride solution (5 g in 100 mL H$_2$O). A violet color indicates a salicylate.

Strychnine. To 5 mL filtered stomach contents or washings add 1 mL concentrated HCl and a small piece of metallic zinc. Boil the mixture for 1 min, cool, and filter. Hold a test tube containing 2 mL concentrated H$_2$SO$_4$ at approximately a 45° angle, and allow 2 mL filtrate to run down the side very slowly. A rose-red ring at the interface of the two liquids indicates strychnine.

Sulfonamides (and other compounds containing a primary aromatic amino group such as a *p*-aminophenol and *p*-aminosalicylate). Place 0.5 mL urine or filtered stomach contents on a piece of newspaper. Add 1 drop concentrated HCl. An intense yellow to orange color is a positive result. Positive and negative control urine specimens must also be carried through the procedure.

Sympathomimetic amines (amphetamine, ephedrine, hydroxyamphetamine, methamphetamine, phenylephrine, synephrine). To 10 mL urine or filtered stomach contents or washings, add sufficient NaOH, 2.0 mol/L, to raise the pH to ~8. Extract with 10 mL chloroform. To the chloroform layer add 2 mL HCl, 5.0 mol/L, and extract. Remove 1 mL of the aqueous layer and add 0.5 mL Sanchez reagent (100 mg dimethylaminobenzaldehyde in 20 mL ethanol acidified with 4 drops concentrated H$_2$SO$_4$). Evaporate to dryness. Note the color of the residue. Dissolve the residue in water and again note the color.

Compound	Residue	Solution
Amphetamine	Yellow-green	Yellow-green
Ephedrine	Dark green	Pale green
Hydroxyamphetamine	Red-brown	Yellow-brown
Methamphetamine	Dark green	Pale green
Phenylephrine	Orange	Pale orange
Synephrine	Yellow-brown	Dark yellow

Note: Certain opium alkaloids will yield a red or violet residue; aqueous solution of the residue will be colorless or pale yellow.

Thallium. To 1 mL urine or filtered stomach contents or washings, add 2 drops concentrated HCl and 5 drops bromine water. Five minutes later, add 5 drops sulfosalicylic acid solution (20 g in 100 mL H_2O), 0.5 mL toluene, and 5 drops aqueous methyl violet solution (0.1 g in 100 mL H_2O) and shake. A blue to blue-green color in the toluene layer is indicative of thallium.

Volatile reducing substances (alcohol, aldehydes, chloral hydrate). Place 1 mL blood, urine, or stomach contents or washings in a small screwcapped test tube. Wipe the sides of the tube dry. Place a loose plug of glass wool about halfway down the tube. Moisten the wool with a drop of reagent (0.5 g potassium dichromate in 100 mL H_2SO_4, 21.6 mol/L). Screw on the cap tightly and immerse the bottom of the tube in hot (almost boiling) water. The reagent will change color from orange to green in less than 10 min if a significant amount of volatile reducing substance is present.

Other lists of spot tests have been compiled by Sunshine,[63] Curry,[15] and Kaye.[41]

Comments

The results of spot tests must be evaluated with a considerable amount of discretion. As a rule, false positives occur much more frequently than do false negatives, since most spot tests lack absolute specificity. On the other hand, a false negative may result from a number of conditions, including low sensitivity of the method used, binding of the agent or its metabolites to protein or other high molecular weight substances, complexation of the agent or its metabolites with highly adsorptive or reactive compounds, formation of metabolites which do not react the same as the parent compound, volatilization of or chemical change in the agent during processing, masking of color by other substances present, outdated or unstable reagents, and inefficient solvent extraction of the compound. Some of these pitfalls can be overcome by carrying positive and negative controls through the test.

Thin-Layer Chromatography (TLC)

A rapid TLC screening method for the detection of a significant number of commonly abused drugs in urine has been presented by Wahl and Rejent.[70] Blood may also be assayed by this method, but urine usually contains a higher concentration of most drugs, and it is therefore the specimen of choice. The method has been modified slightly in the authors' laboratories.

Materials and Reagents

1. TLC plates. Soft-surfaced plates generally give the best resolution in the system; a plate such as the Analtech Silica Gel GF is recommended. Many analysts prefer to use channeled or preadsorbent plates or both. Score the plates horizontally 10 cm above the spotting line to provide a uniform front and to simplify calculation of R_f values.

2. Standards. A composite aqueous standard consists of amphetamine, codeine, meperidine, methadone, methaqualone, morphine, phenobarbital, quinine, and propoxyphene, each in a concentration of 3 mg/L. Prepare fresh monthly; store refrigerated in a brown bottle.

3. Extraction reagents:

   a. *Buffer.* Adjust saturated ammonium chloride solution to pH 9.2 with concentrated ammonium hydroxide solution.

   b. *Solvent.* Mix chloroform:isopropanol in a ratio of 96:4. *Caution*: Mix and use in a fume hood. Chloroform is a demonstrated carcinogen in animals.

4. Developing solvent mixture:

a. Mix ethyl acetate:methanol:concentrated NH_4OH:water in a ratio of 150:18:0.5:10. Keep tightly stoppered.

b. Replace solvent after 2 plates have been developed in the tank, or when R_f values are not reproducible due to loss of ammonia.

5. Spray reagents:

a. Fluorescamine (Fluram, Roche), 20 mg in 100 mL dry acetone. Keep refrigerated. Prepare fresh monthly.

b. Ninhydrin, 100 mg in 100 mL acetone. Keep refrigerated. Prepare fresh monthly.

c. Iodoplatinate, 5 mL of platinic chloride, 5.0 g/dL, and 45 mL of potassium iodide (10 g/dL) added to 100 mL deionized water. Stable for 6 months.

d. Mercuric sulfate. Dissolve 2 g mercuric oxide in 20 mL concentrated sulfuric acid; dilute to 100 mL with deionized water. Stable indefinitely.

e. Diphenylcarbazone, 100 mg in 100 mL of a 50:50 acetone-chloroform solution. *Caution*: Use fume hood; keep refrigerated; prepare fresh monthly.

f. Dragendorff reagent:

(1) Dissolve 2.2 g bismuth subnitrate in a solution of glacial acetic acid:water, 25:100.

(2) Dissolve 50 g potassium iodide in 125 mL water.

(3) Combine 10 mL each of solution (1) and (2) with 20 mL glacial acetic acid and 100 mL water. Stable for one year if kept tightly stoppered in the dark.

Procedure

1. To ~20 mL urine, add 1.5 mL pH 9.2 buffer, mix, and centrifuge. Extract the clear supernatant liquid with 30 mL extraction solvent by shaking for 10 min. Centrifuge and place the lower (solvent) layer into a 100-mL beaker. Evaporate until the solvent is about half evaporated, add 1 drop concentrated HCl (to reduce volatilization of amines such as amphetamine), and continue evaporation to dryness. Carry 1 mL of the standard plus 19 mL water through the extraction and evaporation procedure.

2. Take up the residue in about 20 μL methanol; spot the entire solution 2 cm above the bottom edge of the TLC plate (or in a preabsorbent layer). Do the same with the extract residue of the standard.

3. Immerse the plate in the solvent tank; develop to the 10-cm scored line. Allow the solvent to equilibrate at this line for 10–15 min. Remove the plate from tank and air-dry in the fume hood.

4. Examine the plate under short and long wavelength ultraviolet light (e.g., in a Chromatovue cabinet); several drugs such as quinine or quinidine will exhibit characteristic fluorescence. Others will quench fluorescence if a plate incorporating an inorganic fluorophore is used. With a sharp pencil, outline the spot(s) observed and calculate the R_f value(s) in this and the following steps.

5. Spray the plate with fluorescamine and re-examine under long wavelength ultraviolet light. This will reveal drugs containing primary amine groups, such as amphetamine. Perform this and all subsequent sequential sprays in a fume hood.

6. Spray with ninhydrin; heat the plate in an oven for 10 min at 85 °C, then activate the plate for 3 min under simultaneous long and short wave ultraviolet light in an ultraviolet view box (e.g., a Chromatovue). Primary and secondary amines are seen as pink to pinkish-tan spots (the latter more characteristic of methamphetamine).

7. Spray with mercuric sulfate. Barbiturates, glutethimide, and phenytoin appear as dry or white areas; in this system, all three drugs will have approximately the same R_f values.

8. Spray with diphenylcarbazone for further confirmation of these drugs. Gluthethimide will be blue, for example, and amobarbital will be red on a violet background.

9. Spray the same plate with iodoplatinate; alkaloids particularly, but also many other nitrogen-containing compounds, become evident. Many drugs will have characteristic colors with this reagent.[12]

10. Spray the same plate with Dragendorff reagent; methaqualone is detected as a bright orange spot; benzodiazepines, such as chlordiazepoxide, diazepam, and flurazepam, will appear as less bright spots. Previously visualized spots will usually turn tan or brown.

Interpretation and Comments

Approximate R_f values of a number of commonly encountered drugs appear in Table 18-1. The table is intended to be only a guide, since it is based on results obtained with Analtech GF plates as reported by Wahl and Rejent.[70] It is highly recommended that each laboratory calculate its own R_f values and include results from additional drug standards.

It is important that all positive results be confirmed by at least one independent method of analysis.

Additional thin-layer chromatographic procedures have been described by Sunshine,[64] Bastos et al.,[4] Forney,[25] and Smith and Stewart,[58] among others.

Gas-Liquid Chromatographic (GLC) Screening Procedure for Basic Drugs

Garriott[29] developed a GLC-based screening method for basic drugs. It has been modified to include the detection of acidic and neutral drugs.

Apparatus

1. Syringes, 10 μL, Hamilton Microliter 701, or equivalent.
2. Gas chromatograph equipped with dual columns, flame ionization detectors, linear temperature programmer, and recorder.
3. Columns, 2 m × 2 mm I.D. packed with:
   a. 3% OV-1 on 80/100 mesh Chromosorb W, A/W DMCS.
   b. 3% OV-17 on 80/100 mesh Chromosorb W, A/W DMCS.
   c. 5% Apiezon L, 5% KOH on 80/100 mesh Chromosorb G, A/W DMCS.
   d. 2.5% SE-30 on 80/100 mesh Chromosorb G, A/W DMCS. Column (a) could be substituted for this column.

Chromatographic Conditions

1. Gas flow rates: nitrogen (carrier gas), 30 mL/min; hydrogen, 30 mL/min; air, 240 mL/min.
2. Column temperature:
   Columns 3a and 3b. Program at 8 °C/min from 174 to 270 °C; hold final temperature for 8 min.
   Column 3c. Program at 8 °C/min from 110 to 190 °C; hold initial temperature for 2 min and final temperature for 4 min.
   Column 3d. Program at 8 °C/min from 130 to 190 °C; hold initial and final temperature for 4 min.

Reagents

1. Methanol, GLC Grade.
2. Drug stock solution, 1 mg/mL in methanol. Transfer 25 mg of drug as the free acid or

Table 18-1. R_f VALUES FOR SOME COMMONLY ENCOUNTERED DRUGS ON ANALTECH GF TLC PLATES[70]

Drug	R_f Value
Morphine	0.14
Codeine	0.22
Amphetamine	0.99
Quinine	0.38
Meperidine	0.53
Amitriptyline	0.63
Methadone	0.67
Pentazocine	0.73
Phencyclidine	0.79
Propoxyphene	0.81
Xylocaine	0.82
Phenobarbital	0.77
Methaqualone	0.87

base to a 25-mL volumetric flask and fill to volume with methanol. Some drugs may require a few drops of sodium hydroxide or hydrochloric acid to go into solution. Place the solution into a screw-top test tube and store away from light. Methanol solutions of drugs have been found to be more stable than those in aqueous solutions and were stable at room temperature for as long as three months.

3. Cholestane internal standard solution, 0.05 mg/mL. Dissolve 25 mg cholestane in 25 mL methanol; dilute 0.5 mL of this stock standard to 10 mL with methanol to make a working standard.

4. Acetic anhydride, ACS grade.

Procedure

1. Dissolve the residue from each extract in 20 µL of internal cholestane standard (reagent 3 above).

2. Into columns 3a and 3c inject 3 µL of the solution from step 1. Take up 2 µL of acetic anhydride into the syringe along with 3 µL of the solution from step 1 and inject the entire 5 µL into column 3d. Positive results on column 3a are verified on column 3b. Positive results on column 3c are verified on column 3d as acetylated derivatives.

3. Determine retention times of peaks; compare to table of retention times previously determined.

Comments

Table 18-2 illustrates some typical retention times for a number of drugs.[29] Each laboratory should prepare its own list of retention times using drug standards on each column.

Table 18-2. RETENTION TIMES OF SEVERAL DRUGS
OBSERVED ON COLUMNS a AND b[29]

Drugs	Column a (Minutes)	Column b (Minutes)
Nicotine	1.0	1.5
Phenmetrazine	1.3	2.2
Methyprylon	1.7	3.4
Meperidine	3.0	4.8
Benzphetamine	3.6	5.6
Diphenhydramine	3.9	6.3
Lidocaine	4.1	6.6
Methapyrilene	5.0	8.0
Chlorpheniramine	5.2	8.0
Methadone metabolite	5.3	8.2
Propoxyphene metabolite (early)	6.3	—
Methadone	6.6	9.4
Methaqualone	6.6	11.0
Propoxyphene	7.0	9.8
Amitriptyline	7.0	10.0
Nortriptyline	7.0	10.6
Imipramine	7.3	10.5
Desipramine	7.3	11.1
Doxepin	7.3	11.2
Trihexylphenidyl	7.5	10.6
Pentazocine	7.8	11.1
Promethazine	7.8	11.4
Oxazepam	8.4	12.9
Codeine	8.8	13.4
Propoxyphene metabolites (doublet)	8.8	—
	9.0	12.9
Methaqualone metabolite	9.0	14.6
Diazepam	9.3	14.5
Diazepam metabolite	10.0	16.5
Propoxyphene metabolite (late)	10.4	15.6
Oxycodone	10.0	16.4
Promethazine metabolite	11.2	> 20.0
Flurazepam	12.3	19.9
Cholestane (internal standard)	12.9	15.1

A single-column, temperature-programmed screening approach has been established by Peel and Perrigo.[51] This technique can be extremely useful when a laboratory has a limited capability in gas-liquid chromatography equipment.

Finkle et al.[21] developed a highly comprehensive GLC-based system, using four columns and three liquid phases. This article also lists relative retention data for almost 600 toxic substances and human metabolites.

Screening Test for Metals

Few simple, sensitive, and comprehensive methods exist for screening of toxic metals in clinical specimens. The Reinsch test for detection of antimony, arsenic, bismuth, and mercury has been described in the section on Spot Tests, as has a test for thallium.

A differential procedure for six metals (bismuth, copper, gold, lead, mercury, and zinc) is based upon the fact that ions of each of these metals will result in a specific color when complexed with dithizone (diphenylthiocarbazone) and adjusted to a specific pH.[41] Table 18-3 indicates these colors. This procedure is quite tedious; steps must be taken to remove contaminant metals from most of the reagents employed.

Analyses based on thin-layer chromatography[73] are more convenient. Detection limits are for the most part high (5–100 μg); thus, a 24-h urine collection is generally the specimen of choice. A typical procedure is described below.

Materials, Reagents, and Methods

1. Thin-layer plates with carboxymethyl cellulose adsorbent.
2. Solvent systems:
    a. Sodium acetate, 1.0 mol/L: acetic acid, 1.0 mol/L (50:50).
    b. Sodium acetate, 0.5 mol/L: acetic acid, 0.5 mol/L (50:50).
3. Detection reagents:

Alizarin
    a. Alizarin, saturated solution in ethanol.
    b. Sodium hydroxide, 1 mol/L.
    c. Glacial acetic acid.

Cinchonine–potassium iodide
    Dissolve 1 g cinchonine in hot water and add a few drops of concentrated nitric acid. Cool and add 2 g potassium iodide. Prepare fresh.

Diphenylcarbazide
    a. Diphenylcarbazide, 100 mg in 100 mL ethanol, 95%.
    b. 25 parts ammonium hydroxide: 75 parts H$_2$O.

Dithizone
    a. Dithizone (diphenylthiocarbazone), 100 mg/dL. Dissolve 100 mg dithizone in 100 mL chloroform.
    b. 25 parts ammonium hydroxide: 75 parts H$_2$O.

8-hydroxyquinoline
    a. 8-Hydroxyquinoline. Dissolve 0.5 g in 60 mL absolute ethanol; when dissolved, add 40 mL deionized water.
    b. 25 parts ammonium hydroxide: 75 parts H$_2$O.

Table 18-3. COLORS OF SELECTED METAL-DITHIZONE COMPLEXES[41]

Metal Ion	Dithizone Complex Color	When Adjusted to a pH of
Bismuth	Cherry-red	4.0
Copper	Red	4.0
Gold	Rose-purple	2.0
Lead	Cherry-red	8.0
Mercury	Orange-peach	2.0
Zinc	Orange	5.5

Procedure

1. Prepare digest of specimen as described on page 1697. Take up residue in small amount of deionized water or acetic acid, 1 mol/L. Prepare reagent blank and metal ion standards. Prepare negative control from pooled urine specimens. All are carried through the digestion procedure.

2. Spot on 2 plates; develop one in solvent system (a) and the other in solvent system (b) to a distance of 10 cm from the origin. Dry the plates.

3. Spray with appropriate detection reagent(s).

Alizarin. Spray with solution (a), dry briefly, and spray with solution (b); spray with solution (c) to remove background color. Ag, Al, Ba, Be, Bi, Ca, Ce, Cd, Co, Cr, Cu, Fe, Ga, Hg, In, Li, Mg, Mn, NH$_4$, Ni, Pb, Pd, Pt, rare earths, Sc, Se, Th, U, and Zr are detected as red to violet spots.

Cinchonine–potassium iodide. Bi will appear as an orange spot; Ag, Hg, Pb, Sb, Tl, and V will give yellow spots; Cu will result in a brown spot; Pt will yield a pink spot.

Diphenylcarbazide. Spray with solution (a) followed immediately by solution (b). Ni will appear as a blue spot; Co will result in an orange-brown spot; Ag, Cu, Mn, Pb, and Sn will give brown spots; Zn will yield a purple spot.

Dithizone. Spray with solution (a). Ag, Au, Bi, Cd, Ni, and Sn will give red to violet spots; Hg will yield a pink to yellow spot; Cu will result in a brown spot. Spray with solution (b). Pb will appear as a red to violet spot.

8-Hydroxyquinoline. Spray with solution (a) followed by solution (b). Inspect under ultraviolet light. Al, Be, Ca, Ge, Mg, Sn, Sr, and Zn give brightly fluorescing spots; a number of cations appear as dark spots.

Comments

Results can be semiquantitated by noting sizes and relative intensities of spots.

Table 18-4 describes "typical" R$_f$ values for several metal ions. Each analyst should prepare his/her own table prior to analysis of specimens.

Screening Tests for Volatile Substances

A number of toxic substances, such as the lower alcohols, acetaldehyde, formaldehyde, acetone, lower molecular weight hydrocarbons, inhalation anesthetics, freons, and carbon monoxide, are volatile at room or body temperature. Others, such as cyanide or fluoride salts, can be converted to the gaseous state by pH adjustments.

Screening methods commonly used to detect volatiles include gas-liquid chromatography, osmometry, spectrophotometry, and microdiffusion analyses.

Table 18-4. R$_f$ VALUES OF METAL IONS[73]

Metal	Solvent System a	Solvent System b
Ag	0.49	0.36
Al	0.65	0.18
Ba	0.40	0.26
Be	0.79	0.55
Bi	0.84	0.50
Ca	0.57	0.35
Cd	0.61	0.36
Co	0.62	0.38
Cu	0.55	0.28
Ga	0.51	0.09
Hg	0.81	0.76
In	0.91	0.55
Mg	0.78	0.53
Mn	0.58	0.38
Ni	0.61	0.38
Pb	0.37	0.13
Sr	0.55	0.38
Tl	0.57	0.46
Zn	0.62	0.35

A headspace method for volatiles is described (see p. 1693). Here, blood or urine is incubated with internal standard and the gas above the liquid phase (headspace) is sampled and subjected to gas-liquid chromatography.

Direct on-column injection of the blood or urine specimen can also be used,[39] but many analysts find that syringe needle plugging (due to heat precipitation of proteins or other non-volatiles) may occur.

Some volatiles, particularly those of lower vapor pressure, may be better analyzed by extracting the agents with a low-boiling solvent, which is then injected into the gas chromatograph.

Recently, the purge-and-trap technique has been adapted to the screening of volatiles in biological fluids; 42 such substances can be identified.[22] In this procedure, volatiles are purged from the specimen using an inert gas, trapped on an adsorbent column (Tenax), and repurged into the gas chromatograph. This method requires an additional accessory for the gas chromatograph; it has the advantage of enhanced sensitivity because nearly 100% of the volatile component is recovered. A further advantage is the independence of the method from temperature, salting-out, and specimen manipulations.

Osmometry is also useful in detecting overdose by ethanol and other polar, water-soluble substances.[50] In the absence of conditions causing high serum osmolality, such as diabetic acidosis and coma, renal failure, and hypernatremia, a workable approximation of blood ethanol concentration in mg/dL can be determined by subtracting an average "normal" value of 290 mOsm/kg H_2O from the observed osmolality of the serum specimen and multiplying the difference by 4.24.

When overdose of chloral hydrate or ethchlorvynol is suspected, screening for these substances can be accomplished by spectrophotometry.[26,45] These methods can also be useful in confirming the presence of these drugs when first detected by gas-liquid chromatography.

Organochlorine Pesticide Screen[66]

Screening for organochlorine pesticides or environmental pollutants can be done on serum or whole blood. Since many of these substances are persistent and ubiquitous in the environment, it is not unusual to detect low concentrations in serum of normal patients. It is important, therefore, to discriminate between "normal" and elevated concentrations; the following procedure is helpful.

Principle

Blood or serum is extracted under strongly acidic conditions to minimize interferences. The extract is examined by GLC using an electron-capture detector (ECD) and aldrin as a marker.

Reagents

All solvents and reagents should be the purest available and pesticide grade.

1. Sulfuric acid, 60% (w/w). Add CAUTIOUSLY, with stirring, 60.0 g concentrated H_2SO_4 to 40.0 g distilled water.

2. Hexane-acetone (85:15). Add 15.0 mL acetone to 85 mL of hexane.

3. Iso-octane (2,2,4-trimethylpentane), pesticide quality.

4. Pesticide standards. Aldrin, heptachlor epoxide, toxaphene, hexachlorocyclohexane (BHC), p,p'-dichlorodiphenyltrichloroethane (p,p'-DDT), p,p'-dichlorodiphenyldichloroethylene (p,p'-DDE), dieldrin, endrin, heptachlor, methoxychlor, or others, from EPA Repository of Environmental Chemicals, Research Triangle, NC. Each vial contains ~100 mg. Weigh vial and cap, add iso-octane, cap, and carefully mix. Transfer the solvent to a 100-mL volumetric flask. Repeat rinse three times. Dry vial and cap and reweigh. Difference in weight is the amount of pesticide now dissolved in the flask. Make appropriate dilutions as required. Stock solutions (1 mg/mL) are stable in the dark and refrigerated. Dilutions should be made fresh monthly.

5. Aldrin marker. Prepare aldrin solution, 200 ng/mL, in iso-octane.

Chromatograph Conditions

A variety of column packings have been used in 2 m × 2 mm, glass columns: 3% OV-17, 5% OV-210, 3% OV-101, 4% SE 30/6% QF-1. Isothermal conditions: column temperature, 200 °C; injection port, 300 °C; detector, 300 °C; ECD.

Procedure

All clean glassware should be rinsed with hexane prior to use.
1. Weigh 2 g of specimen into a 50-mL extracting tube.
2. Add 500 µL aldrin marker. Vortex 10 s.
3. Add 1.5 mL 60% H_2SO_4. Vortex 10 s.
4. Add additional 1.5 mL 60% H_2SO_4. Vortex 10 s.
5. Add 2.0 mL 60% H_2SO_4. Vortex 30 s.
6. Allow to cool for several minutes.
7. Add 3.0 mL hexane-acetone (85:15) and vortex 30 s.
8. Centrifuge 10 min at 2000 × g.
9. Transfer top solvent layer to a 15-mL centrifuge tube, using disposable pipet.
10. Repeat steps 7 to 9 two additional times to give a total of 9 mL extract.
11. Evaporate combined solvent in centrifuge tube with a tube heater, preferably an evaporative concentrator (K-569300, Kontes, Vineland, NJ) or similar device, to about 0.5 mL. (Do *not* take to dryness.)
12. Inject 2.0 µL of extract into GLC-ECD.

Results

All retention times are relative to aldrin (3.0 min on 3% OV-17).

Compound	Relative Retention Time (RRT)
α-BHC	0.54
Chlordane	0.60
γ-BHC (Lindane)	0.69
β-BHC	0.80
Heptachlor	0.82
Aldrin	1.00
Oxychlordane	1.38
Heptachlor epoxide	1.54
γ-Chlordane	1.69
Nonachlor	1.76
o,p′-DDE	1.82
α-Chlordane	1.86
p,p′-DDE	2.23
Dieldrin	2.40
o,p′-DDD	2.65
Chlordecone	2.77
Endrin	2.93
o,p′-DDT	3.16
p,p′-DDD	3.48
p,p′-DDT	4.18
Mirex	6.1
Methoxychlor	8.1

Interpretation

Aldrin is rarely encountered as such in biological specimens since it is rapidly metabolized to dieldrin. If a large dieldrin peak is present, the assay should be repeated without adding aldrin as a marker.

Aldrin is added to the specimen at a concentration of 50 ppb. For screening purposes, peaks with an RRT corresponding to a specific pesticide but with peak areas less than half of aldrin can be reported as *estimates* (e.g., p,p′-DDT ~20 ppb) and are probably within "normal" limits. (See Table 18-5.) A test giving an estimated concentration of > 25 ppb should be repeated by a more specific method. It is important to note that this is a screening procedure and thus results

Table 18-5. "NORMAL" CONCENTRATIONS OF
INSECTICIDES IN HUMAN BLOOD [66]

	Mean (ppb)	Range (ppb)
p,p′-DDE	26	2–500
p,p′-DDT	12	2–100
p,p′-TDE	11	2– 70
Dieldrin	4.5	1– 25
Methoxychlor	19.0	10– 40
Lindane	3.5	1– 20
α-BHC	3.5	1– 10
Heptachlor epoxide	4	2– 20

are presumptive only. Absolute characterization or quantitation of a pesticide requires more specific methods.[67]

OTHER SCREENING PROCEDURES

A number of tests, often indirect in nature, can be useful in screening for toxic substances. Many of these are particularly useful when the physician has already made a provisional diagnosis based on clinical signs, symptoms, known exposure to a substance, information from friends or relatives, and medical history. Examples include the determination of cholinesterase activity when poisoning by organophosphate or carbamate pesticide is suspected, and the screening test for ketones when it is likely that the patient has consumed isopropanol. These tests have been described in the section on Spot Tests.

Abnormal results of several tests commonly performed in the clinical chemistry laboratory may indicate the need to pursue more specific procedures to confirm or rule out poisoning. Table 18-6 gives some examples. Extensive descriptions appear in the literature.[9,34,42]

Table 18-6. EXAMPLES OF BIOCHEMICAL ABNORMALITIES
WHICH MAY RESULT FROM TOXIC AGENTS[42]

Acidosis: Methanol, ethylene glycol, iron salts, phenytoin, salicylates, tricyclic antidepressant drugs
Albuminuria: Many toxic substances
Aminoaciduria: Cadmium, phosphorus
Bilirubinemia: Ethylene glycol monomethyl ether
Copper deficiency: Molybdenum
Delta-aminolevulinic dehydratase activity (decreased): Lead
Hematemesis: Acetonitrile, arsenic, boron, tetrachloroethane
Hematuria: Anticoagulant drugs, benzidine, carbon tetrachloride, naphthalene, picric acid, sulfonamides, turpentine
Hemoglobinuria: Acetic acid, arsine, naphthalene, stibine
Hemolysis: Acetic acid, arsine, lead, naphthalene
Hemoptysis: Chlorine
Hypercalciuria: Beryllium, cadmium, phosphorus
Hyperglycemia: Acetone, insulin, isoniazid, phenytoin, salicylates (early stages)
Hyperkalemia: Digitalis glycosides, ethylene glycol, isoniazid
Hypoglycemia: Ethanol, methanol, salicylates (late stages)
Hypokalemia: Tricyclic antidepressant drugs, diuretics
Hypoxia: Aniline, dinitrotoluene
Jaundice: Arsine, chlorinated hydrocarbons, dinitrophenol, gold, naphthalene, nitrobenzene, stibine, sulfonamides
Lactic acid dehydrogenase (increased): Zinc oxide
Liver damage: Many toxic substances
Methemoglobinemia: Aniline, chlorates, mercaptans, nitrated and chlorinated benzenes and toluenes, oxides of nitrogen, phenacetin, pyridium, sulfonamides
Oxygen saturation (decreased): Beryllium
Porphyrinuria: Benzene, chlorinated hydrocarbons, lead, mercury
Proteinuria: Phosphorus, thallium
Renal damage: Many toxic substances
Stools, bloody: Anticoagulant drugs, antimony, arsenic, castor beans, copper, iron, oxalic acid, methaqualone, zinc chloride
Sulfhemoglobinemia: Nitrobenzene, nitrochlorobenzene
Uremia: Diethylene glycol, isopropanol, lead, paraquat

Liquid chromatography (high-performance or high-pressure chromatography, also referred to as HPLC) is emerging as a general screening technique for the detection of drugs in blood.[40] Twenty commonly abused drugs (acetaminophen, theophylline, salicylate, primidone, methyprylon, phenobarbital, butabarbital, ethchlorvynol, butalbital, chlordiazepoxide, pentobarbital, amobarbital, phenytoin, gluthethimide, secobarbital, flurazepam, nitrazepam, methaqualone, N-desmethyldiazepam, and diazepam) can be determined simultaneously in a single run in ~35 min. The primary difficulty in adapting liquid chromatography to a screening method is that no single column packing or solvent system (unlike gas-liquid chromatography) will effectively resolve widely different chemical classes of drugs. Perhaps in the near future ternary (or even quaternary or higher) solvent delivery systems will overcome this problem.

SEPARATION OF TOXIC SUBSTANCES FROM BIOLOGICAL SPECIMENS

In most chemical analytical procedures the analyte must be separated from the biological matrix prior to the application of an instrumental analysis. This critical but frequently neglected step in analytical toxicology deserves some emphasis. Historically, analytical toxicologists have tended to classify toxic substances according to the manner by which they are separated from the biological specimen. For this reason, some of the toxicological reference textbooks have categorized poisons as (1) gases, (2) volatile substances, (3) corrosives, (4) metals, (5) nonmetals, and (6) nonvolatile organic substances. Those compounds which did not fit into any of these categories were grouped into a miscellaneous category. In contemporary analytical toxicology there has been a tendency to combine some categories which naturally overlap, or to eliminate others. The following discussion will include those separation procedures which are commonly utilized, but the discussion will not attempt to deal with all separation procedures.

Methods for the Determination of Volatile Substances

Volatile toxic compounds consist of gases or liquids that have boiling points of 100 °C or lower. For this reason, they can be separated from biological specimens by steam distillation. This procedure, although useful, necessitates assembly of glass apparatus, and particularly for some volatile substances, great care in conducting the separation. As a result, steam distillation is primarily of historic interest except for some specific applications.

The principle of *microdiffusion,* which can also be used to separate many members of this group, has the advantages of small sample size, minimal and inexpensive equipment, and simplicity of operation; it is particularly applicable for separating gases and volatile substances. In the microdiffusion technique, the specimen containing the substance to be separated and a "trapping" solution are placed in separate compartments inside a sealed container. The specimen and the trapping solution share the same vapor space. The substance to be separated, because of its vapor pressure, leaves the specimen and enters the common vapor space, from which it is absorbed by the trapping solution. Thus, by gaseous diffusion the substance to be separated is continuously transferred from the specimen to the trapping solution until an equilibrium is reached. If the trapping solution contains a reagent that converts the separated substance into a different compound, equilibrium does not occur and a quantitative transfer results. The entire operation can be carried out with a small specimen by using a Conway unit. This unit consists of two round, concentric chambers molded into a porcelain, glass, or plastic dish that can be sealed. The trapping solution is placed in the center well and the specimen (and in some cases a "releasing" agent) in the outer compartment (Figure 18-3).

The time required for completion of the diffusion process is variable, depending on the vapor pressure of the substance to be separated, the volume of the specimen solution, the nature of the trapping solution, and the temperature at which the process is conducted. In general, the diffusion time is shortened by higher temperatures, by small volumes of specimen solutions, and by substances of high vapor pressure. It should be emphasized that procedures worked out for a specific diffusion assembly may require different time periods for completion in an assembly of different dimensions. Also, if diffusion is carried out at an elevated temperature, precautions must be taken to ensure against loss of expanding gases, which tend to lift the glass plate. This can

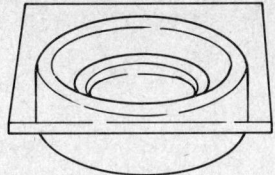

Porcelain diffusion unit

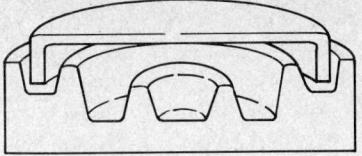

Sealable porcelain diffusion unit, showing cover sealed in liquid.

Figure 18-3. Microdiffusion chambers of the Conway type.

occur even at room temperature if cold solutions are used. It is advisable to place a 300-g weight on the lid at room temperature, and at elevated temperatures the lid should be clamped in place. For a mathematical consideration of these variables, see the publication by Conway.[14]

A number of toxicologically important volatile substances have been measured by this technique. Since most of these procedures are colorimetric in nature, all that is required besides the necessary reagents and the Conway unit is a spectrophotometer. Feldstein and Klendshoj[20] have applied this method to a number of solvents and gases as shown in Table 18-7. The diffusion times shown in this table frequently can be shortened by carrying out the reaction at an elevated temperature.

A second general approach to the separation of volatile substances is by the use of *head-space gas chromatography*. This general approach is applicable to more substances than the microdiffusion procedure and is more convenient for those laboratories equipped with a gas

Table 18-7. MICRODIFFUSION TECHNIQUES*

| Analysis | Outer Compartment | | Inner Compartment | Diffusion Time |
	Sample	*Liberating Agent*		
Acetaldehyde	3 mL blood 5 mL tissue[†]	3–4 drops H_2SO_4, 10%	3.3 mL $NaHSO_3$, 0.15 mol/L	3 h
Acetone	3 mL blood 5 mL tissue[†]	3–4 drops H_2SO_4, 10%	3.3 mL $NaHSO_3$, 0.15 mol/L	3 h
Carbon monoxide	1.0 mL blood	1 mL H_2SO_4, 10%	2.0 mL palladium chloride, 0.005 mol/L, in HCl, 0.01 mol/L	1 h
Cyanide	2–4 mL blood 5 mL tissue[†]	3–4 drops H_2SO_4, 10%	3.3 mL NaOH, 0.1 mol/L	3 h
Ethanol	0.8 mL blood 4 mL tissue[†]	1 mL K_2CO_3, sat. soln.	2 mL acid-dichromate, $K_2Cr_2O_7$, 0.02 mol/L in H_2SO_4, 7.76 mol/L	3 h
Fluoride	1.0 mL blood 1.0 mL urine	$HClO_4$, conc.	Zr SPADNS[‡]	24 h at 60 °C
Formaldehyde	3 mL blood 5 mL tissue[†]	3–4 drops H_2SO_4, 10%	3.3 mL $NaHSO_3$, 0.15 mol/L	3 h
Halogenated hydrocarbons	1–4 mL blood 1–4 mL tissue[†]	—	1.0 mL toluene	3 h
Isopropanol	2.0 mL blood 5 mL tissue[†]	1 mL K_2CO_3, sat. soln.	3.3 mL H_2SO_4, 10%	3 h
Methanol	2.0 mL blood 5 mL tissue[†]	1 mL K_2CO_3, sat. soln.	3.3 mL H_2SO_4, 10%	3 h
Phenols	2–4 mL blood 5 mL tissue[†]	3–4 drops H_2SO_4, 10%	3.3 mL NaOH, 0.1 mol/L	3 h
Sulfide	2–4 mL blood 5 mL tissue[†]	3–4 drops H_2SO_4, 10%	3.3 mL NaOH, 0.1 mol/L	3 h

*After Feldstein, M., and Klendshoj, N.: The determination of volatile substances by microdiffusion analysis. J. Forensic Sci., 2:39-58, 1957.
[†]Tissue homogenate.
[‡]Zirconyl chloride·8 H_2O, 0.4 mmol/L, and 4,5-dihydroxy-3-(*p*-sulfophenylazo)-2,7-naphthalene disulfonic acid trisodium salt, 4.6 mmol/L, in HCl, 2.0 mol/L.

chromatograph. The principle of this method is to decrease the solubility of the volatile substance in the biological sample by saturating the aqueous phase with salt. This step, coupled with the confinement of the specimen in a closed container that can be incubated at a specific temperature, drives the volatile substance from the specimen into the vapor space above the specimen, which can then be easily sampled by a syringe needle. Generally the specimen is measured into a vial fitted with a septum stopper and a quantity of salt is added to the specimen to achieve a saturated solution. After appropriate mixing, the vial is placed in a water bath maintained at a specific temperature and, after equilibration, a measured sample of the vapor space in the vial above the aqueous medium is sampled and injected into an appropriate gas chromatograph. Dubowski has reported a number of column packings which are appropriate for these types of analyses (Table 18-8), together with relative retention times of many volatile substances which can be measured in this fashion (Table 18-9).[18]

Methods for the Determination of Nonvolatile Organic Substances

This is the largest group of substances encountered in the clinical toxicology laboratory and includes most drugs and alkaloids, pesticides, and occupational and environmental hazards and pollutants. The problems associated with the analysis of this group can be complex. Extraction methods utilizing water-immiscible solvents must usually be employed to separate the substance from the specimen. However, extractions are frequently not quantitative or may result in troublesome emulsions. Some toxic substances are rapidly metabolized, excreted, or bound to protein, making their detection difficult. Many substances are chemically similar to naturally occurring compounds and must be differentiated from them by purification steps. Because of the frequent importance of measuring metabolites as well as the parent compound, analysis may require additional separation procedures. Finally, contaminants which may interfere with the subsequent analysis are frequently encountered and must be separated from the analyte of interest.

Those substances which are acids or bases are usually water soluble when they are in the form of salts. Reconversion of the compound to the free acid or base makes it less water soluble but more soluble in solvents used in extraction procedures. For example, in Figure 18-4, note that if the pH of the specimen is made acidic, organic acids are converted to the relatively nonpolar, free acid form that is soluble in an appropriate organic solvent. Conversely, if the specimen is rendered alkaline in its pH, organic bases are converted to the relatively nonpolar free base form that is solvent soluble. Neutral compounds, obviously, exist in the same relatively nonpolar form regardless of the pH of the medium. Generally, these are more soluble in the organic solvent used for extraction. Frequently this group of substances tends to carry along more contaminants than other substances, due to lipid material. The most difficult group of organic substances to extract effectively is amphoteric in nature. As can be seen in Figure 18-4, either acidification or alkalinization may result in polar, water-soluble forms which are not

Table 18-8. ABSOLUTE RETENTION TIMES FOR ETHANOL USING THE GAS-CHROMATOGRAPHIC CONDITIONS DESCRIBED BELOW

Packing	Temperature (°C)	Absolute Retention Time for Ethanol (min)
1. 0.5% Carbowax K-600 and 3.8% Hallcomid 18 on Teflon 6HC, 40- to 60-mesh	100	0.24
2. 0.5% Carbowax K-600 and 3.8% Hallcomid 18 on Teflon 6HC, 40- to 60-mesh*	50	1.8
3. 0.4% Carbowax 1500 on Carbopack A	80	1.47
4. Porapak Q, 80- to 100-mesh	155	1.60
5. Chromosorb 102, 100- to 120-mesh	155	2.16
6. 15% Carbowax 1500 on Celite 525	80	2.58
7. Porapak S, 80- to 100-mesh	155	2.51
8. 30% Carbowax, 2M* on Chromosorb W, AW, 60- to 80-mesh	130	2.12

*6′ × 1/4″ O.D. column, rather than the 6′ × 1/8″ size used for the other supports.
Reprinted with permission, from Dubowski, K.: Organic volatile substances. *In*: Methodology for Analytical Toxicology, I. Sunshine, Ed. Boca Raton, Fla., CRC Press, Inc., 1975.

Table 18–9. RELATIVE RETENTION TIMES OF VOLATILE SUBSTANCES

Compound	Relative Retention Time*	Compound	Relative Retention Time*
Petroleum ether	0.19	Benzene	1.22
Acetaldehyde	0.20	Isopropanol	1.24
n-Pentane	0.21	Methyl cyclohexane	1.32
Diethyl ether	0.25	Allyl ether	1.39
Propionaldehyde	0.36	Butanol, tertiary	1.40
Methylacetate	0.38	n-Propyl acetate	1.63
Acetone	0.40	Chloroform	1.64
Carbon disulfide	0.40	Allyl acetate	1.69
Acrolein	0.41	1-Chloro-3-methyl butane	1.72
n-Hexane	0.44	Trichlorethanol	1.80
Methyl iodide	0.44	1,4-Dioxane	1.81
Isopropyl ether	0.47	Ethylene dichloride	1.84
Diethylamine	0.57	Trichlorethylene	2.10
Methanol	0.60	n-Octane	2.40
Acetonitrile	0.65	n-Propanol	2.50
Methylene chloride	0.69	2,3-Dichloropropane	2.63
Ethyl acetate	0.71	Methyl isobutyl ketone	2.75
Cyclohexane	0.72	Butanol, secondary	3.04
Tetrahydrofuran	0.78	Toluene	3.15
Acrylonitrile	0.80	Paraldehyde	3.45
Butanone	0.84	n-Butylacetate	4.00
n-Heptane	0.98	Isobutanol	4.68
1,1,1-Trichloroethane	0.98	n-Nonane	5.95
Ethanol	1.00	Pyridine	6.08
Carbon tetrachloride	1.01	n-Butanol	6.60
Propionitrile	1.14	Xylene	7.57

*Relative to ethanol, whose absolute retention time is 1.9 ± 0.1 min on a $6' \times 1/4''$ O.D. column packed with 5% Carbowax K-600 and 3% Halcomid 18 on 60- to 80-mesh Teflon 6HC.

Reprinted with permission, from Dubowski, K.: Organic volatile substances. *In*: Methodology for Analytical Toxicology, I. Sunshine, Ed. Boca Raton, Fla., CRC Press, Inc., 1975.

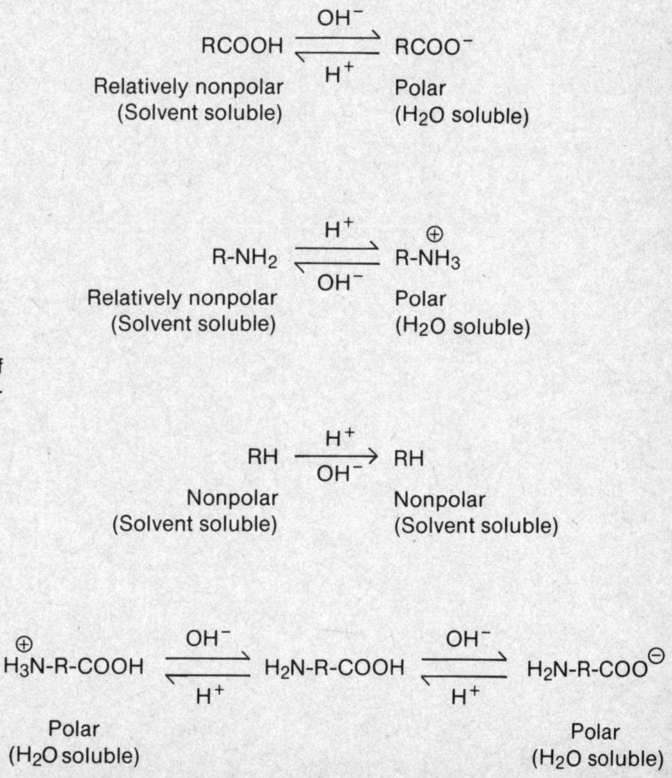

$$RCOOH \underset{H^+}{\overset{OH^-}{\rightleftharpoons}} RCOO^-$$

Relatively nonpolar   Polar
(Solvent soluble)   (H$_2$O soluble)

$$R\text{-}NH_2 \underset{OH^-}{\overset{H^+}{\rightleftharpoons}} R\text{-}NH_3^{\oplus}$$

Relatively nonpolar   Polar
(Solvent soluble)   (H$_2$O soluble)

Figure 18-4. Effects of manipulating pH of solvent for separation of nonvolatile organic substances by solvent extractions.

$$RH \underset{OH^-}{\overset{H^+}{\longrightarrow}} RH$$

Nonpolar   Nonpolar
(Solvent soluble)   (Solvent soluble)

$$\overset{\oplus}{H_3}N\text{-}R\text{-}COOH \underset{H^+}{\overset{OH^-}{\rightleftharpoons}} H_2N\text{-}R\text{-}COOH \underset{H^+}{\overset{OH^-}{\rightleftharpoons}} H_2N\text{-}R\text{-}COO^{\ominus}$$

Polar   Polar
(H$_2$O soluble)   (H$_2$O soluble)

appreciably soluble in organic solvents. For these substances a specific pH must be found at which the compound predominantly exists in the relatively nonpolar, solvent-soluble form.

Utilizing the pH-dependent solubility properties of nonvolatile organic substances, it is possible to separate chemical species in a specimen by differential extraction (Figure 18-5).

Clean-up Procedures

Differential extractions, as described above, constitute a partial removal of interfering substances in the subsequent analysis for the nonvolatile organic substances. Most analytical procedures currently in use that require a preliminary extraction step rely on a chromatographic step in order to remove interferences and quantitate the analyte simultaneously. Thus, in these analytical procedures based on TLC, HPLC, or GLC, additional clean-up procedures may not be necessary.

Since some toxic substances must be measured at extremely low concentrations (in the parts per billion or parts per trillion range), additional cleanup steps are needed. Some nonvolatile, organic substances have become ubiquitous in the environment; they are found as contaminants in reagents and glassware as well as in the patient's specimen. Phthalate esters used as plasticizers, polychlorinated biphenyls, pesticides, and other pollutants in the environment fall into this category. Most of the common cleanup procedures are chromatographic in nature and, indeed, some published methods utilize serial chromatographic separation in order to achieve increased specificity.

Pesticide chemists have long been concerned with minimizing interferences and their experience with cleanup procedures can be effectively applied to the cleanup of extracts, particularly in the measurement of pesticides or environmental pollutants in biological specimens. Since many environmental pollutants are quite nonpolar in nature and would fall into the category of neutral compounds, extracts are generally contaminated with lipid material. Column chromatography, in which the stationary phase may be magnesium silicate (Florisil) or hydrated silica (silica gel), has been widely used in these applications.[67] More recently, gel-permeation chromatography has been found to be a convenient and extremely efficient means of separating certain nonpolar

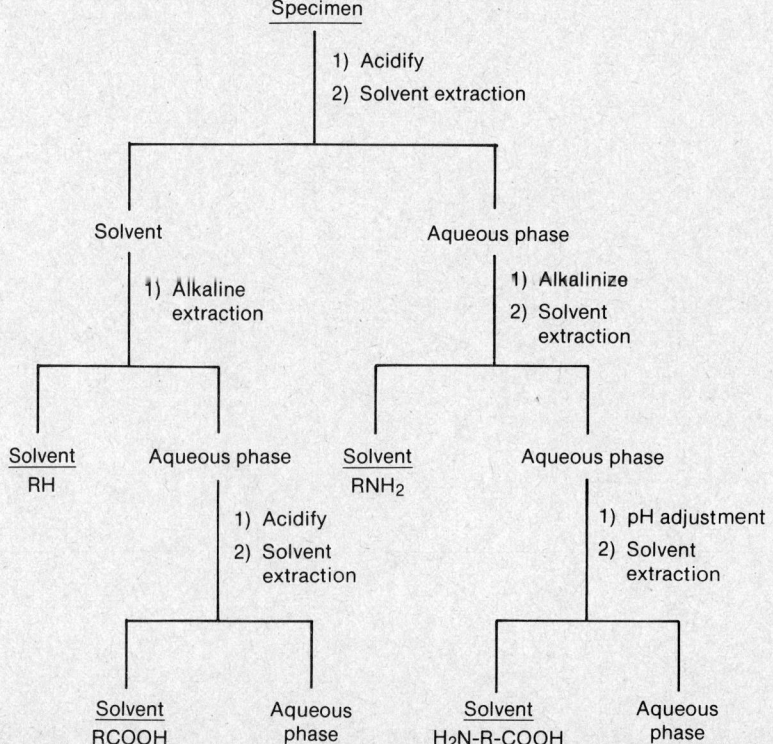

Figure 18-5. A scheme of separation for nonvolatile organic substances by differential solvent extraction.

analytes from specimens with high concentrations of lipids. In fact, separations of these substances from fat-containing samples as well as from milk specimens can be readily achieved.[59] Other purification steps currently in use apply column materials commonly utilized in HPLC, including reverse-phase columns, for effective purification and separation prior to the final analytical step. There is no doubt that the necessity for resorting to purification steps will increase as the need for the detection of small concentrations in limited sample sizes is recognized.

Separation of Inorganic Toxic Agents

For those inorganic toxic substances that are stable at elevated temperatures, separation can be accomplished relatively simply by burning away the organic material in the specimen and measuring the component of interest in the inorganic residue by some standard procedure. The combustion process may be either a wet digestion with strong, oxidizing acids or a dry ashing procedure in a flame or furnace. Some metals and their salts are volatile at high temperatures, and special precautions must be taken to avoid loss during the ashing step. Others are present only in trace quantities and are subject to loss or contamination similar to that which can occur in any trace analysis. Many instrumental procedures, such as atomic absorption spectrometry or anodic stripping voltammetry, do not require extensive specimen manipulation. The method itself accomplishes either destruction of organic material at elevated temperatures or a separation of the analyte from the matrix. Application of atomic absorption is discussed in Chapter 8C. Some specimens of importance in toxicological analysis, however, may require specific preparation procedures. This is particularly true of solid specimens such as hair, nails, bone, or other tissue.

Dry ashing procedures are restricted to those elements which exist in chemical forms that are not volatile at temperatures of ~500 °C. Generally, the specimen is weighed into a silica crucible, treated with appropriate reagents in order to ensure that the element is in the proper oxidation state, and then placed into a cold muffle furnace. The temperature is slowly raised and held at a high level until all organic material is oxidized, leaving a light, inorganic residue. This residue can then be analyzed in a variety of ways.

The *wet digestion procedure* is applicable to those elements that may be more readily volatilized at high temperature. Generally, the specimen is weighed into a quartz or borosilicate glass flask, and treated with a mixture of nitric acid and a small amount of perchloric and sulfuric acids (4:1:1). The mixture is heated slowly to the boiling point of sulfuric acid with occasional addition of oxidizing acids, until no further charring or darkening of the liquid occurs. By connecting the outlet of the flask to a water aspirator pump, acid fumes can safely be diverted to the drain. The resulting solution is neutralized prior to further analysis.

Dialysis and Membrane Filtration

Some toxic substances are difficult to separate by standard procedures. Either their physical properties do not lend themselves to the standard approach or the substance itself may be so labile that it becomes chemically modified during the separation process. Some toxic anions fall into this category and dialysis has been used successfully for these separations. Conventional dialysis across a semipermeable membrane is effective but time consuming. More recently, a technique has been applied that separates low-molecular-weight compounds. The specimen is forced by centrifugation or vacuum filtration through a membrane or special filter that permits passage of substances of limited molecular size. This technique has also been suggested for the separation of protein-bound substances in plasma from those in the free unbound state. The rapidity with which this type of procedure can be carried out makes this means of separation particularly attractive for the clinical laboratory. Few studies have been conducted, however, on the recovery of analytes by this separation procedure. Since many toxic substances of low molecular weight may be associated with macromolecules such as proteins, recovery of the analyte is generally low. Nevertheless, the centrifugal dialysis method of separation is applicable to many analyses, since the analyte is separated from the bulk of naturally occurring interfering substances in the specimen. The essentially clear, aqueous solution resulting from this treatment can be analyzed by a variety of procedures.

Other Separation Procedures

Separation chemistry is a broad field in its own right. Potentially all of the techniques used in this important area can be applied to biological material in order to improve or optimize the separation of a specific toxic substance. For example, electrophoresis, sublimation, countercurrent liquid extraction, and paired-ion extraction techniques are among the methods that have been used. As the challenge to the analytical toxicologist to detect ever lower concentrations of a variety of chemical species continues to increase, the application of new and improved separation techniques will become more important.

IMMUNOASSAY PROCEDURES

The application of immunoassay procedures to the detection of drugs in biological specimens has developed into an important area of toxicological analysis. The remarkable sensitivity and specificity of these procedures frequently permit the direct addition of serum or urine to an appropriate medium containing the necessary components for the reaction, prior to reading the final end point. Thus, a minimum of specimen preparation, frequently with no separation step, is required. Only small volumes of specimens are necessary for these procedures and since few manipulations are involved, the methods are easily adaptable to automation for screening large numbers of patients. The principles of this technique are described more completely in Chapter 1C.

Despite the listed advantages, immunoassay procedures have had only limited application in toxicological analysis. This is primarily due to the following factors:

1. The nature of the toxic substances must be known and an immunoassay kit must be available for such an assay.

2. Although immunoassays are highly specific, frequently there are crossover reactions to structurally similar compounds such as metabolites.

3. When immunoassays are performed on a single specimen, the cost of this assay may be high because standards and controls are included with each assay. If the test is performed infrequently, the limited shelf life of reagents also increases the cost.

For some toxic substances, immunoassays are the only practical approach for the clinical laboratory. This applies to such substances as cardiac glycosides, lysergic acid diethylamide (LSD), and Δ^9-tetrahydrocannabinol (THC). Both LSD and THC are rarely encountered as acute toxic emergencies. In the case of cardiac glycosides, immunoassays are presently the only procedures available. Methods, however, are not always specific. If the patient has ingested a toxic concentration of digitoxin and the immunoassay procedure used cross-reacts with digoxin, the results will be misleading. Generally, chemical or instrumental methods which yield more information than merely the presence or absence of a specific substance are of more practical utility in the clinical toxicology laboratory. Nevertheless, immunoassay procedures are extremely valuable in laboratories engaged in therapeutic drug monitoring (Chapter 17).

DETECTION OF TOXIC SUBSTANCES BY SPECIFIC PROCEDURES

Recommending specific procedures for every toxic substance is impractical, since not only are hundreds of methods available, but for each substance several analytical methodologies may be equally meritorious. In addition, methods may be quickly outdated as new knowledge is gained and technical advancements are made. Therefore, references should be consulted when needed.[2,3,63,65]

The methods outlined in this chapter have the merit that they perform as described and have been used in the authors' laboratories. In addition, they have been selected in order to illustrate a variety of analytical or instrumental applications, thus emphasizing the varied menu of analytical approaches available to the analytical toxicologist.

Screening methods described earlier may or may not be conducted in a quantitative manner.

Generally, they are not sufficiently specific to provide unequivocal evidence for the presence of a toxic substance. The analyst should follow up a diagnostic screening procedure with a reliable quantitative test in order to provide additional information for the clinical management of the patient.

In addition, methods used in therapeutic drug monitoring are generally applicable as specific procedures in toxicology. For this reason, this chapter will describe specific methods not commonly encountered in therapeutic drug monitoring. The reader is also referred to Chapter 17.

CARBON MONOXIDE

Carbon monoxide, the product of incomplete combustion of organic substances, is the most common of the gaseous poisons. It is present in the free state in manufactured gas (coal gas), but not in natural gas. Both types of gas, when used as fuel for stoves, furnaces, and other appliances, release carbon monoxide as one of the combustion products. This is true, of course, of coal, oil, and other types of fuel as well. Malfunctioning or poorly ventilated heating appliances, therefore, are frequently causes of carbon monoxide poisoning. Since the gas is also a component of exhaust fumes from internal combustion engines, accidental poisonings can occur when gasoline-powered tools or outboard motors, as well as automobile engines, are used under conditions of poor ventilation or improper operation.

Carbon monoxide combines reversibly with hemoglobin in a manner almost identical to oxygen, but the bond is about 210 times as strong as that in oxygenated hemoglobin (oxyhemoglobin). As a result, carbon monoxide is not readily displaced from hemoglobin except at high oxygen tension; with prolonged exposure accidental poisonings can occur even at low levels of carbon monoxide in the atmosphere. (See Table 18-10.) In these instances the carbon monoxide level in blood builds up slowly until toxic levels are reached.

The detection and estimation of carbon monoxide in biological specimens can be approached in two general ways: (1) release of the gas from the hemoglobin complex with subsequent direct or indirect measurement of the gas, or (2) estimation of carboxyhemoglobin by its typical color or absorption bands. The first approach can be carried out by gasometric techniques, gas chromatography,[31] microdiffusion (p. 1692), or infrared spectrophotometry.[61] The second approach utilizes spectrophotometric[68] or spectrographic analysis or simple color comparison.

Specimen

Regardless of the analytical method used, the specimen to be analyzed must contain hemoglobin. Relatively little carbon monoxide dissolves in the aqueous or lipid fractions of tissue, compared to that bound to hemoglobin. This rather obvious statement is made since some clinicians request serum, spinal fluid, or even urine carbon monoxide levels. The most satisfactory specimen is whole blood. Clotted blood is less desirable and must be homogenized with a minimum exposure to air before analysis. This can be done using a Ten Broeck hand homogenizer. In those fatalities in which the victim is so badly burned or mutilated that blood is not available, tissue rich in hemoglobin, such as bone marrow or spleen, can be used. In these cases, estimation of carbon monoxide rather than carboxyhemoglobin is done.

Table 18-10. CARBON MONOXIDE TOXICITY*

% (v/v) in Air	Response
0.01	Allowable for an exposure of several hours
0.04–0.05	Can be inhaled for 1 h without appreciable effect
0.06–0.07	Causing a just-noticeable effect after 1 h exposure
0.1–0.12	Causing unpleasant but not dangerous symptoms after 1 h exposure
0.15–0.20	Dangerous for exposure of 1 h
0.4 and above	Fatal in exposure of < 1 h

*From Deichmann, W. B., and Gerarde, H. W.: Symptomatology and Therapy of Toxicological Emergencies. New York, Academic Press, 1964.

Determination of Carbon Monoxide by Spectrophotometry

Hemoglobin and its derivatives have characteristic absorption bands in the visible region that can be utilized to detect carboxyhemoglobin and to measure the quantity present. Oxygenated hemoglobin and carboxyhemoglobin have similar double bands in alkaline solution. The absorption maxima for oxygenated hemoglobin are 576–578 and 540–542 nm; for carboxyhemoglobin they are 568–572 and 538–540 nm (Figure 18-6,*A*). Deoxygenated hemoglobin has a single broad band at 555 nm (Figures 18-6,*B* and *C*).

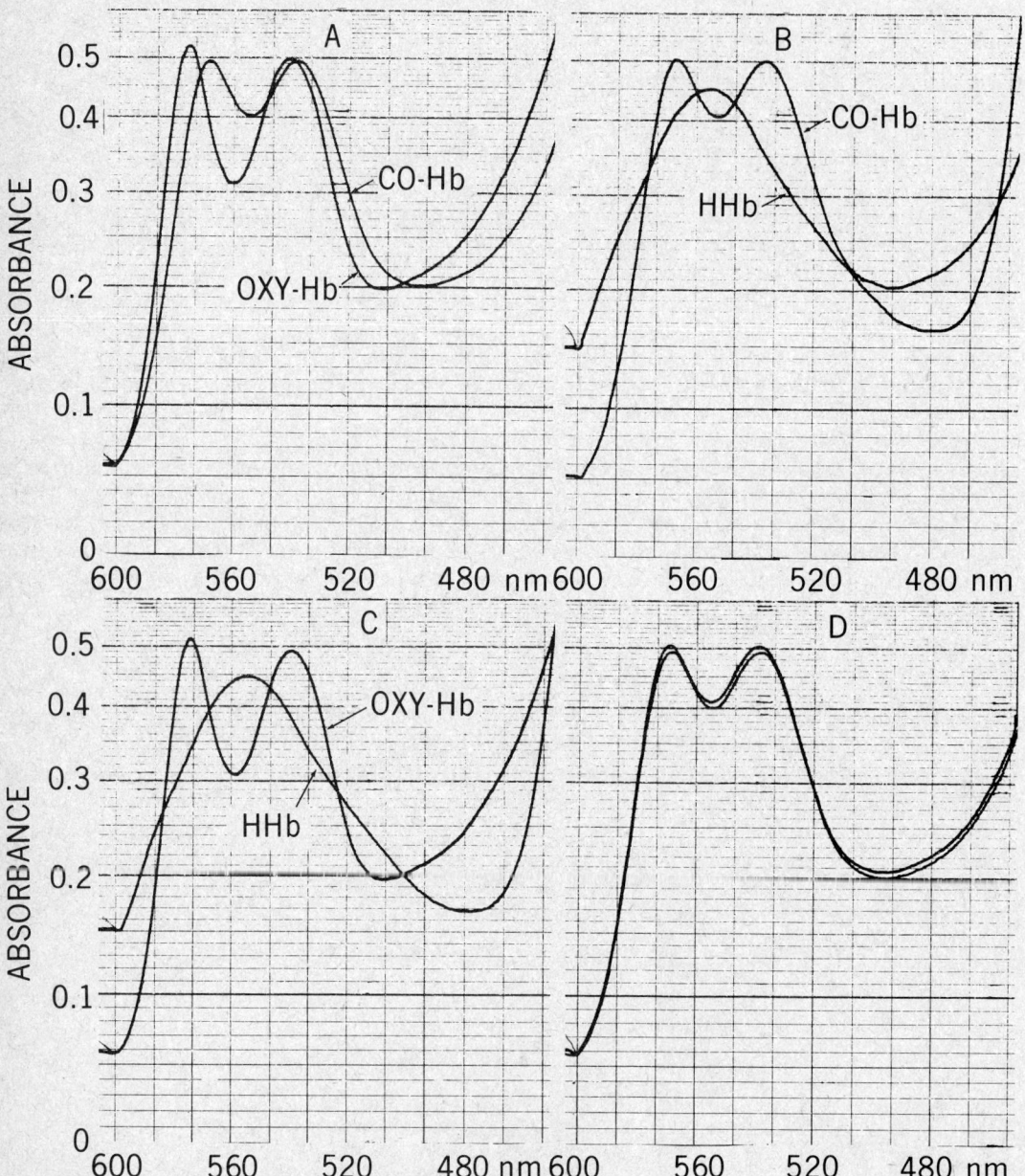

Figure 18-6. Spectral curves (overtraced) of (*A*) 100% oxyhemoglobin (OXY-Hb) and 100% carboxyhemoglobin (CO-Hb) before treatment with sodium dithionite; (*B*) 100% oxyhemoglobin (now HHb) and 100% carboxyhemoglobin after treatment of both with sodium dithionite; (*C*) 100% oxyhemoglobin before and after treatment with sodium dithionite; and (*D*) 100% carboxyhemoglobin before and after treatment with sodium dithionite. (Reproduced by permission of the Institute for Clinical Science, from Tietz, N. W., and Fiereck, E. A.: The spectrophotometric measurement of carboxyhemoglobin. Ann. Clin. Lab. Sci., *3*:36-42, 1973.)

If a weakly alkaline dilution of blood is treated with sodium hydrosulfite, oxygenated hemoglobin (and any methemoglobin present) is converted to deoxygenated hemoglobin (Figure 18-6,*C*). Carboxyhemoglobin is unaffected by such treatment (Figure 18-6,*D*).

$$\begin{matrix} HbO_2 \\ \text{or} \\ MetHb \end{matrix} \quad + \quad Na_2S_2O_4 \longrightarrow Hb$$

$$HbCO \quad + \quad Na_2S_2O_4 \longrightarrow \text{No reaction}$$

This is the basis of several methods for the determination of per cent saturation of hemoglobin by carbon monoxide. The method to be described[68] works satisfactorily with fresh, oxalated blood but is not satisfactory with postmortem blood or specimens containing denatured hemoglobin.

Principle

A dilute hemolysate of blood is treated with sodium dithionite, which reduces methemoglobin and oxyhemoglobin but does not affect carboxyhemoglobin. The absorbance of this solution is measured at 541 and 555 nm, the absorbance ratio A_{541}/A_{555} is calculated, and the per cent carboxyhemoglobin is determined from the standard curve.

Reagents

1. NH$_4$OH, 0.12 mol/L. Dilute 15.9 mL of concentrated NH$_4$OH to 1.0 L with deionized water. This solution is stable.
2. Sodium hydrosulfite (sodium dithionite), reagent grade. Preweigh 10-mg portions of sodium dithionite into individual small tubes. Stopper the test tubes or cover with Parafilm.
3. Carbon monoxide. Lecture bottle (Matheson Gas Products, Division of Will Ross, Inc., East Rutherford, NJ 07073).
4. Oxygen, CP.

Special Apparatus

A narrow bandpass (< 2 nm) spectrophotometer with 10-mm cuvets is required, although the use of a recording spectrophotometer with the same specifications is desirable.

It is imperative that the spectrophotometer be checked regularly for wavelength and spectrophotometric accuracy with appropriate calibrating filters (e.g., NBS Reference Material 930) and with liquid photometric standards (e.g., NBS Reference Material 931).

Procedure

1. Add 100 μL of whole heparinized blood to 25 mL of NH$_4$OH, 0.12 mol/L. Mix the solution and allow it to stand for 2 min. This is the hemolysate.
2. Transfer 3.0 mL of NH$_4$OH and 3.0 mL of the hemolysate into separate 10-mm cuvets. (Analyze the sample in triplicate.)
3. Add 10 mg of sodium dithionite to all cuvets. Cover the cuvets with Parafilm and invert gently 10 times. If a number of hemolysates are analyzed, space the addition of the reducing agent so that each can be read after exactly 5 min.
4. Exactly 5 min after the addition of dithionite to the hemolysates, read the absorbance at 541 and 555 nm against the NH$_4$OH blank.
5. Calculate the ratio of the absorbance at 541 nm to that at 555 nm, A_{541}/A_{555}, and determine the per cent carboxyhemoglobin from the calibration curve. *Note*: For confirmation and for the purpose of record, the sample without and with dithionite (steps 1 and 3, respectively) may be scanned between 450 and 600 nm. (See Figure 18-6.)

Preparation of the Standard Curve

Caution: Use a fume hood when working with carbon monoxide gas.
1. Collect 20 mL of heparinized blood from a healthy person who does not smoke.

2. Transfer a 4.0-mL portion of the fresh, heparinized blood sample into each of two 125-mL separatory funnels. Allow pure oxygen to flow through one funnel and pure carbon monoxide through the other for 15 min, while the funnels are gently rotated. After the addition of the gases, close the separatory funnels and rotate them gently for an additional 15 min. Analyze the fully saturated samples immediately, in triplicate, according to the procedure given above. Use these results for the establishment of the 0 and 100% carboxyhemoglobin calibration points. These samples may not be used to establish the intermediate calibration points.

Plot the ratio of the absorbance at 541 nm to that at 555 nm for the 0% and for the 100% carboxyhemoglobin samples and draw a line between the two points. (See Figure 18-7).

3. Fill the funnel containing the 100% carboxyhemoglobin sample with nitrogen gas and rotate it for 5 min. Treatment with nitrogen removes the physically dissolved CO from the sample, but a small amount of CO will also dissociate from hemoglobin. Determine the exact carboxyhemoglobin content of this sample by the method described, using the standard curve just prepared. Prepare intermediate standards by mixing appropriate proportions of the nitrogen-treated sample with the oxygen-treated sample.

4. Analyze each of the diluted blood samples from step 3 in triplicate, according to the procedure given above.

5. Plot the calculated concentrations against the absorbance ratios obtained. These points should fall on the line drawn for the fully saturated samples, since the curve is linear over the entire range. (See Figure 18-7.)

Interpretation

Normal carbon monoxide levels depend on the degree of exposure to this gas without signs and symptoms of poisoning being produced. For example, in smokers the following levels may occur:

Smokers (one to two packs per day): up to 4–5% saturation of hemoglobin with carbon monoxide.

Heavy smokers (more than two packs per day): up to 8–9% saturation of hemoglobin with carbon monoxide.

Nonsmokers: 0.5–1.5% saturation of hemoglobin with carbon monoxide.

In our experience, patients can survive brief periods of 70–75% saturation. Prolonged periods at these high levels can, of course, be fatal. Interpretation of lethal levels must be related to other factors in each case, i.e., time of exposure, normal hemoglobin level of the patient, age and general health of the patient, degree of activity, and so on.

In patients treated with oxygen, carbon monoxide is fairly rapidly released from hemoglobin. Frequently, a patient is treated with oxygen while being transported to the hospital. By the time a blood sample is drawn and analyzed, the carbon monoxide level may be close to normal. These patients should be kept quiet with good oxygenation to ensure that all tissue-bound carbon monoxide (e.g., with myoglobin and heme-containing enzymes) is dissipated before the patient is discharged. Effects of carbon monoxide toxicity at various carboxyhemoglobin concentrations are shown in Table 18-11.

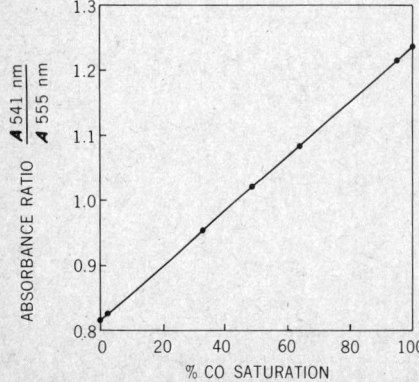

Figure 18-7. Example of a standard curve for conversion of absorbance ratio, A_{541}/A_{555}, to per cent carboxyhemoglobin saturation. (Reproduced by permission of the Institute for Clinical Science, from Tietz, N. W., and Fiereck, E. R.: The spectrophotometric measurement of carboxyhemoglobin. Ann. Clin. Lab. Sci., *3*:36-42, 1973.)

Table 18-11. CARBOXYHEMOGLOBIN EFFECTS*

Carboxyhemoglobin (%)	Response
10	Shortness of breath on vigorous muscular exertion
20	Shortness of breath on moderate exertion, slight headache
30	Decided headache, irritation, ready fatigue, disturbance of judgment
40–50	Headache, confusion, collapse, and fainting on exertion
60–70	Unconsciousness, respiratory failure, and death if exposure is continued
80	Rapidly fatal
Over 80	Immediately fatal

*From Deichmann, W. B., and Gerarde, H. W.: Symptomatology and Therapy of Toxicological Emergencies. New York, Academic Press, 1964.

Reference

Tietz, N. W., and Fiereck, E. A.: The spectrophotometric measurement of carboxyhemoglobin. Ann. Clin. Lab. Sci., 3:36–42, 1973.

DETERMINATION OF VOLATILE SUBSTANCES

Volatile substances include almost all types of chemical compounds; many are solvents commonly used in industry or in household products.

The principle of microdiffusion, used to separate many members of this group, can also be applied in specific and quantitative testing. The Conway unit with its various modifications greatly simplifies this type of separation. Some applications are indicated in Table 18-7.

Volatile substances are ideally suited for analysis by gas chromatography. Numerous methods have been described which are both convenient and accurate. One involves mixing the specimen with potassium carbonate in a tube closed with a stopper having a needle septum. After equilibration, a syringe is used to remove a portion of the "air" above the specimen and to inject it into the gas chromatograph. Some separations which can be achieved are shown in Table 18-9. Quantitative analyses also can be done by the head space technique, but pressure and temperature variables must be carefully controlled. Alternatively, blood or urine can be injected directly. On-column injection of specimens diluted with an internal standard is rapid, accurate, and precise.

In the following procedure, the three most commonly ingested alcohols (ethanol, methanol, and 2-propanol) and acetone, a metabolite of 2-propanol, can be determined simultaneously. Use of this procedure is particularly appropriate, since "alcohol" ingestion may frequently be missed if only ethanol is measured.

Determination of Alcohols by Gas Chromatography

Principle

Poisoning by alcohols almost always results in blood or serum levels that are quite high compared with those of most other toxic agents. This fact, coupled with the low boiling points of alcohols, permits gas chromatographic analysis at low temperatures using specimens which can be injected directly or after simple dilution. Normal constituents of biological specimens do not interfere. For example, if a flame ionization detector is used, the large excess of water in the specimen does not interfere significantly, since it will elicit only a minimal response by the detector.

If large numbers of specimens are to be analyzed, an automatic diluter is useful; occasional specimens can be diluted manually. For greatest accuracy, the use of a recorder with an integrator is recommended. If ±10% accuracy is tolerable, measurement of peak height can be used for quantitation. 1-Propanol is used in this procedure as an internal standard.[63]

Specimen

For specimen collection see page 1706.

Apparatus

Gas chromatograph, equipped with a flame-ionization detector, a glass column, 2 m by 2 mm, packed with 5% Carbowax 20 M on 60/80 Carbopack B. Column temperature, 85 °C; injection port, 100 °C; detector, 125 °C. Place a small plug of glass wool in the end of the column nearest the injection port to trap solids. The plug is easily removed and should be replaced periodically with a clean plug.

Reagents

All chemicals used should be anhydrous, analytical reagent grade.

1. Diluent stock solution, 23.4 g/L. Dilute 30.0 mL of 1-propanol to 1 L with water at 20 °C.

2. Diluent working solution, 234 mg/L. Dilute 10.0 mL of the diluent stock solution to 1.00 L with water.

3. Alcohol stock solutions. Dilute 3.00 mL each of anhydrous methanol, ethanol, and 2-propanol, respectively, to 100.0 mL with water. At 20 °C, this will result in concentrations as follows:

  a. Methanol, 23.7 g/L
  b. Ethanol, 23.7 g/L
  c. 2-Propanol, 23.6 g/L

4. Alcohol standard solutions. Dilute 1.00, 2.00, 4.00, 8.00, and 16.00 mL of each stock solution to 100.0 mL with water. This will result in standard solutions as follows:

  a. Methanol, 237, 474, 948, 1900, and 3790 mg/L
  b. Ethanol, 237, 474, 948, 1900, and 3790 mg/L
  c. 2-Propanol, 236, 472, 944, 1890, and 3780 mg/L

Procedure

1. Dilute the specimen 10-fold with diluent working solution.

2. Inject duplicate 1-μL aliquots of the prepared samples into the gas chromatograph, making sure that the stated operating conditions are met.

3. Inject 1-μL aliquots of each alcohol standard solution, suitably diluted with the diluent working solution.

Calculation

Calculate the ratio of the peak area or peak height of the unknown alcohol to that of the 1-propanol internal standard. Compare this ratio with similar ones obtained from the corresponding alcohol standard solutions. The ratios obtained from the reference solutions are linear functions of their respective concentrations from 0–4000 mg/L.

Accuracy and Precision

This method can easily detect concentrations of 100 mg alcohol/L blood. Lower concentrations can be detected by injecting larger volumes of diluted specimen into the gas chromatograph. The standard deviation of the method is ±30 mg/L when using peak-area ratios.

Interpretation

Under the stated conditions, 1-propanol has a retention time of about 5 min. The retention times for other volatile substances relative to 1-propanol are as follows: acetaldehyde, 0.16; acetone, 0.34; methanol, 0.25; 2-propanol, 0.66; ethanol, 0.5. (See Figure 18-8.)

Reference

Sunshine, I., Ed.: Methodology for Analytical Toxicology. Boca Raton, Fla., CRC Press, 1975, pp. 239-240.

ETHANOL

Ethanol (ethyl alcohol) is the most common toxic substance encountered. Not only is it lethal in its own right, but it is commonly a contributory factor in accidents of all types. In the

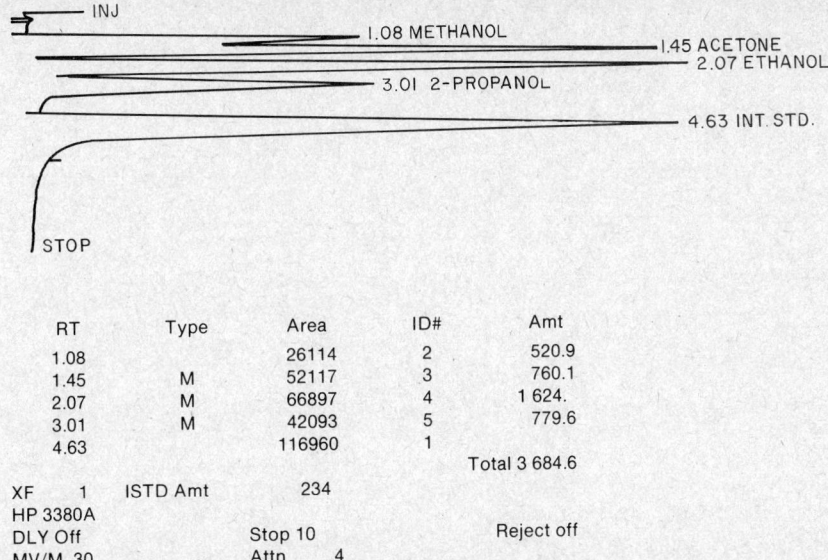

RT	Type	Area	ID#	Amt
1.08		26114	2	520.9
1.45	M	52117	3	760.1
2.07	M	66897	4	1 624.
3.01	M	42093	5	779.6
4.63		116960	1	
				Total 3 684.6

XF	1	ISTD Amt	234	
HP 3380A				
DLY Off		Stop 10		Reject off
MV/M .30		Attn	4	

Figure 18-8. Gas chromatographic separation of some commonly encountered volatile compounds.

case of a patient brought to the hospital in coma, the effect of alcohol, if any is present, must be ruled out in a differential diagnosis of the cause of coma.

There are probably more published methods for the determination of ethanol in blood than for any other toxic substance. In general, they can be divided into methods that are simple but nonspecific and methods that are specific but complex. The ideal method, as far as specificity and rapidity are concerned, is based on the use of gas chromatography. Even this procedure is not absolutely specific by itself, but it is extremely unlikely that an interfering substance would be encountered in biological samples. A commonly used method is an enzymatic one utilizing alcohol dehydrogenase.[44] An ultraviolet spectrophotometer is necessary for this test; however, this method lends itself well to automation.

Determination of Ethanol with Alcohol Dehydrogenase (ADH)

Principle

Ethanol is oxidized in the presence of ADH to acetaldehyde. In the course of this reaction, NAD, a coenzyme, is reduced:

$$C_2H_5OH + NAD^+ \underset{\text{ADH}}{\rightleftharpoons} CH_3CHO + NADH + H^+$$

The increase in NADH can be measured by the increase in absorbance at its absorption maximum of 340 nm. The equilibrium for this reaction lies strongly to the left. At neutral pH and at normal NAD concentrations, <1% of the ethanol present is oxidized to acetaldehyde. However, the reaction can be driven almost completely to the right by maintaining a high pH and removing the acetaldehyde as it is formed by reacting it with semicarbazide.

Both yeast and mammalian liver ADH can be used for this reaction. The Michaelis constant at pH 8 is 30 times greater for yeast than for liver ADH. The turnover number, however, is greater for yeast than for liver ADH. These differences are of only slight importance in analytical work when a large excess of the enzyme is used.

The ADH method is not specific; other alcohols can interfere at high concentrations. Despite this drawback, it is still used in many clinical laboratories because of its convenience and the relatively rare occurrence of other alcohol intoxications.

Many modifications of this method have been reported, and several commercial kits are available. Some vendors claim to have minimized or eliminated interferences due to alcohols other than ethanol (DuPont *aca*).

Specimen

Blood, serum, urine, and saliva are all appropriate specimens for this test. The site of venipuncture, finger puncture, or ear puncture should be cleansed and disinfected with aqueous Zephiran (benzalkonium chloride), aqueous Merthiolate (thimerosal), or other suitable disinfectant. Never use alcohol or other volatile disinfectants. If reusable containers, syringes, and needles are employed, they must not be cleaned or stored with alcohol or other volatile solvents. Fluoride, 5 mg/mL blood, is the best preservative, but citrate, oxalate, and heparin can be used as anticoagulants. All specimens must be stored under refrigeration until they are analyzed.

Procedure

For details, see the insert with the kit to be used.

Comments

This is an extremely sensitive test. Alcohol from outside sources must be carefully avoided as a contaminant.

Temperature and reaction time are not critical since the reaction proceeds to completion. However, if the temperature is lower than 30 °C, the reaction may not be complete after 10 min. This is one reason for checking the standard with each group of tests.

Whole blood is the specimen of choice (see below). Serum or plasma will have much higher ethanol levels than blood. The saliva ethanol concentration is 1.1–1.2 times greater than that of blood, while the urine concentration is quite variable.

Although ethanol is the preferred substrate for this enzyme, the rate of dehydrogenation of 2-propanol is 6%, that for 1-propanol is 36%, and that for 1-butanol is 17.5% relative to that of ethanol. The rate of dehydrogenation of methanol is negligible.

Interpretation

With blood ethanol levels of 500–1000 mg/L, various signs of intoxication may be observed: flushing, loquaciousness, slowing of reflexes, impairment of visual acuity, and so on; however, there is much individual variation in this regard. Individuals with levels >1000 mg/L are considered under the influence of alcohol, and depression of the central nervous system is more apparent. Because of impairment of good judgment and visual acuity as well as slowing of reflexes, driving a motor vehicle or operating machinery is hazardous when an individual is under the influence of alcohol.

With higher blood alcohol levels, central nervous system impairment is more pronounced and coma may appear at levels of 3000 mg/L. Death may occur with levels above 4000 mg/L.

In some areas, state laws require that blood alcohol levels, when measured for legal purposes, be stated in terms of per cent by weight. In these cases, weighing the sample is mandatory. For clinical work, pipeting the sample is more convenient. Interpretation of blood levels is the same in each case. When serum or plasma is the specimen analyzed, concentrations of ethanol average 20% higher than in whole blood, since alcohol distributes into the aqueous phase.

Reference

Lundquist, F.: In: Methods of Biochemical Analysis. D. Glick, Ed. New York, Interscience Publ., Vol. 7, 1959, p. 217.

METHANOL

Methanol (methyl or wood alcohol) is a widely used solvent in paints, varnishes, and paint removers. It is used alone as an antifreeze fluid and with ethanol and soap as a solid canned fuel. Poisonings are usually due to accidental ingestion by children or by alcoholics. In some areas, methanol may be a contaminant in "moonshine."

Methods for the Determination of Methanol

The most convenient and reliable method for methanol determination is gas chromatography. Such a method has already been described. (See p. 1703.)

Methanol can also be measured by a variety of other methods, most of which involve measuring the color intensity after oxidation of methanol to formaldehyde, followed by the development of a color by reacting formaldehyde with chromotropic acid (CTA):

$$CH_3OH + MnO_4^- + 2H^+ \rightarrow CH_2O + MnO_2 + 2\,H_2O$$

$$CH_2O + \text{(chromotropic acid structure)} \rightarrow CH_2O-CTA \text{ (Unknown structure)}$$

Chromotropic acid (CTA)   Violet color

These methods work well since chromotropic acid is a specific reagent for formaldehyde and hence for methanol after oxidation except in conditions mentioned below. The microdiffusion method referred to earlier is useful for the determination of methanol; it also utilizes CTA for color development.

The CTA colorimetric procedure for methanol has two major drawbacks. First, methanol is not quantitatively oxidized to formaldehyde. It is readily apparent that after formation of formaldehyde by the oxidation reaction just noted, the formaldehyde itself can be oxidized to formic acid and further to carbon dioxide as follows:

$$CH_2O + MnO_4^- + H^+ \longrightarrow CO_2 + MnO_2 + 2\,H_2O$$

This means that before a quantitative procedure can be devised, conditions must be chosen such that constant proportions of methanol are oxidized. Thus, the method is empirical; the set conditions must be established and rigidly observed before quantitative results can be achieved.

Second, the presence of reducing substances other than methanol will affect the system so that the procedure can no longer be applied quantitatively. The most common interference in cases of methanol poisoning is ethanol. It is not generally appreciated that the presence of ethanol invalidates a methanol procedure based on oxidation followed by CTA color development, if the calibration curve has been set using pure methanol standards.

The procedure of Hindberg and Wieth[36] obviates both drawbacks. First, the procedure must be carried out identically for standards and unknowns. Second, an excess of ethanol is added to both standards and unknowns. This results in a constant "interference" of a magnitude much greater than would ever be encountered in practice.

A third but minor drawback of any CTA procedure for determining methanol is the use of concentrated sulfuric acid for development of the final color. The dehydrating effect of concentrated sulfuric acid can produce formaldehyde from appropriate organic compounds, resulting in false high values. We have encountered this interference occasionally in patients with severe acidosis. Apparently, some substances may appear in a trichloroacetic acid filtrate, such as glycolic acid, which reacts as follows:

$$H_2C(OH)COOH \xrightarrow[\text{Heat}]{H_2SO_4} HCHO + CO + H_2O$$

Interpretation

Methanol poisioning is considerably more dangerous than poisoning due to ethanol since methanol is metabolized in man to formaldehyde and formic acid. The accumulation of formic and other acids severely reduces the alkali reserve, resulting in a metabolic acidosis. (See Chapter 10.) Furthermore, necrosis of the pancreas and serum amylase elevations have been demonstrated. Therefore, in addition to blood methanol levels, plasma carbon dioxide content, serum amylase determination, and electrolyte studies are useful laboratory tests for determining the severity of the poisoning and for following the progress of treatment.

Metabolites of methyl alcohol can damage the optic nerve, resulting in either temporary or permanent blindness. The mechanism of this effect is not well understood, nor is it a constant finding; nevertheless, prompt treatment of these cases not only may be lifesaving but also may preserve the eyesight.

As little as 2 teaspoonfuls (10 mL) of methanol is considered toxic; fatal results have been reported with dosages between 50 and 200 mL (2 and 8 ounces). A blood level greater than 800 mg/L is dangerous to life.

Treatment is two-fold. First, the acidosis is treated, generally with sodium bicarbonate, given both intravenously and orally. Second, ethanol is administered to saturate the alcohol dehydrogenase enzyme system. Since ethanol is the preferred substrate for this enzyme, this prevents the conversion of methanol to its toxic metabolites.

CYANIDE

Cyanide has a very high affinity for iron in the ferric state and it complexes with the ferric iron of cytochrome oxidase of mitochondria. Thus, cyanide inhibits cellular respiration because of its combination with important respiratory enzymes. This mechanism of action is the same whether cyanide is inhaled as the gas, hydrocyanic acid, or ingested as the potassium or sodium salt or other combined form. Since death follows very quickly if sufficient cyanide is absorbed, the patient rarely survives long enough for treatment. Despite this fact, it is desirable to have a test available for this poison in order to confirm a suspected cyanide death. A microdiffusion method may be used or the following procedure utilizing microdiffusion followed by measurement with an ion-specific electrode.[2]

Determination of Blood Cyanide by Ion-Specific Potentiometry

Principle

Cyanide is isolated from blood by microdiffusion. The concentration of the ion in the sodium hydroxide solution is determined by direct potentiometric measurement with a cyanide-specific electrode.

Reagents

1. Sulfuric acid, 0.5 mol/L. Add 14 mL concentrated H_2SO_4 to 500 mL water.
2. Sodium hydroxide, 0.1 mol/L. Dissolve 2 g NaOH in 500 mL water.
3. Lead acetate solution, 100 g/L. Dissolve 1 g $Pb(C_2H_3O_2)_2$ in 10 mL NaOH, 0.1 mol/L.
4. Stock solution, 1 mg cyanide/mL. Dissolve 25.0 mg KCN in 10 mL water.
5. Cyanide standards in whole blood, 0.5, 1, 2, and 5 mg/L. Prepare fresh by adding stock standard to whole blood.

Instrumental Requirements

pH meter equipped with cyanide-specific electrode (Orion Research).

Procedure

1. Lubricate edges of Conway unit with stopcock grease. Add 3.0 mL NaOH, 0.1 mol/L, to center well.
2. Add 4 mL of known normal blood (blank), blood standards, or specimen to outer compartment.
3. Add 1 mL of H_2SO_4, 0.5 mol/L, to outer compartment. Quickly seal the unit and mix by tilting and swirling gently.
4. Allow to diffuse 1–2 h at 37 °C.
5. Stabilize the cyanide electrode by placing in NaOH, 0.1 mol/L, for 30 min or until a constant voltage reading is attained.
6. Determine the electrode response to the sodium hydroxide solution from the center well of the Conway cell of each of the processed specimens.
7. After recording the initial reading, add 1 drop 10% lead acetate solution to each sodium hydroxide sample. A significant change in electrode potential indicates the presence of sulfide as an interfering substance. In this case record the electrode response to each of the solutions after adding lead acetate solution, and use this value for the calculations.

Calculation

Calculation is based on a response factor derived from a standard curve. This procedure cannot be controlled by the usual methods due to the instability in storage of cyanide-containing blood specimens.

Method Evaluation

Sensitivity: 0.01 mg/L; linearity: 0.1–1.0 mg/L; within-run CV: 5%; relative recovery: 100–109%.

Interferences

Although sulfide is known to interfere in this procedure, the addition of lead acetate to the final solution will confirm its presence or absence. Since the blood used in the preparation of standards will probably contain a low level of cyanide, a blood blank must be analyzed and the result subtracted from those of the standards and the patient specimen.

Reference

Baselt, R. C.: Analytical Procedures for Therapeutic Drug Monitoring and Emergency Toxicology. Davis, Cal., Biomedical Publ., 1980, pp. 94–97.

KEROSENE

Kerosene and other petroleum hydrocarbons are frequently ingested accidentally by children. Any hydrocarbon aspirated into the lungs or absorbed material excreted into the lungs from the general circulation can produce a dangerous chemical pneumonitis. Frequently, the odor of the breath is an indication that this material has been ingested.

Kerosene is used as a fuel for home heating in some areas and is also a common solvent for household products.

The aliphatic hydrocarbons are chemically inert, which makes their detection exceedingly difficult. Generally, one can rely on the characteristic odor of this substance in gastric contents as a positive test. If a sufficient quantity of the hydrocarbon is present and can be physically separated, it can be identified by its physical constants. Some aliphatic hydrocarbon constituents can be detected by head-space gas chromatography (p. 1693).

CORROSIVES

This group includes those strong mineral acids or fixed alkalis that produce chemical burns on contact. There are no good tests that can be carried out on blood, serum, or urine by which the type of acid or alkali can be detected and the ingested quantity estimated.

The only specimen that can be examined profitably is gastric contents. Frequently, this specimen is not available unless the patient has vomited, since gastric lavage is contraindicated in this type of poisoning. If gastric contents are available, the pH should be measured. Ions such as Na^+, K^+, Cl^-, SO_4^{2-}, and PO_4^{3-} can be demonstrated by routine methods used in the clinical laboratory. Obviously, most of the common ions would be present normally in gastric contents. To be of significance in this type of case, a large excess must be present. Since many compounds in the group of corrosives can cause major disturbances in acid-base balance, it is advisable to perform electrolyte studies on blood. Usually the clinician has evidence from lesions in the mouth and esophagus that a corrosive substance has been ingested.

METALS

All metals are toxic if a sufficient quantity is absorbed. Generally they are not encountered in their toxic form in the elemental or free state, but rather in the form of salts. The degree of toxicity of a given metal is dependent on the solubility of the salt; the greater the solubility, the more likely it is that it will be absorbed and the greater will be its toxicity. For example, barium chloride is soluble and extremely toxic, but barium sulfate is insoluble enough to be given orally as a radiopaque medium prior to X-ray examinations of the gastrointestinal tract.

ARSENIC

Arsenic, despite its reputation, is not a common poison. It is still a favorite homicidal poison, but homicidal poisonings are rare. Since arsenic is an ingredient in some herbicides and insecticides, accidental poisonings, both acute and chronic, may still be encountered on occasion.

Clinically, the symptoms of both acute and chronic arsenic poisonings can easily be confused with a variety of other conditions. It is not uncommon, therefore, for a clincian to request that the presence of arsenic be ruled out as an aid in the differential diagnosis. The specimen of choice in this case is urine, even if 2–3 weeks have elapsed after ingestion of the poison. In long-term chronic cases, analysis of hair and nails may be informative, but results are difficult to interpret. (See Interpretation.)

In the older literature, arsenic is frequently described as a "protoplasmic poison." This term is as good as any for describing the mode of action of this element. Arsenic combines readily with proteins because of its great affinity for sulfhydryl groups. This results in the precipitation of proteins, producing gastrointestinal irritation and irreversible inhibition of important enzyme systems, which are important toxic effects of arsenic. The great affinity of arsenic for tissue proteins is also responsible for the rapid removal of arsenic from the blood. Blood, therefore, is not an appropriate specimen except in cases in which a large overdose of arsenic has been ingested.

Arsenic is a trace element that is present in all human tissues. Concentrations may be elevated in specimens from persons who are occupationally or environmentally exposed to this toxic metalloid. Chronic poisoning is best diagnosed by the analysis of hair, whereas urine is the preferred specimen for the early confirmation of acute poisoning.

A colorimetric technique is described which is simple enough to be performed in nearly any laboratory. A more sensitive and specific procedure based on atomic absorption spectrometry is also included. Prior destruction of the organic matrix of the specimen has been described (p. 1697).

Determination of Arsenic by Colorimetry[2]

Principle

A previously ashed specimen is subjected to Zn-HCl arsine generation in a special apparatus (Figure 18-9) following conversion of arsenic to the trivalent form. Hydrogen sulfide formed during the digestion and acidification processes is removed with a glass wool plug soaked in lead acetate solution. The arsine is chelated with diethyldithiocarbamate (DDC) and the absorbance of the chelate measured in a spectrophotometer at 540 nm.

Reagents

1. Stock solution, 1 mg arsenic/mL (Fisher Scientific Co., reference standard SO-A-449). Alternatively dissolve 1.32 g As_2O_3 in 10 mL of NaOH, 400 g/L, and dilute to 1 L with water.
2. Arsenic standards, 0.5, 2, 5, and 10 mg/L in water.

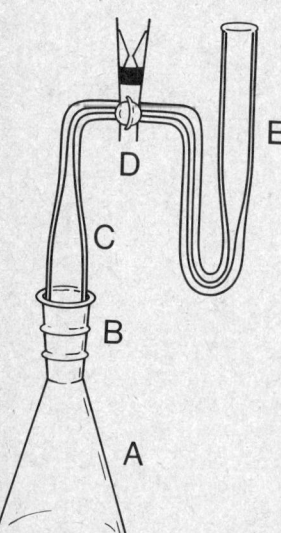

Figure 18-9. Glass apparatus for generating arsine, removing hydrogen sulfide interference, and color development of arsenic-diphenyldithiocarbamate complex. A = generator; B = standard taper neck; C = scrubber; D = ground glass ball-and-socket joint; E = absorber.

3. DDC solution. Dissolve 0.5 g silver diethyldithiocarbamate in 100 mL pyridine. Stable if stored in amber bottle.

4. Hydrochloric acid, concentrated.

5. Stannous chloride solution, 40 g $SnCl_2$ in 100 mL concentrated HCl.

6. Potassium iodide, 150 g/L. Dissolve 15 g KI in 100 mL water.

7. Lead acetate, 100 g/L. Dissolve 10 g $Pb(C_2H_3O_2)_2$ in 100 mL water.

8. Zinc, 20 mesh, low-arsenic content.

9. Arsine generator, Fisher Scientific Co. #1-405.

Procedure

1. Transfer 3.0 mL of the previously ashed specimen to the arsine-generator flask and dilute to about 35 mL with water. A water blank and four aqueous standards, previously carried through the identical ashing procedure as the specimen, are to be analyzed in the same manner.

2. Add 5 mL concentrated HCl; 2 mL KI, 150 g/L; and 8 drops $SnCl_2$ solution. Swirl and allow to stand for 15 min.

3. Wet a pledget of glass wool in the lead acetate solution and place it loosely into the scrubber tube of the arsine generator. Lubricate all joints with stopcock grease. Add 3.0 mL DDC solution to the absorber tube.

4. Add 3 g zinc to the generator flask and immediately insert the scrubber-absorber assembly. Allow arsine evolution to continue for 2 h.

5. Transfer the DDC solution to a 3-mL cuvet and measure the absorbance at 540 nm, using the DDC solution from the water blank as a reference. (This solution may also be analyzed by flameless atomic absorption spectrometry.)

Calculation

Prepare a standard curve of absorbance versus concentration of the aqueous standards. Determine the specimen concentration from this curve.

Method Evaluation

Sensitivity: 0.2 mg/L; linearity: 0.2–10 mg/L; day-to-day CV: 4–9%; relative recovery: 97–108%.

Interferences

Antimony will be detected as arsenic if it is present in sufficiently high concentration, but this is unlikely. The antimony hydride, stibine, forms a chelate which has maximum absorbance at 510 nm. Chromium, copper, and molybdenum may also interfere with the evolution of arsine if present in high concentrations.

Reference

Baselt, R. C.: Analytical Procedures for Therapeutic Drug Monitoring and Emergency Toxicology. Davis, Cal., Biomedical Publ., 1980, pp. 35-36.

Determination of Arsenic by Atomic Absorption Spectrometry[2]

Principle

A previously ashed specimen is treated to convert arsenic to the trivalent form. Arsenic is chelated with diethyldithiocarbamate; the chelate is extracted into chloroform, partitioned back into water, and an aliquot analyzed by flameless atomic absorption spectrometry at 193.7 nm.

Reagents

1. Stock standard solution, 1 mg arsenic/mL (Fisher Scientific Co., reference standard SO-A-449).

2. Arsenic standards, 0.2, 0.5, 1, and 2 mg/L in water.

3. Hydrochloric acid, 8 mol/L.

4. Potassium iodide, saturated.

5. Chloroform, U.S.P.

6. DDC solution, 2 g diethylammonium diethyldithiocarbamate in 100 mL water (prepare fresh).

7. Nitric acid, 2 mol/L.

Instrumental Conditions

1. Atomic absorption spectrometer with graphite furnace.
2. Arsenic electrodeless discharge lamp.
3. Furnace program: dry 20 s at 100 °C; char 20 s at 500 °C; atomize 15 s at 2100 °C.
4. Measure absorption at 193.7 nm.

Procedure

1. Transfer 0.5 mL of a previously ashed specimen to a 15-mL centrifuge tube. A water blank and four aqueous standards, previously carried through the identical ashing procedure as the specimen, are to be analyzed in the same manner. Add 0.3 mL HCl, 8 mol/L, and 0.1 mL saturated KI and vortex.

2. Add 2 mL chloroform and 0.1 mL DDC solution. Vortex for 15 s and allow layers to separate.

3. Discard the upper layer and transfer 1.0 mL of the chloroform layer to a clean tube. Add 1.0 mL HNO$_3$, 2 mol/L, and vortex for 15 s.

4. Introduce 10–30 μL of the water phase into the graphite furnace and analyze.

Calculation

Prepare a standard curve of absorption versus concentration of the aqueous standards. Determine the specimen concentration from this curve.

Method Evaluation

Sensitivity: 0.1 mg/L; linearity: 0.2–2.0 mg/L; day-to-day CV: 10%; relative recovery: not established.

Interferences

None known.

Interpretation

Normal arsenic levels in urine are <0.05 mg/L. In cases of chronic poisoning, arsenic levels in urine will rise to 0.1 mg/L; in acute poisoning, 1.0 mg/L or more may be present.

Since arsenic is readily bound by sulfhydryl groups of protein, considerable arsenic is bound by keratin and subsequently deposited in hair and nails. This phenomenon has led to the analysis of hair and nails in an effort to determine whether a previous exposure to arsenic has occurred. Interpretation of these analyses is difficult because of the problem of differentiating between surface contamination of the hair and endogenous arsenic. If such an examination is required, a minimum of 1.0 g of clean hair (a large handful), clipped close to the scalp, should be collected. For specimen collection, see also Chapter 8C.

Reference

Baselt, R. C.: Analytical Procedures for Therapeutic Drug Monitoring and Emergency Toxicology. Davis, Cal., Biomedical Publ., 1980, pp. 36-37.

LEAD

Lead is still one of the most serious metallic poisons. In adults, inorganic and organic lead compounds may be encountered in industrial exposures. An increasing awareness of this danger has promoted the use of prophylactic measures. Education of workers about the hazards of lead intoxication has also been of help in minimizing industrial poisonings.

Unfortunately, children are particularly sensitive to lead poisoning and the exposure of children to lead-containing paint and plaster, particularly in low income housing, has continued despite regulations, labeling laws, and attempts to educate the public. Severe poisoning in a child can cause lead encephalopathy, which has a high mortality rate. Those children who survive frequently show evidence of permanent central nervous system damage.

The diagnosis of lead poisoning is difficult, and the demonstration of an elevated lead concentration in blood or urine constitutes the most positive indication of absorption of a lead compound. Being a ubiquitous element, lead is normally present in trace amounts in biological material. Analytical procedures are extremely sensitive and must be conducted with great care in order to achieve valid results. This requirement generally makes lead analyses a function of a special laboratory, particularly one which has experience with trace metal analyses and their special problems. Problems which are encountered can be illustrated by some facts relating to lead analysis. An average normal lead concentration in blood is 0.3 mg/L and an amount of 1 mg/L represents a toxic level. Thus, 5 mL of the normal blood specimen contains 1.5 μg, and the abnormal sample contains 5 μg. Obviously any method used must not only be extremely sensitive but also have an excellent accuracy and precision record in order to discriminate between the two extremes of the 3.5-μg range separating the normal and toxic lead levels in blood. In addition, all of the glassware and reagents used in the analysis contain traces of lead. Even after careful selection of reagents and cleaning of glassware, the analyst must still exercise meticulous technique in order to ensure low blank values of lead.

Detection of Lead or Lead Poisoning

The actual analysis may follow one of many techniques. The most reliable methods include colorimetric analysis with diphenylthiocarbazone,[27] polarography,[5] and atomic absorption spectrophotometry.[2]

The clinical laboratory performs two very important functions that aid in the diagnosis of lead poisoning, even if the lead analysis is done by others. First, the specimens to be analyzed must be collected in a valid way, that is, free of contamination. Second, other diagnostic tests can be done for screening purposes or for confirmation. These tests are based on the effects of lead on erythropoiesis. Lead interferes in the biosynthesis of hemoglobin, which results in anemia. Three precursors of hemoglobin that accumulate in lead poisoning are δ-aminolevulinic acid, coproporphyrin III, and protoporphyrin IX. Methods for the detection of these substances and related enzymes are discussed in Chapter 16.

Specimen Collection

For collection of a *24-h urine* specimen, the patient should void directly into a lead-free container (a borosilicate glass or polyethylene container from which surface lead has been removed by washing, then rinsing with hot nitric acid, 1 mol/L, and rinsing twice with metal-free water). A preservative should not be added because it might contaminate the specimen. The entire specimen or a minimum of 100 mL is submitted to the toxicological laboratory for analysis, after recording the total volume. Catheterized specimens should not be used unless it is unavoidable. In this case, the catheter should be cleansed (as just noted) to remove surface lead before sterilization. In some cases we have found that an indwelling catheter through which urine has been flowing freely for 24–48 h is usually free from surface lead. The possibility of contamination always exists when catheterized specimens are submitted for analysis. In an emergency, it may be necessary to analyze a random urine specimen rather than a 24-h urine specimen. In such a case, the specimen must be collected with the same care as just outlined. Interpretation of the result is subject to the same difficulty as discussed next in connection with blood specimens.

Blood specimens can be analyzed as readily as urine, but lead levels may fluctuate widely in different blood specimens from the same patient. We have had the experience of seeing apparently normal lead levels in occasional blood specimens from patients with documented lead poisoning. If the test is to be performed on blood, the exact lead content of anticoagulant or preservative must be known so that proper correction can be made. The needle, syringe, test tube, and stopper should be of lead-free material, cleaned as previously described. Special tubes for blood-lead collection are commercially available. Since most of the lead is in the erythrocytes, a serum lead level is of little value.

As with any analysis of trace substances, the sensitivity of the analysis and the expected level of the analyte determine the amount of specimen to be collected. For example, if a lead method is used that is sensitive to 1 μg of lead and in which the known reagent blank is also 1 μg, and the expected blood level is within normal limits, or about 0.3 mg/L, then a minimum of 10.0 mL of blood must be collected. This quantity of specimen would contain 3 μg of lead, a level that can be differentiated from a blank with some degree of validity. Methods currently available have sensitivities several orders of magnitude less than 1 μg and permit smaller sample sizes to be analyzed.

Reference Ranges

Normal lead levels range up to 0.08 mg/L of urine or 0.3 mg/L of blood. Levels higher than normal indicate increased absorption of lead compounds; levels greater than 0.1 mg/L of urine, or 1 mg/L of blood, are usually associated with signs and symptoms of lead poisoning. Normal blood lead levels in children are 0.15–0.20 mg/L. In this age group, levels of 0.4 mg/L represent an abnormal exposure to lead compounds. Some clinicians prefer urine lead levels to be reported on a per diem basis. It is however, preferable to report these levels in mg/L together with the total volume of the 24-h specimen. This allows the clinician to correlate the 24-h excretion of lead with other factors that may be related to an excessively high or low urinary output.

Determination of Lead by Atomic Absorption Spectrometry

Principle

The valid determination of the lead concentration in a biological specimen is the most definitive diagnostic test for lead poisoning. The analytical problem is particularly great in children, where the need exists for methods requiring small volumes of blood. Classical colorimetric methods in which dithizone is used require very large sample volumes, and even conventional atomic absorption spectrophotometry requires samples of 1 mL or more.

Conventional atomic absorption spectrophotometry necessitates processing a blood specimen by any of a variety of methods prior to actual measurements. This includes ashing the specimen, precipitating protein with trichloroacetic acid followed by direct measurement of lead in the supernatant, or measuring absorption after concentration of lead by solvent extraction. All of these suffer from poor sensitivity, they necessitate rather large blood samples, and they are prone to reagent contamination.

The use of a graphite furnace circumvents many of these problems. Application of this method to 0.5 μL samples of packed red cells or 1 μL of diluted whole blood illustrates the remarkable sensitivity of the technique.

Delves conceived a microsampling atomic absorption method in which a nickel cup and an absorption tube mounted in the flame are used. This technique offers great potential in screening programs and in routine clinical analyses because it accepts 10 μL volumes and is very rapid. Procedures based on this technique require that the analysis be performed by the method of additions to compensate for the matrix effects of blood, and, in the Delves technique, to compensate for poor response to aqueous standards. This increases the time and effort required for the analysis and increases pipeting errors. The "method of additions" involves analysis of the specimen followed by analysis of another specimen to which a known quantity of lead has been added.

Determination of Blood Lead by Graphite Furnace Atomic Absorption Spectrometry[2]

Principle

Whole blood is diluted with a surfactant and injected directly into the graphite furnace of an atomic absorption spectrometer.

Reagents

1. Stock lead standard, 1 mg lead/mL (Fisher Scientific Co., reference standard).
2. Aqueous standards, 0.10, 0.20, 0.40 and 0.80 mg/L. Prepare fresh.
3. Triton X-100 surfactant solution, 0.1% in water.

All glassware used in the procedure should be soaked overnight in 50% nitric acid and rinsed in distilled-deionized water.

Instrumental Conditions

1. Atomic absorption spectrometer with graphite furnace and deuterium background corrector.

2. Lead hollow cathode lamp or electrodeless discharge lamp.
3. Furnace program: dry 30 s at 125 °C; char 40 s at 525 °C; atomize 13 s at 2000 °C.
4. Nitrogen purge gas, 15 mL/min.
5. Measure absorption at 283.3 nm.

Procedure

1. Transfer 50 μL heparinized whole blood to a 12 $\times$ 75 mm tube containing 200 μL surfactant solution. Rinse the pipet tip several times in the solution. Vortex.
2. Inject 15 μL of the solution into the graphite furnace and begin the analysis program.

Calculation

Calculation is based on a standard curve prepared each time specimens are analyzed.

Method Evaluation

Sensitivity: 0.05 mg/L; linearity: 0.10–0.80 mg/L; within-run CV: 2.4–4.1%; relative recovery: 94–104%.

Interferences

Glassware and reagents should be carefully checked for lead contamination by analyzing reagent blanks. Anticoagulants other than oxalate do not interfere. Storage of whole blood specimens beyond several days usually results in poor recovery of lead.

Comments

In any method such as this, when extremely small samples are used and the sensitivity of the method is low, considerable attention must be given to details. The precision of the microscale pipeting is critical to the total precision of the procedure.

Lead standards deteriorate rapidly in polystyrene or borosilicate glass containers exposed to light. This is probably due to adsorption of lead ions on the glass or plastic surface. Aqueous solutions containing lead at 0.5 or 1.0 μg Pb/L are stable for several days in polystyrene or polyethylene containers wrapped in layers of carbon paper. Upon exposure to light, a solution containing 0.5 μg Pb/L lost 20% of the lead within 3 h, 50% within 6 h, and 90% within 24 h. More concentrated solutions saturate the container surface and, after losing about 0.5 μg Pb/L, remain constant for several days. Borosilicate adsorbs more lead than polystyrene. To prevent errors, solutions containing less than 0.2 μg Pb/L should be prepared in a darkened room and analyzed immediately.

No appreciable lead loss occurs in urine stored under conditions similar to those described for aqueous solutions. Urine stored without a preservative in clear polystyrene containers, exposed to light at room temperature, showed no significant lead change for a period of 10 d. These observations of Kopito and Shwachman emphasize the problems associated with trace metal analysis.[43]

Reference

Baselt, R. C.: Analytical Procedure for Therapeutic Drug Monitoring and Emergency Toxicology. Davis, Cal., Biomedical Publ., 1980, pp. 147–148.

Determination of Lead by Anodic Stripping Voltammetry

Electrochemical techniques such as polarography have long been considered important in analysis of inorganic substances. The potential at which a particular species is oxidized or reduced is a qualitative characteristic of the analyte, and the magnitude of the current flowing at that potential is related to the concentration of the analyte.

In anodic stripping voltammetry (ASV), a mercury electrode in contact with the solution to be analyzed is held at a negative potential greater than necessary to reduce the metal ion. This causes the plating of the metal into the mercury of the electrode. The potential is then rapidly and linearly changed to a positive or anodic direction. When the oxidation-reduction potential of a metal/metal ion is reached, the metal will be reoxidized back into solution, resulting in a current flow which is measurable. When the test electrode consists of a thin film of mercury deposited on an inert substrate, a sharp, well-defined current peak will occur.

The potential at which a peak occurs serves to identify the mercury-soluble metals in solution, and the height of the peak is related to the amount of metal in solution. Thus, several different metals may be measured simultaneously at a low concentration. This method has excellent sensitivity; the sample size can be small and sample preparation and manipulation can be minimized.

Application of this procedure to lead analysis in blood is particularly attractive in a clinical setting since the method is rapid, simple, and reproducible.[56] When coupled with a microfluorometric method for estimating protoporphyrin in erythrocytes (Chapter 16), the procedures quickly generate useful data for judging the severity of lead poisoning relative to the clinical state of a patient (Table 18-12).

BROMIDES

Bromides are used in both organic and inorganic forms in medicine, chiefly for the purpose of sedation. These drugs are sometimes abused or may be taken in overdosage accidentally. The nonprescription status of drugs containing bromide makes them easily available to the patient predisposed to drug abuse.

Determination of Bromide in Serum

Principle

The procedure to be described[35] measures free Br^- only; thus, the bromine in most of the organic compounds is not detected. However, when organic bromides are ingested they are metabolized to inorganic bromide. (See Interpretation.)

The bromide anion readily displaces chloride from gold trichloride, forming gold tribromide:

$$AuCl_3 + 3Br^- \longrightarrow AuBr_3 + 3 Cl^-$$

The formation of gold tribromide may also be accompanied by the formation of $AuBrCl_2$ and $AuBr_2Cl$. The resulting brown color is very stable in acid solution and can be read quantitatively in a spectrophotometer at 440 nm.

Reagents

1. Trichloroacetic acid, 100 g/L in water.
2. Gold(III) chloride solution. Wash the contents of a 1.0-g ampule of gold chloride into a 200-mL volumetric flask and dilute to the mark with water. The solution is stable.
3. Trichloroacetic acid (100 g/L)–sodium chloride (0.6 g/L) mixture. Place 0.6 g of NaCl in a 1-L volumetric flask and add 500 mL of water. Add 100 g of trichloroacetic acid and dilute to volume with water.
4. Standards:

   a. Stock, 10 g/L. Weigh exactly 1.000 g of NaBr, reagent grade, dissolve in water and dilute to 100 mL.

Table 18-12. RELATIONSHIP OF BLOOD LEAD AND FREE ERYTHROCYTE PROTOPORPHYRIN (FEP) CONCENTRATIONS

	Interpretation	Blood Concentrations	
		Lead (mg/L)	FEP (mg/L)
I.	Normal	<0.29	<0.59
Ia.	Probable Fe deficiency	<0.29	0.60–1.89
Ib.	Probable transient or declining Pb level	0.30–0.49	<0.59
II.	Minimal elevation of Pb	0.30–0.49	0.60–1.09
III.	Moderate elevation of Pb	0.50–0.79	1.10–1.89
IV.	Extreme elevation of Pb	>0.80	>1.90
	Erythropoietic protoporphyria	<0.29	>1.90
	Probable Pb contamination—repeat Pb assay	>0.50	<1.10

b. Dilute standard, 0.5 g/L. Pipet 10.0 mL of stock standard into a 200 mL volumetric flask and dilute to volume with the trichloroacetic acid–NaCl mixture.

Procedure

1. Prepare a 1:10 trichloroacetic acid filtrate of serum.
2. Pipet 5.0 mL of clear filtrate (sample) into one tube and 5.0 mL of 100 g/L trichloroacetic acid solution (blank) into a second tube.
3. Prepare standards as follows:
   a. Pipet 0.5 mL of dilute standard into a labeled tube and add 4.5 mL of trichloroacetic acid–NaCl mixture. Mix well (corresponds to 500 mg NaBr/L).
   b. Pipet 2.0 mL of dilute standard into a labeled tube and add 3 mL of 100 g/L trichloroacetic acid–NaCl mixture. Mix well (corresponds to 2000 mg NaBr/L).
4. Add 0.5 mL of 5 g/L AuCl$_3$ solution to all tubes. Mix well.
5. Read at 440 nm.

Calculation

$$\frac{A_{\text{unknown}}}{A_{\text{standard}}} \times \text{concentration of standard} = \text{mg NaBr/L}$$

Note: Use the standard whose absorbance is closest to that of the unknown.

Interpretation

Although normal bromide levels in serum are 8–15 mg/L, this method may occasionally give results up to 50 mg/L, even with normal serum. It has been suggested that this is due to a slight turbidity that may at times develop. Therapeutic levels may be in the order of 1000 mg/L, and toxic levels are usually >1500 mg/L. With a single overdose of an organic bromide compound, serum levels of inorganic bromide do not rise above normal levels. After prolonged therapy with these drugs, serum levels of inorganic bromide may increase to >1000 mg/L. At these levels, mental disturbances may be elicited.

Reference

Hepler, O. E.: Manual of Clinical Laboratory Methods. Springfield, Ill., Charles C Thomas, 1963, p. 325.

FLUORIDE

Sodium fluoride is a common ingredient in roach and ant poisons and, as such, it is frequently kept around the house and even in the kitchen. Accidental poisonings have occurred, especially since the white crystalline material can be mistaken for ordinary salt or baking powder. In recent years a blue dye has usually been added to these preparations to avoid this type of accident.

The fatal dose of sodium fluoride is 5–10 g. Once the compound reaches the stomach, the acidity of the gastric contents converts the salt to free hydrofluoric acid, which produces a dark red corrosion of the mucous membrane. For this reason, inorganic fluorides could also be classified with the corrosives.

A number of organic fluoride compounds are extremely toxic, and one of these, sodium fluoroacetate (sometimes called "1080"), has been used as a rat poison. The toxicity of this substance is due to its competition with acetate in the tricarboxylic acid cycle, with the eventual formation of fluorocitric acid. It is estimated that a lethal oral dose in man is ~50 mg.

Despite the marked toxicity of these substances, accidental, suicidal, or homicidal poisonings have not been common in the past. This was fortunate for the analyst because of the difficulty and length of time needed for fluoride analysis. Now, with the aid of ion-specific electrodes, analysis has been greatly simplified. Plastic containers are ideal for collecting specimens as well as for conducting the analysis, since silica in glassware reacts with fluoride to form a volatile product, resulting in loss of fluoride.

Rieders[52] has described a screening test using modified polypropylene Conway cells. The method takes about 1 h for completion.

Normal fluoride concentrations in plasma are 0.1–0.20 mg/L; in urine, 0.2–1.1 mg/L.

Determination of Fluoride in Plasma and Urine by Ion-Specific Potentiometry[2]

Principle

Plasma and urine specimens are diluted with a pH 5 buffer solution. Ionic fluoride is determined by direct ion-specific electrode potentiometry.

Reagents

1. Stock standard solution, 100 mg fluoride/L. Dissolve 221 mg NaF in 1 L of water. Store in a polyethylene container.
2. Aqueous standards, 0.01, 0.05, 0.10, and 0.20 mg/L. Prepare fresh in buffer solution.
3. Buffer solution. Dissolve 6.4 g NaCl in 1 L acetate buffer, pH 5.0, 0.05 mol/L. Store in a polyethylene container.

Instrumental Conditions

Digital pH meter (Orion Research Inc., Cambridge, MA, model 801) with fluoride-specific electrode (Orion Research model 94-09) and KCl reference electrode (Corning Glass Works, Medfield, MA).

Procedure

1. Plasma: Dilute 1 mL plasma with 1 mL buffer solution. Urine: Dilute 1 mL urine with 9 mL buffer solution.
2. Place the electrode into the solution and allow 30-min equilibration prior to recording the ion potential.

Calculation

Calculation is based on a response factor derived from a standard curve. Multiply urine values by 5 to correct for the additional dilution.

Evaluation

Sensitivity: 0.01 mg/L; linearity: 0.01–0.20 mg/L; within-run CV: 5.6%; relative recovery: average 97%.

Interferences

Fluoride contamination by chemicals, water, and glassware may be minimized by using high-purity reagents and polyethylene containers. Specimen containers should be checked for contamination by adding a volume of buffer solution to the containers, allowing equilibration for several hours, and analyzing the solutions for fluoride. Total fluoride concentrations in both plasma and urine are approximately twice those of ionic fluoride, which is measured by this technique; total fluoride may be determined by first treating specimens with an equal volume of perchloric acid, 1 mol/L, and heating at 100 °C for 15 min prior to following this procedure.

Reference

Baselt, R. C.: Analytical Procedures for Therapeutic Drug Monitoring and Emergency Toxicology. Davis, Cal., Biomedical Publ., 1980, pp. 118–119.

BARBITURATES

In cases associated with drug overdose, barbiturates are leading offenders. These drugs are commonly prescribed as treatment for a variety of conditions. Because of their availability, they are frequently the cause of accidental poisoning. Since they are hypnotics (sleep-producing drugs), they are commonly used for suicide or in suicide attempts. There are many individual drugs in this group, but all are chemically similar and produce sleep and, upon overdosage, coma and death. They are chemically characterized by a pyrimidine ring with two substitutions on carbon atom 5. (See Table 18-13.)

Some barbituric acid derivatives may have a methyl group substituted on one of the N atoms of the ring (e.g., mephobarbital and hexobarbital); others have an S atom instead of oxygen at carbon atom 2 (e.g., thiopental and thiamylal). The sulfur derivatives are used as anesthetic agents and are not likely to be encountered in cases of overdosage.

Since these drugs can be readily extracted from blood, serum, or urine, and since they all have good ultraviolet absorption properties, some reliable methods for determining barbiturates in biological materials require the availability of an ultraviolet spectrophotometer, preferably equipped with a recorder.

Another photometric method has been reported in which a complex is formed between mercury and barbituric acid derivatives. The complex is soluble in organic solvents, and the barbiturate can be estimated by using diphenylthiocarbazone to determine the amount of mercury present in the organic solvent. A similar method, which is rapid and does not require a colorimeter, has been described by Curry.[15] This method is useful for blood levels greater than 20 mg/L.

Determination of Barbiturates by Ultraviolet Spectrophotometry

Principle

Ultraviolet spectrophotometric methods for the determination of 5,5-disubstituted barbiturates in biological material are based on the fact that these compounds exist in three forms in solution: a nonionized form in acid solution, with almost no absorption in the range 230–270 nm;

Table 18-13. COMMON BARBITURATES AND THEIR TOXIC BLOOD LEVELS

Generic Name	R₁	R₂	Blood Level When Consciousness Regained (mg/L)
Barbital (Veronal)	$—CH_2—CH_3$	$—CH_2—CH_3$	80
Phenobarbital (Luminal)	$—CH_2—CH_3$	$—C_6H_5$	50
Butabarbital (Butisol)	$—CH_2—CH_3$	$—\overset{CH_3}{\underset{H}{C}}—CH_2—CH_3$	30
Amobarbital (Amytal)	$—CH_2—CH_3$	$—CH_2—CH_2—\overset{CH_3}{\underset{CH_3}{C}}—H$	30
Pentobarbital (Nembutal)	$—CH_2—CH_3$	$—\overset{CH_3}{\underset{H}{C}}—CH_2—CH_2—CH_3$	10
Secobarbital (Seconal)	$—CH_2—CH=CH_2$	$—\overset{CH_3}{\underset{H}{C}}—CH_2—CH_2—CH_3$	10

the first ionized form at pH 9.8–10.5 with an absorption maximum at 240 nm; and the second ionized form at pH 13–14 with an absorption maximum at 252–255 nm and a minimum at 234–237 nm. The three forms of the drug can be represented as follows:

Acid form   *First ionized form*   *Second ionized form*

The 1,5,5-trisubstituted barbiturates exist only in two forms in solution, since they lack one enolizable hydrogen.

Acid form   *Ionized form*

The nonionized form in acid solution has almost no absorption in the range of 230–270 nm; the first and only ionized form at pH 9.8–14 has an absorption maximum at 245 nm.

In the acid form, barbiturates are relatively water insoluble, but they are soluble in organic solvents. In both the first and second ionized forms these drugs are very water soluble but are insoluble in organic solvents. Thus, they can be extracted from blood or serum at physiological pH values or from acidified urine by organic solvents. By washing the organic solvent with an aqueous phosphate buffer of pH 7.4, some interfering impurities, e.g., salicylates, can be removed, although some loss of barbiturates occurs. Shaking the organic solvent with dilute alkaline solution converts the free acid form of barbiturate into its salt, resulting in the transfer of the barbiturate into the aqueous phase. This aqueous extract is used for scanning in the ultraviolet spectrophotometer.

Proper interpretation of a blood or serum barbiturate level cannot be done unless the type of barbiturate present is known. For example, a 10 mg/L concentration of barbital is not too serious, but the same level of secobarbital is close to a lethal level (Table 18-13). Also, 20 mg/L of phenobarbital in blood is a therapeutic level, but the same level of secobarbital can be lethal.

Since the prognosis in a given case of overdosage is influenced by the type of barbiturate involved, it is important to identify the drug or to determine the type of barbiturate present. This is done by treatment of the barbiturate with hot alkaline solution, by thin-layer chromatography, or by GLC (see pp. 1683 and 1724).

About 29 different barbituric acid derivatives are, or have been, used clinically. Of these, only about six or eight are commonly prescribed and available to the general public.

Pharmacologically, the barbiturates can be classified according to their duration of action. Four groups are commonly described: long-, intermediate-, short-, and ultrashort-acting.[32]

1. *Long-acting*: barbital, phenobarbital, mephobarbital, diallylbarbituric acid.
2. *Intermediate-acting*: amobarbital, aprobarbital, butabarbital, hexethal.
3. *Short-acting*: cyclobarbital, pentobarbital, secobarbital.
4. *Ultrashort-acting*: hexobarbital, thiamylal, thiopental.

The ultrashort-acting barbiturates are used exclusively as anesthetic agents and consequently are rarely encountered in cases of accidental or intentional overdosage. Thus, the balance of this discussion is restricted to the classification of the first three groups: long-acting, intermediate-acting, and short-acting barbiturates.

Short-acting barbiturates may be distinguished from long- and intermediate-acting compounds by heat treatment of the alkaline extract (step 9 of the procedure, below). The short-acting compounds are more stable than the others to heat-alkali treatment because they do not hydrolyze to malonic acid and urea as readily. The extent of the decrease of UV absorbance, then, after heat-alkali treatment is evidence of the presence of long-, intermediate-, or short-acting barbiturates.

Reagents

1. Methylene chloride. Prepare a daily supply by washing sufficient CH_2Cl_2 with 1/10 volume of NaOH, 1 mol/L, followed by two washings with 1/10 volume of distilled water. Washing the CH_2Cl_2 may be omitted if a blank determination shows the absence of interfering materials in the CH_2Cl_2.

2. Boric acid, 0.6 mol/L, potassium chloride solution. Dissolve 37.1 g of H_3BO_3 and 44.7 g of KCl in distilled water and make up to the mark in a 1-L volumetric flask. The solution is stable.

3. Sodium hydroxide, 0.45 mol/L. Dissolve 18.0 g of NaOH in distilled water and make up to the mark in a 1-L volumetric flask. This solution need not be standardized, but when equal volumes of NaOH, 0.45 mol/L, and H_3BO_3-KCl, 0.6 mol/L, are mixed, a pH of about 9.9 should result. Check with a pH meter. It may be necessary to add acid or alkali to get the correct pH. The solution is stable.

4. Potassium phosphate, monobasic, 0.5 mol/L. Dissolve 17.0 g of KH_2PO_4 in distilled water and dilute to the mark in a 250-mL volumetric flask. The solution is stable.

5. Sodium phosphate, dibasic, 0.5 mol/L. Dissolve 179 g of $Na_2HPO_4 \cdot 12H_2O$ (or equivalent weight of anhydrous or other hydrated forms) in distilled water and dilute to the mark in a 1-L volumetric flask. The solution is stable.

6. Phosphate buffer, pH 7.5. Mix 19.2 mL of KH_2PO_4, 0.5 mol/L (reagent 4), with 80.0 mL of Na_2HPO_4, 0.5 mol/L (reagent 5). Check the pH with a pH meter and adjust, if necessary, with reagent 4 or 5.

Procedure

For the extractions use separatory funnels with Teflon stopcocks. Stopcock grease should not be used since some of the components of stopcock grease interfere with ultraviolet measurements.

1. Measure 10.0 mL of blood, serum, urine, or gastric contents into a 125-mL separatory funnel. With urine or gastric contents, check with pH with indicator paper and adjust to a pH of 7 or less by the addition of dilute HCl. Specimen volumes other than 10.0 mL can be used with appropriate adjustment of the calculation. If the type of barbiturate need not be identified, 5.0 mL of specimen is sufficient.

2. Add 30 mL of CH_2Cl_2 (reagent 1) and extract the specimen by shaking for 1 min.

3. Draw off the CH_2Cl_2 and filter through Whatman No. 41 filter paper into a second 125-mL separatory funnel. Repeat the extraction two more times, using 30 mL of CH_2Cl_2 each time. If an emulsion occurs at this point, add a few milliliters of CH_2Cl_2 in excess and gently invert the separatory funnel several times. The layers will separate easily with most specimens. If not, centrifuge at 2000 $\times$ g for 5 min.

4. Wash the combined filtered CH_2Cl_2 extracts twice with 5 mL of phosphate buffer. After the second phosphate wash, filter the CH_2Cl_2 through fresh Whatman No. 41 filter paper into a third 125-mL separatory funnel.

5. Extract the CH_2Cl_2 by shaking for 3 min with 10.0 mL NaOH, 0.45 mol/L. Draw off the CH_2Cl_2 and discard it or save it for analysis of other drugs.

6. Transfer the aqueous phase, together with any emulsion present, into a 15-mL centrifuge tube. Centrifuge at 2000 $\times$ g for 5 min. This is the "alkaline extract."

7. Prepare four test tubes as follows:

	Borate Blank	Borate Sample	NaOH Blank	NaOH Sample
H_3BO_3-KCl, 0.6 mol/L	2.0 mL	2.0 mL	—	—
NaOH, 0.45 mol/L	2.0 mL	—	4.0 mL	2.0 mL
Alkaline extract	—	2.0 mL	—	2.0 mL

8. Scan these solutions in the ultraviolet spectrophotometer from 220–300 nm as follows (if only a manual instrument is available, make reading at 5-nm intervals). Read the borate sample against the borate blank as reference. Read the NaOH sample against the NaOH blank as reference.

After completion of the recording of the borate sample, adjust the same recording paper in such a way that the scan of the NaOH sample will exactly overlay the scan of the borate sample. (See Figure 18-10.) If typical barbiturate curves are obtained (see Interpretation) and the type of barbiturate is known, proceed to Calculations. If the type of barbiturate is unknown, proceed as follows:

9. Pipet 5.0 mL of alkaline extract from step 6 into a tube calibrated at 5.0 mL and place into a boiling water bath for exactly 15 min.

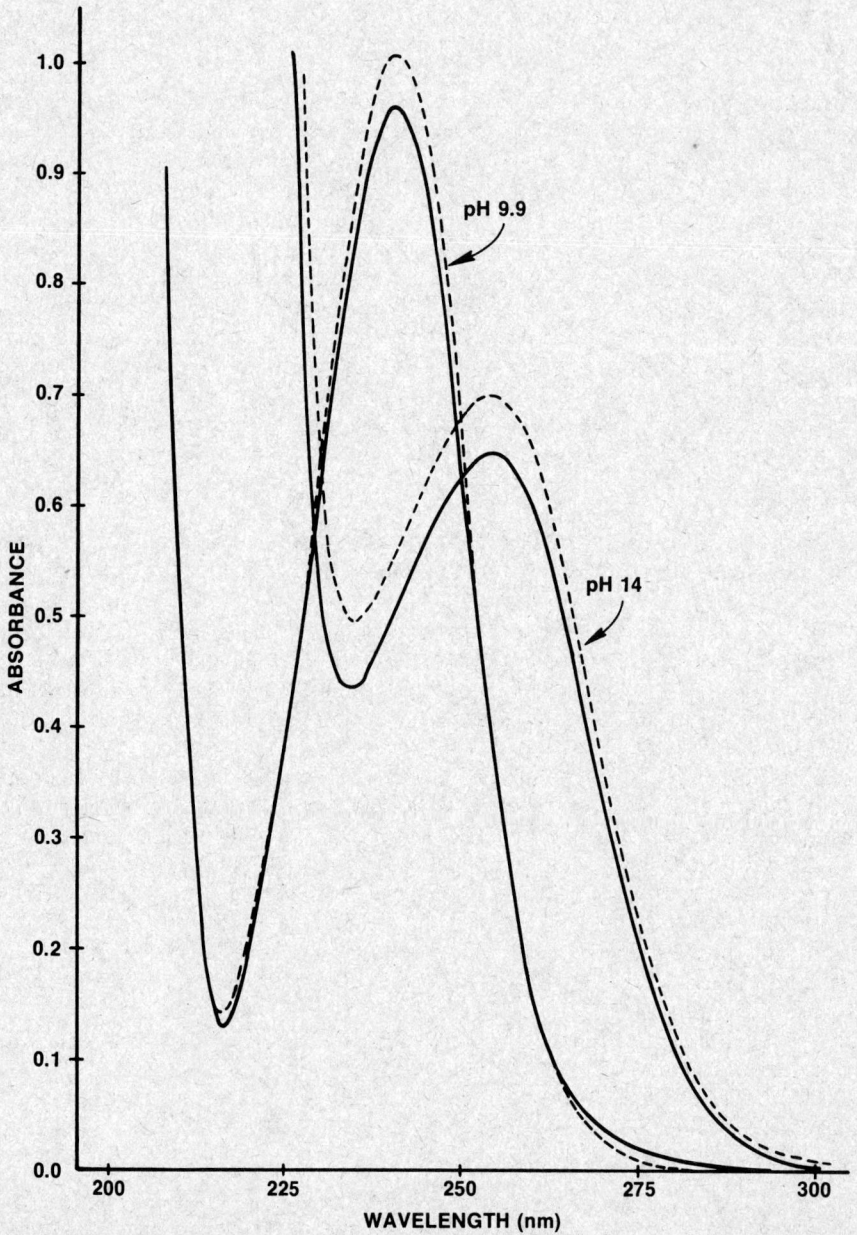

Figure 18-10. Typical ultraviolet absorption curves of secobarbital in alkaline extract before (dashed lines) and after (solid lines) hydrolysis. Note that pH change from 9.9 to 14 shifts absorbance maxima upscale for both the hydrolyzed and unhydrolyzed species.

10. After 15 min, immediately transfer the tube to an ice-water bath. Cool tube to room temperature, adjust the volume to 5.0 mL with distilled water, and mix thoroughly. This is the hydrolyzed alkaline extract.

11. Prepare two test tubes as follows:

	NaOH+ Hydrolyzed Sample	Borate+ Hydrolyzed Sample
H$_3$BO$_3$-KCl, 0.6 mol/L	—	2.0 mL
NaOH, 0.45 mol/L	2.0 mL	—
Hydrolyzed alkaline extract	2.0 mL	2.0 mL

12. Scan these solutions in the ultraviolet spectrophotometer in the same manner as in step 8, using the same blank solutions as reference solutions.

Calculations

Table 18-14 shows the calibration data for 5,5-disubstituted barbiturates commonly encountered. In this table, K_N and K_B are extinction coefficients at 260 nm of the particular barbiturate at a concentration of 1 mg/L in NaOH, 0.45 mol/L, and in borate solution of pH 9.9, respectively. F is a calibration constant and is equal to $1/(K_N - K_B)$. R is the percentage of barbiturate remaining after alkaline hydrolysis for 15 min.

In the calculations to follow, these symbols are used:

A_1 = absorbance of NaOH sample at 260 nm
A_2 = absorbance of borate sample at 260 nm
A_3 = absorbance of NaOH hydrolyzed sample at 260 nm
A_4 = absorbance of borate hydrolyzed sample at 260 nm
C = concentration of barbiturate in sample.

$$(A_1 - A_2) \times F = C \text{ in } \mu g/mL \text{ of solution in cuvet} \tag{a}$$

or

$$(A_1 - A_2) \times F \times 2 = C \text{ in mg/L of specimen} \tag{b}$$

If specimen volumes other than 10.0 mL are used, and the CH$_2$Cl$_2$ is extracted with volumes of NaOH, 0.45 mol/L, other than 10.0 mL, then equation (c) is used:

$$(A_1 - A_2) \times F \times 2 \times \frac{\text{mL NaOH (0.45 mol/L)}}{\text{mL sample}} = C \text{ in mg/L specimen} \tag{c}$$

If the barbiturate is known to be one of the six shown in Table 18-14, the appropriate F value is used. If the specific barbiturate is unknown but it is found by hydrolysis to be long acting, use an F value of 40.6; if intermediate acting, use F = 40.3; and if short acting, use F = 47.8. If the type of barbiturate is unknown and insufficient specimen is available for hydrolysis, an

Table 18-14. CALIBRATION DATA FOR 5,5-DISUBSTITUTED BARBITURATES

Barbiturate	Duration of Action	K_N	K_B	F	R (%)
Phenobarbital	Long	0.0315	0.0086	43.7	31.8
Barbital	Long	0.0329	0.0062	37.5	42.5
Butabarbital	Intermediate	0.0314	0.0056	38.8	49.6
Amobarbital	Intermediate	0.0292	0.0052	41.7	55.8
Pentobarbital	Short	0.0268	0.0060	48.1	98.2
Secobarbital	Short	0.0288	0.0078	47.5	97.6

Modified from Broughton, P.M.G.: Biochem. J., *63*:207-213, 1956.

approximate level can be estimated by using the mean F value of 42.9. In order to determine the type of barbiturate present, the following calculation is made:

$$R = \frac{A_3 - A_4}{A_1 - A_2} \times 100$$

If: R = 30–45, a long-acting barbiturate is present.
R = 45–56, an intermediate-acting barbiturate is present.
R = 90–98, a short-acting barbiturate is present.

Interpretation

The criteria for the identification of barbiturates are as follows (see Figure 18-10):
1. In pH 9.9 solution: a maximum absorbance at 238–240 nm.
2. In pH 14 solution: a maximum absorbance at 252–255 nm.
3. In pH 14 solution: a minimum absorbance at 234–237 nm.
4. Isosbestic points: 227–230 nm and 247–250 nm.
Note: Isosbestic points are points on a spectral curve at which two substances have equal absorbance at the same wavelength.

Dilute solutions of barbiturates may not give all of the characteristic points, and the absorption peaks of salicylates and sulfonamides may obscure some of them; however, the phosphate wash should remove most of these interferences.

The value of R is valid only if a single barbiturate is present. For example, if a mixture of long- and short-acting barbiturates is present, R may indicate an intermediate-acting barbiturate.

Reference

Broughton, P. M. G.: A rapid ultraviolet spectrophotometric method for the detection, estimation and identification of barbiturates in biological material. Biochem. J., *63*:207–213, 1956.

Determination of Barbiturates by GLC

Principle

Drugs are extracted from serum using an internal standard and a single extraction. The extracted residue is dissolved in methylation reagent and injected into a gas chromatograph. Quantitation is performed by calculating peak area ratios and comparing these with a standard curve.[8]

Equipment

Gas chromatograph with FID or NPD. Glass column, 2 m × 2 mm packed with 3% OV-101. Temperature program: 140 °C for 1 min; 10 °C/min to 250 °C; hold 2 min.

Reagents and Standards

1. Butabarbital, USP-NF reference standard.
2. Amobarbital, USP-NF reference standard.
3. Pentobarbital, USP-NF reference standard.
4. Secobarbital, USP-NF reference standard.
5. Phenobarbital, USP-NF reference standard.
Standard solutions (200 µg/mL) of the above are made by dissolving 50.0 mg of each barbiturate (as the free acid) in a 250-mL volumetric flask and diluting to volume with methanol. Refrigerate.
6. Phosphate buffer, 0.5 mol/L, pH 6.4.
   a. Dissolve 17.0 g KH_2PO_4 (M.W. 136) in H_2O and dilute to 250 mL.
   b. Dissolve 17.7 g Na_2HPO_4 (M.W. 142) in H_2O and dilute to 250 mL.
   c. Mix 250 mL KH_2PO_4 solution with 180 mL Na_2HPO_4 solution and check pH.
7. Methylation reagent. Dilute 1 mL TMPAH solution (trimethylphenylammonium hydroxide, Eastman Kodak Co., Rochester, NY) with 2 mL of methanol.

8. Dichloromethane, pesticide grade (Fisher Scientific Co.).

9. Methanol, reagent grade (Baker Chemical Co.).

10. Internal standard, 200 $\mu g/mL$. Place 50.0 mg probarbital extracted and purified from calcium Ipral (E.R. Squibb and Sons, Princeton, NJ) in a 250-mL volumetric flask and dilute to volume with methanol.

Procedure

1. Add 0.5 mL of serum to a 16 $\times$ 125 culture tube with Teflon-lined screw cap.

2. Add 50 μL of internal standard solution (a 200 $\mu g/mL$ solution of probarbital) and 0.5 mL of phosphate buffer.

3. Extract with 5 mL of CH_2Cl_2. Rotate gently for 5 min.

4. Filter the organic layer through Whatman #541 paper into labeled evaporating tubes.

5. Evaporate under a gentle stream of N_2 until 1 mL is left.

6. Transfer the remaining CH_2Cl_2 to a 12 $\times$ 100 test tube and evaporate to dryness under N_2 at 40 °C.

7. Dissolve the residue in 100 μL of methylating reagent and mix.

8. Inject 1 μL into the gas chromatograph.

Calculation

The ratio of the peak area of the barbiturate to the internal standard is calculated. This ratio is compared to a calibration curve for the barbiturate being measured, in which calculated ratios are plotted against the concentration.

Interpretation

Blood barbiturate levels must be interpreted cautiously for the following reasons:

1. There is variation in the response of different individuals to a given dose of any drug.

2. A given concentration of barbiturate has a more profound effect when the blood level is rising than when the blood level is falling.

3. The effects observed in the patient will be more severe if other depressant drugs, particularly alcohol, are present in addition to the barbiturates.

4. Some individuals who are tolerant of or addicted to barbiturates may have high blood levels without obvious effects.

Table 18-13 shows approximate blood barbiturate levels at which coma disappears (falling blood levels). At levels higher than those indicated in Table 18-13, the prognosis is poor, since fatalities have occurred at these levels. Except in epileptics treated with phenobarbital, normal therapeutic blood levels rarely rise higher than 1–2 mg/L.

Figure 18-11 illustrates the advantages in using a nitrogen specific detector for this assay. Since the nitrogen specific detector responds poorly to non-nitrogen containing substances, the resulting chromatogram shows a minimum of nonbarbiturate peaks and thus is a simpler chromatogram.

Reference

Blanke, R. V., and Saady, J. J.: Rapid simultaneous determination of six anti-convulsant drugs by GLC. 26th S.E. Regional Meeting, American Chem. Soc., 1974.

SALICYLATES

Aspirin is responsible for more cases of accidental poisonings in children than any other substance. This extremely effective analgesic is so widely used and readily available (and carelessly handled) that children frequently ingest a toxic quantity by eating the flavored tablets like candy or by mimicking adults. Toxic doses of salicylates initially produce a stimulation of the central nervous system. This may be reflected by hyperventilation, flushing, and fever. Unfortunately, an unrecognized case of salicylate poisoning may be thought to be a case of infection and further aspirin given in a vain attempt to control the fever. Central nervous system stimulation is followed by depression and a complex disturbance of acid-base balance. Initially a respiratory alkalosis occurs caused by severe hyperventilation, but this may be followed, especially in infants, by a

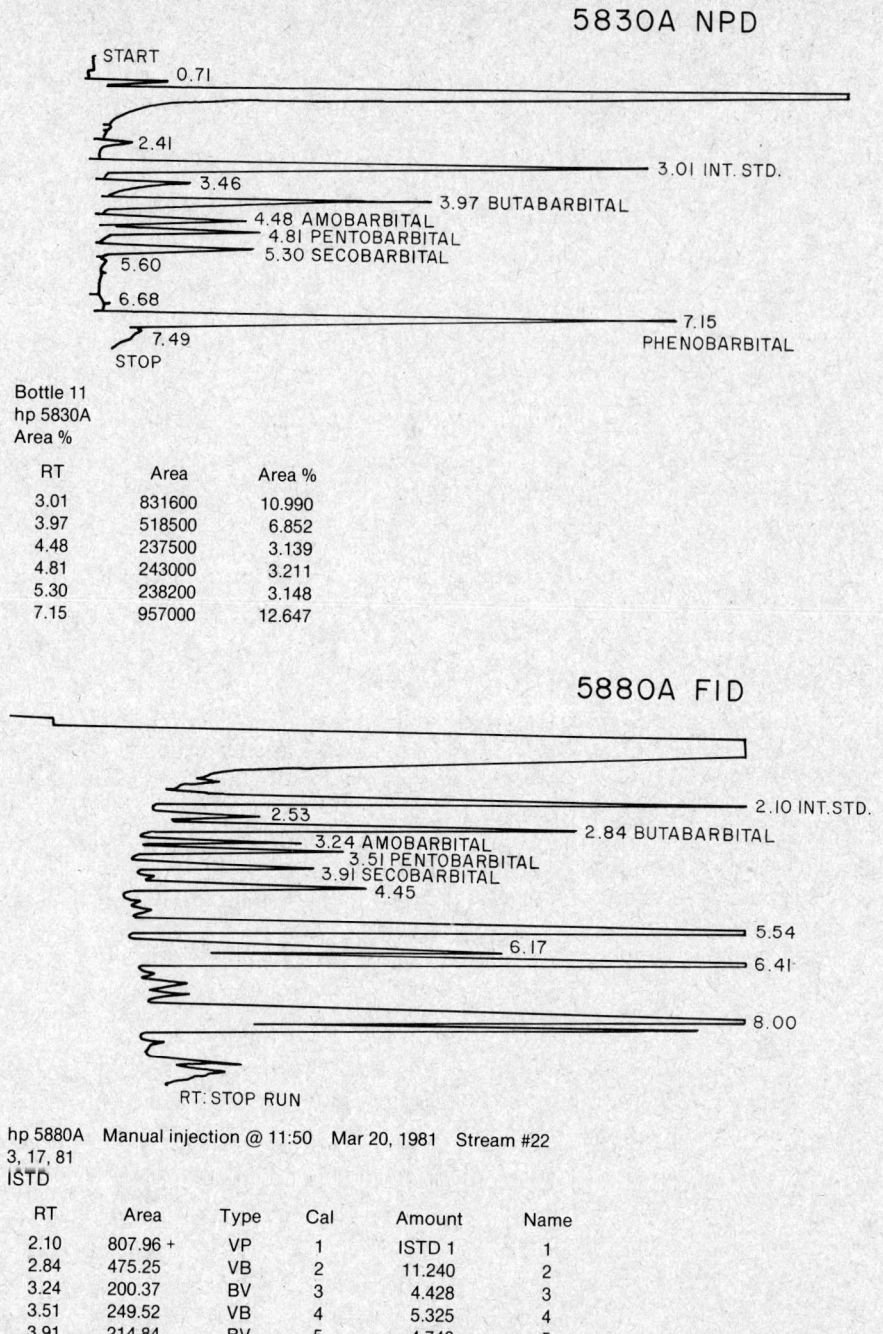

Figure 18-11. GLC tracings of barbiturates in a serum control. Note the cleaner baseline of the tracing obtained with a nitrogen/phosphorus detector (NPD) as compared with that found with a flame ionization detector (FID).

metabolic acidosis due in part to the absorbed acid and in part to accumulation of other metabolic acids. The net effect may be a decrease in blood pH.

Determination of Salicylates in Biological Fluids

A common and convenient procedure for determining salicylate in urine, serum, or other specimen is based on the formation of a violet-colored complex between Fe (III) ions and phenols.

$$\text{Salicylic acid} + Fe^{3+} = \text{Violet color complex}$$

Salicylic acid

This test is not specific for salicylates, but false negative results do not occur.

The method described below is specific and may also be used to detect acetaminophen, another widely used analgesic (see p. 1731). Phenacetin may also be measured by this technique.

Determination of Plasma Salicylate and Acetaminophen by Liquid Chromatography[2]

Principle

Salicylic acid and acetaminophen are extracted into an organic solvent containing an internal standard. The concentrated extract is analyzed by liquid chromatography with ultraviolet absorbance detection at 248 nm.

Reagents

1. Stock standard, 1 mg/mL methanol solutions of salicylic acid and acetaminophen.
2. Plasma standards, 50, 100, and 200 mg salicylate and acetaminophen/L.
3. Extraction solvent. Dissolve 20 mg 8-chlorotheophylline (No. 15816, K & K Labs Division, ICN Pharmaceuticals, Plainview, NY) in chloroform: isopropanol, 1:1, and dilute to 1 L.
4. Sodium chloride, solid.
5. Methanol.
6. Mobile phase. Add 30 mL 2-propanol to 970 mL phosphoric acid, 0.02 mol/L, pH 2.9.

Instrumental Conditions

1. Liquid chromatograph with 248 nm ultraviolet detector.
2. 25 cm × 4 mm I.D. stainless steel column containing ODS-Sil-X-1 (Perkin-Elmer Corp., Norwalk, CN).
3. Column temperature, 40 °C.
4. Solvent flow rate, 1.5 mL/min.

Procedure

1. Transfer 50 µL plasma to a 1.5-mL plastic centrifuge tube. Add 0.5 mL extraction solvent and 0.1–0.2 g solid NaCl and vortex for 2 min.
2. Centrifuge and transfer the supernatant to a glass tube. Evaporate to dryness at 40 °C under a stream of nitrogen.
3. Dissolve the residue in 50 µL methanol and inject 2–5 µL into the chromatograph.

	Retention Time (min)
Acetaminophen	2.2
Salicylate	2.9
Internal standard	4.2

Calculation

Calculation is based on a response factor derived from a standard curve. A quality control specimen containing salicylate and acetaminophen at mid-range concentrations is analyzed daily.

Evaluation

Sensitivity: 10 mg/L for each drug; linearity: 50–1000 mg/L for salicylate; 10–300 mg/L for acetaminophen; day-to-day CV: 3–6%; relative recovery: 96–108% for each drug.

Interferences

Normal plasma constituents do not interfere with the assay. Eight other commonly used drugs do not interfere, although theophylline is found to co-elute with salicylate. Phenacetin has a retention time of 6 min.

Interpretation

Therapeutic levels of salicylic acid rarely rise above 200 mg/L in blood or serum. Above 300 mg/L, toxic symptoms such as headache, tinnitus, flushing, and hyperventilation may be seen. Serum electrolytes should be followed and any imbalance corrected. Lethal salicylate levels are usually greater than 600 mg/L.

Reference

Baselt, R. C.: Analytical Procedures for Therapeutic Drug Monitoring and Emergency Toxicology. Davis, Cal., Biomedical Publ., 1980, pp. 261–262.

Method for the Determination of Propoxyphene

Principle

Propoxyphene exists in two optically active forms. The dextrorotatory form is a narcotic analgesic marketed as Darvon, while the levorotatory form is available as Novrad, an antitussive. The isomers cannot be differentiated by usual chemical means; therefore, levels are reported as propoxyphene. Quantitative methods for assaying blood and tissue levels of this drug have been plagued by many difficulties. Unlike most organic bases, the hydrochloride salt of propoxyphene is soluble in certain organic solvents. Normal metabolism of this drug converts it to norpropoxyphene. Although this metabolite would not be expected to differ greatly from the parent drug, strong alkalinization of norpropoxyphene causes it to rearrange, forming norpropoxyphene amide (Figure 18-12). Once the amide is extracted from an alkaline medium, it cannot be re-extracted from the organic solvent by acid solutions.[46] These factors, plus protein binding of the drug, have contributed to the difficulties associated with assays of propoxyphene in biological specimens.

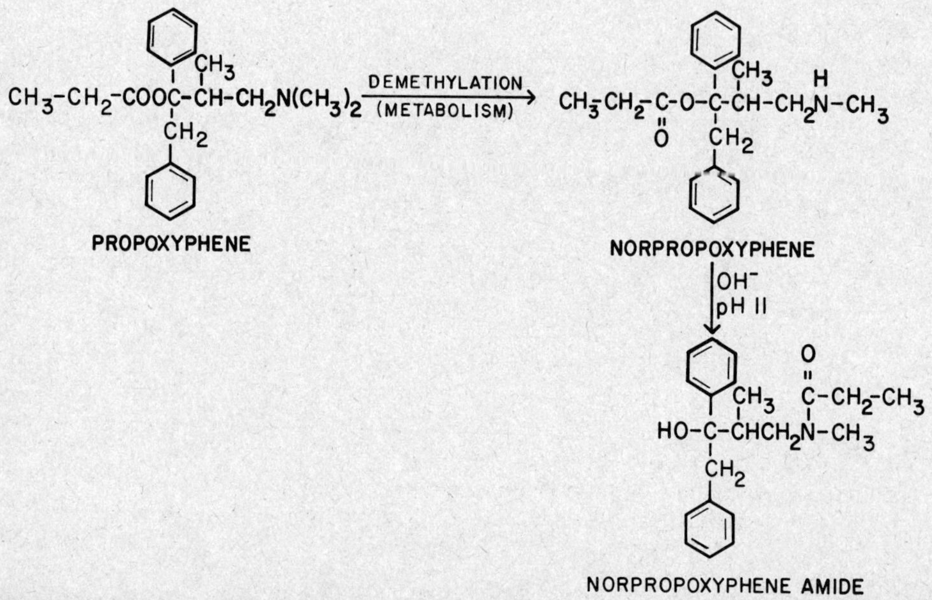

Figure 18-12. Rearrangement of the propoxyphene metabolite, norpropoxyphene, to the amide under alkaline conditions.

Determination of Propoxyphene by GLC[2]

Principle

A three-step extraction is used to extract propoxyphene, norpropoxyphene, and SKF-525A (internal standard). Norpropoxyphene is converted to norpropoxyphene amide with a strong base after back extraction into dilute acid. The final extract is injected into a GC with FID.

Equipment

A gas chromatograph with FID; 2 m $\times$ 2 mm glass column packed with 3% SP-2250 or 3% OV-17.

Reagents

1. Propoxyphene, 200 mg/L in methanol. Weigh out 22.12 mg propoxyphene HCl (USP-NF Reference Standard) and dilute to 100 mL with methanol.
2. Norpropoxyphene, 200 mg/L in methanol. Weigh out 27.13 mg norpropoxyphene maleate (Eli Lilly Co., Indianapolis, IN) and dilute to 100 mL with methanol.
3. SKF-525A 100 mg/L in water. Weigh out 10.0 mg SKF-525A (Smith Kline and French Laboratories, Philadelphia, PA) and dilute to 100 mL with water.
4. Extraction solvent. Hexane:isopropanol (90:10).
5. Phosphate buffer, 1.0 mol/L, pH 11.

    a. Add 50.5 g sodium phosphate, tribasic ($Na_3PO_4 \cdot 12H_2O$, M.W. = 380.12) to 400 mL distilled water. The pH is $\sim$12.5.

    b. Add 16.56 g sodium phosphate, monobasic ($NaH_2PO_4 \cdot H_2O$, M.W. = 137.99) to 100 mL distilled water. The pH is $\sim$4.3.

    c. Add enough monobasic to the tribasic phosphate solution to bring the pH to 11.0.

Extraction

1. To a 15-mL screw top culture tube, add 2 mL of serum or control, 25 μL of the internal standard solution, and 1 mL phosphate buffer, pH 11. Vortex for 5 s.
2. Add 10 mL of hexane:isopropanol (90:10) and mix by gentle rotation for 5 min.
3. Filter the organic layer through filter paper (Whatman #541) into a clean, screw-top culture tube. Add 1 mL of HCl, 0.1 mol/L. Rotate 5 min.
4. Centrifuge and discard the organic layer. To the water layer add 1 drop of NaOH, 500 g/L. Vortex 5 s.
5. Add 1 mL of toluene. Vortex for 1 min. Centrifuge.
6. Transfer the toluene (top layer) to a 5-mL conical centrifuge tube. Evaporate under stream of N_2 at 45 °C.
7. Reconstitute residue with 2 drops of toluene. Vortex. Inject 1–2 μL. (See Figure 18-13.)

Calculations

Calculate the ratio of peak areas and determine concentrations from a standard curve.

Interpretation

Therapeutic concentrations of propoxyphene range from 0.1–0.4 mg/L. The metabolite, norpropoxyphene (measured as norpropoxyphene amide), has about one half of the analgesic potency of propoxyphene, and its therapeutic concentration is 0.1–1.5 mg/L.

Reference

Baselt, R. C.: Analytical Procedures for Therapeutic Drug Monitoring and Emergency Toxicology. Davis, Cal., Biomedical Publ., 1980, pp. 245–247.

AMPHETAMINES

Amphetamine and its homolog methamphetamine are members of a group of phenylethylamine derivatives which share two features: they are potent central nervous system stimulants and they are widely abused. Although amphetamine itself is no longer extensively used thera-

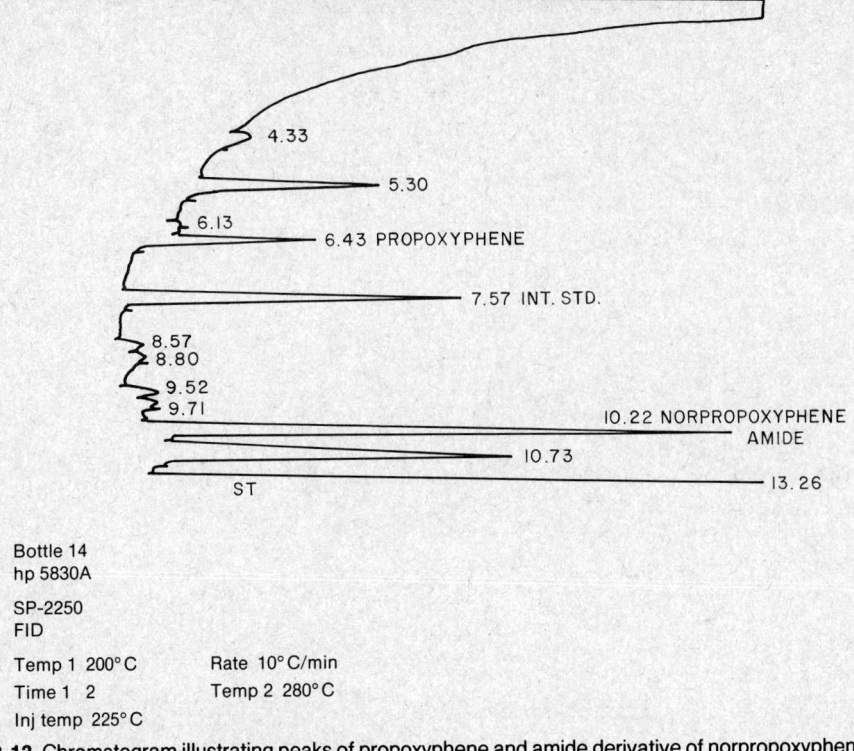

Figure 18-13. Chromatogram illustrating peaks of propoxyphene and amide derivative of norpropoxyphene, obtained by the procedure described (GLC with FID).

peutically, other phenylethylamines are encountered as decongestants, appetite suppressants, and antidepressants. Many are abused for their real or fancied hallucinogenic properties.

Most of these substances can be detected by similar methods. The following procedure, although specifically applied to amphetamine and methamphetamine, will detect most phenylethylamines.

Determination of Amphetamine/Methamphetamine by GLC-ECD[2]

Principle

Amphetamine and methamphetamine are extracted from a basic solution into toluene. Both compounds are derivatized in the toluene using trichloroacetylchloride. The derivatizing reagent is removed with sodium hydroxide. A part of the toluene layer is injected into a gas chromatograph equipped with an electron-capture detector.

Equipment

A gas chromatograph equipped with ECD; 2 m × 2 mm glass column packed with 3% OV-101; column temperature, 180 °C.

Reagents

1. d-Amphetamine sulfate (Smith Kline and French Laboratories, Philadelphia, PA). Prepare a 1.0 g/L solution in methanol by dissolving 17.18 mg of d-amphetamine sulfate in methanol and dilute to 10 mL. Dilute 1 mL of this solution to 100 mL with methanol for a 10 mg/L working standard.

2. d-Methamphetamine HCl (Applied Science, State College, PA). Prepare a 1.0 g/L solution in methanol by dissolving 12.44 mg of d-methamphetamine HCl in methanol and dilute to 10 mL. Dilute 1 mL of this solution to 100 ml with methanol for a 10 mg/L working standard.

3. Methanol, Fisher HPLC grade.

4. Toluene, Fisher HPLC or pesticide grade.

5. Saturated phosphate buffer. Dissolve a sufficient quantity of Na_3PO_4 in water to leave some crystals undissolved in the solution.

6. Sodium hydroxide, 6 mol/L. Dissolve 24 g of NaOH pellets in 100 mL of water.

7. Triethylamine. Add 100 μL of triethylamine (Eastman Kodak, Rochester, NY) to 10 mL of toluene.

8. Trichloroacetylchloride (Fisher Scientific Co., reagent grade).

9. *N*-Propylamphetamine.

a. Dissolve 5 g amphetamine base (liquid) in 200 mL ether and add 10 g propionic anhydride. Let stand for 1 h at room temperature.

b. Wash ether solution with 50 mL of each of the following solutions in sequence: NaOH, 1 mol/L; HCl, 0.5 mol/L; H_2O. Dry ether by shaking with anhydrous Na_2SO_4 or Na_2CO_3.

c. Transfer ether to a refluxing apparatus, add 4 g $LiAlH_4$ slowly, and reflux for 2 h.

d. Cool and add 25 mL H_2O. Extract precipitate with 100 mL ether, discard aqueous phase, and filter ether to remove precipitate.

e. Extract ether with 25 mL HCl, 2 mol/L.

f. Wash HCl layer once with 50 mL ether. Discard ether. Store aqueous layer at 4 °C overnight and collect crystals by filtration. Dry in desiccator. The procedure yields about 4 g of the hydrochloride salt of *N*-propylamphetamine (m.p. 158 °C).

10. Working internal standard, 5 mg/L. Dissolve 5 mg of *N*-propylamphetamine in methanol and dilute to 10 mL to make a 500 mg/L solution. Dilute 1 mL of this solution to 100 mL with methanol.

Extraction

1. To an 8-mL screw-cap extraction tube add 50 μL of internal standard solution, 2 mL serum or plasma, 1 mL saturated phosphate buffer, and 2 mL toluene.

2. Rock for 15 min on hematology-type tube rocker.

3. If an emulsion forms, place in freezer for 15 min.

4. Centrifuge at $2000 \times g$ for 10 min.

5. Transfer 1 mL of toluene layer (upper layer) to a 15-mL screw-cap extraction tube. Add 1 drop of triethylamine solution with a Pasteur pipet. Add 1 drop trichloroacetylchloride with a Pasteur pipet. Mix gently by shaking.

6. Place in 70 °C water bath for 1 h.

7. Add 1.0 mL of NaOH, 6 mol/L. Vortex for 1 min or until the upper toluene layer is completely clear.

8. Inject 2 μL of toluene layer into gas chromatograph. (See Figure 18-14.)

Interpretation

The following levels are seen after a single dose:

	Plasma	*Urine*
Amphetamine	0.02–0.12 mg/L	1–5 mg/L
Methamphetamine	0.01–0.05 mg/L	0.5–4 mg/L

Sensitivity is ∼0.02 mg/L for amphetamine and methamphetamine.

Reference

Baselt, R. C.: Analytical Procedures for Therapeutic Drug Monitoring and Emergency Toxicology. Davis, Cal., Biomedical Publ., 1980, pp. 24-25.

ACETAMINOPHEN

This widely used analgesic is available without a prescription and is frequently recommended when aspirin may present problems to a patient, particularly in pediatrics or after surgery. It is quite safe, but when taken in overdosage may produce liver toxicity due to conversion to a toxic metabolite. Normally, the metabolite is conjugated with glutathione and excreted. If glutathione is depleted for any reason, excess metabolite will bind covalently to hepatic cellular components.

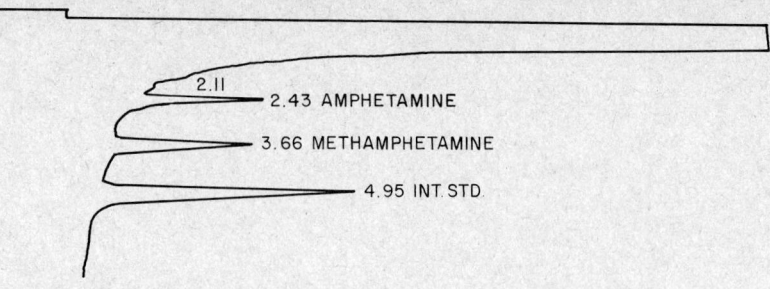

RT	Area	Type	Width	Height	Baseline	Area %
2.43	1785.66	PP	0.16	173.42	358.89	2.466
3.66	2493.18	BB	0.206	189.11	327.53	3.443
4.95	5534.76	BP	0.250	346.69	306.90	7.643

hp 5880A   Manual injection @ 12:02   May 6, 1982
AREA %

Figure 18-14. Chromatogram illustrating peaks of trichloroacetamide derivatives of amphetamine and methamphetamine, obtained by the procedure described (GLC with ECD).

An antidote, *N*-acetylcysteine, can be given to reverse this toxic effect. Analysis not only confirms a suspected overdosage but, by additional analysis of timed specimens, can offer a means of estimating the half-life of the drug. A lengthened half-life indicates that liver toxicity may occur.

The method described is simple and rapid. By utilizing high performance liquid chromatography (HPLC), the less commonly encountered analgesic phenacetin can also be measured.

Determination of Acetaminophen by HPLC

Principle

An internal standard is added to a serum sample and the mixture is deproteinized by trichloroacetic acid. The supernatant is injected into the HPLC. Quantitation is accomplished by calculating the peak height ratio of acetaminophen to the internal standard and reading the sample concentration from the calibration curve.[33]

Equipment

HPLC with UV detector at 250 nm; flow rate, 2.6 mL/min. Column, C18-Spherisorb 10 ODS (Altex).

Reagents

1. Acetaminophen, USP-NF reference standard. Place 100.0 mg acetaminophen into a 100-mL volumetric flask and dilute to volume with methanol (1 g/L).
2. Internal standard, 3-acetamidophenol, 1 g/L (Aldrich Chemical Company, #A720-5). Place 100.0 mg of 3-acetamidophenol into a 100-mL volumetric flask and dilute to volume with methanol.
3. Trichloroacetic acid (TCA), 200 g/L (Baker Chemical Company, Phillipsburg, NJ, #0410, M.W. 163.39). Dissolve 20 g TCA in 100 mL H_2O.
4. Mobile phase, methanol:acetic acid:H_2O, 10:1:89.
Mix 100 mL methanol, 10 mL glacial acetic acid, and 890 mL H_2O.

Procedure

1. Place 500 µL serum into a 10 × 75-mm screw-cap tube containing 50 µL of internal standard and vortex 3 s.
2. Add 500 µL TCA solution. Vortex 5 s and centrifuge 5 min.
3. Aspirate about 500 µL of supernatant into a Hamilton #1001 gastight syringe (Hamilton Co., Reno, NV). Remove any air bubbles.
4. Inject 200 µL through a calibrated loop. Rinse syringe with methanol.
5. Quantitate by calculating the peak height ratio of acetaminophen to the internal standard and reading the concentration from a standard curve. (See Figure 18-15.)

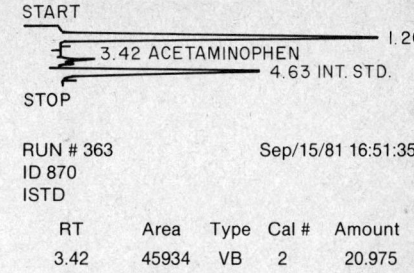

Figure 18-15. Chromatogram and electronic integrated report for acetaminophen obtained with HPLC method described in text.

RUN # 363   Sep/15/81 16:51:35
ID 870
ISTD

RT	Area	Type	Cal #	Amount
3.42	45934	VB	2	20.975
4.63	314940	PB	1&	100.000

Interpretation

Toxic manifestations of acetaminophen are observed if the serum concentration is greater than 100 mg/L or $t_{1/2} > 2$ h. Antidotal treatment is then recommended. Levels as low as 2 mg/L can be measured with this procedure.

The method is linear to at least 200 mg/L. If the patient's concentration is greater, repeat the procedure using 250 μL of patient's serum, and multiply the result by a factor of 2.

Reference

Gotelli, G. W., Kabra, P. M., and Marton, L. J.: Determination of acetaminophen and phenacetin in plasma by HPLC. Clin. Chem., *23*:957-959, 1977.

BENZODIAZEPINES

This class of compounds comprises a therapeutically useful group of structurally related substances that are used as sedative, hypnotic, or anticonvulsant drugs. Therapeutic concentrations are low and many of these drugs are transformed to active metabolites. These in turn are also used as drugs. The method to be described is applicable to many of the currently used drugs of this group.

Determination of Diazepam and Chlordiazepoxide by HPLC

Principle

Serum or urine is extracted under basic conditions into an organic solvent. The solvent is evaporated, and the residue dissolved in methanol and injected into an HPLC. Quantitation is accomplished by calculating the ratio of peak height to an internal standard and by comparing it to a standard curve.[49]

Reagents

1. Diazepam (USP-NF), 100 mg/L. Add 10 mg diazepam to a 100-mL volumetric flask and dilute to mark with methanol.

2. Nordiazepam (Roche Laboratories, Nutley, NJ), 100 mg/L. Add 10 mg nordiazepam to a 100-mL volumetric flask and dilute to mark with methanol.

3. Chlordiazepoxide (USP-NF), 100 mg/L. Add 10 mg chlordiazepoxide to a 100-mL volumetric flask and dilute to mark with methanol. Chlordiazepoxide is light sensitive. Make up fresh with each run.

4. Norchlordiazepoxide, 100 mg/L. Add 10 mg norchlordiazepoxide (Roche Laboratories, Nutley, NJ) to a 100-mL volumetric flask and dilute to mark with methanol.

5. Prazepam (USP-NF), 100 mg/L (internal standard). Add 10 mg prazepam to a 100-mL volumetric flask and dilute to mark with methanol.

6. Borate buffer. Dissolve 14 g sodium borate (Fisher Scientific Co., S-249) in water and dilute to 200 mL. This is a saturated salt solution.

7. Extraction solvent, toluene:hexane:isoamyl alcohol, 78:20:2. Reagents are all HPLC or pesticide grade. (Fisher Scientific Co.)

8. Mobile phase, 70% methanol in water.

   a. Methanol (Fisher Scientific Co., HPLC grade), filter and de-gas.

   b. Water. Add 0.5 mL triethylamine/L and adjust pH to 7 with dropwise addition of glacial acetic acid.

Equipment

1. Gradient HPLC system set for 70% methanol in water and flow rate of 1.0 mL/min.
2. Variable wavelength detector at 240 nm, range 0.05.
3. 5 micron C_{18}-ODS reverse phase column (Altex, Berkeley, CA).
4. Injection loop, 20 μL.

Procedure

1. Rinse all glassware with methanol before use to eliminate any soap residue. If glassware is not rinsed, an extraneous peak may be observed on the chromatogram. (See Figure 18-16.)

2. To a screw-cap culture tube, add: 1 mL serum, 25 μL internal standard solution, 1 mL borate buffer, and 2 mL extraction solvent.

3. Rotate for 10 min at a very slow rate to prevent the formation of emulsions.

4. Centrifuge for 10 min.

5. Transfer 1.5 mL of the upper organic layer to a 15-mL conical centrifuge tube.

6. Evaporate organic solvent to *complete* dryness in heating block at 50 °C under N_2. The organic solvent must be completely removed. Any toluene injected into the HPLC will show on the chromatogram as an interfering peak.

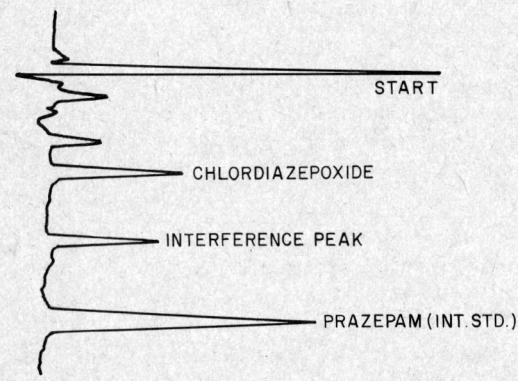

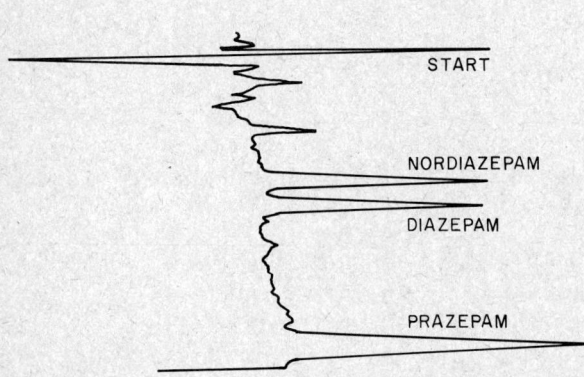

Figure 18-16. Chromatograms showing peaks of benzodiazepines and metabolites. Note interference peak on upper tracing.

Range 0.05
Chart speed 0.5 cm/min

7. Reconstitute the residue in 400 μL of methanol. Inject 20 μL into the HPLC. (See Figure 18-16.)

Interpretation of Results

Since chlordiazepoxide is rather unstable, the standard curve might also be expected to be unstable. The curve should be checked with each sample run.

Drug	Therapeutic Range mg/L	Half-Life
Diazepam	0.5–2.0	21–37 h
Nordiazepam	0.1–0.5	48 h
Chlordiazepoxide	0.5–3.0	7–14 h
Norchlordiazepoxide	0.2–0.7	Not known

Reference

Peat, M. A., and Kopjak, L.: The screening and quantitation of diazepam, fluorazepam, chlordiazepoxide, and their metabolites, in blood and plasma by electron capture gas chromatography, and high pressure liquid chromatography. J. Forensic Sci., 24:46-54, 1979.

MORPHINE, CODEINE, AND HYDROMORPHONE

Opioid analgesics comprise a large group of substances which control pain by a central nervous system depressant effect. Morphine, the prototype compound of this group, was isolated and named by Serturner in 1803. Since that time a large number of naturally derived or synthetic compounds have been described; the clinician therefore has a wide range of choices for use in therapeutic analgesia. Unfortunately, the abuse of these drugs has been a longstanding problem. Fatalities have occurred after overdosage, and adverse effects have also been associated with their abuse. Indeed, it is estimated that in the early 1970's, opiates were responsible for 40–50% of all drug-related fatalities annually.

The isolation, characterization, and quantitation of these drugs have been a continuing challenge. Virtually every analytical technique has been applied to this problem—colorimetry, spectrophotometry, fluorometry, thin-layer chromatography, and gas chromatography. Each of these methods contributed to the state of the art but all suffered from insufficient sensitivity or poor specificity or both. The development of immunoassay techniques provided necessary sensitivity, but immunoassay methods are applicable only when the identity of the drug is known. Positive results must be confirmed by an alternative assay in the presence of metabolites.

Recently, applications of gas liquid chromatography (GLC) with nitrogen detector (NPD) or electron capture detector (ECD), generally after prior derivatization of the drugs, have been reported. These techniques, together with gas chromatography/mass spectroscopy (GC/MS), have been reviewed elsewhere.[23] Here we describe a GC/MS procedure that is rapid, simple, and specific for the simultaneous quantitation of morphine, codeine, and hydromorphone; the procedure uses selected ion monitoring (SIM). Hydromorphone (Dilaudid) was included, since it has proved elusive in many other published assays.

Determination of Opiates

Principle

Serum or urine, previously hydrolyzed to release conjugated opioids, is adjusted to pH of 9.9. Extraction of the drugs together with an added internal standard is followed by derivatization of hydroxyl groups to fluorinated esters (Figure 18-17). GC/MS is carried out while monitoring selected ions characteristic of each compound (Figures 18-18 and 18-19).[54]

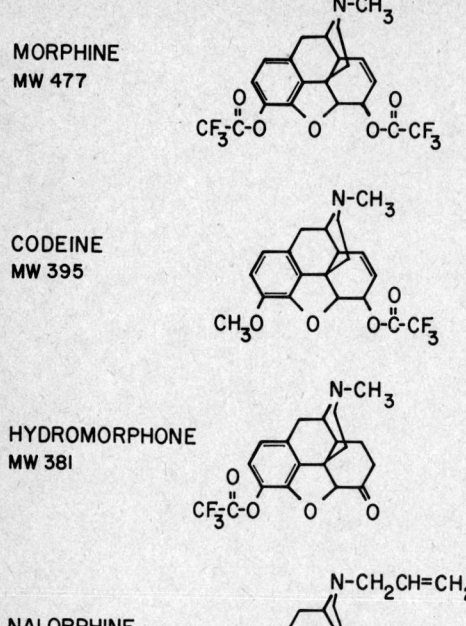

MORPHINE
MW 477

CODEINE
MW 395

Figure 18-17. Trifluroacetyl products formed when MBTFA reacts with opiates. (Reproduced by permission of Preston Publications, Inc., from Saady, J. J., Narasimhachari, N., and Blanke, R.: Rapid, simultaneous quantification of morphine, codeine, and hydromorphone by GC/MS. J. Anal. Toxicol., *6*:235-237, 1982.)

HYDROMORPHONE
MW 381

NALORPHINE
MW 503

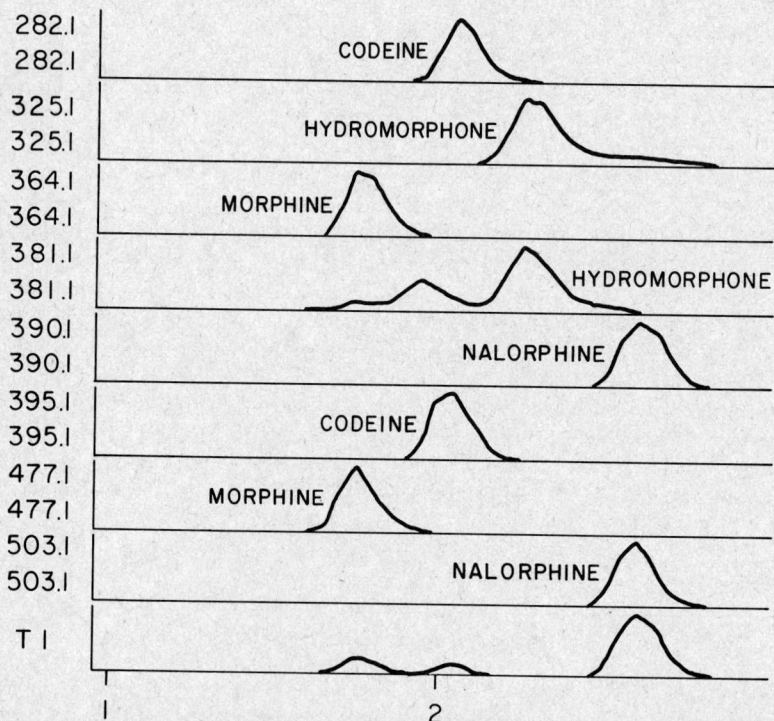

Figure 18-18. Typical selected ion and total ion chromatograms of a plasma extract containing morphine, codeine, hydromorphone, and nalorphine using SIM/GC/MS. (Reproduced by permission of Preston Publications, Inc., from Saady, J. J., Narasimhachari, N., and Blanke, R.: Rapid, simultaneous quantification of morphine, codeine, and hydromorphone by GC/MS. J. Anal. Toxicol., *6*:235-237, 1982.)

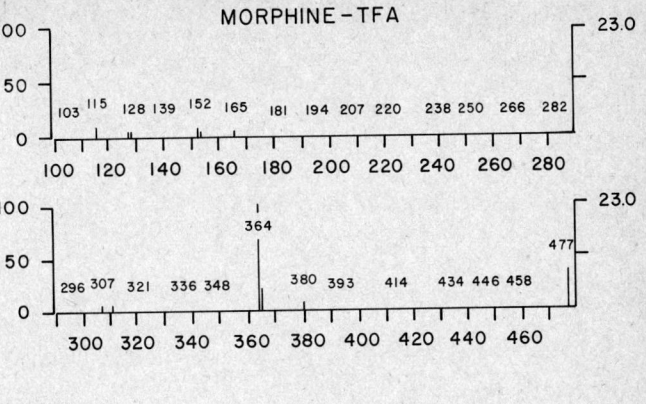

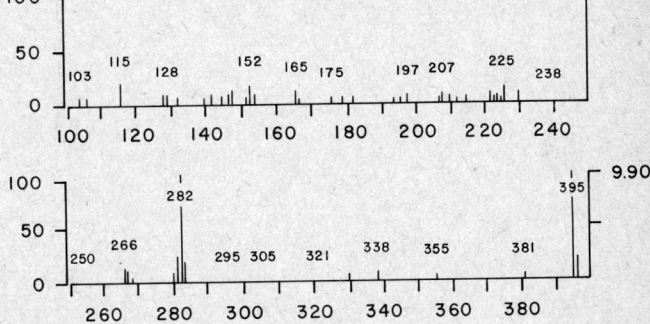

Figure 18-19. Individual mass spectra of MBTFA derivatives of morphine, codeine, hydromorphone, and nalorphine by electron ionization GC/MS. "Relative abundance" is given on the ordinate and *m/z* values on the abscissa. (Reproduced by permission of Preston Publications, Inc., from Saady, J. J., Narasimhachari, N., and Blanke, R.: Rapid, simultaneous quantification of morphine, codeine, and hydromorphone by GC/MS. J. Anal. Toxicol., *6*:235-237, 1982.)

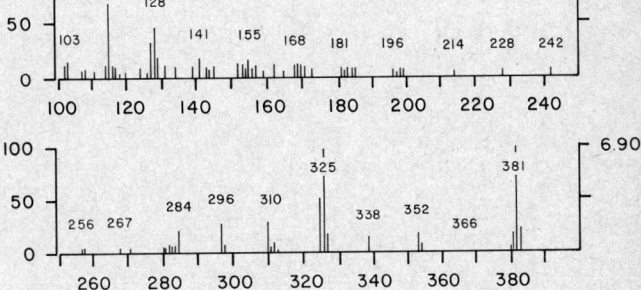

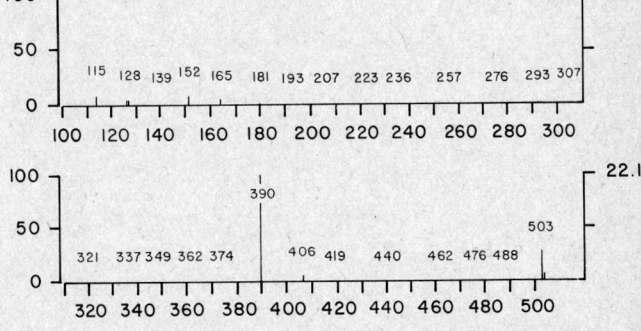

Reagents

1. Stock standards of all drugs are prepared in methanol; each drug is calculated as the free base.

    a. Morphine sulfate (USP-NF reference standard), 230 mg/L.
    b. Codeine phosphate (USP-NF reference standard), 230 mg/L.
    c. Hydromorphone hydrochloride (USP-NF reference standard), 16 mg/L.
    d. Nalorphine HCl (USP-NF reference standard), 1 g/L.

2. *N*-methyl-bis-trifluoroacetamide (MBTFA) (Regis Chemical Company, Morton Grove, IL).

3. Sørenson's phosphate buffer, 1 mol/L, pH 9.9.

4. Toluene, HPLC grade: hexane, pesticide grade; isoamyl alcohol, reagent grade.

5. Extraction solvent, toluene:hexane:isoamyl alcohol, 78:20:2.

Instrumentation

A GC/MS with data system operated in the electron ionization (EI) mode, scanning from m/z 100–550, and capable of monitoring eight ions simultaneously.

Chromatographic column: a 0.9 m × 2 mm (I.D.) glass column packed with 3% OV-101, 100/120 mesh (Supelco, Inc., Bellefonte, PA) operated at an initial temperature of 205 °C for 1 min, then programmed at 12 °C/min to 240 °C; helium flow 35 mL/min.

Procedure

Blood or plasma:

1. Adjust 2 mL of specimen to pH 9.9 with 2 mL phosphate buffer. Verify the pH.
2. Add 20 μg nalorphine (internal standard).
3. Extract the mixture with 4 mL of the extraction solvent for 10 min by gentle rotation.

Urine:

1. Mix 5 mL urine with 20 μL of the internal standard nalorphine.
2. Add 0.5 mL concentrated HCl and hydrolyze in a 100 °C heating block for 1 h. Cool the solution.
3. Add 0.6 mL concentrated NaOH and 2 mL phosphate buffer pH 9.9. Verify the pH and adjust to 9.9 if necessary.
4. Extract the solution with 4 mL extraction solution following the same steps as outlined for blood or plasma.

Derivatization

1. Centrifuge extraction tubes at 2000 × *g* for 10 min.
2. Transfer the organic layer to clean tubes and evaporate to dryness.
3. Add 40 μL of MBTFA to the dry extract.
4. Stopper and heat at 70 °C for 20 min in a water bath.
5. Inject 1 μL into the GC/MS.

Calculation

The ions (m/z) monitored at specific retention times are shown in Table 18-15. Individual standard curves for morphine, codeine, and hydromorphone are established using nalorphine as the internal standard by plotting peak area ratios against concentration. Analytes in unknown specimens are identified on the basis of their retention times and ion characteristics, then quantitated from peak area ratios (Figures 18-18 and 18-19).

Comments

MBTFA is a gentle acetylating agent that will not promote breakdown of the TFA product. It can also be injected directly into a gas chromatographic column; a drying step in the procedure is thereby eliminated. The excess injected reagent is vented to the atmosphere and therefore does not go into the source.

The extraction efficiency for all the opiates has been found to be between 55 and 60%. The quantitative use of the internal standard nalorphine, coupled with the sensitivity of the GC/MS, obviates further extraction.

Table 18-15. GAS CHROMATOGRAPHIC RETENTION DATA FOR THE MBTFA DERIVATIVES OF MORPHINE, CODEINE, HYDROMORPHONE, AND NALORPHINE AND IONS FOR SELECTED ION MONITORING

Compound	RT*	Ions m/z
Morphine-TFA	1.8	477,364
Codeine-TFA	2.1	395,282
Hydromorphone-TFA	2.4	381,325
Nalorphine-TFA	2.7	503,390

*Retention time in minutes.

Reprinted by permission of Preston Publications, Inc., from Saady, J. J., Narasimhachari, N., and Blanke, R.: J. Anal. Toxicol., 6:235-237, 1982.

Monitoring characteristic ions at specific retention times makes selected ion monitoring (SIM) a powerful analytical tool. The use of nalorphine and the sensitivity of response of the system makes this procedure quantitative from <0.02 mg/L to at least 8 mg/L for morphine and codeine. The coefficients of variation of a within-day study were 6.2% for 0.15 mg/L morphine, 5.3% for 0.10 mg codeine/L, and 11% for 0.16 mg hydromorphone/L. The lower limit of detection for hydromorphone is about 0.08 mg/L. Sensitivity is sufficiently great for morphine and codeine that, after assaying a high standard, caution must be exercised to prevent carryover.

Interpretation

Therapeutic concentrations of morphine range from 0.01–0.07 mg/L in serum and from 0.5–10 mg/L in urine. Codeine has similar concentrations in serum, but urine concentrations may range up to 30 mg/L. Hydromorphone is more potent; therapeutic concentrations in serum are <0.03 mg/L but may reach 1 mg/L in urine. Depending on the degree of tolerance, toxic concentrations generally exceed 0.1 mg/L in serum for each of the three opiates.

Reference

Saady, J. J., Narasimhachari, N., and Blanke, R. V.: Rapid simultaneous quantification of morphine, codeine and hydromorphone by GC/MS. J. Anal. Tox., 6:235-237, 1982.

CANNABINOIDS

By definition, marijuana is a mixture of the dried leaves and the flowering tops of the plant *Cannabis sativa L.* The primary active ingredient in marijuana is Δ^9-tetrahydrocannabinol (THC); street marijuana usually contains <5% THC. Hashish, the resinous coating from the flowers and leaves of *Cannabis sativa L.*, contains up to 15% THC and may be smoked as a mixture of the resin and tobacco.

Overdosage of THC is rarely severe enough to be life threatening. This is fortunate for many reasons, not the least of which is the lack of practical methods for the identification of the active ingredient in human serum or urine. Further, THC is almost completely metabolized. The resulting metabolites are held in fat stores and excreted in urine and feces over a long time period. Thus, to distinguish between active and inactive forms in a quantitative manner requires difficult and time-consuming procedures relying on GC/MS.[24]

Nevertheless, for medical-legal reasons or to assess the clinical state of a patient, it may be useful to determine whether a particular individual has used THC in the recent past. A homogeneous enzyme immunoassay procedure can be applied in these cases, provided that results are interpreted cautiously. The method assays the major urinary metabolites of THC and is most sensitive to 11-nor-Δ^9-THC-9-carboxylic acid and to 11-hydroxy-Δ^9-THC. It is designed to detect at a sensitivity of ~20 ng/mL and to discriminate reliably between a negative specimen and one containing 50 ng/mL cannabinoids.

The peak effect of smoking THC occurs in 20–30 min and the duration is 90–120 min after one cigarette. Elevated urinary metabolites are found within hours of exposure and remain detectable for 3–10 d after smoking. Rate of metabolism and, consequently, concentration of urinary metabolites varies widely depending on frequency of use, chronic or single-dose use, and other factors.

Principle

Urine, diluted with a buffer containing malate, is mixed with a reagent containing the coenzyme NAD^+ and antibodies to a THC derivative. Binding occurs to any substance in the urine recognized by the antibody. A second reagent containing a THC derivative labeled with the enzyme malate dehydrogenase is then added. The labeled drug combines with any unfilled antibody binding sites and thereby proportionally reduces the total enzyme activity. The residual enzymatic activity is directly related to the concentration of the drug present in the urine. The active enzyme converts NAD^+ to NADH, resulting in an absorbance change that is measured by a spectrophotometer at 340 nm.[53]

Instrumentation

A digital reading spectrophotometer set at 340 nm, equipped with a flow cell thermostatted at 30 °C and a timer-printer that records and prints out absorbance readings.

A pipetter-diluter capable of sampling 50 μL of liquid and delivering the sample plus 250 μL of buffer with sufficient force to ensure adequate mixing of the components.

Procedure

Refer to the insert in the EMIT-d.a.u. kit for details (Syva, Palo Alto, CA).

Interpretation

Response of an unknown urine specimen equal to or greater than that of the low calibrator is interpreted as "positive"; a response less than that of the low calibrator is "negative." The difference in readings between the negative and low calibrators should be at least 20 *mA* units, and the difference between low and medium calibrators should be at least 35 *mA* units, for satisfactory performance.

Semiquantitative estimates of THC concentrations, in terms of analyte equivalents, can be made by plotting the *A* units for negative, low, and medium calibrators against their respective concentrations (0, 20, 75 ng/mL) on semilog paper. The resulting calibration curve (not linear) can be used to approximate the cannabinoid component concentration in the specimen. For appropriate interpretation the limitations of the method must be fully appreciated.

Interferences

Although a variety of THC derivatives and metabolites will respond to varying degrees in this assay, amphetamine and propoxyphene (< 100 μg/mL), morphine (< 200 μg/mL), benzoyl-ecgonine (< 400 μg/L), methaqualone (< 500 μg/mL), and aspirin, amitriptyline, diazepam, meperidine, phencyclidine, and secobarbital (< 1000 μg/mL) do not significantly interfere. Specimens high in salt content may yield false negative results.

Careful attention to details, frequent blank and control measurements, and good technique are all essential to acceptable results. Positive results must be confirmed by an alternative method specific for Δ^9-THC or its metabolites.

Reference

Rodgers, R., Crowl, C. P., Einstad, W. M., et al.: Homogeneous enzyme immunoassay for cannabinoids in urine. Clin. Chem., *24*:95-100, 1978.

QUALITY ASSURANCE

Quality control is an important and essential component of any laboratory operation. Nevertheless, the implementation of an effective quality control program does not guarantee that a laboratory will report accurate results. What it does assure is that the results reported by a particular laboratory will be consistent. Thus, a laboratory director may spend much time and effort in maintaining a good quality control program, but the end result may be that the laboratory is consistently reporting results of poor quality. Assessment of the accuracy of results is best achieved by participating in an externally administered proficiency testing program.

The quality of a particular laboratory result depends on several factors, but ultimately it depends on the competence of the analyst. It is helpful, of course, for the analyst to have adequate instrumentation and laboratory facilities available so that appropriate methodology can be utilized effectively. Nevertheless, some very competent analytical toxicologists can produce high-quality work even though the laboratory facilities and equipment available to them may be severely limiting. It is important that all laboratories, both small and large, initiate effective quality assurance programs in order for the user to maintain confidence in the laboratory results.

Although the topic of quality assurance is dealt with elsewhere (Chapter 2E), some aspects of considerable importance to the toxicology laboratory should be stressed.

Preventive Maintenance of Measuring Devices

In any laboratory operation in which a measurement is made, the measurement should be considered as a potential variable that must be monitored and documented for future reference. Documentation is important not only for accreditation inspections but also to forestall legal challenges which may question quality control procedures in force at the time a particular analysis was done. Unless documentation is maintained and preserved, it is impossible to demonstrate conclusively that all factors involved in a particular measurement were under control.

Preventive maintenance of instrumentation is also of great importance but at times is more difficult to implement. A gas chromatograph, for example, is subject to variables such as gas flow, temperature settings, programming rates, and column efficiency. Some of these are difficult to monitor with any degree of regularity, yet they are important parameters and cannot be assumed to remain unchanged from day to day. Frequently the manufacturer's operation manual, supplied with the instrument, provides information on recommended preventive maintenance. Using this as a starting point, the analyst should devise a system for periodically checking the variables subject to change. Some variables should be checked on a daily basis while others may be done at extended time periods, either monthly or quarterly.

Standards and Controls

The problem of acquiring reliable and verifiable standards for drugs, metabolites, and other toxic chemicals is a serious one which does not seem to have a simple solution. Some substances are becoming available in a form that can be considered as a primary standard, but generally one must rely on commercial sources for the preparation of standards. The testing of scheduled drugs further requires that the analytical laboratory be licensed by federal and state agencies for the use of these compounds. Some drug standards are available commercially in small quantities which do not require licensure, but these place the analyst in a position of complete reliance on the distributor for authenticity and purity of these substances. Even when drug and chemical standards are acquired in the purest form possible, the analyst must still confirm or determine the state of purity prior to the use of these compounds as standards.[55] Many drugs have a limited shelf life; even when acquired in an extremely pure state, a substance may become modified during storage so that subsequent use would require a reassessment of its purity. One common source of error in this regard is the uncertain amount of water of hydration of very pure crystalline material. Frequently it is not feasible to determine the number of molecules of water associated with the crystalline drug, yet water of hydration can significantly contribute to the total weight of a drug being prepared as a standard.

The most common approach to resolving this problem is to measure a physical property of the standard (such as melting point) and compare it with literature values. The presence of impurities can frequently be detected by a simple TLC or other chromatographic assay of the chemical in question prior to its use as a standard. Methods based on other physical properties can be used, depending upon the circumstances relating to the particular chemical compound in question. Many compounds have significant ultraviolet absorbance properties. Measuring the extinction coefficient after preparing a standard and comparing this measurement with values found in the literature is a useful check as to whether the standard is of adequate purity or has been appropriately prepared. This offers a further advantage in that the standard stock solution can be checked periodically in order to detect any changes that might have occurred due to evaporation or degradation of the compound.[19] In any event, it is of extreme importance to be aware that standard purity is one of the most vulnerable factors in any assay.[10]

Suitable controls for use in clinical toxicology laboratories also present a problem. Ideally, a control is a specimen identical to that which is being assayed routinely but which contains a known quantity of the substance being measured. The control is carried through the entire analytical procedure in the same manner as the unknown specimen and must yield an acceptable result as a check on the validity of the assay being performed.

Laboratories which have a relatively high workload for a particular type of assay can prepare controls by pooling sera or urines and, after an appropriate quantity of material is collected, measure the concentration of the substance in question by repetitive assays of the well-mixed specimen pool. The mean and standard deviation can then be calculated from results obtained from this pool.

Frequently, however, fortifying outdated blood bank plasma with the drug or chemical to be measured and then using this as a control as described above is the only alternative to a true control. This is obviously not identical to a "physiological" specimen, since metabolites or other interferences may be absent from this type of control. When other sources are lacking, however, this artificial control is better than none. Many commercially available controls suffer from this same defect. In addition, many contain a variety of analytes which may pose problems of interferences depending on the assay in use, and thus they must be used cautiously.

The procedures for which it is most difficult to provide adequate control material are screening tests. In these situations, the nature of the analyte may not be known until the screening procedure has been carried out on a specimen from the patient. It is impractical to then obtain, or prepare, a control specimen. Screening procedures must *not* yield negative results, and positive results should be confirmed by specific assays. Hence, it is most important to incorporate a control material which confirms that the test is operating effectively close to its limit of sensitivity. A judicious selection of the most frequently encountered, representative analytes can be added to pooled, analyte-free specimens at concentrations near the sensitivity limit of the test in order to provide control material for these procedures.

Quality control programs for clinical toxicology laboratories are still far from perfect; however, even in an imperfect form, they provide important benefits. These can be achieved only by aggressively implementing, reviewing, and documenting all aspects of the quality control procedure. An important benefit which derives from maintaining good quality control is achieved when the laboratory participates in proficiency testing programs. These are worth the cost of subscription in order to provide the laboratory with an external means of measuring its analytical effectiveness. A good quality control program may enable a laboratory to achieve good reproducibility of reported results, but only when assays are also performed on identical specimens at other similar laboratories is it possible to judge how closely the laboratory is approaching target values of well-prepared specimens.

References

1. Baselt, R. C.: Disposition of Toxic Drugs and Chemicals in Man. 2nd ed. Davis, Cal., Biomedical Publ., 1982.
2. Baselt, R. C: Analytical Procedures for Therapeutic Drug Monitoring and Emergency Toxicology. Davis, Cal., Biomedical Publ., 1980.
3. Baselt, R. C.: Biological Monitoring Methods for Industrial Chemicals. Davis, Cal., Biomedical Publ., 1980.
4. Bastos, M. L., Kananen, G. E., Monforte, J. R. et al.: TLC of basic organic drugs. *In*: Methodology for Analytical Toxicology. I. Sunshine, Ed. Boca Raton, Fla., CRC Press, 1975, pp. 434–442.
5. Blanke, R. V.: Correlation of urinary coproporphyrin III with lead intoxication. J. Forensic Sci., *1*(3):79–87, 1956.
6. Blanke, R. V.: Meaningful toxicology. Medical College of Virginia Quarterly, *9*:302–303, 1973.
7. Blanke, R. V.: Implementation of the clinical toxicology laboratory. *In*: CRC Forum on Drug Analysis in Patient Care. Boca Raton, Fla., CRC Press, 1978.
8. Blanke, R. V., and Saady, J. J.: Rapid simultaneous determination of six anti-convulsant drugs by GLC. Presented at the 26th SE Regional Meeting, Am. Chem. Soc., 1974.
9. Block, J. B.: The Signs and Symptoms of Chemical Exposure. Springfield, Ill., Charles C Thomas, 1980.
10. Booth, E., Drofton, P., and Roberts, L. B.: The influence of standards on interlaboratory quality control programs. Clin. Chim. Acta, *55*:367–375, 1974.
11. Ciba Foundation Symposium 26: The Poisoned Patient: The Role of the Laboratory. New York, Elsevier, 1974.
12. Clarke, E. G. C.: Isolation and Identification of Drugs. London, W. Clous and Sons, 1969.
13. Comstock, E. G.: Treat the patient not the poison. Tex. Med., *65*:25–28, 1969.
14. Conway, E. J.: Microdiffusion Analysis and Volumetric Error. 5th ed. London, Crosby Lockwood & Sons, 1962.
15. Curry, A. S.: Poison Detection in Human Organs. 3rd ed. Springfield, Ill., Charles C Thomas, 1976.
16. Czajka, P., and Duffy, J. P.: Poisoning Emergencies—A Guide for Emergency Medical Personnel. St. Louis, C. V. Mosby, 1980.
17. Decker, W. J., and Treuting, J. J.: Spot tests for rapid diagnosis of poisoning. Clin. Toxicol., *4*:89–97, 1971.

18. Dubowski, K. M.: Organic volatile substances. *In*: Methodology for Analytical Toxicology. I. Sunshine, Ed. Boca Raton, Fla., CRC Press, 1975, pp. 407–411.

19. Elion-Gerritzen, W. E.: The absorbance of the standard: A valuable tool in quality control of spectrophotometric methods. Am. J. Clin. Pathol., *60*:493–498, 1973.

20. Feldstein, M., and Klendshoj, N. C.: The determination of volatile substances by microdiffusion analysis. J. Forensic Sci., *2*:39–58, 1957.

21. Finkle, B. S., Cherry, E. J., and Taylor, D. M.: A GLC based system for the detection of poisons, drugs and human metabolites encountered in forensic toxicology. J. Chromatogr. Sci., *9*:393–419, 1971.

22. Foerster, E. H., and Garriott, J. C.: Analysis for volatile substances in biological samples. J. Anal. Toxicol. *5*:241–244, 1981.

23. Foltz, R. L., Fentiman, A. F., and Foltz, R.: GC/MS assays for abused drugs in body fluids. NIDA Research Monograph 32, DHHS publication number (ADM) 80–1014, 1980.

24. Foltz, R. L., McGinnis, K. M., and Chinn, D. M.: Quantitative measurement of Δ^9-tetrahydrocannabinol and two major metabolites in physiological specimens using capillary column GC-negative ion chemical ionization mass spectrometry. Biomed. Mass Spectrom., *10*:316–323, 1983.

25. Forney, R., Jr.: Modified Davidow TLC drug screening procedure. *In*: Methodology for Analytical Toxicology. I. Sunshine, Ed. Boca Raton, Fla., CRC Press, 1974, pp. 443–446.

26. Frings, C. S., and Cohen, P. S.: Rapid colorimetric method for the quantitative determination of ethchlorvynol (Placidyl) in serum and urine. Am. J. Clin. Pathol., *54*:833–836, 1970.

27. Gant, V. A.: Lead poisoning. Indust. Med., *7*:608–708, 1938.

28. Garrettson, L. K., and Curley, A.: Dieldrin: Studies in a poisoned child. Arch. Environ. Health, *19*:814–822, 1969.

29. Garriott, J. C.: Personal communication, San Antonio, 1973.

30. Gilman, A. G., Goodman, L., and Gilman, A.: The Pharmacological Basis of Therapeutics. 6th ed. New York, Macmillan, 1980.

31. Goldbaum, L. R., Schloegel, E. L., and Dominguez, A. M.: Application of gas chromatography to toxicology. *In*: Progress in Chemical Toxicology, Vol. 1. A. Stohman, Ed. New York, Academic Press, 1963, pp. 11–52.

32. Gosselin, R. E., Hodge, H. C., Smith. R. P., et al.: Clinical Toxicology of Commercial Products. 4th ed. Baltimore, Williams & Wilkins, 1976.

33. Gotelli, G. W., Kabra, P. M., and Marton, L. J.: Determination of acetaminophen and phenacetin in plasma by HPLC. Clin. Chem., *23*:957–959, 1977.

34. Handbook of Common Poisonings in Children. HEW Publication No. (FDA) 76–7004, 1976.

35. Hepler, O. E.: Manual of Clinical Laboratory Methods. Springfield, Ill., Charles C Thomas, 1963, p. 325.

36. Hindberg, J., and Wieth, J. O.: Quantitative determination of methanol in biologic fluids. J. Lab. Clin. Med., *61*:355–362, 1963.

37. In Consultation. Section III—Overdose: No time for heroics. Med. World News, March 24, 1972, pp. 20–26.

38. Ingelfinger, J. A., Isakson, G., Shine, D., et al.: Reliability of the toxic screen in drug overdose. Clin. Pharmacol. Ther., *29*:570–575, 1981.

39. Jain, N. C.: Direct blood injection method for gas chromatographic determination of alcohols and other volatile compounds. Clin. Chem., *17*:82–85, 1971.

40. Kabra, P. M., Stafford, B. E., and Marton, L. J.: Rapid method for screening toxic drugs in serum with liquid chromatography. J. Anal. Toxicol., *5*:177–182, 1981.

41. Kaye, S.: Emergency Toxicology. 5th ed. Springfield, Ill., Charles C Thomas, 1970.

42. Key, M. M., et al., Eds.: Occupational Diseases: A Guide to their Recognition. NIOSH Publication No. 77–181, 1977.

43. Kopito, L., and Shwachman, H.: Determination of lead in urine by atomic absorption spectrometry using coprecipitation with bismuth. J. Lab. Clin. Med., *70*:326–332, 1967.

44. Lundquist, F.: Ethyl alcohol in blood and tissue. *In*: Methods of Biochemical Analysis, Vol. 7. D. Glick, Ed. New York, Interscience Publ., 1959, p. 217.

45. McBay, A. J., and Boling, V. R., Jr.: Spectrophotometric determination of trichloroethanol in chloral hydrate poisoning. J. Anal. Toxicol., *4*:99–101, 1980.

46. McMahon, R. E., Ridolfo, A. J., Culp, H. W., et al.: The fate of radiocarbon-labeled propoxyphene in rat, dog and human. Toxicol. Appl. Pharmacol., *19*:427–444, 1971.

47. Meade, B. W., Widdop, B., Blackmore, D. J., et al.: Simple tests to detect poisons. Ann. Clin. Biochem., *9*:35, 1972.

48. Mulé, S. J.: Routine identification of drugs of abuse in human urine. J. Chromatogr., *55*:255, 1971.

49. Peat, M. A., and Kopjak, L.: The screening and quantitation of diazepam, fluorazepam, chlordiazepoxide and their metabolites in blood and plasma by electron capture gas chromatography and high pressure liquid chromatography. J. Forensic Sci., *24*:46–54, 1979.

50. Pedetski, H. M., Koerner, T. A., Hughes, J. R., et al.: Osmometry in the evaluation of alcohol intoxication. Clin. Toxicol., *5*:343–363, 1972.

51. Peel, J. W., and Perrigo, B. A.: Practical gas chromatographic screening procedure for toxicological analysis. Can. Soc. Forensic Sci. J., *9*:69–74, 1976.

52. Rieders, F.: Fluoride-Type A procedure. *In*: Methodology for Analytical Toxicology. I. Sunshine, Ed. Boca Raton, Fla., CRC Press, 1975, pp. 167–168.

53. Rodgers, R., Crowl, C. P., Einstad, W. M., et al.: Homogeneous enzyme immunoassay for cannabinoids in urine. Clin. Chem., *24*:95–100, 1978.

54. Saady, J. J., Narasimhachari, N., and Blanke, R. V.: Rapid simultaneous quantification of morphine, codeine and hydromorphone by GC/MS. J. Anal. Toxicol., *6*:235–237, 1982.

55. Saady, J. J., Narasimhachari, N., and Friedel, R. O.: Unsuspected impurities in imipramine and desipramine standards and pharmaceutical formulations. Clin. Chem., *27*:343–344, 1981.

56. Searle, B., Chan, W., and Davidow, B.: Determination of lead in blood and urine by anodic stripping voltammetry. Clin. Chem., *19*:76–80, 1973.
57. Sherma, J., and Fried, B.: Thin-layer and paper chromatography. Anal. Chem., *54*:45R, 1982.
58. Smith, R. V., and Stewart, J. T.: Chromatographic methods. *In*: Textbook of Biopharmaceutic Analysis. Philadelphia, Lea & Febiger, 1981, pp. 51–56.
59. Stalling, D. L., Tindle, R. C., and Johnson, J. L.: Cleanup of pesticide and polychlorinated biphenyl residues in fish extracts by gel permeation chromatography. J. Assoc. Off. Anal. Chem., *55*:32–33, 1972.
60. Stewart, R. D.: The use of breath analysis in clinical toxicology. *In*: Essays in Toxicology, Vol. 5. New York, Academic Press, 1974.
61. Stewart, R. D., and Erley, D. S: Detection of toxic compounds in humans and animals by rapid infrared techniques. J. Forensic Sci., *8*:31–45, 1963.
62. Sunshine, I.: Toxicology—a retrospective and prospective look. Clin. Chem., *20*:112–115, 1974.
63. Sunshine, I., Ed.: Methodology for Analytical Toxicology. Boca Raton, Fla., CRC Press, 1975.
64. Sunshine, I.: TLC for weak acids, neutrals and weak bases. *In*: Methodology for Analytical Toxicology. I. Sunshine, Ed. Boca Raton, Fla., CRC Press, 1975, pp. 412–413.
65. Sunshine, I., and Jatlow, P., Eds: Methodology for Analytical Toxicology, Vol. II. Boca Raton, Fla., CRC Press, 1982.
66. Taylor, J. R., Calabrese, R. P., and Blanke, R. V.: Organochlorine and other insecticides. *In*: Intoxications of the Nervous System. Part I. Vol. 36 of Handbook of Clinical Neurology. P. J. Vinken and G. W. Bruyn, Eds. New York, North-Holland Publ. Co., 1979, pp. 391–455.
67. Thompson, J. F., Ed.: Analysis of Pesticide Residues in Human and Environmental Samples. US EPA Health Effects Research Lab., Environmental Toxicology Division, Research Triangle Park, N.C., 1977.
68. Tietz, N. W., and Fiereck, E. A.: The spectrophotometric measurement of carboxyhemoglobin. Ann. Clin. Lab. Sci., *3*:36–42, 1973.
69. Trinder, P.: Rapid determination of salicylates in biological materials. Biochem. J., *57*:301, 1954.
70. Wahl, K., and Rejent, T.: Identification of drugs of abuse in urine using single development thin-layer chromatography. J. Anal. Toxicol., *3*:216–217, 1979.
71. Wiltbank, T. B., Sine, H. E., Brody, B. B., et al.: Are emergency toxicology measurements really used? Clin. Chem., *20*:116–120, 1974.
72. Yodaiken, R. E.: Assessing the risk of occupational disease. *In*: Environmental and Occupational Health Hazards, 35th Annual Stoneburner Lecture Series, CME-MCV, 1982.
73. Zweig, G., and Sherma, J., Eds.: Handbook of Chromatography, Vols. I and II. Boca Raton, Fla., CRC Press, 1972.

Suggested Readings

Quality Control and Proficiency Testing:

Blanke, R. V.: Quality control in the toxicology laboratory. Clin. Toxicol., *13*:141–151, 1978.
Boone, D. J., Guerrant, G. O., and Knouse, R. W.: Proficiency testing in clinical toxicology: Program sponsored by the Center for Disease Control. J. Anal. Toxicol., *1*:147–150, 1977.
Dinovo, E. C.: Forensic toxicology proficiency monitoring: Results, experiences, and comments. J. Anal. Toxicol., *1*:126–129, 1977.
Field, P. H.: Quality assurance and proficiency testing. *In*: Introduction to Forensic Toxicology. R. H. Cravey and R. C. Baselt, Eds. Davis, Cal., Biomedical Publications, 1981.
Flores, A. L.: The National Highway Traffic Safety Administration's blood alcohol proficiency test program. J. Anal. Toxicol., *1*:139–141, 1977.
Jain, N. C.: Blind proficiency testing in urine drug screening: The need for an effective quality control program. J. Anal. Toxicol., *1*:142–146, 1977.
Jatlow, P.: Proficiency testing in clinical toxicology: An overview. J. Anal. Toxicol., *1*:109–110, 1977.
Kelly, R. C.: Some thoughts on proficiency testing in forensic toxicology. J. Anal. Toxicol., *1*:123–125, 1977.
Pippenger, C. E., Paris-Kutt, H., Penry, J. K., et al.: Proficiency testing in determinations of antiepileptic drugs. J. Anal. Toxicol., *1*:118–122, 1977.
Shoemaker, M. J., Klein, M., and Sideman, L.: Drug abuse proficiency testing in Pennsylvania 1972–1976. J. Anal. Toxicol., *1*:130–138, 1977.
Sohn, D.: The College of American Pathologists toxicology program. J. Anal. Toxicol., *1*:111–117, 1977.
Walberg, C. B.: Proficiency assessment programs in toxicology. J. Anal. Toxicol., *1*:105–108, 1977.

Medical-Legal Considerations:

Baselt, R. C., and Cravey, R. H.: Chapter 26, Forensic toxicology. *In*: Casarett and Doull's Toxicology: The Basic Science of Poisons. 2nd ed. New York, Macmillan, 1980.
Cravey, R. H., and Baselt, R. C., Eds.: Introduction to Forensic Toxicology. Davis, Cal., Biomedical Publications, 1981.

CHAPTER 19

BIOCHEMICAL ASPECTS OF PREGNANCY

by Michael F. Greene, M.D., Montserrat deM. Fencl, Ph.D., and Dan Tulchinsky, M.D.

Pregnancy constitutes a special condition in a woman's life that affects various physiologic and endocrinologic systems. Some of the hormones produced during pregnancy are present in much larger amounts than in the nonpregnancy state, whereas other hormones are unique by presenting only during pregnancy. Knowledge of synthesis, transport, and metabolism of these hormones is important not only for understanding their physiologic functions, but also for accurate interpretation of results of analytical measurements of these hormones in biological fluids. These measurements can be utilized to assess placental, fetal, and maternal endocrine systems with respect to fetal well-being and fetal development. (See also section on Reproductive Endocrinology, Chapter 9.)

BIOCHEMICAL CHANGES IN NORMAL PREGNANCY

Hormones produced throughout pregnancy are derived from maternal as well as fetal endocrine glands (see Figure 19-1).[78] In addition, the placenta produces a variety of steroid and protein hormones which are secreted into both the fetal and maternal circulations. Thus, hormones and other substances found in the maternal circulation may originate in the fetal compartment.

The Placenta

The placenta is an incomplete endocrine organ and is only capable of producing hormones by utilizing either fetal or maternal precursors. As Figure 19-2 shows, maternal cholesterol is the main precursor for placental progesterone production. Dehydroepiandrosterone sulfate (DHEAS), derived from both the fetal and maternal adrenals, is the major precursor for placental estrone and estradiol production while fetal 16-α-DHEAS is the main precursor for estriol production. Precursors for placental protein hormone synthesis are derived mainly from the maternal blood supply which nourishes the placenta.

The placenta secretes most of its products into the maternal circulation; only a small portion reaches the fetal circulation. Close proximity of the maternal blood vessels to the site of placental hormone production can explain this preferential secretion of hormones into the maternal blood circulation. Included among the major protein hormones produced by the placenta are human chorionic gonadotropin (hCG) and human placental lactogen (hPL). (See Figure 19-2.) Generally, hormone production by the placenta increases proportionally to the increase in placental mass. Therefore, concentrations of hormones derived from the placenta, such as hPL, increase in maternal peripheral blood as the placenta increases in size toward term; hCG, which is produced maximally at the end of the first trimester (10–12 weeks of pregnancy), is an exception.

Although the placental protein hormones are secreted mainly into the maternal circulation, smaller amounts also reach the fetal circulation where they can exert an effect on both the

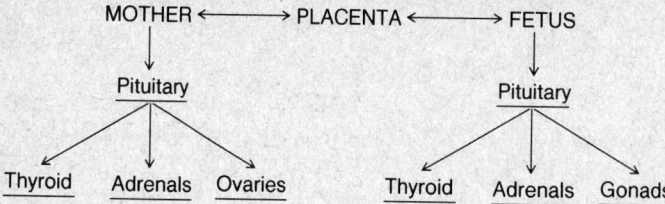

Figure 19-1. Schematic representation of the various major sources of hormones in pregnancy.

maternal and fetal endocrine systems. However, with few exceptions, protein and thyroid hormones of either fetal or maternal origin do not cross the placenta (Table 19-1). In contrast, certain steroid sulfates and unconjugated steroids have the ability to cross from one circulation to the other. One of the exceptions is the steroid conjugate "DHEAS." The extent of the crossover of steroids depends largely on (1) the concentration gradient between the compartments, (2) presence or absence of circulating binding proteins, such as corticosteroid-binding globulin (CBG) in the case of cortisol, and (3) placental metabolism, e.g., cortisol into cortisone.

Maternal Endocrine Changes

The maternal pituitary. Increasing estrogen levels throughout pregnancy will significantly increase the number of pituitary prolactin-producing cells (lactotropes) and cause an increase in prolactin (PRL) secretion up to 10-fold. Conversely, the high estrogen levels during pregnancy will suppress secretion of luteinizing hormone (LH) and follicle stimulating hormone (FSH) to undetectable levels. Baseline levels of other pituitary hormones such as thyroid stimulating hormone (TSH) and adrenocorticotropic stimulating hormone (ACTH) remain unchanged (see Table 19-2), but growth hormone (GH) response to provocative stimuli is blunted.

The maternal thyroid and parathyroid. No clear effect of pregnancy on thyroid function has been demonstrated. Although high levels of thyroxine-binding globulin (TBG) during pregnancy raise the concentration of total thyroxine (T_4) and triiodothyronine (T_3) in blood by ~50%, no change is noted in the free T_4 and T_3 concentrations which represent the active form of these hormones. In contrast, throughout pregnancy there is an increased plasma level of parathyroid hormone (PTH) of ~30–50% with almost no change in plasma ionized calcium, thus suggesting a new "set point" for the secretion of PTH. Calcitonin does not increase predictably during pregnancy, while vitamin 1,25-$(OH)_2D_3$ is increased during pregnancy to promote increased intestinal calcium absorption.

The maternal adrenal. Although some of the increase in plasma cortisol is due to increased CBG and a decreased metabolic clearance rate of cortisol, there also appears to be an increased production of cortisol and a two-fold increase in production of dehydroepiandrosterone sulfate (DHEAS). The diurnal rhythm of cortisol with higher AM than PM levels is maintained, but the absolute plasma level of cortisol is several times greater in pregnancy than in nonpregnancy, mainly because of an increased plasma concentration of sex hormone–binding globulin (SHBG). Increased plasma aldosterone as well as increased deoxycorticosterone concentrations are also observed.

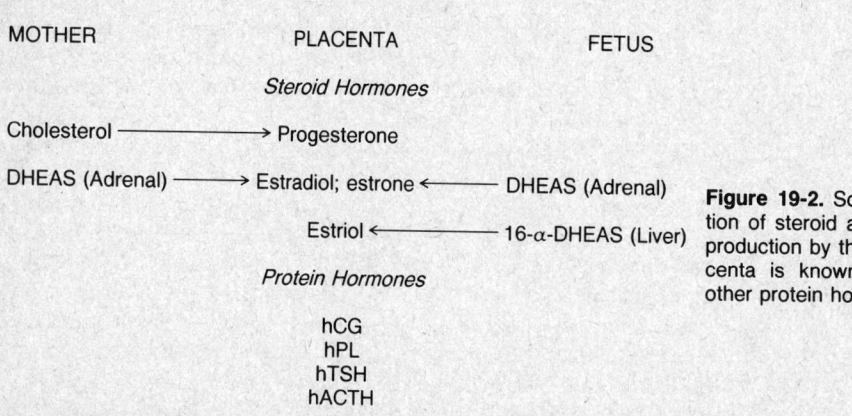

Figure 19-2. Schematic representation of steroid and protein hormone production by the placenta. The placenta is known to produce many other protein hormones.

Table 19-1. THE PLACENTA AS A BARRIER TO HORMONE TRANSPORT

Mother		Placenta (Metabolism)		Fetus
Unconjugated steroids	L	+	L	Unconjugated steroids
Steroid sulfates	L	±	L	Steroid sulfates
Proteins	X	−	X	Proteins
Thyroid hormones	X	−	X	Thyroid hormones

X = lack of placental crossover; L = limited placental crossover.

The maternal ovaries. The maintenance of early pregnancy is dependent on the secretion of adequate amounts of progesterone by the corpus luteum. The continuous function of the corpus luteum of early pregnancy is assured by hCG production and stimulation until enough progesterone is produced by the placenta to assume this function. Experimental removal of the corpus luteum prior to 50 d of pregnancy has resulted in early abortion, whereas its removal after 50 d is without effect because the placenta is already secreting adequate amounts of progesterone.

Changes in the Fetal Endocrine Systems

Both placental and fetal pituitary hormones are required for the normal development of the fetal endocrine system. During the *first trimester,* the fetal pituitary gland is not fully functional and pituitary hypothalamic communications are not well developed. Thus, during the first trimester, placental human chorionic gonadotropin plays a key role in stimulating the fetal testes to produce testosterone which is responsible for the development of the male external genitalia. No similar effect is noted on female gonadal development.

The *second trimester* is characterized by fetal pituitary hormone hypersecretion, presumably as a result of the insensitivity of the hypothalamic-pituitary axis to negative feedback inhibition by circulating hormones. The result is increased pituitary secretion of gonadotropins (LH, FSH), GH, TSH, and ACTH. This increase is illustrated for GH in Figure 19-3. At this stage of development, the secretion of FSH and LH results in oocyte maturation, and the secretion of ACTH and TSH ensures maturation of fetal adrenal and thyroid gland functions.

The *third trimester* of pregnancy is marked by maturation of the fetal hypothalamic-pituitary feedback system. The fetal hypothalamus now recognizes the presence of higher than normal levels of circulating hormones and decreases the secretion of hypothalamic releasing hormones. As a result, the concentrations of most circulating pituitary hormones return to normal.

SCREENING OF THE PREGNANT PATIENT

Effects of Pregnancy on Routine Laboratory Tests

The large amounts of estrogens, progesterone, human placental lactogen, and corticosteroids produced during pregnancy affect various metabolic, physiologic, and endocrinologic systems. The secretion of estrogens and progesterone throughout pregnancy assures appropriate devel-

Figure 19-3. The postulated development of regulatory mechanisms for the control of growth hormone (GH) secretion by the fetal pituitary gland. In the *early-gestation fetus,* the onset of secretion of GH may be autonomous; by *mid-gestation,* when plasma GH concentrations are markedly elevated, the pituitary may be stimulated by unrestrained secretion of GH releasing factor (GRF) by the hypothalamus. In *late gestation,* as the hypothalamus matures, inhibitory influences reduce the secretion of GRF or stimulate the secretion of somatostatin (GIF), or both, with a consequent reduction in pituitary GH secretion. (From Grumbach, M. M., and Kaplan, S. L.: *In:* Foetal and Neonatal Physiology. Cambridge, England, Cambridge University Press, 1973, p. 462.)

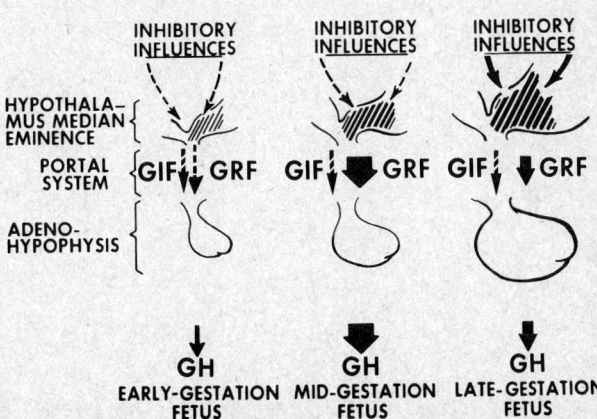

opment of the endometrium, uterine growth, adequate uterine blood supply, and preparation of the uterus for the onset of labor. An increase in resistance to angiotensin, predominance of lipid metabolism over glucose utilization, and increased synthesis by the liver of steroid-binding proteins, fibrinogen, and other proteins are characteristic of pregnancy. As a result of such changes, many of the laboratory reference ranges for the nonpregnant state are not applicable during the pregnant state (Table 19-2).

Hematologic changes. Maternal *blood volume* increases during pregnancy by an average of 45%. Plasma volume increases more rapidly than red cell mass. Therefore, in spite of augmented erythropoiesis, the concentration of *hemoglobin,* the erythrocyte count, and the hematocrit commonly decrease during normal pregnancy. Hemoglobin concentrations at term average 12.1 g/dL compared with 13.3 g/dL for the nonpregnant state. The *leukocyte count* varies considerably during pregnancy from 5000–12 000/μL. During labor and the period immediately after delivery (puerperium) leukocyte counts may be markedly elevated.

The concentrations of several blood *coagulation factors* are increased during pregnancy. Plasma *fibrinogen* increases approximately 50% and ranges from 300–600 mg/dL; the increase contributes to the increase in sedimentation rate. Other clotting factors also increase, including Factor VII, Factor VIII, Factor IX, and Factor X. Prothrombin increases only slightly, while Factors XI and XIII decrease. The Quick one-stage prothrombin time and partial thromboplastin time are both shortened slightly. There does not appear to be a significant decrease in the number of platelets, and the clotting time of whole blood remains unchanged.

Chemical changes. During pregnancy, there is approximately a 40% increase in serum triglycerides, cholesterol, phospholipids, and free fatty acids. Plasma albumin is decreased to an average of 3.0 g/dL in late pregnancy as compared to the mean of 4.3 g/dL in the nonpregnant state; plasma globulins increase slightly. Serum cholinesterase activity is reduced, whereas that of leucine aminopeptidase is markedly elevated. Alkaline phosphatase activity in serum approximately doubles, mainly due to an increase in heat-stable placental alkaline phosphatase.

Renal function. An increase in glomerular filtration rate (GFR) results in an increase in creatinine clearance rate and decreased concentrations of creatinine and urea in serum. The failure of pregnant women to excrete concentrated urine after fluids are withheld does not necessarily indicate renal damage, but may be due to altered fluid distribution. Dye excretion tests (e.g., PSP test) may also give erroneous results. Glucosuria may be present due to increased GFR. Table 19-3 summarizes changes in renal function and associated laboratory findings characteristic of pregnancy.

Table 19-2. PROTEIN, STEROID, AND THYROID HORMONE CHANGES AND ASSOCIATED LABORATORY FINDINGS IN NORMAL PREGNANCY

	Nonpregnant Women (Range)	Pregnant Women at Term (Range)
LH, mIU/mL*	5–25	<2*
FSH, mIU/mL	2–15	<2
PRL, ng/mL	5–25	100–300
GH, ng/mL	<5	<7
TSH, μU/mL	<5	<5
ACTH, pg/mL	7.8–120	12–60
T₄ (total), μg/dL	5–12	7–18
T₃ (total), ng/dL	50–250	120–280
Cortisol, μg/dL	5–25	10–40
Cortisone, μg/dL	1–5	1–10
Deoxycorticosterone, ng/dL	4–16	50–210
Aldosterone, ng/dL	2–10	40–150
Estradiol, ng/mL	0.07–0.3	5–25
Progesterone, ng/mL	1–25	70–250
Parathyroid hormone, μLEq/mL	20–60	20–120†
Calcitonin, pg/mL	50–350	50–450
1,25-(OH)₂ D₃, pg/mL	10–90	20–200

*Since some commercially available LH assays do not distinguish LH from hCG, the "LH" values obtained in the presence of hCG may be extremely high.
†Ranges vary with the laboratory.

Table 19-3. CHANGES IN RENAL FUNCTION AND ASSOCIATED
LABORATORY FINDINGS IN NORMAL PREGNANCY

	Nonpregnant Women	Pregnant Women at Term (Range or Mean)
Renal function		
GFR, mL/min* per 1.73 m^2	100	170
Creatinine clearance, mL/min	90–100	140
Uric acid clearance, mL/min	6–12	12–20
Urine		
Glucose loss, mg/d	Up to 140	Up to 1000
Amino acid loss, mmol/d†	4	15
Protein loss, mg/d	Up to 100	Up to 300
Serum		
Creatinine, mg/dL	0.83	0.53
Urea nitrogen, mg/dL	13	8.7
Uric acid, mg/dL	4–6	3–4
Albumin, g/dL	4.3	3.0

*As determined by inulin infusion.
†Absolute amounts highly variable from one amino acid to another.

EFFECT OF PREGNANCY ON ENDOCRINE TEST RESULTS

HUMAN CHORIONIC GONADOTROPIN (hCG)

Chemistry

Human chorionic gonadotropin is a glycoprotein containing a protein core with branched carbohydrate side chains which usually terminate with sialic acid. hCG is composed of two nonidentical, noncovalently bound glycoprotein subunits, the alpha (α) and the beta (β) subunits. When the dimer is dissociated, most of the hormone activity is lost. However, by equimolar recombination of the two subunits, a major part of the original activity is regained. The molecular weight of the α-subunit is estimated to be 14 900, as calculated from its chemical composition: ~10 200 for the protein and 4700 for the carbohydrate part of the molecule. The amino acid sequence of α-hCG exhibits considerable homology with the α-subunits of human LH, TSH, and FSH. The molecular weight of the β-hCG subunit is estimated to be 23 000; 16 000 for the protein portion and 7000 for the carbohydrate part of the molecule. Some homology exists between the peptide portions of β-hCG and β-LH subunits. The β-subunit of hCG contains 145 amino acid residues, while that of LH contains only 115 amino acid residues. Although 80% of the first 115 amino acids in both β-subunits are similar in sequence, the 30 additional amino acid residues of the β-hCG determine the specificity of the hormone.

Biochemistry and Physiology

Human chorionic gonadotropin appears to be synthesized in the syncytiotrophoblast cells of the placenta. Like other protein hormones, hCG is synthesized on membrane-bound polysomes; its α- and β-subunits are translated from separate mRNA's instead of being synthesized in tandem from one mRNA. Although the proportion of polysomes synthesizing hCG in the placenta declines in later pregnancy, the ratio of α-subunit to β-subunit increases two-fold between the first and third trimesters. This observation coincides with the presence of increasing amounts of free α-chains and decreasing amounts of free β-subunits in maternal plasma as pregnancy progresses. These findings suggest that the synthesis of α- and β-subunits of hCG may be independently controlled, with the β-subunit being the limiting factor in the appearance of complete hCG chains. Many investigators have attempted to elucidate the mechanism that controls the production of hCG and its subunits, and various steroids including progesterone and cortisol have been proposed to play a role. Recent identification of luteinizing hormone–releasing hormone (LRH) and thyrotropin-releasing factors in the placenta raises a possibility that these factors serve as control mechanisms for hCG synthesis. This hypothesis is compatible with the finding of increased release of hCG and cyclic adenosine monophosphate (cAMP) in response to LRH.

From the physiological point of view, hCG plays an important role in maintaining the function of the corpus luteum during the first weeks of pregnancy, i.e., until the luteoplacental shift of progesterone production has occurred. hCG appears also to promote steroidogenesis in the fetoplacental unit and plays an important role in stimulating fetal testicular secretion of testosterone. The human fetal testis has specific binding sites for hCG. During the period of gonadal differentiation, fetal testes in organ culture will produce testosterone in response to stimulation by physiologic levels of hCG. Moreover, the maximal levels of serum testosterone in the fetus occur around the time of peak hCG secretion.

METHODS FOR THE DETERMINATION OF hCG

Several assays involving either biological or immunological techniques are available for the measurement of hCG in biologic fluids (Table 19-4).

Bioassays

The underlying principle of these hCG bioassays is based on the LH-like activity of hCG. Thus, injection of urine containing hCG into experimental animals will produce corpora lutea and corpora hemorrhagica in the ovaries of immature mice (Aschheim and Zondek test), ovulation and corpora lutea formation in female rabbits (Friedman test), and expulsion of spermatozoa in male frogs or toads (Galli-Mainini test). These cumbersome assays were among the first reliable pregnancy tests used, but they have now been replaced by simpler, more rapid, accurate, and specific techniques for the direct measurement of hCG in biologic fluids.

Immunologic Techniques

Rapid agglutination-inhibition tests. For the qualitative detection of hCG, the rapid hemagglutination (or latex particle agglutination) inhibition tests generally suffice. These are presently the most commonly used pregnancy tests and can now even be purchased over the counter and performed by a woman in the privacy of her home. In recent years, a number of commercial kits for pregnancy tests based on hemagglutination (or latex agglutination) have become available, and may be divided into two categories: slide tests (e.g., Gravindex and Pregnosticon) and tube tests (e.g., Pregnosticon and UCG test). The specific protocols for these commercial kit procedures are supplied by the manufacturers, and the following is only a general description of these procedures. The *slide test* is performed on a slide, and the results are obtained within 2 min. Antiserum to hCG is added to the urine, followed by the addition of hCG-coated latex particles. If there is no hCG in the urine (nonpregnant patient), the antiserum will react with the hCG-coated latex particles and agglutination will occur (agglutination = negative test). However, in case of pregnancy, the presence of sufficient quantities of hCG in the urine will neutralize the hCG antibodies and no agglutination will occur within 2 min (positive test). The *tube test* is also based on the hemagglutination-inhibition reaction, but erythrocytes instead of latex particles are coated with hCG. First, urine is mixed with an optimum amount of hCG antiserum in a test tube (Step 1). If hCG is present in the urine specimen (pregnant patient), hCG will combine with and neutralize the hCG antibodies. When hCG-coated erythrocytes are added to the system (Step 2) and left undisturbed for 2 h, no hemagglutination will occur, since all hCG antibodies have been neutralized during step one. As a result, the erythrocytes will settle to the bottom of the tube in a ring pattern (positive test). The size and shape of the ring formation is generally compared with a control tube prepared simultaneously with the urine sample. In the absence of hCG in the urine sample (nonpregnant patient), the hCG antibodies added are not neutralized, and they will react with the hCG-coated erythrocytes and agglutination will occur (negative test) resulting in a mat-like precipitate.

For a semiquantitative assay by hemagglutination-inhibition, the urine is serially diluted until further dilution will not cause inhibition of hemagglutination (end point). A direct comparison with an hCG standard gives the approximate amount of hormone present in the unknown specimen.

Table 19-4. METHODS FOR hCG MEASUREMENTS

Test Type	Sensitivity mIU/mL	Time Required for Assay	Specimen	Comments
1. In vivo bioassay	500–1000	1–5 d	Urine	Measures biological activity of hCG. Does not distinguish LH from hCG. Expensive, time-consuming.
2. Immunologic methods				
Agglutination-inhibition	1000–2000	2 min–2 h	Urine	Nonquantitative, simple, rapid; up to 5% false positives.
β-hCG radioimmunoassay (RIA)	1–10	3–24 h	Serum	Quantitative, specific. Requires RIA equipment and provision for disposal of radioactive waste.
Rapid β-hCG RIA	25–150	2–3 h	Serum	Quantitative, specific. Requires RIA equipment and provision for disposal of radioactive waste.
Enzyme immunoassay (EIA)	5–25	3–5 h	Serum	Quantitative. Nonisotope labeling. Sensitivity depends on antibody and type of enzyme detection.
Immunoradiometric monoclonal antibody (IRMA)	1.5–5	4 h	Serum	Quantitative, high specificity. Requires equipment for counting of radioactivity and provision for disposal of radioactive waste.
3. In vitro bioassay system				
Radioreceptor	100–200	1–2 h	Serum	Quantitative; measures biological activity of hCG. Does not distinguish LH from hCG. Requires equipment for counting and provision for disposal of radioactive waste.
Steroid response in target cells	0.1–0.5	3–5 h	Serum	Quantitative; measures biological activity of hCG. Does not distinguish LH from hCG. Requires RIA equipment and provision for disposal of radioactive waste.
Cytochemical assay	0.5	5 d	Serum	Quantitative; does not distinguish LH from hCG. Too cumbersome for routine clinical use.

Collection of Urine Specimens and Comments

For the qualitative pregnancy test, urine specimens are collected in a clean container at any time of the day; however, the first morning specimen generally contains the greatest concentration of hCG and is therefore preferred. The patient should be instructed to restrict intake of fluids from 2000 h until the morning collection. The urine specimen should be clear; presence of turbidity or urine sediments requires filtration or centrifugation.

False positive results occur in 2–5% of the cases because of the presence in urine of interfering substances such as proteins, drugs, bacteria, erythrocytes, or leukocytes. Cross-reactivity with pituitary gonadotropins may also give false positives (i.e., cases of high LH levels in postmenopausal patients). *False negative* results are common since the test usually does not detect hCG levels at concentrations below 1000–2000 mIU/mL. Therefore, these qualitative tests will not become positive until 8–14 d after the first missed menstrual period; they may also not be sensitive enough for the diagnosis of normal pregnancy after the second trimester and in cases of abnormal or ectopic pregnancies when hCG concentrations are low. Conditions that may cause denaturation of the antiserum to hCG (i.e., temperature, pH, old reagents) may also yield false results. The use of known positive and negative controls and standardization of the reagents are therefore extremely important in order to obtain reliable results. Accuracy for the various commercially available pregnancy tests ranges from 93.8–98.8%. The simplicity and speed with which the results are obtained makes these tests very valuable for the confirmation of pregnancy. It should be remembered, however, that these procedures are not quantitative and do not allow the diagnosis of very early or abnormal pregnancy. For accurate quantitation, more specific immunoassay methods are used. The most common of these tests is the procedure for detection of the hCG-β-subunit. An outline of this method follows.

Determination of hCG by Radioimmunoassay (RIA)

The RIA for hCG provides a very specific assay for the measurements of hCG in blood or urine *if* the antibody used in the RIA has been raised against the β-subunit of hCG or against its carboxyl terminal peptide. This antiserum discriminates between hCG and LH, while most antisera produced against intact hCG do not. Moreover, the β-subunit RIA is also 100- to 200-fold more sensitive than the urine slide test and levels of hCG as low as 1 mIU/mL can be measured in serum (plasma). The sensitivity and specificity of the β-subunit RIA permit the diagnosis of very early pregnancy (6–8 d after conception) and allow the detection of low levels of hCG in patients with abnormal intrauterine or ectopic pregnancy, as well as the monitoring of patients with gestational trophoblastic diseases or hCG-secreting tumors.

The principle of the RIA for the β-subunit of hCG is based upon the ability of a limited quantity of β-subunit antibodies to bind a fixed amount of radiolabeled hCG. Upon addition of an increasing amount of unlabeled hCG, a corresponding increasing fraction of radiolabeled hCG bound to the antibody is displaced. Thus, the percentage of bound radiolabeled hCG decreases as the concentration of unlabeled hCG in the test sample increases. Separation of free from bound hCG is accomplished by techniques such as double antibody, solid-phase systems, polyethylene glycol (PEG), talc, or charcoal, but the double antibody–PEG method is most commonly used. This involves immunoprecipitation with a second antibody directed toward the immunoglobulin present in the β-hCG antiserum and a nonspecific protein precipitation with PEG.

Several commercial kits for the determination of hCG are available. These kits contain β-hCG-antibody, ^{3}H- or ^{125}I-hCG tracer, hCG standards, hCG controls, and precipitating antibodies (or other separating agents), together with detailed instructions. Most of the available kit procedures are similar. Some require a preincubation of unlabeled hormone with antibody to enhance assay sensitivity, followed by an overnight incubation equilibration after labeled hCG is added. Alternatively, a short incubation of 30 min, where all the components of the reaction are incubated simultaneously in a nonequilibrium competitive binding assay, can be used. The choice of a long or short incubation will depend upon the user's requirement for sensitivity, reproducibility, and turnaround time for reporting results.

Levels of hCG < 150 mIU/mL in urine or < 25 mIU/mL in blood are usually not identified by the rapid RIA; however, only a few pregnant patients (< 1%) are expected to have such low levels of hCG at the time of testing. In many instances, it is advisable to use a β-hCG RIA with a sensitivity of 1–10 mIU/mL. This permits detection of very low serum hCG levels in patients with abnormal intrauterine or ectopic pregnancies, as well as the follow-up of serum hCG levels in patients undergoing chemotherapy for hCG-secreting tumors.

Recently, a specific and sensitive commercial kit with two-site *immunoradiometric antibodies* (IRMA) for hCG has been developed (Tandem hCG, Hybritech Inc.). This assay incorporates in a solid phase two different monoclonal antibodies. One antibody is specific for the β-region of hCG, and the second is specific for a sterically removed site on the hCG molecule. The result is an assay which is specific for the β-subunit region and which measures only intact hCG molecules. The sensitivity of the assay claimed by the manufacturer is 1.5 mIU/mL.

Enzyme Immunoassay (EIA) Techniques

EIA methods which utilize enzymes as markers instead of radioactive isotopes have proved to be a suitable alternative for RIA's. The heterogeneous sandwich-type EIA for hCG with double-antibody solid-phase systems is becoming more and more common (e.g., Abbott β-hCG EIA, Abbott Laboratories, North Chicago, IL 60064). This type of EIA for hCG uses an excess of β-hCG antibodies attached to a solid phase to bind the hCG present in the standard or specimen. The bound hCG is then allowed to react with a fixed amount of enzyme-labeled β-hCG antibody. After this second step, the amount of enzyme-labeled β-hCG antibody attached to the solid phase will be proportional to the concentration of hCG in the sample and can be measured indirectly by its enzyme activity after addition of appropriate enzyme substrates. Expensive equipment is not required since enzyme activity is usually determined by photometry. However, in most of these procedures several incubations and washing steps are needed, and it is critical that the solid phase be homogeneously coated with antibody.

Collection of Blood Specimens and Comments

Blood specimens for the hCG immunoassays are collected into suitable tubes without anticoagulants, allowed to clot at room temperature, and centrifuged to obtain clear serum. All specimens not tested within 48 h of collection should be stored at $-20\,°C$. As with most biological materials, repeated freezing and thawing should be avoided. Serum specimens showing gross hemolysis, gross lipemia, or turbidity may give false results.

In assessing the specificity of the β-hCG immunoassay, the determination of the degree of cross-reactivity with LH is of utmost importance. One should assay blood samples with high physiologic LH levels or with LH added to be certain that high levels of this hormone do not significantly influence the hCG results. Assay sensitivity should be set such that small amounts of hCG will be detected, and false positive values due to possible LH presence will be minimized.

Commercially available kits for hCG immunoassay may be calibrated against either the Second International Standard hCG for bioassay (2nd IS-hCG) established in 1964 by the Committee on Biological Standardization of the World Health Organization (WHO) or the most recent WHO International Reference Preparation of hCG (1st IRP-hCG) established in 1974. Because the two standards are not equally pure, laboratories and clinicians must be aware of the implications of using different standards. Results obtained from methods referenced to each of these international standards will not be the same and cannot be directly compared. On a weight basis, the 2nd IS-hCG standard has about one-half the biological potency of the more recently prepared 1st IRP-hCG; that is, 1 ng of 2nd IS-hCG $\simeq$ 5 mIU, and 1 ng of the 1st IRP-hCG $\simeq$ 10 mIU. Therefore, tests standardized with the 2nd IS-hCG standard have about one-half the value (in mIU/mL) of tests standardized with the 1st IRP-hCG. Ultimately, the reported values of hCG will have meaning only in relation to the specificity of the assay system and the reference material used to calibrate it.

DETERMINATION OF hCG WITH IN VITRO BIOASSAY SYSTEMS

In traditional bioassays, the activity of the hormone is analyzed after its administration to an animal. Recently, several simplified assays have been introduced for measuring the biological activity of hCG in cell target tissues using in vitro systems. Among these techniques are radio-receptor and cytochemical assays.

Radioreceptor Assay (RRA)[12]

The principle of this assay is based on the ability of radiolabeled hCG to bind to specific receptors prepared from cell membranes of gonadal tissues and on the inhibition of this binding by unlabeled hCG present in the biological samples (e.g., Biocept G, Wampole Laboratories, Cranbury, NJ 08512). Advantages of RRA are that results can be obtained in less than 2 h and that they reflect the bioactivity of the hormone rather than its immunologic properties. However, the RRA cannot distinguish hCG from hLH. High concentrations of hLH in blood at midcycle and in postmenopausal women (up to 200 mIU/mL) may give false positive results in this assay. To circumvent this problem, the lower limit of this assay is therefore set at 200 mIU/mL. This limits the usefulness of this test for the diagnosis of very early or abnormal pregnancies.

There are other in vitro bioassays for hCG which are based on the ability of hCG to stimulate steroid production by gonadal cell preparations after the interaction of hCG with its cell-membrane receptor (i.e., testosterone production by hCG-stimulated Leydig cells).[25] The steroid produced in vitro is then measured by RIA and is related to the concentration of hCG in the specimen. These methods have high sensitivity and permit measurements of the biological activity of circulating hCG. They do not, however, distinguish between LH and hCG, they are time consuming and expensive to perform, and their sensitivity depends on the preparation of the cells.

Cytochemical Assay

The principle of this method is based on a biochemical response induced by hCG on target cells maintained in vitro (e.g., depletion of ascorbic acid from luteinized ovaries). By measuring such cytochemical reactions it is possible to assess very minute concentrations of hCG (e.g., 0.5 mIU/mL).

Reference

Chayen, J., Daly, J. R., Loveridge, N., et al.: The cytochemical bioassay of hormones. Recent Prog. Horm. Res., *32*:33-72, 1976.

Clinical Applications

Diagnosis of pregnancy. Qualitative tests for hCG in blood or urine are primarily used for the confirmation of pregnancy. Urinary hCG determinations usually suffice to diagnose normal pregnancy when it has progressed beyond the second week following the first missed period. However, a pregnancy test using serum possesses greater sensitivity and may be advantageous for the early diagnosis of pregnancy. Moreover, quantitative and serial determinations of hCG in serum can be helpful whenever abnormalities of pregnancy are suspected and for diagnosis or monitoring treatment in patients with gestational trophoblastic neoplasm (GTN) or other hCG-producing tumors.

In normal pregnancies, the first significant rise in hCG concentration in maternal peripheral blood can be detected between 8 and 10 d after ovulation and soon after ovum implantation. Thereafter, serum and urine concentrations of hCG rise to levels of 30 000–100 000 mIU/mL in serum and 100 000–600 000 mIU/mL in urine after 60–80 days of pregnancy. Subsequently, concentrations of hCG start to decline slowly in serum and urine, and by the end of the second trimester a 90% fall from its peak concentration will usually have occurred (Table 19-5).

Ectopic pregnancy. Measurements of β-hCG in serum can be useful in identifying women with ectopic pregnancies or abnormal intrauterine pregnancies. Patients with ectopic pregnancies often have persistently low serum levels of hCG (50–500 mIU/mL); however, concentrations of hCG depend on the size and viability of the trophoblastic tissue. Not infrequently, high levels of hCG (and a positive urinary pregnancy test) can be associated with ectopic pregnancy. On the other hand, an undetectable hCG serum level (< 1 mIU/mL) in a woman who has missed her period is quite convincing evidence that pregnancy has not occurred. However, relatively low levels of hCG can be associated with a normal early stage or abnormal intrauterine pregnancy. To further distinguish between the various possibilities, repeat measurements of hCG may be needed. In normal intrauterine pregnancy, a rapid increase of hCG will follow, whereas in cases of ectopic pregnancy or abortion, hCG concentrations will fall. However, since hCG is cleared from serum very slowly, it may be detectable in serum or urine up to four weeks after abortion.

Table 19-5. REFERENCE RANGES FOR SERUM hCG IN PREGNANCY

Length of Gestation		
After Fertilization	*After LMP*	mIU/mL
2nd week	4th week	30–100
3rd week	5th week	100–1000
4th week	6th week	1000–10 000
2nd–3rd month	7–14th week	30 000–100 000
2nd trimester	15–26th week	10 000–30 000
3rd trimester	27–40th week	5000–15 000

The samples were assayed using the specific β-hCG assay. The Second International Standard for hCG was used as the reference preparation. LMP = last menstrual period.

For these reasons there are occasional difficulties in interpreting results, particularly when levels of β-hCG range from 5–20 mIU/mL.

In women with amenorrhea secondary to ovarian failure, high gonadotropin levels (high LH) and the small cross reaction of LH in the β-hCG assay (~1% at 50% binding) could account for some false positive results; hCG-like material is also produced by a variety of normal human tissues, and in some patients the appearance of very low levels of "hCG" is a normal phenomenon.

Trophoblastic disease. Serum hCG determinations are very useful for monitoring patients with GTN or other hCG-producing tumors, such as lung carcinoma in male and female patients. These GTN patients characteristically have high hCG levels and low human placental lactogen concentrations. After the tumor is removed and/or treated with chemotherapy or after the uterus has been emptied, hCG should fall to undetectable levels within eight weeks. After serum hCG has been undetectable for three consecutive weeks, monthly follow-up with hCG determinations should be continued for at least six months. The prolonged period of follow-up is required to confirm complete removal of the trophoblastic tumor and to assure that there is no recurrence of the disease.

ESTRIOL

Chemistry

Estriol (E_3), like other natural estrogens, is a derivative of the hydrocarbon estrane with 18 carbon atoms; ring A of the steroid nucleus is aromatic and the hydroxyl group at C-3 is phenolic. E_3 differs from other natural estrogens in having a hydroxyl group at C-16.[14] (See also Chapter 9.)

Biogenesis of Estriol During Pregnancy

The biosynthesis of estrogens during pregnancy differs qualitatively and quantitatively from that of women who are not pregnant. In the nonpregnancy state, the ovaries are the main site of estrogen synthesis; estradiol (E_2) is secreted in microgram quantities and E_3 is only a byproduct of E_2 metabolism. During pregnancy, the placenta is the major source of estrogens; E_3 is produced in milligram quantities and estrone (E_1) and E_2 are produced in microgram amounts. However, the placenta, as opposed to the ovary, cannot accomplish de novo synthesis of estrogens from precursors such as acetate, cholesterol, or progesterone and has to be provided with adrenal C-19 steroid precursors of either maternal or fetal origin.

As shown in Figure 19-4, both the fetus and the placenta cooperate in E_3 biosynthesis in pregnancy. The fetal adrenals provide DHEA-S which is 16α-hydroxylated in the fetal liver. The fetal 16α-OH DHEAS is then hydrolyzed by placental sulfatase, and the 16α-DHEA formed is further metabolized by the placenta via the Δ^5-3β-hydroxysteroid dehydrogenase/isomerase enzyme system to 16α-OH androstenedione (16α-OH Δ^4-adione). The C-19 product is then aromatized by the aromatizing enzyme system (19-hydroxylase, 19-oxidase, and 10,19-desmolase) to estrogens. The 16 OH-estrone thus formed is then converted to estriol by the placental 17β-dehydrogenase which reduces the ketone group at C-17. E_3 is then secreted into maternal and fetal circulations, where its measurement can be used as an indicator of fetoplacental status.

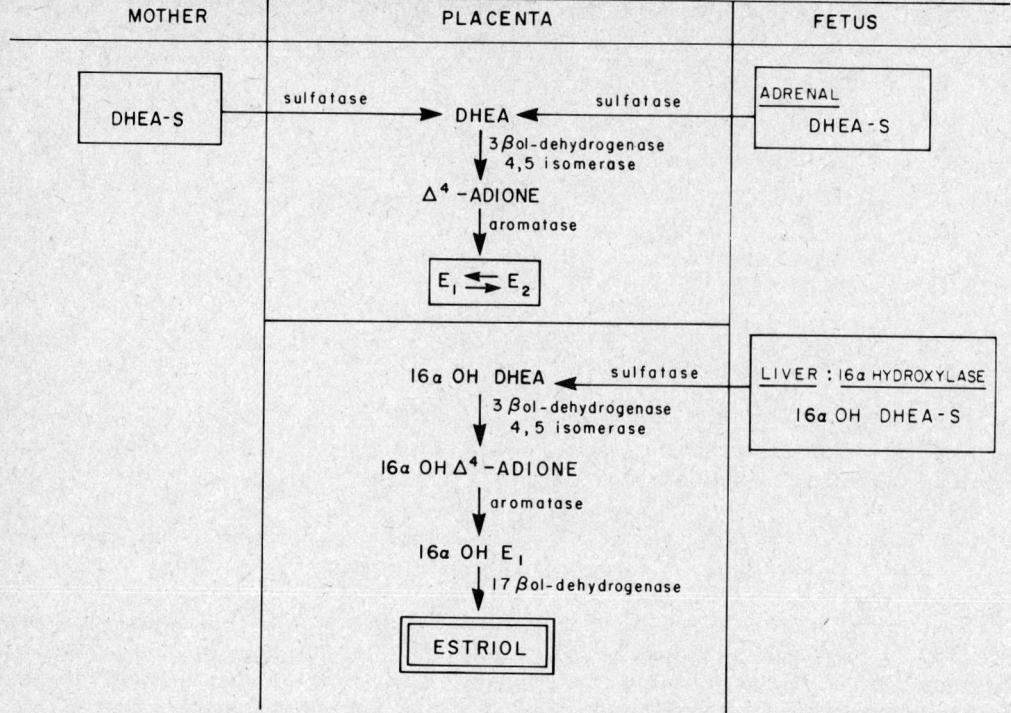

Figure 19-4. Biosynthesis of estrone (E_1), estradiol (E_2), and estriol (E_3) in the fetoplacental unit.

Estriol formed by the placenta reaches the maternal circulation mainly as unconjugated E_3, and it is rapidly conjugated by the maternal liver to various conjugated forms. The clearance rates of these various forms of E_3 are different; it is much slower for estriol sulfates (E_3-3S and E_3-3S-16G) than for E_3 glucuronides (E_3-16G and E_3-3G). The differences in renal clearance and tubular secretion of the various forms of E_3 in pregnant women explain in part the observed differences in their plasma concentrations and their per cent distribution between plasma and urine (Table 19-6).

Clinical Significance

Because the pathways for estriol formation during pregnancy involve both fetus and placenta, E_3 measurements can be sensitive clinical indicators of fetoplacental status. Estriol levels during pregnancy, however, are influenced by many factors other than fetal well-being. These include fetal weight; placental enzyme deficiencies (e.g., sulfatase); primary or secondary fetal adrenal hypoplasia (anencephaly, congenital adrenal hypoplasia, exogenous adrenal hormone therapy); and changes in maternal intestinal flora (antibiotic use), maternal renal excretion (affecting both blood and urine levels), and maternal liver functions. All these potential confounding factors must be considered before an abnormal estriol level can be attributed to deterioration of fetoplacental function. Even in the absence of all these factors, a low E_3 value should always be

Table 19-6. PER CENT DISTRIBUTION OF ESTRIOL (E_3) AND ITS VARIOUS CONJUGATED FORMS IN PLASMA AND URINE IN THE THIRD TRIMESTER OF PREGNANCY

	Steroid	%		Steroid	%
Plasma	E_3	13.8	Urine	E_3	0
	E_3-3S	26.0		E_3-3S	4
	E_3-3S-16G	35.2		E_3-3S-16G	10
	E_3-16G	14.8		E_3-16G	73
	E_3-3G	10.2		E_3-3G	13

G = glucuronide; S = sulfate

evaluated with caution and in conjunction with other independent indices of fetal well-being, such as ultrasound assessment, amniotic fluid evaluation of fetal maturity, oxytocin challenge test, nonstress tests, or other evaluations.

Estriol values during the first and second trimesters of pregnancy are relatively low and are not currently used to any great extent for diagnostic purposes. Estriol determinations during the third trimester (28–40 weeks) are most useful in pregnancies complicated by postmaturity, pre-eclampsia, intrauterine growth retardation (IUGR), and diabetes mellitus.[57] However, they are not helpful in pregnancies with Rh isoimmune disease. (For further details see Complications of Pregnancy.)

The range of normal serum and urinary E_3 levels during pregnancy is very wide. An isolated single estimation of E_3 in a high risk pregnancy has therefore only limited clinical application unless it is definitely subnormal. Serial measurements (i.e., tests performed daily or weekly) to evaluate the trend of E_3 production are more meaningful. In the interpretation of serial E_3 values, a decline of 30% or more from the preceding determination or a continuous fall during three consecutive days generally reflects deteriorating fetoplacental function and is considered a pre-dictor of fetal distress. The turn-around time and reliability of the E_3 assay are therefore of utmost importance. Results should be reported within 4–6 h of specimen collection.

METHODS FOR THE DETERMINATION OF ESTRIOL

Urinary and plasma methods for measurements of estriol have proved equally useful in predicting fetal problems. Day-to-day fluctuations in both urinary and plasma E_3 concentrations are similar. While urinary E_3 determinations have the advantage of measuring the secretion of this hormone over a period of time, plasma specimens reflect concentrations only at a given moment but are not subject to error of collection. Thus, the selection of the method to be used should be based on the availability of a reliable service and should take into consideration the previous experience of the physician and the compliance of the patient.

From the methodological point of view, the estimation of E_3 during pregnancy is relatively easy, due to the large amount of hormone present in blood or urine. A variety of reliable methods, involving colorimetry, fluorometry, gas chromatography, and radioimmunoassays, are available for the analysis of estriol in plasma and urine.

Circulating E_3 is present in plasma in unconjugated and in various conjugated forms (Table 19-6). Most of the plasma methods include organic solvent extraction without prior hydrolysis and therefore measure unconjugated plasma E_3 only. In the urine, E_3 is present mainly in a conjugated form and thus measurement requires prior hydrolysis. In RIA, improved specificity of the antibody for the main E_3 conjugated form in urine (E_3-16-glucuronide) allows its direct measurement without prior hydrolysis. In the following paragraphs the chemical method of Brown et al.[7] for urinary estriol determinations is described, together with more recently used radioimmunoassay methods for urinary and plasma estriol.

The **chemical method** of Brown et al.[7] appears to be suitable for analysis of total E_3 in a clinical laboratory which is not equipped for isotope counting. The principle and main steps are essentially the same as described earlier for the estimation of total estrogens in urine,[15] except that the final measurement is carried out by colorimetry using a modified Kober reaction. (See also Chapter 9.) Because urine in late pregnancy contains more than 500 times the amount of total estrogens that are present in urine of nonpregnant women, only minimal purification of the urinary extract prior to the color reaction is necessary. In acid-hydrolyzed urine, E_3 accounts for some 80% of the total estrogens present during pregnancy. Thus, the method described principally measures estriol. Destruction of estriol by acid hydrolysis or interference with E_3 determination using the Kober reaction has been found to occur in the presence of glucose, mandelamine, chlorothiazide, hydrochlorothiazide, and laxatives containing 1,8 dihydro-oxyanthroquinone. The range and average excretion of urinary total E_3 from 16–42 weeks is given in Table 19-7.

In recent years **radioimmunoassay (RIA)** has become the technique of choice for measuring E_3 in urine or plasma during pregnancy. Compared with colorimetric, fluorometric, or gas chromatographic techniques, RIA's are faster, more sensitive, more specific, and easier to perform. The theoretical and practical aspects of RIA have been discussed in many texts.[1,86] (See also Chapters 1C and 9.)

Table 19-7. URINARY EXCRETION OF TOTAL ESTRIOL AT VARIOUS STAGES OF NORMAL PREGNANCY*

Weeks of Gestation	Range, mg/d	Weeks of Gestation	Range, mg/d
16	1–5	30	8–24
18	1–7	32	8–29
20	2–8	34	9–34
22	3–10	36	10–36
24	4–12	38	11–39
26	5–14	40	12–40+
28	6–20		

*From McLeod, S. C., Brown, J. B., Beischer, N. A., et al.: Aust. N.Z. J. Obstet. Gynaecol., 7:25, 1967.

A wide variety of RIA methods and commercial kits are now available for measuring E_3 during pregnancy. Some of the differences in methodology related to the specificity of the antiserum, the tracer used in the assay (^3H-E_3 vs ^{125}I-E_3) or the method for the separation of free and antibody-bound hormone (e.g., dextran-charcoal, double-antibody separation, or solid-phase system). Methods or commercial test kits using an antiserum specific for unconjugated E_3 can be used to measure total E_3 in urine or plasma after an enzymatic hydrolysis step. In addition, they can measure unconjugated E_3 in plasma after it is selectively extracted with organic solvent (e.g., total E_3 kit, Amersham Corp; E_3 total and/or Free E_3 kit, Clinical Assays, Division of Travenol Lbt. Inc.). Other more simplified and direct methods use an antiserum specific for one or more E_3 conjugates (E_3-16-glucuronide) or specific for unconjugated E_3. The latter methods require no extraction steps.

Regarding the tracer for RIA, undoubtedly ^{125}I is preferred over tritium because of its higher specific activity, rapid counting time, and the ability to dispense with expensive scintillation liquid. Disadvantages include the relatively short useful life of radioiodinated tracer (1–2 months) compared to the long half-life of ^3H (12½ years). Before selecting the tracer for the RIA-E_3, some consideration also should be given to the immunoreactivity of the radioiodinated steroid. One has to be assured that the presence and location of the large iodine atom will not affect the binding of the labeled steroid to the antibody used in the assay.

Estriol Determinations in Urine

Because of the wide availability of commercial kits supplied with a specific protocol, only a general outline of two RIA's for the measurement of total E_3 in pregnancy urine and for unconjugated E_3 in pregnancy plasma will be given here.

The method of Fencl et al.[27] is a simple and sensitive RIA for measuring total E_3 concentrations in highly diluted, unextracted urine samples after a short enzymatic hydrolysis. The prerequisite for such an assay is availability of a specific antiserum for unconjugated E_3. Such antiserum can be raised in rabbits against E_3-6-one-carboximethyloxime coupled to bovine serum albumin, or it can be obtained commercially (Radioassay System Lbt. Inc., Carson, CA). The proper RIA is set up using duplicate aliquots of E_3 standards and of enzyme-hydrolyzed diluted urine together with a fixed amount of ^3H-labeled E_3 and antibody. After a short incubation of 1 h at 4 °C in the assay buffer, separation of free from bound E_3 is accomplished by the dextran-charcoal method;[19] the amount of radioactivity in the bound fraction of all samples is counted in a liquid scintillation counter, and expressed as percentage of that present in the zero standard tube. A standard curve is then constructed in which the percentage of bound radioactivity is plotted versus the log of the E_3 concentrations. The concentration of E_3 in the urine specimen is calculated by interpolation from the standard curve, and by correcting for all urine dilutions. Results are reported in mg E_3/d. The normal distribution of 24-h urinary E_3 values from 16–40 weeks of pregnancy by this RIA are practically identical to those of the colorimetric method that appear in Table 19-7. The RIA is several thousand times more sensitive (25 pg) than the colorimetric method. Substances known to interfere in the colorimetric measurements of E_3, such as glucose, mandelamine, chlorothiazide, and hydrochlorothiazide, do not interfere with measurements of E_3 by the RIA method. Another advantage of the RIA method is that a large number of samples can be simultaneously processed and results can be available in a few hours. Procedural losses are minimal since no extraction and purification steps are included, and there is no need to monitor losses with recovery experiments. Complete enzymatic hydrolysis, however, should be tested by recovery experiments with every new lot of enzyme. Two blank samples (water) and two serum pools (early and late pregnancy) are run with each RIA. Blank values should be below the sensitivity of the method.

Estriol Determinations in Plasma

Plasma samples, as opposed to 24-h urine collections, can be readily obtained and thus permit a rapid assessment of placental E_3 production. However, 85–90% of plasma estriol is conjugated and 10–15% is unconjugated. Since unconjugated E_3 has a shorter $t_{1/2}$ (20 min) in plasma than conjugated estriol, measurement of unconjugated E_3 is considered to be a better reflection of the fetal status. The RIA method of Tulchinsky and Abraham[77] for the measurement of unconjugated E_3 in small volumes of plasma obtained during pregnancy is sensitive, simple, and reliable. Unconjugated E_3 is extracted from plasma with 20 volumes of diethyl ether, using small amounts of tritiated E_3 (1000 cpm) as internal standard for recovery estimation. The ether phase is then transferred to a clean tube and evaporated to dryness. The dried ether extract is then dissolved in assay buffer, and the proper RIA is set up as outlined for total E_3 in pregnancy urine. The concentrations of unconjugated E_3 in the unknown plasma samples are calculated by interpolation from the standard curve and corrections are made for recovery. Results are reported as ng E_3/mL plasma. During pregnancy the mean unconjugated E_3 concentration increases gradually from 1.7 ng/mL at 16–20 weeks to 11.1 ng/mL at term (Table 19-8). Due to individual variations the reference range is very wide, with a five-fold difference between low and high values. Recently, RIA methods have become available which require no prior extraction of estriol from the specimen. These methods are capable of measuring unconjugated estriol directly in serum or plasma without significant interference from conjugated estriol.

HUMAN PLACENTAL LACTOGEN (hPL)

Chemistry

Human placental lactogen (hPL), also known as human chorionic somatomammotropin (hCS), is a single chain polypeptide with a molecular weight of 21 600 and is composed of 190 amino acids with two intramolecular disulfide bridges. When the structure of hPL is compared with that of growth hormone (hGH), their homology is remarkable. Over 80% of their amino acid residues are located in identical positions. Marked similarities in the chemical structure of hPL to that of bovine and human prolactin (PRL) have also become apparent. It is therefore not surprising that hPL shares some growth and lactogenic properties with hGH and PRL.

Biochemistry and Physiology

Human placental lactogen has been localized by immunofluorescence studies in the syncytiotrophoblast cells of the placenta. Evidence of the biosynthesis of hPL by the placenta has been obtained by in vitro studies showing incorporation of labeled amino acids into hPL, by using cultures of placental tissues or cell-free systems containing placental polyribosomes or mRNA. The increase in maternal serum hPL concentration with advancing gestational age can be directly

Table 19-8. PLASMA CONCENTRATION OF UNCONJUGATED ESTRIOL THROUGHOUT PREGNANCY*

Week of Gestation	Mean ± SE (ng/mL)	Range (ng/mL)	Number of Determinations
16–20	1.7 ± 0.25	1–3.2	13
20–24	3.9 ± 0.4	2.1–7.8	14
24–28	6.2 ± 0.4	2.1–10.3	20
28–32	7.5 ± 0.5	4.0–13.6	21
32–36	8.8 ± 0.4	3.6–15.5	55
36–38	9.6 ± 0.6	4.6–18.0	16
38–40	11.1 ± 0.7	5.4–19.8	28

*From Tulchinsky, D., and Abraham, G. E.: J. Clin Endocrinol., *33*:775, 1971.

correlated with the increasing mass of placental tissue and of functional syncytiotrophoblast tissue. The placental secretion of hPL rises steadily during normal pregnancy to 1–2 g/d near term, in contrast to the early rise and subsequent fall of hCG secretion by the placenta. Not much is known about the factors which control the secretion of hPL. It is possible that some maternal factors may influence hPL production and release by the placenta.

From the physiological point of view, hPL has been shown to possess a variety of biological activities, including lactogenic, metabolic, somatotropic, luteotropic, erythropoietic, and aldosterone-stimulating effects.

Lactogenic activity. hPL, either directly or in synergism with prolactin, plays a significant role in preparing the mammary glands for lactation. Animal experiments, both in vivo and in vitro, have clearly demonstrated the lactogenic action of hPL.

Metabolic activity. The many metabolic activities of hPL qualitatively closely resemble those of hGH. These include inhibition of glucose uptake, enhanced lipolysis leading to increased mobilization of free fatty acids, and enhancement of nitrogen retention. Since glucose appears to be a primary energy substrate for the fetus, it has been suggested that the glucose-sparing action of hPL may be a strategy to direct maternal metabolism toward greater utilization of fat for the mother's requirements, thereby sparing maternal glucose for fetal use.

Luteotropic activity. hPL has been shown to have luteotropic activity in laboratory animals, but the importance of this action in human pregnancy is unknown.

Other reported actions of hPL, which include rise in aldosterone levels and erythropoietic effects, are also unclear.

Clinical Applications

The clinical value of measurements of maternal serum hPL in obstetrical complications remains unsettled. hPL in maternal blood has a short half-life of ~15 min and there is a good correlation between its serum concentration and placental weight. It has been suggested that hPL concentrations primarily reflect placental function and fetal distress resulting from placental malfunction. In normal pregnancies, the rise of hPL in maternal peripheral blood begins at the fifth week after the last menstrual period, reaches a plateau at about 34–38 weeks of pregnancy, and falls to undetectable levels within a few hours after delivery (Figure 19-5).

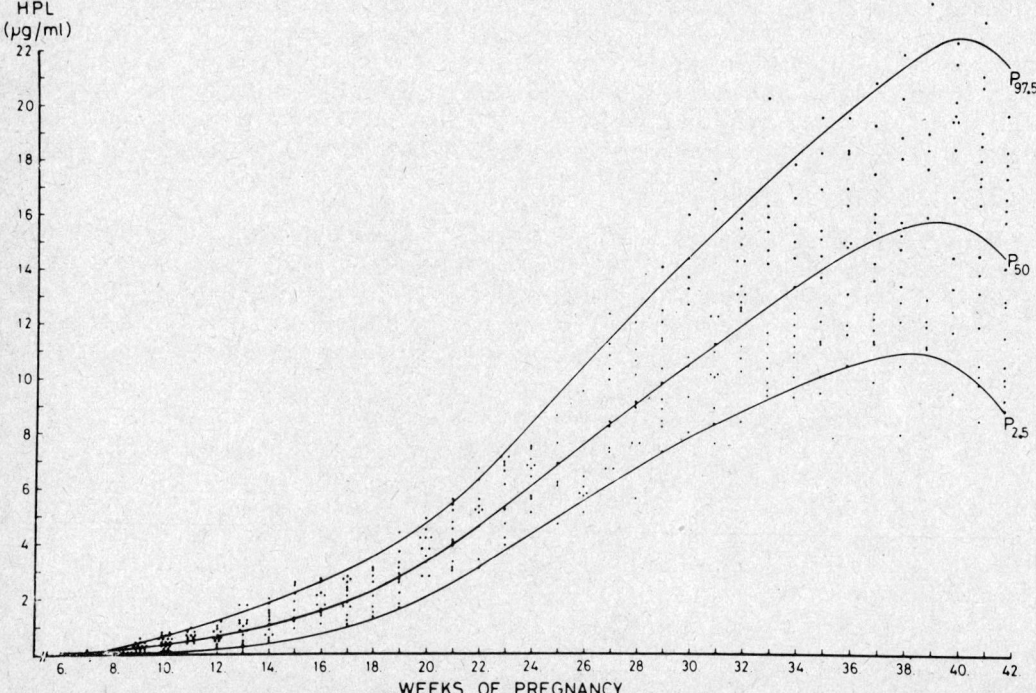

Figure 19-5. Serum hPL levels of 263 women with normal pregnancy followed throughout gestation. The individual values, the median, and the 95 per cent confidence limits are shown. (Olavi Ylikorkala, Acta Obstet. Gynecol. Scand., Suppl. 26, 1973, p. 20.)

Levels of hPL in an individual tend to follow a fixed trend either above or below the population mean, emphasizing the need for serial measurements as opposed to isolated observations. In patients with diabetes mellitus not complicated by hypertensive disease, and in patients with erythroblastosis fetalis where the placenta is larger than normal, high hPL levels are common and should not be interpreted as indicative of good fetal prognosis. Fetal deaths may occur in such patients despite normal or higher than normal hPL levels.

Care should also be taken in interpreting low maternal levels of hPL at term. Isolated cases of extremely low (<1 μg/mL) immunoreactive hPL levels have been reported in normal pregnancies at term. This presumably reflects an alteration in the immunologic property of hPL or a defect in placental synthesis of hPL. Like hCG, hPL has also been detected in nonpregnant patients with certain tumors of the gonads and in some lung or breast cancers.

METHODS FOR THE DETERMINATION OF PLACENTAL LACTOGEN

Radioimmunoassay is usually the method of choice for the quantitative assay of hPL.[3,26] Other methods available are hemagglutination,[72] complement fixation,[83] immunodiffusion,[52] and radioreceptor assay.[68] There is no standard bioassay for hPL measurement, but the pigeon-crop stimulation assay is commonly used.

The Determination of hPL by RIA[26]

A variety of RIA methods have been described for measuring hPL in serum during pregnancy. Because of the wide availability of commercial kits with detailed technical instructions, only a general outline of one RIA method will be given here.

The RIA method of Ermshar and Gusseck[26] utilizes radioiodinated hPL and a specific hPL antiserum and is based upon the ability of a limited quantity of hPL antibodies to bind a fixed amount of hPL. Unlabeled hPL in the test sample competes with radiolabeled hPL for the antibody binding sites. The percentage of bound radiolabeled hPL decreases as the concentration of hPL in the test sample increases. After a 30-min incubation of all reaction components at 37 °C, separation of free from bound hPL is accomplished by precipitation of the bound fraction with polyethylene glycol (PEG). An optimal amount of hPL-free human plasma is added to all assay tubes including standards to facilitate the precipitation of the hPL-antibody-bound complex with PEG while minimizing nonspecific precipitation of free hPL. The amount of radioactivity in the precipitate of all samples is counted in a γ-counter and expressed as a percentage of that present in the zero standard tube after correcting for nonspecific precipitation. A standard curve is then constructed in which the percentage of bound radioactivity is plotted against the log of the standard hPL concentrations. The concentration of hPL in the specimen is calculated by interpolation from the standard curve and using the appropriate dilution factor to correct for any dilutions of the specimen before RIA. Results are reported as ng/mL relative to the hPL reference preparation used.

Specimen Collection and Comments

Collect blood into a suitable tube without anticoagulants. Allow blood to stand at room temperature until clotted and centrifuge to obtain clear serum. Serum specimens are diluted 100-fold with assay buffer to bring the concentrations of hPL within the range of the assay and to minimize additional nonspecific precipitation of immunoglobulin present in the patient's serum.

Specificity of the assay will depend on the specificity of the antiserum used. Rabbit antiserum to hPL cross-reacts to some extent with the structurally related peptide hGH. Usually the importance of this cross reaction is minimal because the affinity of the antiserum to hPL is 1000–10 000 times greater than the affinity to hGH, and the concentration of hPL in pregnancy serum is more than 100-fold the hGH concentration. However, assessment of specificity and titer with every batch of antiserum is necessary. If optimal conditions of temperature, incubation time, and mixture of reagents are carefully controlled in the assay, the precision and reproducibility of this rapid RIA appear to be comparable with more complex procedures. The sensitivity of this method also permits the detection of hPL during early pregnancy.

At present there is no international reference preparation for hPL, although a number of purified preparations can be obtained from research institutes and commercial sources. The quantity of hPL in serum and other fluids should be expressed in terms of weight equivalents of the purified reference preparation used.

HPL can be measured after approximately six weeks of gestation when concentrations are present in the nanogram range only. Thereafter, hPL concentrations gradually rise to peak levels at about 34–38 weeks and then show a slight decline (Figure 19-5). Although there is general agreement on the shape of the secretion curve throughout pregnancy, reported levels vary considerably. At term, reported levels range from 4.0–25 μg/mL.[62] This wide range of levels is partly attributable to the lack of a common reference standard among different authors and to differences in RIA techniques. Each laboratory should establish its own reference range for hPL in normal pregnancy until a common standard reference preparation is available.

COMPLICATIONS OF PREGNANCY

ERYTHROBLASTOSIS FETALIS

Isoimmune hemolytic disease or erythroblastosis fetalis is a disease of the fetus and newborn caused by an incompatibility between fetal and maternal blood. Any of a large number of erythrocyte surface antigens A, B, the antigens Rh, Kell, Duffy, Kidd, and others[84] may be responsible for isoimmune hemolysis. The most common cause of severe disease is sensitization of an Rh negative woman to the D antigen of the Rh system. Commonly used synonyms are "Rh isoimmune disease" or simply "Rh disease."

Pathophysiology

Sensitization, or production of an antibody, may occur in response to any exposure to a foreign antigen. If an Rh negative woman, for example, is inadvertently transfused with Rh positive blood, she may respond by producing anti-D antibodies. With modern transfusion techniques, however, the source for sensitization in women is more likely due to exposure to Rh positive fetal blood from a pregnancy. Although the fetal and maternal blood compartments are generally considered to be separate during normal gestation, it is now known that small numbers of fetal erythrocytes are continually gaining access to the maternal circulation. This antigenic challenge is sufficient in some women to provoke an antibody response. Substantially larger antigenic exposures may result from those disruptions in the integrity of the fetal compartment that accompany spontaneous or induced abortion, ectopic pregnancy, or delivery of an infant. The larger the fetomaternal hemorrhage, the more likely it is that the mother will respond to the challenge with an antibody. Other antigens of the Rh system, C, c, E, e, may also stimulate antibody formation but appear to be only about 1/30th as potent at stimulating a response.

The immunoglobulins produced are usually of the IgG class which cross the placenta quite readily. When the sensitized woman has another Rh positive pregnancy, the antibodies will cross the placenta and cause destruction of the fetal erythrocytes. Furthermore, during pregnancy or at the time of delivery, repeat exposure to Rh positive erythrocytes will cause an augmented response resulting in higher antibody titers. This augmented response is responsible for the frequent observation that subsequent pregnancies seem to be somewhat more severely affected than earlier ones.

The basic pathophysiology of isoimmune hemolysis is the same for all antigens. The differences in severity of disease are due to differences in the expression of the antigen on the surface of the cells, the intrinsic immunogenicity of the antigen, and peculiarities of the immune response of the mother. Destruction of the fetal erythrocytes, which is the central problem, leads to several other problems, which can be summarized as follows: (1) Fetal anemia imposes an extra burden upon the fetal heart to provide an adequate oxygen supply to fetal tissues. (2) Fetal anemia leads to stimulation of the fetal marrow and to extramedullary erythropoiesis in liver and spleen to replace the destroyed erythrocytes. (3) Extramedullary erythropoiesis in the liver destroys hepatocytes and leads to decreased production of serum albumin and decreased oncotic pressure in the intravascular space.

These changes, when severe, lead to congestive heart failure and generalized fetal edema with ascites as well as pleural and pericardial effusions. When the fetal condition has deteriorated to this degree, it is referred to as hydrops fetalis and carries a very grave prognosis. The edema and effusions are readily observable by ultrasound examination. When these changes are observed and there is no therapeutic intervention, intrauterine demise will follow in a relatively short time.

With the fetus in utero, the functions of respiration and removal of bilirubinoid pigments resulting from hemolysis are performed by the placenta and mother. When a fetus is delivered and must assume these functions for itself as a neonate, hydrops is a severe handicap. The lungs are edematous and pleural effusions and ascites physically restrict their ability to expand. The damaged liver is unable to conjugate and excrete bilirubin adequately. When bilirubin accumulates in the blood to excessive concentrations, it passes through the blood-brain barrier to stain the brain and destroy brain cells. This form of brain damage is termed *kernicterus.* Thus, while kernicterus was not a concern in the fetus, it is a significant concern in these sick newborns. A severely erythroblastotic baby can be one of the most challenging problems in the neonatal intensive care unit.

Management of Erythroblastosis Fetalis

The best way to manage any disease is to prevent it. Sensitization can be prevented if the potentially stimulating antigens are removed from the circulation before they have the opportunity to reach and stimulate the maternal immune system. This is the principle upon which the current practice of *prophylaxis* is based. A dose of anti-D Rh immune globulin (RhIg) is administered at the time of any reproductive event which could have exposed the mother to Rh positive fetal erythrocytes (abortion, ectopic gestation, or delivery). In 1968, RhIg was approved for this purpose in the United States by the FDA. Since that time most states have, in fact, passed laws requiring its use when indicated. This use has been primarily responsible for the very dramatic drop in the incidence of pregnancies complicated by Rh disease. More recently, it has been recognized that transplacental fetal-maternal bleeding during an apparently normal pregnancy can lead to antepartum sensitization. This would not be prevented by the immediate postpartum administration of RhIg and therefore administration during the antepartum course, at ∼28 weeks of gestation, has been advocated.

Despite the use of Rh immune prophylaxis, a small number of sensitized pregnancies continue to occur. The main sources of these sensitizations are (1) failure to administer RhIg when indicated postpartum, postabortion, and postectopic (the so-called "utilization gap"); (2) failure to administer RhIg during the antepartum course; and (3) maternal exposure to large volumes of Rh positive cells (in excess of 15 mL of cells or 30 mL of whole blood) which were not entirely neutralized by the standard single dose of RhIg (300 µg).

To care properly for these sensitized women, it is first necessary to distinguish them from the general obstetric population. For this purpose, it is routine at the first prenatal visit to determine each patient's blood group and Rh status as well as to screen the patient's serum for irregular antibodies against a standard panel of erythrocytes. If an antibody which agglutinates erythrocytes is found, it must be identified and titered. If the father possesses the antigen in question, and the maternal antibody is of the IgG class and of sufficiently high titer, then there may be a problem for the fetus. Under these circumstances, further evaluation is required.

Liley noted that the degree of hemolysis in sensitized pregnancies could be assessed by measuring spectrophotometrically the absorbance of bilirubinoid pigments in amniotic fluid, which he named the ΔOD_{450},[44] a term now replaced by ΔA_{450}. Therefore, when the titer indicates the potential for a problem, serial amniocenteses for quantitation of bilirubin in amniotic fluid are indicated. Since the earliest intervention by intrauterine transfusion can be done at 22 weeks of gestation, the first amniocentesis should be performed at about that time. Unless the first amniocentesis shows a very severely affected fetus requiring immediate intervention, the amniocentesis will be repeated at intervals to monitor the pregnancy. The appropriate interval between the amniocenteses is determined by the degree of severity of the disease and the trend with time. The interval may be 3–4 d or 2–3 weeks.

Details of the technique for measuring the ΔA_{450} are given under amniotic fluid analysis. *Interpretation* of the ΔA_{450} depends upon knowing the gestational age of the pregnancy (Figure 19-6). Values which fall into Liley's bottom zone represent an unaffected or very mildly affected fetus. Values in the middle zone are still compatible with a minimally affected fetus, but as values rise within this zone, it is increasingly likely that the fetus is suffering moderate to marked hemolysis. Depending upon the trend with time and the clinical circumstances, some expert

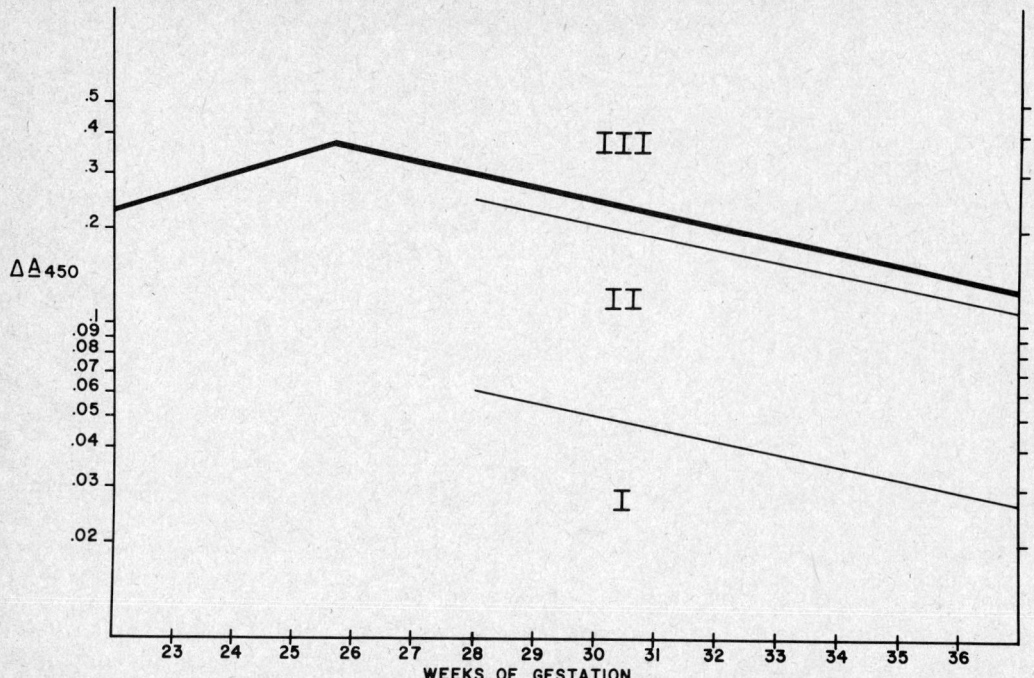

Figure 19-6. Liley's three zone chart (with modification) for interpretation of amniotic fluid ΔA_{450}. Note that the original three zones (divided by the two lighter lines) extended only to 28 weeks of gestation. Uppermost (heavy) line is the upward revision of the "danger line" by Dr. Irving Umansky. For further explanation see text. (Redrawn from Reid, D. E., Ryan, K. J., and Benirschke, K.: Principles and Management of Human Reproduction. Philadelphia, W. B. Saunders, 1972.)

clinicians recommend intervention when the ΔA_{450} has climbed 85% up in the middle zone. Values in the top zone (zone III) indicate severe disease. Without intervention, a fetus whose values fall in the top zone will most likely die.

Therapeutic options for fetuses severely affected by erythroblastosis fetalis are limited. Treatment with promethazine to decrease hemolysis has been recommended, but not convincingly shown to be effective. Plasmapheresis to remove the anti-D globulin has been done, but is impractical because of the frequency with which it must be done over the length of the gestation. Thus, intrauterine transfusion and early delivery are the two most important modes of therapy. The decision to begin intrauterine transfusions is a grave one because these procedures are associated with a considerable potential for fetal morbidity and mortality. Factors which enter into the decision to transfuse in utero include (1) the severity of hemolysis as indicated by the ΔA_{450} and its trend with time, (2) the gestational age of the fetus and pulmonary maturity (tests for fetal lung maturity are discussed in a later section), (3) the presence or absence of hydrops on ultrasound examination, and (4) the mother's past obstetrical history and other factors. An intrauterine transfusion is performed by passing a needle under ultrasound guidance through the maternal abdominal wall and uterus into the fetal peritoneal cavity.[30] Packed erythrocytes of the appropriate group and type (Rh negative, for example, if the antibody involved is anti-D) are then infused either directly through the needle or through a plastic catheter which has been threaded through the needle. The erythrocytes are then absorbed into the fetal circulation across peritoneal surfaces. Once transfusions have begun, they must be continued until the fetus has reached a gestational age which assures a high likelihood for survival in the neonatal intensive care unit. Early delivery and exchange transfusion in the nursery may be a preferable alternative depending upon the particular circumstances. This decision-making process is complex and must be individualized for each patient. It should be noted that at the time of intrauterine transfusion some blood is inevitably spilled into the amniotic fluid. When this blood breaks down, the bilirubin formed will elevate the ΔA_{450}. Thus, the first intrauterine transfusion permanently alters the system and makes further determinations of the ΔA_{450} unreliable.

DIABETES MELLITUS

Diabetes mellitus (DM) is a complex metabolic disturbance caused by inadequate concentrations or inadequate expression of circulating insulin. In patients with untreated DM, hyperglycemia, ketosis, acidosis, and hyperlipidemia may result. Common complications of diabetes include retinopathy, nephropathy, neuropathy, and coronary artery disease. (See also Chapter 6.) Several of the hormonal changes of pregnancy, especially the increased production of glucocorticoids and the production of human placental lactogen, tend to exacerbate some of the pathologic changes in diabetes, including hyperglycemia and the tendency toward ketosis. Following the introduction of insulin therapy for DM, women who survived to reproductive age and were able to conceive then faced significant morbidity, mortality, and fetal wastage during pregnancy. Modern prenatal care with emphasis on scrupulous maintenance of euglycemia has dramatically improved the outlook for pregnant diabetics. Fetuses can now expect a perinatal survival which closely approximates that of the nondiabetic population. Control is established through careful diet and insulin therapy and monitored by capillary blood glucose determinations performed several times daily at home by the patient. Glucose oxidase strips are exposed to blood from a finger stick and the color change of the strip can be read either by eye or in a small portable colorimeter designed specifically for this purpose. (See also Chapter 6.)

A very useful tool to monitor diabetes is the measurement of glycosylated hemoglobin (Hb A_{1C}) which has been introduced in recent years. The theory and practice are reviewed in Chapter 6.

A particularly difficult problem in the care of pregnant diabetics is the high incidence of *congenital anomalies* among the infants of diabetic mothers. Recent work by Miller et al.[49] has shown that an elevated level of glycosylated hemoglobin early in pregnancy, which indicates poor control of DM during organogenesis, correlates significantly with a substantially increased risk for a major congenital anomaly. Patients who had been in good control during the first trimester, however, had the same incidence of major congenital anomalies as the general population. Establishing excellent metabolic control in the first trimester should significantly reduce the risk of major congenital anomalies and associated complications.

The *routine monitoring* and management of pregnant diabetics is somewhat more complex than of nondiabetics. In addition to day-to-day monitoring of capillary blood sugars, a serum and 24-h urine specimen should be collected early in pregnancy to determine baseline creatinine clearance and total protein loss. The diabetic patient population in our institution has an incidence of neural tube defects of 20/1000. Therefore, all of our diabetic patients are screened at 16 weeks' gestation by a quantitative maternal serum α-fetoprotein determination and an ultrasound examination. As discussed below, this will detect approximately 90% of all neural tube defects. Chromosome studies are done only if other standard obstetrical indications are present. As term approaches, pregnancies are monitored with quantitative estriol determinations performed on 24-h urine specimens and with electronic fetal monitoring. Elective deliveries in patients with otherwise uncomplicated diabetes are not done prior to 38 completed weeks of gestation. Fetal lung maturity is checked prior to delivery by determining the lecithin-sphingomyelin (L/S) ratio and the saturated phosphatidylcholine (SPC) concentration in all but emergency cases.

PREMATURE RUPTURE OF MEMBRANES

Premature rupture of the fetal membranes (PROM) is defined in standard texts as rupture of the membranes (i.e., loss of some or all of the amniotic fluid) more than 24 h prior to the onset of spontaneous labor. In recent years, however, PROM has come to take on a new meaning, i.e., rupture of the membranes during the third trimester, but prior to term (35–36 weeks). During this period, a newborn infant may survive with expert neonatal care but it is subject to severe morbidity and mortality due to prematurity. Optimization of neonatal outcome by managing this complication is one of the major challenges in clinical obstetrics and pediatrics today.

Usually, making the diagnosis of PROM is easily done on the basis of history and physical examination. Occasionally, if there is not copious amniotic fluid pouring out of the cervical os, the fluid in the vagina can be tested both for pH with nitrazine paper and for "ferning." The usual vaginal fluid is acidic and will not change nitrazine paper to blue, but amniotic fluid at pH 7.40 will turn the paper blue. When a biological fluid with a sufficiently high salt content is dried on a glass slide, crystals visible under a microscope at low power will grow. These crystals grow in an orderly branching pattern which resembles the fronds of a fern. Thus the phenomenon is termed "ferning." Amniotic fluid will "fern" when dried; vaginal secretions will not.

Once the diagnosis of PROM has been established, a decision must be made to induce early delivery or to continue the pregnancy. Immediate delivery carries with it all of the dangers of prematurity, especially respiratory distress syndrome (RDS), which results from a deficiency of surface active agents (surfactant) in the lungs. Whenever possible, pulmonary indices to estimate the amount of surfactant present should be determined on the amniotic fluid to aid in decision making. Even when the lungs are mature, however, the infant may sustain intracranial hemorrhage or necrotizing enterocolitis which are other common complications of prematurity.

If expectant management is chosen and no effort is made to deliver the patient, infection is an ever-present danger. Intrauterine infection (chorioamnionitis) can be fatal for the fetus and a source of major morbidity (though very rarely fatal) for the mother. The diagnosis of chorioamnionitis can be very difficult since signs and symptoms such as fever, malaise, premature labor, uterine pain, and leukocyte count elevation do not develop until late in the course of the infection. The diagnosis of infection and the decision to deliver are usually made on clinical grounds. However, serum C-reactive protein determinations as well as gas chromatographic techniques for the identification of low molecular weight bacterial metabolic products in amniotic fluid have been proposed to diagnose infection.

PREMATURE LABOR

Our knowledge of the metabolic and endocrine events which sustain labor, including prostaglandin synthesis and release, has advanced dramatically in recent years. In contrast, the mechanism of initiation of parturition at term or preterm in humans is still not entirely clear. Intrauterine infection with or without prior rupture of the fetal membranes or rupture of the membranes with or without infection seems capable of initiating labor. Pyelonephritis, especially when due to a gram-negative organism, and premature separation of a normally implanted placenta can precipitate labor. Regardless of the cause of premature labor, when it is desirable to stop the labor, there are a limited number of therapeutic options. Intravenous ethanol, isoxsuprine, magnesium sulfate, prostaglandin synthetase inhibitors, and betamimetics have all been used and reported to be efficacious in premature labor.

Presently, only betamimetics enjoy widespread acceptance and use. Stimulation of β_2-receptors in myometrium results in increased levels of cAMP and suppression of uterine contractions. The major limitation in the use of betamimetics is the stimulation of cardiac β_1-receptors with resultant tachycardia. Patients being treated simultaneously with betamimetics and high doses of glucocorticoids to stimulate fetal lung maturity are at risk to develop acute pulmonary edema. This is particularly true if patients are too tachycardic, are carrying a multiple gestation, have underlying heart disease (especially mitral valve disease), or receive too much intravenous fluid. Virtually all patients taking betamimetics for premature labor will display changes in the following laboratory results: (1) fall in hematocrit, (2) rise in leukocyte count, and (3) fall in serum K^+. Newborns delivered within 48 h of their mother's receiving betamimetics are at risk for hypoglycemia.

RENAL DISEASE AND HYPERTENSION

Renal disease, chronic hypertension, and pregnancy-induced hypertension (pre-eclampsia) are so tightly interwoven in their clinical presentation, diagnosis, therapy, and implication for the course of the pregnancy that they are discussed together.

Renal disease during pregnancy may originate from all of the same causes as in the nonpregnant patient, i.e., chronic pyelonephritis, chronic glomerulonephritis, diabetic nephropathy, and lupus nephritis. Similarly, the evaluation of renal disease during pregnancy uses the same basic laboratory determinations as in the nongravid patient. The reference ranges for renal function tests in pregnancy, however, are quite different from those in normal nonpregnant women (Table

19-3). Thus, for example, a creatinine clearance which might appear low normal for a nonpregnant patient would be distinctly abnormal in a term pregnancy. In some patients, especially those with diabetic nephropathy, urinary protein loss can be severe (several grams per day), resulting in significant hypoalbuminemia, edema, and weight gain.

The evaluation of *chronic hypertension* during pregnancy includes an assessment of renal function. Regardless of whether a primary renal disease has caused hypertension or vice versa, the implications for the pregnancy are the same. Patients with decreased creatinine clearance, azotemia, and hypertension are at significantly increased risk for premature separation of the placenta (*abruptio placentae*), intrauterine growth retardation, fetal distress, and fetal demise. Therefore, the identification of such high risk pregnancies by close monitoring of blood pressure and of urine for the presence of protein in all pregnant women is of utmost importance. Once such a patient has been identified and evaluated, she should be followed with serial ultrasound examinations to assess fetal growth, with serum or 24-h urine estriol determinations, and with electronic fetal monitoring to assess fetal well-being.

Pre-eclampsia is a syndrome peculiar to pregnancy and consists of hypertension, edema, and proteinuria. It is more common among primigravidas, but women with chronic hypertension also are at risk regardless of their parity. Pre-eclampsia is a systemic disorder whose cause is unknown. It is frequently associated not only with decreased renal function but also with hepatic dysfunction, coagulation abnormalities, and microangiopathic hemolysis. The central nervous system is often irritable and if left untreated, pre-eclampsia may progress to eclampsia with the development of seizures. The clinical diagnosis may occasionally be difficult, especially when changes are gradual or when there is underlying renal disease with pre-existing hypertension and proteinuria.

There is no laboratory test or set of tests which alone establishes or rules out the diagnosis of pre-eclampsia. In the appropriate clinical setting, however, the following laboratory data would help to support the diagnosis: (1) falling creatinine clearance, (2) rising serum creatinine and urea nitrogen, (3) rising serum uric acid, (4) hemoconcentration with rising hematocrit, (5) proteinuria in excess of 400 mg/d, (6) evidence of intravascular hemolysis, such as schistocytes on peripheral smear or hemoglobinuria, (7) evidence of intravascular coagulation, such as falling platelet count or hypofibrinogenemia, and (8) abnormal liver function studies.

The rate at which the disease progresses can vary, and occasionally considerable time can be gained for a pregnancy by vigilant expectant management. Some of the most difficult decisions with respect to delivery occur when there is moderate pre-eclampsia in the early third trimester. Waiting too long to terminate a pre-eclamptic pregnancy can result in fetal demise, maternal cerebrovascular accident, liver rupture, or permanent renal damage.

The time-honored prophylaxis against convulsions in the peripartum period is magnesium sulfate. It can be given either by intermittent intramuscular injection or by continuous intravenous infusion. Under either circumstance serum levels of magnesium should be monitored to maintain therapeutic levels and to avoid toxicity. Therapeutic levels are 3–4 mmol/L while respiratory depression can be seen with levels in excess of 5 mmol/L and complete paralysis at 7–7.5 mmol/L. Intravenous calcium should be available as therapy for magnesium toxicity. Magnesium ions cross the placenta freely, and newborns delivered to mothers who are maintained on magnesium therapy may be lethargic and develop hypocalcemia.

ANTENATAL DIAGNOSIS

The field of antenatal diagnosis has undergone an explosion of knowledge especially within the last 12–15 years. This has been due to the application of known techniques, e.g., cell culture and enzyme determinations, and to the development during this period of new technologies, such as high speed ultrasonic imaging and recombinant DNA techniques. A recent review contains over 180 conditions which have been diagnosed in utero.[71] These conditions range from twins to fatal chromosomal and metabolic disorders.

Ultrasonographic Examination

Basic to all modern antenatal diagnosis is visualization of the uterine contents by the use of ultrasound. Several types of imaging devices are in common use. All operate by placing a transducer in contact with the patient through an acoustic coupling medium of oil or gel. The transducer emits very high frequency sound waves which are transmitted through the tissues and reflected back to the transducer from interfaces of differing tissue densities. The sound waves are emitted in very brief pulses so that the transducer actually spends most of its time "listening" for echoes. The reflected sound waves which are picked up by the transducer are then processed and assembled into an image on a cathode ray tube. This image can then be permanently recorded by any of a number of techniques.

Careful ultrasound examination can establish basic information as to the number of fetuses present, and in addition it can alert the physician to a large number of structural and functional fetal abnormalities. Amniotic fluid volume and fetal activity can also be assessed. Amniocentesis should always be done with the assistance of ultrasound to localize the fetus, placenta, and amniotic fluid. This reduces the incidence of blood-contaminated specimens, multiple needle passes to obtain fluid, and ultimately unsuccessful amniocenteses. It also helps to avoid trauma to the fetus and the mistaken sampling of urine from the maternal bladder.

AMNIOTIC FLUID

Throughout intrauterine life, the fetus lives within a fluid compartment. The amniotic fluid provides a medium in which the fetus can readily move, it cushions the fetus against possible injury, and helps maintain a constant temperature for the fetus. This fluid is a dynamic medium whose volume and chemical composition are controlled within relatively narrow limits.

Volume. The volume of amniotic fluid increases progressively to a mean of 984 mL at the 34th week of gestation, drops off slightly until the 40th week and more sharply toward the 42nd week.[56] Control over volume is established through a complex interrelationship among fluid fluxes and gross fluid movements. Not all of the quantitative relationships are known, and many have been derived from extrapolation from primate models and indirect measurements in humans. Total fluid fluxes between maternal serum and amniotic fluid are large ($\sim$450 mL/h) and result in complete exchange of the amniotic fluid volume every 2–3 h. Gross, unidirectional fluid volume shifts occur into the amniotic cavity by fetal urination and out of the cavity by fetal swallowing. At term, these are about equal at $\sim$450 mL/d. Bidirectional water exchanges which are probably of significant magnitude occur across the following surfaces: (1) placenta (mother-fetus), (2) umbilical vessels through the substance of the umbilical cord (fetus–amniotic fluid), (3) fetal skin (fetus–amniotic fluid), and (4) fetal membranes (amniotic fluid–mother). The fetal tracheobronchial tree is in contact with the amniotic fluid, and measurements in fetal lambs indicate a net flux of 50–80 mL/d of fluid up the trachea from the lungs. This is not significant in terms of the volume of amniotic fluid but is important in releasing surface active agents which reflect lung maturity.

Pathological alterations in fluid volume are encountered fairly frequently in clinical practice. Intrauterine growth retardation and anomalies of the fetal urinary tract, such as bilateral renal agenesis or obstruction of the urethra, are associated with oligohydramnios. Conditions associated with polyhydramnios are as diverse as maternal diabetes mellitus, severe Rh isoimmune disease, fetal esophageal atresia, and anencephaly.

Composition. Early in gestation the composition of the amniotic fluid very closely resembles a complex dialysate of the maternal serum. As the fetus grows, the amniotic fluid changes in several ways (see Table 19-9). Most notably, the Na^+ concentration and osmolality drop, while the concentrations of urea, creatinine, and uric acid rise.[4] The activities of many enzymes in amniotic fluid have been studied with respect to both gestational age and fetal status but have not been found to be clinically useful. The major lipids of interest are the phospholipids whose type and concentrations reflect fetal lung maturity; these will be discussed further below. Numerous steroid and protein hormones are present in amniotic fluid.[85] Levels of androgens and estrogens have been quantitated in attempts to predict fetal sex. The ranges of normal for each group of hormones overlap, however, and such determinations have not been useful. The rare syndrome of congenital adrenal hyperplasia has been diagnosed antenatally by measuring 17-hydroxyprogesterone and pregnanetriol in the amniotic fluid near term.[39] No other diagnostic uses for amniotic fluid hormone measurements have been found. Prostaglandins E_1, E_2, $F_{1\alpha}$, and $F_{2\alpha}$ are all found in small quantities in amniotic fluid. These levels rise gradually during pregnancy and the levels of PGE_2 and $PGF_{2\alpha}$ are very high during active labor.[23] Attempts to demonstrate

Table 19-9. COMPOSITION OF AMNIOTIC FLUID* (MEAN VALUES)

	Gestational Age in Weeks		
	15	*25*	*40*
Na$^+$, mmol/L	136	138	126
K$^+$, mmol/L	3.9	4.0	4.3
Cl$^-$, mmol/L	111	109	103
CO$_2$, mmol/L	16	18	16
Urea N, mg/dL	11	11	18
Creatinine, mg/dL	0.8	0.9	2.2
Glucose, mg/dL	47	39	32
Uric acid, mg/dL	4.0	5.7	10.4
Total protein, g/dL	0.5	0.8	0.3
Bilirubin, mg/dL	0.13	0.14	0.04
Osmolality, mOsmol/kg	272	272	255

*From Benzie, R. J., Doran, T. A., Harkins, J. L., et al.: Am. J. Obstet. Gynecol., *119*:798–810, 1974.

an acute rise in PGE$_2$ or PGF$_{2\alpha}$ immediately before the onset of labor, as the initiation of parturition, have been unsuccessful to date.

Early in pregnancy, there is little or no particulate matter in the amniotic fluid. By 16 weeks of gestation, large numbers of cells are present which have been shed from the surfaces of the amnion, skin, and tracheobronchial tree. They have proved to be of great utility in antenatal diagnosis and will be discussed further below. As pregnancy continues to progress, scalp hair and lanugo (very fine body hair) are also shed into the fluid and contribute to its turbidity.

Determination of ΔA_{450} in Amniotic Fluid

The severity of Rh isoimmune disease can be estimated by determining the amount of bilirubin in the amniotic fluid. This can be done by standard colorimetric techniques (see Chapter 13), but the concentration of bilirubin is generally so low ($\sim$10–30 μg/dL) that the determination of bilirubin can be done more rapidly, accurately, and directly by absorption spectrophotometry. The maximal absorbance of bilirubin is at 450 nm. In the absence of significant amounts of bilirubin, the absorbance curve for the amniotic fluid between 365 and 550 nm is nearly a straight line (see Figure 19-7). The degree to which the curve at 450 nm deviates from a straight line is linearly proportional to the concentration of bilirubin in the amniotic fluid. This is the change in absorbance at 450 nm (ΔA_{450}) formerly and thus occasionally is still referred to in the clinical literature as the ΔOD_{450}. There is normally a small amount of bilirubin in amniotic fluid and this amount changes with gestational age. (See Figure 19-6.) To interpret properly the ΔA_{450} therefore, it is necessary to know the gestational age with reasonable precision.

Specimen

The amniotic fluid specimen is obtained by amniocentesis with a 22 gauge needle under ultrasound guidance, with care not to contaminate the specimen with blood. A total of $\sim$10 mL should be withdrawn. Since bilirubin is unstable in light, the specimen should be protected from light during transport to the laboratory and during storage. This may be done conveniently by wrapping the specimen in aluminum foil. Liley[45] found that the absorbance peak had a $t_{1/2}$ of 10 h in laboratory daylight and 12–18 min in winter sunlight. When stored in the dark, however, the peak is stable for 30 d at room temperature and for at least 9 months under refrigeration. (See also Chapter 3.)

Instrumentation

A recording spectrophotometer is most conveniently used and the fluid is scanned between 350 and 560 nm. The procedure is basically the same whether the instrument records the absorbance or the transmittance on a linear or semilog scale against wavelength. As described by Liley,[44] the original procedure used a semilogarithmic plot of absorbance against wavelength. Subsequently, others have used linear plots with insignificantly different results.[55] If a manual spectrophotometer is used, individual absorbance readings should be taken and recorded on semilog graph paper. Since the major contaminant of concern is oxyhemoglobin with an ab-

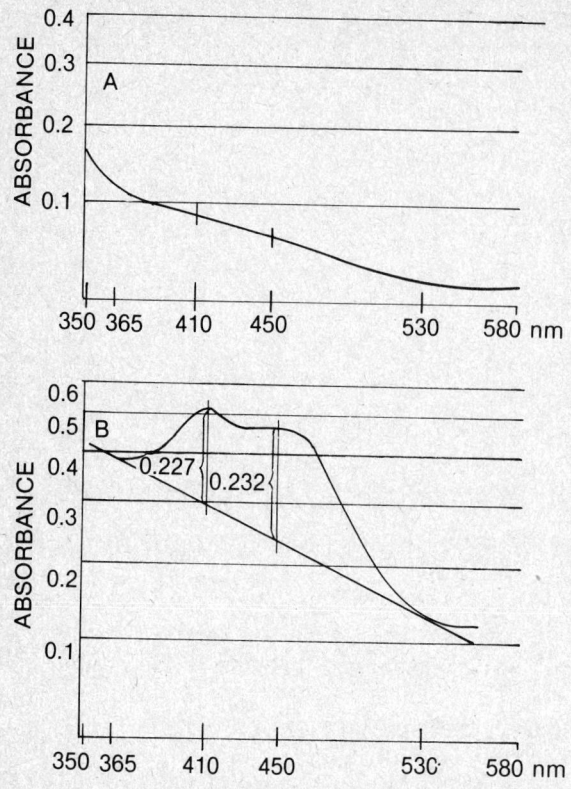

Figure 19-7. *A,* Normal amniotic fluid. Note near linearity of the curve. *B,* Amniotic fluid showing bilirubin peak at 450 nm and oxyhemoglobin peak at approximately 410 nm. Note baseline drawn between linear parts of the curve, from 550 to 365 nm.

sorbance peak at 410 nm, it is important that the instrument have a narrow bandpass, to minimize interference by this compound when making the A_{450} reading.

Procedure

1. Clear the specimen of particulate matter by centrifugation at $\sim$2000 $\times$ g for 5 min.
2. Fill cuvet with undiluted amniotic fluid.
3. Zero the spectrophotometer according to the directions for that particular instrument, using a NaCl solution, 0.15 mol/L.
4a. Scan the fluid from 350–560 nm.
4b. If a manual spectrophotometer is used, make individual absorbance readings at the following wavelengths: 350, 365, 380, 390, 400, 410, 415, 420, 430, 440, 450, 460, 470, 485, 500, 515, 530, 540, 555. Plot the absorbance readings against wavelength on semilog paper with the absorbance values on the semilog scale (y-axis).
5. On the absorbance plot, draw a straight line (the baseline) from the linear area of the curve at 350–375 nm to the linear area of the curve at 525–550 nm.
6. Construct a line perpendicular to the x-axis from the absorbance peak at 450 nm to the baseline. This line represents the ΔA_{450}. (See Figure 19-7.)
7. If there is an oxyhemoglobin peak at 410 nm, determine the ΔA_{410} in a similar fashion.

Calculations

In the absence of other pigment peaks, the ΔA_{450} derived from the graphic method above should be reported directly. If there is oxyhemoglobin present as a peak at 410 nm, then 5% of the ΔA_{410} should be subtracted from the ΔA_{450} and that number reported as the ΔA_{450}.

Sources of Error

Contamination of amniotic fluid with blood should be obvious from the red color and presence of erythrocytes in the button after centrifugation. When a blood-contaminated specimen is obtained, it is important to minimize hemolysis of the erythrocytes since this is the main source of oxyhemoglobin. A bloody sample, therefore, should be refrigerated immediately and processed as soon as possible. Contamination of an amniotic fluid specimen with fetal blood from an affected fetus with a high serum bilirubin could introduce a substantial error, the magnitude of which cannot accurately be determined. Whenever a "bloody tap" is obtained, a Kleihauer-Betke test should be done on the erythrocytes to establish their origin. If they are found to be in large part fetal, this should be kept in mind when interpreting the fluid results. If the amniotic fluid remaining within the amniotic cavity has been contaminated with blood by a traumatic amniocentesis, these erythrocytes will lyse and contaminate the fluid with hemoglobin and bilirubinoid pigments. It will take 2–3 weeks for these pigments to be cleared and for the fluid to return to the original state.

If there are *twins* within separate sacs, the individual sacs must be sampled. Inadvertent sampling of maternal urine is always a possibility, although the risk of this error is reduced with ultrasound guidance.

Occasionally, when following a sensitized pregnancy, delivery will seem imminent and a large dose of glucocorticoids will be administered to stimulate fetal lung maturation. If the patient is not delivered and a repeat amniocentesis is done within 2–4 d of the steroid course, the ΔA_{450} will be lowered relative to the level found immediately before the steroid administration.[11] Further observation and amniocenteses will show that the ΔA_{450} returns to the previous level or higher within a week. It seems doubtful that this phenomenon represents a decrease in hemolysis; instead, it may be an artifact, possibly due to acute changes in amniotic fluid volume.

Meconium is a heterogeneous substance found in the fetal intestine and contains a number of pigments. Meconium staining of amniotic fluid will cause a substantial rise in the ΔA_{450} with a broad and variable peak at 400–415 nm. There is no way to compensate quantitatively for meconium contamination. A single episode of meconium passage into the amniotic fluid will require about 3 weeks to clear.

References

Liley, A.W.: Liquor amnii analysis in management of pregnancy complicated by Rhesus sensitization. Am. J. Obstet. Gynecol., *82*:1359, 1961.

Liley, A.W.: Errors in assessment of hemolytic disease from amniotic fluid. Am. J. Obstet. Gynecol., *86*:485, 1963.

Queenan, J.T.: Amniotic fluid analysis. *In*: Modern Management of the Rh Problem. New York, Harper and Row, 1977, p. 73.

TESTS FOR THE DETERMINATION OF FETAL LUNG MATURITY

Respiratory distress syndrome (RDS) or hyaline membrane disease is the most common immediate complication that a prematurely born infant is likely to encounter. Affected infants require supplemental oxygen or mechanical ventilation to remain properly oxygenated. Their basic problem is a deficiency of surface-active materials produced in the lungs. Normally, when pulmonary surfactant is present in adequate concentrations, it reduces the surface tension in the alveoli so that at end expiration, the walls of the alveoli do not collapse and stick together like a wet piece of paper to a smooth desk top. When surfactant is not present, the alveoli collapse and considerable negative pressure must be generated with each inspiration to reopen them. Depending upon the degree of severity, an infant may not be able to reopen the alveoli, or may do so only with great physical effort from which he may tire. RDS and its therapy may both be the sources of considerable morbidity and mortality; therefore, antenatal prediction of its likely presence at birth is important.

The most common situation in which a test of fetal lung maturity is sought is when early delivery is being anticipated due to some medical or obstetrical indication. Examples are worsening maternal hypertension, severe renal disease, premature rupture of the membranes, intrauterine growth retardation, or premature labor. Results indicating immaturity of the fetal lungs might cause postponement of elective delivery or prompt active intervention to suppress premature labor. If delivery of an infant is inevitable, transfer to a tertiary care center would be appropriate. Alternatively, if the fetal lungs appear to be immature and delivery is likely, administration of a 48-h course of glucocorticoids to the mother to accelerate the rate of fetal lung maturation might be considered.

Pulmonary surfactant is a complex mixture of lipids and proteins with less than 5% carbohydrates. Most of the lipid is phospholipid and the majority of that is lecithin.[2] Up to 85% of the lecithin has saturated fatty acids at both the α- and β-positions, with about 60% of the lecithin as dipalmitoyl lecithin. Other lipids present are phosphatidylglycerol (PG), phosphatidylinositol (PI), phosphatidylethanolamine, and sphingomyelin. The protein fraction may be as high as 20%. Surfactant is synthesized by the Type II pneumocytes in the alveolar epithelium and stored as lamellar bodies. It is released into the alveolar space and makes its way via the tracheal effluent into the amniotic fluid where it can be measured.

Numerous tests of amniotic fluid for "fetal maturity" have been proposed. Some of these, such as the absorbance at 650 nm, creatinine concentration, lipid staining of cells, and "amniocrit," may correlate with gestational age, but either they do not measure lung maturity specifically or they cannot be used in the presence of blood or meconium. Four methods which measure surface active materials directly are presented.

Foam Stability Test (Shake Test) and Foam Stability Index

Principle

When pulmonary surfactant is present in amniotic fluid in sufficient concentration, the fluid is able to form a highly stable surface film which can support the structure of a foam. There are other substances in the fluid such as proteins, bile salts, and salts of free fatty acids which are also capable of forming a stable foam, but these can be excluded from the film by ethanol which competes with other substrates for a position in the surface film. This is the principle upon which the test originally proposed by Clements et al.[17] was based. In that test an effort was made to make the test semiquantitative by using several dilutions of the amniotic fluid. Subsequently, in a series of papers by Sher and Statland,[64-67] a more quantitative test has been devised which they term the foam stability index. This test makes use of the principle that more surfactant activity is necessary to support a stable foam as the fraction of ethanol in the mixture is increased. Thus, a fixed volume of undiluted amniotic fluid is mixed with increasing volumes of ethanol, and the largest fraction of ethanol in which the amniotic fluid is still capable of supporting a foam is noted.

Specimen

Amniotic fluid is obtained by amniocentesis. Fluid may be obtained by transvaginal puncture of the bulging membranes, but it should not be grossly contaminated with vaginal secretions as might occur with aspiration of a vaginal pool after spontaneous rupture of the membranes. Whenever possible, the fluid should be tested immediately. If there is to be a delay of a few hours, the fluid should be centrifuged and refrigerated at 4 °C. If the fluid is not to be tested the same day, it should also be centrifuged and stored frozen at -20 °C. Immediately before testing the fluid, it should be gently inverted several times to obtain a uniform suspension without shaking and creating a foam.

Foam Stability Test by Clements[17]

Procedure

1. Label three 13 × 100 mm round-bottom test tubes 1/1, 1/1.3, 1/2.
2. Pipet 1 mL, 0.75 mL, and 0.50 mL of amniotic fluid into tubes 1/1, 1/1.3, and 1/2, respectively.
3. Pipet 0.25 mL and 0.50 mL of NaCl solution, 0.15 mol/L, into tubes 1/1.3 and 1/2, respectively.
4. Add 1 mL 95% ethanol to each tube.
5. Cap tubes tightly and shake vigorously for 15 s.
6. Place the capped tubes upright into a rack and leave them undisturbed at room temperature for 15 min.
7. At the end of 15 min examine the tubes using a bright overhead light against a flat black background. A test is considered to be positive if there is a ring of small bubbles present at the

air-fluid interface, 360° around the tube. Fetal lungs are judged to be "mature" if all three tubes are positive for a stable foam, and "immature" if none of the tubes will support a stable foam. If tubes at the higher dilutions are negative or equivocal while tubes at the lower dilutions are positive, the test must be interpreted as "intermediate." A more complicated scheme of tube interpretations is given by Schlueter et al.[60]

Comments

1. If delivery occurs within 24 h of obtaining a "mature" test, the risk of RDS is ~0.5%, while an "immature" test is associated with a 77% risk of RDS.

2. Amniotic fluid samples must be mixed well prior to use.

3. Tubes and stoppers must be clean and free of detergents, to avoid false positive readings.

4. The NaCl solution should be prepared from reagent grade NaCl. Saline prepared for clinical use is not satisfactory because it is often packaged in rubber stoppered containers which may contain detergents.

5. The final concentration of the ethanol is critical. A 95% solution of ethanol is hygroscopic; therefore, careful attention must be paid to the preparation of the ethanol reagent from absolute ethanol and water and to its storage in a tightly stoppered container. The 95% ethanol must be accurately pipetted into the amniotic fluid.

6. Clements et al.[17] state that unsaturated phosphatidylcholine will produce a foam in this system but that it is relatively less stable and will break down in a few seconds. In contrast, saturated phosphatidylcholine, which is more important for natural surfactant activity, forms a more stable foam which will last "for several hours" at room temperature. For this reason, they recommend reading the tubes at 15 min. However, if the tubes are left standing too long and the reading is missed, they recommend setting up the test over again rather than trying to shake up the tubes a second time. Keniston et al.[40] examined their shake tests at 1, 5, 10, and 15 min after agitation and did not find a substantial difference in the readings. Statland and Sher[65-67,70] reported reading the tubes 15 s after agitation.

7. The test should be conducted at 20–25 °C. Temperatures substantially below this may retain a relatively unstable foam while higher temperatures may cause the deterioration of an otherwise stable foam.[40]

8. Specimens contaminated with either blood or meconium will tend to give a falsely "mature" test result. Since a significant percentage of fluid samples submitted to a clinical laboratory may be contaminated, this is an important limitation of this test. If the test result is "immature" or "intermediate" in the face of obvious contamination, it is probably reasonable to conclude that the fetal lungs are immature. If possible, the procedure should be repeated on an uncontaminated specimen.

Reference

Clements, J.A., Platzker, A.C.G., Tierney, D.F., et al.: Assessment of the risk of the respiratory distress syndrome by a rapid test for surfactant in amniotic fluid. N. Engl. J. Med., *286*:1077-1081, 1972.

The Foam Stability Index (FSI)[65]

Specimen

See Foam Stability Test.

Procedure

1. Centrifuge the amniotic fluid at $1000 \times g$ for 3 min. Decant the supernatant.

2. Gently mix the supernatant without creating a foam by inverting several times.

3. Set up six 12×75 mm screw-top plastic tubes and label them 0.50, 0.48, 0.47, 0.46, 0.45, and 0.44.

4. Carefully pipet the following volumes of 95% ethanol into the respective tubes: 0.55, 0.51, 0.49, 0.47, 0.45, and 0.43 mL.

5. Add exactly 0.50 mL of the amniotic fluid supernatant to each tube.

6. Cap tubes and shake vigorously for 30 s. Allow to settle for 15 s and then examine.

7. Under bright overhead light against a flat black background examine for the presence of bubbles at the edge of the air-fluid meniscus. A ring of bubbles 360° around the tube indicates a "positive" result. Record the highest concentration of ethanol at which a positive reading is obtained. This is the foam stability index.

Comments

1. In two series, a combined total of 229 fetuses with an FSI of 0.48 or greater were reported, none of whom developed RDS. Of 41 fetuses with an FSI of 0.47, 2 (5%) developed RDS. For 110 fetuses with an FSI of 0.46 or less, 44 (40%) developed RDS.

2. All of the comments made above with reference to the foam stability test regarding the amniotic fluid specimen, reagent purity, glassware cleanliness, and test conditions also pertain to the foam stability index.

3. Recently, a new kit for determining the foam stability index (Lumadex-FSI Test, Beckman Instruments, Brea, CA 92621) has been introduced.[64]

Reference

Sher, G., Statland, B.E., and Freer, D.E.: Clinical evaluation of the quantitative foam stability index test. Obstet. Gynecol., 55:617-620, 1980.

Methods for the Simultaneous Determination of the Lecithin/ Sphingomyelin (L/S) Ratio and Saturated Phosphatidylcholine (SPC)[73]

The major surface active component of the lung surfactant is saturated phosphatidylcholine (SPC) which is a subfraction of the total lecithins. Its concentration in amniotic fluid tends to rise with increasing gestational age. The majority of the SPC in amniotic fluid probably comes from the tracheal effluent, but the degree to which other sources contribute to the amniotic fluid SPC is somewhat conjectural.[2] The concentration of SPC in amniotic fluid can be measured directly and correlated with fetal lung maturity or its concentration can be related to another lipid, sphingomyelin, and that ratio correlated with lung maturity. Sphingomyelin has surface active properties and is found in the lung, but it is ubiquitously distributed in cell membranes, especially those of erythrocytes and in plasma. A substantial percentage of the sphingomyelin in amniotic fluid therefore probably is derived from non-lung sources. Furthermore, diurnal variations in SPC and sphingomyelin concentrations relative to one another indicate that they are mainly derived from different sources. Sphingomyelin is now believed to play no role in the surfactant system in the lung; however, it is used as a convenient marker against which lecithin is measured. The concentration of lecithin relative to sphingomyelin tends to rise with increasing gestational age. This is not a uniform gradual rise; there is a rather sudden rise at 34–36 weeks of gestation which correlates with the development of fetal lung maturity.[33]

The quantitative assay of SPC requires the removal of unsaturated lecithin from the sample. Osmium is a very powerful oxidizing agent which reacts with double bonds in fatty acids and renders lecithins containing unsaturated fatty acids insoluble.[47] This reaction proceeds rapidly to completion and is the principle upon which the present procedure is based.

Specimen

A minimum of 3 mL of amniotic fluid should be obtained, preferably by transabdominal amniocentesis. Specimens obtained from a pool of fluid in the vagina immediately after rupture of the membranes are acceptable, unlike those collected for the shake test. Specimens contaminated with blood or meconium are also acceptable, although the results must be interpreted with care. It has been said that specimens which cannot be analyzed immediately should be either refrigerated or centrifuged and the supernatant frozen, but that uncentrifuged fluid should not be frozen. Recently, van Voorst tot Voorst[79] has shown that the total phospholipid content of amniotic fluid does not change significantly over a week of storage at room temperature, at 4 °C, or when frozen at −20 °C whether it has been centrifuged or not. In a similar study, Schwartz et al.[61] have found the L/S ratio and the concentrations of total lecithin, saturated lecithin, and phosphatidylglycerol to be quite stable over 72 h of storage at room temperature. They noted, however, that the reproducibility of some of their assays was improved by storing the specimens at 4 °C. They conclude that it would seem prudent to refrigerate or to ship on wet or dry ice any specimen which is not to be analyzed within 24 h of collection. Refrigeration of specimens which are not analyzed immediately would seem to be particularly important to slow bacterial metabolism if sterility of the specimen is not maintained.

Materials and Reagents

1. Chromatogram sheets, Eastman 6061 Silica Gel without fluorescent indicator, catalog #13179. Sheets must be activated before use by heating for 24 h in 100 °C oven.

2. Developing solvent. 65 parts chloroform: 25 parts methanol: 4 parts distilled water.

3. Bromothymol blue dye solution. Add 64 mL of sodium hydroxide solution (1 mol/L) to 896 mL distilled water. Dissolve in this solution 10 g boric acid powder. Then add 500 mg bromothymol blue dye and swirl until dissolved (a magnetic stir bar is helpful).

4. Osmium tetroxide

Stock solution: 1 g OsO_4 + 10 mL CCl_4
Working solution: 0.7 mL stock + 9.3 mL CCl_4

5. SPC standard, 100 ng/μL. Mix 9 mL chloroform plus 1 mL of methanol as the solvent, and add 1 mg of SPC. Store in refrigerator.

6. L/S standard. Dissolve 30 mg each of lecithin and sphingomyelin in 10 mL chloroform. Store in refrigerator.

Procedure

Preparation of the Amniotic Fluid Specimen

1. Centrifuge the fluid in a 15-mL tube at 500 × g for 10 min at 4 °C.

2. Pipet 3 mL of the supernatant into a 50-mL tube.

3. Add 3 mL methanol and vortex well.

4. Add 6 mL chloroform and vortex for 30 s.

5. Centrifuge at 500 × g for 5 min.

6. Aspirate off the lower (chloroform) phase and transfer into a 15-mL tube and evaporate to dryness under flowing N_2 in a 60 °C water bath.

7. Rinse down the sides of the tube with a small amount of chloroform and evaporate again as above.

8. Redissolve residue in 150 μL chloroform.

9. Transfer 50 μL to a 12 × 75 mm tube and evaporate tubes containing 100 and 50 μL chloroform extract to dryness. Chill on ice.

Determination of the L/S Ratio

10. Add 2–3 drops of cold acetone to the residue in the 15-mL tube containing 100 μL amniotic fluid residue and agitate to dissolve.

11. Add 0.75 mL cold acetone and let stand on ice for 5 min.

12. Centrifuge at 800 × g for 1 min. Decant off and discard acetone.

13. Dry the tube under flowing N_2 and chill.

14. Redissolve residue in 100 μL, chloroform and spot onto activated chromatogram sheets two 2-μL, two 5-μL, two 10-μL, and two 20-μL bands. Spot a 10-μL and 20-μL band of the standard for reference. Leave blank lanes between the samples so that the sheets can be read in the reflectance spectrodensitometer.

15. Dry sheet in an oven for 2 min.

16. Develop in a tank with 1 cm of solvent and allow the front to migrate to within 0.5–1 cm of top of sheet.

17. Dry in an oven for 2 min and stain by immersion in a tank with bromothymol blue dye. Do not spray.

18. Blot off excess dye and dry in an oven for 5 min.

19. Cool to room temperature and read in a reflectance spectrodensitometer.

Determination of SPC

20. Place the 12 × 75 mm tube containing the 50-μL amniotic-fluid residue prepared above in Step 9 in a rack under a hood.

21. Add 0.5 mL of the OsO_4 working solution and allow to stand at room temperature for 15 min.

22. Evaporate under N_2 in the hood and redissolve the residue in 50 μL chloroform.

23. Place three 10-μL spots onto the TLC sheet, leaving blank lanes between samples.

24. Warm the SPC standard solution to room temperature and place two 10-μL spots (1 μg) onto the TLC sheet.

25. Dry sheet in oven. If spots are particularly dark (usually cases of heavy blood contamination), leave in oven for 10–15 min.

26. Develop, stain, and read chromatograms as above in steps 16–19 for the L/S procedure.

Calculation

$$\text{SPC in sample, } \mu g/dL = \frac{A_{sample} \times 1.0 \times 100}{A_{std} \times 0.2}$$

$$= \frac{A_{sample}}{A_{std}} \times 500$$

where

1.0 = amount of standard (std) used in μg
100 = converts to dL
0.2 = amount of original amniotic fluid in mL

Example. If sample spots give absorbance readings of 33, 35, and 37 ($\bar{x} = 35$) and the standard spots read 20 and 22 ($\bar{x} = 21$), then

$$\text{SPC, } \mu g/dL = \frac{35}{21} \times 500 = 833$$

Interpretation

The L/S ratio indicates mature lungs if it exceeds 2.0/1 and the SPC indicates maturity if it exceeds 500 $\mu g/dL$. Both of these tests predict fetal lung maturity more reliably than they predict immaturity. Less than 2–3% of babies delivered within 24 h of obtaining an L/S ratio of 2.0 or greater can be expected to develop RDS; i.e., > 97–98% of babies predicted to be mature will, in fact, be mature. Almost one-half of the infants with L/S ratios between 1.5 and 2.0, however, will not develop RDS. Thus, in contrast to the high degree of reliability of a mature prediction, only half of the babies predicted to be immature will, in fact, be immature. Corresponding numbers for the SPC are ~1% falsely indicated to be mature and < 20% falsely indicated to be immature. Using a similar OsO_4 and TLC technique, Tsai et al.[75] have found the concentration of SPC, which indicates lung maturity, to be 550 $\mu g/dL$. They have also found the SPC concentration to be more reliable than the L/S ratio measured by one-dimensional TLC.

A number of investigators have reported that infants of diabetic mothers (IDMs) more frequently develop RDS in spite of "mature" L/S ratios of greater than 2.0/1.[20] This has been observed at our institution as well. For this reason, the value for lung "maturity" for an IDM has been raised progressively over the past several years to an L/S ratio of 3.5/1 and an SPC of 1000 $\mu g/dL$. Fetuses with isoimmune hemolytic disease also seem to develop RDS frequently despite "mature" indices. Both the fetuses of diabetic mothers and those affected by isoimmune disease are hyperinsulinemic during intrauterine life. There is thought to be a causal relationship between hyperinsulinemia and "delayed" lung maturity. It is not clear whether or not this is true, however.[31]

Comments

1. In the initial centrifugation of the amniotic fluid to remove particulate matter, it is important to avoid excessive g force and prolonged centrifugation. At g forces in excess of ~500 there is a progressive fall in both the L/S ratio[16] and total phospholipid content.[79] This decline is most rapid up to ~3000 × g but continues more gradually through 10 000 × g. There are widely

varying recommendations in the literature with respect to the centrifugation conditions. The "correct" conditions must be determined empirically by evaluating test results in terms of clinical outcomes. Differences in techniques reported among laboratories make it impossible to state definitively the best centrifugation conditions.[8]

2. It is important in the initial extraction procedure to first create a monophasic system with the addition of methanol, and then to add the chloroform to form the biphasic system. Insoluble proteins will be left at the interface. Failure to mix thoroughly with methanol first will lead to an inadequate extraction and loss of lipid in the protein layer.

3. Not all chromatogram plates are the same. The specified plates have consistently been reliable.

4. While ammonium sulfate charring of lecithin and sphingomyelin spots has been popular, sphingomyelin chars more intensely than an equal weight of dipalmitoyl lecithin.[8] Furthermore, charring is less intense as the degree of saturation of the lipid being examined increases.[69] The phospholipids most important for surfactant activity are saturated lipids and these increase in proportion as gestation advances. Thus the paradox develops that as the lecithin composition "matures," the lecithin will char less intensely. Using very harsh oxidizing conditions (e.g., H_2SO_4/dichromate and high temperatures as opposed to H_2SO_4 alone) may actually drive off some of the carbon as CO_2 and give a less intense color.[8] Staining with bromothymol blue has been shown to be a saturation-insensitive technique. Lecithin which is highly saturated stains with the same intensity as relatively unsaturated lecithins.[69] The major disadvantage of bromothymol blue staining is that it fades rather rapidly, and thus chromatograms should be scanned promptly after staining. Although it has been said that the color developed is not intense enough for reliable densitometric scanning, this has not been a problem in our experience. Dipping the TLC sheet is preferable to spraying because it gives a more uniform color development.

5. The major disadvantage of this one-dimensional technique is that it fails to separate phosphatidylinositol (PI) from lecithin. It thus may slightly overestimate the L/S ratio.

6. Acetic acid may be added to the developing solvent to improve the separation of the PI and PS (phosphatidylserine) from the lecithin. However, this will interfere with color development with bromothymol blue staining.

7. Various reports have found widely different values for the L/S ratio of whole blood; the true value is probably between 1.4/1 and 1.6/1.[73] The values of blood-contaminated specimens will therefore tend to be drawn to that range from either higher or lower values. A high L/S ratio in the presence of even substantial blood contamination can safely be interpreted to predict mature lungs.[32] Similarly, contaminated samples with very low ratios clearly reflect immature lungs. "Borderline" values of bloody specimens must be interpreted with care, as they may represent either very "immature" values drawn up or mature values of just over 2.0 drawn down. Varying effects of meconium contamination have been reported, including some evidence that the L/S ratio may be changed before the degree of contamination is visible to the eye.[82] The L/S ratio of fresh meconium was found in one study to be 1.2/1.[73] When fluids with normal or low L/S ratios were mixed with meconium and their L/S ratios were redetermined, the values were lowered toward 1.2/1.[73]

8. There is essentially no SPC in either blood or meconium. Experimental contamination of amniotic fluid samples with either of these will not change the SPC value.[73] This is for volumes of blood which are relatively small in comparison to the volume of amniotic fluid so that there is no significant dilutional effect from the blood.

9. Vaginal contamination of specimens obtained from a pool of fluid which has been present in the vagina for a relatively brief period does not affect either test.[54]

Reference

Torday, J., Carson, L., and Lawson, E. E.: Saturated phosphatidylcholine in amniotic fluid and prediction of the respiratory-distress syndrome. N. Engl. J. Med., *301:*1013-1018, 1979.

The Determination of the Lung Profile[41,42]

The lung profile, also referred to as the "two-dimensional" lung profile, provides the simultaneous determination of the L/S ratio, PG, and PI as percentages of total phospholipid, and the saturated lecithin as a percentage of total lecithin. By using the two-dimensional system, this technique separates the acidic phospholipids PG and PI from the lecithin fraction so that the lecithin is not overestimated. This procedure depends upon a cold acetone precipitation to precipitate the highly saturated lecithins important for surfactant activity from the soluble, less saturated lecithins which are less important for surfactant activity.

Specimen

Amniotic fluid (5 mL or more) should be obtained by amniocentesis or from a free flow of fluid in the vagina following rupture of the membranes. Specimen handling conditions should be as noted above for the L/S and SPC specimens.

Materials and Reagents

1. TLC plates, Silica Gel H with 5% ammonium sulfate, 20 cm $\times$ 20 cm. Activate before use by storing in 110 °C oven for 24 h.
2. Developing solvents.
   System one: 65 parts chloroform: 25 parts methanol: 4 parts distilled water: 8 parts glacial acetic acid.
   System two: 40 parts tetrahydrofuran: 28.5 parts methylal: 7.8 parts methanol: 4.2 parts ammonium hydroxide (2 mol/L).
3. PG standard, 10 μg/μL. Dissolve 10 mg of PG in 1 mL of solvent (9 parts chloroform:1 part methanol).
4. L/S standard. Dissolve 30 mg each of lecithin and sphingomyelin in 10 mL chloroform.

Procedure

1. Centrifuge the amniotic fluid specimen at 800 $\times$ g for 5 min at 4 °C.
2. Extract the supernatant by adding an equal volume of methanol. Vortex well.
3. Add chloroform (twice the volume of the original sample) and again vortex for 30 s.
4. Centrifuge at 500 $\times$ g for 5 min.
5. Aspirate off the lower, chloroform phase into a 15-mL conical centrifuge tube and evaporate to dryness under flowing N_2 in a 60 °C water bath.
6. Concentrate the specimen to the tip of the tube by rinsing down the sides with several drops of chloroform and evaporate.
7. Place the tube in an ice bath and add two drops of cold acetone with a Pasteur pipet. Agitate slightly. A white precipitate should form.
8. Add 0.75 mL cold acetone and allow to sit in an ice bath for 15 min.
9. Centrifuge at 3000 $\times$ g for 2 min.
10. Transfer the supernatant to a 15-mL conical centrifuge tube. Dry the precipitate under flowing N_2 at 60 °C.
11. Evaporate the supernatant to dryness under flowing N_2 in a 60 °C water bath. Concentrate the specimen down to the tip of the tube by rinsing down the sides with several drops of chloroform and evaporate.
12. Dissolve the acetone-precipitable fraction (from step No. 10) in 30 μL of chloroform. Place 12 μL of this solution onto the activated TLC plate at the bottom right hand corner, 3 cm from the right hand edge. Dry the spot.
13. Dissolve the acetone soluble fraction (from step No. 11) in 30 μL of chloroform. Place 12 μL onto the activated TLC plate at the bottom far left hand corner. Dry the spot.
14. Place 10 μL of the L/S standard and 2 μL of the PG standard in the left hand corner, in separate lanes. Dry the spots.
15. Develop in the first solvent system until the front has run 10 cm from the origin. Dry the plate at approximately 70 °C for ~7 min. Cool to room temperature.
16. Turn the plate 90° so that the original right side is now down. Spot L/S (10 μL) and PG (2 μL) standards on the right hand corner, 3 cm from the bottom of the plate. Dry spots.
17. Develop in the second solvent system until the front has run 10 cm from the origin. (Thus, the acetone-soluble fraction will have been run in one dimension only, while the sample and the L/S and PG standards will have been run in both dimensions.)
18. After drying the plate for ~10 min at room temperature, place on a hot plate pre-heated to 280 °C until thoroughly charred (~15 min).
19. All spot areas are then measured in a reflectance spectrodensitometer.

Calculations

The ratio of the areas under the curves for the saturated lecithin and sphingomyelin is calculated and reported as the L/S ratio to the nearest 0.1.

The areas under the curves for the saturated lecithin and unsaturated lecithin are added. The area for the saturated lecithin is then expressed as a percentage of this total.

The areas under the curves for the total acetone-precipitable fraction (which will include PG, PI, PS, PE, saturated lecithin, and sphingomyelin) are then summed. The areas of the PG and PI spots are then expressed as percentages of this total.

Comments

1. The individual values for each of the determinations within the lung profile are considered when a judgment as to fetal pulmonary maturity is made. The L/S ratio indicates fetal lung maturity if it is 2.0/1 or greater. A phosphatidylglycerol level in excess of 3% of the total acetone-precipitable fraction indicates fetal lung maturity. While there is a trend for the percentage of saturated lecithin, which is acetone precipitable, to rise with increasing L/S ratio and PG percentage, there is no "critical" level cited above in which the fetal lungs would be presumed to be mature. It is also unclear as to just how the low and intermediate PI percentages are to be interpreted. The level of PI in amniotic fluid expressed either as a percentage of total acetone-precipitable phospholipid[42] or as a ratio to sphingomyelin[51] increases from 16 weeks of gestation[35] to a peak at 36 weeks and then falls toward term. Thus, a low or intermediate value for PI could represent a point on the curve at 30–34 weeks with immature lungs on the way up or a point at 38–42 weeks with mature lungs on the way down.

2. The two-dimensional technique separates the acidic phospholipids PG and PI from the lecithin, thus providing a lower and more conservative estimate of the L/S ratio than the one-dimensional technique discussed above. For patients whose pregnancies are uncomplicated by diabetes or isoimmune disease, the rate of false predictions of maturity is <2%.[42] The same criticism of the relatively high (30%) incidence of false predictions of immaturity (L/S ratio <2.0 though no RDS develops) pertains to this L/S ratio technique as well.

3. The precise role of PG in lung surfactant is still unclear.[10] It has been said that the appearance of PG in the amniotic fluid indicates that the "final step" in biochemical maturation of surfactant has occurred.[36,41] It is clear, however, that PG can be found in measurable quantities in amniotic fluid as early as 32 weeks,[28,74] and that its presence in small quantities does not necessarily indicate that the fetal lungs are mature.[21,28,51] The concentration of PG in amniotic fluid rises with gestational age, very closely reflecting the L/S ratio[76] and the SPC concentration.[28] Qualitative techniques which demonstrate the presence of PG in cases of mature fetal lungs and show absence in cases of immature fetal lungs are less sensitive and fail to detect low levels of PG. When sensitive techniques are used, "cut-off" values for PG can be established above which no RDS is seen,[51] but there is still the problem of a high percentage of "false immature" results. Thus, the difference between amniotic fluid specimens reflecting mature lungs and immature lungs is a quantitative difference in PG concentration and not a qualitative difference in its presence or absence. Given these facts, it is not clear that waiting for PG to appear in the amniotic fluid before declaring the lungs to be mature offers any advantage over simply selecting a higher value for the L/S ratio or the SPC concentration. This is also true for pregnancies complicated by maternal diabetes mellitus.[22]

The determination of PG is definitely of value in cases of blood or meconium contamination of the amniotic fluid. Like SPC, PG is not found in either blood or meconium and therefore results are not affected by the presence of either of these contaminants.

4. The acetone precipitation step has been termed "critical" to the procedure since it is said to precipitate specifically the highly saturated lecithins responsible for the majority of the surface activity of surfactant.[34] The ability of acetone to precipitate surfactant lecithin specifically or quantitatively must be seriously questioned, however, in view of the results of a number of investigators. The solubility of lecithin in cold acetone is dependent upon a large number of factors which are difficult or impossible to control, including the precise time and temperature conditions, the water and chloroform content of the acetone, the lipid composition of the original amniotic fluid sample, and the preliminary centrifugation conditions.[8] The precipitation of SPC is not quantitative and may vary from 60–90% among samples.[53] Composition studies have shown that the acetone precipitate may contain both saturated and unsaturated lecithins.[73] These con-

siderations have led Brown and Duck-Chong[8] to conclude that there is no advantage to include the acetone precipitation step in the L/S ratio procedure since it adds variability and procedure time.

5. Comments made with respect to the importance of the centrifugation of the original amniotic fluid specimen, the initial extraction of the fluid with methanol and chloroform, and the charring procedure of the TLC in the one-dimensional technique are pertinent to the two-dimensional technique as well.

References

Kulovich, M. V., and Gluck, L.: The lung profile. II. Complicated pregnancy. Am. J. Obstet. Gynecol., *135*:64-70, 1979.
Kulovich, M. V., Hallman, M. B., and Gluck, L.: The lung profile. I. Normal pregnancy. Am. J. Obstet. Gynecol., *135*:57-63, 1979.

CONGENITAL MALFORMATIONS OF THE CENTRAL NERVOUS SYSTEM

Biology

During embryonic development, the area which is to form the central nervous system (brain and spinal cord) differentiates into a plate of cells. The flat plate then rolls up and its edges fuse into a hollow neural tube which drops into the embryo to develop just under what will become the skin of the back. Neural tube formation is normally complete very early in gestation (by the end of the fourth week following fertilization). Failure of the embryo to complete neural tube fusion successfully will lead to permanent developmental defects of brain or spinal cord such as anencephaly, spina bifida, or myelomeningocele.[38] Approximately 90–95% of these defects are "open," i.e., there is a direct communication between the neural tube and the outside world (which is amniotic fluid for the fetus without intact overlying skin). Under this circumstance, proteins which are normally relatively restricted to the fetal blood and cerebrospinal fluid gain access in large quantities to the amniotic fluid. Furthermore, these proteins in the amniotic fluid gain access in smaller but significant and measurable amounts to the maternal circulation. It is on this principle that present biochemical techniques for the antenatal diagnosis of neural tube defects (NTD's) are based.

ALPHA-FETOPROTEIN

Alpha-fetoprotein (AFP) is a glycoprotein of M.W. ~70 000. It is produced initially by the fetal yolk sac and then by the fetal liver as the yolk sac degenerates. Considerably smaller amounts are probably also produced in the fetal gut and kidneys. Early in embryonic life this protein functions as the major serum protein. During development, the relative concentration of AFP in fetal serum decreases as the liver switches over to production of serum albumin. Maximal concentration in the fetal serum of 2–3 mg/mL is reached at ~16 weeks of gestation and declines steadily to ~0.07 mg/mL by term. (See Figure 19-8.) The rise and fall in concentration of AFP in the amniotic fluid roughly parallels that in the fetal serum, but is two orders of magnitude lower in concentration (~0.015 mg/mL at 16 weeks). The relationship with respect to maternal serum concentration is slightly more complicated due to the additional consideration of fetal membrane and placental permeability. Thus, while the fetal serum concentration of AFP falls progressively from 16 weeks' gestation, the membranes and placenta become increasingly "leaky," leading to a progressive rise in maternal serum AFP from 14 weeks to term.[43] The median maternal serum AFP concentration at 16 weeks' gestation varies markedly among laboratories, the values being from 30–100 ng/mL. (Regarding the method for AFP determination, see Chapter 4 and references 13, 58, 81.)

ACETYLCHOLINESTERASE

Recently, another marker protein in amniotic fluid, which is a useful adjunct in the diagnosis of NTD's, has been described. In the presence of open NTD's, acetylcholinesterase (AChE),

ALPHA-FETOPROTEIN

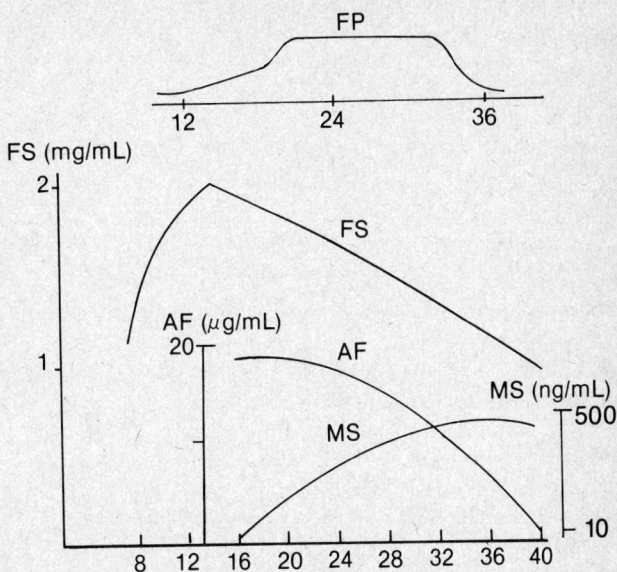

Figure 19-8. Concentrations of AFP in fetal serum (FS), maternal serum (MS), and amniotic fluid (AF). The time-course of fetal production (FP) of AFP is indicated in top curve. (From Lau, H. L., and Linkins, S. E.: Am. J. Obstet. Gynecol., *124*:533, 1976.)

which is of probable central nervous system origin, is secreted into the amniotic fluid in a relatively large quantity. Amniotic fluid from normal pregnancies contains a single AChE isoenzyme which migrates relatively slowly in an electric field. Pregnancies with NTD's (and omphaloceles), however, have also a second isoenzyme which migrates more rapidly. This isoenzyme appears to be relatively specific for neural tissue and is not found in fetal blood.[9] It follows from the discussion above that the few NTD's which are closed lesions, i.e., are covered by intact skin and do not communicate with the amniotic fluid, would not leak this protein and therefore would not be subject to detection by assay for AChE.

TESTING FOR NEURAL TUBE DEFECTS

Any patient who has had a child with an NTD of a type which has significant risk for recurrence should be offered antenatal diagnosis in subsequent pregnancies. Furthermore, the diagnosis should be made by the most sensitive method available, i.e., a combination of ultrasound examination and amniotic fluid AFP analysis. Testing should be done at ~16 weeks' gestation for several reasons: (1) This is a good time for an ultrasound examination because a detailed fetal examination can be done. (2) The AFP concentration in the amniotic fluid is maximal. (3) The amniocentesis is technically relatively easy. (4) If the AFP is elevated there is still time before 24 weeks, the limit for elective termination, to complete the fetal evaluation. When an amniocentesis is done at 16 weeks of gestation for cytogenetic studies, most centers are also performing an AFP assay on the fluid.

If the amniocentesis reveals an elevated AFP (> 3 SD above x̄), and no NTD or other explanation for the elevation (e.g., omphalocele, greater than anticipated gestational age, or intrauterine demise) is found on the initial ultrasound examination, then several further steps are indicated. First, the original fluid sample should be checked to be sure that it was not contaminated with fetal blood. As little as 44 μL of fetal blood in 10 mL of amniotic fluid can cause a doubling of AFP concentration at 16 weeks of gestation.[24] Examination of the button of cells from the centrifuged fluid by Kleihauer-Betke stain alone is not adequate. Immunoelectrophoresis for fetal hemoglobin should also be performed. Having excluded fetal blood contamination as a possible source of error, an AChE assay should be done on the original fluid sample. If the AFP is elevated to > 5 SD above the mean and the fast migrating AChE is present, there is a great chance that a NTD or omphalocele is present.[18,37] A second ultrasound examination should be done 7–14 d following the first examination and every effort made to identify the lesion. Conditions associated with an elevated AFP are listed in Table 19-10. If a lesion is not seen, a second

Table 19-10. CONDITIONS ASSOCIATED WITH ELEVATED AFP

Anencephaly	Nuchal bleb
Congenital nephrosis	Omphalocele
Duodenal atresia	Rh isoimmune disease
Encephalocele	Sacrococcygeal teratoma
Esophageal atresia	Spontaneous abortion
Fetal blood contamination	Trisomy 13
Fetal death	Turner's syndrome with cystic hygroma
Gastroschisis	Twins (AFP increased in maternal serum; amniotic
Hydrocephalus	fluid AFP usually normal)
Meckel's syndrome	Urinary obstruction
Myelomeningocele	
Necrosis of fetal liver secondary to herpesvirus infection	

amniocentesis to repeat AFP and AChE determinations should be performed. Consideration should be given to karyotyping the cells obtained at this amniocentesis.[48] After removing the necessary amount of amniotic fluid (20 mL, see below), a small amount of radiodense dye should be injected into the amniotic cavity and an X-ray taken. The "amniogram" thus obtained will sometimes reveal the deformity not observed by ultrasound.[29] An intermediate degree of AFP elevation, between 3 and 5 SD above the mean with absence of AChE, is sometimes seen with no demonstrable anatomic lesion. These pregnancies should be followed with great care because they are at high risk to result in a low birth weight baby for a variety of reasons.[5,63]

The benefits of making an accurate diagnosis of NTD early in gestation must be weighed against the potential for harming a normal fetus in the course of the evaluation. At present, there are no adverse effects, either immediate or late, known to be associated with diagnostic ultrasound. Amniocentesis is, of course, an invasive procedure and does carry some risk of harming the fetus or causing spontaneous abortion. The risk of harming the fetus short of causing abortion is so small that there are no good estimates of its magnitude. Many estimates of the magnitude of the risk for causing spontaneous abortion were made prior to the routine use of ultrasound for guidance, and were found to be 1/100 to 1/200. However, with the routine use of ultrasound guidance, in experienced hands, in a large medical center setting, this risk may be as low as 1/350.[80]

The fact remains, however, that 90% of all infants with NTD's are delivered to totally unsuspecting parents who have no recognized risk factor for NTD. Thus, it is argued that if a significant impact is to be made on NTD's as a public health issue, a simple, reliable, noninvasive, inexpensive screening test should be made available to all women. Maternal serum AFP screening has been suggested for this purpose and has been shown to be effective in numerous demonstration projects around the world.[6,46] This use of AFP, however, has engendered considerable controversy in this country. The major objections to a maternal serum AFP screening program can be summarized as follows: (1) The incidence of NTD's in the United States population is too low to make screening cost-effective. (2) When used optimally, the screening procedure has a high rate of false positives, which means that ~12.5 unnecessary amniocenteses are performed for each confirmed case; exposing these normal fetuses to the risk of amniocentesis is unacceptable. (3) Quality control of the assay procedure is difficult, and even higher rates of false positives can be expected with commercial kits. (4) Physicians unfamiliar with the appropriate interpretation of the screen results may needlessly terminate normal pregnancies rather than follow the diagnostic schema through to its proper conclusion. (5) Sophisticated ultrasound and genetic counseling support would not be as readily available as a commercial assay kit, again leading to potentially misleading laboratory results and therefore to possible clinical mismanagement. For these reasons, the FDA has been extremely reluctant to approve a commercial RIA kit for AFP.

DIAGNOSTIC USES OF THE CELLULAR ELEMENTS IN AMNIOTIC FLUID

Amniotic fluid contains numerous cells which are shed from various fetal surfaces. The main sources are thought to be skin and amnion with minor contributions from the genitourinary, respiratory, and alimentary tracts. The number of cells per unit volume of fluid tends to rise with increasing gestational age. However, the percentage of these cells which are viable in culture decreases with advancing gestational age. Generally, between 15 and 20 weeks of gestation, the number of cells and their viability are such that cultures can successfully be established with a high degree of reliability. By growing the cells according to standard cell culture techniques, large numbers of cells can be obtained for antenatal diagnostic use.[59] The most important uses for these cells are for the diagnosis of chromosomal abnormalities and metabolic defects.[50] Studies of cell morphology and their adherence to culture surfaces have been said to be of aid in certain situations, but are presently still investigational.

Indications for Chromosomal Studies

The incidence of major and minor chromosomal abnormalities in all live births is approximately 1 in 200. The majority of second trimester diagnostic amniocenteses are performed for the purpose of detecting these major abnormalities. These anomalies are not uniformly distributed among all pregnant women. It is well known that there is a markedly increased tendency toward these problems with increasing maternal age. For Down's syndrome, for example, the risk at maternal age 18 is approximately 1:2000 while at age 40 it is 1:80. Thus, the most common indication for a "genetic amniocentesis" is advanced maternal age.

At what age is a mother to be considered "at risk" by virtue of her age? This is a difficult question with ethical, legal, and practical ramifications. Its present solution reflects the compromise imposed by these considerations. Historically, the risk of having a baby with a chromosomal abnormality has been balanced against the risk of causing loss of the pregnancy as a result of the amniocentesis. Since the risk of having a child with a major chromosomal abnormality is approximately 1:300 at age 35, and this roughly equals the risk of the diagnostic procedure, age 35 or greater has been the generally recommended age for routine cytogenetic studies. If all pregnant women in the United States age 35 or above demanded cytogenetic studies, the present laboratory capacity to perform these studies would be vastly exceeded. In this respect, it is therefore fortunate that this technology is underutilized. In urban areas where cytogenetic studies are heavily utilized, some laboratories refuse to accept specimens from women less than 35 years of age or without some other recognized medical indication. On the other hand, is it fair to equate the birth of a child which may need lifelong custodial care with the loss of an early fetus which the parents never really "knew"? Should this decision be made for patients or should a couple with maternal age 33 be able to choose to have the procedure done and accept the risk? Finally, what might be the legal consequences to a physician or laboratory which refused to perform cytogenetic studies for a couple who then had a profoundly retarded child? Clearly, a full discussion of these questions is beyond the scope of this presentation, but they are worthy of consideration.

Technical Considerations

Cell cultures. Amniotic fluid whose cells are to be cultured should be collected and handled using *sterile* technique. Approximately 20 mL of fluid should be obtained by amniocentesis and the fluid should be kept at body or room temperature and transported to the laboratory as rapidly as possible. It should *never* be frozen or exposed to toxins or fixatives. The specimen may conveniently be stored and transported right in the disposable syringe used for the amniocentesis. Alternatively, it can be transferred into sterile nontoxic plastic tubes. The cells tend to adhere to glass, and therefore glass syringes and tubes should not be used. Upon arrival in the cytogenetics laboratory, samples are centrifuged and the cell pellet is used for preparing cell cultures. For details of preparing the cultures, the reader is referred elsewhere.[59]

In the laboratory, cultures must be observed frequently for growth. If the cells are not growing well, the physician must be notified as soon as possible so that the amniocentesis can be repeated promptly. If failure to grow becomes a frequent problem in a laboratory, possible contamination in the lab with mycoplasma must be considered and tested. Contamination of the sample with maternal cells and overgrowth or combination of maternal and fetal cells in culture is also possible. This is obviously only a problem if the fetus is a female in which case polymorphisms (see below) can be checked between maternal and supposed fetal cells. Finally, any results can pertain only to the fetus from which the fluid was obtained; thus, if there are twins with separate amniotic sacs, each sac must be sampled separately.

Chromosome analysis. Actively growing and dividing amniotic fluid cells in culture are prepared for chromosome analysis by arresting cell division at metaphase. This is done by exposing the cells to an antimitotic agent such as colchicine for several hours. The arrested cells are fixed to a microscope slide and ruptured so that the chromosomes are spread out with little or no overlap of one another. They can then be stained and photographed for analysis. Simple chromosome counting and identification can lead to diagnosis of conditions involving lost chromosomes (45,X Turner's syndrome) or extra chromosomes (47 trisomy 21). More sophisticated staining techniques which demonstrate banding permit the identification of subtle deletions, inversions, and translocations. Making use of minor variations in staining patterns (polymorphisms), an extra chromosome or set of chromosomes can further be traced back to one parent or the other.

Detection of Inborn Errors of Metabolism

Cultured amniotic fluid cells are generally the preferred specimen for studying inborn metabolic errors. For most metabolic studies, e.g., assays of enzyme activity, larger numbers of cells are needed than for cytogenetic studies. Usually, therefore, cultures must be maintained for somewhat longer periods of time. In some instances, enzyme activity assays have been performed directly on amniotic fluid, but this is generally less reliable and cultured cells are the preferred specimens. A list of some of the most common inborn errors of metabolism and their diagnostic tests appears in Table 19-11.

Table 19-11. HEREDITARY DISORDERS FOR WHICH ANTENATAL BIOCHEMICAL DIAGNOSIS HAS BEEN ACCOMPLISHED

Disorder	Diagnosis Made by Demonstration of
Acid phosphatase deficiency	Deficient acid phosphatase (cells)
Acute intermittent porphyria	Deficient uroporphyrinogen I synthetase (cells)
Adenosine deaminase deficiency	Deficient adenosine deaminase (cells)
Argininosuccinic aciduria	Accumulation of argininosuccinic acid (cells)
Congenital adrenal hyperplasia	Elevated 17α-hydroxyprogesterone, pregnanetriol, Δ^4-androstenedione (fluid)
Congenital sialidosis	Deficient neuraminidase activity (cells)
Cystinosis	Accumulation of cystine (cells)
Cystinuria	Accumulation of cystine (fluid)
Duchenne muscular dystrophy	Elevated creatine kinase (fetal blood)
Fabry's disease	Deficient α-galactosidase (cells)
Farber's disease	Deficient ceramidase (cells)
Galactosemia	Deficient galactose-1-phosphate uridyl transferase (cells)
Gangliosidosis GM_1	Deficient β-galactosidase (cells)
Gaucher's disease	Deficient glucocerebrosidase (cells)
Glucose phosphate isomerase deficiency	Decreased glucose phosphate isomerase (cells)
Glutaric acidemia	Deficient glutaryl-CoA dehydrogenase (cells)
Hunter's syndrome	Deficient iduronate sulfatase activity (fluid)
Hurler's syndrome	Deficient α-L-iduronidase activity (cells)
Hypercholesterolemia	Absence of LDL cell surface receptors (cells)
Hypophosphatasia	Deficient alkaline phosphatase (cells)
Krabbe's disease	Deficient cerebroside-β-galactosidase (cells)
Lesch-Nyhan syndrome	Deficient hypoxanthine-guanine phosphoribosyl transferase (cells)
Maple syrup urine disease	Deficient branched-chain ketoacid decarboxylase activity (cells)
Maroteaux-Lamy syndrome	Deficient arylsulfatase (cells)
Menke's disease	Increased incorporation of Cu into cells
Metachromatic leukodystrophy	Deficient arylsulfatase A (cells)
Methylmalonic aciduria	Deficient methylmalonyl-CoA mutase (cells)
Mucolipidosis Type II	Deficient activity of several lysosomal enzymes (cells)
Niemann-Pick disease	Deficient sphingomyelinase (cells)
Osteogenesis imperfecta	Elevated amniotic fluid pyrophosphate
Pompe's disease	Deficient α-1,4-glucosidase (cells)
Propionic acidemia	Deficient propionyl-CoA carboxylase (cells)
Sandhoff's disease	Deficient β-N-acetyl hexosaminidase A + B (cells)
Sanfilippo's syndrome A	Deficient heparin sulfamidase (cells)
Tay-Sachs disease	Deficient β-N-acetyl hexosaminidase A (cells)
Wolman's disease	Deficient acid lipase (cells)
Xeroderma pigmentosa	Absent DNA excision repair synthesis (cells)

Most of these diseases are inherited as autosomal recessive disorders. Thus, in most situations, a first affected individual is born to a couple unexpectedly. It is therefore the subsequent pregnancy which is monitored by amniocentesis for the possible recurrence of the disease. In this case, the parents each carried a relatively rare gene which caused them no harm as heterozygotes, but which, when passed to their offspring in the homozygous condition, could cause serious disease or death. This unexpected carrier state has changed somewhat in recent years with the advent of the ability to test for "carriers" (heterozygotes) among adults. Thus, it is now possible for persons of ethnic groups which are at high risk for carriage of a deleterious gene to be tested for the presence of that gene. The gene for Tay-Sachs disease, for example, is carried by 1 in 30 Ashkenazic Jews. A Jewish couple anticipating having children might be tested in advance, using peripheral blood. If both have normal levels of hexosaminidase A activity, then there is no risk of having an affected child. If either one but not both has an intermediate level of enzyme activity, then they have a 50% chance of having a child who is also a carrier, but no chance of an affected child. If both are carriers, however, then they have a 25% chance of an affected child and should be counseled about the availability of antenatal diagnosis for this disorder. Specific inborn errors of metabolism are discussed in more detail in Chapters 4, 5, and 7 of this book.

Until recently, studies of amniotic fluid cells were limited to metabolic activities which are expressed by the cells. Recombinant DNA technology, however, has permitted the direct examination of the genome without requiring its expression. Antenatal diagnosis of the hemoglobinopathies, which has been reported, is undoubtedly only the first step in the development of this tool, whose power is virtually limitless.

References

1. Abraham, G. E.: Radioimmunoassay of plasma steroid hormones. *In*: Modern Methods in Steroid Analysis. E. Heftman, Ed. New York, Academic Press, 1973, pp. 451-470.
2. Abramovich, D. R., Keeping, J. D., and Thom, H.: The origin of amniotic fluid lecithin. Br. J. Obstet. Gynecol., *82*:204-207, 1975.
3. Beck, P., Parker, M. L., and Daughaday, W. H.: Radioimmunologic measurements of human placental lactogen in plasma by double antibody method during normal and diabetic pregnancies. J. Clin. Endocrinol. Metab., *25*:1457-1462, 1965.
4. Benzie, R. J., Doran, T. A., Harkins, J. L., et al.: Composition of the amniotic fluid and maternal serum in pregnancy. Am. J. Obstet. Gynecol., *119*:798-810, 1974.
5. Brock, D. J. H., Barron, L., Jelen, P., et al.: Maternal serum alpha-fetoprotein measurements as an early indicator of low birth weight. Lancet, *2*:267-270, 1977.
6. Brock, D. J. H., Bolton, A. E., and Monaghan, J. M.: Prenatal diagnosis of anencephaly through maternal serum-alphafetoprotein measurement. Lancet, *2*:923-924, 1973.
7. Brown, J. B., MacLeod, S. C., MacNaughtan, C., et al.: A rapid method for estimating estrogens in urine using a semiautomatic extractor. J. Endocrinol., *42*:5-15, 1968.
8. Brown, L. M., and Duck-Chong, C. G.: Methods of evaluating fetal lung maturity. CRC Crit. Rev. Clin. Lab. Sci., *16*:85-159, 1982.
9. Buamah, P. K., Evans, C., and Ward, A. M.: Amniotic fluid acetylcholinesterase isoenzyme patterns in the diagnosis of neural tube defects. Clin. Chim. Acta, *102*:147-151, 1980.
10. Bustos, R., Kulovich, M. V., Gluck, L., et al.: Significance of phosphatidylglycerol in amniotic fluid in complicated pregnancies. Am. J. Obstet. Gynecol., *133*:899-903, 1979.
11. Caritis, S. N., Mueller-Henbach, E., and Edelstone, D. I.: Effect of betamethasone on analysis of amniotic fluid in the rhesus-sensitized pregnancy. Am. J. Obstet. Gynecol., *127*:529-532, 1977.
12. Catt, K. J., Dufau, M. L., and Tsuruhara, J.: Radioligand-receptor assay of luteinizing hormone and chorionic gonadotropin. J. Clin. Endocrinol., *34*:123-132, 1972.
13. Chard, T.: The assay of alpha-fetoprotein. *In*: Prevention of Neural Tube Defects. The Role of Alpha-fetoprotein. B. F. Crandall and M. A. B. Brazier, Eds. New York, Academic Press, 1978, pp. 141-153.
14. Chattoraj, S. C.: Endocrine function. *In*: Clinical Chemistry, 2nd ed. N. W. Tietz, Ed. Philadelphia, W.B. Saunders, 1976, p. 768.
15. Chattoraj, S. C.: Endocrine function. *In*: Clinical Chemistry, 2nd ed. N. W. Tietz, Ed. Philadelphia: W.B. Saunders, 1976, p. 778.
16. Cherayil, G. D., Wilkinson, E. J., and Borkowf, H. I.: Amniotic fluid lecithin/sphingomyelin ratio changes related to centrifugal force. Obstet. Gynecol., *50*:682-688, 1977.
17. Clements, J. A., Platzker, A. C. G., Tierney, D. F., et al.: Assessment of the risk of the respiratory distress syndrome by a rapid test for surfactant in amniotic fluid. N. Engl. J. Med., *286*:1077-1081, 1972.
18. Collaborative acetylcholinesterase study, Report of the: Amniotic fluid acetylcholinesterase electrophoresis as a secondary test in the diagnosis of anencephaly and open spina bifida in pregnancy. Lancet, *2*:321-324, 1981.
19. Collins, V. P., Bernard, G. J. R., and Hennam, J. F.: Separation techniques. *In*: Steroid Immunoassay. E. H. D. Cameron, S. G. Hillier and K. Griffiths, Eds. Cardiff, Alpha Omega Publishing Ltd., 1975, pp. 223-227.
20. Cruz, A. C., Buhi, W. C., Birk, S. A., et al.: Respiratory distress syndrome with mature lecithin/sphingomyelin ratios: Diabetes mellitus and low Apgar scores. Am. J. Obstet. Gynecol., *126*:78-82, 1976.

21. Cunningham, M. D., Desai, N. S., Thompson, S. A., et al.: Amniotic fluid phosphatidylglycerol in diabetic pregnancies. Am. J. Obstet. Gynecol., 131:719-724, 1978.

22. Cunningham, M. D., McKean, H. E., Gillispie, D. H., et al.: Improved prediction of fetal lung maturity in diabetic pregnancies: A comparison of chromatographic methods. Am. J. Obstet. Gynecol., 142:197-204, 1982.

23. Dawood, M. Y.: Hormones in amniotic fluid. Am. J. Obstet. Gynecol., 128:576-583, 1977.

24. Doran, T. A., Allen, L. C., Pirani, B. B. K., et al.: False positive amniotic fluid alpha fetoprotein levels resulting from contamination with fetal blood: Results of an experiment. Am. J. Obstet. Gynecol., 127:759-762, 1977.

25. Dufau, M. L., Mendelson, C. R., and Catt, K. J.: A highly sensitive in vitro bioassay for luteinizing hormone and chorionic gonadotropin: Testosterone production by dispersed Leydig cells. J. Clin. Endocrinol. Metab., 39:610-613, 1974.

26. Ermshar, C. L., and Gusseck, D. J.: Use of polyethylene glycol in radioimmunoassay of human placental lactogen. Clin. Chem., 24:1767-1769, 1978.

27. Fencl, M. deM., Todd, R., Cohen, J., et al.: Radioimmunoassay of total urinary estriol. Anal. Letters, 14:323-334, 1981.

28. Freer, D. E., Statland, B. E., and Sher, G.: Quantitation of disaturated phosphatidylcholine and phosphatidylglycerol in amniotic fluid by fluorescence diminution: Methodology and clinical results. Clin. Chem., 25:960-968, 1979.

29. Frigoletto, F. D., Jr., and Griscom, N. T.: Amniography for the detection of fetal myelomeningocele. Obstet. Gynecol., 44:286-290, 1974.

30. Frigoletto, F. D., Jr., Umansky, I., Birnholz, J., et al.: Intrauterine fetal transfusion in 365 fetuses during fifteen years. Am. J. Obstet. Gynecol., 139:781-787, 1981.

31. Gabbe, S. G., Lowensohn, R. I., Mestman, J. H., et al.: Lecithin/sphingomyelin ratio in pregnancies complicated by diabetes mellitus. Am. J. Obstet. Gynecol., 128:757-760, 1977.

32. Gibbons, J. M., Jr., Huntley, T. E., and Corral, A. G.: Effect of maternal blood contamination on amniotic fluid analysis. Obstet. Gynecol., 44:657-660, 1974.

33. Gluck, L., and Kulovich, M.: Lecithin/sphingomyelin ratios in amniotic fluid in normal and abnormal pregnancy. Am. J. Obstet. Gynecol., 115:539-546, 1973.

34. Gluck, L., Kulovich, M. V., and Borer, R. C., Jr.: Estimates of fetal lung maturity. Clin. Perinatol., 1:125-139, 1974.

35. Golde, S. H., and Mosley, G. H.: A blind comparison study of the lung phospholipid profile, fluorescence micro-viscosimetry, and the lecithin/sphingomyelin ratio. Am. J. Obstet. Gynecol., 136:222-227, 1980.

36. Hallman, M., Kulovich, M., Kirkpatrick, E., et al.: Phosphatidylinositol and phosphatidylglycerol in amniotic fluid: Indices of lung maturity. Am. J. Obstet. Gynecol., 125:613-617, 1976.

37. Hodgson, A. J., et al.: Combined analysis of acetylcholinesterase and alphafetoprotein improves the accuracy of antenatal diagnosis of neural tube defects. Med. J. Aust., 1:457-460, 1981.

38. Holmes, L. B., Driscoll, S. G., and Atkins, L.: Etiologic heterogeneity of neural tube defects. N. Engl. J. Med., 294:365-369, 1976.

39. Jeffcoate, T. N. A., Fliegner, J. R. H., Russell, S. H., et al.: Diagnosis of adrenogenital syndrome before birth. Lancet, 2:553-555, 1965.

40. Keniston, R. C., Noland, G. L., and Pernoll, M. L.: The effect of blood, meconium and temperature on the rapid surfactant test. Obstet. Gynecol., 48:442-446, 1976.

41. Kulovich, M. V., and Gluck, L.: The lung profile. II. Complicated pregnancy. Am. J. Obstet. Gynecol., 135:64-70, 1979.

42. Kulovich, M. V., Hallman, M. B., and Gluck, L.: The lung profile. I. Normal pregnancy. Am. J. Obstet. Gynecol., 135:57-63, 1979.

43. Lau, H. L., and Linkins, S. E.: Alpha-fetoprotein. Am. J. Obstet. Gynecol., 124:533-554, 1976.

44. Liley, A. W.: Liquor amnii analysis in the management of the pregnancy complicated by rhesus sensitization. Am. J. Obstet. Gynecol., 82:1359-1370, 1961.

45. Liley, A. W.: Errors in the assessment of hemolytic disease from amniotic fluid. Am. J. Obstet. Gynecol., 86:485-494, 1963.

46. Macri, J. N.: Current status of alpha-fetoprotein prenatal testing. In: Perinatal Medicine Today. B. K. Young and A. R. Liss, Eds. New York, A. R. Liss, 1980, pp. 47-63.

47. Mason, R. J., Nellenbogen, J., and Clements, J. A.: Isolation of disaturated phosphatidylcholine with osmium tetroxide. J. Lipid Res., 17:281-284, 1976.

48. Mennuti, M. T., and Zackai, E.: Indications for karyotyping amniotic fluid obtained in follow-up of elevated maternal serum alpha-fetoprotein values. Proceedings of the Second Scarborough Conference. Scarborough, Me., Foundation for Blood Research, 1978, pp. 100-104.

49. Miller, E., Hare, J. W., Cloherty, J. P., et al.: Elevated maternal HbA$_{1C}$ in early pregnancy and major congenital anomalies in infants of diabetic mothers. N. Engl. J. Med., 304:1331-1334, 1981.

50. Milunsky, A.: Genetic Disorders and the Fetus. Diagnosis, Prevention and Treatment. New York, Plenum Press, 1979.

51. Mitnick, M. A., DeMarco, B., and Gibbons, J. M.: Amniotic fluid phosphatidylglycerol and phosphatidylinositol separated by stepwise-development thin-layer chromatography. Clin. Chem., 26:277-281, 1980.

52. Norgaard-Pedersen, B., and Galde, P.: Immunoelectrophoretical quantitation of human placental lactogen (hPL). Scand. J. Immunol., 2(Suppl. 1):184-189, 1973.

53. Penney, L. L., Hagerman, D. D., and Sei, C. A.: Specificity and reproducibility of acetone precipitation in identifying surface-active phosphatidylcholine in amniotic fluid. Clin. Chem., 22:681-682, 1976.

54. Phillippe, M., Acker, D., Torday, J., et al.: The effects of vaginal contamination on two pulmonary phospholipid assays. J. Reprod. Med., 27:283-286, 1982.

55. Queenan, J. T.: Modern Management of the Rh Problem. Hagerstown, Md., Harper and Row, 1977.

56. Queenan, J. T., Thompson, W., Whitfield, C. R., et al.: Amniotic fluid volumes in normal pregnancy. Am. J. Obstet. Gynecol., 114:34-38, 1972.

57. Rothchild, S. B., Tulchinsky, D., Fencl, M., et al.: Estriol determinations in diabetic pregnancies complicated by nephropathy. Am. J. Obstet. Gynecol., *134*:772-775, 1979.
58. Ruoslahti, E., and Seppälä, M.: Studies of carcino-fetal proteins. III. Development of a radioimmunoassay for α-fetoprotein in serum of healthy human adults. Int. J. Cancer, *8*:374-383, 1971.
59. Sandstrom, M. McH., Beauchesne, M. T., Gustashaw, K. M., et al.: Prenatal cytogenetic diagnosis. Meth. Cell. Biol., *26*:35-66, 1982.
60. Schlueter, M. A., Phibbs, R. H., Creasy, R. K., et al.: Antenatal prediction of graduated risk of hyaline membrane disease by amniotic fluid foam test for surfactant. Am. J. Obstet. Gynecol., *134*:761-767, 1979.
61. Schwartz, D. B., Engle, M. J., Brown, D. J., et al.: The stability of phospholipids in amniotic fluid. Am. J. Obstet. Gynecol., *141*:294-298, 1981.
62. Sciarra, J. J., Sherwood, L. M., Varma, A. A., et al.: Human placental lactogen and placental weight. Am. J. Obstet. Gynecol., *101*:413-416, 1968.
63. Seppälä, M., and Ruoslahti, E.: Alpha-fetoprotein in maternal serum: A new marker for detection of fetal distress and intrauterine death. Am. J. Obstet. Gynecol., *115*:48-52, 1973.
64. Sher, G., and Statland, B. E.: Assessment of fetal pulmonary maturity by the Lumadex foam stability index test. Obstet. Gynecol., *61*:444-449, 1983.
65. Sher, G., Statland, B. E., and Freer, D. E.: Clinical evaluation of the quantitative foam stability index tests. Obstet. Gynecol., *55*:617-620, 1980.
66. Sher, G., Statland, B. E., Freer, D. E., et al.: Performance of the amniotic fluid foam stability-50 percent test. A bedside procedure for the prenatal detection of hyaline membrane disease. Am. J. Obstet. Gynecol., *134*:705-707, 1979.
67. Sher, G., Statland, B. E., Freer, D. E., et al.: Assessing fetal lung maturation by the foam stability index test. Obstet. Gynecol., *52*:673-677, 1978.
68. Shiu, R. P., Kelly, P. A., and Friesen, H. G.: Radioreceptor assay for prolactin and other lactogenic hormones. Science, *180*:968-971, 1973.
69. Spillman, T., Cotton, D. B., Lynn, S. C., Jr., et al.: Influence of phospholipid saturation on classical thin-layer chromatographic detection methods and its effect on amniotic fluid lecithin/sphingomyelin ratio determinations. Clin. Chem., *29*:250-255, 1983.
70. Statland, B. E., Sher, G., Freer, D. E., et al.: Evaluation of a modified foam stability (FS-50) test. An assay performed on amniotic fluid to predict fetal pulmonary maturity. Am. J. Clin. Pathol., *69*:514-519, 1978.
71. Stephenson, S. R., and Weaver, D. D.: Prenatal diagnosis—a compilation of diagnosed conditions. Am. J. Obstet. Gynecol., *141*:319-343, 1981.
72. Theppisai, H., Mishell, D. R., Jr., and Nakamura, R. M.: Comparison of a rapid new method of hemagglutination-inhibition assay of HPL with micro-complement fixation and radioimmunoassay. J. Clin. Endocrinol., *32*:382-388, 1971.
73. Torday, J., Carson, L., and Lawson, E. E.: Saturated phosphatidylcholine in amniotic fluid and prediction of the respiratory-distress syndrome. N. Engl. J. Med., *301*:1013-1018, 1979.
74. Torday, J. S., and VanDemottaz Frantz, I. D., III: Prediction of fetal pulmonary maturation in complicated pregnancies. Pediatr. Res., *13*:542a, 1979.
75. Tsai, M. Y., Cain, M., and Josephson, M. W.: Improved thin layer chromatography of disaturated phosphatidylcholine in amniotic fluid. Clin. Chem., *27*:239-242, 1981.
76. Tsai, M. Y., and Marshall, J. G.: Phosphatidylglycerol in 261 samples of amniotic fluid from normal and diabetic pregnancies, as measured by one-dimensional thin layer chromatography. Clin. Chem., *25*:682-685, 1979.
77. Tulchinsky, D., and Abraham, G. E.: Radioimmunoassay of plasma estriol. J. Clin. Endocrinol., *33*:775-782, 1971.
78. Tulchinsky, D., and Ryan, K. J., Eds.: Maternal-Fetal Endocrinology. Philadelphia, W. B. Saunders, 1980.
79. van Voorst tot Voorst, E. J. G. M.: Effects of centrifugation, storage, and contamination of amniotic fluid on its total phospholipid content. Clin. Chem., *26*:232-234, 1980.
80. Verjaal, M., Leschot, N. J., and Treffers, P. E.: Risk of amniocentesis and laboratory findings in a series of 1500 prenatal diagnoses. Prenatal Diagnosis, *1*:173-181, 1981.
81. Vince, J. D., McManus, T. J., Ferguson-Smith, M. A., et al.: A semi-automated serum alpha-fetoprotein radioimmunoassay for prenatal spina bifida screening. Br. J. Obstet. Gynaecol., *82*:718-727, 1975.
82. Wagstaff, T. I., Whyley, G. A., and Freedman, G.: Factors influencing the measurement of the lecithin/sphingomyelin ratio in amniotic fluid. J. Obstet. Gynaecol. Br. Commonw., *81*:264-277, 1974.
83. Wasserman, E., and Levine, L.: Quantitative micro-complement fixation and its use in the study of antigenic structure by specific antigen antibody inhibition. J. Immunol., *87*:290-295, 1961.
84. Weinstein, L.: Irregular antibodies causing hemolytic disease of the newborn. Obstet. Gynecol. Surv., *31*:581-591, 1976.
85. Wu, C. H., Mennuti, M. T., and Mikhail, G.: Free and protein-bound steroids in amniotic fluid of midpregnancy. Am. J. Obstet. Gynecol., *133*:666-672, 1979.
86. Yalow, R. S., and Berson, S. A.: Introduction and general considerations. *In*: Principles of Competitive Protein-Binding Assays. W. D. Odell and W. H. Daughaday, Eds. Philadelphia, J. B. Lippincott Co., 1971, pp. 1-24.

CHAPTER 20

APPENDIX

by Norbert W. Tietz, Ph.D., and Nancy M. Logan, B.A.

CONTENTS

Table 20-1. METRIC UNITS

Prefix Name	Prefix Symbol	Equivalent	Units of Length	Units of Mass*	Units of Capacity
kilo-	k	10^3	kilometer (km)	kilogram (kg)	kiloliter (kL)
		1	meter (m)	gram (g)	liter (L)
deci-	d	10^{-1}	decimeter (dm)	decigram (dg)	deciliter (dL)
centi-	c	10^{-2}	centimeter (cm)	centigram (cg)	centiliter (cL)
milli-	m	10^{-3}	millimeter (mm)	milligram (mg)	milliliter (mL)
micro-	μ	10^{-6}	micrometer (μm)	microgram (μg)	microliter (μL)
nano-	n	10^{-9}	nanometer (nm)	nanogram (ng)	nanoliter (nL)
pico-	p	10^{-12}	picometer (pm)	picogram (pg)	picoliter (pL)

*According to the SI system, mass is the preferred term; weight is more commonly used in the United States.

CONVERSION CHARTS

Table 20-2. UNITS OF LENGTH

Kilometer km	Meter m	Decimeter dm	Centimeter cm	Millimeter mm	Micrometer µm	Nanometer nm	Angstrom Å	Picometer pm	Inch in.
1	10^3	10^4	10^5	10^6	10^9	10^{12}	10^{13}	10^{15}	39.37×10^3
10^{-3}	1	10	10^2	10^3	10^6	10^9	10^{10}	10^{12}	39.37
10^{-4}	10^{-1}	1	10	10^2	10^5	10^8	10^9	10^{11}	39.37×10^{-1}
10^{-5}	10^{-2}	10^{-1}	1	10	10^4	10^7	10^8	10^{10}	39.37×10^{-2}
10^{-6}	10^{-3}	10^{-2}	10^{-1}	1	10^3	10^6	10^7	10^9	39.37×10^{-3}
10^{-9}	10^{-6}	10^{-5}	10^{-4}	10^{-3}	1	10^3	10^4	10^6	39.37×10^{-6}
10^{-12}	10^{-9}	10^{-8}	10^{-7}	10^{-6}	10^{-3}	1	10	10^3	39.37×10^{-9}
10^{-13}	10^{-10}	10^{-9}	10^{-8}	10^{-7}	10^{-4}	10^{-1}	1	10^2	39.37×10^{-10}
10^{-15}	10^{-12}	10^{-11}	10^{-10}	10^{-9}	10^{-6}	10^{-3}	10^{-2}	1	39.37×10^{-12}
2.54×10^{-5}	2.54×10^{-2}	2.54×10^{-1}	2.54	2.54×10	2.54×10^4	2.54×10^7	2.54×10^8	2.54×10^{10}	1

Table 20-3. UNITS OF MASS*

Kilogram kg	Gram g	Decigram dg	Centigram cg	Milligram mg	Microgram µg	Nanogram ng	Picogram pg	Ounce (Av.) oz.	Pound (Av.) lb.
1	10^3	10^4	10^5	10^6	10^9	10^{12}	10^{15}	35.27	2.2
10^{-3}	1	10	10^2	10^3	10^6	10^9	10^{12}	35.27×10^{-3}	2.2×10^{-3}
10^{-4}	10^{-1}	1	10	10^2	10^5	10^8	10^{11}	35.27×10^{-4}	2.2×10^{-4}
10^{-5}	10^{-2}	10^{-1}	1	10	10^4	10^7	10^{10}	35.27×10^{-5}	2.2×10^{-5}
10^{-6}	10^{-3}	10^{-2}	10^{-1}	1	10^3	10^6	10^9	35.27×10^{-6}	2.2×10^{-6}
10^{-9}	10^{-6}	10^{-5}	10^{-4}	10^{-3}	1	10^3	10^6	35.27×10^{-9}	2.2×10^{-9}
10^{-12}	10^{-9}	10^{-8}	10^{-7}	10^{-6}	10^{-3}	1	10^3	35.27×10^{-12}	2.2×10^{-12}
10^{-15}	10^{-12}	10^{-11}	10^{-10}	10^{-9}	10^{-6}	10^{-3}	1	35.27×10^{-15}	2.2×10^{-15}
28.35×10^{-3}	28.35	28.35×10	28.35×10^2	28.35×10^3	28.35×10^6	28.35×10^9	28.35×10^{12}	1	0.0625
0.454	454	454×10	454×10^2	454×10^3	454×10^6	454×10^9	454×10^{12}	16	1

*See footnote for Table 20-1, p. 1790.

Table 20-4. UNITS OF CAPACITY

Kiloliter kL	Liter L	Deciliter dL	Centiliter cL	Milliliter mL	Microliter μL	Nanoliter nL	Picoliter pL	Ounce oz.	Quart qt.
1	10^3	10^4	10^5	10^6	10^9	10^{12}	10^{15}	33.81×10^3	1.06×10^3
10^{-3}	1	10	10^2	10^3	10^6	10^9	10^{12}	33.81	1.06
10^{-4}	10^{-1}	1	10	10^2	10^5	10^8	10^{11}	33.81×10^{-1}	1.06×10^{-1}
10^{-5}	10^{-2}	10^{-1}	1	10	10^4	10^7	10^{10}	33.81×10^{-2}	1.06×10^{-2}
10^{-6}	10^{-3}	10^{-2}	10^{-1}	1	10^3	10^6	10^9	33.81×10^{-3}	1.06×10^{-3}
10^{-9}	10^{-6}	10^{-5}	10^{-4}	10^{-3}	1	10^3	10^6	33.81×10^{-6}	1.06×10^{-6}
10^{-12}	10^{-9}	10^{-8}	10^{-7}	10^{-6}	10^{-3}	1	10^3	33.81×10^{-9}	1.06×10^{-9}
10^{-15}	10^{-12}	10^{-11}	10^{-10}	10^{-9}	10^{-6}	10^{-3}	1	33.81×10^{-12}	1.06×10^{-12}
29.57×10^{-6}	29.57×10^{-3}	29.57×10^{-2}	29.57×10^{-1}	29.57	29.57×10^3	29.57×10^6	29.57×10^9	1	3.125×10^{-2}
0.946×10^{-3}	0.946	0.946×10	0.946×10^2	0.946×10^3	0.946×10^6	0.946×10^9	0.946×10^{12}	32	1

Table 20-5. TEMPERATURE CONVERSIONS (CELSIUS → FAHRENHEIT)

Temp. °C	0	1	2	3	4	5	6	7	8	9
−10	14.0	12.2	10.4	8.6	6.8	5.0	3.2	1.4	−0.4	−2.2
−0	32.0	30.2	28.4	26.6	24.8	23.0	21.2	19.4	17.6	15.8
0	32.0	33.8	35.6	37.4	39.2	41.0	42.8	44.6	46.4	48.2
10	50.0	51.8	53.6	55.4	57.2	59.0	60.8	62.6	64.4	66.2
20	68.0	69.8	71.6	73.4	75.2	77.0	78.8	80.6	82.4	84.2
30	86.0	87.8	89.6	91.4	93.2	95.0	96.8	98.6	100.4	102.2
40	104.0	105.8	107.6	109.4	111.2	113.0	114.8	116.6	118.4	120.2
50	122.0	123.8	125.6	127.4	129.2	131.0	132.8	134.6	136.4	138.2
60	140.0	141.8	143.6	145.4	147.2	149.0	150.8	152.6	154.4	156.2
70	158.0	159.8	161.6	163.4	165.2	167.0	168.8	170.6	172.4	174.2
80	176.0	177.8	179.6	181.4	183.2	185.0	186.8	188.6	190.4	192.2
90	194.0	195.8	197.6	199.4	201.2	203.0	204.8	206.6	208.4	210.2
100	212.0	213.8	215.6	217.4	219.2	221.0	222.8	224.6	226.4	228.2

$°C = 5/9 \times (°F - 32); °F = (9/5 \times °C) + 32.$

MISCELLANEOUS CONSTANTS AND INFORMATION

Table 20-6. GREEK ALPHABET (UPRIGHT AND SLOPING TYPES)

alpha	A	α	A	α	nu	N	ν	N	ν
beta	B	β	B	β	xi	Ξ	ξ	Ξ	ξ
gamma	Γ	γ	Γ	γ	omicron	O	o	O	o
delta	Δ	δ	Δ	δ	pi	Π	π	Π	π
epsilon	E	ε, ε	E	ε, ε	rho	P	ρ	P	ρ
zeta	Z	ζ	Z	ζ	sigma	Σ	σ	Σ	σ
eta	H	η	H	η	tau	T	τ	T	τ
theta	Θ	ϑ, θ	Θ	ϑ, θ	upsilon	Υ	υ	Υ	υ
iota	I	ι	I	ι	phi	Φ	φ	Φ	φ
kappa	K	ϰ, κ	K	ϰ, κ	chi	X	χ	X	χ
lambda	Λ	λ	Λ	λ	psi	Ψ	ψ	Ψ	ψ
mu	M	μ	M	μ	omega	Ω	ω	Ω	ω

Table 20-7. ATOMIC WEIGHTS*

Atomic Number	Name	Symbol	International Atomic Mass	Oxidative States
13	Aluminum	Al	26.98	+3
51	Antimony (Stibium)	Sb	121.8	+3, +5, −3
18	Argon	Ar	39.95	0
33	Arsenic	As	74.92	+3, +5, −3
56	Barium	Ba	137.3	+2
4	Beryllium	Be	9.012	+2
83	Bismuth	Bi	209.0	+3, +5
5	Boron	B	10.81	+3
35	Bromine	Br	79.90	+1, +5, −1
48	Cadmium	Cd	112.4	+2
20	Calcium	Ca	40.08	+2
6	Carbon	C	12.01	+2, +4, −4
58	Cerium	Ce	140.1	+3, +4
55	Cesium	Cs	132.9	+1
17	Chlorine	Cl	35.45	+1, +5, +7, −1
24	Chromium	Cr	52.00	+2, +3, +6
27	Cobalt	Co	58.93	+2, +3
29	Copper	Cu	63.55	+1, +2
9	Fluorine	F	19.00	−1
79	Gold (Aurum)	Au	197.0	+1, +3
2	Helium	He	4.003	0
1	Hydrogen	H	1.008	+1, −1
53	Iodine	I	126.9	+1, +5, +7, −1
26	Iron (Ferrum)	Fe	55.85	+2, +3
57	Lanthanum	La	138.9	+3
82	Lead (Plumbum)	Pb	207.2	+2, +4
3	Lithium	Li	6.941	+1
12	Magnesium	Mg	24.31	+2
25	Manganese	Mn	54.94	+2, +3, +4, +7
80	Mercury	Hg	200.6	+1, +2
42	Molybdenum	Mo	95.94	+6
10	Neon	Ne	20.18	0
28	Nickel	Ni	58.69	+2, +3
7	Nitrogen	N	14.01	+1, +2, +3, +4, +5, −1, −2, −3
8	Oxygen	O	16.00	−2
46	Palladium	Pd	106.4	+2, +4
15	Phosphorus	P	30.97	+3, +5, −3
78	Platinum	Pt	195.1	+2, +4
19	Potassium (Kalium)	K	39.10	+1
34	Selenium	Se	78.96	+4, +6, −2
14	Silicon	Si	28.09	+2, +4, −4
47	Silver (Argentum)	Ag	107.9	+1
11	Sodium (Natrium)	Na	22.99	+1
38	Strontium	Sr	87.62	+2
16	Sulfur	S	32.06	+4, +6, −2
52	Tellurium	Te	127.6	+4, +6, −2
81	Thallium	Tl	204.4	+1, +3
90	Thorium	Th	232.0	+4
50	Tin (Stantium)	Sn	118.7	+2, +4
22	Titanium	Ti	47.88	+2, +3, +4
74	Tungsten (Wolfram)	W	183.8	+6
92	Uranium	U	238.0	+3, +4, +5, +6
23	Vanadium	V	50.94	+2, +3, +4, +5
54	Xenon	Xe	131.3	0
30	Zinc	Zn	65.38	+2

*According to the SI system, mass is the preferred term; weight is more commonly used in the United States. Values have been updated according to Holden, N.E., and Martin, R. L.: Atomic weights of the elements, 1981. Pure and Appl. Chem., 55(7):1011–1118, 1983. Values are based on carbon-12 and rounded to four significant figures.

Table 20-8. BOILING POINTS OF COMMONLY USED SOLVENTS

Name	Alternative Name	Molecular Weight*	Boiling Point[†] (°C)
Acetic acid	Ethanoic acid	60.05	118.5[760]
Acetone	2-Propanone	58.08	56.2
Aniline	Aminobenzene	93.13	184.3[760]
Benzene		78.11	80.1
n-Butanol	1-Butanol	74.12	117.5[760]
Carbon disulfide		76.14	45[760]
Carbon tetrachloride	Tetrachloromethane	153.82	76.8[760]
Chloroform	Trichloromethane	119.38	61.2[760]
Ethanol	Ethyl alcohol	46.07	78.5
Ethyl acetate	Acetic acid ethyl ester	88.11	77.1[760]
Ethyl ether	Diethyl ether	74.12	34.6
Ethylene dichloride	1,2-Dichloroethane	98.96	84[760]
Heptane		100.21	98.4
Isoamyl acetate	Acetic acid 3-methylbutyl ester	130.2	142
Isoamyl alcohol	3-Methyl-1-butanol	88.15	131[760]
Isobutyl alcohol	2-Methyl-1-propanol	74.12	108.4
Isopropyl alcohol	2-Propanol	60.09	82.4
Methanol	Carbinol; Methyl alcohol	32.04	65.0[760]
Methyl isobutyl ketone	4-Methyl-2-propanone	100.16	116.9
Methylene chloride	Dichloromethane	84.93	40
Nitrobenzene		123.11	210.8[760]
Petroleum ether		Varies with fraction	~40–120; varies with fraction
Pyridine		79.10	115.5
Toluene	Methylbenzene	92.13	110.6
m-Xylene	1,3-Dimethylbenzene	106.16	139
o-Xylene	1,2-Dimethylbenzene	106.16	144
p-Xylene	1,4-Dimethylbenzene	106.16	138

*See footnote for Table 20-1, p. 1790.
[†]Superscript indicates the barometric pressure at which the boiling point was measured. If no figure is given, the barometric pressure was measured at approximately 1 atmosphere.

Table 20-9. PRIMARY AND SECONDARY STANDARDS

SRM*	Standard	Formula	Purity,† Assay,† or pH(S) at 25 °C†	Molecular Weight (Also, Gram Quantity Needed for 1 Liter of Solution Containing 1 Mole)
186IIc	Disodium hydrogen phosphate	Na_2HPO_4	pH 7.415	141.98
	Hydrochloric acid	HCl	36.5–38%	36.46
	Nitric acid	HNO_3	69–71%	63.01
	Oxalic acid	$H_2C_2O_4$		90.04
	Oxalic acid, dihydrate	$H_2C_2O_4 \cdot 2H_2O$	pH 6.863	126.07
186Ic	Potassium dihydrogen phosphate	KH_2PO_4	$\geq$ 99.5%	136.09
185e	Potassium hydrogen phthalate	$KHC_8H_4O_4$	pH 4.004	204.22
188	Potassium hydrogen tartrate	$KHC_4H_4O_6$	pH 3.557	188.18
189	Potassium tetroxalate	$KHC_2O_4 \cdot H_2C_2O_4$	pH 1.679	218.16
	Potassium iodate	KIO_3	~99%	214.02
191a	Sodium bicarbonate	$NaHCO_3$	pH 10.011	84.00
192a	Sodium carbonate	N_2CO_3	$\geq$ 99.95%	106.00
40h	Sodium oxalate	$Na_2C_2O_4$	99.972%	134.01
187b	Sodium tetraborate decahydrate (Borax)	$Na_2B_4O_7 \cdot 10H_2O$	pH 9.183	381.27
	Succinic acid	$HOOC—(CH_2)_2—COOH$	$\geq$ 99.0%	118.09
	Sulfuric acid	H_2SO_4	95–98%	98.08
922	Tris (hydroxymethyl)-aminomethane; (tromethamine, THAM or Tris)	$NH_2C(CH_2OH)_3$	99.9%	121.14
923	Tris HCl	$NH_2C(CH_2OH)_3$ HCl	99.7%	157.59

*Standard Reference Material available from the National Bureau of Standards, Washington, D.C.
†May vary with lot or manufacturer.

Table 20-10. pH INDICATORS

Common Name	Chemical Name	pH Range	Color Change	Commonly Used Concentration
Alizarin	1,2-Dihydroxyanthraquinone	5.5–6.8	Yellow to red	0.50 g/dL in ethanol
Bromcresol green (bromocresol green)	3,3'5,5'-Tetrabromo-m-cresol-sulfonphthalein	3.8–5.4	Yellow to blue-green	0.10 g in 7.15 mL NaOH, 0.02 mol/L; dil. to 250 mL with water
Bromcresol purple (bromocresol purple)	5,5'-Dibromo-o-cresolsulfon-phthalein	5.2–6.8	Yellow to purple	0.10 g in 9.25 mL NaOH, 0.02 mol/L; dil. to 250 mL with water
Bromphenol blue (bromophenol blue)	3,3',5,5'-Tetrabromophenol-sulfonphthalein	3.0–4.6	Yellow to purple	0.10 g in 7.45 mL NaOH, 0.02 mol/L; dil. to 250 mL with water
Bromphenol red (bromophenol red)	Dibromophenolsulfonphthalein	5.2–6.8	Yellow to red	0.04 g/dL in water (salt form)
Bromthymol blue (bromothymol blue)	3,3'-Dibromothymolsulfon-phthalein	6.0–7.6	Yellow to blue	0.10 g in 8.0 mL NaOH, 0.02 mol/L; dil. to 250 mL with water
Chlorphenol red (chlorophenol red)	3,3'-Dichlorophenolsulfon-phthalein	5.2–6.8	Yellow to red	0.10 g in 23.6 mL NaOH, 0.01 mol/L; dil. to 250 mL with water
Congo red	Sodium diphenyldiazo-bis-α-naphthylaminesulfonate	3.0–5.0	Blue-violet to red	0.10 g/dL in water
Cresol red	o-Cresolsulfonphthalein; α-hydroxy-α,α-bis(4-hydroxy-m-tolyl)-o-toluenesulfonic acid γ-sultone	7.2–8.8	Yellow to red (orange to amber at pH 2–3)	0.04 g/100 mL of NaOH, 1.1 mmol/L
Litmus	Lacmus; tournesol; turnsole; lacca musica; lacca coerulea (blue coloring matter of various lichens)	4.5–8.3	Red to blue	
Metachrome yellow (alizarine yellow GG)	Sodium m-nitrobenzeneazosali-cylate	10.2–12.0	Colorless to yellow	0.10 g/dL in 50% ethanol
Metacresol purple	m-Cresolsulfonphthalein	1.2–2.8 7.4–9.0	Red to yellow Yellow to purple	0.10 g in 26.2 mL NaOH, 0.01 mol/L + 223.8 mL water
Methyl orange	Sodium p-dimethylaminoazo-benzenesulfonate	3.1–4.4	Red to yellow	0.10 g/dL in water
Methyl red	4'-Dimethylaminobenzene 2-carboxylic acid	4.4–6.2	Red to yellow	0.10 g/dL in ethanol
Methyl violet	Pentamethylbenzyl-p-rosaniline·HCl	0.0–1.6	Yellow to blue	0.01–0.05 g/dL in water
Neutral red	Aminodimethylaminotolu-aminozine hydrochloride	6.8–8.0	Red to yellow	0.10 g/dL in 60% alcohol
Phenol red	Phenolsulfonphthalein; α-hydroxy-α,α-bis(p-hydroxyphenyl)-o-toluenesulfonic acid γ-sultone	6.8–8.4	Yellow to red	0.02–0.05 g/dL in ethanol
Phenolphthalein	3,3-Bis(p-hydroxyphenyl)-phthalide	8.3–10.0	Colorless to red	0.10 g/dL in 95% ethanol
Thymol blue	Thymolsulfonphthalein; α-hydroxy-α,α-bis(5-hydroxycarvacryl)-o-toluenesulfonic acid γ-sultone	1.2–2.8 8.0–9.6	Red to yellow Yellow to blue	If sodium salt is used: 0.04 g/dL in 95% ethanol If acid form is used: 0.04 g/dL NaOH, 1.0 mol/L
Thymolphthalein	5,5'-Diisopropyl-2,2'-dimethyl-phenolphthalein	9.3–10.5	Colorless to blue	0.10 g/dL in 95% ethanol
Toepfer's reagent	Dimethylaminoazobenzene	2.9–4.0	Red to yellow	0.50 g/dL in 95% ethanol

Table 20-11. IONIZATION CONSTANTS K AND pK_a FOR COMMON ACIDS AND BASES IN WATER*

	K	pK_a		K	pK_a
Acetic acid	1.75×10^{-5}	4.76	Imidazole	1.01×10^{-7}	6.95
Acetoacetic acid	2.62×10^{-4}	3.58	Isocitric acid	5.13×10^{-4}	3.29
		(18 °C)		1.99×10^{-5}	4.70
Ammonia	5.6×10^{-10}	9.25		3.98×10^{-7}	6.40
Boric acid†	6.4×10^{-10}	9.19	p-Nitrophenol	7×10^{-8}	7.15
Carbonic acid	4.47×10^{-7}	6.35	Oxalacetic acid	2.75×10^{-3}	2.56
	4.68×10^{-11}	10.34		4.27×10^{-5}	4.37
Citric acid	7.4×10^{-4}	3.13	Oxalic acid	6.5×10^{-2}	1.19
	1.7×10^{-5}	4.77		6.1×10^{-5}	4.21
	4.0×10^{-7}	6.40	Phosphoric acid	7.5×10^{-3}	2.12
Diethylbarbituric acid	3.7×10^{-8}	7.43		6.2×10^{-8}	7.21
(Veronal)				4.8×10^{-13}	12.32
Ethylenediamine	1.4×10^{-7}	6.85	Phosphorous acid	1.0×10^{-2}	2.00
	1.12×10^{-10}	9.93		2.6×10^{-7}	6.59
Ethylenediamine tetra-	1.00×10^{-2}	2.00	Pyruvic acid	3.23×10^{-3}	2.49
acetate	2.16×10^{-3}	2.67	Succinic acid	6.2×10^{-5}	4.21
				2.3×10^{-6}	
	6.92×10^{-7}	6.16	Sulfuric acid	$\gg 1$	—
	5.50×10^{-11}	10.26		1.2×10^{-2}	1.92
Formic acid	1.76×10^{-4}	3.75	Tartaric acid	1.1×10^{-3}	2.96
Glycine	4.5×10^{-3}	2.35		6.9×10^{-5}	4.16
	1.7×10^{-10}	9.77	Triethanolamine	1.26×10^{-8}	7.90
Glycylglycine	7.24×10^{-4}	3.14	Tris(hydroxymethyl)-	8.32×10^{-9}	8.08
	5.62×10^{-9}	8.25	amino methane		
Hydroxylamine	9.1×10^{-9}	8.04			

*Temperature at or near room temperature (25 °C) unless otherwise indicated.
†Boric acid acts as a monotropic acid in aqueous solution.

Table 20-12. COMMON CONCENTRATED ACIDS AND BASES

Acid or Base	Spec. Grav. at 25 °C*	% by Wt. (w/w)*	g/L*	M.W.	Approx. Molarity	Approx. Normality	Approx. mL Required to Make 1 L of 1 N Soln.
Acetic acid (CH_3COOH)	1.06	99.5	1060	60.05	17.6	17.6	57
Ammonium hydroxide (NH_4OH)	0.880	29	252 (NH_3)	17.03 (NH_3)	14.8	14.8	67
Hydrochloric acid (HCl)	1.19	37	440	36.46	12.1	12.1	83
Nitric acid (HNO_3)	1.42	70	990	63.02	15.7	15.7	64
Perchloric acid ($HClO_4$) 60%	1.53	60–62	934	100.46	9.3	9.3	108
" 72%	1.68	70–72	1195	100.46	11.9	11.9	84
Phosphoric acid (H_3PO_4)	1.7	85	1441	98.00	14.7	44.1	23
Potassium hydroxide saturated† (KOH)	1.55	~50	785	56.11	14	14	71
Sodium hydroxide, saturated (NaOH)	1.50	~50	750	40.00	19	19	53
Sulfuric acid (H_2SO_4)	1.84	96	1765	98.08	18	36	28

*May vary with lot or manufacturer.
†Saturated solutions made from the usual C.P. potassium hydroxide will vary in strength, chiefly because of the variable amount of carbonate that such solutions contain.

Table 20-13. % TRANSMISSION-ABSORBANCE CONVERSION CHART

%T	A	%T	A	%T	A	%T	A
1	2.000	1.5	1.824	51	.2924	51.5	.2882
2	1.699	2.5	1.602	52	.2840	52.5	.2798
3	1.523	3.5	1.456	53	.2756	53.5	.2716
4	1.398	4.5	1.347	54	.2676	54.5	.2636
5	1.301	5.5	1.260	55	.2596	55.5	.2557
6	1.222	6.5	1.187	56	.2518	56.5	.2480
7	1.155	7.5	1.126	57	.2441	57.5	.2403
8	1.097	8.5	1.071	58	.2366	58.5	.2328
9	1.046	9.5	1.022	59	.2291	59.5	.2255
10	1.000	10.5	.979	60	.2218	60.5	.2182
11	.959	11.5	.939	61	.2147	61.5	.2111
12	.921	12.5	.903	62	.2076	62.5	.2041
13	.886	13.5	.870	63	.2007	63.5	.1973
14	.854	14.5	.838	64	.1939	64.5	.1905
15	.824	15.5	.810	65	.1871	65.5	.1838
16	.796	16.5	.782	66	.1805	66.5	.1772
17	.770	17.5	.757	67	.1739	67.5	.1707
18	.745	18.5	.733	68	.1675	68.5	.1643
19	.721	19.5	.710	69	.1612	69.5	.1580
20	.699	20.5	.688	70	.1549	70.5	.1518
21	.678	21.5	.668	71	.1487	71.5	.1457
22	.658	22.5	.648	72	.1427	72.5	.1397
23	.638	23.5	.629	73	.1367	73.5	.1337
24	.620	24.5	.611	74	.1308	74.5	.1278
25	.602	25.5	.594	75	.1249	75.5	.1221
26	.585	26.5	.577	76	.1192	76.5	.1163
27	.569	27.5	.561	77	.1135	77.5	.1107
28	.553	28.5	.545	78	.1079	78.5	.1051
29	.538	29.5	.530	79	.1024	79.5	.0996
30	.523	30.5	.516	80	.0969	80.5	.0942
31	.509	31.5	.502	81	.0915	81.5	.0888
32	.495	32.5	.488	82	.0862	82.5	.0835
33	.482	33.5	.475	83	.0809	83.5	.0783
34	.469	34.5	.462	84	.0757	84.5	.0731
35	.456	35.5	.450	85	.0706	85.5	.0680
36	.444	36.5	.438	86	.0655	86.5	.0630
37	.432	37.5	.426	87	.0605	87.5	.0580
38	.420	38.5	.414	88	.0555	88.5	.0531
39	.409	39.5	.403	89	.0505	89.5	.0482
40	.398	40.5	.392	90	.0458	90.5	.0434
41	.387	41.5	.382	91	.0410	91.5	.0386
42	.377	42.5	.372	92	.0362	92.5	.0339
43	.367	43.5	.362	93	.0315	93.5	.0292
44	.357	44.5	.352	94	.0269	94.5	.0246
45	.347	45.5	.342	95	.0223	95.5	.0200
46	.337	46.5	.332	96	.0177	96.5	.0155
47	.328	47.5	.323	97	.0132	97.5	.0110
48	.319	48.5	.314	98	.0088	98.5	.0066
49	.310	49.5	.305	99	.0044	99.5	.0022
50	.301	50.5	.297	100	.0000		

BUFFERS COMMONLY USED IN THE LABORATORY

Table 20-14. SØRENSON'S PHOSPHATE BUFFER, 0.1 MOL/L

Na_2HPO_4 (0.1 mol/L; 14.2 g/L	KH_2PO_4 (0.1 mol/L; 13.6 g/L)	pH at 20 °C*
0.25 mL	9.75 mL	5.29
0.5 mL	9.5 mL	5.59
1.0 mL	9.0 mL	5.91
2.0 mL	8.0 mL	6.24
3.0 mL	7.0 mL	6.47
4.0 mL	6.0 mL	6.64
5.0 mL	5.0 mL	6.81
6.0 mL	4.0 mL	6.98
7.0 mL	3.0 mL	7.17
8.0 mL	2.0 mL	7.38
9.0 mL	1.0 mL	7.73
9.5 mL	0.5 mL	8.04

Mix solutions in the volumes indicated to achieve 10 mL of buffer at the desired pH.
*~0.3 pH unit lower at 37 °C.

Table 20-15. TRIS(HYDROXYMETHYL)AMINOMETHANE BUFFER

Tris(hydroxymethyl)aminomethane (~0.08 mol/L; 10.114 g/L)	HCl, 0.1 mol/L	pH at 23 °C*
50 mL	5.0 mL	9.10
50 mL	7.5 mL	8.92
50 mL	10.0 mL	8.74
50 mL	12.5 mL	8.62
50 mL	15.0 mL	8.50
50 mL	17.5 mL	8.40
50 mL	20.0 mL	8.32
50 mL	22.5 mL	8.23
50 mL	25.0 mL	8.14
50 mL	27.5 mL	8.05
50 mL	30.0 mL	7.96
50 mL	32.5 mL	7.87
50 mL	35.0 mL	7.77
50 mL	37.5 mL	7.66
50 mL	40.0 mL	7.54
50 mL	42.5 mL	7.36
50 mL	45.0 mL	7.20

Mix solutions in the volumes indicated and dilute to 100 mL with deionized water to achieve 100 mL of buffer at the desired pH.
*~0.15 pH unit lower at 37 °C.

Table 20-16. DELORY AND KING'S CARBONATE-BICARBONATE BUFFER

Na$_2$CO$_3$ (0.1 mol/L; 10.6 g/L)	NaCHO$_3$ (0.1 mol/L; 8.4 g/L)	pH at 25 °C*
1.1 mL	8.9 mL	9.1
1.4 mL	8.6 mL	9.2
2.2 mL	7.8 mL	9.4
2.7 mL	7.3 mL	9.5
3.9 mL	6.2 mL	9.7
5.1 mL	4.9 mL	9.9
6.4 mL	3.6 mL	10.1
7.4 mL	2.5 mL	10.3
7.9 mL	2.1 mL	10.4
8.3 mL	1.6 mL	10.5
8.8 mL	1.2 mL	10.6

Mix solutions in the volumes indicated to achieve 10 mL of buffer at the desired pH.
*~0.1 pH unit lower at 37 °C.

Table 20-17. ACETIC ACID–SODIUM ACETATE BUFFER

CH$_3$COOH (0.2 mol/L; Dilute 11.5 mL Reagent Grade Glacial Acetic Acid to 1 L)	CH$_3$COONa (0.2 mol/L; 16.4 g/L)	pH at 25 °C*
9.25 mL	7.5	3.6
88.0 mL	12.0	3.8
82.0 mL	18.0	4.0
73.5 mL	26.5	4.2
63.0 mL	37.0	4.4
52.0 mL	48.0	4.6
41.0 mL	59.0	4.8
30.0 mL	70.0	5.0
21.0 mL	79.0	5.2
14.0 mL	86.0	5.4
9.0 mL	91.0	5.6
6.0 mL	94.0	5.8

Mix solutions in the volumes indicated to achieve 100 mL of buffer at the desired pH.
*~0.05 pH unit lower at 37 °C.

NOMOGRAMS

Table 20-18. NOMOGRAM FOR THE DETERMINATION OF BODY
SURFACE AREA OF CHILDREN*

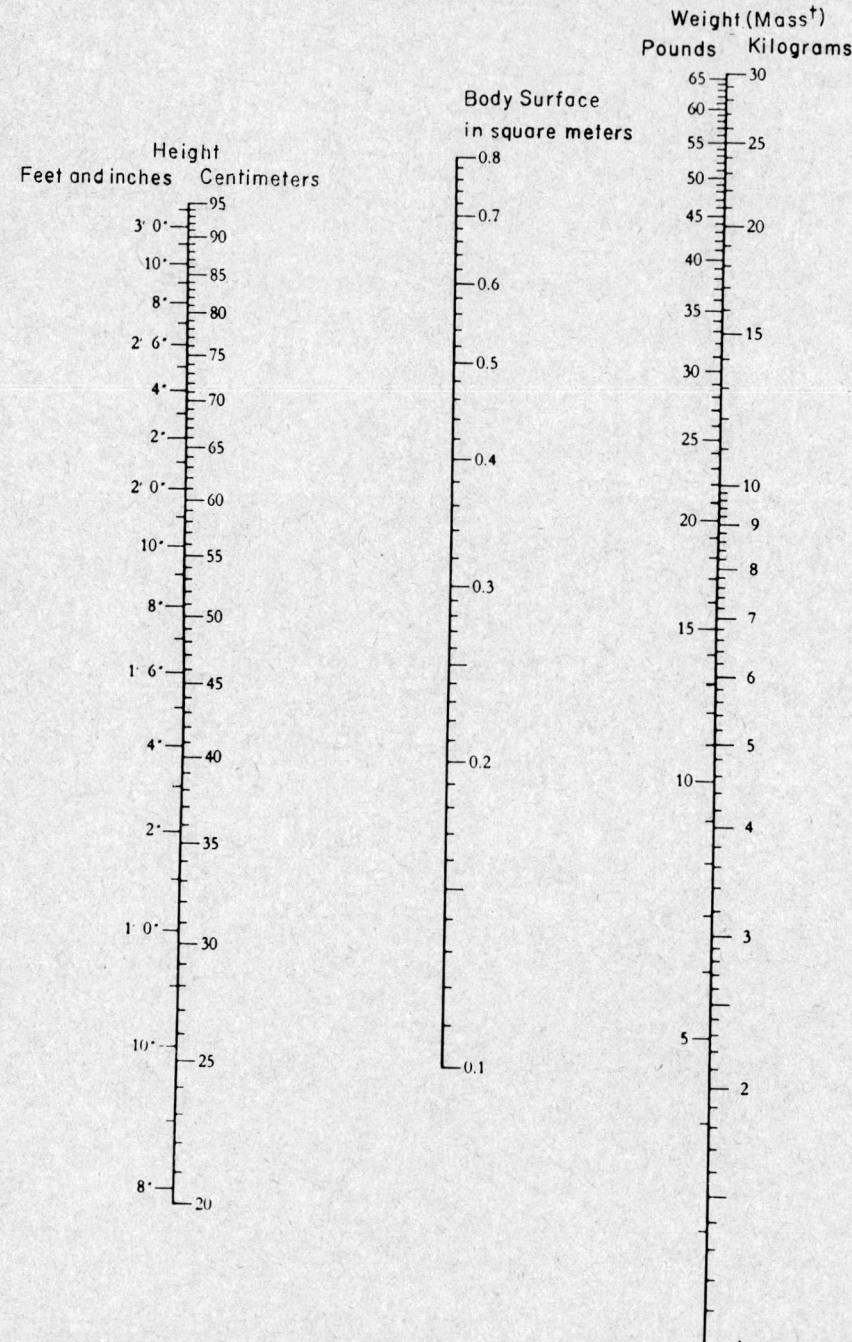

*From DuBois, E.F.: Basal Metabolism in Health and Disease. Philadelphia, Lea & Febiger, 1936.
†See footnote for Table 20-1, p. 1790.

Table 20-19. NOMOGRAM FOR THE DETERMINATION OF BODY SURFACE AREA OF CHILDREN AND ADULTS*

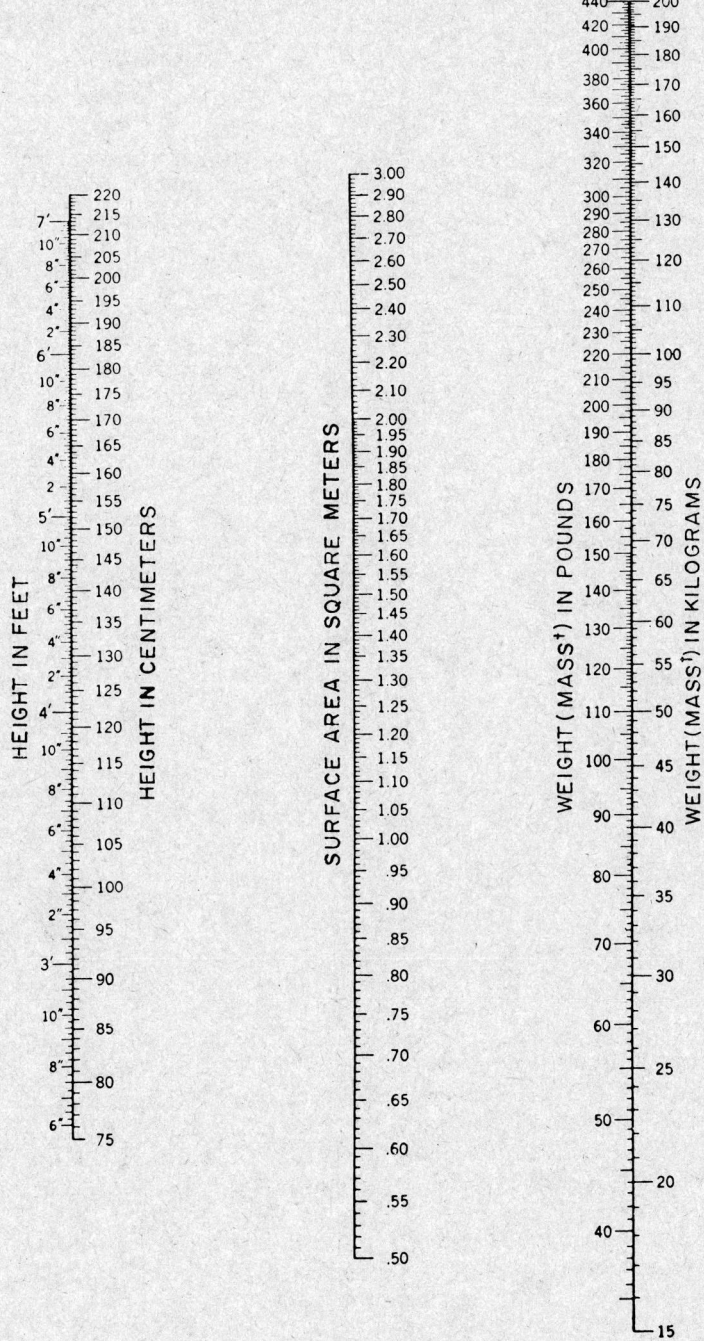

*From Boothby, W. M., and Sandiford, R. B.: Boston M. & S.J. *185*:337, 1921.
†See footnote for Table 20-1, p. 1790.

Table 20-20. NOMOGRAM FOR CALCULATING RELATIVE CENTRIFUGAL FORCE*

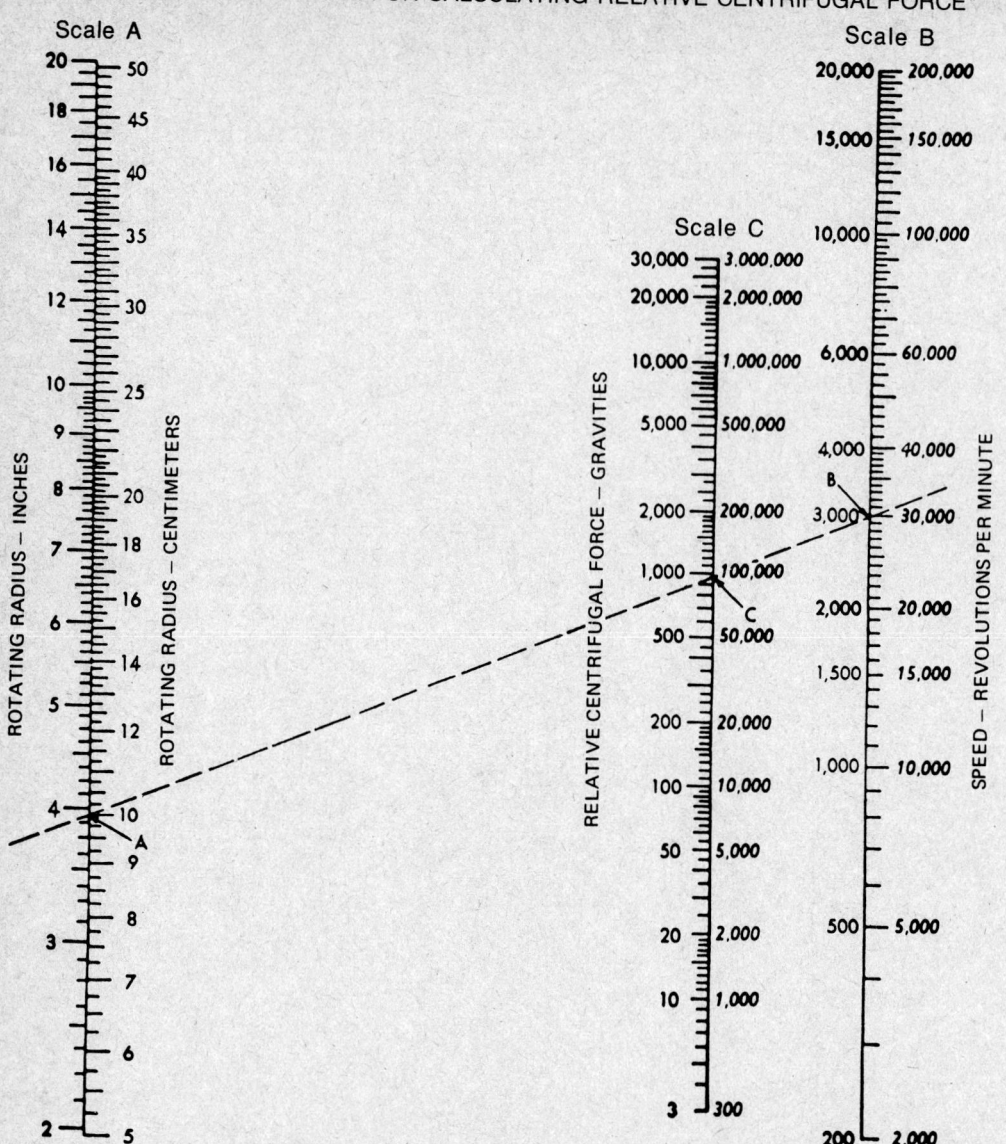

*Modified from the IEC (a division of Damon Corporation, Needham Heights, Mass.) Relative Centrifugal Force Nomograph.

Instructions:

To find, for example, the relative centrifugal force (RCF) at a radial distance of 10 cm from the center of rotation when the centrifuge is operated at a speed of 3000 rpm, place a straightedge on the chart, connecting the 10 cm point on Scale A (rotating radius) with the 3000 rpm point on Scale B (speed). Read the point at which the straightedge intersects Scale C (relative centrifugal force); for this example, that point is 1000 × g. Use the right side of Scales B and C for high-speed and ultracentrifuges.

If the desired RCF is known, the necessary speed for a particular rotating radius may be determined by connecting appropriate points on Scales A and C and reading speed from Scale B.

Equation for calculating RCF:

$$RCF = 1.118 \times 10^{-5} \times r \times n^2$$

where RCF is obtained in gravities (× *g*),
  r is rotating radius in centimeters,
and n is rotating speed in revolutions per min.

STATISTICAL TABLES

Table 20-21. UPPER TAIL AREAS UNDER THE STANDARD GAUSSIAN CURVE FOR POSITIVE VALUES OF z. BECAUSE OF THE SYMMETRY OF THE STANDARD GAUSSIAN CURVE, THE ENTRIES IN THIS TABLE ARE ALSO THE LOWER TAIL AREAS FOR THE NEGATIVE VALUE OF z.

z	0.00	0.01	0.02	0.03	0.04	0.05	0.06	0.07	0.08	0.09
0.0	0.5000	0.4960	0.4920	0.4880	0.4840	0.4801	0.4761	0.4721	0.4681	0.4641
0.1	0.4602	0.4562	0.4522	0.4483	0.4443	0.4404	0.4364	0.4325	0.4286	0.4247
0.2	0.4207	0.4168	0.4129	0.4090	0.4052	0.4013	0.3974	0.3936	0.3897	0.3859
0.3	0.3821	0.3783	0.3745	0.3707	0.3669	0.3632	0.3594	0.3557	0.3520	0.3483
0.4	0.3446	0.3409	0.3372	0.3336	0.3300	0.3264	0.3228	0.3192	0.3156	0.3121
0.5	0.3085	0.3050	0.3015	0.2981	0.2946	0.2912	0.2877	0.2843	0.2810	0.2776
0.6	0.2743	0.2709	0.2676	0.2643	0.2611	0.2578	0.2546	0.2514	0.2483	0.2451
0.7	0.2420	0.2389	0.2358	0.2327	0.2296	0.2266	0.2236	0.2206	0.2177	0.2148
0.8	0.2119	0.2090	0.2061	0.2033	0.2005	0.1977	0.1949	0.1922	0.1894	0.1867
0.9	0.1841	0.1814	0.1788	0.1762	0.1736	0.1711	0.1685	0.1660	0.1635	0.1611
1.0	0.1587	0.1562	0.1539	0.1515	0.1492	0.1469	0.1446	0.1423	0.1401	0.1379
1.1	0.1357	0.1335	0.1314	0.1292	0.1271	0.1251	0.1230	0.1210	0.1190	0.1170
1.2	0.1151	0.1131	0.1112	0.1093	0.1075	0.1056	0.1038	0.1020	0.1003	0.0985
1.3	0.0968	0.0951	0.0934	0.0918	0.0901	0.0885	0.0869	0.0853	0.0838	0.0823
1.4	0.0808	0.0793	0.0778	0.0764	0.0749	0.0735	0.0721	0.0708	0.0694	0.0681
1.5	0.0668	0.0655	0.0643	0.0630	0.0618	0.0606	0.0594	0.0582	0.0571	0.0559
1.6	0.0548	0.0537	0.0526	0.0516	0.0505	0.0495	0.0485	0.0475	0.0465	0.0455
1.7	0.0446	0.0436	0.0427	0.0418	0.0409	0.0401	0.0392	0.0384	0.0375	0.0367
1.8	0.0359	0.0351	0.0344	0.0336	0.0329	0.0322	0.0314	0.0307	0.0301	0.0294
1.9	0.0287	0.0281	0.0274	0.0268	0.0262	0.0256	0.0250	0.0244	0.0239	0.0233
2.0	0.0228	0.0222	0.0217	0.0212	0.0207	0.0202	0.0197	0.0192	0.0188	0.0183
2.1	0.0179	0.0174	0.0170	0.0166	0.0162	0.0158	0.0154	0.0150	0.0146	0.0143
2.2	0.0139	0.0136	0.0132	0.0129	0.0125	0.0122	0.0119	0.0116	0.0113	0.0110
2.3	0.0107	0.0104	0.0102	0.0099	0.0096	0.0094	0.0091	0.0089	0.0087	0.0084
2.4	0.0082	0.0080	0.0078	0.0075	0.0073	0.0071	0.0069	0.0068	0.0066	0.0064
2.5	0.0062	0.0060	0.0059	0.0057	0.0055	0.0054	0.0052	0.0051	0.0049	0.0048
2.6	0.0047	0.0045	0.0044	0.0043	0.0041	0.0040	0.0039	0.0038	0.0037	0.0036
2.7	0.0035	0.0034	0.0033	0.0032	0.0031	0.0030	0.0029	0.0028	0.0027	0.0026
2.8	0.0026	0.0025	0.0024	0.0023	0.0023	0.0022	0.0021	0.0021	0.0020	0.0019
2.9	0.0019	0.0018	0.0018	0.0017	0.0016	0.0016	0.0015	0.0015	0.0014	0.0014
3.0	0.0013	0.0013	0.0013	0.0012	0.0012	0.0011	0.0011	0.0011	0.0010	0.0010
3.1	0.0010	0.0009	0.0009	0.0009	0.0008	0.0008	0.0008	0.0008	0.0007	0.0007
3.2	0.0007	0.0007	0.0006	0.0006	0.0006	0.0006	0.0006	0.0005	0.0005	0.0005
3.3	0.0005	0.0005	0.0005	0.0004	0.0004	0.0004	0.0004	0.0004	0.0004	0.0003
3.4	0.0003	0.0003	0.0003	0.0003	0.0003	0.0003	0.0003	0.0003	0.0003	0.0002
3.5	0.0002	0.0002	0.0002	0.0002	0.0002	0.0002	0.0002	0.0002	0.0002	0.0002
3.6	0.0002	0.0002	0.0001	0.0001	0.0001	0.0001	0.0001	0.0001	0.0001	0.0001
3.7	0.0001	0.0001	0.0001	0.0001	0.0001	0.0001	0.0001	0.0001	0.0001	0.0001
3.8	0.0001	0.0001	0.0001	0.0001	0.0001	0.0001	0.0001	0.0001	0.0001	0.0001
3.9	0.0000	0.0000	0.0000	0.0000	0.0000	0.0000	0.0000	0.0000	0.0000	0.0000

Table 20-22. PROBABILITY POINTS OF THE STUDENT'S t DISTRIBUTION WITH ν DEGREES OF FREEDOM; α IS THE UPPER TAIL PROBABILITY

ν \ α	0.40	0.25	0.10	0.05	0.025	0.01	0.005	0.0005
1	0.325	1.000	3.078	6.314	12.706	31.821	63.657	636.619
2	0.289	0.816	1.886	2.920	4.303	6.965	9.925	31.598
3	0.277	0.765	1.638	2.353	3.182	4.541	5.841	12.941
4	0.271	0.741	1.533	2.132	2.776	3.747	4.604	8.610
5	0.267	0.727	1.476	2.015	2.571	3.365	4.032	6.859
6	0.265	0.718	1.440	1.943	2.447	3.143	3.707	5.959
7	0.263	0.711	1.415	1.895	2.365	2.998	3.499	5.405
8	0.262	0.706	1.397	1.860	2.306	2.896	3.355	5.041
9	0.261	0.703	1.383	1.833	2.262	2.821	3.250	4.781
10	0.260	0.700	1.372	1.812	2.228	2.764	3.169	4.587
11	0.260	0.697	1.363	1.796	2.201	2.718	3.106	4.437
12	0.259	0.695	1.356	1.782	2.179	2.681	3.055	4.318
13	0.259	0.694	1.350	1.771	2.160	2.650	3.012	4.221
14	0.258	0.692	1.345	1.761	2.145	2.624	2.977	4.140
15	0.258	0.691	1.341	1.753	2.131	2.602	2.947	4.073
16	0.258	0.690	1.337	1.746	2.120	2.583	2.921	4.015
17	0.257	0.689	1.333	1.740	2.110	2.567	2.898	3.965
18	0.257	0.688	1.330	1.734	2.101	2.552	2.878	3.922
19	0.257	0.688	1.328	1.729	2.093	2.539	2.861	3.883
20	0.257	0.687	1.325	1.725	2.086	2.528	2.845	3.850
21	0.257	0.686	1.323	1.721	2.080	2.518	2.831	3.819
22	0.256	0.686	1.321	1.717	2.074	2.508	2.819	3.792
23	0.256	0.685	1.319	1.714	2.069	2.500	2.807	3.767
24	0.256	0.685	1.318	1.711	2.064	2.492	2.797	3.745
25	0.256	0.684	1.316	1.708	2.060	2.485	2.787	3.725
26	0.256	0.684	1.315	1.706	2.056	2.479	2.779	3.707
27	0.256	0.684	1.314	1.703	2.052	2.473	2.771	3.690
28	0.256	0.683	1.313	1.701	2.048	2.467	2.763	3.674
29	0.256	0.683	1.311	1.699	2.045	2.462	2.756	3.659
30	0.256	0.683	1.310	1.697	2.042	2.457	2.750	3.646
40	0.255	0.681	1.303	1.684	2.021	2.423	2.704	3.551
60	0.254	0.679	1.296	1.671	2.000	2.390	2.660	3.460
120	0.254	0.677	1.289	1.658	1.980	2.358	2.617	3.373
∞	0.253	0.674	1.282	1.645	1.960	2.326	2.576	3.291

Table 20-23. PROBABILITY POINTS OF THE CHI-SQUARE DISTRIBUTION WITH ν DEGREES OF FREEDOM;
α IS THE UPPER TAIL PROBABILITY

ν \\ α	0.995	0.990	0.975	0.950	0.900	0.500	0.100	0.050	0.025	0.010	0.005
1	0.0000393	0.000157	0.000982	0.00393	0.0158	0.455	2.71	3.84	5.02	6.63	7.88
2	0.0100	0.0201	0.0506	0.103	0.211	1.39	4.61	5.99	7.38	9.21	10.6
3	0.0717	0.115	0.216	0.352	0.584	2.37	6.25	7.81	9.35	11.3	12.8
4	0.207	0.297	0.484	0.711	1.06	3.36	7.78	9.49	11.1	13.3	14.9
5	0.412	0.554	0.831	1.15	1.61	4.35	9.24	11.1	12.8	15.1	16.7
6	0.676	0.872	1.24	1.64	2.20	5.35	10.6	12.6	14.4	16.8	18.5
7	0.989	1.24	1.69	2.17	2.83	6.35	12.0	14.1	16.0	18.5	20.3
8	1.34	1.65	2.18	2.73	3.49	7.34	13.4	15.5	17.5	20.1	22.0
9	1.73	2.09	2.70	3.33	4.17	8.34	14.7	16.9	19.0	21.7	23.6
10	2.16	2.56	3.25	3.94	4.87	9.34	16.0	18.3	20.5	23.2	25.2
11	2.60	3.05	3.82	4.57	5.58	10.3	17.3	19.7	21.9	24.7	26.8
12	3.07	3.57	4.40	5.23	6.30	11.3	18.5	21.0	23.3	26.2	28.3
13	3.57	4.11	5.01	5.89	7.04	12.3	19.8	22.4	24.7	27.7	29.8
14	4.07	4.66	5.63	6.57	7.79	13.3	21.1	23.7	26.1	29.1	31.3
15	4.60	5.23	6.26	7.26	8.55	14.3	22.3	25.0	27.5	30.6	32.8
16	5.14	5.81	6.91	7.96	9.31	15.3	23.5	26.3	28.8	32.0	34.3
17	5.70	6.41	7.56	8.67	10.1	16.3	24.8	27.6	30.2	33.4	35.7
18	6.26	7.01	8.23	9.39	10.9	17.3	26.0	28.9	31.5	34.8	37.2
19	6.84	7.63	8.91	10.1	11.7	18.3	27.2	30.1	32.9	36.2	38.6
20	7.43	8.26	9.59	10.9	12.4	19.3	28.4	31.4	34.2	37.6	40.0
21	8.03	8.90	10.3	11.6	13.2	20.3	29.6	32.7	35.5	38.9	41.4
22	8.64	9.54	11.0	12.3	14.0	21.3	30.8	33.9	36.8	40.3	42.8
23	9.26	10.2	11.7	13.1	14.8	22.3	32.0	35.2	38.1	41.6	44.2
24	9.89	10.9	12.4	13.8	15.7	23.3	33.2	36.4	39.4	43.0	45.6
25	10.5	11.5	13.1	14.6	16.5	24.3	34.4	37.7	40.6	44.3	46.9
26	11.2	12.2	13.8	15.4	17.3	25.3	35.6	38.9	41.9	45.6	48.3
27	11.8	12.9	14.6	16.2	18.1	26.3	36.7	40.1	43.2	47.0	49.6
28	12.5	13.6	15.3	16.9	18.9	27.3	37.9	41.3	44.5	48.3	51.0
29	13.1	14.3	16.0	17.7	19.8	28.3	39.1	42.6	45.7	49.6	52.3
30	13.8	15.0	16.8	18.5	20.6	29.3	40.3	43.8	47.0	50.9	53.7
40	20.7	22.2	24.4	26.5	29.1	39.3	51.8	55.8	59.3	63.7	66.8
50	28.0	29.7	32.4	34.8	37.7	49.3	63.2	67.5	71.4	76.2	79.5
60	35.5	37.5	40.5	43.2	46.5	59.3	74.4	79.1	83.3	88.4	92.0
70	43.3	45.4	48.8	51.7	55.3	69.3	85.5	90.5	95.0	100.4	104.2
80	51.2	53.5	57.2	60.4	64.3	79.3	96.6	101.9	106.6	112.3	116.3
90	59.2	61.8	65.6	69.1	73.3	89.3	107.6	113.1	118.1	124.1	128.3
100	67.3	70.1	74.2	78.0	82.4	99.3	118.5	124.3	129.6	135.8	140.2

Table 20-24. PROBABILITY POINTS OF THE F DISTRIBUTION WITH ν_1 AND ν_2 DEGREES OF FREEDOM

$\alpha = 0.025$ (upper tail)

ν_2 \ ν_1	1	2	3	4	5	6	7	8	9	10	12	15	20	24	30	40	60	120	∞
1	647.8	799.5	864.2	899.6	921.8	937.1	948.2	956.7	963.3	968.6	976.7	984.9	993.1	997.2	1001	1006	1010	1014	1018
2	38.51	39.00	39.17	39.25	39.30	39.33	39.36	39.37	39.39	39.40	39.41	39.43	39.45	39.46	39.46	39.47	39.48	39.49	39.50
3	17.44	16.04	15.44	15.10	14.88	14.73	14.62	14.54	14.47	14.42	14.34	14.25	14.17	14.12	14.08	14.04	13.99	13.95	13.90
4	12.22	10.65	9.98	9.60	9.36	9.20	9.07	8.98	8.90	8.84	8.75	8.66	8.56	8.51	8.46	8.41	8.36	8.31	8.26
5	10.01	8.43	7.76	7.39	7.15	6.98	6.85	6.76	6.68	6.62	6.52	6.43	6.33	6.28	6.23	6.18	6.12	6.07	6.02
6	8.81	7.26	6.60	6.23	5.99	5.82	5.70	5.60	5.52	5.46	5.37	5.27	5.17	5.12	5.07	5.01	4.96	4.90	4.85
7	8.07	6.54	5.89	5.52	5.29	5.12	4.99	4.90	4.82	4.76	4.67	4.57	4.47	4.42	4.36	4.31	4.25	4.20	4.14
8	7.57	6.06	5.42	5.05	4.82	4.65	4.53	4.43	4.36	4.30	4.20	4.10	4.00	3.95	3.89	3.84	3.78	3.73	3.67
9	7.21	5.71	5.08	4.72	4.48	4.32	4.20	4.10	4.03	3.96	3.87	3.77	3.67	3.61	3.56	3.51	3.45	3.39	3.33
10	6.94	5.46	4.83	4.47	4.24	4.07	3.95	3.85	3.78	3.72	3.62	3.52	3.42	3.37	3.31	3.26	3.20	3.14	3.08
11	6.72	5.26	4.63	4.28	4.04	3.88	3.76	3.66	3.59	3.53	3.43	3.33	3.23	3.17	3.12	3.06	3.00	2.94	2.88
12	6.55	5.10	4.47	4.12	3.89	3.73	3.61	3.51	3.44	3.37	3.28	3.18	3.07	3.02	2.96	2.91	2.85	2.79	2.72
13	6.41	4.97	4.35	4.00	3.77	3.60	3.48	3.39	3.31	3.25	3.15	3.05	2.95	2.89	2.84	2.78	2.72	2.66	2.60
14	6.30	4.86	4.24	3.89	3.66	3.50	3.38	3.29	3.21	3.15	3.05	2.95	2.84	2.79	2.73	2.67	2.61	2.55	2.49
15	6.20	4.77	4.15	3.80	3.58	3.41	3.29	3.20	3.12	3.06	2.96	2.86	2.76	2.70	2.64	2.59	2.52	2.46	2.40
16	6.12	4.69	4.08	3.73	3.50	3.34	3.22	3.12	3.05	2.99	2.89	2.79	2.68	2.63	2.57	2.51	2.45	2.38	2.32
17	6.04	4.62	4.01	3.66	3.44	3.28	3.16	3.06	2.98	2.92	2.82	2.72	2.62	2.56	2.50	2.44	2.38	2.32	2.25
18	5.98	4.56	3.95	3.61	3.38	3.22	3.10	3.01	2.93	2.87	2.77	2.67	2.56	2.50	2.44	2.38	2.32	2.26	2.19
19	5.92	4.51	3.90	3.56	3.33	3.17	3.05	2.96	2.88	2.82	2.72	2.62	2.51	2.45	2.39	2.33	2.27	2.20	2.13
20	5.87	4.46	3.86	3.51	3.29	3.13	3.01	2.91	2.84	2.77	2.68	2.57	2.46	2.41	2.35	2.29	2.22	2.16	2.09
21	5.83	4.42	3.82	3.48	3.25	3.09	2.97	2.87	2.80	2.73	2.64	2.53	2.42	2.37	2.31	2.25	2.18	2.11	2.04
22	5.79	4.38	3.78	3.44	3.22	3.05	2.93	2.84	2.76	2.70	2.60	2.50	2.39	2.33	2.27	2.21	2.14	2.08	2.00
23	5.75	4.35	3.75	3.41	3.18	3.02	2.90	2.81	2.73	2.67	2.57	2.47	2.36	2.30	2.24	2.18	2.11	2.04	1.97
24	5.72	4.32	3.72	3.38	3.15	2.99	2.87	2.78	2.70	2.64	2.54	2.44	2.33	2.27	2.21	2.15	2.08	2.01	1.94
25	5.69	4.29	3.69	3.35	3.13	2.97	2.85	2.75	2.68	2.61	2.51	2.41	2.30	2.24	2.18	2.12	2.05	1.98	1.91
26	5.66	4.27	3.67	3.33	3.10	2.94	2.82	2.73	2.65	2.59	2.49	2.39	2.28	2.22	2.16	2.09	2.03	1.95	1.88
27	5.63	4.24	3.65	3.31	3.08	2.92	2.80	2.71	2.63	2.57	2.47	2.36	2.25	2.19	2.13	2.07	2.00	1.93	1.85
28	5.61	4.22	3.63	3.29	3.06	2.90	2.78	2.69	2.61	2.55	2.45	2.34	2.23	2.17	2.11	2.05	1.98	1.91	1.83
29	5.59	4.20	3.61	3.27	3.04	2.88	2.76	2.67	2.59	2.53	2.43	2.32	2.21	2.15	2.09	2.03	1.96	1.89	1.81
30	5.57	4.18	3.59	3.25	3.03	2.87	2.75	2.65	2.57	2.51	2.41	2.31	2.20	2.14	2.07	2.01	1.94	1.87	1.79
40	5.42	4.05	3.46	3.13	2.90	2.74	2.62	2.53	2.45	2.39	2.29	2.18	2.07	2.01	1.94	1.88	1.80	1.72	1.64
60	5.29	3.93	3.34	3.01	2.79	2.63	2.51	2.41	2.33	2.27	2.17	2.06	1.94	1.88	1.82	1.74	1.67	1.58	1.48
120	5.15	3.80	3.23	2.89	2.67	2.52	2.39	2.30	2.22	2.16	2.05	1.94	1.82	1.76	1.69	1.61	1.53	1.43	1.31
∞	5.02	3.69	3.12	2.79	2.57	2.41	2.29	2.19	2.11	2.05	1.94	1.83	1.71	1.64	1.57	1.48	1.39	1.27	1.00

Table 20-24. PROBABILITY POINTS OF THE F DISTRIBUTION WITH ν_1 AND ν_2 DEGREES OF FREEDOM *(Continued)*

$\alpha = 0.05$ (upper tail)

ν_2 \ ν_1	1	2	3	4	5	6	7	8	9	10	12	15	20	24	30	40	60	120	∞
1	161.4	199.5	215.7	224.6	230.2	234.0	236.8	238.9	240.5	241.9	243.9	245.9	248.0	249.1	250.1	251.1	252.2	253.3	254.3
2	18.51	19.00	19.16	19.25	19.30	19.33	19.35	19.37	19.38	19.40	19.41	19.43	19.45	19.45	19.46	19.47	19.48	19.49	19.50
3	10.13	9.55	9.28	9.12	9.01	8.94	8.89	8.85	8.81	8.79	8.74	8.70	8.66	8.64	8.62	8.59	8.57	8.55	8.53
4	7.71	6.94	6.59	6.39	6.26	6.16	6.08	6.04	6.00	5.96	5.91	5.86	5.80	5.77	5.75	5.72	5.69	5.66	5.63
5	6.61	5.79	5.41	5.19	5.05	4.95	4.88	4.82	4.77	4.74	4.68	4.62	4.56	4.53	4.50	4.46	4.43	4.40	4.36
6	5.99	5.14	4.76	4.53	4.39	4.28	4.21	4.15	4.10	4.06	4.00	3.94	3.87	3.84	3.81	3.77	3.74	3.70	3.67
7	5.59	4.74	4.35	4.12	3.97	3.87	3.79	3.73	3.68	3.64	3.57	3.51	3.44	3.41	3.38	3.34	3.30	3.27	3.23
8	5.32	4.46	4.07	3.84	3.69	3.58	3.50	3.44	3.39	3.35	3.28	3.22	3.15	3.12	3.08	3.04	3.01	2.97	2.93
9	5.12	4.26	3.86	3.63	3.48	3.37	3.29	3.23	3.18	3.14	3.07	3.01	2.94	2.90	2.86	2.83	2.79	2.75	2.71
10	4.96	4.10	3.71	3.48	3.33	3.22	3.14	3.07	3.02	2.98	2.91	2.85	2.77	2.74	2.70	2.66	2.62	2.58	2.54
11	4.84	3.98	3.59	3.36	3.20	3.09	3.01	2.95	2.90	2.85	2.79	2.72	2.65	2.61	2.57	2.53	2.49	2.45	2.40
12	4.75	3.89	3.49	3.26	3.11	3.00	2.91	2.85	2.80	2.75	2.69	2.62	2.54	2.51	2.47	2.43	2.38	2.34	2.30
13	4.67	3.81	3.41	3.18	3.03	2.92	2.83	2.77	2.71	2.67	2.60	2.53	2.46	2.42	2.38	2.34	2.30	2.25	2.21
14	4.60	3.74	3.34	3.11	2.96	2.85	2.76	2.70	2.65	2.60	2.53	2.46	2.39	2.35	2.31	2.27	2.22	2.18	2.13
15	4.54	3.68	3.29	3.06	2.90	2.79	2.71	2.64	2.59	2.54	2.48	2.40	2.33	2.29	2.25	2.20	2.16	2.11	2.07
16	4.49	3.63	3.24	3.01	2.85	2.74	2.66	2.59	2.54	2.49	2.42	2.35	2.28	2.24	2.19	2.15	2.11	2.06	2.01
17	4.45	3.59	3.20	2.96	2.81	2.70	2.61	2.55	2.49	2.45	2.38	2.31	2.23	2.19	2.15	2.10	2.06	2.01	1.96
18	4.41	3.55	3.16	2.93	2.77	2.66	2.58	2.51	2.46	2.41	2.34	2.27	2.19	2.15	2.11	2.06	2.02	1.97	1.92
19	4.38	3.52	3.13	2.90	2.74	2.63	2.54	2.48	2.42	2.38	2.31	2.23	2.16	2.11	2.07	2.03	1.98	1.93	1.88
20	4.35	3.49	3.10	2.87	2.71	2.60	2.51	2.45	2.39	2.35	2.28	2.20	2.12	2.08	2.04	1.99	1.95	1.90	1.84
21	4.32	3.47	3.07	2.84	2.68	2.57	2.49	2.42	2.37	2.32	2.25	2.18	2.10	2.05	2.01	1.95	1.92	1.87	1.81
22	4.30	3.44	3.05	2.82	2.66	2.55	2.46	2.40	2.34	2.30	2.23	2.15	2.07	2.03	1.98	1.94	1.89	1.84	1.78
23	4.28	3.42	3.03	2.80	2.64	2.53	2.44	2.37	2.32	2.27	2.20	2.13	2.05	2.01	1.96	1.91	1.86	1.81	1.76
24	4.26	3.40	3.01	2.78	2.62	2.51	2.42	2.36	2.30	2.25	2.18	2.11	2.03	1.98	1.94	1.89	1.84	1.79	1.73
25	4.24	3.39	2.99	2.76	2.60	2.49	2.40	2.34	2.28	2.24	2.16	2.09	2.01	1.96	1.92	1.87	1.82	1.77	1.71
26	4.23	3.37	2.98	2.74	2.59	2.47	2.39	2.32	2.27	2.22	2.15	2.07	1.99	1.95	1.90	1.85	1.80	1.75	1.69
27	4.21	3.35	2.96	2.73	2.57	2.46	2.37	2.31	2.25	2.20	2.13	2.06	1.97	1.93	1.88	1.84	1.79	1.73	1.67
28	4.20	3.34	2.95	2.71	2.56	2.45	2.36	2.29	2.24	2.19	2.12	2.04	1.96	1.91	1.87	1.82	1.77	1.71	1.65
29	4.18	3.33	2.93	2.70	2.55	2.43	2.35	2.28	2.22	2.18	2.10	2.03	1.94	1.90	1.85	1.81	1.75	1.70	1.64
30	4.17	3.32	2.92	2.69	2.53	2.42	2.33	2.27	2.21	2.16	2.09	2.01	1.93	1.89	1.84	1.79	1.74	1.68	1.62
40	4.08	3.23	2.84	2.61	2.45	2.34	2.25	2.18	2.12	2.08	2.00	1.92	1.84	1.79	1.74	1.69	1.64	1.58	1.51
60	4.00	3.15	2.76	2.53	2.37	2.25	2.17	2.10	2.04	1.99	1.92	1.84	1.75	1.70	1.65	1.59	1.53	1.47	1.39
120	3.92	3.07	2.68	2.45	2.29	2.17	2.09	2.02	1.96	1.91	1.83	1.75	1.66	1.61	1.55	1.50	1.43	1.35	1.25
∞	3.84	3.00	2.60	2.37	2.21	2.10	2.01	1.94	1.88	1.83	1.75	1.67	1.57	1.52	1.46	1.39	1.32	1.22	1.00

REFERENCE RANGES

Reference ranges are valuable guidelines for the clinician, but they should not be regarded as absolute indicators of health and disease. Reference ranges should be used with caution since values for "healthy" individuals often overlap significantly with values for persons afflicted with disease. In addition, laboratory values may vary significantly due to methodological differences and mode of standardization. This is especially true for immunological tests, which utilize antibodies that may have different characteristics. As a result, laboratory values in individual institutions may differ from those listed in this appendix. Reference ranges given in other chapters of this book reflect the authors' experience and are specific for the methods discussed. Thus, these values may also differ from those listed here.

The values in this chapter are for adults in the fasting state unless otherwise stated. Values for other age groups, when included, are clearly identified. Most of this material was extracted from *Clinical Guide to Laboratory Tests,* N. W. Tietz, Ed.; Philadelphia, W.B. Saunders Company, 1983. Other sources are listed at the end of the chapter.

All laboratory values are given in conventional and international units. In general, the international units given conform to the SI system (Système International d'Unités). However, in some cases the recommendations of the International Union of Pure and Applied Chemistry (IUPAC) and the Commission on World Standards of the World Association of Societies of Pathology (COWS of WASP) are used, since it is felt that these have found wider acceptance in clinical laboratories and offer advantages over the units recommended in the SI system. If the exact molecular weight of a compound has not been established, e.g., for some proteins, values in the conventional unit are converted to values expressed in mass/L.

Throughout this appendix, we have used the prefixes for units as approved by the CGPM (Conférence Générale des Poids et Mésures), 1964, and the International Union of Pure and Applied Chemistry (IUPAC) and International Federation of Clinical Chemistry (IFCC), Clinica Chimica Acta, 96:157F–183F, 1979. (See Chapter 1A.)

For convenience and to preserve space, we have used standard abbreviations commonly used in laboratory medicine. Less common abbreviations and some nonstandard abbreviations are given below.

Abbreviations

ACD	acid-citrate-dextrose	Ox	oxalate
Amf	amniotic fluid	P	plasma
Ascf	ascitic fluid	Pericf	pericardial fluid
Br	breath	Placent	placental
Cit	citrate	Plf	pleural fluid
CSF	cerebrospinal fluid	Plt	platelets
EDTA	ethylenediaminetetraacetate, edetic acid	post-stim	post-stimulation
		pp	postprandial
F	feces	RBC	red blood cells
F⁻	fluoride	S	serum
F⁻/Ox	fluoride and oxalate	Sal	saliva
Gast Cont	gastric contents	Semf	seminal fluid
Gastf	gastric fluid	Serf	serous fluid
Gast Res	gastric residue	Swt	sweat
H	hair	Synf	synovial fluid
Hep	heparin	T	tissue
Occup. exp.	occupational exposure	U	urine
		WB	whole blood

Table 20-25. CLINICAL CHEMISTRY AND TOXICOLOGY

Test	Specimen	Reference Range		Conversion Factor	Reference Range (International Units)
Acetaldehyde	WB (F⁻/Ox)		*mg/L*		*μmol/L*
			<0.2	× 22.7	<4.5
		Occup. exp.:	<0.5		<11.4
		Toxic:	1–2		22.7–45.4
Acetoacetate					
Semiquant.	S or P	Negative (<3 mg/dL)		× 0.098	Negative (<0.3 mmol/L)
	(F⁻/Ox) U	Negative (<50 mg/dL)			Negative (<4.9 mmol/L)
Acetone					
Semiquant.	S or P	Negative (<3 mg/dL)		× 0.172	Negative (<0.52 mmol/L)
	(F⁻ or Ox)				
			mg/dL		*mmol/L*
Quant.			0.3–2.0	× 0.172	0.05–0.34
		Ketoacidosis:	10–70		1.72–12.04
		Occup. exp.:	<10		<1.72
		Toxic:	>20		>3.44
Semiquant.	U	Negative			Negative
Quant.			0.3	× 0.172	0.05
		Occup. exp.:	>27		>4.64
			mg/L		*μmol/L*
	Br		0.003	× 17.2	0.52
		Occup. exp.:	<0.006		<1.03
Acetonitrile	WB (F⁻/Ox)		*mg/L*		*μmol/L*
		CN⁻:	<0.04	× 38.4	<1.5
		SCN⁻:	<12	× 17.2	<206
		Occup. exp., CN⁻:	<0.1	× 38.4	<3.8
		Toxic,			
		CN⁻:	3–11	× 38.4	115–422
		SCN⁻:	160–230	× 17.2	2752–3956
	U	SCN⁻:	<17		<292
		Occup. exp., SCN⁻:	<20		<344
Acetylcholinesterase in Erythrocytes	WB (ACD, EDTA, or Hep)	Adult:			
		36.9 ± 3.83 U/g Hb (SD)		× 0.0645	2.38 ± 0.23 MU/mol Hb
		1070 ± 111 U/10¹² RBC		× 10⁻³	1.07 ± 0.11 nU/RBC
		12.5 ± 1.30 U/mL RBC		× 1	12.5 ± 1.30 kU/L RBC
		Lower in newborns			
α₁-Acid-Glycoprotein (Orosomucoid)	S	Adult: 55–140 mg/dL		× 10	550–1400 mg/L
	U	0.29–0.68 mg/d			0.29–0.68 mg/d
	CSF	0.41 ± 0.13 mg/dL (SE)		× 10	4.1 ± 1.3 mg/L (SE)
Acrylonitrile	WB (F⁻/Ox)		*mg/L*		*μmol/L*
		CN⁻:	0.04	× 38.4	1.5
		SCN⁻:	0.12	× 17.2	2.1
		Toxic (SCN⁻),			
		Nonsmokers:	>20		>344
		Smokers:	>30		>516
		Toxic (CN⁻):	>0.1	× 38.4	>3.8
	U	Toxic (SCN⁻):	>20	× 17.2	>344
Adrenocorticotropic Hormone (ACTH)	P (EDTA)		*pg/mL*		*ng/L*
		Cord:	143 ± 7.0 (SD)	× 1	143 ± 7.0 (SD)
		1–7 d:	120 ± 8.3		120 ± 8.3
		Adult,			
		0800 h:	25–100	× 1	25–100
		1800 h:	<50		<50
	Amf	10–18 wk:	209 (mean)		209 (mean)
		26–30 wk:	430		430
		35–36 wk:	162 and maintained thereafter		162 and maintained thereafter

Table 20-25. CLINICAL CHEMISTRY AND TOXICOLOGY (*Continued*)

Test	Specimen	Reference Range		Conversion Factor	Reference Range (International Units)
Adrenocorticotropic Hormone–Insulin Stimulation Test *Dose: 0.05–0.15 U insulin/kg, IV, after overnight fast*	S: fasting, 15, 30, and 60 min post-stim	Cortisol: > 20 µg/dL, and 10 µg/dL rise ½ h after max. fall in blood glucose		× 0.028	> 0.56 µmol/L and 0.28 µmol/L rise ½ h after max. fall in blood glucose
		Glucose: 50% below base or signs of hypoglycemia		× 0.01	0.50 (concentration fraction) below base or signs of hypoglycemia
Adrenocorticotropic Hormone Stimulation Test (Prolonged Infusion)	U, 24 h	17-KGS: 17-KS: 17-OHCS:	2–4-fold rise 2-fold rise 2–5-fold rise		Same
Dose: 500 µg Cortrosyn/d × 3	S	Cortisol:	25–50 µg/dL	× 0.028	Cortisol: 0.7–1.4 µmol/L
Adrenocorticotropic Hormone Stimulation Test (Rapid Test) *Dose: 250 µg Cortrosyn, IM*	S: baseline, 30 and 60 min post-stim	Cortisol, Baseline: After Cortrosyn:	> 5.0 µg/dL 2× baseline	× 0.028	Baseline: > 0.14 µmol/L Same
Alanine	S		*mg/dL*		*µmol/L*
		Premature, 1 d: Newborn, 1 d: 1–3 mo: 2–6 mo: 9 mo–2 y: 3–10 y: 6–18 y: Adult:	3.34 ± 0.45 (SD) 2.10–3.65 2.45 ± 0.63 (SD) 1.58–3.68 0.88–2.79 1.22–2.71 1.72–4.85 1.87–5.88	× 112.2	375 ± 50 (SD) 236 – 410 275 ± 71 (SD) 177 – 413 99 – 313 137 – 305 193 – 545 210 – 661
	U, 24 h	10 d–7 wk: 3–12 y: Adult:	*mg/d* 4–10 9–39 8–48	× 11.2	*µmol/d* 45 – 112 101 – 437 90 – 538
Alanine Aminotransferase (ALT, GPT) *Henry, optimized, 30°C*	S	Newborn/Infant: Adult: > 60 y, M: F:	*U/L* 5–28 8–20 7–24 7–16		Same
IFCC, with P-5′-P, 30°C			< 30		
Albumin *Nephelometric, colorimetric*	S	Adult: > 60 y: Avg. ~0.3 g/dL higher in upright individuals	3.5–5.0 g/dL 3.4–4.8 g/dL	× 10	35–50 g/L 34–48 g/L Avg. ~3 g/L higher in upright individuals
Nephelometric, rate	CSF U	10–30 mg/dL At rest: Ambulatory:	< 80 mg/d 50–80 mg/d	× 10	100–300 mg/L Same
Alcohol Dehydrogenase (ADH) *340 nm, 30°C*	S	≤ 2.8 U/L			≤ 2.8 U/L
Aldolase	S	1.0–7.5 U/L (30 °C) 0.3–3.0 U/L at bed rest 1.5–12.0 U/L (37 °C)			Same

Table 20-25. CLINICAL CHEMISTRY AND TOXICOLOGY (*Continued*)

Test	Specimen	Reference Range	Conversion Factor	Reference Range (International Units)
Aldosterone	S or P (Hep, EDTA)	*ng/dL*		*nmol/L*
		Newborn: 5–60	× 0.0277	0.14–1.66
		1 wk–1 y: 1–160		0.03–4.43
		1–3 y: 5–60		0.14–1.66
		3–5 y: <5–80		<0.138–2.22
		5–7 y: <5–50		<0.138–1.39
		7–11 y: 5–70		0.138–1.94
		11–15 y: <5–50		<0.138–1.39
		Adult, *average sodium diet*		
		Supine: 3–10		0.083–0.28
		Upright, F: 5–30		0.138–0.83
		M: 6–22		0.166–0.609
		2–3× higher during pregnancy		
		Adrenal vein: 200–800		5.54–22.16
		Low sodium diet:		
		Increases 2–5-fold		
		Florinef suppression: <4 ng/dL		<0.1 1
		ACTH or angiotensin stimulation, 1 h: 2–5-fold increase		

	U, 24 h	*Total Urinary Na mEq/d*	*Plasma Renin Activity (ng AI*/h)/mL*	*Urinary Aldosterone µg/d*		*Urinary Aldosterone nmol/d*
		<20	5–24	35–80	× 2.77	97–222
		50	2–7	13–33		36–91.4
		100	1–5	5–24		13.8–66.5
		150	0.5–4	3–19		8.3–52.6
		200		1–16		2.7–44.3
		250		1–13		2.7–36

(assuming normal serum Na, K, and extracellular vol)
*Angiotensin I

Test	Specimen	Reference Range	Conversion Factor	Reference Range (International Units)
Aluminum (Al)		*µg/dL*		*µmol/L*
	S	0.4	× 0.371	0.15
	P	0.6–0.7		0.22–0.26
		Patients on Al medication: <3.0 µg/dL		<1.11
	U, random	None detected		None detected
α-Amino Acid Nitrogen (AAN)				
Ninhydrin	P (EDTA)	3.2–5.5 mg/dL	× 0.714	2.28–3.93 mmol/L
Naphthoquinone and FDNB	U, 24 h	50–200 mg/d	× 0.0714	3.57–14.28 mmol/d
δ-Aminolevulinate Dehydratase *Colorimetric, 38 °C*	RBC (Hep)	139–211 U/mL erythrocytes	× 1	139–211 kU/L
δ-Aminolevulinic Acid (δ-ALA)	S	15–23 µg/dL; lower in children	× 0.076	1.14–1.75 µmol/L
	U	1.3–7.0 mg/d	× 7.626	9.9–53.4 µmol/d
Ammonia Nitrogen *Resin or enzymatic*	S or P (NaHep)	*µg N/dL*		*µmol N/L*
		Newborn: 90–150	× 0.714	64–107
		0–2 wk: 79–129		56–92
		>1 mo: 29–70		21–50
		Adult: 15–45		11–32
		40–110		29–79
Conway diffusion		(values for arterial plasma are lower than for venous plasma)		
	U, 24 h	140–1500 mg/d	× 0.0714	10–107 mmol/d

Table 20-25. CLINICAL CHEMISTRY AND TOXICOLOGY (*Continued*)

Test	Specimen	Reference Range	Conversion Factor	Reference Range (International Units)
Ammonium Chloride Loading Test *Dose: 0.1 g ammonium chloride/kg orally at 0800 h after complete midnight fast*	U, hourly between 1000 –1600 h	pH should fall to ≤5.2 between 2 and 8 h after the dose		H^+ concentration should increase to >6 µmol/L between 2 and 8 h after the dose
Amniotic Fluid Analysis ΔA_{450nm} *L/S Ratio, see Lecithin/sphingomyelin Ratio*	Amf	28 wk: 0–0.048 *A* 40 wk: 0–0.02 *A*		Same
Amylase *UV-method, Beckman; BMD*	S U, timed	Adult: 25–125 U/L >70 y: 20–160 U/L 1–17 U/h		Same 1–17 U/h
Amylase/Creatinine Clearance Ratio	S and U, random	1–4%	× 0.01	Clearance fraction: 0.01–0.04
Amyloid-Associated Protein (Amyloid A Protein, AA Protein)	S	0–50 y: <20 mg/dL (Conc. increases after 50 y of age)	× 10	<200 mg/L
Androstenedione	S Amf		× 0.0349 × 3.49	

Androstenedione detail:

		ng/dL (mean ± SE)			nmol/L (mean ± SE)	
		M	F		M	F
Cord:		85 ± 27	93 ± 28		2.97 ± 0.94	3.25 ± 0.98
1–3 mo:		34 ± 11	19 ± 4		1.19 ± 0.38	0.66 ± 0.14
3–5 mo:		24 ± 15	14 ± 5		0.84 ± 0.52	0.49 ± 0.17
Adult:		107 ± 25	151 ± 38		3.73 ± 0.87	5.27 ± 1.33
Midpregnancy,		ng/mL			nmol/L	
Male fetus:		1.0			3.49	
Female fetus:		0.7			2.44	
Term:		0.7–3.5			2.44–12.22	

Test	Specimen	Reference Range	Conversion Factor	Reference Range (International Units)
Angiotensin I	P (KEDTA), periph-vein	11–88 pg/mL	× 1	11–88 ng/L
Angiotensin II	P (KEDTA)	Arterial: 2.4 ± 1.2 ng/dL (SD) Venous: 50–75% of arterial conc	× 10 × 0.01	24 ± 12 ng/L (SD) Fraction of arterial conc.: 0.50–0.75
Anion Gap $[Na - (Cl^- + HCO_3^-)]$	P (Hep)	7–16 mEq/L	× 1	7–16 mmol/L
α_1-Antichymotrypsin (α_1AC)	S	Newborn: ~1 mg/dL Adult: 30–60 mg/dL	× 10	~10 mg/L 300–600 mg/L
Antidiuretic Hormone (hADH, Vasopressin)	P (EDTA)	*Plasma mOsmol/kg* *Plasma ADH pg/mL* 270–280: <1.5 280–285: <2.5 285–290: 1–5 290–295: 2–7 295–300: 4–12	× 1	*Plasma ADH ng/L* <1.5 <2.5 1–5 2–7 4–12
Antimony *NAA, AAS*	P (Hep) U	0.052 ± 0.019 µg/dL (SD) <10 µg/L Toxic: >1 mg/L	× 82.1 × 8.21 × 8.21	4.27 ± 1.56 nmol/L (SD) <82.1 nmol/L >8.21 µmol/L
α_1-Antitrypsin *Nephelometric*	S	Newborn: 145–270 mg/dL Adult: 78–200 mg/dL	× 0.1	1.45–2.70 g/L 0.78–2.00 g/L
Arsenic	WB (Hep) U, 24 h	µg/dL 0.2–6.2 Chronic poisoning: 10–50 Acute poisoning: 60–930 5–50 µg/d	× 0.133 × 0.0133	µmol/L 0.03–0.83 1.33–6.65 7.98–124 0.07–0.67 µmol/d

Table 20-25. CLINICAL CHEMISTRY AND TOXICOLOGY (*Continued*)

Test	Specimen	Reference Range		Conversion Factor	Reference Range (International Units)
Ascorbic Acid, *see Vitamin C*					
Aspartate Aminotransferase (AST, SGOT), *(30 °C)*	S		*U/L*		Same
		Newborn/Infant:	25–75		
		Adult:	8–20		
		> 60y, M:	11–26		
		F:	10–20		
		With P-5′-P:	12–29		
Base Excess	WB (Hep)		*mmol/L*		Same
		Newborn:	(−10)–(−2)		
		Infant:	(−7)–(−1)		
		Child:	(−4)–(+2)		
		Adult:	(−2)–(+3)		
Benzene	S	Negative			Negative
			mg/L		*μmol/L*
		Toxic:	> 1.0	× 12.8	> 13
	U	As phenol:	< 10.0	× 10.6	< 106
		Toxic, as phenol:	> 75.0		> 795
Beryllium *Fluorometric, AAS*	U, 24 h	None detected			None detected
		Toxic:	> 20 μg/L	× 0.111	> 2.22 μmol/L
Bicarbonate (HCO₃⁻)	S	Art.:	21–28 mmol/L		Same
		Ven.:	22–29 mmol/L		
			mEq/L		*mmol/L*
	WB, arterial (Hep; anaerobic)	Newborn:	17.2–23.6	× 1	17.2–23.6
		Infant:	19.0–23.9		19.0–23.9
		2 mo–2 y:	16.3–23.9		16.3–23.9
		Adult:	18–23		18–23
Bile Acid Breath Test *Dose: 300 mL test meal containing 10 μ Ci of ¹⁴C-glycocholic acid and cholyl-2,4-³ H taurine to fasting subjects at 0800 h*	Br	Radioactivity expressed as a % of the original dose; only minute quantities (0.11% ± 0.15%, SD) of radiation are detected at 2 h (0.52 ± 0.09%, SD, cumulative excretion at 4 h)		× 0.01	Fraction of radiation detected at 2 h: 0.0011 ± 0.0015 (SD) Cumulative fraction excreted at 4 h: 0.0052 ± 0.0009 (SD)
Bile Acids, Total *GLC*	S	0.3–2.3 μg/mL		× 2.45	0.74–5.64 μmol/L
	S, 1 h pp	1.8–3.2 μg/mL			4.41–7.84 μmol/L
	F	120–225 mg/d			294–551 μmol/d (Conv. factor based on cholic acid, M.W. 408.6)

Bilirubin
Total

Test	Specimen		Premature *mg/dL*	Full Term *mg/dL*	Conversion Factor	Premature *μmol/L*	Full Term
	S	Cord:	< 2.0	< 2.0	× 17.1	< 34.2	< 34.2
		0–1 d:	< 8.0	< 6.0		< 137	< 103
		1–2 d:	< 12.0	< 8.0		< 205	< 137
		3–5 d:	< 16.0	< 12.0		< 274	< 205
		> 5 d:	< 2.0	0.2–1.0		< 34.2	3.4–17.1
	U	Negative				Negative	
	Amf	28 wk:	0.075 mg/dL (or ΔA_{450} < 0.048)		× 17.1	1.28 μmol/L	
		40 wk:	0.025 mg/dL (or ΔA_{450} < 0.02)			0.43 μmol/L	
Conjugated (direct)	S	0–0.2 mg/dL			× 17.1	0–3.4 μmol/L	

Test	Specimen	Reference Range		Conversion Factor	Reference Range (International Units)
Bismuth	P (Hep)	< 1.0 μg/dL		× 47.85	< 47.9 nmol/L
	U, 24 h	< 20 μg/L		× 4.785	< 95.7 nmol/L
Blood Volume	WB (Hep)	M:	52–83 mL/kg	× 0.001	0.052–0.083 L/kg
		F:	50–75 mL/kg		0.050–0.075 L/kg
Borate	S		*mg/L*		*μmol/L*
		Child:	< 7	× 17.0	< 119
		Adult, M:	< 2		< 34
		Toxic:	> 20		> 340

Table 20-25. CLINICAL CHEMISTRY AND TOXICOLOGY (*Continued*)

Test	Specimen	Reference Range		Conversion Factor	Reference Range (International Units)
Bromosulfophthalein (BSP) Test (5 mg/kg)	S	<6% retention at 45 min		× 0.01	Fraction dye retained: <0.06
C-Peptide	S	Adult: > 60 y, M: F:	*ng/mL* ≤4.0 1.5–5.0 1.4–5.5	× 1	*μg/L* ≤4.0 1.5–5.0 1.4–5.5
C-Reactive Protein	S	Cord blood: Adult:	1–35 μg/dL 6.8–820 μg/dL	× 10	10–350 μg/L 68–8200 μg/L
Cadmium	WB (Hep) U, 24 h	0.1–0.5 μg/dL Toxic: <15 μg/d	10–300 μg/dL	× 8.897 × 0.089 × 0.0089	0.89–4.45 nmol/L 0.89–26.70 μmol/L <0.13 μmol/d
Calcitonin (hCT)	S or P (Hep or EDTA)	Newborn, Term, cord: 48 h: 7 d: Premature, cord: 48 h: 7 d: Adult, M: F: Concentrations decrease with age	*pg/mL* 30–240 91–580 77–293 30–265 108–670 79–570 <100 4 times lower (increases in preg)	× 1	*ng/L* 30–240 91–580 77–293 30–265 108–670 79–570 <100
Calcitonin-Calcium Infusion Stimulation Test *Dose: 15 mg Ca (as gluconate)/kg, IV infusion/4h*	S or P (Hep or EDTA): baseline, 3 and 4 h	Peak hCT level, M: F:	<265 pg/mL <120 pg/mL	× 1	M: <265 ng/L F: <120 ng/L
Calcium, Ionized (Ca$_I$)	S, P, or WB (Hep)	Cord, venous: Newborn, 3–24 h: 24–48 h: Adult: or > 60 y:	*mg/dL* 5.5 ± 0.3 (SD) 4.3–5.1 4.0–4.7 4.48–4.92 2.24–2.46 mEq/L 2.25–2.60 mEq/L	× 0.25 × 0.5	*mmol/L* 1.38 ± 0.08 (SD) 1.08–1.28 1.00–1.18 1.12–1.23 1.12–1.23 1.13–1.30
Calcium, Total	S CSF U, 24 h F	Cord: Newborn, 3–24 h: 24–48 h: 4–7 d: Child: Adult: M, > 60 y: 2.1–2.7 mEq/L or 4.2–5.4 mg/dL *Ca in diet* Free Ca: Low to average: Average (800 mg/d): Average:	*mg/dL* 9.0–11.5 9.0–10.6 7.0–12.0 9.0–10.9 8.8–10.8 8.4–10.2 8.4–10.0 *mg/d* 5–40 50–150 100–300 <0.64 g/d	× 0.25 × 0.50 × 0.25 × 0.025 × 25	*mmol/L* 2.25–2.88 2.25–2.65 1.75–3.0 2.25–2.73 2.20–2.70 2.10–2.55 2.10–2.50 1.05–1.35 mmol/L 1.05–1.35 mmol/L *mmol/d* 0.13–1.00 1.25–3.75 (20 mmol/d): 2.50–7.50 16 mmol/d
Cannabinoids	P (Hep or EDTA) U	Negative; THC after smoking 10 mg: 0.019–0.026 mg/L Negative		× 3.18	0.060–0.083 μmol/L Negative
Carbon Dioxide, Partial Pressure (*p*CO$_2$)	WB, arterial (Hep)	Newborn: Infant: Adult, M: F:	*mm Hg* 27–40 27–41 35–48 32–45	× 0.133	*kPa* 3.59–5.32 3.59–5.45 4.66–6.38 4.26–5.99

Table 20-25. CLINICAL CHEMISTRY AND TOXICOLOGY (*Continued*)

Test	Specimen	Reference Range		Conversion Factor	Reference Range (International Units)
Carbon Dioxide, Total (tCO_2)	S or P (Hep)		*mmol/L*		Same
		Cord:	14–22		
		Newborn:	13–22		
		Premature, 1 wk:	14–27		
		Infant:	20–28		
		Child:	20–28		
		Adult:	22–29		
		> 60 y:	23–31		
Carbon Disulfide *GLC, color (metabolites)*	WB (F⁻/Ox)	Negative			Negative
		Occup. exp.:	<0.3 mg/L	× 13.1	<3.9 µmol/L
	U	Toxic:	>0.5 mg/L		>6.6 µmol/L
		Exposure index:	6–10		6–10
Carbon Monoxide	WB (EDTA)		*% HbCO*		*HbCO Fraction*
		Nonsmokers:	0.5–1.5	× 0.01	0.005–0.015
		Smokers,			
		1–2 pack/d:	4–5		0.04–0.05
		> 2 packs/d:	8–9		0.08–0.09
		Toxic:	> 20		>0.20
		Lethal:	> 50		>0.5
Carboxyhemoglobin, *see Carbon Monoxide*					
Carcinoembryonic Antigen (CEA)	S	Nonsmokers:	0–3.0 ng/mL	× 1	0–3.0 µg/L
		Smokers:	0–5.0 ng/mL		0–5.0 µg/L
β-Carotene *Petroleum ether extraction*	S		*µg/dL*		*µmol/L*
		Infant:	20–70	× 0.0186	0.37–1.30
		Child:	40–130		0.74–2.42
		Adult:	60–200		1.12–3.72
β-Carotene Absorption Test *Dose: 15 000 U carotene orally, 3 × daily with meals for 3 d*	S	Increase by ≥ 35 µg/dL after loading		× 0.0186	Increase by ≥0.65 µmol/L after loading
Catecholamines *HPLC* *Fluorometric*	U, 24 h	<110 µg/d <280 µg/d		× 5.91	<650 nmol/d <1655 nmol/d (Conv. factor based on norepinephrine, M.W. 169.18)
Catecholamines, Fractionated	U, 24 h	Norepinephrine,	*µg/d*		*nmol/d*
		1–4 y:	0–29	× 5.91	0–171
		4–10 y:	8–65		47–384
		10–15 y:	15–80		89–473
		Adult:	0–100		0–591
		Epinephrine,	*µg/d*		*nmol/d*
		1–4 y:	0–6.0	× 5.46	0–32.8
		4–10 y:	0–10.0		0–54.6
		10–15 y:	0.5–20		2.7–109
		Adult:	0–15		0–81.9
		Dopamine,	*µg/d*		*nmol/d*
		1–4 y:	40–260	× 6.53	261–1698
		>4 y:	65–400		425–2610
Catecholamines, Free	P (EDTA and Na meta-bisulfite)		*pg/mL*		*pmol/L*
		Epinephrine, random:	<88	× 5.46	<480
		Norepinephrine, random:	104–548	× 5.91	615–3239
		Dopamine, random:	<136	× 6.53	<888
		Lower values with patient supine			
Cerebrospinal Fluid Volume	CSF	Child:	60–100 mL	× 0.001	0.06–0.10 L
		Adult:	100–160 mL		0.10–0.16 L

Table 20-25. CLINICAL CHEMISTRY AND TOXICOLOGY (*Continued*)

Test	Specimen	Reference Range		Conversion Factor	Reference Range (International Units)
Ceruloplasmin *RID*	S		*mg/dL*	× 10	*μmol/L*
		Newborn:	1–30		10–300
		6 mo–1 y:	15–50		150–500
		1–12 y:	30–65		300–650
		Thereafter:	15–60		150–600
Chenodeoxycholic Acid, Total	S	0.31 ± 0.32 μg/mL		× 2.55	0.79 ± 0.82 μmol/L
Chenodeoxycholyl-glycine, Conjugated	S	0.09 ± 0.01 μg/mL		× 2.22	0.2 ± 0.03 μmol/L
Chloride	S or P (Hep)		*mmol/L*		Same
		Cord:	96–104		
		Newborn:	96–110		
		Thereafter:	98–106		
	CSF	Infant:	111–130		
		Adult:	118–132		
	U, 24 h		*mmol/d*		Same
		Infant:	2–10		
		Child:	15–40		
		Thereafter:	110–250		
		(vary greatly with Cl intake)			
	Swt		*mmol/L*		Same
		Normal:	0–30		
		Cystic fibrosis:	60–200		
	RBC, lysed		49–54		
	Sal		*mmol/L*		Same
		Without stimulation:	5–20		
		After stimulation:	44		
	WB		77–87		
Cholesterol, Total	S or P (EDTA)		*mg/dL*		*mmol/L*
		Cord:	45–100	× 0.0259	1.17–2.59
		Newborn:	53–135		1.37–3.50
		Infant:	70–175		1.81–4.53
		Child:	120–200		3.11–5.18
		Adolescent:	120–210		3.11–5.44
		Adult:	140–310		3.63–8.03
		Recommended (desirable) range for adults:	140–220		3.63–5.70
Cholic Acid, Total	S	0.20 ± 0.17 μg/mL		× 2.45	0.49 ± 0.42 μmol/L
Cholinesterase II (S-Pseudocholinesterase)					
RID	S	0.5–1.5 mg/dL		× 10	5–15 mg/L
DuPont aca		8–18 U/mL			8–18 kU/L
Dibucaine:		83.6 ± 1.3% (SD) inhibition		× 0.01	Fraction of activity inhibited: 0.84 ± 0.013 (SD)
Fluoride:		79.7 ± 1.2% (SD) inhibition			0.80 ± 0.012 (SD)
Cholylglycine, Conjugated	S	0.13 ± 0.01 μg/mL		× 2.15	0.27 ± 0.03 μmol/L
Chorionic Gonadotropin, β-Subunit (β-hCG)	S or P (EDTA)		*mIU/mL*		*IU/L*
		M and nonpregnant F:	<3.0	× 1	<3.0
		F, post-conception,			
		7–10 d:	>3.0		>3.0
		30 d:	100–5000		100–5000
		40 d:	>2000		>2000
		10 wk:	50 000–140 000		50 000–140 000
		14 wk:	10 000–50 000		10 000–50 000
		Trophoblastic disease:	>100 000		>100 000
	U, 24 h	Pregnancy (wk),	*IU/d (mean)*		Same
		6th:	13 000		
		8th:	30 000		
		12–14th:	105 000		
		16th:	46 000		
		Thereafter:	5000–20 000		

Table 20-25. CLINICAL CHEMISTRY AND TOXICOLOGY (*Continued*)

Test	Specimen	Reference Range	Conversion Factor	Reference Range (International Units)
Chromium	WB (Hep)	70 ng/dL (Includes Cr^{6+} bound to RBC)	× 0.192	13.4 nmol/L
	S	14 ng/dL	× 0.192	2.7 nmol/L
	U, 24 h	0.8 μg/d	× 19.23	15.4 nmol/d
	H	0.21 ± 0.14 μg/g (mean ± SD)	× 19.23	4.0 ± 2.7 nmol/g (mean ± SD)
Chymotrypsin	S or P (EDTA and polybrene)	~10 μg/L		~10 μg/L
	F	120–1265 μg/g stool (mean: 290)	× 1	120–1265 mg/kg stool (mean: 290)
Citrulline	S	*μg/dL*		*μmol/L*
		Newborn, 1 d: 158 –508	× 0.057	9 –29
		1–3 mo: 368 ±193 (SD)		21 ±11 (SD)
		2–6 mo: 543 –876		31 –50
		9 mo–10 y: 210 –526		12 –30
		6–18 y: 333 –911		19 –52
		Adult: 210 –964		12 –55
	U, 24 h	Adult: 0.4–8 mg/d	× 5.71	2 –46 μmol/d
		or 2 ± 1 mg/g creatinine (SD)	× 0.65	or 1.3 ± 0.7 mmol/mol creatinine (SD)
Cobalt	WB (no anti-coagulant)	0.20–0.28 μg/dL	× 169.7	33.9–47.5 nmol/L
	U	1–7 μg/L	× 16.97	17.0–118.8 nmol/L
Complement Components				
Total hemolytic complement activity	P (EDTA)	75–160 U/mL or	× 1	75–160 kU/mL
		> 33% of plasma CH50	× 0.01	Fraction of CH50: > 0.33
Total complement decay rate (functional)	P (EDTA)	~10–20%		Fraction decay rate: ~0.10–0.20
		Deficiency: > 50%		> 0.50
Classic pathway components		*mg/dL*		*mg/L*
C1q	S	6.5 ±0.7 (SD)	× 10	65 ±7 (SD)
C1r	S	2.5 –3.8		25 – 38
C1s (C1 esterase)	S	2.5 –3.8		25 – 38
C2	S	2.8 ±0.6 (SD)		28 ±6 (SD)
C3 (β_1 C-globulin)	S	80 –155		800 – 1550
C4 (β_1 E-globulin)	S	13 –37		130 – 370
C5 (β_1 F-globulin)	S	6.4 ±1.3 (SD)		64 ±13 (SD)
C6	S	5.6 ±0.80 (SD)		56 ±8.0 (SD)
C7	S	4.9 –7.0		49 – 70
C8	S	4.3 –6.3		43 – 63
C9	S	4.7 –6.9		47 – 69
Alternative pathway components				
C4 binding protein (reliable assay not presently available)	S	18 –32	× 10	180–320
Factor B (C3 pro-activator)	S	20 –45	× 10	200–450
Properdin	S	2.8 ± 0.4 (SD)	× 10	28 ± 4 (SD)
Regulatory proteins				
β_1 H-globulin (C3b inactivator accelerator)	S	56.1 ± 7.8 (SD)	× 10	561 ± 78 (SD)
C1 inhibitor (Esterase inhibitor)	P (EDTA)	17.4–24.0	× 10	174–240
C1 inhibitor by complement decay rate (functional)	P (EDTA)	~10–20%	× 0.01	Fraction decay rate: ~0.10–0.20
		Deficiency: > 50%		> 0.50
C3b inactivator (KAF)	S	4.0 ± 0.7 (SD) mg/dL	× 10	40 ± 7 (SD) mg/L
S protein	S	41.8–60.0 mg/dL		418–600 mg/L

Table 20-25. CLINICAL CHEMISTRY AND TOXICOLOGY (*Continued*)

Test	Specimen	Reference Range		Conversion Factor	Reference Range (International Units)
Copper	S		*μg/dL*		*μmol/L*
		Birth–6 mo:	20–70	× 0.157	3.14–10.99
		6 y:	90–190		14.13–29.83
		12 y:	80–160		12.56–25.12
		Adult, M:	70–140		10.99–21.98
		F:	80–155		12.56–24.34
		Pregnancy at term:	118–302		18.53–47.41
		Values for blacks ~8–12% higher			Blacks ~8–12% higher
	RBC (Hep)	90–150 μg/dL		× 0.157	14.13–23.55 μmol/L
	U, 24 h	15–30 μg/d		× 0.0157	0.24–0.47 μmol/d
Coproporphyrin	U, 24 h	34–234 μg/d		× 1.5	51–351 nmol/d
	F, 24 h	<30 μg/g dry wt		× 1.5	<45 nmol/g dry wt
		400–1200 μg/d			600–1800 nmol/d
Corticobinding Globulin (CBG), *see Transcortin*					
Corticosterone	S or P (Hep, EDTA, or Ox)	0.13–2.3 μg/dL		× 28.9	3.8–66.5 nmol/L
Cortisol	S or P (Hep)	0800 h:	5–23 μg/dL	× 27.6	138–635 nmol/L
		1600 h:	3–15 μg/dL		83–414 nmol/L
		2000 h:	≤50% of 0800 h	× 0.01	Fraction of 0800 h: ≤0.50
Cortisol, Free	U, 24 h		*μg/d*		*nmol/d*
		Child:	2–27	× 2.76	5.5–74.5
		Adolescent:	5–55		13.8–152
		Adult:	10–100		27.6–276
Creatine	S or P	M: 0.17–0.70 mg/dL		× 76.3	M: 13–53 μmol/L
		F: 0.35–0.93 mg/dL			F: 27–71 μmol/L
	U, 24 h	M: 0–40 mg/d		× 7.63	M: 0–305 μmol/d
		F: 0–80 mg/d			F: 0–609 μmol/d
Creatine Kinase (CK) *Total, 30 °C*	S		*U/L*		Same
		Newborn:	10–200		
		Adult, M:	12–80		
		F:	10–55		
		>60 y, M:	20–110		
		F:	16–80		
		>70 y, M:	22–90		
		F:	16–80		
		Ambulatory, M:	25–90		
		F:	10–70		
		Higher after exercise			
Total, CK-NAC, 30 °C		Newborn:	68–580		
		Adult, M:	25–130		
		F:	10–115		
Isoenzymes		Fraction 2 (MB):	<4–6% of total (method dependent)	× 0.01	Fraction of total: <0.04–0.06
Creatinine *Jaffe, kinetic or enzymatic*	S or P		*mg/dL*		*μmol/L*
		Cord:	0.6–1.2	× 88.4	53–106
		Newborn, 1–4 d:	0.3–1.0		27–88
		Infant:	0.2–0.4		18–35
		Child:	0.3–0.7		27–62
		Adolescent:	0.5–1.0		44–88
		Adult, M:	0.6–1.2		53–106
		F:	0.5–1.1		44–97

Table 20-25. CLINICAL CHEMISTRY AND TOXICOLOGY (*Continued*)

Test	Specimen	Reference Range		Conversion Factor	Reference Range (International Units)
Creatinine (*continued*)					
Jaffe, manual	S or P	0.8–1.5 mg/dL		× 88.4	71–133 μmol/L
	Amf	After 37 wk gestation: > 2.0 mg/dL		× 88.4	After 37 wk gestation: > 177 μmol/L
	U, 24 h		*(mg/d)/kg*	× 8.84	$\mu mol \cdot d^{-1} \cdot kg^{-1}$
		Infant:	8–20		71–180
		Child:	8–22		71–195
		Adolescent:	8–30		71–265
		Adult, M:	14–26		124–230
		F:	11–20		97–177
		Declines with age to (10 mg/kg)/d at age 90 or			
		M:	800–1800 mg/d	× 0.00884	7.1–15.9 mmol/d
		F:	600–1600 mg/d		5.2–14.1 mmol/d
Creatinine Clearance (Endogenous)	S or P, and U	< 40 y, M: (97–137 mL/min)/1.73 m²		× 0.00963	(0.93–1.32 mL/s)/m²
		F: (88–128 mL/min)/1.73 m²			(0.85–1.23 mL/s)/m²
		Decreases (~6.5 mL/min)/1.73 m² per decade			Decreases (~0.06 mL/s)/m² per decade
Cyanide	S		*mg/L*		*μmol/L*
		Nonsmokers:	0.004	× 38.4	0.15
		Smokers:	0.006		0.23
		Nitroprusside therapy:	0.01–0.06		0.38–2.30
		Toxic:	> 0.1		> 3.84
	WB (Ox)	Nonsmokers:	0.016		0.61
		Smokers:	0.041		1.57
		Nitroprusside therapy:	0.05–0.5		1.92–19.20
		Toxic:	> 1		> 38.40
Cyclic AMP	P (EDTA)	M: 5.6–10.9 ng/mL		× 3.04	M: 17–33 nmol/L
		F: 3.6–8.9 ng/mL			F: 11–27 nmol/L
	U, 24 h	< 3.3 mg/d		× 344	< 10 μmol/d
		or < 1.64 mg/g creatinine			or < 565 μmol/mol creatinine
Cystine and Cysteine					
Qual.	U, random	Negative			Negative
Quant.	U, 24 h	10–100 mg/d		× 8.254	83–830 μmol/d Conversion based on cysteine (M.W. 121.16)
Dehydroepiandrosterone (DHEA)					
Total	S		*ng/mL*		*nmol/L*
		Cord:	5.6–20.0	× 3.47	19–69
		Child:	1.0–3.0		3–10
		Adult, M:	1.7–4.2		6–15
		F:	2.0–5.2		7–18
		Pregnancy:	0.5–12.5		2–43
Unconjugated	U, 24 h	Child,	*mg/d*		*μmol/d*
		0–5 y:	< 0.1	× 3.47	< 0.3
		6–9 y:	< 0.2		< 0.7
		10–15 y:	< 0.4		< 1.4
		Adult, M:	0–2.3		0–8.0
		F:	0–1.2		0–4.2
Dehydroepiandrosterone Sulfate (DHEA-SO₄)	S or P (Hep or EDTA)		*μg/mL*		*μmol/L*
		Newborn:	< 300	× 2.6	< 780
		1–4 d:	< 20		< 52
		5–7 d:	< 45		< 117
		Child:	0.60–2.54		1.6–6.6
		Adult, M:	1.99–3.34		5.2–8.7
		F, Premenopausal:	0.82–3.38		2.1–8.8
		Postmenopausal:	0.11–0.61		0.3–1.6
		Pregnancy, term:	0.23–1.17		0.6–3.0
	Amf	10–50 ng/mL		× 2.6	26–130 nmol/L

Table 20-25. CLINICAL CHEMISTRY AND TOXICOLOGY (*Continued*)

Test	Specimen	Reference Range	Conversion Factor	Reference Range (International Units)
Deoxycholic Acid, Total	S	0.22 ± 0.13 μg/mL (SD)	× 2.55	0.56 ± 0.33 μmol/L (SD)
11-Deoxycorticos-terone (DOC)	S or P (Hep, EDTA, or Ox)	Ad lib diet, 0800 h: 4.3–12.3 ng/dL	× 0.0303	0.13–0.37 nmol/L
11-Deoxycortisol (Compound S)	P (Hep, EDTA, or Ox)	*μg/dL* < 1 without metyrapone > 7 after metyrapone	× 0.029	*μmol/L* < 0.029 > 0.203

Dexamethasone Single Dose Overnight Suppression Test
Dose: 1 mg orally at 2300 or 2400 h
	S, 0800 h	Cortisol:
		  Suppression to 5–10 μg/dL
		  or to < 50% of baseline
			× 27.59	Cortisol:
				  Suppression to 138–276 nmol/L or fraction of baseline: < 0.50

Dexamethasone Suppression Test (Standard)
Low dose, adult:
0.5 mg q 6 h × 8
	S, 0800 h
	Control, day 2 and 3
		Cortisol:
		  Suppression on day 3 to < 50% of baseline or to < 5μg/dL
			× 0.01
			× 27.59
				Cortisol:
				  Suppression on day 3; fraction of baseline: < 0.50 or < 138 nmol/L

	U, 24 h Control, day 1, 2, and 3	17-KGS: Suppression on day 2 to < 7.5 mg/d	× 3.467	17-KGS: Suppression on day 2 to < 26 μmol/d
		17-OHCS: Suppression on day 2 to < 4.5 mg/d Free cortisol: < 50% of baseline	×2.76 × 0.01	17-OHCS: Suppression on day 2 to < 12.4 μmol/d Free cortisol: Fraction of baseline: < 0.50

High dose, adult
2.0 mg q 6 h × 8
		Cortisol, 17-KGS, 17-OHCS:
		  Suppression on day 3 to < 50% of baseline
			× 0.01	Cortisol, 17-KGS, 17-OHCS:
				  Suppression on day 3; fraction of baseline: < 0.50

Dihydrotestosterone (DHT) — S

	ng/dL		× 0.0344	*nmol/L*	
Pubertal	M	F		M	F
stage I:	< 10	< 10		< 0.34	< 0.34
II:	< 20	< 15		< 0.69	< 0.52
III:	< 35	< 25		< 1.20	< 0.86
IV–V:	< 75	< 25		< 2.58	< 0.86
Adult:	60–300	10–40		2–10.3	0.34–1.4

Disaccharide Absorption Test — S

	mg/dL	× 0.055	*mmol/L*
Change in glucose from fasting value:	> 30		Change in glucose from fasting value: > 1.67
Inconclusive:	20–30		1.11–1.67
Abnormal:	< 20		< 1.11

Estradiol, (E₂), Total — S or P (Hep or EDTA)

	pg/mL	× 3.67	*pmol/L*
M, Pubertal stage I:	2–8		7–29
II:	11		40
III:	> 20		> 73
Adult, M:	8–36		29–132
F, Pubertal stage I:	0–23		0–84
II:	0–66		0–242
III:	0–105		0–385
IV:	20–300		73–1100
Follicular:	10–90		37–330
Midcycle:	100–500		367–1835
Luteal:	50–240		184–881
(Lower on oral contraceptives)			
Pregnancy,			
6–8 wk:	7.6–11.4		28–42
25–28 wk:	41.8–158		153–580
Term:	127–281		466–1031
Postmenopausal:	10–30		37–110

Table 20-25. CLINICAL CHEMISTRY AND TOXICOLOGY (*Continued*)

Test	Specimen	Reference Range		Conversion Factor	Reference Range (International Units)
Estradiol, (E$_2$), Total (*continued*)					
	U, 24 h		*μg/d*		*nmol/d*
		Adult, M:	0–6	× 3.67	0–22
		F, Follicular:	0–3		0–11
		Ovulatory peak:	4–14		15–51
		Luteal:	4–10		15–37
		Postmenopausal:	0–4		0–15
Estriol (E$_3$), Free	S	Pregnancy (wk),	*μg/L*		*nmol/L*
		25–28:	3.5–12.5	× 3.47	12.1–43.4
		30–32:	4.5–16.0		15.6–55.5
		34:	5.5–18.5		19.1–64.2
		36:	7.0–25.0		24.3–86.8
		37:	8.0–28.0		27.8–97.2
		38:	9.0–32.0		31.2–111.0
		39:	10.0–34.0		34.7–118.0
		40–41:	10.5–25.0		36.4–86.8
	Amf		*ng/mL*		*nmol/L*
		Pregnancy (wk),	*(95% range)*		*(95% range)*
		16–20:	1.0–3.2	× 3.47	3.5–11.1
		20–24:	2.1–7.8		7.3–27.1
		24–28:	2.1–7.8		7.3–27.1
		28–32:	4.0–13.6		13.9–47.2
		32–36:	3.6–15.5		12.5–53.8
		36–38:	4.6–18.0		16.0–62.5
		38–40:	5.4–19.8		18.7–68.7
Estriol (E$_3$), Total	S	Pregnancy (wk),	*ng/mL*		*nmol/L*
		24–28:	30–170	× 3.47	104–590
		28–32:	40–220		139–763
		32–36:	60–280		208–972
		36–40:	80–350		278–1215
		Adult, M and nonpreg F:	<2		<7
	U, 24 h	Pregnancy (wk),	*mg/d*		*μmol/d*
		30:	6–18	× 3.47	21–62
		35:	9–28		31–97
		40:	13–42		45–146
		Decrease of >40% of previous value suggests fetus at risk			Fraction of previous value of <0.60 suggests fetus at risk
	Amf	Pregnancy (wk),	*ng/mL*	× 3.47	*nmol/L*
		<20	1–7		3.5–24.3
		21–32:	5–50		17–174
		33–35:	9–240		31–833
		36–41:	150–213		521–739
Estrogen Receptor Assay (ERA)	T		*fmol/mg Protein*		*nmol/kg Protein*
		Negative:	<3.0	× 1	<3.0
		Borderline positive:	3–10		3–10
		Positive:	>10.0		>10.0
Estrogens, Total	S		*pg/mL*		*ng/L*
		Child:	<30	× 1	<30
		Adult, M:	40–115		40–115
		F, cycle-days,			
		1–10 d:	61–394		61–394
		11–20 d:	122–437		122–437
		21–30 d:	156–350		156–350
		Pregnancy (wk),			
		0–12:	700–8700		700–8700
		36–42:	8000–31 000		8000–31 000
		Prepubertal and postmenopausal:	≤40		≤40
	U, 24 h		*μg/d*		Same
		Child:	1.0 (mean)		
		M, pubertal stage I:	2.5 (mean)		
		II:	5.9 (mean)		
		III:	6.2 (mean)		
		Adult, M:	5–25		
		F, Preovulation:	5–25		
		Ovulation:	28–100		
		Luteal peak:	22–80		
		Pregnancy:	<45 000		
		Postmenopause:	<10		

Table 20-25. CLINICAL CHEMISTRY AND TOXICOLOGY (*Continued*)

Test	Specimen	Reference Range		Conversion Factor	Reference Range (International Units)
Estrone (E$_1$)					
Unconjugated	S		*pg/mL*		*pmol/L*
		M, Pubertal stage I:	11	× 3.70	41
		II:	16		59
		III:	21		78
		Adult:	30–170		111–629
		F, Pubertal stage I:	0–29		0–107
		II:	10–35		37–130
		III:	15–45		56–167
		IV:	20–80		74–296
		Follicular:	20–150		74–555
	U, 24 h		*μg/d*		*nmol/d*
		Adult, M:	3–8	× 3.70	11–30
		F, Ovulatory peak:	11–31		41–115
		Luteal:	10–23		37–85
		Postmenopausal:	1–7		4–26
Conjugated	Amf	38–40 wk:	8.0 ng/mL (mean)	× 2.24	18 nmol/L (mean)
Unconjugated	Amf	Term:	2.5–5.6 ng/mL	× 3.70	9–21 nmol/L
Ethanol	WB (Ox) or S		*mg/dL*		*mmol/L*
		Toxic:	50–100	× 0.217	11–22
		Depression of CNS:	>100		>21.7
		Fatalities reported:	>400		>86.8
Ethylene Glycol	P (EDTA)	0 mg/L		× 16.1	0 μmol/L
	U	0 mg/L		× 16.1	0 μmol/L
Fat, Fecal	F, 72 h		*g/d*		Same
		Infant, breast-fed:	<1		
		0–6 y:	<2		
		Adult:	<7		
		Adult (fat-free diet):	<4		
Fatty Acid Profile	S or P (Hep)		*% of Total Nonesterified Fatty Acids*		*Fraction of Total Nonesterified Fatty Acids*
		Oleic:	26–45	× 0.01	0.26–0.45
		Palmitic:	23–25		0.23–0.25
		Stearic:	10–14		0.10–0.14
		Linoleic:	8–16		0.08–0.16
Fatty Acids, Nonesterified (Free)	S or P (Hep)		*mg/dL*		*mmol/L*
		Adult:	8–25	× 0.0354	0.30–0.90
		Child and obese adults:	<31		<1.10 (Conv. factor based on oleic acid, M.W. 282.47)
Fatty Acids, Total	S	190–420 mg/dL		× 0.0352	7–15 mmol/L
Ferric Chloride Test	U, random	Negative			Negative
Ferritin	S		*ng/mL*		*μg/L*
		Newborn:	25–200	× 1	25–200
		1 mo:	200–600		200–600
		2–5 mo:	50–200		50–200
		6 mo–15 y:	7–140		7–140
		Adult, M:	15–200		15–200
		F:	12–150		12–150
α$_1$-Fetoprotein (AFP)	S	Adult:	<30 ng/mL	× 1	<30 μg/L
		Fetal:	Peak of 200–400 mg/dL in first trimester; falls to 1% of peak (<5 mg/dL) in cord blood	× 0.01	Peak of 2–4 g/L in first trimester; falls to 1% of peak (<0.05 g/L) in cord blood
		1 y:	<30 ng/mL	× 1	<30 μg/L

Table 20-25. CLINICAL CHEMISTRY AND TOXICOLOGY (*Continued*)

Test	Specimen	Reference Range			Conversion Factor	Reference Range (International Units)	
α₁-Fetoprotein (AFP) (*continued*)							
	S, maternal		*µg/dL*				*µg/L*
		Pregnancy (wk)	*10th–90th Percentile*	*Median*		*10th–90th Percentile*	*Median*
		6–9	<1.0–1.5	<1.0	× 10	<10–15	<10
		10–12	<1.0–2.5	<1.0		<10–25	<10
		13–15	<1.0–4.4	1.9		<10–44	19
		16–18	<1.0–7.1	3.6		<10–71	36
		19–21	2.1–9.6	5.9		21–96	59
		22–24	5.1–15.0	7.6		51–150	76
		25–27	5.6–17.0	9.9		56–170	99
		28–30	7.9–19.8	14.0		79–198	140
		31–33	8.4–34.4	14.9		84–344	149
		34–36	6.0–33.5	12.3		60–335	123
		37–40	6.3–16.5	13.5		63–165	135
	Amf		*mg/dL*				*mg/L*
		Pregnancy (wk)	*Median*	*±2 log SD*		*Median*	*±2 log SD*
		11–12	2.4	1.0–5.0	× 10	24	10–50
		13–14	2.3	1.3–4.1		23	13–41
		15–16	1.8	0.9–3.5		18	9–35
		17–18	1.5	0.6–3.3		15	6–33
		19–20	1.0	0.5–2.5		10	5–25
		21–25	0.7	0.4–1.4		7	4–14
		26–30	0.6	0.3–1.0		6	3–10
		31–35	0.2	0.05–0.7		2	0.5–7
		36–40	0.1	0.02–0.3		1	0.2–3
Fibrinogen	P (NaCit)	200–400 mg/dL			× 0.01	2.00–4.00 g/L	
FIGLU *Dose: 5 g histidine q 4 h × 3*	U, 24 h, after initial dose	<35 mg/d			× 5.74	<201 µmol/d	
Fluoride			*µg/mL*			*µmol/L*	
	P (Hep)		0.01–0.2		× 52.6	0.5–10.5	
	U		0.2–1.1			10.5–57.9	
		Occup. exp.:	4–5			210–263	
Folate	S	1.8–9 ng/mL			× 2.265	4.1–20.4 nmol/L	
		>60 y: 2–12 ng/mL				4.5–27.2 nmol/L	
	RBC (EDTA)	150–450 ng/mL packed cells				340–1020 nmol/L packed cells	
		>60 y: 95–500 ng/mL packed cells				215–1132 nmol/L packed cells	
Folate Absorption Test	U, 24 h	45 ± 7% of dose			× 0.01	Fraction of dose: 0.45 ± 0.07	
Follicle Stimulating Hormone (hFSH)	S or P (Hep)		*mIU/mL (IRP-2-hMG)**			*IU/L*	
		Birth–1 y, M:	<1–12		× 1	<1–12	
		F:	<1–20			<1–20	
		1–8 y, M:	<1–6			<1–6	
		F:	<1–4			<1–4	
		9–10 y, M:	<1–10			<1–10	
		F:	2–8			2–8	
		11–12 y, M:	2–12			2–12	
		F:	3–11			3–11	
		13–14 y, M:	3–15			3–15	
		F:	3–15			3–15	
		Adult, M:	4–25			4–25	
		F,					
		Premenopause:	4–30			4–30	
		Midcycle peak:	10–90			10–90	
		Pregnancy: Low to undetectable				Low to undetectable	
		Postmenopause:	40–250			40–250	

*2nd International Reference Preparation of Human Menopausal Gonadotropin

Table 20-25. CLINICAL CHEMISTRY AND TOXICOLOGY (*Continued*)

Test	Specimen	Reference Range	Conversion Factor	Reference Range (International Units)
Follicle Stimulating Hormone (hFSH) (*continued*)	U, 24 h	*IU/d (IRP-2-hMG)**		Same
		Birth–1 y, F: < 0.5–1.4		
		1–8 y, M: < 0.5–4.5		
		F: < 0.5–4.0		
		9–10 y, M: 1–5		
		F: 1–4		
		11–12 y, M: 1.5–5		
		F: 1–8		
		13–14 y, M: 2–12		
		F: 1–10		
		Adult, M: 4–18		
		F: 3–12		
		Higher in males > 60 y		
		*2nd International Reference Preparation of Human Menopausal Gonadotropin		
Formaldehyde	WB (F⁻/Ox)	Occup. exp.: 0.6–4.0 mg/L	× 33.3	20–133 μmol/L
	U	Formic acid: < 17 mg/L	× 21.7	< 369 μmol/L
Free Thyroxine Index (FT$_4$I)	S	*FT$_4$ Index* 1.2–5.0		Same
with normalized T$_3$RU		1–3 d: 9.3–26.6		
		1–4 wk: 7.6–20.8		
		1–4 mo: 7.4–17.9		
		4–12 mo: 6.1–14.5		
		1–6 y: 5.7–13.3		
		> 6 y: 5.5–10.0		
		Borderline low: 4.8		
		Borderline high: 14.0		
Free Triiodothyronine, *see Triiodothyronine, Free*				
Free Triiodothyronine Index (FT$_3$I)	S	1–5 y: 165		Same
		5–10 y: 150		
		10–15 y: 130		
Fructose	S	1–6 mg/dL	× 55.5	55.5–333 μmol/L
	U	< 60 mg/d	× 5.55	< 333 μmol/d
	Semf	> 150 mg/dL	× 0.0555	> 8.33 mmol/L
Fructose Loading Test	S	Peak: 15–25 mg/dL	× 0.0555	0.83–1.39 mmol/L
	U	1–2% of administered dose	× 0.01	Fraction of administered dose: 0.01–0.02
Galactose	S	Newborn: 0–20 mg/dL	× 0.0555	0–1.11 mmol/L
		Thereafter: < 5 mg/dL		< 0.28 mmol/L
	U	Newborn: ≤ 60 mg/dL	× 0.0555	≤ 3.33 mmol/L
		Thereafter: < 14 mg/d	× 0.00555	< 0.08 mmol/d
Galactose Tolerance Test *Dose: 40 g in 250 mL H$_2$O, orally*	WB (Hep with NaF)	60 min: 40–60 mg/dL	× 0.0555	2.22–3.33 mmol/L
		Sum of 30,60,90, 120 min specimens: < 110 mg/dL		< 6.11 mmol/L
	U, 5 h	≤ 3 g/5 h	× 5.55	< 16.7 mmol/5 h
Gastric Acid	Gast Res	*Without Stimulation mmol/L*		Same
		Free: 0–40		
		Combined: 10–20		
		Total: 10–50		
Gastric Content	Gast Res	Volume: 20–100 mL	× 0.001	0.02–0.1 L
				H⁺ concentration:
		pH: 1.5–3.5		32–316 μmol/L
Gastric Secretion Rate	Gast Cont, total; six 15-min spec.	BAO: 0–5 mmol/h		Same
		PAO: 5–20 mmol/h (post Pentagastrin)		
		BAO/PAO: 0.20		

Table 20-25. CLINICAL CHEMISTRY AND TOXICOLOGY (*Continued*)

Test	Specimen	Reference Range		Conversion Factor	Reference Range (International Units)
Gastrin	S		*pg/mL*		*ng/L*
		Cord, venous:	64 ± 12.5 (SEM)	× 1	64 ± 12.5 (SEM)
		4 d:	151 ± 15.8		151 ± 15.8
		<3 y:	32.7 ± 22.3		32.7 ± 22.3
		>3 y:	14.8 ± 14.3		14.8 ± 14.3
		<60 y:	<100		<100
		>60 y:	<100 (15% have values of 100–800 pg/mL)		<100 (15% have values of 100–800 pg/mL)
Gastrin-Calcium Infusion Stimulation Test	S	Gastrin:	Slight or no increase		Same
		Z.E. syndrome:	>450 pg/mL	× 1	Z.E. syndrome: >450 ng/L
Gastrin-Secretin Stimulation Test *IV dose: 5 U secretin/kg*	S: baseline and 15-min intervals for 1 h	Gastrin: No response or slight suppression Z.E. syndrome: Increase >110 pg/mL if baseline is 80–500 pg/mL; increase >1400 pg/mL if baseline is high			Same Z.E. syndrome: Increase >110 ng/L if baseline is 80–500 ng/L; increase >1400 ng/L if baseline is high
Glomerular Selectivity Test (IgG/Albumin Ratio)	U	<0.16 indicates high selectivity			Same
Glucagon	P (Hep or EDTA)		*pg/mL* 30 – 210	× 1	*ng/L* 30 – 210
		Average:	75 ± 4 (SEM)		75 ± 4 (SEM)
		Big glucagon:	113 ± 79		113 ± 79
		Proglucagon:	11 ± 16		11 ± 16
		Glucagon:	31 ± 29		31 ± 29
		Small glucagon:	26 ± 18		26 ± 18
	Amf	(Mostly big glucagon)			
		Midgestation:	43 ± 10 (SE)		43 ± 10 (SE)
		Term:	117 ± 38		117 ± 38
Glucose	S		*mg/dL*		*mmol/L*
		Cord:	45–96	× 0.0555	2.50–5.33
		Premature:	20–60		1.11–3.33
		Neonate:	30–60		1.67–3.33
		Newborn, 1 d:	40–60		2.22–3.33
		>1 d:	50–80		2.78–4.44
		Child:	60–100		3.33–5.55
		Adult:	70–105		3.89–5.83
		>60 y:	80–115		4.44–6.38
		>70 y:	83–110		4.61–6.10
	WB (Hep)	Adult:	65–95		3.61–5.27
	CSF	Infant, child:	60–80		3.33–4.44
		Adult:	40–70		2.22–3.89
Quant., enzymatic	U	<0.5 g/d		× 5.55	<2.8 mmol/d
Qual.	U	Negative			Negative
Glucose-6-Phosphate Dehydrogenase (G-6-PD) in Erythrocytes *WHO and ICSH*	WB (ACD, EDTA, or Hep)	12.1 ± 2.09 U/g Hb (SD) 351 ± 60.6 U/10^{12} RBC 4.11 ± 0.71 U/mL RBC		× 0.0645 × 10^{-3} × 1	0.78 ± 0.13 MU/mol Hb 0.35 ± 0.06 nU/RBC 4.11 ± 0.71 kU/L RBC
Glucose, 2 h Postprandial	S	<120 mg/dL Diabetes: *see Glucose Tolerance Test, Oral*		× 0.0555	<6.7 mmol/L

Test	Specimen	Reference Range *mg/dL*			Conversion Factor	Reference Range (International Units) *mmol/L*	
Glucose Tolerance Test (GTT)		Adults,	*Normal*	*Diabetic*		*Normal*	*Diabetic*
		Fasting:	70–105	>140	× 0.0555	3.9–5.8	>7.8
Oral	S	60 min:	120–170	≥200		6.7–9.4	≥11.1
		90 min:	100–140	≥200		5.6–7.8	≥11.1
		120 min:	70–120	≥140		3.9–6.7	≥7.8
IV	S	5 min:	<250			<13.88	
		90 min: at or below fasting concentration or K = >1.5%					

Table 20-25. CLINICAL CHEMISTRY AND TOXICOLOGY (*Continued*)

Test	Specimen	Reference Range		Conversion Factor	Reference Range (International Units)
Glucose Tolerance Test (GTT) with Cortisone *Dose: 50 mg, 8.5 and 2 h before test*	P (F⁻/Ox): baseline, 1, 1½, 2 h after glucose	Glucose, Fasting: 1 h: 1½ h: 2 h:	*mg/dL* 70–105 <200 <200 <140	× 0.0555	*mmol/L* 3.89–5.83 <11.10 <11.10 <7.77
Glutamine	CSF	Newborn–15 d: 15 d–6 mo: 6–12 mo: 12–30 mo: Adult:	*mg/dL ± SD* 9.7 ± 2.9 8.6 ± 2.7 7.1 ± 2.1 5.4 ± 1.8 6–16	× 0.06842	*mmol/L ± SD* 0.66 ± 0.20 0.59 ± 0.18 0.49 ± 0.14 0.37 ± 0.12 0.41 – 1.10
γ-Glutamyltransferase (GGT) *37°C, aca*	S	M: F:	9–50 U/L 8–40 U/L		Same
Glutathione, Reduced (GSH)	WB (ACD, EDTA, or Hep)	6.57 ± 1.04 μmol/g Hb (SD) 190 ± 30.16 μmol/10¹² RBC 2.23 ± 0.35 μmol/mL RBC		× 0.0645 × 1 × 1	0.42 ± 0.07 mol/mol Hb 190 ± 30 amol/RBC 2.23 ± 0.35 mmol/L RBC (Conv. factor based on hemoglobin, M.W. 64 500)
Glutathione Reductase in Erythrocytes *ICSH, 37°C*	WB (ACD, EDTA, or Hep)	Adult (reaction without added flavin adenine dinucleotide, FAD): 7.18 ± 1.09 U/g Hb (SD) 208 ± 31.6 U/10¹² RBC 2.44 ± 0.37 U/mL RBC Adult (reaction using FAD): 10.4 ± 1.5 U/g Hb (SD) 302 ± 43.5 U/10¹² RBC 3.54 ± 0.51 U/mL RBC		× 0.0645 × 10⁻³ × 1 × 0.0645 × 10⁻³ × 1	0.46 ± 0.07 MU/mol Hb 0.21 ± 0.03 nU/RBC 2.44 ± 0.37 kU/L RBC 0.67 ± 0.10 MU/mol Hb 0.30 ± 0.04 nU/RBC 3.54 ± 0.51 kU/L RBC
Glycerol, Free	P (EDTA)	3–10 y: 11–80 y:	0.56–2.14 mg/dL 0.29–1.72 mg/dL	× 0.1086	0.061–0.232 mmol/L 0.032–0.187 mmol/L
Glycine	U, 24 h	10 d–7 wk: 3–12 y: Adult: or 60 ± 24 mg/g creatinine (SD)	*mg/d* 15–59 12–107 59–294	× 0.01332 × 1.51	*mmol/d* 0.20–0.79 0.16–1.43 0.79–3.92 or 90.6 ± 36.2 mmol/mol creatinine (SD)
Glycolic Acid	U, 24 h	Adult:	15–60 mg/d	× 0.0131	0.20–0.79 mmol/d
Gold	S	<10 μg/dL Therap.:	38–500 μg/dL	× 0.0508	<0.51 μmol/L 1.93–25.4 μmol/L
	U, 24 h	<1 μg/d		× 5.08	<5.08 nmol/d
Gonadotropins, *see Pregnancy Tests and Chorionic Gonadotropin, β-Subunit*					
Growth Hormone (hGH, Somatotropin)	S or P (EDTA or Hep)	Cord: Newborn: Child: (occasional values up to 20) Adult, M: F: >60 y, M: F:	*ng/mL* 10–50 10–40 <1–10 <2 <10 0.4–10 1–14	× 1	*μg/L* 10–50 10–40 <1–10 <2 <10 0.4–10 1–14
	Amf	20 wk: Term:	10 30		10 30
Growth Hormone–Arginine Stimulation Test *Dose, adult: 30 g arginine HCl, IV, within 30 min; child: 0.5 g/kg*	S: baseline, 30-min intervals for 2 h	Fasting: <5 ng/mL; rise to >7 ng/mL during test (peak range 8–35 ng/mL) at 30–60 min		× 1	Fasting: <5 μg/L; rise to >7 μg/L during test (peak range 8–35 μg/L) at 30–60 min

Table 20-25. CLINICAL CHEMISTRY AND TOXICOLOGY (*Continued*)

Test	Specimen	Reference Range	Conversion Factor	Reference Range (International Units)
Growth Hormone–Glucagon Stimulation Test *Dose: 1 mg glucagon, IM or SC*	S: baseline, then hourly for 3–4 h	>7 ng/mL after stimulation or >5 ng/mL rise above baseline	× 1	>7 μg/L after stimulation or >5 μg/L rise above baseline
Growth Hormone–L-Dopa Stimulation Test *Dose, adult: 500 mg L-dopa, orally; child: 10 mg/kg*	S: baseline, 30, 60, 90, 120, and 180 min after L-dopa	Peak: >7 ng/mL or >5 ng/mL rise above baseline	× 1	>7 μg/L or >5 μg/L above baseline
Haptoglobin *RID* *Hemoglobin binding capacity* *Nephelometry*	S	83–267 mg/dL 40–180 mg Hb/dL 26–185 mg/dL	× 10 × 0.155 × 10	830–2670 mg/L 6.20–27.9 μmol Hb/L 260–1850 mg/L

HDL-Cholesterol (HDL-C) — S or P (EDTA)

		mg/dL M	*mg/dL* F		*mmol/L* M	*mmol/L* F
Mean:		45	55	× 0.0259	1.17	1.42
Range,						
Cord blood:		5–50	5–50		0.13–1.30	0.13–1.30
0–14 y:		30–65	30–65		0.78–1.68	0.78–1.68
15–19 y:		30–65	30–70		0.78–1.68	0.78–1.81
20–29 y:		30–70	30–75		0.78–1.81	0.78–1.94
30–39 y:		30–70	30–80		0.78–1.81	0.78–2.07
>40 y:		30–70	30–85		0.78–1.81	0.78–2.20

Values for blacks ~10 mg/dL higher

CHD Risk	*HDL-C,% of Total Chol* M	F		*Fraction HDL-C of Chol* M	F
Dangerous:	<7	<12	× 0.01	<0.07	<0.12
High:	7–15	12–18		0.07–0.15	0.12–0.18
Average:	15–25	18–27		0.15–0.25	0.18–0.27
Below average:	25–37	27–40		0.25–0.37	0.27–0.40
Protection probable:	>37	>40		>0.37	>0.40

Hematocrit — WB (EDTA)

	% Packed Red Cell Volume (V Red Cells/V Whole Blood × 100)		*Volume Fraction (V Red Cells/V Whole Blood)*
Cord blood:	42–60	× 0.01	0.42–0.60
1–3 d (cap):	45–67		0.45–0.67
1 wk:	42–66		0.42–0.66
2 wk:	39–63		0.39–0.63
1 mo:	31–55		0.31–0.55
2 mo:	28–42		0.28–0.42
3–6 mo:	29–41		0.29–0.41
0.5–2 y:	33–39		0.33–0.39
2–6 y:	34–40		0.34–0.40
6–12 y:	35–45		0.35–0.45
12–18 y, M:	37–49		0.37–0.49
F:	36–46		0.36–0.46
18–49 y, M:	41–53		0.41–0.53
F:	36–46		0.36–0.46

Hemoglobin — WB (EDTA)

	g/dL		*mmol/L*
Cord:	13.5–20.0	× 0.155	2.09–3.10
1–3 d (cap):	14.5–22.5		2.25–3.49
1 wk:	13.5–21.5		2.09–3.33
2 wk:	12.5–20.5		1.94–3.18
1 mo:	10.0–18.0		1.55–2.79
2 mo:	9.0–14.0		1.40–2.17
3–6 mo:	9.5–13.5		1.47–2.09
0.5–2 y:	10.5–13.5		1.63–2.09
2–6 y:	11.5–13.5		1.78–2.09
6–12 y:	11.5–15.5		1.78–2.40
12–18 y, M:	13.0–16.0		2.02–2.48
F:	12.0–16.0		1.86–2.48
18–49 y, M:	13.5–17.5		2.09–2.71
F:	12.0–16.0		1.86–2.48

Table 20-25. CLINICAL CHEMISTRY AND TOXICOLOGY (*Continued*)

Test	Specimen	Reference Range	Conversion Factor	Reference Range (International Units)
Hemoglobin (*continued*)				
	P (EDTA, ACD, or Hep)	1–4 mg/dL	× 0.155	0.16–0.62 μmol/L
	S	< 3 mg/dL with butterfly setup and 18-g needle		< 0.47 μmol/L with butterfly setup and 18-g needle
	U, fresh, random	Negative		Negative
Hemoglobin A$_{1c}$ *Electrophoresis Column*	WB (Hep, EDTA, or Ox)	5.6–7.5% of total Hb 6–9% of total Hb	× 0.01	Fraction of Hb: 0.056–0.075 0.06–0.09
Hemoglobin A$_2$ (HbA$_2$)	WB (EDTA or Ox)	Adult: 1.5–3.5% (2 SD) Lower in infants < 1 y β-Thalassemia trait: 3.75–6.5% (2 SD)	× 0.01	0.015–0.035 (2 SD) 0.038–0.065 (2 SD)
Hemoglobin (Hb) Electrophoresis	WB (EDTA, Cit, or Hep)	HbA: > 95% HbA$_2$: 1.5–3.5% HbF: < 2%	× 0.01	Hb Fraction > 0.95 0.015–0.035 < 0.02
Hemoglobin F *Alkali denaturation (White)*	WB (EDTA)	*% HbF* 1 d: 77.0 ± 7.3 5 d: 76.8 ± 5.8 3 wk: 70.0 ± 7.3 6–9 wk: 52.9 ± 11.0 3–4 mo: 23.2 ± 16.0 6 mo: 4.7 ± 2.2 8–11 mo: 1.6 ± 1.0 Adult: < 2.0	× 0.01	*Mass Fraction HbF* 0.77 ± 0.073 0.768 ± 0.058 0.70 ± 0.073 0.529 ± 0.11 0.232 ± 0.16 0.047 ± 0.022 0.016 ± 0.010 < 0.020
Hemoglobin H (HbH) *Isopropanol precipitation*	WB (ACD, EDTA, or Hep)	No precipitation at 40 min		No precipitation at 40 min
Hemopexin	S	Fetus, term: > 30% of mean adult conc. or 18% of maternal conc.	× 0.01	Fraction of adult conc.: > 0.30 Fraction of maternal conc.: 0.18
		Maternal: > 1.5 times that of non-pregnant adults		Same
		Adult: 50–115 mg/dL	× 0.01	0.50–1.15 g/L
	U	Mean: 0.2 mg/d		0.2 mg/d
Hexachlorophene	WB	*mg/L* Infant: < 0.182 Adult: < 0.089 After use: 0.1–0.655	× 2.46	*μmol/L* < 0.448 < 0.219 0.246–1.611
	Milk	< 0.009 mg/L		< 0.022
	Fat	< 0.05 mg/kg	× 2.46	< 0.123 μmol/kg
Homocystine	U, random	Negative		Negative
Homogentisic Acid	U, random	Negative		Negative
Homovanillic Acid (HVA)	U, 24 h	Child: 3–16 μg/mg creatinine	× 0.621	1.86–9.94 mmol/mol creatinine
		Adult: < 15 mg/d	× 5.49	< 82 μmol/d
Hydrogen Sulfide	WB (F$^-$/Ox)	< 0.05 mg/L Toxic: > 0.90 mg/L	× 29.3	< 1.5 μmol/L > 26.4 μmol/L
β-Hydroxybutyric Acid	S or P	Undetectable		Undetectable

Table 20-25. CLINICAL CHEMISTRY AND TOXICOLOGY (*Continued*)

Test	Specimen	Reference Range	Conversion Factor	Reference Range (International Units)
17-Hydroxycorticosteroids (17-OHCS)	U, 24 h	*mg/d*	× 2.76	*μmol/d*
		0–1 y: 0.5–1.0		1.4–2.8
		Child: 1.0–5.6		2.8–15.5
		Adult, M: 3.0–10.0		8.2–27.6
		F: 2.0–8.0		5.5–22
				(Conversion based on hydrocortisone, M.W. 362)
		or 3–7 mg/g creatinine	× 0.312	or 0.9–2.5 mmol/mol creatinine
	Amf	Increases 4–6-fold throughout preg		Same
5-Hydroxyindoleacetic Acid (5-HIAA)				
Qual.	U, random	Negative (<25 mg/d)	× 5.2	Negative (<130 μmol/d)
Quant.	U, 24 h	2–8 mg/d	× 5.2	10.4–41.6 μmol/d
17-Hydroxyprogesterone (17-OHP)	S	*ng/mL*	× 3.03	*nmol/L*
		M, Pub. stage I: 0.1–0.3		0.30–0.91
		Adult: 0.2–1.8		0.61–5.45
		F, Pub. stage I: 0.2–0.5		0.61–1.52
		Follicular: 0.2–0.8		0.61–2.42
		Luteal: 0.8–3.0		2.42–9.10
		Postmenopausal: 0.04–0.5		0.12–1.52
		Values are highest at birth but decrease rapidly during 1st week of life		
Hydroxyproline	S	*mg/dL*	× 76.3	*μmol/L*
		Premature, 1 d: 0.52 ± 0.52		40±40
		6–18 y, M: 0–0.66		0–50
		F: 0–0.58		0–44
		Adult, M: 0–0.55		0–42
		F: 0–0.46		0–34
	U, 24 h	*mg/d*	× 0.0076	*μmol/d*
		1–5 y: 20–65		0.15–0.49
		6–10 y: 35–99		0.27–0.75
		11–14 y: 63–180		0.48–1.37
		18–21 y: 20–55		0.15–0.42
		>21 y: 15–43		0.11–0.33
		Free hydroxyproline: 2–5% of total	× 0.01	Free hydroxyproline fraction of total: 0.02–0.05
Immunoglobulin A (IgA)				
Nephelometric	S	*mg/dL*	× 0.01	*g/L*
		Cord: 0–5		0–0.05
		Newborn: 0–2.2		0–0.02
		4–6 mo: 3–82		0.03–0.82
		6 mo–2 y: 14–108		0.14–1.08
		2–6 y: 23–190		0.23–1.90
		6–12 y: 29–270		0.29–2.70
		12–16 y: 81–232		0.81–2.32
		Adult: 76–390		0.76–3.90
		>60 y, M: 90–410		0.90–4.10
		F: 50–373		0.50–3.73
RIA	CSF	*mg/dL*	× 10	*mg/L*
		15–20 y: 0.07 ± 0.04 (SD)		0.7 ± 0.4 (SD)
		21–40 y: 0.07 ± 0.03		0.7 ± 0.3
		41–60 y: 0.01 ± 0.03		1.0 ± 0.3
		61–87 y: 0.11 ± 0.06		1.1 ± 0.6
Immunoglobulin D (IgD)	S	Newborn: <1.0 mg/dL	× 10	<10 mg/L
		Adult: 0–8 mg/dL		0–80 mg/L
Immunoglobulin E (IgE)	S	*IU/mL*	× 1	*KIU/L*
		Adult: 0–380		0–380
		>60 y, M: 0–250		0–250
		F: 0–175		0–175

Table 20-25. CLINICAL CHEMISTRY AND TOXICOLOGY (*Continued*)

Test	Specimen	Reference Range	Conversion Factor	Reference Range (International Units)
Immunoglobulin G (IgG)		*mg/dL*		*g/L*
Nephelometric	S	Cord: 760–1700	× 0.01	7.6–17.0
		Newborn: 700–1480		7.0–14.8
		½–6 mo: 300–1000		3.0–10.0
		6 mo–2 y: 500–1200		5.0–12.0
		2–6 y: 500–1300		5.0–13.0
		6–12 y: 700–1650		7.0–16.5
		12–16 y: 700–1550		7.0–15.5
		Adult: 650–1500		6.5–15.0
		(higher in blacks)		
RIA	CSF	*mg/dL*		*mg/L*
		15–20 y: 3.5 ± 2.0 (SD)	× 10	35 ± 20 (SD)
		21–40 y: 4.2 ± 1.4		42 ± 14
		41–60 y: 4.7 ± 1.0		47 ± 10
		61–87 y: 5.8 ± 1.6		58 ± 16
Immunoglobulin G/ Albumin Ratio	CSF and S	0.3–0.6		0.3–0.6
Immunoglobulin G Synthesis Rate	CSF and S	−9.9 to +3.3 mg/d		−9.9 to +3.3 mg/d
Immunoglobulin M (IgM)	S	*mg/dL*		*mg/L*
Nephelometric; Std: Beckman; Atlantic Antibodies; WHO		Cord: 4–24	× 10	40–240
		Newborn: 5–30		50–300
		½–6 mo: 15–109		150–1090
		6 mo–2 y: 43–239		430–2390
		2–6 y: 50–199		500–1990
		6–12 y: 50–260		500–2600
		12–16 y: 45–240		450–2400
		Adult: 40–345		400–3450
		Results vary with std. preparation		
	CSF	*mg/dL*		*mg/L*
		15–20 y: 0.02 ± 0.009 (SD)	× 10	0.2 ± 0.09 (SD)
		21–40 y: 0.016 ± 0.003		0.16 ± 0.03
		41–60 y: 0.0017 ± 0.004		0.17 ± 0.04
		61–87 y: 0.0017 ± 0.005		0.17 ± 0.05
Insulin (12 h Fasting)	S	*μIU/mL*		*mIU/L*
		Newborn: 3–20	× 1.0	3–20
		Adult: 6–24		6–24
		> 60 y: 6–35		6–35
	Amf	< 16 wk: Undetectable		Undetectable
		Term: 11.3 (mean)	× 1	11.3 (mean)
Insulin Antibodies	S	Undetectable		Undetectable
Insulin and Glucose Suppression Test	S, every 6–12 h	> 50 mg glucose/dL during a 72-h period of fasting, with values slightly lower in females	× 0.0555	> 2.8 mmol glucose/L during a 72 h period of fasting, with values slightly lower in females
		Insulin: < 4 μIU/mL or undetectable	× 1	Insulin: < 4 mU/L or undetectable
		Normal fasting insulin-glucose ratio: < 0.3		Normal fasting insulin-glucose ratio: < 5.4
Insulin with Oral Glucose Tolerance Test	S	*Min* *Insulin, μIU/mL*		*mIU/L*
		0: 6–24	× 1	6–24
		30: 25–231		25–231
		60: 18–276		18–276
		120: 16–166		16–166
		180: 4–38		4–38

Table 20-25. CLINICAL CHEMISTRY AND TOXICOLOGY (*Continued*)

Test	Specimen	Reference Range	Conversion Factor	Reference Range (International Units)
Insulin Tolerance Test *Dose: 0.1–0.15 U/kg, IV*	S	*Glucose*: Decrease ∼50% of the fasting level by 30 min and return to normal fasting limits by 90–120 min	× 0.01	Fractional decrease in glucose ∼0.50 of the fasting level by 30 min and return to normal fasting limits by 90–120 min
		hGH: Increase of > 5 ng/mL within 60 min of hypoglycemia	× 1	*hGH*: Increase of > 5 μg/L within 60 min of hypoglycemia
		Cortisol: Increase of > 6 μg/dL with peak of > 20 μg/dL	× 27.59	*Cortisol*: Increase of > 165 nmol/L with peak of > 552 nmol/L
Intrinsic Factor, *see Vitamin B$_{12}$ Intrinsic Factor*				

		(mL/min)/1.73 m^2		*(mL/s)/m^2*
Inulin Clearance Test	S and U			
		M F		M F
		20–29 y: 90–174 84–156	× 0.00963	0.87–1.68 0.81–1.50
		30–39 y: 88–168 82–150		0.85–1.62 0.79–1.44
		40–49 y: 78–162 82–146		0.75–1.56 0.79–1.41
		50–59 y: 68–152 66–142		0.65–1.46 0.63–1.37
		60–69 y: 57–137 58–130		0.55–1.32 0.56–1.25
		70–79 y: 42–122 45–121		0.40–1.17 0.43–1.17
		80–89 y: 39–105 39–105		0.38–1.01 0.38–1.01

Test	Specimen	Reference Range	Conversion Factor	Reference Range (International Units)
Iron *ICSH*	S	*μg/dL*		*μmol/L*
		Newborn: 100–250	× 0.179	17.90–44.75
		Infant: 40–100		7.16–17.90
		Child: 50–120		8.95–21.48
		Adult, M: 65–170		11.64–30.43
		F: 50–170		8.95–30.43
		Strongly method dependent		
Iron-Binding Capacity, Total (TIBC)	S	Infant: 100–400 μg/dL	× 0.179	17.90–71.60 μmol/L
		Thereafter: 250–450 μg/dL		44.75–80.55 μmol/L
Iron Saturation	S	20–55%	× 0.01	Fraction saturation 0.20–0.55
Isocitrate Dehydrogenase (ICD), 30 °C	S	1.2–7.0 U/L		1.2–7.0 U/L
Isoleucine	S	*mg/dL*	× 76.3	*μmol/L*
		Premature, 1 d: 0.52 ± 0.26 (SD)		40 ± 20 (SD)
		Newborn, 1 d: 0.35–0.69		27–53
		1–3 mo: 0.77 ± 0.18 (SD)		59 ± 14 (SD)
		2–6 mo: 0.50–1.61		38–123
		9 mo–2 y: 0.34–1.23		26–94
		3–10 y: 0.37–1.10		28–84
		6–18 y: 0.50–1.24		38–95
		Adult: 0.48–1.28		37–98
	U, 24 h	*mg/d*		*μmol/d*
		10 d–7 wk: trace–0.4	× 7.62	trace–3
		3–12 y: 2–7		15–53
		Adult: 2–24		15–183
		or 3 ± 1 mg/g creatinine	× 0.86	or 2.6 ± 0.9 mmol/mol creatinine
17-Ketogenic Steroids (17-KGS)	U, 24 h	*mg/d*	× 3.467	*μmol/d*
		0–1 y: < 1.0		< 3.5
		1–10 y: < 5		< 17
		11–14 y: < 12		< 42
		Adult, M: 5–23		17–80
		F: 3–15		10–52
		> 70 y, M: 3–15		10–52
		F: 3–13		10–45
				(Conversion based on dehydroepiandrosterone, M.W. 288)

Table 20-25. CLINICAL CHEMISTRY AND TOXICOLOGY (*Continued*)

Test	Specimen	Reference Range		Conversion Factor	Reference Range (International Units)
Ketone Bodies					
Qual.	S	Negative			Negative
	U, random	Negative			Negative
Quant.	S	0.5–3.0 mg/dL		× 10	5–30 mg/L
17-Ketosteroids	U, 24 h		*mg/d*		*μmol/d*
(17 KS), Total		14 d–2 y:	≤ 1	× 3.467	≤ 3.5
Zimmerman reaction		2–6 y:	≤ 2		≤ 7
		6–10 y:	1–4		3.5–14
		10–12 y:	1–6		3.5–21
		12–14 y:	3–10		10–35
		14–16 y:	5–12		17–42
		Adult,			
		M, 18–30 y: 9–22			31–76
		M, > 30 y: 8–20			28–70
		F: 6–15			21–52
		Decreases with age			Decreases with age
GLC		Adult, M: 5.0–12.0			17–42
		F: 3.0–10.0			10–35
					(Conversion based on dehydroepiandrosterone, M.W. 288)
17-Ketosteroids (17 KS)					
Fractions	U, 24h		*mg/d*		*μmol/d*
Androsterone		0–1 y:	<0.1	× 3.44	<0.34
		1–5 y:	<0.3		<1.03
		6–9 y:	0.1–1.0		0.34–3.44
		10–15 y, M:	0.2–2.0		0.69–6.88
		F:	0.5–3.0		1.72–10.32
		Adult: M:	2.2–5.0		7.57–17.20
		F:	0.5–2.4		1.72–8.26
Dehydroepiandrosterone		0–5 y:	<0.1	× 3.47	<0.35
		6–9 y:	<0.2		<0.69
		10–15 y:	<0.4		<1.39
		Adult, M:	<2.3		<7.98
		F:	<1.5		<5.21
Etiocholanolone		0–1 y:	<0.1	× 3.44	<0.34
		1–2 y:	<0.4		<1.38
		3–5 y:	<0.7		<2.41
		6–9 y:	0.3–1.0		1.03–3.44
		10–15 y, M:	0.1–1.6		0.34–5.50
		F:	0.7–3.1		2.41–10.66
		Adult, M:	1.9–4.7		6.54–16.17
		F:	1.1–3.5		3.78–12.04
11-β-Hydroxyandrosterone		0–2 y:	<0.3	× 3.26	<0.98
		3–5 y:	<0.4		<1.30
		6–9 y:	0.4–1.0		1.30–3.26
		10–15 y, M:	0.1–1.1		0.33–3.59
		F:	0.2–1.0		0.65–3.26
		Adult, M:	0.5–1.3		1.63–4.24
		F:	0.2–0.6		0.65–1.96
11-β-Hydroxyetiocholanolone		0–2 y:	<0.1	× 3.26	<0.33
		3–5 y:	<0.4		<1.30
		6–9 y:	0.1–0.5		0.33–1.63
		10–15 y, M:	<0.3		<0.98
		F:	0.1–0.5		0.33–1.63
		Adult, M:	0.3–0.7		0.98–2.28
		F:	0.2–0.6		0.65–1.96
11-Ketoandrosterone		≤ 15 y:	<0.1	× 3.27	<0.33
		Adult, M:	0.2–1.0		0.65–3.27
		F:	0.2–0.8		0.65–2.62
11-Ketoetiocholanolone		0–2 y:	<0.1	× 3.27	<0.33
		3–5 y:	<0.4		<1.30
		6–9 y:	0.1–0.7		0.33–2.29
		10–15 y, M:	0.2–0.6		0.65–1.96
		F:	0.1–0.6		0.33–1.96
		Adult, M:	0.2–1.0		0.65–3.27
		F:	0.2–0.8		0.65–2.62

Table 20-25. CLINICAL CHEMISTRY AND TOXICOLOGY (*Continued*)

Test	Specimen	Reference Range	Conversion Factor	Reference Range (International Units)
17-Ketosteroids (17KS) Fractions	U, 24 h			
Beta-Alpha ratio		< 0.2		< 0.2
Alpha-Beta ratio		> 5		> 5
L-Lactate		*mg/dL*		*mmol/L*
	P (NaF)	Venous: 4.5–19.8	× 0.111	0.5–2.2
		Arterial: 4.5–14.4		0.5–1.6
	WB (Hep)	At bed rest,		
		Venous: 8.1–15.3		0.9–1.7
		Arterial: < 11.3		< 1.25
	U, 24 h	496–1982 mg/d	× 0.0111	5.5–22 mmol/d
	CSF	< 25.2 mg/dL	× 0.111	< 2.8 mmol/L
Qual.	Gastf	Negative		Negative
Lactate Dehydrogenase (LDH)				
Total (L → P), 30 °C	S	*U/L*		Same
		Newborn: 160–450		
		Neonate: 300–1500		
		Infant: 100–250		
		Child: 60–170		
		Adult: 45–90		
		> 60 y: 55–100		
Total (P → L), 30 °C		150–320	× 0.01	~0.10 fraction of serum value
	CSF	~10% of serum value		
		% of Total		*Fraction of Total*
Isoenzymes, Agarose	S	Fraction 1: 14–26	× 0.01	0.14–0.26
		Fraction 2: 29–39		0.29–0.39
		Fraction 3: 20–26		0.20–0.26
		Fraction 4: 8–16		0.08–0.16
		Fraction 5: 6–16		0.06–0.16
Lactate Dehydrogenase (LDH) in Erythrocytes	WB (ACD, EDTA, or Hep)	Adult:		
		200 ± 26.5 U/g Hb (SD)	× 0.0645	12.9 ± 1.7 MU/mol Hb
		5800 ± 768 U/10^{12} RBC	× 10^{-3}	5.80 ± 0.77 nU/RBC
		68.0 ± 9.01 U/mL RBC	× 1	68.0 ± 9.01 kU/L RBC (Conv. factor based on hemoglobin, M.W. 64 500)
Lactate/Pyruvate Ratio	WB (Hep)	10/1		10/1
Lactose	S	< 0.5 mg/dL	× 29.21	< 14.6 μmol/L
	U	12–40 mg/dL	× 29.21	350–1168 μmol/L
LDL-Cholesterol (LDL-C)	S or P (EDTA)	*mg/dL*		*mmol/L*
Calculated		M F		M F
		Cord blood: 10–50 10–50	× 0.0259	0.26–1.30 0.26–1.30
		0–19 y: 60–140 60–150		1.55–3.63 1.55–3.89
		20–29 y: 60–175 60–160		1.55–4.53 1.55–4.14
		30–39 y: 80–190 70–170		2.07–4.92 1.81–4.40
		40–49 y: 90–205 80–190		2.33–5.31 2.07–4.92
		50–59 y: 90–205 90–220		2.33–5.31 2.33–5.70
		60–69 y: 90–215 100–235		2.33–5.57 2.59–6.09
		> 70 y: 90–190 95–215		2.33–4.92 2.46–5.57
		Recommended (desirable) range for adults:		
		65–175 mg/dL		1.68–4.53
Lead	WB (Hep)	*μg/dL*		*μmol/L*
		Child: < 30	× 0.0483	< 1.45
		Adult: < 40		< 1.93
		Toxic: ≥ 100		≥ 4.83
	U, 24 h	< 80 μg/L	× 0.00483	< 0.39 μmol/L
Lecithin-Cholesterol Acyltransferase 37 °C	P (Hep, EDTA, or Cit) or S	(92 ± 14.2 μmol/h)/L (SD)	× 0.0167	153 ± 0.24 U/L plasma (SD)
Lecithin/Sphingomyelin (L/S) Ratio	Amf	2.0–5.0 indicates probable fetal lung maturity; > 3.0 in diabetics		Same
Lecithin Phosphorus	Amf	> 0.10 mg/dL indicates probable adequate fetal lung maturity	× 0.3229	> 0.33 mmol/L indicates probable adequate fetal lung maturity

Table 20-25. CLINICAL CHEMISTRY AND TOXICOLOGY (*Continued*)

Test	Specimen	Reference Range	Conversion Factor	Reference Range (International Units)
Leucine		*mg/dL*		*μmol/L*
		Premature, 1 d: 0.92 ± 0.33 (SD)	× 76.3	70 ± 25 (SD)
		Newborn, 1 d: 0.62 ± 1.43		47 ± 109
		1–3 mo: 1.36 ± 0.39		104 ± 30
		9 mo–2 y: 0.59–2.03		45–155
		3–10 y: 0.73–2.33		56–178
		6–18 y: 1.03–2.28		79–174
		Adult: 0.98–2.29		75–175
	U, 24 h	*mg/d*		*μmol/d*
		10 d–7 wk: 0.9–2.0	× 7.624	7–15
		3–12 y: 3–11		23–84
		Adult: 3–70		23–533
		or 4 ± 2 mg/g creatinine	× 0.86	or 3.4 ± 1.7 mmol/mol creatinine
Leukocyte Count (WBC Count)	WB (EDTA)	*Cells × 10³/μL*		*Cells × 10⁹/L*
		Birth: 9.0–30.0	× 10⁶	9.0–30.0
		24 h: 9.4–34.0		9.4–34.0
		1 mo: 5.0–19.5		5.0–19.5
		1–3 y: 6.0–17.5		6.0–17.5
		4–7 y: 5.5–15.5		5.5–15.5
		8–13 y: 4.5–13.5		4.5–13.5
		Adult: 4.5–11.0		4.5–11.0
	CSF	0–5 mononuclear cells/μL		0–5 × 10⁶cells/L
Lipase	S			
Tietz method		< 1.0 unit/mL	× 278	< 278 U/L
BMD turbidimetric		Adult: 10–150 U/L		Same
		> 60 y: 18–180 U/L		
β-Lipoprotein (LDL, Low-Density Lipoprotein)	S, 12–14 h fasting	28–53% of total lipoproteins	× 0.01	Mass fraction of total: 0.28–0.53
Lipoprotein Electrophoresis	S	Distinct β-band; negligible chylomicron and pre-β bands		
Long-Acting Thyroid Stimulating Hormone (LATS)	S	Undetectable		Undetectable
Luteinizing Hormone (hLH)	S or P (Hep); avoid EDTA	*mIU/mL*		*IU/L*
		1–3 mo, M: 22.3 ± 13.4 (SD)	× 1	22.3 ± 13.4 (SD)
		F: 17.4 ± 9.6		17.4 ± 9.6
		3–5 mo, M: 15.5 ± 11.8		15.5 ± 11.8
		F: 13.2 ± 7.6		13.2 ± 7.6
		5–7 mo, M: 17.1 ± 8.0		17.1 ± 8.0
		F: 13.4 ± 8.0		13.4 ± 8.0
		7–12 mo, M: 24.0 ± 18.3		24.0 ± 18.3
		Prepuberty, phase I, M: 3.9 ± 2.1		3.9 ± 2.1
		M, 10–13 y: 4–12		4–12
		12–14 y: 6–12		6–12
		12–17 y: 6–16		6–16
		15–18 y: 7–19		7–19
		Adult: 6–23		–23
		F, 8–12 y: 2.0–11.5		2.0–11.5
		9–14 y: 2.0–14.0		2.0–14.0
		12–18 y: 3.0–29.0		3.0–29.0
		Follicular: 5–30		5–30
		Midcycle: 75–150		75–150
		Luteal: 3–40		3–40
		Postmenopausal: 30–200		30–200
	U	*IU/d*		
		1 wk–2 y: 0.05–3.3		Same
		2–5 y: 0.07–1.2		
		5–8 y: 0.25–2.2		
		8–11 y: 0.19–2.2		
		11–13 y: 0.48–11.28		
		13–15 y: 2.6–27.6		
		15–17 y: 4.6–24.0		
		Adult, M: 13–60		
		F,		
		Follicular: 7.2–23.5		

Table 20-25. CLINICAL CHEMISTRY AND TOXICOLOGY (*Continued*)

Test	Specimen	Reference Range	Conversion Factor	Reference Range (International Units)
Lysozyme (Muramidase)	S or P (EDTA)	5–15 µg/mL	× 1	5–15 mg/L
Macroamylase	S	Present in ~1% of healthy subjects with normal serum amylase activity and in 2.5% of patients with abnormal activity		Same
Magnesium	S	*mEq/L* Newborn, 2–4 d: 1.0–1.8 5 mo–6 y: 1.65 ± 0.23 (2 SD) 6–12 y: 1.56 ± 0.18 12–20 y: 1.56 ± 0.21 Adult: 1.3–2.1 (Higher in females during menses)	× 0.5	*mmol/L* 0.5–0.9 0.83 ±0.12 (2 SD) 0.78 ±0.09 0.78 ±0.11 0.65 –1.05
	U, 24 h	6.0–10.0 mEq/d		3.00–5.00 mmol/d
Melanin	U, random	Negative		Negative
Mercury	WB (EDTA)	<5.0 µg/dL	× 0.0499	<0.25 µmol/L
	U, 24 h	<20 µg/L	× 0.00499	<0.1 µmol/L
		Toxic: >150 µg/L		>0.75 µmol/L
Metanephrine, Total	U, 24 h	*µg/mg creatinine* <1 y: 0.001–4.60 1–2 y: 0.27–5.38 2–5 y: 0.35–2.99 5–10 y: 0.43–2.70 10–15 y: 0.001–1.87 15–18 y: 0.001–0.67 Adult: 0.05–1.20	× 0.574	*mmol/mol creatinine* 0.0006–2.64 0.15–3.09 0.20–1.72 0.25–1.55 0.0006–1.07 0.0006–0.38 0.03–0.69
Methanol	WB (F⁻/Ox)	*mg/L* 1.5	× 0.0312	*mmol/L* 0.05
		Toxic: >200		>6.24
	U	Occup. exp.: <50		<1.56
	Br	<0.8 ppm	× 0.0312	<0.02
		Occup. exp.: <2.5 ppm		<0.08
Methemoglobin (MetHb, Hemiglobin)	WB (EDTA, Hep, or ACD)	0.06–0.24 g/dL or 0.78±0.37% or total Hb (SD)	× 155 × 0.01	9.3–37.2 µmol/L* Mass fraction of total Hb: 0.008 ± 0.0037 (SD) (*Conv. factor based on hemoglobin, M.W. 64 500)
Methionine	S	*mg/dL* Premature, 1 d: 0.52 ± 0.07 (SD) Newborn, 1 d: 0.13 – 0.61 1–3 mo: 0.31 ± 0.13 (SD) 2–6 mo: 0.24 – 0.73 9 mo–2 y: 0.04 – 0.43 3–10 y: 0.16 – 0.24 6–18 y: 0.24 – 0.55 Adult: 0.09 – 0.60	× 67.1	*µmol/L* 35 ±5 (SD) 9 –41 21 ±9 (SD) 16 –49 3 –29 11 –16 16 –37 6 –40
	U, 24 h	*mg/d* 10 d–7 wk: 0.8–2.0 3–12 y: 3 –14 Adult: trace–9.0	× 6.70	*µmol/d* 5 –13 20 –94 trace–60
		or 4.5 ± 2.5 mg/g creatinine (SD)	× 0.76	or 3.4 ± 1.9 mmol/mol creatinine (SD)
Metyrapone (Metopyrone) Stimulation Test *Dose, adult: 750 mg q 4 h × 6; child: 300 mg/m²*	S	11-Deoxycortisol: >7.0 µg/dL	× 28.86	11-Deoxycortisol: >200 nmol/L
		Cortisol: <8 µg/dL	× 27.59	Cortisol: <220 nmol/L
	U, 24	17-KGS: 2.5–3-fold rise, but at least 10 mg/d	× 3.467	17-KGS: 2.5–3-fold rise but at least 35 µmol/d*
		17-KS: >2× base level		Same
		17-OHCS: 3–5× base level		Same *(Conv. factor based on DHEA, M.W. 288)

Table 20-25. CLINICAL CHEMISTRY AND TOXICOLOGY (*Continued*)

Test	Specimen	Reference Range		Conversion Factor	Reference Range (International Units)
Metyrapone (Metopyrone) Stimulation Test (*continued*) *Single Dose Metyrapone Test Dose: 30 mg/kg orally with milk or snack at midnight*	S, 0800 h	11-Deoxycortisol: > 7 µg/dL		× 28.86	> 200 nmol/L
Microsomal Antibodies, Thyroid, *see Thyroid Microsomal Antibodies*					
Molybdenum	S	0.58 ± 0.21 ng/mL (SD)		× 10.42	6.04 ± 2.19 nmol/L (SD)
Mucopolysaccharide Screen (MPS)	U, random	Negative			Negative
Myelin Basic Protein	CSF	< 4 ng/mL		× 1	< 4 µg/L
Myoglobin	S	M: 49 ± 17 µg/L (SD) F: 35 ± 14 µg/L Increases slightly with age			Same
	U, random	Negative			Negative
Niacin	U, 24 h	0.3–1.5 mg/d (mean: 0.6)		× 8.113	2.43–12.17 µmol/d (mean: 4.87)
Nickel	S or P (Hep)		*µg/dL* 0.11 – 0.46	× 170	*nmol/L* 18.7 – 78.2
	WB (Hep)	M: F:	0.45 ± 0.14 (SD) 0.53 ± 0.11		76.5 ± 23.8 (SD) 90.1 ± 18.7
	U, 24 h	M: F:	2.6 ± 1.3 µg/d (SD) 2.2 ± 0.8 µg/d	× 17	44.2 ± 22.1 nmol/d (SD) 37.4 ± 13.6 nmol/d
	F		260 ± 120 µg/d (SD)	× 0.017	4.4 ± 2.1 µmol/d (SD)
	H	M: F:	0.24 ± 0.09 µg/kg 0.19 ± 0.04 µg/kg	× 17	4.1 ± 1.5 nmol/kg 3.2 ± 0.7 nmol/kg
Nitrites	U	Negative Toxic: Nitrites detected			Same
Nitrogen, Total	F	Infants: 0.11–0.52 g N/d Adults: < 2 g N/d		× 71.4	7.9–37 mmol N/d < 143 mmol N/d
Nonprotein Nitrogen (NPN)	WB (Ox) S	< 50 mg/dL < 35 mg/dL		× 0.714	< 35.7 mmol/L < 25.0 mmol/L
Normetanephrine, Total	P (EDTA and Na metabisulfite)	Normotensive: 1.2 ± 0.1 ng/mL (SEM) Primary hypertensive: 2.5 ± 0.2 ng/mL (SEM)		× 5.46	6.55 ± 0.55 nmol/L 13.65 ± 1.09 nmol/L
Occult Blood	F, random	Negative (< 2 mL blood/150 g stool per d)		× 6.67	Negative (< 13.3 mL blood/kg stool per d)
Qual.	U, random	Negative			Negative
Oleic Acid–[125]I Absorption Test *Dose: 50 µCi in milk*	P	> 1.7% of administered dose/L after 4–6 h		× 0.01	Fraction of administered dose: > 0.017
	F, 72 h	Less than 5% of administered dose in 72 h specimen		× 0.01	Fraction of administered dose: < 0.05
Orosomucoid, *see* α_1*-Acid Glycoprotein*					
Osmolality	S	Neonate: May be as low as 266 mOsmol/kg Child, adult: 275–295 mOsmol/kg > 60 y: 280–301 mOsmol/kg			Same
	U, random	50–1400 mOsmol/kg, depending on fluid intake			

Table 20-25. CLINICAL CHEMISTRY AND TOXICOLOGY (*Continued*)

Test	Specimen	Reference Range	Conversion Factor	Reference Range (International Units)
Osmolality (*continued*)	U, random			Same
		After 12 h fluid restriction: > 850 mOsmol/kg		
	U, 24 h	~300–900 mOsmo l/kg		
Osmolality Ratio, Urine/Serum	U and S	1.0–3.0 > 3.0 after 12 h fluid restriction		Same
Oxalate	S	1.0–2.4 µg/mL	× 11.4	11–27 µmol/L
		Ethylene glycol poisoning: > 20 µg/mL		> 228 µmol/L
	U, 24 h	8–40 mg/d	× 11.4	91–456 µmol/d
		Ethylene glycol poisoning: > 150 mg/d		> 1710 µmol/d
Oxygen, P$_{50}$	WB (Hep)	Newborn: 18–24 mm Hg	× 0.133	2.39–3.19 kPa
		Adult, adjusted to pH (P) 7.4: 25–29 mm Hg		3.33–3.86 kPa
Oxygen, Partial Pressure (*p*O$_2$)	WB, arterial (Hep)	*mm Hg*	× 0.133	*kPa*
		Birth: 8–24		1.1–3.2
		5–10 min: 33–75		4.4–10.0
		30 min: 31–85		4.1–11.3
		> 1 h: 55–80		7.3–10.6
		1 d: 54–95		7.2–12.6
		Thereafter: 83–108		11.04–14.36
		(Decreases with age and high altitude)		
Oxygen Saturation	WB, arterial (Hep)		× 0.01	Fraction saturated:
		Newborn: 40–90%		0.40–0.90
		Thereafter: 95–98%		0.95–0.98
Oxytocin	P (EDTA)	< 3.2 µIU/mL	× 1	< 3.2 mIU/L
*p*O$_2$, *see Oxygen, Partial Pressure*				
Palmitic Acid, Total	Amf	Fetal maturity: > 11.5 mg/L	× 3.90	> 44.9 µmol/L
		Fetal maturity: > 8.0 mg/L		> 31.2 µmol/L
Pancreatic Polypeptide	S	*hPP pmol/L*		Same
		Age *mean ± SE*		
		20–29 12.9 ± 1.0		
		30–39 27.4 ± 2.9		
		40–49 39.3 ± 3.1		
		50–59 43.1 ± 6.7		
		60–69 49.3 ± 6.7		
Pantothenic Acid	WB (NaCit) or S	Total: 1.03–1.83 µg/mL	× 4.56	4.70–8.34 µmol/L
Parathyroid Hormone (hPTH)	S	Varies with laboratory,		
		Mayo Clinic, Bioscience:		
		N-terminal 230–630 pg/mL	× 1	230–630 ng/L
		C-terminal 430–1860 pg/mL		430–1860 ng/L
		Nichols Institute:		
		C-terminal 40–100 µLEq/mL		40–100 mLEq/L
		Immuno Nuclear:		
		Midmolecule 0.29–0.85 ng/mL	× 100	29–85 pmol/L
		Higher in pregnancy		
Pentachlorophenol (PCP)	P (EDTA)	*mg/L*	× 3.75	*µmol/L*
		< 1		< 3.8
		Occup. exp.: < 20		< 75.0
		Toxic: > 30		> 112
	U	< 0.6		< 2.3
		Occup. exp.: < 39		< 146
		Toxic: > 60		> 225
Pentoses	U, 24 h	*Total Pentoses*		
		Fruit-free diet: (2–5 mg/kg)/d	× 6.66	(13.3–33.3 µmol/kg)/d
		or 225 mg/d	× 0.00666	1.50 mmol/d
		Slightly higher in children		

Table 20-25. CLINICAL CHEMISTRY AND TOXICOLOGY (*Continued*)

Test	Specimen	Reference Range		Conversion Factor	Reference Range (International Units)
Pentoses (*continued*)	U, 24 h				
		L-Xylulose:	<60 mg/d	× 6.66	<400 μmol/d
		D-Ribose:	<15 mg/d	× 6.66	<100 μmol/d
		D-Ribulose:	traces		
	S	L-Xylulose:	<2 mg/dL	× 66.6	<133.2 μmol/L
Pepsinogen (PG I)	S		*ng/mL ± SEM*		*μg/L ± SEM*
		Premature:	22 ± 2	× 1	22 ± 2
		Cord:	26 ± 2		26 ± 2
		<1 y:	77 ± 5		77 ± 5
		1–2 y:	98 ± 8		98 ± 8
		3–6 y:	92 ± 12		92 ± 12
		7–10 y:	95 ± 8		95 ± 8
		11–14 y:	107 ± 11		107 ± 11
		19–38 y:	133 ± 9		133 ± 9
		Women at delivery:	127 ± 11		127 ± 11
pH (37 °C)	WB, arterial (Hep)	Premature, 48 h:	7.35–7.50		Same
		Birth, full term:	7.11–7.36		
		5–10 min:	7.09–7.30		
		30 min:	7.21–7.38		
		>1 h:	7.26–7.49		
		1 d:	7.29–7.45		
		Thereafter:	7.35–7.45		
		or H^+ conc.:	36–44 nEq/L	× 1	36–44 nmol/L
	U, random	Newborn/neonate:	5–7		Same
		Thereafter:	4.5–8		
		Average	∼6		
	Serf (Plf, Pericf, Ascf)	6.8–7.6			
	Synf	Parallels serum			
	CSF	7.35–7.40			
	Semf	7.2–8.0 (average: 7.8)			
	F	7.0–7.5			
		(May be acid with high lactose intake)			
Phenols	U		*mg/L*		*μmol/L*
		Phenol:	5–8	× 10.6	53–85
		p-Cresol:	20–200	× 9.25	185–1850
		Toxic,			
		Phenol:	>10	× 10.6	>106
		o-Cresol:	>2	× 9.25	>19
		m-Cresol:	>20	× 9.25	>19
		p-Cresol:	>200	× 9.25	>1850
Phenolsulfonphthalein Test (PSP Test) *Dose: 1 mL (6 mg), IV, 30 min after patient has voided and drunk 600 mL water*	U, timed		*% Dose Excreted*		*Fraction of Dose Excreted*:
		15 min:	28–51	× 0.01	0.28–0.51
		30 min:	13–24		0.13–0.24
		60 min:	9–17		0.09–0.17
		120 min:	3–10		0.03–0.10
		Total 2 h:	63–84		0.63–0.84
Phenylalanine	S		*mg/dL*		*mmol/L*
		Premature:	2.0–7.5	× 0.06054	0.12–0.45
		Newborn:	1.2–3.4		0.07–0.21
		Adult:	0.8–1.8		0.05–0.11
	U, 24 h		*mg/d*		*μmol/d*
		10 d–7 wk:	1–2	× 6.05	6–12
		3–12 y:	4–18		24–109
		Adult:	trace–17		trace–103
		or 6 ± 2 mg/g creatinine (SD)		× 0.68	or 4.1 ± 1.4 mmol/mol creatinine (SD)
Phenylpyruvic Acid *Qual.*	U, random	Negative $FeCl_3$ test			Negative by $FeCl_3$ test
Phosphatase, Acid *Prostatic (RIA) Roy, Brower, and Hayden, 37 °C*	S	<3.0 ng/mL 0.11–0.60 U/L		× 1	<3.0 μg/L 0.11–0.60 U/L

Table 20-25. CLINICAL CHEMISTRY AND TOXICOLOGY (*Continued*)

Test	Specimen	Reference Range	Conversion Factor	Reference Range (International Units)
Phosphatase, Alkaline *p-Nitrophenyl phosphate, carbonate buffer, 30°C*	S	*U/L* Infant: 50–165 Child: 20–150 Adult: 20–70 > 60 y: 30–75		Same
Bowers and McComb, 30°C *IFCC, 30°C*		Adult: 25–90 M: 30–90 F: 20–80		
Phosphatidylglycerol (PG)	Amf	Absent: Fetal immaturity Present: Fetal maturity		Same
Phosphofructokinase (PFK) in Erythrocytes *IFCC, 30°C*	WB (ACD, EDTA, or Hep)	9.05 ± 1.89 U/g Hb (SD) 262 ± 55 U/10^{12} RBC 3.08 ± 0.64 U/mL RBC Lower in newborns	$\times$ 0.0645 $\times 10^{-3}$ $\times$ 1	0.58 ± 0.12 MU/mol Hb 0.26 ± 0.05 nU/RBC 3.08 ± 0.64 kU/L RBC
Phospholipids, Total	S or P (EDTA)	*mg/dL* Newborn: 75–170 Infant: 100–275 Child: 180–295 Adult: 125–275 > 65 y: 196–366	$\times$ 0.01	*g/L* 0.75–1.70 1.00–2.75 1.80–2.95 1.25–2.75 1.96–3.66
Phosphorus, Inorganic	S	*mg/dL* Cord: 3.7–8.1 24 h: 3.5–8.6 24–48 h: 5.5–9.5 Infant: 4.5–6.5 Child: 4.5–5.5 Thereafter: 2.7–4.5 > 60 y, M: 2.3–3.7 F: 2.8–4.1	$\times$ 0.323	*mmol/L* 1.20–2.62 1.13–2.78 1.78–3.07 1.45–2.10 1.45–1.78 0.87–1.45 0.74–1.20 0.90–1.32
	U, 24 h	Adults on diet containing 0.9– 1.5 g P and 10 mg Ca/kg: < 1.0 g/d	$\times$ 32.3	Adults on diet containing 29–48 mmol P and 0.25 mmol Ca/kg: < 32.3 mmol/d
		On nonrestricted diet: 0.4–1.3 g/d		On nonrestricted diet: 12.9–42.0 mmol/d
Placental Lactogen (hPL)	S	*μg/mL* Nonpregnant F: < 0.5 Pregnancy (wk): 22: 1.0–3.8 26: 1.5–4.5 30: 2.8–5.8 34: 3.4–6.9 38: 3.6–8.2 42: 3.0–8.0 In diabetic patients, mean values at term: 11.0 ± 1.0	$\times$ 1	*mg/L* < 0.5 1.0–3.8 1.5–4.5 2.8–5.8 3.4–6.9 3.6–8.2 3.0–8.0 11.0 ± 1.0
	Amf	Rises slowly during pregnancy < 20 wk: 0.3–0.4 μg/mL (mean) 36–40 wk: 0.4–0.6 μg/mL (mean)	$\times$ 1 $\times$ 1	0.3–0.4 mg/L (mean) 0.4–0.6 mg/L (mean)
	T, Placent.	300 μg/g	$\times$ 1	300 mg/kg
Plasma Volume	P (Hep)	M: 25–43 mL/kg F: 28–45 mL/kg	$\times$ 0.001	0.025–0.043 L/kg 0.028–0.045 L/kg
Porphobilinogen (PBG) *Quant.* *Qual.*	U, 24 h U, random	0–2.0 mg/d Negative	$\times$ 4.42	0–8.8 μmol/d Negative
Postheparin Lipolytic Activity	P (Hep)	*(μmol FFA/min)/mL* TG < 200 mg/dL: 0.24–0.57 TG > 200 mg/dL: 0.06–0.71	$\times$ 1000	*U/L* 240–570 60–710
Potassium	S	*mEq/L* Premature, Cord: 5.0–10.2 48 h: 3.0–6.0	$\times$ 1	*mmol/L* 5.0–10.2 3.0–6.0

Table 20-25. CLINICAL CHEMISTRY AND TOXICOLOGY (*Continued*)

Test	Specimen	Reference Range		Conversion Factor	Reference Range (International Units)
Potassium (*continued*)		Newborn,	*mEq/L*		*mmol/L*
		Cord:	5.6–12.0	× 1	5.6–12.0
		Newborn:	3.7–5.9		3.7–5.9
		Infant:	4.1–5.3		4.1–5.3
		Child:	3.4–4.7		3.4–4.7
		Thereafter:	3.5–5.1		3.5–5.1
	P (Hep)	M:	3.5–4.5		3.5–4.5
		F:	3.4–4.4		3.4–4.4
	U, 24 h	25–125 mEq/d; varies with diet			25–125 mmol/d; varies with diet
	CSF	70% of plasma level; 2.5–3.2 mEq/L, rises with plasma hyperosmolality		× 0.01 × 1	0.70 plasma level fraction; 2.5–3.2 mmol/L, rises with plasma hyperosmolality
	F, 24 h	~5 mEq/d		× 1	~5 mmol/d
	Gastf	~10 mmol/L; parietal and nonparietal juice have the same conc.		× 1	~10 mmol/L; parietal and nonparietal juice have the same conc.
	RBC	~105 mEq/L		× 1	~105 mmol/L
	Sal	Without stimulation:	19–23 mEq/L	× 1	19–23 mmol/L
		With stimulation:	18–19 mEq/L		18–19 mmol/L
	Swt	5–17 mEq/L		× 1	5–17 mmol/L
Pregnancy Tests *Chorionic Gonadotropin (hCG) Tube Test*					
Qual.	S or U	Negative Positive by 4th–8th d after expected menstrual period			Same
Semiquant.	S or U, 24 h	Peak values up to 120 000 mIU/mL		× 1	Peak values up to 120 000 IU/L
Chorionic Gonadotropin (β-hCG), see Chorionic Gonadotropin, β-Subunit					
Radio Receptor Assay (RRA)					Same
Qual.	S	Negative; pregnancy can be detected < 10 d after conception			
Pregnanediol	U, 24 h		*mg/d*		*μmol/d*
		< 2 y:	< 0.1	× 3.12	< 0.3
		3–5 y:	< 0.3		< 0.9
		6–9 y:	< 0.5		< 1.6
		M, Adult:	0.1–0.7		0.3–2.2
		F, 10–15 y:	0.1–1.2		0.3–3.7
		Adult,			
		Follicular:	< 1.0		< 3.1
		Luteal:	2–7		6.2–21.8
		Postmenopausal:	0.2–1.0		6.2–3.1
		Pregnancy (wk),			
		16:	5–21		16–65
		20:	6–26		19–81
		24:	12–32		37–100
		28:	19–51		59–160
		32:	22–66		69–206
		36:	13–77		41–240
		40:	23–63		72–197
Pregnanetriol	U, 24 h		*mg/d*		*μmol/d*
		2 wk–2 y:	0.02–0.2	× 2.97	0.06–0.59
		2–5 y:	< 0.5		< 1.49
		5–15 y:	< 1.5		< 4.46
		> 15 y:	< 2.0		< 5.94
Pregnenolone	S	Adult:	0.3–2.0 ng/mL	× 3.16	0.95–6.32 nmol/L
Progesterone	S		*ng/mL*		*nmol/L*
		M, Pub. stage I:	0.11–0.26	× 3.18	0.35–0.83
		Adult:	0.12–0.3		0.38–0.95

Table 20-25. CLINICAL CHEMISTRY AND TOXICOLOGY (*Continued*)

Test	Specimen	Reference Range	Conversion Factor	Reference Range (International Units)
Progesterone (*continued*)		*ng/mL*		*nmol/L*
	F, Pub. stage I:	0–0.3	× 3.18	0 –0.95
	II:	0–0.46		0 –1.46
	III:	0–0.6		5 –1.91
	IV:	0.05–13.0		0.16 –41.3
	Follicular:	0.02–0.9		0.06 –2.86
	Luteal:	6.0–30.0		19.1 –95.4
Progesterone Receptor Assay (PRA)	T, tumor	*fmol/mg protein*		*nmol/kg protein*
	Normal, or benign and nonresponsive tumor:	≤ 5	× 1	≤ 5
	Positive:	> 10		> 10
Proinsulin	S	< 30% of total immunoreactive insulin-like material, or < 0.2 ng/mL	× 0.01	Fraction of immuno-reactive insulin-like material: 0.30, or < 0.2 µg/L
Prolactin (hPRL)	S	*ng/mL*		*µg/L*
	Adults, M:	< 20	× 1	< 20
	F, Follicular:	< 23		< 23
	Luteal:	5–40		5–40
	Pregnancy, 1st trimester:	< 80		< 80
	2nd trimester:	< 160		< 160
	3rd trimester:	< 400		< 400
	Newborn:	> 10-fold adult levels		
Prolactin-Insulin Stimulation Test *Dose, adult: 0.5 g arginine HCl/kg, IV, over 30 min*	S	Peak values: 1.4–19.0 × baseline within 35–75 min after stimulation		Same
Properdin	S	*mg/dL ± SD*		*mg/L ± SD*
	Cord:	1.5 ± 0.1	× 10	15 ± 1
	1 mo:	1.4 ± 0.4		14 ± 4
	6 mo:	1.9 ± 0.3		19 ± 3
	Adult:	2.8 ± 0.4		28 ± 4
Prostaglandins, E	P (Hep)	25–200 pg/mL	× 2.82	71–564 pmol/L
F		25–150 pg/mL		71–423 pmol/L
Protein *Total*	S	*g/dL*		*g/L*
	Premature:	3.6–6.0	× 10	36–60
	Newborn:	4.6–7.0		46–70
	Cord:	4.8–8.0		48–80
	1 wk:	4.4–7.6		44–76
	7 mo–1 y:	5.1–7.3		51–73
	1–2 y:	5.6–7.5		56–75
	≥ 3 y:	6.0–8.0		60–80
	Adult, Ambulatory:	6.4–8.3		64–83
	Recumbent:	6.0–7.8		60–78
	> 60 y:	lower by ∼0.2		∼2
Electrophoresis		*Albumin,* *g/dL*		*g/L*
	Adult:	3.5–5.0		35–50
	> 60 y:	3.7–4.7		37–47
	α_1-Globulin, Adult:	0.1–0.3		1–3
	> 60 y:	0.2–0.5		2–5
	α_2-Globulin, Adult:	0.6–1.0		6–10
	> 60 y:	0.5–1.1		5–11
	β-Globulin, Adult:	0.7–1.1		7–11
	> 60 y:	0.5–1.2		5–12
	γ-Globulin, Adult:	0.8–1.6		8–16
	> 60 y:	0.6–1.6		6–16

Table 20-25. CLINICAL CHEMISTRY AND TOXICOLOGY (*Continued*)

Test	Specimen	Reference Range		Conversion Factor	Reference Range (International Units)
Protein (*continued*)					
Total	U, 24 h	1–14 mg/dL		× 10	10–140 mg/L
		50–80 mg/d (at rest)			Same
		<250 mg/d after intense exercise			
Electrophoresis			*Average % of Total Protein*		*Fraction of Total*
		Albumin	37.9	× 0.01	0.379
		α_1-Globulin	27.3		0.273
		α_2-Globulin	19.5		0.195
		β-Globulin	8.8		0.088
		γ-Globulin	3.3		0.033
Total	CSF, lumbar				*mg/L*
Column		8–32 mg/dL		× 10	80–320
Turbidimetry			*mg/dL*		
		Premature:	15–130		150–1300
		Full-term newborn:	40–120		400–1200
		<1 mo:	20–80		200–800
		Thereafter:	15–40		150–400
Electrophoresis			*% of Total*		*Fraction of Total*
		Prealbumin:	2–7	× 0.01	0.02–0.07
		Albumin:	56–76		0.56–0.76
		α_1-Globulin:	2–7		0.02–0.07
		α_2-Globulin:	4–12		0.04–0.12
		β-Globulin:	8–18		0.08–0.18
		γ-Globulin:	3–12		0.03–0.12
Electrophoresis	Synf	Albumin:	63		0.63
		α_1-Globulin:	7		0.07
		α_2-Globulin:	7		0.07
		β-Globulin:	9		0.09
		γ-Globulin:	14		0.14
		Fibrinogen:	0		0
Prothrombin Time	WB (NaCit)				
One-stage (Quick)		In general:	11–15 s (varies with type of thromboplastin)		Same
		Newborn:	Prolonged by 2–3 s		
Two-stage modified (Ware and Seegers)		18–22 s			
Protoporphyrin	WB (Hep or EDTA)	<50 µg/dL RBC		× 0.0178	<0.89 µmol/L RBC
	F, 24 h	≤60 µg/g dry wt or <1500 µg/d		× 0.00178	≤0.11 mmol/kg dry wt or <2.67 µmol/d
Pseudocholinesterase (PCHE), *see* Cholinesterase II					
Pyruvic Acid	WB (Hep)	0.3–0.9 mg/dL		× 0.114	0.03–0.10 mmol/L
	U	88.1 mg/d		× 0.0114	1 mmol/d
	CSF	0.5–1.7 mg/dL		× 0.114	0.06–0.19 mmol/L
Renal Plasma Flow (RPF)	P and U	M:	560–830 mL/min	× 0.01667	9.34–13.84 mL/s
		F:	490–700 mL/min		8.17–11.67 mL/s
		or (390 mL/min)/m² body surface			or (6.50 mL/s)/m² body surface
		>40 y:	decreases ~75 mL/decade		
Renin	P (EDTA)	*Normal sodium diet:*			$\mu g \cdot h^{-1} \cdot L^{-1} \pm SE$
			(ng/h)/mL ± SE		
		Supine:	1.6 ± 1.5	× 1	1.6 ± 1.5
		Standing (4 h):	4.5 ± 2.9		4.5 ± 2.9
		Low sodium:			
		Supine:	3.2 ± 1.1		3.2 ± 1.1
		Standing (4 h):	9.9 ± 4.3		9.9 ± 4.3
Reverse Triiodothyronine (rT$_3$)	S		*ng/dL*		*nmol/L*
		1–5 y:	15–71		0.23–1.09
		5–10 y:	17–79	× 0.0154	0.26–1.22
		10–15 y:	19–88		0.29–1.36
		Adults:	30–80		0.46–1.23

Table 20-25. CLINICAL CHEMISTRY AND TOXICOLOGY (*Continued*)

Test	Specimen	Reference Range		Conversion Factor	Reference Range (International Units)	
Riboflavin (Vitamin B₂)	U, random		*μg/g creatinine*		*μmol/mol creatinine*	
		1–3 y:	500–900	× 0.3	150–270	
		4–6 y:	300–600		90–180	
		7–9 y:	270–500		81–150	
		10–15 y:	200–400		60–120	
		Adult:	80–269		24–81	
		Pregnancy:	90–120		27–36	
Schilling Test (Intrinsic Factor Test) *Dose: 0.5–1.0 μCi* ⁵⁸*Co-Vitamin B₁₂*	U, 24 h	> 7.5% of dose		× 0.01	Fraction of dose: 0.075	
Secretin	S or P (Hep or EDTA)	37 ± 8 pg/mL (SD)		× 1	37 ± 8 ng/L (SD)	
Selenium	WB (Hep)	10–34 μg/dL Serum approx. 20% lower		× 0.127	1.27–4.32 μmol/L	
	U, 24 h	10–100 μg/L		× 0.0127	0.13–1.27 μmol/L	
Serotonin	WB (EDTA)	39–361 ng/mL		× 0.00568	0.22–2.05 μmol/L	
	Plt	~314 ng/10⁹ platelets		× 0.00568	~1.78 amol/platelet	
Sodium	S or P (Hep)		*mEq/L*		*mmol/L*	
		Premature, Cord:	116–140	× 1	116–140	
		48 h:	128–148		128–148	
		Newborn, Cord:	126–166		126–166	
		Newborn:	134–144		134–144	
		Infant:	139–146		139–146	
		Child:	138–145		138–145	
		Thereafter:	136–146		136–146	
	U, 24 h	40–220 mEq/d (diet dependent)		× 1	40–220 mmol/d	
	Swt	10–40 mEq/L		× 1	10–40 mmol/L	
		Cystic fibrosis:	> 70 mEq/L		> 70 mmol/L	
Somatomedin C	P (EDTA)	0.4–2.0 IU/mL		× 1000	400–2000 IU/L	
Specific Gravity	U, random	Newborn:	1.012		Same	
		Infant:	1.002–1.006			
		Adult:	1.002–1.030			
		After 12 h fluid restriction:	> 1.025			
	U, 24 h	1.015–1.025				
Sucrose	S or P	Mean:	0.06 mg/dL	× 29.21	1.75 μmol/L	
	U	Mean:	2.2 mg/dL		64.26 μmol/L	
Sulfhemoglobin	WB (EDTA, Hep or ACD)	≤ 1.0% of total Hb		× 0.01	≤ 0.010 of total Hb (mass fraction)	
Testosterone, Free	S	*ng/dL (mean ± SE)*	*% of Total*		*pmol/L (mean ± SE)*	*Fraction of Total*
		Cord,				
		M: 1.0 ± 0.4	2.9 ± 0.6	× 34.7 and	34.7 ± 13.9	0.029 ± 0.006
		F: 0.89 ± 0.29	3.0 ± 0.5	× 0.01	30.9 ± 10.1	0.03 ± 0.005
		1–15 d,				
		M: 0.8 ± 0.8	1.3 ± 0.2		27.8 ± 27.8	0.013 ± 0.002
		F: 0.14 ± 0.06	1.2 ± 0.2		4.9 ± 2.1	0.012 ± 0.002
		Prepubertal,				
		M: 0.04 ± 0.01	0.7 ± 0.2		1.4 ± 0.35	0.007 ± 0.002
		F: 0.04 ± 0.01	0.7 ± 0.1		1.4 ± 0.35	0.007 ± 0.001
		Adult,				
		M: 7.9 ± 2.3	1.4 ± 0.3		274 ± 80	0.014 ± 0.003
		F: 0.31 ± 0.07	0.9 ± 0.2		10.8 ± 2.4	0.009 ± 0.002
Testosterone, Total	S	*ng/dL ± SD*			*nmol/L ± SD*	
		Cord, M:	39 ± 11	× 0.0347	1.35 ± 0.38	
		F:	30 ± 7		1.04 ± 0.24	
		1–15 d, M:	68 ± 60		2.36 ± 2.08	
		F:	12 ± 6		0.42 ± 0.21	
		1–3 mo, M:	208 ± 68		7.22 ± 2.36	
		F:	9 ± 4		0.31 ± 0.14	

Table 20-25. CLINICAL CHEMISTRY AND TOXICOLOGY (*Continued*)

Test	Specimen	Reference Range		Conversion Factor	Reference Range (International Units)
Testosterone, Total (*continued*)			*ng/dL ± SD*		*nmol/L ± SD*
	S	3–5 mo, M:	95 ± 53		3.30 ± 1.84
		F:	6.7 ± 2.7		0.23 ± 0.09
		5–7 mo, M:	23 ± 18		0.80 ± 0.62
		F:	6.8 ± 3.1		0.24 ± 0.11
		7–12 mo, M:	6.6 ± 4.6		0.23 ± 0.16
		F:	5.5 ± 2.8		0.19 ± 0.10
		Prepubertal, M:	6.6 ± 2.5		0.23 ± 0.09
		F:	6.6 ± 2.5		0.23 ± 0.09
		Adult, M:	572 ± 135		19.85 ± 4.68
		F:	37 ± 10		1.28 ± 0.35
		Pregnancy:	114 ± 38		3.96 ± 1.32
		(Level unrelated to sex of fetus)			
	U, 24 h		*μg/kg* *Body Weight*		*nmol/kg* *Body Weight*
		Pubertal stage I, M: 0.25		× 3.47	0.87
		F: 0.16			0.56
		II, M: 0.34			1.18
		F: 0.16			0.56
		III, M: 0.37			1.28
		F: 0.16			0.56
			μg/d		*nmol/d*
		20–50 y, M: 50–135		× 3.47	173–470
		F: 2–12			7–42
		> 50 y, M: 40–60			139–210
		F: 2–8			7–28
	Amf	*ng/dL*			*nmol/L*
		Fetal age (wk)	*Median* *Range*		*Median* *Range*
		9–12, M: 5.0	2.0–72.6	× 0.0347	0.17 0.07–2.52
		F: 2.7	1.3–4.0		0.09 0.05–0.14
		12–16, M: 25.0	7.0–72.4		0.87 0.24–2.51
		F: 2.6	1.3–10.0		0.09 0.05–0.35
		16–19, M: 19.3	8.4–23.0		0.67 0.29–0.80
		F: 2.9	1.0–9.0		0.10 0.03–0.31
		28–34, M: 12.3	84.0–26.4		0.43 2.91–0.92
		34–40, M: 18.0	2.0–16.0		0.62 0.07–0.56
		F: 3.4	2.2–10.2		0.12 0.08–0.35
Tetrahydrocortisol (THF)	U, 24 h	Adult:	0.5–1.5 mg/d	× 2.72	Adult: 1.4–4.1 μmol/d
Tetrahydrodeoxy-cortisol	U, 24	< 1000 μg/d		× 0.0029	< 2.9 μmol/d
		10-fold increase over baseline after metyrapone			Same
Thallium	WB (NaHep)	0.5 μg/dL		× 48.9	24.5 nmol/L
		Toxic:	10–800 μg/dL	× 0.0489	0.5–39.1 μmol/L
	U, 24 h	< 2.0 μg/L		× 4.89	< 9.78 nmol/L
		Toxic:	1.0–20.0 mg/L		4.9–97.8 μmol/L
Thyroglobulin (Tg)	S	< 50 ng/mL		× 1	< 50 μg/L
Thyroid Antibodies	S	Adult:	≤ 1:10 dilution		Same
		Child:	≤ 1:4		
Thyroid Microsomal Antibodies	S	Nondetectable (hemagglutination) or < 1:10 (IFA)			Same
Thyroid Stimulating Hormone (hTSH)	S or P		*μIU/mL*		*mIU/L*
		Cord:	3–12	× 1	3–12
		Newborn:	3–20		3–20
		Child:	4.5 ± 3.6 (SD)		4.5 ± 3.6 (SD)
		Adult:	< 10		< 10
		> 60 y, M:	2–7.3		2–7.3
		F:	2–16.8		2–16.8
Thyroid Stimulating Hormone—Response to TRH *Dose, adult: 400–500 μg TRH (Thypinone), IV, rapid; child: 7 μg/kg*	S	30 min after stimulation:	*μIU/mL*		*mIU/L*
		Child:	11–35	× 1	11–35
		Adult, M:	15–30		15–30
		F:	20–40		20–40

Table 20-25. CLINICAL CHEMISTRY AND TOXICOLOGY (*Continued*)

Test	Specimen	Reference Range		Conversion Factor	Reference Range (International Units)
Thyroid Uptake of Radioactive Iodine, ^{131}I, ^{125}I, or ^{123}I		2 h: 6 h: 24 h:	<6% 3–20% 8–30%	× 0.01	*Activity Fraction* <0.06 0.03–0.20 0.08–0.30
Thyroid Uptake of $^{99m}TcO_4^-$		Uptake ratio:	<1.15 in euthyroid		<1.15
Thyrotropin Releasing Hormone (hTRH)	P U, 24 h	Newborn, 30 min after delivery: Adult: Adult, M: F:	78 pg/mL (mean), falling to normal at 24 h 5–60 pg/mL 195 ng/d (mean) 119 ng/d (mean)	× 1	78 ng/L (mean) 5–60 ng/L Same
Thyrotropin Releasing Hormone Stimulation Test *Dose, adult: 500 µg TRH, IV*	S	TSH within 30 min, <40 y: >40 y, M: hPRL:	 >6 µIU/mL rise >2 µIU/mL rise 3–5-fold rise above baseline (diminishes with age)		TSH within 30 min, <40 y: >6 mIU/L rise >40 y, M: >2 mIU/L rise hPRL: 3–5-fold rise above baseline (diminishes with age)
Thyroxine Binding Globulin (TBG)	S	 Cord: 1–4 wk: 1–12 mo: 1–5 y: 5–10 y: 10–15 y: Adult: Pregnancy, last 5 mo: As T_4 binding capacity, Adult:	*mg/dL* 1.4–9.4 1.0–9.0 2.0–7.6 2.9–5.4 2.5–5.0 2.1–4.6 1.5–3.4 5.6–10.2 10–25 µg/dL	× 10	*mg/L* 14–94 10–90 20–76 29–54 25–50 21–46 15–34 56–102 As T_4 binding capacity, Adult: 100–250 µg/L
Thyroxine, Free (FT$_4$)	S	0.8–2.4 ng/dL		× 12.9	10.3–31.0 pmol/L
Thyroxine Index, Free, *see Free Thyroxine Index*					
Thyroxine Ratio, Effective (ETR)		0.86–1.13		× 1	0.86–1.13
Thyroxine/TBG Ratio	S	0.2–0.5 T_4 (µg/dL)/TBG (µg/mL)		× 12.9	2.6–6.4 T_4 (nmol/L)/TBG (mg/L)
Thyroxine (T$_4$), Total	S	 Cord: Newborn: Neonate: Infant: 1–5 y: 5–10 y: Adult: >60 y, M: F: Pregnancy, last 5 mo:	*µg/dL* 8–13 11.5–24 9–18 7–15 7.3–15 6.4–13.3 5–12 5–10 5.5–10.5 6.1–17.6	× 12.9	*nmol/L* 103–168 148–310 116–232 90–194 94–194 83–172 65–155 65–129 71–135 79–227
Transcortin	S	 M: F, Follicular: Luteal: Postmenopausal: Pregnancy, 21–28 wk: 33–40 wk:	*mg/dL* 1.5–2.0 1.7–2.0 1.6–2.1 1.7–2.5 4.7–5.4 5.5–7.0	× 10	*mg/L* 15–20 17–20 16–21 17–25 47–54 55–70
Transferrin *Nephelometric*	S	 Newborn: Adult: >60 y:	*mg/dL* 130–275 220–400 180–380	× 0.01	*g/L* 1.30–2.75 2.20–4.00 1.80–3.80

Table 20-25. CLINICAL CHEMISTRY AND TOXICOLOGY (*Continued*)

Test	Specimen	Reference Range	Conversion Factor	Reference Range (International Units)
Transketolase *Ribose/sedoheptulose, 37°C*	WB (Hep)	(9–12 μmol/h)/mL whole blood	× 16.67	150–200 U/L whole blood
		If Hct is abnormal: (2.1–2.4 μmol/h)/10⁹ red cells	× 0.0167	0.035–0.040 nU/red cell

For Triglycerides:

		mg/dL			× 0.0113	*mmol/L*		
Triglycerides (TG) *Fluorometric; colorimetric*	S, ≥12 h		M	F		M		F
		Cord blood:	10–98	10–98		0.11–1.11		0.11–1.11
		0–5 y:	30–86	32–99		0.34–0.97		0.36–1.12
		6–11 y:	31–108	35–114		0.35–1.22		0.40–1.29
		12–15 y:	36–138	41–138		0.41–1.56		0.46–1.56
		16–19 y:	40–163	40–128		0.45–1.84		0.45–1.45
		20–29 y:	44–185	40–128		0.50–2.09		0.45–1.45
		30–39 y:	49–284	38–160		0.55–3.21		0.43–1.81
		40–49 y:	56–298	44–186		0.63–3.37		0.50–2.10
		50–59 y:	62–288	55–247		0.70–3.25		0.62–2.79

Values decrease slightly after age 60
Levels for blacks: 10–20 mg/dL lower

Levels for Blacks: 0.11–0.23 mmol/L lower

Recommended (desirable) levels for adults:

Male:	40–160 mg/dL	0.45–1.81 mmol/L
Female:	35–135 mg/dL	0.40–1.53 mmol/L (Conv. factor based on triolein, M.W. 885)

Test	Specimen	Reference Range	Conversion Factor	Reference Range (International Units)
Triiodothyronine, Free	S			
			mean pg/dL		*mean pmol/L*
		Cord:	130±10 (SE)	× 0.0154	2.0±0.15 (SE)
		1–3 d:	410±20		6.31±0.31
		6 wk:	400±20		6.16±0.31
		Adult (20–50 y):	230–660 pg/dL		3.54–10.16 pmol/L
Triiodothyronine Resin Uptake Test (T₃RU)	S		*% of Total*		*Fraction of Total*
		Newborn:	25–37	× 0.01	0.25–0.37
		Adult:	24–34		0.24–0.34
		>60 y, M:	24–32		0.24–0.32
		F:	22–32		0.22–0.32
Triiodothyronine, Total (T₃-RIA)	S		*ng/dL*		*nmol/L*
		Cord:	30–70	× 0.0154	0.46–1.08
		Newborn:	65–275		0.98–4.26
		1–5 y:	100–260		1.54–4.00
		5–10 y:	90–240		1.39–3.70
		10–15 y:	80–210		1.23–3.23
		Adult:	120–195		1.85–3.00
		>60 y, M:	105–175		1.62–2.69
		F:	108–205		1.66–3.16
Triolein-¹³¹I Absorption Test	P	>1.7% of administered dose/L after 4–6 h	× 0.01	Fraction of administered dose: 0.017/L
Dose: 50 μCi in milk	F, 72 h	<5% of administered dose in 72 h specimen		Fraction of administered dose: <0.50/72 h
Tubular Reabsorption of Phosphate (TRP)	U, 4 h (0800–1200 h) and S	82–95%	× 0.01	Fraction: 0.82–0.95
Tyrosine *Fluorometric*	S		*mg/dL*		*mmol/L*
		Premature:	7.0–24.0	× 55.2	386–1325
		Newborn:	1.6–3.7		88–204
		Adult:	0.8–1.3		44–72
Urea, *see Urea Nitrogen*				
Urea Nitrogen	S		*mg/dL*		*mmol urea/L*
		Cord:	21–40	× 0.166	3.5–6.6
		Premature (1 wk):	3–25		0.5–4.2
		Newborn:	4–12		0.7–2.0
		Infant/child:	5–18		0.8–3.0
		Adult:	7–18		1.2–3.0
		>60 y:	8–21		1.3–3.5
		Higher after protein intake			
	U	12–20 g/d		× 16.66	200–333 mmol urea/d

Table 20-25. CLINICAL CHEMISTRY AND TOXICOLOGY (*Continued*)

Test	Specimen	Reference Range		Conversion Factor	Reference Range (International Units)
Urea Nitrogen/Creatinine Ratio	S	12/1 to 20/1			12/1 to 20/1
Uric Acid					
Phosphotungstate	S		*mg/dL*		*mmol/L*
		Adult, M:	4.5–8.2	× 0.059	0.27–0.48
		F:	3.0–6.5		0.18–0.38
		>60 y, M:	4.2–8.0		0.25–0.47
		F:	3.2–7.3		0.19–0.43
Uricase		Child:	2.0–5.5		0.12–0.32
		Adult, M:	3.5–7.2		0.21–0.42
		F:	2.6–6.0		0.15–0.35
	U, 24 h		*mg/d*		*mmol/d*
		Free purine diet, M:	<420	× 0.0059	<2.48
		F:	slightly lower		slightly lower
		Low purine diet, M:	<480		<2.83
		F:	<400		<2.36
		High purine diet:	<1000		<5.90
		Average diet:	250–750		1.48–4.43
Urine Volume	U, 24 h		*mL/d*		Same
		Newborn (1–2 d):	30–60		
		Infant, 3–10 d:	100–300		
		10–60 d:	250–450		
		60–365 d:	400–500		
		Child, 1–3 y:	500–600		
		3–5 y:	600–700		
		5–8 y:	650–1000		
		8–14 y:	800–1400		
		Adult, M:	800–1800		
		F:	600–1600		
		Older adult:	250–2400		
		(varies with water intake and other factors)			
Urobilinogen	U, 2 h	0.1–0.8 EU*/2 h			Same
	U, 24 h	0.5–4.0 EU/d			
	F	75–275 EU/100 g		× 10	750–2750 EU*/kg
		75–400 EU/d			75–400 EU/d
		40–280 mg/d		× 1.69	67–473 μmol/d
		*Ehrlich Unit			
Uroporphyrin	U, 24 h	<50 μg/d		× 1.2	<60 nmol/d
	F, 24 h	10–40 μg/d			12–48 nmol/d
	RBC (Hep or EDTA)	Negative			Negative
Valine	S		*mg/dL*	× 1	*μmol/L*
		Premature, 1 d:	1.52±0.59 (SD)		130±50 (SD)
		Newborn, 1 d:	0.94–2.88		80–246
		1–3 mo:	2.27±0.57 (SD)		194±49 (SD)
		9 mo–2 y:	0.67–3.07		57–262
		3–10 y:	1.50–3.31		128–283
		6–18 y:	1.83–3.37		156–288
		Adult:	1.65–3.71		141–317
	U, 24 h		*mg/d*		*μmol/d*
		10 d–7 wk:	1–3	× 8.53	9–26
		3–12 y:	2–6		17–51
		Adult:	2–12		17–102
		or 4 ± 1 mg/g creatinine (SD)		× 0.97	or 3.9 ± 1.0 mmol/mol creatinine (SD)
Vanillylmandelic Acid (Vanilmandelic Acid)	U, 24 h		*mg/d*		*μmol/d*
		Newborn:	<1.0	× 5.05	<5.1
		Neonate:	<1.0		<5.1
		Infant:	<2.0		<10.1
		Child:	1–5		5.1–25.3
		Adolescent:	1–5		5.1–25.3
		Thereafter:	2–7		10.1–35.4
		or 1.5–7 μg/mg creatinine		× 0.571	or 0.86–4.00 mmol/mol creatinine

Table 20-25. CLINICAL CHEMISTRY AND TOXICOLOGY (*Continued*)

Test	Specimen	Reference Range	Conversion Factor	Reference Range (International Units)
Vasoactive Intestinal Polypeptide (VIP)	P (Hep)	20–53 pg/mL	× 1	20–53 ng/L
Viscosity	S	1.10–1.22 Centipoise	× 1	1.10–1.22 Centipoise
Vitamin A	S	*μg/dL* Newborn: 35–75 Child: 30–80 Adult: 30–65	× 0.0349	*μmol/L* 1.22–2.62 1.05–2.79 1.05–2.27
Vitamin A Tolerance Test *Dose: 5000 U vit. A in oil/kg orally*	S	3 and/or 6 h: 200–600 μg vit. A/dL	× 0.0349	7–21 μmol/L
Vitamin B$_2$, *see Riboflavin*				
Vitamin B$_6$	P (EDTA)	3.6–18 ng/mL	× 4.046	14.6–72.8 nmol/L
Vitamin B$_{12}$, True	S	100–700 pg/mL > 60 y: 110–800 pg/mL	× 0.738	74–516 pmol/L 81–590 pmol/L
Vitamin B$_{12}$ Intrinsic Factor	Gastf	50–400% enhancement of ^{57}Co-B$_{12}$ uptake by GPIMH* *Guinea pig intestinal mucosal homogenate	× 0.01	Fractional increase in ^{57}Co-B$_{12}$ uptake by GPIMH*: 0.50–4.00
Vitamin C *Dinitrophenylhydrazine, colorimetric*	P (Ox, Hep, or EDTA) Buffy coat (Hep)	0.6–2.0 mg/dL 20–53 μg/10^8 WBC	× 56.78 × 0.568	34–114 μmol/L 11.4–30.1 amol/cell
Vitamin C Saturation Test *Dose: 0.5–2.0 g ascorbate orally over a period of 4 d*	U, 24 h	60–80 % of test dose excreted	× 0.01	Fraction test dose excreted: 0.60–0.80
Vitamin D$_3$, 1,25-dihydroxy	S	25–45 pg/mL	× 2.4	60–108 pmol/L
Vitamin D$_3$, 25-hydroxy	P (Hep)	Summer: 15–80 ng/mL Winter: 14–42 ng/mL	× 2.496	37.4–200 nmol/L 34.9–105 nmol/L
Vitamin E	S	5.0–20 μg/mL	× 2.32	11.6–46.4 μmol/L
Xylene	WB (F/Ox)	Negative Occup. exp.: < 1 mg/L Toxic: > 3 mg/L	× 9.42	Negative < 9.4 μmol/L > 28.3 μmol/L
Xylose Absorption Test	WB (NaF) U, 5 h	*mg/dL* Child, 1 h (5 g dose): > 20 Adult, 2 h (25 g dose): > 25 Child: 16–33% of ingested dose Adult, *g/5 h* 5 g dose: > 1.2 25 g dose: > 4.0 > 65 y: > 3.5	× 0.0666 × 0.01 × 6.66	*mmol/L* > 1.33 > 1.67 Fraction ingested dose: 0.16–0.33 *mmol/5 h* > 8.00 > 26.64 > 23.31
Zinc	S	7–150 μg/dL	× 0.153	10.7–22.9 μmol/L

Table 20-26. DRUGS—THERAPEUTIC AND TOXIC

Doses for drugs listed as part of the protocol for laboratory tests, as well as their therapeutic and toxic ranges, are those commonly accepted on the basis of present knowledge. The constant acquisition of new knowledge makes it necessary to confirm and compare information supplied in this appendix with data furnished by manufacturers; additional experience with these drugs may lead to revised values. It should also be considered that there are significant differences in drug response among individuals. Where no therapeutic or toxic level is given, it was felt by the authors that there are presently no reliable data available.

Data collected for therapeutic monitoring requires careful coordination between the time of sampling and drug administration. Therefore, recommendations have been made regarding the optimal time for sampling of many drugs. Hours in parenthesis indicate the minimum sampling times.

For convenience and to preserve space, we have used standard abbreviations commonly used in laboratory medicine. Less common abbreviations and some nonstandard abbreviations are given below.

Abbreviations

EDTA	ethylenediaminetetraacetate, edetic acid	prem.	premature
Hep	heparin	S	serum
Occup. exp.	occupational exposure	Therap.	therapeutic
P	plasma	U	urine
		WB	whole blood

Drug	Specimen	Reference Range		Conversion Factor	Reference Range (International Units)
Acetaminophen	S or P (Hep or EDTA)	Therap.: Toxic:	10–30 μg/mL >200 μg/mL	× 6.62	66–199 μmol/L >1324 μmol/L
Amikacin	S or P (EDTA)	Therap., Peak: Trough (less severe infection: (life-threatening infection): Toxic, Peak: Trough:	μg/mL 25–35 1–4 4–8 >35–40 >10–15	× 1.71	μmol/L 43–60 1.7–6.8 6.8–13.7 >60–68 >17–26
ε-Aminocaproic Acid	S or P (Hep or EDTA); trough	Therap.:	>130 μg/mL	× 7.62	>991 μmol/L
Amitriptyline	S or P (Hep or EDTA); trough (>12 h after dose)	Therap.: Toxic:	120–250 ng/mL >500 ng/mL	× 3.61	433–903 nmol/L >1805 nmol/L
Amobarbital, quant.	S	Therap.: Toxic:	1–5 μg/mL >10 μg/mL	× 4.42	4–22 μmol/L >44 μmol/L
Amphetamine	S or P (Hep or EDTA)	Therap.: Toxic:	20–30 ng/mL >200 ng/mL	× 7.40	148–222 nmol/L >1480 nmol/L
Bromide	S	Therap.: Toxic:	750–1500 μg/mL >1250 μg/mL	× 0.0125	9.4–18.7 mmol/L >15.6 mmol/L
Caffeine	S or P (Hep or EDTA)	Therap.: Toxic:	3–15 μg/mL >50 μg/mL	× 5.15	15–77 μmol/L >258 μmol/L
Carbamazepine	S or P (Hep or EDTA); trough	Therap.: Toxic:	8–12 μg/mL >15 μg/mL	× 4.23	34–51 μmol/L >63 μmol/L
Carbenicillin	S or P (Hep or EDTA)	Therap: Toxic:	Dependent on minimum inhibition conc. of specific organism >250 μg/mL (neurotoxicity)	× 2.64	Same >660 μmol/L

Table 20-26. DRUGS—THERAPEUTIC AND TOXIC (*Continued*)

Drug	Specimen	Reference Range		Conversion Factor	Reference Range (International Units)
Chloral Hydrate	S	As trichloroethanol: Therap.: 2–12 µg/mL Toxic: > 20 µg/mL		× 6.69	13–80 µmol/L > 134 µmol/L
Chloramphenicol	S or P (Hep or EDTA); trough	Therap.: 10–25 µg/mL Toxic: > 25 µg/mL		× 3.09	31–77 µmol/L > 77 µmol/L
Chlordiazepoxide	S or P (Hep or EDTA); trough	Therap.: 700–1000 ng/mL Toxic: > 5000 ng/mL		× 0.0033	2.34–3.34 µmol/L > 16.7 µmol/L
Chlorpromazine	S or P (Hep or EDTA); trough	Therap.: 50–300 ng/mL Toxic: > 750 ng/mL		× 3.14	157–942 nmol/L > 2355 nmol/L
Cimetidine	S or P (Hep or EDTA); trough	Therap.: > 1.0 µg/mL		× 3.96	> 4.0 µmol/L
Clonazepam	S or P (Hep or EDTA); trough	Therap.: 15–60 ng/mL Toxic: > 80 ng/mL		× 3.17	48–190 nmol/L > 254 nmol/L
Clonidine	S or P (Hep or EDTA)	Therap.: 1.0–2.0 ng/mL		× 4.35	4.4–8.7 nmol/L
Clorazepate	S or P (Hep or EDTA)	As desmethyldiazepam: Therap.: 0.12–1.0 µg/mL		× 3.01	0.36–3.01 µmol/L
Cocaine	S or P (Hep or EDTA); on ice	Therap.: 100–500 ng/mL Toxic: > 1000 ng/mL		× 3.3	330–1650 nmol/L > 3300 nmol/L
Desipramine	S or P (Hep or EDTA); trough (≥ 12 h after dose)	Therap.: 75–160 ng/mL Toxic: > 1000 ng/mL		× 3.75	281–600 nmol/L > 3750 nmol/L
Desmethylmethsux-imide	S	Therap.: 10–40 µg/mL Toxic: > 40 µg/mL		× 5.29	53–212 µmol/L > 212 µmol/L
Diazepam	S or P (Hep or EDTA); trough	Therap.: 100–1000 ng/mL Toxic: > 5000 ng/mL		× 0.0035	0.35–3.51 µmol/L > 17.55 µmol/L
Digitoxin	S or P (Hep or EDTA); ≥ 6 h after dose	Therap.: 20–35 ng/mL Toxic: > 45 ng/mL		× 1.31	26–46 nmol/L > 59 nmol/L
Digoxin	S or P (Hep or EDTA); ≥ 12 h after dose	*ng/mL* Therap., CHF: 0.8–1.5 Arrhyth-mias: 1.5–2.0 Toxic, Adult: > 2.5 Child: > 3.0		× 1.28	*nmol/L* 1.0–1.9 1.9–2.6 > 3.2 > 3.8
Diphenylhydantoin, *see* Phenytoin					

Table 20-26. DRUGS—THERAPEUTIC AND TOXIC (*Continued*)

Drug	Specimen	Reference Range		Conversion Factor	Reference Range (International Units)
Disopyramide	S or P (Hep or EDTA); trough	Therap., Arrhythmias, Atrial: Ventricular: Toxic:	*µg/mL* 2.8–3.2 3.3–7.5 >7	× 2.95	*µmol/L* 8.3–9.4 9.7–22 20.7
Doxepin	S or P (Hep or EDTA); trough (≥12 h after dose)	Therap.: Toxic:	30–150 ng/mL >500 ng/mL	× 3.58	107–537 nmol/L >1790 nmol/L
Ethchlorvynol	S or P (Hep or EDTA)	Therap.: Toxic:	2–8 µg/mL >20 µg/mL	× 6.92	14–55 µmol/L >138 µmol/L
Ethosuximide	S or P (Hep or EDTA); trough	Therap.: Toxic:	40–100 µg/mL >150 µg/mL	× 7.08	283–708 µmol/L >1062 µmol/L
Fenoprofen	P (EDTA)	Therap.:	20–65 µg/mL	× 4.12	82–268 µmol/L
Furosemide	S	Therap.: 30 min after last dose	1–2 µg/mL	× 3.023	3–6 µmol/L
Gentamicin	S or P (EDTA)	Therap., Peak (less severe infection): (severe infection): Trough (less severe infection): (moderate infection): (severe infection): Toxic, Peak: Trough:	*µg/mL* 5–8 8–10 <1 <2 <2–4 >10–12 >2–4	× 2.09	*µmol/L* 10.4–16.7 16.7–20.9 <2.1 <4.2 <4.2–8.4 >21–25 >4.2–8.4
Glutethimide	S	Therap.: Toxic:	2–6 µg/mL >5 µg/mL	× 4.60	9–28 µmol/L >23 µmol/L
Imipramine	S or P (Hep or EDTA); trough (≥12 h after dose)	Therap.: Toxic:	125–250 ng/mL >500 ng/mL	× 3.57	446–893 nmol/L >1785 nmol/L
Isoniazid	S or P (Hep or EDTA)	Therap.: Toxic:	1–7 µg/mL 20–710 µg/mL	× 7.29	7–51 µmol/L 146–5176 µmol/L
Kanamycin	S or P (EDTA)	Therap., Peak: Trough (less severe infection): (life threatening infection): Toxic, Peak: Trough:	*µg/mL* 25–35 1–4 4–8 >35–40 >10–15	× 2.06	*µmol/L* 52–72 2–8 8–16 >72–82 >21–31
Lidocaine	S or P (Hep or EDTA); ≥45 min following bolus dose	Therap.: Toxic, CNS, cardiovascular depression: Seizures, obtundation, decreased cardiac output:	*µg/mL* 1.5–6.0 6–8 >8	× 4.27	*µmol/L* 6.4–26 26–34.2 >34.2

Table 20-26. DRUGS—THERAPEUTIC AND TOXIC (*Continued*)

Drug	Specimen	Reference Range		Conversion Factor	Reference Range (International Units)
Lithium	S or P (Hep or EDTA); ≥ 12 h after last dose	Therap.: Toxic:	0.6–1.2 mEq/L > 2 mEq/L	× 1	0.6–1.2 mmol/L > 2 mmol/L
Lorazepam	S or P (Hep or EDTA)	Therap.:	50–240 ng/mL	× 3.11	156–746 nmol/L
Lysergic Acid Diethylamide	P (EDTA) U	After hallucinogenic dose: 0.005–0.009 µg/mL 0.001–0.050 µg/mL		× 3090	15.5–27.8 nmol/L 3.1–154 nmol/L
Meperidine	S or P (Hep or EDTA)	Therap.: Toxic:	70–500 ng/mL > 1 µg/mL	× 4.04 × 4043	283–2020 nmol/L > 4043 nmol/L
Meprobamate	S	Therap.: Toxic:	6–12 µg/mL > 60 µg/mL	× 4.58	28–55 µmol/L > 275 µmol/L
Methadone	S or P (Hep or EDTA)	Therap.: Toxic:	100–400 ng/mL > 2000 ng/mL	× 0.00323	0.32–1.29 µmol/L > 6.46 µmol/L
Methaqualone	S or P (Hep or EDTA)	Therap.: Toxic:	2–3 µg/mL > 10 µg/mL	× 4.00	8–12 µmol/L > 40 µmol/L
Methotrexate	S or P (Hep or EDTA)	Therap.: Toxic, 1–2 wk after low dose therapy: > 9.1 ng/mL 48 h after high dose therapy:	variable 454 ng/mL	× 2.20	variable > 20 nmol/L > 1000 nmol/L
Methsuximide (N-desmethyl methsuximide)	S	Therap.: Toxic	10–40 µg/mL > 40 µg/mL	× 5.29	53–212 µmol/L > 212 µmol/L
Methyldopa	P (EDTA)	Therap.: Toxic: (not well defined)	1–5 µg/mL > 7 µg/mL	× 4.73	4.7–23.7 µmol/L > 33 µmol/L
Methyprylon	S	Therap.: Toxic:	8–10 µg/mL > 50 µg/mL	× 5.46	43–55 µmol/L > 273 µmol/L
Morphine	S or P (Hep or EDTA)	Surgical anesthesia: Toxic:	65–80 ng/mL 200–5000 ng/mL	× 3.50	227–280 nmol/L 700–17 500 nmol/L
N-acetylprocainamide	S or P (Hep or EDTA); trough	Therap.: Toxic:	5–30 µg/mL > 40 µg/mL	× 3.61	18–108 µmol/L > 144 µmol/L
Nitroprusside	S or P (EDTA)	As thiocyanate: Therap.:	6–29 µg/mL	× 17.2	103–499 µmol/L
Normethsuximide	S	Therap.: Toxic:	10–40 µg/mL > 40 µg/mL	× 5.29	53–212 µmol/L > 212 µmol/L
Nortriptyline	S or P (Hep or EDTA); trough (≥ 12 h after dose)	Therap.: Toxic:	50–150 ng/mL > 500 ng/mL	× 3.80	190–570 nmol/L > 1900 nmol/L
Oxazepam	S or P (Hep or EDTA)	Therap.:	0.2–1.4 µg/mL	× 3.49	0.70–4.9 µmol/L

Table 20-26. DRUGS—THERAPEUTIC AND TOXIC (*Continued*)

Drug	Specimen	Reference Range		Conversion Factor	Reference Range (International Units)
Paraldehyde	S or P (Hep or EDTA)	Therap., Sedation:	$\mu g/mL$ 10–100	× 7.57	$\mu mol/L$ 76–757
		Anesthesia:	> 200		> 1514
		Toxic:	200–400		1514–3028
		Lethal:	> 500		> 3785
			$\mu g/mL$		$\mu mol/L$
Paraquat	WB (EDTA)	Toxic:	0.1–1.6	× 3.89	0.39–6.2
	U	Occup. exp.:	0.3		1.17
		Toxic:	0.9–64		3.50–249
Pentobarbital	S or P (Hep or EDTA); trough	Therap., Hypnotic:	$\mu g/mL$ 1–5	× 4.42	$\mu mol/L$ 4–22
		Therap. coma:	20–50		88–221
		Toxic:	> 10		> 44
Phenacetin	P (EDTA)	Therap.:	1–20 $\mu g/mL$	× 5.58	6–112 $\mu mol/L$
		Toxic:	50–250 $\mu g/mL$		279–1395 $\mu mol/L$
Phencyclidine	S or P (Hep or EDTA)	Toxic:	90–800 ng/mL	× 4.11	370–3288 nmol/L
Phenobarbital	S or P (Hep or EDTA); trough	Therap.:	$\mu g/mL$ 15–40	× 4.31	$\mu mol/L$ 65–170
		Toxic, Slowness, ataxia, nystagmus:	35–80		151–345
		Coma with reflexes:	65–117		280–504
		Coma without reflexes:	> 100		> 430
Phensuximide (both parent and N-desmethyl metabolites)	S or P (Hep or EDTA)	Therap.:	40–60 $\mu g/mL$	× 5.71	228–324 $\mu mol/L$
Phenylbutazone	P (EDTA)	Therap.: (not well defined)	50–100 $\mu g/mL$	× 3.08	162–324 $\mu mol/L$
		Toxic:	> 100 $\mu g/mL$		> 324 $\mu mol/L$
Phenytoin	S or P (Hep or EDTA); trough	Therap.:	10–20 $\mu g/mL$	× 3.96	40–79 $\mu mol/L$
		Toxic:	> 20 $\mu g/mL$		> 79 $\mu mol/L$
Primidone	S or P (Hep or EDTA); trough	Therap.:	5–12 $\mu g/mL$	× 4.58	23–55 $\mu mol/L$
		Toxic:	> 15 $\mu g/mL$		> 69 $\mu mol/L$
Procainamide	S or P (Hep or EDTA); trough	Therap.:	4–10 $\mu g/mL$	× 4.23	17–42 $\mu mol/L$
		Toxic:	> 10–12 $\mu g/mL$		> 42–51 $\mu mol/L$
		Also consider effect of metabolite (NAPA)			
Propoxyphene	P (EDTA)	Therap.:	0.1–0.4 $\mu g/mL$	× 2.95	0.3–1.2 $\mu mol/L$
		Toxic:	> 0.5 $\mu g/mL$		> 1.5 $\mu mol/L$
Propranolol	S or P (Hep or EDTA); trough	Therap.:	50–100 ng/mL	× 3.86	193–386 nmol/L
Protriptyline	S or P (Hep or EDTA); trough (≥ 12 h after dose)	Therap.:	70–250 ng/mL	× 3.80	266–950 nmol/L
		Toxic:	> 500 ng/mL		> 1900 nmol/L

Table 20-26. DRUGS—THERAPEUTIC AND TOXIC (*Continued*)

Drug	Specimen	Reference Range		Conversion Factor	Reference Range (International Units)
Quinidine	S or P (Hep or EDTA); trough	Therap.:	2–5 μg/mL	× 3.08	6–15 μmol/L
		Toxic:	> 6 μg/mL		> 18 μmol/L
Salicylates	S or P (Hep or EDTA); trough	Therap.:	150–300 μg/mL	× 7.24	1086–2172 μmol/L
		Toxic:	> 300 μg/mL		> 2172 μmol/L
Secobarbital	S	Therap.:	1–2 μg/mL	× 4.20	4.2–8.4 μmol/L
		Toxic:	> 5 μg/mL		> 21.0 μmol/L
Theophylline	S or P (Hep or EDTA)	Therap.,	μg/mL		μmol/L
		Bronchodilator:	8–20	× 5.55	44–111
		Prem. apnea:	6–13		33–72
		Toxic:	> 20		> 110
Thiocyanate	S or P (EDTA)		μg/mL		μmol/L
		Nonsmoker:	1–4	× 17.2	17–69
		Smoker:	3–12		52–206
		Therap. after nitroprusside infusion:	6–29		103–499
	U	Nonsmoker:	1–4 mg/d	× 17.2	17–69 μmol/d
		Smoker:	7–17 mg/d		120–292 μmol/d
Thiopental	S or P (Hep or EDTA); trough		μg/mL		μmol/L
		Hypnotic:	1.0–5.0	× 4.13	4.1–20.7
		Coma:	30–100		124–413
		Anesthesia:	7–130		29–536
		Toxic conc.:	> 10		> 41
Thioridazine	S or P (Hep or EDTA)	Therap.:	1.0–1.5 μg/mL	× 2.70	2.7–4.1 μmol/L
		Toxic:	> 10 μg/mL		> 27 μmol/L
Tobramycin	S or P (Hep or EDTA)	Therap.,	μg/mL		μmol/L
		Peak (less severe infection):	5–8	× 2.14	11–17
		(severe infection):	8–10		17–21
		Trough (less severe infection):	< 1		< 2
		(moderate infection):	< 2		< 4
		(severe infection):	< 2–4		< 4–9
		Toxic, Peak:	> 10–12		> 21–26
		Trough:	> 2–4		> 4–9
Valproic Acid	S or P (Hep or EDTA); trough	Therap.:	50–100 μg/mL	× 6.93	347–693 μmol/L
		Toxic:	> 100 μg/mL		> 693 μmol/L
Vancomycin	S or P (Hep or EDTA); trough	Toxic:	> 80–100 μg/mL (not well established)	× 1	> 80–100 mg/L
Warfarin	S or P (Hep or EDTA)	Therap.:	1–10 μg/mL	× 3.24	3–32 μmol/L

References

Beutler, E.: Hemolytic Anemia in Disorders of Red Cell Metabolism. New York. Plenum Publishing Company, 1978.

Brown, S. S., Mitchell, F. L., and Young, D. S., Eds.: Chemical Diagnosis of Disease. Amsterdam, Elsevier/North-Holland Biomedical Press, 1979.

Gilman, A. G., Goodman, L., and Gilman, A., Eds.: The Pharmacological Basis of Therapeutics. 6th ed. New York. Macmillan, 1980.

Mabry, C., and Tietz, N. W.: Tables of normal laboratory values. In Nelson Textbook of Pediatrics. 12th ed. R. E. Behrman, and V. C. Vaughan, III, Eds. Philadelphia, W. B. Saunders Company, 1983.

Tietz, N. W., Ed.: Clinical Guide to Laboratory Tests. Philadelphia, W. B. Saunders Company, 1983.

Tietz, N. W.: Reference ranges and laboratory values of clinical importance. In Cecil Textbook of Medicine. 17th ed. J. B. Wyngaarden and L. H. Smith, Jr., Eds. Philadelphia, W. B. Saunders Company, 1985.

Tietz, N. W., and Blackburn, R. H., Eds.: Reference Ranges and General Information. Clinical Laboratories, A. B. Chandler Medical Center, University of Kentucky, Lexington, Kentucky, 1984.

INDEX

Page numbers in *italic* type indicate figures; page numbers followed by (t) indicate tables.